MACHINERY'S
HANDBOOK

*A Reference Book for the Mechanical Engineer,
Designer, Manufacturing Engineer, Draftsman,
Toolmaker, and Machinist*

MACHINERY'S HANDBOOK
23rd Edition

By ERIK OBERG, FRANKLIN D. JONES
and HOLBROOK L. HORTON

HENRY H. RYFFEL, *Editor*
ROBERT E. GREEN, *Associate Editor*
JAMES H. GERONIMO, *Associate Editor*

INDUSTRIAL PRESS INC.
New York

Library of Congress Cataloging-in-Publication Data

Oberg, Erik, 1881–
  Machinery's handbook.

  Includes index.
  1. Mechanical engineering — Handbooks, manuals, etc. I. Jones, Franklin Day, 1879–
II. Horton, Holbrook Lynedon, 1907–
III. Ryffel, Henry H. IV. Title.
TJ151.O245 1988   621.8′0212   87-31093
ISBN 0-8311-1200-X
ISBN 0-8311-0900-9 (Thumb Indexed)

INDUSTRIAL PRESS INC.
200 Madison Avenue
New York, New York 10016-4078

MACHINERY'S HANDBOOK
Twenty-Third Edition, Second Printing, 1989

Recent composition by Dix Type Inc., Syracuse, New York.
Original composition by Machine Composition Company of Boston.
23RD EDITION printed and bound by
National Publishing Company, Philadelphia, Pennsylvania.

PREFACE

MACHINERY'S HANDBOOK, since the publication of its first edition 75 years ago, has continually increased in popularity throughout the world. The HANDBOOK is now used extensively as a standard work of reference in all countries where machines or other mechanical products are designed and manufactured.

The aim of the publisher is to make each new edition of greater practical value than the preceding one. In the 23rd edition, this objective has been accomplished by rearranging much of the material to make it more accessible to readers looking for information on specific subjects; and by revisions to the HANDBOOK which are carried out as frequently as is practical for a voluminous work that must necessarily be printed in large quantities to meet the constant demand for it both here and abroad.

The new material covers a large variety of subjects that are important to designers, manufacturing engineers, machinists, and builders of everything mechanical. Recent or revised engineering standards are included, together with a large amount of general information and mechanical data representing the latest designing and manufacturing practice.

In selecting material from the almost limitless supply of data pertaining to the mechanical field, the plan is to consider the requirements of the design and production departments of both large and small manufacturing plants as well as the needs of jobbing shops, trade schools, and technical schools.

It should be noted that throughout the HANDBOOK step-by-step worked-out examples are provided where needed to illustrate the use of formulas or to clarify a method of calculation. The formulas themselves are arranged in the best sequence for use with a calculator or computer; wherever useful, tables or charts based on these formulas have been provided.

As in previous editions, increased coverage has been provided for metric standards and data. This expanded coverage recognizes the international nature of manufacturing and the need for common dimensions to serve as the basis for interchangeability of manufactured products. Metric data are provided for the most commonly used elements in mechanical products including: metric module involute splines; M- and MJ-profile screw threads; nineteen types of metric screws, bolts, and nuts; prevailing torque hex nuts; and plain washers. Other areas that have been reviewed and revised include: cams and cam design; continued and conjugate fractions for finding factors for gear ratios; series approximations of functions; derivatives of functions; combined stresses in machine elements; cumulative fatigue damage; strength of perforated metal; geometric dimensioning and tolerancing; conversion coatings of metals; properties and treatment of ferrous and nonferrous materials; polygon shafts; die casting; silent chain; and calculating dimensions for gear replacements.

The ever-increasing use of numerical control in manufacturing requires that information about numerical control be available. For this reason, 48 pages of selected topics, arranged for quick reference, have been added in this edition.

Data and information from many American National Standards Institute (ANSI) Standards will be found in the HANDBOOK and have been extracted with permission of the publisher, the American Society of Mechanical Engineers, United Engineering Center, 345 East 47 Street, New York, New York 10017. These Standards are revised periodically; ASME should be contacted for information concerning the current edition of any particular document. In addition to American and British Standards, data and information from many other sources have been used with permission and the individual organizations acknowledged where this information is cited in the HANDBOOK.

Material from ANSI Standards B7.1-1978, B74.2-1982, B74.3-1974, B74.13-1982, B212.1-1984, B212.4-1986, and B212.5-1986 is reproduced with permission. These Standards are copyrighted by the American National Standards Institute. Copies of these Standards may be purchased from the American National Standards Institute at 1430 Broadway, New York, NY 10018.

On August 24, 1969, the American Standards Association was reconstituted as the United States of America Standards Institute and standards approved as American Standards were designated as USA Standards. There was no change in their index identification or technical content. On October 6, 1969, the name was changed to American National Standards Institute. The present standards' designation is ANSI instead of ASA or USAS. Standards previously adopted by the American Standards Association and not revised are still referred to in the HANDBOOK by the designation ASA.

The editors appreciate the contributions in the 23rd Edition of Mr. Jerome Mogul to the materials and process sections; Professor Edward E. Messal for the three-dimensional stress presentation; Mr. James J. Childs and James Childs Jr. for the new, extensive coverage of numerical control; and Mr. Holbrook L. Horton, Editor Emeritus, for his continuing interest and counsel in the preparation of this new edition.

Directing the editor's attention to a possible defect or to the omission of some matter considered of general value often renders a service to the entire mechanical industry. For this reason and because we desire to perfect the HANDBOOK as far as possible, all criticisms and suggestions about either revisions or the inclusion of new matter are welcome.

Henry H. Ryffel, *Editor*

CONTENTS

Mathematics

Mechanics

Strength of Materials

vii

Strength of Materials (*Continued*)

Properties, Treatment and Testing of Materials

Dimensioning, Gaging and Measuring

Tooling and Toolmaking

Machining Operations

Manufacturing Processes

Manufacturing Processes (*Continued*)

Fasteners

Threads and Threading

Threads and Threading (*Continued*)

Gears, Splines and Cams

Bearings and Other Machine Elements

Measuring Units

Index

Bearings and Other Machine Elements

Measuring Units

Index

MACHINERY'S
HANDBOOK

Mathematical Signs and Commonly Used Abbreviations

$+$	Plus (sign of addition)	π	Pi (3.1416)
$+$	Positive	Σ	Sigma (sign of summation)
$-$	Minus (sign of subtraction)	ω	$\begin{cases}\text{Omega (angles measured}\\ \text{in radians)}\end{cases}$
$-$	Negative		
$\pm\ (\mp)$	Plus or minus (minus or plus)	g	$\begin{cases}\text{Acceleration due to}\\ \text{gravity (32.16 ft. per}\\ \text{sec. per sec.)}\end{cases}$
$\times$	$\begin{cases}\text{Multiplied by (multiplication}\\ \text{sign)}\end{cases}$		
$\cdot$	$\begin{cases}\text{Multiplied by (multiplication}\\ \text{sign)}\end{cases}$	$i\ (\text{or } j)$	$\begin{cases}\text{Imaginary quantity}\\ (\sqrt{-1})\end{cases}$
$\div$	Divided by (division sign)	$\sin$	Sine
$/$	Divided by (division sign)	$\cos$	Cosine
$:$	Is to (in proportion)	$\tan$	Tangent
$=$	Equals	$\cot$	Cotangent
$\neq$	Is not equal to	$\sec$	Secant
$\equiv$	Is identical to	$\csc$	Cosecant
$::$	Equals (in proportion)	vers	Versed sine
$\cong$	$\left.\vphantom{\begin{array}{c}a\\a\end{array}}\right\}$Approximately equals	covers	Coversed sine
$\approx$		$\sin^{-1}a$	$\left.\begin{array}{l}\text{Arc the sine of which}\\ \text{is } a\end{array}\right.$
$>$	Greater than	$\arcsin a$	
$<$	Less than	$(\sin a)^{-1}$	$\begin{cases}\text{Reciprocal of } \sin a\\ 1\div\sin a\end{cases}$
$\geqq$	Greater than or equal to		
$\leqq$	Less than or equal to	$\sin^n x$	nth power of $\sin x$
$\rightarrow$	Approaches as a limit	$\sinh x$	Hyperbolic sine of x
$\propto$	Varies directly as	$\cosh x$	Hyperbolic cosine of x
$\therefore$	Therefore	Δ	Delta (increment of)
$\sqrt{}$	Square root	δ	Delta (variation of)
		d	Differential (in calculus)
$\sqrt[3]{}$	Cube root	∂	$\begin{cases}\text{Partial differentiation}\\ \text{(in calculus)}\end{cases}$
$\sqrt[4]{}$	4th root	$\int$	Integral (in calculus)
$\sqrt[n]{}$	nth root	$\int_b^a$	$\begin{cases}\text{Integral between the}\\ \text{limits } a \text{ and } b\end{cases}$
a^2	a squared (2d power of a)	$!$	$5! = 1\times 2\times 3\times 4\times 5$
a^3	a cubed (3d power of a)	$\angle$	Angle
a^4	4th power of a	$\llcorner$	Right angle
a^n	nth power of a	$\perp$	Perpendicular to
a^{-n}	$1\div a^n$	$\triangle$	Triangle
$\dfrac{1}{n}$	Reciprocal value of n	$\bigcirc$	Circle
		$\square$	Parallelogram
$\log$	Logarithm	$\circ$	$\begin{cases}\text{Degree (circular arc or}\\ \text{temperature)}\end{cases}$
$\log_e$	$\left.\begin{array}{l}\text{Natural or}\\ \text{Napierian logarithm}\end{array}\right.$		
$\ln$		$'$	Minutes or feet
e	$\begin{cases}\text{Base of natural logarithms}\\ (2.71828)\end{cases}$	$''$	Seconds or inches
		a'	a prime
$\lim$	Limit value (of an expression)	a''	a double prime
∞	Infinity	a_1	a sub one
α	Alpha $\left.\vphantom{\begin{array}{c}a\\a\\a\\a\\a\end{array}}\right\}$	a_2	a sub two
β	Beta	a_n	a sub n
γ	Gamma $\quad\begin{array}{l}\text{commonly used}\\ \text{to denote angles}\end{array}$	$(\ )$	Parentheses
θ	Theta	$[\ ]$	Brackets
ϕ	Phi	$\{\ \}$	Braces
μ	Mu (coefficient of friction)		

Prime Numbers and Factors of Numbers

The *factors* of a given number are those numbers which when multiplied together give a product equal to that number; thus, 2 and 3 are factors of 6; and 5 and 7 are factors of 35.

A *prime number* is one which has no factors except itself and 1. Thus, 3, 5, 7, 11, etc., are prime numbers. A factor which is a prime number is called a *prime factor*.

The accompanying "Prime Number and Factor Tables" give the smallest prime factor of all odd numbers from 1 to 9600, and can be used for finding all the factors for numbers up to this limit. For example, find the factors of 931. In the column headed "900," and in the line indicated by "31" in the left-hand column, the smallest prime factor is found to be 7. As this leaves another factor 133 (since $931 \div 7 = 133$), find the smallest prime factor of this number. In the column headed "100" and in the line "33," this is found to be 7, leaving a factor 19. This latter is a prime number; hence, the factors of 931 are $7 \times 7 \times 19$. Where no factor is given for a number in the factor table, it indicates that the number is a prime number.

The last page of the tables lists all prime numbers from 9551 through 18691; and can be used to identify quickly all unfactorable numbers in that range.

For factoring, the following general rules will be found useful:

2 is a factor of any number the right-hand figure of which is an even number or 0. Thus, $28 = 2 \times 14$, and $210 = 2 \times 105$.

3 is a factor of any number the sum of the figures of which is evenly divisible by 3. Thus, 3 is a factor of 1869, because $1 + 8 + 6 + 9 = 24 \div 3 = 8$.

4 is a factor of any number the two right-hand figures of which, considered as one number, are evenly divisible by 4. Thus, 1844 has a factor 4, because $44 \div 4 = 11$.

5 is a factor of any number the right-hand figure of which is 0 or 5. Thus, $85 = 5 \times 17; 70 = 5 \times 14$.

Tables of prime numbers and factors of numbers are particularly useful for calculations involving change-gear ratios for compound gearing, dividing heads, gear-generating machines, and mechanical designs having gear trains.

Example 1: A set of four gears is required in a mechanical design to provide an overall gear ratio of $4104 \div 1200$. Furthermore, no gear in the set is to have more than 120 teeth or less than 24 teeth. Determine the tooth numbers.

First, as explained previously, the factors of 4104 are determined to be: $2 \times 2 \times 2 \times 3 \times 3 \times 57 = 4104$. Next, the factors of 1200 are determined: $2 \times 2 \times 2 \times 2 \times 5 \times 5 \times 3 = 1200$. Therefore, $\dfrac{4104}{1200} = \dfrac{2 \times 2 \times 2 \times 3 \times 3 \times 57}{2 \times 2 \times 2 \times 2 \times 5 \times 5 \times 3} = \dfrac{72 \times 57}{24 \times 50}$. If the factors had been combined differently, say, to give $\dfrac{72 \times 57}{16 \times 75}$, then the 16-tooth gear in the denominator would not satisfy the requirement of no less than 24 teeth.

Example 2: Factor the number 25078 into two numbers neither of which is larger than 200.

The first factor of 25078 is obviously 2, leaving $25078 \div 2 = 12539$ to be factored further. However, from the last table it is seen that 12539 is a prime number; therefore, no solution exists.

Prime Number and Factor Table

From to	0 100	100 200	200 300	300 400	400 500	500 600	600 700	700 800	800 900	900 1000	1000 1100	1100 1200
1	P	P	3	7	P	3	P	P	3	17	7	3
3	P	P	7	3	13	P	3	19	11	3	17	P
5	P	3	5	5	3	5	5	3	5	5	3	5
7	P	P	3	5	11	3	P	7	3	P	19	3
9	3	P	11	3	P	P	3	P	P	3	P	P
11	P	3	P	P	3	7	13	3	P	P	3	11
13	P	P	3	P	7	3	P	23	3	11	P	3
15	3	5	5	3	5	5	3	5	5	3	5	5
17	P	3	7	P	3	11	P	3	19	7	3	P
19	P	7	5	11	P	3	P	P	3	P	P	3
21	3	11	13	3	P	P	3	7	P	3	P	19
23	P	3	P	17	3	P	7	3	P	13	3	P
25	5	5	3	5	5	3	5	5	3	5	5	3
27	3	P	P	3	7	17	3	P	P	3	13	7
29	P	3	P	7	3	23	17	3	P	P	3	P
31	P	P	3	P	P	3	P	17	3	7	P	3
33	3	7	P	3	P	13	3	P	7	3	P	11
35	5	3	5	5	3	5	5	3	5	5	3	5
37	P	P	3	P	19	3	7	11	3	P	17	3
39	3	P	P	3	P	7	3	P	P	3	P	17
41	P	3	P	11	3	P	P	3	29	P	3	7
43	P	11	3	7	P	3	P	P	3	23	7	3
45	3	5	5	3	5	5	3	5	5	3	5	5
47	P	3	13	P	3	P	P	3	7	P	3	31
49	7	P	3	P	P	3	11	7	3	13	P	3
51	3	P	P	3	11	19	3	P	23	3	P	P
53	P	3	11	P	3	7	P	3	P	P	3	P
55	5	5	3	5	5	3	5	5	3	5	5	3
57	3	P	P	3	P	P	3	P	P	3	7	13
59	P	3	7	P	3	13	P	3	P	7	3	19
61	P	7	3	19	P	3	P	P	3	31	P	3
63	3	P	P	3	P	P	3	7	P	3	P	P
65	5	3	5	5	3	5	5	3	5	5	3	5
67	P	P	3	P	P	3	23	13	3	P	11	3
69	3	13	P	3	7	P	3	P	11	3	P	7
71	P	3	P	7	3	P	11	3	13	P	3	P
73	P	P	3	P	11	3	P	P	3	7	29	3
75	3	5	5	3	5	5	3	5	5	3	5	5
77	7	3	P	13	3	P	P	3	P	P	3	11
79	P	P	3	P	P	3	7	19	3	11	13	3
81	3	P	P	3	13	7	3	11	P	3	23	P
83	P	3	P	P	3	11	P	3	P	P	3	7
85	5	5	3	5	5	3	5	5	3	5	5	3
87	3	11	7	3	P	P	3	P	P	3	P	P
89	P	3	17	P	3	19	13	3	7	23	3	29
91	7	P	3	17	P	3	P	7	3	P	P	3
93	3	P	P	3	17	P	3	13	19	3	P	P
95	5	3	5	5	3	5	5	3	5	5	3	5
97	P	P	3	P	7	3	17	P	3	P	P	3
99	3	P	13	3	P	P	3	17	29	3	7	11

Prime Number and Factor Table

From to	1200 1300	1300 1400	1400 1500	1500 1600	1600 1700	1700 1800	1800 1900	1900 2000	2000 2100	2100 2200	2200 2300	2300 2400
1	P	P	3	19	P	3	P	P	3	11	31	3
3	3	P	23	3	7	13	3	11	P	3	P	7
5	5	3	5	5	3	5	5	3	5	5	3	5
7	17	P	3	11	P	3	13	P	3	7	P	3
9	3	7	P	3	P	P	3	23	7	3	47	P
11	7	3	17	P	3	29	P	3	P	P	3	P
13	P	13	3	17	P	3	7	P	3	P	P	3
15	3	5	5	3	5	5	3	5	5	3	5	5
17	P	3	13	37	3	17	23	3	P	29	3	7
19	23	P	3	7	P	3	17	19	3	13	7	3
21	3	P	7	3	P	P	3	17	43	3	P	11
23	P	3	P	P	3	P	P	3	7	11	3	23
25	5	5	3	5	5	3	5	5	3	5	5	3
27	3	P	P	3	P	11	3	41	P	3	17	13
29	P	3	P	11	3	7	31	3	P	P	3	17
31	P	11	3	P	7	3	P	P	3	P	23	3
33	3	31	P	3	23	P	3	P	19	3	7	P
35	5	3	5	5	3	5	5	3	5	5	3	5
37	P	7	3	29	P	3	11	13	3	P	P	3
39	3	13	P	3	11	37	3	7	P	3	P	P
41	17	3	11	23	3	P	7	3	13	P	3	P
43	11	17	3	P	31	3	19	29	3	P	P	3
45	3	5	5	3	5	5	3	5	5	3	5	5
47	29	3	P	7	3	P	P	3	23	19	3	P
49	P	19	3	P	17	3	43	P	3	7	13	3
51	3	7	P	3	13	17	3	P	7	3	P	P
53	7	3	P	P	3	P	17	3	P	P	3	13
55	5	5	3	5	5	3	5	5	3	5	5	3
57	3	23	31	3	P	7	3	19	11	3	37	P
59	P	3	P	P	3	P	11	3	29	17	3	7
61	13	P	3	7	11	3	P	37	3	P	7	3
63	3	29	7	3	P	41	3	13	P	3	31	17
65	5	3	5	5	3	5	5	3	5	5	3	5
67	7	P	3	P	P	3	P	7	3	11	P	3
69	3	37	13	3	P	29	3	11	P	3	P	23
71	31	3	P	P	3	7	P	3	19	13	3	P
73	19	P	3	11	7	3	P	P	3	41	P	3
75	3	5	5	3	5	5	3	5	5	3	5	5
77	P	3	7	19	3	P	P	3	31	7	3	P
79	P	7	3	P	23	3	P	P	3	P	43	3
81	3	P	P	3	41	13	3	7	P	3	P	P
83	P	3	P	P	3	P	7	3	P	37	3	P
85	5	5	3	5	5	3	5	5	3	5	5	3
87	3	19	P	3	7	P	3	P	P	3	P	7
89	P	3	P	7	3	P	P	3	P	11	3	P
91	P	13	3	37	19	3	31	11	3	7	29	3
93	3	7	P	3	P	11	3	P	7	3	P	P
95	5	3	5	5	3	5	5	3	5	5	3	5
97	P	11	3	P	P	3	7	P	3	13	P	3
99	3	P	P	3	P	7	3	P	P	3	11	P

Prime Number and Factor Table

From to	2400 2500	2500 2600	2600 2700	2700 2800	2800 2900	2900 3000	3000 3100	3100 3200	3200 3300	3300 3400	3400 3500	3500 3600
1	7	41	3	37	P	3	P	7	3	P	19	3
3	3	P	19	3	P	P	3	29	P	3	41	31
5	5	3	5	5	3	5	5	5	3	5	3	5
7	29	23	3	P	7	3	31	13	3	P	P	3
9	3	13	P	3	53	P	3	P	P	3	7	11
11	P	3	7	P	3	41	P	3	13	7	3	P
13	19	7	3	P	29	3	23	11	P	P	P	3
15	3	5	5	3	5	5	3	5	5	3	5	5
17	P	3	P	11	3	P	7	3	P	31	3	P
19	41	11	3	P	P	3	P	P	3	P	13	3
21	3	P	P	3	7	23	3	P	P	3	11	7
23	P	3	43	7	3	37	P	3	11	P	3	13
25	5	5	3	5	5	3	5	5	3	5	5	3
27	3	7	37	3	11	P	3	53	7	3	23	P
29	7	3	11	P	3	29	13	3	31	P	3	P
31	11	P	3	P	19	3	7	31	3	P	47	3
33	3	17	P	3	P	7	3	13	53	3	P	P
35	5	3	5	5	3	5	5	3	5	5	3	5
37	P	43	3	7	P	3	P	P	3	47	7	3
39	3	P	7	3	17	P	3	43	41	3	19	P
41	P	3	19	P	3	17	P	3	7	13	3	P
43	7	P	3	13	P	3	17	7	3	P	11	3
45	3	5	5	3	5	5	3	5	5	3	5	5
47	P	3	P	41	3	7	11	3	17	P	3	P
49	31	P	3	P	7	3	P	47	3	17	P	3
51	3	P	11	3	P	13	3	23	P	3	7	53
53	11	3	7	P	3	P	43	3	P	7	3	11
55	5	5	3	5	5	3	5	5	3	5	5	3
57	3	P	P	3	P	P	3	7	P	3	P	P
59	P	3	P	31	3	11	7	3	P	P	3	P
61	23	13	3	11	P	3	P	29	3	P	P	3
63	3	11	P	3	7	P	3	P	13	3	P	7
65	5	3	5	5	3	5	5	3	5	5	3	5
67	P	17	3	P	47	3	P	P	3	7	P	3
69	3	7	17	3	19	P	3	P	7	3	P	43
71	7	3	P	17	3	P	37	3	P	P	3	P
73	P	31	3	47	13	3	7	19	3	P	23	3
75	3	5	5	3	5	5	3	5	5	3	5	5
77	P	3	P	P	3	13	17	3	29	11	3	7
79	37	P	3	7	P	3	P	11	3	31	7	3
81	3	29	7	3	43	11	3	P	17	3	59	P
83	13	3	P	11	3	19	P	3	7	17	3	P
85	5	5	3	5	5	3	5	5	3	5	5	3
87	3	13	P	3	P	29	3	P	19	3	11	17
89	19	3	P	P	3	7	P	3	11	P	3	37
91	47	P	3	P	7	3	11	P	3	P	P	3
93	3	P	P	3	11	41	3	31	37	3	7	P
95	5	3	5	5	3	5	5	3	5	5	3	5
97	11	7	3	P	P	3	19	23	3	43	13	3
99	3	23	P	3	13	P	3	7	P	3	P	59

Prime Number and Factor Table

From to	3600 3700	3700 3800	3800 3900	3900 4000	4000 4100	4100 4200	4200 4300	4300 4400	4400 4500	4500 4600	4600 4700	4700 4800
1	13	P	3	47	P	3	P	11	3	7	43	3
3	3	7	P	3	P	11	3	13	7	3	P	P
5	5	3	5	5	3	5	5	3	5	5	3	5
7	P	11	3	P	P	3	7	59	3	P	17	3
9	3	P	13	3	19	7	3	31	P	3	11	17
11	23	3	37	P	3	P	P	3	11	13	3	7
13	P	47	3	7	P	3	11	19	3	P	7	3
15	3	5	5	3	5	5	3	5	5	3	5	5
17	P	3	11	P	3	23	P	3	7	P	3	53
19	7	P	3	P	P	3	P	7	3	P	31	3
21	3	61	P	3	P	13	3	29	P	P	P	P
23	P	3	P	P	3	7	41	3	P	P	3	P
25	5	5	3	5	5	3	5	5	3	5	5	3
27	3	P	43	3	P	P	3	P	19	3	7	29
29	19	3	7	P	3	P	P	3	43	7	3	P
31	P	7	3	P	29	3	P	61	3	23	11	3
33	3	P	P	3	37	P	3	7	11	3	41	P
35	5	3	5	5	3	5	5	3	5	5	3	5
37	P	37	3	31	11	3	19	P	3	13	P	3
39	3	P	11	3	7	P	3	P	23	3	P	7
41	11	3	23	7	3	41	P	3	P	19	3	11
43	P	19	3	P	13	3	P	43	3	7	P	3
45	3	5	5	3	5	5	3	5	5	3	5	5
47	7	3	P	P	3	11	31	3	P	P	3	47
49	41	23	3	11	P	3	7	P	3	P	P	3
51	3	11	P	3	P	7	3	19	P	3	P	P
53	13	3	P	59	3	P	P	3	61	29	3	7
55	5	5	3	5	5	3	5	5	3	5	5	3
57	3	13	7	3	P	P	3	P	P	3	P	67
59	P	3	17	37	3	P	P	3	7	47	3	P
61	7	P	3	17	31	3	P	7	3	P	59	3
63	3	53	P	3	17	23	3	P	P	3	P	11
65	5	3	5	5	3	5	5	3	5	5	3	5
67	19	P	3	P	7	3	17	11	3	P	13	3
69	3	P	53	3	13	11	3	17	41	3	7	19
71	P	3	7	11	3	43	P	3	17	7	3	13
73	P	7	3	29	P	3	P	P	3	17	P	3
75	3	5	5	3	5	5	3	5	5	3	5	5
77	P	3	P	41	3	P	7	3	11	23	3	17
79	13	P	3	23	P	3	11	29	3	19	P	3
81	3	19	P	3	7	37	3	13	P	3	31	7
83	29	3	11	7	3	47	P	3	P	P	3	P
85	5	5	3	5	5	3	5	5	3	5	5	3
87	3	7	13	3	61	53	3	41	7	3	43	P
89	7	3	P	P	3	59	P	3	67	13	3	P
91	P	17	3	13	P	3	7	P	3	P	P	3
93	3	P	17	3	P	7	3	23	P	3	13	P
95	5	3	5	5	3	5	5	3	5	5	3	5
97	P	P	3	7	17	3	P	P	3	P	7	3
99	3	29	7	3	P	13	3	53	11	3	37	P

Prime Number and Factor Table

From to	4800–4900	4900–5000	5000–5100	5100–5200	5200–5300	5300–5400	5400–5500	5500–5600	5600–5700	5700–5800	5800–5900	5900–6000
1	P	13	3	P	7	3	11	P	3	P	P	3
3	3	P	P	3	11	P	3	P	13	3	7	P
5	5	3	5	5	3	5	5	3	5	5	3	5
7	11	7	3	P	41	3	P	P	3	13	P	3
9	3	P	P	3	P	P	3	7	71	3	37	19
11	17	3	P	19	3	47	7	3	31	P	3	23
13	P	17	3	P	13	3	P	37	3	29	P	3
15	3	5	5	3	5	5	3	5	5	3	5	5
17	P	3	29	7	3	13	P	3	41	P	3	61
19	61	P	3	P	17	3	P	P	3	7	11	3
21	3	7	P	3	23	17	3	P	7	3	P	31
23	7	3	P	47	3	P	11	3	P	59	3	P
25	5	5	3	5	5	3	5	5	3	5	5	3
27	3	13	11	3	P	7	3	P	17	3	P	P
29	11	3	47	23	3	73	61	3	13	17	3	7
31	P	P	3	7	P	3	P	P	3	11	7	3
33	3	P	7	3	P	P	3	11	43	3	19	17
35	5	3	5	5	3	5	5	3	5	5	3	5
37	7	P	3	11	P	3	P	7	3	P	13	3
39	3	11	P	3	13	19	3	29	P	3	P	P
41	47	3	71	53	3	7	P	3	P	P	3	13
43	29	P	3	37	7	3	P	23	3	P	P	3
45	3	5	5	3	5	5	3	5	5	3	5	5
47	37	3	P	P	3	P	13	3	P	7	3	19
49	13	7	3	19	29	3	P	31	3	P	P	3
51	3	P	P	3	59	P	3	7	P	3	P	11
53	23	3	31	P	3	53	7	3	P	11	3	P
55	5	5	3	5	5	3	5	5	3	5	5	3
57	3	P	13	3	7	11	3	P	P	3	P	7
59	43	3	P	7	3	23	53	3	P	13	3	59
61	P	11	3	13	P	3	43	67	3	7	P	3
63	3	7	61	3	19	31	3	P	7	3	11	67
65	5	5	3	5	5	3	5	5	3	5	5	3
67	31	P	3	P	23	3	7	19	3	73	P	3
69	3	P	37	3	11	7	3	P	P	3	P	47
71	P	3	11	P	3	41	P	3	53	29	3	7
73	11	P	3	7	P	3	13	P	3	23	7	3
75	3	5	5	3	5	5	3	5	5	3	5	5
77	P	3	P	31	3	19	P	3	7	53	3	43
79	7	13	3	P	P	3	P	7	3	P	P	3
81	3	17	P	3	P	P	3	P	13	3	P	P
83	19	3	13	71	3	7	P	3	P	P	3	31
85	5	5	3	5	5	3	5	5	3	5	5	3
87	3	P	P	3	17	P	3	37	11	3	7	P
89	P	3	7	P	3	17	11	3	P	7	3	53
91	67	7	3	29	11	3	17	P	3	P	43	3
93	3	P	11	3	67	P	3	7	P	3	71	13
95	5	3	5	5	3	5	5	3	5	5	3	5
97	59	19	3	P	P	3	23	29	3	11	P	3
99	3	P	P	3	7	P	3	11	41	3	17	7

Prime Number and Factor Table

From to	6000 6100	6100 6200	6200 6300	6300 6400	6400 6500	6500 6600	6600 6700	6700 6800	6800 6900	6900 7000	7000 7100	7100 7200
1	17	P	3	P	37	3	7	P	3	67	P	3
3	3	17	P	3	19	7	3	P	P	3	47	P
5	5	3	5	5	3	5	5	3	5	5	3	5
7	P	31	3	7	43	3	P	19	3	P	7	3
9	3	41	7	3	13	23	3	P	11	3	43	P
11	P	3	P	P	3	17	11	3	7	P	3	13
13	7	P	3	59	11	3	17	7	3	31	P	3
15	3	5	5	3	5	5	3	5	5	3	5	5
17	11	3	P	P	3	7	13	3	17	P	3	11
19	13	29	3	71	7	3	P	P	3	11	P	3
21	3	P	P	3	P	P	3	11	19	3	7	P
23	19	3	7	P	3	11	37	3	P	7	3	17
25	5	5	3	5	5	3	5	5	3	5	5	3
27	3	11	13	3	P	61	3	7	P	3	P	P
29	P	3	P	P	3	P	7	3	P	13	3	P
31	37	P	3	13	59	3	19	53	3	29	79	3
33	3	P	23	3	7	47	3	P	P	3	13	7
35	5	3	5	5	3	5	5	3	5	5	3	5
37	P	17	3	P	41	3	P	P	3	7	31	3
39	3	7	17	3	47	13	3	23	7	3	P	11
41	7	3	79	17	3	P	29	3	P	11	3	37
43	P	P	3	P	17	3	7	11	3	53	P	3
45	3	5	5	3	5	5	3	5	5	3	5	5
47	P	3	P	11	3	P	17	3	41	P	3	7
49	23	11	3	7	P	3	61	17	3	P	7	3
51	3	P	7	3	P	P	3	43	13	3	11	P
53	P	3	13	P	3	P	P	3	7	17	3	23
55	5	5	3	5	5	3	5	5	3	5	5	3
57	3	47	P	3	11	79	3	29	P	3	P	17
59	73	3	11	P	3	7	P	3	19	P	3	P
61	11	61	3	P	7	3	P	P	3	P	23	3
63	3	P	P	3	23	P	3	P	P	3	7	13
65	5	3	5	5	3	5	5	3	5	5	3	5
67	P	7	3	P	29	3	59	67	3	P	37	3
69	3	31	P	3	P	P	3	7	P	3	P	67
71	13	3	P	23	3	P	7	3	P	P	3	71
73	P	P	3	P	P	3	P	13	3	19	11	3
75	3	5	5	3	5	5	3	5	5	3	5	5
77	59	3	P	7	3	P	11	3	13	P	3	P
79	P	37	3	P	11	3	P	P	3	7	P	3
81	3	7	11	3	P	P	3	P	7	3	73	43
83	7	3	61	13	3	29	41	3	P	P	3	11
85	5	5	3	5	5	3	5	5	3	5	5	3
87	3	23	P	3	13	7	3	11	71	3	19	P
89	P	3	19	P	3	11	P	3	83	29	3	7
91	P	41	3	7	P	3	P	P	3	P	7	3
93	3	11	7	3	43	19	3	P	61	3	41	P
95	5	3	5	5	3	5	5	3	5	5	3	5
97	7	P	3	P	73	3	37	7	3	P	47	3
99	3	P	P	3	67	P	3	13	P	3	31	23

Prime Number and Factor Table

From to	7200 7300	7300 7400	7400 7500	7500 7600	7600 7700	7700 7800	7800 7900	7900 8000	8000 8100	8100 8200	8200 8300	8300 8400
1	19	7	3	13	11	3	29	P	3	P	59	3
3	3	67	11	3	P	P	3	7	53	3	13	19
5	5	3	5	5	3	5	5	3	5	5	3	5
7	P	P	3	P	P	3	37	P	3	11	29	3
9	3	P	31	3	7	13	3	11	P	3	P	7
11	P	3	P	7	3	11	73	3	P	P	3	P
13	P	71	3	11	23	3	13	41	3	7	43	3
15	3	5	5	3	5	5	3	5	5	3	5	5
17	7	3	P	P	3	P	P	3	P	P	3	P
19	P	13	3	73	19	3	7	P	3	23	P	3
21	3	P	41	3	P	7	3	89	13	3	P	53
23	31	3	13	P	3	P	P	3	71	P	3	7
25	5	5	3	5	5	3	5	5	3	5	5	3
27	3	17	7	3	29	P	3	P	23	3	19	11
29	P	3	17	P	3	59	P	3	7	11	3	P
31	7	P	3	17	13	3	41	7	3	47	P	3
33	3	P	P	3	17	11	3	P	29	3	P	13
35	5	3	5	5	3	5	5	3	5	5	3	5
37	P	11	3	P	7	3	17	P	3	79	P	3
39	3	41	43	3	P	71	3	17	P	3	7	31
41	13	3	7	P	3	P	P	3	11	7	3	19
43	P	7	3	19	P	3	11	13	3	17	P	3
45	3	5	5	3	5	5	3	5	5	3	5	5
47	P	3	11	P	3	61	7	3	13	P	3	17
49	11	P	3	P	P	3	47	P	3	29	73	3
51	3	P	P	3	7	23	3	P	83	3	37	P
53	P	3	29	7	3	P	P	3	P	31	3	P
55	5	5	3	5	5	3	5	5	3	5	5	3
57	3	7	P	3	13	P	3	73	7	3	23	61
59	7	3	P	P	3	P	29	3	P	41	3	13
61	53	17	3	P	47	3	7	19	3	P	11	3
63	3	37	17	3	79	7	3	P	11	3	P	P
65	5	3	5	5	3	5	5	3	5	5	3	5
67	13	53	3	7	11	3	P	31	3	P	7	3
69	3	P	7	3	P	17	3	13	P	3	P	P
71	11	3	31	67	3	19	17	3	7	P	3	11
73	7	73	3	P	P	3	P	7	3	11	P	3
75	3	5	5	3	5	5	3	5	5	3	5	5
77	19	3	P	P	3	7	P	3	41	13	3	P
79	29	47	3	11	7	3	P	79	3	P	17	3
81	3	11	P	3	P	31	3	23	P	3	7	17
83	P	3	7	P	3	43	P	3	59	7	3	83
85	5	5	3	5	5	3	5	5	3	5	5	3
87	3	83	P	3	P	13	3	7	P	3	P	P
89	37	3	P	P	3	P	7	3	P	19	3	P
91	23	19	3	P	P	3	13	61	3	P	P	3
93	3	P	59	3	7	P	3	P	P	3	P	7
95	5	3	5	5	3	5	5	3	5	5	3	5
97	P	13	3	71	43	3	53	11	3	7	P	3
99	3	7	P	3	P	11	3	19	7	3	43	37

Prime Number and Factor Table

From to	8400 8500	8500 8600	8600 8700	8700 8800	8800 8900	8900 9000	9000 9100	9100 9200	9200 9300	9300 9400	9400 9500	9500 9600
1	31	P	3	7	13	3	P	19	3	71	7	3
3	3	11	7	3	P	29	3	P	P	P	P	13
5	5	3	5	5	3	5	5	3	5	5	3	5
7	7	47	3	P	P	3	P	7	3	41	23	3
9	3	67	P	3	23	59	3	P	P	3	97	37
11	13	3	79	31	3	7	P	3	61	P	3	P
13	47	P	3	P	7	3	P	13	3	67	P	3
15	3	5	5	3	5	5	3	5	5	3	5	5
17	19	3	7	23	3	37	71	3	13	7	3	31
19	P	7	P	P	P	3	29	11	3	P	P	3
21	3	P	37	3	P	11	3	7	P	3	P	P
23	P	3	P	11	3	P	7	3	23	P	3	89
25	5	5	3	5	5	3	5	5	3	5	5	3
27	3	P	P	3	7	79	3	P	P	3	11	7
29	P	3	P	7	3	P	P	3	11	19	3	13
31	P	19	3	P	P	3	11	23	3	7	P	3
33	3	7	89	3	11	P	3	P	7	3	P	P
35	5	3	5	5	3	5	5	3	5	5	3	5
37	11	P	3	P	P	3	7	P	3	P	P	3
39	3	P	53	3	P	7	3	13	P	3	P	P
41	23	3	P	P	3	P	P	3	P	P	3	7
43	P	P	3	7	37	3	P	41	3	P	7	3
45	3	5	5	3	5	5	3	5	5	3	5	5
47	P	3	P	P	3	23	83	3	7	13	3	P
49	7	83	3	13	P	3	P	7	3	P	11	3
51	3	17	41	3	53	P	3	P	11	3	13	P
53	79	3	17	P	3	7	11	3	19	47	3	41
55	5	5	3	5	5	3	5	5	3	5	5	3
57	3	43	11	3	17	13	3	P	P	3	7	19
59	11	3	7	19	3	17	P	3	47	7	3	11
61	P	7	3	P	P	3	13	P	3	11	P	3
63	3	P	P	3	P	P	3	7	59	3	P	73
65	5	3	5	5	3	5	5	3	5	5	3	5
67	P	13	3	11	P	3	P	89	3	17	P	3
69	3	11	P	3	7	P	3	53	13	3	17	7
71	43	3	13	7	3	P	47	3	73	P	3	17
73	37	P	3	31	19	3	43	P	3	7	P	3
75	3	5	5	3	5	5	3	5	5	3	5	5
77	7	3	P	67	3	47	29	3	P	P	3	61
79	61	23	3	P	13	3	7	67	3	83	P	3
81	3	P	P	3	83	7	3	P	P	3	19	11
83	17	3	19	P	3	13	31	3	P	11	3	7
85	5	5	3	5	5	3	5	5	3	5	5	3
87	3	31	7	3	P	11	3	P	37	3	53	P
89	13	3	P	11	3	89	61	3	7	41	3	43
91	7	11	3	59	17	3	P	7	3	P	P	3
93	3	13	P	3	P	17	3	29	P	3	11	53
95	5	3	5	5	3	5	5	3	5	5	3	5
97	29	P	3	19	7	3	11	17	3	P	P	3
99	3	P	P	3	11	P	3	P	17	3	7	29

Prime Numbers from 9551 to 18691

9551	10181	10853	11497	12157	12763	13417	14071	14747	15361	16001	16693	17387	18043
9587	10193	10859	11503	12161	12781	13421	14081	14753	15373	16007	16699	17389	18047
9601	10211	10861	11519	12163	12791	13441	14083	14759	15377	16033	16703	17393	18049
9613	10223	10867	11527	12197	12799	13451	14087	14767	15383	16057	16729	17401	18059
9619	10243	10883	11549	12203	12809	13457	14107	14771	15391	16061	16741	17417	18061
9623	10247	10889	11551	12211	12821	13463	14143	14779	15401	16063	16747	17419	18077
9629	10253	10891	11579	12227	12823	13469	14149	14783	15413	16067	16759	17431	18089
9631	10259	10903	11587	12239	12829	13477	14153	14797	15427	16069	16763	17443	18097
9643	10267	10909	11593	12241	12841	13487	14159	14813	15439	16073	16787	17449	18119
9649	10271	10937	11597	12251	12853	13499	14173	14821	15443	16087	16811	17467	18121
9661	10273	10939	11617	12253	12889	13513	14177	14827	15451	16091	16823	17471	18127
9677	10289	10949	11621	12263	12893	13523	14197	14831	15461	16097	16829	17477	18131
9679	10301	10957	11633	12269	12899	13537	14207	14843	15467	16103	16831	17483	18133
9689	10303	10973	11657	12277	12907	13553	14221	14851	15473	16111	16843	17489	18143
9697	10313	10979	11677	12281	12911	13567	14243	14867	15493	16127	16871	17491	18149
9719	10321	10987	11681	12289	12917	13577	14249	14869	15497	16139	16879	17497	18169
9721	10331	10993	11689	12301	12919	13591	14251	14879	15511	16141	16883	17509	18181
9733	10333	11003	11699	12323	12923	13597	14281	14887	15527	16183	16889	17519	18191
9739	10337	11027	11701	12329	12941	13613	14293	14891	15541	16187	16901	17539	18199
9743	10343	11047	11717	12343	12953	13619	14303	14897	15551	16189	16903	17551	18211
9749	10357	11057	11719	12347	12959	13627	14321	14923	15559	16193	16921	17569	18217
9767	10369	11059	11731	12373	12967	13633	14323	14929	15569	16217	16927	17573	18223
9769	10391	11069	11743	12377	12973	13649	14327	14939	15581	16223	16931	17579	18229
9781	10399	11071	11777	12379	12979	13669	14341	14947	15583	16229	16937	17581	18233
9787	10427	11083	11779	12391	12983	13679	14347	14951	15601	16231	16943	17597	18251
9791	10429	11087	11783	12401	13001	13681	14369	14957	15607	16249	16963	17599	18253
9803	10433	11093	11789	12409	13003	13687	14387	14969	15619	16253	16979	17609	18257
9811	10453	11113	11801	12413	13007	13691	14389	14983	15629	16267	16981	17623	18269
9817	10457	11117	11807	12421	13009	13693	14401	15013	15641	16273	16987	17627	18287
9829	10459	11119	11813	12433	13033	13697	14407	15017	15643	16301	16993	17657	18289
9833	10463	11131	11821	12437	13037	13709	14411	15031	15647	16319	17011	17659	18301
9839	10477	11149	11827	12451	13043	13711	14419	15053	15649	16333	17021	17669	18307
9851	10487	11159	11831	12457	13049	13721	14423	15061	15661	16339	17027	17681	18311
9857	10499	11161	11833	12473	13063	13723	14431	15073	15667	16349	17029	17683	18313
9859	10501	11171	11839	12479	13093	13729	14437	15077	15671	16361	17033	17707	18329
9871	10513	11173	11863	12487	13099	13751	14447	15083	15679	16363	17041	17711	18341
9883	10529	11177	11867	12491	13103	13757	14449	15091	15683	16369	17047	17729	18353
9887	10531	11197	11887	12497	13109	13759	14461	15101	15727	16381	17053	17737	18367
9901	10559	11213	11897	12503	13121	13763	14479	15107	15731	16411	17077	17747	18371
9907	10567	11239	11903	12511	13127	13781	14489	15121	15733	16417	17093	17749	18379
9923	10589	11243	11909	12517	13147	13789	14503	15131	15737	16421	17099	17761	18397
9929	10597	11251	11923	12527	13151	13799	14519	15137	15739	16427	17107	17783	18401
9931	10601	11257	11927	12539	13159	13807	14533	15139	15749	16433	17117	17789	18413
9941	10607	11261	11933	12541	13163	13829	14537	15149	15761	16447	17123	17791	18427
9949	10613	11273	11939	12547	13171	13831	14543	15161	15767	16451	17137	17807	18433
9967	10627	11279	11941	12553	13177	13841	14549	15173	15773	16453	17159	17827	18439
9973	10631	11287	11953	12569	13183	13859	14551	15187	15787	16477	17167	17837	18443
10007	10639	11299	11959	12577	13187	13873	14557	15193	15791	16481	17183	17839	18451
10009	10651	11311	11969	12583	13217	13877	14561	15199	15797	16487	17189	17851	18457
10037	10657	11317	11971	12589	13219	13879	14563	15217	15803	16493	17191	17863	18461
10039	10663	11321	11981	12601	13229	13883	14591	15227	15809	16519	17203	17881	18481
10061	10667	11329	11987	12611	13241	13901	14593	15233	15817	16529	17207	17891	18493
10067	10687	11351	12007	12613	13249	13903	14621	15241	15823	16547	17209	17903	18503
10069	10691	11353	12011	12619	13259	13907	14627	15259	15859	16553	17231	17909	18517
10079	10709	11369	12037	12637	13267	13913	14629	15263	15877	16561	17239	17911	18521
10091	10711	11383	12041	12641	13291	13921	14633	15269	15881	16567	17257	17921	18523
10093	10723	11393	12043	12647	13297	13931	14639	15271	15887	16573	17291	17923	18539
10099	10729	11399	12049	12653	13309	13933	14653	15277	15889	16603	17293	17929	18541
10103	10733	11411	12071	12659	13313	13963	14657	15287	15901	16607	17299	17939	18553
10111	10739	11423	12073	12671	13327	13967	14669	15289	15907	16619	17317	17957	18583
10133	10753	11437	12097	12689	13331	13997	14683	15299	15913	16631	17321	17959	18587
10139	10771	11443	12101	12697	13337	13999	14699	15307	15919	16633	17327	17971	18593
10141	10781	11447	12107	12703	13339	14009	14713	15313	15923	16649	17333	17977	18617
10151	10789	11467	12109	12713	13367	14011	14717	15319	15937	16651	17341	17981	18637
10159	10799	11471	12113	12721	13381	14029	14723	15329	15959	16657	17351	17987	18661
10163	10831	11483	12119	12739	13397	14033	14731	15331	15971	16661	17359	17989	18671
10169	10837	11489	12143	12743	13399	14051	14737	15349	15973	16673	17377	18013	18679
10177	10847	11491	12149	12757	13411	14057	14741	15359	15991	16691	17383	18041	18691

Continued Fractions. — In dealing with a cumbersome fraction, or one which does not have satisfactory factors, it may be possible to substitute some other, approximately equal, fraction which is simpler or which can be factored satisfactorily. Continued fractions provide a means of computing a series of fractions each of which is a closer approximation to the original fraction than the one preceding it in the series.

A continued fraction is a proper fraction (one whose numerator is smaller than its denominator) expressed in the form

$$\frac{N}{D} = \cfrac{1}{D_1 + \cfrac{1}{D_2 + \cfrac{1}{D_3 + \dots}}}$$

It is convenient to write the above expression as

$$\frac{N}{D} = \frac{1}{D_1} + \frac{1}{D_2} + \frac{1}{D_3} + \frac{1}{D_4} + \dots$$

The continued fraction is produced from a proper fraction N/D by dividing the numerator N both into itself and into the denominator D. Dividing the numerator into itself gives a result of 1; dividing the numerator into the denominator gives a whole number D_1 plus a remainder fraction R_1. The process is then repeated on the remainder fraction R_1 to obtain D_2 and R_2; then D_3, R_3; etc. until a remainder of zero results. As an example, using $N/D = 2153/9277$,

$$\frac{2153}{9277} = \frac{2153 \div 2153}{9277 \div 2153} = \frac{1}{4 + \cfrac{665}{2153}} = \frac{1}{D_1 + R_1};$$

$$R_1 = \frac{665}{2153} = \frac{1}{3 + \cfrac{158}{665}} = \frac{1}{D_2 + R_2}; \text{ etc.};$$

from which it may be seen that, $D_1 = 4$, $R_1 = 665/2153$; $D_2 = 3$, $R_2 = 158/665$; and, continuing as was explained previously, it would be found that: $D_3 = 4$, $R_3 = 33/158$; . . . ; $D_9 = 2$, $R_9 = 0$. The complete set of continued fraction elements representing 2153/9277 may then be written as

$$\frac{2153}{9277} = \underset{D_1 \dots\dots\dots D_5 \dots\dots\dots D_9}{\frac{1}{4} + \frac{1}{3} + \frac{1}{4} + \frac{1}{4} + \frac{1}{1} + \frac{1}{3} + \frac{1}{1} + \frac{1}{2} + \frac{1}{2}}$$

By following a simple procedure, together with a table organized similar to the one below for the fraction 2153/9277, the denominators D_1, D_2, . . . of the elements of a continued fraction may be used to calculate a series of fractions each of which is a successively closer approximation, called a *convergent,* to the original fraction N/D.

1. The first row of the table contains column numbers numbered from 1 through 2 plus the number of elements, $2 + 9 = 11$ in this example;

2. The second row contains the denominators of the continued fraction elements in sequence but beginning in column 3 instead of column 1 because columns 1 and 2 must be blank in this procedure;

3. The third row contains the convergents to the original fraction as they are calculated and entered. Note that the fractions 1/0 and 0/1 have been inserted into columns 1 and 2. These are two arbitrary convergents, the first equal to infinity, the second to zero, which are used to facilitate the calculations;

4. The convergent in column 3 is now calculated. To find the numerator, multiply the denominator in column 3 by the numerator of the convergent in column 2 and add the numerator of the convergent in column 1. Thus, $4 \times 0 + 1 = 1$;

5. The denominator of the convergent in column 3 is found by multiplying the denominator in column 3 by the denominator of the convergent in column 2 and adding the denominator of the convergent in column 1. Thus, $4 \times 1 + 0 = 4$, and the convergent in column 3 is then $\frac{1}{4}$ as shown in the table; and

6. Finding the remaining successive convergents can be reduced to using the simple equation

$$\text{CONVERGENT}_n = \frac{(D_n)(\text{NUM}_{n-1}) + \text{NUM}_{n-2}}{(D_n)(\text{DEN}_{n-1}) + \text{DEN}_{n-2}}$$

in which n = column number in the table; D_n = denominator in column n; NUM_{n-1} and NUM_{n-2} are numerators and DEN_{n-1} and DEN_{n-2} are denominators of the convergents in the columns indicated by their subscripts; and CONVERGENT_n is the convergent in column n.

Convergents of the Continued Fraction for 2153/9277

Column Number, n	1	2	3	4	5	6	7	8	9	10	11
Denominator, D_n	—	—	4	3	4	4	1	3	1	2	2
Convergent$_n$	$\frac{1}{0}$	$\frac{0}{1}$	$\frac{1}{4}$	$\frac{3}{13}$	$\frac{13}{56}$	$\frac{55}{237}$	$\frac{68}{293}$	$\frac{259}{1116}$	$\frac{327}{1409}$	$\frac{913}{3934}$	$\frac{2153}{9277}$

Notes: The decimal values of the successive convergents in the table are alternately larger and smaller than the value of the original fraction 2153/9277. If the last convergent in the table has the same value as the original fraction 2153/9277, then *all* of the other calculated convergents are correct.

Conjugate Fractions. — In addition to finding approximate ratios by the use of continued fractions and logarithms of ratios, conjugate fractions may be used for the same purpose, independently, or in combination with the other methods.

Two fractions a/b and c/d are said to be conjugate if $ad - bc = \pm 1$. Examples of such pairs are: 0/1 and 1/1; 1/2 and 1/1; and 9/10 and 8/9. Also, *every successive pair of the convergents of a continued fraction are conjugate*. Conjugate fractions have certain properties that are useful for solving ratio problems:

1. No fraction between two conjugate fractions a/b and c/d can have a denominator smaller than either b or d;

2. A new fraction, e/f, conjugate to both fractions of a given pair of conjugate fractions, a/b and c/d, and lying between them, may be created by adding respective numerators, $a + c$, and denominators, $b + d$, so that $e/f = (a + d)/(b + d)$; and

3. The denominator $f = b + d$ of the new fraction e/f is the smallest of any possible fraction lying between a/b and c/d. Thus, 17/19 is conjugate to both 8/9 and 9/10 and no fraction with denominator smaller than 19 lies between them. This property is important if it is desired to minimize the size of the factors of the ratio to be found.

The following example shows the steps to approximate a ratio for a set of gears to any desired degree of accuracy within the limits established for the allowable size of the factors in the ratio.

Example: Find a set of four change gears, ab/cd, to approximate the ratio 2.105399 accurate to within ± 0.0001; no gear is to have more than 120 teeth.

Step 1. Convert the given ratio R to a number r between 0 and 1 by taking its reciprocal: $1/R = 1/2.105399 = 0.4749693 = r$.

Step 2. Select a pair of conjugate fractions a/b and c/d that bracket r. The pair a/b = 0/1 and c/d = 1/1, for example, will bracket 0.4749693.

Step 3. Add the respective numerators and denominators of the conjugates 0/1 and 1/1 to create a new conjugate e/f between 0 and 1: $e/f = (a + c)/(b + d) = (0 + 1)/(1 + 1) = 1/2$.

Step 4. Since 0.4749693 lies between 0/1 and 1/2, e/f must also be between 0/1 and 1/2: $e/f = (0 + 1)/(1 + 2) = 1/3$.

Step 5. Since 0.4749693 now lies between 1/3 and 1/2, e/f must also be between 1/3 and 1/2: $e/f = (1 + 1)/(3 + 2) = 2/5$.

Step 6. Continuing as above to obtain successively closer approximations of e/f to 0.4749693, and using a handheld calculator and a scratch pad to facilitate the process, the fractions below, each of which has factors less than 120, were determined:

Fraction	Numerator Factors	Denominator Factors	Error
19/40	19	$2 \times 2 \times 2 \times 5$	+ .000031
28/59	$2 \times 2 \times 7$	59	− .00039
47/99	47	$3 \times 3 \times 11$	− .00022
104/219	3×41	7×37	− .000066
142/299	2×71	13×23	− .000053
161/339	7×23	3×113	− .000043
218/459	2×109	$3 \times 3 \times 3 \times 17$	− .000024
256/539	$2 \times 2 \times 2 \times 2 \times 2 \times 2 \times 2 \times 2$	$7 \times 7 \times 11$	− .000016
370/779	$2 \times 5 \times 37$	19×41	− .0000014
759/1598	$3 \times 11 \times 23$	$2 \times 17 \times 47$	− .00000059

Factors for the numerators and denominators of the fractions shown above were found with the aid of the Prime Numbers and Factors tables beginning on page 3. Since in Step 1 the desired ratio of 2.105399 was converted to its reciprocal 0.4749693, all of the above fractions should be inverted. Note also that the last fraction, 759/1598, when inverted to become 1598/759, is in error from the desired value by approximately one-half the amount in the example worked out with *Logarithms of Gear Ratios* on page 1677.

Using Continued Fraction Convergents As Conjugates. — Since successive convergents of a continued fraction are also conjugate, they may be used to find a series of additional fractions in between themselves. As an example, the successive convergents 55/237 and 68/293 from the table of convergents for 2153/9277 on page 13 will be used to demonstrate the process for finding the first few in-between ratios.

Step 1. Check the convergents for conjugateness: $55 \times 293 − 237 \times 68 = 16115 − 16116 = − 1$ proving the pair to be conjugate.

Step 2. Set up a table as shown on the next page. The leftmost column of line (1) contains the convergent of lowest value, a/b; the rightmost the higher value, c/d; and the center column the derived value e/f found by adding the respective numerators and denominators of a/b and c/d. The error or difference between e/f and the desired value N/D, error $= N/D − e/f$, is also shown.

Step 3. On line (2) the process used on line (1) is repeated with the e/f value from line (1) becoming the new value of a/b while the c/d value remains unchanged. Had the error in e/f been + instead of −, then e/f would have been the new c/d value and a/b would be unchanged.

Step 4. The process is continued until, as seen on line (4), the error changes sign to + from the previous −. When this occurs, the e/f value becomes the c/d value on the next line instead of a/b as previously and the a/b value remains unchanged.

Desired Fraction $N/D = 2153/9277 = 0.2320793$

a/b	e/f	c/d
(1) 55/237 = .2320675	*123/530 = .2320755 error = − .0000039	68/293 = .2320819
(2) 123/530 = .2320755	191/823 = .2320778 error = − .0000016	68/293 = .2320819
(3) 191/823 = .2320778	*259/1116 = .2320789 error = − .0000005	68/293 = .2320819
(4) 259/1116 = .2320789	327/1409 = .2320795 error = + .0000002	68/293 = .2320819
(5) 259/1116 = .2320789	586/2525 = .2320792 error = − .0000001	327/1409 = .2320795
(6) 586/2525 = .2320792	913/3934 = .2320793 error = − .0000000	327/1409 = .2320795

* Only these ratios had suitable factors below 120.

Positive and Negative Numbers

The degrees on a thermometer scale extending upward from the zero point may be called *positive* and may be preceded by a plus sign; thus +5 degrees means 5 degrees above zero. The degrees below zero may be called *negative* and may be preceded by a minus sign; thus −5 degrees means 5 degrees below zero. In the same way, the ordinary numbers 1, 2, 3, etc., which are larger than 0, are called positive numbers; but numbers can be conceived of as extending in the other direction from 0, numbers that, in fact, are less than 0, and these are called negative. As these numbers must be expressed by the same figures as the positive numbers they are designated by a minus sign placed before them, thus: (−3). A negative number should always be enclosed within parentheses whenever it is written in line with other numbers; for example: $17 + (−13) − 3 × (−0.76)$.

Negative numbers are most commonly met with in the use of logarithms and natural trigonometric functions. The following rules govern calculations with negative numbers.

A negative number can be added to a positive number by subtracting its numerical value from the positive number.

Example: $4 + (−3) = 4 − 3 = 1.$

A negative number can be subtracted from a positive number by adding its numerical value to the positive number.

Example: $4 − (−3) = 4 + 3 = 7.$

A negative number can be added to a negative number by adding the numerical values and making the sum negative.

Example: $(−4) + (−3) = −7.$

A negative number can be subtracted from a larger negative number by subtracting the numerical values and making the difference negative.

Example: $(−4) − (−3) = −1.$

A negative number can be subtracted from a smaller negative number by subtracting the numerical values and making the difference positive.

Example: $(−3) − (−4) = 1.$

If in a subtraction the number to be subtracted is larger than the number from which it is to be subtracted, the calculation can be carried out by subtracting the smaller number from the larger, and indicating that the remainder is negative.

Example: $3 − 5 = −(5 − 3) = −2.$

When a positive number is to be multiplied or divided by a negative number, multiply or divide the numerical values as usual; the product or quotient, respec-

tively, is negative. The same rule is true if a negative number is multiplied or divided by a positive number.

Examples:
$$4 \times (-3) = -12; \quad (-4) \times 3 = -12;$$
$$15 \div (-3) = -5; \quad (-15) \div 3 = -5.$$

When two negative numbers are to be multiplied by each other, the product is positive. When a negative number is divided by a negative number, the quotient is positive.

Examples:
$$(-4) \times (-3) = 12; \quad (-4) \div (-3) = 1.333.$$

The two last rules are often expressed for memorizing as follows: "Equal signs make plus, unequal signs make minus."

Powers, Roots, and Reciprocals

The *square* of a number (or quantity) is the product of that number multiplied by itself. Thus, the square of 9 is $9 \times 9 = 81$. The square of a number is indicated by the *exponent* (^{2}), thus: $9^2 = 9 \times 9 = 81$.

The *cube* or *third power* of a number is the product obtained by using that number as a factor three times. Thus, the cube of 4 is $4 \times 4 \times 4 = 64$, and is written 4^3.

If a number is used as a factor four or five times, respectively, the product is the fourth or fifth power. Thus $3^4 = 3 \times 3 \times 3 \times 3 = 81$, and $2^5 = 2 \times 2 \times 2 \times 2 \times 2 = 32$. A number can be raised to any power by using it as a factor the required number of times.

The *square root* of a given number is that number which, when multiplied by itself, will give a product equal to the given number. The square root of 16 (written $\sqrt{16}$) equals 4, because $4 \times 4 = 16$.

The *cube root* of a given number is that number which, when used as a factor three times, will give a product equal to the given number. Thus, the cube root of 64 (written $\sqrt[3]{64}$) equals 4, because $4 \times 4 \times 4 = 64$.

The fourth, fifth, etc., roots of a given number are those numbers which when used as factors four, five, etc., times, will give as a product the given number. Thus $\sqrt[4]{16} = 2$, because $2 \times 2 \times 2 \times 2 = 16$.

In some formulas there may be such expressions as $(a^2)^3$ and $a^{3/2}$. The first of these, $(a^2)^3$, means that the number a is first to be squared, a^2, and the result then cubed to give a^6. Thus, $(a^2)^3$ is equivalent to a^6 which is obtained by *multiplying* the exponents 2 and 3. Similarly, $a^{3/2}$ may be interpreted as the cube of the square root of a, $(\sqrt{a})^3$, or $(a^{1/2})^3$, so that, for example, $16^{3/2} = (\sqrt{16})^3 = 64$.

The multiplications required for raising numbers to powers and the extracting of roots are greatly facilitated by the use of logarithms. The extracting of the square root and cube root by the regular arithmetical methods is a slow and cumbersome operation, and any roots can be more rapidly found by using logarithms.

When the power to which a number is to be raised is not an integer, say, 1.62, then the use of either logarithms or a scientific calculator becomes the only practical means of solution.

The *reciprocal R* of a number N is obtained by dividing 1 by the number; $R = 1/N$. Reciprocals are useful in some calculations because they avoid the use of negative characteristics as in calculations with logarithms. The tables of Logarithms of Gear Ratios were compiled by using the reciprocals of gear ratios to simplify entering the tables. For example, if the gear ratio is 0.45, the logarithm of the reciprocal, $1/0.45 = 2.22222$, is 0.34679 and this logarithm is found next to the ratio 100:45 in the table on page 1686.

Powers of Ten Notation

Powers of ten notation is used to simplify calculations and insure accuracy, particularly with respect to the position of decimal points, and also simplifies the expression of numbers which are so large or so small as to be unwieldy. For example, the metric (SI) pressure unit pascal is equivalent to 0.00000986923 atmosphere or 0.0001450377 pound/inch^2. In powers of ten notation these figures are 9.86923×10^{-6} atmosphere and 1.450377×10^{-4} pound/inch^2. The notation also facilitates adaptation of numbers for electronic data processing and computer readout.

Expressing Numbers in Powers of Ten Notation. — In this system of notation every number is expressed by two factors, one of which is some integer from 1 to 9 followed by a decimal and the other is some power of 10.

Thus, 10,000 is expressed as 1.0000×10^4 and 10,463 as 1.0463×10^4. The number 43 is expressed 4.3×10 and 568 is expressed 5.68×10^2.

In the case of decimals, the number 0.0001 which as a fraction is $1/10,000$ is expressed as 1×10^{-4} and 0.0001463 is expressed as 1.463×10^{-4}. The decimal 0.498 is expressed as 4.98×10^{-1} and 0.03146 is expressed as 3.146×10^{-2}.

Rules for Converting any Number to Powers of Ten Notation. — Any number can be converted to the powers of ten notation by means of one of two rules.

Rule 1: If the number is a whole number or a whole number and a decimal so that it has digits to the left of the decimal point, the decimal point is moved a sufficient number of places to the *left* to bring it to the immediate right of the first digit. With the decimal point shifted to this position, the number so written comprises the *first* factor when written in powers of ten.

The number of places that the decimal point is moved to the left to bring it immediately to the right of the first digit is the *positive* index or power of 10 that comprises the *second* factor when written in powers of ten notation.

Thus, to write 4639 in this notation, the decimal point is moved three places to the left giving the two factors: 4.639×10^3. Similarly,

$$431.412 = 4.31412 \times 10^2$$
$$986388 = 9.86388 \times 10^5$$

Rule 2: If the number is a decimal, i.e., it has digits entirely to the right of the decimal point, then the decimal point is moved a sufficient number of places to the *right* to bring it immediately to the right of the first digit. With the decimal point shifted to this position, the number so written comprises the *first* factor when written in powers of ten notation.

The number of places that the decimal point is moved to the *right* to bring it immediately to the right of the first digit is the *negative* index or power of 10 that follows the number when written in powers of ten notation.

Thus, to bring the decimal point in 0.005721 to the immediate right of the first digit which is 5, it must be moved *three* places to the right, giving the two factors: 5.721×10^{-3}. Similarly,

$$0.469 = 4.69 \times 10^{-1}$$
$$0.0000516 = 5.16 \times 10^{-5}$$

Multiplying Numbers Written in Powers of Ten Notation. — When multiplying two numbers written in the powers of ten notation together, the procedure is as follows:

1. Multiply the first factor of one number by the first factor of the other to obtain the first factor of the product.

2. Add the index of the second factor (which is some power of 10) of one number

to the index of the second factor of the other number to obtain the index of the second factor (which is some power of 10) in the product. Thus:

$$(4.31 \times 10^{-2}) \times (9.0125 \times 10) =$$

$$(4.31 \times 9.0125) \times 10^{-2+1} = 38.844 \times 10^{-1}$$

$$(5.986 \times 10^4) \times (4.375 \times 10^3) =$$

$$(5.986 \times 4.375) \times 10^{4+3} = 26.189 \times 10^7$$

In the preceding calculations neither of the results shown are in the conventional powers of ten form since the first factor in each case has two digits. In the conventional powers of ten notation the results would be:

$$38.844 \times 10^{-1} = 3.884 \times 10^0 = 3.884$$

since $10^0 = 1$, and

$$26.189 \times 10^7 = 2.619 \times 10^8$$

in each case rounding the first factor off to three decimal places.

When multiplying several numbers written in this notation together, the procedure is the same. All of the first factors are multiplied together to get the first factor of the product and all of the indices of the respective powers of ten are added together, taking into account their respective signs, to get the index of the second factor of the product. Thus $(4.02 \times 10^{-3}) \times (3.987 \times 10) \times (4.863 \times 10^5) = (4.02 \times 3.987 \times 4.863) \times (10^{-3+1+5}) = 77.94 \times 10^3 = 7.79 \times 10^4$ rounding off the first factor to two decimal places.

Dividing Numbers Written in Powers of Ten Notation. — When dividing one number by another when both are written in this notation, the procedure is as follows:

1. Divide the first factor of the dividend by the first factor of the divisor to get the first factor of the quotient.

2. Subtract the index of the second factor of the divisor from the index of the second factor of the dividend, taking into account their respective signs, to get the index of the second factor of the quotient. Thus:

$$(4.31 \times 10^{-2}) \div (9.0125 \times 10) =$$

$$(4.31 \div 9.0125) \times (10^{-2-1}) = 0.4782 \times 10^{-3} = 4.782 \times 10^{-4}$$

It can be seen, then, that where several numbers of different magnitudes are to be multiplied and divided this system of notation is helpful.

Example: Find the quotient of $\dfrac{250 \times 4698 \times 0.00039}{43678 \times 0.002 \times 0.0147}$

Solution: Changing all of these numbers to powers of ten notation and performing the operations indicated:

$$\frac{(2.5 \times 10^2) \times (4.698 \times 10^3) \times (3.9 \times 10^{-4})}{(4.3678 \times 10^4) \times (2 \times 10^{-3}) \times (1.47 \times 10^{-2})}$$

$$= \frac{(2.5 \times 4.698 \times 3.9)\,(10^{2+3-4})}{(4.3678 \times 2 \times 1.47)\,(10^{4-3-2})} = \frac{45.8055 \times 10}{12.8413 \times 10^{-1}}$$

$$= 3.5670 \times 10^{1-(-1)}$$

$$= 3.5670 \times 10^2$$

$$= 356.70$$

Preferred Numbers

Preferred numbers are series of numbers selected to be used for standardization purposes in preference to any other numbers. Their use will lead to simplified practice and they should be employed whenever possible for individual standard sizes and ratings, or for a series, in applications similar to the following:

1. Important or characteristic linear dimensions, such as diameters and lengths, areas, volume, weights, capacities.

2. Ratings of machinery and apparatus in horsepower, kilowatts, kilovolt-amperes, voltages, currents, speeds, power-factors, pressures, heat units, temperatures, gas or liquid-flow units, weight-handling capacities, etc.

3. Characteristic ratios of figures for all kinds of units.

American National Standard for Preferred Numbers. — This ANSI Standard Z17.1-1973 covers basic series of preferred numbers which are independent of any measurement system and therefore can be used with metric or customary units.

The numbers are rounded values of the following five geometric series of numbers: $10^{N/5}$, $10^{N/10}$, $10^{N/20}$, $10^{N/40}$, and $10^{N/80}$, where N is an integer in the series 0, 1, 2, 3, etc. The designations used for the five series are respectively R5, R10, R20, R40, and R80, where R stands for Renard (Charles Renard, originator of the first preferred number system) and the number indicates the root of 10 on which the particular series is based.

The R5 series gives 5 numbers approximately 60 per cent apart, the R10 series gives 10 numbers approximately 25 per cent apart, the R20 series gives 20 numbers approximately 12 per cent apart, the R40 series gives 40 numbers approximately 6 per cent apart, and the R80 series gives 80 numbers approximately 3 per cent apart. The number of sizes for a given purpose can be minimized by using first the R5 series and adding sizes from the R10 and R20 series as needed. The R40 and R80 series are used principally for expressing tolerances in sizes based on preferred numbers. Preferred numbers below 1 are formed by dividing the given numbers by 10, 100, etc., and numbers above 10 are obtained by multiplying the given numbers by 10, 100, etc. Sizes graded according to the system may not be exactly proportional to one another due to the fact that preferred numbers may differ from calculated values by $+1.26$ per cent to -1.01 per cent. Deviations from preferred numbers are used in some instances — for example, where whole numbers are needed, such as 32 instead of 31.5 for the number of teeth in a gear.

Basic Series of Preferred Numbers (ANSI Z17.1-1973)

Series Designation								
R5	R10	R20	R40	R40	R80	R80	R80	R80
Preferred Numbers								
1.00	1.00	1.00	1.00	3.15	1.00	1.80	3.15	5.60
1.60	1.25	1.12	1.06	3.35	1.03	1.85	3.25	5.80
2.50	1.60	1.25	1.12	3.55	1.06	1.90	3.35	6.00
4.00	2.00	1.40	1.18	3.75	1.09	1.95	3.45	6.15
6.30	2.50	1.60	1.25	4.00	1.12	2.00	3.55	6.30
...	3.15	1.80	1.32	4.25	1.15	2.06	3.65	6.50
...	4.00	2.00	1.40	4.50	1.18	2.12	3.75	6.70
...	5.00	2.24	1.50	4.75	1.22	2.18	3.87	6.90
...	6.30	2.50	1.60	5.00	1.25	2.24	4.00	7.10
...	8.00	2.80	1.70	5.30	1.28	2.30	4.12	7.30
...	...	3.15	1.80	5.60	1.32	2.36	4.25	7.50
...	...	3.55	1.90	6.00	1.36	2.43	4.37	7.75
...	...	4.00	2.00	6.30	1.40	2.50	4.50	8.00
...	...	4.50	2.12	6.70	1.45	2.58	4.62	8.25
...	...	5.00	2.24	7.10	1.50	2.65	4.75	8.50
...	...	5.60	2.36	7.50	1.55	2.72	4.87	8.75
...	...	6.30	2.50	8.00	1.60	2.80	5.00	9.00
...	...	7.10	2.65	8.50	1.65	2.90	5.15	9.25
...	...	8.00	2.80	9.00	1.70	3.00	5.20	9.50
...	...	9.00	3.00	9.50	1.75	3.07	5.45	9.75

Principal Algebraic Expressions and Formulas

$a \times a = aa = a^2$

$a \times a \times a = aaa = a^3$

$a \times b = ab$

$a^2b^2 = (ab)^2$

$a^2a^3 = a^{2+3} = a^5$

$a^4 \div a^3 = a^{4-3} = a$

$a^0 = 1$

$a^2 - b^2 = (a + b)(a - b)$

$(a + b)^2 = a^2 + 2ab + b^2$

$(a - b)^2 = a^2 - 2ab + b^2$

$\dfrac{a^3}{b^3} = \left(\dfrac{a}{b}\right)^3$

$\dfrac{1}{a^3} = \left(\dfrac{1}{a}\right)^3 = a^{-3}$

$(a^2)^3 = a^{2 \times 3} = (a^3)^2 = a^6$

$a^3 + b^3 = (a + b)(a^2 - ab + b^2)$

$a^3 - b^3 = (a - b)(a^2 + ab + b^2)$

$(a + b)^3 = a^3 + 3a^2b + 3ab^2 + b^3$

$(a - b)^3 = a^3 - 3a^2b + 3ab^2 - b^3$

$\sqrt{a} \times \sqrt{a} = a$

$\sqrt[3]{a} \times \sqrt[3]{a} \times \sqrt[3]{a} = a$

$(\sqrt[3]{a})^3 = a$

$\sqrt[3]{a^2} = (\sqrt[3]{a})^2 = a^{2/3}$

$\sqrt[4]{\sqrt[3]{a}} = \sqrt[4 \times 3]{a} = \sqrt[3]{\sqrt[4]{a}}$

$\sqrt[3]{ab} = \sqrt[3]{a} \times \sqrt[3]{b}$

$\sqrt[3]{\dfrac{a}{b}} = \dfrac{\sqrt[3]{a}}{\sqrt[3]{b}}$

$\sqrt[3]{\dfrac{1}{a}} = \dfrac{1}{\sqrt[3]{a}} = a^{-1/3}$

$\sqrt{a} + \sqrt{b} = \sqrt{a + b + 2\sqrt{ab}}$

When

$a \times b = x$, then $\log a + \log b = \log x$

$a \div b = x$, then $\log a - \log b = \log x$

$a^3 = x$, then $3 \log a = \log x$

$\sqrt[3]{a} = x$, then $\dfrac{\log a}{3} = \log x$

Equations

An equation is a statement of equality between two expressions, as $5x = 105$. The unknown quantity in an equation is generally designated by the letter x. If there is more than one unknown quantity, the others are designated by letters also selected at the end of the alphabet, as y, z, u, t, etc.

An equation of the first degree is one which contains the unknown quantity only in the first power, as $3x = 9$. A quadratic equation is one which contains the unknown quantity in the second, but no higher, power, as $x^2 + 3x = 10$.

Solving Equations of the First Degree with One Unknown. — Transpose all the terms containing the unknown x to one side of the equals sign, and all the other terms to the other side. Combine and simplify the expressions as far as possible, and divide both sides by the coefficient of the unknown x. (See the rules given for transposition of formulas.)

Example:

$$22x - 11 = 15x + 10$$
$$22x - 15x = 10 + 11$$
$$7x = 21$$
$$x = 3$$

Solution of Equations of the First Degree with Two Unknowns. — The form of the simplified equations is:

$$ax + by = c$$
$$a_1x + b_1y = c_1$$

Then,

$$x = \frac{cb_1 - c_1b}{ab_1 - a_1b} \qquad y = \frac{ac_1 - a_1c}{ab_1 - a_1b}$$

Example:

$$3x + 4y = 17$$
$$5x - 2y = 11$$
$$x = \frac{17 \times (-2) - 11 \times 4}{3 \times (-2) - 5 \times 4} = \frac{-34 - 44}{-6 - 20} = \frac{-78}{-26} = 3.$$

The value of y can now be most easily found by inserting the value of x in one of the equations:

$$5 \times 3 - 2y = 11; \quad 2y = 15 - 11 = 4; \quad y = 2.$$

Solution of Quadratic Equations with One Unknown. — If the form of the equation is $ax^2 + bx + c = 0$, then

$$x = \frac{-b \pm \sqrt{b^2 - 4ac}}{2a}$$

Example: Given the equation, $1x^2 + 6x + 5 = 0$, then $a = 1$, $b = 6$, and $c = 5$.

$$x = \frac{-6 \pm \sqrt{6^2 - 4 \times 1 \times 5}}{2 \times 1} = \frac{(-6) + 4}{2} = -1; \quad \text{or} \quad \frac{(-6) - 4}{2} = -5$$

If the form of the equation is $ax^2 + bx = c$, then

$$x = \frac{-b \pm \sqrt{b^2 + 4ac}}{2a}$$

Example: A right-angle triangle has a hypotenuse 5 inches long and one side which is one inch longer than the other; find the lengths of the two sides.

Let $x =$ one side and $x + 1 =$ other side; then $x^2 + (x + 1)^2 = 5^2$ or $x^2 + x^2 + 2x + 1 = 25$; or $2x^2 + 2x = 24$; or $x^2 + x = 12$. Now referring to the basic formula, $ax^2 + bx = c$, we find, in this case, that $a = 1$; $b = 1$; and $c = 12$; hence

$$x = \frac{-1 \pm \sqrt{1 + 4 \times 1 \times 12}}{2 \times 1} = \frac{(-1) + 7}{2} = 3 \quad \text{or} \quad x = \frac{(-1) - 7}{2} = -4$$

Since the positive value (3) would apply in this case, the lengths of the two sides are $x = 3$ inches and $x + 1 = 4$ inches.

Cubic Equations. — If the given equation has the form: $x^3 + ax + b = 0$, then

$$x = \left(-\frac{b}{2} + \sqrt{\frac{a^3}{27} + \frac{b^2}{4}} \right)^{1/3} + \left(-\frac{b}{2} - \sqrt{\frac{a^3}{27} + \frac{b^2}{4}} \right)^{1/3}$$

The equation $x^3 + px^2 + qx + r = 0$, may be reduced to the form $x_1^3 + ax_1 + b = 0$ by substituting $x_1 - \frac{p}{3}$ for x in the given equation.

Series. — Some hand calculations, as well as computer programs of certain types of mathematical problems, may be facilitated by the use of an appropriate series. For example, in some gear problems, the angle corresponding to a given or calculated involute function is found by using a series together with an iterative procedure such as the Newton-Raphson method described on page 23. The following are those series most commonly used for such purposes. In the series for trigonometric functions, the angles x are in radians (1 radian = $180/\pi$ degrees). The expression $\exp(-x^2)$ means that the base e of the natural logarithm system is raised to the $-x^2$ power; $e =$ 2.7182818.

(1)	$\sin x = x - x^3/3! + x^5/5! - x^7/7! + \cdots$	for all values of x.
(2)	$\cos x = 1 - x^2/2! + x^4/4! - x^6/6! + \cdots$	for all values of x.
(3)	$\tan x = x + x^3/3 + 2x^5/15 + 17x^7/315 + 62x^9/2835 + \cdots$	for $|x| < \pi/2$.
(4)	$\arcsin x = x + x^3/6 + 1\cdot3\cdot x^5/(2\cdot4\cdot5)$ $+ 1\cdot3\cdot5\cdot x^7/(2\cdot4\cdot6\cdot7) + \cdots$	for $|x| \leq 1$.
(5)	$\arccos x = \pi/2 - \arcsin x$	
(6)	$\arctan x = x - x^3/3 + x^5/5 - x^7/7 + \cdots$	for $|x| \leq 1$.
(7)	$e^x = 1 + x + x^2/2! + x^3/3! + \cdots$	for all values of x.
(8)	$\exp(-x^2) = 1 - x^2 + x^4/2! - x^6/3! + \cdots$	for all values of x.
(9)	$a^x = 1 + x\log_e a + (x\log_e a)^2/2! + (x\log_e a)^3/3! + \cdots$	for all values of x.
(10)	$1/(1 + x) = 1 - x + x^2 - x^3 + x^4 - \cdots$	for $|x| < 1$.
(11)	$1/(1 - x) = 1 + x + x^2 + x^3 + x^4 + \cdots$	for $|x| < 1$.
(12)	$1/(1 + x)^2 = 1 - 2x + 3x^2 - 4x^3 + 5x^4 - \cdots$	for $|x| < 1$.
(13)	$1/(1 - x)^2 = 1 + 2x + 3x^2 + 4x^3 + 5x^5 + \cdots$	for $|x| < 1$.
(14)	$\sqrt{(1 + x)} = 1 + x/2 - x^2/(2\cdot4) + 1\cdot3\cdot x^3/(2\cdot4\cdot6)$ $- 1\cdot3\cdot5\cdot x^4/(2\cdot4\cdot6\cdot8) - \cdots$	for $|x| < 1$.
(15)	$1/\sqrt{(1 + x)} = 1 - x/2 + 1\cdot3\cdot x^2/(2\cdot4) - 1\cdot3\cdot5\cdot x^3/(2\cdot4\cdot6) + \cdots$	for $|x| < 1$.
(16)	$(a + x)^n = a^n + na^{n-1}x + n(n - 1)a^{n-2}x^2/2!$ $+ n(n - 1)(n - 2)a^{n-3}x^3/3! + \cdots$	for $x^2 < a^2$.

Derivatives of Functions. — The following are formulas for obtaining the derivatives of basic mathematical functions. In these formulas, the letter a denotes a constant; the letter x denotes a variable; and the letters u and v denote functions of the variable x. The expression d/dx means the derivative with respect to x, and as such applies to whatever expression in parentheses follows it. Thus, d/dx (ax) means the derivative with respect to x of the product (ax) of the constant a and the variable x, as given by formula (3).

To simplify the form of the formulas, the symbol D is used to represent d/dx. Thus, D is equivalent to d/dx and other forms as follows:

$$\mathrm{D}(ax) = d/dx(ax) = \frac{d}{dx}(ax).$$

(1) $\mathrm{D}(a) = 0$ (2) $\mathrm{D}(x) = 1$ (3) $\mathrm{D}(ax) = a\cdot\mathrm{D}(x) = a\cdot1 = a$

(4) $\mathrm{D}(u + v) = \mathrm{D}(u) + \mathrm{D}(v)$ *Example:* $\mathrm{D}(x^4 + 2x^2) = 4x^3 + 4x$

(5) $\mathrm{D}(uv) = v\cdot\mathrm{D}(u) + u\cdot\mathrm{D}(v)$
       Example: $\mathrm{D}(x^2\cdot ax^3) = ax^3\cdot2x + x^2\cdot3ax^2 = 5ax^4$

(6) $\mathrm{D}(u/v) = \dfrac{v\cdot\mathrm{D}(u) - u\cdot\mathrm{D}(v)}{v^2}$
       Example: $\mathrm{D}(ax^2/\sin x) = (2ax\cdot\sin x - ax^2\cdot\cos x)/\sin^2 x$

(7) $\mathrm{D}(x^n) = n\cdot x^{n-1}$ *Example:* $\mathrm{D}(5x^7) = 35x^6$

(8) $\mathrm{D}(e^x) = e^x$

(9) $\mathrm{D}(a^x) = a^x\cdot\log_e a$ *Example:* $\mathrm{D}(11^x) = 11^x\cdot\log_e 11$

(10) $\mathrm{D}(u^v) = v\cdot u^{v-1}\cdot\mathrm{D}(u) + u^v\cdot\log_e u\cdot\mathrm{D}(v)$

(11) $D(\log_e x) = 1/x$

(12) $D(\log_a x) = 1/x \cdot \log_e a = \log_a e / x$

(13) $D(\sin x) = \cos x$ *Example:* $D(a \cdot \sin x) = a \cdot \cos x$

(14) $D(\cos x) = -\sin x$ (15) $D(\tan x) = \sec^2 x$

Solving Numerical Equations Having One Unknown. — The Newton-Raphson method is a procedure for solving various kinds of numerical algebraic and transcendental equations in one unknown. The steps in the procedure are simple and can be used with either a handheld calculator or as a subroutine in a computer program.

Examples of types of equations that can be solved to any desired degree of accuracy by this method are:

$$f(x) = x^2 - 101 = 0; \quad f(x) = x^3 - 2x^2 - 5 = 0; \quad \text{and}$$
$$f(x) = 2.9x - \cos x - 1 = 0.$$

The procedure begins with an estimate, r_1, of the root satisfying the given equation. This estimate is obtained by judgment, inspection, or plotting a rough graph of the equation and observing the value r_1 where the curve crosses the x axis. This value is then used to calculate values $r_2, r_3, \ldots r_n$ progressively closer to the exact value.

Before continuing, it is necessary to calculate the first derivative, $f'(x)$, of the function. In the above examples, $f'(x)$ is, respectively, $2x$, $3x^2 - 4x$, and $2.9 + \sin x$. These values were found by the methods described in *Derivatives* on page 22.

In the steps that follow,

r_1 is the first estimate of the value of the root of $f(x) = 0$;
$f(r_1)$ is the value of $f(x)$ for $x = r_1$;
$f'(x)$ is the first derivative of $f(x)$;
$f'(r_1)$ is the value of $f'(x)$ for $x = r_1$.

The second approximation of the root of $f(x) = 0$, r_2, is calculated from:

$$r_2 = r_1 - [f(r_1)/f'(r_1)];$$

and, to continue further approximations,

$$r_n = r_{n-1} - [f(r_{n-1})/f'(r_{n-1})].$$

Example: Find the square root of 101 using Newton-Raphson methods.
This problem can be restated as an equation to be solved, i.e.,

$$f(x) = x^2 - 101 = 0.$$

Step 1. By inspection, it is evident that $r_1 = 10$ may be taken as the first approximation of the root of this equation.
Then,

$$f(r_1) = f(10) = 10^2 - 101 = -1.$$

Step 2. The first derivative, $f'(x)$, of $x^2 - 101$ is $2x$ as stated previously so that,

$$f'(10) = 2(10) = 20.$$

Then,

$$r_2 = r_1 - f(r_1)/f'(r_1) = 10 - (-1)/20 = 10 + 0.05 = 10.05.$$

Check: $10.05^2 = 101.0025$; error = 0.0025

Step 3. The next, better, approximation is:

$$r_3 = r_2 - [f(r_2)/f'(r_2)] = 10.05 - [f(10.05)/f'(10.05)]$$
$$= 10.05 - [(10.05^2 - 101)/(2(10.05))] = 10.049875.$$

Check: $10.049875^2 = 100.9999875$; error = 0.0000125

Rearrangement and Transposition of Terms in Formulas

A formula is a rule for a calculation expressed by using letters and signs instead of writing out the rule in words; by this means it is possible to condense, in a very small space, the essentials of long and cumbersome rules. The letters used in formulas simply stand in place of the figures which are to be substituted when solving a specific problem.

As an example, the formula for the horsepower transmitted by belting may be written:

$$P = \frac{SVW}{33,000}$$

in which P = horsepower transmitted;
                  S = working stress of belt per inch of width, in pounds;
                  V = velocity of belt in feet per minute;
                  W = width of belt in inches.

If the working stress S, the velocity V, and the width W are known, the horsepower can be found directly from this formula by inserting the given values. Assume $S = 33$; $V = 600$; and $W = 5$. Then:

$$P = \frac{33 \times 600 \times 5}{33,000} = 3.$$

Assume, however, that the horsepower P, the stress S, and the velocity V are known, and that the width of belt, W, is to be found. The formula must then be rearranged so that the symbol W will be on one side of the equals sign and all the known quantities on the other. The rearranged formula is as follows:

$$\frac{P \times 33,000}{SV} = W.$$

The quantities (S and V) that were in the numerator on the right side of the equals sign are moved to the denominator on the left side, and "33,000" which was in the denominator on the right side of the equals sign is moved to the numerator on the other side. Symbols which are not part of a fraction, like "P" in the formula first given, are to be considered as being numerators (having the denominator 1).

Thus, any formula of the form $A = \dfrac{B}{C}$ can be rearranged as below:

$$A \times C = B, \quad \text{and} \quad C = \frac{B}{A}$$

Suppose a formula to be of the form:

$$A = \frac{B \times C}{D}$$

Then: $$D = \frac{B \times C}{A}; \quad \frac{A \times D}{C} = B; \quad \frac{A \times D}{B} = C.$$

The method given is only directly applicable when all the quantities in the numerator or denominator are standing independently or are *factors of a product*. If connected by + or − signs, the entire numerator or denominator must be moved as a unit, thus,

Given: $$\frac{B + C}{A} = \frac{D + E}{F}, \quad \text{to solve for } F$$

then $$\frac{F}{A} = \frac{D + E}{B + C}$$

and $$F = \frac{A(D + E)}{B + C}$$

A quantity preceded by a $+$ or $-$ sign can be transposed to the opposite side of the equals sign by changing its sign; if the sign is $+$, change it to $-$ on the other side; if it is $-$, change it to $+$. This is called *transposition* of terms.

Example:

$$B + C = A - D; \quad \text{then} \quad B + C + D = A;$$
$$B = A - D - C;$$
$$C = A - D - B;$$

Order of Performing Arithmetic Operations

When several numbers or quantities in a formula are connected by signs indicating that additions, subtractions, multiplications, or divisions are to be made, the multiplications and divisions should be carried out first, in the order in which they appear, before the additions or subtractions are performed.

Examples:

$$10 + 26 \times 7 - 2 = 10 + 182 - 2 = 190.$$
$$18 \div 6 + 15 \times 3 = 3 + 45 = 48.$$
$$12 \div 14 \div 2 - 4 = 12 + 7 - 4 = 15.$$

When it is required that certain additions and subtractions should precede multiplications and divisions, use is made of parentheses ( ) and brackets [ ]. These indicate that the calculation inside the parentheses or brackets should be carried out complete by itself before the remaining calculations are commenced. If one bracket is placed inside of another, the one inside is first calculated.

Examples:

$$(6 - 2) \times 5 + 8 = 4 \times 5 + 8 = 20 + 8 = 28.$$
$$6 \times (4 + 7) \div 22 = 6 \times 11 \div 22 = 66 \div 22 = 3.$$
$$2 + [10 \times 6(8 + 2) - 4] \times 2 = 2 + [10 \times 6 \times 10 - 4] \times 2$$
$$= 2 + [600 - 4] \times 2 = 2 + 596 \times 2 = 2 + 1192 = 1194.$$

The parentheses are considered as a sign of multiplication; for example, $6(8 + 2)$ $= 6 \times (8 + 2)$.

The line or bar between the numerator and denominator in a fractional expression is to be considered as a division sign. For example,

$$\frac{12 + 16 + 22}{10} = (12 + 16 + 22) \div 10 = 50 \div 10 = 5.$$

In formulas the multiplication sign ($\times$) is often left out between symbols or letters, the values of which are to be multiplied. Thus

$$AB = A \times B, \quad \text{and} \quad \frac{ABC}{D} = (A \times B \times C) \div D$$

Ratio and Proportion

The *ratio* between two quantities is the quotient obtained by dividing the first quantity by the second. For example, the ratio between 3 and 12 is $\frac{1}{4}$, and the ratio between 12 and 3 is 4. Ratio is generally indicated by the sign (:); thus 12 : 3 indicates the ratio of 12 to 3.

A *reciprocal* or *inverse* ratio is the reciprocal of the original ratio. Thus, the inverse ratio of 5 : 7 is 7 : 5.

In a *compound* ratio each term is the product of the corresponding terms in two or more simple ratios. Thus, when

$$8 : 2 = 4, \qquad 9 : 3 = 3, \qquad 10 : 5 = 2,$$

then the compound ratio is:

$$8 \times 9 \times 10 : 2 \times 3 \times 5 = 4 \times 3 \times 2,$$
$$720 : 30 = 24.$$

Proportion is the equality of ratios. Thus,

$$6 : 3 = 10 : 5, \quad \text{or} \quad 6 : 3 :: 10 : 5.$$

The first and last terms in a proportion are called the *extremes;* the second and third, the *means*. The product of the extremes is equal to the product of the means. Thus,

$$25 : 2 = 100 : 8 \quad \text{and} \quad 25 \times 8 = 2 \times 100.$$

If three terms in a proportion are known, the remaining term may be found by the following rules:

The first term is equal to the product of the second and third terms, divided by the fourth.

The second term is equal to the product of the first and fourth terms, divided by the third.

The third term is equal to the product of the first and fourth terms, divided by the second.

The fourth term is equal to the product of the second and third terms, divided by the first.

Examples: — Let x be the term to be found, then;

$$x : 12 = 3.5 : 21 \qquad x = \frac{12 \times 3.5}{21} = \frac{42}{21} = 2$$

$$\tfrac{1}{4} : x = 14 : 42 \qquad x = \frac{\tfrac{1}{4} \times 42}{14} = \frac{1}{4} \times 3 = \frac{3}{4}$$

$$5 : 9 = x : 63 \qquad x = \frac{5 \times 63}{9} = \frac{315}{9} = 35$$

$$\tfrac{1}{4} : \tfrac{7}{8} = 4 : x \qquad x = \frac{\tfrac{7}{8} \times 4}{\tfrac{1}{4}} = \frac{3\tfrac{1}{2}}{\tfrac{1}{4}} = 14.$$

If the second and third terms are the same, either is said to be the *mean proportional* between the other two. Thus, $8 : 4 = 4 : 2$, and 4 is the mean proportional between 8 and 2. The mean proportional between two numbers may be found by multiplying the numbers together, and extracting the square root of the product. Thus, the mean proportional between 3 and 12 is found as below:

$$3 \times 12 = 36, \quad \text{and} \quad \sqrt{36} = 6,$$

which is the mean proportional.

Practical Examples Involving Simple Proportion. — If it takes 18 days to assemble 4 lathes, how long would it require to assemble 14 lathes?

Let the number of days to be found be x. Then write out the proportion as below:

$$4 : 18 = 14 : x$$
$$\text{(lathes : days = lathes : days)}$$

Find now the fourth term by the rule given:

$$x = \frac{18 \times 14}{4} = 63 \text{ days.}$$

Thirty-four linear feet of bar stock are required for the blanks for 100 clamping bolts. How many feet of stock would be required for 912 bolts?

Let $x =$ total length of stock required for 912 bolts.

$$34 : 100 = x : 912$$
$$\text{(feet : bolts = feet : bolts)}$$

Then, the third term $x = \dfrac{34 \times 912}{100} = 310$ feet, approximately.

Inverse Proportion. — In an inverse proportion, as one of the items involved *increases,* the corresponding item in the proportion *decreases,* or vice versa. For example, a factory employing 270 men completes a given number of typewriters weekly, the number of working hours being 44 per week. How many men would be required for the same production if the working hours were reduced to 40 per week?

The time per week is in an inverse proportion to the number of men employed; the shorter the time, the more men. The inverse proportion is written:

$$270 : x = 40 : 44$$

(men, 44-hour basis: men, 40-hour basis = time, 40-hour basis: time, 44-hour basis) Thus

$$\frac{270}{x} = \frac{40}{44} \quad \text{and} \quad x = \frac{270 \times 44}{40} = 297 \text{ men.}$$

Problems Involving Both Simple and Inverse Proportions. — If two groups of data are related both by direct (simple) and inverse proportions among the various quantities, then a simple mathematical relation that may be used in solving problems is as follows:

$$\frac{\text{Product of all directly proportional items in first group}}{\text{Product of all inversely proportional items in first group}}$$
$$= \frac{\text{Product of all directly proportional items in second group}}{\text{Product of all inversely proportional items in second group}}$$

Example: If a man capable of turning 65 studs in a day of 10 hours is paid $6.50 per hour, how much per hour ought a man be paid who turns 72 studs in a 9-hour day, if compensated in the same proportion?

The first group of data in this problem consists of the number of hours worked by the first man, his hourly wage, and the number of studs which he produces per day; the second group contains similar data for the second man except for his unknown hourly wage which may be indicated by x.

The labor cost per stud, as may be seen, is directly proportional to the number of hours worked and the hourly wage. These quantities, therefore, are used in the numerators of the fractions in the formula. The labor cost per stud is inversely proportional to the number of studs produced per day. (The greater the number of studs produced in a given time the less the cost per stud.) The numbers of studs per day, therefore, are placed in the denominators of the fractions in the formula. Thus,

$$\frac{10 \times 6.50}{65} = \frac{9 \times x}{72}$$

$$x = \frac{10 \times 6.50 \times 72}{65 \times 9} = \$8.00 \text{ per hour}$$

Percentage

If out of 100 pieces made, 12 do not pass inspection, it is said that 12 per cent (12 of the hundred) are rejected. If a quantity of steel is bought for $100 and sold for $140, the profit is 28.6 per cent of the selling price.

The per cent of gain or loss is found by dividing the amount of gain or loss by the *original* number of which the percentage is wanted, and multiplying the quotient by 100.

Examples: — Out of a total output of 280 castings a day, 30 castings are, on an average, rejected. What is the percentage of bad castings?

$$\frac{30}{280} \times 100 = 10.7 \text{ per cent.}$$

If by a new process 100 pieces can be made in the same time as 60 could formerly be made, what is the gain in output of the new process over the old, expressed in per cent?

Original number, 60; gain 100 − 60 = 40. Hence,

$$\frac{40}{60} \times 100 = 66.7 \text{ per cent.}$$

Care should be taken always to use the original number, or the number of which the percentage is wanted, as the divisor in all percentage calculations. In the example just given, it is the percentage of gain over the old output 60 that is wanted and not the percentage with relation to the new output 100. Mistakes are often made by overlooking this important point.

Interest

Interest is the money paid for the use of money lent for a certain time. *Simple* interest is the interest paid on the principal (money lent) only. When simple interest that is due is not paid, and its amount is added to the interest-bearing principal, the interest calculated on this new principal is called *compound* interest. The compounding of the interest into the principal may take place yearly or more often, according to circumstances.

Simple Interest. — The following formulas are applicable to the calculations involving simple interest. Let:

P = principal or amount of money lent;
p = per cent of interest;
r = interest rate = the interest, expressed decimally, on \$1.00 for one year = the per cent of interest divided by 100; thus, if the interest is 6 per cent, the rate $r = {}^{6}/_{100} = 0.06$;
n = the number of years for which interest is calculated;
I = the amount of interest for n years at the given rate;
P_n = principal with interest for n years added, or the total amount after n years.

Then:

Interest for n years, $I = Prn$.
Total amount after n years, $P_n = P + Prn = P(1 + rn)$.
Interest rate $r = I \div Pn$.
Number of years $n = I \div Pr$.
Principal, or amount lent $= I \div rn$.

Example: — Assume that \$250 has been lent for three years at 6 per cent simple interest. Then:

$$P = 250; \ p = 6; \ r = p \div 100 = 0.06; \ n = 3.$$
$$I = Prn = 250 \times 0.06 \times 3 = \$45.$$
$$P_n = P + I = 250 + 45 = \$295.$$

The accurate interest for one day is $^1/_{365}$ of the interest for one year. Banks, however, customarily take the year as composed of 12 months of 30 days, making a total of 360 days to a year.

Compound Interest. — The following formulas are applicable when compound interest is to be computed, using the same notation as for simple interest, and assuming that the interest is compounded annually.

The total amount after n years, $P_n = P(1 + r)^n$.

The principal $P = \dfrac{P_n}{(1 + r)^n}$ 　　　　The rate $r = \sqrt[n]{\dfrac{P_n}{P}} - 1$

The number of years during which the money is lent

$$n = \frac{\log P_n - \log P}{\log (1 + r)}$$

Example: — In what time will $500 become $1000 at 6 per cent interest compounded yearly?

$$P_n = 1000; P = 500; r = 0.06;$$

$$\log (1000) = 3; \log (500) = 2.69897; \log (1.06) = 0.025306.$$

Substituting these values in the formula:

$$n = \frac{3 - 2.69897}{0.025306} = 11.9 \text{ years}$$

This result is the number of years in which any principal will double itself at 6 per cent compound interest.

Interest Compounded More Often than Annually: If the interest is payable q times a year, it will be computed q times during each year, or nq times during n years. The rate for each compounding will be $r \div q$, if r is the annual rate. Hence, at the end of n years the amount due will be:

$$P_n = P\left(1 + \frac{r}{q}\right)^{nq}$$

Thus, if the term be five years, the interest be payable quarterly, and the annual rate be 6 per cent, then, $n = 5$; $q = 4$; $r = 0.06$; $r \div q = 0.06 \div 4 = 0.015$; and $nq = 5 \times 4 = 20$.

Present Value and Discount. — The present value V of a given amount due in a given time is the sum which placed at interest for the given time will produce the given amount. Hence,

At simple interest, $V = \dfrac{P_n}{1 + nr}$

At compound interest, $V = \dfrac{P_n}{(1 + r)^n}$

in which P_n is the amount due in n years time, and r is the rate of simple interest, or the per cent divided by 100.

The *true discount D* is the difference between the amount due at the end of n years and the present value, or

At simple interest, $D = P_n - V = \dfrac{P_n nr}{1 + nr}$

At compound interest, $D = P_n - V = P_n\left[1 - \dfrac{1}{(1 + r)^n}\right]$

These formulas are for interest compounded annually. If the interest is payable and compounded semi-annually, or quarterly, modify the formulas as indicated in the formulas for compound interest.

Example: — Required the present value and discount of $500 due in six months at 6 per cent simple interest. Here, $P_n = 500$; $n = \frac{6}{12}$ years $= \frac{1}{2}$; $r = 0.06$; then,

$$V = \frac{500}{1 + 0.5 \times 0.06} = \$485.44.$$

$$D = 500 - 485.44 = \$14.56.$$

Example: — Required the sum which placed at 5 per cent compound interest, will in three years produce $5000. Here, $P_n = 5000$; $r = 0.05$; $n = 3$. Then,

$$V = \frac{5000}{(1 + 0.05)^3} = 4319.19.$$

Bank discount is calculated at simple interest on the total amount of a promissory note for the term of the note and on the basis of a year of 360 days.

Annuities. — An annuity is a fixed sum paid at regular intervals. In the formulas given below, yearly payments are assumed. It is customary to calculate annuities on the basis of compound interest.

If an annuity A is to be paid out for n consecutive years, the interest rate being then the present value P of the annuity is:

$$P = A \frac{(1+r)^n - 1}{(1+r)^n r}$$

Example: — If an annuity of $200 is to be paid for 10 years, what is the present amount of money that need be deposited if the interest is 5 per cent? Here,

$$A = 200; \quad r = 5 \div 100 = 0.05; \quad n = 10.$$

$$P = 200 \frac{1.05^{10} - 1}{1.05^{10} \times 0.05} = 1544.36.$$

The annuity that a principal P, drawing interest at the rate r, will give for a period of n years, is:

$$A = \frac{Pr(1+r)^n}{(1+r)^n - 1}$$

Example: — A sum of $10,000 is placed at 4 per cent interest. What is the amount of the annuity which can be paid for 20 years out of this sum? Here,

$$P = 10,000; \quad r = 0.04; \quad n = 20.$$

$$A = \frac{10,000 \times 0.04 \times 1.04^{20}}{1.04^{20} - 1} = 735.82.$$

If at the beginning of each year a sum A is set aside at an interest rate r, then the total value of the sum set aside, with interest, will be at the end of n years:

$$P_n = A \frac{(1+r)[(1+r)^n - 1]}{r}$$

If at the end of each year a sum A is set aside at an interest rate r, then the total value of the principal, with interest, at the end of n years will be:

$$P_n = A \frac{(1+r)^n - 1}{r}$$

If a principal P is increased or decreased by a sum A at the end of each year, then the value of the principal after n years will be:

$$P_n = P(1+r)^n \pm A \frac{(1+r)^n - 1}{r}$$

If the sum A by which the principal P is decreased each year is greater than the total yearly interest on the principal, then the principal, with the accumulated interest, will be entirely used up in n years:

$$n = \frac{\log A - \log (A - Pr)}{\log (1 + r)}$$

Sinking Funds.— Amortization is "the extinction of a debt, usually by means of a sinking fund." The sinking fund is created by a fixed investment S placed annually at compound interest for a term of years, and is hence an annuity of sufficient size to produce at the end of the term of years the amount necessary for the repayment of the principal of the debt, or to provide a definite sum for other purposes. Let:

S = the annual investment;

r = rate of interest (the per cent divided by 100);

P = the amount of the sinking fund;

n = the number of years for its creation.

Then:

$$P = S \frac{(1 + r)^{n} - 1}{r}, \quad \text{and} \quad S = \frac{Pr}{(1 + r)^{n} - 1}$$

which formulas correspond to those given above, where a sum A was laid aside at the end of each year.

Example: — If \$2000 is invested annually for 10 years, at 4 per cent compound interest, as a sinking fund, what would be the total amount of the fund at the expiration of the term? Here, $S = 2000$; $n = 10$; $r = 0.04$.

$$P = 2000 \frac{1.04^{10} - 1}{0.04} = 24{,}012.25$$

Cost of Mixture

When an alloy is composed of several metals varying in price, the price per pound of the alloy can be found as in the following example: An alloy is composed of 50 pounds of copper at 14 cents a pound, 10 pounds of tin at 29 cents a pound, 20 pounds of zinc at 5 cents a pound, and 5 pounds of lead at 4 cents a pound. What is the cost of the alloy per pound, no account being taken of the cost of mixing it?

Multiply the number of pounds of each of the ingredients by its price per pound, add these products together, and divide the sum by the total weight of all the ingredients. The quotient is the price per pound of the alloy.

$50 \times 14 + 10 \times 29 + 20 \times 5 + 5 \times 4 = 700 + 290 + 100 + 20 = 1110$.

Total weight of metal in alloy $= 50 + 10 + 20 + 5 = 85$.

Price per pound of alloy $= {}^{1110}\!/_{85} = 13$ cents, approximately.

In general, let a, b, c and d be the weights of each of the ingredients, and w, x, y and z be their respective values per unit weight. Then the average price P per unit weight of the alloy is found by the formula:

$$P = \frac{aw + bx + cy + dz}{a + b + c + d}$$

Example: — Find the average price per pound of an alloy containing 40 pounds of tin at 30 cents per pound, 48 pounds of lead at 4 cents per pound, 10 pounds of antimony at 8 cents per pound, and 2 pounds of copper at 15 cents per pound.

$$P = \frac{40 \times 30 + 48 \times 4 + 10 \times 8 + 2 \times 15}{40 + 48 + 10 + 2} = \frac{1502}{100} = 15.02 \text{ cents.}$$

Arithmetical Progression

An arithmetical progression is a series of numbers in which each consecutive term differs from the preceding one by a fixed amount called the *common difference*, d. Thus, 1, 3, 5, 7, etc., is an arithmetical progression where the difference d is 2. The difference in this case is *added* to the preceding term, and the progression is called increasing. In the series 13, 10, 7, 4, etc., the difference is (-3), and the progression is called decreasing. In any arithmetical progression (or part of progression) let

a = the first term considered;
l = the last term considered;
n = the number of terms;
d = the common difference;
S = the sum of n terms.

Then the general formulas are:

$$l = a + (n-1)d \quad \text{and} \quad S = \frac{a+l}{2} \times n$$

In these formulas d is positive in an increasing and negative in a decreasing progression. When any three of the five quantities above are given, the other two can be found by the formulas in the accompanying table of arithmetical progression.

Example: — In an arithmetical progression, the first term equals 5, and the last term 40. The difference is 7. Find the sum of the progression.

$$S = \frac{a+l}{2d}(l+d-a) = \frac{5+40}{2 \times 7}(40+7-5) = 135.$$

Geometrical Progression

A geometrical progression or a geometrical series is a series in which each term is derived by multiplying the preceding term by a constant multiplier called the *ratio*. When the ratio is greater than 1, the progression is increasing; when smaller than 1, it is decreasing. Thus, 2, 6, 18, 54, etc., is an increasing geometrical progression with a ratio of 3, while 24, 12, 6, etc., is a decreasing progression with a ratio of ½.

In any geometrical progression (or part of progression) let

a = the first term;
l = the last (or nth) term;
n = the number of terms;
r = the ratio of the progression;
S = the sum of n terms.

Then the general formulas are:

$$l = ar^{n-1} \quad \text{and} \quad S = \frac{rl-a}{r-1}$$

When any three of the five quantities above are given, the other two can be found by the formulas tabulated in the accompanying table. Geometrical progressions are used for finding the successive speeds in machine tool drives, in interest calculations, etc.

Example: — The lowest speed of a lathe is 20 R.P.M. The highest speed is 225 R.P.M. There are 18 speeds. Find the ratio between successive speeds.

$$\text{Ratio, } r = \sqrt[n-1]{\frac{l}{a}} = \sqrt[17]{\frac{225}{20}} = \sqrt[17]{11.25} = 1.153.$$

Formulas for Arithmetical Progression

To Find	Given	Use Equation
a	*d* *l* *n*	$a = l - (n-1)\,d$
	d *n* *S*	$a = \dfrac{S}{n} - \dfrac{n-1}{2} \times d$
	d *l* *S*	$a = \dfrac{d}{2} \pm \dfrac{1}{2}\sqrt{(2\,l+d)^2 - 8\,dS}$
	l *n* *S*	$a = \dfrac{2\,S}{n} - l$
d	*a* *l* *n*	$d = \dfrac{l-a}{n-1}$
	a *n* *S*	$d = \dfrac{2\,S - 2\,an}{n\,(n-1)}$
	a *l* *S*	$d = \dfrac{l^2 - a^2}{2\,S - l - a}$
	l *n* *S*	$d = \dfrac{2\,nl - 2\,S}{n\,(n-1)}$
l	*a* *d* *n*	$l = a + (n-1)\,d$
	a *d* *S*	$l = -\dfrac{d}{2} \pm \dfrac{1}{2}\sqrt{8\,dS + (2\,a-d)^2}$
	a *n* *S*	$l = \dfrac{2\,S}{n} - a$
	d *n* *S*	$l = \dfrac{S}{n} + \dfrac{n-1}{2} \times d$
n	*a* *d* *l*	$n = 1 + \dfrac{l-a}{d}$
	a *d* *S*	$n = \dfrac{d - 2\,a}{2\,d} \pm \dfrac{1}{2\,d}\sqrt{8\,dS + (2\,a-d)^2}$
	a *l* *S*	$n = \dfrac{2\,S}{a+l}$
	d *l* *S*	$n = \dfrac{2\,l+d}{2\,d} \pm \dfrac{1}{2\,d}\sqrt{(2\,l+d)^2 - 8\,dS}$
S	*a* *d* *n*	$S = \dfrac{n}{2}\,[2\,a + (n-1)\,d]$
	a *d* *l*	$S = \dfrac{a+l}{2} + \dfrac{l^2 - a^2}{2\,d} = \dfrac{a+l}{2}\,(l+d-a)$
	a *l* *n*	$S = \dfrac{n}{2}\,(a+l)$
	d *l* *n*	$S = \dfrac{n}{2}\,[2\,l - (n-1)\,d]$

Formulas for Geometrical Progression

To Find	Given			Use Equation
a	l	n	r	$a = \dfrac{l}{r^{n-1}}$
	n	r	S	$a = \dfrac{(r-1)\,S}{r^n - 1}$
	l	r	S	$a = lr - (r-1)\,S$
	l	n	S	$a\,(S-a)^{n-1} = l\,(S-l)^{n-1}$
l	a	n	r	$l = ar^{n-1}$
	a	r	S	$l = \dfrac{1}{r}\,[a + (r-1)\,S]$
	a	n	S	$l\,(S-l)^{n-1} = a\,(S-a)^{n-1}$
	n	r	S	$l = \dfrac{S\,(r-1)\,r^{n-1}}{r^n - 1}$
n	a	l	r	$n = \dfrac{\log l - \log a}{\log r} + 1$
	a	r	S	$n = \dfrac{\log\,[a + (r-1)\,S] - \log a}{\log r}$
	a	l	S	$n = \dfrac{\log l - \log a}{\log\,(S-a) - \log\,(S-l)} + 1$
	l	r	S	$n = \dfrac{\log l - \log\,[lr - (r-1)\,S]}{\log r} + 1$
r	a	l	n	$r = \sqrt[n-1]{\dfrac{l}{a}}$
	a	n	S	$r^n = \dfrac{Sr}{a} + \dfrac{a-S}{a}$
	a	l	S	$r = \dfrac{S-a}{S-l}$
	l	n	S	$r^n = \dfrac{Sr^{n-1}}{S-l} - \dfrac{l}{S-l}$
S	a	n	r	$S = \dfrac{a\,(r^n - 1)}{r - 1}$
	a	l	r	$S = \dfrac{lr - a}{r - 1}$
	a	l	n	$S = \dfrac{\sqrt[n-1]{l^n} - \sqrt[n-1]{a^n}}{\sqrt[n-1]{l} - \sqrt[n-1]{a}}$
	l	n	r	$S = \dfrac{l\,(r^n - 1)}{(r-1)\,r^{n-1}}$

Geometrical Propositions

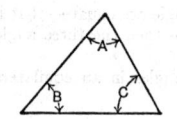

The sum of the three angles in a triangle always equals 180 degrees. Hence, if two angles are known, the third angle can always be found.

$$A + B + C = 180° \qquad A = 180° - (B + C)$$
$$B = 180° - (A + C) \qquad C = 180° - (A + B)$$

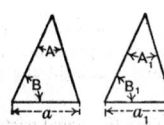

If one side and two angles in one triangle are equal to one side and similarly located angles in another triangle, then the remaining two sides and angle are also equal.

If $a = a_1$, $A = A_1$ and $B = B_1$, then the two other sides and the remaining angle are also equal.

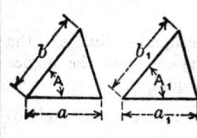

If two sides and the angle between them in one triangle are equal to two sides and a similarly located angle in another triangle, then the remaining side and angles are also equal.

If $a = a_1$, $b = b_1$ and $A = A_1$, then the remaining side and angles are also equal.

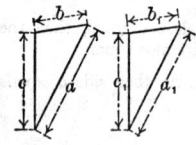

If the three sides in one triangle are equal to the three sides of another triangle, then the angles in the two triangles are also equal.

If $a = a_1$, $b = b_1$ and $c = c_1$, then the angles between the respective sides are also equal.

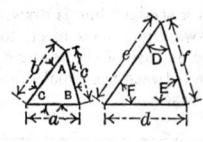

If the three sides of one triangle are proportional to corresponding sides in another triangle, then the triangles are called *similar*, and the angles in the one are equal to the angles in the other.

If $a : b : c = d : e : f$, then $A = D$, $B = E$ and $C = F$.

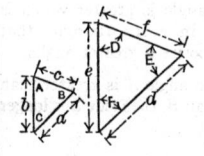

If the angles in one triangle are equal to the angles of another triangle, then the triangles are similar and their corresponding sides are proportional.

If $A = D$, $B = E$ and $C = F$, then $a : b : c = d : e : f$.

Geometrical Propositions

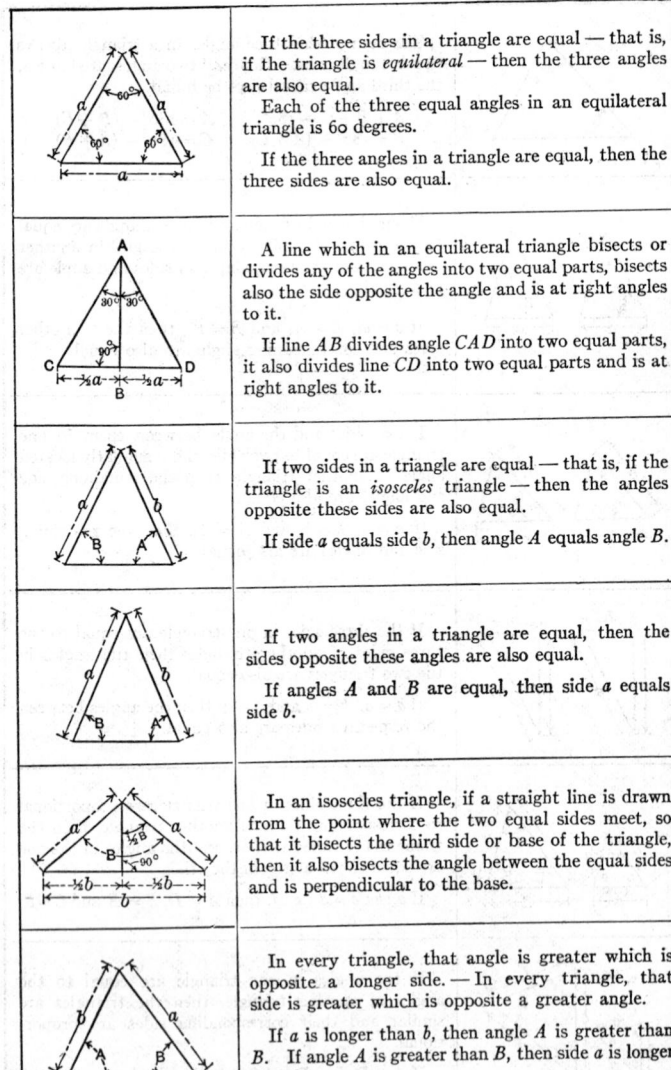

If the three sides in a triangle are equal — that is, if the triangle is *equilateral* — then the three angles are also equal.

Each of the three equal angles in an equilateral triangle is 60 degrees.

If the three angles in a triangle are equal, then the three sides are also equal.

A line which in an equilateral triangle bisects or divides any of the angles into two equal parts, bisects also the side opposite the angle and is at right angles to it.

If line AB divides angle CAD into two equal parts, it also divides line CD into two equal parts and is at right angles to it.

If two sides in a triangle are equal — that is, if the triangle is an *isosceles* triangle — then the angles opposite these sides are also equal.

If side a equals side b, then angle A equals angle B.

If two angles in a triangle are equal, then the sides opposite these angles are also equal.

If angles A and B are equal, then side a equals side b.

In an isosceles triangle, if a straight line is drawn from the point where the two equal sides meet, so that it bisects the third side or base of the triangle, then it also bisects the angle between the equal sides and is perpendicular to the base.

In every triangle, that angle is greater which is opposite a longer side. — In every triangle, that side is greater which is opposite a greater angle.

If a is longer than b, then angle A is greater than B. If angle A is greater than B, then side a is longer than b.

Geometrical Propositions

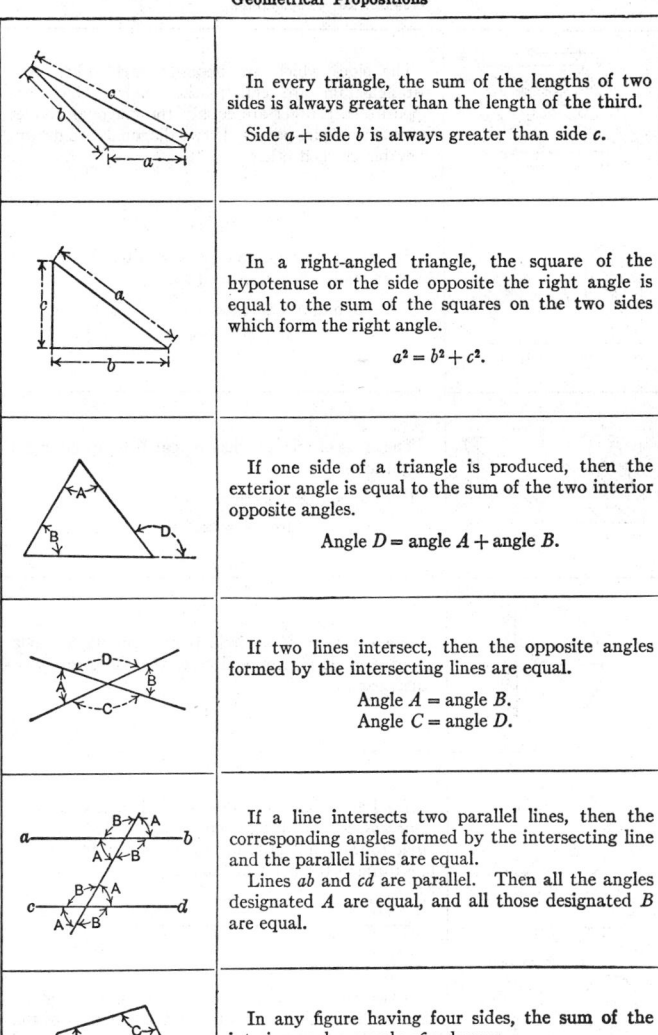

In every triangle, the sum of the lengths of two sides is always greater than the length of the third.

Side a + side b is always greater than side c.

In a right-angled triangle, the square of the hypotenuse or the side opposite the right angle is equal to the sum of the squares on the two sides which form the right angle.

$$a^2 = b^2 + c^2.$$

If one side of a triangle is produced, then the exterior angle is equal to the sum of the two interior opposite angles.

Angle D = angle A + angle B.

If two lines intersect, then the opposite angles formed by the intersecting lines are equal.

Angle A = angle B.
Angle C = angle D.

If a line intersects two parallel lines, then the corresponding angles formed by the intersecting line and the parallel lines are equal.

Lines ab and cd are parallel. Then all the angles designated A are equal, and all those designated B are equal.

In any figure having four sides, the sum of the interior angles equals 360 degrees.

$A + B + C + D$ = 360 degrees.

Geometrical Propositions

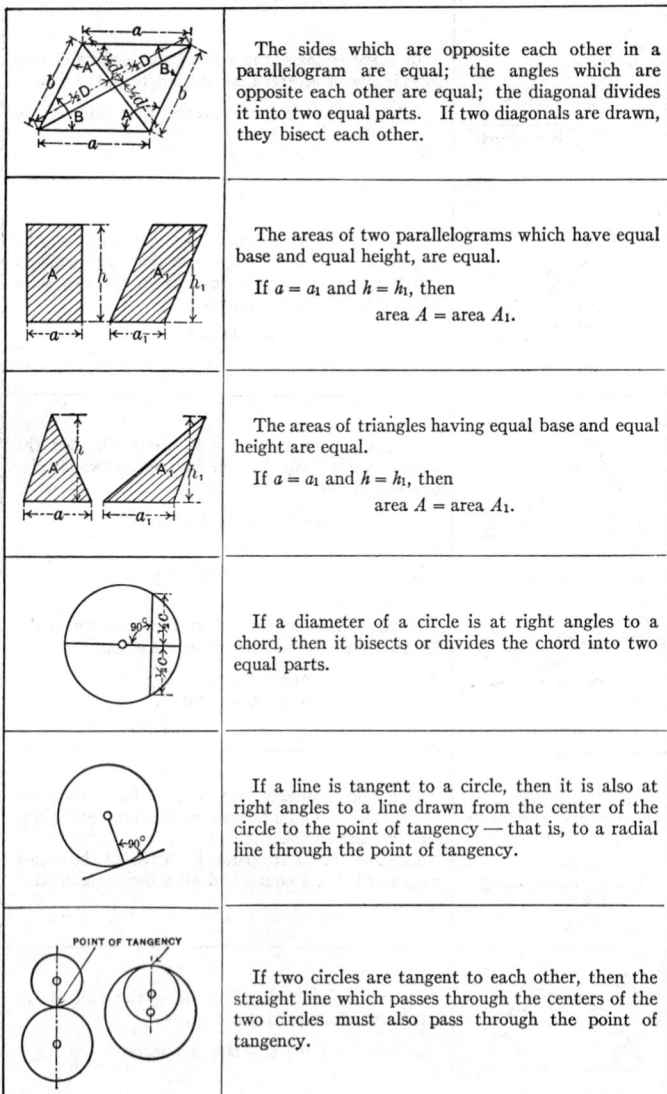

The sides which are opposite each other in a parallelogram are equal; the angles which are opposite each other are equal; the diagonal divides it into two equal parts. If two diagonals are drawn, they bisect each other.

The areas of two parallelograms which have equal base and equal height, are equal.

If $a = a_1$ and $h = h_1$, then

$$\text{area } A = \text{area } A_1.$$

The areas of triangles having equal base and equal height are equal.

If $a = a_1$ and $h = h_1$, then

$$\text{area } A = \text{area } A_1.$$

If a diameter of a circle is at right angles to a chord, then it bisects or divides the chord into two equal parts.

If a line is tangent to a circle, then it is also at right angles to a line drawn from the center of the circle to the point of tangency — that is, to a radial line through the point of tangency.

If two circles are tangent to each other, then the straight line which passes through the centers of the two circles must also pass through the point of tangency.

Geometrical Propositions

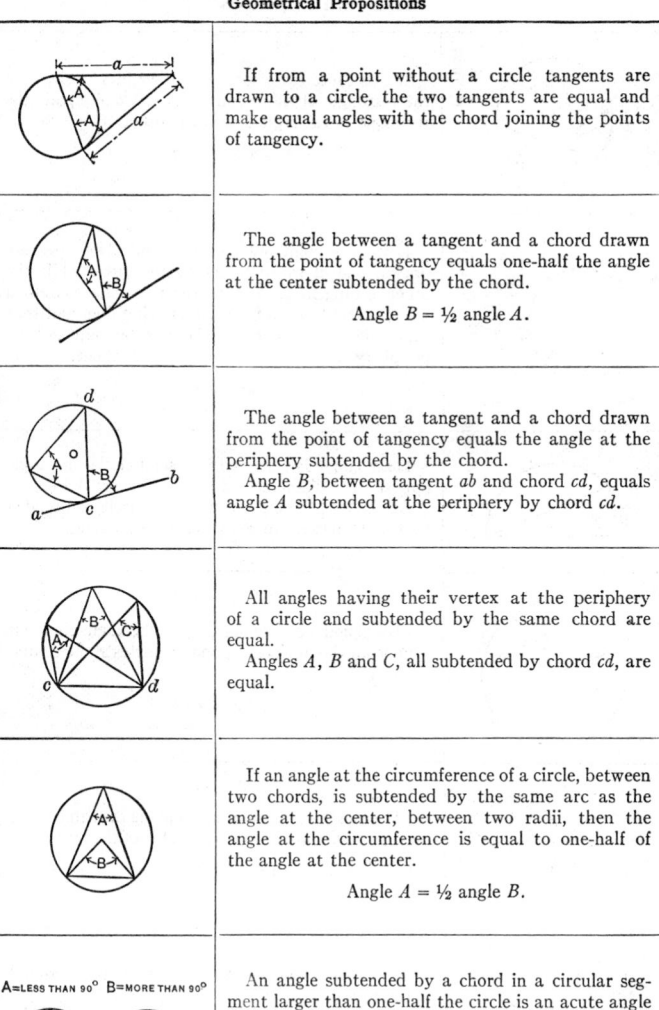

If from a point without a circle tangents are drawn to a circle, the two tangents are equal and make equal angles with the chord joining the points of tangency.

The angle between a tangent and a chord drawn from the point of tangency equals one-half the angle at the center subtended by the chord.

Angle B = ½ angle A.

The angle between a tangent and a chord drawn from the point of tangency equals the angle at the periphery subtended by the chord.

Angle B, between tangent ab and chord cd, equals angle A subtended at the periphery by chord cd.

All angles having their vertex at the periphery of a circle and subtended by the same chord are equal.

Angles A, B and C, all subtended by chord cd, are equal.

If an angle at the circumference of a circle, between two chords, is subtended by the same arc as the angle at the center, between two radii, then the angle at the circumference is equal to one-half of the angle at the center.

Angle A = ½ angle B.

An angle subtended by a chord in a circular segment larger than one-half the circle is an acute angle — an angle less than 90 degrees. An angle subtended by a chord in a circular segment less than one-half the circle is an obtuse angle — an angle greater than 90 degrees.

Geometrical Propositions

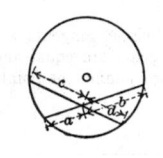

If two chords intersect each other in a circle, then the rectangle of the segments of the one equals the rectangle of the segments of the other.

$$a \times b = c \times d.$$

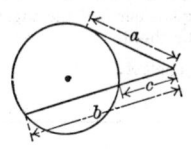

If from a point outside of a circle two lines are drawn, one of which intersects the circle while the other is tangent to it, then the rectangle contained by the total length of the intersecting line, and that part of it which is between the outside point and the periphery, equals the square of the tangent.

$$a^2 = b \times c.$$

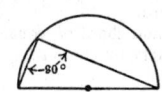

If a triangle is inscribed in a semi-circle, the angle opposite the diameter is a right (90-degree) angle.

All angles at the periphery of a circle, subtended by the diameter, are right (90-degree) angles.

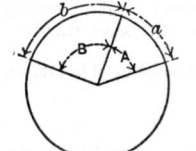

The lengths of circular arcs of the same circle are proportional to the corresponding angles at the center.

$$A : B = a : b.$$

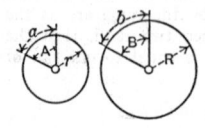

The lengths of circular arcs having the same center angle are proportional to the lengths of the radii.

If $A = B$, then $a : b = r : R$.

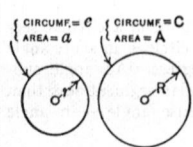

The circumferences of two circles are proportional to their radii.

The areas of two circles are proportional to the squares of their radii.

$$c : C = r : R.$$
$$a : A = r^2 : R^2.$$

Geometrical Constructions

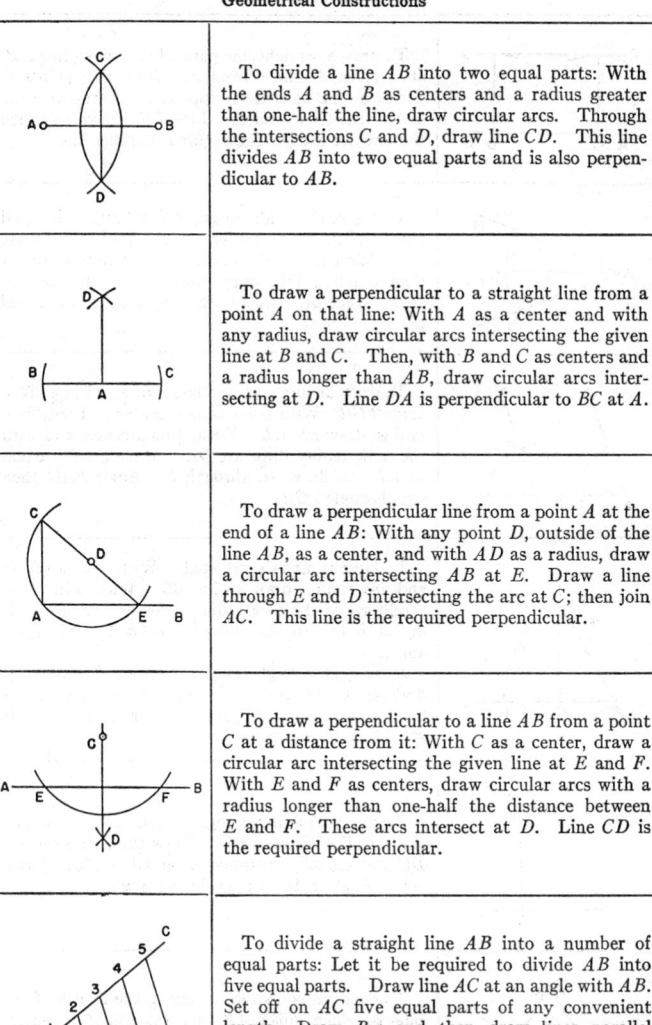

To divide a line AB into two equal parts: With the ends A and B as centers and a radius greater than one-half the line, draw circular arcs. Through the intersections C and D, draw line CD. This line divides AB into two equal parts and is also perpendicular to AB.

To draw a perpendicular to a straight line from a point A on that line: With A as a center and with any radius, draw circular arcs intersecting the given line at B and C. Then, with B and C as centers and a radius longer than AB, draw circular arcs intersecting at D. Line DA is perpendicular to BC at A.

To draw a perpendicular line from a point A at the end of a line AB: With any point D, outside of the line AB, as a center, and with AD as a radius, draw a circular arc intersecting AB at E. Draw a line through E and D intersecting the arc at C; then join AC. This line is the required perpendicular.

To draw a perpendicular to a line AB from a point C at a distance from it: With C as a center, draw a circular arc intersecting the given line at E and F. With E and F as centers, draw circular arcs with a radius longer than one-half the distance between E and F. These arcs intersect at D. Line CD is the required perpendicular.

To divide a straight line AB into a number of equal parts: Let it be required to divide AB into five equal parts. Draw line AC at an angle with AB. Set off on AC five equal parts of any convenient length. Draw $B5$ and then draw lines parallel with $B5$ through the other division points on AC. The points where these lines intersect AB are the required division points.

Geometrical Constructions

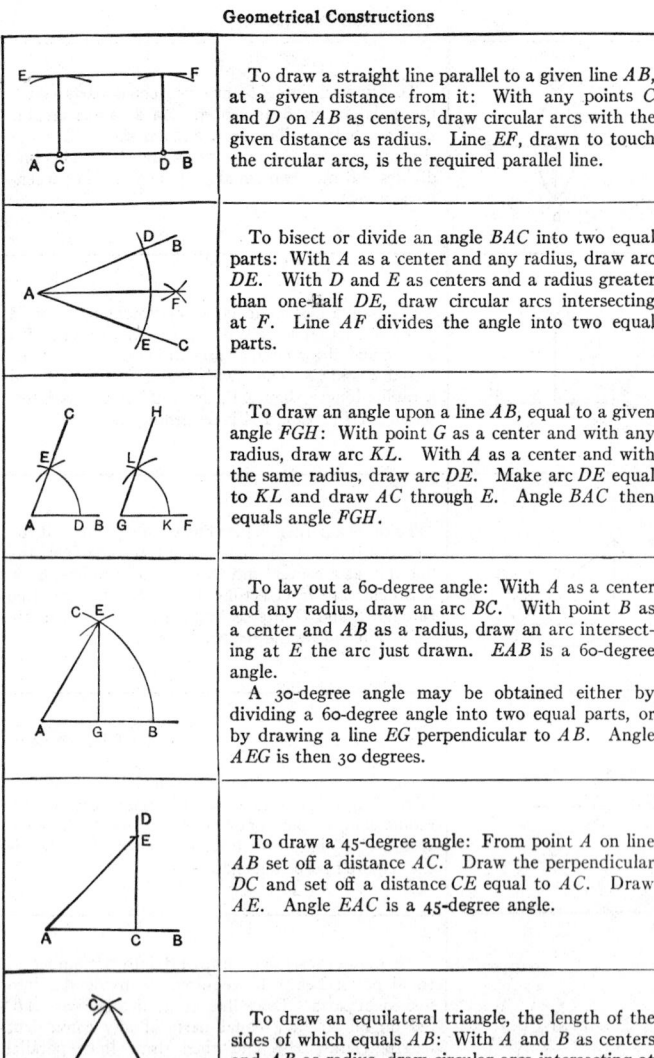

To draw a straight line parallel to a given line *AB*, at a given distance from it: With any points *C* and *D* on *AB* as centers, draw circular arcs with the given distance as radius. Line *EF*, drawn to touch the circular arcs, is the required parallel line.

To bisect or divide an angle *BAC* into two equal parts: With *A* as a center and any radius, draw arc *DE*. With *D* and *E* as centers and a radius greater than one-half *DE*, draw circular arcs intersecting at *F*. Line *AF* divides the angle into two equal parts.

To draw an angle upon a line *AB*, equal to a given angle *FGH*: With point *G* as a center and with any radius, draw arc *KL*. With *A* as a center and with the same radius, draw arc *DE*. Make arc *DE* equal to *KL* and draw *AC* through *E*. Angle *BAC* then equals angle *FGH*.

To lay out a 60-degree angle: With *A* as a center and any radius, draw an arc *BC*. With point *B* as a center and *AB* as a radius, draw an arc intersecting at *E* the arc just drawn. *EAB* is a 60-degree angle.

A 30-degree angle may be obtained either by dividing a 60-degree angle into two equal parts, or by drawing a line *EG* perpendicular to *AB*. Angle *AEG* is then 30 degrees.

To draw a 45-degree angle: From point *A* on line *AB* set off a distance *AC*. Draw the perpendicular *DC* and set off a distance *CE* equal to *AC*. Draw *AE*. Angle *EAC* is a 45-degree angle.

To draw an equilateral triangle, the length of the sides of which equals *AB*: With *A* and *B* as centers and *AB* as radius, draw circular arcs intersecting at *C*. Draw *AC* and *BC*. Then *ABC* is an equilateral triangle.

Geometrical Constructions

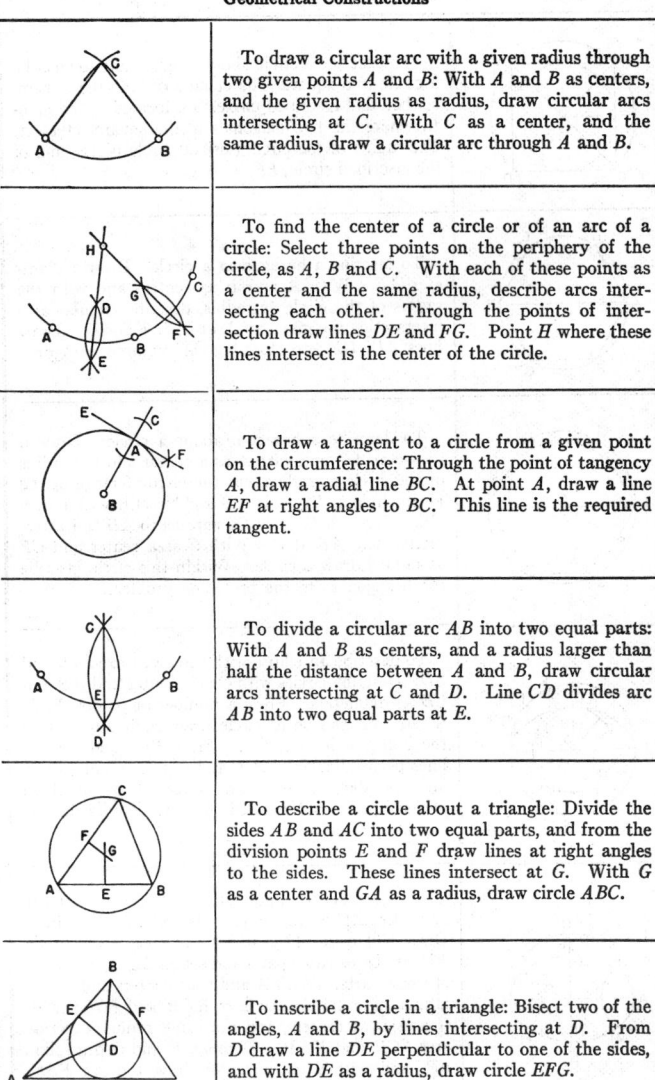

	To draw a circular arc with a given radius through two given points A and B: With A and B as centers, and the given radius as radius, draw circular arcs intersecting at C. With C as a center, and the same radius, draw a circular arc through A and B.
	To find the center of a circle or of an arc of a circle: Select three points on the periphery of the circle, as A, B and C. With each of these points as a center and the same radius, describe arcs intersecting each other. Through the points of intersection draw lines DE and FG. Point H where these lines intersect is the center of the circle.
	To draw a tangent to a circle from a given point on the circumference: Through the point of tangency A, draw a radial line BC. At point A, draw a line EF at right angles to BC. This line is the required tangent.
	To divide a circular arc AB into two equal parts: With A and B as centers, and a radius larger than half the distance between A and B, draw circular arcs intersecting at C and D. Line CD divides arc AB into two equal parts at E.
	To describe a circle about a triangle: Divide the sides AB and AC into two equal parts, and from the division points E and F draw lines at right angles to the sides. These lines intersect at G. With G as a center and GA as a radius, draw circle ABC.
	To inscribe a circle in a triangle: Bisect two of the angles, A and B, by lines intersecting at D. From D draw a line DE perpendicular to one of the sides, and with DE as a radius, draw circle EFG.

Geometrical Constructions

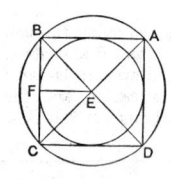

To describe a circle about a square and to inscribe a circle in a square: The centers of both the circumscribed and inscribed circles are located at the point E, where the two diagonals of the square intersect. The radius of the circumscribed circle is AE, and of the inscribed circle, EF.

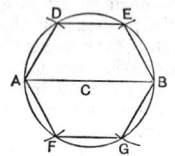

To inscribe a hexagon in a circle: Draw a diameter AB. With A and B as centers and with the radius of the circle as radius, describe circular arcs intersecting the given circle at D, E, F and G. Draw lines AD, DE, etc., forming the required hexagon.

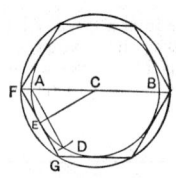

To describe a hexagon about a circle: Draw a diameter AB, and with A as a center and the radius of the circle as radius, cut the circumference of the given circle at D. Join AD and bisect it with radius CE. Through E, draw FG parallel to AD and intersecting line AB at F. With C as a center and CF as radius, draw a circle. Within this circle inscribe the hexagon as in the preceding problem.

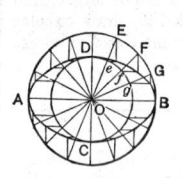

To describe an ellipse with the given axes AB and CD: Describe circles with O as a center and AB and CD as diameters. From a number of points, E, F, G, etc., on the outer circle draw radii intersecting the inner circle at e, f, g. From E, F and G draw lines perpendicular to AB, and from e, f and g draw lines parallel to AB. The intersections of these perpendicular and parallel lines are points on the curve of the ellipse.

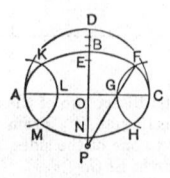

To construct an approximate ellipse by circular arcs: Let AC be the major axis and BN the minor. Draw half circle ADC with O as a center. Divide BD into three equal parts and set off BE equal to one of these parts. With A and C as centers and OE as radius, describe circular arcs KLM and FGH; with G and L as centers, and the same radius, describe arcs FCH and KAM. Through F and G draw line FP, and with P as a center draw the arc FBK. Arc HNM is drawn in the same manner.

Geometrical Constructions

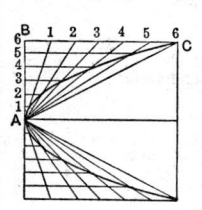

To construct a parabola: Divide line AB into a number of equal parts and divide BC into the same number of parts. From the division points on AB draw horizontal lines. From the division points on BC draw lines to point A. The points of intersection between lines drawn from points numbered alike are points on the parabola.

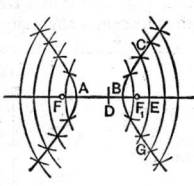

To construct a hyperbola: From focus F lay off a distance FD equal to the transverse axis, or the distance AB between the two branches of the curve. With F as a center and any distance FE greater than FB as a radius, describe a circular arc. Then with F_1 as a center and DE as a radius, describe arcs intersecting at C and G the arc just described. C and G are points on the hyperbola. Any number of points can be found in a similar manner.

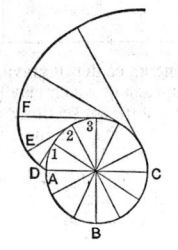

To construct an involute: Divide the circumference of the base circle ABC into a number of equal parts. Through the division points 1, 2, 3, etc., draw tangents to the circle and make the lengths $D\,1$, $E\,2$, $F\,3$, etc., of these tangents equal to the actual length of the arcs $A\,1$, $A\,2$, $A\,3$, etc.

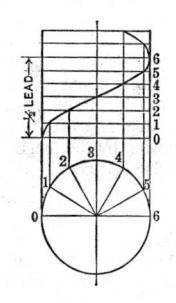

To construct a helix: Divide half the circumference of the cylinder on the surface of which the helix is to be described into a number of equal parts. Divide half the lead of the helix into the same number of equal parts. From the division points on the circle representing the cylinder draw vertical lines, and from the division points on the lead draw horizontal lines as shown. The intersections between lines numbered alike are points on the helix.

The Prismoidal Formula. — The prismoidal formula is a general formula by which the volume of any prism, pyramid or frustum of a pyramid may be found.

A_1 = area at one end of the body;
A_2 = area at the other end;
A_m = area of middle section between the two end surfaces;
h = height of body.

Then, volume V of the body is

$$V = \frac{h}{6}(A_1 + 4A_m + A_2)$$

Pappus or Guldinus Rules. — By means of these rules the area of any surface of revolution and the volume of any solid of revolution may be found. The area of the surface swept out by the revolution of a line *ABC* (see illustration) about the axis *DE* equals the length of the line multiplied by the length of the path of its center of gravity,

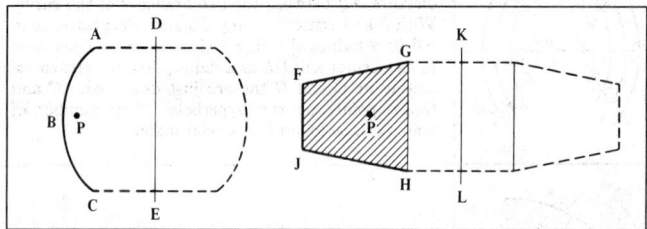

P. If the line is of such a shape that it is difficult to determine its center of gravity, then the line may be divided into a number of short sections, each of which may be considered as a straight line, and the areas swept out by these different sections, as computed by the rule given, may be added to find the total area. The line must lie wholly on one side of the axis of revolution and must be in the same plane.

The volume of a solid body formed by the revolution of a surface *FGHJ* about axis *KL* equals the area of the surface multiplied by the length of the path of its center of gravity. The surface must lie wholly on one side of the axis of revolution and in the same plane.

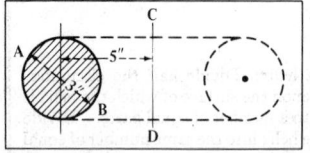

Example: — By means of these rules the area and volume of a cylindrical ring or torus may be found. The torus is formed by a circle *AB* being rotated about axis *CD*. The center of gravity of the circle is at its center. Hence, with the dimensions given in the illustration, the length of the path of the center of gravity of the circle is $3.1416 \times 10 = 31.416$ inches. Multiplying by the length of the circumference of the circle, which is $3.1416 \times 3 = 9.4248$ inches, gives:

$$31.416 \times 9.4248 = 296.089 \text{ square inches}$$

which is the area of the torus.

The volume equals the area of the circle, which is $0.7854 \times 9 = 7.0686$ square inches, multiplied by the path of the center of gravity, which is 31.416, as before; hence,

$$\text{volume} = 7.0686 \times 31.416 = 222.067 \text{ cubic inches.}$$

Example of Approximate Method for Finding the Area of a Surface of Revolution. — The accompanying illustration is shown in order to give an example of the approximate method based on Guldinus' rule, that can be used for finding the area of a symmetrical body. In the illustration, the dimensions in common fractions are the known dimensions; those in decimals are found by actual measurements on a figure drawn to scale. The method for finding the area is as follows: First separate such areas as are cylindrical, conical or spherical, as these can be found by exact formulas. In the illustration *ABCD* is a cylinder, the area of the surface of which can be easily found. The top area *EF* is simply a circular area, and can thus be computed separately. The remainder of the surface generated by rotating line *AF* about the axis *GH*

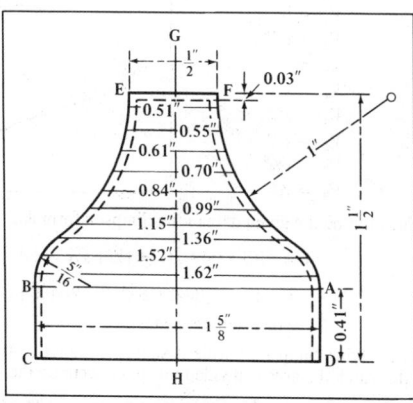

is found by the approximate method explained in the previous section. From point *A*, set off equal distances on line *AF*. In the present case each division indicated is ⅛ inch long. From the central or middle point of each of these parts draw a line at right angles to the axis of rotation *GH*, measure the length of these lines or diameters (the length of each is given in decimals), add all these lengths together and multiply the sum by the length of one division set off on line *AF* (in this case, ⅛ inch), and multiply this product by π. This gives the approximate area of the surface of revolution.

In setting off divisions ⅛ inch long along line *AF*, the last division does not reach exactly to point *F*, but only to a point 0.03 inch below it. The part 0.03 inch high at the top of the cup, can be considered as a cylinder of ½ inch diameter and 0.03 inch height, the area of the cylindrical surface of which is easily computed. By adding the various surfaces together the total surface of the cup is found as below:

Cylinder, 1⅝ inch diameter, 0.41 inch high . . . . 2.093 square inches
Circle, ½ inch diameter . . . . 0.196 square inches
Cylinder, ½ inch diameter, 0.03 inch high . . . . 0.047 square inches
Irregular surface . . . . 3.868 square inches

      Total. . . . . 6.204 square inches

Area of Plane Surfaces of Irregular Outline. — One of the most useful and accurate methods for determining the approximate area of a plane figure or irregular outline is known as *Simpson's Rule*. In applying Simpson's Rule to find an area the work is done in four steps:

1. Divide the area into an *even* number, *N*, of parallel strips of equal width *W*; for example, in the accompanying diagram the area has been divided into 8 strips of equal width;

2. Label the sides of the strips V_0, V_1, V_2, etc. up to V_N;

3. Measure the heights V_0, V_1, V_2, . . . V_N of the sides of the strips;

4. Substitute the heights V_0, V_1, etc. in the following formula to find the area A of the figure:

$$A = \frac{W}{3}[(V_0 + V_N) + 4(V_1 + V_3 + \cdots V_{N-1}) + 2(V_2 + V_4 + \cdots V_{N-2})]$$

Example: The area of the accompanying figure was divided into 8 strips on a full-size drawing and the following data obtained. Calculate the area using Simpson's Rule.

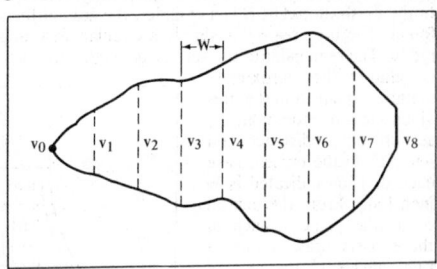

$W = \frac{1}{2}''$
$V_0 = 0''$
$V_1 = \frac{3}{4}''$
$V_2 = 1\frac{1}{4}''$
$V_3 = 1\frac{1}{2}''$
$V_4 = 1\frac{5}{8}''$
$V_5 = 2\frac{1}{4}''$
$V_6 = 2\frac{1}{2}''$
$V_7 = 1\frac{3}{4}''$
$V_8 = \frac{1}{2}''$

Substituting the given data in the Simpson formula,

$$A = \frac{\frac{1}{2}}{3}[(0 + \frac{1}{2}) + 4(\frac{3}{4} + 1\frac{1}{2} + 2\frac{1}{4} + 1\frac{3}{4}) + 2(1\frac{1}{4} + 1\frac{5}{8} + 2\frac{1}{2})]$$

$$= \frac{1}{6}[(\frac{1}{2}) + 4(6\frac{1}{4}) + 2(5\frac{3}{8})] = \frac{1}{6}[36\frac{1}{4}]$$

$$= 6.04 \text{ square inches}$$

In applying Simpson's Rule it should be noted that the larger the number of strips into which the area is divided the more accurate the results obtained.

Areas Enclosed by Cycloidal Curves. — The area between a cycloid and the straight line upon which the generating circle rolls, equals three times the area of the generating circle (see diagram, page 56). The areas between epicycloidal and hypocycloidal curves and the "fixed circle" upon which the generating circle is rolled, may be determined by the following formulas, in which a = radius of the fixed circle upon which the generating circle rolls; b = radius of the generating circle; A = the area for the epicycloidal curve; and A_1 = the area for the hypocycloidal curve.

$$A = \frac{3.1416b^2(3a + 2b)}{a}; \qquad A_1 = \frac{3.1416b^2(3a - 2b)}{a}$$

Find the Contents of Cylindrical Tanks at Different Levels. — In conjunction with the table "Segments of Circles of Radius = 1," presented on pages 70 and 71, the following relations can give a close approximation of the liquid contents, at any level, in a cylindrical tank.

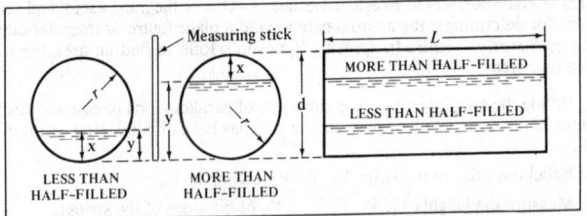

A long measuring rule calibrated in length units or simply a plain stick can be used for measuring contents at a particular level. In turn, the rule or stick can be graduated to serve as a volume gauge for the tank in question. The only requirements are: that the cross section of the tank is circular; the tank's dimensions are known; the gauge rod is inserted vertically through the top center of the tank so that it rests on the exact bottom of the tank; and that consistent English *or* metric units are used throughout the calculations.

(1) Tank Constant = $K = Cr^2L$ (remains the same for any given tank); (2) For a tank that is completely full: $V_T = \pi K$; (3) $V_s = KA$; (4) $V = V_s$, when tank is less than half full; (5) $V = V_T - V_s = V_T - KA$, when tank is more than half full.

Where

C = liquid volume conversion factor, the exact value of which depends on the length and liquid volume units being used during measurement: 0.00433 U.S. gal/in.^{3}; 7.48 U.S. gal/ft^3; 0.00360 U.K. gal/in.^{3}; 6.23 U.K. gal/ft^3; 0.001 liters/cm^3; or 1000 liters/m^3;

V_T = total volume of liquid tank can hold;

V_s = volume formed by segment of circle having depth = x in given tank (see diagram);

V = volume of liquid at particular level in tank;

d = diameter of tank; L = length of tank; r = radius of tank (= ½ diameter);

A = segment area of a corresponding unit circle taken from pages 70 or 71;

y = actual depth of contents in tank as shown on a gauge rod or stick;

x = depth of the segment of a circle to be considered in given tank. As can be seen in above diagram, x is the actual depth of contents (y) when the tank is less than half full, and is the depth of the void $(d - y)$ above the contents when the tank is more than half full. From pages 70 and 71 it can also be seen that h, the height of a segment of a corresponding unit circle, is x/r.

Example: A tank is 20 feet long and 6 feet in diameter. Convert a long inch-stick into a gauge that is graduated at 1000 and 3000 U.S. gallons.

$$L = 20 \times 12 = 240 \text{ in.}; r = \text{⁶⁄₂} \times 12 = 36 \text{ in.}$$

From formula (1): $K = 0.00433(36)^2(240) = 1347.$

From formula (2): $V_T = 3.142 \times 1347 = 4232$ U.S. gal.

The 72-inch mark from the bottom on the inch-stick can be graduated for the rounded full volume "4230"; and the halfway point 36" for 4230/2 or "2115." It can be seen that the 1000 gal mark would be below the halfway mark. From formulas (3) and (4):

$$A_{1000} = \frac{1000}{1347} = 0.7424; \text{ from page 71, } h \text{ can be interpolated as } 0.5724^*; \text{ and}$$

$x = y = 36 \times 0.5724 = 20.61.$

Therefore, 1000 gal mark is graduated 20⅝" from bottom of rod.

It can be seen that the 3000 mark would be above the halfway mark. Therefore, the circular segment considered is the cross section of the void space at the top of the tank. From formulas (3) and (5):

$$A_{3000} = \frac{4230 - 3000}{1347} = 0.9131; h = 0.6648; x = 36 \times 0.6648 = 23.93".$$

Therefore, 3000 gal mark is $72.00 - 23.93 = 48.07$, or at the 48¹⁄₁₆" mark from the bottom.

* If the desired level of accuracy permits, interpolation can be omitted by choosing h directly from the table for the value of A nearest that calculated above.

Areas and Dimensions of Plane Figures

In the following tables are given formulas for the areas of plane figures, together with other formulas relating to their dimensions and properties; the surfaces of solids; and the volumes of solids. The notation used in the formulas is, as far as possible, given in the illustration accompanying them; where this has not been possible, it is given at the beginning of each set of formulas.

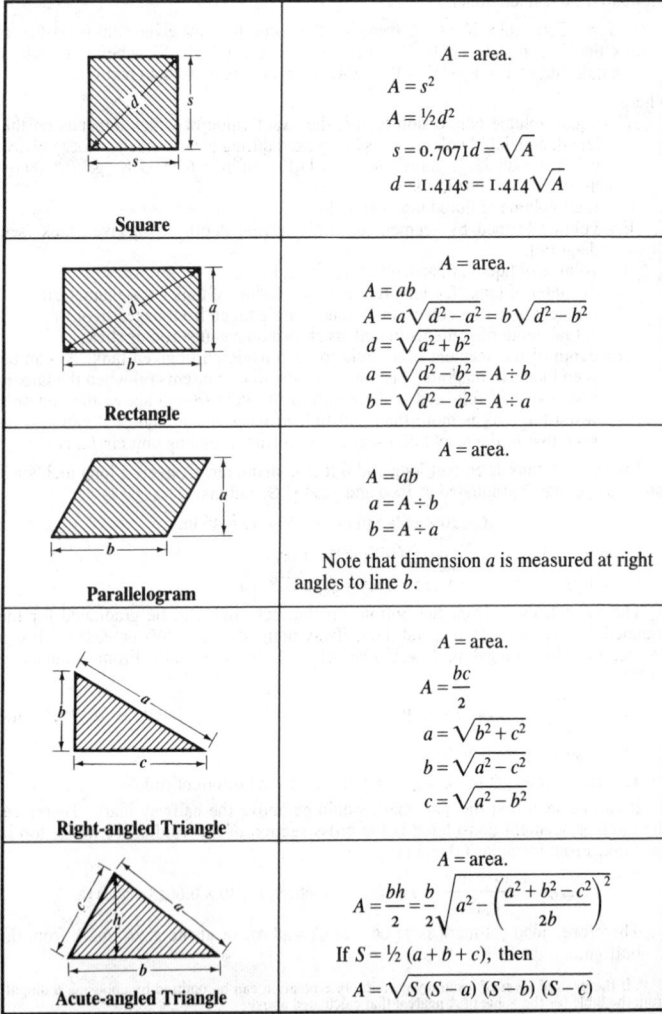

Square	A = area. $A = s^2$ $A = \frac{1}{2}d^2$ $s = 0.7071d = \sqrt{A}$ $d = 1.414s = 1.414\sqrt{A}$
Rectangle	A = area. $A = ab$ $A = a\sqrt{d^2 - a^2} = b\sqrt{d^2 - b^2}$ $d = \sqrt{a^2 + b^2}$ $a = \sqrt{d^2 - b^2} = A \div b$ $b = \sqrt{d^2 - a^2} = A \div a$
Parallelogram	A = area. $A = ab$ $a = A \div b$ $b = A \div a$ Note that dimension a is measured at right angles to line b.
Right-angled Triangle	A = area. $A = \dfrac{bc}{2}$ $a = \sqrt{b^2 + c^2}$ $b = \sqrt{a^2 - c^2}$ $c = \sqrt{a^2 - b^2}$
Acute-angled Triangle	A = area. $A = \dfrac{bh}{2} = \dfrac{b}{2}\sqrt{a^2 - \left(\dfrac{a^2 + b^2 - c^2}{2b}\right)^2}$ If $S = \frac{1}{2}(a + b + c)$, then $A = \sqrt{S(S-a)(S-b)(S-c)}$

Examples of the Use of the Formulas (English and metric units)

Below are given examples, some in English and some in metric units, showing the use of the formulas on the opposite page. Each section corresponds to the opposite section on the previous page, and the illustration on that page should be referred to. The notation used in the illustrations is also used in the examples given.

Square. — Assume that the side s of a square is 15 inches. Find the area and the length of the diagonal.

$$\text{Area} = A = s^2 = 15^2 = 225 \text{ square inches.}$$
$$\text{Diagonal} = d = 1.414s = 1.414 \times 15 = 21.21 \text{ inches.}$$

The area of a square is 625 square inches. Find the length of the side s and the diagonal d.

$$s = \sqrt{A} = \sqrt{625} = 25 \text{ inches.}$$
$$d = 1.414\sqrt{A} = 1.414 \times 25 = 35.35 \text{ inches.}$$

Rectangle. — The side a of a rectangle is 12 centimeters, and the area 70.5 square centimeters. Find the length of the side b, and the diagonal d.

$$b = A \div a = 70.5 \div 12 = 5.875 \text{ centimeters.}$$
$$d = \sqrt{a^2 + b^2} = \sqrt{12^2 + 5.875^2} = \sqrt{178.516} = 13.361 \text{ centimeters.}$$

The sides of a rectangle are 30.5 and 11 centimeters long. Find the area.

$$\text{Area} = a \times b = 30.5 \times 11 = 335.5 \text{ square centimeters.}$$

Parallelogram. — The base b of a parallelogram is 16 feet. The height a is 5.5 feet. Find the area.

$$\text{Area} = A = a \times b = 5.5 \times 16 = 88 \text{ square feet.}$$

The area of a parallelogram is 12 square inches. The height is 1.5 inches. Find the length of the base b.

$$b = A \div a = 12 \div 1.5 = 8 \text{ inches.}$$

Right-angled Triangle. — The sides b and c in a right-angled triangle are 6 and 8 inches. Find side a and the area.

$$a = \sqrt{b^2 + c^2} = \sqrt{6^2 + 8^2} = \sqrt{36 + 64} = \sqrt{100} = 10 \text{ inches.}$$
$$A = \frac{b \times c}{2} = \frac{6 \times 8}{2} = \frac{48}{2} = 24 \text{ square inches.}$$

If $a = 10$ and $b = 6$ had been known, but not c, the latter would have been found as follows:

$$c = \sqrt{a^2 - b^2} = \sqrt{10^2 - 6^2} = \sqrt{100 - 36} = \sqrt{64} = 8 \text{ inches.}$$

Acute-angled Triangle. — If $a = 10$, $b = 9$, and $c = 8$ centimeters, what is the area of the triangle?

$$A = \frac{b}{2}\sqrt{a^2 - \left(\frac{a^2 + b^2 - c^2}{2b}\right)^2} = \frac{9}{2}\sqrt{10^2 - \left(\frac{10^2 + 9^2 - 8^2}{2 \times 9}\right)^2} = 4.5\sqrt{100 - \left(\frac{117}{18}\right)^2}$$

$$= 4.5\sqrt{100 - 42.25} = 4.5\sqrt{57.75} = 4.5 \times 7.60 = 34.20 \text{ square centimeters.}$$

Areas and Dimensions of Plane Figures

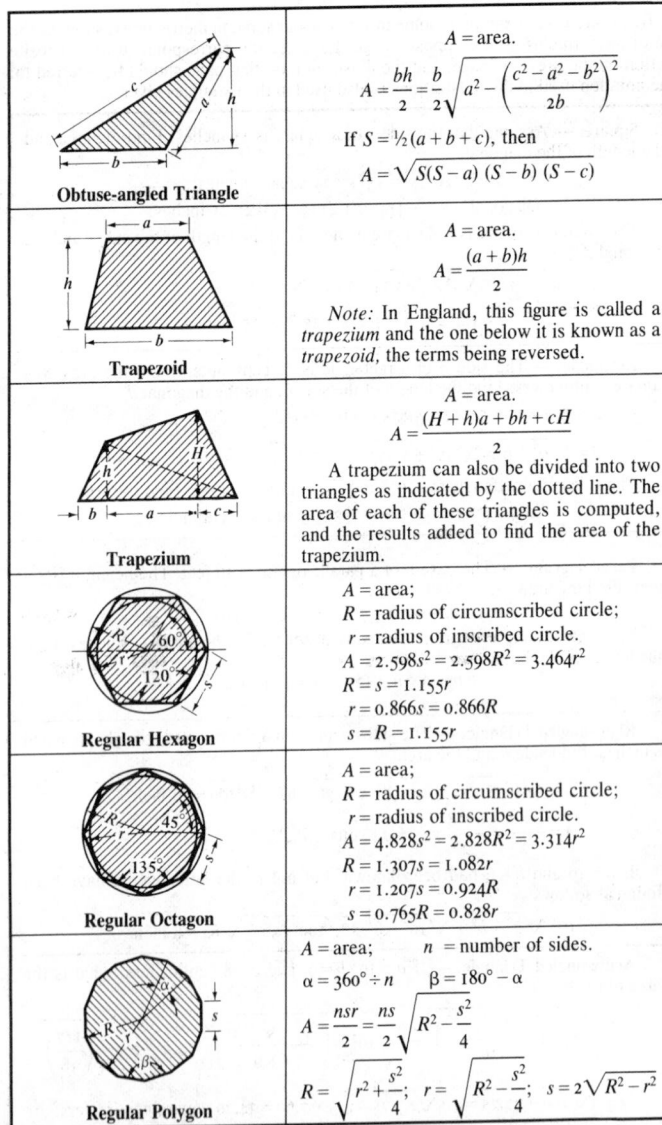

Obtuse-angled Triangle	A = area. $$A = \frac{bh}{2} = \frac{b}{2}\sqrt{a^2 - \left(\frac{c^2 - a^2 - b^2}{2b}\right)^2}$$ If $S = \frac{1}{2}(a + b + c)$, then $$A = \sqrt{S(S - a)(S - b)(S - c)}$$
Trapezoid	A = area. $$A = \frac{(a + b)h}{2}$$ *Note:* In England, this figure is called a *trapezium* and the one below it is known as a *trapezoid,* the terms being reversed.
Trapezium	A = area. $$A = \frac{(H + h)a + bh + cH}{2}$$ A trapezium can also be divided into two triangles as indicated by the dotted line. The area of each of these triangles is computed, and the results added to find the area of the trapezium.
Regular Hexagon	A = area; R = radius of circumscribed circle; r = radius of inscribed circle. $A = 2.598s^2 = 2.598R^2 = 3.464r^2$ $R = s = 1.155r$ $r = 0.866s = 0.866R$ $s = R = 1.155r$
Regular Octagon	A = area; R = radius of circumscribed circle; r = radius of inscribed circle. $A = 4.828s^2 = 2.828R^2 = 3.314r^2$ $R = 1.307s = 1.082r$ $r = 1.207s = 0.924R$ $s = 0.765R = 0.828r$
Regular Polygon	A = area; n = number of sides. $\alpha = 360° \div n$ $\beta = 180° - \alpha$ $$A = \frac{nsr}{2} = \frac{ns}{2}\sqrt{R^2 - \frac{s^2}{4}}$$ $$R = \sqrt{r^2 + \frac{s^2}{4}};\quad r = \sqrt{R^2 - \frac{s^2}{4}};\quad s = 2\sqrt{R^2 - r^2}$$

Examples of the Use of the Formulas (English and metric units)

Obtuse-angled Triangle. — The side $a = 5$, side $b = 4$, and side $c = 8$ inches. Find the area.

$$S = \tfrac{1}{2}(a + b + c) = \tfrac{1}{2}(5 + 4 + 8) = \tfrac{1}{2} \times 17 = 8.5$$

$$A = \sqrt{S(S-a)(S-b)(S-c)} = \sqrt{8.5\,(8.5-5)\,(8.5-4)\,(8.5-8)}$$

$$= \sqrt{8.5 \times 3.5 \times 4.5 \times 0.5} = \sqrt{66.937} = 8.18 \text{ square inches.}$$

Trapezoid. — Side $a = 23$ meters, side $b = 32$ meters, and height $h = 12$ meters. Find the area.

$$A = \frac{(a+b)h}{2} = \frac{(23+32)12}{2} = \frac{55 \times 12}{2} = \frac{660}{2} = 330 \text{ square meters.}$$

Trapezium. — Let $a = 10$, $b = 2$, $c = 3$, $h = 8$, and $H = 12$ inches. Find the area.

$$A = \frac{(H+h)a + bh + cH}{2} = \frac{(12+8)10 + 2 \times 8 + 3 \times 12}{2}$$

$$= \frac{20 \times 10 + 16 + 36}{2} = \frac{252}{2} = 126 \text{ square inches.}$$

Regular Hexagon. — The side s of a regular hexagon is 40 millimeters. Find the area and the radius r of the inscribed circle.

$$A = 2.598s^2 = 2.598 \times 40^2 = 2.598 \times 1600 = 4156.8 \text{ square millimeters.}$$

$$r = 0.866s = 0.866 \times 40 = 34.64 \text{ millimeters.}$$

What is the length of the side of a hexagon that is described about a circle of 50 millimeters radius? — Here $r = 50$. Hence,

$$s = 1.155r = 1.155 \times 50 = 57.75 \text{ millimeters.}$$

Regular Octagon. — Find the area and the length of the side of an octagon that is inscribed in a circle of 12 inches diameter.
Diameter of circumscribed circle = 12 inches; hence, $R = 6$ inches.

$$A = 2.828R^2 = 2.828 \times 6^2 = 2.828 \times 36 = 101.81 \text{ square inches.}$$

$$s = 0.765R = 0.765 \times 6 = 4.590 \text{ inches.}$$

Regular Polygon. — Find the area of a polygon having 12 sides, inscribed in a circle of 8 centimeters radius. The length of the side s is 4.141 centimeters.

$$A = \frac{ns}{2}\sqrt{R^2 - \frac{s^2}{4}} = \frac{12 \times 4.141}{2}\sqrt{8^2 - \frac{4.141^2}{4}} = 24.846\sqrt{59.713}$$

$$= 24.846 \times 7.727 = 191.98 \text{ square centimeters.}$$

Areas and Dimensions of Plane Figures

Circle	A = area; C = circumference. $A = \pi r^2 = 3.1416 r^2 = 0.7854 d^2$ $C = 2\pi r = 6.2832 r = 3.1416 d$ $r = C \div 6.2832 = \sqrt{A \div 3.1416} = 0.564\sqrt{A}$ $d = C \div 3.1416 = \sqrt{A \div 0.7854} = 1.128\sqrt{A}$ Length of arc for center-angle of $1° = 0.008727 d$ Length of arc for center-angle of $n° = 0.008727 nd$
Circular Sector	A = area; l = length of arc; α = angle, in degrees. $l = \dfrac{r \times \alpha \times 3.1416}{180} = 0.01745 r\alpha = \dfrac{2A}{r}$ $A = \tfrac{1}{2} rl = 0.008727 \alpha r^2$ $\alpha = \dfrac{57.296\, l}{r} \qquad r = \dfrac{2A}{l} = \dfrac{57.296\, l}{\alpha}$
Circular Segment	A = area; l = length of arc; α = angle, in degrees. $c = 2\sqrt{h(2r - h)} \qquad A = \tfrac{1}{2}[rl - c(r - h)]^*$ $r = \dfrac{c^2 + 4h^2}{8h} \qquad\qquad l = 0.01745 r\alpha$ $h = r - \tfrac{1}{2}\sqrt{4r^2 - c^2} \qquad \alpha = \dfrac{57.296\, l}{r}$ $h = r[1 - \cos(\alpha/2)] \qquad$ * See also p. 70.
Circular Ring	A = area. $A = \pi(R^2 - r^2) = 3.1416(R^2 - r^2)$ $= 3.1416(R + r)(R - r)$ $= 0.7854(D^2 - d^2) = 0.7854(D + d)(D - d)$
Circular Ring Sector	A = area; α = angle, in degrees. $A = \dfrac{\alpha\pi}{360}(R^2 - r^2) = 0.00873\alpha(R^2 - r^2)$ $= \dfrac{\alpha\pi}{4 \times 360}(D^2 - d^2) = 0.00218\alpha(D^2 - d^2)$
Spandrel or Fillet	A = area. $A = r^2 - \dfrac{\pi r^2}{4} = 0.215 r^2$ $= 0.1075 c^2$

Examples of the Use of the Formulas (English and metric units)

Circle. — Find the area A and circumference C of a circle with a diameter of 2¾ inches.

$A = 0.7854d^2 = 0.7854 \times 2.75^2 = 0.7854 \times 2.75 \times 2.75 = 5.9396$ square inches.

$C = 3.1416d = 3.1416 \times 2.75 = 8.6394$ inches.

The area of a circle is 16.8 square inches. Find its diameter.

$d = 1.128\sqrt{A} = 1.128\sqrt{16.8} = 1.128 \times 4.099 = 4.624$ inches.

Circular Sector. — The radius of a circle is 35 millimeters, and angle α of a sector of the circle is 60 degrees. Find the area of the sector and the length of arc l.

$A = 0.008727\alpha r^2 = 0.008727 \times 60 \times 35^2 = 0.5236 \times 35 \times 35$
$\qquad\qquad\qquad\qquad\qquad\qquad\quad = 641.41$ square millimeters
$\qquad\qquad\qquad\qquad\qquad\qquad\quad = 6.41$ square centimeters.

$l = 0.01745r\alpha = 0.01745 \times 35 \times 60 = 36.645$ millimeters.

Circular Segment. — The radius r of a circular segment is 60 inches and the height h is 8 inches. Find the length of the chord c.

$c = 2\sqrt{h(2r - h)} = 2\sqrt{8 \times (2 \times 60 - 8)} = 2\sqrt{896} = 2 \times 29.93 = 59.86$ inches.

If $c = 16$, and $h = 6$ inches, what is the radius of the circle of which the segment is a part?

$r = \dfrac{c^2 + 4h^2}{8h} = \dfrac{16^2 + 4 \times 6^2}{8 \times 6} = \dfrac{256 + 144}{48} = \dfrac{400}{48} = 8\frac{1}{3}$ inches.

Circular Ring. — Let the outside diameter $D = 12$ centimeters and the inside diameter $d = 8$ centimeters. Find area of ring.

$A = 0.7854\,(D^2 - d^2) = 0.7854\,(12^2 - 8^2) = 0.7854\,(144 - 64) = 0.7854 \times 80$
$\qquad = 62.83$ square centimeters.

By the alternative formula:

$A = 0.7854\,(D + d)\,(D - d) = 0.7854\,(12 + 8)\,(12 - 8) = 0.7854 \times 20 \times 4$
$\qquad = 62.83$ square centimeters.

Circular Ring Sector. — Find the area, if the outside radius $R = 5$ inches, the inside radius $r = 2$ inches, and $\alpha = 72$ degrees.

$A = 0.00873\alpha(R^2 - r^2) = 0.00873 \times 72(5^2 - 2^2)$
$\qquad = 0.6286(25 - 4) = 0.6286 \times 21 = 13.2$ square inches.

Spandrel or Fillet. — Find the area of a spandrel, the radius of which is 0.7 inch.

$A = 0.215r^2 = 0.215 \times 0.7^2 = 0.215 \times 0.7 \times 0.7 = 0.105$ square inch.

If chord c were given as 2.2 inches, what would be the area?

$A = 0.1075c^2 = 0.1075 \times 2.2^2 = 0.1075 \times 4.84 = 0.520$ square inch.

Areas and Dimensions of Plane Figures

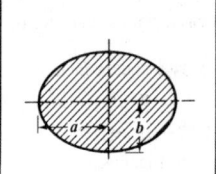 **Ellipse**	A = area; P = perimeter or circumference. $A = \pi ab = 3.1416ab$. An approximate formula for the perimeter is: $$P = 3.1416\sqrt{2(a^2 + b^2)}$$ A closer approximation is: $$P = 3.1416\sqrt{2(a^2 + b^2) - \frac{(a-b)^2}{2.2}}$$
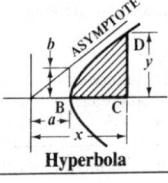 **Hyperbola**	A = area BCD. $$A = \frac{xy}{2} - \frac{ab}{2}\ln\left(\frac{x}{a} + \frac{y}{b}\right)$$
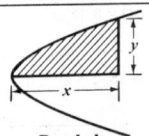 **Parabola**	l = length of arc. $$l = \frac{p}{2}\left[\sqrt{\frac{2x}{p}\left(1 + \frac{2x}{p}\right)} + \ln\left(\sqrt{\frac{2x}{p}} + \sqrt{1 + \frac{2x}{p}}\right)\right]$$ When x is small in proportion to y, the following is a close approximation: $$l = y\left[1 + \frac{2}{3}\left(\frac{x}{y}\right)^2 - \frac{2}{5}\left(\frac{x}{y}\right)^4\right], \text{ or } l = \sqrt{y^2 + \frac{4}{3}x^2}$$
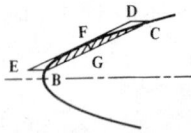 **Parabola**	A = area. $A = \frac{2}{3}xy$ (The area is equal to two-thirds of a rectangle which has x for its base and y for its height.)
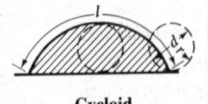 **Segment of Parabola**	A = area. Area $BFC = A = \frac{2}{3}$ area of parallelogram $BCDE$. If FG is the height of the segment, measured at right angles to BC, then: Area of segment $BFC = \frac{2}{3}BC \times FG$
 Cycloid	A = area; l = length of cycloid. $A = 3\pi r^2 = 9.4248r^2 = 2.3562d^2$ $\quad = 3 \times$ area of generating circle $l = 8r = 4d$

Examples of the Use of the Formulas (English and metric units)

Ellipse. — The larger or major axis is 200 millimeters. The smaller or minor axis is 150 millimeters. Find the area and the approximate circumference. Here, then, $a = 100$, and $b = 75$.

$A = 3.1416ab = 3.1416 \times 100 \times 75 = 23,562$ square millimeters
$$= 235.62 \text{ square centimeters.}$$

$$P = 3.1416\sqrt{2(a^2 + b^2)} = 3.1416\sqrt{2\,(100^2 + 75^2)} = 3.1416\sqrt{2 \times 15,625}$$

$$= 3.1416\sqrt{31,250} = 3.1416 \times 176.78 = 555.37 \text{ millimeters}$$
$$= 55.537 \text{ centimeters.}$$

Hyperbola. — The half-axes a and b are 3 and 2 inches, respectively. Find area shown shaded in illustration for $x = 8$ and $y = 5$.
Inserting the known values in the formula:

$$A = \frac{8 \times 5}{2} - \frac{3 \times 2}{2} \times \ln\left(\frac{8}{3} + \frac{5}{2}\right) = 20 - 3 \times \ln 5.167$$

$$= 20 - 3 \times 1.6423 = 20 - 4.927 = 15.073 \text{ square inches.}$$

Parabola. — If $x = 2$ and $y = 24$ feet, what is the approximate length l of the parabolic curve?

$$l = y\left[1 + \frac{2}{3}\left(\frac{x}{y}\right)^2 - \frac{2}{5}\left(\frac{x}{y}\right)^4\right] = 24\left[1 + \frac{2}{3}\left(\frac{2}{24}\right)^2 - \frac{2}{5}\left(\frac{2}{24}\right)^4\right]$$

$$= 24\left[1 + \frac{2}{3} \times \frac{1}{144} - \frac{2}{5} \times \frac{1}{20,736}\right] = 24 \times 1.0046 = 24.11 \text{ feet.}$$

Parabola. — Let the dimension x in the illustration be 15 centimeters, and y, 9 centimeters. Find the area of the shaded portion of the parabola.

$$A = \tfrac{2}{3} \times xy = \tfrac{2}{3} \times 15 \times 9 = 10 \times 9 = 90 \text{ square centimeters.}$$

Segment of Parabola. — The length of the chord $BC = 19.5$ inches. The distance between lines BC and DE, measured at right angles to BC, is 2.25 inches. This is the height of the segment. Find the area.

$$\text{Area} = A = \tfrac{2}{3}BC \times FG = \tfrac{2}{3} \times 19.5 \times 2.25 = 29.25 \text{ square inches.}$$

Cycloid. — The diameter of the generating circle of a cycloid is 6 inches. Find the length l of the cycloidal curve, and the area enclosed between the curve and the base line.

$$l = 4d = 4 \times 6 = 24 \text{ inches.}$$
$$A = 2.3562d^2 = 2.3562 \times 6^2 = 2.3562 \times 36 = 84.82 \text{ square inches.}$$

AREAS AND VOLUMES

Volumes of Solids

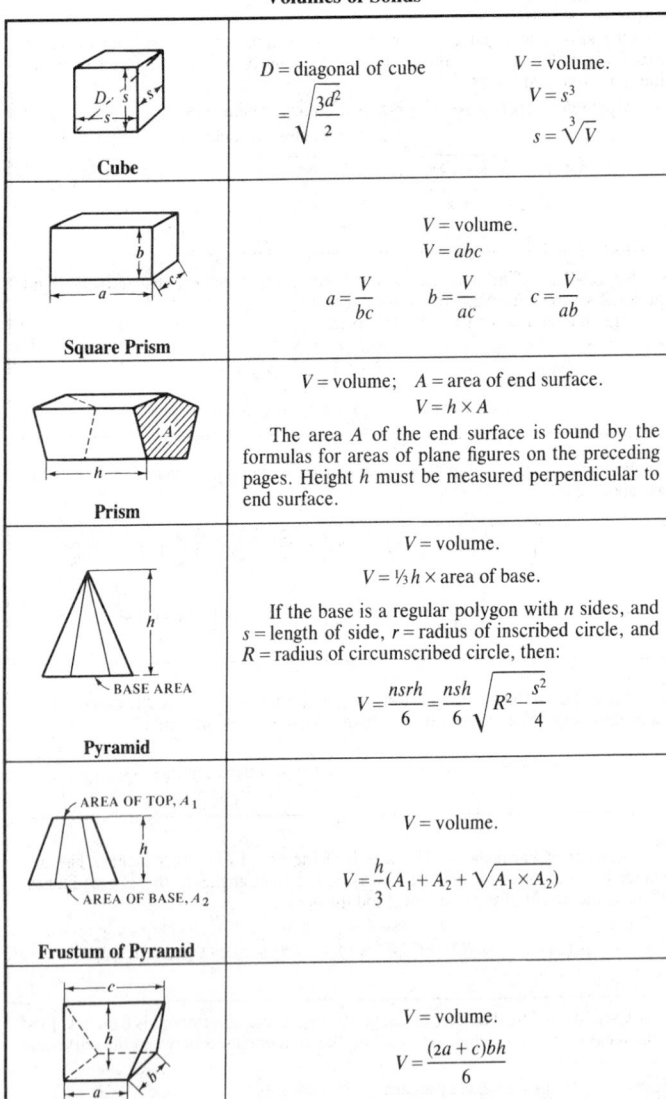

Cube	D = diagonal of cube $\qquad$ V = volume. $= \sqrt{\dfrac{3d^2}{2}}$ $\qquad\qquad$ $V = s^3$ $\qquad\qquad\qquad\qquad\qquad s = \sqrt[3]{V}$
Square Prism	V = volume. $V = abc$ $a = \dfrac{V}{bc} \qquad b = \dfrac{V}{ac} \qquad c = \dfrac{V}{ab}$
Prism	V = volume; $\quad A$ = area of end surface. $V = h \times A$ The area A of the end surface is found by the formulas for areas of plane figures on the preceding pages. Height h must be measured perpendicular to end surface.
Pyramid	V = volume. $V = \frac{1}{3} h \times$ area of base. If the base is a regular polygon with n sides, and s = length of side, r = radius of inscribed circle, and R = radius of circumscribed circle, then: $V = \dfrac{nsrh}{6} = \dfrac{nsh}{6} \sqrt{R^2 - \dfrac{s^2}{4}}$
Frustum of Pyramid	V = volume. $V = \dfrac{h}{3}(A_1 + A_2 + \sqrt{A_1 \times A_2})$
Wedge	V = volume. $V = \dfrac{(2a + c)bh}{6}$

Examples of the Use of the Formulas (English and metric units)

Cube. — The side of a cube equals 9.5 centimeters. Find its volume.

Volume = $V = s^3 = 9.5^3 = 9.5 \times 9.5 \times 9.5 = 857.375$ cubic centimeters.

The volume of a cube is 231 cubic centimeters. What is the length of the side?

$$s = \sqrt[3]{V} = \sqrt[3]{231} = 6.136 \text{ centimeters.}$$

Square Prism. — In a square prism, $a = 6$, $b = 5$, $c = 4$. Find the volume.

$$V = a \times b \times c = 6 \times 5 \times 4 = 120 \text{ cubic inches.}$$

How high should a box be made to contain 25 cubic feet, if it is 4 feet long and 2½ feet wide? Here, $a = 4$, $c = 2.5$, and $V = 25$. Then,

$$b = \text{depth} = \frac{V}{ac} = \frac{25}{4 \times 2.5} = \frac{25}{10} = 2.5 \text{ feet.}$$

Prism. — A prism having for its base a regular hexagon with a side s of 7.5 centimeters, is 25 centimeters high. Find the volume.

Area of hexagon = $A = 2.598s^2 = 2.598 \times 56.25 = 146.14$ square centimeters.

Volume of prism = $h \times A = 25 \times 146.14 = 3653.5$ cubic centimeters.

Pyramid. — A pyramid, having a height of 9 feet, has a base formed by a rectangle, the sides of which are 2 and 3 feet, respectively. Find the volume.

Area of base = $2 \times 3 = 6$ square feet; $h = 9$ feet.

Volume = $V = \frac{1}{3}h \times$ area of base = $\frac{1}{3} \times 9 \times 6 = 18$ cubic feet.

Frustum of Pyramid. — The pyramid in the previous example is cut off 4½ feet from the base, the upper part being removed. The sides of the rectangle forming the top surface of the frustum are, then, 1 and 1½ feet long, respectively. Find the volume of the frustum.

Area of top = $A_1 = 1 \times 1\frac{1}{2} = 1\frac{1}{2}$ sq. ft.   Area of base = $A_2 = 2 \times 3 = 6$ sq. ft.

$$V = \frac{4.5}{3}(1.5 + 6 + \sqrt{1.5 \times 6}) = 1.5(7.5 + \sqrt{9}) = 1.5 \times 10.5 = 15.75 \text{ cubic feet.}$$

Wedge. — Let $a = 4$ inches, $b = 3$ inches, and $c = 5$ inches. The height $h = 4.5$ inches. Find the volume.

$$V = \frac{(2a+c)bh}{6} = \frac{(2 \times 4 + 5) \times 3 \times 4.5}{6} = \frac{(8+5) \times 13.5}{6} = \frac{13 \times 13.5}{6}$$

$$= \frac{175.5}{6} = 29.25 \text{ cubic inches.}$$

Volumes of Solids

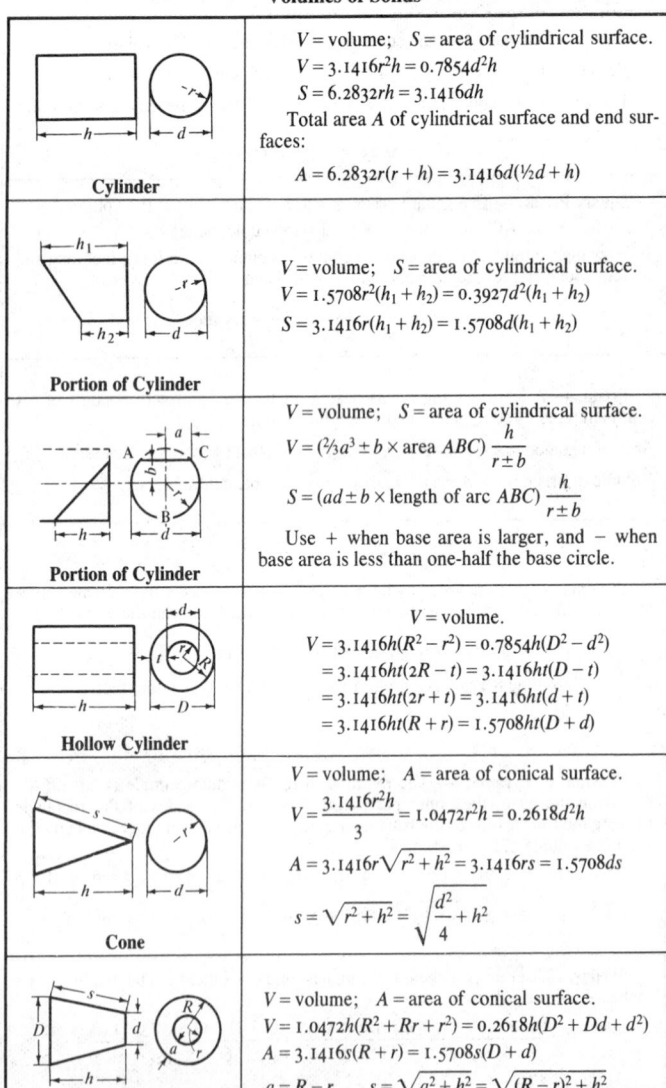

Cylinder

V = volume; S = area of cylindrical surface.
$V = 3.1416r^2h = 0.7854d^2h$
$S = 6.2832rh = 3.1416dh$
    Total area A of cylindrical surface and end surfaces:
$$A = 6.2832r(r + h) = 3.1416d(\tfrac{1}{2}d + h)$$

Portion of Cylinder

V = volume; S = area of cylindrical surface.
$V = 1.5708r^2(h_1 + h_2) = 0.3927d^2(h_1 + h_2)$
$S = 3.1416r(h_1 + h_2) = 1.5708d(h_1 + h_2)$

Portion of Cylinder

V = volume; S = area of cylindrical surface.
$$V = (\tfrac{2}{3}a^3 \pm b \times \text{area } ABC)\,\frac{h}{r \pm b}$$
$$S = (ad \pm b \times \text{length of arc } ABC)\,\frac{h}{r \pm b}$$
    Use + when base area is larger, and − when base area is less than one-half the base circle.

Hollow Cylinder

V = volume.
$V = 3.1416h(R^2 - r^2) = 0.7854h(D^2 - d^2)$
$ = 3.1416ht(2R - t) = 3.1416ht(D - t)$
$ = 3.1416ht(2r + t) = 3.1416ht(d + t)$
$ = 3.1416ht(R + r) = 1.5708ht(D + d)$

Cone

V = volume; A = area of conical surface.
$$V = \frac{3.1416r^2h}{3} = 1.0472r^2h = 0.2618d^2h$$
$$A = 3.1416r\sqrt{r^2 + h^2} = 3.1416rs = 1.5708ds$$
$$s = \sqrt{r^2 + h^2} = \sqrt{\frac{d^2}{4} + h^2}$$

Frustum of Cone

V = volume; A = area of conical surface.
$V = 1.0472h(R^2 + Rr + r^2) = 0.2618h(D^2 + Dd + d^2)$
$A = 3.1416s(R + r) = 1.5708s(D + d)$
$a = R - r \qquad s = \sqrt{a^2 + h^2} = \sqrt{(R - r)^2 + h^2}$

Examples of the Use of the Formulas (English and metric units)

Cylinder. — The diameter of a cylinder is 2.5 inches. The length or height is 20 inches. Find the volume, and the area of the cylindrical surface S.

$V = 0.7854d^2h = 0.7854 \times 2.5^2 \times 20 = 0.7854 \times 6.25 \times 20 = 98.17$ cubic inches.

$S = 3.1416dh = 3.1416 \times 2.5 \times 20 = 157.08$ square inches.

Portion of Cylinder. — A cylinder 125 millimeters in diameter, is cut off at an angle, as shown in the illustration. Dimension $h_1 = 150$, and $h_2 = 100$ mm. Find the volume and the area S of the cylindrical surface.

$V = 0.3927d^2(h_1 + h_2) = 0.3927 \times 125^2 \times (150 + 100)$

$\quad = 0.3927 \times 15,625 \times 250 = 1,533,984$ cubic millimeters $= 1534$ cm^3.

$S = 1.5708d(h_1 + h_2) = 1.5708 \times 125 \times 250$

$\quad = 49,087.5$ square millimeters $= 490.9$ square centimeters.

Portion of Cylinder. — Find the volume of a cylinder so cut off that line AC passes through the center of the base circle — that is, the base area is a half-circle. The diameter of the cylinder $= 5$ inches, and height $h = 2$ inches.

In this case $a = 2.5$; $b = 0$; area $ABC = 0.5 \times 0.7854 \times 5^2 = 9.82$; $r = 2.5$.

$$V = \left(\frac{2}{3} \times 2.5^3 + 0 \times 9.82\right)\frac{2}{2.5 + 0} = \frac{2}{3} \times 15.625 \times 0.8 = 8.33 \text{ cubic inches.}$$

Hollow Cylinder. — A cylindrical shell, 28 centimeters high, is 36 centimeters in outside diameter, and 4 centimeters thick. Find its volume.

$V = 3.1416ht(D - t) = 3.1416 \times 28 \times 4(36 - 4) = 3.1416 \times 28 \times 4 \times 32$

$\quad = 11,259.5$ cubic centimeters.

Cone. — Find the volume and area of conical surface of a cone, the base of which is a circle of 6 inches diameter, and the height of which is 4 inches.

$V = 0.2618d^2h = 0.2618 \times 6^2 \times 4 = 0.2618 \times 36 \times 4 = 37.7$ cubic inches.

$A = 3.1416r\sqrt{r^2 + h^2} = 3.1416 \times 3 \times \sqrt{3^2 + 4^2} = 9.4248 \times \sqrt{25}$

$\quad = 47.124$ square inches.

Frustum of Cone. — Find the volume of a frustum of a cone of the following dimensions: $D = 8$ centimeters; $d = 4$ centimeters; $h = 5$ centimeters.

$V = 0.2618 \times 5(8^2 + 8 \times 4 + 4^2) = 0.2618 \times 5(64 + 32 + 16)$

$\quad = 0.2618 \times 5 \times 112 = 146.61$ cubic centimeters.

Volumes of Solids

Sphere	V = volume; A = area of surface. $$V = \frac{4\pi r^3}{3} = \frac{\pi d^3}{6} = 4.1888r^3 = 0.5236d^3$$ $$A = 4\pi r^2 = \pi d^2 = 12.5664r^2 = 3.1416d^2$$ $$r = \sqrt[3]{\frac{3V}{4\pi}} = 0.6204\sqrt[3]{V}$$
Spherical Sector	V = volume; A = total area of conical and spherical surface. $$V = \frac{2\pi r^2 h}{3} = 2.0944r^2 h$$ $$A = 3.1416r(2h + \tfrac{1}{2}c)$$ $$c = 2\sqrt{h(2r - h)}$$
Spherical Segment	V = volume; A = area of spherical surface. $$V = 3.1416h^2\left(r - \frac{h}{3}\right) = 3.1416h\left(\frac{c^2}{8} + \frac{h^2}{6}\right)$$ $$A = 2\pi rh = 6.2832rh = 3.1416\left(\frac{c^2}{4} + h^2\right)$$ $$c = 2\sqrt{h(2r - h)}; \quad r = \frac{c^2 + 4h^2}{8h}$$
Spherical Zone	V = volume; A = area of spherical surface. $$V = 0.5236h\left(\frac{3c_1^2}{4} + \frac{3c_2^2}{4} + h^2\right)$$ $$A = 2\pi rh = 6.2832rh$$ $$r = \sqrt{\frac{c_2^2}{4} + \left(\frac{c_2^2 - c_1^2 - 4h^2}{8h}\right)^2}$$
Spherical Wedge	V = volume; A = area of spherical surface; α = center angle in degrees. $$V = \frac{\alpha}{360} \times \frac{4\pi r^3}{3} = 0.0116\alpha r^3$$ $$A = \frac{\alpha}{360} \times 4\pi r^2 = 0.0349\alpha r^2$$
Hollow Sphere	V = volume of material used to make a hollow sphere $$V = \frac{4\pi}{3}(R^3 - r^3) = 4.1888(R^3 - r^3)$$ $$= \frac{\pi}{6}(D^3 - d^3) = 0.5236(D^3 - d^3)$$

Examples of the Use of the Formulas (English and metric units)

Sphere. — Find volume and surface of a sphere 6.5 centimeters diam.

$$V = 0.5236d^3 = 0.5236 \times 6.5^3 = 0.5236 \times 6.5 \times 6.5 \times 6.5 = 143.79 \text{ cm}^3.$$

$$A = 3.1416d^2 = 3.1416 \times 6.5^2 = 3.1416 \times 6.5 \times 6.5 = 132.73 \text{ cm}^2.$$

The volume of a sphere is 64 cubic centimeters. Find its radius.

$$r = 0.6204\sqrt[3]{64} = 0.6204 \times 4 = 2.4816 \text{ centimeters.}$$

Spherical Sector. — Find the volume of a sector of a sphere 6 inches in diameter, the height h of the sector being 1.5 inch. Also find length of chord c. Here $r = 3$, and $h = 1.5$.

$$V = 2.0944r^2h = 2.0944 \times 3^2 \times 1.5 = 2.0944 \times 9 \times 1.5 = 28.27 \text{ cubic inches.}$$

$$c = 2\sqrt{h(2r - h)} = 2\sqrt{1.5(2 \times 3 - 1.5)} = 2\sqrt{6.75} = 2 \times 2.598$$
$$= 5.196 \text{ inches.}$$

Spherical Segment. — A segment of a sphere has the following dimensions: $h = 50$ millimeters; $c = 125$ millimeters. Find the volume V and the radius of the sphere of which the segment is a part.

$$V = 3.1416 \times 50 \times \left(\frac{125^2}{8} + \frac{50^2}{6}\right) = 157.08 \times \left(\frac{15,625}{8} + \frac{2500}{6}\right)$$
$$= 157.08 \times 2369.79 = 372,247 \text{ cubic millimeters} = 372 \text{ cm}^3.$$
$$r = \frac{125^2 + 4 \times 50^2}{8 \times 50} = \frac{15,625 + 10,000}{400} = \frac{25,625}{400} = 64 \text{ millimeters.}$$

Spherical Zone. — In a spherical zone, let $c_1 = 3$; $c_2 = 4$; and $h = 1.5$ inch. Find the volume.

$$V = 0.5236 \times 1.5 \times \left(\frac{3 \times 3^2}{4} + \frac{3 \times 4^2}{4} + 1.5^2\right) = 0.5236 \times 1.5 \times \left(\frac{27}{4} + \frac{48}{4} + 2.25\right)$$
$$= 0.5236 \times 1.5 \times 21 = 16.493 \text{ cubic inches.}$$

Spherical Wedge. — Find the area of the spherical surface and the volume of a wedge of a sphere. The diameter of the sphere is 100 millimeters, and the center angle α is 45 degrees.

$$V = 0.0116 \times 45 \times 50^3 = 0.0116 \times 45 \times 125,000$$
$$= 65,250 \text{ cubic millimeters} = 65.25 \text{ cubic centimeters.}$$
$$A = 0.0349 \times 45 \times 50^2 = 3926.25 \text{ square millimeters} = 39.26 \text{ cm}^2.$$

Hollow Sphere. — Find the volume of a hollow sphere, 8 inches in outside diameter, with a thickness of material of 1.5 inch.
Here $R = 4$; $r = 4 - 1.5 = 2.5$.

$$V = 4.1888(4^3 - 2.5^3) = 4.1888(64 - 15.625) = 4.1888 \times 48.375$$
$$= 202.63 \text{ cubic inches.}$$

Volumes of Solids

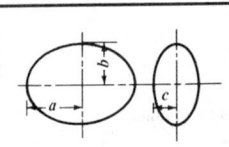 **Ellipsoid**	V = volume. $$V = \frac{4\pi}{3}abc = 4.1888abc$$ In an ellipsoid of revolution, or spheroid, where $c = b$: $$V = 4.1888ab^2$$
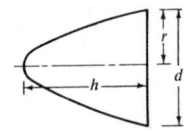 **Paraboloid**	V = volume; $V = \frac{1}{2}\pi r^2 h = 0.3927d^2h$ A = area; $A = \frac{2\pi}{3p}\left[\sqrt{\left(\frac{d^2}{4}+p^2\right)^3} - p^3\right]$ in which $$p = \frac{d^2}{8h}$$
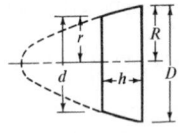 **Paraboloidal Segment**	V = volume. $$V = \frac{\pi}{2}h(R^2 + r^2) = 1.5708h(R^2 + r^2)$$ $$= \frac{\pi}{8}h(D^2 + d^2) = 0.3927h(D^2 + d^2)$$
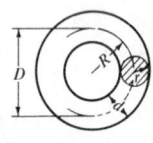 **Torus**	V = volume; A = area of surface. $$V = 2\pi^2 Rr^2 = 19.739Rr^2$$ $$= \frac{\pi^2}{4}Dd^2 = 2.4674Dd^2$$ $$A = 4\pi^2 Rr = 39.478Rr$$ $$= \pi^2 Dd = 9.8696Dd$$
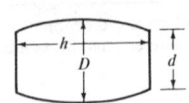 **Barrel**	V = approximate volume. If the sides are bent to the arc of a circle: $$V = \frac{1}{12}\pi h(2D^2 + d^2) = 0.262h(2D^2 + d^2)$$ If the sides are bent to the arc of a parabola: $$V = 0.209h(2D^2 + Dd + \frac{3}{4}d^2)$$
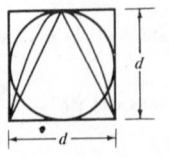	If d = base diameter and height of a cone, a paraboloid and a cylinder, and the diameter of a sphere, then the volumes of these bodies are to each other as below: $$\text{Cone : paraboloid : sphere : cylinder} = \frac{1}{3}:\frac{1}{2}:\frac{2}{3}:1$$

Examples of the Use of the Formulas (English and metric units)

Ellipsoid or Spheroid. — Find the volume of a spheroid in which $a = 5$, and $b = c = 1.5$ inch.

$$V = 4.1888 \times 5 \times 1.5^2 = 4.1888 \times 5 \times 2.25 = 47.124 \text{ cubic inches.}$$

Paraboloid. — Find the volume of a paraboloid in which $h = 300$ millimeters and $d = 125$ millimeters.

$$V = 0.3927 d^2 h = 0.3927 \times 125^2 \times 300 = 0.3927 \times 15,625 \times 300$$
$$= 1,840,781 \text{ cubic millimeters} = 1,840.8 \text{ cubic centimeters.}$$

Segment of Paraboloid. — Find the volume of a segment of a paraboloid in which $D = 5$ inches, $d = 3$ inches, and $h = 6$ inches.

$$V = 0.3927 h(D^2 + d^2) = 0.3927 \times 6 \times (5^2 + 3^2) = 0.3927 \times 6 \times (25 + 9)$$
$$= 0.3927 \times 6 \times 34 = 80.11 \text{ cubic inches.}$$

Torus. — Find the volume and area of surface of a torus in which $d = 1.5$ and $D = 5$ inches.

$$V = 2.4674 \times 5 \times 1.5^2 = 2.4674 \times 5 \times 2.25 = 27.76 \text{ cubic inches.}$$
$$A = 9.8696 \times 5 \times 1.5 = 74.022 \text{ square inches}$$

Barrel. — Find the approximate contents of a barrel, the inside dimensions of which are $D = 60$ centimeters; $d = 50$ centimeters; $h = 120$ centimeters.

$$V = 0.262 h(2D^2 + d^2) = 0.262 \times 120 \times (2 \times 60^2 + 50^2)$$
$$= 0.262 \times 120 \times (7200 + 2500) = 0.262 \times 120 \times 9700$$
$$= 304,968 \text{ cubic centimeters} = 0.305 \text{ cubic meter.}$$

Assume, as an example, that the diameter of the base of a cone, paraboloid, and cylinder is 2 inches, that the height is 2 inches, and that the diameter of a sphere is 2 inches. Then the volumes, written in formula-form, are as below:

Cone	Paraboloid	Sphere	Cylinder	
$\dfrac{3.1416 \times 2^2 \times 2}{12}$	$:\dfrac{3.1416 \times 2^2 \times 2}{8}$	$:\dfrac{3.1416 \times 2^3}{6}$	$:\dfrac{3.1416 \times 2^2 \times 2}{4}$	$= \dfrac{1}{3}:\dfrac{1}{2}:\dfrac{2}{3}:1$

Diameter of Circle Enclosing a Given Number of Smaller Circles. — Four of many possible compact arrangements of circles within a circle are shown at A, B, C, and D in Fig. 1. To determine the diameter of the smallest enclosing circle for a particular number of enclosed circles all of the same size, three factors that influence the size of the enclosing circle should be considered. These are discussed in the paragraphs that follow which are based on the article, *How Many Wires Can Be Packed into a Circular Conduit*, by Jacques Dutka, Machinery, October 1956.

1. Arrangement of Center or Core Circles: The four most common arrangements of center or core circles are shown cross-sectioned in Fig. 1. It may seem, offhand, that the "A" pattern would require the smallest enclosing circle for a given number of enclosed circles but this is not always the case since the most compact arrangement will, in part, depend on the number of circles to be enclosed.

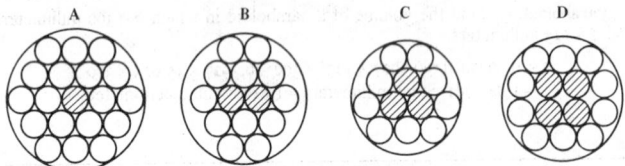

Fig. 1. Arrangements of Circles Within a Circle

2. Diameter of Enclosing Circle When Outer Layer of Circles is Complete: Successive, complete "layers" of circles may be placed around each of the central cores, Fig. 1, of 1, 2, 3, or 4 circles as the case may be. The number of circles contained in arrangements of complete "layers" around a central core of circles, as well as the diameter of the enclosing circle, may be obtained using the data in Table 1. Thus, for example, the "A" pattern in Fig. 1 shows, by actual count, a total of 19 circles arranged in two complete "layers" around a central core consisting of one circle; this agrees with the data shown in the left half of Table 1 for $n = 2$.

To determine the diameter of the enclosing circle, the data in the right half of Table 1 is used. Thus, for $n = 2$ and an "A" pattern, the diameter D is 5 times the diameter d of the enclosed circles.

3. Diameter of Enclosing Circle When Outer Layer of Circles is not Complete: In most cases it is possible to reduce the size of the enclosing circle from that required if the outer layer were complete. Thus, for example, the "B" pattern in Fig. 1 shows that the central core consisting of 2 circles is surrounded by 1 complete layer of 8 circles and 1 partial, outer layer of 4 circles so that the total number of circles enclosed is 14. If the outer layer was complete then (from Table 1) the total number of enclosed circles would be 24 and the diameter of the enclosing circle would be $6d$; however, since the outer layer is composed of only 4 circles out of a possible 14 for a complete second layer, a smaller diameter of enclosing circle may be used. Table 2 shows that for a total of 14 enclosed circles arranged in a "B" pattern with the outer layer of circles incomplete, the diameter for the enclosing circle is $4.606d$.

Table 2 can be used to determine the smallest enclosing circle for a given number of circles to be enclosed by direct comparison of the "A," "B," and "C" columns. For data outside the range of Table 2, use the formulas in Dr. Dutka's article.

Approximate Formula When Number of Enclosed Circles is Large: When a large number of circles are to be enclosed, the arrangement of the center circles has little effect on the diameter of enclosing circle. For numbers of circles greater than 10,000 the diameter of the enclosing circle may be calculated within 2 per cent from the formula: $D = d(1 + \sqrt{N \div 0.907})$. In this formula, D = diameter of enclosing circle; d = diameter of enclosed circles; and N is the number of enclosed circles.

Table 1. Number of Circles Contained in Complete Layers of Circles and Diameter of Enclosing Circle (English or metric units)

No. Complete Layers Over Core, n	Number of Circles in Center Pattern							
	1	2	3	4	1	2	3	4
	Arrangement of Circles in Center Pattern (see Fig. 1)							
	"A"	"B"	"C"	"D"	"A"	"B"	"C"	"D"
	Number of Circles, N, Enclosed				Diameter, D, of Enclosing Circle*			
0	1	2	3	4	d	$2d$	$2.155d$	$2.414d$
1	7	10	12	14	$3d$	$4d$	$4.055d$	$4.386d$
2	19	24	27	30	$5d$	$6d$	$6.033d$	$6.379d$
3	37	44	48	52	$7d$	$8d$	$8.024d$	$8.375d$
4	61	70	75	80	$9d$	$10d$	$10.018d$	$10.373d$
5	91	102	108	114	$11d$	$12d$	$12.015d$	$12.372d$
n	**	**	**	**	**	**	**	**

* Diameter D is given in terms of d, the diameter of the enclosed circles.
** For n complete layers over core, the number of enclosed circles N for "A" center pattern is $3n^2 + 3n + 1$; for "B," $3n^2 + 5n + 2$; for "C," $3n^2 + 6n + 3$; for "D," $3n^2 + 7n + 4$; while the diameter D of the enclosing circle for "A" center pattern is $(2n + 1)d$; for "B," $(2n + 2)d$; for "C," $(1 + 2\sqrt{n^2 + n + \frac{1}{3}})d$; and for "D," $(1 + \sqrt{4n^2 + 5.644n + 2})d$.

Table 2. Factors for Determining Diameter, D, of Smallest Enclosing Circle for Various Numbers, N, of Enclosed Circles* (English or metric units)

No. N	Center Circle Pattern			No. N	Center Circle Pattern			No. N	Center Circle Pattern		
	"A"	"B"	"C"		"A"	"B"	"C"		"A"	"B"	"C"
	Diameter Factor K				Diameter Factor K				Diameter Factor K		
2	3	2	. . .	34	7.001	7.083	7.111	66	9.718	9.545	9.327
3	3	2.733	2.155	35	7.001	7.245	7.111	67	9.718	9.545	9.327
4	3	2.733	3.310	36	7.001	7.245	7.111	68	9.718	9.545	9.327
5	3	3.646	3.310	37	7.001	7.245	7.430	69	9.718	9.661	9.327
6	3	3.646	3.310	38	7.929	7.245	7.430	70	9.718	9.661	10.019
7	3	3.646	4.056	39	7.929	7.558	7.430	71	9.718	9.889	10.019
8	4.465	3.646	4.056	40	7.929	7.558	7.430	72	9.718	9.889	10.019
9	4.465	4	4.056	41	7.929	7.558	7.430	73	9.718	9.889	10.019
10	4.465	4	4.056	42	7.929	7.558	7.430	74	10.166	9.889	10.019
11	4.465	4.606	4.056	43	7.929	8.001	8.024	75	10.166	10	10.019
12	4.465	4.606	4.056	44	8.212	8.001	8.024	76	10.166	10	10.238
13	4.465	4.606	5.164	45	8.212	8.001	8.024	77	10.166	10.540	10.238
14	5	4.606	5.164	46	8.212	8.001	8.024	78	10.166	10.540	10.238
15	5	5.359	5.164	47	8.212	8.001	8.024	79	10.166	10.540	10.452
16	5	5.359	5.164	48	8.212	8.001	8.024	80	10.166	10.540	10.452
17	5	5.359	5.164	49	8.212	8.550	8.572	81	10.166	10.540	10.452
18	5	5.359	5.164	50	8.212	8.550	8.572	82	10.166	10.540	10.452
19	5	5.583	5.619	51	8.212	8.550	8.572	83	10.166	10.540	10.452
20	6.292	5.583	5.619	52	8.212	8.550	8.572	84	10.166	10.540	10.452
21	6.292	5.583	5.619	53	8.212	8.811	8.572	85	10.166	10.644	10.866
22	6.292	5.583	6.034	54	8.212	8.811	8.572	86	11	10.644	10.866
23	6.292	6.001	6.034	55	8.212	8.811	9.083	87	11	10.644	10.866
24	6.292	6.001	6.034	56	9.001	8.811	9.083	88	11	10.644	10.866
25	6.292	6.197	6.034	57	9.001	8.938	9.083	89	11	10.849	10.866
26	6.292	6.197	6.034	58	9.001	8.938	9.083	90	11	10.849	10.866
27	6.292	6.568	6.034	59	9.001	8.938	9.083	91	11	10.849	11.067
28	6.292	6.568	6.774	60	9.001	8.938	9.083	92	11.393	10.849	11.067
29	6.292	6.568	6.774	61	9.001	9.186	9.083	93	11.393	11.149	11.067
30	6.292	6.568	6.774	62	9.718	9.186	9.083	94	11.393	11.149	11.067
31	6.292	7.083	7.111	63	9.718	9.186	9.083	95	11.393	11.149	11.067
32	7.001	7.083	7.111	64	9.718	9.186	9.327	96	11.393	11.149	11.067
33	7.001	7.083	7.111	65	9.718	9.545	9.327	97	11.393	11.441	11.264

See footnote at end of table.

Table 2. (*Concluded*) **Factors for Determining Diameter, D, of Smallest Enclosing Circle for Various Numbers, N, of Enclosed Circles*** (English or metric units)

No.	Center Circle Pattern			No.	Center Circle Pattern			No.	Center Circle Pattern		
	"A"	"B"	"C"		"A"	"B"	"C"		"A"	"B"	"C"
N	Diameter Factor K			N	Diameter Factor K			N	Diameter Factor K		
98	11.584	11.441	11.264	153	14..15	14	14.013	208	16.100	16	16.144
99	11.584	11.441	11.264	154	14.115	14	14.013	209	16.100	16.133	16.144
100	11.584	11.441	11.264	155	14.115	14.077	14.013	210	16.100	16.133	16.144
101	11.584	11.536	11.264	156	14.115	14.077	14.013	211	16.100	16.133	16.144
102	11.584	11.536	11.264	157	14.115	14.077	14.317	212	16.621	16.133	16.144
103	11.584	11.536	12.016	158	14.115	14.077	14.317	213	16.621	16.395	16.144
104	11.584	11.536	12.016	159	14.115	14.229	14.317	214	16.621	16.395	16.276
105	11.584	11.817	12.016	160	14.115	14.229	14.317	215	16.621	16.395	16.276
106	11.584	11.817	12.016	161	14.115	14.229	14.317	216	16.621	16.395	16.276
107	11.584	11.817	12.016	162	14.115	14.229	14.317	217	16.621	16.525	16.276
108	11.584	11.817	12.016	163	14.115	14.454	14.317	218	16.621	16.525	16.276
109	11.584	12	12.016	164	14.857	14.454	14.317	219	16.621	16.525	16.276
110	12.136	12	12.016	165	14.857	14.454	14.317	220	16.621	16.525	16.535
111	12.136	12.270	12.016	166	14.857	14.454	14.317	221	16.621	16.589	16.535
112	12.136	12.270	12.016	167	14.857	14.528	14.317	222	16.621	16.589	16.535
113	12.136	12.270	12.016	168	14.857	14.528	14.317	223	16.621	16.716	16.535
114	12.136	12.270	12.016	169	14.857	14.528	14.614	224	16.875	16.716	16.535
115	12.136	12.358	12.373	170	15	14.528	14.614	225	16.875	16.716	16.535
116	12.136	12.358	12.373	171	15	14.748	14.614	226	16.875	16.716	17.042
117	12.136	12.358	12.373	172	15	14.748	14.614	227	16.875	16.716	17.042
118	12.136	12.358	12.373	173	15	14.748	14.614	228	16.875	16.716	17.042
119	12.136	12.533	12.373	174	15	14.748	14.614	229	16.875	16.716	17.042
120	12.136	12.533	12.373	175	15	14.893	15.048	230	16.875	17.094	17.042
121	12.136	12.533	12.548	176	15	14.893	15.048	231	16.875	17.094	17.042
122	13	12.533	12.548	177	15	14.893	15.048	232	16.875	17.094	17.166
123	13	12.533	12.548	178	15	14.893	15.048	233	16.875	17.094	17.166
124	13	12.533	12.719	179	15	15.107	15.048	234	16.875	17.094	17.166
125	13	12.533	12.719	180	15	15.107	15.048	235	16.875	17.094	17.166
126	13	12.533	12.719	181	15	15.107	15.190	236	17	17.094	17.166
127	13	12.790	12.719	182	15	15.107	15.190	237	17	17.094	17.166
128	13.166	12.790	12.719	183	15	15.178	15.190	238	17	17.094	17.166
129	13.166	12.790	12.719	184	15	15.178	15.190	239	17	17.463	17.166
130	13.166	12.790	13.056	185	15	15.178	15.190	240	17	17.463	17.166
131	13.166	13.125	13.056	186	15	15.178	15.190	241	17	17.463	17.290
132	13.166	13.125	13.056	187	15	15.526	15.469	242	17.371	17.463	17.290
133	13.166	13.125	13.056	188	15.423	15.526	15.469	243	17.371	17.523	17.290
134	13.166	13.125	13.056	189	15.423	15.526	15.469	244	17.371	17.523	17.290
135	13.166	13.125	13.056	190	15.423	15.526	15.469	245	17.371	17.523	17.290
136	13.166	13.125	13.221	191	15.423	15.731	15.469	246	17.371	17.523	17.290
137	13.166	13.289	13.221	192	15.423	15.731	15.469	247	17.371	17.523	17.654
138	13.166	13.289	13.221	193	15.423	15.731	15.743	248	17.371	17.523	17.654
139	13.166	13.289	13.221	194	15.423	15.731	15.743	249	17.371	17.523	17.654
140	13.490	13.289	13.221	195	15.423	15.731	15.743	250	17.371	17.523	17.654
141	13.490	13.530	13.221	196	15.423	15.731	15.743	251	17.371	17.644	17.654
142	13.490	13.530	13.702	197	15.423	15.731	15.743	252	17.371	17.644	17.654
143	13.490	13.530	13.702	198	15.423	15.731	15.743	253	17.371	17.644	17.773
144	13.490	13.530	13.702	199	15.423	15.799	16.012	254	18.089	17.644	17.773
145	13.490	13.768	13.859	200	16.100	15.799	16.012	255	18.089	17.704	17.773
146	13.490	13.768	13.859	201	16.100	15.799	16.012	256	18.089	17.704	17.773
147	13.490	13.768	13.859	202	16.100	15.799	16.012	257	18.089	17.704	17.773
148	13.490	13.768	13.859	203	16.100	15.934	16.012	258	18.089	17.704	17.773
149	13.490	14	13.859	204	16.100	15.934	16.012	259	18.089	17.823	18.010
150	13.490	14	13.859	205	16.100	15.934	16.012	260	18.089	17.823	18.010
151	13.490	14	14.013	206	16.100	15.934	16.012	261	18.089	17.823	18.010
152	14.115	14	14.013	207	16.100	16	16.012	262	18.089	17.823	18.010

* The diameter D of the enclosing circle is equal to the diameter factor, K, multiplied by d, the diameter of the enclosed circles or $D = K \times d$. For example, if the number of circles to be enclosed, N, is 12, and the center circle arrangement is "C," then for $d = 1\frac{1}{2}$ inches, $D = 4.056 \times 1\frac{1}{2} = 6.084$ inches. If $d = 50$ millimeters, then $D = 4.056 \times 50 = 202.9$ millimeters.

Diameters of Circles and Sides of Squares of Equal Area

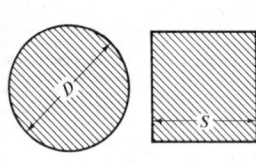

The table below will be found useful for determining the diameter of a circle of an area equal to that of a square, the side of which is known, or for determining the side of a square which has an area equal to that of a circle, the area or diameter of which is known. For example, if the diameter of a circle is 17½ inches, it is found from the table that the side of a square of the same area is 15.51 inches.

Diam. of Circle, D	Side of Square, S	Area of Circle or Square	Diam. of Circle, D	Side of Square, S	Area of Circle or Square	Diam. of Circle, D	Side of Square S	Area of Circle or Square
½	0.44	0.196	20½	18.17	330.06	40½	35.89	1288.25
1	0.89	0.785	21	18.61	346.36	41	36.34	1320.25
1½	1.33	1.767	21½	19.05	363.05	41½	36.78	1352.65
2	1.77	3.142	22	19.50	380.13	42	37.22	1385.44
2½	2.22	4.909	22½	19.94	397.61	42½	37.66	1418.63
3	2.66	7.069	23	20.38	415.48	43	38.11	1452.20
3½	3.10	9.621	23½	20.83	433.74	43½	38.55	1486.17
4	3.54	12.566	24	21.27	452.39	44	38.99	1520.53
4½	3.99	15.904	24½	21.71	471.44	44½	39.44	1555.28
5	4.43	19.635	25	22.16	490.87	45	39.88	1590.43
5½	4.87	23.758	25½	22.60	510.71	45½	40.32	1625.97
6	5.32	28.274	26	23.04	530.93	46	40.77	1661.90
6½	5.76	33.183	26½	23.49	551.55	46½	41.21	1698.23
7	6.20	38.485	27	23.93	572.56	47	41.65	1734.94
7½	6.65	44.179	27½	24.37	593.96	47½	42.10	1772.05
8	7.09	50.265	28	24.81	615.75	48	42.54	1809.56
8½	7.53	56.745	28½	25.26	637.94	48½	42.98	1847.45
9	7.98	63.617	29	25.70	660.52	49	43.43	1885.74
9½	8.42	70.882	29½	26.14	683.49	49½	43.87	1924.42
10	8.86	78.540	30	26.59	706.86	50	44.31	1963.50
10½	9.31	86.590	30½	27.03	730.62	50½	44.75	2002.96
11	9.75	95.033	31	27.47	754.77	51	45.20	2042.82
11½	10.19	103.87	31½	27.92	779.31	51½	45.64	2083.07
12	10.63	113.10	32	28.36	804.25	52	46.08	2123.72
12½	11.08	122.72	32½	28.80	829.58	52½	46.53	2164.75
13	11.52	132.73	33	29.25	855.30	53	46.97	2206.18
13½	11.96	143.14	33½	29.69	881.41	53½	47.41	2248.01
14	12.41	153.94	34	30.13	907.92	54	47.86	2290.22
14½	12.85	165.13	34½	30.57	934.82	54½	48.30	2332.83
15	13.29	176.71	35	31.02	962.11	55	48.74	2375.83
15½	13.74	188.69	35½	31.46	989.80	55½	49.19	2419.22
16	14.18	201.06	36	31.90	1017.88	56	49.63	2463.01
16½	14.62	213.82	36½	32.35	1046.35	56½	50.07	2507.19
17	15.07	226.98	37	32.79	1075.21	57	50.51	2551.76
17½	15.51	240.53	37½	33.23	1104.47	57½	50.96	2596.72
18	15.95	254.47	38	33.68	1134.11	58	51.40	2642.08
18½	16.40	268.80	38½	34.12	1164.16	58½	51.84	2687.83
19	16.84	283.53	39	34.56	1194.59	59	52.29	2733.97
19½	17.28	298.65	39½	35.01	1225.42	59½	52.73	2780.51
20	17.72	314.16	40	35.45	1256.64	60	53.17	2827.43

Segments of Circles for Radius = 1
(English or metric units)

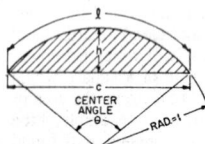

CENTER ANGLE θ RAD = 1

Formulas for segments of circles are given on page 54. When the central angle θ and radius r are known, the tables on these pages can be used to find the length of arc l, height of segment h, chord length c, and segment area A. When angle θ and radius r are not known, but segment height h and chord length c are known or can be measured, the ratio h/c can be used to enter the table and find θ, l, and A by linear interpolation. Radius r is found by the formula on page 54. The value of l is then multiplied by the radius r and the area a by r^2, the square of the radius.

Angle θ can be found thus with an accuracy of about 0.001 degree; arc length l with an error of about 0.02 per cent; and area A with an error ranging from about 0.02 per cent for the highest entry value of h/c to about 1 per cent for values of h/c of about 0.050. For lower values of h/c, and where greater accuracy is required, area A should be found by the formula on page 54.

θ, Deg.	l	h	c	Area A	h/c	θ, Deg.	l	h	c	Area A	h/c
1	0.01745	0.00004	0.01745	0.0000	0.00218	41	0.71558	0.06333	0.70041	0.0298	0.09041
2	0.03491	0.00015	0.03490	0.0000	0.00436	42	0.73304	0.06642	0.71674	0.0320	0.09267
3	0.05236	0.00034	0.05235	0.0000	0.00655	43	0.75049	0.06958	0.73300	0.0342	0.09493
4	0.06981	0.00061	0.06980	0.0000	0.00873	44	0.76794	0.07282	0.74921	0.0366	0.09719
5	0.08727	0.00095	0.08724	0.0001	0.01091	45	0.78540	0.07612	0.76537	0.0391	0.09946
6	0.10472	0.00137	0.10467	0.0001	0.01309	46	0.80285	0.07950	0.78146	0.0418	0.10173
7	0.12217	0.00187	0.12210	0.0002	0.01528	47	0.82030	0.08294	0.79750	0.0445	0.10400
8	0.13963	0.00244	0.13951	0.0002	0.01746	48	0.83776	0.08645	0.81347	0.0473	0.10628
9	0.15708	0.00308	0.15692	0.0003	0.01965	49	0.85521	0.09004	0.82939	0.0503	0.10856
10	0.17453	0.00381	0.17431	0.0004	0.02183	50	0.87266	0.09369	0.84524	0.0533	0.11085
11	0.19199	0.00460	0.19169	0.0006	0.02402	51	0.89012	0.09741	0.86102	0.0565	0.11314
12	0.20944	0.00548	0.20906	0.0008	0.02620	52	0.90757	0.10121	0.87674	0.0598	0.11543
13	0.22689	0.00643	0.22641	0.0010	0.02839	53	0.92502	0.10507	0.89240	0.0632	0.11773
14	0.24435	0.00745	0.24374	0.0012	0.03058	54	0.94248	0.10899	0.90798	0.0667	0.12004
15	0.26180	0.00856	0.26105	0.0015	0.03277	55	0.95993	0.11299	0.92350	0.0704	0.12235
16	0.27925	0.00973	0.27835	0.0018	0.03496	56	0.97738	0.11705	0.93894	0.0742	0.12466
17	0.29671	0.01098	0.29562	0.0022	0.03716	57	0.99484	0.12118	0.95432	0.0781	0.12698
18	0.31416	0.01231	0.31287	0.0026	0.03935	58	1.01229	0.12538	0.96962	0.0821	0.12931
19	0.33161	0.01371	0.33010	0.0030	0.04155	59	1.02974	0.12964	0.98485	0.0863	0.13164
20	0.34907	0.01519	0.34730	0.0035	0.04374	60	1.04720	0.13397	1.00000	0.0906	0.13397
21	0.36652	0.01675	0.36447	0.0041	0.04594	61	1.06465	0.13837	1.01508	0.0950	0.13632
22	0.38397	0.01837	0.38162	0.0047	0.04814	62	1.08210	0.14283	1.03008	0.0996	0.13866
23	0.40143	0.02008	0.39874	0.0053	0.05035	63	1.09956	0.14736	1.04500	0.1043	0.14101
24	0.41888	0.02185	0.41582	0.0061	0.05255	64	1.11701	0.15195	1.05984	0.1091	0.14337
25	0.43633	0.02370	0.43288	0.0069	0.05476	65	1.13446	0.15661	1.07460	0.1141	0.14574
26	0.45379	0.02563	0.44990	0.0077	0.05697	66	1.15192	0.16133	1.08928	0.1192	0.14811
27	0.47124	0.02763	0.46689	0.0086	0.05918	67	1.16937	0.16611	1.10387	0.1244	0.15048
28	0.48869	0.02970	0.48384	0.0096	0.06139	68	1.18682	0.17096	1.11839	0.1298	0.15287
29	0.50615	0.03185	0.50076	0.0107	0.06361	69	1.20428	0.17587	1.13281	0.1353	0.15525
30	0.52360	0.03407	0.51764	0.0118	0.06583	70	1.22173	0.18085	1.14715	0.1410	0.15765
31	0.54105	0.03637	0.53448	0.0130	0.06805	71	1.23918	0.18588	1.16141	0.1468	0.16005
32	0.55851	0.03874	0.55127	0.0143	0.07027	72	1.25664	0.19098	1.17557	0.1528	0.16246
33	0.57596	0.04118	0.56803	0.0157	0.07250	73	1.27409	0.19614	1.18965	0.1589	0.16488
34	0.59341	0.04370	0.58474	0.0171	0.07473	74	1.29154	0.20136	1.20363	0.1651	0.16730
35	0.61087	0.04628	0.60141	0.0186	0.07696	75	1.30900	0.20665	1.21752	0.1715	0.16973
36	0.62832	0.04894	0.61803	0.0203	0.07919	76	1.32645	0.21199	1.23132	0.1781	0.17216
37	0.64577	0.05168	0.63460	0.0220	0.08143	77	1.34390	0.21739	1.24503	0.1848	0.17461
38	0.66323	0.05448	0.65114	0.0238	0.08367	78	1.36136	0.22285	1.25864	0.1916	0.17706
39	0.68068	0.05736	0.66761	0.0257	0.08592	79	1.37881	0.22838	1.27216	0.1986	0.17952
40	0.69813	0.06031	0.68404	0.0277	0.08816	80	1.39626	0.23396	1.28558	0.2057	0.18199

Segments of Circles for Radius = 1
(English or metric units)

θ Deg.	l	h	c	Area A	h/c	θ, Deg.	l	h	c	Area A	h/c
81	1.41372	0.23959	1.29890	0.2130	0.18446	131	2.28638	0.58531	1.81992	0.7658	0.32161
82	1.43117	0.24529	1.31212	0.2205	0.18694	132	2.30383	0.59326	1.82709	0.7803	0.32470
83	1.44862	0.25104	1.32524	0.2280	0.18943	133	2.32129	0.60125	1.83412	0.7950	0.32781
84	1.46608	0.25686	1.33826	0.2358	0.19193	134	2.33874	0.60927	1.84101	0.8097	0.33094
85	1.48353	0.26272	1.35118	0.2437	0.19444	135	2.35619	0.61732	1.84776	0.8245	0.33409
86	1.50098	0.26865	1.36400	0.2517	0.19696	136	2.37365	0.62539	1.85437	0.8395	0.33725
87	1.51844	0.27463	1.37671	0.2599	0.19948	137	2.39110	0.63350	1.86084	0.8546	0.34044
88	1.53589	0.28066	1.38932	0.2682	0.20201	138	2.40855	0.64163	1.86716	0.8697	0.34364
89	1.55334	0.28675	1.40182	0.2767	0.20456	139	2.42601	0.64979	1.87334	0.8850	0.34686
90	1.57080	0.29289	1.41421	0.2854	0.20711	140	2.44346	0.65798	1.87939	0.9003	0.35010
91	1.58825	0.29909	1.42650	0.2942	0.20967	141	2.46091	0.66619	1.88528	0.9158	0.35337
92	1.60570	0.30534	1.43868	0.3032	0.21224	142	2.47837	0.67443	1.89104	0.9314	0.35665
93	1.62316	0.31165	1.45075	0.3123	0.21482	143	2.49582	0.68270	1.89665	0.9470	0.35995
94	1.64061	0.31800	1.46271	0.3215	0.21741	144	2.51327	0.69098	1.90211	0.9627	0.36327
95	1.65806	0.32441	1.47455	0.3309	0.22001	145	2.53073	0.69929	1.90743	0.9786	0.36662
96	1.67552	0.33087	1.48629	0.3405	0.22261	146	2.54818	0.70763	1.91261	0.9945	0.36998
97	1.69297	0.33738	1.49791	0.3502	0.22523	147	2.56563	0.71598	1.91764	1.0105	0.37337
98	1.71042	0.34394	1.50942	0.3601	0.22786	148	2.58309	0.72436	1.92252	1.0266	0.37678
99	1.72788	0.35055	1.52081	0.3701	0.23050	149	2.60054	0.73276	1.92726	1.0428	0.38021
100	1.74533	0.35721	1.53209	0.3803	0.23315	150	2.61799	0.74118	1.93185	1.0590	0.38366
101	1.76278	0.36392	1.54325	0.3906	0.23582	151	2.63545	0.74962	1.93630	1.0753	0.38714
102	1.78024	0.37068	1.55429	0.4010	0.23849	152	2.65290	0.75808	1.94059	1.0917	0.39064
103	1.79769	0.37749	1.56522	0.4117	0.24117	153	2.67035	0.76655	1.94474	1.1082	0.39417
104	1.81514	0.38434	1.57602	0.4224	0.24387	154	2.68781	0.77505	1.94874	1.1247	0.39772
105	1.83260	0.39124	1.58671	0.4333	0.24657	155	2.70526	0.78356	1.95259	1.1413	0.40129
106	1.85005	0.39818	1.59727	0.4444	0.24929	156	2.72271	0.79209	1.95630	1.1580	0.40489
107	1.86750	0.40518	1.60771	0.4556	0.25202	157	2.74017	0.80063	1.95985	1.1747	0.40852
108	1.88496	0.41221	1.61803	0.4669	0.25476	158	2.75762	0.80919	1.96325	1.1915	0.41217
109	1.90241	0.41930	1.62823	0.4784	0.25752	159	2.77507	0.81776	1.96651	1.2084	0.41585
110	1.91986	0.42642	1.63830	0.4901	0.26028	160	2.79253	0.82635	1.96962	1.2253	0.41955
111	1.93732	0.43359	1.64825	0.5019	0.26306	161	2.80998	0.83495	1.97257	1.2422	0.42328
112	1.95477	0.44081	1.65808	0.5138	0.26585	162	2.82743	0.84357	1.97538	1.2592	0.42704
113	1.97222	0.44806	1.66777	0.5259	0.26866	163	2.84489	0.85219	1.97803	1.2763	0.43083
114	1.98968	0.45536	1.67734	0.5381	0.27148	164	2.86234	0.86083	1.98054	1.2934	0.43464
115	2.00713	0.46270	1.68678	0.5504	0.27431	165	2.87979	0.86947	1.98289	1.3105	0.43849
116	2.02458	0.47008	1.69610	0.5629	0.27715	166	2.89725	0.87813	1.98509	1.3277	0.44236
117	2.04204	0.47750	1.70528	0.5755	0.28001	167	2.91470	0.88680	1.98714	1.3449	0.44627
118	2.05949	0.48496	1.71433	0.5883	0.28289	168	2.93215	0.89547	1.98904	1.3621	0.45020
119	2.07694	0.49246	1.72326	0.6012	0.28577	169	2.94961	0.90415	1.99079	1.3794	0.45417
120	2.09440	0.50000	1.73205	0.6142	0.28868	170	2.96706	0.91284	1.99239	1.3967	0.45817
121	2.11185	0.50758	1.74071	0.6273	0.29159	171	2.98451	0.92154	1.99383	1.4140	0.46220
122	2.12930	0.51519	1.74924	0.6406	0.29452	172	3.00197	0.93024	1.99513	1.4314	0.46626
123	2.14675	0.52284	1.75763	0.6540	0.29747	173	3.01942	0.93895	1.99627	1.4488	0.47035
124	2.16421	0.53053	1.76590	0.6676	0.30043	174	3.03687	0.94766	1.99726	1.4662	0.47448
125	2.18166	0.53825	1.77402	0.6813	0.30341	175	3.05433	0.95638	1.99810	1.4836	0.47865
126	2.19911	0.54601	1.78201	0.6950	0.30640	176	3.07178	0.96510	1.99878	1.5010	0.48284
127	2.21657	0.55380	1.78987	0.7090	0.30941	177	3.08923	0.97382	1.99931	1.5184	0.48708
128	2.23402	0.56163	1.79759	0.7230	0.31243	178	3.10669	0.98255	1.99970	1.5359	0.49135
129	2.25147	0.56949	1.80517	0.7372	0.31548	179	3.12414	0.99127	1.99992	1.5533	0.49566
130	2.26893	0.57738	1.81262	0.7514	0.31854	180	3.14159	1.00000	2.00000	1.5708	0.50000

Formulas and Table for Regular Polygons
(English and metric units)

N = number of sides.
S = length of side.
R = radius of circumscribed circle.
r = radius of inscribed circle.
A = area of polygon.
$\alpha = 180° \div N$ = one-half center angle of one side.

Formulas:

$$A = (N \times \cot \alpha \times S^2) \div 4 \qquad R = S \div (2 \sin \alpha) \qquad S = 2R \times \sin \alpha$$
$$A = N \times \sin \alpha \times \cos \alpha \times R^2 \qquad R = r \div \cos \alpha \qquad S = 2r \times \tan \alpha$$
$$A = N \times \tan \alpha \times r^2 \qquad r = R \times \cos \alpha$$
$$r = R \times \cos \alpha$$
$$r = (S \times \cot \alpha) \div 2 \qquad R = \sqrt{A \div (N \sin \alpha \cos \alpha)}^* \qquad S = 2\sqrt{A \times \tan \alpha} \div N^*$$
$$r = \sqrt{A \times \cot \alpha} \div N^*$$

* These formulas may be used to calculate R, S, or r needed to provide a required area A.

Examples of Use of Table. (English and metric units)

A regular hexagon is inscribed in a circle of 6 inches diameter. Find the area and the radius of an inscribed circle. — Here $R = 3$. From the table, area $(A) = 2.5981 R^2 = 2.5981 \times 9 = 23.3829$ square inches. Radius of inscribed circle, $r = 0.866 R = 0.866 \times 3 = 2.598$ inches.

An octagon is inscribed in a circle of 100 millimeters diameter. Thus $R = 50$. Find the area and radius of an inscribed circle. From the table, $A = 2.8284 R^2 = 2.8284 \times 2500 = 7071$ mm$^2 = 70.7$ cm^2. Radius of inscribed circle, $r = 0.9239 R = 0.9239 \times 50 = 46.195$ mm.

Thirty-two bolts are to be equally spaced on the periphery of a bolt-circle, 16 inches in diameter. Find the chordal distance between the bolts. — Chordal distance equals the side (S) of a polygon with 32 sides. $R = 8$. Hence, $S = 0.196 R = 0.196 \times 8 = 1.568$ inch.

Sixteen bolts are to be equally spaced on the periphery of a bolt-circle, 250 millimeters diameter. Find the chordal distance between the bolts. — Chordal distance equals the side (S) of a polygon with 16 sides. $R = 125$. Thus, $S = 0.3902 R = 0.3902 \times 125 = 48.775$ millimeters.

No. of Sides	$A =$	$A =$	$A =$	$R =$	$S =$	$S =$	$R =$	$r =$	$r =$	No. of Sides
3	0.4330 S^2	1.2990 R^2	5.1962 r^2	0.5774 S	1.7321 R	3.4641 r	2.0000 r	0.5000 R	0.2887 S	3
4	1.0000 S^2	2.0000 R^2	4.0000 r^2	0.7071 S	1.4142 R	2.0000 r	1.4142 r	0.7071 R	0.5000 S	4
5	1.7205 S^2	2.3776 R^2	3.6327 r^2	0.8507 S	1.1756 R	1.4531 r	1.2361 r	0.8090 R	0.6882 S	5
6	2.5981 S^2	2.5981 R^2	3.4641 r^2	1.0000 S	1.0000 R	1.1547 r	1.1547 r	0.8660 R	0.8660 S	6
7	3.6339 S^2	2.7364 R^2	3.3710 r^2	1.1524 S	0.8678 R	0.9631 r	1.1099 r	0.9010 R	1.0383 S	7
8	4.8284 S^2	2.8284 R^2	3.3137 r^2	1.3066 S	0.7654 R	0.8284 r	1.0824 r	0.9239 R	1.2071 S	8
9	6.1818 S^2	2.8925 R^2	3.2757 r^2	1.4619 S	0.6840 R	0.7279 r	1.0642 r	0.9397 R	1.3737 S	9
10	7.6942 S^2	2.9389 R^2	3.2492 r^2	1.6180 S	0.6180 R	0.6498 r	1.0515 r	0.9511 R	1.5388 S	10
12	11.196 S^2	3.0000 R^2	3.2154 r^2	1.9319 S	0.5176 R	0.5359 r	1.0353 r	0.9659 R	1.8660 S	12
16	20.109 S^2	3.0615 R^2	3.1826 r^2	2.5629 S	0.3902 R	0.3978 r	1.0196 r	0.9808 R	2.5137 S	16
20	31.569 S^2	3.0902 R^2	3.1677 r^2	3.1962 S	0.3129 R	0.3168 r	1.0125 r	0.9877 R	3.1569 S	20
24	45.575 S^2	3.1058 R^2	3.1597 r^2	3.8306 S	0.2611 R	0.2633 r	1.0086 r	0.9914 R	3.7979 S	24
32	81.225 S^2	3.1214 R^2	3.1517 r^2	5.1011 S	0.1960 R	0.1970 r	1.0048 r	0.9952 R	5.0766 S	32
48	183.08 S^2	3.1326 R^2	3.1461 r^2	7.6449 S	0.1308 R	0.1311 r	1.0021 r	0.9979 R	7.6285 S	48
64	325.69 S^2	3.1365 R^2	3.1441 r^2	10.190 S	0.0981 R	0.0983 r	1.0012 r	0.9988 R	10.178 S	64

Distance Across Corners of Squares and Hexagons
(English and metric units)

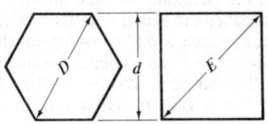

$$D = 1.154701 \, d$$
$$E = 1.414214 \, d$$

d	D	E	d	D	E	d	D	E	d	D	E
1/32	0.0361	0.0442	0.9	1.0392	1.2728	32	36.9504	45.2548	67	77.3649	94.7523
1/16	0.0722	0.0884	29/32	1.0464	1.2816	33	38.1051	46.6690	68	78.5196	96.1665
3/32	0.1083	0.1326	15/16	1.0825	1.3258	34	39.2598	48.0833	69	79.6743	97.5807
0.1	0.1155	0.1414	31/32	1.1186	1.3700	35	40.4145	49.4975	70	80.8290	98.9949
1/8	0.1443	0.1768	1.0	1.1547	1.4142	36	41.5692	50.9117	71	81.9837	100.409
5/32	0.1804	0.2210	2.0	2.3094	2.8284	37	42.7239	52.3259	72	83.1384	101.823
3/16	0.2165	0.2652	3.0	3.4641	4.2426	38	43.8786	53.7401	73	84.2931	103.238
0.2	0.2309	0.2828	4.0	4.6188	5.6569	39	45.0333	55.1543	74	85.4478	104.652
7/32	0.2526	0.3094	5.0	5.7735	7.0711	40	46.1880	56.5685	75	86.6025	106.066
1/4	0.2887	0.3536	6.0	6.9282	8.4853	41	47.3427	57.9828	76	87.7572	107.480
9/32	0.3248	0.3977	7.0	8.0829	9.8995	42	48.4974	59.3970	77	88.9119	108.894
0.3	0.3464	0.4243	8.0	9.2376	11.3137	43	49.6521	60.8112	78	90.0666	110.309
5/16	0.3608	0.4419	9.0	10.3923	12.7279	44	50.8068	62.2254	79	91.2213	111.723
11/32	0.3969	0.4861	10	11.5470	14.1421	45	51.9615	63.6396	80	92.3760	113.137
3/8	0.4330	0.5303	11	12.7017	15.5563	46	53.1162	65.0538	81	93.5307	114.551
0.4	0.4619	0.5657	12	13.8564	16.9706	47	54.2709	66.4680	82	94.6854	115.966
13/32	0.4691	0.5745	13	15.0111	18.3848	48	55.4256	67.8823	83	95.8401	117.380
7/16	0.5052	0.6187	14	16.1658	19.7990	49	56.5803	69.2965	84	96.9948	118.794
15/32	0.5413	0.6629	15	17.3205	21.2132	50	57.7350	70.7107	85	98.1495	120.208
0.5	0.5774	0.7071	16	18.4752	22.6274	51	58.8897	72.1249	86	99.3042	121.622
17/32	0.6134	0.7513	17	19.6299	24.0416	52	60.0444	73.5391	87	100.459	123.037
9/16	0.6495	0.7955	18	20.7846	25.4558	53	61.1991	74.9533	88	101.614	124.451
19/32	0.6856	0.8397	19	21.9393	26.8701	54	62.3538	76.3675	89	102.768	125.865
0.6	0.6928	0.8485	20	23.0940	28.2843	55	63.5085	77.7817	90	103.923	127.279
5/8	0.7217	0.8839	21	24.2487	29.6985	56	64.6632	79.1960	91	105.078	128.693
21/32	0.7578	0.9281	22	25.4034	31.1127	57	65.8179	80.6102	92	106.232	130.108
11/16	0.7939	0.9723	23	26.5581	32.5269	58	66.9726	82.0244	93	107.387	131.522
0.7	0.8083	0.9899	24	27.7128	33.9411	59	68.1273	83.4386	94	108.542	132.936
23/32	0.8299	1.0165	25	28.8675	35.3553	60	69.2820	84.8528	95	109.697	134.350
3/4	0.8660	1.0607	26	30.0222	36.7696	61	70.4367	86.2670	96	110.851	135.765
25/32	0.9021	1.1049	27	31.1769	38.1838	62	71.5914	87.6812	97	112.006	137.179
0.8	0.9238	1.1314	28	32.3316	39.5980	63	72.7461	89.0955	98	113.161	138.593
13/16	0.9382	1.1490	29	33.4863	41.0122	64	73.9008	90.5097	99	114.315	140.007
27/32	0.9743	1.1932	30	34.6410	42.4264	65	75.0555	91.9239	100	115.470	141.421
7/8	1.0104	1.2374	31	35.7957	43.8406	66	76.2102	93.3381	...	...	...

A desired value not given directly in the table can be obtained by the simple addition of two or more values taken directly from the table. Further values can be obtained by shifting the decimal point.

Example 1: Find D when $d = 2\tfrac{5}{16}$ inches. From the table, $2 = 2.3094$, and $\tfrac{5}{16} = 0.3608$. Therefore, $D = 2.3094 + 0.3608 = 2.6702$ inches.

Example 2: Find E when $d = 20.25$ millimeters. From the table, $20 = 28.2843$; $0.2 = 0.2828$; and $0.05 = 0.0707$ (obtained by shifting the decimal point one place to the left at $d = 0.5$). Thus, $E = 28.2843 + 0.2828 + 0.0707 = 28.6378$ millimeters.

SOLUTION OF TRIANGLES

Any figure bounded by three straight lines is called a triangle. Any one of the three lines may be called the base, and the line drawn from the angle opposite the base at right angles to it is called the height or altitude of the triangle.

If all the three sides of a triangle are of equal length, the triangle is called *equilateral*. Each one of the three angles in an equilateral triangle equals 60 degrees. If two sides are of equal length, the triangle is an *isosceles* triangle. If one angle is a right or 90-degree angle, the triangle is a *right* or *right-angled* triangle. The side opposite the right angle is called the *hypotenuse*.

If all the angles are less than 90 degrees, the triangle is called an *acute* or *acute-angled* triangle. If one of the angles is larger than 90 degrees, the triangle is called an *obtuse-angled* triangle. Both acute and obtuse-angled triangles are known under the common name of *oblique-angled* triangles. The sum of the three angles in every triangle is 180 degrees.

The sides and angles of any triangle which are not known can be found when: 1. All the three sides; 2. Two sides and one angle; or 3. One side and two angles, are given. In other words, if a triangle is considered as consisting of six parts, three angles and three sides, the unknown parts can be determined when any three parts are given, provided at least one of the given parts is a side.

Functions of Angles. — The functions of angles used in solving triangles are sine, cosine, tangent, cotangent, secant, and cosecant. These expressions are usually abbreviated as follows:

sin = sine,	cot = cotangent,
cos = cosine,	sec = secant,
ʈan = tangent,	cosec = cosecant.

If in a right-angled triangle (see the illustration in the table below) the lengths of the three sides are represented by a, b, and c, and the angles opposite each of these sides by A, B, and C, then the side c opposite the right angle is the hypotenuse;

Trigonometrical Functions of Angles

The *sine* of an angle equals the opposite side divided by the hypotenuse. Hence, $\sin B = b \div c$, and $\sin A = a \div c$.

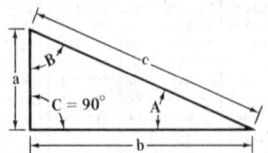

The *cosine* of an angle equals the adjacent side divided by the hypotenuse. Hence, $\cos B = a \div c$, and $\cos A = b \div c$.

The *tangent* of an angle equals the opposite side divided by the adjacent side. Hence, $\tan B = b \div a$, and $\tan A = a \div b$.

The *cotangent* of an angle equals the adjacent side divided by the opposite side. Hence, $\cot B = a \div b$, and $\cot A = b \div a$.

The *secant* of an angle equals the hypotenuse divided by the adjacent side. Hence, $\sec B = c \div a$, and $\sec A = c \div b$.

The *cosecant* of an angle equals the hypotenuse divided by the opposite side. Hence, $\operatorname{cosec} B = c \div b$, and $\operatorname{cosec} A = c \div a$.

It should be noted that the functions of the angles can be found in this manner only when the triangle is right-angled.

side *b* is called the *side adjacent* to angle *A* and is also the *side opposite* to angle *B;* side *a* is the side adjacent to angle *B* and the side opposite to angle *A*. The meanings of the various functions of angles can be explained by the aid of a right-angled triangle.

The following relation exists between the angular functions of the two acute angles in a right-angled triangle: The sine of angle *B* equals the cosine of angle *A;* the tangent of angle *B* equals the cotangent of angle *A*, and *vice versa*. The sum of the two acute angles in a right-angled triangle always equals 90 degrees; hence, when one angle is known, the other can easily be found. When any two angles together make 90 degrees, one is called the *complement* of the other, and in that case the sine of the one equals the cosine of the other, and the tangent of the one equals the cotangent of the other.

The Law of Sines. — In any triangle, any side is to the sine of the angle opposite that side as any other side is to the sine of the angle opposite that side. If *a*, *b*, and *c* are the sides, and *A*, *B*, and *C* their opposite angles, respectively, then:

$$\frac{a}{\sin A} = \frac{b}{\sin B} = \frac{c}{\sin C}, \text{ so that:}$$

$$a = \frac{b \sin A}{\sin B}, \quad \text{or,} \quad a = \frac{c \sin A}{\sin C};$$

$$b = \frac{a \sin B}{\sin A}, \quad \text{or,} \quad b = \frac{c \sin B}{\sin C};$$

$$c = \frac{a \sin C}{\sin A}, \quad \text{or,} \quad c = \frac{b \sin C}{\sin B}$$

The Law of Cosines. — In any triangle, the square of any side is equal to the sum of the squares of the other two sides minus twice their product times the cosine of the included angle; or if *a*, *b* and *c* be the sides and *A*, *B*, and *C* are the opposite angles, respectively, then:

$$a^2 = b^2 + c^2 - 2bc \cos A$$
$$b^2 = a^2 + c^2 - 2ac \cos B$$
$$c^2 = a^2 + b^2 - 2ab \cos C$$

These two laws, together with the proposition that the sum of the three angles equals 180 degrees, are the basis of all formulas relating to the solution of triangles.

Formulas for the solution of right-angled and oblique-angled triangles, arranged in tabular form, are given on the following pages.

Signs of Trigonometric Functions. — On page 81 a diagram, "Signs of Trigonometric Functions," is given. This diagram shows the proper sign (+ or −) for the trigonometric functions of angles in each of the four quadrants, 0 to 90, 90 to 180, 180 to 270, and 270 to 360 degrees. Thus, the cosine of an angle between 90 and 180 degrees is negative; the sine of the same angle is positive.

Trigonometric Identities. — Trigonometric identities are formulas that show the relationship between different trigonometric functions. They may be used to change the form of some trigonometric expressions to simplify calculations. For example, if a formula has a term, $2 \sin A \cos A$, the equivalent but simpler term $\sin 2A$ may be substituted. The identities given below may themselves be combined or rearranged in various ways to form new identities.

1. *Basic:*

$$\tan A = \frac{\sin A}{\cos A} = \frac{1}{\cot A} \qquad \sec A = \frac{1}{\cos A} \qquad \csc A = \frac{1}{\sin A}$$

2. *Negative-Angle:*

$$\sin (-A) = -\sin A \qquad \cos (-A) = \cos A \qquad \tan (-A) = -\tan A$$

3. *Pythagorean:*

$$\sin^2 A + \cos^2 A = 1 \qquad 1 + \tan^2 A = \sec^2 A \qquad 1 + \cot^2 A = \csc^2 A$$

4. *Sum and Difference of Angles:*

$$\tan (A + B) = \frac{\tan A + \tan B}{1 - \tan A \tan B} \qquad \cot (A + B) = \frac{\cot A \cot B - 1}{\cot B + \cot A}$$

$$\tan (A - B) = \frac{\tan A - \tan B}{1 + \tan A \tan B} \qquad \cot (A - B) = \frac{\cot A \cot B + 1}{\cot B - \cot A}$$

$$\sin (A + B) = \sin A \cos B + \cos A \sin B \qquad \cos (A + B) = \cos A \cos B - \sin A \sin B$$

$$\sin (A - B) = \sin A \cos B - \cos A \sin B \qquad \cos (A - B) = \cos A \cos B + \sin A \sin B$$

5. *Double-Angle:*

$$\cos 2A = \cos^2 A - \sin^2 A = 2 \cos^2 A - 1 = 1 - 2 \sin^2 A$$

$$\sin 2A = 2 \sin A \cos A \qquad \tan 2A = \frac{2 \tan A}{1 - \tan^2 A} = \frac{2}{\cot A - \tan A}$$

6. *Half-Angle:*

$$\sin \tfrac{1}{2}A = \sqrt{\tfrac{1}{2}(1 - \cos A)} \qquad \cos \tfrac{1}{2}A = \sqrt{\tfrac{1}{2}(1 + \cos A)}$$

$$\tan \tfrac{1}{2}A = \sqrt{\frac{1 - \cos A}{1 + \cos A}} = \frac{1 - \cos A}{\sin A} = \frac{\sin A}{1 + \cos A}$$

7. *Product-to-Sum:*

$$\sin A \cos B = \tfrac{1}{2}[\sin (A + B) + \sin (A - B)]$$

$$\cos A \cos B = \tfrac{1}{2}[\cos (A + B) + \cos (A - B)]$$

$$\sin A \sin B = \tfrac{1}{2}[\cos (A - B) - \cos (A + B)]$$

$$\tan A \tan B = \frac{\tan A + \tan B}{\cot A + \cot B}$$

8. *Sum and Difference of Functions:*

$$\sin A + \sin B = 2[\sin \tfrac{1}{2} (A + B) \cos \tfrac{1}{2} (A - B)]$$

$$\sin A - \sin B = 2[\sin \tfrac{1}{2} (A - B) \cos \tfrac{1}{2} (A + B)]$$

$$\cos A + \cos B = 2[\cos \tfrac{1}{2} (A + B) \cos \tfrac{1}{2} (A - B)]$$

$$\cos A - \cos B = -2[\sin \tfrac{1}{2} (A + B) \sin \tfrac{1}{2} (A - B)]$$

$$\tan A + \tan B = \frac{\sin (A + B)}{\cos A \cos B} \qquad \cot A + \cot B = \frac{\sin (B + A)}{\sin A \sin B}$$

$$\tan A - \tan B = \frac{\sin (A - B)}{\cos A \cos B} \qquad \cot A - \cot B = \frac{\sin (B - A)}{\sin A \sin B}$$

Solution of Right-angled Triangles

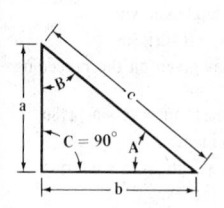

As shown in the illustration, the sides of the right-angled triangle are designated a and b and the hypotenuse, c. The angles opposite each of these sides are designated A and B, respectively.

Angle C, opposite the hypotenuse c is the right angle, and is therefore always one of the known quantities.

Sides and Angles Known	Formulas for Sides and Angles to be Found		
Side a; side b	$c = \sqrt{a^2 + b^2}$	$\tan A = \dfrac{a}{b}$	$B = 90° - A$
Side a; hypotenuse c	$b = \sqrt{c^2 - a^2}$	$\sin A = \dfrac{a}{c}$	$B = 90° - A$
Side b; hypotenuse c	$a = \sqrt{c^2 - b^2}$	$\sin B = \dfrac{b}{c}$	$A = 90° - B$
Hypotenuse c; angle B ...	$b = c \times \sin B$	$a = c \times \cos B$	$A = 90° - B$
Hypotenuse c; angle A ...	$b = c \times \cos A$	$a = c \times \sin A$	$B = 90° - A$
Side b; angle B	$c = \dfrac{b}{\sin B}$	$a = b \times \cot B$	$A = 90° - B$
Side b; angle A	$c = \dfrac{b}{\cos A}$	$a = b \times \tan A$	$B = 90° - A$
Side a; angle B	$c = \dfrac{a}{\cos B}$	$b = a \times \tan B$	$A = 90° - B$
Side a; angle A	$c = \dfrac{a}{\sin A}$	$b = a \times \cot A$	$B = 90° - A$

Examples of the Solution of Right-angled Triangles (English and metric units)

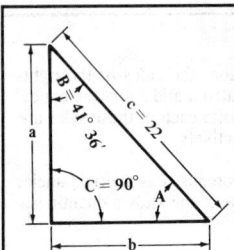

Hypotenuse and one angle known:
$$c = 22 \text{ inches}; \quad B = 41° 36'.$$
Then, by the formulas given on the preceding page:
$$a = c \times \cos B = 22 \times \cos 41° 36' = 22 \times 0.74780$$
$$= 16.4516 \text{ inches.}$$
$$b = c \times \sin B = 22 \times \sin 41° 36' = 22 \times 0.66393$$
$$= 14.6065 \text{ inches.}$$
$$A = 90° - B = 90° - 41° 36' = 48° 24'.$$

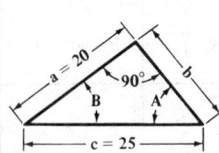

Hypotenuse and one side known:
$$c = 25 \text{ centimeters}; \quad a = 20 \text{ centimeters}.$$
From the formulas on the preceding page:
$$b = \sqrt{c^2 - a^2} = \sqrt{25^2 - 20^2} = \sqrt{625 - 400}$$
$$= \sqrt{225} = 15 \text{ centimeters.}$$
$$\sin A = \frac{a}{c} = \frac{20}{25} = 0.8$$
Hence, $A = 53° 8'.$
$$B = 90° - A = 90° - 53° 8' = 36° 52'.$$

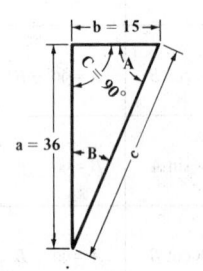

Two sides known:
$$a = 36 \text{ inches}; \quad b = 15 \text{ inches}.$$
Then, by the formulas given on the preceding page:
$$c = \sqrt{a^2 + b^2} = \sqrt{36^2 + 15^2} = \sqrt{1296 + 225}$$
$$= \sqrt{1521} = 39 \text{ inches.}$$
$$\tan A = \frac{a}{b} = \frac{36}{15} = 2.4$$
Hence, $A = 67° 23'.$
$$B = 90° - A = 90° - 67° 23' = 22° 37'.$$

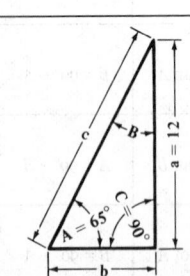

One side and one angle known:
$$a = 12 \text{ meters}; \quad A = 65°.$$
Then, by the formulas given on the preceding page:
$$c = \frac{a}{\sin A} = \frac{12}{\sin 65°} = \frac{12}{0.90631} = 13.2405 \text{ meters.}$$
$$b = a \times \cot A = 12 \times \cot 65° = 12 \times 0.46631$$
$$= 5.5957 \text{ meters.}$$
$$B = 90° - A = 90° - 65° = 25°.$$

Solution of Oblique-angled Triangles

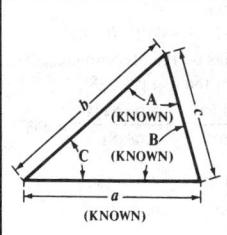

One side and two angles known:

Call the known side a, the angle opposite it A, and the other known angle B. Then:

$C = 180° - (A + B)$; or if angles B and C are given, but not A, then $A = 180° - (B + C)$.

$$C = 180° - (A + B)$$

$$b = \frac{a \times \sin B}{\sin A} \qquad c = \frac{a \times \sin C}{\sin A}$$

$$\text{Area} = \frac{a \times b \times \sin C}{2}$$

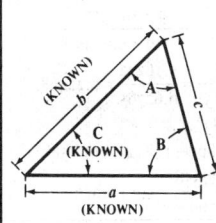

Two sides and the angle between them known:

Call the known sides a and b, and the known angle between them C. Then:

$$\tan A = \frac{a \times \sin C}{b - (a \times \cos C)}$$

$$B = 180° - (A + C) \qquad c = \frac{a \times \sin C}{\sin A}$$

Side c may also be found directly as below:

$$c = \sqrt{a^2 + b^2 - (2ab \times \cos C)}$$

$$\text{Area} = \frac{a \times b \times \sin C}{2}$$

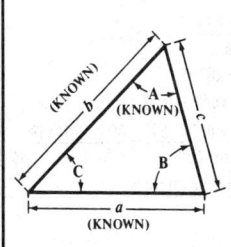

Two sides and the angle opposite one of the sides known:

Call the known angle A, the side opposite it a, and the other known side b. Then:

$$\sin B = \frac{b \times \sin A}{a} \qquad C = 180° - (A + B)$$

$$c = \frac{a \times \sin C}{\sin A} \qquad \text{Area} = \frac{a \times b \times \sin C}{2}$$

If, in the above, angle $B >$ angle A but $< 90°$, then a second solution B_2, C_2, c_2 exists for which: $B_2 = 180° - B$; $C_2 = 180° - (A + B_2)$; $c_2 = (a \times \sin C_2) \div \sin A$; Area $= (a \times b \times \sin C_2) \div 2$. If $a \geqq b$, then the first solution only exists. If $a < b \times \sin A$, then no solution exists.

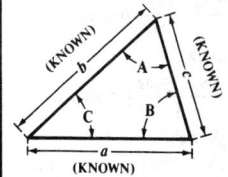

All three sides known:

Call the sides a, b, and c, and the angles opposite them, A, B, and C. Then:

$$\cos A = \frac{b^2 + c^2 - a^2}{2bc} \qquad \sin B = \frac{b \times \sin A}{a}$$

$$C = 180° - (A + B) \qquad \text{Area} = \frac{a \times b \times \sin C}{2}$$

Examples of the Solution of Oblique-angled Triangles (English and metric units)

	Side and angles known: $a = 5$ centimeters; $A = 80°$; $B = 62°$ Then, by the formulas on the preceding page: $C = 180° - (80° + 62°) = 180° - 142° = 38°$. $b = \dfrac{a \times \sin B}{\sin A} = \dfrac{5 \times \sin 62°}{\sin 80°} = \dfrac{5 \times 0.88295}{0.98481} = 4.483$ centimeters. $c = \dfrac{a \times \sin C}{\sin A} = \dfrac{5 \times \sin 38°}{\sin 80°} = \dfrac{5 \times 0.61566}{0.98481} = 3.126$ centimeters.
	Sides and angle known: $a = 9$ inches; $b = 8$ inches; $C = 35°$. $\tan A = \dfrac{a \times \sin C}{b - (a \times \cos C)} = \dfrac{9 \times \sin 35°}{8 - (9 \times \cos 35°)}$ $= \dfrac{9 \times 0.57358}{8 - (9 \times 0.81915)} = \dfrac{5.16222}{0.62765} = 8.22468$. Hence, $A = 83° \, 4'$. $B = 180° - (A + C) = 180° - 118° \, 4' = 61° \, 56'$. $c = \dfrac{a \times \sin C}{\sin A} = \dfrac{9 \times 0.57358}{0.99269} = 5.2$ inches.
	Sides and angle known: $a = 20$ centimeters; $b = 17$ centimeters; $A = 61°$. $\sin B = \dfrac{b \times \sin A}{a} = \dfrac{17 \times \sin 61°}{20}$ $= \dfrac{17 \times 0.87462}{20} = 0.74343$. Hence, $B = 48° \, 1'$. $C = 180° - (A + B) = 180° - 109° \, 1' = 70° \, 59'$. $c = \dfrac{a \times \sin C}{\sin A} = \dfrac{20 \times \sin 70° \, 59'}{\sin 61°} = \dfrac{20 \times 0.94542}{0.87462}$ $= 21.62$ centimeters.
	Sides known: $a = 8$ inches; $b = 9$ inches; $c = 10$ inches. $\cos A = \dfrac{b^2 + c^2 - a^2}{2bc} = \dfrac{9^2 + 10^2 - 8^2}{2 \times 9 \times 10}$ $= \dfrac{81 + 100 - 64}{180} = \dfrac{117}{180} = 0.65000$. Hence, $A = 49° \, 27'$. $\sin B = \dfrac{b \times \sin A}{a} = \dfrac{9 \times 0.75984}{8} = 0.85482$. Hence, $B = 58° \, 44'$. $C = 180° - (A + B) = 180° - 108° \, 11' = 71° \, 49'$.

Tables of Trigonometric Functions. — The numerical values for the natural or trigonometric functions for all angles from 0 to 360 degrees are given in the tables, pages 82 to 126. The chart below shows how to enter the table.

How to Enter Table of Trigonometric Functions

For Angles from	Enter Table for		For Angles from	Enter Table for	
	Degrees and Function	Minutes		Degrees and Function	Minutes
0° to 45°	at top	at left	180° to 225°	at top	at left
45° to 90°	at bottom	at right	225° to 270°	at bottom	at right
90° to 135°	at bottom	at left	270° to 315°	at bottom	at left
135° to 180°	at top	at right	315° to 360°	at top	at right

Examples: The sine of 26° is 0.43837; of 126°, 0.80902; of 226°, −0.71934.

Signs of Trigonometric Functions

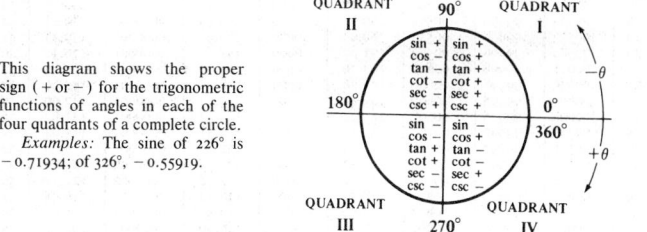

This diagram shows the proper sign ($+$ or $-$) for the trigonometric functions of angles in each of the four quadrants of a complete circle.

Examples: The sine of 226° is −0.71934; of 326°, −0.55919.

Useful Relationships Among Angles

Angle / Function	θ	−θ	90° ± θ	180° ± θ	270° ± θ	360° ± θ
sin	sin θ	− sin θ	+ cos θ	∓ sin θ	− cos θ	± sin θ
cos	cos θ	+ cos θ	∓ sin θ	− cos θ	± sin θ	+ cos θ
tan	tan θ	− tan θ	∓ cot θ	± tan θ	∓ cot θ	± tan θ
cot	cot θ	− cot θ	∓ tan θ	± cot θ	∓ tan θ	± cot θ
sec	sec θ	+ sec θ	∓ csc θ	− sec θ	± csc θ	+ sec θ
csc	csc θ	− csc θ	+ sec θ	∓ csc θ	− sec θ	± csc θ

Examples: cos 270° − θ = − sin θ; tan 90° + θ = − cot θ.

Involute Functions. — Involute functions are used in certain formulas relating to the design and measurement of gear teeth as well as measurement of threads over wires. Included in the trigonometric tables, pages 82 to 126, are values for the involute functions of angles of from 0 to 90 degrees. These involute functions were calculated from the formula: Involute of θ = tan θ − π × θ ÷ 180 when θ is given in degrees.

Sevolute Functions. — Sevolute functions are used in certain spline calculations. They may be computed by subtracting the involute of an angle from the secant of the angle. Thus, sevolute 20° = sec 20° − inv 20° = 1.0642 − 0.014904 = 1.0493.

Versed Sine and Versed Cosine. — These functions are sometimes used in formulas for segments of a circle and may be obtained by using the trigonometric tables together with the relationships: versed sine θ = 1 − cos θ; and versed cosine θ = 1 − sin θ.

0° or 180° **Trigonometric and Involute Functions** **179° or 359°**

M	Sine	Cosine	Tan.	Cotan.	Secant	Cosec.	Involute 0°–1°	READ UP	M
0	0.00000	1.0000	0.00000	Infinite	1.0000	Infinite	.0000000	Infinite	60
1	.00029	.0000	.00029	3437.7	.0000	3437.7	.0000000	3436.2	59
2	.00058	.0000	.00058	1718.9	.0000	1718.9	.0000000	1717.3	58
3	.00087	.0000	.00087	1145.9	.0000	1145.9	.0000000	1144.3	57
4	.00116	.0000	.00116	859.44	.0000	859.44	.0000000	857.87	56
5	0.00145	1.0000	0.00145	687.55	1.0000	687.55	.0000000	685.98	55
6	.00175	.0000	.00175	572.96	.0000	572.96	.0000000	571.39	54
7	.00204	.0000	.00204	491.11	.0000	491.11	.0000000	489.54	53
8	.00233	.0000	.00233	429.72	.0000	429.72	.0000000	428.15	52
9	.00262	.0000	.00262	381.97	.0000	381.97	.0000000	380.40	51
10	0.00291	1.00000	0.00291	343.77	1.0000	343.78	.0000000	342.21	50
11	.00320	.99999	.00320	312.52	.0000	312.52	.0000000	310.95	49
12	.00349	.99999	.00349	286.48	.0000	286.48	.0000000	284.91	48
13	.00378	.99999	.00378	264.44	.0000	264.44	.0000000	262.87	47
14	.00407	.99999	.00407	245.55	.0000	245.55	.0000000	243.99	46
15	0.00436	0.99999	0.00436	229.18	1.0000	229.18	.0000000	227.62	45
16	.00465	.99999	.00465	214.86	.0000	214.86	.0000000	213.29	44
17	.00495	.99999	.00495	202.22	.0000	202.22	.0000000	200.65	43
18	.00524	.99999	.00524	190.98	.0000	190.99	.0000000	189.42	42
19	.00553	.99998	.00553	180.93	.0000	180.93	.0000001	179.37	41
20	0.00582	0.99998	0.00582	171.89	1.0000	171.89	.0000001	170.32	40
21	.00611	.99998	.00611	163.70	.0000	163.70	.0000001	162.14	39
22	.00640	.99998	.00640	156.26	.0000	156.26	.0000001	154.69	38
23	.00669	.99998	.00669	149.47	.0000	149.47	.0000001	147.90	37
24	.00698	.99998	.00698	143.24	.0000	143.24	.0000001	141.67	36
25	0.00727	0.99997	0.00727	137.51	1.0000	137.51	.0000001	135.94	35
26	.00756	.99997	.00756	132.22	.0000	132.22	.0000001	130.66	34
27	.00785	.99997	.00785	127.32	.0000	127.33	.0000002	125.76	33
28	.00814	.99997	.00815	122.77	.0000	122.78	.0000002	121.21	32
29	.00844	.99996	.00844	118.54	.0000	118.54	.0000002	116.98	31
30	0.00873	0.99996	0.00873	114.59	1.0000	114.59	.0000002	113.03	30
31	.00902	.99996	.00902	110.89	.0000	110.90	.0000002	109.33	29
32	.00931	.99996	.00931	107.43	.0000	107.43	.0000003	105.86	28
33	.00960	.99995	.00960	104.17	.0000	104.18	.0000003	102.61	27
34	.00989	.99995	.00989	101.11	.0000	101.11	.0000003	99.546	26
35	0.01018	0.99995	0.01018	98.218	1.0001	98.223	.0000004	96.657	25
36	.01047	.99995	.01047	95.489	.0001	95.495	.0000004	93.929	24
37	.01076	.99994	.01076	92.908	.0001	92.914	.0000004	91.348	23
38	.01105	.99994	.01105	90.463	.0001	90.469	.0000005	88.904	22
39	.01134	.99994	.01135	88.144	.0001	88.149	.0000005	86.584	21
40	0.01164	0.99993	0.01164	85.940	1.0001	85.946	.0000005	84.381	20
41	.01193	.99993	.01193	83.844	.0001	83.849	.0000006	82.285	19
42	.01222	.99993	.01222	81.847	.0001	81.853	.0000006	80.288	18
43	.01251	.99992	.01251	79.943	.0001	79.950	.0000007	78.385	17
44	.01280	.99992	.01280	78.126	.0001	78.133	.0000007	76.568	16
45	0.01309	0.99991	0.01309	76.390	1.0001	76.397	.0000007	74.832	15
46	.01338	.99991	.01338	74.729	.0001	74.736	.0000008	73.172	14
47	.01367	.99991	.01367	73.139	.0001	73.146	.0000009	71.582	13
48	.01396	.99990	.01396	71.615	.0001	71.622	.0000009	70.058	12
49	.01425	.99990	.01425	70.153	.0001	70.160	.0000010	68.597	11
50	0.01454	0.99989	0.01455	68.750	1.0001	68.757	.0000010	67.194	10
51	.01483	.99989	.01484	67.402	.0001	67.409	.0000011	65.846	9
52	.01513	.99989	.01513	66.105	.0001	66.113	.0000012	64.550	8
53	.01542	.99988	.01542	64.858	.0001	64.866	.0000012	63.303	7
54	.01571	.99988	.01571	63.657	.0001	63.665	.0000013	62.102	6
55	0.01600	0.99987	0.01600	62.499	1.0001	62.507	.0000014	60.944	5
56	.01629	.99987	.01629	61.383	.0001	61.391	.0000014	59.828	4
57	.01658	.99986	.01658	60.306	.0001	60.314	.0000015	58.752	3
58	.01687	.99986	.01687	59.266	.0001	59.274	.0000016	57.712	2
59	.01716	.99985	.01716	58.261	.0001	58.270	.0000017	56.708	1
60	0.01745	0.99985	0.01746	57.290	1.0002	57.299	.0000018	55.737	0

M	Cosine	Sine	Cotan.	Tan.	Cosec.	Secant	READ DOWN	89°–90°	M
								Involute	

90° or 270° **89° or 269°**

1° or 181° **Trigonometric and Involute Functions** **178° or 358°**

M	Sine	Cosine	Tan.	Cotan.	Secant	Cosec.	Involute 1°–2°	READ UP	M
0	0.01745	0.99985	0.01746	57.290	1.0002	57.299	.0000018	55.737	60
1	.01774	.99984	.01775	56.351	.0002	56.359	.0000019	54.798	59
2	.01803	.99984	.01804	55.442	.0002	55.451	.0000020	53.889	58
3	.01832	.99983	.01833	54.561	.0002	54.570	.0000021	53.009	57
4	.01862	.99983	.01862	53.709	.0002	53.718	.0000022	52.156	56
5	0.01891	.99982	0.01891	52.882	1.0002	52.892	.0000023	51.330	55
6	.01920	.99982	.01920	52.081	.0002	52.090	.0000024	50.529	54
7	.01949	.99981	.01949	51.303	.0002	51.313	.0000025	49.752	53
8	.01978	.99980	.01978	50.549	.0002	50.558	.0000026	48.997	52
9	.02007	.99980	.02007	49.816	.0002	49.826	.0000027	48.265	51
10	0.02036	0.99979	0.02036	49.104	1.0002	49.114	.0000028	47.553	50
11	.02065	.99979	.02066	48.412	.0002	48.422	.0000029	46.862	49
12	.02094	.99978	.02095	47.740	.0002	47.750	.0000031	46.190	48
13	.02123	.99977	.02124	47.085	.0002	47.096	.0000032	45.536	47
14	.02152	.99977	.02153	46.449	.0002	46.460	.0000033	44.900	46
15	0.02181	.99976	0.02182	45.829	1.0002	45.840	.0000035	44.280	45
16	.02211	.99976	.02211	45.226	.0002	45.237	.0000036	43.677	44
17	.02240	.99975	.02240	44.639	.0003	44.650	.0000037	43.090	43
18	.02269	.99974	.02269	44.066	.0003	44.077	.0000039	42.518	42
19	.02298	.99974	.02298	43.508	.0003	43.520	.0000040	41.960	41
20	0.02327	0.99973	0.02328	42.964	1.0003	42.976	.0000042	41.417	40
21	.02356	.99972	.02357	42.433	.0003	42.445	.0000044	40.886	39
22	.02385	.99972	.02386	41.916	.0003	41.928	.0000045	40.369	38
23	.02414	.99971	.02415	41.411	.0003	41.423	.0000047	39.864	37
24	.02443	.99970	.02444	40.917	.0003	40.930	.0000049	39.371	36
25	0.02472	0.99969	0.02473	40.436	1.0003	40.448	.0000050	38.890	35
26	.02501	.99969	.02502	39.965	.0003	39.978	.0000052	38.420	34
27	.02530	.99968	.02531	39.506	.0003	39.519	.0000054	37.960	33
28	.02560	.99967	.02560	39.057	.0003	39.070	.0000056	37.512	32
29	.02589	.99966	.02589	38.618	.0003	38.631	.0000058	37.073	31
30	0.02618	0.99966	0.02619	38.188	1.0003	38.202	.0000060	36.644	30
31	.02647	.99965	.02648	37.769	.0004	37.782	.0000062	36.224	29
32	.02676	.99964	.02677	37.358	.0004	37.371	.0000064	35.814	28
33	.02705	.99963	.02706	36.956	.0004	36.970	.0000066	35.412	27
34	.02734	.99963	.02735	36.563	.0004	36.576	.0000068	35.019	26
35	0.02763	.99962	0.02764	36.178	1.0004	36.191	.0000070	34.634	25
36	.02792	.99961	.02793	35.801	.0004	35.815	.0000073	34.258	24
37	.02821	.99960	.02822	35.431	.0004	35.445	.0000075	33.889	23
38	.02850	.99959	.02851	35.070	.0004	35.084	.0000077	33.527	22
39	.02879	.99959	.02881	34.715	.0004	34.730	.0000080	33.173	21
40	0.02908	0.99958	0.02910	34.368	1.0004	34.382	.0000082	32.826	20
41	.02938	.99957	.02939	34.027	.0004	34.042	.0000085	32.486	19
42	.02967	.99956	.02968	33.694	.0004	33.708	.0000087	32.152	18
43	.02996	.99955	.02997	33.366	.0004	33.381	.0000090	31.825	17
44	.03025	.99954	.03026	33.045	.0005	33.060	.0000092	31.505	16
45	0.03054	0.99953	0.03055	32.730	1.0005	32.746	.0000095	31.190	15
46	.03083	.99952	.03084	32.421	.0005	32.437	.0000098	30.881	14
47	.03112	.99952	.03114	32.118	.0005	32.134	.0000101	30.578	13
48	.03141	.99951	.03143	31.821	.0005	31.836	.0000103	30.281	12
49	.03170	.99950	.03172	31.528	.0005	31.544	.0000106	29.989	11
50	0.03199	0.99949	0.03201	31.242	1.0005	31.258	.0000109	29.703	10
51	.03228	.99948	.03230	30.960	.0005	30.976	.0000112	29.421	9
52	.03257	.99947	.03259	30.683	.0005	30.700	.0000115	29.145	8
53	.03286	.99946	.03288	30.412	.0005	30.428	.0000118	28.874	7
54	.03316	.99945	.03317	30.145	.0006	30.161	.0000122	28.607	6
55	0.03345	0.99944	0.03346	29.882	1.0006	29.899	.0000125	28.345	5
56	.03374	.99943	.03376	29.624	.0006	29.641	.0000128	28.087	4
57	.03403	.99942	.03405	29.371	.0006	29.388	.0000131	27.834	3
58	.03432	.99941	.03434	29.122	.0006	29.139	.0000135	27.586	2
59	.03461	.99940	.03463	28.877	.0006	28.894	.0000138	27.341	1
60	0.03490	0.99939	0.03492	28.636	1.0006	28.654	.0000142	27.100	0
M	Cosine	Sine	Cotan.	Tan.	Cosec.	Secant	READ DOWN	88°–89° Involute	M

91° or 271°   **88° or 268°**

2° or 182°   Trigonometric and Involute Functions   177° or 357°

M	Sine	Cosine	Tan.	Cotan.	Secant	Cosec.	Involute 2°–3°	READ UP	M
0	0.03490	0.99939	0.03492	28.636	1.0006	28.654	.0000142	27.100	60
1	.03519	.99938	.03521	28.399	.0006	28.417	.0000145	26.864	59
2	.03548	.99937	.03550	28.166	.0006	28.184	.0000149	26.631	58
3	.03577	.99936	.03579	27.937	.0006	27.955	.0000153	26.402	57
4	.03606	.99935	.03609	27.712	.0007	27.730	.0000157	26.177	56
5	0.03635	0.99934	0.03638	27.490	1.0007	27.508	.0000160	25.955	55
6	.03664	.99933	.03667	27.271	.0007	27.290	.0000164	25.737	54
7	.03693	.99932	.03696	27.057	.0007	27.075	.0000168	25.523	53
8	.03723	.99931	.03725	26.845	.0007	26.864	.0000172	25.311	52
9	.03752	.99930	.03754	26.637	.0007	26.655	.0000176	25.103	51
10	.03781	0.99929	.03783	26.432	1.0007	26.451	.0000180	24.899	50
11	.03810	.99927	.03812	26.230	.0007	26.249	.0000185	24.697	49
12	.03839	.99926	.03842	26.031	.0007	26.050	.0000189	24.498	48
13	.03868	.99925	.03871	25.835	.0007	25.854	.0000193	24.303	47
14	.03897	.99924	.03900	25.642	.0007	25.661	.0000198	24.110	46
15	0.03926	0.99923	0.03929	25.452	1.0008	25.471	.0000202	23.920	45
16	.03955	.99922	.03958	25.264	.0008	25.284	.0000207	23.733	44
17	.03984	.99921	.03987	25.080	.0008	25.100	.0000211	23.549	43
18	.04013	.99919	.04016	24.898	.0008	24.918	.0000216	23.367	42
19	.04042	.99918	.04046	24.719	.0008	24.739	.0000220	23.188	41
20	0.04071	0.99917	0.04075	24.542	1.0008	24.562	.0000225	23.012	40
21	.04100	.99916	.04104	24.368	.0008	24.388	.0000230	22.838	39
22	.04129	.99915	.04133	24.196	.0009	24.216	.0000235	22.666	38
23	.04159	.99913	.04162	24.026	.0009	24.047	.0000240	22.497	37
24	.04188	.99912	.04191	23.859	.0009	23.880	.0000245	22.330	36
25	0.04217	0.99911	0.04220	23.695	1.0009	23.716	.0000250	22.166	35
26	.04246	.99910	.04250	23.532	.0009	23.553	.0000256	22.004	34
27	.04275	.99909	.04279	23.372	.0009	23.393	.0000261	21.844	33
28	.04304	.99907	.04308	23.214	.0009	23.235	.0000266	21.686	32
29	.04333	.99906	.04337	23.058	.0009	23.079	.0000272	21.530	31
30	0.04362	0.99905	0.04366	22.904	1.0010	22.926	.0000277	21.377	30
31	.04391	.99904	.04395	22.752	.0010	22.774	.0000283	21.225	29
32	.04420	.99902	.04424	22.602	.0010	22.624	.0000288	21.075	28
33	.04449	.99901	.04454	22.454	.0010	22.476	.0000294	20.928	27
34	.04478	.99900	.04483	22.308	.0010	22.330	.0000300	20.782	26
35	0.04507	0.99898	0.04512	22.164	1.0010	22.187	.0000306	20.638	25
36	.04536	.99897	.04541	22.022	.0010	22.044	.0000312	20.496	24
37	.04565	.99896	.04570	21.881	.0010	21.904	.0000318	20.356	23
38	.04594	.99894	.04599	21.743	.0011	21.766	.0000324	20.218	22
39	.04623	.99893	.04628	21.606	.0011	21.629	.0000330	20.081	21
40	0.04653	0.99892	0.04658	21.470	1.0011	21.494	.0000336	19.946	20
41	.04682	.99890	.04687	21.337	.0011	21.360	.0000343	19.813	19
42	.04711	.99889	.04716	21.205	.0011	21.229	.0000349	19.681	18
43	.04740	.99888	.04745	21.075	.0011	21.098	.0000356	19.551	17
44	.04769	.99886	.04774	20.946	.0011	20.970	.0000362	19.423	16
45	0.04798	0.99885	0.04803	20.819	1.0012	20.843	.0000369	19.296	15
46	.04827	.99883	.04833	20.693	.0012	20.717	.0000376	19.171	14
47	.04856	.99882	.04862	20.569	.0012	20.593	.0000382	19.047	13
48	.04885	.99881	.04891	20.446	.0012	20.471	.0000389	18.925	12
49	.04914	.99879	.04920	20.325	.0012	20.350	.0000396	18.804	11
50	0.04943	0.99878	0.04949	20.206	1.0012	20.230	.0000403	18.684	10
51	.04972	.99876	.04978	20.087	.0012	20.112	.0000411	18.566	9
52	.05001	.99875	.05007	19.970	.0013	19.995	.0000418	18.449	8
53	.05030	.99873	.05037	19.855	.0013	19.880	.0000425	18.334	7
54	.05059	.99872	.05066	19.740	.0013	19.766	.0000433	18.220	6
55	0.05088	0.99870	0.05095	19.627	1.0013	19.653	.0000440	18.107	5
56	.05117	.99869	.05124	19.516	.0013	19.541	.0000448	17.996	4
57	.05146	.99867	.05153	19.405	.0013	19.431	.0000455	17.886	3
58	.05175	.99866	.05182	19.296	.0013	19.322	.0000463	17.777	2
59	.05205	.99864	.05212	19.188	.0014	19.214	.0000471	17.669	1
60	0.05234	0.99863	0.05241	19.081	1.0014	19.107	.0000479	17.563	0
M	Cosine	Sine	Cotan.	Tan.	Cosec.	Secant	READ DOWN	87°–88° Involute	M

Trigonometric and Involute Functions

M	Sine	Cosine	Tan.	Cotan.	Secant	Cosec.	Involute 3°–4°	READ UP	M
0	0.05234	0.99863	0.05241	19.081	1.0014	19.107	.0000479	17.563	60
1	.05263	.99861	.05270	18.976	.0014	19.002	.0000487	17.457	59
2	.05292	.99860	.05299	18.871	.0014	18.898	.0000495	17.353	58
3	.05321	.99858	.05328	18.768	.0014	18.794	.0000503	17.250	57
4	.05350	.99857	.05357	18.666	.0014	18.692	.0000512	17.148	56
5	0.05379	0.99855	0.05387	18.564	1.0014	18.591	.0000520	17.047	55
6	.05408	.99854	.05416	18.464	.0015	18.492	.0000529	16.948	54
7	.05437	.99852	.05445	18.366	.0015	18.393	.0000537	16.849	53
8	.05466	.99851	.05474	18.268	.0015	18.295	.0000546	16.752	52
9	.05495	.99849	.05503	18.171	.0015	18.198	.0000555	16.655	51
10	0.05524	0.99847	0.05533	18.075	1.0015	18.103	.0000563	16.559	50
11	.05553	.99846	.05562	17.980	.0015	18.008	.0000572	16.465	49
12	.05582	.99844	.05591	17.886	.0016	17.914	.0000581	16.371	48
13	.05611	.99842	.05620	17.793	.0016	17.822	.0000591	16.279	47
14	.05640	.99841	.05649	17.702	.0016	17.730	.0000600	16.187	46
15	0.05669	0.99839	0.05678	17.611	1.0016	17.639	.0000609	16.096	45
16	.05698	.99838	.05708	17.521	.0016	17.549	.0000619	16.007	44
17	.05727	.99836	.05737	17.431	.0016	17.460	.0000628	15.918	43
18	.05756	.99834	.05766	17.343	.0017	17.372	.0000638	15.830	42
19	.05785	.99833	.05795	17.256	.0017	17.285	.0000647	15.743	41
20	0.05814	0.99831	0.05824	17.169	1.0017	17.198	.0000657	15.657	40
21	.05844	.99829	.05854	17.084	.0017	17.113	.0000667	15.571	39
22	.05873	.99827	.05883	16.999	.0017	17.028	.0000677	15.487	38
23	.05902	.99826	.05912	16.915	.0017	16.945	.0000687	15.403	37
24	.05931	.99824	.05941	16.832	.0018	16.862	.0000698	15.320	36
25	0.05960	0.99822	0.05970	16.750	1.0018	16.779	.0000708	15.238	35
26	.05989	.99821	.05999	16.668	.0018	16.698	.0000718	15.157	34
27	.06018	.99819	.06029	16.587	.0018	16.618	.0000729	15.077	33
28	.06047	.99817	.06058	16.507	.0018	16.538	.0000739	14.997	32
29	.06076	.99815	.06087	16.428	.0019	16.459	.0000750	14.918	31
30	0.06105	0.99813	0.06116	16.350	1.0019	16.380	.0000761	14.840	30
31	.06134	.99812	.06145	16.272	.0019	16.303	.0000772	14.763	29
32	.06163	.99810	.06175	16.195	.0019	16.226	.0000783	14.686	28
33	.06192	.99808	.06204	16.119	.0019	16.150	.0000794	14.610	27
34	.06221	.99806	.06233	16.043	.0019	16.075	.0000805	14.535	26
35	0.06250	0.99804	0.06262	15.969	1.0020	16.000	.0000817	14.460	25
36	.06279	.99803	.06291	15.895	.0020	15.926	.0000828	14.387	24
37	.06308	.99801	.06321	15.821	.0020	15.853	.0000840	14.313	23
38	.06337	.99799	.06350	15.748	.0020	15.780	.0000851	14.241	22
39	.06366	.99797	.06379	15.676	.0020	15.708	.0000863	14.169	21
40	0.06395	0.99795	0.06408	15.605	1.0021	15.637	.0000875	14.098	20
41	.06424	.99793	.06438	15.534	.0021	15.566	.0000887	14.027	19
42	.06453	.99792	.06467	15.464	.0021	15.496	.0000899	13.958	18
43	.06482	.99790	.06496	15.394	.0021	15.427	.0000911	13.888	17
44	.06511	.99788	.06525	15.325	.0021	15.358	.0000924	13.820	16
45	0.06540	0.99786	0.06554	15.257	1.0021	15.290	.0000936	13.752	15
46	.06569	.99784	.06584	15.189	.0022	15.222	.0000949	13.684	14
47	.06598	.99782	.06613	15.122	.0022	15.155	.0000961	13.617	13
48	.06627	.99780	.06642	15.056	.0022	15.089	.0000974	13.551	12
49	.06656	.99778	.06671	14.990	.0022	15.023	.0000987	13.486	11
50	0.06685	0.99776	0.06700	14.924	1.0022	14.958	.0001000	13.421	10
51	.06714	.99774	.06730	14.860	.0023	14.893	.0001013	13.356	9
52	.06743	.99772	.06759	14.795	.0023	14.829	.0001026	13.292	8
53	.06773	.99770	.06788	14.732	.0023	14.766	.0001040	13.229	7
54	.06802	.99768	.06817	14.669	.0023	14.703	.0001053	13.166	6
55	0.06831	0.99766	0.06847	14.606	1.0023	14.640	.0001067	13.103	5
56	.06860	.99764	.06876	14.544	.0024	14.578	.0001080	13.042	4
57	.06889	.99762	.06905	14.482	.0024	14.517	.0001094	12.980	3
58	.06918	.99760	.06934	14.421	.0024	14.456	.0001108	12.920	2
59	.06947	.99758	.06963	14.361	.0024	14.395	.0001122	12.859	1
60	0.06976	0.99756	0.06993	14.301	1.0024	14.336	.0001136	12.800	0
M	Cosine	Sine	Cotan.	Tan.	Cosec.	Secant	READ DOWN	86°–87° Involute	M

4° or 184° Trigonometric and Involute Functions 175° or 355°

M	Sine	Cosine	Tan.	Cotan.	Secant	Cosec.	Involute 4°–5°	READ UP	M
0	0.06976	.99756	0.06993	14.301	1.0024	14.336	.0001136	12.800	60
1	.07005	.99754	.07022	14.241	.0025	14.276	.0001151	12.740	59
2	.07034	.99752	.07051	14.182	.0025	14.217	.0001165	12.682	58
3	.07063	.99750	.07080	14.124	.0025	14.159	.0001180	12.623	57
4	.07092	.99748	.07110	14.065	.0025	14.101	.0001194	12.566	56
5	0.07121	.99746	0.07139	14.008	1.0025	14.044	.0001209	12.508	55
6	.07150	.99744	.07168	13.951	.0026	13.987	.0001224	12.451	54
7	.07179	.99742	.07197	13.894	.0026	13.930	.0001239	12.395	53
8	.07208	.99740	.07227	13.838	.0026	13.874	.0001254	12.339	52
9	.07237	.99738	.07256	13.782	.0026	13.818	.0001269	12.284	51
10	0.07266	.99736	0.07285	13.727	1.0027	13.763	.0001285	12.229	50
11	.07295	.99734	.07314	13.672	.0027	13.708	.0001300	12.174	49
12	.07324	.99731	.07344	13.617	.0027	13.654	.0001316	12.120	48
13	.07353	.99729	.07373	13.563	.0027	13.600	.0001332	12.066	47
14	.07382	.99727	.07402	13.510	.0027	13.547	.0001347	12.013	46
15	0.07411	.99725	0.07431	13.457	1.0028	13.494	.0001363	11.960	45
16	.07440	.99723	.07461	13.404	.0028	13.441	.0001380	11.908	44
17	.07469	.99721	.07490	13.352	.0028	13.389	.0001396	11.855	43
18	.07498	.99719	.07519	13.300	.0028	13.337	.0001412	11.804	42
19	.07527	.99716	.07548	13.248	.0028	13.286	.0001429	11.753	41
20	0.07556	.99714	0.07578	13.197	1.0029	13.235	.0001445	11.702	40
21	.07585	.99712	.07607	13.146	.0029	13.184	.0001462	11.651	39
22	.07614	.99710	.07636	13.096	.0029	13.134	.0001479	11.601	38
23	.07643	.99707	.07665	13.046	.0029	13.084	.0001496	11.551	37
24	.07672	.99705	.07695	12.996	.0030	13.035	.0001513	11.502	36
25	0.07701	.99703	0.07724	12.947	1.0030	12.985	.0001530	11.453	35
26	.07730	.99701	.07753	12.898	.0030	12.937	.0001548	11.405	34
27	.07759	.99699	.07782	12.850	.0030	12.888	.0001565	11.356	33
28	.07788	.99696	.07812	12.801	.0030	12.840	.0001583	11.309	32
29	.07817	.99694	.07841	12.754	.0031	12.793	.0001601	11.261	31
30	0.07846	.99692	0.07870	12.706	1.0031	12.745	.0001619	11.214	30
31	.07875	.99689	.07899	12.659	.0031	12.699	.0001637	11.167	29
32	.07904	.99687	.07929	12.612	.0031	12.652	.0001655	11.121	28
33	.07933	.99685	.07958	12.566	.0032	12.606	.0001674	11.075	27
34	.07962	.99683	.07987	12.520	.0032	12.560	.0001692	11.029	26
35	0.07991	.99680	0.08017	12.474	1.0032	12.514	.0001711	10.983	25
36	.08020	.99678	.08046	12.429	.0032	12.469	.0001729	10.938	24
37	.08049	.99676	.08075	12.384	.0033	12.424	.0001748	10.894	23
38	.08078	.99673	.08104	12.339	.0033	12.379	.0001767	10.849	22
39	.08107	.99671	.08134	12.295	.0033	12.335	.0001787	10.805	21
40	0.08136	.99668	0.08163	12.251	1.0033	12.291	.0001806	10.761	20
41	.08165	.99666	.08192	12.207	.0034	12.248	.0001825	10.718	19
42	.08194	.99664	.08221	12.163	.0034	12.204	.0001845	10.674	18
43	.08223	.99661	.08251	12.120	.0034	12.161	.0001865	10.632	17
44	.08252	.99659	.08280	12.077	.0034	12.119	.0001885	10.589	16
45	0.08281	.99657	0.08309	12.035	1.0034	12.076	.0001905	10.547	15
46	.08310	.99654	.08339	11.992	.0035	12.034	.0001925	10.505	14
47	.08339	.99652	.08368	11.950	.0035	11.992	.0001945	10.463	13
48	.08368	.99649	.08397	11.909	.0035	11.951	.0001965	10.422	12
49	.08397	.99647	.08427	11.867	.0035	11.909	.0001986	10.381	11
50	0.08426	.99644	0.08456	11.826	1.0036	11.868	.0002007	10.340	10
51	.08455	.99642	.08485	11.785	.0036	11.828	.0002028	10.299	9
52	.08484	.99639	.08514	11.745	.0036	11.787	.0002049	10.259	8
53	.08513	.99637	.08544	11.705	.0036	11.747	.0002070	10.219	7
54	.08542	.99635	.08573	11.664	.0037	11.707	.0002091	10.179	6
55	0.08571	.99632	0.08602	11.625	1.0037	11.668	.0002113	10.140	5
56	.08600	.99630	.08632	11.585	.0037	11.628	.0002134	10.101	4
57	.08629	.99627	.08661	11.546	.0037	11.589	.0002156	10.062	3
58	.08658	.99625	.08690	11.507	.0038	11.551	.0002178	10.023	2
59	.08687	.99622	.08720	11.468	.0038	11.512	.0002200	9.9847	1
60	0.08716	.99619	0.08749	11.430	1.0038	11.474	.0002222	9.9465	0
M	Cosine	Sine	Cotan.	Tan.	Cosec.	Secant	READ DOWN	85°–86° Involute	M

94° or 274° 85° or 265°

5° or 185° **Trigonometric and Involute Functions** **174° or 354°**

M	Sine	Cosine	Tan.	Cotan.	Secant	Cosec.	Involute 5°–6°	READ UP	M
0	0.08716	.99619	0.08749	11.430	1.0038	11.474	.0002222	9.9465	60
1	.08745	.99617	.08778	11.392	.0038	11.436	.0002244	9.9086	59
2	.08774	.99614	.08807	11.354	.0039	11.398	.0002267	9.8710	58
3	.08803	.99612	.08837	11.316	.0039	11.360	.0002289	9.8336	57
4	.08831	.99609	.08866	11.279	.0039	11.323	.0002312	9.7965	56
5	0.08860	.99607	0.08895	11.242	1.0039	11.286	.0002335	9.7596	55
6	.08889	.99604	.08925	11.205	.0040	11.249	.0002358	9.7230	54
7	.08918	.99602	.08954	11.168	.0040	11.213	.0002382	9.6866	53
8	.08947	.99599	.08983	11.132	.0040	11.176	.0002405	9.6504	52
9	.08976	.99596	.09013	11.095	.0041	11.140	.0002429	9.6145	51
10	0.09005	.99594	0.09042	11.059	1.0041	11.105	.0002452	9.5788	50
11	.09034	.99591	.09071	11.024	.0041	11.069	.0002476	9.5433	49
12	.09063	.99588	.09101	10.988	.0041	11.034	.0002500	9.5081	48
13	.09092	.99586	.09130	10.953	.0042	10.998	.0002524	9.4731	47
14	.09121	.99583	.09159	10.918	.0042	10.963	.0002549	9.4383	46
15	0.09150	.99580	0.09189	10.883	1.0042	10.929	.0002573	9.4038	45
16	.09179	.99578	.09218	10.848	.0042	10.894	.0002598	9.3694	44
17	.09208	.99575	.09247	10.814	.0043	10.860	.0002622	9.3353	43
18	.09237	.99572	.09277	10.780	.0043	10.826	.0002647	9.3014	42
19	.09266	.99570	.09306	10.746	.0043	10.792	.0002673	9.2677	41
20	0.09295	.99567	0.09335	10.712	1.0043	10.758	.0002698	9.2342	40
21	.09324	.99564	.09365	10.678	.0044	10.725	.0002723	9.2009	39
22	.09353	.99562	.09394	10.645	.0044	10.692	.0002749	9.1679	38
23	.09382	.99559	.09423	10.612	.0044	10.659	.0002775	9.1350	37
24	.09411	.99556	.09453	10.579	.0045	10.626	.0002801	9.1023	36
25	0.09440	.99553	0.09482	10.546	1.0045	10.593	.0002827	9.0699	35
26	.09469	.99551	.09511	10.514	.0045	10.561	.0002853	9.0376	34
27	.09498	.99548	.09541	10.481	.0045	10.529	.0002879	9.0056	33
28	.09527	.99545	.09570	10.449	.0046	10.497	.0002906	8.9737	32
29	.09556	.99542	.09600	10.417	.0046	10.465	.0002933	8.9421	31
30	0.09585	.99540	0.09629	10.385	1.0046	10.433	.0002959	8.9106	30
31	.09614	.99537	.09658	10.354	.0047	10.402	.0002986	8.8793	29
32	.09642	.99534	.09688	10.322	.0047	10.371	.0003014	8.8482	28
33	.09671	.99531	.09717	10.291	.0047	10.340	.0003041	8.8173	27
34	.09700	.99528	.09746	10.260	.0047	10.309	.0003069	8.7866	26
35	.09729	.99526	.09776	10.229	1.0048	10.278	.0003096	8.7561	25
36	.09758	.99523	.09805	10.199	.0048	10.248	.0003124	8.7257	24
37	.09787	.99520	.09834	10.168	.0048	10.217	.0003152	8.6956	23
38	.09816	.99517	.09864	10.138	.0049	10.187	.0003180	8.6656	22
39	.09845	.99514	.09893	10.108	.0049	10.157	.0003209	8.6358	21
40	0.09874	.99511	0.09923	10.078	1.0049	10.128	.0003237	8.6061	20
41	.09903	.99508	.09952	10.048	.0049	10.098	.0003266	8.5767	19
42	.09932	.99506	.09981	10.019	.0050	10.068	.0003295	8.5474	18
43	.09961	.99503	.10011	9.9893	.0050	10.039	.0003324	8.5183	17
44	.09990	.99500	.10040	9.9601	.0050	10.010	.0003353	8.4893	16
45	0.10019	.99497	0.10069	9.9310	1.0051	9.9812	.0003383	8.4606	15
46	.10048	.99494	.10099	9.9021	.0051	9.9525	.0003412	8.4320	14
47	.10077	.99491	.10128	9.8734	.0051	9.9239	.0003442	8.4035	13
48	.10106	.99488	.10158	9.8448	.0051	9.8955	.0003472	8.3752	12
49	.10135	.99485	.10187	9.8164	.0052	9.8672	.0003502	8.3471	11
50	0.10164	.99482	0.10216	9.7882	.0052	9.8391	.0003532	8.3192	10
51	.10192	.99479	.10246	9.7601	.0052	9.8112	.0003563	8.2914	9
52	.10221	.99476	.10275	9.7322	.0053	9.7834	.0003593	8.2638	8
53	.10250	.99473	.10305	9.7044	.0053	9.7558	.0003624	8.2363	7
54	.10279	.99470	.10334	9.6768	.0053	9.7283	.0003655	8.2090	6
55	0.10308	.99467	0.10363	9.6493	1.0054	9.7010	.0003686	8.1818	5
56	.10337	.99464	.10393	9.6220	.0054	9.6739	.0003718	8.1548	4
57	.10366	.99461	.10422	9.5949	.0054	9.6469	.0003749	8.1280	3
58	.10395	.99458	.10452	9.5679	.0054	9.6200	.0003781	8.1012	2
59	.10424	.99455	.10481	9.5411	.0055	9.5933	.0003813	8.0747	1
60	0.10453	.99452	0.10510	9.5144	1.0055	9.5668	.0003845	8.0483	0
M	Cosine	Sine	Cotan.	Tan.	Cosec.	Secant	READ DOWN	84°–85° Involute	M

95° or 275° **84° or 264°**

6° or 186°   **Trigonometric and Involute Functions**   **173° or 353°**

M	Sine	Cosine	Tan.	Cotan.	Secant	Cosec.	Involute 6°–7°	READ UP	M
0	0.10453	0.99452	0.10510	9.5144	1.0055	9.5668	.0003845	8.0483	60
1	.10482	.99449	.10540	.4878	.0055	.5404	.0003877	8.0220	59
2	.10511	.99446	.10569	.4614	.0056	.5141	.0003909	7.9959	58
3	.10540	.99443	.10599	.4352	.0056	.4880	.0003942	7.9699	57
4	.10569	.99440	.10628	.4090	.0056	.4620	.0003975	7.9441	56
5	0.10597	0.99437	0.10657	9.3831	1.0057	9.4362	.0004008	7.9184	55
6	.10626	.99434	.10687	.3572	.0057	.4105	.0004041	7.8929	54
7	.10655	.99431	.10716	.3315	.0057	.3850	.0004074	7.8675	53
8	.10684	.99428	.10746	.3060	.0058	.3596	.0004108	7.8422	52
9	.10713	.99424	.10775	.2806	.0058	.3343	.0004141	7.8171	51
10	0.10742	0.99421	0.10805	9.2553	1.0058	9.3092	.0004175	7.7921	50
11	.10771	.99418	.10834	.2302	.0059	.2842	.0004209	7.7673	49
12	.10800	.99415	.10863	.2052	.0059	.2593	.0004244	7.7426	48
13	.10829	.99412	.10893	.1803	.0059	.2346	.0004278	7.7180	47
14	.10858	.99409	.10922	.1555	.0059	.2100	.0004313	7.6935	46
15	0.10887	0.99406	0.10952	9.1309	1.0060	9.1855	.0004347	7.6692	45
16	.10916	.99402	.10981	.1065	.0060	.1612	.0004382	7.6450	44
17	.10945	.99399	.11011	.0821	.0060	.1370	.0004417	7.6210	43
18	.10973	.99396	.11040	.0579	.0061	.1129	.0004453	7.5970	42
19	.11002	.99393	.11070	.0338	.0061	.0890	.0004488	7.5732	41
20	0.11031	0.99390	0.11099	9.0098	1.0061	9.0652	.0004524	7.5496	40
21	.11060	.99386	.11128	8.9860	.0062	.0415	.0004560	7.5260	39
22	.11089	.99383	.11158	.9623	.0062	.0179	.0004595	7.5026	38
23	.11118	.99380	.11187	.9387	.0062	8.9944	.0004632	7.4793	37
24	.11147	.99377	.11217	.9152	.0063	.9711	.0004669	7.4561	36
25	0.11176	0.99374	0.11246	8.8918	1.0063	8.9479	.0004706	7.4330	35
26	.11205	.99370	.11276	.8686	.0063	.9248	.0004743	7.4101	34
27	.11234	.99367	.11305	.8455	.0064	.9019	.0004780	7.3873	33
28	.11263	.99364	.11335	.8225	.0064	.8790	.0004817	7.3646	32
29	.11291	.99360	.11364	.7996	.0064	.8563	.0004854	7.3420	31
30	0.11320	0.99357	0.11394	8.7769	1.0065	8.8337	.0004892	7.3195	30
31	.11349	.99354	.11423	.7542	.0065	.8112	.0004930	7.2972	29
32	.11378	.99351	.11452	.7317	.0065	.7888	.0004968	7.2750	28
33	.11407	.99347	.11482	.7093	.0066	.7665	.0005006	7.2528	27
34	.11436	.99344	.11511	.6870	.0066	.7444	.0005045	7.2308	26
35	0.11465	0.99341	0.11541	8.6648	1.0066	8.7223	.0005083	7.2089	25
36	.11494	.99337	.11570	.6427	.0067	.7004	.0005122	7.1871	24
37	.11523	.99334	.11600	.6208	.0067	.6786	.0005161	7.1655	23
38	.11552	.99330	.11629	.5989	.0067	.6569	.0005200	7.1439	22
39	.11580	.99327	.11659	.5772	.0068	.6353	.0005240	7.1225	21
40	0.11609	0.99324	0.11688	8.5555	1.0068	8.6138	.0005280	7.1011	20
41	.11638	.99320	.11718	.5340	.0068	.5924	.0005319	7.0799	19
42	.11667	.99317	.11747	.5126	.0069	.5711	.0005359	7.0587	18
43	.11696	.99313	.11777	.4913	.0069	.5500	.0005400	7.0377	17
44	.11725	.99310	.11806	.4701	.0069	.5289	.0005440	7.0168	16
45	0.11754	0.99307	0.11836	8.4490	1.0070	8.5079	.0005481	6.9960	15
46	.11783	.99303	.11865	.4280	.0070	.4871	.0005522	6.9753	14
47	.11812	.99300	.11895	.4071	.0070	.4663	.0005563	6.9546	13
48	.11840	.99296	.11924	.3863	.0071	.4457	.0005604	6.9341	12
49	.11869	.99293	.11954	.3656	.0071	.4251	.0005645	6.9137	11
50	0.11898	0.99290	0.11983	8.3450	1.0072	8.4047	.0005687	6.8934	10
51	.11927	.99286	.12013	.3245	.0072	.3843	.0005729	6.8732	9
52	.11956	.99283	.12042	.3041	.0072	.3641	.0005771	6.8531	8
53	.11985	.99279	.12072	.2838	.0073	.3439	.0005813	6.8331	7
54	.12014	.99276	.12101	.2636	.0073	.3238	.0005856	6.8132	6
55	0.12043	0.99272	0.12131	8.2434	1.0073	8.3039	.0005898	6.7934	5
56	.12071	.99269	.12160	.2234	.0074	.2840	.0005941	6.7737	4
57	.12100	.99265	.12190	.2035	.0074	.2642	.0005985	6.7540	3
58	.12129	.99262	.12219	.1837	.0074	.2446	.0006028	6.7345	2
59	.12158	.99258	.12249	.1640	.0075	.2250	.0006071	6.7151	1
60	0.12187	0.99255	0.12278	8.1443	1.0075	8.2055	.0006115	6.6957	0
M	Cosine	Sine	Cotan.	Tan.	Cosec.	Secant	READ DOWN	83°–84° Involute	M

96° or 276°   **83° or 263°**

M	Sine	Cosine	Tan.	Cotan.	Secant	Cosec.	Involute 7°–8°	READ UP	M
0	0.12187	0.99255	0.12278	8.1443	1.0075	8.2055	.0006115	6.6957	60
1	.12216	.99251	.12308	.1248	.0075	.1861	.0006159	6.6765	59
2	.12245	.99248	.12338	.1054	.0076	.1668	.0006203	6.6573	58
3	.12274	.99244	.12367	.0860	.0076	.1476	.0006248	6.6383	57
4	.12302	.99240	.12397	.0667	.0077	.1285	.0006292	6.6193	56
5	.12331	.99237	.12426	8.0476	1.0077	8.1095	.0006337	6.6004	55
6	.12360	.99233	.12456	.0285	.0077	.0905	.0006382	6.5816	54
7	.12389	.99230	.12485	.0095	.0078	.0717	.0006427	6.5629	53
8	.12418	.99226	.12515	7.9906	.0078	.0529	.0006473	6.5443	52
9	.12447	.99222	.12544	.9718	.0078	.0342	.0006518	6.5258	51
10	0.12476	0.99219	0.12574	7.9530	1.0079	8.0156	.0006564	6.5073	50
11	.12504	.99215	.12603	.9344	.0079	7.9971	.0006610	6.4890	49
12	.12533	.99211	.12633	.9158	.0079	.9787	.0006657	6.4707	48
13	.12562	.99208	.12662	.8973	.0080	.9604	.0006703	6.4525	47
14	.12591	.99204	.12692	.8789	.0080	.9422	.0006750	6.4344	46
15	0.12620	0.99200	0.12722	7.8606	1.0081	7.9240	.0006797	6.4164	45
16	.12649	.99197	.12751	.8424	.0081	.9059	.0006844	6.3985	44
17	.12678	.99193	.12781	.8243	.0081	.8879	.0006892	6.3806	43
18	.12706	.99189	.12810	.8062	.0082	.8700	.0006939	6.3628	42
19	.12735	.99186	.12840	.7882	.0082	.8522	.0006987	6.3451	41
20	0.12764	0.99182	0.12869	7.7704	1.0082	7.8344	.0007035	6.3275	40
21	.12793	.99178	.12899	.7525	.0083	.8168	.0007083	6.3100	39
22	.12822	.99175	.12929	.7348	.0083	.7992	.0007132	6.2926	38
23	.12851	.99171	.12958	.7171	.0084	.7817	.0007181	6.2752	37
24	.12880	.99167	.12988	.6996	.0084	.7642	.0007230	6.2579	36
25	0.12908	0.99163	0.13017	7.6821	1.0084	7.7469	.0007279	6.2407	35
26	.12937	.99160	.13047	.6647	.0085	.7296	.0007328	6.2236	34
27	.12966	.99156	.13076	.6473	.0085	.7124	.0007378	6.2065	33
28	.12995	.99152	.13106	.6301	.0086	.6953	.0007428	6.1896	32
29	.13024	.99148	.13136	.6129	.0086	.6783	.0007478	6.1727	31
30	0.13053	0.99144	0.13165	7.5958	1.0086	7.6613	.0007528	6.1559	30
31	.13081	.99141	.13195	.5787	.0087	.6444	.0007579	6.1391	29
32	.13110	.99137	.13224	.5618	.0087	.6276	.0007629	6.1224	28
33	.13139	.99133	.13254	.5449	.0087	.6109	.0007680	6.1058	27
34	.13168	.99129	.13284	.5281	.0088	.5942	.0007732	6.0893	26
35	0.13197	0.99125	0.13313	7.5113	1.0088	7.5776	.0007783	6.0729	25
36	.13226	.99122	.13343	.4947	.0089	.5611	.0007835	6.0565	24
37	.13254	.99118	.13372	.4781	.0089	.5446	.0007887	6.0402	23
38	.13283	.99114	.13402	.4615	.0089	.5282	.0007939	6.0240	22
39	.13312	.99110	.13432	.4451	.0090	.5119	.0007991	6.0078	21
40	0.13341	0.99106	0.13461	7.4287	1.0090	7.4957	.0008044	5.9917	20
41	.13370	.99102	.13491	.4124	.0091	.4795	.0008096	5.9757	19
42	.13399	.99098	.13521	.3962	.0091	.4635	.0008150	5.9598	18
43	.13427	.99094	.13550	.3800	.0091	.4474	.0008203	5.9439	17
44	.13456	.99091	.13580	.3639	.0092	.4315	.0008256	5.9281	16
45	0.13485	0.99087	0.13609	7.3479	1.0092	7.4156	.0008310	5.9123	15
46	.13514	.99083	.13639	.3319	.0093	.3998	.0008364	5.8967	14
47	.13543	.99079	.13669	.3160	.0093	.3840	.0008418	5.8811	13
48	.13572	.99075	.13698	.3002	.0093	.3684	.0008473	5.8655	12
49	.13600	.99071	.13728	.2844	.0094	.3527	.0008527	5.8500	11
50	0.13629	0.99067	0.13758	7.2687	1.0094	7.3372	.0008582	5.8346	10
51	.13658	.99063	.13787	.2531	.0095	.3217	.0008638	5.8193	9
52	.13687	.99059	.13817	.2375	.0095	.3063	.0008693	5.8040	8
53	.13716	.99055	.13846	.2220	.0095	.2909	.0008749	5.7888	7
54	.13744	.99051	.13876	.2066	.0096	.2757	.0008805	5.7737	6
55	.13773	.99047	.13906	7.1912	1.0096	7.2604	.0008861	5.7586	5
56	.13802	.99043	.13935	.1759	.0097	.2453	.0008917	5.7436	4
57	.13831	.99039	.13965	.1607	.0097	.2302	.0008974	5.7287	3
58	.13860	.99035	.13995	.1455	.0097	.2152	.0009031	5.7138	2
59	.13889	.99031	.14024	.1304	.0098	.2002	.0009088	5.6990	1
60	0.13917	0.99027	0.14054	7.1154	1.0098	7.1853	.0009145	5.6842	0
M	Cosine	Sine	Cotan.	Tan.	Cosec.	Secant	READ DOWN	82°–83° Involute	M

M	Sine	Cosine	Tan.	Cotan.	Secant	Cosec.	Involute 8°–9°	READ UP	M
0	0.13917	0.99027	0.14054	7.1154	1.0098	7.1853	.0009145	5.6842	60
1	.13946	.99023	.14084	.1004	.0099	.1705	.0009203	5.6695	59
2	.13975	.99019	.14113	.0855	.0099	.1557	.0009260	5.6549	58
3	.14004	.99015	.14143	.0706	.0100	.1410	.0009318	5.6403	57
4	.14033	.99011	.14173	.0558	.0100	.1263	.0009377	5.6258	56
5	0.14061	0.99006	0.14202	7.0410	1.0100	7.1117	.0009435	5.6113	55
6	.14090	.99002	.14232	.0264	.0101	.0972	.0009494	5.5969	54
7	.14119	.98998	.14262	.0117	.0101	.0827	.0009553	5.5826	53
8	.14148	.98994	.14291	6.9972	.0102	.0683	.0009612	5.5683	52
9	.14177	.98990	.14321	.9827	.0102	.0539	.0009672	5.5541	51
10	0.14205	0.98986	0.14351	6.9682	1.0102	7.0396	.0009732	5.5400	50
11	.14234	.98982	.14381	.9538	.0103	.0254	.0009792	5.5259	49
12	.14263	.98978	.14410	.9395	.0103	.0112	.0009852	5.5118	48
13	.14292	.98973	.14440	.9252	.0104	6.9971	.0009913	5.4979	47
14	.14320	.98969	.14470	.9110	.0104	.9830	.0009973	5.4839	46
15	0.14349	0.98965	0.14499	6.8969	1.0105	6.9690	.0010034	5.4701	45
16	.14378	.98961	.14529	.8828	.0105	.9550	.0010096	5.4563	44
17	.14407	.98957	.14559	.8687	.0105	.9411	.0010157	5.4425	43
18	.14436	.98953	.14588	.8547	.0106	.9273	.0010219	5.4288	42
19	.14464	.98948	.14618	.8408	.0106	.9135	.0010281	5.4152	41
20	0.14493	0.98944	0.14648	6.8269	1.0107	6.8998	.0010343	5.4016	40
21	.14522	.98940	.14678	.8131	.0107	.8861	.0010406	5.3881	39
22	.14551	.98936	.14707	.7994	.0108	.8725	.0010469	5.3746	38
23	.14580	.98931	.14737	.7856	.0108	.8589	.0010532	5.3612	37
24	.14608	.98927	.14767	.7720	.0108	.8454	.0010595	5.3478	36
25	0.14637	0.98923	0.14796	6.7584	1.0109	6.8320	.0010659	5.3345	35
26	.14666	.98919	.14826	.7448	.0109	.8186	.0010722	5.3212	34
27	.14695	.98914	.14856	.7313	.0110	.8052	.0010786	5.3080	33
28	.14723	.98910	.14886	.7179	.0110	.7919	.0010851	5.2949	32
29	.14752	.98906	.14915	.7045	.0111	.7787	.0010915	5.2818	31
30	0.14781	0.98902	0.14945	6.6912	1.0111	6.7655	.0010980	5.2687	30
31	.14810	.98897	.14975	.6779	.0112	.7523	.0011045	5.2557	29
32	.14838	.98893	.15005	.6646	.0112	.7392	.0011111	5.2428	28
33	.14867	.98889	.15034	.6514	.0112	.7262	.0011176	5.2299	27
34	.14896	.98884	.15064	.6383	.0113	.7132	.0011242	5.2170	26
35	0.14925	0.98880	0.15094	6.6252	1.0113	6.7003	.0011308	5.2042	25
36	.14954	.98876	.15124	.6122	.0114	.6874	.0011375	5.1915	24
37	.14982	.98871	.15153	.5992	.0114	.6745	.0011441	5.1788	23
38	.15011	.98867	.15183	.5863	.0115	.6618	.0011508	5.1662	22
39	.15040	.98863	.15213	.5734	.0115	.6490	.0011575	5.1536	21
40	0.15069	0.98858	0.15243	6.5606	1.0116	6.6363	.0011643	5.1410	20
41	.15097	.98854	.15272	.5478	.0116	.6237	.0011711	5.1285	19
42	.15126	.98849	.15302	.5350	.0116	.6111	.0011779	5.1161	18
43	.15155	.98845	.15332	.5223	.0117	.5986	.0011847	5.1037	17
44	.15184	.98841	.15362	.5097	.0117	.5861	.0011915	5.0913	16
45	0.15212	0.98836	0.15391	6.4971	1.0118	6.5736	.0011984	5.0790	15
46	.15241	.98832	.15421	.4846	.0118	.5612	.0012053	5.0668	14
47	.15270	.98827	.15451	.4721	.0119	.5489	.0012122	5.0546	13
48	.15299	.98823	.15481	.4596	.0119	.5366	.0012192	5.0424	12
49	.15327	.98818	.15511	.4472	.0120	.5243	.0012262	5.0303	11
50	0.15356	0.98814	0.15540	6.4348	1.0120	6.5121	.0012332	5.0182	10
51	.15385	.98809	.15570	.4225	.0120	.4999	.0012402	5.0062	9
52	.15414	.98805	.15600	.4103	.0121	.4878	.0012473	4.9942	8
53	.15442	.98800	.15630	.3980	.0121	.4757	.0012544	4.9823	7
54	.15471	.98796	.15660	.3859	.0122	.4637	.0012615	4.9704	6
55	0.15500	0.98791	0.15689	6.3737	1.0122	6.4517	.0012687	4.9586	5
56	.15529	.98787	.15719	.3617	.0123	.4398	.0012758	4.9468	4
57	.15557	.98782	.15749	.3496	.0123	.4279	.0012830	4.9350	3
58	.15586	.98778	.15779	.3376	.0124	.4160	.0012903	4.9233	2
59	.15615	.98773	.15809	.3257	.0124	.4042	.0012975	4.9117	1
60	0.15643	0.98769	0.15838	6.3138	1.0125	6.3925	.0013048	4.9000	0

| M | Cosine | Sine | Cotan. | Tan. | Cosec. | Secant | READ DOWN | 81°–82° Involute | M |

M	Sine	Cosine	Tan.	Cotan.	Secant	Cosec.	Involute 9°–10°	READ UP	M
0	0.15643	0.98769	0.15838	6.3138	1.0125	6.3925	.0013048	4.9000	60
1	.15672	.98764	.15868	.3019	.0125	.3807	.0013121	4.8885	59
2	.15701	.98760	.15898	.2901	.0126	.3691	.0013195	4.8769	58
3	.15730	.98755	.15928	.2783	.0126	.3574	.0013268	4.8654	57
4	.15758	.98751	.15958	.2666	.0127	.3458	.0013342	4.8540	56
5	0.15787	0.98746	0.15988	6.2549	1.0127	6.3343	.0013416	4.8426	55
6	.15816	.98741	.16017	.2432	.0127	.3228	.0013491	4.8312	54
7	.15845	.98737	.16047	.2316	.0128	.3113	.0013566	4.8199	53
8	.15873	.98732	.16077	.2200	.0128	.2999	.0013641	4.8086	52
9	.15902	.98728	.16107	.2085	.0129	.2885	.0013716	4.7974	51
10	0.15931	0.98723	0.16137	6.1970	1.0129	6.2772	.0013792	4.7862	50
11	.15959	.98718	.16167	.1856	.0130	.2659	.0013868	4.7751	49
12	.15988	.98714	.16196	.1742	.0130	.2546	.0013944	4.7640	48
13	.16017	.98709	.16226	.1628	.0131	.2434	.0014020	4.7529	47
14	.16046	.98704	.16256	.1515	.0131	.2323	.0014097	4.7419	46
15	0.16074	0.98700	0.16286	6.1402	1.0132	6.2211	.0014174	4.7309	45
16	.16103	.98695	.16316	.1290	.0132	.2100	.0014251	4.7199	44
17	.16132	.98690	.16346	.1178	.0133	.1990	.0014329	4.7090	43
18	.16160	.98686	.16376	.1066	.0133	.1880	.0014407	4.6982	42
19	.16189	.98681	.16405	.0955	.0134	.1770	.0014485	4.6873	41
20	0.16218	0.98676	0.16435	6.0844	1.0134	6.1661	.0014563	4.6765	40
21	.16246	.98671	.16465	.0734	.0135	.1552	.0014642	4.6658	39
22	.16275	.98667	.16495	.0624	.0135	.1443	.0014721	4.6551	38
23	.16304	.98662	.16525	.0514	.0136	.1335	.0014800	4.6444	37
24	.16333	.98657	.16555	.0405	.0136	.1227	.0014880	4.6338	36
25	0.16361	0.98652	0.16685	6.0296	1.0137	6.1120	.0014960	4.6232	35
26	.16390	.98648	.16615	.0188	.0137	.1013	.0015040	4.6126	34
27	.16419	.98643	.16645	.0080	.0138	.0906	.0015120	4.6021	33
28	.16447	.98638	.16674	5.9972	.0138	.0800	.0015201	4.5916	32
29	.16476	.98633	.16704	.9865	.0139	.0694	.0015282	4.5812	31
30	0.16505	0.98629	0.16734	5.9758	1.0139	6.0589	.0015363	4.5708	30
31	.16533	.98624	.16764	.9651	.0140	.0483	.0015445	4.5604	29
32	.16562	.98619	.16794	.9545	.0140	.0379	.0015527	4.5501	28
33	.16591	.98614	.16824	.9439	.0141	.0274	.0015609	4.5398	27
34	.16620	.98609	.16854	.9333	.0141	.0170	.0015691	4.5295	26
35	0.16648	0.98604	0.16884	5.9228	1.0142	6.0067	.0015774	4.5193	25
36	.16677	.98600	.16914	.9124	.0142	5.9963	.0015857	4.5091	24
37	.16706	.98595	.16944	.9019	.0143	.9860	.0015941	4.4990	23
38	.16734	.98590	.16974	.8915	.0143	.9758	.0016024	4.4888	22
39	.16763	.98585	.17004	.8811	.0144	.9656	.0016108	4.4788	21
40	0.16792	0.98580	0.17033	5.8708	1.0144	5.9554	.0016193	4.4687	20
41	.16820	.98575	.17063	.8605	.0145	.9452	.0016277	4.4587	19
42	.16849	.98570	.17093	.8502	.0145	.9351	.0016362	4.4487	18
43	.16878	.98565	.17123	.8400	.0146	.9250	.0016447	4.4388	17
44	.16906	.98561	.17153	.8298	.0146	.9150	.0016533	4.4289	16
45	0.16935	0.98556	0.17183	5.8197	1.0147	5.9049	.0016618	4.4190	15
46	.16964	.98551	.17213	.8095	.0147	.8950	.0016704	4.4092	14
47	.16992	.98546	.17243	.7994	.0148	.8850	.0016791	4.3994	13
48	.17021	.98541	.17273	.7894	.0148	.8751	.0016877	4.3896	12
49	.17050	.98536	.17303	.7794	.0149	.8652	.0016964	4.3799	11
50	0.17078	0.98531	0.17333	5.7694	1.0149	5.8554	.0017051	4.3702	10
51	.17107	.98526	.17363	.7594	.0150	.8456	.0017139	4.3605	9
52	.17136	.98521	.17393	.7495	.0150	.8358	.0017227	4.3509	8
53	.17164	.98516	.17423	.7396	.0151	.8261	.0017315	4.3413	7
54	.17193	.98511	.17453	.7297	.0151	.8164	.0017403	4.3317	6
55	0.17222	0.98506	0.17483	5.7199	1.0152	5.8067	.0017492	4.3222	5
56	.17250	.98501	.17513	.7101	.0152	.7970	.0017581	4.3127	4
57	.17279	.98496	.17543	.7004	.0153	.7874	.0017671	4.3032	3
58	.17308	.98491	.17573	.6906	.0153	.7778	.0017760	4.2938	2
59	.17336	.98486	.17603	.6809	.0154	.7683	.0017850	4.2844	1
60	0.17365	0.98481	0.17633	5.6713	1.0154	5.7588	.0017941	4.2750	0
M	Cosine	Sine	Cotan.	Tan.	Cosec.	Secant	READ DOWN	80°–81° Involute	M

10° or 190°   Trigonometric and Involute Functions   169° or 349°

M	Sine	Cosine	Tan.	Cotan.	Secant	Cosec.	Involute 10°–11°	READ UP	M
0	0.17365	0.98481	0.17633	5.6713	1.0154	5.7588	.0017941	4.2750	60
1	.17393	.98476	.17663	.6617	.0155	.7493	.0018031	4.2657	59
2	.17422	.98471	.17693	.6521	.0155	.7398	.0018122	4.2564	58
3	.17451	.98466	.17723	.6425	.0156	.7304	.0018213	4.2471	57
4	.17479	.98461	.17753	.6329	.0156	.7210	.0018305	4.2378	56
5	0.17508	0.98455	0.17783	5.6234	1.0157	5.7117	.0018397	4.2286	55
6	.17537	.98450	.17813	.6140	.0157	.7023	.0018489	4.2194	54
7	.17565	.98445	.17843	.6045	.0158	.6930	.0018581	4.2103	53
8	.17594	.98440	.17873	.5951	.0158	.6838	.0018674	4.2012	52
9	.17623	.98435	.17903	.5857	.0159	.6745	.0018767	4.1921	51
10	0.17651	0.98430	0.17933	5.5764	1.0160	5.6653	.0018860	4.1830	50
11	.17680	.98425	.17963	.5671	.0160	.6562	.0018954	4.1740	49
12	.17708	.98420	.17993	.5578	.0161	.6470	.0019048	4.1650	48
13	.17737	.98414	.18023	.5485	.0161	.6379	.0019142	4.1560	47
14	.17766	.98409	.18053	.5393	.0162	.6288	.0019237	4.1471	46
15	0.17794	0.98404	0.18083	5.5301	1.0162	5.6198	.0019332	4.1382	45
16	.17823	.98399	.18113	.5209	.0163	.6107	.0019427	4.1293	44
17	.17852	.98394	.18143	.5118	.0163	.6017	.0019523	4.1204	43
18	.17880	.98389	.18173	.5026	.0164	.5928	.0019619	4.1116	42
19	.17909	.98383	.18203	.4936	.0164	.5838	.0019715	4.1028	41
20	0.17937	0.98378	0.18233	5.4845	1.0165	5.5749	.0019812	4.0941	40
21	.17966	.98373	.18263	.4755	.0165	.5660	.0019909	4.0853	39
22	.17995	.98368	.18293	.4665	.0166	.5572	.0020006	4.0766	38
23	.18023	.98362	.18323	.4575	.0166	.5484	.0020103	4.0679	37
24	.18052	.98357	.18353	.4486	.0167	.5396	.0020201	4.0593	36
25	0.18081	0.98352	0.18384	5.4397	1.0168	5.5308	.0020299	4.0507	35
26	.18109	.98347	.18414	.4308	.0168	.5221	.0020398	4.0421	34
27	.18138	.98341	.18444	.4219	.0169	.5134	.0020496	4.0335	33
28	.18166	.98336	.18474	.4131	.0169	.5047	.0020596	4.0250	32
29	.18195	.98331	.18504	.4043	.0170	.4960	.0020695	4.0165	31
30	0.18224	0.98325	0.18534	5.3955	1.0170	5.4874	.0020795	4.0080	30
31	.18252	.98320	.18564	.3868	.0171	.4788	.0020895	3.9995	29
32	.18281	.98315	.18594	.3781	.0171	.4702	.0020995	3.9911	28
33	.18309	.98310	.18624	.3694	.0172	.4617	.0021096	3.9827	27
34	.18338	.98304	.18654	.3607	.0173	.4532	.0021197	3.9743	26
35	0.18367	0.98299	0.18684	5.3521	1.0173	5.4447	.0021298	3.9660	25
36	.18395	.98294	.18714	.3435	.0174	.4362	.0021400	3.9577	24
37	.18424	.98288	.18745	.3349	.0174	.4278	.0021502	3.9494	23
38	.18452	.98283	.18775	.3263	.0175	.4194	.0021605	3.9411	22
39	.18481	.98277	.18805	.3178	.0175	.4110	.0021707	3.9329	21
40	0.18509	0.98272	0.18835	5.3093	1.0176	5.4026	.0021810	3.9247	20
41	.18538	.98267	.18865	.3008	.0176	.3943	.0021914	3.9165	19
42	.18567	.98261	.18895	.2924	.0177	.3860	.0022017	3.9083	18
43	.18595	.98256	.18925	.2839	.0178	.3777	.0022121	3.9002	17
44	.18624	.98250	.18955	.2755	.0178	.3695	.0022226	3.8921	16
45	0.18652	0.98245	0.18986	5.2672	1.0179	5.3612	.0022330	3.8840	15
46	.18681	.98240	.19016	.2588	.0179	.3530	.0022435	3.8759	14
47	.18710	.98234	.19046	.2505	.0180	.3449	.0022541	3.8679	13
48	.18738	.98229	.19076	.2422	.0180	.3367	.0022646	3.8599	12
49	.18767	.98223	.19106	.2339	.0181	.3286	.0022752	3.8519	11
50	0.18795	0.98218	0.19136	5.2257	1.0181	5.3205	.0022859	3.8439	10
51	.18824	.98212	.19166	.2174	.0182	.3124	.0022965	3.8360	9
52	.18852	.98207	.19197	.2092	.0183	.3044	.0023073	3.8281	8
53	.18881	.98201	.19227	.2011	.0183	.2963	.0023180	3.8202	7
54	.18910	.98196	.19257	.1929	.0184	.2883	.0023288	3.8124	6
55	0.18938	0.98190	0.19287	5.1848	1.0184	5.2804	.0023396	3.8045	5
56	.18967	.98185	.19317	.1767	.0185	.2724	.0023504	3.7967	4
57	.18995	.98179	.19347	.1686	.0185	.2645	.0023613	3.7889	3
58	.19024	.98174	.19378	.1606	.0186	.2566	.0023722	3.7812	2
59	.19052	.98168	.19408	.1526	.0187	.2487	.0023831	3.7735	1
60	0.19081	0.98163	0.19438	5.1446	1.0187	5.2408	.0023941	3.7657	0
M	Cosine	Sine	Cotan.	Tan.	Cosec.	Secant	READ DOWN	79°–80° Involute	M

11° or 191° **Trigonometric and Involute Functions** **168° or 348°**

M	Sine	Cosine	Tan.	Cotan.	Secant	Cosec.	Involute 11°–12°	READ UP	M
0	0.19081	0.98163	0.19438	5.1446	1.0187	5.2408	.0023941	3.7657	60
1	.19109	.98157	.19468	.1366	.0188	.2330	.0024051	3.7581	59
2	.19138	.98152	.19498	.1286	.0188	.2252	.0024161	3.7504	58
3	.19167	.98146	.19529	.1207	.0189	.2174	.0024272	3.7428	57
4	.19195	.98140	.19559	.1128	.0189	.2097	.0024383	3.7352	56
5	0.19224	0.98135	0.19589	5.1049	1.0190	5.2019	.0024495	3.7275	55
6	.19252	.98129	.19619	.0970	.0191	.1942	.0024607	3.7200	54
7	.19281	.98124	.19649	.0892	.0191	.1865	.0024719	3.7124	53
8	.19309	.98118	.19680	.0814	.0192	.1789	.0024831	3.7049	52
9	.19338	.98112	.19710	.0736	.0192	.1712	.0024944	3.6974	51
10	0.19366	0.98107	0.19740	5.0658	1.0193	5.1636	.0025057	3.6899	50
11	.19395	.98101	.19770	.0581	.0194	.1560	.0025171	3.6825	49
12	.19423	.98096	.19801	.0504	.0194	.1484	.0025285	3.6750	48
13	.19452	.98090	.19831	.0427	.0195	.1409	.0025399	3.6676	47
14	.19481	.98084	.19861	.0350	.0195	.1333	.0025513	3.6603	46
15	0.19509	0.98079	0.19891	5.0273	1.0196	5.1258	.0025628	3.6529	45
16	.19538	.98073	.19921	.0197	.0197	.1183	.0025744	3.6456	44
17	.19566	.98067	.19952	.0121	.0197	.1109	.0025859	3.6382	43
18	.19595	.98061	.19982	.0045	.0198	.1034	.0025975	3.6309	42
19	.19623	.98056	.20012	4.9969	.0198	.0960	.0026091	3.6237	41
20	0.19652	0.98050	0.20042	4.9894	1.0199	5.0886	.0026208	3.6164	40
21	.19680	.98044	.20073	.9819	.0199	.0813	.0026325	3.6092	39
22	.19709	.98039	.20103	.9744	.0200	.0739	.0026443	3.6020	38
23	.19737	.98033	.20133	.9669	.0201	.0666	.0026560	3.5948	37
24	.19766	.98027	.20164	.9594	.0201	.0593	.0026678	3.5876	36
25	0.19794	0.98021	0.20194	4.9520	1.0202	5.0520	.0026797	3.5805	35
26	.19823	.98016	.20224	.9446	.0202	.0447	.0026916	3.5734	34
27	.19851	.98010	.20254	.9372	.0203	.0375	.0027035	3.5663	33
28	.19880	.98004	.20285	.9298	.0204	.0302	.0027154	3.5592	32
29	.19908	.97998	.20315	.9225	.0204	.0230	.0027274	3.5521	31
30	0.19937	0.97992	0.20345	4.9152	1.0205	5.0159	.0027394	3.5451	30
31	.19965	.97987	.20376	.9078	.0205	.0087	.0027515	3.5381	29
32	.19994	.97981	.20406	.9006	.0206	.0016	.0027636	3.5311	28
33	.20022	.97975	.20436	.8933	.0207	4.9944	.0027757	3.5241	27
34	.20051	.97969	.20466	.8860	.0207	.9873	.0027879	3.5171	26
35	0.20079	0.97963	0.20497	4.8788	1.0208	4.9803	.0028001	3.5102	25
36	.20108	.97958	.20527	.8716	.0209	.9732	.0028123	3.5033	24
37	.20136	.97952	.20557	.8644	.0209	.9662	.0028246	3.4964	23
38	.20165	.97946	.20588	.8573	.0210	.9591	.0028369	3.4895	22
39	.20193	.97940	.20618	.8501	.0210	.9521	.0028493	3.4827	21
40	0.20222	0.97934	0.20648	4.8430	1.0211	4.9452	.0028616	3.4758	20
41	.20250	.97928	.20679	.8359	.0212	.9382	.0028741	3.4690	19
42	.20279	.97922	.20709	.8288	.0212	.9313	.0028865	3.4622	18
43	.20307	.97916	.20739	.8218	.0213	.9244	.0028990	3.4555	17
44	.20336	.97910	.20770	.8147	.0213	.9175	.0029115	3.4487	16
45	0.20364	0.97905	0.20800	4.8077	1.0214	4.9106	.0029241	3.4420	15
46	.20393	.97899	.20830	.8007	.0215	.9037	.0029367	3.4353	14
47	.20421	.97893	.20861	.7937	.0215	.8969	.0029494	3.4286	13
48	.20450	.97887	.20891	.7867	.0216	.8901	.0029620	3.4219	12
49	.20478	.97881	.20921	.7799	.0217	.8833	.0029747	3.4152	11
50	0.20507	0.97875	0.20952	4.7729	1.0217	4.8765	.0029875	3.4086	10
51	.20535	.97869	.20982	.7659	.0218	.8697	.0030003	3.4020	9
52	.20563	.97863	.21013	.7591	.0218	.8630	.0030131	3.3954	8
53	.20592	.97857	.21043	.7522	.0219	.8563	.0030260	3.3888	7
54	.20620	.97851	.21073	.7453	.0220	.8496	.0030389	3.3822	6
55	0.20649	0.97845	0.21104	4.7385	1.0220	4.8429	.0030518	3.3757	5
56	.20677	.97839	.21134	.7317	.0221	.8362	.0030648	3.3692	4
57	.20706	.97833	.21164	.7249	.0222	.8296	.0030778	3.3627	3
58	.20734	.97827	.21195	.7181	.0222	.8229	.0030908	3.3562	2
59	.20763	.97821	.21225	.7114	.0223	.8163	.0031039	3.3497	1
60	0.20791	0.97815	0.21256	4.7046	1.0223	4.8097	.0031171	3.3433	0
M	Cosine	Sine	Cotan.	Tan.	Cosec.	Secant	READ DOWN	78°–79° Involute	M

101° or 281° **78° or 258°**

12° or 192°   **Trigonometric and Involute Functions**   **167° or 347°**

M	Sine	Cosine	Tan.	Cotan.	Secant	Cosec.	Involute 12°–13°	READ UP	M
0	0.20791	0.97815	0.21256	4.7046	1.0223	4.8097	.0031171	3.3433	60
1	.20820	.97809	.21286	.6979	.0224	.8032	.0031302	3.3368	59
2	.20848	.97803	.21316	.6912	.0225	.7966	.0031434	3.3304	58
3	.20877	.97797	.21347	.6845	.0225	.7901	.0031566	3.3240	57
4	.20905	.97791	.21377	.6779	.0226	.7836	.0031699	3.3177	56
5	0.20933	0.97784	0.21408	4.6712	1.0227	4.7771	.0031832	3.3113	55
6	.20962	.97778	.21438	.6646	.0227	.7706	.0031966	3.3050	54
7	.20990	.97772	.21469	.6580	.0228	.7641	.0032100	3.2987	53
8	.21019	.97766	.21499	.6514	.0228	.7577	.0032234	3.2923	52
9	.21047	.97760	.21529	.6448	.0229	.7512	.0032369	3.2861	51
10	0.21076	0.97754	0.21560	4.6382	1.0230	4.7448	.0032504	3.2798	50
11	.21104	.97748	.21590	.6317	.0230	.7384	.0032639	3.2735	49
12	.21132	.97742	.21621	.6252	.0231	.7321	.0032775	3.2673	48
13	.21161	.97735	.21651	.6187	.0232	.7257	.0032911	3.2611	47
14	.21189	.97729	.21682	.6122	.0232	.7194	.0033048	3.2549	46
15	0.21218	0.97723	0.21712	4.6057	1.0233	4.7130	.0033185	3.2487	45
16	.21246	.97717	.21743	.5993	.0234	.7067	.0033322	3.2426	44
17	.21275	.97711	.21773	.5928	.0234	.7004	.0033460	3.2364	43
18	.21303	.97705	.21804	.5864	.0235	.6942	.0033598	3.2303	42
19	.21331	.97698	.21834	.5800	.0236	.6879	.0033736	3.2242	41
20	0.21360	0.97692	0.21864	4.5736	1.0236	4.6817	.0033875	3.2181	40
21	.21388	.97686	.21895	.5673	.0237	.6755	.0034014	3.2120	39
22	.21417	.97680	.21925	.5609	.0238	.6693	.0034154	3.2060	38
23	.21445	.97673	.21956	.5546	.0238	.6631	.0034294	3.1999	37
24	.21474	.97667	.21986	.5483	.0239	.6569	.0034434	3.1939	36
25	0.21502	0.97661	0.22017	4.5420	1.0240	4.6507	.0034575	3.1879	35
26	.21530	.97655	.22047	.5357	.0240	.6446	.0034716	3.1819	34
27	.21559	.97648	.22078	.5294	.0241	.6385	.0034858	3.1759	33
28	.21587	.97642	.22108	.5232	.0241	.6324	.0035000	3.1699	32
29	.21616	.97636	.22139	.5169	.0242	.6263	.0035142	3.1640	31
30	0.21644	0.97630	0.22169	4.5107	1.0243	4.6202	.0035285	3.1581	30
31	.21672	.97623	.22200	.5045	.0243	.6142	.0035428	3.1522	29
32	.21701	.97617	.22231	.4983	.0244	.6081	.0035572	3.1463	28
33	.21729	.97611	.22261	.4922	.0245	.6021	.0035716	3.1404	27
34	.21758	.97604	.22292	.4860	.0245	.5961	.0035860	3.1345	26
35	0.21786	0.97598	0.22322	4.4799	1.0246	4.5901	.0036005	3.1287	25
36	.21814	.97592	.22353	.4737	.0247	.5841	.0036150	3.1229	24
37	.21843	.97585	.22383	.4676	.0247	.5782	.0036296	3.1170	23
38	.21871	.97579	.22414	.4615	.0248	.5722	.0036441	3.1112	22
39	.21899	.97573	.22444	.4555	.0249	.5663	.0036588	3.1055	21
40	0.21928	0.97566	0.22475	4.4494	1.0249	4.5604	.0036735	3.0997	20
41	.21956	.97560	.22505	.4434	.0250	.5545	.0036882	3.0939	19
42	.21985	.97553	.22536	.4373	.0251	.5486	.0037029	3.0882	18
43	.22013	.97547	.22567	.4313	.0251	.5428	.0037177	3.0825	17
44	.22041	.97541	.22597	.4253	.0252	.5369	.0037325	3.0768	16
45	0.22070	0.97534	0.22628	4.4194	1.0253	4.5311	.0037474	3.0711	15
46	.22098	.97528	.22658	.4134	.0253	.5253	.0037623	3.0654	14
47	.22126	.97521	.22689	.4075	.0254	.5195	.0037773	3.0598	13
48	.22155	.97515	.22719	.4015	.0255	.5137	.0037923	3.0541	12
49	.22183	.97508	.22750	.3956	.0256	.5079	.0038073	3.0485	11
50	0.22212	0.97502	0.22781	4.3897	1.0256	4.5022	.0038224	3.0429	10
51	.22240	.97496	.22811	.3838	.0257	.4964	.0038375	3.0373	9
52	.22268	.97489	.22842	.3779	.0258	.4907	.0038527	3.0317	8
53	.22297	.97483	.22872	.3721	.0258	.4850	.0038679	3.0261	7
54	.22325	.97476	.22903	.3662	.0259	.4793	.0038831	3.0206	6
55	0.22353	0.97470	0.22934	4.3604	1.0260	4.4736	.0038984	3.0150	5
56	.22382	.97463	.22964	.3546	.0260	.4679	.0039137	3.0095	4
57	.22410	.97457	.22995	.3488	.0261	.4623	.0039291	3.0040	3
58	.22438	.97450	.23026	.3430	.0262	.4566	.0039445	2.9985	2
59	.22467	.97444	.23056	.3372	.0262	.4510	.0039599	2.9930	1
60	0.22495	0.97437	0.23087	4.3315	1.0263	4.4454	.0039754	2.9876	0
M	Cosine	Sine	Cotan.	Tan.	Cosec.	Secant	READ DOWN	77°–78° Involute	M

13° or 193°   Trigonometric and Involute Functions   166° or 346°

M	Sine	Cosine	Tan.	Cotan.	Secant	Cosec.	Involute 13°–14°	READ UP	M
0	0.22495	0.97437	0.23087	4.3315	1.0263	4.4454	.0039754	2.9876	60
1	.22523	.97430	.23117	.3257	.0264	.4398	.0039909	2.9821	59
2	.22552	.97424	.23148	.3200	.0264	.4342	.0040065	2.9767	58
3	.22580	.97417	.23179	.3143	.0265	.4287	.0040221	2.9713	57
4	.22608	.97411	.23209	.3086	.0266	.4231	.0040377	2.9659	56
5	0.22637	0.97404	0.23240	4.3029	1.0266	4.4176	.0040534	2.9605	55
6	.22665	.97398	.23271	.2972	.0267	.4121	.0040692	2.9551	54
7	.22693	.97391	.23301	.2916	.0268	.4066	.0040849	2.9497	53
8	.22722	.97384	.23332	.2859	.0269	.4011	.0041007	2.9444	52
9	.22750	.97378	.23363	.2803	.0269	.3956	.0041166	2.9390	51
10	0.22778	0.97371	0.23393	4.2747	1.0270	4.3901	.0041325	2.9337	50
11	.22807	.97365	.23424	.2691	.0271	.3847	.0041484	2.9284	49
12	.22835	.97358	.23455	.2635	.0271	.3792	.0041644	2.9231	48
13	.22863	.97351	.23485	.2580	.0272	.3738	.0041804	2.9178	47
14	.22892	.97345	.23516	.2524	.0273	.3684	.0041965	2.9126	46
15	0.22920	0.97338	0.23547	4.2468	1.0273	4.3630	.0042126	2.9073	45
16	.22948	.97331	.23578	.2413	.0274	.3576	.0042288	2.9021	44
17	.22977	.97325	.23608	.2358	.0275	.3522	.0042450	2.8968	43
18	.23005	.97318	.23639	.2303	.0276	.3469	.0042612	2.8916	42
19	.23033	.97311	.23670	.2248	.0276	.3415	.0042775	2.8864	41
20	0.23062	0.97304	0.23700	4.2193	1.0277	4.3362	.0042938	2.8812	40
21	.23090	.97298	.23731	.2139	.0278	.3309	.0043101	2.8761	39
22	.23118	.97291	.23762	.2084	.0278	.3256	.0043266	2.8709	38
23	.23146	.97284	.23793	.2030	.0279	.3203	.0043430	2.8658	37
24	.23175	.97278	.23823	.1976	.0280	.3150	.0043595	2.8606	36
25	0.23203	0.97271	0.23854	4.1922	1.0281	4.3098	.0043760	2.8555	35
26	.23231	.97264	.23885	.1868	.0281	.3045	.0043926	2.8504	34
27	.23260	.97257	.23916	.1814	.0282	.2993	.0044092	2.8453	33
28	.23288	.97251	.23946	.1760	.0283	.2941	.0044259	2.8402	32
29	.23316	.97244	.23977	.1706	.0283	.2889	.0044426	2.8352	31
30	0.23345	0.97237	0.24008	4.1653	1.0284	4.2837	.0044593	2.8301	30
31	.23373	.97230	.24039	.1600	.0285	.2785	.0044761	2.8251	29
32	.23401	.97223	.24069	.1547	.0286	.2733	.0044929	2.8201	28
33	.23429	.97217	.24100	.1493	.0286	.2681	.0045098	2.8150	27
34	.23458	.97210	.24131	.1441	.0287	.2630	.0045267	2.8100	26
35	0.23486	0.97203	0.24162	4.1388	1.0288	4.2579	.0045437	2.8050	25
36	.23514	.97196	.24193	.1335	.0288	.2527	.0045607	2.8001	24
37	.23542	.97189	.24223	.1282	.0289	.2476	.0045777	2.7951	23
38	.23571	.97182	.24254	.1230	.0290	.2425	.0045948	2.7902	22
39	.23599	.97176	.24285	.1178	.0291	.2375	.0046120	2.7852	21
40	0.23627	0.97169	0.24316	4.1126	1.0291	4.2324	.0046291	2.7803	20
41	.23656	.97162	.24347	.1074	.0292	.2273	.0046464	2.7754	19
42	.23684	.97155	.24377	.1022	.0293	.2223	.0046636	2.7705	18
43	.23712	.97148	.24408	.0970	.0294	.2173	.0046809	2.7656	17
44	.23740	.97141	.24439	.0918	.0294	.2122	.0046983	2.7607	16
45	0.23769	0.97134	0.24470	4.0867	1.0295	4.2072	.0047157	2.7558	15
46	.23797	.97127	.24501	.0815	.0296	.2022	.0047331	2.7510	14
47	.23825	.97120	.24532	.0764	.0297	.1973	.0047506	2.7462	13
48	.23853	.97113	.24562	.0713	.0297	.1923	.0047681	2.7413	12
49	.23882	.97106	.24593	.0662	.0298	.1873	.0047857	2.7365	11
50	0.23910	0.97100	0.24624	4.0611	1.0299	4.1824	.0048033	2.7317	10
51	.23938	.97093	.24655	.0560	.0299	.1774	.0048210	2.7269	9
52	.23966	.97086	.24686	.0509	.0300	.1725	.0048387	2.7221	8
53	.23995	.97079	.24717	.0459	.0301	.1676	.0048564	2.7174	7
54	.24023	.97072	.24747	.0408	.0302	.1627	.0048742	2.7126	6
55	0.24051	0.97065	0.24778	4.0358	1.0302	4.1578	.0048921	2.7079	5
56	.24079	.97058	.24809	.0308	.0303	.1529	.0049099	2.7031	4
57	.24108	.97051	.24840	.0257	.0304	.1481	.0049279	2.6984	3
58	.24136	.97044	.24871	.0207	.0305	.1432	.0049458	2.6937	2
59	.24164	.97037	.24902	.0158	.0305	.1384	.0049638	2.6890	1
60	0.24192	0.97030	0.24933	4.0108	1.0306	4.1336	.0049819	2.6843	0
M	Cosine	Sine	Cotan.	Tan.	Cosec.	Secant	READ DOWN	76°–77° Involute	M

14° or 194°   Trigonometric and Involute Functions   165° or 345°

M	Sine	Cosine	Tan.	Cotan.	Secant	Cosec.	Involute 14°–15°	READ UP	M
0	0.24192	0.97030	0.24933	4.0108	1.0306	4.1336	.0049819	2.6843	60
1	.24220	.97023	.24964	.0058	.0307	.1287	.0050000	2.6797	59
2	.24249	.97015	.24995	.0009	.0308	.1239	.0050182	2.6750	58
3	.24277	.97008	.25026	3.9959	.0308	.1191	.0050364	2.6703	57
4	.24305	.97001	.25056	.9910	.0309	.1144	.0050546	2.6657	56
5	0.24333	0.96994	0.25087	3.9861	1.0310	4.1096	.0050729	2.6611	55
6	.24362	.96987	.25118	.9812	.0311	.1048	.0050912	2.6565	54
7	.24390	.96980	.25149	.9763	.0311	.1001	.0051096	2.6519	53
8	.24418	.96973	.25180	.9714	.0312	.0954	.0051280	2.6473	52
9	.24446	.96966	.25211	.9665	.0313	.0906	.0051465	2.6427	51
10	0.24474	0.96959	0.25242	3.9617	1.0314	4.0859	.0051650	2.6381	50
11	.24503	.96952	.25273	.9568	.0314	.0812	.0051835	2.6336	49
12	.24531	.96945	.25304	.9520	.0315	.0765	.0052021	2.6290	48
13	.24559	.96937	.25335	.9471	.0316	.0718	.0052208	2.6245	47
14	.24587	.96930	.25366	.9423	.0317	.0672	.0052395	2.6199	46
15	0.24615	0.96923	0.25397	3.9375	1.0317	4.0625	.0052582	2.6154	45
16	.24644	.96916	.25428	.9327	.0318	.0579	.0052770	2.6109	44
17	.24672	.96909	.25459	.9279	.0319	.0532	.0052958	2.6064	43
18	.24700	.96902	.25490	.9232	.0320	.0486	.0053147	2.6019	42
19	.24728	.96894	.25521	.9184	.0321	.0440	.0053336	2.5975	41
20	0.24756	0.96887	0.25552	3.9136	1.0321	4.0394	.0053526	2.5930	40
21	.24784	.96880	.25583	.9089	.0322	.0348	.0053716	2.5886	39
22	.24813	.96873	.25614	.9042	.0323	.0302	.0053907	2.5841	38
23	.24841	.96866	.25645	.8995	.0324	.0256	.0054098	2.5797	37
24	.24869	.96858	.25676	.8947	.0324	.0211	.0054289	2.5753	36
25	0.24897	0.96851	0.25707	3.8900	1.0325	4.0165	.0054481	2.5709	35
26	.24925	.96844	.25738	.8854	.0326	.0120	.0054674	2.5665	34
27	.24954	.96837	.25769	.8807	.0327	.0075	.0054867	2.5621	33
28	.24982	.96829	.25800	.8760	.0327	.0029	.0055060	2.5577	32
29	.25010	.96822	.25831	.8714	.0328	3.9984	.0055254	2.5533	31
30	0.25038	0.96815	0.25862	3.8667	1.0329	3.9939	.0055448	2.5490	30
31	.25066	.96807	.25893	.8621	.0330	.9894	.0055643	2.5446	29
32	.25094	.96800	.25924	.8575	.0331	.9850	.0055838	2.5403	28
33	.25122	.96793	.25955	.8528	.0331	.9805	.0056034	2.5360	27
34	.25151	.96785	.25986	.8482	.0332	.9760	.0056230	2.5317	26
35	0.25179	0.96778	0.26017	3.8436	1.0333	3.9716	.0056427	2.5274	25
36	.25207	.96771	.26048	.8391	.0334	.9672	.0056624	2.5231	24
37	.25235	.96764	.26079	.8345	.0334	.9627	.0056822	2.5188	23
38	.25263	.96756	.26110	.8299	.0335	.9583	.0057020	2.5145	22
39	.25291	.96749	.26141	.8254	.0336	.9539	.0057218	2.5103	21
40	0.25320	0.96742	0.26172	3.8208	1.0337	3.9495	.0057417	2.5060	20
41	.25348	.96734	.26203	.8163	.0338	.9451	.0057616	2.5018	19
42	.25376	.96727	.26235	.8118	.0338	.9408	.0057817	2.4975	18
43	.25404	.96719	.26266	.8073	.0339	.9364	.0058017	2.4933	17
44	.25432	.96712	.26297	.8028	.0340	.9320	.0058218	2.4891	16
45	0.25460	0.96705	0.26328	3.7983	1.0341	3.9277	.0058420	2.4849	15
46	.25488	.96697	.26359	.7938	.0342	.9234	.0058622	2.4807	14
47	.25516	.96690	.26390	.7893	.0342	.9190	.0058824	2.4765	13
48	.25545	.96682	.26421	.7848	.0343	.9147	.0059027	2.4724	12
49	.25573	.96675	.26452	.7804	.0344	.9104	.0059230	2.4682	11
50	0.25601	0.96667	0.26483	3.7760	1.0345	3.9061	.0059434	2.4640	10
51	.25629	.96660	.26515	.7715	.0346	.9018	.0059638	2.4599	9
52	.25657	.96653	.26546	.7671	.0346	.8976	.0059843	2.4558	8
53	.25685	.96645	.26577	.7627	.0347	.8933	.0060048	2.4516	7
54	.25713	.96638	.26608	.7583	.0348	.8890	.0060254	2.4475	6
55	0.25741	0.96630	0.26639	3.7539	1.0349	3.8848	.0060460	2.4434	5
56	.25769	.96623	.26670	.7495	.0350	.8806	.0060667	2.4393	4
57	.25798	.96615	.26701	.7451	.0350	.8763	.0060874	2.4353	3
58	.25826	.96608	.26733	.7408	.0351	.8721	.0061081	2.4312	2
59	.25854	.96600	.26764	.7364	.0352	.8679	.0061289	2.4271	1
60	0.25882	0.96593	0.26795	3.7321	1.0353	3.8637	.0061498	2.4231	0
M	Cosine	Sine	Cotan.	Tan.	Cosec.	Secant	READ DOWN	75°–76° Involute	M

104° or 284°   75° or 255°

15° or 195°   **Trigonometric and Involute Functions**   **164° or 344°**

M	Sine	Cosine	Tan.	Cotan.	Secant	Cosec.	Involute 15°–16°	READ UP	M
0	0.25882	0.96593	0.26795	3.7321	1.0353	3.8637	.0061498	2.4231	60
1	.25910	.96585	.26826	.7277	.0354	.8595	.0061707	2.4190	59
2	.25938	.96578	.26857	.7234	.0354	.8553	.0061917	2.4150	58
3	.25966	.96570	.26888	.7191	.0355	.8512	.0062127	2.4109	57
4	.25994	.96562	.26920	.7148	.0356	.8470	.0062337	2.4069	56
5	0.26022	0.96555	0.26951	3.7105	1.0357	3.8428	.0062548	2.4029	55
6	.26050	.96547	.26982	.7062	.0358	.8387	.0062760	2.3989	54
7	.26079	.96540	.27013	.7019	.0358	.8346	.0062972	2.3949	53
8	.26107	.96532	.27044	.6976	.0359	.8304	.0063184	2.3909	52
9	.26135	.96524	.27076	.6933	.0360	.8263	.0063397	2.3870	51
10	0.26163	0.96517	0.27107	3.6891	1.0361	3.8222	.0063611	2.3830	50
11	.26191	.96509	.27138	.6848	.0362	.8181	.0063825	2.3791	49
12	.26219	.96502	.27169	.6806	.0363	.8140	.0064039	2.3751	48
13	.26247	.96494	.27201	.6764	.0363	.8100	.0064254	2.3712	47
14	.26275	.96486	.27232	.6722	.0364	.8059	.0064470	2.3672	46
15	0.26303	0.96479	0.27263	3.6680	1.0365	3.8018	.0064686	2.3633	45
16	.26331	.96471	.27294	.6638	.0366	.7978	.0064902	2.3594	44
17	.26359	.96463	.27326	.6596	.0367	.7937	.0065119	2.3555	43
18	.26387	.96456	.27357	.6554	.0367	.7897	.0065337	2.3516	42
19	.26415	.96449	.27388	.6512	.0368	.7857	.0065555	2.3477	41
20	0.26443	0.96440	0.27419	3.6470	1.0369	3.7817	.0065773	2.3439	40
21	.26471	.96433	.27451	.6429	.0370	.7777	.0065992	2.3400	39
22	.26500	.96425	.27482	.6387	.0371	.7737	.0066211	2.3361	38
23	.26528	.96417	.27513	.6346	.0372	.7697	.0066431	2.3323	37
24	.26556	.96410	.27545	.6305	.0372	.7657	.0066652	2.3285	36
25	0.26584	0.96402	0.27576	3.6264	1.0373	3.7617	.0066873	2.3246	35
26	.26612	.96394	.27607	.6222	.0374	.7577	.0067094	2.3208	34
27	.26640	.96386	.27638	.6181	.0375	.7538	.0067316	2.3170	33
28	.26668	.96379	.27670	.6140	.0376	.7498	.0067539	2.3132	32
29	.26696	.96371	.27701	.6100	.0377	.7459	.0067762	2.3094	31
30	0.26724	0.96363	0.27732	3.6059	1.0377	3.7420	.0067985	2.3056	30
31	.26752	.96355	.27764	.6018	.0378	.7381	.0068209	2.3018	29
32	.26780	.96347	.27795	.5978	.0379	.7341	.0068434	2.2981	28
33	.26808	.96340	.27826	.5937	.0380	.7302	.0068659	2.2943	27
34	.26836	.96332	.27858	.5897	.0381	.7263	.0068884	2.2906	26
35	0.26864	0.96324	0.27889	3.5856	1.0382	3.7225	.0069110	2.2868	25
36	.26892	.96316	.27921	.5816	.0382	.7186	.0069337	2.2831	24
37	.26920	.96308	.27952	.5776	.0383	.7147	.0069564	2.2793	23
38	.26948	.96301	.27983	.5736	.0384	.7108	.0069791	2.2756	22
39	.26976	.96293	.28015	.5696	.0385	.7070	.0070019	2.2719	21
40	0.27004	0.96285	0.28046	3.5656	1.0386	3.7032	.0070248	2.2682	20
41	.27032	.96277	.28077	.5616	.0387	.6993	.0070477	2.2645	19
42	.27060	.96269	.28109	.5576	.0388	.6955	.0070706	2.2608	18
43	.27088	.96261	.28140	.5536	.0388	.6917	.0070936	2.2572	17
44	.27116	.96253	.28172	.5497	.0389	.6879	.0071167	2.2535	16
45	0.27144	0.96246	0.28203	3.5457	1.0390	3.6840	.0071398	2.2498	15
46	.27172	.96238	.28234	.5418	.0391	.6803	.0071630	2.2462	14
47	.27200	.96230	.28266	.5379	.0392	.6765	.0071862	2.2425	13
48	.27228	.96222	.28297	.5339	.0393	.6727	.0072095	2.2389	12
49	.27256	.96214	.28329	.5300	.0394	.6689	.0072328	2.2353	11
50	0.27284	0.96206	0.28360	3.5261	1.0394	3.6652	.0072561	2.2316	10
51	.27312	.96198	.28391	.5222	.0395	.6614	.0072796	2.2280	9
52	.27340	.96190	.28423	.5183	.0396	.6576	.0073030	2.2244	8
53	.27368	.96182	.28454	.5144	.0397	.6539	.0073266	2.2208	7
54	.27396	.96174	.28486	.5105	.0398	.6502	.0073501	2.2172	6
55	0.27424	0.96166	0.28517	3.5067	1.0399	3.6465	.0073738	2.2137	5
56	.27452	.96158	.28549	.5028	.0400	.6427	.0073975	2.2101	4
57	.27480	.96150	.28580	.4989	.0400	.6390	.0074212	2.2065	3
58	.27508	.96142	.28612	.4951	.0401	.6353	.0074450	2.2030	2
59	.27536	.96134	.28643	.4912	.0402	.6316	.0074688	2.1994	1
60	0.27564	0.96126	0.28675	3.4874	1.0403	3.6280	.0074927	2.1959	0
M	Cosine	Sine	Cotan.	Tan.	Cosec.	Secant	READ DOWN	74°–75° Involute	M

105° or 285°   **74° or 254°**

16° or 196°   **Trigonometric and Involute Functions**   **163° or 343°**

M	Sine	Cosine	Tan.	Cotan.	Secant	Cosec.	Involute 16°–17°	READ UP	M
0	0.27564	0.96126	0.28675	3.4874	1.0403	3.6280	.0074927	2.1959	60
1	.27592	.96118	.28706	.4836	.0404	.6243	.0075166	2.1923	59
2	.27620	.96110	.28738	.4798	.0405	.6206	.0075406	2.1888	58
3	.27648	.96102	.28769	.4760	.0406	.6169	.0075647	2.1853	57
4	.27676	.96094	.28801	.4722	.0406	.6133	.0075888	2.1818	56
5	0.27704	0.96086	0.28832	3.4684	1.0407	3.6097	.0076130	2.1783	55
6	.27731	.96078	.28864	.4646	.0408	.6060	.0076372	2.1748	54
7	.27759	.96070	.28895	.4608	.0409	.6024	.0076614	2.1713	53
8	.27787	.96062	.28927	.4570	.0410	.5988	.0076857	2.1678	52
9	.27815	.96054	.28958	.4533	.0411	.5951	.0077101	2.1643	51
10	0.27843	0.96046	0.28990	3.4495	1.0412	3.5915	.0077345	2.1609	50
11	.27871	.96037	.29021	.4458	.0413	.5879	.0077590	2.1574	49
12	.27899	.96029	.29053	.4420	.0413	.5843	.0077835	2.1540	48
13	.27927	.96021	.29084	.4383	.0414	.5808	.0078081	2.1505	47
14	.27955	.96013	.29116	.4346	.0415	.5772	.0078327	2.1471	46
15	0.27983	0.96005	0.29147	3.4308	1.0416	3.5736	.0078574	2.1437	45
16	.28011	.95997	.29179	.4271	.0417	.5700	.0078822	2.1402	44
17	.28039	.95989	.29210	.4234	.0418	.5665	.0079069	2.1368	43
18	.28067	.95981	.29242	.4197	.0419	.5629	.0079318	2.1334	42
19	.28095	.95972	.29274	.4160	.0420	.5594	.0079567	2.1300	41
20	0.28123	0.95964	0.29305	3.4124	1.0421	3.5559	.0079817	2.1266	40
21	.28150	.95956	.29337	.4087	.0421	.5523	.0080067	2.1233	39
22	.28178	.95948	.29368	.4050	.0422	.5488	.0080317	2.1199	38
23	.28206	.95940	.29400	.4014	.0423	.5453	.0080568	2.1165	37
24	.28234	.95931	.29432	.3977	.0424	.5418	.0080820	2.1131	36
25	0.28262	0.95923	0.29463	3.3941	1.0425	3.5383	.0081072	2.1098	35
26	.28290	.95915	.29495	.3904	.0426	.5348	.0081325	2.1064	34
27	.28318	.95907	.29526	.3868	.0427	.5313	.0081578	2.1031	33
28	.28346	.95898	.29558	.3832	.0428	.5279	.0081832	2.0998	32
29	.28374	.95890	.29590	.3796	.0429	.5244	.0082087	2.0964	31
30	0.28402	0.95882	0.29621	3.3759	1.0429	3.5209	.0082342	2.0931	30
31	.28429	.95874	.29653	.3723	.0430	.5175	.0082597	2.0898	29
32	.28457	.95865	.29685	.3687	.0431	.5140	.0082853	2.0865	28
33	.28485	.95857	.29716	.3652	.0432	.5106	.0083110	2.0832	27
34	.28513	.95849	.29748	.3616	.0433	.5072	.0083367	2.0799	26
35	0.28541	0.95841	0.29780	3.3580	1.0434	3.5037	.0083625	2.0766	25
36	.28569	.95832	.29811	.3544	.0435	.5003	.0083883	2.0734	24
37	.28597	.95824	.29843	.3509	.0436	.4969	.0084142	2.0701	23
38	.28625	.95816	.29875	.3473	.0437	.4935	.0084401	2.0668	22
39	.28652	.95807	.29906	.3438	.0438	.4901	.0084661	2.0636	21
40	0.28680	0.95799	0.29938	3.3402	1.0439	3.4867	.0084921	2.0603	20
41	.28708	.95791	.29970	.3367	.0439	.4833	.0085182	2.0571	19
42	.28736	.95782	.30001	.3332	.0440	.4799	.0085444	2.0538	18
43	.28764	.95774	.30033	.3297	.0441	.4766	.0085706	2.0506	17
44	.28792	.95766	.30065	.3261	.0442	.4732	.0085969	2.0474	16
45	0.28820	0.95757	0.30097	3.3226	1.0443	3.4699	.0086232	2.0442	15
46	.28847	.95749	.30128	.3191	.0444	.4665	.0086496	2.0410	14
47	.28875	.95740	.30160	.3156	.0445	.4632	.0086760	2.0378	13
48	.28903	.95732	.30192	.3122	.0446	.4598	.0087025	2.0346	12
49	.28931	.95724	.30224	.3087	.0447	.4565	.0087290	2.0314	11
50	0.28959	0.95715	0.30255	3.3052	1.0448	3.4532	.0087556	2.0282	10
51	.28987	.95707	.30287	.3017	.0449	.4499	.0087823	2.0250	9
52	.29015	.95698	.30319	.2983	.0450	.4465	.0088090	2.0219	8
53	.29042	.95690	.30351	.2948	.0450	.4432	.0088358	2.0187	7
54	.29070	.95681	.30382	.2914	.0451	.4399	.0088626	2.0156	6
55	0.29098	0.95673	0.30414	3.2879	1.0452	3.4367	.0088895	2.0124	5
56	.29126	.95664	.30446	.2845	.0453	.4334	.0089164	2.0093	4
57	.29154	.95656	.30478	.2811	.0454	.4301	.0089434	2.0061	3
58	.29182	.95647	.30509	.2777	.0455	.4268	.0089704	2.0030	2
59	.29209	.95639	.30541	.2743	.0456	.4236	.0089975	1.9999	1
60	0.29237	0.95630	0.30573	3.2709	1.0457	3.4203	.0090247	1.9968	0
M	Cosine	Sine	Cotan.	Tan.	Cosec.	Secant	READ DOWN	73°–74° Involute	M

17° or 197°   **Trigonometric and Involute Functions**   **162° or 342°**

M	Sine	Cosine	Tan.	Cotan.	Secant	Cosec.	Involute 17°-18°	READ UP	M
0	0.29237	0.95630	0.30573	3.2709	1.0457	3.4203	.0090247	1.9968	60
1	.29265	.95622	.30605	.2675	.0458	.4171	.0090519	1.9937	59
2	.29293	.95613	.30637	.2641	.0459	.4138	.0090792	1.9906	58
3	.29321	.95605	.30669	.2607	.0460	.4106	.0091065	1.9875	57
4	.29348	.95596	.30700	.2573	.0461	.4073	.0091339	1.9844	56
5	0.29376	0.95588	0.30732	3.2539	1.0462	3.4041	.0091614	1.9813	55
6	.29404	.95579	.30764	.2506	.0463	.4009	.0091889	1.9782	54
7	.29432	.95571	.30796	.2472	.0463	.3977	.0092164	1.9751	53
8	.29460	.95562	.30828	.2438	.0464	.3945	.0092440	1.9721	52
9	.29487	.95554	.30860	.2405	.0465	.3913	.0092717	1.9690	51
10	0.29515	0.95545	0.30891	3.2371	1.0466	3.3881	.0092994	1.9660	50
11	.29543	.95536	.30923	.2338	.0467	.3849	.0093272	1.9629	49
12	.29571	.95528	.30955	.2305	.0468	.3817	.0093551	1.9599	48
13	.29599	.95519	.30987	.2272	.0469	.3785	.0093830	1.9568	47
14	.29626	.95511	.31019	.2238	.0470	.3754	.0094109	1.9538	46
15	0.29654	0.95502	0.31051	3.2205	1.0471	3.3722	.0094390	1.9508	45
16	.29682	.95493	.31083	.2172	.0472	.3691	.0094670	1.9478	44
17	.29710	.95485	.31115	.2139	.0473	.3659	.0094952	1.9448	43
18	.29737	.95476	.31147	.2106	.0474	.3628	.0095234	1.9418	42
19	.29765	.95467	.31178	.2073	.0475	.3596	.0095516	1.9388	41
20	0.29793	0.95459	0.31210	3.2041	1.0476	3.3565	.0095799	1.9358	40
21	.29821	.95450	.31242	.2008	.0477	.3534	.0096083	1.9328	39
22	.29849	.95441	.31274	.1975	.0478	.3502	.0096367	1.9298	38
23	.29876	.95433	.31306	.1943	.0479	.3471	.0096652	1.9269	37
24	.29904	.95424	.31338	.1910	.0480	.3440	.0096937	1.9239	36
25	0.29932	0.95415	0.31370	3.1878	1.0480	3.3409	.0097223	1.9209	35
26	.29960	.95407	.31402	.1845	.0481	.3378	.0097510	1.9180	34
27	.29987	.95398	.31434	.1813	.0482	.3347	.0097797	1.9150	33
28	.30015	.95389	.31466	.1780	.0483	.3317	.0098085	1.9121	32
29	.30043	.95380	.31498	.1748	.0484	.3286	.0098373	1.9092	31
30	0.30071	0.95372	0.31530	3.1716	1.0485	3.3255	.0098662	1.9062	30
31	.30098	.95363	.31562	.1684	.0486	.3224	.0098951	1.9033	29
32	.30126	.95354	.31594	.1652	.0487	.3194	.0099241	1.9004	28
33	.30154	.95345	.31626	.1620	.0488	.3163	.0099532	1.8975	27
34	.30182	.95337	.31658	.1588	.0489	.3133	.0099823	1.8946	26
35	0.30209	0.95328	0.31690	3.1556	1.0490	3.3102	.010012	1.8917	25
36	.30237	.95319	.31722	.1524	.0491	.3072	.010041	1.8888	24
37	.30265	.95310	.31754	.1492	.0492	.3042	.010070	1.8859	23
38	.30292	.95301	.31786	.1460	.0493	.3012	.010099	1.8830	22
39	.30320	.95293	.31818	.1429	.0494	.2981	.010129	1.8801	21
40	0.30348	0.95284	0.31850	3.1397	1.0495	3.2951	.010158	1.8773	20
41	.30376	.95275	.31882	.1366	.0496	.2921	.010188	1.8744	19
42	.30403	.95266	.31914	.1334	.0497	.2891	.010217	1.8715	18
43	.30431	.95257	.31946	.1303	.0498	.2861	.010247	1.8687	17
44	.30459	.95248	.31978	.1271	.0499	.2831	.010277	1.8658	16
45	0.30486	0.95240	0.32010	3.1240	1.0500	3.2801	.010307	1.8630	15
46	.30514	.95231	.32042	.1209	.0501	.2772	.010336	1.8602	14
47	.30542	.95222	.32074	.1178	.0502	.2742	.010366	1.8573	13
48	.30570	.95213	.32106	.1146	.0503	.2712	.010396	1.8545	12
49	.30597	.95204	.32139	.1115	.0504	.2683	.010426	1.8517	11
50	0.30625	0.95195	0.32171	3.1084	1.0505	3.2653	.010456	1.8489	10
51	.30653	.95186	.32203	.1053	.0506	.2624	.010487	1.8461	9
52	.30680	.95177	.32235	.1022	.0507	.2594	.010517	1.8433	8
53	.30708	.95168	.32267	.0991	.0508	.2565	.010547	1.8405	7
54	.30736	.95159	.32299	.0961	.0509	.2535	.010577	1.8377	6
55	0.30763	0.95150	0.32331	3.0930	1.0510	3.2506	.010608	1.8349	5
56	.30791	.95142	.32363	.0899	.0511	.2477	.010638	1.8321	4
57	.30819	.95133	.32396	.0868	.0512	.2448	.010669	1.8293	3
58	.30846	.95124	.32428	.0838	.0513	.2419	.010699	1.8266	2
59	.30874	.95115	.32460	.0807	.0514	.2390	.010730	1.8238	1
60	0.30902	0.95106	0.32492	3.0777	1.0515	3.2361	.010760	1.8210	0
M	Cosine	Sine	Cotan.	Tan.	Cosec.	Secant	READ DOWN	72°-73° Involute	M

M	Sine	Cosine	Tan.	Cotan.	Secant	Cosec.	Involute 18°-19°	READ UP	M
0	0.30902	0.95106	0.32492	3.0777	1.0515	3.2361	.010760	1.8210	60
1	.30929	.95097	.32524	.0746	.0516	.2332	.010791	1.8183	59
2	.30957	.95088	.32556	.0716	.0517	.2303	.010822	1.8155	58
3	.30985	.95079	.32588	.0686	.0518	.2274	.010853	1.8128	57
4	.31012	.95070	.32621	.0655	.0519	.2245	.010884	1.8101	56
5	0.31040	0.95061	0.32653	3.0625	1.0520	3.2217	.010915	1.8073	55
6	.31068	.95052	.32685	.0595	.0521	.2188	.010946	1.8046	54
7	.31095	.95043	.32717	.0565	.0522	.2159	.010977	1.8019	53
8	.31123	.95033	.32749	.0535	.0523	.2131	.011008	1.7992	52
9	.31151	.95024	.32782	.0505	.0524	.2102	.011039	1.7965	51
10	0.31178	0.95015	0.32814	3.0475	1.0525	3.2074	.011071	1.7938	50
11	.31206	.95006	.32846	.0445	.0526	.2045	.011102	1.7911	49
12	.31233	.94997	.32878	.0415	.0527	.2017	.011133	1.7884	48
13	.31261	.94988	.32911	.0385	.0528	.1989	.011165	1.7857	47
14	.31289	.94979	.32943	.0356	.0529	.1960	.011196	1.7830	46
15	0.31316	0.94970	0.32975	3.0326	1.0530	3.1932	.011228	1.7803	45
16	.31344	.94961	.33007	.0296	.0531	.1904	.011260	1.7776	44
17	.31372	.94952	.33040	.0267	.0532	.1876	.011291	1.7750	43
18	.31399	.94943	.33072	.0237	.0533	.1848	.011323	1.7723	42
19	.31427	.94933	.33104	.0208	.0534	.1820	.011355	1.7697	41
20	0.31454	0.94924	0.33136	3.0178	1.0535	3.1792	.011387	1.7670	40
21	.31482	.94915	.33169	.0149	.0536	.1764	.011419	1.7644	39
22	.31510	.94906	.33201	.0120	.0537	.1736	.011451	1.7617	38
23	.31537	.94897	.33233	.0090	.0538	.1708	.011483	1.7591	37
24	.31565	.94888	.33266	.0061	.0539	.1681	.011515	1.7565	36
25	0.31593	0.94878	0.33298	3.0032	1.0540	3.1653	.011547	1.7538	35
26	.31620	.94869	.33330	.0003	.0541	.1625	.011580	1.7512	34
27	.31648	.94860	.33363	2.9974	.0542	.1598	.011612	1.7486	33
28	.31675	.94851	.33395	.9945	.0543	.1570	.011644	1.7460	32
29	.31703	.94842	.33427	.9916	.0544	.1543	.011677	1.7434	31
30	0.31730	0.94832	0.33460	2.9887	1.0545	3.1515	.011709	1.7408	30
31	.31758	.94823	.33492	.9858	.0546	.1488	.011742	1.7382	29
32	.31786	.94814	.33524	.9829	.0547	.1461	.011775	1.7356	28
33	.31813	.94805	.33557	.9800	.0548	.1433	.011807	1.7330	27
34	.31841	.94795	.33589	.9772	.0549	.1406	.011840	1.7304	26
35	0.31868	0.94786	0.33621	2.9743	1.0550	3.1379	.011873	1.7278	25
36	.31896	.94777	.33654	.9714	.0551	.1352	.011906	1.7253	24
37	.31923	.94768	.33686	.9686	.0552	.1325	.011939	1.7227	23
38	.31951	.94758	.33718	.9657	.0553	.1298	.011972	1.7201	22
39	.31979	.94749	.33751	.9629	.0554	.1271	.012005	1.7176	21
40	0.32006	0.94740	0.33783	2.9600	1.0555	3.1244	.012038	1.7150	20
41	.32034	.94730	.33816	.9572	.0556	.1217	.012071	1.7125	19
42	.32061	.94721	.33848	.9544	.0557	.1190	.012105	1.7100	18
43	.32089	.94712	.33881	.9515	.0558	.1163	.012138	1.7074	17
44	.32116	.94702	.33913	.9487	.0559	.1137	.012172	1.7049	16
45	0.32144	0.94693	0.33945	2.9459	1.0560	3.1110	.012205	1.7024	15
46	.32171	.94684	.33978	.9431	.0561	.1083	.012239	1.6998	14
47	.32199	.94674	.34010	.9403	.0563	.1057	.012272	1.6973	13
48	.32227	.94665	.34043	.9375	.0564	.1030	.012306	1.6948	12
49	.32254	.94656	.34075	.9347	.0565	.1004	.012340	1.6923	11
50	0.32282	0.94646	0.34108	2.9319	1.0566	3.0977	.012373	1.6898	10
51	.32309	.94637	.34140	.9291	.0567	.0951	.012407	1.6873	9
52	.32337	.94627	.34173	.9263	.0568	.0925	.012441	1.6848	8
53	.32364	.94618	.34205	.9235	.0569	.0898	.012475	1.6823	7
54	.32392	.94609	.34238	.9208	.0570	.0872	.012509	1.6798	6
55	0.32419	0.94599	0.34270	2.9180	1.0571	3.0846	.012543	1.6774	5
56	.32447	.94590	.34303	.9152	.0572	.0820	.012578	1.6749	4
57	.32474	.94580	.34335	.9125	.0573	.0794	.012612	1.6724	3
58	.32502	.94571	.34368	.9097	.0574	.0768	.012646	1.6699	2
59	.32529	.94561	.34400	.9070	.0575	.0742	.012681	1.6675	1
60	0.32557	0.94552	0.34433	2.9042	1.0576	3.0716	.012715	1.6650	0
M	Cosine	Sine	Cotan.	Tan.	Cosec.	Secant	READ DOWN	71°-72° Involute	M

19° or 199° Trigonometric and Involute Functions 160° or 340°

M	Sine	Cosine	Tan.	Cotan.	Secant	Cosec.	Involute 19°–20°	READ UP	M
0	0.32557	0.94552	0.34433	2.9042	1.0576	3.0716	.012715	1.6650	60
1	.32584	.94542	.34465	.9015	.0577	.0690	.012750	1.6626	59
2	.32612	.94533	.34498	.8987	.0578	.0664	.012784	1.6601	58
3	.32639	.94523	.34530	.8960	.0579	.0638	.012819	1.6577	57
4	.32667	.94514	.34563	.8933	.0580	.0612	.012854	1.6553	56
5	0.32694	0.94504	0.34596	2.8905	1.0582	3.0586	.012888	1.6528	55
6	.32722	.94495	.34628	.8878	.0583	.0561	.012923	1.6504	54
7	.32749	.94485	.34661	.8851	.0584	.0535	.012958	1.6480	53
8	.32777	.94476	.34693	.8824	.0585	.0509	.012993	1.6455	52
9	.32804	.94466	.34726	.8797	.0586	.0484	.013028	1.6431	51
10	0.32832	0.94457	0.34758	2.8770	1.0587	3.0458	.013063	1.6407	50
11	.32859	.94447	.34791	.8743	.0588	.0433	.013098	1.6383	49
12	.32887	.94438	.34824	.8716	.0589	.0407	.013134	1.6359	48
13	.32914	.94428	.34856	.8689	.0590	.0382	.013169	1.6335	47
14	.32942	.94418	.34889	.8662	.0591	.0357	.013204	1.6311	46
15	0.32969	0.94409	0.34922	2.8636	1.0592	3.0331	.013240	1.6287	45
16	.32997	.94399	.34954	.8609	.0593	.0306	.013275	1.6264	44
17	.33024	.94390	.34987	.8582	.0594	.0281	.013311	1.6240	43
18	.33051	.94380	.35020	.8556	.0595	.0256	.013346	1.6216	42
19	.33079	.94370	.35052	.8529	.0597	.0231	.013382	1.6192	41
20	0.33106	0.94361	0.35085	2.8502	1.0598	3.0206	.013418	1.6169	40
21	.33134	.94351	.35118	.8476	.0599	.0181	.013454	1.6145	39
22	.33161	.94342	.35150	.8449	.0600	.0156	.013490	1.6122	38
23	.33189	.94332	.35183	.8423	.0601	.0131	.013526	1.6098	37
24	.33216	.94322	.35216	.8397	.0602	.0106	.013562	1.6075	36
25	0.33244	0.94313	0.35248	2.8370	1.0603	3.0081	.013598	1.6051	35
26	.33271	.94303	.35281	.8344	.0604	.0056	.013634	1.6028	34
27	.33298	.94293	.35314	.8318	.0605	.0031	.013670	1.6004	33
28	.33326	.94284	.35346	.8291	.0606	.0007	.013707	1.5981	32
29	.33353	.94274	.35379	.8265	.0607	2.9982	.013743	1.5958	31
30	0.33381	0.94264	0.35412	2.8239	1.0608	2.9957	.013779	1.5935	30
31	.33408	.94254	.35445	.8213	.0610	.9933	.013816	1.5911	29
32	.33436	.94245	.35477	.8187	.0611	.9908	.013852	1.5888	28
33	.33463	.94235	.35510	.8161	.0612	.9884	.013889	1.5865	27
34	.33490	.94225	.35543	.8135	.0613	.9859	.013926	1.5842	26
35	0.33518	0.94215	0.35576	2.8109	1.0614	2.9835	.013963	1.5819	25
36	.33545	.94206	.35608	.8083	.0615	.9811	.013999	1.5796	24
37	.33573	.94196	.35641	.8057	.0616	.9786	.014036	1.5773	23
38	.33600	.94186	.35674	.8032	.0617	.9762	.014073	1.5750	22
39	.33627	.94176	.35707	.8006	.0618	.9738	.014110	1.5728	21
40	0.33655	0.94167	0.35740	2.7980	1.0619	2.9713	.014148	1.5705	20
41	.33682	.94157	.35772	.7955	.0621	.9689	.014185	1.5682	19
42	.33710	.94147	.35805	.7929	.0622	.9665	.014222	1.5659	18
43	.33737	.94137	.35838	.7903	.0623	.9641	.014259	1.5637	17
44	.33764	.94127	.35871	.7878	.0624	.9617	.014297	1.5614	16
45	0.33792	0.94118	0.35904	2.7852	1.0625	2.9593	.014334	1.5591	15
46	.33819	.94108	.35937	.7827	.0626	.9569	.014372	1.5569	14
47	.33846	.94098	.35969	.7801	.0627	.9545	.014409	1.5546	13
48	.33874	.94088	.36002	.7776	.0628	.9521	.014447	1.5524	12
49	.33901	.94078	.36035	.7751	.0629	.9498	.014485	1.5501	11
50	0.33929	0.94068	0.36068	2.7725	1.0631	2.9474	.014523	1.5479	10
51	.33956	.94058	.36101	.7700	.0632	.9450	.014560	1.5457	9
52	.33983	.94049	.36134	.7675	.0633	.9426	.014598	1.5434	8
53	.34011	.94039	.36167	.7650	.0634	.9403	.014636	1.5412	7
54	.34038	.94029	.36199	.7625	.0635	.9379	.014674	1.5390	6
55	0.34065	0.94019	0.36232	2.7600	1.0636	2.9355	.014713	1.5368	5
56	.34093	.94009	.36265	.7575	.0637	.9332	.014751	1.5346	4
57	.34120	.93999	.36298	.7550	.0638	.9308	.014789	1.5324	3
58	.34147	.93989	.36331	.7525	.0640	.9285	.014827	1.5301	2
59	.34175	.93979	.36364	.7500	.0641	.9261	.014866	1.5279	1
60	0.34202	0.93969	0.36397	2.7475	1.0642	2.9238	.014904	1.5257	0
M	Cosine	Sine	Cotan.	Tan.	Cosec.	Secant	READ DOWN	70°–71° Involute	M

20° or 200°   **Trigonometric and Involute Functions**   **159° or 339°**

M	Sine	Cosine	Tan.	Cotan.	Secant	Cosec.	Involute 20°–21°	READ UP	M
0	0.34202	0.93969	0.36397	2.7475	1.0642	2.9238	.014904	1.5257	60
1	.34229	.93959	.36430	.7450	.0643	.9215	.014943	1.5236	59
2	.34257	.93949	.36463	.7425	.0644	.9191	.014982	1.5214	58
3	.34284	.93939	.36496	.7400	.0645	.9168	.015020	1.5192	57
4	.34311	.93929	.36529	.7376	.0646	.9145	.015059	1.5170	56
5	0.34339	0.93919	0.36562	2.7351	1.0647	2.9122	.015098	1.5148	55
6	.34366	.93909	.36595	.7326	.0649	.9099	.015137	1.5126	54
7	.34393	.93899	.36628	.7302	.0650	.9075	.015176	1.5105	53
8	.34421	.93889	.36661	.7277	.0651	.9052	.015215	1.5083	52
9	.34448	.93879	.36694	.7253	.0652	.9029	.015254	1.5061	51
10	0.34475	0.93869	0.36727	2.7228	1.0653	2.9006	.015293	1.5040	50
11	.34503	.93859	.36760	.7204	.0654	.8983	.015333	1.5018	49
12	.34530	.93849	.36793	.7179	.0655	.8960	.015372	1.4997	48
13	.34557	.93839	.36826	.7155	.0657	.8938	.015411	1.4975	47
14	.34584	.93829	.36859	.7130	.0658	.8915	.015451	1.4954	46
15	0.34612	0.93819	0.36892	2.7106	1.0659	2.8892	.015490	1.4933	45
16	.34639	.93809	.36925	.7082	.0660	.8869	.015530	1.4911	44
17	.34666	.93799	.36958	.7058	.0661	.8846	.015570	1.4890	43
18	.34694	.93789	.36991	.7034	.0662	.8824	.015609	1.4869	42
19	.34721	.93779	.37024	.7009	.0663	.8801	.015649	1.4847	41
20	0.34748	0.93769	0.37057	2.6985	1.0665	2.8779	.015689	1.4826	40
21	.34775	.93759	.37090	.6961	.0666	.8756	.015729	1.4805	39
22	.34803	.93748	.37123	.6937	.0667	.8733	.015769	1.4784	38
23	.34830	.93738	.37157	.6913	.0668	.8711	.015809	1.4763	37
24	.34857	.93728	.37190	.6889	.0669	.8688	.015849	1.4742	36
25	0.34884	0.93718	0.37223	2.6865	1.0670	2.8666	.015890	1.4721	35
26	.34912	.93708	.37256	.6841	.0671	.8644	.015930	1.4700	34
27	.34939	.93698	.37289	.6818	.0673	.8621	.015971	1.4679	33
28	.34966	.93688	.37322	.6794	.0674	.8599	.016011	1.4658	32
29	.34993	.93677	.37355	.6770	.0675	.8577	.016052	1.4637	31
30	0.35021	0.93667	0.37388	2.6746	1.0676	2.8555	.016092	1.4616	30
31	.35048	.93657	.37422	.6723	.0677	.8532	.016133	1.4595	29
32	.35075	.93647	.37455	.6699	.0678	.8510	.016174	1.4575	28
33	.35102	.93637	.37488	.6675	.0680	.8488	.016214	1.4554	27
34	.35130	.93626	.37521	.6652	.0681	.8466	.016255	1.4533	26
35	0.35157	0.93616	0.37554	2.6628	1.0682	2.8444	.016296	1.4513	25
36	.35184	.93606	.37588	.6605	.0683	.8422	.016337	1.4492	24
37	.35211	.93596	.37621	.6581	.0684	.8400	.016379	1.4471	23
38	.35239	.93585	.37654	.6558	.0685	.8378	.016420	1.4451	22
39	.35266	.93575	.37687	.6534	.0687	.8356	.016461	1.4430	21
40	0.35293	0.93565	0.37720	2.6511	1.0688	2.8334	.016502	1.4410	20
41	.35320	.93555	.37754	.6488	.0689	.8312	.016544	1.4389	19
42	.35347	.93544	.37787	.6464	.0690	.8291	.016585	1.4369	18
43	.35375	.93534	.37820	.6441	.0691	.8269	.016627	1.4349	17
44	.35402	.93524	.37853	.6418	.0692	.8247	.016669	1.4328	16
45	0.35429	0.93514	0.37887	2.6395	1.0694	2.8225	.016710	1.4308	15
46	.35456	.93503	.37920	.6371	.0695	.8204	.016752	1.4288	14
47	.35484	.93493	.37953	.6348	.0696	.8182	.016794	1.4268	13
48	.35511	.93483	.37986	.6325	.0697	.8161	.016836	1.4248	12
49	.35538	.93472	.38020	.6302	.0698	.8139	.016878	1.4227	11
50	0.35565	0.93462	0.38053	2.6279	1.0700	2.8117	.016920	1.4207	10
51	.35592	.93452	.38086	.6256	.0701	.8096	.016962	1.4187	9
52	.35619	.93441	.38120	.6233	.0702	.8075	.017004	1.4167	8
53	.35647	.93431	.38153	.6210	.0703	.8053	.017047	1.4147	7
54	.35674	.93420	.38186	.6187	.0704	.8032	.017089	1.4127	6
55	0.35701	0.93410	0.38220	2.6165	1.0705	2.8010	.017132	1.4107	5
56	.35728	.93400	.38253	.6142	.0707	.7989	.017174	1.4087	4
57	.35755	.93389	.38286	.6119	.0708	.7968	.017217	1.4067	3
58	.35782	.93379	.38320	.6096	.0709	.7947	.017259	1.4048	2
59	.35810	.93368	.38353	.6074	.0710	.7925	.017302	1.4028	1
60	0.35837	0.93358	0.38386	2.6051	1.0711	2.7904	.017345	1.4008	0
M	Cosine	Sine	Cotan.	Tan.	Cosec.	Secant	READ DOWN	69°–70° Involute	M

21° or 201°   **Trigonometric and Involute Functions**   **158° or 338°**

M	Sine	Cosine	Tan.	Cotan.	Secant	Cosec.	Involute 21°–22°	READ UP	M
0	0.35837	0.93358	0.38386	2.6051	1.0711	2.7904	.017345	1.4008	60
1	.35864	.93348	.38420	.6028	.0713	.7883	.017388	1.3988	59
2	.35891	.93337	.38453	.6006	.0714	.7862	.017431	1.3969	58
3	.35918	.93327	.38487	.5983	.0715	.7841	.017474	1.3949	57
4	.35945	.93316	.38520	.5961	.0716	.7820	.017517	1.3929	56
5	0.35973	0.93306	0.38553	2.5938	1.0717	2.7799	.017560	1.3910	55
6	.36000	.93295	.38587	.5916	.0719	.7778	.017603	1.3890	54
7	.36027	.93285	.38620	.5893	.0720	.7757	.017647	1.3871	53
8	.36054	.93274	.38654	.5871	.0721	.7736	.017690	1.3851	52
9	.36081	.93264	.38687	.5848	.0722	.7715	.017734	1.3832	51
10	0.36108	0.93253	0.38721	2.5826	1.0723	2.7695	.017777	1.3812	50
11	.36135	.93243	.38754	.5804	.0725	.7674	.017821	1.3793	49
12	.36162	.93232	.38787	.5782	.0726	.7653	.017865	1.3774	48
13	.36190	.93222	.38821	.5759	.0727	.7632	.017908	1.3754	47
14	.36217	.93211	.38854	.5737	.0728	.7612	.017952	1.3735	46
15	0.36244	0.93201	0.38888	2.5715	1.0730	2.7591	.017996	1.3716	45
16	.36271	.93190	.38921	.5693	.0731	.7570	.018040	1.3697	44
17	.36298	.93180	.38955	.5671	.0732	.7550	.018084	1.3677	43
18	.36325	.93169	.38988	.5649	.0733	.7529	.018129	1.3658	42
19	.36352	.93159	.39022	.5627	.0734	.7509	.018173	1.3639	41
20	0.36379	0.93148	0.39055	2.5605	1.0736	2.7488	.018217	1.3620	40
21	.36406	.93137	.39089	.5583	.0737	.7468	.018262	1.3601	39
22	.36434	.93127	.39122	.5561	.0738	.7447	.018306	1.3582	38
23	.36461	.93116	.39156	.5539	.0739	.7427	.018351	1.3563	37
24	.36488	.93106	.39190	.5517	.0740	.7407	.018395	1.3544	36
25	0.36515	0.93095	0.39223	2.5495	1.0742	2.7386	.018440	1.3525	35
26	.36542	.93084	.39257	.5473	.0743	.7366	.018485	1.3506	34
27	.36569	.93074	.39290	.5452	.0744	.7346	.018530	1.3487	33
28	.36596	.93063	.39324	.5430	.0745	.7325	.018575	1.3469	32
29	.36623	.93052	.39357	.5408	.0747	.7305	.018620	1.3450	31
30	0.36650	0.93042	0.39391	2.5386	1.0748	2.7285	.018665	1.3431	30
31	.36677	.93031	.39425	.5365	.0749	.7265	.018710	1.3412	29
32	.36704	.93020	.39458	.5343	.0750	.7245	.018755	1.3394	28
33	.36731	.93010	.39492	.5322	.0752	.7225	.018800	1.3375	27
34	.36758	.92999	.39526	.5300	.0753	.7205	.018846	1.3356	26
35	0.36785	0.92988	0.39559	2.5279	1.0754	2.7185	.018891	1.3338	25
36	.36812	.92978	.39593	.5257	.0755	.7165	.018937	1.3319	24
37	.36839	.92967	.39626	.5236	.0757	.7145	.018983	1.3301	23
38	.36867	.92956	.39660	.5214	.0758	.7125	.019028	1.3282	22
39	.36894	.92945	.39694	.5193	.0759	.7105	.019074	1.3264	21
40	0.36921	0.92935	0.39727	2.5172	1.0760	2.7085	.019120	1.3245	20
41	.36948	.92924	.39761	.5150	.0761	.7065	.019166	1.3227	19
42	.36975	.92913	.39795	.5129	.0763	.7046	.019212	1.3208	18
43	.37002	.92902	.39829	.5108	.0764	.7026	.019258	1.3190	17
44	.37029	.92892	.39862	.5086	.0765	.7006	.019304	1.3172	16
45	0.37056	0.92881	0.39896	2.5065	1.0766	2.6986	.019350	1.3153	15
46	.37083	.92870	.39930	.5044	.0768	.6967	.019397	1.3135	14
47	.37110	.92859	.39963	.5023	.0769	.6947	.019443	1.3117	13
48	.37137	.92849	.39997	.5002	.0770	.6927	.019490	1.3099	12
49	.37164	.92838	.40031	.4981	.0771	.6908	.019536	1.3080	11
50	0.37191	0.92827	0.40065	2.4960	1.0773	2.6888	.019583	1.3062	10
51	.37218	.92816	.40098	.4939	.0774	.6869	.019630	1.3044	9
52	.37245	.92805	.40132	.4918	.0775	.6849	.019676	1.3026	8
53	.37272	.92794	.40166	.4897	.0777	.6830	.019723	1.3008	7
54	.37299	.92784	.40200	.4876	.0778	.6811	.019770	1.2990	6
55	0.37326	0.92773	0.40234	2.4855	1.0779	2.6791	.019817	1.2972	5
56	.37353	.92762	.40267	.4834	.0780	.6772	.019864	1.2954	4
57	.37380	.92751	.40301	.4813	.0782	.6752	.019912	1.2936	3
58	.37407	.92740	.40335	.4792	.0783	.6733	.019959	1.2918	2
59	.37434	.92729	.40369	.4772	.0784	.6714	.020006	1.2900	1
60	0.37461	0.92718	0.40403	2.4751	1.0785	2.6695	.020054	1.2883	0
M	Cosine	Sine	Cotan.	Tan.	Cosec.	Secant	READ DOWN	68°–69° Involute	M

22° or 202°   **Trigonometric and Involute Functions**   **157° or 337°**

M	Sine	Cosine	Tan.	Cotan.	Secant	Cosec.	Involute 22°–23°	READ UP	M
0	0.37461	0.92718	0.40403	2.4751	1.0785	2.6695	.020054	1.2883	60
1	.37488	.92707	.40436	.4730	.0787	.6675	.020101	1.2865	59
2	.37515	.92697	.40470	.4709	.0788	.6656	.020149	1.2847	58
3	.37542	.92686	.40504	.4689	.0789	.6637	.020197	1.2829	57
4	.37569	.92675	.40538	.4668	.0790	.6618	.020244	1.2812	56
5	0.37595	.92664	0.40572	2.4648	1.0792	2.6599	.020292	1.2794	55
6	.37622	.92653	.40606	.4627	.0793	.6580	.020340	1.2776	54
7	.37649	.92642	.40640	.4606	.0794	.6561	.020388	1.2759	53
8	.37676	.92631	.40674	.4586	.0796	.6542	.020436	1.2741	52
9	.37703	.92620	.40707	.4566	.0797	.6523	.020484	1.2723	51
10	.37730	.92609	.40741	2.4545	1.0798	2.6504	.020533	1.2706	50
11	.37757	.92598	.40775	.4525	.0799	.6485	.020581	1.2688	49
12	.37784	.92587	.40809	.4504	.0801	.6466	.020629	1.2671	48
13	.37811	.92576	.40843	.4484	.0802	.6447	.020678	1.2653	47
14	.37838	.92565	.40877	.4464	.0803	.6429	.020726	1.2636	46
15	0.37865	0.92554	0.40911	2.4443	1.0804	2.6410	.020775	1.2619	45
16	.37892	.92543	.40945	.4423	.0806	.6391	.020824	1.2601	44
17	.37919	.92532	.40979	.4403	.0807	.6372	.020873	1.2584	43
18	.37946	.92521	.41013	.4383	.0808	.6354	.020921	1.2567	42
19	.37973	.92510	.41047	.4362	.0810	.6335	.020970	1.2549	41
20	0.37999	0.92499	0.41081	2.4342	1.0811	2.6316	.021019	1.2532	40
21	.38026	.92488	.41115	.4322	.0812	.6298	.021069	1.2515	39
22	.38053	.92477	.41149	.4302	.0814	.6279	.021118	1.2498	38
23	.38080	.92466	.41183	.4282	.0815	.6260	.021167	1.2481	37
24	.38107	.92455	.41217	.4262	.0816	.6242	.021217	1.2463	36
25	0.38134	0.92443	0.41251	2.4242	1.0817	2.6223	.021266	1.2446	35
26	.38161	.92432	.41285	.4222	.0819	.6205	.021316	1.2429	34
27	.38188	.92421	.41319	.4202	.0820	.6186	.021365	1.2412	33
28	.38215	.92410	.41353	.4182	.0821	.6168	.021415	1.2395	32
29	.38241	.92399	.41387	.4162	.0823	.6150	.021465	1.2378	31
30	0.38268	0.92388	0.41421	2.4142	1.0824	2.6131	.021514	1.2361	30
31	.38295	.92377	.41455	.4122	.0825	.6113	.021564	1.2344	29
32	.38322	.92366	.41490	.4102	.0827	.6095	.021614	1.2327	28
33	.38349	.92355	.41524	.4083	.0828	.6076	.021665	1.2310	27
34	.38376	.92343	.41558	.4063	.0829	.6058	.021715	1.2294	26
35	0.38403	0.92332	0.41592	2.4043	1.0830	2.6040	.021765	1.2277	25
36	.38430	.92321	.41626	.4023	.0832	.6022	.021815	1.2260	24
37	.38456	.92310	.41660	.4004	.0833	.6003	.021866	1.2243	23
38	.38483	.92299	.41694	.3984	.0834	.5985	.021916	1.2226	22
39	.38510	.92287	.41728	.3964	.0836	.5967	.021967	1.2210	21
40	0.38537	0.92276	0.41763	2.3945	1.0837	2.5949	.022018	1.2193	20
41	.38564	.92265	.41797	.3925	.0838	.5931	.022068	1.2176	19
42	.38591	.92254	.41831	.3906	.0840	.5913	.022119	1.2160	18
43	.38617	.92243	.41865	.3886	.0841	.5895	.022170	1.2143	17
44	.38644	.92231	.41899	.3867	.0842	.5877	.022221	1.2127	16
45	0.38671	0.92220	0.41933	2.3847	1.0844	2.5859	.022272	1.2110	15
46	.38698	.92209	.41968	.3828	.0845	.5841	.022324	1.2093	14
47	.38725	.92198	.42002	.3808	.0846	.5823	.022375	1.2077	13
48	.38752	.92186	.42036	.3789	.0848	.5805	.022426	1.2060	12
49	.38778	.92175	.42070	.3770	.0849	.5788	.022478	1.2044	11
50	0.38805	0.92164	0.42105	2.3750	1.0850	2.5770	.022529	1.2028	10
51	.38832	.92152	.42139	.3731	.0852	.5752	.022581	1.2011	9
52	.38859	.92141	.42173	.3712	.0853	.5734	.022632	1.1995	8
53	.38886	.92130	.42207	.3692	.0854	.5716	.022684	1.1978	7
54	.38912	.92119	.42242	.3673	.0856	.5699	.022736	1.1962	6
55	0.38939	0.92107	0.42276	2.3654	1.0857	2.5681	.022788	1.1946	5
56	.38966	.92096	.42310	.3635	.0858	.5663	.022840	1.1930	4
57	.38993	.92085	.42345	.3616	.0860	.5646	.022892	1.1913	3
58	.39020	.92073	.42379	.3597	.0861	.5628	.022944	1.1897	2
59	.39046	.92062	.42413	.3578	.0862	.5611	.022997	1.1881	1
60	0.39073	0.92050	0.42447	2.3559	1.0864	2.5593	.023049	1.1865	0
M	Cosine	Sine	Cotan.	Tan.	Cosec.	Secant	READ DOWN	67°–68° Involute	M

23° or 203°   Trigonometric and Involute Functions   156° or 336°

M	Sine	Cosine	Tan.	Cotan.	Secant	Cosec.	Involute 23°–24°	READ UP	M
0	0.39073	0.92050	0.42447	2.3559	1.0864	2.5593	.023049	1.1865	60
1	.39100	.92039	.42482	.3539	.0865	.5576	.023102	1.1849	59
2	.39127	.92028	.42516	.3520	.0866	.5558	.023154	1.1833	58
3	.39153	.92016	.42551	.3501	.0868	.5541	.023207	1.1817	57
4	.39180	.92005	.42585	.3483	.0869	.5523	.023259	1.1800	56
5	0.39207	0.91994	0.42619	2.3464	1.0870	2.5506	.023312	1.1784	55
6	.39234	.91982	.42654	.3445	.0872	.5488	.023365	1.1768	54
7	.39260	.91971	.42688	.3426	.0873	.5471	.023418	1.1752	53
8	.39287	.91959	.42722	.3407	.0874	.5454	.023471	1.1736	52
9	.39314	.91948	.42757	.3388	.0876	.5436	.023524	1.1721	51
10	0.39341	0.91936	0.42791	2.3369	1.0877	2.5419	.023577	1.1705	50
11	.39367	.91925	.42826	.3351	.0878	.5402	.023631	1.1689	49
12	.39394	.91914	.42860	.3332	.0880	.5384	.023684	1.1673	48
13	.39421	.91902	.42894	.3313	.0881	.5367	.023738	1.1657	47
14	.39448	.91891	.42929	.3294	.0883	.5350	.023791	1.1641	46
15	0.39474	0.91879	0.42963	2.3276	1.0884	2.5333	.023845	1.1626	45
16	.39501	.91868	.42998	.3257	.0885	.5316	.023899	1.1610	44
17	.39528	.91856	.43032	.3238	.0887	.5299	.023952	1.1594	43
18	.39555	.91845	.43067	.3220	.0888	.5282	.024006	1.1578	42
19	.39581	.91833	.43101	.3201	.0889	.5264	.024060	1.1563	41
20	0.39608	0.91822	0.43136	2.3183	1.0891	2.5247	.024114	1.1547	40
21	.39635	.91810	.43170	.3164	.0892	.5230	.024169	1.1531	39
22	.39661	.91799	.43205	.3146	.0893	.5213	.024223	1.1516	38
23	.39688	.91787	.43239	.3127	.0895	.5196	.024277	1.1500	37
24	.39715	.91775	.43274	.3109	.0896	.5180	.024332	1.1485	36
25	0.39741	0.91764	0.43308	2.3090	1.0898	2.5163	.024386	1.1469	35
26	.39768	.91752	.43343	.3072	.0899	.5146	.024441	1.1454	34
27	.39795	.91741	.43378	.3053	.0900	.5129	.024495	1.1438	33
28	.39822	.91729	.43412	.3035	.0902	.5112	.024550	1.1423	32
29	.39848	.91718	.43447	.3017	.0903	.5095	.024605	1.1407	31
30	0.39875	0.91706	0.43481	2.2998	1.0904	2.5078	.024660	1.1392	30
31	.39902	.91694	.43516	.2980	.0906	.5062	.024715	1.1377	29
32	.39928	.91683	.43550	.2962	.0907	.5045	.024770	1.1361	28
33	.39955	.91671	.43585	.2944	.0909	.5028	.024825	1.1346	27
34	.39982	.91660	.43620	.2925	.0910	.5012	.024881	1.1331	26
35	0.40008	0.91648	0.43654	2.2907	1.0911	2.4995	.024936	1.1315	25
36	.40035	.91636	.43689	.2889	.0913	.4978	.024992	1.1300	24
37	.40062	.91625	.43724	.2871	.0914	.4962	.025047	1.1285	23
38	.40088	.91613	.43758	.2853	.0915	.4945	.025103	1.1270	22
39	.40115	.91601	.43793	.2835	.0917	.4928	.025159	1.1254	21
40	0.40141	0.91590	0.43828	2.2817	1.0918	2.4912	.025214	1.1239	20
41	.40168	.91578	.43862	.2799	.0920	.4895	.025270	1.1224	19
42	.40195	.91566	.43897	.2781	.0921	.4879	.025326	1.1209	18
43	.40221	.91555	.43932	.2763	.0922	.4862	.025382	1.1194	17
44	.40248	.91543	.43966	.2745	.0924	.4846	.025439	1.1179	16
45	0.40275	0.91531	0.44001	2.2727	1.0925	2.4830	.025495	1.1164	15
46	.40301	.91519	.44036	.2709	.0927	.4813	.025551	1.1149	14
47	.40328	.91508	.44071	.2691	.0928	.4797	.025608	1.1134	13
48	.40355	.91496	.44105	.2673	.0929	.4780	.025664	1.1119	12
49	.40381	.91484	.44140	.2655	.0931	.4764	.025721	1.1104	11
50	0.40408	0.91472	0.44175	2.2637	1.0932	2.4748	.025778	1.1089	10
51	.40434	.91461	.44210	.2620	.0934	.4731	.025834	1.1074	9
52	.40461	.91449	.44244	.2602	.0935	.4715	.025891	1.1059	8
53	.40488	.91437	.44279	.2584	.0936	.4699	.025948	1.1044	7
54	.40514	.91425	.44314	.2566	.0938	.4683	.026005	1.1030	6
55	0.40541	0.91414	0.44349	2.2549	1.0939	2.4667	.026062	1.1015	5
56	.40567	.91402	.44384	.2531	.0941	.4650	.026120	1.1000	4
57	.40594	.91390	.44418	.2513	.0942	.4634	.026177	1.0985	3
58	.40621	.91378	.44453	.2496	.0944	.4618	.026235	1.0971	2
59	.40647	.91366	.44488	.2478	.0945	.4602	.026292	1.0956	1
60	0.40674	0.91355	0.44523	2.2460	1.0946	2.4586	.026350	1.0941	0
M	Cosine	Sine	Cotan.	Tan.	Cosec.	Secant	READ DOWN	66°–67° Involute	M

24° or 204° **Trigonometric and Involute Functions** **155° or 335°**

M	Sine	Cosine	Tan.	Cotan.	Secant	Cosec.	Involute 24°–25°	READ UP	M
0	0.40674	0.91355	0.44523	2.2460	1.0946	2.4586	.026350	1.0941	60
1	.40700	.91343	.44558	.2443	.0948	.4570	.026407	1.0927	59
2	.40727	.91331	.44593	.2425	.0949	.4554	.026465	1.0912	58
3	.40753	.91319	.44627	.2408	.0951	.4538	.026523	1.0897	57
4	.40780	.91307	.44662	.2390	.0952	.4522	.026581	1.0883	56
5	0.40806	0.91295	0.44697	2.2373	1.0953	2.4506	.026639	1.0868	55
6	.40833	.91283	.44732	.2355	.0955	.4490	.026697	1.0854	54
7	.40860	.91272	.44767	.2338	.0956	.4474	.026756	1.0839	53
8	.40886	.91260	.44802	.2320	.0958	.4458	.026814	1.0825	52
9	.40913	.91248	.44837	.2303	.0959	.4442	.026872	1.0810	51
10	0.40939	0.91236	0.44872	2.2286	1.0961	2.4426	.026931	1.0796	50
11	.40966	.91224	.44907	.2268	.0962	.4411	.026989	1.0781	49
12	.40992	.91212	.44942	.2251	.0963	.4395	.027048	1.0767	48
13	.41019	.91200	.44977	.2234	.0965	.4379	.027107	1.0752	47
14	.41045	.91188	.45012	.2216	.0966	.4363	.027166	1.0738	46
15	0.41072	0.91176	0.45047	2.2199	1.0968	2.4348	.027225	1.0724	45
16	.41098	.91164	.45082	.2182	.0969	.4332	.027284	1.0709	44
17	.41125	.91152	.45117	.2165	.0971	.4316	.027343	1.0695	43
18	.41151	.91140	.45152	.2148	.0972	.4300	.027402	1.0681	42
19	.41178	.91128	.45187	.2130	.0974	.4285	.027462	1.0666	41
20	0.41204	0.91116	0.45222	2.2113	1.0975	2.4269	.027521	1.0652	40
21	.41231	.91104	.45257	.2096	.0976	.4254	.027581	1.0638	39
22	.41257	.91092	.45292	.2079	.0978	.4238	.027640	1.0624	38
23	.41284	.91080	.45327	.2062	.0979	.4222	.027700	1.0610	37
24	.41310	.91068	.45362	.2045	.0981	.4207	.027760	1.0596	36
25	0.41337	0.91056	0.45397	2.2028	1.0982	2.4191	.027820	1.0581	35
26	.41363	.91044	.45432	.2011	.0984	.4176	.027880	1.0567	34
27	.41390	.91032	.45467	.1994	.0985	.4160	.027940	1.0553	33
28	.41416	.91020	.45502	.1977	.0987	.4145	.028000	1.0539	32
29	.41443	.91008	.45538	.1960	.0988	.4130	.028060	1.0525	31
30	0.41469	0.90996	0.45573	2.1943	1.0989	2.4114	.028121	1.0511	30
31	.41496	.90984	.45608	.1926	.0991	.4099	.028181	1.0497	29
32	.41522	.90972	.45643	.1909	.0992	.4083	.028242	1.0483	28
33	.41549	.90960	.45678	.1892	.0994	.4068	.028302	1.0469	27
34	.41575	.90948	.45713	.1876	.0995	.4053	.028363	1.0455	26
35	0.41602	0.90936	0.45748	2.1859	1.0997	2.4038	.028424	1.0441	25
36	.41628	.90924	.45784	.1842	.0998	.4022	.028485	1.0427	24
37	.41655	.90911	.45819	.1825	.1000	.4007	.028546	1.0414	23
38	.41681	.90899	.45854	.1808	.1001	.3992	.028607	1.0400	22
39	.41707	.90887	.45889	.1792	.1003	.3977	.028668	1.0386	21
40	0.41734	0.90875	0.45924	2.1775	1.1004	2.3961	.028729	1.0372	20
41	.41760	.90863	.45960	.1758	.1006	.3946	.028791	1.0358	19
42	.41787	.90851	.45995	.1742	.1007	.3931	.028852	1.0345	18
43	.41813	.90839	.46030	.1725	.1009	.3916	.028914	1.0331	17
44	.41840	.90826	.46065	.1708	.1010	.3901	.028976	1.0317	16
45	0.41866	0.90814	0.46101	2.1692	1.1011	2.3886	.029037	1.0303	15
46	.41892	.90802	.46136	.1675	.1013	.3871	.029099	1.0290	14
47	.41919	.90790	.46171	.1659	.1014	.3856	.029161	1.0276	13
48	.41945	.90778	.46206	.1642	.1016	.3841	.029223	1.0262	12
49	.41972	.90766	.46242	.1625	.1017	.3826	.029285	1.0249	11
50	0.41998	0.90753	0.46277	2.1609	1.1019	2.3811	.029348	1.0235	10
51	.42024	.90741	.46312	.1592	.1020	.3796	.029410	1.0222	9
52	.42051	.90729	.46348	.1576	.1022	.3781	.029472	1.0208	8
53	.42077	.90717	.46383	.1560	.1023	.3766	.029535	1.0195	7
54	.42104	.90704	.46418	.1543	.1025	.3751	.029598	1.0181	6
55	0.42130	0.90692	0.46454	2.1527	1.1026	2.3736	.029660	1.0168	5
56	.42156	.90680	.46489	.1510	.1028	.3721	.029723	1.0154	4
57	.42183	.90668	.46525	.1494	.1029	.3706	.029786	1.0141	3
58	.42209	.90655	.46560	.1478	.1031	.3692	.029849	1.0127	2
59	.42235	.90643	.46595	.1461	.1032	.3677	.029912	1.0114	1
60	0.42262	0.90631	0.46631	2.1445	1.1034	2.3662	.029975	1.0100	0
M	Cosine	Sine	Cotan.	Tan.	Cosec.	Secant	READ DOWN	65°–66° Involute	M

25° or 205° Trigonometric and Involute Functions 154° or 334°

M	Sine	Cosine	Tan.	Cotan.	Secant	Cosec.	Involute 25°–26°	READ UP	M
0	0.42262	0.90631	0.46631	2.1445	1.1034	2.3662	.029975	1.0100	60
1	.42288	.90618	.46666	.1429	.1035	.3647	.030039	1.0087	59
2	.42315	.90606	.46702	.1413	.1037	.3633	.030102	1.0074	58
3	.42341	.90594	.46737	.1396	.1038	.3618	.030166	1.0060	57
4	.42367	.90582	.46772	.1380	.1040	.3603	.030229	1.0047	56
5	0.42394	0.90569	0.46808	2.1364	1.1041	2.3588	.030293	1.0034	55
6	.42420	.90557	.46843	.1348	.1043	.3574	.030357	1.0021	54
7	.42446	.90545	.46879	.1332	.1044	.3559	.030420	1.0007	53
8	.42473	.90532	.46914	.1315	.1046	.3545	.030484	0.9994	52
9	.42499	.90520	.46950	.1299	.1047	.3530	.030549	0.9981	51
10	0.42525	0.90507	0.46985	2.1283	1.1049	2.3515	.030613	0.9968	50
11	.42552	.90495	.47021	.1267	.1050	.3501	.030677	0.9954	49
12	.42578	.90483	.47056	.1251	.1052	.3486	.030741	0.9941	48
13	.42604	.90470	.47092	.1235	.1053	.3472	.030806	0.9928	47
14	.42631	.90458	.47128	.1219	.1055	.3457	.030870	0.9915	46
15	0.42657	0.90446	0.47163	2.1203	1.1056	2.3443	.030935	0.9902	45
16	.42683	.90433	.47199	.1187	.1058	.3428	.031000	0.9889	44
17	.42709	.90421	.47234	.1171	.1059	.3414	.031065	0.9876	43
18	.42736	.90408	.47270	.1155	.1061	.3400	.031130	0.9863	42
19	.42762	.90396	.47305	.1139	.1062	.3385	.031195	0.9850	41
20	0.42788	0.90383	0.47341	2.1123	1.1064	2.3371	.031260	0.9837	40
21	.42815	.90371	.47377	.1107	.1066	.3356	.031325	0.9824	39
22	.42841	.90358	.47412	.1092	.1067	.3342	.031390	0.9811	38
23	.42867	.90346	.47448	.1076	.1069	.3328	.031456	0.9798	37
24	.42894	.90334	.47483	.1060	.1070	.3314	.031521	0.9785	36
25	0.42920	0.90321	0.47519	2.1044	1.1072	2.3299	.031587	0.9772	35
26	.42946	.90309	.47555	.1028	.1073	.3285	.031653	0.9759	34
27	.42972	.90296	.47590	.1013	.1075	.3271	.031718	0.9747	33
28	.42999	.90284	.47626	.0997	.1076	.3257	.031784	0.9734	32
29	.43025	.90271	.47662	.0981	.1078	.3242	.031850	0.9721	31
30	0.43051	0.90259	0.47698	2.0965	1.1079	2.3228	.031917	0.9708	30
31	.43077	.90246	.47733	.0950	.1081	.3214	.031983	0.9695	29
32	.43104	.90233	.47769	.0934	.1082	.3200	.032049	0.9683	28
33	.43130	.90221	.47805	.0918	.1084	.3186	.032116	0.9670	27
34	.43156	.90208	.47840	.0903	.1085	.3172	.032182	0.9657	26
35	0.43182	0.90196	0.47876	2.0887	1.1087	2.3158	.032249	0.9644	25
36	.43209	.90183	.47912	.0872	.1089	.3144	.032315	0.9632	24
37	.43235	.90171	.47948	.0856	.1090	.3130	.032382	0.9619	23
38	.43261	.90158	.47984	.0840	.1092	.3115	.032449	0.9606	22
39	.43287	.90146	.48019	.0825	.1093	.3101	.032516	0.9594	21
40	0.43313	0.90133	0.48055	2.0809	1.1095	2.3088	.032583	0.9581	20
41	.43340	.90120	.48091	.0794	.1096	.3074	.032651	0.9569	19
42	.43366	.90108	.48127	.0778	.1098	.3060	.032718	0.9556	18
43	.43392	.90095	.48163	.0763	.1099	.3046	.032785	0.9543	17
44	.43418	.90082	.48198	.0748	.1101	.3032	.032853	0.9531	16
45	0.43445	0.90070	0.48234	2.0732	1.1102	2.3018	.032920	0.9518	15
46	.43471	.90057	.48270	.0717	.1104	.3004	.032988	0.9506	14
47	.43497	.90044	.48306	.0701	.1106	.2990	.033056	0.9493	13
48	.43523	.90032	.48342	.0686	.1107	.2976	.033124	0.9481	12
49	.43549	.90019	.48378	.0671	.1109	.2962	.033192	0.9469	11
50	0.43575	0.90007	0.48414	2.0655	1.1110	2.2949	.033260	0.9456	10
51	.43602	.89994	.48450	.0640	.1112	.2935	.033328	0.9444	9
52	.43628	.89981	.48486	.0625	.1113	.2921	.033397	0.9431	8
53	.43654	.89968	.48521	.0609	.1115	.2907	.033465	0.9419	7
54	.43680	.89956	.48557	.0594	.1117	.2894	.033534	0.9407	6
55	0.43706	0.89943	0.48593	2.0579	1.1118	2.2880	.033602	0.9394	5
56	.43733	.89930	.48629	.0564	.1120	.2866	.033671	0.9382	4
57	.43759	.89918	.48665	.0549	.1121	.2853	.033740	0.9370	3
58	.43785	.89905	.48701	.0533	.1123	.2839	.033809	0.9357	2
59	.43811	.89892	.48737	.0518	.1124	.2825	.033878	0.9345	1
60	0.43837	0.89879	0.48773	2.0503	1.1126	2.2812	.033947	0.9333	0
M	Cosine	Sine	Cotan.	Tan.	Cosec.	Secant	READ DOWN	64°–65° Involute	M

26° or 206° **Trigonometric and Involute Functions** **153° or 333°**

M	Sine	Cosine	Tan.	Cotan.	Secant	Cosec.	Involute 26°–27°	READ UP	M
0	0.43837	0.89879	0.48773	2.0503	1.1126	2.2812	.033947	.93329	60
1	.43863	.89867	.48809	.0488	.1128	.2798	.034016	.93207	59
2	.43889	.89854	.48845	.0473	.1129	.2785	.034086	.93085	58
3	.43916	.89841	.48881	.0458	.1131	.2771	.034155	.92963	57
4	.43942	.89828	.48917	.0443	.1132	.2757	.034225	.92842	56
5	0.43968	0.89816	0.48953	2.0428	1.1134	2.2744	.034294	.92720	55
6	.43994	.89803	.48989	.0413	.1136	.2730	.034364	.92599	54
7	.44020	.89790	.49026	.0398	.1137	.2717	.034434	.92478	53
8	.44046	.89777	.49062	.0383	.1139	.2703	.034504	.92357	52
9	.44072	.89764	.49098	.0368	.1140	.2690	.034574	.92236	51
10	0.44098	0.89752	0.49134	2.0353	1.1142	2.2677	.034644	.92115	50
11	.44124	.89739	.49170	.0338	.1143	.2663	.034714	.91995	49
12	.44151	.89726	.49206	.0323	.1145	.2650	.034785	.91875	48
13	.44177	.89713	.49242	.0308	.1147	.2636	.034855	.91755	47
14	.44203	.89700	.49278	.0293	.1148	.2623	.034926	.91635	46
15	0.44229	0.89687	0.49315	2.0278	1.1150	2.2610	.034996	.91515	45
16	.44255	.89674	.49351	.0263	.1151	.2596	.035067	.91396	44
17	.44281	.89662	.49387	.0248	.1153	.2583	.035138	.91276	43
18	.44307	.89649	.49423	.0233	.1155	.2570	.035209	.91157	42
19	.44333	.89636	.49459	.0219	.1156	.2556	.035280	.91038	41
20	0.44359	0.89623	0.49495	2.0204	1.1158	2.2543	.035352	.90919	40
21	.44385	.89610	.49532	.0189	.1159	.2530	.035423	.90801	39
22	.44411	.89597	.49568	.0174	.1161	.2517	.035494	.90682	38
23	.44437	.89584	.49604	.0160	.1163	.2504	.035566	.90564	37
24	.44464	.89571	.49640	.0145	.1164	.2490	.035637	.90446	36
25	0.44490	0.89558	0.49677	2.0130	1.1166	2.2477	.035709	.90328	35
26	.44516	.89545	.49713	.0115	.1168	.2464	.035781	.90210	34
27	.44542	.89532	.49749	.0101	.1169	.2451	.035853	.90092	33
28	.44568	.89519	.49786	.0086	.1171	.2438	.035925	.89975	32
29	.44594	.89506	.49822	.0072	.1172	.2425	.035997	.89858	31
30	0.44620	0.89493	0.49858	2.0057	1.1174	2.2412	.036069	.89741	30
31	.44646	.89480	.49894	.0042	.1176	.2399	.036142	.89624	29
32	.44672	.89467	.49931	.0028	.1177	.2385	.036214	.89507	28
33	.44698	.89454	.49967	.0013	.1179	.2372	.036287	.89390	27
34	.44724	.89441	.50004	1.9999	.1180	.2359	.036359	.89274	26
35	0.44750	0.89428	0.50040	1.9984	1.1182	2.2346	.036432	.89158	25
36	.44776	.89415	.50076	.9970	.1184	.2333	.036505	.89042	24
37	.44802	.89402	.50113	.9955	.1185	.2320	.036578	.88926	23
38	.44828	.89389	.50149	.9941	.1187	.2308	.036651	.88810	22
39	.44854	.89376	.50185	.9926	.1189	.2295	.036724	.88694	21
40	0.44880	0.89363	0.50222	1.9912	1.1190	2.2282	.036798	.88579	20
41	.44906	.89350	.50258	.9897	.1192	.2269	.036871	.88464	19
42	.44932	.89337	.50295	.9883	.1194	.2256	.036945	.88349	18
43	.44958	.89324	.50331	.9868	.1195	.2243	.037018	.88234	17
44	.44984	.89311	.50368	.9854	.1197	.2230	.037092	.88119	16
45	0.45010	0.89298	0.50404	1.9840	1.1198	2.2217	.037166	.88004	15
46	.45036	.89285	.50441	.9825	.1200	.2205	.037240	.87890	14
47	.45062	.89272	.50477	.9811	.1202	.2192	.037314	.87776	13
48	.45088	.89259	.50514	.9797	.1203	.2179	.037388	.87662	12
49	.45114	.89245	.50550	.9782	.1205	.2166	.037462	.87548	11
50	0.45140	0.89232	0.50587	1.9768	1.1207	2.2153	.037537	.87434	10
51	.45166	.89219	.50623	.9754	.1208	.2141	.037611	.87320	9
52	.45192	.89206	.50660	.9740	.1210	.2128	.037686	.87207	8
53	.45218	.89193	.50696	.9725	.1212	.2115	.037761	.87094	7
54	.45243	.89180	.50733	.9711	.1213	.2103	.037835	.86980	6
55	0.45269	0.89167	0.50769	1.9697	1.1215	2.2090	.037910	.86868	5
56	.45295	.89153	.50806	.9683	.1217	.2077	.037985	.86755	4
57	.45321	.89140	.50843	.9669	.1218	.2065	.038060	.86642	3
58	.45347	.89127	.50879	.9654	.1220	.2052	.038136	.86530	2
59	.45373	.89114	.50916	.9640	.1222	.2039	.038211	.86417	1
60	0.45399	0.89101	0.50953	1.9626	1.1223	2.2027	.038287	.86305	0
M	Cosine	Sine	Cotan.	Tan.	Cosec.	Secant	READ DOWN	63°–64° Involute	M

116° or 296° **63° or 243°**

27° or 207°   Trigonometric and Involute Functions   152° or 332°

M	Sine	Cosine	Tan.	Cotan.	Secant	Cosec.	Involute 27°–28°	READ UP	M
0	0.45399	0.89101	0.50953	1.9626	1.1223	2.2027	.038287	.86305	60
1	.45425	.89087	.50989	.9612	.1225	.2014	.038362	.86193	59
2	.45451	.89074	.51026	.9598	.1227	.2002	.038438	.86082	58
3	.45477	.89061	.51063	.9584	.1228	.1989	.038514	.85970	57
4	.45503	.89048	.51099	.9570	.1230	.1977	.038590	.85858	56
5	0.45529	0.89035	0.51136	1.9556	1.1232	2.1964	.038666	.85747	55
6	.45554	.89021	.51173	.9542	.1233	.1952	.038742	.85636	54
7	.45580	.89008	.51209	.9528	.1235	.1939	.038818	.85525	53
8	.45606	.88995	.51246	.9514	.1237	.1927	.038894	.85414	52
9	.45632	.88981	.51283	.9500	.1238	.1914	.038971	.85303	51
10	0.45658	0.88968	0.51319	1.9486	1.1240	2.1902	.039047	.85193	50
11	.45684	.88955	.51356	.9472	.1242	.1890	.039124	.85082	49
12	.45710	.88942	.51393	.9458	.1243	.1877	.039201	.84972	48
13	.45736	.88928	.51430	.9444	.1245	.1865	.039278	.84862	47
14	.45762	.88915	.51467	.9430	.1247	.1852	.039355	.84752	46
15	0.45787	0.88902	0.51503	1.9416	1.1248	2.1840	.039432	.84643	45
16	.45813	.88888	.51540	.9402	.1250	.1828	.039509	.84533	44
17	.45839	.88875	.51577	.9388	.1252	.1815	.039586	.84424	43
18	.45865	.88862	.51614	.9375	.1253	.1803	.039664	.84314	42
19	.45891	.88848	.51651	.9361	.1255	.1791	.039741	.84205	41
20	0.45917	0.88835	0.51688	1.9347	1.1257	2.1779	.039819	.84096	40
21	.45942	.88822	.51724	.9333	.1259	.1766	.039897	.83987	39
22	.45968	.88808	.51761	.9319	.1260	.1754	.039974	.83879	38
23	.45994	.88795	.51798	.9306	.1262	.1742	.040052	.83770	37
24	.46020	.88782	.51835	.9292	.1264	.1730	.040131	.83662	36
25	0.46046	0.88768	0.51872	1.9278	1.1265	2.1718	.040209	.83554	35
26	.46072	.88755	.51909	.9265	.1267	.1705	.040287	.83446	34
27	.46097	.88741	.51946	.9251	.1269	.1693	.040366	.83338	33
28	.46123	.88728	.51983	.9237	.1270	.1681	.040444	.83230	32
29	.46149	.88715	.52020	.9223	.1272	.1669	.040523	.83123	31
30	0.46175	0.88701	0.52057	1.9210	1.1274	2.1657	.040602	.83015	30
31	.46201	.88688	.52094	.9196	.1276	.1645	.040680	.82908	29
32	.46226	.88674	.52131	.9183	.1277	.1633	.040759	.82801	28
33	.46252	.88661	.52168	.9169	.1279	.1621	.040838	.82694	27
34	.46278	.88647	.52205	.9155	.1281	.1609	.040918	.82587	26
35	0.46304	0.88634	0.52242	1.9142	1.1282	2.1596	.040997	.82480	25
36	.46330	.88620	.52279	.9128	.1284	.1584	.041076	.82374	24
37	.46355	.88607	.52316	.9115	.1286	.1572	.041156	.82267	23
38	.46381	.88593	.52353	.9101	.1288	.1560	.041236	.82161	22
39	.46407	.88580	.52390	.9088	.1289	.1549	.041316	.82055	21
40	0.46433	0.88566	0.52427	1.9074	1.1291	2.1537	.041395	.81949	20
41	.46458	.88553	.52464	.9061	.1293	.1525	.041475	.81844	19
42	.46484	.88539	.52501	.9047	.1294	.1513	.041556	.81738	18
43	.46510	.88526	.52538	.9034	.1296	.1501	.041636	.81632	17
44	.46536	.88512	.52575	.9020	.1298	.1489	.041716	.81527	16
45	0.46561	0.88499	0.52613	1.9007	1.1300	2.1477	.041797	.81422	15
46	.46587	.88485	.52650	.8993	.1301	.1465	.041877	.81317	14
47	.46613	.88472	.52687	.8980	.1303	.1453	.041958	.81212	13
48	.46639	.88458	.52724	.8967	.1305	.1441	.042039	.81107	12
49	.46664	.88445	.52761	.8953	.1307	.1430	.042120	.81003	11
50	0.46690	0.88431	0.52798	1.8940	1.1308	2.1418	.042201	.80898	10
51	.46716	.88417	.52836	.8927	.1310	.1406	.042282	.80794	9
52	.46742	.88404	.52873	.8913	.1312	.1394	.042363	.80690	8
53	.46767	.88390	.52910	.8900	.1313	.1382	.042444	.80586	7
54	.46793	.88377	.52947	.8887	.1315	.1371	.042526	.80482	6
55	0.46819	0.88363	0.52985	1.8873	1.1317	2.1359	.042607	.80378	5
56	.46844	.88349	.53022	.8860	.1319	.1347	.042689	.80275	4
57	.46870	.88336	.53059	.8847	.1320	.1336	.042771	.80172	3
58	.46896	.88322	.53096	.8834	.1322	.1324	.042853	.80068	2
59	.46921	.88308	.53134	.8820	.1324	.1312	.042935	.79965	1
60	0.46947	0.88295	0.53171	1.8807	1.1326	2.1301	.043017	.79862	0
M	Cosine	Sine	Cotan.	Tan.	Cosec.	Secant	READ DOWN	62°–63° Involute	M

117° or 297°   62° or 242°

28° or 208°   Trigonometric and Involute Functions   151° or 331°

M	Sine	Cosine	Tan.	Cotan.	Secant	Cosec.	Involute 28°–29°	READ UP	M
0	0.46947	0.88295	0.53171	1.8807	1.1326	2.1301	.043017	.79862	60
1	.46973	.88281	.53208	.8794	.1327	.1289	.043100	.79759	59
2	.46999	.88267	.53246	.8781	.1329	.1277	.043182	.79657	58
3	.47024	.88254	.53283	.8768	.1331	.1266	.043264	.79554	57
4	.47050	.88240	.53320	.8755	.1333	.1254	.043347	.79452	56
5	.47076	0.88226	0.53358	1.8741	1.1334	2.1242	.043430	.79350	55
6	.47101	.88213	.53395	.8728	.1336	.1231	.043513	.79247	54
7	.47127	.88199	.53432	.8715	.1338	.1219	.043596	.79146	53
8	.47153	.88185	.53470	.8702	.1340	.1208	.043679	.79044	52
9	.47178	.88172	.53507	.8689	.1342	.1196	.043762	.78942	51
10	.47204	0.88158	0.53545	1.8676	1.1343	2.1185	.043845	.78841	50
11	.47229	.88144	.53582	.8663	.1345	.1173	.043929	.78739	49
12	.47255	.88130	.53620	.8650	.1347	.1162	.044012	.78638	48
13	.47281	.88117	.53657	.8637	.1349	.1150	.044096	.78537	47
14	.47306	.88103	.53694	.8624	.1350	.1139	.044180	.78436	46
15	.47332	0.88089	0.53732	1.8611	1.1352	2.1127	.044264	.78335	45
16	.47358	.88075	.53769	.8598	.1354	.1116	.044348	.78234	44
17	.47383	.88062	.53807	.8585	.1356	.1105	.044432	.78134	43
18	.47409	.88048	.53844	.8572	.1357	.1093	.044516	.78033	42
19	.47434	.88034	.53882	.8559	.1359	.1082	.044601	.77933	41
20	.47460	0.88020	0.53920	1.8546	1.1361	2.1070	.044685	.77833	40
21	.47486	.88006	.53957	.8533	.1363	.1059	.044770	.77733	39
22	.47511	.87993	.53995	.8520	.1365	.1048	.044855	.77633	38
23	.47537	.87979	.54032	.8507	.1366	.1036	.044939	.77533	37
24	.47562	.87965	.54070	.8495	.1368	.1025	.045024	.77434	36
25	.47588	0.87951	0.54107	1.8482	1.1370	2.1014	.045110	.77334	35
26	.47614	.87937	.54145	.8469	.1372	.1002	.045195	.77235	34
27	.47639	.87923	.54183	.8456	.1374	.0991	.045280	.77136	33
28	.47665	.87909	.54220	.8443	.1375	.0980	.045366	.77037	32
29	.47690	.87896	.54258	.8430	.1377	.0969	.045451	.76938	31
30	.47716	0.87882	0.54296	1.8418	1.1379	2.0957	.045537	.76839	30
31	.47741	.87868	.54333	.8405	.1381	.0946	.045623	.76741	29
32	.47767	.87854	.54371	.8392	.1383	.0935	.045709	.76642	28
33	.47793	.87840	.54409	.8379	.1384	.0924	.045795	.76544	27
34	.47818	.87826	.54446	.8367	.1386	.0913	.045881	.76446	26
35	.47844	0.87812	0.54484	1.8354	1.1388	2.0901	.045967	.76348	25
36	.47869	.87798	.54522	.8341	.1390	.0890	.046054	.76250	24
37	.47895	.87784	.54560	.8329	.1392	.0879	.046140	.76152	23
38	.47920	.87770	.54597	.8316	.1393	.0868	.046227	.76054	22
39	.47946	.87756	.54635	.8303	.1395	.0857	.046313	.75957	21
40	.47971	0.87743	0.54673	1.8291	1.1397	2.0846	.046400	.75859	20
41	.47997	.87729	.54711	.8278	.1399	.0835	.046487	.75762	19
42	.48022	.87715	.54748	.8265	.1401	.0824	.046575	.75665	18
43	.48048	.87701	.54786	.8253	.1402	.0813	.046662	.75568	17
44	.48073	.87687	.54824	.8240	.1404	.0802	.046749	.75471	16
45	0.48099	0.87673	0.54862	1.8228	1.1406	2.0791	.046837	.75375	15
46	.48124	.87659	.54900	.8215	.1408	.0779	.046924	.75278	14
47	.48150	.87645	.54938	.8202	.1410	.0768	.047012	.75181	13
48	.48175	.87631	.54975	.8190	.1412	.0757	.047100	.75085	12
49	.48201	.87617	.55013	.8177	.1413	.0747	.047188	.74989	11
50	0.48226	0.87603	0.55051	1.8165	1.1415	2.0736	.047276	.74893	10
51	.48252	.87589	.55089	.8152	.1417	.0725	.047364	.74797	9
52	.48277	.87575	.55127	.8140	.1419	.0714	.047452	.74701	8
53	.48303	.87561	.55165	.8127	.1421	.0703	.047541	.74606	7
54	.48328	.87546	.55203	.8115	.1423	.0692	.047630	.74510	6
55	0.48354	0.87532	0.55241	1.8103	1.1424	2.0681	.047718	.74415	5
56	.48379	.87518	.55279	.8090	.1426	.0670	.047807	.74319	4
57	.48405	.87504	.55317	.8078	.1428	.0659	.047896	.74224	3
58	.48430	.87490	.55355	.8065	.1430	.0648	.047985	.74129	2
59	.48456	.87476	.55393	.8053	.1432	.0637	.048074	.74034	1
60	0.48481	0.87462	0.55431	1.8040	1.1434	2.0627	.048164	.73940	0
M	Cosine	Sine	Cotan.	Tan.	Cosec.	Secant	READ DOWN	61°–62° Involute	M

29° or 209°   **Trigonometric and Involute Functions**   **150° or 330°**

M	Sine	Cosine	Tan.	Cotan.	Secant	Cosec.	Involute 29°–30°	READ UP	M
0	0.48481	0.87462	0.55431	1.8040	1.1434	2.0627	.048164	.73940	60
1	.48506	.87448	.55469	.8028	.1435	.0616	.048253	.73845	59
2	.48532	.87434	.55507	.8016	.1437	.0605	.048343	.73751	58
3	.48557	.87420	.55545	.8003	.1439	.0594	.048432	.73656	57
4	.48583	.87406	.55583	.7991	.1441	.0583	.048522	.73562	56
5	0.48608	0.87391	0.55621	1.7979	1.1443	2.0573	.048612	.73468	55
6	.48634	.87377	.55659	.7966	.1445	.0562	.048702	.73374	54
7	.48659	.87363	.55697	.7954	.1446	.0551	.048792	.73280	53
8	.48684	.87349	.55736	.7942	.1448	.0540	.048883	.73186	52
9	.48710	.87335	.55774	.7930	.1450	.0530	.048973	.73093	51
10	0.48735	0.87321	0.55812	1.7917	1.1452	2.0519	.049063	.72999	50
11	.48761	.87306	.55850	.7905	.1454	.0508	.049154	.72906	49
12	.48786	.87292	.55888	.7893	.1456	.0498	.049245	.72813	48
13	.48811	.87278	.55926	.7881	.1458	.0487	.049336	.72720	47
14	.48837	.87264	.55964	.7868	.1460	.0476	.049427	.72627	46
15	0.48862	0.87250	0.56003	1.7856	1.1461	2.0466	.049518	.72534	45
16	.48888	.87235	.56041	.7844	.1463	.0455	.049609	.72441	44
17	.48913	.87221	.56079	.7832	.1465	.0445	.049701	.72349	43
18	.48938	.87207	.56117	.7820	.1467	.0434	.049792	.72256	42
19	.48964	.87193	.56156	.7808	.1469	.0423	.049884	.72164	41
20	0.48989	0.87178	0.56194	1.7796	1.1471	2.0413	.049976	.72072	40
21	.49014	.87164	.56232	.7783	.1473	.0402	.050068	.71980	39
22	.49040	.87150	.56270	.7771	.1474	.0392	.050160	.71888	38
23	.49065	.87136	.56309	.7759	.1476	.0381	.050252	.71796	37
24	.49090	.87121	.56347	.7747	.1478	.0371	.050344	.71704	36
25	0.49116	0.87107	0.56385	1.7735	1.1480	2.0360	.050437	.71613	35
26	.49141	.87093	.56424	.7723	.1482	.0350	.050529	.71521	34
27	.49166	.87079	.56462	.7711	.1484	.0339	.050622	.71430	33
28	.49192	.87064	.56501	.7699	.1486	.0329	.050715	.71339	32
29	.49217	.87050	.56539	.7687	.1488	.0318	.050808	.71248	31
30	0.49242	0.87036	0.56577	1.7675	1.1490	2.0308	.050901	.71157	30
31	.49268	.87021	.56616	.7663	.1491	.0297	.050994	.71066	29
32	.49293	.87007	.56654	.7651	.1493	.0287	.051087	.70975	28
33	.49318	.86993	.56693	.7639	.1495	.0276	.051181	.70885	27
34	.49344	.86978	.56731	.7627	.1497	.0266	.051274	.70794	26
35	0.49369	0.86964	0.56769	1.7615	1.1499	2.0256	.051368	.70704	25
36	.49394	.86949	.56808	.7603	.1501	.0245	.051462	.70614	24
37	.49419	.86935	.56846	.7591	.1503	.0235	.051556	.70524	23
38	.49445	.86921	.56885	.7579	.1505	.0225	.051650	.70434	22
39	.49470	.86906	.56923	.7567	.1507	.0214	.051744	.70344	21
40	0.49495	0.86892	0.56962	1.7556	1.1509	2.0204	.051838	.70254	20
41	.49521	.86878	.57000	.7544	.1510	.0194	.051933	.70165	19
42	.49546	.86863	.57039	.7532	.1512	.0183	.052027	.70075	18
43	.49571	.86849	.57078	.7520	.1514	.0173	.052122	.69986	17
44	.49596	.86834	.57116	.7508	.1516	.0163	.052217	.69897	16
45	0.49622	0.86820	0.57155	1.7496	1.1518	2.0152	.052312	.69808	15
46	.49647	.86805	.57193	.7485	.1520	.0142	.052407	.69719	14
47	.49672	.86791	.57232	.7473	.1522	.0132	.052502	.69630	13
48	.49697	.86777	.57271	.7461	.1524	.0122	.052597	.69541	12
49	.49723	.86762	.57309	.7449	.1526	.0112	.052693	.69452	11
50	0.49748	0.86748	0.57348	1.7437	1.1528	2.0101	.052788	.69364	10
51	.49773	.86733	.57386	.7426	.1530	.0091	.052884	.69275	9
52	.49798	.86719	.57425	.7414	.1532	.0081	.052980	.69187	8
53	.49824	.86704	.57464	.7402	.1533	.0071	.053076	.69099	7
54	.49849	.86690	.57503	.7391	.1535	.0061	.053172	.69011	6
55	0.49874	0.86675	0.57541	1.7379	1.1537	2.0051	.053268	.68923	5
56	.49899	.86661	.57580	.7367	.1539	.0040	.053365	.68835	4
57	.49924	.86646	.57619	.7355	.1541	.0030	.053461	.68748	3
58	.49950	.86632	.57657	.7344	.1543	.0020	.053558	.68660	2
59	.49975	.86617	.57696	.7332	.1545	.0010	.053655	.68573	1
60	0.50000	0.86603	0.57735	1.7321	1.1547	2.0000	.053751	.68485	0
M	Cosine	Sine	Cotan.	Tan.	Cosec.	Secant	READ DOWN	60°–61° Involute	M

30° or 210°   **Trigonometric and Involute Functions**   **149° or 329°**

M	Sine	Cosine	Tan.	Cotan.	Secant	Cosec.	Involute 30°–31°	READ UP	M
0	0.50000	0.86603	0.57735	1.7321	1.1547	2.0000	.053751	.68485	60
1	.50025	.86588	.57774	.7309	.1549	1.9990	.053849	.68398	59
2	.50050	.86573	.57813	.7297	.1551	.9980	.053946	.68311	58
3	.50076	.86559	.57851	.7286	.1553	.9970	.054043	.68224	57
4	.50101	.86544	.57890	.7274	.1555	.9960	.054140	.68137	56
5	0.50126	0.86530	0.57929	1.7262	1.1557	1.9950	.054238	.68050	55
6	.50151	.86515	.57968	.7251	.1559	.9940	.054336	.67964	54
7	.50176	.86501	.58007	.7239	.1561	.9930	.054433	.67877	53
8	.50201	.86486	.58046	.7228	.1563	.9920	.054531	.67791	52
9	.50227	.86471	.58085	.7216	.1565	.9910	.054629	.67705	51
10	.50252	0.86457	0.58124	1.7205	1.1566	1.9900	.054728	.67618	50
11	.50277	.86442	.58162	.7193	.1568	.9890	.054826	.67532	49
12	.50302	.86427	.58201	.7182	.1570	.9880	.054924	.67447	48
13	.50327	.86413	.58240	.7170	.1572	.9870	.055023	.67361	47
14	.50352	.86398	.58279	.7159	.1574	.9860	.055122	.67275	46
15	0.50377	0.86384	0.58318	1.7147	1.1576	1.9850	.055221	.67189	45
16	.50403	.86369	.58357	.7136	.1578	.9840	.055320	.67104	44
17	.50428	.86354	.58396	.7124	.1580	.9830	.055419	.67019	43
18	.50453	.86340	.58435	.7113	.1582	.9821	.055518	.66933	42
19	.50478	.86325	.58474	.7102	.1584	.9811	.055617	.66848	41
20	0.50503	0.86310	0.58513	1.7090	1.1586	1.9801	.055717	.66763	40
21	.50528	.86295	.58552	.7079	.1588	.9791	.055817	.66678	39
22	.50553	.86281	.58591	.7067	.1590	.9781	.055916	.66593	38
23	.50578	.86266	.58631	.7056	.1592	.9771	.056016	.66509	37
24	.50603	.86251	.58670	.7045	.1594	.9762	.056116	.66424	36
25	.50628	0.86237	0.58709	1.7033	1.1596	1.9752	.056217	.66340	35
26	.50654	.86222	.58748	.7022	.1598	.9742	.056317	.66255	34
27	.50679	.86207	.58787	.7011	.1600	.9732	.056417	.66171	33
28	.50704	.86192	.58826	.6999	.1602	.9722	.056518	.66087	32
29	.50729	.86178	.58865	.6988	.1604	.9713	.056619	.66003	31
30	0.50754	0.86163	0.58905	1.6977	1.1606	1.9703	.056720	.65919	30
31	.50779	.86148	.58944	.6965	.1608	.9693	.056821	.65835	29
32	.50804	.86133	.58983	.6954	.1610	.9684	.056922	.65752	28
33	.50829	.86119	.59022	.6943	.1612	.9674	.057023	.65668	27
34	.50854	.86104	.59061	.6932	.1614	.9664	.057124	.65585	26
35	0.50879	0.86089	0.59101	1.6920	1.1616	1.9654	.057226	.65501	25
36	.50904	.86074	.59140	.6909	.1618	.9645	.057328	.65418	24
37	.50929	.86059	.59179	.6898	.1620	.9635	.057429	.65335	23
38	.50954	.86045	.59218	.6887	.1622	.9625	.057531	.65252	22
39	.50979	.86030	.59258	.6875	.1624	.9616	.057633	.65169	21
40	0.51004	0.86015	0.59297	1.6864	1.1626	1.9606	.057736	.65086	20
41	.51029	.86000	.59336	.6853	.1628	.9597	.057838	.65004	19
42	.51054	.85985	.59376	.6842	.1630	.9587	.057940	.64921	18
43	.51079	.85970	.59415	.6831	.1632	.9577	.058043	.64839	17
44	.51104	.85956	.59454	.6820	.1634	.9568	.058146	.64756	16
45	0.51129	0.85941	0.59494	1.6808	1.1636	1.9558	.058249	.64674	15
46	.51154	.85926	.59533	.6797	.1638	.9549	.058352	.64592	14
47	.51179	.85911	.59573	.6786	.1640	.9539	.058455	.64510	13
48	.51204	.85896	.59612	.6775	.1642	.9530	.058558	.64428	12
49	.51229	.85881	.59651	.6764	.1644	.9520	.058662	.64346	11
50	0.51254	0.85866	0.59691	1.6753	1.1646	1.9511	.058765	.64265	10
51	.51279	.85851	.59730	.6742	.1648	.9501	.058869	.64183	9
52	.51304	.85836	.59770	.6731	.1650	.9492	.058973	.64102	8
53	.51329	.85821	.59809	.6720	.1652	.9482	.059077	.64020	7
54	.51354	.85806	.59849	.6709	.1654	.9473	.059181	.63939	6
55	0.51379	0.85792	0.59888	1.6698	1.1656	1.9463	.059285	.63858	5
56	.51404	.85777	.59928	.6687	.1658	.9454	.059390	.63777	4
57	.51429	.85762	.59967	.6676	.1660	.9444	.059494	.63696	3
58	.51454	.85747	.60007	.6665	.1662	.9435	.059599	.63615	2
59	.51479	.85732	.60046	.6654	.1664	.9425	.059704	.63534	1
60	0.51504	0.85717	0.60086	1.6643	1.1666	1.9416	.059809	.63454	0
M	Cosine	Sine	Cotan.	Tan.	Cosec.	Secant	READ DOWN	59°–60° Involute	M

120° or 300°   **59° or 239°**

31° or 211°   **Trigonometric and Involute Functions**   **148° or 328°**

M	Sine	Cosine	Tan.	Cotan.	Secant	Cosec.	Involute 31°–32°	READ UP	M
0	0.51504	0.85717	0.60086	1.6643	1.1666	1.9416	.059809	.63454	60
1	.51529	.85702	.60126	.6632	.1668	.9407	.059914	.63373	59
2	.51554	.85687	.60165	.6621	.1670	.9397	.060019	.63293	58
3	.51579	.85672	.60205	.6610	.1672	.9388	.060124	.63212	57
4	.51604	.85657	.60245	.6599	.1675	.9379	.060230	.63132	56
5	0.51628	0.85642	0.60284	1.6588	1.1677	1.9369	.060335	.63052	55
6	.51653	.85627	.60324	.6577	.1679	.9360	.060441	.62972	54
7	.51678	.85612	.60364	.6566	.1681	.9351	.060547	.62892	53
8	.51703	.85597	.60403	.6555	.1683	.9341	.060653	.62812	52
9	.51728	.85582	.60443	.6545	.1685	.9332	.060759	.62733	51
10	0.51753	0.85567	0.60483	1.6534	1.1687	1.9323	.060866	.62653	50
11	.51778	.85551	.60522	.6523	.1689	.9313	.060972	.62574	49
12	.51803	.85536	.60562	.6512	.1691	.9304	.061079	.62494	48
13	.51828	.85521	.60602	.6501	.1693	.9295	.061186	.62415	47
14	.51852	.85506	.60642	.6490	.1695	.9285	.061292	.62336	46
15	0.51877	0.85491	0.60681	1.6479	1.1697	1.9276	.061400	.62257	45
16	.51902	.85476	.60721	.6469	.1699	.9267	.061507	.62178	44
17	.51927	.85461	.60761	.6458	.1701	.9258	.061614	.62099	43
18	.51952	.85446	.60801	.6447	.1703	.9249	.061721	.62020	42
19	.51977	.85431	.60841	.6436	.1705	.9239	.061829	.61942	41
20	0.52002	0.85416	0.60881	1.6426	1.1707	1.9230	.061937	.61863	40
21	.52026	.85401	.60921	.6415	.1710	.9221	.062045	.61785	39
22	.52051	.85385	.60960	.6404	.1712	.9212	.062153	.61706	38
23	.52076	.85370	.61000	.6393	.1714	.9203	.062261	.61628	37
24	.52101	.85355	.61040	.6383	.1716	.9194	.062369	.61550	36
25	0.52126	0.85340	0.61080	1.6372	1.1718	1.9184	.062478	.61472	35
26	.52151	.85325	.61120	.6361	.1720	.9175	.062586	.61394	34
27	.52175	.85310	.61160	.6351	.1722	.9166	.062695	.61316	33
28	.52200	.85294	.61200	.6340	.1724	.9157	.062804	.61239	32
29	.52225	.85279	.61240	.6329	.1726	.9148	.062913	.61161	31
30	0.52250	0.85264	0.61280	1.6319	1.1728	1.9139	.063022	.61083	30
31	.52275	.85249	.61320	.6308	.1730	.9130	.063131	.61006	29
32	.52299	.85234	.61360	.6297	.1732	.9121	.063241	.60929	28
33	.52324	.85218	.61400	.6287	.1735	.9112	.063350	.60851	27
34	.52349	.85203	.61440	.6276	.1737	.9103	.063460	.60774	26
35	0.52374	0.85188	0.61480	1.6265	1.1739	1.9094	.063570	.60697	25
36	.52399	.85173	.61520	.6255	.1741	.9084	.063680	.60620	24
37	.52423	.85157	.61561	.6244	.1743	.9075	.063790	.60544	23
38	.52448	.85142	.61601	.6234	.1745	.9066	.063901	.60467	22
39	.52473	.85127	.61641	.6223	.1747	.9057	.064011	.60390	21
40	0.52498	0.85112	0.61681	1.6212	1.1749	1.9048	.064122	.60314	20
41	.52522	.85096	.61721	.6202	.1751	.9039	.064232	.60237	19
42	.52547	.85081	.61761	.6191	.1753	.9031	.064343	.60161	18
43	.52572	.85066	.61801	.6181	.1756	.9022	.064454	.60085	17
44	.52597	.85051	.61842	.6170	.1758	.9013	.064565	.60009	16
45	0.52621	0.85035	0.61882	1.6160	1.1760	1.9004	.064677	.59933	15
46	.52646	.85020	.61922	.6149	.1762	.8995	.064788	.59857	14
47	.52671	.85005	.61962	.6139	.1764	.8986	.064900	.59781	13
48	.52696	.84989	.62003	.6128	.1766	.8977	.065012	.59705	12
49	.52720	.84974	.62043	.6118	.1768	.8968	.065123	.59630	11
50	0.52745	0.84959	0.62083	1.6107	1.1770	1.8959	.065236	.59554	10
51	.52770	.84943	.62124	.6097	.1773	.8950	.065348	.59479	9
52	.52794	.84928	.62164	.6087	.1775	.8941	.065460	.59403	8
53	.52819	.84913	.62204	.6076	.1777	.8933	.065573	.59328	7
54	.52844	.84897	.62245	.6066	.1779	.8924	.065685	.59253	6
55	0.52869	0.84882	0.62285	1.6055	1.1781	1.8915	.065798	.59178	5
56	.52893	.84866	.62325	.6045	.1783	.8906	.065911	.59103	4
57	.52918	.84851	.62366	.6034	.1785	.8897	.066024	.59028	3
58	.52942	.84836	.62406	.6024	.1788	.8888	.066137	.58954	2
59	.52967	.84820	.62446	.6014	.1790	.8880	.066250	.58879	1
60	0.52992	0.84805	0.62487	1.6003	1.1792	1.8871	.066364	.58804	0
M	Cosine	Sine	Cotan.	Tan.	Cosec.	Secant	READ DOWN	58°–59° Involute	M

32° or 212° Trigonometric and Involute Functions 147° or 327°

M	Sine	Cosine	Tan.	Cotan.	Secant	Cosec.	Involute 32°–33°	READ UP	M
0	0.52992	0.84805	0.62487	1.6003	1.1792	1.8871	.066364	.58804	60
1	.53017	.84789	.62527	.5993	.1794	.8862	.066478	.58730	59
2	.53041	.84774	.62568	.5983	.1796	.8853	.066591	.58656	58
3	.53066	.84758	.62608	.5972	.1798	.8844	.066705	.58581	57
4	.53091	.84743	.62649	.5962	.1800	.8836	.066819	.58507	56
5	0.53115	0.84728	0.62689	1.5952	1.1803	1.8827	.066934	.58433	55
6	.53140	.84712	.62730	.5941	.1805	.8818	.067048	.58359	54
7	.53164	.84697	.62770	.5931	.1807	.8810	.067163	.58285	53
8	.53189	.84681	.62811	.5921	.1809	.8801	.067277	.58211	52
9	.53214	.84666	.62852	.5911	.1811	.8792	.067392	.58138	51
10	0.53238	0.84650	0.62892	1.5900	1.1813	1.8783	.067507	.58064	50
11	.53263	.84635	.62933	.5890	.1815	.8775	.067622	.57991	49
12	.53288	.84619	.62973	.5880	.1818	.8766	.067738	.57917	48
13	.53312	.84604	.63014	.5869	.1820	.8757	.067853	.57844	47
14	.53337	.84588	.63055	.5859	.1822	.8749	.067969	.57771	46
15	0.53361	0.84573	0.63095	1.5849	1.1824	1.8740	.068084	.57698	45
16	.53386	.84557	.63136	.5839	.1826	.8731	.068200	.57625	44
17	.53411	.84542	.63177	.5829	.1828	.8723	.068316	.57552	43
18	.53435	.84526	.63217	.5818	.1831	.8714	.068432	.57479	42
19	.53460	.84511	.63258	.5808	.1833	.8706	.068549	.57406	41
20	0.53484	0.84495	0.63299	1.5798	1.1835	1.8697	.068665	.57333	40
21	.53509	.84480	.63340	.5788	.1837	.8688	.068782	.57261	39
22	.53534	.84464	.63380	.5778	.1839	.8680	.068899	.57188	38
23	.53558	.84448	.63421	.5768	.1842	.8671	.069016	.57116	37
24	.53583	.84433	.63462	.5757	.1844	.8663	.069133	.57044	36
25	0.53607	0.84417	0.63503	1.5747	1.1846	1.8654	.069250	.56972	35
26	.53632	.84402	.63544	.5737	.1848	.8646	.069367	.56900	34
27	.53656	.84386	.63584	.5727	.1850	.8637	.069485	.56828	33
28	.53681	.84370	.63625	.5717	.1852	.8629	.069602	.56756	32
29	.53705	.84355	.63666	.5707	.1855	.8620	.069720	.56684	31
30	0.53730	0.84339	0.63707	1.5697	1.1857	1.8612	.069838	.56612	30
31	.53754	.84324	.63748	.5687	.1859	.8603	.069956	.56540	29
32	.53779	.84308	.63789	.5677	.1861	.8595	.070075	.56469	28
33	.53804	.84292	.63830	.5667	.1863	.8586	.070193	.56398	27
34	.53828	.84277	.63871	.5657	.1866	.8578	.070312	.56326	26
35	0.53853	0.84261	0.63912	1.5647	1.1868	1.8569	.070430	.56255	25
36	.53877	.84245	.63953	.5637	.1870	.8561	.070549	.56184	24
37	.53902	.84230	.63994	.5627	.1872	.8552	.070668	.56113	23
38	.53926	.84214	.64035	.5617	.1875	.8544	.070788	.56042	22
39	.53951	.84198	.64076	.5607	.1877	.8535	.070907	.55971	21
40	0.53975	0.84182	0.64117	1.5597	1.1879	1.8527	.071026	.55900	20
41	.54000	.84167	.64158	.5587	.1881	.8519	.071146	.55829	19
42	.54024	.84151	.64199	.5577	.1883	.8510	.071266	.55759	18
43	.54049	.84135	.64240	.5567	.1886	.8502	.071386	.55688	17
44	.54073	.84120	.64281	.5557	.1888	.8494	.071506	.55618	16
45	0.54097	0.84104	0.64322	1.5547	1.1890	1.8485	.071626	.55547	15
46	.54122	.84088	.64363	.5537	.1892	.8477	.071747	.55477	14
47	.54146	.84072	.64404	.5527	.1895	.8468	.071867	.55407	13
48	.54171	.84057	.64446	.5517	.1897	.8460	.071988	.55337	12
49	.54195	.84041	.64487	.5507	.1899	.8452	.072109	.55267	11
50	0.54220	0.84025	0.64528	1.5497	1.1901	1.8443	.072230	.55197	10
51	.54244	.84009	.64569	.5487	.1903	.8435	.072351	.55127	9
52	.54269	.83994	.64610	.5477	.1906	.8427	.072473	.55057	8
53	.54293	.83978	.64652	.5468	.1908	.8419	.072594	.54988	7
54	.54317	.83962	.64693	.5458	.1910	.8410	.072716	.54918	6
55	0.54342	0.83946	0.64734	1.5448	1.1912	1.8402	.072838	.54849	5
56	.54366	.83930	.64775	.5438	.1915	.8394	.072959	.54779	4
57	.54391	.83915	.64817	.5428	.1917	.8385	.073082	.54710	3
58	.54415	.83899	.64858	.5418	.1919	.8377	.073204	.54641	2
59	.54440	.83883	.64899	.5408	.1921	.8369	.073326	.54572	1
60	0.54464	0.83867	0.64941	1.5399	1.1924	1.8361	.073449	.54503	0
M	Cosine	Sine	Cotan.	Tan.	Cosec.	Secant	READ DOWN	57°–58° Involute	M

33° or 213°   Trigonometric and Involute Functions   146° or 326°

M	Sine	Cosine	Tan.	Cotan.	Secant	Cosec.	Involute 33°-34°	READ UP	M
0	0.54464	0.83867	0.64941	1.5399	1.1924	1.8361	.073449	.54503	60
1	.54488	.83851	.64982	.5389	.1926	.8353	.073572	.54434	59
2	.54513	.83835	.65024	.5379	.1928	.8344	.073695	.54365	58
3	.54537	.83819	.65065	.5369	.1930	.8336	.073818	.54296	57
4	.54561	.83804	.65106	.5359	.1933	.8328	.073941	.54228	56
5	0.54586	0.83788	0.65148	1.5350	1.1935	1.8320	.074064	.54159	55
6	.54610	.83772	.65189	.5340	.1937	.8312	.074188	.54090	54
7	.54635	.83756	.65231	.5330	.1939	.8303	.074312	.54022	53
8	.54659	.83740	.65272	.5320	.1942	.8295	.074435	.53954	52
9	.54683	.83724	.65314	.5311	.1944	.8287	.074559	.53885	51
10	0.54708	0.83708	0.65355	1.5301	1.1946	1.8279	.074684	.53817	50
11	.54732	.83692	.65397	.5291	.1949	.8271	.074808	.53749	49
12	.54756	.83676	.65438	.5282	.1951	.8263	.074932	.53681	48
13	.54781	.83660	.65480	.5272	.1953	.8255	.075057	.53613	47
14	.54805	.83645	.65521	.5262	.1955	.8247	.075182	.53546	46
15	0.54829	0.83629	0.65563	1.5253	1.1958	1.8238	.075307	.53478	45
16	.54854	.83613	.65604	.5243	.1960	.8230	.075432	.53410	44
17	.54878	.83597	.65646	.5233	.1962	.8222	.075557	.53343	43
18	.54902	.83581	.65688	.5224	.1964	.8214	.075683	.53275	42
19	.54927	.83565	.65729	.5214	.1967	.8206	.075808	.53208	41
20	0.54951	0.83549	0.65771	1.5204	1.1969	1.8198	.075934	.53141	40
21	.54975	.83533	.65813	.5195	.1971	.8190	.076060	.53073	39
22	.54999	.83517	.65854	.5185	.1974	.8182	.076186	.53006	38
23	.55024	.83501	.65896	.5175	.1976	.8174	.076312	.52939	37
24	.55048	.83485	.65938	.5166	.1978	.8166	.076439	.52872	36
25	0.55072	0.83469	0.65980	1.5156	1.1981	1.8158	.076565	.52805	35
26	.55097	.83453	.66021	.5147	.1983	.8150	.076692	.52739	34
27	.55121	.83437	.66063	.5137	.1985	.8142	.076819	.52672	33
28	.55145	.83421	.66105	.5127	.1987	.8134	.076946	.52605	32
29	.55169	.83405	.66147	.5118	.1990	.8126	.077073	.52539	31
30	0.55194	0.83389	0.66189	1.5108	1.1992	1.8118	.077200	.52472	30
31	.55218	.83373	.66230	.5099	.1994	.8110	.077328	.52406	29
32	.55242	.83356	.66272	.5089	.1997	.8102	.077455	.52340	28
33	.55266	.83340	.66314	.5080	.1999	.8094	.077583	.52274	27
34	.55291	.83324	.66356	.5070	.2001	.8086	.077711	.52207	26
35	0.55315	0.83308	0.66398	1.5061	1.2004	1.8078	.077839	.52141	25
36	.55339	.83292	.66440	.5051	.2006	.8070	.077968	.52076	24
37	.55363	.83276	.66482	.5042	.2008	.8062	.078096	.52010	23
38	.55388	.83260	.66524	.5032	.2011	.8055	.078225	.51944	22
39	.55412	.83244	.66566	.5023	.2013	.8047	.078354	.51878	21
40	0.55436	0.83228	0.66608	1.5013	1.2015	1.8039	.078483	.51813	20
41	.55460	.83212	.66650	.5004	.2018	.8031	.078612	.51747	19
42	.55484	.83195	.66692	.4994	.2020	.8023	.078741	.51682	18
43	.55509	.83179	.66734	.4985	.2022	.8015	.078871	.51616	17
44	.55533	.83163	.66776	.4975	.2025	.8007	.079000	.51551	16
45	0.55557	0.83147	0.66818	1.4966	1.2027	1.8000	.079130	.51486	15
46	.55581	.83131	.66860	.4957	.2029	.7992	.079260	.51421	14
47	.55605	.83115	.66902	.4947	.2032	.7984	.079390	.51356	13
48	.55630	.83098	.66944	.4938	.2034	.7976	.079520	.51291	12
49	.55654	.83082	.66986	.4928	.2036	.7968	.079651	.51226	11
50	0.55678	0.83066	0.67028	1.4919	1.2039	1.7960	.079781	.51161	10
51	.55702	.83050	.67071	.4910	.2041	.7953	.079912	.51096	9
52	.55726	.83034	.67113	.4900	.2043	.7945	.080043	.51032	8
53	.55750	.83017	.67155	.4891	.2046	.7937	.080174	.50967	7
54	.55775	.83001	.67197	.4882	.2048	.7929	.080305	.50903	6
55	0.55799	0.82985	0.67239	1.4872	1.2050	1.7922	.080437	.50838	5
56	.55823	.82969	.67282	.4863	.2053	.7914	.080569	.50774	4
57	.55847	.82953	.67324	.4854	.2055	.7906	.080700	.50710	3
58	.55871	.82936	.67366	.4844	.2057	.7898	.080832	.50646	2
59	.55895	.82920	.67409	.4835	.2060	.7891	.080964	.50582	1
60	0.55919	0.82904	0.67451	1.4826	1.2062	1.7883	.081097	.50518	0
M	Cosine	Sine	Cotan.	Tan.	Cosec.	Secant	READ DOWN	56°-57° Involute	M

34° or 214° **Trigonometric and Involute Functions** **145° or 325°**

M	Sine	Cosine	Tan.	Cotan.	Secant	Cosec.	Involute 34°–35°	READ UP	M
0	0.55919	0.82904	0.67451	1.4826	1.2062	1.7883	.081097	.50518	60
1	.55943	.82887	.67493	.4816	.2065	.7875	.081229	.50454	59
2	.55968	.82871	.67536	.4807	.2067	.7868	.081362	.50390	58
3	.55992	.82855	.67578	.4798	.2069	.7860	.081494	.50326	57
4	.56016	.82839	.67620	.4788	.2072	.7852	.081627	.50263	56
5	0.56040	0.82822	0.67663	1.4779	1.2074	1.7844	.081760	.50199	55
6	.56064	.82806	.67705	.4770	.2076	.7837	.081894	.50135	54
7	.56088	.82790	.67748	.4761	.2079	.7829	.082027	.50072	53
8	.56112	.82773	.67790	.4751	.2081	.7821	.082161	.50009	52
9	.56136	.82757	.67832	.4742	.2084	.7814	.082294	.49945	51
10	0.56160	0.82741	0.67875	1.4733	1.2086	1.7806	.082428	.49882	50
11	.56184	.82724	.67917	.4724	.2088	.7799	.082562	.49819	49
12	.56208	.82708	.67960	.4715	.2091	.7791	.082697	.49756	48
13	.56232	.82692	.68002	.4705	.2093	.7783	.082831	.49693	47
14	.56256	.82675	.68045	.4696	.2096	.7776	.082966	.49630	46
15	0.56280	0.82659	0.68088	1.4687	1.2098	1.7768	.083100	.49568	45
16	.56305	.82643	.68130	.4678	.2100	.7761	.083235	.49505	44
17	.56329	.82626	.68173	.4669	.2103	.7753	.083371	.49442	43
18	.56353	.82610	.68215	.4659	.2105	.7745	.083506	.49380	42
19	.56377	.82593	.68258	.4650	.2108	.7738	.083641	.49317	41
20	0.56401	0.82577	0.68301	1.4641	1.2110	1.7730	.083777	.49255	40
21	.56425	.82561	.68343	.4632	.2112	.7723	.083913	.49192	39
22	.56449	.82544	.68386	.4623	.2115	.7715	.084049	.49130	38
23	.56473	.82528	.68429	.4614	.2117	.7708	.084185	.49068	37
24	.56497	.82511	.68471	.4605	.2120	.7700	.084321	.49006	36
25	0.56521	0.82495	0.68514	1.4596	1.2122	1.7693	.084457	.48944	35
26	.56545	.82478	.68557	.4586	.2124	.7685	.084594	.48882	34
27	.56569	.82462	.68600	.4577	.2127	.7678	.084731	.48820	33
28	.56593	.82446	.68642	.4568	.2129	.7670	.084868	.48758	32
29	.56617	.82429	.68685	.4559	.2132	.7663	.085005	.48697	31
30	0.56641	0.82413	0.68728	1.4550	1.2134	1.7655	.085142	.48635	30
31	.56665	.82396	.68771	.4541	.2136	.7648	.085280	.48574	29
32	.56689	.82380	.68814	.4532	.2139	.7640	.085418	.48512	28
33	.56713	.82363	.68857	.4523	.2141	.7633	.085555	.48451	27
34	.56736	.82347	.68900	.4514	.2144	.7625	.085693	.48389	26
35	0.56760	0.82330	0.68942	1.4505	1.2146	1.7618	.085832	.48328	25
36	.56784	.82314	.68985	.4496	.2149	.7610	.085970	.48267	24
37	.56808	.82297	.69028	.4487	.2151	.7603	.086108	.48206	23
38	.56832	.82281	.69071	.4478	.2154	.7596	.086247	.48145	22
39	.56856	.82264	.69114	.4469	.2156	.7588	.086386	.48084	21
40	0.56880	0.82248	0.69157	1.4460	1.2158	1.7581	.086525	.48023	20
41	.56904	.82231	.69200	.4451	.2161	.7573	.086664	.47962	19
42	.56928	.82214	.69243	.4442	.2163	.7566	.086804	.47902	18
43	.56952	.82198	.69286	.4433	.2166	.7559	.086943	.47841	17
44	.56976	.82181	.69329	.4424	.2168	.7551	.087083	.47780	16
45	0.57000	0.82165	0.69372	1.4415	1.2171	1.7544	.087223	.47720	15
46	.57024	.82148	.69416	.4406	.2173	.7537	.087363	.47660	14
47	.57047	.82132	.69459	.4397	.2176	.7529	.087503	.47599	13
48	.57071	.82115	.69502	.4388	.2178	.7522	.087644	.47539	12
49	.57095	.82098	.69545	.4379	.2181	.7515	.087784	.47479	11
50	0.57119	0.82082	0.69588	1.4370	1.2183	1.7507	.087925	.47419	10
51	.57143	.82065	.69631	.4361	.2185	.7500	.088066	.47359	9
52	.57167	.82048	.69675	.4352	.2188	.7493	.088207	.47299	8
53	.57191	.82032	.69718	.4344	.2190	.7485	.088348	.47239	7
54	.57215	.82015	.69761	.4335	.2193	.7478	.088490	.47179	6
55	0.57238	0.81999	0.69804	1.4326	1.2195	1.7471	.088631	.47119	5
56	.57262	.81982	.69847	.4317	.2198	.7463	.088773	.47060	4
57	.57286	.81965	.69891	.4308	.2200	.7456	.088915	.47000	3
58	.57310	.81949	.69934	.4299	.2203	.7449	.089057	.46940	2
59	.57334	.81932	.69977	.4290	.2205	.7442	.089200	.46881	1
60	0.57358	0.81915	0.70021	1.4281	1.2208	1.7434	.089342	.46822	0
M	Cosine	Sine	Cotan.	Tan.	Cosec.	Secant	READ DOWN	55°–56° Involute	M

124° or 304° **55° or 235°**

35° or 215°   Trigonometric and Involute Functions   144° or 324°

M	Sine	Cosine	Tan.	Cotan.	Secant	Cosec.	Involute 35°–36°	READ UP	M
0	0.57358	0.81915	0.70021	1.4281	1.2208	1.7434	.089342	.46822	60
1	.57381	.81899	.70064	.4273	.2210	.7427	.089485	.46762	59
2	.57405	.81882	.70107	.4264	.2213	.7420	.089628	.46703	58
3	.57429	.81865	.70151	.4255	.2215	.7413	.089771	.46644	57
4	.57453	.81848	.70194	.4246	.2218	.7406	.089914	.46585	56
5	0.57477	0.81832	0.70238	1.4237	1.2220	1.7398	.090058	.46526	55
6	.57501	.81815	.70281	.4229	.2223	.7391	.090201	.46467	54
7	.57524	.81798	.70325	.4220	.2225	.7384	.090345	.46408	53
8	.57548	.81782	.70368	.4211	.2228	.7377	.090489	.46349	52
9	.57572	.81765	.70412	.4202	.2230	.7370	.090633	.46291	51
10	0.57596	0.81748	0.70455	1.4193	1.2233	1.7362	.090777	.46232	50
11	.57619	.81731	.70499	.4185	.2235	.7355	.090922	.46173	49
12	.57643	.81714	.70542	.4176	.2238	.7348	.091067	.46115	48
13	.57667	.81698	.70586	.4167	.2240	.7341	.091211	.46057	47
14	.57691	.81681	.70629	.4158	.2243	.7334	.091356	.45998	46
15	0.57715	0.81664	0.70673	1.4150	1.2245	1.7327	.091502	.45940	45
16	.57738	.81647	.70717	.4141	.2248	.7320	.091647	.45882	44
17	.57762	.81631	.70760	.4132	.2250	.7312	.091793	.45824	43
18	.57786	.81614	.70804	.4124	.2253	.7305	.091938	.45766	42
19	.57810	.81597	.70848	.4115	.2255	.7298	.092084	.45708	41
20	0.57833	0.81580	0.70891	1.4106	1.2258	1.7291	.092230	.45650	40
21	.57857	.81563	.70935	.4097	.2260	.7284	.092377	.45592	39
22	.57881	.81546	.70979	.4089	.2263	.7277	.092523	.45534	38
23	.57904	.81530	.71023	.4080	.2265	.7270	.092670	.45476	37
24	.57928	.81513	.71066	.4071	.2268	.7263	.092816	.45419	36
25	0.57952	0.81496	0.71110	1.4063	1.2271	1.7256	.092963	.45361	35
26	.57976	.81479	.71154	.4054	.2273	.7249	.093111	.45304	34
27	.57999	.81462	.71198	.4045	.2276	.7242	.093258	.45246	33
28	.58023	.81445	.71242	.4037	.2278	.7235	.093406	.45189	32
29	.58047	.81428	.71285	.4028	.2281	.7228	.093553	.45132	31
30	0.58070	0.81412	0.71329	1.4019	1.2283	1.7221	.093701	.45074	30
31	.58094	.81395	.71373	.4011	.2286	.7213	.093849	.45017	29
32	.58118	.81378	.71417	.4002	.2288	.7206	.093998	.44960	28
33	.58141	.81361	.71461	.3994	.2291	.7199	.094146	.44903	27
34	.58165	.81344	.71505	.3985	.2293	.7192	.094295	.44846	26
35	0.58189	0.81327	0.71549	1.3976	1.2296	1.7185	.094443	.44789	25
36	.58212	.81310	.71593	.3968	.2299	.7179	.094592	.44733	24
37	.58236	.81293	.71637	.3959	.2301	.7172	.094742	.44676	23
38	.58260	.81276	.71681	.3951	.2304	.7165	.094891	.44619	22
39	.58283	.81259	.71725	.3942	.2306	.7158	.095041	.44563	21
40	0.58307	0.81242	0.71769	1.3934	1.2309	1.7151	.095190	.44506	20
41	.58330	.81225	.71813	.3925	.2311	.7144	.095340	.44450	19
42	.58354	.81208	.71857	.3916	.2314	.7137	.095490	.44393	18
43	.58378	.81191	.71901	.3908	.2317	.7130	.095641	.44337	17
44	.58401	.81174	.71946	.3899	.2319	.7123	.095791	.44281	16
45	0.58425	0.81157	0.71990	1.3891	1.2322	1.7116	.095942	.44225	15
46	.58449	.81140	.72034	.3882	.2324	.7109	.096093	.44169	14
47	.58472	.81123	.72078	.3874	.2327	.7102	.096244	.44113	13
48	.58496	.81106	.72122	.3865	.2329	.7095	.096395	.44057	12
49	.58519	.81089	.72167	.3857	.2332	.7088	.096546	.44001	11
50	0.58543	0.81072	0.72211	1.3848	1.2335	1.7081	.096698	.43945	10
51	.58567	.81055	.72255	.3840	.2337	.7075	.096850	.43889	9
52	.58590	.81038	.72299	.3831	.2340	.7068	.097002	.43833	8
53	.58614	.81021	.72344	.3823	.2342	.7061	.097154	.43778	7
54	.58637	.81004	.72388	.3814	.2345	.7054	.097306	.43722	6
55	0.58661	0.80987	0.72432	1.3806	1.2348	1.7047	.097459	.43667	5
56	.58684	.80970	.72477	.3798	.2350	.7040	.097611	.43611	4
57	.58708	.80953	.72521	.3789	.2353	.7033	.097764	.43556	3
58	.58731	.80936	.72565	.3781	.2355	.7027	.097917	.43501	2
59	.58755	.80919	.72610	.3772	.2358	.7020	.098071	.43446	1
60	0.58779	0.80902	0.72654	1.3764	1.2361	1.7013	.098224	.43390	0

| M | Cosine | Sine | Cotan. | Tan. | Cosec. | Secant | READ DOWN | 54°–55° Involute | M |

36° or 216°   **Trigonometric and Involute Functions**   **143° or 323°**

M	Sine	Cosine	Tan.	Cotan.	Secant	Cosec.	Involute 36°–37°	READ UP	M
0	0.58779	0.80902	0.72654	1.3764	1.2361	1.7013	.098224	.43390	60
1	.58802	.80885	.72699	.3755	.2363	.7006	.098378	.43335	59
2	.58826	.80867	.72743	.3747	.2366	.6999	.098532	.43280	58
3	.58849	.80850	.72788	.3739	.2369	.6993	.098686	.43225	57
4	.58873	.80833	.72832	.3730	.2371	.6986	.098840	.43171	56
5	0.58896	0.80816	0.72877	1.3722	1.2374	1.6979	.098994	.43116	55
6	.58920	.80799	.72921	.3713	.2376	.6972	.099149	.43061	54
7	.58943	.80782	.72966	.3705	.2379	.6966	.099304	.43006	53
8	.58967	.80765	.73010	.3697	.2382	.6959	.099459	.42952	52
9	.58990	.80748	.73055	.3688	.2384	.6952	.099614	.42897	51
10	0.59014	0.80730	0.73100	1.3680	1.2387	1.6945	.099769	.42843	50
11	.59037	.80713	.73144	.3672	.2390	.6939	.099925	.42788	49
12	.59061	.80696	.73189	.3663	.2392	.6932	.10008	.42734	48
13	.59084	.80679	.73234	.3655	.2395	.6925	.10024	.42680	47
14	.59108	.80662	.73278	.3647	.2397	.6918	.10039	.42625	46
15	0.59131	0.80644	0.73323	1.3638	1.2400	1.6912	.10055	.42571	45
16	.59154	.80627	.73368	.3630	.2403	.6905	.10070	.42517	44
17	.59178	.80610	.73413	.3622	.2405	.6898	.10086	.42463	43
18	.59201	.80593	.73457	.3613	.2408	.6892	.10102	.42409	42
19	.59225	.80576	.73502	.3605	.2411	.6885	.10118	.42355	41
20	0.59248	0.80558	0.73547	1.3597	1.2413	1.6878	.10133	.42302	40
21	.59272	.80541	.73592	.3588	.2416	.6871	.10149	.42248	39
22	.59295	.80524	.73637	.3580	.2419	.6865	.10165	.42194	38
23	.59318	.80507	.73681	.3572	.2421	.6858	.10181	.42141	37
24	.59342	.80489	.73726	.3564	.2424	.6852	.10196	.42087	36
25	0.59365	0.80472	0.73771	1.3555	1.2427	1.6845	.10212	.42034	35
26	.59389	.80455	.73816	.3547	.2429	.6838	.10228	.41980	34
27	.59412	.80438	.73861	.3539	.2432	.6832	.10244	.41927	33
28	.59436	.80420	.73906	.3531	.2435	.6825	.10260	.41874	32
29	.59459	.80403	.73951	.3522	.2437	.6818	.10276	.41820	31
30	0.59482	0.80386	0.73996	1.3514	1.2440	1.6812	.10292	.41767	30
31	.59506	.80368	.74041	.3506	.2443	.6805	.10308	.41714	29
32	.59529	.80351	.74086	.3498	.2445	.6799	.10323	.41661	28
33	.59552	.80334	.74131	.3490	.2448	.6792	.10339	.41608	27
34	.59576	.80316	.74176	.3481	.2451	.6785	.10355	.41555	26
35	0.59599	0.80299	0.74221	1.3473	1.2453	1.6779	.10371	.41502	25
36	.59622	.80282	.74267	.3465	.2456	.6772	.10388	.41450	24
37	.59646	.80264	.74312	.3457	.2459	.6766	.10404	.41397	23
38	.59669	.80247	.74357	.3449	.2462	.6759	.10420	.41344	22
39	.59693	.80230	.74402	.3440	.2464	.6753	.10436	.41292	21
40	0.59716	0.80212	0.74447	1.3432	1.2467	1.6746	.10452	.41239	20
41	.59739	.80195	.74492	.3424	.2470	.6739	.10468	.41187	19
42	.59763	.80178	.74538	.3416	.2472	.6733	.10484	.41134	18
43	.59786	.80160	.74583	.3408	.2475	.6726	.10500	.41082	17
44	.59809	.80143	.74628	.3400	.2478	.6720	.10516	.41030	16
45	0.59832	0.80125	0.74674	1.3392	1.2480	1.6713	.10533	.40977	15
46	.59856	.80108	.74719	.3384	.2483	.6707	.10549	.40925	14
47	.59879	.80091	.74764	.3375	.2486	.6700	.10565	.40873	13
48	.59902	.80073	.74810	.3367	.2489	.6694	.10581	.40821	12
49	.59926	.80056	.74855	.3359	.2491	.6687	.10598	.40769	11
50	0.59949	0.80038	0.74900	1.3351	1.2494	1.6681	.10614	.40717	10
51	.59972	.80021	.74946	.3343	.2497	.6674	.10630	.40666	9
52	.59995	.80003	.74991	.3335	.2499	.6668	.10647	.40614	8
53	.60019	.79986	.75037	.3327	.2502	.6661	.10663	.40562	7
54	.60042	.79968	.75082	.3319	.2505	.6655	.10679	.40511	6
55	0.60065	0.79951	0.75128	1.3311	1.2508	1.6649	.10696	.40459	5
56	.60089	.79934	.75173	.3303	.2510	.6642	.10712	.40407	4
57	.60112	.79916	.75219	.3295	.2513	.6636	.10729	.40356	3
58	.60135	.79899	.75264	.3287	.2516	.6629	.10745	.40305	2
59	.60158	.79881	.75310	.3278	.2519	.6623	.10762	.40253	1
60	0.60182	0.79864	0.75355	1.3270	1.2521	1.6616	.10778	.40202	0
M	Cosine	Sine	Cotan.	Tan.	Cosec.	Secant	READ DOWN	53°–54° Involute	M

Trigonometric and Involute Functions

M	Sine	Cosine	Tan.	Cotan.	Secant	Cosec.	Involute 37°–38°	READ UP	M
0	0.60182	0.79864	0.75355	1.3270	1.2521	1.6616	.10778	.40202	60
1	.60205	.79846	.75401	.3262	.2524	.6610	.10795	.40151	59
2	.60228	.79829	.75447	.3254	.2527	.6604	.10811	.40100	58
3	.60251	.79811	.75492	.3246	.2530	.6597	.10828	.40049	57
4	.60274	.79793	.75538	.3238	.2532	.6591	.10844	.39998	56
5	0.60298	0.79776	0.75584	1.3230	1.2535	1.6584	.10861	.39947	55
6	.60321	.79758	.75629	.3222	.2538	.6578	.10878	.39896	54
7	.60344	.79741	.75675	.3214	.2541	.6572	.10894	.39845	53
8	.60367	.79723	.75721	.3206	.2543	.6565	.10911	.39794	52
9	.60390	.79706	.75767	.3198	.2546	.6559	.10928	.39743	51
10	0.60414	0.79688	0.75812	1.3190	1.2549	1.6553	.10944	.39693	50
11	.60437	.79671	.75858	.3182	.2552	.6546	.10961	.39642	49
12	.60460	.79653	.75904	.3175	.2554	.6540	.10978	.39592	48
13	.60483	.79635	.75950	.3167	.2557	.6534	.10995	.39541	47
14	.60506	.79618	.75996	.3159	.2560	.6527	.11011	.39491	46
15	0.60529	0.79600	0.76042	1.3151	1.2563	1.6521	.11028	.39441	45
16	.60553	.79583	.76088	.3143	.2566	.6515	.11045	.39390	44
17	.60576	.79565	.76134	.3135	.2568	.6508	.11062	.39340	43
18	.60599	.79547	.76180	.3127	.2571	.6502	.11079	.39290	42
19	.60622	.79530	.76226	.3119	.2574	.6496	.11096	.39240	41
20	0.60645	0.79512	0.76272	1.3111	1.2577	1.6489	.11113	.39190	40
21	.60668	.79494	.76318	.3103	.2579	.6483	.11130	.39140	39
22	.60691	.79477	.76364	.3095	.2582	.6477	.11146	.39090	38
23	.60714	.79459	.76410	.3087	.2585	.6471	.11163	.39040	37
24	.60738	.79441	.76456	.3079	.2588	.6464	.11180	.38990	36
25	0.60761	0.79424	0.76502	1.3072	1.2591	1.6458	.11197	.38941	35
26	.60784	.79406	.76548	.3064	.2593	.6452	.11215	.38891	34
27	.60807	.79388	.76594	.3056	.2596	.6446	.11232	.38841	33
28	.60830	.79371	.76640	.3048	.2599	.6439	.11249	.38792	32
29	.60853	.79353	.76686	.3040	.2602	.6433	.11266	.38742	31
30	0.60876	0.79335	0.76733	1.3032	1.2605	1.6427	.11283	.38693	30
31	.60899	.79318	.76779	.3024	.2608	.6421	.11300	.38643	29
32	.60922	.79300	.76825	.3017	.2610	.6414	.11317	.38594	28
33	.60945	.79282	.76871	.3009	.2613	.6408	.11334	.38545	27
34	.60968	.79264	.76918	.3001	.2616	.6402	.11352	.38496	26
35	0.60991	0.79247	0.76964	1.2993	1.2619	1.6396	.11369	.38446	25
36	.61015	.79229	.77010	.2985	.2622	.6390	.11386	.38397	24
37	.61038	.79211	.77057	.2977	.2624	.6383	.11403	.38348	23
38	.61061	.79193	.77103	.2970	.2627	.6377	.11421	.38299	22
39	.61084	.79176	.77149	.2962	.2630	.6371	.11438	.38251	21
40	0.61107	0.79158	0.77196	1.2954	1.2633	1.6365	.11455	.38202	20
41	.61130	.79140	.77242	.2946	.2636	.6359	.11473	.38153	19
42	.61153	.79122	.77289	.2938	.2639	.6353	.11490	.38104	18
43	.61176	.79105	.77335	.2931	.2641	.6346	.11507	.38055	17
44	.61199	.79087	.77382	.2923	.2644	.6340	.11525	.38007	16
45	0.61222	0.79069	0.77428	1.2915	1.2647	1.6334	.11542	.37958	15
46	.61245	.79051	.77475	.2907	.2650	.6328	.11560	.37910	14
47	.61268	.79033	.77521	.2900	.2653	.6322	.11577	.37861	13
48	.61291	.79016	.77568	.2892	.2656	.6316	.11595	.37813	12
49	.61314	.78998	.77615	.2884	.2659	.6310	.11612	.37765	11
50	0.61337	0.78980	0.77661	1.2876	1.2661	1.6303	.11630	.37716	10
51	.61360	.78962	.77708	.2869	.2664	.6297	.11647	.37668	9
52	.61383	.78944	.77754	.2861	.2667	.6291	.11665	.37620	8
53	.61406	.78926	.77801	.2853	.2670	.6285	.11682	.37572	7
54	.61429	.78908	.77848	.2846	.2673	.6279	.11700	.37524	6
55	0.61451	0.78891	0.77895	1.2838	1.2676	1.6273	.11718	.37476	5
56	.61474	.78873	.77941	.2830	.2679	.6267	.11735	.37428	4
57	.61497	.78855	.77988	.2822	.2682	.6261	.11753	.37380	3
58	.61520	.78837	.78035	.2815	.2684	.6255	.11771	.37332	2
59	.61543	.78819	.78082	.2807	.2687	.6249	.11788	.37285	1
60	0.61566	0.78801	0.78129	1.2799	1.2690	1.6243	.11806	.37237	0
M	Cosine	Sine	Cotan.	Tan.	Cosec.	Secant	READ DOWN	52°–53°	M
								Involute	

38° or 218°   Trigonometric and Involute Functions   **141° or 321°**

M	Sine	Cosine	Tan.	Cotan.	Secant	Cosec.	Involute 38°–39°	READ UP	M
0	0.61566	0.78801	0.78129	1.2799	1.2690	1.6243	.11806	.37237	60
1	.61589	.78783	.78175	.2792	.2693	.6237	.11824	.37189	59
2	.61612	.78765	.78222	.2784	.2696	.6231	.11842	.37142	58
3	.61635	.78747	.78269	.2776	.2699	.6225	.11859	.37094	57
4	.61658	.78729	.78316	.2769	.2702	.6219	.11877	.37047	56
5	0.61681	.78711	0.78363	1.2761	1.2705	1.6213	.11895	.36999	55
6	.61704	.78694	.78410	.2753	.2708	.6207	.11913	.36952	54
7	.61726	.78676	.78457	.2746	.2710	.6201	.11931	.36905	53
8	.61749	.78658	.78504	.2738	.2713	.6195	.11949	.36858	52
9	.61772	.78640	.78551	.2731	.2716	.6189	.11967	.36810	51
10	0.61795	0.78622	0.78598	1.2723	1.2719	1.6183	.11985	.36763	50
11	.61818	.78604	.78645	.2715	.2722	.6177	.12003	.36716	49
12	.61841	.78586	.78692	.2708	.2725	.6171	.12021	.36669	48
13	.61864	.78568	.78739	.2700	.2728	.6165	.12039	.36622	47
14	.61887	.78550	.78786	.2693	.2731	.6159	.12057	.36575	46
15	0.61909	0.78532	0.78834	1.2685	1.2734	1.6153	.12075	.36529	45
16	.61932	.78514	.78881	.2677	.2737	.6147	.12093	.36482	44
17	.61955	.78496	.78928	.2670	.2740	.6141	.12111	.36435	43
18	.61978	.78478	.78975	.2662	.2742	.6135	.12129	.36388	42
19	.62001	.78460	.79022	.2655	.2745	.6129	.12147	.36342	41
20	0.62024	0.78442	0.79070	1.2647	1.2748	1.6123	.12165	.36295	40
21	.62046	.78424	.79117	.2640	.2751	.6117	.12184	.36249	39
22	.62069	.78405	.79164	.2632	.2754	.6111	.12202	.36202	38
23	.62092	.78387	.79212	.2624	.2757	.6105	.12220	.36156	37
24	.62115	.78369	.79259	.2617	.2760	.6099	.12238	.36110	36
25	0.62138	0.78351	0.79306	1.2609	1.2763	1.6093	.12257	.36063	35
26	.62160	.78333	.79354	.2602	.2766	.6087	.12275	.36017	34
27	.62183	.78315	.79401	.2594	.2769	.6082	.12293	.35971	33
28	.62206	.78297	.79449	.2587	.2772	.6076	.12312	.35925	32
29	.62229	.78279	.79496	.2579	.2775	.6070	.12330	.35879	31
30	0.62251	0.78261	0.79544	1.2572	1.2778	1.6064	.12348	.35833	30
31	.62274	.78243	.79591	.2564	.2781	.6058	.12367	.35787	29
32	.62297	.78225	.79639	.2557	.2784	.6052	.12385	.35741	28
33	.62320	.78205	.79686	.2549	.2787	.6046	.12404	.35695	27
34	.62342	.78188	.79734	.2542	.2790	.6040	.12422	.35649	26
35	0.62365	0.78170	0.79781	1.2534	1.2793	1.6035	.12441	.35604	25
36	.62388	.78152	.79829	.2527	.2796	.6029	.12459	.35558	24
37	.62411	.78134	.79877	.2519	.2799	.6023	.12478	.35512	23
38	.62433	.78116	.79924	.2512	.2802	.6017	.12496	.35467	22
39	.62456	.78098	.79972	.2504	.2804	.6011	.12515	.35421	21
40	0.62479	0.78079	0.80020	1.2497	1.2807	1.6005	.12534	.35376	20
41	.62502	.78061	.80067	.2489	.2810	.6000	.12552	.35330	19
42	.62524	.78043	.80115	.2482	.2813	.5994	.12571	.35285	18
43	.62547	.78025	.80163	.2475	.2816	.5988	.12590	.35240	17
44	.62570	.78007	.80211	.2467	.2819	.5982	.12608	.35194	16
45	0.62592	0.77988	0.80258	1.2460	1.2822	1.5976	.12627	.35149	15
46	.62615	.77970	.80306	.2452	.2825	.5971	.12646	.35104	14
47	.62638	.77952	.80354	.2445	.2828	.5965	.12664	.35059	13
48	.62660	.77934	.80402	.2437	.2831	.5959	.12683	.35014	12
49	.62683	.77916	.80450	.2430	.2834	.5953	.12702	.34969	11
50	0.62706	0.77897	0.80498	1.2423	1.2837	1.5948	.12721	.34924	10
51	.62728	.77879	.80546	.2415	.2840	.5942	.12740	.34879	9
52	.62751	.77861	.80594	.2408	.2843	.5936	.12759	.34834	8
53	.62774	.77843	.80642	.2401	.2846	.5930	.12778	.34790	7
54	.62796	.77824	.80690	.2393	.2849	.5925	.12797	.34745	6
55	0.62819	0.77806	0.80738	1.2386	1.2852	1.5919	.12815	.34700	5
56	.62842	.77788	.80786	.2378	.2855	.5913	.12834	.34656	4
57	.62864	.77769	.80834	.2371	.2859	.5907	.12853	.34611	3
58	.62887	.77751	.80882	.2364	.2862	.5902	.12872	.34567	2
59	.62909	.77733	.80930	.2356	.2865	.5896	.12891	.34522	1
60	0.62932	0.77715	0.80978	1.2349	1.2868	1.5890	.12911	.34478	0
M	Cosine	Sine	Cotan.	Tan.	Cosec.	Secant	READ DOWN	51°–52° Involute	M

128° or 308°   **51° or 231°**

39° or 219°   **Trigonometric and Involute Functions**   **140° or 320°**

M	Sine	Cosine	Tan.	Cotan.	Secant	Cosec.	Involute 39°–40°	READ UP	M
0	0.62932	0.77715	0.80978	1.2349	1.2868	1.5890	.12911	.34478	60
1	.62955	.77696	.81027	.2342	.2871	.5884	.12930	.34434	59
2	.62977	.77678	.81075	.2334	.2874	.5879	.12949	.34389	58
3	.63000	.77660	.81123	.2327	.2877	.5873	.12968	.34345	57
4	.63022	.77641	.81171	.2320	.2880	.5867	.12987	.34301	56
5	0.63045	0.77623	0.81220	1.2312	1.2883	1.5862	.13006	.34257	55
6	.63068	.77605	.81268	.2305	.2886	.5856	.13025	.34213	54
7	.63090	.77586	.81316	.2298	.2889	.5850	.13045	.34169	53
8	.63113	.77568	.81364	.2290	.2892	.5845	.13064	.34125	52
9	.63135	.77550	.81413	.2283	.2895	.5839	.13083	.34081	51
10	0.63158	0.77531	0.81461	1.2276	1.2898	1.5833	.13102	.34037	50
11	.63180	.77513	.81510	.2268	.2901	.5828	.13122	.33993	49
12	.63203	.77494	.81558	.2261	.2904	.5822	.13141	.33949	48
13	.63225	.77476	.81606	.2254	.2907	.5816	.13160	.33906	47
14	.63248	.77458	.81655	.2247	.2910	.5811	.13180	.33862	46
15	0.63271	0.77439	0.81703	1.2239	1.2913	1.5805	.13199	.33818	45
16	.63293	.77421	.81752	.2232	.2916	.5799	.13219	.33775	44
17	.63316	.77402	.81800	.2225	.2919	.5794	.13238	.33731	43
18	.63338	.77384	.81849	.2218	.2923	.5788	.13258	.33688	42
19	.63361	.77366	.81898	.2210	.2926	.5783	.13277	.33645	41
20	0.63383	0.77347	0.81946	1.2203	1.2929	1.5777	.13297	.33601	40
21	.63406	.77329	.81995	.2196	.2932	.5771	.13316	.33558	39
22	.63428	.77310	.82044	.2189	.2935	.5766	.13336	.33515	38
23	.63451	.77292	.82092	.2181	.2938	.5760	.13355	.33471	37
24	.63473	.77273	.82141	.2174	.2941	.5755	.13375	.33428	36
25	0.63496	0.77255	0.82190	1.2167	1.2944	1.5749	.13395	.33385	35
26	.63518	.77236	.82238	.2160	.2947	.5744	.13414	.33342	34
27	.63541	.77218	.82287	.2153	.2950	.5738	.13434	.33299	33
28	.63563	.77199	.82336	.2145	.2953	.5732	.13454	.33256	32
29	.63585	.77181	.82385	.2138	.2957	.5727	.13473	.33213	31
30	0.63608	0.77162	0.82434	1.2131	1.2960	1.5721	.13493	.33171	30
31	.63630	.77144	.82483	.2124	.2963	.5716	.13513	.33128	29
32	.63653	.77125	.82531	.2117	.2966	.5710	.13533	.33085	28
33	.63675	.77107	.82580	.2109	.2969	.5705	.13553	.33042	27
34	.63698	.77088	.82629	.2102	.2972	.5699	.13572	.33000	26
35	0.63720	0.77070	0.82678	1.2095	1.2975	1.5694	.13592	.32957	25
36	.63742	.77051	.82727	.2088	.2978	.5688	.13612	.32915	24
37	.63765	.77033	.82776	.2081	.2981	.5683	.13632	.32872	23
38	.63787	.77014	.82825	.2074	.2985	.5677	.13652	.32830	22
39	.63810	.76996	.82874	.2066	.2988	.5672	.13672	.32787	21
40	0.63832	0.76977	0.82923	1.2059	1.2991	1.5666	.13692	.32745	20
41	.63854	.76959	.82972	.2052	.2994	.5661	.13712	.32703	19
42	.63877	.76940	.83022	.2045	.2997	.5655	.13732	.32661	18
43	.63899	.76921	.83071	.2038	.3000	.5650	.13752	.32618	17
44	.63922	.76903	.83120	.2031	.3003	.5644	.13772	.32576	16
45	0.63944	0.76884	0.83169	1.2024	1.3007	1.5639	.13792	.32534	15
46	.63966	.76866	.83218	.2017	.3010	.5633	.13812	.32492	14
47	.63989	.76847	.83268	.2009	.3013	.5628	.13833	.32450	13
48	.64011	.76828	.83317	.2002	.3016	.5622	.13853	.32408	12
49	.64033	.76810	.83366	.1995	.3019	.5617	.13873	.32366	11
50	0.64056	0.76791	0.83415	1.1988	1.3022	1.5611	.13893	.32324	10
51	.64078	.76772	.83465	.1981	.3026	.5606	.13913	.32283	9
52	.64100	.76754	.83514	.1974	.3029	.5601	.13934	.32241	8
53	.64123	.76735	.83564	.1967	.3032	.5595	.13954	.32199	7
54	.64145	.76717	.83613	.1960	.3035	.5590	.13974	.32158	6
55	.64167	.76698	.83662	1.1953	1.3038	1.5584	.13995	.32116	5
56	.64190	.76679	.83712	.1946	.3041	.5579	.14015	.32075	4
57	.64212	.76661	.83761	.1939	.3045	.5573	.14035	.32033	3
58	.64234	.76642	.83811	.1932	.3048	.5568	.14056	.31992	2
59	.64256	.76623	.83860	.1925	.3051	.5563	.14076	.31950	1
60	0.64279	0.76604	0.83910	1.1918	1.3054	1.5557	.14097	.31909	0
M	Cosine	Sine	Cotan.	Tan.	Cosec.	Secant	READ DOWN	50°–51° Involute	M

40° or 220° Trigonometric and Involute Functions **139° or 319°**

M	Sine	Cosine	Tan.	Cotan.	Secant	Cosec.	Involute 40°–41°	READ UP	M
0	0.64279	0.76604	0.83910	1.1918	1.3054	1.5557	.14097	.31909	60
1	.64301	.76586	.83960	.1910	.3057	.5552	.14117	.31868	59
2	.64323	.76567	.84009	.1903	.3060	.5546	.14138	.31826	58
3	.64346	.76548	.84059	.1896	.3064	.5541	.14158	.31785	57
4	.64368	.76530	.84108	.1889	.3067	.5536	.14179	.31744	56
5	0.64390	0.76511	0.84158	1.1882	1.3070	1.5530	.14200	.31703	55
6	.64412	.76492	.84208	.1875	.3073	.5525	.14220	.31662	54
7	.64435	.76473	.84258	.1868	.3076	.5520	.14241	.31621	53
8	.64457	.76455	.84307	.1861	.3080	.5514	.14261	.31580	52
9	.64479	.76436	.84357	.1854	.3083	.5509	.14282	.31539	51
10	0.64501	0.76417	0.84407	1.1847	1.3086	1.5504	.14303	.31498	50
11	.64524	.76398	.84457	.1840	.3089	.5498	.14324	.31457	49
12	.64546	.76380	.84507	.1833	.3093	.5493	.14344	.31417	48
13	.64568	.76361	.84556	.1826	.3096	.5488	.14365	.31376	47
14	.64590	.76342	.84606	.1819	.3099	.5482	.14386	.31335	46
15	0.64612	0.76323	0.84656	1.1812	1.3102	1.5477	.14407	.31295	45
16	.64635	.76304	.84706	.1806	.3105	.5472	.14428	.31254	44
17	.64657	.76286	.84756	.1799	.3109	.5466	.14448	.31214	43
18	.64679	.76267	.84806	.1792	.3112	.5461	.14469	.31173	42
19	.64701	.76248	.84856	.1785	.3115	.5456	.14490	.31133	41
20	0.64723	0.76229	0.84906	1.1778	1.3118	1.5450	.14511	.31092	40
21	.64746	.76210	.84956	.1771	.3122	.5445	.14532	.31052	39
22	.64768	.76192	.85006	.1764	.3125	.5440	.14553	.31012	38
23	.64790	.76173	.85057	.1757	.3128	.5435	.14574	.30971	37
24	.64812	.76154	.85107	.1750	.3131	.5429	.14595	.30931	36
25	0.64834	0.76135	0.85157	1.1743	1.3135	1.5424	.14616	.30891	35
26	.64856	.76116	.85207	.1736	.3138	.5419	.14638	.30851	34
27	.64878	.76097	.85257	.1729	.3141	.5413	.14659	.30811	33
28	.64901	.76078	.85308	.1722	.3144	.5408	.14680	.30771	32
29	.64923	.76059	.85358	.1715	.3148	.5403	.14701	.30731	31
30	0.64945	0.76041	0.85408	1.1708	1.3151	1.5398	.14722	.30691	30
31	.64967	.76022	.85458	.1702	.3154	.5392	.14743	.30651	29
32	.64989	.76003	.85509	.1695	.3157	.5387	.14765	.30611	28
33	.65011	.75984	.85559	.1688	.3161	.5382	.14786	.30572	27
34	.65033	.75965	.85609	.1681	.3164	.5377	.14807	.30532	26
35	0.65055	0.75946	0.85660	1.1674	1.3167	1.5372	.14829	.30492	25
36	.65077	.75927	.85710	.1667	.3171	.5366	.14850	.30453	24
37	.65100	.75908	.85761	.1660	.3174	.5361	.14871	.30413	23
38	.65122	.75889	.85811	.1653	.3177	.5356	.14893	.30374	22
39	.65144	.75870	.85862	.1647	.3180	.5351	.14914	.30334	21
40	0.65166	0.75851	0.85912	1.1640	1.3184	1.5345	.14936	.30295	20
41	.65188	.75832	.85963	.1633	.3187	.5340	.14957	.30255	19
42	.65210	.75813	.86014	.1626	.3190	.5335	.14979	.30216	18
43	.65232	.75794	.86064	.1619	.3194	.5330	.15000	.30177	17
44	.65254	.75775	.86115	.1612	.3197	.5325	.15022	.30137	16
45	0.65276	0.75756	0.86166	1.1606	1.3200	1.5320	.15043	.30098	15
46	.65298	.75738	.86216	.1599	.3203	.5314	.15065	.30059	14
47	.65320	.75719	.86267	.1592	.3207	.5309	.15087	.30020	13
48	.65342	.75700	.86318	.1585	.3210	.5304	.15108	.29981	12
49	.65364	.75680	.86368	.1578	.3213	.5299	.15130	.29942	11
50	0.65386	0.76661	0.86419	1.1571	1.3217	1.5294	.15152	.29903	10
51	.65408	.75642	.86470	.1565	.3220	.5289	.15173	.29864	9
52	.65430	.75623	.86521	.1558	.3223	.5283	.15195	.29825	8
53	.65452	.75604	.86572	.1551	.3227	.5278	.15217	.29786	7
54	.65474	.75585	.86623	.1544	.3230	.5273	.15239	.29747	6
55	0.65496	0.75566	0.86674	1.1538	1.3233	1.5268	.15261	.29709	5
56	.65518	.75547	.86725	.1531	.3237	.5263	.15282	.29670	4
57	.65540	.75528	.86776	.1524	.3240	.5258	.15304	.29631	3
58	.65562	.75509	.86827	.1517	.3243	.5253	.15326	.29593	2
59	.65584	.75490	.86878	.1510	.3247	.5248	.15348	.29554	1
60	0.65606	0.75471	0.86929	1.1504	1.3250	1.5243	.15370	.29516	0
M	Cosine	Sine	Cotan.	Tan.	Cosec.	Secant	READ DOWN 49°–50°	Involute	M

41° or 221°   **Trigonometric and Involute Functions**   **138° or 318°**

M	Sine	Cosine	Tan.	Cotan.	Secant	Cosec.	Involute 41°–42°	READ UP	M
0	0.65606	0.75471	0.86929	1.1504	1.3250	1.5243	.15370	.29516	60
1	.65628	.75452	.86980	.1497	.3253	.5237	.15392	.29477	59
2	.65650	.75433	.87031	.1490	.3257	.5232	.15414	.29439	58
3	.65672	.75414	.87082	.1483	.3260	.5227	.15436	.29400	57
4	.65694	.75395	.87133	.1477	.3264	.5222	.15458	.29362	56
5	0.65716	0.75375	0.87184	1.1470	1.3267	1.5217	.15480	.29324	55
6	.65738	.75356	.87236	.1463	.3270	.5212	.15503	.29286	54
7	.65759	.75337	.87287	.1456	.3274	.5207	.15525	.29247	53
8	.65781	.75318	.87338	.1450	.3277	.5202	.15547	.29209	52
9	.65803	.75299	.87389	.1443	.3280	.5197	.15569	.29171	51
10	0.65825	0.75280	0.87441	1.1436	1.3284	1.5192	.15591	.29133	50
11	.65847	.75261	.87492	.1430	.3287	.5187	.15614	.29095	49
12	.65869	.75241	.87543	.1423	.3291	.5182	.15636	.29057	48
13	.65891	.75222	.87595	.1416	.3294	.5177	.15658	.29019	47
14	.65913	.75203	.87646	.1410	.3297	.5172	.15680	.28981	46
15	0.65935	0.75184	0.87698	1.1403	1.3301	1.5167	.15703	.28943	45
16	.65956	.75165	.87749	.1396	.3304	.5162	.15725	.28906	44
17	.65978	.75146	.87801	.1389	.3307	.5156	.15748	.28868	43
18	.66000	.75126	.87852	.1383	.3311	.5151	.15770	.28830	42
19	.66022	.75107	.87904	.1376	.3314	.5146	.15793	.28792	41
20	0.66044	0.75088	0.87955	1.1369	1.3318	1.5141	.15815	.28755	40
21	.66066	.75069	.88007	.1363	.3321	.5136	.15838	.28717	39
22	.66088	.75050	.88059	.1356	.3325	.5131	.15860	.28680	38
23	.66109	.75030	.88110	.1349	.3328	.5126	.15883	.28642	37
24	.66131	.75011	.88162	.1343	.3331	.5121	.15905	.28605	36
25	0.66153	0.74992	0.88214	1.1336	1.3335	1.5116	.15928	.28567	35
26	.66175	.74973	.88265	.1329	.3338	.5111	.15950	.28530	34
27	.66197	.74953	.88317	.1323	.3342	.5107	.15973	.28493	33
28	.66218	.74934	.88369	.1316	.3345	.5102	.15996	.28455	32
29	.66240	.74915	.88421	.1310	.3348	.5097	.16019	.28418	31
30	0.66262	0.74896	0.88473	1.1303	1.3352	1.5092	.16041	.28381	30
31	.66284	.74876	.88524	.1296	.3355	.5087	.16064	.28344	29
32	.66306	.74857	.88576	.1290	.3359	.5082	.16087	.28307	28
33	.66327	.74838	.88628	.1283	.3362	.5077	.16110	.28270	27
34	.66349	.74818	.88680	.1276	.3366	.5072	.16133	.28233	26
35	0.66371	0.74799	0.88732	1.1270	1.3369	1.5067	.16156	.28196	25
36	.66393	.74780	.88784	.1263	.3373	.5062	.16178	.28159	24
37	.66414	.74760	.88836	.1257	.3376	.5057	.16201	.28122	23
38	.66436	.74741	.88888	.1250	.3380	.5052	.16224	.28085	22
39	.66458	.74722	.88940	.1243	.3383	.5047	.16247	.28048	21
40	0.66480	0.74703	0.88992	1.1237	1.3386	1.5042	.16270	.28012	20
41	.66501	.74683	.89045	.1230	.3390	.5037	.16293	.27975	19
42	.66523	.74664	.89097	.1224	.3393	.5032	.16317	.27938	18
43	.66545	.74644	.89149	.1217	.3397	.5027	.16340	.27902	17
44	.66566	.74625	.89201	.1211	.3400	.5023	.16363	.27865	16
45	0.66588	0.74606	0.89253	1.1204	1.3404	1.5018	.16386	.27828	15
46	.66610	.74586	.89306	.1197	.3407	.5013	.16409	.27792	14
47	.66632	.74567	.89358	.1191	.3411	.5008	.16432	.27755	13
48	.66653	.74548	.89410	.1184	.3414	.5003	.16456	.27719	12
49	.66675	.74528	.89463	.1178	.3418	.4998	.16479	.27683	11
50	0.66697	0.74509	0.89515	1.1171	1.3421	1.4993	.16502	.27646	10
51	.66718	.74489	.89567	.1165	.3425	.4988	.16525	.27610	9
52	.66740	.74470	.89620	.1158	.3428	.4984	.16549	.27574	8
53	.66762	.74451	.89672	.1152	.3432	.4979	.16572	.27538	7
54	.66783	.74431	.89725	.1145	.3435	.4974	.16596	.27501	6
55	0.66805	0.74412	0.89777	1.1139	1.3439	1.4969	.16619	.27465	5
56	.66827	.74392	.89830	.1132	.3442	.4964	.16642	.27429	4
57	.66848	.74373	.89883	.1126	.3446	.4959	.16666	.27393	3
58	.66870	.74353	.89935	.1119	.3449	.4954	.16689	.27357	2
59	.66891	.74334	.89988	.1113	.3453	.4950	.16713	.27321	1
60	0.66913	0.74314	0.90040	1.1106	1.3456	1.4945	.16737	.27285	0
M	Cosine	Sine	Cotan.	Tan.	Cosec.	Secant	READ DOWN 48°–49°	Involute	M

M	Sine	Cosine	Tan.	Cotan.	Secant	Cosec.	Involute 42°-43°	READ UP	M
0	0.66913	0.74314	0.90040	1.1106	1.3456	1.4945	.16737	.27285	60
1	.66935	.74295	.90093	.1100	.3460	.4940	.16760	.27250	59
2	.66956	.74276	.90146	.1093	.3463	.4935	.16784	.27214	58
3	.66978	.74256	.90199	.1087	.3467	.4930	.16807	.27178	57
4	.66999	.74237	.90251	.1080	.3470	.4925	.16831	.27142	56
5	0.67021	0.74217	0.90304	1.1074	1.3474	1.4921	.16855	.27107	55
6	.67043	.74198	.90357	.1067	.3478	.4916	.16879	.27071	54
7	.67064	.74178	.90410	.1061	.3481	.4911	.16902	.27035	53
8	.67086	.74159	.90463	.1054	.3485	.4906	.16926	.27000	52
9	.67107	.74139	.90516	.1048	.3488	.4901	.16950	.26964	51
10	0.67129	0.74120	0.90569	1.1041	1.3492	1.4897	.16974	.26929	50
11	.67151	.74100	.90621	.1035	.3495	.4892	.16998	.26893	49
12	.67172	.74080	.90674	.1028	.3499	.4887	.17022	.26858	48
13	.67194	.74061	.90727	.1022	.3502	.4882	.17045	.26823	47
14	.67215	.74041	.90781	.1016	.3506	.4878	.17069	.26787	46
15	0.67237	0.74022	0.90834	1.1009	1.3510	1.4873	.17093	.26752	45
16	.67258	.74002	.90887	.1003	.3513	.4868	.17117	.26717	44
17	.67280	.73983	.90940	.0996	.3517	.4863	.17142	.26682	43
18	.67301	.73963	.90993	.0990	.3520	.4859	.17166	.26646	42
19	.67323	.73944	.91046	.0983	.3524	.4854	.17190	.26611	41
20	0.67344	0.73924	0.91099	1.0977	1.3527	1.4849	.17214	.26576	40
21	.67366	.73904	.91153	.0971	.3531	.4844	.17238	.26541	39
22	.67387	.73885	.91206	.0964	.3535	.4840	.17262	.26506	38
23	.67409	.73865	.91259	.0958	.3538	.4835	.17286	.26471	37
24	.67430	.73846	.91313	.0951	.3542	.4830	.17311	.26436	36
25	0.67452	0.73826	0.91366	1.0945	1.3545	1.4825	.17335	.26401	35
26	.67473	.73806	.91419	.0939	.3549	.4821	.17359	.26367	34
27	.67495	.73787	.91473	.0932	.3553	.4816	.17383	.26332	33
28	.67516	.73767	.91526	.0926	.3556	.4811	.17408	.26297	32
29	.67538	.73747	.91580	.0919	.3560	.4807	.17432	.26262	31
30	0.67559	0.73728	0.91633	1.0913	1.3563	1.4802	.17457	.26228	30
31	.67580	.73708	.91687	.0907	.3567	.4797	.17481	.26193	29
32	.67602	.73688	.91740	.0900	.3571	.4792	.17506	.26159	28
33	.67623	.73669	.91794	.0894	.3574	.4788	.17530	.26124	27
34	.67645	.73649	.91847	.0888	.3578	.4783	.17555	.26089	26
35	0.67666	0.73629	0.91901	1.0881	1.3582	1.4778	.17579	.26055	25
36	.67688	.73610	.91955	.0875	.3585	.4774	.17604	.26021	24
37	.67709	.73590	.92008	.0869	.3589	.4769	.17628	.25986	23
38	.67730	.73570	.92062	.0862	.3592	.4764	.17653	.25952	22
39	.67752	.73551	.92116	.0856	.3596	.4760	.17678	.25918	21
40	0.67773	0.73531	0.92170	1.0850	1.3600	1.4755	.17702	.25883	20
41	.67795	.73511	.92224	.0843	.3603	.4750	.17727	.25849	19
42	.67816	.73491	.92277	.0837	.3607	.4746	.17752	.25815	18
43	.67837	.73472	.92331	.0831	.3611	.4741	.17777	.25781	17
44	.67859	.73452	.92385	.0824	.3614	.4737	.17801	.25747	16
45	0.67880	0.73432	0.92439	1.0818	1.3618	1.4732	.17826	.25713	15
46	.67901	.73413	.92493	.0812	.3622	.4727	.17851	.25679	14
47	.67923	.73393	.92547	.0805	.3625	.4723	.17876	.25645	13
48	.67944	.73373	.92601	.0799	.3629	.4718	.17901	.25611	12
49	.67965	.73353	.92655	.0793	.3633	.4713	.17926	.25577	11
50	0.67987	0.73333	0.92709	1.0786	1.3636	1.4709	.17951	.25543	10
51	.68008	.73314	.92763	.0780	.3640	.4704	.17976	.25509	9
52	.68029	.73294	.92817	.0774	.3644	.4700	.18001	.25475	8
53	.68051	.73274	.92872	.0768	.3647	.4695	.18026	.25442	7
54	.68072	.73254	.92926	.0761	.3651	.4690	.18051	.25408	6
55	0.68093	0.73234	0.92980	1.0755	1.3655	1.4686	.18076	.25374	5
56	.68115	.73215	.93034	.0749	.3658	.4681	.18101	.25341	4
57	.68136	.73195	.93088	.0742	.3662	.4677	.18127	.25307	3
58	.68157	.73175	.93143	.0736	.3666	.4672	.18152	.25273	2
59	.68179	.73155	.93197	.0730	.3670	.4667	.18177	.25240	1
60	0.68200	0.73135	0.93252	1.0724	1.3673	1.4663	.18202	.25206	0
M	Cosine	Sine	Cotan.	Tan.	Cosec.	Secant	READ DOWN 47°-48°	Involute	M

43° or 223°   Trigonometric and Involute Functions   136° or 316°

M	Sine	Cosine	Tan.	Cotan.	Secant	Cosec.	Involute 43°-44°	READ UP	M
0	0.68200	0.73135	0.93252	1.0724	1.3673	1.4663	.18202	.25206	60
1	.68221	.73116	.93306	.0717	.3677	.4658	.18228	.25173	59
2	.68242	.73096	.93360	.0711	.3681	.4654	.18253	.25140	58
3	.68264	.73076	.93415	.0705	.3684	.4649	.18278	.25106	57
4	.68285	.73056	.93469	.0699	.3688	.4645	.18304	.25073	56
5	0.68306	0.73036	0.93524	1.0692	1.3692	1.4640	.18329	.25040	55
6	.68327	.73016	.93578	.0686	.3696	.4635	.18355	.25006	54
7	.68349	.72996	.93633	.0680	.3699	.4631	.18380	.24973	53
8	.68370	.72976	.93688	.0674	.3703	.4626	.18406	.24940	52
9	.68391	.72957	.93742	.0668	.3707	.4622	.18431	.24907	51
10	0.68412	0.72937	0.93797	1.0661	1.3711	1.4617	.18457	.24874	50
11	.68434	.72917	.93852	.0655	.3714	.4613	.18482	.24841	49
12	.68455	.72897	.93906	.0649	.3718	.4608	.18508	.24808	48
13	.68476	.72877	.93961	.0643	.3722	.4604	.18534	.24775	47
14	.68497	.72857	.94016	.0637	.3726	.4599	.18559	.24742	46
15	0.68518	0.72837	0.94071	1.0630	1.3729	1.4595	.18585	.24709	45
16	.68539	.72817	.94125	.0624	.3733	.4590	.18611	.24676	44
17	.68561	.72797	.94180	.0618	.3737	.4586	.18637	.24643	43
18	.68582	.72777	.94235	.0612	.3741	.4581	.18662	.24611	42
19	.68603	.72757	.94290	.0606	.3744	.4577	.18688	.24578	41
20	0.68624	0.72737	0.94345	1.0599	1.3748	1.4572	.18714	.24545	40
21	.68645	.72717	.94400	.0593	.3752	.4568	.18740	.24512	39
22	.68666	.72697	.94455	.0587	.3756	.4563	.18766	.24480	38
23	.68688	.72677	.94510	.0581	.3759	.4559	.18792	.24447	37
24	.68709	.72657	.94565	.0575	.3763	.4554	.18818	.24415	36
25	0.68730	0.72637	0.94620	1.0569	1.3767	1.4550	.18844	.24382	35
26	.68751	.72617	.94676	.0562	.3771	.4545	.18870	.24350	34
27	.68772	.72597	.94731	.0556	.3775	.4541	.18896	.24317	33
28	.68793	.72577	.94786	.0550	.3778	.4536	.18922	.24285	32
29	.68814	.72557	.94841	.0544	.3782	.4532	.18948	.24253	31
30	0.68835	0.72537	0.94896	1.0538	1.3786	1.4527	.18975	.24220	30
31	.68857	.72517	.94952	.0532	.3790	.4523	.19001	.24188	29
32	.68878	.72497	.95007	.0526	.3794	.4518	.19027	.24156	28
33	.68899	.72477	.95062	.0519	.3797	.4514	.19053	.24123	27
34	.68920	.72457	.95118	.0513	.3801	.4510	.19080	.24091	26
35	0.68941	0.72437	0.95173	1.0507	1.3805	1.4505	.19106	.24059	25
36	.68962	.72417	.95229	.0501	.3809	.4501	.19132	.24027	24
37	.68983	.72397	.95284	.0495	.3813	.4496	.19159	.23995	23
38	.69004	.72377	.95340	.0489	.3817	.4492	.19185	.23963	22
39	.69025	.72357	.95395	.0483	.3820	.4487	.19212	.23931	21
40	0.69046	0.72337	0.95451	1.0477	1.3824	1.4483	.19238	.23899	20
41	.69067	.72317	.95506	.0470	.3828	.4479	.19265	.23867	19
42	.69088	.72297	.95562	.0464	.3832	.4474	.19291	.23835	18
43	.69109	.72277	.95618	.0458	.3836	.4470	.19318	.23803	17
44	.69130	.72257	.95673	.0452	.3840	.4465	.19344	.23772	16
45	0.69151	0.72236	0.95729	1.0446	1.3843	1.4461	.19371	.23740	15
46	.69172	.72216	.95785	.0440	.3847	.4457	.19398	.23708	14
47	.69193	.72196	.95841	.0434	.3851	.4452	.19424	.23676	13
48	.69214	.72176	.95897	.0428	.3855	.4448	.19451	.23645	12
49	.69235	.72156	.95952	.0422	.3859	.4443	.19478	.23613	11
50	0.69256	0.72136	0.96008	1.0416	1.3863	1.4439	.19505	.23582	10
51	.69277	.72116	.96064	.0410	.3867	.4435	.19532	.23550	9
52	.69298	.72095	.96120	.0404	.3871	.4430	.19558	.23519	8
53	.69319	.72075	.96176	.0398	.3874	.4426	.19585	.23487	7
54	.69340	.72055	.96232	.0392	.3878	.4422	.19612	.23456	6
55	0.69361	0.72035	0.96288	1.0385	1.3882	1.4417	.19639	.23424	5
56	.69382	.72015	.96344	.0379	.3886	.4413	.19666	.23393	4
57	.69403	.71995	.96400	.0373	.3890	.4409	.19693	.23362	3
58	.69424	.71974	.96457	.0367	.3894	.4404	.19720	.23330	2
59	.69445	.71954	.96513	.0361	.3898	.4400	.19747	.23299	1
60	.69466	.71934	.96569	1.0355	1.3902	1.4396	.19774	.23268	0
M	Cosine	Sine	Cotan.	Tan.	Cosec.	Secant	READ DOWN	46°–47° Involute	M

133° or 313°

46° or 226°

44° or 224° Trigonometric and Involute Functions **135° or 315°**

M	Sine	Cosine	Tan.	Cotan.	Secant	Cosec.	Involute 44°–45°	READ UP	M
0	0.69466	0.71934	0.96569	1.0355	1.3902	1.4396	.19774	.23268	60
1	.69487	.71914	.96625	.0349	.3906	.4391	.19802	.23237	59
2	.69508	.71894	.96681	.0343	.3909	.4387	.19829	.23206	58
3	.69529	.71873	.96738	.0337	.3913	.4383	.19856	.23174	57
4	.69549	.71853	.96794	.0331	.3917	.4378	.19883	.23143	56
5	.69570	.71833	.96850	1.0325	1.3921	1.4374	.19910	.23112	55
6	.69591	.71813	.96907	.0319	.3925	.4370	.19938	.23081	54
7	.69612	.71792	.96963	.0313	.3929	.4365	.19965	.23050	53
8	.69633	.71772	.97020	.0307	.3933	.4361	.19992	.23020	52
9	.69654	.71752	.97076	.0301	.3937	.4357	.20020	.22989	51
10	0.69675	0.71732	0.97133	1.0295	1.3941	1.4352	.20047	.22958	50
11	.69696	.71711	.97189	.0289	.3945	.4348	.20075	.22927	49
12	.69717	.71691	.97246	.0283	.3949	.4344	.20102	.22896	48
13	.69737	.71671	.97302	.0277	.3953	.4340	.20130	.22865	47
14	.69758	.71650	.97359	.0271	.3957	.4335	.20157	.22835	46
15	0.69779	0.71630	0.97416	1.0265	1.3961	1.4331	.20185	.22804	45
16	.69800	.71610	.97472	.0259	.3965	.4327	.20212	.22773	44
17	.69821	.71590	.97529	.0253	.3969	.4322	.20240	.22743	43
18	.69842	.71569	.97586	.0247	.3972	.4318	.20268	.22712	42
19	.69862	.71549	.97643	.0241	.3976	.4314	.20296	.22682	41
20	0.69883	0.71529	0.97700	1.0235	1.3980	1.4310	.20323	.22651	40
21	.69904	.71508	.97756	.0230	.3984	.4305	.20351	.22621	39
22	.69925	.71488	.97813	.0224	.3988	.4301	.20379	.22590	38
23	.69946	.71468	.97870	.0218	.3992	.4297	.20407	.22560	37
24	.69966	.71447	.97927	.0212	.3996	.4293	.20435	.22530	36
25	0.69987	0.71427	0.97984	1.0206	1.4000	1.4288	.20463	.22499	35
26	.70008	.71407	.98041	.0200	.4004	.4284	.20490	.22469	34
27	.70029	.71386	.98098	.0194	.4008	.4280	.20518	.22439	33
28	.70049	.71366	.98155	.0188	.4012	.4276	.20546	.22409	32
29	.70070	.71345	.98213	.0182	.4016	.4271	.20575	.22378	31
30	0.70091	0.71325	0.98270	1.0176	1.4020	1.4267	.20603	.22348	30
31	.70112	.71305	.98327	.0170	.4024	.4263	.20631	.22318	29
32	.70132	.71284	.98384	.0164	.4028	.4259	.20659	.22288	28
33	.70153	.71264	.98441	.0158	.4032	.4255	.20687	.22258	27
34	.70174	.71243	.98499	.0152	.4036	.4250	.20715	.22228	26
35	.70195	.71223	.98556	1.0147	1.4040	1.4246	.20743	.22198	25
36	.70215	.71203	.98613	.0141	.4044	.4242	.20772	.22168	24
37	.70236	.71182	.98671	.0135	.4048	.4238	.20800	.22138	23
38	.70257	.71162	.98728	.0129	.4052	.4234	.20828	.22108	22
39	.70277	.71141	.98786	.0123	.4057	.4229	.20857	.22079	21
40	0.70298	0.71121	0.98843	1.0117	1.4061	1.4225	.20885	.22049	20
41	.70319	.71100	.98901	.0111	.4065	.4221	.20914	.22019	19
42	.70339	.71080	.98958	.0105	.4069	.4217	.20942	.21989	18
43	.70360	.71059	.99016	.0099	.4073	.4213	.20971	.21960	17
44	.70381	.71039	.99073	.0093	.4077	.4208	.20999	.21930	16
45	0.70401	0.71019	0.99131	1.0088	1.4081	1.4204	.21028	.21900	15
46	.70422	.70998	.99189	.0082	.4085	.4200	.21056	.21871	14
47	.70443	.70978	.99247	.0076	.4089	.4196	.21085	.21841	13
48	.70463	.70957	.99304	.0070	.4093	.4192	.21114	.21812	12
49	.70484	.70937	.99362	.0064	.4097	.4188	.21142	.21782	11
50	0.70505	0.70916	0.99420	1.0058	1.4101	1.4183	.21171	.21753	10
51	.70525	.70896	.99478	.0052	.4105	.4179	.21200	.21723	9
52	.70546	.70875	.99536	.0047	.4109	.4175	.21229	.21694	8
53	.70567	.70855	.99594	.0041	.4113	.4171	.21257	.21665	7
54	.70587	.70834	.99652	.0035	.4118	.4167	.21286	.21635	6
55	0.70608	0.70813	0.99710	1.0029	1.4122	1.4163	.21315	.21606	5
56	.70628	.70793	.99768	.0023	.4126	.4159	.21344	.21577	4
57	.70649	.70772	.99826	.0017	.4130	.4154	.21373	.21548	3
58	.70670	.70752	.99884	.0012	.4134	.4150	.21402	.21518	2
59	.70690	.70731	.99942	.0006	.4138	.4146	.21431	.21489	1
60	0.70711	0.70711	1.00000	1.0000	1.4142	1.4142	.21460	.21460	0
M	Cosine	Sine	Cotan.	Tan.	Cosec.	Secant	READ DOWN	45°–46° Involute	M

Conversion Tables of Angular Measure. — The accompanying tables of degrees, minutes, and seconds into radians; radians into degrees, minutes, and seconds; radians into degrees and decimals of a degree; and minutes and seconds into decimals of a degree and vice versa facilitate the conversion of measurements.

Example: The Degrees, Minutes, and Seconds into Radians Table is used to find the number of radians in 324 degrees, 25 minutes, 13 seconds as follows:

300 degrees	= 5.235988 radians
20 degrees	= 0.349066 radian
4 degrees	= 0.069813 radian
25 minutes	= 0.007272 radian
13 seconds	= 0.000063 radian
324°25′13″	= 5.662202 radians

Example: The Radians into Degrees and Decimals of a Degree, and Radians into Degrees, Minutes and Seconds Tables are used to find the number of decimal degrees or degrees, minutes and seconds in 0.734 radian as follows:

0.7 radian	= 40.1070 degrees	0.7 radian	= 40° 6′25″
0.03 radian	= 1.7189 degrees	0.03 radian	= 1°43′ 8″
0.004 radian	= 0.2292 degree	0.004 radian	= 0°13′45″
0.734 radian	= 42.0551 degrees	0.734 radian	= 41°62′78″ or 42°3′18″

Degrees, Minutes, and Seconds into Radians
(Based on 180 degrees = π radians)

Degrees into Radians									
Deg.	Rad.	Deg.	Rad.	Deg.	Rad.	Deg.	Rad.	Deg.	Rad.
1000	17.453293	100	1.745329	10	0.174533	1	0.017453	0.1	0.001745	0.01	0.000175
2000	34.906585	200	3.490659	20	0.349066	2	0.034907	0.2	0.003491	0.02	0.000349
3000	52.359878	300	5.235988	30	0.523599	3	0.052360	0.3	0.005236	0.03	0.000524
4000	69.813170	400	6.981317	40	0.698132	4	0.069813	0.4	0.006981	0.04	0.000698
5000	87.266463	500	8.726646	50	0.872665	5	0.087266	0.5	0.008727	0.05	0.000873
6000	104.719755	600	10.471976	60	1.047198	6	0.104720	0.6	0.010472	0.06	0.001047
7000	122.173048	700	12.217305	70	1.221730	7	0.122173	0.7	0.012217	0.07	0.001222
8000	139.626340	800	13.962634	80	1.396263	8	0.139626	0.8	0.013963	0.08	0.001396
9000	157.079633	900	15.707963	90	1.570796	9	0.157080	0.9	0.015708	0.09	0.001571
10000	174.532925	1000	17.453293	100	1.745329	10	0.174533	1.0	0.017453	0.10	0.001745

Minutes into Radians									
Min.	Rad.	Min.	Rad.	Min.	Rad.	Min.	Rad.	Min.	Rad.
1	0.000291	11	0.003200	21	0.006109	31	0.009018	41	0.011926	51	0.014835
2	0.000582	12	0.003491	22	0.006400	32	0.009308	42	0.012217	52	0.015126
3	0.000873	13	0.003782	23	0.006690	33	0.009599	43	0.012508	53	0.015417
4	0.001164	14	0.004072	24	0.006981	34	0.009890	44	0.012799	54	0.015708
5	0.001454	15	0.004363	25	0.007272	35	0.010181	45	0.013090	55	0.015999
6	0.001745	16	0.004654	26	0.007563	36	0.010472	46	0.013381	56	0.016290
7	0.002036	17	0.004945	27	0.007854	37	0.010763	47	0.013672	57	0.016581
8	0.002327	18	0.005236	28	0.008145	38	0.011054	48	0.013963	58	0.016872
9	0.002618	19	0.005527	29	0.008436	39	0.011345	49	0.014254	59	0.017162
10	0.002909	20	0.005818	30	0.008727	40	0.011636	50	0.014544	60	0.017453

Seconds into Radians									
Sec.	Rad.	Sec.	Rad.	Sec.	Rad.	Sec.	Rad.	Sec.	Rad.
1	0.000005	11	0.000053	21	0.000102	31	0.000150	41	0.000199	51	0.000247
2	0.000010	12	0.000058	22	0.000107	32	0.000155	42	0.000204	52	0.000252
3	0.000015	13	0.000063	23	0.000112	33	0.000160	43	0.000208	53	0.000257
4	0.000019	14	0.000068	24	0.000116	34	0.000165	44	0.000213	54	0.000262
5	0.000024	15	0.000073	25	0.000121	35	0.000170	45	0.000218	55	0.000267
6	0.000029	16	0.000078	26	0.000126	36	0.000175	46	0.000223	56	0.000271
7	0.000034	17	0.000082	27	0.000131	37	0.000179	47	0.000228	57	0.000276
8	0.000039	18	0.000087	28	0.000136	38	0.000184	48	0.000233	58	0.000281
9	0.000044	19	0.000092	29	0.000141	39	0.000189	49	0.000238	59	0.000286
10	0.000048	20	0.000097	30	0.000145	40	0.000194	50	0.000242	60	0.000291

Radians into Degrees and Decimals of a Degree
(Based on π radians = 180 degrees)

Rad.	Deg.	Rad	Deg.	Rad.	Deg.	Rad.	Deg.	Rad.	Deg.	Rad.	Deg.
10	572.9578	1	57.2958	0.1	5.7296	0.01	0.5730	0.001	0.0573	0.0001	0.0057
20	1145.9156	2	114.5916	0.2	11.4592	0.02	1.1459	0.002	0.1146	0.0002	0.0115
30	1718.8734	3	171.8873	0.3	17.1887	0.03	1.7189	0.003	0.1719	0.0003	0.0172
40	2291.8312	4	229.1831	0.4	22.9183	0.04	2.2918	0.004	0.2292	0.0004	0.0229
50	2864.7890	5	286.4789	0.5	28.6479	0.05	2.8648	0.005	0.2865	0.0005	0.0286
60	3437.7468	6	343.7747	0.6	34.3775	0.06	3.4377	0.006	0.3438	0.0006	0.0344
70	4010.7046	7	401.0705	0.7	40.1070	0.07	4.0107	0.007	0.4011	0.0007	0.0401
80	4583.6624	8	458.3662	0.8	45.8366	0.08	4.5837	0.008	0.4584	0.0008	0.0458
90	5156.6202	9	515.6620	0.9	51.5662	0.09	5.1566	0.009	0.5157	0.0009	0.0516
100	5729.5780	10	572.9578	1.0	57.2958	0.10	5.7296	0.010	0.5730	0.0010	0.0573

Radians into Degrees, Minutes and Seconds
(Based on π radians = 180 degrees)

Rad.	Angle	Rad.	Angle	Rad.	Angle	Rad.	Angle	Rad.	Angle	Rad.	Angle
10	572°57'28"	1	57°17'45"	0.1	5°43'46"	0.01	0°34'23"	0.001	0°3'26"	0.0001	0°0'21"
20	1145°54'56"	2	114°35'30"	0.2	11°27'33"	0.02	1°8'45"	0.002	0°6'53"	0.0002	0°0'41"
30	1718°52'24"	3	171°53'14"	0.3	17°11'19"	0.03	1°43'8"	0.003	0°10'19"	0.0003	0°1'2"
40	2291°49'52"	4	229°10'59"	0.4	22°55'6"	0.04	2°17'31"	0.004	0°13'45"	0.0004	0°1'23"
50	2864°47'20"	5	286°28'44"	0.5	28°38'52"	0.05	2°51'53"	0.005	0°17'11"	0.0005	0°1'43"
60	3437°44'48"	6	343°46'29"	0.6	34°22'39"	0.06	3°26'16"	0.006	0°20'38"	0.0006	0°2'4"
70	4010°42'16"	7	401°4'14"	0.7	40°6'25"	0.07	4°0'39"	0.007	0°24'4"	0.0007	0°2'24"
80	4583°39'44"	8	458°21'58"	0.8	45°50'12"	0.08	4°35'1"	0.008	0°27'30"	0.0008	0°2'45"
90	5156°37'13"	9	515°39'43"	0.9	51°33'58"	0.09	5°9'24"	0.009	0°30'56"	0.0009	0°3'6"
100	5729°34'41"	10	572°57'28"	1.0	57°17'45"	0.10	5°43'46"	0.010	0°34'23"	0.0010	0°3'26"

Minutes and Seconds into Decimals of a Degree and Vice Versa
(Based on 1 second = 0.00027778 degrees)

Minutes into Decimals of a Degree						Seconds into Decimals of a Degree					
Min.	Deg.	Min.	Deg.	Min.	Deg.	Sec.	Deg.	Sec.	Deg.	Sec.	Deg.
1	0.0167	21	0.3500	41	0.6833	1	0.0003	21	0.0058	41	0.0114
2	0.0333	22	0.3667	42	0.7000	2	0.0006	22	0.0061	42	0.0117
3	0.0500	23	0.3833	43	0.7167	3	0.0008	23	0.0064	43	0.0119
4	0.0667	24	0.4000	44	0.7333	4	0.0011	24	0.0067	44	0.0122
5	0.0833	25	0.4167	45	0.7500	5	0.0014	25	0.0069	45	0.0125
6	0.1000	26	0.4333	46	0.7667	6	0.0017	26	0.0072	46	0.0128
7	0.1167	27	0.4500	47	0.7833	7	0.0019	27	0.0075	47	0.0131
8	0.1333	28	0.4667	48	0.8000	8	0.0022	28	0.0078	48	0.0133
9	0.1500	29	0.4833	49	0.8167	9	0.0025	29	0.0081	49	0.0136
10	0.1667	30	0.5000	50	0.8333	10	0.0028	30	0.0083	50	0.0139
11	0.1833	31	0.5167	51	0.8500	11	0.0031	31	0.0086	51	0.0142
12	0.2000	32	0.5333	52	0.8667	12	0.0033	32	0.0089	52	0.0144
13	0.2167	33	0.5500	53	0.8833	13	0.0036	33	0.0092	53	0.0147
14	0.2333	34	0.5667	54	0.9000	14	0.0039	34	0.0094	54	0.0150
15	0.2500	35	0.5833	55	0.9167	15	0.0042	35	0.0097	55	0.0153
16	0.2667	36	0.6000	56	0.9333	16	0.0044	36	0.0100	56	0.0156
17	0.2833	37	0.6167	57	0.9500	17	0.0047	37	0.0103	57	0.0158
18	0.3000	38	0.6333	58	0.9667	18	0.0050	38	0.0106	58	0.0161
19	0.3167	39	0.6500	59	0.9833	19	0.0053	39	0.0108	59	0.0164
20	0.3333	40	0.6667	60	1	20	0.0056	40	0.0111	60	0.0167

Example 1: Convert 11'37" to decimals of a degree. From the left table, 11' = 0.1833 degree. From the right table, 37" = 0.0103 degree. Adding, 11'37" = 0.1833 + 0.0103 = 0.1936 degree.

Example 2: Convert 0.1234 degree to minutes and seconds. From the left table, 0.1167 degree = 7'. Subtracting 0.1167 from 0.1234 gives 0.0067. From the right table, 0.0067 = 24" so that 0.1234 = 7'24".

MECHANICS

Throughout the Mechanics section in this Handbook, both English and metric SI data and formulas are given to cover the requirements of working in either system of measurement. Except for the passage entitled "The Use of the Metric SI System in Mechanics Calculations," formulas and text relating exclusively to SI are given in bold face type.

Definitions. — The science of mechanics deals with the effects of forces in causing or preventing motion. *Statics* is that branch of mechanics which deals with bodies in equilibrium, i.e., the forces acting on them cause them to remain at rest or to move with uniform velocity. *Dynamics* is that branch of mechanics which deals with bodies not in equilibrium, i.e., the forces acting on them cause them to move with non-uniform velocity. *Kinetics* is that branch of dynamics which deals with both the forces acting on bodies and the motions which they cause. *Kinematics* is that branch of dynamics which deals only with the motions of bodies without reference to the forces that cause them.

Definitions of certain terms and quantities as used in mechanics follow:

A *force* may be defined simply as a push or a pull; the push or pull may result from the force of contact between bodies or from a force, such as magnetism or gravitation, in which no direct contact takes place.

Matter is any substance that occupies space; gases, liquids, solids, electrons, atoms, molecules, etc., all fit this definition.

Inertia is that property of matter which causes it to resist any change in its motion or state of rest.

Mass is a measure of the inertia of a body.

Work, in mechanics, is the product of force times distance and is expressed by a combination of units of force and distance, as foot-pounds, inch-pounds, meter-kilograms, etc. **The metric SI unit of work is the joule, which is the work done when the point of application of a force of one newton is displaced through a distance of one meter in the direction of the force.**

Power, in mechanics, is the product of force times distance divided by time; it measures the performance of a given amount of work in a given time. It is the rate of doing work and as such is expressed in foot-pounds per minute, foot-pounds per second, kilogram-meters per second, etc. **The metric SI unit is the watt, which is one joule per second.**

Horsepower is the unit of power that has been adopted for engineering work. One horsepower is equal to 33,000 foot-pounds per minute or 550 foot-pounds per second. The *kilowatt,* used in electrical work, equals 1.34 horsepower; or 1 horsepower equals 0.746 kilowatt. **However, in the metric SI, the term horsepower is not used, and the basic unit of power is the watt. This unit, and the derived units milliwatt and kilowatt, for example, are the same as those used in electrical work.**

Torque or *moment* of a force is a measure of the tendency of the force to rotate the body upon which it acts about an axis. The magnitude of the moment due to a force acting in a plane perpendicular to some axis is obtained by multiplying the force by the perpendicular distance from the axis to the line of action of the force. (If the axis of rotation is not perpendicular to the plane of the force, then the components of the force in a plane perpendicular to the axis of rotation are used to find the resultant moment of the force by finding the moment of each component and adding these component moments algebraically.) Moment or torque is commonly expressed in pound-feet, pound-inches, kilogram-meters, etc. **The metric SI unit is the newton-meter (N · m).**

Velocity is the time-rate of change of distance and is expressed as distance divided by time, that is, feet per second, miles per hour, centimeters per second, meters per second, etc.

American National Standard Letter Symbols for Mechanics and Time-Related Phenomena (ANSI Y10.3-1984)

Term	Symbol
Acceleration, angular	α (alpha)
Acceleration, due to gravity	g
Acceleration, linear	a
Amplitude*	A
Angle	α (alpha)
	β (beta)
	γ (gamma)
	θ (theta)
	ϕ (phi)
	ψ (psi)
Angle, solid	Ω (omega)
Angular frequency	ω (omega)
Angular momentum	L
Angular velocity	ω (omega)
Arc length	s
Area	A
Axes, through any point*	$X\text{-}X$
	$Y\text{-}Y$
	$Z\text{-}Z$
Bulk modulus	K
Breadth (width)	b
Coefficient of expansion, linear*	α (alpha)
Coefficient of friction	μ (mu)
Concentrated load (same as force)	F
Deflection of beam, max*	δ (delta)
Density	ρ (rho)
Depth	d
	δ (delta)
	t
Diameter	D
	d
Displacement*	u
	v
	w
Distance, linear*	s
Eccentricity of application of load*	e
Efficiency*	η (eta)
Elasticity, modulus of	E
Elasticity, modulus of, in shear	G
Elongation, total*	δ (delta)
Energy, kinetic	E_k
	K
	T
Energy, potential	E_p
	V
	Φ (phi)
Factor of safety*	N
	n
Force or load, concentrated	F
Frequency	f
Gyration, radius of*	k
Height	h

Term	Symbol
Inertia, moment of	I
	J
Inertia, polar (area) moment of*	J
Inertia, product (area) moment of*	I_{xy}
Length	L
	l
Load per unit distance*	q
	w
Load, total*	P
	W
Mass	m
Moment of force, including bending moment	M
Neutral axis, distance to extreme fiber from*	c
Period	T
Poisson's ratio	μ (mu)
	ν (nu)
Power	P
Pressure, normal force per unit area	p
Radius	r
Revolutions per unit of time	n
Second moment of area (second axial moment of area)	I_a
Second polar moment of area	I_p
	J
Section modulus	Z
Shear force in beam section*	V
Spring constant (load per unit deflection)*	k
Statical moment of any area about a given axis*	Q
Strain, normal	ϵ (epsilon)
Strain, shear	γ (gamma)
Stress, concentration factor*	K
Stress, normal	σ (sigma)
Stress, shear	τ (tau)
Temperature, absolute†	T
	θ (theta)
Temperature†	t
	θ (theta)
Thickness	d
	δ (delta)
	t
Time	t
Torque	T
Velocity, linear	v
Volume	V
Wavelength	λ (lambda)
Weight	W
Weight per unit volume	γ (gamma)
Work	W

* Not specified in Standard   † Specified in ANSI Y10.4-1982

Acceleration is defined as the time-rate of change of velocity and is expressed as velocity divided by time or as distance divided by time squared, that is, in feet per second, per second or feet per second squared; inches per second, per second or inches per second squared; centimeters per second, per second or centimeters per second squared; etc. **The metric SI unit is the meter per second squared.**

Unit Systems. — In mechanics calculations, both *absolute* and *gravitational* systems of units are employed. The fundamental units in absolute systems are *length*, *time*, and *mass*, and from these units, the dimension of force is derived. Two absolute systems which have been in use for many years are the cgs (centimeter-gram-second) and the MKS (meter-kilogram-second) systems. Another system, known as MKSA (meter-kilogram-second-ampere), links the MKS system of units of mechanics with electro magnetic units.

The Conference General des Poids et Mesures (CGPM), which is the body responsible for all international matters concerning the metric system, adopted in 1954 a rationalized and coherent system of units based on the four MKSA units and including the kelvin as the unit of temperature, and the candela as the unit of luminous intensity. In 1960, the CGPM formally named this system the 'Systeme International d'Unites,' for which the abbreviation is SI in all languages. In 1971, the 14th CGPM adopted a seventh base unit, the mole, which is the unit of quantity ("amount of substance"). Further details of the SI are given in the Weights and Measures section, and its application in mechanics calculations, contrasted with the use of the English system, is considered on page 132.

The fundamental units in gravitational systems are *length, time,* and *force,* and from these units, the dimension of mass is derived. In the gravitational system most widely used in English measure countries, the units of length, time, and force are, respectively, the foot, the second, and the pound. The corresponding unit of mass, commonly called the *slug*, is equal to 1 pound second^2 per foot and is derived from the formula, $M = W \div g$ in which M = mass in slugs, W = weight in pounds, and g = acceleration due to gravity, commonly taken as 32.16 feet per second^2. A body that weighs 32.16 lbs. on the surface of the earth has, therefore, a mass of one slug.

Many engineering calculations utilize a system of units consisting of the inch, the second, and the pound. The corresponding units of mass are pounds second^2 per inch and the value of g is taken as 386 inches per second^2.

In a gravitational system that has been widely used in metric countries, the units of length, time, and force are, respectively, the meter, the second, and the kilogram. The corresponding units of mass are kilograms second^2 per meter and the value of g is taken as 9.81 meters per second^2.

Acceleration of Gravity g Used in Mechanics Formulas. — The acceleration of a freely falling body has been found to vary according to location on the earth's surface as well as with height, the value at the equator being 32.09 feet per second, per second while at the poles it is 32.26 ft/sec^2. In the United States it is customary to regard 32.16 as satisfactory for most practical purposes in engineering calculations.

Standard Pound Force: For use in defining the magnitude of a standard unit of force, known as the *pound force,* a fixed value of 32.1740 ft/sec^2, designated by the symbol g_0, has been adopted by international agreement. As a result of this agreement, whenever the term mass, M, appears in a mechanics formula and the substitution $M = W/g$ is made, use of the standard value $g_0 = 32.1740$ ft/sec^2 is implied although as stated previously, it is customary to use approximate values for g except in those cases where extreme accuracy is required.

The Use of the Metric SI System in Mechanics Calculations. — The SI system is a development of the traditional metric system based on decimal arithmetic; fractions are avoided. For each physical quantity, units of different sizes are formed by multiplying or dividing a single base value by powers of 10. Thus, changes can be made very simply by adding zeros or shifting decimal points. For example, the meter is the basic unit of length; the kilometer is a multiple (1,000 meters); and the millimeter is a sub-multiple (one-thousandth of a meter).

In the older metric system, the simplicity of a series of units linked by powers of 10 is an advantage for plain quantities such as length, but this simplicity is lost as soon as more complex units are encountered. For example, in different branches of science and engineering, energy may appear as the erg, the calorie, the kilogram-meter, the liter-atmosphere, or the horsepower-hour. In contrast, the SI provides only one basic unit for each physical quantity, and universality is thus achieved.

There are seven base-units, and in mechanics calculations three are used, which are for the basic quantities of length, mass, and time, expressed as the meter (m), the kilogram (kg), and the second (s). The other four base-units are the ampere (A) for electric current, the kelvin (K) for thermodynamic temperature, the candela (cd) for luminous intensity, and the mole (mol) for amount of substance.

The SI is a coherent system. A system of units is said to be coherent if the product or quotient of any two unit quantities in the system is the unit of the resultant quantity. For example, in a coherent system in which the foot is a unit of length, the square foot is the unit of area, whereas the acre is not. Further details of the SI, and definitions of the units, are given at the end of the book.

Other physical quantities are derived from the base-units. For example, the unit of velocity is the meter per second (m/s), which is a combination of the base-units of length and time. The unit of acceleration is the meter per second squared (m/s^2). By applying Newton's second law of motion — force is proportional to mass multiplied by acceleration — the unit of force is obtained, which is the kg·m/s^2. This unit is known as the newton, or N. Work, or force times distance, is the kg·m^2/s^2, which is the joule, (1 joule = 1 newton-meter) and energy is also expressed in these terms. The abbreviation for joule is J. Power, or work per unit time, is the kg·m^2/s^3, which is the watt (1 watt = 1 joule per second = 1 newton-meter per second). The abbreviation for watt is W.

The coherence of SI units has two important advantages. The first, that of uniqueness and therefore universality, has been explained. The second is that it greatly simplifies technical calculations. Equations representing physical principles can be applied without introducing such numbers as 550 in power calculations, which, in the English system of measurement have to be used to convert units. Thus conversion factors largely disappear from calculations carried out in SI units, with a great saving in time and labor.

Mass, weight, force, load. SI is an absolute system (see page 131), and consequently it is necessary to make a clear distinction between mass and weight. The *mass* of a body is a measure of its inertia, whereas the weight of a body is the *force* exerted on it by gravity. In a fixed gravitational field, weight is directly proportional to mass, and the distinction between the two can be easily overlooked. However, if a body is moved to a different gravitational field, for example, that of the moon, its weight alters, but its mass remains unchanged. Since the gravitational field on earth varies from place to place by only a small amount, and weight is proportional to mass, it is practical to use the weight of unit mass as a unit of force, and this procedure is adopted in both the English and older metric systems of measurement. In common usage, they are given the same names, and we say that a mass of 1 pound has a weight of 1 pound. In the former case the pound is being used as a unit of mass, and in the latter case, as a unit of force. This procedure is convenient in some branches of engineering, but leads to confusion in others.

As mentioned earlier, Newton's second law of motion states that force is proportional to mass times acceleration. Because an unsupported body on the earth's surface falls with acceleration g (32 ft/s^2 approximately), the pound (force) is that force which will impart an acceleration of g ft/s^2 to a pound (mass). Similarly, the kilogram (force) is that force which will impart an acceleration of g (9.8 meters per second^2 approximately), to a mass of one kilogram. In the SI, the *newton* is that force which will impart unit acceleration (1 m/s^2) to a mass of one kilogram. It is therefore smaller than the kilogram (force) in the ratio $1:g$ (about 1:9.8). This fact has important consequences in engineering calculations. The factor g now disappears from a wide range of formulas in dynamics, but appears in many formulas in statics where it was formerly absent. It is however not quite the same g, for reasons which will now be explained.

In the article on page 168, the mass of a body is referred to as M, but it is immediately replaced in subsequent formulas by W/g, where W is the weight in pounds (force), which leads to familiar expressions such as $WV^2/2g$ for kinetic energy. In this treatment, the M which appears briefly is really expressed in terms of the slug (page 131), a unit normally used only in aeronautical engineering. In everyday engineers' language, weight and mass are regarded as synonymous and expressions such as $WV^2/2g$ are used without pondering the distinction. Nevertheless, on reflection it seems odd that g should appear in a formula which has nothing to do with gravity at all. In fact the g used here is not the true, local value of the acceleration due to gravity, but an arbitrary standard value which has been chosen as part of the definition of the pound (force) and is more properly designated g_o (page 131). Its function is not to indicate the strength of the local gravitational field, but to convert from one unit to another.

In the SI the unit of mass is the *kilogram*, and the unit of force (and therefore weight) is the *newton*.

The following are typical statements in dynamics expressed in SI units:

A force of R newtons acting on a mass of M kilograms produces an acceleration of R/M meters per second^2. The kinetic energy of a mass of M kg moving with velocity V m/s is $\frac{1}{2} MV^2$ kg (m/s)^2 or $\frac{1}{2} MV^2$ joules. The work done by a force of R newtons moving a distance L meters is RL Nm, or RL joules. If this work were converted entirely into kinetic energy we could write $RL = \frac{1}{2} MV^2$ and it is instructive to consider the units. Remembering that the N is the same as the kg $\cdot$ m/s^2, we have (kg $\cdot$ m/s^2)$^2 \times$ m = kg (m/s)^2, which is obviously correct. It will be noted that g does not appear anywhere in these statements.

In contrast, in many branches of engineering where the weight of a body is important, rather than its mass, using SI units g does appear where formerly it was absent. Thus if a rope hangs vertically supporting a mass of M kilograms the tension in the rope is Mg N. Here g is the acceleration due to gravity, and its units are m/s^2. The ordinary numerical value of 9.81 will be sufficiently accurate for most purposes on earth. The expression is still valid elsewhere, for example, on the moon, provided the proper value of g is used. The maximum tension the rope can safely withstand (and other similar properties) will also be specified in terms of the newton, so that direct comparison may be made with the tension predicted.

Words like load and weight have to be used with greater care. In everyday language we might say "a lift carries a load of five people of average weight 70 kg," but in precise technical language we say that if the average mass is 70 kg, then the average weight is $70g$ N, and the total load (that is force) on the lift is $350g$ N.

If the lift starts to rise with acceleration a m/s^2, the load becomes $350 (g + a)$ N; both g and a have units of m/s^2, the mass is in kg, so the load is in terms of kg $\cdot$ m/s^2, which is the same as the newton.

Pressure and stress. These quantities are expressed in terms of force per unit area. In the SI the unit is the pascal (Pa), which expressed in terms of SI derived

and base units is the newton per meter squared (N/m²). The pascal is very small — it is only equivalent to 0.15×10^{-3} lb/in² — hence the kilopascal (kPa = 1000 pascals), and the megapascal (MPa = 10^6 pascals) may be more convenient multiples in practice. Thus, note: 1 newton per millimeter squared = 1 meganewton per meter squared = 1 megapascal.

In addition to the pascal, the bar, a non-SI unit, is in use in the field of pressure measurement in some countries, including England. Thus, in view of existing practice, the International Committee of Weights and Measures (CIPM) decided in 1969 to retain this unit for a limited time for use with those of SI. The bar = 10^5 pascals and the hectobar = 10^7 pascals.

Scalar and Vector Quantities. — The quantities dealt with in mechanics are of two kinds according to whether magnitude alone or direction as well as magnitude must be known in order to completely specify them. Quantities such as time, volume and density are completely specified when their magnitude is known. Such quantities are called *scalar* quantities. Quantities such as force, velocity, acceleration, moment and displacement which must, in order to be specified completely, have a specific direction as well as magnitude, are called *vector* quantities.

Graphical Representation of Forces. — A force has three characteristics which, when known, determine it. They are *direction, point of application,* and *magnitude.* The direction of a force is the direction in which it tends to move the body upon which it acts. The point of application is the place on the line of action where the force is applied. Forces may conveniently be represented by straight lines and arrow heads. The arrow head indicates the direction of the force, and the length of the line

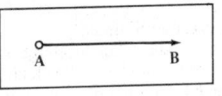

its magnitude to any suitable scale. The point of application may be at any point on the line, but it is generally convenient to assume it to be at one end. In the accompanying illustration, a force is supposed to act along line *AB* in a direction from left to right. The length of line *AB* shows the magnitude of the force. If point *A* is the point of application, the force is exerted as a pull, but if point *B* be assumed to be the point of application, it would indicate that the force is exerted as a push.

Velocities, moments, displacements, etc. may similarly be represented and manipulated graphically because they are all of the same class of quantities called vectors. (*See* Scalar and Vector Quantities.)

Algebraic Composition and Resolution of Force Systems. — The graphical methods shown on pages 136 and 137 are convenient for solving problems involving force systems in which all of the forces lie in the same plane and only a few forces are involved. If many forces are involved, however, or the forces do not lie in the same plane, it is better to use algebraic methods to avoid complicated space diagrams. Systematic procedures for solving force problems by algebraic methods are outlined beginning on page 138. In connection with the use of these procedures, it is necessary to define several terms applicable to force systems in general.

The single force which produces the same effect upon a body as two or more forces acting together is called their *resultant.* The separate forces which can be so combined are called the *components.* Finding the resultant of two or more forces is called the *composition of forces,* and finding two or more components of a given force, the *resolution of forces.* Forces are said to be *concurrent* when their lines of action can be extended to meet at a common point; forces that are *parallel* are, of course, *nonconcurrent.* Two forces having the same line of action are said to be *collinear.* Two forces equal in magnitude, parallel, and in opposite directions constitute

a *couple*. Forces all in the same plane are said to be *coplanar;* if not in the same plane, they are called *noncoplanar* forces.

The *resultant* of a system of forces is the simplest equivalent system that can be determined. It may be a single force, a couple, or a noncoplanar force and a couple. This last type of resultant, a noncoplanar force and a couple, may be replaced, if desired, by two *skewed* forces (forces that are nonconcurrent, nonparallel, and noncoplanar). When the resultant of a system of forces is zero, the system is in equilibrium, that is, the body on which the force system acts remains at rest or continues to move with uniform velocity.

Couples. — If the forces *AB* and *CD* are equal and parallel but act in opposite directions, then the resultant equals o, or, in other words, the two forces have no resultant and are called a couple. A couple tends to produce rotation. The measure of this tendency is called the moment of the couple and is the product of one of the forces multiplied by the distance between the two.

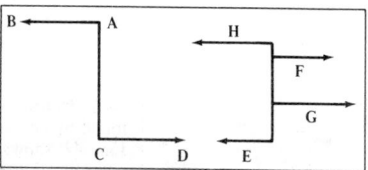

As a couple has no resultant, no single force can balance or counteract the tendency of the couple to produce rotation. To prevent the rotation of a body acted upon by a couple, two other forces are therefore required, forming a second couple. In the illustration, *E* and *F* form one couple and *G* and *H* are the balancing couple. The body on which they act is in equilibrium if the moments of the two couples are equal and tend to rotate the body in opposite directions. A couple may also be represented by a vector in the direction of the axis about which the couple acts. The length of the vector, to some scale, represents the magnitude of the couple, and the direction of the vector is that in which a right-hand screw would advance if it were to be rotated by the couple.

Composition of a Single Force and Couple. — A single force and a couple in the same plane or in parallel planes may be replaced by another single force equal and parallel to the first force, at a distance from it equal to the moment of the couple divided by the magnitude of the force. The new single force is located so that the moment of the resultant about the point of application of the original force is of the same sign as the moment of the couple.

In the figure, with the couple $N - N$ in the position shown, the resultant of P, $-N$, and N is O (which equals P) acting on a line through point c so that $(P - N) \times ac = N \times bc$.

Thus, it follows that,

$$ac = \frac{N(ac + bc)}{P} = \frac{\text{Moment of Couple}}{P}$$

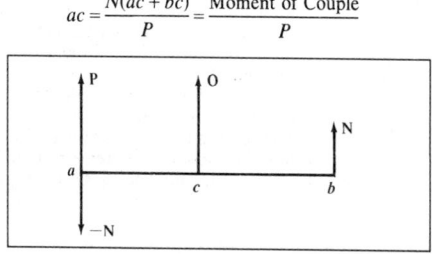

Graphical Composition and Resolution of Forces

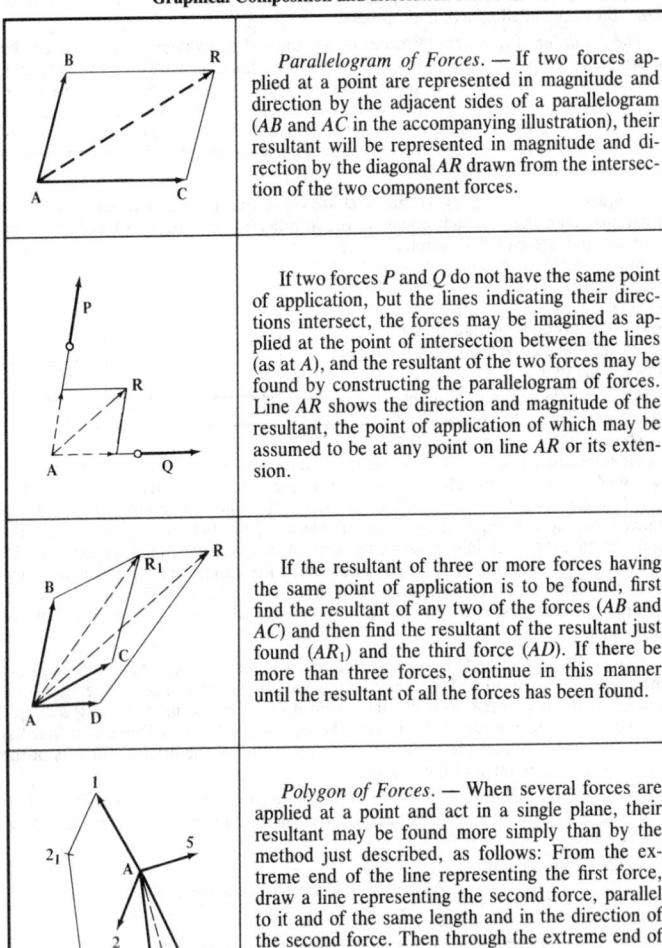

Parallelogram of Forces. — If two forces applied at a point are represented in magnitude and direction by the adjacent sides of a parallelogram (*AB* and *AC* in the accompanying illustration), their resultant will be represented in magnitude and direction by the diagonal *AR* drawn from the intersection of the two component forces.

If two forces *P* and *Q* do not have the same point of application, but the lines indicating their directions intersect, the forces may be imagined as applied at the point of intersection between the lines (as at *A*), and the resultant of the two forces may be found by constructing the parallelogram of forces. Line *AR* shows the direction and magnitude of the resultant, the point of application of which may be assumed to be at any point on line *AR* or its extension.

If the resultant of three or more forces having the same point of application is to be found, first find the resultant of any two of the forces (*AB* and *AC*) and then find the resultant of the resultant just found (*AR_1*) and the third force (*AD*). If there be more than three forces, continue in this manner until the resultant of all the forces has been found.

Polygon of Forces. — When several forces are applied at a point and act in a single plane, their resultant may be found more simply than by the method just described, as follows: From the extreme end of the line representing the first force, draw a line representing the second force, parallel to it and of the same length and in the direction of the second force. Then through the extreme end of this line draw a line parallel to, and of the same length and direction as the third force, and continue this until all the forces have been thus represented. Then draw a line from the point of application of the forces (as *A*) to the extreme point (as 5_1) of the line last drawn. This line (*A* 5_1) is the resultant of the forces.

Graphical Composition and Resolution of Forces

	The resultant of two forces applied at the same point and acting in the same direction, is equal to the sum of the forces. For example, if the two forces AB and AC, one equal to two and the other equal to three pounds, are applied at point A, then their resultant AD equals the sum of these forces, or five pounds.
	If two forces act in opposite directions, then their resultant is equal to their difference, and the direction of the resultant is the same as the direction of the greater of the two forces. For example: AB and AC are both applied at point A; then, if AB equals four and AC equals six pounds, the resultant AD equals two pounds and acts in the direction of AC.
	Parallel Forces. — If two forces are parallel and act in the same direction, then their resultant is parallel to both lines, is located between them, and is equal to the sum of the two components. The point of application of the resultant divides the line joining the points of application of the components inversely as the magnitude of the components. Thus, $AB:CE = CD:AD$.
	The resultant of two parallel and unequal forces acting in opposite directions is parallel to both lines, is located outside of them on the side of the greater of the components, has the same direction as the greater component, and is equal in magnitude to the difference between the two components. The point of application on the line AC produced is found from the proportion: $$AB:CD = CE:AE.$$
Moment of P about A Equals $P \times l$	*Moment of a Force.* — The moment of a force with respect to a point is the product of the force multiplied by the perpendicular distance from the given point to the direction of the force. In the illustration, the moment of the force P with relation to point A is $P \times AB$. The perpendicular distance AB is called the lever-arm of the force. The moment is the measure of the tendency of the force to produce rotation about the given point, which is termed the center of moments. If the force is measured in pounds and the distance in inches, the moment is expressed in inch-pounds. **In metric SI units, the moment is expressed in newton-meters (N·m), or newton-millimeters (N·mm).** The moment of the resultant of any number of forces acting together in the same plane is equal to the algebraic sum of the moments of the separate forces.

Table I. Algebraic Solution of Force Systems — All Forces in the Same Plane

Finding Two Concurrent Components of a Single Force	
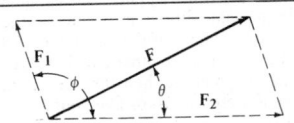	*Case I:* To find two components F_1 and F_2 at angles θ and ϕ, ϕ not being 90°. $$F_1 = \frac{F \sin \theta}{\sin \phi}$$ $$F_2 = \frac{F \sin (\phi - \theta)}{\sin \phi}$$
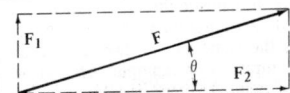	*Case II:* Components F_1 and F_2 form 90° angle. $$F_1 = F \sin \theta$$ $$F_2 = F \cos \theta$$

Finding the Resultant of Two Concurrent Forces	
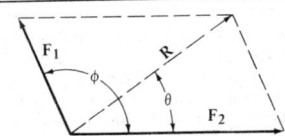	*Case I:* Forces F_1 and F_2 do not form 90° angle. $$R = \frac{F_1 \sin \phi}{\sin \theta}, \quad \text{or,} \quad R = \frac{F_2 \sin \phi}{\sin (\phi - \theta)}, \quad \text{or}$$ $$R = \sqrt{F_1^2 + F_2^2 + 2F_1F_2 \cos \phi}$$ $$\tan \theta = \frac{F_1 \sin \phi}{F_1 \cos \phi + F_2}$$
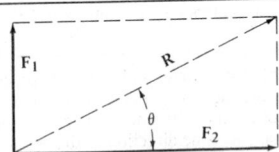	*Case II:* Forces F_1 and F_2 form 90° angle. $$R = \frac{F_2}{\cos \theta}, \quad \text{or,} \quad R = \frac{F_1}{\sin \theta}, \quad \text{or}$$ $$R = \sqrt{F_1^2 + F_2^2}$$ $$\tan \theta = \frac{F_1}{F_2}$$

Finding the Resultant of Three or More Concurrent Forces	
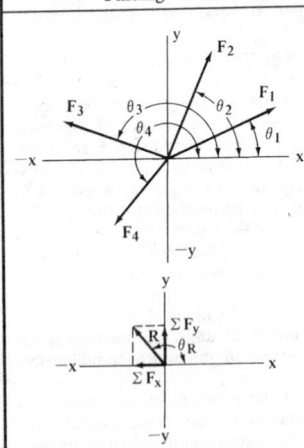	To determine resultant of forces F_1, F_2, F_3, etc. making angles, respectively, of θ_1, θ_2, θ_3, etc. with the x axis, find the x and y components F_x and F_y of each force and arrange in a table similar to that shown below for a system of three forces. Find the algebraic sum of the F_x and F_y components (ΣF_x and ΣF_y) and use these to determine resultant R.

Force	F_x	F_y
F_1	$F_1 \cos \theta_1$	$F_1 \sin \theta_1$
F_2	$F_2 \cos \theta_2$	$F_2 \sin \theta_2$
F_3	$F_3 \cos \theta_3$	$F_3 \sin \theta_3$
	ΣF_x	ΣF_y

$$R = \sqrt{(\Sigma F_x)^2 + (\Sigma F_y)^2}$$

$$\cos \theta_R = \frac{\Sigma F_x}{R}$$

$$\text{or,} \quad \tan \theta_R = \frac{\Sigma F_y}{\Sigma F_x}$$

Table 1 (*Continued*). **Algebraic Solution of Force Systems — All Forces in the Same Plane**

Finding a Force and a Couple Which Together are Equivalent to a Single Force	
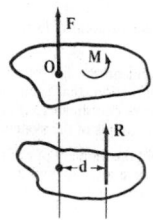	To resolve a single force F into a couple of moment M and a force P passing through any chosen point O at a distance d from the original force F, use the relations $$P = F$$ $$M = F \times d$$ The moment M must, of course, tend to produce rotation about O in the same direction as the original force. Thus, as seen in the diagram, F tends to produce clockwise rotation; hence M is shown clockwise.

Finding the Resultant of a Single Force and a Couple	
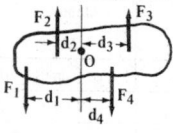	The resultant of a single force F and a couple M is a single force R equal in magnitude and direction to F and parallel to it at a distance d to the left or right of F. $$R = F$$ $$d = M \div R$$ Resultant R is placed to the left or right of point of application O of the original force F depending on which position will give R the same direction of moment about O as the original couple M.

Finding the Resultant of a System of Parallel Forces	
	To find the resultant of a system of coplanar parallel forces, proceed as indicated below.

1. Select any convenient point O from which perpendicular distances d_1, d_2, d_3, etc. to parallel forces F_1, F_2, F_3, etc. can be specified or calculated.
2. Find the algebraic sum of all the forces; this will give the magnitude of the resultant of the system.

$$R = \Sigma F = F_1 + F_2 + F_3 + \cdots$$

3. Find the algebraic sum of the moments of the forces about O; clockwise moments may be taken as negative and counterclockwise moments as positive:

$$\Sigma M_O = F_1 d_1 + F_2 d_2 + \cdots$$

4. Calculate the distance d from O to the line of action of resultant R:

$$d = \Sigma M_O \div R$$

This distance is measured to the left or right from O depending on which position will give the moment of R the same direction of rotation about O as the couple ΣM_O, that is, if ΣM_O is negative, then d is left or right of O depending on which direction will make $R \times d$ negative.

Note Concerning Interpretation of Results: If $R = 0$, then the resultant of the system is a couple ΣM_O; if $\Sigma M_O = 0$ then the resultant is a single force R; if both R and $\Sigma M_O = 0$, then the system is in equilibrium.

Table I (*Continued*). **Algebraic Solution of Force Systems — All Forces in the Same Plane**

Finding the Resultant of Forces Not Intersecting at a Common Point

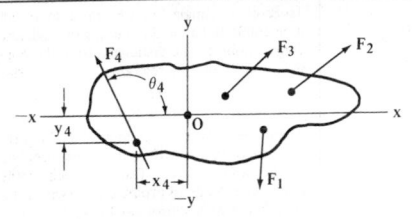

To determine the resultant of a coplanar, nonconcurrent, nonparallel force system as shown in the diagram, proceed as shown below.

1. Draw a set of x and y coordinate axes through any convenient point O in the plane of the forces as shown in the diagram.

2. Determine the x and y coordinates of any convenient point on the line of action of each force and the angle θ, measured in a counterclockwise direction, that each line of action makes with the positive x axis. For example, in the diagram, coordinates x_4, y_4, and θ_4 are shown for F_4. Similar data should be known for each of the forces of the system.

3. Calculate the x and y components (F_x, F_y) of each force and the moment of each component about O. Counterclockwise moments are considered positive and clockwise moments are negative. Tabulate all results in a manner similar to that shown below for a system of three forces and find ΣF_x, ΣF_y, ΣM_O by algebraic addition.

Force	Coordinates of F			Components of F		Moment of F about O
F	x	y	θ	F_x	F_y	$M_O = xF_y - yF_x$
F_1	x_1	y_1	θ_1	$F_1 \cos \theta_1$	$F_1 \sin \theta_1$	$x_1 F_1 \sin \theta_1 - y_1 F_1 \cos \theta_1$
F_2	x_2	y_2	θ_2	$F_2 \cos \theta_2$	$F_2 \sin \theta_2$	$x_2 F_2 \sin \theta_2 - y_2 F_2 \cos \theta_2$
F_3	x_3	y_3	θ_3	$F_3 \cos \theta_3$	$F_3 \sin \theta_3$	$x_3 F_3 \sin \theta_3 - y_3 F_3 \cos \theta_3$
				ΣF_x	ΣF_y	ΣM_O

4. Compute the resultant of the system and the angle θ_R it makes with the x axis by using the formulas:

$$R = \sqrt{(\Sigma F_x)^2 + (\Sigma F_y)^2}$$

$$\cos \theta_R = \Sigma F_x \div R, \quad \text{or,} \quad \tan \theta_R = \Sigma F_y \div \Sigma F_x$$

5. Calculate the distance d from O to the line of action of the resultant R:

$$d = \Sigma M_O \div R$$

Distance d is in such direction from O as will make the moment of R about O have the same sign as ΣM_O.

Note Concerning Interpretation of Results: If $R = 0$, then the resultant is a couple ΣM_O; if $\Sigma M_O = 0$, then R passes through O; if both $R = 0$ and $\Sigma M_O = 0$, then the system is in equilibrium.

Table 1 (*Concluded*). **Algebraic Solution of Force Systems — All Forces in the Same Plane**

Example: Find the resultant of three coplanar nonconcurrent forces for which the following data are given.

$$F_1 = 10 \text{ lbs}; \quad x_1 = 5 \text{ in.}; \quad y_1 = -1 \text{ in.}; \quad \theta_1 = 270°.$$
$$F_2 = 20 \text{ lbs}; \quad x_2 = 4 \text{ in.}; \quad y_2 = 1.5 \text{ in.}; \quad \theta_2 = 50°.$$
$$F_3 = 30 \text{ lbs}; \quad x_3 = 2 \text{ in.}; \quad y_3 = 2 \text{ in.}; \quad \theta_3 = 60°.$$

$$F_{x_1} = 10 \cos 270° = 10 \times 0 = 0 \text{ lbs.}$$
$$F_{x_2} = 20 \cos 50° = 20 \times 0.64279 = 12.86 \text{ lbs.}$$
$$F_{x_3} = 30 \cos 60° = 30 \times 0.5000 = 15.00 \text{ lbs.}$$

$$F_{y_1} = 10 \times \sin 270° = 10 \times (-1) = -10.00 \text{ lbs.}$$
$$F_{y_2} = 20 \times \sin 50° = 20 \times 0.76604 = 15.32 \text{ lbs.}$$
$$F_{y_3} = 30 \times \sin 60° = 30 \times 0.86603 = 25.98 \text{ lbs.}$$

$$M_{O_1} = 5 \times (-10) - (-1) \times 0 = -50 \text{ in. lbs.}$$
$$M_{O_2} = 4 \times 15.32 - 1.5 \times 12.86 = 41.99 \text{ in. lbs.}$$
$$M_{O_3} = 2 \times 25.98 - 2 \times 15 = 21.96 \text{ in. lbs.}$$

Note: When working in metric SI units, pounds are replaced by newtons (N); inches by meters or millimeters, and inch-pounds by newton-meters (N · m) or newton-millimeters (N · mm).

Force F	Coordinates of F			Components of F		Moment of F about O
	x	y	θ	F_x	F_y	
$F_1 = 10$	5	-1	270°	0	-10.00	-50.00
$F_2 = 20$	4	1.5	50°	12.86	15.32	41.99
$F_3 = 30$	2	2	60°	15.00	25.98	21.96
				27.86	31.30	13.95

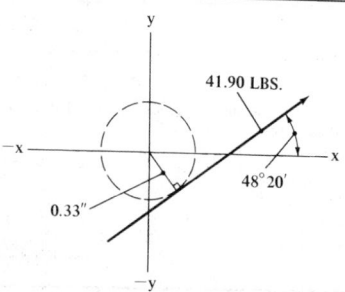

$$R = \sqrt{(27.86)^2 + (31.30)^2}$$
$$= 41.90 \text{ lbs.}$$
$$\tan \theta_R = \frac{31.30}{27.86} = 1.1235$$
$$\theta_R = 48° \, 20'$$
$$d = \frac{13.95}{41.90} = 0.33 \text{ inches}$$

measured as shown on the diagram.

41.90 LBS.

48° 20'

0.33"

Table 2. Algebraic Solution of Force Systems — Forces Not in Same Plane

Resolving a Single Force Into Its Three Rectangular Components

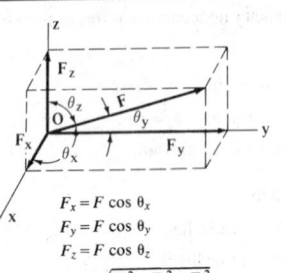

The diagram shows how a force F may be resolved at any point O on its line of action into three concurrent components each of which is perpendicular to the other two.

The x, y, z components F_x, F_y, F_z of force F are determined from the accompanying relations in which θ_x, θ_y, θ_z are the angles which the force F makes with the x, y, z axes.

$$F_x = F \cos \theta_x$$
$$F_y = F \cos \theta_y$$
$$F_z = F \cos \theta_z$$
$$F = \sqrt{F_x^2 + F_y^2 + F_z^2}$$

Finding the Resultant of Any Number of Concurrent Forces

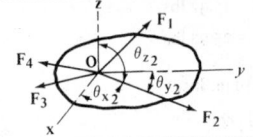

To find the resultant of any number of non-coplanar concurrent forces F_1, F_2, F_3, etc., use the procedure outlined below.

1. Draw a set of x, y, z axes at O, the point of concurrency of the forces. The angles each force makes measured counterclockwise from the positive x, y, and z coordinate axes must be known in addition to the magnitudes of the forces. For force F_2, for example, the angles are θ_{x2}, θ_{y2}, θ_{z2} as indicated on the diagram.
2. Apply the first three formulas given under the heading "Resolving a Single Force Into Its Three Rectangular Components" to each force to find its x, y, and z components. Tabulate these calculations as shown below for a system of three forces. Algebraically add the calculated components to find ΣF_x, ΣF_y, and ΣF_z which are the components of the resultant.

Force	Angles			Components of Forces		
F	θ_x	θ_y	θ_z	F_x	F_y	F_z
F_1	θ_{x1}	θ_{y1}	θ_{z1}	$F_1 \cos \theta_{x1}$	$F_1 \cos \theta_{y1}$	$F_1 \cos \theta_{z1}$
F_2	θ_{x2}	θ_{y2}	θ_{z2}	$F_2 \cos \theta_{x2}$	$F_2 \cos \theta_{y2}$	$F_2 \cos \theta_{z2}$
F_3	θ_{x3}	θ_{y3}	θ_{z3}	$F_3 \cos \theta_{x3}$	$F_3 \cos \theta_{y3}$	$F_3 \cos \theta_{z3}$
				ΣF_x	ΣF_y	ΣF_z

3. Find the resultant of the system from the formula
$$R = \sqrt{(\Sigma F_x)^2 + (\Sigma F_y)^2 + (\Sigma F_z)^2}$$

4. Calculate the angles θ_{xR}, θ_{yR}, and θ_{zR} that the resultant R makes with the respective coordinate axes:
$$\cos \theta_{xR} = \frac{\Sigma F_x}{R}$$
$$\cos \theta_{yR} = \frac{\Sigma F_y}{R}$$
$$\cos \theta_{zR} = \frac{\Sigma F_z}{R}$$

Table 2 (*Continued*). **Algebraic Solution of Force Systems — Forces Not in Same Plane**

Finding the Resultant of Parallel Forces Not in the Same Plane

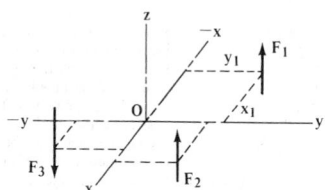

In the diagram, forces F_1, F_2, etc. represent a system of noncoplanar parallel forces. To find the resultant of such systems, use the procedure shown below.

1. Draw a set of x, y, and z coordinate axes through any point O in such a way that one of these axes, say the z axis, is parallel to the lines of action of the forces. The x and y axes then will be perpendicular to the forces.

2. Set the distances of each force from the x and y axes in a table as shown below. For example, x_1 and y_1 are the x and y distances for F_1 shown in the diagram.

3. Calculate the moment of each force about the x and y axes and set the results in the table as shown for a system consisting of three forces. The algebraic sums of the moments ΣM_x and ΣM_y are then obtained. (In taking moments about the x and y axes, assign counterclockwise moments a plus ($+$) sign and clockwise moments a minus ($-$) sign. In deciding whether a moment is counterclockwise or clockwise, look from the positive side of the axis in question toward the negative side.)

Force	Coordinates of Force F		Moments M_x and M_y due to F	
F	x	y	M_x	M_y
F_1	x_1	y_1	$F_1 y_1$	$F_1 x_1$
F_2	x_2	y_2	$F_2 y_2$	$F_2 x_2$
F_3	x_3	y_3	$F_3 y_3$	$F_3 x_3$
ΣF			ΣM_x	ΣM_y

4. Find the algebraic sum ΣF of all the forces; this will be the resultant R of the system.

$$R = \Sigma F = F_1 + F_2 + \cdots$$

5. Calculate x_R and y_R, the moment arms of the resultant:

$$x_R = \Sigma M_y \div R$$
$$y_R = \Sigma M_x \div R$$

These moment arms are measured in such direction along the x and y axes as will give the resultant a moment of the same direction of rotation as ΣM_x and ΣM_y.

Note Concerning Interpretation of Results: If ΣM_x and ΣM_y are both 0, then the resultant is a single force R along the z axis; if R is also 0, then the system is in equilibrium. If R is 0 but ΣM_x and ΣM_y are not both 0, then the resultant is a couple

$$M_R = \sqrt{(\Sigma M_x)^2 + (\Sigma M_y)^2}$$

that lies in a plane parallel to the z axis and making an angle θ_R measured in a counterclockwise direction from the positive x axis and calculated from the following formula:

$$\sin \theta_R = \frac{\Sigma M_x}{M_R}$$

Table 2 (*Continued*). **Algebraic Solution of Force Systems — Forces Not in Same Plane**

Finding the Resultant of Nonparallel Forces Not Meeting at a Common Point

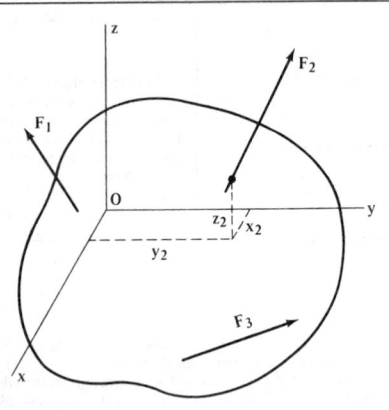

The diagram shows a system of noncoplanar, nonparallel, nonconcurrent forces F_1, F_2, etc. for which the resultant is to be determined. Generally speaking, the resultant will be a noncoplanar force and a couple which may be further combined, if desired, into two forces which are skewed.

Since this is the most general force system that can be devised, each of the other systems so far described represents a special, simpler case of this general force system. The method of solution described below for a system of three forces applies for any number of forces.

1. Select a set of coordinate x, y, and z axes at any desired point O in the body as shown in the diagram.

2. Determine the x, y, and z coordinates of any convenient point on the line of action of each force as shown for F_2. Also determine the angles θ_x, θ_y, θ_z that each force makes with each coordinate axis. These angles are measured counterclockwise from the positive direction of the x, y, and z axes. This data is tabulated, as shown in the table accompanying Step 3, for convenient use in subsequent calculations.

3. Calculate the x, y, and z components of each force using the formulas given in the accompanying table. Add these components algebraically to get ΣF_x, ΣF_y and ΣF_z which are the components of the resultant, R, given by the formula,

$$R = \sqrt{(\Sigma F_x)^2 + (\Sigma F_y)^2 + (\Sigma F_z)^2}$$

Force	Coordinates of Force F						Components of F		
F	x	y	z	θ_x	θ_y	θ_z	F_x	F_y	F_z
F_1	x_1	y_1	z_1	θ_{x1}	θ_{y1}	θ_{z1}	$F_1 \cos \theta_{x1}$	$F_1 \cos \theta_{y1}$	$F_1 \cos \theta_{z1}$
F_2	x_2	y_2	z_2	θ_{x2}	θ_{y2}	θ_{z2}	$F_2 \cos \theta_{x2}$	$F_2 \cos \theta_{y2}$	$F_2 \cos \theta_{z2}$
F_3	x_3	y_3	z_3	θ_{x3}	θ_{y3}	θ_{z3}	$F_3 \cos \theta_{x3}$	$F_3 \cos \theta_{y3}$	$F_3 \cos \theta_{z3}$
							ΣF_x	ΣF_y	ΣF_z

The resultant force R makes angles of θ_{xR}, θ_{yR}, and θ_{zR} with the x, y, and z axes, respectively, and passes through the selected point O. These angles are determined from the formulas,

$$\cos \theta_{xR} = \Sigma F_x \div R$$
$$\cos \theta_{yR} = \Sigma F_y \div R$$
$$\cos \theta_{zR} = \Sigma F_z \div R$$

Table 2 (*Concluded*). **Algebraic Solution of Force Systems — Forces Not in Same Plane**

4. Calculate the moments M_x, M_y, M_z about x, y, and z axes, respectively due to the F_x, F_y, and F_z components of each force and set them in tabular form. The formulas to use are given in the accompanying table.

   In interpreting moments about the x, y, and z axes, consider counterclockwise moments a plus (+) sign and clockwise moments a minus (−) sign. In deciding whether a moment is counterclockwise or clockwise, look from the positive side of the axis in question toward the negative side.

Force	Moments of Components of F (F_x, F_y, F_z) about x, y, z axes		
F	$M_x = yF_z - zF_y$	$M_y = zF_x - xF_z$	$M_z = xF_y - yF_x$
F_1	$M_{x1} = y_1F_{z1} - z_1F_{y1}$	$M_{y1} = z_1F_{x1} - x_1F_{z1}$	$M_{z1} = x_1F_{y1} - y_1F_{x1}$
F_2	$M_{x2} = y_2F_{z2} - z_2F_{y2}$	$M_{y2} = z_2F_{x2} - x_2F_{z2}$	$M_{z2} = x_2F_{y2} - y_2F_{x2}$
F_3	$M_{x3} = y_3F_{z3} - z_3F_{y3}$	$M_{y3} = z_3F_{x3} - x_3F_{z3}$	$M_{z3} = x_3F_{y3} - y_3F_{x3}$
	ΣM_x	ΣM_y	ΣM_z

5. Add the component moments algebraically to get ΣM_x, ΣM_y and ΣM_z which are the components of the resultant couple, M, given by the formula,

$$M = \sqrt{(\Sigma M_x)^2 + (\Sigma M_y)^2 + (\Sigma M_z)^2}$$

The resultant couple M will tend to produce rotation about an axis making angles of β_x, β_y, and β_z with the x, y, z axes, respectively. These angles are determined from the formulas,

$$\cos \beta_x = \frac{\Sigma M_x}{M} \qquad \cos \beta_y = \frac{\Sigma M_y}{M} \qquad \cos \beta_z = \frac{\Sigma M_z}{M}$$

General Method of Locating Resultant When Its Components are Known

To determine the position of the resultant force of a system of forces, proceed as follows:

From the origin, point O, of a set of coordinate axes x, y, z, lay off on the x axis a length A representing the algebraic sum ΣF_x of the x components of all the forces. From the end of line A lay off a line B representing ΣF_y, the algebraic sum of the y components; this line B is drawn in a direction parallel to the y axis. From the end of line B lay off a line C representing ΣF_z. Finally, draw a line R from O to the end of C; R will be the resultant of the system.

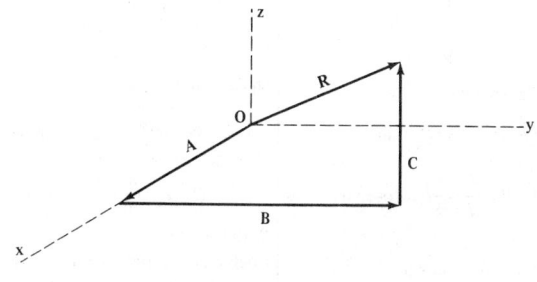

Inclined Plane — Wedge

W = weight of body.

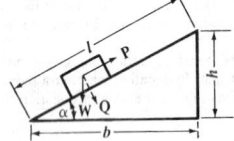

Neglecting friction:

$$P = W \times \frac{h}{l} = W \times \sin \alpha$$

$$W = P \times \frac{l}{h} = \frac{P}{\sin \alpha} = P \times \operatorname{cosec} \alpha$$

$$Q = W \times \frac{b}{l} = W \times \cos \alpha$$

If friction is taken into account, then force P to pull body up is:

$$P = W (\mu \cos \alpha + \sin \alpha)$$

Force P_1 to pull body down is:

$$P_1 = W (\mu \cos \alpha - \sin \alpha)$$

Force P_2 to hold body stationary:

$$P_2 = W (\sin \alpha - \mu \cos \alpha)$$

in which μ is the coefficient of friction.

W = weight of body.

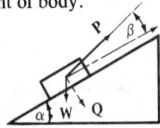

Neglecting friction:

$$P = W \times \frac{\sin \alpha}{\cos \beta}$$

$$W = P \times \frac{\cos \beta}{\sin \alpha}$$

$$Q = W \times \frac{\cos (\alpha + \beta)}{\cos \beta}$$

With friction:

Coefficient of friction $= \mu = \tan \phi$.

$$P = W \times \frac{\sin (\alpha + \phi)}{\cos (\beta - \phi)}$$

W = weight of body.

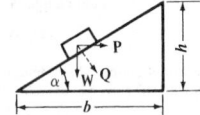

Neglecting friction:

$$P = W \times \frac{h}{b} = W \times \tan \alpha$$

$$W = P \times \frac{b}{h} = P \times \cot \alpha$$

$$Q = \frac{W}{\cos \alpha} = W \times \sec \alpha$$

With friction:

Coefficient of friction $= \mu = \tan \phi$.

$$P = W \tan (\alpha + \phi)$$

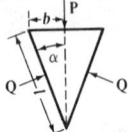

Neglecting friction:

$$P = 2Q \times \frac{b}{l} = 2Q \times \sin \alpha$$

$$Q = P \times \frac{l}{2b} = \frac{1}{2} P \times \operatorname{cosec} \alpha$$

With friction:

Coefficient of friction $= \mu$.

$$P = 2Q (\mu \cos \alpha + \sin \alpha)$$

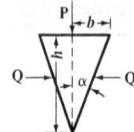

Neglecting friction:

$$P = 2Q \times \frac{b}{h} = 2Q \times \tan \alpha$$

$$Q = P \times \frac{h}{2b} = \frac{1}{2} P \times \cot \alpha$$

With friction:

Coefficient of friction $= \mu = \tan \phi$.

$$P = 2Q \tan (\alpha + \phi)$$

Levers*

Types of Levers	Examples
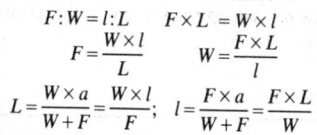 $F:W=l:L \qquad F\times L = W\times l$ $F=\dfrac{W\times l}{L} \qquad W=\dfrac{F\times L}{l}$ $L=\dfrac{W\times a}{W+F}=\dfrac{W\times l}{F}; \quad l=\dfrac{F\times a}{W+F}=\dfrac{F\times L}{W}$	A pull of 80 pounds is exerted at the end of the lever, at W; $l = 12$ inches and $L = 32$ inches. Find the value of force F required to balance the lever. $$F=\frac{80\times 12}{32}=\frac{960}{32}=30 \text{ pounds.}$$ If $F = 20$; $W = 180$; and $l = 3$; how long must L be made to secure equilibrium? $$L=\frac{180\times 3}{20}=27.$$
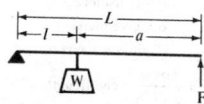 $F:W=l:L \qquad F\times L = W\times l$ $F=\dfrac{W\times l}{L} \qquad W=\dfrac{F\times L}{l}$ $L=\dfrac{W\times a}{W-F}=\dfrac{W\times l}{F}; \quad l=\dfrac{F\times a}{W-F}=\dfrac{F\times L}{W}$	Total length L of a lever is 25 inches. A weight of 90 pounds is supported at W; l is 10 inches. Find the value of F. $$F=\frac{90\times 10}{25}=36 \text{ pounds.}$$ If $F = 100$ pounds, $W = 2200$ pounds, and $a = 5$ feet, what should L equal to secure equilibrium? $$L=\frac{2200\times 5}{2200-100}=5.24 \text{ feet.}$$
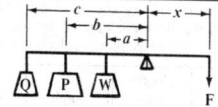 When three or more forces act on lever: $F\times x = W\times a + P\times b + Q\times c$ $x=\dfrac{W\times a+P\times b+Q\times c}{F}$ $F=\dfrac{W\times a+P\times b+Q\times c}{x}$	Let $W = 20$, $P = 30$, and $Q = 15$ pounds; $a = 4$, $b = 7$, and $c = 10$ inches. If $x = 6$ inches, find F. $$F=\frac{20\times 4+30\times 7+15\times 10}{6}=73\tfrac{1}{3}\text{ lbs.}$$ Assuming $F = 20$ in the example above, how long must lever arm x be made? $$x=\frac{20\times 4+30\times 7+15\times 10}{20}=22 \text{ ins.}$$

* The above formulas are valid using metric SI units, with forces expressed in newtons, and lengths in meters. However, it should be noted that the weight of a mass W kilograms is equal to a force of Wg newtons, where g is approximately 9.81 m/s². Thus, supposing that in the first example $l = 0.4$ m, $L = 1.2$ m, and $W = 30$ kg, then the weight of W is $30g$ newtons, so that the force F required to balance the lever is

$$F=\frac{30g\times 0.4}{1.2}=10g=98.1 \text{ newtons.}$$

This force could be produced by suspending a mass of 10 kg at F.

Toggle-joint. — If arms *ED* and *EH* are of unequal length:

$$P = \frac{Fa}{b}$$

The relation between P and F changes constantly as F moves downwards.

If arms *ED* and *EH* are equal:

$$P = \frac{Fa}{2h}$$

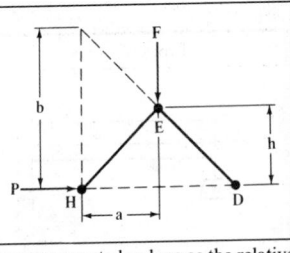

A double toggle-joint does not increase the pressure exerted so long as the relative distances moved by F and P remain the same.

Toggle-joints with Equal Arms

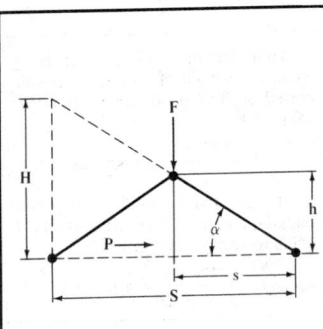

F = force applied;
P = resistance;
α = given angle.

$$2P \sin \alpha = F \cos \alpha;$$

$$\frac{P}{F} = \frac{\cos \alpha}{2 \sin \alpha} = \text{coefficient};$$

or, $P = F \times$ coefficient.

Equivalent expressions:

$$P = \frac{FS}{4h}; \qquad P = \frac{Fs}{H}, \text{ as per diagram.}$$

To use the table, measure angle α, and find the coefficient in the table corresponding to the angle found. The coefficient is the ratio of the resistance to the force applied, and multiplying the force applied by the coefficient gives the resistance, neglecting friction.

Angle	Coefficient	Angle	Coefficient	Angle	Coefficient	Angle	Coefficient
0° 2′	862	0° 50′	34.4	2° 45′	10.4	8° 0′	3.58
0 4	456	0 55	31.2	2 50	10.1	8 30	3.35
0 6	285	1 0	28.6	3 0	9.54	9 0	3.15
0 8	216	1 10	24.6	3 15	8.81	9 30	2.99
0 10	171	1 15	22.9	3 30	8.17	10 0	2.84
0 12	143	1 20	21.5	3 45	7.63	11 0	2.57
0 14	122	1 30	19.1	4 0	7.25	12 0	2.35
0 15	115	1 40	17.2	4 15	6.73	13 0	2.17
0 16	107	1 45	16.4	4 30	6.35	14 0	2.00
0 18	95.4	1 50	15.6	4 45	6.02	15 0	1.87
0 20	85.8	2 0	14.3	5 0	5.71	16 0	1.74
0 25	68.6	2 10	13.2	5 30	5.19	17 0	1.64
0 30	57.3	2 15	12.7	6 0	4.76	18 0	1.54
0 35	49.1	2 20	12.5	6 30	4.39	19 0	1.45
0 40	42.8	2 30	11.5	7 0	4.07	20 0	1.37
0 45	38.2	2 40	10.7	7 30	3.79	...	...

Wheels and Pulleys*

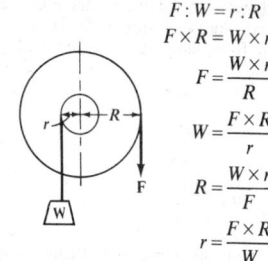

$$F : W = r : R$$

$$F \times R = W \times r$$

$$F = \frac{W \times r}{R}$$

$$W = \frac{F \times R}{r}$$

$$R = \frac{W \times r}{F}$$

$$r = \frac{F \times R}{W}$$

The radius of a drum on which is wound the lifting rope of a windlass is 2 inches. What force will be exerted at the periphery of a gear of 24 inches diameter, mounted on the same shaft as the drum and transmitting power to it, if one ton (2000 pounds) is to be lifted? Here $W = 2000$; $R = 12$; $r = 2$.

$$F = \frac{2000 \times 2}{12} = 333 \text{ pounds.}$$

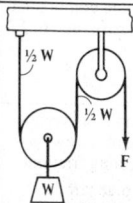

$$F = \frac{1}{2}W$$

The velocity with which weight W will be raised equals one-half the velocity of the force applied at F.

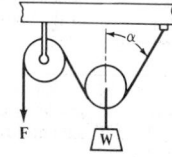

$$F : W = \sec \alpha : 2$$

$$F = \frac{W \times \sec \alpha}{2}$$

$$W = 2F \times \cos \alpha$$

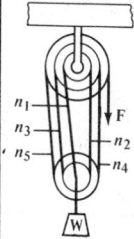

n = number of strands or parts of rope (n_1, n_2, etc.).

$$F = \frac{1}{n} \times W$$

The velocity with which W will be raised equals $\frac{1}{n}$ of the velocity of the force applied at F.

In the illustration is shown a combination of a double and triple block. The pulleys each turn freely on a pin as axis, and are drawn with different diameters, to show the parts of the rope more clearly. There are 5 parts of rope. Therefore, if 200 pounds is to be lifted, the force F required at the end of the rope is:

$$F = \frac{1}{5} \times 200 = 40 \text{ pounds.}$$

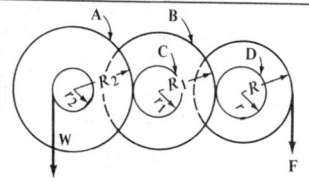

A, B, C and D are the pitch circles of gears.

$$F = \frac{W \times r \times r_1 \times r_2}{R \times R_1 \times R_2}$$

$$W = \frac{F \times R \times R_1 \times R_2}{r \times r_1 \times r_2}$$

Let the pitch diameters of gears A, B, C and D be 30, 28, 12 and 10 inches, respectively. Then $R_2 = 15$; $R_1 = 14$; $r_1 = 6$; and $r = 5$. Let $R = 12$, and $r_2 = 4$. Then the force F required to lift a weight W of 2000 pounds, friction being neglected, is:

$$F = \frac{2000 \times 5 \times 6 \times 4}{12 \times 14 \times 15} = 95 \text{ pounds.}$$

---

* *Note:* The above formulas are valid using metric SI units, with forces expressed in newtons, and lengths in meters or millimeters. (See note on page 147 concerning weight and mass).

Differential Pulley — Screw

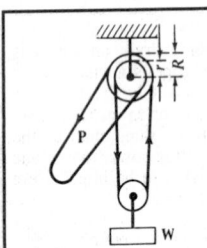

Differential Pulley. — In the differential pulley a chain must be used, engaging sprockets, so as to prevent the chain from slipping over the pulley faces.

$$P \times R = \frac{1}{2} W(R - r)$$

$$P = \frac{W(R - r)}{2R}$$

$$W = \frac{2PR}{R - r}$$

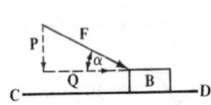

Force Moving Body on Horizontal Plane. — F tends to move B along line CD; Q is the component which actually moves B; P is the pressure, due to F, of the body on CD.

$$Q = F \times \cos \alpha; \qquad P = \sqrt{F^2 - Q^2}$$

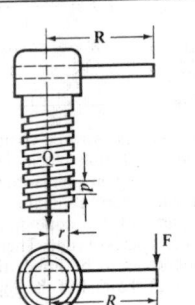

Screw. — F = force at end of handle or wrench; R = lever-arm of F; r = pitch radius of screw; p = lead of thread; Q = load. Then, neglecting friction:

$$F = Q \times \frac{p}{6.2832R} \qquad Q = F \times \frac{6.2832R}{p}$$

If μ is the coefficient of friction, then:

For motion in direction of load Q which *assists* it:

$$F = Q \times \frac{6.2832\mu r - p}{6.2832r + \mu p} \times \frac{r}{R}$$

For motion opposite load Q which *resists* it:

$$F = Q \times \frac{p + 6.2832\mu r}{6.2832r - \mu p} \times \frac{r}{R}$$

Center of Gravity. — The center of gravity of a body, volume, area, or line is that point at which if the body, volume, area, or line were suspended it would be perfectly balanced in all positions. For symmetrical bodies of uniform material it is at the geometric center. The center of gravity of a uniform round rod, for example, is at the center of its diameter halfway along its length; the center of gravity of a sphere is at the center of the sphere. For solids, areas, and arcs that are not symmetrical, the determination of the center of gravity may be made experimentally or may be calculated by the use of formulas.

The tables that follow give such formulas for some of the more important shapes. For more complicated and unsymmetrical shapes the methods outlined on page 156 may be used.

Example: A piece of wire is bent into the form of a semi-circular arc of 10-inch radius. How far from the center of the arc is the center of gravity located?

Accompanying the third diagram on page 151 is a formula for the distance from the center of gravity of an arc to the center of the arc: $a = 2r \div \pi$. Therefore, in this case,

$$a = 2 \times 10 \div 3.1416 = 6.366 \text{ inches.}$$

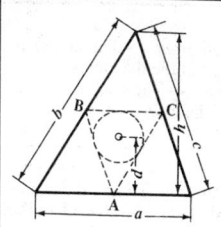

Perimeter of a Triangle. — If A, B and C are the middle points of the sides of the triangle, then the center of gravity is at the center of the circle that can be inscribed in triangle ABC. The distance d of the center of gravity from side a is:

$$d = \frac{h(b + c)}{2(a + b + c)}$$

where h is the height perpendicular to a.

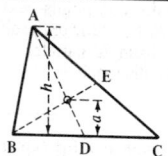

Area of Triangle. — The center of gravity is at the intersection of lines AD and BE, which bisect the sides BC and AC. The perpendicular distance from the center of gravity to any one of the sides is equal to one-third the height perpendicular to that side. Hence, $a = h \div 3$.

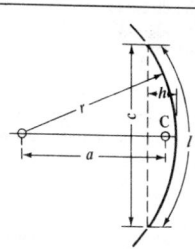

Circular Arc. — The center of gravity is on the line that bisects the arc, at a distance

$$a = \frac{r \times c}{l} = \frac{c(c^2 + 4h^2)}{8lh}$$ from the center of the circle.

For an arc equal to one-half the periphery:

$$a = 2r \div \pi = 0.6366r$$

For an arc equal to one-quarter of the periphery:

$$a = 2r\sqrt{2} \div \pi = 0.9003r$$

For an arc equal to one-sixth of the periphery:

$$a = 3r \div \pi = 0.9549r$$

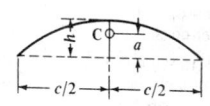

Circular Arc (approximate). —

$$a = \tfrac{2}{3}h$$

This formula is very nearly exact for all arcs less than one-quarter of the periphery. The error is only about one per cent for a quarter circle, and decreases for smaller arcs.

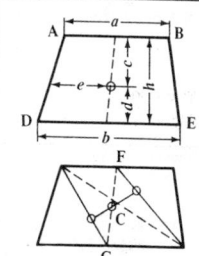

Area of Trapezoid. — The center of gravity is on the line joining the middle points of parallel lines AB and DE.

$$c = \frac{h(a + 2b)}{3(a + b)} \qquad d = \frac{h(2a + b)}{3(a + b)}$$
$$e = \frac{a^2 + ab + b^2}{3(a + b)}$$

The trapezoid can also be divided into two triangles. The center of gravity is at the intersection of the line joining the centers of gravity of the triangles, and the middle line FG.

Center of Gravity

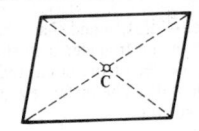

Perimeter of a Parallelogram. — The center of gravity is at the intersection of the diagonals.

Area of a Parallelogram. — The center of gravity is at the intersection of the diagonals.

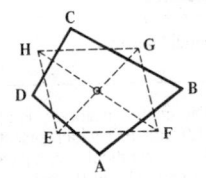

Any Four-sided Figure. — Two cases are possible, as shown in the illustration. To find the center of gravity of the four-sided figure *ABCD*, each of the sides is divided into three equal parts. A line is then drawn through each pair of division points next to the points of intersection *A*, *B*, *C*, and *D* of the sides of the figure. These lines form a parallelogram *EFGH*; the intersection of the diagonals *EG* and *FH* locates center of gravity.

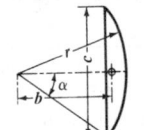

Circle Segment. — The distance of the center of gravity from the center of the circle is:

$$b = \frac{c^3}{12A} = \frac{2}{3} \times \frac{r^3 \sin^3 \alpha}{A}$$

in which A = area of segment.

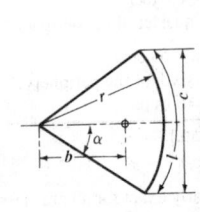

Circle Sector. — Distance b from center of gravity to center of circle is:

$$b = \frac{2rc}{3l} = \frac{r^2 c}{3A} = 38.197 \frac{r \sin \alpha}{\alpha}$$

in which A = area of sector, and α is expressed in degrees.

For the area of a half-circle:
$$b = 4r \div 3\pi = 0.4244r$$
For the area of a quarter circle:
$$b = 4\sqrt{2} \times r \div 3\pi = 0.6002r$$
For the area of a sixth of a circle:
$$b = 2r \div \pi = 0.6366r$$

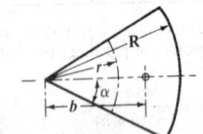

Part of Circle Ring. — Distance b from center of gravity to center of circle is:

$$b = 38.197 \frac{(R^3 - r^3) \sin \alpha}{(R^2 - r^2)\alpha}$$

Angle α is expressed in degrees.

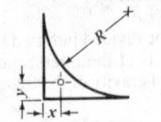

Spandrel or Fillet.

$$\text{Area} = .2146R^2 \qquad \begin{aligned} x &= .2234R \\ y &= .2234R \end{aligned}$$

Center of Gravity

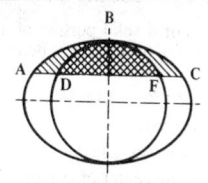

Segment of an Ellipse. — The center of gravity of an elliptic segment *ABC*, symmetrical about one of the axes, coincides with the center of gravity of the segment *DBF* of a circle, the diameter of which is equal to that axis of the ellipse about which the elliptic segment is symmetrical.

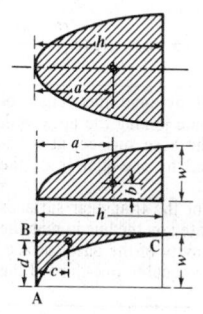

Area of a Parabola. — For the complete parabolic area, the center of gravity is on the center line or axis, and

$$a = \frac{3h}{5}$$

For one-half of the parabola:

$$a = \frac{3h}{5} \quad \text{and} \quad b = \frac{3w}{8}$$

For the complement area *ABC*:

$$c = 0.3h \quad \text{and} \quad d = 0.75w$$

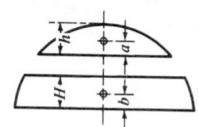

Sperical Surface of Segments and Zones of Spheres. — Distances *a* and *b* which determine the center of gravity, are:

$$a = \frac{h}{2} \qquad b = \frac{H}{2}$$

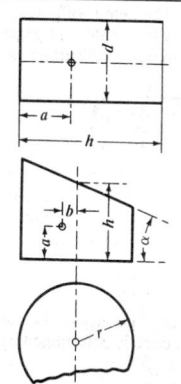

Cylinder. — The center of gravity of a solid cylinder (or prism) with parallel end surfaces, is located at the middle of the line that joins the centers of gravity of the end surfaces.

The center of gravity of a cylindrical surface or shell, with the base or end surface in one end, is found from:

$$a = \frac{2h^2}{4h + d}$$

The center of gravity of a cylinder cut off by an inclined plane is located by:

$$a = \frac{h}{2} + \frac{r^2 \tan^2 \alpha}{8h} \qquad b = \frac{r^2 \tan \alpha}{4h}$$

where α is the angle between the obliquely cut off surface and the base surface.

Center of Gravity

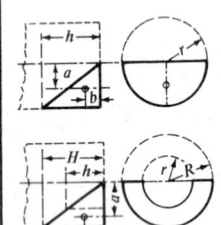

Portion of Cylinder. — For a solid portion of a cylinder, as shown, the center of gravity is determined by:

$$a = \tfrac{3}{16} \times 3.1416r \qquad b = \tfrac{3}{32} \times 3.1416h$$

For the cylindrical surface only:

$$a = \tfrac{1}{4} \times 3.1416r \qquad b = \tfrac{1}{8} \times 3.1416h$$

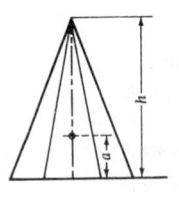

If the cylinder is hollow, the center of gravity of the solid shell is found by:

$$a = \tfrac{3}{16} \times 3.1416 \frac{R^4 - r^4}{R^3 - r^3}; \quad b = \tfrac{3}{32} \times 3.1416 \frac{H^4 - h^4}{H^3 - h^3}$$

Pyramid. — In a solid pyramid the center of gravity is located on the line joining the apex with the center of gravity of the base surface, at a distance from the base equal to one-quarter of the height; or $a = \tfrac{1}{4}h$.

The center of gravity of the triangular surfaces forming the pyramid is located on the line joining the apex with the center of gravity of the base surface, at a distance from the base equal to one-third of the height; or $a = \tfrac{1}{3}h$.

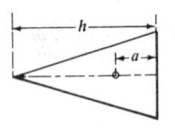

Cone. — The same rules apply as for the pyramid.

For the solid cone:

$$a = \tfrac{1}{4}h$$

For the conical surface:

$$a = \tfrac{1}{3}h$$

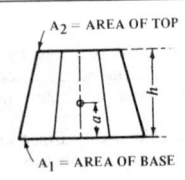

A₂ = AREA OF TOP

A₁ = AREA OF BASE

Frustum of Pyramid. — The center of gravity is located on the line that joins the centers of gravity of the end surfaces. If A_1 = area of base surface, and A_2 area of top surface,

$$a = \frac{h(A_1 + 2\sqrt{A_1 \times A_2} + 3A_2)}{4(A_1 + \sqrt{A_1 \times A_2} + A_2)}$$

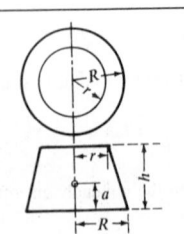

Frustum of Cone. — The same rules apply as for the frustum of a pyramid. For a solid frustum of a circular cone the formula below is also used:

$$a = \frac{h(R^2 + 2Rr + 3r^2)}{4(R^2 + Rr + r^2)}$$

The location of the center of gravity of the conical surface of a frustum of a cone is determined by:

$$a = \frac{h(R + 2r)}{3(R + r)}$$

Center of Gravity

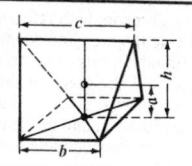	*Wedge.* — The center of gravity is on the line joining the center of gravity of the base with the middle point of the edge, and is located at: $$a = \frac{h(b+c)}{2(2b+c)}$$
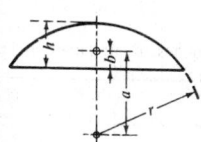	*Spherical Segment.* — The center of gravity of a solid segment is determined by: $$a = \frac{3(2r-h)^2}{4(3r-h)}$$ $$b = \frac{h(4r-h)}{4(3r-h)}$$ For a half-sphere, $a = b = \frac{3}{8}r$
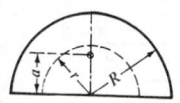	*Half of a Hollow Sphere.* — The center of gravity is located at: $$a = \frac{3(R^4-r^4)}{8(R^3-r^3)}$$
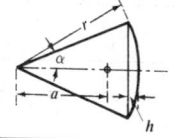	*Spherical Sector.* — The center of gravity of a solid sector is at: $$a = \frac{3}{8}(1 + \cos \alpha)r = \frac{3}{8}(2r-h)$$
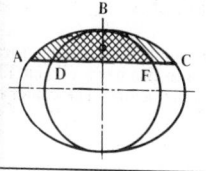	*Segment of Ellipsoid or Spheroid.* — The center of gravity of a solid segment ABC, symmetrical about the axis of rotation, coincides with the center of gravity of the segment DBF of a sphere, the diameter of which is equal to the axis of rotation of the spheroid.
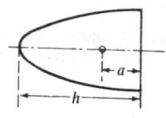	*Paraboloid.* — The center of gravity of a solid paraboloid of rotation is at: $$a = \frac{1}{3}h$$
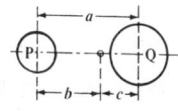	*Center of Gravity of Two Bodies.* — If the weights of the bodies are P and Q, and the distance between their centers of gravity is a, then: $$b = \frac{Qa}{P+Q} \qquad c = \frac{Pa}{P+Q}$$

Center of Gravity of Figures of any Outline. — If the figure is symmetrical about a center line, as in Fig. 1, the center of gravity will be located on that line. To find the exact location on that line, the simplest method is by taking moments with reference to any convenient axis at right angles to this center line. Divide the area into geometrical figures, the centers of gravity of which can be easily found. In this case, divide the figure into three rectangles *KLMN, EFGH* and *OPRS*. Call the areas of these rectangles *A, B* and *C*, respectively, and find the center of gravity of each. Then select any convenient axis, as *XX*, at right angles to the center line *YY*, and determine distances *a, b* and *c*. The distance *y* of the center of gravity of the complete figure from the axis *XX* is then found from the equation:

$$y = \frac{Aa + Bb + Cc}{A + B + C}$$

As an example, assume that the area *A* is 24 square inches, *B*, 14 square inches

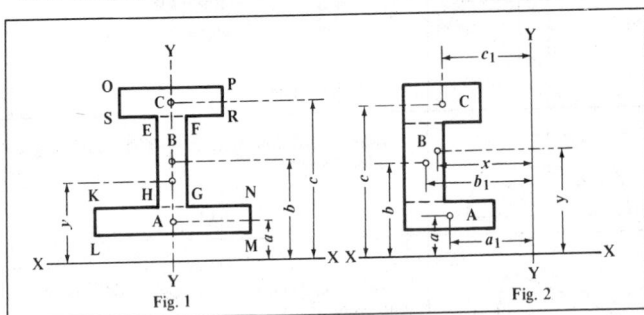

Fig. 1   Fig. 2

and *C*, 16 square inches, and that *a* = 3 inches, *b* = 7.5 inches, and *c* = 12 inches Then:

$$y = \frac{24 \times 3 + 14 \times 7.5 + 16 \times 12}{24 + 14 + 16} = \frac{369}{54} = 6.83 \text{ inches.}$$

If the figure, the center of gravity of which is to be found, is not symmetrical about any axis, then moments must be taken with relation to two axes *XX* and *YY* centers of gravity of which can be easily found, the same as before. The center of gravity is determined by the equations:

$$x = \frac{Aa_1 + Bb_1 + Cc_1}{A + B + C} \qquad y = \frac{Aa + Bb + Cc}{A + B + C}$$

As an example, let *A* = 14 square inches, *B* = 18 square inches, and *C* = 20 square inches. Let *a* = 3 inches, *b* = 7 inches, and *c* = 11.5 inches. Let *a₁* = 6.5 inches, *b₁* = 8.5 inches, and *c₁* = 7 inches. Then:

$$x = \frac{14 \times 6.5 + 18 \times 8.5 + 20 \times 7}{14 + 18 + 20} = \frac{384}{52} = 7.38 \text{ inches.}$$

$$y = \frac{14 \times 3 + 18 \times 7 + 20 \times 11.5}{14 + 18 + 20} = \frac{398}{52} = 7.65 \text{ inches.}$$

In other words, the center of gravity is located at a distance of 7.65 inches from the axis *XX* and 7.38 inches from the axis *YY*.

Moments of Inertia. — An important property of areas and solid bodies is the moment of inertia. Standard formulas are derived by multiplying elementary particles of area or mass by the squares of their distances from reference axes. Moments of inertia, therefore, depend on the location of reference axes. Values are minimum when these axes pass through the centers of gravity.

Three kinds of moments of inertia occur in engineering formulas:

(1) *Moments of inertia of plane area, I,* in which the axis is in the plane of the area, are found in formulas for calculating deflections and stresses in beams. When dimensions are given in inches, the units of I are inches^4. A table of formulas for calculating the I of common areas can be found in the Strength of Materials section.

(2) *Polar moments of inertia of plane areas, J,* in which the axis is at right angles to the plane of the area, occur in formulas for the torsional strength of shafting. When dimensions are given in inches, the units of J are inches^4. If moments of inertia, I, are known for a plane area with respect to both x and y axes, then the polar moment for the z axis may be calculated using the equation,

$$J_z = I_x + I_y$$

A table of formulas for calculating J for common areas can be found in the Shafting section.

When metric SI units are used, the formulas referred to in (1) and (2) above, are valid if the dimensions are given consistently in meters or millimeters. If meters are used, the units of I and J are in meters^4; if millimeters are used, these units are in millimeters^4.

(3) *Polar moments of inertia of masses, J_M*,* appear in dynamics equations in-

* In some books the symbol I denotes the polar moment of inertia of masses; J_M is used in this handbook to avoid confusion with moments of inertia of plane areas.

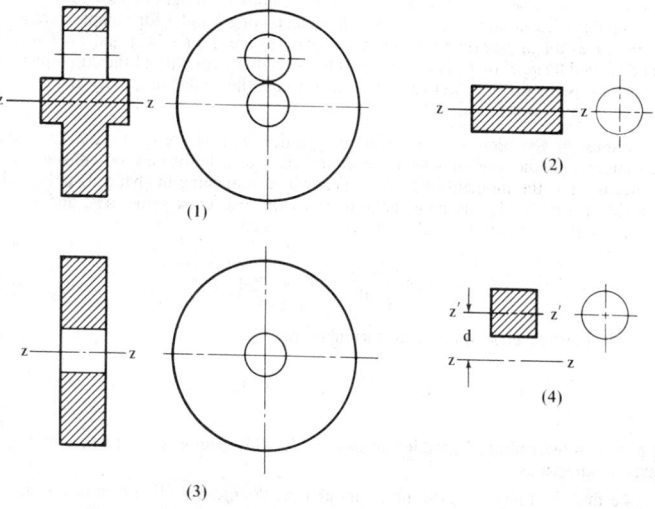

Moments of Inertia of Complex Masses

volving rotational motion. J_M bears the same relationship to angular acceleration as mass does to linear acceleration. If units are in the foot-pound-second system, the units of J_M are ft-lbs-sec^2 or slug-ft^2. (1 slug = 1 pound second^2 per foot.) If units are in the inch-pound-second system, the units of J_M are inch-lbs-sec^2.

If metric SI values are used, the units of J_M are kilogram-meter squared. Formulas for calculating J_M for various bodies are given beginning on page 159. If the polar moment of inertia J is known for the area of a body of constant cross section, J_M may be calculated using the equation,

$$J_M = \frac{\rho L}{g} J$$

where ρ is the density of the material, L the length of the part, and g the gravitational constant. If dimensions are in the foot-pound-second system, ρ is in lbs per ft^3, L is in ft, g is 32.16 ft per sec^2, and J is in ft^4. If dimensions are in the inch-pound-second system, ρ is in lbs per in^3, L in inches, g is 386 inches per sec^2, and J is in inches^4.

Using metric SI units, the above formula becomes $J_M = \rho L J$, where ρ = the density in kilograms/meter^3, L = the length in meters, and J = the polar moment of inertia in meters^4. The units of J_M are kg · m^2.

Moments of inertia of complex areas and masses may be evaluated by the addition and subtraction of elementary areas and masses. For example, the accompanying figure shows a complex mass at (1); its mass polar moment of inertia can be determined by adding together the moments of inertia of the bodies shown at (2) and (3), and subtracting that at (4). Thus, $J_{M1} = J_{M2} + J_{M3} - J_{M4}$. All of these moments of inertia are with respect to the axis of rotation z–z. Formulas for J_{M2} and J_{M3} can be obtained from the tables beginning on page 159. The moment of inertia for the body at (4) can be evaluated by using the following transfer-axis equation: $J_{M4} = J_{M4}' + d^2 M$. The term J_{M4}' is the moment of inertia with respect to axis z'–z'; it may be evaluated using the same equation that applies to J_{M2}. d is the distance between the z–z and the z'–z' axes, and M is the mass of the body (= weight in lbs ÷ g).

Similar calculations can be made when calculating I and J for complex areas. In these cases the appropriate transfer-axis equations are: $I = I' + d^2 A$ and $J = J' + d^2 A$. The primed term, I' or J', is with respect to the center of gravity of the corresponding area A; d is the distance between the axis through the center of gravity and the axis to which I or J is referred.

Radius of Gyration. — The radius of gyration with reference to an axis is that distance from the axis at which the entire mass of a body may be considered as concentrated, the moment of inertia, meanwhile, remaining unchanged. If W is the weight of a body; J_M, its moment of inertia with respect to some axis; and k_o, the radius of gyration with respect to the same axis, then:

$$k_o = \sqrt{\frac{J_M g}{W}} \quad \text{and} \quad J_M = \frac{W k_o^2}{g}$$

When using metric SI units, the formulas are:

$$k_o = \sqrt{\frac{J_M}{M}} \quad \text{and} \quad J_M = M k_o^2,$$

where k_o = the radius of gyration in meters, J_M = kilogram-meter squared, and M = mass in kilograms.

To find the radius of gyration of an area, as the cross-section of a beam, divide the moment of inertia of the area by the area and extract the square root.

(Continued on page 164)

Moments of Inertia, J_M

(J_M = polar moment of inertia of masses, see page 157. M = mass of body.)

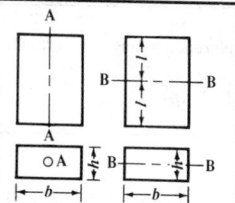	*Prism.* — With reference to axis $A - A$: $$J_M = \frac{M}{12}(h^2 + b^2)$$ With reference to axis $B - B$: $$J_M = M\left(\frac{l^2}{3} + \frac{h^2}{12}\right)$$
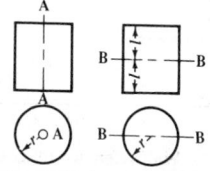	*Cylinder.* — With reference to axis $A - A$: $$J_M = \tfrac{1}{2}Mr^2$$ With reference to axis $B - B$: $$J_M = M\left(\frac{l^2}{3} + \frac{r^2}{4}\right)$$
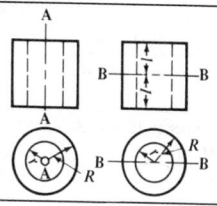	*Hollow Cylinder.* — With reference to axis $A - A$: $$J_M = \tfrac{1}{2}M(R^2 + r^2)$$ With reference to axis $B - B$: $$J_M = M\left(\frac{l^2}{3} + \frac{R^2 + r^2}{4}\right)$$
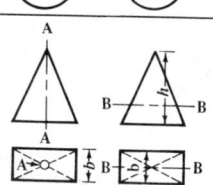	*Pyramid*, rectangular base. — With reference to axis $A - A$: $$J_M = \frac{M}{20}(a^2 + b^2)$$ With reference to axis $B - B$ (through the center of gravity): $$J_M = M\left(\frac{3}{80}h^2 + \frac{b^2}{20}\right)$$
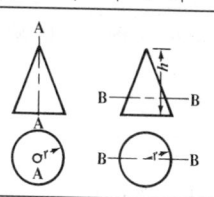	*Cone.* — With reference to axis $A - A$: $$J_M = \frac{3M}{10}r^2$$ With reference to axis $B - B$ (through the center of gravity): $$J_M = \frac{3M}{20}\left(r^2 + \frac{h^2}{4}\right)$$
	*Frustum of Cone.* — With reference to axis $A - A$: $$J_M = \frac{3M}{10}\frac{(R^5 - r^5)}{(R^3 - r^3)}$$

Moments of Inertia, J_M

(J_M = polar moment of inertia of masses, see page 157. M = mass of body.)

	Sphere. — With reference to any axis through the center: $$J_M = \tfrac{2}{5}Mr^2$$
	Spherical Sector. — With reference to axis $A-A$: $$J_M = \frac{M}{5}(3rh - h^2)$$
	Spherical Segment. — With reference to axis $A-A$: $$J_M = M\left(r^2 - \frac{3rh}{4} + \frac{3h^2}{20}\right)\frac{2h}{3r - h}$$
	Ellipsoid. — With reference to axis $A-A$: $$J_M = \frac{M}{5}(b^2 + c^2)$$ With reference to axis $B-B$: $$J_M = \frac{M}{5}(a^2 + c^2)$$ With reference to axis $C-C$: $$J_M = \frac{M}{5}(a^2 + b^2)$$
	Paraboloid. — With reference to axis $A-A$: $$J_M = \tfrac{1}{3}Mr^2$$ With reference to axis $B-B$ (through the center of gravity): $$J_M = M\left(\frac{r^2}{6} + \frac{h^2}{18}\right)$$
	Torus. — With reference to axis $A-A$: $$J_M = M\left(\frac{R^2}{2} + \frac{5r^2}{8}\right)$$ With reference to axis $B-B$: $$J_M = M(R^2 + \tfrac{3}{4}r^2)$$

Radius of Gyration

Bar of Small Diameter. Axis at end. 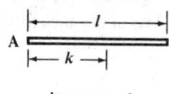 $k = 0.5773l$ $k^2 = \frac{1}{3}l^2$ Axis at center. 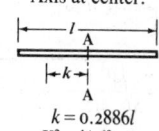 $k = 0.2886l$ $K^2 = \frac{1}{12}l^2$	Thin Circular Disk. Axis through center. Cylinder. Axis through center. 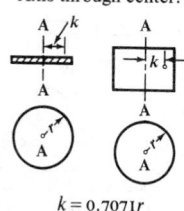 $k = 0.7071r$ $k^2 = \frac{1}{2}r^2$	Cylinder. Axis, diameter at mid-length. 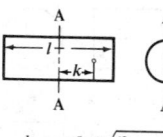 $k = 0.289\sqrt{l^2 + 3r^2}$ $k^2 = \frac{l^2}{12} + \frac{r^2}{4}$
Bar of Small Diameter, bent to Circular Shape. Axis, a diameter of the ring.  $k = 0.7071r$ $k^2 = \frac{1}{2}r^2$	Thin Circular Disk. Axis its diameter. 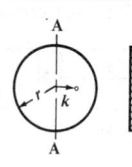 $k = \frac{1}{2}r$ $k^2 = \frac{1}{4}r^2$	Cylinder. Axis, diameter at end. 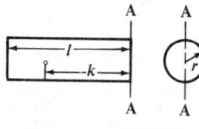 $k = 0.289\sqrt{4l^2 + 3r^2}$ $k^2 = \frac{l^2}{3} + \frac{r^2}{4}$
Bar of Small Diameter, bent to Circular Shape. Axis through center of ring. 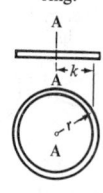 $k = r$; $k^2 = r^2$	Parallelogram (Thin flat plate). Axis at base. 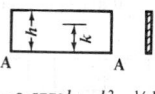 $k = 0.5773h$; $k^2 = \frac{1}{3}h^2$ Axis at mid-height. 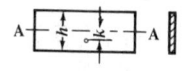 $k = 0.2886h$; $k^2 = \frac{1}{12}h^2$	Thin, Flat, Circular Ring. Axis its diameter. 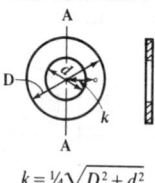 $k = \frac{1}{4}\sqrt{D^2 + d^2}$ $k^2 = \frac{D^2 + d^2}{16}$

Radius of Gyration

Thin Hollow Cylinder. Axis, diameter at mid-length.	Cylinder. Axis at a distance.	Parallelepiped. Axis at distance from end.
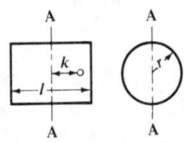	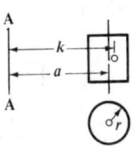	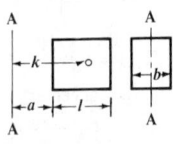
$k = 0.289\sqrt{l^2 + 6r^2}$ $k^2 = \dfrac{l^2}{12} + \dfrac{r^2}{2}$	$k = \sqrt{a^2 + \tfrac{1}{2}r^2}$ $k^2 = a^2 + \tfrac{1}{2}r^2$	$k = \sqrt{\dfrac{4l^2 + b^2}{12} + a^2 + al}$
Hollow Cylinder. Longitudinal Axis.	Rectangular Prism. Axis through center.	Cone. Axis at base.
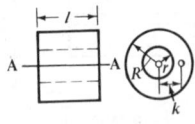	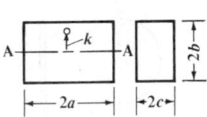	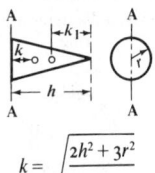
$k = 0.7071\sqrt{R^2 + r^2}$ $k^2 = \tfrac{1}{2}(R^2 + r^2)$	$k = 0.577\sqrt{b^2 + c^2}$ $k^2 = \tfrac{1}{3}(b^2 + c^2)$	$k = \sqrt{\dfrac{2h^2 + 3r^2}{20}}$ Axis at apex. $k_1 = \sqrt{\dfrac{12h^2 + 3r^2}{20}}$
Hollow Cylinder. Axis, diameter at mid-length.	Parallelepiped. Axis at one end, central.	Cone. Axis through its center line.
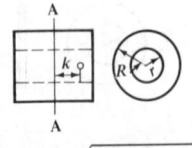	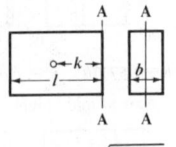	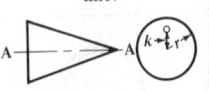
$k = 0.289\sqrt{l^2 + 3(R^2 + r^2)}$ $k^2 = \dfrac{l^2}{12} + \dfrac{R^2 + r^2}{4}$	$k = 0.289\sqrt{4l^2 + b^2}$ $k^2 = \dfrac{4l^2 + b^2}{12}$	$k = 0.5477r$ $k^2 = 0.3r^2$

Radius of Gyration

Frustum of Cone.
Axis at large end.

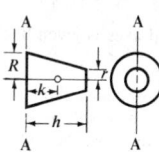

$$k = \sqrt{\frac{h^2}{10}\left(\frac{R^2 + 3Rr + 6r^2}{R^2 + Rr + r^2}\right) + \frac{3}{20}\left(\frac{R^5 - r^5}{R^3 - r^3}\right)}$$

Sphere.
Axis at a distance.

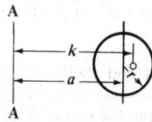

$$k = \sqrt{a^2 + \tfrac{2}{5} r^2}$$
$$k^2 = a^2 + \tfrac{2}{5} r^2$$

Sphere.
Axis its diameter.

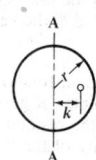

$$k = 0.6325r; \quad k^2 = \tfrac{2}{5} r^2$$

Thin Spherical Shell.

$$k = 0.8165r; \quad k^2 = \tfrac{2}{3} r^2$$

Ellipsoid.
Axis through center.

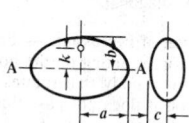

$$k = 0.447\sqrt{b^2 + c^2}$$
$$k^2 = \tfrac{1}{5}(b^2 + c^2)$$

Hollow Sphere.
Axis its diameter.

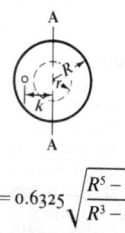

$$k = 0.6325\sqrt{\frac{R^5 - r^5}{R^3 - r^3}}$$
$$k^2 = \frac{2(R^5 - r^5)}{5(R^3 - r^3)}$$

Paraboloid.
Axis through center.

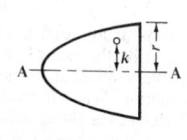

$$k = 0.5773r$$
$$k^2 = \tfrac{1}{3} r^2$$

When the axis, with reference to which the radius of gyration is taken, passes through the center of gravity, the radius of gyration is the least possible and is called the *principal* radius of gyration. If k is the radius of gyration with respect to such an axis passing through the center of gravity of a body, then the radius of gyration, k_o, with respect to a parallel axis at a distance d from the gravity axis is given by:

$$k_o = \sqrt{k^2 + d^2}$$

A table of radii of gyration for various bodies and axes is given beginning on page 161.

Center and Radius of Oscillation. — If a body oscillates about a horizontal axis which does not pass through its center of gravity, there will be a point on the line drawn from the center of gravity, perpendicular to the axis, the motion of which will be the same as if the whole mass were concentrated at that point. This point is called the *center of oscillation.* The *radius of oscillation* is the distance between the center of oscillation and the point of suspension. In a straight line, or in a bar of small diameter, suspended at one end and oscillating about it, the center of oscillation is at two-thirds the length of the rod from the end by which it is suspended.

When the vibrations are perpendicular to the plane of the figure, and the figure is suspended by the vertex of an angle or its uppermost point, the radius of oscillation of an isosceles triangle is equal to ¾ of the height of the triangle; of a circle, ⅝ of the diameter; of a parabola, ⁵⁄₇ of the height.

If the vibrations are in the plane of the figure, then the radius of oscillation of a circle equals ¾ of the diameter; of a rectangle, suspended at the vertex of one angle, ⅔ of the diagonal.

Center of Percussion. — For a body that moves without rotation, the resultant of all the forces acting on the body passes through the center of gravity. On the other hand, for a body that rotates about some *fixed axis,* the resultant of all the forces acting on it does not pass through the center of gravity of the body but through a point called the *center of percussion.* The center of percussion is useful in determining the position of the resultant in mechanics problems involving angular acceleration of bodies about a fixed axis.

Finding the Center of Percussion when the Radius of Gyration and the Location of the Center of Gravity are Known: The center of percussion lies on a line drawn through the center of rotation and the center of gravity. The distance from the axis of rotation to the center of percussion may be calculated from the following formula in which q = distance from the axis of rotation to the center of percussion; k_o = the radius of gyration of the body *with respect to the axis of rotation;* and r = the distance from the axis of rotation to the center of gravity of the body.

$$q = k_o^2 \div r$$

Velocity and Acceleration

Motion is a progressive change of position of a body. Velocity is the rate of motion, that is, the rate of change of position. When the velocity of a body is the same at every moment during which the motion takes place, the latter is called *uniform* motion. When the velocity is variable and constantly increasing, the rate at which it changes is called *acceleration;* that is, acceleration is the rate at which the velocity of a body changes in a unit of time, as the change in feet per second, in one second. When the motion is decreasing instead of increasing, it is called *retarded* motion, and the rate at which the motion is retarded is frequently called the *deceleration.* If the acceleration is uniform, the motion is called *uniformly accelerated* motion. An example of such motion is found in that of falling bodies.

Motion with Constant Velocity. — In the formulas that follow, S = distance moved; V = velocity; t = time of motion, θ = angle of rotation, and ω = angular velocity; the usual units for these quantities are, respectively, feet, feet per second, seconds, radians, and radians per second. Any other consistent set of units may be employed.

Constant Linear Velocity:

$$S = V \times t; \quad V = S \div t; \quad t = S \div V$$

Constant Angular Velocity:

$$\theta = \omega t; \quad \omega = \theta \div t; \quad t = \theta \div \omega$$

Relation between Angular Motion and Linear Motion: The relation between the angular velocity of a rotating body and the linear velocity of a point at a distance r feet from the center of rotation is:

$$V \text{ (ft per sec)} = r \text{ (ft)} \times \omega \text{ (radians per sec)}$$

Similarly, the distance moved by the point during rotation through angle θ is:

$$S \text{ (ft)} = r \text{ (ft)} \times \theta \text{ (radians)}$$

Linear Motion with Constant Acceleration. — The relations between distance, velocity, and time for linear motion with constant or uniform acceleration are given by the formulas in the accompanying table. In these formulas, the acceleration is assumed to be in the same direction as the initial velocity; hence, if the acceleration in a particular problem should happen to be in a direction opposite that of the initial velocity, then a should be replaced by $-a$. Thus, for example, the formula $V_f = V_o + at$ becomes $V_f = V_o - at$ when a and V_o are opposite in direction.

Example: A car is moving at 60 mph when the brakes are suddenly locked and the

Linear Motion with Constant Acceleration

To Find	Known	Formula	To Find	Known	Formula
		Motion Uniformly Accelerated From Rest			
S	a, t	$S = \frac{1}{2}at^2$	t	S, V_f	$t = 2S \div V_f$
	V_f, t	$S = \frac{1}{2}V_f t$		S, a	$t = \sqrt{2S \div a}$
	V_f, a	$S = V_f^2 \div 2a$		a, V_f	$t = V_f \div a$
V_f	a, t	$V_f = at$	a	S, t	$a = 2S \div t^2$
	S, t	$V_f = 2S \div t$		S, V	$a = V_f^2 \div 2S$
	a, S	$V_f = \sqrt{2aS}$		V_f, t	$a = V_f \div t$
		Motion Uniformly Accelerated From Initial Velocity V_o			
S	a, t, V_o	$S = V_o t + \frac{1}{2}at^2$	t	V_o, V_f, a	$t = (V_f - V_o) \div a$
	V_o, V_f, t	$S = (V_f + V_o)t \div 2$		V_o, V_f, S	$t = 2S \div (V_f + V_o)$
	V_o, V_f, a	$S = (V_f^2 - V_o^2) \div 2a$			
	V_f, a, t	$S = V_f t - \frac{1}{2}at^2$	a	V_o, V_f, S	$a = (V_f^2 - V_o^2) \div 2S$
				V_o, V_f, t	$a = (V_f - V_o) \div t$
V_f	V_o, a, t	$V_f = V_o + at$		V_o, S, t	$a = 2(S - V_o t) \div t^2$
	V_o, S, t	$V_f = (2S \div t) - V_o$		V_f, S, t	$a = 2(V_f t - S) \div t^2$
	V_o, a, S	$V_f = \sqrt{V_o^2 + 2aS}$			
	S, a, t	$V_f = (S \div t) + \frac{1}{2}at$			*Meaning of Symbols*
V_o	V_f, a, S	$V_o = \sqrt{V_f^2 - 2aS}$			S = distance moved in feet;
	V_f, S, t	$V_o = (2S \div t) - V_f$			V_f = final velocity, feet per second;
	V_f, a, t	$V_o = V_f - at$			V_o = initial velocity, feet per second;
	S, a, t	$V_o = (S \div t) - \frac{1}{2}at$			a = acceleration, feet per second per second; t = time of acceleration in seconds.

car begins to skid. If it takes 2 seconds to slow the car to 30 mph, at what rate is it being decelerated, how long is it before the car comes to a halt, and how far will it have traveled?

The initial velocity V_o of the car is 60 mph or 88 ft/sec and the acceleration a due to braking is opposite in direction to V_o since the car is slowed to 30 mph or 44 ft/sec.

Since V_o, V_f, and t are known, a can be determined from the formula

$$a = (V_f - V_o) \div t = (44 - 88) \div 2$$
$$a = -22 \text{ ft/sec}^2$$

The time required to stop the car can be determined from the formula

$$t = (V_f - V_o) \div a = (0 - 88) \div -22$$
$$t = 4 \text{ seconds}$$

The distance traveled by the car is obtained from the formula

$$S = (V_f + V_o)t \div 2 = (0 + 88)4 \div 2$$
$$= 176 \text{ feet}$$

Rotary Motion with Constant Acceleration. — The relations among angle of rotation, angular velocity, and time for rotation with constant or uniform acceleration are given in the accompanying table. In these formulas, the acceleration is assumed to be in the same direction as the initial angular velocity; hence, if the acceleration in a particular problem should happen to be in a direction opposite that of the initial angular velocity, then α should be replaced by $-\alpha$. Thus, for example, the formula $\omega_f = \omega_o + \alpha t$ becomes $\omega_f = \omega_o - \alpha t$ when α and ω_o are opposite in direction.

Rotary Motion with Constant Acceleration

To Find	Known	Formula	To Find	Known	Formula
Motion Uniformly Accelerated From Rest					
θ	α, t	$\theta = \frac{1}{2}\alpha t^2$	t	θ, ω_f	$t = 2\theta \div \omega_f$
	ω_f, t	$\theta = \frac{1}{2}\omega_f t$		θ, α	$t = \sqrt{2\theta \div \alpha}$
	ω_f, α	$\theta = \omega_f^2 \div 2\alpha$		α, ω_f	$t = \omega_f \div \alpha$
ω_f	α, t	$\omega_f = \alpha t$	α	θ, t	$\alpha = 2\theta \div t^2$
	θ, t	$\omega_f = 2\theta \div t$		θ, ω_f	$\alpha = \omega_f^2 \div 2\theta$
	α, θ	$\omega_f = \sqrt{2\alpha\theta}$		ω_f, t	$\alpha = \omega_f \div t$
Motion Uniformly Accelerated From Initial Velocity ω_o					
θ	α, t, ω_o	$\theta = \omega_o t + \frac{1}{2}\alpha t^2$	α	$\omega_o, \omega_f, \theta$	$\alpha = (\omega_f^2 - \omega_o^2) \div 2\theta$
	ω_o, ω_f, t	$\theta = (\omega_f + \omega_o)t \div 2$		ω_o, ω_f, t	$\alpha = (\omega_f - \omega\theta) \div t$
	$\omega_o, \omega_f, \alpha$	$\theta = (\omega_f^2 - \omega_o^2) \div 2\alpha$		ω_o, θ, t	$\alpha = 2(\theta - \omega_o t) \div t^2$
	ω_f, α, t	$\theta = \omega_f t - \frac{1}{2}\alpha t^2$		ω_f, θ, t	$\alpha = 2(\omega_f t - \theta) \div t^2$
ω_f	ω_o, α, t	$\omega_f = \omega_o + \alpha t$	*Meaning of Symbols*	
	ω_o, θ, t	$\omega_f = (2\theta \div t) - \omega_o$	θ = angle of rotation, radians;	
	ω_o, α, θ	$\omega_f = \sqrt{\omega_o^2 + 2\alpha\theta}$	ω_f = final angular velocity, radians per second;	
	θ, α, t	$\omega_f = (\theta \div t) + \frac{1}{2}\alpha t$	ω_o = initial angular velocity, radians per second;	
ω_o	ω_f, α, θ	$\omega_o = \sqrt{\omega_f^2 - 2\alpha\theta}$	α = angular acceleration, radians per second, per second;	
	ω_f, θ, t	$\omega_o = (2\theta \div t) - \omega_f$	t = time in seconds.	
	ω_f, α, t	$\omega_o = \omega_f - \alpha t$		
	θ, α, t	$\omega_o = (\theta \div t) - \frac{1}{2}\alpha t$		
t	$\omega_o, \omega_f, \alpha$	$t = (\omega_f - \omega_o) \div \alpha$	1 degree = 0.01745 radians	
	$\omega_o, \omega_f, \theta$	$t = 2\theta \div (\omega_f + \omega_o)$	(See conversion table on page 127)	

Linear Acceleration of a Point on a Rotating Body: A point on a body rotating about a fixed axis has a linear acceleration *a* that is the resultant of two component accelerations. The first component is the centripetal or normal acceleration which is directed from the point *P* toward the axis of rotation; its magnitude is $r\omega^2$ where *r* is the radius from the axis to the point *P* and ω is the angular velocity of the body at the time acceleration *a* is to be determined. The second component of *a* is the tangential acceleration which is equal to $r\alpha$ where α is the angular acceleration of the body.

The acceleration of point *P* is the resultant of $r\omega^2$ and $r\alpha$ and is given by the formula

$$a = \sqrt{(r\omega^2)^2 + (r\alpha)^2}$$

When $\alpha = 0$, this formula reduces to: $a = r\omega^2$

Example: A flywheel on a press rotating at 120 rpm is slowed to 102 rpm during a punching operation that requires ¾ second for the punching portion of the cycle. What angular deceleration does the flywheel experience?

From the table on page 200, the angular velocities corresponding to 120 rpm and 102 rpm, respectively, are 12.57 and 10.68 radians per second. Therefore, using the formula

$$\alpha = (\omega_f - \omega_o) \div t$$

$$\alpha = (10.68 - 12.57) \div \tfrac{3}{4} = -1.89 \div \tfrac{3}{4}$$

$$\alpha = -2.52 \text{ radians per second per second}$$

which is, from the table on page 200, 24 rpm per second, per second. The minus sign in the answer indicates that the acceleration α acts to slow the flywheel, that is, the flywheel is being decelerated.

Force, Work, Energy, and Momentum

Accelerations Resulting from Unbalanced Forces. — In the section describing the resolution and composition of forces it was stated that when the resultant of a system of forces is zero, the system is in equilibrium, that is, the body on which the force system acts remains at rest or continues to move with uniform velocity. If, however, the resultant of a system of forces is not zero, the body on which the forces act will be accelerated in the direction of the unbalanced force. To determine the relation between the unbalanced force and the resulting acceleration, Newton's laws of motion must be applied. These laws may be stated as follows:

First Law: Every body continues in a state of rest or in uniform motion in a straight line, until it is compelled by a force to change its state of rest or motion.

Second Law: Change of motion is proportional to the force applied, and takes place along the straight line in which the force acts. The "force applied" represents the resultant of *all* the forces acting on the body. This law is sometimes worded: An unbalanced force acting on a body causes an acceleration of the body in the direction of the force and of magnitude proportional to the force and inversely proportional to the mass of the body. Stated as a formula, $R = Ma$ where *R* is the resultant of *all* the forces acting on the body, *M* is the mass of the body (mass = weight *W* divided by acceleration due to gravity *g*), and *a* is the acceleration of the body resulting from application of force *R*.

Third Law: To every action there is always an equal reaction, or, in other words, if a force acts to change the state of motion of a body, the body offers a resistance equal and directly opposite to the force.

Newton's second law may be used to calculate linear and angular accelerations of a body produced by unbalanced forces and torques acting on the body; however,

it is necessary first to use the methods described under "Composition and Resolution of Forces" to determine the magnitude and direction of the resultant of *all* forces acting on the body. Then, for a body moving with pure translation,

$$R = Ma = \frac{W}{g}a$$

where R is the resultant force in pounds acting on a body weighing W pounds; g is the gravitational constant, usually taken as 32.16 ft/sec², approximately; and a is the resulting acceleration in ft/sec² of the body due to R and in the same direction as R.

Using metric SI units, the formula is $R = Ma$, where $R =$ force in newtons (N), $M =$ mass in kilograms, and $a =$ acceleration in meters/second squared. It should be noted that the weight of a body of mass M kg is Mg newtons, where g is approximately 9.81 m/s².

Free Body Diagram: In order to correctly determine the effect of forces on the motion of a body it is necessary to resort to what is known as a *free body diagram.* This diagram shows (1) the body removed or isolated from contact with all other bodies that exert force on the body and (2) *all* the forces acting on the body. Thus, for example, in diagram (a) the block being pulled up the plane is acted upon by certain forces; the free body diagram of this block is shown at (b). Note that all forces

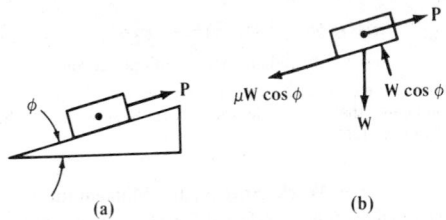

(a)   (b)

acting on the block are indicated. These forces include: (1) the force of gravity (weight); (2) the pull of the cable, P; (3) the normal component, $W \cos \phi$, of the force exerted on the block by the plane; and (4) the friction force, $\mu W \cos \phi$, of the plane on the block.

In preparing a free body diagram, it is important to realize that only those forces exerted *on* the body being considered are shown; forces exerted by the body on other bodies are disregarded. It is this feature that makes the free body diagram an invaluable aid in the solution of problems in mechanics.

Example: A 100-pound body is being hoisted by a winch, the tension in the hoisting cable being kept constant at 110 pounds. At what rate is the body accelerated?

Two forces are acting on the body, its weight, 100 pounds downward, and the pull of the cable, 110 pounds upward. The resultant force R, from a free body diagram, is therefore $110 - 100$. Thus, applying Newton's second law,

$$110 - 100 = \frac{100}{32.16}a$$

$$a = \frac{32.16 \times 10}{100} = 3.216 \text{ ft/sec}^2 \text{ upward}$$

It should be noted that since in this problem the resultant force R was positive ($110 - 100 = +10$), the acceleration a is also positive, that is, a is in the same direction as R, which is in accord with Newton's second law.

Example using SI metric units: **A body of mass 50 kilograms is being hoisted**

by a winch, and the tension in the cable is 600 newtons. What is the acceleration? The weight of the 50 kg body is $50g$ newtons, where g = approximately 9.81 m/s² (see note on page 184). Applying the formula $R = Ma$, the calculation is: $(600 - 50g) = 50a$. Thus,

$$a = \frac{600 - 50g}{50} = \frac{600 - (50 \times 9.81)}{50} = 2.19 \text{ m/s}^2.$$

Formulas Relating Torque and Angular Acceleration: For a body rotating about a fixed axis the relation between the unbalanced torque acting to produce rotation and the resulting angular acceleration may be determined from any one of the following formulas, each based on Newton's second law:

$$T_o = J_M \alpha$$
$$T_o = M k_o^2 \alpha$$
$$T_o = \frac{W k_o^2 \alpha}{g} = \frac{W k_o^2 \alpha}{32.16}$$

where T_o is the unbalanced torque in pounds-feet; J_M in ft-lbs-sec² is the moment of inertia of the body about the axis of rotation; k_o in feet is the radius of gyration of the body with respect to the axis of rotation, and α in radians per second, per second is the angular acceleration of the body.

Example: A flywheel has a diameter of 3 feet and weighs 1000 pounds. What torque must be applied, neglecting bearing friction, to accelerate the flywheel at the rate of 100 revolutions per minute, per second?

From page 159 the moment of inertia of a solid cylinder with respect to a gravity axis at right angles to the circular cross-section is given as ½ Mr^2. From page 200, 100 rpm = 10.47 radians per second, hence an acceleration of 100 rpm per second = 10.47 radians per second, per second. Therefore, using the first of the preceding formulas,

$$T_o = J_M \alpha = \frac{1}{2} \frac{1000}{32.16} \left(\frac{3}{2}\right)^2 \times 10.47$$
$$= 366 \text{ ft-lbs}$$

Using metric SI units, the formulas are: $T_o = J_M \alpha = M k_o^2 \alpha$, where T_o = torque in newton-meters; J_M = the moment of inertia in kg · m², and α = the angular acceleration in radians per second squared.

Example: A flywheel has a diameter of 1.5 m, and a mass of 800 kg. What torque is needed to produce an angular acceleration of 100 revolutions per minute, per second? As in the preceding example, α = 10.47 rad/s². Thus:

$$J_M = \frac{1}{2}Mr^2 = \frac{1}{2} \times 800 \times 0.75^2 = 225 \text{ kg} \cdot \text{m}^2.$$

Therefore: $T_o = J_M \alpha = 225 \times 10.47 = 2356$ N · m.

Energy. — A body is said to possess energy when it is capable of doing work or overcoming resistance. The energy may be either mechanical or non-mechanical, the latter including chemical, electrical, thermal, and atomic energy.

Mechanical energy includes *kinetic energy* (energy possessed by a body because of its motion) and *potential energy* (energy possessed by a body because of its position in a field of force and/or its elastic deformation).

Kinetic Energy: The motion of a body may be one of pure translation, pure rotation, or a combination of rotation and translation. By translation is meant motion in which every line in the body remains parallel to its original position throughout the motion, that is, no rotation is associated with the motion of the body.

The kinetic energy of a translating body is given by the formula

$$\text{Kinetic Energy in ft lbs due to translation} = E_{KT} = \tfrac{1}{2}MV^2 = \frac{Wv^2}{2g} \qquad (a)$$

where M = mass of body ($= W \div g$); V = velocity of the center of gravity of the body in feet per second; W = weight of body in pounds; and g = acceleration due to gravity = 32.16 feet per second, per second.

The kinetic energy of a body rotating about a fixed axis O is expressed by the formula:

$$\text{Kinetic Energy in ft lbs due to rotation} = E_{KR} = \tfrac{1}{2}J_{MO}\omega^2 \qquad (b)$$

where J_{MO} is the moment of inertia of the body about the fixed axis O in pounds-feet-seconds^2, and ω = angular velocity in radians per second.

For a body that is moving with both translation and rotation, the total kinetic energy is given by the following formula as the sum of the kinetic energy due to translation of the center of gravity and the kinetic energy due to rotation about the center of gravity:

$$\text{Total Kinetic Energy in ft lbs} = E_T = \tfrac{1}{2}MV^2 + \tfrac{1}{2}J_{MG}\omega^2$$

$$= \frac{WV^2}{2g} + \tfrac{1}{2}J_{MG}\omega^2$$

$$= \frac{WV^2}{2g} + \tfrac{1}{2}\frac{Wk^2\omega^2}{g}$$

$$= \frac{W}{2g}(V^2 + k^2\omega^2) \qquad (c)$$

where J_{MG} is the moment of inertia of the body about its gravity axis in pounds-feet-seconds^2, k is the radius of gyration in feet with respect to an axis through the center of gravity, and the other quantities are as previously defined.

In the metric SI system, energy is expressed as the joule (J). One joule = 1 newton-meter. The kinetic energy of a translating body is given by the formula $E_{KT} = \tfrac{1}{2}MV^2$, where M = mass in kilograms, and V = velocity in meters per second. Kinetic energy due to rotation is expressed by the formula $E_{KR} = \tfrac{1}{2}J_{MO}\omega^2$, where J_{MO} = moment of inertia in kg · m^2, and ω = the angular velocity in radians per second. Total kinetic energy $E_T = \tfrac{1}{2}MV^2 + \tfrac{1}{2}J_{MO}\omega^2$ joules $= \tfrac{1}{2}M(V^2 + k^2\omega^2)$ joules, where k = radius of gyration in meters.

Potential Energy: The most common example of a body having potential energy because of its position in a field of force is that of a body elevated to some height above the earth. Here the field of force is the gravitational field of the earth and the potential energy E_{PF} of a body weighing W pounds elevated to some height S in feet above the surface of the earth is WS foot-pounds. If the body is permitted to drop from this height its potential energy E_{PF} will be converted to kinetic energy. Thus, after falling through height S the kinetic energy of the body will be WS ft-lbs.

In metric SI units, the potential energy E_{PF} of a body of mass M kilograms elevated to a height of S meters, is MgS joules. After it has fallen a distance S, the kinetic energy gained will thus be MgS joules.

Another type of potential energy is elastic potential energy, such as possessed by a spring that has been compressed or extended. The amount of work in ft lbs done in compressing the spring S feet is equal to $KS^2/2$, where K is the spring constant in pounds per foot. Thus, when the spring is released to act against some resistance, it can perform $KS^2/2$ ft-lbs of work which is the amount of elastic potential energy E_{PE} stored in the spring.

Using metric SI units, the amount of work done in compressing the spring a distance S meters is $KS^2/2$ joules, where K is the spring constant in newtons per meter.

Work Performed by Forces and Couples. — The work U done by a force F in moving an object along some path is the product of the distance S the body is moved and the component $F \cos \alpha$ of the force F in the direction of S.

$$U = FS \cos \alpha$$

where U = work in ft-lbs; S = distance moved in feet; F = force in lbs; and α = angle between line of action of force and the path of S.

If the force is in the same direction as the motion, then $\cos \alpha = \cos 0 = 1$ and this formula reduces to:

$$U = FS$$

Similarly, the work done by a couple T turning an object through an angle θ is:

$$U = T\theta$$

where T = torque of couple in pounds-feet and θ = the angular rotation in radians.

The above formulas can be used with metric SI units: U is in joules; S is in meters; F is in newtons, and T is in newton-meters.

Relation between Work and Energy. — Theoretically, when work is performed on a body and there are no energy losses (such as due to friction, air resistance, etc.), the energy acquired by the body is equal to the work performed on the body; this energy may be either potential, kinetic, or a combination of both.

In actual situations, however, there may be energy losses that must be taken into account. Thus, the relation between work done on a body, energy losses, and the energy acquired by the body can be stated as:

$$\text{Work Performed} - \text{Losses} = \text{Energy Acquired}$$

$$U - \text{Losses} = E_T$$

Example 1: A 12-inch cube of steel weighing 490 pounds is being moved on a horizontal conveyor belt at a speed of 6 miles per hour (8.8 feet per second). What is the kinetic energy of the block?

Since the block is not rotating, Formula (a) for the kinetic energy of a body moving with pure translation applies:

$$\text{Kinetic Energy} = \frac{WV^2}{2g}$$

$$= \frac{490 \times (8.8)^2}{2 \times 32.16} = 590 \text{ ft-lbs}$$

A similar example using metric SI units is as follows: If a cube of mass 200 kg is being moved on a conveyor belt at a speed of 3 meters per second, what is the kinetic energy of the cube? It is:

$$\text{Kinetic Energy} = \tfrac{1}{2}MV^2 = \tfrac{1}{2} \times 200 \times 3^2 = 900 \text{ joules.}$$

Example 2: If the conveyor in Example 1 is brought to an abrupt stop, how long would it take for the steel block to come to a stop and how far along the belt would it slide before stopping if the coefficient of friction μ between the block and the conveyor belt is 0.2 and the block slides without tipping over?

The only force acting to slow the motion of the block is the friction force between the block and the belt. This force F is equal to the weight of the block, W, multiplied by the coefficient of friction; $F = \mu W = 0.2 \times 490 = 98$ lbs.

The time required to bring the block to a stop can be determined from the impulse-momentum formula (c) on page 173.

$$R \times t = \frac{W}{g}(V_f - V_o)$$

$$(-98)t = \frac{490}{32.16} \times (0 - 8.8)$$

$$t = \frac{490 \times 8.8}{98 \times 32.16} = 1.37 \text{ seconds}$$

The distance the block slides before stopping can be determined by equating the kinetic energy of the block and the work done by friction in stopping it:

Kinetic energy of block ($WV^2/2g$) = Work done by friction ($F \times S$)

$$590 = 98 \times S$$

$$S = \frac{590}{98} = 6.0 \text{ feet}$$

If metric SI units are used, the calculation is as follows (for the cube of 200 kg mass): The friction force = μ multiplied by the weight Mg where g = approximately 9.81 m/s². Thus, $\mu M g = 0.2 \times 200g = 392.4$ newtons. The time t required to bring the block to a stop is $(-392.4)t = 200(0-3)$. Therefore,

$$t = \frac{200 \times 3}{392.4} = 1.53 \text{ seconds}$$

The kinetic energy of the block is equal to the work done by friction, that is $392.4 \times S = 900$ joules. Thus, the distance S which the block moves before stopping is

$$S = \frac{900}{392.4} = 2.29 \text{ meters}$$

Force of a Blow. — A body that weighs W pounds and falls S feet from an initial position of rest is capable of doing WS foot-pounds of work. The work performed during its fall may be, for example, that necessary to drive a pile a distance d into the ground. Neglecting losses in the form of dissipated heat and strain energy, the work done in driving the pile is equal to the product of the impact force acting on the pile and the distance d which the pile is driven. Since the impact force is not accurately known, an average value, called the "average force of the blow," may be assumed. Equating the work done on the pile and the work done by the falling body, which in this case is a pile driver:

Average force of blow $\times d = WS$

or,   Average force of blow $= \dfrac{WS}{d}$

where, S = total height in feet through which the driver falls, including the distance d that the pile is driven;

W = weight of driver in pounds;

d = distance in feet which pile is driven.

When using metric SI units, it should be noted that a body of mass M kilograms has a weight of Mg newtons, where $g =$ approximately 9.81 m/s². If the body falls a distance S meters, it can do work equal to MgS joules. The average force of the blow is MgS/d newtons, where d is the distance in meters that the pile is driven.

Example: — A pile driver weighing 200 pounds strikes the top of the pile after having fallen from a height of 20 feet. It forces the pile into the ground a distance of ½ foot. Before the ram is brought to rest, it will then have performed $200 \times (20 + \frac{1}{2})$ = 4100 foot-pounds of work, and as this energy is expended in a distance of one-half foot, the average force of the blow equals $4100 \div \frac{1}{2} = 8200$ pounds.

A similar example using metric SI units is as follows: A pile driver of mass 100 kilograms falls 10 meters and moves the pile a distance of 0.3 meters. The work done = $100g(10 + 0.3)$ joules, and it is expended in 0.3 meters. Thus, the average force is

$$\frac{100g \times 10.3}{0.3} = 33680 \text{ newtons, or } 33.68 \text{ kN.}$$

Impulse and Momentum. — The *linear momentum* of a body is defined as the product of the mass M of the body and the velocity V of the center of gravity of the body:

$$\text{Linear momentum} = MV, \text{ or, since } M = W \div g$$

$$\text{Linear momentum} = \frac{WV}{g} \qquad \text{(a)}$$

It should be noted that linear momentum is a vector quantity, the momentum being in the same direction as V.

Linear impulse is defined as the product of the resultant R of *all* the forces acting on a body and the time t that the resultant acts:

$$\text{Linear Impulse} = Rt \qquad \text{(b)}$$

The change in the linear momentum of a body is numerically equal to the linear impulse that causes the change in momentum:

$$\text{Linear Impulse} = \text{change in Linear Momentum}$$

$$Rt = \frac{W}{g}V_f - \frac{W}{g}V_o = \frac{W}{g}(V_f - V_o) \qquad \text{(c)}$$

where V_f, the final velocity of the body after time t and V_o the initial velocity of the body are both in the same direction as the applied force R. If V_o and V_f are in opposite directions, then the minus sign in the formula becomes a plus sign.

In metric SI units, the formulas are: **Linear Momentum** = MV kg · m/s, where $M =$ mass in kg, and $V =$ velocity in meters per second; and **Linear Impulse** = Rt newton-seconds, where $R =$ force in newtons, and $t =$ time in seconds. In formula (c) above, W/g is replaced by M when SI units are used.

Example: A 1000-pound block is pulled up a 2-degree incline by a cable exerting a constant force F of 600 pounds. If the coefficient of friction μ between the block and the plane is 0.5, how fast will the block be moving up the plane 10 seconds after the pull is applied?

The resultant force R causing the body to be accelerated up the plane is the difference between F, the force acting up the plane, and P, the force acting to resist motion up the plane. This latter force for a body on a plane is given by the formula at the top of page 146 as $P = W (\mu \cos \alpha + \sin \alpha)$ where α is the angle of the incline.

Thus,
$$R = F - P = F - W(\mu \cos \alpha + \sin \alpha)$$
$$= 600 - 1000(0.5 \cos 2° + \sin 2°)$$
$$= 600 - 1000(0.5 \times 0.99939 + 0.03490)$$
$$= 600 - 535$$
$$R = 65 \text{ pounds.}$$

Formula (c) can now be applied to determine the speed at which the body will be moving up the plane after 10 seconds.

$$Rt = \frac{W}{g}V_f - \frac{W}{g}V_o$$

$$65 \times 10 = \frac{1000}{32.2}V_f - \frac{1000}{32.2} \times 0$$

$$V_f = \frac{65 \times 10 \times 32.2}{1000} = 20.9 \text{ ft per sec}$$

$$= 14.3 \text{ miles per hour}$$

A similar example using metric SI units is as follows: A 500 kg block is pulled up a 2 degree incline by a constant force F of 4 kN. The coefficient of friction μ between the block and the plane is 0.5. How fast will the block be moving 10 seconds after the pull is applied?

The resultant force R is:

$$R = F - Mg(\mu \cos \alpha + \sin \alpha)$$
$$= 4000 - 500 \times 9.81(0.5 \times 0.99939 + 0.03490)$$
$$= 1378 \text{ N or } 1.378 \text{ kN.}$$

Formula (c) can now be applied to determine the speed at which the body will be moving up the plane after 10 seconds. Replacing W/g by M in the formula, the calculation is:

$$Rt = MV_f - MV_o$$

$$1378 \times 10 = 500(V_f - 0)$$

$$V_f = \frac{1378 \times 10}{500} = 27.6 \text{ m/s.}$$

Angular Impulse and Momentum: In a manner similar to that for linear impulse and moment, the formulas for angular impulse and momentum for a body rotating about a fixed axis are:

$$\text{Angular momentum} = J_M\omega \qquad \qquad (a)$$
$$\text{Angular impulse} = T_o t \qquad \qquad (b)$$

where J_M is the moment of inertia of the body about the axis of rotation in pounds-feet-seconds^2, ω is the angular velocity in radians per second, T_o is the torque in pounds-feet about the axis of rotation, and t is the time in seconds that T_o acts.

The change in angular momentum of a body is numerically equal to the angular impulse that causes the change in angular momentum:

$$\text{Angular Impulse} = \text{Change in Angular Momentum}$$
$$T_o t = J_M\omega_f - J_M\omega_o = J_M(\omega_f - \omega_o) \qquad \qquad (c)$$

where ω_f and ω_o are the final and initial angular velocities, respectively.

Example: A flywheel having a moment of inertia of 25 lbs-ft-sec² is revolving with an angular velocity of 10 radians per second when a constant torque of 20 lbs-ft is applied to reverse its direction of rotation. For what length of time must this constant torque act to stop the flywheel and bring it up to a reverse speed of 5 radians per second?

Applying formula (c),

$$T_o t = J_M(\omega_f - \omega_o)$$
$$20t = 25(10 - [-5]) = 250 + 125$$
$$t = 375 \div 20 = 18.8 \text{ seconds}$$

A similar example using metric SI units is as follows: A flywheel with a moment of inertia of 20 kilogram-meters² is revolving with an angular velocity of 10 radians per second when a constant torque of 30 newton-meters is applied to reverse its direction of rotation. For what length of time must this constant torque act to stop the flywheel and bring it up to a reverse speed of 5 radians per second? Applying formula (c), the calculation is:

$$T_o t = J_M(\omega_f - \omega_o),$$
$$30t = 20(10 - [-5]).$$

Thus, $t = \dfrac{20 \times 15}{30} = 10$ seconds.

Formulas for Work and Power. — The formulas in the accompanying table may be used to determine work and power in terms of the applied force and the velocity at the point of application of the force.

Formulas for Work and Power

To Find	Known	Formula	To Find	Known	Formula
S	P, t, F	$S = P \times t \div F$	P	F, V	$P = F \times V$
	K, F	$S = K \div F$		F, S, t	$P = F \times S \div t$
	t, F, hp	$S = 550 \times t \times hp \div F$		K, t	$P = K \div t$
				hp	$P = 550 \times hp$
V	P, F	$V = P \div F$	K	F, S	$K = F \times S$
	K, F, t	$V = K \div (F \times t)$		P, t	$K = P \times t$
	F, hp	$V = 550 \times hp \div F$		F, V, t	$K = F \times V \times t$
				t, hp	$K = 550 \times t \times hp$
t	F, S, P	$t = F \times S \div P$	hp	F, S, t	$hp = F \times S \div (550 \times t)$
	K, F, V	$t = K \div (F \times V)$		P	$hp = P \div 550$
	F, S, hp	$t = F \times S \div (550 \times hp)$		F, V	$hp = F \times V \div 550$
				K, t	$hp = K \div (550 \times t)$
F	P, V	$F = P \div V$			
	K, S	$F = K \div S$			
	K, V, t	$F = K \div (V \times t)$			
	V, hp	$F = 550 \times hp \div V$			

Meaning of Symbols: S = distance in feet; V = constant or average velocity in feet per second; t = time in seconds; F = constant or average force in pounds; P = power in foot-pounds per second; K = work in foot-pounds; and hp = horsepower.

Note: The metric SI unit of work is the joule (one joule = 1 newton-meter), and the unit of power is the watt (one watt = 1 joule per second = 1 N · m/s). The term horsepower is not used. Thus, those formulas above which involve horsepower and the factor 550 are not applicable when working in SI units. The remaining formulas can be used, and the units are: S = distance in meters; V = constant or average velocity in meters per second; t = time in seconds; F = force in newtons; P = power in watts; K = work in joules.

Example: — A casting weighing 300 pounds is to be lifted by means of an overhead crane. The casting is lifted 10 feet in 12 seconds. What is the horsepower developed? Here $F = 300$; $S = 10$; $t = 12$.

$$\text{hp} = \frac{F \times S}{550t} = \frac{300 \times 10}{550 \times 12} = 0.45.$$

A similar example using metric SI units, is as follows: A casting of mass 150 kg is lifted 4 meters in 15 seconds by means of a crane. What is the power? Here $F = 150g$ N, $S = 4$ m, and $t = 15$ s. Thus:

$$\text{Power} = \frac{FS}{t} = \frac{150g \times 4}{15} = \frac{150 \times 9.81 \times 4}{15}$$

$$= 392 \text{ watts or } 0.392 \text{ kW.}$$

Centrifugal Force

Centrifugal Force. — When a body rotates about any axis other than one at its center of mass, it exerts an outward radial force called centrifugal force upon the axis or any arm or cord from the axis which restrains it from moving in a straight (tangential) line. In the following formulas:

F = centrifugal force in pounds;

W = weight of revolving body in pounds;

v = velocity at radius R on body in feet per second;

n = number of revolutions per minute;

g = acceleration due to gravity = 32.16 feet per second per second;

R = perpendicular distance from axis of rotation to center of mass, or
 for practical use, to center of gravity of revolving body.

Note: If a body rotates about its own center of mass, R equals zero and v equals zero. This means that the *resultant* of the centrifugal forces of all the elements of the body is equal to zero or, in other words, no centrifugal force is exerted on the axis of rotation. The centrifugal force of any part or element of such a body is found by the equations given below where R is the radius to the center of gravity of the part or element. In case of a flywheel rim, the mean radius of the rim meets practical requirements, as this is the radius to center of gravity of a thin radial section.

$$F = \frac{Wv^2}{gR} = \frac{Wv^2}{32.16R} = \frac{4WR\pi^2 n^2}{60 \times 60g} = \frac{WRn^2}{2933} = 0.000341 WRn^2$$

$$W = \frac{FRg}{v^2} = \frac{2933F}{Rn^2} \qquad v = \sqrt{\frac{FRg}{W}}$$

$$R = \frac{Wv^2}{Fg} = \frac{2933F}{Wn^2} \qquad n = \sqrt{\frac{2933F}{WR}}$$

(If n is the number of revolutions per second instead of per minute, then $F = 1.227WRn^2$.)

If metric SI units are used in the foregoing formulas, W/g is replaced by M, which is the mass in kilograms; F = centrifugal force in newtons; v = velocity in

meters per second; n = number of revolutions per minute; and R = the radius in meters. Thus:

$$F = Mv^2/R = \frac{Mn^2(2\pi R)^2}{60^2 R} = 0.01097\ MRn^2.$$

If the rate of rotation is expressed as n_1 = revolutions per second, then $F = 39.48\ MRn_1^2$; if it is expressed as ω radians per second, then $F = MR\omega^2$.

Calculating Centrifugal Force. — In the ordinary formula for centrifugal force, $F = 0.000341\ WRn^2$; the mean radius R of the flywheel or pulley rim is given in feet. For small dimensions, it is more convenient to have the formula in the form:

$$F = 0.000028416 Wrn^2$$

in which F = centrifugal force, in pounds; W = weight of rim, in pounds; r = mean radius of rim, in inches; n = number of revolutions per minute.

In this formula let $C = 0.000028416n^2$. This, then, is the centrifugal force of one pound, one inch from the axis. The formula can now be written in the form,

$$F = WrC$$

C is calculated for various values of the revolutions per minute n, and the calculated values of C are given in Table 1. To find the centrifugal force in any given case, simply find the value of C in the table and multiply it by the product of W and r, the four multiplications in the original formula given thus having been reduced to two.

Example: A cast-iron flywheel with a mean rim radius of 9 inches, is rotated at a speed of 800 revolutions per minute. If the weight of the rim is 20 pounds, what is the centrifugal force?

From Table 1, for n = 800 revolutions per minute, the value of C is 18.1862.

Thus,

$$F = WrC$$
$$= 20 \times 9 \times 18.1862$$
$$= 3273.52 \text{ pounds}$$

Using metric SI units, $0.01097n^2$ is the centrifugal force acting on a body of 1 kilogram mass rotating at n revolutions per minute at a distance of 1 meter from the axis. If this value is designated C_1, then the centrifugal force of mass M kilograms rotating at this speed at a distance from the axis of R meters, is C_1MR newtons. To simplify calculations, values for C_1 are given in Table 2. If it is required to work in terms of millimeters, the force is $0.001\ C_1MR_1$ newtons, where R_1 is the radius in millimeters.

Example: A steel pulley with a mean rim radius of 120 millimeters is rotated at a speed of 1100 revolutions per minute. If the mass of the rim is 5 kilograms, what is the centrifugal force?

From Table 2, for n = 1100 revolutions per minute, the value of C_1 is 13,269.1.

Thus,

$$F = 0.001\ C_1MR_1$$
$$= 0.001 \times 13,269.1 \times 5 \times 120$$
$$= 7961.50 \text{ newtons}$$

Table I. Factors C for Calculating Centrifugal Force (English units)

n	C	n	C	n	C	n	C
50	0.07104	100	0.28416	470	6.2770	5200	768.369
51	0.07391	101	0.28987	480	6.5470	5300	798.205
52	0.07684	102	0.29564	490	6.8227	5400	828.611
53	0.07982	103	0.30147	500	7.1040	5500	859.584
54	0.08286	104	0.30735	600	10.2298	5600	891.126
55	0.08596	105	0.31328	700	13.9238	5700	923.236
56	0.08911	106	0.31928	800	18.1862	5800	955.914
57	0.09232	107	0.32533	900	23.0170	5900	989.161
58	0.09559	108	0.33144	1000	28.4160	6000	1022.980
59	0.09892	109	0.33761	1100	34.3834	6100	1057.360
60	0.10230	110	0.34383	1200	40.9190	6200	1092.310
61	0.10573	115	0.37580	1300	48.0230	6300	1127.830
62	0.10923	120	0.40921	1400	55.6954	6400	1163.920
63	0.11278	125	0.44400	1500	63.9360	6500	1200.580
64	0.11639	130	0.48023	1600	72.7450	6600	1237.800
65	0.12006	135	0.51788	1700	82.1222	6700	1275.590
66	0.12378	140	0.55695	1800	92.0678	6800	1313.960
67	0.12756	145	0.59744	1900	102.5820	6900	1352.890
68	0.13140	150	0.63936	2000	113.6640	7000	1392.380
69	0.13529	160	0.72745	2100	125.3150	7100	1432.450
70	0.13924	170	0.82122	2200	137.5330	7200	1473.090
71	0.14325	180	0.92067	2300	150.3210	7300	1514.290
72	0.14731	190	1.02590	2400	163.6760	7400	1556.060
73	0.15143	200	1.1367	2500	177.6000	7500	1598.400
74	0.15561	210	1.2531	2600	192.0920	7600	1641.310
75	0.15984	220	1.3753	2700	207.1530	7700	1684.780
76	0.16413	230	1.5032	2800	222.7810	7800	1728.830
77	0.16848	240	1.6358	2900	238.9790	7900	1773.440
78	0.17288	250	1.7760	3000	255.7400	8000	1818.620
79	0.17734	260	1.9209	3100	273.0780	8100	1864.370
80	0.18186	270	2.0715	3200	290.9800	8200	1910.690
81	0.18644	280	2.2278	3300	309.4500	8300	1957.580
82	0.19107	290	2.3898	3400	328.4890	8400	2005.030
83	0.19576	300	2.5574	3500	348.0960	8500	2053.060
84	0.20050	310	2.7308	3600	368.2710	8600	2101.650
85	0.20530	320	2.9098	3700	389.0150	8700	2150.810
86	0.21016	330	3.0945	3800	410.3270	8800	2200.540
87	0.21508	340	3.2849	3900	432.2070	8900	2250.830
88	0.22005	350	3.4809	4000	454.6560	9000	2301.700
89	0.22508	360	3.6823	4100	477.6730	9100	2353.130
90	0.23017	370	3.8901	4200	501.2580	9200	2405.130
91	0.23531	380	4.1032	4300	525.4120	9300	2457.700
92	0.24051	390	4.3220	4400	550.1340	9400	2510.840
93	0.24577	400	4.5466	4500	575.4240	9500	2564.540
94	0.25108	410	4.7767	4600	601.2830	9600	2618.820
95	0.25645	420	5.0126	4700	627.7090	9700	2673.660
96	0.26188	430	5.2541	4800	654.7050	9800	2729.070
97	0.26737	440	5.5013	4900	682.2680	9900	2785.050
98	0.27291	450	5.7542	5000	710.4000	10000	2841.600
99	0.27851	460	6.0128	5100	739.1000		

Table 2. Factors C_1 for Calculating Centrifugal Force (Metric SI units)

n	C_1	n	C_1	n	C_1	n	C_1
50	27.4156	100	109.662	470	2,422.44	5200	296,527
51	28.5232	101	111.867	480	2,526.62	5300	308,041
52	29.6527	102	114.093	490	2,632.99	5400	319,775
53	30.8041	103	116.341	500	2,741.56	5500	331,728
54	31.9775	104	118.611	600	3,947.84	5600	343,901
55	33.1728	105	120.903	700	5,373.45	5700	356,293
56	34.3901	106	123.217	800	7,018.39	5800	368,904
57	35.6293	107	125.552	900	8,882.64	5900	381,734
58	36.8904	108	127.910	1000	10,966.2	6000	394,784
59	38.1734	109	130.290	1100	13,269.1	6100	408,053
60	39.4784	110	132.691	1200	15,791.4	6200	421,542
61	40.8053	115	145.028	1300	18,532.9	6300	435,250
62	42.1542	120	157.914	1400	21,493.8	6400	449,177
63	43.5250	125	171.347	1500	24,674.0	6500	463,323
64	44.9177	130	185.329	1600	28,073.5	6600	477,689
65	46.3323	135	199.860	1700	31,692.4	6700	492,274
66	47.7689	140	214.938	1800	35,530.6	6800	507,078
67	49.2274	145	230.565	1900	39,588.1	6900	522,102
68	50.7078	150	246.740	2000	43,864.9	7000	537,345
69	52.2102	160	280.735	2100	48,361.1	7100	552,808
70	53.7345	170	316.924	2200	53,076.5	7200	568,489
71	55.2808	180	355.306	2300	58,011.3	7300	584,390
72	56.8489	190	395.881	2400	63,165.5	7400	600,511
73	58.4390	200	438.649	2500	68,538.9	7500	616,850
74	60.0511	210	483.611	2600	74,131.7	7600	633,409
75	61.6850	220	530.765	2700	79,943.8	7700	650,188
76	63.3409	230	580.113	2800	85,975.2	7800	667,185
77	65.0188	240	631.655	2900	92,226.0	7900	684,402
78	66.7185	250	685.389	3000	98,696.0	8000	701,839
79	68.4402	260	741.317	3100	105,385	8100	719,494
80	70.1839	270	799.438	3200	112,294	8200	737,369
81	71.9494	280	859.752	3300	119,422	8300	755,463
82	73.7369	290	922.260	3400	126,770	8400	773,777
83	75.5463	300	986.960	3500	134,336	8500	792,310
84	77.3777	310	1,053.85	3600	142,122	8600	811,062
85	79.2310	320	1,122.94	3700	150,128	8700	830,034
86	81.1062	330	1,194.22	3800	158,352	8800	849,225
87	83.0034	340	1,267.70	3900	166,796	8900	868,635
88	84.9225	350	1,343.36	4000	175,460	9000	888,264
89	86.8635	360	1,421.22	4100	184,342	9100	908,113
90	88.8264	370	1,501.28	4200	193,444	9200	928,182
91	90.8113	380	1,583.52	4300	202,766	9300	948,469
92	92.8182	390	1,667.96	4400	212,306	9400	968,976
93	94.8469	400	1,754.60	4500	222,066	9500	989,702
94	96.8976	410	1,843.42	4600	232,045	9600	1,010,650
95	98.9702	420	1,934.44	4700	242,244	9700	1,031,810
96	101.065	430	2,027.66	4800	252,662	9800	1,053,200
97	103.181	440	2,123.06	4900	263,299	9900	1,074,800
98	105.320	450	2,220.66	5000	274,156	10000	1,096,620
99	107.480	460	2,320.45	5100	285,232	...	...

Balancing Rotating Parts

Static Balancing. — There are several methods of testing the standing or static balance of a rotating part. A simple method that is sometimes used for flywheels, etc., is illustrated by the diagram, Fig. 1. An accurate shaft is inserted through the bore of the finished wheel, which is then mounted on carefully leveled "parallels" A. If the wheel is in an unbalanced state, it will turn until the heavy side is downward. When it will stand in any position as the result of counterbalancing and reducing the heavy portions, it is said to be in standing or static balance. Another test which is used for disk-shaped parts is shown in Fig. 2. The disk D is mounted on a vertical arbor attached to an adjustable cross-slide B. The latter is carried by a table C, which is supported by a knife-edged bearing. A pendulum having an adjustable screw-weight W at the lower end is suspended from cross-slide B. To test the static balance of disk D, slide B is adjusted until pointer E of the pendulum coincides with the center of a stationary scale F. Disk D is then turned halfway around without moving the slide, and if the indicator remains stationary, it shows that the disk is in balance for this particular position. The test is then repeated for ten or twelve other positions, and the heavy sides are reduced, usually by drilling out the required amount of metal. There are several other devices for testing the static balance which are designed on this same principle.

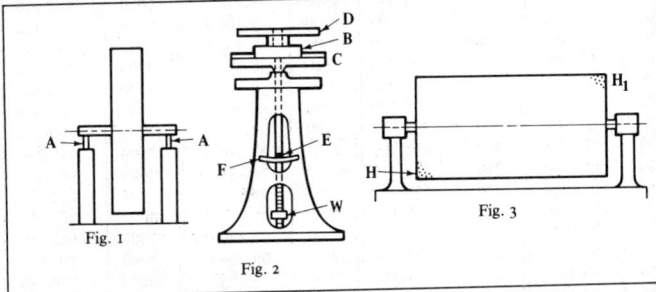

Fig. 1

Fig. 2

Fig. 3

Running or Dynamic Balance. — A cylindrical body may be in perfect static balance and not be in a balanced state when rotating at high speed. If the part is in the form of a thin disk, static balancing, if carefully done, may be accurate enough for high speeds, but if the rotating part is long in proportion to its diameter, and the unbalanced portions are at opposite ends or in different planes, the balancing must be done so as to counteract the centrifugal force of these heavy parts when they are rotating rapidly. This is known as a running balance or dynamic balancing. To illustrate, if a heavy section is located at H (Fig. 3), and another correspondingly heavy section at H_1, one may exactly counterbalance the other when the cylinder is stationary, and this static balance may be sufficient for a part rigidly mounted and rotating at a comparatively slow speed; but when the speed is very high, as in the case of turbine rotors, etc., the heavy masses H and H_1, being in different planes, are in an unbalanced state owing to the effect of centrifugal force, which results in excessive strains and injurious vibrations. Theoretically, to obtain a perfect running balance the exact position of the heavy sections should be located and the balancing effected either by reducing their weight or by adding counterweights opposite each section and in the same plane at the proper radius; but if the rotating part is rigidly mounted on a stiff shaft, a running balance that is sufficiently accurate for practical purposes can be obtained by means of comparatively few counterbalancing weights located with reference to the unbalanced parts.

Balancing Calculations. — As indicated previously, centrifugal forces caused by an unbalanced mass or masses in a rotating machine member cause additional loads on the bearings which are transmitted to the housing or frame and to other machine members. Such dynamically unbalanced conditions can occur even though static balance (balance at zero speed) exists. Dynamic balance can be achieved by the addition of one or two masses rotating about the same axis and at the same speed as the unbalanced masses. A single unbalanced mass can be balanced by one counterbalancing mass located 180 degrees opposite and in the same plane of rotation as the unbalanced mass, if the product of their respective radii and masses are equal; i.e., $M_1 r_1 = M_2 r_2$. Two or more unbalanced masses rotating in the same plane can be balanced by a single mass rotating in the same plane, or by two masses rotating about the same axis in two separate planes. Likewise, two or more unbalanced masses rotating in different planes can be balanced by two masses rotating about the same axis in separate planes. When the unbalanced masses are in separate planes they may be in static balance but not in dynamic balance; i.e., they may be balanced when not rotating but unbalanced when rotating. If a system is in dynamic balance, it will remain in balance at all speeds, although this is not strictly true at the critical speed of the system. (See Critical Speeds.)

In all of the equations that follow the symbol M denotes either mass in kilograms or in slugs, or weight in pounds. Either mass or weight units may be used and the equations may be used with metric or with customary English units without change; however, in a given problem the units must be all metric or all customary English.

Counterbalancing Several Masses Located in a Single Plane. — In all balancing problems, it is the product of counterbalancing mass (or weight) and its radius that is calculated; it is thus necessary to select either the mass or the radius and then calculate the other value from the product of the two quantities. Design considerations usually make this decision self-evident. The angular position of the counterbalancing mass must also be calculated. Referring to Fig. 1:

$$M_B r_B = \sqrt{(\Sigma Mr \cos \theta)^2 + (\Sigma Mr \sin \theta)^2} \qquad (1)$$

$$\tan \theta_B = \frac{-(\Sigma Mr \sin \theta)}{-(\Sigma Mr \cos \theta)} = \frac{y}{x} \qquad (2)$$

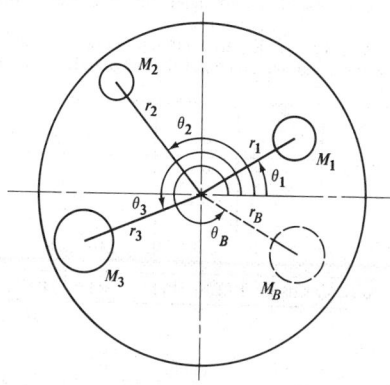

Fig. 1

**Table 1. Relationship of the Signs of the Functions of the Angle
with Respect to the Quadrant in Which They Occur**

		Angle θ			
		0° to 90°	90° to 180°	180° to 270°	270° to 360°
		Signs of the Functions			
tan		$\dfrac{+y}{+x}$	$\dfrac{+y}{-x}$	$\dfrac{-y}{-x}$	$\dfrac{-y}{+x}$
sine		$\dfrac{+y}{+r}$	$\dfrac{+y}{+r}$	$\dfrac{-y}{+r}$	$\dfrac{-y}{+r}$
cosine		$\dfrac{+x}{+r}$	$\dfrac{-x}{+r}$	$\dfrac{-x}{+r}$	$\dfrac{+x}{+r}$

where:

$M_1, M_2, M_3, \ldots M_n =$ any unbalanced mass or weight, kg or lb

$M_B =$ counterbalancing mass or weight, kg or lb

$r =$ radius to center of gravity of any unbalanced mass or weight mm or in.

$r_B =$ radius to center of gravity of counterbalancing mass or weight mm or in.

$θ =$ angular position of r of any unbalanced mass or weight, degrees

$θ_B =$ angular position of r_B of counterbalancing mass or weight degrees

x and $y =$ see Table 1.

Table 1 is helpful in finding the angular position of the counterbalancing mass or weight. It indicates the range of the angles within which this angular position occurs by noting the plus and minus signs of the numerator and the denominator of the terms in Equation 2. In a like manner, Table 1 is helpful in determining the *sign* of the sine or cosine functions for angles ranging from 0 to 360 degrees. Balancing problems are usually solved most conveniently by arranging the arithmetical calculations in a tabular form.

Example: Referring to Fig. 1, the particular values of the unbalanced weights have been entered in the table below. Calculate the magnitude of the counterbalancing weight if its radius is to be 10 inches.

No.	M lb.	r in.	θ deg.	cos θ	sin θ	Mr cos θ	Mr sin θ
1	10	10	30	0.8660	0.5000	86.6	50.0
2	5	20	120	−0.5000	0.8660	−50.0	86.6
3	15	15	200	−0.9397	−0.3420	−211.4	−77.0
						−174.8 = $\Sigma Mr \cos θ$	59.6 = $\Sigma Mr \sin θ$

$$M_B = \frac{\sqrt{(\Sigma Mr \cos θ)^2 + (\Sigma Mr \sin θ)^2}}{r_B} = \frac{\sqrt{(-174.8)^2 + (59.6)^2}}{10}$$

$$M_B = 18.5 \text{ lb}$$

$$\tan θ_B = \frac{-(\Sigma Mr \sin θ)}{-(\Sigma Mr \cos θ)} = \frac{-(59.6)}{-(-174.8)} = \frac{-y}{+x}; \quad θ_B = 341°10'$$

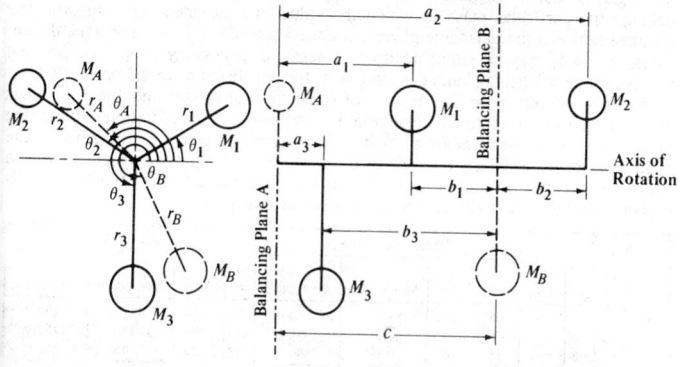

Fig. 2

Counterbalancing Masses Located in Two or More Planes. — Unbalanced masses or weights rotating about a common axis in two separate planes of rotation form a couple, which must be counterbalanced by masses or weights, also located in two separate planes, call them planes A and B, and rotating about the same common axis (see *Couples*, page 135). In addition, they must be balanced in the direction perpendicular to the axis, as before. Since two counterbalancing masses are required, two separate equations are required to calculate the product of each mass or weight and its radius, and two additional equations are required to calculate the angular positions. The planes A and B selected as balancing planes may be any two planes separated by any convenient distance c, along the axis of rotation. In Fig. 2:

For balancing plane A:

$$M_A r_A = \frac{\sqrt{(\Sigma Mrb \cos \theta)^2 + (\Sigma Mrb \sin \theta)^2}}{c} \tag{3}$$

$$\tan \theta_A = \frac{-(\Sigma Mrb \sin \theta)}{-(\Sigma Mrb \cos \theta)} = \frac{y}{x} \tag{4}$$

For balancing plane B:

$$M_B r_B = \frac{\sqrt{(\Sigma Mra \cos \theta)^2 + (\Sigma Mra \sin \theta)^2}}{c} \tag{5}$$

$$\tan \theta_B = \frac{-(\Sigma Mra \sin \theta)}{-(\Sigma Mra \cos \theta)} = \frac{y}{x} \tag{6}$$

Where: M_A and M_B are the mass or weight of the counterbalancing masses in the balancing planes A and B, respectively; r_A and r_B are the radii; and θ_A and θ_B are

the angular positions of the balancing masses in these planes. M, r, and θ are the mass or weight, radius, and angular positions of the unbalanced masses, with the subscripts defining the particular mass to which the values are assigned. The length c, the distance between the balancing planes, is always a positive value. The axial dimensions, a and b, may be either positive or negative, depending upon their position relative to the balancing plane; e.g., in Fig. 2, the dimension b_2 would be negative.

Example: Referring to Fig. 2, a set of values for the masses and dimensions has been selected and put into convenient table form below. The separation of balancing planes, c, is assumed as being 15 inches. If in balancing plane A, the radius of the counterbalancing weight is selected to be 10 inches; calculate the magnitude of the counterbalancing mass and its position. If in balancing plane B, the counterbalancing mass is selected to be 10 lb; calculate its radius and position.

Plane	M lb	r in.	θ deg.	Balancing Plane A				Balancing Plane B			
				b in.	Mrb	$Mrb \cos \theta$	$Mrb \sin \theta$	a in.	Mra	$Mra \cos \theta$	$Mra \sin \theta$
1	10	8	30	6	480	415.7	240.0	9	720	623.5	360.0
2	8	10	135	−6	−480	339.4	−339.4	21	1680	−1187.9	1187.9
3	12	9	270	12	1296	0.0	−1296.0	3	324	0.0	−324.0
A	?	10	?	15*	. . .	755.1 = $\Sigma Mrb \cos \theta$	−1395.4 = $\Sigma Mrb \sin \theta$	0	. . .	−564.4 = $\Sigma Mra \cos \theta$	1223.9 = $\Sigma Mra \sin \theta$
B	10	?	?	0	. . .			15*	. . .		

* 15 inches = distance c between planes A and B.

For balancing plane A:

$$M_A = \frac{\sqrt{(\Sigma Mrb \cos \theta)^2 + (\Sigma Mrb \sin \theta)^2}}{r_A c} = \frac{\sqrt{(755.1)^2 + (-1395.4)^2}}{10\,(15)}$$

$$M_A = 10.6 \text{ lb}$$

$$\tan \theta_A = \frac{-(\Sigma Mrb \sin \theta)}{-(\Sigma Mrb \cos \theta)} = \frac{-(-1395.4)}{-(755.1)} = \frac{+y}{-x}$$

$$\theta_A = 118° \, 25'$$

For balancing plane B:

$$r_B = \frac{\sqrt{(\Sigma Mra \cos \theta)^2 + (\Sigma Mra \sin \theta)^2}}{M_B c} = \frac{\sqrt{(-564.4)^2 + (1223.9)^2}}{10\,(15)}$$

$$= 8.985 \text{ in.}$$

$$\tan \theta_B = \frac{-(\Sigma Mra \sin \theta)}{-(\Sigma Mra \cos \theta)} = \frac{-(1223.9)}{-(-564.4)} = \frac{-y}{+x}$$

$$\theta_B = 294° \, 45'$$

Balancing Lathe Fixtures. — Lathe fixtures rotating at a high speed requir balancing. Often it is assumed that the center of gravity of the workpiece and fixture and of the counterbalancing masses are in the same plane; however, this is not usuall the case. Counterbalancing masses are required in two separate planes to preven excessive vibration or bearing loads at high speeds. Usually a single counterbalancin

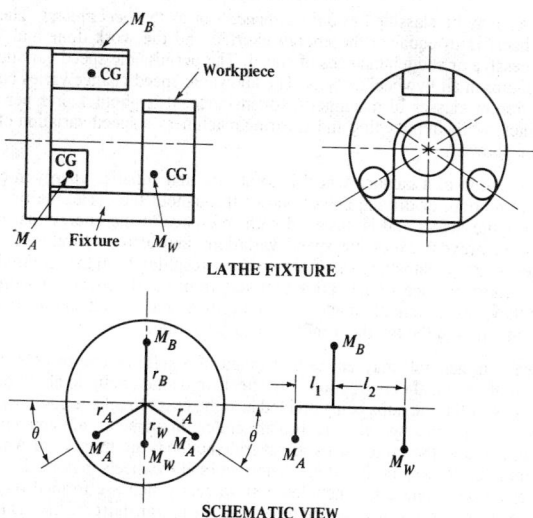

LATHE FIXTURE

SCHEMATIC VIEW

Fig. 3

mass is placed in one plane selected to be 180 degrees directly opposite the combined center of gravity of the workpiece and the fixture. Two equal counterbalancing masses are then placed in the second counterbalancing plane, equally spaced on each side of the fixture. Referring to Fig. 3, the two counterbalancing masses M_A and the two angles θ are equal. For the design in this illustration, the following formulas can be used to calculate the magnitude of the counterbalancing masses. Since their angular positions are fixed by the design, they are not calculated.

$$M_B = \frac{M_w r_w (l_1 + l_2)}{r_B l_1} \tag{7}$$

$$M_A = \frac{M_B r_B - M_w r_w}{2r_A \sin \theta} \tag{8}$$

In these formulas M_w and r_w denote the mass or weight and the radius of the combined center of gravity of the workpiece and the fixture.

Example: In Fig. 3 the combined weight of the workpiece and the fixture is 18.5 lb. The following dimensions were determined from the layout of the fixture and by calculating the centers of gravity: $r_w = 2$ in.; $r_A = 6.25$ in.; $r_B = 6$ in.; $l_1 = 3$ in.; $l_2 = 5$ in.; and $\theta = 30°$. Calculate the weights of the counterbalancing masses.

$$M_B = \frac{M_w r_w (l_1 + l_2)}{r_B l_1} = \frac{18.5 \times 2 \times 8}{6 \times 3} = 16.44 \text{ lb}$$

$$M_A = \frac{M_B r_B - M_w r_w}{2r_A \sin \theta} = \frac{(16.44 \times 6) - (18.5 \times 2)}{(2 \times 6.25) \sin 30°} = 9.86 \text{ lb (each weight).}$$

FLYWHEELS

Flywheels may be classified as *balance wheels* or as *flywheel pulleys*. The object of all flywheels is to equalize the energy exerted and the work done and thereby prevent excessive or sudden changes of speed. The permissible speed variation is an important factor in all flywheel designs. The allowable speed change varies considerably for different classes of machinery; for instance, it is about 1 or 2 per cent in steam engines, while in punching and shearing machinery a speed variation of 20 per cent may be allowed.

As the function of a balance wheel is to absorb and equalize energy in case the resistance to motion, or driving power, varies throughout the cycle, the rim section is generally quite heavy and is designed with reference to the energy that must be stored in it to prevent excessive speed variations and also with reference to the strength necessary to withstand safely the stresses resulting from the required speed. The rims of most balance wheels are either square or nearly square in section, but flywheel pulleys are commonly made wide to accommodate a belt and relatively thin in a radial direction, although this is not an invariable rule.

Flywheels, in general, may either be formed of a solid or one-piece section, or they may be of sectional construction. Flywheels in diameters up to about eight feet are usually cast solid, the hubs being divided in some cases to relieve cooling stresses. Flywheels ranging from, say, eight feet to fifteen feet in diameter, are commonly cast in half sections, and the larger sizes in several sections, the number of which may equal the number of arms in the wheel. The sectional flywheels may be divided into two general classes. One class includes cast wheels which are formed of sections principally because a solid casting would be too large to transport readily. The second class includes wheels of sectional construction which, by reason of the materials used and the special arrangement of the sections, enables much higher peripheral speeds to be obtained safely than would be possible with ordinary sectional wheels of the type not designed especially for high speeds. Various designs have been built to withstand the extreme stresses encountered in some classes of service. The rims in some cases are laminated, being partly or entirely formed of numerous segment-shaped steel plates. Another type of flywheel, which is superior to an ordinary sectional wheel, has a solid cast-iron rim connected to the hub by disk-shaped steel plates instead of cast spokes. Steel wheels may be divided into three distinct types, including (1) those having the center and rim built up entirely of steel plates, (2) those having a cast-iron center and steel rim, and (3) those having a cast-steel center and rim formed of steel plates. Wheels having wire-wound rims have been used to a limited extent when extremely high speeds have been necessary.

When the rim is formed of sections held together by joints it is very important to design these joints properly. The ordinary bolted and flanged rim joints located between the arms average about 20 per cent of the strength of a solid rim and about 25 per cent is the maximum strength obtainable for a joint of this kind. However, by placing the joints at the ends of the arms instead of between them, an efficiency of 50 per cent of the strength of the rim may be obtained. This is due to the fact that the joint is not subjected to the outward bending stresses between the arms but is directly supported by the arm, the end of which is secured to the rim just beneath the joint. When the rim sections of heavy balance wheels are held together by steel links shrunk into place, an efficiency of 60 per cent may be obtained; and by using a rim of box or I-section, a link type of joint connection may have an efficiency of 100 per cent.

Energy Due to Changes of Velocity. — When a flywheel absorbs energy from a variable driving force, as in the case of a steam engine, the velocity increases; and

when this stored energy is given out, the velocity diminishes. When the driven member of a machine encounters a variable resistance in performing its work, as when the punch of a punching machine is passing through a steel plate, the flywheel gives up energy while the punch is at work, and, consequently, the speed of the flywheel is reduced. The total energy that a flywheel would give out if brought to a standstill is given by the formula:

$$E = \frac{Wv^2}{2g} = \frac{Wv^2}{64.32}$$

in which E = total energy of flywheel, in foot-pounds;
  W = weight of flywheel rim, in pounds;
  v = velocity at mean radius of flywheel rim, in feet per second;
  g = acceleration due to gravity = 32.16.

If the velocity of a flywheel changes, the energy it will absorb or give up is proportional to the difference between the squares of its initial and final speeds, and is equal to the difference between the energy which it would give out if brought to a full stop and that which is still stored in it at the reduced velocity. Hence:

$$E_1 = \frac{Wv_1^2}{2g} - \frac{Wv_2^2}{2g} = \frac{W(v_1^2 - v_2^2)}{64.32}$$

in which E_1 = energy in foot-pounds which a flywheel will give out while the speed is reduced from v_1 to v_2;
  W = weight of flywheel rim, in pounds;
  v_1 = velocity at mean radius of flywheel rim before any energy has been given out, in feet per second;
  v_2 = velocity of flywheel rim at end of period during which the energy has been given out, in feet per second.

Ordinarily, the effects of the arms and hub do not enter into flywheel calculations, and only the weight of the rim is considered. In computing the velocity, the mean radius of the rim is commonly used.

Using metric SI units, the formulas are $E = \frac{1}{2}Mv^2$, and $E_1 = \frac{1}{2}M(v_1^2 - v_2^2)$, where E and E_1 are in joules; M = the mass of the rim in kilograms; and v, v_1, and v_2 = velocities in meters per second. Note: In the SI, the unit of mass is the kilogram. If the weight of the flywheel rim is given in kilograms, the value referred to is the mass, M. Should the weight be given in newtons, N, then

$$M = \frac{W \text{ (newtons)}}{g},$$

where g is approximately 9.81 meters per second squared.

General Procedure in Flywheel Design. — The general method of designing a flywheel is to determine first the value of E_1 or the energy the flywheel must either supply or absorb for a given change in velocity, which, in turn, varies for different classes of service. The mean diameter of the flywheel may be assumed, or it may be fixed within certain limits by the general design of the machine. Ordinarily the speed of the flywheel shaft is known, at least approximately; the values of v_1 and v_2 can then be determined, the latter depending upon the allowable percentage of speed variation. When these values are known, the weight of the rim and the cross-sectional area required to obtain this weight may be computed. The general procedure will be illustrated more in detail by considering the design of flywheels for punching and shearing machinery.

Flywheels for Presses, Punches, Shears, Etc. — In these classes of machinery, the work that the machine performs is of an intermittent nature and is done during

a small part of the time required for the driving shaft of the machine to make a complete revolution. In order to distribute the work of the machine over the entire period of revolution of the driving shaft, a heavy-rimmed flywheel is placed on the shaft, giving the belt an opportunity to perform an almost uniform amount of work during the whole revolution. During the greater part of the revolution of the driving shaft, the belt power is used to accelerate the speed of the flywheel. During the part of the revolution when the work is done, the energy thus stored up in the flywheel is given out at the expense of its velocity. The problem is to determine the weight and cross-sectional area of the rim when the conditions affecting the design of the flywheel are known.

Example: — A flywheel is required for a punching machine capable of punching ¾-inch holes through structural steel plates ¾ inch thick. This machine (see accompanying diagram) is of the general type having a belt-driven shaft at the rear which carries a flywheel and a pinion that meshes with a large gear on the main shaft at the top of the machine. It is assumed that the relative speeds of the pinion and large gear are 7 to 1, respectively, and that the slide is to make 30 working strokes per minute. The preliminary lay-out shows that the flywheel should have a mean diameter (see enlarged detail) of about 30 inches. Find the weight of the flywheel and the size of the rim.

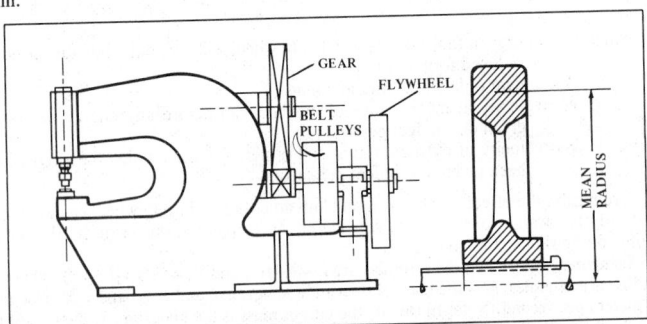

Energy Supplied by Flywheel. — The energy which the flywheel must give up for a given change in velocity, and the weight of rim necessary to supply that energy, must be determined. The maximum force for shearing a ¾-inch hole through ¾-inch structural steel equals approximately the circumference of the hole multiplied by the thickness of the stock multiplied by the tensile strength, which is nearly the same as the shearing resistance of the steel. Thus, in this case, $3.1416 \times ¾ \times ¾ \times 60,000 = 106,000$ pounds. The average force will be much less than the maximum. Some designers assume that the average force is about one-half the maximum, although experiments show that the material is practically sheared off when the punch has entered the sheet a distance equal to about one-third the sheet thickness. On this latter basis, the average energy E_a in foot-pounds is 2200 in this case. Thus:

$$E_a = \frac{106,000 \times \tfrac{1}{3} \times ¾}{12} = \frac{106,000}{4 \times 12} = 2200 \text{ foot-pounds.}$$

If the efficiency of the machine is taken at 85 per cent, the energy required will equal $2200 \div 0.85 = 2600$ foot-pounds nearly. Assume that the energy supplied by the belt while the punch is at work is determined by calculation to equal 175 foot-pounds. Then the flywheel must supply $2600 - 175 = 2425$ foot-pounds $= E_1$.

Dimensions of Flywheels for Punches and Shears

(Maximum number of revolutions per minute given in table should never be exceeded for cast-iron flywheels.)

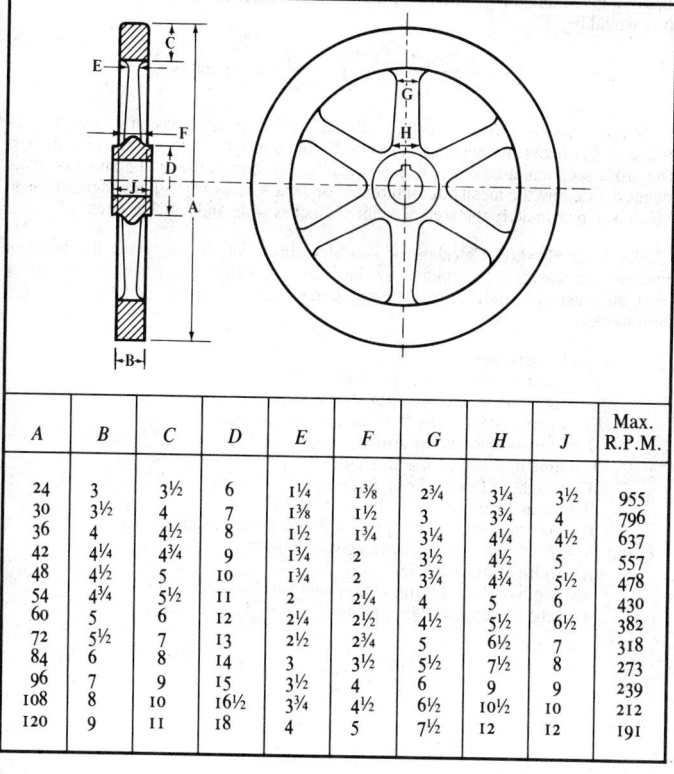

A	B	C	D	E	F	G	H	J	Max. R.P.M.
24	3	3½	6	1¼	1⅜	2¾	3¼	3½	955
30	3½	4	7	1⅜	1½	3	3¾	4	796
36	4	4½	8	1½	1¾	3¼	4¼	4½	637
42	4¼	4¾	9	1¾	2	3½	4½	5	557
48	4½	5	10	1¾	2	3¾	4¾	5½	478
54	4¾	5½	11	2	2¼	4	5	6	430
60	5	6	12	2¼	2½	4½	5½	6½	382
72	5½	7	13	2½	2¾	5	6½	7	318
84	6	8	14	3	3½	5½	7½	8	273
96	7	9	15	3½	4	6	9	9	239
108	8	10	16½	3¾	4½	6½	10½	10	212
120	9	11	18	4	5	7½	12	12	191

Rim Velocity at Mean Radius. — When the mean radius of the flywheel is known, the velocity of the rim at the mean radius, in feet per second, is:

$$v = \frac{2 \times 3.1416 \times R \times n}{60}$$

in which v = velocity at mean radius of flywheel, in feet per second;
            R = mean radius of flywheel rim, in feet;
            n = number of revolutions per minute.

According to the preliminary lay-out the mean diameter in this case should be about 30 inches and the driving shaft is to make 210 R.P.M.; hence,

$$v = \frac{2 \times 3.1416 \times 1.25 \times 210}{60} = 27.5 \text{ feet per second.}$$

Weight of Flywheel Rim. — Assuming that the allowable variation in velocity when punching is about 15 per cent, and values of v_1 and v_2 are respectively 27.5 and 23.4 feet per second (27.5 × 0.85 = 23.4), the weight of a flywheel rim necessary to supply a given amount of energy in foot-pounds while the speed is reduced from v_1 to v_2 would be:

$$W = \frac{E_1 \times 64.32}{v_1^2 - v_2^2} = \frac{2425 \times 64.32}{27.5^2 - 23.4^2} = 750 \text{ pounds.}$$

Size of Rim for Given Weight. — Since 1 cubic inch of cast iron weighs 0.26 pound, a flywheel rim weighing 750 pounds contains 750 ÷ 0.26 = 2884 cubic inches. The cross-sectional area of the rim in square inches equals the total number of cubic inches divided by the mean circumference, or 2884 ÷ 94.25 = 31 square inches nearly, which is approximately the area of a rim 5⅛ inches wide and 6 inches deep.

Simplified Flywheel Calculations. — Calculations for designing the flywheels of punches and shears are simplified by the following formulas and the accompanying table of constants applying to different percentages of speed reduction. In these formulas let:

H.P. = horsepower required;
N = number of strokes per minute;
E = total energy required per stroke, in foot-pounds;
E_1 = energy given up by flywheel, in foot-pounds;
T = time in seconds per stroke;
T_1 = time in seconds of actual cut;
W = weight of flywheel rim, in pounds;
D = mean diameter of flywheel rim, in feet;
R = maximum allowable speed of flywheel in revolutions per minute;
C and C_1 = speed reduction values as given in table;
a = width of flywheel rim;
b = depth of flywheel rim;
y = ratio of depth to width of rim.

$$\text{H.P.} = \frac{EN}{33,000} = \frac{E}{T \times 550} \qquad E_1 = E\left(1 - \frac{T_1}{T}\right)$$

$$W = \frac{E_1}{CD^2R^2} \qquad a = \sqrt{\frac{1.22W}{12Dy}} \qquad b = ay$$

For cast-iron flywheels, with a maximum stress of 1000 pounds per square inch:

$$W = C_1E_1 \qquad R = 1940 \div D$$

Values of C and C_1 in the Previous Formulas

Per Cent Reduction	C	C_1	Per Cent Reduction	C	C_1
2½	0.00000213	0.1250	10	0.00000810	0.0328
5	0.00000426	0.0625	15	0.00001180	0.0225
7½	0.00000617	0.0432	20	0.00001535	0.0173

Example 1: — A hot slab shear is required to cut a slab 4×15 inches which, at a shearing stress of 6000 pounds per square inch, gives a force between the knives of 360,000 pounds. The total energy required for the cut will then be $360,000 \times \frac{4}{12} = $ 120,000 foot-pounds. The shear is to make 20 strokes per minute; the actual cutting time is 0.75 second, and the balance of the stroke is 2.25 seconds.

The flywheel is to have a mean diameter of 6 feet 6 inches and is to run at a speed of 200 R.P.M.; the reduction in speed to be 10 per cent per stroke when cutting.

$$\text{H.P.} = \frac{120,000 \times 20}{33,000} = 72.7 \text{ horsepower;}$$

$$E_1 = 120,000 \times \left(1 - \frac{0.75}{3}\right) = 90,000 \text{ foot-pounds;}$$

$$W = \frac{90,000}{0.0000081 \times 6.5^2 \times 200^2} = 6570 \text{ pounds.}$$

Assuming a ratio of 1.22 between depth and width of rim,

$$a = \sqrt{\frac{6570}{12 \times 6.5}} = 9.18 \text{ inches;}$$

$$b = 1.22 \times 9.18 = 11.2 \text{ inches;}$$

or size of rim, say, $9 \times 11\frac{1}{2}$ inches.

Example 2: — Suppose that the flywheel in Example 1 is to be made with a stress of 1000 pounds, due to centrifugal force, per square inch of rim section.

$$C_1 \text{ for 10 per cent} = 0.0328;$$

$$W = 0.0328 \times 90,000 = 2950 \text{ pounds.}$$

$$R = \frac{1940}{D}. \quad \text{If } D = 6 \text{ feet,} \quad R = \frac{1940}{6} = 323 \text{ R.P.M.}$$

Assuming a ratio of 1.22 between depth and width of rim, as before:

$$a = \sqrt{\frac{2950}{12 \times 6}} = 6.4 \text{ inches;}$$

$$b = 1.22 \times 6.4 = 7.8 \text{ inches;}$$

or size of rim, say, $6\frac{1}{4} \times 8$ inches.

Centrifugal Stresses in Flywheel Rims. — In general, high speed is desirable for flywheels in order to avoid using wheels which are unnecessarily large and heavy. The centrifugal tension or hoop tension stress which tends to rupture a flywheel rim of given area, depends solely upon the rim velocity, and is independent of the rim radius. The bursting velocity of a flywheel, based on hoop stress alone (not considering bending stresses), is related to the tensile stress in the flywheel rim by the following formula which is based on the centrifugal force formula from mechanics.

$$V = \sqrt{10 \times s} \quad \text{or,} \quad s = V^2 \div 10$$

where V = velocity of outside circumference of rim in feet per second, and s is the tensile strength of the rim material in pounds per square inch.

For cast iron having a tensile strength of 19,000 pounds per square inch the bursting speed would be:

$$V = \sqrt{10 \times 19,000} = 436 \text{ feet per second}$$

Built-up Flywheels: Flywheels built up of solid disks of rolled steel plate stacked and bolted together on a through shaft have greater speed capacity than other types. The maximum hoop stress is at the bore and is given by the formula,

$$s = 0.0194V^2[4.333 + (d/D)^2]$$

In this formula, s and V are the stress and velocity as previously defined and d and D are the bore and outside diameters, respectively.

Assuming the plates to be of steel having a tensile strength of 60,000 pounds per square inch and a safe working stress of 24,000 pounds per square inch (using a factor of safety of 2.5 on stress or $\sqrt{2.5}$ on speed) and taking the worst condition (when d approaches D), the safe rim speed for this type of flywheel is 500 feet per second or 30,000 feet per minute.

Combined Stresses in Flywheels. — The bending stresses in the rim of a flywheel may exceed the centrifugal (hoop tension) stress predicted by the simple formula $s = V^2 \div 10$ by a considerable amount. By taking into account certain characteristics of flywheels, relatively simple formulas have been developed to determine the stress due to the combined effect of hoop tension and bending stress. Some of the factors that influence the magnitude of the maximum combined stress acting at the rim of a flywheel are:

1. *The number of spokes.* Increasing the number of spokes decreases the rim span between spokes and hence decreases the bending moment. Thus an eight-spoke wheel can be driven to a considerably higher speed before bursting than a six-spoke wheel having the same rim.

2. *The relative thickness of the spokes.* If the spokes were extremely thin, like wires, they could offer little constraint to the rim in expanding to its natural diameter under centrifugal force, and hence would cause little bending stress. Conversely, if the spokes were extremely heavy in proportion to the rim, they would restrain the rim thereby setting up heavy bending stresses at the junctions of the rim and spokes.

3. *The relative thickness of the rim to the diameter.* If the rim is quite thick (i.e. has a large section modulus in proportion to span), its resistance to bending will be great and bending stress small. Conversely, thin rims with a section modulus small in comparison with diameter or span have little resistance to bending, thus are subject to high bending stresses.

4. *Residual stresses.* These include shrinkage stresses, impact stresses, and stresses caused by operating torques and imperfections in the material. Residual stresses are taken into account by the use of a suitable factor of safety. (See "Factors of Safety for Flywheels.")

The formulas that follow give the maximum combined stress at the rim of flywheels having 6, 8, and 10 spokes. These formulas are for flywheels with *rectangular rim sections* and take into account the first three of the four factors listed as influencing the magnitude of the combined stress in flywheels.

For 6 spokes:

$$s = \frac{V^2}{10}\left[1 + \left(\frac{0.56B - 1.81}{3Q + 3.14}\right)Q\right]$$

For 8 spokes:

$$s = \frac{V^2}{10}\left[1 + \left(\frac{0.42B - 2.53}{4Q + 3.14}\right)Q\right]$$

For 10 spokes:

$$s = \frac{V^2}{10}\left[1 + \left(\frac{0.33B - 3.22}{5Q + 3.14}\right)Q\right]$$

In these formulas, s = maximum combined stress in pounds per square inch; Q = ratio of mean spoke cross-section area to rim cross-section area; B = ratio of outside diameter of rim to rim thickness; and V = velocity of flywheel rim in feet per second.

Thickness of Cast Iron Flywheel Rims. — The mathematical analysis of the stresses in flywheel rims is not conclusive owing to the uncertainty of shrinkage stresses in castings or the strength of the joint in the case of sectional wheels. When a flywheel of ordinary design is revolving at high speed, the tendency of the rim is to bend or bow outward between the arms, and the bending stresses may be serious, especially if the rim is wide and thin and the spokes are rather widely spaced. When the rims are thick, this tendency does not need to be considered, but in the case of a thin rim, running at a high rate of speed, the stress in the middle might become sufficiently great to cause the wheel to fail. The proper thickness of a cast-iron rim to resist this tendency is given for solid rims by Formula 1 and for a jointed rim by Formula 2.

$$t = \frac{0.475d}{n^2 \left(\dfrac{6000}{v^2} - \dfrac{1}{10} \right)} \quad (1); \qquad t = \frac{0.95d}{n^2 \left(\dfrac{6000}{v^2} - \dfrac{1}{10} \right)} \quad (2)$$

In these formulas, t = thickness of rim, in inches; d = diameter of flywheel, in inches; n = number of arms; v = peripheral speed, in feet per second.

Tables of Safe Speeds for Flywheels. — The accompanying Table 1, prepared by T. C. Rathbone of The Fidelity and Casualty Company of New York, gives general recommendations for safe rim speeds for flywheels of various constructions. Table 2 shows the number of revolutions per minute corresponding to the rim speeds in Table 1.

Table 1. Safe Rim Speeds for Flywheels*

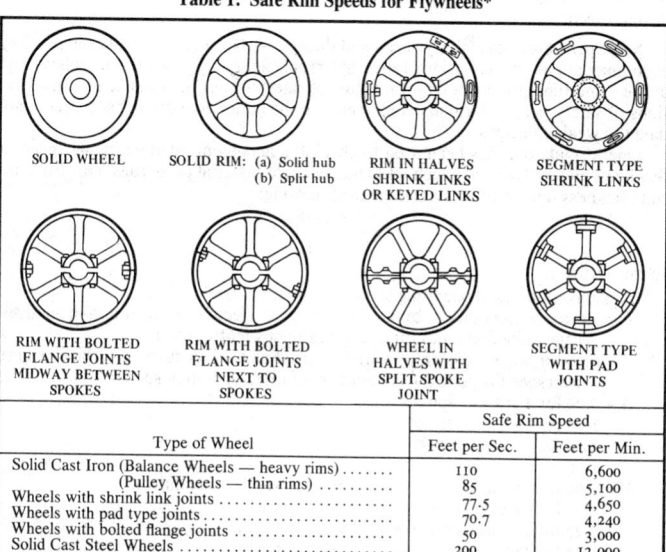

		Safe Rim Speed	
Type of Wheel		Feet per Sec.	Feet per Min.
Solid Cast Iron (Balance Wheels — heavy rims)		110	6,600
(Pulley Wheels — thin rims)		85	5,100
Wheels with shrink link joints		77.5	4,650
Wheels with pad type joints		70.7	4,240
Wheels with bolted flange joints		50	3,000
Solid Cast Steel Wheels		200	12,000
Wheels built up of stacked steel plates		500	30,000

* To find the safe speed in revolutions per minute, divide the safe rim speed in feet per minute by 3.14 times the outside diameter of the flywheel rim in feet. For flywheels up to 15 feet in diameter, see Table 2.

Table 2. Safe Speeds of Rotation for Flywheels*

Outside Diameter of Rim, in Feet	Safe Rim Speed in Feet per Minute (from Table 1)						
	6,600	5,100	4,650	4,240	3,000	12,000	30,000
	Safe Speed of Rotation in Revolutions per Minute						
1	2100	1623	1480	1350	955	3820	9549
2	1050	812	740	676	478	1910	4775
3	700	541	493	450	318	1273	3183
4	525	406	370	338	239	955	2387
5	420	325	296	270	191	764	1910
6	350	271	247	225	159	637	1592
7	300	232	211	193	136	546	1364
8	263	203	185	169	119	478	1194
9	233	180	164	150	106	424	1061
10	210	162	148	135	96	382	955
11	191	148	135	123	87	347	868
12	175	135	123	113	80	318	796
13	162	125	114	104	73	294	735
14	150	116	106	97	68	273	682
15	140	108	99	90	64	255	637

* Safe speeds of rotation are based on safe rim speeds shown in Table 1.

Factors of Safety for Flywheels. — Cast-iron flywheels are commonly designed with a factor of safety of 10 to 13. A factor of safety of 10 applied to the tensile strength of a flywheel material is equivalent to a factor of safety of $\sqrt{10}$ or 3.16 on the speed of the flywheel due to the fact that the stress on the rim of a flywheel increases as the square of the speed. Thus, a flywheel operating at a speed twice that for which it was designed would undergo rim stresses four times as great as at the design speed.

Safe Speed Formulas for Flywheels and Pulleys. — No simple formula can possibly accommodate all of the various types and proportions of flywheels and pulleys and at the same time provide a uniform factor of safety for each. Because of considerations of safety, such a formula would penalize the better constructions to accommodate the weaker designs.

One formula that has been used to check the maximum rated operating speed of flywheels and pulleys and which takes into account material properties, construction, rim thickness, and joint efficiencies is the following:

$$N = \frac{CAMEK}{D}$$

In this formula,

N = maximum rated operating speed in revolutions per minute
C = 1.0 for wheels driven by a constant speed electric motor (i.e. a–c squirrel-cage induction motor or a–c synchronous motor, etc.)
  = 0.90 for wheels driven by variable speed motors, engines or turbines where overspeed is not over 110 per cent of rated operating speed.
A = 0.90 for 4 arms or spokes
  1.00 for 6 arms or spokes
  1.08 for 8 arms or spokes
  1.50 for disc type
M = 1.00 for cast iron of 20,000 psi tensile strength, or unknown
  1.12 for cast iron of 25,000 psi tensile strength
  1.22 for cast iron of 30,000 psi tensile strength
  1.32 for cast iron of 35,000 psi tensile strength
  2.20 for nodular iron of 60,000 psi tensile strength
  2.45 for cast steel of 60,000 psi tensile strength
  2.75 for plate or forged steel of 60,000 psi tensile strength

E = joint efficiency
    1.0 for solid rim
    0.85 for link or prison joints
    0.75 for split rim — bolted joint at arms
    0.70 for split rim — bolted joint between arms
K = 1355 for rim thickness equal to 1 per cent of outside diameter
    1650 for rim thickness equal to 2 per cent of outside diameter
    1840 for rim thickness equal to 3 per cent of outside diameter
    1960 for rim thickness equal to 4 per cent of outside diameter
    2040 for rim thickness equal to 5 per cent of outside diameter
    2140 for rim thickness equal to 7 per cent of outside diameter
    2225 for rim thickness equal to 10 per cent of outside diameter
    2310 for rim thickness equal to 15 per cent of outside diameter
    2340 for rim thickness equal to 20 per cent of outside diameter
D = outside diameter of rim in feet

Example: A six-spoke solid cast iron balance wheel 8 feet in diameter has a rectangular rim 10 inches thick. What is the safe speed, in revolutions per minute, if driven by a constant speed motor?

In this case, $C = 1$; $A = 1$; $M = 1$, since tensile strength is unknown; $E = 1$; $K = 2225$ since the rim thickness is approximately 10 per cent of the wheel diameter; $D = 8$ feet. Thus,

$$N = \frac{1 \times 1 \times 1 \times 2225}{8} = 278 \text{ rpm}$$

(*Note:* This safe speed is slightly greater than the value of 263 rpm obtainable directly from Tables 1 and 2.)

Tests to Determine Flywheel Bursting Speeds. — Tests made by Prof. C. H. Benjamin, to determine the bursting speeds of flywheels, showed the following results:

Cast-iron Wheels with Solid Rims. — Cast-iron wheels having solid rims burst at a rim speed of 395 feet per second, corresponding to a centrifugal tension of about 15,600 pounds per square inch.

Wheels with Jointed Rims. — Four wheels were tested with joints and bolts inside the rim, after the familiar design ordinarily employed for band wheels, but with the joints located at points one-fourth of the distance from one arm to the next, these being the points of least bending moment, and, consequently, the points at which the deflection due to centrifugal force would be expected to have the least effect. The tests, however, did not bear out this conclusion. The wheels burst at a rim speed of 194 feet per second, corresponding to a centrifugal tension of about 3750 pounds per square inch. These wheels, therefore, were only about one-quarter as strong as the wheels with solid rims, and burst at practically the same speed as wheels in a previous series of tests in which the rim joints were midway between the arms.

Bursting Speed for Link Joints. — Another type of wheel with deep rim, fastened together at the joints midway between the arms by links shrunk into recesses, after the manner of flywheels for massive engines, gave much superior results. This wheel burst at a speed of 256 feet per second, indicating a centrifugal tension of about 6600 pounds per square inch.

Wheel having Tie-rods. — Tests were made on a band wheel having joints inside the rim, midway between the arms, and in all respects like others of this design previously tested, except that tie-rods were used to connect the joints with the hub. It burst at a speed of 225 feet per second, showing an increase of strength of from 30 to 40 per cent over similar wheels without the tie-rods.

Wheel Rim of I-section. — Several wheels of special design, not in common use, were also tested, the one giving the greatest strength being an English wheel, with solid rim of I-section, made of high-grade cast iron and with the rim tied to the hub by steel wire spokes. These spokes were adjusted to have a uniform tension. The wheel gave way at a rim speed of 424 feet per second, which is slightly higher than the speed of rupture of the solid rim wheels with ordinary style of spokes.

Tests on Flywheel of Special Construction. — A test was made on a flywheel 49 inches in diameter and weighing about 900 pounds. The rim was 6¾ inches wide and 1⅛ inches thick, and was built of ten segments, the material being cast steel. Each joint was secured by three "prisoners" of an I-section on the outside face, by link prisoners on each edge, and by a dovetailed bronze clamp on the inside, fitting over lugs on the rim. The arms were of phosphor-bronze, twenty in number, ten on each side, and were a cross in section. These arms came midway between the rim joints and were bolted to plane faces on the polygonal hub. The rim was further reinforced by a system of diagonal bracing, each section of the rim being supported at five points on each side, in such a way as to relieve it almost entirely from bending. The braces, like the arms, were of phosphor-bronze, and all bolts and connecting links of steel. This wheel was designed as a model of a proposed 30-foot flywheel. On account of the excessive air resistance the wheel was enclosed at the sides between sheet-metal disks. This wheel burst at 1775 revolutions per minute or at a linear speed of 372 feet per second. The hub and main spokes of the wheel remained nearly in place, but parts of the rim were found two hundred feet away. This sudden failure of the rim casting was unexpected, as it was thought the flange bolts would be the parts to give way first. The tensile strength of the casting at the point of fracture was about four times the strength of the wheel rim at a solid section.

Stresses in Rotating Disks. — When a disk of uniform width is rotated, the max. stress S_t is tangential and at the bore of the hub, and the tangential stress is always greater than the radial stress at the same point on the disk. If S_t = maximum tangential stress in pounds per sq. in.; w = weight of material, lb. per cu. in.; N = rev. per min.; m = Poisson's ratio = 0.3 for steel; R = outer radius of disk, inches; r = inner radius of disk or radius of bore, inches.

$$S_t = 0.0000071 \ wN^2[(3+m)R^2 + (1-m)r^2]$$

Steam Engine Flywheels. — The variable amount of energy during each stroke and the allowable percentage of speed variation are of especial importance in designing steam engine flywheels. The earlier the point of cut-off, the greater the variation in energy and the larger the flywheel that will be required. The weight of the reciprocating parts and the length of the connecting-rod also affect the variation. The following formula is used for computing the weight of the flywheel rim:

Let W = weight of rim in pounds;

       D = mean diameter of rim in feet;

       N = number of revolutions per minute;

       $\dfrac{1}{n}$ = allowable variation in speed (from ⅟₅₀ to ⅟₁₀₀);

       E = excess and deficiency of energy in foot-pounds;

       c = factor of energy excess, from the accompanying table;

   H.P. = indicated horsepower.

Then, if the indicated horsepower is given:

$$W = \frac{387,587,500 \times cn \times \text{H.P.}}{D^2 N^3} \tag{1}$$

If the work in foot-pounds is given, then:

$$W = \frac{11{,}745nE}{D^2 N^2} \qquad (2)$$

In the second formula, E equals the average work in foot-pounds done by the engine in one revolution, multiplied by the decimal given in the accompanying table, "Factors for Engine Flywheel Calculations," which covers both the condensing and non-condensing engines:

Factors for Engine Flywheel Calculations

Condensing Engines						
Fraction of stroke at which steam is cut off	$\frac{1}{3}$	$\frac{1}{4}$	$\frac{1}{5}$	$\frac{1}{6}$	$\frac{1}{7}$	$\frac{1}{8}$
Factor of energy excess	0.163	0.173	0.178	0.184	0.189	0.191
Non-condensing Engines						
Steam cut off at			$\frac{1}{2}$	$\frac{1}{3}$	$\frac{1}{4}$	$\frac{1}{5}$
Factor of energy excess			0.160	0.186	0.209	0.232

Example 1. — A non-condensing engine of 150 indicated horsepower is to make 200 revolutions per minute, with a speed variation of 2 per cent. The average cut-off is to be at one-quarter stroke, and the flywheel is to have a mean diameter of 6 feet. Required, the necessary weight of rim in pounds.

From the table $c = 0.209$, and from the data given H.P. = 150; $N = 200$; $1/n = 1/50$ or $n = 50$; $D = 6$.

Substituting these values in equation (1):

$$W = \frac{387{,}587{,}500 \times 0.209 \times 50 \times 150}{6^2 \times 200^3} = 2110 \text{ pounds, nearly.}$$

Example 2. — A condensing engine, 24×42 inches, cuts off at one-third stroke and has a mean effective pressure of 50 pounds per square inch. The flywheel is to be 18 feet in mean diameter and make 75 revolutions per minute with a variation of 1 per cent. Required, weight of rim.

The work done on the piston in one revolution is equal to the pressure on the piston multiplied by the distance traveled or twice the stroke in feet. The area of the piston in this case is 452.4 square inches, and twice the stroke is 7 feet. The work done on the piston in one revolution is, therefore, $452.4 \times 50 \times 7 = 158{,}340$ foot-pounds. From the table $c = 0.163$, and therefore:

$$E = 158{,}340 \times 0.163 = 25{,}810 \text{ foot-pounds.}$$

From the data given: $n = 100$; $D = 18$; $N = 75$. Substituting these values in equation (2):

$$W = \frac{11{,}745 \times 100 \times 25{,}810}{18^2 \times 75^2} = 16{,}650 \text{ pounds, nearly.}$$

Spokes or Arms of Flywheels. — Flywheel arms are usually of elliptical cross-section. The major axis of the ellipse is in the plane of rotation to give the arms greater resistance to bending stresses and reduce the air resistance which may be considerable at high velocity. The stresses in the arms may be severe, due to the inertia of a heavy rim when sudden load changes occur. The strength of the arms should equal three-fourths the strength of the shaft in torsion.

If W equals the width of the arm at the hub (length of major axis) and D equals the shaft diameter, then W equals $1.3\ D$ for a wheel having 6 arms; and for an 8-arm wheel W equals $1.2\ D$. The thickness of the arm at the hub (length of minor axis) equals one-half the width. The arms usually taper toward the rim. The cross-sectional area at the rim should not be less than two-thirds the area at the hub.

Critical Speeds of Rotating Bodies and Shafts. — If a body or disk mounted upon a shaft rotates about it, the center of gravity of the body or disk must be at the center of the shaft, if a perfect running balance is to be obtained. In most cases, however, the center of gravity of the disk will be slightly removed from the center of the shaft, owing to the difficulty of perfect balancing. Now, if the shaft and disk be rotated, the centrifugal force generated by the heavier side will be greater than that generated by the lighter side geometrically opposite to it, and the shaft will deflect toward the heavier side, causing the center of the disk to rotate in a small circle. A rotating shaft without a body or disk mounted on it can also become dynamically unstable, and the resulting vibrations and deflections that occur can result in damage not only to the shaft but to the machine of which it is a part. These conditions hold true up to a comparatively high speed; but a point is eventually reached (at several thousand revolutions per minute) when momentarily there will be excessive vibration, and then the parts will run quietly again. The speed at which this occurs is called the *critical speed* of the wheel or shaft, and the phenomenon itself for the shaft-mounted disk or body is called the *settling* of the wheel. The explanation of the settling is that at this speed the axis of rotation changes, and the wheel and shaft, instead of rotating about their geometrical center, begin to rotate about an axis through their center of gravity. The shaft itself is then deflected so that for every revolution its geometrical center traces a circle around the center of gravity of the rotating mass.

Critical speeds depend upon the magnitude or location of the load or loads carried by the shaft, the length of the shaft, its diameter and the kind of supporting bearings. The normal operating speed of a machine may or may not be higher than the critical speed. For instance, some steam turbines exceed the critical speed, although they do not run long enough at the critical speed for the vibrations to build up to an excessive amplitude. The practice of the General Electric Co. at Schenectady is to keep below the critical speeds. It is assumed that the maximum speed of a machine may be within 20 per cent high or low of the critical speed without vibration troubles. Thus, in a design of steam turbine sets, critical speed is a factor that determines the size of the shafts, for both the generators and turbines. While a machine may run very close to the critical speed, the alignment and play of the bearings, the balance and construction generally, will require extra care, resulting in a more expensive machine; moreover, while such a machine may run smoothly for a considerable time, any looseness or play that may develop later, causing a slight imbalance, will immediately set up excessive vibrations.

The formulas commonly used to determine critical speeds are sufficiently accurate for general purposes. There are cases, however, where the torque applied to a shaft has an important effect on its critical speed. Investigations have shown that the critical speeds of a uniform shaft are decreased as the applied torque is increased, and that there exist critical torques which will reduce the corresponding critical speed of the shaft to zero. A detailed analysis of the effects of applied torques on critical speeds may be found in a paper, "Critical Speeds of Uniform Shafts under Axial Torque," by Golumb and Rosenberg presented at the First U.S. National Congress of Applied Mechanics in 1951.

Formulas for Critical Speeds. — The critical speed formulas given in the accompanying table (from the paper on Critical Speed Calculation presented

Critical Speed Formulas

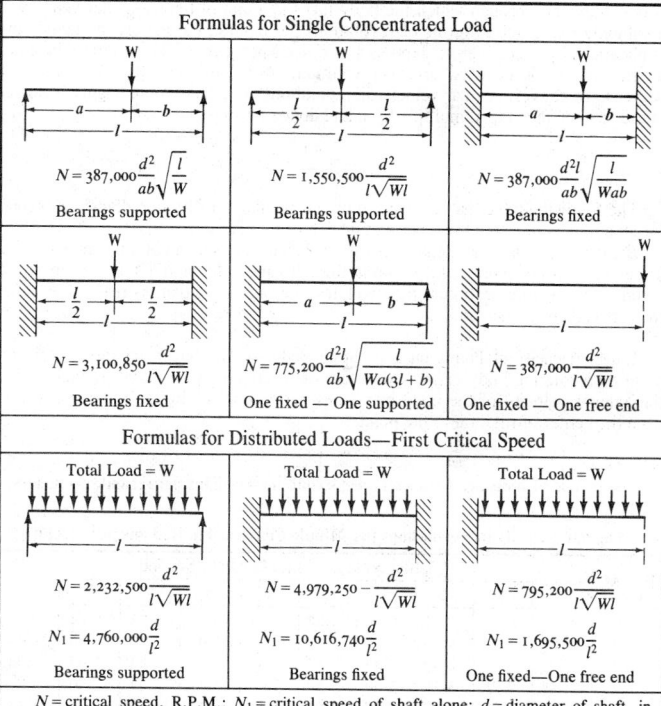

Formulas for Single Concentrated Load

$$N = 387,000 \frac{d^2}{ab}\sqrt{\frac{l}{W}}$$

Bearings supported

$$N = 1,550,500 \frac{d^2}{l\sqrt{Wl}}$$

Bearings supported

$$N = 387,000 \frac{d^2 l}{ab}\sqrt{\frac{l}{Wab}}$$

Bearings fixed

$$N = 3,100,850 \frac{d^2}{l\sqrt{Wl}}$$

Bearings fixed

$$N = 775,200 \frac{d^2 l}{ab}\sqrt{\frac{l}{Wa(3l+b)}}$$

One fixed — One supported

$$N = 387,000 \frac{d^2}{l\sqrt{Wl}}$$

One fixed — One free end

Formulas for Distributed Loads—First Critical Speed

Total Load = W

$$N = 2,232,500 \frac{d^2}{l\sqrt{Wl}}$$

$$N_1 = 4,760,000 \frac{d}{l^2}$$

Bearings supported

Total Load = W

$$N = 4,979,250 \frac{d^2}{l\sqrt{Wl}}$$

$$N_1 = 10,616,740 \frac{d}{l^2}$$

Bearings fixed

Total Load = W

$$N = 795,200 \frac{d^2}{l\sqrt{Wl}}$$

$$N_1 = 1,695,500 \frac{d}{l^2}$$

One fixed—One free end

N = critical speed, R.P.M.; N_1 = critical speed of shaft alone; d = diameter of shaft, in inches; W = load applied to shaft, in pounds; l = distance between centers of bearings, in inches; a and b = distances from bearings to load.

before the A.S.M.E. by S. H. Weaver) apply to (1) shafts with single concentrated loads and (2) shafts carrying uniformly distributed loads. These formulas also cover different conditions as regards bearings. If the bearings are self-aligning or very short, the shaft is considered supported at the ends; whereas, if the bearings are long and rigid, the shaft is considered fixed. These formulas, for both concentrated and distributed loads, apply to vertical shafts as well as horizontal shafts, the critical speeds having the same value in both cases. The data required for the solution of critical speed problems are the same as for shaft deflection. As the shaft is usually of variable diameter and its stiffness is increased by a long hub, an ideal shaft of uniform diameter and equal stiffness must be assumed.

In calculating critical speeds, the weight of the shaft is either neglected or, say, one-half to two-thirds of the weight is added to the concentrated load. The formulas apply to steel shafts having a modulus of elasticity $E = 29,000,000$. While a shaft carrying a number of loads or a distributed load may have an infinite number of critical speeds, ordinarily it is the first critical speed that is of importance in engineering work, which is the speed obtained by the formulas given in the table for distributed loads.

Angular Velocity of Rotating Bodies. — The angular velocity of a rotating body equals the angle through which the body turns in a unit of time. Angular velocity is commonly expressed in terms of revolutions per minute, but in certain engineering applications it is necessary to express it as radians per second. By definition there are 2π radians in 360 degrees, or one revolution, so that one radian $= 360 \div 2\pi = 57.3$ degrees. To convert angular velocity in revolutions per minute, n, to angular velocity in radians per second, ω, multiply by π and divide by 30:

$$\omega = \frac{\pi n}{30} \tag{1}$$

The following table may be used to obtain angular velocity in radians per second for all numbers of revolutions per minute from 1 to 239.

Example: To find the angular velocity in radians per second of a flywheel making 97 revolutions per minute, locate 90 in the left-hand column and 7 at the top of the columns; at the intersection of the two lines, the angular velocity is read off as equal to 10.16 radians per second.

Linear Velocity of Points on a Rotating Body. — The linear velocity, v, of any point on a rotating body expressed in feet per second may be found by multiplying the angular velocity of the body in radians per second, ω, by the radius, r, in feet from the center of rotation to the point:

$$v = \omega r \tag{2}$$

The metric SI units are v = meters per second; ω = radians per second, r = meters.

Angular Velocity in Revolutions per Minute Converted to Radians per Second

R.P.M.	Angular Velocity in Radians per Second									
	0	1	2	3	4	5	6	7	8	9
0	0.00	0.10	0.21	0.31	0.42	0.52	0.63	0.73	0.84	0.94
10	1.05	1.15	1.26	1.36	1.47	1.57	1.67	1.78	1.88	1.99
20	2.09	2.20	2.30	2.41	2.51	2.62	2.72	2.83	2.93	3.04
30	3.14	3.25	3.35	3.46	3.56	3.66	3.77	3.87	3.98	4.08
40	4.19	4.29	4.40	4.50	4.61	4.71	4.82	4.92	5.03	5.13
50	5.24	5.34	5.44	5.55	5.65	5.76	5.86	5.97	6.07	6.18
60	6.28	6.39	6.49	6.60	6.70	6.81	6.91	7.02	7.12	7.23
70	7.33	7.43	7.54	7.64	7.75	7.85	7.96	8.06	8.17	8.27
80	8.38	8.48	8.59	8.69	8.80	8.90	9.01	9.11	9.21	9.32
90	9.42	9.53	9.63	9.74	9.84	9.95	10.05	10.16	10.26	10.37
100	10.47	10.58	10.68	10.79	10.89	11.00	11.10	11.20	11.31	11.41
110	11.52	11.62	11.73	11.83	11.94	12.04	12.15	12.25	12.36	12.46
120	12.57	12.67	12.78	12.88	12.98	13.09	13.19	13.30	13.40	13.51
130	13.61	13.72	13.82	13.93	14.03	14.14	14.24	14.35	14.45	14.56
140	14.66	14.76	14.87	14.97	15.08	15.18	15.29	15.39	15.50	15.60
150	15.71	15.81	15.92	16.02	16.13	16.23	16.34	16.44	16.55	16.65
160	16.75	16.86	16.96	17.07	17.17	17.28	17.38	17.49	17.59	17.70
170	17.80	17.91	18.01	18.12	18.22	18.33	18.43	18.53	18.64	18.74
180	18.85	18.95	19.06	19.16	19.27	19.37	19.48	19.58	19.69	19.79
190	19.90	20.00	20.11	20.21	20.32	20.42	20.52	20.63	20.73	20.84
200	20.94	21.05	21.15	21.26	21.36	21.47	21.57	21.68	21.78	21.89
210	21.99	22.10	22.20	22.30	22.41	22.51	22.62	22.72	22.83	22.93
220	23.04	23.14	23.25	23.35	23.46	23.56	23.67	23.77	23.88	23.98
230	24.09	24.19	24.29	24.40	24.50	24.61	24.71	24.82	24.92	25.03

Pendulums

Types of Pendulums. — A *compound* or *physical* pendulum consists of any rigid body suspended from a fixed horizontal axis about which the body may oscillate in a vertical plane due to the action of gravity.

A *simple* or *mathematical* pendulum is similar to a compound pendulum except that the mass of the body is concentrated at a single point which is suspended from a fixed horizontal axis by a weightless cord. Actually, a simple pendulum cannot be constructed since it is impossible to have either a weightless cord or a body whose mass is entirely concentrated at one point. A good approximation, however, consists of a small, heavy bob suspended by a light, fine wire. If these conditions are not met by the pendulum, it should be considered as a compound pendulum.

A *conical* pendulum is similar to a simple pendulum except that the weight suspended by the cord moves at a uniform speed around the circumference of a circle in a horizontal plane instead of oscillating back and forth in a vertical plane. The principle of the conical pendulum is employed in the Watt fly-ball governor.

A *torsional* pendulum in its simplest form consists of a disk fixed to a slender rod, the other end of which is fastened to a fixed frame. When the disc is twisted through some angle and released, it will then oscillate back and forth about the axis of the rod because of the torque exerted by the rod.

Pendulum Formulas. — From the formulas that follow, the period of vibration or time required for one complete cycle back and forth may be determined for the types of pendulums shown in the accompanying diagram.

Four Types of Pendulum

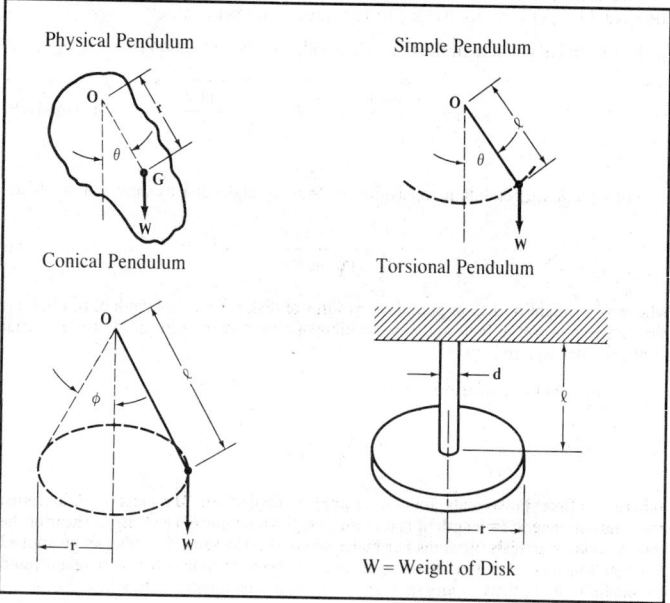

Physical Pendulum

Simple Pendulum

Conical Pendulum

Torsional Pendulum

W = Weight of Disk

For a *simple* pendulum,

$$T = 2\pi \sqrt{\frac{l}{g}} \qquad (1)$$

where T = period in seconds for one complete cycle; g = acceleration due to gravity = 32.17 feet per second per second (approximately); and l is the length of the pendulum in feet as shown on the accompanying diagram.

For a *physical* or *compound* pendulum,

$$T = 2\pi \sqrt{\frac{k_0^2}{gr}} \qquad (2)$$

where k_0 = radius of gyration of the pendulum about the axis of rotation, in feet, and r is the distance from the axis of rotation to the center of gravity, in feet.

The metric SI units that can be used in the two above formulas are T = seconds; g = approximately 9.81 meters per second squared, which is the value for acceleration due to gravity; l = the length of the pendulum in meters; k_0 = the radius of gyration in meters, and r = the distance from the axis of rotation to the center of gravity, in meters.

Formulas (1) and (2) are accurate when the angle of oscillation θ shown in the diagram is very small. For θ equal to 22 degrees, these formulas give results that are too small by 1 per cent; for θ equal to 32 degrees, by 2 per cent.

For a *conical* pendulum, the time in seconds for one revolution is:

$$T = 2\pi \sqrt{\frac{l \cos \phi}{g}}; \quad \text{or,} \quad T = 2\pi \sqrt{\frac{r \cot \phi}{g}} \qquad (3a) \text{ and } (3b)$$

For a *torsional* pendulum consisting of a thin rod and a disk as shown in the figure

$$T = \frac{2}{3} \sqrt{\frac{\pi W r^2 l}{g d^4 G}} \qquad (4)$$

where W = weight of disk in pounds; r = radius of disk in feet; l = length of rod in feet; d = diameter of rod in feet; and G = modulus of elasticity in shear of the rod material in pounds per square inch.

The formula using metric SI units is:

$$T = 8 \sqrt{\frac{\pi M r^2 l}{d^4 G}}$$

Where T = time in seconds for one complete oscillation; M = mass in kilograms; r = radius in meters; l = length of rod in meters; G = modulus of elasticity in shear of the rod material in pascals (newtons per meter squared). The same formula can be applied using millimeters, providing dimensions are expressed in millimeters throughout, and the modulus of elasticity in megapascals (newtons per millimeter squared).

STRENGTH OF MATERIALS

Strength of materials deals with the relations between the external forces applied to elastic bodies and the resulting deformations and stresses. In the design of structures and machines, the application of the principles of strength of materials is necessary if satisfactory materials are to be utilized and adequate proportions obtained to resist functional forces.

Forces are produced by the action of gravity, by accelerations and impacts of moving parts, by gasses and fluids under pressure, by the transmission of mechanical power, etc. In order to analyze the stresses and deflections of a body, the magnitudes, directions and points of application of forces acting on the body must be known. Information given in the Mechanics section provides the basis for evaluating force systems.

The time element in the application of a force on a body is an important consideration. Thus a force may be static or change so slowly that its maximum value can be treated as if it were static; it may be suddenly applied, as in the case of impact; or it may have a repetitive or cyclic behavior.

Also important is the environment in which forces act on a machine or part. Such factors as high and low temperatures; the presence of corrosive gasses, vapors and liquids; radiation; etc. may have a marked effect on how well parts are able to resist stresses.

Throughout the Strength of Materials section in this Handbook, both English and metric SI data and formulas are given to cover the requirements of working in either system of measurement. Formulas and text relating exclusively to SI units are given in bold-face type.

Mechanical Properties of Materials. — Many mechanical properties of materials are determined from tests, some of which give relationships between stresses and strains as shown by the curves in the accompanying figures.

Stress is force per unit area and is usually expressed in pounds per square inch. If the stress tends to stretch or lengthen the material, it is called *tensile* stress; if to compress or shorten the material, a *compressive* stress; and if to shear the material, a *shearing* stress. Tensile and compressive stresses always act at right-angles to (normal to) the area being considered; shearing stresses are always in the plane of the area (at right-angles to compressive or tensile stresses).

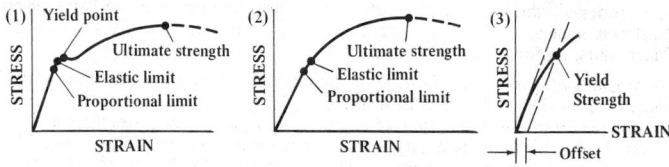

Fig. 1. Stress-strain curves

In the SI, the unit of stress is the pascal (Pa), the newton per meter squared (N/m². The megapascal (newtons per millimeter squared) is often an appropriate sub-multiple for use in practice.

Unit strain is the amount by which a dimension of a body changes when the body is subjected to a load, divided by the original value of the dimension. The simpler term *strain* is often used instead of unit strain.

Proportional limit is the point on a stress-strain curve at which it begins to deviate from the straight-line relationship between stress and strain.

Elastic limit is the maximum stress to which a test specimen may be subjected

and still return to its original length upon release of the load. A material is said to be stressed within the *elastic region* when the working stress does not exceed the elastic limit, and to be stressed in the *plastic region* when the working stress does exceed the elastic limit. The elastic limit for steel is for all practical purposes the same as its proportional limit.

Yield point is a point on the stress-strain curve at which there is a sudden increase in strain without a corresponding increase in stress. Not all materials have a yield point. Some representative values of the yield point (in ksi) are as follows:

Aluminum, wrought, 2014-T6 . . .	60	Titanium, alloy, 5Al, 2.5Sn	110
Aluminum, wrought, 6061-T6 . . .	35	Steel for bridges and buildings, ASTM	
Beryllium copper	140	A7-61T, All shapes	33
Brass, naval.	25–50	Steel, castings, high strength, for	
Cast iron, malleable.	32–45	structural purposes, ASTM A148.60	
Cast iron, nodular	45–65	(seven grades)	40–145
Magnesium, AZ80A-T5	38	Steel, stainless (0.08–0.2 C, 17 Cr, 7	
Titanium, pure	55–70	Ni) ¼ hard	78

Yield strength, S_y, is the maximum stress that can be applied without permanent deformation of the test specimen. This is the value of the stress at the elastic limit for materials for which there is an elastic limit. Because of the difficulty in determining the elastic limit, and since many materials do not have an elastic region, yield strength is often determined by the offset method as illustrated by the accompanying figure at (3). Yield strength in such a case is the stress value on the stress-strain curve corresponding to a definite amount of permanent set or strain, usually 0.1 or 0.2 per cent of the original dimension.

Ultimate strength, S_u, (also called *tensile strength*) is the maximum stress value obtained on a stress-strain curve.

Modulus of elasticity, E, (also called *Young's modulus*) is the ratio of unit stress to unit strain within the proportional limit of a material in tension or compression. Some representative values of Young's modulus (in 10^6 psi) are as follows:

Aluminum, cast, pure	9	Titanium, pure	15.5
Aluminum, wrought, 2014-T6 . .	10.6	Titanium, alloy, 5 Al, 2.5 Sn . . .	17
Beryllium copper	19	Steel for bridges and buildings, ASTM	
Brass, naval.	15	A7-61T, All shapes	29
Bronze, phosphor, ASTM B159 .	15	Steel, castings, high strength, for	
Cast iron, malleable.	26	structural purposes, ASTM A148-60	
Cast iron, nodular	23.5	(seven grades)	29
Magnesium, AZ80A-T5	6.5		

Modulus of elasticity in shear, G, is the ratio of unit stress to unit strain within the proportional limit of a material in shear.

Poisson's ratio, μ, is the ratio of lateral strain to longitudinal strain for a given material subjected to uniform longitudinal stresses within the proportional limit. The term is found in certain equations associated with strength of materials. Values of Poisson's ratio for common materials are as follows:

Aluminum	0.334	Nickel silver	0.322
Beryllium copper	0.285	Phosphor bronze	0.349
Brass	0.340	Rubber	0.500
Cast iron, gray	0.211	Steel, cast.	0.265
Copper	0.340	high carbon	0.295
Inconel	0.290	mild	0.303
Lead	0.431	nickel.	0.291
Magnesium	0.350	Wrought iron	0.278
Monel metal	0.320	Zinc	0.331

Compressive Properties. — From compression tests, *compressive yield strength*, S_{cy}, and *compressive ultimate strength*, S_{cu}, are determined. Ductile materials under compression loading merely swell or buckle without fracture, hence do not have a compressive ultimate strength.

Shear Properties. — The properties of *shear yield strength*, S_{sy}, *shear ultimate strength*, S_{su}, and the *modulus of rigidity*, G, are determined by direct shear and torsional tests. The modulus of rigidity is also known as the modulus of elasticity in shear. It is the ratio of the shear stress, τ, to the shear strain, γ, in radians, within the proportional limit: $G = \tau/\gamma$.

Fatigue Properties. — When a material is subjected to many cycles of stress reversal or fluctuation (variation in magnitude without reversal), failure may occur, even though the maximum stress at any cycle is considerably less than the value at which failure would occcur if the stress were constant. Fatigue properties are determined by subjecting test specimens to stress cycles and counting the number of cycles to failure. From a series of such tests in which maximum stress values are progressively reduced, S–N diagrams can be plotted as illustrated by the accompanying figures. The S–N diagram at (1) shows the behavior of a material for which there is an *endurance limit*, S_{en}. Endurance limit is the stress value at which the number of cycles to failure is infinite. Steels have endurance limits that vary according to hardness, composition, and quality; but many non-ferrous metals do not. The S–N diagram at (2) does not have an endurance limit. For a metal that does not have an endurance limit, it is standard practice to specify fatigue strength as the stress value corresponding to a specific number of stress reversals, usually 100,000,000 or 500,000,000.

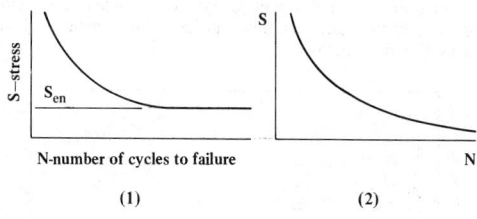

(1)   (2)

Fig. 2. S–N diagrams

The Influence of Mean Stress on Fatigue. — Most published data on the fatigue properties of metals are for completely reversed alternating stresses, that is, the mean stress of the cycle is equal to zero. However, if a structure is subjected to stresses that fluctuate between different values of tension and compression, then the mean stress is not zero.

When fatigue data for a specified mean stress and design life are not available for a material, the influence of nonzero mean stress can be estimated from empirical relationships that relate failure at a given life, under zero mean stress, to failure at the same life under zero mean cyclic stress. One widely used formula is Goodman's linear relationship, which is

$$S_a = S(1 - S_m/S_u),$$

where S_a is the alternating stress associated with some nonzero mean stress, S_m. S is the alternating fatigue strength at zero mean stress. S_u is the ultimate tensile strength.

Goodman's linear relationship is usually represented graphically on a so-called Goodman diagram, as shown below. The alternating fatigue strength or the alternating stress for a given number of endurance cycles is plotted on the ordinate (y-axis) and the static tensile strength is plotted on the abscissa (x-axis). The straight line joining the alternating fatigue strength, S, and the tensile strength, S_u, is the Goodman line.

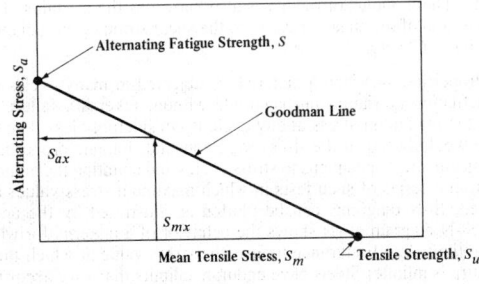

Goodman Diagram

The value of an alternating stress S_{ax} at a known value of mean stress S_{mx} is determined as shown by the dashed lines on the diagram.

For ductile materials, the Goodman law is usually conservative, since approximately 90 per cent of actual test data for most ferrous and nonferrous alloys fall above the Goodman line, even at low endurance values where the yield strength is exceeded. For many brittle materials, however, actual test values can fall below the Goodman line, as illustrated below:

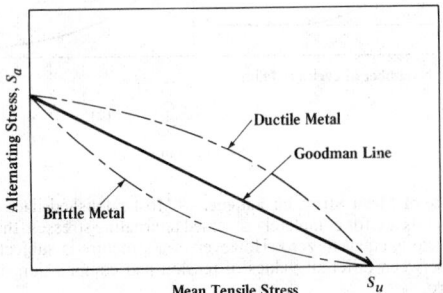

As a rule of thumb, materials having an elongation of less than 5 per cent in a tensile test may be regarded as brittle. Those having an elongation of 5 per cent or more may be regarded as ductile.

Cumulative Fatigue Damage. — Most data are determined from tests at a constant stress amplitude. This is easy to do experimentally, and the data can be presented in a straightforward manner. In actual engineering applications, however, the alternating stress amplitude usually changes in some way during service operation. Such changes, referred to as "spectrum loading," make the direct use of standard S-N fatigue curves inappropriate. A problem exists, therefore, in predicting the fatigue

life under varying stress amplitude from conventional, constant-amplitude S-N fatigue data.

The assumption in predicting spectrum loading effects is that operation at a given stress amplitude and number of cycles will produce a certain amount of permanent fatigue damage and that subsequent operation at different stress amplitude and number of cycles will produce additional fatigue damage and a sequential accumulation of total damage, which at a critical value will cause fatigue failure. While the assumption appears simple, the amount of damage incurred at any stress amplitude and number of cycles has proven difficult to determine, and several "cumulative damage" theories have been advanced.

One of the first and simplest methods for evaluating cumulative damage is known as Miner's law or the linear damage rule, where it is assumed that n_1 cycles at a stress of S_1, for which the average number of cycles to failure is N_1, cause an amount of damage n_1/N_1. Failure is predicted to occur when

$$\Sigma \, n/N = 1.$$

The term n/N is known as the "cycle ratio" or the damage fraction.

The greatest advantages of the Miner rule are its simplicity and prediction reliability, which approximates that of more complex theories. For these reasons it is widely used. It should be noted, however, that it does not account for all influences, and errors are to be expected in failure prediction ability.

Modes of Fatigue Failure. — Several modes of fatigue failure are:

Low/High-Cycle Fatigue: This fatigue process covers cyclic loading in two significantly different domains, with different physical mechanisms of failure. One domain is characterized by relatively low cyclic loads, strain cycles confined largely to the elastic range, and long lives or a high number of cycles to failure; traditionally, this has been called "high-cycle fatigue." The other domain has cyclic loads that are relatively high, significant amounts of plastic strain induced during each cycle, and short lives or a low number of cycles to failure. This has commonly been called "low-cycle fatigue" or cyclic strain-controlled fatigue.

The transition from low- to high-cycle fatigue behavior occurs in the range from approximately 10,000 to 100,000 cycles. Many define low-cycle fatigue as failure that occurs in 50,000 cycles or less.

Thermal Fatigue: Cyclic temperature changes in a machine part will produce cyclic stresses and strains if natural thermal expansions and contractions are either wholly or partially constrained. These cyclic strains produce fatigue failure just as though they were produced by external mechanical loading. When strain cycling is produced by a fluctuating temperature field, the failure process is termed "thermal fatigue."

While thermal fatigue and mechanical fatigue phenomena are very similar, and can be mathematically expressed by the same types of equations, the use of mechanical fatigue results to predict thermal fatigue performance must be done with care. In most cases, for equal values of plastic strain range, the number of cycles to failure is much less by factors up to 2.5 for thermally cycled than for mechanically cycled samples.

Corrosion Fatigue: Corrosion fatigue is a failure mode where cyclic stresses and a corrosion-producing environment combine to initiate and propagate cracks in fewer stress cycles and at lower stress amplitudes than would be required in a more inert environment. The corrosion process forms pits and surface discontinuities that act as stress raisers to accelerate fatigue cracking. The cyclic loads may also cause cracking and flaking of the corrosion layer, baring fresh metal to the corrosive environment. Each process accelerates the other, making the cumulative result more serious.

Surface or Contact Fatigue: Surface fatigue failure is usually associated with rolling surfaces in contact, and results in pitting, cracking, and spalling of the contacting surfaces from cyclic Hertz contact stresses that cause the maximum values of cyclic shear stresses to be slightly below the surface. The cyclic subsurface shear stresses generate cracks that propagate to the contacting surface, dislodging particles in the process.

Combined Creep and Fatigue: This is a failure mode in which all of the conditions for both creep failure and fatigue failure exist simultaneously. Each process influences the other in producing failure; this interaction is not well understood.

Factors of Safety. — There is always a risk that the working stress to which a member is subjected will exceed the strength of its material. The purpose of a factor of safety is to minimize this risk.

Factors of safety can be incorporated into design calculations in many ways. For most calculations the following equation is used:

$$s_w = S_m/f_s \tag{1}$$

f_s is the factor of safety, S_m is the strength of the material in pounds per square inch, and s_w is the allowable working stress, also in pounds per square inch. Since the factor of safety is greater than 1, the allowable working stress will be less than the strength of the material.

In general, S_m is based on yield strength for ductile materials, ultimate strength for brittle materials, and fatigue strength for parts subjected to cyclic stressing. Most strength values are obtained by testing standard specimens at 68°F. in normal atmospheres. If, however, the character of the stress or environment differs significantly from that used in obtaining standard strength data, then special data must be obtained. If special data are not available, standard data must be suitably modified.

General recommendations for values of factors of safety are given in the following table.

f_s	Application
1.3–1.5	For use with highly reliable materials where loading and environmental conditions are not severe, and where weight is an important consideration.
1.5–2	For applications using reliable materials where loading and environmental conditions are not severe.
2–2.5	For use with ordinary materials where loading and environmental conditions are not severe.
2.5–3	For less tried and for brittle materials where loading and environmental conditions are not severe.
3–4	For applications in which material properties are not reliable and where loading and environmental conditions are not severe, or where reliable materials are to be used under difficult loading and environmental conditions.

Working Stress. — Calculated working stresses are the products of calculated nominal stress values and stress concentration factors. Calculated nominal stress values are based on the assumption of idealized stress distributions. Such nominal stresses may be simple stresses, combined stresses, or cyclic stresses. Depending on the nature of the nominal stress, one of the following equations apply:

$$(2)\ s_w = K\sigma \qquad\qquad (4)\ s_w = K\sigma' \qquad\qquad (6)\ s_w = K\sigma_{cy}$$
$$(3)\ s_w = K\tau \qquad\qquad (5)\ s_w = K\tau' \qquad\qquad (7)\ s_w = K\tau_{cy}$$

where: K is a stress concentration factor; σ and τ are, respectively, simple normal (tensile or compressive) and shear stresses; σ' and τ' are combined normal and shear stresses; σ_{cy} and τ_{cy} are cyclic normal and shear stresses.

Where there is uneven stress distribution, as illustrated in the table (on page 214) of simple stresses for Cases 3, 4 and 6, the maximum stress is the one to which the stress concentration factor is applied in computing working stresses. The location of the maximum stress in each case is discussed under the section "Simple Stresses" and the formulas for these maximum stresses are given in the table of simple stresses on page 214.

Stress Concentration Factors. — Stress concentration is related to type of material, the nature of the stress, environmental conditions, and the geometry of parts. When stress concentration factors are not available that specifically match all of the foregoing conditions, then the following equation may be used:

$$K = 1 + q (K_t - 1) \tag{8}$$

K_t is a theoretical stress concentration factor that is a function only of the geometry of a part and the nature of the stress; q is the *index of sensitivity* of the material. If the geometry is such as to provide no theoretical stress concentration, $K_t = 1$.

Curves for evaluating K_t are on pages 210 to 213. For constant stresses in cast iron and in ductile materials, $q = 0$ (hence $K = 1$). For constant stresses in brittle materials such as hardened steel, q may be taken as 0.15; for very brittle materials such as steels that have been quenched but not drawn, q may be taken as 0.25. When stresses are suddenly applied (impact stresses) q ranges from 0.4 to 0.6 for ductile materials; for cast iron it is taken as 0.5, and 1 for brittle materials.

Simple Stresses. — Simple stresses are produced by constant conditions of loading on elements that can be represented as beams, rods, or bars. The table on page 214 summarizes information pertaining to the calculation of simple stresses. Following is an explanation of the symbols used in simple stress formulae: σ = simple normal (tensile or compressive) stress in pounds per square inch; τ = simple shear stress in pounds per square inch; F = external force in pounds; V = shearing force in pounds; M = bending moment in inch-pounds; T = torsional moment in inch-pounds; A = cross-sectional area in square inches; Z = section modulus in inches³; Z_p = polar section modulus in inches³; I = moment of inertia in inches⁴; J = polar moment of inertia in inches⁴; a = area of the web of wide flange and I beams in square inches; y = perpendicular distance from axis through center of gravity of cross-sectional area to stressed fiber in inches; c = radial distance from center of gravity to stressed fiber in inches.

SI metric units can be applied in the calculations in place of the English units of measurement without changes to the formulas. The SI units are the newton (N), which is the unit of force; the meter; the meter squared; the pascal (Pa) which is the newton per meter squared (N/m^2); and the newton-meter ($N \cdot m$) for moment of force. Often in design work using the metric system, the millimeter is employed rather than the meter. In such instances, the dimensions can be converted to meters before the stress calculations are begun. Alternatively, the same formulas can be applied using millimeters in place of the meter, providing the treatment is consistent throughout. In such instances, stress and strength properties must be expressed in megapascals (MPa), which is the same as newtons per millimeter squared (N/mm^2), and moments in newton-millimeters ($N \cdot mm$). Note: $1\ N/mm^2 = 1\ N/10^{-6}m^2 = 10^6\ N/m^2 = 1$ meganewton/$m^2 = 1$ megapascal.

For direct tension and direct compression loading, Cases 1 and 2 in the Table of Simple Stresses on page 214, the force F must act along a line through the center of gravity of the section at which the stress is calculated. The equation for direct compression loading applies only to members for which the ratio of length to least radius of gyration is relatively small, approximately 20, otherwise the member must be treated as a column.

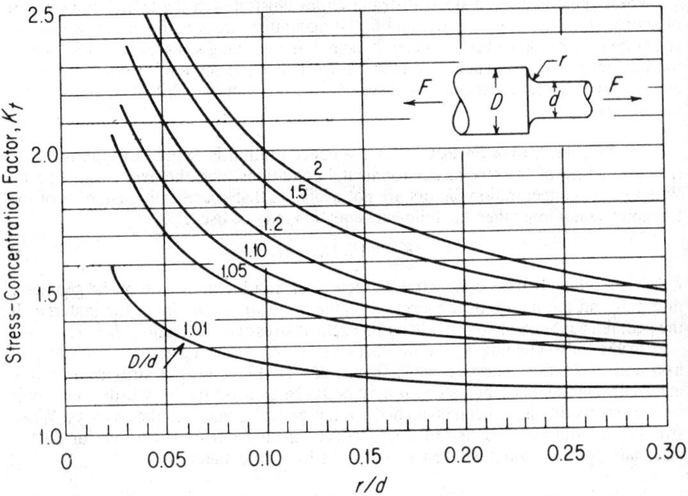

Fig. 3. Stress-concentration factor, K_t, for a filleted shaft in tension*

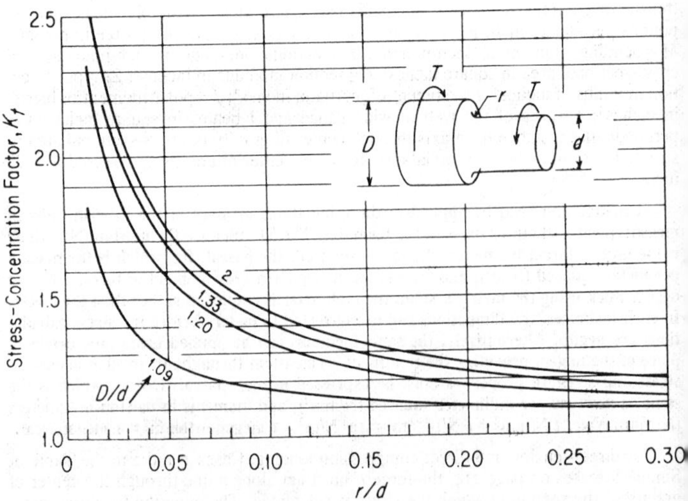

Fig. 4. Stress-concentration factor, K_t, for a filleted shaft in torsion*

* See footnote to Fig. 9.

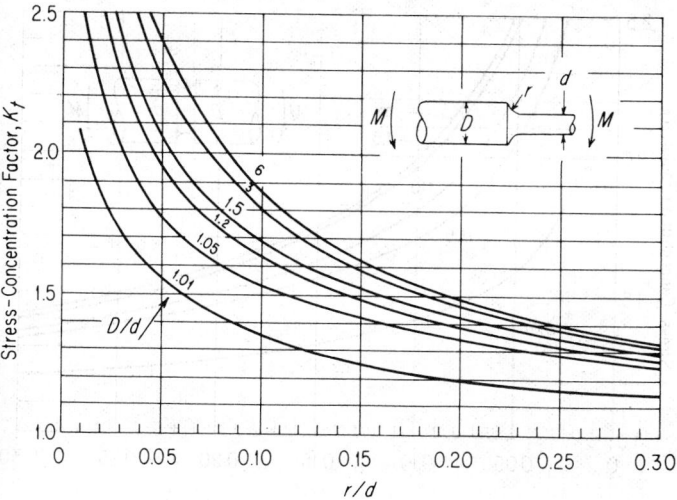

Fig. 5.   Stress-concentration factor, K_t, for a shaft with shoulder fillet in bending*

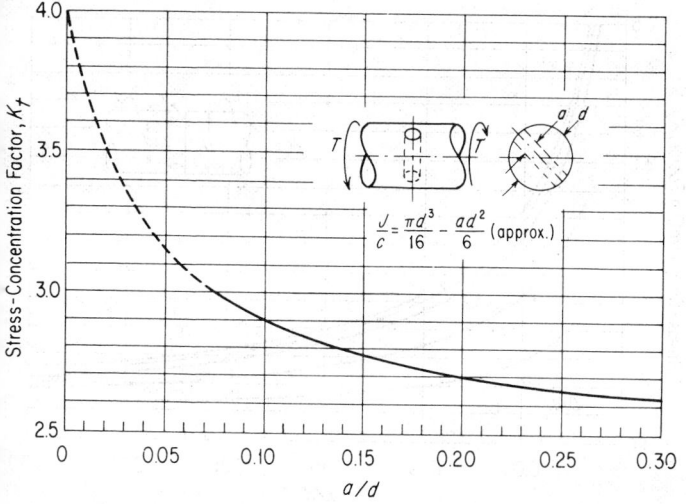

Fig. 6.   Stress-concentration factor, K_t, for a shaft, with a transverse hole, in torsion*

* See footnote to Fig. 9.

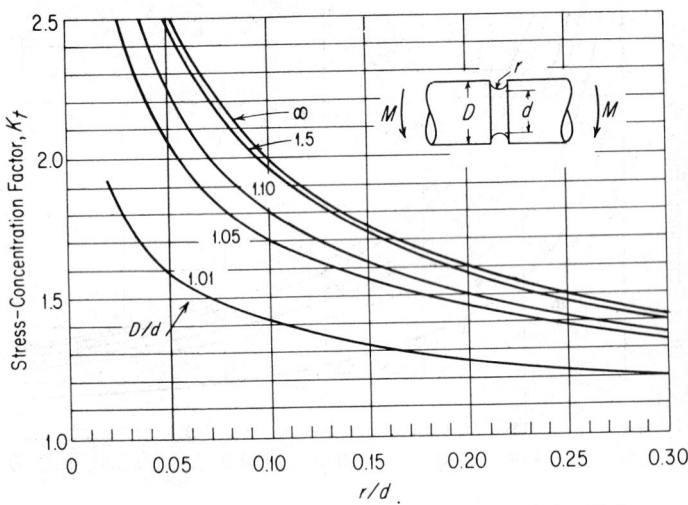

Fig. 7.   Stress-concentration factor, K_t, for a grooved shaft in bending*

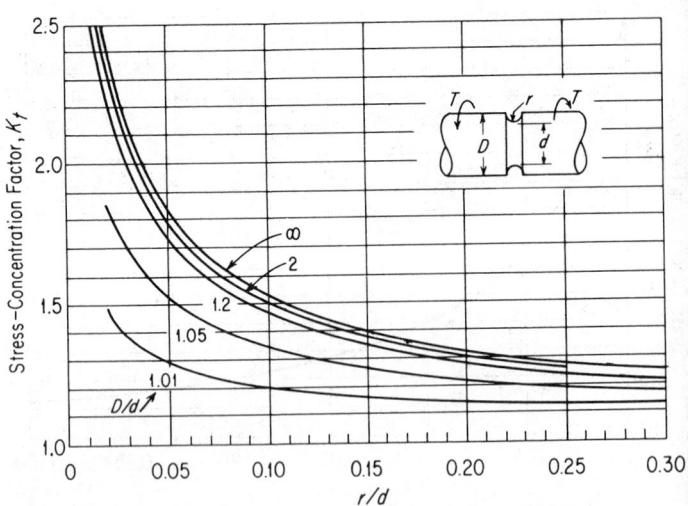

Fig. 8.   Stress-concentration factor, K_t, for a grooved shaft in torsion*

* See footnote to Fig. 9.

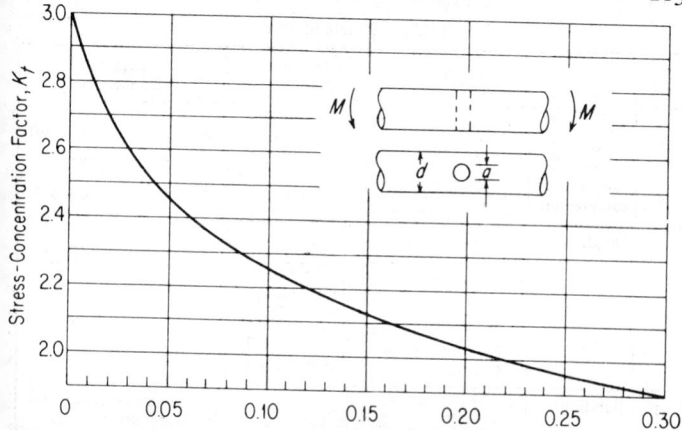

Fig. 9.   Stress-concentration factor, K_t, for a shaft, with a transverse hole, in bending*

The tables on pages 260 to 271 give equations for calculating stresses due to bending for common types of beams and conditions of loading. Where these tables are not applicable, stress may be calculated using Equation (11) in the table on page 214. In using this equation it is necessary to determine the value of the bending moment at the point where the stress is to be calculated. For beams of constant cross-section, stress is ordinarily calculated at the point coinciding with the maximum value of bending moment. Bending loading results in the characteristic stress distribution shown in the table for Case 3. It will be noted that the maximum stress values are at the surfaces farthest from the neutral plane. One of the surfaces is stressed in tension and the other in compression. It is for this reason that the $\pm$ sign is used in Equation (11). Numerous tables for evaluating section moduli are given in the following pages.

Shear stresses caused by bending have maximum values at neutral planes and zero values at the surfaces farthest from the neutral axis, as indicated by the stress distribution diagram shown for Case 4 in the table on page 214. Values for V in Equations (12), (13) and (14) can be determined from shearing force diagrams. The shearing force diagram shown in Case 4 corresponds to the bending moment diagram for Case 3. As shown in this diagram, the value taken for V is represented by the greatest vertical distance from the x axis. The shear stress caused by direct shear loading, Case 5, has a uniform distribution. However, the shear stress caused by torsion loading, Case 6, has a zero value at the axis and a maximum value at the surface farthest from the axis.

Deflections. — For direct tension and direct compression loading on members with uniform cross sections, deflection can be calculated using Eq. (17). For direct tension loading, e is an elongation; for direct compression loading, e is a contraction. Deflection is in inches when the load F is in pounds, the length L over which deflection occurs is in inches, the cross-sectional area A is in square inches, and the modulus of elasticity E is in pounds per square inch. The angular deflection of members with uniform circular cross sections subject to torsion loading can be calculated with Eq. (18).

$$e = FL/AE \quad (17) \qquad\qquad \theta = TL/GJ \quad (18)$$

* Source: R. E. Peterson, Design Factors for Stress Concentration, *Machine Design*, vol. 23, 951. For other stress concentration charts, see Lipson and Juvinall, *The Handbook of Stress and Strength*, The Macmillan Co., 1963.

Table of Simple Stresses*

Case	Type of Loading	Illustration	Stress Distribution	Stress Equations
1	Direct tension		Uniform	$\sigma = \dfrac{F}{A}$ (9)
2	Direct compression		Uniform	$\sigma = -\dfrac{F}{A}$ (10)
3	Bending	Bending moment diagram	Neutral plane	$\sigma = \pm\dfrac{M}{Z} = \pm\dfrac{My}{I}$ (11)
4	Bending	Shearing force diagram	Neutral plane	For beams of rectangular cross-section: $\tau = \dfrac{3V}{2A}$ (12) For beams of solid circular cross-section: $\tau = \dfrac{4V}{3A}$ (13) For wide flange and I beams (approximately): $\tau = \dfrac{V}{a}$ (14)
5	Direct shear		Uniform	$\tau = \dfrac{F}{A}$ (15)
6	Torsion			$\tau = \dfrac{T}{Z_p} = \dfrac{Tc}{J}$ (16)

* See page 209 for explanation of the stress equation symbols, in terms of English and metric SI units of measurement.

The angular deflection θ is in radians when the torsional moment T is in inch-pounds, the length L over which the member is twisted is in inches, the modulus of rigidity G is in pounds per square inch, and the polar moment of inertia J is in inches.[4]

Metric SI units can be used in Equations (17) and (18), where F = force in newtons (N); L = length over which deflection or twisting occurs in meters; A = cross-sectional area in meters squared; E = the modulus of elasticity in (newtons per meter squared); θ = radians; T = the torsional moment in newton-meters (N·m); G = modulus of rigidity, in pascals; and J = the polar moment of inertia in meters[4]. If the load (F) is applied as a weight, it should be noted that the weight of a mass M kilograms is Mg newtons, where g = 9.81 m/s². Millimeters can be used in the calculations in place of meters, providing the treatment is consistent throughout.

Combined Stresses. — A member may be loaded in such a way that a combination of simple stresses acts at a point. Three general cases occur, examples of which are shown in the accompanying illustration.

Superposition of stresses: The figure at (1) illustrates a common situation that results in simple stresses combining by superposition at points a and b. The equal and opposite forces F_1 will cause a compressive stress $\sigma_1 = -F_1/A$. Force F_2 will cause a bending moment M to exist in the plane of points a and b. The resulting stress $\sigma_2 = \pm M/Z$. The combined stress at point a,

$$\sigma_a' = -\frac{F_1}{A} - \frac{M}{Z}; \quad \text{and at } b, \quad \sigma_b' = -\frac{F_1}{A} + \frac{M}{Z} \qquad (19 \text{ and } 20)$$

where the minus sign indicates a compressive stress and the plus sign a tensile stress. Thus, the stress at a will be compressive and at b either tensile or compressive depending on which term in the equation for σ_b' has the greatest value.

Normal stresses at right angles: This is shown in the figure at (2). This combination of stresses occurs, for example, in tanks subjected to internal or external pressure. The principle normal stresses are $\sigma_x = F_1/A_1$, $\sigma_y = F_2/A_2$, and $\sigma_z = 0$ in this plane stress problem. Determine the values of these three stresses with their signs, order them algebraically, and then calculate the maximum shear stress:

$$\tau = (\sigma_{\text{LARGEST}} - \sigma_{\text{SMALLEST}})/2 \qquad (21)$$

Normal and shear stresses: The example in the figure at (3) shows a member subjected to a torsional shear stress, $\tau = T/Z_p$, and a direct compressive stress, $\sigma = -F/A$. At some point a on the member the principal normal stresses are calculated using the equation,

$$\sigma' = \frac{\sigma}{2} \pm \sqrt{\left(\frac{\sigma}{2}\right)^2 + \tau^2} \qquad (22)$$

The maximum shear stress is calculated by using the equation,

$$\tau' = \sqrt{\left(\frac{\sigma}{2}\right)^2 + \tau^2} \qquad (23)$$

The point a should ordinarily be selected where stress is a maximum value. For the example shown in the figure at (3), the point a can be anywhere on the cylindrical surface since the combined stress has the same value anywhere on that surface.

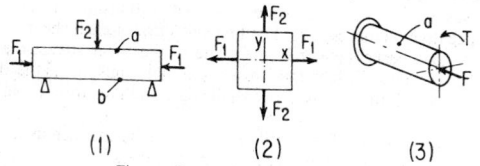

(1)   (2)   (3)

Fig. 10. Types of Combined Loading

Three-Dimensional Stress. — Three-dimensional or triaxial stress occurs in assemblies such as a shaft press-fitted into a gear bore or in pipes and cylinders subjected to internal or external fluid pressure. Triaxial stress also occurs in two-dimensional stress problems if the loads produce normal stresses that are either both tensile or both compressive. In either case the calculated maximum shear stress, based on the corresponding two-dimensional theory, will be less than the true maximum value because of three-dimensional effects. Therefore, if the stress analysis is to be based on the maximum-shear-stress theory of failure, the triaxial stress cubic equation

should first be used to calculate the three principal stresses and from these the true maximum shear stress. The following procedure provides the principal maximum normal tensile and compressive stresses and the true maximum shear stress at any point on a body subjected to any combination of loads.

The basis for the procedure is the stress cubic equation

$$S^3 - AS^2 + BS - C = 0$$

in which:

$$A = S_x + S_y + S_z, \quad B = S_x S_y + S_y S_z + S_z S_x - S_{xy}^2 - S_{yz}^2 - S_{zx}^2;$$
$$C = S_x S_y S_z + 2S_{xy} S_{yz} S_{zx} - S_x S_{yz}^2 - S_y S_{zx}^2 - S_z S_{xy}^2;$$

and S_x, S_y, etc., are as shown in Fig. 1.

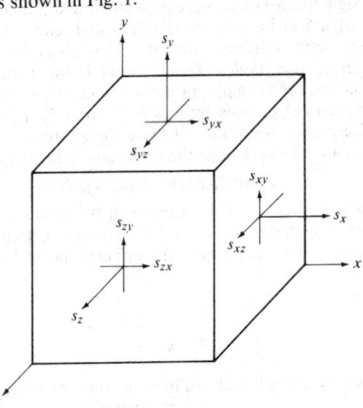

Fig. 1. *XYZ* Coordinate System Showing Positive Directions of Stresses

The coordinate system *XYZ* in Fig. 1 shows the positive directions of the normal and shear stress components on an elementary cube of material. Only six of the nine components shown are needed for the calculations: the normal stresses S_x, S_y, and S_z on three of the faces of the cube; and the three shear stresses S_{xy}, S_{yz}, and S_{zx}. The remaining three shear stresses are known since $S_{yx} = S_{xy}$, $S_{zy} = S_{yz}$, and $S_{xz} = S_{zx}$. The normal stresses S_x, S_y, and S_z are shown as positive (tensile) stresses; the opposite direction is negative (compressive). The first subscript of each shear stress identifies the coordinate axis perpendicular to the plane of the shear stress; the second subscript identifies the axis to which the stress is parallel. Thus, S_{xy} is the shear stress in the *YZ* plane to which the *X* axis is perpendicular, and the stress is parallel to the *Y* axis.

Step 1. Draw a diagram of the hardware to be analyzed, such as the shaft shown in Fig. 2, and show the applied loads *P*, *T*, and any others.

Step 2. For any point at which the stresses are to be analyzed, draw a coordinate diagram similar to Fig. 1 and show the magnitudes of the stresses resulting from the applied loads (these stresses may be calculated by using standard basic equations from strength of materials, and should include any stress concentration factors).

Step 3. Substitute the values of the six stresses S_x, S_y, S_z, S_{xy}, S_{yz}, and S_{zx}, including zero values, into the formulas for the quantities *A* through *K*. The quantities *I*, *J*, and *K* represent the principal normal stresses at the point analyzed. As a check, if the algebraic sum $I + J + K$ equals *A*, within rounding errors, then the calculations up to this point should be correct.

$$D = A^2/3 - B, \quad E = A \times B/3 - C - 2 \times A^3/27$$
$$F = \sqrt{(D^3/27)}, \quad G = \arccos(-E/(2 \times F)), \quad H = \sqrt{(D/3)}$$
$$I = 2 \times H \times \cos(G/3) + A/3$$
$$J = 2 \times H \times [\cos(G/3 + 120°)] + A/3$$
$$K = 2 \times H \times [\cos(G/3 + 240°)] + A/3$$

Step 4. Calculate the true maximum shear stress, $S_{s(max)}$ using the formula

$$S_{s(max)} = 0.5 \times (S_{large} - S_{small})$$

in which S_{large} is equal to the algebraically largest of the calculated principal stresses I, J, or K and S_{small} is algebraically the smallest.

The maximum principal normal stresses and the maximum true shear stress calculated above may be used with any of the various theories of failure.

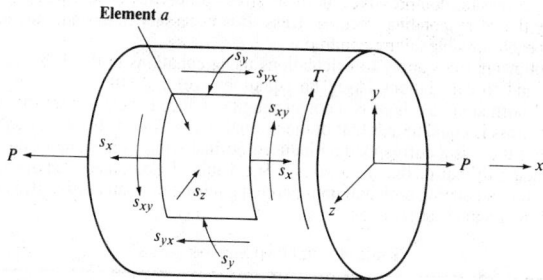

Fig. 2. Example of Triaxial Stress on an Element a of Shaft Surface Caused by Load P, Torque T, and 5000 psi Hydraulic Pressure

Example: A torque T on the shaft in Fig. 2 causes a shearing stress S_{xy} of 8000 psi in the outer fibers of the shaft; while the loads P at the ends of the shaft produce a tensile stress S_x of 4000 psi. The shaft passes through a hydraulic cylinder so that the shaft circumference is subjected to the hydraulic pressure of 5000 psi in the cylinder, causing compressive stresses S_y and S_z of -5000 psi on the surface of the shaft. Find the maximum shear stress at any point A on the surface of the shaft.

From the statement of the problem $S_x = +4000$ psi, $S_y = -5000$ psi, $S_z = -5000$ psi, $S_{xy} = +8000$ psi, $S_{yz} = 0$ psi, and $S_{zx} = 0$ psi.

$A = 4000 - 5000 - 5000 = -6000$
$B = (4000 \times -5000) + (-5000 \times -5000) + (-5000 \times 4000) - 8000^2 - 0^2 - 0^2$
$\quad = -7.9 \times 10^7$
$C = (4000 \times -5000 \times -5000) + 2 \times 8000 \times 0 \times 0 - (4000 \times 0^2)$
$\quad - (-5000 \times 0^2) - (-5000 \times 8000^2) = 4.2 \times 10^{11}$
$D = A^2/3 - B = 9.1 \times 10^7$
$E = A \times B/3 - C - 2 \times A^3/27 = -2.46 \times 10^{11}$
$F = \sqrt{(D^3/27)} = 1.6706 \times 10^{11}$
$G = \arccos(-E/(2 \times F)) = 42.586$ degrees, $H = \sqrt{(D/3)} = 5507.57$
$I = 2 \times H \times \cos(G/3) + A/3 = 8678.8$, say, 8680 psi
$J = 2 \times H \times [\cos(G/3 + 120°)] + A/3 = -9678.78$, say, -9680 psi
$K = 2 \times H \times [\cos(G/3 + 240°)] + A/3 = -5000$ psi

Check: $8680 + (-9680) + (-5000) = -6000$ within rounding error.
$S_{s(max)} = 0.5 \times (8680 - (-9680)) = 9180$ psi

Table of Combined Stresses. — Beginning below, this table lists equations for maximum nominal tensile or compressive (normal) stresses, and maximum nominal shear stresses for common machine elements. These equations were derived using general equations (19), (20), (22), and (23). The equations apply to the critical points indicated on the figures. Cases 1 through 4 are cantilever beams. These may be loaded with a combination of a vertical and horizontal force, or by a single oblique force. If the single oblique force F and the angle θ are given, then horizontal and vertical forces can be calculated using the equations $F_x = F \cos \theta$ and $F_y = F \sin \theta$. In cases 9 and 10 of the table, the equations for $\sigma_a{}'$ can give a tensile and a compressive stress because of the $\pm$ sign in front of the radical. Equations involving direct compression are valid only if machine elements have relatively short lengths with respect to their sections, otherwise column equations apply.

Calculation of worst stress condition: Stress failure can occur at any critical point if either the tensile, compressive, or shear stress properties of the material are exceeded by the corresponding working stress. It is necessary to evaluate the factor of safety for each possible failure condition.

The following rules apply to calculations using equations in the Table of Simple Stresses, and to calculations based on equations (19) and (20). *Rule 1:* For every calculated normal stress there is a corresponding induced shear stress; the value of the shear stress is equal to half that of the normal stress. *Rule 2:* For every calculated shear stress there is a corresponding induced normal stress; the value of the normal stress is equal to that of the shear stress. The Table of Combined Stresses includes equations for calculating both maximum nominal tensile or compressive stresses, and maximum nominal shear stresses.

Table of Combined Stresses — 1

Case	Type of Beam and Loading*	Maximum Nominal Tens. or Comp. Stress	Maximum Nominal Shear Stress
1		$\sigma_a{}' = \dfrac{1.273}{d^2}\left(\dfrac{8LF_y}{d} - F_x\right)$	$\tau_a{}' = 0.5\sigma_a{}'$
		$\sigma_b{}' = -\dfrac{1.273}{d^2}\left(\dfrac{8LF_y}{d} + F_x\right)$	$\tau_b{}' = 0.5\sigma_b{}'$
2		$\sigma_a{}' = \dfrac{1.273}{d^2}\left(F_x + \dfrac{8LF_y}{d}\right)$	$\tau_a{}' = 0.5\sigma_a{}'$
		$\sigma_b{}' = \dfrac{1.273}{d^2}\left(F_x - \dfrac{8LF_y}{d}\right)$	$\tau_b{}' = 0.5\sigma_b{}'$
3		$\sigma_a{}' = \dfrac{1}{bh}\left(\dfrac{6LF_y}{h} - F_x\right)$	$\tau_a{}' = 0.5\sigma_a{}'$
		$\sigma_b{}' = -\dfrac{1}{bh}\left(\dfrac{6LF_y}{h} + F_x\right)$	$\tau_b{}' = 0.5\sigma_b{}'$

* *Case 1:* Circular cantilever beam in direct compression and bending. *Case 2:* Circular cantilever beam in direct tension and bending. *Case 3:* Rectangular cantilever beam in direct compression and bending. See page 215 for combined stress symbols.

Table of Combined Stresses — 2

Case	Type of Beam and Loading*	Maximum Nominal Tens. or Comp. Stress	Maximum Nominal Shear Stress
4		$\sigma_a' = \dfrac{1}{bh}\left(F_x + \dfrac{6LF_y}{h}\right)$	$\tau_a' = 0.5\sigma_a'$
		$\sigma_b' = \dfrac{1}{bh}\left(F_x - \dfrac{6LF_y}{h}\right)$	$\tau_b' = 0.5\sigma_b'$
5		$\sigma_a' = -\dfrac{1.273}{d^2}\left(\dfrac{2LF_y}{d} + F_x\right)$	$\tau_a' = 0.5\sigma_a'$
		$\sigma_b' = \dfrac{1.273}{d^2}\left(\dfrac{2LF_y}{d} - F_x\right)$	$\tau_b' = 0.5\sigma_b'$
6		$\sigma_a' = \dfrac{1.273}{d^2}\left(F_x - \dfrac{2LF_y}{d}\right)$	$\tau_a' = 0.5\sigma_a'$
		$\sigma_b' = \dfrac{1.273}{d^2}\left(F_x + \dfrac{2LF_y}{d}\right)$	$\tau_b' = 0.5\sigma_b'$
7		$\sigma_a' = -\dfrac{1}{bh}\left(\dfrac{3LF_y}{2h} + F_x\right)$	$\tau_a' = 0.5\sigma_a'$
		$\sigma_b' = \dfrac{1}{bh}\left(\dfrac{3LF_y}{2h} - F_x\right)$	$\tau_b' = 0.5\sigma_b'$
8		$\sigma_a' = \dfrac{1}{bh}\left(F_x - \dfrac{3LF_y}{2h}\right)$	$\tau_a' = 0.5\sigma_a'$
		$\sigma_b' = \dfrac{1}{bh}\left(F_x + \dfrac{3LF_y}{2h}\right)$	$\tau_b' = 0.5\sigma_b'$
9	a anywhere on surface	$\sigma_a' = -\dfrac{0.637}{d^2}\left[F \pm \sqrt{F^2 + \left(\dfrac{8T}{d}\right)^2}\right]$	$\tau_a' = -\dfrac{0.637}{d^2}\sqrt{F^2 + \left(\dfrac{8T}{d}\right)^2}$
10	a anywhere on surface	$\sigma_a' = \dfrac{0.637}{d^2}\left[F \pm \sqrt{F^2 + \left(\dfrac{8T}{d}\right)^2}\right]$	$\tau_a' = \dfrac{0.637}{d^2}\sqrt{F^2 + \left(\dfrac{8T}{d}\right)^2}$
11		$\sigma_a' = \dfrac{1.273F}{d^2}\left(1 - \dfrac{8e}{d}\right)$	$\tau_a' = 0.5\sigma_a'$
		$\sigma_b' = \dfrac{1.273F}{d^2}\left(1 + \dfrac{8e}{d}\right)$	$\tau_b' = 0.5\sigma_b'$

* *Case 4:* Rectangular cantilever beam in direct compression and bending. *Case 5:* Circular beam or shaft in direct compression and bending. *Case 6:* Circular beam or shaft in direct tension and bending. *Case 7:* Rectangular beam or shaft in direct compression and bending. *Case 8:* Rectangular beam or shaft in direct tension and bending. *Case 9:* Circular shaft in direct compression and torsion. *Case 10:* Circular shaft in direct tension and torsion. *Case 11:* Offset link, circular cross section, in direct tension. See page 215 for combined stress symbols.

Table of Combined Stresses — 3

Case	Type of Beam and Loading*	Maximum Nominal Tens. or Comp. Stress	Maximum Nominal Shear Stress
12		$\sigma_a' = \dfrac{1.273F}{d^2}\left(\dfrac{8e}{d} - 1\right)$ $\sigma_b' = -\dfrac{1.273F}{d^2}\left(\dfrac{8e}{d} + 1\right)$	$\tau_a' = 0.5\sigma_a'$ $\tau_b' = 0.5\sigma_b'$
13		$\sigma_a' = \dfrac{F}{bh}\left(1 - \dfrac{6e}{h}\right)$ $\sigma_b' = \dfrac{F}{bh}\left(1 + \dfrac{6e}{h}\right)$	$\tau_a' = 0.5\sigma_a'$ $\tau_b' = 0.5\sigma_b'$
14		$\sigma_a' = \dfrac{F}{bh}\left(\dfrac{6e}{h} - 1\right)$ $\sigma_b' = -\dfrac{F}{bh}\left(\dfrac{6e}{h} + 1\right)$	$\tau_a' = 0.5\sigma_a'$ $\tau_b' = 0.5\sigma_b'$

* *Case 12:* Offset link, circular cross section, in direct compression. *Case 13:* Offset link, rectangular section, in direct tension. *Case 14:* Offset link, rectangular section, in direct compression. See page 215 for combined stress symbols.

Formulas from the simple and combined stress tables, as well as tension and shear factors, can be applied without change in calculations using metric SI units. Stresses are given in newtons per meter squared (N/m^2) or in N/mm^2.

Sample Calculations. — The following examples illustrate some typical strength of materials calculations, using both English and metric SI units of measurement.

Example 1(a): A round bar made from SAE 1025 low carbon steel is to support a direct tension load of 50,000 pounds. Using a factor of safety of 4, and assuming that the stress concentration factor $K = 1$, a suitable standard diameter is to be determined. Calculations are to be based on a yield strength of 40,000 psi.

Because the factor of safety and strength of the material are known, the allowable working stress s_w may be calculated using Equation (1): $40,000/4 = 10,000$ psi. The relationship between working stress s_w and nominal stress σ is given by equation (2). Since $K = 1$, $\sigma = 10,000$ psi. Applying Equation (9) in the Table of Simple Stresses, the area of the bar can be solved for: $A = 50,000/10,000$ or 5 square inches. The next largest standard diameter corresponding to this area is 2$\%_{16}$ inches.

Example 1(b): A similar example to that given in 1(a), using metric SI units is as follows. A round steel bar of 300 meganewtons/meter^2 yield strength, is to withstand a direct tension of 200 kilonewtons. Using a safety factor of 4, and assuming that the stress concentration factor $K = 1$, a suitable diameter is to be determined.

Because the factor of safety and the strength of the material are known

the allowable working stress s_w may be calculated using Equation (1): $300/4 = 75$ mega-newtons/meter^2. The relationship between working stress and nominal stress σ is given by Equation (2). Since $K = 1$, $\sigma = 75$ MN/m^2. Applying Equation (9) in the Table of Simple Stresses, the area of the bar can be determined from:

$$A = \frac{200 \text{ kN}}{75 \text{ MN/m}^2} = \frac{200,000 \text{ N}}{75,000,000 \text{ N/m}^2} = 0.00267 \text{ m}^2.$$

The diameter corresponding to this area is 0.058 meters, which, rounded up would be 0.06 m.

Millimeters can be employed in the calculations in place of meters, providing the treatment is consistent throughout. In this instance the diameter would be 60 mm.

Note: If the tension in the bar is produced by hanging a mass of M kilograms from the end of it, the value is Mg newtons, where $g =$ approximately 9.81 meters per second^2.

Example 2(a): What would the total elongation of the bar in Example *1(a)* be if its length were 60 inches? Applying Equation (17),

$$e = \frac{50,000 \times 60}{5.157 \times 30,000,000} = 0.019 \text{ inch}$$

Example 2(b): What would be the total elongation of the bar in Example *1(b)* if its length were 1.5 meters? The problem is solved by applying Equation (17) in which $F = 200$ kilonewtons; $L = 1.5$ meters; $A = \pi 0.06^2/4 = 0.00283$ m^2. Assuming a modulus of elasticity E of 200 giganewtons/meter^2, then the calculation is:

$$e = \frac{200,000 \times 1.5}{0.00283 \times 200,000,000,000} = 0.000530 \text{ m}.$$

The calculation is less unwieldy if carried out using millimeters in place of meters; then $F = 200$ kN; $L = 1500$ mm; $A = 2830$ mm^2, and $E = 200,000$ N/mm^2. Thus:

$$e = \frac{200,000 \times 1500}{2830 \times 200,000} = 0.530 \text{ mm}.$$

Example 3(a): Determine the size for the section of a square bar which is to be held firmly at one end and is to support a load of 3000 pounds at the outer end. The bar is to be 30 inches long and is to be made from SAE 1045 medium carbon steel with a yield point of 60,000 psi. A factor of safety of 3 and a stress concentration factor of 1.3 are to be used.

From Equation (1) the allowable working stress $s_w = 60,000/3 = 20,000$ psi. The applicable equation relating working stress and nominal stress is equation (2); hence, $\sigma = 20,000/1.3 = 15,400$ psi. The member must be treated as a cantilever beam subject to a bending moment of 30×3000 or 90,000 inch-pounds. Solving Equation (11) in the Table of Simple Stresses for section modulus: $Z = 90,000/15,400 = 5.85$ inch^3. Since the section modulus for a square section with neutral axis equidistant from either side is $a^3/6$, where a is the dimension of the square, $a = \sqrt[3]{35.1} = 3.27$ inches. The size of the bar can therefore be $3\frac{5}{16}$ inches.

Example 3(b): A similar example to that given in 3(a), using metric SI units is as follows. Determine the size for the section of a square bar which is to be held firmly at one end and is to support a load of 1600 kilograms at the outer end. The bar is to be 1 meter long, and is to be made from steel with a yield strength of 500

newtons/mm^2. A factor of safety of 3, and a stress concentration factor of 1.3 are to be used. The calculation can be performed using millimeters throughout.

From Equation (1) the allowable working stress $s_w = 500$ N/mm^2/3 = 167 N/mm^2. The formula relating working stress and nominal stress is Equation (2); hence $\sigma = 167/1.3 = 128$ N/mm^2. Since a mass of 1600 kg equals a weight of 1600 g newtons, where $g = 9.81$ meters/second^2, the force acting on the bar is 15,700 newtons. The bending moment on the bar, which must be treated as a cantilever beam, is thus 1000 mm × 15,700 N = 15,700,000 N · mm. Solving Equation (11) in the Table of Simple Stresses for section modulus: $Z = M/\sigma = 15,700,000/128 = 123,000$ mm^3. Since the section modulus for a square section with neutral axis equidistant from either side is $a^3/6$, where a is the dimension of the square,

$$a = \sqrt[3]{6 \times 123,000} = 90.4 \text{ mm.}$$

Example 4(a): Find the working stress in a 2-inch diameter shaft through which a transverse hole ¼ inch in diameter has been drilled. The shaft is subject to a torsional moment of 80,000 inch-pounds and is made from hardened steel so that the index of sensitivity $q = 0.2$.

The polar section modulus is calculated using the equation shown in the stress concentration curve for a Round Shaft in Torsion with Transverse Hole, page 211,

$$\frac{J}{c} = Z_p = \frac{\pi 2^3}{16} - \frac{2^2}{4 \times 6} = 1.4 \text{ inches}^3.$$

The nominal shear stress due to the torsion loading is computed using Equation (16) in the Table of Simple Stresses:

$$\tau = 80,000/1.4 = 57,200 \text{ psi.}$$

Referring to the previously mentioned stress concentration curve on page 211, K_t is 2.82 since d/D is 0.125. The stress concentration factor may now be calculated by means of Equation (8): $K = 1 + 0.2(2.82 - 1) = 1.36$. Working stress calculated with Equation (3) is $s_w = 1.36 \times 57,200 = 77,800$ psi.

Example 4(b): A similar example to that given in 4(a), using metric SI units is as follows. Find the working stress in a 50 mm diameter shaft through which a transverse hole 6 mm in diameter has been drilled. The shaft is subject to a torsional moment of 8000 newton-meters, and has an index of sensitivity of $q = 0.2$. If the calculation is made in millimeters, the torsional moment is 8,000,000 N · mm.

The polar section modulus is calculated using the equation shown in the stress concentration curve for a Round Shaft with Transverse Hole, page 211:

$$\frac{J}{c} = Z_p = \frac{\pi \times 50^3}{16} - \frac{6 \times 50^2}{6}$$
$$= 24,544 - 2500 = 22,044 \text{ mm}^3.$$

The nominal shear stress due to torsion loading is computed using Equation (16) in the Table of Simple Stresses:

$$\tau = 8,000,000/22,044 = 363 \text{ N/mm}^2 = 363 \text{ megapascals.}$$

Referring to the previously mentioned stress concentration curve on page 211, K_t is 2.85, since $a/d = 6/50 = 0.12$. The stress concentration factor may now be calculated by means of Equation (8): $K = 1 + 0.2(2.85 - 1) = 1.37$. From Equation (3), working stress $s_w = 1.37 \times 363 = 497$ N/mm^2 = 497 megapascals.

Example 5(a): For Case 3 in the Table of Combined Stresses, calculate the least factor of safety if: a 5052-H32 aluminum beam is 10 inches long; one inch wide, and 2 inches high. Yield strengths are 23,000 psi tension; 21,000 psi compression; 13,000 psi shear. The stress concentration factor is 1.5; F_y is 600 lbs; F_x 500 lbs.

From Table of Combined Stresses, Case 3:

$$\sigma_b' = -\frac{1}{1 \times 2}\left(\frac{6 \times 10 \times 600}{2} + 500\right) = -9250 \text{ psi (in compression)}$$

The other formulas for Case 3 give $\sigma_a' = 8750$ psi (in tension); $\tau_a' = 4375$ psi, and $\tau_b' = 4625$ psi. Using equation (4) for the nominal compressive stress of 9250 psi: $S_w = 1.5 \times 9250 = 13,900$ psi. From equation (1) $f_s = 21,000/13,900 = 1.51$. Applying equations (1), (4) and (5) in appropriate fashion to the other calculated nominal stress values for tension and shear will show that the factor of safety of 1.51, governed by the compressive stress at b on the beam, is minimum.

Example 5(b): **What maximum F can be applied in Case 3 if the aluminum beam is 200 mm long; 20 mm wide; 40 mm high; $\theta = 30°$; $f_s = 2$, governing for compression, $K = 1.5$, and $S_m = 144N/mm^2$ for compression.**

From equation (1) $S_w = -144/2 = 72N/mm^2$. Therefore, from equation (4), $\sigma_b' = -72/1.5 = -48N/mm^2$. Since $F_x = F \cos 30° = 0.866 F$, and $F_y = F \sin 30° = 0.5 F$:

$$-48 = -\frac{1}{20 \times 40}\left(0.866F + \frac{6 \times 200 \times 0.5F}{40}\right)$$

$$F = 2420 \text{ N}$$

Stresses and Deflections in a Loaded Ring. — For *thin* rings, that is, rings in which the dimension d shown in the accompanying diagram is small compared with D, the maximum stress in the ring is due primarily to bending moments produced by the forces P. The maximum stress due to bending is:

$$S = \frac{PDd}{4\pi I} \qquad (1)$$

For a ring of circular cross section where d is the diameter of the bar from which the ring is made,

$$S = \frac{1.621PD}{d^3}, \quad \text{or,} \quad P = \frac{0.617Sd^3}{D} \qquad (2)$$

The increase in the vertical diameter of the ring due to load P is:

$$\text{Increase in vertical diameter} = \frac{0.0186PD^3}{EI} \text{ inches} \qquad (3)$$

The *decrease* in the horizontal diameter will be about 92% of the increase in the vertical diameter given by Formula (3). In the above formulas, P = load on ring in pounds; D = mean diameter of ring in inches; S = tensile stress in pounds per square inch, I = moment of inertia of section in inches^4; and E = modulus of elasticity of material in pounds per square inch.

Strength of Taper Pins. — The mean diameter of taper pin required to safely transmit a known torque, may be found from the formulas:

$$d = 1.13\sqrt{\frac{T}{DS}} \qquad (1), \quad \text{and} \qquad d = 283\sqrt{\frac{H.P.}{NDS}} \qquad (2)$$

in which formulas T = torque in inch-pounds; S = safe unit stress in pounds per square

inch; H.P. = horsepower transmitted; N = number of revolutions per minute; and d and D denote dimensions shown in the figure.

Formula (1) can be used with metric SI units where d and D denote dimensions shown in the figure in millimeters; T = torque in newton-millimeters (N · mm); and S = safe unit stress in newtons per millimeter² (N/mm²). Formula (2) is replaced by:

$$d = 110.3 \sqrt{\frac{\text{Power}}{NDS}},$$

where d and D denote dimensions shown in the figure in millimeters; S = safe unit stress in N/mm²; N = number of revolutions per minute, and Power = power transmitted in watts.

Examples: — A lever secured to a 2-inch round shaft by a steel tapered pin (dimension $d = ⅜$ inch) has a pull of 50 pounds at a 30-inch radius from shaft center. Find S, the unit working stress on the pin. By rearranging Formula (1):

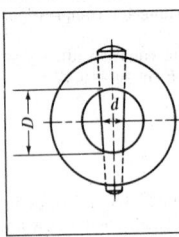

$$S = \frac{1.27T}{Dd^2} = \frac{1.27 \times 50 \times 30}{2 \times \left(\frac{3}{8}\right)^2} = 6770$$

pounds per square inch (nearly), which is a safe unit working stress for machine steel in shear.

Let $P = 50$ pounds, $R = 30$ inches, $D = 2$ inches, and $S = 6000$ pounds unit working stress. Using Formula (1) to find d:

$$d = 1.13 \sqrt{\frac{T}{DS}} = 1.13 \sqrt{\frac{50 \times 30}{2 \times 6000}} = 1.13 \sqrt{\frac{1}{8}} = 0.4 \text{ inch.}$$

A similar example using SI units is as follows: A lever secured to a 50 mm round shaft by a steel tapered pin ($d = 10$ mm) has a pull of 200 newtons at a radius of 800 mm. Find S, the working stress on the pin. By rearranging Formula (1):

$$S = \frac{1.27T}{Dd^2} = \frac{1.27 \times 200 \times 800}{50 \times 10^2} = 40.6 \text{ N/mm}^2 = 40.6 \text{ megapascals.}$$

If a shaft of 50 mm diameter is to transmit power of 12 kilowatts at a speed of 500 rpm, find the mean diameter of the pin for a material having a safe unit stress of 40 N/mm². Using the formula:

$$d = 110.3 \sqrt{\frac{\text{Power}}{NDS}}, \quad \text{then} \quad d = 110.3 \sqrt{\frac{12,000}{500 \times 50 \times 40}}$$

$$= 110.3 \times 0.1096 = 12.09 \text{ mm.}$$

Moment of Inertia of Built-up Sections. — The usual method of calculating the moment of inertia of a built-up section involves the calculations of the moment of inertia for each element of the section about its own neutral axis, and the transferring of this moment of inertia to the previously found neutral axis of the whole built-up section. A much simpler method that can be used in the case of any section which can be divided into rectangular elements bounded by lines parallel and perpendicular to the neutral axis is the so-called tabular method based upon the formula: $I = b(h_1{}^3 - h^3)/3$ in which I = the moment of inertia about axis DE, Fig. 1, and b, h and h_1 are dimensions as given in the same illustration.

The method may be illustrated by applying it to the section shown in Fig. 2, and for simplicity of calculation shown "massed" in Fig. 3. The calculation may then be tabulated as shown in the accompanying table. The distance from the axis DE to the neutral axis xx (which will be designated as d) is found by dividing the sum of the geometrical moments by the area. The moment of inertia about the neutral axis is then found in the usual way by subtracting the area multiplied by d^2 from the moment of inertia about the axis DE.

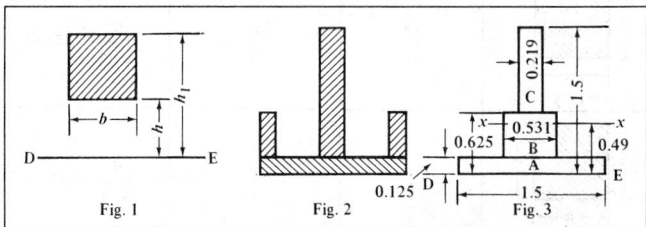

Fig. 1   Fig. 2   Fig. 3

Tabulated Calculation of Moment of Inertia

Section	Breadth b	Height h_1	Area $b(h_1 - h)$	h_1^2	Moment $\dfrac{b(h_1^2 - h^2)}{2}$	h_1^3	I about axis DE $\dfrac{b(h_1^3 - h^3)}{3}$
A	1.500	0.125	0.187	0.016	0.012	0.002	0.001
B	0.531	0.625	0.266	0.391	0.100	0.244	0.043
C	0.219	1.500	0.191	2.250	0.203	3.375	0.228
			$A = 0.644$		$M = 0.315$		$I_{DE} = 0.272$

The distance d from DE, the axis through the base of the configuration, to the neutral axis xx is:

$$d = \frac{M}{A} = \frac{0.315}{0.644} = 0.49$$

The moment of inertia of the entire section with reference to the neutral axis xx is:

$$I_N = I_{DE} - Ad^2$$
$$= 0.272 - 0.644 \times 0.49^2$$
$$= 0.117$$

Formulas for Moments of Inertia, Section Moduli, etc. — On the following pages are given formulas for the moments of inertia and other properties of forty-two different cross-sections. The formulas give the area of the section and the distance y from the neutral axis to the extreme fiber, in each case. In some cases, where the formulas for the section modulus and radius of gyration are very lengthy, the formula for the section modulus, for example, has been simply given as $\dfrac{I}{y}$. The radius of gyration is sometimes given as $\sqrt{\dfrac{I}{A}}$ to save space.

Moments of Inertia, Section Moduli, etc., of Sections

Section A = area y = distance from axis to extreme fiber	Moment of Inertia I	Section Modulus $Z = \dfrac{I}{y}$	Radius of Gyration $k = \sqrt{\dfrac{I}{A}}$
 $A = a^2; \quad y = \frac{1}{2}a$	$\dfrac{a^4}{12}$	$\dfrac{a^3}{6}$	$\dfrac{a}{\sqrt{12}} = 0.289\,a$
 $A = a^2; \quad y = a$	$\dfrac{a^4}{3}$	$\dfrac{a^3}{3}$	$\dfrac{a}{\sqrt{3}} = 0.577\,a$
 $A = a^2$ $y = \dfrac{a}{\sqrt{2}} = 0.707\,a$	$\dfrac{a^4}{12}$	$\dfrac{a^3}{6\sqrt{2}} = 0.118\,a^3$	$\dfrac{a}{\sqrt{12}} = 0.289\,a$
 $A = a^2 - b^2; \quad y = \frac{1}{2}a$	$\dfrac{a^4 - b^4}{12}$	$\dfrac{a^4 - b^4}{6a}$	$\sqrt{\dfrac{a^2 + b^2}{12}}$ $= 0.289\sqrt{a^2 + b^2}$
 $A = a^2 - b^2$ $y = \dfrac{a}{\sqrt{2}} = 0.707\,a$	$\dfrac{a^4 - b^4}{12}$	$\dfrac{\sqrt{2}\,(a^4 - b^4)}{12a}$ $= 0.118\dfrac{a^4 - b^4}{a}$	$\sqrt{\dfrac{a^2 + b^2}{12}}$ $= 0.289\sqrt{a^2 + b^2}$
 $A = bd; \quad y = \frac{1}{2}d$	$\dfrac{bd^3}{12}$	$\dfrac{bd^2}{6}$	$\dfrac{d}{\sqrt{12}} = 0.289\,d$

Moments of Inertia, Section Moduli, etc., of Sections

A = area y = distance from axis to extreme fiber	Moment of Inertia I	Section Modulus $Z = \dfrac{I}{y}$	Radius of Gyration $k = \sqrt{\dfrac{I}{A}}$
 $A = bd; \quad y = d$	$\dfrac{bd^3}{3}$	$\dfrac{bd^2}{3}$	$\dfrac{d}{\sqrt{3}} = 0.577\,d$
 $A = bd - hk$ $y = \frac{1}{2}\,d$	$\dfrac{bd^3 - hk^3}{12}$	$\dfrac{bd^3 - hk^3}{6d}$	$\sqrt{\dfrac{bd^3 - hk^3}{12(bd - hk)}}$ $= 0.289\sqrt{\dfrac{bd^3 - hk^3}{bd - hk}}$
 $A = bd$ $y = \dfrac{bd}{\sqrt{b^2 + d^2}}$	$\dfrac{b^3 d^3}{6(b^2 + d^2)}$	$\dfrac{b^2 d^2}{6\sqrt{b^2 + d^2}}$	$\dfrac{bd}{\sqrt{6(b^2 + d^2)}}$ $= 0.408\dfrac{bd}{\sqrt{b^2 + d^2}}$
 $A = bd$ $y = \frac{1}{2}(d\cos\alpha + b\sin\alpha)$	$\dfrac{bd}{12}(d^2\cos^2\alpha + b^2\sin^2\alpha)$	$\dfrac{bd}{6} \times$ $\left(\dfrac{d^2\cos^2\alpha + b^2\sin^2\alpha}{d\cos\alpha + b\sin\alpha}\right)$	$\sqrt{\dfrac{d^2\cos^2\alpha + b^2\sin^2\alpha}{12}}$ $= 0.289 \times$ $\sqrt{d^2\cos^2\alpha + b^2\sin^2\alpha}$
 $A = \frac{1}{2}\,bd; \quad y = \frac{2}{3}\,d$	$\dfrac{bd^3}{36}$	$\dfrac{bd^2}{24}$	$\dfrac{d}{\sqrt{18}} = 0.236\,d$
 $A = \frac{1}{2}\,bd; \quad y = d$	$\dfrac{bd^3}{12}$	$\dfrac{bd^2}{12}$	$\dfrac{d}{\sqrt{6}} = 0.408\,d$

Moments of Inertia, Section Moduli, etc.

Section	Area of Section, A	Distance from Neutral Axis to Extreme Fiber, y
	$\dfrac{d(a+b)}{2}$	$\dfrac{d(a+2b)}{3(a+b)}$
	$\dfrac{3d^2\tan 30^\circ}{2} = 0.866\,d^2$	$\dfrac{d}{2}$
	$\dfrac{3d^2\tan 30^\circ}{2} = 0.866\,d^2$	$\dfrac{d}{2\cos 30^\circ} = 0.577\,d$
	$2d^2\tan 22\tfrac{1}{2}^\circ = 0.828\,d^2$	$\dfrac{d}{2}$
	$\dfrac{\pi d^2}{4} = 0.7854\,d^2$	$\dfrac{d}{2}$
	$\dfrac{\pi(D^2-d^2)}{4}$ $= 0.7854(D^2-d^2)$	$\dfrac{D}{2}$
	$\dfrac{\pi d^2}{8} = 0.393\,d^2$	$\dfrac{(3\pi-4)\,d}{6\pi}$ $= 0.288\,d$
	$\dfrac{\pi(R^2-r^2)}{2}$ $= 1.5708(R^2-r^2)$	$\dfrac{4(R^3-r^3)}{3\pi(R^2-r^2)}$ $= 0.424\dfrac{R^3-r^3}{R^2-r^2}$

Moments of Inertia, Section Moduli, etc.

Moment of Inertia, I	Section Modulus, $Z = \dfrac{I}{y}$	Radius of Gyration, $k = \sqrt{\dfrac{I}{A}}$
$\dfrac{d^3(a^2+4ab+b^2)}{36(a+b)}$	$\dfrac{d^2(a^2+4ab+b^2)}{12(a+2b)}$	$\sqrt{\dfrac{d^2(a^2+4ab+b^2)}{18(a+b)^2}}$
$\dfrac{A}{12}\left[\dfrac{d^2(1+2\cos^2 30°)}{4\cos^2 30°}\right]$ $=0.06d^4$	$\dfrac{A}{6}\left[\dfrac{d(1+2\cos^2 30°)}{4\cos^2 30°}\right]$ $=0.12d^3$	$\sqrt{\dfrac{d^2(1+2\cos^2 30°)}{48\cos^2 30°}}$ $=0.264d$
$\dfrac{A}{12}\left[\dfrac{d^2(1+2\cos^2 30°)}{4\cos^2 30°}\right]$ $=0.06d^4$	$\dfrac{A}{6.9}\left[\dfrac{d(1+2\cos^2 30°)}{4\cos^2 30°}\right]$ $=0.104d^3$	$\sqrt{\dfrac{d^2(1+2\cos^2 30°)}{48\cos^2 30°}}$ $=0.264d$
$\dfrac{A}{12}\left[\dfrac{d^2(1+2\cos^2 22\frac{1}{2}°)}{4\cos^2 22\frac{1}{2}°}\right]$ $=0.055d^4$	$\dfrac{A}{6}\left[\dfrac{d(1+2\cos^2 22\frac{1}{2}°)}{4\cos^2 22\frac{1}{2}°}\right]$ $=0.109d^3$	$\sqrt{\dfrac{d^2(1+2\cos^2 22\frac{1}{2}°)}{48\cos^2 22\frac{1}{2}°}}$ $=0.257d$
$\dfrac{\pi d^4}{64}=0.049d^4$	$\dfrac{\pi d^3}{32}=0.098d^3$	$\dfrac{d}{4}$
$\dfrac{\pi(D^4-d^4)}{64}$ $=0.049(D^4-d^4)$	$\dfrac{\pi(D^4-d^4)}{32D}$ $=0.098\dfrac{D^4-d^4}{D}$	$\dfrac{\sqrt{D^2+d^2}}{4}$
$\dfrac{(9\pi^2-64)d^4}{1152\pi}$ $=0.007d^4$	$\dfrac{(9\pi^2-64)d^3}{192(3\pi-4)}$ $=0.024d^3$	$\dfrac{\sqrt{(9\pi^2-64)d^2}}{12\pi}$ $=0.132d$
$0.1098(R^4-r^4)$ $-\dfrac{0.283R^2r^2(R-r)}{R+r}$	$\dfrac{I}{y}$	$\sqrt{\dfrac{I}{A}}$

STRENGTH OF MATERIALS

Moments of Inertia, Section Moduli, etc.

Section	Area of Section, A	Distance from Neutral Axis to Extreme Fiber, y
	$\pi ab = 3.1416ab$	a
	$\pi(ab - cd)$ $= 3.1416(ab - cd)$	a
	$dt + 2a(s + n)$	$\dfrac{d}{2}$
	$dt + 2a(s + n)$	$\dfrac{b}{2}$
	$dt + a(s + n)$	$\dfrac{d}{2}$
	$dt + a(s + n)$	$b - [b^2 s + \dfrac{ht^2}{2} + \dfrac{g}{3}(b - t)^2$ $\times (b + 2t)] \div A$ in which g = slope of flange $= \dfrac{h - l}{2(b - t)}$
	$\dfrac{l(T + t)}{2} + Tn + a(s + n)$	$d - [3s^2(b - T)$ $+ 2am(m + 3s) + 3Td^2$ $- l(T - t)(3d - l)] \div 6A$

Moments of Inertia, Section Moduli, etc.

Moment of Inertia, I	Section Modulus, $Z = \dfrac{I}{y}$	Radius of Gyration, $k = \sqrt{\dfrac{I}{A}}$
$\dfrac{\pi a^3 b}{4} = 0.7854\, a^3 b$	$\dfrac{\pi a^2 b}{4} = 0.7854\, a^2 b$	$\dfrac{a}{2}$
$\dfrac{\pi}{4}(a^3 b - c^3 d)$ $= 0.7854(a^3 b - c^3 d)$	$\dfrac{\pi(a^3 b - c^3 d)}{4a}$ $= 0.7854\dfrac{a^3 b - c^3 d}{a}$	$\tfrac{1}{2}\sqrt{\dfrac{a^3 b - c^3 d}{ab - cd}}$
$\tfrac{1}{12}\left[bd^3 - \dfrac{1}{4g}(h^4 - l^4) \right]$ in which g = slope of flange $= (h - l)/(b - t) = \tfrac{1}{6}$ for standard I-beams.	$\dfrac{1}{6d}\left[bd^3 - \dfrac{1}{4g}(h^4 - l^4) \right]$	$\sqrt{\dfrac{\tfrac{1}{12}\left[bd^3 - \dfrac{1}{4g}(h^4 - l^4) \right]}{dt + 2a(s + n)}}$
$\tfrac{1}{12}\left[b^3(d - h) + lt^3 + \dfrac{g}{4}(b^4 - t^4) \right]$ in which g = slope of flange (see above).	$\dfrac{1}{6b}\left[b^3(d - h) + lt^3 + \dfrac{g}{4}(b^4 - t^4) \right]$	$\sqrt{\dfrac{I}{A}}$
$\tfrac{1}{12}\left[bd^3 - \dfrac{1}{8g}(h^4 - l^4) \right]$ in which g = slope of flange $= (h - l)/2(b - t) = \tfrac{1}{6}$ for standard channels.	$\dfrac{1}{6d}\left[bd^3 - \dfrac{1}{8g}(h^4 - l^4) \right]$	$\sqrt{\dfrac{\tfrac{1}{12}\left[bd^3 - \dfrac{1}{8g}(h^4 - l^4) \right]}{dt + a(s + n)}}$
$\tfrac{1}{3}\left[2sb^3 + lt^3 + \dfrac{g}{2}(b^4 - t^4) \right] - A(b - y)^2$ in which g = slope of flange (see above).	$\dfrac{I}{y}$	$\sqrt{\dfrac{I}{A}}$
$\tfrac{1}{12}[l^3(T + 3t) + 4bn^3 - 2am^3] - A(d - y - n)^2$	$\dfrac{I}{y}$	$\sqrt{\dfrac{I}{A}}$

Moments of Inertia, Section Moduli, etc.

Section	Area of Section, A	Distance from Neutral Axis to Extreme Fiber, y
	$\dfrac{l(T+t)}{2} + Tn$ $+ a(s+n)$	$\dfrac{b}{2}$
	$t(2a-t)$	$a - \dfrac{a^2 + at - t^2}{2(2a-t)}$
	$t(2a-t)$	$\dfrac{a^2 + at - t^2}{2(2a-t)\cos 45°}$
	$bd - h(b-t)$	$\dfrac{d}{2}$
	$bd - h(b-t)$	$\dfrac{b}{2}$
	$bd - h(b-t)$	$\dfrac{d}{2}$
	$bd - h(b-t)$	$b - \dfrac{2b^2 s + ht^2}{2bd - 2h(b-t)}$
	$dt + s(b-t)$	$\dfrac{d}{2}$

Moments of Inertia, Section Moduli, etc.

Moment of Inertia, I	Section Modulus, $Z = \dfrac{I}{y}$	Radius of Gyration, $k = \sqrt{\dfrac{I}{A}}$
$\dfrac{sb^3 + mT^3 + lt^3}{12}$ $+ \dfrac{am[2a^2 + (2a + 3T)^2]}{36}$ $+ \dfrac{l(T-t)[(T-t)^2 + 2(T+2t)^2]}{144}$	$\dfrac{I}{y}$	$\sqrt{\dfrac{I}{A}}$
$\frac{1}{3}[ty^3 + a(a-y)^3$ $- (a-t)(a-y-t)^3]$	$\dfrac{I}{y}$	$\sqrt{\dfrac{I}{A}}$
$\dfrac{A}{12}[7(a^2 + b^2) - 12y^2]$ $- 2ab^2(a-b)$ in which $b = (a-t)$	$\dfrac{I}{y}$	$\sqrt{\dfrac{I}{A}}$
$\dfrac{bd^3 - h^3(b-t)}{12}$	$\dfrac{bd^3 - h^3(b-t)}{6d}$	$\sqrt{\dfrac{bd^3 - h^3(b-t)}{12[bd - h(b-t)]}}$
$\dfrac{2sb^3 + ht^3}{12}$	$\dfrac{2sb^3 + ht^3}{6b}$	$\sqrt{\dfrac{2sb^3 + ht^3}{12[bd - h(b-t)]}}$
$\dfrac{bd^3 - h^3(b-t)}{12}$	$\dfrac{bd^3 - h^3(b-t)}{6d}$	$\sqrt{\dfrac{bd^3 - h^3(b-t)}{12[bd - h(b-t)]}}$
$\dfrac{2sb^3 + ht^3}{3} - A(b-y)^2$	$\dfrac{I}{y}$	$\sqrt{\dfrac{I}{A}}$
$\dfrac{td^3 + s^3(b-t)}{12}$	$\dfrac{td^3 + s^3(b-t)}{6d}$	$\sqrt{\dfrac{td^3 + s^3(b-t)}{12[td + s(b-t)]}}$

Moments of Inertia, Section Moduli, etc.

Section	Area of Section, A	Distance from Neutral Axis to Extreme Fiber, y
	$bs + ht + as$	$d - [td^2 + s^2(b-t) + s(a-t)(2d-s)] \div 2A$
	$bs + ht$	$d - \dfrac{d^2 t + s^2(b-t)}{2(bs+ht)}$
	$bs + \dfrac{h(T+t)}{2}$	$d - [3bs^2 + 3ht(d+s) + h(T-t)(h+3s)] \div 6A$
	$t(a+b-t)$	$b - \dfrac{t(2d+a)+d^2}{2(d+a)}$
	$t(a+b-t)$	$a - \dfrac{t(2c+b)+c^2}{2(c+b)}$
	$t[b + 2(a-t)]$	$\dfrac{b}{2}$
	$t[b + 2(a-t)]$	$\dfrac{2a-t}{2}$

Moments of Inertia, Section Moduli, etc.

Moment of Inertia, I	Section Modulus, $Z = \dfrac{I}{y}$	Radius of Gyration, $k = \sqrt{\dfrac{I}{A}}$
$\frac{1}{3}[b(d-y)^3 + ay^3 - (b-t)(d-y-s)^3 - (a-t)(y-s)^3]$	$\dfrac{I}{y}$	$\sqrt{\dfrac{I}{A}}$
$\frac{1}{3}[ty^3 + b(d-y)^3 - (b-t)(d-y-s)^3]$	$\dfrac{I}{y}$	$\sqrt{\dfrac{\mathrm{I}}{3(bs+ht)}[ty^3 + b(d-y)^3 - (b-t)(d-y-s)^3]}$
$\frac{1}{12}[4bs^3 + h^3(3t+T)] - A(d-y-s)^2$	$\dfrac{I}{y}$	$\sqrt{\dfrac{I}{A}}$
$\frac{1}{3}[ty^3 + a(b-y)^3 - (a-t)(b-y-t)^3]$	$\dfrac{I}{y}$	$\sqrt{\dfrac{\mathrm{I}}{3t(a+b-t)}[ty^3 + a(b-y)^3 - (a-t)(b-y-t)^3]}$
$\frac{1}{3}[ty^3 + b(a-y)^3 - (b-t)(a-y-t)^3]$	$\dfrac{I}{y}$	$\sqrt{\dfrac{\mathrm{I}}{3t(a+b-t)}[ty^3 + b(a-y)^3 - (b-t)(a-y-t)^3]}$
$\dfrac{ab^3 - c(b-2t)^3}{12}$	$\dfrac{ab^3 - c(b-2t)^3}{6b}$	$\sqrt{\dfrac{ab^3 - c(b-2t)^3}{12t[b+2(a-t)]}}$
$\dfrac{b(a+c)^3 - 2c^3d - 6a^2cd}{12}$	$\dfrac{b(a+c)^3 - 2c^3d - 6a^2cd}{6(2a-t)}$	$\sqrt{\dfrac{b(a+c)^3 - 2c^3d - 6a^2cd}{12t[b+2(a-t)]}}$

Section Modulus, Area, etc., of Sections for Punch and Shear Frames. — Machine frames cannot be standardized so as to permit tables of section modulus, area, etc., of the sections to be made up in the same way as for standard structural steel sections, but it is possible to arrange a table for punch and shear frames so as to simplify the work of selecting proper sections. A table of these quantities is, therefore, given. To illustrate the use of the table, take as an example the punch frame shown diagrammatically in Fig. 1. The distance from the center line of the punch to the back of the gap is 24 inches. Assume that the maximum pressure P, tending to force the jaws apart, is that due to punching a 1-inch circular hole in soft steel plate 1 inch thick, or say about 157,000 pounds. Consider the section at TX. The action of P is such as to produce a tensile stress on the section to the left of the neutral axis N, with a compressive stress to the right of N, both due to flexure; and,

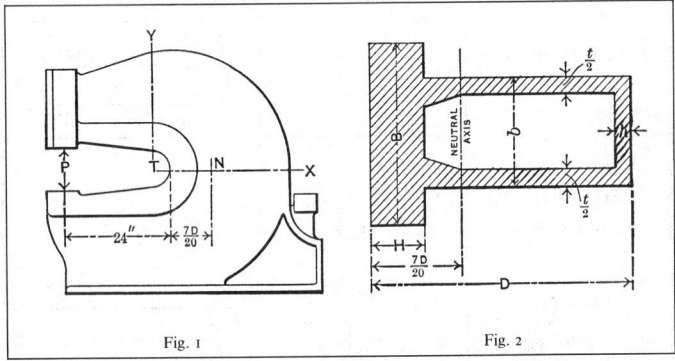

Fig. 1   Fig. 2

besides, there is a tensile stress distributed uniformly over the section. It is usually sufficient to determine the maximum tensile stress to the left of N.

Maximum tensile stress = flexure tensile stress + uniformly distributed tensile stress.

$$\text{Flexure tensile stress} = \frac{\text{Moment of } P \text{ about } N}{\text{Tensile section modulus of section } TX}$$

$$\text{Uniformly distributed tensile stress} = \frac{P}{\text{Area of section } TX}$$

Assume that D, in Fig. 2, is about 30 inches. Then $7D/20 = 10\frac{1}{2}$ inches. If 3000 psi is the allowable fiber stress, and the stress due to flexure only is considered, the required section modulus for tension would be: $157,000 \times (24 + 10\frac{1}{2})/3000 = 1800$, about.

As no allowance has been made for the additional tensile stress uniformly distributed over the section, a section must be selected the section modulus of which is somewhat greater than 1800, say about 2500.

The table of "Properties of Sections for Punch and Shear Frames" gives a great variety of shapes and proportions, the only dimensions common to all being the depth of section, D, and the distance of the neutral axis from the extreme tension fiber. This location of the neutral axis represents average practice and insures an economical distribution of metal. Generous fillets and rounded corners should, of course, be used on the actual section.

Suppose that a deep narrow section is desired, similar to that in Fig. 2, in which the dimensions are as follows: $D = 10$ inches; $B = 7$ inches; $b = 4$ inches; $H = 1.98$ inch; $\frac{1}{2} t = h = \frac{5}{8}$ inch; F (area) = 26.3 square inches; and Z_t (section modulus for tension) = 78.6. This section, taken from the table, is not, of course, large enough, but a similar one which is large enough may be easily found as follows:

If two sections A and B are similar in all respects, then

$$\frac{\text{Area of } A}{\text{Area of } B} = \frac{(\text{Any dimension of } A)^2}{(\text{Corresponding dimension of } B)^2},$$

and

$$\frac{\text{Section modulus of } A}{\text{Section modulus of } B} = \frac{(\text{Any dimension of } A)^3}{(\text{Corresponding dimension of } B)^3};$$

$$\frac{\text{Required section modulus}}{\text{Modulus of section from the table}} = \frac{(\text{Required } D)^3}{(D \text{ of section from the table})^3}$$

Hence, $\dfrac{2500}{78.6} = \dfrac{(\text{Required } D)^3}{(10)^3}$. This last equation solved for D gives $D = 31.7$ inches, nearly. As the large section will be exactly similar to the small one, the area of the large section is found from the equation:

$$\frac{\text{Area of required section}}{26.3} = \frac{(31.7)^2}{(10)^2}$$

and area = 264 square inches, about.

The neutral axis will be $\frac{7}{20} \times 31.7$, or 11.1 inches from the extreme tension fiber.

This trial section may now be tested for the maximum stress on it.

Flexure tensile stress = $157,000 \times (24 + 11.1)/2500 = 2200$ psi, about.

Uniformly distributed tensile stress = $157,000/264 = 600$ psi, about.

The total tensile stress is, therefore, 2800 pounds per square inch.

If this result had not been near enough to the 3000 pounds per square inch assumed, another section could have been selected and worked out in the same way to get a closer result. The depth D of the required section being 31.7 inches as compared with 10 inches of the similar section given in the table, each of the dimensions of the required section will be 3.17 times the corresponding one given in the table. Hence $D = 31.7$, say 32 inches; $B = 22.2$, say 22 inches; $b = 12.7$, say 13 inches; $H = 6.3$, say $6\frac{1}{2}$ inches, $\frac{1}{2} t = h = 1.98$, say 2 inches. The webs thicken gradually from the neutral axis to the tension flange so as to avoid too sudden a change in the section. For selecting the section at TY, Fig. 1, the procedure would be the same except that there would be no uniformly distributed stress to be added as in the case of section TX.

The method here used to determine the maximum stress on the curved section TX is actually correct only in the case of *straight* sections; for curved sections, various investigators have shown that the maximum stress is greater than indicated by straight beam formulas by a factor of up to 3 or more.

Since determination of the dimensions for a curved beam based on the use of theoretically correct formulas is a laborious trial-and-error process, it is common practice to use the method outlined and to apply an ample factor of safety.

A simplified method for the determination of stresses in a curved beam of known proportions is shown on page 282.

Properties of Sections for Punch and Shear Frames

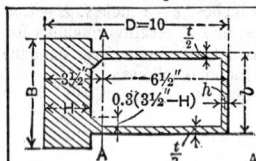

Z_c = Section Modulus for Compression;
Z_t = Section Modulus for Tension;
F = Area of Section;
I = Moment of Inertia about Gravity Axis $A - A$.

All dimensions in inches

B	b	$h = \frac{1}{2}t$	H	F	I	Z_c	Z_t
		¼	0.57	15.36	228.51	35.20	65.40
	10	⅜	1.10	23.43	311.78	47.95	89.10
		½	1.80	31.82	397.83	61.20	113.70
		¼	0.51	14.66	200.77	30.89	57.36
	9	⅜	0.99	21.64	290.32	44.66	82.95
		½	1.61	29.56	371.95	57.22	106.27
		⅝	2.41	38.69	438.24	67.44	125.21
10		¼	0.44	13.87	180.98	27.80	51.50
		⅜	0.88	20.41	272.07	41.90	77.70
	8	½	1.38	27.27	345.47	53.20	98.50
		⅝	2.04	35.20	410.37	63.10	117.00
		¾	3.50	49.63	462.98	71.20	132.00
		¼	0.38	13.15	172.76	26.50	49.30
		⅜	0.77	19.21	261.46	40.30	74.60
	7	½	1.24	25.69	320.52	49.30	91.50
		⅝	1.74	32.24	378.80	58.30	108.00
		¾	2.34	39.42	428.57	65.90	122.60
		¼	0.59	14.66	204.06	31.40	58.30
	9	⅜	1.20	21.57	291.47	44.80	83.40
		½	2.00	30.68	363.50	55.80	103.60
		¼	0.50	13.83	185.91	28.60	53.65
		⅜	1.00	20.34	268.15	41.28	76.65
	8	½	1.70	28.07	338.66	52.15	96.81
		⅝	2.60	37.12	403.27	62.10	115.12
9		¼	0.42	13.00	173.33	26.60	49.80
		⅜	0.89	19.25	250.41	38.40	71.60
	7	½	1.42	25.65	317.24	48.80	90.50
		⅝	2.11	33.03	375.00	57.70	107.10
		¾	3.06	42.13	420.36	64.60	120.10
		¼	0.36	12.40	161.70	24.90	46.20
		⅜	0.75	17.92	226.20	34.80	64.70
	6	½	1.23	23.89	290.50	44.70	83.00
		⅝	1.79	30.22	345.70	53.10	100.60
		¾	2.62	38.26	392.00	60.30	112.00
		¼	0.62	14.09	187.46	28.90	53.55
	8	⅜	1.20	20.50	268.38	41.30	76.70
8		½	2.10	28.79	336.75	51.80	96.20
		¼	0.55	13.69	172.25	26.50	49.20
	7	⅜	1.10	19.50	248.56	38.30	71.10
		½	1.90	27.07	315.01	48.50	90.00

Properties of Sections for Punch and Shear Frames

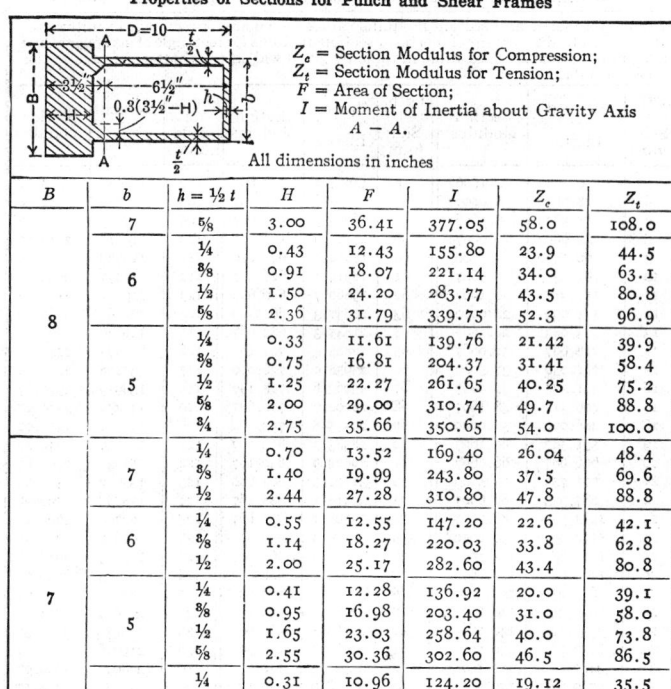

Z_c = Section Modulus for Compression;
Z_t = Section Modulus for Tension;
F = Area of Section;
I = Moment of Inertia about Gravity Axis $A - A$.

All dimensions in inches

B	b	$h = \frac{1}{2}t$	H	F	I	Z_c	Z_t
	7	⅝	3.00	36.41	377.05	58.0	108.0
	6	¼	0.43	12.43	155.80	23.9	44.5
		⅜	0.91	18.07	221.14	34.0	63.1
		½	1.50	24.20	283.77	43.5	80.8
		⅝	2.36	31.79	339.75	52.3	96.9
8	5	¼	0.33	11.61	139.76	21.42	39.9
		⅜	0.75	16.81	204.37	31.41	58.4
		½	1.25	22.27	261.65	40.25	75.2
		⅝	2.00	29.00	310.74	49.7	88.8
		¾	2.75	35.66	350.65	54.0	100.0
	7	¼	0.70	13.52	169.40	26.04	48.4
		⅜	1.40	19.99	243.80	37.5	69.6
		½	2.44	27.28	310.80	47.8	88.8
	6	¼	0.55	12.55	147.20	22.6	42.1
		⅜	1.14	18.27	220.03	33.8	62.8
		½	2.00	25.17	282.60	43.4	80.8
7	5	¼	0.41	12.28	136.92	20.0	39.1
		⅜	0.95	16.98	203.40	31.0	58.0
		½	1.65	23.03	258.64	40.0	73.8
		⅝	2.55	30.36	302.60	46.5	86.5
	4	¼	0.31	10.96	124.20	19.12	35.5
		⅜	0.76	15.71	183.00	28.15	52.3
		½	1.31	20.80	232.20	35.7	66.4
		⅝	1.98	26.30	275.00	42.3	78.6
	6	¼	0.68	12.56	156.25	24.0	44.6
		⅜	1.56	18.79	222.10	32.6	60.5
		½	3.50	30.00	275.00	42.3	78.6
6	5	¼	0.53	11.68	180.83	29.0	51.6
		⅜	1.27	17.25	214.79	33.0	61.4
		½	2.35	24.15	253.85	39.0	73.4
	4	¼	0.38	10.88	125.59	19.3	35.8
		⅜	1.00	15.84	181.36	27.9	51.8
		½	1.80	20.87	228.96	35.2	65.4
	5	¼	0.73	11.76	139.73	21.5	40.0
		⅜	1.70	17.28	196.98	30.2	56.2
	4	¼	0.55	10.97	125.40	19.3	35.9
5		⅜	1.45	16.13	181.52	28.0	52.0
	3	¼	0.84	10.16	110.14	17.25	32.0
		⅜	1.12	14.90	167.70	25.7	47.6
		½	2.10	19.99	198.57	30.5	56.8

Moments of Inertia and Section Moduli for Rectangles (Metric units)

Moment of inertia and section modulus values shown are for rectangles 1 millimeter wide. To obtain moment of inertia or section modulus for rectangle of given length of side, multiply appropriate value in table by given width. (See bottom of page 226 for basic formulas.)

Lgth. of Side (mm)	Moment of Inertia	Section Modulus	Lgth. of Side (mm)	Moment of Inertia	Section Modulus	Lgth. of Side (mm)	Moment of Inertia	Section Modulus
5	10.4167	4.16667	56	14634.7	522.667	107	102087	1908.17
6	18.0000	6.00000	57	15432.8	541.500	108	104976	1944.00
7	28.5833	8.16667	58	16259.3	560.667	109	107919	1980.17
8	42.6667	10.6667	59	17114.9	580.167	110	110917	2016.67
9	60.7500	13.5000	60	18000.0	600.000	111	113969	2053.50
10	83.3333	16.6667	61	18915.1	620.167	112	117077	2090.67
11	110.917	20.1667	62	19860.7	640.667	113	120241	2128.17
12	144.000	24.0000	63	20837.3	661.500	114	123462	2166.00
13	183.083	28.1667	64	21845.3	682.667	115	126740	2204.17
14	228.667	32.6667	65	22885.4	704.167	116	130075	2242.67
15	281.250	37.5000	66	23958.0	726.000	117	133468	2281.50
16	341.333	42.6667	67	25063.6	748.167	118	136919	2320.67
17	409.417	48.1667	68	26202.7	770.667	119	140430	2360.17
18	486.000	54.0000	69	27375.8	793.500	120	144000	2400.00
19	571.583	60.1667	70	28583.3	816.667	121	147630	2440.17
20	666.667	66.6667	71	29825.9	840.167	122	151321	2480.67
21	771.750	73.5000	72	31104.0	864.000	123	155072	2521.50
22	887.333	80.6667	73	32418.1	888.167	124	158885	2562.67
23	1013.92	88.1667	74	33768.7	912.667	125	162760	2604.17
24	1152.00	96.0000	75	35156.3	937.500	126	166698	2646.00
25	1302.08	104.167	76	36581.3	962.667	127	170699	2688.17
26	1464.67	112.667	77	38044.4	988.167	128	174763	2730.67
27	1640.25	121.500	78	39546.0	1014.00	130	183083	2816.67
28	1829.33	130.667	79	41086.6	1040.17	132	191664	2904.00
29	2032.42	140.167	80	42666.7	1066.67	135	205031	3037.50
30	2250.00	150.000	81	44286.8	1093.50	138	219006	3174.00
31	2482.58	160.167	82	45947.3	1120.67	140	228667	3266.67
32	2730.67	170.667	83	47648.9	1148.17	143	243684	3408.17
33	2994.75	181.500	84	49392.0	1176.00	147	264710	3601.50
34	3275.33	192.667	85	51177.1	1204.17	150	281250	3750.00
35	3572.92	204.167	86	53004.7	1232.67	155	310323	4004.17
36	3888.00	216.000	87	54875.3	1261.50	160	341333	4266.67
37	4221.08	228.167	88	56789.3	1290.67	165	374344	4537.50
38	4572.67	240.667	89	58747.4	1320.17	170	409417	4816.67
39	4943.25	253.500	90	60750.0	1350.00	175	446615	5104.17
40	5333.33	266.667	91	62797.6	1380.17	180	486000	5400.00
41	5743.42	280.167	92	64890.7	1410.67	185	527635	5704.17
42	6174.00	294.000	93	67029.8	1441.50	190	571583	6016.67
43	6625.58	308.167	94	69215.3	1472.67	195	617906	6337.50
44	7098.67	322.667	95	71447.9	1504.17	200	666667	6666.67
45	7593.75	337.500	96	73728.0	1536.00	210	771750	7350.00
46	8111.33	352.667	97	76056.1	1568.17	220	887333	8066.67
47	8651.92	368.167	98	78432.7	1600.67	230	1013920	8816.67
48	9216.00	384.000	99	80858.3	1633.50	240	1152000	9600.00
49	9804.08	400.167	100	83333.3	1666.67	250	1302080	10416.7
50	10416.7	416.667	101	85858.4	1700.17	260	1464670	11266.7
51	11054.3	433.5	102	88434.0	1734.00	270	1640250	12150.0
52	11717.3	450.667	103	91060.6	1768.17	280	1829330	13066.7
53	12406.4	468.167	104	93738.7	1802.67	290	2032410	14016.7
54	13122.0	486.000	105	96468.8	1837.5	300	2250000	15000.0
55	13864.6	504.167	106	99251.3	1872.67	...	...	...

Section Moduli for Rectangles

Section modulus values shown are for rectangles 1 inch wide. To obtain section modulus for rectangle of given length of side, multiply value in table by given width.

Length of Side	Section Modulus	Length of Side	Section Modulus	Length of Side	Section Modulus	Length of Side	Section Modulus
1/8	0.0026	2 3/4	1.26	12	24.00	25	104.2
3/16	0.0059	3	1.50	12 1/2	26.04	26	112.7
1/4	0.0104	3 1/4	1.76	13	28.17	27	121.5
5/16	0.0163	3 1/2	2.04	13 1/2	30.38	28	130.7
3/8	0.0234	3 3/4	2.34	14	32.67	29	140.2
7/16	0.032	4	2.67	14 1/2	35.04	30	150
1/2	0.042	4 1/2	3.38	15	37.5	32	171
5/8	0.065	5	4.17	15 1/2	40.0	34	193
3/4	0.094	5 1/2	5.04	16	42.7	36	216
7/8	0.128	6	6.00	16 1/2	45.4	38	241
1	0.167	6 1/2	7.04	17	48.2	40	267
1 1/8	0.211	7	8.17	17 1/2	51.0	42	294
1 1/4	0.260	7 1/2	9.38	18	54.0	44	323
1 3/8	0.315	8	10.67	18 1/2	57.0	46	353
1 1/2	0.375	8 1/2	12.04	19	60.2	48	384
1 5/8	0.440	9	13.50	19 1/2	63.4	50	417
1 3/4	0.510	9 1/2	15.04	20	66.7	52	451
1 7/8	0.586	10	16.67	21	73.5	54	486
2	0.67	10 1/2	18.38	22	80.7	56	523
2 1/4	0.84	11	20.17	23	88.2	58	561
2 1/2	1.04	11 1/2	22.04	24	96.0	60	600

Section Moduli and Moments of Inertia for Round Shafts*

Diam.	Section Modulus	Moment of Inertia	Diam.	Section Modulus	Moment of Inertia	Diam.	Section Modulus	Moment of Inertia
1/8	0.00019	0.00001	27/64	0.0074	0.00155	23/32	0.0364	0.01308
9/64	0.00027	0.00002	7/16	0.0082	0.00180	47/64	0.0388	0.01425
5/32	0.00037	0.00003	29/64	0.0091	0.00207	3/4	0.0413	0.01550
11/64	0.00050	0.00004	15/32	0.0101	0.00237	49/64	0.0440	0.01684
3/16	0.00065	0.00006	31/64	0.0111	0.00270	25/32	0.0467	0.01825
13/64	0.00082	0.00008	1/2	0.0123	0.00306	51/64	0.0496	0.01976
7/32	0.00102	0.00011	33/64	0.0134	0.00346	13/16	0.0526	0.02135
15/64	0.00126	0.00015	17/32	0.0147	0.00390	53/64	0.0557	0.02305
1/4	0.00153	0.00019	35/64	0.0160	0.00438	27/32	0.0588	0.02483
17/64	0.00183	0.00024	9/16	0.0174	0.00491	55/64	0.0622	0.02673
9/32	0.00218	0.00031	37/64	0.0189	0.00547	7/8	0.0656	0.02872
19/64	0.00256	0.00038	19/32	0.0205	0.00609	57/64	0.0692	0.03083
5/16	0.00299	0.00047	39/64	0.0222	0.00676	29/32	0.0728	0.03305
21/64	0.00344	0.00057	5/8	0.0239	0.00748	59/64	0.0767	0.03539
11/32	0.00398	0.00068	41/64	0.0258	0.00825	15/16	0.0807	0.03785
23/64	0.0045	0.00082	21/32	0.0277	0.00909	61/64	0.0849	0.04044
3/8	0.0052	0.00097	43/64	0.0297	0.00999	31/32	0.0891	0.04316
25/64	0.0058	0.00114	11/16	0.0318	0.01095	63/64	0.0934	0.04601
13/32	0.0066	0.00133	45/64	0.0341	0.01198	...	...	...

* The diameters and values given on pages 242 to 246 are valid as English or metric units.

In this and succeeding tables, the *Polar Section Modulus* for a shaft of given diameter can be obtained by multiplying its Section Modulus by 2. Similarly its *Polar Moment of Inertia* can be obtained by multiplying its Moment of Inertia by 2.

STRENGTH OF MATERIALS

Section Moduli and Moments of Inertia for Round Shafts

Diam.	Section Modulus	Moment of Inertia	Diam.	Section Modulus	Moment of Inertia	Diam.	Section Modulus	Moment of Inertia
1.00	0.0981	0.0490	1.50	0.3313	0.2485	2.00	0.7854	0.7854
1.01	0.1011	0.0510	1.51	0.3380	0.2552	2.01	0.7972	0.8012
1.02	0.1041	0.0531	1.52	0.3447	0.2620	2.02	0.8092	0.8172
1.03	0.1072	0.0552	1.53	0.3516	0.2689	2.03	0.8212	0.8335
1.04	0.1104	0.0574	1.54	0.3585	0.2761	2.04	0.8334	0.8501
1.05	0.1136	0.0596	1.55	0.3655	0.2833	2.05	0.8457	0.8669
1.06	0.1169	0.0619	1.56	0.3727	0.2907	2.06	0.8582	0.8839
1.07	0.1202	0.0643	1.57	0.3799	0.2982	2.07	0.8707	0.9012
1.08	0.1236	0.0667	1.58	0.3872	0.3059	2.08	0.8834	0.9188
1.09	0.1271	0.0692	1.59	0.3946	0.3137	2.09	0.8962	0.9366
1.10	0.1307	0.0718	1.60	0.4021	0.3217	2.10	0.9092	0.9547
1.11	0.1342	0.0745	1.61	0.4097	0.3298	2.11	0.9222	0.9729
1.12	0.1379	0.0772	1.62	0.4173	0.3380	2.12	0.9354	0.9915
1.13	0.1416	0.0800	1.63	0.4251	0.3465	2.13	0.9487	1.0103
1.14	0.1454	0.0829	1.64	0.4330	0.3550	2.14	0.9621	1.0295
1.15	0.1493	0.0859	1.65	0.4410	0.3638	2.15	0.9757	1.0488
1.16	0.1532	0.0888	1.66	0.4490	0.3727	2.16	0.9894	1.0685
1.17	0.1572	0.0919	1.67	0.4572	0.3818	2.17	1.0031	1.0884
1.18	0.1613	0.0951	1.68	0.4655	0.3910	2.18	1.0171	1.1086
1.19	0.1654	0.0984	1.69	0.4738	0.4004	2.19	1.0311	1.1291
1.20	0.1696	0.1018	1.70	0.4823	0.4100	2.20	1.0454	1.1499
1.21	0.1739	0.1052	1.71	0.4908	0.4197	2.21	1.0596	1.1709
1.22	0.1782	0.1087	1.72	0.4995	0.4296	2.22	1.0741	1.1923
1.23	0.1826	0.1123	1.73	0.5083	0.4397	2.23	1.0887	1.2139
1.24	0.1871	0.1160	1.74	0.5171	0.4499	2.24	1.1034	1.2358
1.25	0.1917	0.1198	1.75	0.5261	0.4603	2.25	1.1183	1.2580
1.26	0.1963	0.1237	1.76	0.5352	0.4710	2.26	1.1332	1.2806
1.27	0.2011	0.1277	1.77	0.5444	0.4818	2.27	1.1483	1.3034
1.28	0.2058	0.1317	1.78	0.5536	0.4927	2.28	1.1636	1.3265
1.29	0.2107	0.1359	1.79	0.5630	0.5039	2.29	1.1790	1.3499
1.30	0.2157	0.1402	1.80	0.5726	0.5153	2.30	1.1945	1.3737
1.31	0.2207	0.1445	1.81	0.5821	0.5268	2.31	1.2101	1.3977
1.32	0.2258	0.1490	1.82	0.5918	0.5385	2.32	1.2259	1.4234
1.33	0.2309	0.1535	1.83	0.6016	0.5505	2.33	1.2418	1.4468
1.34	0.2362	0.1582	1.84	0.6115	0.5626	2.34	1.2579	1.4718
1.35	0.2415	0.1630	1.85	0.6216	0.5749	2.35	1.2741	1.4971
1.36	0.2469	0.1679	1.86	0.6317	0.5875	2.36	1.2904	1.5227
1.37	0.2524	0.1729	1.87	0.6419	0.6002	2.37	1.3069	1.5487
1.38	0.2580	0.1780	1.88	0.6524	0.6132	2.38	1.3235	1.5750
1.39	0.2636	0.1832	1.89	0.6628	0.6263	2.39	1.3403	1.6016
1.40	0.2694	0.1886	1.90	0.6734	0.6397	2.40	1.3572	1.6286
1.41	0.2752	0.1940	1.91	0.6840	0.6532	2.41	1.3742	1.6559
1.42	0.2811	0.1995	1.92	0.6948	0.6670	2.42	1.3914	1.6836
1.43	0.2870	0.2052	1.93	0.7057	0.6810	2.43	1.4087	1.7116
1.44	0.2931	0.2110	1.94	0.7168	0.6953	2.44	1.4262	1.7399
1.45	0.2993	0.2170	1.95	0.7279	0.7097	2.45	1.4438	1.7686
1.46	0.3055	0.2230	1.96	0.7392	0.7244	2.46	1.4615	1.7977
1.47	0.3118	0.2292	1.97	0.7505	0.7393	2.47	1.4794	1.8271
1.48	0.3182	0.2355	1.98	0.7620	0.7544	2.48	1.4975	1.8526
1.49	0.3247	0.2419	1.99	0.7736	0.7698	2.49	1.5156	1.8870

Section Moduli and Moments of Inertia for Round Shafts

Diam.	Section Modulus	Moment of Inertia	Diam.	Section Modulus	Moment of Inertia	Diam.	Section Modulus	Moment of Inertia
2.50	1.5340	1.9175	3.00	2.6510	3.9761	3.50	4.2090	7.3662
2.51	1.5525	1.9483	3.01	2.6773	4.0293	3.51	4.2455	7.4507
2.52	1.5711	1.9796	3.02	2.7041	4.0831	3.52	4.2818	7.5360
2.53	1.5899	2.0112	3.03	2.7310	4.1375	3.53	4.3184	7.6220
2.54	1.6088	2.0431	3.04	2.7581	4.1924	3.54	4.3552	7.7087
2.55	1.6279	2.0755	3.05	2.7855	4.2478	3.55	4.3922	7.7962
2.56	1.6471	2.1083	3.06	2.8130	4.3038	3.56	4.4294	7.8845
2.57	1.6665	2.1414	3.07	2.8406	4.3604	3.57	4.4669	7.9734
2.58	1.6860	2.1749	3.08	2.8685	4.4175	3.58	4.5045	8.0631
2.59	1.7057	2.2088	3.09	2.8965	4.4751	3.59	4.5424	8.1536
2.60	1.7260	2.2432	3.10	2.9250	4.5333	3.60	4.5804	8.2448
2.61	1.7455	2.2779	3.11	2.9531	4.5921	3.61	4.6187	8.3367
2.62	1.7656	2.3130	3.12	2.9817	4.6514	3.62	4.6572	8.4296
2.63	1.7859	2.3485	3.13	3.0104	4.7113	3.63	4.6959	8.5231
2.64	1.8064	2.3844	3.14	3.0394	4.7718	3.64	4.7347	8.6174
2.65	1.8270	2.4208	3.15	3.0685	4.8330	3.65	4.7740	8.7125
2.66	1.8478	2.4575	3.16	3.0978	4.8946	3.66	4.8133	8.8084
2.67	1.8686	2.4947	3.17	3.1274	4.9568	3.67	4.8529	8.9050
2.68	1.8897	2.5322	3.18	3.1570	5.0197	3.68	4.8926	9.0025
2.69	1.9110	2.5702	3.19	3.1869	5.0832	3.69	4.9325	9.1007
2.70	1.9320	2.6087	3.20	3.2170	5.1472	3.70	4.9730	9.1998
2.71	1.9539	2.6476	3.21	3.2472	5.2119	3.71	5.0133	9.2996
2.72	1.9756	2.6868	3.22	3.2777	5.2771	3.72	5.0540	9.4003
2.73	1.9975	2.7266	3.23	3.3083	5.3430	3.73	5.0948	9.5018
2.74	2.0195	2.7668	3.24	3.3391	5.4094	3.74	5.1359	9.6041
2.75	2.0417	2.8074	3.25	3.3701	5.4765	3.75	5.1771	9.7072
2.76	2.0641	2.8484	3.26	3.4014	5.5442	3.76	5.2187	9.8112
2.77	2.0866	2.8899	3.27	3.4328	5.6126	3.77	5.2605	9.9160
2.78	2.1093	2.9319	3.28	3.4644	5.6815	3.78	5.3024	10.0216
2.79	2.1321	2.9743	3.29	3.4961	5.7511	3.79	5.3444	10.1286
2.80	2.1550	3.0172	3.30	3.5280	5.8214	3.80	5.3870	10.2350
2.81	2.1783	3.0605	3.31	3.5603	5.8923	3.81	5.4297	10.3436
2.82	2.2016	3.1043	3.32	3.5926	5.9638	3.82	5.4726	10.4526
2.83	2.2251	3.1486	3.33	3.6252	6.0363	3.83	5.5156	10.5624
2.84	2.2488	3.1933	3.34	3.6580	6.1088	3.84	5.5590	10.6732
2.85	2.2727	3.2385	3.35	3.6909	6.1823	3.85	5.6025	10.7848
2.86	2.2966	3.2842	3.36	3.7241	6.2564	3.86	5.6462	10.8970
2.87	2.3208	3.3304	3.37	3.7575	6.3312	3.87	5.6903	11.0110
2.88	2.3452	3.3771	3.38	3.7909	6.4067	3.88	5.7345	11.1250
2.89	2.3697	3.4242	3.39	3.8246	6.4829	3.89	5.7789	11.2400
2.90	2.3940	3.4719	3.40	3.8590	6.5597	3.90	5.8240	11.3560
2.91	2.4192	3.5200	3.41	3.8928	6.6372	3.91	5.8685	11.4730
2.92	2.4442	3.5686	3.42	3.9272	6.7154	3.92	5.9137	11.5910
2.93	2.4695	3.6178	3.43	3.9617	6.7943	3.93	5.9590	11.7100
2.94	2.4949	3.6674	3.44	3.9965	6.8739	3.94	6.0046	11.8290
2.95	2.5204	3.7175	3.45	4.0314	6.9542	3.95	6.0505	11.9500
2.96	2.5461	3.7682	3.46	4.0666	7.0352	3.96	6.0966	12.0690
2.97	2.5720	3.8196	3.47	4.1019	7.1168	3.97	6.1429	12.1930
2.98	2.5981	3.8711	3.48	4.1375	7.1976	3.98	6.1894	12.3170
2.99	2.6243	3.9233	3.49	4.1732	7.2824	3.99	6.2361	12.4410

Section Moduli and Moments of Inertia for Round Shafts

Diam.	Section Modulus	Moment of Inertia	Diam.	Section Modulus	Moment of Inertia	Diam.	Section Modulus	Moment of Inertia
4.00	6.2830	12.566	4.50	8.946	20.129	5.00	12.272	30.680
4.01	6.3304	12.692	4.51	9.006	20.308	5.01	12.345	30.926
4.02	6.3779	12.820	4.52	9.066	20.489	5.02	12.420	31.173
4.03	6.4256	12.948	4.53	9.126	20.671	5.03	12.493	31.423
4.04	6.4736	13.077	4.54	9.186	20.854	5.04	12.568	31.673
4.05	6.5217	13.207	4.55	9.247	21.039	5.05	12.644	31.925
4.06	6.5701	13.337	4.56	9.308	21.224	5.06	12.718	32.179
4.07	6.6188	13.469	4.57	9.370	21.411	5.07	12.794	32.434
4.08	6.6677	13.602	4.58	9.431	21.599	5.08	12.870	32.691
4.09	6.7169	13.736	4.59	9.493	21.788	5.09	12.946	32.949
4.10	6.7660	13.871	4.60	9.556	21.979	5.10	13.023	33.209
4.11	6.8159	14.007	4.61	9.618	22.170	5.11	13.099	33.470
4.12	6.8657	14.143	4.62	9.681	22.363	5.12	13.177	33.733
4.13	6.9164	14.281	4.63	9.744	22.557	5.13	13.254	33.997
4.14	6.9663	14.420	4.64	9.807	22.753	5.14	13.332	34.263
4.15	7.0169	14.560	4.65	9.870	22.950	5.15	13.410	34.530
4.16	7.0677	14.701	4.66	9.934	23.148	5.16	13.488	34.799
4.17	7.1188	14.843	4.67	9.998	23.347	5.17	13.567	35.070
4.18	7.1702	14.985	4.68	10.063	23.548	5.18	13.645	35.342
4.19	7.2217	15.129	4.69	10.127	23.750	5.19	13.725	35.615
4.20	7.2740	15.274	4.70	10.193	23.953	5.20	13.804	35.891
4.21	7.3256	15.420	4.71	10.258	24.157	5.21	13.884	36.168
4.22	7.3779	15.568	4.72	10.323	24.363	5.22	13.964	36.446
4.23	7.4305	15.715	4.73	10.389	24.570	5.23	14.045	36.726
4.24	7.4833	15.865	4.74	10.455	24.779	5.24	14.125	37.008
4.25	7.5364	16.015	4.75	10.522	24.989	5.25	14.206	37.291
4.26	7.5898	16.166	4.76	10.588	25.200	5.26	14.287	37.576
4.27	7.6433	16.319	4.77	10.655	25.412	5.27	14.369	37.863
4.28	7.6972	16.472	4.78	10.722	25.626	5.28	14.451	38.151
4.29	7.7513	16.626	4.79	10.790	25.841	5.29	14.534	38.440
4.30	7.8060	16.782	4.80	10.857	26.058	5.30	14.616	38.732
4.31	7.8602	16.938	4.81	10.925	26.275	5.31	14.699	39.025
4.32	7.9149	17.096	4.82	10.994	26.495	5.32	14.782	39.320
4.33	7.9701	17.255	4.83	11.062	26.715	5.33	14.866	39.617
4.34	8.0254	17.415	4.84	11.131	26.937	5.34	14.949	39.915
4.35	8.0810	17.576	4.85	11.200	27.160	5.35	15.034	40.215
4.36	8.1369	17.738	4.86	11.269	27.385	5.36	15.118	40.516
4.37	8.1930	17.902	4.87	11.339	27.611	5.37	15.202	40.819
4.38	8.2494	18.066	4.88	11.409	27.839	5.38	15.288	41.124
4.39	8.3060	18.231	4.89	11.479	28.067	5.39	15.373	41.431
4.40	8.3630	18.398	4.90	11.550	28.298	5.40	15.459	41.739
4.41	8.4200	18.566	4.91	11.621	28.530	5.41	15.545	42.049
4.42	8.4775	18.735	4.92	11.692	28.763	5.42	15.631	42.361
4.43	8.5351	18.905	4.93	11.763	28.997	5.43	15.718	42.674
4.44	8.5930	19.077	4.94	11.835	29.233	5.44	15.805	42.990
4.45	8.6513	19.249	4.95	11.907	29.471	5.45	15.893	43.307
4.46	8.7097	19.423	4.96	11.979	29.710	5.46	15.980	43.626
4.47	8.7685	19.598	4.97	12.052	29.950	5.47	16.068	43.946
4.48	8.8274	19.773	4.98	12.124	30.192	5.48	16.157	44.268
4.49	8.8867	19.950	4.99	12.198	30.435	5.49	16.245	44.592

Section Moduli and Moments of Inertia for Round Shafts

Diam.	Section Modulus	Moment of Inertia	Diam.	Section Modulus	Moment of Inertia	Diam.	Section Modulus	Moment of Inertia
5.5	16.3338	44.9180	30	2650.72	39760.8	54.5	15892.4	433068
6	21.2058	63.6173	30.5	2785.48	42478.5	55	16333.8	449180
6.5	26.9613	87.6241	31	2924.73	45333.2	55.5	16783.4	465738
7	33.6740	117.859	31.5	3068.54	48329.5	56	17241.1	482750
7.5	41.4175	155.316	32	3216.99	51471.9	56.5	17707.0	500223
8	50.2655	201.062	32.5	3370.16	54765.0	57	18181.3	518167
8.5	60.2916	256.239	33	3528.11	58213.8	57.5	18663.9	536588
9	71.5694	322.062	33.5	3690.92	61822.9	58	19155.1	555497
9.5	84.1726	399.820	34	3858.66	65597.2	58.5	19654.8	574901
10	98.1748	490.874	34.5	4031.41	69541.9	59	20163.0	594810
10.5	113.650	596.660	35	4209.24	73661.8	59.5	20680.0	615230
11	130.671	718.688	35.5	4392.23	77962.1	60	21205.8	636173
11.5	149.312	858.541	36	4580.44	82448.0	60.5	21740.3	657645
12	169.646	1017.88	36.5	4773.96	87124.7	61	22283.3	679656
12.5	191.748	1198.42	37	4972.85	91997.7	61.5	22836.3	702215
13	215.690	1401.99	37.5	5177.19	97072.2	62	23397.8	725332
13.5	241.547	1630.44	38	5387.05	102354	62.5	23968.5	749014
14	269.392	1885.74	38.5	5602.50	107848	63	24548.3	773272
14.5	299.298	2169.91	39	5823.63	113561	63.5	25137.4	798114
15	331.340	2485.05	39.5	6050.50	119497	64	25735.9	823550
15.5	365.591	2833.33	40	6283.19	125664	64.5	26343.8	849589
16	402.124	3216.99	40.5	6521.76	132066	65	26961.3	876241
16.5	441.013	3638.36	41	6766.30	138709	65.5	27588.2	903514
17	482.333	4099.83	41.5	7016.88	145600	66	28224.9	931420
17.5	526.155	4603.86	42	7273.57	152745	66.5	28871.2	959967
18	572.555	5153.00	42.5	7536.45	160150	67	29527.3	989166
18.5	621.606	5749.85	43	7805.58	167820	67.5	30193.3	1019030
19	673.381	6397.12	43.5	8081.05	175763	68	30869.3	1049560
19.5	727.954	7097.55	44	8362.92	183984	68.5	31555.3	1080770
20	785.398	7853.98	44.5	8651.27	192491	69	32251.3	1112670
20.5	845.788	8669.33	45	8946.18	201289	69.5	32957.5	1145270
21	909.197	9546.56	45.5	9247.71	210385	70	33674.0	1178590
21.5	975.698	10488.8	46	9555.94	219787	70.5	34400.7	1212630
22	1045.37	11499.0	46.5	9870.95	229500	71	35137.8	1247390
22.5	1118.27	12580.6	47	10192.8	239531	71.5	35885.4	1282900
23	1194.49	13736.7	47.5	10521.6	249887	72	36643.5	1319170
23.5	1274.10	14970.7	48	10857.3	260576	72.5	37412.3	1356190
24	1357.17	16286.0	48.5	11200.2	271604	73	38191.7	1393990
24.5	1443.17	17686.2	49	11550.2	282979	73.5	38981.8	1432580
25	1533.98	19174.8	49.5	11907.4	294707	74	39782.8	1471960
25.5	1627.87	20755.4	50	12271.9	306796	74.5	40594.6	1512150
26	1725.52	22431.8	50.5	12643.7	319253	75	41417.5	1553160
26.5	1827.00	24207.7	51	13023.0	332086	75.5	42251.4	1594990
27	1932.37	26087.1	51.5	13409.8	345302	76	43096.4	1637660
27.5	2041.73	28073.8	52	13804.2	358908	76.5	43952.6	1681190
28	2155.13	30171.9	52.5	14206.2	372913	77	44820.0	1725570
28.5	2272.66	32385.4	53	14616.0	387323	77.5	45698.8	1770830
29	2394.38	34718.6	53.5	15033.5	402147	78	46589.0	1816970
29.5	2520.38	37175.6	54	15459.0	417393	78.5	47490.7	1864010

Section Moduli and Moments of Inertia for Round Shafts

Diam.	Section Modulus	Moment of Inertia	Diam.	Section Modulus	Moment of Inertia	Diam.	Section Modulus	Moment of Inertia
79	48404.0	1911960	103.5	108848	5632890	128	205887	13176790
79.5	49328.9	1960820	104	110433	5742530	128.5	208310	13383890
80	50265.5	2010620	104.5	112034	5853760	129	210751	13593420
80.5	51213.9	2061360	105	113650	5966600	129.5	213211	13805400
81	52174.1	2113050	105.5	115281	6081070	130	215690	14019850
81.5	53146.3	2165710	106	116928	6197170	130.5	218188	14236790
82	54130.4	2219350	106.5	118590	6314930	131	220706	14456230
82.5	55126.7	2273980	107	120268	6434350	131.5	223243	14678200
83	56135.1	2329610	107.5	121962	6555470	132	225799	14902720
83.5	57155.7	2386250	108	123672	6667280	132.5	228375	15129810
84	58188.6	2443920	108.5	125398	6802820	133	230970	15359480
84.5	59233.9	2502630	109	127139	6929080	133.5	233584	15591750
85	60291.6	2562390	109.5	128897	7057100	134	236219	15826650
85.5	61361.8	2623220	110	130671	7186880	134.5	238873	16064200
86	62444.7	2685120	110.5	132461	7318450	135	241547	16304410
86.5	63540.2	2748110	111	134267	7451810	135.5	244241	16547300
87	64648.4	2812210	111.5	136090	7586990	136	246954	16792890
87.5	65769.4	2877410	112	137929	7723990	136.5	249688	17041210
88	66903.4	2943750	112.5	139784	7862850	137	252442	17292280
88.5	68050.3	3011220	113	141656	8003570	137.5	255216	17546100
89	69210.2	3079850	113.5	143545	8146170	138	258010	17802720
89.5	70383.2	3149650	114	145450	8290660	138.5	260825	18062130
90	71569.4	3220620	114.5	147373	8437070	139	263660	18324370
90.5	72768.9	3292790	115	149312	8585410	139.5	266516	18589460
91	73981.7	3366170	115.5	151268	8735700	140	269392	18857410
91.5	75207.9	3440760	116	153241	8887950	140.5	272288	19128250
92	76447.5	3516590	116.5	155231	9042190	141	275206	19401990
92.5	77700.7	3593660	117	157238	9198420	141.5	278144	19678670
93	78967.6	3671990	117.5	159263	9356670	142	281103	19958290
93.5	80248.1	3751600	118	161304	9516950	142.5	284083	20240880
94	81542.4	3832490	118.5	163364	9679280	143	287083	20526460
94.5	82850.5	3914690	119	165440	9843690	143.5	290105	20815050
95	84172.6	3998200	119.5	167534	10010170	144	293148	21106680
95.5	85508.6	4083040	120	169646	10178760	144.5	296213	21401360
96	86858.8	4169220	120.5	171775	10349470	145	299298	21699110
96.5	88223.0	4256760	121	173923	10522320	145.5	302405	21999960
97	89601.5	4345670	121.5	176088	10697320	146	305533	22303930
97.5	90994.2	4435970	122	178271	10874500	146.5	308683	22611030
98	92401.3	4527660	122.5	180471	11053870	147	311854	22921300
98.5	93822.8	4620780	123	182690	11235450	147.5	315047	23234750
99	95258.9	4715320	123.5	184927	11419250	148	318262	23551400
99.5	96709.5	4811300	124	187182	11605310	148.5	321499	23871280
100	98174.8	4908740	124.5	189456	11793620	149	324757	24194410
100.5	99654.8	5007650	125	191748	11984220	149.5	328038	24520800
101	101150	5108050	125.5	194058	12177130	150	331340	24850490
101.5	102659	5209960	126	196387	12372350	.....	....	....
102	104184	5313380	126.5	198734	12569910	.....	....	....
102.5	105724	5418330	127	201100	12769820	.....	....	....
103	107278	5524830	127.5	203484	12972110	.....	....	....

Strength and Stiffness of Perforated Metals. — It is common practice to use perforated metals in equipment enclosures to provide cooling by the flow of air or fluids. If the perforated material is to serve also as a structural member, then calculations of stiffness and strength must be made that take into account the effect of the perforations on the strength of the panels.

The accompanying table provides equivalent or effective values of the yield strength S^*; modulus of elasticity E^*; and Poisson's ratio v^* of perforated metals in terms of the values for solid material. The S^*/S and E^*/E ratios, given in the accompanying table for the standard round hole staggered pattern, can be used to determine the safety margins or deflections for perforated metal use as compared to the unperforated metal for any geometry or loading condition.

Perforated material has different strengths depending on the direction of loading; therefore, values of S^*/S in the table are given for the width (strongest) and length (weakest) directions. Also, the effective elastic constants are for plane stress conditions and apply to the in-plane loading of thin perforated sheets; the bending stiffness is greater. However, since most loading conditions involve a combination of bending and stretching, it is more convenient to use the same effective elastic constants for these combined loading conditions. The plane stress effective elastic constants given in the table can be conservatively used for all loading conditions.

Mechanical Properties of Materials Perforated with Round Holes in IPA[a] Standard Staggered Hole Pattern

IPA No.	Perforation Diam. (in.)	Center Distance (in.)	Open Area (%)	S^*/S^c Width (in.)	S^*/S^c Length (in.)	E^*/E^c	v^{*c}
100	0.020	(625)[b]	20	0.530	0.465	0.565	0.32
106	1/16	1/8	23	0.500	0.435	0.529	0.33
107	5/64	7/64	46	0.286	0.225	0.246	0.38
108	5/64	1/8	36	0.375	0.310	0.362	0.35
109	3/32	5/32	32	0.400	0.334	0.395	0.34
110	3/32	3/16	23	0.500	0.435	0.529	0.33
112	1/10	5/32	36	0.360	0.296	0.342	0.35
113	1/8	3/16	40	0.333	0.270	0.310	0.36
114	1/8	7/32	29	0.428	0.363	0.436	0.33
115	1/8	1/4	23	0.500	0.435	0.529	0.33
116	5/32	7/32	46	0.288	0.225	0.249	0.38
117	5/32	1/4	36	0.375	0.310	0.362	0.35
118	3/16	1/4	51	0.250	0.192	0.205	0.42
119	3/16	5/16	33	0.400	0.334	0.395	0.34
120	1/4	5/16	58	0.200	0.147	0.146	0.47
121	1/4	3/8	40	0.333	0.270	0.310	0.36
122	1/4	7/16	30	0.428	0.363	0.436	0.33
123	1/4	1/2	23	0.500	0.435	0.529	0.33
124	3/8	1/2	51	0.250	0.192	0.205	0.42
125	3/8	9/16	40	0.333	0.270	0.310	0.36
126	3/8	5/8	33	0.400	0.334	0.395	0.34
127	7/16	5/8	45	0.300	0.239	0.265	0.38
128	1/2	11/16	47	0.273	0.214	0.230	0.39
129	9/16	3/4	50	0.250	0.192	0.205	0.42

[a] Industrial Perforators Association. [b] Value in parentheses specifies holes per square inch instead of center distance. [c] S^*/S = ratio of yield strength of perforated to unperforated material; E^*/E = ratio of modulus of elasticity of perforated to unperforated material; v^* = Poisson's ratio for given percentage of open area.

Lengths of Angles Bent to Circular Shape. — To calculate the length of an angle-iron used either inside or outside of a tank or smokestack, the following table of constants may be used: Assume, for example, that a stand-pipe, 20 feet inside diameter, is provided with a 3 by 3 by ⅜ inch angle-iron on the inside at the top. The circumference of a circle 20 feet in diameter is 754 inches. From the table of constants, find the constant for a 3 by 3 by ⅜ inch angle-iron, which is 4.319. The length of the angle then is 754 − 4.319 = 749.681 inches. Should the angle be on the outside, add the constant instead of subtracting it; thus, 754 + 4.319 = 758.319 inches.

Size of Angle	Const.	Size of Angle	Const.	Size of Angle	Const.
¼ × 2 × 2	2.879	5⁄16 × 3 × 3	4.123	½ × 5 × 5	6.804
5⁄16 × 2 × 2	3.076	⅜ × 3 × 3	4.319	⅜ × 6 × 6	7.461
⅜ × 2 × 2	3.272	½ × 3 × 3	4.711	½ × 6 × 6	7.854
¼ × 2½ × 2½	3.403	⅜ × 3½ × 3½	4.843	¾ × 6 × 6	8.639
5⁄16 × 2½ × 2½	3.600	½ × 3½ × 3½	5.235	½ × 8 × 8	9.949
⅜ × 2½ × 2½	3.796	⅜ × 4 × 4	5.366	¾ × 8 × 8	10.734
½ × 2½ × 2½	4.188	½ × 4 × 4	5.758	1 × 8 × 8	11.520
¼ × 3 × 3	3.926	⅜ × 5 × 5	6.414	...	

Standard Designations of Rolled Steel Shapes. — Through a joint effort, the American Iron and Steel Institute (AISI) and the American Institute of Steel Construction (AISC) have changed most of the designations for their hot-rolled structural steel shapes. The present designations, standard for steel producing and fabricating industries, should be used when designing, detailing, and ordering steel. The accompanying table compares the present designations with the previous descriptions.

Hot-Rolled Structural Steel Shape Designations (AISI and AISC)*

Present Designation	Type of Shape	Previous Designation
W 24 × 76	W shape	24 WF 76
W 14 × 26	W shape	14 B 26
S 24 × 100	S shape	24 I 100
M 8 × 18.5	M shape	8 M 18.5
M 10 × 9	M shape	10 JR 9.0
M 8 × 34.3	M shape	8 × 8 M 34.3
C 12 × 20.7	American Standard Channel	12 [20.7
MC 12 × 45	Miscellaneous Channel	12 × 4 [45.0
MC 12 × 10.6	Miscellaneous Channel	12 JR [10.6
HP 14 × 73	HP shape	14 BP 73
L 6 × 6 × ¾	Equal Leg Angle	∠6 × 6 × ¾
L 6 × 4 × ⅝	Unequal Leg Angle	∠6 × 4 × ⅝
WT 12 × 38	Structural Tee cut from W shape	ST 12 WF 38
WT 7 × 13	Structural Tee cut from W shape	ST 7 B 13
ST 12 × 50	Structural Tee cut from S shape	ST 12 I 50
MT 4 × 9.25	Structural Tee cut from M shape	ST 4 M 9.25
MT 5 × 4.5	Structural Tee cut from M shape	ST 5 JR 4.5
MT 4 × 17.15	Structural Tee cut from M shape	ST 4 M 17.15
PL ½ × 18	Plate	PL 18 × ½
Bar 1 ⊡	Square Bar	Bar 1 ⊡
Bar 1¼ φ	Round Bar	Bar 1¼ φ
Bar 2½ × ½	Flat Bar	Bar 2½ × ½
Pipe 4 Std.	Pipe	Pipe 4 Std.
Pipe 4 X–Strong	Pipe	Pipe 4 X–Strong
Pipe 4 XX–Strong	Pipe	Pipe 4 XX–Strong
TS 4 × 4 × .375	Structural Tubing: Square	Tube 4 × 4 × .375
TS 5 × 3 × .375	Structural Tubing: Rectangular	Tube 5 × 3 × .375
TS 3 OD × .250	Structural Tubing: Circular	Tube 3 OD × .250

* Data taken from the "Manual of Steel Construction," 8th Edition, 1980, with permission of the American Institute of Steel Construction.

Steel Wide-Flange Sections — I*

Wide-flange sections are designated, in order, by a section letter, nominal depth of the member in inches, and the nominal weight in pounds per foot; thus:

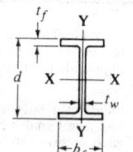

$$W\ 18 \times 64$$

indicates a wide-flange section having a nominal depth of 18 inches, and a nominal weight per foot of 64 pounds. Actual geometry for each section can be obtained from the values below.

Designation	Area, A	Depth, d	Flange Width, b_f	Flange Thickness, t_f	Web Thickness, t_w	Axis X-X I	Axis X-X S	Axis X-X r	Axis Y-Y I	Axis Y-Y S	Axis Y-Y r
	in.²	in.	in.	in.	in.	in.⁴	in.³	in.	in.⁴	in.³	in.
†W 27 × 178	52.3	27.81	14.085	1.190	0.725	6990	502	11.6	555	78.8	3.26
× 161	47.4	27.59	14.020	1.080	0.660	6280	455	11.5	497	70.9	3.24
× 146	42.9	27.38	13.965	0.975	0.605	5630	411	11.4	443	63.5	3.21
× 114	33.5	27.29	10.070	0.930	0.570	4090	299	11.0	159	31.5	2.18
× 102	30.0	27.09	10.015	0.830	0.515	3620	267	11.0	139	27.8	2.15
× 94	27.7	26.92	9.990	0.745	0.490	3270	243	10.9	124	24.8	2.12
× 84	24.8	26.71	9.960	0.640	0.460	2850	213	10.7	106	21.2	2.07
W 24 × 162	47.7	25.00	12.955	1.220	0.705	5170	414	10.4	443	68.4	3.05
× 146	43.0	24.74	12.900	1.090	0.650	4580	371	10.3	391	60.5	3.01
× 131	38.5	24.48	12.855	0.960	0.605	4020	329	10.2	340	53.0	2.97
× 117	34.4	24.26	12.800	0.850	0.550	3540	291	10.1	297	46.5	2.94
× 104	30.6	24.06	12.750	0.750	0.500	3100	258	10.1	259	40.7	2.91
× 94	27.7	24.31	9.065	0.875	0.515	2700	222	9.87	109	24.0	1.98
× 84	24.7	24.10	9.020	0.770	0.470	2370	196	9.79	94.4	20.9	1.95
× 76	22.4	23.92	8.990	0.680	0.440	2100	176	9.69	82.5	18.4	1.92
× 68	20.1	23.73	8.965	0.585	0.415	1830	154	9.55	70.4	15.7	1.87
× 62	18.2	23.74	7.040	0.590	0.430	1550	131	9.23	34.5	9.80	1.38
× 55	16.2	23.57	7.005	0.505	0.395	1350	114	9.11	29.1	8.30	1.34
W 21 × 147	43.2	22.06	12.510	1.150	0.720	3630	329	9.17	376	60.1	2.95
× 132	38.8	21.83	12.440	1.035	0.650	3220	295	9.12	333	53.5	2.93
× 122	35.9	21.68	12.390	0.960	0.600	2960	273	9.09	305	49.2	2.92
× 111	32.7	21.51	12.340	0.875	0.550	2670	249	9.05	274	44.5	2.90
× 101	29.8	21.36	12.290	0.800	0.500	2420	227	9.02	248	40.3	2.89
× 93	27.3	21.62	8.420	0.930	0.580	2070	192	8.70	92.9	22.1	1.84
× 83	24.3	21.43	8.355	0.835	0.515	1830	171	8.67	81.4	19.5	1.83
× 73	21.5	21.24	8.295	0.740	0.455	1600	151	8.64	70.6	17.0	1.81
× 68	20.0	21.13	8.270	0.685	0.430	1480	140	8.60	64.7	15.7	1.80
× 62	18.3	20.99	8.240	0.615	0.400	1330	127	8.54	57.5	13.9	1.77
× 57	16.7	21.06	6.555	0.650	0.405	1170	111	8.36	30.6	9.35	1.35
× 50	14.7	20.83	6.530	0.535	0.380	984	94.5	8.18	24.9	7.64	1.30
× 44	13.0	20.66	6.500	0.450	0.350	843	81.6	8.06	20.7	6.36	1.26
W 18 × 119	35.1	18.97	11.265	1.060	0.655	2190	231	7.90	253	44.9	2.69
× 106	31.1	18.73	11.200	0.940	0.590	1910	204	7.84	220	39.4	2.66
× 97	28.5	18.59	11.145	0.870	0.535	1750	188	7.82	201	36.1	2.65
× 86	25.3	18.39	11.090	0.770	0.480	1530	166	7.77	175	31.6	2.63
× 76	22.3	18.21	11.035	0.680	0.425	1330	146	7.73	152	27.6	2.61
× 71	20.8	18.47	7.635	0.810	0.495	1170	127	7.50	60.3	15.8	1.70
× 65	19.1	18.35	7.590	0.750	0.450	1070	117	7.49	54.8	14.4	1.69
× 60	17.6	18.24	7.555	0.695	0.415	984	108	7.47	50.1	13.3	1.69
× 55	16.2	18.11	7.530	0.630	0.390	890	98.3	7.41	44.9	11.9	1.67
× 50	14.7	17.99	7.495	0.570	0.355	800	88.9	7.38	40.1	10.7	1.65
× 46	13.5	18.06	6.060	0.605	0.360	712	78.8	7.25	22.5	7.43	1.29
× 40	11.8	17.90	6.015	0.525	0.315	612	68.4	7.21	19.1	6.35	1.27
× 35	10.3	17.70	6.000	0.425	0.300	510	57.6	7.04	15.3	5.12	1.22

* Data taken from the "Manual of Steel Construction," 8th Edition, 1980, with permission of the American Institute of Steel Construction. † Consult the AISC Manual, noted above, for W steel shapes having nominal depths greater than 27 in.

Symbols: I = moment of inertia; *S* = section modulus; *r* = radius of gyration.

Steel Wide-Flange Sections — 2*

Wide-flange sections are designated, in order, by a section letter, nominal depth of the member in inches, and the nominal weight in pounds per foot; thus:

W 16 × 78

indicates a wide-flange section having a nominal depth of 16 inches, and a nominal weight per foot of 78 pounds. Actual geometry for each section can be obtained from the values below.

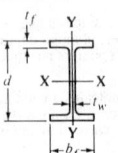

Designation	Area, A	Depth, d	Flange Width, b_f	Flange Thickness, t_f	Web Thickness, t_w	Axis X-X I	Axis X-X S	Axis X-X r	Axis Y-Y I	Axis Y-Y S	Axis Y-Y r
	in.^{2}	in.	in.	in.	in.	in.^{4}	in.^{3}	in.	in.^{4}	in.^{3}	in.
W 16 × 100	29.4	16.97	10.425	0.985	0.585	1490	175	7.10	186	35.7	2.51
× 89	26.2	16.75	10.365	0.875	0.525	1300	155	7.05	163	31.4	2.49
× 77	22.6	16.52	10.295	0.760	0.455	1110	134	7.00	138	26.9	2.47
× 67	19.7	16.33	10.235	0.665	0.395	954	117	6.96	119	23.2	2.46
× 57	16.8	16.43	7.120	0.715	0.430	758	92.2	6.72	43.1	12.1	1.60
× 50	14.7	16.26	7.070	0.630	0.380	659	81.0	6.68	37.2	10.5	1.59
× 45	13.3	16.13	7.035	0.565	0.345	586	72.7	6.65	32.8	9.34	1.57
× 40	11.8	16.01	6.995	0.505	0.305	518	64.7	6.63	28.9	8.25	1.57
× 36	10.6	15.86	6.985	0.430	0.295	448	56.5	6.51	24.5	7.00	1.52
× 31	9.12	15.88	5.525	0.440	0.275	375	47.2	6.41	12.4	4.49	1.17
× 26	7.68	15.69	5.500	0.345	0.250	301	38.4	6.26	9.59	3.49	1.12
W 14 × 730	215.0	22.42	17.890	4.910	3.070	14300	1280	8.17	4720	527	4.69
× 665	196.0	21.64	17.650	4.520	2.830	12400	1150	7.98	4170	472	4.62
× 605	178.0	20.92	17.415	4.160	2.595	10800	1040	7.80	3680	423	4.55
× 550	162.0	20.24	17.200	3.820	2.380	9430	931	7.63	3250	378	4.49
× 500	147.0	19.60	17.010	3.500	2.190	8210	838	7.48	2880	339	4.43
× 455	134.0	19.02	16.835	3.210	2.015	7190	756	7.33	2560	304	4.38
× 426	125.0	18.67	16.695	3.035	1.875	6600	707	7.26	2360	283	4.34
× 398	117.0	18.29	16.590	2.845	1.770	6000	656	7.16	2170	262	4.31
× 370	109.0	17.92	16.475	2.660	1.655	5440	607	7.07	1990	241	4.27
× 342	101.0	17.54	16.360	2.470	1.540	4900	559	6.98	1810	221	4.24
× 311	91.4	17.12	16.230	2.260	1.410	4330	506	6.88	1610	199	4.20
× 283	83.3	16.74	16.110	2.070	1.290	3840	459	6.79	1440	179	4.17
× 257	75.6	16.38	15.995	1.890	1.175	3400	415	6.71	1290	161	4.13
× 233	68.5	16.04	15.890	1.720	1.070	3010	375	6.63	1150	145	4.10
× 211	62.0	15.72	15.800	1.560	0.980	2660	338	6.55	1030	130	4.07
× 193	56.8	15.48	15.710	1.440	0.890	2400	310	6.50	931	119	4.05
× 176	51.8	15.22	15.650	1.310	0.830	2140	281	6.43	838	107	4.02
× 159	46.7	14.98	15.565	1.190	0.745	1900	254	6.38	748	96.2	4.00
× 145	42.7	14.78	15.500	1.090	0.680	1710	232	6.33	677	87.3	3.98

* Data taken from the "Manual of Steel Construction," 8th Edition, 1980, with permission of the American Institute of Steel Construction.

Symbols: I = moment of inertia; S = section modulus; r = radius of gyration.

Steel Wide-Flange Sections — 3*

Wide-flange sections are designated, in order, by a section letter, nominal depth of the member in inches, and the nominal weight in pounds per foot; thus:

$$W\ 14 \times 38$$

indicates a wide-flange section having a nominal depth of 14 inches, and a nominal weight per foot of 38 pounds. Actual geometry for each section can be obtained from the values below.

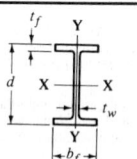

Designation	Area, A	Depth, d	Flange Width, b_f	Flange Thickness, t_f	Web Thickness, t_w	Axis X-X I	S	r	Axis Y-Y I	S	r
	in.²	in.	in.	in.	in.	in.⁴	in.³	in.	in.⁴	in.³	in.
W 14 × 132	38.8	14.66	14.725	1.030	0.645	1530	209	6.28	548	74.5	3.76
× 120	35.3	14.48	14.670	0.940	0.590	1380	190	6.24	495	67.5	3.74
× 109	32.0	14.32	14.605	0.860	0.525	1240	173	6.22	447	61.2	3.73
× 99	29.1	14.16	14.565	0.780	0.485	1110	157	6.17	402	55.2	3.71
× 90	26.5	14.02	14.520	0.710	0.440	999	143	6.14	362	49.9	3.70
× 82	24.1	14.31	10.130	0.855	0.510	882	123	6.05	148	29.3	2.48
× 74	21.8	14.17	10.070	0.785	0.450	796	112	6.04	134	26.6	2.48
× 68	20.0	14.04	10.035	0.720	0.415	723	103	6.01	121	24.2	2.46
× 61	17.9	13.89	9.995	0.645	0.375	640	92.2	5.98	107	21.5	2.45
× 53	15.6	13.92	8.060	0.660	0.370	541	77.8	5.89	57.7	14.3	1.92
× 48	14.1	13.79	8.030	0.595	0.340	485	70.3	5.85	51.4	12.8	1.91
× 43	12.6	13.66	7.995	0.530	0.305	428	62.7	5.82	45.2	11.3	1.89
× 38	11.2	14.10	6.770	0.515	0.310	385	54.6	5.87	26.7	7.88	1.55
× 34	10.0	13.98	6.745	0.455	0.285	340	48.6	5.83	23.3	6.91	1.53
× 30	8.85	13.84	6.730	0.385	0.270	291	42.0	5.73	19.6	5.82	1.49
× 26	7.69	13.91	5.025	0.420	0.255	245	35.3	5.65	8.91	3.54	1.08
× 22	6.49	13.74	5.000	0.335	0.230	199	29.0	5.54	7.00	2.80	1.04
W 12 × 336	98.8	16.82	13.385	2.955	1.775	4060	483	6.41	1190	177	3.47
× 305	89.6	16.32	13.235	2.705	1.625	3550	435	6.29	1050	159	3.42
× 279	81.9	15.85	13.140	2.470	1.530	3110	393	6.16	937	143	3.38
× 252	74.1	15.41	13.005	2.250	1.395	2720	353	6.06	828	127	3.34
× 230	67.7	15.05	12.895	2.070	1.285	2420	321	5.97	742	115	3.31
× 210	61.8	14.71	12.790	1.900	1.180	2140	292	5.89	664	104	3.28
× 190	55.8	14.38	12.670	1.735	1.060	1890	263	5.82	589	93.0	3.25
× 170	50.0	14.03	12.570	1.560	0.960	1650	235	5.74	517	82.3	3.22
× 152	44.7	13.71	12.480	1.400	0.870	1430	209	5.66	454	72.8	3.19
× 136	39.9	13.41	12.400	1.250	0.790	1240	186	5.58	398	64.2	3.16
× 120	35.3	13.12	12.320	1.105	0.710	1070	163	5.51	345	56.0	3.13
× 106	31.2	12.89	12.220	0.990	0.610	933	145	5.47	301	49.3	3.11
× 96	28.2	12.71	12.160	0.900	0.550	833	131	5.44	270	44.4	3.09
× 87	25.6	12.53	12.125	0.810	0.515	740	118	5.38	241	39.7	3.07
× 79	23.2	12.38	12.080	0.735	0.470	662	107	5.34	216	35.8	3.05
× 72	21.1	12.25	12.040	0.670	0.430	597	97.4	5.31	195	32.4	3.04
× 65	19.1	12.12	12.000	0.605	0.390	533	87.9	5.28	174	29.1	3.02
× 58	17.0	12.19	10.010	0.640	0.360	475	78.0	5.28	107	21.4	2.51
× 53	15.6	12.06	9.995	0.575	0.345	425	70.6	5.23	95.8	19.2	2.48
× 50	14.7	12.19	8.080	0.640	0.370	394	64.7	5.18	56.3	13.9	1.96
× 45	13.2	12.06	8.045	0.575	0.335	350	58.1	5.15	50.0	12.4	1.94
× 40	11.8	11.94	8.005	0.515	0.295	310	51.9	5.13	44.1	11.0	1.93
× 35	10.3	12.50	6.560	0.520	0.300	285	45.6	5.25	24.5	7.47	1.54
× 30	8.79	12.34	6.520	0.440	0.260	238	38.6	5.21	20.3	6.24	1.52
× 26	7.65	12.22	6.490	0.380	0.230	204	33.4	5.17	17.3	5.34	1.51
× 22	6.48	12.31	4.030	0.425	0.260	156	25.4	4.91	4.66	2.31	0.847
× 19	5.57	12.16	4.005	0.350	0.235	130	21.3	4.82	3.76	1.88	0.822
× 16	4.71	11.99	3.990	0.265	0.220	103	17.1	4.67	2.82	1.41	0.773
× 14	4.16	11.91	3.970	0.225	0.200	88.6	14.9	4.62	2.36	1.19	0.753

* Data taken from the "Manual of Steel Construction," 8th Edition, 1980, with permission of the American Institute of Steel Construction.

Steel Wide-Flange Sections — 4*

Wide-flange sections are designated, in order, by a section letter, nominal depth of the member in inches, and the nominal weight in pounds per foot; thus:

$$W\ 8 \times 67$$

indicates a wide-flange section having a nominal depth of 8 inches, and a nominal weight per foot of 67 pounds. Actual geometry for each section can be obtained from the values below.

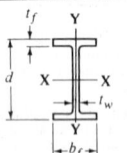

Designation	Area, A	Depth, d	Flange Width, b_f	Flange Thickness, t_f	Web Thickness, t_w	Axis X-X I	Axis X-X S	Axis X-X r	Axis Y-Y I	Axis Y-Y S	Axis Y-Y r
	in.²	in.	in.	in.	in.	in.⁴	in.³	in.	in.⁴	in.³	in.
W10 × 112	32.9	11.36	10.415	1.250	0.755	716	126	4.66	236	45.3	2.68
× 100	29.4	11.10	10.340	1.120	0.680	623	112	4.60	207	40.0	2.65
× 88	25.9	10.84	10.265	0.990	0.605	534	98.5	4.54	179	34.8	2.63
× 77	22.6	10.60	10.190	0.870	0.530	455	85.9	4.49	154	30.1	2.60
× 68	20.0	10.40	10.130	0.770	0.470	394	75.7	4.44	134	26.4	2.59
× 60	17.6	10.22	10.080	0.680	0.420	341	66.7	4.39	116	23.0	2.57
× 54	15.8	10.09	10.030	0.615	0.370	303	60.0	4.37	103	20.6	2.56
× 49	14.4	9.98	10.000	0.560	0.340	272	54.6	4.35	93.4	18.7	2.54
× 45	13.3	10.10	8.020	0.620	0.350	248	49.1	4.32	53.4	13.3	2.01
× 39	11.5	9.92	7.985	0.530	0.315	209	42.1	4.27	45.0	11.3	1.98
× 33	9.71	9.73	7.960	0.435	0.290	170	35.0	4.19	36.6	9.20	1.94
× 30	8.84	10.47	5.810	0.510	0.300	170	32.4	4.38	16.7	5.75	1.37
× 26	7.61	10.33	5.770	0.440	0.260	144	27.9	4.35	14.1	4.89	1.36
× 22	6.49	10.17	5.750	0.360	0.240	118	23.2	4.27	11.4	3.97	1.33
× 19	5.62	10.24	4.020	0.395	0.250	96.3	18.8	4.14	4.29	2.14	0.874
× 17	4.99	10.11	4.010	0.330	0.240	81.9	16.2	4.05	3.56	1.78	0.844
× 15	4.41	9.99	4.000	0.270	0.230	68.9	13.8	3.95	2.89	1.45	0.810
× 12	3.54	9.87	3.960	0.210	0.190	53.8	10.9	3.90	2.18	1.10	0.785
W 8 × 67	19.7	9.00	8.280	0.935	0.570	272	60.4	3.72	88.6	21.4	2.12
× 58	17.1	8.75	8.220	0.810	0.510	228	52.0	3.65	75.1	18.3	2.10
× 48	14.1	8.50	8.110	0.685	0.400	184	43.3	3.61	60.9	15.0	2.08
× 40	11.7	8.25	8.070	0.560	0.360	146	35.5	3.53	49.1	12.2	2.04
× 35	10.3	8.12	8.020	0.495	0.310	127	31.2	3.51	42.6	10.6	2.03
× 31	9.13	8.00	7.995	0.435	0.285	110	27.5	3.47	37.1	9.27	2.02
× 28	8.25	8.06	6.535	0.465	0.285	98.0	24.3	3.45	21.7	6.63	1.62
× 24	7.08	7.93	6.495	0.400	0.245	82.8	20.9	3.42	18.3	5.63	1.61
× 21	6.16	8.28	5.270	0.400	0.250	75.3	18.2	3.49	9.77	3.71	1.26
× 18	5.26	8.14	5.250	0.330	0.230	61.9	15.2	3.43	7.97	3.04	1.23
× 15	4.44	8.11	4.015	0.315	0.245	48.0	11.8	3.29	3.41	1.70	0.876
× 13	3.84	7.99	4.000	0.255	0.230	39.6	9.91	3.21	2.73	1.37	0.843
× 10	2.96	7.89	3.940	0.205	0.170	30.8	7.81	3.22	2.09	1.06	0.841
W 6 × 25	7.34	6.38	6.080	0.455	0.320	53.4	16.7	2.70	17.1	5.61	1.52
× 20	5.87	6.20	6.020	0.365	0.260	41.4	13.4	2.66	13.3	4.41	1.50
× 16	4.74	6.28	4.030	0.405	0.260	32.1	10.2	2.60	4.43	2.20	0.966
× 15	4.43	5.99	5.990	0.260	0.230	29.1	9.72	2.56	9.32	3.11	1.46
× 12	3.55	6.03	4.000	0.280	0.230	22.1	7.31	2.49	2.99	1.50	0.918
× 9	2.68	5.90	3.940	0.215	0.170	16.4	5.56	2.47	2.19	1.11	0.905
W 5 × 19	5.54	5.15	5.030	0.430	0.270	26.2	10.2	2.17	9.13	3.63	1.28
× 16	4.68	5.01	5.000	0.360	0.240	21.3	8.51	2.13	7.51	3.00	1.27
W 4 × 13	3.83	4.16	4.060	0.345	0.280	11.3	5.46	1.72	3.86	1.90	1.00

* Data taken from the "Manual of Steel Construction," 8th Edition, 1980, with permission of the American Institute of Steel Construction.
 Symbols: I = moment of inertia; S = section modulus; r = radius of gyration.

Steel S Sections*

"S" is the section symbol for "I" Beams. S shapes are designated, in order, by their section letter, actual depth in inches, and nominal weight in pounds per foot. Thus:

$$S\ 5 \times 14.75$$

indicates an S shape (or I beam) having a depth of 5 inches and a nominal weight of 14.75 pounds per foot.

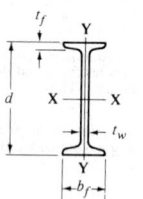

Designation	Area, A in.²	Depth, d in.	Flange Width, b_f in.	Flange Thickness, t_f in.	Web Thickness, t_w in.	Axis X-X I in.⁴	Axis X-X S in.³	Axis X-X r in.	Axis Y-Y I in.⁴	Axis Y-Y S in.³	Axis Y-Y r in.
S 24 × 121	35.6	24.50	8.050	1.090	0.800	3160	258	9.43	83.3	20.7	1.53
× 106	31.2	24.50	7.870	1.090	0.620	2940	240	9.71	77.1	19.6	1.57
× 100	29.3	24.00	7.245	0.870	0.745	2390	199	9.02	47.7	13.2	1.27
× 90	26.5	24.00	7.125	0.870	0.625	2250	187	9.21	44.9	12.6	1.30
× 80	23.5	24.00	7.000	0.870	0.500	2100	175	9.47	42.2	12.1	1.34
S 20 × 96	28.2	20.30	7.200	0.920	0.800	1670	165	7.71	50.2	13.9	1.33
× 86	25.3	20.30	7.060	0.920	0.660	1580	155	7.89	46.8	13.3	1.36
× 75	22.0	20.00	6.385	0.795	0.635	1280	128	7.62	29.8	9.32	1.16
× 66	19.4	20.00	6.255	0.795	0.505	1190	119	7.83	27.7	8.85	1.19
S 18 × 70	20.6	18.00	6.251	0.691	0.711	926	103	6.71	24.1	7.72	1.08
× 54.7	16.1	18.00	6.001	0.691	0.461	804	89.4	7.07	20.8	6.94	1.14
S 15 × 50	14.7	15.00	5.640	0.622	0.550	486	64.8	5.75	15.7	5.57	1.03
× 42.9	12.6	15.00	5.501	0.622	0.411	447	59.6	5.95	14.4	5.23	1.07
S 12 × 50	14.7	12.00	5.477	0.659	0.687	305	50.8	4.55	15.7	5.74	1.03
× 40.8	12.0	12.00	5.252	0.659	0.462	272	45.4	4.77	13.6	5.16	1.06
× 35	10.3	12.00	5.078	0.544	0.428	229	38.2	4.72	9.87	3.89	0.980
× 31.8	9.35	12.00	5.000	0.544	0.350	218	36.4	4.83	9.36	3.74	1.00
S 10 × 35	10.3	10.00	4.944	0.491	0.594	147	29.4	3.78	8.36	3.38	0.901
× 25.4	7.46	10.00	4.661	0.491	0.311	124	24.7	4.07	6.79	2.91	0.954
S 8 × 23	6.77	8.00	4.171	0.426	0.441	64.9	16.2	3.10	4.31	2.07	0.798
× 18.4	5.41	8.00	4.001	0.426	0.271	57.6	14.4	3.26	3.73	1.86	0.831
S 7 × 20	5.88	7.00	3.860	0.392	0.450	42.4	12.1	2.69	3.17	1.64	0.734
× 15.3	4.50	7.00	3.662	0.392	0.252	36.7	10.5	2.86	2.64	1.44	0.766
S 6 × 17.25	5.07	6.00	3.565	0.359	0.465	26.3	8.77	2.28	2.31	1.30	0.675
× 12.5	3.67	6.00	3.332	0.359	0.232	22.1	7.37	2.45	1.82	1.09	0.705
S 5 × 14.75	4.34	5.00	3.284	0.326	0.494	15.2	6.09	1.87	1.67	1.01	0.620
× 10	2.94	5.00	3.004	0.326	0.214	12.3	4.92	2.05	1.22	0.809	0.643
S 4 × 9.5	2.79	4.00	2.796	0.293	0.326	6.79	3.39	1.56	0.903	0.646	0.569
× 7.7	2.26	4.00	2.663	0.293	0.193	6.08	3.04	1.64	0.764	0.574	0.581
S 3 × 7.5	2.21	3.00	2.509	0.260	0.349	2.93	1.95	1.15	0.586	0.468	0.516
× 5.7	1.67	3.00	2.330	0.260	0.170	2.52	1.68	1.23	0.455	0.390	0.522

* Data taken from the "Manual of Steel Construction," 8th Edition, 1980, with permission of the American Institute of Steel Construction.

American Standard Steel Channels*

American Standard Channels are designated, in order, by a section letter, actual depth in inches, and nominal weight per foot in pounds. Thus:

$$C\ 7 \times 14.75$$

indicates an American Standard Channel with a depth of 7 inches and a nominal weight of 14.75 pounds per foot.

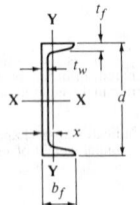

Designation	Area, A	Depth, d	Flange Width, b_f	Flange Thickness, t_f	Web Thickness, t_w	Axis X-X I	S	r	Axis Y-Y I	S	r	x
	in.²	in.	in.	in.	in.	in.⁴	in.³	in.	in.⁴	in.³	in.	in.
C 15 × 50	14.7	15.00	3.716	0.650	0.716	404	53.8	5.24	11.0	3.78	0.867	0.798
× 40	11.8	15.00	3.520	0.650	0.520	349	46.5	5.44	9.23	3.37	0.886	0.777
× 33.9	9.96	15.00	3.400	0.650	0.400	315	42.0	5.62	8.13	3.11	0.904	0.787
C 12 × 30	8.82	12.00	3.170	0.501	0.510	162	27.0	4.29	5.14	2.06	0.763	0.674
× 25	7.35	12.00	3.047	0.501	0.387	144	24.1	4.43	4.47	1.88	0.780	0.674
× 20.7	6.09	12.00	2.942	0.501	0.282	129	21.5	4.61	3.88	1.73	0.799	0.698
C 10 × 30	8.82	10.00	3.033	0.436	0.673	103	20.7	3.42	3.94	1.65	0.669	0.649
× 25	7.35	10.00	2.886	0.436	0.526	91.2	18.2	3.52	3.36	1.48	0.676	0.617
× 20	5.88	10.00	2.739	0.436	0.379	78.9	15.8	3.66	2.81	1.32	0.692	0.606
× 15.3	4.49	10.00	2.600	0.436	0.240	67.4	13.5	3.87	2.28	1.16	0.713	0.634
C 9 × 20	5.88	9.00	2.648	0.413	0.448	60.9	13.5	3.22	2.42	1.17	0.642	0.583
× 15	4.41	9.00	2.485	0.413	0.285	51.0	11.3	3.40	1.93	1.01	0.661	0.586
× 13.4	3.94	9.00	2.433	0.413	0.233	47.9	10.6	3.48	1.76	0.962	0.669	0.601
C 8 × 18.75	5.51	8.00	2.527	0.390	0.487	44.0	11.0	2.82	1.98	1.01	0.599	0.565
× 13.75	4.04	8.00	2.343	0.390	0.303	36.1	9.03	2.99	1.53	0.854	0.615	0.553
× 11.5	3.38	8.00	2.260	0.390	0.220	32.6	8.14	3.11	1.32	0.781	0.625	0.571
C 7 × 14.75	4.33	7.00	2.299	0.366	0.419	27.2	7.78	2.51	1.38	0.779	0.564	0.532
× 12.25	3.60	7.00	2.194	0.366	0.314	24.2	6.93	2.60	1.17	0.703	0.571	0.525
× 9.8	2.87	7.00	2.090	0.366	0.210	21.3	6.08	2.72	0.968	0.625	0.581	0.540
C 6 × 13	3.83	6.00	2.157	0.343	0.437	17.4	5.80	2.13	1.05	0.642	0.525	0.514
× 10.5	3.09	6.00	2.034	0.343	0.314	15.2	5.06	2.22	0.866	0.564	0.529	0.499
× 8.2	2.40	6.00	1.920	0.343	0.200	13.1	4.38	2.34	0.693	0.492	0.537	0.511
C 5 × 9	2.64	5.00	1.885	0.320	0.325	8.90	3.56	1.83	0.632	0.450	0.489	0.478
× 6.7	1.97	5.00	1.750	0.320	0.190	7.49	3.00	1.95	0.479	0.378	0.493	0.484
C 4 × 7.25	2.13	4.00	1.721	0.296	0.321	4.59	2.29	1.47	0.433	0.343	0.450	0.459
× 5.4	1.59	4.00	1.584	0.296	0.184	3.85	1.93	1.56	0.319	0.283	0.449	0.457
C 3 × 6	1.76	3.00	1.596	0.273	0.356	2.07	1.38	1.08	0.305	0.268	0.416	0.455
× 5	1.47	3.00	1.498	0.273	0.258	1.85	1.24	1.12	0.247	0.233	0.410	0.439
× 4.1	1.21	3.00	1.410	0.273	0.170	1.66	1.10	1.17	0.197	0.202	0.404	0.436

* Data taken from the "Manual of Steel Construction," 8th Edition, 1980, with permission of the American Institute of Steel Construction.

Symbols: I = moment of inertia; S = section modulus; r = radius of gyration; x = distance from center of gravity of section to outer face of structural shape.

Steel Angles with Equal Legs†

These angles are commonly designated by section symbol, width of each leg, and thickness, thus:

L 3 × 3 × ¼

indicates a 3 x 3-inch angle of ¼-inch thickness.

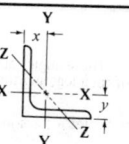

Size	Thickness	Weight per Foot	Area	Axis X-X & Y-Y			Z-Z
				I	r	x or y	r
in.	in.	lb.	in.^{2}	in.^{4}	in.	in.	in.
8 × 8	1⅛	56.9	16.7	98.0	2.42	2.41	1.56
	1	51.0	15.0	89.0	2.44	2.37	1.56
	⅞	45.0	13.2	79.6	2.45	2.32	1.57
	¾	38.9	11.4	69.7	2.47	2.28	1.58
	⅝	32.7	9.61	59.4	2.49	2.23	1.58
	9/16	29.6	8.68	54.1	2.50	2.21	1.59
	½	26.4	7.75	48.6	2.50	2.19	1.59
6 × 6	1	37.4	11.00	35.5	1.80	1.86	1.17
	⅞	33.1	9.73	31.9	1.81	1.82	1.17
	¾	28.7	8.44	28.2	1.83	1.78	1.17
	⅝	24.2	7.11	24.2	1.84	1.73	1.18
	9/16	21.9	6.43	22.1	1.85	1.71	1.18
	½	19.6	5.75	19.9	1.86	1.68	1.18
	7/16	17.2	5.06	17.7	1.87	1.66	1.19
	⅜	14.9	4.36	15.4	1.88	1.64	1.19
	5/16	12.4	3.65	13.0	1.89	1.62	1.20
5 × 5	⅞	27.2	7.98	17.8	1.49	1.57	.973
	¾	23.6	6.94	15.7	1.51	1.52	.975
	⅝	20.0	5.86	13.6	1.52	1.48	.978
	½	16.2	4.75	11.3	1.54	1.43	.983
	7/16	14.3	4.18	10.0	1.55	1.41	.986
	⅜	12.3	3.61	8.74	1.56	1.39	.990
	5/16	10.3	3.03	7.42	1.57	1.37	.994
4 × 4	¾	18.5	5.44	7.67	1.19	1.27	.778
	⅝	15.7	4.61	6.66	1.20	1.23	.779
	½	12.8	3.75	5.56	1.22	1.18	.782
	7/16	11.3	3.31	4.97	1.23	1.16	.785
	⅜	9.8	2.86	4.36	1.23	1.14	.788
	5/16	8.2	2.40	3.71	1.24	1.12	.791
	¼	6.6	1.94	3.04	1.25	1.09	.795
3½ × 3½	½	11.1	3.25	3.64	1.06	1.06	.683
	7/16	9.8	2.87	3.26	1.07	1.04	.684
	⅜	8.5	2.48	2.87	1.07	1.01	.687
	5/16	7.2	2.09	2.45	1.08	.990	.690
	¼	5.8	1.69	2.01	1.09	.968	.694
3 × 3	½	9.4	2.75	2.22	.898	.932	.584
	7/16	8.3	2.43	1.99	.905	.910	.585
	⅜	7.2	2.11	1.76	.913	.888	.587
	5/16	6.1	1.78	1.51	.922	.865	.589
	¼	4.9	1.44	1.24	.930	.842	.592
	3/16	3.71	1.09	.962	.939	.820	.596
2½ × 2½	½	7.7	2.25	1.23	.739	.806	.487
	⅜	5.9	1.73	.984	.753	.762	.487
	5/16	5.0	1.46	.849	.761	.740	.489
	¼	4.1	1.19	.703	.769	.717	.491
	3/16	3.07	.902	.547	.778	.694	.495
2 × 2	⅜	4.7	1.36	.479	.594	.636	.389
	5/16	3.92	1.15	.416	.601	.614	.390
	¼	3.19	.938	.348	.609	.592	.391
	3/16	2.44	.715	.272	.617	.569	.394
	⅛	1.65	.484	.190	.626	.546	.398

† Data taken from the "Manual of Steel Construction," 8th Edition, 1980, with permission of the American Institute of Steel Construction. I = moment of inertia; r = radius of gyration.

Steel Angles with Unequal Legs†

These angles are commonly designated by section symbol, width of each leg, and thickness, thus:

L 7 × 4 × ½

indicates a 7 × 4-inch angle of ½-inch thickness.

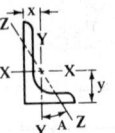

Size	Thick-ness	Weight per Ft.	Area	Axis X-X				Axis Y-Y				Axis Z-Z	Tan A
				I	S	r	y	I	S	r	x	r	
in.	in.	lb.	in.²	in.⁴	in.³	in.	in.	in.⁴	in.³	in.	in.	in.	
9 × 4	⅝	26.3	7.73	64.9	11.5	2.90	3.36	8.32	2.65	1.04	.858	.847	.216
	9/16	23.8	7.00	59.1	10.4	2.91	3.33	7.63	2.41	1.04	.834	.850	.218
	½	21.3	6.25	53.2	9.34	2.92	3.31	6.92	2.17	1.05	.810	.854	.220
8 × 6	1	44.2	13.0	80.8	15.1	2.49	2.65	38.8	8.92	1.73	1.65	1.28	.543
	⅞	39.1	11.5	72.3	13.4	2.51	2.61	34.9	7.94	1.74	1.61	1.28	.547
	¾	33.8	9.94	63.4	11.7	2.53	2.56	30.7	6.92	1.76	1.56	1.29	.551
	⅝	28.5	8.36	54.1	9.87	2.54	2.52	26.3	5.88	1.77	1.52	1.29	.554
	9/16	25.7	7.56	49.3	8.95	2.55	2.50	24.0	5.34	1.78	1.50	1.30	.556
	½	23.0	6.75	44.3	8.02	2.56	2.47	21.7	4.79	1.79	1.47	1.30	.558
	7/16	20.2	5.93	39.2	7.07	2.57	2.45	19.3	4.23	1.80	1.45	1.31	.560
8 × 4	1	37.4	11.0	69.6	14.1	2.52	2.95	11.6	3.94	1.03	1.05	.846	.247
	¾	28.7	8.44	54.9	10.9	2.55	2.95	9.36	3.07	1.05	.953	.852	.258
	9/16	21.9	6.43	42.8	8.35	2.58	2.88	7.43	2.38	1.07	.882	.865	.265
	½	19.6	5.75	38.5	7.49	2.59	2.86	6.74	2.15	1.08	.859	.865	.267
7 × 4	¾	26.2	7.69	37.8	8.42	2.22	2.51	9.05	3.03	1.09	1.01	.860	.324
	⅝	22.1	6.48	32.4	7.14	2.24	2.46	7.84	2.58	1.10	.963	.865	.329
	½	17.9	5.25	26.7	5.81	2.25	2.42	6.53	2.12	1.11	.917	.872	.335
	⅜	13.6	3.98	20.6	4.44	2.27	2.37	5.10	1.63	1.13	.870	.880	.340
6 × 4	⅞	27.2	7.98	27.7	7.15	1.86	2.12	9.75	3.39	1.11	1.12	.857	.421
	¾	23.6	6.94	24.5	6.25	1.88	2.08	8.68	2.97	1.12	1.08	.860	.428
	⅝	20.0	5.86	21.1	5.31	1.90	2.03	7.52	2.54	1.13	1.03	.864	.435
	9/16	18.1	5.31	19.3	4.83	1.90	2.01	6.91	2.31	1.14	1.01	.866	.438
	½	16.2	4.75	17.4	4.33	1.91	1.99	6.27	2.08	1.15	.987	.870	.440
	7/16	14.3	4.18	15.5	3.83	1.92	1.96	5.60	1.85	1.16	.964	.873	.443
	⅜	12.3	3.61	13.5	3.32	1.93	1.94	4.90	1.60	1.17	.941	.877	.446
	5/16	10.3	3.03	11.4	2.79	1.94	1.92	4.18	1.35	1.17	.918	.882	.448
6 × 3½	½	15.3	4.50	16.6	4.24	1.92	2.08	4.25	1.59	.972	.833	.759	.344
	⅜	11.7	3.42	12.9	3.24	1.94	2.04	3.34	1.23	.988	.787	.767	.350
	5/16	9.8	2.87	10.9	2.73	1.95		2.85	1.04	.996	.763	.772	.352
5 × 3½	¾	19.8	5.81	13.9	4.28	1.55	1.75	5.55	2.22	.977	.996	.748	.464
	⅝	16.8	4.92	12.0	3.65	1.56	1.70	4.83	1.90	.991	.951	.751	.472
	½	13.6	4.00	9.99	2.99	1.58	1.66	4.05	1.56	1.01	.906	.755	.479
	7/16	12.0	3.53	8.90	2.64	1.59	1.63	3.63	1.39	1.02	.883	.758	.482
	⅜	10.4	3.05	7.78	2.29	1.60	1.61	3.18	1.21	1.02	.861	.762	.486
	5/16	8.7	2.56	6.60	1.94	1.61	1.59	2.72	1.02	1.03	.838	.766	.489
	¼	7.0	2.06	5.39	1.57	1.62	1.56	2.23	.830	1.04	.814	.770	.492
5 × 3	⅝	15.7	4.61	11.4	3.55	1.57	1.80	3.06	1.39	.815	.796	.644	.349
	½	12.8	3.75	9.45	2.91	1.59	1.75	2.58	1.15	.829	.750	.648	.357
	7/16	11.3	3.31	8.43	2.58	1.60	1.73	2.32	1.02	.837	.727	.651	.361
	⅜	9.8	2.86	7.37	2.24	1.61	1.70	2.04	.888	.845	.704	.654	.364
	5/16	8.2	2.40	6.26	1.89	1.61	1.68	1.75	.753	.853	.681	.658	.368
	¼	6.6	1.94	5.11	1.53	1.62	1.66	1.44	.614	.861	.657	.663	.371
4 × 3½	⅝	14.7	4.30	6.37	2.35	1.22	1.29	4.52	1.84	1.03	1.04	.719	.745
	½	11.9	3.50	5.32	1.94	1.23	1.25	3.79	1.52	1.04	1.00	.722	.750
	7/16	10.6	3.09	4.76	1.72	1.24	1.23	3.40	1.35	1.05	.978	.724	.753
	⅜	9.1	2.67	4.18	1.49	1.25	1.21	2.95	1.17	1.06	.955	.727	.755
	5/16	7.7	2.25	3.56	1.26	1.26	1.18	2.55	.994	1.07	.932	.730	.757
	¼	6.2	1.81	2.91	1.03	1.27	1.16	2.09	.808	1.07	.909	.734	.759

† See footnotes at end of table on page 254.

Steel Angles with Unequal Legs† *(Concluded)*

These angles are commonly designated by section symbol, width of each leg, and thickness, thus:

L 7 × 4 × ½

indicates a 7 × 4-inch angle of ½-inch thickness.

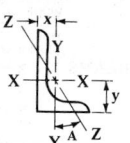

Size	Thickness	Weight per Ft.	Area	Axis X-X				Axis Y-Y				Axis Z-Z	
				I	S	r	y	I	S	r	x	r	Tan A
in.	in.	lb.	in.²	in.⁴	in.³	in.	in.	in.⁴	in.³	in.	in.	in.	
4 × 3	⅝	13.6	3.98	6.03	2.30	1.23	1.37	2.87	1.35	.849	.871	.637	.534
	½	11.1	3.25	5.05	1.89	1.25	1.33	2.42	1.12	.864	.827	.639	.543
	⁷⁄₁₆	9.8	2.87	4.52	1.68	1.25	1.30	2.18	.992	.871	.804	.641	.547
	⅜	8.5	2.48	3.96	1.46	1.26	1.28	1.92	.866	.879	.782	.644	.551
	⁵⁄₁₆	7.2	2.09	3.38	1.23	1.27	1.26	1.65	.734	.887	.759	.647	.554
	¼	5.8	1.69	2.77	1.00	1.28	1.24	1.36	.599	.896	.736	.651	.558
3½ × 3	½	10.2	3.00	3.45	1.45	1.07	1.13	2.33	1.10	.881	.875	.621	.714
	⁷⁄₁₆	9.1	2.65	3.10	1.29	1.08	1.10	2.09	.975	.889	.853	.622	.718
	⅜	7.9	2.30	2.72	1.13	1.09	1.08	1.85	.851	.897	.830	.625	.721
	⁵⁄₁₆	6.6	1.93	2.33	.954	1.10	1.06	1.58	.722	.905	.808	.627	.724
	¼	5.4	1.56	1.91	.776	1.11	1.04	1.30	.589	.914	.785	.631	.727
3½ × 2½	½	9.4	2.75	3.24	1.41	1.09	1.20	1.36	.760	.704	.705	.534	.486
	⁷⁄₁₆	8.3	2.43	2.91	1.26	1.09	1.18	1.23	.677	.711	.682	.535	.491
	⅜	7.2	2.11	2.56	1.09	1.10	1.16	1.09	.592	.719	.660	.537	.496
	⁵⁄₁₆	6.1	1.78	2.19	.927	1.11	1.14	.939	.504	.727	.637	.540	.501
	¼	4.9	1.44	1.80	.755	1.12	1.11	.777	.412	.735	.614	.544	.506
3 × 2½	½	8.5	2.50	2.08	1.04	.913	1.00	1.30	.744	.722	.750	.520	.667
	⁷⁄₁₆	7.6	2.21	1.88	.928	.920	.978	1.18	.664	.729	.728	.521	.672
	⅜	6.6	1.92	1.66	.810	.928	.956	1.04	.581	.736	.706	.522	.676
	⁵⁄₁₆	5.6	1.62	1.42	.688	.937	.933	.898	.494	.744	.683	.525	.680
	¼	4.5	1.31	1.17	.561	.945	.911	.743	.404	.753	.661	.528	.684
	³⁄₁₆	3.39	.996	.907	.430	.954	.888	.577	.310	.761	.638	.533	.688
3 × 2	½	7.7	2.25	1.92	1.00	.924	1.08	.672	.474	.546	.583	.428	.414
	⁷⁄₁₆	6.8	2.00	1.73	.894	.932	1.06	.609	.424	.553	.561	.429	.421
	⅜	5.9	1.73	1.53	.781	.940	1.04	.543	.371	.559	.539	.430	.428
	⁵⁄₁₆	5.0	1.46	1.32	.664	.948	1.02	.470	.317	.567	.516	.432	.435
	¼	4.1	1.19	1.09	.542	.957	.993	.392	.260	.574	.493	.435	.440
	³⁄₁₆	3.07	.902	.842	.415	.966	.970	.307	.200	.583	.470	.439	.446
2½ × 2	⅜	5.3	1.55	.912	.547	.768	.831	.514	.363	.577	.581	.420	.614
	⁵⁄₁₆	4.5	1.31	.788	.466	.776	.809	.446	.310	.584	.559	.422	.620
	¼	3.62	1.06	.654	.381	.784	.787	.372	.254	.592	.537	.424	.626
	³⁄₁₆	2.75	.809	.509	.293	.793	.764	.291	.196	.600	.514	.427	.631

† Data taken from the "Manual of Steel Construction," 8th Edition, 1980, with permission of the American Institute of Steel Construction.

Symbols: I = moment of inertia; S = section modulus; r = radius of gyration; x = distance from center of gravity of section to outer face of structural shape.

Aluminum Association Standard Structural Shapes*

I-BEAMS — Width, Flange Thickness, Web Thickness, Depth, Fillet Radius

CHANNELS — Width, Flange Thickness, Web Thickness, Depth, Fillet Radius

Depth	Width	Weight per Foot	Area	Flange Thickness	Web Thickness	Fillet Radius	Axis X-X			Axis Y-Y			
							I	S	r	I	S	r	x
in.	in.	lb.	in.²	in.	in.	in.	in.⁴	in.³	in.	in.⁴	in.³	in.	in.
I-BEAMS													
3.00	2.50	1.637	1.392	0.20	0.13	0.25	2.24	1.49	1.27	0.52	0.42	.61	...
3.00	2.50	2.030	1.726	0.26	0.15	0.25	2.71	1.81	1.25	0.68	0.54	.63	...
4.00	3.00	2.311	1.965	0.23	0.15	0.25	5.62	2.81	1.69	1.04	0.69	.73	...
4.00	3.00	2.793	2.375	0.29	0.17	0.25	6.71	3.36	1.68	1.31	0.87	.74	...
5.00	3.50	3.700	3.146	0.32	0.19	0.30	13.94	5.58	2.11	2.29	1.31	.85	...
6.00	4.00	4.030	3.427	0.29	0.19	0.30	21.99	7.33	2.53	3.10	1.55	.95	...
6.00	4.00	4.692	3.990	0.35	0.21	0.30	25.50	8.50	2.53	3.74	1.87	.97	...
7.00	4.50	5.800	4.932	0.38	0.23	0.30	42.89	12.25	2.95	5.78	2.57	1.08	...
8.00	5.00	6.181	5.256	0.35	0.23	0.30	59.69	14.92	3.37	7.30	2.92	1.18	...
8.00	5.00	7.023	5.972	0.41	0.25	0.30	67.78	16.94	3.37	8.55	3.42	1.20	...
9.00	5.50	8.361	7.110	0.44	0.27	0.30	102.02	22.67	3.79	12.22	4.44	1.31	...
10.00	6.00	8.646	7.352	0.41	0.25	0.40	132.09	26.42	4.24	14.78	4.93	1.42	...
10.00	6.00	10.286	8.747	0.50	0.29	0.40	155.79	31.16	4.22	18.03	6.01	1.44	...
12.00	7.00	11.672	9.925	0.47	0.29	0.40	255.57	42.60	5.07	26.90	7.69	1.65	...
12.00	7.00	14.292	12.153	0.62	0.31	0.40	317.33	52.89	5.11	35.48	10.14	1.71	...
CHANNELS													
2.00	1.00	0.577	0.491	0.13	0.13	0.10	0.288	0.288	0.766	0.045	0.064	0.303	0.298
2.00	1.25	1.071	0.911	0.26	0.17	0.15	0.546	0.546	0.774	0.139	0.178	0.391	0.471
3.00	1.50	1.135	0.965	0.20	0.13	0.25	1.41	0.94	1.21	0.22	0.22	0.47	0.49
3.00	1.75	1.597	1.358	0.26	0.17	0.25	1.97	1.31	1.20	0.42	0.37	0.55	0.62
4.00	2.00	1.738	1.478	0.23	0.15	0.25	3.91	1.95	1.63	0.60	0.45	0.64	0.65
4.00	2.25	2.331	1.982	0.29	0.19	0.25	5.21	2.60	1.62	1.02	0.69	0.72	0.78
5.00	2.25	2.212	1.881	0.26	0.15	0.30	7.88	3.15	2.05	0.98	0.64	0.72	0.73
5.00	2.75	3.089	2.627	0.32	0.19	0.30	11.14	4.45	2.06	2.05	1.14	0.88	0.95
6.00	2.50	2.834	2.410	0.29	0.17	0.30	14.35	4.78	2.44	1.53	0.90	0.80	0.79
6.00	3.25	4.030	3.427	0.35	0.21	0.30	21.04	7.01	2.48	3.76	1.76	1.05	1.12
7.00	2.75	3.205	2.725	0.29	0.17	0.30	22.09	6.31	2.85	2.10	1.10	0.88	0.84
7.00	3.50	4.715	4.009	0.38	0.21	0.30	33.79	9.65	2.90	5.13	2.23	1.13	1.20
8.00	3.00	4.147	3.526	0.35	0.19	0.30	37.40	9.35	3.26	3.25	1.57	0.96	0.93
8.00	3.75	5.789	4.923	0.41	0.25	0.35	52.69	13.17	3.27	7.13	2.82	1.20	1.22
9.00	3.25	4.983	4.237	0.35	0.23	0.35	54.41	12.09	3.58	4.40	1.89	1.02	0.93
9.00	4.00	6.970	5.927	0.44	0.29	0.35	78.31	17.40	3.63	9.61	3.49	1.27	1.25
10.00	3.50	6.136	5.218	0.41	0.25	0.35	83.22	16.64	3.99	6.33	2.56	1.10	1.02
10.00	4.25	8.360	7.109	0.50	0.31	0.40	116.15	23.23	4.04	13.02	4.47	1.35	1.34
12.00	4.00	8.274	7.036	0.47	0.29	0.40	159.76	26.63	4.77	11.03	3.86	1.25	1.14
12.00	5.00	11.822	10.053	0.62	0.35	0.45	239.69	39.95	4.88	25.74	7.60	1.60	1.61

* Structural sections are available in 6061-T6 aluminum alloy. Data supplied by The Aluminum Association.

Beams

Reaction at the Supports. — When a beam is loaded by vertical loads or forces, the sum of the reactions at the supports equals the sum of the loads. In a simple beam, when the loads are symmetrically placed with reference to the supports, or when the load is uniformly distributed, the reaction at each end will equal one-half of the sum of the loads. When the loads are not symmetrically placed, the reaction at each support may be ascertained from the fact that the algebraic sum of the moments must equal zero. In the accompanying illustration, if moments are taken about the support to the left, then: $R_2 \times 40 - 8000 \times 10 - 10,000 \times 16 - 20,000 \times 20 = 0$; $R_2 = 16,000$ pounds.

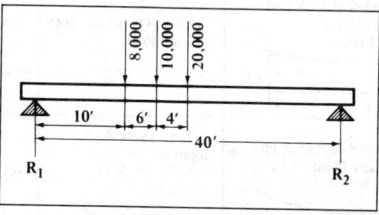

Moments taken about the support at the right will, in the same way, give

$$R_1 = 22,000 \text{ pounds.}$$

The sum of the reactions equals 38,000 pounds, which is also the sum of the loads. If part of the load is uniformly distributed over the beam, this part is first equally divided between the two supports, or the uniform load may be considered as concentrated at its center of gravity.

If metric SI units are used for the calculations, distances may be expressed in meters or millimeters, providing the treatment is consistent, and loads in newtons. *Note:* **If the load is given in kilograms, the value referred to is the mass. A mass of M kilograms has a weight (applies a force) of Mg newtons, where g = approximately 9.81 meters per second².**

Stresses and Deflections in Beams. — On the following pages is given an extensive table of formulas for stresses and deflections in beams, shafts, etc. It is assumed that all the dimensions are in inches, all loads in pounds, and all stresses in pounds per square inch. **The formulas are also valid using metric SI units, with all dimensions in millimeters, all loads in newtons, and stresses and moduli in newtons per millimeter² (N/mm²).** *Note:* **A load due to the weight of a mass of M kilograms is Mg newtons, where g = approximately 9.81 meters per second².** In the tables:

 E = modulus of elasticity of the material;
 I = moment of inertia of the cross-section of the beam;
 Z = section modulus of the cross-section of the beam = $I \div$ distance from neutral axis to extreme fiber;
 W = load on beam;
 s = stress in extreme fiber, or maximum stress in the cross-section considered, due to load W. A positive value of s denotes tension in the upper fibers and compression in the lower ones (as in a cantilever). A negative value of s denotes the reverse (as in a beam supported at the ends). The greatest safe load is that value of W which causes a maxi-

(Continued on page 270)

Stresses and Deflections in Beams

Type of Beam	Stresses	
	General Formula for Stress at any Point	Stresses at Critical Points
Case 1. — Supported at Both Ends, Uniform Load TOTAL LOAD W 	$s = -\dfrac{W}{2Zl}x(l-x)$	Stress at center, $-\dfrac{Wl}{8Z}$ If cross-section is constant, this is the maximum stress.
Case 2. — Supported at Both Ends, Load at Center 	Between each support and load, $s = -\dfrac{Wx}{2Z}$	Stress at center, $-\dfrac{Wl}{4Z}$ If cross-section is constant, this is the maximum stress.
Case 3. — Supported at Both Ends, Load at any Point $a+b=l$	For segment of length a, $s = -\dfrac{Wbx}{Zl}$ For segment of length b, $s = -\dfrac{Wav}{Zl}$	Stress at load, $-\dfrac{Wab}{Zl}$ If cross-section is constant, this is the maximum stress.
Case 4. — Supported at Both Ends, Two Symmetrical Loads 	Between each support and adjacent load, $s = -\dfrac{Wx}{Z}$ Between loads, $s = -\dfrac{Wa}{Z}$	Stress at each load, and at all points between, $-\dfrac{Wa}{Z}$
Case 5. — Both Ends Overhanging Supports Symmetrically, Uniform Load TOTAL LOAD W $L = l + 2c$	Between each support and adjacent end, $s = \dfrac{W}{2ZL}(c-u)^2$ Between supports, $s = \dfrac{W}{2ZL}[c^2 - x(l-x)]$	Stress at each support, $\dfrac{Wc^2}{2ZL}$ Stress at center, $\dfrac{W}{2ZL}(c^2 - \tfrac{1}{4}l^2)$ If cross-section is constant, the greater of these is the maximum stress. If l is greater than $2c$, the stress is zero at points $\sqrt{\tfrac{1}{4}l^2 - c^2}$ on both sides of the center. If cross-section is constant and if $l = 2.828c$, the stresses at supports and center are equal and opposite, and are $\pm\dfrac{WL}{46.62Z}$

Stresses and Deflections in Beams

Deflections (*See footnote*)	
General Formula for Deflection at any Point	Deflections at Critical Points
$$y = \frac{Wx(l-x)}{24EIl}[l^2 + x(l-x)]$$	Maximum deflection, at center, $$\frac{5}{384}\frac{Wl^3}{EI}$$
Between each support and load, $$y = \frac{Wx}{48EI}(3l^2 - 4x^2)$$	Maximum deflection, at load, $\frac{Wl^3}{48EI}$
For segment of length a, $$y = \frac{Wbx}{6EIl}(l^2 - x^2 - b^2)$$ For segment of length b, $$y = \frac{Wav}{6EIl}(l^2 - v^2 - a^2)$$	Deflection at load, $\frac{Wa^2b^2}{3EIl}$ Let a be the length of the shorter segment and b of the longer one. The maximum deflection is in the longer segment, at $$v = b\sqrt{\frac{1}{3} + \frac{2a}{3b}} = v_1, \text{ and is } \frac{Wav_1^3}{3EIl}$$
Between each support and adjacent load, $$y = \frac{Wx}{6EI}[3a(l-a) - x^2]$$ Between loads, $$y = \frac{Wa}{6EI}[3v(l-v) - a^2]$$	Maximum deflection at center, $$\frac{Wa}{24EI}(3l^2 - 4a^2)$$ Deflection at loads $\frac{Wa^2}{6EI}(3l - 4a)$
Between each support and adjacent end, $$y = \frac{Wu}{24EIL}[6c^2(l+u) - u^2(4c-u) - l^3]$$ Between supports, $$y = \frac{Wx(l-x)}{24EIL}[x(l-x) + l^2 - 6c^2]$$	Deflection at ends, $$\frac{Wc}{24EIL}[3c^2(c+2l) - l^3]$$ Deflection at center, $$\frac{Wl^2}{384EIL}(5l^2 - 24c^2)$$ If l is between $2c$ and $2.449c$, there are maximum upward deflections at points $\sqrt{3(\frac{1}{4}l^2 - c^2)}$ on both sides of the center, which are, $-\frac{W}{96EIL}(6c^2 - l^2)^2$

The deflections apply only to cases where the cross section of the beam is constant for its entire length.

Stresses and Deflections in Beams

Type of Beam	Stresses	
	General Formula for Stress at any Point	Stresses at Critical Points
Case 6. — Both Ends Overhanging Supports Unsymmetrically, Uniform Load TOTAL LOAD W $\frac{W}{2l}(l-d+c)$ $\frac{W}{2l}(l+d-c)$	For overhanging end of length c, $$s = \frac{W}{2ZL}(c-u)^2$$ Between supports, $$s = \frac{W}{2ZL}\left\{ c^2\left(\frac{l-x}{l}\right) + d^2\frac{x}{l} - x(l-x) \right\}$$ For overhanging end of length d, $$s = \frac{W}{2ZL}(d-w)^2$$	Stress at support next end of length c, $$\frac{Wc^2}{2ZL}$$ Critical stress between supports is at $$x = \frac{l^2+c^2-d^2}{2l} = x_1$$ and is $\frac{W}{2ZL}(c^2-x_1^2)$ Stress at support next end of length d, $$\frac{Wd^2}{2ZL}$$ If cross-section is constant, the greatest of these three is the maximum stress. If $x_1 > c$, the stress is zero at points $\sqrt{x_1^2-c^2}$ on both sides of $x = x_1$.
Case 7. — Both Ends Overhanging Supports, Load at any Point Between $\frac{Wb}{l}$ $(a+b=l)$ $\frac{Wa}{l}$	Between supports: For segment of length a, $s = -\dfrac{Wbx}{Zl}$ For segment of length b, $s = -\dfrac{Wav}{Zl}$ Beyond supports $s = 0$.	Stress at load, $$-\frac{Wab}{Zl}$$ If cross-section is constant, this is the maximum stress.
Case 8. — Both Ends Overhanging Supports, Single Overhanging Load $\frac{W(c+l)}{l}$ $-\frac{Wc}{l}$	Between load and adjacent support, $$s = \frac{W}{Z}(c-u)$$ Between supports, $$s = \frac{Wc}{Zl}(l-x)$$ Between unloaded end and adjacent support, $s = 0$.	Stress at support adjacent to load, $\dfrac{Wc}{Z}$ If cross-section is constant, this is the maximum stress. Stress is zero at other support.
Case 9. — Both Ends Overhanging Supports, Symmetrical Overhanging Loads	Between each load and adjacent support, $$s = \frac{W}{Z}(c-u)$$ Between supports, $$s = \frac{Wc}{Z}$$	Stress at supports and at all points between, $\dfrac{Wc}{Z}$ If cross-section is constant, this is the maximum stress.

Stresses and Deflections in Beams

Deflections (*See footnote at beginning of Table*)	
General Formula for Deflections at any Point	Deflections at Critical Points

For overhanging end of length c, $$y = \frac{Wu}{24EIL}[2l(d^2+2c^2)$$ $$+6c^2u - u^2(4c-u) - l^3]$$ Between supports, $$y = \frac{Wx(l-x)}{24EIL}\left\{ x(l-x) + l^2 - 2(d^2+c^2) \right.$$ $$\left. -\frac{2}{l}[d^2x + c^2(l-x)] \right\}$$ For overhanging end of length d, $$y = \frac{Ww}{24EIL}[2l(c^2+2d^2)$$ $$+6d^2w - w^2(4d-w) - l^3]$$	Deflection at end c, $$\frac{Wc}{24EIL}[2l(d^2+2c^2)+3c^3-l^3]$$ Deflection at end d, $$\frac{Wd}{24EIL}[2l(c^2+2d^2)+3d^3-l^3]$$ This case is so complicated that convenient general expressions for the critical deflections between supports cannot be obtained.
Between supports, same as Case 3. For overhanging end of length c, $$y = -\frac{Wabu}{6EIl}(l+b)$$ For overhanging end of length d, $$y = -\frac{Wabw}{6EIl}(l+a)$$	Between supports, same as Case 3. Deflection at end c, $-\dfrac{Wabc}{6EIl}(l+b)$ Deflection at end d, $-\dfrac{Wabd}{6EIl}(l+a)$
Between load and adjacent support, $$y = \frac{Wu}{6EI}(3cu - u^2 + 2cl)$$ Between supports, $$y = -\frac{Wcx}{6EIl}(l-x)(2l-x)$$ Between unloaded end and adjacent support, $y = \dfrac{Wclw}{6EI}$	Deflection at load, $\dfrac{Wc^2}{3EI}(c+l)$ Maximum upward deflection is at $$x = 0.42265l, \text{ and is } -\frac{Wcl^2}{15.55EI}$$ Deflection at unloaded end, $\dfrac{Wcld}{6EI}$
Between each load and adjacent support, $$y = \frac{Wu}{6EI}[3c(l+u) - u^2]$$ Between supports, $y = -\dfrac{Wcx}{2EI}(l-x)$	Deflections at loads, $\dfrac{Wc^2}{6EI}(2c+3l)$ Deflection at center, $-\dfrac{Wcl^2}{8EI}$

The above expressions involve the usual approximations of the theory of flexure, and hold only for small deflections. Exact expressions for deflections of any magnitude are as follows:

Between supports the curve is a circle of radius $r = \dfrac{EI}{Wc}$; $y = \sqrt{r^2 - \tfrac{1}{4}l^2} - \sqrt{r^2 - (\tfrac{1}{2}l-x)^2}$

Deflection at center, $\sqrt{r^2 - \tfrac{1}{4}l^2} - r$

Stresses and Deflections in Beams

Type of Beam	Stresses	
	General Formula for Stress at any Point	Stresses at Critical Points
Case 10. — Fixed at One End, Uniform Load TOTAL LOAD W 	$s = \dfrac{W}{2Zl}(l-x)^2$	Stress at support, $\dfrac{Wl}{2Z}$ If cross-section is constant, this is the maximum stress.
Case 11. — Fixed at One End, Load at Other 	$s = \dfrac{W}{Z}(l-x)$	Stress at support, $\dfrac{Wl}{Z}$ If cross-section is constant, this is the maximum stress.
Case 12. — Fixed at One End, Intermediate Load 	Between support and load, $s = \dfrac{W}{Z}(l-x)$ Beyond load, $s = 0$.	Stress at support, $\dfrac{Wl}{Z}$ If cross-section is constant, this is the maximum stress.
Case 13. — Fixed at One End, Supported at the Other, Uniform Load TOTAL LOAD W 	$s = \dfrac{W(l-x)}{2Zl}(\tfrac{1}{4}l - x)$	Maximum stress at point of fixture, $\dfrac{Wl}{8Z}$ Stress is zero at $x = \tfrac{1}{4}l$. Greatest negative stress is at $x = \tfrac{5}{8}l$ and is $-\dfrac{9}{128}\dfrac{Wl}{Z}$
Case 14. — Fixed at One End, Supported at the Other, Load at Center 	Between point of fixture and load, $s = \dfrac{W}{16Z}(3l - 11x)$ Between support and load, $s = -\tfrac{5}{16}\dfrac{Wv}{Z}$	Maximum stress at point of fixture, $\tfrac{3}{16}\dfrac{Wl}{Z}$ Stress is zero at $x = \tfrac{3}{11}l$ Greatest negative stress at center, $-\tfrac{5}{32}\dfrac{Wl}{Z}$

Stresses and Deflections in Beams

Deflections (*See footnote at beginning of Table*)	
General Formula for Deflection at any Point	Deflections at Critical Points
$$y = \frac{Wx^2}{24EIl}[2l^2 + (2l-x)^2]$$	Maximum deflection, at end, $\frac{Wl^3}{8EI}$
$$y = \frac{Wx^2}{6EI}(3l - x)$$	Maximum deflection, at end, $\frac{Wl^3}{3EI}$
Between support and load, $$y = \frac{Wx^2}{6EI}(3l - x)$$ Beyond load, $$y = \frac{Wl^2}{6EI}(3v - l)$$	Deflection at load, $\frac{Wl^3}{3EI}$ Maximum deflection, at end, $$\frac{Wl^2}{6EI}(2l + 3b)$$
$$y = \frac{Wx^2(l-x)}{48EIl}(3l - 2x)$$	Maximum deflection is at $x = 0.5785l$, and is $\frac{Wl^3}{185EI}$ Deflection at center, $\frac{Wl^3}{192EI}$ Deflection at point of greatest negative stress, at $x = \frac{5}{8}l$ is $\frac{Wl^3}{187EI}$
Between point of fixture and load, $$y = \frac{Wx^2}{96EI}(9l - 11x)$$ Between support and load, $$y = \frac{Wv}{96EI}(3l^2 - 5v^2)$$	Maximum deflection is at $v = 0.4472l$, and is $\frac{Wl^3}{107.33EI}$ Deflection at load, $\frac{7}{768}\frac{Wl^3}{EI}$

Stresses and Deflections in Beams

Type of Beam	Stresses	
	General Formula for Stress at any Point	Stresses at Critical Points
Case 15. — Fixed at One End, Supported at the Other, Load at any Point $m = (l+a)(l+b) + al$ $n = al(l+b)$ $\dfrac{Wab(l+b)}{2l^2}$ $W\left[1 - \dfrac{a^2}{2l^3}(3l-a)\right]$ $\dfrac{Wa^2(3l-a)}{2l^3}$	Between point of fixture and load, $$s = \frac{Wb}{2Zl^3}(n - mx)$$ Between support and load, $$s = -\frac{Wa^2v}{2Zl^3}(3l-a)$$	Greatest positive stress, at point of fixture, $$\frac{Wab}{2Zl^2}(l+b)$$ Greatest negative stress, at load, $$-\frac{Wa^2b}{2Zl^3}(3l-a)$$ If $a < 0.5858l$, the first is the maximum stress. If $a = 0.5858l$, the two are equal and are $\pm\dfrac{Wl}{5.83Z}$. If $a > 0.5858l$, the second is the maximum stress. Stress is zero at $x = \dfrac{n}{m}$
Case 16. — Fixed at One End, Free but Guided at the Other, Uniform Load TOTAL LOAD W $\dfrac{Wl}{3}$ $\dfrac{Wl}{6}$ W	$$s = \frac{Wl}{Z}\left\{ \tfrac{1}{3} - \frac{x}{l} + \tfrac{1}{2}\left(\frac{x}{l}\right)^2 \right\}$$	Maximum stress, at support, $\dfrac{Wl}{3Z}$ Stress is zero for $x = 0.4227l$ Greatest negative stress, at free end, $-\dfrac{Wl}{6Z}$
Case 17. — Fixed at One End, Free but Guided at the Other, with Load $\dfrac{Wl}{2}$ W $\dfrac{Wl}{2}$ W	$$s = \frac{W}{Z}(\tfrac{1}{2}l - x)$$	Stress at support, $\dfrac{Wl}{2Z}$ Stress at free end $-\dfrac{Wl}{2Z}$ These are the maximum stresses and are equal and opposite. Stress is zero at center.
Case 18. — Fixed at Both Ends, Uniform Load TOTAL LOAD W $\dfrac{Wl}{12}$ $\dfrac{Wl}{12}$ $\dfrac{W}{2}$ $\dfrac{W}{2}$	$$s = \frac{Wl}{2Z}\left\{ \tfrac{1}{6} - \frac{x}{l} + \left(\frac{x}{l}\right)^2 \right\}$$	Maximum stress, at ends, $\dfrac{WL}{12Z}$ Stress is zero at $x = 0.7887l$ and at $x = 0.2113l$ Greatest negative stress, at center, $-\dfrac{Wl}{24Z}$

Stresses and Deflections in Beams

Deflections (*See footnote at beginning of Table*)	
General Formula for Deflections at any Point	Deflections at Critical Points
Between point of fixture and load, $$y = \frac{Wx^2 b}{12\,EIl^3}(3n - mx)$$ Between support and load, $$y = \frac{Wa^2 v}{12\,EIl^3}[3\,l^2 b - v^2(3l - a)]$$	Deflection at load, $\frac{Wa^3 b^2}{12\,EIl^3}(3l + b)$ If $a < 0.5858l$, maximum deflection is between load and support, at $$v = l\sqrt{\frac{b}{2l + b}} \text{ and is } \frac{Wa^2 b}{6EI}\sqrt{\frac{b}{2l + b}}$$ If $a = 0.5858l$, maximum deflection is at load and is $\frac{Wl^3}{101.9\,EI}$ If $a > 0.5858l$, maximum deflection is between load and point of fixture, at $$x = \frac{2n}{m}, \text{ and is } \frac{Wbn^3}{3\,EIm^2 l^3}$$
$$y = \frac{Wx^2}{24\,EIl}(2l - x)^2$$	Maximum deflection, at free end, $$\frac{Wl^3}{24\,EI}$$
$$y = \frac{Wx^2}{12\,EI}(3l - 2x)$$	Maximum deflection, at free end, $$\frac{Wl^3}{12\,EI}$$
$$y = \frac{Wx^2}{24\,EIl}(l - x)^2$$	Maximum deflection, at center, $$\frac{Wl^3}{384\,EI}$$

Stresses and Deflections in Beams

	Stresses	
Type of Beam	General Formula for Stress at any Point	Stresses at Critical Points
Case 19. — Fixed at Both Ends, Load at Center	Between each end and load, $$s = \frac{W}{2Z}(\tfrac{1}{4}l - x)$$	Stress at ends $\frac{Wl}{8Z}$; at load $-\frac{Wl}{8Z}$ These are the maximum stresses and are equal and opposite. Stress is zero at $x = \frac{1}{4}l$
Case 20. — Fixed at Both Ends, Load at any Point	For segment of length a, $$s = \frac{Wb^2}{Zl^3}[al - x(l + 2a)]$$ For segment of length b, $$s = \frac{Wa^2}{Zl^3}[bl - v(l + 2b)]$$	Stress at end next segment of length a, $$\frac{Wab^2}{Zl^2}$$ Stress at end next segment of length b, $$\frac{Wa^2b}{Zl^2}$$ Maximum stress is at end next shorter segment. Stress is zero for $x = \frac{al}{l + 2a}$ and $v = \frac{bl}{l + 2b}$ Greatest negative stress, at load, $$-\frac{2Wa^2b^2}{Zl^3}$$
Case 21. — Continuous Beam, with Two Equal Spans, Uniform Load TOTAL LOAD ON EACH SPAN, W	$$s = \frac{W(l - x)}{2Zl}(\tfrac{1}{4}l - x)$$	Maximum stress at point A, $\frac{Wl}{8Z}$ Stress is zero at $x = \frac{1}{4}l$. Greatest negative stress is at $x = \frac{5}{8}l$ and is, $-\frac{9}{128}\frac{Wl}{Z}$
Case 22. — Continuous Beam, with Two Unequal Spans, Unequal, Uniform Loads TOTAL LOAD W_1 TOTAL LOAD W_2	Between R_1 and R, $$s = \frac{l_1 - x}{Z}\left\{\frac{(l_1 - x)W_1}{2l_1} - R_1\right\}$$ Between R_2 and R, $$s = \frac{l_2 - u}{Z}\left\{\frac{(l_2 - u)W_2}{2l_2} - R_2\right\}$$	Stress at support R, $$\frac{W_1 l_1^2 + W_2 l_2^2}{8Z(l_1 + l_2)}$$ Greatest stress in the first span is at $x = \frac{l_1}{W_1}(W_1 - R_1)$, and is, $-\frac{R_1^2 l_1}{2ZW_1}$ Greatest stress in the second span is at $u = \frac{l_2}{W_2}(W_2 - R_2)$, and is, $-\frac{R_2^2 l_2}{2ZW_2}$

Stresses and Deflections in Beams

Deflections (*See footnote at beginning of Table*)	
General Formula for Deflections at any Point	Deflections at Critical Points
$$y = \frac{Wx^2}{48EI}(3l - 4x)$$	Maximum deflection, at load, $$\frac{Wl^3}{192EI}$$
For segment of length a, $$y = \frac{Wx^2b^2}{6EIl^3}[2a(l-x)+l(a-x)]$$ For segment of length b, $$y = \frac{Wv^2a^2}{6EIl^3}[2b(l-v)+l(b-v)]$$	Deflection at load, $\dfrac{Wa^3b^3}{3EIl^3}$ Let b be the length of the longer segment and a of the shorter one. The maximum deflection is in the longer segment, at $v = \dfrac{2bl}{l+2b}$, and is $$\frac{2Wa^2b^3}{3EI(l+2b)^2}$$
$$y = \frac{Wx^2(l-x)}{48EIl}(3l-2x)$$	Maximum deflection is at $x = 0.5785l$, and is $\dfrac{Wl^3}{185EI}$ Deflection at center of span, $\dfrac{Wl^3}{192EI}$ Deflection at point of greatest negative stress, at $x = \frac{5}{8}l$ is $\dfrac{Wl^3}{187EI}$
Between R_1 and R, $$y = \frac{x(l_1-x)}{24EI}\left\{(2l_1-x)(4R_1-W_1)\right.$$ $$\left. - \frac{W_1(l_1-x)^2}{l_1}\right\}$$ Between R_2 and R, $$y = \frac{u(l_2-u)}{24EI}\left\{(2l_2-u)(4R_2-W_2)\right.$$ $$\left. - \frac{W_2(l_2-u)^2}{l_2}\right\}$$	This case is so complicated that convenient general expressions for the critical deflections cannot be obtained.

Stresses and Deflections in Beams

Type of Beam	Stresses	
	General Formula for Stress at any Point	Stresses at Critical Points
Case 23. — Continuous Beam, with Two Equal Spans, Equal Loads at Center of Each $$\frac{5}{16}W \qquad \frac{11}{8}W \qquad \frac{5}{16}W$$	Between point A and load, $$s = \frac{W}{16Z}(3l - 11x)$$ Between point B and load, $$s = -\frac{5}{16}\frac{Wv}{Z}$$	Maximum stress at point A, $\dfrac{3}{16}\dfrac{Wl}{Z}$ Stress is zero at $$x = \frac{3}{11}l$$ Greatest negative stress at center of span, $$-\frac{5}{32}\frac{Wl}{Z}$$
Case 24. — Continuous Beam, with Two Unequal Spans, Unequal Loads at any Point of Each $$m = \frac{1}{2(l_1+l_2)}\left(\frac{W_1 a_1 b_1}{l_1}(l_1+a_1) + \frac{W_2 a_2 b_2}{l_2}(l_2+a_2)\right)$$ $$\frac{W_1 b_1 - m}{l_1} \quad \frac{W_1 a_1 + m}{l_1} + \frac{W_2 a_2 + m}{l_2} \quad \frac{W_2 b_2 - m}{l_2}$$ $$= r_1 \qquad\qquad = r \qquad\qquad = r_2$$	Between R_1 and W_1, $$s = -\frac{wr_1}{Z}$$ Between R and W_1, $$s = \frac{1}{l_1 Z}[m(l_1 - u) - W_1 a_1 u]$$ Between R and W_2, $$s = \frac{1}{l_2 Z}[m(l_2 - x) - W_2 a_2 x]$$ Between R_2 and W_2, $$s = -\frac{vr_2}{Z}$$	Stress at load W_1, $$-\frac{a_1 r_1}{Z}$$ Stress at support R, $$\frac{m}{Z}$$ Stress at load W_2, $$-\frac{a_2 r_2}{Z}$$ The greatest of these is the maximum stress.

mum stress equal to, but not exceeding, the greatest safe value of s;

y = deflection measured from the position occupied if the load causing the deflection were removed. A positive value of y denotes deflection below this position; a negative value, deflection upward;

u, v, w, x = variable distances along the beam from a given support to any point.

If there are several kinds of loads, as, for instance, a uniform load and a load at any point, or separate loads at different points, the total stress and the total deflection at any point is found by adding together the various stresses or deflections at the point considered due to each load acting by itself. If the stress or deflection due to any one of the loads is negative, it must be subtracted instead of added.

Remarks Relative to the Use of the Tables. — In the diagrammatical illustrations of the beams and their loading, the values indicated near, but below, the supports are the "reactions" or upward forces at the supports. For Cases 1 to 12, inclusive, the reactions, as well as the formulas for the stresses, are the same whether the beam is of constant or variable cross-section. For the other cases, the reactions and the stresses given are for constant cross-section beams only.

The bending moment at any point in inch-pounds is $s \times Z$ and can be found by omitting the divisor Z in the formula for the stress given in the tables. A positive value of the bending moment denotes tension in the upper fibers and compression in

Stresses and Deflections in Beams

Deflections (*See footnote at beginning of Table*)	
General Formula for Deflections at any Point	Deflections at Critical Points
Between point A and load, $$y = \frac{Wx^2}{96EI}(9l - 11x)$$ Between point B and load, $$y = \frac{Wv}{96EI}(3l^2 - 5v^2)$$	Maximum deflection is at $v = 0.4472l$, and is $$\frac{Wl^3}{107.33EI}$$ Deflection at load, $\dfrac{7}{768}\dfrac{Wl^3}{EI}$
Between R_1 and W_1, $$y = \frac{w}{6EI}\left\{(l_1 - w)(l_1 + w)r_1 - \frac{W_1 b_1^3}{l_1}\right\}$$ Between R and W_1, $$y = \frac{u}{6EIl_1}[W_1 a_1 b_1(l_1 + a_1)$$ $$- W_1 a_1 u^2 - m(2l_1 - u)(l_1 - u)]$$ Between R and W_2, $$y = \frac{x}{6EIl_2}[W_2 a_2 b_2(l_2 + a_2)$$ $$- W_2 a_2 x^2 - m(2l_2 - x)(l_2 - x)]$$ Between R_2 and W_2, $$y = \frac{v}{6EI}\left\{(l_2 - v)(l_2 + v)r_2 - \frac{W_2 b_2^3}{l_2}\right\}$$	Deflection at load W_1, $$\frac{a_1 b_1}{6EIl_1}[2a_1 b_1 W_1 - m(l_1 + a_1)]$$ Deflection at load W_2, $$\frac{a_2 b_2}{6EIl_2}[2a_2 b_2 W_2 - m(l_2 + a_2)]$$ This case is so complicated that convenient general expressions for the maximum deflections cannot be obtained.

the lower ones. A negative value denotes the reverse. The value of W corresponding to a given stress is found by transposition of the formula. For example, in Case 1, the stress at the critical point is $s = -Wl \div 8Z$. From this we find $W = -8Zs \div l$. Of course, the negative sign of W may be ignored.

Deflection of Beam Uniformly Loaded for Part of Its Length. — In the following formulas, lengths are in inches, weights in pounds. W = total load; L = total length between supports; E = modulus of elasticity; I = moment of inertia of beam section;

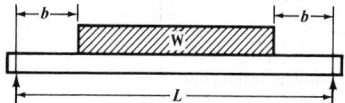

a = fraction of length of beam at each end, that is not loaded = $b \div L$; f = deflection.

$$f = \frac{WL^3}{384EI(1 - 2a)}(5 - 24a^2 + 16a^4)$$

The expression for maximum bending moment is: $M_{max} = \frac{1}{8}WL(1 + 2a)$.

These formulas apply to simple beams resting on supports at the ends.

If the formulas are used with metric SI units, W = total load in newtons; L = total length between supports in millimeters; E = modulus of elasticity in newtons per millimeter^2; I = moment of inertia of beam section in millimeters^4; a = fraction of length

Beams of Uniform Strength Throughout Their Length

(All loads in pounds, all dimensions in inches.)

Type of Beam	Description	Formula*
	Load at one end. Width of beam uniform. Depth of beam decreasing towards loaded end. Outline of beam-shape, parabola with vertex at loaded end.	$P = \dfrac{Sbh^2}{6l}$
	Load at one end. Width of beam uniform. Depth of beam decreasing towards loaded end. Outline of beam, one-half of a parabola with vertex at loaded end. Beam may be reversed so that upper edge is parabolic.	$P = \dfrac{Sbh^2}{6l}$
	Load at one end. Depth of beam uniform. Width of beam decreasing towards loaded end. Outline of beam triangular, with apex at loaded end.	$P = \dfrac{Sbh^2}{6l}$
	Beam of *approximately* uniform strength. Load at one end. Width of beam uniform. Depth of beam decreasing towards loaded end, but not tapering to a sharp point.	$P = \dfrac{Sbh^2}{6l}$
	Uniformly distributed load. Width of beam uniform. Depth of beam decreasing towards outer end. Outline of beam, right-angled triangle.	$P = \dfrac{Sbh^2}{3l}$
	Uniformly distributed load. Depth of beam uniform. Width of beam gradually decreasing towards outer end. Outline of beam is formed by two parabolas which tangent each other at their vertexes at the outer end of the beam.	$P = \dfrac{Sbh^2}{3l}$

* In the formulas, P = load in pounds; S = safe stress in pounds per square inch; and a, b, c, h, and l are in inches. **If metric SI units are used, P is in newtons; S = safe stress in N/mm²; and a, b, c, h, and l are in millimeters.**

Beams of Uniform Strength Throughout Their Length

Type of Beam	Description	Formula*
	Beam supported at both ends. Load concentrated at any point. Depth of beam uniform. Width of beam maximum at point of loading. Outline of beam, two triangles with apexes at points of support.	$P = \dfrac{Sbh^2 l}{6ac}$
	Beam supported at both ends. Load concentrated at any point. Width of beam uniform. Depth of beam maximum at point of loading. Outline of beam is formed by two parabolas with their vertexes at points of support.	$P = \dfrac{Sbh^2 l}{6ac}$
	Beam supported at both ends. Load concentrated in the middle. Depth of beam uniform. Width of beam maximum at point of loading. Outline of beam, two triangles with apexes at points of support.	$P = \dfrac{2 Sbh^2}{3l}$
	Beam supported at both ends. Load concentrated at center. Width of beam uniform. Depth of beam maximum at point of loading. Outline of beam, two parabolas with vertices at points of support.	$P = \dfrac{2 Sbh^2}{3l}$
	Beam supported at both ends. Load uniformly distributed. Depth of beam uniform. Width of beam maximum at center. Outline of beam, two parabolas with vertexes at middle of beam.	$P = \dfrac{4 Sbh^2}{3l}$
	Beam supported at both ends. Load uniformly distributed. Width of beam uniform. Depth of beam maximum at center. Outline of beam one-half of an ellipse.	$P = \dfrac{4 Sbh^2}{3l}$

* For details of English and metric SI units used in the formulas, see previous page.

Rectangular Solid Beams

Style of Loading and Support	Breadth of Beam, b	Height of Beam, h	Stress in Extreme Fibers, f	Length of Beam, l	Total Load, W
	Inches	Inches	Lb./sq. in.	Inches	Pounds
	Millimeters	Millimeters	N/sq. mm.	Millimeters	Newtons
Beam fixed at one end, loaded at the other					
	$\dfrac{6lW}{fh^2}=b$	$\sqrt{\dfrac{6lW}{bf}}=h$	$\dfrac{6lW}{bh^2}=f$	$\dfrac{bfh^2}{6W}=l$	$\dfrac{bfh^2}{6l}=W$
Beam fixed at one end, uniformly loaded					
	$\dfrac{3lW}{fh^2}=b$	$\sqrt{\dfrac{3lW}{bf}}=h$	$\dfrac{3lW}{bh^2}=f$	$\dfrac{bfh^2}{3W}=l$	$\dfrac{bfh^2}{3l}=W$
Beam supported at both ends, single load in middle					
	$\dfrac{3lW}{2fh^2}=b$	$\sqrt{\dfrac{3lW}{2bf}}=h$	$\dfrac{3lW}{2bh^2}=f$	$\dfrac{2bfh^2}{3W}=l$	$\dfrac{2bfh^2}{3l}=W$
Beam supported at both ends, uniformly loaded					
	$\dfrac{3lW}{4fh^2}=b$	$\sqrt{\dfrac{3lW}{4bf}}=h$	$\dfrac{3lW}{4bh^2}=f$	$\dfrac{4bfh^2}{3W}=l$	$\dfrac{4bfh^2}{3l}=W$
Beam supported at both ends, single unsymmetrical load					
	$\dfrac{6Wac}{fh^2l}=b$	$\sqrt{\dfrac{6Wac}{bfl}}=h$	$\dfrac{6Wac}{bh^2l}=f$	$a+c=l$	$\dfrac{bh^2fl}{6ac}=W$
Beam supported at both ends, two symmetrical loads					
	$\dfrac{3Wa}{fh^2}=b$	$\sqrt{\dfrac{3Wa}{bf}}=h$	$\dfrac{3Wa}{bh^2}=f$	l, any length $\dfrac{bh^2f}{3W}=a$	$\dfrac{bh^2f}{3a}=W$

Round Solid Beams

Style of Loading and Support	Diameter of Beam, d	Stress in Extreme Fibers, f	Length of Beam, l	Total Load, W
	Inches	Lb./sq. in.	Inches	Pounds
	Millimeters	N/sq. mm.	Millimeters	Newtons
	Beam fixed at one end, loaded at the other			
	$\sqrt[3]{\dfrac{10.18\,lW}{f}} = d$	$\dfrac{10.18\,lW}{d^3} = f$	$\dfrac{d^3 f}{10.18\,W} = l$	$\dfrac{d^3 f}{10.18\,l} = W$
	Beam fixed at one end, uniformly loaded			
	$\sqrt[3]{\dfrac{5.092\,Wl}{f}} = d$	$\dfrac{5.092\,Wl}{d^3} = f$	$\dfrac{d^3 f}{5.092\,W} = l$	$\dfrac{d^3 f}{5.092\,l} = W$
	Beam supported at both ends, single load in middle			
	$\sqrt[3]{\dfrac{2.546\,Wl}{f}} = d$	$\dfrac{2.546\,Wl}{d^3} = f$	$\dfrac{d^3 f}{2.546\,W} = l$	$\dfrac{d^3 f}{2.546\,l} = W$
	Beam supported at both ends, uniformly loaded			
	$\sqrt[3]{\dfrac{1.273\,Wl}{f}} = d$	$\dfrac{1.273\,Wl}{d^3} = f$	$\dfrac{d^3 f}{1.273\,W} = l$	$\dfrac{d^3 f}{1.273\,l} = W$
	Beam supported at both ends, single unsymmetrical load			
	$\sqrt[3]{\dfrac{10.18\,Wac}{fl}} = d$	$\dfrac{10.18\,Wac}{d^3 l} = f$	$a + c = l$	$\dfrac{d^3 fl}{10.18\,ac} = W$
	Beam supported at both ends, two symmetrical loads			
	$\sqrt[3]{\dfrac{5.092\,Wa}{f}} = d$	$\dfrac{5.092\,Wa}{d^3} = f$	l, any length $\dfrac{d^3 f}{5.092\,W} = a$	$\dfrac{d^3 f}{5.092\,a} = W$

of beam at each end, that is not loaded $= b \div L$; and $f =$ deflection in millimeters. The bending moment M_{max} is in newton-millimeters (N·mm).

Note: A load due to the weight of a mass of M kilograms is Mg newtons, where $g =$ approximately 9.81 meters per second^2.

Beams of Uniform Strength Throughout Their Length. — In nearly all cases, the bending moment in a beam is not uniform throughout its length, but varies. Therefore, a beam of uniform cross-section which is made strong enough at its most strained section, will have an excess of material at every other section. Sometimes it may be desirable to have the cross-section uniform, while in other cases the metal can be more advantageously distributed if the beam is so designed that its cross-section varies from point to point, so that it is at every point just great enough to take care of the bending stresses at that point. A table is given showing beams in which the load is applied in different ways and which are supported by different methods, and the shape of the beam required for uniform strength is indicated. It should be noted that the shape given is the theoretical shape required to resist bending only. It is apparent that sufficient cross-section of beam must also be added either at the points of support (in the case of beams supported at both ends), or at the point of application of the load (in the case of beams loaded at one end), to take care of the vertical shear.

It should be noted that the theoretical shapes of the beams given in the tables on the two following pages are based on the stated assumptions of uniformity of width or depth of cross-section, and unless these are observed in the design, the theoretical outlines do not apply without modifications. For example, in a cantilever with the load at one end, the outline is a parabola only when the width of the beam is uniform. It is not correct to use a strictly parabolic shape when the thickness is not uniform, as, for instance, when the beam is made of an I- or T-section. In such cases, some modification may be necessary; but it is evident that whatever the shape adopted, the correct depth of the section can be obtained by an investigation of the bending moment and the shearing load at a number of points, and then a line can be drawn through the points thus ascertained, which will provide for a beam of practically uniform strength whether the cross-section be of uniform width or not.

Crane Girders with Curved Lower Chords. — An example of a design which makes use of the principles of beams of uniform strength is found in the ordinary fish-belly type of crane girder. When laying out crane girders, the accompanying tables will be found convenient. The figure will explain the use of the tables. A crane girder having a span of 61 feet 3 inches has been assumed as an example. The curved part has a span of 60 feet; one-half of this distance, or 30 feet, is divided for the convenience of the templet makers into ten spaces of 3 feet each. The end ordinate, assumed here to be 1 foot, will be found at the extreme left of the tables under the heading H. The lengths of the remaining nine ordinates follow in order. For short spans, say about 30 feet, it is most convenient to divide the base of the curve into five spaces, as it is the usual practice to give ordinates about every 3 feet. In this case, we would use only every other ordinate in the tables, or, beginning with the left-hand column, the ordinates would be as found in the columns headed H, 8, 6, 4 and 2.

The tables (pages 278 and 279) are calculated from the formula: $X = H \times (M^2 \div N^2)$ in which $H =$ end ordinate; $X =$ required ordinate; $N =$ number of equal spaces into which the base line is divided; $M =$ number of spaces from 0 to the required ordinate. When $N = 10$, as in the case for which the tables are calculated, $N^2 = 100$, and $X = H \times 0.01 M^2$.

Hence, ordinate No. 8 equals $H \times 0.01 \times 64 = 0.64H$. Ordinate No. 4 equals $H \times 0.01 \times 16 = 0.16H$.

Opinions vary considerably as to the allowable working stress in crane girders. Many cranes have girders which are designed for a stress of only 8000 pounds per square inch, while in others the stress will be over 14,000 pounds. However, a

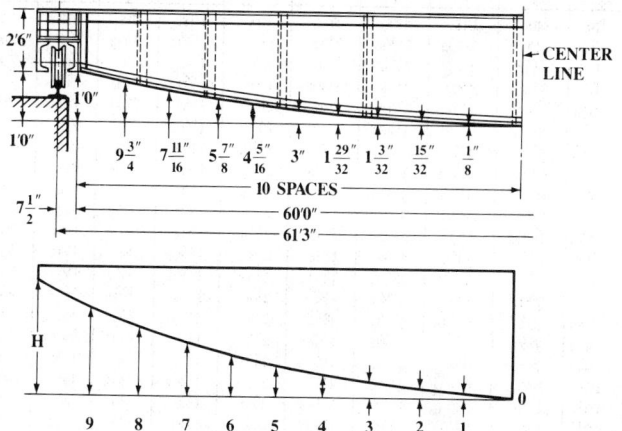

general factor of safety of 5 is the most usual and desirable in crane work, and if that factor of safety is adopted, the working stress should be anywhere from 11,000 to 12,000 pounds per square inch.

Bending Stress Due to an Oblique Transverse Force. — The following illustration shows a beam and a channel being subjected to a transverse force acting at an angle ϕ to the center of gravity.

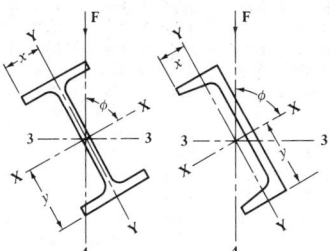

To find the bending stress, the moments of inertia I around axes 3-3 and 4-4 are computed from the following equations:

$$I_3 = I_x \sin^2 \phi + I_y \cos^2 \phi$$
$$I_4 = I_x \cos^2 \phi + I_y \sin^2 \phi$$

The computed bending stress f_b is then found from

$$f_b = M\left(\frac{y}{I_x} \sin \phi + \frac{x}{I_y} \cos \phi\right)$$

where M is the bending moment due to force F.

Ordinates of Parabolas for Crane Girder Design — I

	Ordinates								
H	9	8	7	6	5	4	3	2	1
Ft. Ins.	Ins.	Ins.	Ins.	Ins.	Ins.	Ins.	Ins.	Ins.	Ins.
6	4 7/8	3 27/32	2 15/16	2 5/32	1 1/2	31/32	17/32	1/4	1/16
6 1/4	5 1/16	4	3 1/16	2 1/4	1 9/16	1	9/16	1/4	1/16
6 1/2	5 1/4	4 5/32	3 3/16	2 11/32	1 5/8	1 1/32	19/32	1/4	1/16
6 3/4	5 15/32	4 5/16	3 5/16	2 7/16	1 11/16	1 3/32	19/32	9/32	1/16
7	5 11/16	4 15/32	3 7/16	2 17/32	1 3/4	1 1/8	5/8	9/32	1/16
7 1/4	5 7/8	4 21/32	3 9/16	2 5/8	1 13/16	1 5/32	21/32	9/32	1/16
7 1/2	6 1/16	4 13/16	3 11/16	2 11/16	1 7/8	1 7/32	11/16	5/16	1/16
7 3/4	6 9/32	4 31/32	3 13/16	2 25/32	1 15/16	1 1/4	11/16	5/16	1/16
8	6 15/32	5 1/8	3 15/16	2 7/8	2	1 9/32	23/32	5/16	3/32
8 1/4	6 19/32	5 9/32	4 1/32	2 23/32	2 1/16	1 5/16	3/4	11/32	3/32
8 1/2	6 7/8	5 7/16	4 5/32	3 1/16	2 1/8	1 3/8	3/4	11/32	3/32
8 3/4	7 3/32	5 19/32	4 5/32	3 5/32	2 3/16	1 7/16	25/32	11/32	3/32
9	7 9/32	5 3/4	4 13/32	3 1/4	2 1/4	1 7/16	13/16	3/8	3/32
9 1/4	7 1/2	5 15/16	4 17/32	3 9/32	2 5/16	1 15/32	27/32	3/8	3/32
9 1/2	7 11/16	6 3/32	4 21/32	3 7/16	2 3/8	1 1/2	27/32	3/8	3/32
9 3/4	7 29/32	6 1/4	4 25/32	3 1/2	2 7/16	1 9/16	7/8	13/32	3/32
10	8 3/32	6 13/32	4 29/32	3 19/32	2 1/2	1 19/32	29/32	13/32	3/32
10 1/4	8 5/16	6 9/16	5	3 11/16	2 9/16	1 5/8	15/16	13/32	3/32
10 1/2	8 1/2	6 23/32	5 5/32	3 25/32	2 5/8	1 11/16	15/16	13/32	3/32
10 3/4	8 11/16	6 7/8	5 1/4	3 7/8	2 11/16	1 23/32	31/32	7/16	3/32
11	8 29/32	7 1/32	5 13/32	3 31/32	2 3/4	1 3/4	1	7/16	1/8
11 1/4	9 1/8	7 3/16	5 1/2	4 1/16	2 13/16	1 13/16	1	7/16	1/8
11 1/2	9 5/16	7 3/8	5 5/8	4 5/32	2 7/8	1 27/32	1 1/32	15/32	1/8
11 3/4	9 1/2	7 17/32	5 3/4	4 7/32	2 15/16	1 7/8	1 1/16	15/32	1/8
12	9 3/4	7 11/16	5 7/8	4 5/16	3	1 29/32	1 3/32	15/32	1/8
1 0 1/4	9 15/16	7 27/32	6	4 13/32	3 1/16	1 31/32	1 3/32	1/2	1/8
1 0 1/2	10 1/8	8	6 1/8	4 1/2	3 1/8	2	1 1/8	1/2	1/8
1 0 3/4	10 5/16	8 5/32	6 1/4	4 19/32	3 3/16	2 3/32	1 1/8	1/2	1/8
1 1	10 17/32	8 5/16	6 3/8	4 11/16	3 1/4	2 3/32	1 5/32	17/32	1/8
1 1 1/4	10 3/4	8 15/32	6 1/2	4 25/32	3 5/16	2 1/8	1 3/16	17/32	1/8
1 1 1/2	10 15/16	8 5/8	6 5/8	4 7/8	3 3/8	2 5/32	1 7/32	17/32	1/8
1 1 3/4	11 1/8	8 7/8	6 3/4	4 15/16	3 7/16	2 3/16	1 1/4	9/16	1/8
1 2	11 11/32	8 31/32	6 7/8	5 1/32	3 1/2	2 1/4	1 1/4	9/16	5/32
1 2 1/4	11 17/32	9 1/8	7	5 1/8	3 9/16	2 9/32	1 5/16	9/16	5/32
1 2 1/2	11 3/4	9 3/8	7 3/32	5 7/32	3 5/8	2 5/16	1 5/16	19/32	5/32
1 2 3/4	11 15/16	9 7/16	7 7/32	5 5/16	3 11/16	2 3/8	1 5/16	19/32	5/32
1 3	12 5/32	9 19/32	7 11/32	5 13/32	3 3/4	2 13/32	1 11/32	19/32	5/32
1 3 1/4	12 11/32	9 3/4	7 15/32	5 1/2	3 13/16	2 9/16	1 3/8	19/32	5/32
1 3 1/2	12 9/16	9 15/16	7 19/32	5 19/32	3 7/8	2 1/2	1 13/32	5/8	5/32
1 3 3/4	12 3/4	10 3/32	7 23/32	5 21/32	3 15/16	2 17/32	1 13/32	5/8	5/32
1 4	12 23/32	10 1/4	7 27/32	5 3/4	4	2 9/16	1 7/16	5/8	5/32
1 4 1/4	13 5/32	10 13/32	7 31/32	5 27/32	4 1/16	2 19/32	1 15/32	21/32	5/32
1 4 1/2	13 3/8	10 1/2	8 3/32	5 15/16	4 1/8	2 5/8	1 15/32	21/32	5/32
1 4 3/4	13 9/16	10 23/32	8 5/32	6 1/32	4 3/16	2 11/16	1 1/2	11/16	5/32
1 5	13 25/32	10 7/8	8 11/32	6 1/8	4 1/4	2 23/32	1 17/32	11/16	5/32
1 5 1/4	13 31/32	11 1/32	8 7/16	6 7/32	4 9/16	2 3/4	1 9/16	11/16	3/16
1 5 1/2	14 3/16	11 3/16	8 9/16	6 5/16	4 3/8	2 13/16	1 9/16	11/16	3/16
1 5 3/4	14 3/8	11 3/8	8 11/16	6 13/32	4 7/16	2 27/32	1 19/32	23/32	3/16
1 6	14 19/32	11 17/32	8 13/16	6 15/32	4 1/2	2 7/8	1 5/8	23/32	3/16

Ordinates of Parabolas for Crane Girder Design — 2

	Ordinates								
H	9	8	7	6	5	4	3	2	1
Ft. Ins.	Ft. Ins.	Ft. Ins.	Ft. Ins.	Ins.	Ins.	Ins.	Ins.	Ins.	Ins.
1 6	1 2 19/32	11 17/32	8 13/16	6 15/32	4 1/2	2 7/8	1 5/8	23/32	3/16
1 6 1/4	1 2 25/32	11 11/16	8 15/16	6 9/16	4 9/16	2 15/16	1 21/32	23/32	3/16
1 6 1/2	1 3	11 27/32	9 1/16	6 21/32	4 5/8	2 31/32	1 21/32	3/4	3/16
1 6 3/4	1 3 3/16	1 0	9 3/16	6 3/4	4 11/16	3	1 11/16	3/4	3/16
1 7	1 3 13/32	1 0 5/32	9 5/16	6 27/32	4 3/4	3 1/32	1 23/32	3/4	3/16
1 7 1/4	1 3 19/32	1 0 5/16	9 7/16	6 15/16	4 13/16	3 3/32	1 23/32	3/4	3/16
1 7 1/2	1 3 25/32	1 0 15/32	9 9/16	7	4 7/8	3 1/8	1 3/4	25/32	3/16
1 7 3/4	1 4	1 0 21/32	9 11/16	7 1/8	4 15/16	3 5/32	1 25/32	25/32	3/16
1 8	1 4 3/16	1 0 13/16	9 13/16	7 3/16	5	3 3/16	1 13/16	13/16	3/16
1 8 1/4	1 4 13/32	1 0 31/32	9 15/16	7 9/32	5 1/16	3 1/4	1 13/16	13/16	3/16
1 8 1/2	1 4 19/32	1 1 1/8	10 1/32	7 3/8	5 1/8	3 9/32	1 27/32	13/16	7/32
1 8 3/4	1 4 13/16	1 1 9/32	10 5/32	7 15/32	5 3/16	3 5/16	1 7/8	27/32	7/32
1 9	1 5	1 1 7/16	10 9/32	7 9/16	5 1/4	3 3/8	1 7/8	27/32	7/32
1 9 1/4	1 5 7/32	1 1 19/32	10 13/32	7 21/32	5 5/16	3 13/32	1 29/32	27/32	7/32
1 9 1/2	1 5 13/32	1 1 3/4	10 17/32	7 3/4	5 3/8	3 7/16	1 15/16	7/8	7/32
1 9 3/4	1 5 5/8	1 1 15/16	10 21/32	7 27/32	5 7/16	3 15/32	1 31/32	7/8	7/32
1 10	1 5 13/16	1 2 3/32	10 25/32	7 15/16	5 1/2	3 17/32	1 31/32	7/8	7/32
1 10 1/4	1 6 1/32	1 2 1/4	10 29/32	8	5 9/16	3 9/16	2	29/32	7/32
1 10 1/2	1 6 7/32	1 2 7/16	11 1/32	8 3/32	5 5/8	3 19/32	2 1/32	29/32	7/32
1 10 3/4	1 6 7/16	1 2 9/16	11 5/32	8 3/16	5 11/16	3 5/8	2 1/16	29/32	7/32
1 11	1 6 5/8	1 2 23/32	11 9/32	8 9/32	5 3/4	3 11/16	2 1/16	15/16	7/32
1 11 1/4	1 6 15/16	1 2 7/8	11 13/32	8 3/8	5 13/16	3 23/32	2 3/32	15/16	1/4
1 11 1/2	1 7 1/32	1 3 1/32	11 17/32	8 15/32	5 7/8	3 3/4	2 1/8	15/16	1/4
1 11 3/4	1 7 1/4	1 3 3/16	11 5/8	8 9/16	5 15/16	3 13/16	2 1/8	31/32	1/4
2 0	1 7 7/16	1 3 3/8	11 3/4	8 21/32	6	3 27/32	2 5/32	31/32	1/4
2 0 1/4	1 7 21/32	1 3 17/32	11 7/8	8 23/32	6 1/16	3 7/8	2 3/16	31/32	1/4
2 0 1/2	1 7 27/32	1 3 11/16	1 0	8 13/16	6 1/8	3 15/32	2 7/32	31/32	1/4
2 0 3/4	1 8 1/32	1 3 27/32	1 0 1/8	8 29/32	6 3/16	3 31/32	2 7/32	1	1/4
2 1	1 8 1/4	1 4	1 0 1/4	9	6 1/4	4	2 1/4	1	1/4
2 1 1/4	1 8 15/32	1 4 5/32	1 0 3/8	9 3/32	6 5/16	4 1/32	2 9/32	1	1/4
2 1 1/2	1 8 21/32	1 4 5/16	1 0 1/2	9 3/16	6 3/8	4 3/32	2 9/32	1 1/32	1/4
2 1 3/4	1 8 27/32	1 4 15/32	1 0 5/8	9 9/32	6 7/16	4 1/8	2 5/16	1 1/32	1/4
2 2	1 9 1/16	1 4 21/32	1 0 3/4	9 3/8	6 1/2	4 5/32	2 11/32	1 1/32	1/4
2 2 1/4	1 9 1/4	1 4 13/16	1 0 7/8	9 15/32	6 9/16	4 7/32	2 3/8	1 1/16	1/4
2 2 1/2	1 9 15/32	1 4 31/32	1 1	9 17/32	6 5/8	4 1/4	2 3/8	1 1/16	1/4
2 2 3/4	1 9 21/32	1 5 1/8	1 1 3/32	9 5/8	6 11/16	4 9/32	2 13/32	1 1/16	1/4
2 3	1 9 7/8	1 5 9/32	1 1 1/4	9 23/32	6 3/4	4 5/16	2 7/16	1 3/32	9/32
2 3 1/4	1 10 1/16	1 5 7/16	1 1 11/32	9 13/16	6 13/16	4 3/8	2 15/32	1 3/32	9/32
2 3 1/2	1 10 9/32	1 5 19/32	1 1 15/32	9 29/32	6 7/8	4 13/32	2 15/32	1 3/32	9/32
2 3 3/4	1 10 15/32	1 5 3/4	1 1 19/32	10	6 15/16	4 7/16	2 1/2	1 1/8	9/32
2 4	1 10 11/16	1 5 15/16	1 2 3/32	10 3/32	7	4 15/32	2 17/32	1 1/8	9/32
2 4 1/4	1 10 7/8	1 6 3/32	1 2 7/32	10 3/16	7 1/16	4 17/32	2 17/32	1 1/8	9/32
2 4 1/2	1 11 3/32	1 6 1/4	1 1 31/32	10 1/4	7 1/8	4 9/16	2 9/16	1 5/32	9/32
2 4 3/4	1 11 9/32	1 6 13/32	1 2 3/32	10 11/32	7 3/16	4 19/32	2 19/32	1 5/32	9/32
2 5	1 11 1/2	1 6 9/16	1 2 7/32	10 7/16	7 1/4	4 21/32	2 5/8	1 5/32	9/32
2 5 1/4	1 11 11/16	1 6 23/32	1 2 11/32	10 17/32	7 5/16	4 11/16	2 5/8	1 3/16	9/32
2 5 1/2	1 11 29/32	1 6 7/8	1 2 15/32	10 5/8	7 3/8	4 23/32	2 21/32	1 3/16	5/16
2 5 3/4	2 0 3/32	1 7 1/32	1 2 19/32	10 23/32	7 7/16	4 3/4	2 11/32	1 3/16	5/16
2 6	2 0 5/16	1 7 7/32	1 2 23/32	10 13/16	7 1/2	4 13/16	2 23/32	1 7/32	5/16

Deflection as a Limiting Factor in Beam Design. — For some applications, a beam must be stronger than required by the maximum load it is to support, in order to prevent excessive deflection. Since maximum allowable deflections for such cases vary widely for different classes of service, a general formula for determining them cannot be given. When exceptionally stiff girders are required, one rule is to limit the deflection to 1 inch per 100 feet of span; hence, if l = length of span in inches, deflection = $l \div 1200$. According to another formula, deflection limit = $l \div 360$ where beams are adjacent to materials like plaster which would be broken by excessive beam deflection. Some machine parts of the beam type must be very rigid to maintain alignment under load. For example, the deflection of a punch press column may be limited to 0.010 inch or less. These examples merely illustrate variations in practice. It is impracticable to give general formulas for determining the allowable deflection in any case, because the allowable amount depends on the conditions governing each class of work.

Procedure in Designing for Deflection: Assume that a deflection equal to $l \div 1200$ is to be the limiting factor in selecting a wide-flange (W-shape) beam having a span length of 144 inches. Supports are at both ends and load at center is 15,000 pounds. Deflection y is to be limited to $144 \div 1200 = 0.12$ inch. According to the formula on page 260 (Case 2), in which W = load on beam in pounds, l = length of span in inches, E = modulus of elasticity of material, I = moment of inertia of cross section:

$$\text{Deflection } y = \frac{Wl^3}{48EI} \text{; hence } I = \frac{Wl^3}{48yE} = \frac{15,000 \times 144^3}{48 \times 0.12 \times 29,000,000} = 268.1$$

A structural wide-flange beam having a depth of 12 inches and weighing 36 pounds per foot has a moment of inertia I of 281 and a section modulus (Z or S) of 46.0 (see table, page 251). Checking now for maximum stress s (Case 2, page 260):

$$s = \frac{Wl}{4Z} = \frac{15,000 \times 144}{4 \times 46.0} = 11,740 \text{ lbs. per sq. in.}$$

Although deflection is the limiting factor in this case, the maximum stress is checked to make sure that it is within the allowable limit. As the limiting deflection is decreased, for a given load and length of span, the beam strength and rigidity must be increased, and, consequently, the maximum stress is decreased. Thus, in the preceding example, if the maximum deflection is 0.08 inch instead of 0.12 inch, then the calculated value for the moment of inertia I will be 402; hence a W 12 × 53 beam having an I value of 426 could be used (nearest value above 402). The maximum stress then would be reduced to 7640 pounds per square inch and the calculated deflection is 0.076 inch.

A similar example using metric SI units is as follows. Assume that a deflection equal to $l \div 1000$ millimeters is to be the limiting factor in selecting a W-beam having a span length of 5 meters. Supports are at both ends and the load at the center is 30 kilonewtons. Deflection y is to be limited to $5000 \div 1000 = 5$ millimeters. The formula on page 260 (Case 2) is applied, and W = load on beam in newtons; l = length of span in mm; E = modulus of elasticity (assume 200,000 N/mm² in this example); and I = moment of inertia of cross-section in millimeters⁴. Thus,

$$\text{Deflection } y = \frac{Wl^3}{48EI} \text{;}$$

hence

$$I = \frac{Wl^3}{48yE} = \frac{30,000 \times 5000^3}{48 \times 5 \times 200,000} = 78,125,000 \text{ mm}^4$$

Although deflection is the limiting factor in this case, the maximum stress is checked

STRENGTH OF MATERIALS

Values of the Stress Correction Factor K for Various Curved Beam Sections

Section	R/c	Factor K Inside Fiber	Factor K Outside Fiber	y_0*
Circle/Ellipse (R, c, h)	1.2	3.41	.54	.224R
	1.4	2.40	.60	.151R
	1.6	1.96	.65	.108R
	1.8	1.75	.68	.084R
	2.0	1.62	.71	.069R
	3.0	1.33	.79	.030R
	4.0	1.23	.84	.016R
	6.0	1.14	.89	.0070R
	8.0	1.10	.91	.0039R
	10.0	1.08	.93	.0025R
Rectangle (c, R)	1.2	2.89	.57	.305R
	1.4	2.13	.63	.204R
	1.6	1.79	.67	.149R
	1.8	1.63	.70	.112R
	2.0	1.52	.73	.090R
	3.0	1.30	.81	.041R
	4.0	1.20	.85	.021R
	6.0	1.12	.90	.0093R
	8.0	1.09	.92	.0052R
	10.0	1.07	.94	.0033R
Trapezoid (b, c, 2b, R)	1.2	3.01	.54	.336R
	1.4	2.18	.60	.229R
	1.6	1.87	.65	.168R
	1.8	1.69	.68	.128R
	2.0	1.58	.71	.102R
	3.0	1.33	.80	.046R
	4.0	1.23	.84	.024R
	6.0	1.13	.88	.011R
	8.0	1.10	.91	.0060R
	10.0	1.08	.93	.0039R
Trapezoid (3b, c, 2b, R)	1.2	3.09	.56	.336R
	1.4	2.25	.62	.229R
	1.6	1.91	.66	.168R
	1.8	1.73	.70	.128R
	2.0	1.61	.73	.102R
	3.0	1.37	.81	.046R
	4.0	1.26	.86	.024R
	6.0	1.17	.91	.011R
	8.0	1.13	.94	.0060R
	10.0	1.11	.95	.0039R
Trapezoid (5b, c, 4b, R)	1.2	3.14	.52	
	1.4	2.29	.54	.243R
	1.6	1.93	.62	.179R
	1.8	1.74	.65	.138R
	2.0	1.61	.68	.110R
	3.0	1.34	.76	.050R
	4.0	1.24	.82	.028R
	6.0	1.15	.87	.012R
	8.0	1.12	.91	.0060R
	10.0	1.10	.93	.0039R
Triangle (3/5 b, c, b, R)	1.2	3.26	.44	.361R
	1.4	2.39	.50	.251R
	1.6	1.99	.54	.186R
	1.8	1.78	.57	.144R
	2.0	1.66	.60	.116R
	3.0	1.37	.70	.052R
	4.0	1.27	.75	.029R
	6.0	1.16	.82	.013R
	8.0	1.12	.86	.0060R
	10.0	1.09	.88	.0039R

Section	R/c	Factor K Inside Fiber	Factor K Outside Fiber	y_0*
T-section ($4\frac{1}{2}t$, $3t$, t, $4t$, c, R)	1.2	3.63	.58	.418R
	1.4	2.54	.63	.299R
	1.6	2.14	.67	.229R
	1.8	1.89	.70	.183R
	2.0	1.73	.72	.149R
	3.0	1.41	.79	.069R
	4.0	1.29	.83	.040R
	6.0	1.18	.88	.018R
	8.0	1.13	.91	.010R
	10.0	1.10	.92	.0065R
T-section ($3t$, $2t$, t, $4t$, $6t$, c, R)	1.2	3.55	.67	.409R
	1.4	2.48	.72	.292R
	1.6	2.07	.76	.224R
	1.8	1.83	.78	.178R
	2.0	1.69	.80	.144R
	3.0	1.38	.86	.067R
	4.0	1.26	.89	.038R
	6.0	1.15	.92	.018R
	8.0	1.10	.94	.010R
	10.0	1.08	.95	.0065R
I-section (t, $4t$, t, $3t$, $3t$, c, R)	1.2	2.52	.67	.408R
	1.4	1.90	.71	.285R
	1.6	1.63	.75	.208R
	1.8	1.50	.77	.160R
	2.0	1.41	.79	.127R
	3.0	1.23	.86	.058R
	4.0	1.16	.89	.030R
	6.0	1.10	.92	.013R
	8.0	1.07	.94	.0076R
	10.0	1.05	.95	.0048R
Hollow circle (2d, d, c, R)	1.2	3.28	.58	.269R
	1.4	2.31	.64	.182R
	1.6	1.89	.68	.134R
	1.8	1.70	.71	.104R
	2.0	1.57	.73	.083R
	3.0	1.31	.81	.038R
	4.0	1.21	.85	.020R
	6.0	1.13	.90	.0087R
	8.0	1.10	.92	.0049R
	10.0	1.07	.93	.0031R
Hollow square ($t/2$, $4t$, $4t$, $2t$, t, c, R)	1.2	2.63	.68	.399R
	1.4	1.97	.73	.280R
	1.6	1.66	.76	.205R
	1.8	1.51	.78	.159R
	2.0	1.43	.80	.127R
	3.0	1.23	.86	.058R
	4.0	1.15	.89	.031R
	6.0	1.09	.92	.014R
	8.0	1.07	.94	.0076R
	10.0	1.06	.95	.0048R

Example: The fiber stresses of a curved rectangular beam are calculated as 5000 psi using the straight beam formula, $S = Mc/I$. If the beam is 8 inches deep and its radius of curvature is 12 inches, what are the true stresses? $R/c = 12/4 = 3$. The factors in the table corresponding to $R/c = 3$ are 0.81 and 1.30. Outside fiber stress $= 5000 \times 0.81 = 4050$ psi; inside fiber stress $= 5000 \times 1.30 = 6500$ psi.

*y_0 is the distance from the centroidal axis to the neutral axis of curved beams subjected to pure bending and is measured from the centroidal axis toward the center of curvature.

to make sure that it is within the allowable limit, using the formula from page 260 (Case 2):

$$s = \frac{Wl}{4Z}$$

The units of s are newtons per square millimeter; W is the load in newtons; l is the length in mm; and Z = section modulus of the cross-section of the beam = I ÷ distance in mm from neutral axis to extreme fiber.

Curved Beams. — The formula $S = Mc/I$ used to compute stresses due to bending of beams is based on the assumption that the beams are straight before any loads are applied. In the case of beams having initial curvature, however, the stresses may be considerably higher than predicted by the ordinary straight-beam formula since the effect of initial curvature is to shift the neutral axis of a curved member in from the gravity axis toward the center of curvature (the concave side of the beam). This shift in the position of the neutral axis causes an increase in the stress on the concave side of the beam and decreases the stress at the outside fibers.

Hooks, press frames, and other machine members which as a rule have a rather pronounced initial curvature may have a maximum stress at the inside fibers of up to about 3½ times that predicted by the ordinary straight-beam formula.

Stress Correction Factors for Curved Beams: A simple method for determining the maximum fiber stress due to bending of curved members consists of (1) calculating the maximum stress using the straight-beam formula $S = Mc/I$; and (2) multiplying the calculated stress by a stress correction factor. The table on page 281 gives stress correction factors for some of the common cross-sections and proportions used in the design of curved members.

An example in the application of the method using English units of measurement is given at the bottom of the table. **A similar example using metric SI units is as follows: The fiber stresses of a curved rectangular beam are calculated as 40 newtons per millimeter^2, using the straight beam formula, $S = Mc/I$. If the beam is 150 mm deep and its radius of curvature is 300 mm, what are the true stresses?** $R/c = 300/75 = 4$. From the table on page 281, the K factors corresponding to $R/c = 4$ are 1.20 and 0.85. Thus, the inside fiber stress is $40 \times 1.20 = 48$ N/mm^2 = 48 megapascals; and the outside fiber stress is $40 \times 0.85 = 34$ N/mm^2 = 34 megapascals.

Approximate Formula for Stress Correction Factor: The stress correction factors given in the table on page 281 were determined by Wilson and Quereau and published in the University of Illinois Engineering Experiment Station Circular No. 16, "A Simple Method of Determining Stress in Curved Flexural Members." In this same publication the authors indicate that the following empirical formula may be used to calculate the value of the stress correction factor for the *inside* fibers of sections not covered by the tabular data to within 5 per cent accuracy except in the case of triangular sections where up to 10 per cent deviation may be expected. However, for most engineering calculations, this should prove to be a satisfactory formula for general use in determining the factor for the inside fibers.

$$K = 1.00 + 0.5\frac{I}{bc^2}\left[\frac{1}{R-c} + \frac{1}{R}\right]$$

(Use 1.05 instead of 0.5 in this formula for circular and elliptical sections.)

> I = Moment of inertia of section about centroidal axis;
> b = maximum width of section;
> c = distance from centroidal axis to inside fiber, i.e., to the extreme fiber nearest the center of curvature;
> R = radius of curvature of centroidal axis of beam.

Example: On the accompanying diagram are shown the dimensions of a clamp frame of rectangular cross-section. Determine the maximum stress at points *A* and *B* due to a clamping force of 1000 pounds.

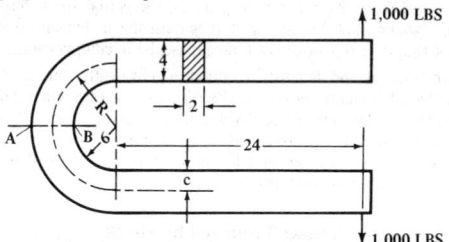

The cross-sectional area $= 2 \times 4 = 8$ square inches; the bending moment at section *AB* is $1000 (24 + 6 + 2) = 32,000$ inch pounds; the distance from the center of gravity of the section at *AB* to point *B* is $c = 2$ inches; and the moment of inertia of the section is, using the formula on page 260, $2 \times (4)^3 \div 12 = 10.667$ inches^4.

Using the straight-beam formula, page 314, the stress at points *A* and *B* due to the bending moment is:

$$S = \frac{Mc}{I} = \frac{32,000 \times 2}{10.667} = 6000 \text{ psi}$$

The stress at *A* is a compressive stress of 6000 psi while that at *B* is a tensile stress of 6000 psi.

These values must be corrected to account for the curvature effect. In the table on page 281 for $R/c = (6 + 2)/(2) = 4$, the value of *K* is found to be 1.20 and 0.85 for points *B* and *A* respectively. Thus, the actual stress due to bending at point *B* is 1.20 $\times$ 6000 = 7200 psi in tension while the stress at point *A* is 0.85 $\times$ 6000 = 5100 psi in compression.

To these stresses at *A* and *B* must be added, algebraically, the direct stress at section *AB* due to the 1000-pound clamping force. The direct stress on section *AB* will be a tensile stress equal to the clamping force divided by the section area. Thus $1000 \div 8 = 125$ psi in tension.

The maximum unit stress at *A* is, therefore, $5100 - 125 = 4975$ psi in compression while the maximum unit stress at *B* is $7200 + 125 = 7325$ psi in tension.

The following is a similar calculation using metric SI units, assuming that it is required to determine the maximum stress at points *A* and *B* due to clamping force of 4 kilonewtons acting on the frame. The frame cross-section is 50 by 100 millimeters, the radius $R = 200$ mm, and the length of the straight portions is 600 mm. Thus, the cross-sectional area $= 50 \times 100 = 5000$ mm^2; the bending moment at *AB* is $4000(600 + 200) = 3,200,000$ newton-millimeters; the distance from the center of gravity of the section at *AB* to point *B* is $c = 50$ mm; and the moment of inertia of the section is, using the formula on page 226, $50 \times (100)^3 \div 12 = 4,170,000$ mm^4.

Using the straight-beam formula, page 280, the stress at points *A* and *B* due to the bending moment is:

$$s = \frac{Mc}{I} = \frac{3,200,000 \times 50}{4,170,000}$$

$$= 38.4 \text{ newtons per millimeter}^2 = 38.4 \text{ megapascals.}$$

The stress at A is a compressive stress of 38.4 N/mm², while that at B is a tensile stress of 38.4 N/mm². These values must be corrected to account for the curvature effect. From the table on page 281, the K factors are 1.20 and 0.85 for points A and B respectively, derived from $R/c = 200/50 = 4$. Thus, the actual stress due to bending at point B is $1.20 \times 38.4 = 46.1$ N/mm² (46.1 megapascals) in tension; and the stress at point A is $0.85 \times 38.4 = 32.6$ N/mm² (32.6 megapascals) in compression.

To these stresses at A and B must be added, algebraically, the direct stress at section AB due to the 4 kN clamping force. The direct stress on section AB will be a tensile stress equal to the clamping force divided by the section area. Thus, $4000/5000 = 0.8$ N/mm². The maximum unit stress at A is, therefore, $32.61 - 0.8 = 31.8$ N/mm² (31.8 megapascals) in compression, and the maximum unit stress at B is $46.1 + 0.8 = 46.9$ N/mm² (46.9 megapascals) in tension.

Stresses Produced by Shocks

Stresses in Beams Produced by Shocks. — Any elastic structure subjected to a shock will deflect until the product of the average resistance, developed by the deflection, and the distance through which it has been overcome, has reached a value equal to the energy of the shock. It follows that for a given shock, the average resisting stresses are inversely proportional to the deflection. If the structure were perfectly rigid, the deflection would be zero, and the stress infinite. The effect of a shock is, therefore, to a great extent dependent upon the elastic property (the springiness) of the structure subjected to the impact.

The energy of a body in motion, such as a falling body, may be spent in each of four ways:

1. In deforming the body struck as a whole.

2. In deforming the falling body as a whole.

3. In partial deformation of both bodies on the surface of contact (most of this energy will be transformed into heat).

4. Part of the energy will be taken up by the supports, if these are not perfectly rigid and inelastic.

How much energy is spent in the last three ways it is in most cases difficult to determine, and for this reason it is safest to figure as if the whole amount were spent as in Case 1. In cases where a reliable judgment is possible, as to what percentage of the energy is spent in other ways than the first, a corresponding fraction of the total energy can be assumed as developing stresses in the body subjected to shocks.

One investigation into the stresses produced by shocks led to the following conclusions: (1) A suddenly applied load will produce the same deflection, and, therefore, the same stress as a static load twice as great. (2) The unit stress p (see formulas in the accompanying table) for a given load producing a shock, varies directly as the square root of the modulus of elasticity E, and inversely as the square root of the length L of the beam and the area of the section. Thus, for instance, if the sectional area of a beam is increased four times, the unit stress will diminish only by half. This is entirely different from the results produced by static loads where the stress would vary inversely with the area, and within certain limits be practically independent of the modulus of elasticity.

In the table, the expression for the approximate value of p, which is applicable whenever the deflection of the beam is small as compared with the total height h through which the body producing the shock is dropped, is always the same for beams supported at both ends and subjected to shock at *any* point between the supports. In the formulas all dimensions are in inches and weights in pounds.

If metric SI units are used, p is in newtons per square millimeter; Q is in newtons;

Stresses Produced in Beams by Shocks

Method of Support and Point Struck by Falling Body	Fiber (Unit) Stress p produced by Weight Q Dropped Through a Distance h	Approximate Value of p
Supported at both ends; struck in center.	$p = \dfrac{QaL}{4I}\left(1 + \sqrt{1 + \dfrac{96hEI}{QL^3}}\right)$	$p = a\sqrt{\dfrac{6QhE}{LI}}$
Fixed at one end; struck at the other.	$p = \dfrac{QaL}{I}\left(1 + \sqrt{1 + \dfrac{6hEI}{QL^3}}\right)$	$p = a\sqrt{\dfrac{6QhE}{LI}}$
Fixed at both ends; struck in center.	$p = \dfrac{QaL}{8I}\left(1 + \sqrt{1 + \dfrac{384hEI}{QL^3}}\right)$	$p = a\sqrt{\dfrac{6QhE}{LI}}$

I = moment of inertia of section; a = distance of extreme fiber from neutral axis; L = length of beam; E = modulus of elasticity.

E = modulus of elasticity in N/mm²; I = moment of inertia of section in millimeters⁴; and h, a, and L in millimeters. *Note:* If Q is given in kilograms, the value referred to is mass. The weight Q of a mass M kilograms is Mg newtons, where g = approximately 9.81 meters per second².

Examples of How Formulas for Stresses Produced by Shocks are Derived: The general formula from which specific formulas for shock stresses in beams, springs, and other machine and structural members are derived is:

$$p = p_s\left(1 + \sqrt{1 + \frac{2h}{y}}\right) \qquad (1)$$

In this formula, p = stress in pounds per square inch due to shock caused by impact of a moving load; p_s = stress in pounds per square inch resulting when moving load is applied statically; h = distance in inches that load falls before striking beam, spring, or other member; y = deflection, in inches, resulting from static load.

As an example of how Formula (1) may be used to obtain a formula for a specific application, suppose that the load W shown applied to the beam in Case 2 on page 260 were dropped on the beam from a height of h inches instead of being gradually applied (static loading). The maximum stress p_s due to load W for Case 2 is given as $Wl \div 4Z$ and the maximum deflection y is given as $Wl^3 \div 48\,EI$. Substituting these values in Formula (1),

$$p = \frac{Wl}{4Z}\left(1 + \sqrt{1 + \frac{2h}{Wl^3 \div 48\,EI}}\right) = \frac{Wl}{4Z}\left(1 + \sqrt{1 + \frac{96\,hEI}{Wl^3}}\right) \qquad (2)$$

If in Formula (2) the letter Q is used in place of W and if Z, the section modulus, is replaced by its equivalent, $I \div$ distance a from neutral axis to extreme fiber of beam, then Formula (2) becomes the first formula given in the accompanying table "Stresses Produced in Beams by Shocks."

Stresses in Helical Springs Produced by Shocks. — A load suddenly applied on a spring will produce the same deflection, and, therefore, also the same unit stress, as a static load twice as great. When the load drops from a height h, the stresses are as given in the accompanying table. The approximate values are applicable when the deflection is small as compared with the height h. The formulas show that the fiber stress for a given shock will be greater in a spring made from a square bar, than in one made from a round bar, if the diameter of coil be the same, and the side of the square bar equals the diameter of the round bar. It is, therefore, more economical to use round stock for springs which must withstand shocks. This is due to the fact that the deflection for the same fiber stress for a square bar spring is smaller than that for a round bar spring, the ratio being as 4 to 5. The round bar spring is therefore capable of storing more energy than a square bar spring for the same stress.

Stresses Produced in Springs by Shocks

Form of Bar from Which Spring is Made	Fiber (Unit) Stress f Produced by Weight Q Dropped a Height h on a Helical Spring	Approximate Value of f
Round	$f = \dfrac{8}{\pi d^3} QD \left(1 + \sqrt{1 + \dfrac{Ghd^4}{4\, QD^3 n}} \right)$	$f = 1.27 \sqrt{\dfrac{QhG}{Dd^2 n}}$
Square	$f = \dfrac{9}{4} \dfrac{QD}{d^3} \left(1 + \sqrt{1 + \dfrac{Ghd^4}{0.9\, \pi QD^3 n}} \right)$	$f = 1.34 \sqrt{\dfrac{QhG}{Dd^2 n}}$

G = modulus of elasticity for torsion; d = diameter or side of bar; D = mean diameter of spring; n = number of coils in spring.

Shocks from Bodies in Motion. — The formulas given can be applied, in general, to shocks from bodies in motion. A body of the weight W moving horizontally with the velocity of v feet per second, has a stored-up energy:

$$E_K = \frac{1}{2} \times \frac{Wv^2}{g} \text{ foot-pounds, or } \frac{6Wv^2}{g} \text{ inch-pounds.}$$

This expression may be substituted for Qh in the tables in the equations for unit stresses containing this quantity, and the stresses produced by the energy of the moving body thereby determined.

The formulas in the tables give the maximum value of the stresses, providing the designer with some definitive guidance even where there may be justification for assuming that only a part of the energy of the shock is taken up by the member under stress.

The formulas can also be applied using metric SI units. The stored-up energy of a body of mass M kilograms moving horizontally with the velocity of v meters per second is:

$$E_K = \frac{1}{2}Mv^2 \text{ newton-meters.}$$

This expression may be substituted for Qh in the appropriate equations in the tables. For calculation in millimeters, $Qh = 1000\, E_K$ newton-millimeters.

Size of Rail Necessary to Carry a Given Load. — The following formulas may be employed for determining the size of rail and wheel suitable for carrying a given load. Let, A = the width of the head of the rail in inches; B = width of the tread of the rail in inches; C = the wheel-load in pounds; D = the diameter of the wheel in inches.

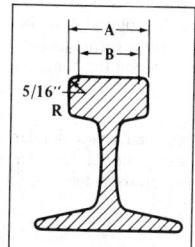

Then the width of the tread of the rail in inches is found from the formula:

$$B = \frac{C}{1250D} \qquad (1)$$

The width A of the head equals $B + \tfrac{5}{8}$ inch. The diameter D of the smallest track wheel that will safely carry the load is found from the formula:

$$D = \frac{C}{A \times K} \qquad (2)$$

in which K = 600 to 800 for steel castings; K = 300 to 400 for cast iron.

As an example, assume that the wheel-load in a given case is 10,000 pounds; the diameter of the wheel is 20 inches; and the material steel casting. Determine the size of rail necessary to carry this load. From Formula (1):

$$B = \frac{10,000}{1250 \times 20} = 0.4 \text{ inch.}$$

Hence the width of the rail required equals $0.4 + \tfrac{5}{8}$ inch = 1.025 inch. Determine also whether a wheel 20 inches in diameter is large enough to safely carry the load. From Formula (2):

$$D = \frac{10,000}{1.025 \times 600} = 16\tfrac{1}{4} \text{ inches.}$$

This is the smallest diameter of track wheel that will safely carry the load; hence a 20-inch wheel is ample.

American Railway Engineering Association Formulas. — The American Railway Engineering Association recommends for safe operation of steel cylinders rolling on steel plates that the allowable load p in pounds per inch of length of the cylinder should not exceed the value calculated from the formula

$$p = \frac{\text{y.s.} - 13,000}{20,000} \, 600 \, d \quad \text{for diameter } d \text{ less than 25 inches.}$$

This formula is based on steel having a yield strength, y.s., of 32,000 pounds per square inch. For roller or wheel diameters of up to 25 inches, the Hertz stress (contact stress) resulting from the calculated load p will be approximately 76,000 pounds per square inch.

Example: For a 10-inch diameter roller the safe load per inch of roller length is

$$p = \frac{32,000 - 13,000}{20,000} \, 600 \times 10 = 5700 \text{ lbs per inch of length.}$$

Therefore, to support a 10,000 pound load the roller or wheel would need to be 10,000/5700 = 1.75 inches wide.

Columns

Strength of Columns or Struts. — Structural members which are subject to compression may be so long in proportion to the diameter or lateral dimensions that failure may be the result (1) of both compression and bending or (2) of bending or buckling to such a degree that compression stress may be ignored. In such cases, the *slenderness ratio* is important. This ratio equals the length l of the column in inches divided by the least radius of gyration r of the cross-section. Various formulas have been used for designing columns which are too slender to be designed for compression only.

Rankine or Gordon Formula. — This formula is generally applied when slenderness ratios range between 20 and 100, and sometimes for ratios up to 120. The notation, in English and metric SI units of measurement, is given on page 290.

$$p = \frac{S}{1 + K\left(\dfrac{l}{r}\right)^2} = \text{ultimate load, lbs. per sq. in.}$$

Factor K may be established by tests with a given material and end condition, and for the probable range of l/r. If determined by calculation, $K = S/C\pi^2 E$. Factor C equals 1 for either rounded or pivoted column ends, 4 for fixed ends, and 1 to 4 for square flat ends. The factors 25,000, 12,500, etc., in the Rankine formulas arranged as on page 290 equal $1/K$, and have been used extensively.

Straight-line Formula. — This general type of formula is often used in designing compression members for buildings, bridges, or similar structural work. It is convenient especially in designing a number of columns which are made of the same material but vary in size, assuming that factor B is known. This factor is determined by tests.

$$p = S_y - B\left(\frac{l}{r}\right) = \text{ultimate load, lbs. per sq. in.}$$

S_y equals yield point, lbs. per square inch, and factor B ranges from 50 to 100. Safe unit stress = p/factor of safety.

Formulas of American Railway Engineering Association. — The formulas which follow apply to structural steel having an ultimate strength of 60,000 to 72,000 pounds per square inch.

For building columns having l/r ratios not greater than 120, allowable unit stress = $17,000 - 0.485\, l^2/r^2$. For columns having l/r ratios greater than 120, allowable unit stress $= \dfrac{18,000}{1 + l^2/18,000 r^2}$

For bridge compression members centrally loaded and with values of l/r not greater than 140:

$$\text{Allowable unit stress, riveted ends} = 15,000 - \frac{1}{4}\frac{l^2}{r^2}$$

$$\text{Allowable unit stress, pin ends} = 15,000 - \frac{1}{3}\frac{l^2}{r^2}$$

Euler Formula. — This formula is for columns which are so slender that bending or buckling action predominates and compressive stresses are not taken into account.

$$P = \frac{C\pi^2 IE}{l^2} = \text{total ultimate load, in pounds}$$

The notation, in English and metric SI units of measurement, is given on page 290. Factors C for different end conditions are included in the Euler formulas at the bottom of the table. According to a series of experiments, Euler formulas should be used if the values of l/r exceed the following ratios: Structural steel and flat ends, 195; hinged ends, 155; round ends, 120; cast iron with flat ends, 120; hinged ends, 100; round ends, 75; oak with flat ends, 130. The *critical slenderness ratio,* which marks the dividing line between the shorter columns and those slender enough to warrant using the Euler formula, depends upon the column material and its end conditions. If the Euler formula is applied when the slenderness ratio is too small, the *calculated* ultimate strength will exceed the yield point of the material and, obviously, will be incorrect.

Eccentrically Loaded Columns. — In the application of the column formulas previously referred to, it is assumed that the action of the load coincides with the axis of the column. If the load is offset relative to the column axis, the column is said to be eccentrically loaded, and its strength is then calculated by using a modification of the Rankine formula, the quantity cz/r^2 being added to the denominator, as shown in the table on the next page. This modified formula is applicable to columns having a slenderness ratio varying from 20 or 30 to about 100.

Machine Elements Subjected to Compressive Loads. — As in the case of structural compression members, an unbraced machine member that is relatively slender, i.e., its length is more than, say, six times the least dimension perpendicular to its longitudinal axis, is usually designed as a column, since failure due to overloading (assuming a compressive load centrally applied in an axial direction) may occur by buckling or a combination of buckling and compression rather than by direct compression alone. In the design of unbraced steel machine "columns" which are to carry compressive loads applied along their longitudinal axes, two formulas are in general use:

(Euler)
$$P_{cr} = \frac{s_y A r^2}{Q} \tag{1}$$

(J. B. Johnson)
$$P_{cr} = A s_y \left(1 - \frac{Q}{4r^2} \right) \quad \text{where} \quad Q = \frac{s_y l^2}{n\pi^2 E} \tag{2 and 3}$$

In these formulas, P_{cr} = critical load in pounds that would result in failure of the column; A = cross-sectional area, square inches; s_y = yield point of material, pounds per square inch; r = least radius of gyration of cross-section, inches; E = modulus of elasticity, pounds per square inch; l = column length, inches; and n = coefficient for end conditions. For both ends fixed, $n = 4$; for one end fixed, one end free, $n = 0.25$; for one end fixed and the other end free but guided, $n = 2$; for round or pinned ends, free but guided, $n = 1$; and for flat ends, $n = 1$ to 4. It should be noted that these values of n represent ideal conditions that are seldom attained in practice; for example, for both ends fixed a value of $n = 3$ to 3.5 may be more realistic than $n = 4$.

If metric SI units are used in these formulas, P_{cr} = critical load in newtons that would result in failure of the column; A = cross-sectional area, square millimeters; s_y = yield point of the material, newtons per square mm; r = least radius of gyration or cross-section, mm; E = modulus of elasticity newtons per square mm; l = column length, mm; and n = a coefficient for end conditions. The coefficients given are valid for calculations in metric units.

Rankine's and Euler's Formulas for Columns

Symbol	Quantity	English Unit	Metric SI Units
p	Ultimate unit load	Lbs./sq. in.	Newtons/sq. mm.
P	Total ultimate load	Pounds	Newtons
S	Ultimate compressive strength of material	Lbs./sq. in.	Newtons/sq. mm.
l	Length of column or strut	Inches	Millimeters
r	Least radius of gyration	Inches	Millimeters
I	Least moment of inertia	Inches^4	Millimeters^4
r^2	Moment of inertia/area of section	Inches^2	Millimeters^2
E	Modulus of elasticity of material	Lbs./sq. in.	Newtons/sq. mm.
c	Distance from neutral axis of cross-section to side under compression	Inches	Millimeters
z	Distance from axis of load to axis coinciding with center of gravity of cross-section	Inches	Millimeters

Rankine's Formulas

Material	Both Ends of Column Fixed	One End Fixed and One End Rounded	Both Ends Rounded
Steel	$p = \dfrac{S}{1 + \dfrac{l^2}{25{,}000\, r^2}}$	$p = \dfrac{S}{1 + \dfrac{l^2}{12{,}500\, r^2}}$	$p = \dfrac{S}{1 + \dfrac{l^2}{6250\, r^2}}$
Cast Iron	$p = \dfrac{S}{1 + \dfrac{l^2}{5000\, r^2}}$	$p = \dfrac{S}{1 + \dfrac{l^2}{2500\, r^2}}$	$p = \dfrac{S}{1 + \dfrac{l^2}{1250\, r^2}}$
Wrought Iron	$p = \dfrac{S}{1 + \dfrac{l^2}{35{,}000\, r^2}}$	$p = \dfrac{S}{1 + \dfrac{l^2}{17{,}500\, r^2}}$	$p = \dfrac{S}{1 + \dfrac{l^2}{8750\, r^2}}$
Timber	$p = \dfrac{S}{1 + \dfrac{l^2}{3000\, r^2}}$	$p = \dfrac{S}{1 + \dfrac{l^2}{1500\, r^2}}$	$p = \dfrac{S}{1 + \dfrac{l^2}{750\, r^2}}$

Formulas Modified for Eccentrically Loaded Columns

Material*	Both Ends of Column Fixed	One End Fixed and One End Rounded	Both Ends Rounded
Steel	$p = \dfrac{S}{1 + \dfrac{l^2}{25{,}000\, r^2} + \dfrac{cz}{r^2}}$	$p = \dfrac{S}{1 + \dfrac{l^2}{12{,}500\, r^2} + \dfrac{cz}{r^2}}$	$p = \dfrac{S}{1 + \dfrac{l^2}{6250\, r^2} + \dfrac{cz}{r^2}}$

* For other materials such as cast iron, etc., use the Rankine formulas given in the upper table and add to the denominator the quantity $\dfrac{cz}{r^2}$.

Euler's Formulas for Slender Columns

Both Ends of Column Fixed	One End Fixed and One End Rounded	Both Ends Rounded	One End Fixed and One End Free
$P = \dfrac{4\pi^2 IE}{l^2}$	$P = \dfrac{2\pi^2 IE}{l^2}$	$P = \dfrac{\pi^2 IE}{l^2}$	$P = \dfrac{\pi^2 IE}{4l^2}$

Allowable Working Loads for Columns: To find the total allowable working load for a given section, divide the total ultimate load P (or $p \times$ area), as found by the appropriate formula above, by a suitable factor of safety.

Allowable Concentric Loads for Steel Pipe Columns†

STANDARD STEEL PIPE								
	Nominal Diameter of Pipe, Inches							
	12	10	8	6	5	4	3½	3
	Wall Thickness of Pipe, Inch							
Effective	.375	.365	.322	.280	.258	.237	.226	.216
Length (KL),	Weight per Foot of Pipe, Pounds							
Feet*	49.56	40.48	28.55	18.97	14.62	10.79	9.11	7.58
	Allowable Concentric Loads in Thousands of Pounds							
6	303	246	171	110	83	59	48	38
7	301	243	168	108	81	57	46	36
8	299	241	166	106	78	54	44	34
9	296	238	163	103	76	52	41	31
10	293	235	161	101	73	49	38	28
11	291	232	158	98	71	46	35	25
12	288	229	155	95	68	43	32	22
13	285	226	152	92	65	40	29	19
14	282	223	149	89	61	36	25	16
15	278	220	145	86	58	33	22	14
16	275	216	142	82	55	29	19	12
17	272	213	138	79	51	26	17	11
18	268	209	135	75	47	23	15	10
19	265	205	131	71	43	21	14	9
20	261	201	127	67	39	19	12	
22	254	193	119	59	32	15	10	
24	246	185	111	51	27	13		
25	242	180	106	47	25	12		
26	238	176	102	43	23			

EXTRA STRONG STEEL PIPE								
	Nominal Diameter of Pipe, Inches							
	12	10	8	6	5	4	3½	3
	Wall Thickness of Pipe, Inch							
Effective	.500	.500	.500	.432	.375	.337	.318	.300
Length (KL),	Weight per Foot of Pipe, Pounds							
Feet*	65.42	54.74	43.39	28.57	20.78	14.98	12.50	10.25
	Allowable Concentric Loads in Thousands of Pounds							
6	400	332	259	166	118	81	66	52
7	397	328	255	162	114	78	63	48
8	394	325	251	159	111	75	59	45
9	390	321	247	155	107	71	55	41
10	387	318	243	151	103	67	51	37
11	383	314	239	146	99	63	47	33
12	379	309	234	142	95	59	43	28
13	375	305	229	137	91	54	38	24
14	371	301	224	132	86	49	33	21
15	367	296	219	127	81	44	29	18
16	363	291	214	122	76	39	25	16
18	353	281	203	111	65	31	20	12
19	349	276	197	105	59	28	18	11
20	344	271	191	99	54	25	16	
21	337	265	185	92	48	22	14	
22	334	260	179	86	44	21		
24	323	248	166	73	37	17		
26	312	236	152	62	32			
28	301	224	137	54	27			

Footnotes appear at bottom of continued table on next page.

Allowable Concentric Loads for Steel Pipe Columns† (Concluded)

	DOUBLE-EXTRA STRONG STEEL PIPE				
	Nominal Diameter of Pipe, Inches				
	8	6	5	4	3
	Wall Thickness of Pipe, Inch				
Effective Length (KL), Feet*	.875	.864	.750	.674	.600
	Weight per Foot of Pipe, Pounds				
	72.42	53.16	38.55	27.54	18.58
	Allowable Concentric Loads in Thousands of Pounds				
6	431	306	216	147	91
7	424	299	209	140	84
8	417	292	202	133	77
9	410	284	195	126	69
10	403	275	187	118	60
11	395	266	178	109	51
12	387	257	170	100	43
13	378	247	160	91	37
14	369	237	151	81	32
15	360	227	141	70	28
16	351	216	130	62	24
17	341	205	119	55	22
18	331	193	108	49	
19	321	181	97	44	
20	310	168	87	40	
22	288	142	72	33	
24	264	119	61		
26	240	102	52		
28	213	88	44		

EFFECTIVE LENGTH FACTORS (K) FOR VARIOUS COLUMN CONFIGURATIONS

	(a)	(b)	(c)	(d)	(e)	(f)
Buckled shape of column is shown by dashed line						
Theoretical K value	0.5	0.7	1.0	1.0	2.0	2.0
Recommended design value when ideal conditions are approximated	0.65	0.80	1.2	1.0	2.10	2.0
End condition code	Rotation fixed and translation fixed					
	Rotation free and translation fixed					
	Rotation fixed and translation free					
	Rotation free and translation free					

† Data from "Manual of Steel Construction," 8th ed., 1980, with permission of the American Institute of Steel Construction.

* With respect to radius of gyration. The effective length (KL) is the actual unbraced length, L, in feet, multiplied by the effective length factor (K) which is dependent upon the restraint at the ends of the unbraced length and the means available to resist lateral movements. K may be determined by referring to the last portion of this table.

Load tables are given for 36 ksi yield stress steel. No load values are given below the heavy horizontal lines, since the Kl/r ratios (where l is the actual unbraced length in inches and r is the governing radius of gyration in inches) would exceed 200.

Application of Euler and Johnson Formulas: To determine whether the Euler or Johnson formula is applicable in any particular case it is necessary to determine the value of the quantity $Q \div r^2$. If $Q \div r^2$ is greater than 2, then the Euler formula (1) should be used; if $Q \div r^2$ is less than 2, then the J. B. Johnson formula is applicable. Most compression members in machine design are in the range of proportions covered by the Johnson formula. For this reason a good procedure is to design machine elements on the basis of the Johnson formula and then as a check calculate $Q \div r^2$ to determine whether the Johnson formula applies or the Euler formula should have been used.

Factor of Safety for Machine Columns: When the conditions of loading and the physical qualities of the material used are accurately known, a factor of safety as low as 1.25 is sometimes used when minimum weight is important. For the usual case, however, a factor of safety of 2 to 2.5 is applied for steady loads. The factor of safety represents the ratio of the critical load P_{cr} to the working load.

Examples of Compression Member Design: A rectangular machine member 24 inches long and ½ × 1 inch in cross-section is to carry a compressive load of 4000 pounds along its axis. What is the factor of safety for this load considering that the material is machinery steel having a yield point of 40,000 pounds per square inch, the load is steady, and each end of the rod has a ball connection so that $n = 1$?

From Formula (3)

$$Q = \frac{40{,}000 \times 24 \times 24}{1 \times 3.1416 \times 3.1416 \times 30{,}000{,}000} = 0.0778$$

(The values 40,000 and 30,000,000 were obtained from the table on page 512.)

The radius of gyration r for a rectangular section (page 226) is $0.289 \times$ the dimension in the direction of bending. In columns, bending is most apt to occur in the direction in which the section is the weakest, the ½-inch dimension in this case. Hence, least radius of gyration $r = 0.289 \times \frac{1}{2} = 0.145$ inch.

$$\frac{Q}{r^2} = \frac{0.0778}{(0.145)^2} = 3.70$$

which is more than 2 so that the Euler formula will be used.

$$P_{cr} = \frac{s_y A r^2}{Q} = \frac{40{,}000 \times \frac{1}{2} \times 1}{3.70}$$

$$= 5400 \text{ pounds so that the factor of safety is } 5400 \div 4000 = 1.35$$

In the preceding example, the column formulas were used to check the adequacy of a column of known dimensions. The more usual problem involves determining what the dimensions should be to resist a specified load. As an example, the previous problem can be reworded as follows:

A 24-inch long bar of rectangular cross-section with width w twice its depth d is to carry a load of 4000 pounds. What must the width and depth be if a factor of safety of 1.35 is to be used?

First determine the critical load P_{cr}:

$$P_{cr} = \text{working load} \times \text{factor of safety}$$

$$= 4000 \times 1.35 = 5400 \text{ pounds.}$$

Next determine Q which, as before, will be 0.0778.
Assume Formula (2) applies:

$$P_{cr} = A s_y \left(1 - \frac{Q}{4r^2} \right)$$

$$5400 = w \times d \times 40,000 \left(1 - \frac{0.0778}{4\ r^2} \right)$$

$$= 2\ d^2 \times 40,000 \left(1 - \frac{0.01945}{r^2} \right)$$

$$\frac{5400}{40,000 \times 2} = d^2 \left(1 - \frac{0.01945}{r^2} \right)$$

As mentioned in the previous example the least radius of gyration r of a rectangle is equal to $0.289 \times$ the least dimension, d, in this case. Therefore, substituting for d the value $r \div 0.289$,

$$\frac{5400}{40,000 \times 2} = \left(\frac{r}{0.289} \right)^2 \left(1 - \frac{0.01945}{r^2} \right)$$

$$\frac{5400 \times 0.289 \times 0.289}{40,000 \times 2} = r^2 - 0.01945$$

$$0.005638 = r^2 - 0.01945$$

$$r^2 = 0.0251$$

Checking to determine if $Q \div r^2$ is greater or less than 2,

$$\frac{Q}{r^2} = \frac{0.0778}{0.0251} = 3.1,$$

therefore Formula (1) should have been used to determine r and dimensions w and d. Using Formula (1),

$$5400 = \frac{40,000 \times 2\ d^2 \times r^2}{Q} = \frac{40,000 \times 2 \times \left(\dfrac{r}{0.289} \right)^2 r^2}{0.0778}$$

$$r^4 = \frac{5400 \times 0.0778 \times 0.289 \times 0.289}{40,000 \times 2}$$

$$= 0.0004386$$

$$r = 0.145 \quad \text{so that}$$

$$d = \frac{0.145}{0.289} = 0.50 \text{ inch}$$

and $w = 2\ d = 1$ inch as in the previous example.

American Institute of Steel Construction. — For main or secondary compression members with l/r ratios up to 120, safe unit stress $= 17,000 - 0.485 l^2/r^2$. For columns and bracing or other secondary members with l/r ratios above 120,

Safe unit stress, psi $= \dfrac{18,000}{1 + l^2/18,000 r^2}$ for bracing and secondary members. For

main members, safe unit stress, psi $= \dfrac{18,000}{1 + l^2/18,000 r^2} \times \left(1.6 - \dfrac{l/r}{200} \right)$

Pipe Columns: Allowable concentric loads for steel pipe columns based on the above formulas are given in the table on page 291.

Plates, Shells and Cylinders

Flat Stayed Surfaces. — In many cases, large flat areas are held against pressure by stays distributed at regular intervals over the surface. In boiler work, these stays are usually screwed into the plate and the projecting end riveted over to insure steam tightness. The U.S. Board of Supervising Inspectors and the American Boiler Makers Association rules give the following formula for flat stayed surfaces:

$$P = \frac{C \times t^2}{S^2}$$

in which P = pressure in pounds per square inch;

C = a constant which equals 112, for plates $7/16$ inch and under; 120, for plates over $7/16$ inch thick; 140, for plates with stays having a nut and bolt on the inside and outside; and 160, for plates with stays having washers of at least one-half the thickness of the plate, and with a diameter at least one-half of the greatest pitch.

t = thickness of plate in 16ths of an inch (thickness = $7/16$, t = 7);

S = greatest pitch of stays in inches.

Strength and Deflection of Flat Plates. — In the majority of cases, the formulas used to determine stresses and deflections in flat plates are based on certain assumptions that can be closely approximated in practice. These assumptions are:

1. the thickness of the plate is not greater than one-quarter the least width of the plate;

2. the greatest deflection when the plate is loaded is less than one-half the plate thickness;

3. the maximum tensile stress resulting from the load does not exceed the elastic limit of the material; and

4. all loads are perpendicular to the plane of the plate.

Plates of ductile materials fail when the maximum stress resulting from deflection under load exceeds the yield strength; for brittle materials, failure occurs when the maximum stress reaches the ultimate tensile strength of the material involved.

Square and Rectangular Flat Plates. — The formulas that follow give the maximum stress and deflection of flat steel plates supported in various ways and subjected to the loading indicated. These formulas are based upon a modulus of elasticity for steel of 30,000,000 pounds per square inch and a value of Poisson's ratio of 0.3. If the formulas for maximum stress, S, are applied without modification to other materials such as cast iron, aluminum, and brass for which the range of Poisson's ratio is about 0.26 to 0.34, the maximum stress calculations will be in error by not more than about 3 per cent. The deflection formulas may also be applied to materials other than steel by substituting in these formulas the appropriate value for E, the modulus of elasticity of the material (see pages 512 and 513). The deflections thus obtained will not be in error by more than about 3 per cent.

In the stress and deflection formulas that follow,

p = uniformly distributed load acting on plate, pounds per square inch;

W = total load on plate, pounds; $W = p \times$ area of plate;

L = distance between supports (length of plate), inches. For rectangular plates, L = long side, l = short side;

t = thickness of plate, inches;

S = maximum tensile stress in plate, pounds per square inch;

d = maximum deflection of plate, inches;

E = modulus of elasticity in tension. E = 30,000,000 pounds per square inch for steel.

If metric SI units are used in the formulas, then,

> W = total load on plate, newtons;
> L = distance between supports (length of plate), millimeters. For
>   rectangular plates, L = long side, l = short side;
> t = thickness of plate, millimeters;
> S = maximum tensile stress in plate, newtons per mm squared;
> d = maximum deflection of plate, mm;
> E = modulus of elasticity, newtons per mm squared.

1. Square flat plate supported at top and bottom of all four edges and a uniformly distributed load over the surface of the plate.

$$S = \frac{0.29W}{t^2} \quad\quad (1) \quad\quad\quad d = \frac{0.0443WL^2}{Et^3} \quad\quad (2)$$

2. Square flat plate supported at the bottom only of all four edges and a uniformly distributed load over the surface of the plate.

$$S = \frac{0.28W}{t^2} \quad\quad (3) \quad\quad\quad d = \frac{0.0443WL^2}{Et^3} \quad\quad (4)$$

3. Square flat plate with all edges firmly fixed and a uniformly distributed load over the surface of the plate.

$$S = \frac{0.31W}{t^2} \quad\quad (5) \quad\quad\quad d = \frac{0.0138WL^2}{Et^3} \quad\quad (6)$$

4. Square flat plate with all edges firmly fixed and a uniform load over small circular area at the center. In equations (7) and (9) r_0 = radius of area to which load is applied. If $r_0 < 1.7t$, use r_s where $r_s = \sqrt{1.6r_0^2 + t^2} - 0.675t$.

$$S = \frac{0.62W}{t^2}\log_e\left(\frac{L}{2r_0}\right) \quad\quad (7) \quad\quad\quad d = \frac{0.0568WL^2}{Et^3} \quad\quad (8)$$

5. Square flat plate with all edges supported above and below, or below only, and a concentrated load at the center. (See Case 4 for definition of r_0).

$$S = \frac{0.62W}{t^2}\left[\log_e\left(\frac{L}{2r_0}\right) + 0.577\right] \quad (9) \quad\quad\quad d = \frac{0.1266WL^2}{Et^3} \quad\quad (10)$$

6. Rectangular plate with all edges supported at top and bottom and a uniformly distributed load over the surface of the plate.

$$S = \frac{0.75W}{t^2\left(\dfrac{L}{l} + 1.61\dfrac{l^2}{L^2}\right)} \quad\quad (11) \quad\quad\quad d = \frac{0.1422W}{Et^3\left(\dfrac{L}{l^3} + \dfrac{2.21}{L^2}\right)} \quad\quad (12)$$

7. Rectangular plate with all edges fixed and a uniformly distributed load over the surface of the plate.

$$S = \frac{0.5W}{t^2\left(\dfrac{L}{l} + \dfrac{0.623l^5}{L^5}\right)} \quad\quad (13) \quad\quad\quad d = \frac{0.0284W}{Et^3\left(\dfrac{L}{l^3} + \dfrac{1.056l^2}{L^4}\right)} \quad\quad (14)$$

Circular Flat Plates. — In the following formulas, R = radius of plate to supporting edge in inches; W = total load in pounds; and other symbols are the same as used for square and rectangular plates.

If metric SI units are used, R = radius of plate to supporting edge in millimeters, and the values of other symbols are the same as those used for square and rectangular plates.

1. Edge supported around the circumference and a uniformly distributed load over the surface of the plate.

$$S = \frac{0.39W}{t^2} \qquad (1) \qquad\qquad d = \frac{0.221WR^2}{Et^3} \qquad (2)$$

2. Edge fixed around circumference and a uniformly distributed load over the surface of the plate.

$$S = \frac{0.24W}{t^2} \qquad (3) \qquad\qquad d = \frac{0.0543WR^2}{Et^3} \qquad (4)$$

3. Edge supported around the circumference and a concentrated load at the center.

$$S = \frac{0.48W}{t^2}\left[1 + 1.3\log_e\frac{R}{0.325t} - 0.0185\frac{t^2}{R^2}\right] \quad (5) \qquad d = \frac{0.55WR^2}{Et^3} \quad (6)$$

4. Edge fixed around circumference and a concentrated load at the center.

$$S = \frac{0.62W}{t^2}\left[\log_e\frac{R}{0.325t} + 0.0264\frac{t^2}{R^2}\right] \quad (7) \qquad\qquad d = \frac{0.22WR^2}{Et^3} \quad (8)$$

Strength of Cylinders Subjected to Internal Pressure. — In designing a cylinder to withstand internal pressure, the choice of formula to be used depends on (1) the kind of material of which the cylinder is made (whether brittle or ductile); (2) the construction of the cylinder ends (whether open or closed); and (3) whether the cylinder is classed as a thin- or a thick-walled cylinder.

A cylinder is considered to be thin-walled when the ratio of wall thickness to inside diameter is 0.1 or less and thick-walled when this ratio is greater than 0.1. Materials such as cast iron, hard steel, cast aluminum are considered to be brittle materials; low-carbon steel, brass, bronze, etc. are considered to be ductile.

In the formulas that follow, p = internal pressure, pounds per square inch; D = inside diameter of cylinder, inches; t = wall thickness of cylinder, inches; μ = Poisson's ratio, = 0.3 for steel, 0.26 for cast iron, 0.34 for aluminum and brass; S = allowable tensile stress, pounds per square inch.

Metric SI units can be used in Formulas (1), (3), (4), and (5), where p = internal pressure in newtons per square millimeter; D = inside diameter of cylinder, millimeters; t = wall thickness, mm; μ = Poisson's ratio, = 0.3 for steel, 0.26 for cast iron, and 0.34 for aluminum and brass; and S = allowable tensile stress, N/mm². For the use of metric SI units in Formula (2), see below.

Thin-walled cylinders:

$$t = \frac{Dp}{2S} \qquad (1)$$

For low-pressure cylinders of cast iron such as are used for certain engine and press applications, a formula in common use is

$$t = \frac{Dp}{2500} + 0.3 \qquad (2)$$

This formula is based on allowable stress of 1250 pounds per square inch and will give a wall thickness 0.3 inch greater than Formula (1) to allow for variations in metal thickness that may result from the casting process.

If metric SI units are used in Formula (2), t = cylinder wall thickness in millimeters; D = inside diameter of cylinder, mm; and the allowable stress is in newtons per square millimeter. The value of 0.3 inches additional wall thickness is 7.62 mm, and the next highest number in preferred metric basic sizes is 8 mm.

Thick-walled cylinders of brittle material; ends open or closed: Lamé's equation is used when cylinders of this type are subjected to internal pressure.

$$t = \frac{D}{2}\left[\sqrt{\frac{S+p}{S-p}} - 1\right] \qquad (3)$$

The table of ratios of outside radius to inside radius of thick cylinders on page 329 is for convenience in calculating the dimensions of cylinders under high internal pressure without the use of Formula 3. As an example of the use of the table, assume that a cylinder of 10 inches inside diameter is to withstand a pressure of 2500 pounds per square inch; the material is cast iron and the allowable stress is 6000 pounds per square inch. To solve the problem, locate the allowable stress per square inch in the left-hand column of the table and the working pressure at the top of the columns. Then find the ratio between the outside and inside radii in the body of the table. In this case, the ratio is 1.558, and hence the outside diameter of the cylinder should be 10 × 1.558, or about 15⅝ inches. The thickness of the cylinder wall will therefore be (15.558 − 10)/2 = 2.779 inches.

Unless very high-grade material is used and sound castings assured, cast iron should not be used for pressures exceeding 2000 pounds per square inch. It is well to leave more metal in the bottom of a hydraulic cylinder than is indicated by the results of calculations, because a hole of some size must be cored in the bottom to permit the entrance of a boring bar when finishing the cylinder, and when this hole is subsequently tapped and plugged it often gives trouble if there is too little thickness.

For steady or gradually applied stresses, the maximum allowable fiber stress S may be assumed to be from 3500 to 4000 pounds per square inch for cast iron; from 6000 to 7000 pounds per square inch for brass; and 12,000 pounds per square inch for steel castings. For intermittent stresses, such as in cylinders for steam and hydraulic work, 3000 pounds per square inch for cast iron; 5000 pounds per square inch for brass; and 10,000 pounds per square inch for steel castings, is ordinarily used. These values give ample factors of safety.

Note: **In metric SI units, 1000 pounds per square inch equals 6.895 newtons per square millimeter.**

Thick-walled cylinders of ductile material; closed ends: Clavarino's equation is used:

$$t = \frac{D}{2}\left[\sqrt{\frac{S+(1-2\mu)p}{S-(1+\mu)p}} - 1\right] \qquad (4)$$

external pressure only. The table thicknesses have been calculated from the formula:

$$t = \frac{[(F \times p) + 1386] D}{86,670}$$

in which D = outside diameter of flue or tube in inches; t = thickness of wall in inches; p = working pressure in pounds per square inch; F = factor of safety. The formula is applicable to working pressures greater than 100 pounds per square inch, to outside diameters from 7 to 18 inches, and to temperatures less than 650° F.

The Formulas (1) and (2) given on the preceding page were determined by Prof. R. T. Stewart, Dean of the Mechanical Engineering Department of the University of Pittsburgh, in a series of experiments carried out at the plant of the National Tube Co., McKeesport, Pa.

The apparent fiber stress under which the different tubes failed varied from about 7000 pounds per square inch for the relatively thinnest to 35,000 pounds per square inch for the relatively thickest walls. Since the average yield point of the material tested was 37,000 pounds and the tensile strength 58,000 pounds per square inch, it is evident that the strength of a tube subjected to external fluid collapsing pressure is not dependent alone upon the elastic limit or ultimate strength of the material from which it is made.

Tubes Subjected to External Pressure

Outside Diameter of Tube, Inches	Working Pressure in Pounds per Square Inch						
	100	120	140	160	180	200	220
	Thickness of Tube in Inches. Safety Factor, 5						
7	0.152	0.160	0.168	0.177	0.185	0.193	0.201
8	0.174	0.183	0.193	0.202	0.211	0.220	0.229
9	0.196	0.206	0.217	0.227	0.237	0.248	0.258
10	0.218	0.229	0.241	0.252	0.264	0.275	0.287
11	0.239	0.252	0.265	0.277	0.290	0.303	0.316
12	0.261	0.275	0.289	0.303	0.317	0.330	0.344
13	0.283	0.298	0.313	0.328	0.343	0.358	0.373
14	0.301	0.320	0.337	0.353	0.369	0.385	0.402
15	0.323	0.343	0.361	0.378	0.396	0.413	0.430
16	0.344	0.366	0.385	0.404	0.422	0.440	0.459
17	0.366	0.389	0.409	0.429	0.448	0.468	0.488
18	0.387	0.412	0.433	0.454	0.475	0.496	0.516

Dimensions and Maximum Allowable Pressure of Tubes Subjected to External Pressure

Outside Diam., Inches	Thickness of Material, Inches	Maximum Pressure Allowed, psi	Outside Diam., Inches	Thickness of Material, Inches	Maximum Pressure Allowed, psi	Outside Diam., Inches	Thickness of Material, Inches	Maximum Pressure Allowed, psi
2	0.095	427	3	0.109	327	4	0.134	303
2¼	0.095	380	3¼	0.120	332	4½	0.134	238
2½	0.109	392	3½	0.120	308	5	0.148	235
2¾	0.109	356	3¾	0.120	282	6	0.165	199

SHAFTS

Torsional Strength of Shafting. — In the formulas that follow,

α = angular deflection of shaft in degrees;
c = distance from center of gravity to extreme fiber;
D = diameter of shaft in inches;
G = torsional modulus of elasticity = 11,500,000 pounds per square inch for steel;
J = polar moment of inertia of shaft cross-section (see table);
l = length of shaft in inches;
N = angular velocity of shaft in revolutions per minute;
P = power transmitted in horsepower;
S_s = allowable torsional shearing stress in pounds per square inch;
T = torsional or twisting moment in inch-pounds;
Z_p = polar section modulus (see table).

The allowable twisting moment for a shaft of any cross-section such as circular, square, etc. is:

$$T = S_s \times Z_p \qquad (1)$$

For a shaft delivering P horsepower at N revolutions per minute the twisting moment T being transmitted is:

$$T = \frac{63,000P}{N} \qquad (2)$$

The twisting moment T as determined by this formula should be less than the value determined by using Formula (1) if the maximum allowable stress S_s is not to be exceeded.

The diameter of a solid circular shaft required to transmit a given torque T is:

$$D = \sqrt[3]{\frac{5.1\,T}{S_s}}; \quad \text{or} \quad D = \sqrt[3]{\frac{321,000\,P}{NS_s}} \qquad \text{(3a) and (3b)}$$

The allowable stresses that are generally used in practice are: 4000 pounds per square inch for main power-transmitting shafts; 6000 pounds per square inch for lineshafts carrying pulleys; and 8500 pounds per square inch for small, short shafts, countershafts, etc. Using these allowable stresses, the horsepower P transmitted by a shaft of diameter D, or the diameter D of a shaft to transmit a given horsepower P may be determined from the following formulas:

For main power-transmitting shafts:

$$P = \frac{D^3N}{80}; \quad \text{or} \quad D = \sqrt[3]{\frac{80\,P}{N}} \qquad \text{(4a) and (4b)}$$

For lineshafts carrying pulleys:

$$P = \frac{D^3N}{53.5}; \quad \text{or} \quad D = \sqrt[3]{\frac{53.5\,P}{N}} \qquad \text{(5a) and (5b)}$$

For small, short shafts:

$$P = \frac{D^3N}{38}; \quad \text{or} \quad D = \sqrt[3]{\frac{38\,P}{N}} \qquad \text{(6a) and (6b)}$$

Shafts which are subjected to shocks, sudden starting and stopping, etc., should be given a greater factor of safety resulting in the use of lower allowable stresses than those just mentioned.

Example: — What would be the diameter of a lineshaft to transmit 10 horsepower if the shaft makes 150 revolutions per minute? Using Formula (5b),

$$D = \sqrt[3]{\frac{53.5 \times 10}{150}} = 1.53, \quad \text{or, say, } 1\%_{16} \text{ inches.}$$

Example: — What horsepower would be transmitted by a short shaft, 2 inches in diameter, carrying but two pulleys close to the bearings if the shaft makes 300 revolutions per minute? Using Formula (6a),

$$P = \frac{2^3 \times 300}{38} = 63 \text{ horsepower.}$$

Torsional Strength of Shafting — Calculations in Metric SI Units. — The allowable twisting moment for a shaft of any cross-section such as circular, square, etc., can be calculated from:

$$T = S_s \times Z_p \tag{1'}$$

where T = torsional or twisting moment in newton-millimeters; S_s = allowable torsional shearing stress in newtons per square millimeter; and Z_p = polar section modulus in millimeters[3].

For a shaft delivering power of P kilowatts at N revolutions per minute, the twisting moment T being transmitted is:

$$T = \frac{9.55 \times 10^6 \, P}{N}; \quad \text{or} \quad T = \frac{10^6 \, P}{\omega} \tag{2'} \text{ and } (2a')$$

where T is in newton-millimeters, and ω = angular velocity in radians per second.

The diameter D of a solid circular shaft required to transmit a given torque T is:

$$D = \sqrt[3]{\frac{5.1T}{S_s}}; \quad \text{or} \quad D = \sqrt[3]{\frac{48.7.7 \times 10^6 P}{N S_s}}; \tag{3a'} \text{ and } (3b')$$

$$\text{or} \quad D = \sqrt[3]{\frac{5.1 \times 10^6 P}{\omega S_s}} \tag{3c'}$$

where D is in millimeters; T is in newton-millimeters; P is power in kilowatts; N = revolutions per minute; S_s = allowable torsional shearing stress in newtons per square millimeter, and ω = angular velocity in radians per second.

If 28 newtons/mm^2 and 59 newtons/mm^2 are taken as the generally allowed stresses for main power-transmitting shafts, and small short shafts respectively, then using these allowable stresses, the power P transmitted by a shaft of diameter D, or the diameter D of a shaft to transmit a given power P may be determined from the following formulas:

For main power-transmitting shafts:

$$P = \frac{D^3 N}{1.77 \times 10^6}; \quad \text{or} \quad D = \sqrt[3]{\frac{1.77 \times 10^6 P}{N}} \tag{4a'} \text{ and } (4b')$$

For small, short shafts:

$$P = \frac{D^3 N}{0.83 \times 10^6}; \quad \text{or} \quad D = \sqrt[3]{\frac{0.83 \times 10^6 P}{N}}, \tag{6a'} \text{ and } (6b')$$

where P is in kilowatts; D is in millimeters, and N = revolutions per minute.

Example: What would be the diameter of a power-transmitting shaft to transmit 150 kW at 500 rpm?

$$D = \sqrt[3]{\frac{1.77 \times 10^6 \times 150}{500}} = 81 \text{ millimeters.}$$

Example: What power would a short shaft, 50 millimeters in diameter, transmit at 400 rpm?

$$P = \frac{50^3 \times 400}{0.83 \times 10^6} = 60 \text{ kilowatts.}$$

Polar Moment of Inertia and Section Modulus. — The *polar moment of inertia, J*, of a cross-section with respect to a polar axis, that is, an axis at right angles to the plane of the cross-section, is defined as the moment of inertia of the cross-section with respect to the point of intersection of the axis and the plane. The polar moment of inertia may be found by taking the sum of the moments of inertia about two perpendicular axes lying in the plane of the cross-section and passing through this point. Thus, for example, the polar moment of inertia of a circular or a square area with respect to a polar axis through the center of gravity is equal to two times the moment of inertia with respect to an axis lying in the plane of the cross-section and passing through the center of gravity.

The polar moment of inertia with respect to a polar axis through the center of gravity is required for problems involving the torsional strength of shafts since this axis is usually the axis about which twisting of the shaft takes place.

The *polar section modulus* (also called section modulus of torsion), Z_p, for *circular* sections may be found by dividing the polar moment of inertia, J, by the distance c from the center of gravity to the most remote fiber. This method may be used to find the *approximate* value of the polar section modulus of sections that are *nearly* round. For other than circular cross-sections, however, the polar section modulus *does not* equal the polar moment of inertia divided by the distance c.

The accompanying table gives formulas for the polar section modulus for several different cross-sections. The polar section modulus multiplied by the allowable torsional shearing stress gives the allowable twisting moment to which a shaft may be subjected (see Formula 1).

Torsional Deflection of Circular Shafts. — In many cases shafting must be proportioned not only to provide the strength required to transmit a given torque, but also to prevent torsional deflection (twisting) through a greater angle than has been found satisfactory for a given type of service.

For a solid circular shaft the torsional deflection in degrees is given by:

$$\alpha = \frac{584Tl}{D^4 G} \tag{7}$$

Example: — Find the torsional deflection for a solid steel shaft 4 inches in diameter and 48 inches long, subjected to a twisting moment of 24,000 inch-pounds. By Formula (7),

$$\alpha = \frac{584 \times 24,000 \times 48}{4^4 \times 11,500,000} = 0.23 \text{ degree.}$$

Formula (7) can be used with metric SI units, where α = angular deflection of shaft in degrees; T = torsional moment in newton-millimeters; l = length of shaft in millimeters; D = diameter of shaft in millimeters; and G = torsional modulus of elasticity in newtons per square millimeter.

Example: Find the torsional deflection of a solid steel shaft, 100 mm in diameter and 1300 mm long, subjected to a twisting moment of 3×10^6 newton-millimeters. The torsional modulus of elasticity is 80,000 newtons/mm². By Formula (7)

$$\alpha = \frac{584 \times 3 \times 10^6 \times 1300}{100^4 \times 80,000} = 0.285 \text{ degree.}$$

Polar Moment of Inertia and Polar Section Modulus

Section	Polar Moment of Inertia J	Polar Section Modulus Z_p
	$\dfrac{a^4}{6} = 0.1667\,a^4$	$0.208\,a^3 = 0.074\,d^3$
	$\dfrac{bd\,(b^2 + d^2)}{12}$	$\dfrac{bd^2}{3 + 1.8\dfrac{d}{b}}$ (*d* is the shorter side)
	$\dfrac{\pi D^4}{32} = 0.098\,D^4$ (see also footnote, p. 241)	$\dfrac{\pi D^3}{16} = 0.196\,D^3$ (see also footnote, p. 241)
	$\dfrac{\pi}{32}\,(D^4 - d^4)$ $= 0.098\,(D^4 - d^4)$	$\dfrac{\pi}{16}\left(\dfrac{D^4 - d^4}{D}\right)$ $= 0.196\left(\dfrac{D^4 - d^4}{D}\right)$
	$\dfrac{5\sqrt{3}}{8}\,s^4 = 1.0825\,s^4$ $= 0.12\,F^4$	$0.20\,F^3$
	$\dfrac{\pi D^4}{32} - \dfrac{s^4}{6}$ $= 0.098\,D^4 - 0.167\,s^4$	$\dfrac{\pi D^3}{16} - \dfrac{s^4}{3\,D}$ $= 0.196\,D^3 - 0.333\,\dfrac{s^4}{D}$
	$\dfrac{\pi D^4}{32} - \dfrac{5\sqrt{3}}{8}\,s^4$ $= 0.098\,D^4 - 1.0825\,s^4$	$\dfrac{\pi D^3}{16} - \dfrac{5\sqrt{3}\,s^4}{4\,D}$ $= 0.196\,D^3 - 2.165\,\dfrac{s^4}{D}$
	$\dfrac{\sqrt{3}\,s^4}{48} = 0.036\,s^4$	$\dfrac{s^3}{20} = 0.05\,s^3$

The diameter of a shaft which is to have a maximum torsional deflection α is given by:

$$D = 4.9 \sqrt[4]{\frac{Tl}{G\alpha}} \qquad (8)$$

Formula (8) can be used with metric SI units, where D = diameter of shaft in millimeters; T = torsional moment in newton-millimeters; l = length of shaft in millimeters; G = torsional modulus of elasticity in newtons per square millimeter; and α = angular deflection of shaft in degrees.

According to some authorities, the allowable twist in steel transmission shafting should not exceed 0.08 degree per foot length of the shaft. The diameter D of a shaft which will permit a maximum angular deflection of 0.08 degree per foot of length for a given torque T or for a given horsepower P can be determined from the formulas:

$$D = 0.29\sqrt[4]{T}; \quad \text{or} \quad D = 4.6\sqrt[4]{\frac{P}{N}} \qquad \text{(9a) and (9b)}$$

Using metric SI units and assuming an allowable twist in steel transmission shafting of 0.26 degree per meter length, Formulas (9a) and (9b) become:

$$D = 2.26\sqrt[4]{T}; \quad \text{or} \quad D = 125.7\sqrt[4]{\frac{P}{N}},$$

where D = diameter of shaft in millimeters; T = torsional moment in newton-millimeters; P = power in kilowatts; and N = revolutions per minute.

Another rule that has been generally used in mill practice limits the deflection to 1 degree in a length equal to 20 times the shaft diameter. For a given torque or horsepower, the diameter of a shaft having this maximum deflection is given by:

$$D = 0.1\sqrt[3]{T}; \quad \text{or} \quad D = 4.0\sqrt[3]{\frac{P}{N}} \qquad \text{(10a) and (10b)}$$

Example: — Find the diameter of a steel lineshaft to transmit 10 horsepower at 150 revolutions per minute with a torsional deflection not exceeding 0.08 degree per foot of length. By Formula (9b),

$$D = 4.6\sqrt[4]{\frac{10}{150}} = 2.35 \text{ inches.}$$

This diameter is larger than that obtained for the same horsepower and rpm in the example given for Formula (5b) in which the diameter was calculated for strength considerations only. The usual procedure in the design of shafting which is to have a specified maximum angular deflection is to compute the diameter first by means of Formula (8), (9a), (9b), (10a), or (10b) and then by means of Formula (3a), (3b), (4b), (5b), or (6b), using the larger of the two diameters thus found.

Linear Deflection of Shafting. — For steel lineshafting, it is considered good practice to limit the linear deflection to a maximum of 0.010 inch per foot of length. The maximum distance in feet between bearings, for average conditions, in order to avoid excessive linear deflection, is determined by the formulas:

$$L = 8.95\sqrt[3]{D^2} \text{ for shafting subject to no bending action except its own weight}$$

$$L = 5.2\sqrt[3]{D^2} \text{ for shafting subject to bending action of pulleys, etc.}$$

in which D = diameter of shaft in inches; L = maximum distance between bearings in feet. Pulleys should be placed as close to the bearings as possible.

Tables of Horsepower Transmitted by Shafting. — The table following, "Horsepower Transmitted by Shafting made from Medium Steel," gives the rela-

Horsepower Transmitted by Shafting made from Medium Steel

Diam. of Shaft, Inches	Transmitting Power, but Subject to No Bending Action Except Its Own Weight						Transmitting Power, and Subject to Bending Action of Pulleys, Belting, Etc.					
	Revolutions per Minute					Max. Distance in Feet Between Bearings	Revolutions per Minute					Max. Distance in Feet Between Bearings
	100	150	200	250	300		100	150	200	250	300	
	Horsepower						Horsepower					
$1\frac{1}{2}$	7	10	14	17	20	11.7	5	7	10	12	14	6.8
$1\frac{5}{8}$	9	13	17	21	26	12.4	6	9	12	15	18	7.2
$1\frac{3}{4}$	11	16	21	26	32	13.0	8	11	15	18	22	7.5
$1\frac{7}{8}$	13	20	26	33	40	13.6	9	14	19	23	28	7.9
2	16	24	32	40	48	14.2	11	17	23	28	34	8.2
$2\frac{1}{8}$	19	29	38	48	58	14.8	14	21	27	34	42	8.6
$2\frac{1}{4}$	23	34	46	57	68	15.4	16	24	33	41	48	8.9
$2\frac{3}{8}$	27	40	54	67	80	16.0	19	29	38	48	58	9.2
$2\frac{1}{2}$	31	47	63	78	94	16.5	22	33	45	55	66	9.6
$2\frac{3}{4}$	42	62	83	102	124	17.6	30	44	59	74	89	10.2
3	54	81	108	134	162	18.6	39	58	77	96	116	10.8
$3\frac{1}{4}$	69	103	137	172	206	19.7	49	74	98	123	148	11.4
$3\frac{1}{2}$	86	129	172	215	258	20.7	61	92	123	153	184	12.0
$3\frac{3}{4}$	105	158	211	264	316	21.6	75	113	151	188	226	12.5
4	128	192	256	320	384	22.6	91	137	183	228	274	13.1

tion between the diameter of shaft, revolutions per minute, horsepower transmitted, and maximum distance in feet between bearings. Assume, for example, that it is required to find the diameter of a shaft for transmitting 40 horsepower at a speed of 250 revolutions per minute. The shaft is not subjected to any bending action except its own weight. From the table, it is found, by locating "40" in the column under 250 revolutions per minute, that the diameter of the shaft required is 2 inches. The maximum permissible distance between the shaft bearings is slightly more than 14 feet. When the exact horsepower cannot be found in the table, it is advisable to take the nearest larger value listed in the table and find the diameter of shafting required to transmit this horsepower. Tables are also given for the horsepower which can be safely transmitted by cold-rolled and turned steel lineshafting. The cold-rolled steel table on page 309 is carried up to 5 inches only, because this diameter is the largest which is commonly cold-rolled. These tables have been used by the transmission department of the Jones & Laughlin Steel Co., and are based on the assumption that bearings are placed at intervals of from 8 to 10 feet and all pulleys are located as near to the bearings as possible. In these tables, the body part in each gives the number of horsepower to be transmitted. For example, assume that a 3-inch cold-rolled steel lineshaft revolves at a speed of 400 revolutions per minute. Find the power that this shaft can safely transmit. By locating 3 inches in the left-hand column and 400 at the top of the vertical columns, and following the vertical column downward until opposite 3 inches, it is found that under the given conditions 154 horsepower may be safely transmitted.

In general, shafting up to three inches in diameter is almost always made from cold-rolled steel. This shafting is true and straight and needs no turning, but if keyways are cut in the shaft, it must, as a rule, be straightened afterwards, as the cutting of the keyways relieves the tension on the surface of the shaft produced by the cold-

SHAFTS

Horsepower Transmitted by Turned Steel Lineshafting

Diam. of Shaft	Number of Revolutions per Minute												
	100	125	150	175	200	225	250	300	350	400	450	500	600
1½	3.7	4.7	5.6	6.6	7.5	8.4	9.4	11.2	13.1	15.0	16.9	18.8	22
1⁹⁄₁₆	4.2	5.3	6.4	7.4	8.5	9.5	10.6	12.7	14.8	17.0	19.0	21	25
1⅝	4.8	5.9	7.1	8.3	9.5	10.7	11.9	14.3	16.6	19.0	21	24	28
1¹¹⁄₁₆	5.3	6.7	8.0	9.3	10.7	12.0	13.4	16.0	18.7	21	24	27	32
1¾	5.9	7.4	8.9	10.4	11.9	13.4	14.9	17.9	21	24	27	30	36
1¹³⁄₁₆	6.6	8.2	9.9	11.5	13.2	14.8	16.5	19.8	23	26	30	33	40
1⅞	7.3	9.1	11.0	12.8	14.7	16.5	18.3	22	26	29	33	37	44
1¹⁵⁄₁₆	8.1	10.0	12.1	14.1	16.1	18.2	20	24	28	32	36	40	48
2	8.9	11.1	13.3	15.6	17.8	20	22	27	31	35	40	44	53
2¹⁄₁₆	9.8	12.3	14.7	17.2	19.6	22	24	29	34	39	44	49	59
2⅛	10.6	13.3	16.0	18.6	21	24	27	32	37	43	48	53	64
2³⁄₁₆	11.6	14.6	17.5	20.0	23	26	29	35	41	47	52	58	70
2¼	12.6	15.8	19.0	22.0	25	28	32	38	44	51	57	63	76
2⁵⁄₁₆	13.7	17.2	21	24	27	31	34	41	48	55	62	69	82
2⅜	14.9	18.6	22	26	30	33	37	45	52	60	67	74	89
2⁷⁄₁₆	16.0	20	24	28	32	36	40	48	56	64	72	80	96
2½	17.4	22	26	30	35	39	43	52	61	69	78	87	104
2⁹⁄₁₆	18.7	23	28	33	37	42	47	56	66	75	84	94	112
2⅝	20	25	30	35	40	45	50	60	71	80	90	100	120
2¹¹⁄₁₆	21	27	32	38	43	48	54	65	76	86	97	108	129
2¾	23	29	35	40	46	52	58	69	81	92	104	115	138
2¹³⁄₁₆	25	31	37	43	49	56	62	74	87	99	111	124	148
2⅞	26	33	40	46	53	59	66	79	92	105	119	132	158
2¹⁵⁄₁₆	28	35	42	49	56	63	70	84	99	113	127	141	169
3	30	37	45	52	60	67	75	90	105	120	135	150	180
3⅛	34	42	51	59	68	76	85	102	119	136	152	170	203
3¼	38	48	57	67	76	86	95	114	134	153	172	191	229
3⅜	43	53	64	75	85	96	107	128	150	171	192	213	256
3½	48	60	72	83	95	107	119	143	167	190	214	238	286
3⅝	53	66	79	93	106	119	132	159	185	211	238	265	317
3¾	59	73	88	103	117	132	146	176	205	234	264	293	351
3⅞	65	81	97	113	129	145	161	194	226	258	291	322	387
4	71	89	107	125	142	160	178	213	249	284	320	356	427
4⅛	78	98	117	136	156	176	195	235	273	312	351	390	468
4¼	85	107	128	149	170	192	213	256	298	341	385	426	511
4⅜	93	116	139	163	186	210	233	279	326	372	419	466	559
4½	102	127	152	178	203	228	253	305	356	405	456	507	610
4⅝	110	138	165	193	220	247	275	330	385	440	495	550	660
4¾	119	149	179	209	238	268	298	357	416	476	537	595	714
4⅞	129	161	193	226	258	290	322	387	452	516	581	646	775
5	139	174	208	244	278	313	347	417	486	557	625	695	835

Horsepower Transmitted by Cold-rolled Steel Lineshafting

Diam. of Shaft	Number of Revolutions per Minute												
	100	125	150	175	200	225	250	300	350	400	450	500	600
1½	4.8	6.0	7.2	8.4	9.6	10.8	12.0	14.4	16.9	19.2	22	24	29
1 9/16	5.5	6.8	8.2	9.5	10.9	12.2	13.6	16.4	19.0	22	25	27	33
1 5/8	6.1	7.6	9.2	10.7	12.2	13.8	15.3	18.4	21	24	28	31	37
1 11/16	6.9	8.6	10.3	12.0	13.7	15.4	17.1	21	24	27	31	34	41
1¾	7.7	9.6	11.5	13.4	15.3	17.2	19.1	23	27	31	34	38	46
1 13/16	8.5	10.6	12.7	14.8	16.9	19.0	21	25	30	34	38	42	51
1 7/8	9.4	11.7	14.1	16.4	18.8	21	23	28	33	38	42	47	57
1 15/16	10.4	13.0	15.6	18.2	21	23	26	31	36	42	47	52	62
2	11.4	14.3	17.2	20	23	26	29	34	40	46	51	57	69
2 1/16	12.6	15.7	18.9	22	25	28	31	38	44	50	56	63	76
2 1/8	13.7	17.1	21	24	27	31	34	41	48	55	61	68	82
2 3/16	15.0	18.7	22	26	30	34	37	45	52	60	67	75	90
2 1/4	16.3	20	24	29	33	37	41	49	57	65	73	81	98
2 5/16	17.7	22	27	31	35	40	44	53	62	71	80	88	106
2 3/8	19.2	24	29	34	38	43	48	57	67	76	86	96	115
2 7/16	20	25	30	36	41	46	51	61	72	81	91	102	122
2 1/2	22	28	33	39	45	50	56	67	78	89	100	112	133
2 9/16	24	30	36	42	48	54	60	72	84	96	108	120	144
2 5/8	26	32	39	45	52	58	64	77	90	104	116	129	155
2 11/16	28	35	42	48	55	62	69	83	97	111	124	138	166
2 3/4	30	37	44	52	59	67	74	89	104	119	133	148	178
2 13/16	32	40	47	55	63	71	79	95	111	127	143	159	190
2 7/8	34	42	51	59	68	76	85	101	119	135	152	169	203
2 15/16	36	45	54	63	72	81	90	108	127	144	162	181	217
3	39	48	58	67	77	87	96	116	135	154	173	192	231
3 1/8	44	54	65	76	87	98	109	131	152	174	196	218	261
3 1/4	49	61	73	86	98	110	122	147	172	196	221	245	294
3 3/8	55	69	83	96	110	124	137	165	192	220	247	275	330
3 1/2	61	77	92	107	123	138	153	184	214	245	276	307	367
3 5/8	68	85	102	119	136	153	170	204	238	272	306	340	408
3 3/4	75	94	113	132	151	170	189	226	264	301	340	377	452
3 7/8	83	104	125	145	166	187	207	249	291	332	379	415	498
4	92	114	137	160	183	206	229	274	320	366	411	457	549
4 1/8	101	125	150	175	201	226	251	300	351	401	451	501	601
4 1/4	110	137	164	192	219	246	273	328	383	438	492	547	657
4 3/8	120	150	180	210	239	268	298	358	418	478	538	597	717
4 1/2	130	163	195	228	261	293	326	391	455	521	586	651	781
4 5/8	141	177	212	247	283	318	354	425	495	566	636	707	848
4 3/4	153	191	230	268	307	344	382	459		613	688	765	919
4 7/8	166	207	249	290	331	372	413	496	580	662	745	827	994
5	179	224	268	313	358	402	447	537	625	715	805	895	1074

Diameters for Finished Shafting (former American Standard ASA B17.1)

Diameters, Inches — Transmission Shafting	Diameters, Inches — Machinery Shafting	Minus Tolerances, Inches*	Diameters, Inches — Transmission Shafting	Diameters, Inches — Machinery Shafting	Minus Tolerances, Inches*	Diameters, Inches — Transmission Shafting	Diameters, Inches — Machinery Shafting	Minus Tolerances, Inches*
	1/2	0.002		1 13/16	0.003		3 3/4	0.004
	9/16	0.002		1 7/8	0.003		3 7/8	0.004
	5/8	0.002	1 15/16	1 15/16	0.003		4	0.004
	11/16	0.002		2	0.003	3 15/16	4 1/4	0.005
	3/4	0.002		2 1/16	0.004		4 1/2	0.005
	13/16	0.002		2 1/8	0.004	4 7/16	4 3/4	0.005
15/16	7/8	0.002	2 3/16	2 3/16	0.004		5	0.005
	15/16	0.002		2 1/4	0.004	4 15/16	5 1/4	0.005
	1	0.002		2 5/16	0.004		5 1/2	0.005
	1 1/16	0.003		2 3/8	0.004	5 7/16	5 3/4	0.005
	1 1/8	0.003	2 7/16	2 7/16	0.004		6	0.005
1 3/16	1 3/16	0.003		2 1/2	0.004	5 15/16	6 1/4	0.006
	1 1/4	0.003		2 5/8	0.004	6 1/2	6 1/2	0.006
	1 5/16	0.003		2 3/4	0.004		6 3/4	0.006
	1 3/8	0.003	2 15/16	2 7/8	0.004	7	7	0.006
1 7/16	1 7/16	0.003		3	0.004		7 1/4	0.006
	1 1/2	0.003		3 1/8	0.004	7 1/2	7 1/2	0.006
	1 9/16	0.003		3 1/4	0.004		7 3/4	0.006
1 11/16	1 5/8	0.003		3 3/8	0.004	8	8	0.006
	1 11/16	0.003	3 7/16	3 1/2	0.004	...	...	...
	1 3/4	0.003		3 5/8	0.004	...	...	...

Note:—These tolerances are *negative* or minus and represent the maximum allowable variation *below* the exact nominal size. For instance the maximum diameter of the 1 15/16 inch shaft is 1.938 inch and its minimum allowable diameter is 1.935 inch. Stock lengths of finished transmission shafting shall be: 16, 20 and 24 feet.

rolling process. Sizes of shafting from three to five inches in diameter may be either cold-rolled or turned, more frequently the latter, while all larger sizes of shafting must be turned, because cold-rolled shafting is not available in diameters larger than five inches.

Design of Transmission Shafting. — The following guidelines for the design of shafting for transmitting a given amount of power under various conditions of loading are based upon formulas given in the former American Standard ASA B17c Code for the Design of Transmission Shafting. These formulas are based on the *maximum-shear theory* of failure which assumes that the elastic limit of a *ductile* ferrous material in shear is practically one-half its elastic limit in tension. This theory agrees, very nearly, with the results of tests on ductile materials and has gained wide acceptance in practice.

The formulas given apply in all cases of shaft design including shafts for special machinery. The limitation of these formulas is that they provide only for the strength of shafting and are not concerned with the torsional or lineal deformation which may, in the case of shafts used in machine design, be the controlling factor (see "Torsional Deflection of Circular Shafts" and "Linear Deflection of Shafting" for deflection considerations). In the formulas that follow,

$B = \sqrt[3]{1 \div (1 - K^4)}$ (see Table 3);

D = outside diameter of shaft in inches;

D_1 = inside diameter of a hollow shaft in inches;

K_m = combined shock and fatigue factor to be applied in every case to the computed bending moment (see Table 1);

K_t = combined shock and fatigue factor to be applied in every case to the computed torsional moment (see Table 1);

M = maximum bending moment in inch-pounds;
N = revolutions per minute;
P = maximum power to be transmitted by shaft in horsepower;
p_t = maximum allowable shearing stress under combined loading conditions in
pounds per square inch (see Table 2);
S = maximum allowable flexural (bending) stress, in either tension or compression in pounds per square inch (see Table 2);
S_s = maximum allowable torsional shearing stress in pounds per square inch
(see Table 2);
T = maximum torsional moment in inch-pounds;
V = maximum transverse shearing load in pounds.

For shafts subjected to pure torsional loads only,

$$D = B \sqrt[3]{\frac{5.1\ K_t T}{S_s}}\ ; \qquad (1a)$$

or

$$D = B \sqrt[3]{\frac{321{,}000\ K_t P}{S_s N}} \qquad (1b)$$

For stationary shafts subjected to bending only,

$$D = B \sqrt[3]{\frac{10.2\ K_m M}{S}} \qquad (2)$$

For shafts subjected to combined torsion and bending,

$$D = B \sqrt[3]{\frac{5.1}{p_t} \sqrt{(K_m M)^2 + (K_t T)^2}} \qquad (3a)$$

or

$$D = B \sqrt[3]{\frac{5.1}{p_t} \sqrt{(K_m M)^2 + \left(\frac{63{,}000\ K_t P}{N}\right)^2}} \qquad (3b)$$

Formulas (1a) to (3b) may be used for solid shafts or for hollow shafts. For solid
shafts the factor B is equal to 1, whereas for hollow shafts the value of B depends on
the value of K which, in turn, depends on the ratio of the inside diameter of the shaft
to the outside diameter ($D_1 \div D = K$). Table 3 gives values of B corresponding to
various values of K.
For short solid shafts subjected only to heavy transverse shear, the diameter of
shaft required is:

$$D = \sqrt{\frac{1.7 V}{S_s}} \qquad (4)$$

**Formulas (1a), (2), (3a) and (4), can be used unchanged with metric SI units. For-
mula (1b) becomes:**

$$D = B \sqrt[3]{\frac{48.7 K_t P}{S_s N}}\ ;$$

312   SHAFTS

and Formula (3b) becomes:

$$D = B \sqrt[3]{\frac{5.1}{p_t} \sqrt{(K_m M)^2 + \left(\frac{9.55 K_t P}{N}\right)^2}}$$

Throughout the formulas, D = outside diameter of shaft in millimeters; T = maximum torsional moment in newton-millimeters; S_s = maximum allowable torsional shearing stress in newtons per millimeter squared (see Table 2); P = maximum power to be transmitted in milliwatts; N = revolutions per minute; M = maximum bending moment in newton-millimeters; S = maximum allowable flexural (bending) stress, either in tension or compression in newtons per millimeter squared (see Table 2); p_t = maximum allowable shearing stress under combined loading conditions in newtons per millimeter squared; and V = maximum transverse shearing load in kilograms. The factors K_m, K_t, and B are unchanged, and D_1 = the inside diameter of a hollow shaft in millimeters.

Table 1. Recommended Values of the Combined Shock and Fatigue Factors for Various Types of Load

Type of Load	Stationary Shafts		Rotating Shafts	
	K_m	K_t	K_m	K_t
Gradually applied and steady	1.0	1.0	1.5	1.0
Suddenly applied, minor shocks only	1.5–2.0	1.5–2.0	1.5–2.0	1.0–1.5
Suddenly applied, heavy shocks	...	...	2.0–3.0	1.5–3.0

Table 2. Recommended Maximum Allowable Working Stresses for Shafts Under Various Types of Load*

Material	Type of Load		
	Simple Bending	Pure Torsion	Combined Stress
"Commercial Steel" shafting without keyways	$S = 16,000$	$S_s = 8000$	$p_t = 8000$
"Commercial Steel" shafting with keyways	$S = 12,000$	$S_s = 6000$	$p_t = 6000$
Steel purchased under definite physical specs.	Note (a)	Note (b)	Note (b)

* If the values in the Table are converted to metric SI units, note that 1000 pounds per square inch = 6.895 newtons per square millimeter.
(a) S = 60 per cent of the elastic limit in tension but not more than 36 per cent of the ultimate tensile strength. (b) S_s and p_t = 30 per cent of the elastic limit in tension but not more than 18 per cent of the ultimate tensile strength.

Table 3. Values of the Factor B Corresponding to Various Values of K for Hollow Shafts*

$K = \dfrac{D_1}{D} =$	0.95	0.90	0.85	0.80	0.75	0.70	0.65	0.60	0.55	0.50
$B = \sqrt[3]{1 \div (1 - K^4)} =$	1.75	1.43	1.28	1.19	1.14	1.10	1.07	1.05	1.03	1.02

* For solid shafts $B = 1$ since $K = 0$. $[B = \sqrt[3]{1 \div (1 - K^4)} = \sqrt[3]{1 \div (1 - 0)} = 1.]$

Effect of Keyways on Shaft Strength. — Keyways cut into a shaft reduce its load carrying ability, particularly when impact loads or stress reversals are involved.

To ensure an adequate factor of safety in the design of a shaft with standard keyway (width, one-quarter, and depth, one-eighth of shaft diameter), the former Code for Transmission Shafting tentatively recommended that shafts with keyways be designed on the basis of a solid circular shaft using not more than 75 per cent of the working stress recommended for the solid shaft. See also page 2234.

Formula for Shafts of Brittle Materials. — The preceding formulas are applicable to ductile materials and are based on the maximum-shear theory of failure which assumes that the elastic limit of a *ductile* material in shear is one-half its elastic limit in tension.

Brittle materials are generally stronger in shear than in tension; therefore, the maximum-shear theory is not applicable. The *maximum-normal-stress theory* of failure is now generally accepted for the design of shafts made from brittle materials. A material may be considered to be brittle if its elongation in a 2-inch gage length is less than 5 per cent. Materials such as cast iron, hardened tool steel, hard bronze, etc., conform to this rule. The diameter of a shaft made of a brittle material may be determined from the following formula which is based on the maximum-normal-stress theory of failure:

$$D = B \sqrt[3]{\frac{5.1}{S_t}[(K_m M) + \sqrt{(K_m M)^2 + (K_t T)^2}]}$$

where S_t is the maximum allowable tensile stress in pounds per square inch and the other quantities are as previously defined.

The formula can be used unchanged with metric SI units, where D = outside diameter of shaft in millimeters; S_t = the maximum allowable tensile stress in newtons per millimeter squared; M = maximum bending moment in newton-millimeters; and T = maximum torsional moment in newton-millimeters. The factors K_m, K_t, and B are unchanged.

Critical Speed of Rotating Shafts. — At certain speeds, a rotating shaft will become dynamically unstable and the resulting vibrations and deflections that occur can result in damage not only to the shaft but to the machine of which it is a part. The speeds at which such dynamic instability occurs are called the critical speeds of the shaft. On pages 235–236 are given formulas for the critical speeds of shafts subject to various conditions of loading and support. A shaft may be safely operated either above or below its critical speed, good practice indicating that the operating speed be at least 20 per cent above or below the critical.

The formulas commonly used to determine critical speeds are sufficiently accurate for general purposes. There are cases, however, where the torque applied to a shaft has an important effect on its critical speed. Investigations have shown that the critical speeds of a uniform shaft are decreased as the applied torque is increased, and that there exist critical torques which will reduce the corresponding critical speed of the shaft to zero. A detailed analysis of the effects of applied torques on critical speeds may be found in a paper, ''Critical Speeds of Uniform Shafts under Axial Torque,'' by Golomb and Rosenberg presented at the First U.S. National Congress of Applied Mechanics in 1951.

Table Giving Comparative Torsional Strengths and Weights of Hollow and Solid Shafting with Same Outside Diameter

(Upper figures in each line give number of per cent decrease in strength; lower figures give per cent decrease in weight.)

Example: — A 4-inch shaft, with a 2-inch hole through it, has a weight 25 per cent less than a solid 4-inch shaft, but its strength is decreased only 6.25 per cent.

Diam. of Solid and Hollow Shaft, Inches	Diameter of Axial Hole in Hollow Shaft, Inches									
	1	1¼	1½	1¾	2	2½	3	3½	4	4½
1½	19.76 44.44	48.23 69.44	.....	.....	.....	.....	.....	.....	.....	.....
1¾	10.67 32.66	26.04 51.02	53.98 73.49	.....	.....	.....	.....	.....	.....	.....
2	6.25 25.00	15.26 39.07	31.65 56.25	58.62 76.54	.....	.....	.....	.....	.....	.....
2¼	3.91 19.75	9.53 30.87	19.76 44.44	36.60 60.49	62.43 79.00	.....	.....	.....	.....	.....
2½	2.56 16.00	6.25 25.00	12.96 36.00	24.01 49.00	40.96 64.00	.....	.....	.....	.....	.....
2¾	1.75 13.22	4.28 20.66	8.86 29.74	16.40 40.48	27.98 52.89	68.30 82.63	.....	.....	.....	.....
3	1.24 11.11	3.01 17.36	6.25 25.00	11.58 34.01	19.76 44.44	48.23 69.44	.....	.....	.....	.....
3¼	0.87 9.46	2.19 14.80	4.54 21.30	8.41 29.00	14.35 37.87	35.02 59.17	72.61 85.22	.....	.....	.....
3½	0.67 8.16	1.63 12.76	3.38 18.36	6.25 25.00	10.67 32.66	26.04 51.02	53.98 73.49	.....	.....	.....
3¾	0.51 7.11	1.24 11.11	2.56 16.00	4.75 21.77	8.09 28.45	19.76 44.44	40.96 64.00	75.89 87.10	.....	.....
4	0.40 6.25	0.96 9.77	1.98 14.06	3.68 19.14	6.25 25.00	15.26 39.07	31.65 56.25	58.62 76.56	.....	.....
4¼	0.31 5.54	0.74 8.65	1.56 12.45	2.89 16.95	4.91 22.15	11.99 34.61	24.83 49.85	46.00 67.83	78.47 88.59	.....
4½	0.25 4.94	0.70 7.72	1.24 11.11	2.29 15.12	3.91 19.75	9.53 30.87	19.76 44.44	36.60 66.49	62.43 79.00	.....
4¾	0.20 4.43	0.50 6.93	1.00 9.97	1.85 13.57	3.15 17.73	7.68 27.70	15.92 39.90	29.48 54.29	50.29 70.91	80.56 89.75
5	0.16 4.00	0.40 6.25	0.81 8.10	1.51 12.25	2.56 16.00	6.25 25.00	12.96 36.00	24.01 49.00	40.96 64.00	65.61 81.00
5½	0.11 3.30	0.27 5.17	0.55 7.43	1.03 10.12	1.75 13.22	4.27 20.66	8.86 29.76	16.40 40.48	27.98 52.89	44.82 66.94
6	0.09 2.77	0.19 4.34	0.40 6.25	0.73 8.50	1.24 11.11	3.02 17.36	6.25 25.00	11.58 34.02	19.76 44.44	31.65 56.25
6½	0.06 2.36	0.14 3.70	0.29 5.32	0.59 7.24	0.90 9.47	2.19 14.79	4.54 21.30	8.41 28.99	14.35 37.87	23.98 47.93
7	0.05 2.04	0.11 3.19	0.22 4.59	0.40 6.25	0.67 8.16	1.63 12.76	3.38 18.36	6.25 25.00	10.67 32.66	17.08 41.33
7½	0.04 1.77	0.08 2.77	0.16 4.00	0.30 5.44	0.51 7.11	1.24 11.11	2.56 16.00	4.75 21.77	8.09 28.45	12.96 36.00
8	0.03 1.56	0.06 2.44	0.13 3.51	0.23 4.78	0.40 6.25	0.96 9.77	1.98 14.06	3.68 19.14	6.25 25.00	10.02 31.64

SPRINGS*

Introduction. — Many advances have been made in the spring industry in recent years. For example: developments in materials permit longer fatigue life at higher stresses; simplified design procedures reduce the complexities of design, and improved methods of manufacture help to speed up some of the complicated fabricating procedures and increase production. New types of testing instruments and revised tolerances also permit higher standards of accuracy. Designers should also consider the possibility of using standard springs now available from stock. They can be obtained from spring manufacturing companies located in different areas, and small shipments usually can be made quickly.

Designers of springs require information in the following order of precedence to simplify design procedures.

1. Spring materials and their applications
2. Allowable spring stresses
3. Spring design data with tables of spring characteristics, tables of formulas, and tolerances.

Only the more commonly used types of springs are covered in detail. Special types and designs rarely used such as torsion bars, volute springs, Belleville washers, constant force, ring and sprial springs and those made from rectangular wire are only described briefly.

Notation. — The following symbols are used in spring equations:

AC = Active coils.
b = Widest width of rectangular wire, inches.
CL = Compressed length, inches.
D = Mean coil diameter, inches = $OD - d$
d = Diameter of wire or side of square, inches.
E = Modulus of elasticity in tension, pounds per square inch.
F = Deflection, for N coils, inches.
$F°$ = Deflection, for N coils, rotary, degrees.
f = Deflection, for one active coil.
FL = Free length, unloaded spring, inches.
G = Modulus of elasticity in torsion, pounds per square inch.
IT = Initial tension, pounds.
K = Curvature stress correction factor.
L = Active length subject to deflection, inches.
N = Number of active coils, total.
P = Load, pounds.
p = pitch, inches.
R = Distance from load to central axis, inches.
S or S_t = Stress, torsional, pounds per square inch.
S_b = Stress, bending, pounds per square inch.
SH = Solid height.
S_{it} = Stress, torsional, due to initial tension, pounds per square inch.
T = Torque = $P \times R$, pound-inches.
TC = Total coils.
t = Thickness, inches.
U = Number of revolutions = $F°/360°$.

* This section was compiled by Harold Carlson, P. E., Consulting Engineer, Lakewood, N.J.

Spring Materials

The spring materials most commonly used include high-carbon spring steels, alloy spring steels, stainless spring steels, copper-base spring alloys, and nickel-base spring alloys.

High-Carbon Spring Steels in Wire Form. — These spring steels are the most commonly used of all spring materials because they are the least expensive, are easily worked, and are readily available. However, they are not satisfactory for springs operating at high or low temperatures or for shock or impact loading. The following wire forms are available:

Music Wire, ASTM A228 (0.80–0.95 per cent carbon): This is the most widely used of all spring materials for small springs operating at temperatures up to about 250 degrees F. It is tough, has a high tensile strength, and can withstand high stresses under repeated loading. The material is readily available in round form in diameters ranging from 0.005 to 0.125 inch and in some larger sizes up to 3/16 inch. It is not available with high tensile strengths in square or rectangular sections. Music wire can be plated easily and is obtainable pretinned or preplated with cadmium, but plating after spring manufacture is usually preferred for maximum corrosion resistance.

Oil-Tempered MB Grade, ASTM A229 (0.60–0.70 per cent carbon): This general-purpose spring steel is commonly used for many types of coil springs where the cost of music wire is prohibitive and in sizes larger than are available in music wire. It is readily available in diameters ranging from 0.125 to 0.500 inch, but both smaller and larger sizes may be obtained. The material should not be used under shock and impact loading conditions, at temperatures above 350 degrees F., or at temperatures in the sub-zero range. Square and rectangular sections of wire are obtainable in fractional sizes. Annealed stock also can be obtained for hardening and tempering after coiling. This material has a heat-treating scale that must be removed before plating.

Oil-Tempered HB Grade, SAE 1080 (0.75–0.85 per cent carbon): This material is similar to the MB Grade except that it has a higher carbon content and a higher tensile strength. It is obtainable in the same sizes and is used for more accurate requirements than the MB Grade, but is not so readily available. In lieu of using this material it may be better to use an alloy spring steel, particularly if a long fatigue life or high endurance properties are needed. Round and square sections are obtainable in the oil-tempered or annealed conditions.

Hard-Drawn MB Grade, ASTM A227 (0.60–0.70 per cent carbon): This grade is used for general-purpose springs where cost is the most important factor. Although increased use in recent years has resulted in improved quality, it is best not to use it where long life and accuracy of loads and deflections are important. It is available in diameters ranging from 0.031 to 0.500 inch and in some smaller and larger sizes also. The material is available in square sections but at reduced tensile strengths. It is readily plated. Applications should be limited to those in the temperature range of 0 to 250 degrees F.

High-Carbon Spring Steels in Flat Strip Form. — Two types of thin, flat, high-carbon spring steel strip are most widely used although several other types are obtainable for specific applications in watches, clocks, and certain instruments. These two compositions are used for over 95 per cent of all such applications. Thin sections of these materials under 0.015 inch having a carbon content of over 0.85 per cent and a hardness of over 47 on the Rockwell C scale are susceptible to hydrogen-embrittlement even though special plating and heating operations are employed. The two types are described as follows:

Cold-Rolled Spring Steel, Blue-Tempered or Annealed, SAE 1074, also 1064, and 1070 (0.60 to 0.80 per cent carbon): This very popular spring steel is available in thicknesses ranging from 0.005 to 0.062 inch and in some thinner and thicker sections. The material is available in the annealed condition for forming in 4-slide machines and in presses, and can readily be hardened and tempered after forming. It is also available in the heat-treated or blue-tempered condition. The steel is obtainable in several finishes such as straw color, blue color, black, or plain. Hardnesses ranging from 42 to 46 Rockwell C are recommended for spring applications. Uses include spring clips, flat springs, clock springs, and motor, power, and spiral springs.

Cold-Rolled Spring Steel, Blue-Tempered Clock Steel, SAE 1095 (0.90 to 1.05 per cent carbon): This popular type should be used principally in the blue-tempered condition. Although obtainable in the annealed condition, it does not always harden properly during heat-treatment as it is a "shallow" hardening type. It is used principally in clocks and motor springs. End sections of springs made from this steel are annealed for bending or piercing operations. Hardnesses usually range from 47 to 51 Rockwell C.

Other materials available in strip form and used for flat springs are brass, phosphor-bronze, beryllium-copper, stainless steels, and nickel alloys.

Alloy Spring Steels. — These spring steels are used for conditions of high stress, and shock or impact loadings. They can withstand both higher and lower temperatures than the high-carbon steels and are obtainable in either the annealed or pretempered conditions.

Chromium Vanadium, ASTM A231: This very popular spring steel is used under conditions involving higher stresses than those for which the high-carbon spring steels are recommended and is also used where good fatigue strength and endurance are needed. It behaves well under shock and impact loading. The material is available in diameters ranging from 0.031 to 0.500 inch and in some larger sizes also. In square sections it is available in fractional sizes. Both the annealed and pretempered types are available in round, square, and rectangular sections. It is used extensively in aircraft-engine valve springs and for springs operating at temperatures up to 425 degrees F.

Silicon Manganese: This alloy steel is quite popular in Great Britain. It is less expensive than chromium-vanadium steel and is available in round, square, and rectangular sections in both annealed and pretempered conditions in sizes ranging from 0.031 to 0.500 inch. It was formerly used for knee-action springs in automobiles. It is used in flat leaf springs for trucks and as a substitute for more expensive spring steels.

Chromium Silicon, ASTM A401: This alloy is used for highly stressed springs requiring long life and which are subjected to shock loading. It can be heat-treated to higher hardnesses than other spring steels so that high tensile strengths are obtainable. The most popular sizes range from 0.031 to 0.500 inch in diameter. Very rarely are square, flat, or rectangular sections used. Hardnesses ranging from 50 to 53 Rockwell C are quite common and the alloy may be used at temperatures up to 475 degrees F. This material is usually ordered specially for each job.

Stainless Spring Steels. — The use of stainless spring steels has increased and several compositions are available all of which may be used for temperatures up to 550 degrees F. They are all corrosion resistant. Only the stainless 18-8 compositions should be used at sub-zero temperatures.

Stainless Type 302, ASTM A313 (18 per cent chromium, 8 per cent nickel): This is a very popular stainless spring steel because it has the highest tensile strength and quite uniform properties. It is cold-drawn to obtain its mechanical properties and cannot be hardened by heat treatment. This material is nonmagnetic only when

fully annealed and becomes slightly magnetic due to the cold-working performed to produce spring properties. It is suitable for use at temperatures up to 550 degrees F. and for sub-zero temperatures. It is very corrosion resistant. The material best exhibits its desirable mechanical properties in diameters ranging from 0.005 to 0.1875 inch although some larger diameters are available. It is also available as hard-rolled flat strip. Square and rectangular sections are available but are infrequently used.

Stainless Type 304, ASTM A313 (18 per cent chromium, 8 per cent nickel): This material is quite similar to Type 302, but has better bending properties and about 5 per cent lower tensile strength. It is a little easier to draw, due to the slightly lower carbon content.

Stainless Type 316, ASTM A313 (18 per cent chromium, 12 per cent nickel, 2 per cent molybdenum): This material is quite similar to Type 302 but is slightly more corrosion resistant because of its higher nickel content. Its tensile strength is 10 to 15 per cent lower than Type 302. It is used for aircraft springs.

Stainless Type 17-7 PH ASTM A313 (17 per cent chromium, 7 per cent nickel): This alloy, which also contains small amounts of aluminum and titanium, is formed in a moderately hard state and then precipitation hardened at relatively low temperatures for several hours to produce tensile strengths nearly comparable to music wire. This material is not readily available in all sizes, and has limited applications due to its high manufacturing cost.

Stainless Type 414, SAE 51414 (12 per cent chromium, 2 per cent nickel): This alloy has tensile strengths about 15 per cent lower than Type 302 and can be hardened by heat-treatment. For best corrosion resistance it should be highly polished or kept clean. It can be obtained hard drawn in diameters up to 0.1875 inch and is commonly used in flat cold-rolled strip for stampings. The material is not satisfactory for use at low temperatures.

Stainless Type 420, SAE 51420 (13 per cent chromium): This is the best stainless steel for use in large diameters above 0.1875 inch and is frequently used in smaller sizes. It is formed in the annealed condition and then hardened and tempered. It does not exhibit its stainless properties until after it is hardened. Clean bright surfaces provide the best corrosion resistance, therefore the heat-treating scale must be removed. Bright hardening methods are preferred.

Stainless Type 431, SAE 51431 (16 per cent chromium, 2 per cent nickel): This spring alloy acquires high tensile properties (nearly the same as music wire) by a combination of heat-treatment to harden the wire plus cold-drawing after heat-treatment. Its corrosion resistance is not equal to Type 302.

Copper-Base Spring Alloys. — Copper-base alloys are important spring materials because of their good electrical properties combined with their good resistance to corrosion. Although these materials are more expensive than the high-carbon and the alloy steels, they nevertheless are frequently used in electrical components and in sub-zero temperatures.

Spring Brass, ASTM B 134 (70 per cent copper, 30 per cent zinc): This material is the least expensive and has the highest electrical conductivity of the copper-base alloys. It has a low tensile strength and poor spring qualities, but is extensively used in flat stampings and where sharp bends are needed. It cannot be hardened by heat-treatment and should not be used at temperatures above 150 degrees F., but is especially good at sub-zero temperatures. Available in round sections and flat strips, this hard-drawn material is usually used in the "spring hard" temper.

Phosphor Bronze, ASTM B 159 (95 per cent copper, 5 per cent tin): This alloy is the most popular of this group because it combines the best qualities of tensile strength, hardness, electrical conductivity, and corrosion resistance with the least cost. It is more expensive than brass, but can withstand stresses 50 per cent higher.

The material cannot be hardened by heat-treatment. It can be used at temperatures up to 212 degrees F. and at sub-zero temperatures. It is available in round sections and flat strip, usually in the "extra-hard" or "spring hard" tempers. It is frequently used for contact fingers in switches because of its low arcing properties. An 8 per cent tin composition is used for flat springs and a superfine grain composition called "Duraflex", has good endurance properties.

Beryllium Copper, ASTM B 197 (98 per cent copper, 2 per cent beryllium): This alloy can be formed in the annealed condition and then precipitation hardened after forming at temperatures around 600 degrees F, for 2 to 3 hours. This produces a high hardness combined with a high tensile strength. After hardening, the material becomes quite brittle and can withstand very little or no forming. It is the most expensive alloy in the group and heat-treating is expensive due to the need for holding the parts in fixtures to prevent distortion. The principal use of this alloy is for carrying electric current in switches and in electrical components. Flat strip is frequently used for contact fingers.

Nickel-Base Spring Alloys. — Nickel-base alloys are corrosion resistant, withstand both elevated and sub-zero temperatures, and their non-magnetic characteristic makes them useful for such applications as gyroscopes, chronoscopes, and indicating instruments. These materials have a high electrical resistance and therefore should not be used for conductors of electrical current.

*Monel** (67 per cent nickel, 30 per cent copper): This material is the least expensive of the nickel-base alloys. It also has the lowest tensile strength but is useful due to its resistance to the corrosive effects of sea water and because it is nearly non-magnetic. The alloy can be subjected to stresses slightly higher than phosphor bronze and nearly as high as beryllium copper. Its high tensile strength and hardness are obtained as a result of cold-drawing and cold-rolling only, since it can not be hardened by heat-treatment. It can be used at temperatures ranging between − 100 to + 425 degrees F. at normal operating stresses and is available in round wires up to ³⁄₁₆ inch in diameter with quite high tensile strengths. Larger diameters and flat strip are available with lower tensile strengths.

*"K" Monel** (66 per cent nickel, 29 per cent copper, 3 per cent aluminum): This material is quite similar to Monel except that the addition of the aluminum makes it a precipitation-hardening alloy. It may be formed in the soft or fairly hard condition and then hardened by a long-time age-hardening heat-treatment to obtain a tensile strength and hardness above Monel and nearly as high as stainless steel. It is used in sizes larger than those usually used with Monel, is non-magnetic and can be used in temperatures ranging from − 100 to + 450 degrees F. at normal working stresses under 45,000 pounds per square inch.

*Inconel** (78 per cent nickel, 14 per cent chromium, 7 per cent iron): This is one of the most popular of the non-magnetic nickel-base alloys because of its corrosion resistance and because it can be used at temperatures up to 700 degrees F. It is more expensive than stainless steel but less expensive than beryllium copper. Its hardness and tensile strength is higher than that of "K" Monel and is obtained as a result of cold-drawing and cold-rolling only. It cannot be hardened by heat treatment. Wire diameters up to ¼ inch have the best tensile properties. It is often used in steam valves, regulating valves, and for springs in boilers, compressors, turbines, and jet engines.

*Inconel "X"** (70 per cent nickel, 16 per cent chromium, 7 per cent iron): This material is quite similar to Inconel but the small amounts of titanium, columbium and aluminum in its composition make it a precipitation-hardening alloy. It can be formed in the soft or partially hard condition and then hardened by holding it

* Trade name of The International Nickel Company.

at 1200 degrees F. for 4 hours. It is non-magnetic and is used in larger sections than Inconel. This alloy is used at temperatures up to 850 degrees F. and at stresses up to 55,000 pounds per square inch.

Duranickel* ("Z" *Nickel*) (98 per cent nickel): This alloy is non-magnetic, corrosion resistant, has a high tensile strength and is hardenable by precipitation hardening at 900 degrees F. for 6 hours. It may be used at the same stresses as Inconel but should not be used at temperatures above 500 degrees F.

Nickel-Base Spring Alloys with Constant Moduli of Elasticity. — Some special nickel alloys have a constant modulus of elasticity over a wide temperature range. These are especially useful where springs undergo temperature changes and must exhibit uniform spring characteristics. These materials have a low or zero thermoelastic coefficient and therefore do not undergo variations in spring stiffness because of modulus changes due to temperature differentials. They also have low hysteresis and creep values which makes them preferred for use in food-weighing scales, precision instruments, gyroscopes, measuring devices, recording instruments and computing scales where the temperature ranges from − 50 to + 150 degrees F. These materials are expensive, none being regularly stocked in a wide variety of sizes. They should not be specified without prior discussion with spring manufacturers since some suppliers may not fabricate springs from these alloys because of the special manufacturing processes required. All of these alloys are used in small wire diameters and in thin strip only and are covered by U.S. patents. They are more specifically described as follows:

Elinvar† (nickel, iron, chromium): This alloy, the first constant-modulus alloy used for hairsprings in watches, is an austenitic alloy hardened only by cold-drawing and cold-rolling. Additions of titanium, tungsten, molybdenum and other alloying elements have brought about improved characteristics and precipitation-hardening abilities. These improved alloys are known by the following trade names: Elinvar Extra, Durinval, Modulvar and Nivarox.

Ni-Span C* (nickel, iron, chromium, titanium): This very popular constant-modulus alloy is usually formed in the 50 per cent cold-worked condition and precipitation-hardened at 900 degrees F. for 8 hours, although heating up to 1250 degrees F. for 3 hours produces hardnesses of 40 to 44 Rockwell C, permitting safe torsional stresses of 60,000 to 80,000 pounds per square inch. This material is ferromagnetic up to 400 degrees F. Above that it becomes non-magnetic.

Iso-Elastic‡ (nickel, iron, chromium, molybdenum): This popular alloy is relatively easy to fabricate and is used at safe torsional stresses of 40,000 to 60,000 pounds per square inch and hardnesses of 30 to 36 Rockwell C. It is used principally in dynamometers, instruments, and food-weighing scales.

Elgiloy§ (nickel, iron, chromium, cobalt): This alloy, also known by the trade names 8J Alloy, Durapower, and Cobenium, is a non-magnetic alloy suitable for sub-zero temperatures and temperatures up to about 1000 degrees F., provided that torsional stresses are kept under 75,000 pounds per square inch. It is precipitation-hardened at 900 degrees F. for 8 hours to produce hardnesses of 48 to 50 Rockwell C. The alloy is used in watch and instrument springs.

Dynavar¶ (nickel, iron, chromium, cobalt): This alloy is a non-magnetic, corrosion-resistant material suitable for sub-zero temperatures and temperatures up to about 750 degrees F., provided that torsional stresses are kept below 75,000 pounds per square inch. It is precipitation-hardened to produce hardnesses of 48 to 50 Rockwell C and is used in watch and instrument springs.

---

* Trade name of the International Nickel Company. † Trade name of Soc. Anon. de Commentry Fourchambault et Decazeville, Paris, France. ‡ Trade name of John Chatillon & Sons. § Trade name of Elgin National Watch Company. ¶ Trade name of Hamilton Watch Company.

Spring Stresses

Allowable Working Stresses for Springs. — The safe working stress for any particular spring depends to a large extent on the following items:

1. Type of spring — whether compression, extension, torsion, etc.;
2. Size of spring — small or large, long or short;
3. Spring material;
4. Size of spring material;
5. Type of service — light, average, or severe;
6. Stress range — low, average, or high;
7. Loading — static, dynamic, or shock;
8. Operating temperature;
9. Design of spring — spring index, sharp bends, hooks.

Consideration should also be given to other factors that affect spring life: corrosion, buckling, friction, and hydrogen embrittlement decrease spring life; manufacturing operations such as high-heat stress-equalizing, presetting, and shot-peening
increase spring life.

Item 5, the type of service to which a spring is subjected, is a major factor in
determining a safe working stress once consideration has been given to type of spring,
kind and size of material, temperature, type of loading, and so on. The types of
service are:

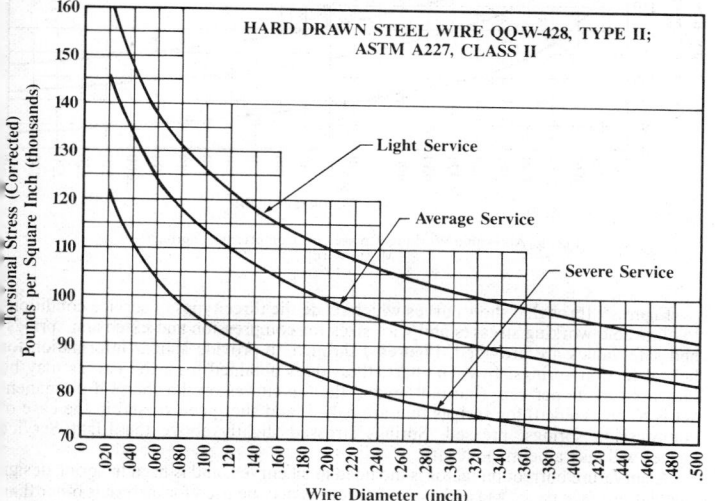

Fig. 1. Allowable Working Stresses for Compression Springs —
Hard Drawn Steel Wire*

* Although these curves are for compression springs, they may also be used for extension
springs; for extension springs, *reduce* the values obtained from the curves by 10 to 15 per cent.

Light Service. This includes springs subjected to static loads or small deflections and seldom-used springs such as those in bomb fuses, projectiles, and safety devices. This service is for 1,000 to 10,000 deflections.

Average Service. This includes springs in general use in machine tools, mechanical products, and electrical components. Normal frequency of deflections not exceeding 18,000 per hour permit such springs to withstand 100,000 to 1,000,000 deflections.

Severe Service. This includes springs subjected to rapid deflections over long periods of time and to shock loading such as in pneumatic hammers, hydraulic controls and valves. This service is for 1,000,000 deflections, and above. Lowering the values 10 per cent permits 10,000,000 deflections.

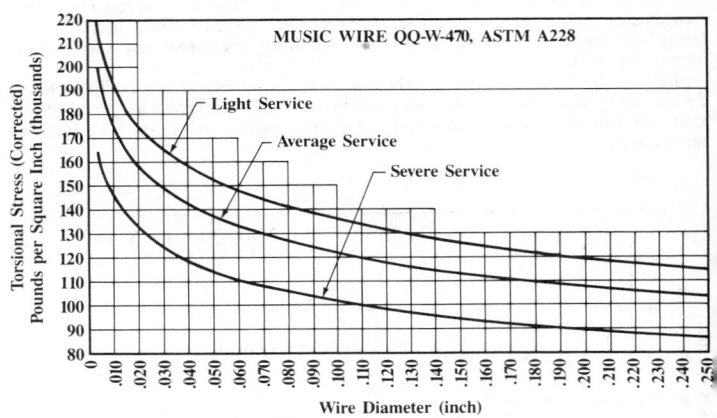

Fig. 2. Allowable Working Stresses for Compression Springs —
Music Wire*

Figures 1 through 6 show curves which relate the three types of service conditions to allowable working stresses and wire sizes for compression and extension springs and safe values are provided. Figures 7 through 10 provide similar information for helical torsion springs. In each chart, the values obtained from the curves may be increased by 20 per cent (but not beyond the top curves on the charts if permanent set is to be avoided) for springs that are baked, and shot-peened, and in the case of compression springs, pressed. Springs stressed slightly above the Light Service curves will take a permanent set.

A curvature correction factor is included in all curves, and is used in spring design calculations (see pages 330 and 332). The curves may be used for materials other than those designated in Figs. 1 through 10, by applying multiplication factors as given in Table 1.

(Continued on page 325)

* Although these curves are for compression springs, they may also be used for extension springs; for extension springs, *reduce* the values obtained from the curves by 10 to 15 per cent.

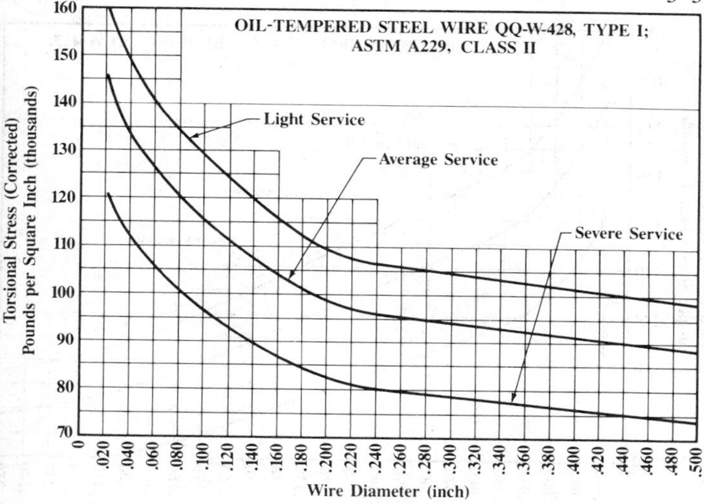

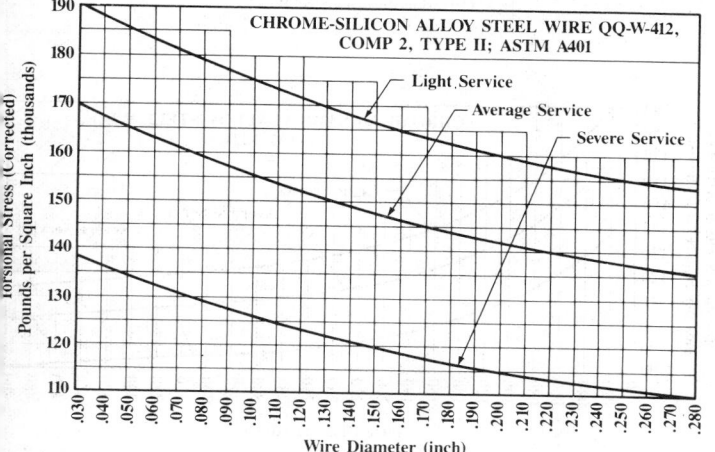

Fig. 3. Allowable Working Stresses for Compression Springs —
Oil-Tempered Steel Wire*

Fig. 4. Allowable Working Stresses for Compression Springs —
Chrome-Silicon Alloy Steel Wire*

* Although these curves are for compression springs, they may also be used for extension springs; for extension springs, *reduce* the values obtained from the curves by 10 to 15 per cent.

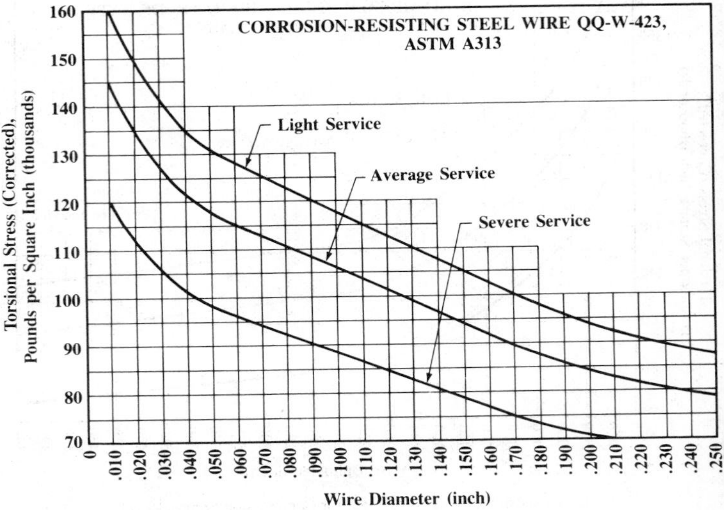

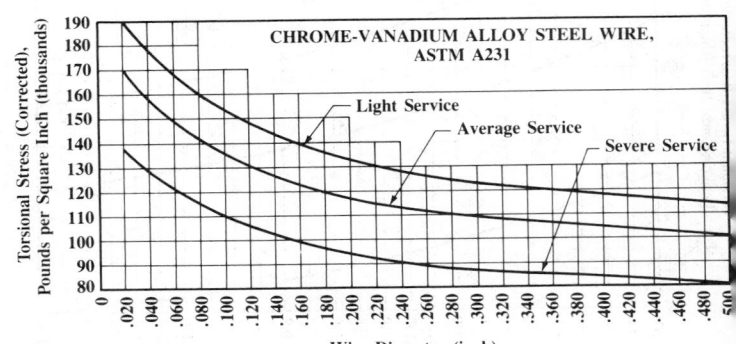

Fig. 5. Allowable Working Stresses for Compression Springs —
Corrosion-Resisting Steel Wire*

Fig. 6. Allowable Working Stresses for Compression Springs —
Chrome-Vanadium Alloy Steel Wire*

* Although these curves are for compression springs, they may also be used for extension springs; for extension springs, *reduce* the values obtained from the curves by 10 to 15 per cent.

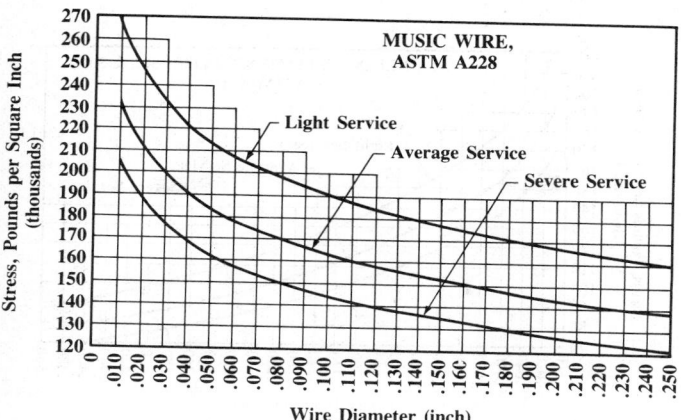

Fig. 7. Recommended Design Stresses in Bending for Helical Torsion Springs — Round Music Wire

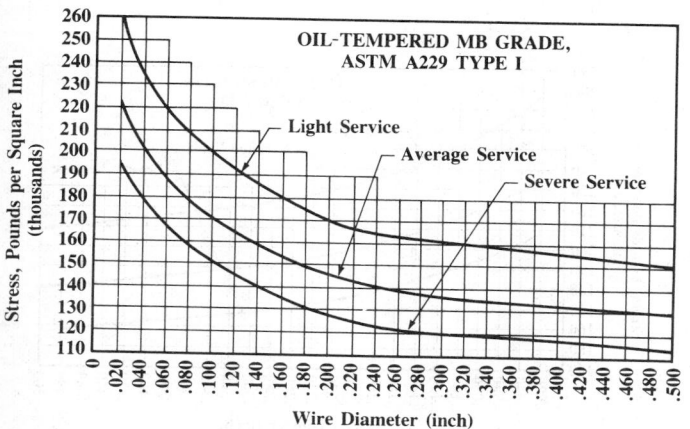

Fig. 8. Recommended Design Stresses in Bending for Helical Torsion Springs — Oil-Tempered MB Round Wire

Endurance Limit for Spring Materials. — When a spring is deflected continually it will become "tired" and fail at a stress far below its elastic limit. This type of failure is called *fatigue failure* and usually occurs without warning. *Endurance limit*

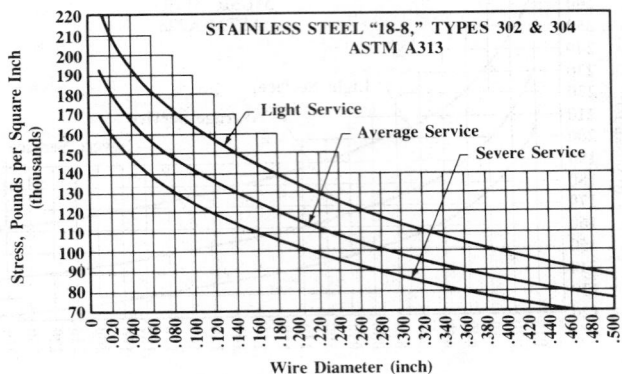

Fig. 9. Recommended Design Stresses in Bending for Helical Torsion Springs —
Stainless Steel Round Wire

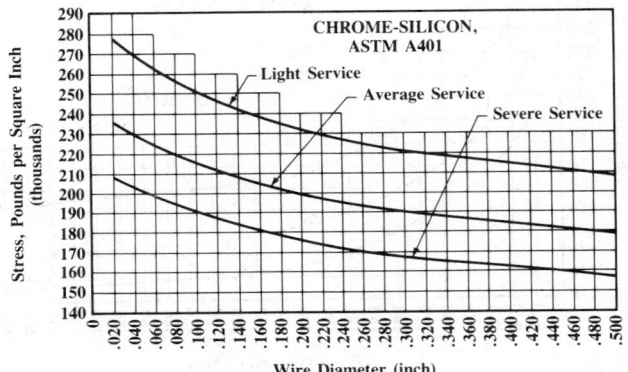

Fig. 10. Recommended Design Stresses in Bending for Helical Torsion Springs —
Chrome-Silicon Round Wire

is the highest stress, or range of stress, in pounds per square inch that can be repeated
indefinitely without failure of the spring. Usually ten million cycles of deflection is
called "infinite life" and is satisfactory for determining this limit.

loss of load may be nearer 10 per cent. Maximum stresses shown in the table are for compression and extension springs and may be increased by 75 per cent for torsion and flat springs. In using the data in Table 2 it should be noted that the values given are for materials in the heat-treated or spring temper condition.

Table 2. Recommended Maximum Working Temperatures and Corresponding Maximum Working Stresses for Springs*

Spring Material	Maximum Working Temperature, Degrees, F.	Maximum Working Stress, Pounds per Square Inch
Brass Spring Wire	150	30,000
Phosphor Bronze	225	35,000
Music Wire	250	75,000
Beryllium-Copper	300	40,000
Hard Drawn Steel Wire	325	50,000
Carbon Spring Steels	375	55,000
Alloy Spring Steels	400	65,000
Monel	425	40,000
K-Monel	450	45,000
Permanickel†	500	50,000
Stainless Steel 18-8	550	55,000
Stainless Chromium 431	600	50,000
Inconel	700	50,000
High Speed Steel	775	70,000
Inconel X	850	55,000
Chromium-Molybdenum-Vanadium	900	55,000
Cobenium, Elgiloy	1000	75,000

* Loss of load at temperatures shown is less than 5 per cent in 48 hours.
† Formerly called Z-Nickel, Type B.

Spring Design Data

This section provides tables of spring characteristics, tables of principal formulas, and other information of a practical nature for designing the more commonly used types of springs.

Standard wire gages for springs: Information on wire gages is given in the section beginning on page 413, and gages in decimals of an inch are given in the table on page 414. It should be noted that the range in this table extends from Number 7/0 through Number 80. However, in spring design, the range most commonly used extends only from Gage Number 4/0 through Number 40. When selecting wire use Steel Wire Gage or Washburn and Moen gage for all carbon steels and alloy steels except music wire; use Brown & Sharpe gage for brass and phosphor bronze wire; use Birmingham gage for flat spring steels, and cold rolled strip; and use piano or music wire gage for music wire.

Spring index: The spring index is the ratio of the mean coil diameter of a spring to the wire diameter (D/d). This ratio is one of the most important considerations in spring design because the deflection, stress, number of coils, and selection of either annealed or tempered material depend to a considerable extent on this ratio. The best proportioned springs have an index of 7 through 9. Indexes of 4 through 7, and 9 through 16 are often used. Springs with values larger than 16 require tolerances

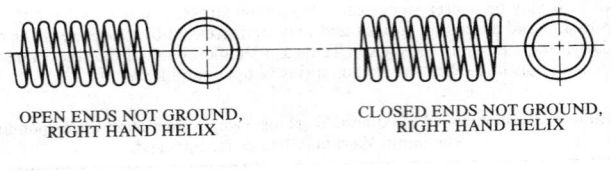

OPEN ENDS NOT GROUND,   CLOSED ENDS NOT GROUND,
RIGHT HAND HELIX   RIGHT HAND HELIX

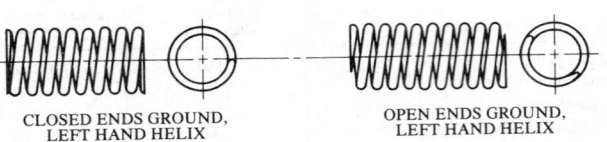

CLOSED ENDS GROUND,   OPEN ENDS GROUND,
LEFT HAND HELIX   LEFT HAND HELIX

Fig. 12. Types of Helical Compression Spring Ends

wider than standard for manufacturing; those with values less than 5 are difficult to coil on automatic coiling machines.

Direction of helix: Unless functional requirements call for a definite hand, the helix of compression and extension springs should be specified as optional. When springs are designed to operate, one inside the other, the helices should be opposite hand to prevent intermeshing. For the same reason, a spring which is to operate freely over a threaded member should have a helix of opposite hand to that of the thread. When a spring is to engage with a screw or bolt, it should, of course, have the same helix as that of the thread.

Helical Compression Spring Design. — After selecting a suitable material and a safe stress value for a given spring, designers should next determine the type of end coil formation best suited for the particular application. Springs with unground ends are less expensive but they do not stand perfectly upright; if this requirement has to be met, closed ground ends are used. Helical compression springs with different types of ends are shown in Fig. 12.

Spring design formulas: Table 3 gives formulas for compression spring dimensional characteristics, and Table 4 gives design formulas for compression and extension springs.

Curvature correction: In addition to the stress obtained from the formulas for load or deflection, there is a direct shearing stress and an increased stress on the inside of the section due to curvature. Therefore the stress obtained by the usual formulas should be multiplied by a factor K taken from the curve in Fig. 13. The corrected stress thus obtained is used only for comparison with the allowable working stress (fatigue strength) curves to determine if it is a safe stress and should not be used in formulas for deflection. The curvature correction factor K is for compression and extension springs made from round wire. For square wire reduce the K value by approximately 4 per cent.

Design procedure: The limiting dimensions of a spring are often determined by the available space in the product or assembly in which it is to be used. The loads and deflections on a spring may also be known or can be estimated, but the wire size and number of coils are usually unknown. Design can be carried out with the aid of the tabular data, which is a simple method, or by calculation alone, as follows:

(Continued on page 332)

Table 3. Formulas for Compression Springs*

Feature	Type of End			
	Open or Plain (not ground)	Open or Plain (with ends ground)	Squared or Closed (not ground)	Closed and Ground
	Formula			
Pitch (p)	$\dfrac{FL-d}{N}$	$\dfrac{FL}{TC}$	$\dfrac{FL-3d}{N}$	$\dfrac{FL-2d}{N}$
Solid Height (SH)	$(TC+1)d$	$TC \times d$	$(TC+1)d$	$TC \times d$
Number of Active Coils (N)	$N=TC$ or $\dfrac{FL-d}{p}$	$N=TC-1$ or $\dfrac{FL}{p}-1$	$N=TC-2$ or $\dfrac{FL-3d}{p}$	$N=TC-2$ or $\dfrac{FL-2d}{p}$
Total Coils (TC)	$\dfrac{FL-d}{p}$	$\dfrac{FL}{p}$	$\dfrac{FL-3d}{p}+2$	$\dfrac{FL-2d}{p}+2$
Free Length (FL)	$(p \times TC)+d$	$p \times TC$	$(p \times N)+3d$	$(p \times N)+2d$

Table 4. Formulas for Compression and Extension Springs*

Feature	Springs made from round wire	Springs made from square wire
	Formula†	
Load (P) Pounds	$\dfrac{0.393\,S\,d^3}{D}$	$\dfrac{0.416\,S\,d^3}{D}$
	$\dfrac{Gd^4F}{8ND^3}$	$\dfrac{Gd^4F}{5.58ND^3}$
Stress, Torsional (S) Pounds per square inch	$\dfrac{GdF}{\pi ND^2}$	$\dfrac{GdF}{2.32ND^2}$
	$\dfrac{PD}{0.393\,d^3}$	$\dfrac{PD}{0.416\,d^3}$
Deflection (F) Inch	$\dfrac{8PND^3}{Gd^4}$	$\dfrac{5.58PND^3}{Gd^4}$
	$\dfrac{\pi SND^2}{Gd}$	$\dfrac{2.32SND^2}{Gd}$
Number of Active Coils (N)	$\dfrac{Gd^4F}{8PD^3}$	$\dfrac{Gd^4F}{5.58PD^3}$
	$\dfrac{GdF}{\pi SD^2}$	$\dfrac{GdF}{2.32SD^2}$
Wire Diameter (d) Inch	$\dfrac{\pi SND^2}{GF}$	$\dfrac{2.32SND^2}{GF}$
	$\sqrt[3]{\dfrac{2.55PD}{S}}$	$\sqrt[3]{\dfrac{PD}{0.416\,S}}$
Stress due to Initial Tension (S_{it})	$\dfrac{S}{P} \times IT$	$\dfrac{S}{P} \times IT$

* The symbol notation is given on page 315.
† Two formulas are given for each feature, and designers can use the one found to be appropriate in a given design instance. The end result from either of any two formulas is the same.

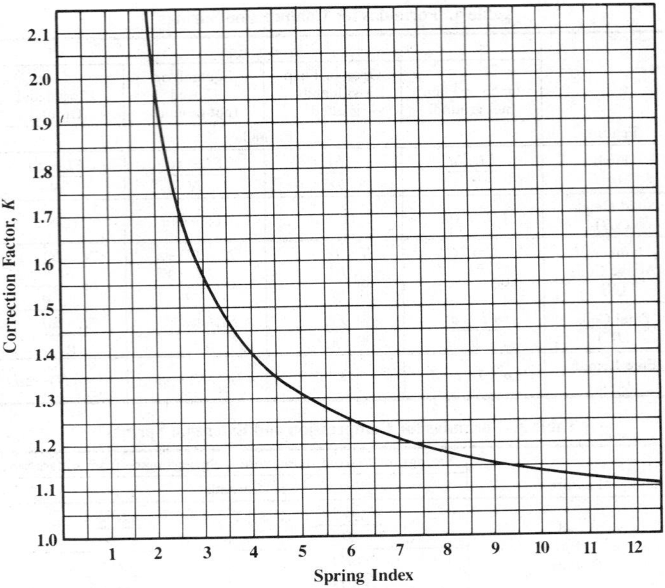

Fig. 13. Compression and Extension Spring Stress Correction for Curvature*

Example: A compression spring with closed and ground ends is to be made from ASTM A229 high carbon steel wire, as shown in Fig. 14. Determine the wire size and number of coils.

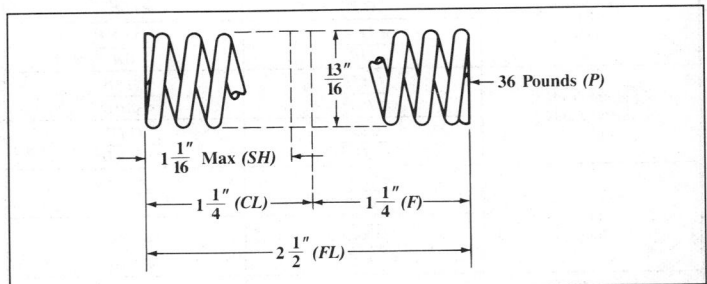

Fig. 14. Compression Spring Design Example

* For springs made from round wire. For springs made from square wire, reduce the K factor values by approximately 4 per cent.

Method 1, using table: Referring to Table 5, page 335, locate the spring outside diameter ($^{13}/_{16}$ inches, from Fig. 14) in the left-hand column. Note from the drawing that the spring load is 36 pounds. Move to the right in the table to the figure nearest this value, which is 41.7 pounds. This is somewhat above the required value but safe. Immediately above the load value, the deflection *f* is given, which in this instance is 0.1594 inch. This is the deflection of 1 coil under a load of 41.7 pounds with an uncorrected torsional stress *S* of 100,000 pounds per square inch. (This is the torsional stress for ASTM A229 oil-tempered MB steel — see page 356). Moving vertically in the table from the load entry, the wire diameter is found to be 0.0915 inch.

The remaining spring design calculations are completed as follows:

Step 1: The stress with a load of 36 pounds is obtained as follows: The 36 pound load is 86.3 per cent of the 41.7 pound load. Therefore, the stress *S* at 36 pounds = $0.863 \times 100,000 = 86,300$ pounds per square inch.

Step 2: The 86.3 per cent figure is also used to determine the deflection per coil *f* at 36 pounds load: $0.863 \times 0.1594 = 0.1375$ inch.

Step 3: The number of active coils $AC = \dfrac{F}{f} = \dfrac{1.25}{0.1375} = 9.1$ (say 9)

Step 4: Total Coils $TC = AC + 2$ (Table 3) $= 9 + 2 = 11$

Therefore a quick answer is: 11 coils of 0.0915 inch diameter wire. However, the design procedure should be completed by carrying out these remaining steps:

Step 5: Solid Height $SH = TC \times d$ (Table 3)

$= 11 \times 0.0915 = 1$ inch

(Total Deflection = $FL - SH = 1.5$ inches)

Step 6: Stress Solid $= \dfrac{86,300}{1.25} \times 1.5 = 103,500$ pounds per square inch

Step 7: Spring Index $= \dfrac{O.D.}{d} - 1 = \dfrac{0.8125}{0.0915} - 1 = 7.9$

Step 8: From Fig. 13, the curvature correction factor $K = 1.185$

Step 9: Total Stress at 36 pounds load = $S \times K = 86,300 \times 1.185 = 102,300$ pounds per square inch. (This is a safe working stress as it is below the 117,000 pounds per square inch permitted for 0.0915 inch wire shown on the middle curve in Fig. 3.)

Step 10: Total Stress at Solid = $103,500 \times 1.185 = 122,800$ pounds per square inch. (This, too, is a safe stress, as it is below the 131,000 pounds per square inch shown on the top curve Fig. 3; therefore the spring will not set.)

Method 2, using formulas: The procedure for design using formulas is as follows (the design example is the same as in Method 1, and the spring is shown in Fig. 14):

Step 1: Select a safe stress *S* below the middle fatigue strength curve Fig. 3 for ASTM A229 steel wire, say 90,000 pounds per square inch. Assume a mean diameter *D* slightly below the $^{13}/_{16}$-inch O.D., say 0.7 inch. Note that the value of *G* is 11,200,000 pounds per square inch (Table 19).

Step 2: A trial wire diameter *d* and other values are found by formulas from

Table 4 as follows: $d = \sqrt[3]{\dfrac{2.55PD}{S}} = \sqrt[3]{\dfrac{2.55 \times 36 \times 0.7}{90,000}}$

$= \sqrt[3]{0.000714} = 0.0894$ inch.

Note: Table 20 can be used to avoid solving the cube root.

Step 3: From the table on page 414, select the nearest wire gauge size, which is 0.0915 inch diameter. Using this value, the mean diameter $D = {}^{13}/_{16}$ inch $- 0.0915 = 0.721$ inch.

(Continued on page 338)

Table 5. Compression and Extension Spring Deflections*

Wire Size or Washburn and Moen Gauge, and Decimal Equivalent†

Deflection f (inch) per coil, at Load P (pounds)‡ — *(cell entries given as "deflection / load")*

Outside Diam. Nom.	Dec.	.010	.012	.014	.016	.018	.020	.022	.024	.026	.028	.030	.032	.034	.036	.038	19 .041	18 .0475	17 .054	16 .0625
7/64	.1094	.0277/.395	.0222/.697	.01824/1.130	.01529/1.722	.01302/2.51	.01121/3.52	.00974/4.79	.00853/6.36	.00751/8.28	.00664/10.59	.00589/13.35								
1/8	.125	.0371/.342	.0299/.600	.0247/.971	.0208/1.475	.01784/2.14	.01548/2.99	.01353/4.06	.01192/5.37	.01050/6.97	.00943/8.89	.00844/11.16	.00758/13.83	.00683/16.95	.00617/20.6					
9/64	.1406	.0478/.301	.0387/.528	.0321/.852	.0272/1.291	.0234/1.868	.0204/2.61	.01794/3.53	.01590/4.65	.01417/6.02	.01271/7.66	.01144/9.58	.01034/11.84	.00937/14.47	.00852/17.51	.00777/21.0				
5/32	.1563	.0600/.268	.0487/.470	.0406/.758	.0345/1.146	.0298/1.656	.0261/2.31	.0230/3.11	.0205/4.10	.01832/5.30	.01640/6.72	.01491/8.39	.01354/10.35	.01234/12.62	.01128/15.23	.01033/18.22	.00909/23.5			
11/64	.1719	.0735/.243	.0598/.424	.0500/.683	.0426/1.031	.0369/1.488	.0324/2.07	.0287/2.79	.0256/3.67	.0230/4.73	.0208/5.99	.01872/7.47	.01716/9.19	.01569/11.19	.01439/13.48	.01324/16.09	.01157/20.8	.00972/33.8		
3/16	.1875	.0884/.221	.0720/.387	.0603/.621	.0516/.938	.0448/1.351	.0394/1.876	.0349/2.53	.0313/3.33	.0281/4.27	.0255/5.40	.0232/6.73	.0212/8.27	.01944/10.05	.01788/12.09	.01650/14.41	.01468/18.47	.01233/30.07	.01155/46.3	
13/64	.2031	.1046/.203	.0854/.355	.0717/.570	.0614/.859	.0534/1.237	.0470/1.716	.0418/2.31	.0375/3.03	.0338/3.90	.0307/4.92	.0280/6.12	.0257/7.52	.0236/9.13	.0218/10.96	.0201/13.05	.01733/16.69	.01468/27.1	.01411/41.5	
7/32	.2188		.1000/.328	.0841/.526	.0721/.793	.0628/1.140	.0555/1.580	.0494/2.13	.0444/2.79	.0401/3.58	.0365/4.52	.0333/5.61	.0306/6.88	.0282/8.35	.0260/10.02	.0241/11.92	.0206/15.22	.01733/24.6	.01690/37.5	.01096/61.3
15/64	.2344		.1156/.305	.0974/.489	.0836/.736	.0730/1.058	.0645/1.465	.0575/1.969	.0518/2.58	.0469/3.21	.0427/4.18	.0391/5.19	.0359/6.35	.0331/7.70	.0307/9.23	.0285/10.97	.0242/13.99	.0206/22.5	.01990/34.3	.01326/55.8
1/4	.250			.1116/.457	.0960/.687	.0839/.987	.0742/1.366	.0663/1.834	.0597/2.40	.0541/3.08	.0494/3.88	.0453/4.82	.0417/5.90	.0385/7.14	.0357/8.56	.0332/10.17	.0299/12.95	.0242/20.8	.0230/31.6	.01578/51.1
9/32	.2813			.1432/.403	.1234/.606	.1080/.870	.0958/1.202	.0857/1.613	.0774/2.11	.0703/2.70	.0643/3.40	.0591/4.22	.0545/5.16	.0505/6.24	.0469/7.47	.0437/8.86	.0395/11.26	.0323/18.01	.0300/26.8	.0215/43.8
5/16	.3125				.1541/.542	.1351/.778	.1200/1.074	.1076/1.440	.0973/1.881	.0886/2.41	.0811/3.03	.0746/3.75	.0690/4.58	.0640/5.54	.0596/6.63	.0556/7.85	.0504/9.97	.0415/15.89	.0385/23.9	.0281/38.3
11/32	.3438					.1633/.703	.1470/.970	.1300/1.300	.1196/1.697	.1090/2.17	.0999/2.73	.0921/3.38	.0852/4.12	.0792/4.98	.0733/5.95	.0690/7.05	.0627/8.94	.0518/14.21	.0436/21.3	.0355/34.1
3/8	.375						.1768/.885	.1589/1.185	.1440/1.546	.1314/1.978	.1206/2.48	.1113/3.07	.1031/3.75	.0960/4.53	.0895/5.40	.0839/6.40	.0764/8.10	.0634/12.85	.0535/19.27	.0438/30.7

* The table is for ASTM A229 oil tempered spring steel with a torsional modulus G of 11,200,000 psi, and an uncorrected torsional stress of 100,000 psi. For other materials use the following factors: stainless steel, multiply f by 1.067; spring brass, multiply f by 2.24; phosphor bronze, multiply f by 1.867; Monel metal, multiply f by 1.244; beryllium copper, multiply f by 1.725; Inconel (non-magnetic), multiply f by 1.045. † Round wire. For square wire, multiply f by 0.707, and multiply P by 1.2. ‡ The upper figure is the deflection and the lower figure the load as read against each spring size. *Note:* Intermediate values can be obtained within reasonable accuracy by interpolation.

Table 5 *(Continued).* **Compression and Extension Spring Deflections***

Header spanning the wire-size columns: **Wire Size or Washburn and Moen Gauge, and Decimal Equivalent**

Data rows give **Deflection f (inch) per coil, at Load P (pounds)** — for each Outside Diameter the upper line is f and the lower line is P.

Nom.	Dec.	1/8 .125	11 .1205	12 .1055	3/32 .0938	13 .0915	14 .080	15 .072	16 .0625	17 .054	18 .0475	19 .041	.038	.036	.034	.032	.030	.028	.026
13/32	.4063	…	…	.0241	.0292	.0304	.0373	.0436	.0531	.0645	.0760	.0913	.1001	.1068	.1143	.1228	.1324	.1434	.1560
		…	…	153.3	103.7	95.6	61.6	43.9	27.9	17.56	11.73	7.41	5.85	4.95	4.15	3.44	2.82	2.28	1.815
7/16	.4375	.0219	.0234	.0293	.0353	.0367	.0450	.0521	.0631	.0764	.0898	.1075	.1178	.1256	.1343	.1441	.1553	.1680	.1827
		245.	217.	138.9	94.3	86.9	56.3	40.1	25.6	16.13	10.79	6.82	5.39	4.56	3.82	3.17	2.60	2.11	1.678
15/32	.4688	.0265	.0282	.0351	.0420	.0437	.0530	.0614	.0741	.0894	.1048	.1252	.1370	.1459	.1560	.1673	.1800	.1947	.212
		223.	197.3	126.9	86.4	79.7	51.7	37.0	23.6	14.91	9.99	6.33	5.00	4.23	3.55	2.94	2.42	1.956	1.559
1/2	.500	.0316	.0335	.0414	.0494	.0512	.0619	.0714	.0859	.1033	.1209	.1441	.1575	.1678	.1792	.1920	.207	.223	.243
		205.	181.1	116.9	80.0	73.6	47.9	34.3	21.9	13.87	9.30	5.90	4.67	3.95	3.31	2.75	2.26	1.826	1.456
17/32	.5313	.0371	.0393	.0482	.0572	.0593	.0714	.0822	.0987	.1183	.1382	.1645	.1796	.1911	.204	.219	.235	.254	.276
		188.8	167.3	108.3	74.1	68.4	44.6	31.9	20.5	12.96	8.70	5.52	4.37	3.70	3.10	2.58	2.12	1.713	1.366
9/16	.5625	.0430	.0455	.0555	.0657	.0680	.0816	.0937	.1122	.1343	.1566	.1861	.203	.216	.230	.247	.265	.286	…
		175.3	155.5	100.9	69.1	63.9	41.7	29.9	19.17	12.16	8.18	5.19	4.11	3.48	2.92	2.42	1.991	1.613	…
19/32	.5938	.0493	.0522	.0634	.0748	.0774	.0926	.1061	.1267	.1514	.1762	.209	.228	.242	.259	.277	.297	…	…
		163.6	145.2	94.4	64.8	60.0	39.1	28.0	18.04	11.46	7.71	4.90	3.88	3.28	2.76	2.29	1.880	…	…
5/8	.625	.0561	.0593	.0718	.0844	.0873	.1041	.1191	.1420	.1693	.1969	.233	.254	.270	.288	.308	.331	…	…
		153.4	136.2	88.7	61.0	56.4	36.9	26.5	17.04	10.83	7.29	4.63	3.67	3.11	2.61	2.17	1.782	…	…
21/32	.6563	.0634	.0668	.0807	.0946	.0978	.1164	.1330	.1582	.1884	.219	.259	.282	.300	.320	.342	…	…	…
		144.3	128.3	83.7	57.6	53.3	34.9	25.1	16.14	10.27	6.92	4.40	3.49	2.95	2.48	2.06	…	…	…
11/16	.6875	.0710	.0748	.0901	.1054	.1089	.1294	.1476	.1753	.208	.242	.286	.311	.331	.352	…	…	…	…
		136.3	121.2	79.2	54.6	50.5	33.1	23.8	15.34	9.76	6.58	4.19	3.32	2.81	2.36	…	…	…	…
23/32	.7188	.0791	.0833	.1000	.1168	.1206	.1431	.1650	.1933	.230	.266	.314	.342	.363	…	…	…	…	…
		129.2	114.9	75.2	51.9	48.0	31.5	22.7	14.61	9.31	6.27	3.99	3.17	2.68	…	…	…	…	…
3/4	.750	.0877	.0923	.1105	.1288	.1329	.1574	.1791	.212	.252	.291	.344	.374	…	…	…	…	…	…
		122.7	109.2	71.5	49.4	45.7	30.0	21.6	13.94	8.89	5.99	3.82	3.03	…	…	…	…	…	…
25/32	.7813	.0967	.1017	.1214	.1413	.1459	.1724	.1960	.232	.275	.318	.375	…	…	…	…	…	…	…
		116.9	104.0	68.2	47.1	43.6	28.7	20.7	13.34	8.50	5.74	3.66	…	…	…	…	…	…	…
13/16	.8125	.1061	.1115	.1329	.1545	.1594	.1881	.214	.253	.299	.346	.407	…	…	…	…	…	…	…
		111.5	99.3	65.2	45.1	41.7	27.5	19.80	12.78	8.15	5.50	3.51	…	…	…	…	…	…	…

* The table is for ASTM A229 oil tempered spring steel with a torsional modulus G of 11,200,000 psi, and an uncorrected torsional stress of 100,000 psi. For other materials, and other important footnotes, see page 334.

Table 5 *(Continued).* **Compression and Extension Spring Deflections***

Wire Size or Washburn and Moen Gauge, and Decimal Equivalent

Deflection *f* (inch) per coil, at Load *P* (pounds)

Each cell is given as deflection *f* / Load *P*.

Outside Diam. Nom.	Dec.	15 (.072)	14 (.080)	13 (.0915)	3/32 (.0938)	12 (.1055)	11 (.1205)	1/8 (.125)	10 (.135)	9 (.1483)	5/32 (.1563)	8 (.162)	7 (.177)	3/16 (.1875)	6 (.192)	5 (.207)	7/32 (.2188)	4 (.2253)
7/8	.875	.251 / 18.26	.222 / 25.3	.1882 / 39.4	.1825 / 41.5	.1574 / 59.9	.1325 / 91.1	.1262 / 102.3	.1138 / 130.5	.0999 / 176.3	.0928 / 209.	.0880 / 234.	.0772 / 312.	.0707 / 377.	.0682 / 407.	.0605 / 521.	.0552 / 626.	.0526 / 691.
29/32	.9063	.271 / 17.57	.239 / 24.3	.204 / 36.9	.1974 / 39.9	.1705 / 57.6	.1438 / 87.5	.1370 / 98.2	.1236 / 125.2	.1087 / 169.0	.1010 / 199.9	.0959 / 224.	.0843 / 299.	.0772 / 360.	.0746 / 389.	.0663 / 498.	.0606 / 598.	.0577 / 660.
15/16	.9375	.292 / 16.94	.258 / 23.5	.219 / 35.6	.213 / 38.4	.1841 / 55.4	.1554 / 84.1	.1479 / 94.4	.1338 / 120.4	.1178 / 162.3	.1096 / 191.9	.1041 / 215.	.0917 / 286.	.0842 / 345.	.0812 / 373.	.0723 / 477.	.0662 / 572.	.0632 / 631.
31/32	.9688	.313 / 16.35	.277 / 22.6	.236 / 34.3	.229 / 37.0	.1982 / 53.4	.1675 / 81.0	.1598 / 90.9	.1445 / 115.9	.1272 / 156.1	.1183 / 184.5	.1127 / 207.	.0994 / 275.	.0913 / 332.	.0882 / 358.	.0786 / 457.	.0721 / 548.	.0688 / 604.
1	1.000	.336 / 15.80	.297 / 21.9	.253 / 33.1	.246 / 35.8	.213 / 51.5	.1801 / 78.1	.1718 / 87.6	.1555 / 111.7	.1372 / 150.4	.1278 / 177.6	.1216 / 198.8	.1074 / 264.	.0986 / 319.	.0954 / 344.	.0852 / 439.	.0783 / 526.	.0747 / 580.
1 1/32	1.031	.359 / 15.28	.317 / 21.1	.271 / 32.0	.263 / 34.6	.228 / 49.8	.1931 / 75.5	.1843 / 84.6	.1669 / 107.8	.1474 / 145.1	.1374 / 171.3	.1308 / 191.6	.1157 / 255.	.1065 / 307.	.1029 / 331.	.0921 / 423.	.0845 / 506.	.0809 / 557.
1 1/16	1.063	.382 / 14.80	.338 / 20.5	.289 / 31.0	.281 / 33.5	.244 / 48.2	.207 / 73.0	.1972 / 81.8	.1788 / 104.2	.1580 / 140.1	.1474 / 165.4	.1404 / 185.0	.1243 / 246.	.1145 / 296.	.1107 / 319.	.0993 / 407.	.0913 / 487.	.0873 / 537.
1 3/32	1.094	.407 / 14.34	.360 / 19.83	.308 / 30.0	.299 / 32.4	.260 / 46.7	.221 / 70.6	.211 / 79.2	.1910 / 100.8	.1691 / 135.5	.1578 / 159.9	.1503 / 178.8	.1332 / 238.	.1229 / 286.	.1188 / 308.	.1066 / 393.	.0982 / 470.	.0939 / 517.
1 1/8	1.125	.432 / 13.92	.383 / 19.24	.328 / 29.1	.318 / 31.4	.277 / 45.2	.235 / 68.4	.224 / 76.7	.204 / 97.6	.1804 / 131.2	.1685 / 154.7	.1604 / 173.0	.1424 / 230.	.1315 / 276.	.1272 / 298.	.1142 / 379.	.1053 / 454.	.1008 / 499.
1 3/16	1.188	.485 / 13.14	.431 / 18.15	.368 / 27.5	.358 / 29.6	.311 / 42.6	.265 / 64.4	.254 / 72.1	.231 / 91.7	.204 / 123.3	.1908 / 145.4	.1812 / 162.4	.1620 / 215.	.1496 / 259.	.1448 / 279.	.1303 / 355.	.1203 / 424.	.1153 / 467.
1 1/4	1.250	.541 / 12.44	.480 / 17.19	.412 / 26.0	.400 / 28.0	.349 / 40.3	.297 / 60.8	.284 / 68.2	.258 / 86.6	.230 / 116.2	.215 / 137.0	.205 / 153.1	.1824 / 203.	.1690 / 244.	.1635 / 263.	.1474 / 334.	.1363 / 399.	.1308 / 438.
1 5/16	1.313	.600 / 11.81	.533 / 16.31	.457 / 24.6	.444 / 26.6	.387 / 38.2	.331 / 57.7	.317 / 64.6	.288 / 82.0	.256 / 110.1	.240 / 129.7	.229 / 144.7	.205 / 191.6	.1894 / 230.	.1836 / 248.	.1657 / 315.	.1535 / 376.	.1472 / 413.
1 3/8	1.375	.662 / 11.25	.588 / 15.53	.506 / 23.4	.491 / 25.3	.429 / 36.3	.367 / 54.8	.351 / 61.4	.320 / 77.9	.285 / 104.4	.267 / 123.0	.255 / 137.3	.227 / 181.7	.211 / 218.	.204 / 235.	.1848 / 298.	.1713 / 356.	.1650 / 391
1 7/16	1.438	.727 / 10.73	.647 / 14.81	.556 / 22.3	.540 / 24.1	.472 / 34.6	.404 / 52.2	.387 / 58.4	.353 / 74.1	.314 / 99.4	.295 / 117.0	.282 / 130.6	.252 / 172.6	.234 / 207.	.227 / 223.	.205 / 283.	.1905 / 337.	.1829 / 371.

* The table is for ASTM A229 oil tempered spring steel with a torsional modulus *G* of 11,200,000 psi, and an uncorrected torsional stress of 100,000 psi. For other materials, and other important footnotes, see page 334.

Table 5 (Concluded). Compression and Extension Spring Deflections*

Wire Size or Washburn and Moen Gauge, and Decimal Equivalent

Deflection f (inch) per coil, at Load P (pounds)

Each cell is given as deflection f (inch) / Load P (pounds).

Outside Diam. Nom.	Dec.	5/16 .3125	0 .3065	9/32 .2813	2 .2625	1/4 .250	3 .2437	4 .2253	7/32 .2188	5 .207	6 .192	3/16 .1875	7 .177	8 .162	5/32 .1563	9 .1483	10 .135	1/8 .125	11 .1205
1½	1.500	.1267/1008.	.1305/947.	.1482/717.	.1612/574.	.1754/499.	.1815/452.	.202/352.	.210/321.	.227/269.	.250/213.	.258/197.1	.277/164.6	.310/124.5	.324/111.5	.350/94.8	.387/70.8	.424/55.8	.443/49.8
1⅝	1.625	.1547/912.	.1592/858.	.1801/650.	.1986/521.	.212/446.	.244/411.	.254/321.	.254/292.	.273/246.	.300/193.9	.309/180.0	.332/150.3	.370/113.9	.387/102.0	.413/86.7	.461/64.8	.505/51.1	.527/45.7
1¾	1.750	.1856/833.	.1908/783.	.215/595.	.237/477.	.253/409.	.261/377.	.290/295.	.301/269.	.323/226.	.355/178.4	.366/165.6	.392/138.5	.437/104.9	.456/94.0	.485/80.0	.542/59.8	.593/47.2	.619/42.2
1⅞	1.875	.219/767.	.225/721.	.253/548.	.278/440.	.296/378.	.306/348.	.339/272.	.351/248.	.377/209.	.414/165.1	.426/153.4	.457/128.2	.508/97.3	.530/87.2	.564/74.2	.629/55.5	.687/43.8	.717/39.2
1 15/16	1.938	.237/737.	.243/693.	.273/528.	.300/425.	.320/364.	.331/335.	.365/262.	.379/239.	.405/201.	.446/159.2	.458/147.9	.492/123.6	.546/93.8	.569/84.2	.605/71.6	.676/53.6	.738/42.3	.769/37.8
2	2.000	.256/710.	.263/668.	.295/509.	.323/409.	.344/351.	.355/324.	.392/253.	.407/231.	.436/194.3	.478/153.7	.492/142.8	.527/119.4	.585/90.6	.610/81.3	.649/69.2	.723/51.8	.789/40.9	.823/36.6
2 1/16	2.063	.275/685.	.282/644.	.316/491.	.346/395.	.369/339.	.381/312.	.421/245.	.436/223.	.467/187.7	.512/148.5	.526/138.1	.564/115.4	.626/87.6	.652/78.5	.693/66.9	.768/50.1	.843/39.6	.878/35.4
2⅛	2.125	.295/661.	.303/622.	.339/474.	.371/381.	.395/327.	.407/302.	.449/236.	.466/216.	.499/181.6	.546/143.8	.562/133.6	.602/111.8	.667/84.9	.696/76.1	.739/64.8	.823/48.5	.898/38.3	.936/34.3
2 3/16	2.188	.316/639.	.324/601.	.362/459.	.396/369.	.421/317.	.435/292.	.479/229.	.497/209.	.532/175.8	.582/139.2	.598/129.5	.641/108.3	.711/82.2	.740/73.8	.786/62.8	.876/47.1	.955/37.2	.995/33.3
2¼	2.250	.337/618.	.346/582.	.387/444.	.423/357.	.449/307.	.463/283.	.511/222.	.529/202.	.566/170.5	.619/135.0	.637/125.5	.681/105.7	.755/79.8	.787/71.6	.835/60.9	.930/45.7	1.013/36.1	1.056/32.3
2 5/16	2.313	.359/599.	.368/564.	.411/430.	.449/347.	.478/298.	.493/275.	.542/215.	.562/196.3	.601/165.4	.657/131.0	.676/121.8	.723/101.9	.801/77.5	.834/69.5	.886/59.2	.986/44.4	1.074/35.1	1.119/31.4
2⅜	2.375	.382/581.	.392/547.	.437/417.	.477/336.	.507/289.	.523/267.	.576/209.	.596/190.7	.637/160.7	.696/127.3	.716/118.3	.763/99.1	.848/75.3	.884/67.6	.938/57.5	1.043/43.1	1.136/34.1	1.184/30.5
2 7/16	2.438	.405/564.	.416/531.	.464/405.	.506/327.	.537/281.	.554/259.	.609/203.	.631/185.1	.674/156.1	.737/123.7	.757/115.1	.810/96.3	.897/73.2	.934/65.7	.991/56.0	1.102/42.0	1.201/33.2	…
2½	2.500	.430/548.	.441/516.	.491/394.	.536/317.	.568/273.	.586/252.	.644/197.5	.667/180.2	.713/151.9	.778/120.4	.800/111.6	.855/93.7	.946/71.3	.986/64.0	1.046/54.5	1.162/40.9	1.266/32.3	…

* The table is for ASTM A229 oil tempered spring steel with a torsional modulus G of 11,200,000 psi, and an uncorrected torsional stress of 100,000 psi. For other materials, and other important footnotes, see page 334.

a close wound extension spring against one another. This force must be overcome before the coils of a spring begin to open up.

Initial tension is wound into extension springs by bending each coil as it is wound away from its normal plane, thereby producing a slight twist in the wire which causes the coil to spring back tightly against the adjacent coil. Initial tension can be wound into cold-coiled extension springs only. Hot-wound springs and springs made from annealed steel are hardened and tempered after coiling, and therefore initial tension cannot be produced. It is possible to make a spring having initial tension only when a high tensile strength, obtained by cold drawing or by heat-treatment, is possessed by the material as it is being wound into springs. Materials that possess the required characteristics for the manufacture of such springs include hard-drawn wire, music wire, pre-tempered wire, 18-8 stainless steel, phosphor-bronze, and many of the hard-drawn copper-nickel, and nonferrous alloys. Permissible torsional stresses resulting from initial tension for different spring indexes are shown in Fig. 16.

Hook failure: The great majority of breakages in extension springs occurs in the hooks. Hooks are subjected to both bending and torsional stresses and have higher stresses than the coils in the spring.

Stresses in regular hooks: The calculations for the stresses in hooks are quite complicated and lengthy. Also, the radii of the bends are difficult to determine and frequently vary between specifications and actual production samples. However, regular hooks are more highly stressed than the coils in the body and are subjected to a bending stress at section B (see Fig. 17). The bending stress S_b at section B should be compared with allowable stresses for torsion springs and with the elastic limit of the material in tension (See Figs. 7 through 10).

Stresses in cross over hooks: Results of tests on springs having a normal average index show that the cross over hooks last longer than regular hooks. These results may not occur on springs of small index or if the cross over bend is made too sharply.

Inasmuch as both types of hooks have the same bending stress, it would appear that the fatigue life would be the same. However, the large bend radius of the regular hooks causes some torsional stresses to coincide with the bending stresses, thus explaining the earlier breakages. If sharper bends were made on the regular hooks, the life should then be the same as for cross over hooks.

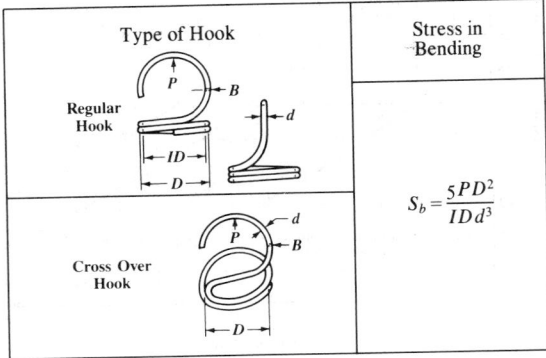

Type of Hook		Stress in Bending
Regular Hook		
Cross Over Hook		$S_b = \dfrac{5PD^2}{ID\,d^3}$

Fig. 17. Formula for the Bending Stress at Section B on Regular and Cross-over Type Hooks

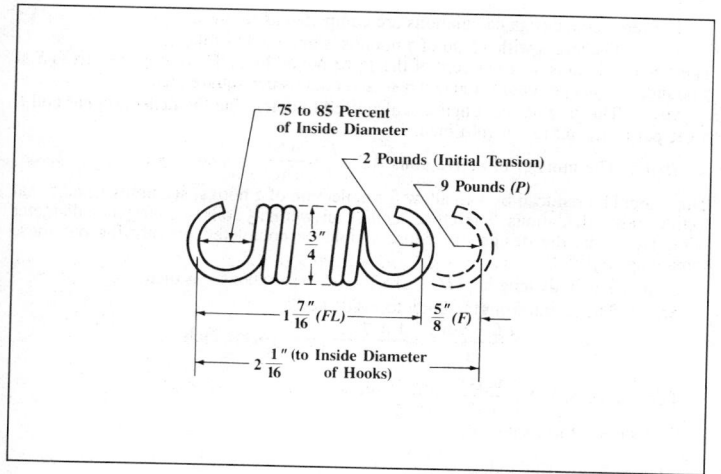

Fig. 18. Extension Spring Design Example

Stresses in half hooks: The formulas for regular hooks can also be used for half hooks, because the smaller bend radius allows for the increase in stress. It will therefore be observed that half hooks have the same stress in bending as regular hooks.

Frequently overlooked facts by many designers are that one full hook deflects an amount equal to ½ a coil and each half hook deflects an amount equal to ¹⁄₁₀th of a coil. Allowances for these deflections should be made when designing springs. Thus, an extension spring, with regular full hooks and having 10 coils, will have a deflection equal to 11 coils, or 10 per cent more than that which may have been calculated.

Extension Spring Design. — The available space in a product or assembly usually determines the limiting dimensions of a spring, but the wire size, number of coils, and initial tension are often unknown.

Example: An extension spring is to be made from spring steel ASTM A229, with regular hooks shown in Fig. 18. Calculate the wire size, number of coils and initial tension.

Note: Allow about 20 to 25 per cent of the 9 pound load for initial tension, say 2 pounds, and then design for a 7 pound load (not 9 pounds) at ⅝ inch deflection. Also use lower stresses than for a compression spring to allow for overstretching during assembly and to obtain a safe stress on the hooks. Proceed as for compression springs, but locate a load in the tables somewhat higher than the 9 pound load.

Method 1, using table: From Table 5 locate ¾ inch outside diameter in the left column and move to the right to locate a load *P* of 13.94 pounds. A deflection *f* of 0.212 inch appears above this figure. Moving vertically from this position to the top of the column a suitable wire diameter of 0.0625 inch is found.

The remaining design calculations are completed as follows:

Step 1: The stress with a load of 7 pounds is obtained as follows:
The 7 pound load is 50.2 per cent of the 13.94 pound load. Therefore, the stress S at 7 pounds = 0.502 per cent × 100,000 = 50,200 pounds per square inch.

Step 2: The 50.2 per cent figure is also used to determine the deflection per coil f:
0.502 per cent × 0.212 = 0.1062 inch.

Step 3: The number of active coils $AC = \dfrac{F}{f} = \dfrac{0.625}{0.1062} = 5.86$ (say 6)

This should be reduced by 1 to allow for deflection of 2 hooks, see notes 1 and 2 that follow these calculations. Therefore, a quick answer is: 5 coils of 0.0625 inch diameter wire. However, the design procedure should be completed by carrying out these remaining steps:

Step 4: The body length = $(TC + 1) \times d = (5 + 1) \times 0.0625 = \frac{3}{8}$ inch.

Step 5: The length from the body to inside hook
$$= \frac{FL - Body}{2} = \frac{1.4375 - 0.375}{2} = 0.531 \text{ inch.}$$

Percentage of I.D. = $\dfrac{0.531}{\text{I.D.}} = \dfrac{0.531}{0.625} = 85$ per cent

(This is satisfactory, see Note 3.)

Step 6: The spring index $= \dfrac{\text{O.D.}}{d} - 1 = \dfrac{0.75}{0.0625} - 1 = 11$

Step 7: The initial tension stress $S_{it} = \dfrac{S \times IT}{P}$

$$= \frac{50,200 \times 2}{7}$$
$$= 14,340 \text{ pounds per square inch.}$$

(Satisfactory, as checked against curve in Fig. 16.)

Step 8: The curvature correction factor $K = 1.12$ (Fig. 13).

Step 9: The total stress = $(50,200 + 14,340) \times 1.12$
$$= 72,285 \text{ pounds per square inch.}$$

(This is less than 106,250 pounds per square inch permitted by the middle curve for 0.0625 inch wire in Fig. 3 and therefore is a safe working stress that permits additional deflection as usually necessary for assembly purposes.)

Step 10: The large majority of hook breakage is due to high stress in bending and should be checked as follows:
From Fig. 17, stress on hook in bending is:
$$S_b = \frac{5PD^2}{\text{I.D.}d^3}$$
$$= \frac{5 \times 9 \times 0.6875^2}{0.625 \times 0.0625^3} = 139,200 \text{ pounds per square inch.}$$

(This is less than the top curve value, Fig. 8, for 0.0625 inch diameter wire, and is therefore safe. See Note 5.)

Note: The following points should be noted when designing extension springs:
1. All coils are active and thus $AC = TC$.
2. Each full hook deflection is approximately equal to ½ coil. Therefore for 2 hooks, reduce the total coils by 1. (Each half hook deflection is nearly equal to 1/10 of a coil.)
3. The distance from the body to the inside of a regular full hook equals 75 to

85 per cent (90 per cent maximum) of the I.D. (For a cross over center hook, it equals the I.D.)

4. Some initial tension should usually be used to hold the spring together. Try not to exceed the maximum curve shown on Fig. 16. Without initial tension, a long spring with many coils will have a different length in the horizontal position than it will when hung vertically.

5. The hooks are stressed in bending, therefore their stress should be less than the maximum bending stress as used for torsion springs — use top fatigue strength curves Figs. 7 through 10.

Method 2, using formulas: The sequence of steps for designing extension springs by formulas is similar to that for compression springs. The formulas for this method are given in Table 3.

Tolerances for Compression and Extension Springs. — Tolerances for coil diameter, free length, squareness, load, and the angle between loop planes for compression and extension springs are given in Tables 6 through 11. To meet the requirements of load, rate, free length, and solid height, it is necessary to vary the number of coils for compression springs by ±5 per cent. For extension springs, the tolerances on the numbers of coils are: for 3 to 5 coils, ±20 per cent; for 6 to 8 coils, ±30 per cent; for 9 to 12 coils, ±40 per cent. For each additional coil, a further 1½ per cent tolerance is added to the extension spring values. Closer tolerances on the number of

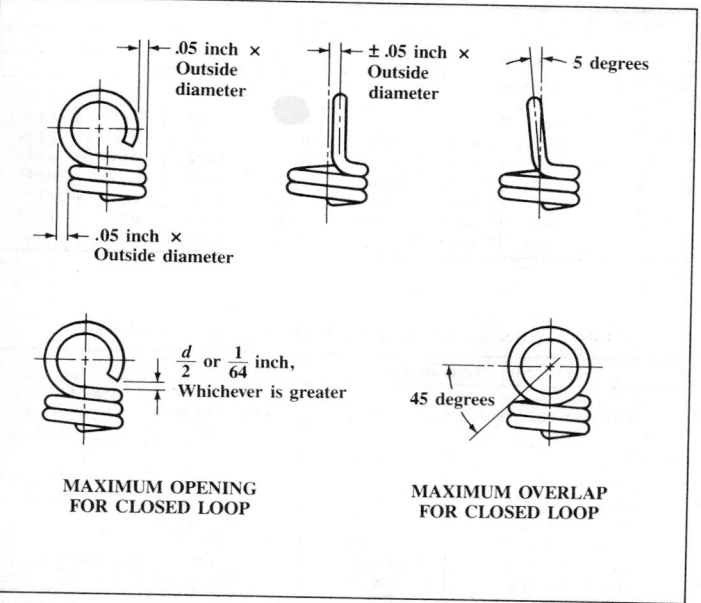

Fig. 19. Maximum Deviations Allowed on Ends and Variation in Alignment of Ends (Loops) for Extension Springs

Table 6. Compression and Extension Spring
Coil Diameter Tolerances*

Wire Diameter, Inch	Spring Index						
	4	6	8	10	12	14	16
	Tolerance, ± inch						
0.015	0.002	0.002	0.003	0.004	0.005	0.006	0.007
0.023	0.002	0.003	0.004	0.006	0.007	0.008	0.010
0.035	0.002	0.004	0.006	0.007	0.009	0.011	0.013
0.051	0.003	0.005	0.007	0.010	0.012	0.015	0.017
0.076	0.004	0.007	0.010	0.013	0.016	0.019	0.022
0.114	0.006	0.009	0.013	0.018	0.021	0.025	0.029
0.171	0.008	0.012	0.017	0.023	0.028	0.033	0.038
0.250	0.011	0.015	0.021	0.028	0.035	0.042	0.049
0.375	0.016	0.020	0.026	0.037	0.046	0.054	0.064
0.500	0.021	0.030	0.040	0.062	0.080	0.100	0.125

Table 7. Compression Spring Normal Free-Length
Tolerances, Squared and Ground Ends*

Number of Active Coils per Inch	Spring Index						
	4	6	8	10	12	14	16
	Tolerance, ± Inch per Inch of Free Length†						
0.5	0.010	0.011	0.012	0.013	0.015	0.016	0.016
1	0.011	0.013	0.015	0.016	0.017	0.018	0.019
2	0.013	0.015	0.017	0.019	0.020	0.022	0.023
4	0.016	0.018	0.021	0.023	0.024	0.026	0.027
8	0.019	0.022	0.024	0.026	0.028	0.030	0.032
12	0.021	0.024	0.027	0.030	0.032	0.034	0.036
16	0.022	0.026	0.029	0.032	0.034	0.036	0.038
20	0.023	0.027	0.031	0.034	0.036	0.038	0.040

Table 8. Extension Spring Normal Free-Length and End Tolerances*

Free-Length Tolerances		End Tolerances	
Spring Free-Length (inch)	Tolerance (inch)	Total Number of Coils	Angle Between Loop Planes (degrees)
Up to 0.5	± 0.020		
Over 0.5 to 1.0	± 0.030	3 to 6	± 25
Over 1.0 to 2.0	± 0.040	7 to 9	± 35
Over 2.0 to 4.0	± 0.060	10 to 12	± 45
Over 4.0 to 8.0	± 0.093	13 to 16	± 60
Over 8.0 to 16.0	± 0.156	Over 16	Random
Over 16.0 to 24.0	± 0.218		

* Courtesy of the Spring Manufacturers Institute
† For springs less than 0.5 inch long, use the tolerances for 0.5 inch long springs. For springs with unground closed ends, multiply the tolerances by 1.7.

Table 9. Compression Spring Squareness Tolerances*

Slenderness Ratio†	Spring Index						
	4	6	8	10	12	14	16
	Squareness Tolerance (± degrees)						
0.5	3.0	3.0	3.5	3.5	3.5	3.5	4.0
1.0	2.5	3.0	3.0	3.0	3.0	3.5	3.5
1.5	2.5	2.5	3.0	3.0	3.0	3.0	3.0
2.0	2.5	2.5	2.5	3.0	3.0	3.0	3.0
3.0	2.0	2.5	2.5	2.5	2.5	3.0	3.0
4.0	2.0	2.0	2.5	2.5	2.5	2.5	3.0
6.0	2.0	2.0	2.0	2.5	2.5	2.5	2.5
8.0	2.0	2.0	2.0	2.0	2.5	2.5	2.5
10.0	2.0	2.0	2.0	2.0	2.0	2.5	2.5
12.0	2.0	2.0	2.0	2.0	2.0	2.0	2.5

Table 10. Compression Spring Normal Load Tolerances

Length Tolerance, ± inch	Deflection (inch)‡							
	0.05	0.10	0.15	0.20	0.25	0.30	0.40	0.50
	Tolerance, ± Per Cent of Load							
0.005	12	7	6	5	...	...	...	...
0.010	...	12	8.5	7	6.5	5.5	...	...
0.020	...	22	15.5	12	10	8.5	7	6
0.030	...	...	22	17	14	12	9.5	8
0.040	...	...	...	22	18	15.5	12	10
0.050	...	...	...	...	22	19	14.5	12
0.060	...	...	...	...	25	22	17	14
0.070	...	...	...	...	...	25	19.5	16
0.080	...	...	...	...	...	...	22	18
0.090	...	...	...	...	...	...	25	20
0.100	...	...	...	...	...	...	...	22
0.200	...	...	...	...	...	...	...	...
0.300	...	...	...	...	...	...	...	...
0.400	...	...	...	...	...	...	...	...
0.500	...	...	...	...	...	...	...	...

Length Tolerance, ± inch	Deflection (inch)‡						
	0.75	1.00	1.50	2.00	3.00	4.00	6.00
	Tolerance, ± Per Cent of Load						
0.005	...	...	...	...	...	...	...
0.010	...	...	...	...	...	...	...
0.020	5	...	...	...	...	...	...
0.030	6	5	...	...	...	...	...
0.040	7.5	6	5	...	...	...	...
0.050	9	7	5.5	...	...	...	...
0.060	10	8	6	5	...	...	...
0.070	11	9	6.5	5.5	...	...	...
0.080	12.5	10	7.5	6	5	...	...
0.090	14	11	8	6	5	...	...
0.100	15.5	12	8.5	7	5.5	...	...
0.200	...	22	15.5	12	8.5	7	5.5
0.300	...	...	22	17	12	9.5	7
0.400	...	...	...	21	15	12	8.5
0.500	...	...	...	25	18.5	14.5	10.5

* Springs with closed and ground ends, in the free position. Squareness tolerances closer than those shown require special process techniques which increase cost. Springs made from fine wire sizes, and with high spring indices, irregular shapes or long free lengths, require special attention in determining appropriate tolerance and feasibility of grinding ends.

† Slenderness Ratio $= \dfrac{FL}{D}$

‡ From free length to loaded position.

Table 11. Extension Spring Normal Load Tolerances

Spring Index	$\frac{FL^*}{F}$	Wire Diameter (inch)										
		0.015	0.022	0.032	0.044	0.062	0.092	0.125	0.187	0.250	0.375	0.437
		Tolerance, ± Per Cent of Load										
4	12	20.0	18.5	17.6	16.9	16.2	15.5	15.0	14.3	13.8	13.0	12.6
	8	18.5	17.5	16.7	15.8	15.0	14.5	14.0	13.2	12.5	11.5	11.0
	6	16.8	16.1	15.5	14.7	13.8	13.2	12.7	11.8	11.2	9.9	9.4
	4.5	15.0	14.7	14.1	13.5	12.6	12.0	11.5	10.3	9.7	8.4	7.9
	2.5	13.1	12.4	12.1	11.8	10.6	10.0	9.1	8.5	8.0	6.8	6.2
	1.5	10.2	9.9	9.3	8.9	8.0	7.5	7.0	6.5	6.1	5.3	4.8
	0.5	6.2	5.4	4.8	4.6	4.3	4.1	4.0	3.8	3.6	3.3	3.2
6	12	17.0	15.5	14.6	14.1	13.5	13.1	12.7	12.0	11.5	11.2	10.7
	8	16.2	14.7	13.9	13.4	12.6	12.2	11.7	11.0	10.5	10.0	9.5
	6	15.2	14.0	12.9	12.3	11.6	10.9	10.7	10.0	9.4	8.8	8.3
	4.5	13.7	12.4	11.5	11.0	10.5	10.0	9.6	9.0	8.3	7.6	7.1
	2.5	11.9	10.8	10.2	9.8	9.4	9.0	8.5	7.9	7.2	6.2	6.0
	1.5	9.9	9.0	8.3	7.7	7.3	7.0	6.7	6.4	6.0	4.9	4.7
	0.5	6.3	5.5	4.9	4.7	4.5	4.3	4.1	4.0	3.7	3.5	3.4
8	12	15.8	14.3	13.1	13.0	12.1	12.0	11.5	10.8	10.2	10.0	9.5
	8	15.0	13.7	12.5	12.1	11.4	11.0	10.6	10.1	9.4	9.0	8.6
	6	14.2	13.0	11.7	11.2	10.6	10.0	9.7	9.3	8.6	8.1	7.6
	4.5	12.8	11.7	10.7	10.1	9.7	9.0	8.7	8.3	7.8	7.2	6.6
	2.5	11.2	10.2	9.5	8.8	8.3	7.9	7.7	7.4	6.9	6.1	5.6
	1.5	9.5	8.6	7.8	7.1	6.9	6.7	6.5	6.2	5.8	4.9	4.5
	0.5	6.3	5.6	5.0	4.8	4.5	4.4	4.2	4.1	3.9	3.6	3.5
10	12	14.8	13.3	12.0	11.9	11.1	10.9	10.5	9.9	9.3	9.2	8.8
	8	14.2	12.8	11.6	11.2	10.5	10.2	9.7	9.2	8.6	8.3	8.0
	6	13.4	12.1	10.8	10.5	9.8	9.3	8.9	8.6	8.0	7.6	7.2
	4.5	12.3	10.8	10.0	9.5	9.0	8.5	8.1	7.8	7.3	6.8	6.4
	2.5	10.8	9.6	9.0	8.4	8.0	7.7	7.3	7.0	6.5	5.9	5.5
	1.5	9.2	8.3	7.5	6.9	6.7	6.5	6.3	6.0	5.6	5.0	4.6
	0.5	6.4	5.7	5.1	4.9	4.7	4.5	4.3	4.2	4.0	3.8	3.7
12	12	14.0	12.3	11.1	10.8	10.1	9.8	9.5	9.0	8.5	8.2	7.9
	8	13.2	11.8	10.7	10.2	9.6	9.3	8.9	8.4	7.9	7.5	7.2
	6	12.6	11.2	10.2	9.7	9.0	8.5	8.2	7.9	7.4	6.9	6.4
	4.5	11.7	10.2	9.4	9.0	8.4	8.0	7.6	7.2	6.8	6.3	5.8
	2.5	10.5	9.2	8.5	8.0	7.8	7.4	7.0	6.6	6.1	5.6	5.2
	1.5	8.9	8.0	7.2	6.8	6.5	6.3	6.1	5.7	5.4	4.8	4.5
	0.5	6.5	5.8	5.3	5.1	4.9	4.7	4.5	4.3	4.2	4.0	3.3
14	12	13.1	11.3	10.2	9.7	9.1	8.8	8.4	8.1	7.6	7.2	7.0
	8	12.4	10.9	9.8	9.2	8.7	8.3	8.0	7.6	7.2	6.8	6.4
	6	11.8	10.4	9.3	8.8	8.3	7.7	7.5	7.2	6.8	6.3	5.9
	4.5	11.1	9.7	8.7	8.2	7.8	7.2	7.0	6.7	6.3	5.8	5.4
	2.5	10.1	8.8	8.1	7.6	7.1	6.7	6.5	6.2	5.7	5.2	5.0
	1.5	8.6	7.7	7.0	6.7	6.3	6.0	5.8	5.5	5.2	4.7	4.5
	0.5	6.6	5.9	5.4	5.2	5.0	4.8	4.6	4.4	4.3	4.2	4.0
16	12	12.3	10.3	9.2	8.6	8.1	7.7	7.4	7.2	6.8	6.3	6.1
	8	11.7	10.0	8.9	8.3	7.8	7.4	7.2	6.8	6.5	6.0	5.7
	6	11.0	9.6	8.5	8.0	7.5	7.1	6.9	6.5	6.2	5.7	5.4
	4.5	10.5	9.1	8.1	7.5	7.2	6.8	6.5	6.2	5.8	5.3	5.1
	2.5	9.7	8.4	7.6	7.0	6.7	6.3	6.1	5.7	5.4	4.9	4.7
	1.5	8.3	7.4	6.6	6.2	6.0	5.8	5.6	5.3	5.1	4.6	4.4
	0.5	6.7	5.9	5.5	5.3	5.1	5.0	4.8	4.6	4.5	4.3	4.1

* $\frac{FL}{F}$ = the ratio of the spring free length (FL) to the deflection (F).

coils for either type of spring, lead to the need for trimming after coiling, and manufacturing time and cost are increased. Figure 19 shows deviations allowed on the ends
of extension springs, and variations in end alignments.

   Torsion Spring Design. — Figure 20 shows the types of ends most commonly used
on torsion springs. To produce them requires only limited tooling. The straight torsion end is the least expensive and should be used whenever possible. After determining the spring load or torque required and selecting the end formations, the
designer usually estimates suitable space or size limitations. However, the space
should be considered approximate until the wire size and number of coils have been
determined. The wire size is dependent principally upon the torque. Design data can
be carried out with the aid of the tabular data, which is a simple method, or by
calculation alone, as shown in the following sections. Also, many other factors affecting the design and operation of torsion springs are covered in the section, Design
Recommendations. Design formulas are shown in Table 12.
   Curvature correction: In addition to the stress obtained from the formulas for load
or deflection, there is a direct shearing stress on the inside of the section due to
curvature. Therefore, the stress obtained by the usual formulas should be multiplied
by the factor K obtained from the curve in Fig. 21. The corrected stress thus obtained
is used only for comparison with the allowable working stress (fatigue strength)
curves to determine if it is a safe value, and should not be used in the formulas for
deflection.
   Torque: Torque is a force applied to a moment arm and tends to produce rotation.
Torsion springs exert torque in a circular arc and the arms are rotated about the
central axis. It should be noted that the stress produced is in bending, not in torsion.
In the spring industry it is customary to specify torque in conjunction with the deflection or with the arms of a spring at a definite position. Formulas for torque are
expressed in pound-inches. If ounce-inches are specified, it is necessary to divide this
value by 16 in order to use the formulas.
   When a load is specified at a distance from a centerline, the torque is, of course,
equal to the load multiplied by the distance. The load can be in pounds or ounces
with the distances in inches or the load can be in grams or kilograms with the dis-

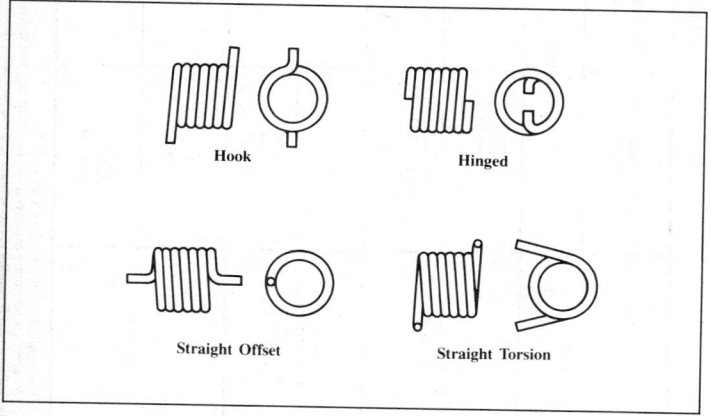

Hook   Hinged

Straight Offset   Straight Torsion

Fig. 20. The Most Commonly Used Types of Ends for Torsion Springs

Table 12. Formulas for Torsion Springs*

Feature	Formula† Springs made from round wire	Formula† Springs made from square wire	Feature	Formula† Springs made from round wire	Formula† Springs made from square wire
$d=$ Wire diameter, Inches	$\sqrt[3]{\dfrac{10.18T}{S_b}}$ $\sqrt[4]{\dfrac{4000TND}{EF°}}$	$\sqrt[3]{\dfrac{6T}{S_b}}$ $\sqrt[4]{\dfrac{2375TND}{EF°}}$	$F°=$ Deflection	$\dfrac{392S_bND}{d}$ $\dfrac{4000TND}{Ed^4}$	$\dfrac{392S_bND}{Ed}$ $\dfrac{2375TND}{Ed^4}$
$S_b=$ Stress, bending pounds per square inch	$\dfrac{10.18T}{d^3}$ $\dfrac{EdF°}{392ND}$	$\dfrac{6T}{d^3}$ $\dfrac{EdF°}{392ND}$	$T=$ Torque Inch lbs. (Also $= P \times R$)	$0.0982S_bd^3$ $\dfrac{Ed^4F°}{4000ND}$	$0.1666S_bd^3$ $\dfrac{Ed^4F°}{2375ND}$
$N=$ Active Coils	$\dfrac{EdF°}{392S_bD}$ $\dfrac{Ed^4F°}{4000TD}$	$\dfrac{EdF°}{392S_bD}$ $\dfrac{Ed^4F°}{2375TD}$	$ID_1=$ Inside Diameter After Deflection, Inches	$\dfrac{N(ID\ \text{free})}{N+\dfrac{F°}{360}}$	$\dfrac{N(ID\ \text{free})}{N+\dfrac{F°}{360}}$
	...	...	...	...	...

* The symbol notation is given on page 315.

† Where two formulas are given for one feature, the designer should use the one found to be appropriate in the given design instance. The end result from either of any two formulas is the same.

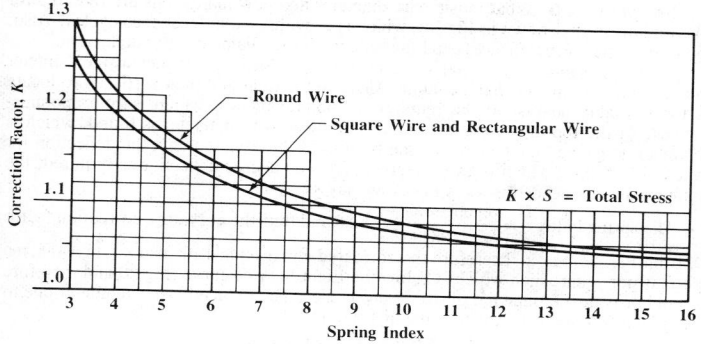

Fig. 21. Torsion Spring Stress Correction for Curvature

tance in centimeters or millimeters, but to use the design formulas, all values must be converted to pounds and inches. Design formulas for torque are based on the tangent to the arc of rotation and presume that a rod is used to support the spring. The stress in bending caused by the moment $P \times R$ is identical in magnitude to the torque T, provided a rod is used.

Theoretically, it makes no difference how or where the load is applied to the arms of torsion springs. Thus, in Fig. 22, the loads shown multiplied by their respective distances produce the same torque; i.e., $20 \times 0.5 = 10$ pounds-inches; $10 \times 1 = 10$ pounds-inches; and $5 \times 2 = 10$ pounds-inches. To further simplify the understanding of torsion spring torque, observe in both Fig. 23 and Fig. 24 that although the turning force is in a circular arc the torque is not equal to P times the radius. The torque in both cases equals $P \times R$ because the spring rests against the support rod at point a.

Torsion spring designs require more effort than other kinds because consideration has to be given to more details such as the proper size of a supporting rod, reduction of the inside diameter, increase in length, deflection of arms, allowance for friction, and method of testing.

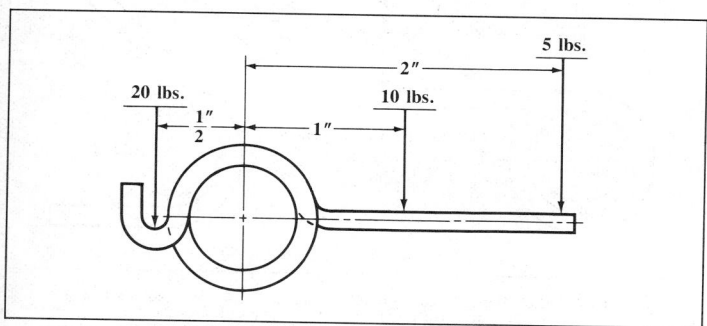

Fig. 22. Right-Hand Torsion Spring

Design example: What music wire diameter and how many coils are required for the torsion spring shown in Fig. 25, which is to withstand at least 1000 cycles? Also, determine the corrected stress and the reduced inside diameter after deflection.

Method 1, using table: From Table 13, page 352, locate the ½ inch inside diameter for the spring in the left-hand column. Move to the right and then vertically to locate a torque value nearest to the required 10 pounds-inches, which is 10.07 pounds-inches. At the top of the same column, the music wire diameter is found, which is Number 31 gauge (0.085 inch). At the bottom of the same column the deflection for one coil is found, which is 15.81 degrees. As a 90-degree deflection is required, the number of coils needed is 90/15.81 = 5.69 (say 5¾ coils).

The spring index $\dfrac{D}{d} = \dfrac{0.500 + 0.085}{0.085} = 6.88$ and thus the curvature correction factor K from Fig. 24 = 1.13. Therefore the corrected stress equals 167,000 × 1.13 = 188,700 pounds per square inch which is below the Light Service curve (Fig. 7) and therefore should provide a fatigue life of over 1,000 cycles. The reduced inside diameter due to deflection is found from the formula in Table 12:

$$\mathrm{ID}_1 = \frac{N(\mathrm{ID\ free})}{N + \dfrac{F}{360}} = \frac{5.75 \times 0.500}{5.75 + \dfrac{90}{360}} = 0.479 \text{ in.}$$

This reduced diameter easily clears a suggested ⅞₁₆ inch diameter supporting rod: 0.479 − 0.4375 = 0.041 inch clearance, and this also allows for the standard tolerance. The overall length of the spring equals the total number of coils plus one, times the wire diameter. Thus, 6¾ × 0.085 = 0.574 inch. If a small space of about 1/64 in. is allowed between the coils to eliminate coil friction, an overall length of 21/32 inch results.

Although this completes the design calculations, other tolerances should be applied in accordance with the Torsion Spring Tolerance Tables shown at the end of this section.

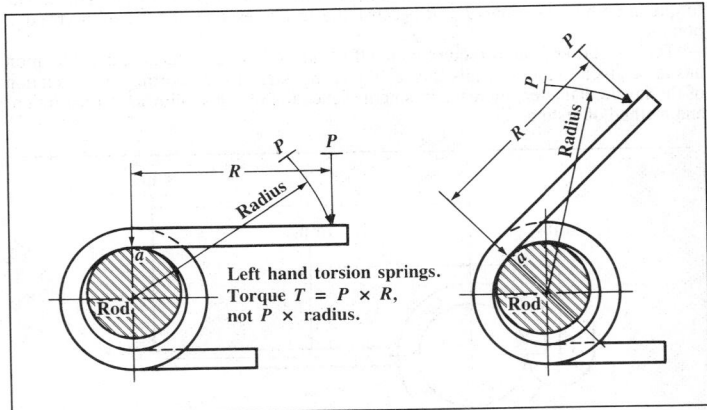

Fig. 23. (*Left*) Left-Hand Torsion Spring. The Torque is $T = P \times R$, Not $P \times$ Radius, because the Spring is Resting Against the Support Rod at Point *a*
Fig. 24. (*Right*) Left-Hand Torsion Spring. As with the Spring in Fig. 23, the Torque is $T = P \times R$, Not $P \times$ Radius, Because the Support Point Is at *a*

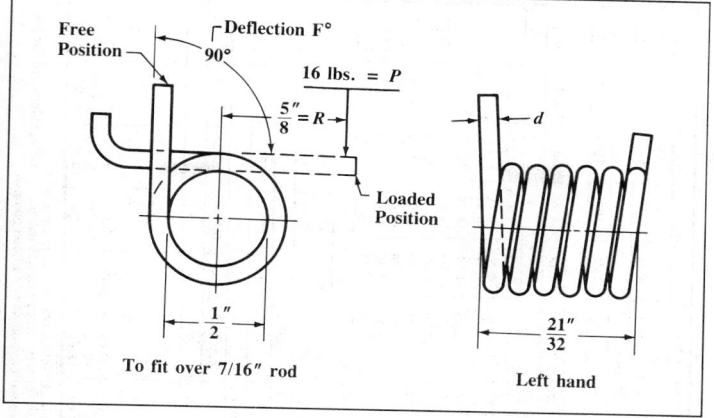

Fig. 25. Torsion Spring Design Example. The Spring is to be Assembled
on a 7/16-Inch Support Rod

Longer fatigue life: If a longer fatigue life is desired, use a slightly larger wire diameter. Usually the next larger gage size is satisfactory. The larger wire will reduce the stress and still exert the same torque, but will require more coils and a longer overall length.

Percentage method for calculating longer life: The spring design can be easily adjusted for longer life as follows:

1. Select the next larger gage size, which is Number 32 (0.090 inch) from Table 13. The torque is 11.88 pound-inches, the design stress is 166,000 pounds per square inch, and the deflection 14.9 degrees per coil. As a percentage the torque is 10/11.88 × 100 = 84 per cent.

2. The new stress is 0.84 × 166,000 = 139,440 pounds per square inch. This value is under the bottom or Severe Service curve, Fig. 7, and thus assures longer life.

3. The new deflection per coil is 0.84 × 14.97 = 12.57 degrees. Therefore, the total number of coils required = 90/12.57 = 7.16 (say 7⅛). The new overall length = 8⅛ × 0.090 = .73 inch (say ¾ inch). A slight increase in the overall length and new arm location are thus necessary.

Method 2, using formulas: When using this method, it is often necessary to solve the formulas several times because assumptions must be made initially either for the stress or for a wire size. The procedure for design using formulas is as follows (the design example is the same as in Method 1, and the spring is shown in Fig. 25):

Step 1: Note from Table 12, page 348 that the wire diameter formula is:

$$d = \sqrt[3]{\frac{10.18\,T}{S_b}}$$

Step 2: Referring to Fig. 7, select a trial stress, say 150,000 pounds per square inch.

(Continued on page 356)

Table 13. Torsion Spring Deflections*

AMW Wire Gauge and Decimal Equivalent†

Inside Diam. Fractional	Decimal	1 (.010)	2 (.011)	3 (.012)	4 (.013)	5 (.014)	6 (.016)	7 (.018)	8 (.020)	9 (.022)	10 (.024)	11 (.026)	12 (.029)	13 (.031)	14 (.033)	15 (.035)	16 (.037)
Design Stress, pounds per sq. in. (thousands)		232	229	226	224	221	217	214	210	207	205	202	199	197	196	194	192
Torque, pounds-inch		.0228	.0299	.0383	.0483	.0596	.0873	.1226	.1650	.2164	.2783	.3486	.4766	.5763	.6917	.8168	.9550
Deflection, degrees per coil																	
1/16	.0625	22.35	20.33	18.64	17.29	16.05	14.15	13.72	11.51	10.56	9.818	9.137	8.343	7.896	…	…	…
5/64	.078125	27.17	24.66	22.55	20.86	19.32	16.96	15.19	13.69	12.52	11.59	10.75	9.768	9.215	…	…	…
3/32	.09375	31.98	28.98	26.47	24.44	22.60	19.78	17.65	15.87	14.47	13.36	12.36	11.19	10.53	10.18	9.646	9.171
7/64	.109375	36.80	33.30	30.38	28.02	25.88	22.60	20.12	18.05	16.43	15.14	13.98	12.62	11.85	11.43	10.82	10.27
1/8	.125	41.62	37.62	34.29	31.60	29.16	25.41	22.59	20.23	18.38	16.91	15.59	14.04	13.17	12.68	11.99	11.36
9/64	.140625	46.44	41.94	38.20	35.17	32.43	28.23	25.06	22.41	20.33	18.69	17.20	15.47	14.49	13.94	13.16	12.46
5/32	.15625	51.25	46.27	42.11	38.75	35.71	31.04	27.53	24.59	22.29	20.46	18.82	16.89	15.81	15.19	14.33	13.56
3/16	.1875	60.89	54.91	49.93	45.91	42.27	36.67	32.47	28.95	26.19	24.01	22.04	19.74	18.45	17.70	16.67	15.75
7/32	.21875	70.52	63.56	57.75	53.06	48.82	42.31	37.40	33.31	30.10	27.55	25.27	22.59	21.09	20.21	19.01	17.94
1/4	.250	80.15	72.20	65.57	60.22	55.38	47.94	42.34	37.67	34.01	31.10	28.49	25.44	23.73	22.72	21.35	20.13

AMW Wire Gauge and Decimal Equivalent†

Inside Diam. Fractional	Decimal	17 (.039)	18 (.041)	19 (.043)	20 (.045)	21 (.047)	22 (.049)	23 (.051)	24 (.055)	25 (.059)	26 (.063)	27 (.067)	28 (.071)	29 (.075)	30 (.080)	31 (.085)
Design Stress, pounds per sq. in. (thousands)		190	188	187	185	184	183	182	180	178	176	174	173	171	169	167
Torque, pounds-inch		1.107	1.272	1.460	1.655	1.876	2.114	2.371	2.941	3.590	4.322	5.139	6.080	7.084	8.497	10.07
Deflection, degrees per coil																
7/64	.109375	9.771	9.320	8.957	…	…	…	…	…	…	…	…	…	…	…	…
1/8	.125	10.80	10.29	9.876	9.447	9.102	8.784	…	…	…	…	…	…	…	…	…
9/64	.140625	11.83	11.26	10.79	10.32	9.929	9.572	9.244	8.654	8.141	…	…	…	…	…	…
5/32	.15625	12.86	12.23	11.71	11.18	10.76	10.36	9.997	9.345	8.778	8.279	7.975	…	…	…	…
3/16	.1875	14.92	14.16	13.55	12.92	12.41	11.94	11.50	10.73	10.05	9.459	9.091	8.663	8.232	7.772	7.364
7/32	.21875	16.97	16.10	15.39	14.66	14.06	13.52	13.01	12.11	11.33	10.64	10.21	9.711	9.212	8.680	8.208
1/4	.250	19.03	18.04	17.22	16.39	15.72	15.09	14.52	13.49	12.60	11.82	11.32	10.76	10.19	9.588	9.053

* For an example in the use of the table, see page 350. Note: Intermediate values may be interpolated within reasonable accuracy.

† For sizes up to 13 gauge, the table values are for music wire with a modulus E of 29,500,000 psi; for sizes from 14 to 26 gauge, the table values are for music wire with a modulus of 29,000,000 psi; and for sizes from 27 to 31 gauge, the values are for music wire with a modulus of 28,000,000 psi.

Table 13 (Continued). Torsion Spring Deflections

AMW Wire Gauge and Decimal Equivalent*

Inside Diam. (Fractional)	Inside Diam. (Decimal)	23 .051	22 .049	21 .047	20 .045	19 .043	18 .041	17 .039	16 .037	15 .035	14 .033	13 .031	12 .029	11 .026	10 .024	9 .022	8 .020
Design Stress, pounds per sq. in. (thousands)		182	183	184	185	187	188	190	192	194	196	197	199	202	205	207	210
Torque, pounds-inch		2.371	2.114	1.876	1.655	1.460	1.272	1.107	.9550	.8168	.6917	.5763	.4766	.3486	.2783	.2164	.1650
		Deflection, degrees per coil															
9/32	.28125	16.03	16.67	17.37	18.13	19.06	19.97	21.09	22.32	23.69	25.23	26.37	28.29	31.72	34.65	37.92	42.03
5/16	.3125	17.53	18.25	19.02	19.87	20.90	21.91	23.15	24.51	26.04	27.74	29.01	31.14	34.95	38.19	41.82	46.39
11/32	.34375	19.04	19.83	20.68	21.60	22.73	23.85	25.21	26.71	28.38	30.25	31.65	33.99	38.17	41.74	45.73	50.75
3/8	.375	20.55	21.40	22.33	23.34	24.57	25.78	27.26	28.90	30.72	32.76	34.28	36.84	41.40	45.29	49.64	55.11
13/32	.40625	22.06	22.98	23.99	25.08	26.41	27.72	29.32	31.09	33.06	35.26	36.92	39.69	44.63	48.85	53.54	59.47
7/16	.4375	23.56	24.56	25.64	26.81	28.25	29.66	31.38	33.28	35.40	37.77	39.56	42.54	47.85	52.38	57.45	63.83
15/32	.46875	25.07	26.14	27.29	28.55	30.08	31.59	33.44	35.47	37.74	40.28	42.20	45.39	51.00	55.93	61.36	68.19
1/2	.500	26.58	27.71	28.95	30.29	31.92	33.53	35.49	37.67	40.08	42.79	44.84	48.24	54.30	59.48	65.27	72.55

AMW Wire Gauge and Decimal Equivalent*

Inside Diam. (Fractional)	Inside Diam. (Decimal)	1/8 .125	37 .118	36 .112	35 .106	34 .100	33 .095	32 .090	31 .085	30 .080	29 .075	28 .071	27 .067	26 .063	25 .059	24 .055
Design Stress, pounds per sq. in. (thousands)		156	158	160	161	163	164	166	167	169	171	173	174	176	178	180
Torque, pounds-inch		29.92	25.49	22.07	18.83	16.00	13.81	11.88	10.07	8.497	7.084	6.080	5.139	4.322	3.590	2.941
		Deflection, degrees per coil														
9/32	.28125	6.973	7.353	7.727	8.090	8.547	8.934	9.418	9.897	10.50	11.17	11.81	12.44	13.00	13.88	14.88
5/16	.3125	7.510	7.929	8.341	8.743	9.248	9.676	10.21	10.74	11.40	12.15	12.85	13.56	14.18	15.15	16.26
11/32	.34375	8.046	8.504	8.955	9.396	9.948	10.42	11.00	11.59	12.31	13.13	13.90	14.67	15.36	16.42	17.64
3/8	.375	8.583	9.080	9.569	10.05	10.65	11.16	11.80	12.43	13.22	14.11	14.95	15.79	16.54	17.70	19.02
13/32	.40625	9.119	9.655	10.18	10.70	11.35	11.90	12.59	13.28	14.13	15.09	15.99	16.90	17.72	18.97	20.40
7/16	.4375	9.655	10.23	10.80	11.35	12.05	12.64	13.38	14.12	15.04	16.07	17.04	18.02	18.90	20.25	21.79
15/32	.46875	10.19	10.81	11.41	12.01	12.75	13.39	14.17	14.96	15.94	17.05	18.09	19.14	20.08	21.52	23.17
1/2	.500	10.73	11.38	12.03	12.66	13.45	14.13	14.97	15.81	16.85	18.03	19.14	20.25	21.26	22.80	24.55

* For sizes up to 13 gauge, the table values are for music wire with a modulus E of 29,500,000 psi; for sizes from 14 to 26 gauge, the table values are for music wire with a modulus of 29,000,000 psi; and for sizes from 27 gauge to 1/8 inch diameter, the values are for music wire with a modulus of 28,500,000 psi.

Table 13 (Continued). Torsion Spring Deflections

AMW Wire Gauge and Decimal Equivalent*

Inside Diam. (Fractional)	Decimal	30 .080	29 .075	28 .071	27 .067	26 .063	25 .059	24 .055	23 .051	22 .049	21 .047	20 .045	19 .043	18 .041	17 .039	16 .037
Design Stress, pounds per sq. in. (thousands)		169	171	173	174	176	178	180	182	183	184	185	187	188	190	192
Torque, pounds-inch		8.497	7.084	6.080	5.139	4.322	3.590	2.941	2.371	2.114	1.876	1.655	1.460	1.272	1.107	.9550
		Deflection, degrees per coil														
17/32	.53125	17.76	19.01	20.18	21.37	22.44	24.07	25.93	28.09	29.29	30.60	32.02	33.76	35.47	37.55	39.86
9/16	.5625	18.67	19.99	21.23	22.49	23.62	25.35	27.32	29.87	30.87	32.25	33.76	35.59	37.40	39.61	42.05
19/32	.59375	19.58	20.97	22.28	23.60	24.80	26.62	28.70	31.10	32.45	33.91	35.50	37.43	39.34	41.67	44.24
5/8	.625	20.48	21.95	23.33	24.72	25.98	27.89	30.08	32.61	34.02	35.56	37.23	39.27	41.28	43.73	46.43
21/32	.65625	21.39	22.93	24.37	25.83	27.16	29.17	31.46	34.12	35.60	37.22	38.97	41.10	43.22	45.78	48.63
11/16	.6875	22.30	23.91	25.42	26.95	28.34	30.44	32.85	35.62	37.18	38.87	40.71	42.94	45.15	47.84	50.82
23/32	.71875	23.21	24.89	26.47	28.07	29.52	31.72	34.33	37.13	38.76	40.52	42.44	44.78	47.09	49.90	53.01
3/4	.750	24.12	25.87	27.52	29.18	30.70	32.99	35.61	38.64	40.33	42.18	44.18	46.62	49.03	51.96	55.20

AMW Wire Gauge and Decimal Equivalent*†

Inside Diam. (Fractional)	Decimal	5 .207	6 .192	3/16 .1875	7 .177	8 .162	9 .1483	10 .135	1/8 .125	37 .118	36 .112	35 .106	34 .100	33 .095	32 .090	31 .085
Design Stress, pounds per sq. in. (thousands)		143	146	149	150	154	158	161	156	158	160	161	163	164	166	167
Torque, pounds-inch		124.6	101.5	96.45	81.68	64.30	50.60	38.90	29.92	25.49	22.07	18.83	16.00	13.81	11.88	10.07
		Deflection, degrees per coil														
17/32	.53125	7.015	7.565	7.856	8.256	9.064	9.441	9.958	11.26	11.96	12.64	13.31	14.15	14.87	15.76	16.65
9/16	.5625	7.312	7.891	8.198	8.620	9.473	9.870	10.42	11.80	12.53	13.25	13.97	14.85	15.61	16.55	17.50
19/32	.59375	7.609	8.218	8.539	8.984	9.882	10.30	10.87	12.34	13.11	13.87	14.62	15.55	16.35	17.35	18.34
5/8	.625	7.906	8.545	8.881	9.348	10.29	10.73	11.33	12.87	13.68	14.48	15.27	16.25	17.10	18.14	19.19
21/32	.65625	8.202	8.872	9.222	9.713	10.70	11.16	11.79	13.41	14.26	15.10	15.92	16.95	17.84	18.93	20.03
11/16	.6875	8.499	9.199	9.564	10.08	11.11	11.59	12.25	13.95	14.83	15.71	16.58	17.65	18.58	19.72	20.88
23/32	.71875	8.796	9.526	9.905	10.44	11.52	12.02	12.71	14.48	15.41	16.32	17.23	18.36	19.32	20.52	21.72
3/4	.750	9.093	9.852	10.25	10.81	11.92	12.44	13.16	15.02	15.99	16.94	17.88	19.06	20.06	21.31	22.56

* For sizes up to 26 gauge, the table values are for music wire with a modulus E of 29,000,000 psi; for sizes from 27 to 1/8 inch diameter the table values are for music wire with a modulus of 28,500,000 psi; for sizes from 10 to 5 gauge are for oil-tempered MB with a modulus of 28,500,000 psi.

† Gauges 31 through 37 are AMW gauges. Gauges 10 through 5 are Washburn and Moen.

Table 13 (Concluded). Torsion Spring Deflections

AMW Wire Gauge and Decimal Equivalent*

Inside Diam. (Fractional)	Inside Diam. (Decimal)	1/8 .125	37 .118	36 .112	35 .106	34 .100	33 .095	32 .090	31 .085	30 .080	29 .075	28 .071	27 .067	26 .063	25 .059	24 .055
Design Stress, pounds per sq. in. (thousands)		156	158	160	161	163	164	166	167	169	171	173	174	176	178	180
Torque, pounds-inch		29.92	25.49	22.07	18.83	16.00	13.81	11.88	10.07	8.497	7.084	6.080	5.139	4.322	3.590	2.941
		Deflection, degrees per coil														
13/16	.8125	16.09	17.14	18.17	19.19	20.46	21.55	22.90	24.25	25.93	27.83	29.61	31.42	33.06	35.54	38.38
7/8	.875	17.17	18.29	19.39	20.49	21.86	23.03	24.58	25.94	27.75	29.79	31.70	33.65	35.42	38.09	41.14
15/16	.9375	18.24	19.44	20.62	21.80	23.26	24.52	26.07	27.63	29.32	31.75	33.80	35.88	37.78	40.64	43.91
1	1.000	19.31	20.59	21.85	23.11	24.66	26.00	27.65	29.32	31.38	33.71	35.89	38.11	40.14	43.19	46.67
1 1/16	1.0625	20.38	21.74	23.08	24.41	26.06	27.48	29.24	31.01	33.20	35.67	37.99	40.35	42.50	45.74	49.44
1 1/8	1.125	21.46	22.89	24.31	25.72	27.46	28.97	30.82	32.70	35.01	37.63	40.08	42.58	44.86	48.28	52.20
1 3/16	1.1875	22.53	24.04	25.53	27.02	28.86	30.45	32.41	34.39	36.83	39.59	42.18	44.81	47.22	50.83	54.97
1 1/4	1.250	23.60	25.19	26.76	28.33	30.27	31.94	33.99	36.08	38.64	41.55	44.27	47.04	49.58	53.38	57.73

Washburn and Moen Gauge or Size and Decimal Equivalent*

Inside Diam. (Fractional)	Inside Diam. (Decimal)	3/8 .375	11/32 .3438	5/16 .3125	9/32 .2813	1/4 .250	3 .2437	4 .2253	7/32 .2188	5 .207	6 .192	3/16 .1875	7 .177	8 .162	5/32 .1563	9 .1483	10 .135
Design Stress, pounds per sq. in. (thousands)		135	136	137	138	139	140	141	142	143	146	149	150	154	156	158	161
Torque, pounds-inch		700.0	542.5	410.6	301.5	213.3	199.0	158.3	146.0	124.6	101.5	96.45	81.68	64.30	58.44	50.60	38.90
		Deflection, degrees per coil															
13/16	.8125	5.880	6.292	6.784	7.382	8.125	8.346	8.933	9.208	9.687	10.51	10.93	11.53	12.74	13.30	14.08	15.54
7/8	.875	6.189	6.632	7.163	7.803	8.603	8.840	9.471	9.766	10.28	11.16	11.61	12.26	13.56	14.16	15.00	16.57
15/16	.9375	6.499	6.972	7.537	8.225	9.081	9.333	10.01	10.32	10.87	11.81	12.30	12.99	14.38	15.02	15.91	17.59
1	1.000	6.808	7.312	7.914	8.647	9.559	9.827	10.55	10.88	11.47	12.47	12.98	13.72	15.19	15.88	16.83	18.62
1 1/16	1.0625	7.118	7.652	8.291	9.069	10.04	10.32	11.09	11.44	12.06	13.12	13.66	14.45	16.01	16.74	17.74	19.64
1 1/8	1.125	7.427	7.993	8.668	9.491	10.52	10.81	11.62	12.00	12.66	13.77	14.35	15.18	16.83	17.59	18.66	20.67
1 3/16	1.1875	7.737	8.333	9.045	9.912	10.99	11.31	12.16	12.56	13.25	14.43	15.03	15.90	17.64	18.45	19.57	21.69
1 1/4	1.250	8.046	8.673	9.422	10.33	11.47	11.80	12.70	13.11	13.84	15.08	15.71	16.63	18.46	19.31	20.49	22.72

* For sizes up to 26 gauge, the table values are for music wire with a modulus E of 29,000,000 psi; for sizes from 27 to 1/8 inch diameter the table values are for music wire with a modulus of 28,500,000 psi; for sizes from 10 gauge to 3/8 inch diameter, the values are for oil-tempered MB with a modulus of 28,500,000 psi.

Table 15. Torsion Spring Tolerance for Angular Relationship of Ends

	Spring Index								
	4	6	8	10	12	14	16	18	20
Number of Coils (N)	Free Angle Tolerance, ± degrees								
1	2	3	3.5	4	4.5	5	5.5	5.5	6
2	4	5	6	7	8	8.5	9	9.5	10
3	5.5	7	8	9.5	10.5	11	12	13	14
4	7	9	10	12	14	15	16	16.5	17
5	8	10	12	14	16	18	20	20.5	21
6	9.5	12	14.5	16	19	20.5	21	22.5	24
8	12	15	18	20.5	23	25	27	28	29
10	14	19	21	24	27	29	31.5	32.5	34
15	20	25	28	31	34	36	38	40	42
20	25	30	34	37	41	44	47	49	51
25	29	35	40	44	48	52	56	60	63
30	32	38	44	50	55	60	65	68	70
50	45	55	63	70	77	84	90	95	100

Table 16. Torsion Spring Coil Diameter Tolerances

	Spring Index						
	4	6	8	10	12	14	16
Wire Diameter, Inch	Coil Diameter Tolerance, ± inch						
0.015	0.002	0.002	0.002	0.002	0.003	0.003	0.004
0.023	0.002	0.002	0.002	0.003	0.004	0.005	0.006
0.035	0.002	0.002	0.003	0.004	0.006	0.007	0.009
0.051	0.002	0.003	0.005	0.007	0.008	0.010	0.012
0.076	0.003	0.005	0.007	0.009	0.012	0.015	0.018
0.114	0.004	0.007	0.010	0.013	0.018	0.022	0.028
0.172	0.006	0.010	0.013	0.020	0.027	0.034	0.042
0.250	0.008	0.014	0.022	0.030	0.040	0.050	0.060

Table 17. Torsion Spring Tolerance on Number of Coils

Number of Coils	Tolerance	Number of Coils	Tolerance
up to 5	± 5°	over 10 to 20	± 15°
over 5 to 10	± 10°	over 20 to 40	± 30°

commercial tolerance on loads is ± 10 percent and is specified with reference to the angular deflection. For example: 100 pounds-inches ± 10 per cent at 45 degrees deflection. One load specified usually suffices. If two loads and two deflections are specified, the manufacturing and testing times are increased. Tolerances smaller than ± 10 per cent require each spring to be individually tested and adjusted, which adds considerably to manufacturing time and cost. Tables 15, 16, and 17 give respectively free angle tolerances, coil diameter tolerances, and tolerances on the number of coils.

Miscellaneous Springs. — This section provides information on various springs, some in common use, some less commonly used.

Conical compression: These springs taper from top to bottom and are useful where an increasing (instead of a constant) load rate is needed, where solid height must be small, and where vibration must be damped. Conical springs with a uniform pitch are easiest to coil. Load and deflection formulas for compression springs can be used — using the average mean coil diameter, and providing the deflection does not cause the largest active coil to lie against the bottom coil. When this happens, each coil must be calculated separately, using the standard formulas for compression springs.

Belleville washers: These washer type springs can sustain relatively large loads with small deflections, but the loads and deflections can be increased by stacking the springs as shown in Fig. 26.

Design data is not given here because the wide variations in ratios of O.D. to I.D., height to thickness, and other factors require too many formulas for convenient use and involve constants obtained from more than 24 curves. It is now practicable to select required sizes from the large stocks carried by several of the larger spring manufacturing companies. Most of these companies also stock curved and wave washers.

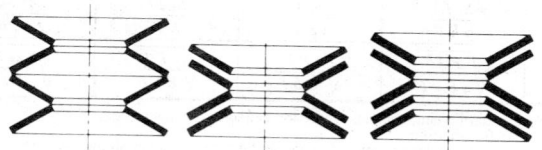

Fig. 26.   Examples of Belleville Spring Combinations

Volute springs: These are often used on army tanks and heavy field artillery, and seldom find additional uses because of their high cost, long production time, difficulties in manufacture, and unavailability of a wide range of materials and sizes. Small volute springs are often replaced with standard compression springs.

Torsion bars: Although the more simple types are often used on motor cars, the more complicated types with specially forged ends are finding fewer applications as time goes on.

Constant force springs: Those made from flat spring steel are finding more applications each year. However, the complicated design procedures can be eliminated by selecting a standard design from thousands now available from several spring manufacturers.

Spiral, clock, and motor springs: Although often used in wind-up type motors for toys and other products, they are difficult to design and results cannot be calculated with precise accuracy. However, many useful designs have been developed and are available from spring manufacturing companies.

Flat springs: These springs are often used to overcome operating space limitations in various products such as electric switches and relays. Table 18 lists formulas for designing flat springs. They are based on standard beam formulas where the deflection is small.

Table 18. Formulas for Flat Springs*†

Feature				
Deflect. f Inches	$\dfrac{PL^3}{4Ebt^3}$ $\dfrac{S_b L^2}{6Et}$	$\dfrac{4PL^3}{Ebt^3}$ $\dfrac{2S_b L^2}{3Et}$	$\dfrac{6PL^3}{Ebt^3}$ $\dfrac{S_b L^2}{Et}$	$\dfrac{5.22\,PL^3}{Ebt^3}$ $\dfrac{0.87\,S_b L^2}{Et}$
Load P Pounds	$\dfrac{2S_b bt^2}{3L}$ $\dfrac{4Ebt^3 F}{L^3}$	$\dfrac{S_b bt^2}{6L}$ $\dfrac{Ebt^3 F}{4L^3}$	$\dfrac{S_b bt^2}{6L}$ $\dfrac{Ebt^3 F}{6L^3}$	$\dfrac{S_b bt^2}{6L}$ $\dfrac{Ebt^3 F}{5.22\,L^3}$
Stress S_b Bending Pounds per sq. inch	$\dfrac{3PL}{2bt^2}$ $\dfrac{6EtF}{L^2}$	$\dfrac{6PL}{bt^2}$ $\dfrac{3EtF}{2L^2}$	$\dfrac{6PL}{bt^2}$ $\dfrac{EtF}{L^2}$	$\dfrac{6PL}{bt^2}$ $\dfrac{EtF}{0.87\,L^2}$
Thickness t Inches	$\dfrac{S_b L^2}{6EF}$ $\sqrt[3]{\dfrac{PL^3}{4EbF}}$	$\dfrac{2S_b L^2}{3EF}$ $\sqrt[3]{\dfrac{4PL^3}{EbF}}$	$\dfrac{S_b L^2}{EF}$ $\sqrt[3]{\dfrac{6PL^3}{EbF}}$	$\dfrac{0.87\,S_b L^2}{EF}$ $\sqrt[3]{\dfrac{5.22\,PL^3}{EbF}}$

* Based on standard beam formulas where the deflection is small
† See page 315 for notation.
Note: Where two formulas are given for one feature, the designer should use the one found to be appropriate for the given design instance. The result from either of any two formulas is the same.

Moduli of Elasticity of Spring Materials. — The modulus of elasticity in tension, denoted by the letter E, and the modulus of elasticity in torsion, denoted by the letter G, are used in formulas relating to spring design. Values of these moduli for various ferrous and nonferrous spring materials are given in Table 19.

General Heat Treating Information for Springs. — The following is general information on the heat treatment of springs, and is applicable to pre-tempered or hard-drawn spring materials only.

Compression springs are baked after coiling (before setting) to relieve residual stresses and thus permit larger deflections before taking a permanent set.

Extension springs also are baked, but heat removes some of the initial tension. Allowance should be made for this loss. Baking at 500 degrees F for 30 minutes removes approximately 50 per cent of the initial tension. The shrinkage in diameter however, will slightly increase the load and rate.

Torsion springs do not actually require baking because coiling causes residual stresses in a direction that is helpful, but such springs frequently are baked so that jarring or handling will not cause them to lose position of ends.

Table 19. Moduli of Elasticity in Torsion and Tension of Spring Materials

FERROUS MATERIALS			NON-FERROUS MATERIALS		
Material (Commercial Name)	Modulus of Elasticity, pounds per square inch		Material (Commercial Name)	Modulus of Elasticity, pounds per square inch	
	In Torsion, G	In Tension, E		In Torsion, G	In Tension, E
Hard Drawn MB			Spring Brass		
Up to 0.032 inch	11,700,000	28,800,000	Type 70-30	5,000,000	15,000,000
0.033 to 0.063 inch	11,600,000	28,700,000	Phosphor Bronze		
0.064 to 0.125 inch	11,500,000	28,600,000	5 per cent tin	6,000,000	15,000,000
0.126 to .625 inch	11,400,000	28,500,000	Beryllium Copper		
Music Wire			Cold Drawn 4 Nos.	7,000,000	17,000,000
Up to 0.032 inch	12,000,000	29,500,000	Pretempered,		
0.033 to 0.063 inch	11,850,000	29,000,000	fully hard	7,250,000	19,000,000
0.064 to 0.125 inch	11,750,000	28,500,000	Inconel† 600	10,500,000	31,000,000*
0.126 to 0.250 inch	11,600,000	28,000,000	Inconel† X 750	10,500,000	31,000,000*
Oil Tempered MB	11,200,000	28,500,000	Mcnel† 400	9,500,000	26,000,000
Chrome-Vanadium	11,200,000	28,500,000	Monel† K 500	9,500,000	26,000,000
Chrome-Silicon	11,200,000	29,500,000	Duranickel† 300	11,000,000	30,000,000
Silicon-Manganese	10,750,000	29,000,000	Permanickel†	11,000,000	30,000,000
Stainless Steel			Ni Span C† 902	10,000,000	27,500,000
Types 302, 304, 316	10,000,000	28,000,000*	Elgiloy‡	12,000,000	29,500,000
Type 17-7 PH	10,500,000	29,500,000	Iso-Elastic**	9,200,000	26,000,000
Type 420	11,000,000	29,000,000			
Type 431	11,400,000	29,500,000			

* May be 2,000,000 pounds per square inch less if material is not fully hard.
† Trade name of the International Nickel Company.   ‡ Trade name of Hamilton Watch Company.   ** Trade name of John Chatillon & Sons.
Note: Modulus G is used for compression and extension springs; modulus E is used for torsion, flat, and spiral springs.

Outside diameters shrink when springs of music wire, pre-tempered MB and other carbon or alloy steels are baked. Baking also slightly increases the free length and these changes cause a little stronger load and increase the rate.

Outside diameters expand when springs of stainless steel (18-8) are baked. The free length also reduces slightly and these changes cause a little lighter load and decrease the rate.

Inconel, Monel, and nickel alloys do not change much when baked.

Beryllium-copper shrinks and deforms when heated. Such springs usually are baked in fixtures or supported on arbors or rods during heating.

Brass and phosphor bronze springs should be given a light heat only. Baking above 450 degrees F will soften the material. Do not heat in salt pots.

Spring brass and phosphor bronze springs which are not too highly stressed and are not subject to severe operating use may receive adequate stress relieving after coiling by immersing them in boiling water for a period of one hour.

Position of loops will change with heat. Parallel hooks may change as much as 45 degrees during baking. Torsion spring arms will alter position considerably. These changes should be allowed for during looping or forming.

Quick heating after coiling either in a high temperature salt pot or by passing a spring through a gas flame is not good practice. Samples heated in this way will not conform with production runs that are properly baked. A small, controlled-temperature oven should be used for samples and for small lot orders.

Plated springs should always be baked before plating to relieve coiling stresses and again after plating to relieve hydrogen embrittlement.

Hardness values fall with high heat — but music wire, hard drawn and stainless steel will increase 2 to 4 points Rockwell C.

Table 20. Squares, Cubes and Fourth Powers of Wire Diameters

Steel Wire Gage (U.S.)	Music or Piano Wire Gage	Diameter Inches	Section Area	Square	Cube	Fourth Power
7-0	..	0.4900	0.1886	0.24010	0.11765	0.05765
6-0	..	0.4615	0.1673	0.21298	0.09829	0.04536
5-0	..	0.4305	0.1456	0.18533	0.07978	0.03435
4-0	..	0.3938	0.1218	0.15508	0.06107	0.02405
3-0	..	0.3625	0.1032	0.13141	0.04763	0.01727
2-0	..	0.331	0.0860	0.10956	0.03626	0.01200
1-0	..	0.3065	0.0738	0.09394	0.02879	0.008825
1	..	0.283	0.0629	0.08009	0.02267	0.006414
2	..	0.2625	0.0541	0.06891	0.01809	0.004748
3	..	0.2437	0.0466	0.05939	0.01447	0.003527
4	..	0.2253	0.0399	0.05076	0.01144	0.002577
5	..	0.207	0.0337	0.04285	0.00887	0.001836
6	..	0.192	0.0290	0.03686	0.00708	0.001359
..	45	0.180	0.0254	0.03240	0.00583	0.001050
7	..	0.177	0.0246	0.03133	0.00555	0.000982
..	44	0.170	0.0227	0.02890	0.00491	0.000835
8	43	0.162	0.0206	0.02624	0.00425	0.000689
..	42	0.154	0.0186	0.02372	0.00365	0.000563
9	..	0.1483	0.0173	0.02199	0.00326	0.000484
..	41	0.146	0.0167	0.02132	0.00311	0.000455
..	40	0.138	0.0150	0.01904	0.00263	0.000363
10	..	0.135	0.0143	0.01822	0.00246	0.000332
..	39	0.130	0.0133	0.01690	0.00220	0.000286
..	38	0.124	0.0121	0.01538	0.00191	0.000237
11	..	0.1205	0.0114	0.01452	0.00175	0.000211
..	37	0.118	0.0109	0.01392	0.00164	0.000194
..	36	0.112	0.0099	0.01254	0.00140	0.000157
..	35	0.106	0.0088	0.01124	0.00119	0.000126
12	..	0.1055	0.0087	0.01113	0.001174	0.0001239
..	34	0.100	0.0078	0.0100	0.001000	0.0001000
..	33	0.095	0.0071	0.00902	0.000857	0.0000815
13	..	0.0915	0.0066	0.00837	0.000766	0.0000701
..	32	0.090	0.0064	0.00810	0.000729	0.0000656
..	31	0.085	0.0057	0.00722	0.000614	0.0000522
14	30	0.080	0.0050	0.0064	0.000512	0.0000410
..	29	0.075	0.0044	0.00562	0.000422	0.0000316
15	..	0.072	0.0041	0.00518	0.000373	0.0000269
..	28	0.071	0.0040	0.00504	0.000358	0.0000254
..	27	0.067	0.0035	0.00449	0.000301	0.0000202
..	26	0.063	0.0031	0.00397	0.000250	0.0000158
16	..	0.0625	0.0031	0.00391	0.000244	0.0000153
..	25	0.059	0.0027	0.00348	0.000205	0.0000121
..	24	0.055	0.0024	0.00302	0.000166	0.00000915
17	..	0.054	0.0023	0.00292	0.000157	0.00000850
..	23	0.051	0.0020	0.00260	0.000133	0.00000677
..	22	0.049	0.00189	0.00240	0.000118	0.00000576
18	..	0.0475	0.00177	0.00226	0.000107	0.00000509
..	21	0.047	0.00173	0.00221	0.000104	0.00000488
..	20	0.045	0.00159	0.00202	0.000091	0.00000410
..	19	0.043	0.00145	0.00185	0.0000795	0.00000342
19	18	0.041	0.00132	0.00168	0.0000689	0.00000283
..	17	0.039	0.00119	0.00152	0.0000593	0.00000231
..	16	0.037	0.00108	0.00137	0.0000507	0.00000187
..	15	0.035	0.00096	0.00122	0.0000429	0.00000150
20	..	0.0348	0.00095	0.00121	0.0000421	0.00000147
..	14	0.033	0.00086	0.00109	0.0000359	0.00000119
21	..	0.0317	0.00079	0.00100	0.0000319	0.00000101
..	13	0.031	0.00075	0.00096	0.0000298	0.000000924
..	12	0.029	0.00066	0.00084	0.0000244	0.000000707
22	..	0.0286	0.00064	0.00082	0.0000234	0.000000669
..	11	0.026	0.00053	0.00068	0.0000176	0.000000457
23	..	0.0258	0.00052	0.00067	0.0000172	0.000000443
..	10	0.024	0.00045	0.00058	0.0000138	0.000000332
24	..	0.023	0.00042	0.00053	0.0000122	0.000000280
..	9	0.022	0.00038	0.00048	0.0000106	0.000000234

Table 21. Causes of Spring Failure*

	Cause	Comments and Recommendations
GROUP 1	High stress	The majority of spring failures are due to high stresses caused by large deflections and high loads. High stresses should be used only for statically loaded springs. Low stresses lengthen fatigue life.
GROUP 1	Hydrogen embrittlement	Improper electro-plating methods and acid cleaning of springs, without proper baking treatment, cause spring steels to become brittle, and is a frequent cause of failure. Non-ferrous springs are immune.
GROUP 1	Sharp bends and holes	Sharp bends on extension, torsion and flat springs and holes or notches in flat springs, cause high concentration of stress resulting in failure. Bend radii should be as large as possible, and tool marks avoided.
GROUP 1	Fatigue	Repeated deflections of springs, especially above 1,000,000 cycles, even with medium stresses, may cause failure. Low stresses should be used if a spring is subject to a very high number of operating cycles conditions.
GROUP 2	Shock loading	Impact, shock, and rapid loading cause far higher stresses than those computed by the regular spring formulas. High carbon spring steels do not withstand shock loading as well as alloy steels.
GROUP 2	Corrosion	Slight rusting or pitting caused by acids, alkali, galvanic corrosion, stress corrosion cracking, or corrosive atmosphere weakens the material and causes higher stresses in the corroded area.
GROUP 2	Faulty heat treatment	Keeping spring materials at the hardening temperature for longer periods than necessary causes an undesirable growth in grain structure resulting in brittleness even though the hardness may be correct.
GROUP 2	Faulty material	Poor material containing inclusions, seams, slivers, and flat material with rough, slit, or torn edges cause early failure. Overdrawn wire, improper hardness, and poor grain structure also result in early failure.
GROUP 3	High temperature	High temperatures reduce spring temper (or hardness), lower the modulus of elasticity thereby causing lower loads, reduce the elastic limit and increase corrosion. Corrosion resisting or nickel alloys should be used.
GROUP 3	Low temperature	Temperatures below −40 degrees F lessen the ability of carbon steels to withstand shock loads. Carbon steels become brittle at −70 degrees F. Corrosion resisting, nickel or non-ferrous alloys should be used.
GROUP 3	Friction	Close fits on rods or in holes result in a wearing away of material and occasional failure. The outside diameters of compression springs expand during deflection but they become smaller on torsion springs.
GROUP 3	Other causes	Enlarged hooks on extension springs increase the stress at the bends. Carrying too much electrical current will cause failure. Welding and soldering frequently destroy the spring temper. Tool marks, nicks, and cuts often raise stresses. Deflecting torsion springs outwardly causes high stresses and winding them tightly causes binding on supporting rods. High speed of deflection, vibration and surging due to operation near natural periods of vibration or their harmonics, cause increased stresses.

* Spring failure may be breakage, high permanent set, or loss of load. The causes are listed in groups in this table. Group 1 covers causes that occur most frequently; Group 2 covers causes that are less frequent; and Group 3 lists causes that occur occasionally.

Table 22. Arbor Diameters for Springs made from Music Wire

Wire Diam. (inch)	Spring Outside Diameter (inch)												
	1/16	3/32	1/8	5/32	3/16	7/32	1/4	9/32	5/16	11/32	3/8	7/16	1/2
	Arbor Diameter (inch)												
.008	.039	.060	.078	.093	.107	.119	.129	.142	.154	.164	…	…	…
.010	.037	.060	.080	.099	.115	.129	.142	.154	.164	…	…	…	…
.012	.034	.059	.081	.101	.119	.135	.150	.163	.177	.189	.200	…	…
.014	.031	.057	.081	.102	.121	.140	.156	.172	.187	.200	.213	.234	…
.016	.028	.055	.079	.102	.123	.142	.161	.178	.194	.209	.224	.250	.271
.018	…	.053	.077	.101	.124	.144	.161	.182	.200	.215	.231	.259	.284
.020	…	.049	.075	.096	.123	.144	.165	.184	.203	.220	.237	.268	.296
.022	…	.046	.072	.097	.122	.145	.165	.186	.206	.224	.242	.275	.305
.024	…	.043	.070	.095	.120	.144	.166	.187	.207	.226	.245	.280	.312
.026	…	…	.067	.093	.118	.143	.166	.187	.208	.228	.248	.285	.318
.028	…	…	.064	.091	.115	.141	.165	.187	.208	.229	.250	.288	.323
.030	…	…	.061	.088	.113	.138	.163	.187	.209	.229	.251	.291	.328
.032	…	…	.057	.085	.111	.136	.161	.185	.209	.229	.251	.292	.331
.034	…	…	…	.082	.109	.134	.159	.184	.208	.229	.251	.292	.333
.036	…	…	…	.078	.106	.131	.156	.182	.206	.229	.250	.294	.333
.038	…	…	…	.075	.103	.129	.154	.179	.205	.227	.251	.293	.335
.041	…	…	…	…	.098	.125	.151	.176	.201	.226	.250	.294	.336
.0475	…	…	…	…	.087	.115	.142	.168	.194	.220	.244	.293	.337
.054	…	…	…	…	…	.103	.132	.160	.187	.212	.245	.287	.336
.0625	…	…	…	…	…	…	.108	.146	.169	.201	.228	.280	.330
.072	…	…	…	…	…	…	…	.129	.158	.186	.214	.268	.319
.080	…	…	…	…	…	…	…	…	.144	.173	.201	.256	.308
.0915	…	…	…	…	…	…	…	…	…	…	.181	.238	.293
.1055	…	…	…	…	…	…	…	…	…	…	…	.215	.271
.1205	…	…	…	…	…	…	…	…	…	…	…	…	.215
.125	…	…	…	…	…	…	…	…	…	…	…	…	.239

Wire Diam. (inch)	Spring Outside Diameter (inches)													
	9/16	5/8	11/16	3/4	13/16	7/8	15/16	1	1 1/8	1 1/4	1 3/8	1 1/2	1 3/4	2
	Arbor Diameter (inches)													
.022	.332	.357	.380	…	…	…	…	…	…	…	…	…	…	…
.024	.341	.367	.393	.415	…	…	…	…	…	…	…	…	…	…
.026	.350	.380	.406	.430	…	…	…	…	…	…	…	…	…	…
.028	.356	.387	.416	.442	.467	…	…	…	…	…	…	…	…	…
.030	.362	.395	.426	.453	.481	.506	…	…	…	…	…	…	…	…
.032	.367	.400	.432	.462	.490	.516	.540	…	…	…	…	…	…	…
.034	.370	.404	.437	.469	.498	.526	.552	.557	…	…	…	…	…	…
.036	.372	.407	.442	.474	.506	.536	.562	.589	…	…	…	…	…	…
.038	.375	.412	.448	.481	.512	.543	.572	.600	.650	…	…	…	…	…
.041	.378	.416	.456	.489	.522	.554	.586	.615	.670	.718	…	…	…	…
.0475	.380	.422	.464	.504	.541	.576	.610	.643	.706	.763	.812	…	…	…
.054	.381	.425	.467	.509	.550	.589	.625	.661	.727	.792	.850	.906	…	…
.0625	.379	.426	.468	.512	.556	.597	.639	.678	.753	.822	.889	.951	…	…
.072	.370	.418	.466	.512	.555	.599	.641	.682	.765	.840	.911	.980	1.11	1.22
.080	.360	.411	.461	.509	.554	.599	.641	.685	.772	.851	.930	1.00	1.13	1.26
.0915	.347	.398	.448	.500	.547	.597	.640	.685	.776	.860	.942	1.02	1.16	1.30
.1055	.327	.381	.433	.485	.535	.586	.630	.683	.775	.865	.952	1.04	1.20	1.35
.1205	.303	.358	.414	.468	.520	.571	.622	.673	.772	.864	.955	1.04	1.22	1.38
.125	.295	.351	.406	.461	.515	.567	.617	.671	.770	.864	.955	1.05	1.23	1.39

STRENGTH AND PROPERTIES OF WIRE ROPE

Wire Rope Construction. — Essentially, a wire rope is made up of a number of strands laid helically about a metallic or non-metallic core. Each strand consists of a number of wires also laid helically about a metallic or non-metallic center. Various types of wire rope have been developed to meet a wide range of uses and operating conditions. These types are distinguished by the kind of core; the number of strands; the number, sizes, and arrangement of the wires in each strand; and the way in which the wires and strands are wound or laid about each other. The following descriptive material is based largely on information supplied by the Bethlehem Steel Co.

Rope Wire Materials: Materials used in the manufacture of rope wire are, in the order of increasing strength: iron, phosphor bronze, traction steel, plow steel, improved plow steel, and bridge rope steel. Iron wire rope is largely used for low-strength applications such as elevator ropes not used for hoisting, and for stationary guy ropes.

Phosphor bronze wire rope is used occasionally for elevator governor-cable rope and for certain marine applications as life lines, clearing lines, wheel ropes and rigging.

Traction steel wire rope is used primarily as a hoist rope for passenger and freight elevators of the traction drive type, an application for which it was specifically designed.

Ropes made of galvanized wire or wire coated with zinc by the electrodeposition process are used in certain applications where additional protection against rusting is required. As will be noted from the tables of wire-rope sizes and strengths, the breaking strength of galvanized wire rope is 10 per cent less than that of ungalvanized (bright) wire rope. Bethanized (zinc-coated) wire rope can be furnished to bright wire rope strength when so specified.

Galvanized carbon steel, tinned carbon steel, and stainless steel are used for small cords and strands ranging in diameter from 1/64 to 3/8 inch and larger.

Marline clad wire rope has each strand wrapped with a layer of tarred marline. This provides hand protection for workers and wear protection for the rope.

Rope Cores: Wire-rope cores are made of fiber, cotton, asbestos, polyvinyl plastic, a small wire rope (independent wire-rope core), a multiple-wire strand (wire-strand core) or a cold-drawn wire-wound spring.

Fiber (manila or sisal) is the type of core most widely used when loads are not too great. It supports the strands in their relative positions and also acts as a cushion to prevent nicking of the wires lying next to the core.

Cotton is used for small ropes such as sash cord and aircraft cord.

Asbestos cores can be furnished for certain special operations where the rope is used in oven operations.

Polyvinyl plastic cores are offered for use where exposure to moisture, acids, or caustics is excessive.

A *wire strand core,* often referred to as WSC, consists of a multiple-wire strand that may be the same as one of the strands of the rope. It is smoother and more solid than the independent wire rope core and provides a better support for the rope strands.

The *independent wire rope core,* often referred to as IWRC, is a small 6×7 wire rope with a wire strand core and is used to provide greater resistance to crushing and distortion of the wire rope. For certain applications it has the advantage over a wire-strand core in that it stretches at a rate closer to that of the rope itself.

Wire ropes with wire-strand cores are, in general, less flexible than wire ropes with independent wire-rope or non-metallic cores.

Ropes with metallic cores are rated 7½ per cent stronger than those with non-metallic cores.

Wire-Rope Lay: The lay of a wire rope is the direction of the helical path in which the strands are laid and, similarly, the lay of a strand is the direction of the helical path in which the wires are laid. If the wires in the strand or the strands in the rope form a helix similar to the threads of a right-hand screw, i.e., they wind around to the right, the lay is called right hand and, conversely, if they wind around to the left, the lay is called left hand. In the *regular lay,* the wires in the strands are laid in the opposite direction to the lay of the strands in the rope. In right-regular lay, the strands are laid to the right and the wires to the left. In left-regular lay, the strands are laid to the left, the wires to the right. In *Lang lay,* the wires and strands are laid in the same direction, i.e., in right Lang lay, both the wires and strands are laid to the right and in left Lang they are laid to the left.

Alternate lay ropes having alternate right and left laid strands are used to resist distortion and prevent clamp slippage, but because other advantages are missing, have limited use.

The regular lay wire rope is most widely used and right regular lay rope is customarily furnished. Regular lay rope has less tendency to spin or untwist when placed under load and is generally selected where long ropes are employed and the loads handled are frequently removed. Lang lay ropes have greater flexibility than regular lay ropes and are more resistant to abrasion and fatigue.

In preformed wire ropes the wires and strands are preshaped into a helical form so that when laid to form the rope they tend to remain in place. In a non-preformed rope, broken wires tend to "wicker out" or protrude from the rope and strands that are not seized tend to spring apart. Preforming also tends to remove locked-in stresses, lengthen service life, and make the rope easier to handle and to spool.

Strand Construction: Various arrangements of wire are used in the construction of wire rope strands. In the simplest arrangement six wires are grouped around a central wire thus making seven wires, all of the same size. Other types of construction known as "filler-wire," Warrington, Seale, etc. make use of wires of different sizes. Their respective patterns of arrangement are shown diagrammatically in the table of wire weights and strengths.

Specifying Wire Rope. — In specifying wire rope the following information will be required: length, diameter, number of strands, number of wires in each strand, type of rope construction, grade of steel used in rope, whether preformed or not preformed, type of center, and type of lay. The manufacturer should be consulted in selecting the best type of wire rope for a new application.

Properties of Wire Rope. — Important properties of wire rope are strength, wear resistance, flexibility, and resistance to crushing and distortion.

Strength: The strength of wire rope depends upon its size, kind of material of which the wires are made and their number, the type of core, and whether the wire is galvanized or not. Strengths of various types and sizes of wire ropes are given in the accompanying tables together with appropriate factors to apply for ropes with steel cores and for galvanized wire ropes.

Wear Resistance: When wire rope must pass back and forth over surfaces which subject it to unusual wear or abrasion, it must be specially constructed to give satisfactory service. Such construction may make use of (1) relatively large outer wires; (2) Lang lay in which wires in each strand are laid in the same direction as the strand; or (3) flattened strands. The object in each case is to provide a greater out-

side surface area to take the wear or abrasion. From the standpoint of material, improved plow steel has not only the highest tensile strength but also the greatest resistance to abrasion in regularly stocked wire rope.

Flexibility: Wire rope which undergoes repeated and severe bending, such as in passing around small sheaves and drums, must have a high degree of flexibility to prevent premature breakage and failure due to fatigue. Greater flexibility in wire rope is obtained by (1) using small wires in larger numbers, (2) using Lang lay, and (3) preforming, that is the wires and strands of the rope are shaped during manufacture to fit the position they will assume in the finished rope.

Resistance to Crushing and Distortion: Where wire rope is to be subjected to transverse loads that may crush or distort it, care should be taken to select a type of construction which will stand up under such treatment. Wire rope designed for such conditions may have (1) large outer wires to spread the load per wire over a greater area and (2) an independent wire core or a high-carbon cold-drawn wound spring core.

Standard Classes of Wire Rope. — Wire rope is commonly designated by two figures, the first indicating the number of strands and the second, the number of wires per strand, as: 6×7, a six-strand rope having seven wires per strand, 8×19, an eight-strand rope having 19 wires per strand, etc. When such numbers are used as designations of standard wire rope classes, the second figure in the designation may be purely nominal in that the number of wires per strand for various ropes in the class may be slightly less or slightly more than the nominal as will be seen from the following brief descriptions. (For ropes with a wire strand core, a second group of two numbers may be used to indicate the construction of the wire core, as 1×21, 1×43, etc.)

6×7 *Class (Standard Coarse Laid Rope):* Wire ropes in this class are for use where resistance to wear, as in dragging over the ground or across rollers, is an important requirement. Heavy hauling, rope transmissions, well drilling are common applications. These wire ropes are furnished in right regular lay and occasionally in Lang lay. The cores may be of fiber, independent wire rope, or wire strand. Since this is a relatively stiff type of construction, ropes in this class should be used with large sheaves and drums. Because of the small number of wires, a larger factor of safety may be called for.

As shown in Table 1, this class includes a 6×7 construction with fiber core: a 6×7 construction with 1×7 wire strand core (sometimes called 7×7); a 6×7 construction with 1×19 wire strand core; and a 6×7 construction with independent wire rope core.

Two special types of wire rope in this class are: aircraft cord, a 6×6 or 7×7 Bethanized wire rope of high tensile strength and sash cord, a 6×7 iron rope used for a variety of purposes where strength is not an important factor.

6×19 *Class (Standard Hoisting Rope):* This is the most popular and widely used class of wire ropes. Ropes in this class are furnished in regular or Lang lay and may be obtained preformed or not preformed. Cores may be of fiber, independent wire rope, or wire strand. As can be seen from Table 2, there are four common types: 6×25 filler wire construction with fiber core (not illustrated), independent wire core, or wire strand core (1×25 or 1×43); 6×19 Warrington construction with fiber core; 6×21 filler wire construction with fiber core; and 6×19, 6×21, and 6×17 Seale construction with fiber core.

6×37 *Class (Extra Flexible Hoisting Rope):* For a given size of rope, the component wires are of smaller diameter than those in the two classes previously described and hence have less resistance to abrasion. Ropes in this class are furnished in regular and Lang lay with fiber core or independent wire rope core, preformed or not preformed.

Table 1. Weights and Strengths of 6 × 7 (Standard Coarse Laid) Wire Ropes, Preformed and Not Preformed

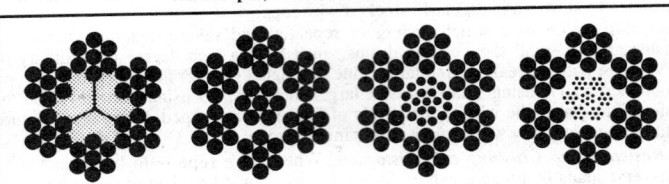

6 × 7 with fiber core   6 × 7 with 1 × 7 WSC   6 × 7 with 1 × 19   6 × 7 with IWRC
(sometimes called 7 × 7)   WSC

Diam., Inches	Approx. Weight per Ft., Pounds	Breaking Strength, Tons of 2000 Lbs.			Diam., Inches	Approx. Weight per Ft., Pounds	Breaking Strength, Tons of 2000 Lbs.		
		Impr. Plow Steel	Plow Steel	Mild Plow Steel			Impr. Plow Steel	Plow Steel	Mild Plow Steel
¼	0.094	2.64	2.30	2.00	¾	0.84	22.7	19.8	17.2
⁵⁄₁₆	0.15	4.10	3.56	3.10	⅞	1.15	30.7	26.7	23.2
⅜	0.21	5.86	5.10	4.43	1	1.50	39.7	34.5	30.0
⁷⁄₁₆	0.29	7.93	6.90	6.00	1⅛	1.90	49.8	43.3	37.7
½	0.38	10.3	8.96	7.79	1¼	2.34	61.0	53.0	46.1
⁹⁄₁₆	0.48	13.0	11.3	9.82	1⅜	2.84	73.1	63.6	55.3
⅝	0.59	15.9	13.9	12.0	1½	3.38	86.2	75.0	65.2

For ropes with steel cores, add 7½ per cent to above strengths.
For galvanized ropes, deduct 10 per cent from above strengths.
Source: Rope diagrams, Bethlehem Steel Co. All data, U.S. Simplified Practice Recommendation 198–50.

As shown in Table 3, there are four common types: 6 × 29 filler wire construction with fiber core and 6 × 36 filler wire construction with independent wire rope core, a special rope for construction equipment; 6 × 35 (two operations) construction with fiber core and 6 × 41 Warrington Seale construction with fiber core, a standard crane rope in this class of rope construction; 6 × 41 filler wire construction with fiber core or independent wire core, a special large shovel rope usually furnished in Lang lay; and 6 × 46 filler wire construction with fiber core or independent wire rope core, a special large shovel and dredge rope.

8 × 19 Class (Special Flexible Hoisting Rope): This is a stable smooth-running rope, especially suitable, because of its flexibility, for high speed operation with reverse bends. Ropes in this class are available in regular lay with fiber core.

As shown in Table 4, there are four common types: 8 × 25 filler wire construction, the most flexible but the least wear resistant rope of the four types; Warrington type in 8 × 19 construction, less flexible than the 8 × 25; 8 × 21 filler wire construction, less flexible than the Warrington; and Seale type in 8 × 19 construction, which has the greatest wear resistance of the four types but is also the least flexible.

Also in this class, but not shown in Table 4 are elevator ropes made of traction steel and iron.

18 × 7 Non-rotating Wire Rope: This is a rope specially designed for use where a minimum of rotating or spinning is called for, especially in the lifting or lowering

Table 2. Weights and Strengths of 6 × 19 (Standard Hoisting) Wire Ropes, Preformed and Not Preformed

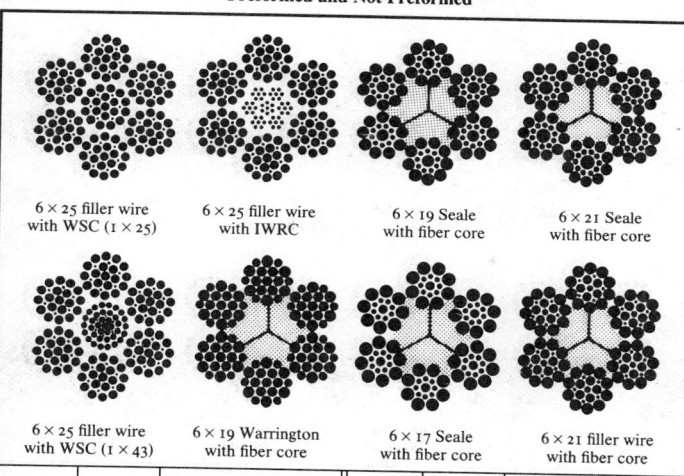

6 × 25 filler wire with WSC (1 × 25)
6 × 25 filler wire with IWRC
6 × 19 Seale with fiber core
6 × 21 Seale with fiber core

6 × 25 filler wire with WSC (1 × 43)
6 × 19 Warrington with fiber core
6 × 17 Seale with fiber core
6 × 21 filler wire with fiber core

Diam., Inches	Approx. Weight per Ft., Pounds	Breaking Strength, Tons of 2000 Lbs.			Diam., Inches	Approx. Weight per Ft., Pounds	Breaking Strength Tons of 2000 Lbs.		
		Impr. Plow Steel	Plow Steel	Mild Plow Steel			Impr. Plow Steel	Plow Steel	Mild Plow Steel
¼	0.10	2.74	2.39	2.07	1¼	2.50	64.6	56.2	48.8
⁵⁄₁₆	0.16	4.26	3.71	3.22	1⅜	3.03	77.7	67.5	58.8
⅜	0.23	6.10	5.31	4.62	1½	3.60	92.0	80.0	69.6
⁷⁄₁₆	0.31	8.27	7.19	6.25	1⅝	4.23	107.	93.4	81.2
½	0.40	10.7	9.35	8.13	1¾	4.90	124.	108.	93.6
⁹⁄₁₆	0.51	13.5	11.8	10.2	1⅞	5.63	141.	123.	107.
⅝	0.63	16.7	14.5	12.6	2	6.40	160.	139.	121.
¾	0.90	23.8	20.7	18.0	2⅛	7.23	179.	156.	...
⅞	1.23	32.2	28.0	24.3	2¼	8.10	200.	174.	...
1	1.60	41.8	36.4	31.6	2½	10.00	244.	212.	...
1⅛	2.03	52.6	45.7	39.8	2¾	12.10	292.	254.	...

The 6 × 25 filler wire with fiber core not illustrated.
For ropes with steel cores, add 7½ per cent to above strengths.
For galvanized ropes, deduct 10 per cent from above strengths.
Source: Rope diagrams, Bethlehem Steel Co. All data, U.S. Simplified Practice Recommendation 198–50.

of free loads with a single-part line. It has an inner layer composed of 6 strands of 7 wires each laid in left Lang lay over a fiber core and an outer layer of 12 strands of 7 wires each laid in right regular lay. The combination of opposing lays tends to prevent rotation when the rope is stretched. However, to avoid any tendency to rotate or spin, loads should be kept to at least one-eighth and preferably one-tenth of the

Table 3. Weights and Strengths of 6 × 37 (Extra Flexible Hoisting) Wire Ropes, Preformed and Not Preformed

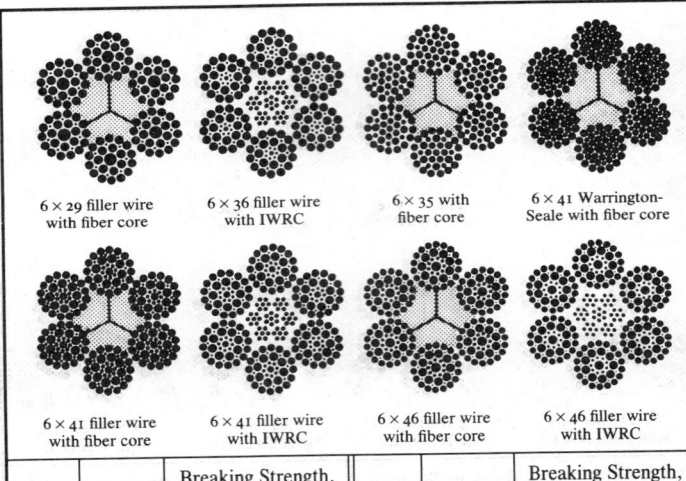

6 × 29 filler wire with fiber core

6 × 36 filler wire with IWRC

6 × 35 with fiber core

6 × 41 Warrington-Seale with fiber core

6 × 41 filler wire with fiber core

6 × 41 filler wire with IWRC

6 × 46 filler wire with fiber core

6 × 46 filler wire with IWRC

Diam., Inches	Approx. Weight per Ft., Pounds	Breaking Strength, Tons of 2000 Lbs.		Diam., Inches	Approx. Weight per Ft., Pounds	Breaking Strength, Tons of 2000 Lbs.	
		Impr. Plow Steel	Plow Steel			Impr. Plow Steel	Plow Steel
¼	0.10	2.59	2.25	1½	3.49	87.9	76.4
5/16	0.16	4.03	3.50	1⅝	4.09	103.	89.3
⅜	0.22	5.77	5.02	1¾	4.75	119.	103.
7/16	0.30	7.82	6.80	1⅞	5.45	136.	118.
½	0.39	10.2	8.85	2	6.20	154.	134.
9/16	0.49	12.9	11.2	2⅛	7.00	173.	150.
⅝	0.61	15.8	13.7	2¼	7.85	193.	168.
¾	0.87	22.6	19.6	2½	9.69	236.	205.
⅞	1.19	30.6	26.6	2¾	11.72	284.	247.
1	1.55	39.8	34.6	3	14.0	335.	291.
1⅛	1.96	50.1	43.5	3¼	16.4	390.	339.
1¼	2.42	61.5	53.5	3½	19.0	449.	390.
1⅜	2.93	74.1	64.5	...	...	...	...

For ropes with steel cores, add 7½ per cent to above strengths.
For galvanized ropes, deduct 10 per cent from above strengths.
Source: Rope diagrams, Bethlehem Steel Co. All data, U.S. Simplified Practice Recommendation 198–50.

breaking strength of the rope. Weights and strengths are shown in Table 5.

Flattened Strand Wire Rope: The wires forming the strands of this type of rope are wound around triangular centers so that a flattened outer surface is provided with a

Table 4. Weights and Strengths of 8 × 19 (Special Flexible Hoisting) Wire Ropes, Preformed and Not Preformed

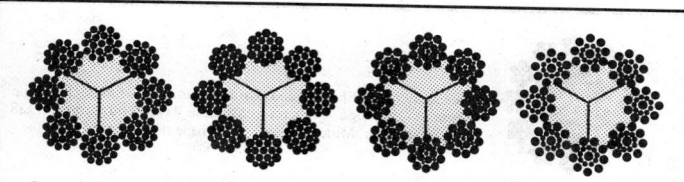

| 8 × 25 filler wire with fiber core | 8 × 19 Warrington with fiber core | 8 × 21 filler wire with fiber core | 8 × 19 Seale with fiber core |

Diam., Inches	Approx. Weight per Ft., Pounds	Breaking Strength, Tons of 2000 Lbs.		Diam., Inches	Approx. Weight per Ft., Pounds	Breaking Strength, Tons of 2000 Lbs.	
		Impr. Plow Steel	Plow Steel			Impr. Plow Steel	Plow Steel
¼	0.09	2.35	2.04	¾	0.82	20.5	17.8
5/16	0.14	3.65	3.18	7/8	1.11	27.7	24.1
3/8	0.20	5.24	4.55	1	1.45	36.0	31.3
7/16	0.28	7.09	6.17	1⅛	1.84	45.3	39.4
½	0.36	9.23	8.02	1¼	2.27	55.7	48.4
9/16	0.46	11.6	10.1	1⅜	2.74	67.1	58.3
5/8	0.57	14.3	12.4	1½	3.26	79.4	69.1

For ropes with steel cores, add 7½ per cent to above strengths.
For galvanized ropes, deduct 10 per cent from above strengths.
Source: Rope diagrams, Bethlehem Steel Co. All data, U.S. Simplified Practice Recommendation 198–50.

greater area than in the regular round rope to withstand severe conditions of abrasion. The triangular shape of the strands also provides superior resistance to crushing. Flattened strand wire rope is usually furnished in Lang lay and may be obtained with fiber core or independent wire rope core. The three types shown in Table 6 are flexible and are designed for hoisting work.

Flat Wire Rope: This type of wire rope is made up of a number of four-strand rope units placed side by side and stitched together with soft steel sewing wire. These four-strand units are alternately right and left lay to resist warping, curling, or rotating in service. Weights and strengths are shown in Table 7.

Simplified Practice Recommendations. — Because the total number of wire rope types is large, manufacturers and users have agreed upon and adopted a U.S. Simplified Practice Recommendation to provide a simplified listing of those kinds and sizes of wire rope which are most commonly used and stocked. These, then, are the types and sizes which are most generally available. Other types and sizes for special or limited uses also may be found in individual manufacturer's catalogs.

Sizes and Strengths of Wire Rope. — The data shown in Tables 1 through 7 have been taken from U.S. Simplified Practice Recommendation 198–50 but do not include those wire ropes shown in that Simplified Practice Recommendation which are intended primarily for marine use.

Table 5. Weights and Strengths of Standard 18 × 7 Nonrotating Wire Rope, Preformed and Not Preformed

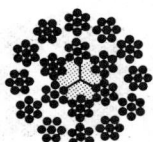

18 × 7 Non-Rotating Rope

Recommended Sheave and Drum Diameters: Single layer on drum . . . 36 rope diameters. Multiple layers on drum . . . 48 rope diameters. Mine service . . . 60 rope diameters.

Diam., Inches	Approx. Weight per Ft., Pounds	Breaking Strength, Tons of 2000 Lbs.		Diam., Inches	Approx. Weight per Ft., Pounds	Breaking Strength, Tons of 2000 Lbs.	
		Impr. Plow Steel	Plow Steel			Impr. Plow Steel	Plow Steel
³⁄₁₆	0.061	1.42	1.24	⅞	1.32	29.5	25.7
¼	0.108	2.51	2.18	1	1.73	38.3	33.3
⁵⁄₁₆	0.169	3.90	3.39	1⅛	2.19	48.2	41.9
⅜	0.24	5.59	4.86	1¼	2.70	59.2	51.5
⁷⁄₁₆	0.33	7.58	6.59	1⅜	3.27	71.3	62.0
½	0.43	9.85	8.57	1½	3.89	84.4	73.4
⁹⁄₁₆	0.55	12.4	10.8	1⅝	4.57	98.4	85.6
⅝	0.68	15.3	13.3	1¾	5.30	114.	98.8
¾	0.97	21.8	19.0	. . .	. . .	. . .	. . .

For galvanized ropes, deduct 10 per cent from above strengths.
Source: Rope diagrams, sheave and drum diameters, and data for ³⁄₁₆, ¼ and ⁵⁄₁₆-inch sizes, Bethlehem Steel Co. All other data, U.S. Simplified Practice Recommendation 198–50.

Wire Rope Diameter: The diameter of a wire rope is the diameter of that circle which will just enclose it, hence when measuring the diameter with calipers, care must be taken to obtain the largest outside dimension, which is taken across the opposite strands, rather than the smallest dimension which is across opposite "valleys" or "flats." It is standard practice for the nominal diameter to be the minimum with all tolerances taken on the plus side. Limits for diameter as well as for minimum breaking strength and maximum pitch are given in Federal Specification for Wire Rope, RR–R—571a.

Wire Rope Strengths: The strength figures shown in the accompanying tables have been obtained by a mathematical derivation based on actual breakage tests of wire rope and represent from 80 to 95 per cent of the total strengths of the individual wires, depending upon the type of rope construction.

Safe Working Loads and Factors of Safety. — The maximum load for which a wire rope is to be used should take into account such associated factors as friction, load caused by bending around each sheave, acceleration and deceleration and, if a long length of rope is to be used for hoisting, the weight of the rope at its maximum

WIRE ROPE

Table 6. Weights and Strengths of Flattened Strand Wire Rope, Preformed and Not Preformed

6 × 25
with fiber core

6 × 30
with fiber core

6 × 27
with fiber core

Diam., Inches	Approx. Weight per Ft., Pounds	Breaking Strength, Tons of 2000 Lbs.		Diam., Inches	Approx. Weight per Ft., Pounds	Breaking Strength, Tons of 2000 Lbs.	
		Impr. Plow Steel	Mild Plow Steel			Impr. Plow Steel	Mild Plow Steel
3/8*	0.25	6.71	...	1 3/8	3.40	85.5	...
1/2*	0.45	11.8	8.94	1 1/2	4.05	101.	...
9/16*	0.57	14.9	11.2	1 5/8	4.75	118.	...
5/8	0.70	18.3	13.9	1 3/4	5.51	136.	...
3/4	1.01	26.2	19.8	2	7.20	176.	...
7/8	1.39	35.4	26.8	2 1/4	9.10	220.	...
1	1.80	46.0	34.8	2 1/2	11.2	269.	...
1 1/8	2.28	57.9	43.8	2 3/4	13.6	321.	...
1 1/4	2.81	71.0	53.7	...	...	...	...

* These sizes in Type B only.
Type H is not in U.S. Simplified Practice Recommendation.
Source: Rope diagrams, Bethlehem Steel Co. All other data, U.S. Simplified Practice Recommendation 198–50.

extension. The condition of the rope — whether new or old, worn or corroded — and type of attachments should also be considered.

Factors of safety for standing rope usually range from 3 to 4; for operating rope, from 5 to 12. Where there is the element of hazard to life or property, higher values are used.

Installing Wire Rope. — The main precaution to be taken in removing and installing wire rope is to avoid kinking which greatly lessens its strength and useful life. Thus, it is preferable when removing wire rope from the reel to have the reel with its axis in a horizontal position and, if possible, mounted so that it will revolve and the wire rope taken off straight. If the rope is in a coil, it should be unwound with the coil in a vertical position as by rolling the coil along the ground. Where a drum is to be used, the rope should be run directly onto it from the reel, taking care to see that it is not bent around the drum in a direction opposite to that on the reel, thus causing it to be subject to reverse bending. On flat or smooth-faced drums it is important that the rope be started from the proper end of the drum. A right lay rope that is being overwound on the drum, that is, it passes over the top of the drum as it is wound on, should be started from the right flange of the drum (looking at the drum from the side that the rope is to come) and a left lay rope from the left flange.

Table 7. Weights and Strengths of Standard Flat Wire Rope, Not Preformed

Flat Wire Rope

This rope consists of a number of 4-strand rope units placed side by side and stitched together with soft steel sewing wire.

Width and Thickness, Inches	No. of Ropes	Approx. Weight per Ft., Pounds	Breaking Strength, Tons of 2000 Lbs.		Width and Thickness, Inches	No. of Ropes	Approx. Weight per Ft., Pounds	Breaking Strength, Tons of 2000 Lbs.	
			Plow Steel	Mild Plow Steel				Plow Steel	Mild Plow Steel
¼ × 1½	7	0.69	16.8	14.6	½ × 4	9	3.16	81.8	71.2
¼ × 2	9	0.88	21.7	18.8	½ × 4½	10	3.82	90.9	79.1
¼ × 2½	11	1.15	26.5	23.0	½ × 5	12	4.16	109.	94.9
¼ × 3	13	1.34	31.3	27.2	½ × 5½	13	4.50	118.	103.
					½ × 6	14	4.85	127.	111.
5⁄16 × 1½	5	0.77	18.5	16.0	½ × 7	16	5.85	145.	126.
5⁄16 × 2	7	1.05	25.8	22.4					
5⁄16 × 2½	9	1.33	33.2	28.8	5⁄8 × 3½	6	3.40	85.8	74.6
5⁄16 × 3	11	1.61	40.5	35.3	5⁄8 × 4	7	3.95	100.	87.1
5⁄16 × 3½	13	1.89	47.9	41.7	5⁄8 × 4½	8	4.50	114.	99.5
5⁄16 × 4	15	2.17	55.3	48.1	5⁄8 × 5	9	5.04	129.	112.
					5⁄8 × 5½	10	5.59	142.	124.
3⁄8 × 2	6	1.25	31.4	27.3	5⁄8 × 6	11	6.14	157.	137.
3⁄8 × 2½	8	1.64	41.8	36.4	5⁄8 × 7	13	7.23	186.	162.
3⁄8 × 3	9	1.84	47.1	40.9	5⁄8 × 8	15	8.32	214.	186.
3⁄8 × 3½	11	2.23	57.5	50.0					
3⁄8 × 4	12	2.44	62.7	54.6	3⁄4 × 5	8	6.50	165.	143.
3⁄8 × 4½	14	2.83	73.2	63.7	3⁄4 × 6	9	7.31	185.	161.
3⁄8 × 5	15	3.03	78.4	68.2	3⁄4 × 7	10	8.13	206.	179.
3⁄8 × 5½	17	3.42	88.9	77.3	3⁄4 × 8	11	9.70	227.	197.
3⁄8 × 6	18	3.63	94.1	81.9					
					7⁄8 × 5	7	7.50	190.	165.
½ × 2½	6	2.13	54.5	47.4	7⁄8 × 6	8	8.56	217.	188.
½ × 3	7	2.47	63.6	55.4	7⁄8 × 7	9	9.63	244.	212.
½ × 3½	8	2.82	72.7	63.3	7⁄8 × 8	10	10.7	271.	236.

Source: Rope diagram, Bethlehem Steel Co.; all data, U.S. Simplified Practice Recommendation 198–50.

When the rope is underwound on the drum, a right lay rope should be started from the left flange and a left lay rope from the right flange. When this is done, the rope will spool evenly and the turns will lie snugly together.

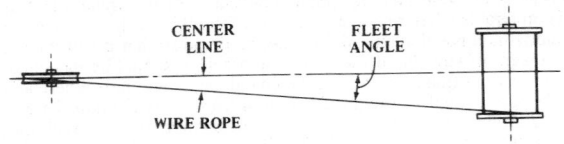

CENTER LINE
FLEET ANGLE
WIRE ROPE

Sheaves and drums should be properly aligned to prevent undue wear. The proper position of the main or lead sheave for the rope as it comes off the drum is governed by what is called the fleet angle or angle between the rope as it stretches from drum

to sheave and an imaginary center-line passing through the center of the sheave groove and a point halfway between the ends of the drum. When the rope is at one end of the drum, this angle should not exceed one and a half to two degrees. With the lead sheave mounted with its groove on this center-line, a safe fleet angle is obtained by allowing 30 feet of lead for each two feet of drum width.

Sheave and Drum Dimensions: Sheaves and drums should be as large as possible to obtain maximum rope life. However, factors such as the need for lightweight equipment for easy transport and use at high speeds, may call for relatively small sheaves with consequent sacrifice in rope life in the interest of over-all economy. No hard and fast rules can be laid down for any particular rope if the utmost in economical performance is to be obtained. Where maximum rope life is of prime importance, the following recommendations of Federal Specification RR–R–571a for minimum sheave or drum diameters D in terms of rope diameter d will be of interest. For 6×7 rope (six strands of 7 wires each) $D = 72d$; for 6×19 rope, $D = 45d$; for 6×25 rope, $D = 45d$; for 6×29 rope, $D = 30d$; for 6×37 rope, $D = 27d$; and for 8×19 rope, $D = 31d$.

Too small a groove for the rope it is to carry will prevent proper seating of the rope in the bottom of the groove and consequently uneven distribution of load on the rope will result. Too large a groove will not give the rope sufficient side support. Federal specifications RR–R–571a recommend that sheave groove diameters be larger than the nominal rope diameters by the following minimum amounts: For ropes of ¼- to ⁵⁄₁₆-inch diameters, ¹⁄₆₄ inch larger; for ⅜- to ¾-inch diameter ropes, ¹⁄₃₂ inch larger; for ¹³⁄₁₆- to 1⅛-inch diameter ropes, ³⁄₆₄ inch larger; for 1³⁄₁₆- to 1½-inch ropes, ¹⁄₁₆ inch larger; for 1⁹⁄₁₆- to 2¼-inch ropes, ³⁄₃₂ inch larger; and for 2⁵⁄₁₆ and larger diameter ropes, ⅛ inch larger. For new or regrooved sheaves these values should be doubled; in other words for ¼- to ⁵⁄₁₆-inch diameter ropes, the groove diameter should be ¹⁄₃₂ inch larger, etc.

Drum or Reel Capacity: The length of wire rope, in feet, that can be spooled onto a drum or reel, is computed by the following formula, where

Table 8. Factors K Used in Calculating Wire Rope Drum and Reel Capacities

Rope Diam., In.	Factor K	Rope Diam., In.	Factor K	Rope Diam., In.	Factor K
³⁄₃₂	23.4	½	0.925	1⅜	0.127
⅛	13.6	⁹⁄₁₆	0.741	1½	0.107
⁹⁄₆₄	10.8	⅝	0.607	1⅝	0.0886
⁵⁄₃₂	8.72	¹¹⁄₁₆	0.506	1¾	0.0770
³⁄₁₆	6.14	¾	0.428	1⅞	0.0675
⁷⁄₃₂	4.59	¹³⁄₁₆	0.354	2	0.0597
¼	3.29	⅞	0.308	2⅛	0.0532
⁵⁄₁₆	2.21	1	0.239	2¼	0.0476
⅜	1.58	1⅛	0.191	2⅜	0.0419
⁷⁄₁₆	1.19	1¼	0.152	2½	0.0380

Note: The values of "K" allow for normal oversize of ropes, and the fact that it is practically impossible to "thread-wind" ropes of small diameter. However, the formula is based on uniform rope winding and will not give correct figures if rope is wound non-uniformly on the reel. The amount of tension applied when spooling the rope will also affect the length. The formula is based on the same number of wraps of rope in each layer, which is strictly correct, but which does not result in appreciable error unless the width (B) of the reel is quite small compared with the flange diameter (H).

A = depth of rope space on drum, inches: $A = (H - D - 2Y) \div 2$
B = width between drum flanges, inches
D = diameter of drum barrel, inches
H = diameter of drum flanges, inches
K = factor from Table 8 for size of line selected
Y = depth not filled on drum or reel where winding is to be less than full capacity
f24L = length of wire rope on drum or reel, feet.

$$L = (A + D) \times A \times B \times K$$

Example: Find the length in feet of %6-inch diameter rope required to fill a drum having the following dimensions: B = 24 inches, D = 18 inches, H = 30 inches,

$$A = (30 - 18 - 0) \div 2 = 6 \text{ inches}$$
$$L = (6 + 18) \times 6 \times 24 \times 0.741 = 2560.0 \text{ or } 2560 \text{ feet}$$

The above formula and factors K allow for normal oversize of ropes but will not give correct figures if rope is wound non-uniformly on the reel.

Load Capacity of Sheave or Drum: To avoid excessive wear and groove corrugation, the radial pressure exerted by the wire rope on the sheave or drum must be kept within certain maximum limits. The radial pressure of the rope is a function of the rope tension, rope diameter, and tread diameter of the sheave and can be determined by the following equation:

$$P = \frac{2T}{D \times d}$$

where
P = Radial pressure in pounds per square inch (see Table 9)
T = Rope tension in pounds
D = Tread diameter of sheave or drum in inches
d = Rope diameter in inches

According to the Bethlehem Steel Co. the radial pressures shown in Table 9 are recommended as maximums according to the material of which the sheave or drum is made.

Table 9. Maximum Radial Pressures for Drums and Sheaves

	Drum or Sheave Material		
	Cast Iron	Cast Steel	Manganese Steel*
Type of Wire Rope	Recommended Maximum Radial Pressures, Pounds per Square Inch		
6 × 7	300†	550†	1500†
6 × 19	500†	900†	2500†
6 × 37	600	1075	3000
6 × 8 Flattened Strand	450	850	2200
6 × 25 Flattened Strand	800	1450	4000
6 × 30 Flattened Strand	800	1450	4000

* 11 to 13 per cent manganese.
† These values are for regular lay rope. For Lang lay rope these values may be increased by 15 per cent.

Rope Loads due to Bending: When a wire rope is bent around a sheave, the resulting bending stress s_b in the outer wire, and equivalent bending load P_b (amount that direct tension load on rope is increased by bending) may be computed by the following formulas: $s_b = Ed_w \div D$; $P_b = s_b A$, where $A = d^2 Q$. E is the modulus of elasticity of the wire rope (this varies with the type and condition of rope from 10,000,000 to 14,000,000. An average value of 12,000,000 is frequently used.) d is the diameter of the wire rope, d_w is the diameter of the component wire (for 6×7 rope, $d_w = 0.106d$; for 6×19 rope, $0.063d$; for 6×37 rope, $0.045d$; and for 8×19 rope, $d_w = 0.050d$), D is the pitch diameter of the sheave in inches, A is the metal cross-sectional area of the rope, and Q is a constant, values for which are: 6×7 (Fiber Core) rope, 0.380; 6×7 (IWRC or WSC), 0.437; 6×19 (Fiber Core), 0.405; 6×19 (IWRC or WSC), 0.475; 6×37 (Fiber Core), 0.400; 6×37 (IWRC), 0.470; 8×19 (Fiber Core), 0.370; and Flattened Strand Rope, 0.440.

Example: Find the bending stress and equivalent bending load due to the bending of a 6×19 (Fiber Core) wire rope of ½-inch diameter around a 24-inch pitch diameter sheave.

$$d_w = 0.063 \times 0.5 = 0.0315 \text{ in.}; \quad A = 0.5^2 \times 0.405 = 0.101 \text{ sq. in.}$$
$$s_b = 12,000,000 \times 0.0315 \div 24 = 15,750 \text{ lbs. per sq. in.}$$
$$P_b = 15,750 \times 0.101 = 1590 \text{ lbs.}$$

Cutting and Seizing of Wire Rope — Wire rope can be cut with mechanical wire rope shears, an abrasive wheel, an electric resistance cutter (used for ropes of smaller diameter only), or an acetylene torch. This last method fuses the ends of the wires in the strands. It is important that the rope be seized on either side of where the cut is to be made. Any annealed low carbon steel wire may be used for seizing, the recommended sizes being as follows: For a wire rope diameter of from ¼- to ¹³⁄₁₆-inch, use a seizing wire of .054-inch (No. 17 Steel Wire Gage); for a rope of 1- to 1⅝-inch diameter, use a .105-inch wire (No. 12); and for rope of 1¾- to 3½-inch diameter, use a .135-inch wire (No. 10). Except for preformed wire ropes, a minimum of two seizings on either side of a cut is recommended. Four seizings should be used on either side of a cut for Lang lay rope, a rope with a steel core, or a non-spinning type of rope.

The following method of seizing is given in Federal specification for wire rope, RR-R-571a. Lay one end of the seizing wire in the groove between two strands of wire rope and wrap the other end tightly in a close helix over the portion in the groove. A seizing iron (round bar ½ to ⅝ inch in diameter by 18 inches long) should be used to wrap the seizing tightly. This bar is placed at right angles to the rope next to the first turn or two of the seizing wire. The seizing wire is brought around the back of the seizing iron so that it now can be wrapped loosely around the wire rope in the opposite direction to that of the seizing coil. As the seizing iron is now rotated around the rope it will carry the seizing wire snugly and tightly into place. When completed, both ends of the seizing should be twisted together tightly.

Maintenance of Wire Rope. — Heavy abrasion, overloading, and bending around sheaves or drums which are too small in diameter are the principal reasons for the rapid deterioration of wire rope. Wire rope in use should be inspected periodically for evidence of wear and damage by corrosion. Such inspection should take place at progressively shorter intervals over the useful life of the rope as wear tends to accelerate with use. Where wear is rapid, the outside of a wire rope will show flattened surfaces in a short time.

If there is any hazard involved in the use of the rope, it may be prudent to estimate the remaining strength and service life. This should be done for the weakest point where the most wear or largest number of broken wires are in evidence. One way to

arrive at a conclusion is to set an arbitrary number of broken wires in a given strand as an indication that the rope should be removed from service and an ultimate strength test run on the worn sample. The arbitrary figure can then be revised and rechecked until a practical working formula is arrived at. A piece of waste rubbed along the wire rope will help to reveal broken wires. The effects of corrosion are not easy to detect since the exterior wires may appear to be only slightly rusty, while the damaging effects of corrosion may be confined to the hidden inner wires where it cannot be seen. To prevent damage by corrosion, the rope should be kept well lubricated. In some cases zinc coated wire rope may be indicated.

Periodic cleaning of wire rope by using a stiff brush and kerosene or with compressed air or live steam and relubricating will help to lengthen rope life and reduce abrasion and wear on sheaves and drums. Before storing after use, wire rope should be cleaned and lubricated.

Lubrication of Wire Rope. — Although wire rope is thoroughly lubricated during manufacture to protect it against corrosion and to reduce friction and wear, this lubrication should be supplemented from time to time. Special lubricants are supplied by wire rope manufacturers. These lubricants vary somewhat with the type of rope application and operating condition. Where the preferred lubricant can not be obtained from the wire rope manufacturer, an adhesive type of lubricant similar to that used for open gearing will often be found suitable. At normal temperatures, some wire rope lubricants may be practically solid and will require thinning before application. Thinning may be done by heating to 160 to 200 degrees F. or by diluting with gasoline or some other fluid which will allow the lubricant to penetrate the rope. The lubricant may be painted on the rope or the rope may be passed through a box or tank filled with the lubricant.

Replacement of Wire Rope. — When an old wire rope is to be replaced, all drums and sheaves should be examined for wear. All evidence of scoring or imprinting of grooves from previous use should be removed and sheaves with flat spots, defective bearings, and broken flanges, should be repaired or replaced. It will frequently be found that the area of maximum wear is located relatively near one end of the rope. By cutting off that portion, the remainder of the rope may be salvaged for continued use. Sometimes the life of a rope can be increased by simply changing it end for end at about one-half the estimated normal life. The worn sections will then no longer come at points which cause the greatest wear.

Wire Rope Slings and Fittings. — A few of the simpler sling arrangements or hitches as they are called, are shown in the accompanying illustration. Normally 6 × 19 Class wire rope is recommended where a diameter in the ¼-inch to 1⅛-inch range is to be used and 6 × 37 Class wire rope where a diameter in the 1¼-inch and larger range is to be used. However, in some cases the 6 × 19 Class may be used even in the larger sizes if resistance to abrasion is of primary importance and the 6 × 37 Class in the smaller sizes if greater flexibility is desired.

The *straight lift hitch,* shown at A, is a straight connector between crane hook and load.

The *basket hitch* may be used with two hooks so that the sides are vertical as shown at B or with a single hook with sides at various angles with the vertical as shown at C, D, and E. As the angle with the vertical increases, a greater tension is placed on the rope so that for any given load, a sling of greater lifting capacity must be used.

The *choker hitch,* shown at F, is widely used for lifting bundles of items such as bars, poles, pipe, and similar objects. The choker hitch holds these items firmly

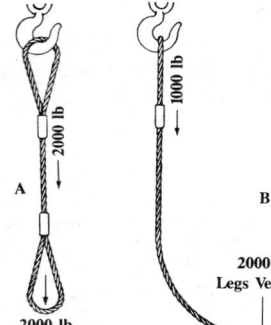

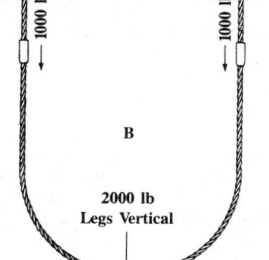

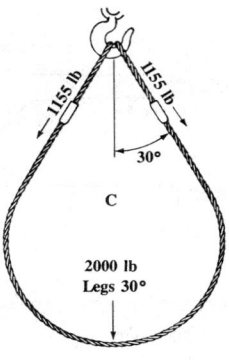

A

**2000 lb
Straight Leg Vertical**

STRAIGHT LIFT

One leg vertical. Load capacity is 100 pct of a single rope.

B

**2000 lb
Legs Vertical**

BASKET HITCH

Two legs vertical. Load capacity is 200 pct of the single rope in the Straight Lift Hitch (A).

C

**2000 lb
Legs 30°**

BASKET HITCH

Two legs at 30 deg with the vertical. Load capacity is 174 pct of the single rope in the Straight Lift Hitch (A).

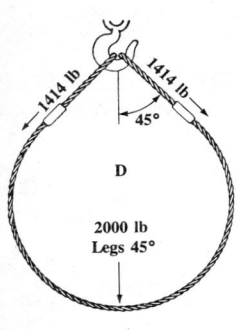

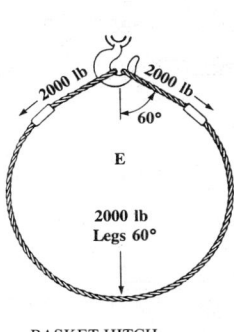

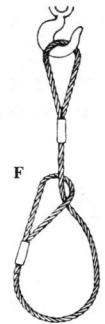

D

**2000 lb
Legs 45°**

BASKET HITCH

Two legs at 45 deg with the vertical. Load capacity is 141 pct of the single rope in the Straight Lift Hitch (A).

E

**2000 lb
Legs 60°**

BASKET HITCH

Two legs at 60 deg with the vertical. Load capacity is 100 pct of the single rope in the Straight Lift Hitch (A).

F

CHOKER HITCH

One leg vertical, with slip-through loop. Rated capacity is 75 pct of the single rope in the Straight Lift Hitch (A).

Wire Rope Slings and Fittings

but the load must be balanced so that it rides safely. Since additional stress is imposed in the rope due to the choking action, the capacity of this type of hitch is 25 per cent less than that of the comparable straight lift. If two choker hitches are used at an angle, these angles must also be taken into consideration as in the basket hitches.

Industrial Types

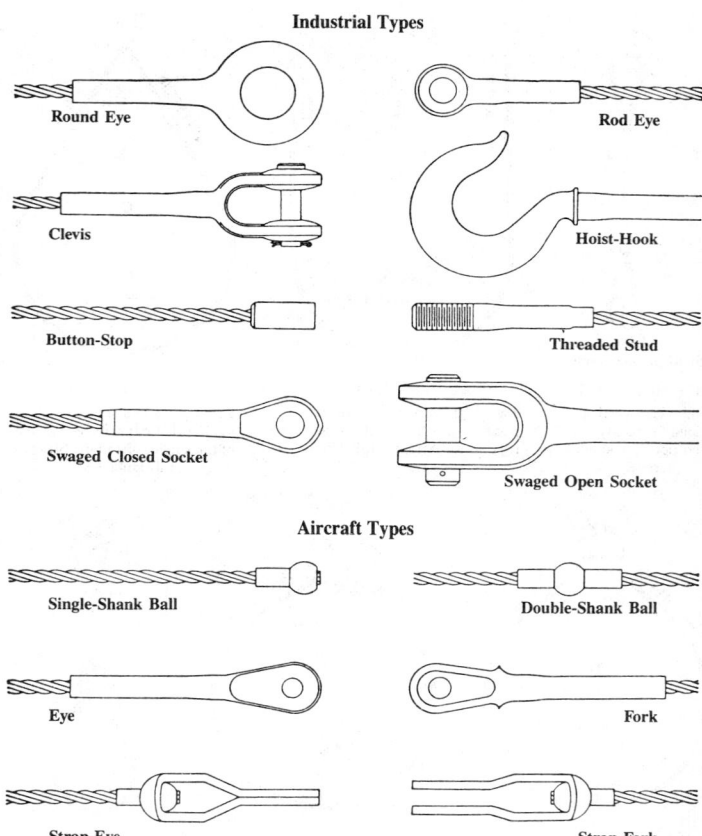

Aircraft Types

Wire Rope Fittings

Wire Rope Fittings. — A wide variety of swaged fittings are available for use with wire rope. A number of industrial and aircraft types are shown in the accompanying illustration. Swaged fittings on wire rope have an efficiency (ability to hold the wire rope) of approximately 100 per cent of the catalogue rope strength. These fittings are applied to the end or body of the wire rope by the application of high pressure through special dies that causes the steel to "flow" around the wires and strands of the rope to form a union that is as strong as the rope itself. The more commonly used type of swaged fittings range from ⅛- to ⅝-inch diameter sizes in the industrial types and from the ¹⁄₁₆- to ⅝-inch sizes in the aircraft types. These fittings are furnished attached to the wire strand, rope, or cable.

Applying Clips and Attaching Sockets. — In attaching U-bolt clips for fastening the end of a wire rope to form a loop, it is essential that the saddle or base of the clip bears against the longer or "live" end of the rope loop and the U-bolt against the shorter or "dead" end. The "U" of the clips should never bear against the live end of the rope because the rope may be cut or kinked. A wire-rope thimble should be used in the loop eye of the rope to prevent kinking when rope clips are used. The strength of a clip fastening is usually less than 80 percent of the strength of the rope. Table 10 gives the proper size, number, and spacing for each size of wire rope.

Table 10. Clips Required for Fastening Wire Rope End

Rope Diam., In.	U-Bolt Diam., In.	Min. No. of Clips	Clip Spacing, In.	Rope Diam., In.	U-Bolt Diam., In.	Min. No. of Clips	Clip Spacing, In.
$3/16$	$11/32$	2	3	$1\frac{1}{8}$	$1\frac{1}{4}$	5	$9\frac{3}{4}$
$1/4$	$7/16$	2	$3\frac{1}{4}$	$1\frac{1}{4}$	$1\frac{7}{16}$	5	$10\frac{3}{4}$
$5/16$	$1/2$	2	$3\frac{1}{4}$	$1\frac{3}{8}$	$1\frac{1}{2}$	6	$11\frac{1}{2}$
$3/8$	$9/16$	2	4	$1\frac{1}{2}$	$1^{23}/32$	6	$12\frac{1}{2}$
$7/16$	$5/8$	2	$4\frac{1}{2}$	$1\frac{5}{8}$	$1\frac{3}{4}$	6	$13\frac{1}{4}$
$1/2$	$11/16$	3	5	$1\frac{3}{4}$	$1^{15}/16$	7	$14\frac{1}{2}$
$5/8$	$3/4$	3	$5\frac{3}{4}$	2	$2\frac{1}{8}$	8	$16\frac{1}{2}$
$3/4$	$7/8$	4	$6\frac{3}{4}$	$2\frac{1}{4}$	$2\frac{5}{8}$	8	$16\frac{1}{2}$
$7/8$	1	4	8	$2\frac{1}{2}$	$2\frac{7}{8}$	8	$17\frac{3}{4}$
1	$1\frac{1}{8}$	4	$8\frac{3}{4}$	...	...	...	...

In attaching commercial sockets of forged steel to wire rope ends, the following procedure is recommended. The wire rope is seized at the end and another seizing is applied at a distance from the end equal to the length of the basket of the socket. As explained in a previous section, soft iron wire is used and particularly for the larger sizes of wire rope, it is important to use a seizing iron to secure a tight winding. For large ropes, the seizing should be several inches long.

The end seizing is now removed and the strands are separated so that the fiber core can be cut back to the next seizing. The individual wires are then untwisted and "broomed out" and for the distance they are to be inserted in the socket are carefully cleaned with benzine, naphtha, or unleaded gasoline. The wires are then dipped into commercial muriatic (hydrochloric) acid and left (usually one to three minutes) until the wires are bright and clean or, if zinc coated, until the zinc is removed. After cleaning, the wires are dipped into a hot soda solution (1 pound of soda to 4 gallons of water at 175 degrees F. minimum) to neutralize the acid. The rope is now placed in a vise. A temporary seizing is used to hold the wire ends together until the socket is placed over the rope end. The temporary seizing is then removed and the socket located so that the ends of the wires are about even with the upper end of the basket. The opening around the rope at the bottom of the socket is now sealed with putty.

A special high grade pure zinc is used to fill the socket. Babbit metal should not be used as it will not hold properly. For proper fluidity and penetration, the zinc is heated to a temperature in the 830- to 900-degree F. range. If a pyrometer is not available to measure the temperature of the molten zinc, a dry soft pine stick dipped into the zinc and quickly withdrawn will show only a slight discoloration and no zinc will adhere to it. If the wood chars, the zinc is too hot. The socket is now permitted to cool and the resulting joint is then ready for use. When properly prepared, its strength should be approximately equal to that of the rope itself.

Rated Capacities for Improved Plow Steel Wire Rope and Wire Rope Slings (in tons of 2,000 lbs) — Independent Wire Rope Core

Rope Diameter (in.)	Vertical A	Vertical B	Vertical C	Choker A	Choker B	Choker C	60° Bridle A	60° Bridle B	60° Bridle C	45° Bridle A	45° Bridle B	45° Bridle C	30° Bridle A	30° Bridle B	30° Bridle C
Single Leg, 6 × 19 Wire Rope															
1/4	0.59	0.56	0.53	0.44	0.42	0.40									
3/8	1.3	1.2	1.1	0.98	0.93	0.86									
1/2	2.3	2.2	2.0	1.7	1.6	1.5									
5/8	3.6	3.4	3.0	2.7	2.5	2.2									
3/4	5.1	4.9	4.2	3.8	3.6	3.1									
7/8	6.9	6.6	5.5	5.2	4.9	4.1									
1	9.0	8.5	7.2	6.7	6.4	5.4									
1 1/8	11	10	9.0	8.5	7.8	6.8									
Single Leg, 6 × 37 Wire Rope															
1 1/4	13	12	10	9.9	9.2	7.9									
1 3/8	16	15	13	12	11	9.6									
1 1/2	19	17	15	14	13	11									
1 3/4	26	24	20	19	18	15									
2	33	30	26	25	23	20									
2 1/4	41	38	33	31	29	25									
Two-Leg Bridle or Basket Hitch, 6 × 19 Wire Rope Sling															
1/4	1.2	1.1	1.0				1.0	0.97	0.92	0.83	0.79	0.75	0.59	0.56	0.53
3/8	2.0	2.5	2.3				2.3	2.1	2.0	1.8	1.8	1.8	1.3	1.2	1.1
1/2	4.0	4.4	3.9				4.0	3.6	3.4	3.2	3.1	2.8	2.3	2.2	2.0
5/8	7.2	6.6	6.0				6.2	5.9	5.2	5.1	4.8	4.2	3.6	3.4	3.0
3/4	10	9.7	8.4				8.9	8.4	7.3	7.2	6.9	5.9	5.1	4.9	4.2
7/8	14	13	11				12	11	9.6	9.8	9.3	7.8	6.9	6.6	5.5
1	18	17	14				15	15	16	13	12	10	9.0	8.5	7.2
1 1/8	23	21	18				19	18	16	16	15	13	11	10	9.0
Two-Leg Bridle or Basket Hitch, 6 × 37 Wire Rope Sling															
1 1/4	26	24	21				23	21	18	19	17	15	13	12	13
1 3/8	32	29	25				28	25	22	22	21	18	16	15	15
1 1/2	38	35	30				33	30	26	27	25	21	19	17	15
1 3/4	51	47	41				44	41	35	36	33	29	26	24	20
2	66	61	53				57	53	46	47	43	37	33	30	26
2 1/4	83	76	66				72	66	67	58	54	47	41	38	33

Rated Capacities for Improved Plow Steel Wire Rope and Wire Rope Slings (in tons of 2,000 lbs) — Fiber Core

Rope Diameter (in.)	Vertical			Choker			60° Bridle			45° Bridle			30° Bridle		
	A	B	C	A	B	C	A	B	C	A	B	C	A	B	C
Single Leg, 6 × 19 Wire Rope															
1/4	0.55	0.51	0.49	0.41	0.38	0.37									
3/8	1.2	1.1	1.1	0.91	0.85	0.80									
1/2	2.1	2.0	1.8	1.6	1.5	1.4									
5/8	3.3	3.1	2.8	2.5	2.3	2.1									
3/4	4.8	4.4	3.9	3.6	3.3	2.9									
7/8	6.4	5.9	5.1	4.8	4.5	3.9									
1	8.4	7.7	6.7	6.3	5.8	5.0									
1 1/8	10	9.5	8.4	7.9	7.1	6.3									
Single Leg, 6 × 37 Wire Rope															
1 1/4	12	11	9.8	9.2	8.3	7.4									
1 3/8	15	13	12	11	10	8.9									
1 1/2	17	16	14	13	12	10									
1 3/4	24	21	19	18	16	14									
2	31	28	25	23	21	18									
Two-Leg Bridle or Basket Hitch, 6 × 19 Wire Rope Sling															
1/4							0.95	0.88	0.85	0.77	0.72	0.70	0.55	0.51	0.49
3/8							2.1	1.9	1.8	1.7	1.6	1.5	1.2	1.1	1.1
1/2							3.7	3.4	3.2	3.0	2.8	2.6	2.1	2.0	1.8
5/8							5.7	5.3	4.8	4.7	4.4	4.0	3.3	3.1	2.8
3/4							8.2	7.6	6.8	6.7	6.2	5.5	4.8	4.4	3.9
7/8							11	10	8.9	9.1	8.4	7.3	6.4	5.9	5.1
1							14	13	11	12	11	9.4	8.4	7.7	6.7
1 1/8							18	16	14	15	13	12	10	9.5	8.4
Two-Leg Bridle or Basket Hitch, 6 × 37 Wire Rope Sling															
1 1/4							21	19	17	17	16	14	12	11	9.8
1 3/8							26	23	20	21	19	17	15	13	12
1 1/2							30	27	24	25	22	20	17	16	14
1 3/4							41	37	33	34	30	27	24	21	19
2							53	43	43	43	39	35	31	26	25

A — socket or swaged terminal attachment; B — mechanical sleeve attachment; C — hand-tucked splice attachment.

Data taken from *Longshoring Industry*, OSHA Safety and Health Standards Digest. OSHA 2232. 1985.

CRANE CHAIN AND HOOKS

Material for Crane Chains.—The best material for crane and hoisting chains is a good grade of wrought iron, in which the percentage of phosphorus, sulfur, silicon, and other impurities is comparatively low. The tensile strength of the best grades of wrought iron does not exceed 46,000 pounds per square inch, whereas mild steel with about 0.15 per cent carbon has a tensile strength nearly double this amount. The ductility and toughness of wrought iron, however, is greater than that of ordinary commercial steel, and for this reason it is preferable for chains subjected to heavy intermittent strains, because wrought iron will always give warning by bending or stretching, before breaking. Another important reason for using wrought iron in preference to steel is that a perfect weld can be effected more easily. Heat-treated alloy steel is also widely used for chains. This steel contains carbon, 0.30 per cent, max; phosphorus, 0.045 per cent, max; and sulfur, 0.045 per cent, max. The selection and amounts of alloying elements are left to the individual manufacturers.

Strength of Chains.—When calculating the strength of chains, it should be observed that the strength of a link subjected to tensile stresses is not equal to twice the strength of an iron bar of the same diameter as the link stock, but is a certain amount less, owing to the bending action caused by the manner in which the load is applied to the link. The strength is also reduced somewhat by the weld. The following empirical formula is commonly used for calculating the breaking load, in pounds, of wrought-iron crane chains:

$$W = 54,000\ D^2$$

in which W = breaking load in pounds and D = diameter of bar (in inches) from which links are made. The working load for chains should not exceed one-third the value of W, and, in many cases, it is one-fourth or one-fifth of the breaking load. When a chain is wound around a casting and severe bending stresses are introduced, a greater factor of safety should be used.

Care of Hoisting and Crane Chains.—Chains used for hoisting heavy loads are subject to deterioration, both apparent and invisible. The links wear, and repeated loading causes localized deformations to form cracks which spread until the links fail. Chain wear can be reduced by occasional lubrication. The life of a wrought-iron chain can be prolonged by frequent annealing or normalizing unless it has been so highly or frequently stressed that small cracks have formed. If this be the case, annealing or normalizing will not "heal" the material, and the links will eventually fracture. To anneal a wrought-iron chain, heat it to cherry-red and allow it to cool slowly. This should be done every six months, and oftener if the chain is subjected to unusually

Maximum Allowable Wear at Any Point of Link

Chain Size (in.)	Maximum Allowable Wear (in.)	Chain Size (in.)	Maximum Allowable Wear (in.)
1/4 (9/32)	3/64	1	3/16
3/8	5/64	1 1/8	7/32
1/2	7/64	1 1/4	1/4
5/8	9/64	1 3/8	3/32
3/4	5/32	1 1/2	5/16
7/8	11/64	1 3/4	11/32

From *Longshoring Industry*, OSHA 2232, 1985.

severe service. Chains should be examined periodically for twists, as a twisted chain will wear rapidly. Any links which have worn excessively should be replaced with new ones, so that every link will do its full share of work during the life of the chain, without exceeding the limit of safety. Chains for hoisting purposes should be made with short links, so that they will wrap closely around the sheaves or drums without bending. The diameter of the winding drums should be not less than 25 or 30 times the diameter of the iron used for the links. The accompanying table lists the maximum allowable wear for various sizes of chains.

Safe Loads for Ropes and Chains. — Safe loads recommended for wire rope or chain slings depend not only upon the strength of the sling but upon the method of applying it to the load, as shown by the accompanying table giving safe loads as prepared by OSHA. The loads recommended in this table are more conservative than those usually specified, in order to provide ample allowance for some unobserved weakness in the sling, or the possibility of excessive strains due to misjudgment or accident.

The working load limit is defined as the maximum load in pounds that should ever be applied to chain, when the chain is new or "in as new" condition, and when the load is uniformly applied in direct tension to a straight length of chain. This limit is also affected by the number of chains used and their configuration. The accompanying table shows the working load limit for various configurations of heat-treated alloy steel chain using a 4 to 1 design factor, which conforms to ISO practice.

Working Load Limit for Heat-Treated Alloy Steel Chain, pounds

Chain Size (in.)	Single Leg	Double Leg				Triple and Quad Leg		
	90°	60°	45°	30°	60°	45°	30°
¼	3,600	6,200	5,050	3,600	9,300	7,600	5,400
⅜	6,400	11,000	9,000	6,400	16,550	13,500	9,500
½	11,400	19,700	16,100	11,400	29,600	24,200	17,100
⅝	17,800	30,800	25,150	17,800	46,250	37,750	26,700
¾	25,650	44,400	36,250	25,650	66,650	54,400	38,450
⅞	34,900	60,400	49,300	34,900	90,650	74,000	52,350

Source: The Crosby Group.

Protection from Sharp Corners: When the load to be lifted has sharp corners or edges, as is often the case with castings, and with structural steel and other similar objects, pads or wooden protective pieces should be applied at these corners, to prevent the slings from being abraded or otherwise damaged where they come in contact with the load. This is especially important when the slings consist of wire cable or fiber rope, although it should also be done even when they are made of chain. Wooden corner-pieces are often provided for use in hoisting loads with sharp angles. If pads of burlap or other soft material are used, they should be thick and heavy enough to sustain the pressure, and distribute it over a considerable area, instead of allowing it to be concentrated directly at the edges of the part to be lifted.

Loads Lifted by Crane Chains. — To find the approximate weight a chain will lift when rove as a tackle, multiply the safe load given in the table "Close-link Hoisting, Sling and Crane Chain" by the number of parts or chains at the movable block, and subtract one-quarter for frictional resistance. To find the size of chain required for lifting a given weight, divide the weight by the number of chains at the movable block, and add one-third for friction; next find in the column headed "Average Safe Working Load" the corresponding load, and then the corresponding size of chain in the column headed "Size." In case of heavy chain or where chain is unusually long, the weight of the chain itself should also be considered.

Safe Working Loads in Pounds for Manila Rope and Chains

Kind of Rope or Chain	Diameter of Rope, or of Rod or Bar for Chain Links, Inch	Rope or Chain Vertical	Sling at 60°	Sling at 45°	Sling at 30°
Manila Rope	1/4	120	204	170	120
	5/16	200	346	282	200
	3/8	270	467	380	270
	7/16	350	605	493	350
	15/32	450	775	635	450
	1/2	530	915	798	530
	9/16	690	1190	973	690
	5/8	880	1520	1240	880
	3/4	1080	1870	1520	1080
	13/16	1300	2250	1830	1300
	7/8	1540	2660	2170	1540
	1	1800	3120	2540	1800
	1 1/16	2000	3400	2800	2000
	1 1/8	2400	4200	3400	2400
	1 1/4	2700	4600	3800	2700
	1 5/16	3000	5200	4200	3000
	1 1/2	3600	6200	5000	3600
	1 5/8	4500	7800	6400	4500
	1 3/4	5200	9000	7400	5200
	2	6200	10,800	8800	6200
	2 1/8	7200	12,400	10,200	7200
Crane Chain (Wrought Iron)	1/4*	1060	1835	1500	1060
	5/16*	1655	2865	2340	1655
	3/8*	2385	4200	3370	2385
	7/16*	3250	5600	4600	3250
	1/2	4200	7400	6000	4200
	9/16*	5400	9200	7600	5400
	5/8	6600	11,400	9400	6600
	3/4	9600	16,600	13,400	9600
	7/8	13,000	22,400	18,400	13,000
	1	17,000	29,400	24,000	17,000
	1 1/8	20,000	34,600	28,400	20,000
	1 1/4	24,800	42,600	35,000	24,800
	1 3/8	30,000	51,800	42,200	30,000
	1 1/2	35,600	61,600	50,400	35,600
	1 5/8	41,800	72,400	59,000	41,800
	1 3/4	48,400	84,000	68,600	48,400
	1 7/8	55,200	95,800	78,200	55,200
	2	63,200	109,600	89,600	63,200
Crane Chain (Alloy Steel)	1/4	3240	5640	4540	3240
	3/8	6600	11,400	9300	6600
	1/2	11,240	19,500	15,800	11,240
	5/8	16,500	28,500	23,300	16,500
	3/4	23,000	39,800	32,400	23,000
	7/8	28,600	49,800	40,600	28,600
	1	38,600	67,000	54,600	38,600
	1 1/8	44,400	77,000	63,000	44,400
	1 1/4	57,400	99,400	81,000	57,400
	1 3/8	67,000	116,000	94,000	67,000
	1 1/2	79,400	137,000	112,000	79,400
	1 5/8	85,000	147,000	119,000	85,000
	1 3/4	95,800	163,000	124,000	95,800

Data taken from *Longshoring Industry*, OSHA Safety and Health Standards Digest, OSHA 2232, 1985.
* These sizes of wrought chain are no longer manufactured in the United States.

Strength of Nylon and Double Braided Nylon Rope

Diam. (in.)	Circumference (in.)	Weight of 100 feet of Rope* (lb)	New Rope Tensile Strength† (lb)	Working Load‡ (lb)	Diam. (in.)	Circumference (in.)	Weight of 100 feet of Rope* (lb)	New Rope Tensile Strength† (lb)	Working Load‡ (lb)
				NYLON ROPE					
3/16	5/8	1.00	900	75	1 5/16	4	45.0	38,800	4,320
1/4	3/4	1.50	1,490	124	1 1/2	4 1/2	55.0	47,800	5,320
5/16	1	2.50	2,300	192	1 5/8	5	66.5	58,500	6,500
3/8	1 1/8	3.50	3,340	278	1 3/4	5 1/2	83.0	70,000	7,800
7/16	1 1/4	5.00	4,500	410	2	6	95.0	83,000	9,200
1/2	1 1/2	6.50	5,750	525	2 1/8	6 1/2	109	95,500	10,600
9/16	1 3/4	8.15	7,200	720	2 1/4	7	129	113,000	12,600
5/8	2	10.5	9,350	935	2 1/2	7 1/2	149	126,000	14,000
3/4	2 1/4	14.5	12,800	1,420	2 5/8	8	168	146,000	16,200
13/16	2 1/2	17.0	15,300	1,700	2 7/8	8 1/2	189	162,000	18,000
7/8	2 3/4	20.0	18,000	2,000	3	9	210	180,000	20,000
1	3	26.4	22,600	2,520	3 1/4	10	264	226,000	25,200
1 1/16	3 1/4	29.0	26,000	2,880	3 1/2	11	312	270,000	30,000
1 1/8	3 1/2	34.0	29,800	3,320	4	12	380	324,000	36,000
1 1/4	3 3/4	40.0	33,800	3,760	...	...	...	...	...
			DOUBLE BRAIDED NYLON ROPE (NYLON COVER — NYLON CORE)						
1/4	3/4	1.56	1,650	150	1 5/16	4	43.1	44,700	5,590
5/16	1	2.44	2,570	234	1 3/8	4 1/4	47.3	49,000	6,130
3/8	1 1/8	3.52	3,700	336	1 1/2	4 1/2	56.3	58,300	7,290
7/16	1 1/2	4.79	5,020	502	1 5/8	5	66.0	68,300	8,540
1/2	1 1/2	6.25	6,550	655	1 3/4	5 1/2	76.6	79,200	9,900
9/16	1 3/4	7.91	8,270	919	2	6	100	103,000	12,900
5/8	2	9.77	10,200	1,130	2 1/8	6 1/2	113	117,000	14,600
3/4	2 1/4	14.1	14,700	1,840	2 1/4	7	127	131,000	18,700
13/16	2 1/2	16.5	17,200	2,150	2 1/2	7 1/2	156	161,000	23,000
7/8	2 3/4	19.1	19,900	2,490	2 5/8	8	172	177,000	25,300
1	3	25.0	26,000	3,250	3	9	225	231,000	33,000
1 1/16	3 1/4	28.2	29,300	3,660	3 1/4	10	264	271,000	38,700
1 1/8	3 1/2	31.6	32,800	4,100	3 1/2	11	329	338,000	48,300
1 1/4	3 3/4	39.1	40,600	5,080	4	12	400	410,000	58,600

Data from Cordage Institute Specifications for nylon rope (three-strand laid and eight-strand plaited, standard construction) and double braided nylon rope.

* Average value is shown. Maximum for nylon rope is 5 per cent higher; tolerance for double braided nylon rope is ±5 per cent.

† Based on tests of new and unused rope of standard construction in accordance with Cordage Institute Standard Test Methods. For double braided nylon rope these values are minimums and are based on a large number of tests by various manufacturers; these values represent results two standard deviations below the mean. The minimum tensile strength is determined by the formula $1057 \times (\text{linear density})^{.995}$.

‡ These values are for rope in good condition with appropriate splices, in noncritical applications, and under normal service conditions. These values should be reduced where life, limb, or valuable property are involved, or for exceptional service conditions such as shock loads or sustained loads.

Close-link Hoisting, Sling and Crane Chain

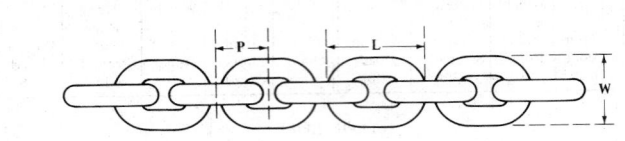

Size	Standard Pitch, P, Inches	Average Weight per Foot in Pounds	Outside Length, L, Inches	Outside Width, W, Inches	Average Safe Working Load in Pounds	Proof Test in Pounds*	Approximate Breaking Load in Pounds
¼	25/32	¾	1⁵/₁₆	⅞	1,200	2,500	5,000
⁵/₁₆	27/32	1	1½	1¹/₁₆	1,700	3,500	7,000
⅜	31/32	1½	1¾	1¼	2,500	5,000	10,000
⁷/₁₆	1⁵/₃₂	2	2¹/₁₆	1⅜	3,500	7,000	14,000
½	1¹¹/₃₂	2½	2⅜	1¹¹/₁₆	4,500	9,000	18,000
⁹/₁₆	1¹⁵/₃₂	3¼	2⅝	1⅞	5,500	11,000	22,000
⅝	1²³/₃₂	4	3	2¹/₁₆	6,700	14,000	27,000
¹¹/₁₆	1¹³/₁₆	5	3¼	2¼	8,100	17,000	32,500
¾	1¹⁵/₁₆	6¼	3½	2½	10,000	20,000	40,000
¹³/₁₆	2¹/₁₆	7	3¾	2¹¹/₁₆	10,500	23,000	42,000
⅞	2³/₁₆	8	4	2⅞	12,000	26,000	48,000
¹⁵/₁₆	2⁷/₁₆	9	4⅜	3¹/₁₆	13,500	29,000	54,000
1	2½	10	4⅝	3¼	15,200	32,000	61,000
1¹/₁₆	2⅝	12	4⅞	3⁵/₁₆	17,200	35,000	69,000
1⅛	2¾	13	5⅛	3¾	19,500	40,000	78,000
1³/₁₆	3¹/₁₆	14½	5⁹/₁₆	3⅞	22,000	46,000	88,000
1¼	3⅛	16	5¾	4⅛	23,700	51,000	95,000
1⁵/₁₆	3⅜	17½	6⅛	4¼	26,000	54,000	104,000
1⅜	3⁹/₁₆	19	6⁷/₁₆	4⁹/₁₆	28,500	58,000	114,000
1⁷/₁₆	3¹¹/₁₆	21½	6¹¹/₁₆	4¾	30,500	62,000	122,000
1½	3⅞	23	7	5	33,500	67,000	134,000
1⁹/₁₆	4	25	7⅜	5⁵/₁₆	35,500	70,500	142,000
1⅝	4¼	28	7¾	5½	38,500	77,000	154,000
1¹¹/₁₆	4½	30	8⅛	5¹¹/₁₆	39,500	79,000	158,000
1¾	4¾	31	8½	5⅞	41,500	83,000	166,000
1¹³/₁₆	5	33	8⅞	6¹/₁₆	44,500	89,000	178,000
1⅞	5¼	35	9¼	6⅜	47,500	95,000	190,000
1¹⁵/₁₆	5½	38	9⅝	6⁹/₁₆	50,500	101,000	202,000
2	5¾	40	10	6¾	54,000	108,000	216,000
2¹/₁₆	6	43	10⅜	6¹⁵/₁₆	57,500	115,000	230,000
2⅛	6¼	47	10¾	7⅛	61,000	122,000	244,000
2³/₁₆	6½	50	11⅛	7⁵/₁₆	64,500	129,000	258,000
2¼	6¾	53	11½	7⅝	68,200	136,500	273,000
2⅜	6⅞	58½	11⅞	8	76,000	152,000	304,000
2½	7	65	12¼	8⅜	84,200	168,500	337,000
2⅝	7⅛	70	12⅝	8¾	90,500	181,000	362,000
2¾	7¼	73	13	9⅛	96,700	193,500	387,000
2⅞	7½	76	13½	9½	103,000	206,000	412,000
3	7¾	86	14	9⅞	109,000	218,000	436,000

* Chains tested to U.S. Government and American Bureau of Shipping requirements.

Dimensions of Crane Hooks

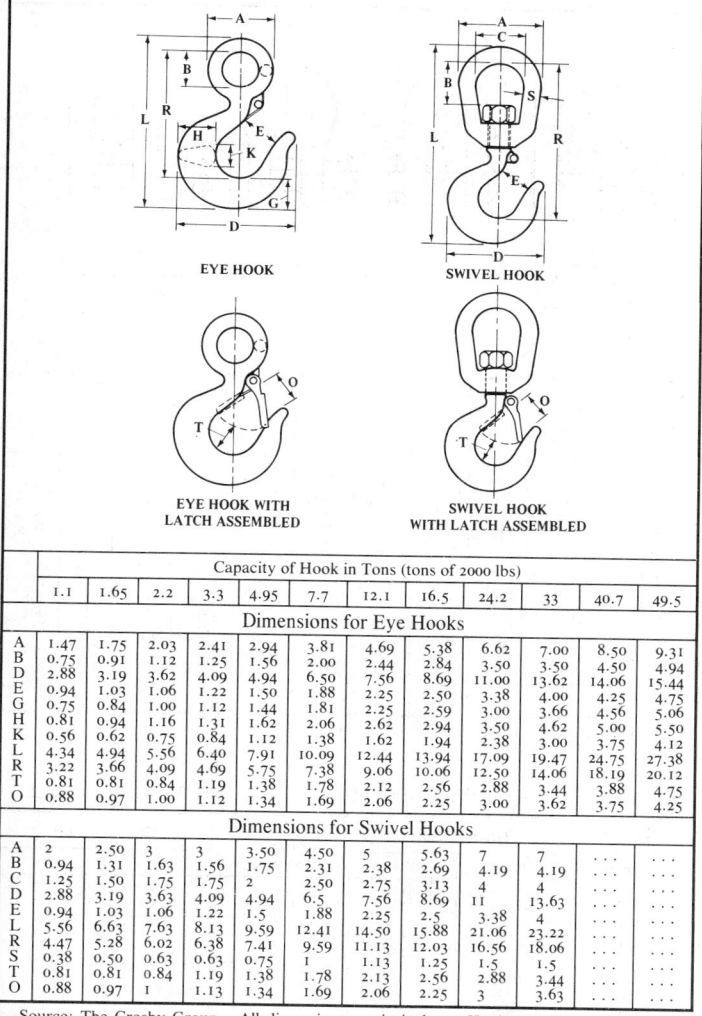

EYE HOOK

SWIVEL HOOK

EYE HOOK WITH
LATCH ASSEMBLED

SWIVEL HOOK
WITH LATCH ASSEMBLED

				Capacity of Hook in Tons (tons of 2000 lbs)							
	1.1	1.65	2.2	3.3	4.95	7.7	12.1	16.5	24.2	33	40.7	49.5

					Dimensions for Eye Hooks							
A	1.47	1.75	2.03	2.41	2.94	3.81	4.69	5.38	6.62	7.00	8.50	9.31
B	0.75	0.91	1.12	1.25	1.56	2.00	2.44	2.84	3.50	3.50	4.50	4.94
D	2.88	3.19	3.62	4.09	4.94	6.50	7.56	8.69	11.00	13.62	14.06	15.44
E	0.94	1.03	1.06	1.22	1.50	1.88	2.25	2.50	3.38	4.00	4.25	4.75
G	0.75	0.84	1.00	1.12	1.44	1.81	2.25	2.59	3.00	3.66	4.56	5.06
H	0.81	0.94	1.16	1.31	1.62	2.06	2.62	2.94	3.50	4.62	5.00	5.50
K	0.56	0.62	0.75	0.84	1.12	1.38	1.62	1.94	2.38	3.00	3.75	4.12
L	4.34	4.94	5.56	6.40	7.91	10.09	12.44	13.94	17.09	19.47	24.75	27.38
R	3.22	3.66	4.09	4.69	5.75	7.38	9.06	10.06	12.50	14.06	18.19	20.12
T	0.81	0.81	0.84	1.19	1.38	1.78	2.12	2.56	2.88	3.44	3.88	4.75
O	0.88	0.97	1.00	1.12	1.34	1.69	2.06	2.25	3.00	3.62	3.75	4.25

					Dimensions for Swivel Hooks							
A	2	2.50	3	3	3.50	4.50	5	5.63	7	7	. . .	. . .
B	0.94	1.31	1.63	1.56	1.75	2.31	2.38	2.69	4.19	4.19	. . .	. . .
C	1.25	1.50	1.75	1.75	2	2.50	2.75	3.13	4	4	. . .	. . .
D	2.88	3.19	3.63	4.09	4.94	6.5	7.56	8.69	11	13.63	. . .	. . .
E	0.94	1.03	1.06	1.22	1.5	1.88	2.25	2.5	3.38	4	. . .	. . .
L	5.56	6.63	7.63	8.13	9.59	12.41	14.50	15.88	21.06	23.22	. . .	. . .
R	4.47	5.28	6.02	6.38	7.41	9.59	11.13	12.03	16.56	18.06	. . .	. . .
S	0.38	0.50	0.63	0.63	0.75	1	1.13	1.25	1.5	1.5	. . .	. . .
T	0.81	0.81	0.84	1.19	1.38	1.78	2.13	2.56	2.88	3.44	. . .	. . .
O	0.88	0.97	1	1.13	1.34	1.69	2.06	2.25	3	3.63	. . .	. . .

Source: The Crosby Group. All dimensions are in inches. Hooks are made of alloy steel, quenched and tempered. For swivel hooks, the data are for a bail of carbon steel. The ultimate load is four times the working load limit (capacity). The swivel hook is a positioning device and is not intended to rotate under load; special load swiveling hooks must be used in such applications.

Dimensions of Forged Round Pin, Screw Pin, and Bolt Type Chain Shackles and Bolt Type Anchor Shackles

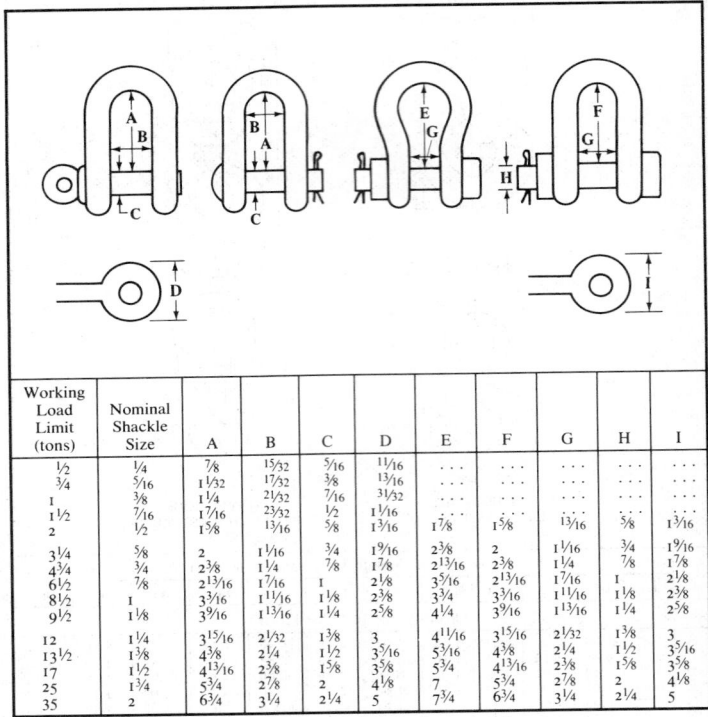

Working Load Limit (tons)	Nominal Shackle Size	A	B	C	D	E	F	G	H	I
1/2	1/4	7/8	15/32	5/16	11/16	...	...	...	...	...
3/4	5/16	1 1/32	17/32	3/8	13/16	...	...	...	...	...
1	3/8	1 1/4	21/32	7/16	31/32	...	...	...	...	...
1 1/2	7/16	1 7/16	23/32	1/2	1 1/16	...	...	...	...	...
2	1/2	1 5/8	13/16	5/8	1 3/16	1 7/8	1 5/8	13/16	5/8	1 3/16
3 1/4	5/8	2	1 1/16	3/4	1 9/16	2 3/8	2	1 1/16	3/4	1 9/16
4 3/4	3/4	2 3/8	1 1/4	7/8	1 7/8	2 13/16	2 3/8	1 1/4	7/8	1 7/8
6 1/2	7/8	2 13/16	1 7/16	1	2 1/8	3 5/16	2 13/16	1 7/16	1	2 1/8
8 1/2	1	3 3/16	1 11/16	1 1/8	2 3/8	3 3/4	3 3/16	1 11/16	1 1/8	2 3/8
9 1/2	1 1/8	3 9/16	1 13/16	1 1/4	2 5/8	4 1/4	3 9/16	1 13/16	1 1/4	2 5/8
12	1 1/4	3 15/16	2 1/32	1 3/8	3	4 11/16	3 15/16	2 1/32	1 3/8	3
13 1/2	1 3/8	4 3/8	2 1/4	1 1/2	3 5/16	5 3/16	4 3/8	2 1/4	1 1/2	3 5/16
17	1 1/2	4 13/16	2 3/8	1 5/8	3 5/8	5 3/4	4 13/16	2 3/8	1 5/8	3 5/8
25	1 3/4	5 3/4	2 7/8	2	4 1/8	7	5 3/4	2 7/8	2	4 1/8
35	2	6 3/4	3 1/4	2 1/4	5	7 3/4	6 3/4	3 1/4	2 1/4	5

All dimensions are in inches. Load limits are in tons of 2000 pounds.
Source: The Crosby Group.

Minimum Sheave- and Drum-Groove Dimensions for Wire Rope Applications-I

Nominal Rope Diameter	Groove Radius		Nominal Rope Diameter	Groove Radius	
	New	Worn		New	Worn
1/4	0.135	0.129	1 1/8	0.605	0.577
5/16	0.167	0.160	1 1/4	0.669	0.639
3/8	0.201	0.190	1 3/8	0.736	0.699
7/16	0.234	0.220	1 1/2	0.803	0.759
1/2	0.271	0.256	1 5/8	0.876	0.833
9/16	0.303	0.288	1 3/4	0.939	0.897
5/8	0.334	0.320	1 7/8	1.003	0.959
3/4	0.401	0.380	2	1.085	1.025
7/8	0.468	0.440	2 1/8	1.137	1.079
1	0.543	0.513	2 1/4	1.210	1.153

See footnotes at end of table.

Minimum Sheave- and Drum-Groove Dimensions for Wire Rope Applications-II

Nominal Rope Diameter	Groove Radius New	Groove Radius Worn	Nominal Rope Diameter	Groove Radius New	Groove Radius Worn
2⅜	1.271	1.199	3¾	1.997	1.918
2½	1.338	1.279	4	2.139	2.050
2⅝	1.404	1.339	4¼	2.264	2.178
2¾	1.481	1.409	4½	2.396	2.298
2⅞	1.544	1.473	4¾	2.534	2.434
3	1.607	1.538	5	2.663	2.557
3⅛	1.664	1.598	5¼	2.804	2.691
3¼	1.731	1.658	5½	2.929	2.817
3⅜	1.807	1.730	5¾	3.074	2.947
3½	1.869	1.794	6	3.198	3.075

All dimensions are in inches. Data taken from *Wire Rope Users Manual*, 2nd ed., American Iron and Steel Institute, Washington, D.C. The values given in this table are applicable to grooves in sheaves and drums and are not generally suitable for pitch design, since other factors may be involved.

Hot Dip Galvanized, Forged Steel Eye-bolts

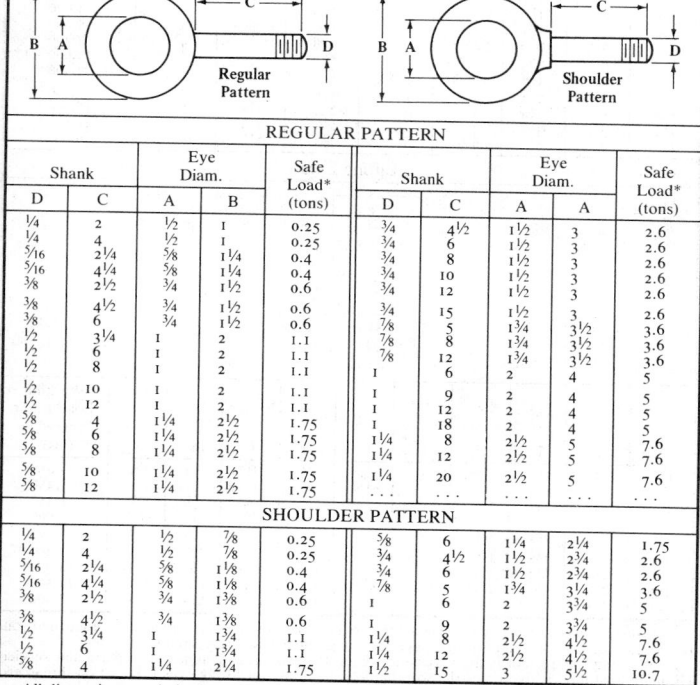

REGULAR PATTERN

D	C	A	B	Safe Load* (tons)	D	C	A	B	Safe Load* (tons)
¼	2	½	1	0.25	¾	4½	1½	3	2.6
¼	4	½	1	0.25	¾	6	1½	3	2.6
5/16	2¼	5/8	1¼	0.4	¾	8	1½	3	2.6
5/16	4¼	5/8	1¼	0.4	¾	10	1½	3	2.6
⅜	2½	¾	1½	0.6	¾	12	1½	3	2.6
⅜	4½	¾	1½	0.6	¾	15	1½	3	2.6
⅜	6	¾	1½	0.6	⅞	5	1¾	3½	3.6
½	3¼	1	2	1.1	⅞	8	1¾	3½	3.6
½	6	1	2	1.1	⅞	12	1¾	3½	3.6
½	8	1	2	1.1	1	6	2	4	5
½	10	1	2	1.1	1	9	2	4	5
½	12	1	2	1.1	1	12	2	4	5
5/8	4	1¼	2½	1.75	1	18	2	4	5
5/8	6	1¼	2½	1.75	1¼	8	2½	5	7.6
5/8	8	1¼	2½	1.75	1¼	12	2½	5	7.6
5/8	10	1¼	2½	1.75	1¼	20	2½	5	7.6
5/8	12	1¼	2½	1.75	...	...	...	...	...

SHOULDER PATTERN

D	C	A	B	Safe Load* (tons)	D	C	A	B	Safe Load* (tons)
¼	2	½	⅞	0.25	5/8	6	1¼	2¼	1.75
¼	4	½	⅞	0.25	¾	4½	1½	2¾	2.6
5/16	2¼	5/8	1⅛	0.4	¾	6	1½	2¾	2.6
5/16	4¼	5/8	1⅛	0.4	⅞	5	1¾	3¼	3.6
⅜	2½	¾	1⅜	0.6	1	6	2	3¾	5
⅜	4½	¾	1⅜	0.6	1	9	2	3¾	5
½	3¼	1	1¾	1.1	1¼	8	2½	4½	7.6
½	6	1	1¾	1.1	1¼	12	2½	4½	7.6
5/8	4	1¼	2¼	1.75	1½	15	3	5½	10.7

All dimensions are in inches. Safe loads are in tons of 2000 pounds.   * The ultimate or breaking load is 5 times the safe working load.   Source: The Crosby Group.

Eye Nuts and Lift Eyes

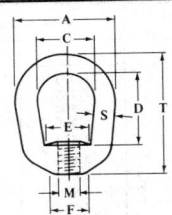

Eye Nuts

The general function of eye nuts is similar to that of eye-bolts. Eye nuts are utilized for a variety of applications in either the swivel or tapped design.

M	A	C	D	E	F	S	T	Working Load Limit (lbs)*
1/4	1 1/4	3/4	1 1/16	21/32	1/2	1/4	1 11/16	520
5/16	1 1/4	3/4	1 1/16	21/32	1/2	5/16	1 11/16	850
3/8	1 5/8	1	1 1/4	3/4	9/16	5/16	2 1/16	1,250
7/16	2	1 1/4	1 1/2	1	13/16	3/8	2 1/2	1,700
1/2	2	1 1/4	1 1/2	1	13/16	3/8	2 1/2	2,250
5/8	2 1/2	1 1/2	2	1 3/16	1	1/2	3 3/4	3,600
3/4	3	1 3/4	2 3/8	1 3/8	1 1/8	5/8	3 7/8	5,200
7/8	3 1/2	2	2 5/8	1 5/8	1 5/16	3/4	4 5/16	7,200
1	4	2 1/4	3 1/16	1 7/8	1 9/16	7/8	5	10,000
1 1/8	4	2 1/4	3 1/16	1 7/8	1 9/16	7/8	5	12,300
1 1/4	4 1/2	2 1/2	3 1/2	1 15/16	1 7/8	1	5 3/4	15,500
1 3/8	5	2 3/4	3 3/4	2	2	1 1/8	6 1/4	18,500
1 1/2	5 5/8	3 1/8	4	2 3/8	2 1/4	1 1/4	6 3/4	22,500
2	7	4	6 1/4	2 3/8	3 3/8	1 1/2	10	40,000

Lifting Eyes

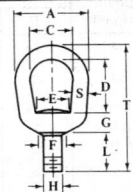

A	C	D	E	F	G	H	L	S	T	Working Load Limit Threaded (lbs)*
1 1/4	3/4	1 1/16	19/32	1/2	3/8	5/16	11/16	1/4	2 3/8	850
1 5/8	1	1 1/4	3/4	9/16	1/2	3/8	15/16	5/16	3	1,250
2	1 1/4	1 1/2	1	13/16	5/8	1/2	1 1/4	3/8	3 3/4	2,250
2 1/2	1 1/2	2	1 3/16	1	11/16	5/8	1 1/2	1/2	4 11/16	3,600
3	1 3/4	2 3/8	1 3/8	1 1/8	7/8	3/4	1 3/4	5/8	5 5/8	5,200
3 1/2	2	2 5/8	1 5/8	1 5/16	15/16	7/8	2	3/4	6 5/16	7,200
4	2 1/4	3 1/16	1 7/8	1 9/16	1 1/16	1	2 1/16	7/8	7 1/16	10,000
4 1/2	2 1/2	3 1/2	1 15/16	1 7/8	1 1/4	1 1/8	2 1/2	1	8 1/4	12,500
5 5/8	3 1/8	4	2 3/8	2 3/8	1 1/4	1 3/8	2 15/16	1 1/4	9 11/16	18,000

All dimensions are in inches. Data for eye nuts are for hot dip galvanized, quenched, and tempered forged steel. Data for lifting eyes are for quenched and tempered forged steel.
* Working Load Limit is the maximum weight the eye nut can support in a particular service. Ultimate load limit is five times the working load limit and for lifting eyes is based on thread size equal to specified shank diameter H.
Source: The Crosby Group.

Winding Drum Scores for Chain

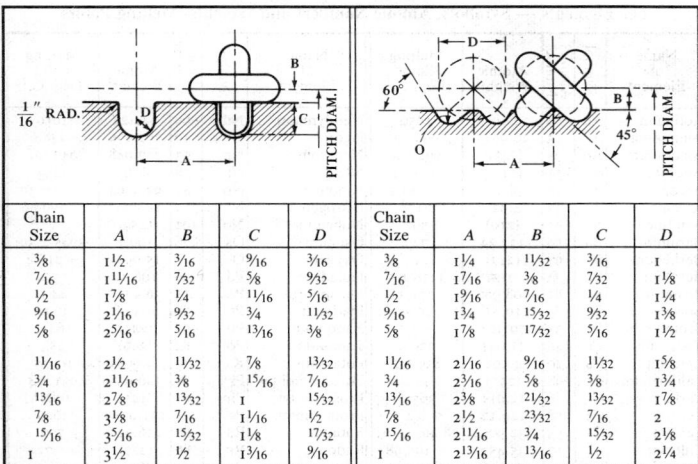

Chain Size	A	B	C	D	Chain Size	A	B	C	D
3/8	1½	3/16	9/16	3/16	3/8	1¼	11/32	3/16	1
7/16	1 11/16	7/32	5/8	9/32	7/16	1 7/16	3/8	7/32	1⅛
½	1⅞	¼	11/16	5/16	½	1 9/16	7/16	¼	1¼
9/16	2 1/16	9/32	¾	11/32	9/16	1¾	15/32	9/32	1⅜
5/8	2 5/16	5/16	13/16	3/8	5/8	1⅞	17/32	5/16	1½
11/16	2½	11/32	7/8	13/32	11/16	2 1/16	9/16	11/32	1⅝
¾	2 11/16	3/8	15/16	7/16	¾	2 3/16	5/8	3/8	1¾
13/16	2⅞	13/32	1	15/32	13/16	2⅜	21/32	13/32	1⅞
7/8	3⅛	7/16	1 1/16	½	7/8	2½	23/32	7/16	2
15/16	3 5/16	15/32	1⅛	17/32	15/16	2 11/16	¾	15/32	2⅛
1	3½	½	1 3/16	9/16	1	2 13/16	13/16	½	2¼

All dimensions are in inches.

Strength of Manila Rope

Diam. (in.)	Circumference (in.)	Weight of 100 feet of Rope* (lb)	New Rope Tensile Strength† (lb)	Working Load‡ (lb)	Diam. (in.)	Circumference (in.)	Weight of 100 feet of Rope* (lb)	New Rope Tensile Strength† (lb)	Working Load‡ (lb)
3/16	5/8	1.50	406	41	1 5/16	4	47.8	13,500	1930
¼	¾	2.00	540	54	1½	4½	60.0	16,700	2380
5/16	1	2.90	900	90	1⅝	5	74.5	20,200	2880
3/8	1⅛	4.10	1220	122	1¾	5½	89.5	23,800	3400
7/16	1¼	5.25	1580	176	2	6	108	28,000	4000
½	1½	7.50	2380	264	2⅛	6½	125	32,400	4620
9/16	1¾	10.4	3100	388	2¼	7	146	37,000	5300
5/8	2	13.3	3960	496	2½	7½	167	41,800	5950
¾	2¼	16.7	4860	695	2⅝	8	191	46,800	6700
13/16	2½	19.5	5850	835	2⅞	8½	215	52,000	7450
7/8	2¾	22.4	6950	995	3	9	242	57,500	8200
1	3	27.0	8100	1160	3¼	10	298	69,500	9950
1 1/16	3¼	31.2	9450	1350	3½	11	366	82,000	11,700
1⅛	3½	36.0	10,800	1540	4	12	434	94,500	13,500
1¼	3¾	41.6	12,200	1740	. . .	. . .	. . .	. . .	. . .

Data from Cordage Institute Rope Specifications for three-strand laid and eight-strand plaited manila rope (standard construction).

* Average value is shown; maximum is 5 per cent higher.

† Based on tests of new and unused rope of standard construction in accordance with Cordage Institute Standard Test Methods.

‡ These values are for rope in good condition with appropriate splices, in noncritical applications, and under normal service conditions. These values should be reduced where life, limb, or valuable property are involved, or for exceptional service conditions such as shock loads or sustained loads.

THE ELEMENTS

The Elements — Symbols, Atomic Numbers and Weights, Melting Points

Name of Element	Symbol	Atomic Number	Atomic Weight[1]	Melting Point Deg. C.[2]	Name of Element	Symbol	Atomic Number	Atomic Weight[1]	Melting Point Deg. C.[2]
Actinium	Ac	89	227.028	1050	Neodymium	Nd	60	144.24	1010
Aluminum	Al	13	26.9815	660.37	Neon	Ne	10	20.1179	−248.67
Americium	Am	95	(243)[1]	994 ± 4	Neptunium	Np	93	237.048	640 ± 1
Antimony	Sb	51	121.75	630.74	Nickel	Ni	28	58.69	1453
Argon	A	18	39.948	− 189.2	Niobium	Nb	41	92.9064	2468 ± 10
Arsenic	As	33	74.9216	817[2]	Nitrogen	N	7	14.0067	− 209.86
Astatine	At	85	(210)	302	Nobelium	No	102	(259)	. . .
Barium	Ba	56	137.33	725	Osmium	Os	76	190.2	3045 ± 30
Berkelium	Bk	97	(247)	. . .	Oxygen	O	8	15.9994	− 218.4
Beryllium	Be	4	9.01218	1278 ± 5	Palladium	Pd	46	106.42	1554
Bismuth	Bi	83	208.980	271.3	Phosphorus	P	15	30.9738	44.1
Boron	B	5	10.81	2079	Platinum	Pt	78	195.08	1772
Bromine	Br	35	79.904	− 7.2	Plutonium	Pu	94	(244)	641
Cadmium	Cd	48	112.41	320.9	Polonium	Po	84	(209)	254
Calcium	Ca	20	40.08	839 ± 2	Potassium	K	19	39.0983	63.25
Californium	Cf	98	(251)	. . .	Praseodymium	Pr	59	140.908	931 ± 4
Carbon	C	6	12.011	3652[3]	Promethium	Pm	61	(145)	1080[4]
Cerium	Ce	58	140.12	798 ± 2	Protactinium	Pa	91	231.0359	1600
Cesium	Cs	55	132.9054	28.40 ± 0.01	Radium	Ra	88	226.025	700
Chlorine	Cl	17	35.453	− 100.98	Radon	Rn	86	(222)	− 71
Chromium	Cr	24	51.996	1857 ± 20	Rhenium	Re	75	186.207	3180
Cobalt	Co	27	51.9332	1495	Rhodium	Rh	45	102.906	1965 ± 3
Copper	Cu	29	63.546	1083.4 ± 0.2	Rubidium	Rb	37	85.4678	38.89
Curium	Cm	96	(247)	1340 ± 40	Ruthenium	Ru	44	101.07	2310
Dysprosium	Dy	66	162.50	1409	Samarium	Sm	62	150.36	1072 ± 5
Einsteinium	Es	99	(252)	. . .	Scandium	Sc	21	44.9559	1539
Erbium	Er	68	167.26	1522	Selenium	Se	34	78.96	217
Europium	Eu	63	151.96	822 ± 5	Silicon	Si	14	28.0855	1410
Fermium	Fm	100	(257)	. . .	Silver	Ag	47	107.868	961.93
Fluorine	F	9	18.9984	− 219.62	Sodium	Na	11	22.9898	97.81 ± 0.03
Francium	Fr	87	(223)	27[4]	Strontium	Sr	38	87.62	769
Gadolinium	Gd	64	157.25	1311 ± 1	Sulfur	S	16	32.06	112.8
Gallium	Ga	31	69.72	29.78	Tantalum	Ta	73	180.9479	2996
Germanium	Ge	32	72.59	937.4	Technetium	Tc	43	(98)	2172
Gold	Au	79	196.967	1064.434	Tellurium	Te	52	127.60	449.5 ± 0.3
Hafnium	Hf	72	178.49	2227 ± 20	Terbium	Tb	65	158.925	1360 ± 4
Helium	He	2	4.00260	− 272.2[5]	Thallium	Tl	81	204.383	303.5
Holmium	Ho	67	164.930	1470	Thorium	Th	90	232.038	1750
Hydrogen	H	1	1.00794	− 259.14	Thulium	Tm	69	168.934	1545 ± 15
Indium	In	49	114.82	156.61	Tin	Sn	50	118.71	231.9681
Iodine	I	53	126.905	113.5	Titanium	Ti	22	47.88	1660 ± 10
Iridium	Ir	77	192.22	2410	Tungsten	W	74	183.85	3410 ± 20
Iron	Fe	26	55.847	1535	(Unnilhexium)	(Unh)	106	(263)	. . .
Krypton	Kr	36	83.80	− 156.6	(Unnilseptium)	(Uns)	107	(262)	. . .
Lanthanum	La	57	138.906	920 ± 5	(Unnilnonium)	(Unn)	109	. . .	. . .
Lawrencium	Lw	103	(260)	. . .	(Unnilpentium)	(Unp)	105	(262)	. . .
Lead	Pb	82	207.2	327.502	(Unnilquadium)	(Unq)	104	(261)	. . .
Lithium	Li	3	6.941	180.54	Uranium	U	92	238.029	1132 ± 0.8
Lutetium	Lu	71	174.967	1656 ± 5	Vanadium	V	23	50.9415	1890 ± 10
Magnesium	Mg	12	24.305	648.8 ± 0.5	Xenon	Xe	54	131.29	− 111.9
Manganese	Mn	25	54.9380	1244 ± 3	Ytterbium	Yb	70	173.04	824 ± 5
Mendelevium	Md	101	(258)	. . .	Yttrium	Y	39	88.9059	1523 ± 8
Mercury	Hg	80	200.59	− 38.87	Zinc	Zn	30	65.39	419.58
Molybdenum	Mo	42	95.94	2617	Zirconium	Zr	40	91.224	1852 ± 2

International Atomic Weights, 1981.   [1] Values in parentheses are for the mass number of the most stable known isotope.   [2] At 28 atm.   [3] Sublimates.   [4] Approximate.   [5] At 26 atm.

Specific gravity is a number indicating how many times a certain volume of a material is heavier than an equal volume of water. As the density of water differs slightly at different temperatures, it is the usual custom to make comparisons on the basis that the water has a temperature of 62 degrees F. The weight of one cubic inch of pure water at 62 degrees F. is 0.0361 pound. If the specific gravity of any material is known, the weight of a cubic inch of the material can, therefore, be found by multiplying its specific gravity by 0.0361.

Example: — The specific gravity of cast iron is 7.2. Find the weight of 5 cubic inches of cast iron.

$$7.2 \times 0.0361 \times 5 = 1.2996 \text{ pound.}$$

To find the weight per cubic foot of a material, the specific gravity of which is known, multiply the specific gravity by 62.355.

If the weight of a cubic inch of a material is known, the specific gravity is found by dividing the weight per cubic inch by 0.0361.

Example: — The weight of a cubic inch of gold is 0.697 pound. Find the specific gravity.

$$0.697 \div 0.0361 = 19.31.$$

If the weight per cubic foot of a material is known, the specific gravity is found by multiplying this weight by 0.01604.

Average Specific Gravity of Miscellaneous Substances

Substance	Sp. Gr.	Weight per Cubic Foot, Lbs.	Substance	Sp. Gr.	Weight per Cubic Foot, Lbs.
Asbestos............	2.4	150	Gypsum............	2.4	150
Asphaltum..........	1.4	87	Ice................	0.9	56
Borax..............	1.8	112	Iron slag	2.7	168
Brick, common.......	1.8	112	Limestone..........	2.6	162
Brick, fire..........	2.3	143	Marble............	2.7	168
Brick, hard..........	2.0	125	Masonry...........	2.4	150
Brick, pressed.......	2.2	137	Mica..............	2.8	175
Brickwork, in mortar...	1.6	100	Mortar............	1.5	94
Brickwork, in cement..	1.8	112	Phosphorus.........	1.8	112
Cement, Portland (set).	3.1	193	Plaster of Paris.......	1.8	112
Chalk..............	2.3	143	Quartz.............	2.6	162
Charcoal...........	0.4	25	Salt, common........	...	48
Coal, anthracite.......	1.5	94	Sand, dry...........	...	100
Coal, bituminous......	1.3	81	Sand, wet...........	...	125
Concrete...........	2.2	137	Sandstone..........	2.3	143
Earth, loose..........	...	75	Slate..............	2.8	175
Earth, rammed.......	...	100	Soapstone..........	2.7	168
Emery..............	4.0	249	Sulphur...........	2.0	125
Glass..............	2.6	162	Tar, bituminous.......	1.2	75
Granite.............	2.7	168	Tile...............	1.8	112
Gravel.............	...	109	Trap rock..........	3.0	187

The weight per cubic foot is calculated on the basis of the specific gravity except for those substances that occur in bulk, heaped or loose form. In these instances, only the weights per cubic foot are given since the voids present in representative samples make the values of the specific gravities insignificant.

Specific Gravity of Liquids. — The specific gravity of liquids is the number which indicates how much a certain volume of the liquid weighs compared with an equal volume of water, the same as in the case of solid bodies. The density of liquids is also often expressed in degrees on the hydrometer, an instrument for determining the density of liquids, provided with graduations made to an arbitrary scale. The hydrometer consists of a glass tube with a bulb at one end containing air, and arranged with a weight at the bottom so as to float in an upright position in the liquid, the density of which is to be measured. The depth to which the hydrometer sinks in the liquid is read off on the graduated scale. The most commonly used hydrometer is the Baumé. The value of the degrees on the Baumé scale differs according to whether the liquid is heavier or lighter than water. The specific gravity for liquids heavier than water equals 145 ÷ (145 − degrees Baumé). For liquids lighter than water, the specific gravity equals 140 ÷ (130 + degrees Baumé).

Specific Gravity of Gases. — The specific gravity of gases is the number which indicates their weight in comparison with that of an equal volume of air. The specific gravity of air is 1, and the comparison is made at 32 degrees F.

Specific Gravity of Gases
(At 32 degrees F.)

Gas	Sp. Gr.	Gas	Sp. Gr.	Gas	Sp. Gr.
Air	1.000	Ether vapor	2.586	Marsh gas	0.555
Acetylene	0.920	Ethylene	0.967	Nitrogen	0.971
Alcohol vapor	1.601	Hydrofluoric acid	2.370	Nitric oxide	1.039
Ammonia	0.592	Hydrochloric acid	1.261	Nitrous oxide	1.527
Carbon dioxide	1.520	Hydrogen	0.069	Oxygen	1.106
Carbon monoxide	0.967	Illuminating gas	0.400	Sulphur dioxide	2.250
Chlorine	2.423	Mercury vapor	6.940	Water vapor	0.623

1 cubic foot of air at 32 degrees F. and atmospheric pressure weighs 0.0807 pound.

Average Weights and Volumes of Fuels

Anthracite coal, 1 cubic foot = 55 to 65 pounds.
        1 ton (2240 pounds) = 34 to 41 cubic feet.
Bituminous coal, 1 cubic foot = 50 to 55 pounds.
        1 ton (2240 pounds) = 41 to 45 cubic feet.
Charcoal, 1 cubic foot = 18 to 18.5 pounds.
        1 ton (2240 pounds) = 120 to 124 cubic feet.
Coke, 1 cubic foot = 28 pounds.
        1 ton (2240 pounds) = 80 cubic feet.

The average weight of a bushel of charcoal is 20 pounds; of a bushel of coke, 40 pounds; of a bushel of anthracite coal, 67 pounds; and of a bushel of bituminous coal, 60 pounds.

Weight of Wood. — The weight of seasoned wood per cord is approximately as follows, assuming about 70 cubic feet of *solid wood* per cord: Beech, 3300 pounds; chestnut, 2600 pounds; elm, 2900 pounds; maple, 3100 pounds; poplar, 2200 pounds; white pine, 2200 pounds; red oak, 3300 pounds; white oak, 3500 pounds.

Weight per Foot of Wood, Board Measure. — The following is the weight in pounds of various kinds of woods, commercially known as dry timber, per foot board measure: White oak, 4.16; white pine, 1.98; Douglas fir, 2.65; short-leaf yellow pine, 2.65; red pine, 2.60; hemlock, 2.08; spruce, 2.08; cypress, 2.39; cedar, 1.93; chestnut, 3.43; Georgia yellow pine, 3.17; California spruce, 2.08.

Specific Gravity of Liquids

Liquid	Sp. Gr.	Liquid	Sp. Gr.	Liquid	Sp. Gr.
Acetic acid...........	1.06	Fluoric acid....	1.50	Petroleum oil....	0.82
Alcohol, commercial...	0.83	Gasoline......	0.70	Phosphoric acid.	1.78
Alcohol, pure.........	0.79	Kerosene......	0.80	Rape oil........	0.92
Ammonia.............	0.89	Linseed oil....	0.94	Sulphuric acid...	1.84
Benzine..............	0.69	Mineral oil....	0.92	Tar............	1.00
Bromine..............	2.97	Muriatic acid..	1.20	Turpentine oil...	0.87
Carbolic acid.........	0.96	Naphtha......	0.76	Vinegar........	1.08
Carbon disulphide.....	1.26	Nitric acid.....	1.50	Water..........	1.00
Cotton-seed oil.......	0.93	Olive oil.......	0.92	Water, sea......	1.03
Ether, sulphuric.......	0.72	Palm oil.......	0.97	Whale oil.......	0.92

Degrees on Baumé's Hydrometer Converted into Specific Gravity

Deg. Baumé	Specific Gravity Liquids Heavier than Water	Specific Gravity Liquids Lighter than Water	Deg. Baumé	Specific Gravity Liquids Heavier than Water	Specific Gravity Liquids Lighter than Water	Deg. Baumé	Specific Gravity Liquids Heavier than Water	Specific Gravity Liquids Lighter than Water
0	1.000	......	27	1.229	0.892	54	1.593	0.761
1	1.007	......	28	1.239	0.886	55	1.611	0.757
2	1.014	......	29	1.250	0.881	56	1.629	0.753
3	1.021	......	30	1.261	0.875	57	1.648	0.749
4	1.028	......	31	1.272	0.870	58	1.667	0.745
5	1.036	......	32	1.283	0.864	59	1.686	0.741
6	1.043	......	33	1.295	0.859	60	1.706	0.737
7	1.051	......	34	1.306	0.854	61	1.726	0.733
8	1.058	......	35	1.318	0.849	62	1.747	0.729
9	1.066	......	36	1.330	0.843	63	1.768	0.725
10	1.074	1.000	37	1.343	0.838	64	1.790	0.721
11	1.082	0.993	38	1.355	0.833	65	1.813	0.718
12	1.090	0.986	39	1.368	0.828	66	1.836	0.714
13	1.099	0.979	40	1.381	0.824	67	1.859	0.710
14	1.107	0.972	41	1.394	0.819	68	1.883	0.707
15	1.115	0.966	42	1.408	0.814	69	1.908	0.704
16	1.124	0.959	43	1.422	0.809	70	1.933	0.700
17	1.133	0.952	44	1.436	0.805	71	1.959	0.696
18	1.142	0.946	45	1.450	0.800	72	1.986	0.693
19	1.151	0.940	46	1.465	0.796	73	2.014	0.689
20	1.160	0.933	47	1.480	0.791	74	2.042	0.686
21	1.169	0.927	48	1.495	0.787	75	2.071	0.683
22	1.179	0.921	49	1.510	0.782	76	2.101	0.679
23	1.189	0.915	50	1.526	0.778	77	2.132	0.676
24	1.198	0.909	51	1.542	0.773	78	2.164	0.673
25	1.208	0.903	52	1.559	0.769	79	2.197	0.669
26	1.219	0.897	53	1.576	0.765	80	2.230	0.666

Specific Gravity and Properties of Metals

Metal or Composition	Chemical Symbol	Specific Gravity	Weight per Cubic Inch, Pound†	Weight per Cubic Foot, Pounds†	Melting Point, Deg. F.	Linear Expansion per Unit Length per Deg. F.	Temp., Deg. F.*	Electric Conductivity Silver = 100
Aluminum	Al	2.70	0.0975	168.5	1220	0.0001244	68	63.0
Antimony	Sb	6.618	0.2390	413.0	1167	0.0000755	68	3.59
Barium	Ba	3.78	0.1365	235.9	1562			30.61
Bismuth	Bi	9.781	0.3532	610.3	520	0.0000077	68	1.40
Boron	B	2.535	0.0916	158.2	4172			
Brass: 80C., 20Z.		8.60	0.3105	536.6	1823			
70C., 30Z.		8.44	0.3048	526.7	1706	0.00001	76	
60C., 40Z.		8.36	0.3018	521.7	1652		to	
50C., 50Z.		8.20	0.2961	511.7	1616		212	
Bronze: 90C., 10T.		8.78	0.3171	547.9	1841	0.00001		
Cadmium	Cd	8.648	0.3123	539.6	610			24.38
Calcium	Ca	1.54	0.0556	96.1	1490			21.77
Chromium	Cr	6.93	0.2502	432.4	2939	0.0000683	68	16.00
Cobalt	Co	8.71	0.3145	543.5	2696	0.0000900	68	16.93
Copper	Cu	8.89	0.3210	554.7	1981	0.0000900	68	97.61
Gold	Au	19.3	0.6969	1204.3	1945	0.0000778	68	76.61
Iridium	Ir	22.42	0.8096	1399.0	4262	0.0000361	68	13.52
Iron, cast	Fe	7.03–7.73	0.254–0.279	438.7–482.4	1990–2300	0.0000055	212	14.57
Iron, wrought		7.80–7.90	0.282–0.285	486.7–493.0	2750	0.0000661	68–212	
Lead	Pb	11.342	0.4096	707.7	621	0.0000163	212	8.42
Magnesium	Mg	1.741	0.0628	108.6	1204	0.0000144	68–212	39.44
Manganese	Mn	7.3	0.2636	455.5	2300			15.75
Mercury (68° F.)	Hg	13.546	0.4892	845.3	−38		68	1.75
Molybdenum	Mo	10.2	0.3683	636.5	4748	0.0000294		17.60
Nickel	Ni	8.8	0.3178	549.1	2651	0.0000700	68	12.89
Platinum	Pt	21.37	0.7717	1333.5	3224	0.0000496	68	14.43
Potassium	K	0.870	0.0314	54.3	144		68	19.62
Silver	Ag	10.42–10.53	0.376–0.380	650.2–657.1	1761	0.0001025		100.00
Sodium	Na	0.9712	0.0351	60.6	207		68	31.98
Steel, Carbon			0.283–0.284	489.0–490.8	2500	0.0000633	68	12.00
Tantalum	Ta	16.6	0.5998	1035.8	5162	0.0000361		54.63
Tellurium	Te	6.25	0.2257	390.0	846		68	0.001
Tin	Sn	7.29	0.2633	454.9	449	0.0000093	104	14.39
Titanium	Ti	4.5	0.1621	280.1	3272	0.0001496	64–212	13.73
Tungsten	W	18.6–19.1	0.672–0.690	1161–1192	6698	0.000049	68–392	14.00
Uranium	U	18.7	0.6753	1166.9	<3362	0.0000239	32–212	16.47
Vanadium	V	5.6	0.2022	349.4	3110	0.000043	32–104	4.95
Zinc	Zn	7.04–7.16	0.254–0.259	439.3–446.8	788	0.00017	68	29.57

*Temperature given for each metal is that at which expansion coefficient shown in previous column was determined   †Avoirdupois

Melting Points of Alloys of Low Fusing Point

Composition in Per Cent				Melting Point, Degrees F.	Composition in Per Cent			Melting Point, Degrees F.
Bismuth	Lead	Tin	Cadmium		Bismuth	Lead	Tin	
50.0	25.0	12.5	12.5	149	20.0	40.0	40.0	293
50.1	26.6	13.3	10.0	158	19.0	38.0	43.0	298
38.4	30.8	15.4	15.4	160	18.1	36.2	45.7	304
27.5	27.5	10.5	34.5	167	17.3	34.6	48.1	311
50.0	34.5	9.3	6.2	171	16.6	33.2	50.2	316
50.0	25.0	25.0	....	187	16.0	36.0	48.0	311
50.0	31.2	18.8	....	201	15.3	38.8	45.9	309
55.6	....	33.3	11.1	203	14.8	40.2	45.0	307
50.0	....	25.0	25.0	203	14.0	43.0	43.0	309
47.0	35.5	17.5	....	208	13.7	44.8	41.5	320
42.1	42.1	15.8	....	226	13.3	46.6	40.1	329
40.0	40.0	20.0	....	235	12.8	49.0	38.2	342
36.5	36.5	27.0	....	243	12.5	50.0	37.5	352
33.3	33.4	33.3	....	253	11.7	46.8	41.5	333
30.8	38.4	30.8	....	266	11.4	45.6	43.0	329
28.5	43.0	28.5	....	270	11.2	44.4	44.4	320
25.0	50.0	25.0	....	300	10.8	43.2	46.0	318
23.5	47.0	29.5	....	304	10.5	42.0	47.5	320
22.2	44.4	33.4	....	289	10.2	41.0	48.8	322
21.0	42.0	37.0	....	289	10.0	40.0	50.0	324

Weights of American Woods, in Pounds per Cubic Foot
(United States Department of Agriculture)

Species	Green	Airdry	Species	Green	Airdry
Alder, red	46	28	Hickory, pecan	62	45
Ash, black	52	34	Hickory, true	63	51
Ash, commercial white	48	41	Honeylocust	61	..
Ash, Oregon	46	38	Larch, western	48	36
Aspen	43	26	Locust, black	58	48
Basswood	42	26	Maple, bigleaf	47	34
Beech	54	45	Maple, black	54	40
Birch	57	44	Maple, red	50	38
Birch, paper	50	38	Maple, silver	45	33
Cedar, Alaska	36	31	Maple, sugar	56	44
Cedar, eastern red	37	33	Oak, red	64	44
Cedar, northern white	28	22	Oak, white	63	47
Cedar, southern white	26	23	Pine, lodgepole	39	29
Cedar, western red	27	23	Pine, northern white	36	25
Cherry, black	45	35	Pine, Norway	42	34
Chestnut	55	30	Pine, ponderosa	45	28
Cottonwood, eastern	49	28	Pines, southern yellow:		
Cottonw'd, northern black	46	24	Pine, loblolly	53	36
Cypress, southern	51	32	Pine, longleaf	55	41
Douglas fir, coast region	38	34	Pine, shortleaf	52	36
Douglas fir, Rocky Mt. reg.	35	30	Pine, sugar	52	25
Elm, American	54	35	Pine, western white	35	27
Elm, rock	53	44	Poplar, yellow	38	28
Elm, slippery	56	37	Redwood	50	28
Fir, balsam	45	25	Spruce, eastern	34	28
Fir, commercial white	46	27	Spruce, Engelmann	39	23
Gum, black	45	35	Spruce, Sitka	33	28
Gum, red	50	34	Sycamore	52	34
Hemlock, eastern	50	28	Tamarack	47	37
Hemlock, western	41	29	Walnut, black	58	38

Table giving Number of Pieces in One Pound, when Weight of One Hundred Pieces is Known

Example: — 100 pieces weigh 11 pounds 5 ounces. From the table, there are then 8.84 pieces in one pound.

Ounces	Pounds								
	0	1	2	3	4	5	6	7	8
0	......	100.00	50.00	33.33	25.00	20.00	16.67	14.29	12.50
1	1600 00	94.12	48.48	32.65	24.61	19.75	16.49	14.16	12.40
2	800.00	88.88	47.06	32.00	24.24	19.51	16.33	14.03	12.31
3	533.33	84.21	45.71	31.37	23.88	19.27	16.16	13.91	12.21
4	400.00	80.00	44.44	30.77	23.52	19.05	16.00	13.79	12.12
5	320.00	76.19	43.24	30.19	23.19	18.82	15.84	13.67	12.03
6	266.66	72.73	42.11	29.63	22.86	18.60	15 69	13.56	11.94
7	228.57	69.57	41.03	29.09	22.53	18.39	15.53	13.44	11.85
8	200.00	66.67	40.00	28.57	22.22	18.18	15.38	13.33	11.76
9	177.78	64.30	39.02	28.07	21.92	17.98	15.24	13.22	11.68
10	160.00	61.54	38.09	27.58	21.62	17.78	15.09	13.11	11.59
11	145.45	59.26	37.21	27.12	21.33	17.58	14.95	13.01	11.51
12	133.33	57.14	36.36	26.67	21.05	17.39	14.81	12.90	11.43
13	123.08	55.17	35.56	26.23	20.78	17.20	14.68	12.80	11.35
14	114.29	53.33	34.78	25.81	20.51	17.02	14.54	12.70	11.27
15	106.67	51.61	34.04	25.40	20.25	16.84	14.41	12.60	11.19

Ounces	Pounds								
	9	10	11	12	13	14	15	16	17
0	11.11	10.00	9.09	8.33	7.69	7.14	6.66	6.25	5.88
1	11.03	9.94	9.04	8.29	7.65	7.11	6.64	6.23	5.86
2	10.96	9.88	8.99	8.25	7.62	7.08	6.61	6.20	5.84
3	10.89	9.82	8.94	8.20	7.58	7.04	6.59	6.17	5.82
4	10.81	9.76	8.89	8.16	7.54	7.01	6.56	6.15	5.80
5	10.74	9.69	8.84	8.12	7.51	6.98	6.53	6.13	5.78
6	10.67	9.64	8.79	8.08	7.47	6.95	6.50	6.11	5.76
7	10.59	9.58	8.74	8.04	7.44	6.92	6.47	6.08	5.74
8	10.53	9.52	8.69	8.00	7.41	6.89	6.45	6.06	5.72
9	10.46	9.47	8.65	7.96	7.37	6.86	6.43	6.04	5.70
10	10.39	9.41	8.60	7.92	7.34	6.84	6.40	6.01	5.68
11	10.32	9.36	8.56	7.88	7.31	6.81	6.37	5.98	5.66
12	10.25	9.30	8.51	7.84	7.27	6.78	6.35	5.96	5.64
13	10.19	9.25	8.46	7.80	7.24	6.75	6.32	5.94	5.62
14	10.13	9.19	8.42	7.76	7.21	6.72	6.30	5.92	5.60
15	10.06	9.14	8.38	7.73	7.17	6.69	6.27	5.90	5.58

Ounces	Pounds								
	18	19	20	21	22	23	24	25	26
0	5.56	5.26	5.00	4.76	4.54	4.35	4.16	4.00	3.84
1	5.54	5.24	4.98	4.74	4.53	4.34	4.15	3.99	3.83
2	5.52	5.23	4.96	4.73	4.52	4.33	4.14	3.98	3.82
3	5.50	5.21	4.95	4.71	4.51	4.32	4.13	3.97	3.81
4	5.48	5.19	4.94	4.70	4.50	4.30	4.12	3.96	3.81
5	5.46	5.18	4.92	4.69	4.49	4.29	4.11	3.95	3.80
6	5.44	5.16	4.90	4.68	4.48	4.28	4.10	3.94	3.79
7	5.42	5.14	4.89	4.66	4.46	4.27	4.09	3.93	3.78
8	5.40	5.12	4.88	4.65	4.44	4.25	4.08	3.92	3.77
9	5.38	5.11	4.86	4.64	4.43	4.24	4.07	3.91	3.76
10	5.36	5.09	4.84	4.63	4.42	4.23	4.06	3.90	3.75
11	5.34	5 08	4.83	4.61	4.41	4.22	4.05	3.89	3.74
12	5.32	5.07	4.82	4.60	4.39	4.21	4.04	3.88	3.73
13	5.31	5.05	4.81	4.59	4.38	4.19	4.03	3.87	3.73
14	5.30	5.04	4.79	4.57	4.37	4.18	4 02	3.86	3.72
15	5.28	5 02	4.78	4.56	4.36	4.17	4.01	3.85	3.71

Theoretical Weights of Round, Square and Hexagon Steel Bars

Thickness or Diameter, Inches	Round Bars		Square Bars		Hexagon Bars	
	Weight, Pounds Per Inch	Weight, Pounds Per Foot	Weight, Pounds Per Inch	Weight, Pounds Per Foot	Weight, Pounds Per Inch	Weight, Pounds Per Foot
1/32	0.0002	0.0026	0.0003	0.0033	0.0002	0.0028
1/16	0.0009	0.0104	0.0011	0.0133	0.0010	0.0115
3/32	0.0020	0.0235	0.0025	0.0299	0.0022	0.0259
1/8	0.0035	0.0417	0.0044	0.0531	0.0038	0.0460
5/32	0.0054	0.0652	0.0069	0.0830	0.0060	0.0719
3/16	0.0078	0.0939	0.0100	0.1195	0.0086	0.1035
7/32	0.0106	0.1278	0.0136	0.1627	0.0117	0.1409
1/4	0.0139	0.1669	0.0177	0.2125	0.0153	0.1840
9/32	0.0176	0.2112	0.0224	0.2689	0.0194	0.2329
5/16	0.0217	0.2608	0.0277	0.3320	0.0240	0.2875
11/32	0.0263	0.3155	0.0335	0.4018	0.0290	0.3479
3/8	0.0313	0.3755	0.0398	0.4781	0.0345	0.4141
13/32	0.0367	0.4407	0.0468	0.5611	0.0405	0.4860
7/16	0.0426	0.5111	0.0542	0.6508	0.0470	0.5636
15/32	0.0489	0.5867	0.0623	0.7471	0.0538	0.6470
1/2	0.0556	0.6676	0.0708	0.8500	0.0613	0.7361
17/32	0.0628	0.7536	0.0800	0.9596	0.0693	0.8310
9/16	0.0704	0.8449	0.0896	1.076	0.0776	0.9317
19/32	0.0785	0.9414	0.0999	1.199	0.0865	1.038
5/8	0.0869	1.043	0.1107	1.328	0.0958	1.150
21/32	0.0958	1.150	0.1220	1.464	0.1057	1.268
11/16	0.1052	1.262	0.1339	1.607	0.1160	1.392
23/32	0.1150	1.380	0.1464	1.756	0.1268	1.521
3/4	0.1252	1.502	0.1594	1.913	0.1380	1.656
25/32	0.1358	1.630	0.1729	2.075	0 1498	1.797
13/16	0.1469	1.763	0.1870	2.245	0.1620	1.944
27/32	0.1584	1.901	0.2017	2.421	0.1747	2.096
7/8	0.1704	2.044	0.2169	2.603	0.1879	2.254
29/32	0.1828	2.193	0.2327	2.792	0.2015	2.418
15/16	0.1956	2.347	0.2490	2.988	0.2157	2.588
31/32	0.2088	2.506	0.2659	3.191	0.2303	2.763
1	0.2225	2.670	0.2833	3.400	0.2454	2.944
1 1/16	0.2512	3.015	0.3199	3.838	0.2770	3.324
1 1/8	0.2816	3.380	0.3586	4.303	0.3106	3.727
1 3/16	0.3138	3.766	0.3995	4.795	0.3460	4.152
1 1/4	0.3477	4.172	0.4427	5.313	0.3834	4.601
1 5/16	0.3833	4.600	0.4881	5.857	0.4227	5.072
1 3/8	0.4207	5.049	0.5357	6.428	0.4639	5.567
1 7/16	0.4598	5.518	0.5855	7.026	0.5070	6.085
1 1/2	0.5007	6.008	0.6375	7.650	0.5521	6.625
1 9/16	0.5433	6.519	0.6917	8.301	0.5991	7.189
1 5/8	0.5876	7.051	0.7482	8.978	0.6479	7.775

Based on a weight of 0.2833 lbs. per cubic in. (489.6 lbs. per cubic ft.).

Theoretical Weights of Round, Square and Hexagon Steel Bars (*Continued*)

Thickness or Diameter, Inches	Round Bars		Square Bars		Hexagon Bars	
	Weight, Pounds Per Inch	Weight, Pounds Per Foot	Weight, Pounds Per Inch	Weight, Pounds Per Foot	Weight, Pounds Per Inch	Weight, Pounds Per Foot
1 11/16	0.6337	7.604	0.8068	9.682	0.6988	8.385
1¾	0.6815	8.178	0.8677	10.41	0.7515	9.018
1 13/16	0.7310	8.773	0.9308	11.17	0.8060	9.67
1⅞	0.7823	9.388	0.9961	11.95	0.8626	10.35
1 15/16	0.8354	10.02	1.064	12.76	0.9211	11.05
2	0.8901	10.68	1.133	13.60	0.9815	11.78
2 1/16	0.9466	11.36	1.205	14.46	1.044	12.53
2⅛	1.005	12.06	1.279	15.35	1.108	13.30
2 3/16	1.065	12.78	1.356	16.27	1.174	14.09
2¼	1.127	13.52	1.434	17.21	1.242	14.91
2 5/16	1.190	14.28	1.515	18.18	1.312	15.75
2⅜	1.255	15.06	1.598	19.18	1.384	16.61
2 7/16	1.322	15.87	1.683	20.20	1.458	17.49
2½	1.391	16.69	1.771	21.25	1.534	18.40
2⅝	1.533	18.40	1.952	23.43	1.691	20.29
2¾	1.683	20.19	2.143	25.71	1.856	22.27
2⅞	1.839	22.07	2.342	28.10	2.028	24.34
3	2.003	24.03	2.550	30.60	2.208	26.50
3⅛	2.173	26.08	2.767	33.20	2.396	28.75
3¼	2.350	28.21	2.993	35.91	2.592	31.10
3⅜	2.535	30.42	3.227	38.73	2.795	33.54
3½	2.726	32.71	3.471	41.65	3.006	36.07
3⅝	2.924	35.09	3.723	44.68	3.224	38.69
3¾	3.129	37.55	3.984	47.81	3.451	41.41
3⅞	3.341	40.10	4.254	51.05	3.684	44.21
4	3.560	42.73	4.533	54.40	3.926	47.11
4⅛	3.786	45.44	4.821	57.85	4.175	50.10
4¼	4.019	48.23	5.118	61.41	4.432	53.18
4⅜	4.259	51.11	5.423	65.08	4.700	56.36
4½	4.506	54.07	5.738	68.85	4.970	59.63
4⅝	4.760	57.12	6.061	72.73	5.248	62.98
4¾	5.021	60.25	6.393	76.71	5.536	66.44
4⅞	5.289	63.46	6.734	80.80	5.831	69.98
5	5.563	66.76	7.083	85.00	6.134	73.61
5⅛	5.845	70.14	7.442	89.30	6.445	77.34
5¼	6.133	73.60	7.809	93.71	6.763	81.16
5⅜	6.429	77.15	8.186	98.23	7.089	85.07
5½	6.732	80.78	8.571	102.85	7.422	89.07
5⅝	7.041	84.49	8.965	107.58	7.763	93.16
5¾	7.357	88.29	9.368	112.41	8.112	97.35
5⅞	7.681	92.17	9.779	117.35	8.470	101.63
6	8.011	96.13	10.200	122.40	8.833	106.00

Based on a weight of 0.2833 lbs. per cubic in. (489.6 lbs. per cubic ft.).
Source: *Steel Products Manual*, American Iron & Steel Institute, with permission.

Weights of Flat Rolled Steel per Lineal Foot in Pounds

U. S. St'd Gage for Plate	Width of Flat Steel, Inches						
	⅛	3/16	¼	5/16	⅜	7/16	½
0000000	0.2126	0.3189	0.4252	0.5315	0.6378	0.7441	0.8504
000000	0.1992	0.2998	0.3984	0.4980	0.5976	0.6972	0.7968
00000	0.1860	0.2790	0.3720	0.4650	0.5580	0.6510	0.7440
0000	0.1728	0.2592	0.3456	0.4320	0.5184	0.6048	0.6912
000	0.1594	0.2391	0.3188	0.3985	0.4782	0.5579	0.6376
00	0.1462	0.2193	0.2924	0.3655	0.4386	0.5117	0.5848
0	0.1328	0.1992	0.2656	0.3320	0.3984	0.4648	0.5312
1	0.1196	0.1794	0.2392	0.2990	0.3588	0.4186	0.4784
2	0.1130	0.1695	0.2260	0.2825	0.3390	0.3955	0.4520
3	0.1062	0.1593	0.2124	0.2655	0.3186	0.3717	0.4248
4	0.0996	0.1494	0.1992	0.2490	0.2988	0.3486	0.3984
5	0.0930	0.1395	0.1860	0.2325	0.2790	0.3255	0.3720
6	0.0864	0.1296	0.1728	0.2160	0.2592	0.3024	0.3456
7	0.0798	0.1197	0.1596	0.1995	0.2394	0.2793	0.3192
8	0.0730	0.1095	0.1460	0.1825	0.2190	0.2555	0.2920
9	0.0664	0.0996	0.1328	0.1660	0.1992	0.2324	0.2656
10	0.0598	0.0897	0.1196	0.1495	0.1794	0.2093	0.2392
11	0.0532	0.0798	0.1064	0.1330	0.1596	0.1862	0.2128
12	0.0466	0.0699	0.0932	0.1165	0.1398	0.1631	0.1864
13	0.0398	0.0597	0.0796	0.0995	0.1194	0.1393	0.1592
14	0.0332	0.0498	0.0664	0.0830	0.0996	0.1162	0.1328
15	0.0298	0.0447	0.0596	0.0745	0.0894	0.1043	0.1192
16	0.0266	0.0399	0.0532	0.0665	0.0798	0.0931	0.1064
17	0.0240	0.0360	0.0480	0.0600	0.0720	0.0840	0.0960
18	0.0212	0.0318	0.0424	0.0530	0.0636	0.0742	0.0848
19	0.0186	0.0279	0.0372	0.0465	0.0558	0.0651	0.0744
20	0.0160	0.0240	0.0320	0.0400	0.0480	0.0560	0.0640
21	0.0146	0.0219	0.0292	0.0365	0.0438	0.0511	0.0584
22	0.0133	0.0201	0.0268	0.0335	0.0402	0.0469	0.0536
23	0.0120	0.0180	0.0240	0.0300	0.0360	0.0420	0.0480
24	0.0106	0.0159	0.0212	0.0265	0.0318	0.0371	0.0424
25	0.0094	0.0141	0.0188	0.0235	0.0282	0.0329	0.0376
26	0.0080	0.0120	0.0160	0.0200	0.0240	0.0280	0.0320
27	0.0074	0.0111	0.0148	0.0185	0.0222	0.0259	0.0296
28	0.0066	0.0099	0.0132	0.0165	0.0198	0.0231	0.0264
29	0.0060	0.0090	0.0120	0.0150	0.0180	0.0210	0.0240
30	0.0054	0.0081	0.0108	0.0135	0.0162	0.0189	0.0212
31	0.0046	0.0069	0.0092	0.0115	0.0138	0.0161	0.0184
32	0.0044	0.0066	0.0088	0.0110	0.0132	0.0154	0.0176
33	0.0040	0.0060	0.0080	0.0100	0.0120	0.0140	0.0160
34	0.0036	0.0054	0.0072	0.0090	0.0108	0.0126	0.0144
35	0.0034	0.0051	0.0068	0.0085	0.0102	0.0119	0.0136
36	0.0030	0.0045	0.0060	0.0075	0.0090	0.0105	0.0120
37	0.0028	0.0042	0.0056	0.0070	0.0084	0.0098	0.0112
38	0.0026	0 0039	0.0052	0.0065	0.0078	0.0091	0.0104

Weights of Flat Rolled Steel Per Lineal Foot in Pounds (*Continued*)

U. S. St'd Gage for Plate	Width of Flat Steel, Inches							
	9/16	5/8	11/16	3/4	13/16	7/8	1	2
0000000	0.9567	1.0630	1.1693	1.2756	1.3819	1.4882	1.7008	3.4016
000000	0.8964	0.9960	1.0956	1.1952	1.2948	1.3944	1.5936	3.1872
00000	0.8370	0.9300	1.0230	1.1160	1.2090	1.3020	1.4880	2.9760
0000	0.7776	0.8640	0.9500	1.0368	1.1232	1.2096	1.3824	2.7648
000	0.7173	0.7970	0.8767	0.9564	1.0361	1.1158	1.2752	2.5504
00	0.6579	0.7310	0.8041	0.8772	0.9505	1.0234	1.1696	2.3392
0	0.5976	0.6640	0.7300	0.7968	0.8632	0.9296	1.0624	2.1248
1	0.5382	0.5980	0.6578	0.7176	0.7774	0.8372	0.9568	1.9136
2	0.5085	0.5650	0.6215	0.6780	0.7345	0.7910	0.9040	1.8080
3	0.4779	0.5310	0.5841	0.6372	0.6903	0.7434	0.8496	1.6992
4	0.4482	0.4980	0.5478	0.5976	0.6474	0.6972	0.7968	1.5936
5	0.4185	0.4650	0.5115	0.5580	0.6045	0.6510	0.7440	1.4880
6	0.3888	0.4320	0.4752	0.5184	0.5616	0.6048	0.6912	1.3824
7	0.3591	0.3990	0.4389	0.4788	0.5187	0.5586	0.6384	1.2768
8	0.3285	0.3650	0.4015	0.4380	0.4745	0.5110	0.5840	1.1680
9	0.2988	0.3320	0.3652	0.3984	0.4316	0.4648	0.5312	1.0624
10	0.2691	0.2990	0.3289	0.3588	0.3887	0.4186	0.4784	0.9568
11	0.2394	0.2660	0.2926	0.3192	0.3458	0.3724	0.4256	0.8512
12	0.2097	0.2330	0.2563	0.2796	0.3029	0.3262	0.3728	0.7456
13	0.1791	0.1990	0.2189	0.2388	0.2587	0.2786	0.3184	0.6368
14	0.1494	0.1660	0.1826	0.1992	0.2158	0.2324	0.2656	0.5312
15	0.1341	0.1490	0.1639	0.1788	0.1937	0.2086	0.2384	0.4768
16	0.1197	0.1330	0.1463	0.1596	0.1729	0.1862	0.2128	0.4256
17	0.1080	0.1200	0.1320	0.1440	0.1560	0.1680	0.1920	0.3840
18	0.0954	0.1060	0.1166	0.1272	0.1378	0.1484	0.1696	0.3392
19	0.0837	0.0930	0.1023	0.1116	0.1209	0.1302	0.1488	0.2976
20	0.0720	0.0800	0.0880	0.0960	0.1040	0.1120	0.1280	0.2560
21	0.0657	0.0730	0.0803	0.0876	0.0949	0.1022	0.1168	0.2336
22	0.0603	0.0670	0.0737	0.0804	0.0871	0.0938	0.1072	0.2144
23	0.0540	0.0600	0.0660	0.0720	0.0780	0.0840	0.0960	0.1920
24	0.0477	0.0530	0.0583	0.0636	0.0689	0.0742	0.0848	0.1696
25	0.0423	0.0470	0.0517	0.0564	0.0611	0.0658	0.0752	0.1504
26	0.0360	0.0400	0.0440	0.0480	0.0520	0.0560	0.0640	0.1280
27	0.0333	0.0370	0.0407	0.0444	0.0481	0.0518	0.0592	0.1184
28	0.0297	0.0330	0.0363	0.0396	0.0429	0.0462	0.0528	0.1056
29	0.0270	0.0300	0.0330	0.0360	0.0390	0.0420	0.0480	0.0960
30	0.0243	0.0270	0.0297	0.0324	0.0351	0.0378	0.0432	0.0864
31	0.0207	0.0230	0.0253	0.0276	0.0299	0.0322	0.0368	0.0736
32	0.0198	0.0220	0.0242	0.0264	0.0286	0.0308	0.0352	0.0704
33	0.0180	0.0200	0.0220	0.0240	0.0260	0.0280	0.0320	0.0640
34	0.0162	0.0180	0.0198	0.0216	0.0234	0.0252	0.0288	0.0576
35	0.0153	0.0170	0.0187	0.0204	0.0221	0.0238	0.0272	0.0544
36	0.0135	0.0150	0.0165	0.0180	0.0195	0.0210	0.0240	0.0480
37	0.0126	0.0140	0.0154	0.0168	0.0182	0.0196	0.0224	0.0448
38	0.0117	0.0130	0.0143	0.0156	0.0169	0.0182	0.0208	0.0416

Weights of Flat Rolled Steel per Lineal Foot in Pounds (*Continued*)

U. S. St'd Gage for Plate	Width of Flat Steel, Inches							
	3	4	5	6	7	8	9	10
0000000	5.1024	6.8032	8.504	10.204	11.905	13.606	15.307	17.008
000000	4.7808	6.3744	7.968	9.561	11.155	12.748	14.342	15.936
00000	4.4640	5.9520	7.440	8.928	10.416	11.904	13.392	14.880
0000	4.1472	5.5296	6.912	8.294	9.676	11.059	12.441	13.824
000	3.8256	5.1008	6.376	7.651	8.926	10.201	11.476	12.752
00	3.5088	4.6784	5.848	7.017	8.187	9.356	10.526	11.696
0	3.1872	4.2496	5.312	6.374	7.436	8.499	9.561	10.624
1	2.8704	3.8272	4.784	5.740	6.697	7.654	8.611	9.568
2	2.7120	3.6160	4.520	5.424	6.328	7.232	8.136	9.040
3	2.5488	3.3984	4.248	5.097	5.947	6.796	7.646	8.496
4	2.3904	3.1872	3.984	4.780	5.577	6.374	7.171	7.968
5	2.2320	2.9760	3.720	4.464	5.208	5.952	6.696	7.440
6	2.0736	2.7648	3.456	4.147	4.838	5.529	6.220	6.912
7	1.9152	2.5536	3.192	3.830	4.468	5.107	5.745	6.384
8	1.7520	2.3360	2.920	3.504	4.088	4.672	5.256	5.840
9	1.5936	2.1248	2.656	3.187	3.718	4.249	4.780	5.312
10	1.4352	1.9136	2.392	2.870	3.348	3.827	4.305	4.784
11	1.2768	1.7024	2.128	2.553	2.979	3.404	3.830	4.256
12	1.1184	1.4912	1.864	2.236	2.609	2.982	3.355	3.728
13	0.9552	1.2736	1.592	1.910	2.228	2.547	2.865	3.184
14	0.7968	1.0624	1.328	1.593	1.859	2.124	2.390	2.656
15	0.7152	0.9536	1.192	1.430	1.668	1.907	2.145	2.384
16	0.6384	0.8512	1.064	1.276	1.489	1.702	1.915	2.128
17	0.5760	0.7680	0.960	1.152	1.344	1.536	1.728	1.920
18	0.5088	0.6786	0.848	1.027	1.187	1.357	1.526	1.696
19	0.4464	0.5952	0.744	0.892	1.041	1.190	1.319	1.488
20	0.3840	0.5120	0.640	0.768	0.896	1.024	1.152	1.280
21	0.3504	0.4672	0.584	0.700	0.817	0.934	1.051	1.168
22	0.3216	0.4288	0.536	0.643	0.750	0.857	0.964	1.072
23	0.2880	0.3840	0.480	0.576	0.672	0.768	0.864	0.960
24	0.2544	0.3392	0.424	0.508	0.593	0.678	0.763	0.848
25	0.2256	0.3008	0.376	0.451	0.526	0.601	0.676	0.752
26	0.1920	0.2560	0.320	0.384	0.448	0.512	0.596	0.640
27	0.1776	0.2368	0.296	0.355	0.414	0.473	0.532	0.592
28	0.1584	0.2112	0.264	0.316	0.369	0.422	0.475	0.528
29	0.1440	0.1920	0.240	0.288	0.336	0.384	0.432	0.480
30	0.1272	0.1768	0.216	0.259	0.302	0.353	0.388	0.432
31	0.1104	0.1472	0.184	0.220	0.257	0.297	0.331	0.368
32	0.1056	0.1408	0.176	0.211	0.246	0.281	0.316	0.352
33	0.0960	0.1280	0.160	0.192	0.224	0.256	0.288	0.320
34	0.0864	0.1152	0.144	0.172	0.201	0.230	0.259	0.288
35	0.0816	0.1088	0.136	0.163	0.190	0.217	0.244	0.272
36	0.0720	0.0960	0.120	0.148	0.168	0.192	0.216	0.240
37	0.0674	0.0896	0.112	0.134	0.156	0.179	0.201	0.224
38	0.0624	0.0832	0.104	0.124	0.145	0.166	0.187	0.208

Weights of Flat Rolled Steel per Lineal Foot in Pounds (*Continued*)

U. S. St'd Gage for Plate	Width of Flat Steel, Inches							
	11	12	13	14	15	16	18	20
0000000	18.708	20.409	22.110	23.811	25.512	27.212	30.614	34.016
000000	17.529	19.123	20.716	22.310	23.904	25.497	28.684	31.872
00000	16.368	17.856	19.344	20.832	22.320	23.808	26.784	29.760
0000	15.206	16.588	17.971	19.353	20.736	22.118	24.883	27.648
000	14.027	15.302	16.577	17.852	19.128	20.403	22.953	25.504
00	12.865	14.035	15.204	16.374	17.544	18.713	21.052	23.392
0	11.686	12.748	13.811	14.873	15.936	16.998	19.123	21.248
1	10.524	11.481	12.438	13.395	14.352	15.308	17.222	19.136
2	9.944	10.848	11.752	12.656	13.560	14.464	16.272	18.080
3	9.345	10.195	11.044	11.894	12.744	13.593	15.292	16.992
4	8.764	9.561	10.358	11.155	11.952	12.748	14.342	15.936
5	8.184	8.928	9.672	10.416	11.160	11.804	13.392	14.880
6	7.603	8.294	8.985	9.676	10.368	11.059	12.441	13.824
7	7.022	7.660	8.299	8.937	9.576	10.214	11.491	12.768
8	6.424	7.008	7.592	8.176	8.760	9.344	10.512	11.680
9	5.843	6.374	6.905	7.436	7.968	8.499	9.561	10.624
10	5.262	5.740	6.219	6.697	7.176	7.654	8.611	9.568
11	4.681	5.107	5.532	5.958	6.384	6.809	7.660	8.512
12	4.100	4.473	4.846	5.219	5.592	5.964	6.710	7.456
13	3.502	3.820	4.139	4.457	4.776	5.094	5.731	6.368
14	2.921	3.187	3.452	3.718	3.984	4.249	4.780	5.312
15	2.622	2.860	3.099	3.337	3.576	3.814	4.291	4.768
16	2.340	2.553	2.766	2.979	3.192	3.404	3.830	4.256
17	2.112	2.304	2.496	2.688	2.880	3.072	3.456	3.840
18	1.865	2.055	2.204	2.374	2.544	2.714	3.056	3.392
19	1.636	1.785	1.934	2.083	2.232	2.380	2.638	2.976
20	1.408	1.536	1.664	1.792	1.920	2.048	2.304	2.560
21	1.284	1.401	1.518	1.635	1.752	1.868	2.102	2.336
22	1.179	1.286	1.393	1.500	1.608	1.715	1.929	2.144
23	1.056	1.152	1.248	1.344	1.440	1.536	1.728	1.920
24	0.932	1.017	1.102	1.187	1.272	1.356	1.526	1.696
25	0.827	0.902	0.977	1.052	1.128	1.203	1.353	1.504
26	0.704	0.768	0.832	0.896	0.960	1.024	1.192	1.380
27	0.651	0.710	0.769	0.828	0.888	0.947	1.065	1.184
28	0.580	0.633	0.686	0.739	0.792	0.844	0.950	1.056
29	0.528	0.576	0.624	0.672	0.720	0.768	0.864	0.960
30	0.475	0.518	0.561	0.604	0.648	0.707	0.777	0.864
31	0.404	0.441	0.478	0.515	0.552	0.594	0.662	0.736
32	0.387	0.422	0.457	0.492	0.528	0.563	0.633	0.714
33	0.352	0.384	0.416	0.448	0.480	0.512	0.576	0.640
34	0.316	0.345	0.374	0.403	0.432	0.460	0.518	0.576
35	0.299	0.326	0.359	0.380	0.408	0.435	0.489	0.544
36	0.264	0.296	0.312	0.336	0.360	0.384	0.432	0.480
37	0.246	0.279	0.291	0.313	0.336	0.358	0.403	0.448
38	0.228	0.249	0.270	0.291	0.312	0.332	0.374	0.416

Weight of Flat Rolled Steel Bars in Pounds per Lineal Foot

(One cubic foot of rolled steel weighs 489.6 pounds.)

Thickness of Bar, Ins.	Width of Bar, Inches																	
	1/4	1/2	3/4	1	1 1/4	1 1/2	1 3/4	2	2 1/4	2 1/2	2 3/4	3	3 1/4	3 1/2	3 3/4	4	4 1/4	4 1/2
1/16	0.053	0.106	0.159	0.21	0.27	0.32	0.37	0.42	0.48	0.53	0.58	0.64	0.69	0.74	0.80	0.85	0.90	0.96
1/8	0.106	0.212	0.319	0.42	0.53	0.64	0.74	0.85	0.96	1.06	1.17	1.27	1.38	1.49	1.59	1.70	1.81	1.91
3/16	0.159	0.319	0.478	0.64	0.80	0.96	1.12	1.28	1.43	1.59	1.75	1.91	2.07	2.23	2.39	2.55	2.71	2.87
1/4	0.213	0.425	0.638	0.85	1.06	1.28	1.49	1.70	1.91	2.13	2.34	2.55	2.76	2.98	3.19	3.40	3.61	3.83
5/16	0.266	0.531	0.797	1.06	1.33	1.59	1.86	2.13	2.39	2.66	2.92	3.19	3.45	3.72	3.98	4.25	4.52	4.78
3/8	0.319	0.638	0.956	1.28	1.59	1.91	2.23	2.55	2.87	3.19	3.51	3.83	4.14	4.46	4.78	5.10	5.42	5.74
7/16	0.372	0.744	1.12	1.49	1.86	2.23	2.60	2.98	3.35	3.72	4.09	4.46	4.83	5.21	5.58	5.95	6.32	6.69
1/2	0.425	0.850	1.28	1.70	2.13	2.55	2.98	3.40	3.83	4.25	4.68	5.10	5.53	5.95	6.38	6.80	7.22	7.65
9/16	0.478	0.956	1.43	1.91	2.39	2.87	3.35	3.83	4.30	4.78	5.26	5.74	6.22	6.69	7.17	7.65	8.13	8.61
5/8	0.531	1.06	1.59	2.13	2.66	3.19	3.72	4.25	4.78	5.31	5.84	6.38	6.91	7.44	7.97	8.50	9.03	9.56
11/16	0.584	1.17	1.75	2.34	2.92	3.51	4.09	4.68	5.26	5.84	6.43	7.01	7.60	8.18	8.77	9.35	9.93	10.52
3/4	0.638	1.28	1.91	2.55	3.19	3.83	4.46	5.10	5.74	6.38	7.01	7.65	8.29	8.93	9.56	10.20	10.84	11.48
13/16	0.691	1.38	2.07	2.76	3.45	4.14	4.83	5.53	6.22	6.91	7.60	8.29	8.98	9.67	10.36	11.05	11.74	12.43
7/8	0.744	1.49	2.23	2.98	3.72	4.46	5.21	5.95	6.69	7.44	8.18	8.93	9.67	10.41	11.16	11.90	12.64	13.39
15/16	0.797	1.59	2.39	3.19	3.98	4.78	5.58	6.38	7.17	7.97	8.77	9.56	10.36	11.16	11.95	12.75	13.55	14.34
1	0.850	1.70	2.55	3.40	4.25	5.10	5.95	6.80	7.65	8.50	9.35	10.20	11.05	11.90	12.75	13.60	14.45	15.30
1 1/16	0.903	1.81	2.71	3.61	4.51	5.42	6.32	7.23	8.13	9.03	9.93	10.84	11.74	12.64	13.55	14.45	15.35	16.26
1 1/8	0.956	1.91	2.87	3.83	4.78	5.74	6.69	7.65	8.61	9.56	10.52	11.48	12.43	13.39	14.34	15.30	16.26	17.21
1 3/16	1.01	2.02	3.03	4.04	5.05	6.06	7.07	8.08	9.08	10.09	11.10	12.11	13.12	14.13	15.14	16.15	17.16	18.17
1 1/4	1.06	2.12	3.19	4.25	5.31	6.38	7.44	8.50	9.56	10.63	11.69	12.75	13.81	14.88	15.94	17.00	18.06	19.13
1 5/16	1.12	2.23	3.34	4.46	5.58	6.69	7.81	8.93	10.04	11.16	12.27	13.39	14.50	15.62	16.73	17.85	18.97	20.08
1 3/8	1.17	2.34	3.50	4.68	5.84	7.01	8.18	9.35	10.52	11.69	12.86	14.03	15.19	16.36	17.53	18.70	19.87	21.04
1 7/16	1.22	2.44	3.66	4.89	6.11	7.33	8.55	9.78	11.00	12.22	13.44	14.66	15.88	17.11	18.33	19.55	20.77	21.99
1 1/2	1.27	2.55	3.82	5.10	6.38	7.65	8.93	10.20	11.48	12.75	14.03	15.30	16.58	17.85	19.13	20.40	21.68	22.95
1 9/16	1.33	2.66	3.98	5.31	6.64	7.97	9.30	10.63	11.95	13.28	14.61	15.92	17.27	18.59	19.92	21.25	22.58	23.91
1 5/8	1.38	2.76	4.14	5.53	6.91	8.29	9.67	11.05	12.43	13.81	15.19	16.58	17.96	19.34	20.72	22.10	23.48	24.86
1 11/16	1.43	2.87	4.30	5.74	7.17	8.61	10.04	11.48	12.91	14.34	15.78	17.21	18.65	20.08	21.52	22.95	24.38	25.82
1 3/4	1.49	2.97	4.46	5.95	7.44	8.93	10.41	11.90	13.39	14.88	16.36	17.85	19.34	20.83	22.31	23.80	25.29	26.78
1 13/16	1.54	3.08	4.62	6.16	7.70	9.24	10.78	12.33	13.87	15.41	16.95	18.49	20.03	21.57	23.11	24.65	26.19	27.73
1 7/8	1.59	3.19	4.78	6.38	7.97	9.56	11.15	12.75	14.34	15.94	17.53	19.13	20.72	22.31	23.91	25.50	27.09	28.69
1 15/16	1.65	3.29	4.94	6.59	8.23	9.88	11.53	13.18	14.82	16.47	18.12	19.76	21.41	23.06	24.70	26.35	28.00	29.64
2	1.70	3.40	5.10	6.80	8.50	10.20	11.90	13.60	15.30	17.00	18.70	20.40	22.10	23.80	25.50	27.20	28.90	30.60

Weight of Flat Rolled Steel Bars in Pounds per Lineal Foot

(One cubic foot of rolled steel weighs 489.6 pounds.)

Thickness of Bar, Ins.	Width of Bar, Inches																	
	4¾	5	5¼	5½	5¾	6	6½	7	7½	8	8½	9	9½	10	10½	11	11½	12
1/16	1.01	1.06	1.11	1.17	1.22	1.27	1.38	1.49	1.59	1.70	1.81	1.91	2.02	2.12	2.23	2.34	2.44	2.55
1/8	2.02	2.12	2.23	2.34	2.44	2.55	2.76	2.97	3.18	3.40	3.61	3.82	4.04	4.25	4.46	4.67	4.89	5.10
3/16	3.03	3.19	3.35	3.51	3.67	3.83	4.14	4.46	4.78	5.10	5.42	5.74	6.06	6.38	6.69	7.01	7.33	7.65
1/4	4.04	4.25	4.46	4.68	4.89	5.10	5.53	5.95	6.38	6.80	7.23	7.65	8.08	8.50	8.93	9.35	9.78	10.20
5/16	5.05	5.31	5.58	5.84	6.11	6.38	6.91	7.44	7.97	8.50	9.03	9.56	10.09	10.63	11.16	11.69	12.22	12.75
3/8	6.06	6.38	6.69	7.01	7.33	7.65	8.29	8.93	9.56	10.20	10.84	11.48	12.11	12.75	13.39	14.03	14.66	15.30
7/16	7.07	7.44	7.81	8.18	8.55	8.93	9.67	10.41	11.16	11.90	12.64	13.39	14.13	14.88	15.62	16.36	17.11	17.85
1/2	8.08	8.50	8.93	9.35	9.78	10.20	11.05	11.90	12.75	13.60	14.45	15.30	16.15	17.00	17.85	18.70	19.55	20.40
9/16	9.08	9.56	10.04	10.52	11.00	11.48	12.43	13.39	14.34	15.30	16.26	17.21	18.17	19.13	20.08	21.04	21.99	22.95
5/8	10.09	10.63	11.16	11.69	12.22	12.75	13.81	14.88	15.94	17.00	18.06	19.13	20.19	21.25	22.31	23.38	24.44	25.50
11/16	11.10	11.69	12.27	12.86	13.44	14.03	15.19	16.36	17.53	18.70	19.87	21.04	22.21	23.38	24.54	25.71	26.88	28.05
3/4	12.11	12.75	13.39	14.03	14.66	15.30	16.58	17.85	19.13	20.40	21.68	22.95	24.23	25.50	26.78	28.05	29.33	30.60
13/16	13.12	13.81	14.50	15.19	15.88	16.58	17.96	19.34	20.72	22.10	23.48	24.86	26.24	27.63	29.01	30.39	31.77	33.15
7/8	14.13	14.88	15.62	16.36	17.11	17.85	19.34	20.83	22.31	23.80	25.29	26.78	28.26	29.75	31.24	32.73	34.21	35.70
15/16	15.14	15.94	16.73	17.53	18.33	19.13	20.72	22.31	23.91	25.50	27.09	28.69	30.28	31.88	33.47	35.06	36.66	38.25
1	16.15	17.00	17.85	18.70	19.55	20.40	22.10	23.80	25.50	27.20	28.90	30.60	32.30	34.00	35.70	37.40	39.10	40.80
1 1/16	17.17	18.06	18.97	19.87	20.77	21.68	23.48	25.29	27.09	28.90	30.71	32.51	34.32	36.13	37.93	39.74	41.54	43.35
1 1/8	18.17	19.13	20.08	21.04	21.99	22.95	24.86	26.78	28.69	30.60	32.51	34.43	36.34	38.25	40.16	42.08	43.99	45.90
1 3/16	19.18	20.19	21.20	22.21	23.22	24.23	26.24	28.26	30.28	32.30	34.32	36.34	38.36	40.38	42.39	44.41	46.43	48.45
1 1/4	20.19	21.25	22.31	23.38	24.44	25.50	27.63	29.75	31.88	34.00	36.13	38.25	40.38	42.50	44.63	46.75	48.88	51.00
1 5/16	21.20	22.31	23.43	24.54	25.66	26.78	29.01	31.24	33.47	35.70	37.93	40.16	42.39	44.63	46.86	49.09	51.32	53.55
1 3/8	22.21	23.38	24.54	25.71	26.88	28.05	30.39	32.73	35.06	37.40	39.74	42.08	44.41	46.75	49.09	51.43	53.76	56.10
1 7/16	23.22	24.44	25.66	26.88	28.10	29.33	31.77	34.21	36.66	39.10	41.54	43.99	46.43	48.88	51.32	53.76	56.21	58.65
1 1/2	24.23	25.50	26.78	28.05	29.33	30.60	33.15	35.70	38.25	40.80	43.35	45.90	48.45	51.00	53.55	56.10	58.65	61.20
1 9/16	25.23	26.56	27.89	29.22	30.55	31.88	34.53	37.19	39.84	42.50	45.16	47.81	50.47	53.13	55.78	58.44	61.09	63.75
1 5/8	26.24	27.63	29.01	30.39	31.77	33.15	35.91	38.68	41.44	44.20	46.96	49.73	52.49	55.25	58.01	60.78	63.54	66.30
1 11/16	27.25	28.69	30.12	31.56	32.99	34.43	37.29	40.16	43.03	45.90	48.77	51.64	54.51	57.38	60.24	63.11	65.98	68.85
1 3/4	28.26	29.75	31.24	32.73	34.21	35.70	38.68	41.65	44.63	47.60	50.58	53.55	56.53	59.50	62.48	65.45	68.43	71.40
1 13/16	29.27	30.81	32.35	33.89	35.43	36.98	40.06	43.14	46.22	49.30	52.38	55.46	58.54	61.63	64.71	67.79	70.87	73.95
1 7/8	30.28	31.88	33.47	35.06	36.66	38.25	41.44	44.63	47.81	51.00	54.19	57.38	60.56	63.75	66.94	70.13	73.31	76.50
1 15/16	31.29	32.94	34.58	36.23	37.88	39.53	42.82	46.11	49.41	52.70	55.99	59.29	62.58	65.88	69.17	72.46	75.76	79.05
2	32.30	34.00	35.70	37.40	39.10	40.80	44.20	47.60	51.00	54.40	57.80	61.20	64.60	68.00	71.40	74.80	78.20	81.60

Areas and Weights of Fillets of Steel, Cast Iron and Brass

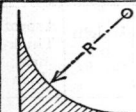

Calculations are based on the following weights:

Steel.........489.6 pounds per cubic foot.
Cast iron.....450   pounds per cubic foot.
Cast brass....504   pounds per cubic foot.

Radius R, Inches	Area, Square Inches	Weight of Steel		Weight of Cast Iron		Weight of Cast Brass	
		Per Foot	Per Inch	Per Foot	Per Inch	Per Foot	Per Inch
¼	0.0134	0.0455	0.0038	0.0418	0.0035	0.0469	0.0040
5⁄16	0.0209	0.0712	0.0059	0.0655	0.0054	0.0733	0.0061
⅜	0.0302	0.1027	0.0085	0.0945	0.0078	0.1058	0.0088
7⁄16	0.0411	0.1397	0.0116	0.1285	0.0107	0.1439	0.0120
½	0.0536	0.1825	0.0152	0.1679	0.0140	0.1880	0.0157
9⁄16	0.0679	0.2310	0.0192	0.2125	0.0177	0.2380	0.0200
⅝	0.0834	0.2847	0.0237	0.2619	0.0218	0.2932	0.0244
11⁄16	0.1014	0.3447	0.0287	0.3171	0.0264	0.3550	0.0300
¾	0.1207	0.4105	0.0342	0.3777	0.0315	0.4228	0.0352
13⁄16	0.1416	0.4817	0.0401	0.4432	0.0369	0.4962	0.0414
⅞	0.1643	0.5580	0.0465	0.5134	0.0428	0.5747	0.0479
15⁄16	0.1886	0.6405	0.0534	0.5893	0.0491	0.6597	0.0550
1	0.2146	0.7300	0.0608	0.6716	0.0559	0.7519	0.0626
1⅛	0.2716	0.9250	0.0771	0.8510	0.0709	0.9527	0.0794
1¼	0.3353	1.140	0.0950	1.049	0.0874	1.174	0.0979
1⅜	0.4057	1.379	0.1150	1.268	0.1057	1.420	0.1183
1½	0.4828	1.642	0.1368	1.511	0.1259	1.691	0.1410
1⅝	0.5668	1.930	0.1608	1.776	0.1479	1.988	0.1657
1¾	0.6572	2.235	0.1862	2.056	0.1713	2.302	0.1920
1⅞	0.7545	2.565	0.2137	2.360	0.1970	2.642	0.2202
2	0.8585	2.917	0.2431	2.684	0.2237	3.005	0.2504
2⅛	0.9692	3.292	0.2743	3.029	0.2502	3.391	0.2826
2¼	1.086	3.695	0.3079	3.399	0.2832	3.806	0.3172
2⅜	1.210	4.115	0.3429	3.786	0.3155	4.238	0.3532
2½	1.341	4.560	0.3800	4.195	0.3496	4.697	0.3914
2⅝	1.478	5.030	0.4192	4.628	0.3857	5.181	0.4317
2¾	1.623	5.507	0.4589	5.066	0.4222	5.672	0.4727
2⅞	1.774	6.027	0.5022	5.545	0.4621	6.208	0.5017
3	1.931	6.565	0.5471	5.940	0.4950	6.762	0.5635
3⅛	2.096	7.125	0.5937	6.555	0.5462	7.339	0.6116
3¼	2.267	7.700	0.6417	7.084	0.5903	7.931	0.6609
3⅜	2.444	8.300	0.6917	7.636	0.6363	8.549	0.7124
3½	2.629	8.925	0.7438	8.211	0.6926	9.193	0.7661
3⅝	2.820	9.575	0.7979	8.809	0.7341	9.862	0.8220
3¾	3.018	10.27	0.8523	9.448	0.7873	10.58	0.8817
3⅞	3.222	10.97	0.9142	10.09	0.8408	11.30	0.9417
4	3.434	11.65	0.9709	10.72	0.8933	12.00	1.000
4¼	3.876	13.15	1.096	12.10	1.008	13.54	1.130
4½	4.346	14.77	1.231	13.59	1.132	15.21	1.270
4¾	4.842	16.45	1.371	15.13	1.261	16.94	1.412
5	5.365	18.25	1.521	16.79	1.400	18.80	1.570

Tin Plate Base Weight and Thickness

Base * Weight, Lbs., and Symbols	Weight per Sq. Ft.	Approx. Thickness, Inch	Base Weight, Lbs., and Symbols	Weight per Sq. Ft.	Approx. Thickness, Inch	Base Weight, Lbs., and Symbols	Weight per Sq. Ft.	Approx. Thickness, Inch
55	0.253	0.006	128–IXL	0.588	0.015	255–7 X	1.125	0.028
60	0.276	0.007	135–IX	0.620	0.015	268–D 4 X	1.230	0.031
65	0 298	0.007	139–DC	0.638	0.016	275–8 X	1.263	0.032
70	0.321	0.008	155–2 X	0.712	0.018	295–9 X	1.355	0.034
75	0.344	0 009	175–3 X	0.804	0.020	315–10 X	1.447	0.036
80	0.367	0.009	180–DX	0.827	0.021	335–11 X	1.539	0.038
85	0.390	0.010	195–4 X	0.895	0.022	355–12 X	1.631	0.041
90	0.413	0.010	210–D 2 X	0.964	0.024	375–13 X	1.722	0.043
95	0.436	0.011	215–5 X	0.988	0.025	395–14 X	1.814	0.045
100–ICL	0.459	0.011	235–6 X	1.08	0.027	415–15 X	1.906	0.048
107–IC	0.491	0.012	240–D 3 X	1.10	0.027	435–16 X	1.998	0.050

* Weight of standard " base box " containing 112 sheets, 14 X 20 inches.

Sheet Zinc Gage
(Matthiessen & Hegeler Zinc Co.)

Gage No.	Thickness, Inches	Gage No.	Thickness, Inches	Gage No.	Thickness, Inches	Gage No.	Thickness, Inches
1	0.002	8	0.016	15	0.040	22	0.090
2	0.004	9	0.018	16	0.045	23	0.100
3	0.006	10	0.020	17	0.050	24	0.125
4	0.008	11	0.024	18	0.055	25	0.250
5	0.010	12	0.028	19	0.060	26	0.375
6	0.012	13	0.032	20	0.070	27	0.500
7	0.014	14	0.036	21	0.080	28	1.000

American "Russia-Iron" Gage

Gage No.	Thickness, Ins.	Gage No.	Thickness, Ins.	Gage No.	Thickness, Ins.	Gage No.	Thickness, Ins.	Gage No.	Thickness, Ins.
7	0.015	9	0.017	11	0.020	13	0.024	15	0.027
8	0.016	10	0.018	12	0.021	14	0.025	16	0.030

Weight in Pounds per Square Foot of Zinc Plate

Gage No.	Weight in Pounds per Sq. Foot	Gage No.	Weight in Pounds per Sq. Foot	Gage No.	Weight in Pounds per Sq. Foot	Gage No.	Weight in Pounds per Sq. Foot
1	0.07	8	0.60	15	1.50	22	3.37
2	0.15	9	0.67	16	1.68	23	3.75
3	0.22	10	0.75	17	1.87	24	4.70
4	0.30	11	0.90	18	2.06	25	9.40
5	0.37	12	1.05	19	2.25	26	14.00
6	0.45	13	1.20	20	2.62	27	18.75
7	0.52	14	1.35	21	3.00	28	37.50

Weights of Non-Ferrous Metal Sheets — 1

American Wire or Brown & Sharpe Gage No.	Thickness, Inch	Approximate Weight, Pounds per Square Foot				
		Copper*	Yellow Brass	Tobin Bronze	5 Per Cent Phosphor-Bronze	Everdur 1010
0000	0.4600	21.33	20.27	20.14	21.20	20.40
000	0.4096	18.99	18.05	17.93	18.88	18.17
00	0.3648	16.92	16.07	16.41	16.81	16.18
0	0.3249	15.06	14.32	14.23	14.98	14.41
1	0.2893	13.41	12.75	12.67	13.33	12.83
2	0.2576	11.94	11.35	11.28	11.87	11.43
3	0.2294	10.64	10.11	10.04	10.57	10.17
4	0.2043	9.473	9.002	8.943	9.414	9.061
5	0.1819	8.434	8.015	7.963	8.382	8.068
6	0.1620	7.512	7.138	7.092	7.465	7.185
7	0.1443	6.691	6.358	6.317	6.649	6.400
8	0.1285	5.958	5.662	5.625	5.921	5.699
9	0.1144	5.304	5.041	5.008	5.272	5.074
10	0.1019	4.725	4.490	4.461	4.696	4.519
11	0.0907	4.206	3.997	3.971	4.180	4.023
12	0.0808	3.747	3.560	3.537	3.723	3.584
13	0.0720	3.338	3.173	3.152	3.318	3.193
14	0.0641	2.972	2.825	2.807	2.954	2.843
15	0.0571	2.648	2.516	2.500	2.631	2.532
16	0.0508	2.355	2.238	2.223	2.341	2.253
17	0.0453	2.100	1.996	1.983	2.087	2.009
18	0.0403	1.869	1.776	1.764	1.857	1.787
19	0.0359	1.665	1.582	1.572	1.654	1.592
20	0.0320	1.484	1.410	1.401	1.475	1.419
21	0.0285	1.321	1.256	1.248	1.314	1.264
22	0.0253	1.178	1.119	1.112	1.170	1.127
23	0.0226	1.048	0.9958	0.9893	1.041	1.002
24	0.0201	0.9320	0.8857	0.8799	0.9263	0.8915
25	0.0179	0.8300	0.7887	0.7836	0.8248	0.7939
26	0.0159	0.7373	0.7006	0.6960	0.7327	0.7052
27	0.0142	0.6584	0.6257	0.6216	0.6544	0.6298
28	0.0126	0.5842	0.5552	0.5516	0.5806	0.5588
29	0.0113	0.5240	0.4979	0.4947	0.5207	0.5012
30	0.0100	0.4637	0.4406	0.4377	0.4608	0.4435
31	0.0089	0.4127	0.3922	0.3897	0.4102	0.3947
32	0.0080	0.3709	0.3525	0.3502	0.3686	0.3548
33	0.0071	0.3292	0.3129	0.3109	0.3272	0.3149
34	0.0063	0.2921	0.2776	0.2758	0.2903	0.2794
35	0.0056	0.2597	0.2468	0.2452	0.2581	0.2484
36	0.0050	0.2318	0.2203	0.2189	0.2304	0.2218
37	0.0045	0.2087	0.1983	0.1970	0.2074	0.1996
38	0.0040	0.1855	0.1763	0.1752	0.1844	0.1774
39	0.0035	0.1623	0.1542	0.1532	0.1613	0.1552
40	0.0031	0.1437	0.1366	0.1357	0.1429	0.1375

* Copper sheets can also be obtained in fractional-inch thicknesses varying by sixteenths of an inch from 1/16 to 2 inches.

Weights of Non-Ferrous Metal Sheets — 2

American Wire or Brown & Sharpe Gage No.	Thickness, Inch	Approximate Weight, Pounds per Square Foot				
		S.A.E. Aluminum Alloys Nos. 26 and 27	S.A.E. Aluminum Alloy No. 28	Aluminum Commercially Pure (99 to 99.4 Per Cent)	Nickel Silver 18%*	Nickel Silver 20%-30%
0000	0.4600	6.680	6.410	6.490	20.93	21.20
000	0.4096	5.950	5.710	5.780	18.64	18.88
00	0.3648	5.290	5.090	5.140	16.60	16.81
0	0.3249	4.720	4.530	4.580	14.78	14.97
1	0.2893	4.200	4.030	4.080	13.16	13.33
2	0.2576	3.738	3.591	3.632	11.72	11.87
3	0.2294	3.329	3.198	3.234	10.44	10.57
4	0.2043	2.964	2.848	2.880	9.296	9.414
5	0.1819	2.640	2.536	2.565	8.277	8.382
6	0.1620	2.351	2.258	2.284	7.372	7.466
7	0.1443	2.094	2.012	2.034	6.566	6.649
8	0.1285	1.865	1.792	1.812	5.847	5.921
9	0.1144	1.660	1.595	1.613	5.206	5.272
10	0.1019	1.479	1.420	1.437	4.637	4.696
11	0.0907	1.316	1.264	1.279	4.127	4.179
12	0.0808	1.172	1.126	1.139	3.677	3.724
13	0.0720	1.045	1.004	1.015	3.276	3.318
14	0.0641	0.930	0.894	0.904	2.917	2.954
15	0.0571	0.829	0.796	0.805	2.598	2.631
16	0.0508	0.737	0.708	0.716	2.312	2.341
17	0.0453	0.657	0.631	0.639	2.061	2.087
18	0.0403	0.585	0.562	0.568	1.834	1.857
19	0.0359	0.5210	0.5010	0.5060	1 634	1.655
20	0.0320	0.4640	0.4460	0.4510	1.456	1.474
21	0.0285	0.4140	0.3970	0.4020	1.297	1.313
22	0.0253	0.3671	0.3527	0.3567	1.156	1.171
23	0.0226	0.3280	0.3150	0.3186	1.028	1.041
24	0.0201	0.2917	0.2802	0.2834	0.9146	0.9262
25	0.0179	0.2597	0.2495	0.2524	0.8145	0.8248
26	0.0159	0.2307	0.2216	0.2242	0.7235	0.7327
27	0.0142	0.2060	0.1980	0.2002	0.6462	0.6544
28	0.0126	0.1828	0.1756	0.1776	0.5734	0.5807
29	0.0113	0.1640	0.1575	0.1593	0.5142	0.5207
30	0.0100	0.1451	0.1394	0.1410	0.4550	0.4608
31	0.0089	0.1296	0.1245	0.1259	0.4050	0.4101
32	0.0080	0.1154	0.1108	0.1121	0.3640	0.3686
33	0.0071	0.1027	0.0987	0.0998	0.3231	0.3272
34	0.0063	0.0914	0.0878	0.0888	0.2867	0.2903
35	0.0056	0.0814	0.0782	0.0791	0.2548	0.2580
36	0.0050	0.0726	0.0697	0.0705	0.2275	0.2304
37	0.0045	0.0646	0.0620	0.0627	0.2048	0.2074
38	0.0040	0.0576	0.0553	0.0560	0.1820	0.1843
39	0.0035	0.0512	0.0492	0.0498	0.1593	0.1613
40	0.0031	0.0456	0.0438	0.0443	0.1411	0.1429

* Multiply weights in this column by 0.9905 for 10 per cent nickel-silver and by 0.9937 for 15 per cent nickel-silver.

WIRE AND SHEET-METAL GAGES

The thicknesses of sheet metals and the diameters of wires conform to various gaging systems. These gage sizes are indicated by numbers, and the following tables give the decimal equivalents of the different gage numbers. Much confusion has resulted from the use of gage numbers, and in ordering materials it is preferable to give the exact dimensions in decimal fractions of an inch. While the dimensions thus specified should conform to the gage ordinarily used for a given class of material, any error in the specification due, for example, to the use of a table having "rounded off" or approximate equivalents, will be apparent to the manufacturer at the time the order is placed. Furthermore, the decimal method of indicating wire diameters and sheet metal thicknesses has the advantage of being self-explanatory, whereas arbitrary gage numbers are not. The decimal system of indicating gage sizes is now being used quite generally, and gage numbers are gradually being discarded. Unfortunately, there is considerable variation in the use of different gages. For example, a gage ordinarily used for copper, brass, and other non-ferrous materials, may at times be used for steel, and vice versa. The gages specified in the following are the ones ordinarily employed for the materials mentioned, but there are in some cases minor exceptions and variations in the different industries.

Wire Gages. — The wire gage system used by practically all of the steel producers in the United States is known by the name Steel Wire Gage or to distinguish it from the Standard Wire Gage (S.W.G.) used in Great Britain it is called the United States Steel Wire Gage. It is the same as the Washburn and Moen, American Steel and Wire Company, and Roebling Wire Gages. The name has the official sanction of the Bureau of Standards at Washington but is not legally effective. The only wire gage which has been recognized in Acts of Congress is the Birmingham Gage (also known as Stub's Iron Wire). The Birmingham Gage is, however, nearly obsolete in both the United States and Great Britain, where it originated. Copper and aluminum wires are specified in decimal fractions. They were formerly universally specified in the United States by the American or Brown & Sharpe Wire Gage. Music spring steel wire, one of the highest quality wires of several types used for mechanical springs, is specified by the piano or music wire gage.

In Great Britain one wire gage has been legalized. This is called the Standard Wire Gage (S.W.G.), formerly called Imperial Wire Gage.

Gages for Rods. — Steel wire rod sizes are designated by fractional or decimal parts of an inch and by the gage numbers of the United States Steel Wire Gage. Copper and aluminum rods are specified by decimal fractions and fractions. Drill rod may be specified in decimal fractions but in the carbon and alloy tool steel grades may also be specified in the Stub's Steel Wire Gage and in the high-speed steel drill rod grade may be specified by the Morse Twist Drill Gage (Manufacturers' Standard Gage for Twist Drills). For gage numbers with corresponding decimal equivalents see the tables of American Standard Straight Shank Twist Drills.

Gages for Wall Thicknesses of Tubing. — At one time the Birmingham or Stub's Iron Wire Gage was used to specify the wall thickness of the following classes of tubing: seamless brass, seamless copper, seamless steel, and aluminum. The Brown & Sharpe Wire Gage was used for brazed brass and brazed copper tubing. Wall thicknesses are now specified by decimal parts of an inch but in the case of steel pressure tubes and steel mechanical tubing the wall thickness may be specified by the Birmingham or Stub's Iron Wire Gage. In Great Britain the Standard Wire Gage (S.W.G.) is used to specify the wall thickness of some kinds of steel tubes.

Table 1. Wire Gages in Approximate Decimals of an Inch

No. of Wire Gage	American Wire or Brown & Sharpe Gage	Steel Wire Gage (U.S.)*	British Standard Wire Gage (Imperial Wire Gage)	Music or Piano Wire Gage	Birmingham or Stub's Iron Wire Gage	Stub's Steel Wire Gage	No. of Wire Gage	Stub's Steel Wire Gage
7/0	...	0.4900	0.5000	...	...	...	51	0.066
6/0	0.5800	0.4615	0.4640	0.004	...	...	52	0.063
5/0	0.5165	0.4305	0.4320	0.005	0.5000	...	53	0.058
4/0	0.4600	0.3938	0.4000	0.006	0.4540	...	54	0.055
3/0	0.4096	0.3625	0.3720	0.007	0.4250	...	55	0.050
2/0	0.3648	0.3310	0.3480	0.008	0.3800	...	56	0.045
1/0	0.3249	0.3065	0.3240	0.009	0.3400	...	57	0.042
1	0.2893	0.2830	0.3000	0.010	0.3000	0.227	58	0.041
2	0.2576	0.2625	0.2760	0.011	0.2840	0.219	59	0.040
3	0.2294	0.2437	0.2520	0.012	0.2590	0.212	60	0.039
4	0.2043	0.2253	0.2320	0.013	0.2380	0.207	61	0.038
5	0.1819	0.2070	0.2120	0.014	0.2200	0.204	62	0.037
6	0.1620	0.1920	0.1920	0.016	0.2030	0.201	63	0.036
7	0.1443	0.1770	0.1760	0.018	0.1800	0.199	64	0.035
8	0.1285	0.1620	0.1600	0.020	0.1650	0.197	65	0.033
9	0.1144	0.1483	0.1440	0.022	0.1480	0.194	66	0.032
10	0.1019	0.1350	0.1280	0.024	0.1340	0.191	67	0.031
11	0.0907	0.1205	0.1160	0.026	0.1200	0.188	68	0.030
12	0.0808	0.1055	0.1040	0.029	0.1090	0.185	69	0.029
13	0.0720	0.0915	0.0920	0.031	0.0950	0.182	70	0.027
14	0.0641	0.0800	0.0800	0.033	0.0830	0.180	71	0.026
15	0.0571	0.0720	0.0720	0.035	0.0720	0.178	72	0.024
16	0.0508	0.0625	0.0640	0.037	0.0650	0.175	73	0.023
17	0.0453	0.0540	0.0560	0.039	0.0580	0.172	74	0.022
18	0.0403	0.0475	0.0480	0.041	0.0490	0.168	75	0.020
19	0.0359	0.0410	0.0400	0.043	0.0420	0.164	76	0.018
20	0.0320	0.0348	0.0360	0.045	0.0350	0.161	77	0.016
21	0.0285	0.0318	0.0320	0.047	0.0320	0.157	78	0.015
22	0.0253	0.0286	0.0280	0.049	0.0280	0.155	79	0.014
23	0.0226	0.0258	0.0240	0.051	0.0250	0.153	80	0.013
24	0.0201	0.0230	0.0220	0.055	0.0220	0.151	...	...
25	0.0179	0.0204	0.0200	0.059	0.0200	0.148	...	...
26	0.0159	0.0181	0.0180	0.063	0.0180	0.146	...	...
27	0.0142	0.0173	0.0164	0.067	0.0160	0.143	...	...
28	0.0126	0.0162	0.0149	0.071	0.0140	0.139	...	...
29	0.0113	0.0150	0.0136	0.075	0.0130	0.134	...	...
30	0.0100	0.0140	0.0124	0.080	0.0120	0.127	...	...
31	0.00893	0.0132	0.0116	0.085	0.0100	0.120	...	...
32	0.00795	0.0128	0.0108	0.090	0.0090	0.115	...	...
33	0.00708	0.0118	0.0100	0.095	0.0080	0.112	...	...
34	0.00630	0.0104	0.0092	0.100	0.0070	0.110	...	...
35	0.00561	0.0095	0.0084	0.106	0.0050	0.108	...	...
36	0.00500	0.0090	0.0076	0.112	0.0040	0.106	...	...
37	0.00445	0.0085	0.0068	0.118	...	0.103	...	...
38	0.00396	0.0080	0.0060	0.124	...	0.101	...	...
39	0.00353	0.0075	0.0052	0.130	...	0.099	...	...
40	0.00314	0.0070	0.0048	0.138	...	0.097	...	...
41	0.00280	0.0066	0.0044	0.146	...	0.095	...	...
42	0.00249	0.0062	0.0040	0.154	...	0.092	...	...
43	0.00222	0.0060	0.0036	0.162	...	0.088	...	...
44	0.00198	0.0058	0.0032	0.170	...	0.085	...	...
45	0.00176	0.0055	0.0028	0.180	...	0.081	...	...
46	0.00157	0.0052	0.0024	...	...	0.079	...	...
47	0.00140	0.0050	0.0020	...	...	0.077	...	...
48	0.00124	0.0048	0.0016	...	...	0.075	...	...
49	0.00111	0.0046	0.0012	...	...	0.072	...	...
50	0.00099	0.0044	0.0010	...	...	0.069	...	...

* Also known as Washburn and Moen, American Steel and Wire Co. and Roebling Wire Gages. A greater selection of sizes is available and is specified by what are known as split gage numbers. They can be recognized by ½ fractions which follow the gage number; i.e., 4½. The decimal equivalents of split gage numbers are in the Steel Products Manual entitled: *Wire and Rods, Carbon Steel* published by the American Iron and Steel Institute, Washington, DC.

Sheet-metal Gages. — The thicknesses of steel sheets now are based upon a weight of 41.82 pounds per square foot per inch thick. This is known as Manufacturers' Standard Gage for Sheet Steel. (See text in table on page 418.) This gage differs from the older United States Standard Gage for iron and steel sheets and plates, established by Congress in 1893, based upon a weight of 40 pounds per square foot per inch of thickness which is the weight of wrought-iron plate.

Thicknesses of aluminum, copper, and copper-base alloys were formerly designated by the American or Brown & Sharpe Wire Gage but now are specified in decimals or fractions of an inch. Copper and copper-base alloy flat products whose thicknesses are below ¼ inch are specified by the 20-series of American Standard Preferred Numbers given in the American Standard B32.1–1952 entitled Preferred Thicknesses for Uncoated Thin Flat Metals (see accompanying Table 2). The thicknesses in this standard are based on the 20- and 40-series of American Standard Preferred Numbers — A.S.A. Z17.1 (see Handbook page 19) and are applicable to uncoated, thin, flat metals and alloys. Although the table given in the American Standard gives only the 20- and 40-series of numbers the statement is made that when intermediate thicknesses are required, selections should be made from thicknesses based on the 80-series of numbers (see Handbook page 19). Each number of the 20-series is approximately 12 per cent greater than the next smaller one and each number of the 40-series is approximately 6 per cent greater than the next smaller one.

Zinc sheets are more often ordered by specifying decimal thickness although a zinc gage exists and is shown in Table 3.

Table 2. Preferred Thicknesses for Uncoated Metals and Alloys — Under 0.250 Inch in Thickness (ASA B32.1-1952)*

Preferred Thickness, Inches		Preferred Thickness, Inches		Preferred Thickness, Inches		Preferred Thickness, Inches	
Based on 20-Series	Based on 40-Series	Based on 20-Series	Based on 40-Series	Based on 20-Series	Based on 40-Series	Based on 20-Series	Based on 40-Series
.....	0.236	0.100	0.100	.....	0.042	0.018	0.018
0.224	0.224	.....	0.095	0.040	0.040	.....	0.017
.....	0.212	0.090	0.090	.....	0.038	0.016	0.016
0.200	0.200	.....	0.085	0.036	0.036	.....	0.015
.....	0.190	0.080	0.080	.....	0.034	0.014	0.014
0.180	0.180	.....	0.075	0.032	0.032	.....	0.013
.....	0.170	0.071	0.071	.....	0.030	0.012	0.012
0.160	0.160	.....	0.067	0.028	0.028	0.011	0.011
.....	0.150	0.063	0.063	.....	0.026	0.010	0.010
0.140	0.140	.....	0.060	0.025	0.025	0.009	0.009
.....	0.132	0.056	0.056	.....	0.024	0.008	0.008
0.125	0.125	.....	0.053	0.022	0.022	0.007	0.007
.....	0.118	0.050	0.050	.....	0.021	0.006	0.006
0.112	0.112	.....	0.048	0.020	0.020	0.005	0.005
.....	0.106	0.045	0.045	.....	0.019	0.004	0.004

* The American Standard ASA B32.1-1952 lists preferred thicknesses which are based on the 20- and 40-series of preferred numbers and states that those based on the 40-series should provide adequate coverage. However, where intermediate thicknesses are required it states that thicknesses be based on the 80-series of preferred numbers (see Handbook page 19).

Most sheet metal products in Great Britain are specified by the British Standard Wire Gage (Imperial Wire Gage).　Black iron and steel sheet and hooping, and galvanized flat and corrugated steel sheet, however, are specified by the Birmingham Gage (B.G.) which was legalized in 1914.　This Birmingham Gage should not be confused with the Birmingham or Stub's Iron Wire Gage mentioned previously.

Table 3.　Sheet-Metal Gages in Approximate Decimals of an Inch

No. of Sheet-Metal Gage	Manufacturers' Standard Gage for Steel*	Birmingham Gage (B.G.) for Sheets, Hoops	Galvanized Sheet Gage	Zinc Gage	No. of Sheet-Metal Gage	Manufacturers' Standard Gage for Steel*	Birmingham Gage (B.G.) for Sheets, Hoops	Galvanized Sheet Gage	Zinc Gage
15/0	...	1.000	...	...	20	0.0359	0.0392	0.0396	0.070
14/0	...	0.9583	...	...	21	0.0329	0.0349	0.0366	0.080
13/0	...	0.9167	...	...	22	0.0299	0.03125	0.0336	0.090
12/0	...	0.8750	...	...	23	0.0269	0.02782	0.0306	0.100
11/0	...	0.8333	...	...	24	0.0239	0.02476	0.0276	0.125
10/0	...	0.7917	...	...	25	0.0209	0.02204	0.0247	...
9/0	...	0.7500	...	...	26	0.0179	0.01961	0.0217	...
8/0	...	0.7083	...	...	27	0.0164	0.01745	0.0202	...
7/0	...	0.6666	...	...	28	0.0149	0.01562	0.0187	...
6/0	...	0.6250	...	...	29	0.0135	0.01390	0.0172	...
5/0	...	0.5883	...	...	30	0.0120	0.01230	0.0157	...
4/0	...	0.5416	...	...	31	0.0105	0.01100	0.0142	...
3/0	...	0.5000	...	...	32	0.0097	0.00980	0.0134	...
2/0	...	0.4452	...	...	33	0.0090	0.00870	...	...
1/0	...	0.3964	...	...	34	0.0082	0.00770	...	...
1	...	0.3532	...	...	35	0.0075	0.00690	...	...
2	...	0.3147	...	...	36	0.0067	0.00610	...	...
3	0.2391	0.2804	...	0.006	37	0.0064	0.00540	...	...
4	0.2242	0.2500	...	0.008	38	0.0060	0.00480	...	...
5	0.2092	0.2225	...	0.010	39	...	0.00430	...	...
6	0.1943	0.1981	...	0.012	40	...	0.00386	...	...
7	0.1793	0.1764	...	0.014	41	...	0.00343	...	...
8	0.1644	0.1570	0.1681	0.016	42	...	0.00306	...	...
9	0.1495	0.1398	0.1532	0.018	43	...	0.00272	...	...
10	0.1345	0.1250	0.1382	0.020	44	...	0.00242	...	...
11	0.1196	0.1113	0.1233	0.024	45	...	0.00215	...	...
12	0.1046	0.0991	0.1084	0.028	46	...	0.00192	...	...
13	0.0897	0.0882	0.0934	0.032	47	...	0.00170	...	...
14	0.0747	0.0785	0.0785	0.036	48	...	0.00152	...	...
15	0.0673	0.0699	0.0710	0.040	49	...	0.00135	...	...
16	0.0598	0.0625	0.0635	0.045	50	...	0.00120	...	...
17	0.0538	0.0556	0.0575	0.050	51	...	0.00107	...	...
18	0.0478	0.0495	0.0516	0.055	52	...	0.00095	...	...
19	0.0418	0.0440	0.0456	0.060	...	...	...	...	...

* For more information and data on Manufacturers' Standard Gage for Steel Sheets see Table 5.

The United States Standard Gage (not shown above) for iron and steel sheets and plates was established by Congress in 1893 and was primarily a *weight* gage rather than a thickness gage.　The equivalent thicknesses were derived from the weight of wrought iron.　The weight per cubic foot was taken at 480 pounds, thus making the weight of a plate 12 inches square and 1 inch thick, 40 pounds.　In converting weight to equivalent thickness, gage tables formerly published contained thicknesses equivalent to the basic weights just mentioned.　For example, a No. 3 U. S. gage represents a wrought-iron plate having a weight of 10 pounds per square foot; hence, if the weight per square foot per inch thick is 40 pounds, the plate thickness for a No. 3 gage = 10 ÷ 40 = 0.25 inch, which was the original thickness equivalent for this gage number.　Since this and the other thickness equivalents were derived from the weight of wrought iron, they are not correct for steel.

Metric Sizes for Flat Metal Products. — American National Standard B32.3M-1984 establishes a preferred series of metric thicknesses, widths, and lengths for flat metal products of rectangular cross section; the thickness and width values are also applicable to base metals that may be coated in later operations. Table 4 lists the preferred thicknesses. Whenever possible, the Preferred Thickness values should be used, with the Second or Third Preference chosen only if no suitable Preferred size is available. Since not all metals and grades are produced in each of the sizes given in Table 4, producers or distributors should be consulted to determine a particular product and size combination's availability.

Table 4. Preferred Metric Thicknesses for All Flat Metal Products (ANSI B32.3M-1984)

Preferred Thickness	Second Preference	Third Preference	Preferred Thickness	Second Preference	Third Preference
0.050	...	...	...	...	3.6
0.060	...	...	...	3.8	...
0.080	...	...	4.0	...	...
0.10	...	...	...	4.2	...
0.12	...	...	...	4.5	...
...	0.14	...	...	4.8	...
0.16	...	...	5.0	...	...
...	0.18	...	...	5.5	...
0.20	...	...	6.0	...	...
...	0.22	...	...	...	6.5
0.25	...	...	...	7.0	...
...	0.28	...	...	...	7.5
0.30	...	...	8.0	...	...
...	0.35	...	...	9.0	...
0.40	...	...	10	...	...
...	0.45	...	...	11	...
0.50	...	...	12	...	...
...	0.55	...	...	14	...
0.60	...	...	16	...	...
...	0.65	...	...	18	...
...	0.70	...	20	...	...
...	...	0.75	...	22	...
0.80	...	...	25	...	...
...	...	0.85	...	28	...
...	0.90	...	30	...	...
...	...	0.95	...	32	...
1.0	...	...	35	...	...
...	...	1.05	...	38	...
1.2	...	...	40	...	...
...	...	...	...	45	...
...	...	1.3	50	...	...
...	1.4	...	...	55	...
...	...	1.5	60	...	...
1.6	...	...	...	70	...
...	...	1.7	80	...	...
...	1.8	...	...	90	...
...	...	1.9	100	...	...
2.0	...	...	...	110	...
...	...	2.1	120	...	...
...	2.2	...	...	130	...
...	...	2.4	140	...	...
2.5	...	...	...	150	...
...	...	2.6	160	...	...
...	2.8	...	180	...	...
3.0	...	...	200	...	...
...	3.2	...	250	...	...
...	...	3.4	300	...	...
3.5	...	...			

All dimensions are in millimeters.

Table 5. Manufacturers' Standard Gage for Sheet Steel

Manufacturers' Standard Gage for Steel Sheets: Although the basic weight of steel used in the manufacture of steel plate, bars, and other steel products is 40.8 pounds per square foot per inch of thickness, the *Manufacturers' Standard Gage for Steel Sheets* is based on a weight of 41.82 pounds per square foot per inch of thickness. This modified figure provides an adjustment for the variation in thickness from the edges to the center of sheets resulting from the rolling process and also for the shearing tolerances which are on the over side. The thicknesses in the table below are based upon this weight of 41.82 pounds and represent standard mill practice. These nominal thicknesses, however, are subject to tolerances or permissible variations as given in the tables on the following pages which include both hot-rolled and cold-rolled sheets.

Standard Gage No.	Ounces per Square Foot	Pounds per Square Foot	Equivalent Thickness, Inch	Standard Gage No.	Ounces per Square Foot	Pounds per Square Foot	Equivalent Thickness, Inch
3	160	10.0000	0.2391	21	22	1.3759	.0329
4	150	9.3760	.2242	22	20	1.2504	.0299
5	140	8.7500	.2092	23	18	1.1250	.0269
6	130	8.1256	.1943	24	16	1.00000	.0239
7	120	7.5000	.1793	25	14	0.87504	.0209
8	110	6.8750	.1644	26	12	.75000	.0179
9	100	6.2521	.1495	27	11	.68600	.0164
10	90	5.6250	.1345	28	10	.62312	.0149
11	80	5.0020	.1196	29	9	.56457	.0135
12	70	4.3740	.1046	30	8	.50184	.0120
13	60	3.7513	.0897	31	7	.43911	.0105
14	50	3.1240	.0747	32	6.5	.40565	.0097
15	45	2.8125	.0673	33	6	.37638	.0090
16	40	2.5008	.0598	34	5.5	.34292	.0082
17	36	2.2500	.0538	35	5	.31365	.0075
18	32	2.0000	.0478	36	4.5	.28019	.0067
19	28	1.7500	.0418	37	4.3	.26765	.0064
20	24	1.5000	.0359	38	4	.25092	.0060

Standard Thickness Tolerances for Steel Sheets — 1
(American Iron and Steel Institute*)

COLD ROLLED ALLOY STEEL (For coils or cut lengths)†								
	Thickness Range and Plus or Minus Thickness Tolerance							
Width	.2299 .1800	.1799 .1420	.1419 .0972	.0971 .0822	.0821 .0710	.0709 .0568	.0567 .0509	.0508 .0314	.0313 .0195
24 to 32	.008	.008	.007	.006	.005	.005	.005	.004	.003
Over 32 to 40	.009	.009	.008	.007	.006	.005	.005	.004	.003
Over 40 to 48	.010	.010	.009	.007	.006	.005	.005	.004	.003
Over 48 to 60	. . .	.010	.010	.008	.006	.006	.005	.004	.003
Over 60 to 70	. . .	.011	.010	.009	.007	.006	.006	.005	. . .
Over 70 to 80	. . .	.012	.011	.009	.007	. . .	. . .	. . .	. . .
Over 80 to 90	. . .	.012	.012	. . .	. . .	. . .	. . .	. . .	. . .
Over 90	. . .	.012	.012	. . .	. . .	. . .	. . .	. . .	. . .

HOT ROLLED ALLOY STEEL (Hand Mill Product)†			
Thickness	Plus or Minus Tolerance	Thickness	Plus or Minus Tolerance
.229 to .188	.015	Under .084 to .073	.007
Under .188 to .146	.014	Under .073 to .059	.006
Under .146 to .131	.012	Under .059 to .041	.005
Under .131 to .115	.010	Under .041 to .027	.004
Under .115 to .099	.009	Under .027 to .019	.003
Under .099 to .084	.008	. . .	. . .

All dimensions are in inches. See end of table for footnotes.

Standard Thickness Tolerances for Steel Sheets — 2
(American Iron and Steel Institute*)

HOT ROLLED ALLOY STEEL SHEETS (Continuous Mill Product — Coils or Cut Lengths)†					
	Thickness Range and Plus or Minus Thickness Tolerance				
Width	.1800 .2299	.0972 .1799	.0822 .0971	.0710 .0821	.0568 .0709
24 to 32	.009	.008	.007	.007	.006
Over 32 to 40	.009	.009	.008	.007	.006
Over 40 to 48	.010	.010	.008	.007	.006
Over 48 to 60	. . .	.010	.008	.007	.007
Over 60 to 72	. . .	.011	.009	.008	.007
Over 72 to 80	. . .	.012	.009	.008	. . .
Over 80 to 90	. . .	.012	.010	. . .	. . .
Over 90	. . .	.012	. . .	. . .	. . .

HOT ROLLED HIGH STRENGTH LOW ALLOY STEEL SHEETS (For Coils and Cut Lengths, Including Pickled Sheets)†§¶				
	Specified Minimum Thickness and Plus Thickness Tolerance			
Width	Over .180 .230	Over .097 .180	Over .082 .097	Over .071 .082
Over 12 to 15	.014	.014	.012	.012
Over 15 to 20	.016	.016	.014	.014
Over 20 to 32	.018	.016	.014	.014
Over 32 to 40	.018	.018	.016	.014
Over 40 to 48	.020	.020	.016	.014
Over 48 to 60	. . .	.020	.016	.014
Over 60 to 72	. . .	.022	.018	.016
Over 72 to 80	. . .	.024	.018	.016

COLD ROLLED HIGH STRENGTH LOW ALLOY STEEL SHEETS (Coils and Cut Lengths)‡¶						
	Specified Minimum Thickness and Plus Thickness Tolerance					
Width	Over .098 .142	Over .071 .098	Over .057 .071	Over .039 .057	Over .019 .039	Over .014 .019
Over 12 to 15	.010	.010	.010	.008	.006	.004
Over 15 to 72	.012	.010	.010	.008	.006	.004
Over 72	.014	.012	.010	.008	.006	. . .

COLD ROLLED HIGH STRENGTH LOW ALLOY STEEL SHEETS (Coils and Cut Lengths)‡¶				
	Specified Minimum Thickness and Plus Thickness Tolerance			
Width	Over .057 .082	Over .039 .057	Over .019 .039	Over .014 .019
2 to 12 (incl.)	.006	.004	.003	.002

All dimensions are in inches.

* *Sheet Steel Data Pocketbook*, AISI, 1980.

† Thickness is measured at any point on the sheet not less than ⅜ inch from a cut edge and not less than ¾ inch from a mill edge.

§ Tolerances listed for this steel do not apply to the uncropped ends of mill edge coils.

¶ The specified thickness ranges also apply when sheet is specified to a nominal thickness, and the tolerances are divided equally, plus and minus.

‡ Thickness is measured at any point on the sheet not less than ⅜ inch from a side edge.

Standard Thickness Tolerances for Carbon Steel Sheet
(American Iron and Steel Institute*)

HOT ROLLED SHEET (For coils and cut lengths, including pickled sheets)†‡

Width	Specified Minimum Thickness and Plus Thickness Tolerance					
	Over .180 .230	Over .098 .180	Over .071 .098	Over .057 .071	Over .051 .057	Over .044 .051
Over 12 to 20	.014	.014	.012	.012	.010	.010
Over 20 to 40	.016	.014	.014	.012	.010	.010
Over 40 to 48	.018	.016	.014	.012	.012	.010
Over 48 to 60	...	.016	.014	.014	.012	...
Over 60 to 72	...	.016	.016	.014	.014	...
Over 72	...	.016	.016	...	...	...

HOT ROLLED SHEET (Heavy thickness, for coils over .180 thick)†‡

Width	Specified Minimum Thickness and Plus Thickness Tolerance			
	Over .375 .500	Over .313 .375	Over .230 .313	Over .180 .230
Over 12 to 20	.028	.024	.020	...
Over 20 to 40	.028	.024	.022	...
Over 40 to 48	.028	.026	.024	...
Over 48 to 60	.030	.028	.024	.020
Over 60 to 72	.032	.030	.026	.022
Over 72	.036	.032	.030	.024

COLD ROLLED SHEET (For coils and cut lengths)‡§

Width	Specified Minimum Thickness and Plus Thickness Tolerance					
	Over .098 .142	Over .071 .098	Over .057 .071	Over .039 .057	Over .019 .039	Over .014 .019
Over 12 to 15	.010	.010	.010	.008	.006	.004
Over 15 to 72	.012	.010	.010	.008	.006	.004
Over 72	.014	.012	.010	.008	.006	...

COLD ROLLED SHEET (For coils and cut lengths)‡§

Width	Specified Minimum Thickness and Plus Thickness Tolerance			
	Over .057 .082	Over .039 .057	Over .019 .039	Over .014 .019
2 to 12 incl.	.010	.008	.006	.004

HOT DIPPED GALVANIZED SHEET (For coils and cut lengths)‡§

Width	Specified Minimum Thickness and Plus Thickness Tolerance					
	Over .101 .187	Over .075 .101	Over .061 .075	Over .043 .061	Over .023 .043	Under .023
To 32	.016	.014	.012	.010	.008	.006
Over 32 to 40	.016	.016	.012	.010	.008	.006
Over 40 to 60	.018	.016	.012	.010	.008	.006
Over 60 to 72	.018	.018	.012	.010	.008	...

ELECTROLYTIC ZINC COATED SHEET (For coils and cut lengths)‡§

Width	Specified Minimum Thickness and Plus Thickness Tolerance				
	Over .057 .100	Over .039 .057	Over .020 .039	Over .014 .020	Under .014
Up to 40	.010	.008	.006	.004	.003
Over 40 to 60	.010	.008	.006	.004	...

All dimensions are in inches.   * *Sheet Steel Data Pocketbook,* AISI, 1980.

  † Thickness is measured at any point across the width not less than ⅜ inch from a cut edge and ¾ inch from a mill edge. Tolerances do not apply to the uncropped ends of mill edge coils.

  ‡ The specified thickness ranges also apply when sheet is specified to a nominal thickness, and the tolerances are divided equally, plus or minus.

  § Thickness is measured at any point across the width not less than ⅜ inch from a side edge.

Permissible Variations in Sizes of Cold-finished and Hot-rolled Bars — 1
(American Iron and Steel Institute*)

HOT ROLLED CARBON STEEL BARS

Size	Tolerance Plus	Tolerance Minus	Out-of-Section†	Size	Tolerance Plus	Tolerance Minus	Out-of-Section†
Rounds, Squares and Round-Cornered Squares							
To 5/16	.005	.005	.008	Over 1½ to 2	1/64	1/64	.023
Over 5/16 to 7/16	.006	.006	.009	Over 2 to 2½	1/32	0	.023
Over 7/16 to 5/8	.007	.007	.010	Over 2½ to 3½	3/64	0	.035
Over 5/8 to 7/8	.008	.008	.012	Over 3½ to 4½	1/16	0	.046
Over 7/8 to 1	.009	.009	.013	Over 4½ to 5½	5/64	0	.058
Over 1 to 1⅛	.010	.010	.015	Over 5½ to 6½	1/8	0	.070
Over 1⅛ to 1¼	.011	.011	.016	Over 6½ to 8¼	5/32	0	.085
Over 1¼ to 1⅜	.012	.012	.018	Over 8¼ to 9½	3/16	0	.100
Over 1⅜ to 1½	.014	.014	.021	Over 9½ to 10	1/4	0	.120
Hexagons							
To ½	.007	.007	.011	Over 1½ to 2	1/32	1/64	1/32
Over ½ to 1	.010	.010	.015	Over 2 to 2½	3/64	1/64	3/64
Over 1 to 1½	.021	.013	.025	Over 2½ to 3½	1/16	1/64	1/16

COLD FINISHED CARBON STEEL BARS

Size	Max. % C Up to .28	Max. % C Over .28 to .55	Max. % C Up to .55‡	Max. % C Over .55§	Width or Diameter	Max. % C Up to .28	Max. % C Over .28 to .55	Max. % C Up to .55‡	Max. % C Over .55§
	Minus Tolerance					Minus Tolerance			
Cold Drawn Rounds¶					**Cold Drawn Flats¶★**				
To 1½	.002	.003	.004	.005	To ¾	.003	.004	.006	.008
Over 1½ to 2½	.003	.004	.005	.006	Over ¾ to 1½	.004	.005	.008	.010
Over 2½ to 4	.004	.005	.006	.007	Over 1½ to 3	.005	.006	.010	.012
Over 4 to 6	.005	.006	.007	.008	Over 3 to 4	.006	.008	.011	.016
Cold Drawn Hexagons¶					Over 4 to 6	.008	.010	.012	.020
To ¾	.002	.003	.004	.006	Over 6	.013	.015	...	...
Over ¾ to 1½	.003	.004	.005	.007	**Turned and Polished Rounds¶**				
Over 1½ to 2½	.004	.005	.006	.008					
Over 2½ to 3⅛	.005	.006	.007	.009	To 1½	.002	.003	.004	.005
Cold Drawn Squares¶					Over 1½ to 2½	.003	.004	.005	.006
					Over 2½ to 4	.004	.005	.006	.007
To ¾	.002	.004	.005	.007	Over 4 to 6	.005	.006	.007	.008
Over ¾ to 1½	.003	.005	.006	.008	Over 6 to 8	.006	.007	.008	.009
Over 1½ to 2½	.004	.006	.007	.009	Over 8 to 9	.007	.008	.009	.010
Over 2½ to 4	.005	.008	.009	.011	Over 9	.008	.009	.010	.011

All dimensions are in inches. * Steel Products Manual — Alloy, Carbon and High Strength Low Alloy Steels: Semifinished for Forging; Hot Rolled and Cold Finished Bars; etc., March, 1986. † Means out-of-round, out-of-square or out-of-hexagon. Out-of-round is the difference between the maximum and minimum diameters of the bar, measured at the same transverse cross section. Out-of-square is the difference in the two dimensions at the same cross section of a square bar between opposite faces. Out-of-hexagon is the greatest difference between any two dimensions at the same cross section between opposite faces. ‡ Stress relieved or annealed after cold finishing. § Or all grades quenched and tempered or normalized and tempered before cold finishing. ¶ Values shown include tolerances for bars that have been annealed, spheroidize annealed, normalized, normalized and tempered or quenched and tempered *before* cold finishing. Values do not include tolerances for bars that are spheroidize annealed, normalized, normalized and tempered or quenched and tempered *after* cold finishing. ★ Width governs the tolerances for both width and thickness of flats. For example, when the maximum of carbon range is .28 per cent or less, for a flat 2 inches wide and 1 inch thick, the width tolerance is .005 inch and the thickness tolerance is the same, namely, .005 inch.

Permissible Variations in Sizes of Cold-finished and Hot-rolled Bars — 2
(American Iron and Steel Institute*)

COLD FINISHED ALLOY STEEL BARS

Size	Max. % C				Width or Diameter	Max. % C			
	Up to .28	Over .28 to .55	Up to .55†	Over .55‡		Up to .28	Over .28 to .55	Up to .55†	Over .55‡
	Minus Tolerance					Minus Tolerance			

Cold Drawn Rounds

Size	Up to .28	Over .28 to .55	Up to .55†	Over .55‡
To 1, in coils	.002	.003	.004	.005
To 1½	.003	.004	.005	.006
Over 1½ to 2½	.004	.005	.006	.007
Over 2½ to 4	.005	.006	.007	.008
Over 4 to 6	.006	.007	.008	.009

Cold Drawn Hexagons

Size	Up to .28	Over .28 to .55	Up to .55†	Over .55‡
To ¾	.003	.004	.005	.007
Over ¾ to 1½	.004	.005	.006	.008
Over 1½ to 2½	.005	.006	.007	.009
Over 2½ to 3⅛	.006	.007	.008	.010
Over 3⅛ to 4	.006	...	...	...

Cold Drawn Squares

Size	Up to .28	Over .28 to .55	Up to .55†	Over .55‡
To ¾	.003	.005	.006	.008
Over ¾ to 1½	.004	.006	.007	.009
Over 1½ to 2½	.005	.007	.008	.010
Over 2½ to 3⅛	.007	.009	.010	.012
Over 4 to 5	.011	...	...	...

Cold Drawn Flats§

Width or Diameter	Up to .28	Over .28 to .55	Up to .55†	Over .55‡
To ¾	.004	.005	.007	.009
Over ¾ to 1½	.005	.006	.009	.011
Over 1½ to 3	.006	.007	.011	.013
Over 3 to 4	.007	.009	.012	.017
Over 4 to 6	.009	.011	.013	.021
Over 6	.014	...	...	...

Turned and Polished Rounds

Width or Diameter	Up to .28	Over .28 to .55	Up to .55†	Over .55‡
To 1½	.003	.004	.005	.006
Over 1½ to 2½	.004	.005	.006	.007
Over 2½ to 4	.005	.006	.007	.008
Over 4 to 6	.006	.007	.008	.009
Over 6 to 8	.007	.008	.009	.010
Over 8 to 9	.008	.009	.010	.011
Over 9	.009	.010	.011	.012

HOT ROLLED TOOL STEEL BARS¶

Size	Tolerance Minus	Tolerance Plus	Size	Tolerance Minus	Tolerance Plus
To ½	.005	.012	Over 2½ to 3	.010	.040
Over ½ to 1	.005	.016	Over 3 to 4	.012	.050
Over 1 to 1½	.006	.020	Over 4 to 5½	.015	.060
Over 1½ to 2	.008	.025	Over 5½ to 6½	.018	.100
Over 2 to 2½	.010	.030	Over 6½ to 8	.020	.150

HOT ROLLED TOOL STEEL FLAT BARS

Thickness Range and Thickness Tolerance

Width	To ¼ −	To ¼ +	Over ¼ to ½ −	Over ¼ to ½ +	Over ½ to 1 −	Over ½ to 1 +	Over 1 to 2 −	Over 1 to 2 +	Over 2 to 3 −	Over 2 to 3 +	Over 3 to 4 −	Over 3 to 4 +
To 1	.006	.010	.008	.012	.010	.016	...	...	...	...	...	...
Over 1 to 2	.006	.014	.008	.016	.010	.020	.020	.024	...	...	...	...
Over 2 to 3	.006	.018	.008	.020	.010	.024	.020	.027	.026	.034	...	...
Over 3 to 4	.008	.020	.010	.022	.013	.024	.024	.030	.032	.042	.040	.048
Over 4 to 5	.010	.020	.012	.024	.015	.030	.027	.035	.032	.042	.042	.050
Over 5 to 6	.012	.020	.014	.030	.018	.030	.030	.035	.036	.046	.044	.054
Over 6 to 7	.014	.027	.016	.032	.018	.035	.030	.040	.036	.048	.046	.056
Over 7 to 10	.018	.030	.020	.035	.024	.040	.035	.045	.040	.054	.052	.064
Over 10 to 12	.020	.035	.025	.040	.030	.045	.040	.050	.046	.060	.056	.072

All dimensions are in inches. * Steel Products Manuals — Alloy, Carbon and High Strength Low Alloy Steels: Semifinished for Forging; Hot Rolled and Cold Finished Bars; etc., March, 1986; and Tool Steels, September, 1981. † Stress relieved or annealed after cold finishing. ‡ With or without stress relieving or annealing after cold finishing. Also, all carbons, quenched and tempered (heat treated), or normalized and tempered before cold finishing. § Width governs the tolerances for both width and thickness of flats. For example: when the maximum of the carbon range is .28 per cent or less, for a flat 2 inches wide and 1 inch thick, the width tolerance is .006 inch and the thickness tolerance is the same, namely, .006 inch. ¶ For rounds (other than high speed steel free of scale and decarburization), squares, octagons, quarter octagons, and hexagons.

Permissible Variations in Sizes of Tool Steel Bars and Flats
(American Iron and Steel Institute*)

TOOL STEEL BARS					
Round Bars, High Speed Steels Free of Scale and Decarburization			Forged Rounds, Squares, Octagons, and Hexagons		
	Tolerance			Tolerance	
Diameter	Minus	Plus	Size	Minus	Plus
¼ to under ⅝	.0015	.0015	Over 1 to 2	.030	.060
⅝ to under 3¹⁄₁₆	.000	.004	Over 2 to 3	.030	.080
3¹⁄₁₆ to under 4¹⁄₁₆	.000	.006	Over 3 to 5	.060	.125
4¹⁄₁₆ to 7	.000	.031	Over 5 to 7	.125	.187
. . .	. . .	. . .	Over 7 to 9	.187	.312

Forged Flats

	Thickness Range and Thickness Tolerance									
	To 1		Over 1 to 3		Over 3 to 5		Over 5 to 7		Over 7 to 9	
Width	Minus	Plus	Minus	Plus	Minus	Plus	Minus	Plus	Minus	Plus
Over 1 to 3	.016	.031	.031	.078	. . .	. . .	. . .	. . .	. . .	. . .
Over 3 to 5	.031	.062	.047	.094	.062	.125	. . .	. . .	. . .	. . .
Over 5 to 7	.047	.094	.062	.125	.078	.156	.125	.187	. . .	. . .
Over 7 to 9	.062	.125	.078	.156	.094	.187	.156	.219	.187	.312

All dimensions are in inches.
* Steel Products Manual — Tool Steels, September, 1981.

Standard Stock Sizes of Machined Tool Steel Bars — Flats and Squares
(U.S. Simplified Practice Recommendation R267-65)

Thickness, Inches	Width, Inches																			
	⅞	1	1⅛	1¼	1⅜	1½	1¾	2	2¼	2½	2¾	3	3½	4	4½	5	6	8	10	12
⅞	X	X	X	X	. . .	X	X	X	. . .	X	. . .	X	X	X	. . .	X	X	X	X	X
1	. . .	X	. . .	X	. . .	X	X	X	X	X	X	X	X	X	X	X	X	X	X	X
1⅛	. . .	. . .	X	X	. . .	X	X	X	X	X	. . .	X	X	X	. . .	X	X	X	X	X
1¼	. . .	. . .	. . .	X	. . .	X	X	X	X	X	X	X	X	X	X	X	X	X	X	X
1⅜	. . .	. . .	. . .	. . .	X	X	X	X	X	X	X	X	X	. . .	X	X	X	X	X	X
1½	. . .	. . .	. . .	. . .	. . .	X	X	X	X	X	X	X	X	X	X	X	X	X	X	X
1¾	. . .	. . .	. . .	. . .	. . .	. . .	X	X	X	X	X	X	X	X	X	X	X	X	X	X
2	. . .	. . .	. . .	. . .	. . .	. . .	. . .	X	X	X	X	X	X	X	X	X	X	X	X	X
2½	. . .	. . .	. . .	. . .	. . .	. . .	. . .	. . .	. . .	X	X	X	X	X	X	X	X	X	X	X
3	. . .	. . .	. . .	. . .	. . .	. . .	. . .	. . .	. . .	. . .	. . .	X	X	X	X	X	X	X	X	X
4	. . .	. . .	. . .	. . .	. . .	. . .	. . .	. . .	. . .	. . .	. . .	. . .	. . .	X	X	X	X	X	X	X
5	. . .	. . .	. . .	. . .	. . .	. . .	. . .	. . .	. . .	. . .	. . .	. . .	. . .	. . .	. . .	X	X	X	X	. . .
6	. . .	. . .	. . .	. . .	. . .	. . .	. . .	. . .	. . .	. . .	. . .	. . .	. . .	. . .	. . .	X	X	X	. . .	. . .

Material: It is recommended that the machined tool steel bars covered by this recommendation shall conform to the chemical and mechanical requirements for the following tool steels as listed in the latest issue of the Tool Steel Manual of the American Iron and Steel Institute: Air-hardening, AISI — A2, A4 and A6; Oil-hardening, AISI — O1, O2 and O6; and High carbon-high chromium, AISI — D2 and D3.

Dimensional tolerances: It is recommended that the sizes in this table be finished in conformity with the following oversize and tolerances: Thickness, 0.015 inch oversize, + 0.020, − 0.000 inch and Width, 0.015 inch oversize, + 0.020, − 0.000 inch.

Example: A 4-inch by 2-inch bar of machined tool steel should measure as follows: 4.015 inch, + 0.020 inch, − 0.000 inch by 2.015 inch, + 0.020 inch, − 0.000 inch.

Surface finish: It is recommended that the surface finish have a roughness height value of 90–110 microinches (see Handbook section Surface Texture). The bars shall be machined on all four sides and be free from surface imperfections such as carburization, decarburization, scale, etc.

Thickness Tolerances for Aluminum Sheet and Plate* (ANSI H35.2-1982) — I

Specified Width — Tolerance — Plus or Minus

Specified Thickness	Up thru 18	Over 18 thru 36	Over 36 thru 48	Over 48 thru 54	Over 54 thru 60	Over 60 thru 66	Over 66 thru 72	Over 72 thru 78	Over 78 thru 84	Over 84 thru 90	Over 90 thru 96	Over 96 thru 132	Over 132 thru 144	Over 144 thru 156	Over 156 thru 168
0.006 to 0.010	.001	.0015	.0025	.0025											
0.011 to 0.017	.0015	.0015	.0025	.0035											
0.018 to 0.028	.0015	.002	.0025	.0035	.004	.004	.004								
0.029 to 0.036	.002	.002	.0025	.004	.004	.005	.005	.006	.006	.007					
0.037 to 0.045	.002	.0025	.003	.004	.005	.005	.005	.006	.006	.007					
0.046 to 0.068	.0025	.003	.004	.005	.006	.006	.006	.007	.007	.008	.009	.013			
0.069 to 0.076	.003	.003	.004	.005	.006	.006	.006	.007	.007	.012	.011	.016			
0.077 to 0.096	.0035	.0035	.004	.005	.006	.006	.006	.007	.007	.012	.012	.016			
0.097 to 0.108	.004	.004	.005	.005	.007	.007	.007	.008	.008	.016	.012	.020			
0.109 to 0.125	.0045	.0045	.005	.005	.007	.007	.007	.008	.008	.016	.012	.020			
0.126 to 0.140	.0045	.0045	.005	.005	.007	.010	.012	.013	.014	.016	.018	.020			
0.141 to 0.172	.006	.006	.008	.008	.009	.012	.014	.015	.016	.017	.019	.023			
0.173 to 0.203	.007	.007	.010	.010	.011	.014	.016	.017	.017	.017	.022	.026			
0.204 to 0.249	.009	.009	.011	.011	.013	.016	.018	.018	.018	.018	.024	.028			
0.250 to 0.320	.013	.013	.013	.013	.015	.018	.020	.020	.020	.020	.025	.030	.035	.042	.053
0.321 to 0.438	.019	.019	.019	.019	.020	.020	.023	.023	.025	.025	.026	.033	.038	.045	.057
0.439 to 0.625	.025	.025	.025	.025	.025	.025	.025	.030	.030	.030	.035	.035	.043	.049	.067
0.626 to 0.875	.030	.030	.030	.030	.030	.030	.030	.037	.037	.037	.045	.045	.054	.059	.077
0.876 to 1.125	.035	.035	.035	.035	.035	.035	.035	.045	.045	.045	.055	.055	.065	.070	.088
1.126 to 1.375	.040	.040	.040	.040	.040	.040	.040	.052	.052	.052	.065	.065	.075	.080	.098
1.376 to 1.625	.045	.045	.045	.045	.045	.045	.045	.060	.060	.060	.075	.075	.085	.090	.108
1.626 to 1.875	.052	.052	.052	.052	.052	.052	.052	.070	.070	.070	.088	.088			
1.876 to 2.250	.060	.060	.060	.060	.060	.060	.060	.080	.080	.080	.100	.100			
2.251 to 2.750	.075	.075	.075	.075	.075	.075	.075	.100	.100	.100	.125	.125			
2.751 to 3.000	.090	.090	.090	.090	.090	.090	.090	.120	.120	.120	.150	.150			
3.001 to 4.000	.110	.110	.110	.110	.110	.110	.110	.140	.140	.140	.160	.160			
4.001 to 5.000	.125	.125	.125	.125	.125	.125	.125	.150	.150	.150	.160				
5.001 to 6.000	.135	.135	.135	.135	.135	.135	.135	.160	.160	.160	.170				

All dimensions are given in inches. * For alloys 2014, 2024, 2036, 2124, 2219, 3004, 5052, 5083, 5086, 5154, 5252, 5254, 5454, 5456, 5652, 6061, 7075, 7178; Alclad alloys, 2014, 2024, 2036, 2124, 2219, 3004, 5052, 5083, 5086, 5154, 5252, 5254, 5454, 5456, 5652, 6061, 7075, 7178; and Brazing Sheet Nos. 11, 12, 21, 22, 23 and 24.

Thickness Tolerances for Aluminum Sheet and Plate* (ANSI H35.2-1982) — 2

Specified Thickness	Up thru 18	Over 18 thru 36	Over 36 thru 54	Over 54 thru 72	Over 72 thru 90	Over 90 thru 102	Over 102 thru 132	Over 132 thru 144	Over 144 thru 156	Over 156 thru 168
					Tolerance — Plus or Minus					
0.006 to 0.007	.001	.001	.002	...	...	...	...	...	...	...
0.008 to 0.010	.001	.0015	.002	...	...	...	...	...	...	...
0.011 to 0.017	.0015	.0015	.002	...	...	...	...	...	...	...
0.018 to 0.028	.0015	.002	.0025	.003	.004	...	...	...	...	...
0.029 to 0.036	.002	.002	.0025	.0035	.005	.006	...	...	...	...
0.037 to 0.045	.002	.0025	.003	.004	.005	.006				
0.046 to 0.068	.0025	.003	.004	.005	.006	.007	.008			
0.069 to 0.076	.0025	.003	.004	.006	.008	.008	.009			
0.077 to 0.096	.003	.003	.004	.006	.008	.009	.010			
0.097 to 0.108	.0035	.004	.005	.007	.009	.010	.012			
0.109 to 0.140	.0045	.0045	.005	.007	.009	.010	.012			
0.141 to 0.172	.006	.006	.008	.009	.011	.012	.015	...		
0.173 to 0.203	.007	.007	.009	.011	.013	.015	.017	...		
0.204 to 0.249	.009	.009	.011	.013	.015	.017	.020	...		
0.250 to 0.320	.013	.013	.013	.015	.017	.020	.023	.032	.040	.050
0.321 to 0.438	.019	.019	.019	.019	.023	.026	.026	.035	.043	.052
0.439 to 0.625	.025	.025	.025	.025	.030	.035	.035	.040	.046	.055
0.626 to 0.875	.030	.030	.030	.030	.037	.045	.045	.050	.056	.064
0.876 to 1.125	.035	.035	.035	.035	.045	.055	.055	.060	.066	.074
1.126 to 1.375	.040	.040	.040	.040	.052	.065	.065	.070	.075	.082
1.376 to 1.625	.045	.045	.045	.045	.060	.075	.075	.080	.085	.092
1.626 to 1.875	.052	.052	.052	.052	.070	.088	.088	...	...	...
1.876 to 2.250	.060	.060	.060	.060	.080	.100	.100	...	...	...
2.251 to 2.750	.075	.075	.075	.075	.100	.125	.125	...	...	...
2.751 to 3.000	.090	.090	.090	.090	.120	.150	.150	...	...	...

All dimensions are given in inches. * For alloys 1060, 1100, 1350, 3003, 3005, 3105, 5005, 5050, 5457, 5657, 1100 Reflector Sheet, Clad 1100 Reflector Sheet, and Clad 3003 Reflector Sheet; Alclad 1060, 1100, 1350, 3003, 3005, 3105, 5005, 5050, 5457, 5657, 1100 Reflector Sheet, and 3003 Reflector Sheet.

Length and Width Tolerances for Aluminum Sheared Flat Sheet and Plate (ANSI H35.2-1982)

Specified Thickness	Up thru 30	Over 30 thru 60	Over 60 thru 120	Over 120 thru 240	Over 240 thru 360	Over 360 thru 480	Over 480 thru 600	Over 600 thru 720
	Specified Length							
	Length Tolerance*							
0.006 to 0.124	± 1/16	± 3/32	± 1/8	± 5/32	± 3/16	± 7/32	± 9/32	...
0.125 to 0.248	± 3/32	± 3/32	± 1/8	± 5/32	± 7/32	± 1/4	± 5/16	...
0.250 to 0.499	+ 1/4	+ 3/8	+ 7/16	+ 1/2	+ 9/16	+ 5/8	+ 11/16	+ 3/4

Specified Thickness	Up thru 6	Over 6 thru 24	Over 24 thru 60	Over 60 thru 96	Over 96 thru 132	Over 132 thru 168
	Specified Width					
	Width Tolerance*					
0.006 to 0.124	± 1/16	± 3/32	± 1/8	± 1/8	± 5/32	...
0.125 to 0.249	± 3/32	± 3/32	± 1/8	± 5/32	± 3/16	...
0.250 to 0.499	+ 1/4	+ 5/16	+ 3/8	+ 3/8	+ 7/16	+ 1/2

All dimensions are given in inches. * Tolerances applicable at ambient mill temperatures; a change in length of 0.013 in. per 100 in. per 10°F (6°C) must be recognized. When a dimension tolerance is specified other than as an equal bilateral tolerance, the value of the standard tolerance is that which applies to the mean of the maximum and minimum dimensions permissible under the tolerance for the dimension under consideration.

Tolerances for Aluminum Sheet and Plate (ANSI H35.2-1982)

SAWED FLAT SHEET AND PLATE

Specified Thickness	Specified Width and Length							
	Up thru 30	Over 30 thru 60	Over 60 thru 120	Over 120 thru 240	Over 240 thru 360	Over 360 thru 480	Over 480 thru 600	Over 600 thru 720
	Tolerance[1]							
0.080 to 0.249		± 1/8	± 3/16	± 1/4	± 1/4	± 5/16	± 3/8	± 7/16
0.250 to 6.000	+ 1/4	+ 5/16	+ 3/8	+ 1/2	+ 9/16	+ 5/8	+ 3/4	+ 7/8

FLAT SHEET AND PLATE

Specified Thickness	Specified Width	Specified Length						
		Up thru 60	Over 60 thru 90	Over 90 thru 120	Over 120 thru 150	Over 150 thru 180	Over 180 thru 210	Over 210 thru 240
		Lateral Bow Tolerance — Allowable Deviation of a Side from a Straight Line						
0.006 to 0.125	Up thru 4	0.250	0.563	1.000	1.563	2.250	3.000	4.000[9]
	Over 4 thru 10	0.094	0.219	0.375	0.563	0.875	1.156	1.500[9]
	Over 10 thru 35	0.063	0.125	0.188	0.250	0.375	0.500	0.750[9]
	Over 35	0.032	0.063	0.125	0.188	0.250	0.375	0.500[9]
0.126 to 0.249	Over 4 thru 15	0.063	0.125	0.250	0.375	0.563	0.750	1.000[9]
	Over 15	0.032	0.063	0.125	0.188	0.250	0.375	0.500[9]
0.250 to 6.000	Up thru 10	0.250	0.563	1.000	1.563	2.250	3.000	4.000[9]
	Over 10 thru 18	0.063	0.125	0.250	0.406	0.594	0.781	1.000[9]
	Over 18	0.032	0.063	0.094	0.125	0.219	0.312	0.438 / 0.562[9]

Specified Length, ft	Specified Width, ft	
	Up thru 3	Over 3
	Squareness Tolerance — Allowable Difference in Length of Diagonals[2]	
Up thru 12	3/32 × Width in feet[3]	5/64 × Width in feet[3]
Over 12	9/64 × Width in feet[3]	7/64 × Width in feet[3]

SAWED OR SHEARED PLATE — All Tempers Except T351, T451, T651, T851, T7351, T7651, O, F, and HX8 and Harder

Specified Thickness	Transverse Flatness Tolerance — Allowable Deviation from Flat[4]		
	Widths over 4 ft thru 6 ft[5]	Widths over 2 ft thru 4 ft[6]	Widths 2 ft and less
0.250 to 0.624	1/2	3/8	Use short-cycle tolerance[8]
0.625 to 1.500	3/8	1/4	
1.501 to 3.000	1/4	3/16	
3.001 to 6.000	1/4	3/16	

SAWED OR SHEARED PLATE — T351, T451, T651, T851, T7351, T7651

Specified Thickness	Transverse Flatness Tolerance — Allowable Deviation from Flat[4]		
	Widths over 4 ft thru 6 ft[5]	Widths over 2 ft thru 4 ft[6]	Widths 2 ft and less[7]
0.250 to 0.624	3/8	5/16	Use short-cycle tolerance[8]
0.625 to 1.500	5/16	5/16	
1.501 to 3.000	3/16	3/16	
3.001 to 6.000	1/8	1/8	

All dimensions are in inches, except where otherwise noted. [1] Tolerances applicable at ambient mill temperatures. A change in length of 0.013 in. per 100 in. per 10°F (6°C) must be recognized. [2] Use values for calculating only. Round result upward to nearest 1/16 in. [3] If specified width is other than an exact multiple of 12 in., the tolerance is determined by using the next largest exact multiple. [4] Allowable deviation from flat with sheet positioned on flat horizontal surface to minimize deviation. [5] For widths over 6 feet, these tolerances apply for any 6 ft of total width. [6] For lengths under 6 feet, the tolerance is 1/8 in. [7] Not applicable to O, F, and HX8 and harder tempers. [8] Short-cycle flatness is the flatness over any 2-ft span in any direction. [9] Also applicable to any 240-in. increment of longer sheet or plate.

Tolerances for Aluminum Wire, Rod and Bar (ANSI H35.2-1982)

ROUND WIRE AND ROD

| Specified Diameter | Diameter Tolerance — Plus or Minus Except as Noted | | | |
| | Allowable Deviation from Specified Diameter | | | |
	Drawn Wire	Cold Finished Rod	Rolled Rod Plus	Minus
Up thru 0.035	.0005	...	...	...
0.036 to 0.064	.001	...	...	...
0.065 to 0.374	.0015	...	...	...
0.375 to 0.500	...	.0015	...	...
0.501 to 1.000	...	.002	...	...
1.001 to 1.500	...	.0025	...	...
1.501 to 2.000	...	.004	.006	.006
2.001 to 3.000	...	.006	.008	.008
3.001 to 3.499	...	.008	.012	.012
3.500 to 5.000	...	.012	.031	.016
5.001 to 8.000	...	...	.062	.031

REDRAW ROD

Specified Diameter	Diameter Tolerance — Plus or Minus. Allowable Deviation from Specified Diameter
0.375	0.020

RECTANGULAR WIRE AND BAR

| Specified Thickness or Width | Tolerance — Plus or Minus | | | |
| | Allowable Deviation from Specified Thickness and Width | | | |
	Drawn Wire and Cold Finished Bar Thickness	Width	Rolled Bar Thickness	Width
Up thru 0.035	.001	...	...	...
0.036 to 0.064	.0015	...	...	...
0.065 to 0.500	.002	.002	.006	...
0.501 to 0.750	.0025	.0025	.008	.016
0.751 to 1.000	.0025	.0025	.012	.016
1.001 to 1.500	.003	.003	.016	.016
1.501 to 2.000	.005	.005	.016	.031
2.001 to 3.000	.008	.008	.020	.031
3.001 to 4.000	...	.010	.020	.031
4.001 to 6.000	...	...	...	.047
6.001 to 10.000	...	...	...	.062

RIVET AND COLD HEADING WIRE AND ROD

| Specified Diameter | Diameter Tolerance | | | |
| | Allowable Deviation from Specified Diameter | | | |
	Rivet Wire Plus	Minus	Rivet Rod Plus	Minus
Up thru 0.061	.0005	.0005	...	...
0.062 to 0.123	.001	.0005	...	...
0.124 to 0.154	.001	.001	...	...
0.155 to 0.374	.002	.001	...	...
0.375 to 0.500	...	...	.002	.001
0.501 to 1.000	...	...	.003	.001

CENTERLESS GROUND ROUND WIRE AND ROD

Specified Diameter	Diameter Tolerance — Plus or Minus. Allowable Deviation from Specified Diameter
0.125 to 0.625	.0005
0.626 to 1.500	.0010
1.501 to 2.000	.0025

SQUARE, HEXAGONAL, AND OCTAGONAL WIRE AND BAR

| Specified Distance Across Flats | Tolerance — Plus or Minus | | |
| | Allowable Deviation from Specified Distance Across Flats | | |
	Drawn Wire	Cold Finished Bar	Rolled Bar
Up thru 0.035	.001	...	...
0.036 to 0.064	.0015	...	...
0.065 to 0.374	.002	...	...
0.375 to 0.500	...	.002	...
0.501 to 1.000	...	.0025	...
1.001 to 1.500	...	.003	...
1.501 to 2.000	...	.005	.016
2.001 to 3.000	...	.008	.020
3.001 to 4.000	...	...	.020

All dimensions are in inches.

Numbering Systems for Metals and Alloys. — Several different numbering systems have been developed for metals and alloys by various trade associations, professional engineering societies, standards organizations, and by private industries for their own use. The numerical code used to identify the metal or alloy may or may not be related to a specification, which is a statement of the technical and commercial requirements that the product must meet. Numbering systems in use include those developed by the American Iron and Steel Institute (AISI), Society of Automotive Engineers (SAE), American Society for Testing and Materials (ASTM), American National Standards Institute (ANSI), Steel Founders Society of America, American Society of Mechanical Engineers (ASME), American Welding Society (AWS), Aluminum Association, Copper Development Association, U.S. Department of Defense (Military Specifications), and the General Accounting Office (Federal Specifications).

The Unified Numbering System (UNS) was developed through a joint effort of the ASTM and the SAE to provide a means of correlating the different numbering systems for metals and alloys that have a commercial standing. This system avoids the confusion caused when more than one identification number is used to specify the same material, or when the same number is assigned to two entirely different materials. It is important to understand that a UNS number is not a specification; it is an identification number for metals and alloys for which detailed specifications are provided elsewhere. There are seventeen series of UNS numbers, which are shown in Table 1. Each UNS number consists of a letter prefix followed by five digits. In some cases the letter is suggestive of the family of metals identified by the series, such as A for aluminum and C for copper. Whenever possible, the numbers in the UNS number groups contain numbering sequences taken directly from other systems in order to facilitate the identification of the material; e.g., the corresponding UNS number for AISI 1020 steel is G10200. The UNS numbers corresponding to the commonly used AISI-SAE numbers that are used to identify plain carbon alloy and tool steels are given in Table 2.

Table 1. Unified Numbering System (UNS) for Metals and Alloys

UNS Series	Metal
Nonferrous Metals and Alloys	
A00001 to A99999	Aluminum and aluminum alloys
C00001 to C99999	Copper and copper alloys
E00001 to E99999	Rare earth and rare earth-like metals and alloys
L00001 to L99999	Low melting metals and alloys
M00001 to M99999	Miscellaneous nonferrous metals and alloys
P00001 to P99999	Precious metals and alloys
R00001 to R99999	Reactive and refractory metals and alloys
Z00001 to Z99999	Zinc and zinc alloys
Ferrous Metals and Alloys	
D00001 to D99999	Specified mechanical property steels
F00001 to F99999	Cast irons
G00001 to G99999	AISI and SAE carbon and alloy steels (except tool steels)
H00001 to H99999	AISI H-steels
J00001 to J99999	Cast steels (except tool steels)
K00001 to K99999	Miscellaneous steels and ferrous alloys
S00001 to S99999	Heat and corrosion resistant (stainless) steels
T00001 to T99999	Tool steels

Table 2. AISI and SAE Numbers and Their Corresponding UNS Numbers for Plain Carbon, Alloy, and Tool Steels

AISI-SAE Numbers	UNS Numbers	AISI-SAE Numbers	UNS Numbers	AISI-SAE Numbers	UNS Numbers	AISI-SAE Numbers	UNS Numbers
Plain Carbon Steels							
1005	G10050	1030	G10300	1070	G10700	1566	G15660
1006	G10060	1035	G10350	1078	G10780	1110	G11100
1008	G10080	1037	G10370	1080	G10800	1117	G11170
1010	G10100	1038	G10380	1084	G10840	1118	G11180
1012	G10120	1039	G10390	1086	G10860	1137	G11370
1015	G10150	1040	G10400	1090	G10900	1139	G11390
1016	G10160	1042	G10420	1095	G10950	1140	G11400
1017	G10170	1043	G10430	1513	G15130	1141	G11410
1018	G10180	1044	G10440	1522	G15220	1144	G11440
1019	G10190	1045	G10450	1524	G15240	1146	G11460
1020	G10200	1046	G10460	1526	G15260	1151	G11510
1021	G10210	1049	G10490	1527	G15270	1211	G12110
1022	G10220	1050	G10500	1541	G15410	1212	G12120
1023	G10230	1053	G10530	1548	G15480	1213	G12130
1025	G10250	1055	G10550	1551	G15510	1215	G12150
1026	G10260	1059	G10590	1552	G15520	12L14	G12144
1029	G10290	1060	G10600	1561	G15610	...	...
Alloy Steels							
1330	G13300	4150	G41500	5140	G51400	8642	G86420
1335	G13350	4161	G41610	5150	G51500	8645	G86450
1340	G13400	4320	G43200	5155	G51550	8655	G86550
1345	G13450	4340	G43400	5160	G51600	8720	G87200
4023	G40230	E4340	G43406	E51100	G51986	8740	G87400
4024	G40240	4615	G46150	E52100	G52986	8822	G88220
4027	G40270	4620	G46200	6118	G61180	9260	G92600
4028	G40280	4626	G46260	6150	G61500	50B44	G50441
4037	G40370	4720	G47200	8615	G86150	50B46	G50461
4047	G40470	4815	G48150	8617	G86170	50B50	G50501
4118	G41180	4817	G48170	8620	G86200	50B60	G50601
4130	G41300	4820	G48200	8622	G86220	51B60	G51601
4137	G41370	5117	G51170	8625	G86250	81B45	G81451
4140	G41400	5120	G51200	8627	G86270	94B17	G94171
4142	G41420	5130	G51300	8630	G86300	94B30	G94301
4145	G41450	5132	G51320	8637	G86370	...	...
4147	G41470	5135	G51350	8640	G86400	...	...
Tool Steels (AISI and UNS Only)							
M1	T11301	T6	T12006	A6	T30106	P4	T51604
M2	T11302	T8	T12008	A7	T30107	P5	T51605
M4	T11304	T15	T12015	A8	T30108	P6	T51606
M6	T11306	H10	T20810	A9	T30109	P20	T51620
M7	T11307	H11	T20811	A10	T30110	P21	T51621
M10	T11310	H12	T20812	D2	T30402	F1	T60601
M3-1	T11313	H13	T20813	D3	T30403	F2	T60602
M3-2	T11323	H14	T20814	D4	T30404	L2	T61202
M30	T11330	H19	T20819	D5	T30405	L3	T61203
M33	T11333	H21	T20821	D7	T30407	L6	T61206
M34	T11334	H22	T20822	O1	T31501	W1	T72301
M36	T11336	H23	T20823	O2	T31502	W2	T72302
M41	T11341	H24	T20824	O6	T31506	W5	T72305
M42	T11342	H25	T20825	O7	T31507	CA2	T90102
M43	T11343	H26	T20826	S1	T41901	CD2	T90402
M44	T11344	H41	T20841	S2	T41902	CD5	T90405
M46	T11346	H42	T20842	S4	T41904	CH12	T90812
M47	T11347	H43	T20843	S5	T41905	CH13	T90813
T1	T12001	A2	T30102	S6	T41906	CO1	T91501
T2	T12002	A3	T30103	S7	T41907	CS5	T91905
T4	T12004	A4	T30104	P2	T51602	...	...
T5	T12005	A5	T30105	P3	T51603	...	...

Standard Steels — Compositions, Applications, and Heat Treatments

Steel is the generic term for a large family of iron–carbon alloys, which are malleable, within some temperature range, immediately after solidification from the molten state. The principal raw materials used in steelmaking are iron ore, coal, and limestone. These materials are converted in a blast furnace into a product known as "pig iron," which contains considerable amounts of carbon, manganese, sulfur, phosphorus, and silicon. Pig iron is hard, brittle, and unsuitable for direct processing into wrought forms. Steelmaking is the process of refining pig iron as well as iron and steel scrap by removing undesirable elements from the melt and then adding desirable elements in predetermined amounts. A primary reaction in most steelmaking is the combination of carbon with oxygen to form a gas. If dissolved oxygen is not removed from the melt prior to or during pouring, the gaseous products continue to evolve during solidification. If the steel is strongly deoxidized by the addition of deoxidizing elements, no gas is evolved, and the steel is termed "killed" because it lies quietly in the molds. Increasing degrees of gas evolution (decreased deoxidation) characterize steels termed "semikilled," "capped," or "rimmed." The degree of deoxidation affects some of the properties of the steel. In addition to oxygen, liquid steel contains measurable amounts of dissolved hydrogen and nitrogen. For some critical steel applications, special deoxidation practices as well as vacuum treatments may be used to reduce and control dissolved gases.

The carbon content of common steel grades ranges from a few hundredths of a per cent to about 1 per cent. All steels also contain varying amounts of other elements, principally manganese, which acts as a deoxidizer and facilitates hot working. Silicon, phosphorus, and sulfur are also always present, if only in trace amounts. Other elements may be present, either as residuals that are not intentionally added, but result from the raw materials or steelmaking practice, or as alloying elements added to effect changes in the properties of the steel.

Steels can be cast to shape, or the cast ingot or strand can be reheated and hot worked by rolling, forging, extrusion, or other processes into a wrought mill shape. Wrought steels are the most widely used of engineering materials, offering a multitude of forms, finishes, strengths, and usable temperature ranges. No other material offers comparable versatility for product design.

Standard Steel Classification. — Wrought steels may be classified systematically into groups based on some common characteristic, such as chemical composition, deoxidation practice, finishing method, or product form. Chemical composition is the most often used basis for identifying and assigning standard designations to wrought steels. While carbon is the principal hardening and strengthening element in steel, no single element controls the steel's characteristics. The combined effect of several elements influences response to heat treatment, hardness, strength, microstructure, corrosion resistance, and formability. The standard steels can be divided broadly into three main groups: carbon steels, alloy steels, and stainless steels.

Carbon Steels: A steel qualifies as a carbon steel when its manganese content is limited to 1.65 per cent (max), silicon to 0.60 per cent (max), and copper to 0.60 per cent (max). With the exception of deoxidizers and boron when specified, no other alloying elements are added intentionally, but they may be present as residuals. If any of these incidental elements are considered detrimental for special applications, maximum acceptable limits may be specified. In contrast to most alloy steels, carbon steels are most often used without a final heat treatment; however, they may be annealed, normalized, case hardened, or quenched and tempered to enhance fabrication or mechanical properties. Carbon steels may be killed, semikilled, capped, or rimmed, and, when necessary, the method of deoxidation may be specified.

Alloy Steels: Alloy steels comprise not only those grades that exceed the element content limits for carbon steel, but also any grade to which different elements than used for carbon steel are added, within specific ranges or specific minimums, to enhance mechanical properties, fabricating characteristics, or any other attribute of the steel. By this definition, alloy steels encompass all steels other than carbon steels; however, by convention, steels containing over 3.99 per cent chromium are considered "special types" of alloy steel, which include the stainless steels and many of the tool steels.

In a technical sense the term alloy steel is reserved for those steels that contain a modest amount of alloying elements (about 1-4 per cent) and generally depend on thermal treatments to develop specific mechanical properties. Alloy steels are always killed, but special deoxidation or melting practices, including vacuum, may be specified for special critical applications. Alloy steels generally require additional care throughout their manufacture, since they are more sensitive to thermal and mechanical operations.

Stainless Steels: Stainless steels are high-alloy steels and have superior corrosion resistance to the carbon and conventional low-alloy steels because they contain relatively large amounts of chromium. Although other elements may also increase corrosion resistance, their usefulness in this respect is limited.

Stainless steels generally contain at least 10 per cent chromium, with or without other elements. It has been customary in the United States, however, to include in the stainless steel classification those steels that contain as little as 4 per cent chromium. Together, these steels form a family known as the stainless and heat-resisting steels, some of which possess very high strength and oxidation resistance. Few, however, contain more than 30 per cent chromium or less than 50 per cent iron.

In the broadest sense, the standard stainless steels can be divided into three groups based on their structures: austenitic, ferritic, and martensitic. In each of the three groups there is one composition that represents the basic, general-purpose alloy. All other compositions are derived from the basic alloy, with specific variations in composition being made to obtain very specific properties.

The *austenitic grades* are nonmagnetic in the annealed condition, although some may become slightly magnetic after cold working. They can be hardened only by cold working, and not by heat treatment, and combine outstanding corrosion and heat resistance with good mechanical properties over a wide temperature range. The austenitic grades are further classified into two subgroups: the chromium–nickel types and the less frequently used chromium–manganese–low-nickel types. The basic composition in the chromium–nickel group is widely known as 18-8 (Cr–Ni) and is the general-purpose austenitic grade. This grade is the basis for over 20 modifications which can be characterized as follows: the chromium–nickel ratio has been modified to change the forming characteristics; the carbon content has been decreased to prevent intergranular corrosion; the elements niobium or titanium have been added to stabilize the structure; or molybdenum has been added or the chromium and nickel contents have been increased to improve corrosion or oxidation resistance.

The standard *ferritic grades* are always magnetic and contain chromium but no nickel. They can be hardened to some extent by cold working, but not by heat treatment, and they combine corrosion and heat resistance with moderate mechanical properties and decorative appeal. The ferritic grades generally are restricted to a narrower range of corrosive conditions than the austenitic grades. The basic ferritic grade contains 17 per cent chromium. In this series there are free machining modifications and grades with increased chromium content to improve scaling resistance. Also in this ferritic group is a 12 per cent chromium steel (the basic composition of the martensitic group) with other elements, such as aluminum or titanium, added to prevent hardening.

Table 1. AISI-SAE System of Designating Carbon and Alloy Steels

AISI-SAE Designation[a]	Type of Steel and Nominal Alloy Content (%)
	Carbon Steels
10XX	Plain Carbon (Mn 1.00% max.)
11XX	Resulfurized
12XX	Resulfurized and Rephosphorized
15XX	Plain Carbon (Max. Mn range 1.00 to 1.65%)
	Manganese Steels
13XX	Mn 1.75
	Nickel Steels
23XX	Ni 3.50
25XX	Ni 5.00
	Nickel–Chromium Steels
31XX	Ni 1.25; Cr 0.65 and 0.80
32XX	Ni 1.75; Cr 1.07
33XX	Ni 3.50; Cr 1.50 and 1.57
34XX	Ni 3.00; Cr 0.77
	Molybdenum Steels
40XX	Mo 0.20 and 0.25
44XX	Mo 0.40 and 0.52
	Chromium–Molybdenum Steels
41XX	Cr 0.50, 0.80, and 0.95; Mo 0.12, 0.20, 0.25, and 0.30
	Nickel–Chromium–Molybdenum Steels
43XX	Ni 1.82; Cr 0.50 and 0.80; Mo 0.25
43BVXX	Ni 1.82; Cr 0.50; Mo 0.12 and 0.35; V 0.03 min.
47XX	Ni 1.05; Cr 0.45; Mo 0.20 and 0.35
81XX	Ni 0.30; Cr 0.40; Mo 0.12
86XX	Ni 0.55; Cr 0.50; Mo 0.20
87XX	Ni 0.55; Cr 0.50; Mo 0.25
88XX	Ni 0.55; Cr 0.50; Mo 0.35
93XX	Ni 3.25; Cr 1.20; Mo 0.12
94XX	Ni 0.45; Cr 0.40; Mo 0.12
97XX	Ni 0.55; Cr 0.20; Mo 0.20
98XX	Ni 1.00; Cr 0.80; Mo 0.25
	Nickel–Molybdenum Steels
46XX	Ni 0.85 and 1.82; Mo 0.20 and 0.25
48XX	Ni 3.50; Mo 0.25
	Chromium Steels
50XX	Cr 0.27, 0.40, 0.50, and 0.65
51XX	Cr 0.80, 0.87, 0.92, 0.95, 1.00, and 1.05
50XXX	Cr 0.50; C 1.00 min.
51XXX	Cr 1.02; C 1.00 min.
52XXX	Cr 1.45; C 1.00 min.
	Chromium–Vanadium Steels
61XX	Cr 0.60, 0.80, and 0.95; V 0.10 and 0.15 min
	Tungsten–Chromium Steels
72XX	W 1.75; Cr 0.75
	Silicon-Manganese Steels
92XX	Si 1.40 and 2.00; Mn 0.65, 0.82, and 0.85; Cr 0.00 and 0.65
	High-Strength Low-Alloy Steels
9XX	Various SAE grades
XXBXX	B denotes boron steels
XXLXX	L denotes leaded steels

AISI	SAE	Stainless Steels
2XX	302XX	Chromium–Manganese–Nickel Steels
3XX	303XX	Chromium–Nickel Steels
4XX	514XX	Chromium Steels
5XX	515XX	Chromium Steels

[a] XX in the last two digits of the carbon and low-alloy designations (but not the stainless steels) indicates that the carbon content (in hundredths of a per cent) is to be inserted.

The standard *martensitic grades* are magnetic and can be hardened by quenching and tempering. They contain chromium and, with two exceptions, no nickel. The basic martensitic grade normally contains 12 per cent chromium. There are over 10 standard compositions in the martensitic series; some are modified to improve machinability and others have small additions of nickel or other elements to improve the mechanical properties or their response to heat treatment. Still others have greatly

Table 2. Compositions of AISI–SAE Standard Carbon Steels

AISI-SAE No.	UNS No.	Composition(%)[1]			
		C	Mn	P(max)[4]	S(max)[4]
Nonresulfurized Grades — 1 per cent Mn (max)					
1005[2]	G10050	0.06 max	0.35 max	0.040	0.050
1006[2]	G10060	0.08 max	0.25–0.40	0.040	0.050
1008	G10080	0.10 max	0.30–0.50	0.040	0.050
1010	G10100	0.08–0.13	0.30–0.60	0.040	0.050
1012	G10120	0.10–0.15	0.30–0.60	0.040	0.050
1015	G10150	0.13–0.18	0.30–0.60	0.040	0.050
1016	G10160	0.13–0.18	0.60–0.90	0.040	0.050
1017	G10170	0.15–0.20	0.30–0.60	0.040	0.050
1018	G10180	0.15–0.20	0.60–0.90	0.040	0.050
1019	G10190	0.15–0.20	0.70–1.00	0.040	0.050
1020	G10200	0.18–0.23	0.30–0.60	0.040	0.050
1021	G10210	0.18–0.23	0.60–0.90	0.040	0.050
1022	G10220	0.18–0.23	0.70–1.00	0.040	0.050
1023	G10230	0.20–0.25	0.30–0.60	0.040	0.050
1025	G10250	0.22–0.28	0.30–0.60	0.040	0.050
1026	G10260	0.22–0.28	0.60–0.90	0.040	0.050
1029	G10290	0.25–0.31	0.60–0.90	0.040	0.050
1030	G10300	0.28–0.34	0.60–0.90	0.040	0.050
1035	G10350	0.32–0.38	0.60–0.90	0.040	0.050
1037	G10370	0.32–0.38	0.70–1.00	0.040	0.050
1038	G10380	0.35–0.42	0.60–0.90	0.040	0.050
1039	G10390	0.37–0.44	0.70–1.00	0.040	0.050
1040	G10400	0.37–0.44	0.60–0.90	0.040	0.050
1042	G10420	0.40–0.47	0.60–0.90	0.040	0.050
1043	G10430	0.40–0.47	0.70–1.00	0.040	0.050
1044	G10440	0.43–0.50	0.30–0.60	0.040	0.050
1045	G10450	0.43–0.50	0.60–0.90	0.040	0.050
1046	G10460	0.43–0.50	0.70–1.00	0.040	0.050
1049	G10490	0.46–0.53	0.60–0.90	0.040	0.050
1050	G10500	0.48–0.55	0.60–0.90	0.040	0.050
1053	G10530	0.48–0.55	0.70–1.00	0.040	0.050
1055	G10550	0.50–0.60	0.60–0.90	0.040	0.050
1059[2]	G10590	0.55–0.65	0.50–0.80	0.040	0.050
1060	G10600	0.55–0.65	0.60–0.90	0.040	0.050
1064[2]	G10640	0.60–0.70	0.50–0.80	0.040	0.050
1065[2]	G10650	0.60–0.70	0.60–0.90	0.040	0.050
1069[2]	G10690	0.65–0.75	0.40–0.70	0.040	0.050
1070	G10700	0.65–0.75	0.60–0.90	0.040	0.050
1078	G10780	0.72–0.85	0.30–0.60	0.040	0.050
1080	G10800	0.75–0.88	0.60–0.90	0.040	0.050
1084	G10840	0.80–0.93	0.60–0.90	0.040	0.050
1086[2]	G10860	0.80–0.93	0.30–0.50	0.040	0.050
1090	G10900	0.85–0.98	0.60–0.90	0.040	0.050
1095	G10950	0.90–1.03	0.30–0.50	0.040	0.050
Nonresulfurized Grades — Over 1 per cent Mn					
1513	G15130	0.10–0.16	1.10–1.40	0.040	0.050
1522	G15220	0.18–0.24	1.10–1.40	0.040	0.050
1524	G15240	0.19–0.25	1.35–1.65	0.040	0.050
1526	G15260	0.22–0.29	1.10–1.40	0.040	0.050
1527	G15270	0.22–0.29	1.20–1.50	0.040	0.050
1541	G15410	0.36–0.44	1.35–1.65	0.040	0.050
1548	G15480	0.44–0.52	1.10–1.40	0.040	0.050
1551	G15510	0.45–0.56	0.85–1.15	0.040	0.050
1552	G15520	0.47–0.55	1.20–1.50	0.040	0.050
1561	G15610	0.55–0.65	0.75–1.05	0.040	0.050
1566	G15660	0.60–0.71	0.85–1.15	0.040	0.050

See footnotes at end of table.

Table 2 (*Concluded*). **Compositions of AISI–SAE Standard Carbon Steels**

AISI-SAE No.	UNS No.	Composition (%)[1]			
		C	Mn	P(max)[4]	S(max)[4]
Free-Machining Grades — Resulfurized					
1110	G11100	0.08–0.13	0.30–0.60	0.040	0.08–0.13
1117	G11170	0.14–0.20	1.00–1.30	0.040	0.08–0.13
1118	G11180	0.14–0.20	1.30–1.60	0.040	0.08–0.13
1137	G11370	0.32–0.39	1.35–1.65	0.040	0.08–0.13
1139	G11390	0.35–0.43	1.35–1.65	0.040	0.13–0.20
1140	G11400	0.37–0.44	0.70–1.00	0.040	0.08–0.13
1141	G11410	0.37–0.45	1.35–1.65	0.040	0.08–0.13
1144	G11440	0.40–0.48	1.35–1.65	0.040	0.24–0.33
1146	G11460	0.42–0.49	0.70–1.00	0.040	0.08–0.13
1151	G11510	0.48–0.55	0.70–1.00	0.040	0.08–0.13
Free-Machining Grades — Resulfurized and Rephosphorized					
1211	G12110	0.13 max	0.60–0.90	0.07–0.12	0.10–0.15
1212	G12120	0.13 max	0.70–1.00	0.07–0.12	0.16–0.23
1213	G12130	0.13 max	0.70–1.00	0.07–0.12	0.24–0.33
1215	G12150	0.09 max	0.75–1.05	0.04–0.09	0.26–0.35
12L14[3]	G12144	0.15 max	0.85–1.15	0.04–0.09	0.26–0.35

[1] The following notes refer to boron, copper, lead, and silicon additions: Boron: Standard killed carbon steels, which are generally fine grain, may be produced with a boron treatment addition to improve hardenability. Such steels are produced to a range of 0.0005–0.003 per cent B. These steels are identified by inserting the letter "B" between the second and third numerals of the AISI or SAE number, e.g., 10B46. Copper: When copper is required, 0.20 per cent (min) is generally specified. Lead: Standard carbon steels can be produced with a lead range of 0.15–0.35 per cent to improve machinability. Such steels are identified by inserting the letter "L" between the second and third numerals of the AISI or SAE number, e.g., 12L15 and 10L45. Silicon: It is not common practice to produce the 12XX series of resulfurized and rephosphorized steels to specified limits for silicon because of its adverse effect on machinability. When silicon ranges or limits are required for resulfurized or nonresulfurized steels, however, these values apply: a range of 0.08 per cent Si for Si max up to 0.15 per cent inclusive, a range of 0.10 per cent Si for Si max over 0.15 to 0.20 per cent inclusive, a range of 0.15 per cent Si for Si max over 0.20 to 0.30 per cent inclusive, and a range of 0.20 per cent Si for Si max over 0.30 to 0.60 per cent inclusive. Example: Si max is 0.25 per cent, range is 0.10–0.25 per cent. [2] Standard grades for wire rods and wire only. [3] 0.15–0.35 per cent Pb. [4] Values given are maximum percentages, except where a range of values is given. Source: *Steel Products Manual*, American Iron and Steel Institute.

increased carbon content, in the tool steel range, and are hardenable to the highest levels of all the stainless steels. The martensitic grades are excellent for service in mild environments such as the atmosphere, freshwater, steam, and weak acids, but are not resistant to severely corrosive solutions.

Standard Steel Numbering System. — The most widely used systems for identifying wrought carbon, low-alloy, and stainless steels are based on chemical composition, and are those of the American Iron and Steel Institute (AISI) and the Society of Automotive Engineers (SAE). These systems are almost identical, but they are carefully coordinated. The standard steels so designated have been developed cooperatively by producers and users and have been found through long experience to cover most of the wrought ferrous metals used in automotive vehicles and related equipment. These designations, however, are not specifications, and should not be used for purchasing unless accompanied by supplementary information necessary to describe commercially the product desired. Engineering societies, associations, and institutes whose members make, specify, or purchase steel products publish standard specifications, many of which have become well known and respected. The most comprehensive and widely used specifications are those published by the American

Table 3. Compositions of AISI-SAE Standard Alloy Steels

AISI-SAE No.	UNS No.	Composition (%)[1,2]							
		C	Mn	P (max)	S (max)	Si	Ni	Cr	Mo
1330	G13300	0.28–0.33	1.60–1.90	0.035	0.040	0.15–0.35	…	…	…
1335	G13350	0.33–0.38	1.60–1.90	0.035	0.040	0.15–0.35	…	…	…
1340	G13400	0.38–0.43	1.60–1.90	0.035	0.040	0.15–0.35	…	…	…
1345	G13450	0.43–0.48	1.60–1.90	0.035	0.040	0.15–0.35	…	…	…
4023	G40230	0.20–0.25	0.70–0.90	0.035	0.040	0.15–0.35			0.20–0.30
4024	G40240	0.20–0.25	0.70–0.90	0.035	0.035–0.050	0.15–0.35			0.20–0.30
4027	G40270	0.25–0.30	0.70–0.90	0.035	0.040	0.15–0.35			0.20–0.30
4028	G40280	0.25–0.30	0.70–0.90	0.035	0.035–0.050	0.15–0.35			0.20–0.30
4037	G40370	0.35–0.40	0.70–0.90	0.035	0.040	0.15–0.35			0.20–0.30
4047	G40470	0.45–0.50	0.70–0.90	0.035	0.040	0.15–0.35			0.20–0.30
4118	G41180	0.18–0.23	0.70–0.90	0.035	0.040	0.15–0.35		0.40–0.60	0.08–0.15
4130	G41300	0.28–0.33	0.40–0.60	0.035	0.040	0.15–0.35		0.80–1.10	0.15–0.25
4137	G41370	0.35–0.40	0.70–0.90	0.035	0.040	0.15–0.35		0.80–1.10	0.15–0.25
4140	G41400	0.38–0.43	0.75–1.00	0.035	0.040	0.15–0.35		0.80–1.10	0.15–0.25
4142	G41420	0.40–0.45	0.75–1.00	0.035	0.040	0.15–0.35		0.80–1.10	0.15–0.25
4145	G41450	0.43–0.48	0.75–1.00	0.035	0.040	0.15–0.35		0.80–1.10	0.15–0.25
4147	G41470	0.45–0.50	0.75–1.00	0.035	0.040	0.15–0.35		0.80–1.10	0.15–0.25
4150	G41500	0.48–0.53	0.75–1.00	0.035	0.040	0.15–0.35		0.80–1.10	0.15–0.25
4161	G41610	0.56–0.64	0.75–1.00	0.035	0.040	0.15–0.35		0.70–0.90	0.25–0.35
4320	G43200	0.17–0.22	0.45–0.65	0.035	0.040	0.15–0.35	1.65–2.00	0.40–0.60	0.20–0.30
4340	G43400	0.38–0.43	0.60–0.80	0.035	0.040	0.15–0.35	1.65–2.00	0.70–0.90	0.20–0.30
E4340[3]	G43406	0.38–0.43	0.65–0.85	0.025	0.025	0.15–0.35	1.65–2.00	0.70–0.90	0.20–0.30
4615	G46150	0.13–0.18	0.45–0.65	0.035	0.040	0.15–0.35	1.65–2.00		0.20–0.30
4620	G46200	0.17–0.22	0.45–0.65	0.035	0.040	0.15–0.35	1.65–2.00		0.20–0.30
4626	G46260	0.24–0.29	0.45–0.65	0.035	0.040	0.15–0.35	0.70–1.00		0.15–0.25
4720	G47200	0.17–0.22	0.50–0.70	0.035	0.040	0.15–0.35	0.90–1.20	0.35–0.55	0.15–0.25
4815	G48150	0.13–0.18	0.40–0.60	0.035	0.040	0.15–0.35	3.25–3.75		0.20–0.30
4817	G48170	0.15–0.20	0.40–0.60	0.035	0.040	0.15–0.35	3.25–3.75		0.20–0.30
4820	G48200	0.18–0.23	0.50–0.70	0.035	0.040	0.15–0.35	3.25–3.75		0.20–0.30
5117	G51170	0.15–0.20	0.70–0.90	0.035	0.040	0.15–0.35		070–0.90	
5120	G51200	0.17–0.22	0.70–0.90	0.035	0.040	0.15–0.35		0.70–0.90	
5130	G51300	0.28–0.33	0.70–0.90	0.035	0.040	0.15–0.35		0.80–1.10	
5132	G51320	0.30–0.35	0.60–0.80	0.035	0.040	0.15–0.35		0.75–1.00	
5135	G51350	0.33–0.38	0.60–0.80	0.035	0.040	0.15–0.35		0.80–1.05	
5140	G51400	0.38–0.43	0.70–0.90	0.035	0.040	0.15–0.35		0.70–0.90	
5150	G51500	0.48–0.53	0.70–0.90	0.035	0.040	0.15–0.35		0.70–0.90	
5155	G51550	0.51–0.59	0.70–0.90	0.035	0.040	0.15–0.35		0.70–0.90	

Table 3 (Continued). Compositions of AISI-SAE Standard Alloy Steels

AISI-SAE No.	UNS No.	Composition (%)[1,2]							
		C	Mn	P (max)	S (max)	Si	Ni	Cr	Mo
5160	G51600	0.56–0.64	0.75–1.00	0.035	0.040	0.15–0.35	…	0.70–0.90	…
E51100[3]	G51986	0.98–1.10	0.25–0.45	0.025	0.025	0.15–0.35	…	0.90–1.15	…
E52100[3]	G52986	0.98–1.10	0.25–0.45	0.025	0.025	0.15–0.35	…	1.30–1.60	…
6118	G61180	0.16–0.21	0.50–0.70	0.035	0.040	0.15–0.35	…	0.50–0.70	0.10–0.15 V
6150	G61500	0.48–0.53	0.70–0.90	0.035	0.040	0.15–0.35	…	0.80–1.10	0.15 V min
8615	G86150	0.13–0.18	0.70–0.90	0.035	0.040	0.15–0.35	0.40–0.70	0.40–0.60	0.15–0.25
8617	G86170	0.15–0.20	0.70–0.90	0.035	0.040	0.15–0.35	0.40–0.70	0.40–0.60	0.15–0.25
8620	G86200	0.18–0.23	0.70–0.90	0.035	0.040	0.15–0.35	0.40–0.70	0.40–0.60	0.15–0.25
8622	G86220	0.20–0.25	0.70–0.90	0.035	0.040	0.15–0.35	0.40–0.70	0.40–0.60	0.15–0.25
8625	G86250	0.23–0.28	0.70–0.90	0.035	0.040	0.15–0.35	0.40–0.70	0.40–0.60	0.15–0.25
8627	G86270	0.25–0.30	0.70–0.90	0.035	0.040	0.15–0.35	0.40–0.70	0.40–0.60	0.15–0.25
8630	G86300	0.28–0.33	0.70–0.90	0.035	0.040	0.15–0.35	0.40–0.70	0.40–0.60	0.15–0.25
8637	G86370	0.35–0.40	0.75–1.00	0.035	0.040	0.15–0.35	0.40–0.70	0.40–0.60	0.15–0.25
8640	G86400	0.38–0.43	0.75–1.00	0.035	0.040	0.15–0.35	0.40–0.70	0.40–0.60	0.15–0.25
8642	G86420	0.40–0.45	0.75–1.00	0.035	0.040	0.15–0.35	0.40–0.70	0.40–0.60	0.15–0.25
8645	G86450	0.43–0.48	0.75–1.00	0.035	0.040	0.15–0.35	0.40–0.70	0.40–0.60	0.15–0.25
8655	G86550	0.51–0.59	0.75–1.00	0.035	0.040	0.15–0.35	0.40–0.70	0.40–0.60	0.15–0.25
8720	G87200	0.18–0.23	0.70–0.90	0.035	0.040	0.15–0.35	0.40–0.70	0.40–0.60	0.20–0.30
8740	G87400	0.38–0.43	0.75–1.00	0.035	0.040	0.15–0.35	0.40–0.70	0.40–0.60	0.20–0.30
8822	G88220	0.20–0.25	0.75–1.00	0.035	0.040	0.15–0.35	0.40–0.70	0.40–0.60	0.30–0.40
9260	G92600	0.56—0.64	0.75–1.00	0.035	0.040	1.80–2.20	…	…	…
Standard Boron Grades[4]									
50B44	G50441	0.43–0.48	0.75–1.00	0.035	0.040	0.15–0.35	…	0.40–0.60	…
50B46	G50461	0.44–0.49	0.75–1.00	0.035	0.040	0.15–0.35	…	0.20–0.35	…
50B50	G50501	0.48–0.53	0.75–1.00	0.035	0.040	0.15–0.35	…	0.40–0.60	…
50B60	G50601	0.56–0.64	0.75–1.00	0.035	0.040	0.15–0.35	…	0.40–0.60	…
51B60	G51601	0.56–0.64	0.75–1.00	0.035	0.040	0.15–0.35	…	0.70–0.90	…
81B45	G81451	0.43–0.48	0.75–1.00	0.035	0.040	0.15–0.35	0.20–0.40	0.35–0.55	0.08–0.15
94B17	G94171	0.15–0.20	0.75–1.00	0.035	0.040	0.15–0.35	0.30–0.60	0.30–0.50	0.08–0.15
94B30	G94301	0.28–0.33	0.75–1.00	0.035	0.040	0.15–0.35	0.30–0.60	0.30–0.50	0.08–0.15

[1] Small quantities of certain elements are present which are not specified or required. These incidental elements may be present to the following maximum amounts: Cu, 0.35 per cent; Ni, 0.25 per cent; Cr, 0.20 per cent; and Mo, 0.06 per cent. [2] Standard alloy steels can also be produced with a lead range of 0.15–0.35 per cent. Such steels are identified by inserting the letter "L" between the second and third numerals of the AISI or SAE number, e.g., 41L40. [3] Electric furnace steel. [4] 0.0005–0.003 per cent B.

Source: Steel Products Manual. American Iron and Steel Institute.

Table 4. Standard Stainless Steels — Typical Compositions

AISI Type (UNS)	Typical Composition (%)	AISI Type (UNS)	Typical Composition (%)
		Austenitic	
201 (S20100)	16–18 Cr, 3.5–5.5 Ni, 0.15 C, 5.5–7.5 Mn, 1.0 Si, 0.060 P, 0.030 S, 0.25 N	310 (S31000)	24–26 Cr, 19–22 Ni, 0.25 C, 2.0 Mn, 1.5 Si, 0.045 P, 0.030 S
202 (S20200)	17–19 Cr, 4–6 Ni, 0.15 C, 7.5–10.0 Mn, 1.0 Si, 0.060 P, 0.030 S, 0.25 N	310S (S31008)	24–26 Cr, 19–22 Ni, 0.08 C, 2.0 Mn, 1.5 Si, 0.045 P, 0.30 S
205 (S20500)	16.5–18 Cr, 1–1.75 Ni, 0.12–0.25 C, 14–15.5 Mn, 0.50 Si, 0.030 P, 0.030 S, 0.32–0.40 N	314 (S31400)	23–26 Cr, 19–22 Ni, 0.25 C, 2.0 Mn, 1.5–3.0 Si, 0.045 P, 0.030 S
301 (S30100)	16–18 Cr, 6–8 Ni, 0.15 C, 2.0 Mn, 1.0 Si, 0.045 P, 0.030 S	316 (S31600)	16–18 Cr, 10–14 Ni, 0.08 C, 2.0 Mn, 1.0 Si, 0.045 P, 0.030 S, 2.0–3.0 Mo
302 (S30200)	17–19 Cr, 8–10 Ni, 0.15 C, 2.0 Mn, 1.0 Si, 0.045 P, 0.030 S	316L (S31603)	16–18 Cr, 10–14 Ni, 0.03 C, 2.0 Mn, 1.0 Si, 0.045 P, 0.030 S, 2.0–3.0 Mo
302B (S30215)	17–19 Cr, 8–10 Ni, 0.15 C, 2.0 Mn, 2.0–3.0 Si, 0.045 P, 0.030 S	316F (S31620)	16–18 Cr, 10–14 Ni, 0.08 C, 2.0 Mn, 1.0 Si, 0.20 P 0.010 S min, 1.75–2.50 Mo
303 (S30300)	17–19 Cr, 8–10 Ni, 0.15 C, 2.0 Mn, 1.0 Si, 0.20 P, 0.015 S min, 0.60 Mo (optional)	316N (S31651)	16–18 Cr, 10–14 Ni, 0.08 C 2.0 Mn, 1.0 Si, 0.045 P, 0.030 S, 2–3 Mo, 0.10–0.16 N
303Se (S30323)	17–19 Cr, 8–10 Ni, 0.15 C, 2.0 Mn, 1.0 Si, 0.20 P, 0.060 S, 0.15 Se min	317 (S31700)	18–20 Cr, 11–15 Ni, 0.08 C, 2.0 Mn, 1.0 Si, 0.045 P, 0.030 S, 3.0–4.0 Mo
304 (S30400)	18–20 Cr, 8–10.50 Ni, 0.08 C, 2.0 Mn, 1.0 Si, 0.045 P, 0.030 S	317L (S31703)	18–20 Cr, 11–15 Ni, 0.03 C, 2.0 Mn, 1.0 Si, 0.045 P, 0.030 S, 3–4 Mo
304L (S30403)	18–20 Cr, 8–12 Ni, 0.03 C, 2.0 Mn, 1.0 Si, 0.045 P, 0.030 S	321 (S32100)	17–19 Cr, 9–12 Ni, 0.08 C, 2.0 Mn, 1.0 Si, 0.045 P, 0.030 S (Ti, 5 × C min)
(S30430)	17–19 Cr, 8–10 Ni, 0.08 C, 2.0 Mn, 1.0 Si, 0.045 P, 0.030 S, 3–4 Cu	329 (S32900)	25–30 Cr, 3–6 Ni, 0.10 C, 2.0 Mn, 1.0 Si, 0.040 P, 0.030 S, 1–2 Mo
304N (S30451)	18–20 Cr, 8–10.5 Ni, 0.08 C, 2.0 Mn, 1.0 Si, 0.045 P, 0.030 S, 0.10–0.16 N	330 (N08330)	17–20 Cr, 34–37 Ni, 0.08 C, 2.0 Mn, 0.75–1.50 Si, 0.040 P, 0.030 S
305 (S30500)	17–19 Cr, 10.50–13 Ni, 0.12 C, 2.0 Mn, 1.0 Si, 0.045 P, 0.030 S	347 (S34700)	17–19 Cr, 9–13 Ni, 0.08 C, 2.0 Mn, 1.0 Si, 0.045 P, 0.030 S (Nb + Ta, 10 × C min)
308 (S30800)	19–21 Cr, 10–12 Ni, 0.08 C, 2.0 Mn, 1.0 Si, 0.045 P, 0.030 S	348 (S34800)	17–19 Cr, 9–13 Ni, 0.08 C, 2.0 Mn, 1.0 Si, 0.045 P, 0.030 S (Nb + Ta, 10 × C min but 0.10 Ta max), 0.20 Ca
309 (S30900)	22–24 Cr, 12–15 Ni, 0.20 C, 2.0 Mn, 1.0 Si, 0.045 P, 0.030 S	384 (S38400)	15–17 Cr, 17–19 Ni, 0.08 C, 2.0 Mn, 1.0 Si, 0.045 P, 0.030 S
309S (S30908)	22–24 Cr, 12–15 Ni, 0.08 C, 2.0 Mn, 1.0 Si, 0.045 P, 0.030 S	. . .	. . .
		. . .	. . .
		Ferritic	
405 (S40500)	11.5–14.5 Cr, 0.08 C, 1.0 Mn, 1.0 Si, 0.040 P, 0.030 S, 0.1–0.3 Al	430FSe (S43023)	16–18 Cr, 0.12 C, 1.25 Mn, 1.0 Si, 0.060 P, 0.060 S, 0.15 Se min

Table 4 (*Concluded*). **Standard Stainless Steels — Typical Compositions**

AISI Type (UNS)	Typical Composition (%)	AISI Type (UNS)	Typical Composition (%)
Ferritic (*Continued*)			
409 (S40900)	10.5–11.75 Cr, 0.08 C, 1.0 Mn, 1.0 Si, 0.045 P, 0.045 S (Ti 6 × C, but with 0.75 max)	434 (S43400)	16–18 Cr, 0.12 C, 1.0 Mn, 1.0 Si, 0.040 P, 0.030 S, 0.75–1.25 Mo
429 (S42900)	14–16 Cr, 0.12 C, 1.0 Mn, 1.0 Si, 0.040 P, 0.030 S	436 (S43600)	16–18 Cr, 0.12 C, 1.0 Mn, 1.0 Si, 0.040 P, 0.030 S, 0.75–1.25 Mo. (Nb + Ta 5 × C min, 0.70 max)
430 (S43000)	16–18 Cr, 0.12 C, 1.0 Mn, 1.0 Si, 0.040 P, 0.030 S	442 (S44200)	18–23 Cr, 0.20 C, 1.0 Mn, 1.0 Si, 0.040 P, 0.030 S
430F (S43020)	16–18 Cr, 0.12 C, 1.25 Mn, 1.0 Si, 0.060 P, 0.15 S min, 0.60 Mo (optional)	446 (S44600)	23–27 Cr, 0.20 C, 1.5 Mn, 1.0 Si, 0.040 P, 0.030 S, 0.25 N
Martensitic			
403 (S40300)	11.5–13.0 Cr, 0.15 C, 1.0 Mn, 0.5 Si, 0.040 P, 0.030 S	420F (S42020)	12–14 Cr, over 0.15 C, 1.25 Mn, 1.0 Si, 0.060 P, 0.15 S min, 0.60 Mo max (optional)
410 (S41000)	11.5–13.5 Cr, 0.15 C, 1.0 Mn, 1.0 Si, 0.040 P, 0.030 S	422 (S42200)	11–13 Cr, 0.50–1.0 Ni, 0.20–0.25 C, 1.0 Mn, 0.75 Si, 0.025 P, 0.025 S, 0.75–1.25 Mo. 0.15–0.30 V, 0.75–1.25 W
414 (S41400)	11.5–13.5 Cr, 1.25–2.50 Ni, 0.15 C, 1.0 Mn, 1.0 Si, 0.040 P, 0.030 S	431 (S43100)	15–17 Cr, 1.25–2.50 Ni, 0.20 C, 1.0 Mn, 1.0 Si, 0.040 P, 0.030 S
416 (S41600)	12–14 Cr, 0.15 C, 1.25 Mn, 1.0 Si, 0.060 P, 0.15 S min, 0.060 Mo (optional)	440A (S44002)	16–18 Cr, 0.60–0.75 C, 1.0 Mn, 1.0 Si, 0.040 P, 0.030 S, 0.75 Mo
416Se (S41623)	12–14 Cr, 0.15 C, 1.25 Mn, 1.0 Si, 0.060 P, 0.060 S, 0.15 Se min	440B (S44003)	16–18 Cr, 0.75–0.95 C, 1.0 Mn, 1.0 Si, 0.040 P, 0.030 S, 0.75 Mo
420 (S42000)	12–14 Cr, 0.15 C min, 1.0 Mn, 1.0 Si, 0.040 P, 0.030 S	440C (S44004)	16–18 Cr, 0.95–1.20 C, 1.0 Mn, 1.0 Si, 0.040 P, 0.030 S, 0.75 Mo
Heat Resisting			
501 (S50100)	4–6 Cr, 0.10 C min, 1.0 Mn, 1.0 Si, 0.040 P, 0.030 S, 0.40–0.65 Mo	502 (S50200)	4–6 Cr, 0.10 C, 1.0 Mn, 1.0 Si, 0.040 P, 0.030 S, 0.40–0.65 Mo

Society for Testing and Materials (ASTM). The U.S. government and various companies also publish their own specification for steel products to serve their own special procurement needs. The Unified Numbering System (UNS) for metals and alloys is also used to designate steels (see pages 428 and 429).

The numerical designation system used by both AISI and SAE for wrought carbon, alloy, and stainless steels is summarized in Table 1. Table 2 lists the compositions of the standard carbon steels; Table 3 lists the standard low-alloy steel compositions; and Table 4 includes the typical compositions of the standard stainless steels.

Thermal Treatment of Steel. — Steel's versatility is due to its response to thermal treatment. While most steel products are used in the as-rolled or un-heat-treated condition, thermal treatment greatly increases the number of properties that can be obtained, because at certain "critical temperatures" iron changes from one type of crystal structure to another. This structural change, known as an allotropic transfor-

mation, is spontaneous and reversible and can be made to occur by simply changing the temperature of the metal.

In steel, the transformation in crystal structure occurs over a range of temperatures, bounded by lower and upper critical points. When heated, most carbon and low-alloy steels have a critical temperature range between 1300 and 1600 degrees F. Steel above this temperature, but below the melting range, has a crystalline structure known as austenite, in which the carbon and alloying elements are dissolved in a solid solution. Below this critical range, the crystal structure changes to a phase known as ferrite, which is capable of maintaining only a very small percentage of carbon in solid solution. The remaining carbon exists in the form of carbides, which are compounds of carbon and iron and certain of the other alloying elements. Depending primarily on cooling rate, the carbides may be present as thin plates alternating with the ferrite (pearlite); as spheroidal globular particles at ferrite grain boundaries or dispersed throughout the ferrite; or as a uniform distribution of extremely fine particles throughout a "ferritelike" phase, which has an acicular (needle-like) appearance, named martensite. In some of the highly alloyed stainless steels the addtion of certain elements stabilizes the austenite structure so that it persists even at very low temperatures (austenitic grades). Other alloying elements can prevent the formation of austenite entirely up to the melting point (ferritic grades).

Fundamentally, all steel heat treatments are intended to either harden or soften the metal. They involve one or a series of operations in which the solid metal is heated and cooled under specified conditions to develop a required structure and properties. In general, there are five major forms of heat treatment for the standard steels that modify properties to suit either fabrication or end use.

Quenching and Tempering: This is the primary hardening treatment for steel and usually consists of three successive operations: heating the steel above the critical range and holding it at these temperatures for a sufficient time to approach a uniform solid solution (austenitizing); cooling the steel rapidly by quenching in oil, water, brine, salt, air, etc., to form a hard, usually brittle, metastable structure known as untempered or white martensite; tempering the steel by reheating it to a temperature below the critical range in order to obtain the proper combination of hardness, strength, ductility, toughness, and structural stability (tempered martensite).

Two well-known modifications of conventional quenching and tempering are "austempering" and "martempering." They involve interrupted quenching techniques (two or more quenching media) that can be utilized for some steels to obtain desired structures and properties while minimizing distortion and cracking problems that may occur in conventional hardening.

Normalizing: In this treatment the steel is heated to a temperature above the critical range, after which it is cooled in still air to produce a generally fine pearlite structure. The purpose is to promote uniformity of structure and properties after a hot-working operation such as forging, extrusion, etc. Steels may be placed in service in the normalized condition, or they may be subjected to additional thermal treatment after subsequent machining or other operations.

Annealing: This treatment consists of heating the steel to a temperature above or within the critical range, then cooling it at a predetermined slow rate (usually in a furnace) to produce a coarse pearlite structure. It is used to soften the steel for improved machinability; to improve or restore ductility for subsequent forming operations; or to eliminate the residual stresses and microstructural effects of cold working.

Spheroidize Annealing: This is a special form of annealing that requires prolonged heating at an appropriate temperature followed by slow cooling in order to produce globular carbides, a structure desirable for machining, cold forming, or cold drawing, or for the effect it will have on subsequent heat treatment.

Stress Relieving: This process reduces internal stresses, caused by machining, cold working, or welding, by heating the steel to a temperature below the critical range and holding it there long enough to equalize the temperature throughout the piece.

See pages 515 to 555 for more information about the heat treatment of steels.

Hardness and Hardenability. — Hardenability is the property of steel that determines the *depth and distribution of hardness* induced by quenching from the austenitizing temperature. Hardenability should not be confused with hardness as such or with maximum hardness. Hardness is a measure of the ability of a metal to resist penetration as determined by any one of a number of standard tests (Brinell, Rockwell, Vickers, etc). The maximum attainable hardness of any steel depends solely on carbon content and is not significantly affected by alloy content. Maximum hardness is realized only when the cooling rate in quenching is rapid enough to ensure full transformation to martensite.

The as-quenched surface hardness of a steel part is dependent on carbon content and cooling rate, but the *depth* to which a certain hardness level is maintained with given quenching conditions is a function of its hardenability. Hardenability is largely determined by the percentage of alloying elements in the steel; however, austenite grain size, time and temperature during austenitizing, and prior microstructure also significantly affect the hardness depth. The hardenability required for a particular part depends on size, design, and service stresses. For highly stressed parts, the best combination of strength and toughness is obtained by through hardening to a martensitic structure followed by adequate tempering. There are applications, however, where through hardening is not necessary or even desirable. For parts that are stressed principally at or near the surface, or in which wear resistance or resistance to shock loading is anticipated, a shallow hardening steel with a moderately soft core may be appropriate.

When through hardening thin sections, carbon steels may be adequate; but as section size increases, alloy steels of increasing hardenability are required. The usual practice is to select the most economical grade that can meet the desired properties consistently. It is not good practice to utilize a higher alloy grade than necessary, since excessive use of alloying elements adds little to the properties and can, in some instances, induce susceptibility to quenching cracks.

Quenching Media: The choice of quenching media is often a critical factor in the selection of the proper hardenability steel for a particular application. Quenching severity can be varied by quenching medium selection, agitation control, and additives that improve the cooling capability of the quenchant. Increasing quenching severity permits the use of less expensive steels of lower hardenability; however, consideration must also be given to the amount of distortion that can be tolerated and the susceptibility to quench cracking. In general, the more severe the quenchant and the less symmetrical the part being quenched, the greater are the size and shape changes that result from quenching and the greater is the risk of quench cracking. Consequently, although water quenching is less costly than oil quenching, and water quenching steels are less expensive than those requiring oil quenching, it is important to know that the parts being hardened can withstand the resulting distortion and the possibility of cracking.

Oil, salt, and synthetic water–polymer quenchants are also used, but they often require steels of higher alloy content and hardenability. A general rule for the selection of steel and quenchant for a particular part is that the steel should have a hardenability not exceeding that required by the severity of the quenchant selected. The carbon content of the steel should also not exceed that required to meet specified hardness and strength, since quench cracking susceptibility increases with carbon content.

While the choice of quenching media is important in hardening, another factor is agitation of the quenching bath. The more rapidly the bath is agitated, the more rapidly heat is removed from the steel and the more effective is the quench.

Hardenability Test Methods: The most commonly used method for determining hardenability is the end-quench test developed by Jominy and Boegehold, and described in detail in both SAE J406 and ASTM A255. In this test a normalized 1-inch-round, approximately 4-inch-long specimen of the steel to be evaluated is heated uniformly to its austenitizing temperature. The specimen is then removed from the furnace, placed in a jig, and immediately end quenched by a jet of room-temperature water. The water is played on the end face of the specimen, without touching the sides, until the entire specimen has cooled. Longitudinal flat surfaces are ground on opposite sides of the piece and Rockwell C scale hardness readings are taken at $\frac{1}{16}$-inch intervals from the quenched end. The resulting data are plotted on graph paper with the hardness values as ordinates (y-axis) and distances from the quenched end as abscissas (x-axis). Representative data have been accumulated for a variety of standard steel grades and are published by SAE and AISI as "H-bands." These data show graphically and in tabular form the high and low limits applicable to each grade. The suffix H following the standard AISI/SAE numerical designation indicates that the steel has been produced to specific hardenability limits.

Experiments have confirmed that the cooling rate at a given point along the Jominy bar corresponds closely to the cooling rate at various locations in round bars of various sizes. In general, when end-quench curves for different steels coincide approximately, similar treatments will produce similar properties in sections of the same size. On occasion it is necessary to predict the end-quench hardenability of a steel not available for testing, and reasonably accurate means of calculating hardness for any Jominy location on a section of steel of known analysis and grain size have been developed.

Tempering: As-quenched steels are in a highly stressed condition and are seldom used without tempering. Tempering imparts plasticity or toughness to the steel, and is inevitably accompanied by a loss in hardness and strength. The loss in strength, however, is only incidental to the very important increase in toughness, which is due to the relief of residual stresses induced during quenching and to precipitation, coalescence, and spheroidization of iron and alloy carbides resulting in a microstructure of greater plasticity.

Alloying slows the tempering rate, so that alloy steel requires a higher tempering temperature to obtain a given hardness than carbon steel of the same carbon content. The higher tempering temperature for a given hardness permits a greater relaxation of residual stress and thereby improves the steel's mechanical properties. Tempering is done in furnaces or in oil or salt baths at temperatures varying from 300 to 1200 degrees F. With most grades of alloy steel, the range between 500 and 700 degrees F. is avoided because of a phenomenon known as "blue brittleness," which reduces impact properties, and tempering the martensitic stainless steels in the range of 800–1100 degrees F. is not recommended because of the low and erratic impact properties and reduced corrosion resistance that result. Maximum toughness is achieved at higher temperatures. It is important to temper parts as soon as possible after quenching, since any delay greatly increases the risk of cracking resulting from the high stress condition in the as-quenched part.

Surface Hardening Treatment (Case Hardening). — Many applications require high hardness or strength primarily at the surface, and complex service stresses frequently require not only a hard, wear resistant surface, but also core strength and toughness to withstand impact stress. To achieve these different properties two general processes are used: (1) The chemical composition of the surface is altered, prior to or after quenching and tempering; the processes used include carburizing, nitrid-

ing, cyaniding, and carbonitriding. (2) Only the surface layer is hardened by the heating and quenching process; the most common processes used are flame hardening and induction hardening.

Carburizing: In this process, carbon is diffused into the part's surface to a controlled depth by heating the part in a carbonaceous medium. The resulting depth of carburization, commonly referred to as case depth, depends on the carbon potential of the medium used and the time and temperature of the carburizing treatment. The steels most suitable for carburizing to enhance toughness are those with sufficiently low carbon contents, usually below 0.03 per cent. Carburizing temperatures range from 1550 to 1750 degrees F., with the temperature and time at temperature adjusted to obtain various case depths. Steel selection, hardenability, and type of quench are determined by section size, desired core hardness, and service requirements.

Three types of carburizing are most often used: (1) *Liquid carburizing* involves heating the steel in molten barium cyanide or sodium cyanide. The case absorbs some nitrogen in addition to carbon, thus enhancing surface hardness. (2) *Gas carburizing* involves heating the steel in a gas of controlled carbon content. When used, the carbon level in the case can be closely controlled. (3) *Pack carburizing* involves sealing both the steel and solid carbonaceous material in a gas-tight container, then heating this combination.

With any of these methods, the part may be either quenched after the carburizing cycle without reheating or air cooled followed by reheating to the austenitizing temperature prior to quenching. The case depth may be varied to suit the conditions of loading in service. Frequently, however, service characteristics require that only selective areas of a part have to be case hardened. Covering the areas not to be cased, with copper plating or a layer of commercial paste, allows the carbon to penetrate only the exposed areas. Another method involves carburizing the entire part, then removing the case in selected areas by machining, prior to quench hardening.

Nitriding: In this process, the steel part is heated to a temperature of 900–1150 degrees F. in an atmosphere of ammonia gas and dissociated ammonia for an extended period of time that depends on the case depth desired. A thin, very hard case results from the formation of nitrides. Strong nitride-forming elements (chromium and molybdenum) are required to be present in the steel, and often special nonstandard grades containing aluminum (a strong nitride former) are used. The major advantage of this process is that parts can be quenched and tempered, then machined, prior to nitriding, because only a little distortion occurs during nitriding.

Cyaniding: This process involves heating the part in a bath of sodium cyanide to a temperature slightly above the transformation range, followed by quenching, to obtain a thin case of high hardness.

Carbonitriding: This process is similar to cyaniding except that the absorption of carbon and nitrogen is accomplished by heating the part in a gaseous atmosphere containing hydrocarbons and ammonia. Temperatures of 1425–1625 degrees F. are used for parts to be quenched, while lower temperatures, 1200–1450 degrees F., may be used where a liquid quench is not required.

Flame Hardening: This process involves rapid heating with a direct high-temperature gas flame, such that the surface layer of the part is heated above the transformation range, followed by cooling at a rate that causes the desired hardening. Steels for flame hardening are usually in the range of 0.30–0.60 per cent carbon, with hardenability appropriate for the case depth desired and the quenchant used. The quenchant is usually sprayed on the surface a short distance behind the heating flame. Immediate tempering is required and may be done in a conventional furnace or by a flame-tempering process, depending on part size and costs.

Induction Hardening: This process is similar in many respects to flame hardening except that the heating is caused by a high-frequency electric current sent through a

coil or inductor surrounding the part. The depth of heating depends on the frequency, the rate of heat conduction from the surface, and the length of the heating cycle. Quenching is usually accomplished with a water spray introduced at the proper time through jets in or near the inductor block or coil. In some instances, however, parts are oil quenched by immersing them in a bath of oil after they reach the hardening temperature.

Quality Variations of Carbon and Alloy Steels. — Carbon steels may be produced with chemical composition (carbon, manganese, phosphorus, sulfur, and silicon) within the specified limits of a given grade and still have characteristics that are dissimilar. Each grade and quality variation thereof has a proper and useful place, depending on the end products to be made and the methods of fabrication.

In all phases of steel production, various practices are employed that determine the quality and types of the finished material.

Quality Classifications: The term "quality" as it technically relates to steel products may be indicative of many conditions such as the degree of internal soundness, relative uniformity of composition, relative freedom from injurious surface imperfections, and finish. Steel quality also relates to general suitability for particular applications. Sheet steel surface requirements may be broadly identified as to the end use by the suffix E for exposed parts requiring a good painted surface and suffix U for unexposed parts for which surface finish is unimportant.

Carbon steel may be obtained in a number of fundamental qualities which reflect various degrees of the quality conditions mentioned above. Some of those qualities may be modified by imposing such requirements as Limited Austenitic Grain Size, Specified Discard, Macroetch Test, Special Heat Treating, Maximum Incidental Alloy Elements, Restricted Chemical Composition, and Nonmetallic Inclusions. In addition, several of the products have special qualities that are intended for specific end uses or fabricating practices.

For complete descriptions of the qualities and supplementary requirements for carbon and alloy steels, reference should be made to the latest applicable American Iron and Steel Institute (AISI) Steel Products Manual Section.

Some of the special quality descriptors for carbon and alloy steels are listed in Table 5.

Applications. — Many factors enter into the selection of a steel for a particular application. These factors include the mechanical and physical properties needed to satisfy the design requirements and service environment; the cost and availability of the material; the cost of processing (machining, heat treatment, welding, etc.); the suitability of available processing equipment or the cost of any new equipment required.

These steel selection considerations require input from designers, metallurgists, manufacturing engineers, service engineers, and procurement specialists, and can be considered proper or optimum when the part is made from the lowest cost material consistent with satisfying engineering and service requirements. Since the factors in selection can vary widely among different organizations, several different steels may be used successfully for similar applications. The best choice of a steel for any application most often results from a balance or trade-offs among the various selection considerations.

The AISI/SAE designated "standard steels" provide a convenient way for engineers and metallurgists to state briefly but clearly the chemical composition and, in some instances, some of the properties desired, and they are widely recognized and used in the United States and in many other countries. There are, however, numerous nonstandard carbon, alloy, and stainless steel grades that are widely used for special applications.

The following sections and tables illustrate the general characteristics and typical applications of most of the standard carbon, alloy, and stainless steel grades.

Table 5. Fundamental Quality Description[a] of Carbon and Alloy Steels[b]

Carbon Steels

SEMIFINISHED FOR FORGING 　Forging Quality 　　Special Hardenability 　　Special Internal Soundness 　　Nonmetallic Inclusion Requirement 　　Special Surface CARBON STEEL STRUCTURAL SECTIONS 　Structural Quality CARBON STEEL PLATES 　Regular Quality 　Structural Quality 　Cold Drawing Quality 　Cold Pressing Quality 　Cold Flanging Quality 　Forging Quality 　Pressure Vessel Quality HOT ROLLED CARBON STEEL BARS 　Merchant Quality 　Special Quality 　　Special Hardenability 　　Special Internal Soundness 　　Nonmetallic Inclusion Requirement 　　Special Surface 　Scrapless Nut Quality 　Axle Shaft Quality 　Cold Extrusion Quality 　Cold Heading and Cold Forging Quality COLD FINISHED CARBON STEEL BARS 　Standard Quality 　　Special Hardenability 　　Special Internal Soundness	HOT ROLLED SHEETS 　Commercial Quality 　Drawing Quality 　Drawing Quality Special Killed 　Structural Quality COLD ROLLED SHEETS 　Commercial Quality 　Drawing Quality 　Drawing Quality Special Killed 　Structural Quality PORCELAIN ENAMELING SHEETS 　Commercial Quality 　Drawing Quality 　Drawing Quality Special Killed LONG TERNE SHEETS 　Commercial Quality 　Drawing Quality 　Drawing Quality Special Killed 　Structural Quality GALVANIZED SHEETS 　Commercial Quality 　Drawing Quality 　Drawing Quality Special Killed 　Lock Forming Quality ELECTROLYTIC ZINC-COATED SHEETS 　Commercial Quality 　Drawing Quality 　Drawing Quality Special Killed 　Structural Quality HOT ROLLED STRIP 　Commercial Quality 　Drawing Quality 　Drawing Quality Special Killed 　Structural Quality COLD ROLLED STRIP	TIN MILL PRODUCTS 　Specific quality descriptions are not applicable to tin mill products. CARBON STEEL WIRE 　Industrial Quality Wire 　Cold Extrusion Wires 　Heading, Forging and Roll Threading Wires 　Mechanical Spring Wires 　Upholstery Spring Construction Wires 　Welding Wire CARBON STEEL FLAT WIRE 　Stitching Wire 　Stapling Wire CARBON STEEL PIPE STRUCTURAL TUBING LINE PIPE OIL COUNTRY TUBULAR GOODS STEEL SPECIALTY TUBULAR PRODUCTS 　Pressure Tubing 　Mechanical Tubing 　Aircraft Tubing HOT ROLLED CARBON STEEL WIRE RODS 　Industrial Quality 　Rods for Manufacture of Wire Intended for Electric Welded Chain 　Rods for Heading, Forging, and Roll Threading Wire 　Rods for Lock Washer Wire 　Rods for Scrapless Nut Wire

Alloy Steels

ALLOY STEEL PLATES 　Drawing Quality 　Pressure Vessel Quality 　Structural Quality 　Aircraft Physical Quality HOT ROLLED ALLOY STEEL BARS 　Regular Quality 　Aircraft Quality or Steel Subject to Magnetic Particle Inspection 　Axle Shaft Quality 　Bearing Quality	ALLOY STEEL WIRE 　Aircraft Quality 　Bearing Quality 　Special Surface Quality COLD FINISHED ALLOY STEEL BARS 　Regular Quality 　Aircraft Quality or Steel Subject to Magnetic Particle Inspection 　Axle Shaft Quality 　Bearing Shaft Quality	LINE PIPE OIL COUNTRY TUBULAR GOODS STEEL SPECIALTY TUBULAR GOODS 　Pressure Tubing 　Mechanical Tubing 　Stainless and Heat Resisting Pipe, Pressure Tubing, and Mechanical Tubing 　Aircraft Tubing

Source: *SAE Handbook,* 1982.

[a] For certain qualities, phosphorus and sulfur are ordinarily furnished to lower limits than the specified maximum. For details, refer to the appropriate AISI manual.

[b] For a detailed description of many of the categories listed, see the appropriate section of the AISI manual.

Carbon Steels. — *SAE steels 1006, 1008, 1010, 1015:* These steels are the lowest carbon steels of the plain carbon type, and are selected where cold formability is the primary requisite of the user. They are produced both as rimmed and killed steels. Rimmed steel is used for sheet, strip, rod, and wire where excellent surface finish or good drawing qualities are required, such as body and fender stock, hoods, lamps, oil pans, and other deep drawn and formed products. It is also used for cold heading wire for tacks, and rivets and low carbon wire products. Killed steel (usually aluminum killed or special killed) is used for difficult stampings or where non-aging properties are needed. Killed steels (usually silicon killed) should be used in preference to rimmed steel for forging or heat treating applications.

These steels have relatively low tensile values and should not be selected where much strength is desired. Within the carbon range of the group, strength and hardness will increase with increase in carbon and/or with cold work, but such increases in strength are at the sacrifice of ductility or the ability to withstand cold deformation. Where cold rolled strip is used the proper temper designation should be specified to obtain the desired properties.

When under 0.15 carbon, the steels are susceptible to serious grain growth, causing brittleness, which may occur as the result of a combination of critical strain (from cold work) followed by heating to certain elevated temperatures. If cold worked parts formed from these steels are to be later heated to temperatures in excess of 1100 degrees F., the user should exercise care to avoid trouble from this cause. When this condition develops it can be overcome by heating the parts to a temperature well in excess of the upper critical point, or at least 1750 degrees F.

Steels in this group, being nearly pure iron or ferritic in structure, do not machine freely and should be avoided for cut screws and operations requiring broaching or smooth finish on turning. The machinability of bar, rod and wire products is improved by cold drawing. Steels in this group are readily welded.

SAE 1016, 1017, 1018, 1019, 1020, 1021, 1022, 1023, 1024, 1025, 1026, 1027, 1030: Steels in this group, due to the carbon range covered, have increased strength and hardness, and reduced cold formability compared to the lowest carbon group. For heat treating purposes they are known as carburizing or case hardening grades. When uniform response to heat treatment is required, or for forgings, killed steel is preferred; for other uses, semi-killed or rimmed steel may be indicated, depending on the combination of properties desired. Rimmed steels can ordinarily be supplied up to 0.25 carbon.

Selection of one of these steels for carburizing applications depends on the nature of the part, the properties desired, and the processing practice preferred. Increase in carbon gives greater core hardness with a given quench, or permits the use of thicker sections. Increase in manganese improves the hardenability of both the core and case; in carbon steels this is the only change in composition that will increase case hardenability. The higher manganese variants also machine much better. For carburizing applications SAE 1016, 1018, and 1019 are widely used for thin sections or water quenched parts. SAE 1022 and 1024 are used for heavier sections or where oil quenching is desired, and SAE 1024 is sometimes used for such parts as transmission and rear axle gears. SAE 1027 is used for parts given a light case to obtain satisfactory core properties without drastic quenching. SAE 1025 and 1030, while not usually regarded as carburizing types, are sometimes used in this manner for larger sections or where greater core hardness is needed.

For cold formed or headed parts the lowest manganese grades (SAE 1017, 1020, and 1025) offer the best formability at their carbon level. SAE 1020 is used for fan blades and some frame members, and SAE 1020 and 1025 are widely used for low strength bolts. The next higher manganese types (SAE 1018, 1021, and 1026) provide increased strength.

Table 6. General Applications of SAE Steels

These applications are intended as a general guide only since the selection may depend upon the exact character of the service, cost of material, machinability when machining is required, or other factors. When more than one steel is recommended for a given application, information on the characteristics of each steel listed will be found in the section beginning on page 445.

Application	SAE No.	Application	SAE No.
Adapters................	1145	Chain pins, transmission ..	4320
Agricultural steel........	1070	"　　"　　"	4815
"　　　　"	1080	"　　"　　"	4820
Aircraft forgings........	4140	Chains, transmission.....	3135
Axles, front or rear......	1040	"　　"　　.....	3140
"　　"　　"	4140	Clutch disks............	1060
Axle shafts.............	1045	"　　"	1070
"　　"	2340	"　　"	1085
"　　"	2345	Clutch springs...........	1060
"　　"	3135	Coil springs.............	4063
"　　"	3140	Cold-headed bolts........	4042
"　　"	3141	Cold-heading steel........	30905
"　　"	4063	Cold-heading wire or rod..	rimmed*
"　　"	4340	"　　"　　" ..	1035
Ball-bearing races........	52100	Cold-rolled steel.........	1070
Balls for ball bearings.....	52100	Connecting-rods.........	1040
Body stock for cars.......	rimmed*	"　　"	3141
Bolts, anchor...........	1040	Connecting-rod bolts.....	3130
Bolts and screws.........	1035	Corrosion resisting.......	51710
Bolts, cold-headed.......	4042	"　　"	30805
Bolts, connecting-rod.....	3130	Covers, transmission......	rimmed*
Bolts, heat-treated.......	2330	Crankshafts............	1045
Bolts, heavy-duty.......	4815	"　　"	1145
"　　"　　"	4820	"　　"	3135
Bolts, steering-arm......	3130	"　　"	3140
Brake levers............	1030	"　　"	3141
"　　"	1040	Crankshafts, Diesel engine.	4340
Bumper bars............	1085	Cushion springs.........	1060
Cams, free-wheeling.....	4615	Cutlery, stainless.........	51335
"　　"	4620	Cylinder studs...........	3130
Camshafts.............	1020	Deep-drawing steel.......	rimmed*
"　　"	1040	"　　"	30905
Carburized parts........	1020	Differential gears........	4023
"　　"　　"	1022	Disks, clutch............	1070
"　　"　　"	1024	"　　"	1060
"　　"　　"	1320	Ductile steel............	30905
"　　"　　"	2317	Fan blades..............	1020
"　　"　　"	2515	Fatigue resisting.........	4340
"　　"　　"	3310	"　　"	4640
"　　"　　"	3115	Fender stock for cars.....	rimmed*
"　　"　　"	3120	Forgings, aircraft........	4140
"　　"　　"	4023	Forgings, carbon steel.....	1040
"　　"　　"	4032	"　　"　　"	1045
"　　"　　"	1117	Forgings, heat-treated.....	3240
"　　"　　"	1118	"　　"　　"	5140

* The "rimmed" and "killed" steels listed are in the SAE 1008, 1010 and 1015 group. See general description of these steels.

Table 6 (*Continued*). General Applications of SAE Steels

These applications are intended as a general guide only since the selection may depend upon the exact character of the service, cost of material, machinability when machining is required, or other factors. When more than one steel is recommended for a given application, information on the characteristics of each steel listed will be found in the section beginning on page 445.

Application	SAE No.	Application	SAE No.
Forgings, heat-treated....	6150	Key stock..............	1030
Forgings, high-duty.......	6150	" "	2330
Forgings, small or medium.	1035	" "	3130
Forgings, large...........	1036	Leaf springs............	1085
Free-cutting carbon steel..	1111	" "	9260
" " " ..	1113	Levers, brake...........	1030
Free-cutting chro.-ni. steel.	30615	" "	1040
Free-cutting mang. steel..	1132	Levers, gear shift........	1030
" " " " ...	1137	Levers, heat-treated......	2330
Gears, carburized........	1320	Lock-washers...........	1060
" "	2317	Mower knives...........	1085
" "	3115	Mower sections..........	1070
" "	3120	Music wire..............	1085
" "	3310	Nuts...................	3130
" "	4119	Nuts, heat-treated.......	2330
" "	4125	Oil-pans, automobile......	rimmed*
" "	4320	Pinions, carburized......	3115
" "	4615	" "	3120
" "	4620	" "	4320
" "	4815	Piston-pins.............	3115
" "	4820	" "	3120
Gears, heat-treated.......	2345	Plow beams.............	1070
Gears, car and truck.....	4027	Plow disks.............	1080
" " " "	4032	Plow shares............	1080
Gears, cyanide-hardening..	5140	Propeller shafts.........	2340
Gears, differential........	4023	" "	2345
Gears, high duty.........	4640	" "	4140
" " "	6150	Races, ball-bearing......	52100
Gears, oil-hardening......	3145	Ring gears.............	3115
" " "	3150	" "	3120
" " "	4340	" "	4119
" " "	5150	Rings, snap.............	1060
Gears, ring..............	1045	Rivets.................	rimmed*
" "	3115	Rod and wire...........	killed*
" "	3120	Rod, cold-heading.......	1035
" "	4119	Roller bearings..........	4815
Gears, transmission.......	3115	Rollers for bearings......	52100
" "	3120	Screws and bolts........	1035
" "	4119	Screw stock, Bessemer....	1111
Gears, truck and bus.....	3310	" " "	1112
" " " "	4320	" " "	1113
Gear shift levers.........	1030	Screw stock, open hearth...	1115
Harrow disks............	1080	Screws, heat-treated......	2330
" "	1095	Seat springs............	1095
Hay-rake teeth..........	1095	Shafts, axle.............	1045

* The "rimmed" and "killed" steels listed are in the SAE 1008, 1010 and 1015 group. See general description of these steels.

Table 6 (*Concluded*). General Applications of SAE Steels

These applications are intended as a general guide only since the selection may depend upon the exact character of the service, cost of material, machinability when machining is required, or other factors. When more than one steel is recommended for a given application, information on the characteristics of each steel listed will be found in the section beginning on page 445.

Application	SAE No.	Application	SAE No.
Shafts, cyanide-hardening..	5140	Steel, cold-heading.......	30905
Shafts, heavy-duty.......	4340	Steel, free-cutting carbon..	11111
" " "	6150	" " " "	1113
" " "	4615	Steel, free-cutting chro.-ni..	30615
" " "	4620	Steel, free-cutting mang...	1132
Shafts, oil-hardening......	5150	" " " " ...	0000
Shafts, propeller.........	2340	Steel, minimum distortion.	4615
" "	2345	" " " .	4620
" "	4140	" " " .	4640
Shafts, transmission......	4140	Steel, soft ductile	30905
Sheets and strips.........	rimmed*	Steering arms...........	4042
Snap rings..............	1060	Steering-arm bolts........	3130
Spline shafts............	1045	Steering knuckles........	3141
" "	1320	Steering-knuckle pins.....	4815
" "	2340	" " "	4820
" "	2345	Studs...................	1040
" "	3115	"	1111
" "	3120	Studs, cold-headed.......	4042
" "	3135	Studs, cylinder..........	3130
" "	3140	Studs, heat-treated.......	2330
" "	4023	Studs, heavy-duty........	4815
Spring clips.............	1060	" " "	4820
Springs, coil.............	1095	Tacks..................	rimmed*
" "	4063	Thrust washers..........	1060
" "	6150	Thrust washers, oil-harden.	5150
Springs, clutch..........	1060	Transmission shafts......	4140
Springs, cushion.........	1060	Tubing.................	1040
Springs, leaf.............	1085	Tubing, front axle........	4140
" "	1095	Tubing, seamless.........	1030
" "	4063	Tubing, welded..........	1020
" "	4068	Universal joints..........	1145
" "	9260	Valve springs............	1060
" "	6150	Washers, lock............	1060
Springs, hard-drawn coiled.	1066	Welded structures........	30705
Springs, oil-hardening.....	5150	Wire and rod............	killed*
Springs, oil-tempered wire.	1066	Wire, cold-heading.......	rimmed*
Springs, seat............	1095	" " "	1035
Springs, valve...........	1060	Wire, hard-drawn spring ..	1045
Spring wire..............	1045	" " " " ...	1055
Spring wire, hard-drawn...	1055	Wire, music..............	1085
Spring wire, oil-tempered..	1055	Wire, oil-tempered spring..	1055
Stainless irons............	51210	Wrist-pins, automobile....	1020
" "	51710	Yokes..................	1145
Steel, cold-rolled.........	1070		

* The "rimmed" and "killed" steels listed are in the SAE 1008, 1010 and 1015 group. See general description of these steels.

All of these steels may be readily welded or brazed by the common commercial methods. SAE 1020 is frequently used for welded tubing. These steels are used for numerous forged parts, the lower carbon grades where high strength is not essential. Forgings from the lower carbon steels usually machine better in the as forged condition without annealing, or after normalizing.

SAE 1030, 1033, 1034, 1035, 1036, 1038, 1039, 1040, 1041, 1042, 1043, 1045, 1046, 1049, 1050, 1052: These steels, of the medium carbon type, are selected for uses where higher mechanical properties are needed and are frequently further hardened and strengthened by heat treatment or by cold work. These grades are ordinarily produced as killed steels.

Steels in this group are suitable for a wide variety of automotive type applications. The particular carbon and manganese level selected is affected by a number of factors. Increase in the mechanical properties required in section thickness, or in depth of hardening, ordinarily indicates either higher carbon or manganese or both. The heat treating practice preferred, particularly the quenching medium, has a great effect on the steel selected. In general, any of the grades over 0.30 carbon may be selectively hardened by induction or flame methods.

The lower carbon and manganese steels in this group find usage for certain types of cold formed parts. SAE 1030 is used for shift and brake levers, SAE 1034 and 1035 are used in the form of wire and rod for cold upsetting such as bolts, and SAE 1038 for bolts and studs. In practically all cases the parts cold formed from these steels are heat treated prior to use. Stampings are usually limited to flat parts or simple bends. The higher carbon SAE 1038, 1040, and 1042 are frequently cold drawn to specified physical properties for use without heat treatment for some applications, such as cylinder head studs.

All of this group of steels are used for forgings, the selection being governed by the section size and the physical properties desired after heat treatment. Thus SAE 1030 and 1035 are used for shifter forks and many small forgings where moderate properties are desired, but the deeper hardening SAE 1036 is used for more critical parts where a higher strength level and more uniformity is essential, such as some front suspension parts. Forgings such as connecting rods, steering arms, truck front axles, axle shafts, and tractor wheels are commonly made from the SAE 1038 to 1045 group. Larger forgings at similar strength levels need more carbon and perhaps more manganese. Examples are crankshafts from SAE 1046 and 1052. These steels are also used for small forgings where high hardness after oil quenching is desired. Suitable heat treatment is necessary on forgings from this group to provide machinability. These steels are also widely used for parts machined from bar stock, the selection following an identical pattern to that described for forgings. They are used both with and without heat treatment, depending on the application and the level of properties needed. As a class they are considered good for normal machining operations. It is also possible to weld these steels by most commercial methods, but precautions should be taken to avoid cracking from too rapid cooling.

SAE 1055, 1060, 1062, 1064, 1065, 1066, 1070, 1074, 1078, 1080, 1085, 1086, 1090, 1095: Steels in this group are of the high carbon type, having more carbon than is required to achieve maximum as quenched hardness. They are used for applications where the higher carbon is needed to improve wear characteristics for cutting edges, to make springs, and for special purposes. Selection of a particular grade is affected by the nature of the part, its end use, and the manufacturing methods available.

In general, cold forming methods are not practical on this group of steels, being limited to flat stampings and springs coiled from small diameter wire. Practically all parts from these steels are heat treated before use, with some variations in heat treating methods to obtain optimum properties for the particular use to which the steel is to be put.

Uses in the spring industry include SAE 1065 for pretempered wire and SAE 10 for cushion springs of hard drawn wire, SAE 1064 may be used for small washe and thin stamped parts, SAE 1074 for light flat springs formed from annealed stoc and SAE 1080 and 1085 for thicker flat springs. SAE 1085 is also used for heavi coil springs. Valve spring wire and music wire are special products.

Due to good wear properties when properly heat treated, the high carbon stee find wide usage in the farm implement industry. SAE 1070 has been used for plo beams, SAE 1074 for plow shares, and SAE 1078 for such parts as rake teet scrapers, cultivator shovels and plow shares. SAE 1085 has been used for scrap blades, disks, and for spring tooth harrows. SAE 1086 and 1090 find use as mow and binder sections, twine holders, and knotter disks.

Free Cutting Steels. — *SAE 1111, 1112, 1113:* This class of steels is intended f those uses where easy machining is the primary requirement. They are characte ized by a higher sulphur content than comparable carbon steels. This results i some sacrifice of cold forming properties, weldability, and forging characteristic In general the uses are similar to those for carbon steels of similar carbon an manganese content.

These steels are commonly known as Bessemer screw stock, and are considere the best machining steels available, machinability improving within the group a sulphur increases. They are used for a wide variety of machined parts. While excellent strength in the cold drawn condition, they have an unfavorable propert of cold shortness and are not commonly used for vital parts. These steels may b cyanided or carburized but when uniform response to heat treating is necessary open hearth steels are recommended.

SAE 1109, 1114, 1115, 1116, 1117, 1118, 1119, 1120, 1126: Steels in this grou are used where a combination of good machinability and more uniform response t heat treatment is needed. The lower carbon varieties are used for small part which are to be cyanided or carbonitrided. SAE 1116, 1117, 1118, and 1119 carr more manganese for better hardenability, permitting oil quenching after cas hardening heat treatments in many instances. The higher carbon SAE 1120 an 1126 provide more core hardness when this is needed.

SAE 1132, 1137, 1138, 1140, 1141, 1144, 1145, 1146, 1151: This group of steel has characteristics comparable to carbon steels of the same carbon level, except fo changes due to higher sulphur as noted previously.

They are widely used for parts where a large amount of machining is necessary or where threads, splines or other operations offer special tooling problems. SAI 1137, for example, is widely used for nuts and bolts and studs with machined thread The higher manganese SAE 1132, 1137, 1141, and 1144 offer greater hardenability the higher carbon types being suitable for oil quenching for many parts. All o these steels may be selectively hardened by induction or flame heating if desired.

Carburizing Grades of Alloy Steels. *Properties of the Case:* The propertie of carburized and hardened cases depend upon the carbon and alloy content, th structure of the case, and the degree and distribution of residual stresses. Th carbon content of the case depends upon the details of the carburizing process and the response of iron and the alloying elements present, to carburization. Th original carbon content of the steel has little or no effect upon the carbon conten produced in the case. The hardenability of the case therefore depends upon th alloy content of the steel and the final carbon content produced by carburizing but not upon the initial carbon content of the steel.

With complete carbide solution the effect of alloying elements upon the harden ability of the case, will in general be the same as the effect of these elements upon th hardenability of the core. As an exception to this, any element which inhibit

carburizing may reduce the hardenability of the case. It is also true that some elements which raise the hardenability of the core may tend to produce more retained austenite and consequently somewhat lower hardness in the case.

Alloy steels are frequently used for case hardening because the required surface hardness can be obtained by moderate speeds of quenching. This may mean less distortion than would be encountered with water quenching. It is usually desirable to select a steel which will attain a minimum surface hardness of 58 or 60 Rockwell C after carburizing and oil quenching. Where section sizes are large, a high hardenability alloy steel may be necessary, while for medium and light sections, low hardenability steels will suffice.

In general, the case hardening alloy steels may be divided into two classes so far as the hardenability of the case is concerned. Only the general type of steel (SAE 3300–4100, etc.) is given. As the original carbon content of the steel has no effect upon the carbon content of the case, the last two digits in the specification numbers are not meaningful so far as the case is concerned.

(a) — *High Hardenability Case.* — *SAE 2500, 3300, 4300, 4800, 9300*

As these are high alloy steels, both the case and the core have high hardenability. These types of steel are used particularly for carburized parts having thick sections, such as bevel drive pinions and heavy gears. Good case properties can be obtained by oil quenching. These steels are likely to have retained austenite in the case after carburizing and quenching, consequently special precautions or treatments, such as refrigeration, may be required.

(b) — *Medium Hardenability Case.* — *SAE 1300, 2300, 4000, 4100, 4600, 5100, 8600, 8700*

Carburized cases of these steels have medium hardenability which means that their hardenability is intermediate between that of plain carbon steel and the higher alloy carburizing steels just described. In general, these steels can be used for average size case hardened automotive parts such as gears, pinions, piston pins, ball studs, universal crosses, crankshafts, etc. Satisfactory case hardness should be produced in most cases by oil quenching.

Core Properties: The core properties of case hardened steels depend upon both carbon and alloy content of the steel. Each of the general types of alloy case hardening steel is usually made with two or more carbon contents so as to produce different hardenability in the core.

The most desirable hardness for the core depends upon the design and functioning of the individual part. In general, where high compressive loads are encountered, relatively high core hardness is beneficial in supporting the case. Low core hardnesses may be desirable where great toughness is essential.

The case hardening steels may be divided into three general classes depending upon hardenability of the core.

(a) — *Low Hardenability Core.* — *SAE 4017, 4023, 4024, 4027[1], 4028[1], 4608, 4615, 4617[1], 8615[1], 8617[1]*

(b) — *Medium Hardenability Core.* — *SAE 1320, 2317, 2512, 2515[1], 3115, 3120, 4032, 4119, 4317, 4620, 4621, 4812, 4815[1], 5115, 5120, 8620, 8622, 8720, 9420*

(c) — *High Hardenability Core.* — *SAE 2517, 3310, 3316, 4320, 4817, 4820, 9310, 9315, 9317*

Heat Treatments: In general, all of the alloy carburizing steels are made fine grain and most are suitable for direct quenching from the carburizing temperature. Several other types of heat treatment involving single and double quenching are also used for most of these steels. (See tables of Typical Heat Treatments for SAE Steels.)

---

[1] Borderline classifications might be considered in the next higher hardenability group.

Directly Hardenable Grades of Alloy Steels. — These steels may be considered in five groups on the basis of approximate mean carbon content of the SAE specification. In general, the last two figures of the specification agree with the mean carbon content. Consequently the heading ".30–.37 Mean Carbon Content of SAE Specification" includes steels such as SAE 1330, 3135, and 4137.

Mean Carbon Content of SAE Specification	Common Applications
(a) — .30–.37 per cent	Heat treated parts requiring moderate strength and great toughness.
(b) — .40–.42 per cent	Heat treated parts requiring higher strength and good toughness.
(c) — .45–.50 per cent	Heat treated parts requiring fairly high hardness and strength with moderate toughness.
(d) — .50–.62 per cent	Springs and hand tools.
(e) — 1.02 per cent	Ball and roller bearings.

It is necessary to deviate from the above plan in the classification of the carbon molybdenum steels. When carbon molybdenum steels are used, it is customary to specify higher carbon content for any given application than would be specified for other alloy steels, due to the low alloy content of these steels. For example, SAE 4063 is used for the same applications as SAE 4140, 4145 and 5150. Consequently in the following discussion, the carbon molybdenum steels have been shown in the groups where they belong on the basis of applications rather than carbon content.

For the present discussion, steels of each carbon content are divided into two or three groups on the basis of hardenability. Transformation ranges and consequently heat treating practices vary somewhat with different alloying elements even though the hardenability is not changed.

.30–.37 Mean Carbon Content of SAE Specification. — These steels are frequently used for water quenched parts of moderate section size and for oil quenched parts of small section size. Typical applications of these steels are connecting rods, steering arms and steering knuckles, axle shafts, bolts, studs, screws, and other parts requiring strength and toughness where section size is small enough to permit obtaining the desired physical properties with the customary heat treatment.

Steels falling in this classification may be subdivided into two groups on the basis of hardenability:

(a) — Low Hardenability: SAE 1330, 1335, 4037, 4042, 4130, 5130, 5132, 8630

(b) — Medium Hardenability: SAE 2330, 3130, 3135, 4137, 5135, 8632, 8635, 8637, 8735, 9437

.40–.42 Mean Carbon Content of SAE Specification. — In general, these steels are used for medium and large size parts requiring high degree of strength and toughness. The choice of the proper steel depends upon the section size and the mechanical properties which must be produced. The low and medium hardenabilty steels are used for average size automotive parts such as steering knuckles, axle shafts, propeller shafts, etc. The high hardenability steels are used particularly for large axles and shafts for large aircraft parts.

These steels are usually considered as oil quenching steels, although some large parts made of the low and medium hardenability classifications may be quenched in water under properly controlled conditions.

These steels may be divided into three groups on the basis of hardenability:

(a) — Low Hardenability: SAE 1340, 4047, 5140, 9440

(b) — Medium Hardenability: SAE 2340, 3140, 3141, 4053, 4063, 4140, 4640, 8640, 8641, 8642, 8740, 8742, 9442

(c) — High Hardenability: SAE 4340, 9840

.45–.50 Mean Carbon Content of SAE Specification. — These steels are used primarily for gears and other parts requiring fairly high hardness as well as strength and toughness. Such parts are usually oil quenched and a minimum of 90 per cent martensite in the as-quenched condition is desirable.

(a) — Low Hardenability: SAE 5045, 5046, 5145, 9747, 9763

(b) — Medium Hardenability: SAE 2345, 3145, 3150, 4145, 5147, 5150, 8645, 8647, 8650, 8745, 8747, 8750, 9445, 9845

(c) — High Hardenability: SAE 4150, 9850

.50–.63 Mean Carbon Content of SAE Specification. — These steels are used primarily for springs and hand tools. The hardenability necessary depends upon the thickness of the material and the quenching practice.

(a) — Medium hardenability: SAE 4068, 5150, 5152, 6150, 8650, 9254, 9255, 9260, 9261

(b) — High hardenability: SAE 8653, 8655, 8660, 9262

1.02 Mean Carbon Content of SAE Specification. — SAE 50100, 51100, 52100 These are straight chromium electric furnace steels used primarily for the races and balls or rollers of anti-friction bearings. They are also used for other parts requiring high hardness and wear resistance. The compositions of the three steels are identical, except for a variation in chromium, with a corresponding variation in hardenability.

(a) — Low Hardenability: SAE 50100

(b) — Medium Hardenability: SAE 51100, 52100

Resulfurized Steel. — Some of the alloy steels, SAE 4024, 4028 and 8641, are made resulfurized so as to give better machinability at a relatively high hardness. In general, increased sulfur results in decreased transverse ductility, notched impact toughness, and weldability.

Typical Mechanical Properties of Steel. — Tables 7–10 provide expected minimum and/or typical mechanical properties of selected standard carbon and alloy steels and stainless steels.

Chromium Nickel Austenitic Steels *(Not capable of heat treatment).* —

SAE 30201: This steel is an austenitic chromium–nickel–manganese stainless steel usually required in flat products. In the annealed condition it exhibits higher strength values than the corresponding chromium–nickel stainless steel (SAE 30301). It is nonmagnetic in the annealed condition, but may be magnetic when cold worked. SAE 30201 is used to obtain high strength by work hardening and is well suited for corrosion-resistant structural members requiring high strength with low weight. It has excellent resistance to a wide variety of corrosive media, showing behavior comparable to stainless grade SAE 30301. It has high ductility and excellent forming properties. Owing to this steel's work hardening rate and yield strength, tools for forming must be designed to allow for a higher springback or recovery rate. It is used for automotive trim, automotive wheel covers, railroad passenger car bodies and structural members, truck trailer bodies.

SAE 30202: Like its corresponding chromium–nickel stainless steel SAE 30302, this steel is a general-purpose stainless steel. It has excellent corrosion resistance and deep drawing qualities. It is nonhardenable by thermal treatments, but may be cold worked to high tensile strengths. In the annealed condition it is nonmagnetic but slightly magnetic when cold worked. Applications for this stainless steel are hub cap, railcar and truck trailer bodies, and spring wire.

SAE 30301: This steel is capable of attaining high tensile strength and ductility by moderate or severe cold working. It is used largely in the cold rolled or cold drawn condition in the form of sheet, strip and wire. Its corrosion resistance is good but not equal to SAE 30302.

SAE 30302: This is the most widely used of the general purpose austenitic chromium nickel stainless steels. It is used for deep drawing largely in the annealed

condition. It can be worked to high tensile strengths but with slightly lower ductility than SAE 30301.

SAE 30303F: This is a free machining type recommended for the manufacture of parts produced on automatic machines. Caution must be used in forging this steel.

SAE 30304: This is similar to SAE 30302 but somewhat superior in corrosion resistance and having superior welding properties for certain types of equipment.

SAE 30305: Similar to SAE 30304 but capable of lower hardness. Has greater ductility with slower work hardening tendency.

SAE 30309: This steel has high heat resisting qualities and is resistant to oxidation at temperatures up to about 1800 deg. F.

SAE 30310: This steel has the highest heat resisting properties of any of the chromium nickel steels listed herewith and is used to resist oxidation at temperatures up to about 1900 deg. F.

SAE 30316: This steel is recommended for use in parts where unusual resistance to chemical or salt water corrosion is necessary. It has superior creep strength at elevated temperatures.

SAE 30317: This steel is similar to SAE 30316 but has the highest corrosion resistance of all these alloys in many environments.

SAE 30321: This steel is recommended for use in the manufacture of welded structures where heat treatment after welding is not feasible. It is also recommended for use where temperatures up to 1600 deg. F. are encountered in service.

SAE 30325: Used for such parts as heat control shafts.

SAE 30347: This steel is similar to SAE 30321 with the following additional statement. This niobium alloy is sometimes preferred to titanium because less niobium is lost in the welding operation.

Stainless Chromium Irons and Steels. —

SAE 51409: This steel is an 11 per cent chromium alloy developed especially for automotive mufflers and tailpipes. Resistance to corrosion and oxidation is very similar to SAE 51410. It is nonhardenable and has good forming and welding characteristics. This alloy is recommended for mildly corrosive applications where surface appearance is not critical.

SAE 51410: This is a general purpose stainless steel capable of heat treatment to show good physical properties. It is used for general stainless applications, both in the heat treated and annealed condition but is not as resistant to corrosion as SAE 51430 in either the annealed or heat treated condition.

SAE 51414: This is a corrosion and heat-resisting nickel-bearing chromium steel with somewhat better corrosion resistance than SAE 51410. It will attain slightly higher mechanical properties when heat treated than SAE 51410. It is used in the form of tempered strip or wire, and in bars and forgings for heat treated parts.

SAE 51416F: This is a free machining grade for the manufacture of parts produced in automatic screw machines.

SAE 51420: This steel is capable of being heat treated to a relatively high hardness. It will harden to a maximum of approximately 500 Brinell. It has its maximum corrosion resisting qualities only in the fully hardened condition. It is used for cutlery, hardened pump shafts, etc.

SAE 51420F: This is similar to SAE 51420 except for its free machining properties.

SAE 51430: This is a steel of high chromium type not capable of heat treatment and is recommended for use in shallow parts requiring moderate draw. Corrosion and heat resistance are superior to SAE 51410.

SAE 51430F: This is similar to SAE 51430 except for its free machining properties.

SAE 51431: This is a nickel-bearing chromium steel designed for heat treatment to high mechanical properties. Its corrosion resistance is superior to other hardenable steels.

(Continued on page 467)

Table 7. Characteristics and Typical Applications of Standard Stainless Steels

AISI Type (UNS)	Characteristics and Typical Applications	AISI Type (UNS)	Characteristics and Typical Applications
201 (S20100)	High work-hardening rate; low-nickel equivalent of type 301. Flatware; automobile wheel covers, trim.	304L (S30403)	Extra-low-carbon modification of type 304 for further restriction of carbide precipitation during welding. Coal hopper linings; tanks for liquid fertilizer and tomato paste.
202 (S20200)	General-purpose low-nickel equivalent of type 302. Kitchen equipment; hub caps; milk handling.		
205 (S20500)	Lower work-hardening rate than type 202; used for spinning and special drawing operations. Nonmagnetic and cryogenic parts.	(S30430)	Lower work-hardening rate than type 305. Severe cold-heading applications.
301 (S30100)	High work-hardening rate; used for structural applications where high strength plus high ductility is required. Railroad cars; trailer bodies; aircraft structurals; fasteners; automobile wheel covers, trim; pole line hardware.	304N (S30451)	Higher nitrogen than type 304 to increase strength with minimum effect on ductility and corrosion resistance, more resistant to increased magnetic permeability. Type 304 applications requiring higher strength.
302 (S30200)	General-purpose austenitic stainless steel. Trim; food handling equipment; aircraft cowlings; antennas; springs; cookware; building exteriors; tanks; hospital, household appliances; jewelry; oil refining equipment; signs.	305 (S30500)	Low work-hardening rate; used for spin forming, severe drawing, cold heading, and forming. Coffee urn tops; mixing bowls; reflectors.
302B (S30215)	More resistant to scale than type 302. Furnace parts; still liners; heating elements; annealing covers; burner sections.	308 (S30800)	Higher-alloy steel having high corrosion and heat resistance. Welding filler metals to compensate for alloy loss in welding; industrial furnaces.
303 (S30300)	Free-machining modification of type 302, for heavier cuts. Screw machine products; shafts; valves; bolts; bushings; nuts.	309 (S30900)	High-temperature strength and scale resistance. Aircraft heaters; heat-treating equipment; annealing covers; furnace parts; heat exchangers; heat-treating trays; oven linings; pump parts.
303Se (S30323)	Free-machining modification of type 302, for lighter cuts; used where hot working or cold heading may be involved. Aircraft fittings; bolts; nuts; rivets; screws; studs.	309S (S30908)	Low-carbon modification of type 309. Welded constructions; assemblies subject to moist corrosion conditions.
304 (S30400)	Low-carbon modification of type 302 for restriction of carbide precipitation during welding. Chemical and food processing equipment; brewing equipment; cryogenic vessels; gutters; downspouts; flashings.	310 (S31000)	Higher elevated temperature strength and scale resistance than type 309. Heat exchangers; furnace parts; combustion chambers; welding filler metals; gas turbine parts; incinerators; recuperators; rolls for roller hearth furnaces.
		310S (S31008)	Low-carbon modification of type 310. Welded constructions; jet engine rings.

Table 7 (*Continued*). **Characteristics and Typical Applications of Standard Stainless Steels**

AISI Type (UNS)	Characteristics and Typical Applications	AISI Type (UNS)	Characteristics and Typical Applications
314 (S31400)	More resistant to scale than type 310. Severe cold heading or forming applications. Annealing and carburizing boxes; heat treating fixtures; radiant tubes.	329 (S32900)	Austenitic–ferritic type with general corrosion resistance similar to type 316 but with better resistance to stress-corrosion cracking; capable of age hardening. Valves; valve fittings; piping; pump parts.
316 (S31600)	Higher corrosion resistance than types 302 and 304; high creep strength. Chemical and pulp handling equipment; photographic equipment; brandy vats; fertilizer parts; ketchup cooking kettles; yeast tubs.	330 (N08330)	Good resistance to carburization and to heat and thermal shock. Heat-treating fixtures.
		347 (S34700)	Similar to type 321 with higher creep strength. Airplane exhaust stacks; welded tank cars for chemicals; jet engine parts.
316L (S31603)	Extra-low-carbon modification of type 316. Welded construction where intergranular carbide precipitation must be avoided. Type 316 applications requiring extensive welding.	348 (S34800)	Similar to type 321; low retentivity. Tubes and pipes for radioactive systems; nuclear energy uses.
316F (S31620)	Higher phosphorus and sulfur than type 316 to improve machining and nonseizing characteristics. Automatic screw machine parts.	384 (S38400)	Suitable for severe cold-heading or cold-forming; lower cold-work hardening rate than type 304. Bolts; rivets; screws; instrument parts.
316N (S31651)	Higher nitrogen than type 316 to increase strength with minimum effect on ductility and corrosion resistance. Type 316 applications requiring extra strength.	403 (S40300)	"Turbine quality" grade. Steam turbine blading and other highly stressed parts including jet engine rings.
		405 (S40500)	Nonhardenable grade for assemblies where air-hardening types such as 410 or 403 are objectionable. Annealing boxes; quenching racks; oxidation resistant partitions.
317 (S31700)	Higher corrosion and creep resistance than type 316. Dyeing and ink manufacturing equipment.	409 (S40900)	General-purpose construction stainless. Automotive exhaust systems; transformer and capacitor cases; dry fertilizer spreaders; tanks for agricultural sprays.
317L (S31703)	Extra-low-carbon modification of type 317 for restriction of carbide precipitation during welding. Welded assemblies.	410 (S41000)	General-purpose heat-treatable type. Machine parts; pump shafts; bolts; bushings; coal chutes; cutlery; hardware; jet engine parts; mining machinery; rifle barrels; screws; valves.
321 (S32100)	Stabilized for weldments subject to severe corrosive conditions, and for service from 800 to 1600 deg. F. Aircraft exhaust manifolds; boiler shells; process equipment; expansion joints; cabin heaters; fire walls; flexible couplings; pressure vessels.	414 (41400)	High hardenability steel. Springs; tempered rules; machine parts, bolts; mining machinery; scissors; ships' bells; spindles; valve seats.

Table 7 (*Concluded*). **Characteristics and Typical Applications of Standard Stainless Steels**

AISI Type (UNS)	Characteristics and Typical Applications	AISI Type (UNS)	Characteristics and Typical Applications
416 (S41600)	Free-machining modification of type 410, for heavier cuts. Aircraft fittings; bolts; nuts; fire extinguisher inserts; rivets; screws.	434 (S43400)	Modification of type 430 designed to resist atmospheric corrosion in the presence of winter road-conditioning and dust-laying compounds. Automotive trim and fasteners.
416Se (S41623)	Free-machining modification of type 410, for lighter cuts. Machined parts requiring hot working or cold heading.	436 (S43600)	Similar to types 430 and 434. Used where low "roping" or "ridging" required. General corrosion and heat-resistant applications such as automobile trim.
420 (S42000)	Higher carbon modification of type 410. Cutlery; surgical instruments; valves; wear-resisting parts; glass molds; hand tools; vegetable choppers.	440A (S44002)	Hardenable to higher hardness than type 420 with good corrosion resistance. Cutlery; bearings; surgical tools.
420F (S42020)	Free-machining modification of type 420. Applications similar to those for type 420 requiring better machinability.	440B (S44003)	Cutlery grade. Cutlery, valve parts; instrument bearings.
422 (S42200)	High strength and toughness at service temperatures up to 1200 deg. F. Steam turbine blades; fasteners.	440C (S44004)	Yields highest hardnesses of hardenable stainless steels. Balls; bearings; races; nozzles; balls and seats for oil well pumps; valve parts.
429 (S42900)	Improved weldability as compared to type 430. Nitric acid and nitrogen-fixation equipment.	442 (S44200)	High-chromium steel, principally for parts that must resist high service temperatures without scaling. Furnace parts; nozzles; combustion chambers.
430 (S43000)	General-purpose nonhardenable chromium type. Decorative trim; nitric acid tanks; annealing baskets; combustion chambers; dishwashers; heaters; mufflers; range hoods; recuperators; restaurant equipment.	446 (S44600)	High-resistance to corrosion and scaling at high temperatures especially for intermittent service; often used in sulfur-bearing atmosphere. Annealing boxes; combustion chambers; glass molds; heaters; pyrometer tubes; recuperators; stirring rods; valves.
430F (S43020)	Free-machining modification of type 430, for heavier cuts. Screw machine parts.		
430FSe (S43023)	Free-machining modification of type 430, for lighter cuts. Machined parts requiring light cold heading or forming.	501 (S50100)	Heat resistance; good mechanical properties at moderately elevated temperatures. Heat exchangers; petroleum refining equipment.
431 (S43100)	Special-purpose hardenable steel used where particularly high mechanical properties are required. Aircraft fittings; beater bars; paper machinery; bolts.	502 (S50200)	More ductility and less strength than type 501. Heat exchangers; petroleum refining equipment; gaskets.

Table 8. Expected Minimum Mechanical Properties, Conventional Practice, of Cold Drawn Carbon Steel Rounds, Squares, and Hexagons

Size, in.	As Cold Drawn					Cold Drawn Followed by Low-Temperature Stress Relief					Cold Drawn Followed by High-Temperature Stress Relief				
	Strength (1000 lb/in.²)		Elongation in 2 in., Per cent	Reduction in Area, Per cent	Hardness, Bhn	Strength (1000 lb/in.²)		Elongation in 2 in., Per cent	Reduction in Area, Per cent	Hardness, Bhn	Strength (1000 lb/in.²)		Elongation in 2 in., Per cent	Reduction in Area, Per cent	Hardness, Bhn
	Tensile	Yield				Tensile	Yield				Tensile	Yield			
AISI 1018 and 1025 Steels															
⅝–⅞	70	60	18	40	143	…	…	…	…	…	65	45	20	45	131
Over ⅞–1¼	65	55	16	40	131	…	…	…	…	…	60	45	20	45	121
Over 1¼–2	60	50	15	35	121	…	…	…	…	…	55	45	16	40	111
Over 2–3	55	45	15	35	111	…	…	…	…	…	50	40	15	40	101
AISI 1117 and 1118 Steels															
⅝–⅞	75	65	15	40	149	70	70	15	40	163	70	50	18	45	143
Over ⅞–1¼	70	60	15	40	143	65	65	15	40	149	65	50	16	45	131
Over 1¼–2	65	55	13	35	131	60	60	13	35	143	60	50	15	40	121
Over 2–3	60	50	12	30	121	55	55	12	35	131	55	45	15	40	111
AISI 1035 Steel															
⅝–⅞	85	75	13	35	170	90	80	13	35	179	80	60	16	45	163
Over ⅞–1¼	80	70	12	35	163	85	75	12	35	170	75	60	15	45	149
Over 1¼–2	75	65	12	35	149	80	70	12	35	163	70	60	15	40	143
Over 2–3	70	60	10	30	143	75	65	10	30	149	65	55	12	35	131
AISI 1040 and 1140 Steels															
⅝–⅞	90	80	12	35	179	95	85	12	35	187	85	65	15	45	170
Over ⅞–1¼	85	75	12	35	170	90	80	12	35	179	80	65	15	45	163
Over 1¼–2	80	70	10	30	163	85	75	10	30	170	75	60	15	40	149
Over 2–3	75	65	10	30	149	80	70	10	30	163	70	55	12	35	143

Source: AISI Committee of Hot Rolled and Cold Finished Bar Producers as published in 1974 DATABOOK issue of the American Society for Metals METAL PROGRESS magazine and used with its permission.

Table 8 (*concluded*). Expected Minimum Mechanical Properties, Conventional Practice, of Cold Drawn Carbon Steel Rounds, Squares, and Hexagons

Size, in.	As Cold Drawn — Strength, Tensile, 1000 lb/in.²	Yield	Elongation in 2 in., Per cent	Reduction in Area, Per cent	Hardness, Bhn	Cold Drawn Followed by Low-Temperature Stress Relief — Strength, Tensile	Yield	Elongation in 2 in., Per cent	Reduction in Area, Per cent	Hardness, Bhn	Cold Drawn Followed by High-Temperature Stress Relief — Strength, Tensile	Yield	Elongation in 2 in., Per cent	Reduction in Area, Per cent	Hardness, Bhn
AISI 1045, 1145, and 1146 Steels															
5/8–7/8	95	85	12	35	187	100	90	12	35	197	90	70	15	45	179
Over 7/8–1¼	90	80	11	30	179	95	85	11	30	187	85	70	15	45	170
Over 1¼–2	85	75	10	30	170	90	80	10	30	179	80	65	15	40	163
Over 2–3	80	70	10	30	163	85	75	10	25	170	75	60	12	35	149
AISI 1050, 1137, and 1151 Steels															
5/8–7/8	100	90	11	35	197	105	95	11	35	212	95	75	15	45	187
Over 7/8–1¼	95	85	11	30	187	100	90	11	30	197	90	75	15	40	179
Over 1¼–2	90	80	10	30	179	95	85	10	25	187	85	70	15	40	170
Over 2–3	85	75	10	30	170	90	80	10	25	179	80	65	12	35	163
AISI 1141 Steel															
5/8–7/8	105	95	11	30	212	110	100	11	30	223	100	80	15	40	197
Over 7/8–1¼	100	90	10	30	197	105	95	10	30	212	95	80	15	40	187
Over 1¼–2	95	85	10	30	187	100	90	10	25	197	85	75	15	40	179
Over 2–3	90	80	10	20	179	95	85	10	20	187	85	70	12	30	170
AISI 1144 Steel															
5/8–7/8	110	100	10	30	223	115	105	10	30	229	105	85	15	40	212
Over 7/8–1¼	105	95	10	30	212	110	100	10	30	223	100	85	15	40	197
Over 1¼–2	100	90	10	25	197	105	95	10	25	212	95	80	15	35	187
Over 2–3	95	85	10	20	187	100	90	10	20	197	90	75	12	30	179

Source: AISI Committee of Hot Rolled and Cold Finished Bar Producers as published in 1974 DATABOOK issue of the American Society for Metals' METAL PROGRESS magazine and used with its permission.

Table 9. Typical Mechanical Properties of Selected Carbon and Alloy Steels
(Hot Rolled, Normalized, and Annealed)

AISI No.*	Treatment	Strength Tensile (lb/in.²)	Strength Yield (lb/in.²)	Elongation, Per cent	Reduction in Area, Per cent	Hardness, Bhn	Impact Strength (Izod), ft-lb
1015	As-rolled	61,000	45,500	39.0	61.0	126	81.5
	Normalized (1700 F)	61,500	47,000	37.0	69.6	121	85.2
	Annealed (1600 F)	56,000	41,250	37.0	69.7	111	84.8
1020	As-rolled	65,000	48,000	36.0	59.0	143	64.0
	Normalized (1600 F)	64,000	50,250	35.8	67.9	131	86.8
	Annealed (1600 F)	57,250	42,750	36.5	66.0	111	91.0
1022	As-rolled	73,000	52,000	35.0	67.0	149	60.0
	Normalized (1700 F)	70,000	52,000	34.0	67.5	143	86.5
	Annealed (1600 F)	65,250	46,000	35.0	63.6	137	89.0
1030	As-rolled	80,000	50,000	32.0	57.0	179	55.0
	Normalized (1700 F)	75,000	50,000	32.0	60.8	149	69.0
	Annealed (1550 F)	67,250	49,500	31.2	57.9	126	51.2
1040	As-rolled	90,000	60,000	25.0	50.0	201	36.0
	Normalized (1650 F)	85,500	54,250	28.0	54.9	170	48.0
	Annealed (1450 F)	75,250	51,250	30.2	57.2	149	32.7
1050	As-rolled	105,000	60,000	20.0	40.0	229	23.0
	Normalized (1650 F)	108,500	62,000	20.0	39.4	217	20.0
	Annealed (1450 F)	92,250	53,000	23.7	39.9	187	12.5
1060	As-rolled	118,000	70,000	17.0	34.0	241	13.0
	Normalized (1650 F)	112,500	61,000	18.0	37.2	229	9.7
	Annealed (1450 F)	90,750	54,000	22.5	38.2	179	8.3
1080	As-rolled	140,000	85,000	12.0	17.0	293	5.0
	Normalized (1650 F)	146,500	76,000	11.0	20.6	293	5.0
	Annealed (1450 F)	89,250	54,500	24.7	45.0	174	4.5
1095	As-rolled	140,000	83,000	9.0	18.0	293	3.0
	Normalized (1650 F)	147,000	72,500	9.5	13.5	293	4.0
	Annealed (1450 F)	95,250	55,000	13.0	20.6	192	2.0
1117	As-rolled	70,600	44,300	33.0	63.0	143	60.0
	Normalized (1650 F)	67,750	44,000	33.5	63.8	137	62.8
	Annealed (1575 F)	62,250	40,500	32.8	58.0	121	69.0
1118	As-rolled	75,600	45,900	32.0	70.0	149	80.0
	Normalized (1700 F)	69,250	46,250	33.5	65.9	143	76.3
	Annealed (1450 F)	65,250	41,250	34.5	66.8	131	78.5
1137	As-rolled	91,000	55,000	28.0	61.0	192	61.0
	Normalized (1650 F)	97,000	57,500	22.5	48.5	197	47.0
	Annealed (1450 F)	84,750	50,000	26.8	53.9	174	36.8
1141	As-rolled	98,000	52,000	22.0	38.0	192	8.2
	Normalized (1650 F)	102,500	58,750	22.7	55.5	201	38.8
	Annealed (1500 F)	86,800	51,200	25.5	49.3	163	25.3
1144	As-rolled	102,000	61,000	21.0	41.0	212	39.0
	Normalized (1650 F)	96,750	58,000	21.0	40.4	197	32.0
	Annealed (1450 F)	84,750	50,250	24.8	41.3	167	48.0

* All grades are fine-grained except those in the 1100 series which are coarse-grained. Austenitizing temperatures are given in parentheses. Heat-treated specimens were oil quenched.

Source: Bethlehem Steel Corp. and Republic Steel Corp. as published in 1974 DATA-BOOK issue of the American Society for Metals' METAL PROGRESS magazine and used with its permission.

Table 9 (*Continued*). **Typical Mechanical Properties of Selected Carbon and Alloy Steels** (Hot Rolled, Normalized, and Annealed)

AISI No.*	Treatment	Strength Tensile lb/in.²	Strength Yield lb/in.²	Elongation, Per cent	Reduction in Area, Per cent	Hardness, Bhn	Impact Strength (Izod), ft-lb
1340	Normalized (1600 F)	121,250	81,000	22.0	62.9	248	68.2
	Annealed (1475 F)	102,000	63,250	25.5	57.3	207	52.0
3140	Normalized (1600 F)	129,250	87,000	19.7	57.3	262	39.5
	Annealed (1500 F)	100,000	61,250	24.5	50.8	197	34.2
4130	Normalized (1600 F)	97,000	63,250	25.5	59.5	197	63.7
	Annealed (1585 F)	81,250	52,250	28.2	55.6	156	45.5
4140	Normalized (1600 F)	148,000	95,000	17.7	46.8	302	16.7
	Annealed (1500 F)	95,000	60,500	25.7	56.9	197	40.2
4150	Normalized (1600 F)	167,500	106,500	11.7	30.8	321	8.5
	Annealed (1500 F)	105,750	55,000	20.2	40.2	197	18.2
4320	Normalized (1640 F)	115,000	67,250	20.8	50.7	235	53.8
	Annealed (1560 F)	84,000	61,625	29.0	58.4	163	81.0
4340	Normalized (1600 F)	185,500	125,000	12.2	36.3	363	11.7
	Annealed (1490 F)	108,000	68,500	22.0	49.9	217	37.7
4620	Normalized (1650 F)	83,250	53,125	29.0	66.7	174	98.0
	Annealed (1575 F)	74,250	54,000	31.3	60.3	149	69.0
4820	Normalized (1580 F)	109,500	70,250	24.0	59.2	229	81.0
	Annealed (1500 F)	98,750	67,250	22.3	58.8	197	68.5
5140	Normalized (1600 F)	115,000	68,500	22.7	59.2	229	28.0
	Annealed (1525 F)	83,000	42,500	28.6	57.3	167	30.0
5150	Normalized (1600 F)	126,250	76,750	20.7	58.7	255	23.2
	Annealed (1520 F)	98,000	51,750	22.0	43.7	197	18.5
5160	Normalized (1575 F)	138,750	77,000	17.5	44.8	269	8.0
	Annealed (1495 F)	104,750	40,000	17.2	30.6	197	7.4
6150	Normalized (1600 F)	136,250	89,250	21.8	61.0	269	26.2
	Annealed (1500 F)	96,750	59,750	23.0	48.4	197	20.2
8620	Normalized (1675 F)	91,750	51,750	26.3	59.7	183	73.5
	Annealed (1600 F)	77,750	55,875	31.3	62.1	149	82.8
8630	Normalized (1600 F)	94,250	62,250	23.5	53.5	187	69.8
	Annealed (1550 F)	81,750	54,000	29.0	58.9	156	70.2
8650	Normalized (1600 F)	148,500	99,750	14.0	40.4	302	10.0
	Annealed (1465 F)	103,750	56,000	22.5	46.4	212	21.7
8740	Normalized (1600 F)	134,750	88,000	16.0	47.9	269	13.0
	Annealed (1500 F)	100,750	60,250	22.2	46.4	201	29.5
9255	Normalized (1650 F)	135,250	84,000	19.7	43.4	269	10.0
	Annealed (1550 F)	112,250	70,500	21.7	41.1	229	6.5
9310	Normalized (1630 F)	131,500	82,750	18.8	58.1	269	88.0
	Annealed (1550 F)	119,000	63,750	17.3	42.1	241	58.0

* All grades are fine-grained except those in the 1100 series which are coarse-grained. Austenitizing temperatures are given in parentheses. Heat-treated specimens were oil quenched.

Source: Bethlehem Steel Corp. and Republic Steel Corp. as published in 1974 DATA-BOOK issue of the American Society for Metals' METAL PROGRESS magazine and used with its permission.

Table 9 (*Continued*). **Typical Mechanical Properties of Selected Carbon and Alloy Steels** (Quenched and Tempered)

AISI No.*	Tempering Temperature, deg F	Strength Tensile	Strength Yield	Elongation, Per cent	Reduction in Area, Per cent	Hardness, Bhn
		1000 lb/in.²				
1030†	400	123	94	17	47	495
	600	116	90	19	53	401
	800	106	84	23	60	302
	1000	97	75	28	65	255
	1200	85	64	32	70	207
1040†	400	130	96	16	45	514
	600	129	94	18	52	444
	800	122	92	21	57	352
	1000	113	86	23	61	269
	1200	97	72	28	68	201
1040	400	113	86	19	48	262
	600	113	86	20	53	255
	800	110	80	21	54	241
	1000	104	71	26	57	212
	1200	92	63	29	65	192
1050†	400	163	117	9	27	514
	600	158	115	13	36	444
	800	145	110	19	48	375
	1000	125	95	23	58	293
	1200	104	78	28	65	235
1050	400	...	...	...	...	...
	600	142	105	14	47	321
	800	136	95	20	50	277
	1000	127	84	23	53	262
	1200	107	68	29	60	223
1060	400	160	113	13	40	321
	600	160	113	13	40	321
	800	156	111	14	41	311
	1000	140	97	17	45	277
	1200	116	76	23	54	229
1080	400	190	142	12	35	388
	600	189	142	12	35	388
	800	187	138	13	36	375
	1000	164	117	16	40	321
	1200	129	87	21	50	255
1095†	400	216	152	10	31	601
	600	212	150	11	33	534
	800	199	139	13	35	388
	1000	165	110	15	40	293
	1200	122	85	20	47	235
1095	400	187	120	10	30	401
	600	183	118	10	30	375
	800	176	112	12	32	363
	1000	158	98	15	37	321
	1200	130	80	21	47	269
1137	400	157	136	5	22	352
	600	143	122	10	33	285
	800	127	106	15	48	262
	1000	110	88	24	62	229
	1200	95	70	28	69	197

* All grades are fine-grained except those in the 1100 series which are coarse-grained. Austenitizing temperatures are given in parentheses. Heat-treated specimens were oil quenched unless otherwise indicated.   † Water quenched.

Source: Bethlehem Steel Corp. and Republic Steel Corp. as published in 1974 DATA-BOOK issue of the American Society for Metals' METAL PROGRESS magazine and used with its permission.

Table 9 (*Continued*). **Typical Mechanical Properties of Selected Carbon and Alloy Steels** (Quenched and Tempered)

AISI No.*	Tempering Temperature, deg F	Strength		Elongation, Per cent	Reduction in Area, Per cent	Hardness, Bhn
		Tensile	Yield			
		1000 lb/in.²				
1137†	400	217	169	5	17	415
	600	199	163	9	25	375
	800	160	143	14	40	311
	1000	120	105	19	60	262
	1200	94	77	25	69	187
1141	400	237	176	6	17	461
	600	212	186	9	32	415
	800	169	150	12	47	331
	1000	130	111	18	57	262
	1200	103	86	23	62	217
1144	400	127	91	17	36	277
	600	126	90	17	40	262
	800	123	88	18	42	248
	1000	117	83	20	46	235
	1200	105	73	23	55	217
1330†	400	232	211	9	39	459
	600	207	186	9	44	402
	800	168	150	15	53	335
	1000	127	112	18	60	263
	1200	106	83	23	63	216
1340	400	262	231	11	35	505
	600	230	206	12	43	453
	800	183	167	14	51	375
	1000	140	120	17	58	295
	1200	116	90	22	66	252
4037	400	149	110	6	38	310
	600	138	111	14	53	295
	800	127	106	20	60	270
	1000	115	95	23	63	247
	1200	101	61	29	60	220
4042	400	261	241	12	37	516
	600	234	211	13	42	455
	800	187	170	15	51	380
	1000	143	128	20	59	300
	1200	115	100	28	66	238
4130†	400	236	212	10	41	467
	600	217	200	11	43	435
	800	186	173	13	49	380
	1000	150	132	17	57	315
	1200	118	102	22	64	245
4140	400	257	238	8	38	510
	600	225	208	9	43	445
	800	181	165	13	49	370
	1000	138	121	18	58	285
	1200	110	95	22	63	230
4150	400	280	250	10	39	530
	600	256	231	10	40	495
	800	220	200	12	45	440
	1000	175	160	15	52	370
	1200	139	122	19	60	290

* All grades are fine-grained except those in the 1100 series which are coarse-grained. Austenitizing temperatures are given in parentheses. Heat-treated specimens were oil quenched unless otherwise indicated. † Water quenched.

Source: Bethlehem Steel Corp. and Republic Steel Corp. as published in 1974 DATA-BOOK issue of the American Society for Metals' METAL PROGRESS magazine and used with its permission.

Table 9 (*Continued*). **Typical Mechanical Properties of Selected Carbon and Alloy Steels** (Quenched and Tempered)

AISI No.*	Tempering Temperature, deg F	Strength Tensile 1000 lb/in.²	Strength Yield 1000 lb/in.²	Elongation, Per cent	Reduction in Area, Per cent	Hardness, Bhn
4340	400	272	243	10	38	520
	600	250	230	10	40	486
	800	213	198	10	44	430
	1000	170	156	13	51	360
	1200	140	124	19	60	280
5046	400	253	204	9	25	482
	600	205	168	10	37	401
	800	165	135	13	50	336
	1000	136	111	18	61	282
	1200	114	95	24	66	235
50B46	400	...	...	...	...	560
	600	258	235	10	37	505
	800	202	181	13	47	405
	1000	157	142	17	51	322
	1200	128	115	22	60	273
50B60	400	...	...	...	...	600
	600	273	257	8	32	525
	800	219	201	11	34	435
	1000	163	145	15	38	350
	1200	130	113	19	50	290
5130	400	234	220	10	40	475
	600	217	204	10	46	440
	800	185	175	12	51	379
	1000	150	136	15	56	305
	1200	115	100	20	63	245
5140	400	260	238	9	38	490
	600	229	210	10	43	450
	800	190	170	13	50	365
	1000	145	125	17	58	280
	1200	110	96	25	66	235
5150	400	282	251	5	37	525
	600	252	230	6	40	475
	800	210	190	9	47	410
	1000	163	150	15	54	340
	1200	117	118	20	60	270
5160	400	322	260	4	10	627
	600	290	257	9	30	555
	800	233	212	10	37	461
	1000	169	151	12	47	341
	1200	130	116	20	56	269
51B60	400	...	...	...	...	600
	600	...	...	...	...	540
	800	237	216	11	36	460
	1000	175	160	15	44	355
	1200	140	126	20	47	290
6150	400	280	245	8	38	538
	600	250	228	8	39	483
	800	208	193	10	43	420
	1000	168	155	13	50	345
	1200	137	122	17	58	282

* All grades are fine-grained except those in the 1100 series which are coarse-grained. Austenitizing temperatures are given in parentheses. Heat-treated specimens were oil quenched.

Source: Bethlehem Steel Corp. and Republic Steel Corp. as published in 1974 DATA-BOOK issue of the American Society for Metals' METAL PROGRESS magazine and used with its permission.

Table 9 (*Concluded*). **Typical Mechanical Properties of Selected Carbon and Alloy Steels** (Quenched and Tempered)

AISI No.*	Tempering Temperature, deg F	Strength		Elongation, Per cent	Reduction in Area, Per cent	Hardness, Bhn
		Tensile	Yield			
		1000 lb/in.²				
81B45	400	295	250	10	33	550
	600	256	228	8	42	475
	800	204	190	11	48	405
	1000	160	149	16	53	338
	1200	130	115	20	55	280
8630	400	238	218	9	38	465
	600	215	202	10	42	430
	800	185	170	13	47	375
	1000	150	130	17	54	310
	1200	112	100	23	63	240
8640	400	270	242	10	40	505
	600	240	220	10	41	460
	800	200	188	12	45	400
	1000	160	150	16	54	340
	1200	130	116	20	62	280
86B45	400	287	238	9	31	525
	600	246	225	9	40	475
	800	200	191	11	41	395
	1000	160	150	15	49	335
	1200	131	127	19	58	280
8650	400	281	243	10	38	525
	600	250	225	10	40	490
	800	210	192	12	45	420
	1000	170	153	15	51	340
	1200	140	120	20	58	280
8660	400	...	...	...	...	580
	600	...	...	...	...	535
	800	237	225	13	37	460
	1000	190	176	17	46	370
	1200	155	138	20	53	315
8740	400	290	240	10	41	578
	600	249	225	11	46	495
	800	208	197	13	50	415
	1000	175	165	15	55	363
	1200	143	131	20	60	302
9255	400	305	297	1	3	601
	600	281	260	4	10	578
	800	233	216	8	22	477
	1000	182	160	15	32	352
	1200	144	118	20	42	285
9260	400	...	...	...	...	600
	600	...	...	...	...	540
	800	255	218	8	24	470
	1000	192	164	12	30	390
	1200	142	118	20	43	295
94B30	400	250	225	12	46	475
	600	232	206	12	49	445
	800	195	175	13	57	382
	1000	145	135	16	65	307
	1200	120	105	21	69	250

* All grades are fine-grained except those in the 1100 series which are coarse-grained. Austenitizing temperatures are given in parentheses. Heat-treated specimens were oil quenched.

Source: Bethlehem Steel Corp. and Republic Steel Corp. as published in 1974 DATA-BOOK issue of the American Society for Metals' METAL PROGRESS magazine and used with its permission.

Table 10. Nominal Mechanical Properties of Selected Standard Stainless Steels

Grade	Condition	Tensile Strength (psi)	0.2 Per Cent Yield Strength (psi)	Elongation in 2 in. (%)	Reduction of Area (%)	Hardness Rockwell	Bhn
Austenitic Steels							
201	Annealed	115,000	55,000	55	...	B90	...
	¼-hard	125,000[a]	75,000[a]	20[a]	...	C25	...
	½-hard	150,000[a]	110,000[a]	10[a]	...	C32	...
	¾-hard	175,000[a]	135,000[a]	5[a]	...	C37	...
	Full-hard	185,000[a]	140,000[a]	4[a]	...	C41	...
202	Annealed	105,000	55,000	55	...	B90	...
	¼-hard	125,000[a]	75,000[a]	12[a]	...	C27	...
301	Annealed	110,000	40,000	60	...	B85	165
	¼-hard	125,000[a]	75,000[a]	25[a]	...	C25	...
	½-hard	150,000[a]	110,000[a]	15[a]	...	C32	...
	¾-hard	175,000[a]	135,000[a]	12[a]	...	C37	...
	Full-hard	185,000	140,000[a]	8[a]	...	C41	...
302	Annealed	90,000	37,000	55	65	B82	155
	¼-hard (sheet, strip)	125,000[a]	75,000[a]	12[a]	...	C25	...
	Cold drawn (bar, wire)[b]	To 350,000	...	...	...	...	...
302B	Annealed	95,000	40,000	50	65	B85	165
303, 303 (Se)	Annealed	90,000	35,000	50	55	B84	160
304	Annealed	85,000	35,000	55	65	B80	150
304L	Annealed	80,000	30,000	55	65	B76	140
305	Annealed	85,000	37,000	55	70	B82	156
308	Annealed	85,000	35,000	55	65	B80	150
309, 309S	Annealed	90,000	40,000	45	65	B85	165
310, 310S	Annealed	95,000	40,000	45	65	B87	170
314	Annealed	100,000	50,000	45	60	B87	170
316	Annealed	85,000	35,000	55	70	B80	150
	Cold drawn (bar, wire)[b]	To 300,000	...	...	...	...	...
316L	Annealed	78,000	30,000	55	65	B76	145
317	Annealed	90,000	40,000	50	55	B85	160
321	Anealed	87,000	35,000	55	65	B80	150
347, 348	Annealed	92,000	35,000	50	65	B84	160
Martensitic Steels							
403, 410 416, 416 (Se)	Annealed	75,000	40,000	30	65	B82	155
	Hardened[c]	...	...	...	...	C43	410
	Tempered at						
	400 deg. F.	190,000	145,000	15	55	C41	390
	600 deg. F.	180,000	140,000	15	55	C39	375
	800 deg. F.	195,000	150,000	17	55	C41	390
	1000 deg. F.	145,000	115,000	20	65	C31	300
	1200 deg. F.	110,000	85,000	23	65	B97	225
	1400 deg. F.	90,000	60,000	30	70	B89	180
414	Annealed	120,000	95,000	17	55	C22	235
	Hardened[c]	...	...	...	...	C44	426
	Tempered at						
	400 deg. F.	200,000	150,000	15	55	C43	415
	600 deg. F.	190,000	145,000	15	55	C41	400
	800 deg. F.	200,000	150,000	16	58	C43	415
	1000 deg. F.	145,000	120,000	20	60	C34	325
	1200 deg. F.	120,000	105,000	20	65	C24	260
420, 420F	Annealed	95,000	50,000	25	55	B92	195
	Hardened[d]	...	...	...	...	C54	540
	Tempered at						
	600 deg. F.	230,000	195,000	8	25	C50	500

Table 10 (*Concluded*). **Nominal Mechanical Properties of Selected Standard Stainless Steels**

Grade	Condition	Tensile Strength (psi)	0.2 Per Cent Yield Strength (psi)	Elongation in 2 in. (%)	Reduction of Area (%)	Hardness Rockwell	Hardness Bhn
			Martensitic Steels (*Continued*)				
431	Annealed	125,000	95,000	20	60	C24	260
	Hardened[d]	...	...	...	...	C45	440
	Tempered at						
	400 deg. F.	205,000	155,000	15	55	C43	415
	600 deg. F.	195,000	150,000	15	55	C41	400
	800 deg. F.	205,000	155,000	15	60	C43	415
	1000 deg. F.	150,000	130,000	18	60	C34	325
	1200 deg. F.	125,000	95,000	20	60	C24	260
440A	Annealed	105,000	60,000	20	45	B95	215
	Hardened[d]	...	...	...	...	C56	570
	Tempered						
	600 deg. F.	260,000	240,000	5	20	C51	510
440B	Annealed	107,000	62,000	18	35	B96	220
	Hardened[d]	...	...	...	...	C58	590
	Tempered						
	600°F	280,000	270,000	3	15	C55	555
440C, 440F	Annealed	110,000	65,000	13	25	B97	230
	Hardened[d]	...	...	...	...	C60	610
	Tempered						
	600 deg. F.	285,000	275,000	2	10	C57	580
501	Annealed	70,000	30,000	28	65	...	160
502	Annealed	70,000	30,000	30	75	B80	150
			Ferritic Steels				
405	Annealed	70,000	40,000	30	60	B80	150
430	Annealed	75,000	45,000	30	60	B82	155
430F, 430F (Se)	Annealed	80,000	55,000	25	60	B86	170
446	Annealed	80,000	50,000	23	50	B86	170

[a] Minimum. [b] Depending on size and amount of cold reduction. [c] Hardening temperature 1800 degrees F., 1-in.-diam. bars. [d] Hardening temperature 1900 degrees F., 1-in.-diam. bars.
Source: *Metals Handbook*, 8th edition, Volume I.

SAE 51440A: A hardenable chromium steel with greater quenched hardness than SAE 51420 and greater toughness than SAE 51440B and 51440C. Maximum corrosion resistance is obtained in the fully hardened and polished condition.

SAE 51440B: A hardenable chromium steel with greater quenched hardness than SAE 51440A. Maximum corrosion resistance is obtained in the fully hardened and polished condition. Capable of hardening to 50–60 Rockwell C depending upon carbon content.

SAE 51440C: This steel has the greatest quenched hardness and wear resistance upon heat treatment of any corrosion- or heat-resistant steel.

SAE 51440F: The same as SAE 51440C, except for its free machining characteristics.

SAE 51442: A corrosion- and heat-resisting chromium steel with corrosion-resisting properties slightly better than SAE 51430 and with good scale resistance up to 1600 deg. F.

SAE 51446: A corrosion- and heat-resisting steel with maximum amount of chromium consistent with commercial malleability. Used principally for parts which must resist high temperatures in service without scaling. Resists oxidation up to 2000 deg. F.

SAE 51501: Used for its heat and corrosion resistance and good mechanical properties at temperatures up to approximately 1000 deg. F.

High-Strength, Low-Alloy Steels. — High-strength, low-alloy (HSLA) steel represents a specific group of steels in which enhanced mechanical properties and, sometimes, resistance to atmospheric corrosion are obtained by the addition of moderate amounts of one or more alloying elements other than carbon. Different types are available, some of which are carbon–manganese steels and others contain further alloy additions, governed by special requirements for weldability, formability, toughness, strength, and economics. These steels may be obtained in the form of sheet, strip, plates, structural shapes, bars, and bar size sections.

HSLA steels are especially characterized by their mechanical properties, obtained in the as-rolled condition. They are not intended for quenching and tempering. For certain applications they are sometimes annealed, normalized, or stress relieved with some influence on mechanical properties.

Where these steels are used for fabrication by welding, care must be exercised in selection of grade and in the details of the welding process. Certain grades may be welded without preheat or postheat.

Because of their high strength-to-weight ratio, abrasion resistance, and, in certain compositions, improved atmospheric corrosion resistance, these steels are adapted particularly for use in mobile equipment and other structures where substantial weight savings are generally desirable. Typical applications are truck bodies, frames, structural members, scrapers, truck wheels, cranes, shovels, booms, chutes, and conveyors.

Grade 942X: A niobium or vanadium treated carbon manganese high-strength steel similar to 945X and 945C except for somewhat improved welding and forming properties.

Grade 945A: A HSLA steel with excellent welding characteristics, both arc and resistance, and the best formability, weldability, and low-temperature notch toughness of the high-strength steels. It is generally used in sheet, strip, and light plate thicknesses.

Grade 945C: A carbon–manganese high-strength steel with satisfactory arc welding properties if adequate precautions are observed. It is similar to grade 950C, except that lower carbon and manganese improve arc welding characteristics, formability and low temperature notch toughness at some sacrifice in strength.

Grade 945X: A niobium or vanadium treated carbon–manganese high-strength steel similar to 945C, except for somewhat improved welding and forming properties.

Grade 950A: A HSLA steel with good weldability, both arc and resistance, with good low-temperature notch toughness, and good formability. It is generally used in sheet, strip, and light plate thicknesses.

Grade 950B: A HSLA steel with satisfactory arc welding properties and fairly good low-temperature notch toughness and formability.

Grade 950C: A carbon–manganese high-strength steel that can be arc welded with special precautions, but is unsuitable for resistance welding. The formability and toughness are fair.

Grade 950D: A HSLA steel with good weldability, both arc and resistance, and fairly good formability. Where low-temperature properties are important, the effect of phosphorus in conjunction with other elements present should be considered.

Grade 950X: A niobium or vanadium treated carbon–manganese high-strength steel similar to 950C, except for somewhat improved welding and forming properties.

Grades 955X, 960X, 965X, 970X, 980X: These are steels similar to 945X and 950X with higher strength obtained by increased amounts of strengthening elements, such as carbon or manganese, or by the addition of nitrogen up to about 0.015 per cent. This increased strength involves reduced formability and usually decreased weldability. Toughness will vary considerably with composition and mill practice.

The composition, minimum mechanical properties, and formability of the HSLA steel grades are shown in Tables 11–15.

Table 11. Chemical Composition[a] Ladle Analysis (max. per cent)

Grade	C	Mn	P
942X	0.21	1.35	0.04
945A	0.15	1.00	0.04
945C	0.23	1.40	0.04
945X	0.22	1.35	0.04
950A	0.15	1.30	0.04
950B	0.22	1.30	0.04
950C	0.25	1.60	0.04
950D	0.15	1.00	0.04
950X	0.23	1.35	0.15
955X	0.25	1.35	0.04
960X	0.26	1.45	0.04
965X	0.26	1.45	0.04
970X	0.26	1.65	0.04
980X	0.26	1.65	0.04

[a] Sulfur, 0.05 per cent max; silicon, 0.90 per cent max.
Source: *SAE Handbook*, 1982.

Table 12. Minimum Mechanical Properties[a]

Grade	Form	Yield Strength[b] (psi min)	Tensile Strength (psi min)	Elongation (% min) 2 in.	8 in.
942X	Plates, shapes, bars to 4 in. incl.	42,000	60,000	24	20
945A, C	Sheet and strip	45,000	60,000	22	...
	Plates, shapes, bars To ½ in. incl.	45,000	65,000	22	18
	½–1½ in. incl.	42,000	62,000	24	19
	1½–3 in. incl.	40,000	62,000	24	19
945X	Sheet and strip	45,000	60,000	25	...
	Plates, shapes, bars To 1½ in. incl.	45,000	60,000	22	19
950A, B, C, D	Sheet and strip	50,000	70,000	22	...
	Plates, shapes, bars To ½ in. incl.	50,000	70,000	22	18
	½–1½ in. inc.	45,000	67,000	24	19
	1½–3 in. incl.	42,000	63,000	24	19
950X	Sheet and strip	50,000	65,000	22	...
	Plates, shapes, bars To 1½ in. incl.	50,000	65,000	...	18
955X	Sheet and strip	55,000	70,000	20	...
	Plates, shapes, bars To 1½ in. incl.	55,000	70,000	...	17
960X	Sheet and strip	60,000	75,000	18	...
	Plates, shapes, bars To 1½ in. incl.	60,000	75,000	...	16
965X	Sheet and strip	65,000	80,000	16	...
	Plates, shapes, bars To ¾ in. incl.	65,000	80,000	...	15
970X	Sheet and strip	70,000	85,000	14	...
	Plates, shapes, bars To ¾ in. incl.	70,000	85,000	...	14
980X	Sheet and strip	80,000	95,000	12	...
	Plates to ⅜ in. incl.	80,000	95,000	...	10

[a] Mechanical properties to be determined in accordance with ASTM A 370.
[b] Yield strength to be measured at 0.2 per cent offset.
Source: *SAE Handbook*, 1982.

Table 13. Steel Grades in Approximate Order of Increasing Excellence

Weldability	Formability	Toughness
980X	980X	980X
970X	970X	970X
965X	965X	965X
960X	960X	960X
955X, 950C, 942X	955X	955X
945C	950C	945C, 950C, 942X
950B, 950X	950D	945X, 950X
945X	950B, 950X, 942X	950D
950D	945C, 945X	950B
950A	950A	950A
945A	945A	945A

Source: *SAE Handbook*, 1982.

Table 14. Typical SAE Heat Treatments for Grades of Chromium–Nickel Austenitic Steels Not Hardenable by Thermal Treatment

SAE Steels	AISI No.	Treatment No.	Annealing[a] Temperature (deg. F)	Quenching Medium
30201	201	I	1850–2050	Water or air
30202	202	I	1850–2050	Water or air
30301	301	I	1800–2100	Water or air
30302	302	I	1800–2100	Water or air
30303F[b]	303	I	1800–2100	Water or air
30304	304	I	1800–2100	Water or air
30305	305	I	1800–2100	Water or air
30309	309	I	1800–2100	Water or air
30310	310	I	1800–2100	Water or air
30316	316	I	1800–2100	Water or air
30317	317	I	1800–2100	Water or air
30321	321	I	1800–2100	Water or air
30325	325	I	1800–2100	Water or air
30330	...	I	2050–2250	Air
30347	347	I	1800–2100	Water or air

[a] Quench to produce full austenitic structure using water or air in accordance with thickness of section. Annealing temperatures given cover process and full annealing as already established and used by industry, the lower end of the range being used for process annealing.

[b] Suffix A denotes steel differing only in carbon content. Suffix F denotes a free machining steel.

Source: *SAE Handbook*, 1982.

Table 15. Typical SAE Heat Treatments for Stainless Chromium Steels

SAE Steels	AISI No.	Normalizing Temperature (deg. F)	Subcritical Annealing Temperature (deg. F)	Full Annealing[a] Temperature (deg. F)	Hardening Temperature (deg. F)	Quenching Medium	Temper
51409	...	...	...	1550–1650	...	Air	...
51410	410	...	1300–1350[b]	1550–1650	1750–1850	Oil or air	To desired hardness
51414	414	...	1200–1250[b]	...	1750–1850	Oil or air	To desired hardness
51416	416	...	1300–1350[b]	1550–1650	1750–1850	Oil or air	To desired hardness
51420	420	...	1350–1450[b]	1550–1650	1800–1850	Oil or air	To desired hardness
51420F[c]	...	...	1350–1450[b]	1550–1650	1800–1850	Oil or air	To desired hardness
51430	430	...	1400–1500[d]	...	...	...	...
51430F[c]	...	...	1250–1500[d]	...	...	...	...
51431	431	...	1150–1225[b]	...	1800–1900	Oil or air	To desired hardness
51434 51436	...	...	1400–1500[d]	...	...	...	...
51440A[c] 51440B[c] 51440C[c] 51440F[c]	440A[c] 440B[c] 440C[c]	...	1350–1440[b]	1550–1650	1850–1950	Oil or air	To desired hardness
51442	442	...	1400–1500[d]	...	...	...	...
51446	446	...	1500–1650[b]	...	...	...	...
51501	501	...	1325–1375[d]	1525–1600	1600–1700	Oil or air	To desired hardness

[a] Cool slowly in furnace.
[b] Usually air cooled but may be furnace cooled.
[c] Suffixes A, B, and C denote three types of steel differing only in carbon content. Suffix F denotes a free machining steel.
[d] Cool rapidly in air.
Source: SAE Handbook, 1982.

TOOL STEELS

Tool steels, as the designation implies, serve primarily for making tools used in manufacturing and in the trades for the working and forming of metals, also wood, plastics and other industrial materials. Tools must withstand high specific loads, often concentrated at exposed areas, may have to operate at elevated or rapidly changing temperatures and in continual contact with abrasive types of work materials, are often subjected to shocks or may have to perform under other varieties of adverse conditions. Nevertheless, the tool when employed under circumstances which are regarded as normal operating conditions, should not suffer major damage, untimely wear resulting in the dulling of the edges, or be susceptible to detrimental metallurgical changes.

Tools for less demanding uses, such as ordinary handtools, including hammers, chisels, files, mining bits, etc., are often made of standard AISI steels which are not considered as belonging to any of the tool steel categories.

The steel for most types of tools must be used in a heat treated state, generally hardened and tempered, in order to provide the properties needed for the particular application. The adaptability to heat treatment with a minimum of harmful effects, which dependably results in the intended beneficial changes in material properties, is still another requirement which tool steels must satisfy.

In order to meet such varied requirements, steel types of different chemical composition, often produced by special metallurgical processes, have been developed. Due to the large number of tool steel types produced by the steel mills, which generally are made available with proprietary designations, it became rather difficult for the user to select those types which are most suitable for any specific application, unless the recommendations of a particular steel producer or producers are obtained.

Substantial clarification has resulted from the development of a classification system which is now widely accepted throughout the industry, on the part of both the producers and the users of tool steels. That system is used in the following as a base for providing concise information on tool steel types, their properties and methods of tool steel selection.

The tool steel classification system establishes seven basic categories of tool and die steels. These categories are associated with the predominant applicational characteristics of the tool steel types they comprise. A few of these categories are composed of several groups to distinguish between families of steel types which, while serving the same general purpose, differ in regard to one or more dominant characteristics.

In order to offer an easily applicable guide for the selection of tool steel types best suited for a particular application, the subsequent discussions and tables are based on the above mentioned application-related categories. As an introduction to the detailed surveys, a concise discussion is presented of the principal tool steel characteristics which govern the suitability for varying service purposes and operational conditions. A brief review of the major steel alloying elements and of the effect of these constituents on the significant characteristics of tool steels is also given in the sections which follow.

The Properties of Tool Steels. — Tool steels must possess certain properties to a higher than ordinary degree, in order to make them adaptable for uses which require the ability of sustaining heavy loads and performing dependably even under adverse conditions.

The extent and the type of loads, the characteristics of the operating condition and the expected performance with regard to both the duration and the level of

consistency, are the principal considerations, in combination with the aspects of cost, which govern the selection of tool steels for specific applications.

While it is not possible to define and to apply exact parameters for measuring the significant tool steel characteristics, certain properties can be determined which may greatly assist in appraising the suitability of various types of tool steels for specific uses.

Because tool steels are generally heat treated to make them adaptable to the intended use by enhancing the desirable properties, *the behavior of the steel during heat treatment* is of prime importance. The behavior of the steel comprises, in this respect, both the resistance to harmful effects and the attainment of the desirable properties. The following are considered the major properties related to heat treatment:

Safety in Hardening: This designation expresses the ability of the steel to withstand the harmful effects of exposure to very high heat and particularly to the sudden temperature changes during quenching without harmful effects. One way of obtaining this property is by means of alloying elements which reduce the critical speed at which the quenching must be carried out, thus permitting the use of less harsh acting quenching media such as oil, salt or just still air.

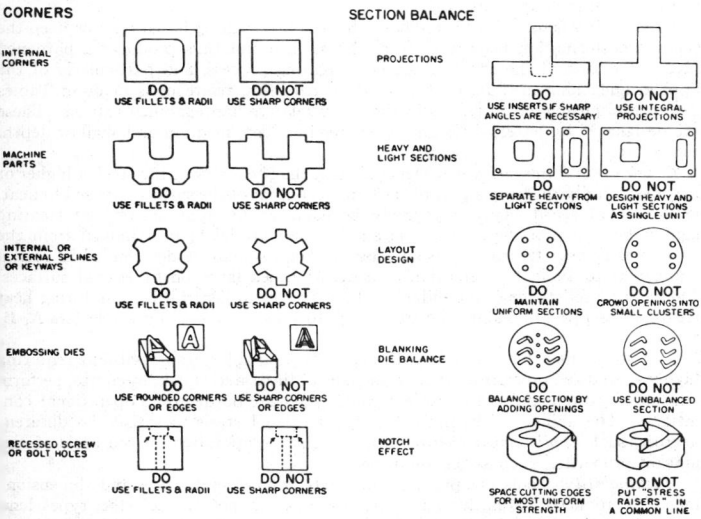

Tool and die design tips to reduce breakage in heat treatment.
Courtesy of Society of Automotive Engineers, Inc.

The most common harm which parts made of tool steel can suffer during heat treat is the development of cracks. In addition to the composition of the steel and the applied heat treating process, the configuration of the part can also affect the sensitivity to cracking. The accompanying figure illustrates a few design characteristics related to cracking and warpage in heat treatment; the observation of these design tips, which call for generous filleting, avoidance of sharp angles and

major changes without transition in the cross section, is particularly advisable when using tool steel types with a low index value for safety in hardening.

In current practice this property of tool steels is rated in the order of decreasing safety (that is increasing sensitivity) as Highest, Very High, High, Medium and Low safety, expressed in the Tables 4–9 by the letters A, B, C, D, and E.

Distortions in Heat Treating: In parts made of tool steels such distortions are often a consequence of inadequate design (see previous illustration) or improper heat treatment (e.g. lack of stress relieving). However, certain types of tool steels display different degrees of sensitivity to distortion. Those which are less stable require safer design of the parts for which they are used, more careful heat treatment including the proper support of long and slender parts, or thin sections, and possibly greater grinding allowance for permitting subsequent correction of the distorted shape. In some cases parts made of a type of steel generally sensitive to distortions can be heat treated with very little damage when the requirements of the part call for a relatively shallow hardened layer over a soft core. However, for intricate shapes and large tools such steel types should be selected which possess superior non-deforming properties. The ratings used in Tables 4–9 express the non-deforming properties (stability of shape in heat treatment) of the steel types and start with the lowest distortion (the best stability) designated as A; the greatest susceptibility to distortion being designated as E.

Depth of Hardening: This is indicated by a relative rating based on how deep the phase transformation penetrates from the surface and thus produces a hardened layer. Because of the effect of the heat treating process, and particularly of the applied quenching medium on the depth of hardness, reference is made in Tables 4–9 to the quench which results in the listed relative hardenability values. These are designated by letters A, B and C, expressing deep, medium and shallow depth, respectively.

Resistance to Decarburization: Depending on the chemistry of the steel, a higher or lower sensitivity to losing a part of the carbon content of the surface exposed to heat exists. That sensitivity can partially be balanced by appropriate heat treating equipment and processes. Also the amount of material to be removed from the surface after heat treatment, usually by grinding, should be determined in such a manner as to avoid the retention of a decarburized layer on functional surfaces. The relative resistance of individual tool steel types to decarburization during heat treatment is rated in Tables 4–9 from High to Low, expressed by the letters A, B and C.

Tool steels must be workable with generally available means, without requiring highly specialized processes. The tools made of these steels must, of course, perform adequately, often under adverse environmental and burdensome operational conditions. The ability of the individual types of tool steels to satisfy, to different degrees, such *applicational requirements* can also be appraised on the basis of significant properties, such as the following:

Machinability: Tools are precision products, whose final shape and dimension must be produced by machining, a process to which not all tool steel types lend themselves equally well. That difference in machinability is particularly evident in tool steels which, depending on their chemical composition, may contain substantial amounts of metallic carbides, beneficial to increased wear resistance, yet detrimental to the service life of tools with which the steel has to be worked. The microstructure of the steel type can also affect the ease of machining and, in some types, certain phase conditions such as those due to low carbon content, may cause difficulties in achieving a fine surface finish. Certain types of tool steels have their machinability improved by the addition of slight amounts of sulfur or lead. In the selection of tool steels machinability may be considered as a factor in the cost of making the tool, and also for reasons of feasibility, particularly in the case of in

tricate tool shapes. The ratings in Tables 4–9, starting with A for the greatest ease of machining, to E for the lowest machinability, refer to working of the steel in an unhardened condition. That characteristic is not necessarily identical with the grindability which expresses how well the steel is adapted to grinding after heat treating. The ease of grinding, however, may become an important consideration in tool steel selection, particularly in the case of tools like cutting tools and dies, which require regular sharpening, involving extensive grinding. AVCO Bay State Abrasives Company has compiled information on the relative grindability of frequently used types of tool steels. A simplified version of that information is presented in Table 1 which assigns the listed tool steel types to one of the following grindability grades: High (A), Medium (B), Low (C) and Very Low (D), expressing decreasinging ratios of volume of metal removed to wheel wear.

Hot Hardness: This property designates the steel's resistance to the softening effect of elevated temperature. This characteristic is related to the tempering temperature of the type of steel which is controlled by various alloying elements, such as tungsten, molybdenum, vanadium, cobalt and chromium.

Table 1. Relative Grindability of Selected Types of Frequently Used Tool Steels

AISI Tool Steel Type	H41	H42	H43	Other H	D2	D3	D5	D7	A Types	O Types	L Types	F Types
Relative Grindability Index	B	B	B	A	B	B	B	C	A	A	A	B

AISI High Speed Tool Steel Type	M1	M2	M3 (1)	M3 (2)	M4	M7	M8	M10	M15	M36	M43	T1	T2	T3	T5	T6	T15
Relative Grindability Index	A	B	C	C	D	B	A	B	D	B	B	A	B	C	B	B	D

Hot hardness is a necessary property of tools used for hot work, like forging, casting and hot extrusion. Hot hardness is also important in cutting tools operated at high speed which generates sufficient heat to raise their temperature well above the level where ordinary steels lose their hardness; hence the designation *high speed steels* which refers to a family of tool steels developed for such uses. Frequently it is the degree of the tool steel's resistance to softening at elevated temperature which governs important process data, such as the applicable cutting speed. In the ratings of Tables 4–9, tool steel types having the highest hot hardness are marked with A, subsequent letters expressing gradually decreasing capacity to endure elevated temperature without losing hardness.

Wear Resistance: The gradual erosion of the tool's operating surface, most conspicuously occurring at the exposed edges, is known as wear. Resistance to wear prolongs the useful life of the tool, by delaying the degradation of its surface through abrasive contact with the work at regular operating temperatures; these vary according to the type of the process. Wear resistance is observable experimentally and measurable by comparison. Certain types of metallic carbides embedded into the steel matrix are considered to be the prime contributing factors to wear resistance, besides the hardness of the heat treated steel material. The ratings of Tables 4–9, starting with A for the best to E for poor, are based on conditions considered as normal in operations for which various types of tool materials are primarily used.

Toughness: In tool steels this property expresses their ability to sustain shocks, suddenly applied and relieved loads, or major impacts without breaking. Steels used for making tools must also be able to absorb such forces with only a minimum of elastic deformation and without permanent deformation to any extent which would interfere with the proper functioning of the tool. Certain types of tool steels, particularly those with high carbon content and without the presence of beneficial

alloying constituents, tend to be the most sensitive to shocks, although they can also be made to act tougher when used for tools which permit a hardened case to be supported by a soft core. Tempering improves toughness, while generally reducing hardness. The rating indexes in Tables 4–9, A for the highest toughness, through E for the types most sensitive to shocks, apply to tools heat treated to hardness values normally used for the particular type of tool steel.

Common Tool Faults and Failures. — While the proper selection of the steel grade used for any particular type of tool is of great importance, it should be recognized that many of the failures experienced in common practice originate from causes other than those related to the tool material.

To permit a better appraisal of the actual causes of failure and possible corrective action, a general, although not complete, list of common tool faults, resulting failures and corrective actions is shown in Table 2. In this list the potential failure causes are grouped into four categories. The possibility of more than a single cause being responsible for the experienced failure should not be excluded.

Finally, it must be remembered that the proper usage of tools is indispensable for obtaining satisfactory performance and tool life. Using the tools properly involves, for example, the avoidance of damage to the tool, overload, excessive speeds and feeds; the application of adequate coolant when called for; assurance of a rigid setup; proper alignment; and firm tool and work holding.

The Effect of Alloying Elements on Tool Steel Properties. — *Carbon (C):* The presence of carbon, usually in excess of 0.60 per cent for non-alloyed types, is essential for assuring the hardenability of steels to the levels needed for tools. Raising the carbon content by different amounts up to a maximum of about 1.3 per cent increases the hardness slightly and the wear resistance considerably. The amount of carbon in tool steels is specified for attaining certain properties (such as in the water-hardening category where higher carbon content may be chosen to improve wear resistance, although to the detriment of toughness) or, in the alloyed types of tool steels, in conformance with the other constituents for producing well balanced metallurgical and performance properties.

Manganese (Mn): In smaller amounts, to about 0.60 per cent, manganese is added for reducing brittleness and to improve forgeability. Larger amounts of manganese improve hardenability, permitting oil quenching for non-alloyed carbon steels, thus reducing deformation, although with regard to several other properties, manganese is not an equivalent replacement for the regular alloying elements.

Silicon (Si): In itself silicon may not be considered an alloying element of tool steels, but it is needed as a deoxidizer and improves the hot-forming properties of the steel. In combination with certain alloying elements the silicon content is sometimes raised up to about 2 per cent for increasing the strength and toughness of steels used for tools which have to sustain shock loads.

Tungsten (W): This is one of the important alloying elements of tool steels, particularly because of two valuable properties: it improves "hot hardness", that is, the resistance of the steel to the softening effect of elevated temperature, and it also forms hard, abrasion resistant carbides, thus improving the wear properties of tool steels.

Vanadium (V): This element contributes to the refinement of the carbide structure and thus improves the forgeability of alloy tool steels. Vanadium has a very strong tendency to form a hard carbide, which improves both the hardness and the wear properties of tool steels, however, a large amount of vanadium carbide makes the grinding of the tool very difficult (causes low grindability).

Molybdenum (Mo): In small amounts molybdenum improves certain metallurgical properties of alloy steels, such as deep hardening and toughness. It is used often

Table 2.   Common Tool Faults, Failures and Cures — 1

Fault Description	Probable Failure	Possible Cure
IMPROPER TOOL DESIGN		
Drastic section changes — widely different thicknesses of adjacent wall sections or protruding elements	In liquid quenching the thin section will cool and then harden more rapidly than the adjacent thicker section, setting up stresses which may exceed the strength of the steel.	Make such parts of two pieces or use an air hardening tool steel which avoids the harsh action of a liquid quench.
Sharp corners on shoulders or in square holes	Cracking can occur, particularly in liquid quenching, due to stress concentrations.	Apply fillets to the corners and/or use an air hardening tool steel.
Sharp cornered keyways	Failure may arise during service, usually considered as caused by fatigue.	The use of round keyways should be preferred when the general configuration of the part makes it prone to failure due to square keyways.
Abrupt section changes in battering tools	Pneumatic tools, due to impact in service, are particularly sensitive to stress concentrations which lead to fatigue failures.	Use taper transitions, which are better than even generous fillets.
Functional inadequacy of tool design — e.g., insufficient guidance for a punch	Excessive wear or breakage in service may occur.	Assure solid support, avoid unnecessary play, adapt travel length to operational conditions (e.g. punch to penetrate to four-fifths of thickness in hard work material.)
Improper tool clearance, such as in blanking and punching tools	Deformed and burred parts may be produced, excessive tool wear or breakage can result.	Adapt clearances to material conditions and dimensions for reducing tool load and also for obtaining clean sheared surfaces.
FAULTY CONDITION OR INADEQUATE GRADE OF TOOL STEEL		
Improper tool steel grade selection	Typical failures: Chipping — insufficient toughness. Wear — poor abrasion resistance. Softening—inadequate "red hardness".	Choose the tool steel grade by following recommendations and improve selection when needed, guided by property ratings.
Material defects — voids, streaks, tears, flakes, surface cooling cracks, etc	When not recognized during material inspection, tools made of defective steel often prove to be useless.	Obtain tool steels from reliable sources and inspect tool material for detectable defects.
Decarburized surface layer ("bark") in rolled tool steel bars	Cracking may originate from the decarburized layer, or it will not harden ("soft skin").	Provide allowance for stock to be removed from all surfaces of hot rolled tool steel. Recommended amounts are listed in tool steel catalogs and vary according to section size, generally about 10 percent for smaller and 5 percent for larger diameters.
Brittleness caused by poor carbide distribution in high alloy tool steels	Excessive brittleness can cause chipping or breakage during the service of the tool.	Bars with large diameter (above about four inches) tend to be prone to nonuniform carbide distribution. Choose upset forged discs instead of large diameter bars.

Table 2.   Common Tool Faults, Failures and Cures — 2

Fault Description	Probable Failure	Possible Cure
FAULTY CONDITION OR INADEQUATE GRADE OF TOOL STEEL		
Unfavorable grain flow	Improper grain flow of the steel used for milling cutters and similar tools can cause the breaking out of teeth.	Upset forged discs made with an upset ratio of about 2 to 1 (starting to upset thickness) display radial grain flow. Highly stressed tools, e.g., gear shaper cutters, may require the cross forging of blanks.
HEAT TREATMENT FAULTS		
Improper preparation for heat treatment. Certain tools may require stress relieving or annealing, and often pre-heating too	Tools highly stressed during machining or forming, unless stress relieved, may aggravate the thermal stresses of heat treatment, thus causing cracks. Excessive temperature gradients developed in non-preheated tools with different section thicknesses can cause warpage.	Stress relieve, when needed before hardening. Anneal prior to heavy machining or cold forming (e.g., hobbing). Pre-heat tools (a) having substantial section thickness variations, or (b) requiring high quenching temperatures, as those made of high speed tool steels.
Overheating during hardening; quenching from too high a temperature	Causes grain coarsening and a sensitivity to cracking which is more pronounced in tools with drastic section changes.	Overheated tools have a characteristic microstructure which aids recognition of the cause of failure and indicates the need for improved temperature control.
Low hardening temperature	The tool may not harden at all, or in its outer portion only, thereby setting up stresses which can lead to cracks.	Controlling both the temperature of the furnace and the time of holding the tool on quenching temperature, will prevent this not too frequent deficiency.
Inadequate composition or condition of the quenching media	Water hardening tool steels are particularly sensitive to inadequate quenching media, which can cause soft spots or even violent cracking.	For water hardening tool steels use water free of dissolved air and contaminants, also assure sufficient quantity and proper agitation of the quench.
Improper handling during and after quenching	Cracking of the tools, particularly of those with sharp corners, during the heat treatment can result from holding the part too long in the quench or incorrectly applied tempering.	Following the steel producer's specifications is a safe way for assuring proper heat treat handling. In general, the tool should be left in the quench until it reaches a temperature of 150 to 200°F, and is then transferred promptly into a warm tempering furnace.
Insufficient tempering	The omission of double tempering for steel types which require it may cause early failure by heat checking in hot work or make the tool abnormally sensitive to grinding checks.	Apply double tempering for highly alloyed tool steel of the high speed, hot work and high chromium categories, for removing stresses caused by martensite formed during the first tempering phase. Second temper also increases hardness of most high speed steels.

Table 2. Common Tool Faults, Failures and Cures — 3

Fault Description	Probable Failure	Possible Cure
HEAT TREATMENT FAULTS		
Decarburization and carburization	Unless hardened in a nuetral atmosphere the original carbon content of the tool surface may be changed: Reduced (decarburization) causing a soft layer which wears rapidly. Increased (carburization) which, when excessive, may cause brittleness.	Heating in neutral atmosphere or well maintained salt bath and controlling the furnace temperature and the time during which the tool is subjected to heating can usually keep the carbon unbalance within acceptable limits.
GRINDING DAMAGES		
Excessive stock removal rate causing the heating of the part surface beyond the applied tempering temperature	Scorched tool surface displaying temper colors varying from yellow to purple, depending on the degree of heat. This causes softening of the ground surface. When coolant is used, a local rehardening can take place, often resulting in cracks.	Prevention: by reducing speed and feed, or using coarser, softer, more open structured grinding wheel, with ample coolant. Correction: by subsequent light stock removal eliminate the discolored layer. Not always a cure, because the effects of abusive grinding may not be corrected.
Improper grinding wheel specifications; grain too fine or bond too hard	Intense localized heating during the grinding may set up surface stresses causing grinding cracks. These are either parallel but at right angles to the direction of grinding or, when more advanced, form a network. May need cold etch or magnetic particle testing to become recognizable.	Prevention: by correcting the grinding wheel specifications. Correction: in the case of shallow (.002 to .004 inch) cracks, by removing the damaged layer, when permitted by the design of the tool, using very light grinding passes.
Incorrectly dressed or loaded grinding wheel	Heating of the work surface can cause scorching or cracking. Incorrect dressing can also cause a poor finish of the ground work surface.	Dress wheel with sharper diamond and faster diamond advance to produce coarser wheel surface. Alternate dressing methods, like crush-dressing, can improve wheel surface conditions. Dress wheel regularly to avoid loading or glazing of the wheel surface.
Inadequate coolant, with regard to composition, amount, distribution and cleanliness	Introducing into the tool surface heat which is not adequately dissipated or absorbed by the coolant, can cause softening, or even the development of cracks.	Improve coolant supply and quality, or reduce stock removal rate in order to generate less heat in the grinding.
Damage caused by abusive abrasive cutoff	The intensive heat developed during this process can cause a hardening of the steel surface, or may even result in cracks.	Reduce rate of advance, adopt wheel specifications better suited for the job. Use ample coolant or, when harmful effect not eliminated, replace abrasive cutoff by some cooler acting stock separation method (e.g., sawing or lathe cutoff) unless damaged surface is being removed by subsequent machining.

Note: Illustrated examples of tool failures from causes such as listed above may be found in "The Tool Steel Trouble Shooter" handbook, published by Bethlehem Steel Corporation.

larger amounts in certain high speed tool steels to replace tungsten, primarily for economic reasons, often with nearly equivalent results.

Cobalt (Co): As an alloying element of tool steels, cobalt increases hot hardness and is used in applications where that property is needed. Substantial addition of cobalt, however, raises the critical quenching temperature of the steel with a tendency to increase the decarburization of the surface, and also reduces toughness.

Chromium (Cr): This element is added in amounts of several per cent to high alloy tool steels, and up to 12 per cent to types in which chromium is the major alloying element. It improves hardenability and, together with high carbon, provides both wear resistance and toughness, a combination valuable in certain tool applications. However, high chromium raises the hardening temperature of the tool steel, and thus can cause proneness to hardening deformations. A high percentage of chromium also affects the grindability of the tool steel.

Nickel (Ni): Generally in combination with other alloying elements, particularly chromium, nickel is used to improve the toughness and, to some extent, the wear resistance of tool steels.

The addition of more than one alloy to a steel often produces what is called a synergistic effect. Thus, the combined effects of two or more alloy elements may be greater than the sum of the individual effects of each alloy.

Classification of Tool Steels. — Steels for tools must satisfy a number of different, often conflicting requirements. The need for specific steel properties arising from widely varying applications has led to the development of many compositions of tool steels, each intended to meet a particular combination of applicational requirements. The resultant diversity of tool steels, their number being continually expanded by the addition of new developments, made it extremely difficult for the user to select the type best suited to his needs, or to find equivalent alternatives for specific types available from particular sources.

As a cooperative industrial effort under the sponsorship of AISI and SAE, a tool classification system has been developed in which the commonly used tool steels are grouped into seven major categories. These categories, several of which contain more than a single group, are listed in the following with the letter symbols used for

(Continued on page 485)

Category Designation	Letter Symbol	Group Designation
High Speed Tool Steels	M T	Molybdenum types Tungsten types
Hot Work Tool Steels	H1–H19 H20–H39 H40–H59	Chromium types Tungsten types Molybdenum types
Cold Work Tool Steels	D A O	High carbon, high chromium types Medium alloy, air hardening types Oil hardening types
Shock Resisting Tool Steels	S	. . . . .
Mold Steels	P	. . . . .
Special Purpose Tool Steels	L F	Low alloy types Carbon tungsten types
Water Hardening Tool Steels	W	. . . . .

Classification, Approximate Compositions, and Properties Affecting Selection of Tool and Die Steels (From SAE Recommended Practice)

Type of Tool Steel	Chemical Composition†								Non-warping Prop.	Safety in Hardening	Toughness	Depth of Hardening	Wear Resistance
	C	Mn	Si	Cr	V	W	Mo	Co					
Water Hardening													
.80 Carbon	.70–.85	*	*	*						Poor	Fair	Good[4]	Shallow	Fair
.90 Carbon	.85–.95	*	*	*						Poor	Fair	Good[4]	Shallow	Fair
1.00 Carbon	.95–1.10	*	*	*						Poor	Fair	Good[4]	Shallow	Good
1.20 Carbon	1.10–1.30	*	*	*						Poor	Fair	Good[4]	Shallow	Good
.90 Carbon-V	.85–.95	*	*	*	.15–.35					Poor	Fair	Good[4]	Shallow	Fair
1.00 Carbon-V	.95–1.10	*	*	*	.15–.35					Poor	Fair	Good	Shallow	Good
1.00 Carbon-VV	.90–1.10	*	*	*	.35–.50					Poor	Fair	Good	Shallow	Good
Oil Hardening													
Low Manganese	.90	1.20	.25	.50	.20[1]				Good	Good	Fair	Deep	Good
High Manganese	.90	1.60	.25	.35[1]	.20[1]				Good	Good	Fair	Deep	Good
High Carbon-High Chromium‡	2.15	.35	.35	12.00	.80[1]		.30[1]		Good	Good	Poor	Through	Best
Chromium	1.00	.35	.25	1.40			.80[1]		Good	Good	Fair	Deep	Good
Molybdenum Graphitic	1.45	.75	1.00				.40		Fair	Good	Fair	Deep	Good
Nickel-Chromium**	.75	.70	.25	.85	.25[1]		.50[1]		Fair	Good	Fair	Deep	Fair
Air Hardening													
High Carbon-High Chromium	1.50	.40	.40	12.00	.80[1]		.90		Best	Best	Fair	Through	Best
5 Per Cent Chromium Air Hard.	1.00	.60	.25	5.25	.40[1]		1.10		Best	Best	Fair	Through	Good
High Carbon-High Chromium-Cobalt	1.50	.40	.40	12.00	.80[1]		.90	.60[1]	Best	Best	Fair	Through	Best
Shock Resisting													
Chromium-Tungsten	.50	.25	.35	1.40	.20	2.25	.40[1]		Fair	Good	Good	Deep	Fair
Silicon-Molybdenum	.50	.40	1.00		.25[1]		.50		Poor[2]	Poor[3]	Best	Deep	Fair
Silicon-Manganese	.55	.80	2.00	.30[1]	.25[1]		.40[1]		Poor[2]	Poor[3]	Best	Deep	Fair
Hot Work													
Chromium-Molybdenum-Tungsten	.35	.30	1.00	5.00	.25[1]	1.25	1.50		Good	Good	Good	Through	Fair
Chromium-Molybdenum-V	.35	.30	1.00	5.00	.40		1.50		Good	Good	Good	Through	Fair
Chromium-Molybdenum-VV	.35	.30	1.00	5.00	.90		1.50		Good	Good	Good	Through	Fair
Tungsten	.32	.30	.20	3.25	.40	9.00			Good	Good	Good	Through	Fair

See footnotes at end of table.

Classification, Approximate Compositions, and Properties Affecting Selection of Tool and Die Steels (Concluded)

Type of Tool Steel	Chemical Composition†								Non-Warping Prop.	Safety in Hardening	Toughness	Depth of Hardening	Wear Resistance
	C	Mn	Si	Cr	V	W	Mo	Co					
High Speed													
Tungsten, 18-4-1	.70	.30	.30	4.10	1.10	18.00	...	...	Good	Good	Poor	Through	Good
Tungsten, 18-4-2	.80	.30	.30	4.10	2.10	18.50	.80	...	Good	Good	Poor	Through	Good
Tungsten, 18-4-3	1.05	.30	.30	4.10	3.25	18.50	.70	...	Good	Good	Poor	Through	Best
Cobalt-Tungsten, 14-4-2-5	.80	.30	.30	4.10	2.00	14.00	.80	5.00	Good	Fair	Poor	Through	Good
Cobalt-Tungsten, 18-4-1-5	.75	.30	.30	4.10	1.00	18.00	.80	5.00	Good	Fair	Poor	Through	Good
Cobalt-Tungsten, 18-4-2-8	.80	.30	.30	4.10	1.75	18.50	.80	8.00	Good	Fair	Poor	Through	Good
Cobalt-Tungsten, 18-4-2-12	.80	.30	.30	4.10	1.75	20.00	.80	12.00	Good	Fair	Poor	Through	Good
Molybdenum, 8-2-1	.80	.30	.30	4.00	1.15	1.50	8.50	...	Good	Fair	Poor	Through	Good
Molybdenum-Tungsten, 6-6-2	.83	.30	.30	4.10	1.90	6.25	5.00	...	Good	Fair	Poor	Through	Good
Molybdenum-Tungsten, 6-6-3	1.15	.30	.30	4.10	3.25	5.75	5.25	...	Good	Fair	Poor	Through	Best
Molybdenum-Tungsten, 6-6-4	1.30	.30	.30	4.25	4.25	5.75	5.25	...	Good	Fair	Poor	Through	Best
Cobalt-Molybdenum-Tungsten, 6-6-2-8	.85	.30	.30	4.10	2.00	6.00	5.00	8.00	Good	Fair	Poor	Through	Good

† C = carbon; Mn = manganese; Si = silicon; Cr = chromium; V = vanadium; W = tungsten; Mo = molybdenum; Co = cobalt.
‡ This steel may have 0.50 per cent nickel as an optional element. The steel has been found to give satisfactory application either with or without the element present.

* Carbon tool steels are usually available in four grades or qualities: *Special (Grade 1)* — The highest quality water-hardening carbon tool steel, controlled for hardenability, chemistry held to closest limits, and subject to rigid tests to insure maximum uniformity in performance; *Extra (Grade 2)* — A high quality water-hardening carbon tool steel, controlled for hardenability, subject to tests to insure good service; *Standard (Grade 3)* — A good quality water-hardening carbon tool steel, not controlled for hardenability, recommended for application where some latitude with respect to uniformity is permissible; *Commercial (Grade 4)* — A commercial quality water-hardening carbon tool steel, not controlled for hardenability, not subject to special tests. On *special* and *extra* grades, limits on manganese, silicon, and chromium are not generally required if Shepherd hardenability limits are specified. For *standard* and *commercial* grades, limits are 0.35 max. each for Mn and Si; 0.15 max. Cr for standard; 0.20 max. Cr for commercial.

** Approximate nickel content of this steel is 1.50%.

Shock Resisting Steels. if chromium, vanadium and molybdenum are not present, then the hardenability will be affected.

1 Optional element. Steels have found satisfactory application either with or without the element present. In the case of silicon manganese steel listed under
2 Poor when water quenched, fair when oil quenched. 3 Poor when water quenched, good when oil quenched.
4 Toughness decreases somewhat with increasing depth of hardening.

Table 3. Quick Reference Guide for Tool Steel Selection — 1

Application Areas	Tool Steel Categories and AISI Letter Symbols						
	High Speed Tool Steels, M and T	Hot Work Tool Steels, H	Cold Work Tool Steels, D, A and O	Shock Resisting Tool Steels, S	Mold Steels, P	Special Purpose Tool Steels, L and F	Water Hardening Tool Steels, W
			Examples of Typical Application				
Cutting Tools Single point types (lathe, planer, boring) Milling cutters Drills Reamers Taps Threading dies Form cutters	General purpose production tools: M2, T1 For increased abrasion resistance: M3, M4, and M10 Heavy-duty work calling for high hot hardness: T5, T15 Heavy-duty work calling for high abrasion resistance: M42, M44		Tools with keen edges (knives, razors) Tools for operations where no high speed is involved, yet stability in heat treatment and substantial abrasion resistance are needed				Uses which do not require hot hardness or high abrasion resistance. Examples with carbon content of applicable group: Taps (1.05/1.10%C) Reamers (1.10/1.15%C) Twist drills (1.20/1.25%C) Files (1.35/1.40%C)
Hot Forging Tools and Dies Dies and inserts Forging machine plungers and pierces	For combining hot hardness with high abrasion resistance: M2, T1	Dies for presses and hammers: H20, H21 For severe conditions over extended service periods: H22 to H26, also H43	Hot trimming dies: D2	Hot trimming dies Blacksmith tools Hot swaging dies			Smith's tools (.65/.70%C) Hot chisels (.70/.75%C) Drop forging dies (.90/1.00%C) Applications limited to short-run production
Hot Extrusion Tools and Dies Extrusion dies and mandrels, Dummy blocks Valve extrusion tools	Brass extrusion dies: T1	Extrusion dies and dummy blocks: H20 to H26 For tools which are exposed to less heat: H10 to H19		Compression molding: S1			

Table 3. Quick Reference Guide for Tool Steel Selection — 2

Application Areas	High Speed Tool Steels, M and T	Hot Work Tool Steels, H	Cold Work Tool Steels, D, A and O	Shock Resisting Tool Steels, S	Mold Steels, P	Special Purpose Tool Steels, L and F	Water Hardening Tool Steels, W
			Tool Steel Categories and AISI Letter Symbols				
			Examples of Typical Application				
Cold Forming Dies	Burnishing tools: M1, T1	Cold heading die casings: H13	Drawing dies: or Coining tools: O1, D2	Hobbing and short-run applications: S1, S7		Blanking, forming, and trimmer dies when toughness has precedence over abrasion resistance: L6	Cold heading dies: W1 or W2 (C ≈ 1.00%) Bending dies: W1 (C ≈ 1.00%)
Bending, forming, drawing, and deep drawing dies and punches			Forming and bending dies: A2 Thread rolling dies: D2	Rivet sets and rivet busters			
Shearing Tools	Special dies for cold and hot work: T1 For work requiring high abrasion resistance: M2, M3	For shearing knives: H11, H12 For severe hot shearing applications: M21, M25	Dies for medium runs: A2, A6 also O1 and O4	Cold and hot shear blades		Knives for work requiring high toughness: L6	Trimming dies (.90/.95%C) Cold blanking and punching dies (1.00%C)
Dies for piercing, punching, and trimming			Dies for long runs: D2, D3	Hot punching and piercing tools			
Shear blades			Trimming dies (also for hot trimming): A2	Boilermaker's tools			
Die Casting and Molding Dies		For aluminum and lead: H11 and H13 For brass: H21	A2 and A6 O1		Plastic molds; P2 to P4, and P20		
Structural Parts for Severe Service Conditions	Roller bearings for high temperature environment: T1 Lathe centers: M2 and T1	For aircraft components (landing gears, arrester hooks, rocket cases): H11	Lathe centers: D2, D3 Arbors: O1 Bushings: A4 Gages: D2	Pawls Clutch parts		Spindles, clutch parts (where high toughness is needed): L6	Spring steel (1.10/1.15%C)
Battering Tools for Hand and Power Tool Use				Pneumatic chisels for cold work: S5 For higher performance: S7			For intermittent use: W1 (.80%C)

their identification. The individual types of tool steels within each category are identified by suffix numbers following the letter symbols.

The subsequent detailed discussion of tool steels will be in agreement with these categories, showing for each type the percentages of the major alloying elements. However, these values are for identification only; tool steels of different producers may, in the mean analysis of the individual types, deviate from the listed percentages.

The Selection of Tool Steels for Particular Applications. While the advice of the specialized steel producer is often sought as a reliable source of information, in many cases the engineer is faced with the task of making his own tool steel selection. It must be realized that frequently the designation of the tool or of the process will not define the particular tool steel type which is best suited for the job. For that reason tool steel selection tables naming a single type for each listed applicational category cannot take into consideration such, often conflicting work factors, as ease of tool fabrication and maintenance (resharpening), productivity, product quality and tooling cost.

When data related to past experience with tool steels for identical or similar applications are not available, a tool steel selection procedure may be followed which is based on information supplied in this Handbook section and comprises these steps:

1. For identifying the AISI category which contains the sought type of steel the Quick Reference Table on pages 483 and 484 should be consulted.
2. Within the defined category
   (a) find from the listed applications of the most frequently used types of tool steels that particular type which corresponds to the job on hand; or
   (b) evaluate from the table of property ratings the best compromise between any conflicting properties (e.g., compromising on wear resistance for obtaining better toughness).

For those willing to refine even further the first choice or to improve on it in the case of not entirely satisfactory experience in one or more meaningful respects, the identifying analyses of the different types of tool steels within each general category may provide additional guidance. In this procedure the general discussion of the effects of different alloying elements on the properties of tools steels, in a previous section, will probably be found useful.

The following two examples illustrate the procedure for refining an original choice with the purpose of adopting a tool steel grade best suited to a particular set of conditions:

Example 1. Workpiece — Trimming Dies: For the manufacture of a type of trimming die the first choice was grade A2, because for the planned medium rate of production the lower material cost was considered an advantage.

A subsequent rise in the production rate indicated the use of a higher alloy tool steel, such as D2, whose increased abrasion resistance would permit longer runs between regrinds.

Then a still further increase in the abrasion resistant properties was sought, which led to the use of D7, the high carbon and high chromium content of which provided an excellent edge retainment, although at the cost of greatly reduced grindability. Finally, it became a matter of economic appraisal, whether the somewhat shorter tool regrind intervals (for D2), or the more expensive tool sharpening (for D7) constituted the lesser burden.

Example 2. Workpiece — Circular form cutter made of high speed tool steel for use on multiple-spindle automatic turning machines: The first choice from the Quick Reference Guide may be the classical tungsten base high speed tool steel T1, because of its good performance and ease of heat treatment, or its alternate in the molybdenum high speed tool steel category, the type M2.

In practice, neither of these grades provided a tool which could hold its edge and profile over the economical tool change time, because of the abrasive properties of the work material and the high cutting speed applied in the cycle. An overrating of the problem resulted in reaching for the top of the scale, making the tool from T15, a high alloy high speed tool steel (high vanadium and high cobalt).

While the performance of the tools made of T15 was excellent, the cost of this steel type was rather high, and the grinding of the tool, both for making it and in the regularly needed re-sharpening, proved to be very time consuming and expensive. Therefore an intermediate tool steel type was tried, the M3 which, while providing added abrasion resistance (due to increased carbon and vanadium content) was less expensive and much easier to grind than the T15.

High Speed Tool Steels. — Their primary application is for tools used for the working of metals at high cutting speeds. Cutting metal at high speed generates heat, the penetration of the cutting tool edge into the work material requires great hardness and strength, and the continued frictional contact of the tool with both the parent material and the detached chips can only be sustained by an abrasion resistant tool edge. Accordingly, the dominant properties of high speed steel are (a) resistance to the softening effect of elevated temperature, (b) great hardness penetrating to substantial depth from the surface, and (c) excellent abrasion resistance.

High speed tool steels are listed in the AISI specifications in two groups: Molybdenum types and Tungsten types, these designations expressing the dominant alloying element of the respective group.

Molybdenum Type High Speed Tool Steels. — In distinction to the traditional tungsten base high speed steels, the tool steels listed in this category are considered to have molybdenum as the principal alloying constituent, this element being also used in the designation of the group. Actually, in several types listed in this category other significant elements like tungsten and cobalt might be present in equal, or even greater amount. The available range of types comprises also high speed tool steels with higher than usual carbon and vanadium content; these alloying elements have been increased to obtain better abrasion resistance although such a change in composition may adversely affect the machinability and the grindability of the steel. The series in whose AISI identification numbers 4 is the first digit were developed for attaining exceptionally high hardness in heat treatment which, for these types, usually comprises triple tempering rather than the double tempering generally applied for high speed tool steels.

Properties and Applications of Frequently Used Molybdenum Types. — *AISI M1:* This was developed as a substitute for the classical T1 to save on the alloying element tungsten by replacing most of it with molybdenum. In most uses this steel is an acceptable substitute, although it requires greater care or more advanced equipment for its heat treatment than the tungsten alloyed type it replaces. Selected for cutting tools like drills, taps, milling cutters, reamers, lathe tools used for lighter cuts, and also for shearing dies.

AISI M2: Similar to M1, yet with substantial tungsten content replacing a part of the molybdenum. It is one of the general-purpose high speed tool steels, combining the economic advantages of the molybdenum type steels with greater ease of hardening, excellent wear resistance and improved toughness. It is a preferred steel type for the manufacture of general-purpose lathe tools; of most categories of multiple-edge cutting tools, like milling cutters, taps, dies, reamers and also for form tools in lathe operations.

AISI M3: A high speed tool steel with increased vanadium content for improved wear resistance, yet still below the level where vanadium would interfere with the

Table 4. Molybdenum High Speed Steels

Identifying Chemical Composition and Typical Heat Treatment Data

AISI Type	M1	M2	M3 Cl.1	M3 Cl.2	M4	M6	M7	M10	M30	M33	M34	M36	M41	M42	M43	M44	M46	M47
Identifying Chemical Elements in Per Cent																		
C	.80	.85; 1.00	1.05	1.20	1.30	.80	1.00	.85; 1.00	.80	.99	.99	.80	1.10	1.10	1.20	1.15	1.25	1.10
W	1.50	6.00	6.00	6.00	5.50	4.00	1.75	…	2.00	1.50	2.00	6.00	6.75	1.50	2.75	5.25	2.00	1.50
Mo	8.00	5.00	5.00	5.00	4.50	5.00	8.75	8.00	8.00	9.50	8.00	5.00	3.75	9.50	8.00	6.25	8.25	9.50
Cr	4.00	4.00	4.00	4.00	4.00	4.00	4.00	4.00	4.00	4.00	4.00	4.00	4.25	3.75	3.75	4.25	4.00	3.75
V	1.00	2.00	2.40	3.00	4.00	1.50	2.00	2.00	1.25	1.15	2.00	2.00	2.00	1.15	1.60	2.25	3.20	1.25
Co	…	…	…	…	…	12.00	…	…	5.00	8.00	8.00	8.00	5.00	8.00	8.25	12.00	8.25	5.00
Heat Treat Data																		
Hardening Temperature Range, °F	2150-2225	2175-2225	2200-2250	2200-2250	2200-2250	2150-2200	2150-2225	2150-2225	2200-2250	2200-2250	2200-2250	2225-2275	2175-2220	2175-2210	2175-2220	2190-2240	2175-2225	2150-2200
Tempering Temperature Range, °F	1000-1100	1000-1160	1000-1100	1000-1100	1000-1100	1000-1100	1000-1100	1000-1100	1000-1100	1000-1100	1000-1100	1000-1100	1000-1100	950-1100	950-1100	1000-1160	975-1100	975-1100
Approx. Tempered Hardness, R_c	65-60	65-60	66-61	66-61	66-61	66-61	66-61	65-60	65-60	65-60	65-60	65-60	70-65	70-65	70-65	70-62	69-67	70-65

Relative Ratings of Properties (A = greatest to E = least)

AISI Type	M1	M2	M3 Cl.1	M3 Cl.2	M4	M6	M7	M10	M30	M33	M34	M36	M41	M42	M43	M44	M46	M47
Characteristics in Heat Treatment																		
Safety in Hardening	D	D	D	D	D	D	D	D	D	D	D	D	D	D	D	D	D	D
Depth of Hardening	A	A	A	A	A	A	A	A	A	A	A	A	A	A	A	A	A	A
Resistance to Decarburization	C	B	B	B	B	C	C	C	C	C	C	C	C	C	C	C	C	C
Stability of Shape in Heat Treatment — Quenching Medium: Air or Salt	C	C	C	C	B	C	C	C	C	C	C	C	C	C	C	C	C	C
Oil	C	B	B	B	B	C	C	C	C	C	C	C	C	C	C	C	C	C
Service Properties																		
Machinability	D	D	D	D/E	D	D	D	D	D	D	D	D	D	D	D	D	D	D
Hot Hardness	D	B	B	B	B	B	B	B	A	B	A	B	A	A	A	A	A	A
Wear Resistance	B	B	B	B	A	A	B	B	B	B	B	B	B	B	B	B	B	B
Toughness	E	E	E	E	E	E	E	E	E	E	E	E	E	E	E	E	E	E

ease of grinding. Preferred for cutting tools requiring the improved wear resistance, like broaches, form tools, milling cutters, chasers, reamers, etc.

AISI M7: The chemical composition of this type is similar to that of M1, except for the higher carbon and vanadium content which raises the cutting efficiency without materially reducing the toughness. Because of its sensitivity to decarburization heat treatment in a salt bath or a controlled atmosphere is advisable. Used for blanking and trimming dies, shear blades, lathe tools and thread rolling dies.

AISI M10: While the relatively high vanadium content assures excellent wear and cutting properties, the only slightly increased carbon does not cause brittleness to an extent which is harmful in many applications. Form cutters and single point lathe tools, broaches, planer tools, punches, blanking dies, shear blades, etc., are examples of typical uses.

AISI M42: In applications where high hardness both at regular and at elevated temperatures is needed, this type of high speed steel with high cobalt content can provide excellent service. Typical applications are tool bits, form tools, shaving tools, fly cutters, roll turning tools, thread rolling dies, etc. Important uses are found for M42, and also for other types of the "M40" group in the working of "difficult-to-machine" type alloys.

Tungsten Type High Speed Tool Steels. — For several decades following their introduction the tungsten base high speed steels were the only types available for cutting operations involving the generation of substantial heat, and are still preferred by users who do not have that kind of advanced heat treating equipment which the efficient hardening of the molybdenum type high speed tool steels requires. Most types of tungsten high speed steel display excellent resistance to decarburization and can be brought to good hardness by means of simple heat treating equipment. However, even in the case of the tungsten type high speed steels, heat treatment by using modern methods and furnaces can appreciably improve the metallurgical qualities of the hardened material and the performance of the cutting tools made of these steels.

Properties and Applications of Frequently Used Tungsten Types. — *AISI T1:* Also mentioned as the 18-4-1 type with reference to the nominal percentage of its principal alloying elements (W-Cr-V), it is considered as the classical type of high speed tool steel. The chemical composition of T1 was developed around the turn of this century, and has since changed very little. T1 is still considered as perhaps the best general-purpose high speed tool steel because of the comparative ease of its machining and heat treatment. It combines a high degree of cutting ability with relative toughness. T1 steel is used for all types of multiple-edge cutting tools like drills, reamers, milling cutters, threading taps and dies, light- and medium-duty lathe tools, also for punches, dies, machine knives, as well as structural parts which are subjected to elevated temperatures, like lathe centers, certain types of antifriction bearings, etc.

AISI T2: Similar to T1 except for somewhat higher carbon content and twice the vanadium contained in the former grade. Its handling ease, both in machining and heat treating, is comparable to that of T1, although it should be held at the quenching temperature slightly longer, particularly when the heating is carried out in a controlled atmosphere furnace. The applications are similar to that of T1, however, because of its increased wear resistance T2 is preferred for tools required for finer cuts, and where the form or size retention of the tool is particularly important, such as for form and finishing tools.

AISI T5: The essential characteristic of this type of high speed steel, its superior red hardness, stems from its substantial cobalt content which, combined with the

Table 5. Tungsten High Speed Tool Steels

			T1	T2	T4	T5	T6	T8	T15
Identifying Chemical Data	Identifying Chemical Elements in Per Cent	AISI Type							
		C	.75	.80	.75	.80	.80	.75	1.50
		W	18.00	18.00	18.00	18.00	20.00	14.00	12.00
		Cr	4.00	4.00	4.00	4.00	4.50	4.00	4.00
		V	1.00	2.00	1.00	2.00	1.50	2.00	5.00
		Co	...	...	5.00	...	...	5.00	5.00
Heat Treat Data		Hardening Temperature Range, °F	2300-2375	2300-2375	2300-2375	2325-2375	2325-2375	2300-2375	2200-2300
		Tempering Temperature Range, °F	1000-1100	1000-1100	1000-1100	1000-1100	1000-1100	1000-1100	1000-1200
		Approx. Tempered Hardness, R_c	65-60	66-61	66-62	65-60	65-60	65-60	68-63
Relative Ratings of Properties (A = greatest to E = least)									
Characteristics in Heat Treatment		Safety in Hardening	C	C	D	D	D	D	D
		Depth of Hardening	A	A	A	A	A	A	A
		Resistance to Decarburization	A	A	B	C	C	B	B
		Stability of Shape in Heat Treatment — Quenching Medium: Air or Salt	C	C	C	C	C	C	C
		Stability of Shape in Heat Treatment — Quenching Medium: Oil	D	D	D	D	D	D	D
Service Properties		Machinability	D	D	D	D	D/E	D	D/E
		Hot Hardness	B	B	A	A	A	A	A
		Wear Resistance	B	B	B	B	B	B	A
		Toughness	E	E	E	E	E	E	E

relatively high amount of vanadium, also produces excellent wear resistance. In heat treatment the tendency for decarburization must be considered and heating in controlled, slightly reducing atmosphere is recommended. This type of high speed tool steel is mainly used for single point tools and inserts, it is well adapted for working at high speeds and feeds, for cutting hard materials and those which produce discontinuous chips, also for non-ferrous metals and, in general, for all kinds of tools needed for hogging (removing great bulks of material).

AISI T15: The performance qualities of this high alloy tool steel surpass most of those found in other grades of high speed tool steels. The high vanadium content, supported by uncommonly high carbon assures superior cutting ability and wear resistance. The addition of high cobalt increases the "hot hardness," and therefore tools made of T15 can sustain cutting speed in excess of those commonly applicable to tools made of steel. The machining and heat treatment of T15 does not cause extraordinary problems, although for best results, heating to high temperature is often applied in its heat treatment, and double, or even triple tempering is recommended. On the other hand, T15 is rather difficult to grind because of the presence of large amounts of very hard metallic carbides, therefore it is considered to have a very low "grindability" index. The main uses are in the field of high speed cutting and the working of hard metallic materials, T15 being often considered to represent in its application a transition from the regular high speed tool steels to cemented carbides. Lathe tool bits, form cutters, solid and inserted blade milling cutters are examples of uses for cutting tools, however the application of this steel type is not confined to this primary area of utilization; excellent results may also be attained for such tools as cold work dies, punches, blanking and forming dies, etc. The low toughness rating of the T15 steel excludes its application for operations which involve shock or sudden variations in load.

Hot Work Tool Steels. — For tools which in their regular service are in contact with hot metals over a shorter or longer period of time, with or without cooling being applied, a family of special tool steels has been developed and these are comprised under the general designation of hot work steels. The essential property of these steels is the capability of sustaining elevated temperature without seriously affecting the usefulness of the tools made of them. Depending on the purpose of the tools for which they were developed, the particular types of hot work tool steels have different dominant properties and are assigned to either of three groups, based primarily on their principal alloying elements.

Hot Work Tool Steels — Chromium Types. — As referred to in the group designation, the chromium content is considered the characteristic element of these tool steels. Their predominant properties are high hardenability, excellent toughness and great ductility, even at the cost of wear resistance. Some members of this family are made with the addition of tungsten, and in one type, cobalt as well. These alloying elements improve the resistance to the softening effect of elevated temperatures, but reduce ductility.

Properties and Applications of Frequently Used Chromium Types. — *AISI H11:* This hot work tool steel of the chromium-molybdenum-vanadium type has excellent ductility, can be machined easily and retains its strength at temperatures up to 1000 degrees F. These properties, combined with relatively good abrasion and shock resistance, account for the varied fields of application of H11, which include the following typical uses: (a) structural applications where high strength is needed at elevated operating temperatures, as for jet engine components and (b) hot work tools, particularly of the kind whose service involves shocks and drastic cooling of the tool, such as in extrusion tools, pierce and draw punches, bolt header dies, etc.

Table 6. Hot Work Tool Steels

Identifying Chemical Composition and Typical Heat Treatment Data

Relative Ratings of Properties (A = greatest to D = least)

		Chromium Types						Tungsten Types						Molybdenum Types		
AISI Group Type		H10	H11	H12	H13	H14	H19	H21	H22	H23	H24	H25	H26	H41	H42	H43
Identifying Elements in %	C	.40	.35	.35	.35	.40	.40	.35	.35	.35	.45	.25	.50	.65	.60	.55
	W	...	...	1.50	...	5.00	4.25	9.00	11.00	12.00	15.00	15.00	18.00	1.50	6.00	...
	Mo	2.50	1.50	1.50	1.50	...	...	...	...	...	...	...	...	8.00	5.00	8.00
	Cr	3.25	5.00	5.00	5.00	5.00	4.25	3.50	2.00	12.00	3.00	4.00	4.00	4.00	4.00	4.00
	V	.40	.40	.40	1.00	...	2.00	...	...	...	...	...	1.00	1.00	2.00	2.00
	Co	...	...	...	...	...	4.25	...	...	...	...	...	...	...	...	...
Heat Treat Data	Hardening Temperature Range °F	1850-1900	1825-1875	1825-1875	1825-1900	1850-1950	2000-2200	2000-2200	2000-2200	2000-2300	2000-2250	2100-2300	2150-2300	2000-2175	2050-2225	2000-2175
	Tempering Temperature Range °F	1000-1200	1000-1200	1000-1200	1000-1200	1100-1200	1000-1300	1100-1250	1100-1250	1200-1500	1050-1200	1050-1250	1050-1250	1050-1200	1050-1200	1050-1200
	Approx. Tempered Hardness, R_c	56-39	54-38	55-38	53-38	47-40	59-40	54-36	52-39	47-30	55-45	44-35	58-43	60-50	60-50	58-45
Characteristics in Heat Treatment	Safety in Hardening	A	A	A	A	A	B	B	B	B	B	B	B	C	C	C
	Depth of Hardening	A	A	A	A	A	A	A	A	A	A	A	A	A	A	A
	Resistance to Decarburization	B	B	B	B	B	A	B	B	B	B	B	B	C	B	C
	Stability of shape in heat treatment — Quenching medium, Air or Salt	B	B	B	B	C	C	C	C	...	C	C	C	C	C	C
	Stability of shape in heat treatment — Quenching medium, Oil	...	...	...	...	...	...	...	...	...	...	...	...	...	...	...
Service Properties	Machinability	C/D	C/D	C/D	C/D	D	D	D	D	D	D	D	D	D	D	D
	Hot Hardness	C	C	C	C	C	C	C	C	B	C	C	B	D	D	D
	Wear Resistance	D	D	D	D	D	C/D	C/D	C/D	C/D	C/D	C	C	B	B	B
	Toughness	C	B	B	B	C	C	C	C	D	D	D	D	C	C	D

AISI H12: The properties of this type are comparable to those of H11, with increased abrasion resistance and hot hardness, resulting from the addition of tungsten, yet in an amount which does not affect the good toughness of this steel type. The applications, based on these properties, are hot work tools which often have to withstand severe impact, such as various punches, bolt header dies, trimmer dies, hot shear blades, also H12 is used to make aluminum extrusion dies and die-casting dies.

AISI H13: This type differs from the preceding ones particularly in properties related to the addition of 1 per cent vanadium, which contributes to increased hot hardness, abrasion resistance and reduced sensitivity for heat checking. Such properties are needed in die casting, particularly of aluminum, where the dies are subjected to drastic cooling at high operating temperatures. Besides die-casting dies, H13 is also widely used for extrusion dies, trimmer dies, hot gripper and header dies, hot shear blades, etc.

AISI H19: This high-alloyed hot work tool steel, containing chromium, tungsten, cobalt and vanadium, has excellent resistance to abrasion and shocks at elevated temperature. It is particularly well adapted to severe hot work uses where the tool, in order to retain its size and shape, must withstand wear and the washing out effect of hot work material. Typical applications include brass extrusion dies and dummy blocks, inserts for forging and valve extrusion dies, press forging dies, hot punches, etc.

Hot Work Tool Steels — Tungsten Types.

Substantial amounts of tungsten, yet very low carbon content characterizes the hot work tool steels of this group. These tool steels have been developed for applications where the tool is in contact with the hot work material over extended periods of time, therefore the resistance of the steel to the softening effect of elevated temperatures is of prime importance, even to the extent of accepting a lower degree of toughness.

Properties and Applications of Frequently used Tungsten Types.

AISI H21: This is a medium tungsten alloyed hot work tool steel with substantially increased abrasion resistance over the chromium alloyed types, yet possessing a degree of toughness which represents a transition between the chromium and the higher alloyed tungsten steel types. The principal applications are for tools subjected to continued abrasion, yet to only a limited amount of shock loads, like tools for the extrusion of brass, both dies and dummy blocks, piercers for forging machines, inserts for forging tools, hot nut tools, etc. Another typical application is dies for the hot extrusion of automobile valves.

AISI H24: The comparatively high tungsten content (about 14 per cent) results in good hardness, great compression strength and excellent abrasion resistance, but makes it sensitive to shock loads. Taking these properties into account, the principal applications include extrusion dies for brass in long run operations, hot forming and gripper dies with shallow impressions, punches which are subjected to great wear yet only to moderate shocks, hot shear blades, etc.

AISI H26: This high alloyed tungsten type hot work steel resembles in its composition the tungsten type high speed steel AISI T1, except for the somewhat lower carbon content for improved toughness. The high amount of tungsten provides the maximum resistance to the softening effect of elevated temperature and assures excellent wear-resistant properties, including the withstanding of the washing-out effect of certain processes. However, this type is less resistant to thermal shocks than the chromium hot work steels. Typical applications comprise extrusion dies for long production runs, extrusion mandrels operated without cooling, hot piercing punches, hot forging dies and inserts. It is also used as special structural steel for springs operating at elevated temperatures.

Hot Work Tool Steels — Molybdenum Types. — These are closely related to certain types of molybdenum high speed steels and possess excellent resistance to the softening effect of elevated temperature but their ductility is rather low. These steel types are generally available on special orders only.

Properties and Applications of Frequently Used Molybdenum Types. — *AISI H43:* The principal constituents of this hot work steel, chromium, molybdenum and vanadium, provide excellent abrasion and wear-resistant properties at elevated temperatures. H43 has a good resistance against the development of heat checks and a toughness adequate for many different applications. These include tools and operations which tend to cause surface wear in high temperature work, like hot headers, punch and die inserts, hot heading and hot nut dies, as well as different kinds of punches operating at high temperature in service involving considerable wear.

Cold Work Tool Steels. — Tool steels of this category are primarily intended for die work, although their use is by no means restricted to that general field. Cold work tool steels are extensively used for tools whose regular service does not involve elevated temperatures. They are available in chemical compositions adjusted to the varying requirements of a wide range of different applications. According to their predominant properties, characterized either by the chemical composition or by the quenching medium in heat treatment, the cold work tool steels are assigned to three different groups.

Cold Work Tool Steels — High Carbon, High Chromium Types. — The chemical composition of tool steels of this family is characterized by the very high chromium content, in the order of 12 to 13 per cent, and the uncommonly high carbon content in the range of about 1.50 to 2.30 per cent. Additional alloying elements which are present in different amounts in some of the steel types of this group are vanadium, molybdenum and cobalt, each of which contributes desirable properties. The predominant properties of the whole group are: (a) excellent dimensional stability in heat treatment where, with one exception, air quench is used; (b) great wear resistance particularly in the types with the highest carbon content; and (c) rather good machinability.

Properties and Applications of Frequently Used High Carbon-High Chromium Types. — *AISI D2:* An air hardening die steel with high carbon, high chromium content with several desirable tool steel properties, such as abrasion resistance, high hardness as well as non-deforming characteristics. The carbon content of this type, although relatively high, is not particularly detrimental to its machining. The ease of working can further be improved by selecting the same basic type with the addition of sulfur. Several steel producers supply the sulfurized version of D2, in which the uniformly distributed sulfide particles substantially improve the machinability and the resulting surface finish. The applications comprise primarily cold working press tools for shearing (blanking and stamping dies, punches, shear blades), for forming (bending, seaming), also for thread rolling dies, solid gages and wear-resistant structural parts. Dies for hot trimming of forgings are also made of D2 when heat treated to a lower hardness for the purpose of increasing toughness.

AISI D3: The high carbon content of this high chromium tool steel type results in excellent resistance to wear and abrasion and also provides superior compressive strength as long as the pressure is gradually applied, without exerting sudden shocks. In hardening, an oil quench is used, without affecting the excellent non-deforming properties of this type. Its deep hardening properties make it particularly suitable for tools which require repeated regrinding during their service life, such as different types of dies and punches. The more important applications

comprise blanking, stamping and trimming dies and punches for long production runs; forming, bending and drawing tools; also structural elements like plug and ring gages, lathe centers, etc., in applications where high wear resistance is important.

Cold Work Tool Steels — Oil Hardening Types. — With a relatively low percentage of alloying elements, yet with a substantial amount of manganese, these less expensive types of tool steels attain good depth of hardness in an oil quench, although at the cost of reduced resistance to deformation. Their good machinability supports general purpose applications, yet because of relatively low wear resistance they are mostly selected for comparatively short-run work.

Properties and Applications of Frequently Used Oil Hardening Types. — *AISI O1:* A low alloy tool steel which is hardened in oil and exhibits only a low tendency to shrinking or warping. It is used for cutting tools, the operation of which does not generate high heat, such as taps and threading dies, reamers, broaches, and for press tools like blanking, trimming, and forming dies in short- or medium-run operations.

AISI O2: Manganese is the dominant alloying element in this type of oil hardening tool steel which has good non-deforming properties, can be machined easily and performs satisfactorily in low volume production. The low hardening temperature results in good safety in hardening, both with regard to form stability and freedom from cracking. The combination of handling ease including free-machining properties, with good wear resistance, makes this type of tool steel adaptable to a wide range of common applications such as cutting tools for low and medium speed operations; forming tools including thread rolling dies; structural parts such as bushings, fixed gages, and also for plastic molding dies.

AISI O6: This oil hardening type of tool steel belongs to a group often designated as graphitic because of the presence of small particles of graphitic carbon which are uniformly dispersed throughout the steel. Usually about one-third of the total carbon is present as free graphite in nodular form, which contributes to the uncommon ease of machining. In the service of parts made of this type of steel the free graphite acts like a lubricant, reducing wear and galling. The ease of hardening is also excellent, requiring only comparatively low quenching temperature. Deep hardness penetration is produced and the oil quench causes very little dimensional change. The principal applications of the O6 tool steel are in the field of structural parts, like arbors, bushings, bodies for inserted tool cutters and shanks for cutting tools, jigs and machine parts and fixed gages like plugs, rings, snap gages, etc. It is also used for blanking, forming, and trimming dies and punches, in applications where the stability of the tool material is more important than high wear resistance.

Cold Work Tool Steels — Medium Alloy Air Hardening Types. — The desirable non-deforming properties of the high chromium types are approached by the members of this family, with substantially lower alloy content which, however, is sufficient to permit hardening by air quenching. The machinability is good, and the comparatively low wear resistance is balanced by relatively high toughness, a property which in certain applications may be considered of prime importance.

Properties and Applications of Frequently Used Medium Alloy Air Hardening Types. — *AISI A2:* The lower chromium content, about 5 per cent, makes this air hardening tool steel less expensive than the high chromium types, without affecting its nondeforming properties. The somewhat reduced wear resistance is balanced by greater toughness, making this type suitable for press work where the process calls for tough tool materials. The machinability is improved by the addition of about 0.12 percent sulfur, offered as a variety of the basic composition by several steel producers. The prime uses of this tool steel type are punches for

Table 7. Cold Work Tool Steels

Identifying Chemical Composition and Typical Heat Treatment Data

Group Types are divided into: **High Carbon High Chromium Types** (D2, D3, D4, D5, D7); **Medium Alloy Air Hardening Types** (A2, A3, A4, A6, A7, A8, A9, A10); **Oil Hardening Types** (O1, O2, O6, O7).

AISI / Property	D2	D3	D4	D5	D7	A2	A3	A4	A6	A7	A8	A9	A10	O1	O2	O6	O7
Identifying Elements in %																	
C	1.50	2.25	2.25	1.50	2.35	1.00	1.25	1.00	.70	2.25	.55	.50	1.35	.90	.90	1.45	1.20
Mn						1.00		2.00	2.00				1.80	1.00	1.60	1.00	
Si												1.00	1.25			1.00	
W										1.00	1.25			.50			1.75
Mo	1.00		1.00	1.00	1.00	1.00	1.00	1.00	1.00	1.00	1.25	1.40	1.50			.25	
Cr	12.00	12.00	12.00	12.00	12.00	5.00	5.00	1.00	1.00	5.25	5.00	5.00		.50			.75
V					4.00		1.00			4.75		1.00					
Co				3.00													
Ni												1.50	1.80				
Heat Treatment Data																	
Hardening Temperature Range °F	1800-1875	1700-1800	1775-1850	1800-1875	1850-1950	1700-1800	1750-1850	1500-1600	1525-1600	1750-1800	1800-1850	1800-1875	1450-1500	1450-1500	1400-1475	1450-1500	1550-1525
Quenching Medium	Air	Oil	Air	Air	Air	Air	Air	Air	Air	Air	Air	Air	Air	Oil	Oil	Oil	Oil
Tempering Temperature Range °F	400-1000	400-1000	400-1000	400-1000	300-1000	350-1000	350-1000	350-800	300-800	300-1000	350-1100	950-1150	350-800	350-500	350-500	350-600	350-550
Approx. Tempered Hardness, R_c	61-54	61-54	61-54	61-54	65-58	62-57	65-57	62-54	60-54	67-57	60-50	56-35	62-55	62-57	62-57	63-58	64-58
Service Properties — Characteristics in Ht. T. *(Relative Ratings of Properties, A = greatest to E = least)*																	
Safety in Hardening	A	C	A	A	A	A	A	A	A	A	A	A	A	B	B	B	B
Depth of Hardening	A	A	A	A	A	A	A	A	A	A	A	A	A	B	B	B	B
Resistance to Decarburization	B	B	B	B	B	B	B	A/B	A/B	B	B	B	A/B	A	A	A	A
Stability of Shape in Heat Treatment	A	B	A	A	A	A	A	A	A	A	A	A	A	B	B	B	B
Machinability	C	E	E	E	E	D	D	D/E	D/E	E	D	D	C/D	C	C	C	C
Hot Hardness	C	C	C	C	C	C	C	D	D	C	C	C	D	E	E	E	E
Wear Resistance	B	B/C	B	A	A	B	B	C/D	C/D	A	C/D	C/D	C	C/D	C/D	C/D	C
Toughness	E	E	E	E	E	D	D	C	C	E	C	C	B	D	D	D	C

blanking and forming, cold and hot trimming dies (the latter heat treated to a lower hardness), thread rolling dies and also plastic molding dies.

AISI A6: The composition of this type of tool steel makes it adaptable to air hardening from a relatively low temperature, which is comparable to that of oil hardening types, yet offering improved stability in the heat treating. Its reduced tendency to heat treat distortions makes this tool steel type well adapted for die work, forming tools, gages, etc., which do not require the highest degree of wear-resistance.

Shock Resisting, Mold and Special-purpose Tool Steels. — There are fields of tool application in which specific properties of the tool steels have dominant significance, determining to a great extent the performance and the service life of the tools which are made of these materials. To meet these requirements special types of tool steels have been developed. These individual types grew into families with members which, while similar in their major characteristics, provide related properties to different degrees. Originally developed for a specific use, the resulting particular properties of some of these tool steels made them desirable for other uses as well. In the tool steel classification system they are shown in three groups.

Shock Resisting Tool Steels. — These are made with low carbon content for increased toughness, even at the expense of wear resistance which is generally low. Each member of this group also contains alloying elements, different in composition and amount, selected to provide properties particularly adjusted to specific applications. Such varying properties are the degree of toughness (generally high in all members), hot hardness, abrasion resistance and machinability.

Properties and Applications of Frequently Used Shock Resisting Types. — *AISI S1:* This chromium-tungsten alloyed tool steel combines, in its hardened state, great toughness with high hardness and strength. Although it has a low carbon content for reasons of good toughness, the carbon forming alloys contribute to deep hardenability and abrasion resistance. When high wear resistance is also required, this property can be improved by carburizing the surface of the tool while still retaining its shock resistant characteristics. Primary uses are for battering tools, including hand and pneumatic chisels. The chemical composition, particularly the silicon and tungsten content, provide good hot hardness too, up to operating temperatures of about 1050 °F, making this tool steel type also adaptable for such hot work tool applications involving shock loads, as headers, piercers, forming tools, drop forge die inserts and heavy shear blades.

AISI S2: This steel type serves primarily for hand chisels and pneumatic tools, although it also has limited applications for hot work. While its wear resistance properties are only moderate, S2 is sometimes used for forming and thread rolling applications, when the resistance to rupturing is more important than extended service life. For hot work applications this steel requires heat treatment in a neutral atmosphere in order to safely avoid either carburization or decarburization of the surface. Such conditions make this tool steel type particularly susceptible to failure in hot work uses.

AISI S5: This is essentially a silicon-manganese type tool steel with small additions of chromium, molybdenum and vanadium for the purpose of improved deep hardening and refinement of the grain structure. The most important properties of this steel are its high elastic limit and good ductility resulting in excellent shock-resisting characteristics, when used at atmospheric temperatures. Its recommended quenching medium is oil, although a water quench may also be applied as long as the design of the tools avoids sharp corners or drastic sectional changes. Typical applications include pneumatic tools in severe service, like chipping chisels,

Table 8. Shock Resisting, Mold, and Special Purpose Tool Steels

Identifying Chemical Compositions and Typical Heat Treatment Data

Column groups: **Shock Resisting Tool Steels** (S1, S2, S5, S7); **Mold Steels** (P2, P3, P4, P5, P6, P20, P21[a]); **Special Purpose Tool Steels** (L2[b], L3[b], L6, F1, F2)

Category / Types	S1	S2	S5	S7	P2	P3	P4	P5	P6	P20	P21[a]	L2[b]	L3[b]	L6	F1	F2
Identifying Elements in Per Cent																
C	.50	.50	.55	.50	.07	.10	.07	.10	.10	.35	.20	.50/1.10	1.00	.70	1.00	1.25
Mn	...	...	.80	...	...	...	...	...	...	...	...	...	...	...	...	...
Si	...	1.00	2.00	...	...	...	...	...	...	...	...	...	...	...	...	...
W	2.50	...	...	...	...	...	...	...	...	...	...	...	...	...	1.25	3.50
Mo	...	.50	.40	1.40	.20	...	.75	...	...	.40	...	...	...	.25	...	...
Cr	1.50	...	...	3.25	2.00	.60	5.00	2.25	1.50	1.25	...	1.00	1.50	.75	...	...
V	...	...	...	...	...	...	...	...	...	...	...	.20	.20	...	...	...
Ni	...	...	...	...	...	1.25	...	...	3.50	...	4.00	...	...	1.50	...	...
Heat Treat Data																
Hardening Temperature	1650-1750	1550-1650	1600-1700	1700-1750	1525-1550*	1475-1525*	1775-1825*	1550-1600*	1450-1500*	1500-1600*	So't'n treat.	1550-1700	1500-1600	1450-1550	1450-1600	1450-1600
Tempering Temp. Range, °F	400-1200	350-800	350-800	400-1150	350-500	350-500	350-900	350-500	350-450	900-1100	Aged	350-1000	350-600	350-1000	350-500	350-500
Approx. Tempered Hardness Rc	58-40	60-50	60-50	57-45	64-58†	64-58†	64-58†	64-58†	61-58†	37-28†	40-30	63-45	63-56	62-45	64-60	65-62

Relative Ratings of Properties (A = greatest to E = least)

Category / Types	S1	S2	S5	S7	P2	P3	P4	P5	P6	P20	P21[a]	L2[b]	L3[b]	L6	F1	F2
Characteristics in Heat Treatment																
Safety in Hardening	C	B	B	B/C	C	C	C		B	C	A	C	D	C	E	E
Depth of Hardening	B	B	B	A	B‡	B‡	B	B‡	B‡	A	A	A	B	B	E	C
Resist. to Decarb.	B	C	C	B	A	A	A	A	A	A	A	B	A	A	A	A
Quench. Med. — Air				A			B				A					
Quench. Med. — Oil	D		D		C	C		C	C	B		D	D	C		
Quench. Med. — Water[e]		E										D/E	D/E		D/E	B
Stability of Shape in Heat Treatment		E														
Service Properties																
Machinability	D	C/D	C/D	C	D	D	D	D	D	C/D	D	C	C	D	C	C
Hot Hardness	E	E	E	C	E	E	D	E	E	E	D	E	E	E	E	E
Wear Resistance	D/E	D/E	D/E	D/E	D	D	C	D	D	D/E	D/E	D/E	D	D/E	D	B/C
Toughness	B	A	A	B	C	C	B	C	C	C	B	C	D	B/C	E	E

*After carburizing   †Carburized case   ‡Solution treated in hardening.

ᵃContains also about 1.20 per cent Al. Solution treated in hardening.

ᵇQuenched in oil.   ᵉSometimes brine is used.

also shear blades, heavy duty punches, bending rolls, etc. Occasionally this steel is also used for structural applications, like shanks for carbide tools and machine parts subject to shocks.

Mold Steels. — These differ from all other types of tool steels by their very low carbon content, generally requiring carburizing for obtaining a hard operating surface. A special property of most steel types in this group is the adaptability to shaping by impression (hobbing) instead of by conventional machining. They also have high resistance to decarburization in heat treatment and dimensional stability, characteristics which obviate the need for grinding following heat treatment. The molding dies for plastic materials require an excellent surface finish, even to the degree of high luster; the generally high chromium content of these types of tool steels greatly aids in meeting this requirement.

Properties and Applications of Frequently Used Mold Steel Types. — *AISI P2 and P4:* Essentially, both types of tool steels were developed for the same special purpose, that is, the making of plastic molding dies. The application conditions of plastic molding dies require high core strength, good wear resistance at elevated temperature and excellent surface finish. Both types are carburizing steels which possess good dimensional stability. Because hobbing, that is, sinking the cavity by pressing into the tool material a punch representing the inverse replica of the cavity, is the process by which many of the plastic molding die cavities are produced, the "hobbability" of the tool steels used for this purpose is an important requirement. The different chemistry of these two types of mold steels is responsible for the high core hardness of the P4, which makes this latter type better suited for applications requiring high strength at elevated temperature.

AISI P6: This nickel-chromium type plastic mold steel has exceptional core strength and develops a deep carburized case. Due to the high nickel-chromium content the cavities of dies made of this steel type are produced by machining rather than by hobbing. An outstanding characteristic of this steel type is the high luster which is produced by the polishing of the hard case surface.

AISI P20: This is a general type mold steel which is adaptable to both through hardening and carburized case hardening. In through hardening an oil quench is used and a relatively lower, yet deeply penetrating hardness is obtained, such as is needed for zinc die-casting dies and injection plastic molds. Carburizing produces, after the direct quenching and tempering, a very hard case and comparatively high core hardness. When thus heat treated this steel is particularly well adapted for making compression, transfer and plunger type plastic molds.

Special-purpose Tool Steels. — These consist of several low alloy types of tool steels which were developed to provide transitional types between the more commonly used basic types of tool steels, and thereby contribute to the balancing of certain conflicting properties such as wear resistance and toughness; to offer intermediate depth of hardening; and to be less expensive than the higher alloyed types of tool steels.

Properties and Applications of Frequently Used Special-purpose Types. — *AISI D6:* This is a low-alloy type special-purpose tool steel. The comparatively safe hardening and the fair non-deforming properties, combined with the service advantage of good toughness in comparison to most other oil hardening types, explains the acceptance of this steel with a rather special chemical composition. The uses of L6 are limited to tools whose toughness requirements prevail over abrasion-resistant properties, such as forming rolls and forming and trimmer dies in applications where a combination of moderate shock and wear-resistant properties

are sought. The areas of use also include structural parts, like clutch members, pawls, knuckle pins, etc., which must withstand shock loads and still display good wear properties.

AISI F2: This carbon-tungsten type is one of the most abrasion resistant of all water hardening tool steels. However, it is sensitive to thermal changes, such as are involved in heat treatment and is also susceptible to distortions. Consequently, its use is limited to tools of simple shape in order to avoid cracking in hardening. The shallow hardening characteristics of F2, which result in a tough core, are desirable properties for certain tool types which, at the same time, require the excellent wear-resistant properties of this tool steel type.

Water Hardening Tool Steels. — Steel types in this category are made without, or with only a minimum amount of alloying elements and need in their heat treatment the harsh quenching action of water or brine, hence the general designation of the category.

Water hardening steels are usually available with different percentages of carbon, to provide properties required for different applications; the classification system lists a carbon range of 0.60 to 1.40 per cent. In practice, however, the steel mills produce these steels in a few varieties of differing carbon content, often giving proprietary designations to each particular group. Typical carbon content limits of frequently used water hardening tool steels are 0.70–0.90, 0.90–1.10, 1.05–1.20 and 1.20–1.30 per cent. The appropriate group should be chosen according to the intended use, as indicated in the steel selection guide for this category, keeping in mind that while higher carbon content results in deeper hardness penetration, it also reduces toughness.

The general system distinguishes the following four grades: (1) special, (2) extra, (3) standard and (4) commercial, listed in the order of decreasing quality. The differences between these grades, which are not offered by all steel mills, are defined in principle only. The distinguishing characteristics are purity and consistency, resulting from different degrees of process refinement and inspection steps applied in making the steel. Higher qualities are selected for assuring dependable uniformity and performance of the tools made of the steel.

Since the groups with higher carbon content are more sensitive to heat treatment defects and are generally used for the more demanding applications, the better grades are usually chosen for the high carbon types and the lower grades for applications where steels with lower carbon content only are needed.

Water hardening tool steels, while being the least expensive, have several drawbacks which, however, are quite acceptable in many types of applications. Such limiting properties are the tendency to deformation in heat treatment due to harsh effects of the applied quenching medium, the sensitivity to heat during the use of the tools made of these steels, the only fair degree of toughness and the shallow penetration of hardness. However, this last mentioned property may prove a desirable characteristic in certain applications, such as, e.g., cold heading dies, because the relatively shallow hard case is supported by tough, although softer core.

The AISI designation for water hardening tool steels is W, followed by a numeral indicating the type, primarily defined by the steel's chemical composition, as shown in the following table.

Recommended Applications of Water Hardening Type W-1 (Plain carbon) Tool Steels. —

Group I (C-0.70 to 0.90%): Relatively tough and therefore preferred for tools which are subjected to shocks or abusive treatment. For such applications as: hand tools — chisels, screwdriver blades, cold punches, nail sets, etc., and fixture elements — vise jaws, anvil faces, chuck jaws, etc.

Table 9. Water Hardening Tool Steels — Identifying Chemical Composition and Heat Treatment Data

	AISI Types		W1	W2	W5
Identifying Elements in Per Cent	C		0.60 to 1.40	0.60 to 1.40	1.10
			Varying carbon content may be available		
	V		...	0.25	...
	Cr		These elements are adjusted		0.50
	Mn		to satisfy the hardening requirements		
	Si				
Heat Treatment Data	Hardening Temperature Ranges °F	0.60 — 0.80%	1450 to 1500		
		0.85 — 1.05 %	1425 to 1550		
	Varying with Carbon Content	1.10 — 1.40%	1400 to 1525		
	Quenching Medium		Brine or Water		
	Tempering Temperature Range, °F		350 to 650		
	Approx. Tempered Hardness, R_c		64 to 50		

Relative Ratings of Properties (A = greatest to E = least)							
Characteristics in Heat Treatment				Service Properties			
Safety in Hardening	Depth of Hardening	Resistance to Decarburization	Stability of Shape in Heat Treatment	Machinability	Hot Hardness	Wear Resistance	Toughness
D	C	A	E	A	E	D/E	C/D

Group II (C-0.90 to 1.10%): Combines greater hardness with fair toughness, resulting in improved cutting capacity and moderate ability to sustain shock loads. For such applications as: hand tools — knives, center punches, pneumatic chisels; cutting tools — reamers, hand taps and threading dies, wood augers; die parts — drawing and heading dies, shear knives, cutting and forming dies; and fixture elements — drill bushings, lathe centers, collets, fixed gages.

Group III (C-1.05 to 1.20%): The higher carbon content increases the depth of hardness penetrations, yet reduces toughness, thus the resistance to shock loads. Preferred for applications where wear resistance and cutting ability are the prime considerations. For such applications as: hand tools — woodworking chisels, paper knives; cutting tools (for low speed applications) — milling cutters, reamers;

planer tools, thread chasers, center drills; and die parts — cold blanking, coining, bending dies.

Group IV (C-1.20 to 1.30%): The high carbon content produces a hard case of considerable depth, with improved wear resistance, yet sensitive to shock and concentrated stresses. Selected for applications where the capacity to withstand abrasive wear is needed, and also where the retention of a keen edge or the original shape of the tool is important. For such applications as: cutting tools — (a) for finishing work, like cutters, reamers, (b) for cutting chilled cast iron and forming tools — for ferrous and nonferrous metals, burnishing tools.

By adding small amounts of alloying elements to W-steel types 2, and 5, certain characteristics which are desirable for specific applications, are improved. The vanadium in type 2 contributes to retaining a greater degree of fine grain structure after heat treating. Chromium in type 5 improves the deep hardening characteristics of the steel, a property being needed for large sections, and also assists in maintaining a keen cutting edge, which is desirable in cutting tools, like broaches, reamers, threading taps and dies.

Mill Production Forms of Tool Steels.

Tool steels are produced in many different forms, although not all those listed in the following are always readily available; certain forms and shapes being made on special orders only.

Hot Finished Bars and Cold Finished Bars: These are the most commonly produced forms of tool steels. Bars can be furnished in many different cross sections, the round shape being the most common. Sizes can vary over a wide range, with a more limited number of standard stock sizes. Various conditions may also be available, however, technological limitations prevent all conditions applying to every size, shape or type of steel. Tool steel bars may be supplied in one of the following conditions and surface finishes:

Conditions: Hot rolled or forged (natural); Hot rolled or forged and annealed; Hot rolled or forged and heat treated; Cold or hot drawn (as drawn); and Cold or Hot drawn and annealed.

Finishes: Hot rolled finish (scale not removed); Pickled or blast cleaned; Cold drawn; Turned or machined; Rough ground; Centerless ground or precision flat ground; and Polished (rounds only).

Other forms in which tool steels are supplied are the following:

Rolled or Forged Special Shapes: These are usually produced on special orders only, for the purpose of reducing material loss and machining time in the large-volume manufacture of certain frequently used types of tools.

Forgings: All types of tool steels may be supplied in the form of forgings, which are usually specified for special shapes and also for dimensions which are beyond the range covered by bars.

Wires: Tool steel wires are produced either by hot or cold drawing and are specified for the purpose of obtaining special shapes, controlled dimensional accuracy, improved surface finish or special mechanical properties. Round wire is commonly produced within an approximate size range of 0.015 to 0.500 inch, which also indicates the limits within which other shapes of tool steel wires, like oval, square, rectangular, etc., may be produced.

Drill Rods: These are produced in round, rectangular, square, hexagonal and octagonal shapes, usually with tight dimensional tolerances in order to eliminate subsequent machining, thereby offering manufacturing economies for the users.

Hot Rolled Plates and Sheets, and Cold Rolled Strips: Such forms of tool steel are generally specified for the high volume production of specific tool types.

Tool Bits: These are semifinished tools and are used by clamping in a tool holder or shank in a manner permitting ready replacement. Tool bits are commonly made of high speed types of tool steels, mostly in square, but also in round, rectangular and

other shapes. Tool bits are made of hot rolled bars and are commonly, yet not exclusively, supplied in hardened and ground form, ready for use after the appropriate cutting edges are ground, usually in the user's plant.

Hollow Bars: These are generally produced by trepanning, boring or drilling of solid round rods and are used for making tools or structural parts of annular shapes, like rolls, ring gages, bushings, etc.

Tolerances of Dimensions. — Such tolerances have been developed and published by the Iron and Steel Institute (AISI) as a compilation of available industry experience which, however, does not exclude the establishment of closer tolerances, particularly for hot rolled products manufactured in large quantities. The tolerances differ for various categories of production processes (e.g., forged, hot rolled, cold drawn, centerless ground) and of general shapes. See Handbook pages 421 and 422.

Allowances for Machining. — These serve to assure freedom from soft spots and defects of the tool surface, thereby preventing failures in the heat treatment or in service. After removing from its surface a layer of specific thickness, known as the allowance, the bar or other form of tool steel material should have a surface without decarburization and other surface defects, such as scale marks or seams. The industry-wide accepted machining allowance values for tool steels in different conditions, shapes and size ranges are spelled out in AISI specifications and are generally also listed in the tool steel catalogs of the producer companies.

Decarburization Limits. — The heating of the steel for its production operation causes the oxidation of the exposed surfaces resulting in the loss of carbon. That condition, termed decarburization, penetrates to a certain depth from the surface, depending on the applied process, the shape and the dimensions of the product. The toleranced values of decarburization must be considered as one of the factors for defining the machining allowances, which must also compensate for expected variations of size and shape, the dimensional effects of heat treatment, and so forth. Decarburization can be present not only in hot rolled and forged, but also in rough turned and cold drawn conditions.

Advances in Tool Steel Making Technology. — Significant advances in processes for tool steel production have been made which offer more homogeneous materials of greater density and higher purity for applications where such extremely high quality is required. Two of these methods of tool steel production are of particular interest.

Vacuum melted tool steels: These are produced by the consumable electrode method involving a remelting of the steel originally produced by conventional processes. Inside a vacuum-tight shell which has been evacuated, the electrode cast of tool steel of the desired chemical analysis is lowered into a water cooled copper mold where it strikes a low voltage-high amperage arc causing the electrode to be consumed by gradual melting. The undesirable gases and volatiles are drawn off by the vacuum, and the inclusions float on the surface of the pool, accumulating on the top of the produced ingot, to be removed later by cropping. In the field of tool steels the consumable electrode vacuum melting (CVM) process is applied primarily to the production of special grades of hot work and high speed tool steels.

High speed tool steels produced by powder metallurgy: The steel produced by conventional methods is reduced to a fine powder by a gas atomization process. The powder is compacted by a hot isostatic method with pressures in the range of 15,000 to 17,000 psi. The compacted billets are hot rolled to the final bar size, yielding a tool steel material which has 100 per cent theoretical density. High speed tool steels produced by the P/M method offer a tool material providing increased tool wear life and high impact strength, of particular advantage in interrupted cuts.

Trade Names of AISI Classified Tool Steels* — I

AISI Type	Al-Tech	Atlas Steels	Bethlehem	Braeburn	Carpenter	Columbia	Crucible	Jessop	Latrobe	Simonds	Teledyne Vasco	Universal-Cyclops
							Producer					
HIGH SPEED TOOL STEELS — TUNGSTEN TYPES												
T1	LXX	Spartan-7	Bethlehem T-1	Vinco	Star Zenith	Clarite	Rex AA	Supremus	Electrite No. 1 XL	Red Streak	Red-Cut Superior	...
T2	ML	...	...	Twinvan	...	Vanite	...	Supremus Extra	Electrite No. 19	Lock Port Special	E.V.M.	...
T4	Panther Special	...	...	Cobalt	...	Acmite	Rex AAA	Purple Label	Electrite Cobalt	Tunco	Red Cut Cobalt	...
T5	Super Panther	Nipigon	...	Bonded Carbide JR	...	Cobite	...	Purple Label Extra	Electrite Super Cobalt	Super Cobalt	Circle C	...
T6	...	...	...	...	...	Cobite II	...	King Cobalt	Electrite Ultra Cobalt	...	...	...
T8	...	...	...	...	...	Maxite	Rex 95	Jessop T8	...	...	...	...
T15	Panther 5	Sabre	...	Braeburn T15	...	Maxite 15	CPM Rex T-15	Jessop T15	Electrite Dynavan	...	Vasco Supreme	...
HIGH SPEED TOOL STEELS — MOLYBDENUM TYPES												
M1	LMW	Mohican-8	Bethlehem M-1	Mocut	Starmax, Starmax FM	Molite 1	Rex TMO	Mogul	Electrite Tatmo	STM	8-N-2	Motung
M2	DBL-2	Sixix	Bethlehem M-2	Braemow, Mocarb	Speed Star, Speed Star FM	Molite 2	Rex M-2, CPM Rex M-2	Mustang	Electrite Double Six M2 XL	Molva T	Vasco M-2	Motung 652
M3 Class 1	DBL-2½	Atlas M-3	...	Braevan	...	Molite 3 Class 1	Rex M-3-1	Jessop M3 Class 1	Electrite Corsair XL	Molva TC1	Van Cut	Unicut
M3 Class 2	DBL-3	...	...	Braevan 2	...	Molite 3 Class 2	Rex M-3-2, CPM Rex M-3-2	Jessop M3 Class 2	Electrite Crusader XL	Molva TC 2	Van Cut Type 2	Unicut 2

* Source: Committee of Tool Steel Producers, American Iron and Steel Institute, 1000 16th St., N.W., Washington, D.C. 20036

Trade Names of AISI Classified Tool Steels — 2

AISI Type	Al-Tech	Atlas Steels	Bethlehem	Braeburn	Carpenter	Columbia	Crucible	Jessop	Latrobe	Simonds	Teledyne Vasco	Universal-Cyclops
					HIGH SPEED TOOL STEELS — MOLYBDENUM TYPES (Continued)							
M4	DBL-4	Atlas M-4	...	Braefour	Four Star	Molite 4	CPM Rex M-4	Jessop M4	Electrite Stark	Molva HC	Neatro	Cyclops M4
M6	...	...	...	Congo	...	...	...	...	...	...	...	...
M7	LMW-V	Atlas M-7	Bethlehem M-7	Motuf	Seven Star	Molite 7	Rex M-7	Jessop M7	Electrite Tatmo V	Molva C	Vasco M-7	Motung CV
M10	VLM	Atlas M-10	Bethlehem M-10	Motemp	Ten Star	Molite 10	Rex VM	Jessop M10	Electrite TNW	Molva	Van Lom	...
M30	Super LMW	...	...	Como	...	Molite 30	...	...	Electrite Lacomo	...	8-N-2 Cobalt	Super Motung
M33	Super LMW Extra	...	...	Braeburn M33	...	Molite 33	Rex M-33	...	Electrite Kelvan	STMCO	8-N-2 Cobalt 8	Super Motung 33
M34	Super LMW Special	Atlas M-34	...	...	...	Molite 34	...	...	Electrite Tatmo Cobalt	...	...	...
M36	Super DBL	...	...	Moco	...	Molite 36	...	...	Electrite CO-6	...	Victory Cobalt	...
M41	...	...	...	...	...	Molite 41	Rex 49	Jessop RC 70	...	...	...	...
M42	Exocut	Atlas M-42	...	Braemax	Super Star	Molite 42	Rex M-42, CPM Rex M-42	...	Electrite Dynamax	...	Hypercut	Cyclops M-42
M43	...	...	...	...	...	...	...	...	Electrite Dynacut	...	...	...
M44	...	...	...	Braecut	...	...	...	...	...	...	...	...
M46	AL-46	...	...	...	...	...	Rex M-46	...	...	...	...	...
M47	Exohard	...	...	...	...	...	...	...	...	...	...	...

Trade Names of AISI Classified Tool Steels — 3

AISI Type	Al-Tech	Atlas Steels	Bethlehem	Braeburn	Carpenter	Columbia	Crucible	Jessop	Latrobe	Simonds	Teledyne Vasco	Universal-Cyclops
HIGH SPEED TOOL STEELS — MOLYBDENUM TYPES (Continued)												
M50	HTB-2	Atlas	...	...	Carpenter M50	...	...	...	MV-1	...	Vasco M50	BHT
M52	Oglala	...	...	Natrona 52	Carpenter M52	...	...	...	MV-2	...	Vasco M52	...
HOT WORK TOOL STEELS — CHROMIUM TYPES												
H10	AL-173	...	...	Pressurdie 6	...	Columbia H10	Peerless 56	...	Dart	...	...	...
H11	Potomac A	Atlas H-11	Cromo-V	Pressurdie 3L	882, 882 FM	Firedie	Nu-Die, Halcomb 218	Dica B (Modified)	Dycast No. 1	Howord A	Hot Form No. 2	Thermold H11
H12	Potomac	Crodi	Cromo-W	Pressurdie 2	345, 345 FM	Alcodie	Chro-Mow	Dica B	LPD	Howord B	Hot Form No. 1	Thermold H12
H13	Potomac M	Dievac	Cromo-High-V	Pressurdie 3	883, 883 FM	Firedie 13	Nu-Die V	Dica B Vanadium	VDC, Viscount 20, Viscount 44	Howord C	Hot Form V	Thermold H13
H14	...	Red Indian	...	Pressurdie 1	...	Firedie 14	...	...	Lumdie	...	...	...
H19	B-47	Atlas H-19	...	Pressurdie C	...	Firedie 19	Halcomb 425	...	Lesco 19	...	W.C.C.	...
HOT WORK TOOL STEELS — TUNGSTEN TYPES												
H21	Atlas A	Seneca	57 HW	T-Alloy A	TK	Formite 21	Peerless A	2-BLC	CLW	...	Marvel	Thermold H21
H22	Atlas B	...	...	T-Alloy	TK (Modified)	...	...	2-BMC	...	...	...	...
H23	...	...	...	HCA	...	Formite 23	...	...	Kalkos	...	W.W. Hot Work	...
H24	...	...	...	T-Alloy B	...	Formite 24	...	2-BHC	CHW	...	SC Special	...

Trade Names of AISI Classified Tool Steels — 4

AISI Type	Al-Tech	Atlas Steels	Bethlehem	Braeburn	Carpenter	Columbia	Crucible	Jessop	Latrobe	Simonds	Teledyne Vasco	Universal-Cyclops
				Producer								
HOT WORK TOOL STEELS — TUNGSTEN TYPES (Continued)												
H25	Mohawk	…	…	T-Alloy C					EHW No. 1		Forge-Die	…
H26	…	Spartan-5	Special HS-55	Vinco Hot Work	Carpenter H26	Clarite HW 26	Rex AA PX		Electrite No. 5	…	Red Cut Superior J	…
HOT WORK TOOL STEELS — MOLYBDENUM TYPE												
H42	…							Mustang L.C.	Electrite No. 7			
COLD WORK TOOL STEELS — HIGH CARBON HIGH CHROMIUM TYPES												
D2	Ontario	FNS	Lehigh H	Superior 3	610, 610 FM	Atmodie	Airdi 150	CNS-1	Olympic FM	CCM	Ohio Die	Ultradie 3
D3	Huron		Lehigh S	Superior 1	Hampden	Superdie	…	CNS-2	GSN	HCCM	…	…
D4		NN		AT 2	…	Atmodie 4	HYCC	CNS-3	GSN + MO	…	Crocar	…
D5	AL-D-5			Superior 2	Carpenter D5	Atmodie 5	…	3 C Special	Cobalt Chrome FM	…	…	…
D7	Huron V			…	Carpenter D7	…	HYCV	Truwear	BR-4 FM	ARS	…	…
SPECIAL PURPOSE TOOL STEELS — LOW ALLOY TYPES												
L2	Albany Caroga	…	Tough M	…	…	Columbia L2	Halvan	ET-6	Crown, Superb	…	Vanadium Type H	Cyclops L2
L6	Tioga	Atlas L-6	Bethalloy	…	R.D.S.	Nicrodie	Champaloy	ET-4	NDS	…	Nikro M	Cyclops L6
COLD WORK TOOL STEELS — MEDIUM ALLOY AIR HARDENING TYPES												
A2	Sagamore	Cromoloy	A-H5	Airque	484, 484 FM	E-Z-Die	Airkool	Windsor	Select B FM	Airtrue	Air Hard	Sparta
A3	…	…	Airque V	…	…	…	…	…	…	…	…	…
A4	…	…	Air-4	…	…	…	…	…	…	…	…	…
A6	Apache	Nutherm	A-6	…	Vega	Uni-Die	CSM 6	Jess-Air	…	…	…	Lo-Air

Trade Names of AISI Classified Tool Steels — 5

AISI Type	Al-Tech	Atlas Steels	Bethlehem	Braeburn	Carpenter	Columbia	Crucible	Jessop	Latrobe	Simonds	Teledyne Vasco	Timken	Universal-Cyclops
COLD WORK TOOL STEELS — MEDIUM ALLOY AIR HARDENING TYPES (Continued)													
A7	Sagamore V	…	…	…	…	E-Z-Die V	Airkool V	BX 3	BR-3	A7W	Chromewear	…	…
A8	AL-158	…	Cromo-W55	Pressurdie 16	Carpenter A8	Columbia A-8	…	…	MGR	Airtrue LC	Hot Form No. 3	…	…
A9	…	…	…	…	Carpenter A9	Formdie	…	…	…	…	…	…	Thermold J
A10	…	…	…	…	…	…	…	…	…	…	…	Graph-Air	…
COLD WORK TOOL STEELS — OIL HARDENING TYPES													
O1	Saratoga	Keewatin	BTR	Kiski	Carpenter O-1	EXL-Die	Ketos	Truform	Badger	Teenax	Colonial No. 6	…	Wando
O2	Deward	…	…	…	Stentor	…	…	Special Oil Hardening	…	…	…	…	…
O6	Oligraph	…	O-6 Graphitic	…	…	Col-Graph	Halgraph	Truglide	…	…	…	Graph-Mo	…
O7	Utica	…	…	…	…	Tapdie	…	…	W Tap	BFD	Red Star Tungsten	…	…
SHOCK RESISTING TOOL STEELS													
S1	Seminole	Falcon-6	67 Chisel	Vibro	Excelo	Buster Alloy	Atha Pneu	Top Notch	XL Chisel	Commando	Par-Exc	…	…
S2	…	…	Imperial	…	Solar	…	…	RTS	…	Havoc	…	…	Venango Special
S5	AL-602, AL-609	Monark-2	Omega	Alloy 10	481 Collet	Silico Alloy	La Belle Silicon #2	259 Grade	Lanark	Orleans	Mosil	…	Cyclops S5
S6	…	…	…	…	…	Columbia S6	La Belle HT	…	…	…	…	…	…
S7	AL-7	…	Bearcat	…	Carpenter S-7	Shock-Die	Crucible S-7	Super Shock 7	…	…	Simoch	…	…

Trade Names of AISI Classified Tool Steels — 6

AISI Type	Al-Tech	Atlas Steels	Bethlehem	Braeburn	Carpenter	Columbia	Crucible	Jessop	Latrobe	Simonds	Teledyne Vasco
					MOLD STEELS						
P2	…	…	Duramold B	…	…	…	…	…	…	…	…
P4	…	…	Duramold A	…	Super Samson	…	…	…	…	…	…
P5					Samson Extra						Vasco Chromold VM
P6	…	Super Impacto PQ	Duramold N	…	No. 158						
P20	Almold-20	Mold Special	Bethlehem P-20				CSM #2	P20	…	…	…
P21	…	…	…	…		…		…	Cascade	…	…
					WATER HARDENING TOOL STEELS						
W1	Pompton	X-10, X-12, Alpha, XX-95	X, XCL, XX, Cold Die	Extra	Comet Green Label, H-9 Double Header, No. 11 Special, Titan, Reading Tap	Columbia Special, Extra, Extra Headerdie, Standard	Sanderson Extra, Labelle Cold Header, Black Diamond	Washington Lion, Lion Extra, New Process Cold Header	Carbon Types	Red Label, Blue Label, Diamond S, Green Label	Colonial No. 14
W2	Python	Asa-10	Best, Superior	Coldie	Nitro Special Vanadium	Vanadium Extra, Standard	Alva Extra	Lion Van., Lion Extra	Carbon Vanadium Types	Red Label Extra	Colonial No. 7
W5	Crow	Atlas "Q"	Bethlehem W-5	Braeburn W-5	U.D.R.	Waterdie Extra, Standard	…	W-5	CFS	…	…

Superhard Tool Materials. — Superhard tool materials are much harder than any tool steel. Some of these materials and cemented carbides have metallic characteristics; others, such as ceramics, are considered nonmetallic. Cemented carbides are the most widely used group for metalworking tools, mining tools, and wear-resistant parts; steel-bonded carbides are used to make forming and stamping tools; ceramics, boron-nitride, and diamonds are used chiefly in cutting tools.

There is no universally accepted system for classifying superhard tool materials. Systems most often used are the SAE Classification (J 1072), the C-Grade Systems, the ISO Classification, and the British Hard Metal Association Systems.

Cemented Carbides: Tungsten carbide with a cobalt binder was the first cemented carbide produced. This original material has been modified mainly by varying the amount of metal used as the binder, by varying the grain size of the carbide, and by substituting other metallic carbides, such as titanium and tantalum carbides, for part of the tungsten carbide. Those cemented carbides containing only tungsten carbide are referred to as "straight grades," while those incorporating other metallic carbides in addition to tungsten carbide are called "complex grades." Straight grades are used for cutting tools, drawing dies, forming die inserts, punches, and other types of tools. Complex grades are used mainly to make cutting tools and cutting tool inserts. Cemented carbides hold their high hardness at temperatures well beyond those where high speed steel begins to soften.

Cemented carbide tools were originally constructed by brazing them to steel shanks. It is now more common for the carbides to be used as small indexable inserts that are mechanically held in a steel holder.

Coated Cemented Carbides: Coated cemented carbides have a thin layer of material, which is even harder than the carbide, deposited on their surfaces by either physical or chemical vapor deposition. The most common coatings are titanium carbide, aluminum oxide, titanium nitride, and multiple coatings consisting of thin layers of titanium carbide, titanium carbonitride, and titanium nitride. Coated cemented carbides are now widely used because they permit a significant increase in cutting speed in high productivity machining operations.

Ceramics: Ceramic cutting tool materials utilize aluminum oxide as the base material and are available as indexable inserts; however, ceramics are more brittle than cemented carbides and require more care. They are generally available in three groups: Group A-1 consists of aluminum oxide with about 10 per cent of other oxides or carbides, mainly those of titanium, magnesium, molybdenum, chromium, nickel, or cobalt. Group A-2 is essentially aluminum oxide. Group A-3 is aluminum oxide with 25–30 per cent of a refractory carbide such as titanium carbide.

Ceramic tools are not subject to oxidation, and they retain their resistance to wear and deformation at much higher temperatures than the best cemented carbides and can be used for long periods of time at speeds much higher than those possible with cemented carbides. They do not require coolants.

Cubic Boron Nitride: A high-pressure, high-temperature process bonds a layer of cubic boron nitride to a cemented carbide substrate. Because of their cost, these tools are usually brazed onto a steel holder and reground, as needed, with diamond wheels for reuse. The tools so produced are primarily used to machine superalloys at speeds several times higher than those possible with cemented carbides.

Diamond: Diamond is the hardest and most scratch-resistant material known, and these characteristics make it very attractive as a tool material. Even in small sizes, however, diamonds are expensive and very brittle, and they start to oxidize rapidly at 1200°F. Diamond is unsatisfactory for machining ferrous alloys, but can be used effectively in machining high-silicon cast aluminum alloys, copper and its alloys, cemented tungsten carbides, high-aluminum ceramics, and many composite materials.

Physical Properties

Physical Properties of Heat-treated Steels. — Steels that have been "fully hardened" to the same hardness when quenched will have the same tensile and yield strengths regardless of composition and alloying elements. When the hardness of such a steel is known, it is also possible to predict its reduction of area and tempering temperature. The accompanying charts illustrating these relationships have been prepared by the Society of Automotive Engineers.

Chart 1 gives the range of Brinell hardnesses that could be expected for any particular tensile strength or it may be used to determine the range of tensile strengths that would correspond to any particular hardness. Chart 2 shows the relationship between the tensile strength or hardness and the yield point. The solid-line curve is the normal-expectancy curve. The dotted-line curves give the range of the variation of scatter of the plotted data. Chart 3 shows the relationship that exists between the tensile strength (or hardness) and the reduction of area. The curve to the left represents the alloy steels and that on the right the carbon steels. Both curves are normal-expectancy curves and the extremities of the perpendicular lines going through them represent the variations from the normal-expectancy curves which may be caused by quality differences and by the magnitude of parasitic stresses induced by quenching. Chart 4 shows the relationship between the hardness (or approximately equivalent tensile strength) and the tempering temperature. Three curves are given, one for fully hardened steels with a carbon content between 0.40 and 0.55 per cent, one for fully hardened steels with a carbon content between 0.30 and 0.40 per cent, and one for steels that are not fully hardened.

Referring to Chart 1 it can be seen that for a tensile strength of say 200,000 pounds per square inch the Brinell hardness could range from something in the order of 375 to 425. Taking 400 as the mean hardness value and using Chart 4, it can be seen that the tempering temperature of fully hardened steels of 0.40 to 0.55 per cent carbon content would be 990 degrees F. and that of fully hardened steels of 0.30 to 0.40 per cent carbon would be 870 degrees F. This chart also shows that the tempering temperature for a steel not fully hardened would approach 520 degrees F. A yield point of $0.9 \times 200,000$ or 180,000 pounds per square inch is indicated (Chart 2) for the fully hardened steel with a tensile strength of 200,000 pounds per square inch. Most alloy steels of 200,000 pounds per square inch tensile strength would probably have a reduction in area of close to 44 per cent (Chart 3) but some would have values in the range of 35 to 53 per cent. Carbon steels of the same tensile strength would probably have a reduction in area of close to 24 per cent but could possibly range from 17 to 31 per cent.

Charts 2 and 3 represent steel in the quenched and tempered condition and Chart 1 represents steel in the hardened and tempered, as-rolled, annealed, and normalized conditions. These charts give a good general indication of mechanical properties; however, more exact information when required should be obtained from tests on samples of the individual heats of steel under consideration.

Strength Data for Ferrous Metals. — The accompanying Table 1 gives ultimate strengths, yield points and moduli of elasticity for various ferrous metals. Values are given as ranges, minimum values, and average values. Ranges of values are due to differences in size and shape of sections, heat-treatments undergone, and composition in those cases where several slightly different materials are listed under one general classification. The values in the table are meant to serve as a guide in the selection of ferrous materials and should not be used to write specifications. More specific data should be obtained from the supplier.

Strength Data for Non-ferrous Metals. — The ultimate tensile, shear, and yield strengths and moduli of elasticity of many non-ferrous metals are given in Table 2. Values for the most part are given in ranges rather than as single values

Physical Property Charts for Heat-Treated Steels (SAE General Information)

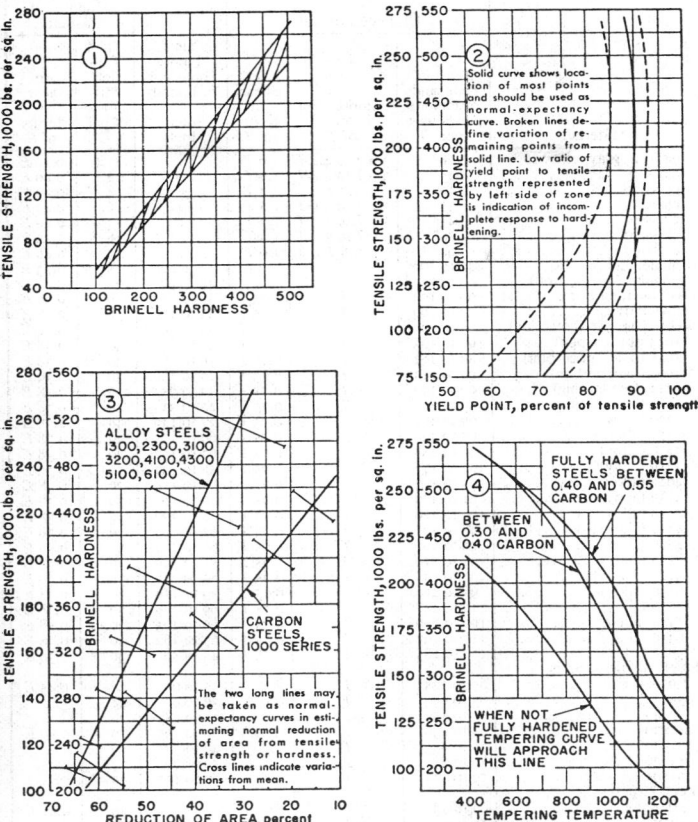

because of differences in composition, forms, sizes and shapes for the aluminum alloys plus differences in heat-treatments undergone for the other non-ferrous metals. The values in the table are meant to serve as a guide, not as specifications. More specific data should be obtained from the supplier.

Effect of Temperature on Strength and Elasticity of Metals. — Most ferrous metals have a maximum strength at approximately 400 degrees F. while the strength of non-ferrous alloys is a maximum at about room temperature. The table on page 514 gives general data for variation in metal strength with temperature.

The modulus of elasticity of metals decreases regularly with increasing temperatures above room temperature until at some elevated temperature it falls off rapidly and reaches zero at the melting point.

PROPERTIES OF MATERIALS

Table 1. Strength Data for Iron and Steel

Material	Ultimate Strength			Yield Point, Thousands of Pounds per Square Inch	Modulus of Elasticity,	
	Tension, Thousands of Pounds per Square Inch, T	Compression, in terms of T	Shear, in terms of T		in Tension, Millions of psi, E	in Shear,[b] in terms of E
Cast iron, gray, class 20 ...	20[a]	3.6 T to 4.4 T	1.6 T	...	11.6	0.40 E
class 25	25[a]	3.6 T to 4.4 T	1.4 T	...	14.2	0.40 E
class 30	30[a]	3.7 T	1.4 T	...	14.5	0.40 E
class 35	35[a]	3.2 T to 3.9 T	1.4 T	...	16.0	0.40 E
class 40	40[a]	3.1 T to 3.4 T	1.3 T	...	17	0.40 E
class 50	50[a]	3.0 T to 3.4 T	1.3 T	...	18	0.40 E
class 60	60[a]	2.8 T	1.0 T	...	19.9	0.40 E
malleable	40 to 100[c]		...	30 to 80[c]	25	0.43 E
nodular (ductile iron) ...	60 to 120[d]	...	...	40 to 90[d]	23	...
Cast steel, carbon	60 to 100	T	0.75 T	30 to 70	30	0.38 E
low alloy	70 to 200	T	0.75 T	45 to 170	30	0.38 E
Steel, SAE 950 (low alloy).	65 to 70	T	0.75 T	45 to 50	30	0.38 E
1025 (low carbon)	60 to 103	T	0.75 T	40 to 90	30	0.38 E
1045 (medium carbon) ..	80 to 182	T	0.75 T	50 to 162	30	0.38 E
1095 (high carbon)......	90 to 213	T	0.75 T	20 to 150	30	0.39 E
1112 (free cutting)*	60 to 100	T	0.75 T	30 to 95	30	0.38 E
1212 (free cutting)	57 to 80	T	0.75 T	25 to 72	30	0.38 E
1330 (alloy)............	90 to 162	T	0.75 T	27 to 149	30	0.38 E
2517 (alloy)[e]	88 to 190	T	0.75 T	60 to 155	30	0.38 E
3140 (alloy)	93 to 188	T	0.75 T	62 to 162	30	0.38 E
3310 (alloy)[e]	104 to 172	T	0.75 T	56 to 142	30	0.38 E
4023 (alloy)[e]	105 to 170	T	0.75 T	60 to 114	30	0.38 E
4130 (alloy)............	81 to 179	T	0.75 T	46 to 161	30	0.38 E
4340 (alloy)	109 to 220	T	0.75 T	68 to 200	30	0.38 E
4640 (alloy)	98 to 192	T	0.75 T	62 to 169	30	0.38 E
4820 (alloy)[e]	98 to 209	T	0.75 T	68 to 184	30	0.38 E
5150 (alloy)	98 to 210	T	0.75 T	51 to 190	30	0.38 E
52100 (alloy)	100 to 238	T	0.75 T	81 to 228	30	0.38 E
6150 (alloy)	96 to 228	T	0.75 T	59 to 210	30	0.38 E
8650 (alloy)	110 to 228	T	0.75 T	69 to 206	30	0.38 E
8740 (alloy)	100 to 179	T	0.75 T	60 to 165	30	0.38 E
9310 (alloy)[e]	117 to 187	T	0.75 T	63 to 162	30	0.38 E
9840 (alloy)	120 to 285	T	0.75 T	45 to 50	30	0.38 E
Steel, stainless, SAE						
30302[f]	85 to 125	T	...	35 to 95	28	0.45 E
30321[f]	85 to 95	T	...	30 to 60	28	...
30347[f]	90 to 100	T	...	35 to 65	28	...
51420[g]	95 to 230	T	...	50 to 195	29	0.40 E
51430[h]	75 to 85	T	...	40 to 70	29	...
51446[h]	80 to 85	T	...	50 to 70	29	...
51501[g]	70 to 175	T	...	30 to 135	29	...
Steel, structural,						
common	60 to 75	T	0.75 T	33[a]	29	0.41 E
rivet.................	52 to 62	T	0.75 T	28[a]	29	...
rivet, high strength	68 to 82	T	0.75 T	38[a]	29	...
Wrought iron	34 to 54	T	0.83 T	23 to 32	28	...

[a] Minimum specified value of the American Society for Testing and Materials. The specifications for the various materials are as follows: Cast iron, ASTM A48; structural steel for bridges and structures, ASTM A7; structural rivet steel, ASTM A141; high-strength structural rivet steel, ASTM A195.

[b] Synonymous in other literature to the modulus of elasticity in torsion and the modulus of rigidity, G.

[c] Range of minimum specified values of the ASTM (ASTM A47, A197, and A220).

[d] Range of minimum specified values of the ASTM (ASTM A339) and the Munitions Board Standards Agency (MIL-I-17166A and MIL-I-11466).

[e] Carburizing grades of steel.

[f] Non-hardenable nickel-chromium and chromium-nickel-manganese steel (austenitic).

[g] Hardenable chromium steel (martensitic).

[h] Non-hardenable chromium steel (ferritic).

Table 2. Strength Data for Non-Ferrous Metals*

Material	Ultimate Strength, Thousands of Pounds per Square Inch		Yield Strength (0.2 per cent offset), Thousands of Pounds per Square Inch	Modulus of Elasticity, Millions of Pounds per Square Inch	
	in Tension	in Shear		in Tension, E	in Shear, G
Aluminum alloys, cast,					
sand cast	19 to 35	14 to 26	8 to 25	10.3	. . .
heat-treated	20 to 48	20 to 34	16 to 40	10.3	. . .
permanent mold cast, ...	23 to 35	16 to 27	9 to 24	10.3	. . .
heat-treated	23 to 48	15 to 36	8.5 to 43	10.3	. . .
die-cast	30 to 46	19 to 29	16 to 27	10.3	. . .
Aluminum alloys, wrought,					
annealed	10 to 42	7 to 26	4 to 22	10.0 to 10.6	. . .
cold-worked	12 to 63	8 to 34	11 to 59	10.0 to 10.3	. . .
heat-treated	22 to 83	14 to 48	13 to 73	10.0 to 11.4	. . .
Aluminum bronze, cast, . . .	62 to 90	. . .	25 to 37	15 to 18	. . .
heat-treated	80 to 110	. . .	32 to 65	15 to 18	. . .
Aluminum bronze, wrought,					
annealed	55 to 80	. . .	20 to 40	16 to 19	. . .
cold-worked	71 to 110	. . .	62 to 66	16 to 19	. . .
heat-treated	101 to 151	. . .	48 to 94	16 to 19	. . .
Brasses, leaded, cast	32 to 40	29 to 31	12 to 15	12 to 14	. . .
flat products, wrought	46 to 85	31 to 45	14 to 62	14 to 17	5.3 to 6.4
wire, wrought	50 to 88	34 to 46	. . .	15	5.6
Brasses, non-leaded,					
flat products, wrought	34 to 99	28 to 48	10 to 65	15 to 17	5.6 to 6.4
wire, wrought	40 to 130	29 to 60	. . .	15 to 17	5.6 to 6.4
Copper, wrought,					
flat products	32 to 57	22 to 29	10 to 53	17	6.4
wire	35 to 66	24 to 33	. . .	17	6.4
Inconel, cast	70 to 95	. . .	30 to 45	23	. . .
flat products, wrought	80 to 170	. . .	30 to 160	31	11
wire, wrought	80 to 185	. . .	25 to 175	31	11
Lead	2.2 to 4.9	. . .	. . .	0.8 to 2.0	. . .
Magnesium, cast,					
sand & permanent mold . .	22 to 40	17 to 22	12 to 23	6.5	2.4
die-cast	33	20	22	6.5	2.4
Magnesium, wrought,					
sheet and plate	35 to 42	21 to 23	20 to 32	6.5	2.4
bars, rods, and shapes . . .	37 to 55	19 to 27	26 to 44	6.5	2.4
Monel, cast	65 to 90	. . .	32 to 40	19	. . .
flat products, wrought	70 to 140	. . .	25 to 130	26	9.5
wire, wrought	70 to 170	. . .	25 to 160	26	9.5
Nickel, cast, ...	45 to 60	. . .	20 to 30	21.5	. . .
flat products, wrought	55 to 130	. . .	15 to 115	30	11
wire, wrought	50 to 165	. . .	10 to 155	30	11
Nickel silver, cast	40 to 50	. . .	24 to 25	. . .	. . .
flat products, wrought	49 to 115	41 to 59	18 to 90	17.5 to 18	6.6 to 6.8
wire, wrought	50 to 145	. . .	25 to 90	17.5 to 18	6.6 to 6.8
Phosphor bronze, wrought,					
flat products	40 to 128	. . .	14 to 80	15 to 17	5.6 to 6.4
wire	50 to 147	. . .	20 to 80	16 to 17	6 to 6.4
Silicon bronze, wrought,					
flat products	56 to 110	42 to 63	21 to 62	15	5.6
wire	50 to 145	36 to 70	25 to 70	15 to 17	5.6 to 6.4
Tin bronze, leaded, cast	21 to 38	23 to 43	15 to 18	10 to 14.5	. . .
Titanium	50 to 135	. . .	40 to 120	15.0 to 16.5	. . .
Zinc, commercial rolled	19.5 to 31	. . .	. . .	. . .	. . .
Zirconium	22 to 83	. . .	. . .	9 to 14.5	4.8

* Consult the index for data on metals not listed and for more data on metals listed.

Average Ultimate Strength of Common Materials other than Metals
(Pounds per square inch)

Material	Compression	Tension
Bricks, best hard..........................	12,000	400
Bricks, light red..........................	1,000	40
Brickwork, common........................	1,000	50
Brickwork, best...........................	2,000	300
Cement, Portland, one month old.............	2,000	400
Cement, Portland, one year old..............	3,000	500
Concrete, Portland.........................	1,000	200
Concrete, Portland, one year old.............	2,000	400
Granite...................................	19,000	700
Limestone and sandstone....................	9,000	300
Trap rock.................................	20,000	800
Slate.....................................	14,000	500
Vulcanized Fiber..........................	39,000	13,000

Influence of Temperature on the Strength of Metals

Material	Degrees Fahrenheit							
	210	400	570	750	930	1100	1300	1475
	Strength in Per Cent of Strength at 70 Degrees F.							
Wrought iron.....	104	112	116	96	76	42	25	15
Cast iron.........	....	100	99	92	76	42	....	....
Steel castings.....	109	125	121	97	57	....	....	....
Structural steel....	103	132	122	86	49	28	....	....
Copper...........	95	85	73	59	42	....	....	....
Bronze...........	101	94	57	26	18	....	....	....

Strength of Copper-zinc-tin Alloys
(U. S. Government Tests)

Copper	Zinc	Tin	Tensile Strength, Lbs. per Sq. In.	Copper	Zinc	Tin	Tensile Strength, Lbs. per Sq. In.	Copper	Zinc	Tin	Tensile Strength, Lbs. per Sq. In.
45	50	5	15,000	60	20	20	10,000	75	20	5	45,000
50	45	5	50,000	65	30	5	50,000	75	15	10	45,000
50	40	10	15,000	65	25	10	42,000	75	10	15	43,000
55	43	2	65,000	65	20	15	30,000	75	5	20	41,000
55	40	5	62,000	65	15	20	18,000	80	15	5	45,000
55	35	10	32,500	65	10	25	12,000	80	10	10	45,000
55	30	15	15,000	70	25	5	45,000	80	5	15	47,500
60	37	3	60,000	70	20	10	44,000	85	10	5	43,500
60	35	5	52,500	70	15	15	37,000	85	5	10	46,500
60	30	10	40,000	70	10	20	30,000	90	5	5	42,000

HEAT TREATMENT OF STANDARD STEELS

Heat-Treating Definitions. — This glossary of heat-treating terms has been adopted by the American Foundrymen's Association, the American Society for Metals, the American Society for Testing and Materials, and the Society of Automotive Engineers. Since it is not intended to be a specification but is strictly a set of definitions, temperatures have purposely been omitted.

Aging: Describes a time-temperature-dependent change in the properties of certain alloys. Except for strain aging and age softening, it is the result of precipitation from a solid solution of one or more compounds whose solubility decreases with decreasing temperature. For each alloy susceptible to aging, there is a unique range of time-temperature combinations to which it will respond.

Annealing: A term denoting a treatment, consisting of heating to and holding at a suitable temperature followed by cooling at a suitable rate, used primarily to soften but also to simultaneously produce desired changes in other properties or in microstructure. The purpose of such changes may be, but is not confined to, improvement of machinability; facilitation of cold working; improvement of mechanical or electrical properties; or increase in stability of dimensions. The time-temperature cycles used vary widely both in maximum temperature attained and in cooling rate employed, depending on the composition of the material, its condition, and the results desired. When applicable, the following more specific process names should be used: Black Annealing, Blue Annealing, Box Annealing, Bright Annealing, Cycle Annealing, Flame Annealing, Full Annealing, Graphitizing, Intermediate Annealing, Isothermal Annealing, Process Annealing, Quench Annealing, and Spheroidizing. When the term is used without qualification, full annealing is implied. When applied only for the relief of stress, the process is properly called stress relieving.

Black Annealing: Box annealing or pot annealing, used mainly for sheet, strip, or wire.

Blue Annealing: Heating hot rolled sheet in an open furnace to a temperature within the transformation range and then cooling in air, to soften the metal. The formation of a bluish oxide on the surface is incidental.

Box Annealing: Annealing in a sealed container under conditions that minimize oxidation. In box annealing the charge is usually heated slowly to a temperature below the transformation range, but sometimes above or within it, and is then cooled slowly; this process is also called "close annealing" or "pot annealing."

Bright Annealing: Annealing in a protective medium to prevent discoloration of the bright surface.

Cycle Annealing: An annealing process employing a predetermined and closely controlled time-temperature cycle to produce specific properties or microstructure.

Flame Annealing: Annealing in which the heat is applied directly by a flame.

Full Annealing: Austenitizing and then cooling at a rate such that the hardness of the product approaches a minimum.

Graphitizing: Annealing in such a way that some or all of the carbon is precipitated as graphite.

Intermediate Annealing: Annealing at one or more stages during manufacture and before final thermal treatment.

Isothermal Annealing: Austenitizing and then cooling to and holding at a temperature at which austenite transforms to a relatively soft ferrite-carbide aggregate.

Process Annealing: An imprecise term to denote various treatments used to improve workability. For the term to be meaningful, the condition of the material and the time-temperature cycle used must be stated.

Quench Annealing: Annealing an austenitic alloy by *Solution Heat Treatment.*

Spheroidizing: Heating and cooling in a cycle designed to produce a spheroidal or globular form of carbide.

Austempering: Quenching from a temperature above the transformation range, in a medium having a rate of heat abstraction high enough to prevent the formation of high-temperature transformation products, and then holding the alloy, until transformation is complete, at a temperature below that of pearlite formation and above that of martensite formation.

Austenitizing: Forming austenite by heating into the transformation range (partial austenitizing) or above the transformation range (complete austenitizing). When used without qualification, the term implies complete austenitizing.

Baking: Heating to a low temperature in order to remove entrained gases.

Bluing: A treatment of the surface of iron-base alloys, usually in the form of sheet or strip, on which, by the action of air or steam at a suitable temperature, a thin blue oxide film is formed on the initially scale-free surface, as a means of improving appearance and resistance to corrosion. This term is also used to denote a heat treatment of springs after fabrication, to reduce the internal stress created by coiling and forming.

Carbon Potential: A measure of the ability of an environment containing active carbon to alter or maintain, under prescribed conditions, the carbon content of the steel exposed to it. In any particular environment, the carbon level attained will depend on such factors as temperature, time, and steel composition.

Carbon Restoration: Replacing the carbon lost in the surface layer from previous processing by carburizing this layer to substantially the original carbon level.

Carbonitriding: A case hardening process in which a suitable ferrous material is heated above the lower transformation temperature in a gaseous atmosphere of such composition as to cause simultaneous absorption of carbon and nitrogen by the surface and, by diffusion, create a concentration gradient. The process is completed by cooling at a rate that produces the desired properties in the workpiece.

Carburizing: A process in which carbon is introduced into a solid iron-base alloy by heating above the transformation temperature range while in contact with a carbonaceous material which may be a solid, liquid, or gas. Carburizing is frequently followed by quenching to produce a hardened case.

Case: (1) The surface layer of an iron-base alloy which has been suitably altered in composition and can be made substantially harder than the interior or core by a process of case hardening. (2) The term case is also used to designate the hardened surface layer of a piece of steel that is large enough to have a distinctly softer core or center.

Cementation: The process of introducing elements into the outer layer of metal objects by means of high-temperature diffusion.

Cold Treatment: Exposing to suitable subzero temperatures for the purpose of obtaining desired conditions or properties, such as dimensional or microstructural stability. When the treatment involves the transformation of retained austenite, it is usually followed by a tempering treatment.

Conditioning Heat Treatment: A preliminary heat treatment used to prepare a material for a desired reaction to a subsequent heat treatment. For the term to be meaningful, the treatment used must be specified.

Controlled Cooling: A term used to describe a process by which a steel object is cooled from an elevated temperature, usually from the final hot forming operation in a predetermined manner of cooling to avoid hardening, cracking, or internal damage.

Core: (1) The interior portion of an iron-base alloy which after case hardening is substantially softer than the surface layer or case. (2) The term core is also used to designate the relatively soft central portion of certain hardened tool steels.

Critical Range or *Critical Temperature Range:* Synonymous with *Transformation Range,* which is preferred.

Cyaniding: A process of case hardening an iron-base alloy by the simultaneous absorption of carbon and nitrogen by heating in a cyanide salt. Cyaniding is usually followed by quenching to produce a hard case.

Decarburization: The loss of carbon from the surface of an iron-base alloy as the result of heating in a medium which reacts with the carbon.

Drawing: Drawing, or drawing the temper, is synonymous with *Tempering,* which is preferable.

Eutectic Alloy: The alloy composition that freezes at constant temperature similar to a pure metal. The lowest melting (or freezing) combination of two or more metals. The alloy structure (homogeneous) of two or more solid phases formed from the liquid eutectically.

Hardenability: In a ferrous alloy, the property that determines the depth and distribution of hardness induced by quenching.

Hardening: Any process of increasing hardness of metal by suitable treatment, usually involving heating and cooling.

Hardening, Age: See *Aging*

Hardening, Case: A process of surface hardening involving a change in the composition of the outer layer of an iron-base alloy followed by appropriate thermal treatment. Typical case-hardening processes are *Carburizing, Cyaniding, Carbonitriding* and *Nitriding.*

Hardening, Flame: A process of heating the surface layer of an iron-base alloy above the transformation temperature range by means of a high-temperature flame, followed by quenching.

Hardening, Precipitation: A process of hardening an alloy in which a constituent precipitates from a supersaturated solid solution. See also *Aging.*

Hardening, Secondary: An increase in hardness following the normal softening that occurs during the tempering of certain alloy steels.

Heating, Differential: A heating process by which the temperature is made to vary throughout the object being heated so that on cooling different portions may have such different physical properties as may be desired.

Heating, Induction: A process of local heating by electrical induction.

Heat Treatment: A combination of heating and cooling operations applied to a metal or alloy in the solid state to obtain desired conditions or properties. Heating for the sole purpose of hot working is excluded from the meaning of this definition.

Heat Treatment, Solution: A treatment in which an alloy is heated to a suitable temperature and held at this temperature for a sufficient length of time to allow a desired constituent to enter into solid solution, followed by rapid cooling to hold the constituent in solution. The material is then in a supersaturated, unstable state, and may subsequently exhibit *Age Hardening.*

Homogenizing: A high-temperature heat-treatment process intended to eliminate or to decrease chemical segregation by diffusion.

Isothermal Transformation: A change in phase at constant temperature.

Malleablizing: A process of annealing white cast iron in which the combined carbon is wholly or in part transformed to graphitic or free carbon and, in some cases, part of the carbon is removed completely. See *Temper Carbon.*

Maraging: A precipitation hardening treatment applied to a special group of iron-base alloys to precipitate one or more intermetallic compounds in a matrix of essentially carbon-free martensite.

Martempering: A hardening procedure in which an austenitized ferrous workpiece is quenched into an appropriate medium whose temperature is maintained substantially at the M_s of the workpiece, held in the medium until its temperature is uniform throughout but not long enough to permit bainite to form, and then cooled in air. The treatment is followed by tempering.

Nitriding: A process of case hardening in which an iron-base alloy of special composition is heated in an atmosphere of ammonia or in contact with nitrogenous material. Surface hardening is produced by the absorption of nitrogen without quenching.

Normalizing: A process in which an iron-base alloy is heated to a temperature above the transformation range and subsequently cooled in still air at room temperature.

Overheated: A metal is said to have been overheated if, after exposure to an unduly high temperature, it develops an undesirably coarse grain structure but is not permanently damaged. The structure damaged by overheating can be corrected by suitable heat treatment or by mechanical work or by a combination of the two. In this respect it differs from a Burnt structure.

Patenting: A process of heat treatment applied to medium or high carbon steel in wire making prior to the wire drawing or between drafts. It consists in heating to a temperature above the transformation range, followed by cooling to a temperature below that range in air or in a bath of molten lead or salt maintained at a temperature appropriate to the carbon content of the steel and the properties required of the finished product.

Preheating: Heating to an appropriate temperature immediately prior to austenitizing when hardening high hardenability constructional steels, many of the tool steels, and heavy sections.

Quenching: Rapid cooling. When applicable, the following more specific terms should be used: Direct Quenching, Fog Quenching, Hot Quenching, Interrupted Quenching, Selective Quenching, Slack Quenching, Spray Quenching, and Time Quenching.

Direct Quenching: Quenching carburized parts directly from the carburizing operation.

Fog Quenching: Quenching in a mist.

Hot Quenching: An imprecise term used to cover a variety of quenching procedures in which a quenching medium is maintained at a prescribed temperature above 160 degrees F. (71 degrees C).

Interrupted Quenching: A quenching procedure in which the workpiece is removed from the first quench at a temperature substantially higher than that of the quenchant and is then subjected to a second quenching system having a different cooling rate than the first.

Selective Quenching: Quenching only certain portions of a workpiece.

Slack Quenching: The incomplete hardening of steel due to quenching from the austenitizing temperature at a rate slower than the critical cooling rate for the particular steel, resulting in the formation of one or more transformation products in addition to martensite.

Spray Quenching: Quenching in a spray of liquid.

Time Quenching: Interrupted quenching in which the duration of holding in the quenching medium is controlled.

Soaking: Prolonged heating of a metal at a selected temperature.

Stabilizing Treatment: A treatment applied to stabilize the dimensions of a workpiece or the structure of a material such as (1) before finishing to final dimensions, heating a workpiece to or somewhat beyond its operating temperature and then cooling to room temperature a sufficient number of times to ensure stability of dimensions in service, (2) transforming retained austenite in those materials which retain substantial amounts when quench hardened (see cold treatment), (3) heating a solution-treated austenitic stainless steel that contains controlled amounts of titanium or niobium plus tantalum to a temperature below the solution-heat-treating temperature to cause precipitation of finely divided, uniformly distributed carbides of those elements, thereby substantially reducing the amount of carbon available for the formation of chromium carbides in the grain boundaries upon subsequent exposure to temperatures in the sensitizing range.

Stress Relieving: A process to reduce internal residual stresses in a metal object by heating the object to a suitable temperature and holding for a proper time at that temperature. This treatment may be applied to relieve stresses induced by casting, quenching, normalizing, machining, cold working or welding.

Temper Carbon: The free or graphitic carbon which comes out of solution usually in the form of rounded nodules in the structure during *Graphitizing* or *Malleablizing.*

Tempering: Heating a quench hardened or normalized ferrous alloy to a temperature below the transformation range to produce desired changes in properties.

> *Double Tempering:* A treatment in which quench hardened steel is given two complete tempering cycles at substantially the same temperature for the purpose of ensuring completion of the tempering reaction and promoting stability of the resulting microstructure.

> *Snap Temper:* A precautionary interim stress-relieving treatment applied to high hardenability steels immediately after quenching to prevent cracking because of delay in tempering them at the prescribed higher temperature.

> *Temper Brittleness:* Brittleness that results when certain steels are held within, or are cooled slowly through, a certain range of temperatures below the transformation range. The brittleness is revealed by notched bar impact tests at or below room temperature.

Transformation Ranges or Transformation Temperature Ranges: Those ranges of temperature within which austenite forms during heating and transforms during cooling. The two ranges are distinct, sometimes overlapping but never coinciding. The limiting temperatures of the ranges depend on the composition of the alloy and on the rate of change of temperature, particularly during cooling.

Transformation Temperature: The temperature at which a change in phase occurs. The term is sometimes used to denote the limiting temperature of a transformation range. The following symbols are used for iron and steels:

Ac_{cm} — In hypereutectoid steel, the temperature at which the solution of cementite in austenite is completed during heating.

Ac_1 — The temperature at which austenite begins to form during heating.

Ac_3 — The temperature at which transformation of ferrite to austenite is completed during heating.

Ac_4 — The temperature at which austenite transforms to delta ferrite during heating.

Ae_1, Ae_3, Ae_{cm}, Ae_4 — The temperatures of phase changes at equilibrium.

Ar_{cm} — In hypereutectoid steel, the temperature at which precipitation of cementite starts during cooling.

Ar_1 — The temperature at which transformation of austenite to ferrite or to ferrite plus cementite is completed during cooling.

Ar_3 — The temperature at which austenite begins to transform to ferrite during cooling.

Ar_4 — The temperature at which delta ferrite transforms to austenite during cooling.

M_s — The temperature at which transformation of austenite to martensite starts during cooling.

M_f — The temperature, during cooling, at which transformation of austenite to martensite is substantially completed.

All these changes except the formation of martensite occur at lower temperatures during cooling than during heating, and depend on the rate of change of temperature.

Structure of Fully Annealed Carbon Steel. — In carbon steel that has been fully annealed, there are normally present, apart from such impurities as phosphorus and sulfur, two constituents: the element iron in a form metallurgically known as *ferrite* and the chemical compound iron carbide in the form metallurgically known as *cementite*. This latter constituent consists of 6.67 per cent carbon and 93.33 per cent iron. A certain proportion of these two constituents will be present as a mechanical mixture. This mechanical mixture, the amount of which depends upon the carbon content of the steel, consists of alternate bands or layers of ferrite and cementite. Under the microscope the matrix frequently has the appearance of mother-of-pearl and hence has been named *pearlite*. Pearlite contains about 0.85 per cent carbon and 99.15 per cent iron, neglecting impurities. A fully annealed steel containing 0.85 per cent carbon would consist entirely of pearlite. Such a steel is known as *eutectoid* steel and has a laminated structure characteristic of a eutectic alloy. Steel which has less than 0.85 per cent carbon (*hypoeutectoid* steel) has an excess of ferrite above that required to mix with the cementite present to form pearlite, hence both ferrite and pearlite are present in the fully annealed state. Steel having a carbon content greater than 0.85 per cent (*hypereutectoid* steel) has an excess of cementite over that required to mix with the ferrite to form pearlite, hence both cementite and pearlite are present in the fully annealed state. The structural constitution of carbon steel in terms of ferrite, cementite, pearlite and austenite for different carbon contents and at different temperatures is shown by the accompanying diagram.

Effect of Heating Fully Annealed Carbon Steel. — When carbon steel in the fully annealed state is heated above the lower critical point, which is some temperature in the range of 1335 to 1355 degrees F. (depending upon the carbon content), the alternate bands or layers of ferrite and cementite which make up the pearlite begin to merge into each other. This process continues until the pearlite is thoroughly "dissolved," forming what is known as *austenite*. If the temperature of the steel continues to rise and there is present, in addition to the pearlite, any excess ferrite or cementite, this also will begin to dissolve into the austenite until finally only austenite will be present. The temperature at which the excess ferrite or cementite is completely dissolved in the austenite is called the *upper critical point*. This temperature varies with the carbon content of the steel much more widely than the lower critical point (see diagram).

Effect of Slow Cooling on Carbon Steel. — If carbon steel which has been heated to the point where it consists entirely of austenite is slowly cooled, the process of transformation which took place during the heating will be reversed but the upper and lower critical points will occur at somewhat lower temperatures than they do on

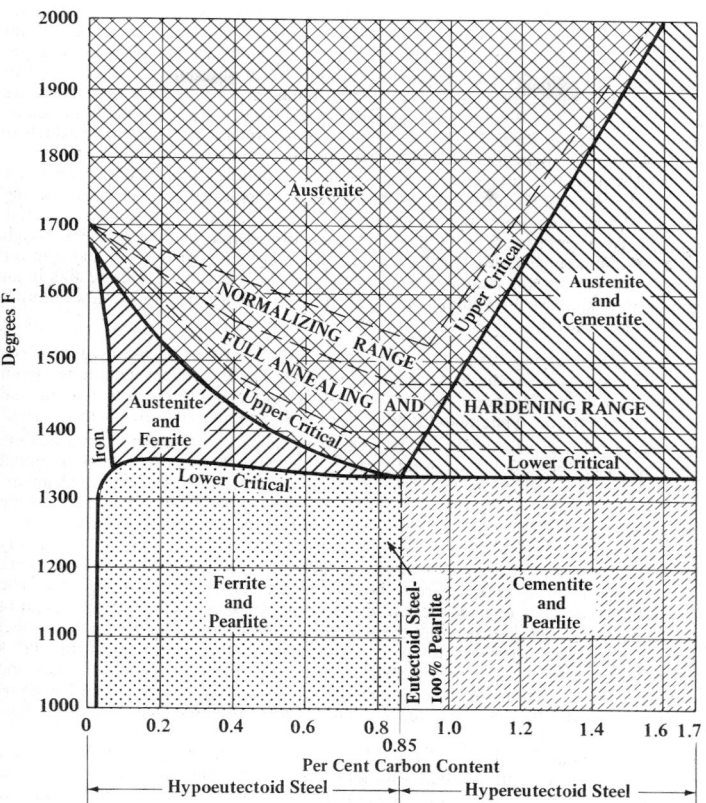

Phase Diagram of Carbon Steel

heating. Assuming that the steel was originally fully annealed, its structure upon returning to atmospheric temperature after slow cooling will be the same as before in terms of the proportions of ferrite or cementite and pearlite present. The austenite will have entirely disappeared.

Effect of Rapid Cooling or Quenching on Carbon Steel. — Observations have shown that as the rate at which carbon steel is cooled from an austenitic state is increased, the temperature at which the austenite begins to change into pearlite drops more and more below the slow cooling transformation temperature of about 1300 degrees F. (For example, a 0.80 per cent carbon steel that is cooled at such a rate that the temperature drops 500 degrees in one second will show transformation of austenite beginning at 930 degrees F.) As the cooling rate is increased, the laminations of the pearlite formed by the transformation of the austenite become finer and finer

up to the point where they cannot be detected under a high power microscope, while the steel itself increases in hardness and tensile strength. As the rate of cooling is still further increased, this transformation temperature suddenly drops to around 500 degrees F. or lower, depending upon the carbon content of the steel. The cooling rate at which this sudden drop in transformation temperature takes place is called the *critical cooling rate*. When a piece of carbon steel is quenched at this rate or faster, a new structure is formed. The austenite is transformed into *martensite* which is characterized by an angular needlelike structure and a very high hardness.

If carbon steel is subjected to a severe quench or to extremely rapid cooling, a small percentage of the austenite, instead of being transformed into martensite during the quenching operation, may be retained. Over a period of time, however, this remaining austenite tends to be gradually transformed into martensite even though the steel is not subjected to further heating or cooling. Since martensite has a lower density than austenite, such a change, or "aging" as it is called, often results in an appreciable increase in volume or "growth" and the setting up of new internal stresses in the steel.

Steel Heat-Treating Furnaces. — Various types of furnaces heated by gas, oil, or electricity are used for the heat-treatment of steel. These furnaces include the oven or box type in various modifications for "in-and-out" or for continuous loading and unloading; the retort type; the pit type; the pot type; and the salt-bath electrode type.

Oven or Box Furnaces: This type of furnace has a box or oven-shaped heating chamber. The "in-and-out" oven furnaces are loaded by hand or by a track-mounted car which, when rolled into the furnace, forms the bottom of the heating chamber. The car type is used where heavy or bulky pieces must be handled. Some oven type furnaces are provided with a full muffle or a semi-muffle which is an enclosed refractory chamber into which the parts to be heated are placed. The full-muffle, being fully enclosed, prevents any flames or burning gases from coming in contact with the work and permits a special atmosphere to be used to protect or condition the work. The semi-muffle, which is open at the top, protects the work from direct impingement of the flame although it does not shut the work off from the hot gases. In the direct-heat type oven furnace, the work is open to the flame. In the electric oven furnace, a retort is provided if gas atmospheres are to be employed to confine the gas and prevent it from attacking the heating elements. Where muffles are used, they must be replaced periodically and a greater amount of fuel is required than in a direct-heat type of oven furnace.

For continuous loading and unloading, there are several types such as rotary hearth car; roller-, furnace belt-, walking-beam or pusher-conveyor; and a continuous-kiln type through which track-mounted cars are run. In the continuous type of furnace, the work may pass through several zones maintained at different temperatures for preheating, heating, soaking, and cooling.

Retort Furnace: This is a vertical type of furnace provided with a cylindrical metal retort into which the parts to be heat-treated are suspended either individually, if large enough, or in a container of some sort. The use of a retort permits special gas atmospheres to be employed for carburizing, nitriding, etc.

Pit Type Furnace: This is a vertical furnace arranged for the loading of parts in a metal basket. The parts are heated by convection, the basket, when lowered into place, fitting into the furnace chamber in such a way as to provide a dead-air space to prevent direct heating.

Pot Type Furnace: This furnace is used for the immersion method of heat-treating small parts. A cast-alloy pot is employed to hold a bath of molten lead or salt in which the parts are placed for heating.

Salt Bath Electrode Furnace: In this type of electric furnace, heating is accom-

plished by means of electrodes suspended directly in the salt bath. The patented grouping and design of electrodes provide an electromagnetic action which results in an automatic stirring action. This stirring tends to produce an even temperature throughout the bath.

Vacuum Furnace: Vacuum heat treatment is a relatively new development in metallurgical processing, with a vacuum substituting for the more commonly used protective gas atmospheres. The most often used furnace is the "cold wall" type, consisting of a water-cooled vessel that is maintained near ambient temperature during operation. During quenching, the chamber is backfilled up to or above atmospheric pressure with an inert gas, which is circulated by an internal fan. When even faster cooling rates are needed, furnaces are available with capability for liquid quenching, performed in an isolated chamber.

Fluidized-Bed Furnace: Fluidized-bed techniques are not new; however, new furnace designs have extended the technology into the temperature ranges required for most common heat treatments. In fluidization, a bed of dry, finely divided particles, typically aluminum oxide, is made to behave like a liquid by feeding gas upward through the bed. An important characteristic of the bed is high-efficiency heat transfer. Applications include continuous or batch-type units for all general heat treatments.

Hardening

Basic Steps in Hardening. — The operation of hardening steel consists fundamentally of two steps. The first step is to heat the steel to some temperature above (usually at least 100 degrees F. above) its transformation point so that it becomes entirely austenitic in structure. The second step is to quench the steel at some rate faster than the critical rate (which depends on the carbon content, the amounts of alloying elements present other than carbon, and the grain size of the austenite) to produce a martensitic structure. The hardness of a martensitic steel depends upon its carbon content and ranges from about 460 Brinell at 0.20 per cent carbon to about 710 Brinell above 0.50 carbon. In comparison, ferrite has a hardness of about 90 Brinell, pearlite about 240 Brinell, and cementite around 550 Brinell.

Critical Points of Decalescence and Recalescence. — The critical or transformation point at which pearlite is transformed into austenite as it is being heated is also called the *decalescence point.* If the temperature of the steel was observed as it passed through the decalescence point, it would be noted that it would continue to absorb heat without appreciably rising in temperature, although the immediate surroundings were hotter than the steel. Similarly, the critical or transformation point at which austenite is transformed back into pearlite upon cooling is called the *recalescence point.* When this point is reached, the steel will give out heat so that its temperature instead of continuing to fall, will momentarily increase.

The recalescence point is lower than the decalescence point by anywhere from 85 to 215 degrees F., and the lower of these points does not manifest itself unless the higher one has first been fully passed. These critical points have a direct relation to the hardening of steel. Unless a temperature sufficient to reach the decalescence point is obtained, so that the pearlite is changed into austenite, no hardening action can take place; and unless the steel is cooled suddenly before it reaches the recalescence point, thus preventing the changing back again from austenite to pearlite, no hardening can take place. The critical points vary for different kinds of steel and must be determined by tests in each case. It is the variation in the critical points that makes it necessary to heat different steels to different temperatures when hardening.

Hardening Temperatures. — The maximum temperature to which a steel is heated before quenching to harden it is called the hardening temperature. Harden-

ing temperatures vary for different steels and different classes of service, although, in general, it may be said that the hardening temperature for any given steel is above the lower critical point of that steel. Just how far above this point the hardening temperature lies for any particular steel depends on three factors: (1) The chemical composition of the steel; (2) the amount of excess ferrite (if the steel has less than 0.85 per cent carbon content) or the amount of excess cementite (if the steel has more than 0.85 per cent carbon content) that is to be dissolved in the austenite; and (3) the maximum grain size permitted, if desired.

The general range of full hardening temperatures for carbon steels is shown by the diagram. This range is merely indicative of general practice and is not intended to represent absolute hardening temperature limits. It can be seen that for steels of less than 0.85 per cent carbon content, the hardening range is above the upper critical point — that is, above the temperature at which all of the excess ferrite has been dissolved in the austenite. On the other hand, for steels of more than 0.85 per cent carbon content, the hardening range lies somewhat below the upper critical point. This indicates that in this hardening range some of the excess cementite still remains undissolved in the austenite. If steel of more than 0.85 per cent carbon content were heated above the upper critical point and then quenched, the resulting grain size would be excessively large.

At one time it was considered desirable to heat steel only to the minimum temperature at which it would fully harden, one of the reasons being to avoid grain growth that takes place at higher temperature. It is now realized that no such rule as this can be applied generally since there are factors other than hardness which must be taken into consideration. For example, in many cases toughness can be impaired by too low a temperature just as much as by too high a temperature. It is true, however, that too high hardening temperatures result in warpage, distortion, increased scale, and decarburization.

Hardening Temperatures for Carbon Tool Steels. — The best hardening temperatures for any given tool steel are dependent upon the type of tool and the intended class of service. Wherever possible, the specific recommendations of the tool steel manufacturer should be followed. General recommendations for hardening temperatures of carbon tool steels based on carbon content are as follows: For steel of 0.65 to 0.80 per cent carbon content, 1450 to 1550 degrees F.; for steel of 0.80 to 0.95 per cent carbon content, 1410 to 1460 degrees F.; for steel of 0.95 to 1.10 per cent carbon content, 1390 to 1430 degrees F.; and for steels of 1.10 per cent and over carbon content, 1380 to 1420 degrees F. For a given hardening temperature range, the higher temperatures tend to produce deeper hardness penetration and increased compressional strength while the lower temperatures tend to result in shallower hardness penetration but increased resistance to splitting or bursting stresses.

Determining Hardening Temperatures. — A hardening temperature can be specified directly or it may be specified indirectly as a certain temperature rise above the lower critical point of the steel. Where the temperature is specified directly, a pyrometer of the type which indicates the furnace temperature or a pyrometer of the type which indicates the work temperature may be employed. If the pyrometer shows furnace temperature, care must be taken to allow sufficient time for the work to reach the furnace temperature after the pyrometer indicates that the required hardening temperature has been attained. If the pyrometer indicates work temperature, then, where the work-piece is large, time must be allowed for the interior of the work to reach the temperature of the surface which is the temperature indicated by the pyrometer.

Where the hardening temperature is specified as a given temperature rise above the critical point of the steel, a pyrometer which indicates the temperature of the work

should be used. The critical point, as well as the given temperature rise, can be more accurately determined with this type of pyrometer. As the work is heated, its temperature, as indicated by the pyrometer, rises steadily until the lower critical or decalescence point of the steel is reached. At this point, the temperature of the work ceases to rise and the pyrometer indicating or recording pointer remains stationary or fluctuates slightly. After a certain elapsed period, depending upon the rate at which heat is furnished to the work, the internal changes in structure of the steel which take place at the lower critical point are completed and the temperature of the work again begins to rise. Since a small fluctuation in temperature may occur in the interval during which structural changes are taking place, for uniform practice the critical point may be considered as the temperature at which the pointer first becomes stationary.

Heating Steel in Liquid Baths. — The liquid baths commonly used for heating steel tools preparatory to hardening are molten lead, sodium cyanide, barium chloride, a mixture of barium and potassium chloride and other metallic salts. The molten substance is retained in a crucible or pot and the heat required may be obtained from gas, oil, or electricity. The principal advantages of heating baths are as follows: No part of the work can be heated to a temperature above that of the bath; the temperature can be easily maintained at whatever degree has proved, in practice, to give the best results; the submerged steel can be heated uniformly, and the finished surfaces are protected against oxidation.

Salt Baths. — Molten baths of various salt mixtures or compounds are used extensively for heat-treating operations such as hardening and tempering; they are also utilized for annealing ferrous and non-ferrous metals. Commercial salt-bath mixtures are available which meet a wide range of temperature and other metallurgical requirements. For example, there are neutral baths for heating tool and die steels without carburizing the surfaces; baths for carburizing the surfaces of low-carbon steel parts; baths adapted for the usual tempering temperatures of, say, 300 to 1100 degrees F.; and baths which may be heated to temperatures up to approximately 2400 degrees F. for hardening high-speed steels. Salt baths are also adapted for local or selective hardening, the type of bath being selected to suit the requirements. For example, a neutral bath may be used for annealing the ends of tubing or other parts, or an activated cyanide bath for carburizing the ends of shafts or other parts. Surfaces which are not to be carburized are protected by copper plating. When the work is immersed, the unplated parts are subjected to the carburizing action.

Baths may consist of a mixture of sodium, potassium, barium, and calcium chlorides or nitrates of sodium, potassium, barium, and calcium in varying proportions, to which sodium carbonate and sodium cyanide are sometimes added to prevent decarburization. Various proportions of these salts provide baths of different properties. Potassium cyanide is seldom used as sodium cyanide costs less. The specific gravity of a salt bath is not as high as that of a lead bath; consequently the work may be suspended in a salt bath and does not have to be held below the surface as in a lead bath.

The Lead Bath. — The lead bath is extensively used, but is not adapted to the high temperatures required for hardening high-speed steel, as it begins to vaporize at about 1190 degrees F. As the temperature increases, the lead volatilizes and gives off poisonous vapors; hence, lead furnaces should be equipped with hoods to carry away the fumes. Lead baths are generally used for temperatures below 1500 or 1600 degrees F. They are often employed for heating small pieces which must be hardened in quantities. It is important to use pure lead that is free from sulphur. The work should be pre-heated before plunging it into the molten lead.

Defects in Hardening. — Uneven heating is the cause of most of the defects in hardening. Cracks of a circular form, from the corners or edges of a tool, indicate uneven heating in hardening. Cracks of a vertical nature and dark-colored fissures indicate that the steel has been burned and should be put on the scrap heap. Tools which have hard and soft places have been either unevenly heated, unevenly cooled, or " soaked," a term used to indicate prolonged heating. A tool not thoroughly moved about in the hardening fluid will show hard and soft places, and have a tendency to crack. Tools which are hardened by dropping them to the bottom of the tank, sometimes have soft places, owing to contact with the floor or sides.

Scale on Hardened Steel. — The formation of scale on the surface of hardened steel is due to the contact of oxygen with the heated steel; hence, to prevent scale, the heated steel must not be exposed to the action of the air. When using an oven heating furnace, the flame should be so regulated that it is not visible in the heating chamber. The heated steel should be exposed to the air as little as possible, when transferring it from the furnace to the quenching bath. An old method of preventing scale and retaining a fine finish on dies used in jewelry manufacture, small taps, etc. is as follows: Fill the die impression with powdered boracic acid and place near the fire until the acid melts; then add a little more acid to insure covering all the surfaces. The die is then hardened in the usual way. If the boracic acid does not come entirely off in the quenching bath, immerse the work in boiling water. Dies hardened by this method are said to be as durable as those heated without the acid.

Hardening or Quenching Baths. — The purpose of a quenching bath is to remove heat from the steel being hardened at a rate that is faster than the critical cooling rate. Generally speaking, the more rapid the rate of heat extraction above the cooling rate, the higher will be the resulting hardness. To obtain the different rates of cooling required by different classes of work, baths of various kinds are used. These include plain or fresh water, brine, caustic soda solutions, oils of various classes, oil-water emulsions, baths of molten salt or lead for high-speed steels and air cooling for some high-speed steel tools when a slow rate of cooling is required. To minimize distortion and cracking where such tendencies are present, without sacrificing depth of hardness penetration, a quenching medium should be selected that will cool rapidly at the higher temperatures and more slowly at the lower temperatures, i.e., below 750 degrees F. Oil quenches in general meet this requirement.

Oil Quenching Baths: Oil is used very extensively as a quenching medium as it results in a good proportion of hardness, toughness, and freedom from warpage when used with standard steels. Oil baths are used extensively for alloy steels. Various kinds of oils are employed such as prepared mineral oils and vegetable, animal and fish oils, either singly or in combination. Prepared mineral quenching oils are widely used because they have good quenching characteristics, are chemically stable, do not have an objectionable odor, and are relatively inexpensive. Special compounded oils of the soluble type are used in many plants instead of such oils as fish oil, linseed oil, cottonseed oil, etc. The soluble properties enable the oil to form an emulsion with water.

Oil cools steel at a slower rate than water, but the rate is fast enough for alloy steel. Oils have different cooling rates, however, and this rate may vary through the initial and final stages of the quenching operation. Faster cooling in the initial stage and slower cooling at lower temperatures is preferable because there is less danger of cracking the steel. The temperature of quenching oil baths should range ordinarily between 90 and 130 degrees F. A fairly constant temperature may be maintained either by circulating the oil through cooling coils or by using a tank provided with a cold water jacket.

A good quenching oil should possess a flash and fire point sufficiently high to be safe under the conditions used and 350 degrees F. should be about the minimum point. The specific heat of the oil regulates the hardness and toughness of the quenched steel; and the greater the specific heat, the higher will be the hardness produced. Specific heats of quenching oils vary from 0.20 to 0.75, the specific heats of fish, animal and vegetable oils usually being from 0.2 to 0.4, and of soluble and mineral oils from 0.5 to 0.7. The efficient temperature range for quenching oil is from 90 to 140 degrees F.

Quenching in Water. — Many carbon tool steels are hardened by immersing them in a bath of fresh water, but water is not an ideal quenching medium. Contact between the water and work and the cooling of the hot steel is impaired by the formation of gas bubbles or an insulating vapor film especially in holes, cavities or pockets. The result is uneven cooling and in some cases excessive strains which may cause the tool to crack; in fact, there is greater danger of cracking in a fresh water bath than in one containing salt water or brine.

In order to secure more even cooling and reduce danger of cracking, either rock salt (8 or 9 per cent) or caustic soda (3 to 5 per cent) may be added to the bath in order to eliminate or prevent the formation of a vapor film or gas pockets, thus promoting rapid early cooling. Brine is commonly used and ¾ pound of rock salt per gallon of water is equivalent to about 8 per cent of salt. Brine is not inherently a more severe or drastic quenching medium than plain water, although it may seem to be because the brine makes better contact with the heated steel and, consequently, cooling is more effective. In still bath quenching, a slow up-and-down movement of the tool is preferable to a violent swishing around.

The temperature of water-base quenching baths should preferably be kept around 70 degrees F., but 70 to 90 or 100 degrees F. is a safe range. The temperature of the hardening bath has a great deal to do with the hardness obtained. The higher the temperature of the quenching water, the more nearly does its effect approach that of oil; and if boiling water is used for quenching, it will have an effect even more gentle than that of oil — in fact, it would leave the steel nearly soft. Parts of irregular shape are sometimes quenched in a water bath that has been warmed somewhat to prevent sudden cooling and cracking.

When water is used, it should be "soft" as unsatisfactory results will be obtained with "hard" water. Any contamination of water-base quenching liquids by soap tends to decrease their rate of cooling. A water bath having 1 or 2 inches of oil on the top is sometimes employed to advantage for quenching tools made of high-carbon steel as the oil through which the work first passes reduces the sudden quenching action of the water.

The bath should be amply large to dissipate the heat rapidly and the temperature should be kept about constant so that successive pieces will be cooled at the same rate. Irregularly shaped parts should be immersed so that the heaviest or thickest section enters the bath first. After immersion, the part to be hardened should be agitated in the bath; the agitation reduces the tendency of the formation of a vapor coating on certain surfaces, and a more uniform rate of cooling is obtained. The work should never be dropped to the bottom of the bath until quite cool.

Flush or Local Quenching by Pressure-spraying: When dies for cold heading, drawing, extruding, etc., or other tools, require a hard working surface and a relatively soft but tough body, the quenching may be done by spraying water under pressure against the interior or other surfaces to be hardened. Special spraying fixtures are used to hold the tool and apply the spray where the hardening is required. The pressure-spray prevents the formation of gas pockets previously referred to in connection with the fresh water quenching bath; hence fresh water is effective for flush quenching and there is no advantage in using brine.

Quenching in Molten Salt Bath. — A molten salt bath may be used in preference to oil for quenching high-speed steel. The object in using a liquid salt bath for quenching (instead of an oil bath) is to obtain maximum hardness with minimum cooling stresses and distortion which might result in cracking expensive tools, especially if there are irregular sections. The temperature of the quenching bath may be around 1100 or 1200 degrees F. Quenching is followed by cooling to room temperature and then the tool is tempered or drawn in a bath having a temperature range of 950 to 1100 degrees F. In many cases, the tempering temperature is about 1050 degrees F.

Tanks for Quenching Baths. — The main point to be considered in a quenching bath is to keep it at a uniform temperature, so that successive pieces quenched will be subjected to the same heat. The next consideration is to keep the bath agitated, so that it will not be of different temperatures in different places; if thoroughly agitated and kept in motion, as is the case with the bath shown in Fig. 1, it is not even necessary to keep the pieces in motion in the bath, as steam will not be likely to form around the pieces quenched. Experience has proved that if a piece is held still in a thoroughly agitated bath, it will come out much straighter than if it has been moved around in an unagitated bath. This is an important consideration, especially when hardening long pieces. It is, besides, no easy matter to keep heavy and long pieces in motion unless it be done by mechanical means.

In Fig. 1 is shown a water or brine tank for quenching baths. Water is forced by a pump or other means through the supply pipe into the intermediate space between the outer and inner tank. From the intermediate space it is forced into the inner tank through holes as indicated. The water returns to the storage tank by overflowing from the inner tank into the outer one and then through the overflow pipe as indicated. In Fig. 3 is shown another water or brine tank of a more common type. In this case the water or brine is pumped from the storage tank and continuously returned to it. If the storage tank contains a large volume of water, there is no need of a special means for cooling. Otherwise, arrangements must be made for cooling the water after it has passed through the tank. The bath is agitated by the force with which the water is pumped into it. The holes at A are drilled at an angle, so as to throw the water toward the center of the tank. In Fig. 2 is shown an oil quenching tank in which water is circulated in an outer surrounding tank for keeping the oil bath cool. Air is forced into the oil bath to keep it agitated. Fig. 4 shows the ordinary type of quenching tank cooled by water forced through a coil of pipe. This can be used for oil, water or brine. Fig. 5 shows a similar type of quenching tank, but with two coils of pipe. Water flows through one of these and steam through the other. By these means it is possible to keep the bath at a constant temperature.

Interrupted Quenching. — *Austempering, martempering,* and *isothermal quenching* are three methods of interrupted quenching that have been developed to obtain greater toughness and ductility for given hardnesses and to avoid the difficulties of quench cracks, internal stresses, and warpage, frequently experienced when the conventional method of quenching steel directly and rapidly from above the transformation point to atmospheric temperature is employed. In each of these three methods, quenching is begun when the work has reached some temperature above the transformation point and is conducted at a rate faster than the critical rate. The rapid cooling of the steel is interrupted, however, at some temperature above that at which martensite begins to form. The three methods differ in the temperature range at which interruption of the rapid quench takes place, the length of time that the steel is held at this temperature, and whether the subsequent cooling to atmospheric temperature is rapid or slow, and is or is not preceded by a tempering operation.

One of the reasons for maintaining the steel at a constant temperature for a definite period of time is to permit the inside sections of the piece to reach the same temper-

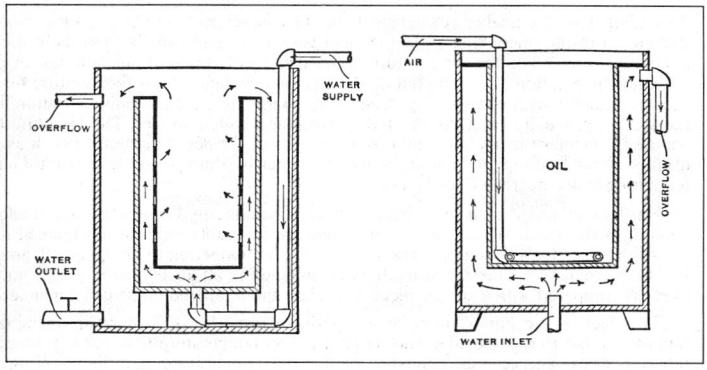

Fig. 1.   Fig. 2

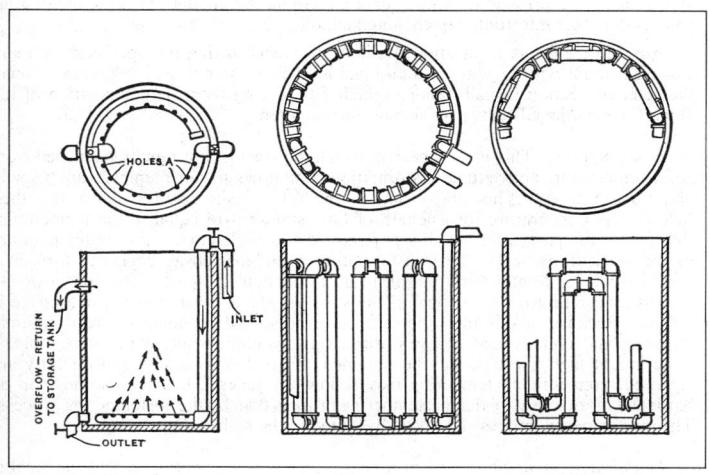

Fig. 3   Fig. 4   Fig. 5

ature as the outer sections so that when transformation of the structure does take place, it will occur at about the same rate and period of time throughout the piece. In order to maintain the constant temperature required in interrupted quenching, a quenching arrangement for absorbing and dissipating a large quantity of heat without increase in temperature is needed. Molten salt baths equipped for water spray or air cooling around the exterior of the bath container have been used for this purpose.

Austempering: This is a heat-treating process in which steels are quenched in a bath maintained at some constant temperature in the range of 350 to 800 degrees F.,

depending upon the analysis of the steel and the characteristics to be obtained. Upon immersion in the quenching bath, the steel is cooled more rapidly than the critical quenching rate. When the temperature of the steel reaches that of the bath, however, the quenching action is interrupted. If the steel is now held at this temperature for a predetermined length of time, say, from 10 to 60 minutes, the austenitic structure of the steel is gradually changed into a new structure, called *bainite*. The structure of bainite is acicular (needlelike) and resembles that of tempered martensite such as is usually obtained by quenching in the usual manner to atmospheric temperature and tempering at 400 degrees F. or higher.

Hardnesses ranging up to 60 Rockwell C, depending upon the carbon and alloy content of the steel, are obtainable and compare favorably wtih those obtained for the respective steels by a conventional quench and tempering to above 400 degrees F. Much greater toughness and ductility is obtained in an austempered piece, however, as compared with a similar piece quenched and tempered in the usual manner.

Two factors are important in austempering. First, the steel must be quenched rapidly enough to the specified sub-transformation temperature to avoid any formation of pearlite, and second, it must be held at this temperature until the transformation from austenite to bainite is completed. Time and temperature transformation curves (called S-curves because of their shape) have been developed for different steels and these provide important data governing the conduct of austempering, as well as the other interrupted quenching methods.

Austempering has been applied chiefly to steels having 0.60 per cent or more carbon content with or without additional low alloy content, and to pieces of small diameter or section, usually under 1 inch but varying with the composition of the steel. Case hardened parts may also be austempered.

Martempering: This is a process in which the steel is first rapidly quenched from some temperature above the transformation point down to some temperature (usually about 400 degrees F.) just above that at which martensite begins to form. It is then held at this temperature for a length of time sufficient to equalize the temperature throughout the part, after which it is removed and cooled in air. As the temperature of the steel drops below the transformation point, martensite begins to form in a matrix of austenite at a fairly uniform rate throughout the piece. The soft austenite acts as a cushion to absorb some of the stresses which develop as the martensite is formed. Because of this fact, the difficulties presented by quench cracks, internal stresses, and dimensional changes are largely avoided, while at the same time, a structure of high hardness can be obtained. If greater toughness and ductility are required, conventional tempering may follow. In general, heavier sections can be hardened more easily by the martempering process than by the austempering process. The martempering process is especially suited to the higher alloyed steels.

Isothermal Quenching: This process resembles austempering in that the steel is first rapidly quenched from above the transformation point down to a temperature which is above that at which martensite begins to form and is held at this temperature until the austenite is completely transformed into bainite. The constant temperature to which the piece is quenched and then maintained is usually 450 degrees F. or above. The process differs from austempering in that after transformation to a bainite structure has been completed, the steel is immersed in another bath and is brought up to some higher temperature, depending upon the characteristics desired, and is maintained at this temperature for a definite period of time, followed by cooling in air. Thus, tempering to obtain the desired toughness or ductility takes place immediately after the structure of the steel has changed to bainite and before it is cooled to atmospheric temperature.

Tempering

The object of *tempering* or *drawing* is to reduce the brittleness in hardened steel and to remove the internal strains caused by the sudden cooling in the quenching bath. The tempering process consists in heating the steel by various means to a certain temperature and then cooling it. When steel is in a fully hardened condition, its structure consists largely of *martensite*. On reheating to a temperature of from about 300 to 750 degrees F., a softer and tougher structure known as *troostite* is formed. If the steel is reheated to a temperature of from 750 to 1290 degrees F., a structure known as *sorbite* is formed which has somewhat less strength than troostite but much greater ductility.

Tempering Temperatures. — If steel is heated in an oxidizing atmosphere a film of oxide forms on the surface which changes color as the temperature increases. These oxide colors (see table) have been used extensively in the past as a means of gaging the correct amount of temper; but since these colors are affected to some extent by the composition of the metal, the method is not dependable.

The availability of reliable pyrometers in combination with tempering baths of oil, salt or lead make it possible to heat the work uniformly and to a given temperature within close limits.

Suggested temperatures for tempering various tools are given in the accompanying table.

Tempering in Oil. — Oil baths are extensively used for tempering tools (especially in quantity), the work being immersed in oil heated to the required temperature, which is indicated by a thermometer. It is important that the oil have a uniform temperature throughout and that the work be immersed long enough to acquire this temperature. Cold steel should not be plunged into a bath heated for tempering, owing to the danger of cracking it. The steel should either be preheated to about 300 degrees F., before placing it in the bath, or the latter should be at a comparatively low temperature before immersing the steel, and then be heated to the required degree. A temperature of from 650 to 700 degrees F. can be obtained with heavy tempering oils; for higher temperatures, either a bath of nitrate salts or a lead bath may be used.

In tempering, the best method is to immerse the pieces to be tempered in the oil before starting to heat the latter. They are then heated with the oil. After the pieces tempered are taken out of the oil bath, they should immediately be dipped in a tank of caustic soda, and after that in a tank of hot water. This will remove all oil which might adhere to the tools. The following tempering oil has given satisfactory results: mineral oil, 94 per cent; saponifiable oil, 6 per cent; specific gravity, 0.920; flash point, 550 degrees F.; fire test, 625 degrees F.

Tempering in Salt Baths. — Molten salt baths may be used for tempering or drawing operations. Nitrate baths are particularly adapted for the usual drawing temperature range of, say, 300 to 1100 degrees F. Tempering in an oil bath usually is limited to temperatures of 500 to 600 degrees, and some heat-treating specialists recommend the use of a salt bath for temperatures above 350 or 400 degrees, as it is considered more efficient and economical. Tempering in a bath (salt or oil) has several advantages, such as ease in controlling the temperature range and maintenance of a uniform temperature. The work is also heated much more rapidly in a molten bath. While a gas- or oil-fired muffle or semi-muffle furnace may be used for tempering, a salt bath or oil bath is preferable. A salt bath is recommended for tempering high-speed steel, although furnaces may also be used. The bath or furnace temperature should be increased gradually, say, from 300 to 400 degrees up to the tempering temperature which may range from 1050 to 1150 degrees F. for high-speed steel.

Tempering in a Lead Bath. — The lead bath is commonly used for heating steel in connection with tempering, as well as for hardening. The bath is first heated to the temperature at which the steel should be tempered; the pre-heated work is then placed in the bath long enough to acquire this temperature, after which it is removed and cooled. As the melting temperature of pure lead is about 620 degrees F., tin is commonly added to it to lower the temperature sufficiently for tempering. Reductions in temperature can be obtained by varying the proportions of lead and tin, as shown by the table, "Temperatures of Lead Bath Alloys."

Temperatures of Lead Bath Alloys

Parts Lead	Parts Tin	Melting Temp., Deg. F.	Parts Lead	Parts Tin	Melting Temp., Deg. F.	Parts Lead	Parts Tin	Melting Temp., Deg. F.
200	8	560	39	8	510	19	8	460
100	8	550	33	8	500	17	8	450
75	8	540	28	8	490	16	8	440
60	8	530	24	8	480	15	8	430
48	8	520	21	8	470	14	8	420

To Prevent Lead from Sticking to Steel. — To prevent hot lead from sticking to parts heated in it, mix common whiting with wood alcohol, and paint the part that is to be heated. Water can be used instead of alcohol, but in that case the paint must be thoroughly dry, as otherwise the moisture will cause the lead to "fly." Another method is to make a thick paste according to the following formula: Pulverized charred leather, 1 pound; fine wheat flour, 1½ pounds; fine table salt, 2 pounds. Coat the tool with this paste and heat slowly until dry, then proceed to harden. Still another method is to heat the work to a blue color, or about 600 degrees F., and then dip it in a strong solution of salt water, prior to heating in the lead bath. The lead is sometimes removed from parts having fine projections or teeth, by using a stiff brush just before immersing in the cooling bath. This is necessary to prevent the formation of soft spots.

Tempering in Sand. — The sand bath is used for tempering certain classes of work. One method is to deposit the sand on an iron plate or in a shallow box which has burners beneath it. With this method of tempering, tools such as boiler punches, etc., can be given a varying temper by placing them endwise in the sand. As the temperature of the sand bath is higher toward the bottom, a tool can be so placed that the color of the lower end will be a deep dark blue when the middle portion is a very dark straw, and the working end or top a light straw color, the hardness gradually increasing from the bottom up.

Double Tempering. — In tempering high-speed steel tools, it is common practice to repeat the tempering operation or "double temper" the steel. This is done by heating the steel to the tempering temperature (say 1050 degrees F.) and holding it at that temperature for two hours. It is then cooled to room temperature, reheated to 1050 degrees F. for another two-hour period, and again cooled to room temperature. After the first tempering operation, some untempered martensite remains in the steel. This martensite is not only tempered by a second tempering operation but is relieved of internal stresses, thus improving the steel for service conditions. The hardening temperature for the higher alloy steels may affect the hardness after tempering. For example, molybdenum high-speed steel when heated to 2100 degrees F. had a hardness of 61 Rockwell C after tempering whereas a temperature of 2250 degrees F. resulted in a hardness of 64.5 Rockwell C after tempering.

Temperatures as Indicated by the Color of Plain Carbon Steel

Degrees Centigrade	Degrees Fahrenheit	Color of Steel	Degrees Centigrade	Degrees Fahrenheit	Color of Steel
221.1	430	Very pale yellow	265.6	510	Spotted red-brown
226.7	440	Light yellow	271.1	520	Brown-purple
232.2	450	Pale straw-yellow	276.7	530	Light purple
237.8	460	Straw-yellow	282.2	540	Full purple
243.3	470	Deep straw-yellow	287.8	550	Dark purple
248.9	480	Dark yellow	293.3	560	Full blue
254.4	490	Yellow-brown	298.9	570	Dark blue
260.0	500	Brown-yellow	337.8	640	Light blue

Tempering Temperatures for Various Plain Carbon Steel Tools

Degrees F.	Class of Tool
495 to 500	Taps ½ inch or over, for use on automatic screw machines.
495 to 500	Nut taps ½ inch and under.
515 to 520	Taps ¼ inch and under, for use on automatic screw machines.
525 to 530	Thread dies to cut thread close to shoulder.
500 to 510	Thread dies for general work.
495	Thread dies for tool steel or steel tube.
525 to 540	Dies for bolt threader threading to shoulder.
460 to 470	Thread rolling dies.
430 to 435	Hollow mills (solid type) for roughing on automatic screw machine.
485	Knurls.
450	Twist drills for hard service.
450	Centering tools for automatic screw machine.
430	Forming tools for automatic screw machine.
430 to 435	Cut-off tools for automatic screw machine.
440 to 450	Profile cutters for milling machine.
430	Formed milling cutters.
435 to 440	Milling cutters.
430 to 440	Reamers.
460	Counterbores and countersinks.
480	Cutters for tube or pipe-cutting machine.
460 to 520	Snaps for pneumatic hammers — harden full length, temper to 460 degrees, then bring point to 520 degrees.

Annealing, Spheroidizing, and Normalizing

Annealing of steel is a heat-treating process in which the steel is heated to some elevated temperature, usually in or near the critical range, is held at this temperature for some period of time, and is then cooled, usually at a slow rate. Spheroidizing and normalizing may be considered as special cases of annealing.

The *full annealing* of carbon steel consists in heating it slightly above the *upper* critical point for hypo-eutectoid steels (steels of less than 0.85 per cent carbon content) and slightly above the *lower* critical point for hyper-eutectoid steels (steels of more than 0.85 per cent carbon content), holding it at this temperature until it is uniformly heated and then slowly cooling it to 1000 degrees F. or below. The resulting structure is layer-like or lamellar in character due to the pearlite which is

formed during the slow cooling. Annealing is employed (1) to soften steel for machining, cutting, stamping, etc., or for some particular service; (2) to alter ductility, toughness, electrical or magnetic characteristics or other physical properties; (3) to refine the crystal structure; (4) to produce grain reorientation; or (5) to relieve stresses and hardness resulting from cold working.

The *spheroidizing* of steel, according to the American Society of Metals, is "any process of heating and cooling that produces a rounded or globular form of carbide." High-carbon steels are spheroidized to improve their machinability especially in continuous cutting operations such as are performed by lathes and screw machines. In low-carbon steels, spheroidizing may be employed to meet certain strength requirements before subsequent heat-treatment. Spheroidizing also tends to increase resistance to abrasion.

The *normalizing* of steel consists in heating it to some temperature above that used for annealing, usually about 100 degrees F. above the upper critical range, and then cooling it in still air at room temperature. Normalizing is intended to put the steel into a uniform, unstressed condition of proper grain size and refinement so that it will properly respond to further heat-treatments. It is particularly important in the case of forgings which are to be later heat-treated. Normalizing may or may not (depending upon the composition) leave steel in a sufficiently soft state for machining with available tools. In some cases, annealing for machinability is preceded by normalizing and the combined treatment — frequently called a *double anneal* — produces a better result than a simple anneal.

Annealing Practice. — For carbon steels, the following annealing temperatures are recommended by the American Society for Testing and Materials: Steels of less than 0.12 per cent carbon content, 1600 to 1700 degrees F.; steels of 0.12 to 0.29 per cent carbon content, 1550 to 1600 degrees F.; steels of 0.30 to 0.49 per cent carbon content, 1500 to 1550 degrees F.; and for 0.50 to 1.00 per cent carbon steels, from 1450 to 1500 degrees F. Slightly lower temperatures are satisfactory for steels having more than 0.75 per cent manganese content. Heating should be uniform to avoid the formation of additional stresses. In the case of large work-pieces, the heating should be slow enough so that the temperature of the interior does not lag too far behind that of the surface.

It has been found that in annealing steel, the higher the temperature to which it is heated to produce an austenitic structure, the greater the tendency of the structure to become lamellar (pearlitic) in cooling. On the other hand, the closer the austenitizing temperature to the critical temperature, the greater is the tendency of the annealed steel to become spheroidal.

Rate of Cooling: After heating the steel to some temperature within the annealing range, it should be cooled slowly enough to permit the development of the desired softness and ductility. In general, the slower the cooling rate, the greater the resulting softness and ductility. Steel of a high carbon content should be cooled more slowly than steel of a low carbon content; also the higher the alloy content, the slower is the cooling rate usually required. Where extreme softness and ductility are not required, the steel may be cooled in the annealing furnace to some temperature well below the critical point, say, to about 1000 degrees F. and then removed and cooled in air.

Annealing by Constant Temperature Transformation. — It has been found that steel which has been heated above the critical point so that it has an austenitic structure can be transformed into a lamellar (pearlitic) or a spheroidal structure by holding it for a definite period of time at some constant sub-critical temperature. In other words, it is feasible to anneal steel by means of a constant-temperature transformation as well as by the conventional continuous cooling method. When th

constant-temperature transformation method is employed, the steel, after being heated to some temperature above the critical and held at this temperature until it is austenitized, is cooled as rapidly as feasible to some relatively high sub-critical transformation temperature. The selection of this temperature is governed by the desired microstructure and hardness required and is taken from a transformation time and temperature curve (often called a TTT curve.) As drawn for a particular steel, such a curve shows the length of time required to transform that steel from an austenitic state at various sub-critical temperatures. After being held at the selected sub-critical temperature for the required length of time, the steel is cooled to room temperature — again, as rapidly as feasible. This rapid cooling down to the selected transformation temperature and then down to room temperature has a negligible effect on the structure of the steel and makes possible in many instances a considerable saving in time over the conventional slow cooling method of annealing.

The softest condition in steel can be developed by heating it to a temperature usually less than 100 degrees F. above the lower critical point and then cooling it to some temperature, usually less than 100 degrees below the critical point, where it is held until the transformation is completed. Certain steels require a very lengthy period of time for transformation of the austenite when held at a constant temperature within this range. For such steels a practical procedure is to allow most of the transformation to take place in this temperature range where a soft product is formed and then to finish the transformation at a lower temperature where the time for the completion of the transformation is short.

Spheroidizing Practice. — A common method of spheroidizing steel consists in heating it to or slightly below the lower critical point, holding it at this temperature for a period of time, and then cooling it slowly to about 1000 degrees F. or below. The length of time for which the steel is held at the spheroidizing temperature largely governs the degree of spheroidization. High-carbon steel may be spheroidized by subjecting it to a temperature that alternately rises and falls between a point within and a point without the critical range. Tool steel may be spheroidized by heating to a temperature slightly above the critical range and then, after being held at this temperature for a period of time, cooling without removal from the furnace.

Normalizing Practice. — When using the lower carbon steels, simple normalizing is often sufficient to place the steel in its best condition for machining and will lessen distortion in carburizing or hardening. In the medium and higher carbon steels, combined normalizing and annealing constitutes the best practice. For unimportant parts, the normalizing may be omitted entirely or annealing practiced only when the steel is otherwise difficult to machine. Both processes are recommended in the following heat-treatments (for S.A.E. steels) as representing the best metallurgical practice. The temperatures recommended for normalizing and annealing have been made indefinite in many instances because of the many different types of furnace used in various plants and the difference in results desired.

Case Hardening

In order to harden low-carbon steel it is necessary to increase the carbon content of the surface of the steel so that a thin outer "case" can be hardened by heating the steel to the hardening temperature and then quenching it. The process, therefore, involves two separate operations. The first is the *carburizing* operation for impregnating the outer surface with sufficient carbon, and the second operation is that of heat-treating the carburized parts so as to obtain a hard outer case and, at the same time, give the "core" the required physical properties. The term "case hardening" is ordinarily used to indicate the complete process of carburizing and hardening.

Carburization. — Carburization is the result of heating iron or steel to a temperature below its melting point in the presence of a solid, liquid, or gaseous material which decomposes so as to liberate carbon when heated to the temperature used. In this way, it is possible to obtain by the gradual penetration, diffusion, or absorption of the carbon by the steel, a "zone" or "case" of higher carbon content at the outer surfaces than that of the original object. When a carburized object is rapidly cooled or quenched in water, oil, brine, etc., from the proper temperature, this case becomes hard, leaving the inside of the piece soft, but of great toughness.

Use of Carbonaceous Mixtures. — When carburizing materials of the solid class are used, the case-hardening process consists in packing steel articles in metal boxes or pots, with a carbonaceous compound surrounding the steel objects. The boxes or pots are sealed and placed in a carburizing oven or furnace maintained usually at a temperature of from about 1650 to 1700 degrees F. for a length of time depending upon the extent of the carburizing action desired. The carbon from the carburizing compound will then be absorbed by the steel on the surfaces desired, and the low-carbon steel is converted into high-carbon steel at these portions, while the internal sections and the insulated parts of the object retain practically their original low-carbon content. The result is a steel of a dual structure, a high-carbon and a low-carbon steel in the same piece. The carburized steel may now be heat-treated by heating and quenching, in much the same way as high-carbon steel is hardened, in order to develop the properties of hardness and toughness; but as the steel is, in reality, two steels in one, one high-carbon and one low-carbon, the correct heat-treatment after carburizing includes two distinct processes, one suitable for the high-carbon portion or the "case," as it is generally termed, and one suitable for the low-carbon portion or core. The method of heat-treatment varies according to the kind of steel used. Usually an initial heating and slow cooling is followed by reheating to 1400–1450 degrees F., quenching in oil or water, and a final tempering. More definite information is given in the following section on S.A.E. steels.

Carburizers: There are many commercial carburizers on the market in which the materials used as the generator may be hard and soft wood charcoal, animal charcoal, coke, coal, beans and nuts, bone and leather, or various combinations of these. The energizers may be barium, cyanogen, and ammonium compounds, various salts, soda ash, or lime and oil hydrocarbons.

Pack-Hardening. — When cutting tools, gages, and other parts made from high-carbon steels are heated for hardening while packed in some carbonaceous material in order to protect delicate edges, corners or finished surfaces, the process usually is known as pack-hardening. Thus, the purpose is to protect the work, prevent scale formation, insure uniform heating, and minimize the danger of cracking and warpage. The work is packed, as in carburizing, and in the same type of receptacle. Common hardwood charcoal often is used, especially if it has had an initial heating to eliminate shrinkage and discharge its more impure gases. The lowest temperature required for hardening should be employed for pack-hardening — usually 1400 to 1450 degrees F. for carbon steels. Pack-hardening has also been applied to high-speed steels, but modern developments in heat-treating salts have made it possible to harden high-speed steel without decarburization, injury to sharp edges, or marring the finished surfaces. See paragraph on Salt Baths.

Cyanide Hardening. — When low-carbon steel requires a very hard outer surface but does not need high shock-resisting qualities, the cyanide hardening process may be employed to produce what is known as superficial hardness. This superficial hardening is the result of carburizing a very thin outer skin (which may be only a few thousandths inch thick) by immersing the steel in a bath containing sodium cyanide. The temperatures usually vary from 1450 to 1650 degrees F. and the percentage of

sodium cyanide in the bath extends over a wide range, depending upon the steel used and properties required.

Nitriding Process. — Nitriding is a process for surface hardening certain alloy steels by heating the steel in an atmosphere of nitrogen (ammonia gas) at approximately 950 degrees F. The steel is then cooled slowly. Finish machined surfaces hardened by nitriding are subject to minimum distortion. The physical properties, such as toughness, high impact strength, etc., can be imparted to the core by previous heat-treatments and are unaffected by drawing temperatures up to 950 degrees F. The "Nitralloy" steels suitable for this process may readily be machined in the heat-treated as well as in the annealed state, and they forge as easily as alloy steels of the same carbon content. Certain heat-treatments must be applied prior to nitriding, the first being annealing to relieve rolling, forging, or machining strains. Parts or sections not requiring heat-treating should be machined or ground to the exact dimensions required. Close tolerances must be maintained in finish machining, and allowances for growth due to adsorption of nitrogen should be made, and this usually amounts to about 0.0005 inch for a case depth of 0.02 inch. Parts requiring heat-treatment for definite physical properties are forged or cut from annealed stock, heat-treated for the desired physical properties, rough machined, normalized, and finish machined. If quenched and drawn parts are normalized afterwards, the drawing and normalizing temperatures should be alike. The normalizing temperature may be below but should never be above the drawing temperature.

Ion Nitriding. — Ion nitriding, also referred to as glow discharge nitriding, is a relatively new process for case hardening of steel parts such as tool spindles, cutting tools, extrusion equipment, forging dies, gears, and crankshafts. An electrical potential ionizes low-pressure nitrogen gas, and the ions produced are accelerated to and impinge upon the workpiece, heating it to the appropriate temperature for diffusion to take place. Therefore, there is no requirement for a supplemental heat source. The inward diffusion of the nitrogen ions forms the iron and alloy nitrides in the case. White layer formation, familiar in conventional gas nitriding, is readily controlled by this process.

Liquid Carburizing. — Activated liquid salt baths are now used extensively for carburizing. Sodium cyanide and other salt baths are used. The salt bath is heated by electrodes immersed in it, the bath itself acting as the conductor and resistor. One or more groups of electrodes, with two or more electrodes per group, may be used. The heating is accompanied by a stirring action to insure uniform temperature and carburizing activity throughout the bath. The temperature may be controlled by a thermocouple immersed in the bath and connecting with a pyrometer designed to provide automatic regulation. The advantages of liquid baths include rapid action; uniform carburization; minimum distortion; and elimination of the packing and unpacking required when carbonaceous mixtures are used. In selective carburizing, the portions of the work which are not to be carburized are copperplated and the entire piece is then immersed in an activated cyanide bath. The copper inhibits any carburizing action on the plated parts, and this method offers a practical solution for selectively carburizing any portion of a steel part.

Gas Carburizing. — When carburizing gases are used, the mixture varies with the type of case and quality of product desired. The gaseous hydrocarbons most widely used are methane (natural gas), propane, and butane. These carbon bearing gases are mixed with air, with manufactured gases of several types, with flue gas, or with other specially prepared "diluent" gases. It is necessary to maintain a continuous fresh stream of carburizing gases to the carburizing retort or muffle, as well as to remove the spent gases from the muffle continuously, in order to obtain the correct mixture

of gases inside the muffle. A slight pressure is maintained on the muffle to exclude foreign gases.

The horizontal rotary type of gas carburizing furnace has a retort or muffle which revolves slowly. This type is adapted to small parts such as ball and roller bearings, chain links, small axles, bolts, etc. With this type of furnace very large pieces such as gears, for example, may be injured by successive shocks due to tumbling within the rotor.

The vertical pit type of gas carburizer has a stationary rotor which is placed vertically in a pit. The work, instead of circulating in the gases as with the rotary type, is stationary and the gases circulate around it. This type is applicable to long large shafts or other parts or shapes that cannot be rolled in a rotary type of furnace. There are three types of continuous gas furnaces which may be designated as (1) direct quench and manually operated, (2) direct quench and mechanically operated, and (3) cooling-zone type. Where production does not warrant using a large continuous type furnace, a horizontal muffle furnace of the batch type may be used, especially if the quantities of work are varied and the production not continuous.

Vacuum Carburizing. — Vacuum carburizing is a high-temperature gas carburizing process that is performed at pressures below atmospheric. The furnace atmosphere usually consists solely of an enriching gas, such as natural gas, pure methane, or propane; nitrogen is sometimes used as a carrier gas. Vacuum carburizing offers a number of advantages such as combining of processing operations and reduced total processing time.

Carburizing Steels. — A low-carbon steel containing, say, from 0.10 to 0.20 per cent of carbon is suitable for carburized case hardening. In addition to straight-carbon steels, the low-carbon alloy steels are employed. The alloys add to case-hardened parts the same advantageous properties which they give to other classes of steel. Various steels for case hardening will be found in the section on S.A.E. steels.

To Clean Work after Case Hardening. — To clean work, especially if knurled where dirt is likely to stick into crevices after case hardening, wash it in caustic soda (1 part soda to 10 parts water). In making this solution, the soda should be put into hot water gradually, and the mixture stirred until the soda is thoroughly dissolved. A still more effective method of cleaning is to dip the work into a mixture of 1 part sulfuric acid and 2 parts water. Leave the pieces in this mixture about three minutes; then wash them off immediately in a soda solution.

Flame Hardening. — This method of hardening is especially applicable to the selective hardening of large steel forgings or castings which must be finish-machined prior to heat-treatment, or which because of size or shape cannot be heat-treated by using a furnace or bath. An oxy-acetylene torch is used to heat quickly the surface to be hardened; this surface is then quenched to secure a hardened layer which may vary in depth from a mere skin to ¼ inch with hardness ranging from 400 to 700 Brinell. A multi-flame torch-head may be equipped with quenching holes or a spray nozzle back of the flame. This is not a carburizing or a case hardening process as the torch is only a heating medium. Most authorities recommend tempering or drawing of the hardened surface at temperatures between 200 and 350 degrees F. This treatment may be done in a standard furnace, an oil bath, or with a gas flame. It should follow the hardening process as closely as possible. Medium-carbon and many low-alloy steels are suitable for hardening. Plain carbon steels ranging from 0.35 to 0.60 per cent carbon will give hardnesses of from 400 to 700 Brinell. Steels in the 0.40 to 0.45 per cent carbon range are preferred, as they have excellent core properties and produce hardnesses of from 400 to 500 Brinell without checking or cracking. Higher carbon steels will give greater hardnesses, but extreme care must be taken to prevent cracking. This requires careful control of the quenching operation.

Spinning Method of Flame Hardening: This method is employed on circular objects that can be rotated or spun past a stationary flame. It may be subdivided according to the speed of rotation, as, first, where the part is rotated slowly in front of a stationary flame and the quench is applied immediately after the flame. This method is used on large circular pieces such as track wheels and bearing surfaces. There will be a narrow band of material with lower hardness between adjacent torches if more than one path of the flame is required to harden the surface. There will also be an area of lower hardness where the flame is extinguished. A second method is applicable to small rollers or pinions. The work is spun at a speed of 50 to 150 R.P.M. in front of the flame until the entire piece has reached the proper temperature; then it is quenched as a unit by a cooling spray or ejecting into a cooling bath.

The Progressive Method: With this method the torch travels along the face of the work while the work remains stationary. It is used to harden lathe ways, gear teeth, and track rails.

The Stationary or Spot-hardening Method: When this method is employed, the work and torch are both stationary. When the spot to be hardened reaches the quenching temperature, the flame is removed and the quench applied.

The Combination Method: This is a combination of the spinning and progressive methods. It is used for long bearing surfaces. The work rotates slowly past the torch as the torch travels longitudinally across the face of the work at the rate of the torch width per revolution of the work.

The equipment for the stationary method of flame-hardening consists merely of an acetylene torch, an oxy-acetylene supply, and a suitable means of quenching; but when the other methods are employed, work-handling tools are essential and specially designed torches are desirable. A lathe is ideally suited for the spinning or combination hardening method, while a planer is adapted for progressive hardening. Production jobs, such as the hardening of gears, require specially designed machines. These machines reduce handling and hardening time, as well as assuring consistent results.

Induction Hardening. — The hardening of steel by means of induction heating and subsequent quenching in either liquid or air, is particularly applicable to parts which require localized hardening or controlled depth of hardening and to irregularly shaped parts, such as cams which require uniform surface hardening around their contour. Advantages offered by induction hardening are: (1) a short heating cycle which may range from a fraction of a second to several seconds (heat energy can be induced in a piece of steel at the rate of 100 to 250 B.T.U. per square inch per minute by induction heating, as compared with a rate of 3 B.T.U. per square inch per minute for the same material at room temperature when placed in a furnace with a wall temperature of 000 degrees F.); (2) absence of tendency to produce oxidation or decarburization; (3) exact control of depth and area of hardening; (4) close regulation of degree of hardness obtained by automatic timing of heating and quenching cycles; (5) minimum amount of warpage or distortion; and (6) possibility of substituting carbon steels for higher cost alloy steels.

The principal advantage of induction hardening to the designer lies in its application to localized zones. Thus, specific areas in a given part can be heat treated separately to the respective hardnesses required. Parts can be designed so that the stresses at any given point in the finished piece can be relieved by local heating. Parts can be designed in which welded or brazed assemblies are built up prior to heat treating with only internal surfaces or projections requiring hardening.

Types of Induction Heating Equipment. — Induction heating is secured by placing the metal part inside or close to an "applicator" coil of one or more turns, through which alternating current is passed. The coil, formed to suit the general class of work

to be heated, is usually made of copper tubing through which water is passed to prevent overheating of the coil itself. In most cases, the work piece is held either in a fixed position or is rotated slowly within or close to the applicator coil. Where the length of work is too great to permit heating in a fixed position, progressive heating may be employed. Thus, a rod or tube of steel may be fed through an applicator coil of one or more turns so that the heating zone travels progressively along the entire length of the work piece.

The frequency of the alternating current used and the type of generator employed to supply this current to the applicator coil depend upon the character of the work to be done. There are three types of equipment used commercially to produce high-frequency current for induction heating: (1) motor generator sets which deliver current at frequencies of approximately 1000, 2000, 3000, and 10,000 cycles; (2) spark gap oscillator units which produce frequencies ranging from 80,000 to 300,000 cycles; and (3) vacuum tube oscillator sets, which produce currents at frequencies ranging from 350,000 to 15,000,000 cycles or more.

Depth of Heat Penetration. — Generally speaking, the higher the frequency used, the shallower the depth of heat penetration. For heating clear through, for deep hardening and for large work pieces, low power concentrations and low frequencies are usually employed. For very shallow and closely controlled depths of heating, as in surface hardening, and in localized heat treating of small work pieces, currents at high frequencies are employed.

For example, a ½-inch round bar of hardenable steel will be heated through its entire structure quite rapidly by an induced current of 2000 cycles. After quenching the bar would show through hardness with a decrease in hardness from surface to center. The same piece of steel could be readily heated and surface hardened to a depth of 0.100 inch with current at 9600 cycles, and to an even shallower depth with current at 100,000 cycles. A ¼-inch bar, however, would not reach a sufficiently high temperature at 2000 cycles to permit hardening but at 9600 cycles through hardening would be accomplished, while current at over 100,000 cycles would be needed for surface hardening.

Types of Steel for Induction Hardening. — Most of the standard types of steel can be hardened by induction heating, providing the carbon content is sufficient to produce the desired degree of hardness by quenching. Thus, low carbon steels with carburized case, medium and high carbon steels (both plain and alloy), and cast iron with a portion of the carbon in combined form, may be used for this purpose. In the case of alloy steels, induction heating should be limited primarily to the shallow hardening type, i.e., those of low alloy content, otherwise the severe quench usually required may result in a highly-stressed surface with consequent reduced load-carrying capacity and danger of cracking.

Through Hardening, Annealing and Normalizing by Induction. — For through hardening, annealing and normalizing by induction, low power concentrations are desirable to prevent too great a temperature differential between surface and interior of the work. A satisfactory rate of heating is obtained when the total power input to the work is slightly greater than the radiation losses at the desired temperature. If possible, as low a frequency should be used as is consistent with good electric coupling. A number of applicator coils may be connected in a series so that several work pieces can be heated simultaneously, thus reducing the power input to each. Widening the spacing between work and applicator coil also will reduce the amount of power delivered to the work.

Induction Surface Hardening. — As indicated in "Depth of Heat Penetration," currents at much higher frequencies are required in induction surface hardening than in through hardening by induction. In general, the smaller the work piece, the thinner

the section, or the shallower the depth to be hardened, the higher will be the frequency required. High power concentrations are also needed to make possible a short heating period so that an undue amount of heat will not be conducted to adjacent or interior areas, where a change in hardness is not desired. Generators of large capacity and applicator coils of but a few turns, or even a single turn, provide the necessary concentration of power in the localized area to be hardened.

Induction heating of internal surfaces, such as the interior of a hollow cylindrical part or the inside of a hole, can be accomplished readily with applicator coils shaped to match the cross-section of the opening, which may be round, square, elliptical, etc. If the internal surface is of short length, a multiturn applicator coil extending along its entire length may be employed. Where the power available is insufficient to heat the entire internal surface at once, progressive heating is used. For this purpose an applicator coil of few turns — often but a single turn — is employed, and either coil or work is moved so that the heated zone passes progressively from one end of the hole or opening to the other. For bores of small diameter, a hairpin shaped applicator, extending the entire length of the hole, may be employed and the work rotated about the axis of the hole to insure even heating.

Quenching After Induction Heating. — After induction heating, quenching may be by immersion in a liquid bath (usually oil), by liquid spray (usually water), or by self-quenching. (The term "self-quenching" is used when there is no quenching medium and hardening of the heated section is due chiefly to rapid absorption of heat by the mass of cool metal adjacent to it.) Quenching by immersion offers the advantage of even cooling and is particularly satisfactory for through heated parts. Spray quenching may be arranged so that the quenching ring and applicator coil are in the same or adjacent units, permitting the quenching cycle to follow immediately the heating cycle without removal of the work from the holding fixture. Automatic timing to a fraction of a second may also be employed for both heating and quenching with this arrangement to secure the exact degree of hardness desired. Self-quenching is applicable only in thin-surface hardening where the mass of adjacent cool metal in the part is great enough to conduct the heat rapidly out of the surface layer which is being hardened. It has been recommended that for adequate self-quenching, the mass of the unheated section should be at least ten times that of the heated shell. It has been found difficult to use the self-quenching technique to produce hardened shells of much more than about 0.060 inch thickness. Close to this limit, self-quenching can only be accomplished with the easily hardenable steels. By using a combination of self-quench and liquid quench, however, it is possible to produced hardened shells on work too thin to self-quench completely. In general, self-quenching is confined chiefly to relatively small parts and simple shapes.

Induction Hardening of Gear Teeth. — Several advantages are claimed for the induction hardening of gear teeth. One advantage is that the gear teeth can be completely machined, including shaving, when in the soft-annealed or normalized condition, and then hardened, since when induction heating is used, distortion is held to a minimum. Another advantage claimed is that bushings and inserts can be assembled in the gears before hardening. A wide latitude in choice of built-up webs and easily machined hubs is afforded since the hardness of neither web nor hub is affected by the induction hardening operation although slight dimensional changes may occur in certain designs. Regular carbon steels can be used in place of alloy steels for a wide variety of gears, and a steel with a higher carbon content can frequently be substituted for a carburizing steel so that the carburizing operation can be eliminated. Another saving in time is the elimination of cleaning after hardening.

In heating spur gear teeth by induction, the gear is usually placed inside a circular unit which combines the applicator coil and quenching ring. An automatic timing

Table 1. Typical Heat Treatments for SAE Carbon Steels

Carburizing Grades								
SAE No.	Normalize Deg. F.	Carburize Deg. F.	Cool *	Reheat Deg. F.	Cool *	2nd Reheat Deg. F.	Cool *	Temper[3] Deg. F.
1010 to 1022	....	1650–1700	A	....	..	....	..	250–400
	....	1650–1700	B	1400–1450	A	....	..	250–400
	....	1650–1700	C	1400–1450	A	....	..	250–400
	....	1650–1700	C	1650–1700	B	1400–1450	A	250–400
	....	1500–1650[1]	B	....	..	....	..	Optional
	....	1350–1575[2]	D	....	..	....	..	Optional
1024	1650–1750[4]	1650–1700	E	....	..	....	..	250–400
	....	1350–1575[2]	D	....	..	....	..	Optional
1025 1026 1027	....	1650–1700	A	....	..	....	..	250–400
	....	1500–1650[1]	B	....	..	....	..	Optional
	....	1350–1575[2]	D	....	..	....	..	Optional
1030	....	1500–1650[1]	B	....	..	....	..	Optional
	....	1350–1575[2]	D	....	..	....	..	Optional
1111 1112 1113	....	1500–1650[1]	B	....	..	....	..	Optional
	....	1350–1575[2]	D	....	..	....	..	Optional
1109 to 1120	....	1650–1700	A	....	..	....	..	250–400
	....	1650–1700	B	1400–1450	A	....	..	250–400
	....	1650–1700	C	1400–1450	A	....	..	250–400
	....	1650–1700	C	1650–1700	B	1400–1450	A	250–400
	....	1500–1650[1]	B	....	..	....	..	Optional
	....	1350–1575[2]	D	....	..	....	..	Optional
1126	....	1500–1650[1]	B	....	..	....	..	Optional
	....	1350–1575[2]	D	....	..	....	..	Optional

Heat Treating Grades					
SAE Number	Normalize Deg. F.	Anneal	Harden Deg. F.	Quench *	Temper Deg. F.
1025 & 1030	....	....	1575–1650	A	
1033 to 1035	....	....	1525–1575	B	
1036	1600–1700	....	1525–1575	B	
	....	....	1525–1575	B	
1038 to 1040	1600–1700	....	1525–1575	B	
	....	....	1525–1575	B	
1041	1600–1700	and/or 1400–1500	1475–1550	E	
1042 to 1050	1600–1700	....	1475–1550	B	To Desired Hardness
1052 & 1055	1550–1650	and/or 1400–1500	1475–1550	E	
1060 to 1074	1550–1650	and/or 1400–1500	1475–1550	E	
1078	....	1400–1500[5]	1450–1500	A	
1080 to 1090	1550–1650	and/or 1400–1500[5]	1450–1500	E[6]	
1095	....	1400–1500[5]	1450–1500	F	
	....	1400–1500[5]	1500–1600	E	
1132 & 1137	1600–1700	and/or 1400–1500	1525–1575	B	
1138 & 1140	....	....	1500–1550	B	
	1600–1700	....	1500–1550	B	
1141 & 1144	....	1400–1500	1475–1550	E	
	1600–1700	1400–1500	1475–1550	E	
1145 to 1151	....	....	1475–1550	B	
	1600–1700	....	1475–1550	B	

* Symbols: A = Water or Brine; B = Water or oil; C = Cool slowly; D = Air or Oil; E = Oil; F = Water, Brine or Oil.
[1] Activated or cyanide baths; may be given refining heat as in other processes.
[2] Carbonitriding atmospheres; may be given refining heat as in other processes.
[3] Even where tempering temperatures are shown, tempering is not mandatory on many applications. It is usually employed for partial stress relief, and improves resistance to grinding cracks.
[4] Normalizing temperatures at least 50 deg. F. above the carburizing temperature are sometimes recommended where minimum heat treat distortion is of vital importance.
[5] Slow cooling produces a spheroidal structure in these high carbon steels which is sometimes required for machining purposes.
[6] May be water or brine quenched by special techniques such as partial immersion or time quenched, otherwise they are subject to quench cracking.

Table 2. Typical Heat Treatments for SAE Alloy Steels

SAE No.	Normalize[1]	Cycle Anneal[3]	Carburized Deg. F.	Cool *	Reheat Deg. F.	Cool *	Temper[8] Deg. F
			Carburizing Grades				
	yes	...	1650-1700	E	1400-1450[6]	E	250-350
	yes	...	1650-1700	E	1475-1525[7]	E	250-350
1320	yes	...	1650-1700	C	1400-1450[6]	E	250-350
	yes	...	1650-1700	C	1500-1550[7]	E	250-350
	yes	...	1650-1700	E[5]	....	..	250-350
	yes	...	1500-1650[4]	E	....	..	250-350
	yes	yes	1650-1700	E	1375-1425[6]	E	250-350
	yes	yes	1650-1700	E	1450-1500[7]	E	250-350
2317	yes	yes	1650-1700	C	1375-1425[6]	E	250-350
	yes	yes	1650-1700	C	1475-1525[7]	E	250-350
	yes	yes	1650-1700	E[5]	....	..	250-350
	yes	yes	1450-1650[4]	E	....	..	250-350
2512 to 2517	yes[2]	...	1650-1700	C	1325-1375[6]	E	250-350
	yes[2]	...	1650-1700	C	1425-1475[7]	E	250-350
	yes	...	1650-1700	E	1400-1450[6]	E	250-350
	yes	...	1650-1700	E	1475-1525[7]	E	250-350
3115 & 3120	yes	...	1650-1700	C	1400-1450[6]	E	250-350
	yes	...	1650-1700	C	1500-1550[7]	E	250-350
	yes	...	1650-1700	E[5]	....	..	250-350
	yes	...	1500-1650[4]	E	....	..	250-350
3310 & 3316	yes[2]	...	1650-1700	E	1400-1450[6]	E	250-350
4017 to 4032	yes[2]	...	1650-1700	C	1475-1500[7]	E	250-350
4119 & 4125	yes	yes	1650-1700	E[5]	....	..	250-350
	yes	...	1650-1700	E[5]	....	..	250-350
	yes	yes	1650-1700	E	1425-1475[6]	E	250-350
	yes	yes	1650-1700	E	1475-1525[7]	E	250-350
4317 & 4320	yes	yes	1650-1700	C	1425-1475[6]	E	250-350
4608 to 4621	yes	yes	1650-1700	C	1475-1525[7]	E	250-350
	yes	yes	1650-1700	E[5]	....	..	250-350
	yes	...	1500-1650[4]	E	....	..	250-350
	yes[2]	yes	1650-1700	E	1375-1425[6]	E	250-350
4812 to 4820	yes[2]	yes	1650-1700	E	1450-1500[7]	E	250-35c
	yes[2]	yes	1650-1700	C	1375-1425[6]	E	250-350
	yes[2]	yes	1650-1700	C	1450-1500[7]	E	250-350
	...	...	1650-1700	E[5]	....	..	250-350
	yes	...	1650-1700	E	1425-1475[6]	E	250-350
	yes	...	1650-1700	E	1500-1550[7]	E	250-350
5115 & 5120	yes	...	1650-1700	C	1425-1475[6]	E	250-350
	yes	...	1650-1700	C	1500-1550[7]	E	250-350
	yes	...	1500-1650[4]	E	....	..	250-350
	yes	yes	1650-1700	E	1475-1525[6]	E	250-350
8615 to 8625	yes	yes	1650-1700	E	1525-1575[7]	E	250-350
8720	yes	yes	1650-1700	C	1475-1525[6]	E	250-350
	yes	yes	1650-1700	C	1525-1575[7]	E	250-350
	yes	yes	1650-1700	E[5]	....	..	250-350
	yes	...	1500-1650[4]	E	....	..	250-350
9310 to 9317	yes[2]	...	1650-1700	E	1400-1450[6]	E	250-350
	yes[2]	...	1650-1700	C	1500-1525	E	250-350

* Symbols: C = Cool slowly; E = Oil.
[1] Normalizing temperatures should not be less than 50 degrees F. higher than the carburizing temperature. Follow by air cooling.
[2] After normalizing, reheat to temperatures of 1000-1200 degrees F. and hold approximately 4 hours.
[3] Where cycle annealing is desired, heat to normalizing temperature — hold for uniformity — cool rapidly to 1000-1250 degrees F.; hold 1 to 3 hours, then air or furnace cool to obtain a structure suitable for machining and finish.
[4] This treatment is for activated or cyanide baths, and parts may be given refining heats as indicated for other heat treating processes.
[5] This treatment applicable to fine-grained steels only. When fine-grained steels are employed, a second reheat is often unnecessary.
[6] This treatment when case hardness only is paramount.
[7] This treatment when higher core hardness is desired.
[8] Tempering treatment is optional. Tempering is generally employed for partial stress relief and improved resistance to cracking from grinding operations.

Table 2 (*Continued*). Typical Heat Treatments for SAE Alloy Steels

		Directly Hardenable Grades			
SAE No.	Normalize Deg. F.	Anneal Deg. F.	Harden Deg. F.	Quench *	Temper Deg. F.
1330	{1600-1700 and/or	1500-1600	1525-1575 / 1525-1575	B / B	To desired hardness / To desired hardness
1335 & 1340	{1600-1700 and/or	1500-1600	1500-1550 / 1525-1575	E / E	To desired hardness / To desired hardness
2330	{1600-1700 and/or	1400-1500	1450-1500 / 1450-1500	E / E	To desired hardness / To desired hardness
2340 & 2345	{1600-1700 and/or	1400-1500	1425-1475 / 1425-1475	E / E	To desired hardness / To desired hardness
3130	1600-1700	....	1500-1550	B	To desired hardness
3135 to 3141	{1600-1700 and/or	1450-1550	1500-1550 / 1500-1550	E / E	To desired hardness / To desired hardness
3145 & 3150	{1600-1700 and/or	1400-1500	1500-1550 / 1500-1550	E / E	To desired hardness / To desired hardness
4037 & 4042	....	1525-1575	1500-1575	E	{Gears, 350-450 / To desired hardness
4047 & 4053	....	1450-1550	1500-1575	E	To desired hardness
4063 & 4068	....	1450-1550	1475-1550	E	To desired hardness
4130	1600-1700 and/or	1450-1550	1600-1650	B	To desired hardness
4137 & 4140	1600-1700 and/or	1450-1550	1550-1600	E	To desired hardness
4145 & 4150	1600-1700 and/or	1450-1550	1500-1600	E	To desired hardness
4340	1600-1700 and draw	1100-1225	1475-1525	E	To desired hardness
4640	{1600-1700 and/or	1450-1550 / 1450-1500	1450-1500 / 1450-1500	E / E	Gears, 350-450 / To desired hardness
5045 & 5046	1600-1700 and/or	1450-1550	1475-1550	G	250-300
5130 & 5132	1650-1750 and/or	1450-1550	1500-1550	E	{To desired hardness / Gears, 350-400
5135 to 5145	1650-1750 and/or	1450-1550	1500-1550	E	{To desired hardness / Gears, 350-400
5147 to 5152	1650-1750 and/or	1450-1550	1475-1550	E	{To desired hardness / Gears, 350-400
50100 51100 52100	{........	1350-1450 / 1350-1450	1425-1475 / 1500-1600	H / E	To desired hardness / To desired hardness
6150	1650-1750 and/or	1550-1650	1600-1650 / 1500-1650	E / E	To desired hardness
9254 to 9262	....	....	1550-1650	B	To desired hardness
8627 to 8632	1600-1700 and/or	1450-1550	1550-1650	B	To desired hardness
8635 to 8641	1600-1700 and/or	1450-1550	1525-1575	E	To desired hardness
8642 to 8653	1600-1700 and/or	1450-1550	1500-1550	E	To desired hardness
8655 & 8660	1650-1750 and/or	1450-1550	1475-1550	E	To desired hardness
8735 & 8740	1600-1700 and/or	1450-1550	1525-1575	E	To desired hardness
8745 & 8750	1600-1700 and/or	1450-1550	1500-1550	E	To desired hardness
9437 & 9440	1600-1700 and/or	1450-1550	1550-1600	E	To desired hardness
9442 to 9747	1600-1700 and/or	1450-1550	1500-1600	E	To desired hardness
9840	1600-1700 and/or	1450-1550	1500-1550	E	To desired hardness
9845 & 9850	1600-1700 and/or	1450-1550	1500-1550	E	To desired hardness

		Heat Treating Grades — Chromium-Nickel Austenitic Steels			
SAE No.	Normalize	Anneal[9]	Harden Deg. F.	Quenching Medium	Temper
30301 to 30347	....	1800-2100	....	Water or Air	....

* Symbols: B = Water or oil; E = Oil; G = Water, caustic solution or oil; H = Water.

[9] Quench to produce full austenitic structure using water or air in accordance with thickness of section. Annealing temperatures given cover process and full annealing as now used by industry, the lower end of the range being used for process annealing.

Table 2 (*Continued*). **Typical Heat Treatments for SAE Alloy Steels**

		Heat Treating Grades — Stainless Chromium Irons and Steels				
SAE No.*	Normalize	Aub-critical Anneal, Deg. F.	Full Anneal Deg. F.	Harden Deg. F.	Quenching Medium	Temper Deg. F.
51410	...	1300–1350[10]	1550–1650[11]	...	Oil or air	To desired hardness
	...	...	...	1750–1850		
51414	...	1200–1250[10]	...	...	Oil or air	To desired hardness
	...	...	...	1750–1850		
51416	...	1300–1350[10]	1550–1650[11]	...	Oil or air	To desired hardness
	...	...	...	1750–1850		
51420 51420F	...	1350–1450[10]	1550–1650[11]	...	Oil or air	To desired hardness
	...	...	...	1800–1850		
51430	...	1400–1500[12]	...	...	...	...
51430F	...	1250–1500[12]	...	...	...	...
51431	...	1150–1225[10]	...	1800–1900	Oil or air	To desired hardness
51440A 51440B 51440C 51440F	...	1350–1440[10]	1550–1650[11]	1850–1950	Oil or air	To desired hardness
51442	...	1400–1500[12]	...	...	...	...
51446	...	1500–1650[12]	...	...	...	...
51501	...	1325–1375[10]	1525–1600[11]	1600–1700	Oil or air	To desired hardness

* Suffixes A, B and C denote three types of steel differing in carbon content only. Suffix F denotes a free machining steel.
[10] Usually air cooled, but may be furnace cooled.
[11] Cool slowly in furnace.
[12] Cool rapidly in air.

device controls both the heating and quenching cycles. During the heating cycle, the gear is rotated at 25 to 35 R.P.M. to insure uniform heating.

In hardening bevel gears, the applicator coil is wound to conform to the face angle of the gear. In some spiral-bevel gears, there is a tendency to obtain more heat on one side of the tooth than on the other. In some sizes of spiral-bevel gears this can be overcome by applying slightly more heat to insure hardening of the concave side. In some forms of spiral-bevel gears, it has been the practice to carburize that part of the gear surface which is to be hardened, after the teeth have been rough-cut. This is followed by the finish-cutting operation, after which the teeth can be induction heated, using a long enough period to heat the entire tooth. When the gear is quenched only the carburized surface will become hardened.

Laser and Electron Beam Surface Hardening. — Industrial lasers and electron beam equipment are now available for surface hardening of steels. The laser and electron beams can generate very intense energy fluxes and steep temperature profiles in the workpiece, so that external quench media are not needed. This self-quenching is due to a cold interior with sufficient mass acting as a large heat sink to rapidly cool the hot surface by conducting heat to the interior of a part. The laser beam is a beam of light and does not require a vacuum for operation. The electron beam is a stream of electrons and processing usually takes place in a vacuum chamber or envelope. Both processes are normally applied to finished machined or ground surfaces, since little distortion results.

Heat-Treating High-Speed Steels

Cobaltcrom Steel. — This is a tungstenless alloy steel or high-speed steel which contains approximately 1.5 per cent carbon, 12.5 per cent chromium, and 3.5 per cent cobalt. Tools such as dies, milling cutters, etc., made from cobaltcrom steel can be cast to shape in suitable molds, the teeth of cutters being formed so that it is necessary only to grind them.

Before the blanks can be machined, they must be annealed; this operation is performed by pack-annealing at the temperature of 1800 degrees F., for a period of from three to six hours, according to the size of the castings being annealed. The following directions are given for the hardening of blanking and trimming dies, milling cutters, and similar tools made from cobaltcrom steel: Heat slowly in a hardening furnace to about 1830 degrees F., and hold the temperature at this point until the tools are thoroughly soaked. Then reduce the temperature about 50 degrees, withdraw the tools from the furance, and allow them to cool in the atmosphere. As soon as the red color disappears from the cooling tool, place it in quenching oil until cold. The slight drop of 50 degrees in temperature while the tool is still in the hardening furnace is highly important in order to obtain proper results. The steel will be injured if the tool is heated above 1860 degrees F. In cooling milling cutters or other rotary tools, it is suggested that they be suspended on a wire to insure a uniform rate of cooling.

Tools that are subjected to shocks or vibration, such as pneumatic rivet sets, shear blades, etc., should be heated slowly to 1650 degrees F., after which the temperature should be reduced to about 1610 degrees F., at which point the tool should be removed from the furnace and permitted to cool in the atmosphere. There is no appreciable scaling present in the hardening of cobaltcrom steel tools.

Preheating Tungsten High-Speed Steel. — Tungsten high-speed steel must be hardened at a very high temperature; consequently, tools made from such steel are seldom hardened without at least one preheating stage to avoid internal strain. This applies especially to milling cutters, taps, and other tools having thin teeth and thick bodies and to forming tools of irregular shape and section. The tools should be heated slowly and carefully to a temperature somewhat below the critical point of the steel, usually in the range of 1500 to 1600 degrees F. By so limiting the preheating temperature, the operation is not unduly sensitive and the tool may be safely left in the furnace until it reaches a uniform temperature throughout its length and cross section.

A single stage of preheating is customary for tools that are simple in form and are not more than from 1 to 1½ inches in thickness. For large, intricate tools, two stages of preheating are frequently used. The first brings the tool up to a temperature of about 1100 to 1200 degrees F., and the second raises its temperature to 1550 to 1600 degrees F. A preheating time of 5 minutes for each ¼ inch in tool thickness has been recommended for a furnace temperature of 1600 degrees F. This is where a single stage of preheating is used and the furnace capacity is sufficient for maintaining practically constant temperature when the tools are changed. To prevent undue chilling, it is common practice to insert a single tool or a small lot in the hardening furnace as often as a tool or lot is removed, rather than to insert a full charge of cold metal at one time.

Preheating is usually done in a simple type of oven furnace heated by gas, electricity, or oil. Atmospheric control is seldom used, although in the case of 18–4–1 steel a slightly reducing atmosphere (2 to 6 per cent carbon monoxide) has been found to produce the least amount of scale and will result in a better surface after final hardening.

Table 1. Relation of Hardening Temperature to Cutting Efficiency

Hardening Temperature, Degrees F.	Typical Analyses of High-Speed Steels			
	18-4-1	14-4-2	18-4-1 Cobalt	14-4-2 Cobalt
2200	0.86	0.83	0.84	0.85
2250	0.88	0.88	0.86	0.88
2300	0.90	0.93	0.90	0.91
2350	0.95	0.98	0.94	0.94
2400	0.99	0.98	0.98	0.98
2450	1.00	. . .	0.99	1.00
2500	0.98	. . .	1.00	0.97

Table 2. Length of Heating Time for Through Hardening

High-Speed Steel Tool Thickness, in Inches	Time in Furnace at High Heat, in Minutes	High-Speed Steel Tool Thickness, in Inches	Time in Furnace at High Heat, in Minutes	High-Speed Steel Tool Thickness, in Inches	Time in Furnace at High Heat, in Minutes
¼	2	1½	7	5	18
½	3	2	8	6	20
¾	4	3	12	8	25
1	5	4	15	10	30

Hardening of Tungsten High-Speed Steel. — All tungsten high-speed steels must be heated to a temperature close to their fusion point to develop their maximum efficiency as metal-cutting tools. This requires a hardening temperature ranging from 2200 to 2500 degrees F. The effect of changes in the hardening temperature on the cutting efficiency of several of the more common high-speed steels are shown in Table 1. The figures given are ratios, the value 1.00 for each steel being assigned to the highest observed cutting speed for that steel. The figures for different steels therefore cannot be directly compared with each other, except to note changes in the point of maximum cutting efficiency.

The figures in the table refer to tools heated in an oven type furnace in which a neutral atmosphere is maintained. The available data indicate that a steel reaches its best cutting qualities at a temperature approximately 50 degrees F. lower than the figures in the table if it is hardened in a bath type furnace. It is, however, desirable, in any case, to use a hardening temperature approximately 50 degrees lower than that giving maximum cutting qualities, in order to avoid possibility of overheating the tool.

Length of Time for Heating: The cutting efficiency of a tool is affected by the time that it is kept at the hardening temperature, almost as much as by the hardening temperature itself. It has been common practice to heat a tool for hardening until a " sweat " appeared on its surface. This sweat is presumably a melting of the oxide film on the surface of a tool heated in an oxidizing atmosphere. It does not show when the tool is heated in an inert atmosphere. This method of determining the proper heating time is at best an approximation and indicates only the temperature on the outside of the tool rather than the condition of the interior. As such, it cannot be relied upon to give consistent results.

The only safe method is to heat the tool for a definite predetermined time, based on the size and the thickness of metal which the heat must penetrate to reach the

interior. The values given in Table 2 are based on a series of experiments to determine the relative cutting efficiency of a group of tools hardened in an identical manner, except for variations in the time the tools were kept at the hardening temperature. The time given is based on that required to harden throughout a tool resting on a conducting hearth; the tool receives heat freely from three sides, on its large top surface and its smaller side surfaces. (The table does not apply to a disk lying flat on the hearth.) In the case of a tool having a projecting cutting edge, such as a tap, the thickness or depth of the projection portion on which the cutting edge is formed should be used in referring to the table.

The time periods given in Table 2 are based on complete penetration of the hardening. For very thick tools, the practical procedure is to harden to a depth sufficient to produce an adequate cutting edge, leaving the interior of the tool relatively soft.

Where atmosphere control is not provided, it will be found impracticable in many cases to use both the temperature for maximum cutting efficiency, given in Table 1, and the heating time, given in Table 2, because abnormal scaling, grain growth, and surface decarburization of the tool will result. The principal value of an accurate control of the furnace atmosphere appears to lie in the fact that its use makes possible the particular heat-treatment that produces the best structure in the tool without destruction of the tool surface or grain.

Quenching Tungsten High-Speed Steel. — High-speed steel is usually quenched in oil. The oil bath offers a convenient quench; it calls for no unusual care in handling and brings about a uniform and satisfactory rate of cooling, which does not vary appreciably with the temperature of the oil. Some authorities believe it desirable to withdraw the tool from the oil bath for a few seconds after it has reached a dull red. It is also believed desirable to move the tool around in the quenching oil, particularly immediately after it has been placed in it, to prevent the formation of a gas film on the tool. Such a film is usually a poor conductor of heat and slows up the rate of cooling.

Salt Bath: Quenching in a lead or salt bath at from 1000 to 1200 degrees F. has the advantage that cooling of the tool from hardening to room temperature is accomplished in two stages, thus reducing the possibility of setting up internal strains which may tend to crack the tool. The quenching temperature is sufficiently below the lower critical point for a tool so quenched to be allowed to cool to room temperature in still air. This type of quench is particularly advantageous for tools of complicated section which would easily develop hardening cracks. The salt quench has the advantage that the tool sinks and requires only a support, while the same tool will float in the lead bath and must be held under the surface. It is believed that the lead quench gives a somewhat higher matrix hardness, and is of advantage for tools that tend to fail by nose abrasion. Tools treated as described are brittle unless given a regular tempering treatment, as the 1100-degree F. quenching temperature is not a substitute for later tempering at the same temperature, after the tool has cooled to room temperature.

Air Cooling: Many high-speed steel tools are quenched in air, either in a stream of dry compressed air or in still air. Small sections harden satisfactorily in still air, but heavier sections should be subjected to air under pressure. One advantage of air cooling is that the tool can be kept straight and free from distortion, although it is likely that there will be more scale on a tool thus quenched than when oil, lead, or salt is used. Thin flat tools, which must be kept straight and flat, can be cooled advantageously between steel plates.

Straightening High-Speed Tools when Quenching. — The final straightness required in a tool must be considered when it is quenched. When a number of

similar tools are to be hardened, a jig can be used to advantage for holding the tools while quenching. When long slender tools are quenched without holders, they frequently warp and must be straightened later. The best time for this straightening is during the first few minutes after the tools have been quenched, as the steel is then quite pliable and may be bent without difficulty. The straightening must be done at once, as the tools become hard in a few minutes.

Anneal Before Rehardening. — Tools that are too soft after hardening must be annealed before rehardening. A quick anneal, such as previously described, is all that is required to put such a tool into the proper condition for rehardening. This treatment is absolutely essential. In the case of milling cutters and forming tools of irregular section, a full anneal should be used.

Tempering or Drawing Tungsten High-Speed Steel. — The tempering or drawing temperature for high-speed steel tools usually varies from 900 to 1200 degrees F. This temperature is higher for turning and planing tools than for such tools as milling cutters, forming tools, etc. If the temperature is below 800 degrees F., the tool is likely to be too brittle. The general idea is to temper tools at the highest temperature likely to occur in actual service. Since this temperature ordinarily would not be known, the general practice is to temper at whatever temperature experience with that particular steel and tool has proved to be the best. The furnace used for tempering usually is kept at a temperature of from 1000 to 1100 degrees F. for ordinary high-speed steels and from 1200 to 1300 degrees F. for steels of the cobalt type. These furnace temperatures apply to tools of the class used on lathes and planers. Such tools, in service, frequently heat to the point of visible redness. Milling cutters, forming tools, or any other tools for lighter duty, may be tempered as low as 850 or 900 degrees F. When the tool has reached the temperature of the furnace it should be held at this temperature for from one to several hours until it has been heated evenly throughout. It should then be allowed to cool gradually in the air and in a place that is dry and free from air drafts. In tempering, the tool should not be quenched, as this tends to produce strains which may result later in cracks.

Annealing Tungsten High-Speed Steel. — The following method of annealing high-speed steel has been used extensively. Use an iron box or pipe of sufficient size to allow at least one-half inch of packing between the pieces of steel to be annealed and the sides of the box or pipe. It is not necessary that each piece of steel be kept separate from every other piece, but only that the steel be prevented from touching the sides of the annealing pipe or box. Pack carefully with powdered charcoal, fine dry lime or mica (preferably charcoal), and cover with an air-tight cap, or lute with fire clay; heat slowly to 1600 to 1650 degrees F. and keep at this heat from 2 to 8 hours, depending upon the size of the pieces to be annealed. A piece 2 by 1 by 8 inches requires about three hours. Cool as slowly as possible, and do not expose to the air until cold as cooling in air is likely to cause partial hardening. A good way is to allow the box or pipe to remain in the furnace until cold.

Hardening Molybdenum High-Speed Steels. — In Table 3 are given the compositions of several molybdenum high-speed steels which are widely used for general commercial tool applications. The general method of hardening molybdenum high-speed steels resembles that used for 18–4–1 tungsten high-speed steel except that the hardening temperatures are lower and more precautions must be taken to avoid decarburization, especially on tools made from Type I or Type II steels, when the surface is not ground after hardening. Either salt baths or atmosphere controlled furnaces are recommended for hardening molybdenum high-speed steel. The usual method is to preheat uniformly in a separate furnace to 1250 to 1550

Table 3. Composition of Molybdenum High-Speed Steels

Element	Molybdenum-Tungsten		Molybdenum-Vanadium	Tungsten-Molybdenum
	Type Ia (Per Cent)	Type Ib* (Per Cent)	Type II (Per Cent)	Type III (Per Cent)
Carbon	0.70–0.85	0.76–0.82	0.70–0.90	0.75–0.90
Tungsten	1.25–2.00	1.60–2.30	....	5.00–6.00
Chromium	3.00–5.00	3.70–4.20	3.00–5.00	3.50–5.00
Vanadium	0.90–1.50	1.05–1.35	1.50–2.25	1.25–1.75
Molybdenum	8.00–9.50	8.00–9.00	7.50–9.50	3.50–5.50
Cobalt	See footnote	4.50–5.50	See footnote	See footnote

*Cobalt may be used in any of these steels in varying amounts up to 9 per cent, and the vanadium content may be as high as 2.25 per cent. When cobalt is used in Type III the vanadium content may be as high as 2.25 per cent. When cobalt is used in Type III steel, this steel becomes susceptible to decarburization. As an illustration of the use of cobalt, Type Ib steel is included. This is steel T10 in the U. S. Navy Specification 46S37, dated November 1, 1939.

degrees F. and then transfer to a high-heat furnace maintained at some temperature in the hardening temperature range given in Table 4. Single point cutting tools, in general, should be hardened at the upper end of the temperature range indicated by Table 4. Slight grain coarsening is not objectionable in such tools when they are properly supported in service and are not subjected to chattering; however, when these tools are used for intermittent cuts, it is better to use the middle of the temperature range. All other cutting tools, such as drills, countersinks, taps, milling cutters, reamers, broaches, form tools, etc., should be hardened in the middle of the range shown. For certain tools, such as slender taps, cold punches, blanking and trimming dies, etc., where greater toughness to resist shocks is required, the lower end of the hardening temperature range should be used.

Table 4. Heat-Treatment of Molybdenum High-Speed Steels

Heat-Treating Operation	Molybdenum-Tungsten - Types Ia and Ib* (Temp., in Deg. F.)	Molybdenum-Vanadium Type II (Temp., in Deg. F.)	Tungsten-Molybdenum Type III (Temp., in Deg. F.)
Forging	1850–2000	1850–2000	1900–2050
Not below	1600	1600	1600
Annealing	1450–1550	1450–1550	1450–1550
Strain Relief	1150–1350	1150–1350	1150–1350
Preheating	1250–1500	1250–1500	1250–1550
Hardening†	2150–2250*	2150–2250	2175–2275
Salt	2150–2225	2150–2225	2150–2250
Tempering	950–1100	950–1100	950–1100

*Under similar conditions Type Ib steel requires a slightly higher hardening heat than Type Ia.
†The higher side of the hardening range should be used for large sections, and the lower side for small sections.

Molybdenum high-speed steels can be pack-hardened following the same practice as is used for tungsten high-speed steels, but keeping on the lower side of the hardening range (approximately 1850 degrees F.). Special surface treatments such as nitriding by immersion in molten cyanide that are used for tungsten high-speed steels are also applicable to molybenum high-speed tools.

When heated in an open fire or in furnaces without atmosphere control, these steels do not sweat like 18–4–1 steels; consequently, determining the proper time in the high-heat chamber is a matter of experience. This time approximates that used with 18–4–1 steels, although it may be slightly longer when the lower part of the hardening range is used. Much can be learned by preliminary hardening of test pieces and checking up on the hardness fracture and structure. It is difficult to give the exact heating time, as this is affected by temperature, type of furnace, size and shape, and furnace atmosphere. Rate of heat transfer is most rapid in salt baths, and slowest in controlled-atmosphere furnaces with high carbon-monoxide content.

Quenching and Tempering of Molybdenum High-speed Tools. — Quenching may be done in oil, air, or molten bath. To reduce the possibility of breakage and undue distortion of intricately shaped tools, it is advisable to quench in a molten bath at approximately 1100 degrees F. The tool also may be quenched in oil and removed while still red, or at approximately 1100 degrees F. The tool is then cooled in air to room temperature, and tempered immediately to avoid cracking.

When straightening is necessary, it should be done after quenching and before cooling to room temperature prior to tempering.

To temper, the tools should be reheated slowly and uniformly to 950 to 1100 degrees F. For general work, 1050 degrees F. is most common. The tools should be held at this temperature at least one hour. Two hours is a safer minimum, and four hours is maximum. The time and temperature depend on the hardness and toughness required. Where tools are subjected to more or less shock, multiple temperings are suggested.

Protective Coatings for Molybdenum Steels. — Borax may be applied by sprinkling it lightly over the steel when the latter is heated in a furnace to a slow temperature (1200 to 1400 degrees F.). Small tools may be rolled in a box of borax before heating. Another method more suitable for finished tools is to apply the borax or boric acid in the form of a supersaturated water solution. In such cases, the tools are immersed in the solution at 180 to 212 degrees F., or it may be applied with a brush or spray. Pieces so treated are heated as usual, taking care in handling to insure good adherence. Special protective coatings or paints, when properly applied, have been found extremely useful. They do not fuse or run at the temperatures used, and therefore do not affect the furnace hearth. When applying these coatings, it is necessary to have a surface free from scale or grease to insure good adherence. They may be sprayed or brushed on, and usually one thin coat is sufficient. Heavy coats tend to pit the surface of the tool and also are difficult to remove. Tools covered with these coatings should be allowed to dry before they are charged into the preheat furnace. After hardening and tempering, the coating can be easily removed by light blasting with sand or steel shot. When tools are lightly ground, these coatings come off immediately.

Nitriding High-speed Steel Tools. — Nitriding as applied to high-speed steel is for the purpose of increasing tool life by producing a very hard skin or case, the thickness of which ordinarily is from 0.001 to 0.002 inch. Nitriding is done after the tool has been fully heat-treated and finish ground. (The process differs entirely from that which is applied to certain alloy steels in order to surface harden them by

heating in an atmosphere of nitrogen or ammonia gas.) The temperature of the high-speed steel nitriding bath, which is a mixture of sodium and potassium cyanides, is equal to or slightly lower than the tempering temperature. For ordinary tools, this temperature usually varies from about 1025 to 1050 degrees F.; but if the tools are exceptionally fragile, the range may be reduced to 950 or 1000 degrees F. Accurate temperature control is essential to prevent exceeding the final tempering temperature. The nitriding time may vary from 10 or 15 minutes to 30 minutes or longer, as determined by experiment. The shorter periods are applied to tools for iron or steel, or any shock-resisting tools, and the longer periods are for tools used in machining non-ferrous metals and plastics. This nitriding process is applied to tools such as hobs, reamers, taps, box-tools, form tools, milling cutters, etc. Nitriding may increase tool life 50 to 200 per cent, or more, but it should always be preceded by correct heat-treatment.

Nitriding Bath Mixtures and Temperatures: A mixture of 60 per cent sodium cyanide and 40 per cent potassium cyanide is commonly used. This mixture has a melting point of 925 degrees F. which is gradually reduced to 800 degrees F. as the cyanate content of the bath increases. A more economical mixture of 70 per cent sodium cyanide and 30 per cent potassium cyanide may be used if the operating temperature of the bath is only 1050 degrees F. Nitriding bath temperature should not exceed 1100 degrees F. because higher temperatures accelerate the formation of carbonate at the expense of the essential cyanide. A third mixture suitable for nitriding consists of 55 per cent sodium cyanide, 25 per cent potassium chloride and 20 per cent sodium carbonate. This mixture melts at 930 degrees F.

Equipment for Hardening High-speed Steel. — Equipment for hardening high-speed steel consists of a hardening furnace capable of maintaining a temperature of 2350 to 2450 degrees F.; a preheating furnace capable of maintaining a temperature of 1700 to 1800 degrees F., and of sufficient size to hold a number of pieces of the work; a tempering (drawing) furnace capable of maintaining a temperature of 1000 to 1200 degrees F. as a general rule; and a water-cooled tank of quenching oil.

High-speed steels usually are heated for hardening either in some type of electric furnace or in a gas-fired furnace of the muffle type. The small furnaces used for high-speed steel seldom are oil-fired. It is desirable to use automatic temperature control and, where an oven type of furnace is employed, a controlled atmosphere is advisable because of the variations in cutting qualities caused by hardening under uncontrolled conditions. Some furnaces in both electric and fuel-fired types are equipped with a salt bath suitable for high-speed steel hardening temperatures. Salt baths have the advantage of providing protection against the atmosphere during the heating period. A type of salt developed for commercial use is water-soluble, so that all deposits from the hardening bath may be removed by immersion in water after quenching in oil or salt, or after air-cooling. One type of electric furnace heats the salt bath internally by electrodes immersed in it. The same type is also applied to various heat-treating operations, such as cyanide hardening, liquid carburizing, tempering, and annealing.

An open-forge fire has many disadvantages, especially in hardening cutters or other tools that cannot be ground all over after hardening. The air blast decarburizes the steel and lack of temperature control makes it impossible to obtain uniform results. Electric and gas furnaces provide continuous uniform heat, and the temperature may be regulated accurately, especially when pyrometers are used. In shops equipped with only one furnace for carbon steel and one for high-speed steel, the tempering can be done in the furnace used for hardening carbon steel after the preheating is finished and the steel has been removed for hardening.

Heating High-speed Steel for Forging. — Care should be taken not to heat high-speed steel for forging too abruptly. In winter, the steel may be extremely cold when brought into the forge shop. If the steel is put directly into the hot forge fire, it is likely to develop cracks which will show up later in the finished tool. It should, therefore, be warmed gradually before heating for forging.

Sub-zero Treatment of Steel

The sub-zero treatment consists in subjecting the steel, after hardening and either before or after tempering, to a sub-zero temperature (which usually ranges from −100 degrees F. to −120 degrees F.) and for a period of time varying with the size or volume of the tool, gage or other part. Commercial equipment is available for obtaining these low temperatures.

The sub-zero treatment is employed by most gage manufacturers in order to stabilize precision gages and prevent subsequent changes in size or form. Sub-zero treatment is also applied to some high-speed steel cutting tools. The object in this case is to increase the durability or life of the tools; however, up to the present time the results of tests by metallurgists and tool engineers often differ considerably and in some instances are contradictory. Methods of procedure also vary, especially in regard to the order and number of operations in the complete heat-treating and cooling cycle.

Changes Resulting From Sub-zero Treatment. — When steel is at the hardening temperature, it contains a solid solution of carbon and iron known as *austenite*. When the steel is hardened by sudden cooling, most of the austenite which is relatively soft, tough and ductile even at room temperatures, is transformed to martensite which is a hard and strong constituent. If all of the austenite were changed to martensite upon reaching room temperature, this would be an ideal hardening operation, but many steels retain some austenite. In general, the higher the carbon and alloy contents and the higher the hardening temperature, the greater the tendency to retain austenite. When steel is cooled to sub-zero temperatures, the stability of the retained austenite is reduced so that it is more readily transformed. To obtain more complete transformation the sub-zero treatment may be repeated. The ultimate transformation of austenite to martensite may take place in carbon steel without the aid of sub-zero treatment, but this natural transformation might require six months or longer, whereas by refrigeration this change occurs in a few hours.

The thorough, uniform heating which is always recommended in heat-treating operations should be accompanied by thorough, uniform cooling when the sub-zero treatment is applied. To insure uniform cooling, the sub-zero cooling period should be increased for the larger tools and it may range from two to six hours. The tool or other part is sometimes surrounded by one or more layers of heavy wrapping or asbestos paper to delay the cooling somewhat and insure uniformity. After the cooling cycle is started, it should continue without interruption.

In some cases, the sub-zero treatment may cause cracking. Normally the austenite in steel provides a cushioning effect which may prevent cracking or breakage resulting from treatments involving temperature and dimensional changes; but, if this cushioning effect is removed, particularly at very low temperatures as in sub-zero treatments, there may be danger of cracking especially in the case of tools having large or irregular sections and sharp corners offering relatively low resistance to stresses. This is one reason why sub-zero treatments may differ in regard to the cooling and tempering cycle.

Stabilizing Dimensions of Gages or Precision Parts by Sub-zero Cooling. — Transformation of austenite into martensite is accompanied by an increase in volume; consequently, the transformation of austenite which may naturally occur over a period of months or years, tends to change the dimensions and form of steel parts, and such changes may be serious in the case of precision gages, close-fitting machine parts, etc. To prevent such changes, the sub-zero treatment has proved effective. Gage-blocks, for example, may be stabilized by hardening followed by repeated cycles of chilling and tempering, to transform a large percentage of the austenite into martensite.

Order of Operations for Stabilizing Precision Gages: If precision gages, sine-bars, etc., are heat-treated in the ordinary manner and then are finished without some stabilizing treatment, dimensional changes and warpage are liable to occur. Sub-zero cooling provides a practical and fairly rapid method of obtaining the necessary stabilization by transforming the austenite into martensite. In stabilization treatments of this kind, tempering is the final operation. One series of treatments which has been recommended after hardening and rough-grinding is as follows:

(a) Cool to -120 degrees F. (This cooling period may require from one to six hours, depending on the size and form of the gage.)

(b) Place gage in boiling water for two hours (oil or salt bath may also be used).

Note: Steps (a) and (b) may be repeated from two to six times, depending on the size and form of the gage. These repeated cycles will eventually transform practically all of the austenite into martensite. Two or three cooling and drawing operations usually are sufficient for such work as thread gages and gage-blocks.

(c) Follow with regular tempering or drawing operation and finish gage by lapping.

Series of Stabilizing Treatments for Chromium Steel: The following series of treatments has proved successful in stabilizing precision gage-blocks made from S.A.E. 52100 chromium steel.

(a) Preheat to 600 degrees F. and then heat to 1575 degrees F. for a period of four minutes.

(b) Quench in oil at 85 degrees F. (Uniform quenching is essential)

(c) Temper at 275 degrees F. for one hour.

(d) Cool in tempering furnace to room temperature.

(e) Continue cooling in atmosphere of industrial refrigerator for six hours with temperature of atmosphere at -120 degrees F.

(f) Allow gage-blocks to return to room temperature and again temper.

Note: The complete treatment consists of six sub-zero cooling periods, each followed by a tempering operation. The transformation to martensite is believed to be complete even after the fifth cooling period. The hardness is about 66 Rockwell C. Transformation is checked by magnetic tests based upon the magnetism of martensite and the non-magnetic qualities of austenite.

Stabilizing Dimensions of Close-fitting Machine Parts. — Sub-zero treatment will always cause an increase in size. Machine parts subjected to repeated and perhaps drastic changes in temperature, as in aircraft, may eventually cause trouble due to growth or warpage as the austenite gradually changes to martensite. In some instances, the sizes of close-fitting moving parts have increased sufficiently to cause seizure. Such treatment, for example, may be applied to precision bearings made from S.A.E. 52100 or alloy carburizing steels for stabilizing or aging them. *Time* aging of 52100 steels after hardening has been found to cause changes as large as .0025 inch in medium size sections. A practical remedy is to apply the sub-zero treatment before the final grinding or other machining operation.

Sub-zero Treatment of Carburized Parts to Improve Physical Properties. — The sub-zero treatment has been applied to carburized machine parts. In the case of carburized gears, for example, the amount of retained austenite may be sufficient to reduce the life of the gears. In one case, the Rockwell hardness was increased from 55 C to 65 C without loss of impact resistance qualities; in fact, impact and fatigue resistance may be increased in some cases.

Application of Sub-zero Treatments to High-speed Steel. — The sub-zero treatment has been applied to such tools as milling cutters, hobs, taps, broaches and drills. It is applicable to different classes of high-speed steels, such as the 18-4-1 tungsten, 18-4-14 cobalt, and the molybdenum high-speed steels. This *cold* treatment is applied preferably in conjunction with the heat-treatment, both being combined in a continuous cycle of operations. The general procedure is either to harden the steel, cool it to a sub-zero temperature, and then temper; or, especially if there is more than one tempering operation, the first one may *precede* sub-zero cooling. The cooling and tempering cycle may be repeated two or more times. The number and order of the operations, or the complete cycle, may be varied to suit the class of work and, to minimize the danger of cracking, particularly if the tool has large or irregular sections, sharp corners or edges, or a high cobalt content. A sub-zero treatment of some kind with a final tempering operation for stress relief, is intended to increase strength and toughness without much loss in hardness; consequently, if there is greater strength at a given hardness, tools subjected to sub-zero treatment can operate with a higher degree of hardness than those heat-treated in the ordinary manner, or, if greater toughness is preferred, this can be obtained by tempering to the original degree of hardness.

Order of Cooling and Tempering Periods for High-speed Steel. — The order or cycle for the cooling and tempering periods has not been standardized. The methods which follow have been applied to high-speed steel tools. They are given as examples of procedure and are subject to possible changes due to subsequent developments. The usual ranges of preheating and hardening temperatures are given; but for a particular steel, the recommended temperatures should be obtained from the manufacturer.

1. *Double Sub-zero Treatment:* (For rugged simple tool forms without irregular sections, sharp corners or edges where cracks might develop during the sub-zero treatment).

(a) Preheat to from 1400 to 1600 degrees F. (double preheating is preferable, the first preheating ranging from 700 to 1000 degrees F.).

(b) Heat to the hardening temperature. (Note: Tests indicate that the effect of sub-zero treatment on high-speed steel may be influenced decidedly by the hardening temperature. If this temperature is near the lower part of the range, the results are unsatisfactory. Effective temperatures for ordinary high-speed steels appear to range from 2300 to 2350 degrees F.)

(c) Quench in oil, salt, lead or air, down to a workpiece temperature of 150–200 degrees F. (Note: One method is to quench in oil; a second method is to quench in oil to about 200–225 degrees F. and then air-cool; a third method is to quench in salt bath at 1050–1100 degrees F. and then air-cool.)

(d) Cool in refrigerating unit to temperature of − 100 to − 120 degrees F. *right after quenching.* (Note: Tests have shown that a delay of one hour has a detrimental effect, and in ten hours the efficiency of the sub-zero treatment is reduced 50 per cent. This is due to the fact that the austenite becomes more and more stabilized when the sub-zero treatment is delayed; consequently, the austenite is more difficult

to transform into martensite.) This refrigerating period usually varies from two to six hours, depending on the size of the tool. Remove tool from refrigerating unit and allow to return to room temperature.

(e) Temper to required hardness for a period of two and one-half to three hours. The tempering temperature usually varies from a minimum of 1000 to 1100 degrees F. for ordinary high-speed steels. Tests indicate that if this first tempering is less than two and one-half hours at 1050 degrees F., there will not be sufficient precipitation of carbides at the tempering temperature to allow complete transformation of the retained austenite on cooling, whereas more than three hours causes some loss in room temperature hardness, hot hardness, strength, and toughness.

(f) Repeat sub-zero treatment, step (d).

(g) Repeat tempering operation, step (e). (Note: The time for the second tempering operation is sometimes reduced to about one-half the time required for the first tempering.)

2. *Single Sub-zero Treatment:* This treatment is the same as procedure No. 1 except that a second sub-zero cooling is omitted; hence the cycle consists of hardening, sub-zero cooling, and double tempering. Procedure No. 3 which follows also has one sub-zero cooling period in the cycle, but this *follows* the first tempering operation.

3. *Tempering Followed by Sub-zero Treatment:* (For tools having irregular sections, sharp corners or edges where cracks might develop if the hardening operation were followed immediately by sub-zero cooling.)

(a) and (b) Preheat and heat for hardening.
(c) Quench as described under Procedure No. 1.
(d) Temper to required hardness.
(e) Cool to sub-zero temperature − 100 to − 120 degrees F. and then allow tool to return to room temperature.
(f) Repeat tempering operation.

Testing the Hardness of Metals

Brinell Hardness Test. — The Brinell test for determining the hardness of metallic materials consists in applying a known load to the surface of the material to be tested through a hardened steel ball of known diameter. The diameter of the resulting permanent impression in the metal is measured and the Brinell Hardness Number (BHN) is then calculated from the following formula in which D = diameter of ball in millimeters, d = measured diameter at the rim of the impression in millimeters, and P = applied load in kilograms.

$$\text{BHN} = \frac{\text{load on indenting tool in kilograms}}{\text{surface area of indentation in sq. mm.}} = \frac{P}{\frac{\pi D}{2}(D - \sqrt{D^2 - d^2})}$$

If the steel ball were not deformed under the applied load and if the impression were truly spherical, then the above formula would be a general one, and any combination of applied load and size of ball could be used. The impression, however, is not quite a spherical surface since there must always be some deformation of the steel ball and some recovery of form of the metal in the impression; hence for a standard Brinell test, the size and characteristics of the ball and the magnitude of the applied load must be standardized. In the standard Brinell test, a ball 10 millimeters in diameter and a load of 3000, 1500, or 500 kilograms is used. It is desirable, although not mandatory, that the test load be of such magnitude that the diameter

of the impression be in the range of 2.50 to 4.75 millimeters. The following test loads and approximate Brinell numbers for this range of impression diameters are: 3000 kg., 160 to 600 BHN; 1500 kg., 80 to 300 BHN; 500 kg., 26 to 100 BHN. In making a Brinell test the load should be applied steadily and without a jerk for at least 15 seconds in the case of iron and steel, and at least 30 seconds in testing other metals. A minimum period of two minutes, for example, has been recommended for magnesium and magnesium alloys. (For the softer metals, loads of 250 kg., 125 kg., or 100 kg., are sometimes used.)

According to the American Society for Testing and Materials Standard E10-66, a steel ball may be used on material having a BHN not over 450, a Hultgren ball on material not over 500, or a carbide ball on material not over 630. The Brinell hardness test is not recommended for material having a BHN over 630.

Rockwell Hardness Test. — The Rockwell hardness tester is essentially a machine that measures hardness by determining the depth of penetration of a penetrator into the specimen under certain fixed conditions of test. The penetrator may be either a steel ball or a diamond sphero-conical penetrator. The hardness number is related to the depth of indentation and the number is higher the harder the material. A minor load of 10 kg. is first applied which causes an initial penetration; the dial is set at zero on the black-figure scale, and the major load is applied. This major load is customarily 60 kg. or 100 kg. when a steel ball is used as a penetrator, but other loads may be used when found necessary. The ball penetrator is $\frac{1}{16}$ inch in diameter normally; but other penetrators of larger diameter, such as $\frac{1}{8}$ inch, may be employed for soft metals. When a diamond sphero-conical penetrator is employed the load usually is 150 kg. Experience decides the best combination of load and penetrator for use. After the major load is applied and removed, according to standard procedure, the reading is taken while the minor load is still applied.

The Rockwell Hardness Scales. — The various Rockwell scales and their applications are shown in the table below. The type of penetrator and load used with each are shown in Tables 1 and 2 which give comparative hardness values for different hardness scales.

Scale	Testing Application
A	For tungsten carbide and other extremely hard materials. Also for thin, hard sheets.
B	For materials of medium hardness such as low and medium carbon steels in the annealed condition.
C	For materials harder than Rockwell B-100.
D	Where somewhat lighter load is desired than on C scale, as on case hardened pieces.
E	For very soft materials such as bearing metals.
F	Same as E scale but using $\frac{1}{16}$-inch ball.
G	For metals harder than tested on B scale.
H & K	For softer metals.
15-N; 30-N; 45-N	Where shallow impression or small area is desired. For hardened steel and hard alloys.
15-T; 30-T; 45-T	Where shallow impression or small area is desired for materials softer than hardened steel.

Shore's Scleroscope. — The scleroscope is an instrument which measures the hardness of the work in terms of elasticity. A diamond-tipped hammer is allowed to drop from a known height on the metal to be tested. As this hammer strikes the metal, it rebounds, and the harder the metal, the greater the rebound. The extreme height of the rebound is recorded, and an average of a number of readings taken on a single piece will give a good indication of the hardness of the work. The surface smoothness of the work affects the reading of the instrument. The readings are also affected by the contour and mass of the work and the depth of the case, in carburized work, the soft core of light-depth carburizing, pack-hardening, or cyanide hardening, absorbing the force of the hammer fall and decreasing the rebound. The hammer weighs about 40 grains, the height of the rebound of hardened steel is in the neighborhood of 100 on the scale, or about 6¼ inches, while the total fall is about 10 inches or 255 millimeters.

Vickers Hardness Test. — The Vickers test is similar in principle to the Brinell test. The standard Vickers penetrator is a square-based diamond pyramid having an included point angle of 136 degrees. The numerical value of the hardness number equals the applied load in kilograms divided by the area of the pyramidal impression. A smooth, firmly supported, flat surface is required. The load, which usually is applied for 30 seconds, may either be 5, 10, 20, 30, 50 or 120 kilograms. The 50-kilogram load is usually employed. The hardness number is based upon the diagonal length of the square impression. The Vickers test, which is considered very accurate, may, with proper load regulation, be applied to thin sheets as well as to larger sections.

Knoop Hardness Numbers. — The Knoop hardness test is applicable to extremely thin metal, plated surfaces, exceptionally hard and brittle materials, very shallow carburized or nitrided surfaces, or whenever the applied load must be kept below 3600 grams. The Knoop indentor is a diamond ground to an elongated pyramidal form and it produces an indentation having long and short diagonals with a ratio of approximately 7 to 1. The longitudinal angle of the indentor is 172 degrees 30 minutes and the transverse angle 130 degrees. The Tukon Tester in which the Knoop indentor is used is fully automatic under electronic control. The Knoop hardness number equals load in kilograms divided by the projected area of indentation in square millimeters. The indentation number corresponding to the long diagonal and for a given load, may be determined from a table computed for a theoretically perfect indentor. The load, which may be varied from 25 to 3600 grams, is applied for a definite period and always normal to the surface tested. Lapped plane surfaces free from scratches are required.

Monotron Hardness Indicator. — With this instrument, a diamond-ball impressor point ¾ mm. in diameter is forced into the material to a depth of 9/5000 inch and the pressure required to produce this constant impression indicates the hardness. One of two dials shows the pressure in kilograms and pounds, and the other shows the depth of the impression in millimeters and inches. Readings in Brinell numbers may be obtained by means of a scale designated as $M - 1$.

Keep's Test. — With this apparatus a standard steel drill is caused to make a definite number of revolutions, while it is pressed with standard force against the specimen to be tested. The hardness is automatically recorded on a diagram on which a dead soft material gives a horizontal line, while a material as hard as the drill itself gives a vertical line, intermediate hardness being represented by the corresponding angle between 0 and 90 degrees.

Table 1. Comparative Hardness Scales for Steel — 1

Rockwell C-Scale Hardness Number	Diamond Pyramid Hardness Number Vickers	Brinell Hardness Number 10-mm. Ball, 3000-kgf Load			Rockwell Hardness Number		Rockwell Superficial Hardness Number Superficial Diam. Penetrator			Shore Scleroscope Hardness Number
		Standard Ball	Hultgren Ball	Tungsten Carbide Ball	A-Scale 60-kgf Load Diam. Penetrator	D-Scale 100-kgf Load Diam. Penetrator	15-N Scale 15-kgf Load	30-N Scale 30-kgf Load	45-N Scale 45-kgf Load	
68	940	...	...	...	85.6	76.9	93.2	84.4	75.4	97
67	900	...	...	...	85.0	76.1	92.9	83.6	74.2	95
66	865	...	...	...	84.5	75.4	92.5	82.8	73.3	92
65	832	...	...	739	83.9	74.5	92.2	81.9	72.0	91
64	800	...	...	722	83.4	73.8	91.8	81.1	71.0	88
63	772	...	...	705	82.8	73.0	91.4	80.1	69.9	87
62	746	...	...	688	82.3	72.2	91.1	79.3	68.8	85
61	720	...	...	670	81.8	71.5	90.7	78.4	67.7	83
60	697	...	613	654	81.2	70.7	90.2	77.5	66.6	81
59	674	...	599	634	80.7	69.9	89.8	76.6	65.5	80
58	653	...	587	615	80.1	69.2	89.3	75.7	64.3	78
57	633	...	575	595	79.6	68.5	88.9	74.8	63.2	76
56	613	...	561	577	79.0	67.7	88.3	73.9	62.0	75
55	595	...	546	560	78.5	66.9	87.9	73.0	60.9	74
54	577	...	534	543	78.0	66.1	87.4	72.0	59.8	72
53	560	...	519	525	77.4	65.4	86.9	71.2	58.6	71
52	544	500	508	512	76.8	64.6	86.4	70.2	57.4	69
51	528	487	494	496	76.3	63.8	85.9	69.4	56.1	68
50	513	475	481	481	75.9	63.1	85.5	68.5	55.0	67
49	498	464	469	469	75.2	62.1	85.0	67.6	53.8	66
48	484	451	455	455	74.7	61.4	84.5	66.7	52.5	64
47	471	442	443	443	74.1	60.8	83.9	65.8	51.4	63
46	458	432	432	432	73.6	60.0	83.5	64.8	50.3	62
45	446	421	421	421	73.1	59.2	83.0	64.0	49.0	60
44	434	409	409	409	72.5	58.5	82.5	63.1	47.8	58
43	423	400	400	400	72.0	57.7	82.0	62.2	46.7	57
42	412	390	390	390	71.5	56.9	81.5	61.3	45.5	56
41	402	381	381	381	70.9	56.2	80.9	60.4	44.3	55
40	392	371	371	371	70.4	55.4	80.4	59.5	43.1	54
39	382	362	362	362	69.9	54.6	79.9	58.6	41.9	52
38	372	353	353	353	69.4	53.8	79.4	57.7	40.8	51
37	363	344	344	344	68.9	53.1	78.8	56.8	39.6	50
36	354	336	336	336	68.4	52.3	78.3	55.9	38.4	49
35	345	327	327	327	67.9	51.5	77.7	55.0	37.2	48
34	336	319	319	319	67.4	50.8	77.2	54.2	36.1	47
33	327	311	311	311	66.8	50.0	76.6	53.3	34.9	46
32	318	301	301	301	66.3	49.2	76.1	52.1	33.7	44
31	310	294	294	294	65.8	48.4	75.6	51.3	32.5	43
30	302	286	286	286	65.3	47.7	75.0	50.4	31.3	42
29	294	279	279	279	64.7	47.0	74.5	49.5	30.1	41
28	286	271	271	271	64.3	46.1	73.9	48.6	28.9	41
27	279	264	264	264	63.8	45.2	73.3	47.7	27.8	40

Note: The values in this table shown in **bold faced** type correspond to those shown in American Society for Testing and Materials Specification E140-67.

Table 1. Comparative Hardness Scales for Steel — 2

Rockwell C-Scale Hardness Number	Diamond Pyramid Hardness Number Vickers	Brinell Hardness Number 10-mm. Ball, 3000-kgf Load			Rockwell Hardness Number		Rockwell Superficial Hardness Number Superficial Brale Penetrator			Shore Scleroscope Hardness Number
		Standard Ball	Hultgren Ball	Tungsten Carbide Ball	A-Scale 60-kgf Load Diam. Penetrator	D-Scale 100-kgf Load Diam. Penetrator	15-N Scale 15-kgf Load	30-N Scale 30-kgf Load	45-N Scale 45-kgf Load	
26	272	258	258	258	63.3	44.6	72.8	46.8	26.7	38
25	266	253	253	253	62.8	43.8	72.2	45.9	25.5	38
24	260	247	247	247	62.4	43.1	71.6	45.0	24.3	37
23	254	243	243	243	62.0	42.1	71.0	44.0	23.1	36
22	248	237	237	237	61.5	41.6	70.5	43.2	22.0	35
21	243	231	231	231	61.0	40.9	69.9	42.3	20.7	35
20	238	226	226	226	60.5	40.1	69.4	41.5	19.6	34
(18)	230	219	219	219	...	...	...	...	...	33
(16)	222	212	212	212	...	...	...	...	...	32
(14)	213	203	203	203	...	...	...	...	...	31
(12)	204	194	194	194	...	...	...	...	...	29
(10)	196	187	187	187	...	...	...	...	...	28
(8)	188	179	179	179	...	...	...	...	...	27
(6)	180	171	171	171	...	...	...	...	...	26
(4)	173	165	165	165	...	...	...	...	...	25
(2)	166	158	158	158	...	...	...	...	...	24
(0)	160	152	152	152	...	...	...	...	...	24

Note: The values in this table shown in **bold faced** type correspond to those shown in American Society for Testing and Materials Specification E140-67.
Values in ( ) are beyond the normal range and are given for information only.

Comparison of Hardness Scales. — Tables 1 and 2 show comparisons of various hardness scales. All such tables are based on the assumption that the metal tested is homogeneous to a depth several times that of the indentation. To the extent that the metal being tested is not homogeneous, errors are introduced because different loads and different shapes of penetrators meet the resistance of metal of varying hardness, depending on the depth of indentation. Another source of error is introduced in comparing the hardness of different materials as measured on different hardness scales. This arises from the fact that in any hardness test, metal that is severely cold-worked actually supports the penetrator and different metals, different alloys, and different analyses of the same type of alloy have different cold-working properties. In spite of the possible inaccuracies introduced by such factors, it is of considerable value to be able to compare hardness values in a general way.

The data shown in Table 1 are based upon extensive tests on carbon and alloy steels mostly in the heat-treated condition, but have been found to be reliable on constructional alloy steels and tool steels in the as-forged, annealed, normalized, quenched and tempered conditions, providing they are homogeneous. These hardness comparisons are not as accurate for special cases such as high manganese steel, 18–8 stainless steel and other austenitic steels, nickel base alloys, as well as constructional alloy steels and nickel base alloys in the cold-worked condition.

The data shown in Table 2 are for hardness measurements of unhardened steel, steel of soft temper, grey and malleable cast iron, and most non-ferrous metals. Again these hardness comparisons are not as accurate for annealed metals of high Rockwell B hardness such as austenitic stainless steel, nickel and high nickel alloys and cold-worked metals of low B-scale hardness such as aluminum and the softer alloys.

Table 2. Comparative Hardness Scales for Unhardened Steel, Soft-temper Steel, Grey and Malleable Cast Iron, and Non-ferrous Alloys* — 1

Rockwell Hardness Number			Rockwell Superficial Hardness Number			Rockwell Hardness Number			Brinell Hardness Number	
Rockwell B scale 1/16" Ball Penetrator — 100 kg. Load	Rockwell F scale 1/16" Ball Penetrator — 60 kg. Load	Rockwell G scale 1/16" Ball Penetrator — 150 kg. Load	Rockwell Superficial 15-T scale 1/16" Ball Penetrator — 15 kg. Load	Rockwell Superficial 30-T scale 1/16" Ball Penetrator — 30 kg. Load	Rockwell Superficial 45-T scale 1/16" Ball Penetrator — 45 kg. Load	Rockwell E scale 1/8" Ball Penetrator — 100 kg. Load	Rockwell K scale 1/8" Ball Penetrator — 150 kg. Load	Rockwell A scale "Brale" Penetrator — 60 kg. Load	Brinell Scale 10 mm Standard Ball — 500 kg. Load	Brinell Scale 10 mm. Standard Ball — 3000 kg. Load
100	....	82.5	93.0	82.0	72.0	....	....	61.5	201	240
99	....	81.0	92.5	81.5	71.0	....	....	61.0	195	234
98	....	79.0	....	81.0	70.0	....	....	60.0	189	228
97	....	77.5	92.0	80.5	69.0	....	....	59.5	184	222
96	....	76.0	....	80.0	68.0	....	....	59.0	179	216
95	....	74.0	91.5	79.0	67.0	....	....	58.0	175	210
94	....	72.5	....	78.5	66.0	....	....	57.5	171	205
93	....	71.0	91.0	78.0	65.5	....	....	57.0	167	200
92	....	69.0	90.5	77.5	64.5	....	100	56.5	163	195
91	....	67.5	....	77.0	63.5	....	99.5	56.0	160	190
90	....	66.0	90.0	76.0	62.5	....	98.5	55.5	157	185
89	....	64.0	89.5	75.5	61.5	....	98.0	55.0	154	180
88	....	62.5	....	75.0	60.5	....	97.0	54.0	151	176
87	....	61.0	89.0	74.5	59.5	....	96.5	53.5	148	172
86	....	59.0	88.5	74.0	58.5	....	95.5	53.0	145	169
85	....	57.5	....	73.5	58.0	....	94.5	52.5	142	165
84	....	56.0	88.0	73.0	57.0	....	94.0	52.0	140	162
83	....	54.0	87.5	72.0	56.0	....	93.0	51.0	137	159
82	....	52.5	....	71.5	55.0	....	92.0	50.5	135	156
81	....	51.0	87.0	71.0	54.0	....	91.0	50.0	133	153
80	....	49.0	86.5	70.0	53.0	....	90.5	49.5	130	150
79	....	47.5	....	69.5	52.0	....	89.5	49.0	128	147
78	....	46.0	86.0	69.0	51.0	....	88.5	48.5	126	144
77	....	44.0	85.5	68.0	50.0	....	88.0	48.0	124	141
76	....	42.5	....	67.5	49.0	....	87.0	47.0	122	139
75	99.5	41.0	85.0	67.0	48.5	....	86.0	46.5	120	137
74	99.0	39.0	....	66.0	47.5	....	85.0	46.0	118	135
73	98.5	37.5	84.5	65.5	46.5	....	84.5	45.5	116	132
72	98.0	36.0	84.0	65.0	45.5	....	83.5	45.0	114	130
71	97.5	34.5	....	64.0	44.5	100	82.5	44.5	112	127
70	97.0	32.5	83.5	63.5	43.5	99.5	81.5	44.0	110	125
69	96.0	31.0	83.0	62.5	42.5	99.0	81.0	43.5	109	123
68	95.5	29.5	....	62.0	41.5	98.0	80.0	43.0	107	121
67	95.0	28.0	82.5	61.5	40.5	97.5	79.0	42.5	106	119
66	94.5	26.5	82.0	60.5	39.5	97.0	78.0	42.0	104	117
65	94.0	25.0	....	60.0	38.5	96.0	77.5	....	102	116
64	93.5	23.5	81.5	59.5	37.5	95.5	76.5	41.5	101	114
63	93.0	22.0	81.0	58.5	36.5	95.0	75.5	41.0	99	112
62	92.0	20.5	....	58.0	35.5	94.5	74.5	40.5	98	110
61	91.5	19.0	80.5	57.0	34.5	93.5	74.0	40.0	96	108
60	91.0	17.5	....	56.5	33.5	93.0	73.0	39.5	95	107
59	90.5	16.0	80.0	56.0	32.0	92.5	72.0	39.0	94	106

* See note at end of table.

Table 2. Comparative Hardness Scales for Unhardened Steel, Soft-temper Steel, Grey and Malleable Cast-iron, and Non-ferrous Alloys* — 2

Rockwell Hardness Number			Rockwell Superficial Hardness Number			Rockwell Hardness Number				Brinell
Rockwell B scale 1/16" Ball Penetrator 100 kg. Load	Rockwell F scale 1/16" Ball Penetrator 60 kg. Load	Rockwell G scale 1/16" Ball Penetrator 150 kg. Load	Rockwell Superficial 15-T scale 1/16" Ball Penetrator 15 kg. Load	Rockwell Superficial 30-T scale 1/16" Ball Penetrator 30 kg. Load	Rockwell Superficial 45-T scale 1/16" Ball Penetrator 45 kg. Load	Rockwell E scale 1/8" Ball Penetrator 100 kg. Load	Rockwell H scale 1/8" Ball Penetrator 60 kg. Load	Rockwell K scale 1/8" Ball Penetrator 150 kg. Load	Rockwell A scale "Brale" Penetrator 60 kg. Load	Brinell Scale 10-mm. Standard Ball — 500 kg. Load
58	90.0	14.5	79.5	55.0	31.0	92.0	....	71.0	38.5	92
57	89.5	13.0	....	54.5	30.0	91.0	....	70.5	38.0	91
56	89.0	11.5	79.0	54.0	29.0	90.5	....	69.5	....	90
55	88.0	10.0	78.5	53.0	28.0	90.0	....	68.5	37.5	89
54	87.5	8.5	....	52.5	27.0	89.5	....	68.0	37.0	87
53	87.0	7.0	78.0	51.5	26.0	89.0	....	67.0	36.5	86
52	86.5	5.5	77.5	51.0	25.0	88.0	....	66.0	36.0	85
51	86.0	4.0	....	50.5	24.0	87.5	....	65.0	35.5	84
50	85.5	2.5	77.0	49.5	23.0	87.0	....	64.5	35.0	83
50	85.5	2.5	77.0	49.5	23.0	87.0	....	64.5	35.0	83
49	85.0	1.0	76.5	49.0	22.0	86.5	....	63.5	....	82
48	84.5	....	....	48.5	20.5	85.5	....	62.5	34.5	81
47	84.0	....	76.0	47.5	19.5	85.0	....	61.5	34.0	80
46	83.0	....	75.5	47.0	18.5	84.0	....	61.0	33.5	..
45	82.5	....	....	46.0	17.5	84.0	....	60.0	33.0	79
44	82.0	....	75.0	45.5	16.5	83.5	....	59.0	32.5	78
43	81.5	....	74.5	45.0	15.5	82.5	....	58.0	32.0	77
42	81.0	....	....	44.0	14.5	82.0	....	57.5	31.5	76
41	80.5	....	74.0	43.5	13.5	81.5	....	56.5	31.0	75
40	79.5	....	73.5	43.0	12.5	81.0	....	55.5	....	..
39	79.0	....	....	42.0	11.0	80.0	....	54.5	30.5	74
38	78.5	....	73.0	41.5	10.0	79.5	....	54.0	30.0	73
37	78.0	....	72.5	40.5	9.0	79.0	....	53.0	29.5	72
36	77.5	....	....	40.0	8.0	78.5	100	52.0	29.0	..
35	77.0	....	72.0	39.5	7.0	78.0	99.5	51.5	28.5	71
34	76.5	....	71.5	38.5	6.0	77.0	99.0	50.5	28.0	70
33	75.5	....	....	38.0	5.0	76.5	....	49.5	....	69
32	75.0	....	71.0	37.5	4.0	76.0	98.5	48.5	27.5	..
31	74.5	....	....	36.5	3.0	75.5	98.0	48.0	27.0	68
30	74.0	....	70.5	36.0	2.0	75.0	....	47.0	26.5	67
29	73.5	....	70.0	35.5	1.0	74.0	97.5	46.0	26.0	..
28	73.0	....	....	34.5	....	73.5	97.0	45.0	25.5	66
27	72.5	....	69.5	34.0	....	73.0	96.5	44.5	25.0	..
26	72.0	....	69.0	33.0	....	72.5	....	43.5	24.5	65
25	71.0	....	....	32.5	....	72.0	96.0	42.5	....	64
24	70.5	....	68.5	32.0	....	71.0	95.5	41.5	24.0	..
23	70.0	....	68.0	31.0	....	70.5	....	41.0	23.5	63
22	69.5	....	....	30.5	....	70.0	95.0	40.0	23.0	..
21	69.0	....	67.5	29.5	....	69.5	94.5	39.0	22.5	62
20	68.5	....	....	29.0	....	68.5	....	38.0	22.0	..
19	68.0	....	67.0	28.5	....	68.0	94.0	37.5	21.5	61
18	67.0	....	66.5	27.5	....	67.5	93.5	36.5	....	..

* See note at end of table.

Table 2. Comparative Hardness Scales for Unhardened Steel, Soft-temper Steel, Grey and Malleable Cast Iron, and Non-ferrous Alloys* — 3

(Compiled by Wilson Mechanical Instrument Co.)

Rockwell Hardness Number			Rockwell Superficial Hardness Number			Rockwell Hardness Number				Brinell
Rockwell B scale 1/16" Ball Penetrator 100 kg. Load	Rockwell F scale 1/16" Ball Penetrator 60 kg. Load	Rockwell G scale 1/16" Ball Penetrator 150 kg. Load	Rockwell Superficial 15-T scale 1/16" Ball Penetrator 15 kg. Load	Rockwell Superficial 30-T scale 1/16" Ball Penetrator 30 kg. Load	Rockwell Superficial 45-T scale 1/16" Ball Penetrator 45 kg. Load	Rockwell E scale 1/8" Ball Penetrator 100 kg. Load	Rockwell H scale 1/8" Ball Penetrator 60 kg. Load	Rockwell K scale 1/8" Ball Penetrator 150 kg. Load	Rockwell A scale "Brale" Penetrator 60 kg. Load	Brinell Scale 10-mm. Standard Ball — 500 kg. Load
17	66.5	....	....	27.0	....	67.0	93.0	35.5	21.0	60
16	66.0	....	66.0	26.0	....	66.5	....	35.0	20.5	..
15	65.5	....	65.5	25.5	....	65.5	92.5	34.0	20.0	59
14	65.0	....	....	25.0	....	65.0	92.0	33.0	....	..
13	64.5	....	65.0	24.0	....	64.5	....	32.0	....	58
12	64.0	....	64.5	23.5	....	64.0	91.5	31.5	....	..
11	63.5	....	....	23.0	....	63.5	91.0	30.5	....	..
10	63.0	....	64.0	22.0	....	62.5	90.5	29.5	....	57
9	62.0	....	....	21.5	....	62.0	....	29.0	....	..
8	61.5	....	63.5	20.5	....	61.5	90.0	28.0	....	..
7	61.0	....	63.0	20.0	....	61.0	89.5	27.0	....	56
6	60.5	....	....	19.5	....	60.5	....	26.0	....	..
5	60.0	....	62.5	18.5	....	60.0	89.0	25.5	....	55
4	59.5	....	62.0	18.0	....	59.0	88.5	24.5	....	..
3	59.0	....	....	17.0	....	58.5	88.0	23.5	....	..
2	58.0	....	61.5	16.5	....	58.0	....	23.0	....	54
1	57.5	....	61.0	16.0	....	57.5	87.5	22.0	....	..
0	57.0	....	....	15.0	....	57.0	87.0	21.0	....	53

* Not applicable to annealed metals of high B-scale hardness such as austenitic stainless steels, nickel and high-nickel alloys nor to cold-worked metals of low B-scale hardness such as aluminum and the softer alloys.

Turner's Sclerometer. — In making this test a weighted diamond point is drawn, once forward and once backward, over the smooth surface of the material to be tested. The hardness number is the weight in grams required to produce a standard scratch.

Mohs's Hardness Scale. — Hardness, in general, is determined by what is known as Mohs's scale, a standard for hardness which is mainly applied to non-metallic elements and minerals. In this hardness scale there are ten degrees or steps, each designated by a mineral, the difference in hardness of the different steps being determined by the fact that any member in the series will scratch any of the preceding members. This scale is as follows:

1. Talc; 2. gypsum; 3. calcite; 4. fluor spar; 5. apatite; 6. orthoclase; 7. quartz; 8. topaz; 9. sapphire or corundum; 10. diamond.

These minerals, arbitrarily selected as standards, are successively harder, from talc, the softest of all minerals, to diamond, the hardest. This scale, which is now universally used for non-metallic minerals, is, however, not applied to metals.

Relation Between Hardness and Tensile Strength. — The approximate relationship between the hardness and tensile strength is shown by the following formula in which B = Brinell hardness number.

Tensile strength = $B \times 515$ (for Brinell numbers up to 175).

Tensile strength = $490 \times B$ (for Brinell numbers larger than 175).

These formulas give the tensile strength in pounds per square inch and apply to steels. This definite relationship between hardness and tensile strength does not apply to non-ferrous metals with the possible exception of certain aluminum alloys.

NON-FERROUS ALLOYS

Copper and Copper Alloys

Pure copper is a reddish, highly malleable metal, and was one of the first to be found and utilized. Copper and its alloys are widely used because of their excellent electrical and thermal conductivities, outstanding resistance to corrosion, ease of fabrication, and broad ranges of obtainable strengths and special properties. There are almost 400 commercial copper and copper-alloy compositions available from mills as wrought products (rod, plate, sheet, strip, tube, pipe, extrusions, foil, forgings, and wire) and from foundries as castings.

Copper alloys are grouped into several general categories according to composition: coppers and high copper alloys; brasses; bronzes; copper nickels; copper–nickel—zinc alloys (nickel silvers); leaded coppers; and special alloys. The designation system originally developed by the U.S. copper and brass industry for identifying copper alloys used a three-digit number preceded by the letters CA. These designations have now been made part of the Unified Numbering System (UNS) simply by expanding the numbers to five digits preceded by the letter C. Because the old numbers are embedded in the new UNS numbers, no confusion results. UNS C10000 to C79999 are assigned to wrought compositions, and UNS C80000 to C99999 are assigned to castings. The designation system is not a specification, but a method for identifying the composition of mill and foundry products. The precise technical and quality assurance requirements to be satisfied are defined in relevant standard specifications issued by the federal government, the military, and the ASTM.

Classification of Copper and Copper Alloys

Family	Principal Alloying Element	UNS Numbers[a]
Coppers, high copper alloys		C1xxxx
Brasses	Zn	C2xxxx, C3xxxx, C4xxxx, C66400 to C69800
Phosphor bronzes	Sn	C5xxxx
Aluminum bronzes	Al	C60600 to C64200
Silicon bronzes	Si	C64700 to C66100
Copper nickels, nickel silvers	Ni	C7xxxx

[a] Wrought alloys.

Cast Copper Alloys. — Generally, casting permits greater latitude in the use of alloying elements than in the fabrication of wrought products, which requires either hot or cold working. The cast compositions of coppers and high copper alloys have a designated minimum copper content and may include other elements to impart special properties. The cast brasses comprise copper–zinc–tin alloys (red, semired, and yellow brasses); manganese bronze alloys (high-strength yellow brasses); leaded manganese bronze alloys (leaded high-strength yellow brasses); and copper–zinc–silicon alloys (silicon brasses and bronzes). The cast bronze alloys have four main families: copper–tin alloys (tin bronzes); copper–tin–lead alloys (leaded and high leaded tin bronzes); copper–tin–nickel alloys (nickel-tin bronzes); and copper–aluminum alloys (aluminum bronzes). The cast copper–nickel alloys containing nickel as the principal alloying element. The leaded coppers are cast alloys containing 20 per cent or more lead.

Table 1 lists the properties and applications of common cast copper alloys.

Table 1. Properties and Applications of Cast Coppers and Copper Alloys

UNS Designation	Nominal Composition (%)	Typical Mechanical Properties, as Cast or Heat Treated†				Typical Applications
		Tensile Strength (ksi)	Yield Strength (ksi)	Elongation in 2 in. (%)	Machinability Rating*	
		Copper Alloys				
C80100	99.95 Cu + Ag min, 0.05 others max	25	9	40	10	Electrical and thermal conductors; corrosion and oxidation resistant applications.
C80300	99.95 Cu + Ag min, 0.034 Ag min, 0.05 others max	25	9	40	10	Electrical and thermal conductors; corrosion and oxidation resistant applications.
C80500	99.75 Cu + Ag min, 0.034 Ag min, 0.02 B max, 0.23 others max	25	9	40	10	Electrical and thermal conductors; corrosion and oxidation resistant applications.
C80700	99.75 Cu + Ag min, 0.02 B max, 0.23 others max	25	9	40	10	Electrical and thermal conductors; corrosion and oxidation resistant applications.
C80900	99.70 Cu + Ag min, 0.034 Ag min, 0.30 others max	25	9	40	10	Electrical and thermal conductors; corrosion and oxidation resistant applications.
C81100	99.70 Cu + Ag min, 0.30 others max	25	9	40	10	Electrical and thermal conductors; corrosion and oxidation resistant applications.
		High-Copper Alloys				
C81300	98.5 Cu min, 0.06 Be, 0.80 Co, 0.40 others max	(53)	(36)	(11)	20	Higher hardness electrical and thermal conductors.
C81400	98.5 Cu min, 0.06 Be, 0.80 Cr, 0.40 others max	(53)	(36)	(11)	20	Higher hardness electrical and thermal conductors.
C81500	98.0 Cu min, 1.0 Cr, 0.50 others max	(51)	(40)	(17)	20	Electrical and/or thermal conductors used as structural members where strength and hardness greater than that of C80100–81100 are required.

* See footnotes at end of table.

Table 1 (*Continued*). **Properties and Applications of Cast Coppers and Copper Alloys**

UNS Designation	Nominal Composition (%)	Typical Mechanical Properties, as Cast or Heat Treated†				Typical Applications
		Tensile Strength (ksi)	Yield Strength (ksi)	Elongation in 2 in. (%)	Machinability Rating*	
		High-Copper Alloys (continued)				
C81700	94.25 Cu min, 1.0 Ag, 0.4 Be, 0.9 Co, 0.9 Ni	(92)	(68)	(8)	30	Electrical and/or thermal conductors used as structural members where strength and hardness greater than that of C80100–81100 are required. Also used in place of C81500 where electrical and/or thermal conductivities can be sacrificed for hardness and strength.
C81800	95.6 Cu min, 1.0 Ag, 0.4 Be, 1.6 Co	50 (102)	25 (75)	20 (8)	20	Resistance welding electrodes, dies.
C82000	96.8 Cu, 0.6 Be, 2.6 Co	50 (100)	20 (75)	20 (8)	20	Current carrying parts, contact and switch blades, bushings and bearings, and soldering iron and re-sistance welding tips.
C82100	97.7 Cu, 0.5 Be, 0.9 Co, 0.9 Ni	(92)	(68)	(8)	30	Electrical and/or thermal conductors used as structural members where strength and hardness greater than that of C80100–81100 are required. Also used in place of C81500 where electrical and/or thermal conductivities can be sacrificed for hardness and strength.
C82200	96.5 Cu min, 0.6 Be, 1.5 Ni	57 (95)	30 (75)	20 (8)	20	Clutch rings, brake drums, seam welder elec-trodes, projection welding dies, spot welding tips, beam welder shapes, bushings, water-cooled holders.
C82400	96.4 Cu min, 1.70 Be, 0.25 Co	72 (150)	37 (140)	20 (1)	20	Safety tools, molds for plastic parts, cams, bush-ings, bearings, valves, pump parts, gears.
C82500	97.2 Cu, 2.0 Be, 0.5 Co, 0.25 Si	80 (160)	45	20 (1)	20	Safety tools, molds for plastic parts, cams, bush-ings, bearings, valves, pump parts.
C82600	95.2 Cu, 2.3 Be, 0.5 Co, 0.25 Si	82 (165)	47 (155)	20 (1)	20	Bearings and molds for plastic parts.

* See footnotes at end of table.

Table 1 (Continued). Properties and Applications of Cast Coppers and Copper Alloys

UNS Designation	Nominal Composition (%)	Tensile Strength (ksi)	Yield Strength (ksi)	Elongation in 2 in. (%)	Machinability Rating*	Typical Applications
		Typical Mechanical Properties, as Cast or Heat Treated†				
High-Copper Alloys (continued)						
C82700	96.3 Cu, 2.45 Be, 1.25 Ni	(155)	(130)	(0)	20	Bearings and molds for plastic parts.
C82800	96.6 Cu, 2.6 Be, 0.5 Co, 0.25 Si	97 (165)	55 (145)	20 (1)	10	Molds for plastic parts, cams, bushings, bearings, valves, pump parts, sleeves.
Red Brasses and Leaded Red Brasses						
C83300	93 Cu, 1.5 Sn, 1.5 Pb, 4 Zn	32	10	35	35	Terminal ends for electrical cables.
C83400	90 Cu, 10 Zn	35	10	30	60	Moderate strength, moderate conductivity castings; rotating bands.
C83600	85 Cu, 5 Sn, 5 Pb, 5 Zn	37	17	30	84	Valves, flanges, pipe fittings, plumbing goods, pump castings, water pump impellers and housings, ornamental fixtures, small gears.
C83800	83 Cu, 4 Sn, 6 Pb, 7 Zn	35	16	25	90	Low-pressure valves and fittings, plumbing supplies and fittings, general hardware, air–gas–water fittings, pump components, railroad catenary fittings.
Semired Brasses and Leaded Semired Brasses						
C84200	80 Cu, 5 Sn, 2.5 Pb, 12.5 Zn	28	15	27	80	Pipe fittings, elbows, T's, couplings, bushings, locknuts, plugs, unions.
C84400	81 Cu, 3 Sn, 7 Pb, 9 Zn	34	15	26	90	General hardware, ornamental castings, plumbing supplies and fixtures, low-pressure valves and fittings.
C84500	78 Cu, 3 Sn, 7 Pb, 12 Zn	35	14	28	90	Plumbing fixtures, cocks, faucets, stops, waste, air and gas fittings, low-pressure valve fittings.
C84800	76 Cu, 3 Sn, 6 Pb, 15 Zn	36	14	30	90	Plumbing fixtures, cocks, faucets, stops, waste, air, and gas, general hardware, and low-pressure valve fittings.

* See footnotes at end of table.

Table I (*Continued*). **Properties and Applications of Cast Coppers and Copper Alloys**

UNS Designation	Nominal Composition (%)	Typical Mechanical Properties, as Cast or Heat Treated[†]				Typical Applications
		Tensile Strength (ksi)	Yield Strength (ksi)	Elongation in 2 in. (%)	Machinability Rating*	
		Yellow Brasses and Leaded Yellow Brasses				
C85200	72 Cu, 1 Sn, 3 Pb, 24 Zn	38	13	35	80	Plumbing fittings and fixtures, ferrules, valves, hardware, ornamental brass, chandeliers, and irons.
C85400	67 Cu, 1 Sn, 3 Pb, 29 Zn	34	12	35	80	General purpose yellow casting alloy not subject to high internal pressure. Furniture hardware, ornamental castings, radiator fittings, ship trimmings, battery clamps, valves and fittings.
C85500	61 Cu, 0.8 Al, bal Zn	60	23	40	80	Ornamental castings.
C85700	63 Cu, 1 Sn, 1 Pb, 34.7 Zn, 0.3 Al	50	18	40	80	Bushings, hardware fittings, ornamental castings.
C85800	58 Cu, 1 Sn, 1 Pb, 40 Zn	55	30	15	80	General purpose die casting alloy having moderate strength.
		Manganese and Leaded Manganese Bronze Alloys				
C86100	67 Cu, 21 Zn, 3 Fe, 5 Al, 4 Mn	95	50	20	30	Marine castings, gears, gun mounts, bushings and bearings, marine racing propellers.
C86200	64 Cu, 26 Zn, 3 Fe, 4 Al, 3 Mn	95	48	20	30	Marine castings, gears, gun mounts, bushings and bearings.
C86300	63 Cu, 25 Zn, 3 Fe, 6 Al, 3 Mn	115	83	15	8	Extra-heavy duty, high-strength alloy. Large valve stems, gears, cams, slow-speed heavy-load bearings, screwdown nuts, hydraulic cylinder parts.
C86400	59 Cu, 1 Pb, 40 Zn	65	25	20	65	Free-machining manganese bronze. Valve stems, marine fittings, lever arms, brackets, light-duty gears.
C86500	58 Cu, 0.5 Sn, 39.5 Zn, 1 Fe, 1 Al	71	28	30	26	Machinery parts requiring strength and toughness, lever arms, valve stems, gears.

Table 1 (*Continued*). **Properties and Applications of Cast Coppers and Copper Alloys**

UNS Designation	Nominal Composition (%)	Typical Mechanical Properties, as Cast or Heat Treated†				Typical Applications
		Tensile Strength (ksi)	Yield Strength (ksi)	Elongation in 2 in. (%)	Machinability Rating*	
		Manganese and Leaded Manganese Bronze Alloys (continued)				
C86700	58 Cu, 1 Pb, 41 Zn	85	42	20	55	High strength, free-machining manganese bronze. Valve stems.
C86800	55 Cu, 37 Zn, 3 Ni, 2 Fe, 3 Mn	82	38	22	30	Marine fittings, marine propellers.
		Silicon Bronzes and Silicon Brasses				
C87200	89 Cu min, 4 Si	55	25	30	40	Bearings, bells, impellers, pump and valve components, marine fittings, corrosion-resistant castings.
C87400	83 Cu, 14 Zn, 3 Si	55	24	30	50	Bearings, gears, impellers, rocker arms, valve stems, clamps.
C87500	82 Cu, 14 Zn, 4 Si	67	30	21	50	Bearings, gears, impellers, rocker arms, valve stems, small boat propellers.
C87600	90 Cu, 5.5 Zn, 4.5 Si	66	32	20	40	Valve stems.
C87800	82 Cu, 14 Zn, 4 Si	85	50	25	40	High-strength, thin-wall die castings; brush holders, lever arms, brackets, clamps, hexagonal nuts.
C87900	65 Cu, 34 Zn, 1 Si	70	35	25	80	General purpose die casting alloy having moderate strength.
		Tin Bronzes				
C90200	93 Cu, 7 Sn	38	16	30	20	Bearings and bushings.
C90300	88 Cu, 8 Sn, 4 Zn	45	21	30	30	Bearings, bushings, pump impellers, piston rings, valve components, seal rings, steam fittings, gears.
C90500	88 Cu, 10 Sn, 2 Zn	45	22	25	30	Bearings, bushings, pump impellers, piston rings, valve components, steam fittings, gears.

* See footnotes at end of table.

Table 1 (*Continued*). **Properties and Applications of Cast Coppers and Copper Alloys**

UNS Designation	Nominal Composition (%)	Typical Mechanical Properties, as Cast or Heat Treated†				Typical Applications
		Tensile Strength (ksi)	Yield Strength (ksi)	Elongation in 2 in. (%)	Machinability Rating*	
		Tin Bronzes (continued)				
C90700	89 Cu, 11 Sn	44 (55)	22 (30)	20 (16)	20	Gears, bearings, bushings.
C90800	87 Cu, 12 Sn	...	...	...	...	...
C90900	87 Cu, 13 Sn	40	20	15	20	Bearings and bushings.
C91000	85 Cu, 14 Sn, 1 Zn	32	25	2	20	Piston rings and bearings.
C91100	84 Cu, 16 Sn	35	25	2	10	Piston rings, bearings, bushings, bridge plates.
C91300	81 Cu, 19 Sn	35	30	0.5	10	Piston rings, bearings, bushings, bridge plates, bells.
C91600	88 Cu, 10.5 Sn, 1.5 Ni	44 (60)	22 (32)	16 (16)	20	Gears.
C91700	86.5 Cu, 12 Sn, 1.5 Ni	44 (60)	22 (32)	16 (16)	20	Gears.
		Leaded Tin Bronzes				
C92200	88 Cu, 6 Sn, 1.5 Pb, 4.5 Zn	40	20	30	42	Valves, fittings, and pressure-containing parts for use up to 550°F.
C92300	87 Cu, 8 Sn, 4 Zn	40	20	25	42	Valves, pipe fittings, and high-pressure steam castings. Superior machinability to C90300.
C92400	88 Cu, 10 Sn, 2 Pb, 2 Zn	...	...	...	...	...
C92500	87 Cu, 11 Sn, 1 Pb, 1 Ni	44	20	20	30	Gears, automotive synchronizer rings.
C92600	87 Cu, 10 Sn, 1 Pb, 2 Zn	44	20	30	40	Bearings, bushings, pump impellers, piston rings, valve components, steam fittings, and gears. Superior machinability to C90500.
C92700	88 Cu, 10 Sn, 2 Pb	42	21	20	45	Bearings, bushings, pump impellers, piston rings, valve components, steam fittings, and gears. Superior machinability to C90500.

* See footnotes at end of table.

Table I (*Continued*). **Properties and Applications of Cast Coppers and Copper Alloys**

UNS Designation	Nominal Composition (%)	Typical Mechanical Properties, as Cast or Heat Treated†				Typical Applications
		Tensile Strength (ksi)	Yield Strength (ksi)	Elongation in 2 in. (%)	Machinability Rating*	
		Leaded Tin Bronzes (continued)				
C92800	79 Cu, 16 Sn, 5 Pb	40	30	1	70	Piston rings.
C92900	84 Cu, 10 Sn, 2.5 Pb, 3.5 Ni	47 (47)	26 (26)	20 (20)	40	Gears, wear plates, guides, cams, parts requiring machinability superior to that of C91600 or 91700.
		High-Leaded Tin Bronzes				
C93200	83 Cu, 7 Sn, 7 Pb, 3 Zn	35	18	20	70	General-utility bearings and bushings.
C93400	84 Cu, 8 Sn, 8 Pb	32	16	20	70	Bearings and bushings.
C93500	85 Cu, 5 Sn, 9 Pb	32	16	20	70	Small bearings and bushings, bronze backing for babbit-lined automotive bearings.
C93700	80 Cu, 10 Sn, 10 Pb	35	18	20	80	Bearings for high speed and heavy pressures, pumps, impellers, corrosion-resistant applications, pressure tight castings.
C93800	78 Cu, 7 Sn, 15 Pb	30	16	18	80	Bearings for general service and moderate pressure, pump impellers and bodies for use in acid mine water.
C93900	79 Cu, 6 Sn, 15 Pb	32	22	7	80	Continuous castings only. Bearings for general service, pump bodies and impellers for mine waters.
C94000	70.5 Cu, 13.0 Sn, 15.0 Pb, 0.50 Zn, 0.75 Ni, 0.25 Fe, 0.05 P, 0.35 Sb	...	...	...	...	...
C94100	70.0 Cu, 5.5 Sn, 18.5 Pb, 3.0 Zn, 1.0 others max.	...	...	...	...	...
C94300	70 Cu, 5 Sn, 25 Pb	27	13	15	80	High-speed bearings for light loads.

* See footnotes at end of table.

Table I (Continued). Properties and Applications of Cast Coppers and Copper Alloys

UNS Designation	Nominal Composition (%)	Typical Mechanical Properties, as Cast or Heat Treated†				Typical Applications
		Tensile Strength (ksi)	Yield Strength (ksi)	Elongation in 2 in. (%)	Machinability Rating*	
High-Leaded Tin Bronzes (continued)						
C94400	81 Cu, 8 Sn, 11 Pb	32	16	18	80	General-utility alloy for bushings and bearings.
C94500	73 Cu, 7 Sn, 20 Pb	25	12	12	80	Locomotive wearing parts, high-speed low-load bearings.
Nickel-Tin Bronzes						
C94700	88 Cu, 5 Sn, 2 Zn, 5 Ni	50 (85)	23 (60)	35 (10)	30	Valve stems and bodies, bearings, wear guides, shift forks, feeding mechanisms, circuit breaker parts, gears, piston cylinders, nozzles.
C94800	87 Cu, 5 Sn, 5 Ni	45 (60)	23 (30)	35 (8)	50	Structural castings, gear components, motion translation devices, machinery parts, bearings.
C94900	80 Cu, 5 Sn, 5 Pb, 5 Zn, 5 Ni	...	...	...	...	...
Aluminum Bronzes						
C95200	88 Cu, 3 Fe, 9 Al	80	27	35	50	Acid-resisting pumps, bearings, gears, valve seats, guides, plungers, pump rods, bushings.
C95300	89 Cu, 1 Fe, 10 Al	75 (85)	27 (42)	25 (15)	55	Pickling baskets, nuts, gears, steel mill slippers, marine equipment, welding jaws.
C95400	85 Cu, 4 Fe, 11 Al	85 (105)	35 (54)	18 (8)	60	Bearings, gears, worms, bushings, valve seats and guides, pickling hooks.
C95410	85 Cu, 4 Fe, 11 Al, 2 Ni	...	...	...	...	...
C95500	81 Cu, 4 Ni, 4 Fe, 11 Al	100 (120)	44 (68)	12 (10)	50	Valve guides and seats in aircraft engines, corrosion-resistant parts, bushings, gears, worms, pickling hooks and baskets, agitators.
C95600	91 Cu, 7 Al, 2 Si	75	34	18	60	Cable connectors, terminals, valve stems, marine hardware, gears, worms, pole-line hardware.

* See footnotes at end of table.

Table 1 (*Continued*). **Properties and Applications of Cast Coppers and Copper Alloys**

UNS Designation	Nominal Composition (%)	Tensile Strength (ksi)	Yield Strength (ksi)	Elongation in 2 in. (%)	Machinability Rating*	Typical Applications
			Aluminum Bronzes (continued)			
C95700	75 Cu, 2 Ni, 3 Fe, 8 Al, 12 Mn	95	45	26	50	Propellers, impellers, stator clamp segments, safety tools, welding rods, valves, pump casings.
C95800	81 Cu, 5 Ni, 4 Fe, 9 Al, 1 Mn	95	38	25	50	Propeller hubs, blades, and other parts in contact with salt water.
			Copper-Nickels			
C96200	88.6 Cu, 10 Ni, 1.4 Fe	45	25	20	10	Components of items being used for seawater corrosion resistance.
C96300	79.3 Cu, 20 Ni, 0.7 Fe	75	55	10	15	Centrifugally cast tailshaft sleeves.
C96400	69.1 Cu, 30 Ni, 0.9 Fe	68	37	28	20	Valves, pump bodies, flanges, elbows used for seawater corrosion resistance.
C96600	68.5 Cu, 30 Ni, 1 Fe, 0.5 Be	(110)	(70)	(7)	20	High-strength constructional parts for seawater corrosion resistance.
C96700	67.6 Cu, 30 Ni, 0.9 Fe, 1.15 Be, 0.15 Zr, 0.15 Ti	(175)	(80)	(10)	40	Corrosion-resistant molds for plastics, high-strength constructional parts for seawater use.
			Nickel Silvers			
C97300	56 Cu, 2 Sn, 10 Pb, 12 Ni, 20 Zn	35	17	20	70	Hardware fittings, valves and valve trim, statuary, ornamental castings.
C97400	59 Cu, 3 Sn, 5 Pb, 17 Ni, 16 Zn	38	17	20	60	Valves, hardware, fittings, ornamental castings.
C97600	64 Cu, 4 Sn, 4 Pb, 20 Ni, 8 Zn	45	24	20	70	Marine castings, sanitary fittings, ornamental hardware, valves, pumps.
C97800	66 Cu, 5 Sn, 2 Pb, 25 Ni, 2 Zn	55	30	15	60	Ornamental and sanitary castings, valves and valve seats, musical instrument components.

* See footnotes at end of table.

Table I (*Concluded*). Properties and Applications of Cast Coppers and Copper Alloys

UNS Designation	Nominal Composition (%)	Typical Mechanical Properties, as Cast or Heat Treated†				Typical Applications
		Tensile Strength (ksi)	Yield Strength (ksi)	Elongation in 2 in. (%)	Machinability Rating*	
Leaded Coppers						
C98200	76.0 Cu, 24.0 Pb	...	...	...	...	...
C98400	70.5 Cu, 28.5 Pb, 1.5 Ag	...	...	...	...	...
C98600	65.0 Cu, 35.0 Pb, 1.5 Ag	...	...	...	...	...
C98800	59.5 Cu, 40.0 Pb, 5.5 Ag	...	...	...	...	...
Special Alloys						
C99300	71.8 Cu, 15 Ni, 0.7 Fe, 11 Al, 1.5 Co	95	55	2	20	Glass making molds, plate glass rolls, marine hardware.
C99400	90.4 Cu, 2.2 Ni, 2.0 Fe, 1.2 Al, 1.2 Si, 3.0 Zn	66 (79)	34 (54)	25	50	Valve stems, marine and other uses requiring resistance to dezincification and dealumification, propeller wheels, electrical parts, mining equipment gears.
C99500	87.9 Cu, 4.5 Ni, 4.0 Fe, 1.2 Al, 1.2 Si, 1.2 Zn	70	40	12	50	Same as C99400, but where higher yield strength is required.
C99600	58 Cu, 2 Al, 40 Mn	81 (81)	36 (44)	34 (27)	...	Damping alloy to reduce noise and vibration.
C99700	56.5 Cu, 1 Al, 1.5 Pb, 12 Mn, 5 Ni, 24 Zn	55	25	25	80	...
C99750	58 Cu, 1 Al, 1 Pb, 20 Mn, 20 Zn	65 (75)	32 (40)	30 (20)	...	...

* Free cutting brass = 100.
† Values in parentheses are for heat-treated condition.
Source: Copper Development Association, New York.

Wrought Copper Alloys. — Wrought copper alloys can be utilized in the annealed, cold worked, stress-relieved, or hardened-by-heat-treatment conditions, depending on composition and end use. The "temper designation" for copper alloys is defined in ASTM Standard Recommended Practice B601, which is applicable to all product forms.

Wrought copper and high copper alloys, like cast alloys, have a designated minimum copper content and may include other elements to impart special properties. Wrought brasses have zinc as the principal alloying element and may have other designated elements. They comprise the copper–zinc alloys; copper–zinc–lead alloys (leaded brasses); and copper–zinc–tin alloys (tin brasses). Wrought bronzes comprise four main groups: copper–tin–phosphorus alloys (phosphor bronze); copper–tin–lead–phosphorus alloys (leaded phosphor bronze); copper–aluminum alloys (aluminum bronzes); and copper–silicon alloys (silicon bronze). Wrought copper–nickel alloys, like the cast alloys, have nickel as the principal alloying element. The wrought copper–nickel–zinc alloys are known as "nickel silvers" because of their color.

Table 2 lists the nominal composition, properties, and applications of common wrought copper alloys.

Table 2. Properties and Applications of Wrought Coppers and Copper Alloys

| Name and Number | Nominal Composition (%) | Mechanical Properties | | Elongation in 2 in. (%) | Machinability Rating[a] | Fabricating Characteristics and Typical Applications |
		Tensile Strength (ksi)	Yield Strength (ksi)			
C10100 Oxygen-free electronic	99.99 Cu	32–66	10–53	55	20	Excellent hot and cold workability; good forgeability. Fabricated by coining, coppersmithing, drawing and upsetting, hot forging and pressing, spinning, swaging, stamping. Uses: busbars, bus conductors, waveguides, hollow conductors, lead-in wires and anodes for vacuum tubes, vacuum seals, transistor components, glass to metal seals, coaxial cables and tubes, klystrons, microwave tubes, rectifiers.
C10200 Oxygen-free copper	99.95 Cu	32–66	10–53	55	20	Fabricating characteristics same as C10100. Uses: busbars, waveguides.
C10300 Oxygen-free, extra-low phosphorus	99.95 Cu, 0.003 P	32–55	10–50	50	20	Fabricating characteristics same as C10100. Uses: busbars, electrical conductors, tubular bus and applications requiring good conductivity and welding or brazing properties.
C10400, C10500, C10700 Oxygen-free, silver-bearing	99.95 Cu	32–66	10–53	55	20	Fabricating characteristics same as C10100. Uses: auto gaskets, radiators, busbars, conductivity wire, contacts, radio parts, winding, switches, terminals, commutator segments; chemical process equipment, printing rolls, clad metals, printed circuit foil.

[a] See footnotes at end of table.

Table 2 (*Continued*). **Properties and Applications of Wrought Coppers and Copper Alloys**

| Name and Number | Nominal Composition (%) | Mechanical Properties | | Elongation in 2 in. (%) | Machinability Rating[a] | Fabricating Characteristics and Typical Applications |
		Tensile Strength (ksi)	Yield Strength (ksi)			
C10800 Oxygen-free, low phosphorus	99.95 Cu, 0.009 P	32–55	10–50	50	20	Fabricating characteristics same as C10100. Uses: refrigerators, air conditioners, gas and heater lines, oil burner tubes, plumbing pipe and tube, brewery tubes, condenser and heat exchanger tubes, dairy and distiller tubes, pulp and paper lines, tanks; air, gasoline, hydraulic and oil lines.
C11000 Electrolytic tough pitch copper	99.90 Cu, 0.04 O	32–66	10–53	55	20	Fabricating characteristics same as C10100. Uses: downspouts, gutters, roofing, gaskets, auto radiators, busbars, nails, printing rolls, rivets, radio parts.
C11000 Electrolytic tough pitch, anneal resistant	99.90 Cu, 0.04 O, 0.01 Cd	66	. . .	. . .	20	Fabricating characteristics same as C10100. Uses: electrical power transmission where resistance to softening under overloads is desired.
C11300, C11400, C11500, C11600 Silver-bearing tough pitch copper	99.90 Cu, 0.04 O, Ag	32–66	10–53	55	20	Fabricating characteristics same as C10100. Uses: gaskets, radiators, busbars, windings, switches, chemical process equipment, clad metals, printed circuit foil.
C12000, C12100	99.9 Cu	32–57	10–53	55	20	Fabricating characteristics same as C10100. Uses: busbars, electrical conductors, tubular bus, and applications requiring welding or brazing.
C12200 Phosphorus deoxidized copper, high residual phosphorus	99.90 Cu, 0.02 P	32–55	10–50	45	20	Fabricating characteristics same as C10100. Uses: gas and heater lines; oil burner tubing; plumbing pipe and tubing; condenser, evaporator, heat exchanger, dairy, and distiller tubing; steam and water lines; air, gasoline, and hydraulic lines.
C12500, C12700, C12800, C12900, C13000 Fire-refined tough pitch with silver	99.88 Cu	32–67	10–53	55	20	Fabricating characteristics same as C10100. Uses: same as C11000.
C14200 Phosphorus deoxidized, arsenical	99.68 Cu, 0.3 As, 0.02 P	32–55	10–50	45	20	Fabricating characteristics same as C10100. Uses: plates for locomotive fireboxes, staybolts, heat exchanger and condenser tubes.

[a] See footnotes at end of table.

Table 2 (*Continued*). **Properties and Applications of Wrought Coppers and Copper Alloys**

Name and Number	Nominal Composition (%)	Mechanical Properties		Elongation in 2 in. (%)	Machinability Rating[a]	Fabricating Characteristics and Typical Applications
		Tensile Strength (ksi)	Yield Strength (ksi)			
C19200	98.97 Cu, 1.0 Fe, 0.03 P	37–77	11–74	40	20	Excellent hot and cold workability. Uses: automotive hydraulic brake lines, flexible hose, electrical terminals, fuse clips, gaskets, gift hollow ware, applications requiring resistance to softening and stress corrosion, air conditioning and heat exchanger tubing.
C14300	99.9 Cu, 0.1 Cd	32–58	11–56	42	20	Fabricating characteristics same as C10100. Uses: anneal resistant electrical applications requiring thermal softening and embrittlement resistance, lead frames, contacts, terminals, solder-coated and solder-fabricated parts, furnace-brazed assemblies and welded components, cable wrap.
C14310	99.8 Cu, 0.2 Cd	32–58	11–56	42	20	Same as C14300.
C14500 Phosphorus deoxidized, tellurium bearing	99.5 Cu, 0.50 Te, 0.008 P	32–56	10–51	50	85	Fabricating characteristics same as C10100. Uses: Forgings and screw machine products, and parts requiring high conductivity, extensive machining, corrosion resistance, copper color, or a combination of these; electrical connectors, motor and switch parts, plumbing fittings, soldering coppers, welding torch tips, transistor bases, and furnace-brazed articles.
C14700 Sulfur-bearing	99.6 Cu, 0.40 S	32–57	10–55	52	85	Fabricating characteristics same as C10100. Uses: screw machine products and parts requiring high conductivity, extensive machining, corrosion resistance, copper color, or a combination of these; electrical connectors, motor and switch components, plumbing fittings, cold headed and machined parts, cold forgings, furnace-brazed articles, screws, soldering coppers, rivets and welding torch tips.
C15000 Zirconium copper	99.8 Cu, 0.15 Zr	29–76	6–72	54	20	Fabricating characteristics same as C10100. Uses: switches, high-temperature circuit breakers, commutators, stud bases for power transmitters, rectifiers, soldering welding tips.
C15500	99.75 Cu, 0.06 P, 0.11 Mg, Ag	40–80	18–72	40	20	Fabricating characteristics same as C10100. Uses: high-conductivity light-duty springs, electrical contacts, fittings, clamps, connectors, diaphragms, electronic components, resistance welding electrodes.

[a] See footnotes at end of table.

Table 2 (*Continued*). **Properties and Applications of Wrought Coppers and Copper Alloys**

Name and Number	Nominal Composition (%)	Mechanical Properties		Elongation in 2 in. (%)	Machinability Rating[a]	Fabricating Characteristics and Typical Applications
		Tensile Strength (ksi)	Yield Strength (ksi)			
C15710	99.8 Cu, 0.2 Al₂O₃	47–105	39–100	20	. . .	Excellent cold workability. Fabricated by extrusion, drawing, rolling, impacting, heading, swaging, bending, machining, blanking, roll threading. Uses: electrical connectors, light-duty current-carrying springs, inorganic insulated wire, thermocouple wire, lead wire, resistance welding electrodes for aluminum, heat sinks.
C15720	99.6 Cu, 0.4 Al₂O₃	67–89	53–85	20	. . .	Excellent cold workability. Fabricated by extrusion, drawing, rolling, impacting, heading, swaging, machining, blanking. Uses: relay and switch springs, lead frames, contact supports, heat sinks, circuit breaker parts, rotor bars, resistance welding electrodes and wheels, connectors, high-strength high-temperature parts.
C15735	99.3 Cu, 0.7 Al₂O₃	70–85	60–82	16	. . .	Excellent cold workability. Fabricated by extrusion, drawing, heading, impacting, machining. Uses: resistance welding electrodes, circuit breakers, feedthrough conductors, heat sinks, motor parts, high-strength high-temperature parts.
C15760	98.9 Cu, 1.1 Al₂O₃	70–94	56–80	20	. . .	Excellent cold workability. Fabricated by extrusion and drawing. Uses: resistance welding electrodes, circuit breakers, electrical connectors, wire feed contact tips, plasma spray nozzles, high-strength high-temperature parts.
C16200 Cadmium copper	99.0 Cu, 1.0 Cd	35–100	7–69	57	20	Excellent cold workability; good hot formability. Uses: trolley wire, heating pad, electric-blanket elements, spring contacts, railbands, high-strength transmission lines, connectors, cable wrap, switch gear components and waveguide cavities.
C16500	98.6 Cu, 0.8 Cd, 0.6 Sn	40–95	14–71	53	20	Fabricating characteristics same as C16200. Uses: electrical springs and contacts, trolley wire, clips, flat cable, resistance welding electrodes.

[a] See footnotes at end of table.

Table 2 (*Continued*). **Properties and Applications of Wrought Coppers and Copper Alloys**

Name and Number	Nominal Composition (%)	Mechanical Properties		Elongation in 2 in. (%)	Machinability Rating[a]	Fabricating Characteristics and Typical Applications
		Tensile Strength (ksi)	Yield Strength (ksi)			
C1700 Beryllium copper	99.5 Cu, 1.7 Be, 0.20 Co	70–190	32–170	45	20	Fabricating characteristics same as C16200. Commonly fabricated by blanking, forming and bending, turning, drilling, tapping. Uses: bellows, Bourdon tubing, diaphragms, fuse clips, fasteners, lock-washers, springs, switch parts, roll pins, valves, welding equipment.
C17200 Beryllium copper	99.5 Cu, 1.9 Be, 0.20 Co	68–212	25–195	48	20	Similar to C17000, particularly for its nonsparking characteristics.
C17300 Beryllium copper	99.5 Cu, 1.9 Be, 0.40 Pb	68–200	25–182	48	50	Combines superior machinability with good fabricating characteristics of C17200.
C17500 Copper-cobalt-beryllium alloy	99.5 Cu, 2.5 Co, 0.6 Be	45–115	25–110	28	...	Fabricating characteristics same as C16200. Uses: fuse clips, fasteners, springs, switch and relay parts, electrical conductors, welding equipment.
C18200, C18400, C18500 Chromium copper	99.5 Cu	34–86	14–77	40	20	Excellent cold workability, good hot workability. Uses: resistance welding electrodes, seam welding wheels, switch gear, electrode holder jaws, cable connectors, current carrying arms and shafts, circuit breaker parts, molds, spot welding tips, flash welding electrodes, electrical and thermal conductors requiring strength, switch contacts.
C18700 Leaded copper	99.0 Cu, 1.0 Pb	32–55	10–50	45	85	Good cold workability; poor hot formability. Uses: connectors, motor and switch parts, screw machine parts requiring high conductivity.
C18900	98.75 Cu, 0.75 Sn, 0.3 Si, 0.20 Mn	38–95	9–52	48	20	Fabricating characteristics same as C10100. Uses: welding rod and wire for inert gas tungsten arc and metal arc welding and oxyacetylene welding of copper.
C19000 Copper-nickel-phosphorus alloy	98.7 Cu, 1.1 Ni, 0.25 P	38–115	20–80	50	30	Fabricating characteristics same as C10100. Uses: springs, clips, electrical connectors, power tube and electron tube components, high-strength electrical conductors, bolts, nails, screws, cotter pins, and parts requiring some combination of high-strength, high-electrical or thermal conductivity, high resistance to fatigue and creep, and good workability.

[a] See footnotes at end of table.

Table 2 (Continued). **Properties and Applications of Wrought Coppers and Copper Alloys**

Name and Number	Nominal Composition (%)	Mechanical Properties		Elongation in 2 in. (%)	Machinability Rating[a]	Fabricating Characteristics and Typical Applications
		Tensile Strength (ksi)	Yield Strength (ksi)			
C19100 Copper-nickel-phosphorus-tellurium alloy	98.15 Cu, 1.1 Ni, 0.50 Te, 0.25 P	36–104	10–92	27	75	Good hot and cold workability. Uses: forgings and screw machine parts requiring high strength, hardenability, extensive machining, corrosion resistance, copper color, good conductivity, or a combination of these; bolts, bushings, electrical connectors, gears, marine hardware, nuts, pinions, tie rods, turnbuckle barrels, welding torch tips.
C19400	97.5 Cu, 2.4 Fe, 0.13 Zn, 0.03 P	45–76	24–73	32	20	Excellent hot and cold workability. Uses: circuit breaker components, contact springs, electrical clamps, electrical springs, electrical terminals, flexible hose, fuse clips, gaskets, gift hollow ware, plug contacts, rivets, and welded condenser tubes.
C19500	97.0 Cu, 1.5 Fe, 0.6 Sn, 0.10 P, 0.80 Co	80–97	65–95	15	20	Excellent hot and cold workability. Uses: electrical springs, sockets, terminals, connectors, clips and other current carrying parts having strength.
C21000 Gilding, 95%	95.0 Cu, 5.0 Zn	34–64	10–58	45	20	Excellent cold workability, good hot workability for blanking, coining, drawing, piercing and punching, shearing, spinning, squeezing and swaging, stamping. Uses: coins, medals, bullet jackets, fuse caps, primers, plaques, jewelry base for gold plate.
C22000 Commercial bronze, 90%	90.0 Cu, 10.0 Zn	37–72	10–62	50	20	Fabricating characteristics same as C21000, plus heading and upsetting, roll threading and knurling, hot forging and pressing. Uses: etching bronze, grillwork, screen cloth, weatherstripping, lipstick cases, compacts, marine hardware, screws, rivets.
C22600 Jewelry bronze, 87.5%	87.5 Cu, 12.5 Zn	39–97	11–62	46	30	Fabricating characteristics same as C21000, plus heading and upsetting, roll threading and knurling. Uses: angles, channels, chain, fasteners, costume jewelry, lipstick cases, compacts, base for gold plate.
C23000 Red brass, 85%	85.0 Cu, 15.0 Zn	39–105	10–63	55	30	Excellent cold workability; good hot formability. Uses: weatherstripping, conduit, sockets, fasteners, fire extinguishers, condenser and heat exchanger tubing, plumbing pipe, radiator cores.

[a] See footnotes at end of table.

Table 2 (*Continued*). **Properties and Applications of Wrought Coppers and Copper Alloys**

Name and Number	Nominal Composition (%)	Mechanical Properties Tensile Strength (ksi)	Yield Strength (ksi)	Elongation in 2 in. (%)	Machinability Rating[a]	Fabricating Characteristics and Typical Applications
C24000 Low brass, 80%	80.0 Cu, 20.0 Zn	42–125	12–65	55	30	Excellent cold workability. Fabricating characteristics same as C23000. Uses: battery caps, bellows, musical instruments, clock dials, pump lines, flexible hose.
C26000 Cartridge brass, 70%	70.0 Cu, 30.0 Zn	44–130	11–65	66	30	Excellent cold workability. Fabricating characteristics same as C23000, except for coining, roll threading, and knurling. Uses: radiator cores and tanks, flashlight shells, lamp fixtures, fasteners, locks, hinges, ammunition components, plumbing accessories, pins, rivets.
C26800, C27000 Yellow brass	65.0 Cu, 35.0 Zn	46–128	14–62	65	30	Excellent cold workability. Fabricating characteristics same as C23000. Uses: same as C26000 except not used for ammunition.
C28000 Muntz metal	60.0 Cu, 40.0 Zn	54–74	21–55	52	40	Excellent hot formability and forgeability for blanking, forming and bending, hot forging and pressing, hot heading and upsetting, shearing. Uses: architectural, large nuts and bolts, brazing rod, condenser plates, heat exchanger and condenser tubing, hot forgings.
C31400 Leaded commercial bronze	89.0 Cu, 1.75 Pb, 9.25 Zn	37–60	12–55	45	80	Excellent machinability. Uses: screws, machine parts, pickling crates.
C31600 Leaded commercial bronze, nickel-bearing	89.0 Cu, 1.9 Pb, 1.0 Ni, 8.1 Zn	37–67	12–59	45	80	Good cold workability; poor hot formability. Uses: electrical connectors, fasteners, hardware, nuts, screws, screw machine parts.
C33000 Low-leaded brass tube	66.0 Cu, 0.5 Pb, 33.5 Zn	47–75	15–60	60	60	Combines good machinability and excellent cold workability. Fabricated by forming and bending, machining, piercing and punching. Uses: pump and power cylinders and liners, ammunition primers, plumbing accessories.
C33200 High-leaded brass tube	66.0 Cu, 1.6 Pb, 32.4 Zn	52–75	20–60	50	80	Excellent machinability. Fabricated by piercing, punching, and machining. Uses: general-purpose screw machine parts.
C33500 Low-leaded brass	65.0 Cu, 0.5 Pb, 34.5 Zn	46–74	14–60	65	60	Similar to C33200. Commonly fabricated by blanking, drawing, machining, piercing and punching, stamping. Uses: butts, hinges, watch backs.

[a] See footnotes at end of table.

Table 2 (*Continued*). **Properties and Applications of Wrought Coppers and Copper Alloys**

| Name and Number | Nominal Composition (%) | Mechanical Properties | | Elongation in 2 in. (%) | Machinability Rating[a] | Fabricating Characteristics and Typical Applications |
		Tensile Strength (ksi)	Yield Strength (ksi)			
C34000 Medium-leaded brass	65.0 Cu, 1.0 Pb, 34.0 Zn	47–88	15–60	60	70	Similar to C33200. Fabricated by blanking, heading and upsetting, machining, piercing and punching, roll threading and knurling, stamping. Uses: butts, gears, nuts, rivets, screws, dials, engravings, instrument plates.
C34200 High-leaded brass	64.5 Cu, 2.0 Pb, 33.5 Zn	49–85	17–62	52	90	Combines excellent machinability with moderate cold workability. Uses: clock plates and nuts, clock and watch backs, gears, wheels and channel plate.
C34900	62.2 Cu, 0.35 Pb, 37.45 Zn	53–68	16–55	72	50	Good cold workability, fair hot workability for bending and forming, heading and upsetting, machining, roll threading and knurling. Uses: building, hardware, rivets and nuts, plumbing goods, and parts requiring moderate cold working combined with some machining.
C35000 Medium-leaded brass	62.5 Cu, 1.1 Pb, 36.4 Zn	45–95	13–70	66	70	Fair cold workability; poor hot formability. Uses: bearing cages, book dies, clock plates, gears, hinges, hose couplings, keys, lock parts, lock tumblers, meter parts, nuts, sink strainers, strike plates, templates, type characters, washers, wear plates.
C35300 High-leaded brass	62.0 Cu, 1.8 Pb, 36.2 Zn	49–85	17–62	52	90	Similar to C34200.
C35600 Extra-high-leaded brass	63.0 Cu, 2.5 Pb, 34.5 Zn	49–74	17–60	50	100	Excellent machinability. Fabricated by blanking, machining, piercing and punching, stamping. Uses: same as C34200 and C35300.
C36000 Free-cutting brass	61.5 Cu, 3.0 Pb, 35.5 Zn	49–68	18–45	53	100	Excellent machinability. Fabricated by machining, roll threading and knurling. Uses: gears, pinions, automatic high-speed screw machine parts.
C36500 to C36800 Leaded Muntz metal	60.0 Cu, 0.6 Pb, 39.4 Zn	54 (As hot rolled)	20	45	60	Combines good machinability with excellent hot formability. Uses: condenser tube plates.
C37000 Free-cutting Muntz metal	60.0 Cu, 1.0 Pb, 39.0 Zn	54–80	20–60	40	70	Fabricating characteristics similar to C36500 to C36800. Uses: automatic screw machine parts.

[a] See footnotes at end of table.

Table 2 (*Continued*). **Properties and Applications of Wrought Coppers and Copper Alloys**

Name and Number	Nominal Composition (%)	Mechanical Properties		Elongation in 2 in. (%)	Machinability Rating[a]	Fabricating Characteristics and Typical Applications
		Tensile Strength (ksi)	Yield Strength (ksi)			
C37700 Forging brass	59.0 Cu, 2.0 Pb, 39.0 Zn	52 (As extruded)	20	45	80	Excellent hot workability. Fabricated by heading and upsetting, hot forging and pressing, hot heading and upsetting, machining. Uses: forgings and pressings of all kinds.
C38500 Architectural bronze	57.0 Cu, 3.0 Pb, 40.0 Zn	60 (As extruded)	20	30	90	Excellent machinability and hot workability. Fabricated by hot forging and pressing, forming, bending, and machining. Uses: architectural extrusions, store fronts, thresholds, trim, butts, hinges, lock bodies and forgings.
C40500	95 Cu, 1 Sn, 4 Zn	39–78	12–70	49	20	Excellent cold workability. Fabricated by blanking, forming and drawing. Uses: meter clips, terminals, fuse clips, contact and relay springs, washers.
C41100	91 Cu, 0.5 Sn, 8.5 Zn	39–106	11–72	13	20	Excellent cold workability, good hot formability. Fabricated by blanking, forming, and drawing. Uses: bushings, bearing sleeves, thrust washers, terminals, connectors, flexible metal hose, electrical conductors.
C41300	90.0 Cu, 1.0 Sn, 9.0 Zn	41–105	12–82	45	20	Excellent cold workability; good hot formability. Fabricated by plater bar for jewelry products, flat springs for electrical switchgear.
C41500	91 Cu, 1.8 Sn, 7.2 Zn	46–81	17–75	44	30	Excellent cold workability. Fabricated by blanking, drawing, bending, forming, shearing, and stamping. Uses: spring applications for electrical switches.
C42200	87.5 Cu, 1.1 Sn, 11.4 Zn	43–88	15–75	46	30	Excellent cold workability; good hot formability. Fabricated by blanking, piercing, forming, and drawing. Uses: sash chains, fuse clips, terminals, spring washers, contact springs, electrical connectors.
C42500	88.5 Cu, 2.0 Sn, 9.5 Zn	45–92	18–76	49	30	Excellent cold workability. Fabricated by blanking, piercing, forming and drawing. Uses: electrical switches, springs, terminals, connectors, fuse clips, pen clips, weather stripping.
C43000	87.0 Cu, 2.2 Sn, 10.8 Zn	46–94	18–73	55	30	Excellent cold workability; good hot formability. Fabricated by blanking, coining, drawing, forming, bending, heading, and upsetting. Uses: same as C42500.

[a] See footnotes at end of table.

Table 2 (*Continued*). **Properties and Applications of Wrought Coppers and Copper Alloys**

Name and Number	Nominal Composi- tion (%)	Mechanical Properties		Elon- gation in 2 in. (%)	Machin- ability Rating[a]	Fabricating Characteristics and Typical Applications
		Tensile Strength (ksi)	Yield Strength (ksi)			
C43400	85.0 Cu, 0.7 Sn, 14.3 Zn	45–88	15–75	49	30	Excellent cold workability. Fabricated by blanking, drawing, bonding, forming, stamping and shearing. Uses: electrical switch parts, blades, relay springs, contacts.
C43500	81.0 Cu, 0.9 Sn, 18.1 Zn	46–80	16–68	46	30	Excellent cold workability for fabrication by forming and bending. Uses: Bourdon tubing and musical instruments.
C44300, C44400, C44500 Inhibited admiralty	71.0 Cu, 28.0 Zn, 1.0 Sn	48–55	18–22	65	30	Excellent cold workability for forming and bending. Uses: condenser, evaporator and heat exchanger tubing, condenser tubing plates, distiller tubing, ferrules.
C46400 to C46700 Naval brass	60.0 Cu, 39.25 Zn, 0.75 Sn	55–88	25–66	50	30	Excellent hot workability and hot forgeability. Fabricated by blanking, drawing, bending, heading and upsetting, hot forging, pressing. Uses: aircraft turnbuckle barrels, balls, bolts, marine hardware, nuts, propeller shafts, rivets, valve stems, condenser plates, welding rod.
C48200 Naval brass, medium-leaded	60.5 Cu, 0.7 Pb, 0.8 Sn, 38.0 Zn	56–75	25–53	43	50	Good hot workability for hot forging, pressing, and machining operations. Uses: marine hardware, screw machine products, valve stems.
C48500 Leaded naval brass	60.0 Cu, 1.75 Pb, 37.5 Zn, 0.75 Sn	55–77	25–53	40	70	Combines excellent hot forgeability and machinability. Fabricated by hot forging and pressing, machining. Uses: marine hardware, screw machine parts, valve stems.
C50500 Phosphor bronze, 1.25% E	98.75 Cu, 1.25 Sn, trace P	40–79	14–50	48	20	Excellent cold workability; good hot formability. Fabricated by blanking, bending, heading and upsetting, shearing and swaging. Uses: electrical contacts, flexible hose, pole-line hardware.
C51000 Phosphor bronze, 5% A	95.0 Cu, 5.0 Sn, trace P	47–140	19–80	64	20	Excellent cold workability. Fabricated by blanking, drawing, bending, heading and upsetting, roll threading and knurling, shearing, stamping. Uses: bellows, Bourdon tubing, clutch discs, cotter pins, diaphragms, fasteners, lock washers, wire brushes, chemical hardware, textile machinery, welding rod.

[a] See footnotes at end of table.

Table 2 (*Continued*). **Properties and Applications of Wrought Coppers and Copper Alloys**

Name and Number	Nominal Composition (%)	Mechanical Properties		Elongation in 2 in. (%)	Machinability Rating[a]	Fabricating Characteristics and Typical Applications
		Tensile Strength (ksi)	Yield Strength (ksi)			
C51100	95.6 Cu, 4.2 Sn, 0.2 P	46–103	50–80	48	20	Excellent cold workability. Uses: bridge bearing plates, locator bars, fuse clips, sleeve bushings, springs, switch parts, truss wire, wire brushes, chemical hardware, perforated sheets, textile machinery, welding rod.
C52100 Phosphor bronze, 8% C	92.0 Cu, 8.0 Sn, trace P	55–140	24–80	70	20	Good cold workability for blanking, drawing, forming and bending, shearing, stamping. Uses: generally for more severe service conditions than C51000.
C52400 Phosphor bronze, 10% D	90.0 Cu, 10.0 Sn, trace P	66–147	28 (Annealed)	70	20	Good cold workability for blanking, forming and bending, shearing. Uses: heavy bars and plates for severe compression, bridge and expansion plates and fittings, articles requiring good spring qualities, resilience, fatigue resistance, good wear and corrosion resistance.
C54400 Free-cutting phosphor bronze	88.0 Cu, 4.0 Pb, 4.0 Zn, 4.0 Sn	44–75	19–63	50	80	Excellent machinability; good cold workability. Fabricated by blanking, drawing, bending, machining, shearing, stamping. Uses: bearings, bushings, gears, pinions, shafts, thrust washers, valve parts.
C60800 Aluminum bronze, 5%	95.0 Cu, 5.0 Al	60	27	55	20	Good cold workability; fair hot formability. Uses: condenser, evaporator and heat exchanger tubes, distiller tubes, ferrules.
C61000	92.0 Cu, 8.0 Al	70–80	30–55	65	20	Good hot and cold workability. Uses: bolts, pump parts, shafts, tie rods, overlay on steel for wearing surfaces.
C61300	92.65 Cu, 0.35 Sn, 7.0 Al	70–85	30–58	42	30	Good hot and cold formability. Uses: nuts, bolts, stringers and threaded members, corrosion-resistant vessels and tanks, structural components, machine parts, condenser tube and piping systems, marine protective sheathing and fastening, munitions mixing troughs and blending chambers.
C61400 Aluminum bronze, D	91.0 Cu, 7.0 Al, 2.0 Fe	76–89	33–60	45	20	Similar to C61300.
C61500	90.0 Cu, 8.0 Al, 2.0 Ni	70–145	22–140	55	30	Good hot and cold workability. Fabricating characteristics similar to C52100. Uses: hardware, decorative metal trim, interior furnishings and other articles requiring high tarnish resistance.

[a] See footnotes at end of table.

Table 2 (*Continued*). **Properties and Applications of Wrought Coppers and Copper Alloys**

| Name and Number | Nominal Composition (%) | Mechanical Properties | | Elongation in 2 in. (%) | Machinability Rating[a] | Fabricating Characteristics and Typical Applications |
		Tensile Strength (ksi)	Yield Strength (ksi)			
C61800	89.0 Cu, 1.0 Fe, 10.0 Al	80–85	39–42.5	28	40	Fabricated by hot forging and hot pressing. Uses: bushings, bearings, corrosion-resistant applications, welding rods.
C61900	86.5 Cu, 4.0 Fe, 9.5 Al	92–152	49–145	30	. . .	Excellent hot formability for fabricating by blanking, forming, bending, shearing, and stamping. Uses: springs, contacts, and switch components.
C62300	87.0 Cu, 3.0 Fe, 10.0 Al	75–98	35–52	35	50	Good hot and cold formability. Fabricated by bending, hot forging, hot pressing, forming, and welding. Uses: bearings, bushings, valve guides, gears, valve seats, nuts, bolts, pump rods, worm gears, and cams.
C62400	86.0 Cu, 3.0 Fe, 11.0 Al	90–105	40–52	18	50	Excellent hot formability for fabrication by hot forging and hot bending. Uses: bushings, gears, cams, wear strips, nuts, drift pins, tie rods.
C62500	82.7 Cu, 4.3 Fe, 13.0 Al	100 (As extruded)	55	1	20	Excellent hot formability for fabrication by hot forging and machining. Uses: guide bushings, wear strips, cams, dies, forming rolls.
C63000	82.0 Cu, 3.0 Fe, 10.0 Al, 5.0 Ni	90–118	50–75	20	30	Good hot formability. Fabricated by hot forming and forging. Uses: nuts, bolts, valve seats, plunger tips, marine shafts, valve guides, aircraft parts, pump shafts, structural members.
C63200	82.0 Cu, 4.0 Fe, 9.0 Al, 5.0 Ni	90–105	45–53	25	30	Good hot formability. Fabricated by hot forming and welding. Uses: nuts, bolts, structural pump parts, shafting requiring corrosion resistance.
C63600	95.5 Cu, 3.5 Al, 1.0 Si	60–84	. . .	64	40	Excellent cold workability; fair hot formability. Fabricated by cold heading. Uses: components for pole line hardware, cold-headed nuts for wire and cable connectors, bolts and screw products.
C63800	99.5 Cu, 2.8 Al, 1.8 Si, 0.40 Co	82–130	54–114	36	. . .	Excellent cold workability and hot formability. Uses: springs, switch parts, contacts, relay springs, glass sealing, and porcelain enameling.
C64200	91.2 Cu, 7.0 Al	75–102	35–68	32	60	Excellent hot formability. Fabricated by hot forming, forging, machining. Uses: valve stems, gears, marine hardware, pole-line hardware, bolts, nuts, valve bodies, and components.

[a] See footnotes at end of table.

Table 2 (*Continued*). **Properties and Applications of Wrought Coppers and Copper Alloys**

Name and Number	Nominal Composition (%)	Mechanical Properties		Elongation in 2 in. (%)	Machinability Rating[a]	Fabricating Characteristics and Typical Applications
		Tensile Strength (ksi)	Yield Strength (ksi)			
C65100 Low-silicon bronze, B	98.5 Cu, 1.5 Si	40–95	15–69	55	30	Excellent hot and cold workability. Fabricated by forming and bending, heading and upsetting, hot forging and pressing, roll threading and knurling, squeezing and swaging. Uses: hydraulic pressure lines, anchor screws, bolts, cable clamps, cap screws, machine screws, marine hardware, nuts, pole-line hardware, rivets, U-bolts, electrical conduits, heat exchanger tubing, welding rod.
C65500 High-silicon bronze, A	97.0 Cu, 3.0 Si	56–145	21–70	63	30	Excellent hot and cold workability. Fabricated by blanking, drawing, forming and bending, heading and upsetting, hot forging and pressing, roll threading and knurling, shearing, squeezing and swaging. Uses: similar to C65100 including propeller shafts.
C66700 Manganese brass	70.0 Cu, 28.8 Zn, 1.2 Mn	45.8–100	12–92.5	60	30	Excellent cold formability. Fabricated by blanking, bending, forming, stamping, welding. Uses: brass products resistance welded by spot, seam, and butt welding.
C67400	58.5 Cu, 36.5 Zn, 1.2 Al, 2.8 Mn, 1.0 Sn	70–92	34–55	28	25	Excellent hot formability. Fabricated by hot forging and pressing, machining. Uses: bushings, gears, connecting rods, shafts, wear plates.
C67500 Manganese bronze, A	58.5 Cu, 1.4 Fe, 39.0 Zn, 1.0 Sn, 0.1 Mn	65–84	30–60	33	30	Excellent hot workability. Fabricated by hot forging and pressing, hot heading and upsetting. Uses: clutch discs, pump rods, shafting, balls, valve stems and bodies.
C68700 Aluminum brass, arsenical	77.5 Cu, 20.5 Zn, 2.0 Al, 0.1 As	60	27	55	30	Excellent cold workability for forming and bending. Uses: condenser, evaporator and heat exchanger tubing, condenser tubing plates, distiller tubing, ferrules.
C68800	73.5 Cu, 22.7 Zn, 3.4 Al, 0.40 Co	82–129	55–114	36	. . .	Excellent hot and cold formability. Fabricated by blanking, drawing, forming and bending, shearing and stamping. Uses: springs, switches, contacts, relays, drawn parts.
C69000	73.3 Cu, 3.4 Al, 0.6 Ni, 22.7 Zn	72–130	50–117	40	. . .	Fabricating characteristics same as C68800. Uses: wiring devices, relays, switches, springs, high-strength shells.

[a] See footnotes at end of table.

Table 2 (*Continued*). **Properties and Applications of Wrought Coppers and Copper Alloys**

| Name and Number | Nominal Composition (%) | Mechanical Properties | | Elongation in 2 in. (%) | Machinability Rating[a] | Fabricating Characteristics and Typical Applications |
		Tensile Strength (ksi)	Yield Strength (ksi)			
C69400 Silicon red brass	81.5 Cu, 14.5 Zn, 4.0 Si	80–100	40–57	25	30	Excellent hot formability for fabrication by forging, screw machine operations. Uses: valve stems where corrosion resistance and high strength are critical.
C70400	92.4 Cu, 1.5 Fe, 5.5 Ni, 0.6 Mn	38–77	40–76	46	20	Excellent cold workability; good hot formability. Fabricated by forming, bending, and welding. Uses: condensers, evaporators, heat exchangers, ferrules, salt water piping, lithium bromide absorption tubing, shipboard condenser intake systems.
C70600 Copper nickel, 10%	88.7 Cu, 1.3 Fe, 10.0 Ni	44–60	16–57	42	20	Good hot and cold workability. Fabricated by forming and bending, welding. Uses: condensers, condenser plates, distiller tubing, evaporator and heat exchanger tubing, ferrules, salt water piping.
C71000 Copper nickel, 20%	79.0 Cu, 21.0 Ni	49–95	13–85	40	20	Good hot and cold formability. Fabricated by blanking, forming and bending, welding. Uses: communication relays, condensers, condenser plates, electrical springs, evaporator and heat exchanger tubes, ferrules, resistors.
C71500 Copper nickel, 30%	70.0 Cu, 30.0 Ni	54–75	20–70	45	20	Similar to C70600.
C71700	67.8 Cu, 0.7 Fe, 31.0 Ni, 0.5 Be	70–200	30–180	40	20	Good hot and cold formability. Uses: high-strength constructional parts for sea water corrosion resistance, hydrophone cases, mooring cable wire, springs, retainer rings, bolts, screws, pins for ocean telephone cable applications.
C72500	88.2 Cu, 9.5 Ni, 2.3 Sn	55–120	22–108	35	20	Excellent cold and hot formability. Fabricated by blanking, brazing, coining, drawing, etching, forming and bending, heading and upsetting, roll threading and knurling, shearing, spinning, squeezing, stamping, and swaging. Uses: relay and switch springs, connectors, brazing alloy, lead frames, control and sensing bellows.
C73500	72.0 Cu, 10.0 Zn, 18.0 Ni	50–110	15–84	37	20	Fabricating characteristics same as C74500. Uses: hollow ware, medallions, jewelry, base for silver plate, cosmetic cases, musical instruments, name plates, contacts.

[a] See footnotes at end of table.

Table 2 (*Concluded*). **Properties and Applications of Wrought Coppers and Copper Alloys**

Name and Number	Nominal Composition (%)	Mechanical Properties		Elongation in 2 in. (%)	Machinability Rating[a]	Fabricating Characteristics and Typical Applications
		Tensile Strength (ksi)	Yield Strength (ksi)			
C74500 Nickel silver, 65–10	65.0 Cu, 25.0 Zn, 10.0 Ni	49–130	18–76	50	20	Excellent cold workability. Fabricated by blanking, drawing, etching, forming and bending, heading and upsetting, roll threading and knurling, shearing, spinning, squeezing, and swaging. Uses: rivets, screws, slide fasteners, optical parts, etching stock, hollow ware, nameplates, platers' bars.
C75200 Nickel silver, 65–18	65.0 Cu, 17.0 Zn, 18.0 Ni	56–103	25–90	45	20	Fabricating characteristics similar to C74500. Uses: rivets, screws, table flatware, truss wire, zippers, bows, camera parts, core bars, temples, base for silver plate, costume jewelry, etching stock, hollow ware, nameplates, radio dials.
C75400 Nickel silver, 65–15	65.0 Cu, 20.0 Zn, 15.0 Ni	53–92	18–79	43	20	Fabricating characteristics similar to C75200. Uses: camera parts, optical equipment, etching stock, jewelry.
C75700 Nickel silver, 65–12	65.0 Cu, 23.0 Zn, 12.0 Ni	52–93	18–79	48	20	Fabricating characteristics similar to C74500. Uses: slide fasteners, camera parts, optical parts, etching stock, nameplates.
C76200	59.0 Cu, 29.0 Zn, 12.0 Ni	57–122	21–110	50	. . .	Fabricating characteristics same as C77000. Uses: electrical terminals, contact springs, release brackets, ornamental bits and spurs, optical parts, surgical instruments, electrical contacts.
C77000 Nickel silver, 55–18	55.0 Cu, 27.0 Zn, 18.0 Ni	60–145	27–90	40	30	Good cold workability. Fabricated by blanking, forming and bending, and shearing. Uses: optical goods, springs, and resistance wire.
C72200	82.0 Cu, 16.0 Ni, 0.5 Cr, 0.8 Fe, 0.5 Mn	46–70	18–66	46	. . .	Good hot and cold workability. Fabricated by forming, bending and welding. Uses: condenser and heat exchanger tubing, salt water piping.
C78200 Leaded nickel silver, 65-8-2	65.0 Cu, 2.0 Pb, 25.0 Zn, 8.0 Ni	53–91	23–76	40	60	Good cold formability. Fabricated by blanking, milling and drilling. Uses: key blanks, watch plates, watch parts.

[a] Free cutting brass = 100.

Source: Copper Development Association, New York.

Aluminum and Aluminum Alloys

Pure aluminum is a silver-white metal characterized by a slightly bluish cast. It has a specific gravity of 2.70, resists the corrosive effects of many chemicals and has a malleability approaching that of gold. When alloyed with other metals numerous properties are obtained which make these alloys useful over a wide range of applications.

Aluminum alloys are light in weight compared to steel, brass, nickel or copper; can be fabricated by all common processes; are available in a wide range of sizes, shapes and forms; resist corrosion; readily accept a wide range of surface finishes; have good electrical and thermal conductivities; and are highly reflective to both heat and light.

Characteristics of Aluminum and Aluminum Alloys. — Aluminum and its alloys lose part of their strength at elevated temperatures, although some alloys retain good strength at temperatures from 400 to 500 degrees F. At subzero temperatures, however, their strength increases without loss of ductility so that aluminum is a particularly useful metal for low-temperature applications.

When aluminum surfaces are exposed to the atmosphere, a thin invisible oxide skin forms immediately which protects the metal from further oxidation. This self-protecting characteristic gives aluminum its high resistance to corrosion. Unless exposed to some substance or condition which destroys this protective oxide coating, the metal remains protected against corrosion. Aluminum is highly resistant to weathering, even in industrial atmospheres. It is also corrosion resistant to many acids. Alkalis are among the few substances that attack the oxide skin and therefore are corrosive to aluminum. Although the metal can safely be used in the presence of certain mild alkalis with the aid of inhibitors, in general, direct contact with alkaline substances should be avoided. Direct contact with certain other metals should be avoided in the presence of an electrolyte; otherwise galvanic corrosion of the aluminum may take place in the vicinity of the contact area. Where other metals must be fastened to aluminum, the use of a bituminous paint coating or insulating tape is recommended.

Aluminum is one of the two common metals having an electrical conductivity high enough for use as an electric conductor. The conductivity of electric-conductor (EC) grade is about 62 per cent that of the International Annealed Copper Standard. Because aluminum has less than one-third the specific gravity of copper, however, a pound of aluminum will go almost twice as far as a pound of copper when used for this purpose. Alloying lowers the conductivity somewhat so that wherever possible the EC grade is used in electric conductor applications. However, aluminum takes a set, which often results in loosening of screwed connectors, leading to arcing and fires. Special clamping designs are therefore required when aluminum is used for electrical wiring, especially in buildings.

Aluminum has nonsparking and nonmagnetic characteristics which make the metal useful for electrical shielding purposes such as in bus bar housings or enclosures for other electrical equipment and for use around inflammable or explosive substances.

Aluminum can be cast by any method known to foundrymen. It can be rolled to any desired thickness down to foil thinner than paper and in sheet form can be stamped, drawn, spun, or roll-formed. The metal also may be hammered or forged. Aluminum wire, drawn from rolled rod, may be stranded into cable of any desired size and type. The metal may be extruded into a variety of shapes. It may be turned, milled, bored, or machined in machines often operating at their maximum speeds. Aluminum rod and bar may readily be employed in the high-speed manufacture of automatic screw-machine parts.

Almost any method of joining is applicable to aluminum — riveting, welding or

brazing. A wide variety of mechanical aluminun fasteners simplifies the assembly of many products. Resin bonding of aluminum parts has been successfully employed, particularly in aircraft components.

For the majority of applications, aluminum needs no protective coating. Mechanical finishes such as polishing, sand blasting or wire brushing meet the majority of needs. When additional protection is desired, chemical, electrochemical, and paint finishes are all used. Vitreous enamels have been developed for aluminum, and the metal may also be electroplated.

Temper Designations for Aluminum Alloys. — The temper designation system adopted by The Aluminum Association and used in industry pertains to all forms of wrought and cast aluminum and aluminum alloys except ingot. It is based on the sequences of basic treatments used to produce the various tempers. The temper designation follows the alloy designation, being separated by a dash.

Basic temper designations consist of letters. Subdivisions of the basic tempers, where required, are indicated by one or more digits following the letter. These designate specific sequences of basic treatments, but only operations recognized as significantly influencing the characteristics of the product are indicated. Should some other variation of the same sequence of basic operations be applied to the same alloy, resulting in different characteristics, then additional digits are added.

The basic temper designations and subdivisions are as follows:

– F *as fabricated:* Applies to products which acquire some temper from shaping processes not having special control over the amount of strain-hardening or thermal treatment. For wrought products, there are no mechanical property limits.

– O *annealed, recrystallized (wrought products only):* Applies to the softest temper of wrought products.

– H *strain-hardened (wrought products only):* Applies to products which have their strength increased by strain-hardening with or without supplementary thermal treatments to produce partial softening.

The – H is always followed by two or more digits. The first digit indicates the specific combination of basic operations, as follows:

– H1 *strain-hardened only:* Applies to products which are strain-hardened to obtain the desired mechanical properties without supplementary thermal treatment. The number following this designation indicates the degree of strain-hardening.

– H2 *strain-hardened and the partially annealed:* Applies to products which are strain-hardened more than the desired final amount and then reduced in strength to the desired level by partial annealing. For alloys that age-soften at room temperature, the – H2 tempers have approximately the same ultimate strength as the corresponding – H3 tempers. For other alloys, the – H2 tempers have approximately the same ultimate strengths as the corresponding – H1 tempers and slightly higher elongations. The number following this designation indicates the degree of strain-hardening remaining after the product has been partially annealed.

– H3 *strain-hardened and then stablilized:* Applies to products which are strain-hardened and then stabilized by a low temperature heating to slightly lower their strength and increase ductility. This designation applies only to the magnesium-containing alloys which, unless stabilized, gradually age-soften at room temperature. The number following this designation indicates the degree of strain-hardening remaining after the product has been strain-hardened a specific amount and then stabilized.

The second digit following the designations – H1, – H2, and – H3 indicates the final degree of strain-hardening. Numeral 8 has been assigned to indicate tempers

having a final degree of strain-hardening equivalent to that resulting from approximately 75 per cent reduction of area. Tempers between − O (annealed) and 8 (full hard) are designated by numerals 1 through 7. Material having an ultimate strength about midway between that of the − O temper and that of the 8 temper is designated by the numeral 4 (half hard); between − O and 4 by the numeral 2 (quarter hard); between 4 and 8 by the numeral 6 (three-quarter hard); etc. (NOTE. For two-digit − H tempers whose second figure is odd, the standard limits for ultimate strength are exactly midway between those for the adjacent two-digit − H tempers whose second figures are even.) Numeral 9 designates extra hard tempers.

The third digit, when used, indicates a variation of a two-digit − H temper. It is used when the degree of control of temper or the mechanical properties are different from but close to those for the two-digit − H temper designation to which it is added. (NOTE. The minimum ultimate strength of a three-digit − H temper is at least as close to that of the corresponding two-digit − H temper as it is to the adjacent two-digit − H tempers.) Numerals 1 through 9 may be arbitrarily assigned and registered with The Aluminum Association for an alloy and product to indicate a specific degree of control of temper or specific mechanical property limits. Zero has been assigned to indicate degrees of control of temper or mechanical property limits negotiated between the manufacturer and purchaser which are not used widely enough to justify registration with The Aluminum Association.

The following three-digit − H temper designations have been assigned for wrought products in all alloys:

− H111   Applies to products which are strain-hardened less than the amount required for a controlled H11 temper.

− H112   Applies to products which acquire some temper from shaping processes not having special control over the amount of strain-hardening or thermal treatment, but for which there are mechanical property limits or mechanical property testing is required.

The following three-digit H temper designations have been assigned for wrought products in alloys containing over a normal 4 per cent magnesium.

− H311   Applies to products which are strain-hardened less than the amount required for a controlled H31 temper.

− H321   Applies to products which are strain-hardened less than the amount required for a controlled H32 temper.

− H323⎫ Applies to products which are specially fabricated to have acceptable resis-
− H343⎭ tance to stress-corrosion cracking.

The following three-digit − H temper designations have been assigned for

Patterned or Embossed Sheet	Fabricated Form
− H114	− O temper
− H124, − H224, − H324	− H11, − H21, − H31 temper, respectively
− H134, − H234, − H334	− H12, − H22, − H32 temper, respectively
− H144, − H244, − H344	− H13, − H23, − H33 temper, respectively
− H154, − H254, − H354	− H14, − H24, − H34 temper, respectively
− H164, − H264, − H364	− H15, − H25, − H35 temper, respectively
− H174, − H274, − H374	− H16, − H26, − H36 temper, respectively
− H184, − H284, − H384	− H17, − H27, − H37 temper, respectively
− H194, − H294, − H394	− H18, − H28, − H38 temper, respectively
− H195, − H395	− H19, − H39 temper, respectively

− W   *solution heat-treated:* An unstable temper applicable only to alloys which spontaneously age at room temperature after solution heat-treatment. This designation is specific only when the period of natural aging is indicated.

−T *thermally treated to produce stable tempers other than* −*F*, −*O*, *or* −*H*: Applies to products which are thermally treated, with or without supplementary strain-hardening, to produce stable tempers.

The −T is always followed by one or more digits. Numerals 2 through 10 have been assigned to indicate specific sequences of basic treatments, as follows:

−T1 *naturally aged to a substantially stable condition:* Applies to products for which the rate of cooling from an elevated temperature shaping process, such as casting or extrusion, is such that their strength is increased by room-temperature aging.

−T2 *annealed (cast products only):* Designates a type of annealing treatment used to improve ductility and increase dimensional stability of castings.

−T3 *solution heat-treated and then cold worked:* Applies to products which are cold worked to improve strength, or in which the effect of cold work in flattening or straightening is recognized in applicable specifications.

−T4 *solution heat-treated and naturally aged to a substantially stable condition:* Applies to products which are not cold worked after solution heat-treatment, or in which the effect of cold work in flattening or straightening may not be recognized in applicable specifications.

−T5 *artificially aged only:* Applies to products which are artificially aged after an elevated-temperature rapid-cool fabrication process, such as casting or extrusion, to improve mechanical properties or dimensional stability, or both.

−T6 *solution heat-treated and then artificially aged:* Applies to products which are not cold worked after solution heat-treatment, or in which the effect of cold work in flattening or straightening may not be recognized in applicable specifications.

−T7 *solution heat-treated and then stabilized:* Applies to products which are stabilized to carry them beyond the point of maximum hardness, providing control of growth or residual stress or both.

−T8 *solution heat-treated, cold worked, and then artificially aged:* Applies to products which are cold worked to improve strength, or in which the effect of cold work in flattening or straightening is recognized in applicable specifications.

−T9 *solution heat-treated, artificially aged, and then cold worked:* Applies to products which are cold worked to improve strength.

−T10 *artificially aged and then cold worked:* Applies to products which are artificially aged after an elevated-temperature rapid-cool fabrication process, such as casting or extrusion, and then cold worked to improve strength.

Additional digits may be added to designations −T1 through −T10 to indicate a variation in treatment which significantly alters the characteristics of the product. These may be arbitrarily assigned and registered with The Aluminum Association for an alloy and product to indicate a specific treatment or specific mechanical property limits.

These additional digits have been assigned for wrought products in all alloys:

−T__51 *stress-relieved by stretching:* Applies to products which are stress-relieved by stretching the following amounts after solution heat-treatment:

Plate	1½ to 3 per cent permanent set
Rod, Bar and Shapes	1 to 3 per cent permanent set
Drawn tube	0.5 to 3 per cent permanent set

Applies directly to plate and rolled or cold-finished rod and bar.

These products receive no further straightening after stretching.

Applies to extruded rod and bar shapes and tube when designated as follows:

- T__510 Products which receive no further straightening after stretching.
- T__511 Products which receive minor straightening after stretching to comply with standard tolerances.

- T__52 *stress-relieved by compressing:* Applies to products which are stress-relieved by compressing after solution heat-treatment, to produce a nominal permanent set of 2½ per cent.

- T__54 *stress-relieved by combined stretching and compressing:* applies to die forgings which are stress relieved by restriking cold in the finish die.

The following two-digit − T temper designations have been assigned for wrought products in all alloys:

- T42   Applies to products solution heat-treated and naturally aged which attain mechanical properties different from those of the − T4 temper.

- T62   Applies to products solution heat-treated and artificially aged which attain mechanical properties different from those of the − T6 temper.

Aluminum Alloy Designation Systems. — Aluminum casting alloys are listed in many specifications of various standardizing agencies. The numbering systems used by each differ and are not always correlatable. Casting alloys are available from producers who use a commercial numbering system and this numbering system is the one used in the tables of aluminum casting alloys which are given further along in this section.

A system of four-digit numerical designations for wrought aluminum and wrought aluminum alloys was adopted by the Aluminum Association in 1954. This system is used by the commercial producers and is similar to the one used by the SAE; the difference being the addition of two prefix letters.

The first digit of the designation identifies the alloy type: 1, indicating an aluminum of 99.00 per cent or greater purity; 2, copper; 3, manganese; 4, silicon; 5, magnesium; 6, magnesium and silicon; 7, zinc; 8, some element other than those aforementioned; 9, unused (not assigned at present). If the second digit in the designation is zero, it indicates that there is no special control on individual impurities; while integers 1 through 9, indicate special control on one or more individual impurities.

In the 1000 series group for aluminum of 99.00 per cent or greater purity, the last two of the four digits indicate to the nearest hundredth the amount of aluminum above 99.00 per cent. Thus designation 1030 indicates 99.30 per cent minimum aluminum. In the 2000 to 8000 series groups the last two of the four digits have no significance but are used to identify different alloys in the group. At the time of adoption of this designation system most of the existing commercial designation numbers were used as these last two digits, as for example, 14S became 2014, 3S became 3003, and 75S became 7075. When new alloys are developed and are commercially used these last two digits are assigned consecutively beginning with − 01, skipping any numbers previously assigned at the time of initial adoption.

Experimental alloys are also designated in accordance with this system but they are indicated by the prefix X. The prefix is dropped upon standardization.

Table 3 lists the nominal composition of commonly used aluminum casting alloys, and Tables 4A and 4B list the typical tensile properties of separately cast bars. Table 5 shows the product forms and nominal compositions of common wrought aluminum alloys, and Table 6 lists their typical mechanical properties.

Heat-treatability of Wrought Aluminum Alloys. — In high-purity form, aluminum is soft and ductile. Most commercial uses, however, require greater strength than pure aluminum affords. This is achieved in aluminum first by the addition of other elements to produce various alloys, which singly or in combination impart strength to the metal. Further strengthening is possible by means which classify the alloys roughly into two categories, non-heat-treatable and heat-treatable.

Non-heat-treatable alloys: The initial strength of alloys in this group depends upon the hardening effect of elements such as manganese, silicon, iron and magnesium, singly or in various combinations. The non-heat-treatable alloys are usually designated, therefore, in the 1000, 3000, 4000, or 5000 series. Since these alloys are work-hardenable, further strengthening is made possible by various degrees of cold working, denoted by the "H" series of tempers. Alloys containing appreciable amounts of magnesium when supplied in strain-hardened tempers are usually given a final elevated-temperature treatment called *stabilizing* for property stability.

Heat-treatable alloys: The initial strength of alloys in this group is enhanced by the addition of alloying elements such as copper, magnesium, zinc, and silicon. Since these elements singly or in various combinations show increasing solid solubility in aluminum with increasing temperature, it is possible to subject them to thermal treatments which will impart pronounced strengthening.

The first step, called *heat-treatment* or *solution heat-treatment,* is an elevated-temperature process designed to put the soluble element in solid solution. This is followed by rapid quenching, usually in water, which momentarily "freezes" the structure and for a short time renders the alloy very workable. It is at this stage that some fabricators retain this more workable structure by storing the alloys at below freezing temperatures until they are ready to form them. At room or elevated temperatures the alloys are not stable after quenching, however, and precipitation of the constituents from the supersaturated solution begins. After a period of several days at room temperature, termed *aging* or *room-temperature precipitation,* the alloy is considerably stronger. Many alloys approach a stable condition at room temperature, but some alloys, particularly those containing magnesium and silicon or magnesium and zinc, continue to age-harden for long periods of time at room temperature.

By heating for a controlled time at slightly elevated temperatures, even further strengthening is possible and properties are stabilized. This process is called *artificial aging* or *precipitation hardening.* By the proper combination of solution heat-treatment, quenching, cold working and artificial aging, the highest strengths are obtained.

Clad Aluminum Alloys. — The heat-treatable alloys in which copper or zinc are major alloying constituents, are less resistant to corrosive attack than the majority of non-heat-treatable alloys. To increase the corrosion resistance of these alloys in sheet and plate form they are often clad with high-purity aluminum, a low magnesium-silicon alloy, or an alloy containing 1 per cent zinc. The cladding, usually from $2\frac{1}{2}$ to 5 per cent of the total thickness on each side, not only protects the composite due to its own inherently excellent corrosion resistance but also exerts a galvanic effect which further protects the core material.

Special composites may be obtained such as clad non-heat-treatable alloys for extra corrosion protection, for brazing purposes, or for special surface finishes. Some alloys in wire and tubular form are clad for similar reasons and on an experimental basis extrusions also have been clad.

Characteristics of Principal Aluminum Alloy Series Groups. — 1000 series: These alloys are characterized by high corrosion resistance, high thermal and electrical conductivity, low mechanical properties and good workability. Moderate

increases in strength may be obtained by strain-hardening. Iron and silicon are the major impurities.

2000 series: Copper is the principal alloying element in this group. These alloys require solution heat-treatment to obtain optimum properties; in the heat-treated condition mechanical properties are similar to, and sometimes exceed, those of mild steel. In some instances artificial aging is employed to further increase the mechanical properties. This treatment materially increases yield strength, with attendant loss in elongation; its effect on tensile (ultimate) strength is not as great. The alloys in the 2000 series do not have as good corrosion resistance as most other aluminum alloys and under certain conditions they may be subject to intergranular corrosion. Therefore, these alloys in the form of sheet are usually clad with a high-purity alloy or a magnesium-silicon alloy of the 6000 series which provides galvanic protection to the core material and thus greatly increases resistance to corrosion. Alloy 2024 is perhaps the best known and most widely used aircraft alloy.

3000 series: Manganese is the major alloying element of alloys in this group, which are generally non-heat-treatable. Because only a limited percentage of manganese, up to about 1.5 per cent, can be effectively added to aluminum, it is used as a major element in only a few instances. One of these, however, is the popular 3003, used for moderate-strength applications requiring good workability.

4000 series: The major alloying element of this group is silicon, which can be added in sufficient quantities to cause substantial lowering of the melting point without producing brittleness in the resulting alloys. For these reasons aluminum-silicon alloys are used in welding wire and as brazing alloys where a lower melting point than that of the parent metal is required. Most alloys in this series are non-heat-treatable, but when used in welding heat-treatable alloys they will pick up some of the alloying constituents of the latter and so respond to heat-treatment to a limited extent. The alloys containing appreciable amounts of silicon become dark gray when anodic oxide finishes are applied, and hence are in demand for architectural applications.

5000 series: Magnesium is one of the most effective and widely used alloying elements for aluminum. When it is used as the major alloying element or with manganese, the result is a moderate to high strength non-heat-treatable alloy. Magnesium is considerably more effective than manganese as a hardener, about 0.8 per cent magnesium being equal to 1.25 per cent manganese, and it can be added in considerably higher quantities. Alloys in this series possess good welding characteristics and good resistance to corrosion in marine atmospheres. However, certain limitations should be placed on the amount of cold work and the safe operating temperatures permissible for the higher magnesium content alloys (over about $3\frac{1}{2}$ per cent for operating temperatures over about 150 deg. F.) to avoid susceptibility to stress corrosion.

6000 series: Alloys in this group contain silicon and magnesium in approximate proportions to form magnesium silicide, thus making them capable of being heat-treated. The major alloy in this series is 6061, one of the most versatile of the heat-treatable alloys. Though less strong than most of the 2000 or 7000 alloys, the magnesium-silicon (or magnesium-silicide) alloys possess good formability and corrosion resistance, with medium strength. Alloys in this heat-treatable group may be formed in the −T4 temper (solution heat-treated but not artificially aged) and then reach full −T6 properties by artificial aging.

7000 series: Zinc is the major alloying element in this group, and when coupled with a smaller percentage of magnesium results in heat-treatable alloys of very high strength. Usually other elements such as copper and chromium are also added in small quantities. Notable member of this group is 7075, which is among the highest strength aluminum alloys available and is used in air-frame structures and for highly stressed parts.

Table 3. Nominal Compositions (in per cent) of Common Aluminum Casting Alloys

Alloy (AA/ANSI Designation)	Product[a]	Si	Fe	Cu	Mn	Mg	Cr	Ni	Zn	Ti	Others Each	Others Total[h]
201.0	S	0.10	0.15	4.0–5.2	0.20–0.50	0.15–0.55	…	…	…	0.15–0.35	0.05[f]	0.10
204.0	S&P	0.20	0.35	4.2–5.0	0.10	0.15–0.35	…	0.05	0.10	0.15–0.30	0.05[g]	0.15
208.0	S&P	2.5–3.5	1.2	3.5–4.5	0.50	0.10	…	0.35	1.0	0.25	…	0.50
222.0	S&P	2.0	1.5	9.2–10.7	0.50	0.15–0.35	…	0.50	0.8	0.25	…	0.35
242.0	S&P	0.7	1.0	3.5–4.5	0.35	1.2–1.8	0.25	1.7–2.3	0.35	0.25	0.05	0.15
295.0	S	0.7–1.5	1.0	4.0–5.0	0.35	0.03	…	…	0.35	0.25	0.05	0.15
308.0	P	5.0–6.0	1.0	4.0–5.0	0.50	0.10	…	…	1.0	0.25	…	0.50
319.0	S&P	5.5–6.5	1.0	3.0–4.0	0.50	0.10	…	0.35	1.0	0.25	…	0.50
328.0	S	7.5–8.5	1.0	1.0–2.0	0.20–0.6	0.20–0.6	0.35	0.25	1.5	0.25	…	0.50
332.0	P	8.5–10.5	1.2	2.0–4.0	0.50	0.50–1.5	…	0.50	1.0	0.25	…	0.50
333.0	P	8.0–10.0	1.0	3.0–4.0	0.50	0.05–0.50	…	0.50	1.0	0.25	…	0.50
336.0	P	11.0–13.0	1.2	0.50–1.5	0.35	0.7–1.3	…	2.0–3.0	0.35	0.25	…	…
355.0	S&P	4.5–5.5	0.6[b]	1.0–1.5	0.50[b]	0.40–0.6	0.25	…	0.35	0.25	0.05	0.15
C355.0	S&P	4.5–5.5	0.20	1.0–1.5	0.10	0.40–0.6	…	…	0.10	0.20	0.05	0.15
356.0	S&P	6.5–7.5	0.6[b]	0.25	0.35[b]	0.20–0.45	…	…	0.35	0.25	0.05	0.15
A356.0	S&P	6.5–7.5	0.20	0.20	0.10	0.25–0.45	…	…	0.10	0.20	0.05	0.15
357.0	S&P	6.5–7.5	0.15	0.05	0.03	0.45–0.6	…	…	0.05	0.20	0.05	0.15
A357.0	S&P	6.5–7.5	0.20	0.20	0.10	0.40–0.7	…	…	0.10	0.04–0.20	0.05[c]	0.15
443.0	S&P	4.5–6.0	0.8	0.6	0.50	0.05	0.25	…	0.50	0.25	0.05	0.35
B443.0	S&P	4.5–6.0	0.8	0.15	0.35	0.05	…	…	0.35	0.25	0.05	0.15
444.0	P	6.5–7.5	0.20	0.10	0.10	0.05	…	…	0.10	0.20	0.05	0.15
512.0	S	1.4–2.2	0.6	0.35	0.8	3.5–4.5	0.25	…	0.35	0.25	0.05	0.35
513.0	P	0.30	0.40	0.10	0.30	3.5–4.5	…	…	1.4–2.2	0.20	0.05	0.15
514.0	S	0.35	0.50	0.15	0.35	3.5–4.5	…	…	0.15	0.25	0.05	0.15
520.0	S	0.25	0.30	0.25	0.15	9.5–10.6	…	…	0.15	0.25	0.05	0.15
705.0	S&P	0.20	0.8	0.20	0.40–0.6	1.4–1.8	0.20–0.40	…	2.7–3.3	0.25	0.05	0.15
707.0	S&P	0.20	0.8	0.20	0.40–0.6	1.8–2.4	0.20–0.40	…	4.0–4.5	0.25	0.05	0.15
710.0	S	0.15	0.50	0.35–0.65	0.05	0.6–0.8	…	…	6.0–7.0	0.25	0.05	0.15
711.0	P	0.30	0.7–1.4	0.35–0.65	0.05	0.25–0.45	…	…	6.0–7.0	0.20	0.05	0.15
712.0	S	0.30	0.50	0.25	0.10	0.50–0.65	0.40–0.6	…	5.0–6.5	0.15–0.25	0.05	0.20
850.0	S&P	0.7	0.7	0.7–1.3	0.10	0.10	…	0.7–1.3	…	0.20	—[e]	0.30
851.0	S&P	2.0–3.0	0.7	0.7–1.3	0.10	0.10	…	0.30–0.7	…	0.20	—[e]	0.30

[a] S = sand cast; P = permanent mold cast. [b] If iron exceeds 0.45 per cent, manganese content should not be less than one-half the iron content. [c] Also contains 0.04–0.07 per cent beryllium. [d] Also contains 0.003–0.007 per cent beryllium, boron 0.005 per cent maximum. [e] Also contains 5.5–7.0 per cent tin. [f] Also contains 0.40–1.0 per cent silver. [g] Also contains 0.05 max. per cent tin. [h] The sum of those "Others" metallic elements 0.010 per cent or more each, expressed to the second decimal before determining the sum. *Source:* Standards for Aluminum Sand and Permanent Mold Castings. Courtesy of the Aluminum Association.

Table 4A. Mechanical Property Limits for Commonly Used Aluminum Sand Casting Alloys[a]

Alloy	Temper[b]	Minimum Properties			Typical Brinell Hardness (500 kgf load, 10-mm ball)
		Tensile Strength (ksi)		Elongation In 2 inches (%)	
		Ultimate	Yield		
201.0	T7	60.0	50.0	3.0	110–140
204.0	T4	45.0	28.0	6.0	. . .
208.0	F	19.0	12.0	1.5	40–70
222.0	O	23.0	. . .	. . .	65–95
222.0	T61	30.0	. . .	. . .	100–130
242.0	O	23.0	. . .	. . .	55–85
242.0	T571	29.0	. . .	. . .	70–100
242.0	T61	32.0	20.0	. . .	90–120
242.0	T77	24.0	13.0	1.0	60–90
295.0	T4	29.0	13.0	6.0	45–75
295.0	T6	32.0	20.0	3.0	60–90
295.0	T62	36.0	28.0	. . .	80–110
295.0	T7	29.0	16.0	3.0	55–85
319.0	F	23.0	13.0	1.5	55–85
319.0	T5	25.0	. . .	. . .	65–95
319.0	T6	31.0	20.0	1.5	65–95
328.0	F	25.0	14.0	1.0	45–75
328.0	T6	34.0	21.0	1.0	65–95
354.0	c	. . .	. . .	. . .	. . .
355.0	T51	25.0	18.0	. . .	50–80
355.0	T6	32.0	20.0	2.0	70–105
355.0	T7	35.0	. . .	. . .	70–100
355.0	T71	30.0	22.0	. . .	60–95
C355.0	T6	36.0	25.0	2.5	75–105
356.0	F	19.0	. . .	2.0	40–70
356.0	T51	23.0	16.0	. . .	45–75
356.0	T6	30.0	20.0	3.0	55–90
356.0	T7	31.0	29.0	. . .	60–90
356.0	T71	25.0	18.0	3.0	45–75
A356.0	T6	34.0	24.0	3.5	70–105
443.0	F	17.0	7.0	3.0	25–55
B443.0	F	17.0	6.0	3.0	25–55
512.0	F	17.0	10.0	. . .	35–65
514.0	F	22.0	9.0	6.0	35–65
520.0	T4[e]	42.0	22.0	12.0	60–90
535.0	F or T5	35.0	18.0	9.0	60–90
705.0	F or T5	30.0	17.0	5.0	50–80
707.0	T5	33.0	22.0	2.0	70–100
707.0	T7	37.0	30.0	1.0	65–95
710.0	F or T5	32.0	20.0	2.0	60–90
712.0	F or T5	34.0	25.0	4.0	60–90
713.0	F or T5	32.0	22.0	3.0	60–90
771.0	T5	42.0	38.0	1.5	85–115
771.0	T51	32.0	27.0	3.0	70–100
771.0	T52	36.0	30.0	1.5	70–100
771.0	T53	36.0	27.0	1.5	. . .
771.0	T6	42.0	35.0	5.0	75–105
771.0	T71	48.0	45.0	2.0	105–135
850.0	T5	16.0	. . .	5.0	30–60
851.0	T5	17.0	. . .	3.0	30–60
852.0	T5	24.0	18.0	. . .	45–75

[a] For separately cast test bars.
[b] F indicates "as cast" condition.
[c] Mechanical properties for these alloys depend on the casting process. For further information consult the individual foundries.
[e] The T4 temper of Alloy 520.0 is unstable; significant room temperature aging occurs within life expectancy of most castings. Elongation may decrease by as much as 80 percent.
Source: Standards for Aluminum Sand and Permanent Mold Castings. Courtesy of the Aluminum Association.

Table 4B. Mechanical Property Limits for Commonly Used Aluminum Permanent Mold Casting Alloys[a]

Alloy	Temper[b]	Minimum Properties			Typical Brinell Hardness (500 kgf load, 10 mm ball)
		Tensile Strength (ksi)		Elongation In 2 inches (%)	
		Ultimate	Yield		
204.0	T4	48.0	29.0	8.0	
208.0	T4	33.0	15.0	4.5	60–90
208.0	T6	35.0	22.0	2.0	75–105
208.0	T7	33.0	16.0	3.0	65–95
222.0	T551	30.0	. . .	. . .	100–130
222.0	T65	40.0	. . .	. . .	125–155
242.0	T571	34.0	. . .	. . .	90–120
242.0	T61	40.0	. . .	. . .	95–125
296.0	T6	35.0	. . .	2.0	75–105
308.0	F	24.0	. . .	. . .	55–85
319.0	F	28.0	14.0	1.5	70–100
319.0	T6	34.0	. . .	2.0	75–105
332.0	T5	31.0	. . .	. . .	90–120
333.0	F	28.0	. . .	. . .	65–100
333.0	T5	30.0	. . .	. . .	70–105
333.0	T6	35.0	. . .	. . .	85–115
333.0	T7	31.0	. . .	. . .	75–105
336.0	T551	31.0	. . .	. . .	90–120
336.0	T65	40.0	. . .	. . .	110–140
354.0	T61	48.0	37.0	3.0	. . .
354.0	T62	52.0	42.0	2.0	. . .
355.0	T51	27.0	. . .	. . .	60–90
355.0	T6	37.0	. . .	1.5	75–105
355.0	T62	42.0	. . .	. . .	90–120
355.0	T7	36.0	. . .	. . .	70–100
355.0	T71	34.0	27.0	. . .	65–95
C355.0	T61	40.0	30.0	3.0	75–105
356.0	F	21.0	. . .	3.0	40–70
356.0	T51	25.0	. . .	. . .	55–85
356.0	T6	33.0	22.0	3.0	65–95
356.0	T7	25.0	. . .	3.0	60–90
356.0	T71	25.0	. . .	3.0	60–90
A356.0	T61	37.0	26.0	5.0	70–100
357.0	T6	45.0	. . .	3.0	75–105
A357.0	T61	45.0	36.0	3.0	85–115
359.0	T61	45.0	34.0	4.0	75–105
359.0	T62	47.0	38.0	3.0	85–115
443.0	F	21.0	7.0	2.0	30–60
B443.0	F	21.0	6.0	2.5	30–60
A444.0	T4	20.0	. . .	20.0	. . .
513.0	F	22.0	12.0	2.5	45–75
535.0	F	35.0	18.0	8.0	60–90
705.0	T5	37.0	17.0	10.0	55–85
707.0	T7	45.0	35.0	3.0	80–110
711.0	T1	28.0	18.0	7.0	55–85
713.0	T5	32.0	22.0	4.0	60–90
850.0	T5	18.0	. . .	8.0	30–60
851.0	T5	17.0	. . .	3.0	30–60
851.0	T6	18.0	. . .	8.0	. . .
852.0	T5	27.0	. . .	3.0	55–85

[a] For separately cast test bars.
[b] F indicates "as cast" condition.
Source: Standards for Aluminum Sand and Permanent Mold Castings. Courtesy of the Aluminum Association.

Table 5. Typical Mechanical Properties of Wrought Aluminum Alloys[a]

Alloy and Temper	Tension				Brinell Hardness Number (500 kg load, 10 mm ball)	Ultimate Shearing Strength (ksi)	Endurance Limit[b] (ksi)
	Strength (ksi)		Elongation in 2 inches (%)				
	Ultimate	Yield	1/16-inch Thick Specimen	1/2-inch Diameter Specimen			
1060-O	10	4	43	...	19	7	3
1060-H12	12	11	16	...	23	8	4
1060-H14	14	13	12	...	26	9	5
1060-H16	16	15	8	...	30	10	6.5
1060-H18	19	18	6	...	35	11	6.5
1100-O	13	5	35	45	23	9	5
1100-H12	16	15	12	25	28	10	6
1100-H14	18	17	9	20	32	11	7
1100-H16	21	20	6	17	38	12	9
1100-H18	24	22	5	15	44	13	9
1350-O	12	4	...	...[d]	...	8	...
1350-H12	14	12	...	...	...	9	...
1350-H14	16	14	...	...	...	10	...
1350-H16	18	16	...	...	...	11	...
1350-H19	27	24	...	...[e]	...	15	7
2011-T3	55	43	...	15	95	32	18
2011-T8	59	45	...	12	100	35	18
2014-O	27	14	...	18	45	18	13
2014-T4, T451	62	42	...	20	105	38	20
2014-T6, T651	70	60	...	13	135	42	18
Alclad 2014-O	25	10	21	...	...	18	...
Alclad 2014-T3	63	40	20	...	...	37	...
Alclad 2014-T4, T451	61	37	22	...	...	37	...
Alclad 2014-T6, T651	68	60	10	...	...	41	...
2017-O	26	10	...	22	45	18	13
2017-T4, T451	62	40	...	22	105	38	18
2018-T61	61	46	...	12	120	39	17
2024-O	27	11	20	22	47	18	13
2024-T3	70	50	18	...	120	41	20
2024-T4, T351	68	47	20	19	120	41	20
2024-T361[c]	72	57	13	...	130	42	18
Alclad 2024-O	26	11	20	...	...	18	...
Alclad 2024-T3	65	45	18	...	...	40	...
Alclad 2024-T4, T351	64	42	19	...	...	40	...
Alclad 2024-T361[c]	67	53	11	...	...	41	...
Alclad 2024-T81, T851	65	60	6	...	...	40	...
Alclad 2024-T861[c]	70	66	6	...	...	42	...
2025-T6	58	37	...	19	110	35	18
2036-T4	49	28	24	...	...	...	18[f]
2117-T4	43	24	...	27	70	28	14
2218-T72	48	37	...	11	95	30	...
2219-O	25	11	18	...	...	...	...
2219-T42	52	27	20	...	...	...	...
2219-T31, T351	52	36	17	...	...	...	...
2219-T37	57	46	11	...	...	...	...
2219-T62	60	42	10	...	...	...	15
2219-T81, T851	66	51	10	...	...	...	15
2219-T87	69	57	10	...	...	...	15
3003-O	16	6	30	40	28	11	7
3003-H12	19	18	10	20	35	12	8
3003-H14	22	21	8	16	40	14	9
3003-H16	26	25	5	14	47	15	10
3003-H18	29	27	4	10	55	16	10

See footnotes at end of table.

Table 5 (Continued). Typical Mechanical Properties of Wrought Aluminum Alloys[a]

Alloy and Temper	Tension Strength (ksi)		Elongation in 2 inches (%)		Brinell Hardness Number (500 kg load, 10 mm ball)	Ultimate Shearing Strength (ksi)	Endurance Limit[b] (ksi)
	Ultimate	Yield	1/16-inch Thick Specimen	1/2-inch Diameter Specimen			
Alclad 3003-O	16	6	30	40	...	11	...
Alclad 3003-H12	19	18	10	20	...	12	...
Alclad 3003-H14	22	21	8	16	...	14	...
Alclad 3003-H16	26	25	5	14	...	15	...
Alclad 3003-H18	29	27	4	10	...	16	...
3004-O	26	10	20	25	45	16	14
3004-H32	31	25	10	17	52	17	15
3004-H34	35	29	9	12	63	18	15
3004-H36	38	33	5	9	70	20	16
3004-H38	41	36	5	6	77	21	16
Alclad 3004-O	26	10	20	25	...	16	...
Alclad 3004-H32	31	25	10	17	...	17	...
Alclad 3004-H34	35	29	9	12	...	18	...
Alclad 3004-H36	38	33	5	9	...	20	...
Alclad 3004-H38	41	36	5	6	...	21	...
3105-O	17	8	24	...	...	12	...
3105-H12	22	19	7	...	...	14	...
3105-H14	25	22	5	...	...	15	...
3105-H16	28	25	4	...	...	16	...
3105-H18	31	28	3	...	...	17	...
3105-H25	26	23	8	...	...	15	...
4032-T6	55	46	...	9	120	38	16
5005-O	18	6	25	...	28	11	...
5005-H12	20	19	10	...	...	14	...
5005-H14	23	22	6	...	...	14	...
5005-H16	26	25	5	...	...	15	...
5005-H18	29	28	4	...	...	16	...
5005-H32	20	17	11	...	36	14	...
5005-H34	23	20	8	...	41	14	...
5005-H36	26	24	6	...	46	15	...
5005-H38	29	27	5	...	51	16	...
5050-O	21	8	24	...	36	15	12
5050-H32	25	21	9	...	46	17	13
5050-H34	28	24	8	...	53	18	13
5050-H36	30	26	7	...	58	19	14
5050-H38	32	29	6	...	63	20	14
5052-O	28	13	25	30	47	18	16
5052-H32	33	28	12	18	60	20	17
5052-H34	38	31	10	14	68	21	18
5052-H36	40	35	8	10	73	23	19
5052-H38	42	37	7	8	77	24	20
5056-O	42	22	...	35	65	26	20
5056-H18	63	59	...	10	105	34	22
5056-H38	60	50	...	15	100	32	22
5083-O	42	21	...	22	...	25	...
5083-H321, H116	46	33	...	16	...	...	23
5086-O	38	17	22	...	...	23	...
5086-H32, H116	42	30	12	...	...	...	...
5086-H34	47	37	10	...	...	27	...
5086-H112	39	19	14	...	...	...	...
5154-O	35	17	27	...	58	22	17
5154-H32	39	30	15	...	67	22	18
5154-H34	42	33	13	...	73	24	19
5154-H36	45	36	12	...	78	26	20
5154-H38	48	39	10	...	60	28	21
5154-H112	35	17	25	...	63	...	17

See footnotes at end of table.

Table 5 *(Continued)*. **Typical Mechanical Properties of Wrought Aluminum Alloys**

Alloy and Temper	Tension				Brinell Hardness Number (500 kg load, 10 mm ball)	Ultimate Shearing Strength (ksi)	Endurance Limit (ksi)
	Strength (ksi)		Elongation in 2 inches (%)				
	Ultimate	Yield	1/16-inch Thick Specimen	1/2-inch Diameter Specimen			
5252-H25	34	25	11	...	68	21	...
5252-H38, H28	41	35	5	...	75	23	...
5254-O	35	17	27	...	58	22	17
5254-H32	39	30	15	...	67	22	18
5254-H34	42	33	13	...	73	24	19
5254-H36	45	36	12	...	78	26	20
5254-H38	48	39	10	...	80	28	21
5254-H112	35	17	25	...	63	...	17
5454-O	36	17	22	...	62	23	...
5454-H32	40	30	10	...	73	24	...
5454-H34	44	35	10	...	81	26	...
5454-H111	38	26	14	...	70	23	...
5454-H112	36	18	18	...	62	23	...
5456-O	45	23	...	24	...	...	...
5456-H112	45	24	...	22	...	...	...
5456-H321, H116	51	37	...	16	90	30	...
5457-O	19	7	22	...	32	12	...
5457-H25	26	23	12	...	48	16	...
5457-H38, H28	30	27	6	...	55	18	...
5652-O	28	13	25	30	47	18	16
5652-H32	33	28	12	18	60	20	17
5652-H34	38	31	10	14	68	21	18
5652-H36	40	35	8	10	73	23	19
5652-H38	42	37	7	8	77	24	20
5657-H25	23	20	12	...	40	14	...
5657-H38, H28	28	24	7	...	50	15	...
6061-O	18	8	25	30	30	12	9
6061-T4, T451	35	21	22	25	65	24	14
6061-T6, T651	45	40	12	17	95	30	14
Alclad 6061-O	17	7	25	...	...	11	...
Alclad 6061-T4, T451	33	19	22	...	...	22	...
Alclad 6061-T6, T651	42	37	12	...	...	27	...
6063-O	13	7	...	...	25	10	8
6063-T1	22	13	20	...	42	14	9
6063-T4	25	13	22	...	...	...	...
6063-T5	27	21	12	...	60	17	10
6063-T6	35	31	12	...	73	22	10
6063-T83	37	35	9	...	82	22	...
6063-T831	30	27	10	...	70	18	...
6063-T832	42	39	12	...	95	27	...
6066-O	22	12	...	18	43	14	...
6066-T4, T451	52	30	...	18	90	29	...
6066-T6, T651	57	52	...	12	120	34	16
6070-T6	55	51	10	...	...	34	14
6101-H111	14	11	...	...	...	...	...
6101-T6	32	28	15	...	71	20	...
6262-T9	58	55	...	10	120	35	13
6463-T1	22	13	20	...	42	14	10
6463-T5	27	21	12	...	60	17	10
6463-T6	35	31	12	...	74	22	10
7001-O	37	22	...	14	60	...	...
7001-T6, T651	98	91	...	9	160	...	22

See footnotes at end of table.

Table 5 (*Concluded*). **Typical Mechanical Properties of Wrought Aluminum Alloys**[a]

Alloy and Temper	Tension				Brinell Hardness Number (500 kg load, 10 mm ball)	Ultimate Shearing Strength (ksi)	Endurance Limit[b] (ksi)
	Strength (ksi)		Elongation in 2 inches (%)				
	Ultimate	Yield	1/16-inch Thick Specimen	1/2-inch Diameter Specimen			
7049-T73	75	66	. . .	12	135	44	. . .
7049-T7352	75	63	. . .	11	135	43	. . .
7050-T73510, T73511	72	63	. . .	12	. . .	. . .	. . .
7050-T7451[g]	76	68	. . .	11	. . .	44	. . .
7050-T7651	80	71	. . .	11	. . .	47	. . .
7075-O	33	15	17	16	60	22	. . .
7075-T6, T651	83	73	11	11	150	48	23
Alclad 7075-O	32	14	17	. . .	. . .	22	. . .
Alclad 7075-T6, T651	76	67	11	. . .	. . .	46	. . .
7178-O	33	15	15	16	. . .	. . .	. . .
7178-T6, T651	88	78	10	11	. . .	. . .	. . .
7178-T76, T7651	83	73	. . .	11	. . .	. . .	. . .
Alclad 7178-O	32	14	16	. . .	. . .	. . .	. . .
Alclad 7178-T6, T651	81	71	10	. . .	. . .	. . .	. . .
8176-H24	17	14	15	. . .	. . .	10	. . .

[a] The data given in this table are intended only as a basis for comparing alloys and tempers and should not be specified as engineering requirements or used for design purposes. The indicated typical mechanical properties for all except O temper material are higher than the specified minimum properties. For O temper products, typical ultimate and yield values are slightly lower than specified (maximum) values.
[b] Based on 500,000,000 cycles of completely reversed stress using the R. R. Moore type of machine and specimen.
[c] Tempers T361 and T861 were formerly designated T36 and T86, respectively.
[d] 1350-O wire should have an elongation of approximately 23 per cent in 10 inches.
[e] 1350-H19 wire should have an elongation of approximately 1.5 per cent in 10 inches.
[f] Based on 10^7 cycles using flexural type testing of sheet specimens.
[g] T7451, although not previously registered, has appeared in the literature and in some specifications as T73651.
Source: Aluminum Standards and Data. Courtesy of the Aluminum Association.

Table 6. Nominal Compositions of Common Wrought Aluminum Alloys

Alloy	Alloying Elements — Aluminum and Normal Impurities Constitute Remainder							
	Si	Cu	Mn	Mg	Cr	Ni	Zn	Ti
1050	. . .	. . .	99.50 per cent minimum aluminum			. . .	. . .	. . .
1060	. . .	. . .	99.60 per cent minimum aluminum			. . .	. . .	. . .
1100	. . .	0.12	99.00 per cent minimum aluminum			. . .	. . .	. . .
1145	. . .	. . .	99.45 per cent minimum aluminum			. . .	. . .	. . .
1175	. . .	. . .	99.75 per cent minimum aluminum			. . .	. . .	. . .
1200	. . .	. . .	99.00 per cent minimum aluminum			. . .	. . .	. . .
1230	. . .	. . .	99.30 per cent minimum aluminum			. . .	. . .	. . .
1235	. . .	. . .	99.35 per cent minimum aluminum			. . .	. . .	. . .
1345	. . .	. . .	99.45 per cent minimum aluminum			. . .	. . .	. . .
1350[f]	. . .	. . .	99.50 per cent minimum aluminum			. . .	. . .	. . .
2011[a]	. . .	5.5	. . .	. . .	. . .	. . .	. . .	. . .
2014	0.8	4.4	0.8	0.50	. . .	. . .	. . .	. . .
2017	0.50	4.0	0.7	0.6	. . .	. . .	. . .	. . .
2018	. . .	4.0	. . .	0.7	. . .	2.0	. . .	. . .
2024	. . .	4.4	0.6	1.5	. . .	. . .	. . .	. . .
2025	0.8	4.4	0.8	. . .	. . .	. . .	. . .	. . .
2036	. . .	2.6	0.25	0.45	. . .	. . .	. . .	. . .
2117	. . .	2.6	. . .	0.35	. . .	. . .	. . .	. . .

Table 6 (*Concluded*). **Nominal Compositions of Common Wrought Aluminum Alloys**

Alloy	Alloying Elements — Aluminum and Normal Impurities Constitute Remainder							
	Si	Cu	Mn	Mg	Cr	Ni	Zn	Ti
2124	...	4.4	0.6	1.5	...	...	...	...
2218	...	4.0	...	1.5	...	2.0	...	...
2219[b]	...	6.3	0.30	...	...	...	...	0.06
2319[b]	...	6.3	0.30	...	...	...	...	0.15
2618[c]	0.18	2.3	...	1.6	...	1.0	...	0.07
3003	...	0.12	1.2	...	...	...	...	...
3004	...	...	1.2	1.0	...	...	...	...
3005	...	...	1.2	0.40	...	...	...	...
4032	12.2	0.9	...	1.0	...	0.9	...	...
4043	5.2	...	...	...	...	...	...	...
4045	10.0	...	...	...	...	...	...	...
4047	12.0	...	...	...	...	...	...	...
4145	10.0	4.0	...	...	...	...	...	...
5005	...	...	...	0.8	...	...	...	...
5050	...	...	...	1.4	...	...	...	...
5052	...	...	...	2.5	0.25	...	...	...
5056	...	...	0.12	5.0	0.12	...	...	...
5083	...	...	0.7	4.4	0.15	...	...	...
5086	...	...	0.45	4.0	0.15	...	...	...
5183	...	...	0.8	4.8	0.15	...	...	...
5252	...	...	...	2.5	...	...	...	...
5254	...	...	...	3.5	0.25	...	...	...
5356	...	...	0.12	5.0	0.12	...	...	0.13
5456	...	...	0.8	5.1	0.12	...	...	...
5457	...	...	0.30	1.0	...	...	...	...
5554	...	...	0.8	2.7	0.12	...	...	0.12
5556	...	...	0.8	5.1	0.12	...	...	0.12
5652	...	...	...	2.5	0.25	...	...	...
5654	...	...	...	3.5	0.25	...	...	0.10
6003	0.7	...	...	1.2	...	...	...	...
6005	0.8	...	...	0.50	...	...	...	...
6053	0.7	...	...	1.2	0.25	...	...	...
6061	0.6	0.28	...	1.0	0.20	...	...	...
6066	1.4	1.0	0.8	1.1	...	...	...	...
6070	1.4	0.28	0.7	0.8	...	...	...	...
6101	0.50	...	...	0.6	...	...	...	...
6105	0.8	...	...	0.62	...	...	...	...
6151	0.9	...	...	0.6	0.25	...	...	...
6201	0.7	...	...	0.8	...	...	...	...
6253	0.7	...	...	1.2	0.25	...	2.0	...
6262[e]	0.6	0.28	...	1.0	0.09	...	...	...
6351	1.0	...	0.6	0.6	...	...	...	...
6463	0.40	...	...	0.7	...	...	...	...
7001	...	2.1	...	3.0	0.26	...	7.4	...
7005[d]	...	...	0.45	1.4	0.13	...	4.5	0.04
7008	...	...	...	1.0	0.18	...	5.0	...
7049	...	1.6	...	2.4	0.16	...	7.7	...
7050[g]	...	2.3	...	2.2	6.2	...	...	...
7072	...	...	...	...	...	...	1.0	...
7075	...	1.6	...	2.5	0.23	...	5.6	...
7108[h]	...	...	...	1.0	...	...	5.0	...
8017[i]	...	0.15	...	0.03	...	...	...	...
8030[j]	...	0.22	...	...	...	...	...	...
8177[k]	...	...	...	0.08	...	...	...	...

[a] Lead and bismuth, 0.40 per cent each.
[b] Vanadium 0.10 per cent; zirconium 0.18 per cent.
[c] Iron 1.1 per cent.
[d] Zirconium 0.14 per cent.
[e] Lead and bismuth, 0.6 per cent each.
[f] Formerly designated EC.
[g] Zirconium 0.12 per cent.
[h] Zirconium 0.18 per cent.
[i] Iron 0.7 per cent.
[j] Boron 0.02 per cent.
[k] Iron 0.35 per cent.
Source: Aluminum Standards and Data. Courtesy of the Aluminum Association.

Magnesium Alloys

Magnesium is the lightest of all structural metals. Pure magnesium is a relatively soft, silver-white metal. The unalloyed metal has little engineering value because of its low strength, poor corrosion and oxidation resistances, difficulty in cold working, and high cost. Its use is restricted to those applications that take advantage of its chemical affinity for oxygen, for instance, as getters in vacuum-tube manufacture and in flashlights and pyrotechnics. However, small amounts of alloying elements improve magnesium's properties considerably, providing a combination of low density and good mechanical strength resulting in a high strength to weight ratio.

Alloys of magnesium are easy to machine and weld, and can be fabricated by most metalworking processes, including die casting. Cold forming, however, is limited to mild deformation or roll bending, since, at room temperature, the alloys work harden quite rapidly. Because of their low modulus of elasticity, magnesium alloys can absorb energy elastically, which in combination with moderate strength provides high damping capacity. Magnesium alloys have good fatigue resistance; however, the metal is particularly sensitive to stress concentrations. Some alloys are heat treated, by solution treating and aging, to improve their properties.

Applications of Magnesium Alloys. — Magnesium alloys are used in a wide variety of structural applications including industrial, materials handling, commercial, and aerospace equipment. In industrial machinery the alloys are used for parts that operate at high speeds and, therefore, must be light in weight to minimize inertial forces. Materials handling equipment includes dockboards, grain shovels, and gravity conveyors. Commercial applications include luggage and ladders. Magnesium alloys' good strength to weight ratios make them valuable for a variety of aircraft structures, where resistance to seawater corrosion is not critical.

Magnesium parts too intricate to fabricate economically by other methods can be produced by casting, including sand, permanent mold, die, investment, and shell molding. Wrought magnesium alloys are produced as bars, billets, and extruded shapes, wire, sheet, plate, and forgings.

Alloy and Temper Designation. — Magnesium alloys are designated by a standard four-part system established by the ASTM, and now also used by the SAE, that indicates both chemical composition and temper. Designations begin with two letters representing the two alloying elements that are specified in the greatest amount; these letters are arranged in order of decreasing percentage of alloying elements or alphabetically if they are present in equal amounts. The letters are followed by digits representing the respective composition percentages, rounded off to whole numbers, and then by a serial letter indicating some variation in composition of minor constituents. The final part, separated by a hyphen, consists of a letter followed by a number, indicating the temper condition.

The letters that designate the more common alloying elements are A, aluminum; E, rare earths; H, thorium; K, zirconium; M, manganese; Q, silver; S, silicon; T, tin; Z, zinc.

The letters and numbers that indicate the temper designation are:

F, as fabricated	T4, solution heat treated
O, annealed	T5, artificially aged
H10, H11, strain hardened	T6, solution heat treated and artificially
H23, H24, H26, strain hardened and	aged
annealed	T8, solution heat treated, cold worked,
	and artificially aged

The nominal composition and typical properties of magnesium alloys are listed in Table 7.

Table 7. Nominal Compositions and Typical Room-Temperature Mechanical Properties of Magnesium Alloys

Alloy	Al	Mn[a]	Th	Zn	Zr	Others	Tensile Strength (ksi)	Yield Strength, Tensile (ksi)	Yield Strength, Compressive (ksi)	Bearing (ksi)	Elongation in 2 in. (%)	Shear Strength (ksi)	Hardness, Rockwell B[b]
Sand and Permanent Mold Castings													
AM100A-T61	10.0	0.1					40	22	22		1		69
AZ63A-T6	6.0	0.15		3.0			40	19	19	52	5	21	73
AZ81A-T4	7.6	0.13		0.7			40	12	12	44	15	18	55
AZ91C-T6	8.7	0.13		0.7			40	21	21	52	6	21	66
EZ33A-T5				2.7	0.6	3.3 RE	23	16	16	40	2	21	50
HK31A-T6			3.3		0.7		32	15	15	40	8	21	55
HZ32A-T5			3.3	2.1	0.7		27	13	13	37	4	20	57
K1A-F					0.7		26	8		18	19	8	
QE22A-T6					0.7	2.5 Ag, 2.1 RE	38	28	28		3		80
QH21A-T6			60		0.7	2.5 Ag, 1.0 RE	40	30			4		
ZE41A-T5				4.2	0.7	1.2 RE	30	20	20	51	3.5	23	62
ZE63A-T6				5.8	0.7	2.6 RE	44	28	28		10		60-85
ZH62A-T5			1.8	5.7	0.7		35	25	25	49	4	24	70
Die Castings													
AM60A-F	6.0	0.13					30	17	17		6		
AS41A-F[d]	4.3	0.35				1.0 Si	32	22	22		4		
AZ91A and B-F[c]	9.0	0.13		0.7			33	22	24		3	20	63
Extruded Bars and Shapes													
AZ10A-F	1.2	0.2		0.4			35	21	10		10		
AZ31 B and C-F[e]	3.0			1.0			38	29	14	33	15	19	49
AZ61A-F	6.5			1.0			45	33	19	41	16	20	60
AZ80A-T5	8.5			0.5			55	40	35		7	24	82
HM31A-F		1.2	3.0				42	33	27	50	10	22	
M1A-F		1.2					37	26	12	28	12	18	44
ZK60A-T5				5.5	0.45[d]		53	44	36	59	11	26	88
Sheet and Plate													
AZ31B-H24	3.0			1.0			42	32	26	47	15	23	73
HK31A-H24			3.0		0.6		33	29	23	41	9	20	68
HM21A-T8		0.6	2.0				34	25	19	39	11	18	

[a] Minimum. [b] 500-kg load, 10-mm ball. [c] A and B are identical except that 0.30 per cent max residual Cu is allowable in AZ91B. [d] For battery applications. [e] Properties of B and C are identical, but AZ31C has 0.15 min Mn, 0.1 max Cu and 0.03 max Ni.

Nickel and Nickel Alloys

Nickel is a white metal, similar in some respects to iron but with good oxidation and corrosion resistances. Nickel and its alloys are used in a variety of applications, usually requiring specific corrosion resistance or high strength at high temperature. Some nickel alloys exhibit very high toughness; others have very high strength, high proportional limits, and high moduli compared to steel. Commercially, pure nickel has good electrical, magnetic, and magnetostrictive properties. Nickel alloys are strong, tough, and ductile at cryogenic temperatures, while several of the so-called nickel-based superalloys have good strength at temperatures up to 2000 degrees F.

Most wrought nickel alloys can be hot and cold worked, machined, and welded successfully; an exception is the most highly alloyed nickel compound — forged nickel-based superalloys — in which these operations are more difficult. The casting alloys can be machined or ground, and many can be welded and brazed.

There are five categories into which the common nickel-based metals and alloys can be separated: the pure nickel and high nickel (over 94 per cent Ni) alloys; the nickel–molybdenum and nickel–molybdenum–chromium superalloys, which are specifically for corrosive or high-temperature, high-strength service; the nickel–molybdenum–copper alloys, which are also specified for corrosion applications; the nickel–copper (Monel) alloys, which are used in actively corrosive environments; and the nickel–chromium and nickel–chromium–iron superalloys, which are noted for their strength and corrosion resistance at high temperatures.

Descriptions and compositions of some commonly used nickel and high nickel alloys are shown in Table 8.

Titanium and Titanium Alloys

Titanium is a gray, light metal with a better strength to weight ratio than any other metal at room temperature, and is used in corrosive environments or in applications that take advantage of its light weight, good strength, and nonmagnetic properties. Titanium is available commercially in many alloys, but many requirements can be met by a single grade of the commercially pure metal. The alloys of titanium are of three metallurgical types: alpha, alpha–beta, and beta, with these designations referring to the predominant phases present in the microstructure.

Titanium has a strong affinity for hydrogen, oxygen, and nitrogen gases, which tend to embrittle the material; carbon is another embrittling agent. Titanium is outstanding in its resistance to strongly oxidizing acids, aqueous chloride solutions, moist chlorine gas, sodium hypochlorite, and seawater and brine solutions. Nearly all nonaircraft applications take advantage of this corrosion resistance. Its uses in aircraft engine compressors and in airframe structures are based on both its high corrosion resistance and high strength to weight ratio.

Forming titanium is similar to forming stainless steel. Titanium and its alloys can be machined and abrasive ground; however, sharp tools and continuous feed are required to prevent work hardening. Tapping is difficult because the metal galls.

Titanium castings can be produced by investment or graphite mold methods; however, because of the highly reactive nature of the metal in the presence of oxygen, casting must be done in a vacuum furnace.

Generally, titanium is welded by gas-tungsten arc or plasma arc techniques, and the key to successful welding lies in proper cleaning and shielding. The alpha–beta titanium alloys can be heat treated for higher strength, but they are not easily welded. Beta and alpha–beta alloys are designed for formability; they are formed in the soft state, and then heat treated for high strength.

The properties of some wrought titanium alloys are shown in Table 9.

Table 8. Common Cast and Wrought Nickel and High Nickel Alloys — Designations, Composition, Typical Properties, and Uses

UNS Designation	Description and Common Name	Nominal Composition (Weight %)	Typical Room-Temperature Properties			Form	Typical Uses
			Tensile (ksi)	0.2% Yield (ksi)	Elong. (%)		
N02200	Commercially pure Ni (Nickel 200)	99.5 Ni	67	22	47	Wrought	Food processing and chemical equipment.
N04400	Nickel–copper alloy (Monel 400)	65 Ni; 32 Cu, 2 Fe	79	30	48	Wrought	Valves, pumps, shafts, marine fixtures and fasteners, electrical and petroleum refining equipment.
N05500	Age-hardened Ni–Cu alloy (Monel K 500)	65 Ni; 30 Cu, 2 Fe, 3 Al + Ti	160	111	24	Wrought	Pump shafts, impellers, springs, fasteners, and electronic and oil well components.
N06002	Ni–Cr Alloy (Hastelloy X)	60 Ni; 22 Cr, 19 Fe, 9 Mo, 0.6 W	114	52	43	Wrought	Turbine and furnace parts, petrochemical equipment.
N06003	Ni–Cr alloy (Nichrome V)	80 Ni; 20 Cr	100	60	30	Wrought	Heating elements, resistors, electronic parts.
N06333	Ni–Cr alloy (RA 333)	48 Ni; 25 Cr, 18 Fe, 3 Mo, 3 W, 3 Co	100	50	50	Wrought	Turbine and furnace parts.
N06600	Ni–Cr alloy (Inconel 600)	75 Ni; 15 Cr, 10 Fe	90	36	47	Wrought	Chemical, electronic, food processing and heat treating equipment; nuclear steam generator tubing.
N06625	Ni–Cr alloy (Inconel 625)	61 Ni; 21 Cr, 2 Fe, 9 Mo, 4 Nb	142	86	42	Wrought	Turbine parts, marine and chemical equipment.
N07001	Age-hardened Ni–Cr alloy (Waspalloy)	58 Ni; 20 Cr, 14 Co, 4 Mo, 3 Al, 1.3 Ti, B, Zr	185	115	25	Wrought	Turbine parts.
N07500	Age-hardened Ni–Cr alloy (Udimet 500)	52 Ni; 18 Cr, 19 Co, 4 Mo, 3 Al, 3 Ti, B, Zr	176	110	16	Wrought & Cast	Turbine parts.
N07750	Age-hardened Ni–Cr alloy (Inconel X-750)	73 Ni; 16 Cr, 7 Fe, 2.5 Ti, 1 Al, 1 Nb	185	130	20	Wrought	Turbine parts, nuclear reactor springs, bolts, extrusion dies, forming tools.

Table 8 (Concluded). **Common Cast and Wrought Nickel and High Nickel Alloys — Designations, Composition, Typical Properties, and Uses**

UNS Designation	Description and Common Name	Nominal Composition (Weight %)	Typical Room-Temperature Properties			Form	Typical Uses
			Tensile (ksi)	0.2% Yield (ksi)	Elong. (%)		
N08800	Ni–Cr–Fe alloy (Incoloy 800)	32 Ni, 21 Cr, 46 Fe, 0.4 Ti, 0.4 Al	87	42	44	Wrought	Heat exchangers, furnace parts, chemical and power plant piping.
N08825	Ni–Cr–Fe alloy (Incoloy 825)	42 Ni, 22 Cr, 30 Fe, 3 Mo, 2 Cu, 1 Ti, Al	91	35	50	Wrought	Heat treating and chemical handling equipment.
N09901	Age-hardened Ni–Cr–Fe alloy (Incoloy 901)	43 Ni, 12 Cr, 36 Fe, 6 Mo, 3 Ti+ Al, B	175	130	14	Wrought	Turbine parts.
N10001	Ni–Mo alloy (Hastelloy B)	67 Ni, 28 Mo, 5 Fe	121	57	63	Wrought	Chemical handling equipment.
N10004	Ni–Cr–Mo alloy (Hastelloy W)	59 Ni, 5 Cr, 25 Mo, 5 Fe, 0.6 V	123	53	55	Wrought	Weld wire for joining dissimilar metals, engine repair and maintenance.
N10276	Ni–Cr–Mo alloy (Hastelloy C-276)	57 Ni, 15 Cr, 16 Mo, 5 Fe, 4 W, 2 Co	116	52	60	Wrought	Chemical handling equipment.
N13100	Ni–Co alloy (IN 100)	60 Ni, 10 Cr, 15 Co, 3 Mo, 5.5 Al, 5 Ti, 1 V, B, Zr	147	123	9	Cast	Turbine parts.

Table 9. Mechanical Properties of Wrought Titanium Alloys

Nominal Composition (%)	Condition	Tensile Strength (ksi)	Yield Strength (ksi)	Elonga-tion (%)	Reduc-tion in Area (%)
			Room Temperature		
Commercially Pure					
99.5 Ti	Annealed	48	35	30	55
99.2 Ti	Annealed	63	50	28	50
99.1 Ti	Annealed	75	65	25	45
99.0 Ti	Annealed	96	85	20	40
99.2 Ti[a]	Annealed	63	50	28	50
98.9[b]	Annealed	75	65	25	42
Alpha Alloys					
5 Al, 2.5 Sn	Annealed	125	117	16	40
5 Al, 2.5 Sn (low O₂)	Annealed	117	108	16	...
Near Alpha Alloys					
8 Al, 1 Mo, 1 V	Duplex annealed	145	138	15	28
11 Sn, 1 Mo, 2.25 Al, 5.0 Zr, 1 Mo, 0.2 Si	Duplex annealed	160	144	15	35
6 Al, 2 Sn, 4 Zr, 2 Mo	Duplex annealed	142	130	15	35
5 Al, 5 Sn, 2 Zr, 2 Mo, 0.25 Si	975°C (1785°F) (½ h), AC + 595°C (1100°F) (2 h), AC	152	140	13	...
6 Al, 2 Nb, 1 Ta, 1 Mo	As rolled 2.5 cm (1 in.) plate	124	110	13	34
6 Al, 2 Sn, 1.5 Zr, 1 Mo, 0.35 Bi, 0.1 Si	Beta forge + duplex anneal	147	137	11	...
Alpha–Beta Alloys					
8 Mn	Annealed	137	125	15	32
3 Al, 2.5 V	Annealed	100	85	20	...
6 Al, 4 V	Annealed	144	134	14	30
	Solution + age	170	160	10	25
6 Al, 4 V (low O₂)	Annealed	130	120	15	35
6 Al, 6 V, 2 Sn	Annealed	155	145	14	30
	Solution + age	185	170	10	20
7 Al, 4 Mo	Solution + age	160	150	16	22
6 Al, 2 Sn, 4 Zr, 6 Mo	Solution + age	184	170	10	23
6 Al, 2 Sn, 2 Zr, 2 Mo, 2 Cr, 0.25 Si	Solution + age	185	165	11	33
10 V, 2 Fe, 3 Al	Solution + age	185	174	10	19
Beta Alloys					
13 V, 11 Cr, 3 Al	Solution + age	177	170	8	...
	Solution + age	185	175	8	...
8 Mo, 8 V, 2 Fe, 3 Al	Solution + age	190	180	8	...
3 Al, 8 V, 6 Cr, 4 Mo, 4 Zr	Solution + age	210	200	7	...
	Annealed	128	121	15	...
11.5 Mo, 6 Zr, 4.5 Sn	Solution + age	201	191	11	...

[a] Also contains 0.2 Pd.   [b] Also contains 0.8 Ni and 0.3 Mo.
Source: Titanium Metals Corp. of America and RMI Co.

Copper–Silicon and Copper–Beryllium Alloys

Everdur. — This copper–silicon alloy is available in five slightly different nominal compositions for applications which require high strength, good fabricating and fusing qualities, immunity to rust, free-machining and a corrosion resistance equivalent to copper. The following table gives the nominal compositions and tensile strengths, yield strengths, and per cent elongations for various tempers and forms.

Uses: (1010) Hot-rolled-and-annealed plates for unfired pressure vessels, and rods for hot forging, hot upsetting, and machining. (1015) Cold-headed-and-roll-threaded bolts and cold-drawn seamless tubes for electrical metallic tubing and rigid conduit. (1012) Screw machine products. (1000) Castings. (1014) Hot forgings and for free machining applications; not for cold working or welding.

Table 10. Nominal Composition and Properties of Everdur

Desig. No.	Nominal Composition					Temper[a]	Strength		Elongation (%)
	Cu	Si	Mn	Pb	Al		Tensile (ksi)	Yield (ksi)	
655	95.80	3.10	1.10	. . .	. . .	A	52	15	35[b]
						HRA	50	18	40
						CRA	52	18	35
						CRHH	71	40	10
						CRH	87	60	3
						H	70 to 85	38 to 50	17 to 8[b]
651	98.25	1.50	0.25	. . .	. . .	AP	38	10	35
						HP	50	40	7
						XHB	75 to 85	45 to 55	8 to 6[b]
661	95.60	3.00	1.00	0.40	. . .	A	52	15	35[b]
						H	85	50	13 to 8[b]
6552	94.90	4.00	1.10	. . .	. . .	AC	45	. . .	15
637	90.75	2.00	. . .	. . .	7.25	A	75 to 90	37.5 to 45	12 to 9[b]

Designation numbers are those of The American Brass Co.

The values given for the tensile strength, yield strength and elongation are all minimum values. Where ranges are shown, the first values given are for the largest diameter or largest size specimens. Yield strength values were determined at 0.50 per cent elongation under load.

[a] Symbols used are: HRA for hot-rolled and annealed tank plates; CRA for cold-rolled sheets and strips; CRHH for cold-rolled half hard strips; and CRH for cold-rolled hard strips. For round, square, hexagonal, and octagonal rods: A for annealed; H for hard; and XHB for extra-hard bolt temper (in coils for cold-heading). For pipe and tube: AP for annealed; and HP for hard. For castings: AC for as cast.

[b] Per cent elongation in 4 times the diameter or thickness of the specimen. All other values are per cent elongation in 2 inches.

Beryllium Copper. — These alloys which contain copper, beryllium, cobalt and in the case of one alloy, silver, fall into two groups. One group whose beryllium content is greater than one per cent is characterized by its high strength and hardness and the other, whose beryllium content is less than 1 per cent, by its high electrical and thermal conductivity. The alloys have many applications in the electrical and aircraft industries or wherever strength, corrosion resistance, conductivity, non-magnetic and non-sparking properties are essential. Beryllium copper is obtainable in the form of strips, rods and bars, wire, platers bars, billets, tubes, and casting ingots.

Composition and Properties: Table 11 lists some of the more common wrought alloys and gives some of their mechanical properties.

Table 11. Wrought Beryllium–Copper Properties

Alloy[a]	Form	Temper[b]	Heat Treatment	Tensile Strength (ksi)	Yield Strength 0.2% Offset (ksi)	Elongation in 2 in. (%)
25	Rod, Bar, and Plate	A	. . .	60–85	20–30	35–60
		½ H or H	. . .	85–130	75–105	10–20
		AT	3 hr at 600°F or mill heat treated	165–190	145–175	3–10
		½ HT or HT	2 hr at 600°F or mill heat treated	175–215	150–200	2–5
	Wire	A	. . .	58–78	20–35	35–55
		¼ H	. . .	90–115	70–95	10–35
		½ H	. . .	110–135	90–110	5–10
		¾ H	. . .	130–155	110–135	2–8
		AT	3 hr at 600°F	165–190	145–175	3–8
		¼ HT	2 hr at 600°F	175–205	160–190	2–5
		½ HT	2 hr at 600°F	190–215	175–200	1–3
		¾ HT	2 hr at 600°F	195–220	180–205	1–3
		XHT	Mill heat treated	115–165	95–145	2–8
165	Rod, Bar, and Plate	A	. . .	60–85	20–30	35–60
		½ H or H	. . .	85–130	75–105	10–20
		AT	3 hr at 650°F or mill heat treated	150–180	125–155	4–10
		½ HT or HT	2 hr at 650°F or mill heat treated	165–200	135–165	2–5
10	Rod, Bar, and Plate	A	. . .	35–55	20–30	20–35
		½ H or H	. . .	65–80	55–75	10–15
		AT	3 hr at 900°F or mill heat treated	100–120	80–100	10–25
		½ HT or HT	2 hr at 900°F or mill heat treated	110–130	100–120	8–20
50	Rod, Bar, and Plate	A	. . .	35–55	20–30	20–35
		½ H or H	. . .	65–80	55–75	10–15
		AT	3 hr at 900°F or mill heat treated	100–120	80–100	10–25
		½ HT or HT	2 hr at 900°F or mill heat treated	110–130	100–120	8–20
35	Rod, Bar, and Plate	A	. . .	35–55	20–30	20–35
		½ H or H	. . .	65–80	55–75	10–15
		AT	3 hr at 900°F or mill heat treated	100–120	80–100	10–25
		½ HT or HT	2 hr at 900°F or mill heat treated	110–130	100–120	8–20

[a] Composition (in per cent) of alloys is as follows: alloy 25: 1.80–2.05 Be, 0.20–0.35 Co, balance Cu; alloy 165: 1.6–1.8 Be, 0.20–0.35 Co, balance Cu; alloy 10: 0.4–0.7 Be, 2.35–2.70 Co, balance Cu; alloy 50, 0.25–0.50 Be, 1.4–1.7 Co, 0.9–1.1 Ag, balance Cu; alloy 35, 0.25–0.50 Be, 1.4–1.◼ Ni, balance Cu.
[b] Temper symbol designations: A, solution annealed; H, hard; HT, heat treated from hard; AT heat treated from solution annealed.

Heat-treatment of Non-Ferrous Alloys

The solution and precipitation methods of heat-treatment may be applied to certain non-ferrous alloys such as wrought aluminum and also to some of the magnesium or Dowmetal alloys (see page 606).

Wrought Aluminum Alloys. — The wrought alloys of aluminum may be divided into two classes depending upon the manner in which their harder tempers are produced. One class comprises the alloys in which strain-hardening, by definite amounts of cold work following the last annealing operation, produces the varying degrees of strength and hardness. The alloys in the other class depend primarily upon heat-treatment processes to develop their higher mechanical properties. While there is a wide range of tensile properties in both classes of alloys, the highest combinations of strength and ductility available in the widest range of products are to be found in the heat-treated alloys. In the aluminum alloys which respond to heat-treatment, the alloying constituents which give the increased strength and hardness are substances which are more soluble in solid aluminum at high temperatures than at low temperatures.

Solution Heat-treatment. — The first step in heat-treatment, frequently called the "solution heat-treatment," consists in heating the alloy to some temperature below the melting point, usually in the range of 900 to 1000 degrees F. for aluminum alloys, to put as much as possible of the alloying constituent into solid solution. The alloy is held at this temperature for some period of time, usually from 20 to 60 minutes for aluminum alloys, according to the thickness of the piece. This permits the entire piece to reach a uniform temperature and the dissolving of the alloying elements in the solid solution to take place throughout. In effect, the alloying constituent has been dissolved in the aluminum and dispersed as completely as when sugar is dissolved in water. The alloy is then quenched and — in contrast to steel after quenching — is in a relatively soft condition.

Precipitation Heat-treatment. — After quenching, the alloy undergoes an aging process which, if carried out at elevated temperatures, is called a "precipitation heat-treatment," because during this stage some of the alloying constituent which is held in solid solution precipitates from the solid solution in the form of extremely fine particles. This precipitation may occur spontaneously at room temperature, as is the case in the so-called "natural aging" of certain alloys, or it may require a "precipitation heat-treatment" or "artificial aging" at about 300 degrees F., in the case of certain other alloys to produce increased hardness and tensile strength.

Heat Treatment of Copper Alloys. — Precipitation hardening of copper alloys is useful in producing materials which have high strength and high electrical conductivity. In the case of beryllium copper alloys the most favorable properties are obtained through the use of both the solution and precipitation methods of heat-treatment. The solution heat-treatment is generally accomplished by heating and keeping the alloy at temperatures ranging from 1450 to 1650 degrees F. in a circulating air furnace for a certain length of time and then water quenching it. The temperature and length of time depend upon the composition of the alloy and physical dimensions of the part being treated. The precipitation heat-treatment may then be accomplished by heating and holding the alloy at a temperature from 600 to 900 degrees F. in a circulating air furnace (salt baths are not recommended) for 3 hours followed by an uncontrolled air cooling. The temperature to be used depends upon the composition of the alloy. Where special physical properties are desired, the times and temperatures may be varied.

American National Standard Drafting Practices

Several American National Standards for use in preparing engineering drawings and related documents are referred to for use.

Sizes of Drawing Sheets. — Recommended trimmed sheet sizes, based on ANSI Y14.1-1980, are shown in the following table.

Size, Inches				Metric Size, mm.			
A	8½ × 11	D	22 × 34	A0	841 × 1189	A3	297 × 420
B	11 × 17	E	34 × 44	A1	594 × 841	A4	210 × 297
C	17 × 22	F	28 × 40	A2	420 × 594		

The standard sizes shown by the left-hand section of the table are based on the dimensions of the commercial letter head, 8½ × 11 inches, in general use in the United States. The use of the basic sheet size 8½ × 11 inches and its multiples permits filing of small tracings and folded blueprints in commercial standard letter files with or without correspondence. These sheet sizes also cut without unnecessary waste from the present 36 inch rolls of paper and cloth.

For drawings made in the metric system of units or for foreign correspondence it is recommended that the metric standard trimmed sheet sizes be used. (Right-hand section of table.) These sizes are based on the width to length ratio of 1 to $\sqrt{2}$.

Line Conventions and Drawings. — American National Standard Y14.2M-1979 establishes line and lettering practices for engineering drawings. The line conventions and the symbols for section lining are as shown on pages 615 and 616.

Surface Texture Symbols. — A detailed explanation of the use of surface texture symbols from American National Standard Y14.36-1978 begins on page 707.

Geometric Dimensioning and Tolerancing. — American National Standard Y14.5M-1982, "Dimensioning and Tolerancing," covers dimensioning, tolerancing, and similar practices for engineering drawings and related documents; excluded are practices related to architectural and civil engineering.

Geometric tolerancing provides a comprehensive system for the accurate transmission of design specifications. Several techniques introduced in this Standard have been accepted by the ISO. (A comparison of the symbols used in ISO standards and Y14.5M is given on page 618.) These techniques include projected tolerance zone, three-plane datum concept, total runout tolerance, multiple datums, and datum targets. Although this Standard follows ISO practice closely, there are still differences between ISO and U.S. practice. One major area of disagreement is the ISO "principle of interdependency" versus the "Taylor principle." Standard Y14.5M and U.S. practice both follow the Taylor principle, in which a boundary of perfect form at MMC (maximum material condition) is prescribed to control variations as well as the size of individual features.

The International System of Units (SI) is featured in Y14.5M, but customary units may be used without violating any principles. On drawings where all dimensions are either in millimeters or in inches, individual identification of linear units is not required. However, the drawing should contain a note stating UNLESS OTHERWISE SPECIFIED, ALL DIMENSIONS ARE IN MILLIMETERS (or IN INCHES, as applicable). Angular units are expressed in degrees and decimals of a degree ($35.4°$) or in degrees (°), minutes ('), and seconds (") ($35° 25' 10''$).

A 90 degree angle is implied where center lines and depicting features are shown on a drawing at right angles and no angle is specified. A 90 degree BASIC angle

American National Standard for Engineering Drawings (ANSI Y14.2M-1979)

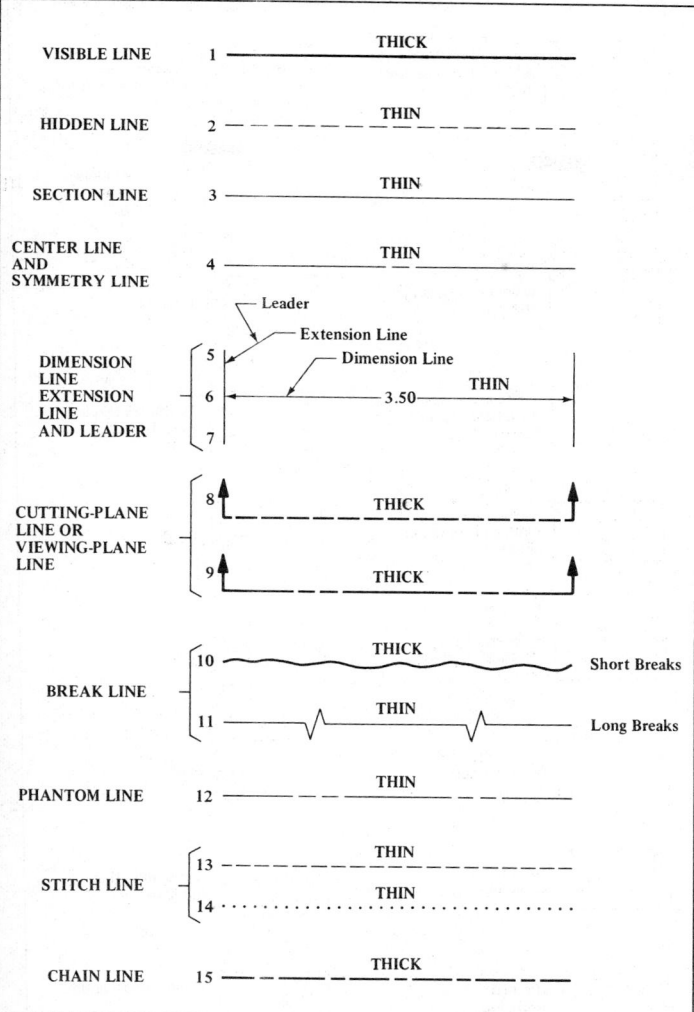

Approximate width of THICK lines for metric drawings, 0.7 mm and for inch drawings, 0.032 inch. Approximate width of THIN lines for metric drawings, 0.35 mm and for inch drawings, 0.016 inch. These approximate line widths are intended to differentiate between THICK and THIN lines and are not values for control of acceptance or rejection of the drawings.

American National Standard Symbols for Section Lining (ANSI Y14.2M-1979)

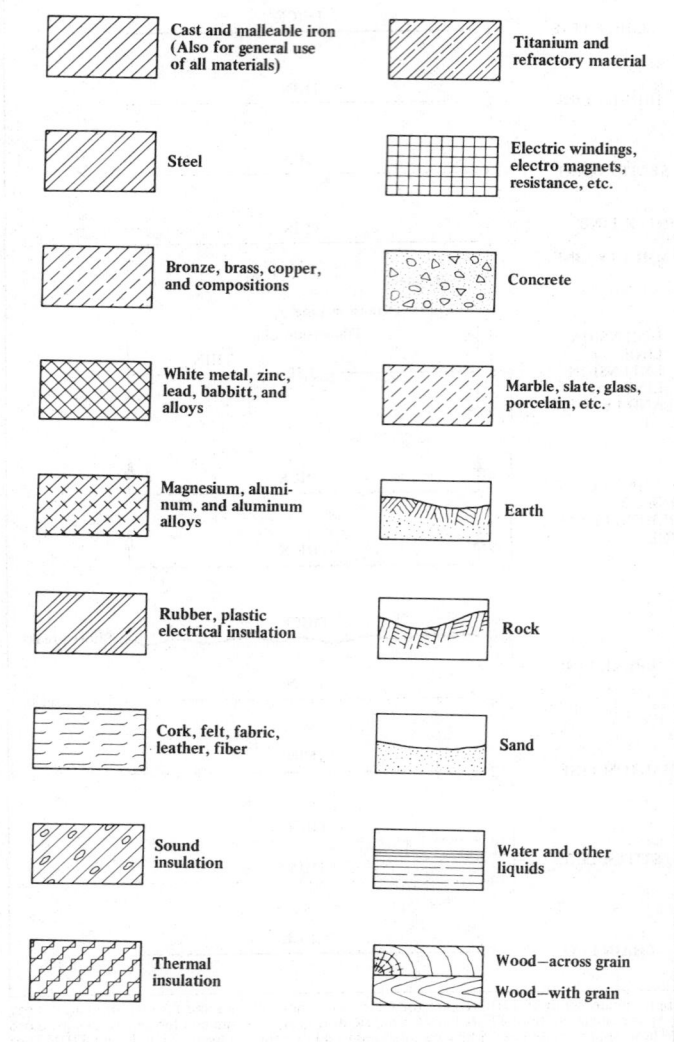

Cast and malleable iron (Also for general use of all materials)

Titanium and refractory material

Steel

Electric windings, electro magnets, resistance, etc.

Bronze, brass, copper, and compositions

Concrete

White metal, zinc, lead, babbitt, and alloys

Marble, slate, glass, porcelain, etc.

Magnesium, aluminum, and aluminum alloys

Earth

Rubber, plastic electrical insulation

Rock

Cork, felt, fabric, leather, fiber

Sand

Sound insulation

Water and other liquids

Thermal insulation

Wood–across grain
Wood–with grain

applies where center lines of features in a pattern or surface shown at right angles on a drawing are located or defined by basic dimensions and no angle is specified.

According to this Standard, all dimensions are applicable at a temperature of 20°C (68°F) unless otherwise specified. Compensation may be made for measurements taken at other temperatures.

Definitions. — The following terms are defined as their use applies to Y14.5M.

Actual Size: The measured size.

Basic Dimension: A numerical value used to describe the theoretically exact size, profile, orientation, or location of a feature or datum target. It is the basis from which permissible variations are established by tolerances on other dimensions, in notes, or in feature control frames.

Datum: A theoretically exact point, axis, or plane derived from the true geometric counterpart of a specified datum feature. A datum is the origin from which the location or geometric characteristics of features of a part are established.

Datum Feature: An actual feature of a part that is used to establish a datum.

Datum Target: A specified point, line, or area on a part used to establish a datum.

Feature: The general term applied to a physical portion of a part, such as a surface, hole, or slot.

Feature of Size: One cylindrical or spherical surface, or a set of two plane parallel surfaces, each of which is associated with a size dimension.

Geometric Tolerance: The general term applied to the category of tolerances used to control form, profile, orientation, location, and runout.

Least Material Condition (LMC): The condition in which a feature of size contains the least amount of material within the stated limits of size — for example, maximum hole diameter, minimum shaft diameter.

Maximum Material Condition (MMC): The condition in which a feature of size contains the maximum amount of material with the stated limits of size — for example, minimum hole diameter, maximum shaft diameter.

Reference Dimension: A dimension, usually without tolerance, used for information purposes only. It is considered auxiliary information and does not govern production or inspection operations. A reference dimension is a repeat of a dimension or is derived from other values shown on the drawing or on related drawings.

Regardless of Feature Size (RFS): The term used to indicate that a geometric tolerance or datum reference applies at any increment of size of the feature within its size tolerance.

True Position: The theoretically exact location of a feature established by basic dimensions.

Virtual Condition: The boundary generated by the collective effects of the specified maximum material condition limit of size of a feature and any applicable geometric tolerances.

Datum Referencing. — A datum indicates the origin of a dimensional relationship between a toleranced feature and a designated feature or features on a part. The designated feature serves as a datum feature, whereas its true geometric counterpart establishes the datum.

Since measurements cannot be made from a true geometric counterpart that is theoretical, a datum is assumed to exist in and be simulated by the associated processing equipment.

For example, machine tables and surface plates, although not true planes, are of such quality that they are used to simulate the datums from which measurements are taken and dimensions are verified. When magnified, flat surfaces of manufactured parts are seen to have irregularities; contact is made with a datum plane at a number of surface extremities or high points.

Comparison of American National Standard Geometric Characteristic and Modifying Symbols with ISO Standard Symbols (ANSI Y14.5M-1982)

Symbol for	ANSI Y14.5	ISO
Straightness	⎯	⎯
Flatness	▱	▱
Circularity	○	○
Cylindricity	⌭	⌭
Profile of a Line	⌒	⌒
Profile of a Surface	⌓	⌓
All Around—Profile	⌾	**NONE**
Angularity	∠	∠
Perpendicularity	⊥	⊥
Parallelism	//	//
Position	⊕	⊕
Concentricity/Coaxiality	◎	◎
Symmetry	**NONE**	≡
Circular Runout	* ↗	↗
Total Runout	* ↗↗	↗↗
At Maximum Material Condition	Ⓜ	Ⓜ
At Least Material Condition	Ⓛ	**NONE**
Regardless of Feature Size	Ⓢ	**NONE**
Projected Tolerance Zone	Ⓟ	Ⓟ
Diameter	∅	∅
Basic Dimension	50	50
Reference Dimension	(50)	(50)
Datum Feature	-A-	* OR * A
Datum Target	Ⓐ⑥/Ⓐ₁	Ⓐ⑥/Ⓐ₁
Target Point	✕	✕
Dimension Origin	⊕→	**NONE**
Feature Control Frame	⊕ ∅ 0.5 Ⓜ A B C	⊕ ∅ 0.5 Ⓜ A B C
Conical Taper	▷	▷
Slope	◿	◿
Counterbore/Spotface	⌴	**NONE**
Countersink	⌵	**NONE**
Depth/Deep	↧	**NONE**
Square (Shape)	□	□
Dimension Not to Scale	15	15
Number of Times/Places	8X	8X
Arc Length	⏜105	**NONE**
Radius	R	R
Spherical Radius	SR	**NONE**
Spherical Diameter	S∅	**NONE**

* Arrowheads may be filled in.

Some American National Standard Symbols for Datum Referencing in Engineering Drawings (ANSI Y14.5M-1982)

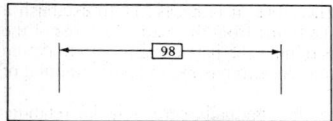

Fig. 1. Basic Dimension Symbol

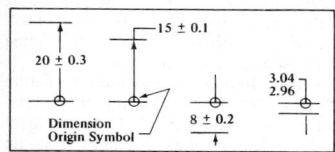

Fig. 2. Dimension Origin Symbol

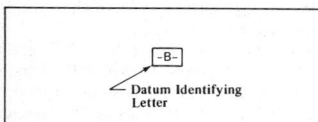

Fig. 3. Datum Feature Symbol

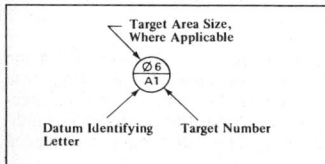

Fig. 4. Datum Target Symbol

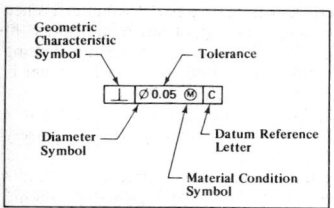

Fig. 5. Feature Control Frame Incorporating a Datum Reference

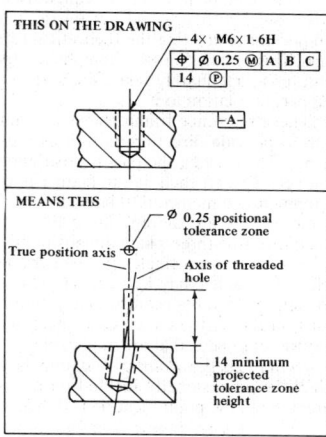

Fig. 6. Feature Control Frame with a Projected Tolerance Zone

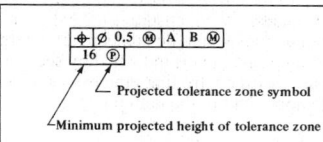

Fig. 7. Projected Tolerance Zone Specified

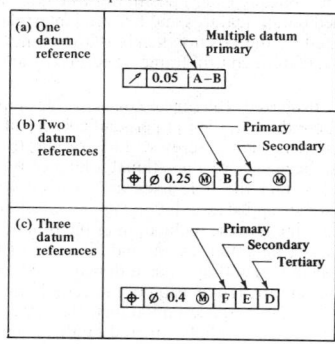

Fig. 8. Order of Precedence of Datum References

Datum Reference Frame: Sufficient datum features, those most important to the design of the part, are chosen to position the part in relation to a set of three mutually perpendicular planes, jointly called a datum reference frame. This reference frame exists only in theory and not on the part. Therefore, it is necessary to establish a method for simulating the theoretical reference frame from the actual features of the part. This simulation is accomplished by positioning the part on appropriate datum features to adequately relate the part to the reference frame and to restrict motion of the part in relation to it.

These reference frame planes are simulated in a mutually perpendicular relationship to provide direction as well as the origin for related dimensions and measurements. Thus, when the part is positioned on the datum reference frame (by physical contact between each datum feature and its counterpart in the associated processing equipment), dimensions related to the datum reference frame by a feature control frame or note are thereby mutually perpendicular. This theoretical reference frame constitutes the three-plane dimensioning system used for datum referencing.

In some cases a single reference frame will suffice. In other cases additional datum reference frames may be necessary where physical separation or the functional relationship of features require that datum reference frames be applied at specific locations on the part. In such cases each feature control frame must contain the datum feature references that are applicable.

Datum Target Points: A datum target point is indicated by the symbol "X," which is dimensionally located on a direct view of the surface. Where there is no direct view, the point location is dimensioned on two adjacent views.

Datum Target Lines: A datum target line is indicated by the symbol "X" on an edge view of the surface, a phantom line on the direct view, or both. Where the length of the datum target line must be controlled, its length and location are dimensioned.

Datum Target Areas: Where it is determined that an area or areas of flat contact are necessary to ensure establishment of the datum (that is, where spherical or pointed pins would be inadequate), a target area of the desired shape is specified. The datum target area is indicated by section lines inside a phantom outline of the desired shape, with controlling dimensions added. The diameter of circular areas is given in the upper half of the datum target symbol. (See Fig. 4.)

Positional Tolerance. — A positional tolerance defines a zone within which the center, axis, or center plane of a feature of size is permitted to vary from true (theoretically exact) position. Basic dimensions establish the true position from specified datum features and between interrelated features. A positional tolerance is indicated by the position symbol, a tolerance, and appropriate datum references placed in a feature control frame. (See Figs. 5 and 8.)

Projected Tolerance Zone. — The application of this concept is recommended where the variation in perpendicularity of the threaded or press-fit holes could cause fasteners such as screws, studs, or pins to interfere with mating parts. An interference can occur where a positional tolerance is applied to the depth of threaded or press-fit holes, and the hole axes are inclined within allowable limits. Unlike the floating fastener application involving clearance holes only, the attitude of a fixed fastener is restrained by the inclination of the produced hole into which it assembles.

The minimum extent and direction of the projected tolerance zone is shown in a drawing view (Fig. 7) as a dimensioned value with a heavy chain line drawn closely adjacent to an extension of the center line of the hole.

Owing to space limitations, the above are only a few of the numerous concepts and related symbols covered by this extensive standard. ANSI Y14.5M-1982 should be referred to for a complete discussion and many examples of the application of geometric dimensioning and tolerancing principles.

Checking Drawings. — In order that the drawings may have a high standard of excellence, a set of instructions, as given in the following, has been issued to the checkers, and also to the draftsmen and tracers in the engineering department of a well-known machine-building company.

Inspecting a New Design: In case a new design is involved, first inspect the lay-outs carefully to see that the parts function correctly under all conditions, that they have the proper relative proportions, that the general design is correct in the matters of strength, rigidity, bearing areas, appearance, convenience of assembly, and direction of motion of the parts, and that there are no interferences. Consider the design as a whole to see if any improvements can be made. If the design appears to be unsatisfactory in any particular, or improvements appear to be possible, call the matter to the attention of the chief engineer.

Checking for Strength: Inspect the design of the part being checked for strength, rigidity, and appearance by comparing it with other parts for similar service whenever possible, giving preference to the later designs in such comparison, unless the later designs are known to be unsatisfactory. If there is any question regarding the matter, compute the stresses and deformations or find out whether the chief engineer has approved the stresses or deformations that will result from the forces applied to the part in service. In checking parts that are to go on a machine of increased size, be sure that standard parts used in similar machines and proposed for use on the larger machine, have ample strength and rigidity under the new and more severe service to which they will be put.

Materials Specified. — Consider the kind of material required for the part and the various possibilities of molding, forging, welding, or otherwise forming the rough part from this material. Then consider the machining operations to see whether changes in form or design will reduce the number of operations or the cost of machining.

See that parts are designed with reference to the economical use of material, and whenever possible, utilize standard sizes of stock and material readily obtainable from local dealers. In the case of alloy steel, special bronze, and similar materials, be sure that the material can be obtained in the size required.

Method of Making Drawing. — Inspect the drawing to see that the projections and sections are made in such a way as to show most clearly the form of the piece and the work to be done on it. Make sure that any worker looking at the drawing will understand what the shape of the piece is and how it is to be molded or machined. Make sure that the delineation is correct in every particular, and that the information conveyed by the drawing as to the form of the piece is complete.

Checking Dimensions: — Check all dimensions to see that they are correct. Scale all dimensions and see that the drawing is to scale. See that the dimensions on the drawing agree with the dimensions scaled from the lay-out. Wherever any dimension is out of scale, see that the dimension is so marked. Investigate any case where the dimension, the scale of the drawing, and the scale of the lay-out do not agree. All dimensions not to scale must be underlined on the tracing. In checking dimensions, note particularly the following points:

See that all figures are correctly formed and that they will print clearly, so that the workers can easily read them correctly.

See that the over-all dimensions are given.

See that all witness lines go to the correct part of the drawing.

See that all arrow points go to the correct witness lines.

See that proper allowance is made for all fits.

See that the tolerances are correctly given where necessary.

See that all dimensions given agree with the corresponding dimensions of adjacent parts.

Be sure that the dimensions given on a drawing are those that the machinist will use, and that the worker will not be obliged to do addition or subtraction in order to obtain the necessary measurements for machining or checking his work.

Avoid strings of dimensions where errors can accumulate. It is generally better to give a number of dimensions from the same reference surface or center line.

When holes are to be located by boring on a horizontal spindle boring machine or other similar machine, give dimensions to centers of bored holes in rectangular coordinates and from the center lines of the first hole to be bored, so that the operator will not be obliged to add measurements or transfer gages.

Checking Assembly. — See that the part can readily be assembled with the adjacent parts. If necessary, provide tapped holes for eyebolts and cored holes for tongs, lugs, or other methods of handling.

Make sure that, in being assembled, the piece will not interfere with other pieces already in place and that the assembly can be taken apart without difficulty.

Check the sum of a number of tolerances; this sum must not be great enough to permit two pieces that should not be in contact to come together.

Checking Castings. — In the case of castings, study the form of the pattern, the methods of molding, the method of supporting and venting the cores, and the effect of draft and rough molding on clearances.

Avoid undue metal thickness, and especially avoid thick and thin sections in the same casting.

Indicate all metal thicknesses, so that the molder will know what chaplets to use for supporting the cores.

See that ample fillets are provided, and that they are properly dimensioned.

See that the cores can be assembled in the mold without crushing or interference.

See that swelling, shrinkage, or misalignment of cores will not make trouble in machining.

See that the amount of extra material allowed for finishing is indicated.

See that there is sufficient extra material for finishing on large castings to permit them to be "cleaned up," even though they warp. In the case of such castings, make sure that the metal thickness will be sufficient after finishing, even though the castings do warp.

Make sure that sufficient sections are shown so that the patternmakers and molders will not be compelled to make assumptions about the form of any part of the casting. These details are particularly important when a number of sections of the casting are similar in form, while others differ slightly.

Checking Machined Parts. — Study the sequences of operations in machining and see that all finish marks are indicated.

See that the finish marks are placed on the lines to which dimensions are given.

See that methods of machining are indicated where necessary.

Give all drill, reamer, tap, and rose bit sizes.

See that jig and gage numbers are indicated at the proper places.

See that all necessary bosses, lugs, and openings are provided for lifting, handling, clamping, and machining the piece.

See that adequate wrench room is provided for all nuts and bolt heads.

Avoid special tools, such as taps, drills, reamers, etc., unless such tools are specially authorized.

Where parts are right- and left-hand, be sure that the hand is correctly designated. When possible, mark parts symmetrical, so as to avoid having them right- and left-hand, but do not sacrifice correct design or satisfactory operation on this account.

When heat-treatment is required, the heat-treatment should be specified.

Check the title, size of machine, the scale, and the drawing number on both the drawing and the drawing record card.

ALLOWANCES AND TOLERANCES FOR FITS

Limits and Fits. — Fits between cylindrical parts or, briefly cylindrical fits, govern the proper assembly and performance of countless mechanisms. Clearance fits permit relative freedom of motion between a shaft and a hole — axially, radially, or both. Interference fits secure a certain amount of tightness between parts, whether these are meant to remain permanently assembled or to be taken apart from time to time. Or again, two parts may be required to fit together snugly — without apparent tightness or looseness. The designer's problem is to specify these different types of fits in such a way that the shop can produce them. This involves the adoption of two manufacturing limits for the hole and two for the shaft, and, hence, the adoption of a manufacturing tolerance on each part.

In selecting and specifying limits and fits for various applications, it is essential in the interests of interchangeable manufacturing that (1) standard definitions of terms relating to limits and fits be used; (2) preferred basic sizes be selected wherever possible to reduce material and tooling costs; (3) limits be based upon a series of preferred tolerances and allowances; and (4) a uniform system of applying tolerances (preferably unilateral) be used. These principles have been incorporated in both the American and British standards for limits and fits. Information about these standards is given beginning on page 630.

Basic Dimensions. — The basic size of a screw thread or machine part is the theoretical or nominal standard size from which variations are made. For example, a shaft may have a *basic* diameter of 2 inches, but a maximum variation of minus 0.010 inch may be permitted. The minimum hole should be of basic size in all cases where the use of standard tools represents the greatest economy. The maximum shaft should be of basic size in all cases where the use of standard purchased material, without further machining, represents the greatest economy, even though special tools are required to machine the mating part.

Tolerances. — Tolerance is the amount of variation permitted on dimensions or surfaces of machine parts. The tolerance is equal to the difference between the maximum and minimum limits of any specified dimension. For example, if the maximum limit for the diameter of a shaft is 2.000 inches and its minimum limit 1.990 inches, the tolerance for this diameter is 0.010 inch. By determining the maximum and minimum clearances required on operating surfaces, the extent of these tolerances is established. As applied to the fitting of machine parts, the word tolerance means the amount that duplicate parts are allowed to vary in size in connection with manufacturing operations, owing to unavoidable imperfections of workmanship. Tolerance may also be defined as the amount that duplicate parts are permitted to vary in size in order to secure sufficient accuracy without unnecessary refinement. The terms "tolerance" and "allowance" are often used interchangeably, but, according to common usage, *allowance* is a difference in dimensions prescribed in order to secure various classes of fits between different parts.

Unilateral and Bilateral Tolerances. — The term "unilateral tolerance" means that the total tolerance, as related to a basic dimension, is in *one* direction only. For example, if the basic dimension were 1 inch and the tolerance were expressed as 1.00 − 0.002, or as 1.00 + 0.002, these would be unilateral tolerances, since the total tolerance in each case is in one direction. On the contrary, if the tolerance were divided, so as to be partly plus and partly minus, it would be classed as "bilateral." Thus, $1.00 \begin{array}{l} + 0.001 \\ - 0.001 \end{array}$ is an example of bilateral tolerance, because the total tolerance of 0.002 is given in two directions — plus and minus.

When unilateral tolerances are used, one of the three following methods should be used to express them:

(1) Specify limiting dimensions only as

    Diameter of hole: 2.250, 2.252
    Diameter of shaft: 2.249, 2.247

(2) One limiting size may be specified with its tolerances as

    Diameter of hole: 2.250 + 0.002, − 0.000
    Diameter of shaft: 2.249 + 0.000, − 0.002

(3) The nominal size may be specified for both parts, with a notation showing both allowance and tolerance, as

    Diameter of hole: 2¼ + 0.002, − 0.000
    Diameter of shaft: 2¼ − 0.001, − 0.003

Bilateral tolerances should be specified as such, usually with plus and minus tolerances of equal amount. Example of the expression of bilateral tolerances follow:

$$2 \pm 0.001 \quad \text{or} \quad 2 \begin{array}{l} + 0.001 \\ - 0.001 \end{array}$$

Application of Tolerances. — According to common practice, tolerances are applied in such a way as to show the permissible amount of dimensional variation in the direction that is less dangerous. When a variation in either direction is equally dangerous, a bilateral tolerance should be given. When a variation in one direction is more dangerous than a variation in another, a unilateral tolerance should be given in the less dangerous direction.

For non-mating surfaces, or atmospheric fits, the tolerances may be bilateral, or unilateral, depending entirely upon the nature of the variations that develop in manufacture. On mating surfaces, with but few exceptions, the tolerances should be unilateral.

Where tolerances are required on the distances between holes, usually they should be bilateral, as variation in either direction is usually equally dangerous. The variation in the distance between shafts carrying gears, however, should always be unilateral and plus; otherwise the gears might run too tight. A slight increase in the backlash between gears is seldom of much importance.

One exception to the use of unilateral tolerances on mating surfaces occurs when tapers are involved. In such cases either bilateral or unilateral tolerances may prove advisable, depending upon conditions. These should be determined in the same manner as the tolerances on the distances between holes. When a variation either in or out of the position of the mating taper surfaces is equally dangerous, the tolerances should be bilateral. When a variation in one direction is of less danger than a variation in the opposite direction, the tolerance should be unilateral and in the less dangerous direction.

Locating Tolerance Dimensions — Only one dimension in the same straight line can be controlled within fixed limits. That is the distance between the cutting surface of the tool and the locating or registering surface of the part being machined. Therefore, it is incorrect to locate any point or surface with tolerances from more than one point in the same straight line.

Every part of a mechanism must be located in each plane. Every operating part must be located with proper operating allowances. After such requirements of location are met, all other surfaces should have liberal clearances. Dimensions should be given between those points or surfaces that it is essential to hold in a specific relation to each other. This applies particularly to those surfaces in each plane which control the location of other component parts. Many dimensions are relatively unimportant in this respect. It is good practice in such cases to establish a common locating-point in each plane and give, so far as possible all such dimensions from these common locating-points. The locating points on the drawing, the locat-

ing or registering points used for machining the surfaces and the locating points for measuring should all be identical.

The initial dimensions placed on component drawings should be the exact dimensions that would be used if it were possible to work without tolerances. Tolerances should be given in that direction in which variations will cause the least harm or danger. When a variation in either direction is equally dangerous, the tolerances should be of equal amount in both directions, or bilateral. The initial clearance, or allowance, between operating parts should be as small as the operation of the mechanism will permit. The maximum clearance should be as great as the proper functioning of the mechanism will permit.

Direction of Tolerances on Gages. — The extreme sizes for all plain limit gages shall not exceed the extreme limits of the part to be gaged. All variations in the gages, whatever their cause or purpose, shall bring these gages within these extreme limits.

The data for gage tolerances on page 658 cover gages to inspect workpieces held to tolerances in the American National Standard ANSI B4.2-1978.

Allowance for Forced Fits. — The allowance per inch of diameter usually ranges from 0.001 inch to 0.0025 inch, 0.0015 being a fair average. Ordinarily the allowance per inch decreases as the diameter increases; thus the total allowance for a diameter of 2 inches might be 0.004 inch, whereas for a diameter of 8 inches the total allowance might not be over 0.009 or 0.010 inch. The parts to be assembled by forced fits are usually made cylindrical, although sometimes they are slightly tapered. The advantages of the taper form are that the possibility of abrasion of the fitted surfaces is reduced; that less pressure is required in assembling; and that the parts are more readily separated when renewal is required. On the other hand, the taper fit is less reliable, because if it loosens, the entire fit is free with but little axial movement. Some lubricant, such as white lead and lard oil mixed to the consistency of paint, should be applied to the pin and bore before assembling, to reduce the tendency of abrasion.

Pressure for Forced Fits. — The pressure required for assembling cylindrical parts depends not only upon the allowance for the fit, but also upon the area of the fitted surfaces, the pressure increasing in proportion to the distance that the inner member is forced in. The approximate ultimate pressure in tons can be determined by the use of the following formula in conjunction with the accompanying

Pressure Factors

Diameter, Inches	Pressure Factor	Diameter, Inches	Pressure Factor	Diameter, Inches	Pressure Factor	Diameter, Inches	Pressure Factor	Diameter, Inches	Pressure Factor
1	500	3½	132	6	75	9	48.7	14	30.5
1¼	395	3¾	123	6¼	72	9½	46.0	14½	29.4
1½	325	4	115	6½	69	10	43.5	15	28.3
1¾	276	4¼	108	6¾	66	10½	41.3	15½	27.4
2	240	4½	101	7	64	11	39.3	16	26.5
2¼	212	4¾	96	7¼	61	11½	37.5	16½	25.6
2½	189	5	91	7½	59	12	35.9	17	24.8
2¾	171	5¼	86	7¾	57	12½	34.4	17½	24.1
3	156	5½	82	8	55	13	33.0	18	23.4
3¼	143	5¾	78	8½	52	13½	31.7	. . .	. . .

table of "Pressure Factors." Assuming that A = area of surface in contact in "fit"; a = total allowance in inches; P = ultimate pressure required, in tons; F = pressure factor based upon assumption that the diameter of the hub is twice the diameter of the bore, that the shaft is of machine steel, and that the hub is of cast iron:

$$P = \frac{A \times a \times F}{2}$$

Allowance for Given Pressure. — By transposing the preceding formula, the approximate allowance for a required ultimate tonnage can be determined. Thus, $a = \dfrac{2P}{AF}$. The average ultimate pressure in tons commonly used ranges from 7 to 10 times the diameter in inches.

Expansion Fits. — In assembling certain classes of work requiring a very tight fit, the inner member is contracted by sub-zero cooling to permit insertion into the outer member and a tight fit is obtained as the temperature rises and the inner part expands. In order to obtain the sub-zero temperature, solid carbon dioxide or "dry ice" has been used but its temperature of about 109 degrees F. below zero will not contract some parts sufficiently to permit insertion in holes or recesses. Greater contraction may be obtained by using high purity liquid nitrogen which has a temperature of about 320 degrees F. below zero. During a temperature reduction from 75 degrees F. to —321 degrees F., the shrinkage per inch of diameter varies from about 0.002 to 0.003 inch for steel; 0.0042 inch for aluminum alloys; 0.0046 inch for magnesium alloys; 0.0033 inch for copper alloys; 0.0023 inch for monel metal; and 0.0017 inch for cast iron (not alloyed). The cooling equipment may vary from an insulated bucket to a special automatic unit, depending upon the kind and quantity of work. One type of unit is so arranged that parts are precooled by vapors from the liquid nitrogen before immersion. With another type, cooling is entirely by the vapor method.

Shrinkage Fits. — General practice seems to favor a smaller allowance for shrinkage fits than for forced fits, although in many shops the allowances are practically the same in each case, and for some classes of work, shrinkage allowances exceed those for forced fits. In any case, the shrinkage allowance varies to a great extent with the form and construction of the part which has to be shrunk into place. The thickness or amount of metal around the hole is the most important factor. The way in which the metal is distributed also has an influence on the results. Shrinkage allowances for locomotive driving wheel tires adopted by the American Railway Master Mechanics Association are as follows:

Center diameter, inches	38	44	50	56	62	66
Allowance, inches	0.040	0.047	0.053	0.060	0.066	0.070

Whether parts are to be assembled by forced or shrinkage fits depends upon conditions. For example, to press a tire over its wheel center, without heating, would ordinarily be a rather awkward and difficult job. On the other hand, pins, etc., are easily and quickly forced into place with a hydraulic press and there is the additional advantage of knowing the exact pressure required in assembling, whereas there is more or less uncertainty connected with a shrinkage fit, unless the stresses are calculated. Tests to determine the difference in the quality of shrinkage and forced fits showed that the resistance of a shrinkage fit to slippage was, for an axial pull, 3.66 times greater than that of a forced fit, and in rotation or torsion, 3.2 times greater. In each comparative test, the dimensions and allowances were the same.

Allowances for Shrinkage Fits. — The most important point to consider when calculating shrinkage fits is the stress in the hub at the bore, which depends chiefly upon the shrinkage allowance. If the allowance is excessive, the elastic limit of the material will be exceeded and permanent set will occur, or, in extreme cases, the ultimate strength of the metal will be exceeded and the hub will burst. The intensity of the grip of the fit and the resistance to slippage depends mainly upon the thickness of the hub; the greater the thickness, the stronger the grip, and *vice versa*. Assuming the modulus of elasticity for steel to be 30,000,000, and for cast iron, 15,000,000, the shrinkage allowance per inch of nominal diameter can be determined by the following formula, in which A = allowance per inch of diameter; T = true tangential tensile stress at inner surface of outer member; C = factor taken from one of the accompanying tables, "Factors for Calculating Shrinkage Fit Allowances."

For a cast-iron hub and steel shaft:

$$A = \frac{T(2 + C)}{30,000,000} \qquad (1)$$

When both hub and shaft are of steel:

$$A = \frac{T(1 + C)}{30,000,000} \qquad (2)$$

If the shaft is solid, the factor C is taken from Table 1; if it is hollow and the hub is of steel, factor C is taken from Table 2; if it is hollow and the hub is of cast iron, the factor is taken from Table 3.

Table 1. Factors for Calculating Shrinkage Fit Allowances

Values of factor C for solid steel shafts of nominal diameter D_1, and hubs of steel or cast iron of nominal external and internal diameters D_2 and D_1, respectively.

Ratio of Diameters $\frac{D_2}{D_1}$	Steel Hub	Cast-iron Hub	Ratio of Diameters $\frac{D_2}{D_1}$	Steel Hub	Cast-iron Hub
1.5	0.227	0.234	2.8	0.410	0.432
1.6	0.255	0.263	3.0	0.421	0.444
1.8	0.299	0.311	3.2	0.430	0.455
2.0	0.333	0.348	3.4	0.438	0.463
2.2	0.359	0.377	3.6	0.444	0.471
2.4	0.380	0.399	3.8	0.450	0.477
2.6	0.397	0.417	4.0	0.455	0.482

Example 1: A steel crank web 15 inches outside diameter is to be shrunk on a 10-inch solid steel shaft. Required the allowance per inch of shaft diameter to produce a maximum tensile stress in the crank of 25,000 pounds per square inch, assuming the stresses in the crank to be equivalent to those in a ring of the diameter given.

The ratio of the external to the internal diameters equals $15 \div 10 = 1.5$; $T =$ 25,000 pounds; from Table 1, $C = 0.227$. Substituting in Formula (2):

$$A = \frac{25,000 \times (1 + 0.227)}{30,000,000} = 0.001 \text{ inch.}$$

Example 2: Find the allowance per inch of diameter for a 10-inch shaft having a 5-inch axial hole through it, other conditions being the same as in Example 1.

The ratio of external to internal diameters of the hub equals $15 \div 10 = 1.5$, as before, and the ratio of external to internal diameters of the shaft equals $10 \div 5 = 2$. From Table 2, we find that factor $C = 0.455$: $T = 25,000$ pounds. Substituting these values in Formula (2):

$$A = \frac{25,000 \times (1 + 0.455)}{30,000,000} = 0.0012 \text{ inch.}$$

The increase in allowance, as compared with Example 1, is due to the fact that the hollow shaft is more compressible.

Table 2. Factors for Calculating Shrinkage Fit Allowances

Values of factor C for hollow steel shafts of external and internal diameters D_1 and D_0, respectively, and steel hubs of nominal external diameter D_2.

$\frac{D_2}{D_1}$	$\frac{D_1}{D_0}$	C	$\frac{D_2}{D_1}$	$\frac{D_1}{D_0}$	C	$\frac{D_2}{D_1}$	$\frac{D_1}{D_0}$	C
1.5	2.0	0.455	2.4	2.0	0.760	3.4	2.0	0.876
	2.5	0.357		2.5	0.597		2.5	0.689
	3.0	0.313		3.0	0.523		3.0	0.602
	3.5	0.288		3.5	0.481		3.5	0.555
1.6	2.0	0.509	2.6	2.0	0.793	3.6	2.0	0.888
	2.5	0.400		2.5	0.624		2.5	0.698
	3.0	0.350		3.0	0.546		3.0	0.611
	3.5	0.322		3.5	0.502		3.5	0.562
1.8	2.0	0.599	2.8	2.0	0.820	3.8	2.0	0.900
	2.5	0.471		2.5	0.645		2.5	0.707
	3.0	0.412		3.0	0.564		3.0	0.619
	3.5	0.379		3.5	0.519		3.5	0.570
2.0	2.0	0.667	3.0	2.0	0.842	4.0	2.0	0.909
	2.5	0.524		2.5	0.662		2.5	0.715
	3.0	0.459		3.0	0.580		3.0	0.625
	3.5	0.422		3.5	0.533		3.5	0.576
2.2	2.0	0.718	3.2	2.0	0.860		.....	......
	2.5	0.565		2.5	0.676		.....	......
	3.0	0.494		3.0	0.591	.....	.....	......
	3.5	0.455		3.5	0.544		.....	......

Example 3: If the crank web in Example 1 is of cast iron and 4000 pounds per square inch is the maximum tensile stress in the hub, what is the allowance per inch of diameter?

$$\frac{D_2}{D_1} = 1.5; \quad T = 4000.$$

In Table 1, we find that $C = 0.234$. Substituting in Formula (1), for cast-iron hubs, $A = 0.0003$ inch, which, owing to the lower tensile strength of cast iron, is

about one-third the shrinkage allowance in Example 1, although the stress is two-thirds of the elastic limit.

Temperatures for Shrinkage Fits. — The temperature to which the outer member in a shrinkage fit should be heated for clearance in assembling the parts depends on the total expansion required and on the coefficient α of linear expansion of the metal (that is, the increase in length of any section of the metal in any direction for an increase in temperature of 1 degree F.). The total expansion in diameter which is required consists of the total allowance for shrinkage and an added amount for clearance. The value of the coefficient α is, for nickel-steel, 0.000007; for steel

Table 3. Factors for Calculating Shrinkage Fit Allowances

Values of factor C for hollow steel shafts and cast-iron hubs. Notation as in Table 2.

$\frac{D_2}{D_1}$	$\frac{D_1}{D_0}$	C	$\frac{D_2}{D_1}$	$\frac{D_1}{D_0}$	C	$\frac{D_2}{D_1}$	$\frac{D_1}{D_0}$	C
1.5	2.0	0.468	2.4	2.0	0.798	3.4	2.0	0.926
	2.5	0.368		2.5	0.628		2.5	0.728
	3.0	0.322		3.0	0.549		3.0	0.637
	3.5	0.296		3.5	0.506		3.5	0.587
1.6	2.0	0.527	2.6	2.0	0.834	3.6	2.0	0.941
	2.5	0.414		2.5	0.656		2.5	0.740
	3.0	0.362		3.0	0.574		3.0	0.647
	3.5	0.333		3.5	0.528		3.5	0.596
1.8	2.0	0.621	2.8	2.0	0.864	3.8	2.0	0.953
	2.5	0.488		2.5	0.679		2.5	0.749
	3.0	0.427		3.0	0.594		3.0	0.656
	3.5	0.393		3.5	0.547		3.5	0.603
2.0	2.0	0.696	3.0	2.0	0.888	4.0	2.0	0.964
	2.5	0.547		2.5	0.698		2.5	0.758
	3.0	0.479		3.0	0.611		3.0	0.663
	3.5	0.441		3.5	0.562		3.5	0.610
2.2	2.0	0.753	3.2	2.0	0.909	.....	.....	......
	2.5	0.592		2.5	0.715			......
	3.0	0.518		3.0	0.625			......
	3.5	0.477		3.5	0.576			......

in general, 0.0000065; for cast iron, 0.0000062. As an example, take an outer member of steel to be expanded 0.005 inch per inch of internal diameter, 0.001 being the shrinkage allowance and the remainder for clearance. Then:

$$\alpha \times t^\circ = 0.005$$

$$t = \frac{0.005}{0.0000065} = 769 \text{ degrees F.}$$

The value t is the number of degrees F. which the temperature of the member must be raised above that of the room temperature.

ANSI Standard Limits and Fits (ANSI B4.1-1967, R1979). — This American National Standard for Preferred Limits and Fits for Cylindrical Parts presents definitions of terms applying to fits between plain (non-threaded) cylindrical parts and makes recommendations on preferred sizes, allowances, tolerances, and fits for use wherever they are applicable. This standard is in accord with the recommendations of American-British-Canadian (ABC) conferences up to a diameter of 20 inches. Experimental work is being carried on with the objective of reaching agreement in the range above 20 inches. The recommendations in the Standard are presented for guidance and for use where they might serve to improve and simplify products, practices, and facilities. They should have application for a wide range of products.

As revised in 1967, and reaffirmed in 1979, the definitions in ANSI B4.1 have been expanded and some of the limits in certain classes have been changed.

Factors Affecting Selection of Fits. — Many factors, such as length of engagement, bearing load, speed, lubrication, temperature, humidity, and materials, must be taken into consideration in the selection of fits for a particular application, and modifications in the ANSI recommendations may be required to satisfy extreme conditions. Subsequent adjustments may also be found desirable as a result of experience in a particular application to suit critical functional requirements or to permit optimum manufacturing economy.

Definitions. — The following terms are defined in this standard:

Nominal Size: The nominal size is the designation which is used for the purpose of general identification.

Dimension: A dimension is a geometrical characteristic such as diameter, length, angle, or center distance.

Size: Size is a designation of magnitude. When a value is assigned to a dimension, it is referred to as the size of that dimension. (It is recognized that the words "dimension" and "size" are both used at times to convey the meaning of magnitude.)

Allowance: An allowance is a prescribed difference between the maximum material limits of mating parts. (See definition of *Fit*). It is a minimum clearance (positive allowance) or maximum interference (negative allowance) between such parts.

Tolerance: A tolerance is the total permissible variation of a size. The tolerance is the difference between the limits of size.

Basic Size: The basic size is that size from which the limits of size are derived by the application of allowances and tolerances.

Design Size: The design size is the basic size with allowance applied, from which the limits of size are derived by the application of tolerances. Where there is no allowance, the design size is the same as the basic size.

Actual Size: An actual size is a measured size.

Limits of Size: The limits of size are the applicable maximum and minimum sizes.

Maximum Material Limit: A maximum material limit is that limit of size that provides the maximum amount of material for the part. Normally it is the maximum limit of size of an external dimension or the minimum limit of size of an internal dimension.*

Minimum Material Limit: A minimum material limit is that limit of size that provides the minimum amount of material for the part. Normally it is the minimum limit of size of an external dimension or the maximum limit of size of an internal dimension.*

---

* An example of exceptions: an exterior corner radius where the maximum radius is the minimum material limit and the minimum radius is the maximum material limit.

Tolerance Limit: A tolerance limit is the variation, positive or negative, by which size is permitted to depart from the design size.

Unilateral Tolerance: A unilateral tolerance is a tolerance in which variation is permitted only in one direction from the design size.

Bilateral Tolerance: A bilateral tolerance is a tolerance in which variation is permitted in both directions from the design size.

Unilateral Tolerance System: A design plan which uses only unilateral tolerances is known as a Unilateral Tolerance System.

Bilateral Tolerance System: A design plan which uses only bilateral tolerances is known as a Bilateral Tolerance System.

Fit: Fit is the general term used to signify the range of tightness which may result from the application of a specific combination of allowances and tolerances in the design of mating parts.

Actual Fit: The actual fit between two mating parts is the relation existing between them with respect to the amount of clearance or interference which is present when they are assembled. (Fits are of three general types: clearance, transition, and interference.)

Clearance Fit: A clearance fit is one having limits of size so specified that a clearance always results when mating parts are assembled.

Interference Fit: An interference fit is one having limits of size so specified that an interference always results when mating parts are assembled.

Transition Fit: A transition fit is one having limits of size so specified that either a clearance or an interference may result when mating parts are assembled.

Basic Hole System: A basic hole system is a system of fits in which the design size of the hole is the basic size and the allowance, if any, is applied to the shaft.

Basic Shaft System: A basic shaft system is a system of fits in which the design size of the shaft is the basic size and the allowance, if any, is applied to the hole.

Preferred Basic Sizes. — In specifying fits, the basic size of mating parts shall be chosen from the decimal series or the fractional series in the following table. All dimensions are given in inches.

Preferred Basic Sizes

Decimal			Fractional					
0.010	2.00	8.50	1/64	0.015625	2¼	2.2500	9½	9.5000
0.012	2.20	9.00	1/32	0.03125	2½	2.5000	10	10.0000
0.016	2.40	9.50	1/16	0.0625	2¾	2.7500	10½	10.5000
0.020	2.60	10.00	3/32	0.09375	3	3.0000	11	11.0000
0.025	2.80	10.50	1/8	0.1250	3¼	3.2500	11½	11.5000
0.032	3.00	11.00	5/32	0.15625	3½	3.5000	12	12.0000
0.040	3.20	11.50	3/16	0.1875	3¾	3.7500	12½	12.5000
0.05	3.40	12.00	¼	0.2500	4	4.0000	13	13.0000
0.06	3.60	12.50	5/16	0.3125	4¼	4.2500	13½	13.5000
0.08	3.80	13.00	3/8	0.3750	4½	4.5000	14	14.0000
0.10	4.00	13.50	7/16	0.4375	4¾	4.7500	14½	14.5000
0.12	4.20	14.00	½	0.5000	5	5.0000	15	15.0000
0.16	4.40	14.50	9/16	0.5625	5¼	5.2500	15½	15.5000
0.20	4.60	15.00	5/8	0.6250	5½	5.5000	16	16.0000
0.24	4.80	15.50	11/16	0.6875	5¾	5.7500	16½	16.5000
0.30	5.00	16.00	¾	0.7500	6	6.0000	17	17.0000
0.40	5.20	16.50	7/8	0.8750	6½	6.5000	17½	17.5000
0.50	5.40	17.00	1	1.0000	7	7.0000	18	18.0000
0.60	5.60	17.50	1¼	1.2500	7½	7.5000	18½	18.5000
0.80	5.80	18.00	1½	1.5000	8	8.0000	19	19.0000
1.00	6.00	18.50	1¾	1.7500	8½	8.5000	19½	19.5000
1.20	6.50	19.00	2	2.0000	9	9.0000	20	20.0000
1.40	7.00	19.50	.....	....	...	....	....	....
1.60	7.50	20.00						
1.80	8.00	.....			All dimensions are given in inches.			

Preferred Series of Tolerances and Allowances (In thousandths of an inch)

0.1	1	10	100	0.3	3	30	...
...	1.2	12	125	...	3.5	35	...
0.15	1.4	14	...	0.4	4	40	...
...	1.6	16	160	...	4.5	45	...
...	1.8	18	...	0.5	5	50	...
0.2	2	20	200	0.6	6	60	...
...	2.2	22	...	0.7	7	70	...
0.25	2.5	25	250	0.8	8	80	...
...	2.8	28	...	0.9	9	...	...

Standard Tolerances. — The series of standard tolerances shown in Table 1 are so arranged that for any one grade they represent approximately similar production difficulties throughout the range of sizes. This table provides a suitable range from which appropriate tolerances for holes and shafts can be selected. This enables the use of standard gages. The tolerances shown in Table 1 have been used in the succeeding tables for different classes of fits.

Table 1. ANSI Standard Tolerances (ANSI B4.1-1967, R1979)

Nominal Size, Inches Over	To	Grade 4	5	6	7	8	9	10	11	12	13
		Tolerances in thousandths of an inch*									
0–	0.12	0.12	0.15	0.25	0.4	0.6	1.0	1.6	2.5	4	6
0.12–	0.24	0.15	0.20	0.3	0.5	0.7	1.2	1.8	3.0	5	7
0.24–	0.40	0.15	0.25	0.4	0.6	0.9	1.4	2.2	3.5	6	9
0.40–	0.71	0.2	0.3	0.4	0.7	1.0	1.6	2.8	4.0	7	10
0.71–	1.19	0.25	0.4	0.5	0.8	1.2	2.0	3.5	5.0	8	12
1.19–	1.97	0.3	0.4	0.6	1.0	1.6	2.5	4.0	6	10	16
1.97–	3.15	0.3	0.5	0.7	1.2	1.8	3.0	4.5	7	12	18
3.15–	4.73	0.4	0.6	0.9	1.4	2.2	3.5	5	9	14	22
4.73–	7.09	0.5	0.7	1.0	1.6	2.5	4.0	6	10	16	25
7.09–	9.85	0.6	0.8	1.2	1.8	2.8	4.5	7	12	18	28
9.85–	12.41	0.6	0.9	1.2	2.0	3.0	5.0	8	12	20	30
12.41–	15.75	0.7	1.0	1.4	2.2	3.5	6	9	14	22	35
15.75–	19.69	0.8	1.0	1.6	2.5	4	6	10	16	25	40
19.69–	30.09	0.9	1.2	2.0	3	5	8	12	20	30	50
30.09–	41.49	1.0	1.6	2.5	4	6	10	16	25	40	60
41.49–	56.19	1.2	2.0	3	5	8	12	20	30	50	80
56.19–	76.39	1.6	2.5	4	6	10	16	25	40	60	100
76.39–100.9		2.0	3	5	8	12	20	30	50	80	125
100.9–131.9		2.5	4	6	10	16	25	40	60	100	160
131.9–171.9		3	5	8	12	20	30	50	80	125	200
171.9–200		4	6	10	16	25	40	60	100	160	250

* All tolerances above heavy line are in accordance with American-British-Canadian (ABC) agreements.

Relation of Machining Processes to Tolerance Grades

This chart may be used as a general guide to determine the machining processes that will, under normal conditions, produce work within the tolerance grades indicated.

(See also Relation of Surface Roughness to Dimensional Tolerances in the Surface Texture Section, page 704.)

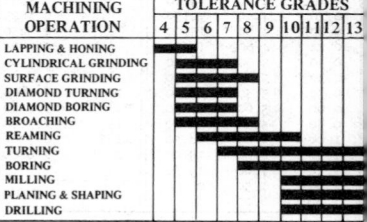

MACHINING OPERATION	TOLERANCE GRADES 4 5 6 7 8 9 10 11 12 13
LAPPING & HONING	
CYLINDRICAL GRINDING	
SURFACE GRINDING	
DIAMOND TURNING	
DIAMOND BORING	
BROACHING	
REAMING	
TURNING	
BORING	
MILLING	
PLANING & SHAPING	
DRILLING	

ANSI Standard Fits. — Tables 2 to 6 inclusive show a series of standard types and classes of fits on a unilateral hole basis, such that the fit produced by mating parts in any one class will produce approximately similar performance throughout the range of sizes. These tables prescribe the fit for any given size, or type of fit; they also prescribe the standard limits for the mating parts which will produce the fit. The fits listed in these tables contain all of those which appear in the approved American-British-Canadian proposal.

Selection of Fits: In selecting limits of size for any application, the type of fit is determined first, based on the use or service required from the equipment being designed; then the limits of size of the mating parts are established, to insure that the desired fit will be produced.

Theoretically, an infinite number of fits could be chosen, but the number of standard fits shown in the accompanying tables should cover most applications.

Designation of Standard Fits: Standard fits are designated by means of the following symbols which facilitate reference to classes of fit for educational purposes. The symbols are not intended to be shown on manufacturing drawings; instead, sizes should be specified on drawings.

The letter symbols used are as follows:

RC   Running or Sliding Clearance Fit
LC   Locational Clearance Fit
LT   Transition Clearance or Interference Fit
LN   Locational Interference Fit
FN   Force or Shrink Fit

These letter symbols are used in conjunction with numbers representing the class of fit; thus FN 4 represents a Class 4, force fit.

Each of these symbols (two letters and a number) represents a complete fit for which the minimum and maximum clearance or interference and the limits of size for the mating parts are given directly in the tables.

Description of Fits. — The classes of fits are arranged in three general groups: running and sliding fits, locational fits, and force fits.

Running and Sliding Fits (RC): Running and sliding fits, for which limits of clearance are given in Table 2, are intended to provide a similar running performance, with suitable lubrication allowance, throughout the range of sizes. The clearances for the first two classes, used chiefly as slide fits, increase more slowly with the diameter than for the other classes, so that accurate location is maintained even at the expense of free relative motion.

These fits may be described as follows:

RC 1 *Close sliding fits* are intended for the accurate location of parts which must assemble without perceptible play.

RC 2 *Sliding fits* are intended for accurate location, but with greater maximum clearance than class RC 1. Parts made to this fit move and turn easily but are not intended to run freely, and in the larger sizes may seize with small temperature changes.

RC 3 *Precision running fits* are about the closest fits which can be expected to run freely, and are intended for precision work at slow speeds and light journal pressures, but are not suitable where appreciable temperature differences are likely to be encountered.

RC 4 *Close running fits* are intended chiefly for running fits on accurate machinery with moderate surface speeds and journal pressures, where accurate location and minimum play is desired.

RC 5 and RC 6 *Medium running fits* are intended for higher running speeds, or heavy journal pressures, or both.

RC 7 *Free running fits* are intended for use where accuracy is not essential, or where large temperature variations are likely to be encountered, or under both these conditions.

RC 8 and RC 9 *Loose running fits* are intended for use where wide commercial tolerances may be necessary, together with an allowance, on the external member.

Locational Fits (LC, LT, and LN): Locational fits are fits intended to determine only the location of the mating parts; they may provide rigid or accurate location, as with interference fits, or provide some freedom of location, as with clearance fits. Accordingly, they are divided into three groups: clearance fits (LC), transition fits (LT), and interference fits (LN).

These are described as follows:

LC   *Locational clearance fits* are intended for parts which are normally stationary, but which can be freely assembled or disassembled. They range from snug fits for parts requiring accuracy of location, through the medium clearance fits for parts such as spigots, to the looser fastener fits where freedom of assembly is of prime importance.

LT   *Locational transition fits* are a compromise between clearance and interference fits, for application where accuracy of location is important, but either a small amount of clearance or interference is permissible.

LN   *Locational interference fits* are used where accuracy of location is of prime importance, and for parts requiring rigidity and alignment with no special requirements for bore pressure. Such fits are not intended for parts designed to transmit frictional loads from one part to another by virtue of the tightness of fit, as these conditions are covered by force fits.

Force Fits (FN): Force or shrink fits constitute a special type of interference fit, normally characterized by maintenance of constant bore pressures throughout the range of sizes. The interference therefore varies almost directly with diameter, and the difference between its minimum and maximum value is small, to maintain the resulting pressures within reasonable limits.

These fits are described as follows:

FN 1   *Light drive fits* are those requiring light assembly pressures, and produce more or less permanent assemblies. They are suitable for thin sections or long fits, or in cast-iron external members.

FN 2   *Medium drive fits* are suitable for ordinary steel parts, or for shrink fits on light sections. They are about the tightest fits that can be used with high-grade cast-iron external members.

FN 3   *Heavy drive fits* are suitable for heavier steel parts or for shrink fits in medium sections.

FN 4 and FN 5   *Force fits* are suitable for parts which can be highly stressed, or for shrink fits where the heavy pressing forces required are impractical.

Graphical Representation of Limits and Fits. — A visual comparison of the hole and shaft tolerances and the clearances or interferences provided by the various types and classes of fits can be obtained from the diagrams on page 635. These diagrams have been drawn to scale for a nominal diameter of 1 inch.

Use of Standard Fit Tables. — *Example 1:* A Class RC 1 fit is to be used in assembling a mating hole and shaft of 2-inch nominal diameter. This class of fit was selected because the application required accurate location of the parts with no perceptible play (see description of RC 1 close sliding fits). From the data in Table 2, establish the limts of size and clearance of the hole and shaft.

Maximum hole = 2 + 0.0005 = 2.0005; minimum hole = 2 inches

(*Continued on page* 644)

Graphical Representation of ANSI Standard Limits and Fits*

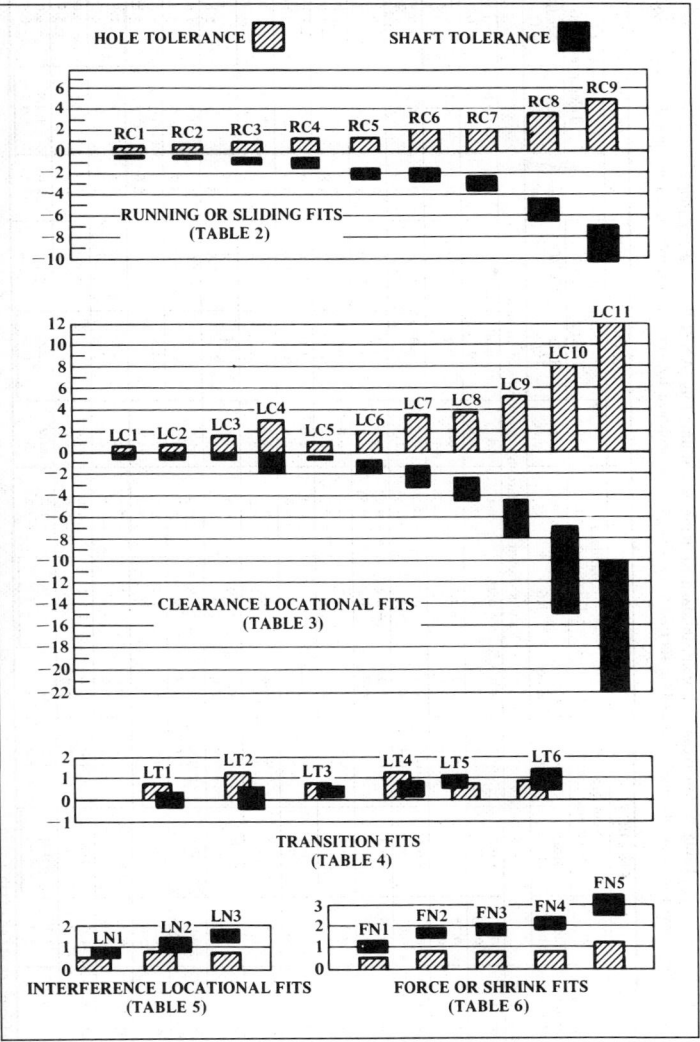

* Diagrams show disposition of hole and shaft tolerances (in thousandths of an inch) with respect to basic size (0) for a diameter of 1 inch.

Table 2. American National Standard Running and Sliding Fits (ANSI B4.1-1967, R1979)

Tolerance limits given in body of table are added or subtracted to basic size (as indicated by + or − sign) to obtain maximum and minimum sizes of mating parts.

Values shown below are in thousandths of an inch

Nominal Size Range, Inches Over – To	Class RC 1			Class RC 2			Class RC 3			Class RC 4		
	Clearance*	Hole H5	Shaft g4	Clearance*	Hole H6	Shaft g5	Clearance*	Hole H7	Shaft f6	Clearance*	Hole H8	Shaft f7
0– 0.12	0.1 / 0.45	+0.2 / 0	−0.1 / −0.25	0.1 / 0.55	+0.25 / 0	−0.1 / −0.3	0.3 / 0.95	+0.4 / 0	−0.3 / −0.55	0.3 / 1.3	+0.6 / 0	−0.3 / −0.7
0.12– 0.24	0.15 / 0.5	+0.2 / 0	−0.15 / −0.3	0.15 / 0.65	+0.3 / 0	−0.15 / −0.35	0.4 / 1.12	+0.5 / 0	−0.4 / −0.7	0.4 / 1.6	+0.7 / 0	−0.4 / −0.9
0.24– 0.40	0.2 / 0.6	+0.25 / 0	−0.2 / −0.35	0.2 / 0.85	+0.4 / 0	−0.2 / −0.45	0.5 / 1.5	+0.6 / 0	−0.5 / −0.9	0.5 / 2.0	+0.9 / 0	−0.5 / −1.1
0.40– 0.71	0.25 / 0.75	+0.3 / 0	−0.25 / −0.45	0.25 / 0.95	+0.4 / 0	−0.25 / −0.55	0.6 / 1.7	+0.7 / 0	−0.6 / −1.0	0.6 / 2.3	+1.0 / 0	−0.6 / −1.3
0.71– 1.19	0.3 / 0.95	+0.4 / 0	−0.3 / −0.55	0.3 / 1.2	+0.5 / 0	−0.3 / −0.7	0.8 / 2.1	+0.8 / 0	−0.6 / −1.3	0.8 / 2.8	+1.2 / 0	−0.8 / −1.6
1.19– 1.97	0.4 / 1.1	+0.4 / 0	−0.4 / −0.7	0.4 / 1.4	+0.6 / 0	−0.4 / −0.8	1.0 / 2.6	+1.0 / 0	−1.0 / −1.6	1.0 / 3.6	+1.6 / 0	−1.0 / −2.0
1.97– 3.15	0.4 / 1.2	+0.5 / 0	−0.4 / −0.7	0.4 / 1.6	+0.7 / 0	−0.4 / −0.9	1.2 / 3.1	+1.2 / 0	−1.2 / −1.9	1.2 / 4.2	+1.8 / 0	−1.2 / −2.4
3.15– 4.73	0.5 / 1.5	+0.6 / 0	−0.5 / −0.9	0.5 / 2.0	+0.9 / 0	−0.5 / −1.1	1.4 / 3.7	+1.4 / 0	−1.4 / −2.3	1.4 / 5.0	+2.2 / 0	−1.4 / −2.8
4.73– 7.09	0.6 / 1.8	+0.7 / 0	−0.6 / −1.1	0.6 / 2.3	+1.0 / 0	−0.6 / −1.3	1.6 / 4.2	+1.6 / 0	−1.6 / −2.6	1.6 / 5.7	+2.5 / 0	−1.6 / −3.2
7.09– 9.85	0.6 / 2.0	+0.8 / 0	−0.6 / −1.2	0.6 / 2.6	+1.2 / 0	−0.6 / −1.4	2.0 / 5.0	+1.8 / 0	−2.0 / −3.2	2.0 / 6.6	+2.8 / 0	−2.0 / −3.8
9.85–12.41	0.8 / 2.3	+0.9 / 0	−0.8 / −1.4	0.8 / 2.9	+1.2 / 0	−0.8 / −1.7	2.5 / 5.7	+2.0 / 0	−2.5 / −3.7	2.5 / 7.5	+3.0 / 0	−2.5 / −4.5
12.41–15.75	1.0 / 2.7	+1.0 / 0	−1.0 / −1.7	1.0 / 3.4	+1.4 / 0	−1.0 / −2.0	3.0 / 6.6	+2.2 / 0	−3.0 / −4.4	3.0 / 8.7	+3.5 / 0	−3.0 / −5.2
15.75–19.69	1.2 / 3.0	+1.0 / 0	−1.2 / −2.0	1.2 / 3.8	+1.6 / 0	−1.2 / −2.2	4.0 / 8.1	+2.5 / 0	−4.0 / −5.6	4.0 / 10.5	+4.0 / 0	−4.0 / −6.5

See footnotes at end of table.

Table 2 (Concluded). American National Standard Running and Sliding Fits (ANSI B4.1-1967, R1979)

Values shown below are in thousandths of an inch

Nominal Size Range, Inches Over	To	Class RC 5 Clearance*	RC 5 Hole H8	RC 5 Shaft e7	Class RC 6 Clearance*	RC 6 Hole H9	RC 6 Shaft e8	Class RC 7 Clearance*	RC 7 Hole H9	RC 7 Shaft d8	Class RC 8 Clearance*	RC 8 Hole H10	RC 8 Shaft c9	Class RC 9 Clearance*	RC 9 Hole H11	RC 9 Shaft
0–	0.12	0.6 / 1.6	+0.6 / 0	−0.6 / −1.0	0.6 / 2.2	+1.0 / 0	−0.6 / −1.2	1.0 / 2.6	+1.0 / 0	−1.0 / −1.6	2.5 / 5.1	+1.6 / 0	−2.5 / −3.5	4.0 / 8.1	+2.5 / 0	−4.0 / −5.6
0.12–	0.24	0.8 / 2.0	+0.7 / 0	−0.8 / −1.3	0.8 / 2.7	+1.2 / 0	−0.8 / −1.5	1.2 / 3.1	+1.2 / 0	−1.2 / −1.9	2.8 / 5.8	+1.8 / 0	−2.8 / −4.0	4.5 / 9.0	+3.0 / 0	−4.5 / −6.0
0.24–	0.40	1.0 / 2.5	+0.9 / 0	−1.0 / −1.6	1.0 / 3.3	+1.4 / 0	−1.0 / −1.9	1.6 / 3.9	+1.4 / 0	−1.6 / −2.5	3.0 / 6.6	+2.2 / 0	−3.0 / −4.4	5.0 / 10.7	+3.5 / 0	−5.0 / −7.2
0.40–	0.71	1.2 / 2.9	+1.0 / 0	−1.2 / −1.9	1.2 / 3.8	+1.6 / 0	−1.2 / −2.2	2.0 / 4.6	+1.6 / 0	−2.0 / −3.0	3.5 / 7.9	+2.8 / 0	−3.5 / −5.1	6.0 / 12.8	+4.0 / 0	−6.0 / −8.8
0.71–	1.19	1.6 / 3.6	+1.2 / 0	−1.6 / −2.4	1.6 / 4.8	+2.0 / 0	−1.6 / −2.8	2.5 / 5.7	+2.0 / 0	−2.5 / −3.7	4.5 / 10.0	+3.5 / 0	−4.5 / −6.5	7.0 / 15.5	+5.0 / 0	−7.0 / −10.5
1.19–	1.97	2.0 / 4.6	+1.6 / 0	−2.0 / −3.0	2.0 / 6.1	+2.5 / 0	−2.0 / −3.6	3.0 / 7.1	+2.5 / 0	−3.0 / −4.6	5.0 / 11.5	+4.0 / 0	−5.0 / −7.5	8.0 / 18.0	+6.0 / 0	−8.0 / −12.0
1.97–	3.15	2.5 / 5.5	+1.8 / 0	−2.5 / −3.7	2.5 / 7.3	+3.0 / 0	−2.5 / −4.3	4.0 / 8.8	+3.0 / 0	−4.0 / −5.8	6.0 / 13.5	+4.5 / 0	−6.0 / −9.0	9.0 / 20.5	+7.0 / 0	−9.0 / −13.5
3.15–	4.73	3.0 / 6.6	+2.2 / 0	−3.0 / −4.4	3.0 / 8.7	+3.5 / 0	−3.0 / −5.2	5.0 / 10.7	+3.5 / 0	−5.0 / −7.2	7.0 / 15.5	+5.0 / 0	−7.0 / −10.5	10.0 / 24.0	+9.0 / 0	−10.0 / −15.0
4.73–	7.09	3.5 / 7.6	+2.5 / 0	−3.5 / −5.1	3.5 / 10.0	+4.0 / 0	−3.5 / −6.0	6.0 / 12.5	+4.0 / 0	−6.0 / −8.5	8.0 / 18.0	+6.0 / 0	−8.0 / −12.0	12.0 / 28.0	+10.0 / 0	−12.0 / −18.0
7.09–	9.85	4.0 / 8.6	+2.8 / 0	−4.0 / −5.8	4.0 / 11.3	+4.5 / 0	−4.0 / −6.8	7.0 / 14.3	+4.5 / 0	−7.0 / −9.8	10.0 / 21.5	+7.0 / 0	−10.0 / −14.5	15.0 / 34.0	+12.0 / 0	−15.0 / −22.0
9.85–	12.41	5.0 / 10.0	+3.0 / 0	−5.0 / −7.0	5.0 / 13.0	+5.0 / 0	−5.0 / −8.0	8.0 / 16.0	+5.0 / 0	−8.0 / −11.0	12.0 / 25.0	+8.0 / 0	−12.0 / −17.0	18.0 / 38.0	+12.0 / 0	−18.0 / −26.0
12.41–	15.75	6.0 / 11.7	+3.5 / 0	−6.0 / −8.2	6.0 / 15.5	+6.0 / 0	−6.0 / −9.5	10.0 / 19.5	+6.0 / 0	−10.0 / −13.5	14.0 / 29.0	+9.0 / 0	−14.0 / −20.0	22.0 / 45.0	+14.0 / 0	−22.0 / −31.0
15.75–	19.69	8.0 / 14.5	+4.0 / 0	−8.0 / −10.5	8.0 / 18.0	+6.0 / 0	−8.0 / −12.0	12.0 / 22.0	+6.0 / 0	−12.0 / −16.0	16.0 / 32.0	+10.0 / 0	−16.0 / −22.0	25.0 / 51.0	+16.0 / 0	−25.0 / −35.0

All data above heavy lines are in accord with ABC agreements. Symbols H5, g4, etc. are hole and shaft designations in ABC system. Limits for sizes above 19.69 inches are also given in the ANSI Standard.

* Pairs of values shown represent minimum and maximum amounts of clearance resulting from application of standard tolerance limits.

Table 3. American National Standard Clearance Locational Fits (ANSI B4.1-1967, R1979)

Tolerance limits given in body of table are added or subtracted to basic size (as indicated by + or − sign) to obtain maximum and minimum sizes of mating parts.

Values shown below are in thousandths of an inch

Nominal Size Range, Inches Over To	Class LC 1 Clearance*	Hole H6	Shaft h5	Class LC 2 Clearance*	Hole H7	Shaft h6	Class LC 3 Clearance*	Hole H8	Shaft h7	Class LC 4 Clearance*	Hole H10	Shaft h9	Class LC 5 Clearance*	Hole H7	Shaft g6
0– 0.12	0 / 0.45	+0.25 / 0	0 / −0.2	0 / 0.65	+0.4 / 0	0 / −0.25	0 / 1	+0.6 / 0	0 / −0.4	0 / 2.6	+1.6 / 0	0 / −1.0	0.1 / 0.75	+0.4 / 0	−0.1 / −0.35
0.12– 0.24	0 / 0.5	+0.3 / 0	0 / −0.2	0 / 0.8	+0.5 / 0	0 / −0.3	0 / 1.2	+0.7 / 0	0 / −0.5	0 / 3.0	+1.8 / 0	0 / −1.2	0.15 / 0.95	+0.5 / 0	−0.15 / −0.45
0.24– 0.40	0 / 0.65	+0.4 / 0	0 / −0.25	0 / 1.0	+0.6 / 0	0 / −0.4	0 / 1.5	+0.9 / 0	0 / −0.6	0 / 3.6	+2.2 / 0	0 / −1.4	0.2 / 1.2	+0.6 / 0	−0.2 / −0.6
0.40– 0.71	0 / 0.7	+0.4 / 0	0 / −0.3	0 / 1.1	+0.7 / 0	0 / −0.4	0 / 1.7	+1.0 / 0	0 / −0.6	0 / 4.4	+2.8 / 0	0 / −1.6	0.25 / 1.35	+0.7 / 0	−0.25 / −0.65
0.71– 1.19	0 / 0.9	+0.5 / 0	0 / −0.4	0 / 1.3	+0.8 / 0	0 / −0.5	0 / 2	+1.2 / 0	0 / −0.7	0 / 5.5	+3.5 / 0	0 / −2.0	0.3 / 1.6	+0.8 / 0	−0.3 / −0.8
1.19– 1.97	0 / 1.0	+0.6 / 0	0 / −0.4	0 / 1.6	+1.0 / 0	0 / −0.6	0 / 2.6	+1.6 / 0	0 / −0.8	0 / 6.5	+4.0 / 0	0 / −2.5	0.4 / 2.0	+1.0 / 0	−0.4 / −1.0
1.97– 3.15	0 / 1.2	+0.7 / 0	0 / −0.5	0 / 1.9	+1.2 / 0	0 / −0.7	0 / 3	+1.8 / 0	0 / −1	0 / 7.5	+4.5 / 0	0 / −3	0.4 / 2.3	+1.2 / 0	−0.4 / −1.1
3.15– 4.73	0 / 1.5	+0.9 / 0	0 / −0.6	0 / 2.3	+1.4 / 0	0 / −0.9	0 / 3.6	+2.2 / 0	0 / −1	0 / 8.5	+5.0 / 0	0 / −3.5	0.5 / 2.8	+1.4 / 0	−0.5 / −1.4
4.73– 7.09	0 / 1.7	+1.0 / 0	0 / −0.7	0 / 2.6	+1.6 / 0	0 / −1.0	0 / 4.1	+2.5 / 0	0 / −1.2	0 / 10.0	+6.0 / 0	0 / −4	0.6 / 3.2	+1.6 / 0	−0.6 / −1.6
7.09– 9.85	0 / 2.0	+1.2 / 0	0 / −0.8	0 / 3.0	+1.8 / 0	0 / −1.2	0 / 4.6	+2.8 / 0	0 / −1.4	0 / 11.5	+7.0 / 0	0 / −4.5	0.6 / 3.6	+1.8 / 0	−0.6 / −1.8
9.85–12.41	0 / 2.1	+1.2 / 0	0 / −0.9	0 / 3.2	+2.0 / 0	0 / −1.2	0 / 5	+3.0 / 0	0 / −1.6	0 / 13.0	+8.0 / 0	0 / −5	0.7 / 3.9	+2.0 / 0	−0.7 / −1.9
12.41–15.75	0 / 2.4	+1.4 / 0	0 / −1.0	0 / 3.6	+2.2 / 0	0 / −1.4	0 / 5.7	+3.5 / 0	0 / −1.8	0 / 15.0	+9.0 / 0	0 / −6	0.7 / 4.3	+2.2 / 0	−0.7 / −2.1
15.75–19.69	0 / 2.6	+1.6 / 0	0 / −1.0	0 / 4.1	+2.5 / 0	0 / −1.6	0 / 6.5	+4 / 0	0 / −2.0	0 / 16.0	+10.0 / 0	0 / −6	0.8 / 4.9	+2.5 / 0	−0.8 / −2.4

See footnotes at end of table.

Table 3 (Concluded). ANSI Standard Clearance Locational Fits (ANSI B4.1-1967, R1979)

Values shown below are in thousandths of an inch

Nominal Size Range, Inches Over — To	Class LC 6			Class LC 7			Class LC 8			Class LC 9			Class LC 10			Class LC 11		
	Clear-ance*	Std. Tol. Hole H9	Std. Tol. Shaft f8	Clear-ance*	Std. Tol. Hole H10	Std. Tol. Shaft e9	Clear-ance*	Std. Tol. Hole H10	Std. Tol. Shaft d9	Clear-ance*	Std. Tol. Hole H11	Std. Tol. Shaft c10	Clear-ance*	Std. Tol. Hole H12	Std. Tol. Shaft	Clear-ance*	Std. Tol. Hole H13	Std. Tol. Shaft
0– 0.12	0.3 / 1.9	+1.0 / 0	−0.3 / −0.9	0.6 / 3.2	+1.6 / 0	−0.6 / −1.6	1.0 / 2.0	+1.6 / 0	−1.0 / −2.0	2.5 / 6.6	+2.5 / 0	−2.5 / −4.1	4 / 12	+4 / 0	−4 / −8	5 / 17	+6 / 0	−5 / −11
0.12– 0.24	0.4 / 2.3	+1.2 / 0	−0.4 / −1.1	0.8 / 3.8	+1.8 / 0	−0.8 / −2.0	1.2 / 4.2	+1.8 / 0	−1.2 / −2.4	2.8 / 7.6	+3.0 / 0	−2.8 / −4.6	4.5 / 14.5	+5 / 0	−4.5 / −9.5	6 / 20	+7 / 0	−6 / −13
0.24– 0.40	0.5 / 2.8	+1.4 / 0	−0.5 / −1.4	1.0 / 4.6	+2.2 / 0	−1.0 / −2.4	1.6 / 5.2	+2.2 / 0	−1.6 / −3.0	3.0 / 8.7	+3.5 / 0	−3.0 / −5.2	5 / 17	+6 / 0	−5 / −11	7 / 25	+9 / 0	−7 / −16
0.40– 0.71	0.6 / 3.2	+1.6 / 0	−0.6 / −1.6	1.2 / 5.6	+2.8 / 0	−1.2 / −2.8	2.0 / 6.4	+2.8 / 0	−2.0 / −3.6	3.5 / 10.3	+4.0 / 0	−3.5 / −6.3	6 / 20	+7 / 0	−6 / −13	8 / 28	+10 / 0	−8 / −18
0.71– 1.19	0.8 / 4.0	+2.0 / 0	−0.8 / −2.0	1.6 / 7.1	+3.5 / 0	−1.6 / −3.6	2.5 / 8.0	+3.5 / 0	−2.5 / −4.5	4.5 / 13.0	+5.0 / 0	−4.5 / −8.0	7 / 23	+8 / 0	−7 / −15	10 / 34	+12 / 0	−10 / −22
1.19– 1.97	1.0 / 5.1	+2.5 / 0	−1.0 / −2.6	2.0 / 8.5	+4.0 / 0	−2.0 / −4.5	3.6 / 9.5	+4.0 / 0	−3.0 / −5.5	5.0 / 15.0	+6 / 0	−5.0 / −9.0	8 / 28	+10 / 0	−8 / −18	12 / 44	+16 / 0	−12 / −28
1.97– 3.15	1.2 / 6.0	+3.0 / 0	−1.2 / −3.0	2.5 / 10.0	+4.5 / 0	−2.5 / −5.5	4.0 / 11.5	+4.5 / 0	−4.0 / −7.0	6.0 / 17.5	+7 / 0	−6.0 / −10.5	10 / 34	+12 / 0	−10 / −22	14 / 50	+18 / 0	−14 / −32
3.15– 4.73	1.4 / 7.1	+3.5 / 0	−1.4 / −3.6	3.0 / 11.5	+5.0 / 0	−3.0 / −6.5	5.0 / 13.5	+5.0 / 0	−5.0 / −8.5	7 / 21	+9 / 0	−7 / −12	11 / 39	+14 / 0	−11 / −25	16 / 60	+22 / 0	−16 / −38
4.73– 7.09	1.6 / 8.1	+4.0 / 0	−1.6 / −4.1	3.5 / 13.5	+6.0 / 0	−3.5 / −7.5	8 / 16	+6 / 0	−6 / −10	8 / 24	+10 / 0	−8 / −14	12 / 44	+16 / 0	−12 / −28	18 / 68	+25 / 0	−18 / −43
7.09– 9.85	2.0 / 9.3	+4.5 / 0	−2.0 / −4.8	4.0 / 15.5	+7.0 / 0	−4.0 / −8.5	7 / 18.5	+7 / 0	−7 / −11.5	10 / 29	+12 / 0	−10 / −17	16 / 52	+18 / 0	−16 / −34	22 / 78	+28 / 0	−22 / −50
9.85–12.41	2.2 / 10.2	+5.0 / 0	−2.2 / −5.2	4.5 / 17.5	+8.0 / 0	−4.5 / −9.5	7 / 20	+8 / 0	−7 / −12	12 / 32	+12 / 0	−12 / −20	20 / 60	+20 / 0	−20 / −40	28 / 88	+30 / 0	−28 / −58
12.41–15.75	2.5 / 12.0	+6.0 / 0	−2.5 / −6.0	5.0 / 20.0	+9.0 / 0	−5 / −11	8 / 23	+9 / 0	−8 / −14	14 / 37	+14 / 0	−14 / −23	22 / 66	+22 / 0	−22 / −44	30 / 100	+35 / 0	−30 / −65
15.75–19.69	2.8 / 12.8	+6.0 / 0	−2.8 / −6.8	5.0 / 21.0	+10.0 / 0	−5 / −11	9 / 25	+10 / 0	−9 / −15	16 / 42	+16 / 0	−16 / −26	25 / 75	+25 / 0	−25 / −50	35 / 115	+40 / 0	−35 / −75

All data above heavy lines are in accordance with American-British-Canadian (ABC) agreements. Symbols H6, H7, s6, etc. are hole and shaft designations in ABC system. Limits for sizes above 19.69 inches are not covered by ABC agreements but are given in the ANSI Standard.
* Pairs of values shown represent minimum and maximum amounts of interference resulting from application of standard tolerance limits.

Table 4. ANSI Standard Transition Locational Fits (ANSI B4.1-1967, R1979)

Values shown below are in thousandths of an inch

Nominal Size Range, Inches Over — To	Class LT 1			Class LT 2			Class LT 3			Class LT 4			Class LT 5			Class LT 6		
		Std. Tolerance Limits			Std. Tolerance Limits			Std. Tolerance Limits			Std. Tolerance Limits			Std. Tolerance Limits			Std. Tolerance Limits	
	Fit*	Hole H7	Shaft js6	Fit*	Hole H8	Shaft js7	Fit*	Hole H7	Shaft k6	Fit*	Hole H8	Shaft k7	Fit*	Hole H7	Shaft n6	Fit*	Hole H7	Shaft n7
0– 0.12	−0.12 +0.52	+0.4 0	+0.12 −0.12	−0.2 +0.8	+0.6 0	+0.2 −0.2							−0.5 +0.15	+0.4 0	+0.5 +0.25	−0.65 +0.15	+0.4 0	+0.65 +0.25
0.12– 0.24	−0.15 +0.65	+0.5 0	+0.15 −0.15	−0.25 +0.95	+0.7 0	+0.25 −0.25							−0.6 +0.2	+0.5 0	+0.6 +0.3	−0.8 +0.2	+0.5 0	+0.8 +0.3
0.24– 0.40	−0.2 +0.8	+0.6 0	+0.2 −0.2	−0.3 +1.2	+0.9 0	+0.3 −0.3	−0.5 +0.5	+0.6 0	+0.5 +0.1	−0.7 +0.8	+0.9 0	+0.7 +0.1	−0.8 +0.2	+0.6 0	+0.8 +0.4	−1.0 +0.2	+0.6 0	+1.0 +0.4
0.40– 0.71	−0.2 +0.9	+0.7 0	+0.2 −0.2	−0.35 +1.35	+1.0 0	+0.35 −0.35	−0.5 +0.6	+0.7 0	+0.5 +0.1	−0.8 +0.9	+1.0 0	+0.8 +0.1	−0.9 +0.2	+0.7 0	+0.9 +0.5	−1.2 +0.2	+0.7 0	+1.2 +0.5
0.71– 1.19	−0.25 +1.05	+0.8 0	+0.25 −0.25	−0.4 +1.6	+1.2 0	+0.4 −0.4	−0.6 +0.7	+0.8 0	+0.6 +0.1	−0.9 +1.1	+1.2 0	+0.9 +0.1	−1.1 +0.2	+0.8 0	+1.1 +0.6	−1.4 +0.2	+0.8 0	+1.4 +0.6
1.19– 1.97	−0.3 +1.3	+1.0 0	+0.3 −0.3	−0.5 +2.1	+1.6 0	+0.5 −0.5	−0.7 +0.9	+1.0 0	+0.7 +0.1	−1.1 +1.5	+1.6 0	+1.1 +0.1	−1.3 +0.3	+1.0 0	+1.3 +0.7	−1.7 +0.3	+1.0 0	+1.7 +0.7
1.97– 3.15	−0.3 +1.5	+1.2 0	+0.3 −0.3	−0.6 +2.4	+1.8 0	+0.6 −0.6	−0.8 +1.1	+1.2 0	+0.8 +0.1	−1.3 +1.7	+1.8 0	+1.3 +0.1	−1.5 +0.4	+1.2 0	+1.5 +0.8	−2.0 +0.4	+1.2 0	+2.0 +0.8
3.15– 4.73	−0.4 +1.8	+1.4 0	+0.4 −0.4	−0.7 +2.9	+2.2 0	+0.7 −0.7	−1.0 +1.3	+1.4 0	+1.0 +0.1	−1.5 +2.1	+2.2 0	+1.5 +0.1	−1.9 +0.4	+1.4 0	+1.9 +1.0	−2.4 +0.4	+1.4 0	+2.4 +1.0
4.73– 7.09	−0.5 +2.1	+1.6 0	+0.5 −0.5	−0.8 +3.3	+2.5 0	+0.8 −0.8	−1.1 +1.5	+1.6 0	+1.1 +0.1	−1.7 +2.4	+2.5 0	+1.7 +0.1	−2.2 +0.4	+1.6 0	+2.2 +1.2	−2.8 +0.4	+1.6 0	+2.8 +1.2
7.09– 9.85	−0.6 +2.4	+1.8 0	+0.6 −0.6	−0.9 +3.7	+2.8 0	+0.9 −0.9	−1.4 +1.6	+1.8 0	+1.4 +0.2	−2.0 +2.6	+2.8 0	+2.0 +0.2	−2.6 +0.4	+1.8 0	+2.6 +1.4	−3.2 +0.4	+1.8 0	+3.2 +1.4
9.85–12.41	−0.6 +2.6	+2.0 0	+0.6 −0.6	−1.0 +4.0	+3.0 0	+1.0 −1.0	−1.4 +1.8	+2.0 0	+1.4 +0.2	−2.2 +2.8	+3.0 0	+2.2 +0.2	−2.6 +0.6	+2.0 0	+2.6 +1.4	−3.4 +0.6	+2.0 0	+3.4 +1.4
12.41–15.75	−0.7 +2.9	+2.2 0	+0.7 −0.7	−1.0 +4.5	+3.5 0	+1.0 −1.0	−1.6 +2.0	+2.2 0	+1.6 +0.2	−2.4 +3.3	+3.5 0	+2.4 +0.2	−3.0 +0.6	+2.2 0	+3.0 +1.6	−3.8 +0.6	+2.2 0	+3.8 +1.6
15.75–19.69	−0.8 +3.3	+2.5 0	+0.8 −0.8	−1.2 +5.2	+4.0 0	+1.2 −1.2	−1.8 +2.3	+2.5 0	+1.8 +0.2	−2.7 +3.8	+4.0 0	+2.7 +0.2	−3.4 +0.7	+2.5 0	+3.4 +1.8	−4.3 +0.7	+2.5 0	+4.3 +1.8

All data above heavy lines are in accord with ABC agreements. Symbols H7, js6, etc. are hole and shaft designations in ABC system.
* Pairs of values shown represent maximum amount of interference (−) and maximum amount of clearance (+) resulting from application of standard tolerance limits.

Table 5. ANSI Standard Interference Locational Fits (ANSI B4.1-1967, R1979)

Tolerance limits given in body of table are added or subtracted to basic size (as indicated by + or − sign) to obtain maximum and minimum sizes of mating parts.

Nominal Size Range, Inches		Class LN 1			Class LN 2			Class LN 3		
		Limits of Inter-ference	Standard Limits		Limits of Inter-ference	Standard Limits		Limits of Inter-ference	Standard Limits	
Over	To		Hole H6	Shaft n5		Hole H7	Shaft p6		Hole H7	Shaft r6
		Values shown below are given in thousandths of an inch								
0−	0.12	0 / 0.45	+0.25 / 0	+0.45 / +0.25	0 / 0.65	+0.4 / 0	+0.65 / +0.4	0.1 / 0.75	+0.4 / 0	+0.75 / +0.5
0.12−	0.24	0 / 0.5	+0.3 / 0	+0.5 / +0.3	0 / 0.8	+0.5 / 0	+0.8 / +0.5	0.1 / 0.9	+0.5 / 0	+0.9 / +0.6
0.24−	0.40	0 / 0.65	+0.4 / 0	+0.65 / +0.4	0 / 1.0	+0.6 / 0	+1.0 / +0.6	0.2 / 1.2	+0.6 / 0	+1.2 / +0.8
0.40−	0.71	0 / 0.8	+0.4 / 0	+0.8 / +0.4	0 / 1.1	+0.7 / 0	+1.1 / +0.7	0.3 / 1.4	+0.7 / 0	+1.4 / +1.0
0.71−	1.19	0 / 1.0	+0.5 / 0	+1.0 / +0.5	0 / 1.3	+0.8 / 0	+1.3 / +0.8	0.4 / 1.7	+0.8 / 0	+1.7 / +1.2
1.19−	1.97	0 / 1.1	+0.6 / 0	+1.1 / +0.6	0 / 1.6	+1.0 / 0	+1.6 / +1.0	0.4 / 2.0	+1.0 / 0	+2.0 / +1.4
1.97−	3.15	0.1 / 1.3	+0.7 / 0	+1.3 / +0.8	0.2 / 2.1	+1.2 / 0	+2.1 / +1.4	0.4 / 2.3	+1.2 / 0	+2.3 / +1.6
3.15−	4.73	0.1 / 1.6	+0.9 / 0	+1.6 / +1.0	0.2 / 2.5	+1.4 / 0	+2.5 / +1.6	0.6 / 2.9	+1.4 / 0	+2.9 / +2.0
4.73−	7.09	0.2 / 1.9	+1.0 / 0	+1.9 / +1.2	0.2 / 2.8	+1.6 / 0	+2.8 / +1.8	0.9 / 3.5	+1.6 / 0	+3.5 / +2.5
7.09−	9.85	0.2 / 2.2	+1.2 / 0	+2.2 / +1.4	0.2 / 3.2	+1.8 / 0	+3.2 / +2.0	1.2 / 4.2	+1.8 / 0	+4.2 / +3.0
9.85−	12.41	0.2 / 2.3	+1.2 / 0	+2.3 / +1.4	0.2 / 3.4	+2.0 / 0	+3.4 / +2.2	1.5 / 4.7	+2.0 / 0	+4.7 / +3.5
12.41−	15.75	0.2 / 2.6	+1.4 / 0	+2.6 / +1.6	0.3 / 3.9	+2.2 / 0	+3.9 / +2.5	2.3 / 5.9	+2.2 / 0	+5.9 / +4.5
15.75−	19.69	0.2 / 2.8	+1.6 / 0	+2.8 / +1.8	0.3 / 4.4	+2.5 / 0	+4.4 / +2.8	2.5 / 6.6	+2.5 / 0	+6.6 / +5.0

All data in this table are in accordance with American-British-Canadian (ABC) agreements. Limits for sizes above 19.69 inches are not covered by ABC agreements but are given in the ANSI Standard.

Symbols H7, p6, etc. are hole and shaft designations in ABC system.

* Pairs of values shown represent minimum and maximum amounts of interference resulting from application of standard tolerance limits.

Table 6. ANSI Standard Force and Shrink Fits (ANSI B4.1-1967, R1979)

Values shown below are in thousandths of an inch

Nominal Size Range, Inches Over / To	Class FN 1 Interference*	Class FN 1 Hole H6	Class FN 1 Shaft	Class FN 2 Interference*	Class FN 2 Hole H7	Class FN 2 Shaft s6	Class FN 3 Interference*	Class FN 3 Hole H7	Class FN 3 Shaft t6	Class FN 4 Interference*	Class FN 4 Hole H7	Class FN 4 Shaft u6	Class FN 5 Interference*	Class FN 5 Hole H8	Class FN 5 Shaft x7
0–0.12	0.05 / 0.5	+0.25 / 0	+0.5 / +0.3	0.2 / 0.85	+0.4 / 0	+0.85 / +0.6				0.3 / 0.95	+0.4 / 0	+0.95 / +0.7	0.3 / 1.3	+0.6 / 0	+1.3 / +0.9
0.12–0.24	0.1 / 0.6	+0.3 / 0	+0.6 / +0.4	0.2 / 1.0	+0.5 / 0	+1.0 / +0.7				0.4 / 1.2	+0.5 / 0	+1.2 / +0.9	0.5 / 1.7	+0.7 / 0	+1.7 / +1.2
0.24–0.40	0.1 / 0.75	+0.4 / 0	+0.75 / +0.5	0.4 / 1.4	+0.6 / 0	+1.4 / +1.0				0.6 / 1.6	+0.6 / 0	+1.6 / +1.2	0.5 / 2.0	+0.9 / 0	+2.0 / +1.4
0.40–0.56	0.1 / 0.8	+0.4 / 0	+0.8 / +0.5	0.5 / 1.6	+0.7 / 0	+1.6 / +1.2				0.7 / 1.8	+0.7 / 0	+1.8 / +1.4	0.6 / 2.3	+1.0 / 0	+2.3 / +1.6
0.56–0.71	0.2 / 0.9	+0.4 / 0	+0.9 / +0.6	0.5 / 1.6	+0.7 / 0	+1.6 / +1.2				0.7 / 1.8	+0.7 / 0	+1.8 / +1.4	0.8 / 2.5	+1.0 / 0	+2.5 / +1.8
0.71–0.95	0.2 / 1.1	+0.5 / 0	+1.1 / +0.7	0.6 / 1.9	+0.8 / 0	+1.9 / +1.4				0.8 / 2.1	+0.8 / 0	+2.1 / +1.6	1.0 / 3.0	+1.2 / 0	+3.0 / +2.2
0.95–1.19	0.3 / 1.2	+0.5 / 0	+1.2 / +0.8	0.6 / 1.9	+0.8 / 0	+1.9 / +1.4	0.8 / 2.1	+0.8 / 0	+2.1 / +1.6	1.0 / 2.3	+0.8 / 0	+2.3 / +1.8	1.3 / 3.3	+1.2 / 0	+3.3 / +2.5
1.19–1.58	0.3 / 1.3	+0.6 / 0	+1.3 / +0.9	0.8 / 2.4	+1.0 / 0	+2.4 / +1.8	1.0 / 2.6	+1.0 / 0	+2.6 / +2.0	1.5 / 3.1	+1.0 / 0	+3.1 / +2.5	1.4 / 4.0	+1.6 / 0	+4.0 / +3.0
1.58–1.97	0.4 / 1.4	+0.6 / 0	+1.4 / +1.0	0.8 / 2.4	+1.0 / 0	+2.4 / +1.8	1.2 / 2.8	+1.0 / 0	+2.8 / +2.2	1.8 / 3.4	+1.0 / 0	+3.4 / +2.8	2.4 / 5.0	+1.6 / 0	+5.0 / +4.0
1.97–2.56	0.6 / 1.8	+0.7 / 0	+1.8 / +1.3	0.8 / 2.7	+1.2 / 0	+2.7 / +2.0	1.3 / 3.2	+1.2 / 0	+3.2 / +2.5	2.3 / 4.2	+1.2 / 0	+4.2 / +3.5	3.2 / 6.2	+1.8 / 0	+6.2 / +5.0
2.56–3.15	0.7 / 1.9	+0.7 / 0	+1.9 / +1.4	1.0 / 2.9	+1.2 / 0	+2.9 / +2.2	1.8 / 3.7	+1.2 / 0	+3.7 / +3.0	2.8 / 4.7	+1.2 / 0	+4.7 / +4.0	4.2 / 7.2	+1.8 / 0	+7.2 / +6.0
3.15–3.94	0.9 / 2.4	+0.9 / 0	+2.4 / +1.8	1.4 / 3.7	+1.4 / 0	+3.7 / +2.8	2.1 / 4.4	+1.4 / 0	+4.4 / +3.5	3.6 / 5.9	+1.4 / 0	+5.9 / +5.0	4.8 / 8.4	+2.2 / 0	+8.4 / +7.0
3.94–4.73	1.1 / 2.6	+0.9 / 0	+2.6 / +2.0	1.6 / 3.9	+1.4 / 0	+3.9 / +3.0	2.6 / 4.9	+1.4 / 0	+4.9 / +4.0	4.6 / 6.9	+1.4 / 0	+6.9 / +6.0	5.8 / 9.4	+2.2 / 0	+9.4 / +8.0

See footnotes at end of table.

Table 6 (Concluded). ANSI Standard Force and Shrink Fits (ANSI B4.1-1967, R1979)

Values shown below are in thousandths of an inch

Nominal Size Range, Inches Over–To	Class FN 1 Inter-ference*	Class FN 1 Hole H6	Class FN 1 Shaft	Class FN 2 Inter-ference*	Class FN 2 Hole H7	Class FN 2 Shaft s6	Class FN 3 Inter-ference*	Class FN 3 Hole H7	Class FN 3 Shaft t6	Class FN 4 Inter-ference*	Class FN 4 Hole H7	Class FN 4 Shaft u6	Class FN 5 Inter-ference*	Class FN 5 Hole H8	Class FN 5 Shaft x7
4.73– 5.52	1.2 / 2.9	+1.0 / 0	+2.9 / +2.2	1.9 / 4.5	+1.6 / 0	+4.5 / +3.5	3.4 / 6.0	+1.6 / 0	+6.0 / +5.0	5.4 / 8.0	+1.6 / 0	+8.0 / +7.0	7.5 / 11.6	+2.5 / 0	+11.6 / +10.0
5.52– 6.30	1.5 / 3.2	+1.0 / 0	+3.2 / +2.5	2.4 / 5.0	+1.6 / 0	+5.0 / +4.0	3.4 / 6.0	+1.6 / 0	+6.0 / +5.0	5.4 / 8.0	+1.6 / 0	+8.0 / +7.0	9.5 / 13.6	+2.5 / 0	+13.6 / +12.0
6.30– 7.09	1.8 / 3.5	+1.0 / 0	+3.5 / +2.8	2.9 / 5.5	+1.6 / 0	+5.5 / +4.5	4.4 / 7.0	+1.6 / 0	+7.0 / +6.0	6.4 / 9.0	+1.6 / 0	+9.0 / +8.0	9.5 / 13.6	+2.5 / 0	+13.6 / +12.0
7.09– 7.88	1.8 / 3.8	+1.2 / 0	+3.8 / +3.0	3.2 / 6.2	+1.8 / 0	+6.2 / +5.0	5.2 / 8.2	+1.8 / 0	+8.2 / +7.0	7.2 / 10.2	+1.8 / 0	+10.2 / +9.0	11.2 / 15.8	+2.8 / 0	+15.8 / +14.0
7.88– 8.86	2.3 / 4.3	+1.2 / 0	+4.3 / +3.5	3.2 / 6.2	+1.8 / 0	+6.2 / +5.0	5.2 / 8.2	+1.8 / 0	+8.2 / +7.0	8.2 / 11.2	+1.8 / 0	+11.2 / +10.0	13.2 / 17.8	+2.8 / 0	+17.8 / +16.0
8.86– 9.85	2.3 / 4.3	+1.2 / 0	+4.3 / +3.5	4.2 / 7.2	+1.8 / 0	+7.2 / +6.0	6.2 / 9.2	+1.8 / 0	+9.2 / +8.0	10.2 / 13.2	+1.8 / 0	+13.2 / +12.0	13.2 / 17.8	+2.8 / 0	+17.8 / +16.0
9.85–11.03	2.8 / 4.9	+1.2 / 0	+4.9 / +4.0	4.0 / 7.2	+2.0 / 0	+7.2 / +6.0	7.0 / 10.2	+2.0 / 0	+10.2 / +9.0	10.0 / 13.2	+2.0 / 0	+13.2 / +12.0	15.0 / 20.0	+3.0 / 0	+20.0 / +18.0
11.03–12.41	2.8 / 4.9	+1.2 / 0	+4.9 / +4.0	5.0 / 8.2	+2.0 / 0	+8.2 / +7.0	7.0 / 10.2	+2.0 / 0	+10.2 / +9.0	12.0 / 15.2	+2.0 / 0	+15.2 / +14.0	17.0 / 22.0	+3.0 / 0	+22.0 / +20.0
12.41–13.98	3.1 / 5.5	+1.4 / 0	+5.5 / +4.5	5.8 / 9.4	+2.2 / 0	+9.4 / +8.0	7.8 / 11.4	+2.2 / 0	+11.4 / +10.0	13.8 / 17.4	+2.2 / 0	+17.4 / +16.0	18.5 / 24.2	+3.5 / 0	+24.2 / +22.0
13.98–15.75	3.6 / 6.1	+1.4 / 0	+6.1 / +5.0	5.8 / 9.4	+2.2 / 0	+9.4 / +8.0	9.8 / 13.4	+2.2 / 0	+13.4 / +12.0	15.8 / 19.4	+2.2 / 0	+19.4 / +18.0	21.5 / 27.2	+3.5 / 0	+27.2 / +25.0
15.75–17.72	4.4 / 7.0	+1.6 / 0	+7.0 / +6.0	6.5 / 10.6	+2.5 / 0	+10.6 / +9.0	9.5 / 13.6	+2.5 / 0	+13.6 / +12.0	17.5 / 21.6	+2.5 / 0	+21.6 / +20.0	24.0 / 30.5	+4.0 / 0	+30.5 / +28.0
17.72–19.69	4.4 / 7.0	+1.6 / 0	+7.0 / +6.0	7.5 / 11.6	+2.5 / 0	+11.6 / +10.0	11.5 / 15.6	+2.5 / 0	+15.6 / +14.0	19.5 / 23.6	+2.5 / 0	+23.6 / +22.0	26.0 / 32.5	+4.0 / 0	+32.5 / +30.0

All data above heavy lines are in accordance with American-British-Canadian (ABC) agreements. Symbols H6, H7, s6, etc. are hole and shaft designations in ABC system. Limits for sizes above 19.69 inches are not covered by ABC agreements but are given in the ANSI standard.
* Pairs of values shown represent minimum and maximum amounts of interference resulting from application of standard tolerance limits.

Maximum shaft = 2 − 0.0004 = 1.9996; minimum shaft = 2 − 0.0007 = 1.9993 inches
Minimum clearance = 0.0004; maximum clearance = 0.0012 inch

Example 2: Establish the limits for a Class LT 1 fit for a 2-inch diameter.
Maximum hole = 2 + 0.0012 = 2.0012; minimum hole = 2 inches
Maximum shaft = 2 + 0.0003 = 2.0003; minimum shaft = 2 − 0.0003 = 1.9997 inches
Maximum resulting *interference* = 0.0003; maximum resulting *clearance* = 0.0015 inch

Modified Standard Fits. — Fits having the same limits of clearance or interference as those shown in Tables 2 to 6 may sometimes have to be produced by using holes or shafts having limits of size other than those shown in these tables. This may be accomplished by using either a *Bilateral Hole (System B)* or a *Basic Shaft System (Symbol S)*. Both methods will result in non-standard holes and shafts.

Bilateral Hole Fits (Symbol B): The common case is where holes are produced with fixed tools, such as drills or reamers; to provide a longer wear life for such tools a bilateral tolerance is desired.

The symbols used for these fits are identical with those used for standard fits except that they are followed by the letter B. Thus, LC 4B is a clearance locational fit, Class 4, except that it is produced with a bilateral hole.

The limits of clearance or interference are identical with those shown in Tables 2 to 6 for the corresponding fits.

The hole tolerance, however, is changed so that the plus limit is that for one grade finer than the value shown in the tables and the minus limit equals the amount by which the plus limit was lowered. The shaft limits are both lowered by the same amount as the lower limit of size of the hole. The finer grade of tolerance required to make these modifications may be obtained from Table 1. For example, an LC 4B fit for a 6-inch diameter hole would have tolerance limits of + 4.0, − 2.0 (+ .0040 inch, − .0020 inch); the shaft would have tolerance limits of − 2.0, − 6.0 (− .0020 inch, − .0060 inch).

Basic Shaft Fits (Symbol S): For these fits the maximum size of the shaft is basic. The limits of clearance or interference are identical with those shown in Tables 2 to 6 for the corresponding fits and the symbols used for these fits are identical with those used for standard fits except that they are followed by the letter S. Thus, LC 4S is a clearance locational fit, Class 4, except that it is produced on a basic shaft basis.

The limits for hole and shaft as given in Tables 2 to 6 are increased for clearance fits (*decreased* for transition or interference fits) by the value of the upper shaft limit; that is, by the amount required to change the maximum shaft to the basic size.

American National Standard Preferred Metric Limits and Fits. — This standard (ANSI B4.2-1978) describes the ISO system of metric limits and fits for mating parts as approved for general engineering usage in the United States. It establishes: (1) the designation symbols used to define dimensional limits on drawings, material stock, related tools, gages, etc.; (2) the preferred basic sizes (first and second choices); (3) the preferred tolerance zones (first, second and third choices); (4) the preferred limits and fits for sizes (first choice only) up to and including 500 millimeters; and (5) the definitions of related terms.

The general terms "hole" and "shaft" can also be taken to refer to the space containing or contained by two parallel faces of any part, such as the width of a slot, the thickness of a key, etc.

Definitions. — The most important terms relating to limits and fits are shown in Fig. 1 and are defined as follows:

Basic Size: The size to which limits of deviation are assigned. The basic size is the same for both members of a fit. For example, it is designated by the numbers 40 in 40H7.

Deviation: The algebraic difference between a size and the corresponding basic size.

Upper Deviation: The algebraic difference between the maximum limit of size and the corresponding basic size.

Lower Deviation: The algebraic difference between the minimum limit of size and the corresponding basic size.

Fundamental Deviation: That one of the two deviations closest to the basic size. For example, it is designated by the letter H in 40H7.

Tolerance: The difference between the maximum and minimum size limits on a part.

Tolerance Zone: A zone representing the tolerance and its position in relation to the basic size.

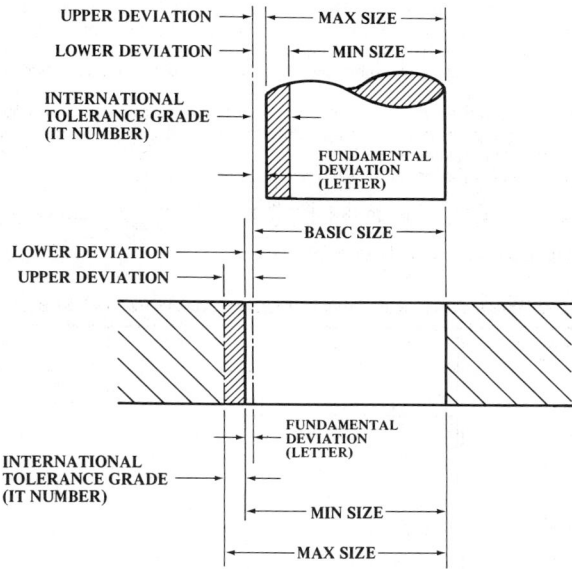

Fig. 1. Illustration of Definitions

International Tolerance Grade (IT): A group of tolerances which vary depending upon the basic size, but which provide the same relative level of accuracy within a given grade. For example, it is designated by the number 7 in 40H7 or as IT7.

Hole Basis: The system of fits where the minimum hole size is basic. The fundamental deviation for a hole basis system is H.

Shaft Basis: The system of fits where the maximum shaft size is basic. The fundamental deviation for a shaft basis system is h.

Clearance Fit: The relationship between assembled parts when clearance occurs under all tolerance conditions.

Interference Fit: The relationship between assembled parts when interference occurs under all tolerance conditions.

Transition Fit: The relationship between assembled parts when either a clearance or an interference fit can result, depending on the tolerance conditions of the mating parts.

Tolerances Designation. — An "International Tolerance grade" establishes the magnitude of the tolerance zone or the amount of part size variation allowed for external and internal dimensions alike (see Fig. 11). Tolerances are expressed in grade numbers which are consistent with International Tolerance grades identified by the prefix IT, such as IT6, IT11, etc. A smaller grade number provides a smaller tolerance zone.

A fundamental deviation establishes the position of the tolerance zone with respect to the basic size (see Fig. 1). Fundamental deviations are expressed by tolerance position letters. Capital letters are used for internal dimensions and lower case or small letters for external dimensions.

Symbols. — By combining the IT grade number and the tolerance position letter, the tolerance symbol is established which identifies the actual maximum and minimum limits of the part. The toleranced size is thus defined by the basic size of the part followed by a symbol composed of a letter and a number, such as 40H7, 40f7, etc.

A fit is indicated by the basic size common to both components, followed by a symbol corresponding to each component, the internal part symbol preceding the external part symbol, such as 40H8/f7.

Some methods of designating tolerances on drawings are:

$$\text{a. } 40H8 \qquad \text{b. } 40H8 \left(\begin{matrix} 40.039 \\ 40.000 \end{matrix} \right) \qquad \text{c. } \left(\begin{matrix} 40.039 \\ 40.000 \end{matrix} \right) 40H8$$

The values in parentheses indicate reference only.

Table 1. American National Standard Preferred Metric Sizes (ANSI B4.2-1978)

Basic Size, mm		Basic Size, mm		Basic Size, mm		Basic Size, mm	
1st Choice	2nd Choice	1st Choice	2nd Choice	1st Choice	2nd Choice	1st Choice	2nd Choice
1.0	. . .	. . .	7.0	25	. . .	. . .	90
. . .	1.1	8.0	. . .	. . .	26	100	. . .
1.2	. . .	. . .	9.0	. . .	28	. . .	110
. . .	1.4	10	. . .	30	. . .	120	. . .
1.6	. . .	. . .	11	. . .	32	. . .	130
. . .	1.8	12	. . .	35	. . .	140	. . .
2.0	. . .	. . .	13	. . .	38	. . .	150
. . .	2.2	14	. . .	40	. . .	160	. . .
2.5	. . .	. . .	15	. . .	42	. . .	170
. . .	2.8	16	. . .	45	. . .	180	. . .
3.0	. . .	. . .	17	. . .	48	. . .	190
. . .	3.5	18	. . .	50	. . .	200	. . .
4.0	. . .	. . .	19	. . .	55	. . .	220
. . .	4.5	20	. . .	60	. . .	250	. . .
5.0	. . .	. . .	21	. . .	65	. . .	280
. . .	5.5	22	. . .	70	300	. . .
6.0	. . .	. . .	23	. . .	75	. . .	320
. . .	6.5	. . .	24	80	. . .	. . .	. . .

Preferred Metric Sizes. — American National Standard ANSI B32.4M presents series of preferred metric sizes for round, square, rectangular, and hexagonal metal products. Table 1 on the previous page gives preferred metric diameters from 1 to 320 millimeters for round metal products. Wherever possible, sizes should be selected from the Preferred Series shown in the table. A Second Preference series is also shown. A Third Preference Series not shown in the table is: 1.3, 2.1, 2.4, 2.6, 3.2, 3.8, 4.2, 4.8, 7.5, 8.5, 9.5, 36, 85, and 95.

Most of the Preferred Series of sizes are derived from the American National Standard "10 series" of preferred numbers (See Preferred Numbers in Index). Most of the Second Preference Series are derived from the "20 series" of preferred numbers. Third Preference sizes are generally from the "40 series" of preferred numbers.

For preferred metric diameters less than 1 millimeter, for preferred across flat metric sizes of square and hexagon metal products, preferred across flat metric sizes of rectangle metal products, and preferred metric lengths of metal products, reference should be made to the Standard.

Preferred Fits. — First choice tolerance zones are used to establish preferred fits in the Standard for Preferred Metric Limits and Fits ANSI B4.2, as shown in Figs. 2 and 3. A complete listing of first, second and third choice tolerance zones is given in the Standard.

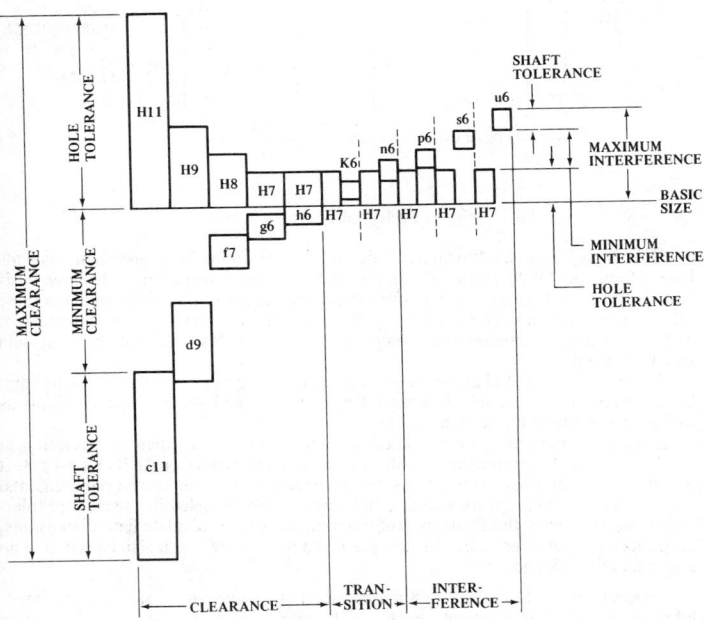

Fig. 2. Preferred Hole Basis Fits

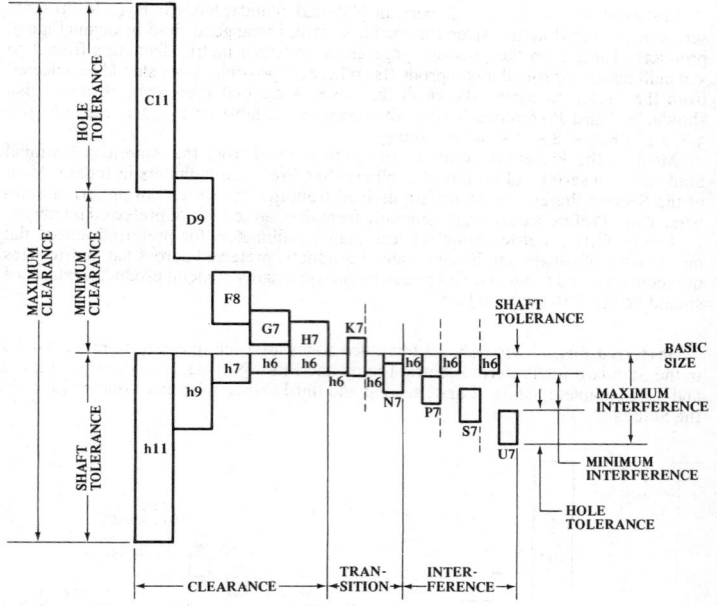

Fig. 3. Preferred Shaft Basis Fits

Hole basis fits have a fundamental deviation of H on the hole, and shaft basis fits have a fundamental deviation of h on the shaft and are shown in Fig. 2 for hole basis and Fig. 3 for shaft basis fits. A description of both types of fits which have the same relative fit condition is given in Fig. 4. Normally the hole basis system is preferred; however, when a common shaft mates with several holes, the shaft basis system should be used.

The hole basis and shaft basis fits shown in Fig. 4 are combined with the first choice sizes shown in Table 1 to form Tables 2, 3, 4, and 5 where specific limits as well as the resultant fits are tabulated.

If the required size is not tabulated in Tables 2 to 5, incl., then the preferred fit can be calculated from numerical values given in an appendix of ANSI B4.2-1978. It is anticipated that other fit conditions may be necessary to meet special requirements, and a preferred fit can be loosened or tightened simply by selecting a standard tolerance zone as given in the Standard. Information on how to calculate limit dimensions, clearances, and interferences, for non-preferred fits and sizes can also be found in an appendix of this Standard.

Applications. — Many factors such as length of engagement, bearing load, speed, lubrication, operating temperatures, humidity, surface texture, and materials must be taken into account in fit selections for a particular application. Choice of other than

(Concluded on page 659)

← More Clearance — More Interference →

ISO SYMBOL		DESCRIPTION	
Hole Basis	Shaft Basis		
H11/c11	C11/h11	*Loose running* fit for wide commercial tolerances or allowances on external members.	Clearance Fits
H9/d9	D9/h9	*Free running* fit not for use where accuracy is essential, but good for large temperature variations, high running speeds, or heavy journal pressures.	Clearance Fits
H8/f7	F8/h7	*Close running* fit for running on accurate machines and for accurate location at moderate speeds and journal pressures.	Clearance Fits
H7/g6	G7/h6	*Sliding* fit not intended to run freely, but to move and turn freely and locate accurately.	Clearance Fits
H7/h6	H7/h6	*Locational clearance* fit provides snug fit for locating stationary parts; but can be freely assembled and disassembled.	Clearance Fits
H7/k6	K7/h6	*Locational transition* fit for accurate location, a compromise between clearance and interference.	Transition Fits
H7/n6	N7/h6	*Locational transition* fit for more accurate location where greater interference is permissible.	Transition Fits
H7/p6[1]	P7/h6	*Locational interference* fit for parts requiring rigidity and alignment with, prime accuracy of location but without special bore pressure requirements.	Interference Fits
H7/s6	S7/h6	*Medium drive* fit for ordinary steel parts or shrink fits on light sections, the tightest fit usable with cast iron.	Interference Fits
H7/u6	U7/h6	*Force* fit suitable for parts which can be highly stressed or for shrink fits where the heavy pressing forces required are impractical.	Interference Fits

[1] Transition fit for basic sizes in range from 0 through 3 mm.

Fig. 4. Description of preferred fits.

Table 2. American National Standard Preferred Hole Basis Metric Clearance Fits (ANSI B4.2-1978)

Basic Size*		Loose Running			Free Running			Close Running			Sliding			Locational Clearance		
		Hole H11	Shaft c11	Fit†	Hole H9	Shaft d9	Fit†	Hole H8	Shaft f7	Fit†	Hole H7	Shaft g6	Fit†	Hole H7	Shaft h6	Fit†
1	Max	1.060	0.940	0.180	1.025	0.980	0.070	1.014	0.994	0.030	1.010	0.998	0.018	1.010	1.000	0.016
	Min	1.000	0.880	0.060	1.000	0.955	0.020	1.000	0.984	0.006	1.000	0.992	0.002	1.000	0.994	0.000
1.2	Max	1.260	1.140	0.180	1.225	1.180	0.070	1.214	1.194	0.030	1.210	1.198	0.018	1.210	1.200	0.016
	Min	1.200	1.080	0.060	1.200	1.155	0.020	1.200	1.184	0.006	1.200	1.192	0.002	1.200	1.194	0.000
1.6	Max	1.660	1.540	0.180	1.625	1.580	0.070	1.614	1.594	0.030	1.610	1.598	0.018	1.610	1.600	0.016
	Min	1.600	1.480	0.060	1.600	1.555	0.020	1.600	1.584	0.006	1.600	1.592	0.002	1.600	1.594	0.000
2	Max	2.060	1.940	0.180	2.025	1.980	0.070	2.014	1.994	0.030	2.010	1.998	0.018	2.010	2.000	0.016
	Min	2.000	1.880	0.060	2.000	1.955	0.020	2.000	1.984	0.006	2.000	1.992	0.002	2.000	1.994	0.000
2.5	Max	2.560	2.440	0.180	2.525	2.480	0.070	2.514	2.494	0.030	2.510	2.498	0.018	2.510	2.500	0.016
	Min	2.500	2.380	0.060	2.500	2.455	0.020	2.500	2.484	0.006	2.500	2.492	0.002	2.500	2.494	0.000
3	Max	3.060	2.940	0.180	3.025	2.980	0.070	3.014	2.994	0.030	3.010	2.998	0.018	3.010	3.000	0.016
	Min	3.000	2.880	0.060	3.000	2.955	0.020	3.000	2.984	0.006	3.000	2.992	0.002	3.000	2.994	0.000
4	Max	4.075	3.930	0.220	4.030	3.970	0.090	4.018	3.990	0.040	4.012	3.996	0.024	4.012	4.000	0.020
	Min	4.000	3.855	0.070	4.000	3.940	0.030	4.000	3.978	0.010	4.000	3.988	0.004	4.000	3.992	0.000
5	Max	5.075	4.930	0.220	5.030	4.970	0.090	5.018	4.990	0.040	5.012	4.996	0.024	5.012	5.000	0.020
	Min	5.000	4.855	0.070	5.000	4.940	0.030	5.000	4.978	0.010	5.000	4.988	0.004	5.000	4.992	0.000
6	Max	6.075	5.930	0.220	6.030	5.970	0.090	6.018	5.990	0.040	6.012	5.996	0.024	6.012	6.000	0.020
	Min	6.000	5.855	0.070	6.000	5.940	0.030	6.000	5.978	0.010	6.000	5.988	0.004	6.000	5.992	0.000
8	Max	8.090	7.920	0.260	8.036	7.960	0.112	8.022	7.987	0.050	8.015	7.995	0.029	8.015	8.000	0.024
	Min	8.000	7.830	0.080	8.000	7.924	0.040	8.000	7.972	0.013	8.000	7.986	0.005	8.000	7.991	0.000
10	Max	10.090	9.920	0.260	10.036	9.960	0.112	10.022	9.987	0.050	10.015	9.995	0.029	10.015	10.000	0.024
	Min	10.000	9.830	0.080	10.000	9.924	0.040	10.000	9.972	0.013	10.000	9.986	0.005	10.000	9.991	0.000
12	Max	12.110	11.905	0.315	12.043	11.956	0.136	12.027	11.984	0.061	12.018	11.994	0.035	12.018	12.000	0.029
	Min	12.000	11.795	0.095	12.000	11.907	0.050	12.000	11.966	0.016	12.000	11.983	0.006	12.000	11.989	0.000
16	Max	16.110	15.905	0.315	16.043	15.950	0.136	16.027	15.984	0.061	16.018	15.994	0.035	16.018	16.000	0.029
	Min	16.000	15.795	0.095	16.000	15.907	0.050	16.000	15.966	0.016	16.000	15.983	0.006	16.000	15.989	0.000
20	Max	20.130	19.890	0.370	20.052	19.935	0.169	20.033	19.980	0.074	20.021	19.993	0.041	20.021	20.000	0.034
	Min	20.000	19.760	0.110	20.000	19.883	0.065	20.000	19.959	0.020	20.000	19.980	0.007	20.000	19.987	0.000

All dimensions are in millimeters.

*The sizes shown are first choice basic sizes (see Table 1). Preferred fits for other sizes can be calculated from data given in ANSI B4.2-1978.

† All fits shown in this table have clearance.

Table 2 (Concluded). American National Standard Preferred Hole Basis Metric Clearance Fits (ANSI B4.2-1978)

Basic Size*		Loose Running			Free Running			Close Running			Sliding			Locational Clearance		
		Hole H11	Shaft c11	Fit†	Hole H9	Shaft d9	Fit†	Hole H8	Shaft f7	Fit†	Hole H7	Shaft g6	Fit†	Hole H7	Shaft h6	Fit†
25	Max	25.130	24.890	0.370	25.052	24.935	0.169	25.033	24.980	0.074	25.021	24.993	0.041	25.021	25.000	0.034
	Min	25.000	24.760	0.110	25.000	24.883	0.065	25.000	24.959	0.020	25.000	24.980	0.007	25.000	24.987	0.000
30	Max	30.130	29.890	0.370	30.052	29.935	0.169	30.033	29.980	0.074	30.021	29.993	0.041	30.021	30.000	0.034
	Min	30.000	29.760	0.110	30.000	29.883	0.065	30.000	29.959	0.020	30.000	29.980	0.007	30.000	29.987	0.000
40	Max	40.160	39.880	0.440	40.062	39.920	0.204	40.039	39.975	0.089	40.025	39.991	0.050	40.025	40.000	0.041
	Min	40.000	39.720	0.120	40.000	39.858	0.080	40.000	39.950	0.025	40.000	39.975	0.009	40.000	39.984	0.000
50	Max	50.160	49.870	0.450	50.062	49.920	0.204	50.039	49.975	0.089	50.025	49.991	0.050	50.025	50.000	0.041
	Min	50.000	49.710	0.130	50.000	49.858	0.080	50.000	49.950	0.025	50.000	49.975	0.009	50.000	49.984	0.000
60	Max	60.190	59.860	0.520	60.074	59.900	0.248	60.046	59.970	0.106	60.030	59.990	0.059	60.030	60.000	0.049
	Min	60.000	59.670	0.140	60.000	59.826	0.100	60.000	59.940	0.030	60.000	59.971	0.010	60.000	59.981	0.000
80	Max	80.190	79.850	0.530	80.074	79.900	0.248	80.046	79.970	0.106	80.030	79.990	0.059	80.030	80.000	0.049
	Min	80.000	79.660	0.150	80.000	79.826	0.100	80.000	79.940	0.030	80.000	79.971	0.010	80.000	79.981	0.000
100	Max	100.220	99.830	0.610	100.087	99.880	0.294	100.054	99.964	0.125	100.035	99.988	0.069	100.035	100.000	0.057
	Min	100.000	99.610	0.170	100.000	99.793	0.120	100.000	99.929	0.036	100.000	99.966	0.012	100.000	99.978	0.000
120	Max	120.220	119.820	0.620	120.087	119.880	0.294	120.054	119.964	0.125	120.035	119.988	0.069	120.035	120.000	0.057
	Min	120.000	119.600	0.180	120.000	119.793	0.120	120.000	119.929	0.036	120.000	119.966	0.012	120.000	119.978	0.000
160	Max	160.250	159.790	0.710	160.100	159.855	0.345	160.063	159.957	0.146	160.040	159.986	0.079	160.040	160.000	0.065
	Min	160.000	159.540	0.210	160.000	159.755	0.145	160.000	159.917	0.043	160.000	159.961	0.014	160.000	159.975	0.000
200	Max	200.290	199.760	0.820	200.115	199.830	0.400	200.072	199.950	0.168	200.046	199.985	0.090	200.046	200.000	0.075
	Min	200.000	199.470	0.240	200.000	199.715	0.170	200.000	199.904	0.050	200.000	199.956	0.015	200.000	199.971	0.000
250	Max	250.290	249.720	0.860	250.115	249.830	0.400	250.072	249.950	0.168	250.046	249.985	0.090	250.046	250.000	0.075
	Min	250.000	249.430	0.280	250.000	249.715	0.170	250.000	249.904	0.050	250.000	249.956	0.015	250.000	249.971	0.000
300	Max	300.320	299.670	0.970	300.130	299.810	0.450	300.081	299.944	0.189	300.052	299.983	0.101	300.052	300.000	0.084
	Min	300.000	299.350	0.330	300.000	299.680	0.190	300.000	299.892	0.056	300.000	299.951	0.017	300.000	299.968	0.000
400	Max	400.360	399.600	1.120	400.140	399.790	0.490	400.089	399.938	0.208	400.057	399.982	0.111	400.057	400.000	0.093
	Min	400.000	399.240	0.400	400.000	399.650	0.210	400.000	399.881	0.062	400.000	399.946	0.018	400.000	399.964	0.000
500	Max	500.400	499.520	1.280	500.155	499.770	0.540	500.097	499.932	0.228	500.063	499.980	0.123	500.063	500.000	0.103
	Min	500.000	499.120	0.480	500.000	499.615	0.230	500.000	499.869	0.068	500.000	499.940	0.020	500.000	499.960	0.000

All dimensions are in millimeters.
*The sizes shown are first choice basic sizes (see Table 1). Preferred fits for other sizes can be calculated from data given in ANSI B4.2-1978.
†All fits shown in this table have clearance.

Table 3. American National Standard Preferred Hole Basis Metric Transition and Interference Fits (ANSI B4.2-1978)

Basic Size*		Locational Transition			Locational Transition			Locational Interference			Medium Drive			Force		
		Hole H7	Shaft k6	Fit†	Hole H7	Shaft n6	Fit†	Hole H7	Shaft p6	Fit†	Hole H7	Shaft s6	Fit†	Hole H7	Shaft u6	Fit†
1	Max	1.010	1.006	+0.010	1.010	1.010	+0.006	1.010	1.012	+0.004	1.010	1.020	-0.004	1.010	1.024	-0.008
	Min	1.000	1.000	-0.006	1.000	1.004	-0.010	1.000	1.006	-0.012	1.000	1.014	-0.020	1.000	1.018	-0.024
1.2	Max	1.210	1.206	+0.010	1.210	1.210	+0.006	1.210	1.212	+0.004	1.210	1.220	-0.004	1.210	1.224	-0.008
	Min	1.200	1.200	-0.006	1.200	1.204	-0.010	1.200	1.206	-0.012	1.200	1.214	-0.020	1.200	1.218	-0.024
1.6	Max	1.610	1.606	+0.010	1.610	1.610	+0.006	1.610	1.612	+0.004	1.610	1.620	-0.004	1.610	1.624	-0.008
	Min	1.600	1.600	-0.006	1.600	1.604	-0.010	1.600	1.606	-0.012	1.600	1.614	-0.020	1.600	1.618	-0.024
2	Max	2.010	2.006	+0.010	2.010	2.010	+0.006	2.010	2.012	+0.004	2.010	2.020	-0.004	2.010	2.024	-0.008
	Min	2.000	2.000	-0.006	2.000	2.004	-0.010	2.000	2.006	-0.012	2.000	2.014	-0.020	2.000	2.018	-0.024
2.5	Max	2.510	2.506	+0.010	2.510	2.510	+0.006	2.510	2.512	+0.004	2.510	2.520	-0.004	2.510	2.524	-0.008
	Min	2.500	2.500	-0.006	2.500	2.504	-0.010	2.500	2.506	-0.012	2.500	2.514	-0.020	2.500	2.518	-0.024
3	Max	3.010	3.006	+0.010	3.010	3.010	+0.006	3.010	3.012	+0.004	3.010	3.020	-0.004	3.010	3.024	-0.008
	Min	3.000	3.000	-0.006	3.000	3.004	-0.010	3.000	3.006	-0.012	3.000	3.014	-0.020	3.000	3.018	-0.024
4	Max	4.012	4.009	+0.011	4.012	4.016	+0.004	4.012	4.020	0.000	4.012	4.027	-0.007	4.012	4.031	-0.011
	Min	4.000	4.001	-0.009	4.000	4.008	-0.016	4.000	4.012	-0.020	4.000	4.019	-0.027	4.000	4.023	-0.031
5	Max	5.012	5.009	+0.011	5.012	5.016	+0.004	5.012	5.020	0.000	5.012	5.027	-0.007	5.012	5.031	-0.011
	Min	5.000	5.001	-0.009	5.000	5.008	-0.016	5.000	5.012	-0.020	5.000	5.019	-0.027	5.000	5.023	-0.031
6	Max	6.012	6.009	+0.011	6.012	6.016	+0.004	6.012	6.020	0.000	6.012	6.027	-0.007	6.012	6.031	-0.011
	Min	6.000	6.001	-0.009	6.000	6.008	-0.016	6.000	6.012	-0.020	6.000	6.019	-0.027	6.000	6.023	-0.031
8	Max	8.015	8.010	+0.014	8.015	8.019	+0.005	8.015	8.024	0.000	8.015	8.032	-0.008	8.015	8.037	-0.013
	Min	8.000	8.001	-0.010	8.000	8.010	-0.019	8.000	8.015	-0.024	8.000	8.023	-0.032	8.000	8.028	-0.037
10	Max	10.015	10.010	+0.014	10.015	10.019	+0.005	10.015	10.024	0.000	10.015	10.032	-0.008	10.015	10.037	-0.013
	Min	10.000	10.001	-0.010	10.000	10.010	-0.019	10.000	10.015	-0.024	10.000	10.023	-0.032	10.000	10.028	-0.037
12	Max	12.018	12.012	+0.017	12.018	12.023	+0.006	12.018	12.029	0.000	12.018	12.039	-0.010	12.018	12.044	-0.015
	Min	12.000	12.001	-0.012	12.000	12.012	-0.023	12.000	12.018	-0.029	12.000	12.028	-0.039	12.000	12.033	-0.044
16	Max	16.018	16.012	+0.017	16.018	16.023	+0.006	16.018	16.029	0.000	16.018	16.039	-0.010	16.018	16.044	-0.015
	Min	16.000	16.001	-0.012	16.000	16.012	-0.023	16.000	16.018	-0.029	16.000	16.028	-0.039	16.000	16.033	-0.044
20	Max	20.021	20.015	+0.019	20.021	20.028	+0.006	20.021	20.035	-0.001	20.021	20.048	-0.014	20.021	20.054	-0.020
	Min	20.000	20.002	-0.015	20.000	20.015	-0.028	20.000	20.022	-0.035	20.000	20.035	-0.048	20.000	20.041	-0.054

All dimensions are in millimeters.

*The sizes shown are first choice basic sizes (see Table 1). Preferred fits for other sizes can be calculated from data given in ANSI B4.2-1978.

†A plus sign indicates clearance; a minus sign indicates interference.

Table 3 (Concluded). American National Standard Preferred Hole Basis Metric Transition and Interference Fits (ANSI B4.2-1978)

Basic Size*		Locational Transition			Locational Transition			Locational Interference			Medium Drive			Force		
		Hole H7	Shaft k6	Fit†	Hole H7	Shaft n6	Fit†	Hole H7	Shaft p6	Fit†	Hole H7	Shaft s6	Fit†	Hole H7	Shaft u6	Fit†
25	Max	25.021	25.015	+0.019	25.021	25.028	+0.006	25.021	25.035	-0.001	25.021	25.048	-0.014	25.021	25.061	-0.027
	Min	25.000	25.002	-0.015	25.000	25.015	-0.028	25.000	25.022	-0.035	25.000	25.035	-0.048	25.000	25.048	-0.061
30	Max	30.021	30.015	+0.019	30.021	30.028	+0.006	30.021	30.035	-0.001	30.021	30.048	-0.014	30.021	30.061	-0.027
	Min	30.000	30.002	-0.015	30.000	30.015	-0.028	30.000	30.022	-0.035	30.000	30.035	-0.048	30.000	30.048	-0.061
40	Max	40.025	40.018	+0.023	40.025	40.033	+0.008	40.025	40.042	-0.001	40.025	40.059	-0.018	40.025	40.076	-0.035
	Min	40.000	40.002	-0.018	40.000	40.017	-0.033	40.000	40.026	-0.042	40.000	40.043	-0.059	40.000	40.060	-0.076
50	Max	50.025	50.018	+0.023	50.025	50.033	+0.008	50.025	50.042	-0.001	50.025	50.059	-0.018	50.025	50.086	-0.045
	Min	50.000	50.002	-0.018	50.000	50.017	-0.033	50.000	50.026	-0.042	50.000	50.043	-0.059	50.000	50.070	-0.086
60	Max	60.030	60.021	+0.028	60.030	60.039	+0.010	60.030	60.051	-0.002	60.030	60.072	-0.023	60.030	60.106	-0.057
	Min	60.000	60.002	-0.021	60.000	60.020	-0.039	60.000	60.032	-0.051	60.000	60.053	-0.072	60.000	60.087	-0.106
80	Max	80.030	80.021	+0.028	80.030	80.039	+0.010	80.030	80.051	-0.002	80.030	80.078	-0.029	80.030	80.121	-0.072
	Min	80.000	80.002	-0.021	80.000	80.020	-0.039	80.000	80.032	-0.051	80.000	80.059	-0.078	80.000	80.102	-0.121
100	Max	100.035	100.025	+0.032	100.035	100.045	+0.012	100.035	100.059	-0.002	100.035	100.093	-0.036	100.035	100.146	-0.089
	Min	100.000	100.003	-0.025	100.000	100.023	-0.045	100.000	100.037	-0.059	100.000	100.071	-0.093	100.000	100.124	-0.146
120	Max	120.035	120.025	+0.032	120.035	120.045	+0.012	120.035	120.059	-0.002	120.035	120.101	-0.044	120.035	120.166	-0.109
	Min	120.000	120.003	-0.025	120.000	120.023	-0.045	120.000	120.037	-0.059	120.000	120.079	-0.101	120.000	120.144	-0.166
160	Max	160.040	160.028	+0.037	160.040	160.052	+0.013	160.040	160.068	-0.003	160.040	160.125	-0.060	160.040	160.215	-0.150
	Min	160.000	160.003	-0.028	160.000	160.027	-0.052	160.000	160.043	-0.068	160.000	160.100	-0.125	160.000	160.190	-0.215
200	Max	200.046	200.033	+0.042	200.046	200.060	+0.015	200.046	200.079	-0.004	200.046	200.151	-0.076	200.046	200.265	-0.190
	Min	200.000	200.004	-0.033	200.000	200.031	-0.060	200.000	200.050	-0.079	200.000	200.122	-0.151	200.000	200.236	-0.265
250	Max	250.046	250.033	+0.042	250.046	250.060	+0.015	250.046	250.079	-0.004	250.046	250.169	-0.094	250.046	250.313	-0.238
	Min	250.000	250.004	-0.033	250.000	250.031	-0.060	250.000	250.050	-0.079	250.000	250.140	-0.169	250.000	250.284	-0.313
300	Max	300.052	300.036	+0.048	300.052	300.066	+0.018	300.052	300.088	-0.004	300.052	300.202	-0.118	300.052	300.382	-0.298
	Min	300.000	300.004	-0.036	300.000	300.034	-0.066	300.000	300.056	-0.088	300.000	300.170	-0.202	300.000	300.350	-0.382
400	Max	400.057	400.040	+0.053	400.057	400.073	+0.020	400.057	400.098	-0.005	400.057	400.244	-0.151	400.057	400.471	-0.378
	Min	400.000	400.004	-0.040	400.000	400.037	-0.073	400.000	400.062	-0.098	400.000	400.208	-0.244	400.000	400.435	-0.471
500	Max	500.063	500.045	+0.058	500.063	500.080	+0.023	500.063	500.108	-0.005	500.063	500.292	-0.189	500.063	500.580	-0.477
	Min	500.000	500.005	-0.045	500.000	500.040	-0.080	500.000	500.068	-0.108	500.000	500.252	-0.292	500.000	500.540	-0.580

All dimensions are in millimeters.

*The sizes shown are first choice basic sizes (see Table 1). Preferred fits for other sizes can be calculated from data given in ANSI B4.2-1978.

†A plus sign indicates clearance; a minus sign indicates interference.

Table 4. American National Standard Preferred Shaft Basis Metric Clearance Fits (ANSI B4.2-1978)

Basic Size*		Loose Running			Free Running			Close Running			Sliding			Locational Clearance		
		Hole C11	Shaft h11	Fit†	Hole D9	Shaft h9	Fit†	Hole F8	Shaft h7	Fit†	Hole G7	Shaft h6	Fit†	Hole H7	Shaft h6	Fit†
1	Max	1.120	1.000	0.180	1.045	1.000	0.070	1.020	1.000	0.030	1.012	1.000	0.018	1.010	1.000	0.016
	Min	1.060	0.940	0.060	1.020	0.975	0.020	1.006	0.990	0.006	1.002	0.994	0.002	1.000	0.994	0.000
1.2	Max	1.320	1.200	0.180	1.245	1.200	0.070	1.220	1.200	0.030	1.212	1.200	0.018	1.210	1.200	0.016
	Min	1.260	1.140	0.060	1.220	1.175	0.020	1.206	1.190	0.006	1.202	1.194	0.002	1.200	1.194	0.000
1.6	Max	1.720	1.600	0.180	1.645	1.600	0.070	1.620	1.600	0.030	1.612	1.600	0.018	1.610	1.600	0.016
	Min	1.660	1.540	0.060	1.620	1.575	0.020	1.606	1.590	0.006	1.602	1.594	0.002	1.600	1.594	0.000
2	Max	2.120	2.000	0.180	2.045	2.000	0.070	2.020	2.000	0.030	2.012	2.000	0.018	2.010	2.000	0.016
	Min	2.060	1.940	0.060	2.020	1.975	0.020	2.006	1.990	0.006	2.002	1.994	0.002	2.000	1.994	0.000
2.5	Max	2.620	2.500	0.180	2.545	2.500	0.070	2.520	2.500	0.030	2.512	2.500	0.018	2.510	2.500	0.016
	Min	2.560	2.440	0.060	2.520	2.475	0.020	2.506	2.490	0.006	2.502	2.494	0.002	2.500	2.494	0.000
3	Max	3.120	3.000	0.180	3.045	3.000	0.070	3.020	3.000	0.030	3.012	3.000	0.018	3.010	3.000	0.016
	Min	3.060	2.940	0.060	3.020	2.975	0.020	3.006	2.990	0.006	3.002	2.994	0.002	3.000	2.994	0.000
4	Max	4.145	4.000	0.220	4.060	4.000	0.090	4.028	4.000	0.040	4.016	4.000	0.024	4.012	4.000	0.020
	Min	4.070	3.925	0.070	4.030	3.970	0.030	4.010	3.988	0.010	4.004	3.992	0.004	4.000	3.992	0.000
5	Max	5.145	5.000	0.220	5.060	5.000	0.090	5.028	5.000	0.040	5.016	5.000	0.024	5.012	5.000	0.020
	Min	5.070	4.925	0.070	5.030	4.970	0.030	5.010	4.988	0.010	5.004	4.992	0.004	5.000	4.992	0.000
6	Max	6.145	6.000	0.220	6.060	6.000	0.090	6.028	6.000	0.040	6.016	6.000	0.024	6.012	6.000	0.020
	Min	6.070	5.925	0.070	6.030	5.970	0.030	6.010	5.988	0.010	6.004	5.992	0.004	6.000	5.992	0.000
8	Max	8.170	8.000	0.260	8.076	8.000	0.112	8.035	8.000	0.050	8.020	8.000	0.029	8.015	8.000	0.024
	Min	8.080	7.910	0.080	8.040	7.964	0.040	8.013	7.985	0.013	8.005	7.991	0.005	8.000	7.991	0.000
10	Max	10.170	10.000	0.260	10.076	10.000	0.112	10.035	10.000	0.050	10.020	10.000	0.029	10.015	10.000	0.024
	Min	10.080	9.910	0.080	10.040	9.964	0.040	10.013	9.985	0.013	10.005	9.991	0.005	10.000	9.991	0.000
12	Max	12.205	12.000	0.315	12.093	12.000	0.136	12.043	12.000	0.061	12.024	12.000	0.035	12.018	12.000	0.029
	Min	12.095	11.890	0.095	12.050	11.957	0.050	12.016	11.982	0.016	12.006	11.989	0.006	12.000	11.989	0.000
16	Max	16.205	16.000	0.315	16.093	16.000	0.136	16.043	16.000	0.061	16.024	16.000	0.035	16.018	16.000	0.029
	Min	16.095	15.890	0.095	16.050	15.957	0.050	16.016	15.982	0.016	16.006	15.989	0.006	16.000	15.989	0.000
20	Max	20.240	20.000	0.370	20.117	20.000	0.169	20.053	20.000	0.074	20.028	20.000	0.041	20.021	20.000	0.034
	Min	20.110	19.870	0.110	20.065	19.948	0.065	20.020	19.979	0.020	20.007	19.987	0.007	20.000	19.987	0.000

All dimensions are in millimeters.
*The sizes shown are first choice basic sizes (see Table 1). Preferred fits for other sizes can be calculated from data given in ANSI B4.2-1978.
†All fits shown in this table have clearance.

Table 4 (*Concluded*). American National Standard Preferred Shaft Basis Metric Clearance Fits (ANSI B4.2-1978)

Basic Size*		Loose Running			Free Running			Close Running			Sliding			Locational Clearance		
		Hole C11	Shaft h11	Fit†	Hole D9	Shaft h9	Fit†	Hole F8	Shaft h7	Fit†	Hole G7	Shaft h6	Fit†	Hole H7	Shaft h6	Fit†
25	Max	25.240	25.000	0.370	25.117	25.000	0.169	25.053	25.000	0.074	25.028	25.000	0.041	25.021	25.000	0.034
	Min	25.110	24.870	0.110	25.065	24.948	0.065	25.020	24.979	0.020	25.007	24.987	0.007	25.000	24.987	0.000
30	Max	30.240	30.000	0.370	30.117	30.000	0.169	30.053	30.000	0.074	30.028	30.000	0.041	30.021	30.000	0.034
	Min	30.110	29.870	0.110	30.065	29.948	0.065	30.020	29.979	0.020	30.007	29.987	0.007	30.000	29.987	0.000
40	Max	40.280	40.000	0.440	40.142	40.000	0.204	40.064	40.000	0.089	40.034	40.000	0.050	40.025	40.000	0.041
	Min	40.120	39.840	0.120	40.080	39.938	0.080	40.025	39.975	0.025	40.009	39.984	0.009	40.000	39.984	0.000
50	Max	50.290	50.000	0.450	50.142	50.000	0.204	50.064	50.000	0.089	50.034	50.000	0.050	50.025	50.000	0.041
	Min	50.130	49.840	0.130	50.080	49.938	0.080	50.025	49.975	0.025	50.009	49.984	0.009	50.000	49.984	0.000
60	Max	60.330	60.000	0.520	60.174	60.000	0.248	60.076	60.000	0.106	60.040	60.000	0.049	60.030	60.000	0.049
	Min	60.140	59.810	0.140	60.100	59.926	0.100	60.030	59.970	0.030	60.010	59.981	0.010	60.000	59.981	0.000
80	Max	80.340	80.000	0.530	80.174	80.000	0.248	80.076	80.000	0.106	80.040	80.000	0.049	80.030	80.000	0.049
	Min	80.150	79.810	0.150	80.100	79.926	0.100	80.030	79.970	0.030	80.010	79.981	0.010	80.000	79.981	0.000
100	Max	100.390	100.000	0.610	100.207	100.000	0.294	100.090	100.000	0.125	100.047	100.000	0.069	100.035	100.000	0.057
	Min	100.170	99.780	0.170	100.120	99.913	0.120	100.036	99.965	0.036	100.012	99.978	0.012	100.000	99.978	0.000
120	Max	120.400	120.000	0.620	120.207	120.000	0.294	120.090	120.000	0.125	120.047	120.000	0.069	120.035	120.000	0.057
	Min	120.180	119.780	0.180	120.120	119.913	0.120	120.036	119.965	0.036	120.012	119.978	0.012	120.000	119.978	0.000
160	Max	160.460	160.000	0.710	160.245	160.000	0.345	160.106	160.000	0.146	160.054	160.000	0.079	160.040	160.000	0.065
	Min	160.210	159.750	0.210	160.145	159.900	0.145	160.043	159.960	0.043	160.014	159.975	0.014	160.000	159.975	0.000
200	Max	200.530	200.000	0.820	200.285	200.000	0.400	200.122	200.000	0.168	200.061	200.000	0.090	200.046	200.000	0.075
	Min	200.240	199.710	0.240	200.170	199.885	0.170	200.050	199.954	0.050	200.015	199.971	0.015	200.000	199.971	0.000
250	Max	250.570	250.000	0.860	250.285	250.000	0.400	250.122	250.000	0.168	250.061	250.000	0.090	250.046	250.000	0.075
	Min	250.280	249.710	0.280	250.170	249.885	0.170	250.050	249.954	0.050	250.015	249.971	0.015	250.000	249.971	0.000
300	Max	300.650	300.000	0.970	300.320	300.000	0.450	300.137	300.000	0.189	300.069	300.000	0.101	300.052	300.000	0.084
	Min	300.330	299.680	0.330	300.190	299.870	0.190	300.056	299.948	0.056	300.017	299.968	0.017	300.000	299.968	0.000
400	Max	400.760	400.000	1.120	400.350	400.000	0.490	400.151	400.000	0.208	400.075	400.000	0.111	400.057	400.000	0.093
	Min	400.400	399.640	0.400	400.210	399.860	0.210	400.062	399.943	0.062	400.018	399.964	0.018	400.000	399.964	0.000
500	Max	500.880	500.000	1.280	500.385	500.000	0.540	500.165	500.000	0.228	500.083	500.000	0.123	500.063	500.000	0.103
	Min	500.480	499.600	0.480	500.230	499.845	0.230	500.068	499.937	0.068	500.020	499.960	0.020	500.000	499.960	0.000

All dimensions are in millimeters.

*The sizes shown are first choice basic sizes (see Table 1). Preferred fits for other sizes can be calculated from data given in ANSI B4.2-1978.

†All fits shown in this table have clearance.

Table 5. American National Standard Preferred Shaft Basis Metric Transition and Interference Fits (ANSI B4.2-1978)

Basic Size*		Locational Transition			Locational Transition			Locational Interference			Medium Drive			Force		
		Hole K7	Shaft h6	Fit†	Hole N7	Shaft h6	Fit†	Hole P7	Shaft h6	Fit†	Hole S7	Shaft h6	Fit†	Hole U7	Shaft h6	Fit†
1	Max	1.000	1.000	+0.006	0.996	1.000	+0.002	0.994	1.000	0.000	0.986	1.000	-0.008	0.982	1.000	-0.012
	Min	0.990	0.994	-0.010	0.986	0.994	-0.014	0.984	0.994	-0.016	0.976	0.994	-0.024	0.972	0.994	-0.028
1.2	Max	1.200	1.200	+0.006	1.196	1.200	+0.002	1.194	1.200	0.000	1.186	1.200	-0.008	1.182	1.200	-0.012
	Min	1.190	1.194	-0.010	1.186	1.194	-0.014	1.184	1.194	-0.016	1.176	1.194	-0.024	1.172	1.194	-0.028
1.6	Max	1.600	1.600	+0.006	1.596	1.600	+0.002	1.594	1.600	0.000	1.586	1.600	-0.008	1.582	1.600	-0.012
	Min	1.590	1.594	-0.010	1.586	1.594	-0.014	1.584	1.594	-0.016	1.576	1.594	-0.024	1.572	1.594	-0.028
2	Max	2.000	2.000	+0.006	1.996	2.000	+0.002	1.994	2.000	0.000	1.986	2.000	-0.008	1.982	2.000	-0.012
	Min	1.990	1.994	-0.010	1.986	1.994	-0.014	1.984	1.994	-0.016	1.976	1.994	-0.024	1.972	1.994	-0.028
2.5	Max	2.500	2.500	+0.006	2.496	2.500	+0.002	2.494	2.500	0.000	2.486	2.500	-0.008	2.482	2.500	-0.012
	Min	2.490	2.494	-0.010	2.486	2.494	-0.014	2.484	2.494	-0.016	2.476	2.494	-0.024	2.472	2.494	-0.028
3	Max	3.000	3.000	+0.006	2.996	3.000	+0.002	2.994	3.000	0.000	2.986	3.000	-0.008	2.982	3.000	-0.012
	Min	2.990	2.994	-0.010	2.986	2.994	-0.014	2.984	2.994	-0.016	2.976	2.994	-0.024	2.972	2.994	-0.028
4	Max	4.003	4.000	+0.011	3.996	4.000	+0.004	3.992	4.000	0.000	3.985	4.000	-0.007	3.981	4.000	-0.011
	Min	3.991	3.992	-0.009	3.984	3.992	-0.016	3.980	3.992	-0.020	3.973	3.992	-0.027	3.969	3.992	-0.031
5	Max	5.003	5.000	+0.011	4.996	5.000	+0.004	4.992	5.000	0.000	4.985	5.000	-0.007	4.981	5.000	-0.011
	Min	4.991	4.992	-0.009	4.984	4.992	-0.016	4.980	4.992	-0.020	4.973	4.992	-0.027	4.969	4.992	-0.031
6	Max	6.003	6.000	+0.011	5.996	6.000	+0.004	5.992	6.000	0.000	5.985	6.000	-0.007	5.981	6.000	-0.011
	Min	5.991	5.992	-0.009	5.984	5.992	-0.016	5.980	5.992	-0.020	5.973	5.992	-0.027	5.969	5.992	-0.031
8	Max	8.005	8.000	+0.014	7.996	8.000	+0.005	7.991	8.000	0.000	7.983	8.000	-0.008	7.978	8.000	-0.013
	Min	7.990	7.991	-0.010	7.981	7.991	-0.019	7.976	7.991	-0.024	7.968	7.991	-0.032	7.963	7.991	-0.037
10	Max	10.005	10.000	+0.014	9.996	10.000	+0.005	9.991	10.000	0.000	9.983	10.000	-0.008	9.978	10.000	-0.013
	Min	9.990	9.991	-0.010	9.981	9.991	-0.019	9.976	9.991	-0.024	9.968	9.991	-0.032	9.963	9.991	-0.037
12	Max	12.006	12.000	+0.017	11.995	12.000	+0.006	11.989	12.000	0.000	11.979	12.000	-0.010	11.974	12.000	-0.015
	Min	11.988	11.989	-0.012	11.977	11.989	-0.023	11.971	11.989	-0.029	11.961	11.989	-0.039	11.956	11.989	-0.044
16	Max	16.006	16.000	+0.017	15.995	16.000	+0.006	15.989	16.000	0.000	15.979	16.000	-0.010	15.974	16.000	-0.015
	Min	15.988	15.989	-0.012	15.977	15.989	-0.023	15.971	15.989	-0.029	15.961	15.989	-0.039	15.956	15.989	-0.044
20	Max	20.006	20.000	+0.019	19.993	20.000	+0.006	19.986	20.000	-0.001	19.973	20.000	-0.014	19.967	20.000	-0.020
	Min	19.985	19.987	-0.015	19.972	19.987	-0.028	19.965	19.987	-0.035	19.952	19.987	-0.048	19.946	19.987	-0.054

All dimensions are in millimeters.
*The sizes shown are first choice basic sizes (see Table 1). Preferred fits for other sizes can be calculated from data given in ANSI B4.2-1978.
†A plus sign indicates clearance and a minus sign indicates interference.

Table 5 (Concluded). American National Standard Preferred Shaft Basis Metric Transition and Interference Fits (ANSI B4.2-1978)

Basic Size*		Locational Transition			Locational Transition			Locational Interference			Medium Drive			Force		
		Hole K7	Shaft h6	Fit†	Hole N7	Shaft h6	Fit†	Hole P7	Shaft h6	Fit†	Hole S7	Shaft h6	Fit†	Hole U7	Shaft h6	Fit†
25	Max	25.006	25.000	+0.019	24.993	25.000	+0.006	24.986	25.000	-0.001	24.973	25.000	-0.014	24.960	25.000	-0.027
	Min	24.985	24.987	-0.015	24.972	24.987	-0.028	24.965	24.987	-0.035	24.952	24.987	-0.048	24.939	24.987	-0.061
30	Max	30.006	30.000	+0.019	29.993	30.000	+0.006	29.986	30.000	-0.001	29.973	30.000	-0.014	29.960	30.000	-0.027
	Min	29.985	29.987	-0.015	29.972	29.987	-0.028	29.965	29.987	-0.035	29.952	29.987	-0.048	29.939	29.987	-0.061
40	Max	40.007	40.000	+0.023	39.992	40.000	+0.008	39.983	40.000	-0.001	39.966	40.000	-0.018	39.949	40.000	-0.035
	Min	39.982	39.984	-0.018	39.967	39.984	-0.033	39.958	39.984	-0.042	39.941	39.984	-0.059	39.924	39.984	-0.076
50	Max	50.007	50.000	+0.023	49.992	50.000	+0.008	49.983	50.000	-0.001	49.966	50.000	-0.018	49.939	50.000	-0.045
	Min	49.982	49.984	-0.018	49.967	49.984	-0.033	49.958	49.984	-0.042	49.941	49.984	-0.059	49.914	49.984	-0.086
60	Max	60.009	60.000	+0.028	59.991	60.000	+0.010	59.979	60.000	-0.002	59.958	60.000	-0.023	59.924	60.000	-0.087
	Min	59.979	59.981	-0.021	59.961	59.981	-0.039	59.949	59.981	-0.051	59.928	59.981	-0.072	59.894	59.981	-0.106
80	Max	80.009	80.000	+0.028	79.991	80.000	+0.010	79.979	80.000	-0.002	79.952	80.000	-0.029	79.909	80.000	-0.072
	Min	79.979	79.981	-0.021	79.961	79.981	-0.039	79.949	79.981	-0.051	79.922	79.981	-0.078	79.879	79.981	-0.121
100	Max	100.010	100.000	+0.032	99.990	100.000	+0.012	99.976	100.000	-0.002	99.942	100.000	-0.036	99.889	100.000	-0.089
	Min	99.975	99.978	-0.025	99.955	99.978	-0.045	99.941	99.978	-0.059	99.907	99.978	-0.093	99.854	99.978	-0.146
120	Max	120.010	120.000	+0.032	119.990	120.000	+0.012	119.976	120.000	-0.002	119.934	120.000	-0.044	119.869	120.000	-0.109
	Min	119.975	119.978	-0.025	119.955	119.978	-0.045	119.941	119.978	-0.059	119.899	119.978	-0.101	119.834	119.978	-0.166
160	Max	160.012	160.000	+0.037	159.988	160.000	+0.013	159.972	160.000	-0.003	159.915	160.000	-0.060	159.825	160.000	-0.150
	Min	159.972	159.975	-0.028	159.948	159.975	-0.052	159.932	159.975	-0.068	159.875	159.975	-0.125	159.785	159.975	-0.215
200	Max	200.013	200.000	+0.042	199.986	200.000	+0.015	199.967	200.000	-0.004	199.895	200.000	-0.076	199.781	200.000	-0.190
	Min	199.967	199.971	-0.033	199.940	199.971	-0.060	199.921	199.971	-0.079	199.849	199.971	-0.151	199.735	199.971	-0.265
250	Max	250.013	250.000	+0.042	249.986	250.000	+0.015	249.967	250.000	-0.004	249.877	250.000	-0.094	249.733	250.000	-0.238
	Min	249.967	249.971	-0.033	249.940	249.971	-0.060	249.921	249.971	-0.079	249.831	249.971	-0.169	249.687	249.971	-0.313
300	Max	300.016	300.000	+0.048	299.986	300.000	+0.018	299.964	300.000	-0.004	299.850	300.000	-0.118	299.670	300.000	-0.298
	Min	299.964	299.968	-0.036	299.934	299.968	-0.066	299.912	299.968	-0.088	299.798	299.968	-0.202	299.618	299.968	-0.382
400	Max	400.017	400.000	+0.053	399.984	400.000	+0.020	399.959	400.000	-0.005	399.813	400.000	-0.151	399.586	400.000	-0.378
	Min	399.960	399.964	-0.040	399.927	399.964	-0.073	399.902	399.964	-0.098	399.756	399.964	-0.244	399.529	399.964	-0.471
500	Max	500.018	500.000	+0.058	499.983	500.000	+0.023	499.955	500.000	-0.005	499.771	500.000	-0.189	499.483	500.000	-0.477
	Min	499.955	499.960	-0.045	499.920	499.960	-0.080	499.892	499.960	-0.108	499.708	499.960	-0.292	499.420	499.960	-0.580

All dimensions are in millimeters.
*The sizes shown are first choice basic sizes (see Table 1). Preferred fits for other sizes can be calculated from data given in ANSI B4.2-1978.
†A plus sign indicates clearance and a minus sign indicates interference.

American National Standard Recommended Gage Usage (ANSI B4.4M-1981)

Gagemakers Tolerance			Workpiece Tolerance	
	Class	ISO Symbol*	IT Grade	Recommended Gage Usage
Rejection of Good Parts Increase / Gage Cost Increase	ZM	0.05 IT11	IT11	Low precision gages recommended used to inspect workpieces held to internal (hole) tolerances C11 and H11 and to external (shaft) tolerances c11 and h11.
	YM	0.05 IT9	IT9	Gages recommended used to inspect workpieces held to internal (hole) tolerances D9 and H9 and to external (shaft) tolerances d9 and h9.
	XM	0.05 IT8	IT8	Precision gages recommended used to inspect workpieces held to internal (hole) tolerances F8 and H8.
	XXM	0.05 IT7	IT7	Recommended used for gages to inspect workpieces held to internal (hole) tolerances G7, H7, K7, N7, P7, S7 and U7 and to external (shaft) tolerances f7 and h7.
	XXXM	0.05 IT6	IT6	High precision gages recommended used to inspect workpieces held to external (shaft) tolerances g6, h6, k6, n6, p6, s6 and u6.

* Gagemakers tolerance is equal to 5 per cent of workpiece tolerance or 5 per cent of applicable IT grade value (see table below).
For workpiece tolerance class values, see previous tables 2 to 5, incl.

American National Standard Gagemakers Tolerances (ANSI B4.4M-1981)

Basic Size		Class ZM (0.05 IT11)	Class YM (0.05 IT9)	Class XM (0.05 IT8)	Class XXM (0.05 IT7)	Class XXXM (0.05 IT6)
Over	To					
0	3	0.0030	0.0012	0.0007	0.0005	0.0003
3	6	0.0037	0.0015	0.0009	0.0006	0.0004
6	10	0.0045	0.0018	0.0011	0.0007	0.0005
10	18	0.0055	0.0021	0.0013	0.0009	0.0006
18	30	0.0065	0.0026	0.0016	0.0010	0.0007
30	50	0.0080	0.0031	0.0019	0.0012	0.0008
50	80	0.0095	0.0037	0.0023	0.0015	0.0010
80	120	0.0110	0.0043	0.0027	0.0017	0.0011
120	180	0.0125	0.0050	0.0031	0.0020	0.0013
180	250	0.0145	0.0057	0.0036	0.0023	0.0015
250	315	0.0160	0.0065	0.0040	0.0026	0.0016
315	400	0.0180	0.0070	0.0044	0.0028	0.0018
400	500	0.0200	0.0077	0.0048	0.0031	0.0020

All dimensions are in millimeters. For closer gagemakers tolerance classes than Class XXXM, specify 5 per cent of IT5, IT4, or IT3 and use the designation 0.05 IT5, 0.05 IT4, etc.

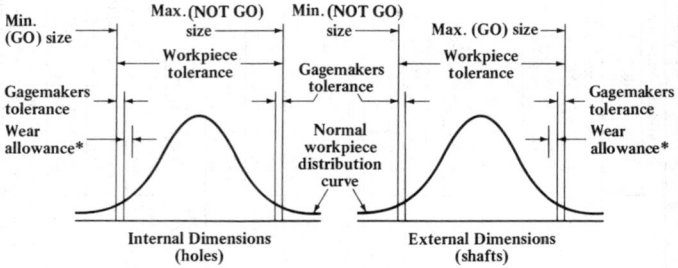

Fig. 5. Relation of Gagemakers Tolerance and Wear Allowance to Workpiece Tolerance

(*Continued from page* 648)

the preferred fits might be considered necessary to satisfy extreme conditions. Subsequent adjustments might also be desired as the result of experience in a particular application to suit critical functional requirements or to permit optimum manufacturing economy. Selection of a departure from these recommendations will depend upon consideration of the engineering and economic factors that might be involved; however, the benefits to be derived from the use of preferred fits should not be overlooked.

A general guide to machining processes which may normally be expected to produce work within the tolerances indicated by the IT grades given in ANSI B4.2-1978 is shown in the chart in Fig. 6.

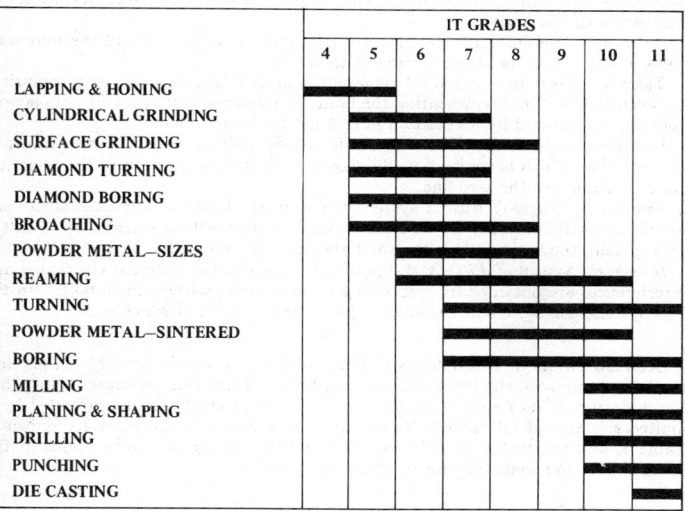

Fig. 6. Relation of Machining Processes to IT Tolerance Grades

British Standard for Metric ISO Limits and Fits. — Based on ISO Recommendation R286, this British Standard (BS 4500:1969) is intended to provide a comprehensive range of metric limits and fits for engineering purposes, and meets the requirements of metrication in the United Kingdom. Sizes up to 3,150 mm are covered by the Standard, but the condensed information presented here embraces dimensions up to 500 mm only. The system is based on a series of tolerances graded to suit all classes of work from the finest to the most coarse, and the different types of fits that can be obtained range from coarse clearance to heavy interference. In the Standard, only cylindrical parts, designated holes and shafts are referred to explicitly, but it is emphasized that the recommendations apply equally well to other sections, and the general term *hole* or *shaft* can be taken to mean the space contained by or containing two parallel faces or tangent planes of any part, such as the width of a slot, or the thickness of a key. It is also strongly emphasized that the grades series of tolerances are intended for the most general application, and should be used wherever possible whether the features of the component involved are members of a fit or not.

Definitions. — The definitions given in the Standard include the following:

Limits of Size: The maximum and minimum sizes permitted for a feature.

Basic Size: The size by reference to which the limits of size are fixed. The basic size is the same for both members of a fit.

Upper Deviation: The algebraical difference between the maximum limit of size and the corresponding basic size. It is designated as ES for a hole, and as es for a shaft, which stands for the French term *écart supérieur*.

Lower Deviation: The algebraical difference between the minimum limit of size and the corresponding basic size. It is designated as EI for a hole, and as ei for a shaft, which stands for the French term *écart inférieur*.

Zero Line: In a graphical representation of limits and fits, the straight line to which the deviations are referred. The zero line is the line of zero deviation and represents the basic size.

Tolerance: The difference between the maximum limit of size and the minimum limit of size. It is an absolute value without sign.

Tolerance Zone: In a graphical representation of tolerances, the zone comprised between the two lines representing the limits of tolerance and defined by its magnitude (tolerance) and by its position in relation to the zero line.

Fundamental Deviation: That one of the two deviations, being the one nearest to the zero line, which is conventionally chosen to define the position of the tolerance zone in relation to the zero line.

Shaft-basis System of Fits: A system of fits in which the different clearances and interferences are obtained by associating various holes with a single shaft. In the ISO system, the basic shaft is the shaft the upper deviation of which is zero.

Hole-basis System of Fits: A system of fits in which the different clearances and interferences are obtained by associating various shafts with a single hole. In the ISO system, the basic hole is the hole the lower deviation of which is zero.

Selected Limits of Tolerance, and Fits. — The number of fit combinations that can be built up with the ISO system is very large. However, experience shows that the majority of fits required for usual engineering products can be provided by a limited selection of tolerances. Limits of tolerance for selected holes are shown in Table 1, and for shafts, in Table 2. Selected fits, based on combinations of the selected hole and shaft tolerances, are given in Table 3.

Tolerances and Fundamental Deviations. — There are 18 tolerance grades intended to meet the requirements of different classes of work, and they are designated IT 01, IT 02, and IT 1 to IT 16. (IT stands for ISO series of tolerances.) Table 4 shows the standardized numerical values for the 18 tolerance grades, which are known as standard tolerances. The system provides 27 fundamental deviations for sizes up to and including 500 mm, and Tables 5 and 6 contain the values for shafts and holes respectively. Upper case (capital) letters designate hole deviations, and the same letters in lower case designate shaft deviations. The deviation j_s (J_s for holes) is provided to meet the need for symmetrical bilateral tolerances. In this instance there is no fundamental deviation, and the tolerance zone, of whatever magnitude, is equally disposed about the zero line.

Calculated Limits of Tolerance. — The deviations and fundamental tolerances provided by the ISO system can be combined in any way that appears necessary to give a required fit. Thus, for example, the deviations H (basic hole) and f (clearance shaft) could be associated, and with each of these deviations any one of the tolerance grades IT 01 to IT 16 could be used. All the limits of tolerance of which

the system is capable of providing for sizes up to and including 500 mm can be calculated from the standard tolerances given in Table 4, and the fundamental deviations given in Tables 5 and 6. The range includes limits of tolerance for shafts and holes used in small high-precision work and horology.

The system provides for the use of either hole-basis or shaft-basis fits, and the Standard includes details of procedures for converting from one type of fit to the other.

The limits of tolerance for a shaft or hole are designated by the appropriate letter indicating the fundamental deviation, followed by a suffix number denoting the tolerance grade. This suffix number is the numerical part of the tolerance grade designation. Thus, a hole tolerance with deviation H, and tolerance grade IT7 is designated H7. Likewise, a shaft with deviation p, and tolerance grade IT 6 is designated p6. The limits of size of a component feature are defined by the basic size, say 45 mm, followed by the appropriate tolerance designation, for example 45 H7 or 45 p6. A fit is indicated by combining the basic size common to both features with the designation appropriate to each of them, for example 45 H7-p6 or 45 H7/p6.

When calculating the limits of size for a shaft, the upper deviation es, or the lower deviation ei, is first obtained from Table 5, depending on the particular letter designation, and nominal dimension. If an upper deviation has been determined, the lower deviation ei = es − IT. The IT value is obtained from Table 4 for the particular tolerance grade being applied. If a lower deviation has been obtained from Table 5, the upper deviation es = ei + IT. When the upper deviation ES has been determined for a hole from Table 6, the lower deviation EI = ES − IT. If a lower deviation EI has been obtained from Table 6, then the upper deviation ES = EI + IT.

The upper deviations for holes K, M, and N with tolerance grades up to and including IT8, and for holes P to ZC with tolerance grades up to and including IT7, must be calculated by adding the delta (Δ) values given in Table 6 as indicated.

Examples of Calculations: The limits of size for a part of 133 mm basic size with a tolerance designation g9 are derived as follows:

From Table 5, the upper deviation (es) is − 0.014 mm. From Table 4, the tolerance grade (IT9) is 0.100 mm. The lower deviation (ei) = es − IT = −0.114 mm, and the limits of size are thus 132.986 and 132.886 mm.

The limits of size for a part 20 mm in size, with tolerance designation D3, are derived as follows: From Table 6, the lower deviation (EI) is + 0.065 mm. From Table 4, the tolerance grade (IT3) is 0.004 mm. The upper deviation (ES) = EI + IT = 0.069 mm, and thus the limits of size for the part are 20.069 and 20.065 mm.

The limits of size for a part 32 mm in size, with tolerance designation M5, which involves a delta value, are obtained as follows: From Table 6, the upper deviation ES is − 0.009 mm + Δ = −0.005 mm. (The delta value given at the end of this table for this size and grade IT 5, is 0.004 mm.) From Table 4, the tolerance grade (IT5) is 0.011 mm. The lower deviation (EI) = ES − IT = −0.016 mm, and thus the limits of size for the part are 31.995 and 31.984 mm.

Where the designations h and H or j_s and J_s are used, it is only necessary to refer to Table 4. For h and H, the fundamental deviation is always zero, and the disposition of the tolerance is always negative(−) for a shaft, and positive (+) for a hole. Thus, the limits for a part 40 mm in size, designated h8 are derived as follows: From Table 4, the tolerance grade (IT 8) is 0.039 mm, and the limits are therefore 40.000 and 39.961 mm.

The limits for a part 60 mm in size, designated j_s7 or J_s7 are derived as follows: From Table 1, the tolerance grade (IT 7) is 0.030 mm, and this value is divided equally about the basic size to give limits of 60.015 and 59.985 mm.

Table 1. British Standard Limits of Tolerance for Selected Holes
(Upper and Lower Deviations) (BS 4500:1969)

| Nominal Sizes, mm | | H7 | | H8 | | H9 | | H11 | |
Over	Up to and Including	ES +	EI	ES +	EI	ES +	EI	ES +	EI
...	3	10	0	14	0	25	0	60	0
3	6	12	0	18	0	30	0	75	0
6	10	15	0	22	0	36	0	90	0
10	18	18	0	27	0	43	0	110	0
18	30	21	0	33	0	52	0	130	0
30	50	25	0	39	0	62	0	160	0
50	80	30	0	46	0	74	0	190	0
80	120	35	0	54	0	87	0	220	0
120	180	40	0	63	0	100	0	250	0
180	250	46	0	72	0	115	0	290	0
250	315	52	0	81	0	130	0	320	0
315	400	57	0	89	0	140	0	360	0
400	500	63	0	97	0	155	0	400	0

ES = Upper deviation.   EI = Lower deviation

The dimensions are given in 0.001 mm, except for the nominal sizes, which are in millimeters.

Table 2. British Standard Limits of Tolerance for Selected Shafts
(Upper and Lower Deviations) (BS 4500:1969)

| Nominal Sizes, mm | | c11 | | d10 | | e9 | | f7 | | g6 | | h6 | | k6 | | n6 | | p6 | | s6 | |
Over	Up to and Incl.	es −	ei −	es −	ei −	es −	ei −	es −	ei −	es −	ei −	es −	ei −	es +	ei +	es +	ei +	es +	ei +	es +	ei +
...	3	60	120	20	60	14	39	6	16	2	8	0	6	6	0	10	4	12	6	20	14
3	6	70	145	30	78	20	50	10	22	4	12	0	8	9	1	16	8	20	12	27	19
6	10	80	170	40	98	25	61	13	28	5	14	0	9	10	1	19	10	24	15	32	23
10	18	95	205	50	120	32	75	16	34	6	17	0	11	12	1	23	12	29	18	39	28
18	30	110	240	65	149	40	92	20	41	7	20	0	13	15	2	28	15	35	22	48	35
30	40	120	280	80	180	50	112	25	50	9	25	0	16	18	2	33	17	42	26	59	43
40	50	130	290	80	180	50	112	25	50	9	25	0	16	18	2	33	17	42	26	59	43
50	65	140	330	100	220	60	134	30	60	10	29	0	19	21	2	39	20	51	32	72	53
65	80	150	340	100	220	60	134	30	60	10	29	0	19	21	2	39	20	51	32	78	59
80	100	170	390	120	260	72	159	36	71	12	34	0	22	25	3	45	23	59	37	93	71
100	120	180	400	120	260	72	159	36	71	12	34	0	22	25	3	45	23	59	37	101	79
120	140	200	450	145	305	85	185	43	83	14	39	0	25	28	3	52	27	68	43	117	92
140	160	210	460	145	305	85	185	43	83	14	39	0	25	28	3	52	27	68	43	125	100
160	180	230	480	145	305	85	185	43	83	14	39	0	25	28	3	52	27	68	43	133	108
180	200	240	530	170	355	100	215	50	96	15	44	0	29	33	4	60	31	79	50	151	122
200	225	260	550	170	355	100	215	50	96	15	44	0	29	33	4	60	31	79	50	159	130
225	250	280	570	170	355	100	215	50	96	15	44	0	29	33	4	60	31	79	50	169	140
250	280	300	620	190	400	110	240	56	108	17	49	0	32	36	4	66	34	88	56	190	158
280	315	330	650	190	400	110	240	56	108	17	49	0	32	36	4	66	34	88	56	202	170
315	355	360	720	210	440	125	265	62	119	18	54	0	36	40	4	73	37	98	62	226	190
355	400	400	760	210	440	125	265	62	119	18	54	0	36	40	4	73	37	98	62	244	208
400	450	440	840	230	480	135	290	68	131	20	60	0	40	45	5	80	40	108	68	272	232
450	500	480	880	230	480	135	290	68	131	20	60	0	40	45	5	80	40	108	68	292	252

es = upper deviation.   ei = lower deviation

The dimensions are given in 0.001 mm, except for the nominal sizes, which are in millimeters.

British Standards Selected Fits. ... mm and Maximum Clearances (BS 4500:1969)

Nominal Sizes, mm		H11-c11		H9-d10		H9-e9		H8-f7		H7-g6		H7-h6		H7-k6		H7-n6		H7-p6		H7-s6	
Over	Up to and Incl.	Min	Max	Min	Max	Min	Max	Min	Max	Min	Max	Min	Max	Min	Max	Min	Max	Min	Max	Min	Max
...	3	60	180	20	85	14	64	6	30	2	18	0	16	−6	+10	−10	+6	−12	+4	−20	−4
3	6	70	220	30	108	20	80	10	40	4	24	0	20	−9	+11	−16	+4	−20	0	−27	−7
6	10	80	260	40	134	25	97	13	50	5	29	0	24	−10	+14	−19	+5	−24	0	−32	−8
10	18	95	315	50	163	32	118	16	61	6	35	0	29	−12	+17	−23	+6	−29	0	−39	−10
18	30	110	370	65	201	40	144	20	74	7	41	0	34	−15	+19	−28	+6	−35	−1	−48	−14
30	40	120	440	80	242	50	174	25	89	9	50	0	41	−18	+23	−33	+8	−42	−1	−59	−18
40	50	130	450	80	242	50	174	25	89	9	50	0	41	−18	+23	−33	+8	−42	−1	−59	−18
50	65	140	520	100	294	60	208	30	106	10	59	0	49	−21	+28	−39	+10	−51	−2	−72	−23
65	80	150	530	100	294	60	208	30	106	10	59	0	49	−21	+28	−39	+10	−51	−2	−78	−29
80	100	170	610	120	347	72	246	36	125	12	69	0	57	−25	+32	−45	+12	−59	−2	−93	−36
100	120	180	620	120	347	72	246	36	125	12	69	0	57	−25	+32	−45	+12	−59	−2	−101	−44
120	140	200	700	145	405	85	285	43	146	14	79	0	65	−28	+37	−52	+13	−68	−3	−117	−52
140	160	210	710	145	405	85	285	43	146	14	79	0	65	−28	+37	−52	+13	−68	−3	−125	−60
160	180	230	730	145	405	85	285	43	146	14	79	0	65	−28	+37	−52	+13	−68	−3	−133	−68
180	200	240	820	170	470	100	330	50	168	15	90	0	75	−33	+42	−60	+15	−79	−4	−151	−76
200	225	260	840	170	470	100	330	50	168	15	90	0	75	−33	+42	−60	+15	−79	−4	−159	−84
225	250	280	860	170	470	100	330	50	168	15	90	0	75	−33	+42	−60	+15	−79	−4	−169	−94
250	280	300	940	190	530	110	370	56	189	17	101	0	84	−36	+48	−66	+18	−88	−4	−190	−126
280	315	330	970	190	530	110	370	56	189	17	101	0	84	−36	+48	−66	+18	−88	−4	−202	−112
315	355	360	1080	210	580	125	405	62	208	18	111	0	93	−40	−53	−73	+20	−98	−5	−226	−133
355	400	400	1120	210	580	125	405	62	208	18	111	0	93	−40	−53	−73	+20	−98	−5	−244	−151
400	450	440	1240	230	635	135	445	68	228	20	123	0	103	−45	+58	−80	+23	−108	−5	−272	−169
450	500	480	1280	230	635	135	445	68	228	20	123	0	103	−45	+58	−80	+23	−108	−5	−292	−189

The dimensions are given in 0.001 mm, except for the nominal sizes, which are in millimeters.
Minus (−) sign indicates negative clearance, i.e., interference.

Table 4.　British Standard Limits and Fits (BS 4500:1969)

Nominal Sizes, mm		Tolerance Grades									
Over	To	IT 01	IT 0	IT 1	IT 2	IT 3	IT 4	IT 5	IT 6	IT 7	IT 8
...	3	0.3	0.5	0.8	1.2	2	3	4	6	10	14
3	6	0.4	0.6	1	1.5	2.5	4	5	8	12	18
6	10	0.4	0.6	1	1.5	2.5	4	6	9	15	22
10	18	0.5	0.8	1.2	2	3	5	8	11	18	27
18	30	0.6	1	1.5	2.5	4	6	9	13	21	33
30	50	0.6	1	1.5	2.5	4	7	11	16	25	39
50	80	0.8	1.2	2	3	5	8	13	19	30	46
80	120	1	1.5	2.5	4	6	10	15	22	35	54
120	180	1.2	2	3.5	5	8	12	18	25	40	63
180	250	2	3	4.5	7	10	14	20	29	46	72
250	315	2.5	4	6	8	12	16	23	32	52	81
315	400	3	5	7	9	13	18	25	36	57	89
400	500	4	6	8	10	15	20	27	40	63	97

Nominal Sizes, mm		Tolerance Grades							
Over	To	IT 9	IT 10	IT 11	IT 12	IT 13	IT 14†	IT 15†	IT 16†
...	3	25	40	60	100	140	250	400	600
3	6	30	48	75	120	180	300	480	750
6	10	36	58	90	150	220	360	580	900
10	18	43	70	110	180	270	430	700	1100
18	30	52	84	130	210	330	520	840	1300
30	50	62	100	160	250	390	620	1000	1600
50	80	74	120	190	300	460	740	1200	1900
80	120	87	140	220	350	540	870	1400	2200
120	180	100	160	250	400	630	1000	1600	2500
180	250	115	185	290	460	720	1150	1850	2900
250	315	130	210	320	520	810	1300	2100	3200
315	400	140	230	360	570	890	1400	2300	3600
400	500	155	250	400	630	970	1550	2500	4000

† Not applicable to sizes below 1 mm.

The dimensions are given in 0.001 mm, except for the nominal sizes which are in millimeters.

Table 5. British Standard Fundamental Deviations for Shafts (BS 4500:1969)

Nominal Size, mm		Grade																
		Fundamental (Upper) Deviation es — 01 to 16												Fundamental (Lower)			Dev'n ei	
Over	To	a*	b*	c	cd	d	e	ef	f	fg	g	h	js†	j (5-6)	j (7)	j (8)	k (4-7)	k (≤3 >7)
...	3	−270	−140	−60	−34	−20	−14	−10	−6	−4	−2	0	±IT/2	−2	−4	−6	0	0
3	6	−270	−140	−70	−46	−30	−20	−14	−10	−6	−4	0		−2	−4		+1	0
6	10	−280	−150	−80	−56	−40	−25	−18	−13	−8	−5	0		−2	−5		+1	0
10	14	−290	−150	−95		−50	−32		−16		−6	0		−3	−6		+1	0
14	18	−290	−150	−95		−50	−32		−16		−6	0		−3	−6		+1	0
18	24	−300	−160	−110		−65	−40		−20		−7	0		−4	−8		+2	0
24	30	−300	−160	−110		−65	−40		−20		−7	0		−4	−8		+2	0
30	40	−310	−170	−120		−80	−50		−25		−9	0		−5	−10		+2	0
40	50	−320	−180	−130		−80	−50		−25		−9	0		−5	−10		+2	0
50	65	−340	−190	−140		−100	−60		−30		−10	0		−7	−12		+2	0
65	80	−360	−200	−150		−100	−60		−30		−10	0		−7	−12		+2	0
80	100	−380	−220	−170		−120	−72		−36		−12	0		−9	−15		+3	0
100	120	−410	−240	−180		−120	−72		−36		−12	0		−9	−15		+3	0
120	140	−460	−260	−200		−145	−85		−43		−14	0		−11	−18		+3	0
140	160	−520	−280	−210		−145	−85		−43		−14	0		−11	−18		+3	0
160	180	−580	−310	−230		−145	−85		−43		−14	0		−11	−18		+3	0
180	200	−660	−340	−240		−170	−100		−50		−15	0		−13	−21		+4	0
200	225	−740	−380	−260		−170	−100		−50		−15	0		−13	−21		+4	0
225	250	−820	−420	−280		−170	−100		−50		−15	0		−13	−21		+4	0
250	280	−920	−480	−300		−190	−110		−56		−17	0		−16	−26		+4	0
280	315	−1050	−540	−330		−190	−110		−56		−17	0		−16	−26		+4	0
315	355	−1200	−600	−360		−210	−125		−62		−18	0		−18	−28		+4	0
355	400	−1350	−680	−400		−210	−125		−62		−18	0		−18	−28		+4	0
400	450	−1500	−760	−440		−230	−135		−68		−20	0		−20	−32		+5	0
450	500	−1650	−840	−480		−230	−135		−68		−20	0		−20	−32		+5	0

The dimensions are in 0.001 mm, except the nominal sizes, which are in millimeters.

* Not applicable to sizes up to 1 mm.   † In grades 7 to 11, the two symmetrical deviations ±IT/2 should be rounded if the IT value in micro-meters is an odd value by replacing it with the even value immediately below. For example, if IT = 175, replace it by 174.

Table 5. *(Continued).* **British Standard Fundamental Deviations for Shafts** (BS 4500:1969)

Nominal Size, mm		Grade 01 to 16 — Fundamental (Lower) Deviation ei													
Over	To	m	n	p	r	s	t	u	v	x	y	z	za	zb	zc
..	3	+2	+4	+6	+10	+14		+18		+20		+26	+32	+40	+60
3	6	+4	+8	+12	+15	+19		+23		+28		+35	+42	+50	+80
6	10	+6	+10	+15	+19	+23		+28		+34		+42	+52	+67	+97
10	14	+7	+12	+18	+23	+28		+33		+40		+50	+64	+90	+130
14	18	+7	+12	+18	+23	+28		+33	+39	+45		+60	+77	+108	+150
18	24	+8	+15	+22	+28	+35		+41	+47	+54	+63	+73	+98	+136	+188
24	30	+8	+15	+22	+28	+35	+41	+48	+55	+64	+75	+88	+118	+160	+218
30	40	+9	+17	+26	+34	+43	+48	+60	+68	+80	+94	+112	+148	+200	+274
40	50	+9	+17	+26	+34	+43	+54	+70	+81	+97	+114	+136	+180	+242	+325
50	65	+11	+20	+32	+41	+53	+66	+87	+102	+122	+144	+172	+226	+300	+405
65	80	+11	+20	+32	+43	+59	+75	+102	+120	+146	+174	+210	+274	+360	+480
80	100	+13	+23	+37	+51	+71	+91	+124	+146	+178	+214	+258	+335	+445	+585
100	120	+13	+23	+37	+54	+79	+104	+144	+172	+210	+254	+310	+400	+525	+690
120	140	+15	+27	+43	+63	+92	+122	+170	+202	+248	+300	+365	+470	+620	+800
140	160	+15	+27	+43	+65	+100	+134	+190	+228	+280	+340	+415	+535	+700	+900
160	180	+15	+27	+43	+68	+108	+146	+210	+252	+310	+380	+465	+600	+780	+1000
180	200	+17	+31	+50	+77	+122	+166	+236	+284	+350	+425	+520	+670	+880	+1150
200	225	+17	+31	+50	+80	+130	+180	+258	+310	+385	+470	+575	+740	+960	+1250
225	250	+17	+31	+50	+84	+140	+196	+284	+340	+425	+520	+640	+820	+1050	+1350
250	280	+20	+34	+56	+94	+158	+218	+315	+385	+475	+580	+710	+920	+1200	+1550
280	315	+20	+34	+56	+98	+170	+240	+350	+425	+525	+650	+790	+1000	+1300	+1700
315	355	+21	+37	+62	+108	+190	+268	+390	+475	+590	+730	+900	+1150	+1500	+1900
355	400	+21	+37	+62	+114	+208	+294	+435	+530	+660	+820	+1000	+1300	+1650	+2100
400	450	+23	+40	+68	+126	+232	+330	+490	+595	+740	+920	+1100	+1450	+1850	+2400
450	500	+23	+40	+68	+132	+252	+360	+540	+660	+820	+1000	+1250	+1600	+2100	+2600

Nominal Size, mm		Grade																				
		01 to 16												6	7	8	>8	≤8	>8	≤8†	≤8	>8§
		Fundamental (Lower) Deviation EI												Fundamental (Upper) Deviation ES								
Over	To	A*	B*	C	CD	D	E	EF	F	FG	G	H	Jst	J			K††		M††		N††	
..	3	+270	+140	+60	+34	+20	+14	+10	+6	+4	+2	0		+2 +4 +6			0	0	−2	△	−4	−4
3	6	+270	+140	+70	+46	+30	+20	+14	+10	+6	+4	0		+5 +6 +10			−1+△	△	−4+△	△	−8+△	0
6	10	+280	+150	+80	+56	+40	+25	+18	+13	+8	+5	0		+5 +8 +12			−1+△	△	−6+△	△	−10+△	0
10	14	+290	+150	+95	...	+50	+32	...	+16	...	+6	0		+6 +10 +15			−1+△	△	−7+△	△	−12+△	0
14	18	+290	+150	+95	...	+50	+32	...	+16	...	+6	0		+6 +10 +15			−1+△	△	−7+△	△	−12+△	0
18	24	+300	+160	+110	...	+65	+40	...	+20	...	+7	0	±IT/2	+8 +12 +20			−2+△	△	−8+△	△	−15+△	0
24	30	+300	+160	+110	...	+65	+40	...	+20	...	+7	0		+8 +12 +20			−2+△	△	−8+△	△	−15+△	0
30	40	+310	+170	+120	...	+80	+50	...	+25	...	+9	0		+10 +14 +24			−2+△	△	−9+△	△	−17+△	0
40	50	+320	+180	+130	...	+80	+50	...	+25	...	+9	0		+10 +14 +24			−2+△	△	−9+△	△	−17+△	0
50	65	+340	+190	+140	...	+100	+60	...	+30	...	+10	0		+13 +18 +28			−2+△	△	−11+△	△	−20+△	0
65	80	+360	+200	+150	...	+100	+60	...	+30	...	+10	0		+13 +18 +28			−2+△	△	−11+△	△	−20+△	0
80	100	+380	+220	+170	...	+120	+72	...	+36	...	+12	0		+16 +22 +34			−3+△	△	−13+△	△	−23+△	0
100	120	+410	+240	+180	...	+120	+72	...	+36	...	+12	0		+16 +22 +34			−3+△	△	−13+△	△	−23+△	0
120	140	+460	+260	+200	...	+145	+85	...	+43	...	+14	0		+18 +26 +41			−3+△	△	−15+△	△	−27+△	0
140	160	+520	+280	+210	...	+145	+85	...	+43	...	+14	0		+18 +26 +41			−3+△	△	−15+△	△	−27+△	0
160	180	+580	+310	+230	...	+145	+85	...	+43	...	+14	0		+18 +26 +41			−3+△	△	−15+△	△	−27+△	0
180	200	+660	+340	+240	...	+170	+100	...	+50	...	+15	0		+22 +30 +47			−4+△	△	−17+△	△	−31+△	0
200	225	+740	+380	+260	...	+170	+100	...	+50	...	+15	0		+22 +30 +47			−4+△	△	−17+△	△	−31+△	0
225	250	+820	+420	+280	...	+170	+100	...	+50	...	+15	0		+22 +30 +47			−4+△	△	−17+△	△	−31+△	0
250	280	+920	+480	+300	...	+190	+110	...	+56	...	+17	0		+25 +36 +55			−4+△	△	−20+△	△	−34+△	0
280	315	+1050	+540	+330	...	+190	+110	...	+56	...	+17	0		+25 +36 +55			−4+△	△	−20+△	△	−34+△	0
315	355	+1200	+600	+360	...	+210	+125	...	+62	...	+18	0		+29 +39 +60			−4+△	△	−21+△	△	−37+△	0
355	400	+1350	+680	+400	...	+210	+125	...	+62	...	+18	0		+29 +39 +60			−4+△	△	−21+△	△	−37+△	0
400	450	+1500	+760	+440	...	+230	+135	...	+68	...	+20	0		+33 +43 +66			−5+△	△	−23+△	△	−40+△	0
450	500	+1650	+840	+480	...	+230	+135	...	+68	...	+20	0		+33 +43 +66			−5+△	△	−23+△	△	−40+△	0

The dimensions are given in 0.001 mm, except the nominal sizes which are in millimeters. *Not applicable to sizes up to 1 mm. †In grades 7 to 11, the two symmetrical deviations ±IT/2 should be rounded if the IT value in micrometers is an odd value, by replacing it with the even value below. For example, if IT = 175, replace it by 174.

†† When calculating deviations for holes K, M, and N with tolerance grades up to and including IT 8, and holes F to ZC with tolerance grades up to and including IT 7, the delta (△) values are added to the upper deviation ES. For example, for 25 P7, ES = −0.022 + 0.008 = −0.014 mm.

‡ Special case: for M6, ES = −9 for sizes from 250 to 315 mm, instead of −11.   § Not applicable to sizes up to 1 mm.

Table 6. (Continued). British Standard Fundamental Deviations for Holes (BS 4500:1969)

The "≤7 / P to ZC" column carries the note: ∇ Same deviation as for grades above 7 increased by Δ.

Nominal Size (Over)	To	P	R	S	T	U	V	X	Y	Z	ZA	ZB	ZC	Δ3	Δ4	Δ5	Δ6	Δ7	Δ8
…	3	-6	-10	-14	…	-18	…	-20	…	-26	-32	-40	-60	0	0	0	0	0	0
3	6	-12	-15	-19	…	-23	…	-28	…	-35	-42	-50	-80	1	1.5	1	3	4	6
6	10	-15	-19	-23	…	-28	…	-34	…	-42	-52	-67	-97	1	1.5	2	3	6	7
10	14	-18	-23	-28	…	-33	…	-40	…	-50	-64	-90	-130	1	2	3	3	7	9
14	18	-18	-23	-28	…	-33	-39	-45	…	-60	-77	-108	-150	1	2	3	3	7	9
18	24	-22	-28	-35	…	-41	-47	-54	-63	-73	-98	-136	-188	1.5	2	3	4	8	12
24	30	-22	-28	-35	-41	-48	-55	-64	-75	-88	-118	-160	-218	1.5	2	3	4	8	12
30	40	-26	-34	-43	-48	-60	-68	-80	-94	-112	-148	-200	-274	1.5	3	4	5	9	14
40	50	-26	-34	-43	-54	-70	-81	-97	-114	-136	-180	-242	-325	1.5	3	4	5	9	14
50	65	-32	-41	-53	-66	-87	-102	-122	-144	-172	-226	-300	-405	2	3	5	6	11	16
65	80	-32	-43	-59	-75	-102	-120	-146	-174	-210	-274	-360	-480	2	3	5	6	11	16
80	100	-37	-51	-71	-91	-124	-146	-178	-214	-258	-335	-445	-585	2	4	5	7	13	19
100	120	-37	-54	-79	-104	-144	-172	-210	-254	-310	-400	-525	-690	2	4	5	7	13	19
120	140	-43	-63	-92	-122	-170	-202	-248	-300	-365	-470	-620	-800	3	4	6	7	15	23
140	160	-43	-65	-100	-134	-190	-228	-280	-340	-415	-535	-700	-900	3	4	6	7	15	23
160	180	-43	-68	-108	-146	-210	-252	-310	-380	-465	-600	-780	-1000	3	4	6	7	15	23
180	200	-50	-77	-122	-166	-236	-284	-350	-425	-520	-670	-880	-1150	3	4	6	9	17	26
200	225	-50	-80	-130	-180	-258	-310	-385	-470	-575	-740	-960	-1250	3	4	6	9	17	26
225	250	-50	-84	-140	-196	-284	-340	-425	-520	-640	-820	-1050	-1350	3	4	6	9	17	26
250	280	-56	-94	-158	-218	-315	-385	-475	-580	-710	-920	-1200	-1550	4	4	7	9	20	29
280	315	-56	-98	-170	-240	-350	-425	-525	-650	-790	-1000	-1300	-1700	4	4	7	9	20	29
315	355	-62	-108	-190	-268	-390	-475	-590	-730	-900	-1150	-1500	-1800	4	5	7	11	21	32
355	400	-62	-114	-208	-294	-435	-530	-660	-820	-1000	-1300	-1650	-2100	4	5	7	11	21	32
400	450	-68	-126	-232	-330	-490	-595	-740	-920	-1100	-1450	-1850	-2400	5	5	7	13	23	34
450	500	-68	-132	-252	-360	-540	-660	-820	-1000	-1250	-1600	-2100	-2600	5	5	7	13	23	34

Grade ≤7 (P to ZC): Same deviation as for grades above 7 increased by Δ. Grade >7 columns (P to ZC) give the Fundamental (Upper) Deviation ES. Δ values apply to Grades 3, 4, 5, 6, 7, 8.

The dimensions are given in 0.001 mm, except the nominal sizes which are in millimeters.

†† When calculating deviations for holes K, M, and N with tolerance grades up to and including IT 8, and holes P to ZC with tolerance grades up

British Standard Preferred Numbers and Preferred Sizes. — This British Standard (PD 6481:1977) gives recommendations for the use of preferred numbers and preferred sizes for functional characteristics and dimensions of various products.

The preferred number system is internationally standardized in ISO 3. It is also referred to as the Renard or R series (see Handbook page 19).

The series in the preferred number system are geometric series, i.e., in which there is a constant ratio between each figure and the succeeding one, within a decimal framework. Thus the R5 series has five steps between 1 and 10, the R10 series has 10 steps between 1 and 10, the R20 series, 20 steps and the R40 series, 40 steps, giving increases between steps of approximately 60, 25, 12 and 6 per cent, respectively.

The preferred size series have been developed from the preferred number series by rounding off the inconvenient numbers in the basic series and adjusting for linear measurement in millimeters. These series are shown in the table below.

Taking all normal considerations into account it is recommended that: (a) For ranges of values of the primary *functional* characteristics (outputs and capacities) of a series of products, the preferred number series R5 to R40 (see page 19) should be used and (b) whenever linear sizes are concerned, the preferred sizes as given in the table below should be used. The presentation of preferred sizes gives designers and users a logical selection and the benefits of rational variety reduction.

The second choice size given should only be used when it is not possible to employ the first choice, and the third choice should be applied only if a size from the second choice cannot be selected. With this procedure, common usage will tend to be concentrated on a limited range of sizes, and a contribution is thus made to variety reduction. However, the decision to use a particular size cannot be taken on the basis that one is first choice and the other not. Account must be taken of the effect on the design, the availability of tools, and other relevant factors.

British Standard Preferred Sizes (PD 6481:1977)

Choice			Choice			Choice			Choice			Choice			Choice		
1st	2nd	3rd	1st	2nd	3rd	1st	2nd	3rd	1st	2nd	3rd	1st	2nd	3rd	1st	2nd	3rd
1					5.2			23	65					122			188
	1.1			5.5				24			66	125			190		
1.2					5.8	25				68				128			192
		1.3	6					26	70			130				195	
	1.4				6.2		28			72				132			198
		1.5		6.5		30					74		135		200		
1.6					6.8		32		75					138			205
		1.7	7					34			76	140				210	
	1.8				7.5	35				78				142			215
		1.9	8				36		80				145		220		
2					8.5		38				82			148			225
		2.1	9			40				85		150				230	
	2.2				9.5		42				88			152			235
		2.4	10					44	90				155		240		
2.5			11			45					92			158			245
		2.6	12					46		95		160				250	
	2.8				13		48				98			162			255
3			14			50			100				165		260		
		3.2			15		52				102			168			265
	3.5		16					54		105		170				270	
		3.8			17	55					108			172			275
4			18					56	110				175		280		
		4.2			19		58				112			178			285
	4.5		20			60				115		180				290	
		4.8			21		62				118			182			295
5			22					64	120				185		300		

For dimensions above 300, each series continues in a similar manner, i.e., the intervals between each series number are the same as between 200 and 300.

Finding Length Differences Due to Temperature Changes. — The tables on this page and the next page give changes in length for changes in temperature from the standard reference temperature of 68 deg F (20 deg C). Thus, for example, in a steel bar with a coefficient of thermal expansion of .0000063 inch per inch per deg F, the increase in length at 73 deg F is 25 + 5 + 1.5 = 31.5 microinches per inch of length.

Length Differences per Centimeter for Change from Standard Temperature of 20 Degrees Celsius (Centigrade)

Temper- ature Deg. C.	Coefficient of thermal expansion of material per degree C. $\times$ 10^6								
	1	2	3	4	5	10	15	20	25	30
	Total change in length from standard, hundredths of microns (microns/100) per centimeter of length†									
0	−20	−40	−60	−80	−100	−200	−300	−400	−500	−600
1	−19	−38	−57	−76	− 95	−190	−285	−380	−475	−570
2	−18	−36	−54	−72	− 90	−180	−270	−360	−450	−540
3	−17	−34	−51	−68	− 85	−170	−255	−340	−425	−510
4	−16	−32	−48	−64	− 80	−160	−240	−320	−400	−480
5	−15	−30	−45	−60	− 75	−150	−225	−300	−375	−450
6	−14	−28	−42	−56	− 70	−140	−210	−280	−350	−420
7	−13	−26	−39	−52	− 65	−130	−195	−260	−325	−390
8	−12	−24	−36	−48	− 60	−120	−180	−240	−300	−360
9	−11	−22	−33	−44	− 55	−110	−165	−220	−275	−330
10	−10	−20	−30	−40	− 50	−100	−150	−200	−250	−300
11	− 9	−18	−27	−36	− 45	− 90	−135	−180	−225	−270
12	− 8	−16	−24	−32	− 40	− 80	−120	−160	−200	−240
13	− 7	−14	−21	−28	− 35	− 70	−105	−140	−175	−210
14	− 6	−12	−18	−24	− 30	− 60	− 90	−120	−150	−180
15	− 5	−10	−15	−20	− 25	− 50	− 75	−100	−125	−150
16	− 4	− 8	−12	−16	− 20	− 40	− 60	− 80	−100	−120
17	− 3	− 6	− 9	−12	− 15	− 30	− 45	− 60	− 75	− 90
18	− 2	− 4	− 6	− 8	− 10	− 20	− 30	− 40	− 50	− 60
19	− 1	− 2	− 3	− 4	− 5	− 10	− 15	− 20	− 25	− 30
20	0	0	0	0	0	0	0	0	0	0
21	1	2	3	4	5	10	15	20	25	30
22	2	4	6	8	10	20	30	40	50	60
23	3	6	9	12	15	30	45	60	75	90
24	4	8	12	16	20	40	60	80	100	120
25	5	10	15	20	25	50	75	100	125	150
26	6	12	18	24	30	60	90	120	150	180
27	7	14	21	28	35	70	105	140	175	210
28	8	16	24	32	40	80	120	160	200	240
29	9	18	27	36	45	90	135	180	225	270
30	10	20	30	40	50	100	150	200	250	300
31	11	22	33	44	55	110	165	220	275	330
32	12	24	36	48	60	120	180	240	300	360
33	13	26	39	52	65	130	195	260	325	390
34	14	28	42	56	70	140	210	280	350	420
35	15	30	45	60	75	150	225	300	375	450
36	16	32	48	64	80	160	240	320	400	480
37	17	34	51	68	85	170	255	340	425	510
38	18	36	54	72	90	180	270	360	450	540
39	19	38	57	76	95	190	285	380	475	570
40	20	40	60	80	100	200	300	400	500	600

For intermediate coefficients add appropriate listed values. For example, a length change for a coefficient of 7 is the sum of values in the 5 and 2 columns. Fractional interpolation may be similarly calculated.

* Or hundredths of micron (microns/100) per centimeter (see table on preceding page).
† Or microinches per inch.

Length Differences per Inch for Change from Standard Temperature of 68 Degrees F

Temperature Deg. F.	Coefficient of thermal expansion of material per degree F. $\times 10^6$									
	1	2	3	4	5	10	15	20	25	30
	Total change in length from standard, microinches per inch of length*									
40	−28	−56	−84	−112	−140	−280	−420	−560	−700	−840
41	−27	−54	−81	−108	−135	−270	−405	−540	−675	−810
42	−26	−52	−78	−104	−130	−260	−390	−520	−650	−780
43	−25	−50	−75	−100	−125	−250	−375	−500	−625	−750
44	−24	−48	−72	−96	−120	−240	−360	−480	−600	−720
45	−23	−46	−69	−92	−115	−230	−345	−460	−575	−690
46	−22	−44	−66	−88	−110	−220	−330	−440	−550	−660
47	−21	−42	−63	−84	−105	−210	−315	−420	−525	−630
48	−20	−40	−60	−80	−100	−200	−300	−400	−500	−600
49	−19	−38	−57	−76	−95	−190	−285	−380	−475	−570
50	−18	−36	−54	−72	−90	−180	−270	−360	−450	−540
51	−17	−34	−51	−68	−85	−170	−255	−340	−425	−510
52	−16	−32	−48	−64	−80	−160	−240	−320	−400	−480
53	−15	−30	−45	−60	−75	−150	−225	−300	−375	−450
54	−14	−28	−42	−56	−70	−140	−210	−280	−350	−420
55	−13	−26	−39	−52	−65	−130	−195	−260	−325	−390
56	−12	−24	−36	−48	−60	−120	−180	−240	−300	−360
57	−11	−22	−33	−44	−55	−110	−165	−220	−275	−330
58	−10	−20	−30	−40	−50	−100	−150	−200	−250	−300
59	−9	−18	−27	−36	−45	−90	−135	−180	−225	−270
60	−8	−16	−24	−32	−40	−80	−120	−160	−200	−240
61	−7	−14	−21	−28	−35	−70	−105	−140	−175	−210
62	−6	−12	−18	−24	−30	−60	−90	−120	−150	−180
63	−5	−10	−15	−20	−25	−50	−75	−100	−125	−150
64	−4	−8	−12	−16	−20	−40	−60	−80	−100	−120
65	−3	−6	−9	−12	−15	−30	−45	−60	−75	−90
66	−2	−4	−6	−8	−10	−20	−30	−40	−50	−60
67	−1	−2	−3	−4	−5	−10	−15	−20	−25	−30
68	0	0	0	0	0	0	0	0	0	0
69	1	2	3	4	5	10	15	20	25	30
70	2	4	6	8	10	20	30	40	50	60
71	3	6	9	12	15	30	45	60	75	90
72	4	8	12	16	20	40	60	80	100	120
73	5	10	15	20	25	50	75	100	125	150
74	6	12	18	24	30	60	90	120	150	180
75	7	14	21	28	35	70	105	140	175	210
76	8	16	24	32	40	80	120	160	200	240
77	9	18	27	36	45	90	135	180	225	270
78	10	20	30	40	50	100	150	200	250	300
79	11	22	33	44	55	110	165	220	275	330
80	12	24	36	48	60	120	180	240	300	360
81	13	26	39	52	65	130	195	260	325	390
82	14	28	42	56	70	140	210	280	350	420
83	15	30	45	60	75	150	225	300	375	450
84	16	32	48	64	80	160	240	320	400	480
85	17	34	51	68	85	170	255	340	425	510
86	18	36	54	72	90	180	270	360	450	540
87	19	38	57	76	95	190	285	380	475	570
88	20	40	60	80	100	200	300	400	500	600
89	21	42	63	84	105	210	315	420	525	630
90	22	44	66	88	110	220	330	440	550	660
91	23	46	69	92	115	230	345	460	575	690
92	24	48	72	96	120	240	360	480	600	720
93	25	50	75	100	125	250	375	500	625	750·
94	26	52	78	104	130	260	390	520	650	780
95	27	54	81	108	135	270	405	540	675	810

* See footnotes at end of table on the preceding page.

MEASURING INSTRUMENTS
AND INSPECTION METHODS

Reading a Vernier. — A general rule for taking readings with a vernier scale is as follows: Note the number of inches and sub-divisions of an inch that the zero mark of the vernier scale has moved along the true scale, and then add to this reading as many thousandths, or hundredths, or whatever fractional part of an inch the vernier reads to, as there are spaces between the vernier zero and that line on the vernier which coincides with one on the true scale. For example, if the zero line of a vernier which reads to thousandths is slightly beyond the 0.5 inch division on the main or true scale, as shown in Fig. 1, and graduation line 10 on the vernier exactly coincides

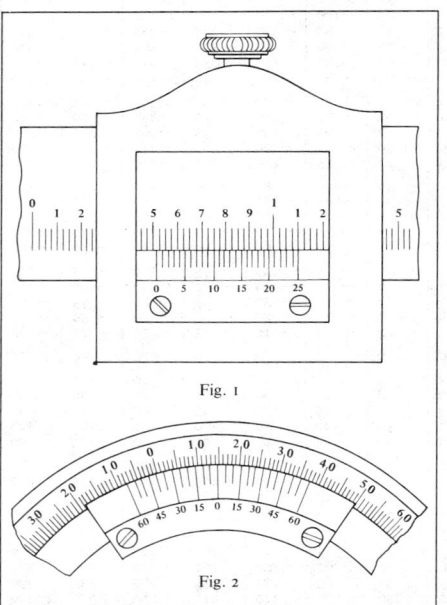

Fig. 1

Fig. 2

with one on the true scale, the reading is 0.5 + 0.010 or 0.510 inch. In order to determine the reading or fractional part of an inch that can be obtained by a vernier, multiply the denominator of the finest sub-division given on the true scale by the total number of divisions on the vernier. For example, if one inch on the true scale is divided into 40 parts or fortieths (as in Fig. 1), and the vernier into twenty-five parts, the vernier will read to thousandths of an inch, as 25 × 40 = 1000. Similarly, if there are sixteen divisions to the inch on the true scale and a total of eight on the vernier, the latter will enable readings to be taken within one-hundred-twenty-eighths of an inch, as 8 × 16 = 128.

If the vernier is on a protractor, note the whole number of degrees passed by the vernier zero mark and then count the spaces between

the vernier zero and that line which coincides with a graduation on the protractor scale. If the vernier indicates angles within five minutes or one-twelfth degree (as in Fig. 2), the number of spaces multiplied by 5 will, of course, give the number of minutes to be added to the whole number of degrees. The reading of the protractor set as illustrated would be 14 whole degrees (the number passed by the zero mark on the vernier) plus 30 minutes, as the graduation 30 on the vernier is the only one to the right of the vernier zero which exactly coincides with a line on the protractor scale. It will be noted that there are duplicate scales on the vernier, one being to the right and the other to the left of zero. The left-hand scale is used when the vernier zero is moved to the left of the zero of the protractor scale, whereas the right-hand graduations are used when the movement is to the right.

Reading a Metric Vernier. — The smallest graduation on the bar (true or main scale) of the metric vernier gage shown in Fig. 1, is 0.5 millimeter. The scale is numbered at each twentieth division, and thus increments of 10, 20, 30, 40 millimeters, etc., are indicated. There are 25 divisions on the vernier scale, occupying the same length as 24 divisions on the bar, which is 12 millimeters. Therefore, one division on the vernier scale equals one twenty-fifth of 12 millimeters = 0.04 × 12 = 0.48 millimeter. Thus, the difference between one bar division (0.50 mm) and one vernier division (2.48 mm) is 0.50 − 0.48 = 0.02 millimeter, which is the minimum measuring increment that the gage provides. To permit direct readings, the vernier scale has graduations to represent tenths of a millimeter (0.1 mm) and fiftieths of a millimeter (0.02 mm).

To read a vernier gage, first note how many millimeters the zero line on the vernier is from the zero line on the bar. Next, find the graduation on the vernier

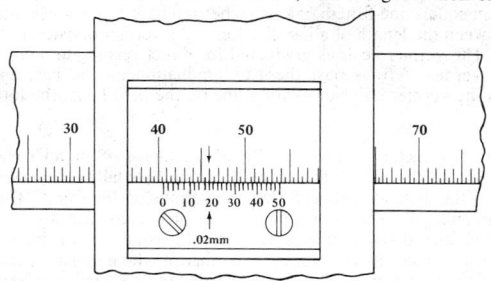

Fig. 1.

scale which exactly coincides with a graduation line on the bar, and note the value of the vernier scale graduation. This value is added to the value obtained from the bar, and the result is the total reading.

In the example shown in Fig. 1, the vernier zero is just past the 40.5 millimeters graduation on the bar. The 0.18 millimeter line on the vernier coincides with a line on the bar, and the total reading is therefore 40.5 + 0.18 = 40.68 mm.

Dual Metric-Inch Vernier. — The vernier gage shown in Fig. 1 has separate metric and inch 50-division vernier scales to permit measurements in either system.

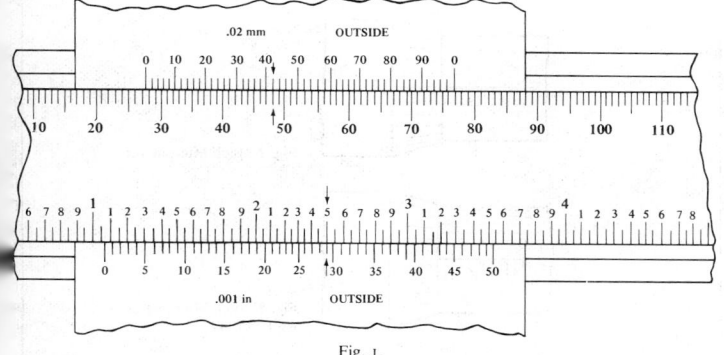

Fig. 1.

A 50-division vernier has more widely spaced graduations than the 25-division vernier shown on the previous pages, and is thus easier to read. On the bar, the smallest metric graduation is 1 millimeter, and the 50 divisions of the vernier occupy the same length as 49 divisions on the bar, which is 49 mm. Therefore, one division on the vernier scale equals one-fiftieth of 49 millimeters = 0.02 × 49 = 0.98 mm. Thus, the difference between one bar division (1.0 mm) and one vernier division (0.98 mm) is 0.02 mm, which is the minimum measuring increment the gage provides.

The vernier scale is graduated for direct reading to 0.02 mm. In the figure, the vernier zero is just past the 27 mm graduation on the bar, and the 0.42 mm graduation on the vernier coincides with a line on the bar. The total reading is therefore 27.42 mm.

The smallest inch graduation on the bar is 0.05 inch, and the 50 vernier divisions occupy the same length as 49 bar divisions, which is 2.45 inches. Therefore, one vernier division equals one-fiftieth of 2.45 inches = 0.02 × 2.45 = 0.049 inch. Thus, the difference between the length of a bar division and a vernier division is 0.050 − 0.049 = 0.001 inch. The vernier scale is graduated for direct reading to 0.001 inch. In the example, the vernier zero is past the 1.05 graduation on the bar, and the 0.029 graduation on the vernier coincides with a line on the bar. Thus, the total reading is 1.079 inches.

Reading a Micrometer. — The spindle of an inch-system micrometer has 40 threads per inch, so that one turn moves the spindle axially 0.025 inch (1 ÷ 40 = 0.025), equal to the distance between two graduations on the frame. The 25 graduations on the thimble allow the 0.025 inch to be further divided, so that turning the thimble through one division moves the spindle axially 0.001 inch (0.025 ÷ 25 = 0.001). To read a micrometer, count the number of whole divisions that are visible on the scale of the frame, multiply this number by 25 (the number of thousandths of an inch that each division represents) and add to the product the number of that division on the thimble which coincides with the axial zero line on the frame. The result will be the diameter expressed in thousandths of an inch. As the numbers 1, 2, 3, etc., opposite every fourth sub-division on the frame, indicate hundreds of thousandths, the reading can easily be taken mentally. Suppose the thimble were screwed out so that graduation 2, and three additional sub-divisions, were visible (as shown in Fig. 1), and that graduation 10 on the thimble coincided with the axial line on the frame. The reading then would be 0.200 + 0.075 + 0.010, or 0.285 inch.

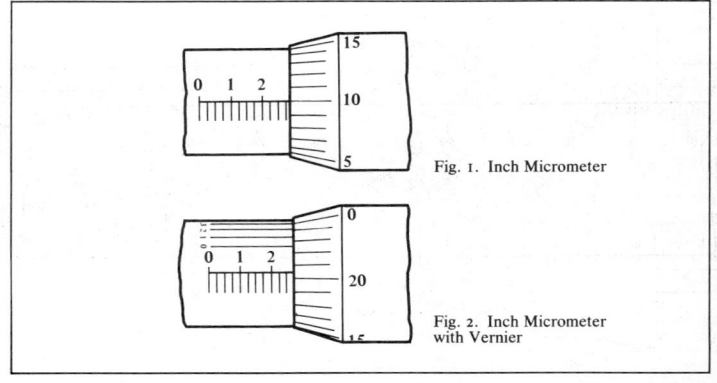

Fig. 1. Inch Micrometer

Fig. 2. Inch Micrometer with Vernier

Some micrometers have a vernier scale on the frame in addition to the regular graduations, so that measurements within 0.0001 part of an inch can be taken. Micrometers of this type are read as follows: First determine the number of thousandths, as with an ordinary micrometer, and then find a line on the vernier scale that exactly coincides with one on the thimble; the number of this line represents the number of ten-thousandths to be added to the number of thousandths obtained by the regular graduations. The reading shown in the illustration, Fig. 2, is 0.270 + 0.0003 = 0.2703 inch.

Micrometers graduated according to the English system of measurement ordinarily have a table of decimal equivalents stamped on the sides of the frame, so that fractions such as sixty-fourths, thirty-seconds, etc., can readily be converted into decimals.

Reading a Metric Micrometer. — The spindle of an ordinary metric micrometer has 2 threads per millimeter, and thus one complete revolution moves the spindle through a distance of 0.5 millimeter. The longitudinal line on the frame is graduated with 1 millimeter divisions and 0.5 millimeter sub-divisions. The thimble has 50 graduations, each being 0.01 millimeter (one-hundredth of a millimeter).

To read a metric micrometer, note the number of millimeter divisions visible on the scale of the sleeve, and add the total to the particular division on the thimble which coincides with the axial line on the sleeve. Suppose that the thimble were screwed out so that graduation 5, and one additional 0.5 sub-division were visible (as shown in Fig. 3), and that graduation 28 on the thimble coincided with the axial line on the sleeve. The reading then would be 5.00 + 0.5 + 0.28 = 5.78 mm.

Some micrometers are provided with a vernier scale on the sleeve in addition to the regular graduations to permit measurements within 0.002 millimeter to be made. Micrometers of this type are read as follows: First determine the number of whole millimeters (if any) and the number of hundredths of a millimeter, as with an ordinary micrometer, and then find a line on the sleeve vernier scale which exactly coincides

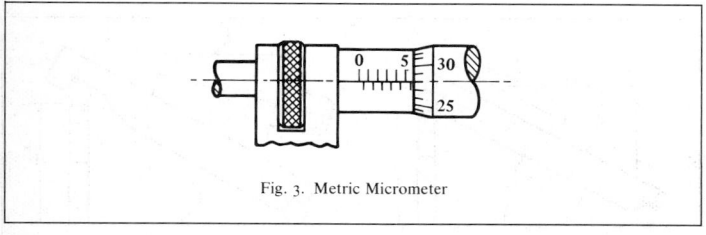

Fig. 3. Metric Micrometer

with one on the thimble. The number of this coinciding vernier line represents the number of two-thousandths of a millimeter to be added to the reading already obtained. Thus, for example, a measurement of 2.958 millimeters would be obtained by reading 2.5 millimeters on the sleeve, adding 0.45 millimeter read from the thimble, and then adding 0.008 millimeter as determined by the vernier.

Note: 0.01 millimeter = 0.000393 inch, and 0.002 millimeter = 0.000078 inch (78 millionths). Therefore, metric micrometers provide smaller measuring increments than comparable inch unit micrometers — the smallest graduation of an ordinary inch reading micrometer is 0.001 inch; the vernier type has graduations down to 0.0001 inch. When using either a metric or inch micrometer, without a vernier, smaller readings than those graduated may of course be obtained by visual interpolation between graduations.

Sine-bar. — The sine-bar is used either for very accurate angular measurements or for locating work at a given angle as, for example, in surface grinding templets, gages, etc. The sine-bar is especially useful in measuring or checking angles when the limit of accuracy is 5 minutes or less. Some bevel protractors are equipped with verniers which read to 5 minutes but the setting depends upon the alignment of graduations whereas a sine-bar usually is located by positive contact with precision gage-blocks selected for whatever dimension is required for obtaining a given angle.

Types of Sine-bars. — A sine-bar consists of a hardened, ground and lapped steel bar which has very accurate cylindrical plugs of equal diameter attached to or near each end. The form illustrated by Fig. 1 has notched ends for receiving the cylindrical plugs which are held firmly against both faces of the notch. The standard center-to-center distance C between the plugs is either 5 or 10 inches. The upper and lower sides of sine-bars are parallel to the center line of the plugs within very close limits. The body of the sine-bar ordinarily has several holes through it to reduce the weight. In the making of the sine-bar shown in Fig. 2, if too much material is removed from one locating notch, regrinding the shoulder at the opposite end would make it possible to obtain the correct center distance. That is the reason for this change in form. The type of sine-bar illustrated by Fig. 3 has the cylindrical disks or plugs attached to one side. These differences in form or arrangement do not, of course, affect the principle governing the use of the sine-bar. An accurate surface plate or master flat is always used in conjunction with a sine-bar in order to form the base from which the vertical measurements are made.

Setting 5-inch Sine-bar to Given Angle. — Since many sine-bars have a length of 5 inches, the accompanying table of constants is based upon that length. These constants represent the vertical distances H for setting a 5-inch sine-bar to the required angle. Assume that the angle is 31° 20'; the table shows that height H

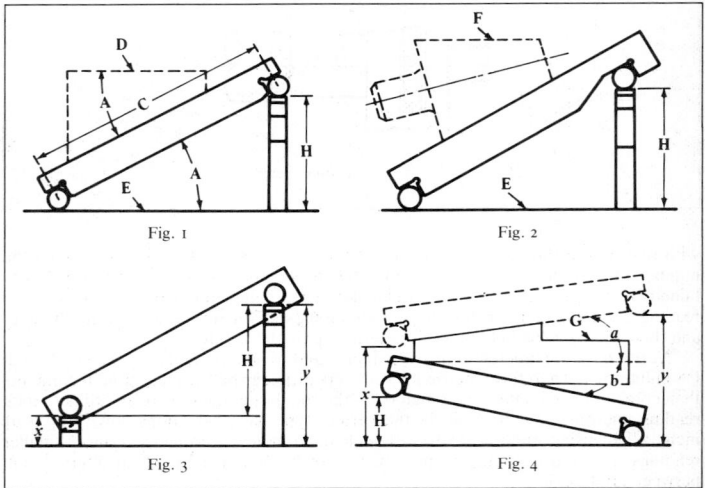

Fig. 1

Fig. 2

Fig. 3

Fig. 4

(Figs. 1, 2 and 3) should equal 2.6001 inches. Note: The constants in the table equal five times sine of the angle; thus the sine of 31° 20′ (see table of trigonometric functions) is 0.52002, and 0.52002 × 5 = 2.6001 inches.

Finding Angle when Height H of Sine-bar is Known. — In finding the angle equivalent to a given height H, the table of constants is used in reverse order. To illustrate, if the height H is 1.4061 inches, the angle to which the sine-bar is set is 16° 20′. (Note: In using the regular table of sines, divide height H by length of sine-bar, find sine equal to quotient and its angle; thus 1.4061 ÷ 5 = 0.2812. Table of sines shows that this is the sine of 16° 20′.)

Checking Angle of Templet or Gage by Using Sine-bar. — Place templet or gage on sine-bar as indicated by dotted lines, Fig. 1. Clamps may be used to hold work in place. Place upper end of sine-bar on gage-blocks having total height H corresponding to the required angle. If upper edge D of work is parallel with surface plate E, then angle A of work equals angle A to which sine-bar is set. Parallelism between edge D and surface plate may be tested by checking the height at each end with a dial gage or some indicating type of comparator.

Measuring Angle of Templet or Gage with Sine-bar. — Adjust height of gage-blocks and sine-bar until edge D, Fig. 1, of gage or templet is parallel with surface plate E; then find angle corresponding to height H of gage-blocks. For example, if height H is 2.5939 inches when D and E are parallel, the table of sine-bar constants shows that angle A of work is 31° 15′.

Checking Taper per Foot with Sine-bar. — As an example, assume that plug gage, Fig. 2, is supposed to have a taper of 6⅛ inches per foot and taper is to be checked by using a 5-inch sine-bar. The table on page 687, Tapers per Foot and Corresponding Angles, shows that the included angle for a taper of 6⅛ inches per foot is 28° 38′ 1″ or practically 28° 38′. The table of sine-bar constants shows that height H of sine-bar should be 2.3960 inches; hence, if the upper surface F of the gage is parallel to surface E when sine-bar is at the height given, the angle corresponds to a taper of 6⅛ inches per foot.

Setting Sine-bar which has Plugs Attached to Side. — If lower plug does not rest directly upon the surface plate (see Fig. 3), note that height H for setting the sine-bar is the difference between heights x and y or the difference between the heights of the plugs; otherwise the procedure in setting the sine-bar and checking angles is the same as previously described.

Checking Templet Having Two Angles. — Assume that angle a of templet, Fig. 4, is 9 degrees, angle b 12 degrees, and that edge G is parallel to surface plate. Table shows that height H equals 1.03956 inches for an angle b of 12 degrees. For an angle a of 9 degrees, the table shows that the difference between measurements x and y when sine-bar is in contact with the upper edge of the templet should equal 0.78217 inch.

Setting 10-inch Sine-bar to Given Angle. — A 10-inch sine-bar may be preferable in some cases because of its longer working surface or because the longer center distance is conducive to greater precision. To obtain the vertical distances H for setting a 10-inch sine bar, first find the setting for a 5-inch sine bar in the following table and then multiply this setting by 2.

Example: As shown in the table, the setting for 39 degrees is 3.14660, hence the vertical height H for setting the 10-inch sine bar is 6.2932 inches.

SINE-BAR CONSTANTS

Constants for Setting a 5-inch Sine-Bar — I

Min.	0°	1°	2°	3°	4°	5°	6°	7°
0	0.00000	0.08726	0.17450	0.26168	0.34878	0.43578	0.52264	0.60935
1	0.00145	0.08872	0.17595	0.26313	0.35023	0.43723	0.52409	0.61079
2	0.00291	0.09017	0.17740	0.26458	0.35168	0.43868	0.52554	0.61223
3	0.00436	0.09162	0.17886	0.26604	0.35313	0.44013	0.52698	0.61368
4	0.00582	0.09308	0.18031	0.26749	0.35459	0.44157	0.52843	0.61512
5	0.00727	0.09453	0.18177	0.26894	0.35604	0.44302	0.52987	0.61656
6	0.00873	0.09599	0.18322	0.27039	0.35749	0.44447	0.53132	0.61801
7	0.01018	0.09744	0.18467	0.27185	0.35894	0.44592	0.53277	0.61945
8	0.01164	0.09890	0.18613	0.27330	0.36039	0.44737	0.53421	0.62089
9	0.01309	0.10035	0.18758	0.27475	0.36184	0.44882	0.53566	0.62234
10	0.01454	0.10180	0.18903	0.27620	0.36329	0.45027	0.53710	0.62378
11	0.01600	0.10326	0.19049	0.27766	0.36474	0.45171	0.53855	0.62522
12	0.01745	0.10471	0.19194	0.27911	0.36619	0.45316	0.54000	0.62667
13	0.01891	0.10617	0.19339	0.28056	0.36764	0.45461	0.54144	0.62811
14	0.02036	0.10762	0.19485	0.28201	0.36909	0.45606	0.54289	0.62955
15	0.02182	0.10907	0.19630	0.28346	0.37054	0.45751	0.54433	0.63099
16	0.02327	0.11053	0.19775	0.28492	0.37199	0.45896	0.54578	0.63244
17	0.02473	0.11198	0.19921	0.28637	0.37344	0.46040	0.54723	0.63388
18	0.02618	0.11344	0.20066	0.28782	0.37489	0.46185	0.54867	0.63532
19	0.02763	0.11489	0.20211	0.28927	0.37634	0.46330	0.55012	0.63677
20	0.02909	0.11634	0.20357	0.29072	0.37779	0.46475	0.55156	0.63821
21	0.03054	0.11780	0.20502	0.29218	0.37924	0.46620	0.55301	0.63965
22	0.03200	0.11925	0.20647	0.29363	0.38069	0.46765	0.55445	0.64109
23	0.03345	0.12071	0.20793	0.29508	0.38214	0.46909	0.55590	0.64254
24	0.03491	0.12216	0.20938	0.29653	0.38360	0.47054	0.55734	0.64398
25	0.03636	0.12361	0.21083	0.29798	0.38505	0.47199	0.55879	0.64542
26	0.03782	0.12507	0.21228	0.29944	0.38650	0.47344	0.56024	0.64686
27	0.03927	0.12652	0.21374	0.30089	0.38795	0.47489	0.56168	0.64830
28	0.04072	0.12798	0.21519	0.30234	0.38940	0.47633	0.56313	0.64975
29	0.04218	0.12943	0.21664	0.30379	0.39085	0.47778	0.56457	0.65119
30	0.04363	0.13088	0.21810	0.30524	0.39230	0.47923	0.56602	0.65263
31	0.04509	0.13234	0.21955	0.30669	0.39375	0.48068	0.56746	0.65407
32	0.04654	0.13379	0.22100	0.30815	0.39520	0.48212	0.56891	0.65551
33	0.04800	0.13525	0.22246	0.30960	0.39665	0.48357	0.57035	0.65696
34	0.04945	0.13670	0.22391	0.31105	0.39810	0.48502	0.57180	0.65840
35	0.05090	0.13815	0.22536	0.31250	0.39954	0.48647	0.57324	0.65984
36	0.05236	0.13961	0.22681	0.31395	0.40099	0.48791	0.57469	0.66128
37	0.05381	0.14106	0.22827	0.31540	0.40244	0.48936	0.57613	0.66272
38	0.05527	0.14252	0.22972	0.31686	0.40389	0.49081	0.57758	0.66417
39	0.05672	0.14397	0.23117	0.31831	0.40534	0.49226	0.57902	0.66561
40	0.05818	0.14542	0.23263	0.31976	0.40679	0.49370	0.58046	0.66705
41	0.05963	0.14688	0.23408	0.32121	0.40824	0.49515	0.58191	0.66849
42	0.06109	0.14833	0.23553	0.32266	0.40969	0.49660	0.58335	0.66993
43	0.06254	0.14979	0.23699	0.32411	0.41114	0.49805	0.58480	0.67137
44	0.06399	0.15124	0.23844	0.32556	0.41259	0.49949	0.58624	0.67281
45	0.06545	0.15269	0.23989	0.32702	0.41404	0.50094	0.58769	0.67425
46	0.06690	0.15415	0.24134	0.32847	0.41549	0.50239	0.58913	0.67570
47	0.06836	0.15560	0.24280	0.32992	0.41694	0.50383	0.59058	0.67714
48	0.06981	0.15705	0.24425	0.33137	0.41839	0.50528	0.59202	0.67858
49	0.07127	0.15851	0.24570	0.33282	0.41984	0.50673	0.59346	0.68002
50	0.07272	0.15996	0.24715	0.33427	0.42129	0.50818	0.59491	0.68146
51	0.07417	0.16141	0.24861	0.33572	0.42274	0.50962	0.59635	0.68290
52	0.07563	0.16287	0.25006	0.33717	0.42419	0.51107	0.59780	0.68434
53	0.07708	0.16432	0.25151	0.33863	0.42564	0.51252	0.59924	0.68578
54	0.07854	0.16578	0.25296	0.34008	0.42708	0.51396	0.60068	0.68722
55	0.07999	0.16723	0.25442	0.34153	0.42853	0.51541	0.60213	0.68866
56	0.08145	0.16868	0.25587	0.34298	0.42998	0.51686	0.60357	0.69010
57	0.08290	0.17014	0.25732	0.34443	0.43143	0.51830	0.60502	0.69154
58	0.08435	0.17159	0.25877	0.34588	0.43288	0.51975	0.60646	0.69298
59	0.08581	0.17304	0.26023	0.34733	0.43433	0.52120	0.60790	0.69443
60	0.08726	0.17450	0.26168	0.34878	0.43578	0.52264	0.60935	0.69587

Constants for Setting a 5-inch Sine-Bar — 2

Min.	8°	9°	10°	11°	12°	13°	14°	15°
0	0.69587	0.78217	0.86824	0.95404	1.03956	1.12476	1.20961	1.29410
1	0.69731	0.78361	0.86967	0.95547	1.04098	1.12617	1.21102	1.29550
2	0.69875	0.78505	0.87111	0.95690	1.04240	1.12759	1.21243	1.29690
3	0.70019	0.78648	0.87254	0.95833	1.04383	1.12901	1.21384	1.29831
4	0.70163	0.78792	0.87397	0.95976	1.04525	1.13042	1.21525	1.29971
5	0.70307	0.78935	0.87540	0.96118	1.04667	1.13184	1.21666	1.30112
6	0.70451	0.79079	0.87683	0.96261	1.04809	1.13326	1.21808	1.30252
7	0.70595	0.79223	0.87827	0.96404	1.04951	1.13467	1.21949	1.30393
8	0.70739	0.79366	0.87970	0.96546	1.05094	1.13609	1.22090	1.30533
9	0.70883	0.79510	0.88113	0.96689	1.05236	1.13751	1.22231	1.30673
10	0.71027	0.79653	0.88256	0.96832	1.05378	1.13892	1.22372	1.30814
11	0.71171	0.79797	0.88399	0.96974	1.05520	1.14034	1.22513	1.30954
12	0.71314	0.79941	0.88542	0.97117	1.05662	1.14175	1.22654	1.31095
13	0.71458	0.80084	0.88686	0.97260	1.05805	1.14317	1.22795	1.31235
14	0.71602	0.80228	0.88829	0.97403	1.05947	1.14459	1.22936	1.31375
15	0.71746	0.80371	0.88972	0.97545	1.06089	1.14600	1.23077	1.31516
16	0.71890	0.80515	0.89115	0.97688	1.06231	1.14742	1.23218	1.31656
17	0.72034	0.80658	0.89258	0.97830	1.06373	1.14883	1.23359	1.31796
18	0.72178	0.80802	0.89401	0.97973	1.06515	1.15025	1.23500	1.31937
19	0.72322	0.80945	0.89544	0.98116	1.06657	1.15166	1.23640	1.32077
20	0.72466	0.81089	0.89687	0.98258	1.06799	1.15308	1.23781	1.32217
21	0.72610	0.81232	0.89830	0.98401	1.06941	1.15449	1.23922	1.32357
22	0.72754	0.81376	0.89973	0.98544	1.07084	1.15591	1.24063	1.32498
23	0.72898	0.81519	0.90117	0.98686	1.07226	1.15732	1.24204	1.32638
24	0.73042	0.81663	0.90260	0.98829	1.07368	1.15874	1.24345	1.32778
25	0.73185	0.81806	0.90403	0.98971	1.07510	1.16015	1.24486	1.32918
26	0.73329	0.81950	0.90546	0.99114	1.07652	1.16157	1.24627	1.33058
27	0.73473	0.82093	0.90689	0.99256	1.07794	1.16298	1.24768	1.33199
28	0.73617	0.82237	0.90832	0.99399	1.07936	1.16440	1.24908	1.33339
29	0.73761	0.82380	0.90975	0.99541	1.08078	1.16581	1.25049	1.33479
30	0.73905	0.82524	0.91118	0.99684	1.08220	1.16723	1.25190	1.33619
31	0.74049	0.82667	0.91261	0.99826	1.08362	1.16864	1.25331	1.33759
32	0.74192	0.82811	0.91404	0.99969	1.08504	1.17006	1.25472	1.33899
33	0.74336	0.82954	0.91547	1.00112	1.08646	1.17147	1.25612	1.34040
34	0.74480	0.83098	0.91690	1.00254	1.08788	1.17288	1.25753	1.34180
35	0.74624	0.83241	0.91833	1.00396	1.08930	1.17430	1.25894	1.34320
36	0.74768	0.83384	0.91976	1.00539	1.09072	1.17571	1.26035	1.34460
37	0.74911	0.83528	0.92119	1.00681	1.09214	1.17712	1.26175	1.34600
38	0.75055	0.83671	0.92262	1.00824	1.09355	1.17854	1.26316	1.34740
39	0.75199	0.83815	0.92405	1.00966	1.09497	1.17995	1.26457	1.34880
40	0.75343	0.83958	0.92547	1.01109	1.09639	1.18136	1.26598	1.35020
41	0.75487	0.84101	0.92690	1.01251	1.09781	1.18278	1.26738	1.35160
42	0.75630	0.84245	0.92833	1.01394	1.09923	1.18419	1.26879	1.35300
43	0.75774	0.84388	0.92976	1.01536	1.10065	1.18560	1.27020	1.35440
44	0.75918	0.84531	0.93119	1.01678	1.10207	1.18702	1.27160	1.35580
45	0.76062	0.84675	0.93262	1.01821	1.10349	1.18843	1.27301	1.35720
46	0.76205	0.84818	0.93405	1.01963	1.10491	1.18984	1.27442	1.35860
47	0.76349	0.84961	0.93548	1.02106	1.10632	1.19125	1.27582	1.36000
48	0.76493	0.85105	0.93691	1.02248	1.10774	1.19267	1.27723	1.36140
49	0.76637	0.85248	0.93834	1.02390	1.10916	1.19408	1.27863	1.36280
50	0.76780	0.85391	0.93976	1.02533	1.11058	1.19549	1.28004	1.36420
51	0.76924	0.85535	0.94119	1.02675	1.11200	1.19690	1.28145	1.36560
52	0.77068	0.85678	0.94262	1.02817	1.11341	1.19832	1.28285	1.36700
53	0.77211	0.85821	0.94405	1.02960	1.11483	1.19973	1.28426	1.36840
54	0.77355	0.85965	0.94548	1.03102	1.11625	1.20114	1.28566	1.36980
55	0.77499	0.86108	0.94691	1.03244	1.11767	1.20255	1.28707	1.37119
56	0.77643	0.86251	0.94833	1.03387	1.11909	1.20396	1.28847	1.37259
57	0.77786	0.86394	0.94976	1.03529	1.12050	1.20538	1.28988	1.37399
58	0.77930	0.86538	0.95119	1.03671	1.12192	1.20679	1.29129	1.37539
59	0.78074	0.86681	0.95262	1.03814	1.12334	1.20820	1.29269	1.37679
60	0.78217	0.86824	0.95404	1.03956	1.12476	1.20961	1.29410	1.37819

Constants for Setting a 5-inch Sine-Bar — 3

Min.	16°	17°	18°	19°	20°	21°	22°	23°
0	1.37819	1.46186	1.54508	1.62784	1.71010	1.79184	1.87303	1.95366
1	1.37958	1.46325	1.54647	1.62922	1.71147	1.79320	1.87438	1.95499
2	1.38098	1.46464	1.54785	1.63059	1.71283	1.79456	1.87573	1.95633
3	1.38238	1.46603	1.54923	1.63197	1.71420	1.79591	1.87708	1.95767
4	1.38378	1.46742	1.55062	1.63334	1.71557	1.79727	1.87843	1.95901
5	1.38518	1.46881	1.55200	1.63472	1.71693	1.79863	1.87977	1.96035
6	1.38657	1.47020	1.55338	1.63609	1.71830	1.79998	1.88112	1.96169
7	1.38797	1.47159	1.55476	1.63746	1.71966	1.80134	1.88247	1.96302
8	1.38937	1.47298	1.55615	1.63884	1.72103	1.80270	1.88382	1.96436
9	1.39076	1.47437	1.55753	1.64021	1.72240	1.80405	1.88516	1.96570
10	1.39216	1.47576	1.55891	1.64159	1.72376	1.80541	1.88651	1.96704
11	1.39356	1.47715	1.56029	1.64296	1.72513	1.80677	1.88786	1.96837
12	1.39496	1.47854	1.56167	1.64433	1.72649	1.80812	1.88920	1.96971
13	1.39635	1.47993	1.56306	1.64571	1.72786	1.80948	1.89055	1.97105
14	1.39775	1.48132	1.56444	1.64708	1.72922	1.81083	1.89190	1.97238
15	1.39915	1.48271	1.56582	1.64845	1.73059	1.81219	1.89324	1.97372
16	1.40054	1.48410	1.56720	1.64983	1.73195	1.81355	1.89459	1.97506
17	1.40194	1.48549	1.56858	1.65120	1.73331	1.81490	1.89594	1.97639
18	1.40333	1.48687	1.56996	1.65257	1.73468	1.81626	1.89728	1.97773
19	1.40473	1.48826	1.57134	1.65394	1.73604	1.81761	1.89863	1.97906
20	1.40613	1.48965	1.57272	1.65532	1.73741	1.81897	1.89997	1.98040
21	1.40752	1.49104	1.57410	1.65669	1.73877	1.82032	1.90132	1.98173
22	1.40892	1.49243	1.57548	1.65806	1.74013	1.82168	1.90266	1.98307
23	1.41031	1.49382	1.57687	1.65943	1.74150	1.82303	1.90401	1.98440
24	1.41171	1.49520	1.57825	1.66081	1.74286	1.82438	1.90535	1.98574
25	1.41310	1.49659	1.57963	1.66218	1.74422	1.82574	1.90670	1.98707
26	1.41450	1.49798	1.58101	1.66355	1.74559	1.82709	1.90804	1.98841
27	1.41589	1.49937	1.58238	1.66492	1.74695	1.82845	1.90939	1.98974
28	1.41729	1.50075	1.58376	1.66629	1.74831	1.82980	1.91073	1.99108
29	1.41868	1.50214	1.58514	1.66766	1.74967	1.83115	1.91207	1.99241
30	1.42008	1.50353	1.58652	1.66903	1.75104	1.83251	1.91342	1.99375
31	1.42147	1.50492	1.58790	1.67041	1.75240	1.83386	1.91476	1.99508
32	1.42287	1.50630	1.58928	1.67178	1.75376	1.83521	1.91610	1.99641
33	1.42426	1.50769	1.59066	1.67315	1.75512	1.83657	1.91745	1.99775
34	1.42565	1.50908	1.59204	1.67452	1.75649	1.83792	1.91879	1.99908
35	1.42705	1.51046	1.59342	1.67589	1.75785	1.83927	1.92013	2.00041
36	1.42844	1.51185	1.59480	1.67726	1.75921	1.84062	1.92148	2.00175
37	1.42984	1.51324	1.59617	1.67863	1.76057	1.84197	1.92282	2.00308
38	1.43123	1.51462	1.59755	1.68000	1.76193	1.84333	1.92416	2.00441
39	1.43262	1.51601	1.59893	1.68137	1.76329	1.84468	1.92550	2.00574
40	1.43402	1.51739	1.60031	1.68274	1.76465	1.84603	1.92685	2.00707
41	1.43541	1.51878	1.60169	1.68411	1.76601	1.84738	1.92819	2.00841
42	1.43680	1.52017	1.60306	1.68548	1.76737	1.84873	1.92953	2.00974
43	1.43820	1.52155	1.60444	1.68685	1.76873	1.85009	1.93087	2.01107
44	1.43959	1.52294	1.60582	1.68821	1.77010	1.85144	1.93221	2.01240
45	1.44098	1.52432	1.60720	1.68958	1.77146	1.85279	1.93355	2.01373
46	1.44237	1.52571	1.60857	1.69095	1.77282	1.85414	1.93490	2.01506
47	1.44377	1.52709	1.60995	1.69232	1.77418	1.85549	1.93624	2.01640
48	1.44516	1.52848	1.61133	1.69369	1.77553	1.85684	1.93758	2.01773
49	1.44655	1.52986	1.61271	1.69506	1.77689	1.85819	1.93892	2.01906
50	1.44794	1.53125	1.61408	1.69643	1.77825	1.85954	1.94026	2.02039
51	1.44934	1.53263	1.61546	1.69779	1.77961	1.86089	1.94160	2.02172
52	1.45073	1.53401	1.61683	1.69916	1.78097	1.86224	1.94294	2.02305
53	1.45212	1.53540	1.61821	1.70053	1.78233	1.86359	1.94428	2.02438
54	1.45351	1.53678	1.61959	1.70190	1.78369	1.86494	1.94562	2.02571
55	1.45490	1.53817	1.62096	1.70327	1.78505	1.86629	1.94696	2.02704
56	1.45629	1.53955	1.62234	1.70463	1.78641	1.86764	1.94830	2.02837
57	1.45769	1.54093	1.62371	1.70600	1.78777	1.86899	1.94964	2.02970
58	1.45908	1.54232	1.62509	1.70737	1.78912	1.87034	1.95098	2.03103
59	1.46047	1.54370	1.62647	1.70873	1.79048	1.87168	1.95232	2.03235
60	1.46186	1.54508	1.62784	1.71010	1.79184	1.87303	1.95366	2.03368

Constants for Setting a 5-inch Sine-Bar — 4

Min.	24°	25°	26°	27°	28°	29°	30°	31°
0	2.03368	2.11309	2.19186	2.26995	2.34736	2.42405	2.50000	2.57519
1	2.03501	2.11441	2.19316	2.27125	2.34864	2.42532	2.50126	2.57644
2	2.03634	2.11573	2.19447	2.27254	2.34993	2.42659	2.50252	2.57768
3	2.03767	2.11705	2.19578	2.27384	2.35121	2.42786	2.50378	2.57893
4	2.03900	2.11836	2.19708	2.27513	2.35249	2.42913	2.50504	2.58018
5	2.04032	2.11968	2.19839	2.27643	2.35378	2.43041	2.50630	2.58142
6	2.04165	2.12100	2.19970	2.27772	2.35506	2.43168	2.50755	2.58267
7	2.04298	2.12231	2.20100	2.27902	2.35634	2.43295	2.50881	2.58391
8	2.04431	2.12363	2.20231	2.28031	2.35763	2.43422	2.51007	2.58516
9	2.04563	2.12495	2.20361	2.28161	2.35891	2.43549	2.51133	2.58640
10	2.04696	2.12626	2.20492	2.28290	2.36019	2.43676	2.51259	2.58765
11	2.04829	2.12758	2.20622	2.28420	2.36147	2.43803	2.51384	2.58889
12	2.04962	2.12890	2.20753	2.28549	2.36275	2.43930	2.51510	2.59014
13	2.05094	2.13021	2.20883	2.28678	2.36404	2.44057	2.51636	2.59138
14	2.05227	2.13153	2.21014	2.28808	2.36532	2.44184	2.51761	2.59262
15	2.05359	2.13284	2.21144	2.28937	2.36660	2.44311	2.51887	2.59387
16	2.05492	2.13416	2.21275	2.29066	2.36788	2.44438	2.52013	2.59511
17	2.05625	2.13547	2.21405	2.29196	2.36916	2.44564	2.52138	2.59635
18	2.05757	2.13679	2.21536	2.29325	2.37044	2.44691	2.52264	2.59760
19	2.05890	2.13810	2.21666	2.29454	2.37172	2.44818	2.52389	2.59884
20	2.06022	2.13942	2.21796	2.29583	2.37300	2.44945	2.52515	2.60008
21	2.06155	2.14073	2.21927	2.29712	2.37428	2.45072	2.52640	2.60132
22	2.06287	2.14205	2.22057	2.29842	2.37556	2.45198	2.52766	2.60256
23	2.06420	2.14336	2.22187	2.29971	2.37684	2.45325	2.52891	2.60381
24	2.06552	2.14468	2.22318	2.30100	2.37812	2.45452	2.53017	2.60505
25	2.06685	2.14599	2.22448	2.30229	2.37940	2.45579	2.53142	2.60629
26	2.06817	2.14730	2.22578	2.30358	2.38068	2.45705	2.53268	2.60753
27	2.06949	2.14862	2.22708	2.30487	2.38196	2.45832	2.53393	2.60877
28	2.07082	2.14993	2.22839	2.30616	2.38324	2.45959	2.53519	2.61001
29	2.07214	2.15124	2.22969	2.30745	2.38452	2.46085	2.53644	2.61125
30	2.07347	2.15256	2.23099	2.30874	2.38579	2.46212	2.53769	2.61249
31	2.07479	2.15387	2.23229	2.31003	2.38707	2.46338	2.53894	2.61373
32	2.07611	2.15518	2.23359	2.31132	2.38835	2.46465	2.54020	2.61497
33	2.07744	2.15649	2.23489	2.31261	2.38963	2.46591	2.54145	2.61621
34	2.07876	2.15781	2.23619	2.31390	2.39090	2.46718	2.54270	2.61745
35	2.08008	2.15912	2.23749	2.31519	2.39218	2.46844	2.54396	2.61869
36	2.08140	2.16043	2.23880	2.31648	2.39346	2.46971	2.54521	2.61993
37	2.08273	2.16174	2.24010	2.31777	2.39474	2.47097	2.54646	2.62117
38	2.08405	2.16305	2.24140	2.31906	2.39601	2.47224	2.54771	2.62241
39	2.08537	2.16436	2.24270	2.32035	2.39729	2.47350	2.54896	2.62364
40	2.08669	2.16567	2.24400	2.32163	2.39857	2.47477	2.55021	2.62488
41	2.08801	2.16698	2.24530	2.32292	2.39984	2.47603	2.55146	2.62612
42	2.08934	2.16830	2.24659	2.32421	2.40112	2.47729	2.55271	2.62736
43	2.09066	2.16961	2.24789	2.32550	2.40239	2.47856	2.55397	2.62860
44	2.09198	2.17092	2.24919	2.32679	2.40367	2.47982	2.55522	2.62983
45	2.09330	2.17223	2.25049	2.32807	2.40494	2.48108	2.55647	2.63107
46	2.09462	2.17354	2.25179	2.32936	2.40622	2.48235	2.55772	2.63231
47	2.09594	2.17485	2.25309	2.33065	2.40749	2.48361	2.55896	2.63354
48	2.09726	2.17616	2.25439	2.33193	2.40877	2.48487	2.56021	2.63478
49	2.09858	2.17746	2.25569	2.33322	2.41004	2.48613	2.56146	2.63601
50	2.09990	2.17877	2.25698	2.33451	2.41132	2.48739	2.56271	2.63725
51	2.10122	2.18008	2.25828	2.33579	2.41259	2.48866	2.56396	2.63849
52	2.10254	2.18139	2.25958	2.33708	2.41386	2.48992	2.56521	2.63972
53	2.10386	2.18270	2.26088	2.33836	2.41514	2.49118	2.56646	2.64096
54	2.10518	2.18401	2.26217	2.33965	2.41641	2.49244	2.56771	2.64219
55	2.10650	2.18532	2.26347	2.34093	2.41769	2.49370	2.56895	2.64343
56	2.10782	2.18663	2.26477	2.34222	2.41896	2.49496	2.57020	2.64466
57	2.10914	2.18793	2.26606	2.34350	2.42023	2.49622	2.57145	2.64590
58	2.11045	2.18924	2.26736	2.34479	2.42150	2.49748	2.57270	2.64713
59	2.11177	2.19055	2.26866	2.34607	2.42278	2.49874	2.57394	2.64836
60	2.11309	2.19186	2.26995	2.34736	2.42405	2.50000	2.57519	2.64960

Constants for Setting a 5-inch Sine-Bar — 5

Min.	32°	33°	34°	35°	36°	37°	38°	39°
0	2.64960	2.72320	2.79596	2.86788	2.93893	3.00908	3.07831	3.14660
1	2.65083	2.72441	2.79717	2.86907	2.94010	3.01024	3.07945	3.14773
2	2.65206	2.72563	2.79838	2.87026	2.94128	3.01140	3.08060	3.14886
3	2.65330	2.72685	2.79958	2.87146	2.94246	3.01256	3.08174	3.14999
4	2.65453	2.72807	2.80079	2.87265	2.94363	3.01372	3.08289	3.15112
5	2.65576	2.72929	2.80199	2.87384	2.94481	3.01488	3.08403	3.15225
6	2.65699	2.73051	2.80319	2.87503	2.94598	3.01604	3.08518	3.15338
7	2.65822	2.73173	2.80440	2.87622	2.94716	3.01720	3.08632	3.15451
8	2.65946	2.73295	2.80560	2.87741	2.94833	3.01836	3.08747	3.15564
9	2.66069	2.73416	2.80681	2.87860	2.94951	3.01952	3.08861	3.15676
10	2.66192	2.73538	2.80801	2.87978	2.95068	3.02068	3.08976	3.15789
11	2.66315	2.73660	2.80921	2.88097	2.95185	3.02184	3.09090	3.15902
12	2.66438	2.73782	2.81042	2.88216	2.95303	3.02300	3.09204	3.16015
13	2.66561	2.73903	2.81162	2.88335	2.95420	3.02415	3.09318	3.16127
14	2.66684	2.74025	2.81282	2.88454	2.95538	3.02531	3.09433	3.16240
15	2.66807	2.74147	2.81402	2.88573	2.95655	3.02647	3.09547	3.16353
16	2.66930	2.74268	2.81523	2.88691	2.95772	3.02763	3.09661	3.16465
17	2.67053	2.74390	2.81643	2.88810	2.95889	3.02878	3.09775	3.16578
18	2.67176	2.74511	2.81763	2.88929	2.96007	3.02994	3.09890	3.16690
19	2.67299	2.74633	2.81883	2.89048	2.96124	3.03110	3.10004	3.16803
20	2.67422	2.74754	2.82003	2.89166	2.96241	3.03226	3.10118	3.16915
21	2.67545	2.74876	2.82123	2.89285	2.96358	3.03341	3.10232	3.17028
22	2.67668	2.74997	2.82243	2.89403	2.96475	3.03457	3.10346	3.17140
23	2.67791	2.75119	2.82363	2.89522	2.96592	3.03572	3.10460	3.17253
24	2.67913	2.75240	2.82484	2.89641	2.96709	3.03688	3.10574	3.17365
25	2.68036	2.75362	2.82603	2.89759	2.96826	3.03803	3.10688	3.17478
26	2.68159	2.75483	2.82723	2.89878	2.96944	3.03919	3.10802	3.17590
27	2.68282	2.75605	2.82843	2.89996	2.97061	3.04034	3.10916	3.17702
28	2.68404	2.75726	2.82963	2.90115	2.97178	3.04150	3.11030	3.17815
29	2.68527	2.75847	2.83083	2.90233	2.97294	3.04265	3.11143	3.17927
30	2.68650	2.75968	2.83203	2.90351	2.97411	3.04381	3.11257	3.18039
31	2.68772	2.76090	2.83323	2.90470	2.97528	3.04496	3.11371	3.18151
32	2.68895	2.76211	2.83443	2.90588	2.97645	3.04611	3.11485	3.18264
33	2.69018	2.76332	2.83563	2.90707	2.97762	3.04727	3.11599	3.18376
34	2.69140	2.76453	2.83682	2.90825	2.97879	3.04842	3.11712	3.18488
35	2.69263	2.76575	2.83802	2.90943	2.97996	3.04957	3.11826	3.18600
36	2.69385	2.76696	2.83922	2.91061	2.98112	3.05073	3.11940	3.18712
37	2.69508	2.76817	2.84042	2.91180	2.98229	3.05188	3.12053	3.18824
38	2.69630	2.76938	2.84161	2.91298	2.98346	3.05303	3.12167	3.18936
39	2.69753	2.77059	2.84281	2.91416	2.98463	3.05418	3.12281	3.19048
40	2.69875	2.77180	2.84401	2.91534	2.98579	3.05533	3.12394	3.19160
41	2.69998	2.77301	2.84520	2.91652	2.98696	3.05648	3.12508	3.19272
42	2.70120	2.77422	2.84640	2.91771	2.98813	3.05764	3.12621	3.19384
43	2.70243	2.77543	2.84759	2.91889	2.98929	3.05879	3.12735	3.19496
44	2.70365	2.77664	2.84879	2.92007	2.99046	3.05994	3.12848	3.19608
45	2.70487	2.77785	2.84998	2.92125	2.99162	3.06109	3.12962	3.19720
46	2.70610	2.77906	2.85118	2.92243	2.99279	3.06224	3.13075	3.19831
47	2.70732	2.78027	2.85237	2.92361	2.99395	3.06339	3.13189	3.19943
48	2.70854	2.78148	2.85357	2.92479	2.99512	3.06454	3.13302	3.20055
49	2.70976	2.78269	2.85476	2.92597	2.99628	3.06568	3.13415	3.20167
50	2.71099	2.78389	2.85596	2.92715	2.99745	3.06683	3.13529	3.20278
51	2.71221	2.78510	2.85715	2.92833	2.99861	3.06798	3.13642	3.20390
52	2.71343	2.78631	2.85834	2.92950	2.99977	3.06913	3.13755	3.20502
53	2.71465	2.78752	2.85954	2.93068	3.00094	3.07028	3.13868	3.20613
54	2.71587	2.78873	2.86073	2.93186	3.00210	3.07143	3.13982	3.20725
55	2.71709	2.78993	2.86192	2.93304	3.00326	3.07257	3.14095	3.20836
56	2.71831	2.79114	2.86311	2.93422	3.00443	3.07372	3.14208	3.20948
57	2.71953	2.79235	2.86431	2.93540	3.00559	3.07487	3.14321	3.21059
58	2.72076	2.79355	2.86550	2.93657	3.00675	3.07601	3.14434	3.21171
59	2.72198	2.79476	2.86669	2.93775	3.00791	3.07716	3.14547	3.21282
60	2.72320	2.79596	2.86788	2.93893	3.00908	3.07831	3.14660	3.21394

Constants for Setting a 5-inch Sine-Bar — 6

Min.	40°	41°	42°	43°	44°	45°	46°	47°
0	3.21394	3.28030	3.34565	3.40999	3.47329	3.53553	3.59670	3.65677
1	3.21505	3.28139	3.34673	3.41106	3.47434	3.53656	3.59771	3.65776
2	3.21617	3.28249	3.34781	3.41212	3.47538	3.53759	3.59872	3.65875
3	3.21728	3.28359	3.34889	3.41318	3.47643	3.53862	3.59973	3.65974
4	3.21839	3.28468	3.34997	3.41424	3.47747	3.53965	3.60074	3.66073
5	3.21951	3.28578	3.35105	3.41531	3.47852	3.54067	3.60175	3.66172
6	3.22062	3.28688	3.35213	3.41637	3.47956	3.54170	3.60276	3.66271
7	3.22173	3.28797	3.35321	3.41743	3.48061	3.54273	3.60376	3.66370
8	3.22284	3.28907	3.35429	3.41849	3.48165	3.54375	3.60477	3.66469
9	3.22395	3.29016	3.35537	3.41955	3.48270	3.54478	3.60578	3.66568
10	3.22507	3.29126	3.35645	3.42061	3.48374	3.54580	3.60679	3.66667
11	3.22618	3.29235	3.35753	3.42168	3.48478	3.54683	3.60779	3.66766
12	3.22729	3.29345	3.35860	3.42274	3.48583	3.54785	3.60880	3.66865
13	3.22840	3.29454	3.35968	3.42380	3.48687	3.54888	3.60981	3.66964
14	3.22951	3.29564	3.36076	3.42486	3.48791	3.54990	3.61081	3.67063
15	3.23062	3.29673	3.36183	3.42591	3.48895	3.55093	3.61182	3.67161
16	3.23173	3.29782	3.36291	3.42697	3.48999	3.55195	3.61283	3.67260
17	3.23284	3.29892	3.36399	3.42803	3.49104	3.55297	3.61383	3.67359
18	3.23395	3.30001	3.36506	3.42909	3.49208	3.55400	3.61484	3.67457
19	3.23506	3.30110	3.36614	3.43015	3.49312	3.55502	3.61584	3.67556
20	3.23617	3.30219	3.36721	3.43121	3.49416	3.55604	3.61684	3.67655
21	3.23728	3.30329	3.36829	3.43227	3.49520	3.55707	3.61785	3.67753
22	3.23838	3.30438	3.36936	3.43332	3.49624	3.55809	3.61885	3.67852
23	3.23949	3.30547	3.37044	3.43438	3.49728	3.55911	3.61986	3.67950
24	3.24060	3.30656	3.37151	3.43544	3.49832	3.56013	3.62086	3.68049
25	3.24171	3.30765	3.37259	3.43649	3.49936	3.56115	3.62186	3.68147
26	3.24281	3.30874	3.37366	3.43755	3.50039	3.56217	3.62286	3.68245
27	3.24392	3.30983	3.37473	3.43861	3.50143	3.56319	3.62387	3.68344
28	3.24503	3.31092	3.37581	3.43966	3.50247	3.56421	3.62487	3.68442
29	3.24613	3.31201	3.37688	3.44072	3.50351	3.56523	3.62587	3.68540
30	3.24724	3.31310	3.37795	3.44177	3.50455	3.56625	3.62687	3.68639
31	3.24835	3.31419	3.37902	3.44283	3.50558	3.56727	3.62787	3.68737
32	3.24945	3.31528	3.38010	3.44388	3.50662	3.56829	3.62887	3.68835
33	3.25056	3.31637	3.38117	3.44494	3.50766	3.56931	3.62987	3.68933
34	3.25166	3.31746	3.38224	3.44599	3.50869	3.57033	3.63087	3.69031
35	3.25277	3.31854	3.38331	3.44704	3.50973	3.57135	3.63187	3.69130
36	3.25387	3.31963	3.38438	3.44810	3.51077	3.57236	3.63287	3.69228
37	3.25498	3.32072	3.38545	3.44915	3.51180	3.57338	3.63387	3.69326
38	3.25608	3.32181	3.38652	3.45020	3.51284	3.57440	3.63487	3.69424
39	3.25718	3.32289	3.38759	3.45126	3.51387	3.57541	3.63587	3.69522
40	3.25829	3.32398	3.38866	3.45231	3.51491	3.57643	3.63687	3.69620
41	3.25939	3.32507	3.38973	3.45336	3.51594	3.57745	3.63787	3.69718
42	3.26049	3.32615	3.39080	3.45441	3.51697	3.57846	3.63886	3.69816
43	3.26159	3.32724	3.39187	3.45546	3.51801	3.57948	3.63986	3.69913
44	3.26270	3.32832	3.39294	3.45651	3.51904	3.58049	3.64086	3.70011
45	3.26380	3.32941	3.39400	3.45757	3.52007	3.58151	3.64185	3.70109
46	3.26490	3.33049	3.39507	3.45862	3.52111	3.58252	3.64285	3.70207
47	3.26600	3.33158	3.39614	3.45967	3.52214	3.58354	3.64385	3.70305
48	3.26710	3.33266	3.39721	3.46072	3.52317	3.58455	3.64484	3.70402
49	3.26820	3.33375	3.39827	3.46177	3.52420	3.58557	3.64584	3.70500
50	3.26930	3.33483	3.39934	3.46281	3.52523	3.58658	3.64683	3.70598
51	3.27040	3.33591	3.40041	3.46386	3.52627	3.58759	3.64783	3.70695
52	3.27150	3.33700	3.40147	3.46491	3.52730	3.58861	3.64882	3.70793
53	3.27260	3.33808	3.40254	3.46596	3.52833	3.58962	3.64982	3.70890
54	3.27370	3.33916	3.40360	3.46701	3.52936	3.59063	3.65081	3.70988
55	3.27480	3.34025	3.40467	3.46806	3.53039	3.59164	3.65181	3.71085
56	3.27590	3.34133	3.40573	3.46910	3.53142	3.59266	3.65280	3.71183
57	3.27700	3.34241	3.40680	3.47015	3.53245	3.59367	3.65379	3.71280
58	3.27810	3.34349	3.40786	3.47120	3.53348	3.59468	3.65478	3.71378
59	3.27920	3.34457	3.40893	3.47225	3.53451	3.59569	3.65578	3.71475
60	3.28030	3.34565	3.40999	3.47329	3.53553	3.59670	3.65677	3.71572

Constants for Setting a 5-inch Sine-Bar — 7

Min.	48°	49°	50°	51°	52°	53°	54°	55°
0	3.71572	3.77355	3.83022	3.88573	3.94005	3.99318	4.04508	4.09576
1	3.71670	3.77450	3.83116	3.88664	3.94095	3.99405	4.04594	4.09659
2	3.71767	3.77546	3.83209	3.88756	3.94184	3.99493	4.04679	4.09743
3	3.71864	3.77641	3.83303	3.88847	3.94274	3.99580	4.04765	4.09826
4	3.71961	3.77736	3.83396	3.88939	3.94363	3.99668	4.04850	4.09909
5	3.72059	3.77831	3.83489	3.89030	3.94453	3.99755	4.04936	4.09993
6	3.72156	3.77927	3.83583	3.89122	3.94542	3.99842	4.05021	4.10076
7	3.72253	3.78022	3.83676	3.89213	3.94631	3.99930	4.05106	4.10159
8	3.72350	3.78117	3.83769	3.89304	3.94721	4.00017	4.05191	4.10242
9	3.72447	3.78212	3.83862	3.89395	3.94810	4.00104	4.05277	4.10325
10	3.72544	3.78307	3.83955	3.89487	3.94899	4.00191	4.05362	4.10409
11	3.72641	3.78402	3.84049	3.89578	3.94988	4.00279	4.05447	4.10492
12	3.72738	3.78498	3.84142	3.89669	3.95078	4.00366	4.05532	4.10575
13	3.72835	3.78593	3.84235	3.89760	3.95167	4.00453	4.05617	4.10658
14	3.72932	3.78688	3.84328	3.89851	3.95256	4.00540	4.05702	4.10741
15	3.73029	3.78782	3.84421	3.89942	3.95345	4.00627	4.05787	4.10823
16	3.73126	3.78877	3.84514	3.90033	3.95434	4.00714	4.05872	4.10906
17	3.73222	3.78972	3.84607	3.90124	3.95523	4.00801	4.05957	4.10989
18	3.73319	3.79067	3.84700	3.90215	3.95612	4.00888	4.06042	4.11072
19	3.73416	3.79162	3.84793	3.90306	3.95701	4.00975	4.06127	4.11155
20	3.73513	3.79257	3.84886	3.90397	3.95790	4.01062	4.06211	4.11238
21	3.73609	3.79352	3.84978	3.90488	3.95878	4.01148	4.06296	4.11320
22	3.73706	3.79446	3.85071	3.90579	3.95967	4.01235	4.06381	4.11403
23	3.73802	3.79541	3.85164	3.90669	3.96056	4.01322	4.06466	4.11486
24	3.73899	3.79636	3.85257	3.90760	3.96145	4.01409	4.06550	4.11568
25	3.73996	3.79730	3.85349	3.90851	3.96234	4.01495	4.06635	4.11651
26	3.74092	3.79825	3.85442	3.90942	3.96322	4.01582	4.06720	4.11733
27	3.74189	3.79919	3.85535	3.91032	3.96411	4.01669	4.06804	4.11816
28	3.74285	3.80014	3.85627	3.91123	3.96500	4.01755	4.06889	4.11898
29	3.74381	3.80109	3.85720	3.91214	3.96588	4.01842	4.06973	4.11981
30	3.74478	3.80203	3.85812	3.91304	3.96677	4.01928	4.07058	4.12063
31	3.74574	3.80297	3.85905	3.91395	3.96765	4.02015	4.07142	4.12145
32	3.74671	3.80392	3.85997	3.91485	3.96854	4.02101	4.07227	4.12228
33	3.74767	3.80486	3.86090	3.91576	3.96942	4.02188	4.07311	4.12310
34	3.74863	3.80581	3.86182	3.91666	3.97031	4.02274	4.07395	4.12392
35	3.74959	3.80675	3.86274	3.91756	3.97119	4.02361	4.07480	4.12475
36	3.75056	3.80769	3.86367	3.91847	3.97207	4.02447	4.07564	4.12557
37	3.75152	3.80863	3.86459	3.91937	3.97296	4.02533	4.07648	4.12639
38	3.75248	3.80958	3.86551	3.92027	3.97384	4.02619	4.07732	4.12721
39	3.75344	3.81052	3.86644	3.92118	3.97472	4.02706	4.07817	4.12803
40	3.75440	3.81146	3.86736	3.92208	3.97560	4.02792	4.07901	4.12885
41	3.75536	3.81240	3.86828	3.92298	3.97649	4.02878	4.07985	4.12967
42	3.75632	3.81334	3.86920	3.92388	3.97737	4.02964	4.08069	4.13049
43	3.75728	3.81428	3.87012	3.92478	3.97825	4.03050	4.08153	4.13131
44	3.75824	3.81522	3.87104	3.92568	3.97913	4.03136	4.08237	4.13213
45	3.75920	3.81616	3.87196	3.92658	3.98001	4.03222	4.08321	4.13295
46	3.76016	3.81710	3.87288	3.92748	3.98089	4.03308	4.08405	4.13377
47	3.76112	3.81804	3.87380	3.92838	3.98177	4.03394	4.08489	4.13459
48	3.76207	3.81898	3.87472	3.92928	3.98265	4.03480	4.08572	4.13540
49	3.76303	3.81992	3.87564	3.93018	3.98353	4.03566	4.08656	4.13622
50	3.76399	3.82086	3.87656	3.93108	3.98441	4.03652	4.08740	4.13704
51	3.76495	3.82180	3.87748	3.93198	3.98529	4.03738	4.08824	4.13785
52	3.76590	3.82273	3.87840	3.93288	3.98616	4.03823	4.08908	4.13867
53	3.76686	3.82367	3.87931	3.93378	3.98704	4.03909	4.08991	4.13949
54	3.76782	3.82461	3.88023	3.93468	3.98792	4.03995	4.09075	4.14030
55	3.76877	3.82554	3.88115	3.93557	3.98880	4.04081	4.09158	4.14112
56	3.76973	3.82648	3.88207	3.93647	3.98967	4.04166	4.09242	4.14193
57	3.77068	3.82742	3.88298	3.93737	3.99055	4.04252	4.09326	4.14275
58	3.77164	3.82835	3.88390	3.93826	3.99143	4.04337	4.09409	4.14356
59	3.77259	3.82929	3.88481	3.93916	3.99230	4.04423	4.09493	4.14437
60	3.77355	3.83022	3.88573	3.94005	3.99318	4.04508	4.09576	4.14519

Measuring Tapers with Vee-Block and Sine-Bar. — The taper on a conical part may be checked or found by placing the part in a vee-block which rests on the surface of a sine-plate or sine-bar as shown in the accompanying diagram. The advantage of this method is that the axis of the vee-block may be aligned with the sides of the sine-bar. Thus when the tapered part is placed in the vee-block it will be aligned perpendicular to the transverse axis of the sine-bar.

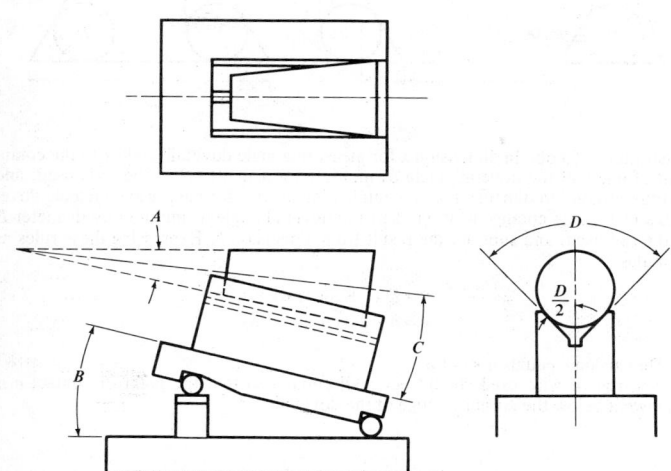

The sine-bar is set to angle $B = (C + A/2)$ where $A/2$ is one-half the included angle of the tapered part. If D is the included angle of the precision vee-block, the angle C is calculated from the formula:

$$\sin C = \frac{\sin A/2}{\sin D/2}$$

If dial indicator readings show no change across all points along the top of the taper surface, then this checks that the angle A of the taper is correct.

If the indicator readings vary, proceed as follows to find the actual angle of taper: (1) Adjust the angle of the sine-bar until the indicator reading is constant. Then find the new angle B' as explained on page 677 in the paragraph "Measuring Angle of Templet or Gage with Sine-bar." (2) Using the angle B' calculate the actual half-angle $A'/2$ of the taper from the formula:

$$\tan \frac{A'}{2} = \frac{\sin B'}{\csc \dfrac{D}{2} + \cos B'}$$

The taper per foot corresponding to certain half-angles of taper may be found in the table on page 687.

Measuring Dovetail Slides. — Dovetail slides which must be machined accurately to a given width are commonly gaged by using pieces of cylindrical rod or wire and measuring as indicated by the dimensions x and y of the accompanying

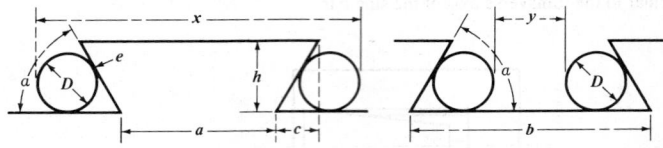

illustrations. To obtain dimension x for measuring male dovetails, add 1 to the cotangent of one-half the dovetail angle α, multiply by diameter D of the rods used, and add the product to dimension a. To obtain dimension y for measuring a female dovetail, add 1 to the cotangent of one-half the dovetail angle α, multiply by diameter D of the rod used, and subtract the result from dimension b. Expressing these rules as formulas:

$$x = D\,(1 + \cot \tfrac{1}{2}\,\alpha) + a.$$
$$y = b - D\,(1 + \cot \tfrac{1}{2}\,\alpha).$$

Dimension c equals $h \times \cot \alpha$.

The rod or wire used should be small enough so that the point of contact e is somewhat below the corner or edge of the dovetail.

Rules for Figuring Tapers

Given	To Find	Rule
The taper per foot.	The taper per inch.	Divide the taper per foot by 12.
The taper per inch.	The taper per foot.	Multiply the taper per inch by 12.
End diameters and length of taper in inches.	The taper per foot.	Subtract small diameter from large; divide by length of taper, and multiply quotient by 12.
Large diameter and length of taper in inches, and taper per foot.	Diameter at small end in inches.	Divide taper per foot by 12; multiply by length of taper, and subtract result from large diameter.
Small diameter and length of taper in inches, and taper per foot.	Diameter at large end in inches.	Divide taper per foot by 12; multiply by length of taper, and add result to small diameter.
The taper per foot and two diameters in inches.	Distance between two given diameters in inches.	Subtract small diameter from large; divide remainder by taper per foot, and multiply quotient by 12.
The taper per foot.	Amount of taper in a certain length given in inches.	Divide taper per foot by 12; multiply by given length of tapered part.

Tapers per Foot and Corresponding Angles*

Taper per Foot	Included Angle	Angle with Center Line	Taper per Foot	Included Angle	Angle with Center Line
1/64	0° 4′ 29″	0° 2′ 14″	1 7/8	8° 56′ 4″	4° 28′ 2″
1/32	0 8 57	0 4 29	1 15/16	9 13 51	4 36 56
1/16	0 17 54	0 8 57	2	9 31 38	4 45 49
3/32	0 26 51	0 13 26	2 1/8	10 7 11	5 3 36
1/8	0 35 49	0 17 54	2 1/4	10 42 42	5 21 21
5/32	0 44 46	0 22 23	2 3/8	11 18 11	5 39 5
3/16	0 53 43	0 26 51	2 1/2	11 53 37	5 56 49
7/32	1 2 40	0 31 20	2 5/8	12 29 2	6 14 31
1/4	1 11 37	0 35 49	2 3/4	13 4 24	6 32 12
9/32	1 20 34	0 40 17	2 7/8	13 39 43	6 49 52
5/16	1 29 31	0 44 46	3	14 15 0	7 7 30
11/32	1 38 28	0 49 14	3 1/8	14 50 14	7 25 7
3/8	1 47 25	0 53 43	3 1/4	15 25 26	7 42 43
13/32	1 56 22	0 58 11	3 3/8	16 0 34	8 0 17
7/16	2 5 19	1 2 40	3 1/2	16 35 39	8 17 50
15/32	2 14 16	1 7 8	3 5/8	17 10 42	8 35 21
1/2	2 23 13	1 11 37	3 3/4	17 45 41	8 52 50
17/32	2 32 10	1 16 5	3 7/8	18 20 36	9 10 18
9/16	2 41 7	1 20 33	4	18 55 29	9 27 44
19/32	2 50 4	1 25 2	4 1/8	19 30 17	9 45 9
5/8	2 59 1	1 29 30	4 1/4	20 5 3	10 2 31
21/32	3 7 57	1 33 59	4 3/8	20 39 44	10 19 52
11/16	3 16 54	1 38 27	4 1/2	21 14 22	10 37 11
23/32	3 25 51	1 42 55	4 5/8	21 48 55	10 54 28
3/4	3 34 47	1 47 24	4 3/4	22 23 25	11 11 42
25/32	3 43 44	1 51 52	4 7/8	22 57 50	11 28 55
13/16	3 52 41	1 56 20	5	23 32 12	11 46 6
27/32	4 1 37	2 0 49	5 1/8	24 6 29	12 3 14
7/8	4 10 33	2 5 17	5 1/4	24 40 41	12 20 21
29/32	4 19 30	2 9 45	5 3/8	25 14 50	12 37 25
15/16	4 28 26	2 14 13	5 1/2	25 48 53	12 54 27
31/32	4 37 23	2 18 41	5 5/8	26 22 52	13 11 26
1	4 46 19	2 23 9	5 3/4	26 56 47	13 28 23
1 1/16	5 4 11	2 32 6	5 7/8	27 30 36	13 45 18
1 1/8	5 22 3	2 41 2	6	28 4 21	14 2 10
1 3/16	5 39 55	2 49 57	6 1/8	28 38 1	14 19 0
1 1/4	5 57 47	2 58 53	6 1/4	29 11 35	14 35 48
1 5/16	6 15 38	3 7 49	6 3/8	29 45 5	14 52 32
1 3/8	6 33 29	3 16 44	6 1/2	30 18 29	15 9 15
1 7/16	6 51 19	3 25 40	6 5/8	30 51 48	15 25 54
1 1/2	7 9 10	3 34 35	6 3/4	31 25 2	15 42 31
1 9/16	7 27 0	3 43 30	6 7/8	31 58 11	15 59 5
1 5/8	7 44 49	3 52 25	7	32 31 13	16 15 37
1 11/16	8 2 38	4 1 19	7 1/8	33 4 11	16 32 5
1 3/4	8 20 27	4 10 14	7 1/4	33 37 3	16 48 31
1 13/16	8 38 16	4 19 8	7 3/8	34 9 49	17 4 54

* For conversions into decimal degrees and radians see pages 127 and 128.

Accurate Measurement of Angles and Tapers. — When great accuracy is required in the measurement of angles, or when originating tapers, disks are commonly used. The principle of the disk method of taper measurement is that if two disks of unequal diameters are placed either in contact or a certain distance apart, lines tangent to their peripheries will represent an angle or taper, the degree of which depends upon the diameters of the two disks and the distance between them. The

gage shown in the accompanying illustration, which is a form commonly used for originating tapers or measuring angles accurately, is set by means of disks. This gage consists of two adjustable straight-edges A and A_1, which are in contact with disks B and B_1. The angle α or the taper between the straight-edges depends, of course, upon the diameters of the disks and the center distance C, and as

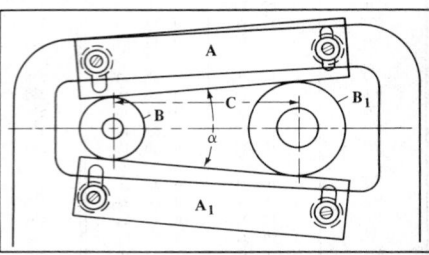

these three dimensions can be measured accurately, it is possible to set the gage to a given angle within very close limits. Moreover, if a record of the three dimensions is kept, the exact setting of the gage can be reproduced quickly at any time. The following rules may be used for adjusting a gage of this type, and cover all problems likely to arise in practice. Disks are also occasionally used for the setting of parts in angular positions when they are to be machined accurately to a given angle; the rules will be found applicable to these conditions also.

To Find Angle for Given Taper per Foot. — When the taper in inches per foot is known, and the corresponding angle α is required. *Rule:* Divide the

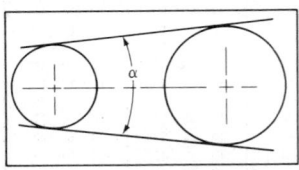

taper in inches per foot by 24; find the angle corresponding to the quotient, in a table of tangents, and double this angle.

Example: What angle α is equivalent to a taper of 1½ inch per foot?

$\dfrac{1.5}{24} = 0.0625$. The angle whose tangent is 0.0625 equals 3 degrees 35 minutes, nearly; then, 3 deg. 35 min. × 2 = 7 deg. 10 min.

To Find Angle for Given Disk Dimensions. — When the diameters D and d of the large and small disks and the center distance are given, to determine

the angle α. *Rule:* Divide the difference between the disk diameters by twice the center distance; find the angle corresponding to the quotient, in a table of sines, and double the angle.

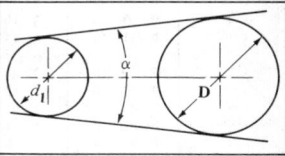

Example: If the disk diameters are 1 and 1.5 inch, respectively, and the center distance is 5 inches, find the included angle α.

$\dfrac{1.5 - 1}{2 \times 5} = 0.05$. The angle whose sine is 0.05 equals 2 degrees 52 minutes; then, 2 deg. 52 min. × 2 = 5 deg. 44 min. = angle α.

To Find the Taper per Foot. — When the diameters D and d of the large and small disks and the center distance C are given, to determine the taper per foot (measured at right angles to line through disk centers).

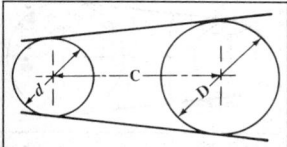

Rule: Divide the difference between the disk diameters by twice the center distance; find the angle corresponding to the quotient, in a table of sines; then find the tangent corresponding to this angle, and multiply the tangent by 24.

Example: If disk diameters are 1 and 1.5 inch, respectively, and center distance is 5 inches, find the taper per foot.

$$\frac{1.5 - 1}{2 \times 5} = 0.05.$$ The angle whose sine is 0.05 equals 2 degrees 52 minutes;

$$\tan 2° 52' = 0.05007; \qquad 0.05007 \times 24 = 1.2017 \text{ inch taper per foot.}$$

Taper Measured at Right Angles to One Side. — When one side is taken as a base line, and the taper is measured at right angles to that side, use the following rule for determining the taper per foot. *Rule:* Divide the difference between the disk diameters D and d by twice the center distance C; find the angle corresponding to the quotient, in a table of sines; double this angle and find the corresponding tangent; then multiply the tangent by 12.

Example: If the disk diameters are 2 and 3 inches, respectively, and the center distance is 5 inches, what is the taper per foot measured at right angles to one side?

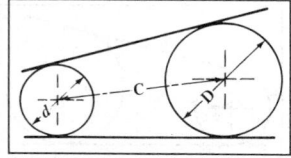

$$\frac{3 - 2}{2 \times 5} = 0.1.$$ The angle whose sine is 0.1 equals 5 degrees 44 minutes 21 seconds;

then, $$2 \times 5° 44' 21'' = 11° 28' 42''; \qquad \tan 11° 28' 42'' = 0.20306;$$

$$0.20306 \times 12 = 2.4367 \text{ inches taper per foot.}$$

To Find Center Distance for a Given Taper. — When the taper, in inches per foot, is given, to determine center distance x. *Rule:* Divide the taper by 24 and find the angle corresponding to the quotient in a table of tangents; then find the sine corresponding to this angle and divide the difference between the disk diameters by twice the sine.

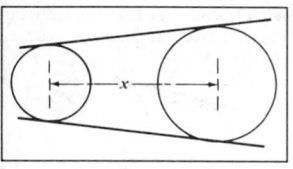

Example: Gage is to be set to ¾ inch per foot, and disk diameters are 1.25 and 1.5 inch, respectively. Find the required center distance for the disks.

$$\frac{0.75}{24} = 0.03125.$$ The angle whose tangent is 0.03125 equals 1 degree 47.4 minutes;

$$\sin 1° 47.4' = 0.03123; \qquad 1.50 - 1.25 = 0.25 \text{ inch;}$$

$$\frac{0.25}{2 \times 0.03123} = 4.002 \text{ inches} = \text{center distance } x.$$

To Find Center Distance for a Given Angle. — When straight-edges must be set to a given angle α, to determine center distance x between disks of known diameter. *Rule:* Find the sine of half the angle α in a table of sines; divide the difference between the disk diameters by double this sine.

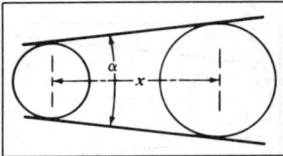

Example: If an angle α of 20 degrees is required, and the disks are 1 and 3 inches in diameter, respectively, find the required center distance x.

$$\frac{20}{2} = 10 \text{ degrees}; \qquad \sin 10° = 0.17365;$$

$$\frac{3 - 1}{2 \times 0.17365} = 5.759 \text{ inches} = \text{center distance } x.$$

Center Distance when Taper is Measured from One Side. — When taper is measured at right angles to one side, use the following rule for determining the center

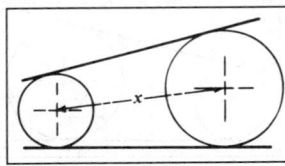

distance x. *Rule:* Divide the taper in inches per foot by 12; find the angle corresponding to the quotient, in a table of tangents; find the sine of one-half this angle, and then divide the difference between the disk diameters by double the sine.

Example: If taper measured at right angles to one side is 6.9 inches per foot, and the disks are 2 and 5 inches in diameter, respectively, what is center distance x?

$$\frac{6.9}{12} = 0.575.$$ The angle whose tangent is 0.575 equals 29 degrees 54 minutes;

then, $$\frac{29 \text{ deg. } 54 \text{ min.}}{2} = 14 \text{ deg. } 57 \text{ min}; \qquad \sin 14°57' = 0.25798;$$

$$\frac{5 - 2}{2 \times 0.25798} = 5.814 \text{ inches, center distance.}$$

Angular Measurements with Disks in Contact. — When the two disks are to be in contact and the diameter of the small disk is known, the diameter D of the large disk for a given angle α can be obtained as follows. *Rule:* Multiply twice the diameter of the small disk by the sine of one-half the required angle; divide this product by 1 minus the sine of one-half the required angle; add the quotient to the diameter of the small disk to obtain the diameter of the large disk.

Example: The required angle α is 15 degrees. Find diameter of large disk, to be in contact with a standard 1-inch reference disk.

$$\sin 7° 30' = 0.13053. \qquad 2 \times 1 \times 0.13053 = 0.26106;$$

$$\frac{0.26106}{1 - 0.13053} = 0.3002. \qquad 1 + 0.3002 = 1.3002 = \text{diameter of large disk.}$$

Compound Angles. — Three types of compound angles are illustrated by Figs. 1–6. The first type is shown in Figs. 1, 2 and 3; the second type in Fig. 4; and the third type in Figs. 5 and 6.

In Fig. 1 is shown what might be considered as a thread-cutting tool without front clearance. A is a known angle in plane y–y of the top surface. C is the corresponding angle in plane x–x which is at some given angle B with plane y–y. Thus, angles A and B are components of the compound angle C.

Problem Referring to Fig. 1: The angle $2A$ in plane y–y is known, as is also the angle B between planes x–x and y–y. It is required to find the compound angle $2C$ in plane x–x.

Solution: Let $2A = 60°$ and $B = 15°$

Then
$$\tan C = \tan A \cos B$$
$$\tan C = \tan 30° \cos 15°$$
$$\tan C = 0.57735 \times 0.96592$$
$$\tan C = 0.55767$$
$$C = 29° 8.8' \qquad 2C = 58° 17.6'$$

Fig. 2 shows a thread-cutting tool with front clearance angle B. Angle A equals one-half the angle between the cutting edges in plane y–y of the top surface and compound angle C is one-half the angle between the cutting edges in a plane x–x at right angles to the inclined front edge of the tool. The angle between planes y–y and x–x is, therefore, equal to clearance angle B.

Problem Referring to Fig. 2: Find the angle $2C$ between the front faces of a thread-cutting tool having a known clearance angle B, which will permit the grinding of these faces so that their top edges will form the desired angle $2A$ for cutting the thread.

Solution: Let $2A = 60°$ and $B = 15°$

Then
$$\tan C = \frac{\tan A}{\cos B} = \frac{\tan 30°}{\cos 15°} = \frac{0.57735}{0.96592}$$
$$\tan C = 0.59772$$
$$C = 30° 52' \qquad 2C = 61° 44'$$

In Fig. 3 is shown a form-cutting tool in which A is one-half the angle between the cutting edges in plane y–y of the top surface; B is the front clearance angle; and C is one-half the angle between the cutting edges in plane x–x at right angles to the front edges of the tool. The formula for finding angle C when angles A and B are known is the same as that for Fig. 2.

Problem Referring to Fig. 3: Find the angle $2C$ between the front faces of a form-cutting tool having a known clearance angle B which will permit the grinding of these faces so that their top edges will form the desired angle $2A$ for form cutting.

Solution: Let $2A = 46°$ and $B = 12°$

Then
$$\tan C = \frac{\tan A}{\cos B} = \frac{\tan 23°}{\cos 12°} = \frac{0.42447}{0.97815}$$
$$\tan C = 0.43395$$
$$C = 23° 27.5' \qquad 2C = 46° 55'$$

In Fig. 4 is shown a wedge-shaped block, the top surface of which is inclined at compound angle C with the base in a plane at right angles with the base and at angle R with the front edge. Angle A in the vertical plane of the front of the plate

Formulas for Compound Angles

C = compound angle in plane x-x and is the resultant of angles A and B

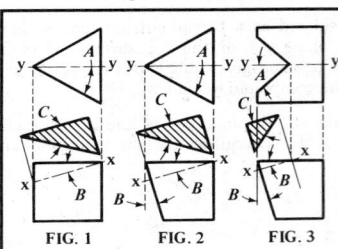

FIG. 1   FIG. 2   FIG. 3

For given angles A and B, find the resultant angle C in plane x-x. Angle B is measured in vertical plane y-y of midsection.

(Fig. 1) $\tan C = \tan A \times \cos B$

(Fig. 2) $\tan C = \dfrac{\tan A}{\cos B}$

(Fig. 3) (Same formula as for Fig. 2)

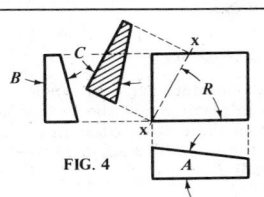

FIG. 4

Fig. 4. In machining plate to angles A and B, it is held at angle C in plane x-x. Angle of rotation R in plane parallel to base (or complement of R) is for locating plate so that plane x-x is perpendicular to axis of pivot on angle-plate or work-holding vise.

$$\tan R = \frac{\tan B}{\tan A} \; ; \; \tan C = \frac{\tan A}{\cos R}$$

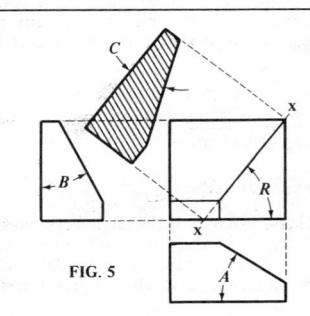

FIG. 5

Fig. 5. Angle R in horizontal plane parallel to base is angle from plane x-x to side having angle A.

$$\tan R = \frac{\tan A}{\tan B}$$

$$\tan C = \tan A \cos R = \tan B \sin R$$

Compound angle C is angle in plane x-x from base to corner formed by intersection of planes inclined to angles A and B. This formula for C may be used to find cot of complement of C_1, Fig. 6.

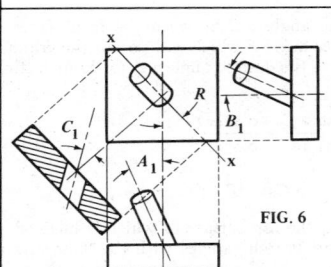

FIG. 6

Fig. 6. Angles A_1 and B_1 are measured in vertical planes of front and side elevations. Plane x-x is located by angle R from centerline or from plane of angle B_1.

$$\tan R = \frac{\tan A_1}{\tan B_1}$$

$$\tan C_1 = \frac{\tan A_1}{\sin R} = \frac{\tan B_1}{\cos R}$$

The resultant angle C_1 would be required in drilling hole for pin.

and angle B in the vertical plane of one side which is at right angles to the front are components of angle C.

Problem Referring to Fig. 4: Find the compound angle C of a wedge-shaped block having known component angles A and B in sides at right angles to each other.

Solution:
$$\text{Let } A = 47° \ 14' \text{ and } B = 38° \ 10'$$
$$\tan R = \frac{\tan B}{\tan A} = \frac{\tan 38° \ 10'}{\tan 47° \ 14'}$$
$$\tan R = \frac{0.78598}{1.0812} = 0.72695$$
$$R = 36° \ 0.9'$$
$$\tan C = \frac{\tan A}{\cos R} = \frac{\tan 47° \ 14'}{\cos 36° \ 0.9'}$$
$$\tan C = \frac{1.0812}{0.80887} = 1.3367$$
$$C = 53° \ 12'$$

In Fig. 5 is shown a four-sided block, two sides of which are at right angles to each other and to the base of the block. The other two sides are inclined at an oblique angle with the base. Angle C is a compound angle formed by the intersection of these two inclined sides and the intersection of a vertical plane passing through x–x, and the base of the block. The components of angle C are angles A and B and angle R is the angle in the base plane of the block between the plane of angle C and the plane of angle A.

Problem Referring to Fig. 5: Find the angles C and R in the block shown in Fig. 5 when angles A and B are known.

Solution: Let angle $A = 27°$ and $B = 36°$
$$\tan R = \frac{\cot B}{\cot A} = \frac{\cot 36°}{\cot 27°} = \frac{1.3764}{1.9626}$$
$$\tan R = 0.70131 \qquad R = 35° \ 2.5'$$
$$\cot C = \sqrt{\cot^2 A + \cot^2 B} = \sqrt{(1.9626)^2 + (1.3764)^2}$$
$$\cot C = \sqrt{5.74627572} = 2.3971$$
$$C = 22° \ 38.6'$$

Problem Referring to Fig. 6: A rod or pipe is inserted into a rectangular block at an angle. Angle C_1 is the compound angle of inclination (measured from the vertical) in a plane passing through the center line of the rod or pipe and at right angles to the top surface of the block. Angles A_1 and B_1 are the angles of inclination of the rod or pipe when viewed respectively in the front and side planes of the block. Angle R is the angle between the plane of angle C_1 and the plane of angle B_1. Find angles C_1 and R when a rod or pipe is inclined at known angles A_1 and B_1.

Solution: Let $A_1 = 39°$ and $B_1 = 34°$
$$\text{Then} \quad \tan C_1 = \sqrt{\tan^2 A_1 + \tan^2 B_1} = \sqrt{(0.80978)^2 + (0.67451)^2}$$
$$\tan C_1 = \sqrt{1.1107074} = 1.0539$$
$$C_1 = 46° \ 30.2'$$
$$\tan R = \frac{\tan A_1}{\tan B_1} = \frac{0.80978}{0.67451}$$
$$\tan R = 1.2005 \qquad R = 50° \ 12.4'$$

Checking a V-shaped Groove by Measurement Over Pins. — In checking a groove of the shape shown in Fig. 7, it is necessary to measure the dimension X over the pins of radius R. If values for the radius R, dimension Z, and the angles α and β are known, the problem is to determine the distance Y, to arrive at the required overall dimension for X. If a line AC is drawn from the bottom of the V to the center of the pin at the left in Fig. 7, and a line CB from the center of this pin to its point of tangency with the side of the V, a right-angled triangle is formed in which one side, CB, is known and one angle CAB, can be determined. A line drawn from the center of a circle to the point of intersection of two tangents to the circle bisects the angle made by the tangent lines, and angle CAB therefore equals $\frac{1}{2}(\alpha + \beta)$. The length AC and the angle DAC can now be found, and with AC known in the right-angled triangle ADC, AD, which is equal to Y, can be found.

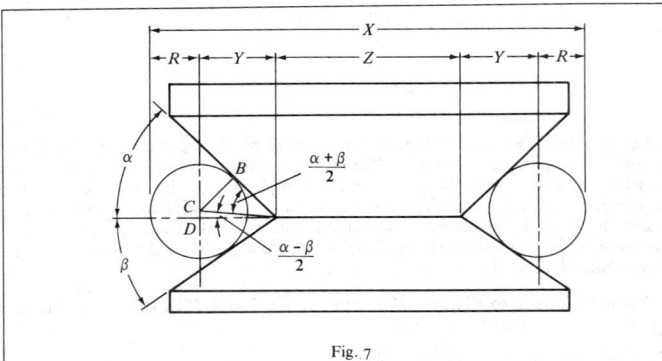

Fig. 7

The value for X can be obtained from the formula

$$X = Z + 2R \left(\csc\frac{\alpha + \beta}{2} \cos\frac{\alpha - \beta}{2} + 1 \right)$$

For example, if $R = 0.500$, $Z = 1.824$, $\alpha = 45$ degrees, and $\beta = 35$ degrees,

$$X = 1.824 + (2 \times 0.5) \left(\csc\frac{45° + 35°}{2} \cos\frac{45° - 35°}{2} + 1 \right)$$

$$X = 1.824 + \csc 40° \cos 5° + 1$$

$$X = 1.824 + 1.5557 \times 0.99619 + 1$$

$$X = 1.824 + 1.550 + 1 = 4.374$$

Checking Radius of Arc by Measurement Over Rolls. — The radius R of large-radius concave and convex gages of the type shown in Fig. 8 can be checked by measurement L over two rolls with the gage resting on the rolls as shown. If the diameter of the rolls D, the length L, and the height H of the top of the arc above the surface plate (for the concave gage, Fig. 8a) are known or can be measured, the radius R of the workpiece to be checked can be calculated trigonometrically, as follows.

Referring to Fig. 8a for the concave gage, if L and D are known, cb can be found and if H and D are known, ce can be found. With cb and ce known, ab can be found by means of a diagram as shown in Fig. 8c. In this diagram cb and ce are shown a

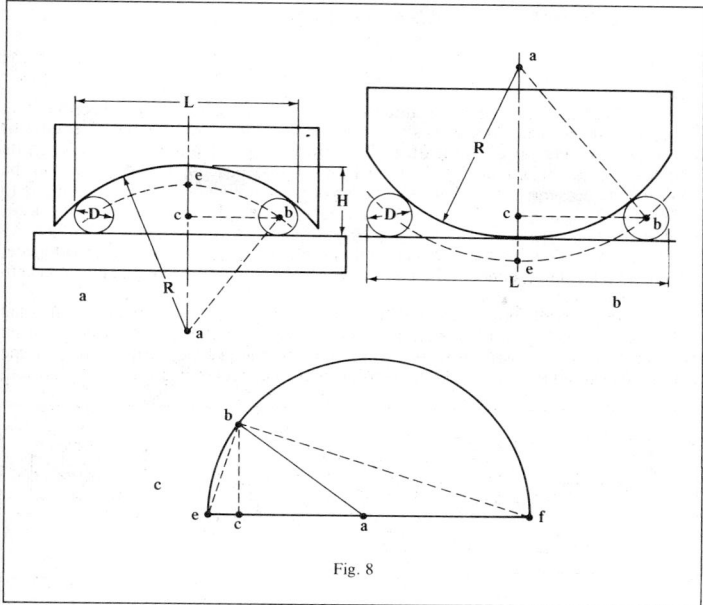

Fig. 8

right angles as in Fig. 8a. A line is drawn connecting points b and e and line ce is extended to the right. A line is now drawn from point b perpendicular to be and intersecting the extension of ce at point f. A semicircle can now be drawn through points b, e, and f with point a as the center. Triangles bce and bcf are similar and have a common side. Thus $ce:bc::bc:cf$. With ce and bc known, cf can be found from this proportion and hence ef which is the diameter of the semicircle and radius ab. Then $R = ab + D/2$.

The procedure for the convex gage is similar. The distances cb and ce are readily found and from these two distances ab is computed on the basis of similar triangles as before. Radius R is then readily found.

The derived formulas for concave and convex gages are as follows:

Formulas: $R = \dfrac{(L-D)^2}{8(H-D)} + \dfrac{H}{2}$ (Concave gage Fig. 8a)

$R = \dfrac{(L-D)^2}{8D}$ (Convex gage Fig. 8b)

For example: For Fig. 8a, let $L = 17.8$, $D = 3.20$, and $H = 5.72$, then

$$R = \frac{(17.8 - 3.20)^2}{8(5.72 - 3.20)} + \frac{5.72}{2} = \frac{(14.60)^2}{8 \times 2.52} + 2.86$$

$$R = \frac{213.16}{20.16} + 2.86 = 13.43$$

For Fig. 8b, let $L = 22.28$ and $D = 3.40$, then

$$R = \frac{(22.28 - 3.40)^2}{8 \times 3.40} = \frac{356.45}{27.20} = 13.1$$

Checking for Various Shaft Conditions. — An indicating height gage, together with V-blocks can be used to check shafts for ovality, taper, straightness (bending or curving), and concentricity of features (as shown exaggerated in Fig. 9). If a shaft on which work has been completed shows lack of concentricity, it may be due to the shaft having become bent or bowed because of mishandling or oval or tapered due to poor machine conditions. In checking for concentricity, the first step is to check for ovality, or out-of-roundness, as in Fig. 9a. The shaft is supported in a suitable V-block on a surface table and the dial indicator plunger is placed over the workpiece, which is then rotated beneath the plunger to obtain readings of the amount of eccentricity.

This procedure (sometimes called clocking, owing to the resemblance of the dial indicator to a clock face) is repeated for other shaft diameters as necessary, and, in addition to making a written record of the measurements, the positions of extreme conditions should be marked on the workpiece for later reference.

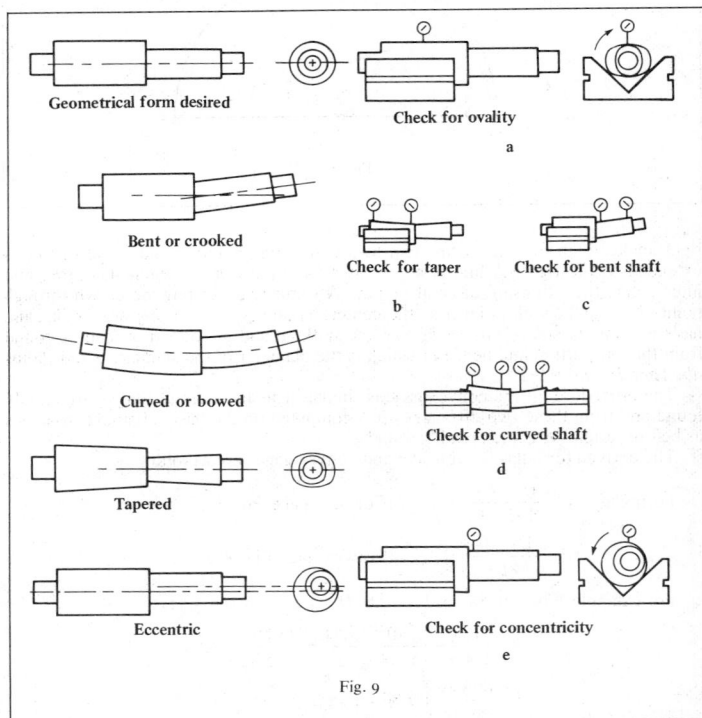

Geometrical form desired

Check for ovality

a

Bent or crooked

Check for taper

Check for bent shaft

b

c

Curved or bowed

Check for curved shaft

d

Tapered

Eccentric

Check for concentricity

e

Fig. 9

To check for taper, the shaft is supported in the V-block and the dial indicator is used to measure the maximum height over the shaft at various positions along its length, as shown in Fig. 9b, without turning the workpiece. Again, the shaft should be marked with the reading positions and values, also the direction of the taper, and a written record should be made of the amount and direction of any taper discovered.

Checking for a bent shaft requires that the shaft be clocked at the shoulder and at the farther end, as shown in Fig. 9c. For a second check the shaft is rotated only 90° or a quarter turn. When the recorded readings are compared with those from the ovality and taper checks, the three conditions can be distinguished.

To detect a curved or bowed condition, the shaft should be suspended in two V-blocks with only about ⅛ inch of each end in each vee. Alternatively, the shaft can be placed between centers. The shaft is then clocked at several points, as shown in Fig. 9d, but preferably not at those locations used for the ovality, taper, or crooked-ness checks. If the single element due to curvature is to be distinguished from the effects of ovality, taper, and crookedness, and its value assessed, great care must be taken to differentiate between the conditions detected by the measurements.

Finally, the amount of eccentricity between one shaft diameter and another may be tested by the setup shown in Fig. 9e. With the indicator plunger in contact with the smaller diameter, close to the shoulder, the shaft is rotated in the V-block and the indicator needle position is monitored to find the maximum and minimum readings.

Curvature, ovality, or crookedness conditions may tend to cancel each other, as shown in Fig. 10, and one or more of these degrees of defectiveness may add them-selves to the true eccentricity readings, depending on their angular positions. Fig. 10a shows, for instance, how crookedness and ovality tend to cancel each other, and also shows their effect in falsifying the reading for eccentricity. As the same shaft is turned in the V-block to the position shown in Fig. 10b, the maximum curvature reading could tend to cancel or reduce the maximum eccentricity reading. Where maximum readings for ovality, curvature, or crookedness occur at the same angular position, their values should be subtracted from the eccentricity reading to arrive at a true picture of the shaft condition. Confirmation of eccentricity readings may be obtained by reversing the shaft in the V-block, as shown in Fig. 10c, and clocking the larger diameter of the shaft.

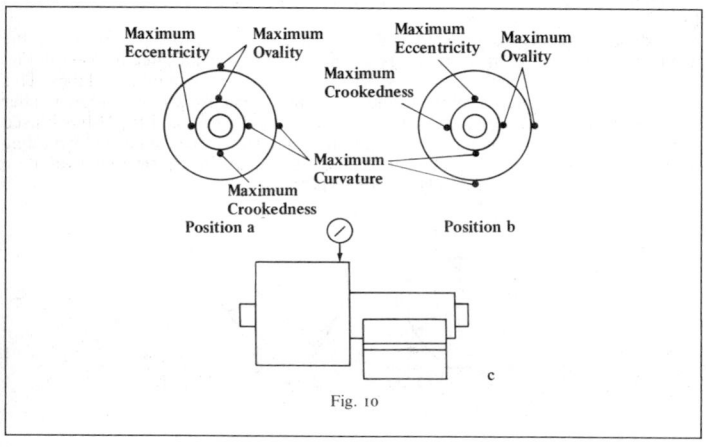

Fig. 10

Out-of-Roundness — Lobing. — With the imposition of finer tolerances and the development of improved measurement methods, it has become apparent that no hole, cylinder, or sphere can be produced with a perfectly symmetrical round shape. Some of the conditions are diagrammed in Fig. 11, where Fig. 11a shows simple ovality and Fig. 11b shows ovality occurring in two directions. From the observation of such conditions have come the terms lobe and lobing. Fig. 11c shows the three-lobed shape common with centerless-ground components, and Fig. 11d is typical of multi-lobed shapes. In Fig. 11e are shown surface waviness, surface roughness, and out-of-roundness, which often are combined with lobing.

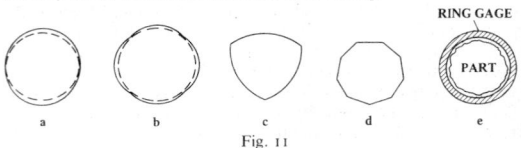

RING GAGE

PART

a   b   c   d   e

Fig. 11

In Figs. 11a through 11d the cylinder (or hole) diameters are shown at full size but the lobes are magnified some 10,000 times to make them visible. In precision parts, the deviation from the round condition is usually only in the range of millionths of an inch, although it occasionally can be 0.0001 inch, 0.0002 inch, or more. For instance, a 3-inch-diameter part may have a lobing condition amounting to an inaccuracy of only 30 millionths (0.000030 inch). Even if the distortion (ovality, waviness, roughness) is small, it may cause hum, vibration, heat buildup, and wear, possibly leading to eventual failure of the component or assembly.

Plain elliptical out-of-roundness (two lobes), or any even number of lobes, can be detected by rotating the part on a surface plate under a dial indicator of adequate resolution, or by using an indicating caliper or snap gage. However, supporting such a part in a V-block during measurement will tend to conceal roundness errors. Ovality in a hole can be detected by a dial-type bore gage or internal measuring machine. Parts with odd numbers of lobes require an instrument that can measure the envelope or complete circumference. Plug and ring gages will tell whether a shaft can be assembled into a bearing, but not whether there will be a good fit, as illustrated in Fig. 11e.

A standard, 90-degree included-angle V-block can be used to detect and count the number of lobes, but to measure the exact amount of lobing indicated by $R - r$ in Fig. 12 requires a V-block with an angle α, which is related to the number of lobes. This angle α can be calculated from the formula $2\alpha = 180° - 360°/N$, where N is the number of lobes. Thus, for a three-lobe form, α becomes 30 degrees, and the V-block used should have a 60-degree included angle. The distance M, which is obtained by rotating the part under the comparator plunger, is converted to a value for the radial variation in cylinder contour by the formula $M = (R - r)(1 + \csc \alpha)$.

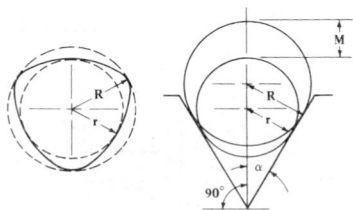

Fig. 12

Using a V-block (even of appropriate angle) for parts with odd numbers of lobes will give exaggerated readings when the distance $R - r$ (Fig. 12) is used as the measure of the amount of out-of-roundness. The accompanying table shows the appropriate V-block angles for various odd numbers of lobes, and the factors $(1 + \csc \alpha)$ by which the readings are increased over the actual out-of-roundness values.

Table of Lobes, V-block Angles and Exaggeration Factors in Measuring Out-of-round Conditions in Shafts

Number of Lobes	Included Angle of V-block (deg)	Exaggeration Factor $(1 + \csc \alpha)$
3	60	3.00
5	108	2.24
7	128.57	2.11
9	140	2.06

Measurement of a complete circumference requires special equipment, often incorporating a precision spindle running true within two millionths (0.000002) inch. A stylus attached to the spindle is caused to traverse the internal or external cylinder being inspected, and its divergences are processed electronically to produce a polar chart similar to the wavy outline in Fig. 11e. The electronic circuits provide for the variations due to surface effects to be separated from those of lobing and other departures from the "true" cylinder traced out by the spindle.

Measuring by Light-wave Interference Bands. — Surface variations as small as two millionths (0.000002) inch can be detected by light-wave interference methods, using an optical flat. An optical flat is a transparent block, usually of plate glass, clear fused quartz, or borosilicate glass, the faces of which are finished to extremely fine limits (of the order of 1 to 8 millionths [0.000001 to 0.000008] inch, depending on the application) for flatness. When an optical flat is placed on a "flat" surface, as shown in Fig. 13, any small departure from flatness will result in formation of a wedge-shaped layer of air between the work surface and the underside of the flat.

Light rays reflected from the work surface and the underside of the flat either interfere with or reinforce each other. Interference of two reflections results when the air gap measures exactly half the wavelength of the light used, and produces a dark band across the work surface when viewed perpendicularly, under monochromatic helium light. A light band is produced halfway between the dark bands when the rays reinforce each other. With the 0.0000232-inch-wavelength helium light used, the dark bands occur where the optical flat and the work surface are separated by 11.6 millionths (0.0000116) inch, or multiples thereof.

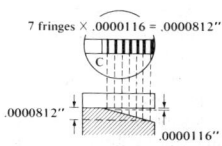

Fig. 13

For instance, at a distance of seven dark bands from the point of contact, as shown in Fig. 13, the underface of the optical flat is separated from the work surface by a distance of 7×0.0000116 inch or 0.0000812 inch. The bands are separated more widely and the indications become increasingly distorted as the viewing angle departs from the perpendicular. If the bands appear straight, equally spaced and parallel with each other, the work surface is flat. Convex or concave surfaces cause the bands to curve correspondingly, and a cylindrical tendency in the work surface will produce unevenly spaced, straight bands.

Control and Production of Surface Texture. — Surface characteristics should not be controlled on a drawing or specification unless such control is essential to functional performance or appearance of the product. Imposition of such restrictions when unnecessary may increase production costs and in any event will serve to lessen the emphasis on the control specified for important surfaces.

Smoothness and roughness are relative, i.e., surfaces may be either smooth or rough for the purpose intended; what is smooth for one purpose may be rough for another purpose. In the mechanical field comparatively few surfaces require any control of surface texture beyond that afforded by the processes required to obtain the necessary dimensional characteristics.

Working surfaces such as on bearings, pistons, and gears are typical of surfaces for which optimum performance may require control of the surface characteristics. Non-working surfaces such as on the walls of transmission cases, crankcases, or housings seldom require any surface control. Experimentation or experience with surfaces performing similar functions is the best criterion on which to base selection of optimum surface characteristics.

Determination of required characteristics for working surfaces may involve consideration of such conditions as the area of contact, the load, speed, direction of motion, type and amount of lubricant, temperature, material and physical characteristics of component parts. Variations in any one of the conditions also may require a change in the specified surface characteristics.

Production: Surface texture is a result of the processing method. The surface obtained from casting, forging, or burnishing is the result of plastic deformation; if machined or ground, lapped or honed, the surface obtained is the result of action of cutting tools, abrasives or other forces. It is important to understand that surfaces with similar roughness average ratings may not have the same performance, due to tempering, sub-surface effects, different profile waveforms, etc.

American National Standard Surface Texture (Surface Roughness, Waviness and Lay). — American National Standard ANSI B46.1-1985 is concerned with the geometric irregularities of surfaces of solid materials, physical specimens for gaging roughness, and the characteristics of stylus instrumentation for measuring roughness. It defines surface texture and its constituents: roughness, waviness, lay and flaws. A set of symbols for drawings, specifications and reports is established. In order to ensure a uniform basis for measurements, the standard also provides specifications for Precision Reference Specimens, and Roughness Comparison Specimens, and establishes requirements for stylus type instruments. The standard is not concerned with luster, appearance, color, corrosion resistance, wear resistance, hardness, subsurface microstructure, surface integrity and many other characteristics which may be governing considerations in specific applications.

The standard does not define the degrees of surface roughness and waviness or type of lay suitable for specific purposes, nor does it specify the means by which any degree of such irregularities may be obtained or produced. However, criteria for selection of surface qualities and information on instrument techniques and methods of producing, controlling and inspecting surfaces are included in Appendixes attached to the standard. The Appendix sections are not considered a part of the standard; they are included for clarification or information purposes only.

Surfaces, in general, are very complex in character. The standard deals only with the height, width, and direction of surface irregularities since these are of practical importance in specific applications. Surface texture designations as delineated in this standard may not be a sufficient index to performance. Other part characteristics such as dimensional and geometrical relationships, material, metallurgy, and stress must also be controlled.

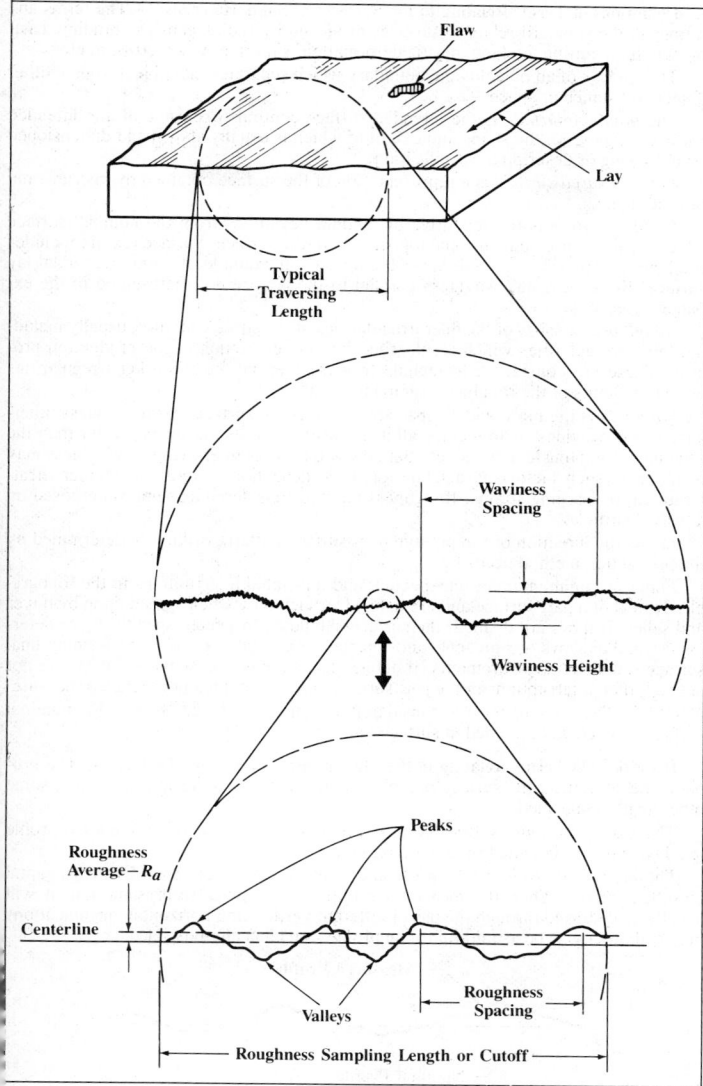

Fig. 1. Pictorial Display of Surface Characteristics

Definitions of Terms Relating to the Surfaces of Solid Materials. — The terms and ratings in the standard relate to surfaces produced by such means as abrading, casting, coating, cutting, etching, plastic deformation, sintering, wear, erosion, etc.

The *surface* of an object is the boundary which separates that object from another object, substance or space.

The *nominal surface* is the intended surface contour (exclusive of any intended surface roughness), the shape and extent of which is usually shown and dimensioned on a drawing or descriptive specification.

The *measured surface* is a representation of the surface obtained by instrumental or other means.

Surface texture is the repetitive or random deviations from the nominal surface which form the three-dimensional topography of the surface. Surface texture includes roughness, waviness, lay and flaws. Figure 1 is an example of a uni-directional lay surface. Roughness and waviness parallel to the lay are not represented in the expanded views.

Roughness consists of the finer irregularities of the surface texture, usually including those irregularities which result from the inherent action of the production process. These are considered to include traverse feed marks and other irregularities within the limits of the roughness sampling length.

Waviness is the more widely spaced component of surface texture. Unless otherwise noted, waviness is to include all irregularities whose spacing is greater than the roughness sampling length and less than the waviness sampling length. Waviness may result from such factors as machine or work deflections, vibration, chatter, heat-treatment or warping strains. Roughness may be considered as being superposed on a 'wavy' surface.

Lay is the direction of the predominant surface pattern, ordinarily determined by the production method used.

Flaws are unintentional, unexpected, and unwanted interruptions in the topography typical of a part surface and are defined as such only when agreed upon by buyer and seller. If flaws are defined, the surface should be inspected specifically to determine whether flaws are present, and rejected or accepted prior to performing final surface roughness measurements. If defined flaws are not present, or if flaws are not defined, then interruptions in the part surface may be included in roughness measurements.   The *error of form* is considered as being that deviation from the nominal surface which is not included in surface texture.

Definitions of Terms Relating to the Measurement of Surface Texture. — The *profile* is the contour of the surface in a plane perpendicular to the surface, unless some other angle is specified.

The *nominal profile* is a profile of the nominal surface; it is the intended profile (exclusive of any intended roughness profile). See Fig. 2.

The *measured profile* is a representation of the profile obtained by instrumental or other means. When the measured profile is a graphical representation, it will usually be distorted through the use of different vertical and horizontal magnifications but shall otherwise be as faithful to the profile as technically possible.

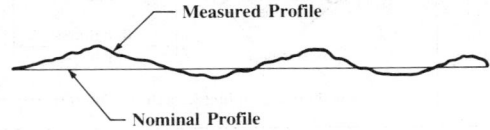

Fig. 2. Nominal and Measured Profiles

The *modified profile* is a measured profile where filter mechanisms (including the instrument datum) are used to minimize certain surface texture characteristics and emphasize others.

The *graphical centerline* is the line about which roughness is measured and is a line parallel to the general direction of the profile within the limits of the sampling length, such that the sums of the areas contained between it and those parts of the profile that lie on either side are equal. (See Fig. 3.)

The *electrical mean line* is the centerline established by the selected cutoff and its associated circuitry in an electronic roughness average measuring instrument.

A *peak* is the point of maximum height on that portion of a profile that lies above the centerline and between two intersections of the profile with the centerline.

A *valley* is the point of maximum depth on that portion of a profile that lies below the centerline and between two intersections of the profile with the centerline.

The *spacing* is the distance between specified points on the profile measured parallel to the nominal profile.

The *roughness spacing* is the average spacing between adjacent peaks of the measured profile within the roughness sampling length.

The *waviness spacing* is the average spacing between adjacent peaks of the measured profile within the waviness sampling length.

The *sampling length* is the nominal spacing within which a surface characteristic is determined.

The *roughness sampling length* is the sampling length within which the roughness average is determined. This length is chosen, or specified, to separate the profile irregularities which are designated as roughness from those irregularities designated as waviness.

The *cutoff* is the electrical response characteristic of the roughness measuring instrument which is selected to limit the spacing of the surface irregularities to be included in the assessment of roughness average. The cutoff is rated in millimeters. In most electrical averaging instruments, the cutoff can be selected. It is a characteristic of the instrument rather than of the surface being measured. In specifying the cutoff, care must be taken to choose a value which will include all the surface irregularities which it is desired to assess.

The *waviness sampling length* is the sampling length within which the waviness height is determined.

The *traversing length* is the length of profile which is traversed by the stylus to establish a representative measurement. See Section 5.2.5 of the standard for values which must be used for different type measurements.

Height is considered to be those measurements of the profile in a direction normal to the nominal profile.

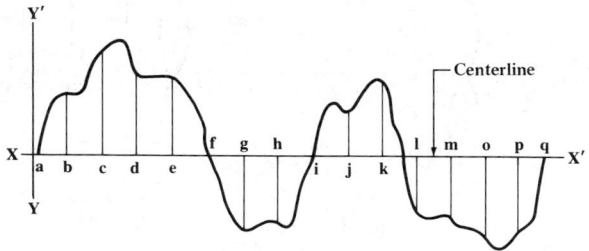

Fig. 3. Short Section of Hypothetical Profile Divided into Increments

Roughness average (R_a), also known as arithmetic average (AA) and centerline average (CLA), is the arithmetic average of the absolute values of the measured profile height deviations taken within the sampling length and measured from the graphical centerline. For graphical determinations of roughness average, the height deviations are measured normal to the chart centerline, as shown in Fig. 3. Roughness average is expressed in micrometers. A micrometer is one millionth of a meter (0.000001 meter). For written specifications or references as to surface requirements, micrometer can be abbreviated as μm. A microinch is one millionth of an inch (0.000001 inch). For written specifications or reference to surface roughness requirements, microinch may be abbreviated as μin. One microinch equals 0.0254 micrometer (1 μin. = 0.0254 μm).

Roughness average value (R_a) from continuously averaging meter readings. So that uniform interpretation may be made of readings from stylus-type instruments of the continuously averaging type, it should be understood that the reading which is considered significant is the mean reading around which the needle tends to dwell or fluctuate with a small amplitude.

The *peak-to-valley height* is the maximum excursion above the centerline plus the maximum excursion below the centerline within the sampling length. This value is typically 3 or more times the roughness average.

The *waviness height* is the peak-to-valley height of the modified profile from which roughness and flaws have been removed by filtering, smoothing, or other means. The measurement is to be taken normal to the nominal profile within the limits of the waviness sampling length.

Relation of Surface Roughness to Tolerances. — Since the measurement of surface roughness involves the determination of the average linear deviation of the measured surface from the nominal surface, there is a direct relationship between the dimensional tolerance on a part and the permissible surface roughness. It is evident that a requirement for the accurate measurement of a dimension is that the variations introduced by surface roughness should not exceed the tolerance placed on the dimension. If this is not the case, the measurement of the dimension will be subject to an uncertainty greater than the required tolerance as illustrated in Fig. 4. The standard

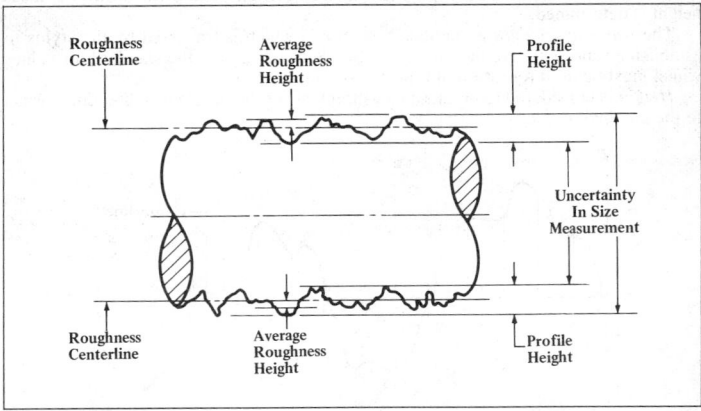

Fig. 4. Uncertainty in Dimensional Measurement due to Surface Roughness (ASA B46.1-1962)

thod of measuring surface roughness involves the determination of the average viation from the mean surface. On most surfaces the total profile height of the rface roughness (peak-to-valley height) will be approximately four times the measured average surface roughness in microinches. This factor will vary somewhat with e character of the surface under consideration, but the value of four may be used establish approximate profile heights.

From these considerations it follows that if the arithmetical average value of rface roughness specified on a part exceeds one eighth of the dimensional tolerance, e whole tolerance will be taken up by the roughness height. In most cases, a smaller ughness specification than this will be found; but on parts where very small dimenonal tolerances are given, it is necessary to specify a suitably small surface roughss in order that useful dimensional measurements can be made. The tables on pages 2 and 659 show the relation between machining processes and working tolerances. lues for surface roughness produced by common processing methods are

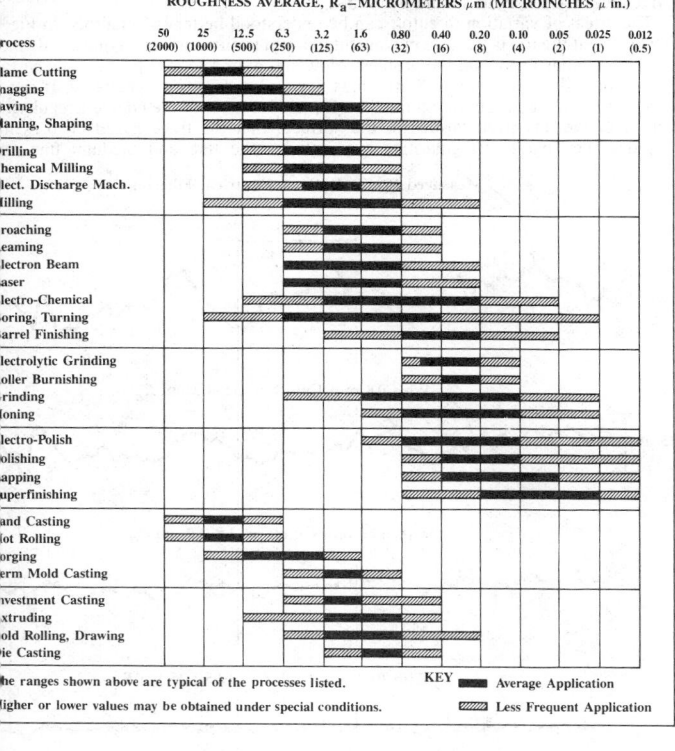

Fig. 5. Surface Roughness Produced by Common Production Methods

shown in Fig. 5. The ability of a processing operation to produce a specific surfa
roughness depends on many factors. For example, in surface grinding, the final su
face depends on the peripheral speed of the wheel, the speed of the traverse, the ra
of feed, the grit size, bonding material and state of dress of the wheel, the amou
and type of lubrication at the point of cutting, and the mechanical properties of th
piece being ground. A small change in any of the above factors can have a mark
effect on the surface produced.

Selecting Cutoff for Roughness Measurements. — In general, surfaces will conta
irregularities with a large range of widths. Stylus-type instruments are designed
respond only to irregularity spacings less than a given value, called cutoff. In son
cases, such as surfaces in which actual contact area with a mating surface is impo
tant, the largest convenient cutoff will be used. In other cases, such as surfac
subject to fatigue failure, only the irregularities of small width will be important, a
more significant values will be obtained when a short cutoff is used. In still oth
cases, such as identifying chatter marks on machined surfaces, information is need
on only the widely spaced irregularities. For such measurements, a large cutoff val
and a larger radius stylus should be used.

The effect of variation in cutoff can be understood better by reference to Fig.
The profile at the top is the true movement of a stylus on a surface having a roughne
spacing of about 1 mm and the profiles below are interpretations of the same surfa
with cutoff value settings of 0.8 mm, 0.25 mm and 0.08 mm, respectively. It can b
seen that the trace based on 0.8 mm cutoff includes most of the coarse irregulariti
and all of the fine irregularities of the surface; that the trace based on 0.25 m
excludes the coarser irregularities but includes the fine and medium fine; an

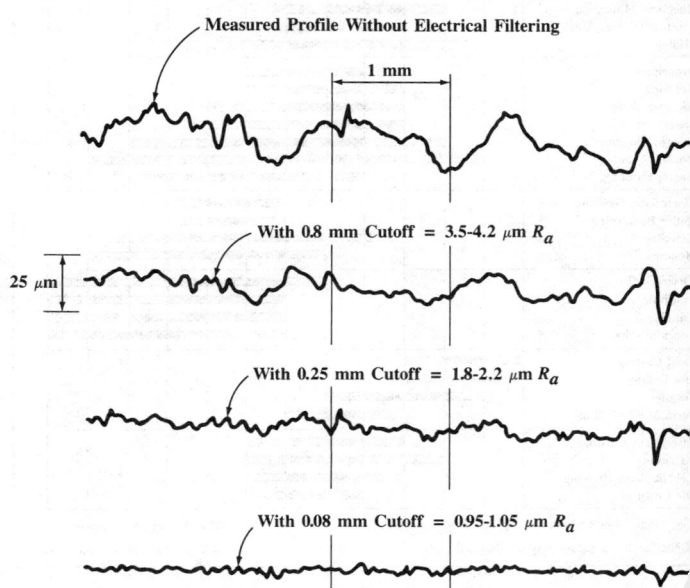

Fig. 6. Effects of Various Cutoff Values

that the trace based on 0.08 mm cutoff includes only the very fine irregularities. In this example the effect of reducing the cutoff has been to reduce the roughness average indication. However, had the surface been made up only of irregularities as fine as those of the bottom trace, the roughness average values would have been the same for all three cutoff settings.

In other words, all irregularities having a spacing less than the value of the cutoff used are included in a measurement. Obviously, if the cutoff value is too small to include coarser irregularities of a surface, the measurements will not agree with those taken with a larger cutoff. For this reason, care must be taken to choose a cutoff value which will include all of the surface irregularities it is desired to assess.

To become proficient in the use of continuously averaging stylus-type instruments the inspector or machine operator must realize that for uniform interpretation, the reading which is considered significant is the mean reading around which the needle tends to dwell or fluctuate under small amplitude.

Drawing Practices for Surface Texture Symbols. — American National Standard ANSI Y14.36-1978 establishes the method to designate controls for surface texture of solid materials. It includes methods for controlling roughness, waviness, and lay, and provides a set of symbols for use on drawings, specifications, or other documents. The units (metric or nonmetric) shall be consistent with the other units used on the drawing or documents. The numerical values expressed in the standard are stated in metric units and are to be regarded as standard. Approximate nonmetric equivalents are shown for reference.

Surface Texture Symbol. — The symbol used to designate control of surface irregularities is shown in Fig. 1(a). Where surface texture values other than roughness average are specified, the symbol must be drawn with the horizontal extension as shown in Fig. 1(e).

Symbol		Meaning
(a)	√	Basic Surface Texture Symbol. Surface may be produced by any method except when the bar or circle (Figure 1b or 1d) is specified.
(b)	▽	Material Removal By Machining Is Required. The horizontal bar indicates that material removal by machining is required to produce the surface and that material must be provided for that purpose.
(c) 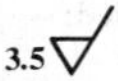	3.5 ▽	Material Removal Allowance. The number indicates the amount of stock to be removed by machining in millimeters (or inches). Tolerances may be added to the basic value shown or in a general note.
(d) 	◁	Material Removal Prohibited. The circle in the vee indicates that the surface must be produced by processes such as casting, forging, hot finishing, cold finishing, die casting, powder metallurgy or injection molding without subsequent removal of material.
(e) 	√	Surface Texture Symbol. To be used when any surface characteristics are specified above the horizontal line or to the right of the symbol. Surface may be produced by any method except when the bar or circle (Figure 1b and 1d) is specified.
(f)		

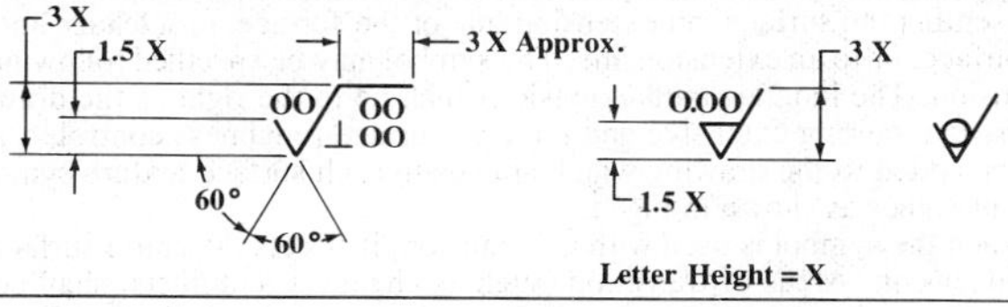

Letter Height = X

Fig. 1. Surface Texture Symbols and Construction

Use of Surface Texture Symbols: When required from a functional standpoint, the desired surface characteristics should be specified. Where no surface texture control is specified, the surface produced by normal manufacturing methods is satisfactory provided it is within the limits of size (and form) specified in accordance with ANSI Y14.5-1973. Dimensioning and Tolerancing. This is not viewed as good practice; there should always be some maximum value, either specifically or by default (for example, in the manner of the note shown in Fig. 2).

Material Removal Required or Prohibited: The surface texture symbol is modified when necessary to require or prohibit removal of material. When it is necessary to indicate that a surface must be produced by removal of material by machining, specify the symbol shown in Fig. 1(b). When required, the amount of material to be removed is specified as shown in Fig. 1(c), in millimeters for metric drawings and in inches for nonmetric drawings. Tolerance for material removal may be added to the basic value shown or specified in a general note. When it is necessary to indicate that a surface must be produced without material removal, specify the machining prohibited symbol as shown in Fig. 1(d).

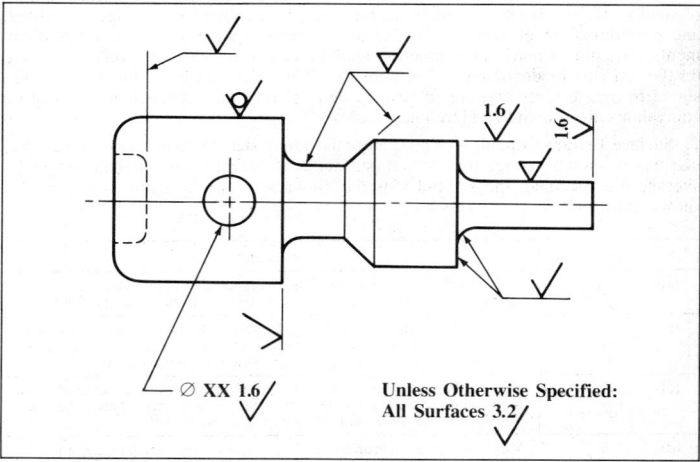

Fig. 2. Application of Surface Texture Symbols

Proportions of Surface Texture Symbols: The recommended proportions for drawing the surface texture symbol are shown in Fig. 1(f). The letter height and line width should be the same as that for dimensions and dimension lines.

Applying Surface Texture Symbols. — The point of the symbol should be on a line representing the surface, an extension line of the surface, or a leader line directed to the surface, or to an extension line. The symbol may be specified following a diameter dimension. The long leg (and extension) shall be to the right as the drawing is read. For parts requiring extensive and uniform surface roughness control, a general note may be added to the drawing which applies to each surface texture symbol specified without values as shown in Fig. 2.

When the symbol is used with a dimension, it affects the entire surface defined by the dimension. Areas of transition, such as chamfers and fillets, shall conform with the roughest adjacent finished area unless otherwise indicated.

Table 1. Preferred Series Roughness Average Values (R_a)

μm	μin	μm	μin
0.012	0.5	1.25	50
0.025*	1*	1.60*	63*
0.050*	2*	2.0	80
0.075*	3	2.5	100
0.10*	4*	3.2*	125*
0.125	5	4.0	160
0.15	6	5.0	200
0.20*	8*	6.3*	250*
0.25	10	8.0	320
0.32	13	10.0	400
0.40*	16*	12.5*	500*
0.50	20	15	600
0.63	25	20	800
0.80*	32*	25*	1000*
1.00	40	. . .	. . .

* Recommended

Surface texture values, unless otherwise specified, apply to the complete surface. Drawings or specifications for plated or coated parts shall indicate whether the surface texture values apply before plating, after plating or both before and after plating.

Include in the symbol only those values required to specify and verify the required texture characteristics. Values should be in metric units for metric drawings and nonmetric units for nonmetric drawings.

Roughness and waviness measurements, unless otherwise specified, apply in a direction which gives the maximum reading; generally across the lay.

Table 2. Standard Roughness Sampling Length (Cutoff) Values

mm	in.	mm	in.
0.08	0.003	2.5	0.1
0.25	0.010	8.0	0.3
0.80	0.030	25.0	1.0

Roughness Average (R_a): The preferred series of specified roughness average values is given in Table 1.

Cutoff or Roughness Sampling Length: Standard values are listed in Table 2. When no value is specified, the value 0.8 mm (0.030 in.) applies.

Waviness Height: The preferred series of maximum waviness height values is listed in Table 3. Waviness is not currently shown in ISO Standards. It is included here to follow present industry practice in the United States.

Table 3. Preferred Series Maximum Waviness Height Values

mm	in.	mm	in.	mm	in.
0.0005	0.00002	0.008	0.0003	0.12	0.005
0.0008	0.00003	0.012	0.0005	0.20	0.008
0.0012	0.00005	0.020	0.0008	0.25	0.010
0.0020	0.00008	0.025	0.001	0.38	0.015
0.0025	0.0001	0.05	0.002	0.50	0.020
0.005	0.0002	0.08	0.003	0.80	0.030

Lay Symbol	Meaning	Example Showing Direction of Tool Marks
—	Lay approximately parallel to the line representing the surface to which the symbol is applied.	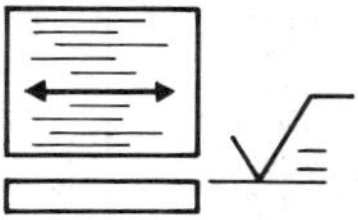
⊥	Lay approximately perpendicular to the line representing the surface to which the symbol is applied.	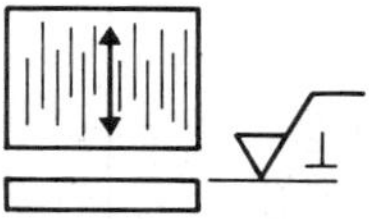
X	Lay angular in both directions to line representing the surface to which the symbol is applied.	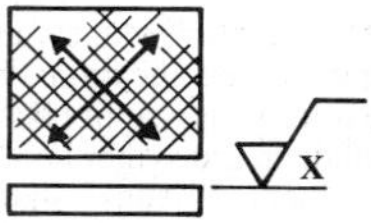
M	Lay multidirectional.	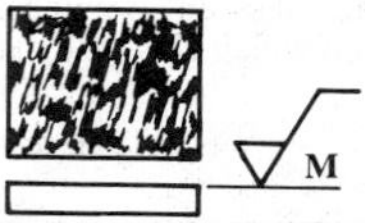
C	Lay approximately circular relative to the center of the surface to which the symbol is applied.	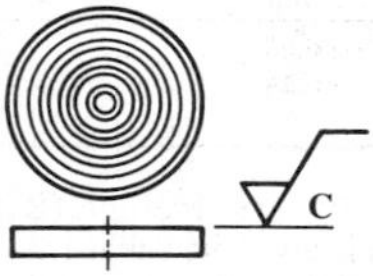
R	Lay approximately radial relative to the center of the surface to which the symbol is applied.	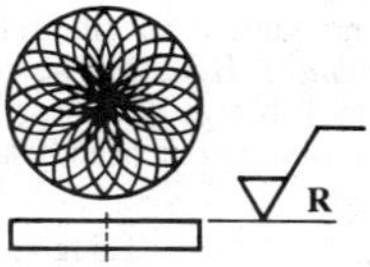
P	Lay particulate, non-directional, or protuberant.	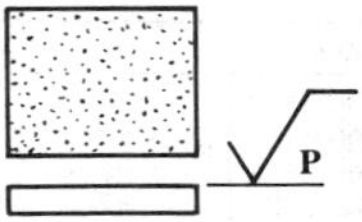

Fig. 3. Lay Symbols

Lay: Symbols for designating the direction of lay are shown and interpreted in Fig. 3.

Example Designations. — Figure 4 illustrates examples of designations of roughness, waviness, and lay by insertion of values in appropriate positions relative to the symbol. Where surface roughness control of several operations is required within a given area, or on a given surface, surface qualities may be designated, as in Fig. 5(a). If a surface must be produced by one particular process or a series of processes, they should be specified as shown in Fig. 5(b). Where special requirements are needed

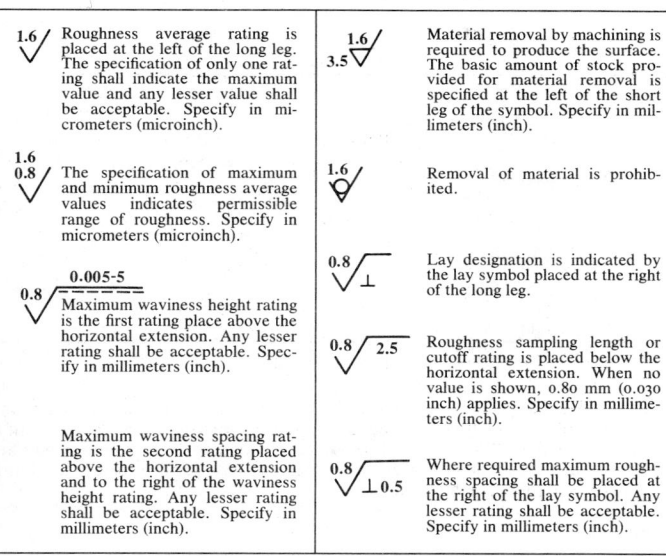

Fig. 4. Application of Surface Texture Values to Symbol

on a designated surface, a note should be added at the symbol giving the requirements and the area involved. An example is illustrated in Fig. 5(c).

Surface Texture of Castings. — Surface characteristics should not be controlled on a drawing or specification unless such control is essential to functional performance or appearance of the product. Imposition of such restrictions when unnecessary may increase production costs and in any event will serve to lessen the emphasis on the control specified for important surfaces. Surface characteristics of castings should never be considered on the same basis as machined surfaces. Castings are characterized by random distribution of non-directional deviations from the nominal surface.

Surfaces of castings rarely need control beyond that provided by the production method necessary to meet dimensional requirements. Comparison specimens are frequently used for evaluating surfaces having specific functional requirements. Surface texture control should not be specified unless required for appearance or function of the surface. Specification of such requirements may increase cost to the user.

Engineers should recognize that different areas of the same castings may have different surface textures. It is recommended that specifications of the surface be

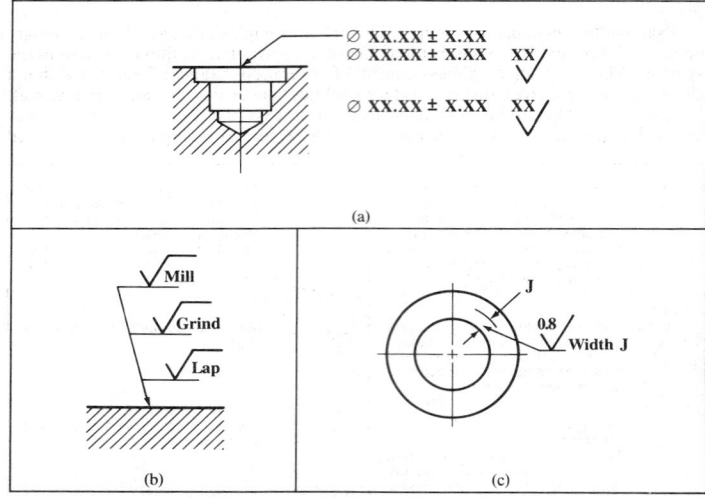

Fig. 5. Examples of Special Designations

limited to defined areas of the casting. Practicality of, and methods of determining that a casting's surface texture meets the specification shall be coordinated with the producer. The Society of Automotive Engineers standard J435 "Automotive Steel Castings" describes methods of evaluating steel casting surface texture used in the automotive and related industries.

Metric Dimensions on Drawings. — The length units of the metric system that are most generally used in connection with any work relating to mechanical engineering are the meter (39.37 inches) and the millimeter (0.03937 inch). One meter equals 1000 millimeters. On mechanical drawings, all dimensions are generally given in millimeters, no matter how large the dimensions may be. In fact, dimensions of such machines as locomotives and large electrical apparatus are given exclusively in millimeters. This practice is adopted to avoid mistakes due to misplacing decimal points, or mis-reading dimensions as when other units are used as well. When dimensions are given in millimeters, many of them can be given without resorting to decimal points, as a millimter is only a little more than 1/32 inch. Only dimensions of precision need be given in decimals of a millimeter; such dimensions are generally given in hundredths of a millimeter — for example, 0.02 millimeter, which is equal to 0.0008 inch. As 0.01 millimeter is equal to 0.0004 inch, it is seldom that dimensions would be given with greater accuracy than to hundredths of a millimeter.

Scales of Metric Drawings. — Drawings made to the metric system are not made to scales of ½, ¼, ⅛, etc., as in the case of drawings made to the English system. If the object cannot be drawn full size, it may be drawn ½.5, ⅕, 1/10, 1/20, 1/50, 1/100, 1/200, 1/500, or 1/1000 size. If the object is too small and has to be drawn larger it is drawn 2, 5, or 10 times its actual size.

Precision Gage Blocks. — Precision gage blocks are usually purchased in sets comprising a specific number of blocks of different sizes. The nominal gage lengths of individual blocks in a set are determined mathematically so that particular desired lengths can be obtained by combining selected blocks. They are made to several different tolerance grades which categorize them as master blocks, calibration blocks, inspection blocks, and workshop blocks. *Master blocks* are employed as basic reference standards; *calibration blocks* are used for high precision gaging work and calibrating inspection blocks; *inspection blocks* are used as toolroom standards and for checking and setting limit and comparator gages, for example. The *workshop blocks* are working gages used as shop standards for a variety of direct precision measurements and gaging applications, including sine bar settings.

Federal Specification GGG-G-15B, Gage Blocks (see below), lists typical sets, and gives details of materials, design, and manufacturing requirements, and tolerance grades. When there is in a set no single block of the exact size that is wanted, two or more blocks are combined by "wringing" them together. This is achieved by first placing one block crosswise on the other and applying some pressure. Then a swiveling motion is applied and the blocks are twisted to a parallel position. This causes them to adhere firmly to one another.

When combining blocks for a given dimension, the object is to use as few blocks as possible to obtain the dimension. The procedure for selecting blocks is based on successively eliminating the right-hand figure of the desired dimension.

Example. Referring to gage block set number 1 in Table 1, determine the blocks required to obtain 3.6742 inches. *Step 1:* Eliminate .0002 by selecting a .1002 block. Subtract .1002 from 3.6742 = 3.5740. *Step 2:* Eliminate .004 by selecting a .124 block. Subtract .124 from 3.5740 = 3.450. *Step 3:* Eliminate .450 with a block this size. Subtract .450 from 3.450 = 3.000. *Step 4:* Select a 3.000 inch block. The combined blocks are .1002 + .124 + .450 + 3.000 = 3.6742 inches.

Federal Specification for Gage Blocks, Inch and Metric Sizes. — This Specification, GGG-G-15B, November 6, 1970, which supersedes GGG-G-15a, September 22, 1964, covers design, manufacturing, and purchasing details for precision gage blocks in inch and metric sizes up to and including 20 inches and 500 millimeters gage lengths. The shapes of blocks are designated Style 1, which is rectangular; Style 2, which is square with a center accessory hole, and Style 3, which defines other shapes as may be specified by the purchaser. Blocks may be made from steel, chromium-plated steel, chromium carbide, tungsten carbide, and other materials to specification. There are four tolerance grades, which are designated Grade 0.5 (formerly Grade AAA in the GGG-G-15a issue of the Specification); Grade 1 (formerly Grade AA); Grade 2 (formerly Grade A+); and Grade 3 (a compromise between former Grades A and B). Grade 0.5 blocks are special reference gages used for extremely high precision gaging work, and are not recommended for general use. Grade 1 blocks are laboratory reference standards used for calibrating inspection gage blocks and high precision gaging work. Grade 2 blocks are used as inspection and toolroom standards, and Grade 3 blocks are used as shop standards.

Inch and metric sizes of blocks in specific sets are given in Tables 1 and 2. It is not a complete list of available sizes, and it should be noted that some gage blocks must be ordered as specials, some may not be available in all materials, and some may not be available from all manufacturers. Gage block set number 4 (88 blocks), listed in the Specification, is not given in Table 1. It is the same as set number 1 (81 blocks) but contains seven additional blocks measuring 0.0625, 0.078125, 0.093750, 0.100025, 0.100050, 0.100075, and 0.109375 inch. In Table 2, gage block set number 3M (112 blocks) is not given. It is similar to set number 2M (88 blocks), and the chief difference is the inclusion of a larger number of blocks in the 0.5 millimeter increment series up to 24.5 mm.

Table 1.   Gage Block Sets* — Inch Sizes (Federal Specification GGG-G-15B†)

SET NUMBER 1 (81 BLOCKS)

First Series: 0.0001 Inch Increments (9 Blocks)

.1001	.1002	.1003	.1004	.1005	.1006	.1007	.1008	.1009

Second Series: 0.001 Inch Increments (49 Blocks)

.101	.102	.103	.104	.105	.106	.107	.108	.109	.110
.111	.112	.113	.114	.115	.116	.117	.118	.119	.120
.121	.122	.123	.124	.125	.126	.127	.128	.129	.130
.131	.132	.133	.134	.135	.136	.137	.138	.139	.140
.141	.142	.143	.144	.145	.146	.147	.148	.149	

Third Series: 0.50 Inch Increments (19 Blocks)

.050	.100	.150	.200	.250	.300	.350	.400	.450	.500
.550	.600	.650	.700	.750	.800	.850	.900	.950	

Fourth Series: 1.000 Inch Increments (4 Blocks)

1.000	2.000	3.000	4.000

SET NUMBER 5 (21 BLOCKS)

First Series: 0.0001 Inch Increments (9 Blocks)

.0101	.0102	.0103	.0104	.0105	.0106	.0107	.0108	.0109

Second Series: 0.001 Inch Increments (11 Blocks)

.010	.011	.012	.013	.014	.015	.016	.017	.018	.019	.020

One Block 0.01005 Inch

SET NUMBER 6 (28 BLOCKS)

First Series: 0.0001 Inch Increments (9 Blocks)

.0201	.0202	.0203	.0204	.0205	.0206	.0207	.0208	.0209

Second Series: 0.001 Inch Increments (9 Blocks)

.021	.022	.023	.024	.025	.026	.027	.028	.029

Third Series: 0.010 Inch Increments (9 Blocks)

.010	.020	.030	.040	.050	.060	.070	.080	.090

One Block 0.02005 Inch

LONG GAGE BLOCK SET NUMBER 7 (8 BLOCKS)

Whole Inch Series (8 Blocks)

5	6	7	8	10	12	16	20

SET NUMBER 8 (36 BLOCKS)

First Series: 0.0001 Inch Increments (9 Blocks)

.1001	.1002	.1003	.1004	.1005	.1006	.1007	.1008	.1009

Second Series: 0.001 Inch Increments (11 Blocks)

.100	.101	.102	.103	.104	.105	.106	.107	.108	.109	.110

Third Series: 0.010 Inch Increments (8 Blocks)

.120	.130	.140	.150	.160	.170	.180	.190

Fourth Series: 0.100 Inch Increments (4 Blocks)

.200	.300	.400	.500

Whole Inch Series (3 Blocks)

1	2	4

One Block 0.050 Inch

SET NUMBER 9 (20 BLOCKS)

First Series: 0.0001 Inch Increments (9 Blocks)

.0501	.0502	.0503	.0504	.0505	.0506	.0507	.0508	.0509

Second Series: 0.001 Inch Increments (10 Blocks)

.050	.051	.052	.053	.054	.055	.056	.057	.058	.059

One Block 0.05005 Inch

* Set number 4 is not shown, and the Specification does not list a set 2 or 3.
† Arranged here in incremental series for convenience of use.

Table 2. Gage Block Sets* — Metric Sizes (Federal Specification GGG-G-15B†)

SET NUMBER 1M (47 BLOCKS)								
First Series: 0.01 Millimeter Increments (9 Blocks)								
1.01	1.02	1.03	1.04	1.05	1.06	1.07	1.08	1.09
Second Series: 0.10 Millimeter Increments (9 Blocks)								
1.10	1.20	1.30	1.40	1.50	1.60	1.70	1.80	1.90

Third Series: 1.0 Millimeter Increments (24 Blocks)

	1.0	2.0	3.0	4.0	5.0	6.0	7.0	8.0	9.0	10.0	
11	12	13	14	15	16	17	18	19	20	21	22	23	24

Fourth Series: 20 Millimeter Increments (3 Blocks)
40   60   80

One Block 1.005 mm   One Block 100 mm

SET NUMBER 2M (88 BLOCKS)									
First Series: 0.001 Millimeter Increments (9 Blocks)									
1.001	1.002	1.003	1.004	1.005	1.006	1.007	1.008	1.009	
Second Series: 0.01 Millimeter Increments (49 Blocks)									
1.01	1.02	1.03	1.04	1.05	1.06	1.07	1.08	1.09	1.10
1.11	1.12	1.13	1.14	1.15	1.16	1.17	1.18	1.19	1.20
1.21	1.22	1.23	1.24	1.25	1.26	1.27	1.28	1.29	1.30
1.31	1.32	1.33	1.34	1.35	1.36	1.37	1.38	1.39	1.40
1.41	1.42	1.43	1.44	1.45	1.46	1.47	1.48	1.49	
Third Series: 0.50 Millimeter Increments (19 Blocks)									
0.5	1.0	1.5	2.0	2.5	3.0	3.5	4.0	4.5	5.0
5.5	6.0	6.5	7.0	7.5	8.0	8.5	9.0	9.5	

Fourth Series: 10 Millimeter Increments (10 Blocks)

10	20	30	40	50	60	70	80	90	100

One Block 1.0005 mm

SET NUMBER 4M (64 BLOCKS)									
First Series: 0.001 Millimeter Increments (9 Blocks)									
2.001	2.002	2.003	2.004	2.005	2.006	2.007	2.008	2.009	
Second Series: 0.01 Millimeter Increments (49 Blocks)									
2.01	2.02	2.03	2.04	2.05	2.06	2.07	2.08	2.09	2.10
2.11	2.12	2.13	2.14	2.15	2.16	2.17	2.18	2.19	2.20
2.21	2.22	2.23	2.24	2.25	2.26	2.27	2.28	2.29	2.30
2.31	2.32	2.33	2.34	2.35	2.36	2.37	2.38	2.39	2.40
2.41	2.42	2.43	2.44	2.45	2.46	2.47	2.48	2.49	
Third Series: 0.10 Millimeter Increments (5 Blocks)									
	2.50		2.60		2.70		2.80		2.90

One Block 2.00 mm

SET NUMBER 5M (24 BLOCKS)								
First Series: 0.01 Millimeter Increments (9 Blocks)								
.41	.42	.43	.44	.45	.46	.47	.48	.49
Second Series: 0.05 Millimeter Increments (5 Blocks)								
	.20		.25		.30		.35	.40
Third Series: 0.05 Millimeter Increments (10 Blocks)								
.50	.55	.60	.65	.70	.75	.80	.85	.90	.95

LONG GAGE BLOCK SET NUMBER 6M (8 BLOCKS)							
Whole Millimeter Series (8 Blocks)							
125	150	175	200	250	300	400	500

Note: Gage blocks measuring 1.09 millimeters and under in set number 1M, and blocks measuring 1.5 millimeters and under in set number 2M, are not available in tolerance grade 0.5.

* Set number 3M is not listed.

† Arranged here in incremental series for convenience of use.

CUTTING TOOLS

Tool Contour. — Tools for turning, planing, etc., are made in straight, bent, off-set, and other forms to place the cutting edges in convenient positions for operating on differently located surfaces. The contour or shape of the cutting edge may also be varied to suit different classes of work. Tool shapes, however, are not only related to the kind of operation, but, in roughing tools particularly, the contour may have a decided effect upon the cutting efficiency of the tool. To illustrate, an increase in the side cutting-edge angle of a roughing tool, or in the nose radius, tends to permit higher cutting speeds because the chip will be thinner for a given feed rate. Such changes, however, may result in chattering or vibrations unless the work and the machine are rigid; hence, the most desirable contour may be a compromise between the ideal form and one that is needed to meet practical requirements.

Terms and Definitions. — The terms and definitions relating to single-point tools vary somewhat in different plants, but the following are in general use.

Single-point Tool: This term is applied to tools for turning, planing, boring, etc., which have a cutting edge at one end. This cutting edge may be formed on one end of a solid piece of steel, or the cutting part of the tool may consist of an insert or tip which is held to the body of the tool by brazing, welding, or mechanical means.

Shank: The shank is the main body of the tool. If the tool is an inserted cutter type, the shank supports the cutter or bit. (See diagram, Fig. 1.)

Nose: A general term sometimes used to designate the cutting end but usually relating more particularly to the rounded tip of the cutting end.

Face: The surface against which the chips bear, as they are severed in turning or planing operations, is called the face.

Flank: The flank is that end surface adjacent to the cutting edge and below it when the tool is in a horizontal position as for turning.

Base: The base is the surface of the tool shank that bears against the supporting tool-holder or block.

Side Cutting Edge: The side cutting edge is the cutting edge on the side of the tool. Tools such as shown in Fig. 1 do the bulk of the cutting with this cutting edge and are, therefore, sometimes called side cutting edge tools.

End Cutting Edge: The end cutting edge is the cutting edge at the end of the tool.

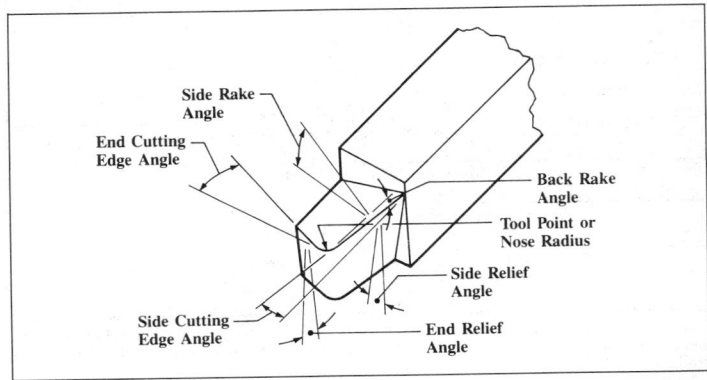

Fig. 1. Terms Applied to Single-point Turning Tools

On side cutting edge tools, the end cutting edge can be used for light plunging and facing cuts. Cut-off tools and similar tools have only one cutting edge located on the end. These tools and other tools that are intended to cut primarily with the end cutting edge are sometimes called end cutting edge tools.

Rake: A metal-cutting tool is said to have rake when the tool face or surface against which the chips bear as they are being severed, is inclined for the purpose of either increasing or diminishing the keenness or bluntness of the edge. The magnitude of the rake is most conveniently measured by two angles called the back rake angle and the side rake angle. The tool shown in Fig. 1 has rake. If the face of the tool did not incline but was parallel to the base, there would be no rake; the rake angles would be zero.

Positive Rake: If the inclination of the tool face is such as to make the cutting edge keener or more acute than when the rake angle is zero, the rake angle is defined as positive.

Negative Rake: If the inclination of the tool face makes the cutting edge less keen or more blunt than when the rake angle is zero, the rake is defined as negative.

Back Rake: The back rake is the inclination of the face toward or away from the end or the end cutting edge of the tool. When the inclination is away from the end cutting edge, as shown in Fig. 1, the back rake is positive. If the inclination is downward toward the end cutting edge the back rake is negative.

Side Rake: The side rake is the inclination of the face toward or away from the side cutting edge. When the inclination is away from the side cutting edge, as shown in Fig. 1, the side rake is positive. If the inclination is toward the side cutting edge the side rake is negative.

Relief: The flanks below the side cutting edge and the end cutting edge must be relieved to allow these cutting edges to penetrate into the workpiece when taking a cut. If the flanks are not provided with relief, the cutting edges will rub against the workpiece and be unable to penetrate in order to form the chip. Relief is also provided below the nose of the tool to allow it to penetrate into the workpiece. The relief at the nose is usually a blend of the side relief and the end relief.

End Relief Angle: The end relief angle is a measure of the relief below the end cutting edge.

Side Relief Angle: The side relief angle is a measure of the relief below the side cutting edge.

Back Rake Angle: The back rake angle is a measure of the back rake. It is measured in a plane that passes through the side cutting edge and is perpendicular to the base. Thus, the back rake angle can be defined by measuring the inclination of the side cutting edge with respect to a line or plane that is parallel to the base. The back rake angle may be positive, negative, or zero depending upon the magnitude and direction of the back rake.

Side Rake Angle: The side rake angle is a measure of the side rake. This angle is always measured in a plane that is perpendicular to the side cutting edge and perpendicular to the base. Thus, the side rake angle is the angle of inclination of the face perpendicular to the side cutting edge with reference to a line or a plane that is parallel to the base.

End Cutting Edge Angle: The end cutting edge angle is the angle made by the end cutting edge with respect to a plane perpendicular to the axis of the tool shank. It is provided to allow the end cutting edge to clear the finish machined surface on the workpiece.

Side Cutting Edge Angle: The side cutting edge angle is the angle made by the side cutting edge and a plane that is parallel to the side of the shank.

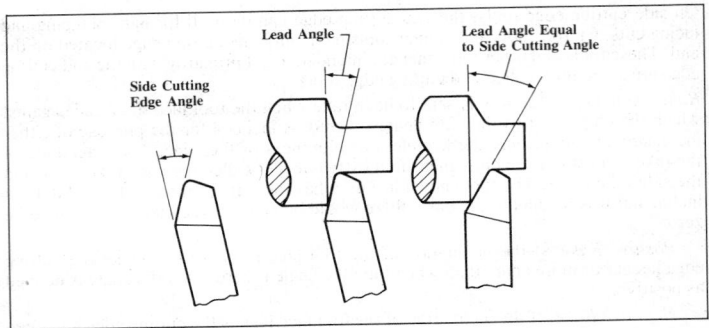

Fig. 2. Lead Angle on Single-point Turning Tool

Nose Radius: The nose radius is the radius of the nose of the tool. The performance of the tool, in part, is influenced by nose radius so that it must be carefully controlled.

Lead Angle: The lead angle, shown in Fig. 2, is not ground on the tool. It is a tool setting angle which has a great influence on the performance of the tool. The lead angle is bounded by the side cutting edge and a plane perpendicular to the workpiece surface when the tool is in position to cut; or, more exactly, the lead angle is the angle between the side cutting edge and a plane perpendicular to the direction of the feed travel.

Solid Tool: A solid tool is a cutting tool made from one piece of tool material.

Brazed Tool: A brazed tool is a cutting tool having a blank of cutting-tool material permanently brazed to a steel shank.

Blank: A blank is an unground piece of cutting-tool material from which a brazed tool is made.

Tool Bit: A tool bit is a relatively small cutting tool that is clamped in a holder in such a way that it can readily be removed and replaced. It is intended primarily to be reground when dull and not indexed.

Tool-bit Blank: The tool-bit blank is an unground piece of cutting-tool material from which a tool bit can be made by grinding. It is available in standard sizes and shapes.

Tool-bit Holder: Usually made from forged steel, the tool-bit holder is used to hold the tool bit, to act as an extended shank for the tool bit, and to provide a means for clamping in the tool post.

Straight-shank Tool-bit Holder: A straight-shank tool-bit holder has a straight shank when viewed from the top. The axis of the tool bit is held parallel to the axis of the shank.

Offset-shank Tool-bit Holder: An offset-shank tool-bit holder has the shank bent to the right or left, as seen in Fig. 3. The axis of the tool bit is held at an angle with respect to the axis of the shank.

Side cutting Tool: A side cutting tool has its major cutting edge on the side of the cutting part of the tool. The major cutting edge may be parallel or at an angle with respect to the axis of the tool.

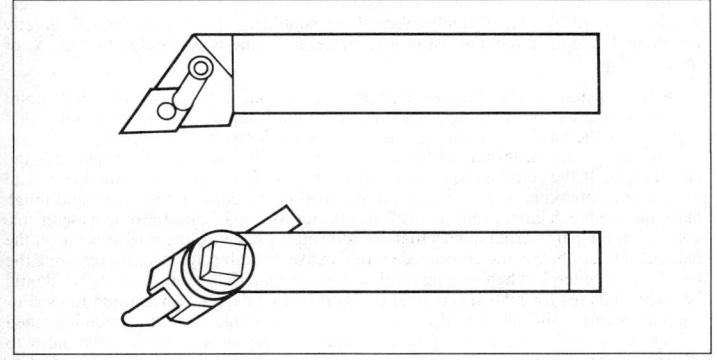

Fig. 3. Top: Right-hand Offset-shank, Indexable Insert Holder
Bottom: Right-hand Offset-shank Tool-bit Holder

Indexable Inserts: An indexable insert is a relatively small piece of cutting-tool material that is geometrically shaped to have two or several cutting edges that are used until dull. The insert is then indexed on the holder to apply a sharp cutting edge. When all the cutting edges have been dulled, the insert is discarded. The insert is held in a pocket or against other locating surfaces on an indexable insert holder by means of a mechanical clamping device that can be tightened or loosened easily.

Indexable Insert Holder: Made of steel, an indexable insert holder is used to hold indexable inserts. It is equipped with a mechanical clamping device that holds the inserts firmly in a pocket or against other seating surfaces.

Straight-shank Indexable Insert Holder: A straight-shank indexable insert tool-holder is essentially straight when viewed from the top, although the cutting edge of the insert may be oriented parallel, or at an angle, to the axis of the holder.

Offset-shank Indexable Insert Holder: An offset-shank indexable insert holder has the head end, or the end containing the insert pocket, offset to the right or left, as shown in Fig. 3.

End cutting Tool: An end cutting tool has its major cutting edge on the end of the cutting part of the tool. The major cutting edge may be perpendicular or at an angle, with respect to the axis of the tool.

Curved Cutting-edge Tool: A curved cutting-edge tool has a continuously variable side cutting edge angle. The cutting edge is usually in the form of a smooth, continuous curve along its entire length, or along a large portion of its length.

Right-hand Tool: A right-hand tool has the major, or working, cutting edge on the right-hand side when viewed from the cutting end with the face up. As used in a lathe, such a tool is usually fed into the work from right to left, when viewed from the shank end.

Left-hand Tool: A left-hand tool has the major or working cutting edge on the left-hand side when viewed from the cutting end with the face up. As used in a lathe, the tool is usually fed into the work from left to right, when viewed from the shank end.

Neutral-hand Tool: A neutral-hand tool is a tool to cut either left to right or right to left; or the cut may be parallel to the axis of the shank as when plunge cutting.

Chipbreaker: A groove formed in or on a shoulder on the face of a turning tool back of the cutting edge to break up the chips and prevent the formation of long,

continuous chips which would be dangerous to the operator and also bulky and cumbersome to handle. A chipbreaker of the shoulder type may be formed directly on the tool face or it may consist of a separate piece that is held either by brazing or by clamping.

Relief Angles. — The end relief angle and the side relief angle on single-point cutting tools are usually, though not invariably, made equal to each other. The relief angle under the nose of the tool is a blend of the side and end relief angles.

The size of the relief angles has a pronounced effect on the performance of the cutting tool. If the relief angles are too large, the cutting edge will be weakened and in danger of breaking when a heavy cutting load is placed on it by a hard and tough material. On finish cuts, rapid wear of the cutting edge may cause problems with size control on the part. Relief angles that are too small will cause the rate of wear on the flank of the tool below the cutting edge to increase, thereby significantly reducing the tool life. In general, when cutting hard and tough materials, the relief angles should be 6 to 8 degrees for high-speed steel tools and 5 to 7 degrees for carbide tools. For medium steels, mild steels, cast iron, and other average work the recommended values of the relief angles are 8 to 12 degrees for high-speed steel tools and 5 to 10 degrees for carbides. Ductile materials having a relatively low modulus of elasticity should be cut using larger relief angles. For example, the relief angles recommended for turning copper, brass, bronze, aluminum, ferritic malleable iron, and similar metals are 12 to 16 degrees for high-speed steel tools and 8 to 14 degrees for carbides.

Larger relief angles generally tend to produce a better finish on the finish machined surface because less surface of the worn flank of the tool rubs against the workpiece. For this reason, single-point thread-cutting tools should be provided with relief angles that are as large as circumstances will permit. Problems encountered when machining stainless steel may be overcome by increasing the size of the relief angle. The relief angles used should never be smaller than necessary.

Rake Angles. — Machinability tests have confirmed that when the rake angle along which the chip slides, called the true rake angle, is made larger in the positive direction, the cutting force and the cutting temperature will decrease. Also, the tool life for a given cutting speed will increase with increases in the true rake angle up to an optimum value, after which it will decrease again. For turning tools which cut primarily with the side cutting edge, the true rake angle corresponds rather closely with the side rake angle except when taking shallow cuts. Increasing the side rake angle in the positive direction lowers the cutting force and the cutting temperature, while at the same time it results in a longer tool life or a higher permissible cutting speed up to an optimum value of the side rake angle. After the optimum value is exceeded, the cutting force and the cutting temperature will continue to drop; however, the tool life and the permissible cutting speed will decrease.

As an approximation, the magnitude of the cutting force will decrease about one per cent per degree increase in the side rake angle. While not exact, this rule of thumb does correspond approximately to test results and can be used to make rough estimates. Of course, the cutting force also increases about one per cent per degree decrease in the side rake angle. The limiting value of the side rake angle for optimum tool life or cutting speed depends upon the work material and the cutting tool material. In general, it will occur at lower values for hard and tough work materials. Cemented carbides are harder and more brittle than high-speed steel; therefore, the rake angles usually used for cemented carbides are larger than for high-speed steel.

Negative rake angles cause the face of the tool to slope in the opposite direction from positive rake angles and, as might be expected, they have an opposite effect. For side cutting edge tools, increasing the side rake angle in a negative direction will result in an increase in the cutting force and an increase in the cutting temperature o

approximately one per cent per degree change in rake angle. For example, if the side rake angle is changed from 5 degrees positive to 5 degrees negative, the cutting force will be about 10 per cent larger. Usually the tool life will also decrease when negative side rake angles are used, although the tool life will sometimes increase when the negative rake angle is not too large and when a fast cutting speed is used.

Negative side rake angles are usually used in combination with negative back rake angles on single-point cutting tools. The negative rake angles strengthen the cutting edges enabling them to sustain heavier cutting loads and shock loads. They are recommended for turning very hard materials and for heavy interrupted cuts. There is also an economic advantage in favor of using negative rake indexable inserts and tool holders inasmuch as the cutting edges provided on both the top and bottom of the insert can be used.

On turning tools that cut primarily with the side cutting edge, the effect of the back rake angle alone is much less than the effect of the side rake angle although the direction of the change in cutting force, cutting temperature, and tool life is the same. The effect that the back rake angle has can be ignored unless, of course, extremely large changes in this angle are made. A positive back rake angle does improve the performance of the nose of the tool somewhat and is helpful in taking light finishing cuts. A negative back rake angle strengthens the nose of the tool and is helpful when interrupted cuts are taken. The back rake angle has a very significant effect on the performance of end cutting edge tools, such as cut-off tools. For these tools, the effect of the back rake angle is very similar to the effect of the side rake angle on side cutting edge tools.

Side Cutting Edge and Lead Angles. — These angles are considered together because the side cutting edge angle is usually designed to provide the desired lead angle when the tool is being used. The side cutting edge angle and the lead angle will be equal when the shank of the cutting tool is positioned perpendicular to the workpiece, or, more correctly, perpendicular to the direction of the feed. When the shank is not perpendicular, the lead angle is determined by the side cutting edge and an imaginary line perpendicular to the feed direction.

The flow of the chips over the face of the tool is approximately perpendicular to the side cutting edge except when shallow cuts are taken. The thickness of the undeformed chip is measured perpendicular to the side cutting edge. As the lead angle is increased, the length of chip in contact with the side cutting edge is increased, and the chip will become longer and thinner. This effect is the same as increasing the depth of cut and decreasing the feed, although the actual depth of cut and feed remain the same and the same amount of metal is removed. The effect of lengthening and thinning the chip by increasing the lead angle is very beneficial as it increases the tool life for a given cutting speed or that speed can be increased. Increasing the cutting speed while the feed and the tool life remain the same leads to faster production.

However, an adverse effect must be considered. Chatter can be caused by a cutting edge that is oriented at a high lead angle when turning and sometimes, when turning long and slender shafts, even a small lead angle can cause chatter. In fact, an unsuitable lead angle of the side cutting edge is one of the principal causes of chatter. When chatter occurs, often simply reducing the lead angle will cure it. Sometimes, very long and slender shafts can be turned successfully with a tool having a zero degree lead angle (and having a small nose radius). Boring bars, being usually somewhat long and slender, are also susceptible to chatter if a large lead angle is used. The lead angle for boring bars should be kept small, and for very long and slender boring bars a zero degree lead angle is recommended. It is impossible to provide a rule that will determine when chatter caused by a lead angle will occur and when it will not. In making a judgment, the first consideration is the length to diameter ratio of the part

to be turned, or of the boring bar. Then the method of holding the workpiece must be considered — a part that is firmly held is less apt to chatter. Finally, the overall condition and rigidity of the machine must be considered because they may be the real cause of chatter.

Although chatter can be a problem, the advantages gained from high lead angles are such that the lead angle should be as large as possible at all times.

End Cutting Edge Angle. — The size of the end cutting edge angle is important when tool wear by cratering occurs. Frequently, the crater will enlarge until it breaks through the end cutting edge just behind the nose, and tool failure follows shortly. Reducing the size of the end cutting edge angle tends to delay the time of crater breakthrough. When cratering takes place, the recommended end cutting edge angle is 8 to 15 degrees. If there is no cratering, the angle can be made larger. Larger end cutting edge angles may be required to enable profile turning tools to plunge into the work without interference from the end cutting edge.

Nose Radius. — The tool nose is a very critical part of the cutting edge since it cuts the finished surface on the workpiece. If the nose is made to a sharp point, the finish machined surface will usually be unacceptable and the life of the tool will be short. Thus, a nose radius is required to obtain an acceptable surface finish and tool life. The surface finish obtained is determined by the feed rate and by the nose radius if other factors such as the work material, the cutting speed, and cutting fluids are not considered. A large nose radius will give a better surface finish and will permit a faster feed rate to be used.

Machinability tests have demonstrated that increasing the nose radius will also improve the tool life or allow a faster cutting speed to be used. For example, high-speed steel tools were used to turn an alloy steel in one series of tests where complete or catastrophic tool failure was used as a criterion for the end of tool life. The cutting speed for a 60-minute tool life was found to be 125 fpm when the nose radius was 1/16 inch and 160 fpm when the nose radius was 1/4 inch.

A very large nose radius can often be used but a limit is sometimes imposed because the tendency for chatter to occur is increased as the nose radius is made larger. A nose radius that is too large can cause chatter and when it does, a smaller nose radius must be used on the tool. It is always good practice to make the nose radius as large as is compatible with the operation being performed.

Chipbreakers. — Many steel turning tools are equipped with chipbreaking devices to prevent the formation of long continuous chips in connection with the turning of steel at the high speeds made possible by high-speed steel and especially cemented carbide tools. Long steel chips are dangerous to the operator, cumbersome to handle, and they may twist around the tool and cause damage. Broken chips not only occupy less space, but permit a better flow of coolant to the cutting edge. Several different forms of chipbreakers are illustrated in Fig. 4.

Angular Shoulder Type: The angular shoulder type shown at *A* is one of the commonly used forms. As the enlarged sectional view shows, the chipbreaking shoulder is located back of the cutting edge. The angle *a* between the shoulder and cutting edge may vary from 6 to 15 degrees or more, 8 degrees being a fair average. The ideal angle, width *W* and depth *G*, depend upon the speed and feed, the depth of cut, and the material. As a general rule, width *W*, at the end of the tool, varies from 3/32 to 7/32 inch, and the depth *G* may range from 1/64 to 1/16 inch. The shoulder radius equals depth *G*. If the tool has a large nose radius, the corner of the shoulder at the nose end may be beveled off, as illustrated at *B*, to prevent it from coming into contact with the work. The width *K* for type *B* should equal approximately 1.5 times the nose radius.

Parallel Shoulder Type: Diagram C shows a design with a chipbreaking shoulder that is parallel with the cutting edge. With this form, the chips are likely to come off in short curled sections. The parallel form may also be applied to straight tools which do not have a side cutting-edge angle. The tendency with this parallel shoulder form is to force the chips against the work and damage it.

Groove Type: This type (diagram D) has a groove in the face of the tool produced by grinding. Between the groove and the cutting edge, there is a land *L*. Under ideal conditions, this width *L*, the groove width *W*, and the groove depth *G*, would be varied to suit the feed, depth of cut and material. For average use, *L* is about $\frac{1}{32}$ inch; *G*, $\frac{1}{32}$ inch; and, *W*, $\frac{1}{16}$ inch. There are differences of opinion concerning the relative merits of the groove type and the shoulder type. Both types have proved satisfactory when properly proportioned for a given class of work.

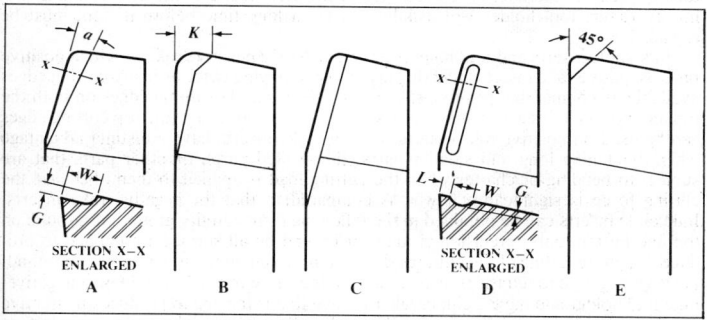

Fig. 4. Different Forms of Chipbreakers for Turning Tools

Chipbreaker for Light Cuts: Diagram E illustrates a form of chipbreaker that is sometimes used on tools for finishing cuts having a maximum depth of about $\frac{1}{32}$ inch. This chipbreaker is a shoulder type having an angle of 45 degrees and a maximum width of about $\frac{1}{16}$ inch. It is important in grinding all chipbreakers to give the chip-bearing surfaces a fine finish, such as would be obtained by honing. This finish greatly increases the life of the tool.

Planing Tools. — Many of the principles which govern the shape of turning tools also apply in the grinding of tools for planing. The amount of rake depends upon the hardness of the material, and the direction of the rake should be away from the *working part* of the cutting edge. The angle of clearance should be about 4 or 5 degrees for planer tools, which is less than for lathe tools. This small clearance is allowable because a planer tool is held about square with the platen, whereas a lathe tool, the height and inclination of which can be varied, may not always be clamped in the same position.

Carbide Tools: Carbide tools for planing usually have negative rake. Round-nose and square-nose end-cutting tools should have a "negative back rake" (or front rake) of 2 or 3 degrees. Side cutting tools may have a negative back rake of 10 degrees, a negative side rake of 5 degrees, and a side cutting-edge angle of 8 degrees.

Indexable Inserts. — A large proportion of the cemented carbide, single-point cutting tools are indexable inserts and indexable insert tool holders. Dimensional specifications for solid sintered carbide indexable inserts are given in

American National Standard ANSI B94.25-1982. Samples of the many insert shapes are shown in Table 3. Most modern, cemented carbide, face milling cutters are of the indexable insert type. Larger size end milling cutters, side milling or slotting cutters, boring tools, and a wide variety of special tools are made to use indexable inserts. These inserts are primarily made from cemented carbide, although most of the cemented oxide cutting tools are also indexable inserts.

The objective of this type of tooling is to provide an insert with several cutting edges. When an edge is worn, the insert is indexed in the tool holder until all the cutting edges are used up, after which it is discarded. The insert is not intended to be reground. The advantages are that the cutting edges on the tool can be rapidly changed without removing the tool holder from the machine, tool-grinding costs are eliminated, and the cost of the insert is less than the cost of a similar, brazed carbide tool. Of course, the cost of the tool holder must be added to the cost of the insert; however, one tool holder will usually last for a long time before it, too, must be replaced.

Indexable inserts and tool holders are made with a negative rake or with a positive rake. Negative rake inserts have the advantage of having twice as many cutting edges available as comparable positive rake inserts, because the cutting edges on both the top and bottom of negative rake inserts can be used, while only the top cutting edges can be used on positive rake inserts. Positive rake inserts have a distinct advantage when machining long and slender parts, thin-walled parts, or other parts that are subject to bending or chatter when the cutting load is applied to them, because the cutting force is significantly lower as compared to that for negative rake inserts. Indexable inserts can be obtained in the following forms: utility ground, or ground on top and bottom only; precision ground, or ground on all surfaces; prehoned to produce a slight rounding of the cutting edge; and precision molded, which are unground. Positive-negative rake inserts also are available. These inserts are held on a negative-rake tool holder and have a chipbreaker groove that is formed to produce an effective positive-rake angle while cutting. Cutting edges may be available on the top surface only, or on both top and bottom surfaces. The positive-rake chipbreaker surface may be ground or precision molded on the insert.

Many materials, such as gray cast iron, form a discontinuous chip. For these materials an insert that has plain faces without chipbreaker grooves should always be used. Steels and other ductile materials form a continuous chip that must be broken into small segments when machined on lathes and planers having single-point, cemented-carbide and cemented-oxide cutting tools; otherwise, the chips can cause injury to the operator. In this case a chipbreaker must be used. Some inserts are made with chipbreaker grooves molded or ground directly on the insert. When inserts with plain faces are used, a cemented-carbide plate-type chipbreaker is clamped on top of the insert.

Identification System for Indexable Inserts. — The size of indexable inserts is determined by the diameter of an inscribed circle (I.C.), except for rectangular and parallelogram inserts where the length and width dimensions are used. To describe an insert in its entirety, a standard (ANSI B212.4-1986) identification system is used where each position number designates a feature of the insert. The ANSI Standard includes items now commonly used and facilitates identification of items not in common use. Identification consists of up to ten positions; each position defines a characteristic of the insert as shown below:

1 2 3 4 5 6 7 8* 9 10*
T N M G 5 4 3 A

* Eighth and Tenth Positions are only used when required.

1. *Shape:* The shape of an insert is designated by a letter: **R** for round; **S**, square; **T**, triangle; **A**, 85° parallelogram; **B**, 82° parallelogram; **C**, 80° diamond; **D**, 55° diamond; **E**, 75° diamond; **H**, hexagon; **K**, 55° parallelogram; **L**, rectangle; **M**, 86° diamond; **O**, octagon; **P**, pentagon; **V**, 35° diamond; and **W**, 80° trigon.

2. *Relief Angle (Clearances):* The second position is a letter denoting the relief angles; N for 0°; **A**, 3°; **B**, 5°; **C**, 7°; **P**, 11°; **D**, 15°; **E**, 20°; **F**, 25°; **G**, 30°; **H**, 0° & 11°*; **J**, 0° & 14°*; **K**, 0° & 17°*; **L**, 0° & 20°*; **M**, 11° & 14°*; **R**, 11° & 17°*; **S**, 11° & 20°*. When mounted on a holder, the actual relief angle may be different from that on the insert.

* Second angle is secondary facet angle, which may vary by ± 1°.

3. *Tolerances:* The third position is a letter and indicates the tolerances which control the indexability of the insert. Tolerances specified do not imply the method of manufacture.

Symbol	Tolerance (± from nominal)		Symbol	Tolerance (± from nominal)	
	Inscribed Circle, Inch	Thickness, Inch		Inscribed Circle, Inch	Thickness, Inch
A	0.001	0.001	H	0.0005	0.001
B	0.001	0.005	J	0.002–0.005	0.001
C	0.001	0.001	K	0.002–0.005	0.001
D	0.001	0.005	L	0.002–0.005	0.001
E	0.001	0.001	M	0.002–0.004†	0.005
F	0.0005	0.001	U	0.005–0.010†	0.005
G	0.001	0.005	N	0.002–0.004†	0.001

† Exact tolerance is determined by size of insert. (See ANSI B94.25.)

4. *Type:* The type of insert is designated by a letter. **A**, with hole; **B**, with hole and countersink; **C**, with hole and two countersinks; **F**, chip grooves both surfaces, no hole; **G**, same as F but with hole; **H**, with hole, one countersink, and chip groove on one rake surface; **J**, with hole, two countersinks and chip grooves on two rake surfaces; **M**, with hole and chip groove on one rake surface; **N**, without hole; **Q**, with hole and two countersinks; **R**, without hole but with chip groove on one rake surface; **T**, with hole, one countersink, and chip groove on one rake face; **U**, with hole, two countersinks, and chip grooves on two rake faces; and **W**, with hole and one countersink. *Note:* a dash may be used after position 4 to separate the shape-describing portion from the following dimensional description of the insert and is not to be considered a position in the standard description.

5. *Size:* The size of the insert is designated by a one- or a two-digit number. For regular polygons and diamonds, it is the number of eighths of an inch in the nominal size of the inscribed circle, and will be a one- or two-digit number when the number of eighths is a whole number. It will be a two-digit number, including one decimal place, when it is not a whole number. Rectangular and parallelogram inserts require two digits: the first digit indicates the number of eighths of an inch width and the second digit, the number of quarters of an inch length.

6. *Thickness:* The thickness is designated by a one- or two-digit number, which indicates the number of sixteenths of an inch in the thickness of the insert. It is a one-digit number when the number of sixteenths is a whole number; it is a two-digit number carried to one decimal place when the number of sixteenths of an inch is not a whole number.

7. *Cutting Point Configuration:* The cutting point, or nose radius, is designated by a number representing 1/64ths of an inch; a flat at the cutting point or nose, is designated by a letter: **0** for sharp corner; **1**, 1/64 inch radius; **2**, 1/32 inch radius; **3**, 3/64

inch radius; **4,** $\frac{1}{16}$ inch radius; **5,** $\frac{5}{64}$ inch radius; **6,** $\frac{3}{32}$ inch radius; **7,** $\frac{7}{64}$ inch radius; **8,** $\frac{1}{8}$ inch radius; **A,** square insert with 45° chamfer; **D,** square insert with 30° chamfer; **E,** square insert with 15° chamfer; **F,** square insert with 3° chamfer; **K,** square insert with 30° double chamfer; **L,** square insert with 15° double chamfer; **M,** square insert with 3° double chamfer; **N,** truncated triangle insert; and **P,** flatted corner triangle insert.

8. *Special Cutting Point Definition:* The eighth position, if it follows a letter in the 7th position, is a number indicating the number of $\frac{1}{64}$ths of an inch measured parallel to the edge of the facet.

9. *Hand:* **R,** right; **L,** left; to be used when required in ninth position.

10. *Other Conditions:* The tenth position defines special conditions (such as edge treatment, surface finish) as follows: **A,** honed, 0.0005 inch to less than 0.003 inch; **B,** honed, 0.003 inch to less than 0.005 inch; **C,** honed, 0.005 inch to less than 0.007 inch; **J,** polished, 4 microinch arithmetic average (AA) on rake surfaces only; **T,** chamfered, manufacturer's standard negative land, rake face only.

Indexable Insert Tool Holders. — Indexable insert tool holders are made from a good grade of steel which is heat treated to a hardness of 44 to 48 Rc for most normal applications. Accurate pockets that serve to locate the insert in position and to provide surfaces against which the insert can be clamped are machined in the ends of tool holders. A cemented carbide seat usually is provided, and is held in the bottom of the pocket by a screw or by the clamping pin, if one is used. The seat is necessary to provide a flat bearing surface upon which the insert can rest and, in so doing, it adds materially to the ability of the insert to withstand the cutting load. The seating surface of the holder may provide a positive-, negative-, or a neutral-rake orientation to the insert when it is in position on the holder. Holders, therefore, are classified as positive, negative, or neutral rake.

Four basic methods are used to clamp the insert on the holder: 1. Clamping, usually top clamping; 2. Pin-lock clamping; 3. Multiple clamping using a clamp, usually a top clamp, and a pin lock; and 4. Clamping the insert with a machine screw. All top clamps are actuated by a screw that forces the clamp directly against the insert. When required, a cemented-carbide, plate-type chipbreaker is placed between the clamp and the insert. Pin-lock clamps require an insert having a hole: the pin acts against the walls of the hole to clamp the insert firmly against the seating surfaces of the holder. Multiple or combination clamping, simultaneously using both a pin-lock and a top clamp, is recommended when taking heavier or interrupted cuts. Holders are available on which all the above-mentioned methods of clamping may be used. Other holders are made with only a top clamp or a pin lock. Screw-on type holders use a machine screw to hold the insert in the pocket. Most standard indexable insert holders are either straight-shank or offset-shank, although special holders are made having a wide variety of configurations.

The common shank sizes of indexable insert tool holders are shown in Table 1. Not all styles are available in every shank size. Positive- and negative-rake tools are also not available in every style or shank size. Some manufacturers provide additional shank sizes for certain tool holder styles. For more complete details the manufacturers' catalogs must be consulted.

Identification System for Indexable Insert Holders. —The following identification system conforms to the American National Standard, ANSI B212.5-1986, Metric Holders for Indexable Inserts.

Table 1. Standard Shank Sizes for Indexable Insert Holders

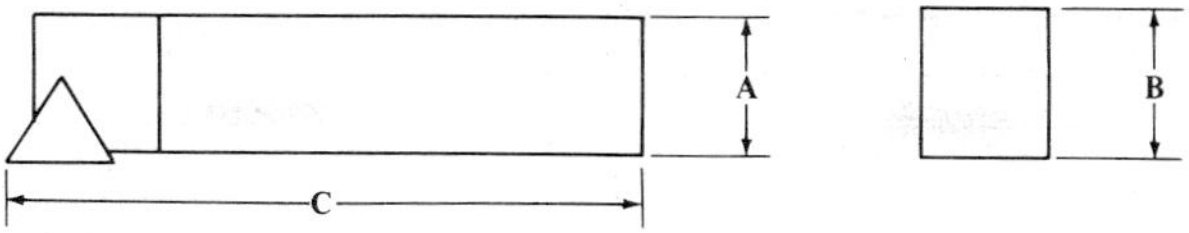

Basic Shank Size	Shank Dimensions for Indexable Insert Holders					
	A		B		C*	
	In.	mm	In.	mm	In.	mm
½ × ½ × 4½	.500	12.70	.500	12.70	4.500	114.30
⅝ × ⅝ × 4½	.625	15.87	.625	15.87	4.500	114.30
⅝ × 1¼ × 6	.625	15.87	1.250	31.75	6.000	152.40
¾ × ¾ × 4½	.750	19.05	.750	19.05	4.500	114.30
¾ × 1 × 6	.750	19.05	1.000	25.40	6.000	152.40
¾ × 1¼ × 6	.750	19.05	1.250	31.75	6.000	152.40
1 × 1 × 6	1.000	25.40	1.000	25.40	6.000	152.40
1 × 1¼ × 6	1.000	25.40	1.250	31.75	6.000	152.40
1 × 1½ × 6	1.000	25.40	1.500	38.10	6.000	152.40
1¼ × 1¼ × 7	1.250	31.75	1.250	31.75	7.000	177.80
1¼ × 1½ × 8	1.250	31.75	1.500	38.10	8.000	203.20
1⅜ × 2¹⁄₁₆ × 6⅜	1.375	34.92	2.062	52.37	6.380	162.05
1½ × 1½ × 7	1.500	38.10	1.500	38.10	7.000	177.80
1¾ × 1¾ × 9½	1.750	44.45	1.750	44.45	9.500	241.30
2 × 2 × 8	2.000	50.80	2.000	50.80	8.000	203.20

* Holder length; may vary by manufacturer. Actual shank length depends on holder style.

Each position in the system designates a feature of the holder in the following sequence:

1	2	3	4	5	—	6	—	7	—	8*	—	9	—	10*
C	**T**	**N**	**A**	**R**	—	**85**	—	**25**	—	**D**	—	**16**	—	**Q**

1. *Method of Holding Horizontally Mounted Insert:* The method of holding or clamping is designated by a letter: **C**, top clamping, insert without hole; **M**, top and hole clamping, insert with hole; **P**, hole clamping, insert with hole; **S**, screw clamping through hole, insert with hole; **W**, wedge clamping.

2. *Insert Shape:* The insert shape is identified by a letter: **H**, hexagonal; **O**, octagonal; **P**, pentagonal; **S**, square; **T**, triangular; **C**, rhombic, 80° included angle; **D**, rhombic, 55° included angle; **E**, rhombic, 75° included angle; **M**, rhombic, 86° included angle; **V**, rhombic, 35° included angle; **W**, hexagonal, 80° included angle; **L**, rectangular; **A**, parallelogram, 85° included angle; **B**, parallelogram, 82° included angle; **K**, parallelogram, 55° included angle; **R**, round. The included angle is always the smaller angle.

3. *Holder Style:* The holder style designates the shank style and the side cutting edge angle, or end cutting edge angle, or the purpose for which the holder is used. It is designated by a letter: **A** for straight shank with 0° side cutting edge angle; **B**, straight shank with 15° side cutting edge angle; **C**, straight-shank end cutting tool with 0° end cutting edge angle; **D**, straight shank with 45° side cutting edge angle; **E**, straight shank with 30° side cutting edge angle; **F**, offset shank with 0° end cutting edge angle; **G**, offset shank with 0° side cutting edge angle; **J**, offset shank with negative 3° side cutting edge angle; **K**, offset shank with 15° end cutting edge angle; **L**, offset shank with negative 5° side cutting edge angle and 5° end cutting edge angle; **M**, straight shank with 40° side cutting edge angle; **N**, straight shank with 27° side cutting edge angle; **R**, offset shank with 15° side cutting edge angle; **S**, offset shank with 45° side cutting edge angle; **T**, offset shank with 30° side cutting edge angle; **U**,

Table 2. Letter Symbols for Qualification of Tool Holders — Position 10
(ANSI B212.5-1986)

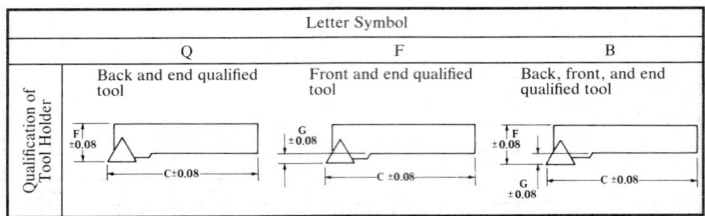

offset shank with negative 3° end cutting edge angle; **V**, straight shank with 17½° side cutting edge angle; **W**, offset shank with 30° end cutting edge angle; **Y**, offset shank with 5° end cutting edge angle.

4. *Normal Clearances:* The normal clearances of inserts are identified by letters: **A**, 3°; **B**, 5°; **C**, 7°; **D**, 15°; **E**, 20°; **F**, 25°; **G**, 30°; **N**, 0°; **P**, 11°.

5. *Hand of tool:* The hand of the tool is designated by a letter: **R** for right-hand; **L**, left-hand; and **N**, neutral, or either hand.

6. *Tool Height for Rectangular Shank Cross Sections:* The tool height for tool holders with a rectangular shank cross section and the height of cutting edge equal to shank height is given as a two-digit number representing this value in millimeters. For example, a height of 32 mm would be encoded as 32; 8 mm would be encoded as 08, where the one-digit value is preceded by a zero.

7. *Tool Width for Rectangular Shank Cross Sections:* The tool width for tool holders with a rectangular shank cross section is given as a two-digit number representing this value in millimeters. For example, a width of 25 mm would be encoded as 25; 8 mm would be encoded as 08, where the one-digit value is preceded by a zero.

8. *Tool Length:* The tool length is designated by a letter: **A**, 32 mm; **B**, 40 mm; **C**, 50 mm; **D**, 60 mm; **E**, 70 mm; **F**, 80 mm; **G**, 90 mm; **H**, 100 mm; **J**, 110 mm; **K**, 125 mm; **L**, 140 mm; **M**, 150 mm; **N**, 160 mm; **P**, 170 mm; **Q**, 180 mm; **R**, 200 mm; **S**, 250 mm; **T**, 300 mm; **U**, 350 mm; **V**, 400 mm; **W**, 450 mm; **X**, special length to be specified; **Y**, 500 mm.

9. *Indexable Insert Size:* The size of indexable inserts is encoded as follows: For insert shapes C, D, E, H, M, O, P, R, S, T, V, the side length (the diameter for R inserts) in millimeters is used as a two-digit number, with decimals being disregarded. For example, the symbol for a side length of 16.5 mm is 16. For insert shapes A, B, K, L, the length of the main cutting edge or of the longer cutting edge in millimeters is encoded as a two-digit number, disregarding decimals. If the symbol obtained has only one digit, then it should be preceded by a zero. For example, the symbol for a main cutting edge of 19.5 mm is 19; for an edge of 9.52 mm, the symbol is 09.

10. *Special Tolerances:* Special tolerances are indicated by a letter: **Q**, back and end qualified tool; **F**, front and end qualified tool; **B**, back, front, and end qualified tool. A qualified tool is one that has tolerances of ±0.08 mm for dimensions *F*, *G*, and *C*. (See Table 2.)

Selecting Indexable Insert Holders. — A guide for selecting indexable insert holders is provided by Table 3B. Some operations such as deep grooving, cut-off, and threading are not given in this table. However, tool holders designed specifically for these operations are available. The boring operations listed in Table 3B refer primarily to larger holes, into which the holders will fit. Smaller holes are bored using boring bars. An examination of this table shows that several tool-holder styles can be used and frequently are used for each operation. Selection of the best holder for a given

job depends largely on the job and there are certain basic facts that should be considered in making the selection.

Rake Angle: A negative-rake insert has twice as many cutting edges available as a comparable positive-rake insert. Sometimes the tool life obtained when using the second face may be less than that obtained on the first face because the tool wear on the cutting edges of the first face may reduce the insert strength. Nevertheless, the advantage of negative-rake inserts and holders is such that they should be considered first in making any choice. Positive-rake holders should be used where lower cutting forces are required, as when machining slender or small-diameter parts, when chatter may occur, and for machining some materials, such as aluminum, copper, and certain grades of stainless steel, when positive-negative rake inserts can sometimes be used to advantage. These inserts are held on negative-rake holders that have their rake surfaces ground or molded to form a positive-rake angle.

Insert Shape: The configuration of the workpiece, the operation to be performed, and the lead angle required often determine the insert shape. When these factors need not be considered, the insert shape should be selected on the basis of insert strength and the maximum number of cutting edges available. Thus, a round insert is the strongest and has a maximum number of available cutting edges. It can be used with heavier feeds while producing a good surface finish. Round inserts are limited by their tendency to cause chatter, which may preclude their use. The square insert is the next most effective shape, providing good corner strength and more cutting edges than all other inserts except the round insert. The only limitation of this insert shape is that it must be used with a lead angle. Therefore, the square insert cannot be used for turning square shoulders or for back facing. Triangle inserts are the most versatile and can be used to perform more operations than any other insert shape. The 80-degree diamond insert is designed primarily for heavy turning and facing operations, using the 100-degree corners, and for turning and back-facing square shoulders using the 80-degree corners. The 55- and 35-degree diamond inserts are intended primarily for tracing.

Lead Angle: Tool holders should be selected to provide the largest possible lead angle, although limitations are sometimes imposed by the nature of the job. For example, when turning and back-facing a shoulder, a negative lead angle must be used. Slender or small-diameter parts may deflect, causing difficulties in holding size, or chatter when the lead angle is too large.

End Cutting Edge Angle: When tracing or contour turning, the plunge angle is determined by the end cutting edge angle. A 2-deg minimum clearance angle should be provided between the workpiece surface and the end cutting edge of the insert. Table 3A provides the maximum plunge angle for holders commonly used to plunge when tracing where insert shape identifiers are S = square; T = triangle; D = 55-deg diamond, V = 35-deg diamond. When severe cratering cannot be avoided, an insert having a small, end cutting edge angle is desirable to delay the crater breakthrough behind the nose. For very heavy cuts a small, end cutting edge angle will strengthen the corner of the tool. Tool holders for numerical control machines are discussed in the NC section.

Table 3A. Maximum Plunge Angle for Tracing or Contour Turning

Tool Holder Style	Insert Shape	Maximum Plunge Angle	Tool Holder Style	Insert Shape	Maximum Plunge Angle
E	T	58°	J	D	30°
D and S	S	43°	J	V	50°
H	D	71°	N	T	55°
J	T	25°	N	D	58°–60°

Table 3B. Indexable Insert Holder Application Guide

Tool	Tool Holder Style	Insert Shape	Rake N-Negative P-Positive	Turn	Face	Turn and Face	Turn and Backface	Trace	Groove	Chamfer	Bore	Plane
0°	A	T	N	•	•						•	
			P	•	•						•	
0°	A	T	N	•	•			•				
			P	•	•			•				
	A	R	N	•	•	•						•
	A	R	N	•	•	•		•				•
15°	B	T	N	•	•						•	
			P	•	•						•	
15°	B	T	N	•	•			•			•	
			P	•	•			•			•	
15°	B	S	N	•	•						•	
			P	•	•						•	
5° 15°	B	C	N	•	•	•					•	•
	C	T	N	•	•				•	•		
			P	•	•				•	•		
45°	D	S	N	•	•	•		•		•	•	•
			P	•	•	•		•		•	•	•
30°	E	T	N	•	•			•	•	•		
			P	•	•			•	•	•		
	F	T	N	•	•						•	
			P	•	•						•	
0°	G	T	N	•	•						•	
			P	•	•						•	

Table 3B (*Concluded*). **Indexable Insert Holder Application Guide**

Tool	Tool Holder Style	Insert Shape	Rake N-Negative P-Positive	Application								
				Turn	Face	Turn and Face	Turn and Backface	Trace	Groove	Chamfer	Bore	Plane
	G	R	N	•	•	•						
	G	C	N	•	•	•						
			P	•	•	•						
	H	D	N	•	•			•				
	J	T	N				•	•				
			P				•	•				
	J	D	N				•	•				
	J	V	N				•	•				
	K	S	N	•	•						•	
			P	•	•						•	
	K	C	N	•	•						•	
	L	C	N			•	•					
	N	T	N	•	•			•				
			P	•	•			•				
	N	D	N	•	•			•				
	S	S	N	•	•	•		•		•	•	•
			P	•	•	•		•		•	•	•
	W	S	N	•	•							

Sintered Carbide Blanks and Cutting Tools. — As shown in Table 4, American National Standard ANSI B212.1-1984 provides standard sizes and designations for eight styles of sintered carbide blanks. These blanks are the unground solid carbide from which either solid or tipped cutting tools are made. Tipped cutting tools are made by brazing a blank onto a shank to produce the cutting tool; these tools differ from carbide *insert* cutting tools which consist of a carbide insert held mechanically in a tool holder. A typical single-point carbide-tipped cutting tool is shown in the diagram on page 733.

Table 4. American National Standard Sizes and Designations for Carbide Blanks* (ANSI B212.1-1984)

T	W	L	Style 1000	Style 2000
			Blank Designation	
1/16	1/8	5/8	1010	2010
1/16	5/32	1/4	1015	2015
1/16	3/16	1/4	1020	2020
1/16	1/4	1/4	1025	2025
1/16	1/4	5/16	1030	2030
3/32	1/8	3/4	1035	2035
3/32	3/16	5/16	1040	2040
3/32	3/16	1/2	1050	2050
3/32	1/4	3/8	1060	2060
3/32	1/4	1/2	1070	2070
3/32	5/16	3/8	1080	2080
3/32	3/8	3/8	1090	2090
3/32	3/8	1/2	1100	2100
3/32	7/16	1/2	1105	2105
1/8	3/16	3/4	1110	2110
1/8	1/4	1/2	1120	2120
1/8	1/4	5/8	1130	2130
1/8	1/4	3/4	1140	2140
1/8	5/16	7/16	1150	2150
1/8	5/16	1/2	1160	2160
1/8	5/16	5/8	1170	2170
1/8	3/8	1/2	1180	2180
1/8	3/8	3/4	1190	2190
1/8	1/2	1/2	1200	2200
1/8	1/2	3/4	1210	2210
1/8	3/4	3/4	1215	2215
5/32	3/8	9/16	1220	2220
5/32	3/8	3/4	1230	2230
5/32	5/8	5/8	1240	2240
3/16	5/16	7/16	1250	2250
3/16	5/16	5/8	1260	2260
3/16	3/8	1/2	1270	2270
3/16	3/8	5/8	1280	2280
3/16	3/8	3/4	1290	2290
3/16	7/16	5/8	1300	2300
3/16	7/16	13/16	1310	2310
3/16	1/2	1/2	1320	2320
3/16	1/2	3/4	1330	2330
3/16	3/4	3/4	1340	2340

T	W	L	Style 0000	Style 1000	Style 3000	Style 4000
			Blank Designation			
1/4	3/8	9/16	0350	1350	3350	4350
1/4	3/8	3/4	0360	1360	3360	4360
1/4	7/16	5/8	0370	1370	3370	4370
1/4	1/2	3/4	0380	1380	3380	4380
1/4	9/16	1	0390	1390	3390	4390
1/4	5/8	5/8	0400	1400	3400	4400
1/4	3/4	3/4	0405	1405	3405	4405
1/4	3/4	1	0410	1410	3410	4410
1/4	1	1	0415	1415	3415	4415
5/16	7/16	5/8	0420	1420	3420	4420
5/16	7/16	15/16	0430	1430	3430	4430
5/16	1/2	3/4	0440	1440	3440	4440
5/16	1/2	1	0450	1450	3450	4450
5/16	5/8	1	0460	1460	3460	4460
5/16	3/4	3/4	0470	1470	3470	4470
5/16	3/4	1	0475	1475	3475	4475
5/16	3/4	1 1/4	0480	1480	3480	4480
3/8	1/2	3/4	0490	1490	3490	4490
3/8	1/2	1	0500	1500	3500	4500
3/8	5/8	1	0510	1510	3510	4510
3/8	5/8	1 1/4	0515	1515	3515	4515
3/8	3/4	1 1/4	0520	1520	3520	4520
3/8	3/4	1 1/2	0525	1525	3525	4525
1/2	3/4	1	0530	1530	3530	4530
1/2	3/4	1 1/4	0540	1540	3540	4540
1/2	3/4	1 1/2	0550	1550	3550	4550

T	W	L	F	Style 5000	Style 6000	Style 7000
1/16	1/4	5/16	...	5030	...	...
3/32	1/4	3/8	1/16	...	...	7060
3/32	5/16	3/8		5080	6080	...
3/32	3/8	1/2		5100	6100	...
3/32	7/16	1/2		5105	...	...
1/8	5/16	5/8	3/32	...	...	7170
1/8	1/2	1/2		5200	6200	...
5/32	3/8	3/4	1/8	...	...	7230
5/32	3/8	5/8		5240	6240	...
3/16	3/4	3/4		5340	6340	...
1/4	1	3/4		5410	...	...

* See diagram on page 733.

All dimensions are in inches.

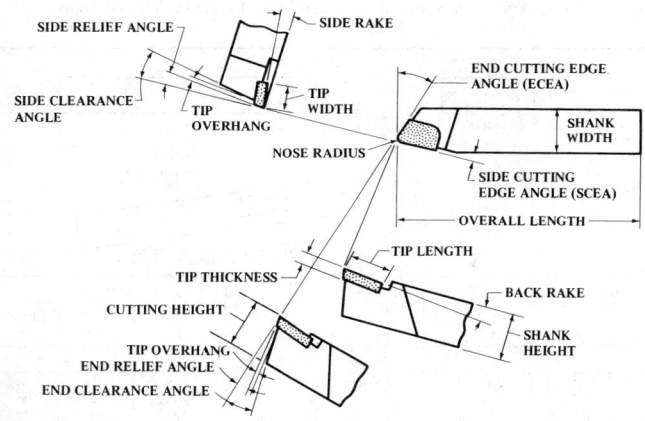

A typical single-point carbide tipped cutting tool. The side rake, side relief, and the clearance angles are *normal* to the side-cutting edge, rather than the shank, to facilitate its being ground on a tilting-table grinder. The end-relief and clearance angles are *normal* to the end-cutting edge. The back-rake angle is parallel to the side-cutting edge.

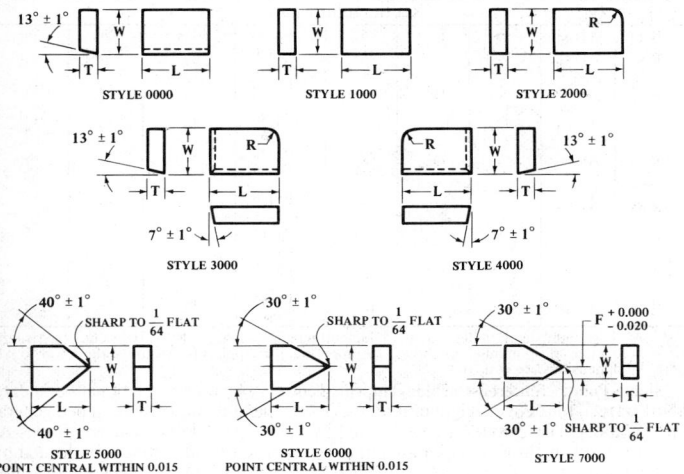

Eight styles of sintered carbide blanks. Standard dimensions for these blanks are given in Table 4.

Table 5. American National Standard Style A Carbide Tipped Tools
(ANSI B212.1-1984)

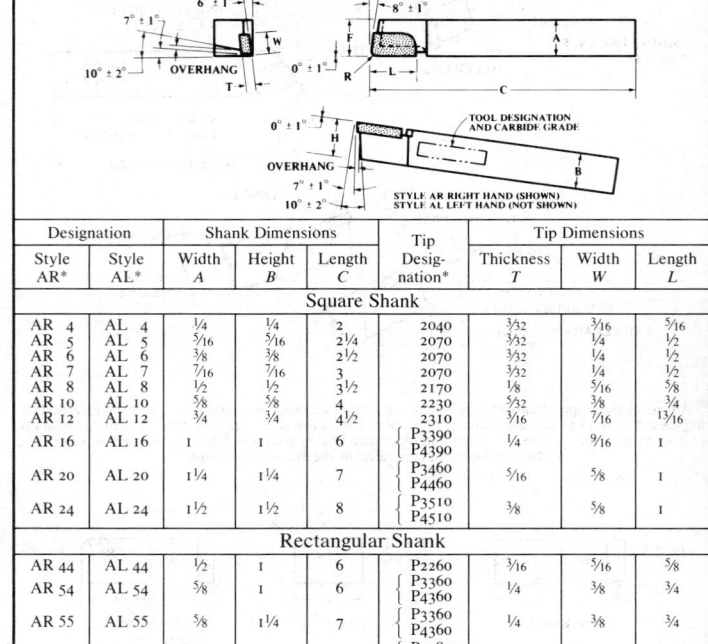

Designation		Shank Dimensions			Tip Desig-nation*	Tip Dimensions		
Style AR*	Style AL*	Width A	Height B	Length C		Thickness T	Width W	Length L
Square Shank								
AR 4	AL 4	1/4	1/4	2	2040	3/32	3/16	5/16
AR 5	AL 5	5/16	5/16	2 1/4	2070	3/32	1/4	1/2
AR 6	AL 6	3/8	3/8	2 1/2	2070	3/32	1/4	1/2
AR 7	AL 7	7/16	7/16	3	2070	3/32	1/4	1/2
AR 8	AL 8	1/2	1/2	3 1/2	2170	1/8	5/16	5/8
AR 10	AL 10	5/8	5/8	4	2230	5/32	3/8	3/4
AR 12	AL 12	3/4	3/4	4 1/2	2310	3/16	7/16	13/16
AR 16	AL 16	1	1	6	{ P3390 { P4390	1/4	9/16	1
AR 20	AL 20	1 1/4	1 1/4	7	{ P3460 { P4460	5/16	5/8	1
AR 24	AL 24	1 1/2	1 1/2	8	{ P3510 { P4510	3/8	5/8	1
Rectangular Shank								
AR 44	AL 44	1/2	1	6	P2260	3/16	5/16	5/8
AR 54	AL 54	5/8	1	6	{ P3360 { P4360	1/4	3/8	3/4
AR 55	AL 55	5/8	1 1/4	7	{ P3360 { P4360	1/4	3/8	3/4
AR 64	AL 64	3/4	1	6	{ P3380 { P4380	1/4	1/2	3/4
AR 66	AL 66	3/4	1 1/2	8	{ P3430 { P4430	5/16	7/16	15/16
AR 85	AL 85	1	1 1/4	7	{ P3460 { P4460	5/16	5/8	1
AR 86	AL 86	1	1 1/2	8	{ P3510 { P4510	3/8	5/8	1
AR 88	AL 88	1	2	10	{ P3510 { P4510	3/8	5/8	1
AR 90	AL 90	1 1/2	2	10	{ P3540 { P4540	1/2	3/4	1 1/4

* "A" is straight shank, o deg., SCEA (side-cutting-edge angle). '"R" is right-cut. "L" is left-cut. Where a pair of tip numbers is shown, the upper number applies to AR tools, the lower to AL tools. All dimensions are in inches.

Single-Point, Sintered-Carbide-Tipped Tools. — American National Standard ANSI B212.1-1984 covers eight different styles of single-point, carbide-tipped general purpose tools. These styles are designated by the letters A to G inclusive. Styles A, B, F, G, and E with offset point are either right- or left-hand cutting as indicated by the letters R or L. Dimensions of tips and shanks are given in Tables 5 to 11. For dimensions and tolerances not shown, see the Standard. Also see the Standard for the identification system, dimensions, and tolerances of sintered carbide boring tools.

Table 6. American National Standard Style B Carbide Tipped Tools with 15-degree Side-cutting-edge Angle (ANSI B212.1-1984)

Designation		Shank Dimensions			Tip Desig- nation*	Tip Dimensions		
Style BR	Style BL	Width A	Height B	Length C		Thickness T	Width W	Length L
Square Shank								
BR 4	BL 4	1/4	1/4	2	2015	1/16	5/32	1/4
BR 5	BL 5	5/16	5/16	2 1/4	2040	3/32	3/16	5/16
BR 6	BL 6	3/8	3/8	2 1/2	2070	3/32	1/4	1/2
BR 7	BL 7	7/16	7/16	3	2070	3/32	1/4	1/2
BR 8	BL 8	1/2	1/2	3 1/2	2070	3/32	5/16	5/8
BR 10	BL 10	5/8	5/8	4	2170	1/8	5/16	5/8
BR 12	BL 12	3/4	3/4	4 1/2	2230	5/32	3/8	3/4
					2310	3/16	7/16	13/16
BR 16	BL 16	1	1	6	3390 / 4390	1/4	9/16	1
BR 20	BL 20	1 1/4	1 1/4	7	3460 / 4460	5/16	5/8	1
BR 24	BL 24	1 1/2	1 1/2	8	3510 / 4510	3/8	5/8	1
Rectangular Shank								
BR 44	BL 44	1/2	1	6	2260	3/16	5/16	5/8
BR 54	BL 54	5/8	1	6	3360 / 4360	1/4	3/8	3/4
BR 55	BL 55	5/8	1 1/4	7	3360 / 4360	1/4	3/8	3/4
BR 64	BL 64	3/4	1	6	3380 / 4380	1/4	1/2	3/4
BR 66	BL 66	3/4	1 1/2	8	3430 / 4430	5/16	7/16	15/16
BR 85	BL 85	1	1 1/4	7	3460 / 4460	5/16	5/8	1
BR 86	BL 86	1	1 1/2	8	3510 / 4510	3/8	5/8	1
BR 88	BL 88	1	2	10	3510 / 4510	3/8	5/8	1
BR 90	BL 90	1 1/2	2	10	3540 / 4540	1/2	3/4	1 1/4

* Where a pair of tip numbers is shown, the upper number applies to BR tools, the lower to BL tools. All dimensions are in inches.

A number follows the letters of the tool style and hand designation and for square shank tools, represents the number of sixteenths of an inch of width, W, and height, H. With rectangular shanks, the first digit of the number indicates the number of eighths of an inch in the shank width, W, and the second digit the number of quarters of an inch in the shank height, H. One exception is the 1 1/2 × 2-inch size which has been arbitrarily assigned the number 90.

Table 7. American National Standard Style C Carbide Tipped Tools
(ANSI B212.1-1984)

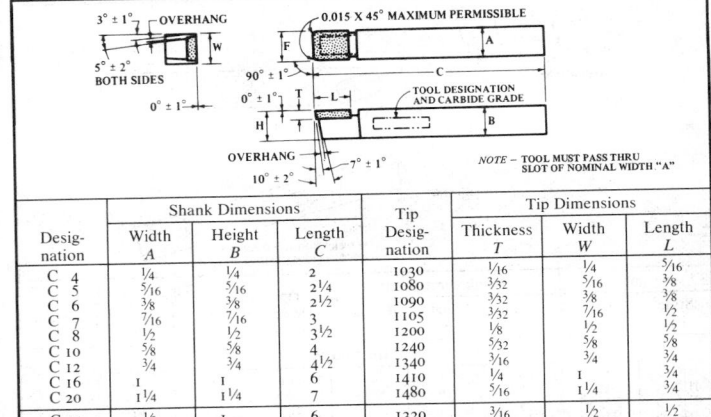

Desig-nation	Shank Dimensions			Tip Desig-nation	Tip Dimensions		
	Width A	Height B	Length C		Thickness T	Width W	Length L
C 4	¼	¼	2	1030	¹⁄₁₆	¼	⁵⁄₁₆
C 5	⁵⁄₁₆	⁵⁄₁₆	2¼	1080	³⁄₃₂	⁵⁄₁₆	³⁄₈
C 6	³⁄₈	³⁄₈	2½	1090	³⁄₃₂	³⁄₈	³⁄₈
C 7	⁷⁄₁₆	⁷⁄₁₆	3	1105	³⁄₃₂	⁷⁄₁₆	½
C 8	½	½	3½	1200	⅛	½	½
C 10	⅝	⅝	4	1240	⁵⁄₃₂	⅝	⅝
C 12	¾	¾	4½	1340	³⁄₁₆	¾	¾
C 16	1	1	6	1410	¼	1	¾
C 20	1¼	1¼	7	1480	⁵⁄₁₆	1¼	¾
C 44	½	1	6	1320	³⁄₁₆	½	½
C 54	⅝	1	6	1400	¼	⅝	⅝
C 55	⅝	1¼	7	1400	¼	⅝	⅝
C 64	¾	1	6	1405	¼	¾	¾
C 66	¾	1½	8	1470	⁵⁄₁₆	¾	¾
C 86	1	1½	8	1475	⁵⁄₁₆	1	¾

All dimensions are in inches.   Square shanks above horizontal line; rectangular below.

Table 8. American National Standard Style D, 80-degree Nose-angle
Carbide Tipped Tools (ANSI B212.1-1984)

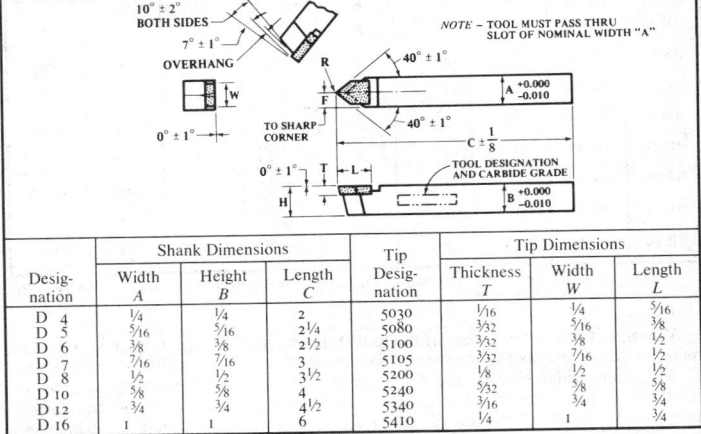

Desig-nation	Shank Dimensions			Tip Desig-nation	Tip Dimensions		
	Width A	Height B	Length C		Thickness T	Width W	Length L
D 4	¼	¼	2	5030	¹⁄₁₆	¼	⁵⁄₁₆
D 5	⁵⁄₁₆	⁵⁄₁₆	2¼	5080	³⁄₃₂	⁵⁄₁₆	³⁄₈
D 6	³⁄₈	³⁄₈	2½	5100	³⁄₃₂	³⁄₈	½
D 7	⁷⁄₁₆	⁷⁄₁₆	3	5105	³⁄₃₂	⁷⁄₁₆	½
D 8	½	½	3½	5200	⅛	½	½
D 10	⅝	⅝	4	5240	⁵⁄₃₂	⅝	⅝
D 12	¾	¾	4½	5340	³⁄₁₆	¾	¾
D 16	1	1	6	5410	¼	1	¾

All dimensions are in inches.

**Table 9a. American National Standard Style E, 60-degree Nose-angle, Carbide
Tipped Tools** (ANSI B212.1-1984)

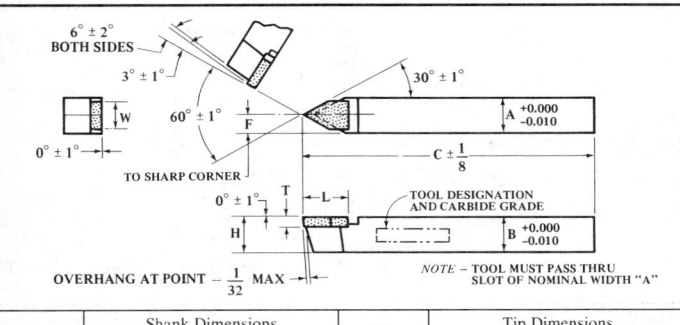

Desig- nation	Shank Dimensions			Tip Desig- nation	Tip Dimensions		
	Width A	Height B	Length C		Thickness T	Width W	Length L
E 4	1/4	1/4	2	6030	1/16	1/4	5/16
E 5	5/16	5/16	2 1/4	6080	3/32	5/16	3/8
E 6	3/8	3/8	2 1/2	6100	3/32	3/8	1/2
E 8	1/2	1/2	3 1/2	6200	1/8	1/2	1/2
E 10	5/8	5/8	4	6240	5/32	5/8	5/8
E 12	3/4	3/4	4 1/2	6340	3/16	3/4	3/4

All dimensions are in inches.

**Table 9b. American National Standard Styles ER and EL, 60-degree Nose-angle,
Carbide Tipped Tools with Offset Point** (ANSI B212.1-1984)

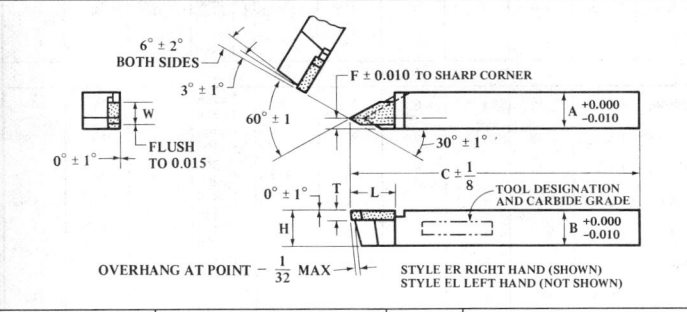

Designation		Shank Dimensions			Tip Desig- nation	Tip Dimensions		
Style ER	Style EL	Width A	Height B	Length C		Thick. T	Width W	Length L
ER 4	EL 4	1/4	1/4	2	1020	1/16	3/16	1/4
ER 5	EL 5	5/16	5/16	2 1/4	7060	3/32	1/4	3/8
ER 6	EL 6	3/8	3/8	2 1/2	7060	3/32	1/4	3/8
ER 8	EL 8	1/2	1/2	3 1/2	7170	1/8	5/16	5/8
ER 10	EL 10	5/8	5/8	4	7170	1/8	5/16	5/8
ER 12	EL 12	3/4	3/4	4 1/2	7230	5/32	3/8	3/4

All dimensions are in inches.

Table 10. American National Standard Style F, Offset, End-cutting Carbide Tipped Tools (ANSI B212.1-1984)

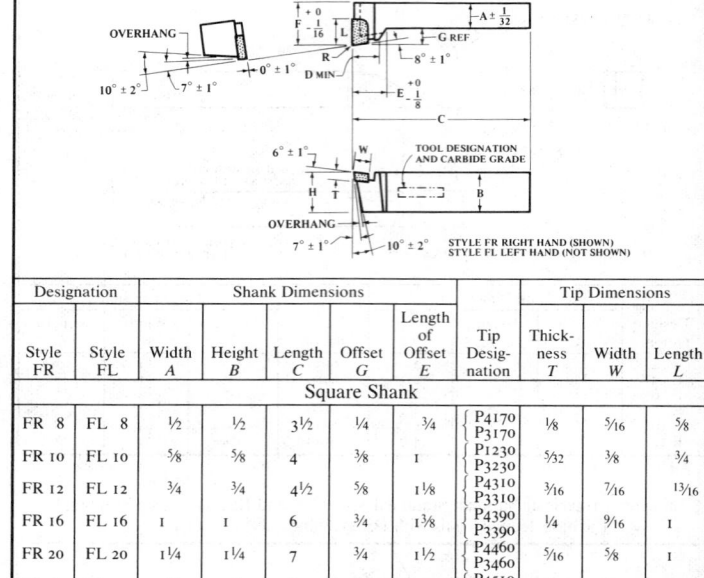

Designation		Shank Dimensions					Tip Dimensions			
Style FR	Style FL	Width A	Height B	Length C	Offset G	Length of Offset E	Tip Designation	Thickness T	Width W	Length L
Square Shank										
FR 8	FL 8	½	½	3½	¼	¾	P4170 / P3170	⅛	5/16	⅝
FR 10	FL 10	⅝	⅝	4	⅜	1	P1230 / P3230	5/32	⅜	¾
FR 12	FL 12	¾	¾	4½	⅝	1⅛	P4310 / P3310	3/16	7/16	13/16
FR 16	FL 16	1	1	6	¾	1⅜	P4390 / P3390	¼	9/16	1
FR 20	FL 20	1¼	1¼	7	¾	1½	P4460 / P3460	5/16	⅝	1
FR 24	FL 24	1½	1½	8	¾	1½	P4510 / P3510	⅜	⅝	1
Rectangular Shank										
FR 44	FL 44	½	1	6	½	⅞	P4260 / P1260	3/16	5/16	⅝
FR 55	FL 55	⅝	1¼	7	⅝	1⅛	P4360 / P3360	¼	⅜	¾
FR 64	FL 64	¾	1	6	⅝	1 3/16	P4380 / P3380	¼	½	¾
FR 66	FL 66	¾	1½	8	¾	1¼	P4430 / P3430	5/16	7/16	15/16
FR 85	FL 85	1	1¼	7	¾	1½	P4460 / P3460	5/16	⅝	1
FR 86	FL 86	1	1½	8	¾	1½	P4510 / P3510	⅜	⅝	1
FR 90	FL 90	1½	2	10	¾	1⅝	P4540 / P3540	½	¾	1¼

All dimensions are in inches. Where a pair of tip numbers is shown, the upper number applies to FR tools, the lower number to FL tools.

Single-point Tool Nose Radii. — The tool nose radii recommended in the American National Standard are as follows: For square-shank tools up to and including ⅜-inch square tools, 1/64 inch; for those over ⅜-inch square through 1¼-inches square, 1/32 inch; and for those above 1¼-inches square, 1/16 inch. For rectangular-shank tools with shank section of ½ × 1 inch through 1 × 1½ inches, the nose radii are 1/32 inch, and for 1 × 2 and 1½ × 2 inch shanks, the nose radius is 1/16 inch.

Single-point Tool Angle Tolerances. — The tool angles shown on the diagrams

Table 11. American National Standard Style G, Offset, Side-cutting, Carbide Tipped Tools (ANSI B212.1-1984)

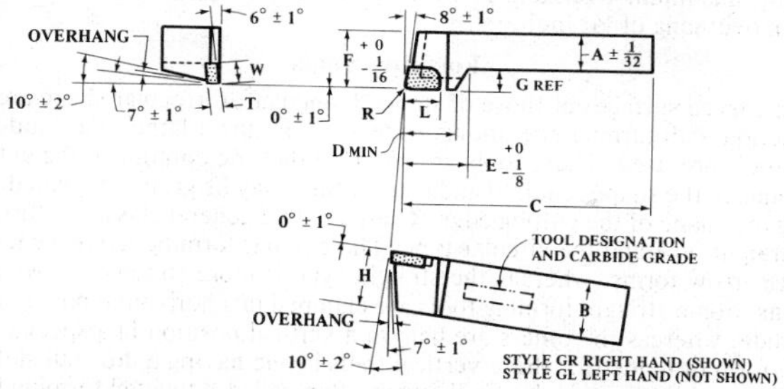

Designation		Shank Dimensions					Tip Dimensions			
Style GR	Style GL	Width A	Height B	Length C	Offset G	Length of Offset E	Tip Designation	Thickness T	Width W	Length L
Square Shank										
GR 8	GL 8	½	½	3½	¼	1 1/16	P3170 P4170	⅛	5/16	⅝
GR 10	GL 10	⅝	⅝	4	⅜	1⅜	P3230 P4230	5/32	⅜	¾
GR 12	GL 12	¾	¾	4½	⅜	1½	P3310 P2310	3/16	7/16	13/16
GR 16	GL 16	1	1	6	½	1 11/16	P3390 P4390	¼	9/16	1
GR 20	GL 20	1¼	1¼	7	¾	1 13/16	P3460 P4460	5/16	⅝	1
GR 24	GL 24	1½	1½	8	¾	1 13/16	P3510 P4510	⅜	⅝	1
Rectangular Shank										
GR 44	GL 44	½	1	6	¼	1 1/16	P3260 P4260	3/16	5/16	⅝
GR 55	GL 55	⅝	1¼	7	⅜	1⅜	P3360 P4360	¼	⅜	¾
GR 64	GL 64	¾	1	6	½	1 7/16	P3380 P4380	¼	½	¾
GR 66	GL 66	¾	1½	8	½	1⅝	P3430 P4430	5/16	7/16	15/16
GR 85	GL 85	1	1¼	7	½	1 11/16	P3460 P4460	5/16	⅝	1
GR 86	GL 86	1	1½	8	½	1 11/16	P3510 P4510	⅜	⅝	1
GR 90	GL 90	1½	2	10	¾	2 1/16	P3540 P4540	½	¾	1¼

All dimensions are in inches. Where a pair of tip numbers is shown, the upper number applies to GR tools, the lower number to GL tools.

in the Tables 5 through 11 are general recommendations. Tolerances applicable to these angles are ± 1 degree on all angles except end and side clearance angles; for these the tolerance is ± 2 degrees.

Single-point Tool Tip Overhang. — As shown on the diagram, page 733, the tip of the brazed carbide blank overhangs the shank of the tool by a small amount. The amount of overhang is either 1/32 or 1/16 inch depending on the size of the tool.

For tools shown in Tables 5, 6, 7, 8, 10, and 11, the maximum overhang is ¹⁄₃₂ inch for shank size numbers 4, 5, 6, 7, 8, 10, 12, and 44; for other shank sizes in these tables, the maximum overhang is ¹⁄₁₆ inch. In Tables 9a and 9b all tools have a maximum overhang of ¹⁄₃₂ inch.

Forming Tools

When curved surfaces or those of stepped, angular or irregular shape are required in connection with turning operations, especially on turret lathes and "automatics", forming tools are used. These tools are so made that the contour of the cutting edge corresponds to the shape required and usually they may be ground repeatedly without changing the shape of the cutting edge. There are two general classes of forming tools — the straight type and the circular type. The circular forming tool is generally used on small narrow forms, whereas the straight type is more suitable for wide forming operations. Some straight forming tools are clamped in a horizontal position upon the cut-off slide, whereas the others are held in a vertical position in a special holder. A common form of holder for these vertical tools is one having a dovetail slot in which the forming tool is clamped; hence they are often called "dovetail forming tools." In many cases, two forming tools are used, especially when a very smooth surface is required, one being employed for roughing and the other for finishing.

There was an American standard for forming tool blanks. This former standard (to be presented later in this section) covers both straight or dovetailed and the circular forms. Dimensions of the finished blanks are given excepting, of course, the formed part which must be shaped to suit whatever job the tool is to be used for. This former standard includes the important dimensions of holders for both straight and circular tools.

Dimensions of Steps on Straight or Dovetail Forming Tools. — The diagrams at the top of the accompanying table, illustrate a straight or "dovetail" forming tool. The upper or cutting face lies in the same plane as the center of the work and there is no rake. (Many forming tools have rake to increase the cutting efficiency, and this type will be referred to later.) In making a forming tool, the various steps measured perpendicular to the front face (as at d) must be proportioned so as to obtain the required radial dimensions on the work. For example, if D equals the difference between two radial dimensions on the work, then:

$$\text{Step } d = D \times \text{cosine front clearance angle.}$$

Angles on Straight Forming Tools. — In making forming tools to the required shape or contour, any angular surfaces (like the steps referred to in the previous paragraph) are affected by the clearance angle. For example, assume that angle A on the work (see diagram at top of accompanying table) is 20 degrees. The angle on the tool in plane x-x, in that case, will be slightly less than 20 degrees. In making the tool, this modified or reduced angle is required because of the convenience in machining and measuring the angle square to the front face of the tool or in the plane x-x.

If the angle on the work is measured from a line parallel to the axis (as at A in diagram), then the reduced angle on the tool as measured square to the front face (or in plane x-x) is found as follows:

$$\text{tan reduced angle on tool} = \text{tan } A \times \text{cos front clearance angle.}$$

If angle A on the work is larger than, say, 45 degrees, it may be given on the drawing as indicated at B. In this case, the angle is measured from a plane perpendicular to the axis of the work. When the angle is so specified, the angle on the tool in plane x-x may be found as follows:

$$\text{tan reduced angle on tool} = \frac{\text{tan } B}{\text{cos clearance angle}}$$

Dimensions of Steps and Angles on Straight Forming Tools.

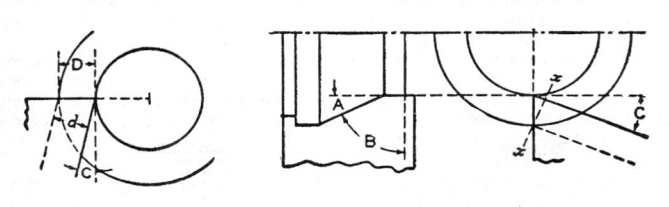

Upper section of table gives depth d of step on forming tool for a given dimension D which equals the actual depth of the step on the work, measured radially and along the cutting face of the tool (See diagram at left). First, locate depth D required on work; then find depth d on tool under tool clearance angle C. Depth d is measured perpendicular to front face of tool.

Radial Depth of Step D	Depth d of step on tool			Radial Depth of Step D	Depth d of step on tool		
	When $C = 10°$	When $C = 15°$	When $C = 20°$		When $C = 10°$	When $C = 15°$	When $C = 20°$
0.001	0.00098	0.00096	0.00094	0.040	0.03939	0.03863	0.03758
0.002	0.00197	0.00193	0.00187	0.050	0.04924	0.04829	0.04698
0.003	0.00295	0.00289	0.00281	0.060	0.05908	0.05795	0.05638
0.004	0.00393	0.00386	0.00375	0.070	0.06893	0.06761	0.06577
0.005	0.00492	0.00483	0.00469	0.080	0.07878	0.07727	0.07517
0.006	0.00590	0.00579	0.00563	0.090	0.08863	0.08693	0.08457
0.007	0.00689	0.00676	0.00657	0.100	0.09848	0.09659	0.09396
0.008	0.00787	0.00772	0.00751	0.200	0.19696	0.19318	0.18793
0.009	0.00886	0.00869	0.00845	0.300	0.29544	0.28977	0.28190
0.010	0.00984	0.00965	0.00939	0.400	0.39392	0.38637	0.37587
0.020	0.01969	0.01931	0.01879	0.500	0.49240	0.48296	0.46984
0.030	0.02954	0.02897	0.02819	....	....	....	....

Section of table below gives angles as measured in plane x-x perpendicular to front face of forming tool (See diagram on right). Find in first column the angle A required on work; then find reduced angle in plane x-x under given clearance angle C.

Angle A in Plane of Tool Cutting Face	Angle on tool in plane x-x			Angle A in Plane of Tool Cutting Face	Angle on tool in plane x-x		
	When $C = 10°$	When $C = 15°$	When $C = 20°$		When $C = 10°$	When $C = 15°$	When $C = 20°$
5°	4° 55′	4° 50′	4° 42′	50°	49° 34′	49° 1′	48° 14′
10	9 51	9 40	9 24	55	54 35	54 4	53 18
15	14 47	14 31	14 8	60	59 37	59 8	58 26
20	19 43	19 22	18 53	65	64 40	64 14	63 36
25	24 40	24 15	23 40	70	69 43	69 21	68 50
30	29 37	29 9	28 29	75	74 47	74 30	74 5
35	34 35	34 4	33 20	80	79 51	79 39	79 22
40	39 34	39 1	38 15	85	84 55	84 49	84 41
45	44 34	44 0	43 13	....	....	....	....

Table Giving Step Dimensions and Angles on Straight or Dovetailed Forming Tools. — The accompanying table "Dimensions of Steps and Angles on Straight Forming Tools" gives the required dimensions and angles within its range, direct or without calculation.

To Find Dimension of Step: The upper section of the table is used in determining the dimensions of steps. The radial depth of the step or the actual cutting depth D (see left-hand diagram) is given in the first column of the table. The columns which follow give the corresponding depths d for a front clearance angle of either

10, 15 or 20 degrees, as the case may be. To illustrate the use of the table, suppose a tool is required for turning the part shown in Fig. 1 which has diameters of 0.75, 1.25 and 1.75 inch, respectively. The difference between the largest and the smallest radius is 0.5 inch, which is the depth of one step. Assume that the clearance angle is 15 degrees. First, locate 0.5 in the column headed "Radial Depth of Step D"; then find depth d in the column headed "when $C = 15°$." As will be seen, this depth is 0.48296 inch. Prac-

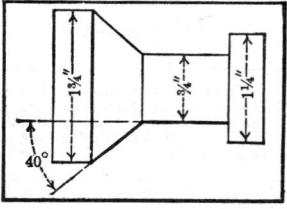

Fig. 1

tically the same procedure is followed in determining the depth of the second step on the tool. The difference in the radii in this case equals 0.25. Since this value is not given directly in the table, we first find the depth equivalent to 0.200 and then add to it the depth equivalent to 0.050. Thus, we have in this case 0.19138 + 0.04829 = 0.24147. In using this table, it is assumed that the top face of the tool is set at the height of the axis of the work.

To Find Angle: The lower section of the table applies to angles when they are measured relative to the axis of the work. The application of the table will again be illustrated by using the part shown in Fig. 1. The angle in this case is 40 degrees (which is also the angle in the plane of the cutting face of the tool). If the clearance angle is 15 degrees, then the angle measured in plane x-x square to the face of the tool is shown by the table to be 39° 1′ — a reduction of practically one degree.

Straight Forming Tools With Rake. — If a straight forming tool has rake, the depth x of each step (see Fig. 2), measured perpendicular to the front or clearance face, is affected not only by the clearance angle but by the rake angle F and the

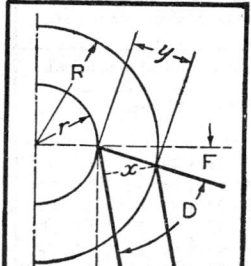

radii R and r of the steps on the work. First it is necessary to find three angles. These angles, which will be designated as A, B and C, are not shown on the drawing.

Angle $A = 180° -$ rake angle F; $\sin B = \dfrac{r \sin A}{R}$;

angle $C = 180° - (A + B)$

$y = \dfrac{R \sin C}{\sin A}$; angle D of tool $= 90° - (E + F)$;

depth $x = y \sin D$

If the work has two or more shoulders, the depth x for other steps on the tool may be determined for each radius r. If the work has curved or angular forms, it is more practical to use a tool without rake because its profile, in the plane of the cutting face, duplicates that of the work.

Fig. 2

Example: Assume that radius R equals 0.625 and radius r equals 0.375 inch so that the step on the work has a radial depth of 0.25 inch. The tool has a rake angle F of 10 degrees and a clearance angle E of 15 degrees. Then angle $A = 180 - 10 = 170$ degrees.

$$\sin B = \frac{0.375 \times 0.17365}{0.625} = 0.10419$$

Angle $B = 5°\ 59'$ nearly. Angle $C = 180 - (170° + 5°\ 59') = 4°\ 1'$

$$\text{Dimension } y = \frac{0.625 \times 0.07005}{0.17365} = 0.25212$$

Angle $D = 90° - (15 + 10) = 65$ degrees.

Depth x of step $= 0.25212 \times 0.90631 = 0.2285$ inch.

Circular Forming Tools. — To provide sufficient peripheral clearance on circular forming tools, the cutting face is off-set with relation to the center of the tool a distance C as shown in Fig. 3. Whenever a circular tool has two or more diameters,

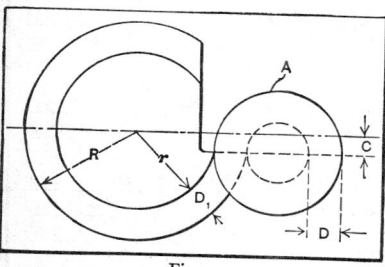

the difference in the radii of the steps on the tool will, therefore, not correspond exactly to the difference in the steps on the work. The form produced with the tool also changes, although the change is very slight, unless the amount of off-set C is considerable. Assume that a circular tool is required to produce the piece A having two diameters as shown. If the difference D_1 between the large and small radii of the tool were made equal to dimension D required on the work, D would be a certain amount over-size,

Fig. 3

depending upon the off-set C of the cutting edge. The following formulas can be used to determine the radii of circular forming tools for turning parts to different diameters:

Let R = largest radius of tool in inches; D = difference in radii of steps on work; C = amount cutting edge is off-set from center of tool; r = required radius in inches; then:

$$r = \sqrt{(\sqrt{R^2 - C^2} - D)^2 + C^2}. \qquad (1)$$

If the small radius r is given and the large radius R is required, then:

$$R = \sqrt{(\sqrt{r^2 - C^2} + D)^2 + C^2}. \qquad (2)$$

To illustrate, if D (Fig. 3) is to be 1/8 inch, the large radius R is 1 1/8 inch, and C is 5/32 inch, what radius r would be required to compensate for the off-set C of the cutting edge? Inserting these values in Formula (1):

$$r = \sqrt{\left(\sqrt{(1\tfrac{1}{8})^2 - (\tfrac{5}{32})^2} - \tfrac{1}{8}\right)^2 + (\tfrac{5}{32})^2} = 1.0014 \text{ inch}.$$

The value of r is thus found to be 1.0014 inch; hence the diameter $= 2 \times 1.0014 = 2.0028$ inches instead of 2 inches, as would have been the case if the cutting edge had been exactly on the center-line. Formulas for circular tools used on different makes of screw machines can be simplified when the values R and C are constant for each size of machine. The accompanying table "Formulas for Circular Form-

Formulas for Circular Forming Tools
(For notation, see Fig. 3)

Make of Machine	Size of Machine	Radius R, Inches	Offset C, Inches	Radius r, Inches
Brown & Sharpe	No. 00	0.875	0.125	$r = \sqrt{(0.8660 - D)^2 + 0.0156}$
	No. 0	1.125	0.15625	$r = \sqrt{(1.1141 - D)^2 + 0.0244}$
	No. 2	1.50	0.250	$r = \sqrt{(1.4790 - D)^2 + 0.0625}$
	No. 6	2.00	0.3125	$r = \sqrt{(1.975 - D)^2 + 0.0976}$
Acme	No. 51	0.75	0.09375	$r = \sqrt{(0.7441 - D)^2 + 0.0088}$
	No. 515	0.75	0.09375	$r = \sqrt{(0.7441 - D)^2 + 0.0088}$
	No. 52	1.0	0.09375	$r = \sqrt{(0.9956 - D)^2 + 0.0088}$
	No. 53	1.1875	0.125	$r = \sqrt{(1.1809 - D)^2 + 0.0156}$
	No. 54	1.250	0.15625	$r = \sqrt{(1.2402 - D)^2 + 0.0244}$
	No. 55	1.250	0.15625	$r = \sqrt{(1.2402 - D)^2 + 0.0244}$
	No. 56	1.50	0.1875	$r = \sqrt{(1.4882 - D)^2 + 0.0352}$
Cleveland	¼″	0.625	0.03125	$r = \sqrt{(0.6242 - D)^2 + 0.0010}$
	⅜″	0.84375	0.0625	$r = \sqrt{(0.8414 - D)^2 + 0.0039}$
	⅝″	1.15625	0.0625	$r = \sqrt{(1.1546 - D)^2 + 0.0039}$
	⅞″	1.1875	0.0625	$r = \sqrt{(1.1859 - D)^2 + 0.0039}$
	1¼″	1.375	0.0625	$r = \sqrt{(1.3736 - D)^2 + 0.0039}$
	2″	1.375	0.0625	$r = \sqrt{(1.3736 - D)^2 + 0.0039}$
	2¼″	1.625	0.125	$r = \sqrt{(1.6202 - D)^2 + 0.0156}$
	2¾″	1.875	0.15625	$r = \sqrt{(1.8685 - D)^2 + 0.0244}$
	3¼″	1.875	0.15625	$r = \sqrt{(1.8685 - D)^2 + 0.0244}$
	4¼″	2.50	0.250	$r = \sqrt{(2.4875 - D)^2 + 0.0625}$
	6″	2.625	0.250	$r = \sqrt{(2.6131 - D)^2 + 0.0625}$

ing Tools" gives the standard values of R and C for circular tools used on different automatics. The formulas for determining the radius r (see column at right-hand side of table) contain a constant which represents the value of the expression $\sqrt{R^2 - C^2}$ in formula (1).

The table "Constants for Determining Diameters of Circular Forming Tools" has been compiled to facilitate proportioning tools of this type. It gives constants for computing the various diameters of forming tools, when the cutting face of the tool is either ⅛, 3/16, ¼ or 5/16 inch below the horizontal center-line. As there is no standard distance for the location of the cutting face, the table has been prepared to correspond with distances commonly used. As an example, suppose the tool is required for a part having three diameters of 1.75, 0.75 and 1.25 inch, respectively, as shown in Fig. 1, and that the largest diameter of the tool is 3 inches and its cutting face is ¼ inch below the horizontal center-line. The first step would be to determine approximately the respective diameters of the forming tool and then correct these diameters by the use of the table. To produce the three diameters shown in Fig. 1, with a 3-inch forming tool, the tool diameters would be approximately 2, 3 and 2.5 inches, respectively. The first dimension (2 inches) is 1 inch less in diameter than that of the tool, and the necessary correction should be given

in the column "Correction for Difference in Diameter"; but as the table is only extended to half-inch differences, it will be necessary to obtain this particular correction in two steps. On the line for 3-inch diameter and under corrections for ½ inch, we find 0.0085; then in line with 2½ and under the same heading, we find 0.0129; hence the total correction would be 0.0085 + 0.0129 = 0.0214 inch. This correction is added to the approximate diameter, making the exact diameter of the first step 2 + 0.0214 = 2.0214 inches. The next step would be computed in the same way, by noting on the 3-inch line the correction for ½ inch and adding it to the approximate diameter of the second step, giving an exact diameter of 2.5 + 0.0085 = 2.5085 inches. Therefore, to produce the part shown in Fig. 1, the tool should have three steps of 3, 2.0214 and 2.5085 inches, respectively, provided the cutting face is ¼ inch below the center. All diameters are computed in this way, from the largest diameter of the tool.

The tables "Corrected Diameters of Circular Forming Tools" are especially applicable to tools used on Brown & Sharpe automatic screw machines. Directions for using these tables are given at the end of Table 4.

Circular Tools Having Top Rake. — Circular forming tools without top rake are satisfactory for brass, but tools for steel or other tough metals cut better when there is a rake angle of 10 or 12 degrees. For such tools the small radius r (see Fig. 3) for an outside radius R may be found by the formula:

$$r = \sqrt{P^2 + R^2 - 2PR\cos\theta}$$

To find the value of P proceed as follows: sin ϕ = small rad. on work × sin rake angle ÷ large rad. on work. Angle β = rake angle − ϕ. P = large rad. on work × sin β ÷ sin rake angle. Angle θ = rake angle + δ. Sin δ = vertical height C from center of tool to center of work ÷ R. It is assumed that the point of tool is to be set at same height as the work center.

Dimensions for Circular Cut-off Tools

Diam. of Stock	Soft Brass, Copper		Norway Iron, Machine Steel		Drill Rod, Tool Steel	
	a = 23 Deg.		a = 15 Deg.		a = 12 Deg.	
	T	x	T	x	T	x
1/16	0.031	0.013	0.039	0.010	0.043	0.009
1/8	0.044	0.019	0.055	0.015	0.062	0.013
3/16	0.052	0.022	0.068	0.018	0.076	0.016
1/4	0.062	0.026	0.078	0.021	0.088	0.019
5/16	0.069	0.029	0.087	0.023	0.098	0.021
3/8	0.076	0.032	0.095	0.025	0.107	0.023
7/16	0.082	0.035	0.103	0.028	0.116	0.025
1/2	0.088	0.037	0.110	0.029	0.124	0.026
9/16	0.093	0.039	0.117	0.031	0.131	0.028
5/8	0.098	0.042	0.123	0.033	0.137	0.029
11/16	0.103	0.044	0.129	0.035	0.145	0.031
3/4	0.107	0.045	0.134	0.036	0.152	0.032
13/16	0.112	0.047	0.141	0.038	0.158	0.033
7/8	0.116	0.049	0.146	0.039	0.164	0.035
15/16	0.120	0.051	0.151	0.040	0.170	0.036
1	0.124	0.053	0.156	0.042	0.175	0.037

The length of the blade equals radius of stock $R + x + r + \frac{1}{32}$ inch (for notation see illustration above); r = ¹⁄₁₆ inch for ³⁄₈- to ¾-inch stock, and ³⁄₃₂ inch for ¾- to 1-inch stock.

Constants for Determining Diameters of Circular Forming Tools

Diam. of Tool	Radius of Tool	Cutting Face 1/8 Inch Below Center			Cutting Face 3/16 Inch Below Center			Cutting Face 1/4 Inch Below Center			Cutting Face 5/16 Inch Below Center		
		Correction for 1/8 Inch Difference in Diam.	Correction for 1/4 Inch Difference in Diam.	Correction for 1/2 Inch Difference in Diam.	Correction for 1/8 Inch Difference in Diam.	Correction for 1/4 Inch Difference in Diam.	Correction for 1/2 Inch Difference in Diam.	Correction for 1/8 Inch Difference in Diam.	Correction for 1/4 Inch Difference in Diam.	Correction for 1/2 Inch Difference in Diam.	Correction for 1/8 Inch Difference in Diam.	Correction for 1/4 Inch Difference in Diam.	Correction for 1/2 Inch Difference in Diam.
1	0.500												
1⅛	0.5625	0.0036			0.0086			0.0167			0.0298		
1¼	0.625	0.0028	0.0065		0.0067	0.0154		0.0128	0.0296		0.0221	0.0519	
1⅜	0.6875	0.0023			0.0054			0.0102			0.0172		
1½	0.750	0.0019	0.0042	0.0107	0.0045	0.0099	0.0253	0.0083	0.0185	0.0481	0.0138	0.0310	0.0829
1⅝	0.8125	0.0016			0.0037			0.0069			0.0114		
1¾	0.875	0.0014	0.0030		0.0032	0.0069		0.0058	0.0128		0.0095	0.0210	
1⅞	0.9375	0.0012			0.0027			0.0050			0.0081		
2	1.000	0.0010	0.0022	0.0052	0.0024	0.0051	0.0121	0.0044	0.0094	0.0223	0.0070	0.0152	0.0362
2⅛	1.0625	0.0009			0.0021			0.0038			0.0061		
2¼	1.125	0.0008	0.0017		0.0018	0.0040		0.0034	0.0072		0.0054	0.0116	
2⅜	1.1875	0.0007			0.0016			0.0029			0.0048		
2½	1.250	0.0006	0.0014	0.0031	0.0015	0.0031	0.0071	0.0027	0.0057	0.0129	0.0043	0.0092	0.0208
2⅝	1.3125	0.0006			0.0013			0.0024			0.0038		
2¾	1.375	0.0005	0.0011		0.0012	0.0026		0.0022	0.0046		0.0035	0.0073	
2⅞	1.4375	0.0005			0.0011			0.0020			0.0032		
3	1.500	0.0004	0.0009	0.0021	0.0010	0.0021	0.0047	0.0018	0.0038	0.0085	0.0029	0.0061	0.0135
3⅛	1.5625	0.0004			0.0009			0.0017			0.0027		
3¼	1.625	0.0003	0.0008		0.0008	0.0018		0.0015	0.0032		0.0024	0.0051	
3⅜	1.6875	0.0003			0.0008			0.0014			0.0023		
3½	1.750	0.0003	0.0007	0.0015	0.0007	0.0015	0.0033	0.0013	0.0028	0.0060	0.0021	0.0044	0.0095
3⅝	1.8125	0.0003			0.0007			0.0012			0.0019		
3¾	1.875	0.0002	0.0006		0.0006	0.0013		0.0011	0.0024		0.0018	0.0038	

Corrected Diameters of Circular Forming Tools — 1

Length c on Tool	Number of B. & S. Automatic Screw Machine			Length c on Tool	Number of B. & S. Automatic Screw Machine		
	No. ∞	No. 0	No. 2		No. ∞	No. 0	No. 2
0.001	1.7480	2.2480	2.9980	0.058	1.6353	2.1352	2.8857
0.002	1.7460	2.2460	2.9961	0.059	1.6333	2.1332	2.8837
0.003	1.7441	2.2441	2.9941	0.060	1.6313	2.1312	2.8818
0.004	1.7421	2.2421	2.9921	0.061	1.6294	2.1293	2.8798
0.005	1.7401	2.2401	2.9901	0.062	1.6274	2.1273	2.8778
0.006	1.7381	2.2381	2.9882	1/16	1.6264	2.1263	2.8768
0.007	1.7362	2.2361	2.9862	0.063	1.6254	2.1253	2.8759
0.008	1.7342	2.2341	2.9842	0.064	1.6234	2.1233	2.8739
0.009	1.7322	2.2321	2.9823	0.065	1.6215	2.1213	2.8719
0.010	1.7302	2.2302	2.9803	0.066	1.6195	2.1194	2.8699
0.011	1.7282	2.2282	2.9783	0.067	1.6175	2.1174	2.8680
0.012	1.7263	2.2262	2.9763	0.068	1.6155	2.1154	2.8660
0.013	1.7243	2.2243	2.9744	0.069	1.6136	2.1134	2.8640
0.014	1.7223	2.2222	2.9724	0.070	1.6116	2.1115	2.8621
0.015	1.7203	2.2203	2.9704	0.071	1.6096	2.1095	2.8601
1/64	1.7191	2.2191	2.9692	0.072	1.6076	2.1075	2.8581
0.016	1.7184	2.2183	2.9685	0.073	1.6057	2.1055	2.8561
0.017	1.7164	2.2163	2.9665	0.074	1.6037	2.1035	2.8542
0.018	1.7144	2.2143	2.9645	0.075	1.6017	2.1016	2.8522
0.019	1.7124	2.2123	2.9625	0.076	1.5997	2.0996	2.8503
0.020	1.7104	2.2104	2.9606	0.077	1.5978	2.0976	2.8483
0.021	1.7085	2.2084	2.9586	0.078	1.5958	2.0956	2.8463
0.022	1.7065	2.2064	2.9566	5/64	1.5955	2.0954	2.8461
0.023	1.7045	2.2045	2.9547	0.079	1.5938	2.0937	2.8443
0.024	1.7025	2.2025	2.9527	0.080	1.5918	2.0917	2.8424
0.025	1.7005	2.2005	2.9507	0.081	1.5899	2.0897	2.8404
0.026	1.6986	2.1985	2.9488	0.082	1.5879	2.0877	2.8384
0.027	1.6966	2.1965	2.9468	0.083	1.5859	2.0857	2.8365
0.028	1.6946	2.1945	2.9448	0.084	1.5839	2.0838	2.8345
0.029	1.6926	2.1925	2.9428	0.085	1.5820	2.0818	2.8325
0.030	1.6907	2.1906	2.9409	0.086	1.5800	2.0798	2.8306
0.031	1.6887	2.1886	2.9389	0.087	1.5780	2.0778	2.8286
1/32	1.6882	2.1881	2.9384	0.088	1.5760	2.0759	2.8266
0.032	1.6867	2.1866	2.9369	0.089	1.5740	2.0739	2.8247
0.033	1.6847	2.1847	2.9350	0.090	1.5721	2.0719	2.8227
0.034	1.6827	2.1827	2.9330	0.091	1.5701	2.0699	2.8207
0.035	1.6808	2.1807	2.9310	0.092	1.5681	2.0679	2.8187
0.036	1.6788	2.1787	2.9290	0.093	1.5661	2.0660	2.8168
0.037	1.6768	2.1767	2.9271	3/32	1.5647	2.0645	2.8153
0.038	1.6748	2.1747	2.9251	0.094	1.5642	2.0640	2.8148
0.039	1.6729	2.1727	2.9231	0.095	1.5622	2.0620	2.8128
0.040	1.6709	2.1708	2.9211	0.096	1.5602	2.0600	2.8109
0.041	1.6689	2.1688	2.9192	0.097	1.5582	2.0581	2.8089
0.042	1.6669	2.1668	2.9172	0.098	1.5563	2.0561	2.8069
0.043	1.6649	2.1649	2.9152	0.099	1.5543	2.0541	2.8050
0.044	1.6630	2.1629	2.9133	0.100	1.5523	2.0521	2.8030
0.045	1.6610	2.1609	2.9113	0.101	1.5503	2.0502	2.8010
0.046	1.6590	2.1589	2.9093	0.102	1.5484	2.0482	2.7991
3/64	1.6573	2.1572	2.9076	0.103	1.5464	2.0462	2.7971
0.047	1.6570	2.1569	2.9073	0.104	1.5444	2.0442	2.7951
0.048	1.6550	2.1549	2.9054	0.105	1.5425	2.0422	2.7932
0.049	1.6531	2.1529	2.9034	0.106	1.5405	2.0403	2.7912
0.050	1.6511	2.1510	2.9014	0.107	1.5385	2.0383	2.7892
0.051	1.6491	2.1490	2.8995	0.108	1.5365	2.0363	2.7873
0.052	1.6471	2.1470	2.8975	0.109	1.5346	2.0343	2.7853
0.053	1.6452	2.1451	2.8955	7/64	1.5338	2.0336	2.7846
0.054	1.6432	2.1431	2.8936	0.110	1.5326	2.0324	2.7833
0.055	1.6412	2.1411	2.8916	0.111	1.5306	2.0304	2.7814
0.056	1.6392	2.1391	2.8896	0.112	1.5287	2.0284	2.7794
0.057	1.6373	2.1372	2.8877	0.113	1.5267	2.0264	2.7774

Corrected Diameters of Circular Forming Tools — 2

Length c on Tool	Number of B. & S. Automatic Screw Machine			Length c on Tool	Number of B. & S. Automatic Screw Machine		
	No. 00	No. 0	No. 2		No. 00	No. 0	No. 2
0.113	1.5267	2.0264	2.7774	0.171	1.4124	1.9119	2.6634
0.114	1.5247	2.0245	2.7755	11/64	1.4107	1.9103	2.6617
0.115	1.5227	2.0225	2.7735	0.172	1.4104	1.9099	2.6614
0.116	1.5208	2.0205	2.7715	0.173	1.4084	1.9080	2.6595
0.117	1.5188	2.0185	2.7696	0.174	1.4065	1.9060	2.6575
0.118	1.5168	2.0166	2.7676	0.175	1.4045	1.9040	2.6556
0.119	1.5148	2.0146	2.7656	0.176	1.4025	1.9021	2.6536
0.120	1.5129	2.0126	2.7637	0.177	1.4006	1.9001	2.6516
0.121	1.5109	2.0106	2.7617	0.178	1.3986	1.8981	2.6497
0.122	1.5089	2.0087	2.7597	0.179	1.3966	1.8961	2.6477
0.123	1.5070	2.0067	2.7578	0.180	1.3947	1.8942	2.6457
0.124	1.5050	2.0047	2.7558	0.181	1.3927	1.8922	2.6438
0.125	1.5030	2.0027	2.7538	0.182	1.3907	1.8902	2.6418
0.126	1.5010	2.0008	2.7519	0.183	1.3888	1.8882	2.6398
0.127	1.4991	1.9988	2.7499	0.184	1.3868	1.8863	2.6379
0.128	1.4971	1.9968	2.7479	0.185	1.3848	1.8843	2.6359
0.129	1.4951	1.9948	2.7460	0.186	1.3829	1.8823	2.6339
0.130	1.4932	1.9929	2.7440	0.187	1.3809	1.8804	2.6320
0.131	1.4912	1.9909	2.7420	3/16	1.3799	1.8794	2.6310
0.132	1.4892	1.9889	2.7401	0.188	1.3789	1.8784	2.6300
0.133	1.4872	1.9869	2.7381	0.189	1.3770	1.8764	2.6281
0.134	1.4853	1.9850	2.7361	0.190	1.3750	1.8744	2.6261
0.135	1.4833	1.9830	2.7342	0.191	1.3730	1.8725	2.6241
0.136	1.4813	1.9810	2.7322	0.192	1.3711	1.8705	2.6222
0.137	1.4794	1.9790	2.7302	0.193	1.3691	1.8685	2.6202
0.138	1.4774	1.9771	2.7282	0.194	1.3671	1.8665	2.6182
0.139	1.4754	1.9751	2.7263	0.195	1.3652	1.8646	2.6163
0.140	1.4734	1.9731	2.7243	0.196	1.3632	1.8626	2.6143
9/64	1.4722	1.9719	2.7231	0.197	1.3612	1.8606	2.6123
0.141	1.4715	1.9711	2.7224	0.198	1.3592	1.8587	2.6104
0.142	1.4695	1.9692	2.7204	0.199	1.3573	1.8567	2.6084
0.143	1.4675	1.9672	2.7184	0.200	1.3553	1.8547	2.6064
0.144	1.4655	1.9652	2.7165	0.201	....	1.8527	2.6045
0.145	1.4636	1.9632	2.7145	0.202	....	1.8508	2.6025
0.146	1.4616	1.9613	2.7125	0.203	....	1.8488	2.6006
0.147	1.4596	1.9593	2.7106	13/64	....	1.8486	2.6003
0.148	1.4577	1.9573	2.7086	0.204	....	1.8468	2.5986
0.149	1.4557	1.9553	2.7066	0.205	....	1.8449	2.5966
0.150	1.4537	1.9534	2.7047	0.206	....	1.8429	2.5947
0.151	1.4517	1.9514	2.7027	0.207	....	1.8409	2.5927
0.152	1.4498	1.9494	2.7007	0.208	....	1.8390	2.5908
0.153	1.4478	1.9474	2.6988	0.209	....	1.8370	2.5888
0.154	1.4458	1.9455	2.6968	0.210	....	1.8350	2.5868
0.155	1.4439	1.9435	2.6948	0.211	....	1.8330	2.5849
0.156	1.4419	1.9415	2.6929	0.212	....	1.8311	2.5829
5/32	1.4414	1.9410	2.6924	0.213	....	1.8291	2.5809
0.157	1.4399	1.9395	2.6909	0.214	....	1.8271	2.5790
0.158	1.4380	1.9376	2.6889	0.215	....	1.8252	2.5770
0.159	1.4360	1.9356	2.6870	0.216	....	1.8232	2.5751
0.160	1.4340	1 9336	2.6850	0.217	....	1.8212	2.5731
0.161	1.4321	1.9317	2.6830	0.218	....	1.8193	2.5711
0.162	1.4301	1.9297	2.6811	7/32	....	1.8178	2.5697
0.163	1.4281	1.9277	2.6791	0.219	....	1.8173	2.5692
0.164	1.4262	1.9257	2.6772	0.220	....	1.8153	2.5672
0.165	1.4242	1.9238	2.6752	0.221	....	1.8133	2.5653
0.166	1.4222	1.9218	2.6732	0.222	....	1.8114	2.5633
0.167	1.4203	1.9198	2.6713	0.223	....	1.8094	2.5613
0.168	1.4183	1.9178	2.6693	0.224	....	1.8074	2.5594
0.169	1.4163	1.9159	2.6673	0.225	....	1.8055	2.5574
0.170	1.4144	1.9139	2.6654	0.226	....	1.8035	2.5555

Corrected Diameters of Circular Forming Tools — 3

Length c on Tool	No. of B. & S. Machine No. 0	No. 2	Length c on Tool	No. of B. & S. Machine No. 0	No. 2	Length c on Tool	No. 2 B. & S Machine
0.227	1.8015	2.5535	0.284	1.6894	2.4418	0.341	2.3303
0.228	1.7996	2.5515	0.285	1.6874	2.4398	0.342	2.3284
0.229	1.7976	2.5496	0.286	1.6854	2.4378	0.343	2.3264
0.230	1.7956	2.5476	0.287	1.6835	2.4359	11/32	2.3250
0.231	1.7936	2.5456	0.288	1.6815	2.4340	0.344	2.3245
0.232	1.7917	2.5437	0.289	1.6795	2.4320	0.345	2.3225
0.233	1.7897	2.5417	0.290	1.6776	2.4300	0.346	2.3206
0.234	1.7877	2.5398	0.291	1.6756	2.4281	0.347	2.3186
15/64	1.7870	2.5390	0.292	1.6736	2.4261	0.348	2.3166
0.235	1.7858	2.5378	0.293	1.6717	2.4242	0.349	2.3147
0.236	1.7838	2.5358	0.294	1.6697	2.4222	0.350	2.3127
0.237	1.7818	2.5339	0.295	1.6677	2.4203	0.351	2.3108
0.238	1.7799	2.5319	0.296	1.6658	2.4183	0.352	2.3088
0.239	1.7779	2.5300	19/64	1.6641	2.4166	0.353	2.3069
0.240	1.7759	2.5280	0.297	1.6638	2.4163	0.354	2.3049
0 241	1.7739	2.5260	0.298	1.6618	2.4144	0.355	2.3030
0.242	1.7720	2.5241	0.299	1.6599	2.4124	0.356	2.3010
0.243	1.7700	2.5221	0.300	1.6579	2.4105	0.357	2.2991
0.244	1.7680	2.5201	0.301	....	2.4085	0.358	2.2971
0.245	1.7661	2.5182	0.302	....	2.4066	0.359	2.2952
0.246	1.7641	2.5162	0.303	....	2.4046	23/64	2.2945
0.247	1.7621	2.5143	0.304	....	2.4026	0.360	2.2932
0.248	1.7602	2.5123	0.305	....	2.4007	0.361	2.2913
0.249	1.7582	2.5104	0.306	....	2.3987	0.362	2.2893
0.250	1.7562	2.5084	0.307	....	2.3968	0.363	2.2874
0.251	1.7543	2.5064	0.308	....	2.3948	0.364	2.2854
0.252	1.7523	2.5045	0.309	....	2.3929	0.365	2.2835
0.253	1.7503	2.5025	0.310	....	2.3909	0.366	2.2815
0.254	1.7484	2.5005	0.311	....	2.3890	0.367	2.2796
0.255	1.7464	2.4986	0.312	....	2.3870	0.368	2.2776
0.256	1.7444	2.4966	5/16	....	2.3860	0.369	2.2757
0.257	1.7425	2.4947	0.313	....	2.3851	0.370	2.2737
0.258	1.7405	2.4927	0.314	....	2.3831	0.371	2.2718
0.259	1.7385	2.4908	0.315	....	2.3811	0.372	2.2698
0.260	1.7366	2.4888	0.316	....	2.3792	0.373	2.2679
0.261	1.7346	2.4868	0.317	....	2.3772	0.374	2.2659
0.262	1.7326	2.4849	0.318	....	2.3753	0.375	2.2640
0.263	1.7306	2.4829	0.319	....	2.3733	0.376	2.2620
0.264	1.7287	2.4810	0.320	....	2.3714	0.377	2.2601
0.265	1.7267	2.4790	0.321	....	2.3694	0.378	2.2581
17/64	1.7255	2.4778	0.322	....	2.3675	0.379	2.2562
0.266	1.7248	2.4770	0.323	....	2.3655	0.380	2.2542
0.267	1.7228	2.4751	0.324	....	2.3636	0.381	2.2523
0.268	1.7208	2.4731	0.325	....	2.3616	0.382	2.2503
0.269	1.7189	2.4712	0.326	....	2.3596	0.383	2.2484
0.270	1.7169	2.4692	0.327	....	2.3577	0.384	2.2464
0.271	1.7149	2.4673	0.328	....	2.3557	0.385	2.2445
0.272	1.7130	2.4653	21/64	....	2.3555	0.386	2.2425
0.273	1.7110	2.4633	0.329	....	2.3538	0.387	2.2406
0.274	1.7090	2.4614	0.330	....	2.3518	0.388	2.2386
0.275	1.7071	2.4594	0.331	....	2.3499	0.389	2.2367
0.276	1.7051	2.4575	0.332	....	2.3479	0.390	2.2347
0.277	1.7031	2.4555	0.333	....	2.3460	25/64	2.2335
0.278	1.7012	2.4535	0.334	....	2.3440	0.391	2.2328
0.279	1.6992	2.4516	0.335	....	2.3421	0.392	2.2308
0.280	1.6972	2.4496	0.336	....	2.3401	0.393	2.2289
0.281	1.6953	2.4477	0.337	....	2.3381	0.394	2.2269
9/32	1.6948	2.4472	0.338	....	2.3362	0.395	2.2250
0.282	1.6933	2.4457	0.339	....	2.3342	0.396	2.2230
0.283	1.6913	2.4438	0.340	....	2.3323	0.397	2.2211

Corrected Diameters of Circular Forming Tools — 4

Length c on Tool	No. 2 B. & S. Machine	Length c on Tool	No. 2 B. & S. Machine	Length c on Tool	No. 2 B. & S. Machine	Length c on Tool	No. 2 B. & S. Machine
0.398	2.2191	0.423	2.1704	0.449	2.1199	0.474	2.0713
0.399	2.2172	0.424	2.1685	0.450	2.1179	0.475	2.0694
0.400	2.2152	0.425	2.1666	0.451	2.1160	0.476	2.0674
0.401	2.2133	0.426	2.1646	0.452	2.1140	0.477	2.0655
0.402	2.2113	0.427	2.1627	0.453	2.1121	0.478	2.0636
0.403	2.2094	0.428	2.1607	29/64	2.1118	0.479	2.0616
0.404	2.2074	0.429	2.1588	0.454	2.1101	0.480	2.0597
0.405	2.2055	0.430	2.1568	0.455	2.1082	0.481	2.0577
0.406	2.2035	0.431	2.1549	0.456	2.1063	0.482	2.0558
13/32	2.2030	0.432	2.1529	0.457	2.1043	0.483	2.0538
0.407	2.2016	0.433	2.1510	0.458	2.1024	0.484	2.0519
0.408	2.1996	0.434	2.1490	0.459	2.1004	0.485	2.0500
0.409	2.1977	0.435	2.1471	0.460	2.0985	0.486	2.0480
0.410	2.1957	0.436	2.1452	0.461	2.0966	0.487	2.0461
0.411	2.1938	0.437	2.1432	0.462	2.0946	0.488	2.0441
0.412	2.1919	7/16	2.1422	0.463	2.0927	0.489	2.0422
0.413	2.1899	0.438	2.1413	0.464	2.0907	0.490	2.0403
0.414	2.1880	0.439	2.1393	0.465	2.0888	0.491	2.0383
0.415	2.1860	0.440	2.1374	0.466	2.0868	0.492	2.0364
0.416	2.1841	0.441	2.1354	0.467	2.0849	0.493	2.0344
0.417	2.1821	0.442	2.1335	0.468	2.0830	0.494	2.0325
0.418	2.1802	0.443	2.1315	15/32	2.0815	0.495	2.0306
0.419	2.1782	0.444	2.1296	0.469	2.0810	0.496	2.0286
0.420	2.1763	0.445	2.1276	0.470	2.0791	0.497	2.0267
0.421	2.1743	0.446	2.1257	0.471	2.0771	0.498	2.0247
27/64	2.1726	0.447	2.1237	0.472	2.0752	0.499	2.0228
0.422	2.1724	0.448	2.1218	0.473	2.0733	0.500	2.0209

Method of Using Tables for "Corrected Diameters of Circular Forming Tools." — These tables are especially applicable to the Brown & Sharpe automatic screw machines. The maximum diameter D of forming tools for these machines should be as follows: For No. oo machine, 1¾ inch; for No. o machine, 2¼ inches; for No. 2 machine, 3 inches. To find the other diameters of the tool for any piece

Dimensions of Forming Tools for B. & S. Automatic Screw Machines

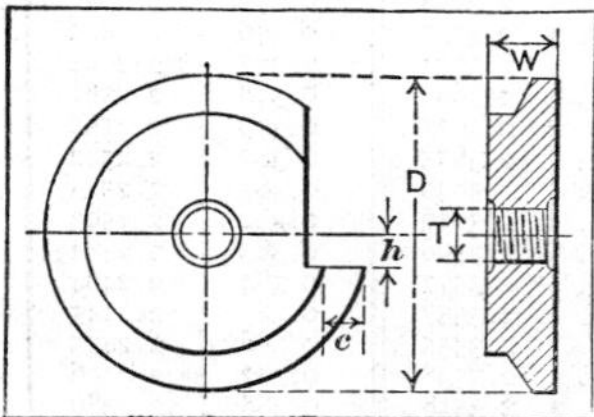

	No. of Machine	Max. Diam., D	h	T	W
	oo	1¾	⅛	⅜–16	¼
	o	2¼	5/32	½–14	5/16
	2	3	¼	⅝–12	⅜
	6	4	5/16	¾–12	⅜

to be formed, proceed as follows: Subtract the smallest diameter of the work from that diameter of the work which is to be formed by the required tool diameter; divide the remainder by 2; locate the quotient obtained in the column headed "Length c on Tool," and opposite the figure thus located and in the column headed by the number of the machine used, read off directly the diameter to which the tool is to be made. The quotient obtained, which is located in the column headed "Length c on Tool," is the length c as shown in the illustration above.

Example: — A piece of work is to be formed on a No. o machine to two diameters, ne being ¼ inch and one 0.550 inch; find the diameters of the tool. The maximum ool diameter is 2¼ inches. This will be the diameter which will cut the ¼ inch iameter of the work. To find the other diameter, proceed according to the rule iven: 0.550 − ¼ = 0.300; 0.300 ÷ 2 = 0.150. In Table 2, opposite 0.150, we nd that the required tool diameter is 1.9534 inch. These tables are for tools vithout rake

Arrangement of Circular Tools. — When applying circular tools to automatic crew machines, their arrangement has an important bearing on the results obtained. he various ways of arranging the circular tools, with relation to the rotation of he spindle, are shown at *A*, *B*, *C* and *D*, in the illustration. These diagrams epresent the view obtained when looking towards the chuck. The arrangement hown at *A* gives good results on long forming operations on brass and steel for the eason that the pressure of the cut on the front tool is downward; the support is nore rigid than when the forming tool is turned upside down on the front slide s shown at *B*; here the stock, turning up towards the tool, has a tendency to lift

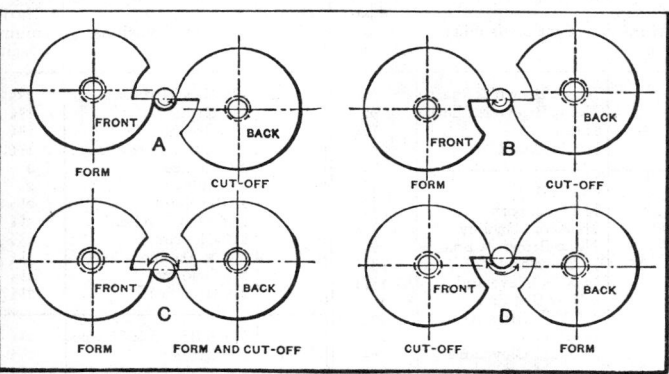

he cross-slide, causing chattering; therefore, the arrangement shown at *A* is ecommended when a high finish is desired. The arrangement at *B* works satis-actorily for short steel pieces which do not require a high finish; it allows the hips to drop clear of the work, and is especially advantageous when making crews, when the forming and cut-off tools operate after the die, as no time is lost n reversing the spindle. The arrangement at *C* is recommended for heavy cutting n large work, when both tools are used for forming the piece; a rigid support is hen necessary for both tools and a good supply of oil is also required. The arrange-nent at *D* is objectionable and should be avoided; it is used only when a left-hand hread is cut on the piece and when the cut-off tool is used on the front slide, leaving he heavy cutting to be performed from the rear slide. In all "cross-forming" work, t is essential that the spindle be kept in good condition, and that the collet or huck have a parallel contact upon the bar which is being formed.

Feeds and Speeds for Forming Tools. — Approximate feeds and speeds for orming tools are given in the table beginning on page 1020. The feeds and speeds re average values, and if the job at hand has any features out of the ordinary the gures given should be altered accordingly.

Former ASA Standard Forming Tool Blanks (ASA B5.7-1954). — The standar covers both circular and dovetail forming tool blanks such as are commonly used c hand screw machines and on automatic screw machines. It also includes the moun ing and clamping elements of the tool-holders used on these machines. The purpo of the standard is to provide interchangeability and also to permit a reduction i the number of forming tool blanks required. In order to establish a minimu number of blank sizes, the machines have been classified for reference purposes int six different groups of comparable stock capacities as shown by the accompanyir Table 1. Each group of machines takes a definite size tool and the holders a provided with suitable mounting or clamping devices. The machine group numbe have been assigned arbitrarily to identify the size of the tool with the machines c which that size may be used. This standard is not intended to provide interchange ability of holders for various makes of machines except in connection with tl mounting or clamp details.

Table 1.   Machine Classifications for Former ASA Standard Forming Tool Blanks

Group Number	Type of Machine	Maximum Capacity*	Group Number	Type of Machine	Maximum Capacity
1	No. 00 Brown & Sharpe	3/8	4	2 5/8 New Britain	2 5/8
	No. 19 Brown & Sharpe	3/8		1 3/4 Gridley	1 3/4
	Index "0"	7/16		1 3/4 Greenlee	1 3/4
	3/8 Cleveland	5/8		No. 4 Brown & Sharpe	1 7/8
				2 Greenlee	2
2	3/8 Gridley	3/8		2 Gridley	2
	1/2 Davenport	7/8		2 Cleveland	2 1/2
	9/16 Acme Gridley	9/16		2 × 2 3/4 Cleveland	3 1/4
	No. 0 Brown & Sharpe	5/8		2 1/4 Cleveland	2 1/2
	5/8 Cleveland	3/4		2 1/4 × 2 3/4 Cleveland	3 1/4
	5/8 × 7/8 Cleveland	1 1/16		2 1/4 Gridley	2 1/4
	1 New Britain	1		2 1/4 Greenlee	2 1/4
	1 3/8 New Britain	1 3/8			
	7/8 Greenlee	1	5	No. 6 Brown & Sharpe	2 5/8
	7/8 × 1 1/4 Cleveland	7/8		2 5/8 Gridley	2 5/8
				2 3/4 × 3 3/4 Cleveland	3 1/4
3	7/8 Gridley	7/8		2 3/4 × 4 Cleveland	2 3/4
	1 Acme Gridley	1		3 Gridley	3
	No. 2 Brown & Sharpe	1 1/8		3 9/16 Gridley	3 9/16
	1 1/4 Gridley	1 1/4			
	1 1/4 Cleveland	1 1/4	6	3 1/4 Gridley	3 1/4
	1 1/4 Cleveland	1 3/8		3 1/2 Gridley	3 1/2
	1 1/4 × 1 1/2 Cleveland	1 3/4		4 Gridley	4
	1 1/4 × 1 1/2 Cleveland	1 1/2		4 Cleveland	4
	1 1/4 Greenlee	1 1/4		4 1/4 Cleveland	4 1/4
	1 3/8 Gridley	1 3/8		4 1/2 Cleveland	4 1/2
	1 1/2 Greenlee	1 1/2		4 1/4 Gridley	4 1/4
	1 5/8 Gridley	1 5/8		4 1/2 Gridley	4 1/2
				5 Gridley	5
4	1 5/8 Gridley	1 5/8		5 1/2 Cleveland	5 1/2
	1 5/8 Acme Gridley	1 5/8		6 3/4 Cleveland	6 3/4
	1 5/8 New Britain	1 5/8		7 3/4 Cleveland	7 3/4
	2 1/4 New Britain	2 1/4			

* The group classification numbers apply to all machine models of the respective make listed having the maximum capacities indicated. Dimensions in inches.

Table 2. Former ASA Standard Dimensions of Finished Blanks for Circular Forming Tools With Threaded Mounting Hole

The circular tool blanks in Table 2, which are provided with threaded mounting holes, are used for tools having nominal outside diameters of 1¾, 2¼, and 3 inches. For the larger blank diameters listed in Table 3, the mounting holes are unthreaded and counterbored. This method of mounting is for tool blanks having nominal outside diameters of 3½, 4, and 5 inches.

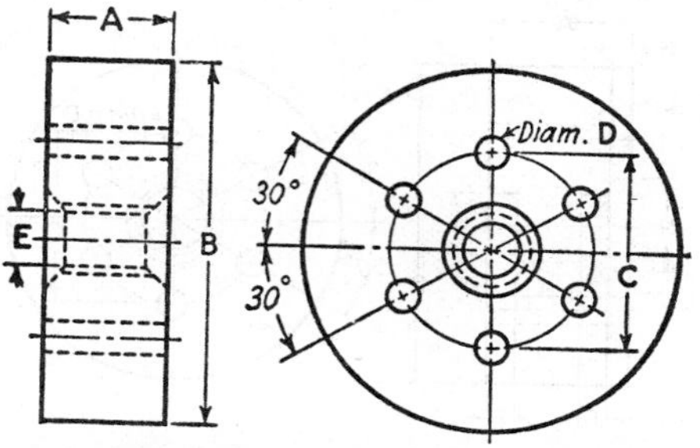

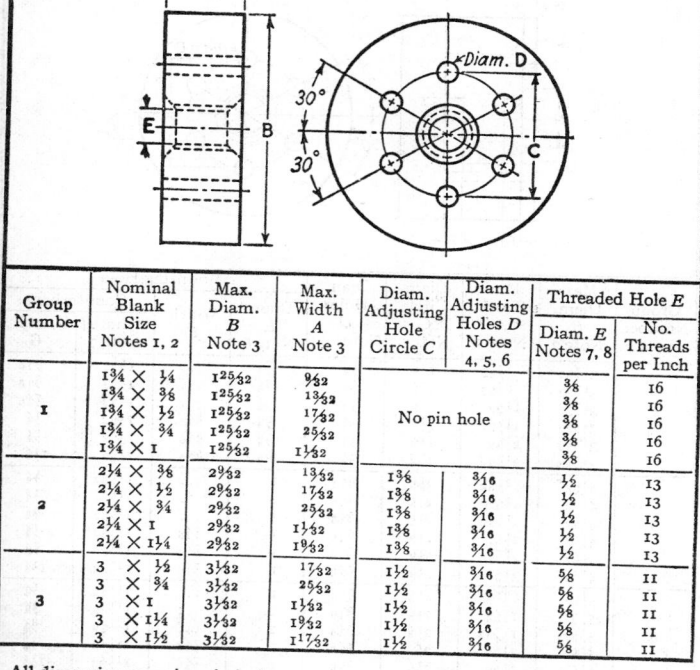

Group Number	Nominal Blank Size Notes 1, 2	Max. Diam. *B* Note 3	Max. Width *A* Note 3	Diam. Adjusting Hole Circle *C*	Diam. Adjusting Holes *D* Notes 4, 5, 6	Threaded Hole *E*	
						Diam. *E* Notes 7, 8	No. Threads per Inch
1	1¾ × ¼	1 25/32	9/32	No pin hole		3/8	16
	1¾ × 3/8	1 25/32	13/32			3/8	16
	1¾ × ½	1 25/32	17/32			3/8	16
	1¾ × ¾	1 25/32	25/32			3/8	16
	1¾ × 1	1 25/32	1 1/32			3/8	16
2	2¼ × 3/8	2 9/32	13/32	1 3/8	3/16	½	13
	2¼ × ½	2 9/32	17/32	1 3/8	3/16	½	13
	2¼ × ¾	2 9/32	25/32	1 3/8	3/16	½	13
	2¼ × 1	2 9/32	1 1/32	1 3/8	3/16	½	13
	2¼ × 1¼	2 9/32	1 9/32	1 3/8	3/16	½	13
3	3 × ½	3 1/32	17/32	1½	3/16	5/8	11
	3 × ¾	3 1/32	25/32	1½	3/16	5/8	11
	3 × 1	3 1/32	1 1/32	1½	3/16	5/8	11
	3 × 1¼	3 1/32	1 9/32	1½	3/16	5/8	11
	3 × 1½	3 1/32	1 17/32	1½	3/16	5/8	11

All dimensions are given in inches.

(1) Blanks are designated by giving the nominal outside diameter and width.

(2) Blanks made of high-speed steel shall be stamped H.S.

(3) The tolerance on diameter *B* and width *A* is −1/64 inch.

(4) Six adjusting holes shall have a minimum depth of ¼ inch, greater depth or through holes being optional. Adjusting holes shall be equally spaced on the circle *C* and be slightly chamfered.

(5) Diameter of adjusting holes *D* shall be ± 0.002.

(6) Diameter *C* of location circle shall be ± 0.002.

(7) The threaded hole *E* shall be made to Unified and American Standard, Class 2B. Both ends of threaded hole *E* shall be chamfered 35 degrees.

(8) Commercial blanks may not, in all cases, have the standard number of threads per inch listed in this table.

Table 3. Former ASA Standard Dimensions for Finished Blanks for Circular Forming Tools With Counterbored Mounting Hole

In obtaining circular forming tool blanks from commercial sources, the blank size is designated by giving the nominal outside diameter and nominal width. In the table below, the nominal diameter and width is 1/32 inch less than the maximum diameter and width. For example, a forming tool having a maximum diameter of 3 17/32 inches and maximum width of 17/32 inch would be designated as 3 1/2 × 1/2.

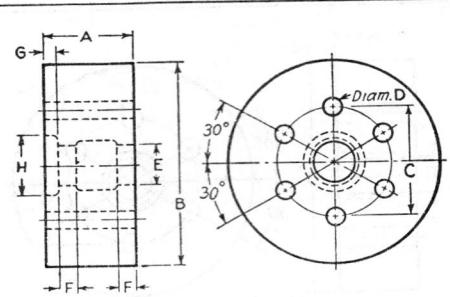

Group Number	Max. Diam. B Note 2	Max. Width A Note 2	Diam. Circle C Note 5	Diam. Holes D Notes 3, 4	Mounting Hole			
					Diam. E Note 6	Width F	Counterbore Diam. H Note 7	Depth G
4	3 17/32	17/32	1 7/8	1/4	3/4	...	1 5/32	3/16
	3 17/32	25/32	1 7/8	1/4	3/4	...	1 5/32	3/16
	3 17/32	1 5/32	1 7/8	1/4	3/4	...	1 5/32	7/16
	3 17/32	1 17/32	1 7/8	1/4	3/4	5/16	1 5/32	1/4
	3 17/32	2 5/32	1 7/8	1/4	3/4	5/16	1 5/32	3/4
	3 17/32	2 17/32	1 7/8	1/4	3/4	5/16	1 5/32	1 1/4
5	4 5/32	2 1/32	2 1/8	5/16	1	...	1 11/32	1/4
	4 5/32	1 5/32	2 1/8	5/16	1	...	1 11/32	1/4
	4 5/32	1 17/32	2 1/8	5/16	1	...	1 11/32	3/4
	4 5/32	2 5/32	2 1/8	5/16	1	7/16	1 11/32	1/4
	4 5/32	2 17/32	2 1/8	5/16	1	7/16	1 11/32	3/4
	4 5/32	3 5/32	2 1/8	5/16	1	7/16	1 11/32	1 1/4
6	5 5/32	2 1/32	2 1/8	5/16	1	...	1 11/32	1/4
	5 5/32	1 5/32	2 1/8	5/16	1	...	1 11/32	1/4
	5 5/32	1 17/32	2 1/8	5/16	1	...	1 11/32	3/4
	5 5/32	2 5/32	2 1/8	5/16	1	7/16	1 11/32	1/4
	5 5/32	3 5/32	2 1/8	5/16	1	7/16	1 11/32	1 1/4
	5 5/32	4 5/32	2 1/8	5/16	1	7/16	1 11/32	2 1/4

All dimensions are given in inches.

(1) Blanks made of high-speed steel shall be stamped H.S.

(2) The tolerance on diameter B and width A is −1/64 inch.

(3) Six adjusting holes shall have a minimum depth of 1/4 inch, greater depth or through holes being optional. Adjusting holes shall be equally spaced on the circle C and be slightly chamfered. Holes D, if blind, to be on face opposite one containing counterbore H.

(4) Diameter of adjusting holes D shall be ± 0.002.

(5) Diameter C of location circle shall be ± 0.002.

(6) Bolt hole E has a tolerance of +0.003, −0.000 with 45 deg. chamfer at both ends.

(7) Diameter H shall have a slight radius at the bottom of counterbore.

Table 4. Former ASA Standard Finished Blanks for Dovetailed Forming Tools

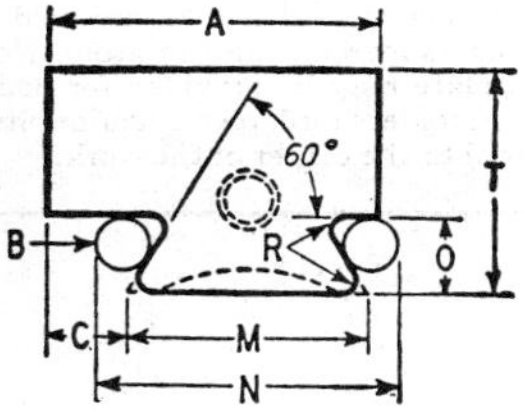

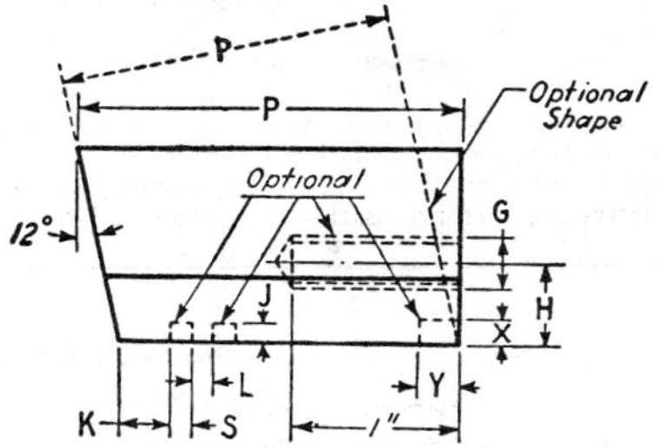

Group No.	A +0 −1/64	T +0 −1/64	P ±.010	M ±.001	O ±.010	C ±.010	N† ±.003	B ±.0003	G	H ±.010	Y, X ±.010	No. of Adjusting Slots	J, L, S *	K ±.010
1	1 1/32	29/32	1 1/2	0.732	9/32	1/4	0.834	0.156	5/16–18	3/8	1/8	2	1/8	5/16
	1 9/32	29/32	1 1/2	0.951	19/64	19/64	1.035	0.156	5/16–18	1/2	3/16	2	1/8	5/16
2	1 9/32	29/32	2	0.951	19/64	19/64	1.035	0.156	5/16–18	1/2	3/16	3	1/8	1 1/32
	1 17/32	29/32	2	0.951	19/64	19/64	1.035	0.156	5/16–18	1/2	3/16	...	...	...
	1 25/32	1 5/32	2	1.250	13/32	7/16	1.464	0.250	5/16–18	1/2	1/4	3	1/8	1 1/32
3	1 25/32	1 5/32	2 7/16	1.250	13/32	7/16	1.464	0.250	5/16–18	1/2	1/4	3	1/8	1 1/32
	2 9/32	1 5/32	2 7/16	1.250	13/32	7/16	1.464	0.250	5/16–18	1/2	1/4	...	...	...
	2 25/32	1 5/32	2 7/16	1.250	13/32	7/16	1.464	0.250	5/16–18	1/4	1/4	...	...	...
4	2 25/32	2 17/32	2 5/8	1.614	35/64	1/2	2.349	0.500	5/16–18	1/2	1/4	...	...	...
	2 25/32	2 17/32	2 5/8	1.882	35/64	1/2	2.617	0.500	7/16–14	3/8	1/4	...	...	...
	3 1/32	2 17/32	2 5/8	1.882	35/64	1/2	2.617	0.500	7/16–14	3/8	1/4	...	...	...
5	2 25/32	1 9/32	2 7/16	2.000	33/64	1/2	2.771	0.500	7/16–14	3/8	5/16	...	...	...
	2 25/32	3 1/32	3	2.000	33/64	1/2	2.771	0.500	7/16–14	3/8	5/16	...	...	...
	3 1/32	3 1/32	3	2.000	33/64	1/2	2.771	0.500	7/16–14	3/8	5/16	...	...	...
	3 9/32	3 1/32	3	2.000	33/64	1/2	2.771	0.500	7/16–14	3/8	5/16	...	...	...
6	3 9/32	3 17/32	4	2.238	35/64	9/16	2.973	0.500	7/16–14	3/8	5/16	...	...	...
	3 17/32	3 17/32	4	2.883	43/64	5/8	3.815	0.625	7/16–14	3/8	5/16	...	...	...
	4 1/32	3 17/32	4	2.883	43/64	5/8	3.815	0.625	7/16–14	3/8	5/16	...	...	...
	4 17/32	3 17/32	4	2.883	43/64	5/8	3.815	0.625	7/16–14	3/8	5/16	...	...	...

All dimensions are given in inches. Blanks made of high-speed steel shall be stamped H. S. Radius *R* is 1/32 inch for blank widths up to 1 17/32 inch inclusive and 1/16 inch for larger widths.

* *Tolerances:* On *J* and *L*, ±0.010; on *S*, +0.003, −0.000.

† The method of measuring the dovetail is optional. It may be measured with roll plugs *B*, as shown above, or it may be measured with conventional micrometers through the use of small 60-degree angle blocks having a known distance from the vertex of the angle to a surface contacted by the micrometer as shown in the diagram on the right. This method would eliminate the necessity of grinding the shoulder under the dovetail.

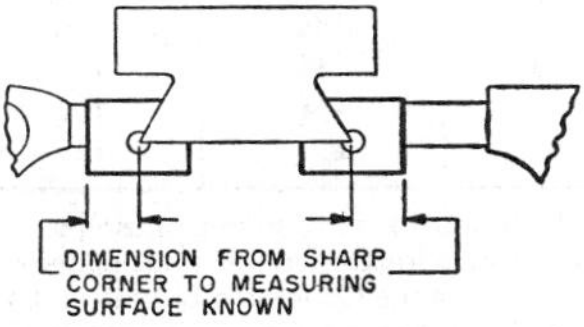

Table 5. Typical Circular Forming Tool-holder for Machines in Group No. 1

The circular forming tools for machines in group Nos. 1 to 3, inclusive (Tables 5 and 6), are held by a stud which enters a threaded hole in the center of the blank, and by an auxiliary clamp of the hook-bolt type as shown by plan views of the diagrams above the tables. Circular tools for machine group Nos. 4 to 6, inclusive (Table 7), are held by a bolt and nut in conjunction with a hook-bolt. Blanks for machines in group No. 1 (Table 5) have no adjusting pin-holes but a friction plate may be provided for use in adjusting the cutting edge to the center of the work. Holders for larger machines have positive means for adjusting the cutting edge of the tool to the center of the work.

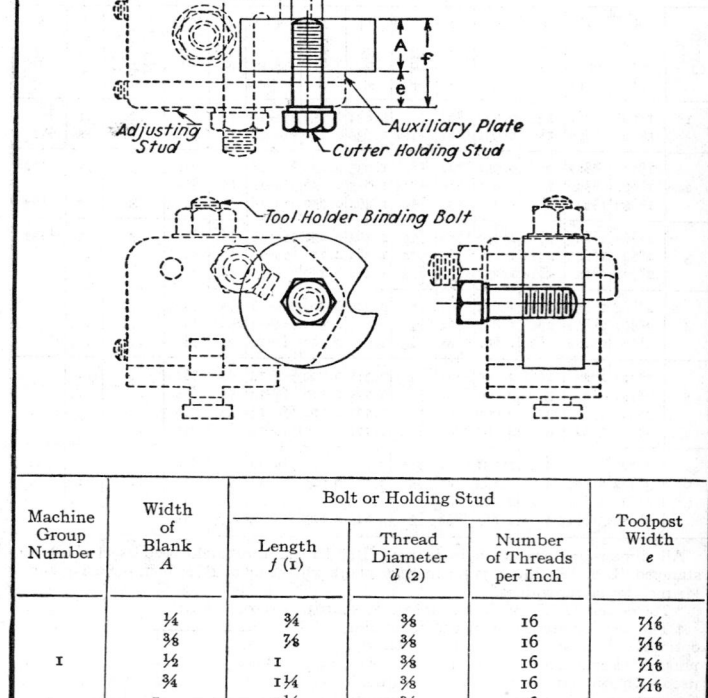

| Machine Group Number | Width of Blank A | Bolt or Holding Stud | | | Toolpost Width e |
		Length f (1)	Thread Diameter d (2)	Number of Threads per Inch	
	¼	¾	⅜	16	7⁄16
	⅜	⅞	⅜	16	7⁄16
1	½	1	⅜	16	7⁄16
	¾	1¼	⅜	16	7⁄16
	1	1½	⅜	16	7⁄16

All dimensions are given in inches.
Tolerance for dimensions not otherwise specified shall be held to ±0.010.
(1) Tolerance for length of bolt is ±1⁄32.
(2) Screw thread shall be made to Unified and American Standard, Class 3A.

Table 6. Typical Circular Forming Tool-holder for Machines in Groups Nos. 2 and 3

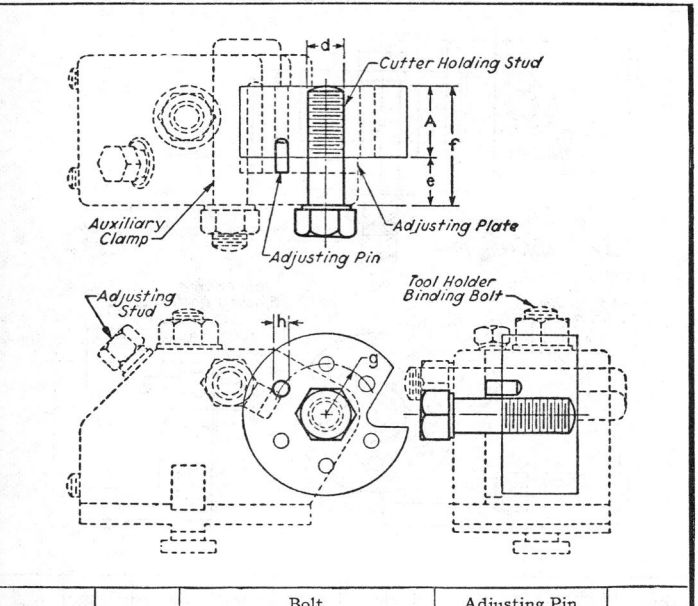

Machine Group Number	Width of Blank A	Bolt				Adjusting Pin		Tool Post e
		Length f Note 1	Thread Diam. d (2)	No. of Threads per Inch (3)		Radius g (4)	Diam. h (5)	
2	$\frac{3}{8}$	$1\frac{1}{8}$	$\frac{1}{2}$	13		$1\frac{1}{16}$	0.185	$\frac{3}{4}$
	$\frac{1}{2}$	$1\frac{1}{4}$	$\frac{1}{2}$	13		$1\frac{1}{16}$	0.185	$\frac{3}{4}$
	$\frac{3}{4}$	$1\frac{1}{2}$	$\frac{1}{2}$	13		$1\frac{1}{16}$	0.185	$\frac{3}{4}$
	1	$1\frac{3}{4}$	$\frac{1}{2}$	13		$1\frac{1}{16}$	0.185	$\frac{3}{4}$
	$1\frac{1}{4}$	2	$\frac{1}{2}$	13		$1\frac{1}{16}$	0.185	$\frac{3}{4}$
3	$\frac{1}{2}$	$1\frac{3}{8}$	$\frac{5}{8}$	11		$\frac{3}{4}$	0.185	$1\frac{3}{16}$
	$\frac{3}{4}$	$1\frac{5}{8}$	$\frac{5}{8}$	11		$\frac{3}{4}$	0.185	$1\frac{3}{16}$
	1	$1\frac{7}{8}$	$\frac{5}{8}$	11		$\frac{3}{4}$	0.185	$1\frac{3}{16}$
	$1\frac{1}{4}$	$2\frac{1}{8}$	$\frac{5}{8}$	11		$\frac{3}{4}$	0.185	$1\frac{3}{16}$
	$1\frac{1}{2}$	$2\frac{3}{8}$	$\frac{5}{8}$	11		$\frac{3}{4}$	0.185	$1\frac{3}{16}$

All dimensions are given in inches.
Tolerances for dimensions not otherwise specified shall be held to ±0.010
(1) Tolerance for length of bolt is ±$\frac{1}{32}$.
(2) Screw thread d shall be made to Unified and American Standard, Class 3A.
(3) Commercial blanks may not, in all cases, have the standard number of threads per inch listed in this table.
(4) Tolerance for the radius of the adjusting pin hole location g is ±0.001.
(5) Tolerance for the diameter of the pin is +0.000, − 0.001.

Table 7. Typical Circular Forming Tool-holder for Machines in Groups Nos. 4, 5, and 6

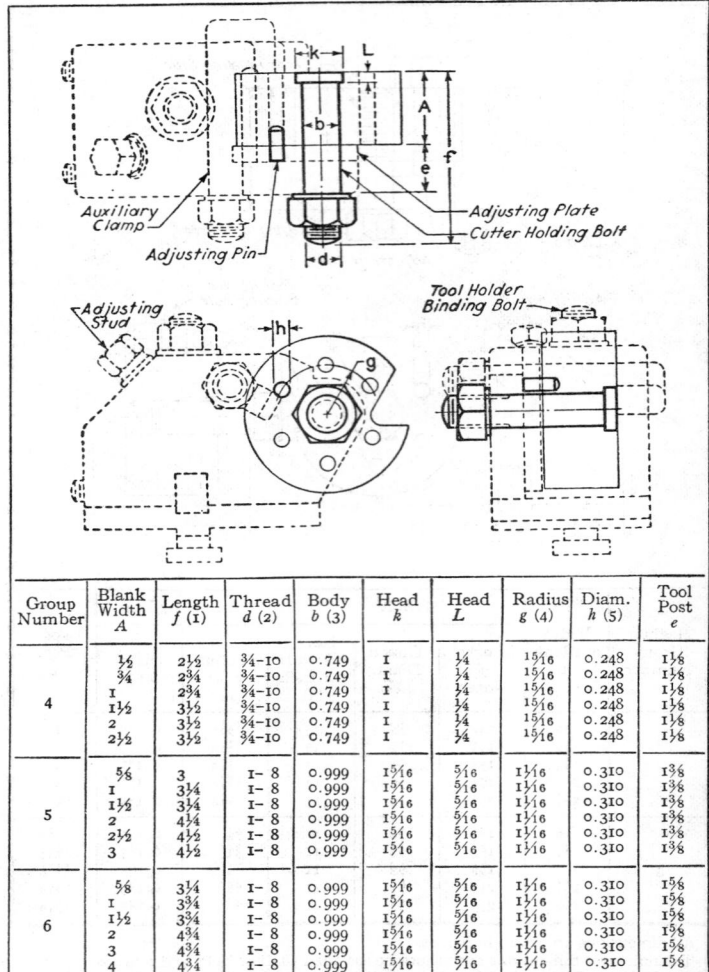

Group Number	Blank Width A	Length f (1)	Thread d (2)	Body b (3)	Head k	Head L	Radius g (4)	Diam. h (5)	Tool Post e
4	1/2	2 1/2	3/4-10	0.749	1	1/4	15/16	0.248	1 1/8
	3/4	2 3/4	3/4-10	0.749	1	1/4	15/16	0.248	1 1/8
	1	2 3/4	3/4-10	0.749	1	1/4	15/16	0.248	1 1/8
	1 1/2	3 1/2	3/4-10	0.749	1	1/4	15/16	0.248	1 1/8
	2	3 1/2	3/4-10	0.749	1	1/4	15/16	0.248	1 1/8
	2 1/2	3 1/2	3/4-10	0.749	1	1/4	15/16	0.248	1 1/8
5	5/8	3	1-8	0.999	1 5/16	5/16	1 1/16	0.310	1 3/8
	1	3 1/4	1-8	0.999	1 5/16	5/16	1 1/16	0.310	1 3/8
	1 1/2	3 3/4	1-8	0.999	1 5/16	5/16	1 1/16	0.310	1 3/8
	2	4 1/4	1-8	0.999	1 5/16	5/16	1 1/16	0.310	1 3/8
	2 1/2	4 1/2	1-8	0.999	1 5/16	5/16	1 1/16	0.310	1 3/8
	3	4 1/2	1-8	0.999	1 5/16	5/16	1 1/16	0.310	1 3/8
6	5/8	3 1/4	1-8	0.999	1 5/16	5/16	1 1/16	0.310	1 5/8
	1	3 3/4	1-8	0.999	1 5/16	5/16	1 1/16	0.310	1 5/8
	1 1/2	3 3/4	1-8	0.999	1 5/16	5/16	1 1/16	0.310	1 5/8
	2	4 3/4	1-8	0.999	1 5/16	5/16	1 1/16	0.310	1 5/8
	3	4 3/4	1-8	0.999	1 5/16	5/16	1 1/16	0.310	1 5/8
	4	4 3/4	1-8	0.999	1 5/16	5/16	1 1/16	0.310	1 5/8

Tolerance for dimensions not otherwise specified shall be held to ±0.010.
(1) Tolerance for length of bolt f is ±1/32.
(2) The screw thread d shall be made to Unified and American Standard, Class 3A.
(3) The body of the screw b shall have tolerance of +0.000, −0.001.
(4) Tolerance for the radius of the adjusting pin hole location g is ±0.001.
(5) Tolerance for the diameter of the adjusting pin h is +0.000, −0.001.

Table 8. Former ASA Standard Typical Dovetail Forming Tool-holders

Tool-holder with Collar-head Screw Adjustment

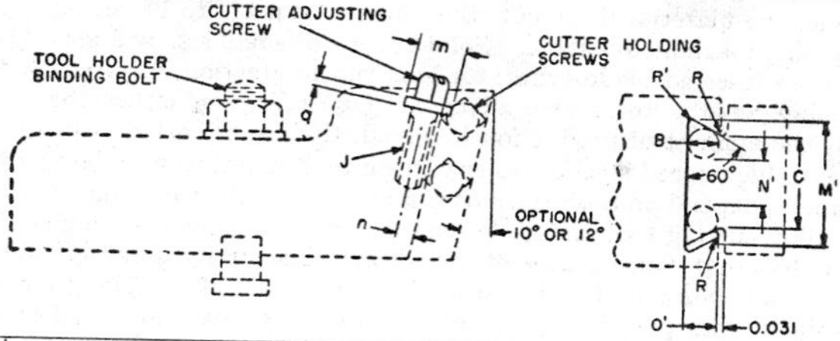

M' +.010 −.000	C +.010 −.000	N' +.002 −.000	O' ±.003	B ±.0003	R +.010 −.000	R' +.010 −.000	n +.010 −.000	m +.010 −.000	q +.000 −.002	J UNF-3A
0.732	0.489	0.305	¼	0.156	¹⁄₃₂	¹⁄₆₄	1¹⁄₆₄	⁹⁄₁₆	0.122	¼ −28
0.951	0.690	0.524	¹⁷⁄₆₄	0.156	¹⁄₃₂	¹⁄₆₄	1⁵⁄₆₄	1¹⁄₁₆	0.122	⁵⁄₁₆-24
1.250	0.9085	0.567	⅜	0.250	¹⁄₁₆	¹⁄₃₂	1⁵⁄₆₄	1¹⁄₁₆	0.122	⁵⁄₁₆-24
1.614	1.110	0.248	³³⁄₆₄	0.500	¹⁄₁₆	¹⁄₃₂				
1.882	1.378	0.516	³³⁄₆₄	0.500	¹⁄₁₆	¹⁄₃₂				
2.000	1.532	0.634	³¹⁄₆₄	0.500	¹⁄₁₆	¹⁄₃₂	These dimensions			
2.238	1.734	0.872	³³⁄₆₄	0.500	¹⁄₁₆	¹⁄₃₂	are not given in			
2.883	2.235	1.175	⁴¹⁄₆₄	0.625	¹⁄₁₆	¹⁄₃₂	the standard.			

Tool-holder with Hook-bolt Adjustment

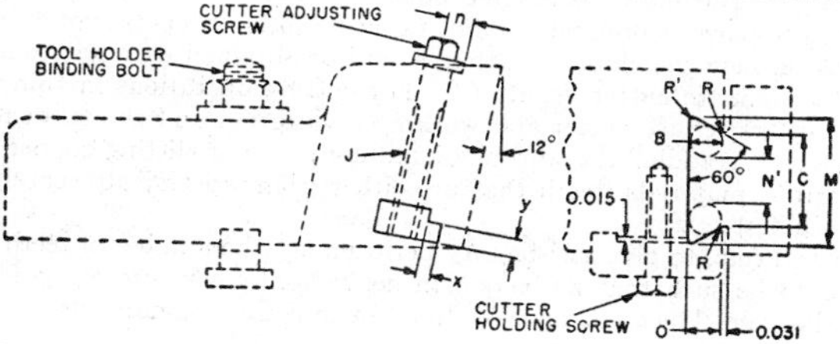

M' +.010 −.000	C +.010 −.000	N' +.002 −.000	O' ±.003	B ±.0003	R +.010 −.000	R' +.010 −.000	n +.010 −.000	x +.010 −.000	y +.010 −.000	J *
0.732	0.489	0.305	¼	0.156	¹⁄₃₂	¹⁄₆₄	⁷⁄₃₂	⅛	⅛	¼ −28
0.951	0.690	0.524	¹⁷⁄₆₄	0.156	¹⁄₃₂	¹⁄₆₄	¼	³⁄₁₆	³⁄₁₆	⁵⁄₁₆-24
1.250	0.9085	0.567	⅜	0.250	¹⁄₁₆	¹⁄₃₂	¼	¼	¼	⁵⁄₁₆-24
1.614	1.110	0.248	³³⁄₆₄	0.500	¹⁄₁₆	¹⁄₃₂	⁹⁄₃₂	¼	¼	⅜ −24
1.882	1.378	0.516	³³⁄₆₄	0.500	¹⁄₁₆	¹⁄₃₂	⁹⁄₃₂	⁵⁄₁₆	⁵⁄₁₆	⅜ −24
2.000	1.532	0.634	³¹⁄₆₄	0.500	¹⁄₁₆	¹⁄₃₂	⁵⁄₁₆	⁵⁄₁₆	⁵⁄₁₆	⁷⁄₁₆-20
2.238	1.734	0.872	³³⁄₆₄	0.500	¹⁄₁₆	¹⁄₃₂	⁵⁄₁₆	⁵⁄₁₆	⁵⁄₁₆	⁷⁄₁₆-20
2.883	2.235	1.175	⁴¹⁄₆₄	0.625	¹⁄₁₆	¹⁄₃₂	⁵⁄₁₆	⁵⁄₁₆	⁵⁄₁₆	⁷⁄₁₆-20

All dimensions are given in inches.
* The adjusting screw thread is UNF-3A; adjusting nut thread is UNF-2B.

MILLING CUTTERS

Selection of Milling Cutters. — The most suitable type of milling cutter for a particular milling operation will depend upon, among other factors, the kind of cut to be made, the material to be cut, the number of parts to be machined, and the type of milling machine available. Solid cutters of small size will usually cost less, initially, than inserted blade types; for long-run production, inserted-blade cutters will probably have a lower over-all cost. Depending on either the material to be cut, or the amount of production involved, the use of carbide tipped cutters in preference to high-speed steel or other cutting tool materials may be justified.

Rake angles depend on both the cutter material and the work material. Carbide and cast alloy cutting tool materials generally have smaller rake angles than high-speed steel tool materials because of their lower edge strength and greater abrasion resistance. Soft work materials permit higher radial rake angles than hard materials; thin cutters permit zero or practically zero axial rake angles; and wide cutters operate smoother with high axial rake angles. (See page 1621.)

Cutting edge relief or clearance angles are usually from 3 to 6 degrees for hard or tough materials, 4 to 7 degrees for average materials, and 6 to 12 degrees for easily machined materials. (See page 790.)

The number of teeth in the milling cutter is also a factor that should be given consideration as explained in the next paragraph.

Number of Teeth in Milling Cutters. — In determining the number of teeth a milling cutter should have for optimum performance, there is no universal rule that covers every case. There are, however, two factors which should be considered in making a choice: (1) The number of teeth should never be so great as to reduce the chip space between the teeth to a point where a free flow of chips is prevented; (2) The chip space should be smooth and without sharp corners that would cause clogging of the chips in the space.

For milling ductile materials which produce a continuous and curled chip, a cutter with large chip spaces is preferable. Such coarse tooth cutters permit an easier flow of the chips through the chip space than would be obtained with fine tooth cutters, and help to eliminate cutter "chatter." For cutting operations in thin materials, fine tooth cutters reduce cutter and workpiece vibration and the tendency for the cutter teeth to "straddle" the workpiece and dig in. For slitting copper and other soft non-ferrous materials, teeth that are either chamfered or alternately flat and V-shaped are best.

As a general rule, to give satisfactory performance the number of teeth in milling cutters should be such that *no more than two teeth at a time are engaged in the cut.* Based on this rule, The Cincinnati Milling Machine Co. recommends the use of the following formulas:

For face milling cutters,

$$T = \frac{6.3\,D}{W} \tag{1}$$

For peripheral milling cutters,

$$T = \frac{12.6\,D\cos A}{D + 4d} \tag{2}$$

where T = number of teeth in cutter; D = cutter diameter in inches; W = width of cut in inches; d = depth of cut in inches; and A = helix angle of cutter.

To find the number of teeth that a cutter should have when other than two teeth in the cut at the same time is desired, Formulas (1) and (2) should be divided by two and the result multiplied by the number of teeth desired in the cut.

Example: Determine the required number of teeth in a face mill where $D = 6$ inches and $W = 4$ inches. Using Formula (1),

$$T = \frac{6.3 \times 6}{4} = 10 \text{ teeth, approximately}$$

Example: Determine the required number of teeth in a plain milling cutter where $D = 4$ inches and $d = \frac{1}{4}$ inch. Using Formula (2),

$$T = \frac{12.6 \times 4 \times \cos 0°}{4 + (4 \times \frac{1}{4})} = 10 \text{ teeth, approximately}$$

In *high speed milling* with sintered carbide, high-speed steel, and cast non-ferrous cutting tool materials, a formula that permits full use of the power available at the cutter but prevents overloading of the motor driving the milling machine is:

$$T = \frac{K \times H}{F \times N \times d \times W} \qquad (3)$$

where T = number of cutter teeth; H = horsepower available at the cutter; F = feed per tooth in inches; N = revolutions per minute of cutter; d = depth of cut in inches; W = width of cut in inches; and K = a constant which may be taken as 0.65 for average steel, 1.5 for cast iron, and 2.5 for aluminum. These values are conservative and take into account dulling of the cutter in service.

Example: Determine the required number of teeth in a sintered carbide tipped face mill for high speed milling of 200 Brinell hardness alloy steel if $H = 10$ horsepower; $F = 0.008$ inch; $N = 272$ rpm; $d = 0.125$ inch; $W = 6$ inches; and K for alloy steel is 0.65. Using Formula (3),

$$T = \frac{0.65 \times 10}{0.008 \times 272 \times 0.125 \times 6} = 4 \text{ teeth, approximately}$$

American National Standard Milling Cutters. — According to American National Standard ANSI B94.19-1985, milling cutters may be classified in two general ways which are given as follows:

By Type of Relief on Cutting Edges: Milling cutters may be described on the basis of one of two methods of providing relief for the cutting edges. *Profile sharpened* cutters are those on which relief is obtained and which are resharpened by grinding a narrow land back of the cutting edges. Profile sharpened cutters may produce flat, curved, or irregular surfaces. *Form relieved* cutters are those which are so relieved that by grinding only the faces of the teeth the original form is maintained throughout the life of the cutters. Form relieved cutters may produce flat, curved or irregular surfaces.

By Method of Mounting: Milling cutters may be described by one of two methods used to mount the cutter. *Arbor type* cutters are those which have a hole for mounting on an arbor and usually have a keyway to receive a driving key. These are sometimes called *Shell type*. *Shank type* cutters are those which have a straight or tapered shank to fit the machine tool spindle or adapter.

Explanation of the "Hand" of Milling Cutters. — In the ANSI Standard the terms "right hand" and "left hand" are used to describe hand of rotation, hand of cutter and hand of flute helix.

Hand of Rotation or *Hand of Cut* is described as either "right hand" if the cutter revolves counterclockwise as it cuts when viewed from a position in front of a horizontal milling machine and facing the spindle or "left hand" if the cutter revolves clockwise as it cuts when viewed from the same position.

American National Standard Plain Milling Cutters (ANSI B94.19-1985)

Cutter Diameter			Range of Face Widths Nom.*	Hole Diameter		
Nom.	Max.	Min.		Nom.	Max.	Min.
Light-duty Cutters[1]						
2½	2.515	2.485	3/16, 1/4, 5/16, 3/8, 1/2, 5/8, 3/4, 1, 1½, 2 and 3	1	1.00075	1.0000
3	3.015	2.985	3/16, 1/4, 5/16, 3/8, 5/8, 3/4, and 1½	1	1.00075	1.0000
3	3.015	2.985	1/2, 5/8, 3/4, 1, 1¼, 1½, 2 and 3	1¼	1.2510	1.2500
4	4.015	3.985	1/4, 5/16 and 3/8	1	1.00075	1.0000
4	4.015	3.985	3/8, 1/2, 5/8, 3/4, 1, 1½, 2, 3 and 4	1¼	1.2510	1.2500
Heavy-duty Cutters[2]						
2½	2.515	2.485	2	1	1.00075	1.0000
2½	2.515	2.485	4	1	1.0010	1.0000
3	3.015	2.985	2, 2½, 3, 4 and 6	1¼	1.2510	1.2500
4	4.015	3.985	2, 3, 4 and 6	1½	1.5010	1.5000
High-helix Cutters[3]						
3	3.015	2.985	4 and 6	1¼	1.2510	1.2500
4	4.015	3.985	8	1½	1.5010	1.5000

All dimensions are in inches. All cutters are high-speed steel. Plain milling cutters are of cylindrical shape, having teeth on the peripheral surface only.

* *Tolerances on Face Widths:* Up to 1 inch, inclusive, ± 0.001 inch; over 1 to 2 inches, inclusive, + 0.010, − 0.000 inch; over 2 inches, + 0.020, − 0.000 inch.

[1] Light-duty plain milling cutters with face widths under ¾ inch have straight teeth. Cutters with ¾-inch face and wider have helix angles of not less than 15 degrees nor greater than 25 degrees.

[2] Heavy-duty plain milling cutters have a helix angle of not less than 25 degrees nor greater than 45 degrees.

[3] High-helix plain milling cutters have a helix angle of not less than 45 degrees nor greater than 52 degrees.

Hand of Cutter: Some types of cutters require special consideration when referring to their hand. These are principally cutters with unsymmetrical forms, face type cutters, or cutters with threaded holes. *Symmetrical* cutters may be reversed on the arbor in the same axial position and rotated in the cutting direction without altering the contour produced on the work-piece, and may be considered as either right or left hand. *Unsymmetrical* cutters reverse the contour produced on the work-piece when reversed on the arbor in the same axial position and rotated in the cutting direction. A *single-angle* cutter is considered to be a right-hand cutter if it revolves counterclockwise, or a left-hand cutter if it revolves clockwise, when cutting as viewed from the side of the larger diameter. The *hand of rotation* of a single angle milling cutter need not necessarily be the same as its *hand of cutter.* A *single corner rounding* cutter is considered to be a right-hand cutter if it revolves counterclockwise, or a left-hand cutter if it revolves clockwise, when cutting as viewed from the side of the smaller diameter.

American National Standard Side Milling Cutters (ANSI B94.19-1985)

Cutter Diameter			Range of Face Widths Nom.*	Hole Diameter		
Nom.	Max.	Min.		Nom.	Max.	Min.
Side Cutters[1]						
2	2.015	1.985	3/16, 1/4, 3/8	5/8	0.62575	0.6250
2 1/2	2.515	2.485	1/4, 3/8, 1/2	7/8	0.87575	0.8750
3	3.015	2.985	1/4, 5/16, 3/8, 7/16, 1/2	1	1.00075	1.0000
4	4.015	3.985	1/4, 3/8, 1/2, 5/8, 3/4, 7/8	1	1.00075	1.0000
4	4.015	3.985	1/2, 5/8, 3/4	1 1/4	1.2510	1.2500
5	5.015	4.985	1/2, 5/8, 3/4	1	1.00075	1.0000
5	5.015	4.985	1/2, 5/8, 3/4, 1	1 1/4	1.2510	1.2500
6	6.015	5.985	1/2	1	1.00075	1.0000
6	6.015	5.985	1/2, 5/8, 3/4, 1	1 1/4	1.2510	1.2500
7	7.015	6.985	3/4	1 1/4	1.2510	1.2500
7	7.015	6.985	3/4	1 1/2	1.5010	1.5000
8	8.015	7.985	3/4, 1	1 1/4	1.2510	1.2500
8	8.015	7.985	3/4, 1	1 1/2	1.5010	1.5000
Staggered-tooth Side Cutters[2]						
2 1/2	2.515	2.485	1/4, 5/16, 3/8, 1/2	7/8	0.87575	0.8750
3	3.015	2.985	3/16, 1/4, 5/16, 3/8	1	1.00075	1.0000
3	3.015	2.985	1/2, 5/8, 3/4	1 1/4	1.2510	1.2500
4	4.015	3.985	1/4, 5/16, 3/8, 7/16, 1/2, 5/8, 3/4 and 7/8	1 1/4	1.2510	1.2500
5	5.015	4.985	1/2, 5/8, 3/4	1 1/4	1.2510	1.2500
6	6.015	5.985	3/8, 1/2, 5/8, 3/4, 7/8, 1	1 1/4	1.2510	1.2500
8	8.015	7.985	3/8, 1/2, 5/8, 3/4, 1	1 1/2	1.5010	1.5000
Half Side Cutters[3]						
4	4.015	3.985	3/4	1 1/4	1.2510	1.2500
5	5.015	4.985	3/4	1 1/4	1.2510	1.2500
6	6.015	5.985	3/4	1 1/4	1.2510	1.2500

All dimensions are in inches. All cutters are high-speed steel. Side milling cutters are of cylindrical shape, having teeth on the periphery and on one or both sides.

* *Tolerances on Face Widths:* For side cutters, up to 3/4 inch face width, inclusive, +0.000 −0.0005 inch, and over 3/4 to 1 inch, inclusive, +0.000 −0.0010 inch; and for half side cutters, +0.015, −0.000 inch.

[1] Side milling cutters have straight peripheral teeth and side teeth on both sides.

[2] Staggered-tooth side milling cutters have peripheral teeth of alternate right- and left-hand helix and alternate side teeth.

[3] Half side milling cutters have side teeth on one side only. The peripheral teeth are helical of the same hand as the cut. Made either with right-hand or left-hand cut.

Hand of Flute Helix: Milling cutters may have *straight flutes* which means that their cutting edges are in planes parallel to the cutter axis. Milling cutters with flute helix in one direction only are described as having a right-hand helix if the flutes twist away from the observer in a clockwise direction when viewed from either end of the cutter or as having a left-hand helix if the flutes twist away from the observer in a counterclockwise direction when viewed from either end of the cutter. *Staggered tooth cutters* are milling cutters with every other flute of opposite (right and left hand) helix.

An illustration describing the various milling cutter elements of both a profile cutter and a form-relieved cutter is given on page 766.

American National Standard Staggered Teeth, T-Slot Milling Cutters with Brown & Sharpe Taper and Weldon Shanks (ANSI B94.19-1985)

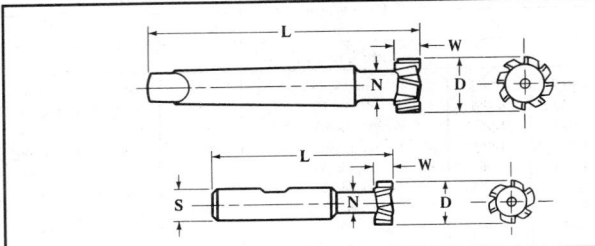

Bolt Size	Cutter Diam., D	Face Width, W	Neck Diam., N	With B. & S. Taper*		With Weldon Shank	
				Length, L	Taper No.	Length, L	Diam., S
1/4	9/16	15/64	17/64	. . .	. . .	2 19/32	1/2
5/16	21/32	17/64	21/64	. . .	. . .	2 11/16	1/2
3/8	25/32	21/64	13/32	. . .	. . .	3 1/4	3/4
1/2	31/32	25/64	17/32	5	7	3 7/16	3/4
5/8	1 1/4	31/64	21/32	5 1/4	7	3 15/16	1
3/4	1 15/32	5/8	25/32	6 7/8	9	4 7/16	1
1	1 27/32	53/64	1 1/32	7 1/4	9	4 13/16	1 1/4

All dimensions are in inches. All cutters are high-speed steel and only right-hand cutters are standard.
   * For dimensions of Brown & Sharpe taper shanks, see information given on page 901.
   Tolerances: On D, + 0.000, − 0.010 inch; on W, + 0.000, − 0.005 inch; on N, + 0.000, − 0.005 inch; on L, ± 1/16 inch; on S, − 0.0001 to − 0.0005 inch.

American National Standard Form Relieved Corner Rounding Cutters with Weldon Shanks (ANSI B94.19-1985)

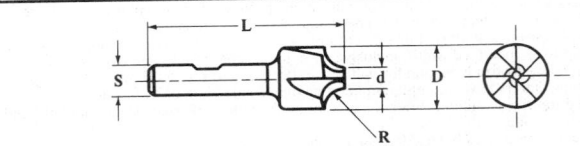

Rad., R	Diam., D	Diam., d	S	L	Rad., R	Diam., D	Diam., d	S	L
1/16	7/16	1/4	3/8	2 1/2	3/8	1 1/4	3/8	1/2	3 1/2
3/32	1/2	1/4	3/8	2 1/2	3/16	7/8	5/16	3/4	3 1/8
1/8	5/8	1/4	1/2	3	1/4	1	3/8	3/4	3 1/4
5/32	3/4	5/16	1/2	3	5/16	1 1/8	3/8	7/8	3 1/2
3/16	7/8	5/16	1/2	3	3/8	1 1/4	3/8	7/8	3 3/4
1/4	1	3/8	1/2	3	7/16	1 3/8	3/8	1	4
5/16	1 1/8	3/8	1/2	3 1/4	1/2	1 1/2	3/8	1	4 1/8

All dimensions are in inches. All cutters are high-speed steel. Right-hand cutters are standard.
   Tolerances: On D, ± 0.010 inch; on diameter of circle, $2R$, ± 0.001 inch for cutters up to and including 1/8-inch radius; + 0.002, − 0.001 inch for cutters over 1/8-inch radius; on S, − 0.0001 to − 0.0005 inch; and on L, ± 1/16 inch.

American National Standard Metal Slitting Saws (ANSI B94.19-1985)

Cutter Diameter			Range of Face Widths Nom.*	Hole Diameter		
Nom.	Max.	Min.		Nom.	Max.	Min.
Plain Metal Slitting Saws¹						
2½	2.515	2.485	⅟₃₂, ³⁄₆₄, ¹⁄₁₆, ³⁄₃₂, ⅛	⅞	0.87575	0.8750
3	3.015	2.985	⅟₃₂, ³⁄₆₄, ¹⁄₁₆, ³⁄₃₂, ⅛ and ⁵⁄₃₂	1	1.00075	1.0000
4	4.015	3.985	⅟₃₂, ³⁄₆₄, ¹⁄₁₆, ³⁄₃₂, ⅛, ⁵⁄₃₂ and ³⁄₁₆	1	1.00075	1.0000
5	5.015	4.985	¹⁄₁₆, ³⁄₃₂, ⅛	1	1.00075	1.0000
5	5.015	4.985	⅛	1¼	1.2510	1.2500
6	6.015	5.985	¹⁄₁₆, ³⁄₃₂, ⅛	1	1.00075	1.0000
6	6.015	5.985	⅛, ³⁄₁₆	1¼	1.2510	1.2500
8	8.015	7.985	⅛	1	1.00075	1.0000
8	8.015	7.985	⅛	1¼	1.2510	1.2500
Metal Slitting Saws with Side Teeth²						
2½	2.515	2.485	¹⁄₁₆, ³⁄₃₂, ⅛	⅞	0.87575	0.8750
3	3.015	2.985	¹⁄₁₆, ³⁄₃₂, ⅛, ⁵⁄₃₂	1	1.00075	1.0000
4	4.015	3.985	¹⁄₁₆, ³⁄₃₂, ⅛, ⁵⁄₃₂, ³⁄₁₆	1	1.00075	1.0000
5	5.015	4.985	¹⁄₁₆, ³⁄₃₂, ⅛, ⁵⁄₃₂, ³⁄₁₆	1	1.00075	1.0000
5	5.015	4.985	⅛	1¼	1.2510	1.2500
6	6.015	5.985	¹⁄₁₆, ³⁄₃₂, ⅛, ³⁄₁₆	1	1.00075	1.0000
6	6.015	5.985	⅛, ³⁄₁₆	1¼	1.2510	1.2500
8	8.015	7.985	⅛	1	1.00075	1.0000
8	8.015	7.985	⅛, ³⁄₁₆	1¼	1.2510	1.2500
Metal Slitting Saws with Staggered Peripheral and Side Teeth³						
3	3.015	2.985	³⁄₁₆	1	1.00075	1.0000
4	4.015	3.985	³⁄₁₆	1	1.00075	1.0000
5	5.015	4.985	³⁄₁₆, ¼	1	1.00075	1.0000
6	6.015	5.985	³⁄₁₆, ¼	1	1.00075	1.0000
6	6.015	5.985	³⁄₁₆, ¼	1¼	1.2510	1.2500
8	8.015	7.985	³⁄₁₆, ¼	1¼	1.2510	1.2500
10	10.015	9.985	³⁄₁₆, ¼	1¼	1.2510	1.2500
12	12.015	11.985	¼, ⁵⁄₁₆	1½	1.5010	1.5000

All dimensions are in inches. All saws are high-speed steel. Metal slitting saws are similar to plain or side milling cutters but are relatively thin.

* Tolerances on face widths are plus or minus 0.001 inch.

¹ Plain metal slitting saws are relatively thin plain milling cutters having peripheral teeth only. They are furnished with or without hub and their sides are concaved to the arbor hole or hub.

² Metal slitting saws with side teeth are relatively thin side milling cutters having both peripheral and side teeth.

³ Metal slitting saws with staggered peripheral and side teeth are relatively thin staggered tooth milling cutters having peripheral teeth of alternate right- and left-hand helix and alternate side teeth.

Milling Cutter Terms

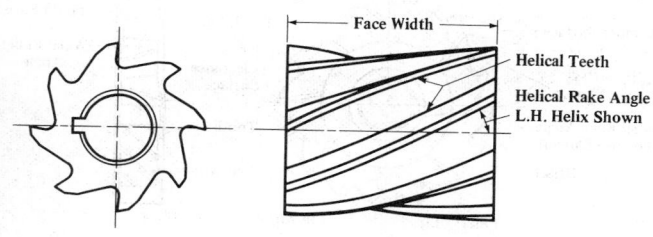

Face Width

Helical Teeth

Helical Rake Angle
L.H. Helix Shown

American National Standard Single- and Double-angle Milling Cutters
(ANSI B94.19-1985)

Cutter Diameter				Hole Diameter		
Nom.	Max.	Min.	Nominal Face Width*	Nom.	Max.	Min.
Single-angle Cutters[1]						
†1¼	1.265	1.235	⁷⁄₁₆	⅜-24 UNF-2B RH ⅜-24 UNF-2B LH ½-20 UNF-2B RH		
†1⅝	1.640	1.610	⁹⁄₁₆			
2¾	2.765	2.735	½	1	1.00075	1.0000
3	3.015	2.985	½	1¼	1.2510	1.2500
Double-angle Cutters[2]						
2¾	2.765	2.735	½	1	1.00075	1.0000

All dimensions are in inches. All cutters are high-speed steel.

* Face width tolerances are plus or minus 0.015 inch.

† These cutters have threaded holes, the sizes of which are given under "Hole Diameter."

[1] Single-angle milling cutters have peripheral teeth, one cutting edge of which lies in a conical surface and the other in the plane perpendicular to the cutter axis. There are two types: one has a plain keywayed hole and has an included tooth angle of either 45 or 60 degrees plus or minus 10 minutes; the other has a threaded hole and has an included tooth angle of 60 degrees plus or minus 10 minutes. Cutters with a right-hand threaded hole have a right-hand hand of rotation and a right-hand hand of cutter. Cutters with a left-hand threaded hole have a left-hand hand of rotation and a left-hand hand of cutter. Cutters with plain keywayed holes are standard as either right-hand or left-hand cutters.

[2] Double-angle milling cutters have symmetrical peripheral teeth both sides of which lie in conical surfaces. They are designated by the included angle, which may be 45, 60 or 90 degrees. Tolerances are plus or minus 10 minutes for the half angle on each side of the center.

Milling Cutter Terms (*Continued*)

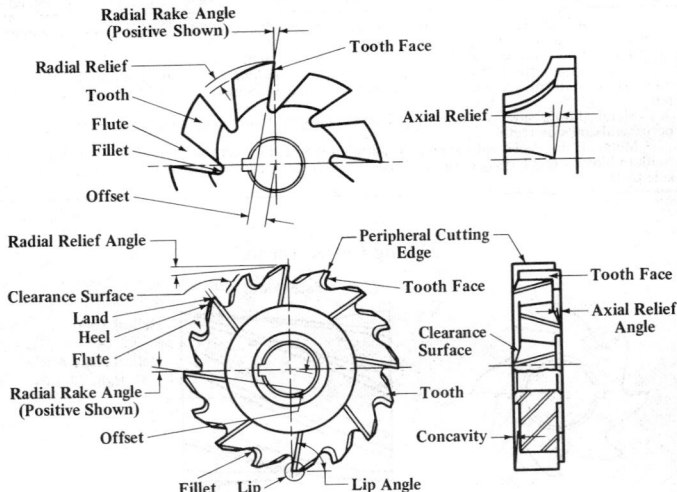

American National Standard Shell Mills (ANSI B94.19-1985)

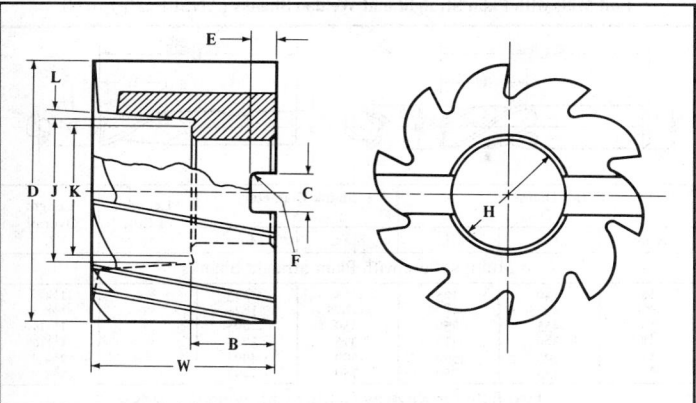

Diam., D	Width, W	Diam., H	Length, B	Width, C	Depth, E	Radius, F	Diam., J	Diam., K	Angle, L
inches	inches	inches	inches	inches	inches	inches	inches	inches	degrees
1¼	1	½	⅝	¼	5⁄32	1⁄64	11⁄16	⅝	0
1½	1⅛	½	⅝	¼	5⁄32	1⁄64	11⁄16	⅝	0
1¾	1¼	¾	¾	5⁄16	3⁄16	1⁄32	15⁄16	⅞	0
2	1⅜	¾	¾	5⁄16	3⁄16	1⁄32	15⁄16	⅞	0
2¼	1½	1	¾	⅜	7⁄32	1⁄32	1¼	1 3⁄16	0
2½	1⅝	1	¾	⅜	7⁄32	1⁄32	1⅜	1 3⁄16	0
2¾	1⅝	1	¾	⅜	7⁄32	1⁄32	1½	1 3⁄16	5
3	1¾	1¼	¾	½	9⁄32	1⁄32	1 21⁄32	1½	5
3½	1⅞	1¼	¾	½	9⁄32	1⁄32	1 11⁄16	1½	5
4	2¼	1½	1	⅝	⅜	1⁄16	2 1⁄32	1⅞	5
4½	2¼	1½	1	⅝	⅜	1⁄16	2 1⁄16	1⅞	10
5	2¼	1½	1	⅝	⅜	1⁄16	2 9⁄16	1⅞	10
6	2¼	2	1	¾	7⁄16	1⁄16	2 13⁄16	2½	15

All cutters are high-speed steel. Right-hand cutters with right-hand helix and square corners are standard.

Tolerances: On D, + 1⁄64 inch; on W, ± 1⁄64 inch; on H, + 0.0005 inch; on B, + 1⁄64 inch; on C, at least + 0.008 but not more than + 0.012 inch; on E, + 1⁄64 inch; on J, ± 1⁄64 inch; on K, ± 1⁄64 inch.

End Mill Terms

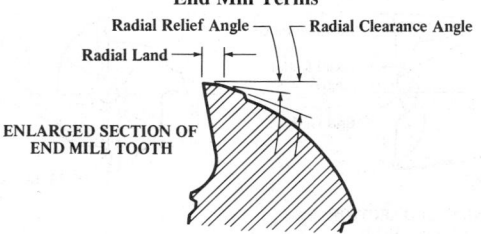

American National Standard Multiple- and Two-flute Single End Helical End Mills with Plain Straight and Weldon Shanks (ANSI B94.19-1985)

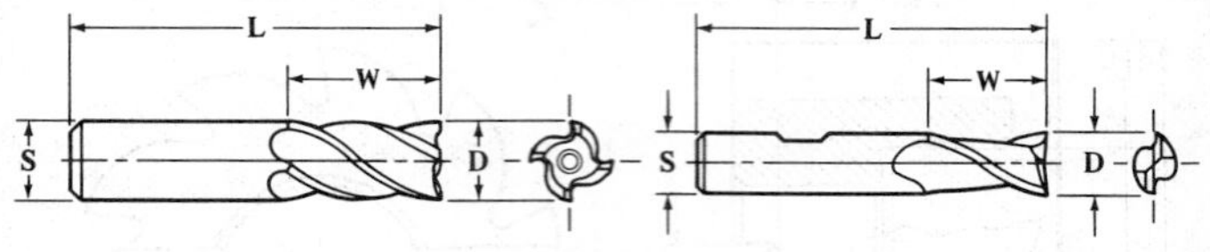

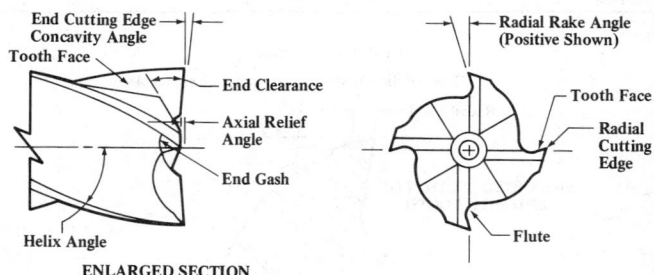

Cutter Diameter, D			Shank Diameter, S		Length of Cut, W	Length Overall, L
Nom.	Max.	Min.	Max.	Min.		
Multiple-flute with Plain Straight Shanks						
1/8	.130	.125	.125	.1245	5/16	1 1/4
3/16	.1925	.1875	.1875	.1870	1/2	1 3/8
1/4	.255	.250	.250	.2495	5/8	1 11/16
3/8	.380	.375	.375	.3745	3/4	1 13/16
1/2	.505	.500	.500	.4995	15/16	2 1/4
3/4	.755	.750	.750	.7495	1 1/4	2 5/8
Two-flute for Keyway Cutting with Weldon Shanks						
1/8	.125	.1235	.375	.3745	3/8	2 5/16
3/16	.1875	.1860	.375	.3745	7/16	2 5/16
1/4	.250	.2485	.375	.3745	1/2	2 5/16
5/16	.3125	.3110	.375	.3745	9/16	2 5/16
3/8	.375	.3735	.375	.3745	9/16	2 5/16
1/2	.500	.4985	.500	.4995	1	3
5/8	.625	.6235	.625	.6245	1 5/16	3 7/16
3/4	.750	.7485	.750	.7495	1 5/16	3 9/16
7/8	.875	.8735	.875	.8745	1 1/2	3 3/4
1	1.000	.9985	1.000	.9995	1 5/8	4 1/8
1 1/4	1.250	1.2485	1.250	1.2495	1 5/8	4 1/8
1 1/2	1.500	1.4985	1.250	1.2495	1 5/8	4 1/8

All dimensions are in inches. All cutters are high-speed steel. Right-hand cutters with right-hand helix are standard.

The helix angle is not less than 10 degrees for multiple-flute cutters with plain straight shanks; the helix angle is optional with the manufacturer for two-flute cutters with Weldon shanks.

Tolerances: On W, ± 1/32 inch; on L, ± 1/16 inch.

End Mill Terms (*Continued*)

End Cutting Edge —
Concavity Angle
Tooth Face
— End Clearance
— Axial Relief Angle
— End Gash
Helix Angle
ENLARGED SECTION OF END MILL

— Radial Rake Angle (Positive Shown)
— Tooth Face
— Radial Cutting Edge
— Flute

American National Standard Regular-, Long-, and Extra Long-length, Multiple-flute Medium Helix Single-end End Mills with Weldon Shanks (ANSI B94.19-1985)

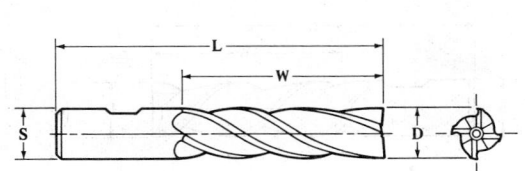

AS INDICATED BY THE DIMENSIONS GIVEN BELOW, SHANK DIAMETER S MAY BE LARGER, SMALLER, OR THE SAME AS THE CUTTER DIAMETER D.

Cutter Diam., D	Regular Mills				Long Mills				Extra Long Mills			
	S	W	L	N*	S	W	L	N*	S	W	L	N*
⅛†	3/8	3/8	2⁵⁄₁₆	4	...	...	...	...	...	...	...	...
³⁄₁₆†	3/8	1/2	2³⁄₈	4	...	...	...	...	...	...	...	...
¼†	3/8	5/8	2⁷⁄₁₆	4	3/8	1¼	3¹⁄₁₆	4	3/8	1¾	3⁹⁄₁₆	4
⁵⁄₁₆†	3/8	3/4	2½	4	3/8	1³⁄₈	3⅛	4	3/8	2	3¾	4
⅜†	3/8	3/4	2½	4	3/8	1½	3¼	4	3/8	2½	4¼	4
⁷⁄₁₆	3/8	1	2¹¹⁄₁₆	4	1/2	1¾	3¾	4	...	...	...	...
½	3/8	1	2¹¹⁄₁₆	4	1/2	2	4	4	1/2	3	5	4
½†	1/2	1¼	3¼	4	...	...	...	...	...	...	...	...
⁹⁄₁₆	1/2	1³⁄₈	3⅜	4	...	...	...	...	...	...	...	...
⅝	1/2	1³⁄₈	3⅜	4	5/8	2½	4⅝	4	5/8	4	6⅛	4
¹¹⁄₁₆	1/2	1⅝	3⅝	4	...	...	...	...	...	...	...	...
¾	1/2	1⅝	3⅝	4	3/4	3	5¼	4	3/4	4	6¼	4
⅝†	5/8	1⅝	3¾	4	...	...	...	...	...	...	...	...
¹¹⁄₁₆	5/8	1⅝	3¾	4	...	...	...	...	...	...	...	...
¾†	5/8	1⅝	3¾	4	...	...	...	...	...	...	...	...
¹³⁄₁₆	5/8	1⅞	4	6	...	...	...	...	...	...	...	...
⅞	5/8	1⅞	4	6	7/8	3½	5¾	4	7/8	5	7¼	4
1	5/8	1⅞	4	6	1	4	6½	4	1	6	8½	4
⅞	7/8	1⅞	4⅛	4	...	...	...	...	...	...	...	...
1	7/8	1⅞	4⅛	4	...	...	...	...	...	...	...	...
1⅛	7/8	2	4¼	6	1	4	6½	6	...	...	...	...
1¼	7/8	2	4¼	6	1	4	6½	6	1¼	6	8½	6
1	1	2	4½	4	...	...	...	...	...	...	...	...
1⅛	1	2	4½	6	...	...	...	...	...	...	...	...
1¼	1	2	4½	6	...	...	...	...	...	...	...	...
1⅜	1	2	4½	6	...	...	...	...	...	...	...	...
1½	1	2	4½	6	1	4	6½	6	...	...	...	...
1¼	1¼	2	4½	6	1¼	4	6½	6	...	...	...	...
1½	1¼	2	4½	6	1¼	4	6½	6	1¼	8	10½	6
1¾	1¼	2	4½	6	1¼	4	6½	6	...	...	...	...
2	1¼	2	4½	8	1¼	4	6½	8	...	...	...	...

All dimensions are in inches. All cutters are high-speed steel. Helix angle is greater than 19 degrees but not more than 39 degrees. Right-hand cutters with right-hand helix are standard.

Tolerances: On D, + 0.003 inch; on S, − 0.0001 to − 0.0005 inch; on W, ± ¹⁄₃₂ inch; on L, ± ¹⁄₁₆ inch.

* N = Number of flutes.

† In this size of regular mill a left-hand cutter with left-hand helix is also standard.

American National Standard Two-flute, High Helix, Regular-, Long-, and Extra Long-length, Single-end End Mills with Weldon Shanks (ANSI B94.19-1985)

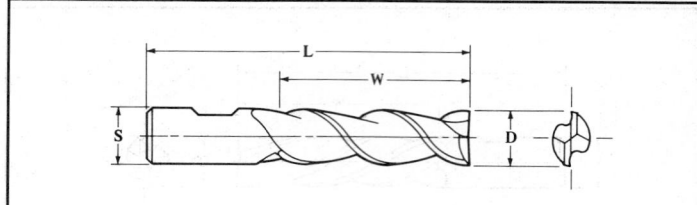

Cutter Diam., D	Regular Mill			Long Mill			Extra Long Mill		
	S	W	L	S	W	L	S	W	L
1/4	3/8	5/8	2 7/16	3/8	1 1/4	3 1/16	3/8	1 3/4	3 9/16
5/16	3/8	3/4	2 1/2	3/8	1 3/8	3 1/8	3/8	2	3 3/4
3/8	3/8	3/4	2 1/2	3/8	1 1/2	3 1/4	3/8	2 1/2	4 1/4
7/16	3/8	1	2 11/16	1/2	1 3/4	3 3/4	...	...	...
1/2	1/2	1 1/4	3 1/4	1/2	2	4	1/2	3	5
5/8	5/8	1 5/8	3 3/4	5/8	2 1/2	4 5/8	5/8	4	6 1/8
3/4	3/4	1 5/8	3 7/8	3/4	3	5 1/4	3/4	4	6 1/4
7/8	7/8	1 7/8	4 1/8	...	...	...	...	...	...
1	1	2	4 1/2	1	4	6 1/2	1	6	8 1/2
1 1/4	1 1/4	2	4 1/2	1 1/4	4	6 1/2	1 1/4	6	8 1/2
1 1/2	1 1/4	2	4 1/2	1 1/4	4	6 1/2	1 1/4	8	10 1/2
2	1 1/4	2	4 1/2	1 1/4	4	6 1/2	...	...	...

All dimensions are in inches. All cutters are high-speed steel. Right-hand cutters with right-hand helix are standard. Helix angle is greater than 39 degrees.

Tolerances: On D, + 0.003 inch; on S, − 0.0001 to − 0.0005 inch; on W, ± 1/32 inch; and on L, ± 1/16 inch.

American National Standard Combination Shanks* for End Mills (ANSI B94.19-1985)

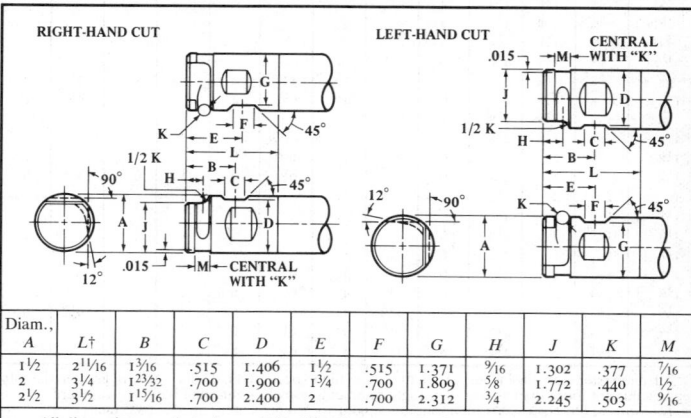

Diam., A	L†	B	C	D	E	F	G	H	J	K	M
1 1/2	2 11/16	1 3/16	.515	1.406	1 1/2	.515	1.371	9/16	1.302	.377	7/16
2	3 1/4	1 23/32	.700	1.900	1 3/4	.700	1.809	5/8	1.772	.440	1/2
2 1/2	3 1/2	1 15/16	.700	2.400	2	.700	2.312	3/4	2.245	.503	9/16

All dimensions are in inches.   * Modified for use as Weldon or Pin Drive shank.
† Length of shank.

American National Standard Roughing, Single-End End Mills with Weldon Shanks, High-Speed Steel (ANSI B94.19-1985)

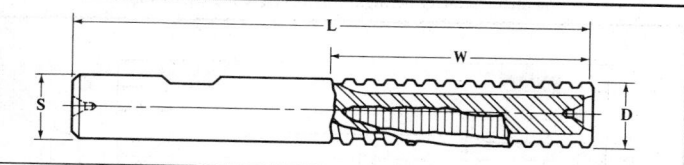

Diameter		Length		Diameter		Length	
Cutter	Shank	Cut	Overall	Cutter	Shank	Cut	Overall
D	*S*	*W*	*L*	*D*	*S*	*W*	*L*
½	½	1	3	2	2	2	5¾
½	½	1¼	3¼	2	2	3	6¾
½	½	2	4	2	2	4	7¾
⅝	⅝	1¼	3⅜	2	2	5	8¾
⅝	⅝	1⅝	3¾	2	2	6	9¾
⅝	⅝	2½	4⅝	2	2	7	10¾
¾	¾	1½	3¾	2	2	8	11¾
¾	¾	1⅝	3⅞	2	2	10	13¾
¾	¾	3	5¼	2	2	12	15¾
1	1	2	4½	2½	2	4	7¾
1	1	4	6½	2½	2	6	9¾
1¼	1¼	2	4½	2½	2	8	11¾
1¼	1¼	4	6½	2½	2	10	13¾
1½	1¼	2	4½	3	2½	4	7¾
1½	1¼	4	6½	3	2½	6	9¾
1¾	1¼	2	4½	3	2½	8	11¾
1¾	1¼	4	6½	3	2½	10	13¾

All dimensions are in inches. Right-hand cutters with right-hand helix are standard.
Tolerances: Outside diameter, + 0.025, − 0.005 inch; length of cut, + ⅛, − ¹⁄₃₂ inch.

American National Standard Heavy Duty, Medium Helix Single-End End Mills, 2½-inch Combination Shank, High-Speed Steel (ANSI B94.19-1985)

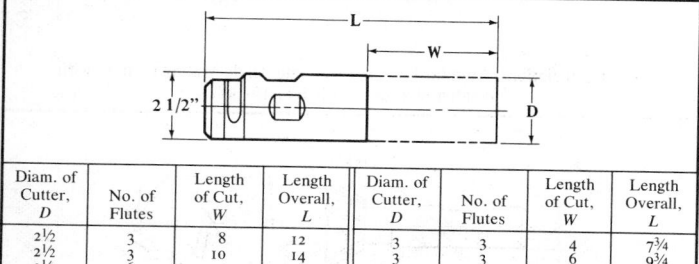

Diam. of Cutter, *D*	No. of Flutes	Length of Cut, *W*	Length Overall, *L*	Diam. of Cutter, *D*	No. of Flutes	Length of Cut, *W*	Length Overall, *L*
2½	3	8	12	3	3	4	7¾
2½	3	10	14	3	3	6	9¾
2½	6	4	8	3	3	8	11¾
2½	6	6	10	3	8	4	7¾
2½	6	8	12	3	8	6	9¾
2½	6	10	14	3	8	8	11¾
2½	6	12	16	3	8	10	13¾
3	2	4	7¾	3	8	12	15¾
3	2	6	9¾	. . .	. . .	. . .	. . .

All dimensions are in inches. For shank dimensions see page 770. Right-hand cutters with right-hand helix are standard. Helix angle is greater than 19 degrees but not more than 39 degrees.
Tolerances: On *D*, + 0.005 inch; on *W*, ± ¹⁄₃₂ inch; on *L*, ± ¹⁄₁₆ inch.

American National Standard Stub-, Regular-, and Long-length, Four-flute, Medium Helix, Plain-end, Double-end Miniature End Mills with 3/16-inch Diameter Straight Shanks (ANSI B94.19-1985)

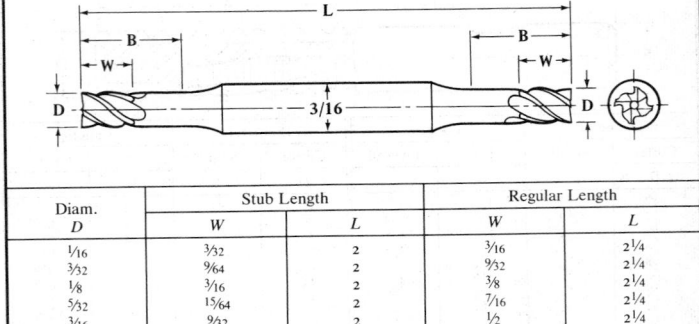

Diam.	Stub Length		Regular Length	
D	W	L	W	L
1/16	3/32	2	3/16	2 1/4
3/32	9/64	2	9/32	2 1/4
1/8	3/16	2	3/8	2 1/4
5/32	15/64	2	7/16	2 1/4
3/16	9/32	2	1/2	2 1/4

Diam.	Long Length		
D	B	W	L
1/16	3/8	7/32	2 1/2
3/32	1/2	9/32	2 5/8
1/8	3/4	3/4	3 1/8
5/32	7/8	7/8	3 1/4
3/16	1	1	3 3/8

All dimensions are in inches. All cutters are high-speed steel. Right-hand cutters with right-hand helix are standard. Helix angle is greater than 19 degrees but not more than 39 degrees.
Tolerances: On D, + 0.003 inch (if the shank is the same diameter as the cutting portion, however, then the tolerance on the cutting diameter is − 0.0025 inch.); on W, + 1/32, − 1/64 inch; and on L, ± 1/16 inch.

American National Standard 60-degree Single-Angle Milling Cutters with Weldon Shanks (ANSI B94.19-1985)

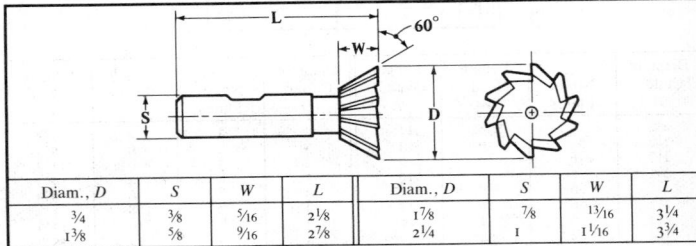

Diam., D	S	W	L	Diam., D	S	W	L
3/4	3/8	5/16	2 1/8	1 7/8	7/8	13/16	3 1/4
1 3/8	5/8	9/16	2 7/8	2 1/4	1	1 1/16	3 3/4

All dimensions are in inches. All cutters are high-speed steel. Right-hand cutters are standard.
Tolerances: On D, ± 0.015 inch; on S, − 0.0001 to − 0.0005 inch; on W, ± 0.015 inch; and on L, ± 1/16 inch.

American National Standard Stub-, Regular-, and Long-length, Two-flute, Medium Helix, Plain- and Ball-end, Double-end Miniature End Mills with 3/16-inch Diameter Straight Shanks (ANSI B94.19-1985)

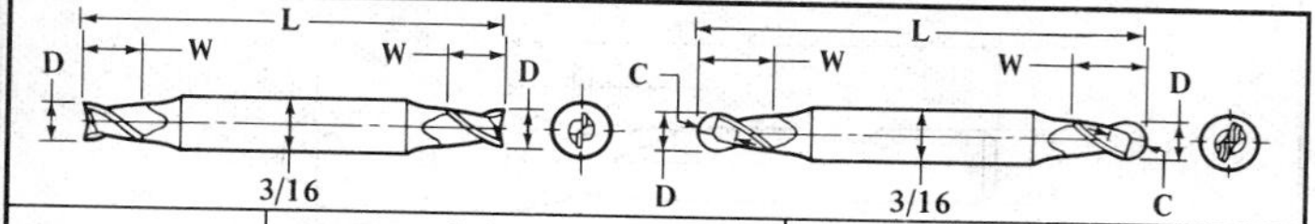

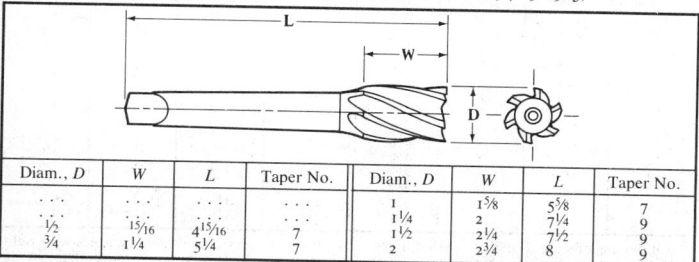

Diam., C and D	Stub Length				Regular Length			
	Plain End		Ball End		Plain End		Ball End	
	W	L	W	L	W	L	W	L
1/32	3/64	2	...	...	3/32	2¼	...	...
3/64	1/16	2	...	...	9/64	2¼	...	...
1/16	3/32	2	3/32	2	3/16	2¼	3/16	2¼
5/64	1/8	2	...	...	15/64	2¼	...	...
3/32	9/64	2	9/64	2	9/32	2¼	9/32	2¼
7/64	5/32	2	...	...	21/64	2¼	...	...
1/8	3/16	2	3/16	2	3/8	2¼	3/8	2¼
9/64	7/32	2	...	...	13/32	2¼	...	...
5/32	15/64	2	15/64	2	7/16	2¼	7/16	2¼
11/64	1/4	2	...	...	1/2	2¼	...	...
3/16	9/32	2	9/32	2	1/2	2¼	1/2	2¼

Diam., D	Long Length		
	Plain End		
	B*	W	L
1/16	3/8	7/32	2½
3/32	1/2	9/32	2⅝
1/8	3/4	3/4	3⅛
5/32	7/8	7/8	3¼
3/16	1	1	3⅜

All dimensions are in inches. All cutters are high-speed steel. Right-hand cutters with right-hand helix are standard. Helix angle is greater than 19 degrees but not more than 39 degrees.

Tolerances: On C and D, − 0.0015 inch for stub and regular length; + 0.003 inch for long length (if the shank is the same diameter as the cutting portion, however, then the tolerance on the cutting diameter is − 0.0025 inch.); on W, + 1/32, − 1/64 inch; and on L, ± 1/16 inch.

* B is the length below the shank.

American National Standard Multiple Flute, Helical Series End Mills with Brown & Sharpe Taper Shanks* (ANSI B94.19-1985)

Diam., D	W	L	Taper No.	Diam., D	W	L	Taper No.
...	...	...	...	1	1⅝	5⅝	7
1/2	15/16	4 15/16	7	1¼	2	7¼	9
3/4	1¼	5¼	7	1½	2¼	7½	9
				1¾	2¾	8	9

All dimensions are in inches. All cutters are high-speed steel. Right-hand cutters with right-hand helix are standard. Helix angle is not less than 10 degrees.

No. 5 taper is standard without tang; Nos. 7 and 9 are standard with tang only.

Tolerances: On D, + 0.005 inch; on W, ± 1/32 inch; and on L ± 1/16 inch.

* For dimensions of B & S taper shanks, see information given on page 901.

American National Standard Stub- and Regular-length, Two-flute, Medium Helix, Plain- and Ball-end, Single-end End Mills with Weldon Shanks* (ANSI B94.19-1985)

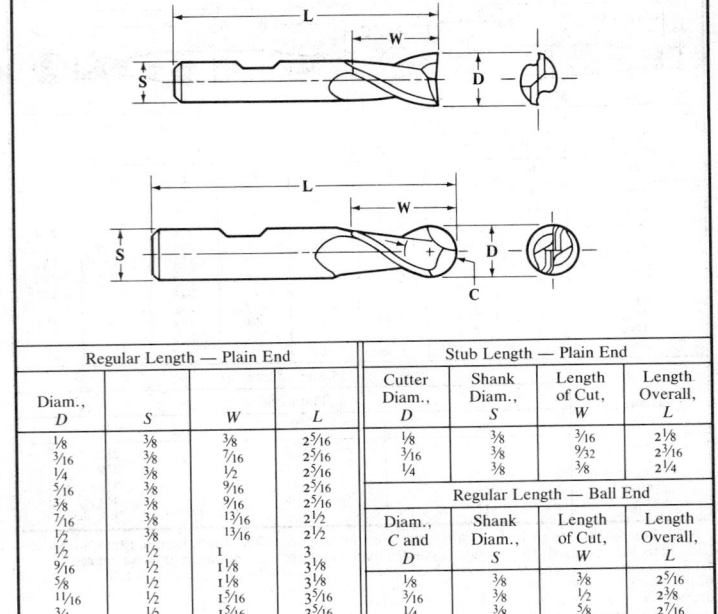

Regular Length — Plain End

Diam., D	S	W	L
1/8	3/8	3/8	2 5/16
3/16	3/8	7/16	2 5/16
1/4	3/8	1/2	2 5/16
5/16	3/8	9/16	2 5/16
3/8	3/8	9/16	2 5/16
7/16	3/8	13/16	2 1/2
1/2	3/8	13/16	2 1/2
1/2	1/2	1	3
9/16	1/2	1 1/8	3 1/8
5/8	1/2	1 1/8	3 1/8
11/16	1/2	1 5/16	3 5/16
3/4	1/2	1 5/16	3 5/16
5/8	5/8	1 5/16	3 7/16
11/16	5/8	1 5/16	3 7/16
3/4	5/8	1 5/16	3 7/16
13/16	5/8	1 1/2	3 5/8
7/8	5/8	1 1/2	3 5/8
1	5/8	1 1/2	3 5/8
7/8	7/8	1 1/2	3 3/4
1	7/8	1 1/2	3 3/4
1 1/8	7/8	1 3/8	3 7/8
1 1/4	7/8	1 5/8	3 7/8
1	1	1 5/8	4 1/8
1 1/8	1	1 5/8	4 1/8
1 1/4	1	1 5/8	4 1/8
1 3/8	1	1 5/8	4 1/8
1 1/2	1	1 5/8	4 1/8
1 1/4	1 1/4	1 5/8	4 1/8
1 1/2	1 1/4	1 5/8	4 1/8
1 3/4	1 1/4	1 5/8	4 1/8
2	1 1/4	1 5/8	4 1/8

Stub Length — Plain End

Cutter Diam., D	Shank Diam., S	Length of Cut, W	Length Overall, L
1/8	3/8	3/16	2 1/8
3/16	3/8	9/32	2 3/16
1/4	3/8	3/8	2 1/4

Regular Length — Ball End

Diam., C and D	Shank Diam., S	Length of Cut, W	Length Overall, L
1/8	3/8	3/8	2 5/16
3/16	3/8	1/2	2 3/8
1/4	3/8	5/8	2 7/16
5/16	3/8	3/4	2 1/2
3/8	3/8	3/4	2 1/2
7/16	1/2	1	3
1/2	1/2	1	3
9/16	1/2	1 1/8	3 1/8
5/8	1/2	1 1/8	3 1/8
5/8	5/8	1 3/8	3 1/2
3/4	1/2	1 5/16	3 5/16
3/4	3/4	1 5/8	3 7/8
7/8	7/8	2	4 1/4
1	1	2 1/4	4 3/4
1 1/8	1	2 1/4	4 3/4
1 1/4	1 1/4	2 1/2	5
1 1/2	1 1/4	2 1/2	5

All dimensions are in inches. All cutters are high-speed steel. Right-hand cutters with right-hand helix are standard. Helix angle is greater than 19 degrees but not more than 39 degrees. * The following single-end end mills are available in premium high speed steel: ball end, two flute, with D ranging from 1/8 to 1 1/2 inches; ball end, multiple flute, with D ranging from 1/8 to 1 inch; and plain end, two flute, with D ranging from 1/8 to 1 1/2 inches.

Tolerances: On C and D, − 0.0015 inch for stub-length mills; + 0.003 inch for regular-length mills; on S, − 0.0001 to − 0.0005 inch; on W, ± 1/32 inch; and on L, ± 1/16 inch.

American National Standard Long Length Single-end and Stub-, and Regular-length, Double-end, Plain- and Ball-end, Medium Helix, Two-flute End Mills with Weldon Shanks (ANSI B94.19-1985)

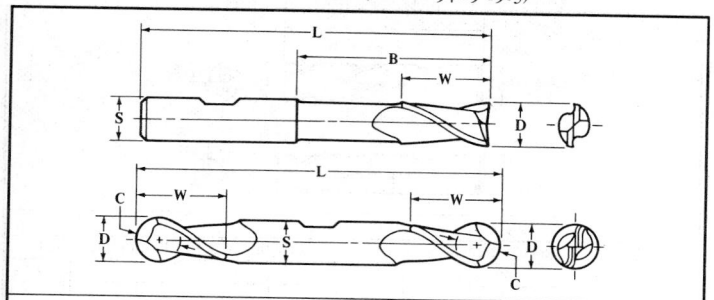

Single End

Diam., C and D	Long Length — Plain End				Long Length — Ball End			
	S	B*	W	L	S	B*	W	L
1/8	...	...	...	...	3/8	13/16	1/2	2 3/8
3/16	...	...	...	...	3/8	1 1/8	1/2	2 11/16
1/4	3/8	1 1/2	5/8	3 1/16	3/8	1 1/2	5/8	3 1/16
5/16	3/8	1 3/4	3/4	3 5/16	3/8	1 3/4	3/4	3 5/16
3/8	3/8	1 3/4	3/4	3 5/16	3/8	1 3/4	3/4	3 5/16
7/16	...	...	...	...	1/2	1 7/8	1	3 11/16
1/2	1/2	2 7/32	1	4	1/2	2 1/4	1	4
5/8	5/8	2 23/32	1 3/8	4 5/8	5/8	2 3/4	1 3/8	4 5/8
3/4	3/4	3 11/32	1 5/8	5 3/8	3/4	3 3/8	1 5/8	5 3/8
1	1	4 31/32	2 1/2	7 1/4	1	5	2 1/2	7 1/4
1 1/4	1 1/4	4 31/32	3	7 1/4	...	...	...	...

Double End

Diam., C and D	Stub Length — Plain End			Regular Length — Plain End			Regular Length — Ball End		
	S	W	L	S	W	L	S	W	L
1/8	3/8	3/16	2 3/4	3/8	3/8	3 1/16	3/8	3/8	3 1/16
5/32	3/8	15/64	2 3/4	3/8	7/16	3 1/8	...	...	...
3/16	3/8	9/32	2 3/4	3/8	7/16	3 1/8	3/8	7/16	3 1/8
7/32	3/8	21/64	2 7/8	3/8	1/2	3 1/8	...	...	...
1/4	3/8	3/8	2 7/8	3/8	1/2	3 1/8	3/8	1/2	3 1/8
9/32	...	...	...	3/8	9/16	3 1/8	...	...	...
5/16	...	...	...	3/8	9/16	3 1/8	3/8	9/16	3 1/8
11/32	...	...	...	3/8	9/16	3 1/8	...	...	...
3/8	...	...	...	3/8	9/16	3 1/8	3/8	9/16	3 1/8
13/32	...	...	...	1/2	13/16	3 3/4	...	...	...
7/16	...	...	...	1/2	13/16	3 3/4	1/2	13/16	3 3/4
15/32	...	...	...	1/2	13/16	3 3/4	...	...	...
1/2	...	...	...	1/2	13/16	3 3/4	1/2	13/16	3 3/4
9/16	...	...	...	5/8	1 1/8	4 1/2	...	...	...
5/8	...	...	...	5/8	1 1/8	4 1/2	5/8	1 1/8	4 1/2
11/16	...	...	...	3/4	1 5/16	5	...	...	...
3/4	...	...	...	3/4	1 5/16	5	3/4	1 5/16	5
7/8	...	...	...	7/8	1 9/16	5 1/2	...	...	...
1	...	...	...	1	1 5/8	5 7/8	1	1 5/8	5 7/8

All dimensions are in inches. All cutters are high-speed steel. Right-hand cutters with right-hand helix are standard. Helix angle is greater than 19 degrees but not more than 39 degrees.

Tolerances: On C and D, + 0.003 inch for single-end mills, − 0.0015 inch for double-end mills; on S, − 0.0001 to − 0.0005 inch; on W, ± 1/32 inch; and on L, ± 1/16 inch.

* B is the length below the shank.

American National Standard Regular-, Long-, and Extra Long-length, Three- and Four-flute, Medium Helix, Center Cutting, Single-end End Mills with Weldon Shanks* (ANSI B94.19-1985)

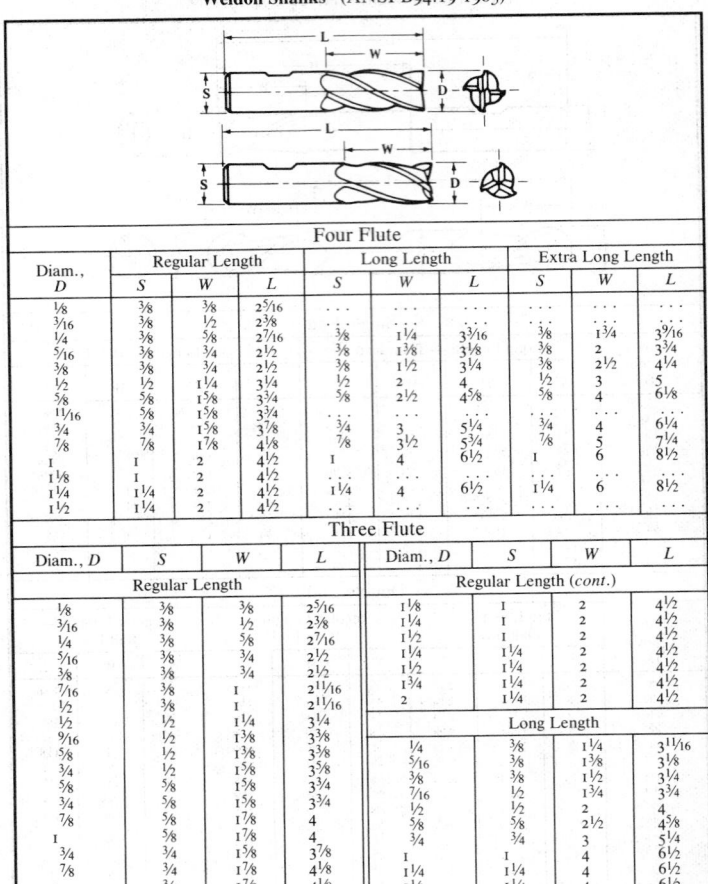

Four Flute

Diam., D	Regular Length			Long Length			Extra Long Length		
	S	W	L	S	W	L	S	W	L
1/8	3/8	3/8	2 5/16	...	...	...	...	...	...
3/16	3/8	1/2	2 3/8	...	...	...	...	...	...
1/4	3/8	5/8	2 7/16	3/8	1 1/4	3 3/16	3/8	1 3/4	3 9/16
5/16	3/8	3/4	2 1/2	3/8	1 3/8	3 1/8	3/8	2	3 3/4
3/8	3/8	3/4	2 1/2	3/8	1 1/2	3 1/4	3/8	2 1/2	4 1/4
1/2	1/2	1 1/4	3 1/4	1/2	2	4	1/2	3	5
5/8	5/8	1 5/8	3 3/4	5/8	2 1/2	4 5/8	5/8	4	6 1/8
11/16	5/8	1 5/8	3 3/4	...	...	...	...	...	...
3/4	3/4	1 5/8	3 7/8	3/4	3	5 1/4	3/4	4	6 1/4
7/8	7/8	1 7/8	4 1/8	7/8	3 1/2	5 3/4	7/8	5	7 1/4
1	1	2	4 1/2	1	4	6 1/2	1	6	8 1/2
1 1/8	1	2	4 1/2	...	...	...	...	...	...
1 1/4	1 1/4	2	4 1/2	1 1/4	4	6 1/2	1 1/4	6	8 1/2
1 1/2	1 1/4	2	4 1/2	...	...	...	...	...	...

Three Flute

Regular Length

Diam., D	S	W	L	Diam., D	S	W	L
1/8	3/8	3/8	2 5/16	1 1/8	1	2	4 1/2
3/16	3/8	1/2	2 3/8	1 1/4	1	2	4 1/2
1/4	3/8	5/8	2 7/16	1 1/2	1	2	4 1/2
5/16	3/8	3/4	2 1/2	1 1/4	1 1/4	2	4 1/2
3/8	3/8	3/4	2 1/2	1 1/2	1 1/4	2	4 1/2
7/16	3/8	1	2 11/16	1 3/4	1 1/4	2	4 1/2
1/2	3/8	1	2 11/16	2	1 1/4	2	4 1/2
1/2	1/2	1 1/4	3 1/4				
9/16	1/2	1 3/8	3 3/8				
5/8	1/2	1 3/8	3 3/8				
3/4	1/2	1 5/8	3 5/8				
5/8	5/8	1 5/8	3 3/4				
3/4	5/8	1 5/8	3 3/4				
7/8	5/8	1 7/8	4				
1	5/8	1 7/8	4				
3/4	3/4	1 5/8	3 7/8				
7/8	3/4	1 7/8	4 1/8				
1	3/4	1 7/8	4 1/8				
1	7/8	1 7/8	4 1/8				
1	1	2	4 1/2				

Long Length

Diam., D	S	W	L
1/4	3/8	1 1/4	3 11/16
5/16	3/8	1 3/8	3 1/8
3/8	3/8	1 1/2	3 1/4
7/16	1/2	1 3/4	3 3/4
1/2	1/2	2	4
5/8	5/8	2 1/2	4 5/8
3/4	3/4	3	5 1/4
1	1	4	6 1/2
1 1/4	1 1/4	4	6 1/2
1 1/2	1 1/4	4	6 1/2
1 3/4	1 1/4	4	6 1/2
2	1 1/4	4	6 1/2

All dimensions are in inches. All cutters are high-speed steel. Right-hand cutters with right-hand helix are standard. Helix angle is greater than 19 degrees but not more than 39 degrees.

* The following center-cutting, single-end end mills are available in premium high speed steel: regular length, multiple flute, with D ranging from 1/8 to 1 1/2 inches; long length, multiple flute, with D ranging from 3/8 to 1 1/4 inches; and extra long-length, multiple flute, with D ranging from 3/8 to 1 1/4 inches.

Tolerances: On D, + 0.003 inch; on S, − 0.0001 to − 0.0005 inch; on W, ± 1/32 inch; and on L, ± 1/16 inch.

American National Standard Stub- and Regular-length, Four-flute, Medium Helix, Double-end End Mills with Weldon Shanks (ANSI B94.19-1985)

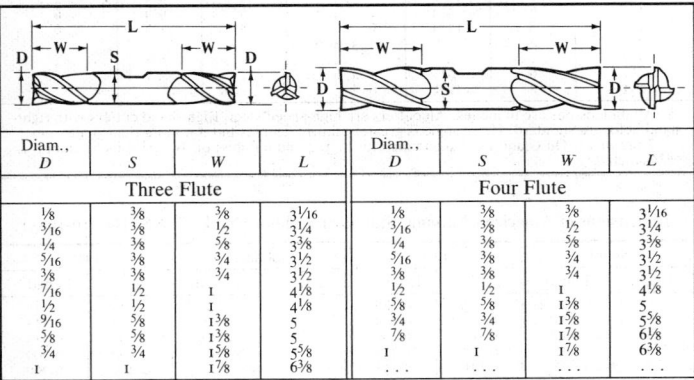

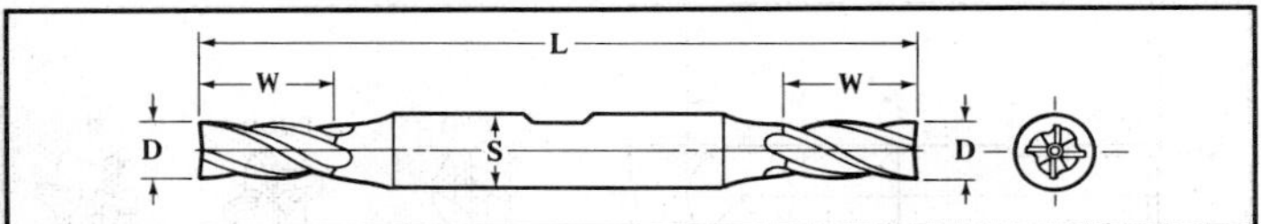

Diam., D	S	W	L	Diam., D	S	W	L	Diam., D	S	W	L
Stub Length											
1/8	3/8	3/16	2¾	3/16	3/8	9/32	2¾	1/4	3/8	3/8	2⅞
5/32	3/8	15/64	2¾	7/32	3/8	21/64	2⅞	...	...	...	...
Regular Length											
1/8*	3/8	3/8	3 1/16	11/32	3/8	3/4	3½	5/8*	5/8	1 3/8	5
5/32*	3/8	7/16	3 1/8	3/8*	3/8	3/4	3½	11/16	3/4	1 5/8	5⅝
3/16*	3/8	1/2	3¼	13/32	1/2	1	4 1/8	3/4*	3/4	1 5/8	5⅝
7/32	3/8	9/16	3¼	7/16	1/2	1	4 1/8	13/16	7/8	1 7/8	6 1/8
1/4*	3/8	5/8	3 3/8	15/32	1/2	1	4 1/8	7/8	7/8	1 7/8	6 1/8
9/32	3/8	11/16	3 3/8	1/2*	1/2	1	4 1/8	1	1	1 7/8	6 3/8
5/16*	3/8	3/4	3½	9/16	5/8	1 3/8	5	...	...	...	...

All dimensions are in inches. All cutters are high-speed steel. Right-hand cutters with right-hand helix are standard. Helix angle is greater than 19 degrees but not more than 39 degrees.

Tolerances: On D, + 0.003 inch (if the shank is the same diameter as the cutting portion, however, then the tolerance on the cutting diameter is − 0.0025 inch); on S, − 0.0001 to − 0.0005 inch; on W, ± 1/32 inch; and on L, ± 1/16 inch.

* In this size of regular mill a left-hand cutter with a left-hand helix is also standard.

American National Standard Three- and Four-flute, Regular Length, Medium Helix, Center Cutting, Double-end End Mills with Weldon Shanks (ANSI B94.19-1985)

Diam., D	S	W	L	Diam., D	S	W	L
Three Flute				Four Flute			
1/8	3/8	3/8	3 1/16	1/8	3/8	3/8	3 1/16
3/16	3/8	1/2	3¼	3/16	3/8	1/2	3¼
1/4	3/8	5/8	3 3/8	1/4	3/8	5/8	3 3/8
5/16	3/8	3/4	3½	5/16	3/8	3/4	3½
3/8	3/8	3/4	3½	3/8	3/8	3/4	3½
7/16	1/2	1	4 1/8	1/2	1/2	1	4 1/8
1/2	1/2	1	4 1/8	5/8	5/8	1 3/8	5
9/16	5/8	1 3/8	5	3/4	3/4	1 5/8	5⅝
5/8	5/8	1 3/8	5	7/8	7/8	1 7/8	6 1/8
3/4	3/4	1 5/8	5⅝	1	1	1 7/8	6 3/8
1	1	1 7/8	6 3/8	...	...	...	...

All dimensions are in inches. All cutters are high-speed steel. Right-hand cutters with right-hand helix are standard. Helix angle is greater than 19 degrees but not more than 39 degrees.

Tolerances: On D, + 0.0015 inch; on S, − 0.0001 to − 0.0005 inch; on W, ± 1/32 inch; and on L, ± 1/16 inch.

American National Standard Plain- and Ball-end, Heavy Duty, Medium Helix, Single-end End Mills with 2-inch Diameter Shanks (ANSI B94.19-1985)

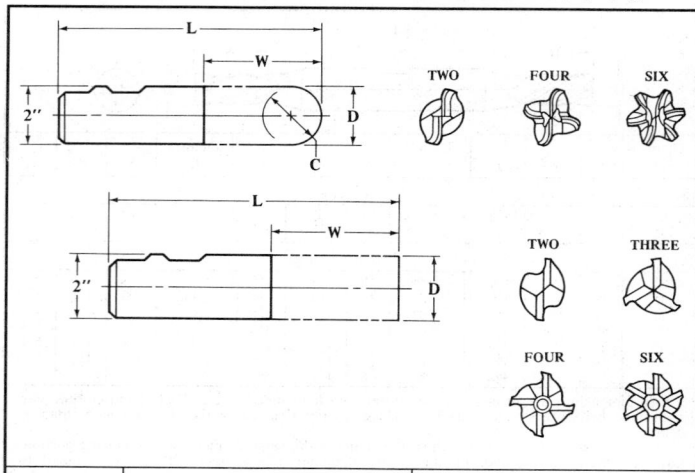

Diam., C and D	Plain End			Ball End		
	W	L	No. of Flutes	W	L	No. of Flutes
2	2	5¾	2, 4, 6	...	...	...
2	3	6¾	2, 3	...	...	...
2	4	7¾	2, 3, 4, 6	4	7¾	6
2	...	...	...	5	8¾	2, 4
2	6	9¾	2, 3, 4, 6	6	9¾	6
2	8	11¾	6	8	11¾	6
2½	4	7¾	2, 3, 4, 6	...	...	...
2½	...	...	...	5	8¾	4
2½	6	9¾	2, 4, 6	...	...	...
2½	8	11¾	6	...	...	...

All dimensions are in inches. All cutters are high-speed steel. Right-hand cutters with right-hand helix are standard. Helix angle is greater than 19 degrees but not more than 39 degrees.

Tolerances: On C and D, + 0.005 inch for 2, 3, 4 and 6 flutes; on W, ±¹⁄₁₆ inch; and on L, ±¹⁄₁₆ inch.

Dimensions of American National Standard Weldon Shanks (ANSI B94.19-1985)

Shank		Flat		Shank		Flat	
Diam.	Length	X†	Length*	Diam.	Length	X†	Length*
⅜	1⁹⁄₁₆	0.325	0.280	1	2⁹⁄₃₂	0.925	0.515
½	1²⁵⁄₃₂	0.440	0.330	1¼	2⁹⁄₃₂	1.156	0.515
⅝	1²⁹⁄₃₂	0.560	0.400	1½	2¹¹⁄₁₆	1.406	0.515
¾	2¹⁄₃₂	0.675	0.455	2	3¼	1.900	0.700
⅞	2¹⁄₃₂	0.810	0.455	2½	3½	2.400	0.700

All dimensions are in inches.

Centerline of flat is at half-length of shank except for 1½-, 2- and 2½-inch shanks where it is 1³⁄₁₆, 1²⁷⁄₃₂ and 1¹⁵⁄₁₆ from shank end, respectively.

* Minimum.   † X is distance from bottom of flat to opposite side of shank.

Tolerance on shank diameter, − 0.0001 to − 0.0005 inch.

American National Standard Form Relieved, Concave, Convex, and Corner-rounding Arbor-type Cutters* (ANSI B94.19-1985)

Concave   Convex   Corner–rounding

Diameter C or Radius R			Cutter Diam. D[1]	Width W ± .010[2]	Diameter of Hole H		
Nom.	Max.	Min.			Nom.	Max.	Min.
Concave Cutters[3]							
1/8	0.1270	0.1240	2 1/4	1/4	1	1.00075	1.00000
3/16	0.1895	0.1865	2 1/4	3/8	1	1.00075	1.00000
1/4	0.2520	0.2490	2 1/2	7/16	1	1.00075	1.00000
5/16	0.3145	0.3115	2 3/4	9/16	1	1.00075	1.00000
3/8	0.3770	0.3740	2 3/4	5/8	1	1.00075	1.00000
7/16	0.4395	0.4365	3	3/4	1	1.00075	1.00000
1/2	0.5040	0.4980	3	13/16	1	1.00075	1.00000
5/8	0.6290	0.6230	3 1/2	1	1 1/4	1.251	1.250
3/4	0.7540	0.7480	3 3/4	1 3/16	1 1/4	1.251	1.250
7/8	0.8790	0.8730	4	1 3/8	1 1/4	1.251	1.250
1	1.0040	0.9980	4 1/4	1 9/16	1 1/4	1.251	1.250
Convex Cutters[3]							
1/8	0.1270	0.1230	2 1/4	1/8	1	1.00075	1.00000
3/16	0.1895	0.1855	2 1/4	3/16	1	1.00075	1.00000
1/4	0.2520	0.2480	2 1/2	1/4	1	1.00075	1.00000
5/16	0.3145	0.3105	2 3/4	5/16	1	1.00075	1.00000
3/8	0.3770	0.3730	2 3/4	3/8	1	1.00075	1.00000
7/16	0.4395	0.4355	3	7/16	1	1.00075	1.00000
1/2	0.5020	0.4980	3	1/2	1	1.00075	1.00000
5/8	0.6270	0.6230	3 1/2	5/8	1 1/4	1.251	1.250
3/4	0.7520	0.7480	3 3/4	3/4	1 1/4	1.251	1.250
7/8	0.8770	0.8730	4	7/8	1 1/4	1.251	1.250
1	1.0020	0.9980	4 1/4	1	1 1/4	1.251	1.250
Corner-rounding Cutters[4]							
1/8	0.1260	0.1240	2 1/2	1/4	1	1.00075	1.00000
1/4	0.2520	0.2490	3	13/32	1	1.00075	1.00000
3/8	0.3770	0.3740	3 3/4	9/16	1 1/4	1.251	1.250
1/2	0.5020	0.4990	4 1/4	3/4	1 1/4	1.251	1.250
5/8	0.6270	0.6240	4 1/4	15/16	1 1/4	1.251	1.250

All dimensions in inches. All cutters are high-speed steel and are form relieved.
Right-hand corner rounding cutters are standard, but left-hand cutter for 1/4-inch size is also standard.

* For key and keyway dimensions for these cutters, see page 1614. [1] Tolerances on cutter diameter are + 1/16, − 1/16 inch for all sizes. [2] Tolerance does not apply to convex cutters. [3] Size of cutter is designated by specifying diameter C of circular form. [4] Size of cutter is designated by specifying radius R of circular form.

American National Standard Roughing and Finishing Gear Milling Cutters for Gears with 14½-degree Pressure Angles (ANSI B94.19-1985)

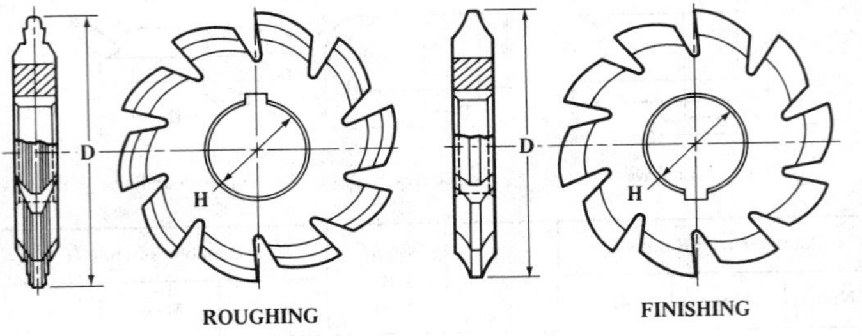

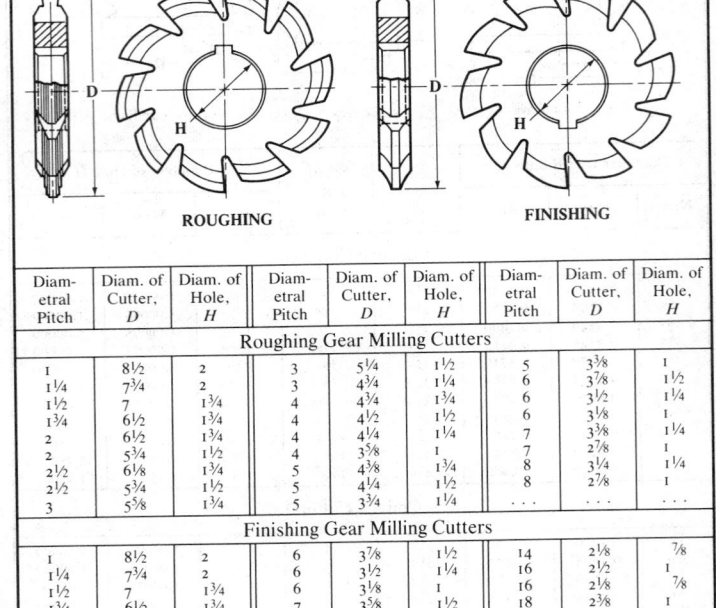

Diametral Pitch	Diam. of Cutter, *D*	Diam. of Hole, *H*	Diametral Pitch	Diam. of Cutter, *D*	Diam. of Hole, *H*	Diametral Pitch	Diam. of Cutter, *D*	Diam. of Hole, *H*
Roughing Gear Milling Cutters								
1	8½	2	3	5¼	1½	5	3⅜	1
1¼	7¾	2	3	4¾	1¼	6	3⅞	1½
1½	7	1¾	4	4¾	1¾	6	3½	1¼
1¾	6½	1¾	4	4½	1½	6	3⅛	1
2	6½	1¾	4	4¼	1¼	7	3⅜	1¼
2	5¾	1½	4	3⅝	1	7	2⅞	1
2½	6⅛	1¾	5	4⅜	1¾	8	3¼	1¼
2½	5¾	1½	5	4¼	1½	8	2⅞	1
3	5⅝	1¾	5	3¾	1¼	...	...	...
Finishing Gear Milling Cutters								
1	8½	2	6	3⅞	1½	14	2⅛	⅞
1¼	7¾	2	6	3½	1¼	16	2½	1
1½	7	1¾	6	3⅛	1	16	2⅛	⅞
1¾	6½	1¾	7	3⅝	1½	18	2⅜	1
2	6½	1¾	7	3⅜	1¼	18	2	⅞
2	5¾	1½	7	2⅞	1	20	2⅜	1
2½	6⅛	1¾	8	3½	1½	20	2	⅞
2½	5¾	1½	8	3¼	1¼	22	2¼	1
3	5⅝	1¾	8	2⅞	1	22	2	⅞
3	5¼	1½	9	3⅛	1¼	24	2¼	1
3	4¾	1¼	9	2¾	1	24	1¾	⅞
4	4¾	1¾	10	3	1¼	26	1¾	⅞
4	4½	1½	10	2¾	1	28	1¾	⅞
4	4¼	1¼	10	2⅜	⅞	30	1¾	⅞
4	3⅝	1	11	2⅝	1	32	1¾	⅞
5	4⅜	1¾	11	2⅜	⅞	36	1¾	⅞
5	4¼	1½	12	2⅞	1¼	40	1¾	⅞
5	3¾	1¼	12	2⅝	1	48	1¾	⅞
5	3⅜	1	12	2¼	⅞	...	...	...
6	4¼	1¾	14	2½	1	...	...	...

All dimensions are in inches.

All gear milling cutters are high-speed steel and are form relieved.

For keyway dimensions see page 783.

Tolerances: On outside diameter, +1/16, −1/16 inch; on hole diameter, through 1-inch hole diameter, +0.00075 inch, over 1-inch and through 2-inch hole diameter, +0.0010 inch.

For cutter number relative to numbers of gear teeth, see page 1794. Roughing cutters are made with No. 1 cutter form only.

American National Standard Gear Milling Cutters for Mitre and Bevel Gears with 14½-degree Pressure Angles (ANSI B94.19-1985)

Diametral Pitch	Diameter of Cutter, D	Diameter of Hole, H	Diametral Pitch	Diameter of Cutter, D	Diameter of Hole, H
3	4	1¼	10	2⅜	⅞
4	3⅝	1¼	12	2¼	⅞
5	3⅜	1¼	14	2⅛	⅞
6	3⅛	1	16	2⅛	⅞
7	2⅞	1	20	2	⅞
8	2⅞	1	24	1¾	⅞

All dimensions are in inches.
All cutters are high-speed steel and are form relieved.
For keyway dimensions see page 783. For cutter selection see page 1871.
  Tolerances: On outside diameter, + ¹⁄₁₆, − ¹⁄₁₆ inch; on hole diameter, through 1-inch hole diameter, + 0.00075 inch, for 1¼-inch hole diameter, + 0.0010 inch.
  To select the cutter number for bevel gears with the axis at any angle, double the back cone radius and multiply the result by the diametral pitch. This procedure gives the number of equivalent spur gear teeth and is the basis for selecting the cutter number from the table on page 1794.

American National Standard Roller Chain Sprocket Milling Cutters* (ANSI B94.19-1985)

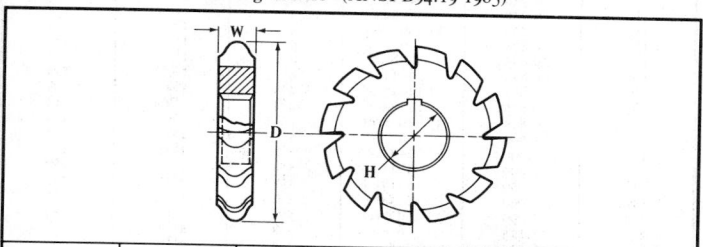

Chain Pitch	Diam. of Roll	No. of Teeth in Sprocket	Diam. of Cutter, D	Width of Cutter, W	Diam. of Hole, M
¼	0.130	6	2¾	⁵⁄₁₆	1
¼	0.130	7–8	2¾	⁵⁄₁₆	1
¼	0.130	9–11	2¾	⁵⁄₁₆	1
¼	0.130	12–17	2¾	⁵⁄₁₆	1
¼	0.130	18–34	2¾	⁹⁄₃₂	1
¼	0.130	35 and over	2¾	⁹⁄₃₂	1
⅜	0.200	6	2¾	¹⁵⁄₃₂	1
⅜	0.200	7–8	2¾	¹⁵⁄₃₂	1
⅜	0.200	9–11	2¾	¹⁵⁄₃₂	1
⅜	0.200	12–17	2¾	⁷⁄₁₆	1
⅜	0.200	18–34	2¾	⁷⁄₁₆	1
⅜	0.200	35 and over	2¾	¹³⁄₃₂	1
½	0.313	6	3	¾	1
½	0.313	7–8	3	¾	1
½	0.313	9–11	3⅛	¾	1
½	0.313	12–17	3⅛	¾	1
½	0.313	18–34	3⅛	²³⁄₃₂	1
½	0.313	35 and over	3⅛	¹¹⁄₁₆	1
⅝	0.400	6	3⅛	¾	1
⅝	0.400	7–8	3⅛	¾	1
⅝	0.400	9–11	3¼	¾	1
⅝	0.400	12–17	3¼	¾	1

All dimensions are in inches.
All cutters are high-speed steel and are form relieved.
* For tooth form, keyway dimensions and tolerances, see end of table.

(Concluded) American National Standard Roller Chain Sprocket Milling Cutters* (ANSI B94.19-1985)

Chordal Pitch	Diam. of Roll	No. of Teeth in Sprocket	Diam. of Cutter, D	Width of Cutter, D	Diam. of Hole, H
5/8	0.400	18–34	3¼	23/32	1
5/8	0.400	35 and over	3¼	11/16	1
3/4	0.469	6	3¼	29/32	1
3/4	0.469	7–8	3¼	29/32	1
3/4	0.469	9–11	3⅜	29/32	1
3/4	0.469	12–17	3⅜	7/8	1
3/4	0.469	18–34	3⅜	27/32	1
3/4	0.469	35 and over	3⅜	13/16	1
1	0.625	6	3⅞	1½	1¼
1	0.625	7–8	4	1½	1¼
1	0.625	9–11	4⅛	1 15/32	1¼
1	0.625	12–17	4⅛	1 15/32	1¼
1	0.625	18–34	4¼	1 13/32	1¼
1	0.625	35 and over	4¼	1 11/32	1¼
1¼	0.750	6	4¼	1 13/16	1¼
1¼	0.750	7–8	4⅜	1 13/16	1¼
1¼	0.750	9–11	4½	1 25/32	1¼
1¼	0.750	12–17	4½	1¾	1¼
1¼	0.750	18–34	4⅝	1 11/16	1¼
1¼	0.750	35 and over	4⅝	1⅝	1¼
1½	0.875	6	4⅜	1 13/16	1¼
1½	0.875	7–8	4½	1 13/16	1¼
1½	0.875	9–11	4⅝	1 25/32	1¼
1½	0.875	12–17	4⅝	1¾	1¼
1½	0.875	18–34	4¾	1 11/16	1¼
1½	0.875	35 and over	4¾	1⅝	1¼
1¾	1.000	6	5	2 3/32	1½
1¾	1.000	7–8	5⅛	2 3/32	1½
1¾	1.000	9–11	5¼	2 1/16	1½
1¾	1.000	12–17	5⅜	2 1/32	1½
1¾	1.000	18–34	5½	1 31/32	1½
1¾	1.000	35 and over	5½	1⅞	1½
2	1.125	6	5⅜	2 13/32	1½
2	1.125	7–8	5½	2 13/32	1½
2	1.125	9–11	5⅝	2⅜	1½
2	1.125	12–17	5¾	2 5/16	1½
2	1.125	18–34	5⅞	2¼	1½
2	1.125	35 and over	5⅞	2 5/32	1½
2¼	1.406	6	5⅞	2 11/16	1½
2¼	1.406	7–8	6	2 11/16	1½
2¼	1.406	9–11	6¼	2 21/32	1½
2¼	1.406	12–17	6⅜	2 19/32	1½
2¼	1.406	18–34	6½	2 15/32	1½
2¼	1.406	35 and over	6½	2 13/32	1½
2½	1.563	6	6⅜	3	1¾
2½	1.563	7–8	6⅝	3	1¾
2½	1.563	9–11	6¾	2 15/16	1¾
2½	1.563	12–17	6⅞	2 29/32	1¾
2½	1.563	18–34	7	2¾	1¾
2½	1.563	35 and over	7⅛	2 11/16	1¾
3	1.875	6	7½	3 19/32	2
3	1.875	7–8	7¾	3 19/32	2
3	1.875	9–11	7⅞	3 17/32	2
3	1.875	12–17	8	3 15/32	2
3	1.875	18–34	8	3 11/32	2
3	1.875	35 and over	8¼	3 7/32	2

All dimensions are in inches.

All cutters are high-speed steel and are form relieved.

* For tooth form, see ANSI sprocket tooth form table on page 2301.

For keyway dimensions see page 783.

Tolerances: Outside diameter, + 1/16, − 1/16 inch; hole diameter, through 1-inch diameter, + 0.00075 inch, above 1-inch diameter and through 2-inch diameter, + 0.0010 inch.

American National Standard Keys and Keyways for Milling Cutters and Arbors (ANSI B94.19-1985)

ARBOR AND KEYSEAT — CUTTER HOLE AND KEYWAY — ARBOR AND KEY

Nom. Arbor and Cutter Hole Diam.	Nom. Size Key (Square)	Arbor and Keyseat				Hole and Keyway					Arbor and Key			
		A Max.	A Min.	B Max.	B Min.	C Max.	C Min.	D† Min.	H Nom.	Corner Radius	E Max.	E Min.	F Max.	F Min.
1/2	3/32	0.0947	0.0937	0.4531	0.4481	0.106	0.099	0.5578	3/64	0.020	0.0932	0.0927	0.5468	0.5408
5/8	1/8	0.1260	0.1250	0.5625	0.5575	0.137	0.130	0.6985	1/16	1/32	0.1245	0.1240	0.6875	0.6815
3/4	1/8	0.1260	0.1250	0.6875	0.6825	0.137	0.130	0.8225	1/16	1/32	0.1245	0.1240	0.8125	0.8065
7/8	1/8	0.1260	0.1250	0.8125	0.8075	0.137	0.130	0.9475	1/16	1/32	0.1245	0.1240	0.9375	0.9315
1	1/4	0.2510	0.2500	0.8438	0.8388	0.262	0.255	1.1040	3/32	1/32	0.2495	0.2490	1.0940	1.0880
1 1/4	5/16	0.3135	0.3125	1.0630	1.0580	0.343	0.318	1.3850	1/8	3/64	0.3120	0.3115	1.3750	1.3690
1 1/2	3/8	0.3760	0.3750	1.2810	1.2760	0.410	0.385	1.6660	1/8	1/16	0.3745	0.3740	1.6560	1.6500
1 3/4	7/16	0.4385	0.4375	1.5000	1.4950	0.473	0.448	1.9480	5/32	1/16	0.4370	0.4365	1.9380	1.9320
2	1/2	0.5010	0.5000	1.6870	1.6820	0.535	0.510	2.1980	3/16	1/16	0.4995	0.4990	2.1880	2.1820
2 1/2	5/8	0.6260	0.6250	2.0940	2.0890	0.660	0.635	2.7330	7/32	1/16	0.6245	0.6240	2.7180	2.7120
3	3/4	0.7510	0.7500	2.5000	2.4950	0.785	0.760	3.2650	1/4	3/32	0.7495	0.7490	3.2500	3.2440
3 1/2	7/8	0.8760	0.8750	3.0000	2.9950	0.910	0.885	3.8900	3/8	3/32	0.8745	0.8740	3.8750	3.8690
4	1	1.0010	1.0000	3.3750	3.3700	1.035	1.010	4.3900	3/8	3/32	0.9995	0.9990	4.3750	4.3690
4 1/2	1 1/8	1.1260	1.1250	3.8130	3.8080	1.160	1.135	4.9530	7/16	1/8	1.1245	1.1240	4.9380	4.9320
5	1 1/4	1.2510	1.2500	4.2500	4.2450	1.285	1.260	5.5150	1/2	1/8	1.2495	1.2490	5.5000	5.4940

All dimensions given in inches.
† D max. is 0.010 inch larger than D min.

American National Standard Woodruff Keyseat Cutters† — Shank-type Straight-teeth and Arbor-type Staggered-teeth (ANSI B94.19-1985)

Shank-type Cutters

Cutter Number	Nom. Diam. of Cutter, D	Width of Face, W	Length Overall, L	Cutter Number	Nom. Diam. of Cutter, D	Width of Face, W	Length Overall, L	Cutter Number	Nom. Diam. of Cutter, D	Width of Face, W	Length Overall, L
202	1/4	1/16	2 1/16	506	3/4	5/32	2 5/32	809	1 1/8	1/4	2 1/4
202 1/2	5/16	1/16	2 1/16	606	3/4	3/16	2 3/16	1009	1 1/8	5/16	2 5/16
302 1/2	5/16	3/32	2 3/32	806	3/4	1/4	2 1/4	610	1 1/4	3/16	2 3/16
203	3/8	1/16	2 1/16	507	7/8	5/32	2 5/32	710	1 1/4	7/32	2 7/32
303	3/8	3/32	2 3/32	607	7/8	3/16	2 3/16	810	1 1/4	1/4	2 1/4
403	3/8	1/8	2 1/8	707	7/8	7/32	2 7/32	1010	1 1/4	5/16	2 5/16
204	1/2	1/16	2 1/16	807	7/8	1/4	2 1/4	1210	1 1/4	3/8	2 3/8
304	1/2	3/32	2 3/32	608	1	3/16	2 3/16	811	1 3/8	1/4	2 1/4
404	1/2	1/8	2 1/8	708	1	7/32	2 7/32	1011	1 3/8	5/16	2 5/16
305	5/8	3/32	2 3/32	808	1	1/4	2 1/4	1211	1 3/8	3/8	2 3/8
405	5/8	1/8	2 1/8	1008	1	5/16	2 5/16	812	1 1/2	1/4	2 1/4
505	5/8	5/32	2 5/32	1208	1	3/8	2 3/8	1012	1 1/2	5/16	2 5/16
605	5/8	3/16	2 3/16	609	1 1/8	3/16	2 3/16	1212	1 1/2	3/8	2 3/8
406	3/4	1/8	2 1/8	709	1 1/8	7/32	2 7/32	...	...	...	...

Arbor-type Cutters

Cutter Number	Nom. Diam. of Cutter, D	Width of Face, W	Diam. of Hole, H	Cutter Number	Nom. Diam. of Cutter, D	Width of Face, W	Diam. of Hole, H	Cutter Number	Nom. Diam. of Cutter, D	Width of Face, W	Diam. of Hole, H
617	2 1/8	3/16	3/4	1022	2 3/4	5/16	1	1628	3 1/2	1/2	1
817	2 1/8	1/4	3/4	1222	2 3/4	3/8	1	1828	3 1/2	9/16	1
1017	2 1/8	5/16	3/4	1422	2 3/4	7/16	1	2028	3 1/2	5/8	1
1217	2 1/8	3/8	3/4	1622	2 3/4	1/2	1	2428	3 1/2	3/4	1
822	2 3/4	1/4	1	1228	3 1/2	3/8	1	...	...	...	...

All dimensions are given in inches. All cutters are high-speed steel.

Shank type cutters are standard with right-hand cut and straight teeth. All sizes have 1/2-inch diameter straight shank.

Arbor type cutters have staggered teeth.

For Woodruff key and key-slot dimensions, see pages 2241 through 2244.

Tolerances: Face width W for shank type cutters: 1/16- to 5/32-inch face, + 0.0000, − 0.0005; 3/16 to 7/32, − 0.0002, − 0.0007; 1/4, − 0.0003, − 0.0008; 5/16, − 0.0004, − 0.0009; 3/8, − 0.0005, − 0.0010 inch. Face width W for arbor type cutters; 3/16 inch face, − 0.0002, − 0.0007; 1/4, − 0.0003, − 0.0008; 5/16, − 0.0004, − 0.0009; 3/8 and over, − 0.0005, − 0.0010 inch. Hole size H: + 0.00075, − 0.0000 inch. Diameter D for shank type cutters: 1/4- through 3/4-inch diameter, + 0.010, + 0.015, 7/8 through 1 1/8, + 0.012, + 0.017; 1 1/4 through 1 1/2, + 0.015, + 0.020 inch. These tolerances include an allowance for sharpening. For arbor type cutters diameter D is furnished 1/32 inch larger than listed and a tolerance of ± 0.002 inch applies to the oversize diameter.

Setting-angles for Milling Straight Teeth of Uniform Land Width in End Mills, Angular Cutters, and Taper Reamers. — The accompanying tables give setting-angles for the dividing head when straight teeth, having a land of uniform width throughout their length, are to be milled using single-angle fluting cutters. These setting-angles depend upon three factors: the number of teeth to be cut; the angle of the blank in which the teeth are to be cut; and the angle of the fluting cutter. Setting-angles for various combinations of these three factors are given in the tables. For example, assume that 12 teeth are to be cut on the end of an end mill using a 60-degree cutter. By following the horizontal line from 12 teeth, read in the column under 60 degrees that the dividing head should be set to an angle of 70 degrees and 32 minutes.

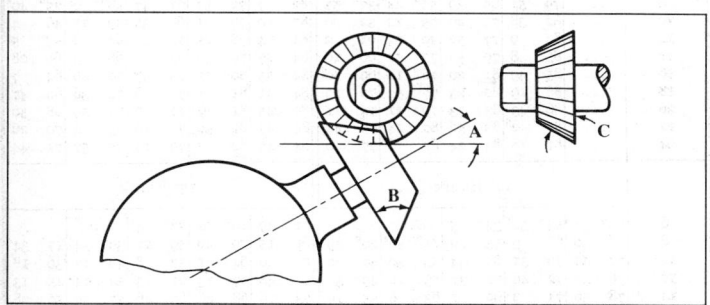

The following formulas, which were used to compile these tables, may be used to calculate the setting-angles for combinations of number of teeth, blank angle, and cutter angle not covered by the tables. In these formulas, A = setting-angle for dividing head, B = angle of blank in which teeth are to be cut, C = angle of fluting cutter, N = number of teeth to be cut, and D and E are angles not shown on the accompanying diagram and which are used only to simplify calculations.

$$\tan D = \cos (360°/N) \times \cot B \qquad (1)$$
$$\sin E = \tan (360°/N) \times \cot C \times \sin D \qquad (2)$$
$$\text{Setting-angle } A = D - E \qquad (3)$$

Example: Suppose 9 teeth are to be cut in a 35-degree blank using a 55-degree single-angle fluting cutter. In this case, $N = 9$, $B = 35°$, and $C = 55°$.

$$\tan D = \cos (360°/9) \times \cot 35° = 0.76604 \times 1.4281 = 1.0940; \text{ and } D = 47°34'$$
$$\sin E = \tan (360°/9) \times \cot 55° \times \sin 47°34' = 0.83910 \times 0.70021 \times 0.73806$$
$$= 0.43365; \text{ and } E = 25°42'$$

$$\text{Setting-angle } A = 47°34' - 25°42' = 21°52'$$

In the case of end mills and side mills the angle of the blank B is 0 degrees and the following simplified formula may be used to find the setting angle A

$$\cos A = \tan (360°/N) \times \cot C \qquad (4)$$

Example: If in the previous example the blank angle was 0 degrees,
$$\cos A = \tan (360°/9) \times \cot 55°$$
$$= 0.83910 \times 0.70021 = 0.58755; \text{ and setting-angle } A = 54°1'$$

Angles of Elevation for Milling Straight Teeth in 0-, 5-, 10-, 15-, 20-, 25-, 30-, and 35-degree Blanks Using Single-angle Fluting Cutters

0° Blank (End Mill) and 5° Blank

No. of Teeth	| 0° Blank (End Mill) | 90°	80°	70°	60°	50°	5° Blank 90°	80°	70°	60°	50°
6	...	72° 13'	50° 55'	...	...	80° 4'	62° 34'	41° 41'	...	...
8	...	79° 51'	68° 39'	54° 44'	32° 57'	82° 57'	72° 52'	61° 47'	48° 0'	25° 40'
10	...	82° 38'	74° 40'	65° 12'	52° 26'	83° 50'	76° 31'	68° 35'	59° 11'	46° 4'
12	...	84° 9'	77° 52'	70° 32'	61° 2'	84° 14'	78° 25'	72° 10'	64° 52'	55° 5'
14	...	85° 8'	79° 54'	73° 51'	66° 10'	84° 27'	79° 36'	74° 24'	68° 23'	60° 28'
16	...	85° 49'	81° 20'	76° 10'	69° 40'	84° 35'	80° 25'	75° 57'	70° 49'	64° 7'
18	...	86° 19'	82° 23'	77° 52'	72° 13'	84° 41'	81° 1'	77° 2'	72° 36'	66° 47'
20	...	86° 43'	83° 13'	79° 11'	74° 11'	84° 45'	81° 29'	77° 59'	73° 59'	68° 50'
22	...	87° 2'	83° 52'	80° 14'	75° 44'	84° 47'	81° 50'	78° 40'	75° 4'	70° 26'
24	...	87° 18'	84° 24'	81° 6'	77° 0'	84° 49'	82° 7'	79° 15'	75° 57'	71° 44'

10° Blank and 15° Blank

No. of Teeth	10° Blank 90°	80°	70°	60°	50°	15° Blank 90°	80°	70°	60°	50°
6	70° 34'	53° 50'	34° 5'	...	...	61° 49'	46° 12'	28° 4'	...	...
8	76° 0'	66° 9'	55° 19'	41° 56'	20° 39'	69° 15'	59° 46'	49° 21'	36° 34'	17° 34'
10	77° 42'	70° 31'	62° 44'	53° 30'	40° 42'	71° 40'	64° 41'	57° 8'	48° 12'	36° 18'
12	78° 30'	72° 46'	66° 37'	59° 26'	49° 50'	72° 48'	67° 13'	61° 13'	54° 14'	45° 12'
14	78° 56'	74° 9'	69° 2'	63° 6'	55° 19'	73° 26'	68° 46'	63° 46'	57° 59'	50° 38'
16	79° 12'	75° 5'	70° 41'	65° 37'	59° 1'	73° 50'	69° 49'	65° 30'	60° 33'	54° 20'
18	79° 22'	75° 45'	71° 53'	67° 27'	61° 43'	74° 5'	70° 33'	66° 46'	62° 26'	57° 0'
20	79° 30'	76° 16'	72° 44'	68° 52'	63° 47'	74° 16'	71° 6'	67° 44'	63° 52'	59° 3'
22	79° 35'	76° 40'	73° 33'	69° 59'	65° 25'	74° 24'	71° 32'	68° 29'	65° 0'	60° 40'
24	79° 39'	76° 59'	74° 9'	70° 54'	66° 44'	74° 30'	71° 53'	69° 6'	65° 56'	61° 59'

20° Blank and 25° Blank

No. of Teeth	20° Blank 90°	80°	70°	60°	50°	25° Blank 90°	80°	70°	60°	50°
6	53° 57'	39° 39'	23° 18'	...	...	47° 0'	34° 6'	19° 33'	...	...
8	62° 46'	53° 45'	43° 53'	31° 53'	14° 31'	56° 36'	48° 8'	38° 55'	27° 47'	11° 33'
10	65° 47'	59° 4'	51° 50'	43° 18'	32° 1'	60° 2'	53° 40'	46° 47'	38° 43'	27° 47'
12	67° 12'	61° 49'	56° 2'	49° 18'	40° 40'	61° 42'	56° 33'	51° 2'	44° 38'	36° 10'
14	68° 0'	63° 29'	58° 39'	53° 4'	46° 0'	62° 38'	58° 19'	53° 41'	48° 20'	41° 22'
16	68° 30'	64° 36'	60° 26'	55° 39'	49° 38'	63° 13'	59° 29'	55° 29'	50° 53'	44° 57'
18	68° 50'	65° 24'	61° 44'	57° 32'	52° 17'	63° 37'	60° 19'	56° 48'	52° 48'	47° 34'
20	69° 3'	65° 59'	62° 43'	58° 58'	54° 18'	63° 53'	60° 56'	57° 47'	54° 11'	49° 33'
22	69° 14'	66° 28'	63° 30'	60° 7'	55° 55'	64° 5'	61° 25'	58° 34'	55° 19'	51° 9'
24	69° 21'	66° 49'	64° 7'	61° 2'	57° 12'	64° 14'	61° 47'	59° 12'	56° 13'	52° 26'

30° Blank and 35° Blank

No. of Teeth	30° Blank 90°	80°	70°	60°	50°	35° Blank 90°	80°	70°	60°	50°
6	40° 54'	29° 22'	16° 32'	...	...	35° 32'	25° 19'	14° 3'	...	...
8	50° 46'	42° 55'	34° 24'	24° 12'	10° 14'	45° 17'	38° 5'	30° 18'	21° 4'	8° 41'
10	54° 29'	48° 30'	42° 3'	34° 31'	24° 44'	49° 7'	43° 33'	37° 35'	30° 38'	21° 40'
12	56° 18'	51° 26'	46° 14'	40° 12'	32° 32'	51° 3'	46° 30'	41° 39'	36° 2'	28° 55'
14	57° 21'	53° 15'	48° 52'	43° 49'	37° 27'	52° 9'	48° 19'	44° 12'	39° 28'	33° 33'
16	58° 0'	54° 27'	50° 39'	46° 19'	40° 52'	52° 50'	49° 20'	45° 56'	41° 51'	36° 45'
18	58° 26'	55° 18'	51° 57'	48° 7'	43° 20'	53° 18'	50° 21'	47° 12'	43° 36'	39° 8'
20	58° 44'	55° 55'	52° 56'	49° 30'	45° 15'	53° 38'	50° 59'	48° 10'	44° 57'	40° 57'
22	58° 57'	56° 24'	53° 42'	50° 36'	46° 46'	53° 53'	51° 29'	48° 56'	46° 1'	42° 24'
24	59° 8'	56° 48'	54° 20'	51° 30'	48° 0'	54° 4'	51° 53'	49° 32'	46° 52'	43° 35'

Angles of Elevation for Milling Straight Teeth in 40-, 45-, 50-, 55-, 60-, 65-, 70-, and 75-degree Blanks Using Single-angle Fluting Cutters

40° Blank | 45° Blank

No. of Teeth	90°	80°	70°	60°	50°	90°	80°	70°	60°	50°
6	30° 48'	21° 48'	11° 58'	...	...	26° 34'	18° 43'	10° 11'	...	...
8	40 7	33 36	26 33	18° 16'	7° 23'	35 16	29 25	23 8	15° 48'	5° 58'
10	43 57	38 51	33 32	27 3	18 55	38 58	34 21	29 24	23 40	16 10
12	45 54	41 43	37 14	32 3	25 33	40 54	37 5	33 0	28 18	22 13
14	47 3	43 29	39 41	35 19	29 51	42 1	38 46	35 17	31 18	26 9
16	47 45	44 39	41 21	37 33	32 50	42 44	39 54	36 52	33 24	28 57
18	48 14	45 29	42 34	39 13	35 5	43 13	40 42	38 1	34 56	30 1
20	48 35	46 7	43 30	40 30	36 47	43 34	41 18	38 53	36 8	32 37
22	48 50	46 36	44 13	41 30	38 8	43 49	41 46	39 34	37 5	34 53
24	49 1	46 58	44 48	42 19	39 15	44 0	42 7	40 7	37 50	35 55

50° Blank | 55° Blank

No. of Teeth	90°	80°	70°	60°	50°	90°	80°	70°	60°	50°
6	22° 45'	15° 58'	8° 38'	...	...	19° 17'	13° 30'	7° 15'	...	...
8	30 41	25 31	19 59	13° 33'	5° 20'	26 21	21 52	17 3	11° 30'	4° 17'
10	34 10	30 2	25 39	20 32	14 9	29 32	25 55	22 3	17 36	11 52
12	36 0	32 34	28 53	24 42	19 27	31 14	28 12	24 59	21 17	16 32
14	37 5	34 9	31 1	27 26	22 58	32 15	29 39	26 53	23 43	19 40
16	37 47	35 13	32 29	29 22	25 30	32 56	30 38	28 12	25 26	21 54
18	38 15	35 58	33 33	30 46	27 21	33 21	31 20	29 10	26 43	23 35
20	38 35	36 32	34 21	31 52	28 47	33 40	31 51	29 54	27 42	24 53
22	38 50	36 58	34 59	32 44	29 57	33 54	32 15	30 29	28 28	25 55
24	39 1	37 19	35 30	33 25	30 52	34 5	32 34	30 57	29 7	26 46

60° Blank | 65° Blank

No. of Teeth	90°	80°	70°	60°	50°	90°	80°	70°	60°	50°
6	16° 6'	11° 12'	6° 2'	...	...	13° 7'	9° 8'	4° 53'	...	...
8	22 13	18 24	14 9	9° 37'	3° 44'	18 15	15 6	11 42	7° 50'	3° 1'
10	25 2	21 56	18 37	14 49	10 5	20 40	18 4	15 19	12 9	8 15
12	26 34	23 57	21 10	17 59	14 13	21 59	19 48	17 28	14 49	11 32
14	27 29	25 14	22 50	20 6	16 44	22 48	20 55	18 54	16 37	13 48
16	28 5	26 7	24 1	21 37	18 40	23 18	21 39	19 53	17 53	15 24
18	28 29	26 44	24 52	22 44	20 6	23 40	22 11	20 37	18 50	16 37
20	28 46	27 11	25 30	23 35	21 14	23 55	22 35	21 10	19 33	17 34
22	29 0	27 34	26 2	24 17	22 8	24 6	22 53	21 36	20 8	18 20
24	29 9	27 50	26 26	24 50	22 52	24 15	23 8	21 57	20 36	18 57

70° Blank | 75° Blank

No. of Teeth	90°	80°	70°	60°	50°	90°	80°	70°	60°	50°
6	10° 18'	7° 9'	3° 48'	...	...	7° 38'	5° 19'	2° 50'	...	...
8	14 26	11 55	9 14	6° 9'	2° 21'	10 44	8 51	6 51	4° 34'	1° 45'
10	16 25	14 21	12 8	9 37	6 30	12 14	10 40	9 1	7 8	4 49
12	17 30	15 45	13 53	11 45	9 8	13 4	11 45	10 21	8 45	6 47
14	18 9	16 38	15 1	13 11	10 55	13 34	12 26	11 13	9 50	8 7
16	18 35	17 15	15 50	14 13	12 13	13 54	12 54	11 50	10 37	9 7
18	18 53	17 42	16 24	14 59	13 13	14 8	13 14	12 17	11 12	9 51
20	19 6	18 1	16 53	15 35	13 59	14 18	13 29	12 38	11 39	10 27
22	19 15	18 16	17 15	16 3	14 35	14 25	13 41	12 53	12 0	10 54
24	19 22	18 29	17 33	16 25	15 5	14 31	13 50	13 7	12 18	11 18

Angles of Elevation for Milling Straight Teeth in 80- and 85-degree Blanks Using Single-angle Fluting Cutters

No. of Teeth	Angle of Fluting Cutter									
	90°	80°	70°	60°	50°	90°	80°	70°	60°	50°
	80° Blank					85° Blank				
6	5° 2′	3° 30′	1° 52′	. . .	. . .	2° 30′	1° 44′	0° 55′	. . .	. . .
8	7 6	5 51	4 31	3° 2′	1° 8′	3 32	2 55	2 15	1° 29′	0° 34′
10	8 7	7 5	5 59	4 44	3 11	4 3	3 32	2 59	2 21	1 35
12	8 41	7 48	6 52	5 48	4 29	4 20	3 53	3 25	2 53	2 15
14	9 2	8 16	7 28	6 32	5 24	4 30	4 7	3 43	3 15	2 42
16	9 15	8 35	7 51	7 3	6 3	4 37	4 17	3 56	3 30	3 1
18	9 24	8 48	8 10	7 26	6 33	4 42	4 24	4 5	3 43	3 16
20	9 31	8 58	8 24	7 44	6 56	4 46	4 29	4 12	3 52	3 28
22	9 36	9 6	8 35	7 59	7 15	4 48	4 33	4 18	3 59	3 37
24	9 40	9 13	8 43	8 11	7 30	4 50	4 36	4 22	4 5	3 45

Spline-Shaft Milling Cutter. — The most efficient method of forming splines on shafts is by hobbing, but special milling cutters may also be used. Since the cutter forms the space between adjacent splines, it must be made to suit the number of splines and the root diameter of the shaft. The cutter angle B equals 360 degrees divided by the number of splines. The following formulas are for determining the chordal width C at the root of the splines or the chordal width across the concave

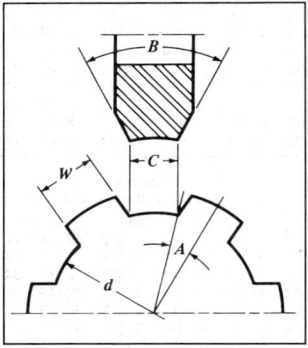

edge of the cutter. In these formulas, A = angle between center line of spline and a radial line passing through the intersection of the root circle and one side of the spline; W = width of spline; d = root diameter of splined shaft; C = chordal width at root circle between adjacent splines; N = number of splines.

$$\text{Sin } A = \frac{W}{d} \; ; \; C = d \times \sin\left(\frac{180}{N} - A\right)$$

Splines of involute form are often used in preference to the straight-sided type. Dimensions of the American Standard involute splines and hobs are given in the section on splines.

Cutter Grinding

Wheels for Sharpening Milling Cutters. — Milling cutters may be sharpened either by using the periphery of a disk wheel or the face of a cup wheel. The latter grind the lands of the teeth flat, whereas the periphery of a disk wheel leaves the teeth slightly concave back of the cutting edges. The concavity produced by disk wheel reduces the effective clearance angle on the teeth, the effect being more pronounced for wheels of small diameter than for wheels of large diameter. For this reason, large diameter wheels are preferred when sharpening milling cutters with disk type wheels. Irrespective of what type of wheel is used to sharpen a milling cutter, any burr resulting from grinding should be carefully removed by a hand stoning operation. Stoning also helps to reduce the roughness of grinding marks and improves the quality of the finish produced on the surface being machined.

Specifications of Grinding Wheels for Sharpening Milling Cutters

Cutter Material	Operation	Grinding Wheel			
		Abrasive Material	Grain Size	Grade	Bond
Carbon Tool Steel	Roughing Finishing	Aluminum Oxide Aluminum Oxide	46–60 100	K H	Vitrified Vitrified
High-speed Steel:					
18-4-1 {	Roughing Finishing	Aluminum Oxide Aluminum Oxide	60 100	K,H H	Vitrified Vitrified
18-4-2 {	Roughing Finishing	Aluminum Oxide Aluminum Oxide	80 100	F,G,H H	Vitrified Vitrified
Cast Non-Ferrous Tool Material	Roughing Finishing	Aluminum Oxide Aluminum Oxide	46 100–120	H,K,L,N H	Vitrified Vitrified
Sintered Carbide	Roughing after Brazing	Silicon Carbide	60	G	Vitrified
	Roughing Finishing	Diamond Diamond	100 Up to 500	* *	Resinoid Resinoid
Carbon Tool Steel and High-Speed Steel**	Roughing Finishing	Cubic Boron Nitride Cubic Boron Nitride	80–100 100–120	R,P S,T	Resinoid Resinoid

* Not indicated in diamond wheel markings.
** For hardnesses above Rockwell C 56.

Wheel Speeds and Feeds for Sharpening Milling Cutters. — Relatively low cutting speeds should be used when sharpening milling cutters to avoid tempering and heat checking. Dry grinding is recommended in all cases except when diamond wheels are employed. The surface speed of grinding wheels should be in the range of 4500 to 6500 feet per minute for grinding milling cutters of high-speed steel or cast non-ferrous tool material. For sintered carbide cutters, 5000 to 5500 feet per minute should be used.

The maximum stock removed per pass of the grinding wheel should not exceed about 0.0004 inch for sintered carbide cutters; 0.003 inch for large high-speed steel and cast non-ferrous tool material cutters; and 0.0015 inch for narrow saws and slotting cutters of high-speed steel or cast non-ferrous tool material. The stock removed per pass of the wheel may be increased for backing-off operations such as the grinding of secondary clearance behind the teeth since there is usually a sufficient body of metal to carry off the heat.

Clearance Angles for Milling Cutter Teeth. — The clearance angle provided on the cutting edges of milling cutters has an important bearing on cutter performance, cutting efficiency, and cutter life between sharpenings. It is desirable in all cases to use a clearance angle as small as possible so as to leave more metal back of the cutting edges for better heat dissipation and to provide maximum support. Excessive clearance angles not only weaken the cutting edges, but also increase the likelihood of "chatter" which will result in poor finish on the machined surface and reduce the life of the cutter. According to The Cincinnati Milling Machine Co., milling cutters used for general purpose work and having diameters from ⅛ to 3 inches should have clearance angles from 13 to 5 degrees, respectively, decreasing proportionately as the diameter increases. General purpose cutters over 3 inches

in diameter should be provided with a clearance angle of 4 to 5 degrees. The land width is usually ¼₄, ¹⁄₃₂, and ¹⁄₁₆ inch, respectively, for small, medium, and large cutters.

The primary clearance or relief angle for best results varies according to the material being milled about as follows: low carbon, high carbon, and alloy steels, 3 to 5 degrees; cast iron and medium and hard bronze, 4 to 7 degrees; brass, soft bronze, aluminum, magnesium, plastics, etc., 10 to 12 degrees. When milling cutters are resharpened, it is customary to grind a secondary clearance angle of 3 to 5 degrees behind the primary clearance angle to reduce the land width to its original value and thus avoid interference with the surface to be milled. A general formula for plain milling cutters, face mills, and form relieved cutters which gives the clearance angle C, in degrees, necessitated by the feed per revolution F, in inches, the width of land L, in inches, the depth of cut d, in inches, the cutter diameter D, in inches, and the Brinell hardness number B of the work being cut is:

$$C = \frac{45860}{DB}\left(1.5\,L + \frac{F}{\pi D}\,\sqrt{d\,(D-d)}\,\right)$$

Rake Angles for Milling Cutters. — In peripheral milling cutters, the rake angle is generally defined as the angle in degrees that the tooth face deviates from a radial line to the cutting edge. In face milling cutters, the teeth are inclined with respect to both the radial and axial lines. These angles are called *radial* and *axial* rake, respectively. The radial and axial rake angles may be positive, zero, or negative.

Positive rake angles should be used whenever possible for all types of high-speed steel milling cutters. For sintered carbide tipped cutters, zero and negative rake angles are frequently employed to provide more material back of the cutting edge to resist shock loads.

Rake Angles for High-speed Steel Cutters: Positive rake angles of 10 to 15 degrees are satisfactory for milling steels of various compositions with plain milling cutters. For softer materials such as magnesium and aluminum alloys, the rake angle may be 25 degrees or more. Metal slitting saws for cutting alloy steel usually have rake angles from 5 to 10 degrees, whereas zero and sometimes negative rake angles are used for saws to cut copper and other soft non-ferrous metals to reduce the tendency to "hog in." Form relieved cutters usually have rake angles of 0, 5, or 10 degrees. Commercial face milling cutters usually have 10 degrees positive radial and axial rake angles for general use in milling cast iron, forged and alloy steel, brass, and bronze; for milling castings and forgings of magnesium and free-cutting aluminum and their alloys, the rake angles may be increased to 25 degrees positive or more, depending on the operating conditions; a smaller rake angle is used for abrasive or difficult to machine aluminum alloys.

Cast Non-ferrous Tool Material Milling Cutters: Positive rake angles are generally provided on milling cutters using cast non-ferrous tool materials although negative rake angles may be used advantageously for some operations such as those where shock loads are encountered or where it is necessary to eliminate vibration when milling thin sections.

Sintered Carbide Milling Cutters: Peripheral milling cutters such as slab mills, slotting cutters, saws, etc., tipped with sintered carbide, generally have negative radial rake angles of 5 degrees for soft low carbon steel and 10 degrees or more for alloy steels. Positive axial rake angles of 5 and 10 degrees, respectively, may be provided, and for slotting saws and cutters, 0 degree axial rake may be used. On soft materials such as free-cutting aluminum alloys, positive rake angles of 10 to 20 degrees are used. For milling abrasive or difficult to machine aluminum alloys, small positive or even negative rake angles are used.

Eccentric Type Radial Relief. — When the radial relief angles on peripheral teeth of milling cutters are ground with a disc type grinding wheel in the conventional manner the ground surfaces on the lands are slightly concave, conforming approximately to the radius of the wheel. A flat land is produced when the radial relief angle is ground with a cup wheel. Another entirely different method of grinding the radial angle is by the eccentric method, which produces a slightly convex surface on the land. If the radial relief angle at the cutting edge is equal for all of the three types of land mentioned, it will be found that the land with the eccentric relief will drop away from the cutting edge a somewhat greater distance for a given distance around the land than will the others. This is evident from a study of Table 2 entitled, "Indicator Drops for Checking Radial Relief Angles on Peripheral Teeth." It is claimed that this feature is an advantage of the eccentric type relief.

The setup for grinding an eccentric relief is shown in Fig. 2. In this setup the point of contact between the cutter and the tooth rest must be in the same plane as the centers, or axes, of the grinding wheel and the cutter. A wide face is used on the grinding wheel, which is trued and dressed at an angle with respect to the axis of the cutter. An alternate method is to tilt the wheel at this angle. Then as the cutter is traversed and rotated past the grinding wheel while in contact with the tooth rest, an eccentric relief will be generated by the angular face of the wheel. This type of relief can only be ground on the peripheral teeth on milling cutters having helical flutes because the combination of the angular wheel face and the twisting motion of the cutter is required to generate the eccentric relief. Therefore, an eccentric relief cannot be ground on the peripheral teeth of straight fluted cutters.

Table 1 is a table of wheel angles for grinding an eccentric relief for different combinations of relief angles and helix angles. When angles are required that cannot be found in this table, the wheel angle, W, can be calculated by using the following formula, in which R is the radial relief angle and H is the helix angle of the flutes on the cutter.

$$\tan W = \tan R \times \tan H$$

Indicator Drop Method of Checking Relief and Rake Angles. — The most convenient and inexpensive method of checking the relief and rake angles on milling cutters is by the indicator drop method. Three tables, Tables 2, 3 and 4, of indicator

(*Continued on page* 794)

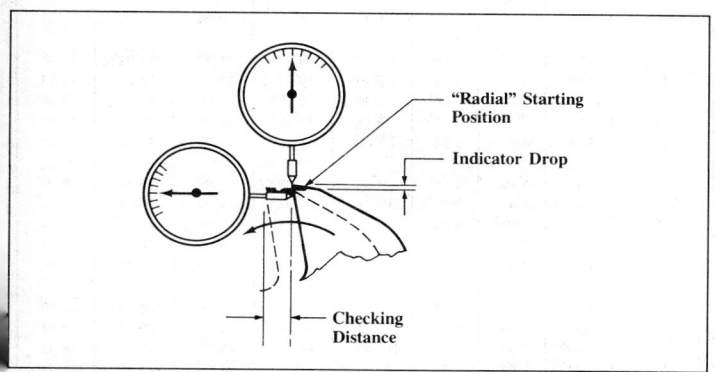

Fig. 1. Setup for Checking the Radial Relief Angle by Indicator Drop Method

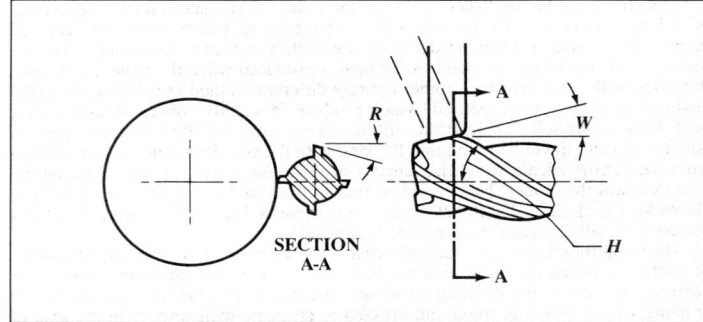

Fig. 2. Setup for Grinding Eccentric Type Radial Relief Angle

Table 1. Grinding Wheel Angles for Grinding Eccentric Type Radial Relief Angle

Radial Relief Angle, R, Degrees	Helix Angle of Cutter Flutes, H, Degrees							
	12	18	20	30	40	45	50	52
	Wheel Angle, W, Degrees							
1	0°13′	0°19′	0°22′	0°35′	0°50′	1°00′	1°12′	1°17′
2	0°26′	0°39′	0°44′	1°09′	1°41′	2°00′	2°23′	2°34′
3	0°38′	0°59′	1°06′	1°44′	2°31′	3°00′	3°34′	3°50′
4	0°51′	1°18′	1°27′	2°19′	3°21′	4°00′	4°46′	5°07′
5	1°04′	1°38′	1°49′	2°53′	4°12′	5°00′	5°57′	6°23′
6	1°17′	1°57′	2°11′	3°28′	5°02′	6°00′	7°08′	7°40′
7	1°30′	2°17′	2°34′	4°03′	5°53′	7°00′	8°19′	8°56′
8	1°43′	2°37′	2°56′	4°38′	6°44′	8°00′	9°30′	10°12′
9	1°56′	2°57′	3°18′	5°13′	7°34′	9°00′	10°41′	11°28′
10	2°09′	3°17′	3°40′	5°49′	8°25′	10°00′	11°52′	12°43′
11	2°22′	3°37′	4°03′	6°24′	9°16′	11°00′	13°03′	13°58′
12	2°35′	3°57′	4°25′	7°00′	10°07′	12°00′	14°13′	15°13′
13	2°49′	4°17′	4°48′	7°36′	10°58′	13°00′	15°23′	16°28′
14	3°02′	4°38′	5°11′	8°11′	11°49′	14°00′	16°33′	17°42′
15	3°16′	4°59′	5°34′	8°48′	12°40′	15°00′	17°43′	18°56′
16	3°29′	5°19′	5°57′	9°24′	13°32′	16°00′	18°52′	20°09′
17	3°43′	5°40′	6°21′	10°01′	14°23′	17°00′	20°01′	21°22′
18	3°57′	6°02′	6°45′	10°37′	15°15′	18°00′	21°10′	22°35′
19	4°11′	6°23′	7°09′	11°15′	16°07′	19°00′	22°19′	23°47′
20	4°25′	6°45′	7°33′	11°52′	16°59′	20°00′	23°27′	24°59′
21	4°40′	7°07′	7°57′	12°30′	17°51′	21°00′	24°35′	26°10′
22	4°55′	7°29′	8°22′	13°08′	18°44′	22°00′	25°43′	27°21′
23	5°09′	7°51′	8°47′	13°46′	19°36′	23°00′	26°50′	28°31′
24	5°24′	8°14′	9°12′	14°25′	20°29′	24°00′	27°57′	29°41′
25	5°40′	8°37′	9°38′	15°04′	21°22′	25°00′	29°04′	30°50′

Table 2. Indicator Drops for Checking the Radial Relief Angle on Peripheral Teeth

Cutter Diameter, Inch	Recommended Range of Radial Relief Angles, Degrees	Checking Distance, Inch	Indicator Drops, Inches				Recommended Max. Primary Land Width, Inch
			For Flat and Concave Relief		For Eccentric Relief		
			Min.	Max.	Min.	Max.	
1/16	20–25	.005	.0014	.0019	.0020	.0026	.007
3/32	16–20	.005	.0012	.0015	.0015	.0019	.007
1/8	15–19	.010	.0018	.0026	.0028	.0037	.015
5/32	13–17	.010	.0017	.0024	.0024	.0032	.015
3/16	12–16	.010	.0016	.0023	.0022	.0030	.015
7/32	11–15	.010	.0015	.0022	.0020	.0028	.015
1/4	10–14	.015	.0017	.0028	.0027	.0039	.015
9/32	10–14	.015	.0018	.0029	.0027	.0039	.020
5/16	10–13	.015	.0019	.0027	.0027	.0035	.020
11/32	10–13	.015	.0020	.0028	.0027	.0035	.020
3/8	10–13	.015	.0020	.0029	.0027	.0035	.020
13/32	10–13	.015	.0020	.0029	.0027	.0035	.020
7/16	9–12	.020	.0022	.0032	.0032	.0044	.025
15/32	9–12	.020	.0022	.0033	.0032	.0043	.025
1/2	9–12	.020	.0023	.0034	.0032	.0043	.025
9/16	9–12	.020	.0024	.0034	.0032	.0043	.025
5/8	8–11	.020	.0022	.0032	.0028	.0039	.025
11/16	8–11	.030	.0029	.0045	.0043	.0059	.035
3/4	8–11	.030	.0030	.0046	.0043	.0059	.035
13/16	8–11	.030	.0031	.0047	.0043	.0059	.035
7/8	8–11	.030	.0032	.0048	.0043	.0059	.035
15/16	7–10	.030	.0027	.0043	.0037	.0054	.035
1	7–10	.030	.0028	.0044	.0037	.0054	.035
1 1/8	7–10	.030	.0029	.0045	.0037	.0054	.035
1 1/4	7–10	.030	.0024	.0040	.0037	.0053	.035
1 3/8	6–9	.030	.0025	.0041	.0032	.0048	.035
1 1/2	6–9	.030	.0026	.0041	.0032	.0048	.035
1 5/8	6–9	.030	.0026	.0042	.0032	.0048	.035
1 3/4	6–9	.030	.0026	.0042	.0032	.0048	.035
1 7/8	6–9	.030	.0027	.0043	.0032	.0048	.035
2	6–9	.030	.0027	.0043	.0032	.0048	.035
2 1/4	5–8	.030	.0022	.0038	.0026	.0042	.040
2 1/2	5–8	.030	.0023	.0039	.0026	.0042	.040
2 3/4	5–8	.030	.0023	.0039	.0026	.0042	.040
3	5–8	.030	.0023	.0039	.0026	.0042	.040
3 1/2	5–8	.030	.0024	.0040	.0026	.0042	.047
4	5–8	.030	.0024	.0040	.0026	.0042	.047
5	4–7	.030	.0019	.0035	.0021	.0037	.047
6	4–7	.030	.0019	.0035	.0021	.0037	.047
7	4–7	.030	.0020	.0036	.0021	.0037	.060
8	4–7	.030	.0020	.0036	.0021	.0037	.060
10	4–7	.030	.0020	.0036	.0021	.0037	.060
12	4–7	.030	.0020	.0036	.0021	.0037	.060

Table 3. Indicator Drops for Checking Relief Angles on Side Teeth and End Teeth

Checking Distance, Inch	Given Relief Angle								
	1°	2°	3°	4°	5°	6°	7°	8°	9°
	Indicator Drop, inch								
.005	.00009	.00017	.00026	.00035	.0004	.0005	.0006	.0007	.0008
.010	.00017	.00035	.00052	.0007	.0009	.0011	.0012	.0014	.0016
.015	.00026	.0005	.00079	.0010	.0013	.0016	.0018	.0021	.0024
.031	.00054	.0011	.0016	.0022	.0027	.0033	.0038	.0044	.0049
.047	.00082	.0016	.0025	.0033	.0041	.0049	.0058	.0066	.0074
.062	.00108	.0022	.0032	.0043	.0054	.0065	.0076	.0087	.0098

Table 4. Indicator Drops for Checking Rake Angles on Milling Cutter Face

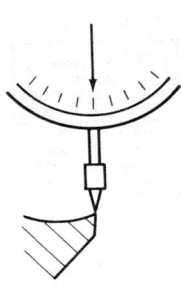

Set indicator to read zero
on horizontal plane passing
through cutter axis. Zero
cutting edge against indicator.

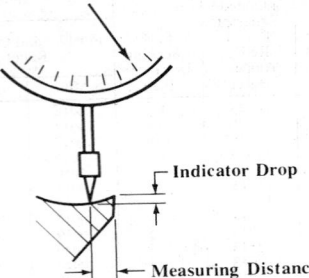

├ Indicator Drop

← Measuring Distance

Move cutter or indicator
measuring distance.

Rate Angle, Deg.	Measuring Distance, inch				Rate Angle, Deg.	Measuring Distance, inch			
	.031	.062	.094	.125		.031	.062	.094	.125
	Indicator Drop, inch					Indicator Drop, inch			
1	.0005	.0011	.0016	.0022	11	.0060	.0121	.0183	.0243
2	.0011	.0022	.0033	.0044	12	.0066	.0132	.0200	.0266
3	.0016	.0032	.0049	.0066	13	.0072	.0143	.0217	.0289
4	.0022	.0043	.0066	.0087	14	.0077	.0155	.0234	.0312
5	.0027	.0054	.0082	.0109	15	.0083	.0166	.0252	.0335
6	.0033	.0065	.0099	.0131	16	.0089	.0178	.0270	.0358
7	.0038	.0076	.0115	.0153	17	.0095	.0190	.0287	.0382
8	.0044	.0087	.0132	.0176	18	.0101	.0201	.0305	.0406
9	.0049	.0098	.0149	.0198	19	.0107	.0213	.0324	.0430
10	.0055	.0109	.0166	.0220	20	.0113	.0226	.0342	.0455

drops are provided in this section, for checking radial relief angles on the peripheral teeth, relief angles on side and end teeth, and rake angles on the tooth faces.

The setup for checking the radial relief angle is illustrated in Fig. 1. Two dial test indicators are required, one of which should have a sharp pointed contact point. This indicator is positioned so that the axis of its spindle is vertical, passing through the axis of the cutter. The cutter may be held by its shank in the spindle of a tool and cutter grinder workhead, or between centers while mounted on a mandrel. The cutter is rotated to the position where the vertical indicator contacts a cutting edge. The second indicator is positioned with its spindle axis horizontal and with the contact point touching the tool face just below the cutting edge. With both indicators adjusted to read zero, the cutter is rotated a distance equal to the checking distance, as determined by the reading on the second indicator. Then the indicator drop is read on the vertical indicator and checked against the values in the tables. The indicator drops for radial relief angles ground by a disc type grinding wheel and those ground with a cup wheel are so nearly equal that the values are listed together; values for the eccentric type relief are listed separately, since they are larger. A similar procedure is used to check the relief angles on the side and end teeth of milling cutters; however, only one indicator is used. Also, instead of rotating the cutter, the indicator or the cutter must be moved a distance equal to the checking distance in a straight line.

Various Set-ups Used in Grinding the Clearance Angle on Milling Cutter Teeth

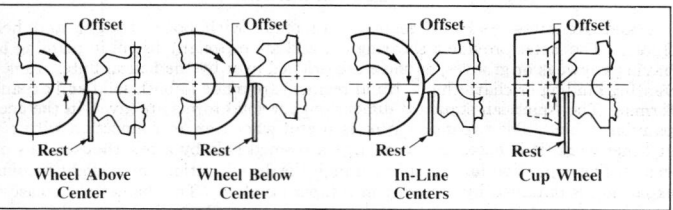

Wheel Above Center   Wheel Below Center   In-Line Centers   Cup Wheel

Distance to Set Center of Wheel Above the Cutter Center (Disk Wheel)

Diam. of Wheel, Inches	Desired Clearance Angle, Degrees											
	1	2	3	4	5	6	7	8	9	10	11	12
	*Distance to Offset Wheel Center Above Cutter Center, Inches											
3	.026	.052	.079	.105	.131	.157	.183	.209	.235	.260	.286	.312
4	.035	.070	.105	.140	.174	.209	.244	.278	.313	.347	.382	.416
5	.044	.087	.131	.174	.218	.261	.305	.348	.391	.434	.477	.520
6	.052	.105	.157	.209	.261	.314	.366	.417	.469	.521	.572	.624
7	.061	.122	.183	.244	.305	.366	.427	.487	.547	.608	.668	.728
8	.070	.140	.209	.279	.349	.418	.488	.557	.626	.695	.763	.832
9	.079	.157	.236	.314	.392	.470	.548	.626	.704	.781	.859	.936
10	.087	.175	.262	.349	.436	.523	.609	.696	.782	.868	.954	1.040

* Calculated from the formula: Offset = Wheel Diameter × ½ × Sine of Clearance Angle.

Distance to Set Center of Wheel Below the Cutter Center (Disk Wheel)

Diam. of Cutter, Inches	Desired Clearance Angle, Degrees											
	1	2	3	4	5	6	7	8	9	10	11	12
	*Distance to Offset Wheel Center Below Cutter Center, Inches											
2	.017	.035	.052	.070	.087	.105	.122	.139	.156	.174	.191	.208
3	.026	.052	.079	.105	.131	.157	.183	.209	.235	.260	.286	.312
4	.035	.070	.105	.140	.174	.209	.244	.278	.313	.347	.382	.416
5	.044	.087	.131	.174	.218	.261	.305	.348	.391	.434	.477	.520
6	.052	.105	.157	.209	.261	.314	.366	.417	.469	.521	.572	.624
7	.061	.122	.183	.244	.305	.366	.427	.487	.547	.608	.668	.728
8	.070	.140	.209	.279	.349	.418	.488	.557	.626	.695	.763	.832
9	.079	.157	.236	.314	.392	.470	.548	.626	.704	.781	.859	.936
10	.087	.175	.262	.349	.436	.523	.609	.696	.782	.868	.954	1.040

* Calculated from the formula: Offset = Cutter Diameter × ½ × Sine of Clearance Angle.

Distance to Set Tooth Rest Below Center Line of Wheel and Cutter. — When the clearance angle is ground with a disk type wheel by keeping the center line of the wheel in line with the center line of the cutter, the tooth rest should be lowered by an amount given by the following formula:

$$\text{Offset} = \frac{\text{Wheel Diam.} \times \text{Cutter Diam.} \times \text{Sine of One-half the Clearance Angle}}{\text{Wheel Diam.} + \text{Cutter Diam.}}$$

Distance to Set Tooth Rest Below Cutter Center When Cup Wheel is Used. — When the clearance is ground with a cup wheel, the tooth rest is set below the center of the cutter the same amount as given in the table for "Distance to Set Center of Wheel Below the Cutter Center (Disk Wheel)."

REAMERS

Hand Reamers. — Hand reamers are made with both straight and helical flutes. The latter provide a shearing cut and are especially useful in reaming holes having keyways or grooves, as these are bridged over by the helical flutes, thus preventing binding or chattering. Hand reamers are made in both solid and expansion forms. The American standard dimensions for solid forms are given in the accompanying table. The expansion type is useful whenever, in connection with repair or other work, it is necessary to enlarge a reamed hole by a few thousandths of an inch. The expansion form is split through the fluted section and a slight amount of expansion is obtained by screwing in a tapering plug. The diameter increase may vary from 0.005 to 0.008 inch for reamers up to about 1 inch diameter and from 0.010 to 0.012 inch for diameters between 1 and 2 inches. Hand reamers are tapered slightly on the end to facilitate starting them properly. The actual diameter of the shanks of commercial reamers may be from 0.002 to 0.005 inch under the reamer size. That part of the shank which is squared should be turned smaller in diameter than the shank itself, so that, when applying a wrench, no burr may be raised which may mar the reamed hole if the reamer is passed clear through it.

When fluting reamers, the cutter is so set with relation to the center of the reamer blank that the tooth gets a slight negative rake; that is, the cutter should be set *ahead* of the center, as shown in the illustration accompanying the table giving the amount to set the cutter ahead of the radial line. The amount is so selected that a tangent to the circumference of the reamer at the cutting point makes an angle of approximately 95 degrees with the front face of the cutting edge.

Amount to Set Cutter Ahead of Radial Line to Obtain Negative Front Rake

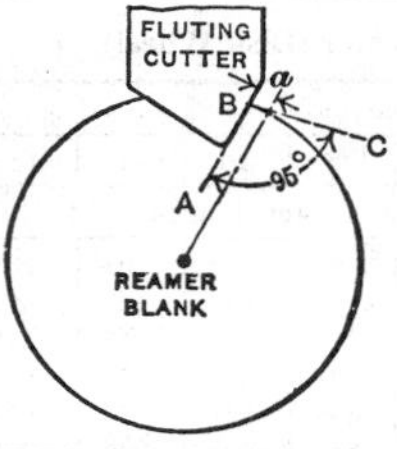

	Size of Reamer	Dimension *a*, Inches	Size of Reamer	Dimension *a*, Inches	Size of Reamer	Dimension *a*, Inches
	¼	0.011	⅞	0.038	2	0.087
	⅜	0.016	1	0.044	2¼	0.098
	½	0.022	1¼	0.055	2½	0.109
	⅝	0.027	1½	0.066	2¾	0.120
	¾	0.033	1¾	0.076	3	0.131

When fluting reamers, it is necessary to "break up the flutes"; that is, to space the cutting edges unevenly around the reamer. The difference in spacing should be very slight and need not exceed two degrees one way or the other. The manner in which the breaking up of the flutes is usually done is to move the index head to which the reamer is fixed a certain amount more or less than would be the case if the spacing were regular. A table is given showing the amount of this additional movement of the index crank for reamers with different numbers of flutes. When a reamer is provided with helical flutes, the angle of spiral should be such that the cutting edges make an angle of about 10 or at most 15 degrees with the axis of the reamer.

The relief of the cutting edges should be comparatively slight. An eccentric relief, that is, one where the land back of the cutting edge is convex, rather than flat, is used by one or two manufacturers, and is preferable for finishing reamers as the reamer will hold its size longer. When hand reamers are used merely for removing stock, or simply for enlarging holes, the flat relief is better, because the

reamer has a keener cutting edge. The width of the land of the cutting edges should be about 1/32 inch for a 1/4-inch, 1/16 inch for a 1-inch, and 3/32 inch for a 3-inch reamer.

Irregular Spacing of Teeth in Reamers

Number of flutes in reamer	4	6	8	10	12	14	16
Index circle to use	39	39	39	39	39	49	20
Before cutting	Move Spindle the Number of Holes below More or Less than for Regular Spacing						
2d flute	8 less	4 less	3 less	2 less	4 less	3 less	2 less
3d flute	4 more	5 more	5 more	3 more	4 more	2 more	2 more
4th flute	6 less	7 less	2 less	5 less	1 less	2 less	1 less
5th flute	.....	6 more	4 more	2 more	3 more	4 more	2 more
6th flute	.....	5 less	6 less	2 less	4 less	1 less	2 less
7th flute	.....	.....	2 more	3 more	4 more	3 more	1 more
8th flute	.....	.....	3 less	2 less	3 less	2 less	2 less
9th flute	.....	.....	.....	5 more	2 more	1 more	2 more
10th flute	.....	.....	.....	1 less	2 less	3 less	2 less
11th flute	.....	.....	.....	.....	3 more	3 more	1 more
12th flute	.....	.....	.....	.....	4 less	2 less	2 less
13th flute	.....	.....	.....	.....	.....	2 more	2 more
14th flute	.....	.....	.....	.....	.....	3 less	1 less
15th flute	.....	.....	.....	.....	.....	.....	2 more
16th flute	.....	.....	.....	.....	.....	.....	2 less

Threaded-end Hand Reamers. — Hand reamers are sometimes provided with a thread at the extreme point in order to give them a uniform feed when reaming. The diameter on the top of this thread at the point of the reamer is slightly smaller than the reamer itself, and the thread tapers upward until it reaches a dimension of from 0.003 to 0.008 inch, according to size, below the size of the reamer; at this point the thread stops and a short neck about 1/16 inch wide separates the threaded portion from the actual reamer which is provided with a short taper from 3/16 to 7/16 inch long up to where the standard diameter is reached. The length of the threaded portion and the number of threads per inch for reamers of this kind are given in the accompanying table. The thread employed is a sharp V-thread.

Dimensions for Threaded-end Hand Reamers

Sizes of Reamers	Length of Threaded Part	No. of Threads per Inch	Diam. of Thread at Point of Reamer	Sizes of Reamers	Length of Threaded Part	No. of Threads per Inch	Diam. of Thread at Point of Reamer
			Full diameter				Full diameter
1/8–5/16	3/8	32	−0.006	1 1/32 –1 1/2	9/16	18	−0.010
11/32–1/2	7/16	28	−0.006	1 17/32–2	9/16	18	−0.012
17/32–3/4	1/2	24	−0.008	2 1/32 –2 1/2	9/16	18	−0.015
25/32–1	9/16	18	−0.008	2 17/32–3	9/16	18	−0.020

Fluted Chucking Reamers. — Reamers of this type are used in turret lathes. screw machines, etc., for enlarging holes and finishing them smooth and to the required size. The best results are obtained with a floating type of holder which permits a reamer to align itself with the hole being reamed. These reamers are intended for removing a small amount of metal, 0.005 to 0.010 inch being common allowances. Fluted chucking reamers are provided either with a straight shank or a standard taper shank. (See table for standard dimensions.)

Rose Chucking Reamers. — The rose type of reamer is used for enlarging cored or other holes. The cutting edges at the end are ground to a 45-degree bevel. This type of reamer will remove considerable metal in one cut. The cylindrical part of the reamer has no cutting edges, but merely grooves cut for the full length of the reamer body, providing a way for the chips to escape and a channel for lubricant to reach the cutting edges. There is no relief on the cylindrical surface of the body part, but it is slightly back-tapered so that the diameter at the point with the beveled cutting edges is slightly larger than the diameter further back. The back-taper should not exceed 0.001 inch per inch. This form of reamer usually

Fluting Cutters for Reamers

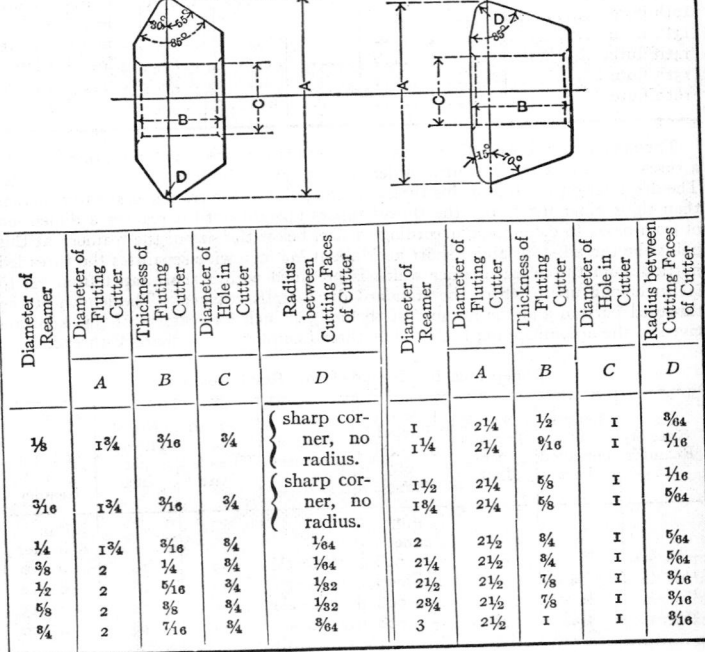

Diameter of Reamer	Diameter of Fluting Cutter	Thickness of Fluting Cutter	Diameter of Hole in Cutter	Radius between Cutting Faces of Cutter	Diameter of Reamer	Diameter of Fluting Cutter	Thickness of Fluting Cutter	Diameter of Hole in Cutter	Radius between Cutting Faces of Cutter
	A	B	C	D		A	B	C	D
$\frac{1}{8}$	$1\frac{3}{4}$	$\frac{3}{16}$	$\frac{3}{4}$	sharp corner, no radius.	1	$2\frac{1}{4}$	$\frac{1}{2}$	1	$\frac{3}{64}$
					$1\frac{1}{4}$	$2\frac{1}{4}$	$\frac{9}{16}$	1	$\frac{1}{16}$
$\frac{3}{16}$	$1\frac{3}{4}$	$\frac{3}{16}$	$\frac{3}{4}$	sharp corner, no radius.	$1\frac{1}{2}$	$2\frac{1}{4}$	$\frac{5}{8}$	1	$\frac{1}{16}$
					$1\frac{3}{4}$	$2\frac{1}{4}$	$\frac{5}{8}$	1	$\frac{5}{64}$
$\frac{1}{4}$	$1\frac{3}{4}$	$\frac{3}{16}$	$\frac{3}{4}$	$\frac{1}{64}$	2	$2\frac{1}{2}$	$\frac{3}{4}$	1	$\frac{5}{64}$
$\frac{3}{8}$	2	$\frac{1}{4}$	$\frac{3}{4}$	$\frac{1}{64}$	$2\frac{1}{4}$	$2\frac{1}{2}$	$\frac{3}{4}$	1	$\frac{5}{64}$
$\frac{1}{2}$	2	$\frac{5}{16}$	$\frac{3}{4}$	$\frac{1}{32}$	$2\frac{1}{2}$	$2\frac{1}{2}$	$\frac{7}{8}$	1	$\frac{3}{16}$
$\frac{5}{8}$	2	$\frac{3}{8}$	$\frac{3}{4}$	$\frac{1}{32}$	$2\frac{3}{4}$	$2\frac{1}{2}$	$\frac{7}{8}$	1	$\frac{3}{16}$
$\frac{3}{4}$	2	$\frac{7}{16}$	$\frac{3}{4}$	$\frac{3}{64}$	3	$2\frac{1}{2}$	1	1	$\frac{3}{16}$

Dimensions of Formed Reamer Fluting Cutters

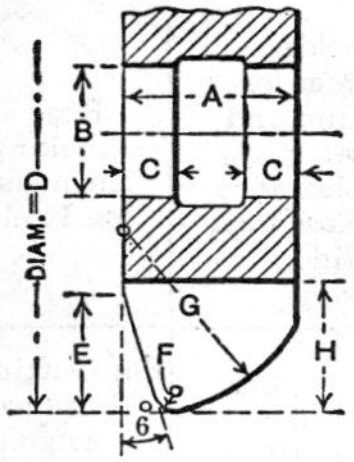

The making and maintenance of cutters of the formed type involves greater expense than the use of the angular cutters of which dimensions are given on the previous page; but the form of flute produced by the formed type of cutter is preferred by many reamer users. The claims made for the formed type of flute are that the chips can be more readily removed from the reamer, and that the reamer has greater strength and is less likely to crack or spring out of shape in hardening.

Size of Reamers used for	Number of Teeth in Reamer	Diam. of Cutter	Width of Cutter	Diam. of Hole	Width of Bearing	Length of Bevel	Radius	Radius	Depth of Tooth	Number of Teeth in Cutter
		D	*A*	*B*	*C*	*E*	*F*	*G*	*H*	
1/8–3/16	6	1 3/4	3/16	7/8	....	0.125	0.016	7/32	0.21	14
1/4–5/16	6	1 3/4	1/4	7/8	....	0.152	0.022	9/32	0.25	13
3/8–7/16	6	1 7/8	3/8	7/8	1/8	0.178	0.029	1/2	0.28	12
1/2–11/16	6–8	2	7/16	7/8	1/8	0.205	0.036	9/16	0.30	12
3/4–1	8	2 1/8	1/2	7/8	5/32	0.232	0.042	11/16	0.32	12
1 1/16–1 1/2	10	2 1/4	9/16	7/8	5/32	0.258	0.049	3/4	0.38	11
1 9/16–2 1/8	12	2 3/8	5/8	7/8	3/16	0.285	0.056	27/32	0.40	11
2 1/4 –3	14	2 5/8	11/16	7/8	3/16	0.312	0.062	7/8	0.44	10

produces holes slightly larger than its size and it is, therefore, always made from 0.005 to 0.010 inch smaller than its nominal size, so that it may be followed by a fluted reamer for finishing. The grooves on the cylindrical portion are cut by a convex cutter having a width equal to from one-fifth to one-fourth the diameter of the rose reamer itself. The depth of the groove should be from one-eighth to one-sixth the diameter of the reamer. The teeth at the end of the reamer are milled with a 75-degree angular cutter; the width of the land of the cutting edge should be about one-fifth the distance from tooth to tooth. If an angular cutter is preferred to a convex cutter for milling the grooves on the cylindrical portion, because of the higher cutting speed possible when milling, an 80-degree angular cutter slightly rounded at the point may be used.

Cutters for Fluting Rose Chucking Reamers. — The cutters used for fluting rose chucking reamers on the end are 80-degree angular cutters for 1/4 and 5/16 inch diameter reamers; 75-degree angular cutters for 3/8 and 7/16 inch reamers; and 70-degree angular cutters for all larger sizes. The grooves on the cylindrical portion are milled with convex cutters of approximately the following sizes for given diameters of reamers: 5/32-inch convex cutter for 1/2-inch reamers; 5/16-inch cutter for 1-inch reamers; 3/8-inch cutter for 1 1/2-inch reamers; 13/32-inch cutters for 2-inch reamers; and 15/32-inch cutters for 2 1/2-inch reamers. The smaller sizes of reamers, from 1/4 to 3/8 inch in diameter, are often milled with regular double-angle reamer fluting cutters having a radius of 1/64 inch for 1/4-inch reamer, and 1/32 inch for 5/16- and 3/8-inch sizes.

Vertical Adjustment of Tooth-rest for Grinding Clearance on Reamers

Size of Reamer	Hand Reamer for Steel. Cutting Clearance Land 0.006 inch Wide		Hand Reamer for Cast Iron and Bronze. Cutting Clearance Land 0.025 inch Wide		Chucking Reamer for Cast Iron and Bronze. Cutting Clearance Land 0.025 inch Wide		Rose Chucking Reamers for Steel
	For Cutting Clearance	For Second Clearance	For Cutting Clearance	For Second Clearance	For Cutting Clearance	For Second Clearance	For Cutting Clearance on Angular Edge at End
½	0.012	0.052	0.032	0.072	0.040	0.080	0.080
⅝	0.012	0.062	0.032	0.072	0.040	0.090	0.090
¾	0.012	0.072	0.035	0.095	0.040	0.100	0.100
⅞	0.012	0.082	0.040	0.120	0.045	0.125	0.125
1	0.012	0.092	0.040	0.120	0.045	0.125	0.125
1⅛	0.012	0.102	0.040	0.120	0.045	0.125	0.125
1¼	0.012	0.112	0.045	0.145	0.050	0.160	0.160
1⅜	0.012	0.122	0.045	0.145	0.050	0.160	0.175
1½	0.012	0.132	0.048	0.168	0.055	0.175	0.175
1⅝	0.012	0.142	0.050	0.170	0.060	0.200	0.200
1¾	0.012	0.152	0.052	0.192	0.060	0.200	0.200
1⅞	0.012	0.162	0.056	0.196	0.060	0.200	0.200
2	0.012	0.172	0.056	0.216	0.064	0.224	0.225
2⅛	0.012	0.172	0.059	0.219	0.064	0.224	0.225
2¼	0.012	0.172	0.063	0.223	0.064	0.224	0.225
2⅜	0.012	0.172	0.063	0.223	0.068	0.228	0.230
2½	0.012	0.172	0.065	0.225	0.072	0.232	0.230
2⅝	0.012	0.172	0.065	0.225	0.075	0.235	0.235
2¾	0.012	0.172	0.065	0.225	0.077	0.237	0.240
2⅞	0.012	0.172	0.070	0.230	0.080	0.240	0.240
3	0.012	0.172	0.072	0.232	0.080	0.240	0.240
3⅛	0.012	0.172	0.075	0.235	0.083	0.240	0.240
3¼	0.012	0.172	0.078	0.238	0.083	0.243	0.245
3⅜	0.012	0.172	0.081	0.241	0.087	0.247	0.245
3½	0.012	0.172	0.084	0.244	0.090	0.250	0.250
3⅝	0.012	0.172	0.087	0.247	0.093	0.253	0.250
3¾	0.012	0.172	0.090	0.250	0.097	0.257	0.255
3⅞	0.012	0.172	0.093	0.253	0.100	0.260	0.255
4	0.012	0.172	0.096	0.256	0.104	0.264	0.260
4⅛	0.012	0.172	0.096	0.256	0.104	0.264	0.260
4¼	0.012	0.172	0.096	0.256	0.106	0.266	0.265
4⅜	0.012	0.172	0.096	0.256	0.108	0.268	0.265
4½	0.012	0.172	0.100	0.260	0.108	0.268	0.265
4⅝	0.012	0.172	0.100	0.260	0.110	0.270	0.270
4¾	0.012	0.172	0.104	0.264	0.114	0.274	0.275
4⅞	0.012	0.172	0.106	0.266	0.116	0.276	0.275
5	0.012	0.172	0.110	0.270	0.118	0.278	0.275

Reamer Difficulties. — There are certain frequently occurring problems in reaming for which it is necessary to apply remedial measures. These difficulties include the production of oversize holes, bellmouth holes and holes with a poor finish. The following is taken from suggestions for correction of these difficulties by the National Twist Drill and Tool Co. and Winter Brothers Co.*

Oversize Holes: The cutting of a hole oversize from the start of the reaming operations usually indicates a mechanical defect in the setup or reamer. Thus, the wrong reamer for the work-piece material may have been used or there may be inadequate work-piece support, inadequate or worn guide bushings, or misalignment of the spindles, bushings or work-piece or runout of the spindle or reamer holder. The reamer itself may be defective due to chamfer runout or runout of the cutting end due to a bent or non-concentric shank.

When reamers gradually start to cut oversize, it is due to pickup or galling, principally on the reamer margins. This condition is partly due to the work-piece material. Mild steels, certain cast irons and some aluminum alloys are particularly troublesome in this respect.

Corrective measures include reducing the reamer margin widths to about 0.005 to 0.010 inch, use of hard case surface treatments on high speed steel reamers, either alone or in combination with black oxide treatments, and the use of a high grade finish on the reamer faces, margins, and chamfer relief surfaces.

Bellmouth Holes: The cutting of a hole that becomes oversize at the entry end with the oversize decreasing gradually along its length always reflects misalignment of the cutting portion of the reamer with respect to the hole. The obvious solution is to provide improved guiding of the reamer by the use of accurate bushings and pilot surfaces. If this is not feasible, and the reamer is cutting in a vertical position, a flexible element may be employed to hold the reamer in such a way that it has both radial and axial float with the hope that the reamer will follow the original hole and prevent the bellmouth condition.

In horizontal setups where the reamer is held fixed and the work-piece rotated, any misalignment exerts a sideways force on the reamer as it is fed to depth, resulting in the formation of a tapered hole. This type of bellmouthing can frequently be reduced by shortening the bearing length of the cutting portion of the reamer. One way to do this is to reduce the reamer diameter by 0.010 to 0.030 inch, depending on size and length, behind a short full-diameter section, $\frac{1}{8}$ to $\frac{1}{2}$ inch long according to length and size, following the chamfer. The second method is to grind a high back taper, 0.008 to 0.015 inch per inch, behind the short full-diameter section. Either of these modifications reduces the length of the reamer tooth which can cause the bellmouth condition.

Poor Finish: The most obvious step towards producing a good finish is to reduce the reamer feed per revolution. Feeds as low as 0.0002 to 0.0005 inch per tooth have been used successfully. However, reamer life will be better if the maximum feasible feed is used.

The minimum practical amount of reaming stock allowance will often improve finish by reducing the volume of chips and the resulting heat generated on the cutting portion of the chamfer. Too little reamer stock, however, can be troublesome in that the reamer teeth may not cut freely but will actually deflect the work material out of the way. When this happens, excessive heat, poor finish and rapid reamer wear can occur.

Because of their superior abrasion resistance, carbide reamers are often used when fine finishes are required. When properly conditioned, carbide reamers can produce a large number of good quality holes. Careful honing of the carbide reamer edges is very important.

* *Metal Cuttings,* "Some Aspects of Reamer Design and Operation," April 1963.

Dimensions of Centers for Reamers and Arbors

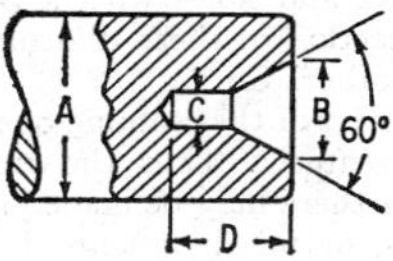

Diameter of Arbor A	Largest Diameter of Center B	Number of Drill C	Depth of Hole D	Diameter of Arbor A	Largest Diameter of Center B	Number of Drill C	Depth of Hole D	Diameter of Arbor A	Largest Diameter of Center B	Letter of Drill C	Depth of Hole D
				3/4	3/8	25	7/16	2 1/2	11/16	J	27/32
				13/16	13/32	20	1/2	2 5/8	45/64	K	7/8
				7/8	7/16	17	17/32	2 3/4	23/32	L	29/32
				15/16	15/32	12	9/16	2 7/8	47/64	M	29/32
				1	1/2	8	19/32	3	3/4	N	15/16
				1 1/8	33/64	5	5/8	3 1/8	49/64	N	31/32
				1 1/4	17/32	3	21/32	3 1/4	25/32	O	31/32
				1 3/8	35/64	2	21/32	3 3/8	51/64	O	1
				1 1/2	9/16	1	11/16	3 1/2	13/16	P	1
1/4	1/8	55	5/32	....	....	Letter	...	3 5/8	53/64	Q	1 1/16
5/16	5/32	52	3/16	1 5/8	37/64	A	23/32	3 3/4	27/32	R	1 1/16
3/8	3/16	48	7/32	1 3/4	19/32	B	23/32	3 7/8	55/64	R	1 1/16
7/16	7/32	43	1/4	1 7/8	39/64	C	3/4	4	7/8	S	1 1/8
1/2	1/4	39	5/16	2	5/8	E	3/4	4 1/4	29/32	T	1 1/8
9/16	9/32	33	11/32	2 1/8	41/64	F	25/32	4 1/2	15/16	V	1 3/16
5/8	5/16	30	3/8	2 1/4	21/32	G	13/16	4 3/4	31/32	W	1 1/4
11/16	11/32	29	13/32	2 3/8	43/64	H	27/32	5	1	X	1 1/4

Straight Shank Center Reamers and Machine Countersinks (ANSI B94.2-1983)

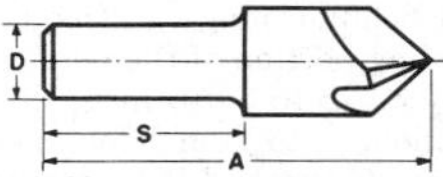

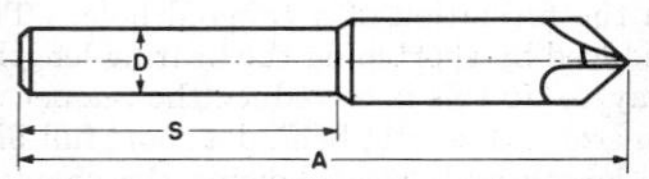

Center Reamers (Short Countersinks)				Machine Countersinks			
Diam. of Cut	Approx. Length Overall A	Length of Shank S	Diam. of Shank D	Diam. of Cut	Approx. Length Overall A	Length of Shank S	Diam. of Shank D
1/4	1 1/2	3/4	3/16	1/2	3 7/8	2 1/4	1/2
3/8	1 3/4	7/8	1/4	5/8	4	2 1/4	1/2
1/2	2	1	3/8	3/4	4 1/8	2 1/4	1/2
5/8	2 1/4	1	3/8	7/8	4 1/4	2 1/4	1/2
3/4	2 5/8	1 1/4	1/2	1	4 3/8	2 1/4	1/2

All dimensions are given in inches. Material is high-speed steel. Reamers and countersinks have 3 or 4 flutes. Center reamers are standard with either 60, 82, 90 or 100 degrees included angle. Machine countersinks are standard with either 60 or 82 degrees included angle.

Tolerances: On length overall *A* the tolerance is ±1/8 inch for center reamers in a size range of from 1/4 to 3/8 inch, incl., and machine countersinks in a size range of from 1/2 to 5/8 inch, incl.; ±3/16 inch for center reamers, 1/2 to 3/4 inch, incl. and machine countersinks, 3/4 to 1 inch, incl. On shank diameter *D* the tolerance is −.0005 to −.002 inch. On shank length *S*, the tolerance is ±1/16 inch.

Expansion Chucking Reamers — Straight and Taper Shanks* (ANSI B94.2-1983)

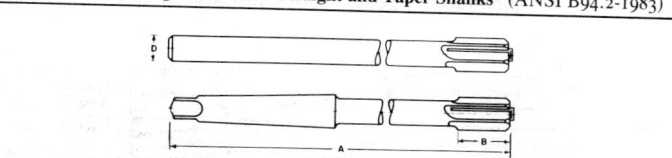

Diam. of Reamer	Length, A	Flute Length, B	Shank Diam., D Max.	Shank Diam., D Min.	Diam. of Reamer	Length, A	Flute Length, B	Shank Diam., D Max.	Shank Diam., D Min.
3/8	7	3/4	0.3105	0.3095	1 3/32	10 1/2	1 5/8	0.8745	0.8730
13/32	7	3/4	0.3105	0.3095	1 1/8	11	1 3/4	0.8745	0.8730
7/16	7	7/8	0.3730	0.3720	1 5/32	11	1 3/4	0.8745	0.8730
15/32	7	7/8	0.3730	0.3720	1 3/16	11	1 3/4	0.9995	0.9980
1/2	8	1	0.4355	0.4345	1 7/32	11	1 3/4	0.9995	0.9980
17/32	8	1	0.4355	0.4345	1 1/4	11 1/2	1 7/8	0.9995	0.9980
9/16	8	1 1/8	0.4355	0.4345	1 5/16	11 1/2	1 7/8	0.9995	0.9980
19/32	8	1 1/8	0.4355	0.4345	1 3/8	12	2	0.9995	0.9980
5/8	9	1 1/4	0.4355	0.4345	1 7/16	12	2	1.2495	1.2480
21/32	9	1 1/4	0.5620	0.5605	1 1/2	12 1/2	2 1/8	1.2495	1.2480
11/16	9	1 1/4	0.5620	0.5605	1 9/16†	12 1/2	2 1/8	1.2495	1.2480
23/32	9	1 1/4	0.5620	0.5605	1 5/8	13	2 1/4	1.2495	1.2480
3/4	9 1/2	1 3/8	0.5620	0.5605	1 11/16†	13	2 1/4	1.2495	1.2480
25/32	9 1/2	1 3/8	0.6245	0.6230	1 3/4	13 1/2	2 3/8	1.2495	1.2480
13/16	9 1/2	1 3/8	0.6245	0.6230	1 13/16†	13 1/2	2 3/8	1.4995	1.4980
27/32	9 1/2	1 3/8	0.6245	0.6230	1 7/8	14	2 1/2	1.4995	1.4980
7/8	10	1 1/2	0.6245	0.6230	1 15/16†	14	2 1/2	1.4995	1.4980
29/32	10	1 1/2	0.7495	0.7480	2	14	2 1/2	1.4995	1.4980
15/16	10	1 1/2	0.7495	0.7480	2 1/8‡	14 1/2	2 3/4	. . .	. . .
31/32	10	1 1/2	0.7495	0.7480	2 1/4‡	14 1/2	2 3/4	. . .	. . .
1	10 1/2	1 5/8	0.7495	0.7480	2 3/8‡	15	3	. . .	. . .
1 1/32	10 1/2	1 5/8	0.8745	0.8730	2 1/2‡	15	3	. . .	. . .
1 1/16	10 1/2	1 5/8	0.8745	0.8730	. . .				

All dimensions in inches. Material is high-speed steel. The number of flutes is as follows: 3/8- to 15/32-inch sizes, 4 to 6; 1/2- to 31/32-inch sizes, 6 to 8; 1- to 1 11/16-inch sizes, 8 to 10; 1 3/4- to 1 15/16-inch sizes, 8 to 12; 2- to 2 1/4-inch sizes, 10 to 12; 2 3/8- and 2 1/2-inch sizes, 10 to 14. The expansion feature of these reamers provides a means of adjustment which is important in reaming holes to close tolerances. When worn undersize, they may be expanded and reground to the original size.

* Taper is Morse taper: No. 1 for sizes 3/8 to 19/32 incl.; No. 2 for sizes 5/8 to 29/32 incl.; No. 3 for sizes 15/16 to 1 7/32, incl.; No. 4 for sizes 1 1/4 to 1 5/8, incl.; and No. 5 for sizes 1 3/4 to 2 1/2, incl. For amount of taper see page 898.

† Straight shank only.   ‡ Taper shank only.

Tolerances: On reamer diameter, 3/8- to 1-inch sizes, incl., +.0001 to +.0005 inch; over 1-inch size, +.0002 to +.0006 inch. On length A and flute length B, 3/8- to 1-inch sizes, incl., ±1/16 inch; 1 1/32- to 2-inch sizes, incl., ±3/32 inch; over 2-inch sizes, ±1/8 inch.

Illustration of Terms Applying to Reamers — 1

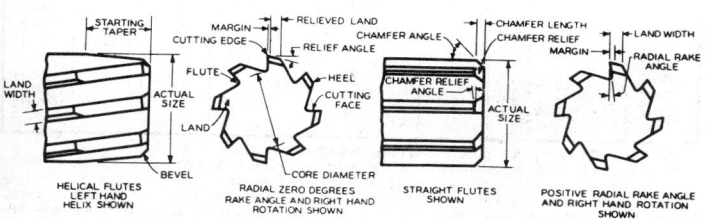

HAND REAMER   MACHINE REAMER

Illustration of Terms Applying to Reamers — 2

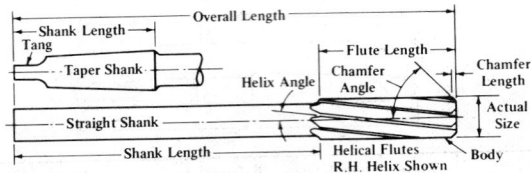

Chucking Reamer, Straight and Taper Shank

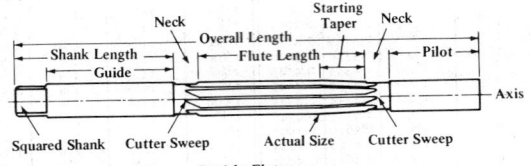

Straight Flutes

Hand Reamer, Pilot and Guide

American National Standard Fluted Taper Shank Chucking Reamers — Straight and Helical Flutes, Fractional Sizes (ANSI B94.2-1983)

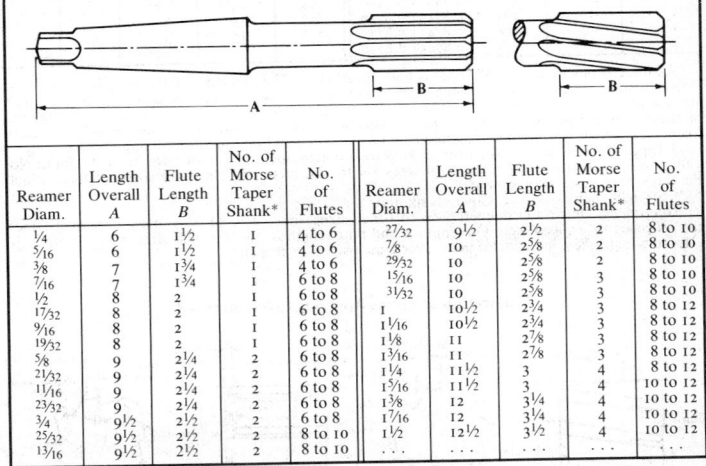

Reamer Diam.	Length Overall A	Flute Length B	No. of Morse Taper Shank*	No. of Flutes	Reamer Diam.	Length Overall A	Flute Length B	No. of Morse Taper Shank*	No. of Flutes
1/4	6	1 1/2	1	4 to 6	27/32	9 1/2	2 1/2	2	8 to 10
5/16	6	1 1/2	1	4 to 6	7/8	10	2 5/8	2	8 to 10
3/8	7	1 3/4	1	4 to 6	29/32	10	2 5/8	2	8 to 10
7/16	7	1 3/4	1	6 to 8	15/16	10	2 5/8	3	8 to 10
1/2	8	2	1	6 to 8	31/32	10	2 5/8	3	8 to 10
17/32	8	2	1	6 to 8	1	10 1/2	2 3/4	3	8 to 12
9/16	8	2	1	6 to 8	1 1/16	10 1/2	2 3/4	3	8 to 12
19/32	8	2	1	6 to 8	1 1/8	11	2 7/8	3	8 to 12
5/8	9	2 1/4	2	6 to 8	1 3/16	11	2 7/8	3	8 to 12
21/32	9	2 1/4	2	6 to 8	1 1/4	11 1/2	3	4	8 to 12
11/16	9	2 1/4	2	6 to 8	1 5/16	11 1/2	3	4	10 to 12
23/32	9	2 1/4	2	6 to 8	1 3/8	12	3 1/4	4	10 to 12
3/4	9 1/2	2 1/2	2	6 to 8	1 7/16	12	3 1/4	4	10 to 12
25/32	9 1/2	2 1/2	2	8 to 10	1 1/2	12 1/2	3 1/2	4	10 to 12
13/16	9 1/2	2 1/2	2	8 to 10	...	...	...	...	...

All dimensions are given in inches. Material is high-speed steel.

* American National Standard self-holding tapers (see Table 5, p. 903).

Helical flute reamers with right-hand helical flutes are standard.

Tolerances: On reamer diameter, 1/4-inch size, + .0001 to + .0004 inch; over 1/4- to 1-inch size, + .0001 to + .0005 inch; over 1-inch size, + .0002 to + .0006 inch. On length overall A and flute length B, 1/4- to 1-inch size, incl., ± 1/16 inch; 1 1/16- to 1 1/2-inch size, incl., 3/32 inch.

Hand Reamers — Straight and Helical Flutes (ANSI B94.2-1983)

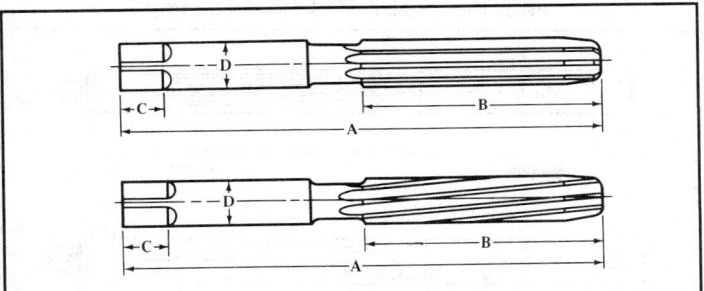

Reamer Diameter			Length Overall A	Flute Length B	Square Length C	Size of Square	No. of Flutes
Straight Flutes	Helical Flutes	Decimal Equivalent					
1/8	. . .	0.1250	3	1½	5/32	0.095	4 to 6
5/32	. . .	0.1562	3¼	1⅝	7/32	0.115	4 to 6
3/16	. . .	0.1875	3½	1¾	7/32	0.140	4 to 6
7/32	. . .	0.2188	3¾	1⅞	¼	0.165	4 to 6
¼	¼	0.2500	4	2	¼	0.185	4 to 6
9/32	. . .	0.2812	4¼	2⅛	¼	0.210	4 to 6
5/16	5/16	0.3125	4½	2¼	5/16	0.235	4 to 6
11/32	. . .	0.3438	4¾	2⅜	5/16	0.255	4 to 6
⅜	⅜	0.3750	5	2½	⅜	0.280	4 to 6
13/32	. . .	0.4062	5¼	2⅝	⅜	0.305	6 to 8
7/16	7/16	0.4375	5½	2¾	7/16	0.330	6 to 8
15/32	. . .	0.4688	5¾	2⅞	7/16	0.350	6 to 8
½	½	0.5000	6	3	½	0.375	6 to 8
17/32	. . .	0.5312	6¼	3⅛	½	0.400	6 to 8
9/16	9/16	0.5625	6½	3¼	9/16	0.420	6 to 8
19/32	. . .	0.5938	6¾	3⅜	9/16	0.445	6 to 8
⅝	⅝	0.6250	7	3½	⅝	0.470	6 to 8
21/32	. . .	0.6562	7⅜	3 11/16	⅝	0.490	6 to 8
11/16	11/16	0.6875	7¾	3⅞	11/16	0.515	6 to 8
23/32	. . .	0.7188	8⅛	4 1/16	11/16	0.540	6 to 8
¾	¾	0.7500	8⅜	4 3/16	¾	0.560	6 to 8
. . .	13/16	0.8125	9⅛	4 9/16	13/16	0.610	8 to 10
⅞	⅞	0.8750	9¾	4⅞	⅞	0.655	8 to 10
. . .	15/16	0.9375	10¼	5⅛	15/16	0.705	8 to 10
1	1	1.0000	10⅞	5 7/16	1	0.750	8 to 10
1⅛	1⅛	1.1250	11⅝	5 13/16	1	0.845	8 to 10
1¼	1¼	1.2500	12¼	6⅛	1	0.935	8 to 12
1⅜	1⅜	1.3750	12⅝	6 5/16	1	1.030	10 to 12
1½	1½	1.5000	13	6½	1⅛	1.125	10 to 14

All dimensions in inches. Material is high-speed steel. The nominal shank diameter D is the same as the reamer diameter. Helical-flute hand reamers with left-hand helical flutes are standard. Reamers are tapered slightly on the end to facilitate proper starting.

Tolerances: On diameter of reamer, up to ¼-inch size, incl., + .0001 to + .0004 inch; over ¼- to 1-inch size, incl., + .0001 to + .0005 inch; over 1-inch size, + .0002 to + .0006 inch. On length overall A and flute length B, ⅛- to 1-inch size, incl., ± 1/16 inch; 1⅛- to 1½-inch size, incl., ± 3/32 inch. On length of square C, ⅛- to 1 inch size, incl., ± 1/32 inch; 1⅛- to 1½-inch size, incl., ± 1/16 inch. On shank diameter D, ⅛- to 1-inch size, incl., − .001 to − .005 inch; 1⅛- to 1½-inch size, incl., − .0015 to − .006 inch. On size of square, ⅛- to ½-inch size, incl., − .004 inch; 17/32- to 1-inch size, incl., − .006 inch; 1⅛- to 1½-inch size, incl., − .008 inch.

American National Standard Expansion Hand Reamers — Straight and Helical Flutes, Squared Shank (ANSI B94.2-1983)

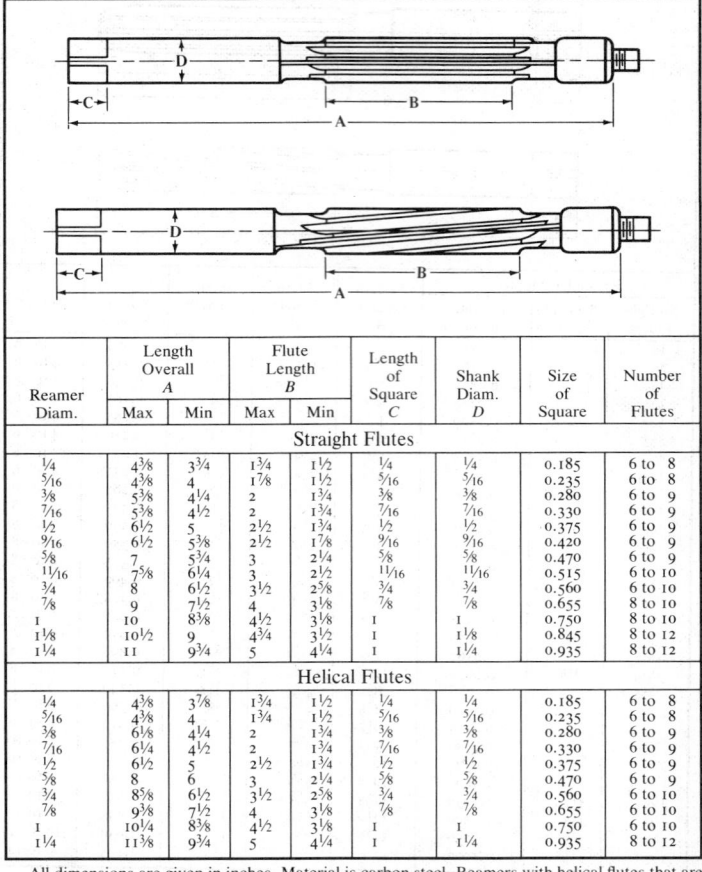

Reamer Diam.	Length Overall A		Flute Length B		Length of Square C	Shank Diam. D	Size of Square	Number of Flutes
	Max	Min	Max	Min				
Straight Flutes								
¼	4⅜	3¾	1¾	1½	¼	¼	0.185	6 to 8
⁵⁄₁₆	4⅜	4	1⅞	1½	⁵⁄₁₆	⁵⁄₁₆	0.235	6 to 8
⅜	5⅜	4¼	2	1¾	⅜	⅜	0.280	6 to 9
⁷⁄₁₆	5⅜	4½	2	1¾	⁷⁄₁₆	⁷⁄₁₆	0.330	6 to 9
½	6½	5	2½	1¾	½	½	0.375	6 to 9
⁹⁄₁₆	6½	5⅜	2½	1⅞	⁹⁄₁₆	⁹⁄₁₆	0.420	6 to 9
⅝	7	5¾	3	2¼	⅝	⅝	0.470	6 to 9
¹¹⁄₁₆	7⅝	6¼	3	2½	¹¹⁄₁₆	¹¹⁄₁₆	0.515	6 to 10
¾	8	6½	3½	2⅝	¾	¾	0.560	6 to 10
⅞	9	7½	4	3⅛	⅞	⅞	0.655	8 to 10
1	10	8⅜	4½	3⅛	1	1	0.750	8 to 10
1⅛	10½	9	4¾	3½	1	1⅛	0.845	8 to 12
1¼	11	9¾	5	4¼	1	1¼	0.935	8 to 12
Helical Flutes								
¼	4⅜	3⅞	1¾	1½	¼	¼	0.185	6 to 8
⁵⁄₁₆	4⅜	4	1¾	1½	⁵⁄₁₆	⁵⁄₁₆	0.235	6 to 8
⅜	6⅛	4¼	2	1¾	⅜	⅜	0.280	6 to 9
⁷⁄₁₆	6¼	4½	2	1¾	⁷⁄₁₆	⁷⁄₁₆	0.330	6 to 9
½	6½	5	2½	1¾	½	½	0.375	6 to 9
⅝	8	6	3	2¼	⅝	⅝	0.470	6 to 9
¾	8⅝	6½	3½	2⅝	¾	¾	0.560	6 to 10
⅞	9⅜	7½	4	3⅛	⅞	⅞	0.655	6 to 10
1	10¼	8⅜	4½	3⅛	1	1	0.750	6 to 10
1¼	11⅜	9¾	5	4¼	1	1¼	0.935	8 to 12

All dimensions are given in inches. Material is carbon steel. Reamers with helical flutes that are left hand are standard. Expansion hand reamers are primarily designed for work where it is necessary to enlarge reamed holes by a few thousandths. The pilots and guides on these reamers are ground undersize for clearance. The maximum expansion on these reamers is as follows: .006 inch for the ¼- to ⁷⁄₁₆-inch sizes. .010 inch for the ½- to ⅞-inch sizes and .012 inch for the 1- to 1¼-inch sizes.

Tolerances: On length overall A and flute length B, ±¹⁄₁₆ inch for ¼- to 1-inch sizes, ±³⁄₃₂ inch for 1⅛- to 1¼-inch sizes; on length of square C, ±¹⁄₃₂ inch for ¼- to 1-inch sizes, ±¹⁄₁₆ inch for 1⅛- to 1¼-inch sizes; on shank diameter D, −.001 to −.005 inch for ¼- to 1-inch sizes, −.0015 to −.006 inch for 1⅛- to 1¼-inch sizes; on size of square, −.004 inch for ¼- to ½-inch sizes, −.006 inch for ⁹⁄₁₆- to 1-inch sizes, and −.008 inch for 1⅛- to 1¼-inch sizes.

Taper Shank Jobbers Reamers — Straight Flutes (ANSI B94.2-1983)

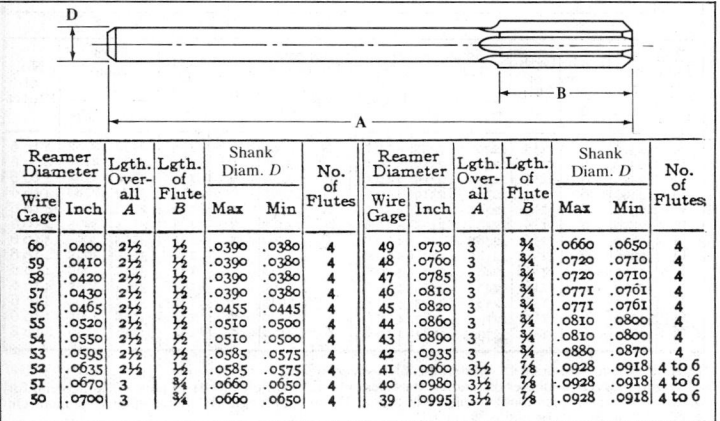

Reamer Diameter		Length Overall *A*	Length of Flute *B*	No. of Morse Taper Shank*	No. of Flutes
Fractional	Dec. Equiv.				
1/4	0.2500	5 3/16	2	1	6 to 8
5/16	0.3125	5 1/2	2 1/4	1	6 to 8
3/8	0.3750	5 13/16	2 1/2	1	6 to 8
7/16	0.4375	6 1/8	2 3/4	1	6 to 8
1/2	0.5000	6 7/16	3	1	6 to 8
9/16	0.5625	6 3/4	3 1/4	1	6 to 8
5/8	0.6250	7 9/16	3 1/2	2	6 to 8
11/16	0.6875	8	3 7/8	2	8 to 10
3/4	0.7500	8 3/8	4 3/16	2	8 to 10
13/16	0.8125	8 13/16	4 9/16	2	8 to 10
7/8	0.8750	9 3/16	4 7/8	2	8 to 10
15/16	0.9375	10	5 1/8	3	8 to 10
1	1.0000	10 3/8	5 7/16	3	8 to 10
1 1/16	1.0625	10 5/8	5 5/8	3	8 to 10
1 1/8	1.1250	10 7/8	5 13/16	3	8 to 10
1 3/16	1.1875	11 1/8	6	3	8 to 12
1 1/4	1.2500	12 9/16	6 1/8	4	8 to 12
1 3/8	1.3750	12 13/16	6 5/16	4	10 to 12
1 1/2	1.5000	13 1/8	6 1/2	4	10 to 12

All dimensions in inches. Material is high-speed steel.
* American National Standard self-holding tapers (see Table 5, p. 903).
Tolerances: On reamer diameter, 1/4-inch size, + .0001 to + .0004 inch; over 1/4- to 1-inch size, incl., + .0001 to + .0005 inch; over 1-inch size, + .0002 to + .0006 inch. On overall length *A* and length of flute *B*, 1/4- to 1-inch size, incl., ± 1/16 inch; and 1 1/16- to 1 1/2-inch size, incl., ± 3/32 inch.

Straight Shank Chucking Reamers — Straight Flutes, Wire Gage Sizes — 1
(ANSI B94.2-1983)

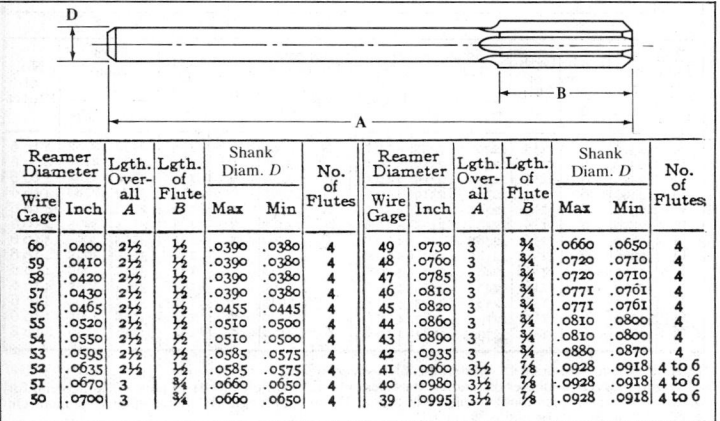

Reamer Diameter		Lgth. Over-all *A*	Lgth. of Flute *B*	Shank Diam. *D*		No. of Flutes	Reamer Diameter		Lgth. Over-all *A*	Lgth. of Flute *B*	Shank Diam. *D*		No. of Flutes
Wire Gage	Inch			Max	Min		Wire Gage	Inch			Max	Min	
60	.0400	2 1/2	1/2	.0390	.0380	4	49	.0730	3	3/4	.0660	.0650	4
59	.0410	2 1/2	1/2	.0390	.0380	4	48	.0760	3	3/4	.0720	.0710	4
58	.0420	2 1/2	1/2	.0390	.0380	4	47	.0785	3	3/4	.0720	.0710	4
57	.0430	2 1/2	1/2	.0390	.0380	4	46	.0810	3	3/4	.0771	.0761	4
56	.0465	2 1/2	1/2	.0455	.0445	4	45	.0820	3	3/4	.0771	.0761	4
55	.0520	2 1/2	1/2	.0510	.0500	4	44	.0860	3	3/4	.0810	.0800	4
54	.0550	2 1/2	1/2	.0510	.0500	4	43	.0890	3	3/4	.0810	.0800	4
53	.0595	2 1/2	1/2	.0585	.0575	4	42	.0935	3	3/4	.0880	.0870	4
52	.0635	2 1/2	1/2	.0585	.0575	4	41	.0960	3 1/2	7/8	.0928	.0918	4 to 6
51	.0670	3	3/4	.0660	.0650	4	40	.0980	3 1/2	7/8	.0928	.0918	4 to 6
50	.0700	3	3/4	.0660	.0650	4	39	.0995	3 1/2	7/8	.0928	.0918	4 to 6

Straight Shank Chucking Reamers — Straight Flutes, Wire Gage Sizes — 2
(ANSI B94.2-1983)

Reamer Diameter		Lgth. Over-all A	Lgth. of Flute B	Shank Diam. D		No. of Flutes	Reamer Diameter		Lgth. Over-all A	Lgth. of Flute B	Shank Diam. D		No. of Flutes
Wire Gage	Inch			Max	Min		Wire Gage	Inch			Max	Min	
38	.1015	3½	⅞	.0950	.0940	4 to 6	19	.1660	4½	1⅛	.1595	.1585	4 to 6
37	.1040	3½	⅞	.0950	.0940	4 to 6	18	.1695	4½	1⅛	.1595	.1585	4 to 6
36	.1065	3½	⅞	.1030	.1020	4 to 6	17	.1730	4½	1⅛	.1645	.1635	4 to 6
35	.1100	3½	⅞	.1030	.1020	4 to 6	16	.1770	4½	1⅛	.1704	.1694	4 to 6
34	.1110	3½	⅞	.1055	.1045	4 to 6	15	.1800	4½	1⅛	.1755	.1745	4 to 6
33	.1130	3½	⅞	.1055	.1045	4 to 6	14	.1820	4½	1⅛	.1755	.1745	4 to 6
32	.1160	3½	⅞	.1120	.1110	4 to 6	13	.1850	4½	1⅛	.1805	.1795	4 to 6
31	.1200	3½	⅞	.1120	.1110	4 to 6	12	.1890	4½	1⅛	.1805	.1795	4 to 6
30	.1285	3½	⅞	.1190	.1180	4 to 6	11	.1910	5	1¼	.1860	.1850	4 to 6
29	.1360	4	1	.1275	.1265	4 to 6	10	.1935	5	1¼	.1860	.1850	4 to 6
28	.1405	4	1	.1350	.1340	4 to 6	9	.1960	5	1¼	.1895	.1885	4 to 6
27	.1440	4	1	.1350	.1340	4 to 6	8	.1990	5	1¼	.1895	.1885	4 to 6
26	.1470	4	1	.1430	.1420	4 to 6	7	.2010	5	1¼	.1945	.1935	4 to 6
25	.1495	4	1	.1430	.1420	4 to 6	6	.2040	5	1¼	.1945	.1935	4 to 6
24	.1520	4	1	.1460	.1450	4 to 6	5	.2055	5	1¼	.2016	.2006	4 to 6
23	.1540	4	1	.1460	.1450	4 to 6	4	.2090	5	1¼	.2016	.2006	4 to 6
22	.1570	4	1	.1510	.1500	4 to 6	3	.2130	5	1¼	.2075	.2065	4 to 6
21	.1590	4½	1⅛	.1530	.1520	4 to 6	2	.2210	6	1½	.2173	.2163	4 to 6
20	.1610	4½	1⅛	.1530	.1520	4 to 6	1	.2280	6	1½	.2173	.2163	4 to 6

All dimensions in inches.   Material is high-speed steel.
Tolerances: On diameter of reamer, plus .0001 to plus .0004 inch.   On overall length *A*, plus or minus 1/16 inch.   On length of flute *B*, plus or minus 1/16 inch.

Straight Shank Chucking Reamers — Straight Flutes, Letter Sizes (ANSI B94.2-1983)

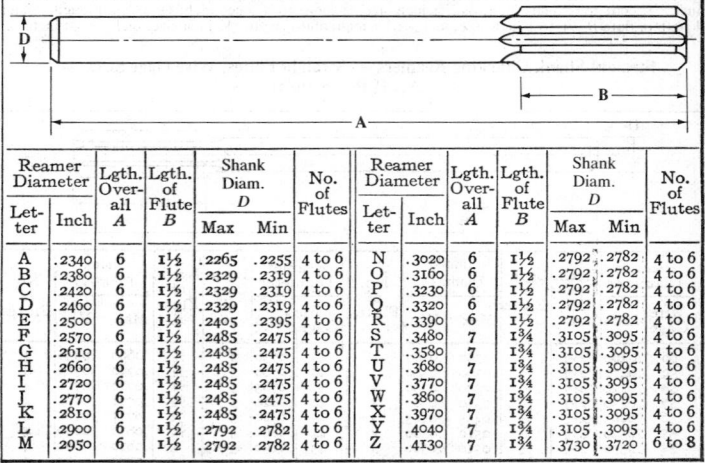

Reamer Diameter		Lgth. Over-all A	Lgth. of Flute B	Shank Diam. D		No. of Flutes	Reamer Diameter		Lgth. Over-all A	Lgth. of Flute B	Shank Diam. D		No. of Flutes
Let-ter	Inch			Max	Min		Let-ter	Inch			Max	Min	
A	.2340	6	1½	.2265	.2255	4 to 6	N	.3020	6	1½	.2792	.2782	4 to 6
B	.2380	6	1½	.2329	.2319	4 to 6	O	.3160	6	1½	.2792	.2782	4 to 6
C	.2420	6	1½	.2329	.2319	4 to 6	P	.3230	6	1½	.2792	.2782	4 to 6
D	.2460	6	1½	.2329	.2319	4 to 6	Q	.3320	6	1½	.2792	.2782	4 to 6
E	.2500	6	1½	.2405	.2395	4 to 6	R	.3390	6	1½	.2792	.2782	4 to 6
F	.2570	6	1½	.2485	.2475	4 to 6	S	.3480	7	1¾	.3105	.3095	4 to 6
G	.2610	6	1½	.2485	.2475	4 to 6	T	.3580	7	1¾	.3105	.3095	4 to 6
H	.2660	6	1½	.2485	.2475	4 to 6	U	.3680	7	1¾	.3105	.3095	4 to 6
I	.2720	6	1½	.2485	.2475	4 to 6	V	.3770	7	1¾	.3105	.3095	4 to 6
J	.2770	6	1½	.2485	.2475	4 to 6	W	.3860	7	1¾	.3105	.3095	4 to 6
K	.2810	6	1½	.2485	.2475	4 to 6	X	.3970	7	1¾	.3105	.3095	4 to 6
L	.2900	6	1½	.2792	.2782	4 to 6	Y	.4040	7	1¾	.3105	.3095	4 to 6
M	.2950	6	1½	.2792	.2782	4 to 6	Z	.4130	7	1¾	.3730	.3720	6 to 8

All dimensions in inches.   Material is high-speed steel.
Tolerances: On diameter of reamer, for sizes A to E, incl., plus .0001 to plus .0004 inch and for sizes F to Z, incl., plus .0001 to plus .0005 inch. On overall length *A*, plus or minus 1/16 inch. On length of flute *B*, plus or minus 1/16 inch.

Straight Shank Chucking Reamers — Straight Flutes, Decimal Sizes (ANSI B94.2-1983)

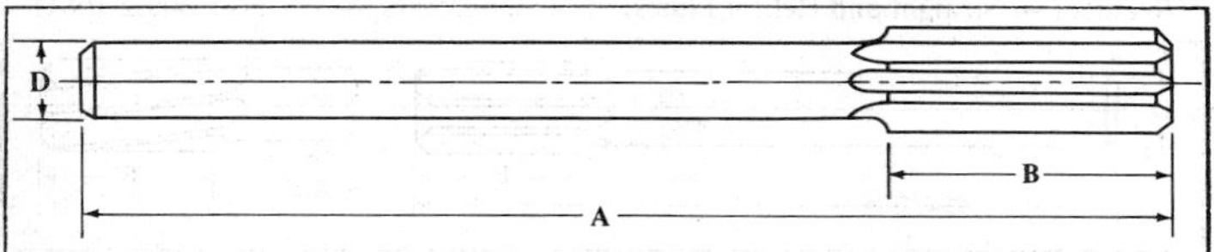

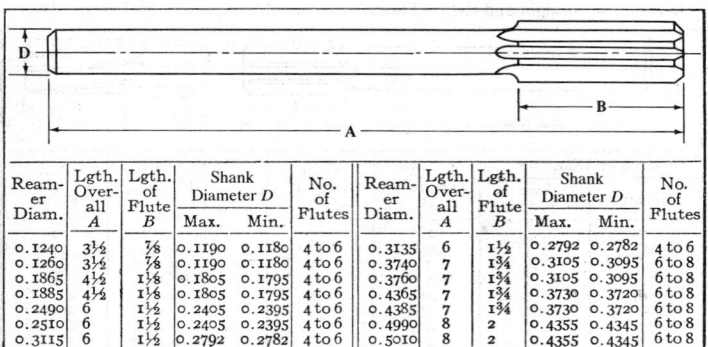

Reamer Diam.	Lgth. Overall A	Lgth. of Flute B	Shank Diameter D Max.	Shank Diameter D Min.	No. of Flutes	Reamer Diam.	Lgth. Overall A	Lgth. of Flute B	Shank Diameter D Max.	Shank Diameter D Min.	No. of Flutes
0.1240	3½	⅞	0.1190	0.1180	4 to 6	0.3135	6	1½	0.2792	0.2782	4 to 6
0.1260	3½	⅞	0.1190	0.1180	4 to 6	0.3740	7	1¾	0.3105	0.3095	6 to 8
0.1865	4½	1⅛	0.1805	0.1795	4 to 6	0.3760	7	1¾	0.3105	0.3095	6 to 8
0.1885	4½	1⅛	0.1805	0.1795	4 to 6	0.4365	7	1¾	0.3730	0.3720	6 to 8
0.2490	6	1½	0.2405	0.2395	4 to 6	0.4385	7	1¾	0.3730	0.3720	6 to 8
0.2510	6	1½	0.2405	0.2395	4 to 6	0.4990	8	2	0.4355	0.4345	6 to 8
0.3115	6	1½	0.2792	0.2782	4 to 6	0.5010	8	2	0.4355	0.4345	6 to 8

All dimensions in inches. Material is high-speed steel.

Tolerances: On diameter of reamer, for 0.124 to 0.249-inch sizes, plus .0001 to plus .0004 inch and for 0.251 to 0.501-inch sizes, plus .0001 to plus .0005 inch. On overall length *A*, plus or minus 1/16 inch. On length of flute *B*, plus or minus 1/16 inch.

Stub Screw Machine Reamers — Helical Flutes (ANSI B94.2-1983)

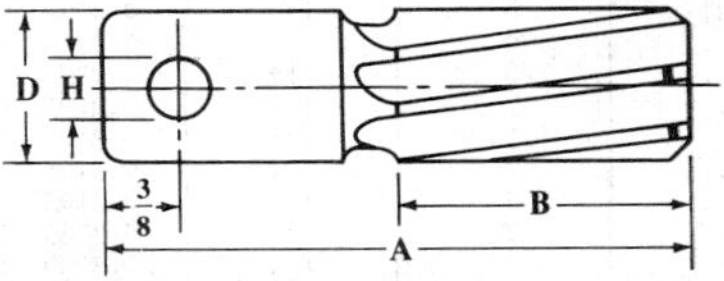

Series No.	Diameter Range	Length Overall A	Length of Flute B	Diam. of Shank D	Size of Hole H	Flute No.	Series No.	Diameter Range	Length Overall A	Length of Flute B	Diam. of Shank D	Size of Hole H	Flute No.
00	.0600-.066	1¾	½	⅛	1/16	4	12	.3761- .407	2½	1¼	½	3/16	6
0	.0661-.074	1¾	½	⅛	1/16	4	13	.4071- .439	2½	1¼	½	3/16	6
1	.0741-.084	1¾	½	⅛	1/16	4	14	.4391- .470	2½	1¼	½	3/16	6
2	.0841-.096	1¾	½	⅛	1/16	4	15	.4701- .505	2½	1¼	½	3/16	6
3	.0961-.126	2	¾	⅛	1/16	4	16	.5051- .567	3	1½	⅝	¼	6
4	.1261-.158	2¼	1	¼	3/32	4	17	.5671- .630	3	1½	⅝	¼	6
5	.1581-.188	2¼	1	¼	3/32	4	18	.6301- .692	3	1½	⅝	¼	6
6	.1881-.219	2¼	1	¼	3/32	6	19	.6921- .755	3	1½	¾	5/16	8
7	.2191-.251	2¼	1	¼	3/32	6	20	.7551- .817	3	1½	¾	5/16	8
8	.2511-.282	2¼	1	⅜	⅛	6	21	.8171- .880	3	1½	¾	5/16	8
9	.2821-.313	2¼	1	⅜	⅛	6	22	.8801- .942	3	1½	¾	5/16	8
10	.3131-.344	2½	1¼	⅜	⅛	6	23	.9421-1.010	3	1½	¾	5/16	8
11	.3441-.376	2½	1¼	⅜	⅛	6	...	...	...	...	...	...	...

All dimensions in inches. Material is high-speed steel.

These reamers are standard with right-hand cut and left-hand helical flutes within the size ranges shown.

Tolerances: On diameter of reamer, for sizes 00 to 7, incl., plus .0001 to plus .0004 inch and for sizes 8 to 23, incl., plus .0001 to plus .0005 inch. On overall length *A*, plus or minus 1/16 inch. On length of flute *B*, plus or minus 1/16 inch. On diameter of shank *D*, minus .0005 to minus .002 inch.

American National Standard Straight Shank Rose Chucking and Chucking Reamers — Straight and Helical Flutes,* Fractional Sizes — 1 (ANSI B94.2-1983)

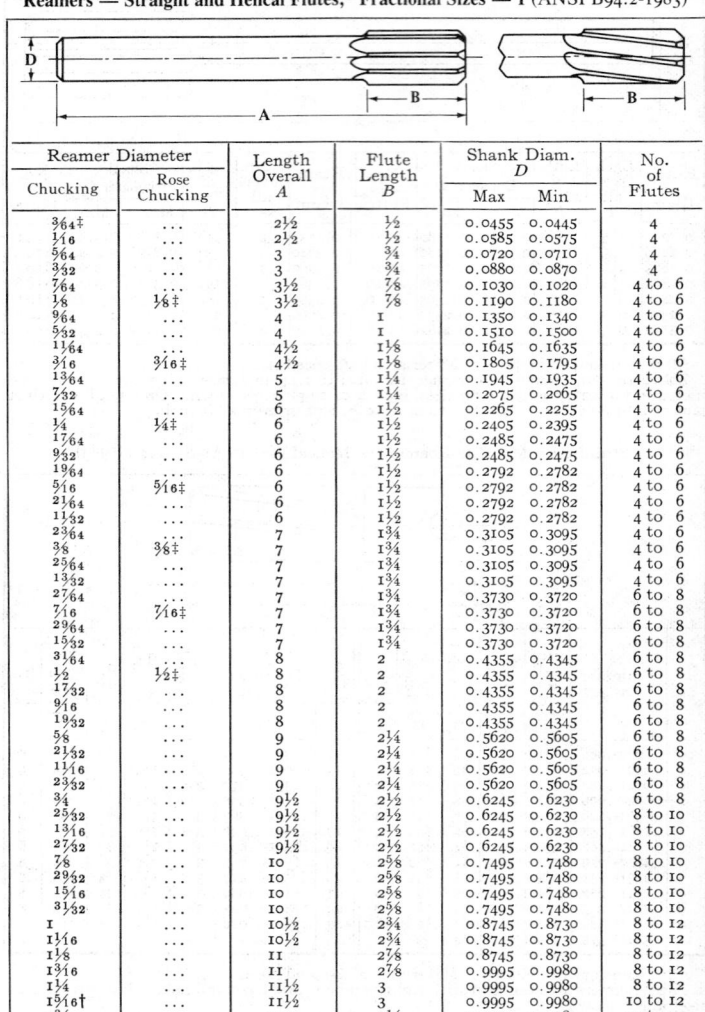

Reamer Diameter — Chucking	Reamer Diameter — Rose Chucking	Length Overall A	Flute Length B	Shank Diam. D — Max	Shank Diam. D — Min	No. of Flutes
3/64‡	...	2½	½	0.0455	0.0445	4
1/16	...	2½	½	0.0585	0.0575	4
5/64	...	3	¾	0.0720	0.0710	4
3/32	...	3	¾	0.0880	0.0870	4
7/64	...	3½	⅞	0.1030	0.1020	4 to 6
1/8	1/8‡	3½	⅞	0.1190	0.1180	4 to 6
9/64	...	4	1	0.1350	0.1340	4 to 6
5/32	...	4	1	0.1510	0.1500	4 to 6
11/64	...	4½	1⅛	0.1645	0.1635	4 to 6
3/16	3/16‡	4½	1⅛	0.1805	0.1795	4 to 6
13/64	...	5	1¼	0.1945	0.1935	4 to 6
7/32	...	5	1¼	0.2075	0.2065	4 to 6
15/64	...	6	1½	0.2265	0.2255	4 to 6
1/4	1/4‡	6	1½	0.2405	0.2395	4 to 6
17/64	...	6	1½	0.2485	0.2475	4 to 6
9/32	...	6	1½	0.2485	0.2475	4 to 6
19/64	...	6	1½	0.2792	0.2782	4 to 6
5/16	5/16‡	6	1½	0.2792	0.2782	4 to 6
21/64	...	6	1½	0.2792	0.2782	4 to 6
11/32	...	6	1½	0.2792	0.2782	4 to 6
23/64	...	7	1¾	0.3105	0.3095	4 to 6
3/8	3/8‡	7	1¾	0.3105	0.3095	4 to 6
25/64	...	7	1¾	0.3105	0.3095	4 to 6
13/32	...	7	1¾	0.3105	0.3095	4 to 6
27/64	...	7	1¾	0.3730	0.3720	6 to 8
7/16	7/16‡	7	1¾	0.3730	0.3720	6 to 8
29/64	...	7	1¾	0.3730	0.3720	6 to 8
15/32	...	7	1¾	0.3730	0.3720	6 to 8
31/64	...	8	2	0.4355	0.4345	6 to 8
1/2	1/2‡	8	2	0.4355	0.4345	6 to 8
17/32	...	8	2	0.4355	0.4345	6 to 8
9/16	...	8	2	0.4355	0.4345	6 to 8
19/32	...	8	2	0.4355	0.4345	6 to 8
5/8	...	9	2¼	0.5620	0.5605	6 to 8
21/32	...	9	2¼	0.5620	0.5605	6 to 8
11/16	...	9	2¼	0.5620	0.5605	6 to 8
23/32	...	9	2¼	0.5620	0.5605	6 to 8
3/4	...	9½	2½	0.6245	0.6230	6 to 8
25/32	...	9½	2½	0.6245	0.6230	8 to 10
13/16	...	9½	2½	0.6245	0.6230	8 to 10
27/32	...	9½	2½	0.6245	0.6230	8 to 10
7/8	...	10	2⅝	0.7495	0.7480	8 to 10
29/32	...	10	2⅝	0.7495	0.7480	8 to 10
15/16	...	10	2⅝	0.7495	0.7480	8 to 10
31/32	...	10	2⅝	0.7495	0.7480	8 to 10
1	...	10½	2¾	0.8745	0.8730	8 to 12
1 1/16	...	10½	2¾	0.8745	0.8730	8 to 12
1 1/8	...	11	2⅞	0.8745	0.8730	8 to 12
1 3/16	...	11	2⅞	0.9995	0.9980	8 to 12
1 1/4	...	11½	3	0.9995	0.9980	8 to 12
1 5/16†	...	11½	3	0.9995	0.9980	10 to 12
1 3/8	...	12	3¼	0.9995	0.9980	10 to 12
1 7/16†	...	12	3¼	1.2495	1.2480	10 to 12
1 1/2	...	12½	3½	1.2495	1.2480	10 to 12

For footnote see next page.

American National Standard Straight Shank Rose Chucking and Chucking Reamers — Straight and Helical Flutes*, Fractional Sizes — 2 (ANSI B94.2-1983)

All dimensions are given in inches. Material is high-speed steel. Chucking reamers are end cutting on the chamfer and the relief for the outside diameter is ground in back of the margin for the full length of land. Lands of rose chucking reamers are not relieved on the periphery but have a relatively large amount of back taper.

* Helical flutes are right- or left-hand helix, right-hand cut, except sizes 1 1/16 through 1 1/2 inches, which are right-hand helix only.

‡ Reamer with straight flutes is standard only.   † Reamer with helical flutes is standard only.

Tolerances: On reamer diameter, up to 1/4-inch size, incl., +.0001 to +.0004 inch; over 1/4- to 1-inch size, incl., +.0001 to +.0005 inch; over 1-inch size, +.0002 to +.0006 inch. On length overall *A* and flute length *B*, up to 1-inch size, incl., ±1/16 inch; 1 1/16- to 1 1/2-inch size, incl., ±3/32 inch.

Shell Reamers — Straight and Helical Flutes (ANSI B94.2-1983)

Diameter of Reamer	Length Overall A	Flute Length B	Hole Diameter Large End H	Fitting Arbor No.	Number of Flutes
3/4	2 1/4	1 1/2	0.375	4	8 to 10
7/8	2 1/2	1 3/4	0.500	5	8 to 10
15/16*	2 1/2	1 3/4	0.500	5	8 to 10
1	2 1/2	1 3/4	0.500	5	8 to 10
1 1/16	2 3/4	2	0.625	6	8 to 12
1 1/8	2 3/4	2	0.625	6	8 to 12
1 3/16	2 3/4	2	0.625	6	8 to 12
1 1/4	2 3/4	2	0.625	6	8 to 12
1 5/16	3	2 1/4	0.750	7	8 to 12
1 3/8	3	2 1/4	0.750	7	8 to 12
1 7/16	3	2 1/4	0.750	7	8 to 12
1 1/2	3	2 1/4	0.750	7	10 to 14
1 9/16	3	2 1/4	0.750	7	10 to 14
1 5/8	3	2 1/4	0.750	7	10 to 14
1 11/16	3 1/2	2 1/2	1.000	8	10 to 14
1 3/4	3 1/2	2 1/2	1.000	8	12 to 14
1 13/16	3 1/2	2 1/2	1.000	8	12 to 14
1 7/8	3 1/2	2 1/2	1.000	8	12 to 14
1 15/16	3 1/2	2 1/2	1.000	8	12 to 14
2	3 1/2	2 1/2	1.000	8	12 to 14
2 1/16*	3 3/4	2 3/4	1.250	9	12 to 16
2 1/8	3 3/4	2 3/4	1.250	9	12 to 16
2 3/16*	3 3/4	2 3/4	1.250	9	12 to 16
2 1/4	3 3/4	2 3/4	1.250	9	12 to 16
2 3/8*	3 3/4	2 3/4	1.250	9	14 to 16
2 1/2*	3 3/4	2 3/4	1.250	9	14 to 16

All dimensions are given in inches. Material is high-speed steel. Helical flute shell reamers with left-hand helical flutes are standard. Shell reamers are designed as a sizing or finishing reamer and are held on an arbor provided with driving lugs. The holes in these reamers are ground with a taper of 1/8 inch per foot.   * Helical flutes only.

Tolerances: On diameter of reamer, 3/4- to 1-inch size, incl., +.0001 to +.0005 inch; over 1-inch size, +.0002 to +.0006 inch. On length overall *A* and flute length *B*, 3/4- to 1-inch size, incl., ±1/16 inch; 1 1/16- to 2-inch size, incl., ±3/32 inch; 2 1/16- to 2 1/2-inch size, incl., ±1/8 inch.

American National Standard Arbors for Shell Reamers — Straight and Taper Shanks (ANSI B94.2-1983)

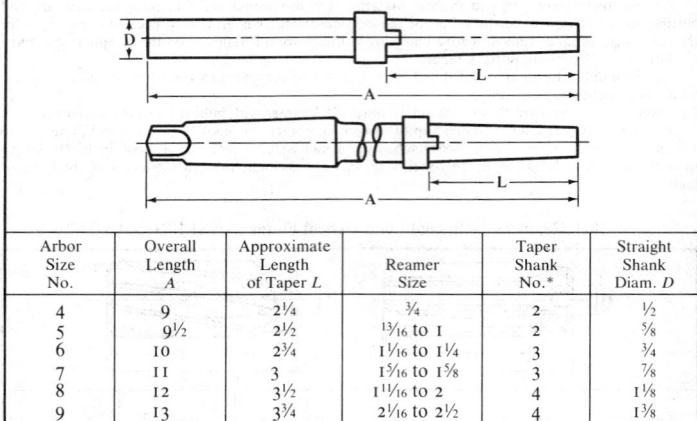

Arbor Size No.	Overall Length A	Approximate Length of Taper L	Reamer Size	Taper Shank No.*	Straight Shank Diam. D
4	9	2¼	¾	2	½
5	9½	2½	13⁄16 to 1	2	⅝
6	10	2¾	1 1⁄16 to 1¼	3	¾
7	11	3	1 5⁄16 to 1⅝	3	⅞
8	12	3½	1 11⁄16 to 2	4	1⅛
9	13	3¾	2 1⁄16 to 2½	4	1⅜

All dimensions are given in inches. These arbors are designed to fit standard shell reamers (see table). End which fits reamer has taper of ⅛ inch per foot.

* American National Standard self-holding tapers (see Table 5, p. 903).

Tolerances: On overall length A, in a range of arbor size numbers 4 to 5, incl., ± 1⁄16 inch; arbor size nos. 6 to 8, incl., ± 3⁄32 inch; arbor size no. 9, ± ⅛ inch. On diameter of shank D, − .0005 to − .002 inch.

American National Standard Driving Slots and Lugs for Shell Reamers or Shell Reamer Arbors (ANSI B94.2-1983)

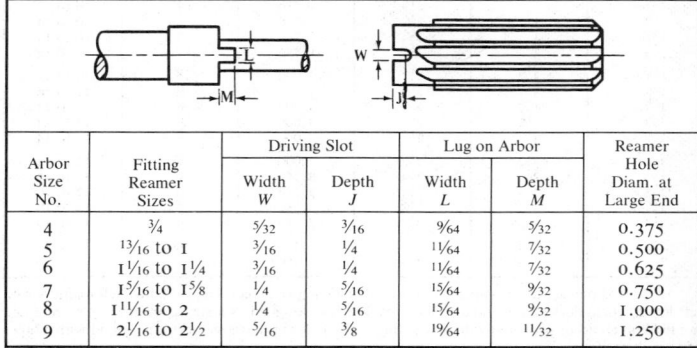

Arbor Size No.	Fitting Reamer Sizes	Driving Slot		Lug on Arbor		Reamer Hole Diam. at Large End
		Width W	Depth J	Width L	Depth M	
4	¾	5⁄32	3⁄16	9⁄64	5⁄32	0.375
5	13⁄16 to 1	3⁄16	¼	11⁄64	7⁄32	0.500
6	1 1⁄16 to 1¼	3⁄16	¼	11⁄64	7⁄32	0.625
7	1 5⁄16 to 1⅝	¼	5⁄16	15⁄64	9⁄32	0.750
8	1 11⁄16 to 2	¼	5⁄16	15⁄64	9⁄32	1.000
9	2 1⁄16 to 2½	5⁄16	⅜	19⁄64	11⁄32	1.250

All dimensions are given in inches. The hole in shell reamers has a taper of ⅛ inch per foot, with arbors tapered to correspond. Shell reamer arbor tapers are made to permit a driving fit with the reamer.

American National Standard Morse Taper Finishing Reamers (ANSI B94.2-1983)

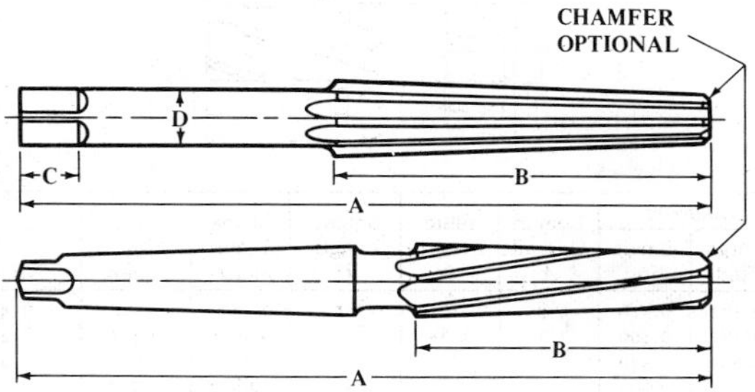

Straight Flutes and Squared Shank

Taper No.*	Small End Diam. (Ref.)	Large End Diam. (Ref.)	Length Overall A	Flute Length B	Square Length C	Shank Diam. D	Square Size
0	0.2503	0.3674	3¾	2¼	5/16	5/16	0.235
I	0.3674	0.5170	5	3	7/16	7/16	0.330
2	0.5696	0.7444	6	3½	5/8	5/8	0.470
3	0.7748	0.9881	7¼	4¼	7/8	.7/8	0.655
4	1.0167	1.2893	8½	5¼	I	1⅛	0.845
5	1.4717	1.8005	9¾	6¼	1⅛	1½	1.125

Straight and Spiral Flutes and Taper Shank						Squared and Taper Shank	

Taper No.*	Small End Diam. (Ref.)	Large End Diam. (Ref.)	Length Overall A	Flute Length B	Taper Shank No.*		Number of Flutes
0	0.2503	0.3674	5 11/32	2¼	0		4 to 6 incl.
I	0.3674	0.5170	6 5/16	3	I		6 to 8 incl.
2	0.5696	0.7444	7⅜	3½	2		6 to 8 incl.
3	0.7748	0.9881	8⅞	4¼	3		8 to 10 incl.
4	1.0167	1.2893	10⅞	5¼	4		8 to 10 incl.
5	1.4717	1.8005	13⅛	6¼	5		10 to 12 incl.

All dimensions are given in inches. Material is high-speed steel. The chamfer on the cutting end of the reamer is optional. Squared shank reamers are standard with straight flutes. Tapered shank reamers are standard with straight or spiral flutes. Spiral flute reamers are standard with left-hand spiral flutes.   * Morse. For amount of taper see Table 1A, page 898.

Tolerances: On overall length *A* and flute length *B*, in taper numbers 0 to 3, incl., ± 1/16 inch, in taper numbers 4 and 5, ± 3/32 inch. On length of square *C*, in taper numbers 0 to 3, incl., ± 1/32 inch; in taper numbers 4 and 5, ± 1/16 inch. On shank diameter *D*, − .0005 to − .002 inch. On size of square, in taper numbers 0 and 1, − .004 inch; in taper numbers 2 and 3, − .006 inch; in taper numbers 4 and 5, − .008 inch.

Taper Pipe Reamers — Spiral Flutes (ANSI B94.2-1983)

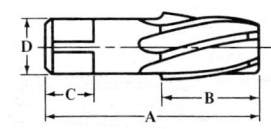

| Nom. Size | Diameter | | Length Overall *A* | Flute Length *B* | Square Length *C* | Shank Diameter *D* | Size of Square | No. of Flutes |
	Large End	Small End						
⅛	0.362	0.316	2⅛	¾	⅜	0.4375	0.328	4 to 6
¼	0.472	0.406	2⁷⁄₁₆	1¹⁄₁₆	⁷⁄₁₆	0.5625	0.421	4 to 6
⅜	0.606	0.540	2⁹⁄₁₆	1¹⁄₁₆	½	0.7000	0.531	4 to 6
½	0.751	0.665	3⅛	1⅜	⅝	0.6875	0.515	4 to 6
¾	0.962	0.876	3¼	1⅜	¹¹⁄₁₆	0.9063	0.679	6 to 10
1	1.212	1.103	3¾	1¾	¹³⁄₁₆	1.1250	0.843	6 to 10
1¼	1.553	1.444	4	1¾	¹⁵⁄₁₆	1.3125	0.984	6 to 10
1½	1.793	1.684	4¼	1¾	1	1.5000	1.125	6 to 10
2	2.268	2.159	4½	1¾	1⅛	1.8750	1.406	8 to 12

All dimensions are given in inches. These reamers are tapered ¾ inch per foot and are intended for reaming holes to be tapped with American National Standard Taper Pipe Thread taps. Material is high-speed steel. Reamers are standard with left-hand spiral flutes.

Tolerances: On length overall *A* and flute length *B*, ⅛- to ¾-inch size, incl., ± ¹⁄₁₆ inch; 1- to 1½-inch size, incl., ± ³⁄₃₂ inch; 2-inch size, ± ⅛ inch. On length of square *C*, ⅛- to ¾-inch size, incl., ± ¹⁄₃₂ inch; 1- to 2-inch size, incl., ± ¹⁄₁₆ inch. On shank diameter *D*, ⅛-inch size, − .0015 inch; ¼- to 1-inch size, incl., − .002 inch; 1¼- to 2-inch size, incl., − .003 inch. On size of square, ⅛-inch size, − .004 inch; ¼- to ¾-inch size, incl., − .006 inch; 1- to 2-inch size, incl., − .008 inch.

B & S Taper Reamers — Straight and Spiral Flutes, Squared Shank

Taper No.*	Diam., Small End	Diam., Large End	Overall Length	Square Length	Flute Length	Diam. of Shank	Size of Square	No. of Flutes
1	0.1974	0.3176	4¾	¼	2⅞	⁹⁄₃₂	0.210	4 to 6
2	0.2474	0.3781	5⅛	⁵⁄₁₆	3⅛	¹¹⁄₃₂	0.255	4 to 6
3	0.3099	0.4510	5½	⅜	3⅜	¹³⁄₃₂	0.305	4 to 6
4	0.3474	0.5017	5⅞	⁷⁄₁₆	3¹¹⁄₁₆	⁷⁄₁₆	0.330	4 to 6
5	0.4474	0.6145	6⅜	½	4	⁹⁄₁₆	0.420	4 to 6
6	0.4974	0.6808	6⅞	⅝	4⅜	⅝	0.470	4 to 6
7	0.5974	0.8011	7½	¾	4⅞	¾	0.560	6 to 8
8	0.7474	0.9770	8⅛	¹³⁄₁₆	5½	¹³⁄₁₆	0.610	6 to 8
9	0.8974	1.1530	8⅞	⅞	6⅛	1	0.750	6 to 8
10	1.0420	1.3376	9¾	1	6⅞	1⅛	0.845	6 to 8

These reamers are no longer ANSI Standard.

All dimensions are given in inches. Material is high-speed steel. The chamfer on the cutting end of the reamer is optional. All reamers are finishing reamers. Spiral flute reamers are standard with left-hand spiral flutes. B & S taper reamers are designed for use in reaming out Brown & Sharpe standard taper sockets.

*For taper per foot, see page 901.

Tolerances: On length overall *A* and flute length *B*, taper nos. 1 to 7, incl., ± ¹⁄₁₆ inch; taper nos. 8 to 10, incl., ± ³⁄₃₂ inch. On length of square *C*, taper nos. 1 to 9, incl., ± ¹⁄₃₂ inch; taper no. 10, ± ¹⁄₁₆ inch. On shank diameter *D*, − .0005 to − .002 inch. On size of square, taper nos. 1 to 3, incl., − .004 inch; taper nos. 4 to 9, incl., − .006 inch; taper no. 10, − .008 inch.

American National Standard Die-Maker's Reamers (ANSI B94.2-1983)

Letter Size	Diameter Small End	Diameter Large End	Length A	Length B	Letter Size	Diameter Small End	Diameter Large End	Length A	Length B	Letter Size	Diameter Small End	Diameter Large End	Length A	Length B
AAA	0.055	0.070	2¼	1⅛	G	0.135	0.158	3	1¾	O	0.250	0.296	5	3½
AA	0.065	0.080	2¼	1⅛	H	0.145	0.169	3¼	1⅞	P	0.275	0.327	5½	4
A	0.075	0.090	2¼	1⅛	I	0.160	0.184	3¼	1⅞	Q	0.300	0.358	6	4½
B	0.085	0.103	2⅜	1⅜	J	0.175	0.199	3¼	1⅞	R	0.335	0.397	6½	4¾
C	0.095	0.113	2½	1⅜	K	0.190	0.219	3½	2¼	S	0.370	0.435	6¾	5
D	0.105	0.126	2⅝	1⅝	L	0.205	0.234	3½	2¼	T	0.405	0.473	7	5¼
E	0.115	0.136	2¾	1⅝	M	0.220	0.252	4	2½	U	0.440	0.511	7¼	5½
F	0.125	0.148	3	1¾	N	0.235	0.274	4½	3	…	…..	…..	…	…

All dimensions in inches. Material is high-speed steel. These reamers are designed for use in diemaking, have a taper of ¾ degree included angle or 0.013 inch per inch, and have 2 or 3 flutes. Reamers are standard with left-hand spiral flutes. * May have conical end.

Tolerances: On length overall *A* and flute length *B*, ±1/16 inch.

Taper Pin Reamers — Straight and Left-Hand Spiral Flutes, Squared Shank; and Left-Hand High-Spiral Flutes, Round Shank (ANSI B94.2-1983)

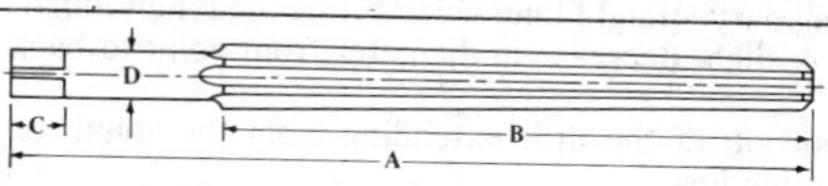

No. of Taper Pin Reamer	Diameter at Large End of Reamer (Ref.)	Diameter at Small End of Reamer (Ref.)	Overall Length of Reamer *A*	Length of Flute *B*	Length of Square *C*†	Diameter of Shank *D*	Size of Square†
8/0*	0.0514	0.0351	1⅝	25/32	…	1/16	…..
7/0	0.0666	0.0497	1 13/16	13/16	5/32	5/64	0.060
6/0	0.0806	0.0611	1 15/16	15/16	5/32	3/32	0.070
5/0	0.0966	0.0719	2 3/16	1 3/16	5/32	7/64	0.080
4/0	0.1142	0.0869	2 5/16	1 5/16	5/32	1/8	0.095
3/0	0.1302	0.1029	2 5/16	1 5/16	5/32	9/64	0.105
2/0	0.1462	0.1137	2 9/16	1 9/16	7/32	5/32	0.115
0	0.1638	0.1287	2 15/16	1 11/16	7/32	11/64	0.130
1	0.1798	0.1447	2 15/16	1 11/16	7/32	3/16	0.140
2	0.2008	0.1605	3 3/16	1 15/16	1/4	13/64	0.150
3	0.2294	0.1813	3 11/16	2 5/16	1/4	15/64	0.175
4	0.2604	0.2071	4 1/16	2 9/16	1/4	17/64	0.200
5	0.2994	0.2409	4 5/16	2 13/16	5/16	5/16	0.235
6	0.3540	0.2773	5 7/16	3 11/16	3/8	23/64	0.270
7	0.4220	0.3297	6 5/16	4 7/16	3/8	13/32	0.305
8	0.5050	0.3971	7 3/16	5 3/16	7/16	7/16	0.330
9	0.6066	0.4805	8 5/16	6 1/16	9/16	9/16	0.420
10	0.7216	0.5799	9 5/16	6 13/16	5/8	5/8	0.470

All dimensions in inches. Reamers have a taper of ¼ inch per foot and are made of high-speed steel. Straight flute reamers of carbon steel are also standard. The number of flutes is as follows: 3 or 4, for 7/0 to 4/0 sizes; 4 to 6, for 3/0 to 0 sizes; 5 or 6, for 1 to 5 sizes; 6 to 8, for 6 to 9 sizes; 7 or 8, for the 10 size in the case of straight- and spiral-flute reamers; and 2 or 3, for 8/0 to 8 sizes; 2 to 4, for the 9 and 10 sizes in the case of high-spiral flute reamers.

* Not applicable to straight and left-hand spiral fluted, squared shank reamers. † Not applicable to high-spiral flute reamers.

Tolerances: On length overall *A* and flute length *B*, ± 1/16 inch. On length of square *C*, ± 1/32 inch. On shank diameter *D*, − .001 to − .005 inch for straight- and spiral-flute reamers and − .0005 to − .002 inch for high-spiral flute reamers. On size of square, − .004 inch for 7/0 to 7 sizes and − .006 inch for 8 to 10 sizes.

TWIST DRILLS AND COUNTERBORES

Twist drills are made with either straight or tapered shanks. The former are by far the more popular type in the smaller sizes because the price is comparatively low. The smaller sizes are parallel throughout their length whereas the larger sizes are ground with a "back taper" thus slightly reducing the drill body diameter nearest the shank. This feature is introduced to prevent binding when the drill is worn.

Straight Shank Drills: Straight shank drills have cylindrical shanks which may be of the same or of a different diameter than the body diameter of the drill and may be made with or without driving flats, tang, or grooves.

Taper Shank Drills: Taper shank drills are preferable to the straight shank type for drilling medium and large size holes. The taper on the shank conforms to one of the tapers in the American Standard (Morse) Series.

American National Standard. — American National Standard B94.11M-1979 covers nomenclature, definitions, sizes and tolerances for High Speed Steel Straight and Taper Shank Drills and Combined Drills and Countersinks, Plain and Bell types. It covers both inch and metric sizes. Dimensional tables from the Standard will be found on the following pages.

Definitions of Twist Drill Terms. — The following definitions are included in the Standard.

Axis: The imaginary straight line which forms the longitudinal center of the drill.

Back Taper: A slight decrease in diameter from point to back in the body of the drill.

Body: The portion of the drill extending from the shank or neck to the outer corners of the cutting lips.

Body Diameter Clearance: That portion of the land that has been cut away so it will not rub against the wall of the hole.

Chisel Edge: The edge at the ends of the web that connects the cutting lips.

Chisel Edge Angle: The angle included between the chisel edge and the cutting lip as viewed from the end of the drill.

Clearance Diameter: The diameter over the cutaway portion of the drill lands.

Drill Diameter: The diameter over the margins of the drill measured at the point.

Flutes: Helical or straight grooves cut or formed in the body of the drill to provide cutting lips, to permit removal of chips, and to allow cutting fluid to reach the cutting lips.

Helix Angle: The angle made by the leading edge of the land with a plane containing the axis of the drill.

Land: The peripheral portion of the drill body between adjacent flutes.

Land Width: The distance between the leading edge and the heel of the land measured at a right angle to the leading edge.

Lips — Two Flute Drill: The cutting edges extending from the chisel edge to the periphery.

Lips — Three or Four Flute Drill (Core Drill): The cutting edges extending from the bottom of the chamfer to the periphery.

Lip Relief: The axial relief on the drill point.

Lip Relief Angle: The axial relief angle at the outer corner of the lip. It is measured by projection into a plane tangent to the periphery at the outer corner of the lip. (Lip relief angle is usually measured across the margin of the twist drill.)

Margin: The cylindrical portion of the land which is not cut away to provide clearance.

Neck: The section of reduced diameter between the body and the shank of a drill.

Overall Length: The length from the extreme end of the shank to the outer corners of the cutting lips. It does not include the conical shank end often used on straight

shank drills, nor does it include the conical cutting point used on both straight and taper shank drills. (For core drills with an external center on the cutting end it is the same as for two-flute drills. For core drills with an internal center on the cutting end, the overall length is to the extreme ends of the tool.)

Point: The cutting end of a drill made up of the ends of the lands, the web, and the lips. In form, it resembles a cone, but departs from a true cone to furnish clearance behind the cutting lips.

Point Angle: The angle included between the lips projected upon a plane parallel to the drill axis and parallel to the cutting lips.

Shank: The part of the drill by which it is held and driven.

Tang: The flattened end of a taper shank, intended to fit into a driving slot in the socket.

Tang Drive: Two opposite parallel driving flats on the end of a straight shank.

Web: The central portion of the body that joins the end of the lands. The end of the web forms the chisel edge on a two-flute drill.

Web Thickness: The thickness of the web at the point unless another specific location is indicated.

Web Thinning: The operation of reducing the web thickness at the point to reduce drilling thrust.

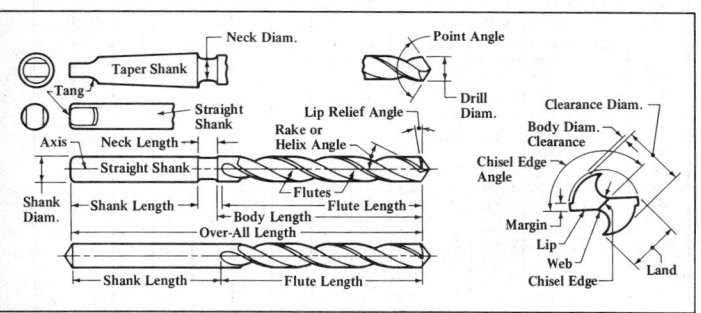

ANSI Standard Twist Drill Nomenclature

Types of Drill. — Drills may be classified based on the type of shank, number of flutes or hand of cut.

Straight Shank Drills: Those having cylindrical shanks which may be the same or different diameter than the body of the drill. The shank may be with or without driving flats, tang, grooves, or threads.

Taper Shank Drills: Those having conical shanks suitable for direct fitting into tapered holes in machine spindles, driving sleeves, or sockets. Tapered shanks generally have a driving tang.

Two-Flute Drills: The conventional type of drill used for originating holes.

Three-Flute Drills (Core Drills): Drill commonly used for enlarging and finishing drilled, cast or punched holes. They will not produce original holes.

Four-Flute Drills (Core Drills): Used interchangeably with three-flute drills. They are of similar construction except for the number of flutes.

Right-Hand Cut: When viewed from the cutting point, the counterclockwise rotation of a drill in order to cut.

Left-Hand Cut: When viewed from the cutting point, the clockwise rotation of a drill in order to cut.

Table 1. American National Standard Straight Shank Twist Drills — Jobbers Length through 17.5 mm, Taper Length through 12.7 mm, and Screw Machine Length through 25.4 mm Diameter (ANSI B94.11M-1979)

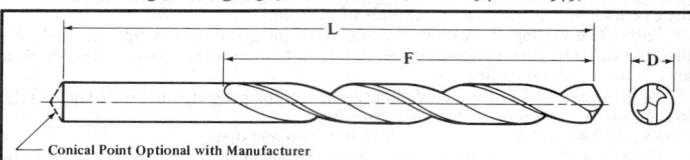

— Conical Point Optional with Manufacturer

Drill Diameter, D*		Equivalent		Jobbers Length				Taper Length†				Screw Machine Length†			
Frac., No. or Ltr.	mm	Deci. In.	mm	Flute F In.	F mm	Overall L In.	L mm	Flute F In.	F mm	Overall L In.	L mm	Flute F In.	F mm	Overall L In.	L mm
97	0.15	0.0059	0.150	1/16	1.5	3/4	19	...	...	...	...	...	...	...	...
96	0.16	0.0063	0.160	1/16	1.5	3/4	19	...	...	...	...	...	...	...	...
95	0.17	0.0067	0.170	1/16	1.5	3/4	19	...	...	...	...	...	...	...	...
94	0.18	0.0071	0.180	1/16	1.5	3/4	19	...	...	...	...	...	...	...	...
93	0.19	0.0075	0.190	1/16	1.5	3/4	19	...	...	...	...	...	...	...	...
92	0.20	0.0079	0.200	1/16	1.5	3/4	19	...	...	...	...	...	...	...	...
91		0.0083	0.211	5/64	2.0	3/4	19	...	...	...	...	...	...	...	...
90	0.22	0.0087	0.221	5/64	2.0	3/4	19	...	...	...	...	...	...	...	...
89		0.0091	0.231	5/64	2.0	3/4	19	...	...	...	...	...	...	...	...
88		0.0095	0.241	5/64	2.0	3/4	19	...	...	...	...	...	...	...	...
	0.25	0.0098	0.250	5/64	2.0	3/4	19	...	...	...	...	...	...	...	...
87		0.0100	0.254	5/64	2.0	3/4	19	...	...	...	...	...	...	...	...
86		0.0105	0.267	3/32	2.5	3/4	19	...	...	...	...	...	...	...	...
85	0.28	0.0110	0.280	3/32	2.5	3/4	19	...	...	...	...	...	...	...	...
84		0.0115	0.292	3/32	2.5	3/4	19	...	...	...	...	...	...	...	...
	0.30	0.0118	0.300	3/32	2.5	3/4	19	...	...	...	...	...	...	...	...
83		0.0120	0.305	3/32	2.5	3/4	19	...	...	...	...	...	...	...	...
82		0.0125	0.318	3/32	2.5	3/4	19	...	...	...	...	...	...	...	...
	0.32	0.0126	0.320	3/32	2.5	3/4	19	...	...	...	...	...	...	...	...
81		0.0130	0.330	3/32	2.5	3/4	19	...	...	...	...	...	...	...	...
80		0.0135	0.343	1/8	3	3/4	19	...	...	...	...	...	...	...	...
	0.35	0.0138	0.350	1/8	3	3/4	19	...	...	...	...	...	...	...	...
79		0.0145	0.368	1/8	3	3/4	19	...	...	...	...	...	...	...	...
	0.38	0.0150	0.380	3/16	5	3/4	19	...	...	...	...	...	...	...	...
1/64		0.0156	0.396	3/16	5	3/4	19	...	...	...	...	...	...	...	...
	0.40	0.0157	0.400	3/16	5	3/4	19	...	...	...	...	...	...	...	...
78		0.0160	0.406	3/16	5	7/8	22	...	...	...	...	...	...	...	...
	0.42	0.0165	0.420	3/16	5	7/8	22	...	...	...	...	...	...	...	...
	0.45	0.0177	0.450	3/16	5	7/8	22	...	...	...	...	...	...	...	...
77		0.0180	0.457	3/16	5	7/8	22	...	...	...	...	...	...	...	...
	0.48	0.0189	0.480	3/16	5	7/8	22	...	...	...	...	...	...	...	...
	0.50	0.0197	0.500	3/16	5	7/8	22	...	...	...	...	...	...	...	...
76		0.0200	0.508	3/16	5	7/8	22	...	...	...	...	...	...	...	...
75		0.0210	0.533	1/4	6	1	25	...	...	...	...	...	...	...	...
	0.55	0.0217	0.550	1/4	6	1	25	...	...	...	...	...	...	...	...
74		0.0225	0.572	1/4	6	1	25	...	...	...	...	...	...	...	...
	0.60	0.0236	0.600	5/16	8	1 1/8	29	...	...	...	...	...	...	...	...
73		0.0240	0.610	5/16	8	1 1/8	29	...	...	...	...	...	...	...	...
72		0.0250	0.635	5/16	8	1 1/8	29	...	...	...	...	...	...	...	...
	0.65	0.0256	0.650	3/8	10	1 1/4	32	...	...	...	...	...	...	...	...

* Fractional inch, number, letter, and metric sizes.
† Dimensions for taper and screw machine lengths begin on following page.

Table 1 (*Continued*). **Straight Shank Twist Drills — Jobbers Length through 17.5 mm, Taper Length through 12.7 mm, and Screw Machine Length through 25.4 mm Diameter** (ANSI B94.11M-1979)

Drill Diameter, *D**		Equivalent		Jobbers Length				Taper Length				Screw Machine Length			
				Flute *F*		Overall *L*		Flute *F*		Overall *L*		Flute *F*		Overall *L*	
Frac., No. or Ltr.	mm	Deci. In.	mm	In.	mm	In.	mm	In.	mm	In.	mm	In.	mm	In.	mm
71		0.0260	0.660	⅜	10	1¼	32	...							
	0.70	0.0276	0.700	⅜	10	1¼	32	...							
70		0.0280	0.711	⅜	10	1¼	32	...							
69		0.0292	0.742	½	13	1⅜	35	...							
	0.75	0.0295	0.750	½	13	1⅜	35	...							
68		0.0310	0.787	½	13	1⅜	35	...							
1/32		0.0312	0.792	½	13	1⅜	35	...							
	0.80	0.0315	0.800	½	13	1⅜	35	...							
67		0.0320	0.813	½	13	1⅜	35	...							
66		0.0330	0.838	½	13	1⅜	35	...							
	0.85	0.0335	0.850	⅝	16	1½	38	...							
65		0.0350	0.889	⅝	16	1½	38	...							
	0.90	0.0354	0.899	⅝	16	1½	38	...							
64		0.0360	0.914	⅝	16	1½	38	...							
63		0.0370	0.940	⅝	16	1½	38	...							
	0.95	0.0374	0.950	⅝	16	1½	38	...							
62		0.0380	0.965	⅝	16	1½	38	...							
61		0.0390	0.991	11/16	17	1⅝	41	...							
	1.00	0.0394	1.000	11/16	17	1⅝	41	1⅛	29	2¼	57	½	13	1⅜	35
60		0.0400	1.016	11/16	17	1⅝	41	1⅛	29	2¼	57	½	13	1⅜	35
59		0.0410	1.041	11/16	17	1⅝	41	1⅛	29	2¼	57	½	13	1⅜	35
	1.05	0.0413	1.050	11/16	17	1⅝	41	1⅛	29	2¼	57	½	13	1⅜	35
58		0.0420	1.067	11/16	17	1⅝	41	1⅛	29	2¼	57	½	13	1⅜	35
57		0.0430	1.092	¾	19	1¾	44	1⅛	29	2¼	57	½	13	1⅜	35
	1.10	0.0433	1.100	¾	19	1¾	44	1⅛	29	2¼	57	½	13	1⅜	35
	1.15	0.0453	1.150	¾	19	1¾	44	1⅛	29	2¼	57	½	13	1⅜	35
56		0.0465	1.181	¾	19	1¾	44	1⅛	29	2¼	57	½	13	1⅜	35
3/64		0.0469	1.191	¾	19	1¾	44	1⅛	29	2¼	57	½	13	1⅜	35
	1.20	0.0472	1.200	⅞	22	1⅞	48	1¾	44	3	76	⅝	16	1⅝	41
	1.25	0.0492	1.250	⅞	22	1⅞	48	1¾	44	3	76	⅝	16	1⅝	41
	1.30	0.0512	1.300	⅞	22	1⅞	48	1¾	44	3	76	⅝	16	1⅝	41
55		0.0520	1.321	⅞	22	1⅞	48	1¾	44	3	76	⅝	16	1⅝	41
	1.35	0.0531	1.350	⅞	22	1⅞	48	1¾	44	3	76	⅝	16	1⅝	41
54		0.0550	1.397	⅞	22	1⅞	48	1¾	44	3	76	⅝	16	1⅝	41
	1.40	0.0551	1.400	⅞	22	1⅞	48	1¾	44	3	76	⅝	16	1⅝	41
	1.45	0.0571	1.450	⅞	22	1⅞	48	1¾	44	3	76	⅝	16	1⅝	41
	1.50	0.0591	1.500	⅞	22	1⅞	48	1¾	44	3	76	⅝	16	1⅝	41
53		0.0595	1.511	⅞	22	1⅞	48	1¾	44	3	76	⅝	16	1⅝	41
	1.55	0.0610	1.550	⅞	22	1⅞	48	1¾	44	3	76	⅝	16	1⅝	41
1/16		0.0625	1.588	⅞	22	1⅞	48	1¾	44	3	76	⅝	16	1⅝	41
	1.60	0.0630	1.600	⅞	22	1⅞	48	2	51	3¾	95	11/16	17	1 11/16	43
52		0.0635	1.613	⅞	22	1⅞	48	2	51	3¾	95	11/16	17	1 11/16	43
	1.65	0.0650	1.650	1	25	2	51	2	51	3¾	95	11/16	17	1 11/16	43
	1.70	0.0669	1.700	1	25	2	51	2	51	3¾	95	11/16	17	1 11/16	43
51		0.0670	1.702	1	25	2	51	2	51	3¾	95	11/16	17	1 11/16	43

* Fractional inch, number, letter, and metric sizes.

Table 1 (*Continued*). Straight Shank Twist Drills — Jobbers Length through 17.5 mm, Taper Length through 12.7 mm, and Screw Machine Length through 25.4 mm Diameter (ANSI B94.11M-1979)

Drill Diameter, D*				Jobbers Length				Taper Length				Screw Machine Length			
Frac., No. or Ltr.	mm	Equivalent Deci. In.	mm	Flute F In.	mm	Overall L In.	mm	Flute F In.	mm	Overall L In.	mm	Flute F In.	mm	Overall L In.	mm
	1.75	0.0689	1.750	1	25	2	51	2	51	3¾	95	11/16	17	1 11/16	43
50		0.0700	1.778	1	25	2	51	2	51	3¾	95	11/16	17	1 11/16	43
	1.80	0.0709	1.800	1	25	2	51	2	51	3¾	95	11/16	17	1 11/16	43
	1.85	0.0728	1.850	1	25	2	51	2	51	3¾	95	11/16	17	1 11/16	43
49		0.0730	1.854	1	25	2	51	2	51	3¾	95	11/16	17	1 11/16	43
	1.90	0.0748	1.900	1	25	2	51	2	51	3¾	95	11/16	17	1 11/16	43
48		0.0760	1.930	1	25	2	51	2	51	3¾	95	11/16	17	1 11/16	43
	1.95	0.0768	1.950	1	25	2	51	2	51	3¾	95	11/16	17	1 11/16	43
5/64		0.0781	1.984	1	25	2	51	2	51	3¾	95	11/16	17	1 11/16	43
47		0.0785	1.994	1	25	2	51	2	51	3¾	95	11/16	17	1 11/16	43
	2.00	0.0787	2.000	1	25	2	51	2¼	57	4¼	108	11/16	17	1 11/16	43
	2.05	0.0807	2.050	1⅛	29	2⅛	54	2¼	57	4¼	108	¾	19	1¾	44
46		0.0810	2.057	1⅛	29	2⅛	54	2¼	57	4¼	108	¾	19	1¾	44
45		0.0820	2.083	1⅛	29	2⅛	54	2¼	57	4¼	108	¾	19	1¾	44
	2.10	0.0827	2.100	1⅛	29	2⅛	54	2¼	57	4¼	108	¾	19	1¾	44
	2.15	0.0846	2.150	1⅛	29	2⅛	54	2¼	57	4¼	108	¾	19	1¾	44
44		0.0860	2.184	1⅛	29	2⅛	54	2¼	57	4¼	108	¾	19	1¾	44
	2.20	0.0866	2.200	1¼	32	2¼	57	2¼	57	4¼	108	¾	19	1¾	44
	2.25	0.0886	2.250	1¼	32	2¼	57	2¼	57	4¼	108	¾	19	1¾	44
43		0.0890	2.261	1¼	32	2¼	57	2¼	57	4¼	108	¾	19	1¾	44
	2.30	0.0906	2.300	1¼	32	2¼	57	2¼	57	4¼	108	¾	19	1¾	44
	2.35	0.0925	2.350	1¼	32	2¼	57	2¼	57	4¼	108	¾	19	1¾	44
42		0.0935	2.375	1¼	32	2¼	57	2¼	57	4¼	108	¾	19	1¾	44
3/32		0.0938	2.383	1¼	32	2¼	57	2¼	57	4¼	108	¾	19	1¾	44
	2.40	0.0945	2.400	1⅜	35	2⅜	60	2½	64	4⅝	117	13/16	21	1 13/16	46
41		0.0960	2.438	1⅜	35	2⅜	60	2½	64	4⅝	117	13/16	21	1 13/16	46
	2.45	0.0965	2.450	1⅜	35	2⅜	60	2½	64	4⅝	117	13/16	21	1 13/16	46
40		0.0980	2.489	1⅜	35	2⅜	60	2½	64	4⅝	117	13/16	21	1 13/16	46
	2.50	0.0984	2.500	1⅜	35	2⅜	60	2½	64	4⅝	117	13/16	21	1 13/16	46
39		0.0995	2.527	1⅜	35	2⅜	60	2½	64	4⅝	117	13/16	21	1 13/16	46
38		0.1015	2.578	1 7/16	37	2½	64	2½	64	4⅝	117	13/16	21	1 13/16	46
	2.60	0.1024	2.600	1 7/16	37	2½	64	2½	64	4⅝	117	13/16	21	1 13/16	46
37		0.1040	2.642	1 7/16	37	2½	64	2½	64	4⅝	117	13/16	21	1 13/16	46
	2.70	0.1063	2.700	1 7/16	37	2½	64	2½	64	4⅝	117	13/16	21	1 13/16	46
36		0.1065	2.705	1 7/16	37	2½	64	2½	64	4⅝	117	13/16	21	1 13/16	46
7/64		0.1094	2.779	1½	38	2⅝	67	2½	64	4⅝	117	13/16	21	1 13/16	46
35		0.1100	2.794	1½	38	2⅝	67	2¾	70	5⅛	130	⅞	22	1⅞	48
	2.80	0.1102	2.800	1½	38	2⅝	67	2¾	70	5⅛	130	⅞	22	1⅞	48
34		0.1110	2.819	1½	38	2⅝	67	2¾	70	5⅛	130	⅞	22	1⅞	48
33		0.1130	2.870	1½	38	2⅝	67	2¾	70	5⅛	130	⅞	22	1⅞	48
	2.90	0.1142	2.900	1⅝	41	2¾	70	2¾	70	5⅛	130	⅞	22	1⅞	48
32		0.1160	2.946	1⅝	41	2¾	70	2¾	70	5⅛	130	⅞	22	1⅞	48
	3.00	0.1181	3.000	1⅝	41	2¾	70	2¾	70	5⅛	130	⅞	22	1⅞	48
31		0.1200	3.048	1⅝	41	2¾	70	2¾	70	5⅛	130	⅞	22	1⅞	48
	3.10	0.1220	3.100	1⅝	41	2¾	70	2¾	70	5⅛	130	⅞	22	1⅞	48

* Fractional inch, number, letter, and metric sizes.

Table 1 (*Continued*). **Straight Shank Twist Drills — Jobbers Length through 17.5 mm, Taper Length through 12.7 mm, and Screw Machine Length through 25.4 mm Diameter (ANSI B94.11M-1979)**

Drill Diameter, D*				Jobbers Length				Taper Length				Screw Machine Length			
Frac., No. or Ltr.	mm	Equivalent		Flute F		Overall L		Flute F		Overall L		Flute F		Overall L	
		Deci. In.	mm	In.	mm	In.	mm	In.	mm	In.	mm	In.	mm	In.	mm
⅛		0.1250	3.175	1⅝	41	2¾	70	2¾	70	5⅛	130	⅞	22	1⅞	48
	3.20	0.1260	3.200	1⅝	41	2¾	70	3	76	5⅜	137	15/16	24	1 15/16	49
30		0.1285	3.264	1⅝	41	2¾	70	3	76	5⅜	137	15/16	24	1 15/16	49
	3.30	0.1299	3.300	1¾	44	2⅞	73	3	76	5⅜	137	15/16	24	1 15/16	49
	3.40	0.1339	3.400	1¾	44	2⅞	73	3	76	5⅜	137	15/16	24	1 15/16	49
29		0.1360	3.454	1¾	44	2⅞	73	3	76	5⅜	137	15/16	24	1 15/16	49
	3.50	0.1378	3.500	1¾	44	2⅞	73	3	76	5⅜	137	15/16	24	1 15/16	49
28		0.1405	3.569	1¾	44	2⅞	73	3	76	5⅜	137	15/16	24	1 15/16	49
9/64		0.1406	3.571	1¾	44	2⅞	73	3	76	5⅜	137	15/16	24	1 15/16	49
	3.60	0.1417	3.600	1⅞	48	3	76	3	76	5⅜	137	1	25	2 1/16	52
27		0.1440	3.658	1⅞	48	3	76	3	76	5⅜	137	1	25	2 1/16	52
	3.70	0.1457	3.700	1⅞	48	3	76	3	76	5⅜	137	1	25	2 1/16	52
26		0.1470	3.734	1⅞	48	3	76	3	76	5⅜	137	1	25	2 1/16	52
25		0.1495	3.797	1⅞	48	3	76	3	76	5⅜	137	1	25	2 1/16	52
	3.80	0.1496	3.800	1⅞	48	3	76	3	76	5⅜	137	1	25	2 1/16	52
24		0.1520	3.861	2	51	3⅛	79	3	76	5⅜	137	1	25	2 1/16	52
	3.90	0.1535	3.900	2	51	3⅛	79	3	76	5⅜	137	1	25	2 1/16	52
23		0.1540	3.912	2	51	3⅛	79	3	76	5⅜	137	1	25	2 1/16	52
5/32		0.1562	3.967	2	51	3⅛	79	3	76	5⅜	137	1	25	2 1/16	52
22		0.1570	3.988	2	51	3⅛	79	3⅜	86	5¾	146	1 1/16	27	2⅛	54
	4.00	0.1575	4.000	2⅛	54	3¼	83	3⅜	86	5¾	146	1 1/16	27	2⅛	54
21		0.1590	4.039	2⅛	54	3¼	83	3⅜	86	5¾	146	1 1/16	27	2⅛	54
20		0.1610	4.089	2⅛	54	3¼	83	3⅜	86	5¾	146	1 1/16	27	2⅛	54
	4.10	0.1614	4.100	2⅛	54	3¼	83	3⅜	86	5¾	146	1 1/16	27	2⅛	54
	4.20	0.1654	4.200	2⅛	54	3¼	83	3⅜	86	5¾	146	1 1/16	27	2⅛	54
19		0.1660	4.216	2⅛	54	3¼	83	3⅜	86	5¾	146	1 1/16	27	2⅛	54
	4.30	0.1693	4.300	2⅛	54	3¼	83	3⅜	86	5¾	146	1 1/16	27	2⅛	54
18		0.1695	4.305	2⅛	54	3¼	83	3⅜	86	5¾	146	1 1/16	27	2⅛	54
11/64		0.1719	4.366	2⅛	54	3¼	83	3⅜	86	5¾	146	1 1/16	27	2⅛	54
17		0.1730	4.394	2 3/16	56	3⅜	86	3⅜	86	5¾	146	1⅛	29	2 3/16	56
	4.40	0.1732	4.400	2 3/16	56	3⅜	86	3⅜	86	5¾	146	1⅛	29	2 3/16	56
16		0.1770	4.496	2 3/16	56	3⅜	86	3⅜	86	5¾	146	1⅛	29	2 3/16	56
	4.50	0.1772	4.500	2 3/16	56	3⅜	86	3⅜	86	5¾	146	1⅛	29	2 3/16	56
15		0.1800	4.572	2 3/16	56	3⅜	86	3⅜	86	5¾	146	1⅛	29	2 3/16	56
	4.60	0.1811	4.600	2 3/16	56	3⅜	86	3⅜	86	5¾	146	1⅛	29	2 3/16	56
14		0.1820	4.623	2 3/16	56	3⅜	86	3⅜	86	5¾	146	1⅛	29	2 3/16	56
13	4.70	0.1850	4.700	2 5/16	59	3½	89	3⅜	86	5¾	146	1⅛	29	2 3/16	56
3/16		0.1875	4.762	2 5/16	59	3½	89	3⅜	86	5¾	146	1⅛	29	2 3/16	56
12	4.80	0.1890	4.800	2 5/16	59	3½	89	3⅝	92	6	152	1 3/16	30	2¼	57
11		0.1910	4.851	2 5/16	59	3½	89	3⅝	92		152	1 3/16	30	2¼	57
	4.90	0.1929	4.900	2 7/16	62	3⅝	92	3⅝	92	6	152	1 3/16	30	2¼	57
10		0.1935	4.915	2 7/16	62	3⅝	92	3⅝	92	6	152	1 3/16	30	2¼	57
9		0.1960	4.978	2 7/16	62	3⅝	92	3⅝	92	6	152	1 3/16	30	2¼	57
	5.00	0.1969	5.000	2 7/16	62	3⅝	92	3⅝	92	6	152	1 3/16	30	2¼	57
8		0.1990	5.054	2 7/16	62	3⅝	92	3⅝	92	6	152	1 3/16	30	2¼	57

* Fractional inch, number, letter, and metric sizes.

Table 1 (Continued). Straight Shank Twist Drills — Jobbers Length through 17.5 mm, Taper Length through 12.7 mm, and Screw Machine Length through 25.4 mm Diameter (ANSI B94.11M-1979)

Drill Diameter, D*				Jobbers Length				Taper Length				Screw Machine Length			
Frac., No. or Ltr.	mm	Equivalent		Flute		Overall		Flute		Overall		Flute		Overall	
		Deci. In.	mm	F In.	mm	L In.	mm	F In.	mm	L In.	mm	F In.	mm	L In.	mm
	5.10	0.2008	5.100	2 7/16	62	3 5/8	92	3 5/8	92	6	152	1 3/16	30	2 1/4	57
7		0.2010	5.105	2 7/16	62	3 5/8	92	3 5/8	92	6	152	1 3/16	30	2 1/4	57
13/64		0.2031	5.159	2 7/16	62	3 5/8	92	3 5/8	92	6	152	1 3/16	30	2 1/4	57
6		0.2040	5.182	2 1/2	64	3 3/4	95	3 5/8	92	6	152	1 1/4	32	2 3/8	60
	5.20	0.2047	5.200	2 1/2	64	3 3/4	95	3 5/8	92	6	152	1 1/4	32	2 3/8	60
5		0.2055	5.220	2 1/2	64	3 3/4	95	3 5/8	92	6	152	1 1/4	32	2 3/8	60
	5.30	0.2087	5.300	2 1/2	64	3 3/4	95	3 5/8	92	6	152	1 1/4	32	2 3/8	60
4		0.2090	5.309	2 1/2	64	3 3/4	95	3 5/8	92	6	152	1 1/4	32	2 3/8	60
	5.40	0.2126	5.400	2 1/2	64	3 3/4	95	3 5/8	92	6	152	1 1/4	32	2 3/8	60
3		0.2130	5.410	2 1/2	64	3 3/4	95	3 5/8	92	6	152	1 1/4	32	2 3/8	60
	5.50	0.2165	5.500	2 1/2	64	3 3/4	95	3 5/8	92	6	152	1 1/4	32	2 3/8	60
7/32		0.2188	5.558	2 1/2	64	3 3/4	95	3 5/8	92	6	152	1 1/4	32	2 3/8	60
	5.60	0.2205	5.600	2 5/8	67	3 7/8	98	3 3/4	95	6 1/8	156	1 5/16	33	2 7/16	62
2		0.2210	5.613	2 5/8	67	3 7/8	98	3 3/4	95	6 1/8	156	1 5/16	33	2 7/16	62
	5.70	0.2244	5.700	2 5/8	67	3 7/8	98	3 3/4	95	6 1/8	156	1 5/16	33	2 7/16	62
1		0.2280	5.791	2 5/8	67	3 7/8	98	3 3/4	95	6 1/8	156	1 5/16	33	2 7/16	62
	5.80	0.2283	5.800	2 5/8	67	3 7/8	98	3 3/4	95	6 1/8	156	1 5/16	33	2 7/16	62
	5.90	0.2323	5.900	2 5/8	67	3 7/8	98	3 3/4	95	6 1/8	156	1 5/16	33	2 7/16	62
A		0.2340	5.944	2 5/8	67	3 7/8	98	...	...	...	...	1 5/16	33	2 7/16	62
15/64		0.2344	5.954	2 5/8	67	3 7/8	98	3 3/4	95	6 1/8	156	1 5/16	33	2 7/16	62
	6.00	0.2362	6.000	2 3/4	70	4	102	3 3/4	95	6 1/8	156	1 3/8	35	2 1/2	64
B		0.2380	6.045	2 3/4	70	4	102	...	...	...	...	1 3/8	35	2 1/2	64
	6.10	0.2402	6.100	2 3/4	70	4	102	3 3/4	95	6 1/8	156	1 3/8	35	2 1/2	64
C		0.2420	6.147	2 3/4	70	4	102	...	...	...	...	1 3/8	35	2 1/2	64
	6.20	0.2441	6.200	2 3/4	70	4	102	3 3/4	95	6 1/8	156	1 3/8	35	2 1/2	64
D		0.2460	6.248	2 3/4	70	4	102	...	...	...	...	1 3/8	35	2 1/2	64
	6.30	0.2480	6.300	2 3/4	70	4	102	3 3/4	95	6 1/8	156	1 3/8	35	2 1/2	64
E, 1/4		0.2500	6.350	2 3/4	70	4	102	3 3/4	95	6 1/8	156	1 3/8	35	2 1/2	64
	6.40	0.2520	6.400	2 7/8	73	4 1/8	105	3 7/8	98	6 1/4	159	1 7/16	37	2 5/8	67
	6.50	0.2559	6.500	2 7/8	73	4 1/8	105	3 7/8	98	6 1/4	159	1 7/16	37	2 5/8	67
F		0.2570	6.528	2 7/8	73	4 1/8	105	...	...	...	...	1 7/16	37	2 5/8	67
	6.60	0.2598	6.600	2 7/8	73	4 1/8	105	...	...	...	...	1 7/16	37	2 5/8	67
G		0.2610	6.629	2 7/8	73	4 1/8	105	...	...	...	...	1 7/16	37	2 5/8	67
	6.70	0.2638	6.700	2 7/8	73	4 1/8	105	...	...	...	...	1 7/16	37	2 5/8	67
17/64		0.2656	6.746	2 7/8	73	4 1/8	105	3 7/8	98	6 1/4	159	1 7/16	37	2 5/8	67
H		0.2660	6.756	2 7/8	73	4 1/8	105	...	...	...	...	1 1/2	38	2 11/16	68
	6.80	0.2677	6.800	2 7/8	73	4 1/8	105	3 7/8	98	6 1/4	159	1 1/2	38	2 11/16	68
	6.90	0.2717	6.900	2 7/8	73	4 1/8	105	...	...	...	...	1 1/2	38	2 11/16	68
I		0.2720	6.909	2 7/8	73	4 1/8	105	...	...	...	...	1 1/2	38	2 11/16	68
	7.00	0.2756	7.000	2 7/8	73	4 1/8	105	3 7/8	98	6 1/4	159	1 1/2	38	2 11/16	68
J		0.2770	7.036	2 7/8	73	4 1/8	105	...	...	...	...	1 1/2	38	2 11/16	68
	7.10	0.2795	7.100	2 15/16	75	4 1/4	108	...	...	...	...	1 1/2	38	2 11/16	68
K		0.2810	7.137	2 15/16	75	4 1/4	108	...	...	...	...	1 1/2	38	2 11/16	68
9/32		0.2812	7.142	2 15/16	75	4 1/4	108	3 7/8	98	6 1/4	159	1 1/2	38	2 11/16	68
	7.20	0.2835	7.200	2 15/16	75	4 1/4	108	4	102	6 3/8	162	1 9/16	40	2 3/4	70

* Fractional, number, letter, and metric sizes.

Table 1 (*Continued*). **Straight Shank Twist Drills — Jobbers Length through 17.5 mm, Taper Length through 12.7 mm, and Screw Machine Length through 25.4 mm Diameter** (ANSI B94.11M-1979)

Frac., No. or Ltr.	mm	Deci. In.	mm	Flute F In.	mm	Overall L In.	mm	Flute F In.	mm	Overall L In.	mm	Flute F In.	mm	Overall L In.	mm
	7.30	0.2874	7.300	2¹⁵⁄₁₆	75	4¼	108	…	…	…	…	1⁹⁄₁₆	40	2¾	70
L		0.2900	7.366	2¹⁵⁄₁₆	75	4¼	108	…	…	…	…	1⁹⁄₁₆	40	2¾	70
	7.40	0.2913	7.400	3¹⁄₁₆	78	4⅜	111	…	…	…	…	1⁹⁄₁₆	40	2¾	70
M		0.2950	7.493	3¹⁄₁₆	78	4⅜	111	…	…	…	…	1⁹⁄₁₆	40	2¾	70
	7.50	0.2953	7.500	3¹⁄₁₆	78	4⅜	111	4	102	6⅜	162	1⁹⁄₁₆	40	2¾	70
19⁄64		0.2969	7.541	3¹⁄₁₆	78	4⅜	111	4	102	6⅜	162	1⁹⁄₁₆	40	2¾	70
	7.60	0.2992	7.600	3¹⁄₁₆	78	4⅜	111	…	…	…	…	1⅝	41	2¹³⁄₁₆	71
N		0.3020	7.671	3¹⁄₁₆	78	4⅜	111	…	…	…	…	1⅝	41	2¹³⁄₁₆	71
	7.70	0.3031	7.700	3³⁄₁₆	81	4½	114	…	…	…	…	1⅝	41	2¹³⁄₁₆	71
	7.80	0.3071	7.800	3³⁄₁₆	81	4½	114	4	102	6⅜	162	1⅝	41	2¹³⁄₁₆	71
	7.90	0.3110	7.900	3³⁄₁₆	81	4½	114	…	…	…	…	1⅝	41	2¹³⁄₁₆	71
5⁄16		0.3125	7.938	3³⁄₁₆	81	4½	114	4	102	6⅜	162	1⅝	41	2¹³⁄₁₆	71
	8.00	0.3150	8.000	3³⁄₁₆	81	4½	114	4⅛	105	6½	165	1¹¹⁄₁₆	43	2¹⁵⁄₁₆	75
O		0.3160	8.026	3³⁄₁₆	81	4½	114	…	…	…	…	1¹¹⁄₁₆	43	2¹⁵⁄₁₆	75
	8.10	0.3189	8.100	3⁵⁄₁₆	84	4⅝	117	…	…	…	…	1¹¹⁄₁₆	43	2¹⁵⁄₁₆	75
	8.20	0.3228	8.200	3⁵⁄₁₆	84	4⅝	117	4⅛	105	6½	165	1¹¹⁄₁₆	43	2¹⁵⁄₁₆	75
P		0.3230	8.204	3⁵⁄₁₆	84	4⅝	117	…	…	…	…	1¹¹⁄₁₆	43	2¹⁵⁄₁₆	75
	8.30	0.3268	8.300	3⁵⁄₁₆	84	4⅝	117	…	…	…	…	1¹¹⁄₁₆	43	2¹⁵⁄₁₆	75
21⁄64		0.3281	8.334	3⁵⁄₁₆	84	4⅝	117	4⅛	105	6½	165	1¹¹⁄₁₆	43	2¹⁵⁄₁₆	75
	8.40	0.3307	8.400	3⁷⁄₁₆	87	4¾	121	…	…	…	…	1¹¹⁄₁₆	43	3	76
Q		0.3320	8.433	3⁷⁄₁₆	87	4¾	121	…	…	…	…	1¹¹⁄₁₆	43	3	76
	8.50	0.3346	8.500	3⁷⁄₁₆	87	4¾	121	4⅛	105	6½	165	1¹¹⁄₁₆	43	3	76
	8.60	0.3386	8.600	3⁷⁄₁₆	87	4¾	121	…	…	…	…	1¹¹⁄₁₆	43	3	76
R		0.3390	8.611	3⁷⁄₁₆	87	4¾	121	…	…	…	…	1¹¹⁄₁₆	43	3	76
	8.70	0.3425	8.700	3⁷⁄₁₆	87	4¾	121	…	…	…	…	1¹¹⁄₁₆	43	3	76
11⁄32		0.3438	8.733	3⁷⁄₁₆	87	4¾	121	4⅛	105	6½	165	1¹¹⁄₁₆	43	3	76
	8.80	0.3465	8.800	3½	89	4⅞	124	4¼	108	6¾	171	1¾	44	3¹⁄₁₆	78
S		0.3480	8.839	3½	89	4⅞	124	…	…	…	…	1¾	44	3¹⁄₁₆	78
	8.90	0.3504	8.900	3½	89	4⅞	124	4¼	108	6¾	171	1¾	44	3¹⁄₁₆	78
	9.00	0.3543	9.000	3½	89	4⅞	124	…	…	…	…	1¾	44	3¹⁄₁₆	78
T		0.3580	9.093	3½	89	4⅞	124	…	…	…	…	1¾	44	3¹⁄₁₆	78
	9.10	0.3583	9.100	3½	89	4⅞	124	…	…	…	…	1¾	44	3¹⁄₁₆	78
23⁄64		0.3594	9.129	3½	89	4⅞	124	4¼	108	6¾	171	1¾	44	3¹⁄₁₆	78
	9.20	0.3622	9.200	3⅝	92	5	127	4¼	108	6¾	171	1¹³⁄₁₆	46	3⅛	79
	9.30	0.3661	9.300	3⅝	92	5	127	…	…	…	…	1¹³⁄₁₆	46	3⅛	79
U		0.3680	9.347	3⅝	92	5	127	…	…	…	…	1¹³⁄₁₆	46	3⅛	79
	9.40	0.3701	9.400	3⅝	92	5	127	…	…	…	…	1¹³⁄₁₆	46	3⅛	79
	9.50	0.3740	9.500	3⅝	92	5	127	4¼	108	6¾	171	1¹³⁄₁₆	46	3⅛	79
⅜		0.3750	9.525	3⅝	92	5	127	4¼	108	6¾	171	1¹³⁄₁₆	46	3⅛	79
V		0.3770	9.576	3⅝	92	5	127	…	…	…	…	1⅞	48	3¼	83
	9.60	0.3780	9.600	3¾	95	5⅛	130	…	…	…	…	1⅞	48	3¼	83
	9.70	0.3819	9.700	3¾	95	5⅛	130	…	…	…	…	1⅞	48	3¼	83
	9.80	0.3858	9.800	3¾	95	5⅛	130	4⅜	111	7	178	1⅞	48	3¼	83
W		0.3860	9.804	3¾	95	5⅛	130	…	…	…	…	1⅞	48	3¼	83
	9.90	0.3898	9.900	3¾	95	5⅛	130	…	…	…	…	1⅞	48	3¼	83

* Fractional, number, letter, and metric sizes.

Table 1 (*Continued*). **Straight Shank Twist Drills — Jobbers Length through 17.5 mm, Taper Length through 12.7 mm, and Screw Machine Length through 25.4 mm Diameter** (ANSI B94.11M-1979)

Drill Diameter, D*				Jobbers Length				Taper Length				Screw Machine Length			
Frac., No. or Ltr.	mm	Equivalent		Flute F		Overall L		Flute F		Overall L		Flute F		Overall L	
		Deci. In.	mm	In.	mm	In.	mm	In.	mm	In.	mm	In.	mm	In.	mm
25/64		0.3906	9.921	3 3/4	95	5 1/8	130	4 3/8	111	7	178	1 7/8	48	3 1/4	83
	10.00	0.3937	10.000	3 3/4	95	5 1/8	130	4 3/8	111	7	178	1 15/16	49	3 3/16	84
X		0.3970	10.084	3 3/4	95	5 1/8	130	...	...	...	...	1 15/16	49	3 3/16	84
	10.20	0.4016	10.200	3 7/8	98	5 1/8	133	4 3/8	111	7	178	1 15/16	49	3 3/16	84
Y		0.4040	10.262	3 7/8	98	5 1/4	133					1 15/16	49	3 3/16	84
13/32		0.4062	10.317	3 7/8	98	5 1/4	133	4 3/8	111	7	178	1 15/16	49	3 3/16	84
Z		0.4130	10.490	3 7/8	98	5 1/4	133	...	...	...	...	2	51	3 3/8	86
	10.50	0.4134	10.500	3 7/8	98	5 1/4	133	4 5/8	117	7 1/4	184	2	51	3 3/8	86
27/64		0.4219	10.716	3 15/16	100	5 3/8	137	4 5/8	117	7 1/4	184	2	51	3 3/8	86
	10.80	0.4252	10.800	4 1/16	103	5 1/2	140	4 5/8	117	7 1/4	184	2 1/16	52	3 7/16	87
	11.00	0.4331	11.000	4 1/16	103	5 1/2	140	4 5/8	117	7 1/4	184	2 1/16	52	3 7/16	87
7/16		0.4375	11.112	4 1/16	103	5 1/2	140	4 5/8	117	7 1/4	184	2 1/16	52	3 7/16	87
	11.20	0.4409	11.200	4 3/16	106	5 5/8	143	4 3/4	121	7 1/2	190	2 1/8	54	3 9/16	90
	11.50	0.4528	11.500	4 3/16	106	5 5/8	143	4 3/4	121	7 1/2	190	2 1/8	54	3 9/16	90
29/64		0.4531	11.509	4 3/16	106	5 5/8	143	4 3/4	121	7 1/2	190	2 1/8	54	3 9/16	90
	11.80	0.4646	11.800	4 5/16	110	5 3/4	146	4 3/4	121	7 1/2	190	2 1/8	54	3 5/8	92
15/32		0.4688	11.908	4 5/16	110	5 3/4	146	4 3/4	121	7 1/2	190	2 1/8	54	3 5/8	92
	12.00	0.4724	12.000	4 3/8	111	5 7/8	149	4 3/4	121	7 3/4	197	2 3/16	56	3 11/16	94
	12.20	0.4803	12.200	4 3/8	111	5 7/8	149	4 3/4	121	7 3/4	197	2 3/16	56	3 11/16	94
31/64		0.4844	12.304	4 3/8	111	5 7/8	149	4 3/4	121	7 3/4	197	2 3/16	56	3 11/16	94
	12.50	0.4921	12.500	4 1/2	114	6	152	4 3/4	121	7 3/4	197	2 1/4	57	3 3/4	95
1/2		0.5000	12.700	4 1/2	114	6	152	4 3/4	121	7 3/4	197	2 1/4	57	3 3/4	95
	12.80	0.5039	12.800	4 1/2	114	6	152	...				2 3/8	60	3 7/8	98
	13.00	0.5118	13.000	4 1/2	114	6	152	...				2 3/8	60	3 7/8	98
33/64		0.5156	13.096	4 13/16	122	6 5/8	168	...				2 3/8	60	3 7/8	98
	13.20	0.5197	13.200	4 13/16	122	6 5/8	168	...				2 3/8	60	3 7/8	98
17/32		0.5312	13.492	4 13/16	122	6 5/8	168	...				2 3/8	60	3 7/8	98
	13.50	0.5315	13.500	4 13/16	122	6 5/8	168	...				2 3/8	60	3 7/8	98
	13.80	0.5433	13.800	4 13/16	122	6 5/8	168	...				2 1/2	64	4	102
35/64		0.5469	13.891	4 13/16	122	6 5/8	168	...				2 1/2	64	4	102
	14.00	0.5512	14.000	4 13/16	122	6 5/8	168	...				2 1/2	64	4	102
	14.25	0.5610	14.250	4 13/16	122	6 5/8	168	...				2 1/2	64	4	102
9/16		0.5625	14.288	4 13/16	122	6 5/8	168	...				2 1/2	64	4	102
	14.50	0.5709	14.500	4 13/16	122	6 5/8	168	...				2 5/8	67	4 1/8	105
37/64		0.5781	14.684	4 13/16	122	6 5/8	168	...				2 5/8	67	4 1/8	105
	14.75	0.5807	14.750	5 3/16	132	7 1/8	181	...				2 5/8	67	4 1/8	105
	15.00	0.5906	15.000	5 3/16	132	7 1/8	181	...				2 5/8	67	4 1/8	105
19/32		0.5938	15.083	5 3/16	132	7 1/8	181	...				2 5/8	67	4 1/8	105
	15.25	0.6004	15.250	5 3/16	132	7 1/8	181	...				2 3/4	70	4 1/4	108
39/64		0.6094	15.479	5 3/16	132	7 1/8	181	...				2 3/4	70	4 1/4	108
	15.50	0.6102	15.500	5 3/16	132	7 1/8	181	...				2 3/4	70	4 1/4	108
	15.75	0.6201	15.750	5 3/16	132	7 1/8	181	...				2 3/4	70	4 1/4	108
5/8		0.6250	15.875	5 3/16	132	7 1/8	181	...				2 3/4	70	4 1/4	108
	16.00	0.6299	16.000	5 3/16	132	7 1/8	181	...				2 7/8	73	4 1/2	114
	16.25	0.6398	16.250	5 3/16	132	7 1/8	181	...				2 7/8	73	4 1/2	114

* Fractional, number, letter, and metric sizes.

Table 1 (*Concluded*). Straight Shank Twist Drills — Jobbers Length through 17.5 mm, Taper Length through 12.7 mm, and Screw Machine Length through 25.4 mm Diameter (ANSI B94.11M-1979)

Frac., No. or Ltr.	mm	Deci. In.	mm	Flute F In.	mm	Overall L In.	mm	Flute F In.	mm	Overall L In.	mm	Flute F In.	mm	Overall L In.	mm
41/64		0.6406	16.271	5³/₁₆	132	7⅛	181	. . .	. . .	. . .	. . .	2⅞	73	4½	114
	16.50	0.6496	16.500	5³/₁₆	132	7⅛	181	. . .	. . .	. . .	. . .	2⅞	73	4½	114
21/32		0.6562	16.669	5³/₁₆	132	7⅛	181	. . .	. . .	. . .	. . .	2⅞	73	4½	114
	16.75	0.6594	16.750	5⅝	143	7⅝	194	. . .	. . .	. . .	. . .	2⅞	73	4½	114
	17.00	0.6693	17.000	5⅝	143	7⅝	194	. . .	. . .	. . .	. . .	2⅞	73	4½	114
43/64		0.6719	17.066	5⅝	143	7⅝	194	. . .	. . .	. . .	. . .	2⅞	73	4½	114
	17.25	0.6791	17.250	5⅝	143	7⅝	194	. . .	. . .	. . .	. . .	2⅞	73	4½	114
11/16		0.6875	17.462	5⅝	143	7⅝	194	. . .	. . .	. . .	. . .	2⅞	73	4½	114
	17.50	0.6890	17.500	5⅝	143	7⅝	194	. . .	. . .	. . .	. . .	3	76	4¾	121
45/64		0.7031	17.859	. . .	. . .	. . .	. . .	. . .	. . .	. . .	. . .	3	76	4¾	121
	18.00	0.7087	18.000	. . .	. . .	. . .	. . .	. . .	. . .	. . .	. . .	3	76	4¾	121
23/32		0.7188	18.258	. . .	. . .	. . .	. . .	. . .	. . .	. . .	. . .	3	76	4¾	121
	18.50	0.7283	18.500	. . .	. . .	. . .	. . .	. . .	. . .	. . .	. . .	3⅛	79	5	127
47/64		0.7344	18.654	. . .	. . .	. . .	. . .	. . .	. . .	. . .	. . .	3⅛	79	5	127
	19.00	0.7480	19.000	. . .	. . .	. . .	. . .	. . .	. . .	. . .	. . .	3⅛	79	5	127
¾		0.7500	19.050	. . .	. . .	. . .	. . .	. . .	. . .	. . .	. . .	3⅛	79	5	127
49/64		0.7656	19.446	. . .	. . .	. . .	. . .	. . .	. . .	. . .	. . .	3¼	83	5⅛	130
	19.50	0.7677	19.500	. . .	. . .	. . .	. . .	. . .	. . .	. . .	. . .	3¼	83	5⅛	130
25/32		0.7812	19.845	. . .	. . .	. . .	. . .	. . .	. . .	. . .	. . .	3¼	83	5⅛	130
	20.00	0.7879	20.000	. . .	. . .	. . .	. . .	. . .	. . .	. . .	. . .	3⅜	86	5¼	133
51/64		0.7969	20.241	. . .	. . .	. . .	. . .	. . .	. . .	. . .	. . .	3⅜	86	5¼	133
	20.50	0.8071	20.500	. . .	. . .	. . .	. . .	. . .	. . .	. . .	. . .	3⅜	86	5¼	133
13/16		0.8125	20.638	. . .	. . .	. . .	. . .	. . .	. . .	. . .	. . .	3⅜	86	5¼	133
	21.00	0.8268	21.000	. . .	. . .	. . .	. . .	. . .	. . .	. . .	. . .	3½	89	5⅜	137
53/64		0.8281	21.034	. . .	. . .	. . .	. . .	. . .	. . .	. . .	. . .	3½	89	5⅜	137
27/32		0.8438	21.433	. . .	. . .	. . .	. . .	. . .	. . .	. . .	. . .	3½	89	5⅜	137
	21.50	0.8465	21.500	. . .	. . .	. . .	. . .	. . .	. . .	. . .	. . .	3½	89	5⅜	137
55/64		0.8594	21.829	. . .	. . .	. . .	. . .	. . .	. . .	. . .	. . .	3½	89	5⅜	137
	22.00	0.8661	22.000	. . .	. . .	. . .	. . .	. . .	. . .	. . .	. . .	3½	89	5⅜	137
⅞		0.8750	22.225	. . .	. . .	. . .	. . .	. . .	. . .	. . .	. . .	3½	89	5⅜	137
	22.50	0.8858	22.500	. . .	. . .	. . .	. . .	. . .	. . .	. . .	. . .	3⅝	92	5⅝	143
57/64		0.8906	22.621	. . .	. . .	. . .	. . .	. . .	. . .	. . .	. . .	3⅝	92	5⅝	143
	23.00	0.9055	23.000	. . .	. . .	. . .	. . .	. . .	. . .	. . .	. . .	3⅝	92	5⅝	143
29/32		0.9062	23.017	. . .	. . .	. . .	. . .	. . .	. . .	. . .	. . .	3⅝	92	5⅝	143
59/64		0.9219	23.416	. . .	. . .	. . .	. . .	. . .	. . .	. . .	. . .	3¾	95	5¾	146
	23.50	0.9252	23.500	. . .	. . .	. . .	. . .	. . .	. . .	. . .	. . .	3¾	95	5¾	146
15/16		0.9375	23.812	. . .	. . .	. . .	. . .	. . .	. . .	. . .	. . .	3¾	95	5¾	146
	24.00	0.9449	24.000	. . .	. . .	. . .	. . .	. . .	. . .	. . .	. . .	3⅞	98	5⅞	149
61/64		0.9531	24.209	. . .	. . .	. . .	. . .	. . .	. . .	. . .	. . .	3⅞	98	5⅞	149
	24.50	0.9646	24.500	. . .	. . .	. . .	. . .	. . .	. . .	. . .	. . .	3⅞	98	5⅞	149
31/32		0.9688	24.608	. . .	. . .	. . .	. . .	. . .	. . .	. . .	. . .	3⅞	98	5⅞	149
	25.00	0.9843	25.000	. . .	. . .	. . .	. . .	. . .	. . .	. . .	. . .	4	102	6	152
63/64		0.9844	25.004	. . .	. . .	. . .	. . .	. . .	. . .	. . .	. . .	4	102	6	152
1		1.0000	25.400	. . .	. . .	. . .	. . .	. . .	. . .	. . .	. . .	4	102	6	152

* Fractional, number, letter, and metric sizes.

Table 2. American National Standard Straight Shank Twist Drills — Taper Length — Over ½ in. (12.7 mm) Diam., Fractional and Metric Sizes (ANSI B94.11M-1979)

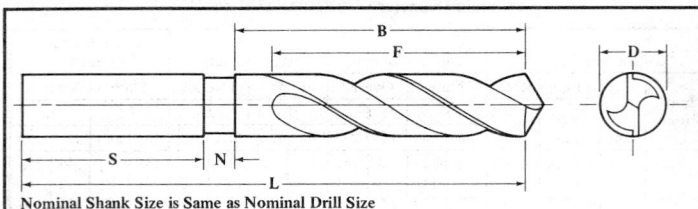

Nominal Shank Size is Same as Nominal Drill Size

Diameter of Drill				Flute Length		Overall Length		Length of Body		Minimum Length of Shk.		Maximum Length of Neck	
D		Decimal Inch Equiv.	Milli-meter Equiv.	F		L		B		S		N	
Frac.	mm			Inch	mm	Inch	mm	Inch	mm	Inch	mm	Inch	mm
	12.80	0.5039	12.800	4¾	121	8	203	4⅞	124	2⅝	66	½	13
	13.00	0.5117	13.000	4¾	121	8	203	4⅞	124	2⅝	66	½	13
33/64		0.5156	13.096	4¾	121	8	203	4⅞	124	2⅝	66	½	13
	13.20	0.5197	13.200	4¾	121	8	203	4⅞	124	2⅝	66	½	13
17/32		0.5312	13.492	4¾	121	8	203	4⅞	124	2⅝	66	½	13
	13.50	0.5315	13.500	4¾	121	8	203	4⅞	124	2⅝	66	½	13
	13.80	0.5433	13.800	4⅞	124	8¼	210	5	127	2¾	70	½	13
35/64		0.5419	13.891	4⅞	124	8¼	210	5	127	2¾	70	½	13
	14.00	0.5512	14.000	4⅞	124	8¼	210	5	127	2¾	70	½	13
	14.25	0.5610	14.250	4⅞	124	8¼	210	5	127	2¾	70	½	13
9/16		0.5625	14.288	4⅞	124	8¼	210	5	127	2¾	70	½	13
	14.50	0.5709	14.500	4⅞	124	8¾	222	5	127	3⅛	79	⅝	16
37/64		0.5781	14.684	4⅞	124	8¾	222	5	127	3⅛	79	⅝	16
	14.75	0.5807	14.750	4⅞	124	8¾	222	5	127	3⅛	79	⅝	16
	15.00	0.5906	15.000	4⅞	124	8¾	222	5	127	3⅛	79	⅝	16
19/32		0.5938	15.083	4⅞	124	8¾	222	5	127	3⅛	79	⅝	16
	15.25	0.6004	15.250	4⅞	124	8¾	222	5	127	3⅛	79	⅝	16
39/64		0.6094	15.479	4⅞	124	8¾	222	5	127	3⅛	79	⅝	16
	15.50	0.6102	15.500	4⅞	124	8¾	222	5	127	3⅛	79	⅝	16
	15.75	0.6201	15.750	4⅞	124	8¾	222	5	127	3⅛	79	⅝	16
5/8		0.6250	15.875	4⅞	124	8¾	222	5	127	3⅛	79	⅝	16
	16.00	0.6299	16.000	5⅛	130	9	228	5¼	133	3⅛	79	⅝	16
	16.25	0.6398	16.250	5⅛	130	9	228	5¼	133	3⅛	79	⅝	16
41/64		0.6406	16.271	5⅛	130	9	228	5¼	133	3⅛	79	⅝	16
	16.50	0.6496	16.500	5⅛	130	9	228	5¼	133	3⅛	79	⅝	16
21/32		0.6562	16.667	5⅛	130	9	228	5¼	133	3⅛	79	⅝	16
	16.75	0.6594	16.750	5⅜	137	9¼	235	5½	140	3⅛	79	⅝	16
	17.00	0.6693	17.000	5⅜	137	9¼	235	5½	140	3⅛	79	⅝	16
43/64		0.6719	17.066	5⅜	137	9¼	235	5½	140	3⅛	79	⅝	16
	17.25	0.6791	17.250	5⅜	137	9¼	235	5½	140	3⅛	79	⅝	16
11/16		0.6875	17.462	5⅜	137	9¼	235	5½	140	3⅛	79	⅝	16
	17.50	0.6890	17.500	5⅝	143	9½	241	5¾	146	3⅛	79	⅝	16
45/64		0.7031	17.859	5⅝	143	9½	241	5¾	146	3⅛	79	⅝	16
	18.00	0.7087	18.000	5⅝	143	9½	241	5¾	146	3⅛	79	⅝	16
23/32		0.7188	18.258	5⅝	143	9½	241	5¾	146	3⅛	79	⅝	16
	18.50	0.7283	18.500	5⅞	149	9¾	247	6	152	3⅛	79	⅝	16
47/64		0.7344	18.654	5⅞	149	9¾	247	6	152	3⅛	79	⅝	16
	19.00	0.7480	19.000	5⅞	149	9¾	247	6	152	3⅛	79	⅝	16
3/4		0.7500	19.050	5⅞	149	9¾	247	6	152	3⅛	79	⅝	16
49/64		0.7656	19.446	6	152	9⅞	251	6⅛	156	3⅛	79	⅝	16

Table 2 (*Continued*). Straight Shank Twist Drills — Taper Length —
Over ½ in. (12.7 mm) Diam. (ANSI B94.11M-1979)

Diameter of Drill				Flute Length		Overall Length		Length of Body		Minimum Length of Shk.		Maximum Length of Neck	
D		Decimal Inch Equiv.	Milli-meter Equiv.	F		L		B		S		N	
Frac.	mm			Inch	mm	Inch	mm	Inch	mm	Inch	mm	Inch	mm
	19.50	0.7677	19.500	6	152	9⅞	251	6⅛	156	3⅛	79	⅝	16
25/32		0.7812	19.842	6	152	9⅞	251	6⅛	156	3⅛	79	⅝	16
	20.00	0.7874	20.000	6⅛	156	10	254	6¼	159	3⅛	79	⅝	16
51/64		0.7969	20.241	6⅛	156	10	254	6¼	159	3⅛	79	⅝	16
	20.50	0.8071	20.500	6⅛	156	10	254	6¼	159	3⅛	79	⅝	16
13/16		0.8125	20.638	6⅛	156	10	254	6¼	159	3⅛	79	⅝	16
	21.00	0.8268	21.000	6⅛	156	10	254	6¼	159	3⅛	79	⅝	16
53/64		0.8281	21.034	6⅛	156	10	254	6¼	159	3⅛	79	⅝	16
27/32		0.8438	21.433	6⅛	156	10	254	6¼	159	3⅛	79	⅝	16
	21.50	0.8465	21.500	6⅛	156	10	254	6¼	159	3⅛	79	⅝	16
55/64		0.8594	21.829	6⅛	156	10	254	6¼	159	3⅛	79	⅝	16
	22.00	0.8661	22.000	6⅛	156	10	254	6¼	159	3⅛	79	⅝	16
7/8		0.8750	22.225	6⅛	156	10	254	6¼	159	3⅛	79	⅝	16
	22.50	0.8858	22.500	6⅛	156	10	254	6¼	159	3⅛	79	⅝	16
57/64		0.8906	22.621	6⅛	156	10	254	6¼	159	3⅛	79	⅝	16
	23.00	0.9055	23.000	6⅛	156	10	254	6¼	159	3⅛	79	⅝	16
29/32		0.9062	23.017	6⅛	156	10	254	6¼	159	3⅛	79	⅝	16
59/64		0.9219	23.416	6⅛	156	10¾	273	6¼	159	3⅞	98	⅝	16
	23.50	0.9252	23.500	6⅛	156	10¾	273	6¼	159	3⅞	98	⅝	16
15/16		0.9375	23.812	6⅛	156	10¾	273	6¼	159	3⅞	98	⅝	16
	24.00	0.9449	24.000	6⅜	162	11	279	6½	165	3⅞	98	⅝	16
61/64		0.9531	24.209	6⅜	162	11	279	6½	165	3⅞	98	⅝	16
	24.50	0.9646	24.500	6⅜	162	11	279	6½	165	3⅞	98	⅝	16
31/32		0.9688	24.608	6⅜	162	11	279	6½	165	3⅞	98	⅝	16
	25.00	0.9843	25.000	6⅜	162	11	279	6½	165	3⅞	98	⅝	16
63/64		0.9844	25.004	6⅜	162	11	279	6½	165	3⅞	98	⅝	16
1		1.0000	25.400	6⅜	162	11	279	6½	165	3⅞	98	⅝	16
	25.50	1.0039	25.500	6½	165	11⅛	282	6⅝	168	3⅞	98	⅝	16
1 1/64		1.0156	25.796	6½	165	11⅛	282	6⅝	168	3⅞	98	⅝	16
	26.00	1.0236	26.000	6½	165	11⅛	282	6⅝	168	3⅞	98	⅝	16
1 1/32		1.0312	26.192	6½	165	11⅛	282	6⅝	168	3⅞	98	⅝	16
	26.50	1.0433	26.560	6⅝	168	11¼	286	6¾	172	3⅞	98	⅝	16
1 3/64		1.0469	26.591	6⅝	168	11¼	286	6¾	172	3⅞	98	⅝	16
1 1/16		1.0625	26.988	6⅝	168	11¼	286	6¾	172	3⅞	98	⅝	16
	27.00	1.0630	27.000	6⅝	168	11¼	286	6¾	172	3⅞	98	⅝	16
1 5/64		1.0781	27.384	6⅞	175	11½	292	7	178	3⅞	98	⅝	16
	27.50	1.0827	27.500	6⅞	175	11½	292	7	178	3⅞	98	⅝	16
1 3/32		1.0938	27.783	6⅞	175	11½	292	7	178	3⅞	98	⅝	16
	28.00	1.1024	28.000	7⅛	181	11¾	298	7¼	184	3⅞	98	⅝	16
1 7/64		1.1094	28.179	7⅛	181	11¾	298	7¼	184	3⅞	98	⅝	16
	28.50	1.1220	28.500	7⅛	181	11¾	298	7¼	184	3⅞	98	⅝	16
1 1/8		1.1250	28.575	7⅛	181	11¾	298	7¼	184	3⅞	98	⅝	16
1 9/64		1.1406	28.971	7¼	184	11⅞	301	7⅜	187	3⅞	98	⅝	16
	29.00	1.1417	29.000	7¼	184	11⅞	301	7⅜	187	3⅞	98	⅝	16
1 5/32		1.1562	29.367	7¼	184	11⅞	301	7⅜	187	3⅞	98	⅝	16
	29.50	1.1614	29.500	7⅜	187	12	305	7½	191	3⅞	98	⅝	16
1 11/64		1.1719	29.766	7⅜	187	12	305	7½	191	3⅞	98	⅝	16
	30.00	1.1811	30.000	7⅜	187	12	305	7½	191	3⅞	98	⅝	16
1 3/16		1.1875	30.162	7⅜	187	12	305	7½	191	3⅞	98	⅝	16
	30.50	1.2008	30.500	7½	190	12⅛	308	7⅝	194	3⅞	98	⅝	16

Table 2 (*Concluded*). **Straight Shank Twist Drills — Taper Length — Over ½ in. (12.7 mm) Diam.** (ANSI B94.11M-1979)

Diameter of Drill				Flute Length		Overall Length		Length of Body		Minimum Length of Shk.		Maximum Length of Neck	
D		Decimal Inch Equiv.	Millimeter Equiv.	*F*		*L*		*B*		*S*		*N*	
Frac.	mm			Inch	mm	Inch	mm	Inch	mm	Inch	mm	Inch	mm
1 13/64		1.2031	30.559	7½	190	12⅛	308	7⅝	194	3⅞	98	⅝	16
1 7/32		1.2188	30.958	7½	190	12⅛	308	7⅝	194	3⅞	98	⅝	16
	31.00	1.2205	31.000	7⅞	200	12½	317	8	203	3⅞	98	⅝	16
1 15/64		1.2344	31.354	7⅞	200	12½	317	8	203	3⅞	98	⅝	16
	31.50	1.2402	31.500	7⅞	200	12½	317	8	203	3⅞	98	⅝	16
1¼		1.2500	31.750	7⅞	200	12½	317	8	203	3⅞	98	⅝	16
	32.00	1.2598	32.000	8½	216	14⅛	359	8⅝	219	4⅞	124	⅝	16
	32.50	1.2795	32.500	8½	216	14⅛	359	8⅝	219	4⅞	124	⅝	16
1 9/32		1.2812	32.542	8½	216	14⅛	359	8⅝	219	4⅞	124	⅝	16
	33.00	1.2992	33.000	8⅝	219	14¼	362	8¾	222	4⅞	124	⅝	16
1 5/16		1.3125	33.338	8⅝	219	14¼	362	8¾	222	4⅞	124	⅝	16
	33.50	1.3189	33.500	8¾	222	14⅜	365	8⅞	225	4⅞	124	⅝	16
	34.00	1.3386	34.000	8¾	222	14⅜	365	8⅞	225	4⅞	124	⅝	16
1 11/32		1.3438	34.133	8¾	222	14⅜	365	8⅞	225	4⅞	124	⅝	16
	34.50	1.3583	34.500	8⅞	225	14½	368	9	229	4⅞	124	⅝	16
1⅜		1.3750	34.925	8⅞	225	14½	368	9	229	4⅞	124	⅝	16
	35.00	1.3780	35.000	9	229	14⅝	372	9⅛	232	4⅞	124	⅝	16
	35.50	1.3976	35.500	9	229	14⅝	372	9⅛	232	4⅞	124	⅝	16
1 13/32		1.4062	35.717	9	229	14⅝	372	9⅛	232	4⅞	124	⅝	16
	36.00	1.4173	36.000	9⅛	232	14¾	375	9¼	235	4⅞	124	⅝	16
	36.50	1.4370	36.500	9⅛	232	14¾	375	9¼	235	4⅞	124	⅝	16
1 7/16		1.4375	36.512	9⅛	232	14¾	375	9¼	235	4⅞	124	⅝	16
	37.00	1.4567	37.000	9¼	235	14⅞	378	9⅜	238	4⅞	124	⅝	16
1 15/32		1.4688	37.308	9¼	235	14⅞	378	9⅜	238	4⅞	124	⅝	16
	37.50	1.4764	37.500	9⅜	238	15	381	9½	241	4⅞	124	⅝	16
	38.00	1.4961	38.000	9⅜	238	15	381	9½	241	4⅞	124	⅝	16
1½		1.5000	38.100	9⅜	238	15	381	9½	241	4⅞	124	⅝	16
1 9/16		1.5625	39.688	9⅝	244	15¼	387	9¾	247	4⅞	124	⅝	16
1⅝		1.6250	41.275	9⅞	251	15⅝	397	10	254	4⅞	124	¾	19
1¾		1.7500	44.450	10½	267	16¼	413	10⅝	270	4⅞	124	¾	19

Table 3. American National Standard Tangs for Straight Shank Drills* (ANSI B94.11M-1979)

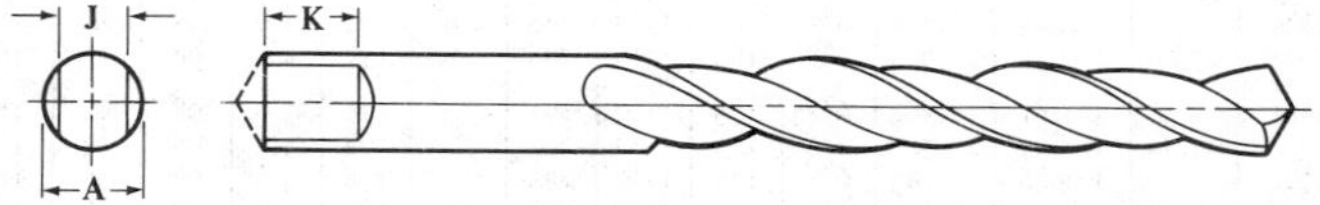

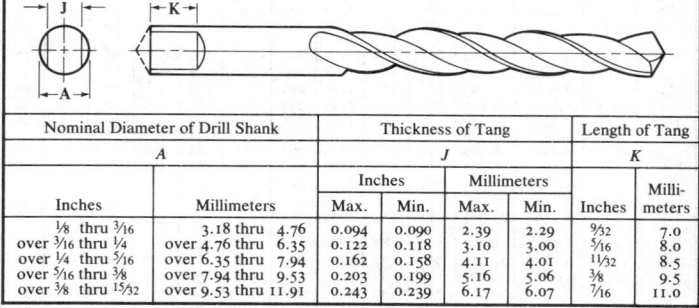

Nominal Diameter of Drill Shank		Thickness of Tang				Length of Tang	
A		*J*				*K*	
		Inches		Millimeters			
Inches	Millimeters	Max.	Min.	Max.	Min.	Inches	Millimeters
⅛ thru 3/16	3.18 thru 4.76	0.094	0.090	2.39	2.29	9/32	7.0
over 3/16 thru ¼	over 4.76 thru 6.35	0.122	0.118	3.10	3.00	5/16	8.0
over ¼ thru 5/16	over 6.35 thru 7.94	0.162	0.158	4.11	4.01	11/32	8.5
over 5/16 thru ⅜	over 7.94 thru 9.53	0.203	0.199	5.16	5.06	⅜	9.5
over ⅜ thru 15/32	over 9.53 thru 11.91	0.243	0.239	6.17	6.07	7/16	11.0

* To fit split sleeve collet type drill drivers. See page 839.

Table 3 (*Concluded*). **Tangs for Straight Shank Drills** (ANSI B94.11-1979)

Nominal Diameter of Drill Shank		Thickness of Tang				Length of Tang	
A		J				K	
		Inches		Millimeters			
Inches	Millimeters	Max.	Min.	Max.	Min.	Inches	Milli-meters
over 15/32 thru 9/16	over 11.91 thru 14.29	0.303	0.297	7.70	7.55	1/2	12.5
over 9/16 thru 21/32	over 14.29 thru 16.67	0.373	0.367	9.47	9.32	9/16	14.5
over 21/32 thru 3/4	over 16.67 thru 19.05	0.443	0.437	11.25	11.10	5/8	16.0
over 3/4 thru 7/8	over 19.05 thru 22.23	0.514	0.508	13.06	12.90	11/16	17.5
over 7/8 thru 1	over 22.23 thru 25.40	0.609	0.601	15.47	15.27	3/4	19.0
over 1 thru 13/16	over 25.40 thru 30.16	0.700	0.692	17.78	17.58	13/16	20.5
over 13/16 thru 13/8	over 30.16 thru 34.93	0.817	0.809	20.75	20.55	7/8	22.0

Table 4. American National Standard Straight Shank Twist Drills — Screw Machine Length — Over 1 in. (25.4 mm) Diam. (ANSI B94.11M-1979)

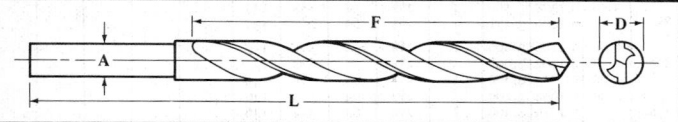

Diameter of Drill				Flute Length		Overall Length		Shank Diameter	
D				F		L		A	
Frac.	mm	Decimal Inch Equivalent	Millimeter Equivalent	Inch	mm	Inch	mm	Decimal Inch	mm
	25.50	1.0039	25.500	4	102	6	152	0.9843	25.00
	26.00	1.0236	26.000	4	102	6	152	0.9843	25.00
1 1/16		1.0625	26.988	4	102	6	152	1.0000	25.40
	28.00	1.1024	28.000	4	102	6	152	0.9843	25.00
1 1/8		1.1250	28.575	4	102	6	152	1.0000	25.40
	30.00	1.1811	30.000	4 1/4	108	6 5/8	168	0.9843	25.00
1 3/16		1.1875	30.162	4 1/4	108	6 5/8	168	1.0000	25.40
1 1/4		1.2500	31.750	4 3/8	111	6 3/4	171	1.0000	25.40
	32.00	1.2598	32.000	4 3/8	111	7	178	1.2402	31.50
1 5/16		1.3125	33.338	4 3/8	111	7	178	1.2500	31.75
	34.00	1.3386	34.000	4 1/2	114	7 1/8	181	1.2402	31.50
1 3/8		1.3750	34.925	4 1/2	114	7 1/8	181	1.2500	31.75
	36.00	1.4173	36.000	4 3/4	121	7 3/8	187	1.2402	31.50
1 7/16		1.4375	36.512	4 3/4	121	7 3/8	187	1.2500	31.75
	38.00	1.4961	38.000	4 7/8	124	7 1/2	190	1.2402	31.50
1 1/2		1.5000	38.100	4 7/8	124	7 1/2	190	1.2500	31.75
1 9/16		1.5625	39.688	4 7/8	124	7 3/4	197	1.5000	38.10
	40.00	1.5748	40.000	4 7/8	124	7 3/4	197	1.4961	38.00
1 5/8		1.6250	41.275	4 7/8	124	7 3/4	197	1.5000	38.10
	42.00	1.6535	42.000	5 1/8	130	8	203	1.4961	38.00
1 11/16		1.6875	42.862	5 1/8	130	8	203	1.5000	38.10
	44.00	1.7323	44.000	5 1/8	130	8	203	1.4961	38.00
1 3/4		1.7500	44.450	5 1/8	130	8	203	1.5000	38.10
	46.00	1.8110	46.000	5 3/8	137	8 1/4	210	1.4961	38.00
1 13/16		1.8125	46.038	5 3/8	137	8 1/4	210	1.5000	38.10
1 7/8		1.8750	47.625	5 3/8	137	8 1/4	210	1.5000	38.10
	48.00	1.8898	48.000	5 5/8	143	8 1/2	216	1.4961	38.00
1 15/16		1.9375	49.212	5 5/8	143	8 1/2	216	1.5000	38.10
	50.00	1.9685	50.000	5 5/8	143	8 1/2	216	1.4961	38.00
2		2.0000	50.800	5 5/8	143	8 1/2	216	1.5000	38.10

Table 5. American National Standard Taper Shank Twist Drills — Fractional and Metric Sizes (ANSI B94.11M-1979)

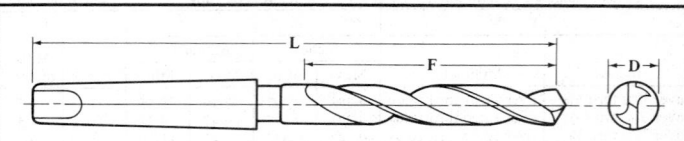

Fraction	mm	Deci. Inch	mm	Morse Taper No.	F Inch	F mm	L Inch	L mm	Morse Taper No.	F Inch	F mm	L Inch	L mm
	3.00	0.1181	3.000	I	1⅞	48	5⅛	130	...	...	...	...	...
⅛		0.1250	3.175	I	1⅞	48	5⅛	130	...	...	...	...	...
	3.20	0.1260	3.200	I	2⅛	54	5⅜	137	...	...	...	...	...
	3.50	0.1378	3.500	I	2⅛	54	5⅜	137	...	...	...	...	...
9/64		0.1406	3.571	I	2⅛	54	5⅜	137	...	...	...	...	...
	3.80	0.1496	3.800	I	2⅛	54	5⅜	137	...	...	...	...	...
5/32		0.1562	3.967	I	2⅛	54	5⅜	137	...	...	...	...	...
	4.00	0.1575	4.000	I	2½	64	5¾	146	...	...	...	...	...
	4.20	0.1654	4.200	I	2½	64	5¾	146	...	...	...	...	...
11/64		0.1719	4.366	I	2½	64	5¾	146	...	...	...	...	...
	4.50	0.1772	4.500	I	2½	64	5¾	146	...	...	...	...	...
3/16		0.1875	4.762	I	2½	64	5¾	146	...	...	...	...	...
	4.80	0.1890	4.800	I	2¾	70	6	152	...	...	...	...	...
	5.00	0.1969	5.000	I	2¾	70	6	152	...	...	...	...	...
13/64		0.2031	5.159	I	2¾	70	6	152	...	...	...	...	...
	5.20	0.2047	5.200	I	2¾	70	6	152	...	...	...	...	...
	5.50	0.2165	5.500	I	2¾	70	6	152	...	...	...	...	...
7/32		0.2183	5.558	I	2¾	70	6	152	...	...	...	...	...
	5.80	0.2223	5.800	I	2⅞	73	6⅛	156	...	...	...	...	...
15/64		0.2344	5.954	I	2⅞	73	6⅛	156	...	...	...	...	...
	6.00	0.2362	6.000	I	2⅞	73	6⅛	156	...	...	...	...	...
	6.20	0.2441	6.200	I	2⅞	73	6⅛	156	...	...	...	...	...
¼		0.2500	6.350	I	2⅞	73	6⅛	156	...	...	...	...	...
	6.50	0.2559	6.500	I	3	76	6¼	159	...	...	...	...	...
17/64		0.2656	6.746	I	3	76	6¼	159	...	...	...	...	...
	6.80	0.2677	6.800	I	3	76	6¼	159	...	...	...	...	...
	7.00	0.2756	7.000	I	3	76	6¼	159	...	...	...	...	...
9/32		0.2812	7.142	I	3	76	6¼	159	...	...	...	...	...
	7.20	0.2835	7.200	I	3⅛	79	6⅜	162	...	...	...	...	...
	7.50	0.2953	7.500	I	3⅛	79	6⅜	162	...	...	...	...	...
19/64		0.2969	7.541	I	3⅛	79	6⅜	162	...	...	...	...	...
	7.80	0.3071	7.800	I	3⅛	79	6⅜	162	...	...	...	...	...
5/16		0.3125	7.938	I	3⅛	79	6⅜	162	...	...	...	...	...
	8.00	0.3150	8.000	I	3¼	83	6½	165	...	...	...	...	...
	8.20	0.3228	8.200	I	3¼	83	6½	165	...	...	...	...	...
21/64		0.3281	8.334	I	3¼	83	6½	165	...	...	...	...	...
	8.50	0.3346	8.500	I	3¼	83	6½	165	...	...	...	...	...
11/32		0.3438	8.733	I	3¼	83	6½	165	...	...	...	...	...
	8.80	0.3465	8.800	I	3½	89	6¾	171	...	...	...	...	...
	9.00	0.3543	9.000	I	3½	89	6¾	171	...	...	...	...	...
23/64		0.3594	9.129	I	3½	89	6¾	171	...	...	...	...	...
	9.20	0.3622	9.200	I	3½	89	6¾	171	...	...	...	...	...
	9.50	0.3740	9.500	I	3½	89	6¾	171	...	...	...	...	...

* Larger or smaller than regular shank. See following pages.

Table 5 *(Continued).* **Taper Shank Twist Drills — Fractional and Metric Sizes**
(ANSI B94.11M-1979)

Drill Diameter, D		Equivalent		Regular Shank					Larger or Smaller Shank*				
		Deci.		Morse Taper No.	Flute Length F		Overall Length L		Morse Taper No.	Flute Length F		Overall Length L	
Fraction	mm	Inch	mm		Inch	mm	Inch	mm		Inch	mm	Inch	mm
⅜		0.3750	9.525	1	3½	89	6¾	171	2	3½	89	7⅜	187
	9.80	0.3858	9.800	1	3⅝	92	7	178					
25/64		0.3906	9.921	1	3⅝	92	7	178	2	3⅝	92	7½	190
	10.00	0.3937	10.000	1	3⅝	92	7	178					
	10.20	0.4016	10.200	1	3⅝	92	7	178					
13/32		0.4062	10.320	1	3⅝	92	7	178	2	3⅝	92	7½	190
	10.50	0.4134	10.500	1	3⅞	98	7¼	184					
27/64		0.4219	10.716	1	3⅞	98	7¼	184	2	3⅞	98	7¾	197
	10.80	0.4252	10.800	1	3⅞	98	7¼	184					
	11.00	0.4331	11.000	1	3⅞	98	7¼	184					
7/16		0.4375	11.112	1	3⅞	98	7¼	184	2	3⅞	98	7¾	197
	11.20	0.4409	11.200	1	4⅛	105	7½	190					
	11.50	0.4528	11.500	1	4⅛	105	7½	190					
29/64		0.4531	11.509	1	4⅛	105	7½	190	2	4⅛	105	8	203
	11.80	0.4646	11.800	1	4⅛	105	7½	190					
15/32		0.4688	11.906	1	4⅛	105	7½	190	2	4⅛	105	8	203
	12.00	0.4724	12.000	2	4⅜	111	8¼	210	1	4⅜	111	7¾	197
	12.20	0.4803	12.200	2	4⅜	111	8¼	210	1	4⅜	111	7¾	197
31/64		0.4844	12.304	2	4⅜	111	8¼	210	1	4⅜	111	7¾	197
	12.50	0.4921	12.500	2	4⅜	111	8¼	210	1	4⅜	111	7¾	197
½		0.5000	12.700	2	4⅜	111	8¼	210	1	4⅜	111	7¾	197
	12.80	0.5034	12.800	2	4⅝	117	8½	216	1	4⅝	117	8	203
	13.00	0.5118	13.000	2	4⅝	117	8½	216	1	4⅝	117	8	203
33/64		0.5156	13.096	2	4⅝	117	8½	216	1	4⅝	117	8	203
	13.20	0.5197	13.200	2	4⅝	117	8½	216	1	4⅝	117	8	203
17/32		0.5312	13.492	2	4⅝	117	8½	216	1	4⅝	117	8	203
	13.50	0.5315	13.500	2	4⅝	117	8½	216	1	4⅝	117	8	203
	13.80	0.5433	13.800	2	4⅞	124	8¾	222	1	4⅞	124	8¼	210
35/64		0.5469	13.891	2	4⅞	124	8¾	222	1	4⅞	124	8¼	210
	14.00	0.5512	14.000	2	4⅞	124	8¾	222	1	4⅞	124	8¼	210
	14.25	0.5610	14.250	2	4⅞	124	8¾	222	1	4⅞	124	8¼	210
9/16		0.5625	14.288	2	4⅞	124	8¾	222	1	4⅞	124	8¼	210
	14.50	0.5709	14.500	2	4⅞	124	8¾	222					
37/64		0.5781	14.684	2	4⅞	124	8¾	222					
	14.75	0.5807	14.750	2	4⅞	124	8¾	222					
	15.00	0.5906	15.000	2	4⅞	124	8¾	222					
19/32		0.5938	15.083	2	4⅞	124	8¾	222					
	15.25	0.6004	15.250	2	4⅞	124	8¾	222					
39/64		0.6094	15.479	2	4⅞	124	8¾	222					
	15.50	0.6102	15.500	2	4⅞	124	8¾	222					
	15.75	0.6201	15.750	2	4⅞	124	8¾	222					
⅝		0.6250	15.875	2	4⅞	124	8¾	222					
	16.00	0.6299	16.000	2	5⅛	130	9	229					
	16.25	0.6398	16.250	2	5⅛	130	9	229					
41/64		0.6406	16.271	2	5⅛	130	9	229	3	5⅛	130	9¾	248
	16.50	0.6496	16.500	2	5⅛	130	9	229					
21/32		0.6562	16.667	2	5⅛	130	9	229	3	5⅛	130	9¾	248
	16.75	0.6594	16.750	2	5⅜	137	9¼	235					
	17.00	0.6693	17.000	2	5⅜	137	9¼	235					
43/64		0.6719	17.066	2	5⅜	137	9¼	235	3	5⅜	137	10	254
	17.25	0.6791	17.250	2	5⅜	137	9¼	235					
11/16		0.6875	17.462	2	5⅜	137	9¼	235	3	5⅜	137	10	254

* Larger or smaller than regular shank.

Table 5 (*Continued*). **Taper Shank Twist Drills — Fractional and Metric Sizes** (ANSI B94.11M-1979)

Drill Diameter, *D*				Regular Shank					Larger or Smaller Shank*				
		Equivalent			Flute Length		Overall Length			Flute Length		Overall Length	
				Morse Taper	*F*		*L*		Morse Taper	*F*		*L*	
Fraction	mm	Deci. Inch	mm	No.	Inch	mm	Inch	mm	No.	Inch	mm	Inch	mm
	17.50	0.6880	17.500	2	5⅝	143	9½	241	...	...	...	...	...
45/64		0.7031	17.859	2	5⅝	143	9½	241	3	5⅝	143	10¼	260
	18.00	0.7087	18.000	2	5⅝	143	9½	241	...	...	...	...	...
23/32		0.7188	18.258	2	5⅝	143	9½	241	3	5⅝	143	10¼	260
	18.50	0.7283	18.500	2	5⅞	149	9¾	248	...	...	...	...	...
47/64		0.7344	18.654	2	5⅞	149	9¾	248	3	5⅞	149	10½	267
	19.00	0.7480	19.000	2	5⅞	149	9¾	248	...	...	...	...	...
¾		0.7500	19.050	2	5⅞	149	9¾	248	3	5⅞	149	10½	267
49/64		0.7656	19.446	2	6	152	9⅞	251	3	6	152	10⅝	270
	19.50	0.7677	19.500	2	6	152	9⅞	251	...	...	...	...	...
25/32		0.7812	19.843	2	6	152	9⅞	251	3	6	152	10⅝	270
	20.00	0.7821	20.000	3	6⅛	156	10¾	273	2	6⅛	156	10	254
51/64		0.7969	20.241	3	6⅛	156	10¾	273	2	6⅛	156	10	254
	20.50	0.8071	20.500	3	6⅛	156	10¾	273	2	6⅛	156	10	254
13/16		0.8125	20.638	3	6⅛	156	10¾	273	2	6⅛	156	10	254
	21.00	0.8268	21.000	3	6⅛	156	10¾	273	2	6⅛	156	10	254
53/64		0.8281	21.034	3	6⅛	156	10¾	273	2	6⅛	156	10	254
27/32		0.8438	21.433	3	6⅛	156	10¾	273	2	6⅛	156	10	254
	21.50	0.8465	21.500	3	6⅛	156	10¾	273	2	6⅛	156	10	254
55/64		0.8594	21.829	3	6⅛	156	10¾	273	2	6⅛	156	10	254
	22.00	0.8661	22.000	3	6⅛	156	10¾	273	2	6⅛	156	10	254
⅞		0.8750	22.225	3	6⅛	156	10¾	273	2	6⅛	156	10	254
	22.50	0.8858	22.500	3	6⅛	156	10¾	273	2	6⅛	156	10	254
57/64		0.8906	22.621	3	6⅛	156	10¾	273	2	6⅛	156	10	254
	23.00	0.9055	23.000	3	6⅛	156	10¾	273	2	6⅛	156	10	254
29/32		0.9062	23.017	3	6⅛	156	10¾	273	2	6⅛	156	10	254
59/64		0.9219	23.416	3	6⅛	156	10¾	273	...	...	...	...	...
	23.50	0.9252	23.500	3	6⅛	156	10¾	273	...	...	...	...	...
15/16		0.9375	23.813	3	6⅛	156	10¾	273	...	...	...	...	...
	24.00	0.9449	24.000	3	6⅜	162	11	279	...	...	...	...	...
61/64		0.9531	24.209	3	6⅜	162	11	279	...	...	...	...	...
	24.50	0.9646	24.500	3	6⅜	162	11	279	...	...	...	...	...
31/32		0.9688	24.608	3	6⅜	162	11	279	...	...	...	...	...
	25.00	0.9843	25.000	3	6⅜	162	11	279	...	...	...	...	...
63/64		0.9844	25.004	3	6⅜	162	11	279	...	...	...	...	...
1		1.0000	25.400	3	6⅜	162	11	279	4	6⅜	162	12	305
	25.50	1.0039	25.500	3	6½	165	11⅛	283	...	...	...	...	...
1 1/64		1.0156	25.796	3	6½	165	11⅛	283	...	...	...	...	...
	26.00	1.0236	26.000	3	6½	165	11⅛	283	...	...	...	...	...
1 1/32		1.0312	26.192	3	6½	165	11⅛	283	4	6½	165	12⅛	308
	26.50	1.0433	26.500	3	6⅝	168	11¼	286	...	...	...	...	...
1 3/64		1.0469	26.591	3	6⅝	168	11¼	286	...	...	...	...	...
1 1/16		1.0625	26.988	3	6⅝	168	11¼	286	4	6⅝	168	12¼	311
	27.00	1.0630	27.000	3	6⅝	168	11¼	286	...	...	...	...	...
1 5/64		1.0781	27.384	4	6⅞	175	12½	318	3	6⅞	175	11½	292
	27.50	1.0827	27.500	4	6⅞	175	12½	318	3	6⅞	175	11½	292
1 3/32		1.0938	27.783	4	6⅞	175	12½	318	3	6⅞	175	11½	292
	28.00	1.1024	28.000	4	7⅛	181	12¾	324	3	7⅛	181	11¾	298
1 7/64		1.1094	28.179	4	7⅛	181	12¾	324	3	7⅛	181	11¾	298
	28.50	1.1220	28.500	4	7⅛	181	12¾	324	3	7⅛	181	11¾	298
1⅛		1.1250	28.575	4	7⅛	181	12¾	324	3	7⅛	181	11¾	298
1 9/64		1.1406	28.971	4	7¼	184	12⅞	327	3	7¼	184	11⅞	302

* Larger or smaller than regular shank.

Table 5 (*Continued*). **Taper Shank Twist Drills — Fractional and Metric Sizes** (ANSI B94.11M-1979)

Drill Diameter, *D*				Regular Shank					Larger or Smaller Shank*				
		Equivalent		Morse Taper No.	Flute Length *F*		Overall Length *L*		Morse Taper No.	Flute Length *F*		Overall Length *L*	
Fraction	mm	Deci. Inch	mm		Inch	mm	Inch	mm		Inch	mm	Inch	mm
	29.00	1.1417	29.000	4	7 1/4	184	12 7/8	327	3	7 1/4	184	11 7/8	302
1 5/32		1.1562	29.367	4	7 1/4	184	12 7/8	327	3	7 1/4	184	11 7/8	302
	29.50	1.1614	29.500	4	7 3/8	187	13	330	3	7 3/8	187	12	305
1 11/64		1.1719	29.797	4	7 3/8	187	13	330	3	7 3/8	187	12	305
	30.00	1.1811	30.000	4	7 3/8	187	13	330	3	7 3/8	187	12	305
1 3/16		1.1875	30.162	4	7 3/8	187	13	330	3	7 3/8	187	12	305
	30.50	1.2008	30.500	4	7 1/2	190	13 1/8	333	3	7 1/2	190	12 1/8	308
1 13/64		1.2031	30.559	4	7 1/2	190	13 1/8	333	3	7 1/2	190	12 1/8	308
1 7/32		1.2188	30.958	4	7 1/2	190	13 1/8	333	3	7 1/2	190	12 1/8	308
	31.00	1.2205	31.000	4	7 7/8	200	13 1/2	343	3	7 7/8	200	12 1/2	318
1 15/64		1.2344	31.354	4	7 7/8	200	13 1/2	343	3	7 7/8	200	12 1/2	318
	31.50	1.2402	31.500	4	7 7/8	200	13 1/2	343	3	7 7/8	200	12 1/2	318
1 1/4		1.2500	31.750	4	7 7/8	200	13 1/2	343	3	7 7/8	200	12 1/2	318
	32.00	1.2598	32.000	4	8 1/2	216	14 1/8	359	...	...	...	...	...
1 17/64		1.2656	32.146	4	8 1/2	216	14 1/8	359	...	...	...	...	...
	32.50	1.2795	32.500	4	8 1/2	216	14 1/8	359	...	...	...	...	...
1 9/32		1.2812	32.542	4	8 1/2	216	14 1/8	359	...	...	...	...	...
1 19/64		1.2969	32.941	4	8 5/8	219	14 1/4	362	...	...	...	...	...
	33.00	1.2992	33.000	4	8 5/8	219	14 1/4	362	...	...	...	...	...
1 5/16		1.3125	33.338	4	8 5/8	219	14 1/4	362	...	...	...	...	...
	33.50	1.3189	33.500	4	8 3/4	222	14 3/8	365	...	...	...	...	...
1 21/64		1.3281	33.734	4	8 3/4	222	14 3/8	365	...	...	...	...	...
	34.00	1.3386	34.000	4	8 3/4	222	14 3/8	365	...	...	...	...	...
1 11/32		1.3438	34.133	4	8 3/4	222	14 3/8	365	...	...	...	...	...
	34.50	1.3583	34.500	4	8 7/8	225	14 1/2	368	...	...	...	...	...
1 23/64		1.3594	34.529	4	8 7/8	225	14 1/2	368	...	...	...	...	...
1 3/8		1.3750	34.925	4	8 7/8	225	14 1/2	368	...	...	...	...	...
	35.00	1.3780	35.000	4	9	229	14 5/8	371	...	...	...	...	...
1 25/64		1.3906	35.321	4	9	229	14 5/8	371	...	...	...	...	...
	35.50	1.3976	35.500	4	9	229	14 5/8	371	...	...	...	...	...
1 13/32		1.4062	35.717	4	9	229	14 5/8	371	...	...	...	...	...
	36.00	1.4173	36.000	4	9 1/8	232	14 3/4	375	...	...	...	...	...
1 27/64		1.4219	36.116	4	9 1/8	232	14 3/4	375	...	...	...	...	...
	36.50	1.4370	36.500	4	9 1/8	232	14 3/4	375	...	...	...	...	...
1 7/16		1.4375	36.512	4	9 1/8	232	14 3/4	375	...	...	...	...	...
1 29/64		1.4531	36.909	4	9 1/4	235	14 7/8	378	...	...	...	...	...
	37.00	1.4567	37.000	4	9 1/4	235	14 7/8	378	...	...	...	...	...
1 15/32		1.4688	37.308	4	9 1/4	235	14 7/8	378	...	...	...	...	...
	37.50	1.4764	37.500	4	9 3/8	238	15	381	...	...	...	...	...
1 31/64		1.4844	37.704	4	9 3/8	238	15	381	...	...	...	...	...
	38.00	1.4961	38.000	4	9 3/8	238	15	381	...	...	...	...	...
1 1/2		1.5000	38.100	4	9 3/8	238	15	381	...	...	...	...	...
1 33/64		1.5156	38.496	...	...	...	...	...	4	9 3/4	238	15	381
1 17/32		1.5312	38.892	5	9 3/8	238	16 3/8	416	4	9 3/8	238	15	381
	39.00	1.5354	39.000	5	9 5/8	244	16 5/8	422	4	9 5/8	244	15 1/4	387
1 35/64		1.5469	39.291	...	...	...	...	...	4	9 5/8	244	15 1/4	387
1 9/16		1.5625	39.688	5	9 5/8	244	16 5/8	422	4	9 5/8	244	15 1/4	387
	40.00	1.5748	40.000	5	9 7/8	251	16 7/8	429	4	9 7/8	251	15 1/2	394
1 37/64		1.5781	40.084	...	...	...	...	...	4	9 7/8	251	15 1/2	394
1 19/32		1.5938	40.483	5	9 7/8	251	16 7/8	429	4	9 7/8	251	15 1/2	394
1 39/64		1.6094	40.879	...	...	...	...	...	4	10	254	15 5/8	397
	41.00	1.6142	41.000	5	10	254	17	432	4	10	254	15 5/8	397
1 5/8		1.6250	41.275	5	10	254	17	432	4	10	254	15 5/8	397

* Larger or smaller than regular shank.

Table 5 (Continued). Taper Shank Twist Drills — Fractional and Metric Sizes
(ANSI B94.11M-1979)

Drill Diameter, D				Regular Shank					Larger or Smaller Shank*				
		Equivalent			Flute Length		Overall Length			Flute Length		Overall Length	
				Morse Taper	F		L		Morse Taper	F		L	
Fraction	mm	Deci. Inch	mm	No.	Inch	mm	Inch	mm	No.	Inch	mm	Inch	mm
1 41/64		1.6406	41.671	...	...	...	...	...	4	10⅛	257	15¾	400
	42.00	1.6535	42.000	5	10⅛	257	17⅛	435	4	10⅛	257	15¾	400
1 21/32		1.6562	42.067	5	10⅛	257	17⅛	435	4	10⅛	257	15¾	400
1 43/64		1.6719	42.466	...	...	...	...	...	4	10⅛	257	15¾	400
1 11/16		1.6875	42.862	5	10⅛	257	17⅛	435	4	10⅛	257	15¾	400
	43.00	1.6929	43.000	5	10⅛	257	17⅛	435	4	10⅛	257	15¾	400
1 45/64		1.7031	43.259	...	...	...	...	...	4	10⅛	257	15¾	400
1 23/32		1.7188	43.658	5	10⅛	257	17⅛	435	4	10⅛	257	15¾	400
	44.00	1.7323	44.000	5	10⅛	257	17⅛	435	4	10⅜	264	16¼	413
1 47/64		1.7344	44.054	...	...	...	...	...	4	10⅜	264	16¼	413
1 3/4		1.7500	44.450	5	10⅛	257	17⅛	435	4	10¾	264	16¼	413
	45.00	1.7717	45.000	5	10⅛	257	17⅛	435	4	10⅜	264	16¼	413
1 25/32		1.7812	45.242	5	10⅛	257	17⅛	435	4	10⅜	264	16¼	413
	46.00	1.8110	46.000	5	10⅛	257	17⅛	435	4	10⅜	264	16¼	413
1 13/16		1.8125	46.038	5	10⅛	257	17⅛	435	4	10⅜	264	16¼	413
1 27/32		1.8438	46.833	5	10⅛	257	17⅛	435	4	10⅜	264	16¼	413
	47.00	1.8504	47.000	5	10⅜	264	17⅜	441	4	10½	267	16½	419
1 7/8		1.8750	47.625	5	10⅜	264	17⅜	441	4	10½	267	16½	419
	48.00	1.8898	48.000	5	10⅜	264	17⅜	441	4	10½	267	16½	419
1 29/32		1.9062	48.417	5	10⅜	264	17⅜	441	4	10½	267	16½	419
	49.00	1.9291	49.000	5	10⅜	264	17⅜	441	4	10⅝	270	16⅝	422
1 15/16		1.9375	49.212	5	10⅜	264	17⅜	441	4	10⅝	270	16⅝	422
	50.00	1.9625	50.000	5	10⅜	264	17⅜	441	4	10⅝	270	16⅝	422
1 31/32		1.9688	50.008	5	10⅜	264	17⅜	441	4	10⅝	270	16⅝	422
2		2.0000	50.800	5	10⅜	264	17⅜	441	4	10⅝	270	16⅝	422
	51.00	2.0079	51.000	5	10⅜	264	17⅜	441	...	...	...	...	...
2 1/32		2.0312	51.592	5	10⅜	264	17⅜	441	...	...	...	...	...
	52.00	2.0472	52.000	5	10¼	260	17⅜	441	...	...	...	...	...
2 1/16		2.0625	52.388	5	10¼	260	17⅜	441	...	...	...	...	...
	53.00	2.0866	53.000	5	10¼	260	17⅜	441	...	...	...	...	...
2 3/32		2.0938	53.183	5	10¼	260	17⅜	441	...	...	...	...	...
2 1/8		2.1250	53.975	5	10¼	260	17⅜	441	...	...	...	...	...
	54.00	2.1260	54.000	5	10¼	260	17⅜	441	...	...	...	...	...
2 5/32		2.1562	54.767	5	10¼	260	17⅜	441	...	...	...	...	...
	55.00	2.1654	55.000	5	10¼	260	17⅜	441	...	...	...	...	...
2 3/16		2.1875	55.563	5	10¼	260	17¾	441	...	...	...	...	...
	56.00	2.2000	56.000	5	10⅛	257	17⅜	441	...	...	...	...	...
2 7/32		2.2188	56.358	5	10⅛	257	17⅜	441	...	...	...	...	...
	57.00	2.2441	57.000	5	10⅛	257	17⅜	441	...	...	...	...	...
2 1/4		2.2500	57.150	5	10⅛	257	17⅜	441	...	...	...	...	...
	58.00	2.2835	58.000	5	10⅛	257	17⅜	441	...	...	...	...	...
2 5/16		2.3125	58.738	5	10⅛	257	17⅜	441	...	...	...	...	...
	59.00	2.3228	59.000	5	10⅛	257	17⅜	441	...	...	...	...	...
	60.00	2.3622	60.000	5	10⅛	257	17⅜	441	...	...	...	...	...
2 3/8		2.3750	60.325	5	10⅛	257	17⅜	441	...	...	...	...	...
	61.00	2.4016	61.000	5	11¼	286	18¾	476	...	...	...	...	...
2 7/16		2.4375	61.912	5	11¼	286	18¾	476	...	...	...	...	...
	62.00	2.4409	62.000	5	11¼	286	18¾	476	...	...	...	...	...
	63.00	2.4803	63.000	5	11¼	286	18¾	476	...	...	...	...	...
2 1/2		2.5000	63.500	5	11¼	286	18¾	476	...	...	...	...	...
	64.00	2.5197	64.000	5	11⅞	302	19½	495	...	...	...	...	...
	65.00	2.5591	65.000	5	11⅞	302	19½	495	...	...	...	...	...

* Larger or smaller than regular shank.

Table 5 (*Concluded*). **Taper Shank Twist Drills — Fractional and Metric Sizes** (ANSI B94.11M-1979)

Fraction	mm	Deci. Inch	mm	Morse Taper No.	Inch	mm	Inch	mm	Morse Taper No.	Inch	mm	Inch	mm
2⁹⁄₁₆		2.5625	65.088	5	11⁷⁄₈	302	19½	495	...	...	...	...	...
	66.00	2.5984	66.000	5	11⁷⁄₈	302	19½	495	...	...	...	...	...
2⅝		2.6250	66.675	5	11⁷⁄₈	302	19½	495	...	...	...	...	...
	67.00	2.6378	67.000	5	12¾	324	20⅜	518	...	...	...	...	...
	68.00	2.6772	68.000	5	12¾	324	20⅜	518	...	...	...	...	...
2¹¹⁄₁₆		2.6875	68.262	5	12¾	324	20⅜	518	...	...	...	...	...
	69.00	2.7165	69.000	5	12¾	324	20⅜	518	...	...	...	...	...
2¾		2.7500	69.850	5	12¾	324	20⅜	518	...	...	...	...	...
	70.00	2.7559	70.000	5	13⅜	340	21⅛	537	...	...	...	...	...
	71.00	2.7953	71.000	5	13⅜	340	21⅛	537	...	...	...	...	...
2¹³⁄₁₆		2.8125	71.438	5	13⅜	340	21⅛	537	...	...	...	...	...
	72.00	2.8346	72.000	5	13⅜	340	21⅛	537	...	...	...	...	...
	73.00	2.8740	73.000	5	13⅜	340	21⅛	537	...	...	...	...	...
2⅞		2.8750	73.025	5	13⅜	340	21⅛	537	...	...	...	...	...
	74.00	2.9134	74.000	5	14	356	21¾	552	...	...	...	...	...
2¹⁵⁄₁₆		2.9375	74.612	5	14	356	21¾	552	...	...	...	...	...
	75.00	2.9528	75.000	5	14	356	21¾	552	...	...	...	...	...
	76.00	2.9921	76.000	5	14	356	21¾	552	...	...	...	...	...
3		3.0000	76.200	5	14	356	21¾	552	...	...	...	...	...
	77.00	3.0315	77.000	6	14⅝	371	24½	622	5	14¼	362	22	559
	78.00	3.0709	78.000	6	14⅝	371	24½	622	5	14¼	362	22	559
3⅛		3.1250	79.375	6	14⅝	371	24½	622	5	14¼	362	22	559
3¼		3.2500	82.550	6	15½	394	25½	648	5	15¼	387	23	584
3½		3.5000	88.900	...	...	...	...	...	5	16¼	413	24	610

* Larger or smaller than regular shank.

Table 6. **American National Standard Combined Drills and Countersinks — Plain and Bell Types** (ANSI B94.11M-1979)

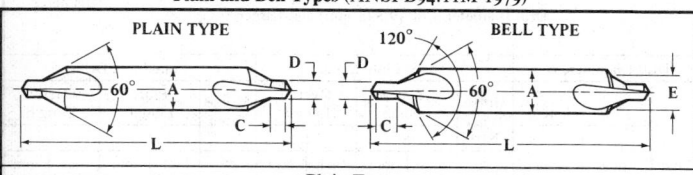

Size Designation	Body Diameter A Inches	Body Diameter A Millimeters	Drill Diameter D Inches	Drill Diameter D Millimeters	Drill Length C Inches	Drill Length C Millimeters	Overall Length L Inches	Overall Length L Millimeters
00	⅛	3.18	.025	0.64	.030	0.76	1⅛	29
0	⅛	3.18	1⁄32	0.79	.038	0.97	1⅛	29
1	⅛	3.18	3⁄64	1.19	3⁄64	1.19	1¼	32
2	3⁄16	4.76	5⁄64	1.98	5⁄64	1.98	1⅞	48
3	¼	6.35	7⁄64	2.78	7⁄64	2.78	2	51
4	5⁄16	7.94	⅛	3.18	⅛	3.18	2⅛	54
5	7⁄16	11.11	3⁄16	4.76	3⁄16	4.76	2¾	70
6	½	12.70	7⁄32	5.56	7⁄32	5.56	3	76
7	⅝	15.88	¼	6.35	¼	6.35	3¼	83
8	¾	19.05	5⁄16	7.94	5⁄16	7.94	3½	89

Table 6 (*Concluded*). **Combined Drills and Countersinks —
Plain and Bell Types** (ANSI B94.11M-1979)

	Bell Type									
Size Designation	Body Diameter *A*		Drill Diameter *D*		Bell Diameter *E*		Drill Length *C*		Overall Length *L*	
	Inches	mm	Inches	mm	Inches	mm	Inches	mm	Inches	mm
11	1/8	3.18	3/64	1.19	0.10	2.5	3/64	1.19	1 1/4	32
12	3/16	4.76	1/16	1.59	0.15	3.8	1/16	1.59	1 7/8	48
13	1/4	6.35	3/32	2.38	0.20	5.1	3/32	2.38	2	51
14	5/16	7.94	7/64	2.78	0.25	6.4	7/64	2.78	2 1/8	54
15	7/16	11.11	5/32	3.97	0.35	8.9	5/32	3.97	2 3/4	70
16	1/2	12.70	3/16	4.76	0.40	10.2	3/16	4.76	3	76
17	5/8	15.88	7/32	5.56	0.50	12.7	7/32	5.56	3 1/4	83
18	3/4	19.05	1/4	6.35	0.60	15.2	1/4	6.35	3 1/2	89

**Table 7. American National Standard Three- and Four-Flute Taper Shank Core
Drills — Fractional Sizes Only** (ANSI B94.11M-1979)

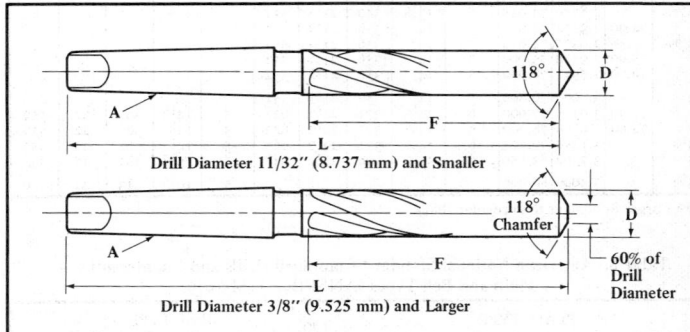

Drill Diameter, *D*			Three-Flute Drills					Four-Flute Drills				
	Equivalent		Morse Taper No.	Flute Length *F*		Overall Length *L*		Morse Taper No.	Flute Length *F*		Overall Length *L*	
Inch	Deci. Inch	mm	A	Inch	mm	Inch	mm	A	Inch	mm	Inch	mm
1/4	0.2500	6.350	1	2 7/8	73	6 1/8	156	...	...	...	...	...
9/32	0.2812	7.142	1	3	76	6 1/4	159	...	...	...	...	...
5/16	0.3175	7.938	1	3 1/8	79	6 3/8	162	...	...	...	...	...
11/32	0.3438	8.733	1	3 1/4	83	6 1/2	165	...	...	...	...	...
3/8	0.3750	9.525	1	3 1/2	89	6 3/4	171	...	...	...	...	...
13/32	0.4062	10.319	1	3 5/8	92	7	178	...	...	...	...	...
7/16	0.4375	11.112	1	3 7/8	98	7 1/4	184	...	...	...	...	...
15/32	0.4688	11.908	1	4 1/8	105	7 1/2	190	...	...	...	...	...
1/2	0.5000	12.700	2	4 3/8	111	8 1/4	210	2	4 3/8	111	8 1/4	210
17/32	0.5312	13.492	2	4 5/8	117	8 1/2	216	2	4 5/8	117	8 1/2	216
9/16	0.5625	14.288	2	4 7/8	124	8 3/4	222	2	4 7/8	124	8 3/4	222
19/32	0.5938	15.083	2	4 7/8	124	8 3/4	222	2	4 7/8	124	8 3/4	222
5/8	0.6250	15.815	2	4 7/8	124	8 3/4	222	2	4 7/8	124	8 3/4	222
21/32	0.6562	16.668	2	5 1/8	130	9	229	2	5 1/8	130	9	229
11/16	0.6875	17.462	2	5 3/8	137	9 1/4	235	2	5 3/8	137	9 1/4	235

Table 7 (*Concluded*). **Three- and Four-Flute Taper Shank Core Drills —**
Fractional Sizes Only (ANSI B94.11M-1979)

Drill Diameter, D			Three-Flute Drills					Four-Flute Drills				
	Equivalent		Morse Taper No.	Flute Length		Overall Length		Morse Taper No.	Flute Length		Overall Length	
	Deci.			F		L			F		L	
Inch	Inch	mm	A	Inch	mm	Inch	mm	A	Inch	mm	Inch	mm
23/32	0.7188	18.258	2	5⅝	143	9½	241	2	5⅝	143	9½	241
3/4	0.7500	19.050	2	5⅞	149	9¾	248	2	5⅞	149	9¾	248
25/32	0.7812	19.842	2	6	152	9⅞	251	2	6	152	9⅞	251
13/16	0.8125	20.638	3	6⅛	156	10¾	273	3	6⅛	156	10¾	273
27/32	0.8438	21.433	3	6⅛	156	10¾	273	3	6⅛	156	10¾	273
7/8	0.8750	22.225	3	6⅛	156	10¾	273	3	6⅛	156	10¾	273
29/32	0.9062	23.019	3	6⅛	156	10¾	273	3	6⅛	156	10¾	273
15/16	0.9375	23.812	3	6⅛	156	10¾	273	3	6⅛	156	10¾	273
31/32	0.9688	24.608	3	6⅜	162	11	279	3	6⅜	162	11	279
1	1.0000	25.400	3	6⅜	162	11	279	3	6⅜	162	11	279
1 1/32	1.0312	26.192	3	6½	165	11⅛	283	3	6½	165	11⅛	283
1 1/16	1.0625	26.988	3	6⅝	168	11¼	286	3	6⅝	168	11¼	286
1 3/32	1.0938	27.783	4	6⅞	175	12½	318	4	6⅞	175	12½	318
1 1/8	1.1250	28.575	4	7⅛	181	12¾	324	4	7⅛	181	12¾	324
1 5/32	1.1562	29.367	4	7¼	184	12⅞	327	4	7¼	184	12⅞	327
1 3/16	1.1875	30.162	4	7⅜	187	13	330	4	7⅜	187	13	330
1 7/32	1.2188	30.958	4	7½	190	13⅛	333	4	7½	190	13⅛	333
1 1/4	1.2500	31.750	4	7⅞	200	13½	343	4	7⅞	200	13½	343
1 9/32	1.2812	32.542	...	...	...	...	...	4	8½	216	14⅛	359
1 5/16	1.3125	33.338	...	...	...	...	...	4	8⅝	219	14¼	362
1 11/32	1.3438	34.133	...	...	...	...	...	4	8¾	222	14⅜	365
1 3/8	1.3750	34.925	...	...	...	...	...	4	8⅞	225	14½	368
1 13/32	1.4062	35.717	...	...	...	...	...	4	9	229	14⅝	371
1 7/16	1.4375	36.512	...	...	...	...	...	4	9⅛	232	14¾	375
1 15/32	1.4688	37.306	...	...	...	...	...	4	9¼	235	14⅞	378
1 1/2	1.5000	38.100	...	...	...	...	...	4	9⅜	238	15	381
1 17/32	1.5312	38.892	...	...	...	...	...	5	9⅜	238	16⅜	416
1 9/16	1.5675	39.688	...	...	...	...	...	5	9⅝	244	16⅝	422
1 19/32	1.5938	40.483	...	...	...	...	...	5	9⅞	251	16⅞	429
1 5/8	1.6250	41.275	...	...	...	...	...	5	10	254	17	432
1 21/32	1.6562	42.067	...	...	...	...	...	5	10⅛	257	17⅛	435
1 11/16	1.6875	42.862	...	...	...	...	...	5	10⅛	257	17⅛	435
1 23/32	1.7188	43.658	...	...	...	...	...	5	10⅛	257	17⅛	435
1 3/4	1.7500	44.450	...	...	...	...	...	5	10⅛	257	17⅛	435
1 25/32	1.7812	45.244	...	...	...	...	...	5	10⅛	257	17⅛	435
1 13/16	1.8125	46.038	...	...	...	...	...	5	10⅛	257	17⅛	435
1 27/32	1.8438	46.833	...	...	...	...	...	5	10⅛	257	17⅛	435
1 7/8	1.8750	47.625	...	...	...	...	...	5	10⅜	264	17⅜	441
1 29/32	1.9062	48.417	...	...	...	...	...	5	10⅜	264	17⅜	441
1 15/16	1.9375	49.212	...	...	...	...	...	5	10⅜	264	17⅜	441
1 31/32	1.9688	50.008	...	...	...	...	...	5	10⅜	264	17⅜	441
2	2.0000	50.800	...	...	...	...	...	5	10⅜	264	17⅜	441
2 1/8	2.1250	53.975	...	...	...	...	...	5	10¼	260	17⅜	441
2 1/4	2.2500	57.150	...	...	...	...	...	5	10⅛	257	17⅜	441
2 3/8	2.3750	60.325	...	...	...	...	...	5	10⅛	257	17⅜	441
2 1/2	2.5000	63.500	...	...	...	...	...	5	11¼	286	18¾	476

British Standard Combined Drills and Countersinks (Center Drills). — BS 328: Part 2: 1972 provides dimensions of combined drills and countersinks for center holes. Three types of drill and countersink combinations are shown in this standard but are not given here. These three types will produce center holes without protecting chamfers, with protecting chamfers, and with protecting chamfers of radius form.

Table 8. American National Standard Three- and Four-Flute Straight Shank Core Drills — Fractional Sizes Only (ANSI B94.11M-1979)

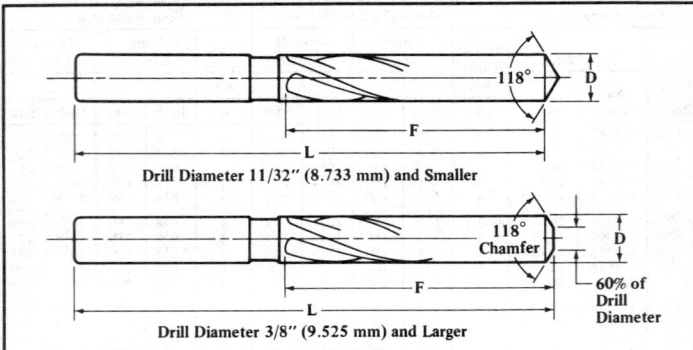

Drill Diameter 11/32″ (8.733 mm) and Smaller

Drill Diameter 3/8″ (9.525 mm) and Larger

Nominal Shank Size is same as Nominal Drill Size

Drill Diameter, D			Three-Flute Drills				Four-Flute Drills			
	Equivalent		Flute Length		Overall Length		Flute Length		Overall Length	
			F		L		F		L	
Inch	Deci. Inch	mm	Inch	mm	Inch	mm	Inch	mm	Inch	mm
¼	0.2500	6.350	3¾	95	6⅛	156	...	...	...	...
9/32	0.2812	7.142	3⅞	98	6¼	159	...	...	...	...
5/16	0.3125	7.938	4	102	6⅜	162	...	...	...	...
11/32	0.3438	8.733	4⅛	105	6½	165	...	...	...	...
⅜	0.3750	9.525	4⅛	105	6¾	171	...	...	...	...
13/32	0.4062	10.317	4⅜	111	7	178	...	...	...	...
7/16	0.4375	11.112	4⅝	117	7¼	184	...	...	...	...
15/32	0.4688	11.908	4¾	121	7½	190	...	...	...	...
½	0.5000	12.700	4¾	121	7¾	197	4¾	121	7¾	197
17/32	0.5312	13.492	4¾	121	8	203	4¾	121	8	203
9/16	0.5625	14.288	4⅞	124	8¼	210	4⅞	124	8¼	210
19/32	0.5938	15.083	4⅞	124	8¾	222	4⅞	124	8¾	222
⅝	0.6250	15.875	4⅞	124	8¾	222	4⅞	124	8¾	222
21/32	0.6562	16.667	5⅛	130	9	229	5⅛	130	9	229
11/16	0.6875	17.462	5⅜	137	9¼	235	5⅜	137	9¼	235
23/32	0.7188	18.258	...	...	...	...	5⅝	143	9½	241
¾	0.7500	19.050	5⅞	149	9¾	248	5⅞	149	9¾	248
25/32	0.7812	19.842	...	...	...	...	6	152	9⅞	251
13/16	0.8125	20.638	...	...	...	...	6⅛	156	10	254
27/32	0.8438	21.433	...	...	...	...	6⅛	156	10	254
⅞	0.8750	22.225	...	...	...	...	6⅛	156	10	254
29/32	0.9062	23.017	...	...	...	...	6⅛	156	10	254
15/16	0.9375	23.812	...	...	...	...	6⅛	156	10¾	273
31/32	0.9688	24.608	...	...	...	...	6⅜	162	11	279
1	1.0000	25.400	...	...	...	...	6⅜	162	11	279
1 1/32	1.0312	26.192	...	...	...	...	6½	165	11⅛	283
1 1/16	1.0625	26.988	...	...	...	...	6⅝	168	11¼	286
1 3/32	1.0938	27.783	...	...	...	...	6⅞	175	11½	292
1⅛	1.1250	28.575	...	...	...	...	7⅛	181	11¾	298
1¼	1.2500	31.750	...	...	...	...	7⅞	200	12½	318

American National Standard Drill Drivers — Split-Sleeve, Collet Type
(ANSI B94.35-1972)

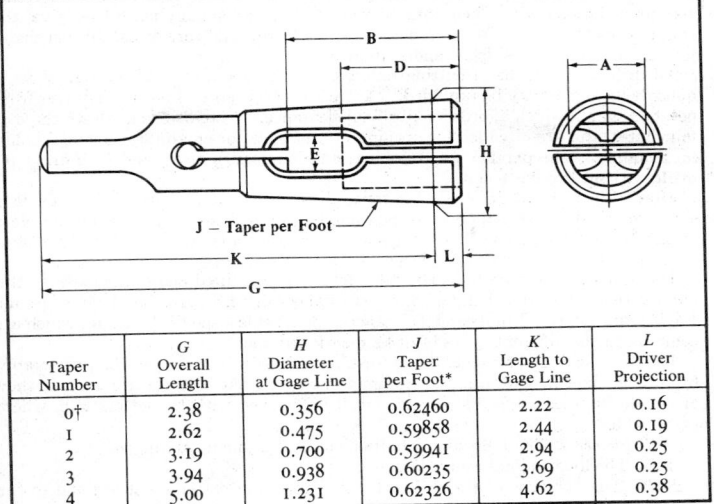

Taper Number	G Overall Length	H Diameter at Gage Line	J Taper per Foot*	K Length to Gage Line	L Driver Projection
0†	2.38	0.356	0.62460	2.22	0.16
1	2.62	0.475	0.59858	2.44	0.19
2	3.19	0.700	0.59941	2.94	0.25
3	3.94	0.938	0.60235	3.69	0.25
4	5.00	1.231	0.62326	4.62	0.38

All dimensions are in inches.
* Taper rate in accordance with ANSI B5.10-1981, Machine Tapers.
† Size 0 is not an American National Standard but is included here to meet special needs.

Drill Drivers — Split-Sleeve, Collet Type. — American National Standard ANSI B94.35-1972 covers split-sleeve, collet type drivers for driving straight shank drills, reamers, and similar tools, without tangs from 0.0390-inch through 0.1220-inch diameter, and with tangs from 0.1250-inch through 0.7500-inch diameter, including metric sizes.

For sizes 0.0390 through 0.0595 inch, the standard taper number is 1 and the optional taper number is 0. For sizes 0.0610 through 0.1875 inch, the standard taper number is 1, first optional taper number is 0, and second optional taper number is 2. For sizes 0.1890 through 0.2520 inch, the standard taper number is 1, first optional taper number is 2, and second optional taper number is 0. For sizes 0.2570 through 0.3750 inch, the standard taper number is 1 and the optional taper number is 2. For sizes 0.3860 through 0.5625 inch, the standard taper number is 2 and the optional taper number is 3. For sizes 0.5781 through 0.7500 inch, the standard taper number is 3 and the optional taper number is 4.

The depth B that the drill enters the driver is 0.44 inch for sizes 0.0390 through 0.0781 inch; 0.50 inch for sizes 0.0785 through 0.0938 inch; 0.56 inch for sizes 0.0960 through 0.1094 inch; 0.62 inch for sizes 0.1100 through 0.1220 inch; 0.75 inch for sizes 0.1250 through 0.1875 inch; 0.88 inch for sizes 0.1890 through 0.2500 inch; 1.00 inch for sizes 0.2520 through 0.3125 inch; 1.12 inches for sizes 0.3160 through 0.3750 inch; 1.25 inches for sizes 0.3860 through 0.4688 inch; 1.31 inches for sizes 0.4844 through 0.5625 inch; 1.47 inches for sizes 0.5781 through 0.6562 inch; and 1.62 inches for sizes 0.6719 through 0.7500 inch.

British Standard Metric Twist Drills. — BS 328:Part 1:1959 (incorporating amendments issued March 1960 and March 1964) covers twist drills made to inch and metric dimensions which are intended for general engineering purposes. ISO recommendations are taken into account. The accompanying tables give the standard metric sizes of Morse taper shank twist drills and core drills, parallel shank jobbing and long series drills, and stub drills.

All drills are right-hand cutting unless otherwise specified, and normal, slow or quick helix angles may be provided. A 'back taper' is ground on the diameter from point to shank, which affords longitudinal clearance. Core drills may have three or four flutes, and are intended for opening up cast holes or enlarging machined holes, for example. The parallel shank jobber, and long series drills, and stub drills are made without driving tenons.

Morse taper shank drills with oversize dimensions are also listed, and Table 7 shows metric drill sizes superseding gage and letter size drills, which are now obsolete in the United Kingdom. To meet special requirements, the Standard lists non-standard sizes for the various types of drills.

The limits of tolerance on cutting diameters, as measured across the lands at the outer corners of a drill, shall be h8, in accordance with BS 1916, Limits and Fits for Engineering (Part 1, Limits and Tolerances), and Table 3 shows the values which are common to the different types of drills mentioned above.

The drills shall be permanently and legibly marked whenever possible, preferably by rolling, showing the size, and the manufacturer's name or trademark. If they are made from high speed steel, they shall be marked with the letters H.S. where practicable.

Drill Elements: The following definitions of drill elements are given.

Axis: The longitudinal centre-line.

Body: That portion of the drill extending from the extreme cutting end to the commencement of the shank.

Shank: That portion of the drill by which it is held and driven.

Flutes: The grooves in the body of the drill which provide lips and permit the removal of chips and allow cutting fluid to reach the lips.

Web (Core): The central portion of the drill situated between the roots of the flutes and extending from the point end towards the shank; the point end of the web or core forms the chisel edge.

Lands: The cylindrical-ground surfaces on the leading edges of the drill flutes. The width of the land is measured at right angles to the flute helix.

Body Clearance: That portion of the body surface reduced in diameter to provide diametral clearance.

Heel: The edge formed by the intersection of the flute surface and the body clearance.

Point: The sharpened end of the drill, consisting of all that part of the drill which is shaped to produce lips, faces, flanks and chisel edge.

Face: That portion of the flute surface adjacent to the lip on which the chip impinges as it is cut from the work.

Flank: That surface on a drill point which extends behind the lip to the following flute.

Lip: (cutting edge): The edge formed by the intersection of the flank and face.

Relative Lip Height: The relative position of the lips measured at the outer corners in a direction parallel to the drill axis.

Outer Corner: The corner formed by the intersection of the lip and the leading edge of the land.

Chisel Edge: The edge formed by the intersection of the flanks.

Chisel Edge Corner: The corner formed by the intersection of a lip and the chisel edge.

Table 1. British Standard Morse Taper Shank Twist Drills and Core Drills — Standard Metric Sizes* (BS 328: Part 1: 1959)

Diameter	Flute Length	Overall Length	Diameter	Flute Length	Overall Length	Diameter	Flute Length	Overall Length
3.00	33	114	16.25	125	223	29.50	175	296
3.20	36	117	16.50			29.75		
3.50	39	120	16.75			30.00		
3.80	43	123	17.00			30.25	180	301
4.00			17.25	130	228	30.50		
4.20			17.50			30.75		
4.50	47	128	17.75			31.00		
4.80	52	133	18.00			31.25		
5.00			18.25	135	233	31.50		
5.20			18.50			31.75	185	306
5.50	57	138	18.75			32.00	185	334
5.80			19.00			32.50		
6.00			19.25	140	238	33.00		
6.20	63	144	19.50			33.50		
6.50			19.75			34.00	190	339
6.80	69	150	20.00			34.50		
7.00			20.25	145	243	35.00		
7.20			20.50			35.50		
7.50			20.75			36.00	195	344
7.80	75	156	21.00			36.50		
8.00			21.25	150	248	37.00		
8.20			21.50			37.50		
8.50			21.75			38.00	200	349
8.80	81	162	22.00			38.50		
9.00			22.25			39.00		
9.20			22.50	155	253	39.50		
9.50			22.75			40.00		
9.80	87	168	23.00			40.50	205	354
10.00			23.25	155	276	41.00		
10.20			23.50			41.50		
10.50			23.75	160	281	42.00		
10.80	94	175	24.00			42.50		
11.00			24.25			43.00	210	359
11.20			24.50			43.50		
11.50			24.75			44.00		
11.80			25.00			44.50		
12.00	101	182	25.25	165	286	45.00		
12.20			25.50			45.50	215	364
12.50			25.75			46.00		
12.80			26.00			46.50		
13.00			26.25			47.00		
13.20			26.50			47.50		
13.50	108	189	26.75	170	291	48.00	220	369
13.80			27.00			48.50		
14.00			27.25			49.00		
14.25	114	212	27.50			49.50		
14.50			27.75			50.00		
14.75			28.00			50.50	225	374
15.00			28.25	175	296	51.00	225	412
15.25	120	218	28.50			52.00		
15.50			28.75			53.00		
15.75			29.00			54.00	230	417
16.00			29.25			55.00		

For footnotes see end of table.

Table 1. (Continued). British Standard Morse Taper Shank Twist Drills and Core Drills — Standard Metric Sizes* (BS 328: Part 1: 1959)

Diameter	Flute Length	Overall Length	Diameter	Flute Length	Overall Length	Diameter	Flute Length	Overall Length
56.00	230	417	71.00	250	437	86.00		
						87.00		
57.00			72.00			88.00	270	524
58.00	235	422	73.00	255	442	89.00		
59.00			74.00			90.00		
60.00			75.00					
61.00			76.00	260	477	91.00		
62.00	240	427				92.00		
63.00			77.00			93.00	275	529
			78.00			94.00		
64.00			79.00	260	514	95.00		
65.00	245	432	80.00					
66.00								
67.00			81.00			96.00		
			82.00			97.00		
68.00			83.00	265	519	98.00	280	534
69.00	250	437	84.00			99.00		
70.00			85.00			100.00		

All dimensions are in millimeters. Tolerances on diameters are given in table below.
* The Morse taper shanks of these twist and core drills are as follows: 3.00 to 14.00 mm diameter, M.T. No. 1; 14.25 to 23.00 mm diameter, M.T. No. 2; 23.25 to 31.50 mm diameter, M.T. No. 3; 31.75 to 50.50 mm diameter, M.T. No. 4; 51.00 to 76.00 mm diameter, M.T. No. 5; 77.00 to 100.00 mm diameter, M.T. No. 6.
Table 2, below, shows twist drills that may be supplied with the shank and length oversize, but they should be regarded as non-preferred.

Table 2. British Standard Morse Taper Shank Twist Drills — Metric Oversize Shank and Length Series† (BS 328: Part 1: 1959)

Diam. Range	Overall Length	M. T. No.	Diam. Range	Overall Length	M. T. No.	Diam. Range	Overall Length	M. T. No.
12.00 to 13.20	199	2	22.50 to 23.00	276	3	45.50 to 47.50	402	5
13.50 to 14.00	206	2	26.75 to 28.00	319	4	48.00 to 50.00	407	5
18.25 to 19.00	256	3	29.00 to 30.00	324	4	50.50	412	5
19.25 to 20.00	251	3	30.25 to 31.50	329	4	64.00 to 67.00	499	6
20.25 to 21.00	266	3	40.50 to 42.50	392	5	68.00 to 71.00	504	6
21.25 to 22.25	271	3	43.00 to 45.00	397	5	72.00 to 75.00	509	6

Diameters and lengths are given in millimeters. For the individual sizes within the diameter ranges given, see Table 1.
† This series of drills should be regarded as non-preferred.

Table 3. British Standard Limits of Tolerance on Diameter for Twist Drills and Core Drills — Metric Series (BS 328: Part 1: 1959)

Drill Size (Diameter measured across lands at outer corners)	Tolerance (h8)
0 to 1 inclusive	Plus 0.000 to Minus 0.014
Over 1 to 3 inclusive	Plus 0.000 to Minus 0.014
Over 3 to 6 inclusive	Plus 0.000 to Minus 0.018
Over 6 to 10 inclusive	Plus 0.000 to Minus 0.022
Over 10 to 18 inclusive	Plus 0.000 to Minus 0.027
Over 18 to 30 inclusive	Plus 0.000 to Minus 0.033
Over 30 to 50 inclusive	Plus 0.000 to Minus 0.039
Over 50 to 80 inclusive	Plus 0.000 to Minus 0.046
Over 80 to 120 inclusive	Plus 0.000 to Minus 0.054

All dimensions are given in millimeters.

**Table 4. British Standard Parallel Shank Jobber Series Twist Drills —
Standard Metric Sizes (BS 328: Part 1: 1959)**

Diameter	Flute Length	Overall Length
0.20	2.5	19
0.22		
0.25	3.0	19
0.28		
0.30	4.0	19
0.32	4	19
0.35		
0.38		
0.40	5	20
0.42		
0.45		
0.48		
0.50	6	22
0.52		
0.55	7	24
0.58		
0.60		
0.62	8	26
0.65		
0.68	9	28
0.70		
0.72		
0.75		
0.78	10	30
0.80		
0.82		
0.85		
0.88	11	32
0.90		
0.92		
0.95		
0.98	12	34
1.00		
1.05		
1.10	14	36
1.15		
1.20	16	38
1.25		
1.30		
1.35	18	40
1.40		
1.45		
1.50		
1.55	20	43
1.60		
1.65		
1.70		

Diameter	Flute Length	Overall Length
1.75	22	46
1.80		
1.85		
1.90		
1.95	24	49
2.00		
2.05		
2.10		
2.15	27	53
2.20		
2.25		
2.30		
2.35		
2.40	30	57
2.45		
2.50		
2.55		
2.60		
2.65	33	61
2.70		
2.75		
2.80		
2.85		
2.90		
2.95		
3.00		
3.10	36	65
3.20		
3.30		
3.40	39	70
3.50	39	70
3.60		
3.70	43	75
3.80		
3.90		
4.00		
4.10		
4.20		
4.30	47	80
4.40		
4.50		
4.60		
4.70		
4.80	52	86
4.90		
5.00		
5.10		
5.20		
5.30		

Diameter	Flute Length	Overall Length
5.40	57	93
5.50		
5.60		
5.70		
5.80		
5.90		
6.00		
6.10	63	101
6.20		
6.30		
6.40		
6.50		
6.60		
6.70		
6.80	69	109
6.90		
7.00		
7.10		
7.20		
7.30		
7.40		
7.50		
7.60	75	117
7.70		
7.80		
7.90		
8.00		
8.10		
8.20		
8.30		
8.40		
8.50		
8.60	81	125
8.70		
8.80		
8.90		
9.00		
9.10		
9.20		
9.30		
9.40		
9.50		
9.60	87	133
9.70		
9.80		
9.90		
10.00		
10.10		

Diameter	Flute Length	Overall Length
10.20	87	133
10.30		
10.40		
10.50		
10.60		
10.70	94	142
10.80		
10.90		
11.00		
11.10		
11.20		
11.30		
11.40		
11.50		
11.60		
11.70		
11.80		
11.90	101	151
12.00		
12.10		
12.20		
12.30		
12.40		
12.50		
12.60		
12.70		
12.80		
12.90		
13.00		
13.10		
13.20		
13.30	108	160
13.40		
13.50		
13.60		
13.70		
13.80		
13.90		
14.00		
14.25	114	169
14.50		
14.75		
15.00		
15.25	120	178
15.50		
15.75		
16.00		

All dimensions are in millimeters. Tolerances on diameters are given in Table 3.

TWIST DRILLS

Table 5. British Standard Parallel Shank Long Series Twist Drills — Standard Metric Sizes (BS 328: Part 1: 1959)

Diameter	Flute Length	Overall Length	Diameter	Flute Length	Overall Length	Diameter	Flute Length	Overall Length
2.00	56	85	6.80	102	156	12.70	134	205
2.05			6.90			12.80		
2.10			7.00			12.90		
2.15	59	90	7.10			13.00		
2.20			7.20			13.10		
2.25			7.30			13.20		
2.30			7.40			13.30	140	214
2.35			7.50			13.40		
2.40	62	95	7.60	109	165	13.50		
2.45			7.70			13.60		
2.50			7.80			13.70		
2.55			7.90			13.80		
2.60			8.00			13.90		
2.65			8.10			14.00		
2.70	66	100	8.20			14.25	144	220
2.75			8.30			14.50		
2.80			8.40			14.75		
2.85			8.50			15.00		
2.90			8.60	115	175	15.25	149	227
2.95			8.70			15.50		
3.00			8.80			15.75		
3.10	69	106	8.90			16.00		
3.20			9.00			16.25	154	235
3.30			9.10			16.50		
3.40	73	112	9.20			16.75		
3.50			9.30			17.00		
3.60			9.40			17.25	158	241
3.70			9.50			17.50		
3.80	78	119	9.60	121	184	17.75		
3.90			9.70			18.00		
4.00			9.80			18.25	162	247
4.10			9.90			18.50		
4.20			10.00			18.75		
4.30	82	126	10.10			19.00		
4.40			10.20			19.25	166	254
4.50			10.30			19.50		
4.60			10.40			19.75		
4.70			10.50			20.00		
4.80	87	132	10.60			20.25	171	261
4.90			10.70	128	195	20.50		
5.00			10.80			20.75		
5.10			10.90			21.00		
5.20			11.00			21.25	176	268
5.30			11.10			21.50		
5.40	91	139	11.20			21.75		
5.50			11.30			22.00		
5.60			11.40			22.25		
5.70			11.50			22.50	180	275
5.80			11.60			22.75		
5.90			11.70			23.00		
6.00			11.80			23.25		
6.10	97	148	11.90	134	205	23.50		
6.20			12.00			23.75	185	282
6.30			12.10			24.00		
6.40			12.20			24.25		
6.50			12.30			24.50		
6.60			12.40			24.75		
6.70			12.50			25.00		
			12.60					

All dimensions are in millimeters. Tolerances on diameters are given in Table 3.

Table 6. British Standard Stub Drills — Metric Sizes (BS 328: Part 1: 1959)

Diameter	Flute Length	Overall Length	Diameter	Flute Length	Overall Length	Diameter	Flute Length	Overall Length	Diameter	Flute Length	Overall Length
0.50	3	20	5.00	26	62	9.50	40	84	14.00	54	107
0.80	5	24	5.20	26	62	9.80	43	89	14.50	56	111
1.00	6	26	5.50	28	66	10.00	43	89	15.00	56	111
1.20	8	30	5.80	28	66	10.20	43	89	15.50	58	115
1.50	9	32	6.00	28	66	10.50	43	89	16.00	58	115
1.80	11	36	6.20	31	70	10.80	47	95	16.50	60	119
2.00	12	38	6.50	31	70	11.00	47	95	17.00	60	119
2.20	13	40	6.80	34	74	11.20	47	95	17.50	62	123
2.50	14	43	7.00	34	74	11.50	47	95	18.00	62	123
2.80	16	46	7.20	34	74	11.80	47	95	18.50	64	127
3.00	16	46	7.50	34	74	12.00	51	102	19.00	64	127
3.20	18	49	7.80	37	79	12.20	51	102	19.50	66	131
3.50	20	52	8.00	37	79	12.50	51	102	20.00	66	131
3.80	22	55	8.20	37	79	12.80	51	102	21.00	68	136
4.00	22	55	8.50	37	79	13.00	51	102	22.00	70	141
4.20	22	55	8.80	40	84	13.20	51	102	23.00	72	146
4.50	24	58	9.00	40	84	13.50	54	107	24.00	75	151
4.80	26	62	9.20	40	84	13.80	54	107	25.00	75	151

All dimensions are given in millimeters. Tolerances on diameters are given in Table 3.

Table 7. British Standard Drills — Metric Sizes Superseding Gauge and Letter Sizes* (BS 328: Part 1: 1959 Appendix B)

Obsolete Drill Size	Recommended Metric Size (mm)	Obsolete Drill Size	Recommended Metric Size (mm)	Obsolete Drill Size	Recommended Metric Size (mm)	Obsolete Drill Size	Recommended Metric Size (mm)	Obsolete Drill Size	Recommended Metric Size (mm)
80	0.35	58	1.05	36	2.70	14	4.60	I	6.90
79	0.38	57	1.10	35	2.80	13	4.70	J	7.00
78	0.40	56	3/64 in.	34	2.80	12	4.80	K	9/32 in.
77	0.45	55	1.30	33	2.85	11	4.90	L	7.40
76	0.50	54	1.40	32	2.95	10	4.90	M	7.50
75	0.52	53	1.50	31	3.00	9	5.00	N	7.70
74	0.58	52	1.60	30	3.30	8	5.10	O	8.00
73	0.60	51	1.70	29	3.50	7	5.10		
72	0.65	50	1.80	28	9/64 in.	6	5.20	P	8.20
71	0.65	49	1.85	27	3.70			Q	8.40
70	0.70	48	1.95	26	3.70	5	5.20	R	8.60
69	0.75	47	2.00	25	3.80	4	5.30	S	8.80
68	1/32 in.	46	2.05	24	3.90	3	5.40	T	9.10
67	0.82	45	2.10	23	3.90	2	5.60		
66	0.85	44	2.20	22	4.00	1	5.80	U	9.30
65	0.90	43	2.25	21	4.00	A	15/64 in.	V	3/8 in.
64	0.92	42	3/32 in.	20	4.10	B	6.00	W	9.80
63	0.95	41	2.45	19	4.20	C	6.10	X	10.10
62	0.98	40	2.50	18	4.30	D	6.20	Y	10.30
61	1.00	39	2.55	17	4.40	E	1/4 in.	Z	10.50
60	1.00	38	2.60	16	4.50	F	6.50	...	...
59	1.05	37	2.65	15	4.60	G	6.60	...	...
						H	17/64 in.	...	...

* Gauge and letter size drills are now obsolete in the United Kingdom and should not be used in the production of new designs. The table is given to assist users in changing over to the recommended standard sizes.

Steels for Twist Drills. — *Carbon Steel:* If the conditions are such that carbon steel drill speeds are sufficient for the purpose, high-speed-steel drills are not economical to use because the difference in performance as compared with the carbon steel tool does not compensate for the difference in price.

High-Speed Steel: For high surface speed drilling operations where carbon steels would fail by tempering, the properties of red hardness and abrasion resistance favor the use of high-speed steel.

Cobalt High-Speed Steel: These high-speed-steel drills are capable of withstanding cutting speeds beyond the range of conventional high-speed-steel drills and have superior resistance to abrasion but are not to be compared with tungsten-carbide tipped tools.

Accuracy of Drilled Holes. — Normally the diameter of drilled holes is not given a tolerance; the size of the hole is expected to be as close to the drill size as can be obtained. The accuracy of holes drilled with a two-fluted twist drill is influenced by many factors, which include: the accuracy of the drill point; the size of the drill; length and shape of the chisel edge; whether or not a bushing is used to guide the drill; the work material; length of the drill; runout of the spindle and the chuck; rigidity of the machine tool, workpiece, and the setup; also the cutting fluid used, if any. Usually when drilling most materials the diameter of the drilled holes will be oversize. The table below provides the results of tests reported by The Metal Cutting Tool Institute in which the diameter of over 2800 holes drilled in steel and cast iron were measured. The values in this table indicate what might be expected under average shop conditions; however, when the drill point is accurately ground and the other machining conditions are correct, the resulting hole size is more likely to be between the mean and average minimum values given in this table. If the drill is ground and used incorrectly, holes that are even larger than the average maximum values can result.

Oversize Diameters in Drilling

Drill Diam., Inch	Amount Oversize, Inch			Drill Diam., Inch	Amount Oversize, Inch		
	Average Max.	Mean	Average Min.		Average Max	Mean	Average Min.
1/16	.002	.0015	.001	1/2	.008	.005	.003
1/8	.0045	.003	.001	3/4	.008	.005	.003
1/4	.0065	.004	.0025	1	.009	.007	.004

Courtesy of The Metal Cutting Institute

There are some conditions which will cause the drilled hole to be undersize. For example, holes drilled in light metals and in other materials having a high coefficient of thermal expansion may contract to a size that is smaller than the diameter of the drill as the metal surrounding the hole is cooled after having been heated by the drilling. The elastic action of the metal surrounding the hole may also cause the drilled hole to be undersize when drilling high strength materials with a drill that is dull at its outer corner.

The accuracy of the drill point has a great effect on the accuracy of the drilled hole. An inaccurately ground twist drill will produce holes that are excessively oversize. The drill point must be symmetrical; i.e., the point angles must be equal, as well as the lip lengths and the axial height of the lips. Any alteration to the lips or to the chisel edge, such as thinning the web, must be done with care to preserve the symmetry of the drill point. An adequate relief should be provided behind the chisel edge to prevent heel drag. On conventionally ground drill points this can be estimated by the chisel edge angle.

When drilling a hole, as the drill point starts to enter the workpiece, the drill will be unstable and will tend to wander. Then as the body of the drill enters the hole the drill will tend to stabilize. The result of this action is a tendency to drill a bellmouth shape in the hole at the entrance and perhaps beyond. Factors contributing to bellmouthing are: an unsymmetrically ground drill point; a large chisel edge length; inadequate relief behind the chisel edge; runout of the spindle and the chuck; using a slender drill that will bend easily; and lack of rigidity of the machine tool, workpiece, or the setup. Correcting these conditions as required will reduce the tendency for bellmouthing to occur and improve the accuracy of the hole diameter and its straightness. Starting the hole with a short stiff drill, such as a center drill, will quickly stabilize the drill that follows and reduce or eliminate bellmouthing; this procedure should always be used when drilling in a lathe, where the work is rotating. Bellmouthing can also be eliminated almost entirely and the accuracy of the hole improved by using a close fitting drill jig bushing placed close to the workpiece. Although specific recommendations cannot be made, many cutting fluids will help to increase the accuracy of the diameters of drilled holes. Double margin twist drills, available in the smaller sizes, will drill a more accurate hole than conventional twist drills having only a single margin at the leading edge of the land. The second land, located on the trailing edge of each land, provides greater stability in the drill bushing and in the hole. These drills are especially useful in drilling intersecting off-center holes. Single and double margin step drills, also available in the smaller sizes, will produce very accurate drilled holes, which are usually less than .002 inch larger than the drill size.

Counterbores. — Counterbores for screw holes are generally made in sets. Each set contains three counterbores: one with the body of the size of the screw head and the pilot the size of the hole to admit the body of the screw; one with the body the size of the head of the screw and the pilot the size of the tap drill; and

Counterbores With Interchangeable Cutters and Guides

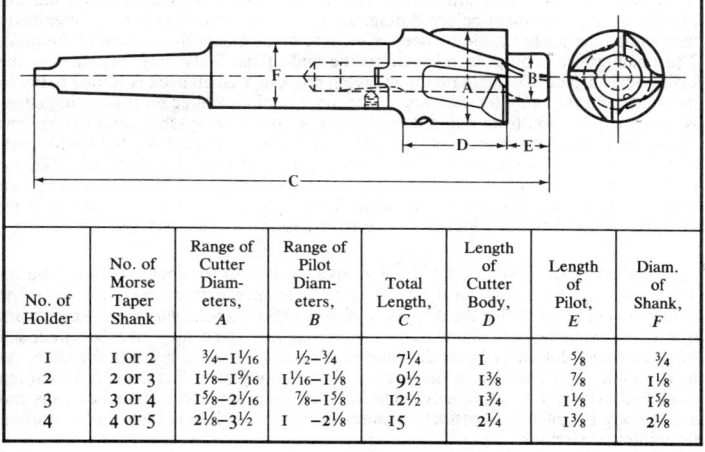

No. of Holder	No. of Morse Taper Shank	Range of Cutter Diameters, A	Range of Pilot Diameters, B	Total Length, C	Length of Cutter Body, D	Length of Pilot, E	Diam. of Shank, F
1	1 or 2	¾–1¹⁄₁₆	½–¾	7¼	1	⅝	¾
2	2 or 3	1⅛–1⁹⁄₁₆	1¹⁄₁₆–1⅛	9½	1⅜	⅞	1⅛
3	3 or 4	1⅝–2¹⁄₁₆	⅞–1⅝	12½	1¾	1⅛	1⅝
4	4 or 5	2⅛–3½	1 –2⅛	15	2¼	1⅜	2⅛

Dimensions of Solid Counterbores

A	B	C	D	E	F	G	A	B	C	D	E	F	G
1/4	11/32	3/16	2 1/2	1/4	7/32	4 7/8	1 5/16	1 3/4	63/64	3 9/16	1 5/16	1 9/32	12 5/16
5/16	13/32	15/64	2 9/16	5/16	9/32	5 5/16	1 3/8	1 27/32	1 1/32	3 5/8	1 3/8	1 11/32	12 3/4
3/8	1/2	9/32	2 5/8	3/8	11/32	5 3/4	1 7/16	1 29/32	1 5/64	3 11/16	1 7/16	1 13/32	13 3/16
7/16	19/32	21/64	2 11/16	7/16	13/32	6 3/16	1 1/2	2	1 1/8	3 3/4	1 1/2	1 7/16	13 3/8
1/2	21/32	3/8	2 3/4	1/2	15/32	6 5/8	1 5/8	2 3/32	1 7/32	3 15/16	1 17/32	1 15/32	13 3/4
9/16	3/4	27/64	2 13/16	9/16	17/32	7 1/16	1 3/4	2 3/16	1 5/16	4 1/8	1 19/32	1 17/32	14 1/8
5/8	27/32	15/32	2 7/8	5/8	19/32	7 1/2	1 7/8	2 9/32	1 13/32	4 5/16	1 5/8	1 9/16	14 1/2
11/16	29/32	33/64	2 15/16	11/16	21/32	7 15/16	2	2 3/8	1 1/2	4 1/2	1 21/32	1 19/32	14 7/8
3/4	1	9/16	3	3/4	23/32	8 3/8	2 1/8	2 15/32	1 19/32	4 11/16	1 23/32	1 21/32	15 1/4
13/16	1 3/32	39/64	3 1/16	13/16	25/32	8 13/16	2 1/4	2 9/16	1 11/16	4 7/8	1 3/4	1 11/16	15 5/8
7/8	1 5/32	21/32	3 1/8	7/8	27/32	9 1/4	2 3/8	2 21/32	1 25/32	5 1/16	1 25/32	1 23/32	16
15/16	1 1/4	45/64	3 3/16	15/16	29/32	9 11/16	2 1/2	2 3/4	1 7/8	5 1/4	1 27/32	1 25/32	16 3/8
1	1 11/32	3/4	3 1/4	1	31/32	10 1/8	2 5/8	2 27/32	1 31/32	5 7/16	1 7/8	1 13/16	16 3/4
1 1/16	1 13/32	51/64	3 5/16	1 1/16	1 1/32	10 9/16	2 3/4	2 15/16	2 1/16	5 5/8	1 29/32	1 27/32	17 1/8
1 1/8	1 1/2	27/32	3 3/8	1 1/8	1 3/32	11	2 7/8	3 1/32	2 5/32	5 13/16	1 31/32	1 29/32	17 1/2
1 3/16	1 19/32	57/64	3 7/16	1 3/16	1 5/32	11 7/16	3	3 1/8	2 1/4	6	2	1 15/16	17 7/8
1 1/4	1 21/32	15/16	3 1/2	1 1/4	1 7/32	11 7/8	...	...	...	...	...	...	...

the third with the body the size of the body of the screw and the pilot the size of the tap drill. Counterbores are usually provided with four flutes cut on a right-hand spiral. The angle of the spiral is 15 degrees with the center line of the counterbore, which corresponds to a lead of the flute equal to about twelve times the diameter of the body of the counterbore. Counterbores for brass are fluted straight.

Small counterbores are often made with three flutes, but should then have the size plainly stamped on them before fluting, as they cannot afterwards be conveniently measured. The flutes should be deep enough to come below the surface of the pilot. The counterbore should be relieved on the end of the body only, and not on the cylindrical surface. To facilitate the relieving process, a small neck is turned between the guide and the body for clearance. The amount of clearance on the cutting edges is, for general work, from 4 to 5 degrees. The accompanying table gives dimensions for straight shank counterbores. The same dimensions, except for the shank part, may be used for Morse taper shank counterbores. The number of shank used for counterbores with bodies of different diameters is usually as follows: Up to 1/2 inch diameter body, No. 1 Morse taper shank; from 9/16 to 7/8 inch, No. 2; from 15/16 to 1 3/8 inch, No. 3; from 1 7/16 to 2 inches, No. 4; from 2 1/16 to 3 inches, No. 5 shank.

Lathe Arbors. — Arbors are usually tapered about 0.006 inch per foot. The diameter or nominal size D in the table is at a distance F from the small end. The diameter G of the drills for the centers conforms to Stub's steel wire gage. The "width of flat," listed in the last column, is for the driving dog. The centers of arbors intended for very heavy duty may be made somewhat larger than those given in the table. As to hardening, the practice at the present time, among manufacturers, is to harden arbors all over, but for extremely accurate work, an arbor having hardened ends and a soft body is generally considered superior, as there is less tendency of distortion from internal stresses.

Proportions of Solid Lathe Arbors

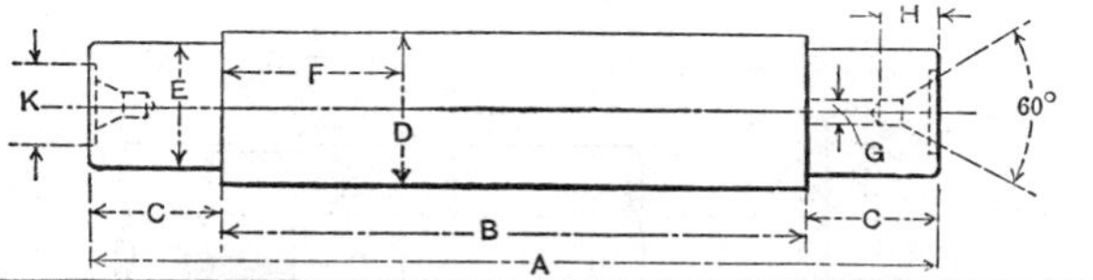

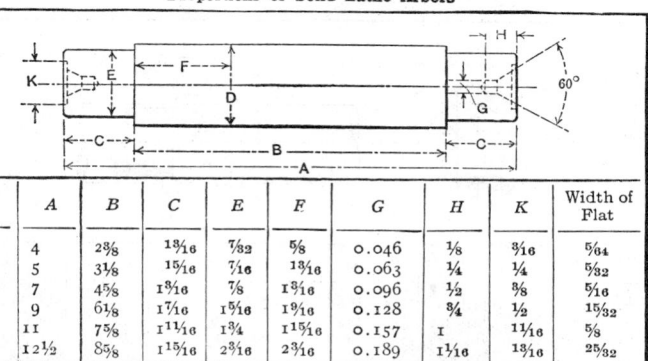

D	*A*	*B*	*C*	*E*	*F*	*G*	*H*	*K*	Width of Flat
¼	4	2⅜	13/16	7/32	⅝	0.046	⅛	3/16	5/64
½	5	3⅛	15/16	7/16	13/16	0.063	¼	¼	5/32
1	7	4⅝	1 3/16	⅞	1 3/16	0.096	½	⅜	5/16
1½	9	6⅛	1 7/16	1 5/16	1 9/16	0.128	¾	½	15/32
2	11	7⅝	1 11/16	1¾	1 15/16	0.157	1	1 1/16	⅝
2½	12½	8⅝	1 15/16	2 3/16	2 3/16	0.189	1 1/16	1 3/16	25/32

Boring-bar Couplings

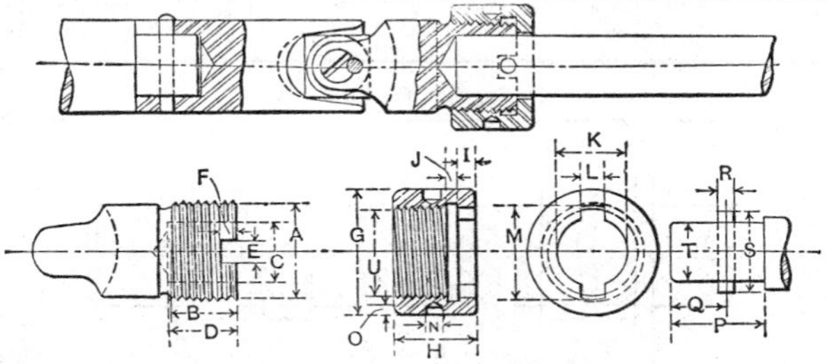

A	*B*	*C*	*D*	*E*	*F*	*G*	*H*	*I*	*J*	*K*	*L*	*M*
¾	½	½	7/16	3/16	5/32	1	⅝	⅛	3/32	17/32	3/16	11/16
1	½	⅝	9/16	¼	7/32	1⅜	⅝	5/32	3/32	21/32	¼	15/16
1¼	⅝	¾	⅝	5/16	9/32	1⅝	13/16	3/16	⅛	25/32	5/16	1 5/32
1½	13/16	1	⅞	7/16	13/32	2	1 1/16	¼	⅛	1 1/32	7/16	1 13/32
2½	⅞	1½	1⅜	9/16	17/32	3	1⅛	¼	⅛	1 17/32	9/16	2⅜
3	⅞	1⅞	1¾	11/16	21/32	3½	1¼	¼	⅛	1 29/32	11/16	2⅞

N	*O*	*P*	*Q*	*R*	*S*	*T*	*U*, Bore	Threads per Inch	Used for Bars
9/64	⅛	¾	⅜	⅛	21/32	0.499	0.6610	16	⅜ to ½
11/64	5/32	⅞	15/32	3/16	⅞	0.624	0.9110	16	½ to ¾
13/64	5/32	1⅛	½	¼	1 1/16	0.749	1.1615	12	¾ to 1
15/64	7/32	1⅜	11/16	⅜	1 5/16	0.999	1.4115	12	1 to 1½
19/64	7/32	1⅞	1⅛	½	2¼	1.4985	2.4115	12	1½ to 2
21/64	7/32	2 5/16	1 7/16	⅝	2¾	1.8735	2.9115	12	2 and over

Boring-bar Cutters — 1

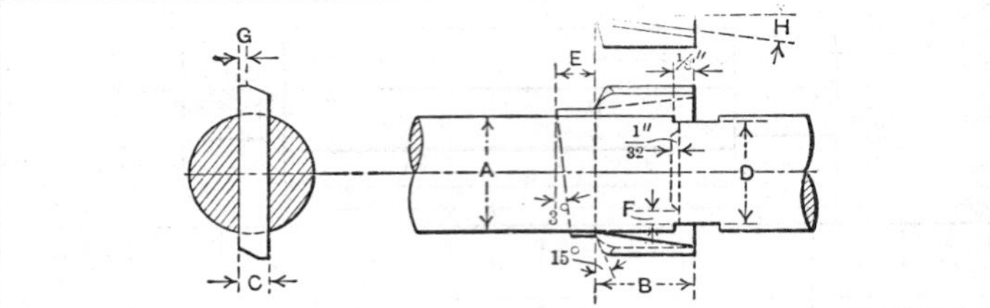

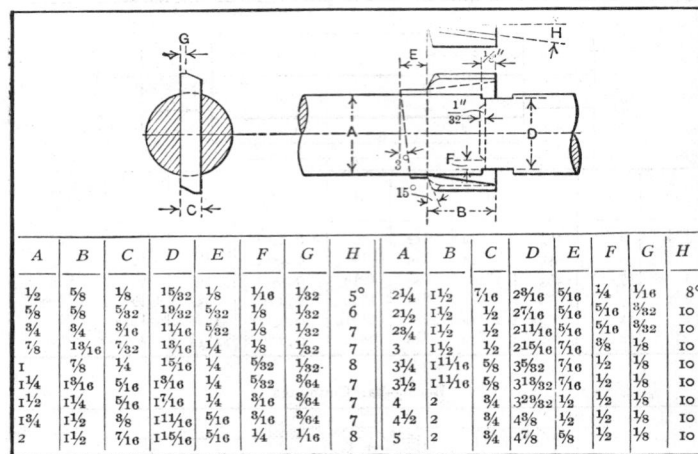

A	B	C	D	E	F	G	H	A	B	C	D	E	F	G	H
1/2	5/8	1/8	15/32	1/8	1/16	1/32	5°	2 1/4	1 1/2	7/16	2 3/16	5/16	1/4	1/16	8°
5/8	5/8	5/32	19/32	5/32	1/8	1/32	6	2 1/2	1 1/2	1/2	2 7/16	5/16	5/16	3/32	10
3/4	3/4	3/16	11/16	5/32	1/8	1/32	7	2 3/4	1 1/2	1/2	2 11/16	5/16	5/16	3/32	10
7/8	13/16	7/32	13/16	1/4	1/8	1/32	7	3	1 1/2	1/2	2 15/16	7/16	3/8	1/8	10
1	7/8	1/4	15/16	1/4	5/32	1/32	8	3 1/4	1 11/16	5/8	3 5/32	7/16	1/2	1/8	10
1 1/4	1 3/16	5/16	1 3/16	1/4	5/32	3/64	7	3 1/2	1 11/16	5/8	3 13/32	7/16	1/2	1/8	10
1 1/2	1 1/4	5/16	1 7/16	1/4	3/16	3/64	7	4	2	3/4	3 29/32	1/2	1/2	1/8	10
1 3/4	1 1/2	3/8	1 11/16	5/16	3/16	3/64	7	4 1/2	2	3/4	4 3/8	1/2	1/2	1/8	10
2	1 1/2	7/16	1 15/16	5/16	1/4	1/16	8	5	2	3/4	4 7/8	5/8	1/2	1/8	10

Boring-bar Cutters — 2

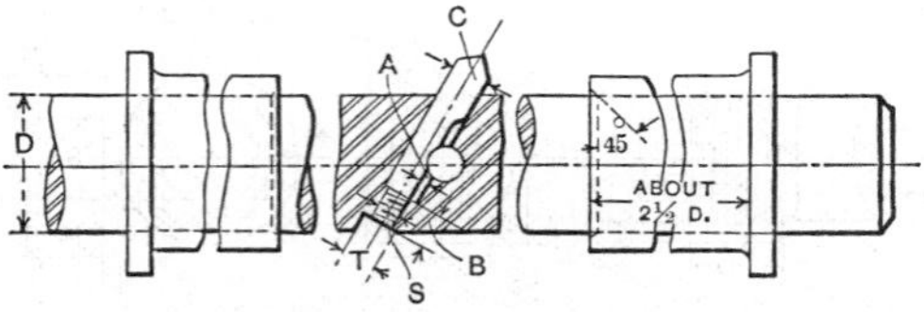

Diameter of Bar, D	Diameter of Cutter, C	Diameter of Pin	Depth of Flat, A	Diameter of Screw, S	Diameter T of Counter-bore	Length B of Thread	Diameter of Bar, D	Diameter of Cutter, C	Diameter of Pin	Depth of Flat, A	Diameter of Screw, S	Diameter T of Counterbore	Length B of Thread
3/8	1/8	1/16	1/64	1/16	1/8	5/64	1 1/2	3/8	9/32	3/64	5/16	3/8	3/8
7/16	1/8	3/32	1/64	1/16	1/8	5/64	1 5/8	7/16	5/16	3/64	5/16	3/8	3/8
1/2	1/8	3/32	1/64	1/16	1/8	5/64	1 3/4	7/16	5/16	3/64	3/8	7/16	1/2
9/16	1/8	1/8	1/32	1/8	5/32	1/8	1 7/8	1/2	11/32	1/16	3/8	7/16	1/2
5/8	3/16	1/8	1/32	1/8	5/32	1/8	2	1/2	3/8	1/16	7/16	1/2	9/16
11/16	3/16	5/32	1/32	1/8	3/16	5/32	2 1/8	9/16	13/32	1/16	7/16	1/2	9/16
3/4	3/16	5/32	1/32	1/8	3/16	5/32	2 1/4	9/16	13/32	1/16	1/2	9/16	5/8
13/16	1/4	5/32	1/32	3/16	1/4	3/16	2 3/8	5/8	7/16	5/64	1/2	9/16	5/8
7/8	1/4	5/32	1/32	3/16	1/4	3/16	2 1/2	5/8	15/32	5/64	9/16	5/8	11/16
15/16	1/4	3/16	1/32	3/16	1/4	1/4	2 5/8	11/16	1/2	5/64	9/16	5/8	11/16
1	1/4	3/16	1/32	3/16	1/4	1/4	2 3/4	11/16	1/2	5/64	5/8	11/16	3/4
1 1/8	5/16	7/32	1/32	1/4	5/16	5/16	2 7/8	3/4	17/32	3/32	5/8	3/4	7/8
1 1/4	5/16	1/4	3/64	1/4	5/16	5/16	3	3/4	9/16	3/32	5/8	3/4	7/8
1 3/8	3/8	1/4	3/64	1/4	5/16	5/16	...	...	...	...	...	...	...

Sintered Carbide Boring Tools. — Industrial experience has shown that the shapes of tools used for boring operations need to be different from those of single-point tools ordinarily used for general applications such as lathe work. Accordingly, Section 5 of American National Standard ANSI B94.5-1974 gives standard sizes, styles and designations for four basic types of sintered carbide boring tools, namely: solid carbide square; carbide-tipped square; solid carbide round; and carbide-tipped round

**Table 1. American National Standard Sintered Carbide Boring Tools —
Style Designations (ANSI B94.5-1974)**

Side Cutting Edge Angle E		Boring Tool Styles			
Degrees	Designation	Solid Square (SS)	Tipped Square (TS)	Solid Round (SR)	Tipped Round (TR)
0	A		TSA		
10	B		TSB		
30	C	SSC	TSC	SRC	TRC
40	D		TSD		
45	E	SSE	TSE	SRE	TRE
55	F		TSF		
90 (0° Rake)	G				TRG
90 (10° Rake)	H				TRH

**Table 2. American National Standard Solid Carbide Square Boring Tools — Style
SSC for 60° Boring Bar and Style SSE for 45° Boring Bar (ANSI B94.5-1974)**

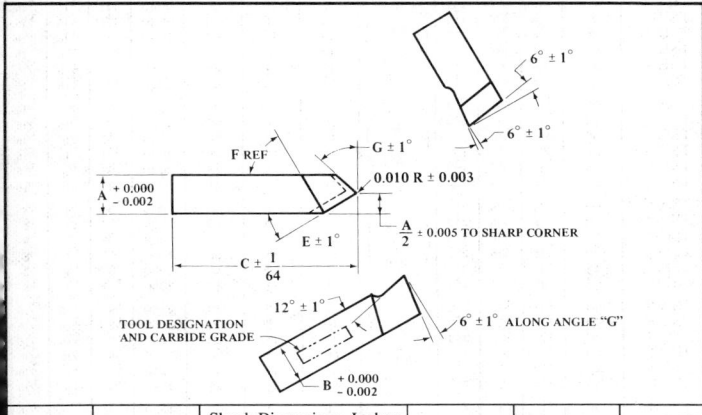

Tool Desig-nation	Boring Bar Angle, Deg. From Axis	Shank Dimensions, Inches			Side Cutting Edge Angle E, Deg.	End Cutting Edge Angle G, Deg.	Shoulder Angle F, Deg.
		Width A	Height B	Length C			
SSC-58	60	5/32	5/32	1	30	38	60
SSE-58	45				45	53	45
SSC-610	60	3/16	3/16	1¼	30	38	60
SSE-610	45				45	53	45
SSC-810	60	¼	¼	1¼	30	38	60
SSE-810	45				45	53	45
SSC-1012	60	5/16	5/16	1½	30	38	60
SSE-1012	45				45	53	45

Table 3. American National Standard Carbide-Tipped Square Boring Tools — Styles TSA and TSB for 90° Boring Bar, Styles TSC and TSD for 60° Boring Bar, and Styles TSE and TSF for 45° Boring Bar (ANSI B94.5-1974)

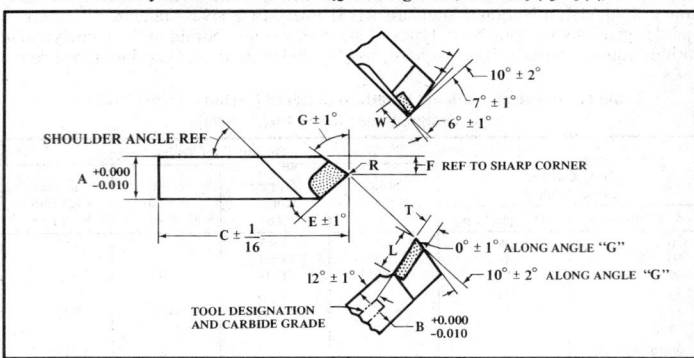

Tool Desig-nation	Bor. Bar Angle From Axis, Deg.	Shank Dimensions, Inches				Side Cut. Edge Angle E, Deg.	End Cut. Edge Angle G, Deg.	Shoulder Angle F, Deg.	Tip No.	Tip Dimensions, Inches		
		A	B	C	R					T	W	L
TSA-5	90	5/16	5/16	1 1/2		0	8	90	2040	3/32	3/16	5/16
TSB-5	90	5/16	5/16	1 1/2		10	8	90	2040	3/32	3/16	5/16
TSC-5	60	5/16	5/16	1 1/2	1/64	30	38	60	2040	3/32	3/16	5/16
TSD-5	60	5/16	5/16	1 1/2	±	40	38	60	2040	3/32	3/16	5/16
TSE-5	45	5/16	5/16	1 1/2	.005	45	53	45	2040	3/32	3/16	5/16
TSF-5	45	5/16	5/16	1 1/2		55	53	45	2040	3/32	3/16	5/16
TSA-6	90	3/8	3/8	1 3/4		0	8	90	2040	3/32	3/16	5/16
TSB-6	90	3/8	3/8	1 3/4		10	8	90	2040	3/32	3/16	5/16
TSC-6	60	3/8	3/8	1 3/4	1/64	30	38	60	2040	3/32	3/16	5/16
TSD-6	60	3/8	3/8	1 3/4	±	40	38	60	2040	3/32	3/16	5/16
TSE-6	45	3/8	3/8	1 3/4	.005	45	53	45	2040	3/32	3/16	5/16
TSF-6	45	3/8	3/8	1 3/4		55	53	45	2040	3/32	3/16	5/16
TSA-7	90	7/16	7/16	2 1/2		0	8	90	2060	3/32	1/4	3/8
TSB-7	90	7/16	7/16	2 1/2		10	8	90	2060	3/32	1/4	3/8
TSC-7	60	7/16	7/16	2 1/2	1/32	30	38	60	2060	3/32	1/4	3/8
TSD-7	60	7/16	7/16	2 1/2	±	40	38	60	2060	3/32	1/4	3/8
TSE-7	45	7/16	7/16	2 1/2	.010	45	53	45	2060	3/32	1/4	3/8
TSF-7	45	7/16	7/16	2 1/2		55	53	45	2060	3/32	1/4	3/8
TSA-8	90	1/2	1/2	2 1/2		0	8	90	2150	1/8	5/16	7/16
TSB-8	90	1/2	1/2	2 1/2		10	8	90	2150	1/8	5/16	7/16
TSC-8	60	1/2	1/2	2 1/2	1/32	30	38	60	2150	1/8	5/16	7/16
TSD-8	60	1/2	1/2	2 1/2	±	40	38	60	2150	1/8	5/16	7/16
TSE-8	45	1/2	1/2	2 1/2	.010	45	53	45	2150	1/8	5/16	7/16
TSF-8	45	1/2	1/2	2 1/2		55	53	45	2150	1/8	5/16	7/16
TSA-10	90	5/8	5/8	3		0	8	90	2220	5/32	3/8	9/16
TSB-10	90	5/8	5/8	3		10	8	90	2220	5/32	3/8	9/16
TSC-10	60	5/8	5/8	3	1/32	30	38	60	2220	5/32	3/8	9/16
TSD-10	60	5/8	5/8	3	±	40	38	60	2220	5/32	3/8	9/16
TSE-10	45	5/8	5/8	3	.010	45	53	45	2220	5/32	3/8	9/16
TSF-10	45	5/8	5/8	3		55	53	45	2220	5/32	3/8	9/16
TSA-12	90	3/4	3/4	3 1/2		0	8	90	2300	3/16	7/16	5/8
TSB-12	90	3/4	3/4	3 1/2		10	8	90	2300	3/16	7/16	5/8
TSC-12	60	3/4	3/4	3 1/2	1/32	30	38	60	2300	3/16	7/16	5/8
TSD-12	60	3/4	3/4	3 1/2	±	40	38	60	2300	3/16	7/16	5/8
TSE-12	45	3/4	3/4	3 1/2	.010	45	53	45	2300	3/16	7/16	5/8
TSF-12	45	3/4	3/4	3 1/2		55	53	45	2300	3/16	7/16	5/8

Table 4. American National Standard Solid Carbide Round Boring Tools — Style SRC for 60° Boring Bar and Style SRE for 45° Boring Bar (ANSI B94.5-1974)

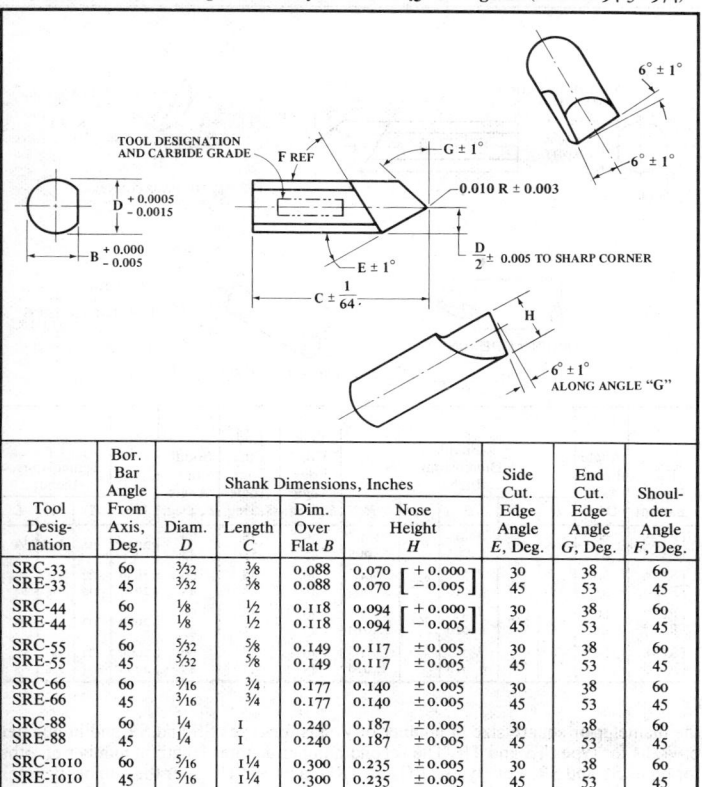

Tool Desig-nation	Bor. Bar Angle From Axis, Deg.	Shank Dimensions, Inches				Side Cut. Edge Angle E, Deg.	End Cut. Edge Angle G, Deg.	Shoul-der Angle F, Deg.
		Diam. D	Length C	Dim. Over Flat B	Nose Height H			
SRC-33	60	3/32	3/8	0.088	0.070 [+0.000 / −0.005]	30	38	60
SRE-33	45	3/32	3/8	0.088	0.070 [+0.000 / −0.005]	45	53	45
SRC-44	60	1/8	1/2	0.118	0.094 [+0.000 / −0.005]	30	38	60
SRE-44	45	1/8	1/2	0.118	0.094 [+0.000 / −0.005]	45	53	45
SRC-55	60	5/32	5/8	0.149	0.117 ±0.005	30	38	60
SRE-55	45	5/32	5/8	0.149	0.117 ±0.005	45	53	45
SRC-66	60	3/16	3/4	0.177	0.140 ±0.005	30	38	60
SRE-66	45	3/16	3/4	0.177	0.140 ±0.005	45	53	45
SRC-88	60	1/4	1	0.240	0.187 ±0.005	30	38	60
SRE-88	45	1/4	1	0.240	0.187 ±0.005	45	53	45
SRC-1010	60	5/16	1 1/4	0.300	0.235 ±0.005	30	38	60
SRE-1010	45	5/16	1 1/4	0.300	0.235 ±0.005	45	53	45

boring tools. In addition to these ready-to-use standard boring tools, solid carbide round and square unsharpened boring tool bits are provided.

Style Designations for Carbide Boring Tools: Table 1 shows designations used to specify the styles of American Standard sintered carbide boring tools. The first letter denotes solid (S) or tipped (T). The second letter denotes square (S) or round (R). The side cutting edge angle is denoted by a third letter (A through H) to complete the style designation. Solid square and round bits with the mounting surfaces ground but the cutting edges unsharpened (Table 7) are designated using the same system except that the third letter indicating side cutting edge angle is omitted.

Size Designation of Carbide Boring Tools: Specific sizes of boring tools are identified by the addition of numbers after the style designation. The first number denotes

Table 5. American National Standard Carbide-Tipped Round Boring Tools — Style TRC for 60° Boring Bar and Style TRE for 45° Boring Bar (ANSI B94.5-1974)

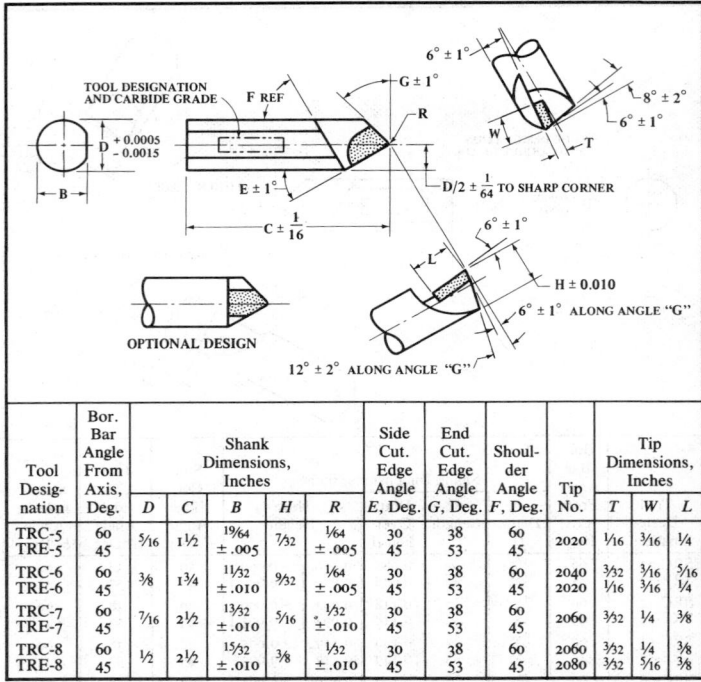

Tool Designation	Bor. Bar Angle From Axis, Deg.	Shank Dimensions, Inches					Side Cut. Edge Angle E, Deg.	End Cut. Edge Angle G, Deg.	Shoulder Angle F, Deg.	Tip No.	Tip Dimensions, Inches		
		D	C	B	H	R					T	W	L
TRC-5	60	5⁄16	1½	19⁄64 ± .005	7⁄32	1⁄64 ± .005	30	38	60	2020	1⁄16	3⁄16	1⁄4
TRE-5	45						45	53	45				
TRC-6	60	3⁄8	1¾	11⁄32 ± .010	9⁄32	1⁄64 ± .005	30	38	60	2040	3⁄32	3⁄16	5⁄16
TRE-6	45						45	53	45	2020	1⁄16	3⁄16	1⁄4
TRC-7	60	7⁄16	2½	13⁄32 ± .010	5⁄16	1⁄32 ± .010	30	38	60	2060	3⁄32	1⁄4	3⁄8
TRE-7	45						45	53	45				
TRC-8	60	1⁄2	2½	15⁄32 ± .010	3⁄8	1⁄32 ± .010	30	38	60	2060	3⁄32	1⁄4	3⁄8
TRE-8	45						45	53	45	2080	3⁄32	5⁄16	3⁄8

the diameter or square size in number of $\frac{1}{32}$nds for types SS and SR and in number of $\frac{1}{16}$ths for types TS and TR. The second number denotes length in number of $\frac{1}{8}$ths for types SS and SR. For styles TRG and TRH, a letter "U" after the number denotes a semi-finished tool (cutting edges unsharpened). Complete designations for the various standard sizes of carbide boring tools are given in Tables 2 through 7. In the diagrams in the tables, angles shown without tolerance are ± 1°.

    Examples of Tool Designation: The designation TSC-8 indicates: a carbide-tipped tool (T); square cross section (S); 30-degree side cutting edge angle (C); and $\frac{9}{16}$ or $\frac{1}{2}$ inch square size (8).

    The designation SRE-66 indicates: a solid carbide tool (S); round cross section (R); 45 degree side cutting edge angle (E); $\frac{6}{32}$ or $\frac{3}{16}$ inch diameter (6); and $\frac{6}{8}$ or $\frac{3}{4}$ inch long (6).

    The designation SS-610 indicates: a solid carbide tool (S); square cross section (S); $\frac{6}{32}$ or $\frac{3}{16}$ inch square size (6); $\frac{10}{8}$ or $1\frac{1}{4}$ inches long (10).

    It should be noted in this last example that the absence of a third letter (from A to H) indicates that the tool has its mounting surfaces ground but that the cutting edges are unsharpened.

Table 6. American National Standard Carbide-Tipped Round General-Purpose Square-End Boring Tools — Style TRG with 0° Rake and Style TRH with 10° Rake (ANSI B94.5-1974)

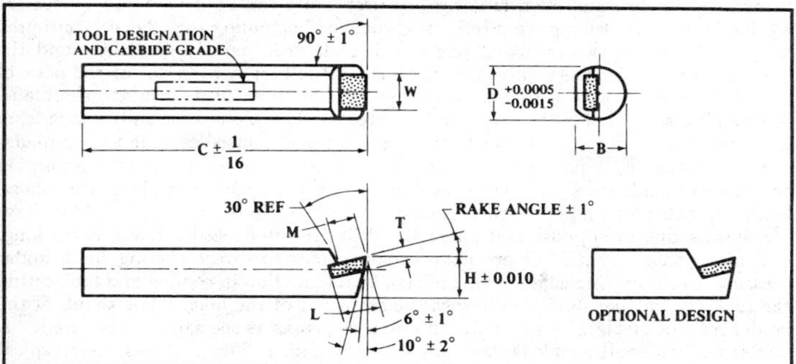

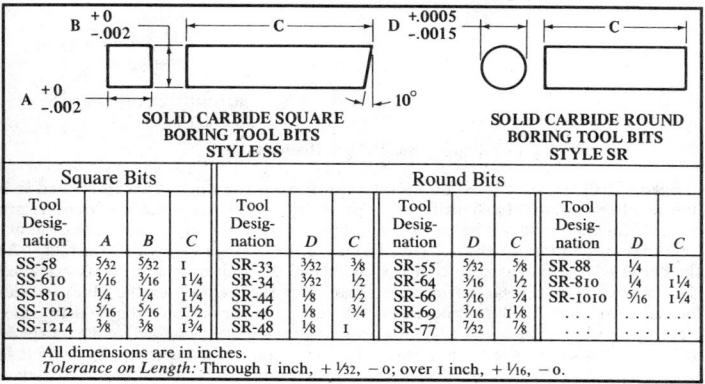

Tool Designation		Shank Dimensions, Inches							Tip Dimensions, Inches		
Finished	Semi-finished *	Diam. *D*	Length *C*	Dim. Over Flat *B*	Nose Height *H*	Set-back *M* (Min)	Rake Angle Deg.	Tip No.	*T*	*W*	*L*
TRG-5 TRH-5	TRG-5U TRH-5U	5/16	1 1/2	19/64 ±.005	3/16 7/32	3/16 3/16	0 10	1025	1/16	1/4	1/4
TRG-6 TRH-6	TRG-6U TRH-6U	3/8	1 3/4	11/32 ±.010	7/32 1/4	3/16	0 10	1030	1/16	5/16	1/4
TRG-7 TRH-7	TRG-7U TRH-7U	7/16	2 1/2	13/32 ±.010	1/4 5/16	3/16	0 10	1080	3/32	5/16	3/8
TRG-8 TRH-8	TRG-8U TRH-8U	1/2	2 1/2	15/32 ±.010	9/32 11/32	1/4	0 10	1090	3/32	3/8	3/8

* Semi-finished tool will be without Flat (B) and carbide unground on the end.

Table 7. Solid Carbide Square and Round Boring Tool Bits

SOLID CARBIDE SQUARE BORING TOOL BITS STYLE SS

SOLID CARBIDE ROUND BORING TOOL BITS STYLE SR

Square Bits				Round Bits								
Tool Designation	*A*	*B*	*C*	Tool Designation	*D*	*C*	Tool Designation	*D*	*C*	Tool Designation	*D*	*C*
SS-58	5/32	5/32	1	SR-33	3/32	3/8	SR-55	5/32	5/8	SR-88	1/4	1
SS-610	3/16	3/16	1 1/4	SR-34	3/32	1/2	SR-64	3/16	1/2	SR-810	1/4	1 1/4
SS-810	1/4	1/4	1 1/4	SR-44	1/8	1/2	SR-66	3/16	3/4	SR-1010	5/16	1 1/4
SS-1012	5/16	5/16	1 1/2	SR-46	1/8	3/4	SR-69	3/16	1 1/8	...	...	...
SS-1214	3/8	3/8	1 3/4	SR-48	1/8	1	SR-77	7/32	7/8	...	...	...

All dimensions are in inches.
Tolerance on Length: Through 1 inch, + 1/32, − 0; over 1 inch, + 1/16, − 0.

Spade Drills and Drilling

Spade drills are used to produce holes ranging in size from about one inch to 6 inches diameter, and even larger. Very deep holes can be drilled and blades are available for core drilling, counterboring, and for bottoming to a flat or contoured shape. There are two principal parts to a spade drill, namely, the blade and the holder. The holder has a slot into which the blade fits; a wide slot at the back of the blade engages with a tongue in the holder slot to accurately locate the blade. A retaining screw holds the two parts together. The blade is usually made from high speed steel, although cast nonferrous metal and cemented carbide-tipped blades are also available. Spade drill holders are classified by a letter symbol designating the range of blade sizes that can be held and by their length. Standard stub, short, long, and extra long holders are available; for very deep holes, special holders having wear strips to support and guide the drill are often used. Long, extra long, and many short length holders have coolant holes to direct cutting fluid, under pressure, to the cutting edges. In addition to its function in cooling and lubricating the tool, the cutting fluid also flushes the chips out of the hole. The shank of the holder may be straight or tapered; automotive shanks as specials are also used. A holder and different shank designs are shown in Fig. 1; Fig. 2 shows some typical blades.

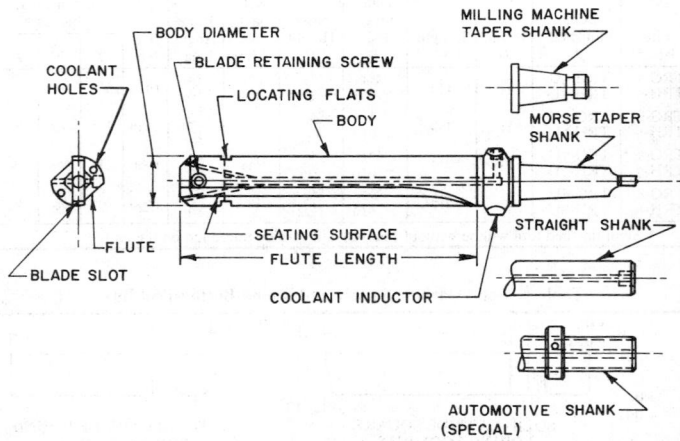

Fig. 1. Spade Drill Blade Holder

Spade Drill Geometry. — Metal separation from the work is accomplished in a like manner by both twist drills and spade drills, and the same mechanisms are involved in each case. The two cutting lips separate the metal by a shearing action that is identical to that of chip formation by a single-point cutting tool. At the chisel edge a much more complex condition exists. Here the metal is extruded sideways while at the same time it is sheared by the rotation of the blunt wedge-formed chisel edge. This accounts for the very high thrust force required to penetrate the work. The chisel edge of a twist drill is slightly rounded, while on spade drills it is a straight edge. For this reason it is likely that with spade drills it is more difficult for the extruded metal to escape from the region of the chisel edge.

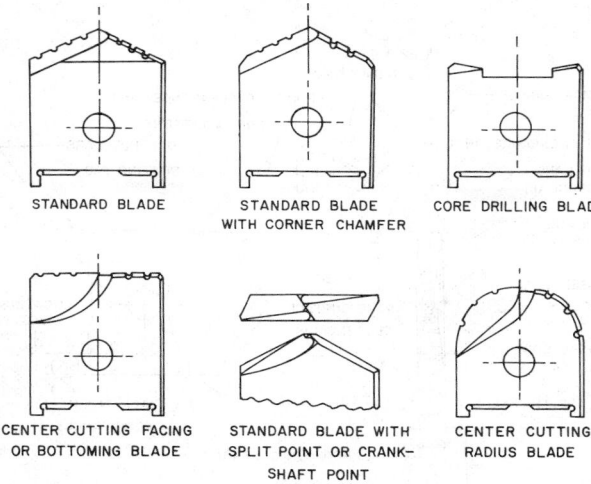

STANDARD BLADE   STANDARD BLADE   CORE DRILLING BLADE
                        WITH CORNER CHAMFER

CENTER CUTTING FACING   STANDARD BLADE WITH   CENTER CUTTING
  OR BOTTOMING BLADE   SPLIT POINT OR CRANK-   RADIUS BLADE
                              SHAFT POINT

Fig. 2. Typical Spade Drill Blades

owever, the edge is shorter in length than on twist drills and the thrust for spade
rilling is less.

Basic spade drill geometry is shown in Fig. 3. Normally, the point angle of a stand-
rd tool is 130 degrees and the lip clearance angle is 18 degrees, resulting in a chisel
lge angle of 108 degrees. The web thickness, is usually about ¼ to ⁵⁄₁₆ as thick
s the blade thickness. Usually the cutting angle is selected to provide this
eb thickness and to provide the necessary strength along the entire length of the
atting lip. A further reduction of the chisel edge length is sometimes desirable to
duce the thrust force in drilling. This can be accomplished by grinding a sec-
dary rake surface at the center, or by grinding a split point, or crankshaft point,
1 the point of the drill.

The larger point angle of a standard spade drill — 130 degrees as compared with
8 degrees on a twist drill — causes the chips to flow more toward the periphery of
ie drill, thereby allowing the chips to enter the flutes of the holder more readily.
he rake angle facilitates the formation of the chip along the cutting lips. For
rilling materials of average hardness the rake angle should be 10 to 12 degrees;
r hard or tough steels it should be 5 to 7 degrees, and for soft and ductile ma-
rials it can be increased to 15 to 20 degrees. The rake surface may be flat or
ounded, and the latter design is called radial rake. Radial rake is usually ground
 that the rake angle is maximum at the periphery and decreases uniformly toward
ie center to provide greater cutting edge strength at the center. A flat rake sur-
ce is recommended for drilling hard and tough materials in order to reduce the
ndency to chipping and to reduce heat damage.

A most important feature of the cutting edge is the chip splitters, which are also
lled chip breaker grooves. Functionally these grooves are chip dividers; instead
 forming a single wide chip along the entire length of the cutting edge, these grooves
use several chips to form that can be readily disposed of through the flutes of the
older. Chip splitters must be carefully ground to prevent the chips from packing in

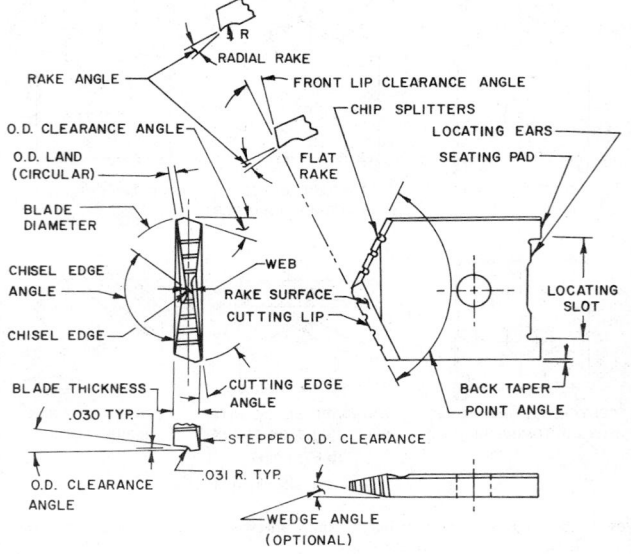

Fig. 3. Spade Drill Blade

the grooves which greatly reduces their effectiveness. They should be ground pe[r]
pendicular to the cutting lip and parallel to the surface formed by the clearanc[e]
angle.   The grooves on the two cutting lips must not overlap when measured r[a]
dially along the cutting lip.   Figure 4 and the accompanying table show the groo[ve]
form and dimensions.

On spade drills, the front lip clearance angle provides the relief.   It may be groun[d]
on a drill grinding machine but usually it is ground flat.   The normal front li[p]
clearance angle is 8 degrees; in some instances a secondary relief angle of abou[t]
14 degrees is ground below the primary clearance.   The wedge angle on the blad[e]
is optional.   It is generally ground on thicker blades having a larger diameter t[han]

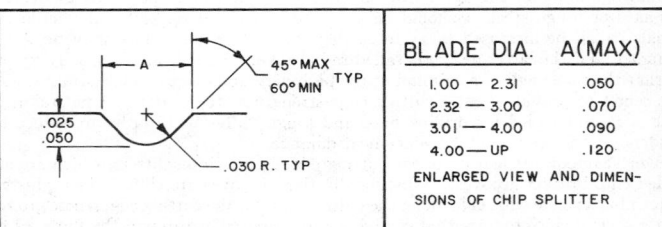

BLADE DIA.	A(MAX)
1.00 — 2.31	.050
2.32 — 3.00	.070
3.01 — 4.00	.090
4.00 — UP	.120
ENLARGED VIEW AND DIMEN-	
SIONS OF CHIP SPLITTER	

Fig. 4.   Spade Drill Chip Splitter Dimensions

revent heel dragging below the cutting lip and to reduce the chisel edge length. The outside diameter land is circular, serving to support and guide the blade in the hole. Usually it is ground to have a back taper of .001 to .002 inch per inch er side. The width of the land is approximately 20 to 25 per cent of the blade thickness. Normally the outside diameter clearance angle behind the land is 7 to 10 degrees. On many spade drill blades the outside diameter clearance surface s stepped about .030 inch below the land.

Spade Drilling. — Spade drills are used on drilling machines and other machine tools where the cutting tool rotates; they are also used on turning machines where the work and not the tool rotates. Although there are some slight operational differences, the method of using the spade drill is basically the same in each case. An adequate supply of cutting fluid must be used, which serves to cool and lubricate the cutting edges; to cool the chips, thus making them brittle and more easily broken; and to flush chips out of the hole. Flood cooling from outside the hole can be used for drilling relatively shallow holes, of about one or two and one-half times the diameter in depth. For deeper holes the cutting fluid should be injected through the holes in the drill. When drilling very deep holes it is often helpful to blow compressed air through the drill in addition to the cutting fluid to facilitate ejection of the chips. Air at full shop pressure is throttled down to a pressure that provides the most efficient ejection. The cutting fluids used are light and medium cutting oils, water soluble oils, and synthetics and the type selected depends on the work material.

Starting a spade drill in the workpiece needs special attention. The straight chisel edge on the spade drill has a tendency to wander as it starts to enter the work, especially if the feed is too light. This can result in a mispositioned hole and possible breakage of the drill point. The best method of starting the hole is to use a stub or short length spade drill holder and a blade of full size which should penetrate at least 1/8 inch at full diameter. The holder is then changed for a longer one as required to complete the hole to depth. Difficulties can be encountered if spotting with a center drill or starting drill is employed because the angles on these drills do not match the 130 degree point angle of the spade drill. Longer spade drills can be started without this starting procedure if the drill is guided by a jig bushing and if the holder is provided with wear strips.

Chip formation warrants the most careful attention as success in spade drilling is dependent on producing short, well-broken chips that can be easily ejected from the hole. Straight, stringy chips or chips that are wound like a clock spring cannot be ejected properly; they tend to pack around the blade, which may result in blade failure. The chip splitters must be functioning to produce a series of narrow chips along each cutting edge. Each chip must be broken and for drilling ductile materials they should be formed into a "C" or "figure 9" shape. Such chips will readily enter the flutes on the holder and flow out of the hole.

Proper chip formation is dependent on the work material, the spade drill geometry, and the cutting conditions under which the machine is operating. Brittle materials such as gray cast iron seldom pose a problem because they produce a discontinuous chip, but austenitic stainless steels and very soft and ductile materials require much attention to obtain satisfactory chip control. Thinning the web or grinding a split point on the blade will sometimes be helpful in obtaining better chip control as these modifications allow use of a heavier feed. Reducing the rake angle to obtain a tighter curl on the chip and grinding a corner chamfer on the tool will sometimes help to produce more manageable chips.

In most instances it is not necessary to experiment with the spade drill blade geometry to obtain satisfactory chip control. This can usually be accomplished by adjusting the cutting conditions; i.e., the cutting speed and the feed rate.

Normally the cutting speed for spade drilling should be 10 to 15 per cent low than that for an equivalent twist drill, although the same speed can be used if lower tool life is acceptable. The recommended cutting speeds for twist drills pages 997–1004 can be used as a starting point; however, they should be decreas by the percentage just given. It is essential to use a heavy feed rate when spa drilling to produce a thick chip and to force the chisel edge into the work. In ducti materials a light feed will produce a thin chip that is very difficult to break. T thick chip on the other hand, which often contains many rupture planes, will cu and break readily. Table 1 gives suggested feed rates for different drill sizes a materials. These should be used as a starting point and some adjustments m be necessary as experience is gained.

Table 1. Feed Rates for Spade Drilling

| Material | Hardness, Bhn | Feed — Inches per Revolution | | | | | |
| | | Spade Drill Diameter — Inches | | | | | |
		1–1¼	1¼–2	2–3	3–4	4–5	5–8
Free Machining Steel	100–240	.014	.016	.018	.022	.025	.03
	240–325	.010	.014	.016	.020	.022	.02
Plain Carbon Steels	100–225	.012	.015	.018	.022	.025	.03
	225–275	.010	.013	.015	.018	.020	.02
	275–325	.008	.010	.013	.015	.018	.02
Free Machining Alloy Steels	150–250	.014	.016	.018	.022	.025	.03
	250–325	.012	.014	.016	.018	.020	.02
	325–375	.010	.010	.014	.016	.018	.02
Alloy Steels	125–180	.012	.015	.018	.022	.025	.03
	180–225	.010	.012	.016	.018	.022	.02
	225–325	.009	.010	.013	.015	.018	.02
	325–400	.006	.008	.010	.012	.014	.01
Tool Steels							
Water Hardening	150–250	.012	.014	.016	.018	.020	.02
Shock Resisting	175–225	.012	.014	.015	.016	.017	.02
Cold Work	200–250	.007	.008	.009	.010	.011	.01
Hot Work	150–250	.012	.013	.015	.016	.018	.02
Mold	150–200	.010	.012	.014	.016	.018	.01
Special Purpose	150–225	.010	.012	.014	.016	.016	.01
High Speed Steel	200–240	.010	.012	.013	.015	.017	.0.
Gray Cast Iron	110–160	.020	.022	.026	.028	.030	.0
	160–190	.015	.018	.020	.024	.026	.0
	190–240	.012	.014	.016	.018	.020	.0
	240–320	.010	.012	.016	.018	.018	.0
Ductile or Nodular Iron	140–190	.014	.016	.018	.020	.022	.0
	190–250	.012	.014	.016	.018	.018	.0
	250–300	.010	.012	.016	.018	.018	.0
Malleable Iron							
Ferritic	110–160	.014	.016	.018	.020	.022	.0
Pearlitic	160–220	.012	.014	.016	.018	.020	.0
	220–280	.010	.012	.014	.016	.018	.0
Free Machining Stainless Steel							
Ferritic	.....	.016	.018	.020	.024	.026	.c
Austenitic	.....	.016	.018	.020	.022	.024	.c
Martensitic	.....	.012	.014	.016	.016	.018	.c
Stainless Steel							
Ferritic	.....	.012	.014	.018	.020	.020	.c
Austenitic	.....	.012	.014	.016	.018	.020	.c
Martensitic	.....	.010	.012	.012	.014	.016	.c
Aluminum Alloys	.....	.020	.022	.024	.028	.030	.c
Copper Alloys (Soft)		.016	.018	.020	.026	.028	.c
(Hard)		.010	.012	.014	.016	.018	.c
Titanium Alloys	.....	.008	.010	.012	.014	.014	.c
High Temperature Alloys	.....	.008	.010	.012	.012	.014	.c

Power Consumption and Thrust for Spade Drilling — In each individual setup there are factors and conditions influencing power consumption that cannot be accounted for in a simple equation; however, those given below will enable the user to estimate power consumption and thrust accurately enough for most practical purposes. They are based on experimentally derived values of unit horsepower, as given in Table 2. As a word of caution, these values are for sharp tools. In spade drilling, it is reasonable to estimate that a dull tool will increase the power consumption and the thrust by 25 to 50 per cent. The unit horsepower values in the table are for the power consumed at the cutting edge, to which must be added the power required to drive the machine tool itself, in order to obtain the horsepower required by the machine tool motor. This is calculated by dividing the horsepower at the cutter by a mechanical efficiency factor, e_m. This factor can be estimated to be .90 for a direct spindle drive with a belt, .75 for a back gear drive, and .70 to .80 for geared head drives. Thus, for spade drilling the formulas are:

$$hp_c = uhp \left(\frac{\pi D^2}{4} \right) fN$$

$$B_s = 148,500 \; uhp \; f \; D$$

$$hp_m = \frac{hp_c}{e_m}$$

$$f = \frac{f_m}{N}$$

Where: hp_c = Horsepower at the cutter
  hp_m = Horsepower at the motor
  B_s = Thrust for spade drilling in pounds
  uhp = Unit horsepower
  D = Drill diameter in inches
  f = Feed in inches per revolution
  f_m = Feed in inches per minute
  N = Spindle speed in revolutions per minute
  e_m = Mechanical efficiency factor

Table 2.   Unit Horsepower for Spade Drilling

Material	Hardness	Uhp	Material	Hardness	Uhp
Plain Carbon and Alloy Steel	85–200 Bhn	.79	Titanium Alloys	250–375 Bhn	.72
	200–275	.94	High Temp Alloys	200–360 Bhn	1.44
	275–375	1.00	Aluminum Alloys	……	.22
	375–425	1.15	Magnesium Alloys	……	.16
	45–52 R_c	1.44	Copper Alloys	20–80 R_B	.43
Cast Irons	110–200 Bhn	.5		80–100 R_B	.72
	200–300	1.08			
Stainless Steels	135–275 Bhn	.94			
	30–45 R_c	1.08			

Example: Estimate the horsepower and thrust required to drive a 2 inch diameter spade drill in AISI 1045 steel that is quenched and tempered to a hardness of 240 BHN. From the table (from Table 1.) on page 860, the cutting speed, V, for drilling this material with a twist drill is 50 feet per minute. This value is reduced by 10 per cent for spade drilling and the speed selected is thus .9 × 50 = 45 feet per minute. The feed rate is .016 ipr and the unit horsepower is .94 (from Table 2). The machine efficiency factor is estimated to be .80 and it will be assumed that a 50 per cent increase in the unit horsepower must be allowed for dull tools.

Step 1. Calculate the spindle speed from the following formula:

$$N = \frac{12V}{\pi D}$$

Where: N = Spindle speed in revolutions per minute
V = Cutting speed in feet per minute
D = Drill diameter in inches

Thus: $N = \frac{12 \times 45}{\pi \times 2} = 86$ revolutions per minute

Step 2. Calculate the horsepower at the cutter:

$$hp_c = uhp\left(\frac{\pi D^2}{4}\right)fN = .94\left(\frac{\pi \times 2^2}{4}\right).016 \times 86 = 4$$

Step 3. Calculate the horsepower at the motor and provide for a 50 per cent power increase for the dull tool:

$$hp_m = \frac{hp_c}{e_m} = \frac{4}{.80} = 5 \text{ horsepower}$$

hp_m (with dull tool) $= 1.5 \times 5 = 7.5$ horsepower

Step 4. Estimate the spade drill thrust:

$$B_s = 148,500 \; uhp \; f \; D = 148,500 \times .94 \times .016 \times 2$$
$$= 4467 \text{ lbs (for sharp tool)}$$
$$B_s = 1.5 \times 4467$$
$$= 6700 \text{ lbs (for dull tool)}$$

Trepanning. — Cutting a groove in the form of a circle or boring or cutting a hole by removing the center or core in one piece is called trepanning. Shallow trepanning, also called face grooving, can be performed on a lathe using a single-point tool that is similar to a grooving tool but has a curved blade. Generally, the minimum outside diameter that can be cut by this method is about three inches and the maximum groove depth is about two inches. Trepanning is probably the most economical method of producing deep holes that are two inches, and larger, in diameter. Fast production rates can be achieved. The tool consists of a hollow bar, or stem, and a hollow cylindrical head to which a carbide or high-speed steel, single-point cutting tool is attached. Usually only one cutting tool is used although for some applications a multiple cutter head must be used; e.g., heads used to start the hole have multiple tools. In operation, the cutting tool produces a circular groove and a residue core which enters the hollow stem after passing through the head. On outside diameter exhaust trepanning tools the cutting fluid is applied through the stem and the chips are flushed around the outside of the tool; inside diameter exhaust tools flush the chips out through the stem with the cutting fluid applied from the outside. For starting the cut a tool that cuts a starting groove in the work must be used, or the trepanning tool must be guided by a bushing. For holes less than about five diameters deep, a machine that rotates the trepanning tool can be used. Often, an ordinary drill press is satisfactory; deeper holes should be machined on a lathe with the work rotating. A hole diameter tolerance of $\pm$ .010 inch can easily be obtained by trepanning and in some cases a tolerance of $\pm$ .001 inch has been held. Hole runout can be held to $\pm$ .003 inch per foot and, at times, to $\pm$ .001 inch per foot. On heat-treated metal a surface finish of 125 to 150 AA can be obtained and on annealed metals 100 to 250 AA is common.

TAPS AND THREADING DIES

General dimensions and tap markings given in the ANSI Standard B94.9-1979 for hand taps, machine screw taps, spiral pointed taps, spiral fluted taps, nut taps, tapper taps, pulley taps, pipe taps, and metric taps of ISO, metric screw thread form are shown in the tables on the pages which follow. This Standard also gives the thread limits for taps with cut threads and ground threads. The thread limits for cut thread and ground thread taps for screw threads are given in Tables 13 through 16 and 19a and 19b; thread limits for cut thread and ground thread taps for pipe threads are given in Tables 17a through 18c. Taps recommended for various classes of Unified and American Standard screw threads are given in Tables 4a through 6 in numbered sizes and Table 7 for nuts in fractional sizes.

Hand Taps and Machine Screw Taps. — Regular hand taps are taps with thread and shank approximately the same length, with a square to accommodate a driving mechanism. Hand taps were first identified as such because they were used by hand; however, for many years they have been generally machine driven.

Regular hand taps in 2, 3, 4 and 6 flutes are furnished in taper, plug or bottoming chamfers. Taper taps are chamfered 7 to 10 threads. Plug taps are chamfered 3 to 5 threads. Bottoming taps are chamfered approximately 1 to 2 threads. For general dimensions see Tables 1a, 1b and 2.

Regular Machine Screw Taps: These are similar to regular hand taps, excepting that they are made in the numbered or machine screw sizes. See Tables 4a and 4b.

Spiral-Pointed Taps: Regular hand and machine screw taps having shallower flutes and wider lands. The cutting face of the first few threads is ground at an angle to force the chips ahead to prevent clogging in the flutes. See Tables 2 and 5.

Spiral-Pointed Short-Flute Taps: Regular hand and machine screw taps made with spiral point only. The balance of the threaded section is left unfluted. They are used for tapping thin material. See Tables 3 and 6.

Regular Spiral-Fluted Taps: Regular hand and machine screw taps having right-hand spiral flutes with a helix angle of from 25 to 35 degrees which are designed to help draw chips from the hole or bridge a keyway. See Tables 3 and 6.

Fast Spiral-Fluted Taps: These are similar to regular spiral-fluted taps except that the helix angle is from 45 to 60 degrees. See Tables 3 and 6.

Nut Taps. — Designed for nut tapping on a low-production basis. Approximately one-half to three quarters of the threaded portion is chamfered. This distributes the cutting load over a great number of teeth and facilitates the entering of the hole by the tap. See Table 7.

Bent-Shank Tapper Taps: These taps are designed for use in an automatic nut tapping machine. The tapped nuts pass over the bent shank and are ejected, so that continuous production is obtained without stopping or reversal. These taps are made in both fractional and machine screw sizes. See Standard.

Pulley Taps. — These taps are exceptionally long hand taps; the extra length being in the shank. The diameter of the shank is the same as the full diameter of the thread. These taps are made available in various lengths. See Table 8.

Pipe Taps. — *Taper Pipe Taps:* For tapping standard taper pipe threads. See Table 10.

Straight Pipe Taps: For producing standard straight pipe threads. See Table 10.

Metric Taps. — Taps for threads with ISO metric thread form are available in three styles, all with straight flutes. See Table 9a.

Spiral-Pointed Metric Taps: These taps for threads of ISO metric screw thread form have shallower flutes and wider lands than regular metric taps. See Table 9b.

Definitions of Tap Terms. — The definitions which follow are taken from ANSI B94.9 but include only the more important terms. Some tap terms are the same as screw thread terms; therefore, see also definitions on pages 1475–1479.

Back Taper: A gradual decrease in the diameter of the thread form on a tap from the chamfered end of the land towards the back which creates a slight radial relief in the threads.

Base of Thread: That which coincides with the cylindrical or conical surface from which the thread projects.

Chamfer: The tapering of the threads at the front end of each land or chaser of a tap by cutting away and relieving the crest of the first few teeth to distribute the cutting action over several teeth.

Chamfer Angle: The angle formed between the chamfer and the axis of the tap measured in an axial plane at the cutting edge.

Chamfer Relief Angle: The complement of the angle formed between a tangent to the relieved surface at the cutting edge and a radial line to the same point on the cutting edge.

Class of Thread: The designation of the class which determines the specification of the size, allowance, and tolerance to which a given threaded product is to be manufactured. It is not applicable to the tools used for threading.

Tap Terms

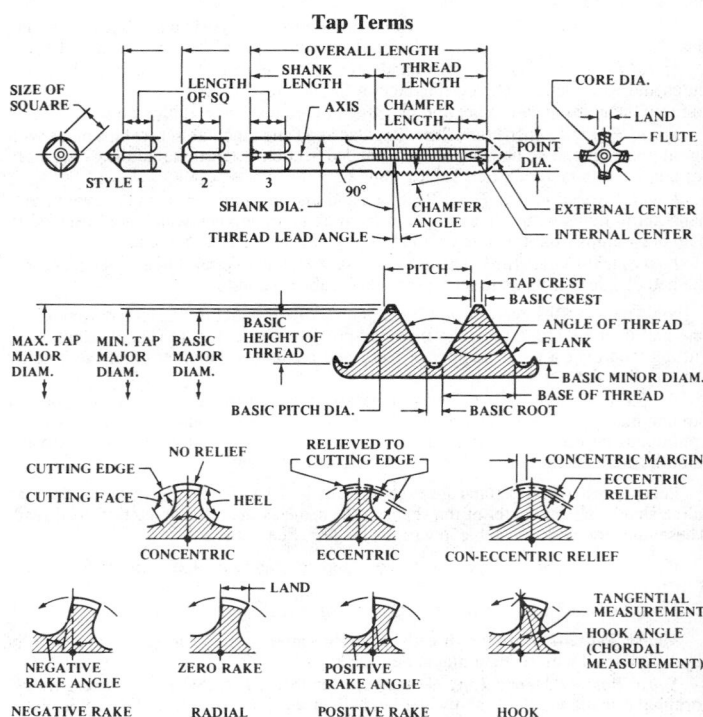

Core Diameter: The diameter of a circle which is tangent to the bottom of the flutes at a given point on the axis.

Crest Clearance: The radial distance between the root of the internal thread and the crest of the external thread of the coaxially assembled design forms of mating threads.

First Full Thread: The first full thread on the cutting edge back of the chamfer. It is at this point that rake, hook, and thread elements are measured.

Flank Angle: The flank angle is the angle between the individual flank and the perpendicular to the axis of the thread, measured in an axial plane. A flank angle of a symmetrical thread is commonly termed the "half-angle of thread."

Flank-Leading: (1) The flank of a thread facing toward the chamfered end of a threading tool. (2) The leading flank of a thread is the one which, when the thread is about to be assembled with a mating thread, faces the mating thread.

Flank-Trailing: The trailing flank of a thread is the one opposite the leading flank.

Flutes: The longitudinal channels formed in a tap to create cutting edges on the thread profile and to provide chip spaces and cutting fluid passages. On a parallel or straight thread tap they may be straight, angular or helical; on a taper thread tap they may be straight, angular or spiral.

Flute–Angular: A flute lying in a plane intersecting the tool axis at an angle.

Flute–Helical: A flute with uniform axial lead and constant helix in a helical path around the axis of a cylindrical tap.

Flute–Spiral: A flute with uniform axial lead in a spiral path around the axis of a conical tap.

Flute Lead Angle: The angle at which a helical or spiral cutting edge at a given point makes with an axial plane through the same point.

Flute–Straight: A flute which forms a cutting edge lying in an axial plane.

Front Taper: A gradual increase in the diameter of the thread form on a tap from the leading end of the tool toward the back.

Heel: The edge of the land opposite the cutting edge.

Hook Angle: The inclination of a concave cutting face, usually specified either as Chordal Hook or Tangential Hook.

Hook–Chordal Angle: The angle between the chord passing through the root and crest of a thread form at the cutting face, and a radial line through the crest at the cutting edge.

Hook–Tangential Angle: The angle between a line tangent to a hook cutting face at the cutting edge and a radial line to the same point.

Interrupted Thread Tap: A tap having an odd number of lands with alternate teeth in the thread helix removed. In some cases alternate teeth are removed only for a portion of the thread length.

Land: One of the threaded sections between the flutes of a tap.

Lead: The distance a screw thread advances axially in one complete turn.

Lead Error: The deviation from prescribed limits.

Lead Deviation: The deviation from the basic nominal lead. *Progressive Lead Deviation:* (1) On a straight thread the deviation from a true helix where the thread helix advances uniformly. (2) On a taper thread the deviation from a true spiral where the thread spiral advances uniformly.

Length of Thread: The length of the thread of the tap includes the chamfered threads and the full threads but does not include an external center. It is indicated by the letter "B" in the illustrations at the heads of the tables.

Limits: The limits of size are the applicable maximum and minimum sizes.

Major Diameter: On a straight thread the major diameter is that of the major cylinder. On a taper thread the major diameter at a given position on the thread axis is that of the major cone at that position.

Minor Diameter: On a straight thread the minor diameter is that of the minor cylinder. On a taper thread the minor diameter at a given position on the thread axis is that of the minor cone at that position.

Pitch Diameter (Simple Effective Diameter): On a straight thread, the pitch diameter is the diameter of the imaginary coaxial cylinder, the surface of which would pass through the thread profiles at such points as to make the width of the groove equal to one-half the basic pitch. On a perfect thread this occurs at the point where the widths of the thread and groove are equal. On a taper thread, the pitch diameter at a given position on the thread axis is the diameter of the pitch cone at that position.

Point Diameter: The diameter at the cutting edge of the leading end of the chamfered section.

Rake: The angular relationship of the straight cutting face of a tooth with respect to a radial line through the crest of the tooth at the cutting edge. Positive rake means that the crest of the cutting face is angularly ahead of the balance of the cutting face of the tooth. Negative rake means that the crest of the cutting face is angularly behind the balance of the cutting face of the tooth. Zero rake means that the cutting face is directly on a radial line.

Relief: The removal of metal behind the cutting edge to provide clearance between the part being threaded and the threaded land.

Relief–Center: Clearance produced on a portion of the tap land by reducing the diameter of the entire thread form between cutting edge and heel.

Relief–Chamfer: The gradual decrease in land height from cutting edge to heel on the chamfered portion of the land on a tap to provide radial clearance for the cutting edge.

Relief–Con-eccentric Thread: Radial relief in the thread form starting back of a concentric margin.

Relief–Double Eccentric Thread: The combination of a slight radial relief in the thread form starting at the cutting edge and continuing for a portion of the land width, and a greater radial relief for the balance of the land.

Relief–Eccentric Thread: Radial relief in the thread form starting at the cutting edge and continuing to the heel.

Relief–Flatted Land: Clearance produced on a portion of the tap land by truncating the thread between cutting edge and heel.

Relief–Grooved Land: Clearance produced on a tap land by forming a longitudinal groove in the center of the land.

Relief–Radial: The clearance produced by removal of metal from behind the cutting edge. Taps should have the chamfer relieved and should have back taper, but may or may not have relief in the angle and on the major diameter of the threads. When the thread angle is relieved, starting at the cutting edge and continuing to the heel, the tap is said to have "eccentric" relief. If the thread angle is relieved back of a concentric margin (usually one-third of land width) the tap is said to have "con-eccentric" relief.

Size–Actual: The measured size of an element on an individual part.

Size–Basic: That size from which the limits of size are derived by the application of allowances and tolerances.

Size–Functional: The functional diameter of an external or internal thread is the pitch diameter of the enveloping thread of perfect pitch, lead and flank angles, having full depth of engagement but clear at crests and roots, and of a specified length of engagement. It may be derived by adding to the pitch diameter in the case of an external thread, or subtracting from the pitch diameter in the case of an internal thread, the cumulative effects of deviations from specified profile, including variations in lead and flank angle over a specified length of engagement. The effects of

(Continued on page 872)

Table 1a. ANSI Standard Ground Thread Regular Hand Taps (ANSI B94.9-1979)

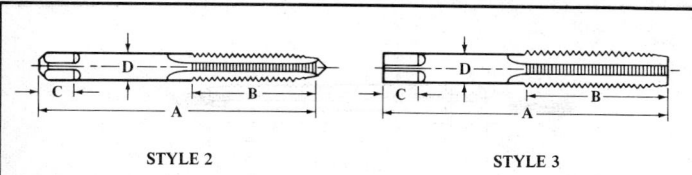

STYLE 2   STYLE 3

Diam. of Tap	NC UNC	NF UNF	No. of Flutes	H_1	H_2	H_3	H_4	H_5	Length Overall, A	Length of Thread, B	Length of Square, C	Diam. of Shank, D	Size of Square, E
1/4	20	...	4	TPB	TPB	TPB	0	PB	2½	1	5/16	0.255	0.191
1/4	...	28	4	PB	PB	PB	PB	...	2½	1	5/16	0.255	0.191
5/16	18	...	4	PB	PB	TPB	...	PB	2 23/32	1 1/8	3/8	0.318	0.238
5/16	...	24	4	PB	P	TPB	PB	...	2 23/32	1 1/8	3/8	0.318	0.238
3/8	16	...	4	PB	PB	TPB	...	PB	2 15/16	1 1/4	7/16	0.381	0.286
3/8	...	24	4	PB	PB	TPB	PB	...	2 15/16	1 1/4	7/16	0.381	0.286
7/16	14	20	4	...	...	TPB	...	PB	3 5/32	1 7/16	13/32	0.323	0.242
1/2	13	...	4	P	...	TPB	...	PB	3 3/8	1 21/32	7/16	0.367	0.275
1/2	...	20	4	PB	...	TPB	...	P	3 3/8	1 21/32	7/16	0.367	0.275
9/16	12	...	4	...	...	TPB	...	P	3 19/32	1 21/32	1/2	0.429	0.322
9/16	...	18	4	...	P	TPB	...	P	3 19/32	1 21/32	1/2	0.429	0.322
5/8	11	...	4	...	P	TPB	...	PB	3 13/16	1 13/16	9/16	0.480	0.360
5/8	...	18	4	...	P	TPB	...	PB	3 13/16	1 13/16	9/16	0.480	0.360
11/16†	...	...	4	...	...	TPB	...	...	4 1/32	1 13/16	5/8	0.542	0.406
3/4	10	...	4	...	P	TPB	...	PB	4 1/4	2	11/16	0.590	0.442
3/4	...	16	4	P	P	TPB	...	PB	4 1/4	2	11/16	0.590	0.442
7/8§	9	...	4	...	...	...	TPB	...	4 11/16	2 7/32	3/4	0.697	0.523
7/8	...	14	4	...	P	...	TPB	...	4 11/16	2 7/32	3/4	0.697	0.523
1§	8	...	4	...	...	...	TPB	...	5 1/8	2 1/2	13/16	0.800	0.600
1	...	12	4	...	...	...	TPB	...	5 1/8	2 1/2	13/16	0.800	0.600
1‡	...	...	4	...	...	...	TPB	...	5 1/8	2 1/2	13/16	0.800	0.600
1 1/8	7	12	4	...	...	...	TPB	...	5 7/16	2 9/16	7/8	0.896	0.672
1 1/4	7	12*	4	...	...	...	TPB	...	5 3/4	2 9/16	1	1.021	0.766
1 3/8	6	12*	4	...	...	...	TPB	...	6 1/16	3	1 1/16	1.108	0.831
1 1/2	6	12*	4	...	...	...	TPB	...	6 3/8	3	1 1/8	1.233	0.925

Tolerances for General Dimensions

Element	Diameter Range	Tolerance	Element	Diameter Range	Tolerance
Length Overall, A	1/4 to 1 incl	±1/32	Diameter of Shank, D	1/4 to 5/8 incl	−0.0015
	1 1/8 to 1 1/2 incl	±1/16		11/16 to 1 1/2 incl	−0.002
Length of Thread, B	1/4 to 1/2 incl	±1/16	Size of Square, E	1/4 to 1/2 incl	−0.004
	9/16 to 1 1/2 incl	±3/32		9/16 to 1 incl	−0.006
Length of Square, C	1/4 to 1 incl	±1/32		1 1/8 to 1 1/2 incl	−0.008
	1 1/8 to 1 1/2 incl	±1/16			

All dimensions are given in inches.

These taps are standard in high-speed steel.

* In these sizes NF-UNF thread taps have six flutes.

† This size has 11 or 16 threads per inch NS.

‡ This size has 14 threads per inch NS.

§ These sizes are also available with plug chamfer in H6 pitch diameter limits.

Chamfer designations are: T = taper, P = plug, and B = bottoming.

Style 2 taps, 3/8 inch and smaller, have external center on thread end (may be removed on bottoming taps) and external partial cone center on shank end with length of cone approximately one-quarter of diameter of shank.

Style 3 taps, larger than 3/8 inch, have internal center in thread and shank ends.

For standard thread limits see Table 14. For eccentricity tolerances see Table 20.

Table 1b. ANSI Standard Cut Thread Regular Hand Taps (ANSI B94.9-1979)

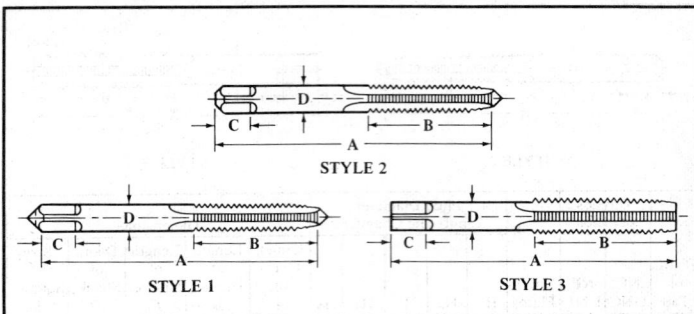

STYLE 2

STYLE 1   STYLE 3

Diam. of Tap	Threads Per Inch					No. of Flutes	Dimensions				
	Carbon Steel			HS Steel			Length Overall, A	Length of Thread, B	Length of Square, C	Diam. of Shank, D	Size of Square, E
	NC UNC	NF UNF	NS	NC UNC	NF UNF						
1/8	...	...	40	...	...	3	1 15/16	5/8	3/16	0.141	0.110
5/32	...	...	32	...	...	4	2 1/8	3/4	1/4	0.168	0.131
3/16	...	...	24, 32	...	...	4	2 3/8	7/8	1/4	0.194	0.152
1/4	20	28	...	20	28	4	2 1/2	1	5/16	0.255	0.191
5/16	18	24	...	18	24	4	2 23/32	1 1/8	3/8	0.318	0.238
3/8	16	24	...	16	24	4	2 15/16	1 1/4	7/16	0.381	0.286
7/16	14	20	...	14	20	4	3 5/32	1 7/16	13/32	0.323	0.242
1/2	13	20	...	13	20	4	3 3/8	1 21/32	7/16	0.367	0.275
9/16	12	18	...	12	...	4	3 19/32	1 21/32	1/2	0.429	0.322
5/8	11	18	...	11	18	4	3 13/16	1 13/16	9/16	0.480	0.360
3/4	10	16	...	10	16	4	4 1/4	2	11/16	0.590	0.442
7/8	9	14	...	9	14	4	4 11/16	2 7/32	3/4	0.697	0.523
1	8	...	14*	8	...	4	5 1/8	2 1/2	13/16	0.800	0.600
1 1/8	7	12	...	...	...	4	5 7/16	2 9/16	7/8	0.896	0.672
1 1/4	7	12†	...	...	...	4	5 3/4	2 9/16	1	1.021	0.766
1 3/8	6*	12†*	...	...	...	4	6 1/16	3	1 1/16	1.108	0.831
1 1/2	6	12†*	...	...	...	4	6 3/8	3	1 1/8	1.233	0.925
1 3/4	5*	...	...	...	...	6	7	3 3/16	1 1/4	1.430	1.072
2	4 1/2*	...	...	...	...	6	7 5/8	3 9/16	1 3/8	1.644	1.233

Tolerances for General Dimensions

Elements	Range	Tolerance	Elements	Range	Tolerance
Length Overall, A	1/16 to 1	± 1/32	Diameter of Shank, D	1/16 to 3/16	− 0.004
	1 1/8 to 2	± 1/16		1/4 to 1	− 0.005
Length of Thread, B	1/16 to 3/16	± 3/64		1 1/8 to 2	− 0.007
	1/4 to 1/2	± 1/16			
	9/16 to 1 1/2	± 3/32	Size of Square, E	1/16 to 1/2	− 0.004
	1 5/8 to 2	± 1/8		9/16 to 1	− 0.006
Length of Square, C	1/16 to 1	± 1/32		1/8 to 2	− 0.008
	1 1/8 to 2	± 1/16			

All dimensions are given in inches.
* Standard in plug chamfer only.
† In these sizes NF-UNF thread taps have six flutes.
These taps are standard in carbon steel and high-speed steel.
Except where indicated, these taps are standard with taper, plug, or bottoming chamfer.
Cut thread taps, sizes 3/8 inch and smaller have optional style center on thread and shank ends;
sizes larger than 3/8 inch have internal centers in thread and shank ends.
For standard thread limits see Table 13. For eccentricity tolerances see Table 20.

Table 2. ANSI Standard Two- and Three-Fluted and Spiral-Pointed Hand Taps (ANSI B94.9-1979)

STYLE 2 · STYLE 3
TWO- AND THREE- FLUTED HAND TAPS
SPIRAL - POINTED HAND TAPS

Diam. of Tap	Threads per Inch NC UNC	NF UNF	No. of Flutes	Pitch Diameter Limits and Chamfers*† H1	H2	H3	H4	H5	Length Over-all, A	Length of Thread, B	Length of Square, C	Diam. of Shank, D	Size of Square, E
Ground Thread High-Speed Steel Two- and Three-Fluted Hand Taps													
1/4	20	...	2			PB			2 1/2	1	5/16	0.255	0.191
1/4	20	...	3	P	P	PB	...	P	2 1/2	1	5/16	0.255	0.191
1/4	...	28	2, 3			PB			2 1/2	1	5/16	0.255	0.191
5/16	18	...	2			PB			2 23/32	1 1/8	3/8	0.318	0.238
5/16	18	...	3			PB			2 23/32	1 1/8	3/8	0.318	0.238
5/16	...	24	3			PB			2 23/32	1 1/8	3/8	0.318	0.238
3/8	16	...	3			PB			2 15/16	1 1/4	7/16	0.381	0.286
3/8	...	24	3			PB			2 15/16	1 1/4	7/16	0.381	0.286
7/16	14	...	3			P			3 5/32	1 7/16	13/32	0.323	0.242
7/16	...	20	3			P			3 5/32	1 7/16	13/32	0.323	0.242
1/2	13	...	3			PB			3 3/8	1 21/32	7/16	0.367	0.275
1/2	...	20	3			P			3 3/8	1 21/32	7/16	0.367	0.275
Ground Thread High-Speed Steel and Cut Thread High-Speed Steel Spiral-Pointed Hand Taps													
1/4	20	...	2	P	P	PB		P	2 1/2	1	5/16	0.255	0.191
1/4*	20	...	3			P		P	2 1/2	1	5/16	0.255	0.191
1/4	...	28	2	P	P	PB	P		2 1/2	1	5/16	0.255	0.191
1/4*	...	28	3	P	P	P	P		2 1/2	1	5/16	0.255	0.191
5/16	18	...	2	P	P	PB		P	2 23/32	1 1/8	3/8	0.318	0.238
5/16*	18	...	3			P		P	2 23/32	1 1/8	3/8	0.318	0.238
5/16	...	24	2	P	P	PB	P		2 23/32	1 1/8	3/8	0.318	0.238
5/16*	...	24	3	P	P	P	P		2 23/32	1 1/8	3/8	0.318	0.238
3/8	16	...	3	P	P	P		P	2 15/16	1 1/4	7/16	0.381	0.286
3/8	...	24	3	P	P	P	P		2 15/16	1 1/4	7/16	0.381	0.286
7/16*	14	20	3	...	P**	P	...	P	3 5/32	1 7/16	13/32	0.323	0.242
1/2	13	20*	3	P	P	P		P	3 3/8	1 21/32	7/16	0.367	0.275
5/8*	11	...	3			P		P	3 13/16	1 13/16	9/16	0.480	0.360
3/4*	10	...	3			P		P	4 1/4	2	11/16	0.590	0.442

Tolerances for General Dimensions

Element	Diameter Range	Tolerance Ground Thread	Cut Thread	Element	Diameter Range	Tolerance Ground Thread	Cut Thread
Overall Length, A	1/4 to 3/4	± 1/32	± 1/32	Shank Diameter, D	1/4 to 5/8	− 0.0015	− 0.005
Thread Length, B	1/4 to 1/2	± 1/16	± 1/16		3/4	− 0.0020	...
	5/8 to 3/4	± 1/32	...	Size of Square, E	1/4 to 1/2	− 0.0040	− 0.004
Square Length, C	1/4 to 3/4	± 1/32	± 1/32		5/8 to 3/4	− 0.0060	...

All dimensions are given in inches. Chamfer designations as follows: P = plug and B = bottoming. Ground thread taps — Style 2, 3/8 inch and smaller, have external center on thread end (may be removed on bottoming taps) and external partial cone center on shank end, with length of cone approximately 1/4 of shank diameter. Ground thread taps — Style 3, larger than 3/8 inch, have internal center in thread and shank ends. Cut thread taps, 3/8 inch and smaller have optional style center on thread and shank ends; sizes larger than 3/8 inch have internal centers in thread and shank ends. For standard thread limits see Tables 13 and 14. * Applies only to ground thread high-speed steel taps. † Cut thread high-speed steel taps are standard with plug chamfer only. ** Applies only to 7/16-14 tap. For eccentricity tolerances see Table 20.

Table 3. Other Types of ANSI Standard Hand Taps (ANSI B94.9-1979)

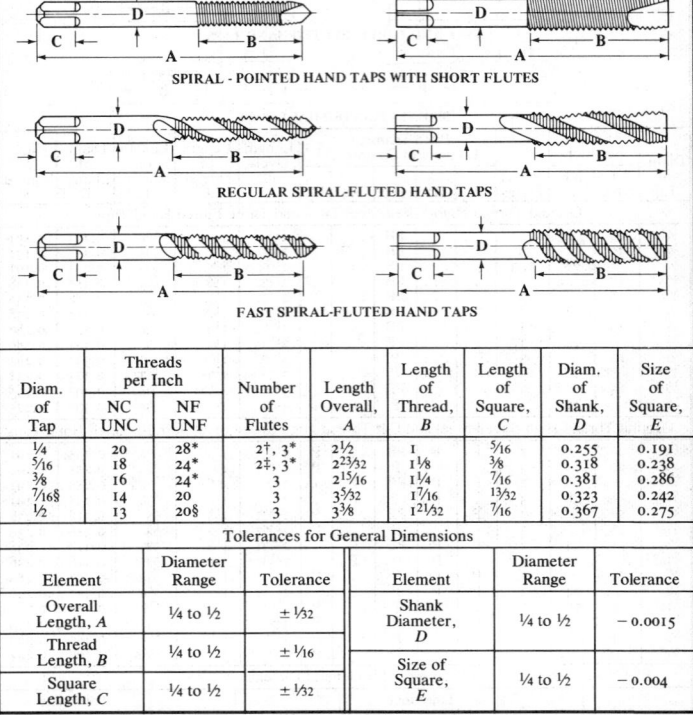

Diam. of Tap	Threads per Inch		Number of Flutes	Length Overall, A	Length of Thread, B	Length of Square, C	Diam. of Shank, D	Size of Square, E
	NC UNC	NF UNF						
¼	20	28*	2†, 3*	2½	1	5⁄16	0.255	0.191
5⁄16	18	24*	2‡, 3*	2²³⁄32	1⅛	⅜	0.318	0.238
⅜	16	24*	3	2¹⁵⁄16	1¼	7⁄16	0.381	0.286
7⁄16§	14	20	3	3⁵⁄32	1⁷⁄16	13⁄32	0.323	0.242
½	13	20§	3	3⅜	1²¹⁄32	7⁄16	0.367	0.275

Tolerances for General Dimensions

Element	Diameter Range	Tolerance	Element	Diameter Range	Tolerance
Overall Length, A	¼ to ½	± 1⁄32	Shank Diameter, D	¼ to ½	− 0.0015
Thread Length, B	¼ to ½	± 1⁄16	Size of Square, E	¼ to ½	− 0.004
Square Length, C	¼ to ½	± 1⁄32			

All dimensions are given in inches. These standard taps are made of high-speed steel with ground threads. For standard thread limits see Table 14.

Spiral-Pointed Hand Taps with Short Flutes: These taps are standard with plug chamfer only in H3 limit and have right-hand spiral flutes with a helix angle of from 25 to 35 degrees. They are provided with spiral point only. The balance of the threaded section is left unfluted.

Regular Spiral-Fluted Hand Taps: These taps are standard with plug or bottoming chamfer in H3 limit and have right-hand spiral flutes with a helix angle of from 45 to 60 degrees.

Fast Spiral-Fluted Hand Taps: These taps are standard with plug or bottoming chamfer in H3 limit and have right-hand spiral flutes with a helix angle of from 45 to 60 degrees.

Style 2 taps, ⅜ inch and smaller, have external center on thread end (may be removed on bottoming taps) and external partial cone center on shank end with cone length approximately ¼ shank diameter.

Style 3 taps larger than ⅜ inch have internal center in thread and shank ends.

* Does not apply to spiral-pointed hand taps with short flutes.

† Does not apply to fast spiral-fluted hand taps or to regular spiral-fluted hand taps with 28 threads per inch.

‡ Applies only to spiral-pointed hand taps with short flutes.

§ Applies only to fast spiral-fluted hand taps.

For standard thread limits see Table 14. For eccentricity tolerances see Table 20.

Table 4a. ANSI Standard Ground Thread Regular Machine Screw Taps
(ANSI B94.9-1979)

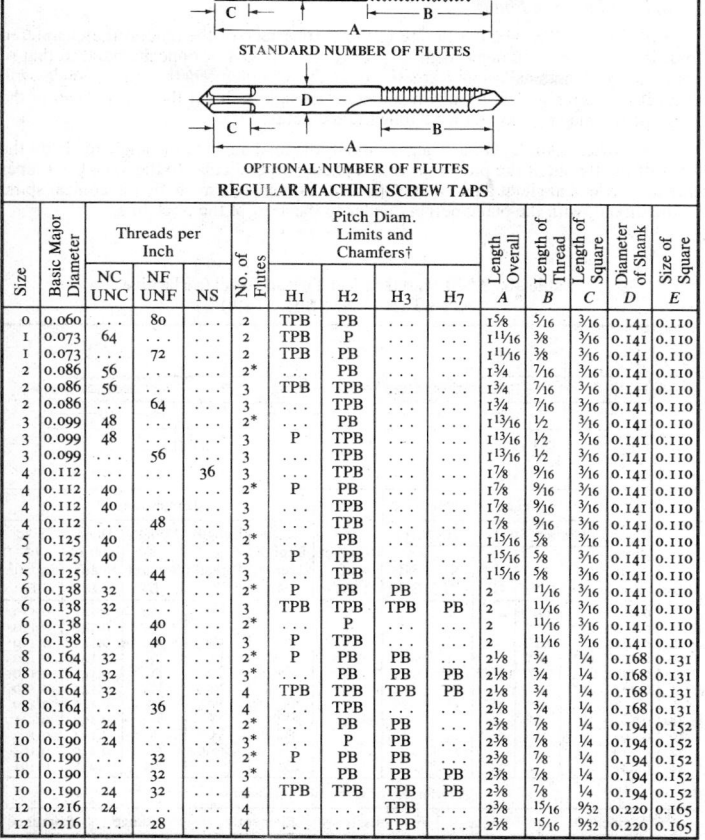

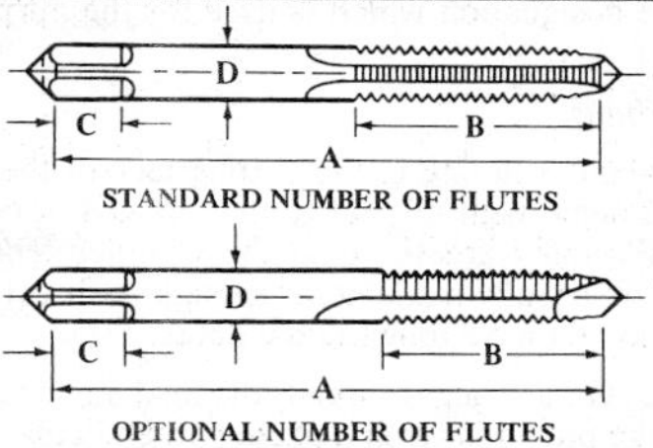

STANDARD NUMBER OF FLUTES

OPTIONAL NUMBER OF FLUTES
REGULAR MACHINE SCREW TAPS

Size	Basic Major Diameter	NC UNC	NF UNF	NS	No. of Flutes	H1	H2	H3	H7	Length Overall A	Length of Thread B	Length of Square C	Diameter of Shank D	Size of Square E
0	0.060	...	...	80	2	TPB	P	...	...	1⅝	5/16	3/16	0.141	0.110
1	0.073	64	...	...	2	TPB	P	...	...	1¹¹/16	⅜	3/16	0.141	0.110
1	0.073	...	72	...	2	TPB	PB	...	...	1¹¹/16	⅜	3/16	0.141	0.110
2	0.086	56	...	...	2*	...	PB	...	...	1¾	7/16	3/16	0.141	0.110
2	0.086	56	...	...	3	TPB	TPB	...	...	1¾	7/16	3/16	0.141	0.110
2	0.086	...	64	...	3	...	TPB	...	...	1¾	7/16	3/16	0.141	0.110
3	0.099	48	...	...	2*	...	PB	...	...	1¹³/16	½	3/16	0.141	0.110
3	0.099	48	...	...	3	P	TPB	...	...	1¹³/16	½	3/16	0.141	0.110
3	0.099	...	56	...	3	...	TPB	...	...	1¹³/16	½	3/16	0.141	0.110
4	0.112	...	...	36	3	...	TPB	...	...	1⅞	9/16	3/16	0.141	0.110
4	0.112	40	...	...	2*	P	PB	...	...	1⅞	9/16	3/16	0.141	0.110
4	0.112	40	...	...	3	...	TPB	...	...	1⅞	9/16	3/16	0.141	0.110
4	0.112	...	48	...	3	...	TPB	...	...	1⅞	9/16	3/16	0.141	0.110
5	0.125	40	...	...	2*	...	PB	...	...	1¹⁵/16	⅝	3/16	0.141	0.110
5	0.125	40	...	...	3	P	TPB	...	...	1¹⁵/16	⅝	3/16	0.141	0.110
5	0.125	...	44	...	3	...	TPB	...	...	1¹⁵/16	⅝	3/16	0.141	0.110
6	0.138	32	...	...	2*	P	PB	PB	...	2	11/16	3/16	0.141	0.110
6	0.138	32	...	...	3	TPB	TPB	TPB	PB	2	11/16	3/16	0.141	0.110
6	0.138	...	40	...	2*	...	P	...	...	2	11/16	3/16	0.141	0.110
6	0.138	...	40	...	3	P	TPB	...	...	2	11/16	3/16	0.141	0.110
8	0.164	32	...	...	2*	P	PB	PB	...	2⅛	¾	¼	0.168	0.131
8	0.164	32	...	...	3*	...	PB	PB	PB	2⅛	¾	¼	0.168	0.131
8	0.164	32	...	...	4	TPB	TPB	TPB	PB	2⅛	¾	¼	0.168	0.131
8	0.164	...	36	...	4	...	TPB	...	...	2⅛	¾	¼	0.168	0.131
10	0.190	24	...	...	2*	...	PB	PB	...	2⅜	⅞	¼	0.194	0.152
10	0.190	24	...	...	3*	...	P	PB	...	2⅜	⅞	¼	0.194	0.152
10	0.190	...	32	...	2*	P	PB	...	...	2⅜	⅞	¼	0.194	0.152
10	0.190	...	32	...	3*	...	PB	PB	...	2⅜	⅞	¼	0.194	0.152
10	0.190	24	32	...	4	TPB	TPB	TPB	PB	2⅜	⅞	¼	0.194	0.152
12	0.216	24	...	...	4	...	...	TPB	...	2⅜	15/16	9/32	0.220	0.165
12	0.216	...	28	...	4	...	...	TPB	...	2⅜	15/16	9/32	0.220	0.165

All dimensions are given in inches.

These taps are standard as high-speed steel taps with ground threads, with standard and optional number of flutes and pitch diameter limits and chamfers as given in the table.

* Optional number of flutes.

Chamfer designations are: T = taper, P = plug, and B = bottoming.

These are style 1 taps and have external centers on thread and shank ends (may be removed on thread end of bottoming taps).

For standard thread limits see Table 15. For eccentricity tolerances see Table 20.

Tolerances: No. 0 to 12 size range — A, ±1/32; B, ±3/64; C, ±1/32; D, −0.0015 for ground threads, −0.004 for cut threads; E, −0.004.

taper, out-of-roundness, and surface defects may be positive or negative on either external or internal threads.

Size–Nominal: The designation which is used for the purpose of general identification.

Spiral Flute: See *Flutes.*

Spiral Point: The angular fluting in the cutting face of the land at the chamfered end. It is formed at an angle with respect to the tap axis of opposite hand to that of rotation. Its length is usually greater than the chamfer length and its angle with respect to the tap axis is usually made great enough to direct the chips ahead of the tap. The tap may or may not have longitudinal flutes.

Thread Lead Angle: On a straight thread, the lead angle is the angle made by the helix of the thread at the pitch line with a plane perpendicular to the axis. On a taper thread, the lead angle at a given axial position is the angle made by the conical spiral of the thread, with the plane perpendicular to the axis, at the pitch line.

Table 4b. ANSI Standard Cut Thread Regular Machine Screw Taps (ANSI B94.9-1979)

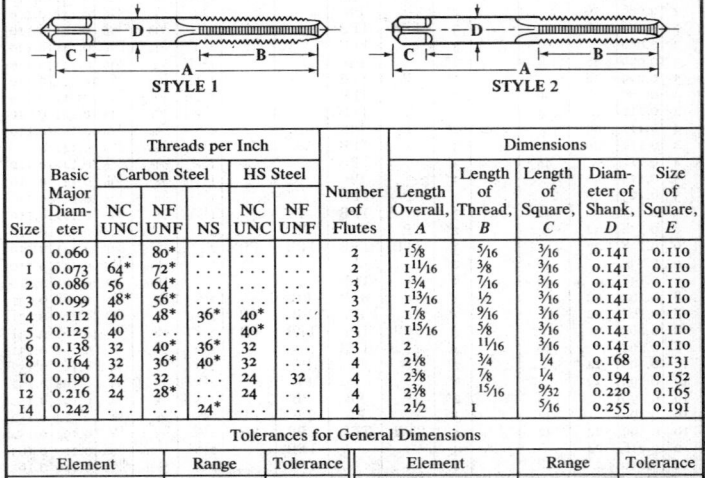

	Basic Major Diameter	Threads per Inch						Dimensions				
		Carbon Steel			HS Steel							
		NC UNC	NF UNF	NS	NC UNC	NF UNF	Number of Flutes	Length Overall, A	Length of Thread, B	Length of Square, C	Diameter of Shank, D	Size of Square, E
Size												
0	0.060	. . .	80*	. . .	. . .	. . .	2	1⅝	⁵⁄₁₆	³⁄₁₆	0.141	0.110
1	0.073	64*	72*	. . .	. . .	. . .	2	1¹¹⁄₁₆	⅜	³⁄₁₆	0.141	0.110
2	0.086	56	64*	. . .	. . .	. . .	3	1¾	⁷⁄₁₆	³⁄₁₆	0.141	0.110
3	0.099	48*	56*	. . .	. . .	. . .	3	1¹³⁄₁₆	½	³⁄₁₆	0.141	0.110
4	0.112	40	48*	36*	40*	. . .	3	1⅞	⁹⁄₁₆	³⁄₁₆	0.141	0.110
5	0.125	40	. . .	. . .	40*	. . .	3	1¹⁵⁄₁₆	⅝	³⁄₁₆	0.141	0.110
6	0.138	32	40*	36*	32	. . .	3	2	¹¹⁄₁₆	³⁄₁₆	0.141	0.110
8	0.164	32	36*	40*	32	. . .	4	2⅛	¾	¼	0.168	0.131
10	0.190	24	32	. . .	24	32	4	2³⁄₈	⅞	¼	0.194	0.152
12	0.216	24	28*	. . .	24	. . .	4	2³⁄₈	¹⁵⁄₁₆	⁹⁄₃₂	0.220	0.165
14	0.242	. . .	. . .	24*	. . .	. . .	4	2½	1	⁵⁄₁₆	0.255	0.191

Tolerances for General Dimensions						
Element	Range	Tolerance		Element	Range	Tolerance
Length Overall, A	0 to 14 incl	± ¹⁄₃₂		Diameter of Shank, D	0 to 12 incl	− 0.004
Length of Thread, B	0 to 12 incl 14	± ³⁄₆₄ ± ¹⁄₁₆			14	− 0.005
Length of Square, C	0 to 14 incl	± ¹⁄₃₂		Size of Square, E	0 to 14 incl	− 0.004

All dimensions are given in inches.
  * These taps are standard with plug chamfer only. All others are standard with taper, plug or bottoming chamfer.
  Styles 1 and 2 cut thread taps have optional style centers on thread and shank ends.
  For standard thread limits see Table 16. For eccentricity tolerances see Table 20.

Table 5. ANSI Standard Spiral-Pointed Machine Screw Taps (ANSI B94.9-1979)

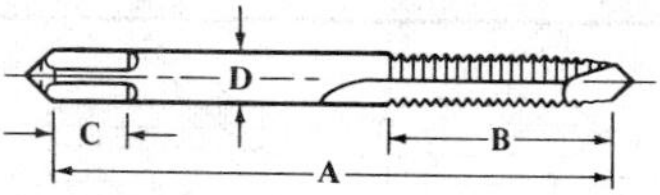

STYLE 1

High-Speed Steel Taps with Ground Threads

Size	Basic Major Diameter	Threads per Inch NC UNC	NF UNF	NS	No. of Flutes	Pitch Diam. Limits and Chamfers† H1	H2	H3	H7	Length Overall A	Length of Thread B	Length of Square C	Diameter of Shank D	Size of Square E
0	0.060	..	80	..	2	PB	PB	..	..	1⅝	5/16	3/16	0.141	0.110
1	0.073	64	72	..	2	P	P	..	..	1 11/16	3/8	3/16	0.141	0.110
2	0.086	56	..	..	2	PB	PB	..	..	1¾	7/16	3/16	0.141	0.110
2	0.086	..	64	..	2	..	P	..	..	1¾	7/16	3/16	0.141	0.110
3	0.099	48	..	..	2	..	PB	..	..	1 13/16	1/2	3/16	0.141	0.110
3	0.099	..	56	..	2	P	P	..	..	1 13/16	1/2	3/16	0.141	0.110
4	0.112	..	..	36	2	..	P	..	..	1⅞	9/16	3/16	0.141	0.110
4	0.112	40	..	..	2	P	PB	..	..	1⅞	9/16	3/16	0.141	0.110
4	0.112	..	48	..	2	P	PB	..	..	1⅞	9/16	3/16	0.141	0.110
5	0.125	40	..	..	2	P	PB	..	..	1 15/16	5/8	3/16	0.141	0.110
5	0.125	..	44	..	2	..	P	..	..	1 15/16	5/8	3/16	0.141	0.110
6	0.138	32	..	..	2	P	PB	PB	PB	2	11/16	3/16	0.141	0.110
6	0.138	..	40	..	2	..	PB	..	..	2	11/16	3/16	0.141	0.110
8	0.164	32	..	..	2	P	PB	PB	PB	2⅛	3/4	1/4	0.168	0.131
8	0.164	..	36	..	2	P	P	..	..	2⅛	3/4	1/4	0.168	0.131
10	0.190	24	..	..	2	P	PB	PB	P	2⅜	7/8	1/4	0.194	0.152
10	0.190	..	32	..	2	PB	PB	PB	P	2⅜	7/8	1/4	0.194	0.152
12	0.216	24	..	..	2	..	..	PB	..	2⅜	15/16	9/32	0.220	0.165
12	0.216	..	28	..	2	..	..	P	..	2⅜	15/16	9/32	0.220	0.165

High-Speed and Carbon Steel Taps with Cut Threads

Size	Basic Major Diameter	Carbon Steel NC UNC	Carbon Steel NF UNF	HS Steel NC UNC	HS Steel NF UNF	No. of Flutes	Length Overall, A	Length of Thread, B	Length of Square, C	Diameter of Shank, D	Size of Square, E
4	0.112	..	..	40	..	2	1⅞	9/16	3/16	0.141	0.110
5	0.125	..	..	40	..	2	1 15/16	5/8	3/16	0.141	0.110
6	0.138	32	..	32	..	2	2	11/16	3/16	0.141	0.110
8	0.164	32	..	32	..	2	2⅛	3/4	1/4	0.168	0.131
10	0.190	24	32	24	32	2	2⅜	7/8	1/4	0.194	0.152
12	0.216	..	..	24	..	2	2⅜	15/16	9/32	0.220	0.165

Tolerances for General Dimensions

Element	Size Range	Tolerance Ground Thread	Cut Thread	Element	Size Range	Tolerance Ground Thread	Cut Thread
Overall Length, A	0 to 12	±1/32	±1/32	Shank Diameter, D	0 to 12	−0.0015	−0.004
Thread Length, B	0 to 12	±3/64	±3/64	Size of Square, E	0 to 12	−0.004	−0.004
Square Length, C	0 to 12	±1/32	±1/32				

All dimensions are given in inches. Chamfer designations are: P = plug and B = bottoming. The cut thread taps are standard with plug chamfer only. The Style 1 ground thread taps have external centers on thread and shank ends (may be removed on thread end of bottoming taps). The Style 1 cut thread taps have optional style centers on thread and shank ends. Standard thread limits for ground threads are given in Table 15 and for cut threads in Table 16. For eccentricity tolerances see Table 20.

Table 6. ANSI Standard Spiral-Pointed Machine Screw Taps with Short Flutes and Regular and Fast Spiral-Fluted Machine Screw Taps (ANSI B94.9-1979)

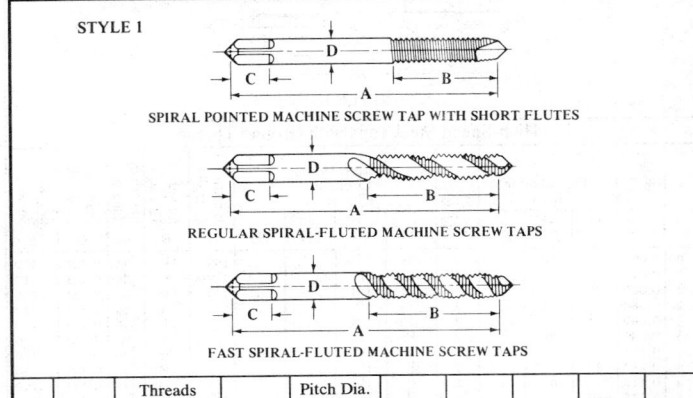

STYLE 1

SPIRAL POINTED MACHINE SCREW TAP WITH SHORT FLUTES

REGULAR SPIRAL-FLUTED MACHINE SCREW TAPS

FAST SPIRAL-FLUTED MACHINE SCREW TAPS

Size	Basic Major Diam-eter	Threads per Inch NC UNC	Threads per Inch NF UNF	No. of Flutes	Pitch Dia. Limits & Chamfers* H2	Pitch Dia. Limits & Chamfers* H3	Length Overall, A	Length of Thread, B	Length of Square, C	Diam-eter of Shank, D	Size of Square, E
3†	0.099	48	...	2	PB	...	1¹³⁄₁₆	½	³⁄₁₆	0.141	0.110
4	0.112	40	...	2	PB	...	1⅞	⁹⁄₁₆	³⁄₁₆	0.141	0.110
5	0.125	40	...	2	PB	...	1¹⁵⁄₁₆	⅝	³⁄₁₆	0.141	0.110
6	0.138	32	...	2	...	PB	2	¹¹⁄₁₆	³⁄₁₆	0.141	0.110
8	0.164	32	...	2‡, 3†	...	PB	2⅛	¾	¼	0.168	0.131
10	0.190	24	32	2‡, 3†	...	PB	2⅜	⅞	¼	0.194	0.152
12§	0.216	24	...	2‡, 3†	...	PB	2⅜	¹⁵⁄₁₆	⁹⁄₃₂	0.220	0.165

Tolerances for General Dimensions

Element	Size Range	Tolerance	Element	Size Range	Tolerance
Overall Length, A	3 to 12	± ¹⁄₃₂	Shank Diameter, D	3 to 12	− 0.0015
Thread Length, B	3 to 12	± ³⁄₆₄	Size of Square, E	3 to 12	− 0.004
Square Length, C	3 to 12	± ¹⁄₃₂			

All dimensions are given in inches. These standard taps are made of high-speed steel with ground threads. For standard thread limits see Table 15. For eccentricity tolerances see Table 20.

Spiral-Pointed Machine Screw Taps with Short Flutes: These taps are standard with plug chamfer only. They are provided with a spiral point only; the balance of the threaded section is left unfluted. These Style 1 taps have external centers on thread and shank ends.

Regular Spiral-Fluted Machine Screw Taps: These taps have right-hand spiral flutes with a helix angle of from 25 to 35 degrees.

Fast Spiral-Fluted Machine Screw Taps: These taps have right-hand spiral flutes with a helix angle of from 45 to 60 degrees.

Both regular and fast spiral-fluted Style 1 taps have external centers on thread and shank ends (may be removed on thread end of bottoming taps).

Chamfer designations: P = plug and B = bottoming.

* Bottom chamfer applies only to regular and fast spiral-fluted machine screw taps.

† Applies only to fast spiral-fluted machine screw taps.

‡ Does not apply to fast spiral-fluted machine screw taps.

§ Does not apply to regular spiral-fluted machine screw taps.

Table 7. ANSI Standard Nut Taps (ANSI B94.9-1979)

Diam. of Tap	Threads per Inch NC UNC	Number of Flutes	Length Overall, A	Length of Thread, B	Length of Square, C	Diameter of Shank, D	Size of Square, E
1/4	20	4	5	1 5/8	9/16	0.185	0.139
5/16	18	4	5 1/2	1 13/16	5/8	0.240	0.180
3/8	16	4	6	2	11/16	0.294	0.220
1/2	13	4	7	2 1/2	7/8	0.400	0.300

Tolerances for General Dimensions

Element	Diameter Range	Tolerance	Element	Diameter Range	Tolerance
Overall Length, A	1/4 to 1/2	± 1/16	Shank Diameter, D	1/4 to 1/2	− 0.005
Thread Length, B	1/4 to 1/2	± 1/16			
Square Length, C	1/4 to 1/2	± 1/32	Size of Square, E	1/4 to 1/2	− 0.004

All dimensions are given in inches. These ground thread high-speed steel taps are standard in H3 limit only. All taps have an internal center in thread end. For standard limits see Table 14.
* Chamfer J is made 1/2 to 3/4 the thread length B.

Table 8. ANSI Standard Pulley Taps (ANSI B94.9-1979)

Diam. of Tap	Threads per Inch NC UNC	Number of Flutes	Length Overall, A	Length of Thread, B	Length of Square, C	Diam. of Shank, D	Length of Close Tolerance, T*	Size of Square, E	Length of Neck, K**
1/4	20	4	6,8	1	5/16	0.255	1 1/2	0.191	3/8
5/16	18	4	6,8	1 1/8	3/8	0.318	1 9/16	0.238	3/8
3/8	16	4	6,8,10	1 1/4	7/16	0.381	1 5/8	0.286	3/8
7/16	14	4	6	1 7/16	1/2	0.444	1 11/16	0.333	7/16
1/2	13	4	6,8,10,12	1 21/32	9/16	0.507	1 11/16	0.380	1/2
5/8	11	4	6,8,10	1 13/16	11/16	0.633	2	0.475	5/8
3/4	10	4	10	2	3/4	0.759	2 1/4	0.569	3/4

Tolerances for General Dimensions

Element	Diameter Range	Tolerance	Element	Diameter Range	Tolerance
Overall Length, A	1/4 to 3/4	± 1/16	Shank Diameter, D	1/4 to 3/4	− 0.005
Thread Length, B	1/4 to 3/4	± 1/16			
Square Length, C	1/4 to 3/4	± 1/32	Size of Square, E	1/4 to 1/2 5/8 to 3/4	− 0.004 − 0.006

All dimensions are given in inches. These ground thread high-speed steel taps are standard with plug chamfer in H3 limit only. All taps have an internal center in thread end. For standard thread limits see Table 14. For eccentricity tolerances see Table 20.
* T is minimum length of shank which is held to eccentricity tolerances.
** K neck optional with manufacturer.

Table 9a. ANSI Standard Ground Thread Metric Taps
ISO Metric Thread Form (ANSI B94.9-1979)

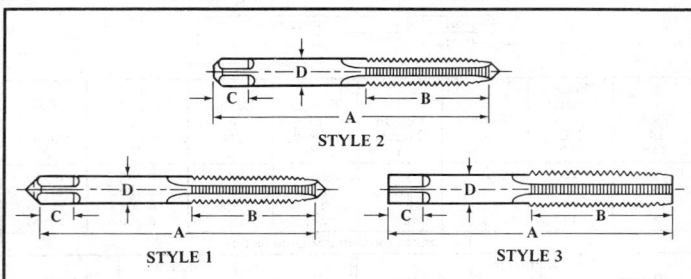

STYLE 2

STYLE 1   STYLE 3

Nom. Diam. mm	Pitch mm	No. of Flutes	Pitch Diameter Limits and Chamfers							Length Overall A	Length of Thread B	Length of Square C	Diam. of Square D	Size of Square E
			D3	D4	D5	D6	D7	D8	D9					
1.6	0.35	2	PB	...	...	...	...	...	...	1⅝	⁵⁄₁₆	³⁄₁₆	.141	.110
2	0.4	3	PB	...	...	...	...	...	...	1¾	⁷⁄₁₆	³⁄₁₆	.141	.110
2.5	0.45	3	PB	...	...	...	...	...	...	1¹³⁄₁₆	½	³⁄₁₆	.141	.110
3	0.5	3	PB	...	...	...	...	...	...	1¹⁵⁄₁₆	⅝	³⁄₁₆	.141	.110
3.5	0.6	3	...	PB	...	...	...	...	...	2	¹¹⁄₁₆	³⁄₁₆	.141	.110
4	0.7	4	...	PB	...	...	...	...	...	2⅛	¾	¼	.168	.131
5	0.8	4	...	PB	...	...	...	...	...	2⅜	⅞	¼	.194	.152
6	1	4	...	...	PB	...	...	...	...	2½	1	⁵⁄₁₆	.255	.198
8	1.25	4	...	...	PB	...	...	...	...	2²³⁄₃₂	1⅛	⅜	.318	.238
10	1.5	4	...	...	...	PB	...	...	...	2¹⁵⁄₁₆	1¼	⁷⁄₁₆	.381	.286
12	1.75	4	...	...	...	PB	...	...	...	3⅜	1²¹⁄₃₂	⁷⁄₁₆	.367	.275
14	2	4	...	...	...	...	PB	...	...	3¹⁹⁄₃₂	1²¹⁄₃₂	½	.429	.322
16	2	4	...	...	...	...	PB	...	...	3¹³⁄₁₆	1¹³⁄₁₆	⁹⁄₁₆	.480	.360
20	2.5	4	...	...	...	...	PB	...	...	4¹⁵⁄₃₂	2	¹¹⁄₁₆	.652	.489
24	3	4	...	...	...	...	...	PB	...	4²⁹⁄₃₂	2⁷⁄₃₂	¾	.760	.570
30	3.5	4	...	...	...	...	...	...	PB	5⁷⁄₁₆	2⁹⁄₁₆	1	1.021	.766
36	4	4	...	...	...	...	...	...	PB	6¹⁄₁₆	3	1⅛	1.233	.925

Tolerances

Element	Nom. Diam. Range, mm	Toler., Inch	Element	Nom. Diam. Range, mm	Toler., Inch
Overall Length, A	M1.6 to M24, incl.	± ¹⁄₃₂	Shank Diameter, D		
	M30 and M36	± ¹⁄₁₆		M1.6 to M14, incl.	− 0.0015
Thread Length, B	M1.6 to M5, incl.	± ³⁄₆₄		M16 to M36	− 0.002
	M6 to M12 incl.	± ¹⁄₁₆	Size of Square, E		
	M14 to M36	± ³⁄₃₂		M1.6 to M12, incl.	− 0.004
Square Length, C	M1.6 to M24, incl.	± ¹⁄₃₂		M14 to M24, incl.	− 0.006
	M30 and M36	± ¹⁄₁₆		M30 and M36	− 0.008

All dimensions are in inches except where otherwise stated.

Chamfer Designation: P — Plug, B — Bottoming. These taps are high-speed steel.

Style 1 taps, sizes M1.6 through M5, have external center on thread and shank ends (may be removed on thread end of bottoming taps).

Style 2 taps, sizes M6, M8 and M10, have external center on thread end (may be removed on bottoming taps) and external partial cone center on shank end with length of cone approximately ¼ of diameter of shank.

Style 3 taps, sizes larger than M10 have external center on thread and shank ends.

For standard thread limits see Tables 19a and 19b.

For eccentricity tolerances of tap elements see Table 20.

Table 9b. ANSI Standard Spiral Pointed Ground Thread Metric Taps
ISO Metric Thread Form (ANSI B94.9-1979)

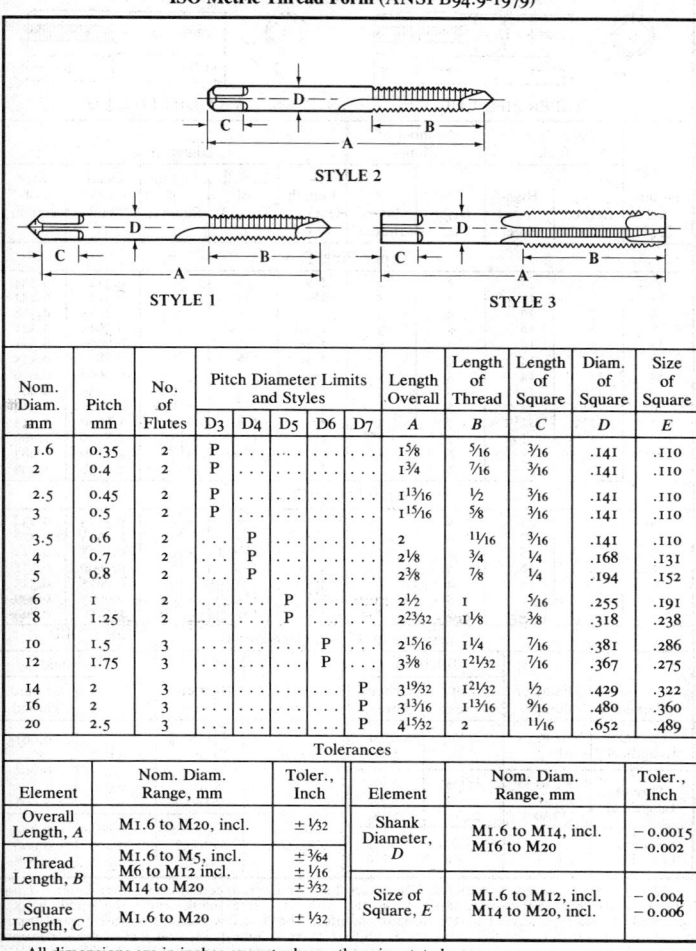

STYLE 2

STYLE 1

STYLE 3

Nom. Diam. mm	Pitch mm	No. of Flutes	Pitch Diameter Limits and Styles					Length Overall	Length of Thread	Length of Square	Diam. of Square	Size of Square
			D3	D4	D5	D6	D7	A	B	C	D	E
1.6	0.35	2	P	...	...	...	...	1⅝	5⁄16	3⁄16	.141	.110
2	0.4	2	P	...	...	...	...	1¾	7⁄16	3⁄16	.141	.110
2.5	0.45	2	P	...	...	...	...	1¹³⁄16	½	3⁄16	.141	.110
3	0.5	2	P	...	...	...	...	1¹⁵⁄16	5⁄8	3⁄16	.141	.110
3.5	0.6	2	...	P	...	...	...	2	11⁄16	3⁄16	.141	.110
4	0.7	2	...	P	...	...	...	2⅛	¾	¼	.168	.131
5	0.8	2	...	P	...	...	...	2⅜	7⁄8	¼	.194	.152
6	1	2	...	...	P	...	...	2½	1	5⁄16	.255	.191
8	1.25	2	...	...	P	...	...	2²³⁄32	1⅛	3⁄8	.318	.238
10	1.5	3	...	...	...	P	...	2¹⁵⁄16	1¼	7⁄16	.381	.286
12	1.75	3	...	...	...	P	...	3⅜	1²¹⁄32	7⁄16	.367	.275
14	2	3	...	...	...	...	P	3¹⁹⁄32	1²¹⁄32	½	.429	.322
16	2	3	...	...	...	...	P	3¹³⁄16	1¹³⁄16	9⁄16	.480	.360
20	2.5	3	...	...	...	...	P	4¹⁵⁄32	2	11⁄16	.652	.489

Tolerances						
Element	Nom. Diam. Range, mm	Toler., Inch	Element	Nom. Diam. Range, mm		Toler., Inch
Overall Length, A	M1.6 to M20, incl.	± 1⁄32	Shank Diameter, D	M1.6 to M14, incl. M16 to M20		− 0.0015 − 0.002
Thread Length, B	M1.6 to M5, incl. M6 to M12 incl. M14 to M20	± 3⁄64 ± 1⁄16 ± 3⁄32	Size of Square, E	M1.6 to M12, incl. M14 to M20, incl.		− 0.004 − 0.006
Square Length, C	M1.6 to M20	± 1⁄32				

All dimensions are in inches except where otherwise stated.

Chamfer Designation: P — Plug. These taps are high-speed steel.

Style 1 taps, sizes M1.6 through M5, have external center on thread and shank ends.

Style 2 taps, sizes M6, M8 and M10, have external center on thread end and external partial cone center on shank end with length of cone approximately ¼ of diameter of shank.

Style 3 taps, sizes larger than M10 have internal center on thread and shank ends.

For standards thread limits see Table 19a and 19b.

For eccentricity tolerances of tap elements see Table 20.

Table 10. ANSI Standard Taper and Straight Pipe Taps (ANSI B94.9-1979)

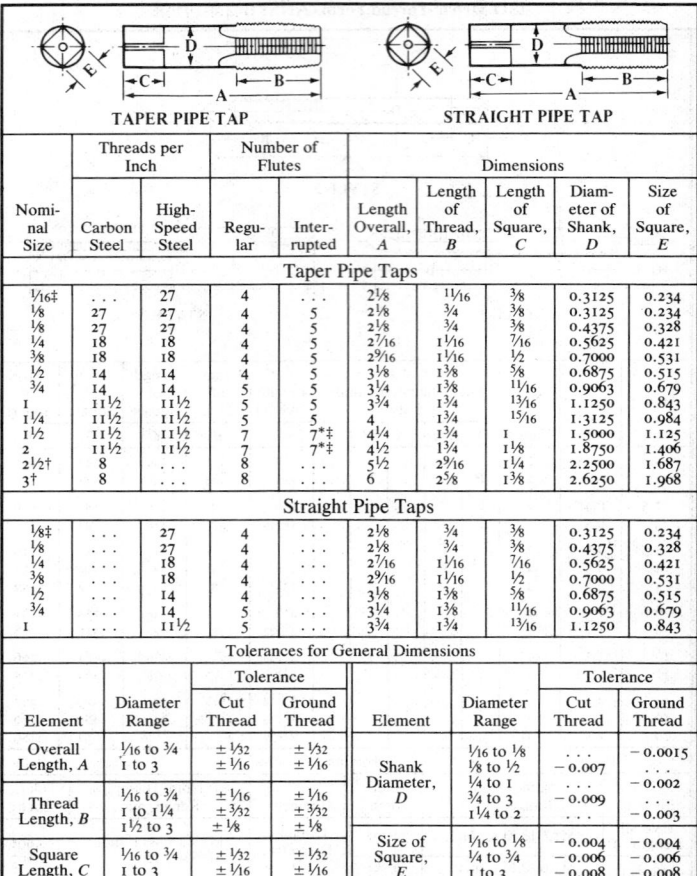

TAPER PIPE TAP   STRAIGHT PIPE TAP

Nominal Size	Threads per Inch		Number of Flutes		Dimensions				
	Carbon Steel	High-Speed Steel	Regular	Interrupted	Length Overall, A	Length of Thread, B	Length of Square, C	Diameter of Shank, D	Size of Square, E
Taper Pipe Taps									
1/16‡	. . .	27	4	. . .	2⅛	11/16	3/8	0.3125	0.234
1/8	27	27	4	5	2⅛	3/4	3/8	0.3125	0.234
1/8	27	27	4	5	2⅛	3/4	3/8	0.4375	0.328
1/4	18	18	4	5	2 7/16	1 1/16	7/16	0.5625	0.421
3/8	18	18	4	5	2 9/16	1 1/16	1/2	0.7000	0.531
1/2	14	14	4	5	3⅛	1 3/8	5/8	0.6875	0.515
3/4	14	14	5	5	3¼	1 3/8	11/16	0.9063	0.679
1	11½	11½	5	5	3¾	1¾	13/16	1.1250	0.843
1¼	11½	11½	5	5	4	1¾	15/16	1.3125	0.984
1½	11½	11½	7	7*‡	4¼	1¾	1	1.5000	1.125
2	11½	11½	7	7*‡	4½	1¾	1⅛	1.8750	1.406
2½†	8	. . .	8	. . .	5½	2 9/16	1¼	2.2500	1.687
3†	8	. . .	8	. . .	6	2 5/8	1 3/8	2.6250	1.968
Straight Pipe Taps									
1/8‡	. . .	27	4	. . .	2⅛	3/4	3/8	0.3125	0.234
1/8	. . .	27	4	. . .	2⅛	3/4	3/8	0.4375	0.328
1/4	. . .	18	4	. . .	2 7/16	1 1/16	7/16	0.5625	0.421
3/8	. . .	18	4	. . .	2 9/16	1 1/16	1/2	0.7000	0.531
1/2	. . .	14	4	. . .	3⅛	1 3/8	5/8	0.6875	0.515
3/4	. . .	14	5	. . .	3¼	1 3/8	11/16	0.9063	0.679
1	. . .	11½	5	. . .	3¾	1¾	13/16	1.1250	0.843

Tolerances for General Dimensions

Element	Diameter Range	Tolerance		Element	Diameter Range	Tolerance	
		Cut Thread	Ground Thread			Cut Thread	Ground Thread
Overall Length, A	1/16 to 3/4	± 1/32	± 1/32	Shank Diameter, D	1/16 to 1/8	. . .	− 0.0015
	1 to 3	± 1/16	± 1/16		1/8 to 1/2	− 0.007	. . .
Thread Length, B	1/16 to 3/4	± 1/16	± 1/16		1/4 to 1	. . .	− 0.002
	1 to 1¼	± 3/32	± 3/32		3/4 to 3	− 0.009	. . .
	1½ to 3	± 1/8	± 1/8		1¼ to 2	. . .	− 0.003
Square Length, C	1/16 to 3/4	± 1/32	± 1/32	Size of Square, E	1/16 to 1/8	− 0.004	− 0.004
	1 to 3	± 1/16	± 1/16		1/4 to 3/4	− 0.006	− 0.006
					1 to 3	− 0.008	− 0.008

All dimensions are given in inches. These taps have an internal center in the thread end. *Taper Pipe Threads:* The 1/8-inch pipe tap is furnished with large size shank unless the small shank is specified. These taps have 2 to 3½ threads chamfer. The first few threads on interrupted thread pipe taps are left full. The following styles and sizes are standard: 1/16 to 2 inches regular ground thread, NPT, NPTF, and ANPT; 1/8 to 2 inches interrupted ground thread, NPT, NPTF and ANPT; 1/8 to 3 inches carbon steel regular cut thread, NPT; 1/8 to 2 inches high-speed steel, regular cut thread, NPT; 1/8 to 1¼ inches high-speed steel interrupted cut thread, NPT. For standard thread limits see Table 17. *Straight Pipe Threads:* The 1/8-inch pipe tap is furnished with large size shank unless the small size is specified. These taps are standard with plug chamfer only. The following styles and sizes are standard: ground threads — 1/8 to 1 inch, NPSC and NPSM; 1/8 to 3/4 inch, NPSF; cut threads — 1/8 to 1 inch, NPSC and NPSM. For standard thread limits see Tables 18a, 18b, and 18c. For eccentricity tolerances see Table 20. * Standard in NPT form of thread only. † Cut thread taps only. ‡ Ground thread taps only.

Table 11. Taps Recommended for Classes 1B,* 2B,† and 3B Unified Screw Threads — Numbered and Fractional Sizes (ANSI B94.9-1979)

Size	Threads per Inch NC UNC	Threads per Inch NF UNF	Recommended Tap For Class of Thread Class 2B†	Recommended Tap For Class of Thread Class 3B	Pitch Diameter Limits For Class of Thread Min All Classes (Basic)	Pitch Diameter Limits For Class of Thread Max Class 2B	Pitch Diameter Limits For Class of Thread Max Class 3B
			Machine Screw Numbered Size Taps				
0	. . .	80	G H2	G H1	0.0519	0.0542	0.0536
1	64	. . .	G H2	G H1	0.0629	0.0655	0.0648
1	. . .	72	G H2	G H1	0.0640	0.0665	0.0659
2	56	. . .	G H2	G H1	0.0744	0.0772	0.0765
2	. . .	64	G H2	G H1	0.0759	0.0786	0.0779
3	48	. . .	G H2	G H1	0.0855	0.0885	0.0877
3	. . .	56	G H2	G H1	0.0874	0.0902	0.0895
4	40	. . .	G H2	G H2	0.0958	0.0991	0.0982
4	. . .	48	G H2	G H1	0.0985	0.1016	0.1008
5	40	. . .	G H2	G H2	0.1088	0.1121	0.1113
5	. . .	44	G H2	G H1	0.1102	0.1134	0.1126
6	32	. . .	G H3	G H2	0.1177	0.1214	0.1204
6	. . .	40	G H2	G H2	0.1218	0.1252	0.1243
8	32	. . .	G H3	G H2	0.1437	0.1475	0.1465
8	. . .	36	G H2	G H2	0.1460	0.1496	0.1487
10	24	. . .	G H3	G H3	0.1629	0.1672	0.1661
10	. . .	32	G H3	G H2	0.1697	0.1736	0.1726
12	24	. . .	G H3	G H3	0.1889	0.1933	0.1922
12	. . .	28	G H3	G H3	0.1928	0.1970	0.1959
			Fractional Size Taps				
1/4	20	. . .	G H5	G H3	0.2175	0.2224	0.2211
1/4	. . .	28	G H4	G H3	0.2268	0.2311	0.2300
5/16	18	. . .	G H5	G H3	0.2764	0.2817	0.2803
5/16	. . .	24	G H4	G H3	0.2854	0.2902	0.2890
3/8	16	. . .	G H5	G H3	0.3344	0.3401	0.3387
3/8	. . .	24	G H4	G H3	0.3479	0.3528	0.3516
7/16	14	. . .	G H5	G H3	0.3911	0.3972	0.3957
7/16	. . .	20	G H5	G H3	0.4050	0.4104	0.4091
1/2	13	. . .	G H5	G H3	0.4500	0.4565	0.4548
1/2	. . .	20	G H5	G H3	0.4675	0.4731	0.4717
9/16	12	. . .	G H5	G H3	0.5084	0.5152	0.5135
9/16	. . .	18	G H5	G H3	0.5264	0.5323	0.5308
5/8	11	. . .	G H5	G H3	0.5660	0.5732	0.5714
5/8	. . .	18	G H5	G H3	0.5889	0.5949	0.5934
3/4	10	. . .	G H5	G H5	0.6850	0.6927	0.6907
3/4	. . .	16	G H5	G H3	0.7094	0.7159	0.7143
7/8	9	. . .	G H6	G H4	0.8028	0.8110	0.8089
7/8	. . .	14	G H6**	G H4	0.8286	0.8356	0.8339
1	8	. . .	G H6	G H4	0.9188	0.9276	0.9254
1	. . .	12	G H6**	G H4	0.9459	0.9535	0.9516
1		14NS	G H6**	G H4	0.9536	0.9609	0.9590
1 1/8	7	. . .	G H8**	G H4	1.0322	1.0416	1.0393
1 1/8	. . .	12	G H6**	G H4	1.0709	1.0787	1.0768
1 1/4	7	. . .	G H8**	G H4	1.1572	1.1668	1.1644
1 1/4	. . .	12	G H6**	G H4	1.1959	1.2039	1.2019
1 3/8	6	. . .	G H8**	G H4	1.2667	1.2771	1.2745
1 3/8	. . .	12	G H6**	G H4	1.3209	1.3291	1.3270
1 1/2	6	. . .	G H8**	G H4	1.3917	1.4022	1.3996
1 1/2	. . .	12	G H6**	G H4	1.4459	1.4542	1.4522

All dimensions are given in inches.
* 1B tapped holes can be produced with cut thread taps.   † Cut thread taps in all fractional sizes and in numbered sizes 3 to 12 NC and NF may be used under normal conditions and in average materials to produce tapped holes in this classification.   ** Standard G H4 taps are also suitable for this class of thread.

The above recommended taps normally produce the class of thread indicated in average materials when used with reasonable care. However, if the tap specified does not give a satisfactory gage fit in the work, a choice of some other limit tap will be necessary.

**Table 12. ANSI Standard Ground Thread Spark Plug Hand Taps —
Metric Sizes (ANSI B94.9-1979)**

Tap Diameter, mm	Pitch, mm	Number of Flutes	Overall Length, In. A	Thread Length, In. B	Square Length, In. C	Shank Diam., In. D	Square Size, In. E
14	1.25	4	3¹⁹⁄₃₂	1²¹⁄₃₂	½	0.429	0.322
18	1.50	4	4¹⁄₃₂	1¹³⁄₁₆	⅝	0.542	0.406

These are high-speed steel Style 3 taps and have internal center in thread and shank ends. They are standard with plug chamfer only, right-hand threads with 60-degree form of thread.

Tolerances: Overall length, ± ¹⁄₃₂ inch; thread length, ± ³⁄₃₂ inch; square length, ± ¹⁄₃₂ inch; shank diameter, 14 mm, − 0.0015 inch, 18 mm, − 0.0020 inch; and size of square, − 0.0040 inch.

**Table 13. ANSI Standard Fraction-size Taps —
Cut Thread Limits (ANSI B94.9-1979)**

Tap Size	NC UNC	NF UNF	NS or UN	Major Diameter Basic	Min.	Max.	Pitch Diameter Basic	Min.	Max.
⅛	...	...	40	0.1250	0.1266	0.1286	0.1088	0.1090	0.1105
⁵⁄₃₂	...	...	32	0.1563	0.1585	0.1605	0.1360	0.1365	0.1380
³⁄₁₆	...	...	24	0.1875	0.1903	0.1923	0.1604	0.1609	0.1624
³⁄₁₆	...	...	32	0.1875	0.1897	0.1917	0.1672	0.1677	0.1692
¼	20	...	...	0.2500	0.2532	0.2557	0.2175	0.2180	0.2200
¼	...	28	...	0.2500	0.2524	0.2549	0.2268	0.2273	0.2288
⁵⁄₁₆	18	...	...	0.3125	0.3160	0.3185	0.2764	0.2769	0.2789
⁵⁄₁₆	...	24	...	0.3125	0.3153	0.3178	0.2854	0.2859	0.2874
⅜	16	...	...	0.3750	0.3789	0.3814	0.3344	0.3349	0.3369
⅜	...	24	...	0.3750	0.3778	0.3803	0.3479	0.3484	0.3499
⁷⁄₁₆	14	...	...	0.4375	0.4419	0.4449	0.3911	0.3916	0.3941
⁷⁄₁₆	...	20	...	0.4375	0.4407	0.4437	0.4050	0.4055	0.4075
½	13	...	...	0.5000	0.5047	0.5077	0.4500	0.4505	0.4530
½	...	20	...	0.5000	0.5032	0.5062	0.4675	0.4680	0.4700
⁹⁄₁₆	12	...	...	0.5625	0.5675	0.5705	0.5084	0.5089	0.5114
⁹⁄₁₆	...	18	...	0.5625	0.5660	0.5690	0.5264	0.5269	0.5289
⅝	11	...	...	0.6250	0.6304	0.6334	0.5660	0.5665	0.5690
⅝	...	18	...	0.6250	0.6285	0.6315	0.5889	0.5894	0.5914
¾	10	...	...	0.7500	0.7559	0.7599	0.6850	0.6855	0.6885
¾	...	16	...	0.7500	0.7539	0.7579	0.7094	0.7099	0.7124
⅞	9	...	...	0.8750	0.8820	0.8860	0.8028	0.8038	0.8068
⅞	...	14	...	0.8750	0.8799	0.8839	0.8286	0.8296	0.8321
1	8	...	...	1.0000	1.0078	1.0118	0.9188	0.9198	0.9228
1	...	...	14	1.0000	1.0049	1.0089	0.9536	0.9546	0.9571
1⅛	7	...	...	1.1250	1.1337	1.1382	1.0322	1.0332	1.0367
1⅛	...	12	...	1.1250	1.1305	1.1350	1.0709	1.0719	1.0749
1¼	7	...	...	1.2500	1.2587	1.2632	1.1572	1.1582	1.1617
1¼	...	12	...	1.2500	1.2555	1.2600	1.1959	1.1969	1.1999
1⅜	6	...	...	1.3750	1.3850	1.3895	1.2667	1.2677	1.2712
1⅜	...	12	...	1.3750	1.3805	1.3850	1.3209	1.3219	1.3249
1½	6	...	...	1.5000	1.5100	1.5145	1.3917	1.3927	1.3962
1½	...	12	...	1.5000	1.5055	1.5100	1.4459	1.4469	1.4499
1¾	5	...	...	1.7500	1.7602	1.7657	1.6201	1.6216	1.6256
2	4½	...	...	2.0000	2.0111	2.0166	1.8557	1.8572	1.8612

All dimensions are given in inches.

Lead Tolerance: Plus or minus 0.003 inch max. per inch of thread.

Angle Tolerance: Plus or minus 35 min. in half angle or 53 min. in full angle for 4½ to 5½ thds. per in.; 40 min. half angle and 60 min. full angle for 6 to 9 thds.; 45 min. half angle and 68 min. full angle for 10 to 28 thds.; 60 min. half angle and 90 min. full angle for 30 to 64 thds. per in.

Table 14. ANSI Standard Fractional-size Taps — Ground Thread Limits (ANSI B94.9-1979)

Size	NC UNC	NF UNF	NS	Major Dia. Basic	Major Dia. Min.	Major Dia. Max.	Basic Pitch Diam.	H1 Min.	H1 Max.	H2 Min.	H2 Max.	H3 & H4* Min.	H3 & H4* Max.	H4,* H5† & H6§ Min.	H4,* H5† & H6§ Max.
1/4	20			0.2500	0.2540	0.2550	0.2175	0.2175	0.2180	0.2180	0.2185	0.2185	0.2190	0.2195†	0.2200†
1/4		28		0.2500	0.2525	0.2535	0.2268	0.2268	0.2273	0.2273	0.2278	0.2278	0.2283	0.2283*	0.2288*
5/16	18			0.3125	0.3170	0.3180	0.2764	0.2764	0.2769	0.2769	0.2774	0.2774	0.2779	0.2784†	0.2789†
5/16		24		0.3125	0.3155	0.3165	0.2854	0.2854	0.2859	0.2859	0.2864	0.2864	0.2869	0.2869*	0.2874*
3/8	16			0.3750	0.3800	0.3810	0.3344	0.3344	0.3349	0.3349	0.3354	0.3354	0.3359	0.3364†	0.3369†
3/8		24		0.3750	0.3780	0.3790	0.3479	0.3479	0.3484	0.3484	0.3489	0.3489	0.3494	0.3494*	0.3499*
7/16	14			0.4375	0.4435	0.4445	0.3911			0.3916	0.3921	0.3921	0.3926	0.3931†	0.3936†
7/16		20		0.4375	0.4415	0.4425	0.4050					0.4060	0.4065	0.4070†	0.4075†
1/2	13			0.5000	0.5065	0.5075	0.4500	0.4500	0.4505	0.4505	0.4510	0.4510	0.4515	0.4520†	0.4525†
1/2		20		0.5000	0.5040	0.5050	0.4675	0.4675	0.4680	0.4680	0.4685	0.4685	0.4690	0.4695†	0.4700†
9/16	12			0.5625	0.5690	0.5700	0.5084					0.5094	0.5099	0.5104†	0.5109†
9/16		18		0.5625	0.5670	0.5680	0.5264			0.5269	0.5274	0.5274	0.5279	0.5284†	0.5289†
5/8	11			0.6250	0.6320	0.6330	0.5660			0.5665	0.5670	0.5670	0.5675	0.5680†	0.5685†
5/8		18		0.6250	0.6295	0.6305	0.5889			0.5894	0.5899	0.5899	0.5904	0.5909†	0.5914†
11/16			11	0.6875	0.6945	0.6955	0.6285					0.6295	0.6300		
11/16			16	0.6875	0.6925	0.6935	0.6469					0.6479	0.6484		
3/4	10			0.7500	0.7575	0.7590	0.6850			0.6855	0.6860	0.6860	0.6865	0.6870†	0.6875†
3/4		16		0.7500	0.7550	0.7560	0.7094	0.7094	0.7099	0.7099	0.7104	0.7104	0.7109	0.7114†	0.7119†
7/8	9			0.8750	0.8835	0.8850	0.8028					0.8043*	0.8048*	0.8053§	0.8058§
7/8		14		0.8750	0.8810	0.8820	0.8286			0.8291	0.8296	0.8301*	0.8306*		
1	8			1.0000	1.0095	1.0110	0.9188					0.9203*	0.9208*	0.9213§	0.9218§
1		12		1.0000	1.0065	1.0075	0.9459					0.9474*	0.9479*		
1			14	1.0000	1.0060	1.0070	0.9536					0.9551*	0.9556*		

All dimensions are given in inches.
* H4 limit value.
† H5 limit value.
§ H6 limit value.

Table 14 (*Concluded*). ANSI Standard Fractional-size Taps — Ground Thread Limits (ANSI B94.9-1979)

| Size | Threads per Inch | | | Major Diameter | | | Pitch Diameter Limits | | |
	NC UNC	NF UNF	NS	Basic	Min.	Max.	Basic Pitch Diam.	H4 Limit Min.	H4 Limit Max.
1⅛	7	...	...	1.1250	1.1350	1.1370	1.0322	1.0332	1.0342
1⅛	...	12	...	1.1250	1.1315	1.1325	1.0709	1.0719	1.0729
1¼	7	...	...	1.2500	1.2600	1.2620	1.1572	1.1582	1.1592
1¼	...	12	...	1.2500	1.2565	1.2575	1.1959	1.1969	1.1979
1⅜	6	...	...	1.3750	1.3870	1.3890	1.2667	1.2677	1.2687
1⅜	...	12	...	1.3750	1.3815	1.3825	1.3209	1.3219	1.3229
1½	6	...	...	1.5000	1.5120	1.5140	1.3917	1.3927	1.3937
1½	...	12	...	1.5000	1.5065	1.5075	1.4459	1.4469	1.4479

All dimensions are given in inches.

Lead Tolerance: Plus or minus 0.0005 inch within any two threads not farther apart than one inch.

Angle Tolerance: Plus or minus 25 min. in half angle for 6 to 9 threads per inch; plus or minus 30 min. in half angle for 10 to 28 threads per inch.

For an explanation of the significance of the H4 limit value range see page 889.

Table 15. ANSI Standard Machine Screw Taps — Ground Thread Limits (ANSI B94.9-1979)

| Size | Threads per Inch | | | Major Diameter | | | Pitch Diameter Limits* | | | | | | |
	NC UNC	NF UNF	NS	Basic	Min.	Max.	Basic Pitch Diam.	H1 Limit Min.	H1 Limit Max.	H2 Limit Min.	H2 Limit Max.	H3 Limit Min.	H3 Limit Max.
0	...	80	...	0.0600	0.0605	0.0615	0.0519	0.0519	0.0524	0.0524	0.0529	...	...
1	64	...	...	0.0730	0.0735	0.0745	0.0629	0.0629	0.0634	0.0634	0.0639	...	...
1	...	72	...	0.0730	0.0735	0.0745	0.0640	0.0640	0.0645	0.0645	0.0650	...	...
2	56	...	...	0.0860	0.0865	0.0875	0.0744	0.0744	0.0749	0.0749	0.0754	...	...
2	...	64	...	0.0860	0.0865	0.0875	0.0759	...	...	0.0764	0.0769	...	...
3	48	...	...	0.0990	0.1000	0.1010	0.0855	0.0855	0.0860	0.0860	0.0865	...	...
3	...	56	...	0.0990	0.0995	0.1005	0.0874	0.0874	0.0879	0.0879	0.0884	...	...
4	...	...	36	0.1120	0.1135	0.1145	0.0940	...	...	0.0945	0.0950	...	...
4	40	...	...	0.1120	0.1135	0.1145	0.0958	0.0958	0.0963	0.0963	0.0968	...	...
4	...	48	...	0.1120	0.1130	0.1140	0.0985	0.0985	0.0990	0.0990	0.0995	...	...
5	40	...	...	0.1250	0.1265	0.1275	0.1088	0.1088	0.1093	0.1093	0.1098	...	...
5	...	44	...	0.1250	0.1260	0.1270	0.1102	...	...	0.1107	0.1112	...	...
6	32	...	...	0.1380	0.1400	0.1410	0.1177	0.1177	0.1182	0.1182	0.1187	0.1187	0.1192
6	...	40	...	0.1380	0.1395	0.1405	0.1218	0.1218	0.1223	0.1223	0.1228	...	...
8	32	...	...	0.1640	0.1660	0.1670	0.1437	0.1437	0.1442	0.1442	0.1447	0.1447	0.1452
8	...	...	36	0.1640	0.1655	0.1665	0.1460	...	...	0.1465	0.1470	...	...
10	24	...	...	0.1900	0.1930	0.1940	0.1629	0.1629	0.1634	0.1634	0.1639	0.1639	0.1644
10	...	32	...	0.1900	0.1920	0.1930	0.1697	0.1697	0.1702	0.1702	0.1707	0.1707	0.1712
12	24	...	...	0.2160	0.2190	0.2200	0.1889	...	...	...	...	0.1899	0.1904
12	...	28	...	0.2160	0.2185	0.2195	0.1928	...	...	...	...	0.1938	0.1943

All dimensions are given in inches.

Lead Tolerance: Plus or minus 0.0005 inch within any two threads not farther apart than one inch.

Angle Tolerance: Plus or minus 30 min. in half angle for 20 to 80 threads per inch.

For an explanation of the significance of the limit value ranges see page 889.

* H7 limits (formerly designated as G) apply to same threads as H3 limits with the exception of the 12-24 and 12-28 threads. H7 limits have minimum and maximum major diameters 0.0020 inch larger than shown and minimum and maximum pitch diameters 0.0020 inch larger than shown for H3 limits.

Table 16. ANSI Standard Machine Screw Taps — Cut Threads Limits (ANSI B9.9-1979)

Size	Threads per Inch			Major Diameter			Pitch Diameter		
	NC UNC	NF UNF	NS or UN	Basic	Min.	Max.	Basic	Min.	Max.
0	. . .	80	. . .	0.0600	0.0609	0.0624	0.0519	0.0521	0.0531
1	64	. . .	. . .	0.0730	0.0740	0.0755	0.0629	0.0631	0.0641
1	. . .	72	. . .	0.0730	0.0740	0.0755	0.0640	0.0642	0.0652
2	56	. . .	. . .	0.0860	0.0872	0.0887	0.0744	0.0746	0.0756
2	. . .	64	. . .	0.0860	0.0870	0.0885	0.0759	0.0761	0.0771
3	48	. . .	. . .	0.0990	0.1003	0.1018	0.0855	0.0857	0.0867
3	. . .	56	. . .	0.0990	0.1002	0.1017	0.0874	0.0876	0.0886
4	. . .	. . .	36	0.1120	0.1137	0.1157	0.0940	0.0942	0.0957
4	40	. . .	. . .	0.1120	0.1136	0.1156	0.0958	0.0960	0.0975
4	. . .	48	. . .	0.1120	0.1133	0.1153	0.0985	0.0987	0.1002
5	40	. . .	. . .	0.1250	0.1266	0.1286	0.1088	0.1090	0.1105
6	32	. . .	. . .	0.1380	0.1402	0.1422	0.1177	0.1182	0.1197
6	. . .	. . .	36	0.1380	0.1397	0.1417	0.1200	0.1202	0.1217
6	. . .	40	. . .	0.1380	0.1396	0.1416	0.1218	0.1220	0.1235
8	32	. . .	. . .	0.1640	0.1662	0.1682	0.1437	0.1442	0.1457
8	. . .	36	. . .	0.1640	0.1657	0.1677	0.1460	0.1462	0.1477
8	. . .	. . .	40	0.1640	0.1656	0.1676	0.1478	0.1480	0.1495
10	24	. . .	. . .	0.1900	0.1928	0.1948	0.1629	0.1634	0.1649
10	. . .	32	. . .	0.1900	0.1922	0.1942	0.1697	0.1702	0.1717
12	24	. . .	. . .	0.2160	0.2188	0.2208	0.1889	0.1894	0.1909
12	. . .	28	. . .	0.2160	0.2184	0.2204	0.1928	0.1933	0.1948
14	. . .	. . .	24	0.2420	0.2448	0.2473	0.2149	0.2154	0.2174

All dimensions are given in inches.
Lead Tolerance: Plus or minus 0.003 inch per inch of thread.   *Angle Tolerance:* Plus or minus 45 min. in half angle and 68 min. in full angle for 20 to 28 threads per inch; plus or minus 60 min. in half angle and 90 min. in full angle for 30 or more threads per inch.

Table 17a. ANSI Standard Taper Pipe Taps — Cut Thread Tolerances for NPT and Ground Thread Tolerances for NPT, NPTF, and ANPT (ANSI B94.9-1979)

Nominal Size	Threads per Inch NPT, NPTF, or ANPT	Gage Measurement*			Taper per Foot, Inches			
		Projection Inches	Tolerance Plus or Minus		Cut Thread		Ground Thread	
			Cut Thread	Ground Thread	Min.	Max.	Min.	Max.
1/16	27	0.312	1/16	1/16	23/32	27/32	23/32	25/32
1/8	27	0.312	1/16	1/16	23/32	27/32	23/32	25/32
1/4	18	0.459	1/16	1/16	23/32	27/32	23/32	25/32
3/8	18	0.454	1/16	1/16	23/32	27/32	23/32	25/32
1/2	14	0.579	1/16	1/16	23/32	13/16	23/32	25/32
3/4	14	0.565	1/16	1/16	23/32	13/16	23/32	25/32
1	11½	0.678	3/32	3/32	23/32	13/16	23/32	25/32
1¼	11½	0.686	3/32	3/32	23/32	13/16	23/32	25/32
1½	11½	0.699	3/32	3/32	23/32	13/16	23/32	25/32
2	11½	0.667	3/32	3/32	23/32	13/16	23/32	25/32
2½	8	0.925	3/32	3/32	47/64	51/64	47/64	25/32
3	8	0.925	3/32	3/32	47/64	51/64	47/64	25/32

All dimensions are given in inches.
* Distance that small end of tap projects through L1 thread ring gage (see ANSI B1.20.3).
Lead Tolerance: Plus or minus 0.003 inch per inch of cut thread and plus or minus 0.0005 inch per inch of ground thread.   *Angle Tolerance:* Plus or minus 40 min. in half angle and 60 min. in full angle for 8 cut threads per inch; plus or minus 45 min. in half angle and 68 min. in full angle for 11½ to 27 cut threads per inch; plus or minus 25 min. in half angle for 8 ground threads per inch; and plus and minus 30 min. in half angle for 11½ to 27 ground threads per inch.

Table 17b. ANSI Standard Taper Pipe Thread — Widths of Flats at Tap Crests and Roots for Cut Thread NPT and Ground Thread NPT, ANPT, and NPTF (ANSI B94.9-1979)

Threads per Inch	Tap Flat Width at	Column I NPT — Cut and Ground Thread ANPT — Ground Thread		Column II NPTF — Ground Thread	
		Minimum*	Maximum*	Minimum	Maximum
27	Major Diameter	0.0014	0.0041	0.0040	0.0055
	Minor Diameter	. . .	0.0041	. . .	0.0040
18	Major Diameter	0.0021	0.0057	0.0050	0.0065
	Minor Diameter	. . .	0.0057	. . .	0.0050
14	Major Diameter	0.0027	0.0064	0.0050	0.0065
	Minor Diameter	. . .	0.0064	. . .	0.0050
11½	Major Diameter	0.0033	0.0073	0.0060	0.0083
	Minor Diameter	. . .	0.0073	. . .	0.0060
8	Major Diameter	0.0048	0.0090	0.0080	0.0103
	Minor Diameter	. . .	0.0090	. . .	0.0080

All dimensions are given in inches.
* Minimum minor diameter flats are not specified. May be sharp as practicable.
Note: Cut Thread taps made to Column I are marked NPT but are not recommended for ANPT applications. Ground Thread taps made to Column I are marked NPT and may be used for NPT and ANPT applications. Ground Thread taps made to Column II are marked NPTF and used for Dryseal application.

Table 18a. ANSI Standard Straight Pipe Taps (NPSF — Dryseal) — Ground Thread Limits (ANSI B94.9-1979)

Nominal Size, Inches	Threads per Inch	Major Diameter		Pitch Diameter			Minor* Diam. Flat, Max.
		Min. G	Max. H	Plug at Gaging Notch E	Min. K	Max. L	
⅛	27	0.3932	0.3942	0.3736	0.3696	0.3701	0.004
¼	18	0.5239	0.5249	0.4916	0.4859	0.4864	0.005
⅜	18	0.6593	0.6603	0.6270	0.6213	0.6218	0.005
½	14	0.8230	0.8240	0.7784	0.7712	0.7717	0.005
¾	14	1.0335	1.0345	0.9889	0.9817	0.9822	0.005

Formulas For American Dryseal (NPSF) Ground Thread Taps

Nominal Size, Inches	Major Diameter		Pitch Diameter		Max. Minor Diam.
	Min. G	Max. H	Min. K	Max. L	
⅛	$H - 0.0010$	$K + Q - 0.0005$	$L - 0.0005$	$E - F$	$M - Q$
¼	$H - 0.0010$	$K + Q - 0.0005$	$L - 0.0005$	$E - F$	$M - Q$
⅜	$H - 0.0010$	$K + Q - 0.0005$	$L - 0.0005$	$E - F$	$M - Q$
½	$H - 0.0010$	$K + Q - 0.0005$	$L - 0.0005$	$E - F$	$M - Q$
¾	$H - 0.0010$	$K + Q - 0.0005$	$L - 0.0005$	$E - F$	$M - Q$

Values to use in Formulas

Threads per Inch	E	F	M	Q
27	Pitch diameter of plug at gaging notch	0.0035	Actual measured pitch diameter	0.0251
18		0.0052		0.0395
14		0.0067		0.0533

All dimensions are given in inches.   * As specified or sharper.
Note: The major diameter of standard taper pipe plug gages and the minor diameter of standard taper pipe ring gages used for gaging dryseal threads are truncated .20p minimum to .25p maximum for all pitches.
Lead Tolerance: Plus or minus 0.0005 inch within any two threads not farther apart than one inch. *Angle Tolerance:* Plus or minus 30 min. in half angle for 14 to 27 threads per inch.

Table 18b. ANSI Standard Straight Pipe Taps (NPS) — Cut Thread Limits
(ANSI B94.9-1979)

Nominal Size	Threads per Inch, NPS, NPSC	Size at Gaging Notch	Pitch Diameter		Values to use in Formulas		
			Min.	Max.	A	B	C
1/8	27	0.3736	0.3721	0.3751	0.0267	0.0296	0.0257
1/4	18	0.4916	0.4908	0.4938	} 0.0408	0.0444	0.0401
3/8	18	0.6270	0.6257	0.6292			
1/2	14	0.7784	0.7776	0.7811	} 0.0535	0.0571	0.0525
3/4	14	0.9889	0.9876	0.9916			
1	11½	1.2386	1.2372	1.2412	0.0658	0.0696	0.0647

The following are approximate formulas, in which M = measured pitch diameter in inches:

Major diam., min. $= M + A$
Major diam., max. $= M + B$ Minor diam., max. $= M - C$

All dimensions are given in inches.

Lead Tolerance: Plus or minus 0.003 inch per inch of thread.   *Angle Tolerance:* All pitches, plus or minus 45 min. in half angle and 68 min. in full angle. Taps made to these specifications are to be marked NPS and used for NPSC thread form.

Table 18c. ANSI Standard Straight Pipe Taps (NPS) — Ground Thread Limits
(ANSI B94.9-1979)

Nominal Size, Inches	Threads per Inch, NPS, NPSC, NPSM	Major Diameter			Pitch Diameter		
		Plug at Gaging Notch	Min. G	Max. H	Plug at Gaging Notch E	Min. K	Max. L
1/8	27	0.3983	0.4022	0.4032	0.3736	0.3746	0.3751
1/4	18	0.5286	0.5347	0.5357	0.4916	0.4933	0.4938
3/8	18	0.6640	0.6701	0.6711	0.6270	0.6287	0.6292
1/2	14	0.8260	0.8347	0.8357	0.7784	0.7086	0.7811
3/4	14	1.0364	1.0447	1.0457	0.9889	0.9906	0.9916
1	11½	1.2966	1.3062	1.3077	1.2386	1.2402	1.2412

Formulas for NPS Ground Thread Taps*

Nominal Size	Major Diameter		Minor Diam.	Threads per Inch	A	B
	Min. G	Max. H	Max.			
				27	0.0296	0.0257
1/8	$H - 0.0010$	$(K + A) - 0.0010$	$M - B$	18	0.0444	0.0401
1/4 to 3/4	$H - 0.0010$	$(K + A) - 0.0020$	$M - B$	14	0.0571	0.0525
1	$H - 0.0015$	$(K + A) - 0.0021$	$M - B$	11½	0.0696	0.0647

The maximum Pitch Diameter of tap is based upon an allowance deducted from the maximum product pitch diameter of NPSC or NPSM, whichever is smaller.

The minimum Pitch Diameter of tap is derived by subtracting the ground thread pitch diameter tolerance for actual equivalent size.

All dimensions are given in inches.   * In the formulas, M equals the actual measured pitch diameter. *Lead tolerance:* Plus or minus 0.0005 inch within any two threads not farther apart than one inch.   *Angle Tolerance:* All pitches, plus or minus 30 min. in half angle. Taps made to these specifications are to be marked NPS and used for NPSC and NPSM.

Standard Marks or Symbols for Identifying Threads on Taps. — All taps are marked with the nominal size, number of threads per inch and the proper symbol to identify the thread form. Taps having multiple threads are marked with diameter, number of threads per inch, form of thread and lead designated in fractions, also double, triple, etc. For example: A 1″-8 double thread special tap with National form of thread will be marked as follows: 1″-8NS Double ¼″ Lead.

Left-hand taps are marked "Left Hand" or "LH" as follows: 1″-8NS Double LH ¼″ Lead.

**Table 19a. ANSI Standard Metric Tap Ground Thread Limits in Inches —
ISO Metric Thread Form (ANSI B94.9-1979)**

Nominal Diam, mm	Pitch, mm	Major Diameter (Inches)			Pitch Diameter (Inches)		
		Basic	Min	Max	Basic	Min	Max
1.6	0.35	0.0630	.0641	.0651	0.0541	0.0550	0.0556
2	0.4	0.0788	.0801	.0811	0.0686	0.0695	0.0701
2.5	0.45	0.0985	.0999	.1009	0.0870	0.0879	0.0885
3	0.5	0.1182	.1198	.1208	0.1054	0.1063	0.1069
3.5	0.6	0.1378	.1397	.1407	0.1225	0.1237	0.1245
4	0.7	0.1575	.1597	.1613	0.1396	0.1408	0.1416
5	0.8	0.1969	.1994	.2010	0.1764	0.1776	0.1784
6	1	0.2363	.2395	.2411	0.2107	0.2122	0.2132
8	1.25	0.3150	.3189	.3214	0.2830	0.2843	0.2855
10	1.5	0.3937	.3985	.4010	0.3554	0.3572	0.3584
12	1.75	0.4725	.4780	.4805	0.4277	0.4295	0.4307
14	2	0.5512	.5575	.5600	0.5001	0.5020	0.5036
16	2	0.6300	.6363	.6388	0.5788	0.5807	0.5823
20	2.5	0.7875	.7954	.7979	0.7235	0.7254	0.7270
24	3	0.9449	.9544	.9583	0.8682	0.8706	0.8722
30	3.5	1.1812	1.1922	1.1901	1.0917	1.0942	1.0962
36	4	1.4174	1.4300	1.4339	1.3151	1.3176	1.3196

Basic pitch diameter is the same as minimum pitch diameter of internal thread, Class 6H as shown in table on pages 1538 and 1539.

* Pitch diameter limits are designated in the Standard as D3 for 1.6 to 3 mm diameter sizes, incl.; D4 for 3.5 to 5 mm sizes, incl.; D5 for 6 and 8 mm sizes; D6 for 10 and 12 mm sizes; D7 for 14 to 20 mm sizes, incl.; D8 for 24 mm size; and D9 for 30 and 36 mm sizes.

Angle tolerances are plus or minus 30 minutes in half angle for pitches ranging from 0.35 through 2.5 mm, incl. and plus or minus 25 minutes in half angle for pitches ranging from 3 to 4 mm, incl.

A maximum deviation of plus or minus 0.0005 inch within any two threads not farther apart than one inch is permitted.

Metric Thread Pitch Diameter Limit Designations. — Where the tap pitch diameter is over or under basic thread pitch diameter by even multiples of 0.0005 inch, the tap will be marked with the letter "D" or "DU", respectively, followed by a limit number. The limit number is determined as follows:

D Limit Number   Amount tap PD High Limit is over Basic PD, divided by 0.0005 inch.

DU Limit Number   Amount tap PD Low Limit is under Basic PD, divided by 0.0005 inch.

Examples: M1.6 × 0.35 — for D3 limit, max tap PD = Basic plus 0.0015 in. Tap PD tolerance = minus 0.0006 in.

M12 × 1.75 — for D6 limit, max tap PD = Basic plus 0.0030 in. Tap PD tolerance = minus 0.0012 in.

M6 × 1 — for DU 4 limit, min tap PD = Basic minus 0.0020 in. Tap PD tolerance = plus 0.0010 in.

**Table 19b. ANSI Standard Metric Tap Ground Thread Limits in Millimeters —
ISO Metric Thread Form (ANSI B94.9-1979)**

Nominal Diam, mm	Pitch, mm	Major Diameter (mm)			Pitch Diameter (mm)		
		Basic	Min	Max	Basic	Min	Max
1.6	0.35	1.600	1.628	1.653	1.373	1.397	1.412
2	0.4	2.000	2.032	2.057	1.740	1.764	1.779
2.5	0.45	2.500	2.536	2.561	2.208	2.232	2.247
3	0.5	3.000	3.040	3.065	2.675	2.699	2.714
3.5	0.6	3.500	3.548	3.573	3.110	3.142	3.162
4	0.7	4.000	4.056	4.097	3.545	3.577	3.597
5	0.8	5.000	5.064	5.105	4.480	4.512	4.532
6	1	6.000	6.080	6.121	5.350	5.390	5.415
8	1.25	8.000	8.100	8.164	7.188	7.222	7.253
10	1.5	10.000	10.120	10.184	9.026	9.073	9.104
12	1.75	12.000	12.140	12.204	10.863	10.910	10.941
14	2	14.000	14.160	14.224	12.701	12.751	12.792
16	2	16.000	16.160	16.224	14.701	14.751	14.792
20	2.5	20.000	20.200	20.264	18.376	18.426	18.467
24	3	24.000	24.240	24.340	22.051	22.114	22.155
30	3.5	30.000	30.280	30.380	27.727	27.792	27.844
36	4	36.000	36.320	36.420	33.402	33.467	33.519

Basic pitch diameter is the same as minimum pitch diameter of internal thread, Class 6H as shown in table on pages 1538 and 1539.

Pitch diameter limits are designated in the Standard as D3 for 1.6 to 3 mm diameter sizes, incl.: D4 for 3.5 to 5 mm sizes, incl.; D5 for 6 and 8 mm sizes; D6 for 10 and 12 mm sizes; D7 for 14 to 20 mm sizes, incl.; D8 for 24 mm size; and D9 for 30 and 36 mm sizes.

Angle tolerances are plus or minus 30 minutes in half angle for pitches ranging from 0.35 through 2.5 mm, incl. and plus or minus 25 minutes in half angle for pitches ranging from 3 to 4 mm, incl.

A maximum lead deviation of plus or minus 0.013 mm within any two threads not farther apart than 25 mm is permitted.

**Table 20. ANSI Standard Eccentricity Tolerances of Tap Elements
When Tested on Dead Centers (ANSI B94.9-1979)**

Element	Range Sizes are Inclusive			Cut Thread		Ground Thread	
	Hand, Mch. Screw	Metric	Pipe	Eccentricity	tiv*	Eccentricity	tiv*
Square	#0–½"	M1.6–M12	1⁄16–1⁄8"	0.0030	0.0060	0.0030	0.0060
(at central point)	17⁄32–4"	M14–M100	1⁄4–4"	0.0040	0.0080	0.0040	0.0080
Shank	#0–5⁄16"	M1.6–M8	1⁄16"	0.0030	0.0060	0.0005	0.0010
	11⁄32–4"	M10–M100	1⁄8–4"	0.0040	0.0080	0.0008	0.0016
Major Diameter	#0–5⁄16"	M1.6–M8	1⁄16"	0.0025	0.0050	0.0005	0.0010
	11⁄32–4"	M10–M100	1⁄8–4"	0.0040	0.0080	0.0008	0.0016
Pitch Diameter	#0–5⁄16"	M1.6–M8	1⁄16"	0.0025	0.0050	0.0005	0.0010
(at first full thread)	11⁄32–4"	M10–M100	1⁄8–4"	0.0040	0.0080	0.0008	0.0016
Chamfer**	#0–½"	M1.6–M12	1⁄16–1⁄8"	0.0020	0.0040	0.0010	0.0020
	17⁄32–4"	M14–M100	1⁄4–4"	0.0030	0.0060	0.0015	0.0030

All dimensions are given in inches.

* tiv = total indicator variation. Figures are given for both eccentricity and total indicator variation to avoid misunderstanding.

** Chamfer should preferably be inspected by light projection to avoid errors due to indicator contact points dropping into the thread groove.

High-speed steel taps are marked HS while carbon steel taps need not be marked with steel designation. Example of marking a ground thread tap: ¼-20 NC GH3 HS

Metric taps are marked with a capital "M" followed by the nominal size in millimeters and the pitch in millimeters separated by the sign "x". For example: M1.6 × 0.35; M6 × 1; M10 × 1.5

Symbols Used for Standard Threads

M	Metric Screw Thread Designation
NC	American National Coarse Thread Series
*UNC	Unified Coarse Thread Series
NF	American National Fine Thread Series
*UNF	Unified Fine Thread Series
NEF	American National Extra-Fine Thread Series
*UNEF	Unified Extra-Fine Thread Series
N	American National 8, 12 and 16 Thread Series (8N, 12N, 16N)
*UN	Unified Constant Pitch Thread Series
NS	American National Thread — Special
*UNS	Unified Thread — Special
UNM	Unified Miniature Thread Series
NR	American National Thread with a 0.108p to 0.144p Controlled Root Radius
UNR	Unified Constant Pitch Thread Series with a 0.108p to 0.144p Controlled Root Radius
UNRC	Unified Coarse Thread Series with a 0.108p to 0.144p Controlled Root Radius
UNRF	Unified Fine Thread Series with a 0.108p to 0.144p Controlled Root Radius
UNJ	Unified Thread Series with a 0.15011p to 0.18042p Controlled Root Radius
UNJC	Unified Coarse Series with a 0.15011p to 0.18042p Controlled Root Radius
UNJF	Unified Fine Series with a 0.15011p to 0.18042p Controlled Root Radius
NH	American National Hose Coupling & Fire Hose Coupling Threads
NPS	For Tap Marking Only (See NPSC and NPSM)
NPSC	†American National Standard Straight Pipe Thread in Pipe Couplings (Tap Marked NPS)
NPSF	†Dryseal American National Std. Fuel Internal Straight Pipe Thread
NPSH	†American National Standard Straight Pipe Threads for Loose Fitting Mechanical Joints for Hose Couplings
NPSI	†Dryseal American National Standard Intermediate Internal Straight Pipe Thread
NPSL	†American National Standard Straight Pipe Thread for Loose-Fitting Mechanical Joints with Locknuts
NPSM	†American National Standard Straight Pipe Threads for Free-Fitting Mechanical Joints (Ground Thread Tap Marked NPS)
ANPT	Aeronautical National Form Taper Pipe Thread (Ground Thread Tap Marked NPT)
NPT	†American National Standard Taper Pipe Thread
NPTF	†Dryseal American National Standard Taper Pipe Thread
NPTR	†American National Standard Taper Thread for Railing Joints (Tap Marked NPT)
NGO	National Gas Outlet Thread
NGS	National Gas Straight Thread
NGT	National Gas Taper Thread (See also "SGT")
PTF-SAE-SHORT	Dryseal SAE Short Taper Pipe Thread
ACME-C	Acme Thread Centralizing
ACME-G	Acme Thread-General Purpose
STUB ACME	Stub Acme Thread
AMO	American Standard Microscope Objective Thread
N BUTT	American Buttress Thread
V	A 60-degree "V" Thread with Truncated Crest and Root. The theoretical "V" form is usually flatted to the user's specifications.
SB	Manufacturers' Stovebolt Standard Thread
STI	Special Thread for Helical Coil Wire Screw Thread Inserts
SGT	Special Gas Taper Thread
SPL-PTF	Dryseal Special Taper Pipe Thread

* Taps are not marked with "U" but with corresponding American Standard Form Symbol.
† Formerly designated USA (American).

Designations for Ground Thread Taps. — Designations for ground threads provide a wide selection of ground thread taps to secure the classes of thread tolerance desired with maximum wear life. They also allow the user to select a tap to limits which in general will suit the individual work and equipment conditions that he encounters.

While the changes from previous standards are major in character because they affect nomenclature, as well as pitch diameter limits in certain ranges, they permit an easy and economical transition from previous standards without obsolescence of existing inventory.

All standard ground thread taps were formerly designated as Commercial Ground, Commercial Ground High, or Precision Ground. Precision Ground thread taps were made to pitch diameter tolerances of .0005 inch, whereas Commercial Ground thread taps were made to pitch diameter tolerances varying from .001 to .0018 inch depending upon the diameter-pitch combination.

The ANSI Standard establishes only one classification of ground thread taps designated as "Ground Thread." The classification previously identified as Commercial Ground, Commercial Ground High, and Precision Ground no longer applies to ground taps.

The ANSI Standard for ground thread taps establishes pitch diameter tolerances for machine screw and hand taps, in size range No. 0 to 1 inch diameter in 0.0005 inch increments. The pitch diameter tolerances for taps over 1 inch to 1½ inch diameter, inclusive, are in 0.001 inch increments.

The pitch diameter limit designations for standard tolerance ranges, described in the following paragraphs, permit selection of the proper tap for the class of thread tolerance desired and maximum tap life.

Ground Thread taps made to pitch diameter tolerances above basic are designated "High" taps, and those made to tolerances below basic, "Low" taps and are identified by the letter "H" and "L" respectively. The tolerance ranges are indicated by a numeral after the "H" and "L" identification. This numeral discloses the number of half-thousandths (.0005) larger than basic of the maximum "H" tap or smaller than basic of the minimum "L" tap. The tolerance equals .0005 inch on taps 1-inch diameter and smaller and .001 inch on taps over 1-inch diameter to 1½-inch diameter, inclusive.

Example 1: A ¼"–20 NC Ground Thread Tap marked G H3 is identified as follows:

    G = the symbol for Ground Thread;
    H = the symbol for above basic; and
    3 = the symbol for pitch diameter limits which in this case are .001" to .0015" above basic.

Example 2: A ¼"–20 NC Ground Thread tap marked G L1 is identified as follows:

    G = symbol for Ground Thread;
    L = symbol for below basic; and
    1 = symbol for pitch diameter limits which in this case are .0000" to .0005" under basic.

Example 3: A 1¼"–7 NC Ground Thread tap marked G H4 is identified as follows:

    G = symbol for Ground Thread;
    H = symbol for above basic; and
    4 = symbol for pitch diameter limits which in this case are .001" to .002" above basic.

It will be noted in these examples that the tolerance range numeral divided by 2 establishes, in thousandths of an inch, the amount that the maximum tap pitch

diameter is above basic in the H series and the amount that the minimum tap pitch diameter is under basic in the L series.

The pitch diameter limit numbers and the corresponding limits for Ground Thread taps are as follows, H indicating above basic (High), and L indicating below basic (Low).

Pitch Diameter Limit Numbers for Taps to 1" Diameter, Inclusive:

$$L1 = \text{Basic to basic minus } .0005''$$
$$H1 = \text{Basic to basic plus } .0005''$$
$$H2 = \text{Basic plus } .0005'' \text{ to basic plus } .001''$$
$$H3 = \text{Basic plus } .001'' \text{ to basic plus } .0015''$$
$$H4 = \text{Basic plus } .0015'' \text{ to basic plus } .002''$$
$$H5 = \text{Basic plus } .002'' \text{ to basic plus } .0025''$$
$$H6 = \text{Basic plus } .0025'' \text{ to basic plus } .003''$$
$$H7 = \text{Basic plus } .003'' \text{ to basic plus } .0035''$$

Pitch Diameter Limit Numbers for Taps Over 1" to 1½" Diameter, Inclusive:

$$H4 = \text{Basic plus } .001'' \text{ to basic plus } .002''.$$

Table 11 presents a guide to assist in the selection of the proper tap to produce a desired class of thread tolerance under normal conditions.

Ground thread tap limits for standard metric taps are given in Tables 9a and 9b.

Acme and Square-threaded Taps. — These taps are usually made in sets, three taps in a set undoubtedly being the most common. For very fine pitches, two taps in a set will be found sufficient, while as many as five taps in a set are used for coarse pitches. Tables are given herewith for proportioning both Acme and square-threaded taps when made in sets. One leading tap maker in cutting the threads of square-threaded taps makes them according to the following rules: The width of the groove between two threads is made equal to one-half the pitch of the thread, less 0.004 inch. This makes the width of the thread itself equal to one-half of the pitch, plus 0.004 inch. The depth of the thread is made equal to 0.45 times the pitch, plus 0.0025 inch. This latter rule produces a thread which for all the ordinarily used pitches for square-threaded taps has a depth less than the generally accepted standard depth, this latter depth being equal to one-half the pitch. The object of this shallow thread is to insure that if the hole to be threaded by the tap is not bored out so as to provide clearance at the bottom of the thread, the tap will cut its own clearance. The hole should, however, always be drilled out large enough so that the cutting of the clearance is not required of the tap.

Another maker follows under ordinary conditions the dimensions given in the accompanying tables, making the diameter at the end of the chamfer of the first tap equal to the root diameter of the thread, plus 0.010 inch. The diameter at the end of the chamfer of the second and third taps is made equal to the diameter of the straight portion of the next previous tap, minus 0.005 inch.

For Acme thread taps, this manufacturer makes the actual root diameter on the first tap 0.010 inch, and on the second tap 0.005 inch less than the standard root diameter. The finishing tap is made with standard root diameter, and a standard thread tool is used for all three taps in a set.

The table, "Dimensions of Acme Thread Taps in Sets of Three Taps" may be used for the length dimensions for Acme taps. The dimensions in this table apply to single-threaded taps. For multiple-threaded taps or taps with very coarse pitch, relative to the diameter, the length of the chamfered part of the thread may be increased. Square-threaded taps are made to the same table as Acme taps, with the exception of the figures in column K, which for square-threaded taps should be equal to the nominal diameter of the tap, no oversize allowance

being customary in these taps. The first tap in a set of Acme taps (not square-threaded taps) should be turned taper in bottom of the thread for a distance of about one-quarter of the length of the threaded part. The taper should be so selected that the root diameter is about 1/32 inch smaller at the point than the proper root diameter of the tap. The first tap should preferably be provided with a short pilot at the point. For very coarse pitches, the first tap may be provided with spiral flutes at right angles to the angle of the thread. Acme and square-threaded taps should be relieved or backed off on the top of the thread of the chamfered portion on all of the taps in the set. When the taps are used as machine taps, rather than as hand taps, they should be relieved in the angle of the thread, as well as on the top, for the whole length of the chamfered portion. Acme taps should also always be relieved on the front side of the thread to within 1/32 inch of the cutting edge.

Table for Making Acme Thread Taps in Sets of Three Taps

No. of Threads per Inch	Amount in Inches to be Added to Root Diameter of Tap to Obtain Diameter of Straight Part of Thread of		No. of Threads per Inch	Amount in Inches to be Added to Root Diameter of Tap to Obtain Diameter of Straight Part of Thread of	
	1st Tap	2d Tap		1st Tap	2d Tap
1	0.468	0.832	5	0.108	0.192
1½	0.318	0.566	5½	0.100	0.178
2	0.243	0.432	6	0.093	0.166
2½	0.198	0.352	7	0.082	0.146
3	0.168	0.298	8	0.074	0.132
3½	0.147	0.261	9	0.068	0.121
4	0.130	0.232	10	0.063	0.112
4½	0.118	0.210	12	0.055	0.098

Table for Making Square-threaded Taps in Sets of Three Taps

No. of Threads per Inch	Amount in Inches to be Added to Root Diameter of Tap to Obtain Diameter of Straight Part of Thread of		No. of Threads per Inch	Amount in Inches to be Added to Root Diameter of Tap to Obtain Diameter of Straight Part of Thread of	
	1st Tap	2d Tap		1st Tap	2d Tap
1	0.410	0.800	5	0.082	0.160
1½	0.273	0.533	5½	0.075	0.146
2	0.205	0.400	6	0.068	0.133
2½	0.164	0.320	7	0.059	0.114
3	0.137	0.267	8	0.051	0.100
3½	0.117	0.229	9	0.046	0.089
4	0.102	0.200	10	0.041	0.080
4½	0.091	0.178	12	0.034	0.067

Proportions of Acme and Square-threaded Taps Made in Sets

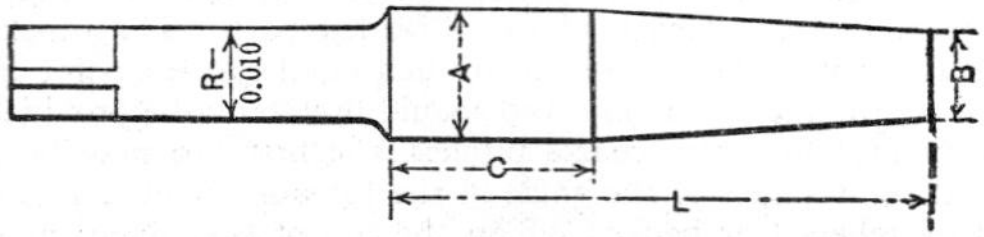

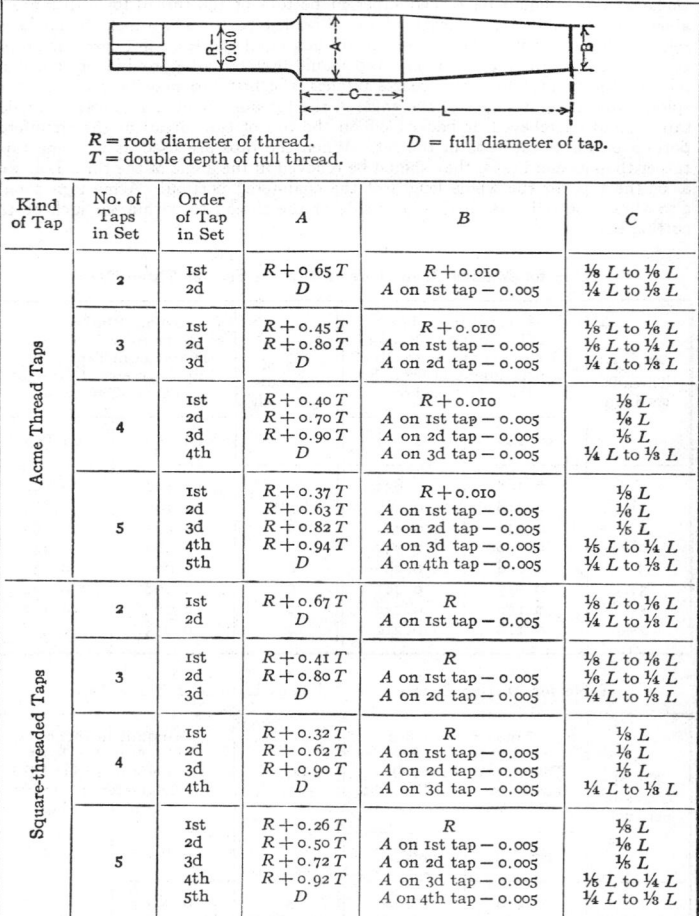

R = root diameter of thread. D = full diameter of tap.
T = double depth of full thread.

Kind of Tap	No. of Taps in Set	Order of Tap in Set	*A*	*B*	*C*
Acme Thread Taps	2	1st	$R + 0.65\,T$	$R + 0.010$	$\frac{1}{8}L$ to $\frac{1}{6}L$
		2d	D	A on 1st tap − 0.005	$\frac{1}{4}L$ to $\frac{1}{3}L$
	3	1st	$R + 0.45\,T$	$R + 0.010$	$\frac{1}{8}L$ to $\frac{1}{6}L$
		2d	$R + 0.80\,T$	A on 1st tap − 0.005	$\frac{1}{6}L$ to $\frac{1}{4}L$
		3d	D	A on 2d tap − 0.005	$\frac{1}{4}L$ to $\frac{1}{3}L$
	4	1st	$R + 0.40\,T$	$R + 0.010$	$\frac{1}{8}L$
		2d	$R + 0.70\,T$	A on 1st tap − 0.005	$\frac{1}{6}L$
		3d	$R + 0.90\,T$	A on 2d tap − 0.005	$\frac{1}{5}L$
		4th	D	A on 3d tap − 0.005	$\frac{1}{4}L$ to $\frac{1}{3}L$
	5	1st	$R + 0.37\,T$	$R + 0.010$	$\frac{1}{8}L$
		2d	$R + 0.63\,T$	A on 1st tap − 0.005	$\frac{1}{6}L$
		3d	$R + 0.82\,T$	A on 2d tap − 0.005	$\frac{1}{5}L$
		4th	$R + 0.94\,T$	A on 3d tap − 0.005	$\frac{1}{5}L$ to $\frac{1}{4}L$
		5th	D	A on 4th tap − 0.005	$\frac{1}{4}L$ to $\frac{1}{3}L$
Square-threaded Taps	2	1st	$R + 0.67\,T$	R	$\frac{1}{8}L$ to $\frac{1}{6}L$
		2d	D	A on 1st tap − 0.005	$\frac{1}{4}L$ to $\frac{1}{3}L$
	3	1st	$R + 0.41\,T$	R	$\frac{1}{8}L$ to $\frac{1}{6}L$
		2d	$R + 0.80\,T$	A on 1st tap − 0.005	$\frac{1}{6}L$ to $\frac{1}{4}L$
		3d	D	A on 2d tap − 0.005	$\frac{1}{4}L$ to $\frac{1}{3}L$
	4	1st	$R + 0.32\,T$	R	$\frac{1}{8}L$
		2d	$R + 0.62\,T$	A on 1st tap − 0.005	$\frac{1}{6}L$
		3d	$R + 0.90\,T$	A on 2d tap − 0.005	$\frac{1}{5}L$
		4th	D	A on 3d tap − 0.005	$\frac{1}{4}L$ to $\frac{1}{3}L$
	5	1st	$R + 0.26\,T$	R	$\frac{1}{8}L$
		2d	$R + 0.50\,T$	A on 1st tap − 0.005	$\frac{1}{6}L$
		3d	$R + 0.72\,T$	A on 2d tap − 0.005	$\frac{1}{5}L$
		4th	$R + 0.92\,T$	A on 3d tap − 0.005	$\frac{1}{5}L$ to $\frac{1}{4}L$
		5th	D	A on 4th tap − 0.005	$\frac{1}{4}L$ to $\frac{1}{3}L$

Adjustable Taps. — Many adjustable taps are now used, especially for accurate work. Some taps of this class are made of a solid piece of steel which is split and provided with means of expanding sufficiently to compensate for wear. Most of the larger adjustable taps have inserted blades or chasers which are rigidly held, but capable of radial adjustment. The use of taps of this general class enables standard sizes to be maintained readily.

Advantages of Collapsing Taps. — Collapsing taps are similar in principle to self-opening dies, except that the action is reversed, the tap chasers moving inward to permit the rapid removal of the tap from the hole. This collapsing action may be due to the engagement of a collar gage-plate or lever on the tap with the surface of the work or with a fixed stop.

Combination Taps and Dies. — Combination tools arranged for cutting an external thread and tapping a hole at the same time may be used to advantage in some cases. If the tap of a combination tool is used for cutting threads of different pitch, the difference in the rate at which they advance is compensated for by providing a floating movement for either the tap or the die.

Dimensions of Acme Thread Taps in Sets of Three Taps

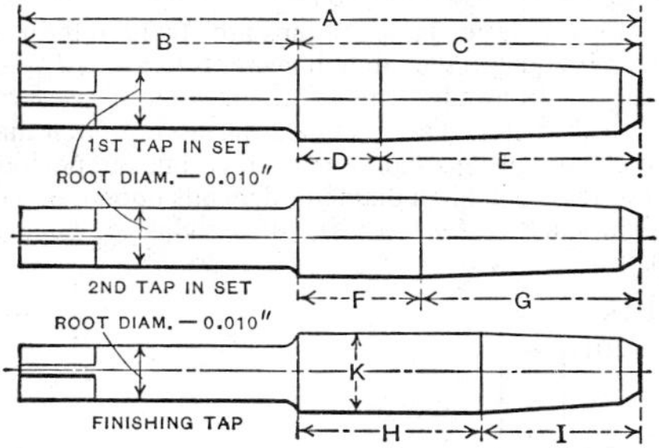

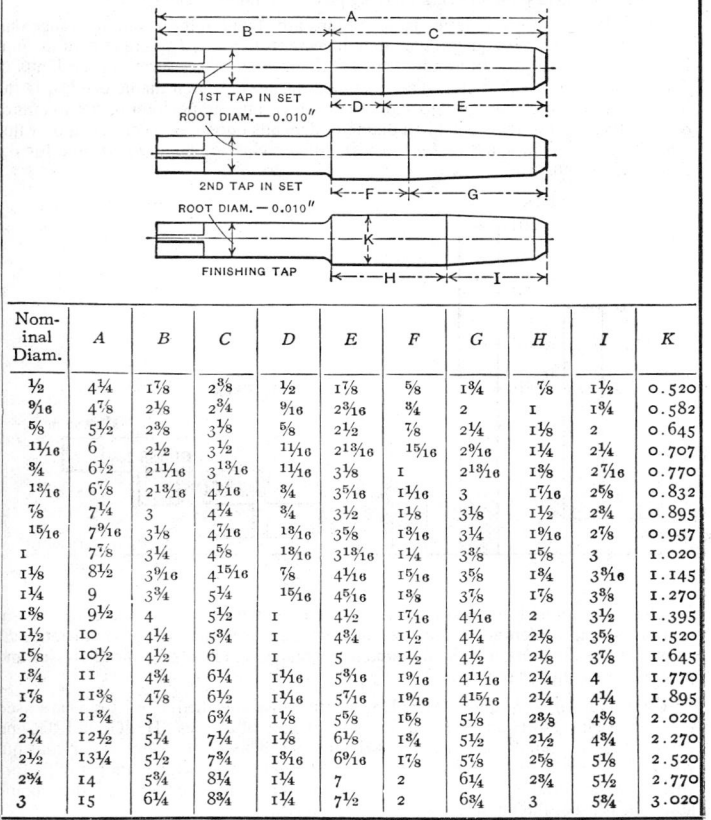

Nominal Diam.	*A*	*B*	*C*	*D*	*E*	*F*	*G*	*H*	*I*	*K*
½	4¼	1⅞	2⅜	½	1⅞	⅝	1¾	⅞	1½	0.520
9/16	4⅞	2⅛	2¾	9/16	2 3/16	¾	2	1	1¾	0.582
⅝	5½	2⅜	3⅛	⅝	2½	⅞	2¼	1⅛	2	0.645
11/16	6	2½	3½	11/16	2 13/16	15/16	2 9/16	1¼	2¼	0.707
¾	6½	2 11/16	3 13/16	11/16	3⅛	1	2 13/16	1⅜	2 7/16	0.770
13/16	6⅞	2 13/16	4 1/16	¾	3 5/16	1 1/16	3	1 7/16	2⅝	0.832
⅞	7¼	3	4¼	¾	3½	1⅛	3⅛	1½	2¾	0.895
15/16	7 9/16	3⅛	4 7/16	13/16	3⅝	1 3/16	3¼	1 9/16	2⅞	0.957
1	7⅞	3¼	4⅝	13/16	3 13/16	1¼	3⅜	1⅝	3	1.020
1⅛	8½	3 9/16	4 15/16	⅞	4 1/16	1 5/16	3⅝	1¾	3 3/16	1.145
1¼	9	3¾	5¼	15/16	4 5/16	1⅜	3⅞	1⅞	3⅜	1.270
1⅜	9½	4	5½	1	4½	1 7/16	4 1/16	2	3½	1.395
1½	10	4¼	5¾	1	4¾	1½	4¼	2⅛	3⅝	1.520
1⅝	10½	4½	6	1	5	1½	4½	2⅛	3⅞	1.645
1¾	11	4¾	6¼	1 1/16	5 3/16	1 9/16	4 11/16	2¼	4	1.770
1⅞	11⅜	4⅞	6½	1 1/16	5 7/16	1 9/16	4 15/16	2¼	4¼	1.895
2	11¾	5	6¾	1⅛	5⅝	1⅝	5⅛	2⅜	4⅜	2.020
2¼	12½	5¼	7¼	1⅛	6⅛	1¾	5½	2½	4¾	2.270
2½	13¼	5½	7¾	1 3/16	6 9/16	1⅞	5⅞	2⅝	5⅛	2.520
2¾	14	5¾	8¼	1¼	7	2	6¼	2¾	5½	2.770
3	15	6¼	8¾	1¼	7½	2	6¾	3	5¾	3.020

British Standard Screwing Taps for ISO Metric Threads. — BS 949: Part 1: 1976 provides dimensions and tolerances for screwing taps for ISO metric coarse-pitch series threads in accordance with BS 3643: Part 2; and for metric fine-pitch series threads in accordance with BS 3643: Part 3.

Table 1 provides dimensional data for the cutting portion of cut-thread taps for coarse-series threads of ISO metric sizes. The sizes shown were selected from the first-choice combinations of diameter and pitch listed in BS 3643: Part 1: 1981. Table 2 provides similar data for ground-thread taps for both coarse- and fine-pitch series threads of ISO metric sizes.

Tolerance Classes of Taps: Three tolerance classes (class 1, class 2, and class 3) are used for the designation of taps used for the production of nuts of the following classes:

nut classes 4H, 5H, 6H, 7H, and 8H, all having zero minimum clearance;

nut classes 4G, 5G, and 6G, all having positive minimum clearance.

The tolerances for the three classes of taps are stated in terms of a tolerance unit t, the value of which is equal to the pitch diameter tolerance, T_{D2}, grade 5, of the nut. Thus, $t = T_{D2}$, grade 5, of the nut. Taps of the different classes vary in the limits of size of the tap pitch diameter. The tolerance on the tap pitch diameter, T_{d2}, is the same for all three classes of taps (20 percent of t), but the position of the tolerance zone with respect to the basic pitch diameter depends upon the lower deviation value Em which is: for tap class 1, $Em = +0.1t$; for tap class 2, $Em = +0.3t$; and for tap class 3, $Em = +0.5t$.

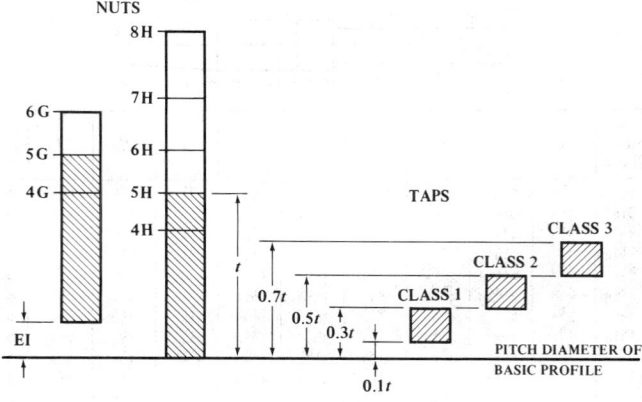

The disposition of the tolerances described is shown in the accompanying illustration of nut class tolerances compared against tap class tolerances. The distance EI shown in this illustration is the minimum clearance, which is zero for H classes and positive for G classes of nuts.

Choice of Tap Tolerance Class: Unless otherwise specified, class 1 taps are used for nuts of classes 4H and 5H; class 2 taps for nuts of classes 6H, 4G, and 5G; and class 3 taps for nuts of classes 7H, 8H, and 6G. This relationship of tap and nut classes is a general one, since the accuracy of tapping varies with a number of factors such as the material being tapped, the condition of the machine tool used, the tapping attachment used, the tapping speed, and the lubricant.

**Table 1. British Standard Screwing Taps for ISO Metric Threads;
Dimensional Limits for the Threaded Portion of Cut Taps —
Coarse Pitch Series* (BS 949: Part 1: 1976)**

Designation	Pitch	Major Diameter Minimum*	Pitch Diameter Basic	Max.	Min.	Tolerance on Thread Angle, Degrees
M1	0.25	1.030	0.838	0.875	0.848	4.0
M1.2	0.25	1.230	1.038	1.077	1.048	4.0
M1.6	0.35	1.636	1.373	1.417	1.385	3.4
M2	0.40	2.036	1.740	1.786	1.752	3.2
M2.5	0.45	2.539	2.208	2.259	2.221	3.0
M3	0.50	3.042	2.675	2.730	2.689	2.9
M4	0.70	4.051	3.545	3.608	3.562	2.4
M5	0.80	5.054	4.480	4.547	4.498	2.3
M6	1.00	6.060	5.350	5.424	5.370	2.0
M8	1.25	8.066	7.188	7.270	7.210	1.8
M10	1.50	10.072	9.026	9.116	9.050	1.6
M12	1.75	12.078	10.863	10.961	10.889	1.5
M16	2.00	16.084	14.701	14.811	14.729	1.4
M20	2.50	20.093	18.376	18.497	18.407	1.3
M24	3.00	24.102	22.051	22.183	22.085	1.2
M30	3.50	30.111	27.727	27.874	27.764	1.1
M36	4.00	36.117	33.402	33.563	33.441	1.0

* See footnotes under Table 2.

Tap Major Diameter: Except when a screwed connection has to be tight against gaseous or liquid pressure, it is undesirable for the mating threads to bear on the roots and crests. By avoiding contact in these regions of the threads, the opposite flanks of the two threads are allowed to make proper load bearing contact when the connection is tightened. In general, the desired clearance between crests and roots of mating threads is obtained by increasing the major and minor diameters of the internal thread. Such an increase in the minor diameter is already provided on threads such as the ISO metric thread, in which there is a basic clearance between the crests of minimum size nuts and the roots of maximum size bolts. For this reason, and the fact that taps are susceptible to wear on the crests of their threads, a minimum size is specified for the major diameter of new taps which provides a reasonable margin for the wear of their crests and at the same time provides the desired clearance at the major diameter of the hole. These minimum major diameters for taps are shown in Tables 1 and 2. The maximum tap major diameter is not specified and is left to the manufacturer to take advantage of this concession to produce taps with as liberal a margin possible for wear on the major diameter.

Tapping Square Threads. — If it is necessary to tap square threads, this should be done by using a set of taps that will form the thread by a progressive cutting action, the taps varying in size in order to distribute the work, especially for threads of comparatively coarse pitch. From three to five taps may be required in a set, depending upon the pitch. Each tap should have a pilot to steady it. The pilot of the first tap has a smooth cylindrical end from 0.003 to 0.005 inch smaller than the hole, and the pilots of following taps should have teeth.

896 TAPS AND THREADING DIES

Table 2. British Standard Screwing Taps for ISO Metric Threads; Dimensional Limits for the Threaded Portion of Ground Taps — Coarse- and Fine-Pitch
(BS 949: Part 1: 1976)

Thread			All Classes of Taps		Class 1 Taps		Class 2 Taps		Class 3 Taps		
	Nominal Major Diam. (basic)				Pitch Diameter						Tolerance on ½ Thd Angle
Designation	d	Pitch p	Min. Major Diam. d_{min}*	Basic Pitch Diam. d_2	$d_2$min	$d_2$max	$d_2$min	$d_2$max	$d_2$min	$d_2$max	
COARSE-PITCH THREAD SERIES											
M1	1	0.25	1.022	0.838	0.844	0.855	...	...	...	...	±60'
M1.2	1.2	0.25	1.222	1.038	1.044	1.055	...	...	...	...	±60'
M1.6	1.6	0.35	1.627	1.373	1.380	1.393	1.393	1.407	...	...	±50'
M2	2	0.40	2.028	1.740	1.747	1.761	1.761	1.776	...	...	±40'
M2.5	2.5	0.45	2.530	2.208	2.216	2.231	2.231	2.246	...	...	±38'
M3	3	0.50	3.032	2.675	2.683	2.699	2.699	2.715	2.715	2.731	±36'
M4	4	0.70	4.038	3.545	3.555	3.574	3.574	3.593	3.593	3.612	±30'
M5	5	0.80	5.040	4.480	4.490	4.510	4.510	4.530	4.530	4.550	±26'
M6	6	1.00	6.047	5.350	5.362	5.385	5.385	5.409	5.409	5.433	±24'
M8	8	1.25	8.050	7.188	7.201	7.226	7.226	7.251	7.251	7.276	±22'
M10	10	1.50	10.056	9.026	9.040	9.068	9.068	9.096	9.096	9.124	±20'
M12	12	1.75	12.064	10.863	10.879	10.911	10.911	10.943	10.943	10.975	±19'
M16	16	2.00	16.068	14.701	14.718	14.752	14.752	14.786	14.786	14.820	±18'
M20	20	2.50	20.072	18.376	18.394	18.430	18.430	18.466	18.466	18.502	±16'
M24	24	3.00	24.085	22.051	22.072	22.115	22.115	22.157	22.157	22.199	±14'
M30	30	3.50	30.090	27.727	27.749	27.794	27.794	27.839	27.839	27.884	±13'
M36	36	4.00	36.094	33.402	33.426	33.473	33.473	33.520	33.520	33.567	±12'
FINE-PITCH THREAD SIZES											
M1 × 0.2	1	0.20	1.020	0.870	0.875	0.885	...	...	...	...	±70'
M1.2 × 0.2	1.2	0.20	1.220	1.070	1.075	1.085	...	...	...	...	±70'
M1.6 × 0.2	1.6	0.20	1.621	1.470	1.475	1.485	...	...	...	...	±70'
M2 × 0.25	2	0.25	2.024	1.838	1.844	1.856	...	...	...	...	±60'
M2.5 × 0.35	2.5	0.35	2.527	2.273	2.280	2.293	2.293	2.307	...	...	±50'
M3 × 0.35	3	0.35	3.028	2.773	2.780	2.794	2.794	2.809	...	...	±50'
M4 × 0.5	4	0.50	4.032	3.675	3.683	3.699	3.699	3.715	3.715	3.731	±36'
M5 × 0.5	5	0.50	5.032	4.675	4.683	4.699	4.699	4.715	4.715	4.731	±36'
M6 × 0.75	6	0.75	6.042	5.513	5.524	5.545	5.545	5.566	5.566	5.587	±28'
M8 × 1	8	1.00	8.047	7.350	7.362	7.385	7.385	7.409	7.409	7.433	±24'
M10 × 1.25	10	1.25	10.050	9.188	9.201	9.226	9.226	9.251	9.251	9.276	±22'
M12 × 1.25	12	1.25	12.056	11.188	11.202	11.230	11.230	11.258	11.258	11.286	±22'
M16 × 1.5	16	1.50	16.060	15.026	15.041	15.071	15.071	15.101	15.101	15.131	±20'
M20 × 1.5	20	1.50	20.060	19.026	19.041	19.071	19.071	19.101	19.101	19.131	±20'
M24 × 2	24	2.00	24.072	22.701	22.719	22.755	22.755	22.791	22.791	22.827	±18'
M30 × 2	30	2.00	30.072	28.701	28.719	28.755	28.755	28.791	28.791	28.827	±18'

All dimensions are in millimeters. The thread sizes in the table have been selected from the preferred series shown in BS 3643: Part 1: 1981. For other sizes, and for second and third choice combinations of diameters and pitches, see the Standard.

* The maximum tap major diameter, d max, is not specified and is left to the manufacturer's discretion.

STANDARD TAPERS

Certain types of small tools and machine parts, such as twist drills, end mills, arbors, lathe centers, etc., are provided with taper shanks which fit into spindles or sockets of corresponding taper, thus providing not only accurate alignment between the tool or other part and its supporting member, but also more or less frictional resistance for driving the tool. There are several standards for "self-holding" tapers, but the American National, Morse, and the Brown & Sharpe are the standards most widely used by American manufacturers.

The name *self-holding* has been applied to the smaller tapers — like the Morse and the Brown & Sharpe — because, where the angle of the taper is only 2 or 3 degrees, the shank of a tool is so firmly seated in its socket that there is considerable frictional resistance to any force tending to turn or rotate the tool relative to the socket. The term "self-holding" is used to distinguish relatively small tapers from the larger or *self-releasing* type. A milling machine spindle having a taper of 3½ inches per foot is an example of a self-releasing taper. The included angle in this case is over 16 degrees and the tool or arbor requires a positive locking device to prevent slipping, but the shank may be released or removed more readily than one having a smaller taper of the self-holding type.

Morse Taper. — Dimensions relating to Morse standard taper shanks and sockets may be found in an accompanying table. The taper for different numbers of Morse tapers is slightly different, but it is approximately ⅝ inch per foot in most cases. The table gives the actual tapers, accurate to five decimal places. Morse taper shanks are used on a variety of tools, and exclusively on the shanks of twist drills. Dimensions for Morse Stub Taper Shanks are given in Table 1B.

Brown & Sharpe Taper. — This standard taper is used for taper shanks on tools such as end mills and reamers, the taper being approximately ½ inch per foot for all sizes except for taper No. 10, where the taper is 0.5161 inch per foot. Brown & Sharpe taper sockets are used for many arbors, collets, and machine tool spindles, especially milling machines and grinding machines. In many cases there are a number of different lengths of sockets corresponding to the same number of taper; all these tapers, however, are of the same diameter at the small end.

Jarno Taper. — The Jarno taper was originally proposed by Oscar J. Beale of the Brown & Sharpe Mfg. Co. This taper is based on such simple formulas that practically no calculations are required when the number of taper is known. The taper per foot of all Jarno taper sizes is 0.600 inch on the diameter. The diameter at the large end is as many eighths, the diameter at the small end is as many tenths, and the length as many half inches as are indicated by the number of the taper. For example, a No. 7 Janor taper is ⅞ inch in diameter at the large end; ⁷⁄₁₀, or 0.700 inch at the small end; and ⅞, or 3½ inches long; hence, diameter at large end = No. of taper ÷ 8; diameter at small end = No. of taper ÷ 10; length of taper = No. of taper ÷ 2. The Jarno taper is used on various machine tools, especially profiling machines and die-sinking machines. It has also been used for the headstock and tailstock spindles of some lathes.

American National Standard Machine Tapers. — This standard includes a self-holding series (Table 5, 7, 8, 9 and 10) and a steep taper series, Table 6. The self-holding taper series consists of 22 sizes which are listed in Table 5. The reference gage for the self-holding tapers is a plug gage. Table 11 gives the dimensions and tolerances for both plug and ring gages applying to this series. Tables 7 to 10 inclusive give the dimensions for self-holding taper shanks and sockets which are classified as to (1) means of transmitting torque from spindle to the tool shank, and (2) means of retaining the shank in the socket. The steep machine tapers consist of a preferred

Table 1A. Morse Standard Taper Shanks

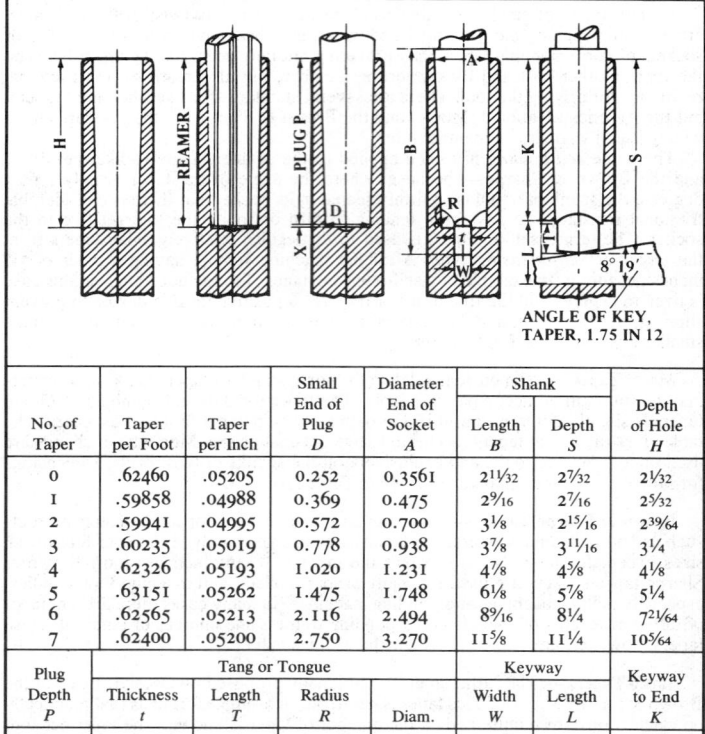

ANGLE OF KEY,
TAPER, 1.75 IN 12

No. of Taper	Taper per Foot	Taper per Inch	Small End of Plug D	Diameter End of Socket A	Shank Length B	Shank Depth S	Depth of Hole H
0	.62460	.05205	0.252	0.3561	2¹¹/₃₂	2⁷/₃₂	2¹/₃₂
1	.59858	.04988	0.369	0.475	2⁹/₁₆	2⁷/₁₆	2⁵/₃₂
2	.59941	.04995	0.572	0.700	3⅛	2¹⁵/₁₆	2³⁹/₆₄
3	.60235	.05019	0.778	0.938	3⅞	3¹¹/₁₆	3¼
4	.62326	.05193	1.020	1.231	4⅞	4⅝	4⅛
5	.63151	.05262	1.475	1.748	6⅛	5⅞	5¼
6	.62565	.05213	2.116	2.494	8⁹/₁₆	8¼	7²¹/₆₄
7	.62400	.05200	2.750	3.270	11⅝	11¼	10⁵/₆₄

Plug Depth P	Tang or Tongue Thickness t	Tang or Tongue Length T	Tang or Tongue Radius R	Diam.	Keyway Width W	Keyway Length L	Keyway to End K
2	.1562	¼	⁵/₃₂	.235	¹¹/₆₄	⁹/₁₆	1¹⁵/₁₆
2⅛	.2031	⅜	³/₁₆	.343	0.218	¾	2¹/₁₆
2⁹/₁₆	.2500	⁷/₁₆	¼	¹⁷/₃₂	0.266	⅞	2½
3³/₁₆	.3125	⁹/₁₆	⁹/₃₂	²³/₃₂	0.328	1³/₁₆	3¹/₁₆
4¹/₁₆	.4687	⅝	⁵/₁₆	³¹/₃₂	0.484	1¼	3⅞
5³/₁₆	.6250	¾	⅜	1¹³/₃₂	0.656	1½	4¹⁵/₁₆
7¼	.7500	1⅛	½	2	0.781	1¾	7
10	1.1250	1⅜	¾	2⅝	1.156	2⅝	9½

series (bold-face type, Table 6) and an intermediate series (light-face type). A self-holding taper is defined as "a taper with an angle small enough to hold a shank in place ordinarily by friction without holding means. (Sometimes referred to as slow taper.)" A steep taper is defined as "a taper having an angle sufficiently large to insure the easy or self-releasing feature." The term "gage line" indicates the basic diameter at or near the large end of the taper.

Table 1B. Morse Stub Taper Shanks

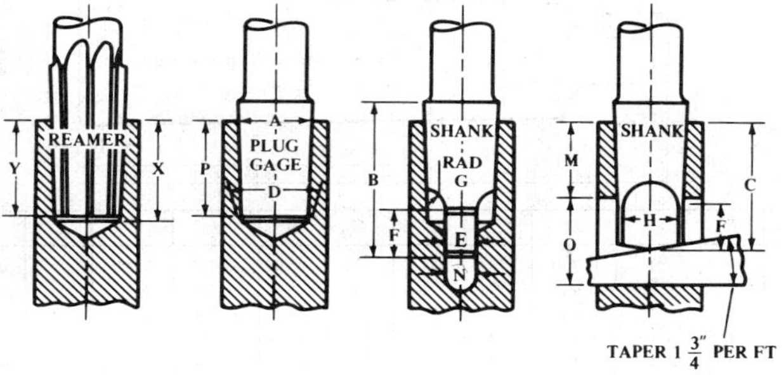

No. of Taper	Taper per Foot*	Taper per Inch†	Small End of Plug,† D	Diam. End of Socket,* A	Shank: Total Length, B	Shank: Depth, C	Tang: Thickness, E	Tang: Length, F
1	.59858	.049882	.4314	.475	1 5/16	1 1/8	13/64	5/16
2	.59941	.049951	.6469	.700	1 11/16	1 7/16	19/64	7/16
3	.60235	.050196	.8753	.938	2	1 3/4	25/64	9/16
4	.62326	.051938	1.1563	1.231	2 3/8	2 1/16	33/64	11/16
5	.63151	.052626	1.6526	1.748	3	2 11/16	3/4	15/16

No. of Taper	Tang: Radius of Mill, G	Tang: Diameter, H	Socket: Plug Depth, P	Socket: Min. Depth of Tapered Hole, Drilled X	Socket: Min. Depth of Tapered Hole, Reamed Y	Socket: Socket End to Tang Slot, M	Tang Slot: Width, N	Tang Slot: Length, O
1	3/16	13/32	7/8	5/16	29/32	25/32	7/32	23/32
2	7/32	39/64	1 1/16	1 5/32	1 7/64	15/16	5/16	15/16
3	9/32	13/16	1 1/4	1 3/8	1 5/16	1 1/16	13/32	1 1/8
4	3/8	1 3/32	1 7/16	1 9/16	1 1/2	1 3/16	17/32	1 3/8
5	9/16	1 19/32	1 13/16	1 15/16	1 7/8	1 7/16	25/32	1 3/4

All dimensions in inches.
* These are basic dimensions.
† These dimensions are calculated for reference only.
Radius *J* is 3/64, 1/16, 5/64, 3/32, and 1/8 inch respectively for Nos. 1, 2, 3, 4, and 5 tapers.

British Standard Tapers. — British Standard 1660: 1972, "Machine Tapers, Reduction Sleeves, and Extension Sockets," contains dimensions for self-holding and self-releasing tapers, reduction sleeves, extension sockets, and turret sockets for tools having Morse and metric 5 per cent taper shanks. Adapters for use with 7/24 tapers and dimensions for spindle noses and tool shanks with self-release tapers and cotter slots are included in this Standard.

Table 2. Dimensions of Morse Taper Sleeves

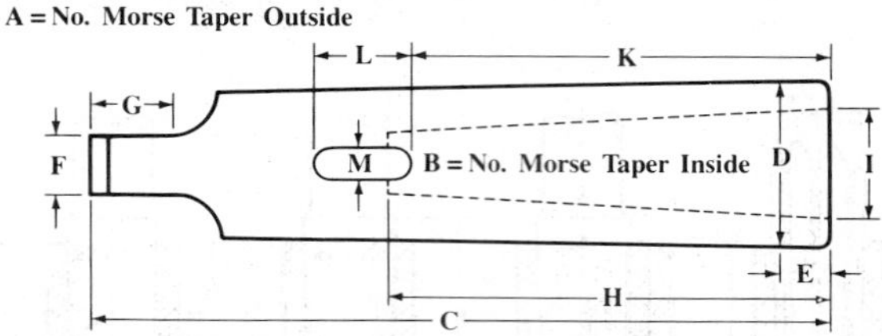

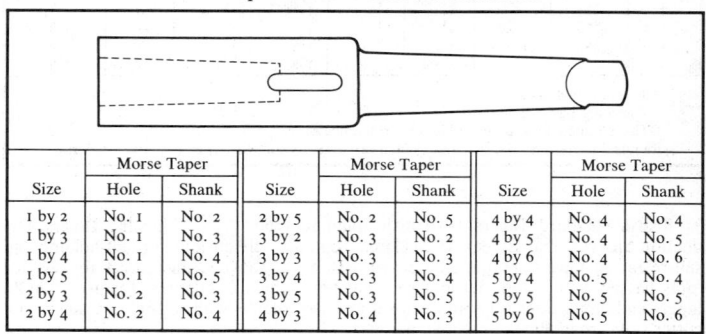

A	B	C	D	E	F	G	H	I	K	L	M
2	1	3⁹⁄₁₆	0.700	⅝	¼	⁷⁄₁₆	2³⁄₁₆	0.475	2¹⁄₁₆	¾	0.213
3	1	3¹⁵⁄₁₆	0.938	¼	⁵⁄₁₆	⁹⁄₁₆	2³⁄₁₆	0.475	2¹⁄₁₆	¾	0.213
3	2	4⁷⁄₁₆	0.938	¾	⁵⁄₁₆	⁹⁄₁₆	2⅝	0.700	2½	⅞	0.260
4	1	4⅞	1.231	¼	¹⁵⁄₃₂	⅝	2³⁄₁₆	0.475	2¹⁄₁₆	¾	0.213
4	2	4⅞	1.231	¼	¹⁵⁄₃₂	⅝	2⅝	0.700	2½	⅞	0.260
4	3	5⅜	1.231	¾	¹⁵⁄₃₂	⅝	3¼	0.938	3¹⁄₁₆	1³⁄₁₆	0.322
5	1	6⅛	1.748	¼	⅝	¾	2³⁄₁₆	0.475	2¹⁄₁₆	¾	0.213
5	2	6⅛	1.748	¼	⅝	¾	2⅝	0.700	2½	⅞	0.260
5	3	6⅛	1.748	¼	⅝	¾	3¼	0.938	3¹⁄₁₆	1³⁄₁₆	0.322
5	4	6⅝	1.748	¾	⅝	¾	4⅛	1.231	3⅞	1¼	0.478
6	1	8⅝	2.494	⅜	¾	1⅛	2³⁄₁₆	0.475	2¹⁄₁₆	¾	0.213
6	2	8⅝	2.494	⅜	¾	1⅛	2⅝	0.700	2½	⅞	0.260
6	3	8⅝	2.494	⅜	¾	1⅛	3¼	0.938	3¹⁄₁₆	1³⁄₁₆	0.322
6	4	8⅝	2.494	⅜	¾	1⅛	4⅛	1.231	3⅞	1¼	0.478
6	5	8⅝	2.494	⅜	¾	1⅛	5¼	1.748	4¹⁵⁄₁₆	1½	0.635
7	3	11⅝	3.270	⅜	1⅛	1⅜	3¼	0.938	3¹⁄₁₆	1³⁄₁₆	0.322
7	4	11⅝	3.270	⅜	1⅛	1⅜	4⅛	1.231	3⅞	1¼	0.478
7	5	11⅝	3.270	⅜	1⅛	1⅜	5¼	1.748	4¹⁵⁄₁₆	1½	0.635
7	6	12½	3.270	1¼	1⅛	1⅜	7⅜	2.494	7	1¾	0.760

Morse Taper Sockets — Hole and Shank Sizes

Size	Morse Taper Hole	Morse Taper Shank	Size	Morse Taper Hole	Morse Taper Shank	Size	Morse Taper Hole	Morse Taper Shank
1 by 2	No. 1	No. 2	2 by 5	No. 2	No. 5	4 by 4	No. 4	No. 4
1 by 3	No. 1	No. 3	3 by 2	No. 3	No. 2	4 by 5	No. 4	No. 5
1 by 4	No. 1	No. 4	3 by 3	No. 3	No. 3	4 by 6	No. 4	No. 6
1 by 5	No. 1	No. 5	3 by 4	No. 3	No. 4	5 by 4	No. 5	No. 4
2 by 3	No. 2	No. 3	3 by 5	No. 3	No. 5	5 by 5	No. 5	No. 5
2 by 4	No. 2	No. 4	4 by 3	No. 4	No. 3	5 by 6	No. 5	No. 6

Table 3. Brown & Sharpe Taper Shanks

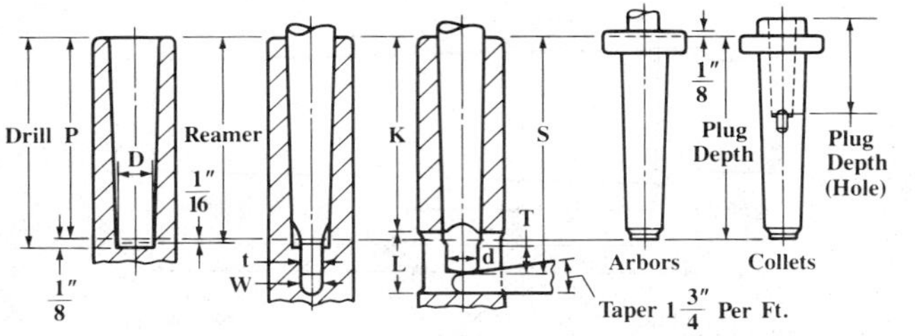

Number of Taper	Taper per Foot (inch)	Diam. of Plug at Small End	Plug Depth, P			Keyway from End of Spindle	Shank Depth	Length of Keyway†	Width of Keyway	Length of Arbor Tongue	Diameter of Arbor Tongue	Thickness of Arbor Tongue
			B & S** Standard	Mill. Mach. Standard	Miscell.							
		D				*K*	*S*	*L*	*W*	*T*	*d*	*t*
*1	.50200	.20000	15/16	. . .	. . .	15/16	1 3/16	3/8	.135	3/16	.170	1/8
*2	.50200	.25000	1 3/16	. . .	. . .	1 11/64	1 1/2	1/2	.166	1/4	.220	5/32
*3	.50200	.31250	1 1/2	. . . .	. . . 1 3/4 2	1 15/32 1 23/32 1 31/32	1 7/8 2 1/8 2 3/8	5/8 5/8 5/8	.197 .197 .197	5/16 5/16 5/16	.282 .282 .282	3/16 3/16 3/16
4	.50240	.35000	. . . 1 11/16	1 1/4 . . .	. . . . . .	1 13/64 1 41/64	1 21/32 2 3/32	11/16 11/16	.228 .228	11/32 11/32	.320 .320	7/32 7/32
5	.50160	.45000	. . . . . . 2 1/8	1 3/4	. . . 2 . . .	1 11/16 1 15/16 2 1/16	2 3/16 2 7/16 2 9/16	3/4 3/4 3/4	.260 .260 .260	3/8 3/8 3/8	.420 .420 .420	1/4 1/4 1/4
6	.50329	.50000	2 3/8	. . .	. . .	2 19/64	2 7/8	7/8	.291	7/16	.460	9/32
7	.50147	.60000	. . . 2 7/8 . . .	 3	2 1/2	2 13/32 2 25/32 2 29/32	3 1/32 3 13/32 3 17/32	15/16 15/16 15/16	.322 .322 .322	15/32 15/32 15/32	.560 .560 .560	5/16 5/16 5/16
8	.50100	.75000	3 9/16	. . .	. . .	3 29/64	4 1/8	1	.353	1/2	.710	11/32
9	.50085	.90010	. . . 4 1/4	4 . . .	. . . . . .	3 7/8 4 1/8	4 5/8 4 7/8	1 1/8 1 1/8	.385 .385	9/16 9/16	.860 .860	3/8 3/8
10	.51612	1.04465	5	. . . 5 11/16 . . .	. . . . . . 6 7/32	4 27/32 5 17/32 6 1/16	5 23/32 6 13/32 6 15/16	1 5/16 1 5/16 1 5/16	.447 .447 .447	21/32 21/32 21/32	1.010 1.010 1.010	7/16 7/16 7/16
11	.50100	1.24995	5 15/16 . . .	. . . 6 3/4	. . . . . .	5 25/32 6 19/32	6 21/32 7 15/32	1 5/16 1 5/16	.447 .447	21/32 21/32	1.210 1.210	7/16 7/16
12	.49973	1.50010	7 1/8 . . .	7 1/8 . . .	. . . 6 1/4	6 15/16 . . .	7 15/16 . . .	1 1/2 . . .	.510 . . .	3/4 . . .	1.460 . . .	1/2 . . .
13	.50020	1.75005	7 3/4	. . .	. . .	7 9/16	8 9/16	1 1/2	.510	3/4	1.710	1/2
14	.50000	2.00000	8 1/4	8 1/4	. . .	8 1/32	9 5/32	1 11/16	.572	27/32	1.960	9/16
15	.50000	2.25000	8 3/4	. . .	. . .	8 17/32	9 21/32	1 11/16	.572	27/32	2.210	9/16
16	.50000	2.50000	9 1/4	. . .	. . .	9	10 1/4	1 7/8	.635	15/16	2.450	5/8
17	.50000	2.75000	9 3/4	. . .	. . .	. . .	. . .	. . .	. . .	. . .	. . .	. . .
18	.50000	3.00000	10 1/4	. . .	. . .	. . .	. . .	. . .	. . .	. . .	. . .	. . .

* Adopted by American Standards Association. ** "B & S Standard" Plug Depths are not used in all cases. † Special lengths of keyway are used instead of standard lengths in some places. Standard lengths need not be used when keyway is for driving only and not for admitting key to force out tool.

Table 4. Jarno Taper Shanks

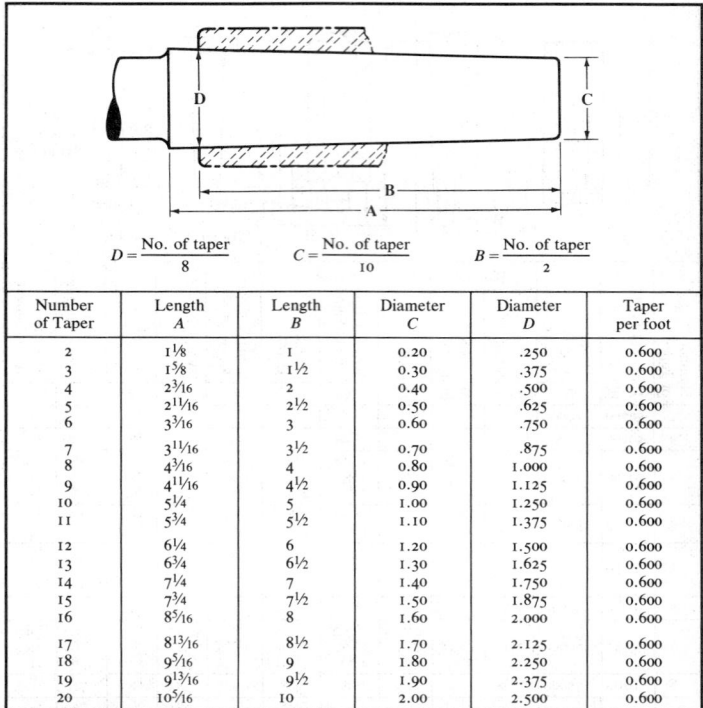

$$D = \frac{\text{No. of taper}}{8} \qquad C = \frac{\text{No. of taper}}{10} \qquad B = \frac{\text{No. of taper}}{2}$$

Number of Taper	Length A	Length B	Diameter C	Diameter D	Taper per foot
2	1⅛	1	0.20	.250	0.600
3	1⅝	1½	0.30	.375	0.600
4	2³⁄₁₆	2	0.40	.500	0.600
5	2¹¹⁄₁₆	2½	0.50	.625	0.600
6	3³⁄₁₆	3	0.60	.750	0.600
7	3¹¹⁄₁₆	3½	0.70	.875	0.600
8	4³⁄₁₆	4	0.80	1.000	0.600
9	4¹¹⁄₁₆	4½	0.90	1.125	0.600
10	5¼	5	1.00	1.250	0.600
11	5¾	5½	1.10	1.375	0.600
12	6¼	6	1.20	1.500	0.600
13	6¾	6½	1.30	1.625	0.600
14	7¼	7	1.40	1.750	0.600
15	7¾	7½	1.50	1.875	0.600
16	8⁵⁄₁₆	8	1.60	2.000	0.600
17	8¹³⁄₁₆	8½	1.70	2.125	0.600
18	9⁵⁄₁₆	9	1.80	2.250	0.600
19	9¹³⁄₁₆	9½	1.90	2.375	0.600
20	10⁵⁄₁₆	10	2.00	2.500	0.600

Tapers for Machine Tool Spindles. — Various standard tapers have been used for the taper holes in the spindles of machine tools requiring a taper hole for receiving either the shank of a cutter, an arbor, a center, or any tool or accessory requiring a tapering seat. The spindles of drilling machines and the taper shanks of twist drills are made to fit the Morse taper. For lathes, the Morse taper is generally used, but some lathes have either the Jarno, Brown & Sharpe, or a special taper. The practice of 33 lathe manufacturers is as follows: 20 use the Morse taper; 5, the Jarno; 3 use special tapers of their own; 2 use modified Morse (longer than the standard but the same taper); 2 use Reed (which is a short Jarno); 1 uses the Brown & Sharpe standard. For grinding machine centers Jarno, Morse and Brown & Sharpe tapers are used. Ten grinding machine manufacturers were divided as follows: 3 use Brown & Sharpe; 3 use Morse, and 4 use Jarno. The Brown & Sharpe taper has been extensively used for milling machine and dividing head spindles. The standard milling machine spindle adopted in 1927 by the milling machine manufacturers of the National Machine Tool Builders' Association, has a taper of 3½ inches per foot. This comparatively steep taper was adopted to insure easy release of arbors.

Table 5. American National Standard Self-holding Tapers — Basic Dimensions (ANSI B5.10-1981)

No. of Taper	Taper per Foot	Diam. at Gage Line* A	Means of Driving and Holding*				Origin of Series
.239	0.50200	0.23922					Brown & Sharpe Taper Series
.299	0.50200	0.29968					
.375	0.50200	0.37525					
1	0.59858	0.47500					
2	0.59941	0.70000					
3	0.60235	0.93800					
4	0.62326	1.23100					Morse Taper Series
4½	0.62400	1.50000	Tang Drive With Shank Held in by Friction (*See Table 7*)	Tang Drive With Shank Held in by Key (*See Table 8*)	Key Drive With Shank Held in by Key (*See Table 9*)	Key Drive With Shank Held in by Draw-bolt (*See Table 10*)	
5	0.63151	1.74800					
6	0.62565	2.49400					
7	0.62400	3.27000					
200	0.750	2.000					¾ Inch per Foot Taper Series
250	0.750	2.500					
300	0.750	3.000					
350	0.750	3.500					
400	0.750	4.000					
450	0.750	4.500					
500	0.750	5.000					
600	0.750	6.000					
800	0.750	8.000					
1000	0.750	10.000					
1200	0.750	12.000					

All dimensions given in inches.
* See illustrations above Tables 7, 8, 9 and 10.

Table 6. ANSI Standard Steep Machine Tapers (ANSI B5.10-1981)

No. of Taper	Taper per Foot (1)	Diam. at Gage Line (2)	Length Along Axis	No. of Taper	Taper per Foot (1)	Diam. at Gage Line (2)	Length Along Axis
5	3.500	0.500	0.6875	35	3.500	1.500	2.2500
10	3.500	0.625	0.8750	40	3.500	1.750	2.5625
15	3.500	0.750	1.0625	45	3.500	2.250	3.3125
20	3.500	0.875	1.3125	50	3.500	2.750	4.0000
25	3.500	1.000	1.5625	55	3.500	3.500	5.1875
30	3.500	1.250	1.8750	60	3.500	4.250	6.3750

All dimensions given in inches.
(1) This taper corresponds to an included angle of 16°, 35′, 39.4″.
The tapers numbered 10, 20, 30, 40, 50, and 60 that are printed in heavy-faced type are designated as the "Preferred Series." The tapers numbered 5, 15, 25, 35, 45, and 55 that are printed in light-faced type are designated as the "Intermediate Series."
(2) The basic diameter at gage line is at large end of taper.

Table 7. American National Standard Taper Drive with Tang, Self-Holding Tapers (ANSI B5.10-1981)

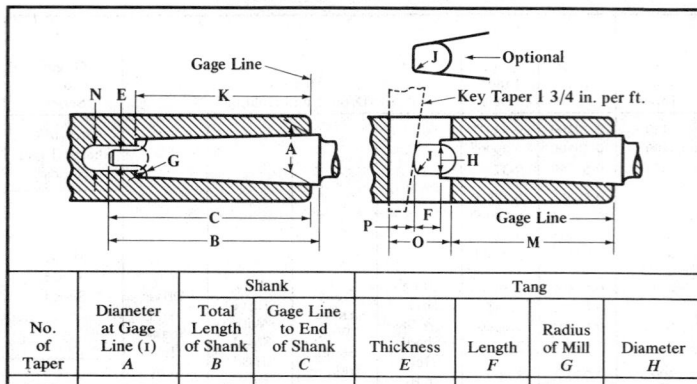

No. of Taper	Diameter at Gage Line (1) A	Shank		Tang			
		Total Length of Shank B	Gage Line to End of Shank C	Thickness E	Length F	Radius of Mill G	Diameter H
0.239	0.23922	1.28	1.19	0.125	0.19	0.19	0.18
0.299	0.29968	1.59	1.50	0.156	0.25	0.19	0.22
0.375	0.37525	1.97	1.88	0.188	0.31	0.19	0.28
1	0.47500	2.56	2.44	0.203	0.38	0.19	0.34
2	0.70000	3.13	2.94	0.250	0.44	0.25	0.53
3	0.93800	3.88	3.69	0.312	0.56	0.22	0.72
4	1.23100	4.88	4.63	0.469	0.63	0.31	0.97
4½	1.50000	5.38	5.13	0.562	0.69	0.38	1.20
5	1.74800	6.12	5.88	0.625	0.75	0.38	1.41
6	2.49400	8.25	8.25	0.750	1.13	0.50	2.00

No. of Taper	Radius J	Socket		Tang Slot			
		Min. Depth of Hole K		Gage Line to Tang Slot M	Width N	Length O	Shank End to Back of Tang Slot P
		Drilled	Reamed				
0.239	0.03	1.06	1.00	0.94	0.141	0.38	0.13
0.299	0.03	1.31	1.25	1.17	0.172	0.50	0.17
0.375	0.05	1.63	1.56	1.47	0.203	0.63	0.22
1	0.05	2.19	2.16	2.06	0.218	0.75	0.38
2	0.06	2.66	2.61	2.50	0.266	0.88	0.44
3	0.08	3.31	3.25	3.06	0.328	1.19	0.56
4	0.09	4.19	4.13	3.88	0.484	1.25	0.50
4½	0.13	4.62	4.56	4.31	0.578	1.38	0.56
5	0.13	5.31	5.25	4.94	0.656	1.50	0.56
6	0.16	7.41	7.33	7.00	0.781	1.75	0.50

All dimensions are in inches. (1) See Table 11 for plug and ring gage dimensions.

Tolerances: For shank diameter A at gage line, $+0.002-0.000$; for hole diameter A, $+0.000-0.002$. For tang thickness E up to No. 5 inclusive, $+0.000-0.006$; No. 6, $+0.000$ -0.008. For width N of tang slot up to No. 5 inclusive, $+0.006-0.000$; No. 6, $+0.008$ -0.000. For centrality of tang E with center line of taper, .0025 (.005 total indicator variation). These centrality tolerances also apply to the tang slot N. On rate of taper, all sizes 0.002 per foot. This tolerance may be applied on *shanks* only in the direction which *increases* the rate of taper and on *sockets* only in the direction which *decreases* the rate of taper. Tolerances for two-decimal dimensions are plus or minus 0.010, unless otherwise specified.

Table 8. American National Standard Taper Drive with Keeper Key Slot, Self-Holding Tapers (ANSI B5.10-1981)

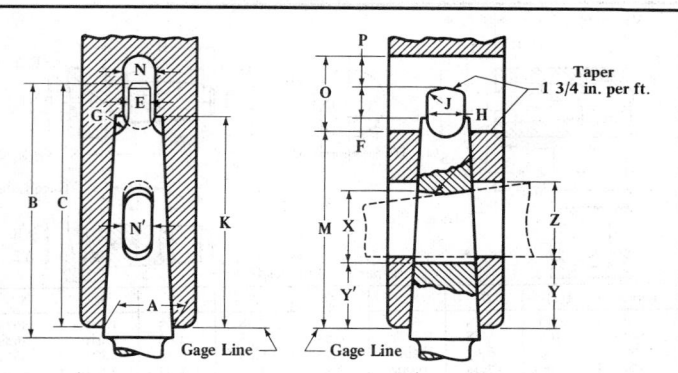

No. of Taper	Diam. at Gage Line (1) A	Shank			Tang					Socket		
		Total Length B	Gage Line to End C	Thickness E	Length F	Radius of Mill G	Diameter H	Radius J	Min. Depth of Hole K		Gage Line to Tang Slot M
									Drill	Ream	
3	0.938	3.88	3.69	0.312	0.56	0.28	0.78	0.08	3.31	3.25	3.06
4	1.231	4.88	4.63	0.469	0.63	0.31	0.97	0.09	4.19	4.13	3.88
4½	1.500	5.38	5.13	0.562	0.69	0.38	1.20	0.13	4.63	4.56	4.32
5	1.748	6.13	5.88	0.625	0.75	0.38	1.41	0.13	5.31	5.25	4.94
6	2.494	8.56	8.25	0.750	1.13	0.50	2.00	0.16	7.41	7.33	7.00
7	3.270	11.63	11.25	1.125	1.38	0.75	2.63	0.19	10.16	10.08	9.50

No. of Taper	Tang Slot			Keeper Slot in Shank			Keeper Slot in Socket		
	Width N	Length O	Shank End to Back of Slot P	Gage Line to Bottom of Slot Y'	Length X	Width N'	Gage Line to Front of Slot Y	Length Z	Width N'
3	0.328	1.19	0.56	1.03	1.13	0.266	1.13	1.19	0.266
4	0.484	1.25	0.50	1.41	1.19	0.391	1.50	1.25	0.391
4½	0.578	1.38	0.56	1.72	1.25	0.453	1.81	1.38	0.453
5	0.656	1.50	0.56	2.00	1.38	0.516	2.13	1.50	0.516
6	0.781	1.75	0.50	2.13	1.63	0.641	2.25	1.75	0.641
7	1.156	2.63	0.88	2.50	1.69	0.766	2.63	1.81	0.766

All dimensions are in inches. (1) See Table 11 for plug and ring gage dimensions.

Tolerances: For shank diameter A at gage line, $+0.002$, -0; for hole diameter A, $+0$, -0.002. For tang thickness E up to No. 5 inclusive, $+0$, -0.006; larger than No. 5, $+0$, -0.008. For width of slots N and N' up to No. 5 inclusive, $+0.006$, -0; larger than No. 5, $+0.008$, -0. For centrality of tang E with center line of taper .0025 (.005 total indicator variation). These centrality tolerances also apply to slots N and N'. On rate of taper, see footnote in Table 7. Tolerances for two-decimal dimensions are ±0.010 unless otherwise specified.

Table 9. American National Standard Nose Key Drive with Keeper Key Slot, Self-Holding Tapers (ANSI B5.10-1981)

Taper 1 3/4 in. per ft.

Taper	A(1)	B'	C	Q	I'	I	R	S
200	2.000	5.13		0.25	1.38	1.63	1.010	.562
250	2.500	5.88		0.25	1.38	2.06	1.010	.562
300	3.000	6.63	Min	0.25	1.63	2.50	2.010	.562
350	3.500	7.44	0.003	0.31	2.00	2.94	2.010	.562
400	4.000	8.19	Max	0.31	2.13	3.31	2.010	.562
450	4.500	9.00	0.035	0.38	2.38	3.81	3.010	.812
500	5.000	9.75	for	0.38	2.50	4.25	3.010	.812
600	6.000	11.31	all	0.44	3.00	5.19	3.010	.812
800	8.000	14.38	sizes	0.50	3.50	7.00	4.010	1.062
1000	10.000	17.44		0.63	4.50	8.75	4.010	1.062
1200	12.000	20.50		0.75	5.38	10.50	4.010	1.062

Taper	D	D'*	W	X	N'	R'	S'	T
200	1.41	0.375	3.44	1.56	0.656	1.000	0.50	4.75
250	1.66	0.375	3.69	1.56	0.781	1.000	0.50	5.50
300	2.25	0.375	4.06	1.56	1.031	2.000	0.50	6.25
350	2.50	0.375	4.88	2.00	1.031	2.000	0.50	6.94
400	2.75	0.375	5.31	2.25	1.031	2.000	0.50	7.69
450	3.00	0.500	5.88	2.44	1.031	3.000	0.75	8.38
500	3.25	0.500	6.44	2.63	1.031	3.000	0.75	9.13
600	3.75	0.500	7.44	3.00	1.281	3.000	0.75	10.56
800	4.75	0.500	9.56	4.00	1.781	4.000	1.00	13.50
1000	...	...	11.50	4.75	2.031	4.000	1.00	16.31
1200	...	...	13.75	5.75	2.031	4.000	1.00	19.00

Taper	U	V	M	N	O	P	Y	Z
200	1.81	1.00	4.50	0.656	1.56	0.94	2.00	1.69
250	2.25	1.00	5.19	0.781	1.94	1.25	2.25	1.69
300	2.75	1.00	5.94	1.031	2.19	1.50	2.63	1.69
350	3.19	1.25	6.75	1.031	2.19	1.50	3.00	2.13
400	3.63	1.25	7.50	1.031	2.19	1.50	3.25	2.38
450	4.19	1.50	8.00	1.031	2.75	1.75	3.63	2.56
500	4.63	1.50	8.75	1.031	2.75	1.75	4.00	2.75
600	5.50	1.75	10.13	1.281	3.25	2.06	4.63	3.25
800	7.38	2.00	12.88	1.781	4.25	2.75	5.75	4.25
1000	9.19	2.50	15.75	2.031	5.00	3.31	7.00	5.00
1200	11.00	3.00	18.50	2.531	6.00	4.00	8.25	6.00

All dimensions are in inches. AE is 0.005 greater than one-half of A.
Width of drive key R'' is 0.001 less than width R' of keyway.
* Thread is UNF-2B for hole; UNF-2A for screw. (1) See Table 11 for plug and ring gage dimensions.

Tolerances: For diameter A of hole at gage line, $+0$, -0.002; for diameter A of shank at gage line, $+0.002$, -0; for width of slots N and N', $+0.008$, -0; for width of drive keyway R' in socket, $+0$, -0.001; for width of drive keyway R in shank, 0.010, -0; for centrality of slots N and N' with center line of spindle, 0.007; for centrality of keyway with spindle center line: for R, 0.004 and for R', 0.002 T.I.V. On rate of taper, see footnote in Table 7. Two-decimal dimensions, ± 0.010 unless otherwise specified.

Table 10. American National Standard Nose Key Drive with Drawbolt, Self-Holding Tapers (ANSI B5.10-1981)

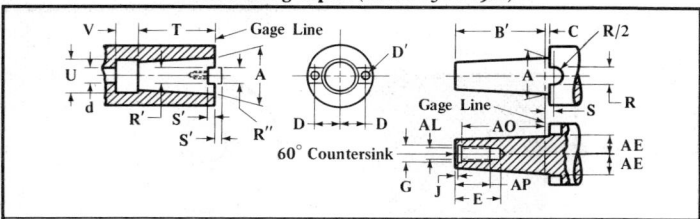

Sockets

No. of Taper	Diam. at Gage Line A*	Drive Key			Drive Keyway			Gage Line to Front of Relief T	Diam. of Relief U	Depth of Relief V	Diam. of Draw Bolt Hole d
		Screw Holes									
		Center Line to Center of Screw D	UNF 2B Hole UNF 2A Screw D'	Width R''	Width R'	Depth S'					
200	2.000	1.41	0.38	0.999	1.000	0.50	4.75	1.81	1.00	1.00
250	2.500	1.66	0.38	0.999	1.000	0.50	5.50	2.25	1.00	1.00
300	3.000	2.25	0.38	1.999	2.000	0.50	6.25	2.75	1.00	1.13
350	3.500	2.50	0.38	1.999	2.000	0.50	6.94	3.19	1.25	1.13
400	4.000	2.75	0.38	1.999	2.000	0.50	7.69	3.63	1.25	1.63
450	4.500	3.00	0.50	2.999	3.000	0.75	8.38	4.19	1.50	1.63
500	5.000	3.25	0.50	2.999	3.000	0.75	9.13	4.63	1.50	1.63
600	6.000	3.75	0.50	2.999	3.000	0.75	10.56	5.50	1.75	2.25
800	8.000	4.75	0.50	3.999	4.000	1.00	13.50	7.38	2.00	2.25
1000	10.000	...	...	3.999	4.000	1.00	16.31	9.19	2.50	2.25
1200	12.000	...	...	3.999	4.000	1.00	19.00	11.00	3.00	2.25

Shanks

No. of Taper	Length from Gage Line B'	Drawbar Hole						Drive Keyway		Center Line to Bottom of Keyway AE
		Diam. UNC-2B AL	Depth of Drilled Hole E	Depth of Thread AP	Diam. of Counter Bore G	Gage Line to First Thread AO	Depth of 60° Chamfer J	Width R	Depth S	
200	5.13	7/8–9	2.44	1.75	0.91	4.78	0.13	1.010	0.562	1.005
250	5.88	7/8–9	2.44	1.75	0.91	5.53	0.13	1.010	0.562	1.255
300	6.63	1–8	2.75	2.00	1.03	6.19	0.19	2.010	0.562	1.505
350	7.44	1–8	2.75	2.00	1.03	7.00	0.19	2.010	0.562	1.755
400	8.19	1 1/2–6	4.00	3.00	1.53	7.50	0.31	2.010	0.562	2.005
450	9.00	1 1/2–6	4.00	3.00	1.53	8.31	0.31	3.010	0.812	2.255
500	9.75	1 1/2–6	4.00	3.00	1.53	9.06	0.31	3.010	0.812	2.505
600	11.31	2–4 1/2	5.31	4.00	2.03	10.38	0.50	3.010	0.812	3.005
800	14.38	2–4 1/2	5.31	4.00	2.03	13.44	0.50	4.010	1.062	4.005
1000	17.44	2–4 1/2	5.31	4.00	2.03	16.50	0.50	4.010	1.062	5.005
1200	20.50	2–4 1/2	5.31	4.00	2.03	19.56	0.50	4.010	1.062	6.005

All dimensions in inches.   * See Table 11 for plug and ring gage dimensions.

Exposed length C is 0.003 minimum and 0.035 maximum for all sizes.

Drive key D' screw sizes are 3/8–24 UNF-2A up to taper No. 400 inclusive and 1/2–20 UNF-2A for larger tapers.

Tolerances: For diameter A of hole at gage line, + 0.000, − 0.002 for all sizes; for diameter A of shank at gage line, + 0.002, − 0.000 for all sizes; for width of drive keyway R' in socket, + 0.000, − 0.001; for width of drive keyway R in shank, + 0.010, − 0.000; for centrality of drive keyway R with center line of shank, 0.004 total indicator variation, and for drive keyway R', with center line of spindle, 0.002. On rate of taper, see footnote in Table 7. Tolerances for two-decimal dimensions are ± 0.010 unless otherwise specified.

Table 11. American National Standard Plug and Ring Gages for the Self-Holding Taper Series (ANSI B5.10-1981)

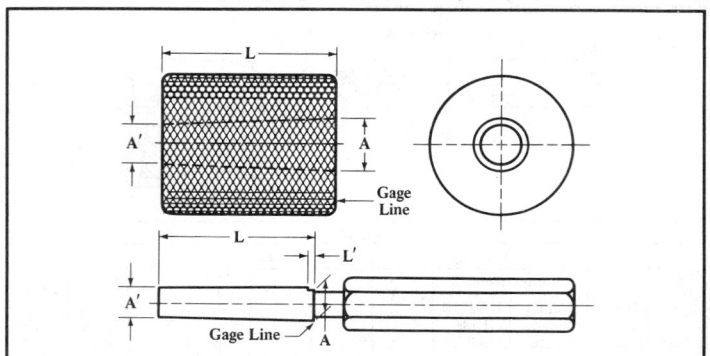

No. of Taper	Taper[1] per Foot	Diameter[1] at Gage Line A	Tolerances for Diameter A[2]			Diameter at Small End A'	Length Gage Line to End L	Depth of Gaging Notch, Plug Gage L'
			Class X Gage	Class Y Gage	Class Z Gage			
.239	0.50200	0.23922	0.00004	0.00007	0.00010	0.20000	0.94	0.048
.299	0.50200	0.29968	0.00004	0.00007	0.00010	0.25000	1.19	0.048
.375	0.50200	0.37525	0.00004	0.00007	0.00010	0.31250	1.50	0.048
1	0.59858	0.47500	0.00004	0.00007	0.00010	0.36900	2.13	0.040
2	0.59941	0.70000	0.00004	0.00007	0.00010	0.57200	2.56	0.040
3	0.60235	0.93800	0.00006	0.00009	0.00012	0.77800	3.19	0.040
4	0.62326	1.23100	0.00006	0.00009	0.00012	1.02000	4.06	0.038
4½	0.62400	1.50000	0.00006	0.00009	0.00012	1.26600	4.50	0.038
5	0.63151	1.74800	0.00008	0.00012	0.00016	1.47500	5.19	0.038
6	0.62565	2.49400	0.00008	0.00012	0.00016	2.11600	7.25	0.038
7	0.62400	3.27000	0.00010	0.00015	0.00020	2.75000	10.00	0.038
200	0.750	2.000	0.00008	0.00012	0.00016	1.703	4.75	0.032
250	0.750	2.500	0.00008	0.00012	0.00016	2.156	5.50	0.032
300	0.750	3.000	0.00010	0.00015	0.00020	2.609	6.25	0.032
350	0.750	3.500	0.00010	0.00015	0.00020	3.063	7.00	0.032
400	0.750	4.000	0.00010	0.00015	0.00020	3.516	7.75	0.032
450	0.750	4.500	0.00010	0.00015	0.00020	3.969	8.50	0.032
500	0.750	5.000	0.00013	0.00019	0.00025	4.422	9.25	0.032
600	0.750	6.000	0.00013	0.00019	0.00025	5.328	10.75	0.032
800	0.750	8.000	0.00016	0.00024	0.00032	7.141	13.75	0.032
1000	0.750	10.000	0.00020	0.00030	0.00040	8.953	16.75	0.032
1200	0.750	12.000	0.00020	0.00030	0.00040	10.766	19.75	0.032

All dimensions are in inches.

[1] The taper per foot and diameter A at gage line are basic dimensions. Dimensions in Column A' are calculated for reference only.

[2] Tolerances for diameter A are plus for plug gages and minus for ring gages.

The amount of taper deviation for Class X, Class Y, and Class Z gages are the same, respectively, as the amounts shown for tolerances on diameter A. Taper deviation is the permissible allowance from true taper at any point of diameter in the length of the gage. On taper *plug* gages, this deviation may be applied only in the direction which *decreases* the rate of taper. On taper *ring* gages, this deviation may be applied only in the direction which *increases* the rate of taper. Tolerances on two-decimal dimensions are ±0.010.

Table 12. American National Standard Plug and Ring Gages for Steep Machine Tapers (ANSI B5.10-1981)

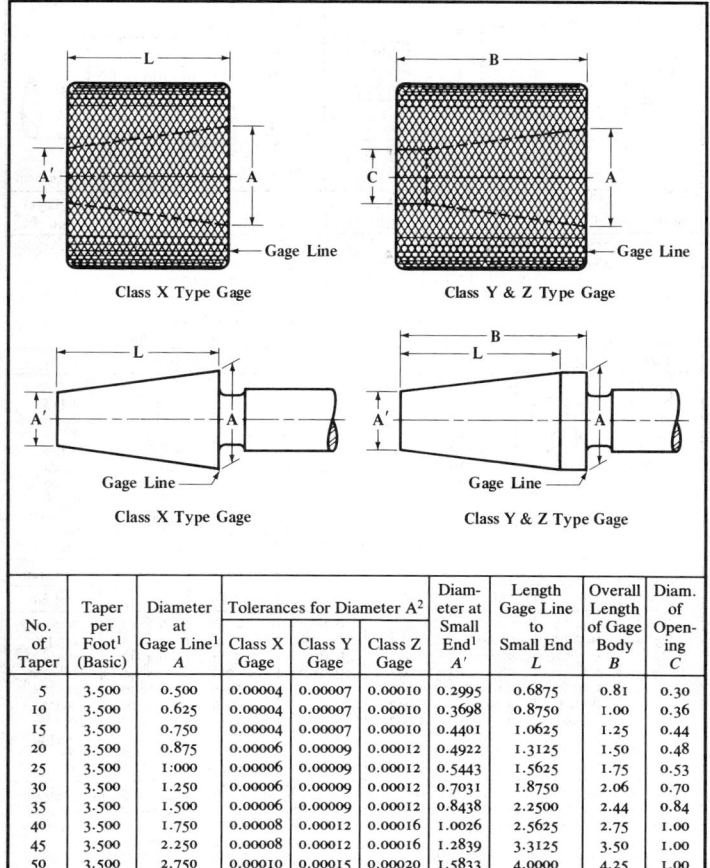

Class X Type Gage   Class Y & Z Type Gage

Class X Type Gage   Class Y & Z Type Gage

No. of Taper	Taper per Foot[1] (Basic)	Diameter at Gage Line[1] A	Tolerances for Diameter A[2]			Diameter at Small End[1] A'	Length Gage Line to Small End L	Overall Length of Gage Body B	Diam. of Opening C
			Class X Gage	Class Y Gage	Class Z Gage				
5	3.500	0.500	0.00004	0.00007	0.00010	0.2995	0.6875	0.81	0.30
10	3.500	0.625	0.00004	0.00007	0.00010	0.3698	0.8750	1.00	0.36
15	3.500	0.750	0.00004	0.00007	0.00010	0.4401	1.0625	1.25	0.44
20	3.500	0.875	0.00006	0.00009	0.00012	0.4922	1.3125	1.50	0.48
25	3.500	1.000	0.00006	0.00009	0.00012	0.5443	1.5625	1.75	0.53
30	3.500	1.250	0.00006	0.00009	0.00012	0.7031	1.8750	2.06	0.70
35	3.500	1.500	0.00006	0.00009	0.00012	0.8438	2.2500	2.44	0.84
40	3.500	1.750	0.00008	0.00012	0.00016	1.0026	2.5625	2.75	1.00
45	3.500	2.250	0.00008	0.00012	0.00016	1.2839	3.3125	3.50	1.00
50	3.500	2.750	0.00010	0.00015	0.00020	1.5833	4.0000	4.25	1.00
55	3.500	3.500	0.00010	0.00015	0.00020	1.9870	5.1875	5.50	1.00
60	3.500	4.250	0.00010	0.00015	0.00020	2.3906	6.3750	6.75	2.00

All dimensions are in inches.

[1] The taper per foot and diameter A at gage line are basic dimensions. Dimensions in Column A' are calculated for reference only.

[2] Tolerances for diameter A are plus for plug gages and minus for ring gages.

The amounts of taper deviation for Class X, Class Y, and Class Z gages are the same, respectively, as the amounts shown for tolerances on diameter A. Taper deviation is the permissible allowance from true taper at any point of diameter in the length of the gage. On taper *plug* gages, this deviation may be applied only in the direction which *decreases* the rate of taper. On taper *ring* gages, this deviation may be applied only in the direction which *increases* the rate of taper. Tolerances on two-decimal dimensions are ±0.010.

Jacobs Tapers and Threads for Drill Chucks and Spindles

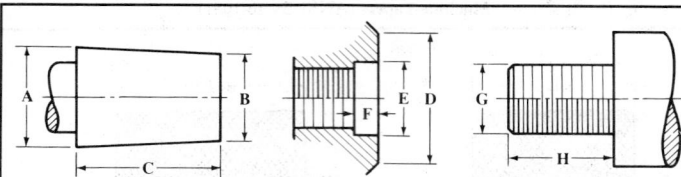

American Standard Thread Form

Taper Series	A	B	C	Taper per Ft.	Taper Series	A	B	C	Taper per Ft.
No. 0	.2500	.22844	.43750	.59145	No. 4	1.1240	1.0372	1.6563	.62886
No. 1	.3840	.33341	.65625	.92508	No. 5	1.4130	1.3161	1.8750	.62010
No. 2	.5590	.48764	.87500	.97861	No. 6	0.6760	0.6241	1.0000	.62292
No. 2ᵃ	.5488	.48764	.75000	.97861	No. 33	0.6240	0.5605	1.0000	.76194
No. 3	.8110	.74610	1.21875	.63898	. . .	. . .	. . .	. . .	. . .

Thread Size	Diameter D		Diameter E		Dimension F	
	Max.	Min.	Max.	Min.	Max.	Min.
5/16–24	0.531	0.516	0.3245	0.3195	0.135	0.115
5/16–24	0.633	0.618	0.3245	0.3195	0.135	0.115
3/8–24	0.633	0.618	0.385	0.380	0.135	0.115
1/2–20	0.860	0.845	0.510	0.505	0.135	0.115
5/8–11	1.125	1.110	0.635	0.630	0.166	0.146
5/8–16	1.125	1.110	0.635	0.630	0.166	0.146
45/64–16	1.250	1.235	0.713	0.708	0.166	0.146
3/4–16	1.250	1.235	0.760	0.755	0.166	0.146
1–8	1.437	1.422	1.036	1.026	0.281	0.250
1–10	1.437	1.422	1.036	1.026	0.281	0.250
1½–8	1.871	1.851	1.536	1.526	0.343	0.312

Threadᵇ Size	G		Hᶜ	Plug Gage Pitch Diam.		Ring Gage Pitch Diam.	
	Max	Min		Go	Not Go	Go	Not Go
5/16–24	0.3114	0.3042	0.437ᵈ	0.2854	0.2902	0.2843	0.2806
3/8–24	0.3739	0.3667	0.562ᵉ	0.3479	0.3528	0.3468	0.3430
1/2–20	0.4987	0.4906	0.562	0.4675	0.4731	0.4662	0.4619
5/8–11	0.6234	0.6113	0.687	0.5660	0.5732	0.5644	0.5589
5/8–16	0.6236	0.6142	0.687	0.5844	0.5906	0.5830	0.5782
45/64–16	0.7016	0.6922	0.687	0.6625	0.6687	0.6610	0.6561
3/4–16	0.7485	0.7391	0.687	0.7094	0.7159	0.7079	0.7029
1–8	1.000	0.9848	1.000	0.9188	0.9242	0.9188	0.9134
1–10	1.000	0.9872	1.000	0.9350	0.9395	0.9350	0.9305
1½–8	1.500	1.4848	1.000	1.4188	1.4242	1.4188	1.4134

Usual Chuck Capacities for Different Taper Series Numbers: No. 0 taper, drill diameters, 0–5/32 inch; No. 1, 0–¼ inch; No. 2, 0–½ inch; No. 2 "Short," 0–5/16 inch; No. 3, 0–½, 1/8–5/8, 3/16–¾, or ¼–13/16 inch; No. 4, 1/8–¾ inch; No. 5, 3/8–1 inch; No. 6, 0–½ inch; No. 33, 0–½ inch.

Usual Chuck Capacities for Different Thread Sizes: Size 5/16–24, drill diameters 0–¼ inch; size 3/8–24, drill diameters 0–3/8, 1/16–3/8, or 5/64–½ inch; size ½–20, drill diameters 0–½, 1/16–3/8, or 5/64–½ inch; size 5/8–11, drill diameters 0–½ inch; size 5/8–16, drill diameters 0–½, 1/8–5/8, or 3/16–¾ inch; size 45/64–16, drill diameters 0–½ inch; size ¾–16, drill diameters 0–½ or 3/16–¾.

ᵃ These dimensions are for the No. 2 "short" taper.
ᵇ Except for 1-8, 1-10, 1½-8 all threads are now manufactured to the American National Standard Unified Screw Thread System, Internal Class 2B, External Class 2A. Effective date 1976.
ᶜ Tolerances for dimension H are as follows: 0.030 inch for thread sizes 5/16–24 to ¾–16, inclusive and 0.125 inch for thread sizes 1–8 to 1½–8, inclusive.
ᵈ Length for Jacobs 0B5/16 chuck is 0.375 inch, length for 1B5/16 chuck is 0.437 inch.
ᵉ Length for Jacobs No. 1BS chuck is 0.437 inch.

Table 1. Essential Dimensions of American National Standard Spindle Noses for Milling Machines (ANSI B5.18-1972)

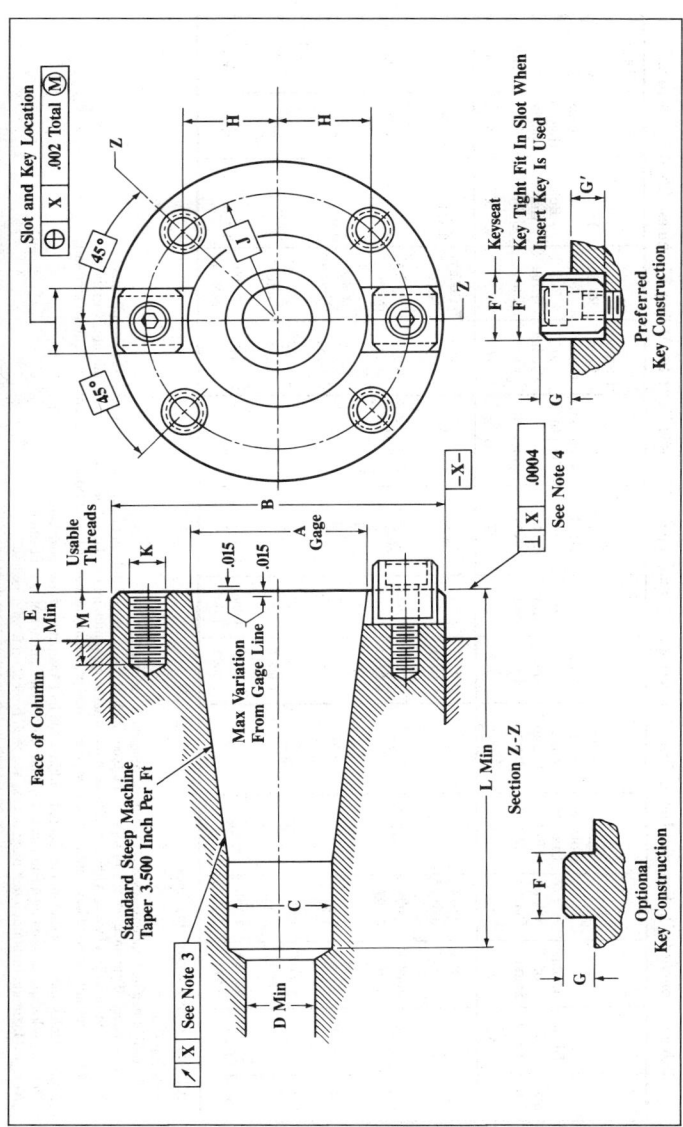

Table I (Concluded). **Essential Dimensions of American National Standard Spindle Noses for Milling Machines (ANSI B5.18-1972)**

Size No.	Gage Diam. of Taper A	Diam. of Spindle B	Pilot Diam. C	Clearance Hole for Draw-in Bolt Min. D	Minimum Dimension Spindle End to Column E	Width of Driving Key F	Width of Keyseat F'	Maximum Height of Driving Key G	Minimum Depth of Keyseat G'	Distance from Center to Driving Keys H	Radius of Bolt Hole Circle J	Size of Threads for Bolt Holes UNC-2B K	Full Depth of Arbor Hole in Spindle Min. L	Depth of Usable Thread for Bolt Hole M
30	1.250	2.7493 2.7488	0.692 0.685	0.66	0.50	0.6255 0.6252	0.624 0.625	0.31	0.31	0.660 0.654	1.0625 (Note 1)	0.375 — 16	2.88	0.62
40	1.750	3.4993 3.4988	1.005 0.997	0.66	0.62	0.6255 0.6252	0.624 0.625	0.31	0.31	0.910 0.904	1.3125 (Note 1)	0.500 — 13	3.88	0.81
45	2.250	3.9993 3.9988	1.286 1.278	0.78	0.62	0.7505 0.7502	0.749 0.750	0.38	0.38	1.160 1.154	1.500 (Note 1)	0.500 — 13	4.75	0.81
50	2.750	5.0618 5.0613	1.568 1.559	1.06	0.75	1.0006 1.0002	0.999 1.000	0.50	0.50	1.410 1.404	2.000 (Note 2)	0.625 — 11	5.50	1.00
60	4.250	8.7180 8.7175	2.381 2.371	1.38	1.50	1.0006 1.0002	0.999 1.000	0.50	0.50	2.420 2.414	3.500 (Note 2)	0.750 — 10	8.62	1.25

All dimensions are given in inches.

Tolerances:

Two-digit decimal dimensions ±0.010 unless otherwise specified.

A —Taper: Tolerance on rate of taper to be 0.001 inch per foot applied only in direction which decreases rate of taper.

F —Centrality of keyway with axis of taper 0.002 total at maximum material condition. (0.002 Total indicator variation)

F' —Centrality of solid key with axis of taper 0.002 total at maximum material condition. (0.002 Total indicator variation)

Note 1: Holes spaced as shown and located within 0.006 inch diameter of true position.

Note 2: Holes spaced as shown and located within 0.010 inch diameter of true position.

Note 3: Maximum turnout on test plug: 0.0004 at 1 inch projection from gage line.

          : 0.0010 at 12 inch projection from gage line.

Note 4: Squareness of mounting face measured near mounting bolt hole circle.

Table 2. Essential Dimensions of American National Standard Tool Shanks for Milling Machines (ANSI B5.18-1972)

Size No.	Gage Diam. of Taper N	Tap Drill Size for Draw-in Thread O	Diam. of Neck P	Size of Thread for Draw-in Bolt UNC-2B M	Pilot Diam. R	Length of Pilot S	Minimum Length of Usable Thread T	Minimum Depth of Clearance Hole U
30	1.250	0.422 0.432	0.66 0.65	0.500 — 13	0.675 0.670	0.81	1.00	2.00
40	1.750	0.531 0.541	0.94 0.93	0.625 — 11	0.987 0.980	1.00	1.12	2.25
45	2.250	0.656 0.666	1.19 1.18	0.750 — 10	1.268 1.260	1.00	1.50	2.75
50	2.750	0.875 0.885	1.50 1.49	1.000 — 8	1.550 1.540	1.00	1.75	3.50
60	4.250	1.109 1.119	2.28 2.27	1.250 — 7	2.360 2.350	1.75	2.25	4.25

Size No.	Distance from Rear of Flange to End of Arbor V	Clearance of Flange from Gage Diameter W	Tool Shank Center-line to Driving Slot X	Width of Driving Slot Y	Distance from Gage Line to Bottom of C'bore Z	Depth of 60° Center K	Diameter of C'bore L
30	2.75	0.045 0.075	0.640 0.625	0.635 0.645	2.50	0.05 0.07	0.525 0.530
40	3.75	0.045 0.075	0.890 0.875	0.635 0.645	3.50	0.05 0.07	0.650 0.655
45	4.38	0.105 0.135	1.140 1.125	0.760 0.770	4.06	0.05 0.07	0.775 0.780
50	5.12	0.105 0.135	1.390 1.375	1.010 1.020	4.75	0.05 0.12	1.025 1.030
60	8.25	0.105 0.135	2.400 2.385	1.010 1.020	7.81	0.05 0.12	1.307 1.312

All dimensions are given in inches.

Tolerances: Two digit decimal dimensions ±0.010 inch unless otherwise specified.
M—Permissible for Class 2B "NoGo" gage to enter five threads before interference.
N—Taper tolerance on rate of taper to be 0.001 inch per foot applied only in direction which increases rate of taper.
Y—Centrality of drive slot with axis of taper shank 0.004 inch at maximum material condition. (0.004 inch total indicator variation)

Table 3. American National Standard Draw-in Bolt Ends (ANSI B5.18-1972)

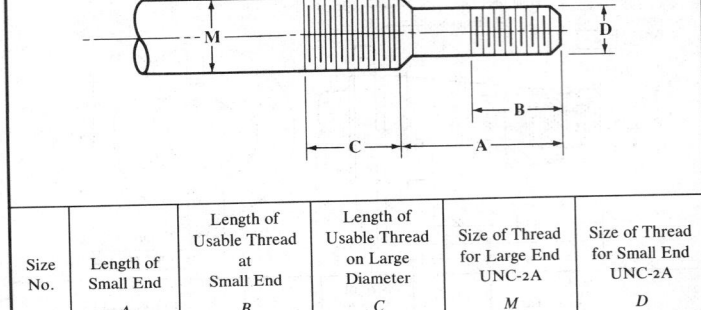

Size No.	Length of Small End	Length of Usable Thread at Small End	Length of Usable Thread on Large Diameter	Size of Thread for Large End UNC-2A	Size of Thread for Small End UNC-2A
	A	*B*	*C*	*M*	*D*
30	1.06	0.75	0.75	0.500 — 13	0.375 — 16
40	1.25	1.00	1.12	0.625 — 11	0.500 — 13
45	1.50	1.12	1.25	0.750 — 10	0.625 — 11
50	1.50	1.25	1.38	1.000 — 8	0.625 — 11
60	1.75	1.37	2.00	1.250 — 7	1.000 — 8

All dimensions are given in inches.

American National Standard Pilot Lead on Centering Plugs for Flatback Milling Cutters (ANSI B5.18-1972)

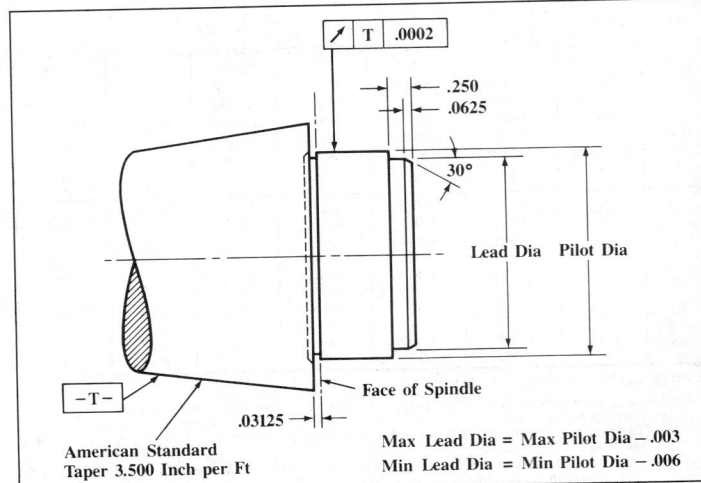

Table 4. Essential Dimensions for American National Standard Spindle Nose with Large Flange (ANSI B5.18-1972)

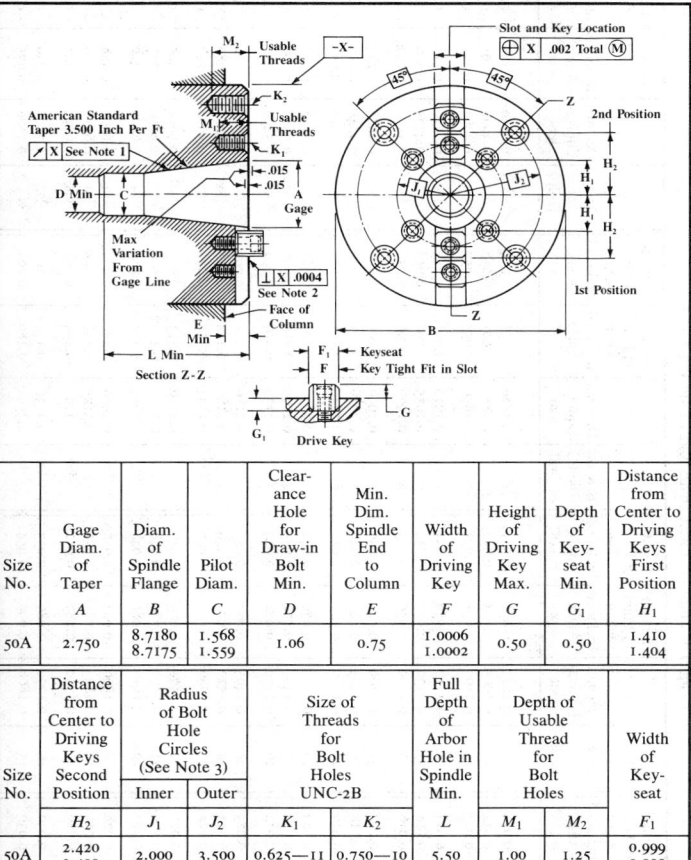

Size No.	Gage Diam. of Taper	Diam. of Spindle Flange	Pilot Diam.	Clear-ance Hole for Draw-in Bolt Min.	Min. Dim. Spindle End to Column	Width of Driving Key	Height of Driving Key Max.	Depth of Key-seat Min.	Distance from Center to Driving Keys First Position
	A	B	C	D	E	F	G	G_1	H_1
50A	2.750	8.7180 8.7175	1.568 1.559	1.06	0.75	1.0006 1.0002	0.50	0.50	1.410 1.404

Size No.	Distance from Center to Driving Keys Second Position	Radius of Bolt Hole Circles (See Note 3) Inner	Radius of Bolt Hole Circles (See Note 3) Outer	Size of Threads for Bolt Holes UNC-2B	Size of Threads for Bolt Holes UNC-2B	Full Depth of Arbor Hole in Spindle Min.	Depth of Usable Thread for Bolt Holes	Depth of Usable Thread for Bolt Holes	Width of Key-seat
	H_2	J_1	J_2	K_1	K_2	L	M_1	M_2	F_1
50A	2.420 2.410	2.000	3.500	0.625—11	0.750—10	5.50	1.00	1.25	0.999 1.000

All dimensions are given in inches.

Tolerances: Two-digit decimal dimensions ±0.010 unless otherwise specified.

A—Tolerance on rate of taper to be 0.001 inch per foot applied only in direction which decreases rate of taper.

F—Centrality of solid key with axis of taper 0.002 inch total at maximum material condition. (0.002 inch Total indicator variation)

F_1—Centrality of keyseat with axis of taper 0.002 inch total at maximum material condition. (0.002 inch Total indicator variation)

Note 1: Maximum runout on test plug: 0.0004 at 1 inch projection from gage line.
: 0.0010 at 12 inch projection from gage line.

Note 2: Squareness of mounting face measured near mounting bolt hole circle.

Note 3: Holes located as shown and within 0.010 inch diameter of true position.

Length of Point on Twist Drills and Centering Tools

Size of Drill	Decimal Equivalent	Length of Point when Included Angle = 90°	Length of Point when Included Angle = 118°	Size of Drill	Decimal Equivalent	Length of Point when Included Angle = 90°	Length of Point when Included Angle = 118°	Size or Diam. of Drill	Decimal Equivalent	Length of Point when Included Angle = 90°	Length of Point when Included Angle = 118°	Diam. of Drill	Decimal Equivalent	Length of Point when Included Angle = 90°	Length of Point when Included Angle = 118°
60	0.0400	0.020	0.012	37	0.1040	0.052	0.031	14	0.1820	0.091	0.055	3/8	0.3750	0.188	0.113
59	0.0410	0.021	0.012	36	0.1065	0.054	0.032	13	0.1850	0.093	0.056	25/64	0.3906	0.195	0.117
58	0.0420	0.021	0.013	35	0.1100	0.055	0.033	12	0.1890	0.095	0.057	13/32	0.4063	0.203	0.122
57	0.0430	0.022	0.013	34	0.1110	0.056	0.033	11	0.1910	0.096	0.057	27/64	0.4219	0.211	0.127
56	0.0465	0.023	0.014	33	0.1130	0.057	0.034	10	0.1935	0.097	0.058	7/16	0.4375	0.219	0.131
55	0.0520	0.026	0.016	32	0.1160	0.058	0.035	9	0.1960	0.098	0.059	29/64	0.4531	0.227	0.136
54	0.0550	0.028	0.017	31	0.1200	0.060	0.036	8	0.1990	0.100	0.060	15/32	0.4688	0.234	0.141
53	0.0595	0.030	0.018	30	0.1285	0.065	0.039	7	0.2010	0.101	0.060	31/64	0.4844	0.242	0.145
52	0.0635	0.032	0.019	29	0.1360	0.068	0.041	6	0.2040	0.102	0.061	1/2	0.5000	0.250	0.150
51	0.0670	0.034	0.020	28	0.1405	0.070	0.042	5	0.2055	0.103	0.062	33/64	0.5156	0.258	0.155
50	0.0700	0.035	0.021	27	0.1440	0.072	0.043	4	0.2090	0.105	0.063	17/32	0.5313	0.266	0.159
49	0.0730	0.037	0.022	26	0.1470	0.074	0.044	3	0.2130	0.107	0.064	35/64	0.5469	0.273	0.164
48	0.0760	0.038	0.023	25	0.1495	0.075	0.045	2	0.2210	0.111	0.067	9/16	0.5625	0.281	0.169
47	0.0785	0.040	0.024	24	0.1520	0.076	0.046	1	0.2280	0.114	0.068	37/64	0.5781	0.289	0.173
46	0.0810	0.041	0.024	23	0.1540	0.077	0.046	15/64	0.2344	0.117	0.070	19/32	0.5938	0.297	0.178
45	0.0820	0.041	0.025	22	0.1570	0.079	0.047	1/4	0.2500	0.125	0.075	39/64	0.6094	0.305	0.183
44	0.0860	0.043	0.026	21	0.1590	0.080	0.048	17/64	0.2656	0.133	0.080	5/8	0.6250	0.313	0.188
43	0.0890	0.045	0.027	20	0.1610	0.081	0.048	9/32	0.2813	0.141	0.084	41/64	0.6406	0.320	0.192
42	0.0935	0.047	0.028	19	0.1660	0.083	0.050	19/64	0.2969	0.148	0.089	21/32	0.6563	0.328	0.197
41	0.0960	0.048	0.029	18	0.1695	0.085	0.051	5/16	0.3125	0.156	0.094	43/64	0.6719	0.336	0.202
40	0.0980	0.049	0.029	17	0.1730	0.087	0.052	21/64	0.3281	0.164	0.098	11/16	0.6875	0.344	0.206
39	0.0995	0.050	0.030	16	0.1770	0.089	0.053	11/32	0.3438	0.171	0.103	23/32	0.7188	0.359	0.216
38	0.1015	0.051	0.030	15	0.1800	0.090	0.054	23/64	0.3594	0.180	0.108	3/4	0.7500	0.375	0.225

BROACHES AND BROACHING

The Broaching Process. — The broaching process may be applied in machining holes or other internal surfaces and also to many flat or other external surfaces. Internal broaching is applied in forming either symmetrical or irregular holes, grooves, or slots in machine parts, especially when the size or shape of the opening, or its length in proportion to diameter or width, make other machining processes impracticable. Broaching originally was utilized for such work as cutting keyways, machining round holes into square, hexagonal, or other shapes, forming splined holes, and for a large variety of other internal operations. The development of broaching machines and broaches finally resulted in extensive application of the process to external, flat, and other surfaces. Most external or surface broaching is done on machines of vertical design, but horizontal machines are also used for some classes of work. The broaching process is very rapid, accurate, and it leaves a finish of good quality. It is employed extensively in automotive and other plants where duplicate parts must be produced in large quantities and frequently to given dimensions within small tolerances.

Types of Broaches. — A number of typical broaches and the operations for which they are intended are shown by the diagrams, Fig. 1. Broach A produces a round-cornered, square hole. Prior to broaching square holes, it is usually the practice to drill a round hole having a diameter d somewhat larger than the width of the square. Hence, the sides are not completely finished, but this unfinished part is not objectionable in most cases. In fact, this clearance space is an advantage during the broaching operation in that it serves as a channel for the broaching lubricant; moreover, the broach has less metal to remove. Broach B is for finishing round holes. Broaching is superior to reaming for some classes of work, because the broach will hold its size for a much longer period, thus insuring greater accuracy. Broaches C and D are for cutting single and double keyways, respectively. Broach C is of rectangular section and, when in use, slides through a guiding bushing which is inserted in the hole. Broach E is for forming four integral splines in a hub. The broach at F is for producing hexagonal holes. Rectangular holes are finished by broach G. The teeth on the sides of this broach are inclined in opposite directions, which has the following advantages: The broach is stronger than it would be if the teeth were opposite and parallel to each other; thin work cannot drop between the inclined teeth, as it tends to do when the teeth are at right angles, because at least two teeth are always cutting; the inclination in opposite directions neutralizes the lateral thrust. The teeth on the edges are staggered, the teeth on one side being midway between the teeth on the other edge, as shown by the dotted line. A double cut broach is shown at H. This type is for finishing, simultaneously, both sides f of a slot, and for similar work. Broach I is the style used for forming the teeth in internal gears. It is practically a series of gear-shaped cutters, the outside diameters of which gradually increase toward the finishing end of the broach. Broach J is for round holes but differs from style B in that it has a continuous helical cutting edge. Some prefer this form because it gives a shearing cut. Broach K is for cutting a series of helical grooves in a hub or bushing. In helical broaching, either the work or the broach is rotated to form the helical grooves as the broach is pulled through.

In addition to the typical broaches shown in Fig. 1, many special designs are now in use for performing more complex operations. Two surfaces on opposite sides of a casting or forging are sometimes machined simultaneously by twin broaches and, in other cases, three or four broaches are drawn through a part at the same time, for finishing as many duplicate holes or surfaces. Notable developments have been made in the design of broaches for external or "surface" broaching.

Pitch of Broach Teeth. — The pitch of broach teeth depends upon the depth of cut or chip thickness, length of cut, the cutting force required and power of the broaching

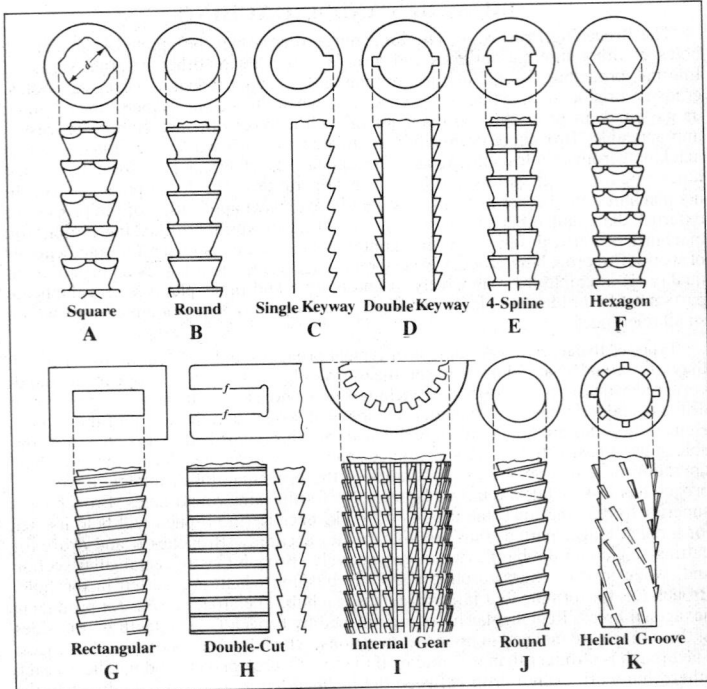

Fig. 1. Types of Broaches

machine. In the pitch formulas which follow

L = length, in inches, of layer to be removed by broaching

d = depth of cut per tooth as shown by Table 1 (For internal broaches, d = depth of cut as measured on one side of broach or one-half difference in diameters of successive teeth in case of a round broach)

F = a factor. (For brittle types of material, F = 3 or 4 for roughing teeth, and 6 for finishing teeth. For ductile types of material, F = 4 to 7 for roughing teeth and 8 for finishing teeth.)

b = width of inches, of layer to be removed by broaching

P = pressure required in tons per square inch, of an area equal to depth of cut times width of cut, in inches (Table 2)

T = usable capacity, in tons, of broaching machine = 70 per cent of maximum tonnage

The minimum pitch shown by Formula (1) is based upon the receiving capacity of the chip space. The minimum, however, should not be less than 0.2 inch unless a smaller pitch is required for exceptionally short cuts to provide at least two teeth in contact simultaneously, with the part being broached. A reduction below 0.2 inch is seldom required in surface broaching but it may be necessary in connection with internal broaching.

Table 1. Designing Data for Surface Broaches

Material to be Broached	Depth of Cut per Tooth, Inch		Face Angle or Rake, Degrees	Clearance Angle, Degrees	
	Roughing*	Finishing		Roughing	Finishing
Steel, High Tensile Strength	0.0015–0.002	0.0005	10–12	1.5–3	0.5–1
Steel, Medium Tensile Strength	0.0025–0.005	0.0005	14–18	1.5–3	0.5–1
Cast Steel	0.0025–0.005	0.0005	10	1.5–3	0.5
Malleable Iron	0.0025–0.005	0.0005	7	1.5–3	0.5
Cast Iron, Soft	0.006 –0.010	0.0005	10–15	1.5–3	0.5
Cast Iron, Hard	0.003 –0.005	0.0005	5	1.5–3	0.5
Zinc Die Castings	0.005 –0.010	0.0010	12**	5	2
Cast Bronze	0.010 –0.025	0.0005	8	0	0
Wrought Aluminum Alloys	0.005 –0.010	0.0010	15**	3	1
Cast Aluminum Alloys . . .	0.005 –0.010	0.0010	12**	3	1
Magnesium Die Castings . .	0.010 –0.015	0.0010	20**	3	1

* The lower depth-of-cut values for roughing are recommended when work is not very rigid, the tolerance is small, a good finish is required, or length of cut is comparatively short.
** In broaching these materials, smooth surfaces for tooth and chip spaces are especially recommended.

Table 2. Broaching Pressure P for Use in Pitch Formula (2)

Material to be Broached	Depth d of Cut per Tooth, Inch					Pressure P, Side-cutting Broaches
	0.024	0.010	0.004	0.002	0.001	
	Pressure P in Tons per Square Inch					
Steel, High Ten. Strength . . .	. . .	. . .	. . .	250	312	200–.004″ cut
Steel, Med. Ten. Strength . . .	. . .	. . .	158	185	243	143–.006″ cut
Cast Steel	. . .	. . .	128	158	. . .	115–.006″ cut
Malleable Iron	. . .	. . .	108	128	. . .	100–.006″ cut
Cast Iron	. . .	115	115	143	. . .	115–.020″ cut
Cast Brass	. . .	50	50	. . .	. . .	. . . .
Brass, Hot Pressed	. . .	85	85	. . .	. . .	. . . .
Zinc Die Castings	. . .	70	70	. . .	. . .	. . . .
Cast Bronze	35	35	. . .	. . .	. . .	. . . .
Wrought Aluminum	. . .	70	70	. . .	. . .	. . . .
Cast Aluminum	. . .	85	85	. . .	. . .	. . . .
Magnesium Alloy	35	35	. . .	. . .	. . .	. . . .

$$\text{Minimum pitch} = 3\sqrt{LdF} \qquad (1)$$

Whether the minimum pitch may be used or not depends upon the power of the available machine. The factor F in the formula provides for the increase in volume as the material is broached into chips. If a broach has adjustable inserts for the finishing teeth, the pitch of the finishing teeth may be smaller than the pitch of the roughing teeth because of the smaller depth d of the cut. The higher value of F for finishing teeth prevents the pitch from becoming too small, so that the spirally curled chips will not be crowded into too small a space. The pitch of the roughing and finishing

teeth should be equal for broaches without separate inserts (notwithstanding the different values of d and F) so that some of the finishing teeth may be ground into roughing teeth after wear makes this necessary.

$$\text{Allowable pitch} = \frac{dLbP}{T} \qquad (2)$$

If the pitch obtained by Formula (2) is larger than the minimum obtained by Formula (1), this larger value should be used because it is based upon the usable power of the machine. As the notation indicates, 70 per cent of the maximum tonnage T is taken as the usable capacity. The 30 per cent reduction is to provide a margin for the increase in broaching load resulting from the gradual dulling of the cutting edges. The procedure in calculating both minimum and allowable pitches will be illustrated by an example.

Example: Determine pitch of broach for cast iron when $L = 9$ inches; $d = 0.004$; and $F = 4$.

$$\text{Minimum pitch} = 3\sqrt{9 \times 0.004 \times 4} = 1.14$$

Next, apply Formula (2). Assume that $b = 3$ and $T = 10$; for cast iron and depth d of 0.004, $P = 115$ (Table 2). Then,

$$\text{Allowable pitch} = \frac{0.004 \times 9 \times 3 \times 115}{10} = 1.24$$

This pitch is safely above the minimum. If in this case the usable tonnage of an available machine were, say, 8 tons instead of 10 tons, the pitch as shown by Formula (2) might be increased to about 1.5 inches, thus reducing the number of teeth cutting simultaneously and, consequently, the load on the machine; or the cut per tooth might be reduced instead of increasing the pitch, especially if only a few teeth are in cutting contact, as might be the case with a short length of cut. If the usable tonnage in the preceding example were, say, 15, then a pitch of 0.84 would be obtained by Formula (2); hence the pitch in this case should not be less than the minimum of approximately 1.14 inches.

Depth of Cut per Tooth. — The term "depth of cut" as applied to surface or external broaches means the difference in the heights of successive teeth. This term, as applied to internal broaches for round, hexagonal or other holes, may indicate the total increase in the diameter of successive teeth; however, to avoid confusion, the term as here used means in all cases and regardless of the type of broach, the depth of cut as measured on one side.

In broaching free cutting steel, the Broaching Tool Institute recommends 0.003 to 0.006 inch depth of cut for surface broaching; 0.002 to 0.003 inch for multi-spline broaching; and 0.0007 to 0.0015 inch for round hole broaching. The accompanying table contains data from a German source and applies specifically to surface broaches. All data relating to depth of cut are intended as a general guide only. While depth of cut is based primarily upon the machinability of the material, some reduction from the depth thus established may be required particularly when the work supporting fixture in surface broaching is not sufficiently rigid to resist the thrust from the broaching operation. In some cases, the pitch and cutting length may be increased to reduce the thrust force. Another possible remedy in surface broaching certain classes of work is to use a side-cutting broach instead of the ordinary depth cutting type. A broach designed for side cutting takes relatively deep narrow cuts which extend nearly to the full depth required. The side cutting section is followed by teeth arranged for depth cutting to obtain the required size and surface finish on the work. In

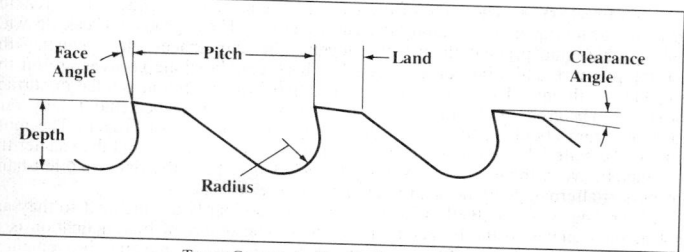

Terms Commonly Used in Broach Design

general, small tolerances in surface broaching require a reduced cut per tooth to minimize work deflection resulting from the pressure of the cut. See pages 981–984 for broaching speeds.

Face Angle or Rake. — The face angle (see diagram) of broach teeth affects the chip flow and varies considerably for different materials. While there are some variations in practice, even for the same material, the angles given in the accompanying table are believed to represent commonly used values. Some broach designers increase the rake angle for finishing teeth in order to improve the finish on the work.

Clearance Angle. — The clearance angle (see illustration) for roughing steel varies from 1.5 to 3 degrees and for finishing steel from 0.5 to 1 degree. Some recommend the same clearance angles for cast iron and others, larger clearance angles varying from 2 to 4 or 5 degrees. Additional data will be found in Table 1.

Land Width. — The width of the land usually is about 0.25 × pitch. It varies, however, from about one-fourth to one-third of the pitch. The land width is selected so as to obtain the proper balance between tooth strength and chip space.

Depth of Broach Teeth. — The tooth depth as established experimentally and on the basis of experience, usually varies from about 0.37 to 0.40 of the pitch. This depth is measured radially from the cutting edge to the bottom of the tooth fillet.

Radius of Tooth Fillet. — The "gullet" or bottom of the chip space between the teeth should have a rounded fillet to strengthen the broach, facilitate curling of the chips, and safeguard against cracking in connection with the hardening operation. One rule is to make the radius equal to one-fourth the pitch. Another is to make it equal 0.4 to 0.6 the tooth depth. A third method preferred by some broach designers is to make the radius equal one-third of the sum obtained by adding together the land width, one-half the tooth depth, and one-fourth of the pitch.

Total Length of Broach. — After the depth of cut per tooth has been determined, the total amount of material to be removed by a broach is divided by this decimal to ascertain the number of cutting teeth required. This number of teeth multiplied by the pitch gives the length of the active portion of the broach. By adding to this dimension the distance over three or four straight teeth, the length of a pilot to be provided at the finishing end of the broach, and the length of a shank which must project through the work and the faceplate of the machine to the draw-head, the over-all length of the broach is found. This calculated length is often greater than the stroke of the machine, or greater than is practical for a broach of the diameter required. In such cases, a set of broaches must be used.

Chip Breakers. — The teeth of broaches frequently have rounded chip-breaking grooves located at intervals along the cutting edges. These grooves break up wide curling chips and prevent them from clogging the chip spaces, thus reducing the cutting pressure and strain on the broach. These chip-breaking grooves are on the roughing teeth only. They are staggered and applied to both round and flat or surface broaches. The grooves are formed by a round edged grinding wheel and usually vary in width from about $\frac{1}{32}$ to $\frac{3}{32}$ inch depending upon the size of broach. The more ductile the material, the wider the chip breaker grooves should be and the smaller the distance between them. Narrow slotting broaches may have the right- and left-hand corners of alternate teeth beveled to obtain chip-breaking action.

Shear Angle. — The teeth of surface broaches ordinarily are inclined so they are not at right angles to the broaching movement. The object of this inclination is to obtain a shearing cut which results in smoother cutting action and an improvement in surface finish. The shearing cut also tends to eliminate troublesome vibration. Shear angles for surface broaches are not suitable for broaching slots or any profiles that resist the outward movement of the chips. When the teeth are inclined, the fixture should be designed to resist the resulting thrusts unless it is practicable to incline the teeth of right- and left-hand sections in opposite directions to neutralize the thrust. The shear angle usually varies from 10 to 25 degrees.

Types of Broaching Machines. — Broaching machines may be divided into horizontal and vertical designs, and they may be classified further according to the method of operation, as, for example, whether a broach in a vertical machine is pulled up or pulled down in forcing it through the work. Horizontal machines usually pull the broach through the work in internal broaching but short rigid broaches may be pushed through. External surface broaching is also done on some machines of horizontal design, but usually vertical machines are employed for flat or other external broaching. Although parts usually are broached by traversing the broach itself, some machines are designed to hold the broach or broaches stationary during the actual broaching operation. This principle has been applied both to internal and surface broaching.

Vertical Duplex Type: The vertical duplex type of surface broaching machine has two slides or rams which move in opposite directions and operate alternately. While the broach connected to one slide is moving downward on the cutting stroke, the other broach and slide is returning to the starting position, and this returning time is utilized for reloading the fixture on that side; consequently, the broaching operation is practically continuous. Each ram or slide may be equipped to perform a separate operation on the same part when two operations are required.

Pull-up Type: Vertical hydraulically operated machines which pull the broach or broaches up through the work are used for internal broaching of holes of various shapes, for broaching bushings, splined holes, small internal gears, etc. A typical machine of this kind is so designed that all broach handling is done automatically.

Pull-down Type: The various movements in the operating cycle of a hydraulic pull-down type of machine equipped with an automatic broach-handling slide, are the reverse of the pull-up type. The broaches for a pull-down type of machine have shanks on each end, there being an upper one for the broach-handling slide and lower one for pulling through the work.

Hydraulic Operation: Modern broaching machines, as a general rule, are operated hydraulically rather than by mechanical means. Hydraulic operation is efficient, flexible in the matter of speed adjustments, low in maintenance cost, and the "smooth" action required for fine precision finishing may be obtained. The hydraulic pressures required, which frequently are 800 to 1000 pounds per square inch, are obtained from a motor-driven pump forming part of the machine. The cutting speeds of broaching machines frequently are between 20 and 30 feet per minute, and the return speed often are double the cutting speed, or higher, to reduce the idle period.

Causes of Broaching Difficulties

Broaching Difficulty	Possible Causes
Stuck broach	Insufficient machine capacity; dulled teeth; clogged chip gullets; failure of power during cutting stroke. To remove a stuck broach, workpiece and broach are removed from the machine as a unit; never try to back out broach by reversing machine. If broach does not loosen by tapping workpiece lightly and trying to slide it off its starting end, mount workpiece and broach in a lathe and turn down workpiece to the tool surface. Workpiece may be sawed longitudinally into several sections in order to free the broach. Check broach design, perhaps tooth relief (back off) angle is too small or depth of cut per tooth is too great.
Galling and pickup	Lack of homogeneity of material being broached — uneven hardness, porosity; improper or insufficient coolant; poor broach design, mutilated broach; dull broach; improperly sharpened broach; improperly designed or outworn fixtures. Good broach design will do away with possible chip build-up on tooth faces and excessive heating. Grinding of teeth should be accurate so that the correct gullet contour is maintained. Contour should be fair and smooth.
Broach breakage	Overloading; broach dullness; improper sharpening; interrupted cutting stroke; backing up broach with workpiece in fixture; allowing broach to pass entirely through guide hole; ill fitting and/or sharp edged key; crooked holes; untrue locating surface; excessive hardness of workpiece; insufficient clearance angle; sharp corners on pull end of broach. When grinding bevels on pull end of broach use wheel that is not too pointed.
Chatter	Too few teeth in cutting contact simultaneously; excessive hardness of material being broached; loose or poorly constructed tooling; surging of ram due to load variations. Chatter can be alleviated by changing the broaching speed, by using shear cutting teeth instead of right angle teeth, and by changing the coolant and the face and relief angles of the teeth.
Drifting or misalignment of tool during cutting stroke	Lack of proper alignment when broach is sharpened in grinding machine, which may be caused by dirt in the female center of the broach; inadequate support of broach during the cutting stroke, on a horizontal machine especially; body diameter too small; cutting resistance variable around I.D. of hole due to lack of symmetry of surfaces to be cut; variations in hardness around I.D. of hole; too few teeth in cutting contact.
Streaks in broached surface	Lands too wide; presence of forging, casting or annealing scale; metal pickup; presence of grinding burrs and grinding and cleaning abrasives.
Rings in the broached hole	Due to surging resulting from uniform pitch of teeth; presence of sharpening burrs on broach; tooth clearance angle too large; locating face not smooth or square; broach not supported for all cutting teeth passing through the work. The use of differential tooth spacing or shear cutting teeth helps in preventing surging. Sharpening burrs on a broach may be removed with a wood block.

Broaching Difficulties. — The accompanying table has been compiled from information supplied by the National Broach and Machine Co. and presents some of the common broaching difficulties, their causes and means of correction.

Tool Wear

Metal cutting tools wear constantly when they are being used. A normal amount of wear should not be a cause for concern until the size of the worn region has reached the point where the tool should be replaced. Normal wear cannot be avoided and should be differentiated from abnormal tool breakage or excessively fast wear. Tool breakage and an excessive rate of wear indicate that the tool is not operating correctly and steps should be taken to correct this situation.

There are several basic mechanisms that cause tool wear. It is generally understood that tools wear as a result of abrasion which is caused by hard particles of work material plowing over the surface of the tool. Wear is also caused by diffusion or alloying between the work material and the tool material. In regions where the conditions of contact are favorable, the work material reacts with the tool material causing an attrition of the tool material. The rate of this attrition is dependent upon the temperature in the region of contact and the reactivity of the tool and the work materials with each other. Diffusion or alloying also occurs where particles of the work material are welded to the surface of the tool. These welded deposits are often quite visible in the form of a built-up edge, as particles or a layer of work material inside a crater or as small mounds attached to the face of the tool. The diffusion or alloying occurring between these deposits and the tool weakens the tool material below the weld. Frequently these deposits are again rejoined to the chip by welding or they are simply broken away by the force of collision with the passing chip. When this happens, a small amount of the tool material may remain attached to the deposit and be plucked from the surface of the tool, to be carried away with the chip. This mechanism can cause chips to be broken from the cutting edge and the formation of small craters on the tool face called pull-outs. It can also contribute to the enlargement of the larger crater that sometimes forms behind the cutting edge. Among the other mechanisms that can cause tool wear are severe thermal gradients and thermal shocks, which cause cracks to form near the cutting edge, ultimately leading to tool failure. This condition can be caused by improper tool grinding procedures, heavy interrupted cuts, or by the improper application of cutting fluids when machining at high cutting speeds. Chemical reactions between the active constituents in some cutting fluids sometimes accelerate the rate of tool wear. Oxidation of the heated metal near the cutting edge also contributes to tool wear, particularly when fast cutting speeds and high cutting temperatures are encountered. Breakage of the cutting edge caused by overloading, heavy shock loads, or improper tool design is not normal wear and should be corrected.

The wear mechanisms described bring about visible manifestations of wear on the tool which should be understood so that the proper corrective measures can be taken when required. These visible signs of wear are described in the following paragraphs and the corrective measures that might be required are given in the accompanying Tool Trouble-Shooting Check List. The best procedure when trouble shooting is to try to correct only one condition at a time. When a correction has been made it should be checked. After one condition has been corrected, work can then start to correct the next condition.

Flank Wear: Tool wear occurring on the flank of the tool below the cutting edge is called flank wear. Flank wear always takes place and cannot be avoided. It should not give rise to concern unless the rate of flank wear is too fast or the flank wear land becomes too large in size. The size of the flank wear can be measured as the distance between the top of the cutting edge and the bottom of the flank wear land. In practice, a visual estimate is usually made instead of a precise measurement, although in many instances flank wear is ignored and the tool wear is "measured" by the loss of size on the part. The best measure of tool wear, however, is flank wear. When it becomes too large, the rubbing action of the wear land

against the workpiece increases and the cutting edge must be replaced. Because conditions vary, it is not possible to give an exact amount of flank wear at which the tool should be replaced. Although there are many exceptions, as a rough estimate, high-speed steel tools should be replaced when the width of the flank wear land reaches .005 to .010 inch for finish turning and .030 to .060 inch for rough turning; and for cemented carbides .005 to .010 inch for finish turning and .020 to .040 inch for rough turning.

Under ideal conditions which, surprisingly, occur quite frequently, the width of the flank wear land will be very uniform along its entire length. When the depth of cut is uneven, such as when turning out-of-round stock, the bottom edge of the wear land may become somewhat slanted, the wear land being wider toward the nose. A jagged-appearing wear land usually is evidence of chipping at the cutting edge. Sometimes only one or two sharp depressions of the lower edge of the wear land will appear, to indicate that the cutting edge has chipped above these depressions. A deep notch will sometimes occur at the "depth of cut line," or that part of the cutting opposite the original surface of the work. This can be caused by a hard surface scale on the work, by a work-hardened surface layer on the work, or when machining high-temperature alloys. Often the size of the wear land is enlarged at the nose of the tool. This can be a sign of crater breakthrough near the nose or of chipping in this region. Under certain conditions, when machining with carbides, it can be an indication of deformation of the cutting edge in the region of the nose.

When a sharp tool is first used, the initial amount of flank wear is quite large in relation to the subsequent total amount. Under normal operating conditions, the width of the flank wear land will increase at a uniform rate until it reaches a critical size after which the cutting edge breaks down completely. This is called catastrophic failure and the cutting edge should be replaced before this occurs. When cutting at slow speeds with high-speed steel tools, there may be long periods when no increase in the flank wear can be observed. For a given work material and tool material, the rate of flank wear is primarily dependent on the cutting speed and then the feed rate.

Cratering: A deep crater will sometimes form on the face of the tool which is easily recognizable. The crater forms at a short distance behind the side cutting edge leaving a small shelf between the cutting edge and the edge of the crater. This shelf is sometimes covered with the built-up edge and at other times it is uncovered. Often the bottom of the crater is obscured with work material that is welded to the tool in this region. Under normal operating conditions, the crater will gradually enlarge until it breaks through a part of the cutting edge. Usually this occurs on the end cutting edge just behind the nose. When this takes place, the flank wear at the nose increases rapidly and complete tool failure follows shortly. Sometimes cratering cannot be avoided and a slow increase in the size of the crater is considered normal. However, if the rate of crater growth is rapid, leading to a short tool life, corrective measures must be taken.

Cutting Edge Chipping: Small chips are sometimes broken from the cutting edge which accelerates tool wear but does not necessarily cause immediate tool failure. Chipping can be recognized by the appearance of the cutting edge and the flank wear land. A sharp depression in the lower edge of the wear land is a sign of chipping and if this edge of the wear land has a jagged appearance it indicates that a large amount of chipping has taken place. Often the vacancy or cleft in the cutting edge that results from chipping is filled up with work material that is tightly welded in place. This occurs very rapidly when chipping is caused by a built-up edge on the face of the tool. In this manner the damage to the cutting edge is healed; however, the width of the wear land below the chip is usually increased and the tool life is shortened.

Deformation: Deformation occurs on carbide cutting tools when taking a very heavy cut using a slow cutting speed and a high feed rate. A large section of the cutting edge then becomes very hot and the heavy cutting pressure compresses the nose of the cutting edge, thereby lowering the face of the tool in the area of the nose. This reduces the relief under the nose, increases the width of the wear land in this region, and shortens the tool life.

Surface Finish: The finish on the machined surface does not necessarily indicate poor cutting tool performance unless there is a rapid deterioration. A good surface finish is, however, sometimes a requirement. The principal cause of a poor surface finish is the built-up edge which forms along the edge of the cutting tool. The elimination of the built-up edge will always result in an improvement of the surface finish. The most effective way to eliminate the built-up edge is to increase the cutting speed. When the cutting speed is increased beyond a certain critical cutting speed, there will be a rather sudden and large improvement in the surface finish. Cemented carbide tools can operate successfully at higher cutting speeds, where the built-up edge does not occur and where a good surface finish is obtained. Whenever possible, cemented carbide tools should be operated at cutting speeds where a good surface finish will result. There are times when such speeds are not possible. Also, high-speed tools cannot be operated at the speed where the built-up edge does not form. In these conditions the most effective method of obtaining a good surface finish is to employ a cutting fluid that has active sulphur or chlorine additives.

Cutting tool materials that do not alloy readily with the work material are also effective in obtaining an improved surface finish. Straight titanium carbide and diamond are the two principal tool materials that fall into this category.

The presence of feed marks can mar an otherwise good surface finish and attention must be paid to the feed rate and the nose radius of the tool if a good surface finish is desired. Changes in the tool geometry can also be helpful. A small "flat," or secondary cutting edge, ground on the end cutting edge behind the nose will sometimes provide the desired surface finish. When the tool is in operation, the flank wear should not be allowed to become too large, particularly in the region of the nose where the finished surface is produced.

Sharpening Twist Drills. — Twist drills are cutting tools designed to perform concurrently several functions, such as penetrating directly into solid material, ejecting the removed chips outside the cutting area, maintaining the essentially straight direction of the advance movement and controlling the size of the drilled hole. The geometry needed for these multiple functions is incorporated into the design of the twist drill in such a manner that it can be retained even after repeated sharpening operations. Twist drills are resharpened many times during their service life, with the practically complete restitution of their original operational characteristics. However, in order to assure all the benefits which the design of the twist drill is capable of providing, the surfaces generated in the sharpening process must agree with the original form of the tool's operating surfaces, unless a change of shape is required for use on a different work material.

The principal elements of the tool geometry which are essential for the adequate cutting performance of twist drills are shown in Fig. 1. The generally used values for these dimensions are the following:

Point angle: Commonly 118°, except for high strength steels, 118° to 135°; aluminum alloys, 90° to 140°; and magnesium alloys, 70° to 118°.

Helix angle: Commonly 24° to 32°, except for magnesium and copper alloys, 10° to 30°.

Lip relief angle: Commonly 10° to 15°, except for high strength or tough steels, 7° to 12°. The lower values of these angle ranges are used for drills of larger diameter, the higher values for the smaller diameters. For drills of diameters less than ¼ inch

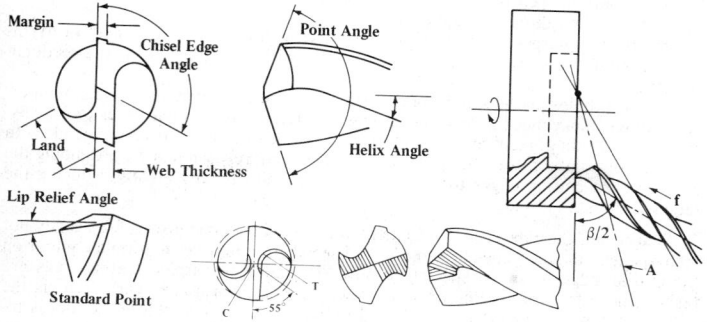

Fig. 1 Top and Bottom Left. The principal elements of tool geometry on twist drills. Fig. 2 Top Right. In grinding the face of the twist drill the tool is swung around the axis *A* of an imaginary cone, while resting in a support tilted by half of the point angle β with respect to the face of the grinding wheel. Feed *f* for stock removal is in the direction of the drill axis.
Fig. 3 Bottom Center. The chisel edge *C* after thinning the web by grinding off area *T*.
Fig. 4 Bottom Right. Split point or "crankshaft" type web thinning.

the lip relief angles are increased beyond the listed max. values up to 24°. For soft and free machining materials, 12° to 18° except for diameters less than ¼ inch, 20° to 26°.

Relief Grinding of the Tool Flanks. — In sharpening twist drills the tool flanks containing the two cutting edges are ground. Each flank consists of a curved surface which provides the relief needed for the easy penetration and free cutting of the tool edges. In grinding the flanks, Fig. 2, the drill is swung around the axis *A* of an imaginary cone while resting in a support which holds the drill at one-half the point angle *B* with respect to the face of the grinding wheel. Feed *f* for stock removal is in the direction of the drill axis. The relief angle is usually measured at the periphery of the twist drill and is also specified by that value. It is not a constant but should increase toward the center of the drill.

The relief grinding of the flank surfaces will generate the chisel angle on the web of the twist drill. The value of that angle, typically 55°, which can be measured, for example, with the protractor of an optical projector, is indicative of the correctness of the relief grinding.

Drill Point Thinning. — The chisel edge is the least efficient operating surface element of the twist drill because it does not cut, but actually squeezes or extrudes the work material. To improve the inefficient cutting conditions caused by the chisel edge, the point width is often reduced in a drill-point thinning operation, resulting in a condition such as that shown in Fig. 3. Point thinning is particularly desirable on larger size drills and also on those which become shorter in usage, because the thickness of the web increases toward the shaft of the twist drill, thereby adding to the length of the chisel edge. The extent of point thinning is limited by the minimum strength of the web needed to avoid splitting of the drill point under the influence of cutting forces.

Both sharpening operations — the relieved face grinding and the point thinning — should be carried out in special drill grinding machines or with twist drill grinding fixtures mounted on general-purpose tool grinding machines, designed to assure the essential accuracy of the required tool geometry. Off-hand grinding may be used

for the important web thinning when a special machine is not available; however, such operation requires skill and experience.

Improperly sharpened twist drills, e.g. those with unequal edge length or asymmetrical point angle, will tend to produce holes with poor diameter and directional control.

For deep holes and also drilling into stainless steel, titanium alloys, high temperature alloys, nickel alloys, very high strength materials and in some cases tool steels, split point grinding, resulting in a "crankshaft" type drill point, is recommended. In this type of pointing, see Fig. 4, the chisel edge is entirely eliminated, extending the positive rake cutting edges to the center of the drill, thereby greatly reducing the required thrust in drilling.

Sharpening Carbide Tools. — Cemented carbide indexable inserts are usually not resharpened but sometimes they require a special grind in order to form a contour on the cutting edge to suit a special purpose. Brazed type carbide cutting tools are resharpened after the cutting edge has become worn. On brazed carbide tools the cutting-edge wear should not be allowed to become excessive before the tool is resharpened. One method of determining when brazed carbide tools need resharpening is by periodic inspection of the flank wear and the condition of the face. Another method is to determine the amount of production which is normally obtained before excessive wear has taken place, or to determine the equivalent period of time. One disadvantage of this method is that slight variations in the work material will often cause the wear rate not to be uniform and the number of parts machined before regrinding will not be the same each time. Usually, sharpening should not require the removal of more than .005 to .010 inch of carbide.

General Procedure in Carbide Tool Grinding: The general procedure depends upon the kind of grinding operation required. If the operation is to resharpen a dull tool, a diamond wheel of 100 to 120 grain size is recommended although a finer wheel — up to 150 grain size — is sometimes used to obtain a better finish. If the tool is new or is a "standard" design and changes in shape are necessary, a 100-grit diamond wheel is recommended for roughing and a finer grit diamond wheel can be used for finishing. Some shops prefer to rough grind the carbide with a vitrified silicon carbide wheel, the finish grinding being done with a diamond wheel. A final operation commonly designated as lapping may or may not be employed for obtaining an extra-fine finish.

Wheel Speeds: — The speed of silicon carbide wheels usually is about 5000 feet per minute. The speeds of diamond wheels generally range from 5000 to 6000 feet per minute; yet lower speeds (550 to 3000 fpm) can be effective.

Offhand Grinding: — In grinding single-point tools (excepting chip breakers) the common practice is to hold the tool by hand, press it against the wheel face and traverse it continuously across the wheel face while the tool is supported on the machine rest or table which is adjusted to the required angle. This is known as "offhand grinding" to distinguish it from the machine grinding of cutters as in regular cutter grinding practice. The selection of wheels adapted to carbide tool grinding is very important.

Silicon Carbide Wheels. — The green colored silicon carbide wheels generally are preferred to the dark gray or gray-black variety, although the latter are sometimes used.

Grain or Grit Sizes: — For roughing, a grain size of 60 is very generally used. For finish grinding with silicon carbide wheels, a finer grain size of 100 or 120 is common. A silicon carbide wheel such as C60-I-7V may be used for grinding both the steel shank and carbide tip. However, for under-cutting steel shanks up to the carbide tip

it may be advantageous to use an aluminum oxide wheel suitable for high speed steel grinding.

Grade: — According to the standard system of marking, different grades from soft to hard are indicated by letters from A to Z. For carbide tool grinding fairly soft grades such as G, H, I and J are used. The usual grades for roughing are I or J and for finishing H, I and J. The grade should be such that a sharp free-cutting wheel will be maintained without excessive grinding pressure. Harder grades than those indicated tend to overheat and crack the carbide.

Structure: — The common structure numbers for carbide tool grinding are 7 and 8. The larger cup-wheels (10 to 14 inches) may be of the porous type and be designated as 12P. The standard structure numbers range from 1 to 15 with progressively higher numbers indicating less density and more open wheel structure.

Diamond Wheels. — Wheels with diamond-impregnated grinding faces are fast and cool cutting and have a very low rate of wear. They are used extensively both for resharpening and for finish grinding of carbide tools when preliminary roughing is required. Diamond wheels are also adapted for sharpening multi-tooth cutters such as milling cutters, reamers, etc., which are ground in a cutter grinding machine.

Resinoid bonded wheels are commonly used for grinding chip breakers, milling cutters, reamers or other multi-tooth cutters. They are also applicable to precision grinding of carbide dies, gages, and various external, internal and surface grinding operations. Fast, cool cutting action is characteristic of these wheels.

Metal bonded wheels are often used for offhand grinding of single-point tools especially when durability or long life and resistance to grooving of the cutting face, are considered more important than the rate of cutting. *Vitrified bonded* wheels are used both for roughing of chipped or very dull tools and for ordinary resharpening and finishing. They provide rigidity for precision grinding, a porous structure for fast cool cutting, sharp cutting action and durability.

Diamond Wheel Grit Sizes. — For roughing with diamond wheels a grit size of 100 is the most common both for offhand and machine grinding. Grit sizes of 120 and 150 are frequently used in offhand grinding of single point tools (1) for resharpening, (2) for a combination roughing and finishing wheel and (3) for chip-breaker grinding. Grit sizes of 220 or 240 are used for ordinary finish grinding all types of tools (offhand and machine) and also for cylindrical, internal and surface finish grinding. Grits of 320 and 400 are used for "lapping" to obtain very fine finishes, and for hand hones. A grit of 500 is for lapping to a mirror finish on such work as carbide gages and boring or other tools for exceptionally fine finishes.

Diamond Wheel Grades. — Diamond wheels are made in several different grades to better adapt them to different classes of work. The grades vary for different types and shapes of wheels. Standard Norton grades are H, J and L for resinoid bonded wheels, grade N for metal bonded wheels and grades J, L, N and P for vitrified wheels. Harder and softer grades than standard may at times be used to advantage.

Diamond Concentration. — The relative amount (by carat weight) of diamond in the diamond section of the wheel is known as the "diamond concentration." Concentrations of 100 (high), 50 (medium) and 25 (low) ordinarily are supplied. A concentration of 50 represents one-half the diamond content of 100 (if the depth of the diamond is the same in each case) and 25 equals one-fourth the content of 100 or one-half the content of 50 concentration.

100 Concentration: — Recommended (especially in grit sizes up to about 220) for general machine grinding of carbides, and for grinding cutters and chip breakers. Vitrified and metal bonded wheels usually have 100 concentration.

50 Concentration: — In the finer grit sizes of 220, 240, 320, 400 and 500, a 50 concentration is recommended for offhand grinding with resinoid bonded cup-wheels.

25 Concentration: — A low concentration of 25 is recommended for offhand grinding with resinoid bonded cup-wheels with grit sizes of 100, 120 and 150.

Depth of Diamond Section: — The radial depth of the diamond section usually varies from 1/16 to 1/4 inch. The depth varies somewhat according to the wheel size and type of bond.

Dry Versus Wet Grinding of Carbide Tools. — In using silicon carbide wheels, grinding should be done either absolutely dry or with enough coolant to flood the wheel and tool. Satisfactory results may be obtained either by the wet or dry method. However, dry grinding is the most prevalent usually because, in wet grinding, operators tend to use an inadequate supply of coolant to obtain better visibility of the grinding operation and avoid getting wet; hence checking or cracking in many cases is more likely to occur in wet grinding than in dry grinding.

Wet Grinding with Silicon Carbide Wheels: — One advantage commonly cited in connection with wet grinding is that an ample supply of coolant permits using wheels about one grade harder than in dry grinding thus increasing the wheel life. Plenty of coolant also prevents thermal stresses and the resulting cracks, and there is less tendency for the wheel to load. A dust exhaust system also is unnecessary.

Wet Grinding with Diamond Wheels: — In grinding with diamond wheels the general practice is to use a coolant to keep the wheel face clean and promote free cutting. The amount of coolant may vary from a small stream to a coating applied to the wheel face by a felt pad.

Coolants for Carbide Tool Grinding. — In grinding either with silicon carbide or diamond wheels a coolant that is used extensively consists of water plus a small amount either of soluble oil, sal soda, or soda ash to prevent corrosion. One prominent manufacturer recommends for silicon carbide wheels about 1 ounce of soda ash per gallon of water and for diamond wheels kerosene. The use of kerosene is quite general for diamond wheels and usually it is applied to the wheel face by a felt pad. Another coolant recommended for diamond wheels consists of 80 per cent water and 20 per cent soluble oil.

Peripheral Versus Flat Side Grinding. — In grinding single point carbide tools with silicon carbide wheels, the roughing preparatory to finishing with diamond wheels may be done either by using the flat face of a cup-shaped wheel (side grinding) or the periphery of a "straight" or disk-shaped wheel. Even where side grinding is preferred, the periphery of a straight wheel may be used for heavy roughing

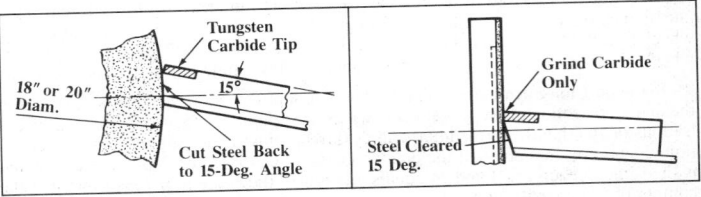

as in grinding back chipped or broken tools (see left-hand diagram). Reasons for preferring peripheral grinding include faster cutting with less danger of localized heating and checking especially in grinding broad surfaces. The advantages usually claimed for side grinding are that proper rake or relief angles are easier to obtain and

the relief or land is ground flat. The diamond wheels used for tool sharpening are designed for side grinding. (See right-hand diagram.)

Lapping Carbide Tools. — Carbide tools may be finished by lapping, especially if an exceptionally fine finish is required on the work as, for example, tools used for precision boring or turning non-ferrous metals. If the finishing is done by using a diamond wheel of very fine grit (such as 240, 320 or 400), the operation is often called "lapping." A second lapping method is by means of a power-driven lapping disk charged with diamond dust, Norbide powder, or silicon carbide finishing compound. A third method is by using a hand lap or hone usually of 320 or 400 grit. In many plants the finishes obtained with carbide tools meet requirements without a special lapping operation. In all cases any feather edge which may be left on tools should be removed and it is good practice to bevel the edges of roughing tools at 45 degrees to leave a chamfer 0.005 to 0.010 inch wide. This is done by hand honing and the object is to prevent crumbling or flaking off at the edges when hard scale or heavy chip pressure is encountered.

Hand Honing: The cutting edge of carbide tools, and tools made from other tool materials, is sometimes hand honed before it is used in order to strengthen the cutting edge. When interrupted cuts or heavy roughing cuts are to be taken, or when the grade of carbide is slightly too hard, hand honing is beneficial because it will prevent chipping, or even possibly, breakage of the cutting edge. Whenever chipping is encountered, hand honing the cutting edge before use will be helpful. It is important, however, to hone the edge lightly and only when necessary. Heavy honing will always cause a reduction in tool life. Normally, removing .002 to .004 inch from the cutting edge is sufficient. When indexable inserts are used, the use of pre-honed inserts is preferred to hand honing although sometimes an additional amount of honing is required. Hand honing of carbide tools in between cuts is sometimes done to defer grinding or to increase the life of a cutting edge on an indexable insert. If correctly done, so as not to change the relief angle, this procedure is sometimes helpful. If improperly done, it can result in a reduction in tool life.

Chip Breaker Grinding. — For this operation a straight diamond wheel is used on a universal tool and cutter grinder, a small surface grinder, or a special chipbreaker grinder. A resinoid bonded wheel of the grade J or N commonly is used and the tool is held rigidly in an adjustable holder or vise. The width of the diamond wheel usually varies from ⅛ to ¼ inch. A vitrified bond may be used for wheels as thick as ¼ inch, and a resinoid bond for relatively narrow wheels.

Summary of Miscellaneous Points. — In grinding a single-point carbide tool, traverse it across the wheel face continuously to avoid localized heating. This traverse movement should be quite rapid in using silicon carbide wheels and comparatively slow with diamond wheels. A hand traversing and feeding movement, whenever practicable, is generally recommended because of greater sensitivity. In grinding, maintain a constant, moderate pressure. Never cool a hot tool by dipping it in a liquid, as this may crack the tip. Wheel rotation should preferably be *against* the cutting edge or from the front face toward the back. If the grinder is driven by a reversing motor, opposite sides of a cup wheel can be used for grinding right- and left-hand tools and with rotation against the cutting edge. If it is necessary to grind the top face of a single-point tool, this should precede the grinding of the side and front relief, and top-face grinding should be minimized to maintain the tip thickness. In machine grinding with a diamond wheel, limit the feed per traverse to 0.001 inch for 100 to 120 grit; 0.0005 inch for 150 to 240 grit; and 0.0002 inch for 320 grit and finer.

JIGS AND FIXTURES

Material for Jig Bushings. — Bushings are generally made of a good grade of tool steel to ensure hardening at a fairly low temperature and to lessen the danger of fire cracking. They can also be made from machine steel, which will answer all practical purposes, provided the bushings are properly casehardened to a depth of about $\frac{1}{16}$ inch. Sometimes, bushings for guiding tools may be made of cast iron, but only when the cutting tool is of such a design that no cutting edges come within the bushing itself. For example, bushings used simply to support the smooth surface of a boring-bar or the shank of a reamer might, in some instances, be made of cast iron, but hardened steel bushings should always be used for guiding drills, reamers, taps, etc., when the cutting edges come in direct contact with the guiding surfaces. If the outside diameter of the bushing is very large, as compared with the diameter of the cutting tool, the cost of the bushing can sometimes be reduced by using an outer cast-iron body and inserting a hardened tool steel bushing.

When tool steel bushings are made and hardened, it is recommended that A-2 steel be used. The furnace should be set to 1750°F and the bushing placed in the furnace and held there approximately 20 minutes after the furnace reaches temperature. Remove the bushing and cool in still air. After the part cools to 100-150°F, immediately place in a tempering furnace that has been heated to 300°F. Remove the bushing after one hour and cool in still air. If an atmospherically controlled furnace is unavailable, the part should be wrapped in stainless foil to prevent scaling and oxidation at the 1750°F temperature.

American National Standard Jig Bushings. — Specifications for the following types of jig bushings are given in American National Standard B94.33-1974. Head Type Press Fit Wearing Bushings, Type H (See Tables 1 and 3); Headless Type Press Fit Wearing Bushings, Type P (See Tables 2 and 3); Slip Type Renewable Wearing Bushings, Type S (See Tables 4 and 5); Fixed Type Renewable Wearing Bushings, Type F (See Tables 5 and 6); Headless Type Liner Bushings, Type L (See Table 7); and Head Type Liner Bushings, Type HL (See Table 8). Specifications for locking mechanisms are also given (See Table 9).

Jig Bushing Definitions. — *Renewable bushings:* Renewable wearing bushings to guide the tool are for use in liners which in turn are installed in the jig. They are used where the bushing will wear out or become obsolete before the jig or where several bushings are to be interchangeable in one hole. Renewable wearing bushings are divided into two classes, "Fixed" and "Slip." Fixed renewable bushings are installed in the liner with the intention of leaving them in place until worn out. Slip renewable bushings are interchangeable in a given size of liner and, to facilitate removal, they are usually made with a knurled head. They are most frequently used where two or more operations requiring different inside diameters are performed in a single jig, such as where drilling is followed by reaming, tapping, spot facing, counterboring or some other secondary operation.

Press Fit Bushings: Press fit wearing bushings to guide the tool are for installation directly in the jig without the use of a liner and are employed principally where the bushings are used for short production runs and will not require replacement. They are intended also for short center distances.

Liner Bushings: Liner bushings are provided with and without heads and are permanently installed in a jig to receive the renewable wearing bushings. They are sometimes called master bushings.

Jig Plate Thickness. — The standard length of the press fit portion of jig bushings as established are based on standardized uniform jig plate thicknesses of $\frac{5}{16}$, $\frac{3}{8}$, $\frac{1}{2}$, $\frac{3}{4}$, 1, $1\frac{3}{8}$, $1\frac{1}{4}$, $2\frac{1}{8}$, $2\frac{1}{2}$, and 3 inches.

Jig Bushing Designation System. — *Inside Diameter:* The inside diameter of the hole is specified by a decimal dimension.

Type Bushing: The type of bushing is specified by a letter: S for Slip Renewable, F for Fixed Renewable, L for Headless Liner, HL for Head Liner, P for Headless Press Fit, and H for Head Press Fit.

Body Diameter: The body diameter is specified in multiples of 0.0156 inch. For example, a .500-inch body diameter = .500/.0156 = 32.

Body Length: The effective or body length is specified in multiples of 0.0625 inch. For example, a .500-inch length = .500/.0625 = 8.

Unfinished Bushings: All bushings with grinding stock on the body diameter are designated by the letter U following the number.

Example: A slip renewable bushing having a hole diameter of 0.5000 inch, a body diameter of 0.750 inch, and a body length of 1.000 inch would be designated as .5000-S-48-16.

Table 1. American National Standard Head Type Press Fit Wearing Bushings — Type H (ANSI B94.33-1974)

Range of Hole Sizes A	Body Diameter B					Body Length C	Radius D	Head Diam. E Max	Head Thickness F Max	Number
	Nom	Unfinished		Finished						
		Max	Min	Max	Min					
0.0135 up to and including 0.0625	0.156	0.166	0.161	0.1578	0.1575	0.250 0.312 0.375 0.500	0.016	0.250	0.094	H-10-4 H-10-5 H-10-6 H-10-8
0.0630 to 0.0995	0.203	0.213	0.208	0.2046	0.2043	0.250 0.312 0.375 0.500 0.750	0.016	0.312	0.094	H-13-4 H-13-5 H-13-6 H-13-8 H-13-12
0.1015 to 0.1405	0.250	0.260	0.255	0.2516	0.2513	0.250 0.312 0.375 0.500 0.750	0.016	0.375	0.094	H-16-4 H-16-5 H-16-6 H-16-8 H-16-12
0.1406 to 0.1875	0.312	0.327	0.322	0.3141	0.3138	0.250 0.312 0.375 0.500 0.750 1.000	0.031	0.438	0.125	H-20-4 H-20-5 H-20-6 H-20-8 H-20-12 H-20-16
0.189 to 0.2500	0.406	0.421	0.416	0.4078	0.4075	0.250 0.312 0.375 0.500 0.750 1.000 1.375 1.750	0.031	0.531	0.156	H-26-4 H-26-5 H-26-6 H-26-8 H-26-12 H-26-16 H-26-22 H-26-28

All dimensions are in inches.

Table 1 (Concluded). American National Standard Head Type Press Fit Wearing Bushings — Type H (ANSI B94.33-1974)

Range of Hole Sizes A	Body Diameter B					Body Length C	Radius D	Head Diam. E Max	Head Thickness F Max	Number
	Nom	Unfinished		Finished						
		Max	Min	Max	Min					
0.2570 to 0.3125	0.500	0.520	0.515	0.5017	0.5014	0.312 0.375 0.500 0.750 1.000 1.375 1.750	0.047	0.625	0.219	H-32-5 H-32-6 H-32-8 H-32-12 H-32-16 H-32-22 H-32-28
0.3160 to 0.4219	0.625	0.645	0.640	0.6267	0.6264	0.312 0.375 0.500 0.750 1.000 1.375 1.750 2.125	0.047	0.812	0.219	H-40-5 H-40-6 H-40-8 H-40-12 H-40-16 H-40-22 H-40-28 H-40-34
0.4375 to 0.5000	0.750	0.770	0.765	0.7518	0.7515	0.500 0.750 1.000 1.375 1.750 2.125	0.062	0.938	0.219	H-48-8 H-48-12 H-48-16 H-48-22 H-29-28 H-48-34
0.5156 to 0.6250	0.875	0.895	0.890	0.8768	0.8765	0.500 0.750 1.000 1.375 1.750 2.125 2.500	0.062	1.125	0.250	H-56-8 H-56-12 H-56-16 H-56-22 H-56-28 H-56-34 H-56-40
0.6406 to 0.7500	1.000	1.020	1.015	1.0018	1.0015	0.500 0.750 1.000 1.375 1.750 2.125 2.500	0.094	1.250	0.312	H-64-8 H-64-12 H-64-16 H-64-22 H-64-28 H-64-34 H-64-40
0.7656 to 1.0000	1.375	1.395	1.390	1.3772	1.3768	0.750 1.000 1.375 1.750 2.125 2.500	0.094	1.625	0.375	H-88-12 H-88-16 H-88-22 H-88-28 H-88-34 H-88-40
1.0156 to 1.3750	1.750	1.770	1.765	1.7523	1.7519	1.000 1.375 1.750 2.125 2.500 3.000	0.094	2.000	0.375	H-112-16 H-112-22 H-112-28 H-112-34 H-112-40 H-112-48
1.3906 to 1.7500	2.250	2.270	2.265	2.2525	2.2521	1.000 1.375 1.750 2.125 2.500 3.000	0.094	2.500	0.375	H-144-16 H-144-22 H-144-28 H-144-34 H-144-40 H-144-48

All dimensions are in inches.
See also Table 3 for additional specifications.

Table 2. American National Standard Headless Type Press Fit Wearing Bushings — Type P (ANSI B94.33-1974)

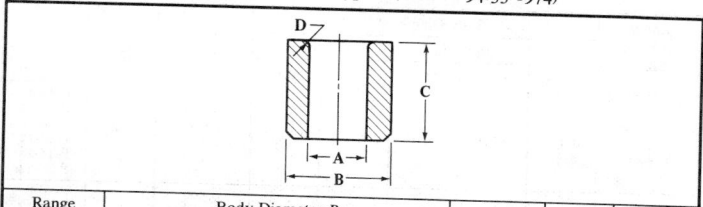

Range of Hole Sizes A	Body Diameter B					Body Length C	Radius D	Number
	Nom	Unfinished		Finished				
		Max	Min	Max	Min			
0.0135 up to and including 0.0625	0.156	0.166	0.161	0.1578	0.1575	0.250 0.312 0.375 0.500	0.016	P-10-4 P-10-5 P-10-6 P-10-8
0.0630 to 0.0995	0.203	0.213	0.208	0.2046	0.2043	0.250 0.312 0.375 0.500 0.750	0.016	P-13-4 P-13-5 P-13-6 P-13-8 P-13-12
0.1015 to 0.1405	0.250	0.260	0.255	0.2516	0.2513	0.250 0.312 0.375 0.500 0.750	0.016	P-16-4 P-16-5 P-16-6 P-16-8 P-16-12
0.1406 to 0.1875	0.312	0.327	0.322	0.3141	0.3138	0.250 0.312 0.375 0.500 0.750 1.000	0.031	P-20-4 P-20-5 P-20-6 P-20-8 P-20-12 P-20-16
0.1890 to 0.2500	0.406	0.421	0.416	0.4078	0.4075	0.250 0.312 0.375 0.500 0.750 1.000 1.375 1.750	0.031	P-26-4 P-26-5 P-26-6 P-26-8 P-26-12 P-26-16 P-26-22 P-26-28
0.2570 to 0.3125	0.500	0.520	0.515	0.5017	0.5014	0.312 0.375 0.500 0.750 1.000 1.375 1.750	0.047	P-32-5 P-32-6 P-32-8 P-32-12 P-32-16 P-32-22 P-32-28
0.3160 to 0.4219	0.625	0.645	0.640	0.6267	0.6264	0.312 0.375 0.500 0.750 1.000 1.375 1.750 2.125	0.047	P-40-5 P-40-6 P-40-8 P-40-12 P-40-16 P-40-22 P-40-28 P-40-34

All dimensions are in inches.

Table 2 (*Concluded*). American National Standard Headless Type Press Fit Wearing Bushings — Type P (ANSI B94.33-1974)

Range of Hole Sizes A	Body Diameter B					Body Length C	Radius D	Number
		Unfinished		Finished				
	Nom	Max	Min	Max	Min			
0.4375 to 0.5000	0.750	0.770	0.765	0.7518	0.7515	0.500 0.750 1.000 1.375 1.750 2.125	0.062	P-48-8 P-48-12 P-48-16 P-48-22 P-48-28 P-48-34
0.5156 to 0.6250	0.875	0.895	0.890	0.8768	0.8765	0.500 0.750 1.000 1.375 1.750 2.125 2.500	0.062	P-56-8 P-56-12 P-56-16 P-56-22 P-56-28 P-56-34 P-56-40
0.6406 to 0.7500	1.000	1.020	1.015	1.0018	1.0015	0.500 0.750 1.000 1.375 1.750 2.125 2.500	0.062	P-64-8 P-64-12 P-64-16 P-64-22 P-64-28 P-64-34 P-64-40
0.7656 to 1.0000	1.375	1.395	1.390	1.3772	1.3768	0.750 1.000 1.375 1.750 2.125 2.500	0.094	P-88-12 P-88-16 P-88-22 P-88-28 P-88-34 P-88-40
1.0156 to 1.3750	1.750	1.770	1.765	1.7523	1.7519	1.000 1.375 1.750 2.125 2.500 3.000	0.094	P-112-16 P-112-22 P-112-28 P-112-34 P-112-40 P-112-48
1.3906 to 1.7500	2.250	2.270	2.265	2.2525	2.2521	1.000 1.375 1.750 2.125 2.500 3.000	0.094	P-144-16 P-144-22 P-144-28 P-144-34 P-144-40 P-144-48

All dimensions are in inches. See Table 3 for additional specifications.

Table 3. Specifications for Head Type H and Headless Type P Press Fit Wearing Bushings (ANSI B94.33-1974)

All dimensions given in inches. Tolerance on dimensions where not otherwise specified shall be ±0.010 inch.

Size and type of chamfer on lead end to be manufacturer's option.

The length, C, is the overall length for the headless type and length underhead for the head type.

The head design shall be in accordance with the manufacturer's practice.

Diameter A must be concentric to diameter B within 0.0005 T.I.V. on finish ground bushings.

The body diameter, B, for unfinished bushings is larger than the nominal diameter in order to provide grinding stock for fitting to jig plate holes. The grinding allowance is:

        0.005 to 0.010 in. for sizes 0.156, 0.203 and 0.250 in.

        0.010 to 0.015 in. for sizes 0.312 and 0.406 in.

        0.015 to 0.020 in. for sizes 0.500 in. and up.

Table 3 (Concluded). Specifications for Head Type H and Headless Type P Press Fit Wearing Bushings (ANSI B94.33-1974)

Hole sizes are in accordance with American National Standard Twist Drill Sizes.
The maximum and minimum values of the hole size, A, shall be as follows:

Nominal Size of Hole	Maximum	Minimum
Above 0.0135 to 0.2500 in., incl.	Nominal + 0.0004 in.	Nominal + 0.0001 in.
Above 0.2500 to 0.7500 in., incl.	Nominal + 0.0005 in.	Nominal + 0.0001 in.
Above 0.7500 to 1.5000 in., incl.	Nominal + 0.0006 in.	Nominal + 0.0002 in.
Above 1.5000 in.	Nominal + 0.0007 in.	Nominal + 0.0003 in.

Bushings in the size range from 0.0135 through 0.3125 will be counterbored to provide for lubrication and chip clearance.

Bushings without counterbore are optional and will be furnished upon request.

The size of the counterbore shall be inside diameter of the bushing + 0.031 inch.

The included angle at the bottom of the counterbore shall be 118 deg, ±2 deg.

The depth of the counterbore shall be in accordance with the table below to provide adequate drill bearing.

	Drill Bushing Hole Size										
	0.0135 to 0.0625		0.0630 to 0.0995		0.1015 to 0.1405		0.1406 to 0.1875		0.1890 to 0.2500		0.2570 to 0.3125	
Body Length	P	H	P	H	P	H	P	H	P	H	P	H
	Minimum Drill Bearing Length — Inch										
0.250	X	0.250	X	X	X	X	X	X	X	X	X	X
0.312	X	0.250	X	X	X	X	X	X	X	X	X	X
0.375	0.250	0.250	X	X	X	X	X	X	X	X	X	X
0.500	0.250	0.250	X	0.312	X	0.312	X	X	X	X	X	X
0.750	+	+	0.375	0.375	0.375	0.375	X	0.375	X	X	X	X
1.000	+	+	+	+	0.375	0.375	X	0.375	X	X	X	X
1.375	+	+	+	+	+	+	0.625	0.625	0.625	0.625	0.625	0.625
1.750	+	+	+	+	+	+	+	+	0.625	0.625	0.625	0.625
							+	+	0.625	0.625	0.625	0.625

All dimensions are in inches.
X indicates no counterbore. + indicates not American National Standard length.

Table 4. American National Standard Slip Type Renewable Wearing Bushings — Type S (ANSI B94.33-1974)

Range of Hole Sizes A	Body Diameter B			Length Under Head C	Radius D	Head Diam. E Max	Head Thickness F Max	Number
	Nom	Max	Min					
0.0135 up to and including 0.0469	0.188	0.1875	0.1873	0.250 0.312 0.375 0.500	0.031	0.312	0.188	S-12-4 S-12-5 S-12-6 S-12-8
0.0492 to 0.1562	0.312	0.3125	0.3123	0.312 0.500 0.750 1.000	0.047	0.562	0.375	S-20-5 S-20-8 S-20-12 S-20-16

All dimensions are in inches.

Table 4 (*Concluded*). American National Standard Slip Type Renewable Wearing Bushings — Type S (ANSI B94.33-1974)

Range of Hole Sizes A	Body Diameter B			Length Under Head C	Radius D	Head Diam. E Max	Head Thickness F Max	Number
	Nom	Max	Min					
0.1570 to 0.3125	0.500	0.5000	0.4998	0.312 0.500 0.750 1.000 1.375 1.750	0.047	0.812	0.438	S-32-5 S-32-8 S-32-12 S-32-16 S-32-22 S-32-28
0.3160 to 0.5000	0.750	0.7500	0.7498	0.500 0.750 1.000 1.375 1.750 2.125	0.094	1.062	0.438	S-48-8 S-48-12 S-48-16 S-48-22 S-48-28 S-48-34
0.5156 to 0.7500	1.000	1.0000	0.9998	0.500 0.750 1.000 1.375 1.750 2.125 2.500	0.094	1.438	0.438	S-64-8 S-64-12 S-64-16 S-64-22 S-64-28 S-64-34 S-64-40
0.7656 to 1.0000	1.375	1.3750	1.3747	0.750 1.000 1.375 1.750 2.125 2.500	0.094	1.812	0.438	S-88-12 S-88-16 S-88-22 S-88-28 S-88-34 S-88-40
1.0156 to 1.3750	1.750	1.7500	1.7497	1.000 1.375 1.750 2.125 2.500 3.000	0.125	2.312	0.625	S-112-16 S-112-22 S-112-28 S-112-34 S-112-40 S-112-48
1.3906 to 1.7500	2.250	2.2500	2.2496	1.000 1.375 1.750 2.125 2.500 3.000	0.125	2.812	0.625	S-144-16 S-144-22 S-144-28 S-144-34 S-144-40 S-144-48

All dimensions are in inches. See also Table 5 for additional specifications.

Table 5. Specifications for Slip Type S and Fixed Type F Renewable Wearing Bushings (ANSI B94.33-1974)

Tolerance on dimensions where not otherwise specified shall be plus or minus 0.010 inch.
Hole sizes are in accordance with the American Standard Twist Drill Sizes.
The maximum and minimum values of hole size, A, shall be as follows:

Nominal Size of Hole	Maximum	Minimum
Above 0.0135 to 0.2500 in. incl.	Nominal + 0.0004 in.	Nominal + 0.0001 in.
Above 0.2500 to 0.7500 in. incl.	Nominal + 0.0005 in.	Nominal + 0.0001 in.
Above 0.7500 to 1.5000 in. incl.	Nominal + 0.0006 in.	Nominal + 0.0002 in.
Above 1.5000	Nominal + 0.0007 in.	Nominal + 0.0003 in.

Table 5 (*Concluded*). Specifications for Slip Type S and Fixed Type F Renewable Wearing Bushings (ANSI B94.33-1974)

The head design shall be in accordance with the manufacturer's practice.

Head of slip type is usually knurled.

When renewable wearing bushings are used with liner bushings of the head type, the length under the head will still be equal to the thickness of the jig plate, because the head of the liner bushing will be countersunk into the jig plate.

Diameter *A* must be concentric to diameter *B* within 0.0005 T.I.R. on finish ground bushings.

Size and type of chamfer on lead end to be manufacturer's option.

Bushings in the size range from 0.0135 through 0.3125 will be counterbored to provide for lubrication and chip clearance.

Bushings without counterbore are optional and will be furnished upon request.

The size of the counterbore shall be inside diameter of the bushings plus 0.031 inch.

The included angle at the bottom of the counterbore shall be 118 deg., plus or minus 2 deg.

The depth of the counterbore shall be in accordance with the table below to provide adequate drill bearing.

	Drill Bearing Hole Size											
	0.0135 to 0.0625		0.0630 to 0.0995		0.1015 to 0.1405		0.1406 to 0.1875		0.1890 to 0.2500		0.2500 to 0.3125	
Body Length	S	F	S	F	S	F	S	F	S	F	S	F
	Minimum Drill Bearing Length											
0.250	0.250	0.250	0.375	0.375	X	0.375	X	0.375	X	0.375	X	X
0.312	0.250	0.250	0.375	0.375	X	0.375	X	0.375	X	0.375	X	X
0.375	0.250	0.250	0.375	0.375	0.375	0.375	0.375	0.375	0.375	0.375	X	X
0.500	0.250	0.250	0.375	0.375	0.375	0.375	0.375	0.375	0.375	0.375	X	X
0.750	0.250	0.250	0.375	0.375	0.375	0.375	0.375	0.375	0.375	0.375	X	X
1.000	0.312	0.312	0.375	0.375	0.375	0.375	0.625	0.625	0.625	0.625	0.625	0.625
1.375	+	+	+	+	+	+	0.625	0.625	0.625	0.625	0.625	0.625
1.750	+	+	+	+	+	+	0.625	0.625	0.625	0.625	0.625	0.625

All dimensions are in inches.
X indicates no counterbore.   + indicates not American National Standard length.

Table 6. American National Standard Fixed Type Renewable Wearing Bushings — Type F (ANSI B94.33-1974)

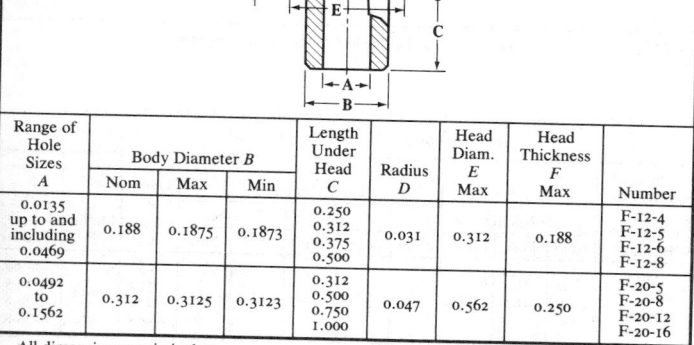

Range of Hole Sizes A	Body Diameter B			Length Under Head C	Radius D	Head Diam. E Max	Head Thickness F Max	Number
	Nom	Max	Min					
0.0135 up to and including 0.0469	0.188	0.1875	0.1873	0.250 0.312 0.375 0.500	0.031	0.312	0.188	F-12-4 F-12-5 F-12-6 F-12-8
0.0492 to 0.1562	0.312	0.3125	0.3123	0.312 0.500 0.750 1.000	0.047	0.562	0.250	F-20-5 F-20-8 F-20-12 F-20-16

All dimensions are in inches.

Table 6 (Concluded). American National Standard Fixed Type Renewable Wearing Bushings — Type F (ANSI B94.33-1974)

Range of Hole Sizes A	Body Diameter B			Length Under Head C	Radius D	Head Diam. E Max	Head Thickness F Max	Number
	Nom	Max	Min					
0.1570 to 0.3125	0.500	0.5000	0.4998	0.312 0.500 0.750 1.000 1.375 1.750	0.047	0.812	0.250	F-32-5 F-32-8 F-32-12 F-32-16 F-32-22 F-32-28
0.3160 to 0.5000	0.750	0.7500	0.7498	0.500 0.750 1.000 1.375 1.750 2.125	0.094	1.062	0.250	F-48-8 F-48-12 F-48-16 F-48-22 F-48-28 F-48-34
0.5156 to 0.7500	1.000	1.0000	0.9998	0.500 0.750 1.000 1.375 1.750 2.125 2.500	0.094	1.438	0.375	F-64-8 F-64-12 F-64-16 F-64-22 F-64-28 F-64-34 F-64-40
0.7656 to 1.0000	1.375	1.3750	1.3747	0.750 1.000 1.375 1.750 2.125 2.500	0.094	1.812	0.375	F-88-12 F-88-16 F-88-22 F-88-28 F-88-34 F-88-40
1.0156 to 1.3750	1.750	1.7500	1.7497	1.000 1.375 1.750 2.125 2.500 3.000	0.125	2.312	0.375	F-112-16 F-112-22 F-112-28 F-112-34 F-112-40 F-112-48
1.3906 to 1.7500	2.250	2.2500	2.2496	1.000 1.375 1.750 2.125 2.500 3.000	0.125	2.812	0.375	F-144-16 F-144-22 F-144-28 F-144-34 F-144-40 F-144-48

All dimensions are in inches. See also Table 5 for additional specifications.

Table 7. American National Standard Headless Type Liner Bushings — Type L (ANSI B94.33-1974)

Range of Hole Sizes in Renewable Bushings	Inside Diameter A			Body Diameter B					Over-all Length C	Radius D	Number
				Nom	Unfinished		Finished				
	Nom	Max	Min		Max	Min	Max	Min			
0.0135 up to and including 0.0469	0.188	0.1879	0.1876	0.312	0.3341	0.3288	0.3141	0.3138	0.250 0.312 0.375 0.500	0.031	L-20-4 L-20-5 L-20-6 L-20-8

All dimensions are in inches. For illustration of bushing, see following page.

Table 7 (*Concluded*). **American National Standard Headless Type Liner Bushings — Type L** (ANSI B94.33-1974)

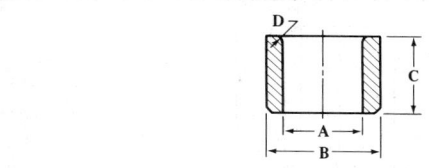

Range of Hole Sizes in Renewable Bushings	Inside Diameter A			Body Diameter B					Over-all Length C	Radius D	Number
				Nom	Unfinished		Finished				
	Nom	Max	Min		Max	Min	Max	Min			
0.0492 to 0.1562	0.312	0.3129	0.3126	0.500	0.520	0.515	0.5017	0.5014	0.312 0.500 0.750 1.000	0.047	L-32-5 L-32-8 L-32-12 L-32-16
0.1570 to 0.3125	0.500	0.5005	0.5002	0.750	0.770	0.765	0.7518	0.7515	0.312 0.500 0.750 1.000 1.375 1.750	0.062	L-48-5 L-48-8 L-48-12 L-48-16 L-48-22 L-48-28
0.3160 to 0.5000	0.750	0.7506	0.7503	1.000	1.020	1.015	1.0018	1.0015	0.500 0.750 1.000 1.375 1.750 2.125	0.062	L-64-8 L-64-12 L-64-16 L-64-22 L-64-28 L-64-34
0.5156 to 0.7500	1.000	1.0007	1.0004	1.375	1.395	1.390	1.3772	1.3768	0.500 0.750 1.000 1.375 1.750 2.125 2.500	0.094	L-88-8 L-88-12 L-88-16 L-88-22 L-88-28 L-88-34 L-88-40
0.7656 to 1.0000	1.375	1.3760	1.3756	1.750	1.770	1.765	1.7523	1.7519	0.750 1.000 1.375 1.750 2.125 2.500	0.094	L-112-12 L-112-16 L-112-22 L-112-28 L-112-34 L-112-40
1.0156 to 1.3750	1.750	1.7512	1.7508	2.250	2.270	2.265	2.2525	2.2521	1.000 1.375 1.750 2.125 2.500 3.000	0.094	L-144-16 L-144-22 L-144-28 L-144-34 L-144-40 L-144-48
1.3906 to 1.7500	2.250	2.2515	2.2510	2.750	2.770	2.765	2.7526	2.7522	1.000 1.375 1.750 2.125 2.500 3.000	0.125	L-176-16 L-176-22 L-176-28 L-176-34 L-176-40 L-176-48

All dimensions are in inches.
Tolerances on dimensions where otherwise not specified are ± 0.010 in.
The body diameter, B for unfinished bushings is 0.015 to 0.020 in. larger than the nominal diameter in order to provide grinding stock for fitting to jig plate holes.
Diameter A must be concentric to diameter B within 0.0005 T.I.R. on finish ground bushings.

Table 8. American National Standard Head Type Liner Bushings — Type HL
(ANSI B94.33-1974)

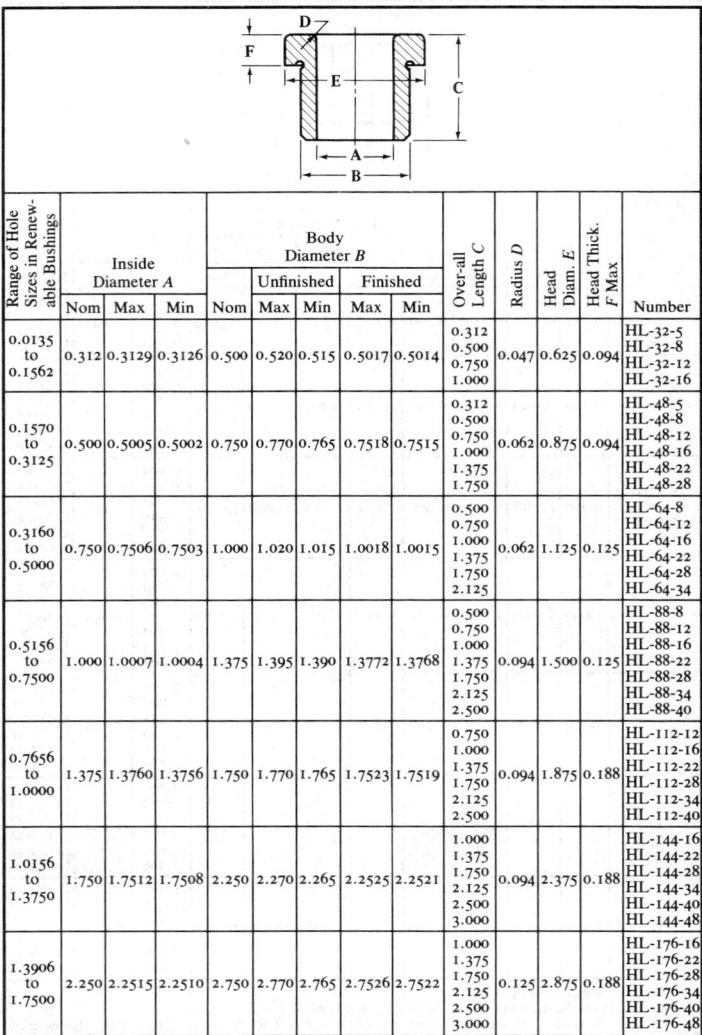

Range of Hole Sizes in Renewable Bushings	Inside Diameter A			Body Diameter B					Over-all Length C	Radius D	Head Diam. E	Head Thick. F Max	Number
					Unfinished		Finished						
	Nom	Max	Min	Nom	Max	Min	Max	Min					
0.0135 to 0.1562	0.312	0.3129	0.3126	0.500	0.520	0.515	0.5017	0.5014	0.312 0.500 0.750 1.000	0.047	0.625	0.094	HL-32-5 HL-32-8 HL-32-12 HL-32-16
0.1570 to 0.3125	0.500	0.5005	0.5002	0.750	0.770	0.765	0.7518	0.7515	0.312 0.500 0.750 1.000 1.375 1.750	0.062	0.875	0.094	HL-48-5 HL-48-8 HL-48-12 HL-48-16 HL-48-22 HL-48-28
0.3160 to 0.5000	0.750	0.7506	0.7503	1.000	1.020	1.015	1.0018	1.0015	0.500 0.750 1.000 1.375 1.750 2.125	0.062	1.125	0.125	HL-64-8 HL-64-12 HL-64-16 HL-64-22 HL-64-28 HL-64-34
0.5156 to 0.7500	1.000	1.0007	1.0004	1.375	1.395	1.390	1.3772	1.3768	0.500 0.750 1.000 1.375 1.750 2.125 2.500	0.094	1.500	0.125	HL-88-8 HL-88-12 HL-88-16 HL-88-22 HL-88-28 HL-88-34 HL-88-40
0.7656 to 1.0000	1.375	1.3760	1.3756	1.750	1.770	1.765	1.7523	1.7519	0.750 1.000 1.375 1.750 2.125 2.500	0.094	1.875	0.188	HL-112-12 HL-112-16 HL-112-22 HL-112-28 HL-112-34 HL-112-40
1.0156 to 1.3750	1.750	1.7512	1.7508	2.250	2.270	2.265	2.2525	2.2521	1.000 1.375 1.750 2.125 2.500 3.000	0.094	2.375	0.188	HL-144-16 HL-144-22 HL-144-28 HL-144-34 HL-144-40 HL-144-48
1.3906 to 1.7500	2.250	2.2515	2.2510	2.750	2.770	2.765	2.7526	2.7522	1.000 1.375 1.750 2.125 2.500 3.000	0.125	2.875	0.188	HL-176-16 HL-176-22 HL-176-28 HL-176-34 HL-176-40 HL-176-48

All dimensions are in inches.
See also footnotes to Table 7.

Table 9. American National Standard Locking Mechanisms for Jig Bushings (ANSI B94.33-1974)

Lock Screw for Use With Slip or Fixed Renewable Bushings

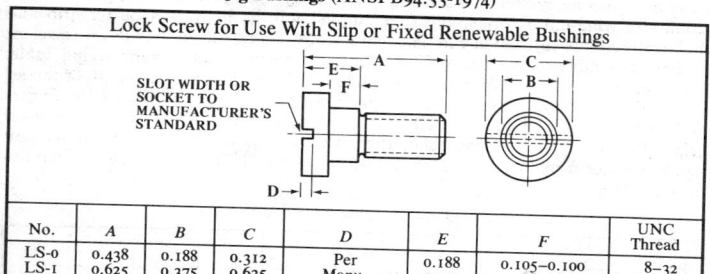

SLOT WIDTH OR SOCKET TO MANUFACTURER'S STANDARD

No.	A	B	C	D	E	F	UNC Thread
LS-0	0.438	0.188	0.312	Per Manufacturer's Standard	0.188	0.105–0.100	8–32
LS-1	0.625	0.375	0.625		0.250	0.138–0.132	5/16–18
LS-2	0.875	0.375	0.625		0.375	0.200–0.194	5/16–18
LS-3	1.000	0.438	0.750		0.375	0.200–0.194	3/8–16

Round Clamp Optional Only For Use With Fixed Renewable Bushing Only

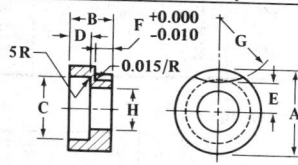

NOTE: F DIMENSION ALLOWS FOR CLAMPING. MATERIAL AND HARDNESS TO MANUFACTURER'S STANDARD. TO CHANGE TO THE ROUND CLAMP IN OLD FIXTURES, REMOVE THE CONVENTIONAL SCREW AND USE THE SAME TAPPED HOLE TO SECURE THE NEW CLAMP WITH STANDARD SOCKET HEAD SCREW.

Number	A	B	C	D	E	F	G	H	Use With Socket Head Screw
RC-1	0.625	0.312	0.484	0.150	0.203	0.125	0.531	0.328	5/16–18
RC-2	0.625	0.438	0.484	0.219	0.187	0.188	0.906	0.328	5/16–18
RC-3	0.750	0.500	0.578	0.281	0.188	0.188	1.406	0.391	3/8–16

Locking Mechanism Dimensions of Slip and Fixed Renewable Bushings

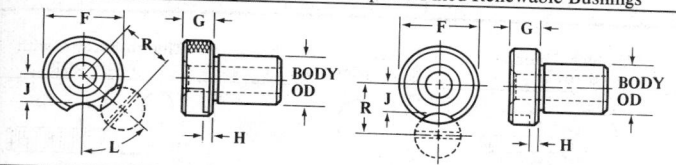

Body OD	Max Diam. F When Used With Locking Device	G Head Thickness Slip	G Head Thickness Fixed	H ±0.005	J	L Max	R	Locking Dim. of Lock Screw (Slip or Fixed)	Locking Dim. of Clamp (Fixed Only)	Max Head Diam. of Mating Liner Used to Clear Locking Device	Clamp or Screw LS or RC
0.188	0.312	0.188	0.188	0.094	0.094	55°	0.266	0.105–0.100	...	...	0
0.312	0.562	0.375	0.250	0.125	0.172	65°	0.500	0.138–0.132	0.125–0.115	0.625	1
0.500	0.812	0.438	0.250	0.125	0.297	65°	0.625	0.138–0.132	0.125–0.115	0.875	1
0.750	1.062	0.438	0.250	0.125	0.422	50°	0.750	0.138–0.132	0.125–0.115	1.125	1
1.000	1.438	0.438	0.375	0.188	0.594	35°	0.922	0.200–0.194	0.187–0.177	1.500	2
1.375	1.812	0.438	0.375	0.188	0.781	30°	1.109	0.200–0.194	0.187–0.177	1.875	2
1.750	2.312	0.625	0.375	0.188	1.000	30°	1.391	0.200–0.194	0.187–0.177	2.375	3
2.250	2.812	0.625	0.375	0.188	1.250	25°	1.641	0.200–0.194	0.187–0.177	2.875	3

All dimensions are in inches.

Wing-nuts for Jigs. — Star Handwheels. — Wing-nuts are used on hook bolts or swiveling eye-bolts, when a comparatively light pressure is required. The thumb- or wing-nut is preferable to a knurled nut, as it gives a better grip and makes it possible to tighten the bolt more firmly. The dimensions of an excellent design of handwheel for use on jigs, etc., are given in an accompanying table. These wheels have a rather long stem or hub which provides a good length of thread and brings the grip or handle far enough from the jig body to prevent the fingers or knuckles from striking it. The "star" design of handle also permits a good grip. By having the casting solid, these handwheels can be tapped out for any size thread, or a plain hole can be drilled when it is desired to attach the handles to round stock.

Dimensions of Wing or Thumb Nuts

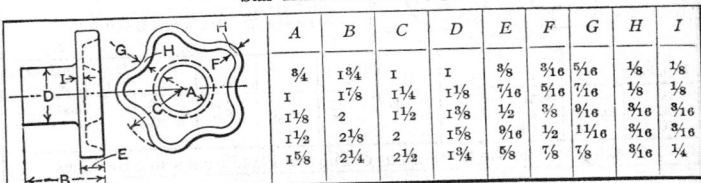

	A	B	C	D	E	F	G
	3/16	5/8	1 3/16	5/16	3/8	7/16	1/8
	1/4	3/4	1 1/2	15/32	1/2	17/32	5/32
	5/16	3/4	1 1/2	15/32	1/2	17/32	5/32
	3/8	13/16	1 3/4	17/32	9/16	5/8	5/32
	7/16	7/8	2	21/32	5/8	11/16	3/16
	1/2	1 1/16	2 1/2	3/4	13/16	7/8	3/16

Star Handwheels for Jigs

	A	B	C	D	E	F	G	H	I
	3/4	1 3/4	1	1	3/8	3/16	5/16	1/8	1/8
	1	1 7/8	1 1/4	1 1/8	7/16	5/16	7/16	1/8	1/8
	1 1/8	2	1 1/2	1 3/8	1/2	3/8	9/16	3/16	3/16
	1 1/2	2 1/8	2	1 5/8	9/16	1/2	11/16	3/16	3/16
	1 5/8	2 1/4	2 1/2	1 3/4	5/8	7/8	7/8	3/16	1/4

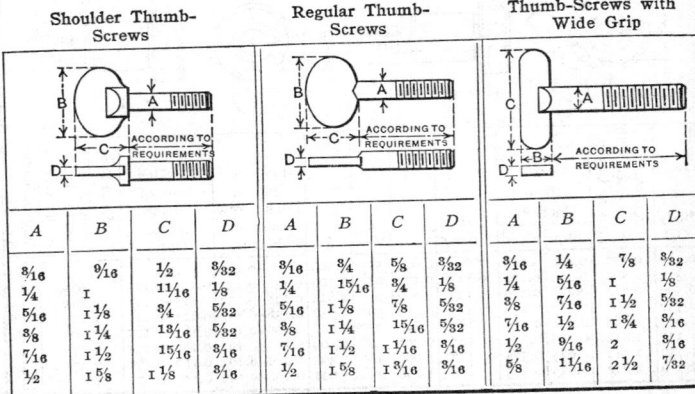

Shoulder Thumb-Screws — Regular Thumb-Screws — Thumb-Screws with Wide Grip

A	B	C	D	A	B	C	D	A	B	C	D
3/16	9/16	1/2	3/32	3/16	3/4	5/8	3/32	3/16	1/4	7/8	3/32
1/4	1	11/16	1/8	1/4	15/16	3/4	1/8	1/4	5/16	1	1/8
5/16	1 1/8	3/4	5/32	5/16	1 1/8	7/8	5/32	3/8	7/16	1 1/2	5/32
3/8	1 1/4	13/16	5/32	3/8	1 1/4	15/16	5/32	7/16	1/2	1 3/4	3/16
7/16	1 1/2	15/16	3/16	7/16	1 1/2	1 1/16	3/16	1/2	9/16	2	3/16
1/2	1 5/8	1 1/8	3/16	1/2	1 5/8	1 3/16	3/16	5/8	11/16	2 1/2	7/32

Collar-head Screws

A	B	C	D	E	F	Diameter G and Threads per Inch
5/8	1/16	3/16	5/16	3/16	1/2	No. 10 — 32
7/8	1/16	3/16	5/16	3/16	5/8	No. 10 — 32
1 1/8	1/16	3/16	5/16	3/16	7/8	No. 10 — 32
1 3/8	1/16	3/16	5/16	3/16	1	No. 10 — 32
1 5/8	1/16	3/16	5/16	3/16	1 1/4	No. 10 — 32
7/8	3/32	1/4	7/16	1/4	5/8	No. 14 — 24
1 1/8	3/32	1/4	7/16	1/4	7/8	No. 14 — 24
1 3/8	3/32	1/4	7/16	1/4	1	No. 14 — 24
1 5/8	3/32	1/4	7/16	1/4	1 1/4	No. 14 — 24
1 7/8	3/32	1/4	7/16	1/4	1 7/16	No. 14 — 24
7/8	1/8	5/16	9/16	5/16	5/8	5/16 — 18
1 1/4	1/8	5/16	9/16	5/16	1	5/16 — 18
1 5/8	1/8	5/16	9/16	5/16	1 1/4	5/16 — 18
2	1/8	5/16	9/16	5/16	1 1/2	5/16 — 18
1	1/8	3/8	11/16	3/8	3/4	3/8 — 16
1 3/4	1/8	3/8	11/16	3/8	1 5/16	3/8 — 16
2 1/2	1/8	3/8	11/16	3/8	1 7/8	3/8 — 16
1 3/8	3/16	7/16	3/4	7/16	1	7/16 — 14
2 1/8	3/16	7/16	3/4	7/16	1 5/8	7/16 — 14
2 1/2	3/16	7/16	3/4	7/16	1 7/8	7/16 — 14
1 3/4	3/16	1/2	7/8	1/2	1 5/16	1/2 — 13
2 1/2	3/16	1/2	7/8	1/2	1 7/8	1/2 — 13
3 1/4	3/16	1/2	7/8	1/2	2 3/8	1/2 — 13

Clamping Screws, Screw Bushings and Studs. — Collar-head screws are used on jigs and fixtures in conjunction with clamps, straps and latches for clamping purposes (see table for dimensions). Rocking collar-screws are used with clamps for rough work, since they adapt themselves to any irregularities of the work and give a full bearing on the clamps in any position the work may assume.

Shoulder-screws are used for fastening clamping blocks, latches, or any parts that must move through a limited distance while still remaining permanently fastened to the tool.

Quarter-turn thumb-screws may be rapidly manipulated and are especially of use in box jigs. An objection to this type of screw is that the wear takes place on the boss on which it acts. Half-turn thumb-screws are also used in box jigs when the quarter-turn thumb-screw cannot be used on account of the work or bushing protruding through the end of the jig. These screws are used in pairs, one on each side of the jig cover.

Screw bushings are generally avoided when accurate work is required, as a threaded bushing is likely to be out of true. Sometimes, however, no other type of bushing is adapted for the work in hand. (See table of "Aligning Screw Bushings.") Studs are used in jig design for locating work with holes in it. The accompanying table shows a recommended form of collar stud.

Rocking Collar-screws

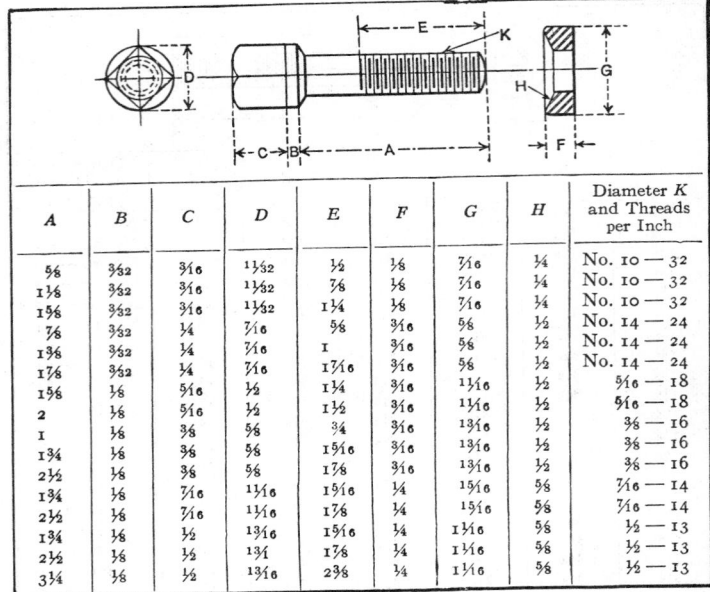

A	B	C	D	E	F	G	H	Diameter K and Threads per Inch
5/8	3/32	3/16	11/32	1/2	1/8	7/16	1/4	No. 10 — 32
1 1/8	3/32	3/16	11/32	7/8	1/8	7/16	1/4	No. 10 — 32
1 5/8	3/32	3/16	11/32	1 1/4	1/8	7/16	1/4	No. 10 — 32
7/8	3/32	1/4	7/16	5/8	3/16	5/8	1/2	No. 14 — 24
1 3/8	3/32	1/4	7/16	1	3/16	5/8	1/2	No. 14 — 24
1 7/8	3/32	1/4	7/16	1 7/16	3/16	5/8	1/2	No. 14 — 24
1 5/8	1/8	5/16	1/2	1 1/4	3/16	11/16	1/2	5/16 — 18
2	1/8	5/16	1/2	1 1/2	3/16	11/16	1/2	5/16 — 18
1	1/8	3/8	5/8	3/4	3/16	13/16	1/2	3/8 — 16
1 3/4	1/8	3/8	5/8	15/16	3/16	13/16	1/2	3/8 — 16
2 1/2	1/8	3/8	5/8	1 7/8	3/16	13/16	1/2	3/8 — 16
1 3/4	1/8	7/16	11/16	1 9/16	1/4	15/16	5/8	7/16 — 14
2 1/2	1/8	7/16	11/16	1 7/8	1/4	15/16	5/8	7/16 — 14
1 3/4	1/8	1/2	13/16	1 9/16	1/4	1 1/16	5/8	1/2 — 13
2 1/2	1/8	1/2	1 3/4	1 7/8	1/4	1 1/16	5/8	1/2 — 13
3 1/4	1/8	1/2	1 13/16	2 3/8	1/4	1 1/16	5/8	1/2 — 13

Shoulder-screws

A	B	C	D		E	Diameter F and Threads per inch
0.249	1/2	1/8	1/2	to 3/4	7/16	No. 10 — 32
0.249	1/2	1/8	3/16	to 7/16	9/16	No. 10 — 32
0.3115	5/8	5/32	3/16	to 1/2	3/8	No. 14 — 24
0.3115	5/8	5/32	9/16	to 7/8	1/2	No. 14 — 24
0.374	3/4	5/32	1/4	to 7/16	3/8	No. 14 — 24
0.374	3/4	5/32	1/2	to 11/16	1/2	No. 14 — 24
0.4365	13/16	5/32	3/8	to 9/16	1/2	5/16 — 18
0.4365	13/16	5/32	7/8	to 1 1/8	3/4	5/16 — 18
0.499	7/8	3/16	3/8	to 9/16	1/2	3/8 — 16
0.499	7/8	3/16	15/16	to 1 1/4	3/4	3/8 — 16
0.5615	1	3/16	1/2	to 3/4	5/8	7/16 — 14
0.5615	1	3/16	1 1/4	to 1 1/2	7/8	7/16 — 14
0.6235	1 1/8	1/4	1/2	to 5/8	5/8	1/2 — 13
0.6235	1 1/8	1/4	1 1/8	to 1 5/8	7/8	1/2 — 13
0.686	1 1/4	1/4	1 1/8	to 1 3/4	1	1/2 — 13
0.7485	1 3/8	5/16	3/4	to 1 3/8	7/8	5/8 — 11
0.7485	1 3/8	5/16	1 1/4	to 2	1 1/8	5/8 — 11

Quarter-turn Thumb-screws

	A	B	C	D	E	Diameter F and Threads per Inch
	7/16	1/2	11/16	3/16	5/16	No. 10 — 32
	5/8	1/2	11/16	3/16	1/2	No. 10 — 32
	7/8	1/2	11/16	3/16	5/8	No. 10 — 32
	1 3/8	1/2	11/16	3/16	1	No. 10 — 32
	5/8	9/16	13/16	1/4	1/2	No. 14 — 24
	1 1/8	9/16	13/16	1/4	7/8	No. 14 — 24
	1 5/8	9/16	13/16	1/4	1 1/4	No. 14 — 24
	5/8	5/8	1	5/16	1/2	5/16 — 18
	1 1/4	5/8	1	9/16	1	5/16 — 18
	2	5/8	1	9/16	1 1/2	5/16 — 18
	3/4	11/16	1 3/16	3/8	9/16	3/8 — 16
	1	11/16	1 3/16	3/8	3/4	3/8 — 16
	1 3/4	11/16	1 3/16	3/8	1 5/16	3/8 — 16
	2 1/2	11/16	1 3/16	3/8	1 7/8	3/8 — 16
	1	3/4	1 5/16	7/16	3/4	7/16 — 14
	1 3/4	3/4	1 5/16	7/16	1 5/16	7/16 — 14
	2 1/2	3/4	1 5/16	7/16	1 1/4	7/16 — 14
	1 3/4	3/4	1 5/16	1/2	1 5/16	1/2 — 13
	2 1/2	3/4	1 5/16	1/2	1 1/4	1/2 — 13
	3 1/4	3/4	1 5/16	1/2	1 5/8	1/2 — 13

Half-turn Thumb-screws

	A	B	C	D	E	F	Diameter G and Threads per Inch
	7/16	1/2	3/16	1 1/16	5/16	3/4	No. 10 — 32
	5/8	1/2	3/16	1 1/16	1/2	3/4	No. 10 — 32
	7/8	1/2	3/16	1 1/16	5/8	3/4	No. 10 — 32
	1 3/8	1/2	3/16	1 1/16	1	3/4	No. 10 — 32
	5/8	9/16	3/16	1 3/16	1/2	7/8	No. 14 — 24
	1 1/8	9/16	3/16	1 3/16	7/8	7/8	No. 14 — 24
	1 5/8	9/16	3/16	1 3/16	1 1/4	7/8	No. 14 — 24
	5/8	5/8	3/16	1	1/2	1	5/16 — 18
	1 1/4	5/8	3/16	1	1	1	5/16 — 18
	2	5/8	3/16	1	1 1/2	1	5/16 — 18
	3/4	11/16	1/4	1 3/16	9/16	1 1/4	3/8 — 16
	1	11/16	1/4	1 3/16	3/4	1 1/4	3/8 — 16
	1 3/4	11/16	1/4	1 3/16	1 5/16	1 1/4	3/8 — 16
	2 1/2	11/16	1/4	1 3/16	1 7/8	1 1/4	3/8 — 16
	1	3/4	1/4	1 5/16	3/4	1 3/8	7/16 — 14
	1 3/4	3/4	1/4	1 5/16	1 5/16	1 3/8	7/16 — 14
	2 1/2	3/4	1/4	1 5/16	1 1/4	1 3/8	7/16 — 14
	1 3/4	3/4	1/4	1 5/16	1 5/16	1 1/2	1/2 — 13
	2 1/2	3/4	1/4	1 5/16	1 1/4	1 1/2	1/2 — 13
	3 1/4	3/4	1/4	1 5/16	1 5/8	1 1/2	1/2 — 13

Aligning Screw Bushings

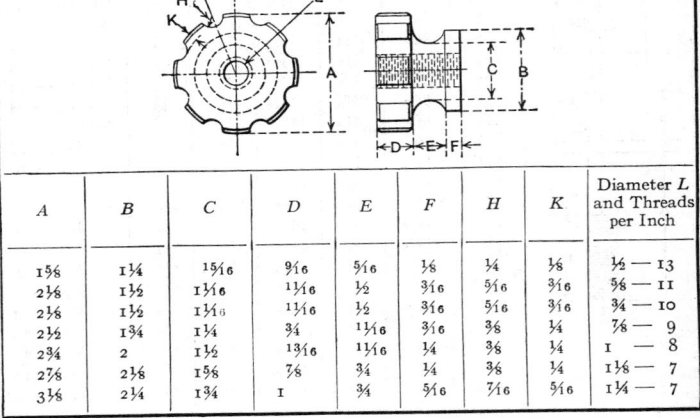

	B	D	E	F	Diameter G and Threads per Inch
	7/8	1/8	5/16	1 1/16	1/2 — 13
	1	5/32	5/16	7/8	5/8 — 11
	1 1/4	3/16	5/16	1	3/4 — 10
	1 1/2	3/16	7/16	1 1/4	1 — 14
	1 7/8	1/4	1/2	1 5/8	1 1/4 — 12
	2 1/4	1/4	5/8	2	1 1/2 — 12
	2 3/4	1/4	3/4	2 3/8	1 3/4 — 8
	3 1/4	1/4	7/8	2 3/4	2 — 8

A and C, according to requirements

Collar Studs — Hardened and Ground

A	B	C	D	E	F
0.251	0.249	1/2	1/2	3/16	1/4 — 5/8
0.3135	0.3115	9/16	1/2	3/16	5/16 — 11/16
0.376	0.374	5/8	5/8	3/16	3/8 — 3/4
0.4385	0.4365	11/16	5/8	3/16	7/16 — 13/16
0.501	0.499	3/4	5/8	1/4	1/2 — 7/8
0.5635	0.5615	7/8	5/8	1/4	1/2 — 7/8
0.626	0.624	1	3/4	1/4	1/2 — 1 1/16
0.6885	0.6865	1 1/8	3/4	1/4	5/8 — 1 3/16
0.751	0.749	1 1/4	7/8	1/4	5/8 — 1 3/8

Hand Nuts

A	B	C	D	E	F	H	K	Diameter L and Threads per Inch
1 5/8	1 1/4	15/16	9/16	5/16	1/8	1/4	1/8	1/2 — 13
2 1/8	1 1/2	1 1/16	11/16	1/2	3/16	5/16	3/16	5/8 — 11
2 1/8	1 1/2	1 1/16	11/16	1/2	3/16	5/16	3/16	3/4 — 10
2 1/2	1 3/4	1 1/4	3/4	11/16	3/16	3/8	1/4	7/8 — 9
2 3/4	2	1 1/2	13/16	11/16	1/4	3/8	1/4	1 — 8
2 7/8	2 1/8	1 5/8	7/8	3/4	1/4	3/8	1/4	1 1/8 — 7
3 1/8	2 1/4	1 3/4	1	3/4	5/16	7/16	5/16	1 1/4 — 7

Dimensions of Jig-Screw Latches

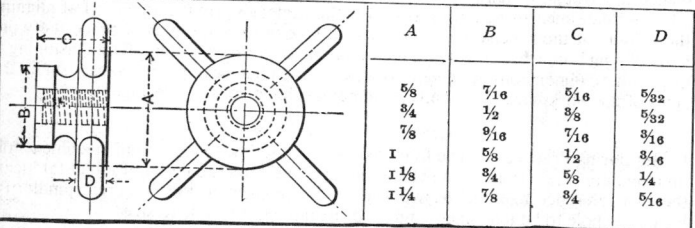

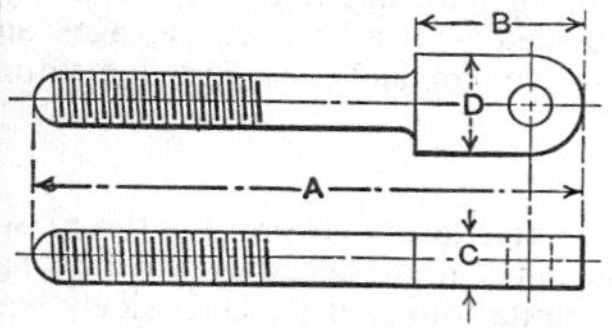

	A	B	C	D
	1 ¼	⅜	⅛	5/16
	1 ¾	⅝	5/32	⅜
	2 ⅜	¾	3/16	7/16
	2 ⅞	⅞	¼	½
	3 ½	1	5/16	⅝
	4 ⅛	1 ¼	⅜	¾

Dimensions of Latch Nuts

	A	B	C	D
	⅝	7/16	5/16	5/32
	¾	½	⅜	5/32
	⅞	9/16	7/16	3/16
	1	⅝	½	3/16
	1 ⅛	¾	⅝	¼
	1 ¼	⅞	¾	5/16

Standard Jig Feet

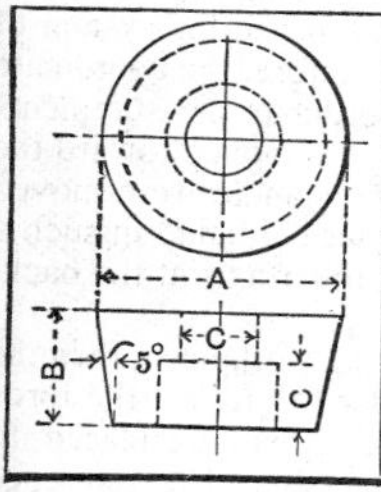

A	B	C	A	B	C
⅜	3/16	⅛	13/16	13/32	7/32
7/16	7/32	9/64	¾	⅜	¼
½	¼	5/32	⅞	7/16	9/32
9/16	9/32	11/64	1	½	5/16
⅝	5/16	3/16	......	......	......

Screws for Jig Feet

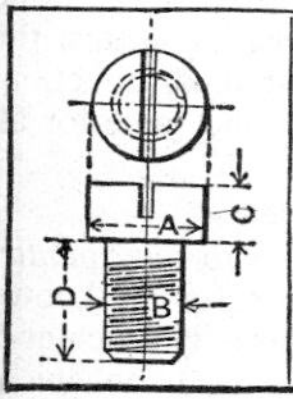

A	B	C	D	A	B	C	D
0.160	⅛	0.110	9/32	0.299	7/32	0.192	7/16
0.191	9/64	0.123	5/16	0.343	¼	0.219	15/32
0.213	5/32	0.137	11/32	0.386	9/32	0.246	½
0.233	11/64	0.150	⅜	0.426	5/16	0.273	17/32
0.256	3/16	0.164	13/32	.....	....	.....	....

Definition of Jig and Fixture. — The distinction between a jig and fixture is not easy to define, but, as a general rule, it is as follows: A jig either holds or is held on the work, and, at the same time, contains guides for the various cutting tools, whereas a fixture holds the work while the cutting tools are in operation, but

does not contain any special arrangements for guiding the tools. A fixture, therefore, must be securely held or fixed to the machine on which the operation is performed — hence the name. A fixture is sometimes provided with a number of gages and stops, but not with bushings or other devices for guiding and supporting the cutting tools.

Jig Borers. — Jig borers are used for precision hole-location work. For this reason the coordinate measuring systems on these machines are designed to provide longitudinal and transverse movements that are accurate to 0.0001 in. One widely used method of obtaining this accuracy utilizes ultraprecision lead screws. Another measuring system employs precision end measuring rods and a micrometer head that are placed in a trough which is parallel to the table movement. However, the purpose of all coordinate measuring systems used is the same: to provide a method of aligning the spindle at the precise location where a hole is to be produced. Since the work table of a jig borer moves in two directions, the coordinate system of dimensioning is used, where dimensions are given from two perpendicular reference axes, usually the sides of the workpiece, frequently its upper left-hand corner. See Fig. 1C.

Jig-Boring Practice. — The four basic steps to follow to locate and machine a hole on a jig borer are: (1) align and clamp the workpiece on the jig-borer table; (2) locate the two reference axes of the workpiece with respect to the jig-borer spindle; (3) locate the hole to be machined; and (4) drill and bore the hole to size.

Align and Clamp the Workpiece: The first consideration in placing the workpiece on the jig-borer table should be the relation of the coordinate measuring system of the jig borer to the coordinate dimensions on the drawing. Therefore, the coordinate measuring system is designed so that the readings of the coordinate measurements are direct when the table is moved toward the left and when it is moved toward the column of the jig borer. The result would be the same if the spindle were moved toward the right and away from the column, with the workpiece situated in such a position that one reference axis is located at the left and the other axis at the back, toward the column.

If the holes to be bored are to pass through the bottom of the workpiece, then the workpiece must be placed on precision parallel bars. In order to prevent the force exerted by the clamps from bending the workpiece the parallel bars are placed directly under the clamps, which hold the workpiece on the table. The reference axes of the workpiece must also be aligned with respect to the transverse and longitudinal table movements before it is firmly clamped. This alignment can be done with a dial-test indicator held in the spindle of the jig borer and bearing against the longitudinal reference edge. As the table is traversed in the longitudinal direction the workpiece is adjusted until the dial-test indicator readings are the same for all positions.

Locate the Two Reference Axes of the Workpiece with Respect to the Spindle: The jig-borer table is now moved to position the workpiece in a precise and known location from where it can be moved again to the location of the holes to be machined. Since all the holes are dimensioned from the two reference axes, the most convenient position to start from is where the axis of the jig-borer spindle and the intersection of the two workpiece reference axes are aligned. This is called the starting position, which is similar to a zero reference position. When so positioned, the longitudinal and transverse measuring systems of the jig borer are set to read zero. Occasionally the reference axes are located outside the body of the workpiece: a convenient edge

or hole on the workpiece is picked up as the starting position, and the dimensions from this point to the reference axes are set on the positioning measuring system.

Locate the Hole: Precise coordinate table movements are used to position the workpiece so that the spindle axis is located exactly where the hole is to be machined. When the measuring system has been set to zero at the starting position, the coordinate readings at the hole location will be the same as the coordinate dimensions of the hole center.

The movements to each hole must be made in one direction for both the transverse and longitudinal directions, to eliminate the effect of any backlash in the leadscrew. The usual table movements are toward the left and toward the column.

The most convenient sequence on machines using micrometer dials as position indicators (machines with lead screws) is to machine the hole closest to the starting position first and then the next closest, and so on. On jig borers using end measuring rods, the opposite sequence is followed: the farthest hole is machined first and then the next farthest, and so on, since it is easier to remove end rods and replace them with shorter rods.

Drill and Bore Hole to Size: The sequence of operations used to produce a hole on a jig borer is as follows: (1) a short, stiff drill, such as a center drill, that will not deflect when cutting should be used to spot a hole when the work and the axis of the machine tool spindle are located at the exact position where the hole is wanted; (2) the initial hole is made by a twist drill; (3) a single-point boring tool that is set to rotate about the axis of the machine tool spindle is then used to generate a cut surface that is concentric to the axis of rotation.

Heat will be generated by the drilling operation, so it is good practice to drill all the holes first, and then allow the workpiece to cool before the holes are bored to size.

Transfer of Tolerances. — All of the dimensions that must be accurately held on precision machines and engine parts are usually given a tolerance. And when such dimensions are changed from the conventional to the coordinate system of dimensioning, the tolerances must also be included. Because of their importance, the transfer of the tolerances must be done with great care, keeping in mind that the sum of the tolerances of any pair of dimensions in the coordinate system must not be larger than the tolerance of the dimension that they replaced in the conventional system. An example is given in Fig. 1.

The first step in the procedure is to change the tolerances given in Fig. 1A to equal, bilateral tolerances given in Fig. 1B. For example, the dimension $2.125^{+.003}_{-.001}$ has a total tolerance of 0.004. The equal, bilateral tolerance would be plus or minus one-half of this value, or ± .002. Then to keep the limiting dimensions the same, the basic dimension must be changed to 2.126, in order to give the required values of 2.128 and 2.124. When changing to equal, bilateral tolerances, if the upper tolerance is decreased (as in this example), the basic dimension must be increased by a like amount. The upper tolerance was decreased by .003 − .002 = .001; therefore, the basic dimension was increased by .001 to 2.126. Conversely, if the upper tolerance is increased, the basic dimension is decreased.

The next step is to transfer the revised basic dimension to the coordinate dimensioning system. To transfer the 2.126 dimension, the distance of the applicable holes from the left reference axis must be determined. The first holes to the right are 0.8750 from the reference axis. The second hole is 2.126 to the right of the first holes. Therefore, the second hole is 0.8750 + 2.126 = 3.0010 to the right of the reference axis. This value is then the coordinate dimension for the second hole, while the 0.8750 value is the coordinate dimension of the first two, vertically aligned holes. This procedure is followed for all the holes to find their distances from the two reference axes. These values are given in Fig. 1C.

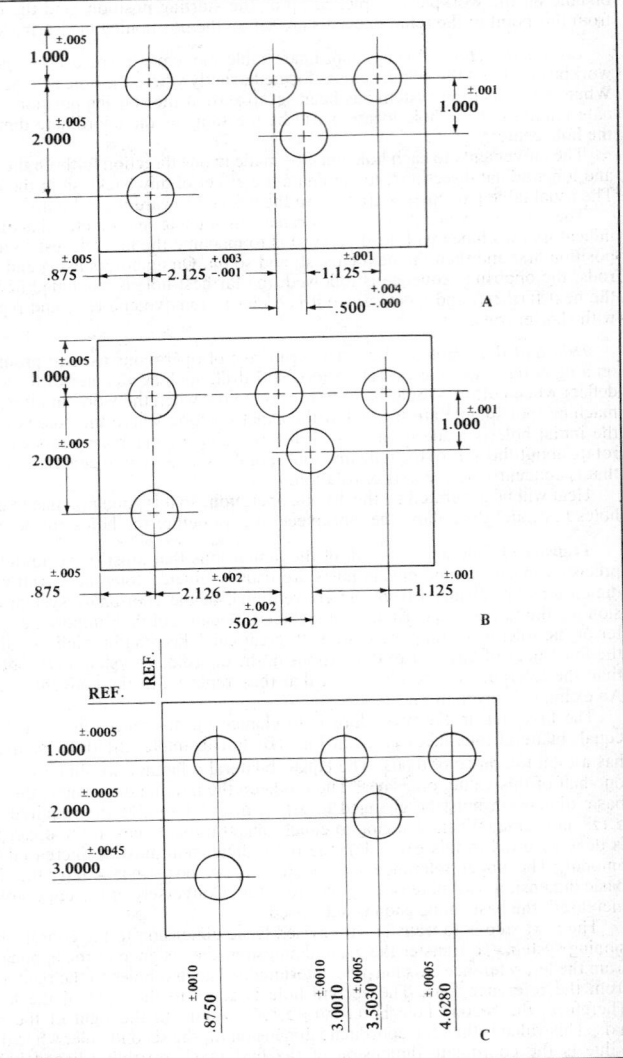

Fig. 1. (A) Conventional Dimensions, Mixed Tolerances; (B) Conventional Dimensions, All Equal, Bilateral Tolerances; and (C) Coordinate Dimensions

The final step is to transfer the tolerances. The 2.126 value in Fig. 1B has been replaced by the 0.8750 and 3.0010 values in Fig. 1C. The 2.126 value has an available tolerance of ±.002. Dividing this amount equally between the two replacement values gives 0.8750 ± .001 and 3.0010 ± .001. The sum of these tolerances is .002, and, as required, does not exceed the tolerance that was replaced. Next transfer the tolerance of the 0.502 dimension. Divide the available tolerance, ±.002, equally between the two replacement values to yield 3.0010 ± .001 and 3.5030 ± .001. The sum of these two tolerances equals the replaced tolerance, as required. However, the 1.125 value of the last hole to the right (coordinate dimension 4.6280 in.) has a tolerance of only ±.001. Therefore, the sum of the tolerances on the 3.5030 and 4.6280 values cannot be larger than .001. Dividing this tolerance equally would give 3.5030 ± .0005 and 4.6280 ± .0005. This new, smaller tolerance replaces the ±.001 tolerance on the 3.5030 value in order to satisfy all tolerance sum requirements. This example shows how the tolerance of a coordinate value is affected by more than one other dimensional requirement.

The following discussion will summarize the various tolerances listed in Fig. 1C. For the 0.8750 ± .0010 dimension, the ±.0010 tolerance together with the ±.0010 tolerance on the 3.0010 dimension is required to maintain the ±.002 tolerance of the 2.126 dimension. The ±.0005 tolerances on the 3.5030 and 4.2680 dimensions are required to maintain the ±.001 tolerance of the 1.125 dimension, at the same time as the sum of the ±.0005 tolerance on the 3.5030 dimension and the ±.001 tolerance on the 3.0010 dimension does not exceed the ±.002 tolerance on the replaced 0.503 dimension. The ±.0005 tolerances on the 1.0000 and 2.0000 values maintain the ±.001 tolerance on the 1.0000 value given at the right in Fig. 1A. The ±.0045 tolerance on the 3.0000 dimension together with the ±.0005 tolerance on the 1.0000 value maintains the ±.005 tolerance on the 2.0000 dimension of Fig. 1A. It should be noted that the 2.000 ± .005 dimension in Fig. 1A was replaced by the 1.0000 and 3.0000 dimensions in Fig. 1C. Each of these values could have had a tolerance of ±.0025, except that the tolerance on the 1.0000 dimension on the left in Fig. 1A is also bound by the ±.001 tolerance on the 1.0000 dimension on the right, thus the ±.0005 tolerance value is used. This procedure requires the tolerance on the 3.0000 value to be increased to ±.0045.

Hole Coordinate Dimension Factors for Jig Boring. — Tables of hole coordinate dimension factors for use in jig boring are given on pages 954 through 961. The coordinate axes shown in the figure accompanying each table are used to reference the tool path; the values listed in each table are for the end points of the tool path. In this machine coordinate system, a positive Y value indicates that the effective motion of the tool with reference to the work is toward the front of the jig borer (the actual motion of the jig borer table is toward the column). Similarly, a positive X value indicates that the effective motion of the tool with respect to the work is toward the right (the actual motion of the jig borer table is toward the left). When entering data into most computer-controlled jig borers, current practice is to use the more familiar Cartesian coordinate axis system in which the positive Y direction is "up" (i.e., pointing toward the column of the jig borer). The computer will automatically change the signs of the entered Y values to the signs that they would have in the machine coordinate system. Therefore, before applying the coordinate dimension factors given in the tables, it is important to determine the coordinate system to be used. If a Cartesian coordinate system is to be used for the tool path, then the sign of the Y values in the tables must be changed, from positive to negative and from negative to positive. For example, when programming for a three-hole type A circle using Cartesian coordinates, the Y values from Table 3 would be $y1 = +0.50000$, $y2 = -0.25000$, and $y3 = -0.25000$.

Table 1. Hole Coordinate Dimension Factors for Jig Boring — Type "A" Hole Circles
(English or metric units)

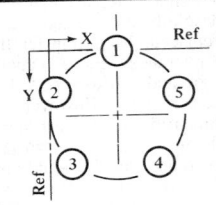

The diagram shows a type "A" circle for a 5-hole circle. Coordinates x, y are given in the table for hole circles of from 3 to 28 holes. Dimensions are for holes numbered in a counterclockwise direction (as shown). Dimensions given are based upon a hole circle of unit diameter. For a hole circle of, say, 3-inch or 3-centimeter diameter, multiply table values by 3.

3 Holes
x1 0.50000
y1 0.00000
x2 0.06699
y2 0.75000
x3 0.93301
y3 0.75000

4 Holes
x1 0.50000
y1 0.00000
x2 0.00000
y2 0.50000
x3 0.50000
y3 1.00000
x4 1.00000
y4 0.50000

5 Holes
x1 0.50000
y1 0.00000
x2 0.02447
y2 0.34549
x3 0.20611
y3 0.90451
x4 0.79389
y4 0.90451
x5 0.97553
y5 0.34549

6 Holes
x1 0.50000
y1 0.00000
x2 0.06699
y2 0.25000
x3 0.06699
y3 0.75000
x4 0.50000
y4 1.00000
x5 0.93301
y5 0.75000
x6 0.93301
y6 0.25000

7 Holes
x1 0.50000
y1 0.00000
x2 0.10908
y2 0.18826
x3 0.01254
y3 0.61126
x4 0.28306
y4 0.95048
x5 0.71694
y5 0.95048
x6 0.98746
y6 0.61126

x7 0.89092
y7 0.18826

8 Holes
x1 0.50000
y1 0.00000
x2 0.14645
y2 0.14645
x3 0.00000
y3 0.50000
x4 0.14645
y4 0.85355
x5 0.50000
y5 1.00000
x6 0.85355
y6 0.85355
x7 1.00000
y7 0.50000
x8 0.85355
y8 0.14645

9 Holes
x1 0.50000
y1 0.00000
x2 0.17861
y2 0.11698
x3 0.00760
y3 0.41318
x4 0.06699
y4 0.75000
x5 0.32899
y5 0.96985
x6 0.67101
y6 0.96985
x7 0.93301
y7 0.75000
x8 0.99240
y8 0.41318
x9 0.82139
y9 0.11698

10 Holes
x1 0.50000
y1 0.00000
x2 0.20611
y2 0.09549
x3 0.02447
y3 0.34549
x4 0.02447
y4 0.65451
x5 0.20611
y5 0.90451
x6 0.50000
y6 1.00000
x7 0.79389
y7 0.90451

x8 0.97553
y8 0.65451
x9 0.97553
y9 0.34549
x10 0.79389
y10 0.09549

11 Holes
x1 0.50000
y1 0.00000
x2 0.22968
y2 0.07937
x3 0.04518
y3 0.29229
x4 0.00509
y4 0.57116
x5 0.12213
y5 0.82743
x6 0.35913
y6 0.97975
x7 0.64087
y7 0.97975
x8 0.87787
y8 0.82743
x9 0.99491
y9 0.57116
x10 0.95482
y10 0.29229
x11 0.77032
y11 0.07937

12 Holes
x1 0.50000
y1 0.00000
x2 0.25000
y2 0.06699
x3 0.06699
y3 0.25000
x4 0.00000
y4 0.50000
x5 0.06699
y5 0.75000
x6 0.25000
y6 0.93301
x7 0.50000
y7 1.00000
x8 0.75000
y8 0.93301
x9 0.93301
y9 0.75000
x10 1.00000
y10 0.50000
x11 0.93301
y11 0.25000
x12 0.75000
y12 0.06699

13 Holes
x1 0.50000
y1 0.00000
x2 0.26764
y2 0.05727
x3 0.08851
y3 0.21597
x4 0.00365
y4 0.43973
x5 0.03249
y5 0.67730
x6 0.16844
y6 0.87426
x7 0.38034
y7 0.98547
x8 0.61966
y8 0.98547
x9 0.83156
y9 0.87426
x10 0.96751
y10 0.67730
x11 0.99635
y11 0.43973
x12 0.91149
y12 0.21597
x13 0.73236
y13 0.05727

14 Holes
x1 0.50000
y1 0.00000
x2 0.28306
y2 0.04952
x3 0.10908
y3 0.18826
x4 0.01254
y4 0.38874
x5 0.01254
y5 0.61126
x6 0.10908
y6 0.81174
x7 0.28306
y7 0.95048
x8 0.50000
y8 1.00000
x9 0.71694
y9 0.95048
x10 0.89092
y10 0.81174
x11 0.98746
y11 0.61126

x12 0.98746
y12 0.38874
x13 0.89092
y13 0.18826
x14 0.71694
y14 0.04952

15 Holes
x1 0.50000
y1 0.00000
x2 0.29663
y2 0.04323
x3 0.12843
y3 0.16543
x4 0.02447
y4 0.34549
x5 0.00274
y5 0.55226
x6 0.06699
y6 0.75000
x7 0.20611
y7 0.90451
x8 0.39604
y8 0.98907
x9 0.60396
y9 0.98907
x10 0.79389
y10 0.90451
x11 0.93301
y11 0.75000
x12 0.99726
y12 0.55226
x13 0.97553
y13 0.34549
x14 0.87157
y14 0.16543
x15 0.70337
y15 0.04323

16 Holes
x1 0.50000
y1 0.00000
x2 0.30866
y2 0.03806
x3 0.14645
y3 0.14645
x4 0.03806
y4 0.30866
x5 0.00000
y5 0.50000
x6 0.03806
y6 0.69134
x7 0.14645
y7 0.85355

x8 0.30866
y8 0.96194
x9 0.50000
y9 1.00000
x10 0.69134
y10 0.96194
x11 0.85355
y11 0.85355
x12 0.96194
y12 0.69134
x13 1.00000
y13 0.50000
x14 0.96194
y14 0.30866
x15 0.85355
y15 0.14645
x16 0.69134
y16 0.03806

17 Holes
x1 0.50000
y1 0.00000
x2 0.31938
y2 0.03376
x3 0.16315
y3 0.13050
x4 0.05242
y4 0.27713
x5 0.00021
y5 0.45387
x6 0.01909
y6 0.63683
x7 0.10099
y7 0.80132
x8 0.23678
y8 0.92551
x9 0.40813
y9 0.99149
x10 0.59187
y10 0.99149
x11 0.76322
y11 0.92551
x12 0.89091
y12 0.80132
x13 0.98091
y13 0.63683
x14 0.99979
y14 0.45387
x15 0.94758
y15 0.27713
x16 0.83685
y16 0.13050
x17 0.68062
y17 0.03376

18 Holes
x1 0.50000
y1 0.00000
x2 0.32899
y2 0.03015
x3 0.17861
y3 0.11698
x4 0.06699
y4 0.25000
x5 0.00760
y5 0.41318
x6 0.00760
y6 0.58682
x7 0.06699
y7 0.75000
x8 0.17861
y8 0.88302
x9 0.32899
y9 0.96985
x10 0.50000
y10 1.00000
x11 0.67101
y11 0.96985
x12 0.82139
y12 0.88302
x13 0.93301
y13 0.75000
x14 0.99240
y14 0.58682
x15 0.99240
y15 0.41318
x16 0.93301
y16 0.25000
x17 0.82139
y17 0.11698
x18 0.67101
y18 0.03015

19 Holes
x1 0.50000
y1 0.00000
x2 0.33765
y2 0.02709
x3 0.19289
y3 0.10543
x4 0.08142
y4 0.22653
x5 0.01530
y5 0.37726
x6 0.00171
y6 0.54129
x7 0.04211
y7 0.70085
... .

Table 1 (*Concluded*). **Hole Coordinate Dimension Factors for Jig Boring — Type "A" Hole Circles** (English or metric units)

x_8 0.13214	**21 Holes**	x_{12} 0.50000	x_{23} 0.63490	x_8 0.00886	x_{16} 0.73236	x_{23} 0.95911
y_8 0.83864	x_1 0.50000	y_{12} 1.00000	y_{23} 0.01854	y_8 0.59369	y_{16} 0.94273	y_{23} 0.30196
x_9 0.26203	y_1 0.00000	x_{13} 0.64087	**24 Holes**	x_9 0.04759	x_{17} 0.83156	x_{24} 0.90106
y_9 0.93974	x_2 0.35262	y_{13} 0.97975	x_1 0.50000	y_9 0.71289	y_{17} 0.87426	y_{24} 0.20142
x_{10} 0.41770	y_2 0.02221	x_{14} 0.77032	y_1 0.00000	x_{10} 0.11474	x_{18} 0.91149	x_{25} 0.82139
y_{10} 0.99318	x_3 0.21834	y_{14} 0.92063	x_2 0.37059	y_{10} 0.81871	y_{18} 0.78403	y_{25} 0.11698
x_{11} 0.58230	y_3 0.08688	x_{15} 0.87787	y_2 0.01704	x_{11} 0.20611	x_{19} 0.96751	x_{26} 0.72440
y_{11} 0.99318	x_4 0.10908	y_{15} 0.82743	x_3 0.25000	y_{11} 0.90451	y_{19} 0.67730	y_{26} 0.05318
x_{12} 0.73797	y_4 0.18826	x_{16} 0.95482	y_3 0.06699	x_{12} 0.31594	x_{20} 0.99635	x_{27} 0.61531
y_{12} 0.93974	x_5 0.03456	y_{16} 0.70771	x_4 0.14645	y_{12} 0.96489	y_{20} 0.56027	y_{27} 0.01348
x_{13} 0.86786	y_5 0.31733	x_{17} 0.99491	y_4 0.14645	x_{13} 0.43733	x_{21} 0.99635	**28 Holes**
y_{13} 0.83864	x_6 0.00140	y_{17} 0.57116	x_5 0.06699	y_{13} 0.99606	y_{21} 0.43973	x_1 0.50000
x_{14} 0.95789	y_6 0.46263	x_{18} 0.99491	y_5 0.25000	x_{14} 0.56267	x_{22} 0.96751	y_1 0.00000
y_{14} 0.70085	x_7 0.01254	y_{18} 0.42884	x_6 0.01704	y_{14} 0.99606	y_{22} 0.32270	x_2 0.38874
x_{15} 0.99829	y_7 0.61126	x_{19} 0.95482	y_6 0.37059	x_{15} 0.68406	x_{23} 0.91149	y_2 0.01254
y_{15} 0.54129	x_8 0.06699	y_{19} 0.29229	x_7 0.00000	y_{15} 0.96489	y_{23} 0.21597	x_3 0.28306
x_{16} 0.98470	y_8 0.75000	x_{20} 0.87787	y_7 0.50000	x_{16} 0.79389	x_{24} 0.83156	y_3 0.04952
y_{16} 0.37726	x_9 0.15991	y_{20} 0.17257	x_8 0.01704	y_{16} 0.90451	y_{24} 0.12574	x_4 0.18826
x_{17} 0.91858	y_9 0.86653	x_{21} 0.77032	y_8 0.62941	x_{17} 0.88526	x_{25} 0.73236	y_4 0.10908
y_{17} 0.22658	x_{10} 0.28306	y_{21} 0.07937	x_9 0.06699	y_{17} 0.81871	y_{25} 0.05727	x_5 0.10908
x_{18} 0.80711	y_{10} 0.95048	x_{22} 0.64087	y_9 0.75000	x_{18} 0.95241	x_{26} 0.61966	y_5 0.18826
y_{18} 0.10543	x_{11} 0.42548	y_{22} 0.02025	x_{10} 0.14645	y_{18} 0.71289	y_{26} 0.01453	x_6 0.04952
x_{19} 0.66235	y_{11} 0.99442	**23 Holes**	y_{10} 0.85355	x_{19} 0.99114	**27 Holes**	y_6 0.28306
y_{19} 0.02709	x_{12} 0.57452	x_1 0.50000	x_{11} 0.25000	y_{19} 0.59369	x_1 0.50000	x_7 0.01254
20 Holes	y_{12} 0.99442	y_1 0.00000	y_{11} 0.93301	x_{20} 0.99901	y_1 0.00000	y_7 0.38874
x_1 0.50000	x_{13} 0.71694	x_2 0.36510	x_{12} 0.37059	y_{20} 0.46860	x_2 0.38469	x_8 0.00000
y_1 0.00000	y_{13} 0.95048	y_2 0.01854	y_{12} 0.98296	x_{21} 0.97553	y_2 0.01348	y_8 0.50000
x_2 0.34549	x_{14} 0.84009	x_3 0.24021	x_{13} 0.50000	y_{21} 0.34549	x_3 0.27560	x_9 0.01254
y_2 0.02447	y_{14} 0.86653	y_3 0.07279	y_{13} 1.00000	x_{22} 0.92216	y_3 0.05318	y_9 0.61126
x_3 0.20611	x_{15} 0.93301	x_4 0.13458	x_{14} 0.62941	y_{22} 0.23209	x_4 0.17861	x_{10} 0.04952
y_3 0.09549	y_{15} 0.75000	y_4 0.15872	y_{14} 0.98296	x_{23} 0.84227	y_4 0.11698	y_{10} 0.71694
x_4 0.09549	x_{16} 0.98746	x_5 0.05606	x_{15} 0.75000	y_{23} 0.13552	x_5 0.09894	x_{11} 0.10908
y_4 0.20611	y_{16} 0.61126	y_5 0.26997	y_{15} 0.93301	x_{24} 0.74088	y_5 0.20142	y_{11} 0.81174
x_5 0.02447	x_{17} 0.99860	x_6 0.01046	x_{16} 0.85355	y_{24} 0.06185	x_6 0.04089	x_{12} 0.18826
y_5 0.34549	y_{17} 0.46263	y_6 0.39827	y_{16} 0.85355	x_{25} 0.62434	y_6 0.30196	y_{12} 0.89092
x_6 0.00000	x_{18} 0.96544	x_7 0.00117	x_{17} 0.93301	y_{25} 0.01571	x_7 0.00760	x_{13} 0.28306
y_6 0.50000	y_{18} 0.31733	y_7 0.53412	y_{17} 0.75000	**26 Holes**	y_7 0.41318	y_{13} 0.95048
x_7 0.02447	x_{19} 0.89092	x_8 0.02887	x_{18} 0.98296	x_1 0.50000	x_8 0.00085	x_{14} 0.38874
y_7 0.65451	y_{19} 0.18826	y_8 0.66744	y_{18} 0.62941	y_1 0.00000	y_8 0.52907	y_{14} 0.98746
x_8 0.09549	x_{20} 0.78166	x_9 0.09152	x_{19} 1.00000	x_2 0.38034	x_9 0.02101	x_{15} 0.50000
y_8 0.79389	y_{20} 0.08688	y_9 0.78834	y_{19} 0.50000	y_2 0.01453	y_9 0.64340	y_{15} 1.00000
x_9 0.20611	x_{21} 0.64738	x_{10} 0.18446	x_{20} 0.98296	x_3 0.26764	x_{10} 0.06699	x_{16} 0.61126
y_9 0.90451	y_{21} 0.02221	y_{10} 0.88786	y_{20} 0.37059	y_3 0.05727	y_{10} 0.75000	y_{16} 0.98746
x_{10} 0.34549	**22 Holes**	x_{11} 0.30080	x_{21} 0.93301	x_4 0.16844	x_{11} 0.13631	x_{17} 0.71694
y_{10} 0.97553	x_1 0.50000	y_{11} 0.95861	y_{21} 0.25000	y_4 0.12574	y_{11} 0.84312	y_{17} 0.95048
x_{11} 0.50000	y_1 0.00000	x_{12} 0.43192	x_{22} 0.85355	x_5 0.08851	x_{12} 0.22525	x_{18} 0.81174
y_{11} 1.00000	x_2 0.35913	y_{12} 0.99534	y_{22} 0.14645	y_5 0.21597	y_{12} 0.91774	y_{18} 0.89092
x_{12} 0.65451	y_2 0.02025	x_{13} 0.56808	x_{23} 0.75000	x_6 0.03249	x_{13} 0.32899	x_{19} 0.89092
y_{12} 0.97553	x_3 0.22968	y_{13} 0.99534	y_{23} 0.06699	y_6 0.32270	y_{13} 0.96985	y_{19} 0.81174
x_{13} 0.79389	y_3 0.07937	x_{14} 0.69920	x_{24} 0.62941	x_7 0.00365	x_{14} 0.44195	x_{20} 0.95048
y_{13} 0.90451	x_4 0.12213	y_{14} 0.95861	y_{24} 0.01704	y_7 0.43973	y_{14} 0.99662	y_{20} 0.71694
x_{14} 0.90451	y_4 0.17257	x_{15} 0.81554	**25 Holes**	x_8 0.00365	x_{15} 0.55805	x_{21} 0.98746
y_{14} 0.79389	x_5 0.04518	y_{15} 0.88786	x_1 0.50000	y_8 0.56027	y_{15} 0.99662	y_{21} 0.61126
x_{15} 0.97553	y_5 0.29229	x_{16} 0.90848	y_1 0.00000	x_9 0.03249	x_{16} 0.67101	x_{22} 1.00000
y_{15} 0.65451	x_6 0.00509	y_{16} 0.78834	x_2 0.37566	y_9 0.67730	y_{16} 0.96985	y_{22} 0.50000
x_{16} 1.00000	y_6 0.42884	x_{17} 0.97113	y_2 0.01571	x_{10} 0.08851	x_{17} 0.77475	x_{23} 0.98746
y_{16} 0.50000	x_7 0.00509	y_{17} 0.66744	x_3 0.25912	y_{10} 0.78403	y_{17} 0.91774	y_{23} 0.38874
x_{17} 0.97553	y_7 0.57116	x_{18} 0.99883	y_3 0.06185	x_{11} 0.16844	x_{18} 0.86369	x_{24} 0.95048
y_{17} 0.34549	x_8 0.04518	y_{18} 0.53412	x_4 0.15773	y_{11} 0.87426	y_{18} 0.84312	y_{24} 0.28306
x_{18} 0.90451	y_8 0.70771	x_{19} 0.98954	y_4 0.13552	x_{12} 0.26764	x_{19} 0.93301	x_{25} 0.89092
y_{18} 0.20611	x_9 0.12213	y_{19} 0.39827	x_5 0.07784	y_{12} 0.94273	y_{19} 0.75000	y_{25} 0.18826
x_{19} 0.79389	y_9 0.82743	x_{20} 0.94394	y_5 0.23209	x_{13} 0.38034	x_{20} 0.97899	x_{26} 0.81174
y_{19} 0.09549	x_{10} 0.22968	y_{20} 0.26997	x_6 0.02447	y_{13} 0.98547	y_{20} 0.64340	y_{26} 0.10908
x_{20} 0.65451	y_{10} 0.92063	x_{21} 0.86542	y_6 0.34549	x_{14} 0.50000	x_{21} 0.99915	x_{27} 0.71694
y_{20} 0.02447	x_{11} 0.35913	y_{21} 0.15872	x_7 0.00099	y_{14} 1.00000	y_{21} 0.52907	y_{27} 0.04952
.. ....	y_{11} 0.97975	x_{22} 0.75979	y_7 0.46860	x_{15} 0.61966	x_{22} 0.99240	x_{28} 0.61126
.. ....	.. ....	y_{22} 0.07279	.. ....	y_{15} 0.98547	y_{22} 0.41318	y_{28} 0.01254

Table 2. Hole Coordinate Dimension Factors for Jig Boring — Type "B" Hole Circles
(English or metric units)

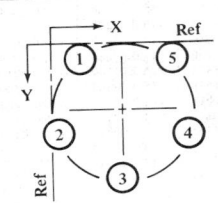

The diagram shows a type "B" circle for a 5-hole circle. Coordinates x, y are given in the table for hole circles of from 3 to 28 holes. Dimensions are for holes numbered in a counterclockwise direction (as shown). Dimensions given are based upon a hole circle of unit diameter. For a hole circle of, say, 3-inch or 3-centimeter diameter, multiply table values by 3.

3 Holes
x1 0.06699, y1 0.25000
x2 0.50000, y2 1.00000
x3 0.93301, y3 0.25000

4 Holes
x1 0.14645, y1 0.14645
x2 0.85355, y2 0.85355
x3 0.85355, y3 0.85355
x4 0.14645

5 Holes
x1 0.20611, y1 0.09549
x2 0.02447, y2 0.65451
x3 0.50000, y3 1.00000
x4 0.97553, y4 0.65451
x5 0.79389, y5 0.09549

6 Holes
x1 0.25000, y1 0.06699
x2 0.00000, y2 0.50000
x3 0.25000, y3 0.93301
x4 0.75000, y4 0.93301
x5 1.00000, y5 0.50000
x6 0.75000, y6 0.06699

7 Holes
x1 0.28306, y1 0.04952
x2 0.01254, y2 0.38874
x3 0.10908, y3 0.81174
x4 0.50000, y4 1.00000
x5 0.89092, y5 0.81174
x6 0.98746, y6 0.38874
x7 0.71694, y7 0.04952

8 Holes
x1 0.30866, y1 0.03806
x2 0.03806, y2 0.30866
x3 0.03806, y3 0.69134
x4 0.30866, y4 0.96194
x5 0.69134, y5 0.96194
x6 0.96194, y6 0.69134
x7 0.96194, y7 0.30866
x8 0.69134, y8 0.03806

9 Holes
x1 0.32899, y1 0.03015
x2 0.06699, y2 0.25000
x3 0.00760, y3 0.58682
x4 0.17861, y4 0.88302
x5 0.50000, y5 1.00000
x6 0.82139, y6 0.88302
x7 0.99240, y7 0.58682
x8 0.93301, y8 0.25000
x9 0.67101, y9 0.03015

10 Holes
x1 0.34549, y1 0.02447
x2 0.09549, y2 0.20611
x3 0.00000, y3 0.50000
x4 0.09549, y4 0.79389
x5 0.34549, y5 0.97553
x6 0.65451, y6 0.97553
x7 0.90451, y7 0.79389
x8 1.00000, y8 0.50000
x9 0.90451, y9 0.20611
x10 0.65451, y10 0.02447

11 Holes
x1 0.35913, y1 0.02025
x2 0.12213, y2 0.17257
x3 0.00509, y3 0.42884
x4 0.04518, y4 0.70771
x5 0.22968, y5 0.92063
x6 0.50000, y6 1.00000
x7 0.77032, y7 0.92063
x8 0.95482, y8 0.70771
x9 0.99491, y9 0.42884
x10 0.87787, y10 0.17257
x11 0.64087, y11 0.02025

12 Holes
x1 0.37059, y1 0.01704
x2 0.14645, y2 0.14645
x3 0.01704, y3 0.37059
x4 0.01704, y4 0.62941
x5 0.14645, y5 0.85355
x6 0.37059, y6 0.98296
x7 0.62941, y7 0.98296
x8 0.85355, y8 0.85355
x9 0.98296, y9 0.62941
x10 0.98296, y10 0.37059
x11 0.85355, y11 0.14645
x12 0.62941, y12 0.01704

13 Holes
x1 0.38034, y1 0.01453
x2 0.16844, y2 0.12574
x3 0.03249, y3 0.32270
x4 0.00365, y4 0.56027
x5 0.08851, y5 0.78403
x6 0.26764, y6 0.94273
x7 0.50000, y7 1.00000
x8 0.73236, y8 0.94273
x9 0.91149, y9 0.78403
x10 0.99635, y10 0.56027
x11 0.96751, y11 0.32270
x12 0.83156, y12 0.12574
x13 0.61966, y13 0.01453

14 Holes
x1 0.38874, y1 0.01254
x2 0.18826, y2 0.10908
x3 0.04952, y3 0.28306
x4 0.00000, y4 0.50000
x5 0.04952, y5 0.71694
x6 0.18826, y6 0.89092
x7 0.38874, y7 0.98746
x8 0.61126, y8 0.98746
x9 0.81174, y9 0.89092
x10 0.95048, y10 0.71694
x11 1.00000, y11 0.50000
x12 0.95048, y12 0.28306
x13 0.81174, y13 0.10908
x14 0.61126, y14 0.01254

15 Holes
x1 0.39604, y1 0.01093
x2 0.20611, y2 0.09549
x3 0.06699, y3 0.25000
x4 0.00274, y4 0.44774
x5 0.02447, y5 0.65451
x6 0.12843, y6 0.83457
x7 0.29663, y7 0.95067
x8 0.50000, y8 1.00000
x9 0.70337, y9 0.95067
x10 0.87157, y10 0.83457
x11 0.97553, y11 0.65451
x12 0.99726, y12 0.44774
x13 0.93301, y13 0.25000
x14 0.79389, y14 0.09549
x15 0.60396, y15 0.01093

16 Holes
x1 0.40245, y1 0.00961
x2 0.22221, y2 0.08427
x3 0.08427, y3 0.22221
x4 0.00961, y4 0.40245
x5 0.00961, y5 0.59755
x6 0.08427, y6 0.77779
x7 0.22221, y7 0.91573
x8 0.40245, y8 0.99039
x9 0.59755, y9 0.99039
x10 0.77779, y10 0.91573
x11 0.91573, y11 0.77779
x12 0.99039, y12 0.59755
x13 0.99039, y13 0.40245
x14 0.91573, y14 0.22221
x15 0.77779, y15 0.08427
x16 0.59755, y16 0.00961

17 Holes
x1 0.40813, y1 0.00851
x2 0.23678, y2 0.07489
x3 0.10099, y3 0.19868
x4 0.01909, y4 0.36317
x5 0.00213, y5 0.54613
x6 0.05242, y6 0.72287
x7 0.16315, y7 0.86950
x8 0.31938, y8 0.96624
x9 0.50000, y9 1.00000
x10 0.68062, y10 0.96624
x11 0.83685, y11 0.86950
x12 0.94758, y12 0.72287
x13 0.99787, y13 0.54613
x14 0.98091, y14 0.36317
x15 0.89901, y15 0.19868
x16 0.76322, y16 0.07489
x17 0.59187, y17 0.00851

18 Holes
x1 0.41318, y1 0.00760
x2 0.25000, y2 0.06699
x3 0.11698, y3 0.17861
x4 0.03015, y4 0.32899
x5 0.00000, y5 0.50000
x6 0.03015, y6 0.67101
x7 0.11698, y7 0.82139
x8 0.25000, y8 0.93301
x9 0.41318, y9 0.99240
x10 0.58682, y10 0.99240
x11 0.75000, y11 0.93301
x12 0.88302, y12 0.82139
x13 0.96985, y13 0.67101
x14 1.00000, y14 0.50000
x15 0.96985, y15 0.32899
x16 0.88302, y16 0.17861
x17 0.75000, y17 0.06699
x18 0.58682, y18 0.00760

19 Holes
x1 0.41779, y1 0.00682
x2 0.26203, y2 0.06026
x3 0.13214, y3 0.16136
x4 0.04211, y4 0.29915
x5 0.00171, y5 0.45871
x6 0.01530, y6 0.62274
x7 0.08142, y7 0.77347
... .

Table 2 (*Concluded*). **Hole Coordinate Dimension Factors for Jig Boring — Type "B"
Hole Circles** (English or metric units)

(continuation, 19 Holes, points 8–19)

n	x	y
8	0.19289	0.89457
9	0.33765	0.97291
10	0.50000	1.00000
11	0.66235	0.97291
12	0.80711	0.89457
13	0.91858	0.77347
14	0.98470	0.62274
15	0.99829	0.45891
16	0.95789	0.29915
17	0.86786	0.16136
18	0.73797	0.06026
19	0.58230	0.00682

20 Holes

n	x	y
1	0.42178	0.00616
2	0.27300	0.05450
3	0.14645	0.14645
4	0.05450	0.27300
5	0.00616	0.42178
6	0.00616	0.57822
7	0.05450	0.72700
8	0.14645	0.85355
9	0.27300	0.94550
10	0.42178	0.99384
11	0.57822	0.99384
12	0.72700	0.94550
13	0.85355	0.85355
14	0.94550	0.72700
15	0.99384	0.57822
16	0.99384	0.42178
17	0.94550	0.27300
18	0.85355	0.14645
19	0.72700	0.05450
20	0.57822	0.00616

21 Holes

n	x	y
1	0.42548	0.00558
2	0.28306	0.04952
3	0.15991	0.13347
4	0.06699	0.25000
5	0.01254	0.38874
6	0.00104	0.53737
7	0.03456	0.68267
8	0.10908	0.81174
9	0.21834	0.91312
10	0.35262	0.97779
11	0.50000	1.00000
12	0.64738	0.97779
13	0.78166	0.91312
14	0.89092	0.81174
15	0.96544	0.68267
16	0.99896	0.53737
17	0.98746	0.38874
18	0.93301	0.25000
19	0.84009	0.13347
20	0.71694	0.04952
21	0.57452	0.00558

22 Holes

n	x	y
1	0.42884	0.00509
2	0.29229	0.04518
3	0.17257	0.12213
4	0.07937	0.22968
5	0.02025	0.35913
6	0.00000	0.50000
7	0.02025	0.64087
8	0.07937	0.77032
9	0.17257	0.87787
10	0.29229	0.95482
11	0.42884	0.99491
12	0.57116	0.99491
13	0.70771	0.95482
14	0.82743	0.87787
15	0.92063	0.77032
16	0.97975	0.64087
17	1.00000	0.50000
18	0.97975	0.35913
19	0.92063	0.22968
20	0.82743	0.12213
21	0.70771	0.04518
22	0.57116	0.00509

23 Holes

n	x	y
1	0.43192	0.00466
2	0.30080	0.04139
3	0.18446	0.11214
4	0.09152	0.21166
5	0.02887	0.33256
6	0.00117	0.46588
7	0.01046	0.60173
8	0.05606	0.73003
9	0.13458	0.84128
10	0.24021	0.92721
11	0.36510	0.98146
12	0.50000	1.00000
13	0.63490	0.98146
14	0.75979	0.92721
15	0.86542	0.84128
16	0.94394	0.73003
17	0.98954	0.60173
18	0.99883	0.46588
19	0.97113	0.33256
20	0.90848	0.21166
21	0.81554	0.11214
22	0.69920	0.04139
23	0.56808	0.00466

24 Holes

n	x	y
1	0.43474	0.00428
2	0.30866	0.03806
3	0.19562	0.10332
4	0.10332	0.19562
5	0.03806	0.30866
6	0.00428	0.43474
7	0.00428	0.56526
8	0.03806	0.69134
9	0.10332	0.80438
10	0.19562	0.89668
11	0.30866	0.96194
12	0.43474	0.99572
13	0.56526	0.99572
14	0.69134	0.96194
15	0.80438	0.89668
16	0.89668	0.80438
17	0.96194	0.69134
18	0.99572	0.56526
19	0.99572	0.43474
20	0.96194	0.30866
21	0.89668	0.19562
22	0.80438	0.10332
23	0.69134	0.03806
24	0.56526	0.00428

25 Holes

n	x	y
1	0.43733	0.00394
2	0.31594	0.03511
3	0.20611	0.09549
4	0.11474	0.18129
5	0.04759	0.28711
6	0.00886	0.40631
7	0.00099	0.53140
8	0.02447	0.65451
9	0.07784	0.76791
10	0.15773	0.86448
11	0.25912	0.93815
12	0.37566	0.98429
13	0.50000	1.00000
14	0.62434	0.98429
15	0.74088	0.93815
16	0.84227	0.86448
17	0.92216	0.76791
18	0.97553	0.65451
19	0.99901	0.53140
20	0.99114	0.40631
21	0.95241	0.28711
22	0.88526	0.18129
23	0.79389	0.09549
24	0.68406	0.03511
25	0.56267	0.00394

26 Holes

n	x	y
1	0.43973	0.00365
2	0.32270	0.03249
3	0.21597	0.08851
4	0.12574	0.16844
5	0.05727	0.26764
6	0.01453	0.38034
7	0.00000	0.50000
8	0.01453	0.61966
9	0.05727	0.73236
10	0.12574	0.83156
11	0.21597	0.91149
12	0.32270	0.96751
13	0.43973	0.99635
14	0.56027	1.00000
15	0.67730	0.99635
16	0.78403	0.91149
17	0.87426	0.83156
18	0.94273	0.73236
19	0.98547	0.61966
20	1.00000	0.50000
21	0.98547	0.38034
22	0.94273	0.26764
23	0.87426	0.16844
24	0.78403	0.08851
25	0.67730	0.03249
26	0.56027	0.00365

27 Holes

n	x	y
1	0.44195	0.00338
2	0.32899	0.03015
3	0.22525	0.08226
4	0.13631	0.15688
5	0.06699	0.25000
6	0.02101	0.35660
7	0.00085	0.47093
8	0.00760	0.58682
9	0.04089	0.69804
10	0.09894	0.79858
11	0.17861	0.88302
12	0.27560	0.94682
13	0.38469	0.98652
14	0.50000	1.00000
15	0.61531	0.98652
16	0.72440	0.94682
17	0.82139	0.88302
18	0.90106	0.79858
19	0.95911	0.69804
20	0.99240	0.58682
21	0.99915	0.47093
22	0.97899	0.35660
23	0.93301	0.25000
24	0.86369	0.15688
25	0.77475	0.08226
26	0.67101	0.03015
27	0.55805	0.00338

28 Holes

n	x	y
1	0.44402	0.00314
2	0.33486	0.02806
3	0.23398	0.07664
4	0.14645	0.14645
5	0.07664	0.23398
6	0.02806	0.33486
7	0.00314	0.44402
8	0.00314	0.55598
9	0.02806	0.66514
10	0.07664	0.76602
11	0.14645	0.85355
12	0.23398	0.92336
13	0.33486	0.97194
14	0.44402	0.99686
15	0.55598	0.99686
16	0.66514	0.97194
17	0.76602	0.92336
18	0.85355	0.85355
19	0.92336	0.76602
20	0.97194	0.66514
21	0.99686	0.55598
22	0.99686	0.44402
23	0.97194	0.33486
24	0.92336	0.23398
25	0.85355	0.14645
26	0.76602	0.07664
27	0.66514	0.02806
28	0.55598	0.00314

Table 3. Hole Coordinate Dimension Factors for Jig Boring — Type "A" Hole Circles, Central Coordinates (English or metric units)

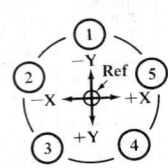

The diagram shows a type "A" circle for a 5-hole circle. Coordinates x, y are given in the table for hole circles of from 3 to 28 holes. Dimensions are for holes numbered in a counterclockwise direction (as shown). Dimensions given are based upon a hole circle of unit diameter. For a hole circle of, say, 3-inch or 3-centimeter diameter, multiply table values by 3.

3 Holes

Hole	x	y
1	0.00000	−0.50000
2	−0.43301	+0.25000
3	+0.43301	+0.25000

4 Holes

Hole	x	y
1	0.00000	−0.50000
2	−0.50000	0.00000
3	0.00000	+0.50000
4	+0.50000	0.00000

5 Holes

Hole	x	y
1	0.00000	−0.50000
2	−0.47553	−0.15451
3	−0.29389	+0.40451
4	+0.29389	+0.40451
5	+0.47553	−0.15451

6 Holes

Hole	x	y
1	0.00000	−0.50000
2	−0.43301	−0.25000
3	−0.43301	+0.25000
4	0.00000	+0.50000
5	+0.43301	+0.25000
6	+0.43301	−0.25000

7 Holes

Hole	x	y
1	0.00000	−0.50000
2	−0.39092	−0.31174
3	−0.48746	+0.11126
4	−0.21694	+0.45048
5	+0.21694	+0.45048
6	+0.48746	+0.11126
7	+0.39092	−0.31174

8 Holes

Hole	x	y
1	0.00000	−0.50000
2	−0.35355	−0.35355
3	−0.50000	0.00000
4	−0.35355	+0.35355
5	0.00000	+0.50000
6	+0.35355	+0.35355
7	+0.50000	0.00000
8	+0.35355	−0.35355

9 Holes

Hole	x	y
1	0.00000	−0.50000
2	−0.32139	−0.38302
3	−0.49240	−0.08682
4	−0.43301	+0.25000
5	−0.17101	+0.46985
6	+0.17101	+0.46985
7	+0.43301	+0.25000
8	+0.49240	−0.08682
9	+0.32139	−0.38302

10 Holes

Hole	x	y
1	0.00000	−0.50000
2	−0.29389	−0.40451
3	−0.47553	−0.15451
4	−0.47553	+0.15451
5	−0.29389	+0.40451
6	0.00000	+0.50000
7	+0.29389	+0.40451
8	+0.47553	+0.15451
9	+0.47553	−0.15451
10	+0.29389	−0.40451

11 Holes

Hole	x	y
1	0.00000	−0.50000
2	−0.27032	−0.42063
3	−0.45482	−0.20771
4	−0.49491	+0.07116
5	−0.37787	+0.32743
6	−0.14087	+0.47975
7	+0.14087	+0.47975
8	+0.37787	+0.32743
9	+0.49491	+0.07116
10	+0.45482	−0.20771
11	+0.27032	−0.42063

12 Holes

Hole	x	y
1	0.00000	−0.50000
2	−0.25000	−0.43301
3	−0.43301	−0.25000
4	−0.50000	0.00000
5	−0.43301	+0.25000
6	−0.25000	+0.43301
7	0.00000	+0.50000
8	+0.25000	+0.43301
9	+0.43301	+0.25000
10	+0.50000	0.00000
11	+0.43301	−0.25000
12	+0.25000	−0.43301

13 Holes

Hole	x	y
1	0.00000	−0.50000
2	−0.23236	−0.44273
3	−0.41149	−0.28403
4	−0.49635	−0.06027
5	−0.46751	+0.17730
6	−0.33156	+0.37426
7	−0.11966	+0.48547
8	+0.11966	+0.48547
9	+0.33156	+0.37426
10	+0.46751	+0.17730
11	+0.49635	−0.06027
12	+0.41149	−0.28403
13	+0.23236	−0.44273

14 Holes

Hole	x	y
1	0.00000	−0.50000
2	−0.21694	−0.45048
3	−0.39092	−0.31174
4	−0.48746	−0.11126
5	−0.48746	+0.11126
6	−0.39092	+0.31174
7	−0.21694	+0.45048
8	0.00000	+0.50000
9	+0.21694	+0.45048
10	+0.39092	+0.31174
11	+0.48746	+0.11126
12	+0.48746	−0.11126
13	+0.39092	−0.31174
14	+0.21694	−0.45048

15 Holes

Hole	x	y
1	0.00000	−0.50000
2	−0.20337	−0.45677
3	−0.37157	−0.33457
4	−0.47553	−0.15451
5	−0.49726	+0.05226
6	−0.43301	+0.25000
7	−0.29389	+0.40451
8	−0.10396	+0.48907
9	+0.10396	+0.48907
10	+0.29389	+0.40451
11	+0.43301	+0.25000
12	+0.49726	+0.05226
13	+0.47553	−0.15451
14	+0.37157	−0.33457
15	+0.20337	−0.45677

16 Holes

Hole	x	y
1	0.00000	−0.50000
2	−0.19134	−0.46194
3	−0.35355	−0.35355
4	−0.46194	−0.19134
5	−0.50000	0.00000
6	−0.46194	+0.19134
7	−0.35355	+0.35355
8	−0.19134	+0.46194
9	0.00000	+0.50000
10	+0.19134	+0.46194
11	+0.35355	+0.35355
12	+0.46194	+0.19134
13	+0.50000	0.00000
14	+0.46194	−0.19134
15	+0.35355	−0.35355
16	+0.19134	−0.46194

17 Holes

Hole	x	y
1	0.00000	−0.50000
2	−0.18062	−0.46624
3	−0.33685	−0.36950
4	−0.44758	−0.22287
5	−0.49787	−0.04613
6	−0.48091	+0.13683
7	−0.39901	+0.30132
8	−0.26322	+0.42511
9	−0.09187	+0.49149
10	+0.09187	+0.49149
11	+0.26322	+0.42511
12	+0.39901	+0.30132
13	+0.48091	+0.13683
14	+0.49787	−0.04613
15	+0.44758	−0.22287
16	+0.33685	−0.36950
17	+0.18062	−0.46624

18 Holes

Hole	x	y
1	0.00000	−0.50000
2	−0.17101	−0.46985
3	−0.32139	−0.38302
4	−0.43301	−0.25000
5	−0.49240	−0.08682
6	−0.49240	+0.08682
7	−0.43301	+0.25000
8	−0.32139	+0.38302
9	−0.17101	+0.46985
10	0.00000	+0.50000
11	+0.17101	+0.46985
12	+0.32139	+0.38302
13	+0.43301	+0.25000
14	+0.49240	+0.08682
15	+0.49240	−0.08682
16	+0.43301	−0.25000
17	+0.32139	−0.38302
18	+0.17101	−0.46985

19 Holes

Hole	x	y
1	0.00000	−0.50000
2	−0.16235	−0.47291
3	−0.30711	−0.39457
4	−0.41858	−0.27347
5	−0.48470	−0.12274
6	−0.49829	+0.04129
7	−0.45789	+0.20085
...	.....	

Table 3 (*Concluded*). **Hole Coordinate Dimension Factors for Jig Boring — Type "A" Hole Circles, Central Coordinates** (English or metric units)

(19 Holes — continued)

Point	x	y
8	-0.36786	+0.33864
9	-0.23797	+0.43974
10	-0.08230	+0.49318
11	+0.08230	+0.49318
12	+0.23797	+0.43974
13	+0.36786	+0.33864
14	+0.45789	+0.20085
15	+0.49829	+0.04129
16	+0.48470	-0.12274
17	+0.41858	-0.27347
18	+0.30711	-0.39457
19	+0.16235	-0.47291

20 Holes

Point	x	y
1	0.00000	-0.50000
2	-0.15451	-0.47553
3	-0.29389	-0.40451
4	-0.40451	-0.29389
5	-0.47553	-0.15451
6	-0.50000	0.00000
7	-0.47553	+0.15451
8	-0.40451	+0.29389
9	-0.29389	+0.40451
10	-0.15451	+0.47553
11	0.00000	+0.50000
12	+0.15451	+0.47553
13	+0.29389	+0.40451
14	+0.40451	+0.29389
15	+0.47553	+0.15451
16	+0.50000	0.00000
17	+0.47553	-0.15451
18	+0.40451	-0.29389
19	+0.29389	-0.40451
20	+0.15451	-0.47553

21 Holes

Point	x	y
1	0.00000	-0.50000
2	-0.14738	-0.47779
3	-0.28166	-0.41312
4	-0.39092	-0.31174
5	-0.46544	-0.18267
6	-0.49860	-0.03737
7	-0.48746	+0.11126
8	-0.43301	+0.25000
9	-0.34009	+0.36653
10	-0.21694	+0.45048
11	-0.07452	+0.49442
12	+0.07452	+0.49442
13	+0.21694	+0.45048
14	+0.34009	+0.36653
15	+0.43301	+0.25000
16	+0.48746	+0.11126
17	+0.49860	-0.03737
18	+0.46544	-0.18267
19	+0.39092	-0.31174
20	+0.28166	-0.41312
21	+0.14738	-0.47779

22 Holes

Point	x	y
1	0.00000	-0.50000
2	-0.14087	-0.47975
3	-0.27032	-0.42063
4	-0.37787	-0.32743
5	-0.45482	-0.20771
6	-0.49491	-0.07116
7	-0.49491	+0.07116
8	-0.45482	+0.20771
9	-0.37787	+0.32743
10	-0.27032	+0.42063
11	-0.14087	+0.47975
12	0.00000	+0.50000
13	+0.14087	+0.47975
14	+0.27032	+0.42063
15	+0.37787	+0.32743
16	+0.45482	+0.20771
17	+0.49491	+0.07116
18	+0.49491	-0.07116
19	+0.45482	-0.20771
20	+0.37787	-0.32743
21	+0.27032	-0.42063
22	+0.14087	-0.47975

23 Holes

Point	x	y
1	0.00000	-0.50000
2	-0.13490	-0.48146
3	-0.25979	-0.42721
4	-0.36542	-0.34128
5	-0.44394	-0.23003
6	-0.48954	-0.10173
7	-0.49883	+0.03412
8	-0.47113	+0.16744
9	-0.40848	+0.28834
10	-0.31554	+0.38786
11	-0.19920	+0.45861
12	-0.06808	+0.49534
13	+0.06808	+0.49534
14	+0.19920	+0.45861
15	+0.31554	+0.38786
16	+0.40848	+0.28834
17	+0.47113	+0.16744
18	+0.48983	+0.03412
19	+0.48954	-0.10173
20	+0.44394	-0.23003
21	+0.36542	-0.34128
22	+0.25979	-0.42721
23	+0.13490	-0.48146

24 Holes

Point	x	y
1	0.00000	-0.50000
2	-0.12941	-0.48296
3	-0.25000	-0.43301
4	-0.35355	-0.35355
5	-0.43301	-0.25000
6	-0.48296	-0.12941
7	-0.50000	0.00000
8	-0.48296	+0.12941
9	-0.43301	+0.25000
10	-0.35355	+0.35355
11	-0.25000	+0.43301
12	-0.12941	+0.48296
13	0.00000	+0.50000
14	+0.12941	+0.48296
15	+0.25000	+0.43301
16	+0.35355	+0.35355
17	+0.43301	+0.25000
18	+0.48296	+0.12941
19	+0.50000	0.00000
20	+0.48296	-0.12941
21	+0.43301	-0.25000
22	+0.35355	-0.35355
23	+0.25000	-0.43301
24	+0.12941	-0.48296

25 Holes

Point	x	y
1	0.00000	-0.50000
2	-0.12434	-0.48429
3	-0.24088	-0.43815
4	-0.34227	-0.36448
5	-0.42216	-0.26791
6	-0.47553	-0.15451
7	-0.49901	-0.03140
8	-0.49114	+0.09369
9	-0.45241	+0.21289
10	-0.38526	+0.31871
11	-0.29389	+0.40451
12	-0.18406	+0.46489
13	-0.06267	+0.49606
14	+0.06267	+0.49606
15	+0.18406	+0.46489
16	+0.29389	+0.40451
17	+0.38526	+0.31871
18	+0.45241	+0.21289
19	+0.49114	+0.09369
20	+0.49901	-0.03140
21	+0.47553	-0.15451
22	+0.42216	-0.26791
23	+0.34227	-0.36448
24	+0.24088	-0.43815
25	+0.12434	-0.48429

26 Holes

Point	x	y
1	0.00000	-0.50000
2	-0.11966	-0.48547
3	-0.23236	-0.44273
4	-0.33156	-0.37426
5	-0.41149	-0.28403
6	-0.46751	-0.17730
7	-0.49635	-0.06027
8	-0.49635	+0.06027
9	-0.46751	+0.17730
10	-0.41149	+0.28403
11	-0.33156	+0.37426
12	-0.23236	+0.44273
13	-0.11966	+0.48547
14	0.00000	+0.50000
15	+0.11966	+0.48547
16	+0.23236	+0.44273
17	+0.33156	+0.37426
18	+0.41149	+0.28403
19	+0.46751	+0.17730
20	+0.49635	+0.06027
21	+0.49635	-0.06027
22	+0.46751	-0.17730
23	+0.41149	-0.28403
24	+0.33156	-0.37426
25	+0.23236	-0.44273
26	+0.11966	-0.48547

27 Holes

Point	x	y
1	0.00000	-0.50000
2	-0.11531	-0.48652
3	-0.22440	-0.44682
4	-0.32139	-0.38302
5	-0.40106	-0.29858
6	-0.45911	-0.19804
7	-0.49240	-0.08682
8	-0.49915	+0.02907
9	-0.47899	+0.14340
10	-0.43301	+0.25000
11	-0.36369	+0.34312
12	-0.27475	+0.41774
13	-0.17101	+0.46985
14	-0.05805	+0.49662
15	+0.05805	+0.49662
16	+0.17101	+0.46985
17	+0.27475	+0.41774
18	+0.36369	+0.34312
19	+0.43301	+0.25000
20	+0.47899	+0.14340
21	+0.49915	+0.02907
22	+0.49240	-0.08682
23	+0.45911	-0.19804
24	+0.40106	-0.29858
25	+0.32139	-0.38302
26	+0.22440	-0.44682
27	+0.11531	-0.48652

28 Holes

Point	x	y
1	0.00000	-0.50000
2	-0.11126	-0.48746
3	-0.21694	-0.45048
4	-0.31174	-0.39092
5	-0.39092	-0.31174
6	-0.45048	-0.21694
7	-0.48746	-0.11126
8	-0.50000	0.00000
9	-0.48746	+0.11126
10	-0.45048	+0.21694
11	-0.39092	+0.31174
12	-0.31174	+0.39092
13	-0.21694	+0.45048
14	-0.11126	+0.48746
15	0.00000	+0.50000
16	+0.11126	+0.48746
17	+0.21694	+0.45048
18	+0.31174	+0.39092
19	+0.39092	+0.31174
20	+0.45048	+0.21694
21	+0.48746	+0.11126
22	+0.50000	0.00000
23	+0.48746	-0.11126
24	+0.45048	-0.21694
25	+0.39092	-0.31174
26	+0.31174	-0.39092
27	+0.21694	-0.45048
28	+0.11126	-0.48746

Table 4. Hole Coordinate Dimension Factors for Jig Boring — Type "B" Hole Circles Central Coordinates (English or metric units)

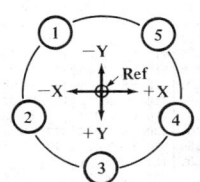

The diagram shows a type "B" circle for a 5-hole circle. Coordinates x, y are given in the table for hole circles of from 3 to 28 holes. Dimensions are for holes numbered in a counterclockwise direction (as shown). Dimensions given are based upon a hole circle of unit diameter. For a hole circle of, say, 3-inch or 3-centimeter diameter, multiply table values by 3.

3 Holes

	x	y
1	−0.43301	−0.25000
2	0.00000	+0.50000
3	+0.43301	−0.25000

4 Holes

	x	y
1	−0.35355	−0.35355
2	−0.35355	+0.35355
3	+0.35355	+0.35355
4	+0.35355	−0.35355

5 Holes

	x	y
1	−0.29389	−0.40451
2	−0.47553	+0.15451
3	0.00000	+0.50000
4	+0.47553	+0.15451
5	+0.29389	−0.40451

6 Holes

	x	y
1	−0.25000	−0.43301
2	−0.50000	0.00000
3	−0.25000	+0.43301
4	+0.25000	+0.43301
5	+0.50000	0.00000
6	+0.25000	−0.43301

7 Holes

	x	y
1	−0.21694	−0.45048
2	−0.48746	−0.11126
3	−0.39092	+0.31174
4	0.00000	+0.50000
5	+0.39092	+0.31174
6	+0.48746	−0.11126
7	+0.21694	−0.45048

8 Holes

	x	y
1	−0.19134	−0.46194
2	−0.46194	−0.19134
3	−0.46194	+0.19134
4	−0.19134	+0.46194
5	+0.19134	+0.46194
6	+0.46194	+0.19134
7	+0.46194	−0.19134
8	+0.19134	−0.46194

9 Holes

	x	y
1	−0.17101	−0.46985
2	−0.43301	−0.25000
3	−0.49240	+0.08682
4	−0.32139	+0.38302
5	0.00000	+0.50000
6	+0.32139	+0.38302
7	+0.49240	+0.08682
8	+0.43301	−0.25000
9	+0.17101	−0.46985

10 Holes

	x	y
1	−0.15451	−0.47553
2	−0.40451	−0.29389
3	−0.50000	0.00000
4	−0.40451	+0.29389
5	−0.15451	+0.47553
6	+0.15451	+0.47553
7	+0.40451	+0.29389
8	+0.50000	0.00000
9	+0.40451	−0.29389
10	+0.15451	−0.47553

11 Holes

	x	y
1	−0.14087	−0.47975
2	−0.37787	−0.32743
3	−0.49491	−0.07116
4	−0.45482	+0.20771
5	−0.27032	+0.42063
6	0.00000	+0.50000
7	+0.27032	+0.42063
8	+0.45482	+0.20771
9	+0.49491	−0.07116
10	+0.37787	−0.32743
11	+0.14087	−0.47975

12 Holes

	x	y
1	−0.12941	−0.48296
2	−0.35355	−0.35355
3	−0.48296	−0.12941
4	−0.48296	+0.12941
5	−0.35355	+0.35355
6	−0.12941	+0.48296
7	+0.12941	+0.48296
8	+0.35355	+0.35355
9	+0.48296	+0.12941
10	+0.48296	−0.12941
11	+0.35355	−0.35355
12	+0.12941	−0.48296

13 Holes

	x	y
1	−0.11966	−0.48547
2	−0.33156	−0.37426
3	−0.46751	−0.17730
4	−0.49635	+0.06027
5	−0.41149	+0.28403
6	−0.23236	+0.44273
7	0.00000	+0.50000
8	+0.23236	+0.44273
9	+0.41149	+0.28403
10	+0.49635	+0.06027
11	+0.46751	−0.17730
12	+0.33156	−0.37426
13	+0.11966	−0.48547

14 Holes

	x	y
1	−0.11126	−0.48746
2	−0.31174	−0.39092
3	−0.45048	−0.21694
4	−0.50000	0.00000
5	−0.45048	+0.21694
6	−0.31174	+0.39092
7	−0.11126	+0.48746
8	+0.11126	+0.48746
9	+0.31174	+0.39092
10	+0.45048	+0.21694
11	+0.50000	0.00000
12	+0.45048	−0.21694
13	+0.31174	−0.39092
14	+0.11126	−0.48746

15 Holes

	x	y
1	−0.10396	−0.48907
2	−0.29389	−0.40451
3	−0.43301	−0.25000
4	−0.49726	−0.05226
5	−0.47553	+0.15451
6	−0.37157	+0.33457
7	−0.20337	+0.45677
8	0.00000	+0.50000
9	+0.20337	+0.45677
10	+0.37157	+0.33457
11	+0.47553	+0.15451
12	+0.49726	−0.05226
13	+0.43301	−0.25000
14	+0.29389	−0.40451
15	+0.10396	−0.48907

16 Holes

	x	y
1	−0.09755	−0.49039
2	−0.27779	−0.41573
3	−0.41573	−0.27779
4	−0.49039	−0.09755
5	−0.49039	+0.09755
6	−0.41573	+0.27779
7	−0.27779	+0.41573
8	−0.09755	+0.49039
9	+0.09755	+0.49039
10	+0.27779	+0.41573
11	+0.41573	+0.27779
12	+0.49039	+0.09755
13	+0.49039	−0.09755
14	+0.41573	−0.27779
15	+0.27779	−0.41573
16	+0.09755	−0.49039

17 Holes

	x	y
1	−0.09187	−0.49149
2	−0.26322	−0.42511
3	−0.39991	−0.30132
4	−0.48091	−0.13683
5	−0.49787	+0.04613
6	−0.44758	+0.22287
7	−0.33685	+0.36950
8	−0.18062	+0.46624
9	0.00000	+0.50000
10	+0.18062	+0.46624
11	+0.33685	+0.36950
12	+0.44758	+0.22287
13	+0.49787	+0.04613
14	+0.48091	−0.13683
15	+0.39991	−0.30132
16	+0.26322	−0.42511
17	+0.09187	−0.49149

18 Holes

	x	y
1	−0.08682	−0.49240
2	−0.25000	−0.43301
3	−0.38302	−0.32139
4	−0.46985	−0.17101
5	−0.50000	0.00000
6	−0.46985	+0.17101
7	−0.38302	+0.32139
8	−0.25000	+0.43301
9	−0.08682	+0.49240
10	+0.08682	+0.49240
11	+0.25000	+0.43301
12	+0.38302	+0.32139
13	+0.46985	+0.17101
14	+0.50000	0.00000
15	+0.46985	−0.17101
16	+0.38302	−0.32139
17	+0.25000	−0.43301
18	+0.08682	−0.49240

19 Holes

	x	y
1	−0.08230	−0.49318
2	−0.23797	−0.43974
3	−0.36786	−0.33864
4	−0.45789	−0.20085
5	−0.49829	−0.04129
6	−0.48470	+0.12274
7	−0.41858	+0.27347
...	.....	.....

Table 4 (*Concluded*). **Hole Coordinate Dimension Factors for Jig Boring — Type "B" Hole Circles, Central Coordinates** (English or metric units)

x8	−0.30711	y8	+0.39457
x9	−0.16235	y9	+0.47291
x10	0.00000	y10	+0.50000
x11	+0.16235	y11	+0.47291
x12	+0.30711	y12	+0.39457
x13	+0.41858	y13	+0.27347
x14	+0.48470	y14	+0.12274
x15	+0.49829	y15	−0.04129
x16	+0.45789	y16	−0.20085
x17	+0.36786	y17	−0.33864
x18	+0.23797	y18	−0.43974
x19	+0.08230	y19	−0.49318

20 Holes

x1	−0.07822	y1	−0.49384
x2	−0.22700	y2	−0.44550
x3	−0.35355	y3	−0.35355
x4	−0.44550	y4	−0.22700
x5	−0.49384	y5	−0.07822
x6	−0.49384	y6	+0.07822
x7	−0.44550	y7	+0.22700
x8	−0.35355	y8	+0.35355
x9	−0.22700	y9	+0.44550
x10	−0.07822	y10	+0.49384
x11	+0.07822	y11	+0.49384
x12	+0.22700	y12	+0.44550
x13	+0.35355	y13	+0.35355
x14	+0.44550	y14	+0.22700
x15	+0.49384	y15	+0.07822
x16	+0.49384	y16	−0.07822
x17	+0.44550	y17	−0.22700
x18	+0.35355	y18	−0.35355
x19	+0.22700	y19	−0.44550
x20	+0.07822	y20	−0.49384
..	....	..	....

21 Holes

x1	−0.07452	y1	−0.49442
x2	−0.21694	y2	−0.45048
x3	−0.34009	y3	−0.36653
x4	−0.43301	y4	−0.25000
x5	−0.48746	y5	−0.11126
x6	−0.49860	y6	+0.03737
x7	−0.46544	y7	+0.18267
x8	−0.39092	y8	+0.31174
x9	−0.28166	y9	+0.41312
x10	−0.14738	y10	+0.47779
x11	0.00000	y11	+0.50000
x12	+0.14738	y12	+0.47779
x13	+0.28166	y13	+0.41312
x14	+0.39092	y14	+0.31174
x15	+0.46544	y15	+0.18267
x16	+0.49860	y16	+0.03737
x17	+0.48746	y17	−0.11126
x18	+0.43301	y18	−0.25000
x19	+0.34009	y19	−0.36653
x20	+0.21694	y20	−0.45048
x21	+0.07452	y21	−0.49442

22 Holes

x1	−0.07116	y1	−0.49491
x2	−0.20771	y2	−0.45482
x3	−0.32743	y3	−0.37787
x4	−0.42063	y4	−0.27032
x5	−0.47975	y5	−0.14087
x6	−0.50000	y6	0.00000
x7	−0.47975	y7	+0.14087
x8	−0.42063	y8	+0.27032
x9	−0.32743	y9	+0.37787
x10	−0.20771	y10	+0.45482
x11	−0.07116	y11	+0.49491
..	....		
x12	+0.07116	y12	+0.49491
x13	+0.20771	y13	+0.45482
x14	+0.32743	y14	+0.37787
x15	+0.42063	y15	+0.27032
x16	+0.47975	y16	+0.14087
x17	+0.50000	y17	0.00000
x18	+0.47975	y18	−0.14087
x19	+0.42063	y19	−0.27032
x20	+0.32743	y20	−0.37787
x21	+0.20771	y21	−0.45482
x22	+0.07116	y22	−0.49491

23 Holes

x1	−0.06808	y1	−0.49534
x2	−0.19920	y2	−0.45861
x3	−0.31554	y3	−0.38786
x4	−0.40848	y4	−0.28834
x5	−0.47113	y5	−0.16744
x6	−0.49883	y6	−0.03412
x7	−0.48954	y7	+0.10173
x8	−0.44394	y8	+0.23003
x9	−0.36542	y9	+0.34128
x10	−0.25979	y10	+0.42721
x11	−0.13490	y11	+0.48146
x12	0.00000	y12	+0.50000
x13	+0.13490	y13	+0.48146
x14	+0.25979	y14	+0.42721
x15	+0.36542	y15	+0.34128
x16	+0.44394	y16	+0.23003
x17	+0.48954	y17	+0.10173
x18	+0.49883	y18	−0.03412
x19	+0.47113	y19	−0.16744
x20	+0.40848	y20	−0.28834
x21	+0.31554	y21	−0.38786
x22	+0.19920	y22	−0.45861
x23	+0.06808	y23	−0.49534

24 Holes

x1	−0.06526	y1	−0.49572
x2	−0.19134	y2	−0.46194
x3	−0.30438	y3	−0.39668
x4	−0.39668	y4	−0.30438
x5	−0.46194	y5	−0.19134
x6	−0.49572	y6	−0.06526
x7	−0.49572	y7	+0.06526
x8	−0.46194	y8	+0.19134
x9	−0.39668	y9	+0.30438
x10	−0.30438	y10	+0.39668
x11	−0.19134	y11	+0.46194
x12	−0.06526	y12	+0.49572
x13	+0.06526	y13	+0.49572
x14	+0.19134	y14	+0.46194
x15	+0.30438	y15	+0.39668
x16	+0.39668	y16	+0.30438
x17	+0.46194	y17	+0.19134
x18	+0.49572	y18	+0.06526
x19	+0.49572	y19	−0.06526
x20	+0.46194	y20	−0.19134
x21	+0.39668	y21	−0.30438
x22	+0.30438	y22	−0.39668
x23	+0.19134	y23	−0.46194
x24	+0.06526	y24	−0.49572

25 Holes

x1	−0.06267	y1	−0.49606
x2	−0.18406	y2	−0.46489
x3	−0.29389	y3	−0.40451
x4	−0.38526	y4	−0.31871
x5	−0.45241	y5	−0.21289
x6	−0.49114	y6	−0.09369
x7	−0.49901	y7	+0.03140
..	....		
x8	−0.47553	y8	+0.15451
x9	−0.42216	y9	+0.26791
x10	−0.34227	y10	+0.36448
x11	−0.24088	y11	+0.43815
x12	−0.12434	y12	+0.48429
x13	0.00000	y13	+0.50000
x14	+0.12434	y14	+0.48429
x15	+0.24088	y15	+0.43815
x16	+0.34227	y16	+0.36448
x17	+0.42216	y17	+0.26791
x18	+0.47553	y18	+0.15451
x19	+0.49901	y19	+0.03140
x20	+0.49114	y20	−0.09369
x21	+0.45241	y21	−0.21289
x22	+0.38526	y22	−0.31871
x23	+0.29389	y23	−0.40451
x24	+0.18406	y24	−0.46489
x25	+0.06267	y25	−0.49606

26 Holes

x1	−0.06027	y1	−0.49635
x2	−0.17730	y2	−0.46751
x3	−0.28403	y3	−0.41149
x4	−0.37426	y4	−0.33156
x5	−0.44273	y5	−0.23236
x6	−0.48547	y6	−0.11966
x7	−0.50000	y7	0.00000
x8	−0.48547	y8	+0.11966
x9	−0.44273	y9	+0.23236
x10	−0.37426	y10	+0.33156
x11	−0.28403	y11	+0.41149
x12	−0.17730	y12	+0.46751
x13	−0.06027	y13	+0.49635
x14	+0.06027	y14	+0.49635
x15	+0.17730	y15	+0.46751
x16	+0.28403	y16	+0.41149
x17	+0.37426	y17	+0.33156
x18	+0.44273	y18	+0.23236
x19	+0.48547	y19	+0.11966
x20	+0.50000	y20	0.00000
x21	+0.48547	y21	−0.11966
x22	+0.44273	y22	−0.23236
x23	+0.37426	y23	−0.33156
x24	+0.28403	y24	−0.41149
x25	+0.17730	y25	−0.46751
x26	+0.06027	y26	−0.49635

27 Holes

x1	−0.05805	y1	−0.49662
x2	−0.17101	y2	−0.46985
x3	−0.27475	y3	−0.41774
x4	−0.36369	y4	−0.34312
x5	−0.43301	y5	−0.25000
x6	−0.47899	y6	−0.14340
x7	−0.49915	y7	−0.02907
x8	−0.49240	y8	+0.08682
x9	−0.45911	y9	+0.19804
x10	−0.40106	y10	+0.29858
x11	−0.32139	y11	+0.38302
x12	−0.22440	y12	+0.44682
x13	−0.11531	y13	+0.48652
x14	0.00000	y14	+0.50000
x15	+0.11531	y15	+0.48652
x16	+0.22440	y16	+0.44682
x17	+0.32139	y17	+0.38302
x18	+0.40106	y18	+0.29858
x19	+0.45911	y19	+0.19804
x20	+0.49240	y20	+0.08682
x21	+0.49915	y21	−0.02907
x22	+0.47899	y22	−0.14340
x23	+0.43301	y23	−0.25000
x24	+0.36369	y24	−0.34312
x25	+0.27475	y25	−0.41774
x26	+0.17101	y26	−0.46985
x27	+0.05805	y27	−0.49662

28 Holes

x1	−0.05598	y1	−0.49686
x2	−0.16514	y2	−0.47194
x3	−0.26602	y3	−0.42336
x4	−0.35355	y4	−0.35355
x5	−0.42336	y5	−0.26602
x6	−0.47194	y6	−0.16514
x7	−0.49686	y7	−0.05598
x8	−0.49686	y8	+0.05598
x9	−0.47194	y9	+0.16514
x10	−0.42336	y10	+0.26602
x11	−0.35355	y11	+0.35355
x12	−0.26602	y12	+0.42336
x13	−0.16514	y13	+0.47194
x14	−0.05598	y14	+0.49686
x15	+0.05598	y15	+0.49686
x16	+0.16514	y16	+0.47194
x17	+0.26602	y17	+0.42336
x18	+0.35355	y18	+0.35355
x19	+0.42336	y19	+0.26602
x20	+0.47194	y20	+0.16514
x21	+0.49686	y21	+0.05598
x22	+0.49686	y22	−0.05598
x23	+0.47194	y23	−0.16514
x24	+0.42336	y24	−0.26602
x25	+0.35355	y25	−0.35355
x26	+0.26602	y26	−0.42336
x27	+0.16514	y27	−0.47194
x28	+0.05598	y28	−0.49686

Lengths of Chords for Spacing Off the Circumferences of Circles

On the following pages are given tables of the lengths of chords for spacing off the circumferences of circles. The object of these tables is to make possible the division of the periphery into a number of equal parts without trials with the dividers. The first table is calculated for circles having a diameter equal to 1. For circles of other diameters, the length of chord given in the table should be multiplied by the diameter of the circle. This first table may be used by tool-makers when setting "buttons" in circular formation. Assume that it is required to divide the periphery of a circle of 20 inches diameter into thirty-two equal parts. From the table the length of the chord is found to be 0.098017 inch, if the diameter of the circle were 1 inch. With a diameter of 20 inches the length of the chord for one division would be 20 × 0.098017 = 1.9603 inches. Another example in metric units: For a 100 millimeter diameter requiring 5 equal divisions, the length of the chord for one division would be 100 × 0.587785 = 58.7785 millimeters.

The two following pages give an additional table for the spacing off of circles, the table, in this case, being worked out for diameters from $\frac{1}{16}$ inch to 14 inches. As an example, assume that it is required to divide a circle having a diameter of 6½ inches into seven equal parts. Find first, in the column headed "6" and in line with 7 divisions, the length of the chord for a 6-inch circle, which is 2.603 inches. Then find the length of the chord for a ½-inch diameter circle, 7 divisions, which is 0.217. The sum of these two values, 2.603 + 0.217 = 2.820 inches, is the length of the chord required for spacing off the circumference of a 6½-inch circle into seven equal divisions.

As another example, assume that it is required to divide a circle having a diameter of 9²³⁄₃₂ inches into 15 equal divisions. First find the length of the chord for a 9-inch circle, which is 1.871 inch. The length of the chord for a ²³⁄₃₂-inch circle can easily be estimated from the table by taking the value that is exactly between those given for ¹¹⁄₁₆ and ¾ inch. The value for ¹¹⁄₁₆ inch is 0.143, and for ¾ inch, 0.156. For ²³⁄₃₂, the value would be 0.150. Then, 1.871 + 0.150 = 2.021 inches.

Lengths of Chords for Spacing Off the Circumferences of Circles with a Diameter Equal to 1

For circles of other diameters multiply length given in table by diameter of circle. (English or metric units)

No. of Spaces	Length of Chord	No. of Spaces	Length of Chord	No. of Spaces	Length of Chord	No. of Spaces	Length of Chord
3	0.866025	22	0.142315	41	0.076549	60	0.052336
4	0.707107	23	0.136167	42	0.074730	61	0.051479
5	0.587785	24	0.130526	43	0.072995	62	0.050649
6	0.500000	25	0.125333	44	0.071339	63	0.049846
7	0.433884	26	0.120537	45	0.069756	64	0.049068
8	0.382683	27	0.116093	46	0.068242	65	0.048313
9	0.342020	28	0.111964	47	0.066793	66	0.047582
10	0.309017	29	0.108119	48	0.065403	67	0.046872
11	0.281733	30	0.104528	49	0.064070	68	0.046183
12	0.258819	31	0.101168	50	0.062791	69	0.045515
13	0.239316	32	0.098017	51	0.061561	70	0.044865
14	0.222521	33	0.095056	52	0.060378	71	0.044233
15	0.207912	34	0.092268	53	0.059241	72	0.043619
16	0.195090	35	0.089639	54	0.058145	73	0.043022
17	0.183750	36	0.087156	55	0.057089	74	0.042441
18	0.173648	37	0.084806	56	0.056070	75	0.041876
19	0.164505	38	0.082579	57	0.055088	76	0.041325
20	0.156434	39	0.080467	58	0.054139	77	0.040789
21	0.149042	40	0.078459	59	0.053222	78	0.040266

Table for Spacing Off the Circumferences of Circles
(See page 962 for explanatory matter.)

No. of Divisions	Degrees in Arc	Diameter of Circle to be Spaced Off														
		Length of Chord														
		1/16	1/8	3/16	1/4	5/16	3/8	7/16	1/2	9/16	5/8	11/16	3/4	13/16	7/8	15/16
3	120	0.054	0.108	0.162	0.217	0.271	0.325	0.379	0.433	0.487	0.541	0.595	0.650	0.704	0.758	0.812
4	90	0.044	0.088	0.133	0.177	0.221	0.265	0.309	0.354	0.398	0.442	0.486	0.530	0.575	0.619	0.663
5	72	0.037	0.073	0.110	0.147	0.184	0.220	0.257	0.294	0.331	0.367	0.404	0.441	0.478	0.514	0.551
6	60	0.031	0.063	0.094	0.125	0.156	0.188	0.219	0.250	0.281	0.313	0.344	0.375	0.406	0.438	0.469
7	51 3/7	0.027	0.054	0.081	0.108	0.136	0.163	0.190	0.217	0.244	0.271	0.298	0.325	0.353	0.380	0.407
8	45	0.024	0.048	0.072	0.096	0.120	0.144	0.167	0.191	0.215	0.239	0.263	0.287	0.311	0.335	0.359
9	40	0.021	0.043	0.064	0.086	0.107	0.128	0.150	0.171	0.192	0.214	0.235	0.257	0.278	0.299	0.321
10	36	0.019	0.039	0.058	0.077	0.097	0.116	0.135	0.155	0.174	0.193	0.212	0.232	0.251	0.270	0.290
11	32 8/11	0.018	0.035	0.053	0.070	0.088	0.106	0.123	0.141	0.158	0.176	0.194	0.211	0.229	0.247	0.264
12	30	0.016	0.032	0.049	0.065	0.081	0.097	0.113	0.129	0.146	0.162	0.178	0.194	0.210	0.226	0.243
13	27 9/13	0.015	0.030	0.045	0.060	0.075	0.090	0.105	0.120	0.135	0.150	0.165	0.179	0.194	0.209	0.224
14	25 5/7	0.014	0.028	0.042	0.056	0.070	0.083	0.097	0.111	0.125	0.139	0.153	0.167	0.181	0.195	0.209
15	24	0.013	0.026	0.039	0.052	0.065	0.078	0.091	0.104	0.117	0.130	0.143	0.156	0.169	0.182	0.195
16	22 1/2	0.012	0.024	0.037	0.049	0.061	0.073	0.085	0.098	0.110	0.122	0.134	0.146	0.159	0.171	0.183
17	21 3/17	0.011	0.023	0.034	0.046	0.057	0.069	0.080	0.092	0.103	0.115	0.126	0.138	0.149	0.161	0.172
18	20	0.011	0.022	0.033	0.043	0.054	0.065	0.076	0.087	0.098	0.109	0.119	0.130	0.141	0.152	0.163
19	18 18/19	0.010	0.021	0.031	0.041	0.051	0.062	0.072	0.082	0.093	0.103	0.113	0.123	0.134	0.144	0.154
20	18	0.010	0.020	0.029	0.039	0.049	0.059	0.068	0.078	0.088	0.098	0.108	0.117	0.127	0.137	0.147
21	17 1/7	0.009	0.019	0.028	0.037	0.047	0.056	0.065	0.075	0.084	0.093	0.102	0.112	0.121	0.130	0.140
22	16 4/11	0.009	0.018	0.027	0.036	0.044	0.053	0.062	0.071	0.080	0.089	0.098	0.107	0.116	0.125	0.133
23	15 15/23	0.009	0.017	0.026	0.034	0.043	0.051	0.060	0.068	0.077	0.085	0.094	0.102	0.111	0.119	0.128
24	15	0.008	0.016	0.024	0.033	0.041	0.049	0.057	0.065	0.073	0.082	0.090	0.098	0.106	0.114	0.122
25	14 2/5	0.008	0.016	0.023	0.031	0.039	0.047	0.055	0.063	0.070	0.078	0.086	0.094	0.102	0.110	0.117
26	13 11/13	0.008	0.015	0.023	0.030	0.038	0.045	0.053	0.060	0.068	0.075	0.083	0.090	0.098	0.105	0.113
28	12 6/7	0.007	0.014	0.021	0.028	0.035	0.042	0.049	0.056	0.063	0.070	0.077	0.084	0.091	0.098	0.105
30	12	0.007	0.013	0.020	0.026	0.033	0.039	0.046	0.052	0.059	0.065	0.072	0.078	0.085	0.091	0.098
32	11 1/4	0.006	0.012	0.018	0.025	0.031	0.037	0.043	0.049	0.055	0.061	0.067	0.074	0.080	0.086	0.092

Table for Spacing Off the Circumferences of Circles

No. of Divisions	Degrees in Arc	Diameter of Circle to be Spaced Off													
		Length of Chord													
		1	2	3	4	5	6	7	8	9	10	11	12	13	14
3	120	0.866	1.732	2.598	3.464	4.330	5.196	6.062	6.928	7.794	8.660	9.526	10.392	11.258	12.124
4	90	0.707	1.414	2.121	2.828	3.536	4.243	4.950	5.657	6.364	7.071	7.778	8.485	9.192	9.899
5	72	0.588	1.176	1.763	2.351	2.939	3.527	4.114	4.702	5.290	5.878	6.466	7.053	7.641	8.229
6	60	0.500	1.000	1.500	2.000	2.500	3.000	3.500	4.000	4.500	5.000	5.500	6.000	6.500	7.000
7	51 3/7	0.434	0.868	1.302	1.736	2.169	2.603	3.037	3.471	3.905	4.339	4.773	5.207	5.640	6.074
8	45	0.383	0.765	1.148	1.531	1.913	2.296	2.679	3.061	3.444	3.827	4.210	4.592	4.975	5.358
9	40	0.342	0.684	1.026	1.368	1.710	2.052	2.394	2.736	3.078	3.420	3.762	4.104	4.446	4.788
10	36	0.309	0.618	0.927	1.236	1.545	1.854	2.163	2.472	2.781	3.090	3.399	3.708	4.017	4.326
11	32 8/11	0.282	0.563	0.845	1.127	1.409	1.690	1.972	2.254	2.536	2.817	3.099	3.381	3.663	3.944
12	30	0.259	0.518	0.776	1.035	1.294	1.553	1.812	2.071	2.329	2.588	2.847	3.106	3.365	3.623
13	27 9/13	0.239	0.479	0.718	0.957	1.197	1.436	1.675	1.915	2.154	2.393	2.632	2.872	3.111	3.350
14	25 5/7	0.223	0.445	0.668	0.890	1.113	1.335	1.558	1.780	2.003	2.225	2.448	2.670	2.893	3.115
15	24	0.208	0.416	0.624	0.832	1.040	1.247	1.455	1.663	1.871	2.079	2.287	2.495	2.703	2.911
16	22 1/2	0.195	0.390	0.585	0.780	0.975	1.171	1.366	1.561	1.756	1.951	2.146	2.341	2.536	2.731
17	21 3/17	0.184	0.367	0.551	0.735	0.919	1.102	1.286	1.470	1.654	1.837	2.021	2.205	2.389	2.572
18	20	0.174	0.347	0.521	0.695	0.868	1.042	1.216	1.389	1.563	1.736	1.910	2.084	2.257	2.431
19	18 18/19	0.165	0.329	0.494	0.658	0.823	0.988	1.152	1.317	1.481	1.646	1.811	1.975	2.140	2.304
20	18	0.156	0.313	0.469	0.626	0.782	0.939	1.095	1.251	1.408	1.564	1.721	1.877	2.034	2.190
21	17 1/7	0.149	0.298	0.447	0.596	0.745	0.894	1.043	1.192	1.341	1.499	1.639	1.789	1.938	2.087
22	16 4/11	0.142	0.285	0.427	0.569	0.712	0.854	0.996	1.139	1.281	1.423	1.565	1.708	1.850	1.992
23	15 15/23	0.136	0.272	0.408	0.545	0.681	0.817	0.953	1.089	1.225	1.362	1.498	1.634	1.770	1.906
24	15	0.131	0.261	0.392	0.522	0.653	0.783	0.914	1.044	1.175	1.305	1.436	1.566	1.697	1.827
25	14 2/5	0.125	0.251	0.376	0.501	0.627	0.752	0.877	1.003	1.128	1.253	1.379	1.504	1.629	1.755
26	13 11/13	0.121	0.241	0.362	0.482	0.603	0.723	0.844	0.964	1.085	1.205	1.326	1.446	1.567	1.688
28	12 6/7	0.112	0.224	0.336	0.448	0.560	0.672	0.784	0.896	1.008	1.120	1.233	1.344	1.456	1.568
30	12	0.105	0.209	0.314	0.418	0.523	0.627	0.732	0.836	0.941	1.045	1.150	1.254	1.359	1.463
32	11 1/4	0.098	0.196	0.294	0.392	0.490	0.588	0.686	0.784	0.882	0.980	1.078	1.176	1.274	1.372

CUTTING SPEEDS AND FEEDS

Work Materials. — The large number of work materials that are commonly machined vary greatly in their basic structure and the ease with which they can be machined. Yet it is possible to group together certain materials having similar machining characteristics, for the purpose of recommending the cutting speed at which they can be cut. Most materials that are machined are metals and it has been found that the most important single factor influencing the ease with which a metal can be cut is its microstructure, followed by any cold work that may have been done to the metal, which increases its hardness. Metals that have a similar, but not necessarily the same microstructure, will tend to have similar machining characteristics. Thus, the grouping of the metals in the accompanying tables has been done on the basis of their microstructure.

With the exception of a few soft and gummy metals, experience has shown that harder metals are more difficult to cut than softer metals. Furthermore, any given metal is more difficult to cut when it is in a harder form than when it is softer. It is more difficult to penetrate the harder metal and more power is required to cut it. This in turn will generate a higher cutting temperature at any given cutting speed thereby making it necessary to use a slower speed, for the cutting temperature must always be kept within the limits that can be sustained by the cutting tool without failure. Hardness, then, is an important property that must be considered when machining a given metal. Hardness alone, however, cannot be used as a measure of cutting speed. For example, if pieces of AISI 11L17 and AISI 1117 steel both have a hardness of 150 Bhn, their recommended cutting speed for high speed steel tools will be 140 fpm and 130 fpm respectively. In some metals two entirely different microstructures can produce the same hardness. As an example, a fine pearlite microstructure and a tempered martensite microstructure can result in the same hardness in a steel. These microstructures will not machine alike. For practical purposes, however, information on hardness is usually easier to obtain than information on microstructure; thus, hardness alone is usually used to differentiate between different cutting speeds for machining a metal.

In some situations the hardness of a metal to be machined is not known. When this occurs, the material condition, which is listed separately in the tables, can be used as a guide. Lacking more specific information on the hardness of the metal to be machined, a knowledge of its material condition is helpful in selecting the cutting speed.

The surface of ferrous metal castings has a scale that is more difficult to machine than the metal below. Some scale is more difficult to machine than others, depending on the foundry sand used, the casting process, the method of cleaning the casting, and the type of metal cast. Special electrochemical treatments can be used in some cases that almost entirely eliminate the effect of the scale on machining, although castings so treated are not frequently encountered. Usually when casting scale is encountered, the cutting speed is reduced approximately 5 or 10 per cent. Difficult-to-machine surface scale can also be encountered when machining hot rolled or forged steel bars.

Metallurgical differences that affect machining characteristics are often found within a single piece of metal. The occurrence of hard spots in castings is an example. Different microstructures and hardness levels may occur within a casting as a result of variations in the cooling rate in different parts of the casting. Such variations are less severe in castings that have been heat treated. Steel bar stock is usually harder toward the outside than toward the center of the bar. Sometimes there are slight metallurgical differences along the length of a bar which can affect its cutting characteristics.

Cutting Tool Materials. — The recommended cutting speeds in the accompanying tables are given for high speed steel and for cemented carbides because these materials are the most commonly used and more data are available for them than for others. Other

materials that are used to make cutting tools are cemented oxides or ceramics, cermets, cast non-ferrous alloys (stellite), single crystal diamonds, polycrystalline diamonds, and cubic boron nitride.

High Speed Steel designates a number of steels that have several properties which enhance their value as cutting tool material. They can be hardened to a high initial or room temperature hardness ranging from 63 R_C to 65 R_C for ordinary high speed steels and up to 70 R_C for the so-called super-high speed steels. They can retain sufficient hardness at temperatures up to 1,000 to 1,100 F to enable them to cut at cutting speeds that will generate these tool temperatures, and they will return to their original hardness when cooled to room temperature. They harden very deeply enabling high speed steels to be ground to the tool shape from solid stock and to be reground many times without sacrificing hardness at the cutting edge. High speed steels can be made soft by annealing so that they can be machined into complex cutting tools such as drills, reamers, and milling cutters and then hardened.

The principal alloying elements of high speed steels are tungsten (W), molybdenum (Mo), chromium (Cr), vanadium (V), together with carbon (C). There are a number of grades of high speed steel which are divided into two types: tungsten high speed steels and molybdenum high speed steels. Tungsten high speed steels are designated by the prefix T before the number which designates the grade and molybdenum high speed steels are designated by the prefix letter M. Tungsten high speed steels were developed first and are still used. A common steel of this type is the T1 which has the following alloy composition: W − 18%, Cr − 4%, V − 1%, and C − .75%. Molybdenum high speed steels were developed because of a shortage in the supply of tungsten. They are less expensive than comparable grades of tungsten high speed steel and are, therefore, widely used at the present time. Molybdenum is used as a substitute for a part of the tungsten as shown by the following composition of the widely used M2 high speed steel: Mo − 5%, W − 6%, Cr − 4%, V − 2%, and C − .85%. Production experience and extensive laboratory tests have shown no significant superiority in either direction between comparable grades of tungsten or molybdenum high speed steel.

The addition of 5 to 12 per cent cobalt to high speed steel increases its hardness at the temperatures encountered in cutting thereby improving its wear resistance and cutting efficiency. Cobalt does slightly increase the brittleness of high speed steel making it susceptible to chipping at the cutting edge. For this reason, cobalt high speed steels are primarily made into single point cutting tools which are used to take heavy roughing cuts in abrasive materials and through rough abrasive surface scales.

The M40 series and T15 are a group of high hardness or so-called super-high-speed steels that can be hardened to 70 R_C; however, they tend to be brittle and difficult to grind. For cutting applications they are usually heat treated to 67-68 R_C to reduce their brittleness and tendency to chip. The M40 series is appreciably easier to grind than T15. A typical composition is M42, having: Mo-9.6%, W-1.6%, Cr-3.75%, V-1.15%, Co-8.25%, C-1.08%. They are recommended for machining tough die steels and other difficult-to-cut materials; they are not recommended for applications where conventional high-speed steels perform well. High-speed steels made by the particle-metallurgy process are tougher and have an improved grindability when compared to similar grades made by the customary process. Conventional high-speed-steel alloys are made by this process as well as grades having a very high alloy content. The particle-metallurgy process enables high-speed steels to attain a higher alloy content than can be economically produced by the customary method. Tools made of these steels can be hardened about 1 R_C higher without a sacrifice in toughness than comparable high-speed steels made by the customary process. They are particularly useful on applications involving intermittent cutting and where tool life is limited by chipping. All of these steels augment rather than replace the conventional high-speed steels.

Cemented Carbides are also called sintered carbides or simply carbides. They are harder than high-speed steels and have excellent wear resistance. They retain a very

high degree of hardness at temperatures up to 1400 F and even higher; therefore, very fast cutting speeds can be used. When used at fast cutting speeds, they produce good surface finishes on the workpiece. Carbides are more brittle than high-speed steel and must, therefore, be used with more care.

Several hundred grades of carbides have been produced and attempts to classify these grades by area of application have not been entirely successful. There are four distinct types of carbides: 1) straight tungsten carbides; 2) crater resistant carbides; 3) titanium carbide; and 4) coated carbides. *Straight tungsten carbide* usually consists of approximately 94 to 97 per cent tungsten carbide with the remainder being cobalt. It is the most abrasion resistant cemented carbide and is used to machine gray cast iron, most non-ferrous metals, and non-metallic materials, where abrasion resistance is the primary criterion. When used to machine steel, straight tungsten carbide will rapidly form a crater on the face of the tool, which has a very adverse effect on the life of the tool. Titanium carbide is added to the tungsten carbide in order to counteract the rapid formation of the crater. In addition, tantalum carbide is also usually added to prevent the cutting edge from deforming when subject to the intense heat and pressure generated in taking a heavy cut. *Crater Resisting Carbides* containing titanium and tantalum carbide additions to the tungsten carbide are used to cut steels, alloy cast irons, and other materials that have a strong tendency to form a crater.

Titanium Carbides, available in several grades, are made entirely from titanium carbide plus small amounts of nickel and molybdenum. These carbides have an excellent resistance to cratering and to heat. Their high hot hardness enables them to operate at higher cutting speeds. They are, however, more brittle and less resistant to mechanical and thermal shocks; therefore, they are not recommended for taking heavy cuts and interrupted cuts. The fundamental abrasion resistance of titanium carbides is also lower and they are not recommended for cutting through scale or oxide films on steel. Although the resistance to cratering of titanium carbides is excellent, failure caused by crater formation can sometimes occur because the chip tends to curl very close to the cutting edge, thereby forming a small crater in this region which may break through.

Coated Carbides are available only as indexable inserts because the coating would be removed by grinding. The principal types of coating material are titanium carbide (TiC), titanium nitride (TiN), and aluminum oxide (Al_2O_3). A very thin layer (approximately .0002 in.) of coating material is deposited over a cemented carbide insert; the material below the coating is called the substrate. The overall performance of the coated carbide is limited by the substrate, which provides the required toughness and resistance to deformation and thermal shock. Although coated carbides are less resistant to mechanical shock than uncoated carbides, they have been used successfully on interrupted cuts and for milling. While both coated and uncoated carbides can fail as a result of thermal shock, certain grades of coated carbides have a better thermal shock resistance than many uncoated carbide grades. By having the right combination of coating material and substrate, and by having the proper cutting edge preparation, coated carbides will exhibit superior performance characteristics. Proper honing of the cutting edge by the carbide producer is very important in the application of coated carbide inserts. With an equal tool life, coated carbides can operate at a higher cutting speed than uncoated carbides. This may be 20 to 30 per cent and in some cases up to 50 per cent faster. Titanium carbide and titanium nitride coated carbides usually operate in the medium (200-800 fpm) cutting speed range, while aluminum oxide coated carbides are used in the higher (800-1600 fpm) cutting speed range.

Carbide Grade Selection: The selection of the best grade of carbide for a particular application is very important. An improper grade of carbide will result in a poor performance—in some cases it may even cause the cutting edge to fail before any significant amount of cutting has been done. Because of the many grades and the many variables that are involved, the carbide producers should be consulted to obtain

recommendations for the application of their grades of carbide. A few general guidelines can be given which are useful to form an orientation. Metal cutting carbides usually range in hardness from about 89.5 HRA (Rockwell A Scale) to 93.0 HRA with the exception of titanium carbide which has a hardness range of 90.5 HRA to 93.5 HRA. Generally the harder carbides are more wear resistant and more brittle while the softer carbides are less wear resistant but tougher. A choice of hardness must be made to suit the given application. The very hard carbides are generally used for taking light finishing cuts. For other applications, select the carbide that has the highest hardness with sufficient strength to prevent chipping or breakage. A grade of straight tungsten carbide should always be used unless cratering is encountered. Straight tungsten carbides are used to machine gray cast iron, ferritic malleable iron, austenitic stainless steel, high temperature alloys, copper, brass, bronze, aluminum alloys, zinc alloy die castings, and plastics. Crater resistant carbides should be used to machine plain carbon steel, alloy steel, tool steel, pearlitic malleable iron, nodular iron, other highly alloyed cast irons, ferritic stainless steel, martensitic stainless steel, and certain high temperature alloys. Titanium carbides are recommended for taking high speed finishing and semi-finishing cuts on steel, especially the low carbon, low alloy steels, which are less abrasive and have a strong tendency to form a crater. They are also used to take light cuts on alloy cast iron and on some high nickel alloys. Non-ferrous materials which are essentially non-abrasive may also be machined with titanium carbides, such as some aluminum alloys and brass. Abrasive materials and others that should not be machined with titanium carbides include gray cast iron, titanium alloys, cobalt and nickel base superalloys, stainless steel, bronze, many aluminum alloys, fiberglass, plastics, and graphite. For materials that can successfully be machined with titanium carbide the cutting speed should be increased 30 to 60 per cent over the values given in the accompanying tables. The feed rate used should not exceed about .020 inch per revolution.

Coated carbides can be used to take cuts ranging from light finishing to heavy roughing on most materials that can be cut with these carbides. For titanium carbide and titanium nitride coatings, the cutting speed should be increased 20 to 30 per cent over the values in the tables; experience may show that in some cases an increase up to 50 per cent is possible. Aluminum oxide coated carbide should be operated at even higher cutting speeds. The coated carbides are recommended for machining all free machining steels, all plain carbon and alloy steels, tool steels, martensitic and ferritic stainless steels, precipitation hardening stainless steels, alloy cast iron, pearlitic and martensitic malleable iron, and nodular iron. They are also recommended for taking light finishing and roughing cuts on austenitic stainless steels. Coated carbides should not be used to machine nickel and cobalt base superalloys, titanium and titanium alloys, brass, bronze, aluminum alloys, pure metals, refractory metals, and non-metals such as fiberglass, graphite, and plastics.

Ceramic Cutting Tool Materials are made from finely powdered aluminum oxide particles sintered into a hard dense structure without a binder material. Aluminum oxide is also combined with titanium carbide to form a composite, which is called a cermet. These materials have a very high hot hardness enabling very high cutting speeds to be used. For example, ceramic cutting tools have been used to cut AISI 1040 steel at a cutting speed of 18,000 fpm with a satisfactory tool life. However, much lower cutting speeds, in the range of 1000 to 4000 fpm and lower are more common because of limitations placed by the machine tool, cutters, and chucks. Although most applications of ceramic and cermet cutting tool materials are for turning, they have also been used successfully for milling. They are relatively brittle and a special cutting edge preparation is required to prevent chipping or edge breakage. This consists of honing or grinding a narrow flat land, .002 to .006 inch wide, on the cutting edge which is made about 30 degrees with respect to the tool face. For some heavy duty applications a wider land is used. The setup should be as rigid as possible and the feed rate should not

normally exceed .020 inch, although in some cases .030 inch has been used successfully. Ceramics and cermets are recommended for roughing and finishing operations on all cast irons, plain carbon and alloy steels, and the stainless steels. Materials up to a hardness of 60 HRC (Rockwell C Scale) can be cut with ceramic and cermet cutting tools. They should not be used to machine aluminum and aluminum alloys, magnesium alloys, titanium, and titanium alloys.

Cast Nonferrous Alloys are made from tungsten, tantalum, chromium, and cobalt plus carbon. Other alloying elements are also used to produce a high temperature resistant and wear resistant material. These alloys cannot be softened by heat treatment and must be cast and ground to shape. The room temperature hardness of cast nonferrous alloys is lower than for high-speed steel, but the hardness and wear resistance is retained to a higher temperature. They are generally marketed under trade names such as Stellite, Crobalt, and Tantung. As a starting point, the cutting speed for cast nonferrous tools can be 20 to 50 per cent greater than the recommended cutting speed for high-speed steel as given in the accompanying tables.

Diamond cutting tools are available in two forms: 1. single crystal natural diamonds shaped to a cutting edge and mounted on a tool holder on a boring bar; and, 2. polycrystalline diamond indexable inserts made from synthetic or natural diamond powders which have been compacted and sintered into a solid mass. Single crystal and polycrystalline diamond cutting tools are very wear resistant. They are, therefore, recommended for machining abrasive materials which cause other cutting tool materials to wear rapidly. Typical of the abrasive materials machined with single crystal and polycrystalline diamond tools are the following: fiberglass, 300 to 1000 fpm; fused silica, 900 to 950 fpm; reinforced melamine plastics, 350 to 1000 fpm; reinforced phenolic plastics, 350 to 1000 fpm; thermosetting plastics, 300 to 2000 fpm; teflon, 600 fpm; nylon, 200 to 300 fpm; mica, 300 to 1000 fpm; graphite, 200 to 2000 fpm; babbitt bearing metal, 700 fpm; and aluminum-silicon alloys, 1000 to 2000 fpm. Another important application of diamond cutting tools is to produce fine surface finishes on soft nonferrous metals that are difficult to finish by other methods. Surface finishes of 1 to 2 microinches can readily be obtained with single crystal diamond tools, and finishes down to 10 microinches can be obtained with polycrystalline diamond tools. In addition to babbitt and the aluminum-silicon alloys, other metals finished with diamond tools include: soft aluminum, 1000 to 2000 fpm; all wrought and cast aluminum alloys, 600 to 1500 fpm; copper, 1000 fpm; brass, 500 to 1000 fpm; bronze, 300 to 600 fpm; oilite bearing metal, 500 fpm; silver, gold, and platinum, 300 to 2500 fpm; and zinc, 1000 fpm. Ferrous alloys, such as cast iron and steel, should not be machined with diamond cutting tools because the high cutting temperatures generated will cause the diamond to transform into carbon.

Cubic Boron Nitride (CBN) is also known by its trade name, Borazon,® and when used as a cutting tool material it is designated by the symbol BZN.® Next to the diamond, it is the hardest known material. It will retain its hardness at a temperature of 1800 F and higher, making it an ideal cutting tool material for machining very hard and tough materials at cutting speeds beyond those possible with other cutting tool materials. Indexable inserts and cutting tool blanks made from this material consist of a layer, approximately .020 inch thick, of polycrystalline cubic boron nitride firmly bonded to the top of a cemented carbide substrate. Cubic boron nitride is recommended for rough and finish turning hardened plain carbon and alloy steels, hardened tool steels, hard cast irons, and particularly the superalloys. As a class, the superalloys are not as hard as hardened steel; however, their combination of high strength and tendency to deform plastically under the pressure of the cut, or gumminess, places them in the class of hard-to-machine materials. Conventional materials that can readily be machined with other cutting tool materials should not be machined with cubic boron nitride. Table 10 provides a representative sample of the materials machined with cubic boron nitride and gives the cutting conditions recommended for rough turning these

materials. Finish turning operations on these materials with cubic boron nitride cutting tools produce excellent surface finishes. Round indexable inserts are recommended when taking severe cuts on these materials in order to provide maximum strength to the insert. When using square or triangular inserts, a large lead angle should be used. Normally the lead angle should be 15 degrees, and whenever possible, 45 degrees. A negative rake angle should always be used, which for most applications is negative 5 degrees. The relief angle should be 5 to 9 degrees. Although cubic boron nitride cutting tools can be used without a coolant, flooding the tool with a water soluble type coolant is recommended.

Carbon Tool Steel is used primarily to make the less expensive drills, taps, and reamers. It is seldom used to make single point cutting tools. Carbon steels are very shallow hardening although some have a small amount of vanadium and chromium added to improve their hardening quality. The cutting speed to use for plain carbon tool steel should be approximately one-half of the recommended speed for high-speed steel.

Cutting Speed, Feed, Depth of Cut. — The cutting conditions that determine the rate of metal removal are the cutting speed, the feed rate, and the depth of cut. These cutting conditions and the nature of the material to be cut determine the power required to take the cut. The cutting conditions must be adjusted to stay within the power available on the machine tool to be used.

The cutting conditions must also be considered in relation to the tool life. Tool life can be defined as the length of time that a cutting tool will cut before becoming dull or before it must be replaced. The end of tool life is defined as a given amount of wear on the flank of the tool by the ANSI Standard Specification For Tool Life Testing With Single-Point Tools — ANSI B94.55M-1985. This standard is followed when making scientific machinability tests with single-point cutting tools in order to achieve uniformity in testing procedure so that results from different machinability laboratories can readily be compared. It is not practicable or necessary to follow this standard in the shop; however, it should be understood that the cutting conditions and tool life are related. For example, the cutting speed and the feed may be increased if a shorter tool life is accepted. Furthermore, the decrease in the tool life will be proportionately greater than the increase in the cutting speed or the feed. Conversely, if the cutting speed or the feed is decreased, the increase in the tool life will be proportionately greater than the decrease in the cutting speed or the feed.

Tool life is influenced most by cutting speed, then by the feed rate, and least by the depth of cut. After the depth of cut is about 10 times greater than the feed rate, a further increase in the depth of cut will have no significant effect on the tool life. This characteristic of the performance of cutting tools is very important in determining the operating or cutting conditions for machining metals.

The first step in selecting the cutting conditions is to select the depth of cut. The depth of cut will be limited by the amount of metal that is to be machined from the workpiece, by the power available on the machine tool, by the rigidity of the workpiece and the cutting tool, and by the rigidity of the setup. Since the depth of cut has the least effect upon the tool life, always use the heaviest depth of cut that is possible.

The second step is to select the feed rate. In selecting the feed rate, consideration must be given to the power available on the machine tool, to the rigidity of the workpiece and the cutting tool, to the rigidity of the setup, and to the surface finish required on the finished workpiece. The available power must be considered in relation to the depth of cut previously selected in considering the feed. Select the maximum feed possible; however, it must not be greater than that which will produce an acceptable surface finish.

The third step is to select the cutting speed. The accompanying tables provide recommended cutting speeds. If previous experience has been had in machining a certain material, this may form the basis for selecting the cutting speed. In either case,

however, the depth of cut should be selected first, followed by the feed, and the last to be selected should be the cutting speed.

Use of Cutting Speed Tables. — On the following pages tables of recommended cutting speeds are provided. The values in these tables are for average conditions and serve as a basis from which to start. In many cases they will be found to be the most satisfactory cutting speeds, while in others, modifications may offer advantages as experience is gained on a particular job. It is not possible to specify a single optimum cutting speed for a material that will fit every situation. Many factors unique to each job may make a modification desirable. These factors include: the size, type, model, and make of the machine tool; its power, rigidity, and the foundation on which it is standing; the workpiece configuration; the rigidity of the workpiece setup, or fixturing; safety aspects of the setup; the particular grade of cemented carbide used; the influence of a cutting fluid; and, the tool life desired. Except for certain difficult-to-machine materials, most materials can be cut successfully over a rather wide range of cutting speeds with a particular type of cutting tool material; however, not just any speed within this range should be used. A cutting speed that is too slow will result in a loss of production and in an increase in the cost of the part. Likewise, a cutting speed that is too fast can have the same result because the tool life will be too short and the production must frequently be interrupted to change tools. Moreover, the cost of sharpening or replacing the cutting tool must be considered. There is usually a narrow range of cutting speeds within which the most economical results will be obtained. When making a modification to the cutting speed the fundamental behaviour of cutting tools must be kept in mind; i.e., increasing the cutting speed will result in a proportionately larger reduction in the tool life; reducing the cutting speed will result in a proportionately larger increase in the tool life.

Feed and Depth of Cut Factors: Factors used to modify the cutting speeds in compensation for different feed rates and depths of cut are given in Table 5. These factors should be used only with the cutting speeds listed in Tables 1 through 4. Moreover, they do not apply when the hardness of a material exceeds approximately 350 to 400 HB. They should, however, be used in the other hardness ranges, at which most materials are machined. Experience in production and scientific investigations conducted in machinability laboratories have shown that a change in the feed or in the depth of cut will require a compensating change in the cutting speed if the tool life is to remain unchanged. For this reason a modification in the cutting speed should be made if the feed or depth of cut is changed significantly when machining the materials listed in the other tables.

Cutting Tool Material: The cutting speeds in the tables listed under cemented carbide are based on the assumption that a correct grade of uncoated straight tungsten carbide or an uncoated crater resisting carbide is used. Most carbide producers will be able to recommend a grade of carbide that can be used at the cutting speed given. An incorrect grade of carbide may, however, require a modification to the cutting speed, and in some instances it may result in an extremely short tool life. High speed steels are less sensitive in this respect; i.e., any type of high speed steel can cut most of the materials listed in the tables at the recommended cutting speed for high speed steel. It is true, however, that certain types of high speed steels do offer advantages in some applications, such as a longer tool life. Very hard and tough materials, such as the superalloys, should be cut with T15 high speed steel, or with one of the M-40 types, as for example M42. These premium high speed steels should not be used to cut most other materials, since they can be cut as well by other types of high speed steel.

Coated Carbides: A faster cutting speed can be used with coated carbide cutting tools than with uncoated tools when machining materials that can be cut by the coated grades. When using titanium carbide or titanium nitride coated cutting tools, the recommended cutting speed in the tables should be increased by 20 to 30 per cent;

Summary of Principal Tables in the Speeds and Feeds Section

Recommended Cutting Speeds for Turning:
- Table 1. Plain carbon and alloy steels
- Table 2. Tool steels
- Table 3. Stainless steels
- Table 4. Ferrous cast metals
- Table 5. Cutting speed feed and depth of cut factors for use with Tables 1-4
- Table 6. Light metals (For milling also)
- Table 7. Copper alloys (For milling also)
- Table 8. Titanium and titanium alloys (For milling and drilling also)
- Table 9. Superalloys (For milling and drilling also)
- Table 10. Hard-to-machine alloys with cubic boron nitride cutting tools

Recommended Cutting Speeds for Milling:
- Table 11. Plain carbon and alloy steels
- Table 12. Tool steels
- Table 13. Stainless steels
- Table 14. Ferrous cast metals
- Table 15. Feed rates using high speed steel cutters
- Table 16. Feed rates using cemented carbide cutters

Recommended Cutting Speeds for Drilling and Reaming:
- Table 17. Plain carbon and alloy steels
- Table 18. Tool steels
- Table 19. Stainless steels
- Table 20. Ferrous cast metals
- Table 21. Light metals
- Table 22. Copper alloys

Machining Power:
- Table 23. Power constants for wrought steels
- Table 24. Power constants for ferrous cast metals
- Table 25. Power constants for high temperature alloys, tool steels, stainless steel and non-ferrous metals
- Table 26. Feed factors for power constants
- Table 27. Tool wear factors
- Table 28. Machine tool efficiency factors
- Table 29. Formulas for calculating the metal removal rate
- Table 30. Work material factors for drilling
- Table 31. Feed factors for drilling
- Table 32. Drill diameter factors for thrust and torque
- Table 33. Chisel edge factors for thrust and torque

in some cases an increase up to 50 per cent is possible. The cutting speed should be increased even more when using aluminum oxide coated carbides.

Titanium Carbides: The cutting speed should be increased by 30 to 60 per cent over the values given in the tables when using titanium carbide cutting tools.

Milling: The recommended cutting speed for milling given in the tables can be used for all face milling operations, slab milling operations, slab milling type cuts taken with end milling cutters, and for shallow slotting cuts taken with either end milling cutters or with side milling cutters. When milling deep slots with end milling cutters or with side milling cutters, the cutting speed should be reduced approximately 10 per

cent. Likewise, the cutting speed should be reduced about 10 per cent when taking wide side milling cuts with side milling cutters; no reduction in the cutting speed is required if the width of the side milling cut is small.

Planing and Shaping: The cutting speeds in Tables 1 through 4, and Tables 6 through 9 can be used for planing and shaping. The feed and depth of cut factors in Table 5 should also be used as explained previously. Very often other factors relating to the machine or the setup will require a reduction in the actual cutting speed used on the job.

Surface Scale: Certain heavy and abrasive surface scales encountered on castings and on some wrought metals require a reduction in the cutting speed, especially when drilling.

Cutting Fluids: Many cutting fluids permit a somewhat higher cutting speed to be used. It is not possible, however, to provide specific recommendations because each proprietary cutting fluid exhibits its own characteristics.

Metric Units: In metric practice the units used for the cutting speed is meters per minute, abbreviated m/min. or mpm. The cutting speeds in the tables are given in inch units, or feet per minute, which is abbreviated fpm. These units can be converted as follows: to obtain meters per minute, multiply feet per minute by 0.3; to obtain feet per minute, multiply meters per minute by 3.28.

Cutting Speed Formulas: Most machining operations are conducted on machine tools having a rotating spindle, and the cutting speed in feet or meters per minute must be converted to a spindle speed, or to revolutions per minute; this is accomplished by use of the following formulas:

For inch units only:

$$N = \frac{12V}{\pi D}$$

For metric units only:

$$N = \frac{1000V}{\pi D}$$

Where: N = Spindle speed; rpm

V = Cutting speed; fpm, or m/min

D = Diameter; in., or mm (For turning, D is the outside diameter of the workpiece. For milling, drilling and reaming, D is the diameter of the cutter.)

π = 3.14

The feed and depth-of-cut factors in Table 5 are only to be used together with Tables 1 through 4. While the values in Table 5 are for inch units, they can be used with metric units by converting the metric feed and depth of cut into inch units. First select the cutting speed from the appropriate table and then apply the factors from Table 5, using the formula given below:

$$V = V_o F_f F_d$$

Where: V = Cutting speed to be used; fpm, or m/min.

V_o = Cutting speed from tables; fpm, or m/min.

F_f = Feed Factor   (From Table 5)

F_d = Depth-of-cut factor   (From Table 5)

Example: Using both the inch and the metric formulas, calculate the spindle speed for

turning a 1¼ inch (31.75 mm) bar of 200-220 HB AISI 1040 steel using depth of cut of .100 in. (2.54 mm) and a feed rate of .015 in. (0.38 mm/rev.). (A small insignificant difference in the answers may occur, which is caused by the conversion of the units and the rounding off of numbers.)

From Table 1: $V_o = 85$ fpm.

From Table 5: $F_f = .91$;
$\qquad\qquad F_d = 1.03$

$$V = V_o F_f F_d$$
$$= 85 \times .91 \times 1.03$$
$$= 80 \text{ fpm}$$

$$N = \frac{12V}{\pi D}$$
$$= \frac{12 \times 80}{\pi \times 1.250}$$
$$= 244 \text{ rpm}$$

$$V_o = 85 \times .3$$
$$= 25.5 \text{ m/min}$$

$$V = V_o F_f F_d$$
$$= 25.5 \times .91 \times 1.03$$
$$= 24 \text{ m/min}$$

$$N = \frac{1000V}{\pi D}$$
$$= \frac{1000 \times 24}{\pi \times 31.75}$$
$$= 241 \text{ rpm}$$

It is often necessary to calculate the cutting speed in feet per minute or in meters per minute, when the diameter of the workpiece or of the cutting tool and the spindle speed are known. In this event, the following formulas are used.

For inch units only:

$$V = \frac{\pi D N}{12} \text{ , where } D \text{ is in inches}$$

For metric units only:

$$V = \frac{\pi D N}{1000} \text{ , where } D \text{ is in millimeters}$$

Example: Calculate the cutting speed in feet per minute and in meters per minute when the spindle speed of a ¾ inch (19.05 mm) drill is 400 rpm.

$$V = \frac{\pi D N}{12} = \frac{\pi \times .75 \times 400}{12} = 78.5 \text{ fpm}$$

$$V = \frac{\pi D N}{1000} = \frac{\pi \times 19.05 \times 400}{1000} = 24 \text{ m/min}$$

Table 1. Recommended Cutting Speeds in Feet per Minute for Turning Plain Carbon and Alloy Steels

Material AISI and SAE Steels	Hardness, HB*	Material Condition*	Cutting Speed, fpm HSS	Carbide
Free Machining Plain Carbon Steels (Resulphurized), 1212, 1213, 1215	100-150	HR, A	150	600
	150-200	CD	160	625
1108, 1109, 1115, 1117, 1118, 1120 1126, 1211	100-150	HR, A	130	500
	150-200	CD	120	525
1132, 1137, 1139, 1140, 1144, 1146, 1151	175-225	HR, A, N, CD	120	400
	275-325	Q and T	75	300
	325-375	Q and T	50	225
	375-425	Q and T	40	200
Free Machining Plain Carbon Steels (Leaded), 11L17, 11L18, 12L13, 12L14	100-150	HR, A, N, CD	140	550
	150-200	HR, A, N, CD	145	560
	200-250	N, CD	110	400
Plain Carbon Steels, 1006, 1008, 1009, 1010, 1012, 1015, 1016, 1017, 1018, 1019, 1020, 1021, 1022, 1023, 1024, 1025, 1026, 1513, 1514	100-125	HR, A, N, CD	120	450
	125-175	HR, A, N, CD	110	400
	175-225	HR, N, CD	90	350
	225-275	CD	70	300
1027, 1030, 1033, 1035, 1036, 1037, 1038, 1039, 1040, 1041, 1042, 1043, 1045, 1046, 1048, 1049, 1050, 1052, 1524, 1526, 1527, 1541	125-175	HR, A, N, CD	100	375
	175-225	HR, A, N, CD	85	325
	225-275	N, CD, Q and T	70	225
	275-325	Q and T	60	200
	325-375	Q and T	40	160
	375-425	Q and T	30	140
1055, 1060, 1064, 1065, 1070, 1074, 1078, 1080, 1084, 1086, 1090, 1095, 1548, 1551, 1552, 1561, 1566	125-175	HR, A, N, CD	100	370
	175-225	HR, A, N, CD	80	320
	225-275	N, CD, Q and T	65	220
	275-325	Q and T	50	180
	325-375	Q and T	35	150
	375-425	Q and T	30	130
Free Machining Alloy Steels (Resulphurized), 4140, 4150	175-200	HR, A, N, CD	110	400
	200-250	HR, N, CD	90	350
	250-300	Q and T	65	300
	300-375	Q and T	50	225
	375-425	Q and T	40	165
Free Machining Alloy Steels (Leaded), 41L30, 41L40, 41L47, 41L50, 43L47, 51L32, 52L100, 86L20, 86L40	150-200	HR, A, N, CD	120	430
	200-250	HR, N, CD	100	380
	250-300	Q and T	75	275
	300-375	Q and T	55	220
	375-425	Q and T	50	200
Alloy Steels, 4012, 4023, 4024, 4028, 4118, 4320, 4419, 4422, 4427, 4615, 4620, 4621, 4626, 4718, 4720, 4815, 4817, 4820, 5015, 5117, 5120, 6118, 8115, 8615, 8617, 8620, 8622, 8625, 8627, 8720, 8822, 94B17	125-175	HR, A, N, CD	100	400
	175-225	HR, N, CD	90	350
	225-275	CD, N, Q and T	70	300
	275-325	Q and T	60	250
	325-375	Q and T	50	200
	375-425	Q and T	35	175

Based on a feed rate of .012 in. per rev. and a depth of cut of .125 in.
* Abbreviations designate: HR, hot rolled; CD, cold drawn; A, annealed; N, normalized; Q and T, quenched and tempered; and HB, Brinell hardness number.

Table 1 (*Concluded*). **Recommended Cutting Speeds in Feet per Minute for Turning Plain Carbon and Alloy Steels**

Material AISI and SAE Steels	Hardness, HB*	Material Condition*	Cutting Speed, fpm	
			HSS	Carbide
Alloy Steels, 1330, 1335, 1340, 1345 4032, 4037, 4042, 4047, 4130, 4135, 4137, 4140, 4142, 4145, 4147, 4150, 4161, 4337, 4340, 50B44, 50B46, 50B50, 50B60, 5130, 5132, 5140, 5145, 5147, 5150, 5160, 51B60, 6150, 81B45, 8630, 8635, 8637, 8640, 8642, 8645, 8650, 8655, 8660, 8740, 9254, 9255, 9260, 9262, 94B30	175-225 225-275 275-325 325-375 375-425	HR, A, N, CD N, CD, Q and T N, Q and T N, Q and T Q and T	85 70 60 40 30	325 275 230 200 150
Alloy Steels, E51100, E52100	175-225 225-275 275-325 325-375 375-425	HR, A, CD N, CD, Q and T N, Q and T N, Q and T Q and T	70 65 50 30 20	310 260 220 180 140
Ultra High Strength Steels (Not AISI) AMS 6421 (98B37 Mod.), AMS 6422 (98BV40), AMS 6424, AMS 6427, AMS 6428, AMS 6430, AMS 6432, AMS 6433, AMS 6434, AMS 6436, AMS 6442, 300M, D6ac	220-300 300-350 350-400 43-48 HRC 48-52 HRC	A N N Q and T Q and T	65 50 35 25 10	270 200 150 120 80
Maraging Steels (Not AISI) 18% Ni Grade 200 18% Ni Grade 250 18% Ni Grade 300 18% Ni Grade 350	250-325 50-52 HRC	A Maraged	60 10	300 80
Nitriding Steels (Not AISI) Nitralloy 125 Nitralloy 135 Nitralloy 135 Mod. Nitralloy 225 Nitralloy 230 Nitralloy N Nitralloy EZ Nitrex 1	200-250 300-350	A N, Q and T	70 30	300 225

Based on a feed rate of .012 in. per rev. and a depth of cut of .125 in.

* Abbreviations designate: HR, hot rolled; CD, cold drawn; A, annealed; N, normalized; Q and T, quenched and tempered; HB, Brinell hardness number; and HRC, Rockwell C scale hardness number.

Cutting Time for Turning, Boring, and Facing. — The time required to turn a length of metal can be determined by the following formula in which T = time in minutes, L = length of cut in inches, f = feed in inches per revolution, and N = lathe spindle speed in revolutions per minute.

$$T = \frac{L}{fN}$$

When making job estimates, the time required to load and to unload the workpiece on the machine, and the machine handling time must be added to the cutting time for each length cut to obtain the floor-to-floor time.

Table 2.   Recommended Cutting Speeds in Feet per Minute for Turning Tool Steels

Material Tool Steels (AISI Types)	Hardness, HB*	Material Condition*	Cutting Speed, fpm	
			HSS	Carbide
Water Hardening W1, W2, W5	150-200	A	100	325
Shock Resisting S1, S2, S5, S6, S7	175-225	A	70	300
Cold Work, Oil Hardening O1, O2, O6, O7	175-225	A	70	250
Cold Work, High Carbon High Chromium D2, D3, D4, D5, D7	200-250	A	45	175
Cold Work, Air Hardening A2, A3, A8, A9, A10	200-250	A	70	250
A4, A6	200-250	A	55	200
A7	225-275	A	45	175
Hot Work, Chromium Type H10, H11, H12, H13, H14, H19	150-200	A	80	300
	200-250	A	65	225
	325-375	Q and T	50	175
	48-50 HRC	Q and T	20	95
	50-52 HRC	Q and T	10	80
	52-54 HRC	Q and T	—	60
	54-56 HRC	Q and T	—	40
Hot Work, Tungsten Type H21, H22, H23, H24, H25, H26	150-200	A	60	250
	200-250	A	50	200
Hot Work, Molybdenum Type H41, H42, H43	150-200	A	55	225
	200-250	A	45	175
Special Purpose, Low Alloy L2, L3, L6	150-200	A	75	325
Mold P2, P3, P4, P5, P6	100-150	A	90	400
P20, P21	150-200	A	80	350
High Speed Steel M1, M2, M6, M10, T1, T2, T6	200-250	A	65	225
M3-1, M4, M7, M30, M33, M34, M36, M41, M42, M43, M44, M46, M47, T5, T8	225-275	A	55	200
T15, M3-2	225-275	A	45	170

Based on a feed rate of .012 in. per rev. and a depth of cut of .125 in.
* Abbreviations designate: A, annealed; Q and T, quenched and tempered; HB, Brinell hardness number; and HRC, Rockwell C scale hardness number.

Cutting Speed for Tapping. — A table of cutting speeds for tapping is not given. Several factors, singly or in combination, can cause very great differences in the permissible tapping speed. The principal factors affecting the tapping speed are the pitch of the thread, the chamfer length on the tap, the percentage of full thread to be cut, the length of the hole to be tapped, the cutting fluid used, whether the threads are straight or tapered, the machine tool used to perform the operation, and the material to be tapped.

The cutting speed for coarse pitch taps must be slower than for fine pitch taps

Table 3. Recommended Cutting Speeds in Feet per Minute for Turning Stainless Steels

Material Stainless Steels	Hardness, HB*	Material Condition*	Cutting Speed, fpm	
			HSS	Carbide
Free Machining Stainless Steels				
(Ferritic), 430F, 430F Se	135-185	A	110	400
(Austenitic), 203EZ, 303, 303Se, 303MA, 303Pb, 303Cu, 303 Plus X	135-185	A	100	350
	225-275	CD	80	325
(Martensitic); 416, 416Se, 416 Plus X, 420F, 420F Se, 440F, 440F Se	135-185	A	110	400
	185-240	A, CD	100	350
	275-325	Q and T	60	250
	375-425	Q and T	30	125
Stainless Steels				
(Ferritic), 405, 409, 429, 430, 434, 436, 442, 446, 502	135-185	A	90	300
(Austenitic), 201, 202, 301, 302, 304, 304L, 305, 308, 321, 347, 348	135-185	A	75	225
	225-275	CD	65	200
(Austenitic), 302B, 309, 309S, 310, 310S, 314, 316, 316L, 317, 330	135-185	A	70	225
(Martensitic), 403, 410, 420, 501	135-175	A	95	350
	175-225	A	85	300
	275-325	Q and T	55	200
	375-425	Q and T	35	125
(Martensitic), 414, 431, Greek Ascoloy	225-275	A	60	250
	275-325	Q and T	50	200
	375-425	Q and T	30	125
(Martensitic), 440A, 440B, 440C	225-275	A	55	200
	275-325	Q and T	45	150
	375-425	Q and T	30	125
(Precipitation Hardening) 15-5PH, 17-4PH, 17-7PH, AF-71, 17-14CuMo, AFC-77, AM-350, AM-355, AM-362, Custom 455, HNM, PH13-8, PH14-8Mo, PH15-7Mo, Stainless W	150-200	A	60	225
	275-325	H	50	200
	325-375	H	40	130
	375-450	H	25	90

Based on a feed rate of .012 in. per rev. and a depth of cut of .125 in.
* Abbreviations designate: A, annealed; CD, cold drawn; Q and T, quenched and tempered; H, precipitation hardened; and HB, Brinell hardness number.

with the same diameter. Usually the difference in pitch becomes more pronounced as the diameter of the tap becomes larger and slight differences in the pitch of smaller diameter taps have little significant effect on the cutting speed. Unlike all other cutting tools, the feed per revolution of a tap cannot be independently adjusted—it is always equal to the lead of the thread and is always greater for coarse pitches than for fine pitches. Furthermore, the thread form of a coarse pitch thread is larger than that of a fine pitch thread; therefore, it is necessary to remove more metal when cutting a coarse pitch thread.

Taps with a long chamfer, such as starting or taper taps, can cut faster in a short hole than short chamfer taps, such as plug taps. In deep holes, however, short chamfer or plug taps can run faster than long chamfer taps. Bottoming taps must be run more slowly than either starting or plug taps. The chamfer helps to start the tap in the hole. It also functions to involve more threads, or thread form cutting edges, on the tap in cutting the thread in the hole. This reduces the cutting load on

Table 4. Recommended Cutting Speeds in Feet per Minute for Turning Ferrous Cast Metals

Material Ferrous Cast Metals	Hardness, HB*	Material Condition*	Cutting Speed, fpm	
			HSS	Carbide
Gray Cast Iron				
ASTM Class 20	120-150	A	120	450
ASTM Class 25	160-200	AC	90	350
ASTM Class 30, 35, and 40	190-220	AC	80	275
ASTM Class 45 and 50	220-260	AC	60	200
ASTM Class 55 and 60	250-320	AC, HT	35	125
ASTM Type 1, 1b, 5 (Ni Resist)	100-215	AC	70	225
ASTM Type 2, 3, 6 (Ni Resist)	120-175	AC	65	210
ASTM Type 2b, 4 (Ni Resist)	150-250	AC	50	200
Malleable Iron				
(Ferritic), 32510, 35018	110-160	MHT	130	500
(Pearlitic), 40010, 43010, 45006, 45008, 48005, 50005	160-200	MHT	95	400
	200-240	MHT	75	275
(Martensitic), 53004, 60003, 60004	200-255	MHT	70	250
(Martensitic), 70002, 70003	220-260	MHT	60	225
(Martensitic), 80002	240-280	MHT	50	140
(Martensitic), 90001	250-320	MHT	30	125
Nodular (Ductile) Iron				
(Ferritic), 60-40-18, 65-45-12	140-190	A	100	450
(Ferritic-Pearlitic), 80-55-06	190-225	AC	80	350
	225-260	AC	65	210
(Pearlitic-Martensitic), 100-70-03	240-300	HT	45	175
(Martensitic), 120-90-02	270-330	HT	30	100
	330-400	HT	15	50
Cast Steels				
(Low Carbon), 1010, 1020	100-150	AC, A, N	110	400
(Medium Carbon), 1030, 1040, 1050	125-175	AC, A, N	100	400
	175-225	AC, A, N	90	350
	225-300	AC, HT	70	300
(Low Carbon Alloy), 1320, 2315, 2320 4110, 4120, 4320, 8020, 8620	150-200	AC, A, N	90	350
	200-250	AC, A, N	80	325
	250-300	AC, HT	60	250
(Medium Carbon Alloy), 1330, 1340, 2325, 2330, 4125, 4130, 4140, 4330, 4340, 8030, 80B30, 8040, 8430, 8440, 8630, 8640, 9525, 9530, 9535	175-225	AC, A, N	80	325
	225-250	AC, A, N	70	280
	250-300	AC, HT	55	250
	300-350	AC, HT	45	220
	350-400	HT	30	150

Based on a feed rate of .012 inch per revolution and a depth of cut of .125 inch.
* Abbreviations designate: A, annealed; AC, as cast; N, normalized; HT, heat treated; MHT, malleablizing heat treatment; and HB, Brinell hardness number.

Table 5.   Cutting Speed Feed and Depth of Cut Factors for Turning*

Feed, in./rev.	Feed Factor, F_f	Depth of Cut, in.	Depth of Cut Factor, F_d
.002	1.50	.005	1.50
.003	1.50	.010	1.42
.004	1.50	.016	1.33
.005	1.44	.031	1.21
.006	1.34	.047	1.15
.007	1.25	.062	1.10
.008	1.18	.078	1.07
.009	1.12	.094	1.04
.010	1.08	.100	1.03
.011	1.04	.125	1.00
.012	1.00	.150	.97
.013	.97	.188	.94
.014	.94	.200	.93
.015	.91	.250	.91
.016	.88	.312	.88
.018	.84	.375	.86
.020	.80	.438	.84
.022	.77	.500	.82
.025	.73	.625	.80
.028	.70	.688	.78
.030	.68	.750	.77
.032	.66	.812	.76
.035	.64	.938	.75
.040	.60	1.000	.74
.045	.57	1.125	.73
.050	.55	1.250	.72
.060	.50	1.375	.71

* For use in conjunction with Tables 1, 2, 3, and 4 only.

any one set of thread form cutting edges. In so doing, more chips and thinner chips are produced which are difficult to remove from deeper holes. Shortening the chamfer length causes fewer thread form cutting edges to cut, thereby producing fewer and thicker chips that can be easily disposed of. Only one or two sets of thread form cutting edges are cut on bottoming taps which cause these cutting edges to assume a heavy cutting load and produce very thick chips.

Spiral pointed taps can operate at a faster cutting speed than taps with normal flutes. These taps are made with supplementary angular flutes on the end which push the chips ahead of the tap and prevent the tapped hole from becoming clogged with chips. They are used primarily to tap open or through holes although some are made with shorter supplementary flutes for tapping blind holes.

The tapping speed must be reduced as the percentage of full thread to be cut is increased. Experiments have shown that the torque required to cut a 100 per cent thread form is more than twice that required to cut a 50 per cent thread form. An increase in the percentage of full thread will also produce a greater volume of chips.

The tapping speed must be lowered as the length of the hole to be tapped is increased. More friction must be overcome in turning the tap and more chips accumulate in the hole. It will be more difficult to apply the cutting fluid at the cutting edges and to lubricate the tap in order to reduce friction. This is especially true when the hole is being tapped in a horizontal position.

Cutting fluids have a very great effect on the cutting speed for tapping. While other operating conditions when tapping frequently cannot be changed, a free selection of the cutting fluid can usually be made. When planning the tapping operation, the selection of a cutting fluid warrants a very careful consideration and perhaps an investigation.

Taper threaded taps, such as pipe taps, must be operated at a slower speed than

Table 6. Recommended Cutting Speeds in Feet per Minute for Turning and Milling
Light Metals

Material, Light Metals	Material Condition*	Cutting Speed, fpm	
		HSS	Carbide
All Wrought Aluminum Alloys {	CD	600	1200
	ST and A	500	1100
All Aluminum Sand and Permanent Mold Casting Alloys {	AC	750	1400
	ST and A	600	1200
All Aluminum Die Casting Alloys† {	AC	125	550
	ST and A	100	450
†except Alloys 390.0 and 392.0 {	AC	80	500
	ST and A	60	425
All Wrought Magnesium Alloys	A, CD, ST, and A	800	2000
All Cast Magnesium Alloys	A, AC, ST and A	800	2000

* Abbreviations designate: A, annealed; AC, as cast; CD, cold drawn; and ST and A, solution treated and aged.

straight thread taps with a comparable diameter. All of the thread form cutting edges of a taper threaded tap that are engaged in the work cut and produce a chip, while only those cutting edges along the chamfer length cut on straight thread taps. In many cases the pipe tap is required to cut the tapered thread from a straight hole which adds to the cutting burden.

The machine tool on which the tapping operation is performed must be considered in selecting the tapping speed. Tapping machines and other machines that are able to feed the tap at a rate of advance equal to the lead of the tap and which have provisions for quickly reversing the spindle can be operated at high cutting speeds. On machines where the feed of the tap is controlled manually—such as on drill presses and turret lathes—the tapping speed must be reduced to allow the operator to maintain a safe control of the operation.

There are other special considerations in selecting the tapping speed. Very accurate threads are usually tapped more slowly than threads with a commercial grade of accuracy. Thread forms that require deep threads for which a large amount of metal must be removed, producing a large volume of chips, require special techniques and slower cutting speeds. Acme, buttress, and square threads are, therefore, generally cut at lower speeds.

Not the least important consideration is the material to be tapped. Like other metal cutting operations, tapping is affected by the basic structure of the metal being tapped and its hardness. Because of all of the factors mentioned above, it is not practical to tabulate the cutting speeds for tapping as has been done for other operations. On a relative basis, the cutting speeds for tapping can be compared to the cutting speed for drilling with high speed steel drills. The speeds may be used as a starting point but normally must be reduced, and must not be applied as such without weighing every factor that affects the tapping operation.

Cutting Speed for Broaching. — Broaching offers many advantages in manufacturing metal parts, including high production rates, excellent surface finishes, and close dimensional tolerances. These advantages are not derived from the use of high cutting speeds; they are derived from the large number of cutting teeth that can be applied in consecutive order in a given period of time, from their configuration and precise dimensions, and from the width or diameter of the surface that can be machined in a single stroke. Most broaching cutters are expensive in their initial

Table 7.   Recommended Cutting Speeds in Feet per Minute for Turning
and Milling Copper Alloys

Material Copper Alloys (Copper Alloy Nos. as per the Copper Development Assn. Inc.)	Material Condition*	Cutting Speed, fpm	
		HSS	Carbide
314 Leaded Commercial Bronze 332 High Leaded Brass 340 Medium Leaded Brass 342 High Leaded Brass 353 High Leaded Brass 356 Extra High Leaded Brass 360 Free Cutting Brass 370 Free Cutting Muntz Metal 377 Forging Brass 385 Architectural Bronze 485 Leaded Naval Brass 544 Free Cutting Phosphor Bronze	A CD	300 350	650 600
226 Jewelry Bronze 230 Red Brass 240 Low Brass 260 Cartridge Brass 70% 268 Yellow Brass 280 Muntz Metal 335 Low Leaded Brass 365 Leaded Muntz Metal 368 Leaded Muntz Metal 443 Admiralty Brass (inhibited) 445 Admiralty Brass (inhibited) 651 Low Silicon Bronze 655 High Silicon Bronze 675 Manganese Bronze 687 Aluminum Brass 770 Nickel Silver 796 Leaded Nickel Silver	A CD	200 250	500 550
102 Oxygen Free Copper 110 Electrolytic Tough Pitch Copper 122 Phosphorus Deoxidized Copper 170 Beryllium Copper 172 Beryllium Copper 175 Beryllium Copper 210 Gilding, 95% 220 Commercial Bronze 502 Phosphor Bronze 1.25% 510 Phosphor Bronze 5% 521 Phosphor Bronze 8% 524 Phosphor Bronze 10% 614 Aluminum Bronze 706 Copper Nickel 10% 715 Copper Nickel 30% 745 Nickel Silver 752 Nickel Silver 754 Nickel Silver 757 Nickel Silver	A CD	100 110	200 225

* Abbreviations used in this column are as follows: A, annealed; CD, cold drawn.

Table 8. Recommended Cutting Speeds in Feet per Minute for Turning, Milling, and Drilling Titanium and Titanium Alloys

Material Titanium and Titanium Alloys	Hardness, HB*	Material Condition*	Cutting Speed, fpm	
			HSS	Carbide
Commercially Pure				
99.5Ti	110-150	A	110	400
99.1Ti, 99.2Ti	180-240	A	90	300
99.0% Ti	250-275	A	70	250
Low Alloyed				
99.5Ti-.15Pd	110-150	A	100	350
99.2Ti-.15Pd, 98.9Ti-.8Ni-.3Mo	180-250	A	85	280
Alpha Alloys and Alpha-Beta Alloys				
5Al-2.5Sn, 8Mn, 2Al-11Sn-5Zr-1Mo, 4Al-3Mo-1V, 5Al-6Sn-2Zr-1Mo, 6Al-2Sn-4Zr-2Mo, 6Al-2Sn-4Zr-6Mo, 6Al-2Sn-4Zr-2Mo-.25Si	300-350	A	50	200
6Al-4V	310-350	A	40	125
6Al-6V-2Sn, 7Al-4Mo, 8Al-1Mo-1V	320-370	A	30	100
8V-5Fe-1Al	320-380	A	20	90
6Al-4V, 6Al-2Sn-4Zr-2Mo, 6Al-2Sn-4Zr-6Mo, 6Al-2Sn-4Zr-2Mo-.25Sn	320-380	ST and A	40	100
4Al-3Mo-1V, 6Al-6V-2Sn, 7Al-4Mo	375-420	ST and A	20	80
1Al-8V-5Fe	375-440	ST and A	20	75
Beta Alloys				
13V-11Cr-3Al, 8Mo-8V-2Fe-3Al, 3Al-8V-6Cr-4Mo-4Zr, 11.5Mo-6Zr-4.5Sn	275-350	A, ST	25	100
	350-440	ST and A	20	60

* Abbreviations designate: A, annealed; ST, solution treated; ST and A, solution treated and aged; and HB, Brinell hardness number.

cost and are expensive to sharpen. For these reasons a long tool life is desirable, and to obtain a long tool life relatively slow cutting speeds are used. In many instances slower cutting speeds are used because of the limitations of the machine in accelerating and stopping heavy broaching cutters. At other times the available power on the machine places a limit on the cutting speed that can be used; i.e., the cubic inches of metal removed per minute must be within the power capacity of the machine.

The cutting speeds for high speed steel broaches range from 3 to 50 feet per minute, although faster speeds have been used. In general, the harder and more difficult to machine materials are cut at a slower cutting speed and those that are easier to machine are cut at a faster speed. Some typical recommendations for high speed steel broaches are: AISI 1040, 10 to 30 fpm; AISI 1060, 10 to 25 fpm; AISI 4140, 10 to 25 fpm; AISI 41L40, 20 to 30 fpm; 201 austenitic stainless steel, 10 to 20 fpm; Class 20 gray cast iron, 20 to 30 fpm; Class 40 gray cast iron, 15 to 25 fpm; aluminum and magnesium alloys, 30 to 50 fpm; copper alloys, 20 to 30 fpm; commercially pure titanium, 20 to 25 fpm; alpha and beta titanium alloys, 5 fpm; and the superalloys, 3 to 10 fpm. Surface broaching operations on gray iron castings have been conducted at a cutting speed of 150 fpm, using indexable insert cemented carbide broaching cutters. In selecting the speed for broaching, the cardinal principle of the performance of all metal cutting tools should be kept in mind;

Table 9. Recommended Cutting Speeds in Feet per Minute for Turning, Milling, and Drilling* Superalloys

Material	Roughing HSS	Roughing Carbide	Finishing HSS	Finishing Carbide
A-286	30-35	120-145	35-40	145-155
AF2-1DA	8-10	35-45	10-15	40-50
Air Resist 213	15-20	55-65	20-25	70-85
Air Resist 13, and 215	10-12	35-40	10-15	45-55
Astroloy	5-10	25-50	5-15	50-75
B-1900	8-10	30-35	8-10	35-50
CW-12M	8-12	55-65	10-15	65-85
Discalloy	15-35	100-150	35-40	140-180
FSX-H14	10-12	35-40	10-15	45-55
GMR-235, and 235D	8-10	30-35	8-10	40-50
Hastelloy B, C, G, and X (wrought)	15-20	60-90	20-25	80-100
Hastelloy B, and C (cast)	8-12	55-65	10-15	75-85
Haynes 25, and 188	15-20	55-65	20-25	70-95
Haynes 36, and 151	10-12	35-40	10-15	45-55
HS 6, 21, 25, 31(X40), 36, and 151	10-12	35-40	10-15	45-55
IN 100, and 738	8-10	30-35	8-10	35-50
Incoloy 800, 801, and 802	30-35	120-160	35-40	145-180
Incoloy 804, and 825	15-20	60-90	20-25	80-100
Incoloy 901	10-20	30-60	20-35	40-80
Inconel 625, 702, 706, 718 (wrought), 721, 722, X750, 751, 901, 600, and 604	15-20	35-60	20-25	60-90
Inconel 700, and 702	10-12	40-65	12-15	65-70
Inconel 713C, and 718 (cast)	8-10	30-35	8-10	40-50
J1300	15-25	80-100	20-30	100-125
J1570	15-20	55-65	20-25	70-85
M252 (wrought)	15-20	65-75	20-25	75-85
M252 (cast)	8-10	30-35	8-10	40-50
Mar-M200, M246, M421, and M432	8-10	30-35	10-12	35-50
Mar-M905, and M918	15-20	55-65	20-25	70-85
Mar-M302, M322, and M509	10-12	35-40	10-15	45-55
N-12M	8-12	55-65	10-15	75-85
N-155	15-20	50-70	15-25	55-75
Nasa Co-W-Re	10-12	35-40	10-15	45-55
Nimonic 75, and 80	15-20	65-75	20-25	75-85
Nimonic 90 and 95	10-12	55-65	12-15	65-75
Refractaloy 26	15-20	60-90	20-25	80-100
René 41	10-15	35-60	12-20	55-80
René 80, and 95	8-10	30-45	10-15	40-50
S-590	10-20	60-90	15-30	80-100
S-816	10-15	45-65	15-20	50-75
TD-Nickel	70-80	250-290	80-100	300-350
Udimet 500, 700, and 710	10-15	30-50	12-20	40-60
Udimet 630	10-20	30-80	20-25	80-100
Unitemp 1753	8-10	35-45	10-15	40-50
V-36	10-15	45-65	15-20	50-75
V-57	30-35	120-160	35-40	145-180
W-545	25-35	110-155	30-40	140-175
W1-52	10-12	35-40	10-15	45-55
Waspaloy	10-30	30-60	25-35	50-95
X-45	10-12	35-40	10-15	45-55
16-25-6	30-35	120-160	35-40	145-180
19-9DL	25-35	110-150	30-40	140-180

* For milling and drilling, use the cutting speeds recommended under roughing.

i.e., increasing the cutting speed results in a proportionately larger reduction in tool life, and conversely, reducing the cutting speed results in a proportionately larger increase in the tool life. When broaching most materials, a suitable cutting fluid should be used to obtain a good surface finish and a better tool life. Gray cast iron can be broached without using a cutting fluid although some shops prefer to use a soluble oil.

Table 10.   Representative Cutting Conditions for Rough Turning Hard-to-Machine
Materials with Single Point Cubic Boron Nitride (CBN) Cutting Tools.

Material	Hardness, HRC	Cutting Speed, fpm	Feed, in./rev.	Depth of Cut, in.
Hardened Ferrous Alloys				
AISI 8620	63	250	.005	.060
AISI 52100	70	270	.020	.050
A2, A6, Cold Work Tool Steel	58	250	.008	.060
D2, Cold Work Tool Steel	54	250	.008	.060
H10, Hot Work Tool Steel	56	200	.005	.060
S5, Shock Resisting Tool Steel	60	400	.008	.060
O1, Oil Hardening Tool Steel	58	250	.008	.060
M2, High Speed Steel	62	250	.008	.060
Chilled Gray Iron	60	400	.010	.200
Meehanite Cast Iron	56	600	.008	.250
Superalloys				
Colmonoy	...	600	.006,	.125
Incoloy 901	...	800	.006	.125
Inconel 600	...	600	.006	.125
Inconel 718	...	600	.006	.125
K-Monel	...	600	.006	.125
René 41	...	600	.006	.125
René 77	...	500	.006	.015
René 95 (Hot Isostatic Pressed)	...	900	.005	.125
René 95 (Forged)	...	450	.005	.125
Stellite	...	600	.006	.125
Waspaloy	...	600	.003	.060

Thread Cutting with Single Point Cutting Tools. — Whenever possible the cutting speed recommended for turning should be used to cut internal and external threads with single point thread cutting tools. This cutting speed can frequently be used on numerically controlled lathes, using either cemented carbide or high speed steel tools. There are occasions, however, when a slower cutting speed must be used as a result of the workpiece configuration, the setup, or when cutting certain difficult-to-machine threads, such as coarse pitch Acme threads. A slightly reduced cutting speed is sometimes used on numerically controlled lathes to obtain a longer tool life.

Thread cutting on an engine lathe is not necessarily a slow speed operation, although on these machines the operation is controlled manually. However, there must never be a compromise with safety; the operator must always be sure that he has control over the machine to the extent that he or others will not be injured, and that the machine or the workpiece will not be damaged. On some jobs a skilled operator can safely manipulate a lathe with such skill that a fast spindle speed can be used to cut the thread, in which case the cutting speed may be equal to that recommended for turning with high speed steel, or with cemented carbide in the case of the more difficult-to-machine materials. Other jobs require using a slower cutting speed, even when the thread cutting operation is performed by a highly skilled operator. Some of the reasons for cutting threads at a lower speed have been given in the previous paragraph. Other reasons involve the ability to safely manipulate the machine such as when cutting a thread close to a large shoulder, or when cutting an internal thread with the cutting tool feeding into the bore.

Cutting Speed for Thread Chasing. — Cutting threads with a self-opening die head is called thread chasing. The die head contains a set of thread chasers that cut the thread and feed the die head in a nut-and-screw-like action. The feed rate is determined entirely by the lead or pitch of the thread. Since the feed of the die head must not be

Tool Trouble-Shooting Check List

Problem	Tool Material	Remedy
Excessive flank wear — Tool life too short	Carbide	1. Change to harder, more wear-resistant grade 2. Reduce the cutting speed 3. Reduce the cutting speed and increase the feed to maintain production 4. Reduce the feed 5. For work hardenable materials — increase the feed 6. Increase the lead angle 7. Increase the relief angles
	HSS	1. Use a coolant 2. Reduce the cutting speed 3. Reduce the cutting speed and increase the feed to maintain production 4. Reduce the feed 5. For work hardenable materials — increase the feed 6. Increase the lead angle 7. Increase the relief angle
Excessive cratering	Carbide	1. Use a crater-resistant grade 2. Use a harder, more wear-resistant grade 3. Reduce the cutting speed 4. Reduce the feed 5. Widen the chip breaker groove
	HSS	1. Use a coolant 2. Reduce the cutting speed 3. Reduce the feed 4. Widen the chip breaker groove
Cutting edge chipping	Carbide	1. Increase the cutting speed 2. Lightly hone the cutting edge 3. Change to a tougher grade 4. Use negative rake tools 5. Increase the lead angle 6. Reduce the feed 7. Reduce the depth of cut 8. Reduce the relief angles 9. If low cutting speed must be used — use a high additive EP cutting fluid
	HSS	1. Use a high additive EP cutting fluid 2. Lightly hone the cutting edge before using 3. Increase the lead angle 4. Reduce the feed 5. Reduce the depth of cut 6. Use a negative rake angle 7. Reduce the relief angles
	Carbide and HSS	1. Check the setup for cause if chatter occurs 2. Check the grinding procedure for tool over-heating 3. Reduce the tool overhang
Cutting edge deformation	Carbide	1. Change to a grade containing more tantalum 2. Reduce the cutting speed 3. Reduce the feed

Tool Trouble-Shooting Check List (*Continued*).

Problem	Tool Material	Remedy
Poor surface finish	Carbide	1. Increase the cutting speed 2. If low cutting speed must be used — use a high additive EP cutting fluid 3. For light cuts — use straight titanium carbide grade 4. Increase the nose radius 5. Reduce the feed 6. Increase the relief angles 7. Use positive rake tools
	HSS	1. Use high additive EP cutting fluid 2. Increase the nose radius 3. Reduce the feed 4. Increase the relief angles 5. Increase the rake angles
	Diamond	1. Use diamond tool for soft materials
Notching at the depth of cut line	Carbide and HSS	1. Increase the lead angle 2. Reduce the feed

too rapid, the thread lead and pitch place a limit on the spindle speed, and thereby on the cutting speed. Other factors affecting the cutting speed are the work material, the type and size of the thread, the thread tolerance, and the finish required. A cutting fluid should be used in most cases, which may also have an effect on the cutting speed. Much slower cutting speeds are recommended for thread chasing, as compared to turning. A cutting speed that is too fast will reduce the life of the thread chasers and may cause the threads cut to be rough or torn. Some typical cutting speeds recommended by one manufacturer of self-opening die heads are given below. These cutting speeds may have to be modified somewhat to suit existing conditions on each job.

Material	Threads per Inch 3-7½	8-15	16-24	25-Up
	Cutting Speed (fpm) for Threads per Inch			
AISI 1010-1035 Steel	20	30	40	50
AISI 1112-1340 Steel	20	30	40	50
AISI 1040-1095 Steel	15	20	25	30
AISI 4130-4820 Steel	8	10	15	20
AISI 5120-52100 Steel	8	10	15	20
Stainless Steel	8	10	15	20
Gray Cast Iron	25	40	50	80
Aluminum Alloys	50	100	150	200
Brass Bar Stock	50	100	150	200
Phosphor Bronze	40	80	100	150
Zinc Die Castings	50	100	150	200

Feed Rate for Milling. — Whenever the power feed is to be used to perform a milling operation, the table feed rate, in inches per minute, should always be calculated in order to achieve the best results. The table feed rate governs the production rate. Failure to calculate the table feed rate may result in overloading the milling cutter which can have serious consequences. Aside from the possible

Table 11. Cutting Speeds in Feet per Minute for Milling Plain Carbon and Alloy Steels

Material AISI and SAE Steels	Hardness, HB*	Material Condition*	Cutting Speed, fpm	
			HSS	Carbide
Free Machining Plain Carbon Steels (Resulphurized), 1212, 1213, 1215	100-150	HR, A	140	600
	150-200	CD	130	550
1108, 1109, 1115, 1117, 1118, 11120, 1126, 1211	100-150	HR, A	130	550
	150-200	CD	115	500
1132, 1137, 1139, 1140, 1144, 1146 1151	175-225	HR, A, N, CD	115	450
	275-325	Q and T	70	290
	325-375	Q and T	45	200
	375-425	Q and T	35	170
Free Machining Plain Carbon Steels (Leaded), 11L17, 11L18, 12L13, 12L14	100-150	HR, A, N, CD	140	600
	150-200	HR, A, N, CD	130	625
	200-250	N, CD	110	400
Plain Carbon Steels, 1006, 1008, 1009, 1010, 1012, 1015, 1016, 1017, 1018, 1019, 1020, 1021, 1022, 1023, 1024, 1025, 1026, 1513, 1514	100-125	HR, A, N, CD	110	425
	125-175	HR, A, N, CD	110	400
	175-225	HR, N, CD	90	350
	225-275	CD	65	250
1027, 1030, 1033, 1035, 1036, 1037, 1038, 1039, 1040, 1041, 1042, 1043, 1045, 1046, 1048, 1049, 1050, 1052, 1524, 1526, 1527, 1541	125-175	HR, A, N, CD	100	375
	175-225	HR, A, N, CD	85	325
	225-275	N, CD, Q and T	70	225
	275-325	Q and T	55	200
	325-375	Q and T	35	160
	375-425	Q and T	25	140
1055, 1060, 1064, 1065, 1070, 1074, 1078, 1080, 1084, 1086, 1090, 1095, 1548, 1551, 1552, 1561, 1566	125-175	HR, A, N, CD	90	350
	175-225	HR, A, N, CD	75	300
	225-275	N, CD, Q and T	60	200
	275-325	Q and T	45	160
	325-375	Q and T	30	145
	375-425	Q and T	15	125
Free Machining Alloy Steels (Resulphurized), 4140, 4150	175-200	HR, A, N, CD	100	400
	200-250	HR, N, CD	90	350
	250-300	Q and T	60	280
	300-375	Q and T	45	220
	375-425	Q and T	35	160
Free Machining Alloy Steels (Leaded), 41L30, 41L40, 41L47, 41L50, 43L47, 51L32, 52L100, 86L20, 86L40	150-200	HR, A, N, CD	115	425
	200-250	HR, N, CD	95	375
	250-300	Q and T	70	260
	300-375	Q and T	50	210
	375-425	Q and T	40	180
Alloy Steels, 4012, 4023, 4024, 4028, 4118, 4320, 4419, 4422, 4427, 4615, 4620, 4621, 4626, 4718, 4720, 4815, 4817, 4820, 5015, 5117, 5120, 6118, 8115, 8615, 8617, 8620, 8622, 8625, 8627, 8720, 8822, 94B17	125-175	HR, A, N, CD	100	400
	175-225	HR, N, CD	90	350
	225-275	CD, N, Q and T	60	250
	275-325	Q and T	50	200
	325-375	Q and T	40	175
	375-425	Q and T	25	150

* Abbreviations designate: HR, hot rolled; CD, cold drawn; A, annealed; N, normalized; Q and T, quenched and tempered; HB, Brinell hardness number; and HRC, Rockwell C scale hardness number.

Table 11 *(Concluded)*. **Cutting Speeds in Feet per Minute for Milling Plain Carbon and Alloy Steels**

Material AISI and SAE Steels	Hardness, HB*	Material Condition*	Cutting Speed, fpm	
			HSS	Carbide
Alloy Steels, 1330, 1335, 1340, 1345, 4032, 4037, 4042, 4047, 4130, 4135, 4137, 4140, 4142, 4145, 4147, 4150, 4161, 4337, 4340, 50B44, 50B46, 50B50, 50B60, 5130, 5132, 5140, 5145, 5147, 5150, 5160, 51B60, 6150, 81B45, 8630, 8635, 8637, 8640, 8642, 8645, 8650, 8655, 8660, 8740, 9254, 9255, 9260, 9262, 94B30	175-225 225-275 275-325 325-375 375-425	HR, A, N, CD N, CD, Q and T N, Q and T N, Q and T Q and T	75 60 50 35 20	310 260 210 180 140
Alloy Steels, E51100, E52100	175-225 225-275 275-325 325-375 375-425	HR, A, CD N, CD, Q and T N, Q and T N, Q and T Q and T	65 60 40 30 20	300 250 130 100 60
Ultra High Strength Steels (Not AISI) AMS 6421 (98B37 Mod.), AMS 6422 (98BV40), AMS 6424, AMS 6427, AMS 6428, AMS 6430, AMS 6432, AMS 6433, AMS 6434, AMS 6436, AMS 6442, 300M, D6 ac	220-300 300-350 350-400 43-48 HRC 48-52 HRC	A N N Q and T Q and T	60 45 20	250 180 130 100 60
Maraging Steels (Not AISI) 18% Ni Grade 200 18% Ni Grade 250 18% Ni Grade 300 18% Ni Grade 350	250-325 50-52 HRC	A Maraged	50 ..	250 60
Nitriding Steels (Not AISI) Nitralloy 125 Nitralloy 135 Nitralloy 135 (Mod.) Nitralloy 225 Nitralloy 230 Nitralloy N Nitralloy EZ Nitrex 1	200-250 300-350	A N, Q and T	60 25	280 200

* Abbreviations designate: HR, hot rolled; CD, cold drawn; A, annealed; N, normalized; Q and T, quenched and tempered; HB, Brinell hardness number; and HRC, Rockwell C scale hardness number.

breakage of equipment, overloading the cutter will cause the tool life of the cutter to decrease. The tool life can also be decreased if the feed rate is too slow, which, in addition, certainly leads to a loss of production.

The basic feed rate for milling cutters is the feed per tooth (f) which is expressed in inches per tooth. There are many factors to consider in selecting the feed per tooth and no formula is available to resolve these factors. Among the factors to consider are: 1) the cutting tool material; 2) the work material and its hardness; 3) the width and the depth of the cut to be taken; 4) the type of milling cutter to be used and its size; 5) the surface finish to be produced; 6) the power available on the milling machine; 7) the rigidity of the milling machine, the workpiece, the setup of the workpiece, the milling cutter, and the cutter mounting.

As a guide to help in the selection of the feed rate, two tables are given; Table 15 is for high-speed steel cutters, and Table 16 is for cemented carbide cutters.

As a cardinal principle, always use the maximum feed rate that conditions will

Table 12. Cutting Speed in Feet per Minute for Milling Tool Steels

Material Tool Steels (AISI Types)	Hardness, HB*	Material Condition*	Cutting Speed, fpm	
			HSS	Carbide
Water Hardening W1, W2, W5	150-200	A	85	250
Shock Resisting S1, S2, S5, S6, S7	175-225	A	55	215
Cold Work, Oil Hardening O1, O2, O6, O7	175-225	A	50	200
Cold Work, High Carbon High Chromium D2, D3, D4, D5, D7	200-250	A	40	150
Cold Work, Air Hardening				
A2, A3, A8, A9, A10	200-250	A	50	200
A4, A6	200-250	A	45	160
A7	225-275	A	40	140
Hot Work, Chromium Type H10, H11, H12, H13, H14, H19	150-200	A	60	250
	200-250	A	50	200
	325-375	Q and T	30	150
	48-50 HRC	Q and T	—	80
	50-52 HRC	Q and T	—	60
	52-54 HRC	Q and T	—	40
	54-56 HRC	Q and T	—	20
Hot Work, Tungsten Type H21, H22, H23, H24, H25, H26	150-200	A	55	200
	200-250	A	45	170
Hot Work, Molybdenum Type H41, H42, H43	150-200	A	55	180
	200-250	A	45	140
Special Purpose, Low Alloy L2, L3, L6	150-200	A	65	300
Mold				
P2, P3, P4, P5, P6	100-150	A	75	350
P20, P21	150-200	A	60	300
High-Speed Steel				
M1, M2, M6, M10, T1, T2, T6	200-250	A	50	175
M3-1, M4, M7, M30, M33, M34, M36, M41, M42, M43, M44, M46, M47, T5, T8	225-275	A	40	150
T15, M3-2	225-275	A	30	130

* Abbreviations designate: A, annealed; Q and T, quenched and tempered; and HB, Brinell hardness number.

permit. Avoid, if possible, using a feed rate that is less than .001 inch per tooth because this will result in a decrease in the tool life of the cutter. When milling hard materials with small diameter end mills, such small feed rates may be necessary, but otherwise use as much feed as possible. Harder materials in general will require lower feed rates than softer materials. The width and the depth of cut also affect the feed rate; wider and deeper cuts must be fed somewhat more slowly than narrow and shallow cuts. A slower feed rate will result in a better surface finish; however, always use the heaviest feed rate that will produce the surface finish

Table 13. Recommended Cutting Speeds in Feet per Minute for Milling Stainless Steels

Material Stainless Steels	Hardness, HB*	Material Condition*	Cutting Speed, fpm	
			HSS	Carbide
Free Machining Stainless Steels				
(Ferritic), 430F, 430F Se	135-185	A	95	375
(Austenitic), 203EZ, 303, 303 Se, 303MA, 303Pb, 303Cu, 303 Plus X	135-185	A	90	325
	225-275	CD	75	300
(Martensitic), 416, 416 Se, 416 Plus X, 420F, 420F Se, 440F, 440F Se	135-185	A	95	375
	185-240	CD	80	325
	275-325	Q and T	50	225
	375-425	Q and T	20	100
Stainless Steels				
(Ferritic), 405, 409, 429, 430, 434, 436, 442, 446, 502	135-185	A	75	275
(Austenitic), 201, 202, 301, 302, 304, 304L, 305, 308, 321, 347, 348	135-185	A	60	200
	225-275	CD	50	180
(Austenitic), 302B, 309, 309S, 310, 310S, 314, 316, 316L, 317, 330	135-185	A	50	200
(Martensitic), 403, 410, 420, 501	135-175	A	75	325
	175-225	A	65	275
	275-325	Q and T	40	175
	375-425	Q and T	25	100
(Martensitic), 414, 431, Greek Ascoloy	225-275	A	55	225
	275-325	Q and T	45	180
	375-425	Q and T	25	100
(Martensitic), 440A, 440B, 440C	225-275	A	50	180
	275-325	Q and T	40	140
	375-425	Q and T	20	100
(Precipitation Hardening) 15-5PH, 17-4PH, 17-7PH, AF-71, 17-14Cu Mo, AFC-77, AM-350, AM-355, AM-362, Custom 455, HNM, PH13-8, PH14-8Mo, PH15-7Mo, Stainless W	150-200	A	60	200
	275-325	H	50	180
	325-375	H	40	110
	375-450	H	25	75

* Abbreviations designate: A, annealed; CD, cold drawn; Q and T, quenched and tempered; H, precipitation hardened; and HB, Brinell hardness number.

desired. Fine chips produced by fine feeds are dangerous when milling magnesium because spontaneous combustion can occur. Thus, when milling magnesium, a fast feed that will produce a relatively thick chip should be used. Cutting stainless steel produces a work hardened layer on the surface that has been cut. When milling this material, the feed should be large enough to allow each cutting edge on the cutter to penetrate below the work hardened layer produced by the previous cutting edge. The heavy feeds recommended for face milling cutters are to be used primarily by larger cutters on milling machines having an adequate amount of power. For smaller face milling cutters, start with the slower feeds and increase the feed as indicated by the performance of the cutter and the machine.

When planning a milling operation that is to entail the use of a high cutting speed and a fast feed rate, always check to determine if the power required to take the cut is within the capacity of the milling machine. Such cutting conditions are often encountered when milling with cemented carbide cutters. The large metal removal

(Continued on page 996)

Table 14. Recommended Cutting Speeds in Feet per Minute for Milling Ferrous Cast Metals

Material Ferrous Cast Metals	Hardness, HB*	Material Condition*	Cutting Speed, fpm	
			HSS	Carbide
Gray Cast Iron				
ASTM Class 20	120-150	A	100	425
ASTM Class 25	160-200	AC	80	325
ASTM Class 30, 35, and 40	190-220	AC	70	250
ASTM Class 45 and 50	220-260	AC	50	190
ASTM Class 55 and 60	250-260	AC, HT	30	110
ASTM Type 1, 1b, 5 (Ni-Resist)	100-215	AC	50	200
ASTM Type 2, 3, 6 (Ni-Resist)	120-175	AC	40	190
ASTM Type 2b, 4 (Ni-Resist)	150-250	AC	30	180
Malleable Iron				
(Ferritic), 32510, 35018	110-160	MHT	110	475
(Pearlitic), 40010, 43010, 45006,	160-200	MHT	80	375
45008, 48005, 50005	200-240	MHT	65	250
(Martensitic), 53004, 60003, 60004	200-255	MHT	55	225
(Martensitic), 70002, 70003	220-260	MHT	50	200
(Martensitic), 80002	240-280	MHT	45	130
(Martensitic), 90001	250-320	MHT	25	110
Nodular (Ductile) Iron				
(Ferritic), 60-40-18, 65-45-12	140-190	A	75	425
(Ferritic-Pearlitic), 80-55-06	190-225	AC	60	325
	225-260	AC	50	200
(Pearlitic-Martensitic), 100-70-03	240-300	HT	40	160
(Martensitic), 120-90-02	270-330	HT	25	90
	330-400	HT	—	30
Cast Steels				
(Low Carbon), 1010, 1020	100-150	AC, A, N	100	375
(Medium Carbon), 1030, 1040, 1050	125-175	AC, A, N	95	375
	175-225	AC, A, N	80	325
	225-300	AC, HT	60	250
(Low Carbon Alloy), 1320, 2315, 2320,	150-200	AC, A, N	85	325
4110, 4120, 4320, 8020, 8620	200-250	AC, A, N	75	300
	250-300	AC, HT	50	225
(Medium Carbon Alloy), 1330, 1340,	175-225	AC, A, N	70	300
2325, 2330, 4125, 4130, 4140, 4330,	225-250	AC, A, N	65	250
4340, 8030, 80B30, 8040, 8430, 8440,	250-300	AC, HT	50	200
8630, 8640, 9525, 9530, 9535	300-350	AC, HT	30	180
	350-400	HT	..	125

* Abbreviations designate: A, annealed; AC, as cast; N, normalized; HT, heat treated; MHT, malleablizing heat treatment; and HB, Brinell hardness number.

SPEEDS AND FEEDS

Table 15. Recommended Feed in Inches per Tooth (f_t) for Milling with High Speed Steel Cutters

Material	Hardness, HB	End Mills Depth of Cut, .250 in. Cutter Diam., in. 1/2	3/4	1 and up	End Mills Depth of Cut, .050 in. Cutter Diam., in. 1/4	1/2	3/4	1 and up	Plain or Slab Mills	Form Relieved Cutters	Face Mills and Shell End Mills	Slotting and Side Mills
					Feed per Tooth, inch							
Free Machining Plain Carbon Steels	100-185	.001	.003	.004	.001	.002	.003	.004	.003-.008	.005	.004-.012	.002-.008
Plain Carbon Steels, AISI 1006 to 1030; 1513 to 1522	100-150	.001	.003	.004	.001	.002	.003	.004	.003-.008	.004	.004-.012	.002-.008
	150-200	.001	.002	.003	.001	.002	.002	.003	.003-.008	.004	.004-.012	.002-.008
AISI 1033 to 1095; 1524 to 1566	120-180	.001	.003	.003	.001	.002	.003	.003	.003-.008	.004	.003-.012	.002-.008
	180-220	.001	.002	.003	.001	.002	.002	.003	.003-.008	.004	.003-.012	.002-.008
	220-300	.001	.002	.002	.001	.001	.002	.003	.002-.006	.003	.002-.008	.002-.006
Alloy Steels having less than 3% Carbon. Typical examples: AISI 4012, 4023, 4027, 4118, 4320, 4419, 4615, 4620, 4626, 4720, 4820, 5015, 5120, 6118, 8115, 8620, 8627, 8720, 8822, 9310, 93B17	125-175	.001	.003	.003	.001	.002	.003	.003	.003-.008	.004	.004-.012	.002-.008
	175-225	.001	.002	.003	.001	.002	.003	.003	.003-.008	.004	.003-.012	.002-.008
	225-275	.001	.002	.003	.001	.001	.002	.003	.002-.006	.003	.003-.008	.002-.006
	275-335	.001	.002	.002	.001	.001	.001	.002	.002-.005	.003	.003-.008	.002-.005
Alloy Steels have 3% Carbon or more. Typical examples: AISI 1330, 1340, 4032, 4037, 4130, 4140, 4150, 4340, 50B40, 50B60, 5130, 51B60, 6150, 81B45, 8630, 8640, 86B45, 8660, 8740, 94B30	175-225	.001	.002	.003	.001	.002	.003	.004	.003-.008	.004	.003-.012	.002-.008
	225-275	.001	.002	.003	.001	.001	.002	.003	.002-.006	.003	.003-.010	.002-.006
	275-335	.001	.002	.002	.001	.001	.002	.003	.002-.005	.003	.002-.008	.002-.005
	325-375	.001	.002	.002	.001	.001	.002	.002	.002-.004	.002	.002-.008	.002-.005
Tool Steel	150-200	.001	.002	.002	.001	.001	.002	.003	.003-.008	.004	.003-.010	.002-.006
	200-250	.001	.002	.002	.001	.001	.002	.003	.002-.006	.003	.003-.008	.002-.005
Gray Cast Iron	120-180	.001	.003	.004	.002	.003	.004	.004	.004-.012	.005	.005-.016	.002-.010
	180-225	.001	.002	.003	.001	.002	.003	.003	.003-.010	.004	.004-.012	.002-.008
	225-300	.001	.002	.002	.001	.001	.002	.002	.002-.006	.003	.002-.008	.002-.005
Ferritic Malleable Iron	110-160	.001	.003	.004	.002	.003	.003	.004	.003-.010	.005	.005-.016	.002-.010
Pearlitic-Martensitic Malleable Iron	160-200	.001	.003	.004	.002	.003	.003	.004	.003-.010	.004	.004-.012	.002-.008
	200-240	.001	.002	.003	.001	.002	.003	.003	.003-.007	.004	.003-.010	.002-.006
	240-300	.001	.002	.002	.001	.001	.002	.002	.002-.006	.003	.002-.008	.002-.005

Table 15 (Concluded). Recommended Feed in Inches per Tooth (f_t) for Milling with High Speed Steel Cutters

Material	Hardness, HB	End Mills							Plain or Slab Mills	Form Relieved Cutters	Face Mills and Shell End Mills	Slotting and Side Mills
		Depth of Cut, .250 in.			Depth of Cut, .050 in.							
		Cutter Diam., in.			Cutter Diam., in.							
		1/2	3/4	1 and up	1/4	1/2	3/4	1 and up				
		Feed per Tooth, inch										
Cast Steel	100-180	.001	.003	.003	.001	.002	.003	.004	.003-.008	.004	.003-.012	.002-.008
	180-240	.001	.002	.003	.001	.002	.003	.003	.003-.008	.004	.003-.010	.002-.006
	240-300	.001	.002	.002	.0005	.002	.002	.002	.002-.006	.003	.003-.008	.002-.005
Zinc Alloys (Die Castings)	...	.002	.003	.004	.001	.003	.004	.006	.003-.010	.005	.004-.015	.002-.012
Copper Alloys (Brasses & Bronzes)	100-150	.002	.004	.005	.002	.003	.005	.006	.003-.015	.004	.004-.020	.002-.010
	150-250	.002	.003	.004	.001	.003	.004	.005	.003-.015	.004	.003-.012	.002-.008
Free Cutting Brasses & Bronzes	80-100	.002	.004	.005	.002	.003	.005	.006	.003-.015	.004	.004-.015	.002-.010
Cast Aluminum Alloys—As Cast	...	.003	.004	.005	.002	.004	.005	.006	.005-.016	.006	.005-.020	.004-.012
Cast Aluminum Alloys—Hardened	...	.003	.004	.005	.002	.003	.004	.005	.004-.012	.005	.005-.020	.004-.012
Wrought Aluminum Alloys—Cold Drawn	...	.003	.004	.005	.002	.003	.004	.005	.004-.014	.005	.005-.020	.004-.012
Wrought Aluminum Alloys—Hardened	...	.002	.003	.004	.001	.002	.003	.004	.003-.012	.004	.005-.020	.004-.012
Magnesium Alloys	135-185	.003	.004	.005	.003	.004	.005	.007	.005-.016	.006	.008-.020	.005-.012
Ferritic Stainless Steel	135-185	.001	.002	.003	.001	.002	.003	.003	.002-.006	.004	.004-.008	.002-.007
Austenitic Stainless Steel	135-185	.001	.002	.003	.001	.002	.003	.003	.003-.007	.004	.005-.008	.002-.007
	185-275	.001	.002	.003	.001	.002	.003	.002	.003-.006	.003	.004-.006	.002-.007
Martensitic Stainless Steel	135-185	.001	.002	.003	.001	.002	.003	.003	.003-.006	.004	.004-.010	.002-.007
	185-225	.001	.002	.002	.001	.002	.002	.003	.003-.006	.004	.003-.008	.002-.007
	225-300	.0005	.002	.002	.0005	.001	.002	.002	.002-.005	.003	.002-.006	.002-.005
Monel	100-160	.001	.003	.004	.001	.003	.003	.004	.002-.006	.004	.002-.008	.002-.006

Table 16. Recommended Feed in Inch per Tooth (f_t) for Milling with Cemented Carbide Cutters

Material	Hardness, HB	Face Mills	Slotting and Side Mills
		Feed per Tooth, inch	
Free Machining Plain Carbon Steels	100-185	.008-.020	.003-.010
Plain Carbon Steels, AISI 1006 to 1030, 1513 to 1522	100-150	.008-.020	.003-.010
	150-200	.008-.020	.003-.010
Plain Carbon Steels, AISI 1033 to 1095, 1524 to 1566	120-180	.005-.020	.003-.010
	180-220	.005-.020	.003-.010
	220-300	.003-.012	.003-.008
Alloy Steels having less than .3% Carbon content. Typical examples: AISI 4012, 4023, 4027, 4118, 4320, 4422, 4427, 4615, 4620, 4626, 4720, 4820, 5015, 5120, 6118, 8115, 8620, 8627, 8720, 8822, 9310, 93B17	125-175	.006-.020	.003-.010
	175-225	.006-.020	.003-.010
	225-275	.006-.016	.003-.010
	275-325	.004-.012	.003-.008
	325-375	.003-.008	.003-.007
Alloy Steels having .3% Carbon content, or more. Typical examples: AISI 1330, 1340, 4032, 4037, 4130, 4140, 4150, 4340, 50B40, 50B60, 5130, 51B60, 6150, 81B45, 8630, 8640, 86B45, 8660, 8740, 94B30	175-225	.005-.020	.003-.010
	225-275	.004-.012	.003-.008
	275-325	.003-.010	.003-.008
	325-375	.003-.008	.003-.007
Tool Steels	200-275	.004-.012	.003-.007
	275-325	.003-.010	.003-.006
	36-45 HRC	.003-.006	.002-.005
	45-55 HRC	.003-.005	.002-.003
Ferritic Stainless Steels	110-160	.005-.015	.003-.010
Austenitic Stainless Steels	135-185	.005-.012	.003-.010
	185-275	.005-.010	.003-.008
Martensitic Stainless Steel	135-185	.005-.015	.003-.010
	185-225	.005-.010	.003-.008
	225-300	.004-.008	.003-.007
Precipitation Hardening Stainless Steels	Annealed	.004-.012	.003-.010
	275-350	.003-.008	.002-.005
	350-450	.002-.005	.002-.004
Cast Steel	100-180	.008-.020	.003-.010
	180-240	.005-.016	.003-.010
	240-300	.004-.012	.003-.008
Gray Cast Iron	140-185	.008-.020	.005-.012
	185-225	.008-.016	.005-.010
	225-300	.005-.012	.004-.008
Ferritic Malleable Iron	110-160	.005-.020	.004-.012
Pearlitic-Martensitic Malleable Iron	160-200	.005-.020	.003-.010
	200-240	.005-.016	.003-.010
	240-300	.004-.010	.003-.008
Nodular (Ductile) Iron	140-200	.008-.020	.003-.010
	200-275	.006-.014	.003-.008
	275-325	.005-.012	.003-.007
	325-400	.003-.008	.002-.004
Copper Alloys (Brasses and Bronzes)	100-150	.005-.020	.003-.012
	150-250	.004-.014	.003-.010

Table 16 (*Concluded*).   **Recommended Feed in Inch per Tooth (f_t) for Milling with Cemented Carbide Cutters**

Material	Hardness, HB	Face Mills	Slotting and Side Mills
		Feed per Tooth, inch	
Wrought and Cast Aluminum Alloys	....	.005-.020	.005-.020
Wrought and Cast Magnesium Alloys	....	.005-.020	.005-.020
Superalloys	....	.003-.010	.002-.006
Titanium Alloys	....	.003-.010	.002-.006
Nickel Alloys	....	.003-.010	.002-.006
Monel	....	.003-.010	.002-.006
Plastics, Hard Rubber, etc.	....	.003-.015	.003-.012

rates that can be attained require a high horsepower output. An example of this type of calculation is given in the section on milling under "Horsepower for Machining." If the size of the cut must be reduced in order to stay within the power capacity of the machine, start by reducing the cutting speed rather than the feed rate in inches per tooth.

The formula for calculating the table feed rate, when the feed in inches per tooth is known, is given below:

$$f_m = f_t\, n_t\, N \qquad (4)$$

Where: f_m = Milling machine table feed rate in inches per minute (ipm)
f_t = Feed rate in inch per tooth (ipt)
n_t = Number of teeth in the milling cutter
N = Spindle speed of the milling machine in revolutions per minute (rpm)

Example: Calculate the feed rate for milling a piece of AISI 1040 steel having a hardness of 160 Bhn. The cutter is a 3-inch diameter high speed steel plain or slab milling cutter with 8 teeth. The width of the cut is 2 inches, the depth of cut is .062 inch and the cutting speed is 100 fpm. From the Table 16, the feed rate selected is .008 inch per tooth.

$$N = \frac{12\,V}{\pi\,D} = \frac{12 \times 100}{3.14 \times 3} = 127 \text{ rpm} \qquad (1)$$

$$f_m = f_t\, n_t\, N = .008 \times 8 \times 127$$
$$= 8 \text{ ipm (approximately)}$$

Feed Rates for Drilling. — The feed rate for drilling is governed primarily by the size of the drill and by the material to be drilled. Other factors that also affect the feed rate that can be used are the workpiece configuration, the rigidity of the machine tool and the workpiece setup, and the length of the chisel edge. A chisel edge that is too long will result in a very significant increase in the thrust force, which may cause large deflections to occur on the machine tool and drill breakage. For ordinary twist drills the feed rate used is .001 to .003 in./rev. for drills smaller than ⅛ in.; .002 to .006 in./rev. for ⅛ to ¼ in. drills; .004 to .010 in./rev. for ¼ to ½ in. drills; .007 to .015 in./rev. for ½ to 1 in. drills; and, .010 to .025 in./rev. for drills larger than 1 inch. The lower values in the feed ranges should be used for hard materials such as tool steels, superalloys, and work hardening stainless steels; the higher values in the feed ranges should be used to drill soft materials such as aluminum and brass.

Table 17. Recommended Cutting Speeds in Feet per Minute for Drilling and Reaming Plain Carbon and Alloy Steels

Material AISI and SAE Steels	Hardness, HB*	Material Condition*	Drilling HSS	Reaming HSS	Reaming Carbide
Free Machining Plain Carbon Steels (Resulphurized) 1212, 1213, 1214	100-150	HR, A	120	80	400
	150-200	CD	125	80	350
1108, 1109, 1115, 1117, 1118, 1120, 1126, 1211	100-150	HR, A	110	75	375
	150-200	CD	120	80	350
1132, 1137, 1139, 1140, 1144, 1146, 1151	175-225	HR, A, N, CD	100	65	350
	275-325	Q and T	70	45	250
	325-375	Q and T	45	30	175
	375-425	Q and T	35	20	100
Free Machining Plain Carbon Steels (Leaded) 11L17, 11L18, 12L13, 12L14	100-150	HR, A, N, CD	130	85	400
	150-200	HR, A, N, CD	120	80	375
	200-250	N, CD	90	60	275
Plain Carbon Steels, 1006, 1008, 1009, 1010, 1012, 1015, 1016, 1017, 1018, 1019, 1020, 1021, 1022, 1023, 1024, 1025, 1026, 1513, 1514	100-125	HR, A, N, CD	100	65	300
	125-175	HR, A, N, CD	90	60	275
	175-225	HR, N, CD	70	45	200
	225-275	CD	60	40	175
1027, 1030, 1033, 1035, 1036, 1037, 1038, 1039, 1040, 1041, 1042, 1043, 1045, 1046, 1048, 1049, 1050, 1052, 1524, 1526, 1527, 1541	125-175	HR, A, N, CD	90	60	250
	175-225	HR, A, N, CD	75	50	200
	225-275	N, CD, Q and T	60	40	150
	275-325	Q and T	50	30	120
	325-375	Q and T	35	20	100
	375-425	Q and T	25	15	80
1055, 1060, 1064, 1065, 1070, 1074, 1078, 1080, 1084, 1086, 1090, 1095, 1548, 1551, 1552, 1561, 1566	125-175	HR, A, N, CD	85	55	250
	175-225	HR, A, N, CD	70	45	200
	225-275	N, CD, Q and T	50	30	140
	275-325	Q and T	40	25	110
	325-375	Q and T	30	20	90
	375-425	Q and T	15	10	70
Free Machining Alloy Steels (Resulphurized), 4140, 4150	175-200	HR, A, N, CD	90	60	250
	200-250	HR, N, CD	80	50	225
	250-300	Q and T	55	30	200
	300-375	Q and T	40	25	150
	375-425	Q and T	30	15	100
Free Machining Alloy Steels (Leaded), 41L30, 41L40, 41L47, 41L50, 43L47, 51L32, 52L100, 86L20, 86L40	150-200	HR, A, N, CD	100	65	285
	200-250	HR, N, CD	90	60	250
	250-300	Q and T	65	40	200
	300-375	Q and T	45	30	150
	375-425	Q and T	30	15	110
Alloy Steels, 4012, 4023, 4024, 4028, 4118, 4120, 4419, 4422, 4427, 4615, 4620, 4621, 4626, 4718, 4720, 4815, 4817, 4820, 5015, 5017, 5020, 6118, 8115, 8615, 8617, 8620, 8622, 8625, 8627, 8620, 8822, 94B17	125-175	HR, A, N, CD	85	55	250
	175-225	HR, N, CD	70	45	225
	225-275	CD, N, Q and T	55	35	200
	275-325	Q and T	50	30	150
	325-375	Q and T	35	25	125
	375-425	Q and T	25	15	90

* Abbreviations designate: A, annealed; HR, hot rolled; CD, cold drawn; N, normalized; Q and T, quenched and tempered; and HB, Brinell hardness number.

Table 17 (*Concluded*). **Recommended Cutting Speeds in Feet per Minute for Drilling and Reaming Plain Carbon and Alloy Steels**

Material AISI and SAE Steels	Hardness, HB*	Material Condition*	Cutting Speed, fpm		
			Drilling	Reaming	
			HSS	HSS	Carbide
Alloy Steels, 1330, 1335, 1340, 1345, 4032, 4037, 4042, 4047, 4130, 4135, 4137, 4140, 4142, 4145, 4147, 4150, 4160, 4337, 4340, 50B44, 50B46, 50B50, 50B60, 5130, 5132, 5140, 5145, 5147, 5150, 5160, 51B60, 6150, 81B45, 8630, 8635, 8637, 8640, 8642, 8645, 8650, 8655, 8660, 8740, 9254, 9255, 9260, 9262, 94B30	175-225	HR, A, N, CD	75	50	200
	225-275	N, CD, Q and T	60	40	175
	275-325	N, Q and T	45	30	150
	325-375	N, Q and T	30	15	100
	375-425	Q and T	20	15	80
Alloy Steels, E51100, E52100	175-225	HR, A, CD	60	40	200
	225-275	N, CD, Q and T	50	30	125
	275-325	N, Q and T	35	25	100
	325-375	N, Q and T	30	20	80
	375-425	Q and T	20	10	50
Ultra High Strength Steels (Not AISI), AMS6424, AMS6421 (98B37 Mod.), AMS6422 (98BV40), AMS6427, AMS6428, AMS6430, AMS6432, AMS6433, AMS6434, AMS6436, AMS6442, 300M, D6ac	220-300	A	50	30	180
	300-350	N	35	20	125
	350-400	N	20	10	90
Maraging Steels (Not AISI) 18% Nickel Grade 200 18% Nickel Grade 250 18% Nickel Grade 300 18% Nickel Grade 350	250-225	A	50	30	175
Nitriding Steels Nitralloy 125 Nitralloy 135 Nitralloy 135 (Mod.) Nitralloy 225 Nitralloy 230 Nitralloy N Nitralloy EZ Nitrex 1	200-250	A	60	40	175
	250-300	N, Q and T	35	20	125

* Abbreviations designate: A, annealed; HR, hot rolled; CD, cold drawn; N, normalized; Q and T, quenched and tempered; and HB, Brinell hardness number.

Drilling Difficulties. — A drill split up the web is evidence of too much feed or insufficient lip clearance at the center due to improper grinding. The rapid wearing away of the extreme outer corners of the cutting edges indicates that the speed is too high. A drill chipping or breaking out at the cutting edges indicates that either the feed is too heavy or the drill has been ground with too much lip clearance. Nothing will "check" a high-speed drill quicker than to turn a stream of cold water on it after it has been heated while in use. It is equally bad to plunge it in cold water after the point has been heated in grinding. the small checks or cracks resulting from this practice will eventually chip out and cause rapid wear or breakage. Insufficient speed in drilling small holes with hand feed greatly increases the risk of breakage, especially at the moment the drill is breaking through the farther side of the work. This is due to the operator's inability to gage the feed when the drill is running too slowly.

Table 18. Recommended Cutting Speeds in Feet per Minute for Drilling and Reaming Tool Steels

Material Tool Steels	Hardness, HB*	Material Condition*	Cutting Speed, fpm		
			Drilling	Reaming	
			HSS	HSS	Carbide
Water Hardening W1, W2, W5	150-200	A	85	55	200
Shock Resisting S1, S2, S5, S6, S7	175-225	A	50	35	175
Cold Work, Oil Hardening O1, O2, O6, O7	175-225	A	45	30	150
Cold Work, High Carbon High Chromium, D2, D3, D4, D5, D7	200-250	A	30	20	80
Cold Work, Air Hardening A2, A3, A8, A9, A10	200-250	A	50	35	150
A4, A6	200-250	A	45	30	125
A7	225-275	A	30	20	100
Hot Work, Chromium Type H10, H11, H12, H13, H14, H19	150-200	A	60	40	200
	200-250	A	50	30	150
	325-375	Q and T	30	20	100
Hot Work, Tungsten Type H21, H22, H23, H24, H25, H26	150-200	A	55	35	150
	200-250	A	40	25	125
Hot Work, Molybdenum Type H41, H42, H43	150-200	A	45	30	150
	200-250	A	35	20	125
Special Purpose, Low Alloy L2, L3, L6	150-200	A	60	40	200
Mold P2, P3, P4, P5, P6	100-150	A	75	50	225
P20, P21	150-200	A	60	40	200
High-Speed Steel M1, M2, M6, M10, T1, T2, T6	200-250	A	45	30	150
M3-1, M4, M7, M30, M33, M34, M36, M41, M42, M43, M44, M46, M47, T5, T8	225-275	A	35	20	100
T15, M3-2	225-275	A	25	15	80

* Abbreviations designate: A, annealed; Q and T, quenched and tempered; and HB, Brinell Hardness Number.

Small drills have heavier webs and smaller flutes in proportion to their size than do larger drills, and breakage due to clogging of chips in the flutes is more likely to occur. When drilling holes more than three times the diameter of the drill, it is advisable to withdraw the drill at intervals to remove the chips and permit coolant to reach the tip of the drill.

Drilling Holes in Glass. — There are several methods of drilling holes in glass. For holes of medium and large size, use brass or copper tubing, having an outside diameter equal to the size of hole required. Revolve the tube at a peripheral speed of about 100 feet per minute, and use carborundum (80 to 100 grit) and light machine oil between the

Table 19. Recommended Cutting Speeds in Feet per Minute for Drilling and Reaming Stainless Steels

Material Stainless Steels	Hardness, HB*	Material Condition*	Cutting Speed, fpm		
			Drilling	Reaming	
			HSS	HSS	Carbide
Free Machining Stainless Steels					
(Ferritic), 430F, 430F Se	135-185	A	90	60	250
(Austenitic), 203EZ, 303, 303 Se, 303 MA, 303 Pb, 303 Cu, 303 Plus X	135-185	A	85	55	225
	225-275	CD	70	45	200
(Martensitic), 416, 416 Se, 416 Plus X, 420F, 420F Se, 440F, 440F Se	135-185	A	90	60	250
	185-240	CD	70	45	200
	275-325	Q and T	40	25	150
	375-425	Q and T	20	10	80
Stainless Steels					
(Ferritic), 405, 409, 429, 430, 434, 436, 442, 446, 502	135-185	A	65	45	200
(Austenitic), 201, 202, 301, 302, 304, 304L, 305, 308, 321, 347, 348	135-185	A	55	35	150
	225-275	CD	50	30	125
(Austenitic), 302B, 309, 309S, 310, 310S, 314, 316, 316L, 317, 330	135-185	A	50	30	150
(Martensitic), 403, 410, 420, 501	135-175	A	75	50	225
	175-225	A	65	45	200
	275-325	Q and T	40	25	125
	375-425	Q and T	25	15	80
(Martensitic), 414, 431 Greek Ascoloy	225-275	A	50	30	150
	275-325	Q and T	40	25	125
	375-425	Q and T	25	15	80
(Martensitic), 440A, 440B, 440C	225-275	A	45	30	125
	275-325	Q and T	40	25	100
	375-425	Q and T	20	10	75
(Precipitation Hardening) 15-5PH, 17-4PH, 17-7Ph, 17-14Cu Mo, AF-71, AFC-77, AM-350, AM-355, AM-362, Custom 455, HNM, PH13-8, PH14-8Mo, PH15-7Mo, Stainless W	150-200	A	50	30	150
	275-325	H	45	25	125
	325-375	H	35	20	75
	375-425	H	20	10	50

* Abbreviations designate: A, annealed; CD, cold drawn; Q and T, quenched and tempered; H, precipitation hardened; and HB, Brinell hardness number.

end of the pipe and the glass. Insert the abrasive under the drill with a thin piece of soft wood, to avoid scratching the glass. The glass should be supported by a felt or rubber cushion, not much larger than the hole to be drilled. If practicable, it is well to drill about halfway through and then turn the glass over and drill down to meet the first cut. Any fin that may be left in the hole can be removed with a round second-cut file wet with turpentine.

Smaller diameter holes are generally drilled with triangular shaped cemented carbide drills that can be purchased in standard sizes. The end of the drill is shaped into a long tapering triangular point. The other end of the cemented carbide bit is brazed on to a

Table 20. Recommended Cutting Speeds in Feet per Minute for Drilling and Reaming Ferrous Cast Metals

Material Ferrous Cast Metals	Hardness, HB*	Material Condition*	Cutting Speed, fpm		
			Drilling	Reaming	
			HSS	HSS	Carbide
Gray Cast Iron					
ASTM Class 20	120-150	A	100	65	300
ASTM Class 25	160-200	AC	90	60	225
ASTM Class 30, 35, and 40	190-220	AC	80	55	180
ASTM Class 45 and 50	220-260	AC	60	40	125
ASTM Class 55 and 60	250-320	AC, HT	30	20	80
ASTM Type 1, 1b, 5 (Ni-Resist)	100-215	AC	50	30	150
ASTM Type 2, 3, 6 (Ni-Resist)	120-175	AC	40	25	140
ASTM Type 2b, 4 (Ni-Resist)	150-250	AC	30	20	125
Malleable Iron					
(Ferritic), 32510, 35018	110-160	MHT	110	75	325
(Pearlitic), 40010, 43010, 45006, 45008, 48005, 50005	160-200	MHT	80	55	250
	200-240	MHT	70	45	180
(Martensitic), 53004, 60003, 60004	200-255	MHT	55	35	160
(Martensitic), 70002, 70003	220-260	MHT	50	30	150
(Martensitic), 80002	240-280	MHT	45	30	100
(Martensitic), 90001	250-320	MHT	25	15	80
Nodular (Ductile) Iron					
(Ferritic), 60-40-18, 65-45-12	140-190	A	100	65	300
(Ferritic-Pearlitic), 80-55-06	190-225	AC	70	45	225
	225-260	AC	50	30	140
(Pearlitic-Martensitic), 100-70-03	240-300	HT	40	25	115
(Martensitic), 120-90-02	270-330	HT	25	15	65
	330-400	HT	10	5	30
Cast Steels					
(Low Carbon), 1010, 1020	100-150	AC, A, N	100	65	250
(Medium Carbon), 1030, 1040, 1050	125-175	AC, A, N	90	60	240
	175-225	AC, A, N	70	45	220
	225-300	AC, HT	55	35	200
(Low Carbon Alloy), 1320, 2315, 2320, 4110, 4120, 4320, 8020, 8620	150-200	AC, A, N	75	50	220
	200-250	AC, A, N	65	40	200
	250-300	AC, HT	50	30	150
(Medium Carbon Alloy), 1330, 1340, 2325, 2330, 4125, 4130, 4140, 4330, 4340, 8030, 80B30, 8040, 8430, 8440, 8630, 8640, 9525, 9530, 9535	175-225	AC, A, N	70	45	200
	225-250	AC, A, N	60	35	180
	250-300	AC, HT	45	30	150
	300-350	AC, HT	30	20	140
	350-400	HT	20	10	100

* Abbreviations designate: A, annealed; AC, as cast; N, normalized; HT, heat treated; MHT, Malleablizing heat treatment; and HB, Brinell hardness number.

Table 21.   Recommended Cutting Speeds in Feet per Minute for Drilling
and Reaming Light Metals

Material Light Metals	Material Condition*	Cutting Speed, fpm		
		Drilling	Reaming	
		HSS	HSS	Carbide
All Wrought Aluminum Alloys {	CD	400	400	800
	ST and A	350	350	750
All Aluminum Sand and Permanent Mold { Casting Alloys	AC	500	500	900
	ST and A	350	350	750
All Aluminum Die Casting Alloys† {	AC	300	300	500
	ST and A	70	70	200
†Except Alloys 390.0 and 392.0 {	AC	125	100	250
	ST and A	45	40	200
All Wrought Magnesium Alloys	A, CD, ST and A	500	500	1000
All Cast Magnesium Alloys	A, AC, ST and A	450	450	1000

* Abbreviations designate: A, annealed; AC, as cast; CD, cold drawn; and ST and A, solution treated and aged.

steel shank. A glass drill can be made to the same shape from hardened drill rod or an old three-cornered file. The location at which the hole is to be drilled is marked on the workpiece. A dam of putty or glazing compound is built up on the work surface to contain the cutting fluid, which can be either kerosene or turpentine mixed with camphor. Chipping on the back edge of the hole can be prevented by placing a scrap plate of glass behind the area to be drilled and drilling into the back-up glass. This procedure also provides additional support to the workpiece and is essential for drilling very thin plates. The hole is usually drilled with an electric hand drill. When the hole is being produced the drill should be given a small circular motion using the point as a fulcrum, thereby providing a clearance for the drill in the hole.

Very small round or intricately shaped holes and narrow slots can be cut in glass by the ultrasonic machining process or by the abrasive jet cutting process.

Estimating Planer Cutting Speeds. — While most planers of modern design have a means of indicating the speed at which the table is traveling, or cutting, many older planers do not. Thus, the following formulas are useful for planers that do not have a means of indicating the table or cutting speed. It is not practicable to provide a formula for calculating the exact cutting speed at which a planer is operating because the time to stop and start the table when reversing varies greatly. The formulas below will, however, provide a reasonable estimate.

$$V_c \cong S_c L$$

$$S_c \cong \frac{V_c}{L}$$

Where: V_c = Cutting speed; fpm, or m/min
S_c = Number of cutting strokes per minute of planer table
L = Length of table cutting stroke; ft, or m

Table 22.   **Recommended Cutting Speeds in Feet per Minute for Drilling and Reaming Copper Alloys**

Material Copper Alloys (Copper Alloy Nos. as per the Copper Development Assn. Inc.)	Material Condition*	Cutting Speed, fpm		
		Drilling	Reaming	
		HSS	HSS	Carbide
314 Leaded Commercial Bronze 332 High Leaded Brass 340 Medium Leaded Brass 342 High Leaded Brass 353 High Leaded Brass 356 Extra High Leaded Brass 360 Free Cutting Brass 370 Free Cutting Muntz Metal 377 Forging Brass 385 Architectural Bronze 485 Leaded Naval Brass 544 Free Cutting Phosphor Bronze	A CD	160 175	160 175	320 360
226 Jewelry Bronze 230 Red Brass 240 Low Brass 260 Cartridge Brass 70% 268 Yellow Brass 280 Muntz Metal 335 Low Leaded Brass 365 Leaded Muntz Metal 368 Leaded Muntz Metal 443 Admiralty Brass (inhibited) 445 Admiralty Brass (inhibited) 651 Low Silicon Bronze 655 High Silicon Bronze 675 Manganese Bronze 687 Aluminum Brass 770 Nickel Silver 796 Leaded Nickel Silver	A CD	120 140	110 120	250 275
102 Oxygen Free Copper 110 Electrolytic Tough Pitch Copper 122 Phosphorus Deoxidized Copper 170 Beryllium Copper 172 Beryllium Copper 175 Beryllium Copper 210 Guilding, 95% 220 Commercial Bronze 502 Phosphor Bronze 1.25% 510 Phosphor Bronze 5% 521 Phosphor Bronze 8% 524 Phosphor Bronze 10% 614 Aluminum Bronze 706 Copper Nickel 10% 715 Copper Nickel 30% 745 Nickel Silver 752 Nickel Silver 754 Nickel Silver 757 Nickel Silver	A CD	60 65	50 60	180 200

* Abbreviations used in this column are as follows: A, annealed; CD, cold drawn

Cutting Speeds and Equivalent R.P.M. for Drills of Number and Letter Sizes

Size No.	Cutting Speed, Feet per Minute										
	30'	40'	50'	60'	70'	80'	90'	100'	110'	130'	150'
	Revolutions per Minute for Number Sizes										
1	503	670	838	1005	1173	1340	1508	1675	1843	2179	2513
2	518	691	864	1037	1210	1382	1555	1728	1901	2247	2593
4	548	731	914	1097	1280	1462	1645	1828	2010	2376	2741
6	562	749	936	1123	1310	1498	1685	1872	2060	2434	2809
8	576	768	960	1151	1343	1535	1727	1919	2111	2495	2879
10	592	790	987	1184	1382	1579	1777	1974	2171	2566	2961
12	606	808	1010	1213	1415	1617	1819	2021	2223	2627	3032
14	630	840	1050	1259	1469	1679	1889	2099	2309	2728	3148
16	647	863	1079	1295	1511	1726	1942	2158	2374	2806	3237
18	678	904	1130	1356	1582	1808	2034	2260	2479	2930	3380
20	712	949	1186	1423	1660	1898	2135	2372	2610	3084	3559
22	730	973	1217	1460	1703	1946	2190	2433	2676	3164	3649
24	754	1005	1257	1508	1759	2010	2262	2513	2764	3267	3769
26	779	1039	1299	1559	1819	2078	2338	2598	2858	3378	3898
28	816	1088	1360	1631	1903	2175	2447	2719	2990	3534	4078
30	892	1189	1487	1784	2081	2378	2676	2973	3270	3864	4459
32	988	1317	1647	1976	2305	2634	2964	3293	3622	4281	4939
34	1032	1376	1721	2065	2409	2753	3097	3442	3785	4474	5162
36	1076	1435	1794	2152	2511	2870	3228	3587	3945	4663	5380
38	1129	1505	1882	2258	2634	3010	3387	3763	4140	4892	5645
40	1169	1559	1949	2339	2729	3118	3508	3898	4287	5067	5846
42	1226	1634	2043	2451	2860	3268	3677	4085	4494	5311	6128
44	1333	1777	2221	2665	3109	3554	3999	4442	4886	5774	6662
46	1415	1886	2358	2830	3301	3773	4244	4716	5187	6130	7074
48	1508	2010	2513	3016	3518	4021	4523	5026	5528	6534	7539
50	1637	2183	2729	3274	3820	4366	4911	5457	6002	7094	8185
52	1805	2406	3008	3609	4211	4812	5414	6015	6619	7820	9023
54	2084	2778	3473	4167	4862	5556	6251	6945	7639	9028	10417
Size	Revolutions per Minute for Letter Sizes										
A	491	654	818	982	1145	1309	1472	1636	1796	2122	2448
B	482	642	803	963	1124	1284	1445	1605	1765	2086	2407
C	473	631	789	947	1105	1262	1420	1578	1736	2052	2368
D	467	622	778	934	1089	1245	1400	1556	1708	2018	2329
E	458	611	764	917	1070	1222	1375	1528	1681	1968	2292
F	446	594	743	892	1040	1189	1337	1486	1635	1932	2229
G	440	585	732	878	1024	1170	1317	1463	1610	1903	2195
H	430	574	718	862	1005	1149	1292	1436	1580	1867	2154
I	421	562	702	842	983	1123	1264	1404	1545	1826	2106
J	414	552	690	827	965	1103	1241	1379	1517	1793	2068
K	408	544	680	815	951	1087	1223	1359	1495	1767	2039
L	395	527	659	790	922	1054	1185	1317	1449	1712	1976
M	389	518	648	777	907	1036	1166	1295	1424	1683	1942
N	380	506	633	759	886	1012	1139	1265	1391	1644	1897
O	363	484	605	725	846	967	1088	1209	1330	1571	1813
P	355	473	592	710	828	946	1065	1183	1301	1537	1774
Q	345	460	575	690	805	920	1035	1150	1266	1496	1726
R	338	451	564	676	789	902	1014	1127	1239	1465	1690
S	329	439	549	659	769	878	988	1098	1207	1427	1646
T	320	426	533	640	746	853	959	1066	1173	1387	1600
U	311	415	519	623	727	830	934	1038	1142	1349	1557
V	304	405	507	608	709	810	912	1013	1114	1317	1520
W	297	396	495	594	693	792	891	989	1088	1286	1484
X	289	385	481	576	672	769	865	962	1058	1251	1443
Y	284	378	473	567	662	756	851	945	1040	1229	1418
Z	277	370	462	555	647	740	832	925	1017	1202	1387

For fractional drill sizes, use table on page 1005.

Revolutions per Minute for Various Cutting Speeds and Diameters

Diameter, Inches	Cutting Speed, Feet per Minute											
	40	50	60	70	80	90	100	120	140	160	180	200
	Revolutions per Minute											
1/4	611	764	917	1070	1222	1376	1528	1834	2139	2445	2750	3056
5/16	489	611	733	856	978	1100	1222	1466	1711	1955	2200	2444
3/8	408	509	611	713	815	916	1018	1222	1425	1629	1832	2036
7/16	349	437	524	611	699	786	874	1049	1224	1398	1573	1748
1/2	306	382	459	535	611	688	764	917	1070	1222	1375	1528
9/16	272	340	407	475	543	611	679	813	951	1086	1222	1358
5/8	245	306	367	428	489	552	612	736	857	979	1102	1224
11/16	222	273	333	389	444	500	555	666	770	888	999	1101
3/4	203	254	306	357	408	458	508	610	711	813	914	1016
13/16	190	237	284	332	379	427	474	569	664	758	853	948
7/8	175	219	262	306	349	392	438	526	613	701	788	876
15/16	163	204	244	285	326	366	407	488	570	651	733	814
1	153	191	229	267	306	344	382	458	535	611	688	764
1 1/16	144	180	215	251	287	323	359	431	503	575	646	718
1 1/8	136	170	204	238	272	306	340	408	476	544	612	680
1 3/16	129	161	193	225	258	290	322	386	451	515	580	644
1 1/4	123	153	183	214	245	274	306	367	428	490	551	612
1 5/16	116	146	175	204	233	262	291	349	407	466	524	582
1 3/8	111	139	167	195	222	250	278	334	389	445	500	556
1 7/16	106	133	159	186	212	239	265	318	371	424	477	530
1 1/2	102	127	153	178	204	230	254	305	356	406	457	508
1 9/16	97.6	122	146	171	195	220	244	293	342	390	439	488
1 5/8	93.9	117	141	165	188	212	234	281	328	374	421	468
1 11/16	90.4	113	136	158	181	203	226	271	316	362	407	452
1 3/4	87.3	109	131	153	175	196	218	262	305	349	392	436
1 7/8	81.5	102	122	143	163	184	204	244	286	326	367	408
2	76.4	95.5	115	134	153	172	191	229	267	306	344	382
2 1/8	72.0	90.0	108	126	144	162	180	216	252	288	324	360
2 1/4	68.0	85.5	102	119	136	153	170	204	238	272	306	340
2 3/8	64.4	80.5	96.6	113	129	145	161	193	225	258	290	322
2 1/2	61.2	76.3	91.7	107	122	138	153	184	213	245	275	306
2 5/8	58.0	72.5	87.0	102	116	131	145	174	203	232	261	290
2 3/4	55.6	69.5	83.4	97.2	111	125	139	167	195	222	250	278
2 7/8	52.8	66.0	79.2	92.4	106	119	132	158	185	211	238	264
3	51.0	63.7	76.4	89.1	102	114	127	152	178	203	228	254
3 1/8	48.8	61.0	73.2	85.4	97.6	110	122	146	171	195	219	244
3 1/4	46.8	58.5	70.2	81.9	93.6	105	117	140	164	188	211	234
3 3/8	45.2	56.5	67.8	79.1	90.4	102	113	136	158	181	203	226
3 1/2	43.6	54.5	65.5	76.4	87.4	98.1	109	131	153	174	196	218
3 5/8	42.0	52.5	63.0	73.5	84.0	94.5	105	126	147	168	189	210
3 3/4	40.8	51.0	61.2	71.4	81.6	91.8	102	122	143	163	184	205
3 7/8	39.4	49.3	59.1	69.0	78.8	88.6	98.5	118	138	158	177	197
4	38.2	47.8	57.3	66.9	76.4	86.0	95.6	115	134	153	172	191
4 1/4	35.9	44.9	53.9	62.9	71.8	80.8	89.8	108	126	144	162	180
4 1/2	34.0	42.4	51.0	59.4	67.9	76.3	84.8	102	119	136	153	170
4 3/4	32.2	40.2	48.2	56.3	64.3	72.4	80.4	96.9	113	129	145	161
5	30.6	38.2	45.9	53.5	61.1	68.8	76.4	91.7	107	122	145	161
5 1/4	29.1	36.4	43.6	50.9	58.2	65.4	72.7	87.2	102	116	131	145
5 1/2	27.8	34.7	41.7	48.6	55.6	62.5	69.4	83.3	97.2	111	125	139
5 3/4	26.6	33.2	39.8	46.5	53.1	59.8	66.4	80.0	93.0	106	120	133
6	25.5	31.8	38.2	44.6	51.0	57.2	63.6	76.3	89.0	102	114	127
6 1/4	24.4	30.6	36.7	42.8	48.9	55.0	61.1	73.3	85.5	97.7	110	122
6 1/2	23.5	29.4	35.2	41.1	47.0	52.8	58.7	70.4	82.2	93.9	106	117
6 3/4	22.6	28.3	34.0	39.6	45.3	50.9	56.6	67.9	79.2	90.6	102	113
7	21.8	27.3	32.7	38.2	43.7	49.1	54.6	65.5	76.4	87.4	98.3	109
7 1/4	21.1	26.4	31.6	36.9	42.2	47.4	52.7	63.2	73.8	84.3	94.9	105
7 1/2	20.4	25.4	30.5	35.6	40.7	45.8	50.9	61.1	71.0	81.4	91.6	102
7 3/4	19.7	24.6	29.5	34.4	39.4	44.3	49.2	59.0	68.9	78.7	88.6	98.4
8	19.1	23.9	28.7	33.4	38.2	43.0	47.8	57.4	66.9	76.5	86.0	95.6

Revolutions per Minute for Various Cutting Speeds and Diameters

Diameter, Inches	Cutting Speed, Feet per Minute											
	225	250	275	300	325	350	375	400	425	450	500	550
	Revolutions per Minute											
1/4	3438	3820	4202	4584	4966	5348	5730	6112	6493	6875	7639	8403
5/16	2750	3056	3362	3667	3973	4278	4584	4889	5195	5501	6112	6723
3/8	2292	2546	2801	3056	3310	3565	3820	4074	4329	4584	5093	5602
7/16	1964	2182	2401	2619	2837	3056	3274	3492	3710	3929	4365	4802
1/2	1719	1910	2101	2292	2483	2675	2866	3057	3248	3439	3821	4203
9/16	1528	1698	1868	2037	2207	2377	2547	2717	2887	3056	3396	3736
5/8	1375	1528	1681	1834	1987	2139	2292	2445	2598	2751	3057	3362
11/16	1250	1389	1528	1667	1806	1941	2084	2223	2362	2501	2779	3056
3/4	1146	1273	1401	1528	1655	1783	1910	2038	2165	2292	2547	2802
13/16	1058	1175	1293	1410	1528	1646	1763	1881	1998	2116	2351	2586
7/8	982	1091	1200	1310	1419	1528	1637	1746	1855	1965	2183	2401
15/16	917	1019	1120	1222	1324	1426	1528	1630	1732	1834	2038	2241
1	859	955	1050	1146	1241	1337	1432	1528	1623	1719	1910	2101
1 1/16	809	899	988	1078	1168	1258	1348	1438	1528	1618	1798	1977
1 1/8	764	849	933	1018	1103	1188	1273	1358	1443	1528	1698	1867
1 3/16	724	804	884	965	1045	1126	1206	1287	1367	1448	1609	1769
1 1/4	687	764	840	917	993	1069	1146	1222	1299	1375	1528	1681
1 5/16	654	727	800	873	946	1018	1091	1164	1237	1309	1455	1601
1 3/8	625	694	764	833	903	972	1042	1111	1181	1250	1389	1528
1 7/16	598	664	730	797	863	930	996	1063	1129	1196	1329	1461
1 1/2	573	636	700	764	827	891	955	1018	1082	1146	1273	1400
1 9/16	550	611	672	733	794	855	916	978	1039	1100	1222	1344
1 5/8	528	587	646	705	764	822	881	940	999	1057	1175	1293
1 11/16	509	566	622	679	735	792	849	905	962	1018	1132	1245
1 3/4	491	545	600	654	709	764	818	873	927	982	1091	1200
1 13/16	474	527	579	632	685	737	790	843	895	948	1054	1159
1 7/8	458	509	560	611	662	713	764	815	866	917	1019	1120
1 15/16	443	493	542	591	640	690	739	788	838	887	986	1084
2	429	477	525	573	620	668	716	764	811	859	955	1050
2 1/8	404	449	494	539	584	629	674	719	764	809	899	988
2 1/4	382	424	468	509	551	594	636	679	721	764	849	933
2 3/8	362	402	442	482	522	563	603	643	683	724	804	884
2 1/2	343	382	420	458	496	534	573	611	649	687	764	840
2 5/8	327	363	400	436	472	509	545	582	618	654	727	800
2 3/4	312	347	381	416	451	486	520	555	590	625	694	763
2 7/8	299	332	365	398	431	465	498	531	564	598	664	730
3	286	318	350	381	413	445	477	509	541	572	636	700
3 1/8	274	305	336	366	397	427	458	488	519	549	611	672
3 1/4	264	293	323	352	381	411	440	470	499	528	587	646
3 3/8	254	283	311	339	367	396	424	452	481	509	566	622
3 1/2	245	272	300	327	354	381	409	436	463	490	545	600
3 5/8	237	263	289	316	342	368	395	421	447	474	527	579
3 3/4	229	254	280	305	331	356	382	407	433	458	509	560
3 7/8	221	246	271	295	320	345	369	394	419	443	493	542
4	214	238	262	286	310	334	358	382	405	429	477	525
4 1/4	202	224	247	269	292	314	337	359	383	404	449	494
4 1/2	191	212	233	254	275	297	318	339	360	382	424	466
4 3/4	180	201	221	241	261	281	301	321	341	361	402	442
5	171	191	210	229	248	267	286	305	324	343	382	420
5 1/4	163	181	199	218	236	254	272	290	308	327	363	399
5 1/2	156	173	190	208	225	242	260	277	294	312	347	381
5 3/4	149	166	182	199	215	232	249	265	282	298	332	365
6	143	159	174	190	206	222	238	254	270	286	318	349
6 1/4	137	152	168	183	198	213	229	244	259	274	305	336
6 1/2	132	146	161	176	190	205	220	234	249	264	293	322
6 3/4	127	141	155	169	183	198	212	226	240	254	283	311
7	122	136	149	163	177	190	204	218	231	245	272	299
7 1/4	118	131	144	158	171	184	197	210	223	237	263	289
7 1/2	114	127	139	152	165	178	190	203	216	229	254	279
7 3/4	111	123	135	148	160	172	185	197	209	222	246	271
8	107	119	131	143	155	167	179	191	203	215	238	262

Revolutions per Minute for Various Cutting Speeds and Diameters
(Metric units)

Diam., mm	Cutting Speed, Metres per Minute											
	5	6	8	10	12	16	20	25	30	35	40	45
	Revolutions per Minute											
5	318	382	509	637	764	1019	1273	1592	1910	2228	2546	2865
6	265	318	424	530	637	849	1061	1326	1592	1857	2122	2387
8	199	239	318	398	477	637	796	995	1194	1393	1592	1790
10	159	191	255	318	382	509	637	796	955	1114	1273	1432
12	133	159	212	265	318	424	531	663	796	928	1061	1194
16	99.5	119	159	199	239	318	398	497	597	696	796	895
20	79.6	95.5	127	159	191	255	318	398	477	557	637	716
25	63.7	76.4	102	127	153	204	255	318	382	446	509	573
30	53.1	63.7	84.9	106	127	170	212	265	318	371	424	477
35	45.5	54.6	72.8	90.9	109	145	182	227	273	318	364	409
40	39.8	47.7	63.7	79.6	95.5	127	159	199	239	279	318	358
45	35.4	42.4	56.6	70.7	84.9	113	141	177	212	248	283	318
50	31.8	38.2	51	63.7	76.4	102	127	159	191	223	255	286
55	28.9	34.7	46.3	57.9	69.4	92.6	116	145	174	203	231	260
60	26.6	31.8	42.4	53.1	63.7	84.9	106	133	159	186	212	239
65	24.5	29.4	39.2	49	58.8	78.4	98	122	147	171	196	220
70	22.7	27.3	36.4	45.5	54.6	72.8	90.9	114	136	159	182	205
75	21.2	25.5	34	42.4	51	68	84.9	106	127	149	170	191
80	19.9	23.9	31.8	39.8	47.7	63.7	79.6	99.5	119	139	159	179
90	17.7	21.2	28.3	35.4	42.4	56.6	70.7	88.4	106	124	141	159
100	15.9	19.1	25.5	31.8	38.2	51	63.7	79.6	95.5	111	127	143
110	14.5	17.4	23.1	28.9	34.7	46.2	57.9	72.3	86.8	101	116	130
120	13.3	15.9	21.2	26.5	31.8	42.4	53.1	66.3	79.6	92.8	106	119
130	12.2	14.7	19.6	24.5	29.4	39.2	49	61.2	73.4	85.7	97.9	110
140	11.4	13.6	18.2	22.7	27.3	36.4	45.5	56.8	68.2	79.6	90.9	102
150	10.6	12.7	17	21.2	25.5	34	42.4	53.1	63.7	74.3	84.9	95.5
160	9.9	11.9	15.9	19.9	23.9	31.8	39.8	49.7	59.7	69.6	79.6	89.5
170	9.4	11.2	15	18.7	22.5	30	37.4	46.8	56.2	65.5	74.9	84.2
180	8.8	10.6	14.1	17.7	21.2	28.3	35.4	44.2	53.1	61.9	70.7	79.6
190	8.3	10	13.4	16.8	20.1	26.8	33.5	41.9	50.3	58.6	67	75.4
200	8	9.5	12.7	15.9	19.1	25.5	31.8	39.8	47.7	55.7	63.7	71.6
220	7.2	8.7	11.6	14.5	17.4	23.1	28.9	36.2	43.4	50.6	57.9	65.1
240	6.6	8	10.6	13.3	15.9	21.2	26.5	33.2	39.8	46.4	53.1	59.7
260	6.1	7.3	9.8	12.2	14.7	19.6	24.5	30.6	36.7	42.8	49	55.1
280	5.7	6.8	9.1	11.4	13.6	18.2	22.7	28.4	34.1	39.8	45.5	51.1
300	5.3	6.4	8.5	10.6	12.7	17	21.2	26.5	31.8	37.1	42.4	47.7
350	4.5	5.4	7.3	9.1	10.9	14.6	18.2	22.7	27.3	31.8	36.4	40.9
400	4	4.8	6.4	8	9.5	12.7	15.9	19.9	23.9	27.9	31.8	35.8
450	3.5	4.2	5.7	7.1	8.5	11.3	14.1	17.7	21.2	24.8	28.3	31.8
500	3.2	3.8	5.1	6.4	7.6	10.2	12.7	15.9	19.1	22.3	25.5	28.6

Revolutions per Minute for Various Cutting Speeds and Diameters
(Metric units)

Diam., mm	Cutting Speed, Metres per Minute											
	50	55	60	65	70	75	80	85	90	95	100	200
	Revolutions per Minute											
5	3183	3501	3820	4138	4456	4775	5093	5411	5730	6048	6366	12,732
6	2653	2918	3183	3448	3714	3979	4244	4509	4775	5039	5305	10,610
8	1989	2188	2387	2586	2785	2984	3183	3382	3581	3780	3979	7958
10	1592	1751	1910	2069	2228	2387	2546	2706	2865	3024	3183	6366
12	1326	1459	1592	1724	1857	1989	2122	2255	2387	2520	2653	5305
16	995	1094	1194	1293	1393	1492	1591	1691	1790	1890	1989	3979
20	796	875	955	1034	1114	1194	1273	1353	1432	1512	1592	3183
25	637	700	764	828	891	955	1019	1082	1146	1210	1273	2546
30	530	584	637	690	743	796	849	902	955	1008	1061	2122
35	455	500	546	591	637	682	728	773	819	864	909	1818
40	398	438	477	517	557	597	637	676	716	756	796	1592
45	354	389	424	460	495	531	566	601	637	672	707	1415
50	318	350	382	414	446	477	509	541	573	605	637	1273
55	289	318	347	376	405	434	463	492	521	550	579	1157
60	265	292	318	345	371	398	424	451	477	504	530	1061
65	245	269	294	318	343	367	392	416	441	465	490	979
70	227	250	273	296	318	341	364	387	409	432	455	909
75	212	233	255	276	297	318	340	361	382	403	424	849
80	199	219	239	259	279	298	318	338	358	378	398	796
90	177	195	212	230	248	265	283	301	318	336	354	707
100	159	175	191	207	223	239	255	271	286	302	318	637
110	145	159	174	188	203	217	231	246	260	275	289	579
120	133	146	159	172	186	199	212	225	239	252	265	530
130	122	135	147	159	171	184	196	208	220	233	245	490
140	114	125	136	148	159	171	182	193	205	216	227	455
150	106	117	127	138	149	159	170	180	191	202	212	424
160	99.5	109	119	129	139	149	159	169	179	189	199	398
170	93.6	103	112	122	131	140	150	159	169	178	187	374
180	88.4	97.3	106	115	124	133	141	150	159	168	177	354
190	83.8	92.1	101	109	117	126	134	142	151	159	167	335
200	79.6	87.5	95.5	103	111	119	127	135	143	151	159	318
220	72.3	79.6	86.8	94	101	109	116	123	130	137	145	289
240	66.3	72.9	79.6	86.2	92.8	99.5	106	113	119	126	132	265
260	61.2	67.3	73.4	79.6	85.7	91.8	97.9	104	110	116	122	245
280	56.8	62.5	68.2	73.9	79.6	85.3	90.9	96.6	102	108	114	227
300	53.1	58.3	63.7	69	74.3	79.6	84.9	90.2	95.5	101	106	212
350	45.5	50	54.6	59.1	63.7	68.2	72.8	77.3	81.8	99.1	91	182
400	39.8	43.8	47.7	51.7	55.7	59.7	63.7	67.6	71.6	75.6	79.6	159
450	35.4	38.9	42.4	46	49.5	53.1	56.6	60.1	63.6	67.2	70.7	141
500	31.8	35	38.2	41.4	44.6	47.7	50.9	54.1	57.3	60.5	63.6	127

Planing Time. — The approximate time required to plane a surface can be determined from the following formula in which T = time in minutes, L = length of stroke in feet, V_c = cutting speed in feet per minute, V_r = return speed in feet per minute, W = width of surface to be planed in inches, F = feed in inches, and 0.025 = approximate reversal time factor per stroke in minutes for most planers.

$$T = \frac{W}{F}\left[L \times \left(\frac{1}{V_c} + \frac{1}{V_r} \right) + 0.025 \right]$$

Speeds for Metal-cutting Saws. — The following speeds and feeds for metal-cutting saws are recommended by Henry Disston & Sons, Inc. Speeds and feeds vary over a wide range depending upon the kind of machine as well as the size, shape, and kind of metal to be cut; hence, the following speeds are intended as a general guide only.

Solid-tooth Circular Saws. — Low-carbon steel: Speed, 60–90 feet per minute; feed, 3–6 inches per minute. Tool and alloy steel: Speed, 40–60 feet per minute; feed, $\frac{1}{32}$–$2\frac{1}{2}$ inches per minute. Cast iron: Speed, 60–70 feet per minute; feed, 5 inches per minute. Brass: Speed, 500–800 feet per minute; feed, 80 inches per minute. Copper: Speed, 600–700 feet per minute; feed, 60 inches per minute. Aluminum: Speed, 800–1000 feet per minute; feed, 90 inches per minute.

"Hot saws" for sawing hot metal (structural shapes, rails, billets, etc., in rolling mills) should operate at speeds of 22,000 to 25,000 feet per minute.

Inserted-tooth Metal-cutting Saws. — Generally speaking, an inserted-tooth saw is not recommended smaller than 18 inches in diameter, although smaller sizes are made. Small saws usually are applied to work requiring fine or relatively fine teeth which are impossible when the teeth are of the inserted type. The following speeds are not suitable for all compositions but represent general practice. Steel: Speed, 40 feet per minute; feed, $2\frac{1}{2}$–10 inches per minute. Cast iron: Speed, 40 feet per minute; feed, 5 inches per minute. Brass: Speed, 500 feet per minute; feed, 80 inches per minute. Copper: Speed, 750 feet per minute; feed, 60 inches per minute. Aluminum: Speed, 1200 feet per minute; feed, 90 inches per minute.

Metal-cutting Band Saws. — The speeds which follow apply to Disston "hard-edge" saws which have milled teeth. Aluminum sheets: Speed of blade, 1000–3000 feet per minute. Bakelite: Speed, 800–1000 feet per minute. Brass sheets and tubing: Speed, 700–1500 feet per minute. Carbon tool steel: Speed, 100–150 feet per minute. Cast iron: Speed, 100–150 feet per minute. Cold-rolled steel: Speed, 150–200 feet per minute. High-speed steel: Speed, 90–125 feet per minute. Malleable iron: Speed, 150–200 feet per minute. Hard rubber: Speed, 150–200 feet per minute. Slate: Speed, 100–150 feet per minute. The number of saw teeth per inch recommended for these different materials ranges from 8 to 14, and the saw user should be guided by the recommendations of the manufacturer.

Hacksaw Speeds. — The following hacksaw speeds are given on the authority of a leading manufacturer: The average rate of travel of the cutting blade in feet per minute, including forward and return strokes, should be as follows: For mild steel, 130; for annealed tool steel, 90; and for unannealed tool steel, 60 feet per minute. Thus in the case of a 6-inch stroke, for example, the revolutions per minute of the driving crank should be 130 for mild steel, 90 for annealed tool steel, and 60 for unannealed tool steel. All of these steels are cut with the use of a cutting compound. Bronze can ordinarily be cut at the same speed as mild steel when a suitable compound is used. Brass heats the blade very rapidly if cut dry, and must be cut with a cooling compound adapted to brass. It also fills up the teeth of the saw if not used with the right kind of compound, but with a suitable compound, it may be cut at the same speed as machine steel.

Speeds for Turning Unusual Materials. — *Slate,* on account of its peculiarly strat-ified formation, is rather difficult to turn, but if handled carefully, can be machined in an ordinary lathe. The cutting speed should be about the same as for cast iron. A sheet of fiber or pressed paper should be interposed between the chuck or steadyrest jaws and the slate, to protect the latter. Slate rolls must not be centered and run on the tailstock. A satisfactory method of supporting a slate roll having journals at the ends is to bore a piece of lignum vitae to receive the turned end of the roll, and center it for the tailstock spindle.

Rubber can be turned at a peripheral speed of 200 feet per minute, although it is much easier to grind it with an abrasive wheel that is porous and soft. For cutting a rubber roll in two, the ordinary parting tool should not be used, but a tool shaped like a knife; such a tool severs the rubber without removing any material.

Gutta percha can be turned as easily as wood, but the tools must be sharp and a good soap-and-water lubricant used.

Copper can be turned easily at 200 feet per minute.

Lime-stone such as is used in the construction of pillars for balconies, etc., can be turned at 150 feet per minute, and the formation of ornamental contours is quite easy. *Marble* is a treacherous material to turn. It should be cut with a tool such as would be used for brass, but at a speed suitable for cast iron. It must be handled very carefully to prevent flaws in the surface.

The foregoing speeds are for high-speed steel tools. Tools tipped with tungsten carbide are adapted for cutting various non-metallic products which cannot be ma-chined readily with steel tools, such as slate, marble, synthetic plastic materials, etc. In drilling slate and marble, use flat drills; and for plastic materials, tungsten-carbide-tipped twist drills. Cutting speeds ranging from 75 to 150 feet per minute have been used for drilling slate (without coolant) and a feed of 0.025 inch per revolution for drills ¾ and 1 inch in diameter.

Estimating Machining Power. — Knowledge of the power required to perform machining operations is useful when planning new machining operations, for optim-izing existing machining operations, and to develop specifications for new machine tools that are to be acquired. The available power on any machine tool places a limit on the size of the cut that it can take. When much metal must be removed from the workpiece it is advisable to estimate the cutting conditions that will utilize the maxi-mum power on the machine. Many machining operations require only light cuts to be taken for which the machine obviously has ample power; in this event, estimating the power required is a wasteful effort. Since conditions in different shops may vary and machine tools are not all designed alike, some variations between the estimated results and those obtained on the job are to be expected; however, by using the methods provided in this section a reasonable estimate of the power required can be made, which will suffice in most practical situations.

The measure of power in customary inch units is the horsepower; in SI metric units it is the kilowatt, which is used for both mechanical and electrical power. The power required to cut a material is dependent upon the rate at which the material is being cut and upon an experimentally determined power constant, K_p, which is also called the unit horsepower, unit power, or specific power consumption. The power constant is equal to the horsepower required to cut a material at a rate of one cubic inch per minute; in SI metric units the power constant is equal to the power in kilowatts required to cut a material at a rate of one cubic centimeter per second, or 1000 cubic millimeters per second ($1 \text{ cm}^3 = 1000 \text{ mm}^3$). Different values of the power constant are required for inch and for metric units, which are related as follows: to obtain the SI metric power constant, multiply the inch power constant by 2.73; to obtain the inch power constant, divide the SI metric power constant by 2.73. Values of the power constant in Tables 23, 24, and 25 can be used for all machining opera-tions except drilling and grinding. Values given are for sharp tools.

Table 23. Power Constants, K_p, for Wrought Steels, Using Sharp Cutting Tools

Material	Brinell Hardness Number	K_p Inch Units	K_p SI Metric Units
Plain Carbon Steels			
All Plain Carbon Steels	80-100	.63	1.72
	100-120	.66	1.80
	120-140	.69	1.88
	140-160	.74	2.02
	160-180	.78	2.13
	180-200	.82	2.24
	200-220	.85	2.32
	220-240	.89	2.43
	240-260	.92	2.51
	260-280	.95	2.59
	280-300	1.00	2.73
	300-320	1.03	2.81
	320-340	1.06	2.89
	340-360	1.14	3.11
Free Machining Steels			
AISI 1108, 1109, 1110, 1115, 1116, 1117, 1118, 1119, 1120, 1125, 1126, 1132	100-120	.41	1.12
	120-140	.42	1.15
	140-160	.44	1.20
	160-180	.48	1.31
	180-200	.50	1.36
AISI 1137, 1138, 1139, 1140, 1141, 1144, 1145, 1146, 1148, 1151	180-200	.51	1.39
	200-220	.55	1.50
	220-240	.57	1.56
	240-260	.62	1.69
Alloy Steels			
AISI 4023, 4024, 4027, 4028, 4032, 4037, 4042, 4047, 4137, 4140, 4142, 4145, 4147, 4150, 4340, 4640, 4815, 4817, 4820, 5130, 5132, 5135, 5140, 5145, 5150, 6118, 6150, 8637, 8640, 8642, 8645, 8650, 8740	140-160	.62	1.69
	160-180	.65	1.77
	180-200	.69	1.88
	200-220	.72	1.97
	220-240	.76	2.07
	240-260	.80	2.18
	260-280	.84	2.29
	280-300	.87	2.38
	300-320	.91	2.48
	320-340	.96	2.62
	340-360	1.00	2.73
AISI 4130, 4320, 4615, 4620, 4626, 5120, 8615, 8617, 8620, 8622, 8625, 8630, 8720	140-160	.56	1.53
	160-180	.59	1.61
	180-200	.62	1.69
	200-220	.65	1.77
	220-240	.70	1.91
	240-260	.74	2.02
	260-280	.77	2.10
	280-300	.80	2.18
	300-320	.83	2.27
	320-340	.89	2.43
AISI 1330, 1335, 1340, E52100	160-180	.79	2.16
	180-200	.83	2.27
	200-220	.87	2.38
	220-240	.91	2.48
	240-260	.95	2.59
	260-280	1.00	2.73

Table 24.   Power Constants, K_p, for Ferrous Cast Metals,
Using Sharp Cutting Tools

Material	Brinell Hardness Number	K_p Inch Units	K_p SI Metric Units	Material	Brinell Hardness Number	K_p Inch Units	K_p SI Metric Units
Gray Cast Iron	100-120	.28	0.76	Malleable Iron Ferritic			
	120-140	.35	0.96		150-175	.42	1.15
	140-160	.38	1.04		175-200	.57	1.56
	160-180	.52	1.42	Pearlitic	200-250	.82	2.24
	180-200	.60	1.64		250-300	1.18	3.22
	200-220	.71	1.94				
	220-240	.91	2.48	Cast Steel	150-175	.62	1.69
Alloy Cast Iron	150-175	.30	0.82		175-200	.78	2.13
	175-200	.63	1.72		200-250	.86	2.35
	200-250	.92	2.51	...	...	...	...
				...	...	...	...

The value of the power constant is essentially unaffected by the cutting speed, the depth of cut, and by the cutting tool material. There are, however, factors that do affect the value of the power constant and thereby the power required to cut a material. These factors include the hardness and microstructure of the work material, the feed rate, the rake angle of the cutting tool, and the condition of the cutting edge of the tool, whether it is sharp or dull. Values in the power constant tables are given for different material hardness levels, whenever this information is available. Feed factors for the power constant are given in Table 26. All metal cutting tools wear as they are used and are expected to continue to cut as the cutting edge becomes worn. A worn cutting edge requires more power to cut than a sharp cutting edge. Factors to provide for tool wear are given in Table 27. In this table the extra-heavy duty category for milling and turning occurs only on operations where the tool is allowed to wear more than a normal amount before it is replaced, such as roll turning. In most instances the effect of the rake angle can be disregarded. The rake angle for which most of the data in the power constant tables are given is positive 14 degrees. Only when the deviation from this angle is large is it necessary to make an adjustment. Using a rake angle that is more positive reduces the power required approximately one per cent per degree; using a rake angle that is more negative increases the power required; again approximately one per cent per

Table 25.   Power Constant, K_p, for High Temperature Alloys, Tool Steel,
Stainless Steel, and Nonferrous Metals, Using Sharp Cutting Tools

Material	Brinell Hardness Number	K_p Inch Units	K_p SI Metric Units	Material	Brinell Hardness Number	K_p Inch Units	K_p SI Metric Units
High Temperature Alloys					150-175	.60	1.64
A286	165	.82	2.24	Stainless Steel	175-200	.72	1.97
A286	285	.93	2.54		200-250	.88	2.40
Chromoloy	200	.78	3.22	Zinc Die Cast Alloys	...	.25	0.68
Chromoloy	310	1.18	3.00	Copper (pure)	...	.91	2.48
Inco 700	330	1.12	3.06	Brass			
Inco 702	230	1.10	3.00	Hard	...	.83	2.27
Hastelloy-B	230	1.10	3.00	Medium	...	.50	1.36
M-252	230	1.10	3.00	Soft	...	.25	0.68
M-252	310	1.20	3.28	Leaded	...	.30	0.82
Ti-150A	340	.65	1.77	Bronze			
U-500	375	1.10	3.00	Hard	...	.91	2.48
				Medium	...	.50	1.36
Monel Metal	...	1.00	2.73	Soft	...	.33	0.90
Tool Steel	175-200	.75	2.05	Aluminum			
	200-250	.88	2.40	Cast	...	.25	0.68
	250-300	.98	2.68	Rolled (hard)	...	.33	0.90
	300-350	1.20	3.28	Magnesium Alloys	...	.10	0.27
	350-400	1.30	3.55				

Table 26. Feed Factor, C, for Power Constants

Inch Units				SI Metric Units			
Feed in.*	C	Feed in.*	C	Feed mm†	C	Feed mm†	C
.001	1.60	.014	.97	0.02	1.70	0.35	.97
.002	1.40	.015	.96	0.05	1.40	0.38	.95
.003	1.30	.016	.94	0.07	1.30	0.40	.94
.004	1.25	.018	.92	0.10	1.25	0.45	.92
.005	1.19	.020	.90	0.12	1.20	0.50	.90
.006	1.15	.022	.88	0.15	1.15	0.55	.88
.007	1.11	.025	.86	0.18	1.11	0.60	.87
.008	1.08	.028	.84	0.20	1.08	0.70	.84
.009	1.06	.030	.83	0.22	1.06	0.75	.83
.010	1.04	.032	.82	0.25	1.04	0.80	.82
.011	1.02	.035	.80	0.28	1.01	0.90	.80
.012	1.00	.040	.78	0.30	1.00	1.00	.78
.013	.98	.060	.72	0.33	.98	1.50	.72

* Turning–in./rev; Milling–in./tooth; Planing and Shaping–in./stroke; Broaching–in./tooth.
†Turning–mm/rev; Milling–mm/tooth; Planing and Shaping–mm/stroke; Broaching–mm/tooth.

degree. Many indexable insert cutting tools have a pressed-in chip breaker or other cutting edge modifications, which have the effect of reducing the power required to cut a material. The extent of this effect cannot be predicted without a test of each design. Cutting fluids will also usually reduce the power required, when operating in the lower range of cutting speeds. Again, the extent of this effect cannot be predicted because each cutting fluid exhibits its own characteristics.

The machine tool transmits the power from the driving motor to the workpiece, where it is used to cut the material. How efficiently this is done is measured by the machine tool efficiency factor, E. Average values of this factor are given in Table 28. Formulas for calculating the metal removal rate, Q, for different machining operations are given in Table 29. These formulas are used together with those given below. The following formulas can be used with either customary inch or with SI metric units.

$$P_c = K_p C Q W \qquad (1)$$

$$P_m = \frac{P_c}{E} = \frac{K_p C Q W}{E} \qquad (2)$$

Where: P_c = Power at the cutting tool; hp, or kW
P_m = Power at the motor; hp, or kW
K_p = Power constant   (See Tables 23, 24, and 25)
Q = Metal removal rate; in.³/min, or cm³/s   (See Table 29)

Table 27. Tool Wear Factors, W

Type of Operation		W
For all operations with sharp cutting tools		1.00
Turning:	Finish turning (lightcuts)	1.10
	Normal rough and semi-finish turning	1.30
	Extra-heavy duty rough turning	1.60-2.00
Milling:	Slab milling	1.10
	End milling	1.10
	Light and medium face milling	1.10-1.25
	Extra-heavy duty face milling	1.30-1.60
Drilling:	Normal drilling	1.30
	Drilling hard-to-machine materials and drilling with a very dull drill	1.50
Broaching:	Normal broaching	1.05-1.10
	Heavy duty surface broaching	1.20-1.30

For planing and shaping, use values given for turning.

Table 28. Machine Tool Efficiency Factors, E

Type of Drive	E	Type of Drive	E
Direct Belt Drive	.90	Geared Head Drive	.70–.80
Back Gear Drive	.75	Oil-Hydraulic Drive	.60–.90

C = Feed factor for power constant   (See Table 26)
W = Tool wear factor   (See Table 27)
E = Machine tool efficiency factor   (See Table 28)
V = Cutting speed, fpm, or m/min
N = Cutting speed, rpm
f = Feed rate for turning; in./rev. or mm/rev
f = Feed rate for planing and shaping; in./stroke, or mm/stroke
f_t = Feed per tooth; in./tooth, or mm/tooth
f_m = Feed rate; in./min, or mm/min
d_t = Maximum depth of cut per tooth; in., or mm
d = Depth of cut; in., or mm
n_t = Number of teeth on milling cutter
n_c = Number of teeth engaged in work
w = width of cut; in. or mm

Example: A 180-200 Bhn AISI shaft is to be turned on a cutting a cutting speed of 350 fpm (107 m/min), a feed rate of .016 in./rev (0.40 mm/rev), and a depth of cut of .100 inch (2.54 mm). Estimate the power at the cutting tool and at the motor, using both the inch and metric data.

Inch units: K_p = .62 (From Table 23) C = .94 (From Table 26)
           W = 1.30 (From Table 27) E = .80 (From Table 28)

$$Q = 12\,Vfd = 12 \times 350 \times .016 \times .100 \quad \text{(From Table 29)}$$
$$= 6.72 \text{ in.}^3/\text{min}$$

$$P_c = K_p C Q W = .62 \times .94 \times 6.72 \times 1.30 = 5 \text{ hp}$$

$$P_m = \frac{P_c}{E} = \frac{5}{.80} = 6.25 \text{ hp}$$

SI metric units: K_p = 1.69 (From Table 23) C = .94 (From Table 26)
            W = 1.30 (From Table 27) E = .80 (From Table 28)

$$Q = \frac{V}{60} fd = \frac{107}{60} \times 0.40 \times 2.54 \quad \text{(From Table 29)}$$
$$= 1.81 \text{ cm}^3/\text{s}$$

$$P_c = K_p C Q W = 1.69 \times .94 \times 1.81 \times 1.30 = 3.74 \text{ kW}$$

$$P_m = \frac{P_c}{E} = \frac{3.74}{.80} = 4.675 \text{ kW}$$

Table 29. Formulas for Calculating the Metal Removal Rate, Q

Operation	Metal Removal Rate	
	For Inch Units Only Q = in.^{3}/min	For SI Metric Units Only Q = cm^3/s
Single Point Tools (Turning, Planing, and Shaping)	$12\,Vfd$	$\dfrac{V}{60} fd$
Milling	$f_m wd$	$\dfrac{f_m wd}{60,000}$
Surface Broaching	$12\,Vwn_c d_t$	$\dfrac{V}{60} wn_c d_t$

Whenever possible the maximum power available on a machine tool should be used when heavy cuts must be taken. The cutting conditions for utilizing the maximum power should be selected in the following order: 1. select the maximum depth of cut that can be used; 2. select the maximum feed rate that can be used; and, 3. estimate the cutting speed that will utilize the maximum power available on the machine. This order is based on obtaining the longest tool life of the cutting tool while at the same time obtaining as much production as possible from the machine. *The life of a cutting tool is most affected by the cutting speed, then by the feed rate, and least of all by the depth of cut.* The maximum metal removal rate that a given machine is capable of machining from a given material is used as the basis for estimating the cutting speed that will utilize all of the power available on the machine.

Example: A .125 inch deep cut is to be taken on a 200-210 Bhn AISI 1050 steel part using a 10 hp geared head lathe. The feed rate selected for this job is .018 in./rev. Estimate the cutting speed that will utilize the maximum power available on the lathe.

$$K_p = .85 \quad \text{(From Table 23)}$$
$$C = .92 \quad \text{(From Table 26)}$$
$$W = 1.30 \quad \text{(From Table 27)}$$
$$E = .80 \quad \text{(From Table 28)}$$

$$Q_{max} = \frac{P_m E}{K_p CW} = \frac{10 \times .80}{.85 \times .92 \times 1.30} \qquad \left(P_m = \frac{K_p CQW}{E} \right)$$

$$= 7.87 \text{ in.}^3/\text{min}$$

$$V = \frac{Q_{max}}{12 \, fd} = \frac{7.87}{12 \times .018 \times .125} \qquad (Q = 12 \, Vfd)$$

$$= 290 \text{ fpm}$$

Example: A 160-180 Bhn gray iron casting that is 6 inches wide is to have ⅛ inch stock removed on a 10 hp milling machine, using an 8 inch diameter, 10 tooth, indexable insert cemented carbide face milling cutter. The feed rate selected for this cutter is .012 in./tooth, and all of the stock (.125 in.) will be removed in one cut. Estimate the cutting speed that will utilize the maximum power available on the machine.

$$K_p = .52 \quad \text{(From Table 24)}$$
$$C = 1.00 \quad \text{(From Table 26)}$$
$$W = 1.20 \quad \text{(From Table 27)}$$
$$E = .80 \quad \text{(From Table 28)}$$

$$Q_{max} = \frac{P_m E}{K_p CW} = \frac{10 \times .80}{.52 \times 1.00 \times 1.20} = 12.82 \text{ in.}^3/\text{min} \qquad \left(P_m = \frac{K_p CQW}{E} \right)$$

$$f_m = \frac{Q_{max}}{wd} = \frac{12.82}{6 \times .125} = 17 \text{ in./min.} \qquad (Q = f_m wd)$$

$$N = \frac{f_{max}}{f_t n_t} = \frac{17}{.012 \times 10} = 140 \text{ rpm} \qquad (f_m = f_t n_t N)$$

$$V = \frac{\pi DN}{12} = \frac{\pi \times 8 \times 140}{12} = 293 \text{ fpm} \qquad \left(N = \frac{12V}{\pi D} \right)$$

Table 30. Work Material Factor, K_d, for Drilling with a Sharp Drill

Work Material	Work Material Constant, K_d
AISI 1117 (Resulfurized free machining mild steel)	12,000
Steel, 200 Bhn	24,000
Steel, 300 Bhn	31,000
Steel, 400 Bhn	34,000
Cast Iron, 150 Bhn	14,000
Most Aluminum Alloys	7,000
Most Magnesium Alloys	4,000
Most Brasses	14,000
Leaded Brass	7,000
Austenitic Stainless Steel (Type 316)	24,000* for Torque 35,000* for Thrust
Titanium Alloy T16A 4V 40R$_c$	18,000* for Torque 29,000* for Thrust
René 41 40R$_c$	40,000*† min.
Hastelloy-C	30,000* for Torque 37,000* for Thrust

* Values based upon a limited number of tests.   † Will increase with rapid wear.

Estimating Drilling Thrust, Torque, and Power. — Although the lips of a drill cut metal and produce a chip in the same manner as the cutting edges of other metal cutting tools, the chisel edge removes the metal by means of a very complex combination of extrusion and cutting. For this reason a separate method must be used to estimate the power required for drilling. Also, it is often desirable to know the magnitude of the thrust and the torque required to drill a hole. The formulas and tabular data provided in this section are based on information supplied by the National Twist Drill Division of Lear Siegler, Inc. The values in Tables 30 through 33 are for sharp drills and the tool wear factors are given in Table 27. For most ordinary drilling operations 1.30 can be used as the tool wear factor. When drilling most difficult-to-machine materials and when the drill is allowed to become very dull, 1.50 should be used as the value of this factor. It is usually more convenient to measure the web thickness at the drill point than the length of the chisel edge; for this reason, the approximate w/d ratio corresponding to each c/d ratio for a correctly ground drill is provided in Table 33. For most standard twist drills the c/d ratio is .18, unless the drill has been ground short or the web has been thinned. The c/d ratio of split point drills is .03. The formulas given below can be used for spade drills, as well as for twist drills. Separate formulas are required for use with customary inch units and for SI metric units.

For inch units only:

$$T = 2K_dF_fF_TBW + K_dd^2JW \tag{3}$$

$$M = K_dF_fF_MAW \tag{4}$$

$$P_c = \frac{MN}{63.025} \tag{5}$$

For SI metric units only:

$$T = 0.05\ K_dF_fF_TBW + 0.007\ K_dd^2JW \tag{6}$$

$$M = \frac{K_dF_fF_MAW}{40,000} = 0.000025\ K_dF_fF_MAW \tag{7}$$

$$P_c = \frac{MN}{9550} \tag{8}$$

Use with either inch or metric units:

$$P_m = \frac{P_c}{E} \qquad\qquad (9)$$

Where: P_c = Power at the cutter; hp, or kW
$\quad\quad\; P_m$ = Power at the motor; hp, or kW
$\quad\quad\;\; M$ = Torque; in. lb, or N·m
$\quad\quad\;\;\; T$ = Thrust; lb, or N
$\quad\quad\; K_d$ = Work material factor   (See Table 30)
$\quad\quad\; F_f$ = Feed factor   (See Table 31)
$\quad\quad\; F_T$ = Thrust factor for drill diameter   (See Table 32)
$\quad\quad\; F_M$ = Torque factor for drill diameter   (See Table 32)
$\quad\quad\;\; A$ = Chisel edge factor for torque   (See Table 33)
$\quad\quad\;\; B$ = Chisel edge factor for thrust   (See Table 33)
$\quad\quad\;\; J$ = Chisel edge factor for thrust   (See Table 33)
$\quad\quad\; W$ = Tool wear factor   (See Table 27)
$\quad\quad\;\; N$ = Spindle speed; rpm
$\quad\quad\;\; E$ = Machine tool efficiency factor   (See Table 28)
$\quad\quad\;\; D$ = Drill diameter; in., or mm
$\quad\quad\;\;\; c$ = Chisel edge length; in., or mm   (See Table 33)
$\quad\quad\; w$ = Web thickness at drill point; in., or mm   (See Table 33)

Example: A standard ⅞ inch drill is to drill steel parts having a hardness of 200 Bhn on a drilling machine having an efficiency of .80. The spindle speed to be used is 350 rpm and the feed rate will be .008 in./rev. Calculate the thrust, torque, and power required to drill these holes:

$\quad\quad K_d = 24,000$ (From Table 30)
$\quad\quad F_f = .021$ (From Table 31)
$\quad\quad F_T = .899$ (From Table 32)
$\quad\quad F_M = .786$ (From Table 32)
$\quad\quad\;\; A = 1.085$ (From Table 33)
$\quad\quad\;\; B = 1.355$ (From Table 33)
$\quad\quad\;\;\; J = .030$ (From Table 33)
$\quad\quad\; W = 1.30$ (From Table 27)

$T = 2K_dF_fF_TBW + K_dd^2JW$

$\quad = 2 \times 24,000 \times .021 \times .899 \times 1.355 \times 1.30 + 24,000 \times .875^2 \times .030 \times 1.30$

$\quad = 2313$ lb

$M = K_dF_fF_MAW$

$\quad = 24,000 \times .021 \times .786 \times 1.085 \times 1.30 = 559$ in. lb

$$P_c = \frac{MN}{63,025} = \frac{559 \times 350}{63,025} = 3.1 \text{ hp} \qquad\qquad P_m = \frac{P_c}{E} = \frac{3.1}{.80} = 3.9 \text{ hp}$$

Twist drills are generally the most highly stressed of all metal cutting tools. They must not only resist the cutting forces on the lips, but also the drill torque resulting from these forces and the very large thrust force required to push the drill through the hole. Therefore, in many instances when drilling smaller holes, the twist drill places a limit on the power used while for very large holes, the machine may limit the power.

Table 31. Feed Factors, F_f, for Drilling

Inch Units				SI Metric Units			
Feed, in./rev	F_f	Feed, in./rev	F_f	Feed, mm/rev	F_f	Feed, mm/rev	F_f
.0005	.0023	.012	.029	0.01	.025	0.30	.382
.001	.004	.013	.031	0.03	.060	0.35	.432
.002	.007	.015	.035	0.05	.091	0.40	.480
.003	.010	.018	.040	0.08	.133	0.45	.528
.004	.012	.020	.044	0.10	.158	0.50	.574
.005	.014	.022	.047	0.12	.183	0.55	.620
.006	.017	.025	.052	0.15	.219	0.65	.708
.007	.019	.030	.060	0.18	.254	0.75	.794
.008	.021	.035	.068	0.20	.276	0.90	.919
.009	.023	.040	.076	0.22	.298	1.00	1.000
.010	.025	.050	.091	0.25	.330	1.25	1.195

Table 32. Drill Diameter Factors: F_T for Thrust; F_M for Torque

Inch Units						SI Metric Units					
Drill Diam., in.	F_T	F_M	Drill Diam., in.	F_T	F_M	Drill Diam., mm	F_T	F_M	Drill Diam., mm	F_T	F_M
.063	.110	.007	.875	.899	.786	1.60	1.46	2.33	22.00	11.86	260.8
.094	.151	.014	.938	.950	.891	2.40	2.02	4.84	24.00	12.71	305.1
.125	.189	.024	1.000	1.000	1.000	3.20	2.54	8.12	25.50	13.34	340.2
.156	.226	.035	1.063	1.050	1.116	4.00	3.03	12.12	27.00	13.97	377.1
.188	.263	.049	1.125	1.099	1.236	4.80	3.51	16.84	28.50	14.58	415.6
.219	.297	.065	1.250	1.195	1.494	5.60	3.97	22.22	32.00	16.00	512.0
.250	.330	.082	1.375	1.290	1.774	6.40	4.42	28.26	35.00	17.19	601.6
.281	.362	.102	1.500	1.383	2.075	7.20	4.85	34.93	38.00	18.36	697.6
.313	.395	.124	1.625	1.475	2.396	8.00	5.28	42.22	42.00	19.89	835.3
.344	.426	.146	1.750	1.565	2.738	8.80	5.96	50.13	45.00	21.02	945.8
.375	.456	.171	1.875	1.653	3.100	9.50	6.06	57.53	48.00	22.13	1062
.438	.517	.226	2.000	1.741	3.482	11.00	6.81	74.90	50.00	22.86	1143
.500	.574	.287	2.250	1.913	4.305	12.50	7.54	94.28	58.00	25.75	1493
.563	.632	.355	2.500	2.081	5.203	14.50	8.49	123.1	64.00	27.86	1783
.625	.687	.429	2.750	2.246	6.177	16.00	9.19	147.0	70.00	29.93	2095
.688	.741	.510	3.000	2.408	7.225	17.50	9.87	172.8	76.00	31.96	2429
.750	.794	.596	3.500	2.724	9.535	19.00	10.54	200.3	90.00	36.53	3293
.813	.847	.689	4.000	3.031	12.13	20.00	10.98	219.7	100.00	39.81	3981

Table 33. Chisel Edge Factors for Torque and Thrust

c/d	Approx. w/d	Torque Factor A	Thrust Factor B	Thrust Factor J	c/d	Approx. w/d	Torque Factor A	Thrust Factor B	Thrust Factor J
.03	.025	1.000	1.100	.001	.18	.155	1.085	1.355	.030
.05	.045	1.005	1.140	.003	.20	.175	1.105	1.380	.040
.08	.070	1.015	1.200	.006	.25	.220	1.155	1.445	.065
.10	.085	1.020	1.235	.010	.30	.260	1.235	1.500	.090
.13	.110	1.040	1.270	.017	.35	.300	1.310	1.575	.120
.15	.130	1.050	1.310	.022	.40	.350	1.395	1.620	.160

For drills of standard design, use $c/d = .18$.
For split point drills, use $c/d = .03$.
c/d = Length of Chisel Edge ÷ Drill Diameter.
w/d = Web Thickness at Drill Point ÷ Drill Diameter.

SCREW MACHINE FEEDS AND SPEEDS

Feeds and Speeds for Automatic Screw Machine Tools. — Approximate feeds and speeds for standard screw machine tools are given in the accompanying table.

Knurling in Automatic Screw Machines. — When knurling is done from the cross slide, it is good practice to feed the knurl gradually to the center of the work, starting to feed when the knurl touches the work and then passing off the center of the work with a quick rise of the cam. The knurl should also dwell for a certain number of revolutions, depending on the pitch of the knurl and the kind of material being knurled. See also Knurls and Knurling on pages 1112–1116.

When two knurls are employed for spiral and diamond knurling from the turret, the knurls can be operated at a higher rate of feed for producing a spiral than they can for producing a diamond pattern. The reason for this is that in the first case the knurls work in the same groove, whereas in the latter case they work independently of each other.

Revolutions Required for Top Knurling. — The depth of the teeth and the feed per revolution govern the number of revolutions required for top knurling from the cross slide. If R is the radius of the stock, d is the depth of the teeth, c is the distance the knurl travels from the point of contact to the center of the work at the feed required for knurling, and r is the radius of the knurl; then

$$c = \sqrt{(R + r)^2 - (R + r - d)^2}$$

For example, if the stock radius R is $\frac{5}{32}$ inch, depth of teeth d is 0.0156 inch, and radius of knurl r is 0.3125 inch, then

$$c = \sqrt{(0.1562 + 0.3125)^2 - (0.1562 + 0.3125 - 0.0156)^2}$$
$$= 0.120 \text{ inch} = \text{cam rise required.}$$

Assume that it is required to find the number of revolutions to knurl a piece of brass $\frac{5}{16}$ inch in diameter using a 32 pitch knurl. The included angle of the teeth for brass is 90 degrees, the circular pitch is 0.03125 inch, and the calculated tooth depth is 0.0156 inch. The distance c (as determined in the previous example) is 0.120 inch. Referring to the accompanying table of feeds and speeds, the feed for top knurling brass is 0.005 inch per revolution. The number of revolutions required for knurling is, therefore, $0.120 \div 0.005 = 24$ revolutions. If conditions permit, the higher feed of 0.008 inch per revolution given in the table may be used, in which case 15 revolutions are required for knurling.

Cams for Threading. — The table "Spindle Revolutions and Cam Rise for Threading" on page 1022 gives the revolutions required for threading various lengths and pitches and the corresponding rise for the cam lobe. To illustrate the use of this table, suppose a set of cams is required for threading a screw to the length of $\frac{3}{8}$ inch in a Brown & Sharpe machine. Assume that the spindle speed is 2400 revolutions per minute; the number of revolutions to complete one piece, 400; time required to make one piece, 10 seconds; pitch of the thread, $\frac{1}{32}$ inch or 32 threads per inch. By referring to the table, under 32 threads per inch, and opposite $\frac{3}{8}$ inch (length of threaded part), the number of revolutions required is found to be 15 and the rise required for the cam, 0.413 inch.

Cams of this type are often cut on a circular milling attachment. When this method is employed, the number of minutes the attachment should be revolved for each 0.001 inch rise, is first determined. As 15 revolutions are required for threading and 400 for completing one piece, that part of the cam surface required for the actual threading operation equals $15 \div 400 = 0.0375$, which is equivalent to 810 minutes of the circumference. As the total rise, in this case, through an arc of 810 minutes is 0.413 inch,

Approximate Cutting Speeds and Feeds for Standard Automatic Screw Machine Tools — Brown and Sharpe

Tool	Cut — Width or Depth, Inches	Cut — Diam. of Hole, Inches	Brass† Feed, Inches per Rev.	Mild or Soft Steel Feed, Inches per Rev.	Mild or Soft Steel Surface Speed, Feet per Min. — Carbon Tools	Mild or Soft Steel Surface Speed, Feet per Min. — H.S.S. Tools	Tool Steel, .80–1.00% C Feed, Inches per Rev.	Tool Steel Surface Speed, Feet per Min. — Carbon Tools	Tool Steel Surface Speed, Feet per Min. — H.S.S. Tools
Boring tools	0.005		.008	.008	50	110	.004	30	60
Box tools, roller rest	1/32		.012	.010	70	150	.005	40	75
	1/16		.010	.008	70	150	.004	40	75
	1/8		.008	.007	70	150	.003	40	75
	3/16		.008	.006	70	150	.002	40	75
Single chip finishing	1/4		.006	.005	70	150	.0015	40	75
			.010	.010			.006		
Finishing	0.005		.003	.0015	50	110	.001	30	75
Center drills		Under 1/8	.006	.0035	50	110	.002	30	75
		Over 1/8							
Cutoff tools — Angular	3/64–1/8		.0015	.0006	80	150	.0004	50	85
Cutoff tools — Circular	1/16–1/8		.0035	.0015	80	150	.001	50	85
Cutoff tools — Straight			.0035	.0015	80	150	.001	50	85
Stock diameter under 1/8 in.			.002	.0008	80	150	.0005	50	85
Dies — Button					30			14	
Dies — Chaser					30	40		16	20
Drills, twist cut		0.02	.0014	.001	40	60	.0006	30	45
		0.04	.002	.0014	40	60	.0008	30	45
		1/16	.004	.002	40	60	.0012	30	45
		3/32	.006	.0025	40	60	.0016	30	45
		1/8	.009	.0035	40	75	.002	30	60
		3/16	.012	.004	40	75	.003	30	60
		1/4	.014	.005	40	75	.003	30	60
		5/16	.016	.005	40	85	.0035	30	60
		3/8–5/8	.016	.006	40		.004	30	
Form tools, circular	1/8		.002	.0009	80	150	.0006	50	85
	1/4		.002	.0008	80	150	.0005	50	85
	3/8		.0015	.0007	80	150	.0004	50	85
	1/2		.0012	.0006	80	150	.0004	50	85
	5/8		.001	.0005	80	150	.0003	50	85
	3/4		.001	.0005	80	150	.0003	50	85
	1		.001	.0004	80	150			85

See footnotes at end of table.

Compiled by Brown and Sharpe.

Approximate Cutting Speeds and Feeds for Standard Automatic Screw Machine Tools — Brown and Sharpe (Concluded)

Tool	Width or Depth, Inches	Diam. of Hole, Inches	Brass† Feed, Inches per Rev.	Mild or Soft Steel — Feed, Inches per Rev.	Mild or Soft Steel — Carbon Tools Surface Speed, Feet per Min.	Mild or Soft Steel — H.S.S. Tools Surface Speed	Tool Steel, .80–1.00% C — Feed, Inches per Rev.	Tool Steel — Carbon Tools Surface Speed, Feet per Min.	Tool Steel — H.S.S. Tools Surface Speed
Hollow mills and balance turning tools — Turned diam. under 5/32 in.	1/32		.012	.010	70	150	.008	40	85
	1/16		.010	.009	70	150	.006	40	85
Turned diam. over 5/32 in.	1/32		.017	.014	70	150	.010	40	85
	1/16		.015	.012	70	150	.008	40	85
	1/8		.012	.010	70	150	.008	40	85
	3/16		.010	.008	70	150	.006	40	85
	1/4		.009	.007	70	150	.0045	40	85
Knee tools	1/32			.010	70	150	.008	40	85
Knurling tools — Turret	On		.020	.015	150		.010	105	
Turret	Off		.040	.030	150		.025	105	
Side or swing			.004	.002	150		.002	105	
Side or swing			.006	.004	150		.003	105	
Top			.005	.003	150		.002	105	
Top			.008	.006	150		.004	105	
Pointing and facing tools			.001	.0008	70	150	.0005	40	80
Pointing and facing tools			.0025	.002	70	150	.0008	40	80
Reamers and bits	.003–.004	1/8 or less	.010–.007	.008–.006	70	105	.006–.004	40	60
Reamers and bits	.004–.008	1/8 or over	.010	.010	70	105	.006–.008	40	60
Recessing tools — End cut			.001	.0006	70	150	.0004	40	75
End cut			.005	.003	70	150	.002	40	75
Inside cut		1/16–1/8	.0025	.002	70	105	.0015	40	60
Inside cut			.0008	.0006	70	105	.0004	40	60
Swing tools, forming	1/8		.002	.0007	70	150	.0005	40	85
	1/4		.0012	.0005	70	150	.0003	40	85
	3/8		.001	.0004	70	150	.0002	40	85
	1/2		.0008	.0003	70	150	.0002	40	85
Turning, straight and taper*	1/32		.008	.006	70	150	.0035	40	85
	1/16		.006	.004	70	150	.003	40	85
	1/8		.005	.003	70	150	.002	40	85
	3/16		.004	.0025	70	150	.0015	40	85
Taps					25	30		12	15

† Use maximum spindle speed on machine.   * For taper turning use feed slow enough for greatest depth of cut.

Spindle Revolutions and Cam Rise for Threading

Number of Threads per Inch

First Line: Revolutions of Spindle for Threading.   Second Line: Rise on Cam for Threading, Inch

Length of Threaded Portion, Inch	14	16	18	20	24	28	30	32	36	40	48	56	64	72	80
1/16	…	…	…	…	3.00 / 0.106	5.00 / 0.157	5.00 / 0.147	5.00 / 0.138	5.50 / 0.134	5.50 / 0.121	6.00 / 0.110	8.00 / 0.129	8.50 / 0.120	9.00 / 0.113	9.50 / 0.107
1/8	…	3.50 / 0.186	3.50 / 0.165	4.00 / 0.170	4.50 / 0.159	6.50 / 0.204	7.00 / 0.205	7.00 / 0.193	7.00 / 0.171	8.00 / 0.176	9.00 / 0.165	11.50 / 0.185	12.50 / 0.176	13.50 / 0.169	14.50 / 0.163
3/16	4.00 / 0.243	4.50 / 0.239	5.00 / 0.236	5.50 / 0.234	6.00 / 0.213	8.50 / 0.267	8.50 / 0.249	9.00 / 0.248	10.00 / 0.244	10.50 / 0.231	12.00 / 0.220	15.00 / 0.241	16.50 / 0.232	18.00 / 0.225	19.50 / 0.219
1/4	5.00 / 0.304	5.50 / 0.292	6.00 / 0.283	6.50 / 0.276	7.50 / 0.266	10.00 / 0.314	10.50 / 0.308	11.00 / 0.303	12.00 / 0.293	13.00 / 0.286	15.00 / 0.275	18.50 / 0.297	20.50 / 0.288	23.50 / 0.294	24.50 / 0.276
5/16	6.00 / 0.364	6.50 / 0.345	7.00 / 0.330	8.00 / 0.340	9.00 / 0.319	12.00 / 0.377	12.50 / 0.367	13.00 / 0.358	14.50 / 0.354	15.50 / 0.341	18.00 / 0.340	22.00 / 0.354	24.50 / 0.345	27.00 / 0.338	29.50 / 0.332
3/8	7.00 / 0.425	7.50 / 0.398	8.50 / 0.401	9.00 / 0.383	10.50 / 0.372	13.50 / 0.424	14.50 / 0.425	15.00 / 0.413	16.50 / 0.403	18.00 / 0.396	21.00 / 0.385	25.50 / 0.410	28.50 / 0.401	31.50 / 0.394	34.50 / 0.388
7/16	7.50 / 0.455	8.50 / 0.451	9.50 / 0.448	10.50 / 0.446	12.00 / 0.425	15.50 / 0.487	16.00 / 0.469	17.00 / 0.468	19.00 / 0.464	20.50 / 0.451	24.00 / 0.440	29.00 / 0.466	32.50 / 0.457	36.00 / 0.450	39.50 / 0.444
1/2	8.50 / 0.516	9.50 / 0.504	10.50 / 0.496	11.50 / 0.489	13.50 / 0.478	17.00 / 0.534	18.00 / 0.528	19.00 / 0.523	21.00 / 0.513	23.00 / 0.506	27.00 / 0.495	32.50 / 0.522	36.50 / 0.513	40.50 / 0.506	44.50 / 0.501
9/16	9.50 / 0.577	10.50 / 0.558	11.50 / 0.543	13.00 / 0.553	15.00 / 0.531	19.00 / 0.597	20.00 / 0.587	21.00 / 0.578	23.50 / 0.574	25.50 / 0.561	30.00 / 0.550	36.00 / 0.579	40.50 / 0.570	45.00 / 0.563	49.50 / 0.559
5/8	10.50 / 0.637	11.50 / 0.611	13.00 / 0.614	14.00 / 0.595	16.50 / 0.584	20.50 / 0.644	22.00 / 0.645	23.00 / 0.633	25.50 / 0.623	28.00 / 0.616	33.00 / 0.605	39.50 / 0.635	44.50 / 0.626	49.50 / 0.619	54.50 / 0.613
11/16	11.00 / 0.668	12.50 / 0.664	14.00 / 0.661	15.50 / 0.659	18.00 / 0.638	22.50 / 0.707	23.50 / 0.689	25.00 / 0.688	28.00 / 0.684	30.50 / 0.671	36.00 / 0.660	43.00 / 0.691	48.50 / 0.682	54.00 / 0.675	59.50 / 0.679
3/4	12.00 / 0.728	13.50 / 0.717	15.00 / 0.708	16.50 / 0.701	19.50 / 0.691	24.00 / 0.754	25.50 / 0.748	27.00 / 0.743	30.00 / 0.733	33.00 / 0.726	39.00 / 0.715	46.50 / 0.747	52.50 / 0.738	58.50 / 0.731	64.50 / 0.726

the number of minutes for each 0.001 inch rise equals 810 ÷ 413 = 1.96 or, approximately, two minutes. If the attachment is graduated to read to five minutes, the cam will be fed laterally 0.0025 inch each time it is turned five minutes.

Practical Points on Cam and Tool Design. — The following general rules are given to aid in designing cams and special tools for automatic screw machines, and apply particularly to Brown and Sharpe machines:

1. Use the highest speeds recommended for the material used that the various tools will stand.

2. Use the arrangement of circular tools best suited for the class of work.

3. Decide on the quickest and best method of arranging the operations before designing the cams.

4. Do not use turret tools for forming when the cross-slide tools can be used to better advantage.

5. Make the shoulder on the circular cutoff tool large enough so that the clamping screw will grip firmly.

6. Do not use too narrow a cutoff blade.

7. Allow 0.005 to 0.010 inch for the circular tools to approach the work and 0.003 to 0.005 inch for the cutoff tool to pass the center.

8. When cutting off work, the feed of the cutoff tool should be decreased near the end of the cut where the piece breaks off.

9. When a thread is cut up to a shoulder, the piece should be grooved or necked to make allowance for the lead on the die. This requires an extra projection on the forming tool and also an extra amount of rise on the cam.

10. Allow sufficient clearance for tools to pass one another.

11. Always make a diagram of the cross-slide tools in position on the work when difficult operations are to be performed; it is also necessary to make a diagram of the tools held in the turret.

12. Do not drill a hole the depth of which is more than 3 times the diameter of the drill, but rather use two or more drills as required. If there are not sufficient holes in the turret for the extra drills needed, make provision for withdrawing the drill clear of the hole and then advancing it into the hole again.

13. Do not run drills at low speeds. Feeds and speeds recommended in the table of feeds and speeds on pages 1020 and 1021 should be followed as far as is practicable.

14. When the turret tools operate farther in than the face of the chuck, see that they will clear the chuck when the turret is revolved.

15. See that the body of all turret tools will clear the side of the chute when the turret is revolved.

16. Use a balance turning tool or a hollow mill for roughing cuts.

17. The rise on the thread lobe should be reduced so that the spindle will reverse when the tap or die holder is drawn out.

18. When bringing another tool into position after a threading operation, allow clearance before revolving the turret.

19. Make provision to revolve the turret rapidly, especially when pieces are being made in from three to five seconds and when only a few tools are used in the turret. It is sometimes convenient to use two sets of tools.

20. When using a belt-shifting attachment for threading, clearance should be allowed, as it requires extra time to shift the belt.

21. When laying out a set of cams for operating on a piece which requires to be slotted, cross-drilled or burred, allowance should be made on the lead cam so that the transferring arm can descend and ascend to and from the work without coming in contact with any of the turret tools.

22. Always allow a vacant hole in the turret when it is necessary to use the transferring arm.

23. When designing special tools allow as much clearance as possible. Do not make them so that they will just clear each other, as a slight inaccuracy in the dimensions will then often cause trouble.

24. When designing special tools having intricate movements, avoid springs as much as possible, and use positive actions.

Stock for Screw Machine Products. — The amount of stock required for the production of 1000 pieces on the automatic screw machine can be obtained directly from the table "Stock required for Screw Machine Products." To use this table, add to the length of the work the width of the cut-off tool blade; then the number of feet of material required for 1000 pieces can be found opposite the figure thus obtained, in the column headed "Feet per 1000 Parts." Screw machine stock usually comes in bars 10 feet long, and in compiling this table an allowance was made for chucking on each bar.

The table can be extended by using the following formula, in which F = number of feet required for 1000 pieces; L = length of piece in inches; W = width of cut-off tool blade in inches.

$$F = (L + W) \times 84.$$

The amount to add to the length of the work, or the width of the cut-off tool, is given in the following, which is standard in a number of machine shops:

Diameter of Stock, Inches	Width of Cut-off Tool Blade, Inches
0.000–0.250	0.045
0.251–0.375	0.062
0.376–0.625	0.093
0.626–1.000	0.125
1.001–1.500	0.156

It is sometimes convenient to know the weight of a certain number of pieces, when estimating the price. The weight of round bar stock can be found by means of the following formulas, in which W = weight in pounds; D = diameter of stock in inches; F = length in feet:

For brass stock: $W = D^2 \times 2.86 \times F.$

For steel stock: $W = D^2 \times 2.675 \times F.$

For iron stock: $W = D^2 \times 2.65 \times F.$

Stock Required for Screw Machine Products

The table gives the amount of stock, in feet, required for 1000 pieces, when the length of the finished part plus the thickness of the cut-off tool blade is known. Allowance has been made for chucking. To illustrate, if length of cut-off tool and work equals 0.140 inch, 11.8 feet of stock is required for the production of 1000 parts.

Length of Piece and Cut-off Tool	Feet per 1000 Parts	Length of Piece and Cut-off Tool	Feet per 1000 Parts	Length of Piece and Cut-off Tool	Feet per 1000 Parts	Length of Piece and Cut-off Tool	Feet per 1000 Parts
0.050	4.2	0.430	36.1	0.810	68.1	1.380	116.0
0.060	5.0	0.440	37.0	0.820	68.9	1.400	117.6
0.070	5.9	0.450	37.8	0.830	69.7	1.420	119.3
0.080	6.7	0.460	38.7	0.840	70.6	1.440	121.0
0.090	7.6	0.470	39.5	0.850	71.4	1.460	122.7
0.100	8.4	0.480	40.3	0.860	72.3	1.480	124.4
0.110	9.2	0.490	41.2	0.870	73.1	1.500	126.1
0.120	10.1	0.500	42.0	0.880	73.9	1.520	127.7
0.130	10.9	0.510	42.9	0.890	74.8	1.540	129.4
0.140	11.8	0.520	43.7	0.900	75.6	1.560	131.1
0.150	12.6	0.530	44.5	0.910	76.5	1.580	132.8
0.160	13.4	0.540	45.4	0.920	77.3	1.600	134.5
0.170	14.3	0.550	46.2	0.930	78.2	1.620	136.1
0.180	15.1	0.560	47.1	0.940	79.0	1.640	137.8
0.190	16.0	0.570	47.9	0.950	79.8	1.660	139.5
0.200	16.8	0.580	48.7	0.960	80.7	1.680	141.2
0.210	17.6	0.590	49.6	0.970	81.5	1.700	142.9
0.220	18.5	0.600	50.4	0.980	82.4	1.720	144.5
0.230	19.3	0.610	51.3	0.990	83.2	1.740	146.2
0.240	20.2	0.620	52.1	1.000	84.0	1.760	147.9
0.250	21.0	0.630	52.9	1.020	85.7	1.780	149.6
0.260	21.8	0.640	53.8	1.040	87.4	1.800	151.3
0.270	22.7	0.650	54.6	1.060	89.1	1.820	152.9
0.280	23.5	0.660	55.5	1.080	90.8	1.840	154.6
0.290	24.4	0.670	56.3	1.100	92.4	1.860	156.3
0.300	25.2	0.680	57.1	1.120	94.1	1.880	158.0
0.310	26.1	0.690	58.0	1.140	95.8	1.900	159.7
0.320	26.9	0.700	58.8	1.160	97.5	1.920	161.3
0.330	27.7	0.710	59.7	1.180	99.2	1.940	163.0
0.340	28.6	0.720	60.5	1.200	100.8	1.960	164.7
0.350	29.4	0.730	61.3	1.220	102.5	1.980	166.4
0.360	30.3	0.740	62.2	1.240	104.2	2.000	168.1
0.370	31.1	0.750	63.0	1.260	105.9	2.100	176.5
0.380	31.9	0.760	63.9	1.280	107.6	2.200	184.9
0.390	32.8	0.770	64.7	1.300	109.2	2.300	193.3
0.400	33.6	0.780	65.5	1.320	110.9	2.400	201.7
0.410	34.5	0.790	66.4	1.340	112.6	2.500	210.1
0.420	35.3	0.800	67.2	1.360	114.3	2.600	218.5

Cutting Fluids for Machining

The goal in all conventional metal-removal operations is to raise productivity and reduce costs by machining at the highest practical speed consistent with long tool life, fewest rejects, and minimum downtime, and with the production of surfaces of satisfactory accuracy and finish. While many machining operations can be performed "dry," the proper application of a cutting fluid generally makes possible: higher cutting speeds, higher feed rates, greater depths of cut, lengthened tool life, decreased surface roughness, increased dimensional accuracy, and reduced power consumption. Selecting the proper cutting fluid for a specific machining situation requires knowledge of fluid functions, properties, and limitations. Cutting fluid selection deserves as much attention as the choice of machine tool, tooling, speeds, and feeds.

In order to understand the action of a cutting fluid, it is important to realize that almost all the energy expended in cutting metal is transformed into heat, primarily by the deformation of the metal into the chip and, to a lesser degree, by the friction of the chip sliding against the tool face. With these factors in mind, it becomes clear that the primary functions of any cutting fluid are: cooling of the tool, workpiece, and chip; reducing friction at the sliding contacts; and reducing or preventing welding or adhesion at the contact surfaces, which forms the "built-up edge" on the tool. Two other functions of cutting fluids are flushing away chips from the cutting zone and protecting the workpiece and tool from corrosion.

The relative importance of the functions is dependent on the material being machined, the cutting tool and conditions, and the finish and accuracy required on the part. For example, cutting fluids with greater lubricity are generally used in low-speed machining and on most difficult-to-cut materials. Cutting fluids with greater cooling ability are generally used in high-speed machining on easier-to-cut materials.

Types of Cutting and Grinding Fluids. — In recent years a wide range of cutting fluids has been developed to satisfy the requirements of new materials of construction and new tool materials and coatings.

There are four basic types of cutting fluids; each has distinctive features, as well as advantages and limitations. Selection of the right fluid is made more complex because the dividing line between types is not always clear. Most machine shops try to use as few different fluids as possible and prefer fluids that have long life, do not require constant changing or modifying, have reasonably pleasant odors, do not smoke or fog in use, and, most important, are neither toxic nor cause irritation to the skin. Other issues in selection are the cost and ease of disposal.

The major divisions and subdivisions used in classifying cutting fluids are:

Cutting Oils, including straight and compounded mineral oils plus additives.

Water-Miscible Fluids, including emulsifiable oils; chemical or synthetic fluids; and semichemical fluids.

Gases.

Paste and Solid Lubricants.

Since the cutting oils and water-miscible types are the most commonly used cutting fluids in machine shops, discussion will be limited primarily to these types. It should be noted, however, that compressed air and inert gases, such as carbon dioxide, nitrogen, and Freon are sometimes used in machining. Paste, waxes, soaps, graphite, and molybdenum disulfide may also be used, either applied directly to the workpiece or as an impregnant in the tool, such as in a grinding wheel.

Cutting Oils. — Cutting oils are generally compounds of mineral oil with the addition of animal, vegetable, or marine oils to improve the wetting and lubricating properties. Sulfur, chlorine, and phosphorous compounds, sometimes called extreme pressure (EP) additives, provide for even greater lubricity. In general, these cutting oils do not cool as well as water-miscible fluids.

Water-Miscible Fluids. — *Emulsions or soluble oils* are a suspension of oil droplets in water. These suspensions are made by blending the oil with emulsifying agents (soap and soaplike materials) and other materials. These fluids combine the lubricating and rust-prevention properties of oil with water's excellent cooling properties. Their properties are affected by the emulsion concentration, with "lean" concentrations providing better cooling but poorer lubrication and with "rich" concentrations having the opposite effect. Additions of sulfur, chlorine, and phosphorus, as with cutting oils, yield "extreme pressure" (EP) grades.

Chemical fluids are true solutions composed of organic and inorganic materials dissolved in water. Inactive types are usually clear fluids combining high rust inhibition, high cooling, and low lubricity characteristics with high surface tension. Surface-active types include wetting agents and possess moderate rust inhibition, high cooling, and moderate lubricating properties with low surface tension. They may also contain chlorine and/or sulfur compounds for extreme pressure properties.

Semichemical fluids are combinations of chemical fluids and emulsions. These fluids have a lower oil content but a higher emulsifier and surface-active-agent content than emulsions, producing oil droplets of much smaller diameter. They possess low surface tension, moderate lubricity and cooling properties, and very good rust inhibition. Sulfur, chlorine, and phosphorus also are sometimes added.

Selection of Cutting Fluids for Different Materials and Operations. — The choice of a cutting fluid depends on many complex interactions including the machinability of the metal; the severity of the operation; the cutting tool material; metallurgical, chemical, and human compatibility; fluid properties, reliability, and stability; and finally cost. Other factors affect results. Some shops standardize on a few cutting fluids which have to serve all purposes. In other cases, one cutting fluid must be used for all the operations performed on a machine. Sometimes, in such cases, a very severe operating condition may be alleviated by applying the "right" cutting fluid manually while the machine supplies the cutting fluid for other operations through its coolant system. Several voluminous textbooks are available with specific recommendations for the use of particular cutting fluids for almost every combination of machining operation and workpiece and tool material. In general, when experience is lacking, it is wise to consult the material supplier and/or any of the many suppliers of different cutting fluids for advice and recommendations. Another excellent source is the Machinability Data Center, one of the many information centers supported by the U.S. Department of Defense. While the following recommendations represent good practice, they are to serve as a guide only, and it is not intended to say that other cutting fluids will not, in certain specific cases, also be effective.

Steels: Caution should be used when using a cutting fluid on steel that is being turned at a high cutting speed with cemented carbide cutting tools. See "The Application of Cutting Fluids to Carbides" later. Frequently this operation is performed dry. If a cutting fluid is used, it should be a soluble oil mixed to a consistency of about 1 part oil to 20 to 30 parts water. A sulfurized mineral oil is recommended for reaming with carbide tipped reamers although a heavy-duty soluble oil has also been used successfully.

The cutting fluid recommended for machining steel with high speed cutting tools depends largely on the severity of the operation. For ordinary turning, boring, drilling, and milling on medium and low strength steels, use a soluble oil having a consistency of 1 part oil to 10 to 20 parts water. For tool steels and tough alloy steels, a heavy-duty soluble oil having a consistency of 1 part oil to 10 parts water is recommended for turning and milling. For drilling and reaming these materials, a light

Cutting Fluids Recommended for Machining Operations — 1

Soluble Oils. — Types of oils or paste compounds which form emulsions when mixed with water: Soluble oils are used extensively in machining both ferrous and non-ferrous metals when the cooling quality is paramount and the chip-bearing pressure is not excessive. Care should be taken in selecting the proper soluble oil for precision grinding operations. Grinding coolants should be free from fatty materials that tend to load the wheel, thus affecting the finish on the machined part. Soluble coolants should contain rust preventive constituents to prevent corrosion.

Base Oils. — Various types of highly sulfurized and chlorinated oils containing inorganic, animal, or fatty materials. This "base stock" usually is "cut back" or blended with a lighter oil, unless the chip-bearing pressures are high, as in cutting alloy steel, in which case the base stock may be used straight. Base oils usually have a viscosity range of from 300 to 900 seconds at 100 degrees F.

Mineral Oils. — This group includes all types of oils extracted from petroleum such as paraffin oil, mineral seal oil, and kerosene. Mineral oils are often blended with base stocks, but they are generally used in the original form for light machining operations on both free-machining steels and non-ferrous metals. The coolants in this class should be of a type that has a relatively high flash point. Care should be taken to see that they are nontoxic, so that they will not be injurious to the operator. The heavier mineral oils (paraffin oils) usually have a viscosity of about 100 seconds at 100 degrees F. Mineral seal oil and kerosene have a viscosity of 35 to 60 seconds at 100 degrees F.

Material to be Cut	Turning	Milling
Aluminum (Note 1–See notes on next page)	Mineral Oil with 10 Percent Fat (or) Soluble Oil	Soluble Oil (96 Per Cent Water) (or) Mineral Seal Oil (or) Mineral Oil
Alloy Steels (Note 2)	25 Per Cent Sulfur base Oil* with 75 Per Cent Mineral Oil	10 Per Cent Lard Oil with 90 Per Cent Mineral Oil
Brass	Mineral Oil with 10 Per Cent Fat	Soluble Oil (96 Per Cent Water)
Tool Steels and Low-carbon Steels	25 Per Cent Lard Oil with 75 Per Cent Mineral Oil	Soluble Oil
Copper	Soluble Oil	Soluble Oil
Monel Metal	Soluble Oil	Soluble Oil
Cast Iron (Note 3)	Dry	Dry
Malleable Iron	Soluble Oil	Soluble Oil
Bronze	Soluble Oil	Soluble Oil
Magnesium (Note 5)	10 Per Cent Lard Oil with 90 Per Cent Mineral Oil	Mineral Seal Oil

The Application of Cutting Fluids to Carbides. — Carbide turning, boring, and similar operations on lathes are performed dry or with the help of soluble oil or chemical cutting fluids. The effectiveness of cutting fluids in improving tool life or by permitting higher cutting speeds to be used, is less with carbides than with high-speed steel tools. Furthermore, the effectiveness of the cutting fluid is lessened as the cutting speed is increased. Cemented carbides are very sensitive to sudden changes in temperature and to temperature gradients within the carbide. Thermal shocks to the carbide will cause thermal cracks to form near the cutting edge which are a prelude to tool failure. An unsteady or interrupted flow of the coolant reaching the cutting edge will generally cause thermal cracks to form near the cutting edge which results in a rapid failure of the tool. The flow of the chip over the face of the tool can cause an interruption to the flow of the coolant reaching the cutting edge even though a steady stream of coolant is directed at the tool. When a cutting fluid is used and frequent tool breakage is encountered, it is often best to cut dry. When a cutting fluid must be used to keep the workpiece cool for size control or to allow it to be handled by the operator, special precautions must be used. Sometimes applying the coolant from the front and the side of the tool simultaneously is helpful. On lathes equipped with overhead shields, it is very effective to apply the coolant from below the tool into the space between the shoulder of the work and the tool flank, in addition to applying the coolant from the top. Another method is not to direct the coolant stream at the cutting tool at all but to direct it at the workpiece above or behind the cutting tool.

The danger of thermal cracking is great when milling with carbide cutters. The nature of the milling operation itself tends to promote thermal cracking because the cutting edge is constantly heated to a high temperature and rapidly cooled as it enters and leaves the workpiece. For this reason, carbide milling operations should be performed dry.

Lower cutting-edge temperatures diminish the danger of thermal cracking. The cutting-edge temperatures usually encountered when reaming with solid carbide or carbide tipped reamers are generally such that thermal cracking is not apt to occur except when reaming certain difficult-to-machine metals. Therefore, cutting fluids are very effective when used on carbide reamers. Practically every kind of cutting fluid has been used, depending on the job material encountered. For difficult surface-finish problems in holes, heavy duty soluble oils, sulfurized mineral-fatty oils, and sulfo-chlorinated mineral-fatty oils have been used successfully. In some cases, the grade and the hardness of the carbide also have an effect on the surface finish of the hole.

Application of Cutting Fluids. — Cutting fluids should be applied where the cutting action is taking place and at the highest possible velocity without causing splashing. As a general rule, it is preferable to supply from 3 to 5 gallons per minute for each single-point tool on a machine such as a turret lathe or automatic. The temperature of the cutting fluid should be kept below 110 degrees F. If the volume of fluid used is not sufficient to maintain the proper temperature, means of cooling it should be provided.

Cutting Fluids for Machining Magnesium. — In machining magnesium, it is the general but not invariable practice in the United States to use a cutting fluid, whereas, in England, magnesium usually is machined dry except in cases where heat generated by high cutting speeds would not be dissipated rapidly enough without a cutting fluid. This condition may exist when, for example, small tools without much heat-conducting capacity are employed on automatics.

The cutting fluid for magnesium should be an anhydrous oil having, at most, a very low acid content. Various mineral-oil cutting fluids are used for magnesium.

MACHINING NONFERROUS METALS AND PLASTICS

Machining Aluminum. — Some of the alloys of aluminum have been machined successfully without any lubricant or cutting compound, but in order to obtain the best results, some form of lubricant is desirable. For many purposes, a soluble cutting oil is good.

Tools for aluminum and aluminum alloys should have larger relief and rake angles than tools for cutting steel. For high-speed steel turning tools the following angles are recommended: relief angles, 14 to 16 degrees; back rake angle, 5 to 20 degrees; side rake angle, 15 to 35 degrees. For very soft alloys even larger side rake angles are sometimes used. High silicon aluminum alloys and some others have a very abrasive effect on the cutting tool. While these alloys can be cut successfully with high-speed-steel tools, cemented carbides are recommended because of their superior abrasion resistance. The tool angles recommended for cemented carbide turning tools are: relief angles, 12 to 14 degrees; back rake angle, 0 to 15 degrees; side rake angle, 8 to 30 degrees.

Cut-off tools and necking tools for machining aluminum and its alloys should have from 12 to 20 degrees back rake angle and the end relief angle should be from 8 to 12 degrees. Excellent threads can be cut with single-point tools in even the softest aluminum. Experience seems to vary somewhat regarding the rake angle for single-point thread cutting tools. Some prefer to use a rather large back and side rake angle although this requires a modification in the included angle of the tool to produce the correct thread contour. When both rake angles are zero, the included angle of the tool is ground equal to the included angle of the thread. Excellent threads have been cut in aluminum with zero rake angle thread-cutting tools using large relief angles, which are 16 to 18 degrees opposite the front side of the thread and 12 to 14 degrees opposite the back side of the thread. In either case, the cutting edges should be ground and honed to a keen edge. It is sometimes advisable to give the face of the tool a few strokes with a hone between cuts when chasing the thread in order to remove any built-up edge on the cutting edge.

Fine surface finishes are often difficult to obtain on aluminum and aluminum alloys, particularly the softer metals. When a fine finish is required, the cutting tool should be honed to a keen edge and the surfaces of the face and the flank will also benefit by being honed smooth. Tool wear is inevitable; however, it should not be allowed to progress too far before the tool is changed or sharpened. A sulphurized mineral oil or a heavy-duty soluble oil will sometimes be helpful in obtaining a satisfactory surface finish. For best results, however, a diamond cutting tool is recommended. Excellent surface finishes can be obtained on even the softest aluminum and aluminum alloys with these tools.

Although ordinary milling cutters can be used successfully in shops where aluminum parts are only machined occasionally, the best results are obtained with coarse-tooth, large helix-angle cutters having large rake and clearance angles. Clearance angles up to 10 to 12 degrees are recommended. When slab milling and end milling a profile, using the peripheral teeth on the end mill, climb milling (also called down milling) will generally produce a better finish on the machined surface than conventional (or up) milling. Face milling cutters should have a large axial rake angle. Standard twist drills can be used without difficulty in drilling aluminum and aluminum alloys although high helix-angle drills are preferred. The wide flutes and high helix-angle in these drills helps to clear the chips. In some cases the use of split-point drills is preferred. Carbide tipped twist drills can be used for drilling aluminum and its alloys which may afford advantages in some production applications. Ordinary hand and machine taps can be used to tap aluminum and its alloys although spiral-fluted ground thread taps give superior results. Experience has shown that such taps should have a right-hand ground flute when intended to cut right-hand threads and the helix angle should be similar to that used in an ordinary twist drill.

Machining Magnesium. — Magnesium alloys are readily machined and with relatively low power consumption per cubic inch of metal removed. The usual practice is to employ high cutting speeds with relatively coarse feeds and deep cuts. Exceptionally fine finishes can be obtained so that grinding to improve the finish usually is unnecessary. The horsepower normally required in machining magnesium varies from 0.15 to 0.30 per cubic inch per minute. While this value is low, especially in comparison with power required for cast iron and steel, the total amount of power for machining magnesium usually is high because of the exceptionally rapid rate at which metal is removed.

Carbide tools are recommended for maximum efficiency, although high-speed steel frequently is employed. Tools should be designed so as to dispose of chips readily or without excessive friction, by employing polished chip-bearing surfaces, ample chip spaces, large clearances, and small contact areas. *Keen-edged tools should always be used.*

Feeds and Speeds for Magnesium: Speeds ordinarily range up to 5000 feet per minute for rough- and finish-turning, up to 3000 feet per minute for rough-milling, and up to 9000 feet per minute for finish-milling. For rough-turning, the following combinations of speed in feet per minute, feed per revolution, and depth of cut are recommended: Speed 300 to 600 feet per minute — feed 0.030 to 0.100 inch, depth of cut 0.5 inch; speed 600 to 1000 — feed 0.020 to 0.080, depth of cut 0.4; speed 1000 to 1500 — feed 0.010 to 0.060, depth of cut 0.3; speed 1500 to 2000 — feed 0.010 to 0.040, depth of cut 0.2; speed 2000 to 5000 — feed 0.010 to 0.030, depth of cut 0.15.

Lathe Tool Angles for Magnesium: The true or actual rake angle resulting from back and side rakes usually varies from 10 to 15 degrees. Back rake varies from 10 to 20, and side rake from 0 to 10 degrees. Reduced back rake may be employed to obtain better chip breakage. The back rake may also be reduced to from 2 to 8 degrees on form tools or other broad tools to prevent chatter.

Parting Tools: For parting tools, the back rake varies from 15 to 20 degrees, the front end relief 8 to 10 degrees, the side relief measured perpendicular to the top face 8 degrees, the side relief measured in the plane of the top face from 3 to 5 degrees.

Milling Magnesium: In general, the coarse-tooth type of cutter is recommended. The number of teeth or cutting blades may be one-third to one-half the number normally used; however, the two-blade fly-cutter has proved to be very satisfactory. As a rule, the land relief or primary peripheral clearance is 10 degrees followed by secondary clearance of 20 degrees. The lands should be narrow, the width being about ³⁄₆₄ to ¹⁄₁₆ inch. The rake, which is positive, is about 15 degrees.

For rough-milling and speeds in feet per minute up to 900 — feed, inch per tooth, 0.005 to 0.025, depth of cut up to 0.5; for speeds 900 to 1500 — feed 0.005 to 0.020, depth of cut up to 0.375; for speeds 1500 to 3000 — feed 0.005 to 0.010, depth of cut up to 0.2.

Drilling Magnesium: If the depth of a hole is less than five times the drill diameter, an ordinary twist drill with highly polished flutes may be used. The included angle of the point may vary from 70 degrees to the usual angle of 118 degrees. The relief angle is about 12 degrees. The drill should be kept sharp and the outer corners rounded to produce a smooth finish and prevent burr formation. For deep hole drilling, use a drill having a helix angle of 40 to 45 degrees with large polished flutes of uniform cross-section throughout the drill length to facilitate the flow of chips. A pyramid-shaped "spur" or "pilot point" at the tip of the drill will reduce the "spiraling or run-off."

Drilling speeds vary from 300 to 2000 feet per minute with feeds per revolution ranging from 0.015 to 0.050 inch.

Reaming Magnesium: Reamers up to 1 inch in diameter should have four flutes; larger sizes, six flutes. These flutes may be either parallel with the axis

or have a negative helix angle of 10 degrees. The positive rake angle varies from 5 to 8 degrees, the relief angle from 4 to 7 degrees, and the clearance angle from 15 to 20 degrees.

Tapping Magnesium: Standard taps may be used unless Class 3B tolerances are required, in which case the tap should be designed for use in magnesium. A high-speed steel concentric type with a ground thread is recommended. The concentric form, which eliminates the radial thread relief, prevents jamming of chips while the tap is being backed out of the hole. The positive rake angle at the front may vary from 10 to 25 degrees and the "heel rake angle" at the back of the tooth from 3 to 5 degrees. The chamfer extends over two to three threads. For holes up to ¼ inch in diameter, two-fluted taps are recommended; for sizes from ½ to ¾ inch, three flutes; and for larger holes, four flutes. Tapping speeds ordinarily range from 75 to 200 feet per minute, and mineral oil cutting fluid should be used.

Threading Dies for Magnesium: Threading dies for use on magnesium should have about the same cutting angles as taps. Narrow lands should be used to provide ample chip space. Either solid or self-opening dies may be used. The latter type is recommended when maximum smoothness is required. Threads may be cut at speeds up to 1000 feet per minute.

Grinding Magnesium: As a general rule, magnesium is ground dry. The highly inflammable dust should be formed into a sludge by means of a spray of water or low-viscosity mineral oil. Accumulations of dust or sludge should be avoided. For surface grinding, when a fine finish is desirable, a low-viscosity mineral oil may be used.

Machining Zinc Alloy Die-Castings. — Machining of zinc alloy die-castings is mostly done without a lubricant. For particular work, especially deep drilling and tapping, a lubricant such as lard oil and kerosene (about half and half) or a 50–50 mixture of kerosene and machine oil may be used to advantage. A mixture of turpentine and kerosene has been been found effective on certain difficult jobs.

Reaming: In reaming, tools with six straight flutes are commonly used, although tools with eight flutes irregularly spaced have been found to yield better results by one manufacturer. Many standard reamers have a land that is too wide for best results. A land about 0.015 inch wide is recommended but this may often be ground down to around 0.007 or even 0.005 inch to obtain freer cutting, less tendency to loading, and reduced heating.

Turning: Tools of high-speed steel are commonly employed although the application of Stellite and carbide tools, even on short runs, is feasible. For steel or Stellite, a positive top rake of from 0 to 20 degrees and an end clearance of about 15 degrees is commonly recommended. Where side cutting is involved, a side clearance of about 4 degrees minimum is recommended. With carbide tools, the end clearance should not exceed 6 to 8 degrees and the top rake should be from 5 to 10 degrees positive. For boring, facing, and other lathe operations, rake and clearance angles are about the same as for tools used in turning.

Machining Monel and Nickel Alloys. — These alloys are machined with high-speed steel and with cemented carbide cutting tools. High-speed steel lathe tools usually have a back rake of 6 to 8 degrees, a side rake of 10 to 15 degrees, and relief angles of 8 to 12 degrees. Broad-nose finishing tools have a back rake of 20 to 25 degrees and an end relief angle of 12 to 15 degrees. In most instances, standard commercial cemented-carbide tool holders and tool shanks can be used which provide an acceptable tool geometry. Honing the cutting edge lightly will help if chipping is encountered.

The most satisfactory tool materials for machining Monel and the softer nickel

alloys, such as Nickel 200 and Nickel 230, are M2 and T5 for high-speed steel and crater resistant grades of cemented carbides. For the harder nickel alloys such as K Monel, Permanickel, Duranickel, and Nitinol alloys, the recommended tool materials are T15, M41, M42, M43, and for high-speed steel, M42. For carbides, a grade of crater resistant carbide is recommended when the hardness is less than 300 Bhn, and when the hardness is more than 300 Bhn, a grade of straight tungsten carbide will often work best, although some crater resistant grades will also work well.

A sulfurized oil or a water-soluble oil is recommended for rough and finish turning. A sulfurized oil is also recommended for milling, threading, tapping, reaming, and broaching. Recommended cutting speeds for Monel and the softer nickel alloys are 70 to 100 fpm for high-speed steel tools and 200 to 300 fpm for cemented carbide tools. For the harder nickel alloys, the recommended speed for high-speed steel is 40 to 70 fpm for a hardness up to 300 Bhn and for a higher hardness, 10 to 20 fpm; for cemented carbides, 175 to 225 fpm when the hardness is less than 300 Bhn and for a higher hardness, 30 to 70 fpm.

Nickel alloys have a high tendency to work harden. To minimize work hardening caused by machining, the cutting tools should be provided with adequate relief angles and positive rake angles. Furthermore, the cutting edges should be kept sharp and replaced when dull to prevent burnishing of the work surface. The depth of cut and feed should be sufficiently large to ensure that the tool penetrates the work without rubbing.

Machining Plastics. — Molded plastic parts do not require machining, as a general rule, unless the tolerances are exceptionally small or there are undercuts, angular holes, or other openings difficult or impracticable to reproduce in a mold. It is common practice, however, to machine laminated phenolic plastics and also cast phenolic plastics, as well as sheet, bar, and tube stock such as is commonly used for making parts such as pinions or gears. The machining characteristics of different plastics vary somewhat so that the general recommendations given may require some modification to obtain the best results for a given class of work. Although plastics are poor conductors of heat, they usually are machined dry or without a cutting fluid. In some cases, either an air jet, water, or a soap solution is used. The maximum speed at which a cutting tool can operate without excessive heating should be determined by actual test.

Turning and Boring Plastics: Tools of high-speed steel, Stellite and carbide are commonly used in machining plastics. The general practice in turning and boring is similar to that for brass so far as the feed and speed are concerned. Speeds usually vary from 250 to 500 feet per minute for high-speed steel tools and from 500 to 1500 for Stellite and carbide tools. According to the Haynes Stellite Co., hot-set molded parts require tools having less clearance and more rake than those used for steel or other metals. The cold-set acetate and polystyrene molded parts should be turned with tools having no rake and plenty of clearance. Parts molded of "Vinylite" plastic can usually be machined satisfactorily with ordinary metal-cutting tools, provided the front and side clearances are about double those required for machine steel.

Cutting Off Plastics: Tools for cutting off should have greater front and side clearances than for steel. The cutting speed should be about half that employed for a turning operation.

Drilling Plastics: Standard twist drills may be used for small holes up to 1/8 or 3/16 inch, but for larger sizes particularly, it is preferable to use the commercial high-speed steel drills designed especially for plastics. These drills are made both in wire gage and fractional sizes. They have relatively large flutes to provide greater space for chips, and polished flutes are preferable. For large-quantity production, carbide-tipped drills are recommended. Extra clearance back of the edges of the flutes tends

to reduce friction and heating. Frequent removal of the drill may be necessary, especially in drilling comparatively deep holes. A feed of 0.007 to 0.015 inch per revolution is a common range. Drilling speeds usually vary from 150 to 300 feet per minute. With carbide-tipped drills, the speeds may be as high as 12,000 to 15,000 rpm. The drill point should have an included angle of 55 to 60 degrees for thin sections and 90 degrees for the thicker sections. The clearance angle usually is 15 degrees. To avoid excessive heating and aid in chip removal, an air jet may be directed into the hole. In some cases, a soap solution is effective as a cutting fluid. If the drill-spindle movement is cam-operated, design the cam to advance the drill tip slowly at the beginning or for about 0.010 inch, then continue at the full rate of speed and allow a dwell of about 2 to 5 revolutions at the bottom of the hole; then withdraw the drill rapidly.

Tapping and Threading Plastics: For tapping small holes in hot- or cold-set molded parts, a high-speed steel nitrided chromium-plated tap is recommended. The rake may vary from 0 to 5 degrees negative. The size of the hole should allow for about three-fourths of the standard thread depth. Small holes generally are tapped dry. If a coolant is used, water is preferable to oil. Tapping speeds usually vary from 40 to 55 per minute. Threaded brass inserts or bushings are often used, in which case they may be tapped either before or after insertion. In cutting a 60-degree thread, such as the Unified Standard, use a tool that is ground to cut on one side only and feed it in at an angle of 30 degrees by setting the compound rest at this angle.

Drill Jigs for Plastics: In the design of jigs for drilling, close-fitting drill bushings should be avoided. They may increase not only the friction on the drill but also the tendency of the chips to plug up the drill flutes. If the operation is such that a drill bushing is absolutely essential, a floating leaf or templet should be employed. When using a templet, the hole should be spotted with the templet in place, using the drill size corresponding to the final hole size; then the templet should be removed and the hole completed. Pilot holes should be avoided, except in special instances when the hole is to be reamed or counterbored.

Sawing Hot-set Molded Parts: The sawing of molded hot-set parts is done chiefly on circular and band saws. Band saws are to be recommended at times for straight cutting because they run cooler than circular saws. Band saw manufacturers advocate saw teeth set to clear, some advocating one-half the thickness of the blade on each side so that saws give a width of cut double their thickness. Narrower saws and more set are needed for cutting curves than for straight cuts. Band saws just soft enough to permit filing are recommended, but saws must be kept sharp. Dull saws cause chipping and might result in saw breakage. Sawing is usually done dry but some recommend water for cooling. Saw teeth should have little set and eight to nine teeth per inch. The speed is 1800 to 2500 feet per minute.

Sawing Cold-set Molded Parts: Sawing of cold-set polystyrene or cellulose acetate can be done with circular saws having nine to twelve teeth per inch for thin sheets and six teeth per inch for thickness over ¼ inch. Saws six to nine inches in diameter are run at speeds of 3000 to 3600 rpm and should be hollow ground. They usually are $\frac{1}{32}$ to $\frac{1}{16}$ inch thick. The use of a water spray gives a cleaner cut. One large saw manufacturer recommends that pieces be cut with a stream of water running in the kerf while the saw is cutting. This applies to both circular and band saws; otherwise, the cold-set type of material will fuse. The circular saws recommended are 14 inches, 12 and 9 gage, 130 teeth, 10 degrees rake to be operated at 3000 rpm and made of a special alloy steel stock.

For band sawing, manufacturers suggest a band saw which is 19 to 20 gage, having twenty points to the inch and hardened and tempered. The saw should be operated at 4000 to 4500 feet per minute.

Milling Hot-set and Cold-set Molded Parts: Milling of molded parts is not as a rule feasible, but where it is required, milling speeds and feeds of the range used for brass are recommended. A speed of 400 fpm with carbon steel cutters and 1200–1600 fpm is recommended with carbide cutters. Single- and double-bladed fly cutters are sometimes used at high speed with fine cuts. Where little material has to be removed, a high-speed woodworking shaper with a carbide-tipped tool can be used to advantage. It is desirable and may be necessary to use an air blast to assure proper chip removal from the milling cutter. Wherever possible, it is recommended that spiral milling cutters be used, and that the number of teeth in the cutter head be such that at least two of them are in contact with the work at all times.

The same general rules that apply to turning, facing, and boring operations also hold for milling Vinyl molded parts. Standard cutters can be used, but higher speeds are feasible if extra clearances are ground on the cutter blades.

Punching Laminated Phenolics: Most grades of laminated phenolics can be punched either hot or cold. The die must be kept sharp, however, in order to produce good results. The minimum clearance between individual punchings, and also between punchings and the edge of the material strips, should be about three times the thickness of the material. For hot punching or shearing, the material is heated in a steam or electric oven, designed to give a uniform temperature throughout the heating chamber. The material is left in the oven just long enough to be uniformly heated to oven temperature. Further heating causes brittleness. Temperatures of 100 to 120 degrees C (212 to 248 degrees F) are recommended. The heating time ranges from five minutes for $\frac{1}{16}$-inch material to thirty minutes for $\frac{1}{4}$-inch material.

Dies for punching laminated plastics are designed the same as for punching metal, except that smaller clearances are allowed between punch and die. In cold punching, this clearance is small, approaching a "sliding fit." The strippers are close fitting and backed with strong springs.

Because these plastic materials expand after being compressed in a die, blanks will be larger than the die diameter and holes will be smaller. On hot punchings, allowance should be made for shrinkage of the material after punching. This shrinkage varies with the grade of material, thickness of piece, and the temperature of the material during the punching operation. For very small holes and blanks, allowances for shrinkage are often neglected, while for large pieces and accurate work, they must be carefully considered.

Shearing Laminated Plastics: Shears suitable for thin metal sheet are used for cutting laminated plastics. The knife must be kept sharp. In trimming paper-base grades cold, clean-cut edges are obtained with thicknesses of $\frac{1}{32}$ inch and under, and when trimmed hot, up to $\frac{1}{8}$ inch. Fabric-base grades are trimmed cold up to $\frac{1}{16}$ inch, and hot up to $\frac{1}{8}$ inch. Greater thicknesses can be sheared if the condition of the edge is not important.

Sawing Plastics: Material up to 1 inch thick may be cut with a 12- to 16-inch circular saw at about 3000 rpm, and material 1 inch thick and over, with a 16-inch saw at about 2400 rpm. A saw used for roughing cuts has bevel teeth, seven to the inch, while a smooth saw, with no set — similar to that for metal — is used for finishing cuts. For use on all thick material, the saws should be hollow-ground to prevent binding. The smaller the projection of the saw above the material on the sawing table, the better will be the sawed edges. A thin sheet of plastic or other material placed under the piece to be sawed, is of advantage when extreme smoothness of cut is desired.

A band saw is used for sawing round blanks from plate stock. The usual band saw is of the bevel-tooth type, with some set, and has three to seven teeth per inch. It is run at 3000 feet per minute.

GRINDING AND OTHER ABRASIVE PROCESSES

Processes and equipment discussed under this heading use abrasive grains for shaping workpieces by means of machining or related methods. Abrasive grains are hard crystals either found in nature or manufactured. The most commonly used materials are aluminum oxide, silicon carbide and diamond. Other materials such as garnet, zirconia, glass and even walnut shells are used for some applications. Abrasive products are used in three basic forms by industry:

a. *Bonded* to form a solid shaped tool such as disks (the basic shape of grinding wheels), cylinders, rings, cups, segments or sticks to name a few.

b. *Coated* on backings made of paper or cloth, in the form of sheets, strips or belts.

c. *Loose,* held in some liquid or solid carrier, (for lapping, polishing, tumbling) or propelled by centrifugal force, air or water pressure against the work surface (blast cleaning).

The applications for abrasive processes are multiple and varied. They include:

a. *Cleaning* of surfaces, also the coarse removal of excess material — such as rough off-hand grinding in foundries to remove gates and risers.

b. *Shaping,* such as in form grinding and tool sharpening.

c. *Sizing,* a general objective, but of primary importance in precision grinding.

d. *Surface finish improvement,* either primarily as in lapping, honing and polishing or as a secondary objective in other types of abrasive processes.

e. *Separating,* as in cut-off or slicing operations.

The main field of application of abrasive processes is in metalworking, because of the capacity of abrasive grains to penetrate into even the hardest metals and alloys. However, the great hardness of the abrasive grains makes the process also preferred for working other hard materials, such as stones, glass and certain types of plastics. Abrasive processes are also chosen for working relatively soft materials, such as wood, rubber, etc., for such reasons as high stock removal rates, long-lasting cutting ability, good form control, and fine finish of the worked surface.

Grinding Wheels

Abrasive Materials. — In earlier times, only natural abrasives were available. From about the beginning of this century, however, manufactured abrasives, primarily silicon carbide and aluminum oxide, have replaced the natural materials; even natural diamonds have been almost completely supplanted by synthetics. Superior and controllable properties, and dependable uniformity characterize the manufactured abrasives.

Both silicon carbide and aluminum oxide abrasives are very hard and brittle. This brittleness, called friability, is controllable for different applications. Friable abrasives break easily, thus forming sharp edges. This decreases the force needed to penetrate into the work material and the heat generated during cutting. Friable abrasives are most commonly used for precision and finish grinding. Tough abrasives resist fracture and last longer. They are used for rough grinding, snagging and off-hand grinding.

As a general rule, although subject to variation:

1. Aluminum oxide abrasives are used for grinding plain and alloyed steel in a soft or hardened condition.

2. Silicon carbide abrasives are selected for cast iron, nonferrous metals and nonmetallic materials.

3. Diamond is the best type of abrasive for grinding cemented carbides. It is also used for grinding glass, ceramics, and in some cases, hardened tool steel.

4. Borazon, the trade name for cubic boron nitride, is a man-made abrasive particularly effective for grinding hardened tool steels. It produces a superior surface

integrity on the ground surface, close tolerances are easier to maintain, and the grinding wheel lasts longer.

Bond Properties and Grinding Wheel Grades. — The four main types of bonds used for grinding wheels are the vitrified, resinoid, rubber and metal.

Vitrified bonds are used for more than half of all grinding wheels made, and are preferred because of their strength and other desirable qualities. Being inert, glass-like materials, vitrified bonds are not affected by water or by the chemical composition of different grinding fluids. Vitrified bonds also withstand the high temperatures generated during normal grinding operations. The structure of vitrified wheels can be controlled over a wide range of strength and porosity. Vitrified wheels, however, are more sensitive to impact than those made with organic bonds.

Resinoid bonds are selected for wheels subjected to impact, or sudden loads or very high operating speeds. They are preferred for snagging, portable grinder uses or roughing operations. The higher flexibility of this type of bond — essentially a filled thermosetting plastic — helps it withstand rough treatment.

Rubber bonds are even more flexible than the resinoid type, and for that reason are used for producing a high finish and for resisting sudden rises in load. Rubber bonded wheels are commonly used for wet cut-off wheels because of the nearly burr-free cuts they produce, and for centerless grinder regulating wheels to provide a stronger grip and more reliable workpiece control.

Metal bonds are used in diamond wheels for grinding carbides. Their very high strength and toughness help to retain the costly diamond abrasive firmly bonded throughout its entire useful life. Metal bonded wheels — either with diamond or aluminum oxide abrasive — are also used in electrical discharge grinding and electro-chemical grinding where an electrically conductive wheel is needed.

In addition to the basic properties of the various bond materials, each can also be applied in different proportions, thereby controlling the grade of the grinding wheel.

Grinding wheel grades commonly associated with hardness, express the amount of bond material in a grinding wheel, and hence the strength by which the bond retains the individual grains.

During grinding, the forces generated when cutting the work material tend to dislodge the abrasive grains. As the grains get dull and if they don't fracture to resharpen themselves, the cutting forces will eventually tear the grains from their supporting bond. For a "soft" wheel the cutting forces will dislodge the abrasive grains before they have an opportunity to fracture. When a "hard" wheel is used, the situation is reversed. Because of the extra bond in the wheel the grains are so firmly held that they never break loose and the wheel becomes glazed. During most grinding operations it is desirable to have an intermediate wheel where there is a continual slow wearing process composed of both grain fracture and dislodgement.

The grades of the grinding wheels are designated by capital letters used in alphabetical order to express increasing "hardness" from A to Z.

Grinding Wheel Structure. — The individual grains, which are encased and held together by the bond material, do not fill the entire volume of the grinding wheel; the intermediate open space is needed for several functional purposes such as heat dissipation, coolant application and particularly for the temporary storage of chips. It follows that the spacing of the grains must be greater for coarse grains which cut thicker chips and for large contact areas within which the chips have to be retained on the surface of the wheel before being disposed of. On the other hand, a wide spacing reduces the number of grains which contact the work surface within a given advance distance, thereby producing a coarser finish.

In general, denser structures are specified for grinding hard materials, for high speed grinding operations, when the contact area is narrow, and also for producing

fine finishes and/or accurate forms. Wheels with open structure are used for tough materials, high stock removal rates, and extended contact areas, such as grinding with the face of the wheel. There are, however, several exceptions to these basic rules, an important one being the grinding of parts made by powder metallurgy, such as cemented carbides; although they represent one of the hardest industrial materials, they require wheels with an open structure.

Most kinds of general grinding operations when carried out with the periphery of the wheel call for medium spacing of the grains. The structure of the grinding wheels is expressed by numerals from 1 to 16, ranging from dense to open. In some cases, "induced porosity" is used with open structure wheels. This term means that the grinding wheel manufacturer has placed filler material (which later burns out when the wheel is fired to vitrify the bond) in the grinding wheel mix. These fillers create large "pores" between grain clusters without changing the total volume of the "pores" in the grinding wheel. Thus, an A46-H12V wheel and an A46H12VP wheel will contain the same amounts of bond, abrasive, and air space. In the former a large number of relatively small pores will be distributed throughout the wheel. The latter will have a smaller number of larger pores.

American National Standard Grinding Wheel Markings. — ANSI Standard B74.13-1982, "Markings for Identifying Grinding Wheels and Other Bonded Abrasives," applies to grinding wheels and other bonded abrasives, segments, bricks, sticks, hones, rubs, and other shapes which are for removing material, or producing a desired surface or dimension. It does not apply to specialities such as sharpening stones and provides only a standard system of markings. Wheels having the same standard markings but made by different wheel manufacturers may not — and probably will not — produce exactly the same grinding action. This desirable result cannot be obtained because of the impossibility of closely correlating any measurable physical properties of bonded abrasive products in terms of their grinding action.

Symbols for designating diamond and cubic boron wheel compositions are given on page 1065.

Sequence of Markings. — The accompanying illustration taken from ANSI B74.13-1982 shows the makeup of a typical wheel or bonded abrasive marking.

Prefix	1 Abrasive Type	2 Grain Size	3 Grade	4 Structure	5 Bond Type	6 Manufacturer's Record
51 –	A –	36 –	L –	5 –	V –	23

The meaning of each letter and number in this or other markings is indicated by the following complete list.

1. *Abrasive Letters:* The letter (A) is used for aluminum oxide and (C) for silicon carbide. The manufacturer may designate some particular type in either of these broad classes, by using his own symbol as a prefix (Example, 51).

2. *Grain Size:* The grain sizes commonly used and varying from coarse to very fine are indicated by the following numbers: 8, 10, 12, 14, 16, 20, 24, 30, 36, 46, 54, 60,

70, 80, 90, 100, 120, 150, 180, and 220. The following additional sizes are used occasionally: 240, 280, 320, 400, 500, and 600. The wheel manufacturer may add to the regular grain number an additional symbol to indicate a special grain combination.

3. *Grade:* Grades are indicated by letters of the alphabet from A to Z in all bonds or processes. Wheel grades from A to Z range from soft to hard.

4. *Structure:* The use of a structure symbol is optional. The structure is indicated by Nos. 1 to 16 (or higher, if necessary) with progressively higher numbers indicating a progressively wider grain spacing (more open structure).

5. *Bond or Process:* Bonds are indicated by the following letters: V, vitrified; S, silicate; E, shellac or elastic; R, rubber; RF, rubber reinforced; B, resinoid (synthetic resins); BF, resinoid reinforced; O, oxychloride.

6. *Manufacturer's Record:* The sixth position may be used for manufacturer's private factory records; this is optional.

American National Standard Shapes and Sizes of Grinding Wheels. — The ANSI Standard B74.2-1982, which includes shapes and sizes of grinding wheels, gives a wide variety of grinding wheel shape and size combinations. These are suitable for the majority of applications. Although grinding wheels can be manufactured to shapes and dimensions different from those listed, it is advisable, for reasons of cost and inventory control, to avoid using special shapes and sizes, unless technically warranted.

Standard shapes and size ranges as given in this Standard together with typical applications are shown in Table 1A for inch dimensions and in Table 1B for metric dimensions.

The operating surface of the grinding wheel is often referred to as the wheel face. In the majority of cases it is the periphery of the grinding wheel which, when not specified otherwise, has a straight profile. However, other face shapes can also be supplied by the grinding wheel manufacturers, and also reproduced during usage by appropriate truing. ANSI B74.2-1982 offers 13 different shapes for grinding wheel faces, which are shown in Table 2.

The Selection of Grinding Wheels. — In selecting a grinding wheel, the determining factors are the composition of the work material, the type of grinding machine, the size range of the wheels used, and the expected grinding results, in this approximate order.

The Norton Company has developed, as the result of extensive test series, a method of grinding wheel recommendation that is more flexible and also better adapted to taking into consideration pertinent factors of the job, than are listings based solely on workpiece categories. This approach is the basis for Tables 3–6, inclusive. Tool steels and constructional steels are considered in the detailed recommendations presented in these tables.

Table 3 assigns most of the standardized tool steels to five different grindability groups. The AISI-SAE tool steel designations are used.

After having defined the grindability group of the tool steel to be ground, the operation to be carried out is found in the first column of Table 4. The second column in this table distinguishes between different grinding wheel size ranges, because wheel size is a factor in determining the contact area between wheel and work, thus affecting the apparent hardness of the grinding wheel. Distinction is also made between wet and dry grinding.

Finally, the last two columns define the essential characteristics of the recommended types of grinding wheels under the headings of first and second choice, respectively. Where letters are used *preceding* A, the standard designation for aluminum oxide, they indicate a degree of friability different from the regular, thus:

(Continued on page 1049)

Table 1A. Standard Shapes and Inch Size Ranges of Grinding Wheels
(ANSI B74.2-1982)

Applications	Size Ranges of Principal Dimensions, Inches		
	D = Diam.	T = Thick.	H = Hole

Type I. Straight Wheel
For peripheral grinding.

Applications	D = Diam.	T = Thick.	H = Hole
CUTTING OFF. (Organic bonds only)	1 to 48	1/64 to 3/8	1/16 to 6
CYLINDRICAL GRINDING. Between centers	12 to 48	1/2 to 6	5 to 20
CYLINDRICAL GRINDING. Centerless grinding wheels	14 to 30	1 to 20	5 or 12
CYLINDRICAL GRINDING. Centerless regulating wheels	8 to 14	1 to 12	3 to 6
INTERNAL GRINDING.	1/4 to 4	1/4 to 2	3/32 to 7/8
OFFHAND GRINDING. Grinding on the periphery — General purpose — For wet tool grinding only	6 to 36 / 30 or 36	1/2 to 4 / 3 or 4	1/2 to 3 / 20
SAW GUMMING. (F-type face)	6 to 12	1/4 to 1 1/2	1/2 to 1 1/4
SNAGGING. Floor stand machines	12 to 24	1 to 3	1 1/4 to 2 1/2
SNAGGING. Floor stand machines (Organic bond, wheel speed over 6500 sfpm)	20 to 36	2 to 4	6 or 12
SNAGGING. Mechanical grinders (Organic bond, wheel speed up to 16,500 SFPM)	24	2 to 3	12
SNAGGING. Portable machines	3 to 8	1/4 to 1	3/8 to 5/8
SNAGGING. Portable machines (Reinforced organic bond, 17,000 sfpm)	6 or 8	3/4 or 1	1
SNAGGING. Swing frame machines	12 to 24	2 to 3	3 1/2 to 12
SURFACE GRINDING. Horizontal spindle machines	6 to 24	1/2 to 6	1 1/4 to 12
TOOL GRINDING. Broaches, cutters, mills, reamers, taps, etc.	6 to 10	1/4 to 1/2	5/8 to 5

Type 2. Cylindrical Wheel
Side grinding wheel — mounted on the diameter; may also be mounted in a chuck or on a plate.

Applications	D = Diam.	T = Thick.	W = Wall
SURFACE GRINDING. Vertical spindle machines	8 to 20	4 or 5	1 to 4

Throughout table large open-head arrows indicate grinding surfaces.

Table 1A (*Continued*). **Standard Shapes and Inch Size Ranges of Grinding Wheels**
(ANSI B74.2-1982)

Applications	Size Ranges of Principal Dimensions, Inches		
	D = Diam.	T = Thick.	H = Hole

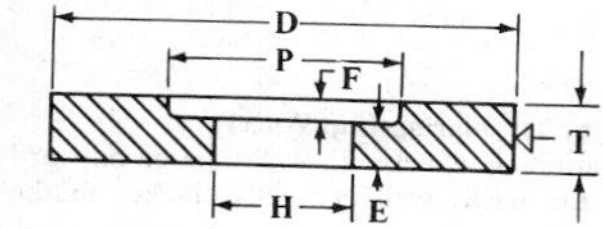

Type 5. Wheel, recessed one side
For peripheral grinding. Allows wider faced wheels than the available mounting thickness, also grinding clearance for the nut and flange.

Applications	D = Diam.	T = Thick.	H = Hole
CYLINDRICAL GRINDING. Between centers	12 to 36	1½ to 4	5 or 12
CYLINDRICAL GRINDING. Centerless regulating wheel	8 to 14	3 to 6	3 or 5
INTERNAL GRINDING.	⅜ to 4	⅜ to 2	⅛ to ⅞
SURFACE GRINDING. Horizontal spindle machines	7 to 24	¾ to 6	1¼ to 12

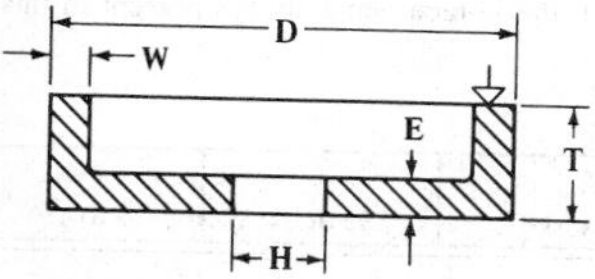

Type 6. Straight-Cup Wheel
Side grinding wheel, in whose dimensioning the wall thickness (W) takes precedence over the diameter of the recess. Hole is ⅝-11 UNC-2B threaded for the snagging wheels and ½ or 1¼" for the tool grinding wheels.

Applications	D = Diam.	T = Thick.	W = Wall
SNAGGING. Portable machines, organic bond only.	4 to 6	2	¾ to 1½
TOOL GRINDING. Broaches, cutters, mills, reamers, taps, etc.	2 to 6	1¼ to 2	5⁄16 or ⅜

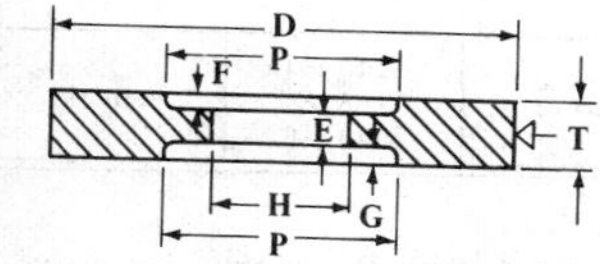

Type 7. Wheel, recessed two sides
Peripheral grinding. Recesses allow grinding clearance for both flanges and also narrower mounting thickness than overall thickness.

Applications	D = Diam.	T = Thick.	H = Hole
CYLINDRICAL GRINDING. Between centers	12 to 36	1½ to 4	5 or 12
CYLINDRICAL GRINDING. Centerless regulating wheel	8 to 14	4 to 20	3 to 6
SURFACE GRINDING. Horizontal spindle machines	12 to 24	2 to 6	5 to 12

Table 1A (*Continued*). Standard Shapes and Inch Size Ranges of Grinding Wheels
(ANSI B74.2-1982)

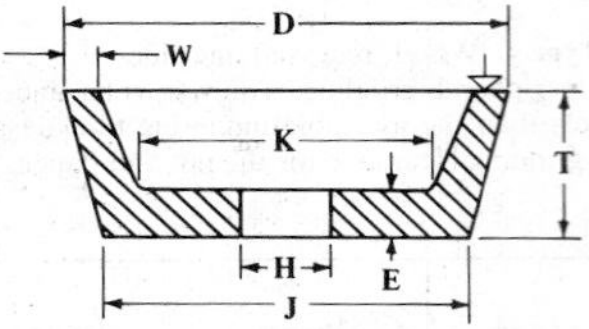

Applications	Size Ranges of Principal Dimensions, Inches		
	D = Diam.	T = Thick.	H = Hole

Type 11. Flaring-Cup Wheel
Side grinding wheel with wall tapered outward from the back; wall generally thicker in the back.

Applications	D = Diam.	T = Thick.	H = Hole
SNAGGING. Portable machines Organic bonds only, threaded hole	4 to 6	2	⅝-11 UNC-2B
TOOL GRINDING. Broaches, cutters, mills, reamers, taps, etc.	2 to 5	1¼ to 2	½ to 1¼

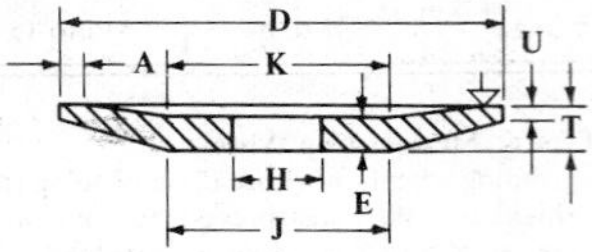

Type 12. Dish Wheel
Grinding on the side or on the U-face of the wheel, the U-face being always present in this type.

Applications	D = Diam.	T = Thick.	H = Hole
TOOL GRINDING. Broaches, cutters, mills, reamers, taps, etc.	3 to 8	½ or ¾	½ to 1¼

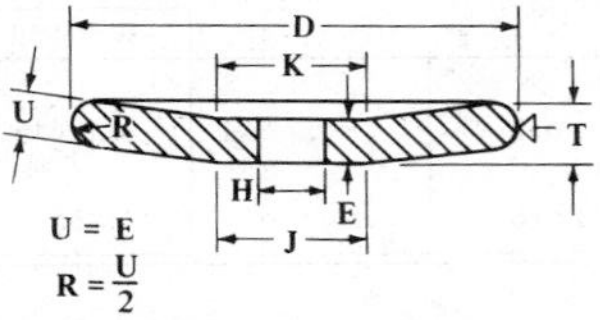

$U = E$
$R = \dfrac{U}{2}$

Type 13. Saucer Wheel
Peripheral grinding wheel, resembling the shape of a saucer, with cross section equal throughout.

Applications	D = Diam.	T = Thick.	H = Hole
SAW GUMMING. Saw tooth shaping and sharpening	8 to 12	½ to 1¾ U & E ¼ to 1½	¾ to 1¼

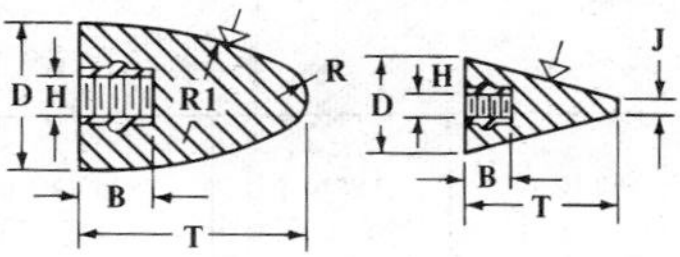

Type 16. Cone, Curved Side

Type 17. Cone, Straight Side, Square Tip

Type 17R. Cone, Straight Side, Round Tip
(Tip Radius $R = J/2$)

Applications	D = Diam.	T = Thick.	H = Hole
SNAGGING. Portable machine Threaded holes	1¼ to 3	2 to 3½	⅜-24 UNF-2B to ⅝-11 UNC-2B

Table 1A (*Continued*). **Standard Shapes and Inch Size Ranges of Grinding Wheels**
(ANSI B74.2-1982)

Applications	Size Ranges of Principal Dimensions, Inches		
	D = Diam.	T = Thick.	H = Hole
Type 18. Plug, Square End Type 18R. Plug, Round End $R = D/2$			
Type 19. Plugs, Conical End, Square Tip Type 19R. Plugs, Conical End, Round Tip (Tip Radius $R = J/2$)			
SNAGGING. Portable machine Threaded holes	1¼ to 3	2 to 3½	⅜-24UNF-2B to ⅝-11UNC-2B
Type 20. Wheel, Relieved One Side Peripheral grinding wheel, one side flat, the other side relieved to a flat.			
CYLINDRICAL GRINDING. Between centers	12 to 36	¾ to 4	5 to 20
Type 21. Wheel, Relieved Two Sides Both sides relieved to a flat.			
Type 22. Wheel, Relieved One Side, Recessed Other Side One side relieved to a flat.			
Type 23. Wheel, Relieved and Recessed Same Side The other side is straight.			
CYLINDRICAL GRINDING. Between centers, with wheel periphery	20 to 36	2 to 4	12 or 20

Table 1A (*Concluded*). **Standard Shapes and Inch Size Ranges of Grinding Wheels**
(ANSI B74.2-1982)

Applications	Size Ranges of Principal Dimensions, Inches		
	D = Diam.	T = Thick.	H = Hole

Type 24. Wheel, Relieved and Recessed One Side, Recessed Other Side
One side recessed, the other side is relieved to a recess.

Type 25. Wheel, Relieved and Recessed One Side, Relieved Other Side
One side relieved to a flat, the other side relieved to a recess.

Type 26. Wheel, Relieved and Recessed Both Sides

Applications	D	T	H
CYLINDRICAL GRINDING. Between centers, with the periphery of the wheel	20 to 36	2 to 4	12 or 20

TYPES 27 & 27A. Wheel, Depressed Center
27. *Portable Grinding:* Grinding normally done by contact with work at approx. a 15° angle with face of the wheel.
27A. *Cutting-off:* Using the periphery as grinding face.

Applications	D	T	H
CUTTING OFF. Reinforced organic bonds only	16 to 30	$U = E = $ 5/32 to 1/4	1 or 1½
SNAGGING. Portable machine	3 to 9	U = Uniform thick. 1/8 to 3/8	3/8 or 7/8

Type 28. Wheel, Depressed Center (Saucer Shaped Grinding Face)
Grinding at approx. 15° angle with wheel face.

Applications	D	T	H
SNAGGING. Portable machine	7 or 9	U = Uniform thickness 1/4	7/8

Table 1B. Standard Shapes and Metric Size Ranges of Grinding Wheels*
(ANSI B74.2-1982)

Applications	Size Ranges of Principal Dimensions, Millimeters		
	D = Diam.	T = Thick.	H = Hole
Type 1. Straight Wheel†			
CUTTING OFF. (nonreinforced and reinforced organic bonds only)	150 to 1250	0.8 to 10	16 to 152.4
CYLINDRICAL GRINDING. Between centers	300 to 1250	20 to 160	127 to 508
CYLINDRICAL GRINDING. Centerless grinding wheels	350 to 750	25 to 500	127 or 304.8
CYLINDRICAL GRINDING. Centerless regulating wheels	200 to 350	25 to 315	76.2 to 152.4
INTERNAL GRINDING.	6 to 100	6 to 50	2.5 to 25
OFFHAND GRINDING. Grinding on the periphery — General purpose — For wet tool grinding only	150 to 900 750 or 900	13 to 100 80 or 100	20 to 76.2 508
SAW GUMMING. (F-type face)	150 to 300	6 to 40	32
SNAGGING. Floor stand machines	300 to 600	25 to 80	32 to 76.2
SNAGGING. Floor stand machines (organic bond, wheel speed over 33 meters per second)	500 to 900	50 to 100	152.4 or 304.8
SNAGGING. Mechanical grinders (organic bond, wheel speed up to 84 meters per second)	600	50 to 80	304.8
SNAGGING. Portable machines	80 to 200	6 to 25	10 to 16
SNAGGING. Swing frame machines (organic bond)	300 to 600	50 to 80	88.9 to 304.8
SURFACE GRINDING. Horizontal spindle machines	150 to 600	13 to 160	32 to 304.8
TOOL GRINDING. Broaches, cutters, mills, reamers, taps, etc.	150 to 250	6 to 20	32 to 127
Type 2. Cylindrical Wheel†			
			W = Wall
SURFACE GRINDING. Vertical spindle machines	200 to 500	100 or 125	25 to 100

* All dimensions are in millimeters.
† See Table 1A for diagrams and descriptions of each wheel type.

Table 1B (*Concluded*). **Standard Shapes and Metric Size Ranges of Grinding Wheels***
(ANSI B74.2-1982)

Applications	Size Ranges of Principal Dimensions, Millimeters		
	D = Diam.	T = Thick.	H = Hole
Type 5. Wheel, recessed one side†			
CYLINDRICAL GRINDING. Between centers	300 to 900	40 to 100	127 or 304.8
CYLINDRICAL GRINDING. Centerless regulating wheels	200 to 350	80 to 160	76.2 or 127
INTERNAL GRINDING.	10 to 100	10 to 50	3.18 to 25
Type 6. Straight-Cup Wheel†			
			W = Wall
SNAGGING. Portable machines, organic bond only (hole is ⅝-11UNC-2B)	100 to 150	50	20 to 40
TOOL GRINDING. Broaches, cutters, mills, reamers, taps, etc. (Hole is 13 to 32 mm)	50 to 150	32 to 50	8 or 10
Type 7. Wheel, recessed two sides†			
CYLINDRICAL GRINDING. Between centers	300 to 900	40 to 100	127 or 304.8
CYLINDRICAL GRINDING. Centerless regulating wheels	200 to 350	100 to 500	76.2 to 152.4
Type 11. Flaring-Cup Wheel†			
SNAGGING. Portable machines, organic bonds only, threaded hole	100 to 150	50	⅝-11UNC-2B
TOOL GRINDING. Broaches, cutters, mills, reamers, taps, etc.	50 to 125	32 to 50	13 to 32
Type 12. Dish Wheel†			
TOOL GRINDING. Broaches, cutters, mills, reamers, taps, etc.	80 to 200	13 or 20	13 to 32
Types 27 and 27A. Wheel, depressed center†			
CUTTING OFF. Reinforced organic bonds only	400 to 750	U = E = 6	25.4 or 38.1
SNAGGING. Portable machines	80 to 230	U = E = 3.2 to 10	9.53 or 22.23

* All dimensions in millimeters.
† See Table 1A for diagrams and descriptions of each wheel type.

Table 2. Standard Shapes of Grinding Wheel Faces (ANSI B74.2-1982)

Recommendations, similar in principle, yet somewhat less discriminating have been developed by the Norton Company for *constructional steels*. These materials can be ground either in their original state (soft) or in their after-hardened state (directly or following carburization). Constructional steels must be distinguished from structural steels which are used primarily by the building industry in mill shapes, without or with a minimum of machining.

Table 3. Classification of Tool Steels by their Relative Grindability

Relative Grindability Group	AISI-SAE Designation of Tool Steels
GROUP 1 — Any area of work surface High grindability tool and die steels (Grindability index greater than 12)	W1, W2, W4, W5 S1, S2, S4, S5 O1, O2, O6, O7 A2, A4, A5, A6 H11, H12, H13, H14, H20, H21, H22, H23, H24, H25, H26 L1, L2, L3, L6, L7
GROUP 2 — Small area of work surface (as found in tools) Medium grindability tool and die steels (Grindability index 3 to 12)	H41, H42, H43 T1, T2, T4, T7, T8 M1, M2, M8, M10, M30 F1, F2, F3 D1, D2, D3, D4, D5, D6
GROUP 3 — Small area of work surface (as found in tools) Low grindability tool and die steels (Grindability index between 1.0 and 3)	T5, T6 M3, M6, M7, M34, M35, M36, M41, M43, M44 D7
GROUP 4 — Large area of work surface (as found in dies) Medium and low grindability tool and die steels (Grindability index between 1.0 and 12)	All steels found in Groups 2 and 3
GROUP 5 — Any area of work surface Very low grindability tool and die steels (Grindability index less than 1.0)	T9, T15 M4, M15

Constructional steels are either plain carbon or alloy type steels assigned in the AISI-SAE specifications to different groups, according to the predominant types of alloying elements. In the following recommendations no distinction is made because of different compositions since that factor generally, has a minor effect on grinding wheel choice in the case of constructional steels. However, separate recommendations are made for soft (Table 5) and hardened (Table 6) constructional steels. For the relatively rare case where the use of a single type of wheel for both soft and hardened steel materials is considered more important than the selection of the best suited types for each condition of the work materials, Table 5 lists "All-Around" wheels in its last column.

For applications where cool cutting properties of the wheel are particularly important, Table 6 lists, as a second alternative, porous-type wheels. The sequence of choices as presented in these tables does not necessarily represent a second, or third best; it can also apply to conditions where the first choice did not provide optimum results and by varying slightly the composition of the grinding wheel, as indicated in the subsequent choices, the performance experience by the first choice might actually be improved.

(Continued on page 1056)

Table 4. Grinding Wheel Recommendations for Hardened Tool Steels According to their Grindability

Operation	Wheel or Rim Diameter, Inches	First Choice Specifications	Second Choice Specifications
GROUP 1 STEELS			
Surfacing			
Surfacing wheels	14 and smaller	Wet FA46-I8V	SFA46-G12VP
	14 and smaller	Dry FA46-H8V	FA46-F12VP
	Over 14	Wet FA36-I8V	SFA36-I8V
Segments or Cylinders	1½ rim or less	Wet FA30-H8V	FA30-F12VP
Cups	¾ rim or less	Wet FA36-H8V	FA46-F12VP
	(for rims wider than 1½ inches, go one grade softer in available specifications)		
Cutter sharpening			
Straight wheel	....	Wet FA46-K8V	FA60-K8V
	....	Dry FA46-J8V	FA46-H12VP
Dish shape	....	Dry FA60-J8V	FA60-H12VP
Cup shape	....	Dry FA46-L8V	FA60-H12VP
	....	Wet SFA46-L5V	SFA60-L5V
Form tool grinding	8 and smaller	Wet FA60-L8V to	FA100-M7V
	8 and smaller	Dry FA60-K8V to	FA100-L8V
	10 and larger	Wet FA60-L8V to	FA80-M6V
Cylindrical	14 and smaller	Wet SFA60-L5V	....
	16 and larger	Wet SFA60-M5V	....
Centerless	....	Wet SFA60-M5V	....
Internal			
Production grinding	Under ½	Wet SFA80-N6V	SFA80-N7V
	½ to 1	Wet SFA60-M5V	SFA60-M6V
	Over 1 to 3	Wet SFA54-L5V	SFA54-L6V
	Over 3	Wet SFA46-L5V	SFA46-K5V
Tool room grinding	Under ½	Dry FA80-L6V	SFA80-L7V
	½ to 1	Dry FA70-K7V	SFA70-K7V
	Over 1 to 3	Dry FA60-J8V	FA60-H12VP
	Over 3	Dry FA46-J8V	FA54-H12VP
GROUP 2 STEELS			
Surfacing			
Straight wheels	14 and smaller	Wet FA46-I8V	FA46-G12VP
	14 and smaller	Dry FA46-H8V	FA46-F12VP
	Over 14	Wet FA46-H8V	SFA46-I8V
Segments or Cylinders	1½ rim or less	Wet FA30-G8V	FA36-E12VP
Cups	¾ rim or less	Wet FA36-H8V	FA46-F12VP
	(for rims wider than 1½ inches, go one grade softer in available specifications)		

Table 4. *(Continued)* **Grinding Wheel Recommendations for Hardened Tool Steels According to their Grindability**

Operation	Wheel or Rim Diameter, Inches	First Choice Specifications	Second Choice Specifications
GROUP 2 STEELS (Continued)			
Cutter sharpening Straight wheel	....	Wet FA46-L5V	FA60-K8V
	....	Dry FA46-J8V	FA60-H12VP
Dish shape	....	Dry FA60-J5V	FA60-G12VP
Cup shape	....	Dry FA46-K5V	FA60-G12VP
	....	Wet FA46-L5V	FA60-J8V
Form tool grinding	8 and smaller	Wet FA60-K8V to	FA120-L8V
	8 and smaller	Dry FA80-K8V to	FA150-K8V
	10 and larger	Wet FA60-K8V to	FA120-L8V
Cylindrical	14 and less	Wet FA60-L5V	SFA60-L5V
	16 and larger	Wet FA60-K5V	SFA60-K5V
Centerless	....	Wet FA60-M5V	SFA60-M5V
Internal Production grinding	Under ½	Wet FA80-L6V	SFA80-L6V
	½ to 1	Wet FA70-K5V	SFA70-K5V
	Over 1 to 3	Wet FA60-J8V	SFA60-J7V
	Over 3	Wet FA54-J8V	SFA54-J8V
			SFA80-K7V
Tool room grinding	Under ½	Dry FA80-I8V	
	½ to 1	Dry FA70-J8V	SFA70-J7V
	Over 1 to 3	Dry FA60-I8V	FA60-G12VP
	Over 3	Dry FA54-I8V	FA54-G12VP
GROUP 3 STEELS			
Surfacing Straight wheels	14 and smaller	Wet FA60-I8V	FA60-G12VP
	14 and smaller	Dry FA60-H8V	FA60-F12VP
	Over 14	Wet FA60-H8V	SFA60-I8V
Segments or Cylinders Cups	1½ rim or less	Wet FA46-G8V	FA46-E12VP
	¾ rim or less	Wet FA46-G8V	FA46-E12VP
	(for rims wider than 1½ inches, go one grade softer in available specifications)		
Cutter sharpening Straight wheel	....	Wet FA46-J8V	FA60-J8V
	....	Dry FA46-I8V	FA46-G12VP
Dish shape	....	Dry FA60-H8V	FA60-F12VP
Cup shape	....	Dry FA46-I8V	FA60-F12VP
	....	Wet FA46-J8V	FA60-J8V
Form tool grinding	8 and smaller	Wet FA80-K8V to	FA150-L9V
	8 and smaller	Dry FA100-J8V to	FA150-K8V
	10 and larger	Wet FA80-J8V to	FA150-J8V

Table 4. (*Continued*) **Grinding Wheel Recommendations for Hardened Tool Steels According to their Grindability**

Operation	Wheel or Rim Diameter, Inches	First Choice Specifications	Second Choice Specifications
GROUP 3 STEELS (Continued)			
Cylindrical	14 and less	Wet FA80-L5V	SFA80-L6V
	16 and larger	Wet FA60-L6V	SFA60-K5V
Centerless	. . . .	Wet FA60-L5V	SFA60-L5V
Internal			
Production grinding	Under ½	Wet FA90-L6V	SFA90-L6V
	½ to 1	Wet FA80-L6V	SFA80-L6V
	Over 1 to 3	Wet FA70-K5V	SFA70-K5V
	Over 3	Wet FA60-J5V	SFA60-J5V
Tool room grinding	Under ½	Dry FA90-K8V	SFA90-K7V
	½ to 1	Dry FA80-J8V	SFA80-J7V
	Over 1 to 3	Dry FA70-I8V	FA70-G12VP
	Over 3	Dry FA60-I8V	FA60-G12VP
GROUP 4 STEELS			
Surfacing			
Straight wheels	14 and smaller	Wet FA60-I8V	C60-JV
	14 and smaller	Wet FA60-H8V	C60-IV
	Over 14	Wet FA46-H8V	C60-HV
Segments	1½ rim or less	Wet FA46-G8V	C46-HV
Cylinders	1½ rim or less	Wet FA46-G8V	C60-HV
Cups	¾ rim or less	Wet FA46-G8V	C60-IV
	(for rims wider than 1½ inches, go one grade softer in available specifications)		
Form tool grinding	8 and smaller	Wet FA60-J8V to FA150-K8V	
	8 and smaller	Dry FA80-I8V to FA180-J8V	
	10 and larger	Wet FA60-J8V to FA150-K8V	
Cylindrical	14 and less	Wet FA80-K8V	C60-KV
	16 and larger	Wet FA60-J8V	C60-KV
Internal			
Production grinding	Under ½	Wet FA90-L6V	C90-LV
	½ to 1	Wet FA80-K5V	C80-KV
	Over 1 to 3	Wet FA70-J8V	C70-JV
	Over 3	Wet FA60-I8V	C60-IV
Tool room grinding	Under ½	Dry FA90-K8V	C90-KV
	½ to 1	Dry FA80-J8V	C80-JV
	Over 1 to 3	Dry FA70-I8V	C70-IV
	Over 3	Dry FA60-H8V	C60-HV

Table 4. *(Concluded)* Grinding Wheel Recommendations for Hardened Tool Steels According to their Grindability

Operation	Wheel (or Rim) Diameter, Inches	First Choice Specifications	Second Choice Specifications	Third Choice Specifications
		GROUP 5 STEELS		
Surfacing				
Straight wheels	14 and smaller	Wet SFA60-H8V	FA60-E12VP	C60-IV
	14 and smaller	Dry SFA80-H8V	FA80-E12VP	C80-HV
	Over 14	Wet SFA60-H8V	FA60-E12VP	C60-HV
Segments or Cylinders	1½ rim or less	Wet SFA46-G8V	FA46-E12VP	C46-GV
Cups	¾ rim or less	Wet SFA60-G8V	FA60-E12VP	C60-GV
	(for rims wider than 1½ inches, go one grade softer in available specifications)			
Cutter sharpening				
Straight wheels	...	Wet SFA60-I8V	SFA60-G12VP	...
	...	Dry SFA60-H8V	SFA80-F12VP	...
Dish shape	...	Dry SFA80-H8V	SFA80-F12VP	...
Cup shape	...	Wet SFA60-I8V	SFA60-G12VP	...
	...	Wet SFA60-J8V	SFA60-H12VP	...
Form tool grinding	8 and smaller	Wet FA80-J8V	FA180-J9V	FA80-H12VP
	8 and smaller	Dry FA100-I8V	to FA220-J9V	FA80-G12VP
	10 and larger	Wet FA80-J8V	to FA180-J9V	...
Cylindrical	14 and less	Wet FA80-J8V	C80-KV	...
	16 and larger	Wet FA80-I8V	C80-KV	...
Centerless	...	Wet FA80-J5V	C80-LV	...
Internal				
Production grinding	Under ½	Wet FA100-L8V	C90-MV	...
	½ to 1	Wet FA90-K8V	C80-LV	...
	Over 1 to 3	Wet FA80-J8V	C70-KV	...
	Over 3	Wet FA70-I8V	C60-IV	...
Tool room grinding	Under ½	Dry FA100-K8V	C90-KV	...
	½ to 1	Dry FA90-J8V	C80-JV	...
	Over 1 to 3	Dry FA80-I8V	C70-IV	FA80-G12VP
	Over 3	Dry FA70-I8V	C60-IV	FA70-G12VP

Table 5. Grinding Wheel Recommendations for Constructional Steels (Soft)

Grinding Operation	Wheel or Rim Diameter, Inches	First Choice	Alternate Choice (Porous type)	All-Around Wheel
Surfacing Straight wheels	14 and smaller	Wet FA46-J8V	FA46-H12VP	FA46-J8V
	14 and smaller	Dry FA46-I8V	FA46-H12VP	FA46-I8V
	Over 14	Wet FA36-J8V	FA36-H12VP	FA36-J8V
Segments Cylinders Cups	1½ rim or less	Wet FA24-H8V	FA30-F12VP	FA24-H8V
	1½ rim or less	Wet FA24-I8V	FA30-G12VP	FA24-H8V
	¾ rim or less	Wet FA24-H8V	FA30-F12VP	FA30-H8V
	(For wider rims, go one grade softer)			
Cylindrical	14 and smaller	Wet SFA60-M5V	....	SFA60-L5V
	16 and larger	Wet SFA54-M5V	....	SFA54-L5V
Centerless	....	Wet SFA54-N5V	....	SFA60-M5V
Internal	Under ½	Wet SFA60-M5V	....	SFA80-L6V
	½ to 1	Wet SFA60-L5V	....	SFA60-K5V
	Over 1 to 3	Wet SFA54-K5V	....	SFA54-J5V
	Over 3	Wet SFA46-K5V	....	SFA46-J5V

Table 6. Grinding Wheel Recommendations for Constructional Steels (Hardened or Carburized)

Grinding Operation	Wheel or Rim Diameter, Inches	First Choice	Alternate Choice (Porous Type)
Surfacing Straight wheels	14 and smaller	Wet FA46-I8V	FA46-G12VP
	14 and smaller	Dry FA46-H8V	FA46-F12VP
	Over 14	Wet FA36-I8V	FA36-G12VP
Segments or Cylinders Cups	1½ rim or less	Wet FA30-H8V	FA36-F12VP
	¾ rim or less	Wet FA36-H8V	FA46-F12VP
	(For wider rims, go one grade softer)		
Forms and Radius Grinding	8 and smaller	Wet FA60-L7V to FA100-M8V	
	8 and smaller	Dry FA60-K8V to FA100-L8V	
	10 and larger	Wet FA60-L7V to FA 80-M7V	
Cylindrical Work diameter			
1″ and smaller	14 and smaller	Wet SFA80-L6V	....
Over 1″	14 and smaller	Wet SFA80-K5V	....
1″ and smaller	16 and larger	Wet SFA60-L5V	....
Over 1″	16 and larger	Wet SFA60-L5V	....
Centerless	....	Wet SFA80-M6V	....
Internal	Under ½	Wet SFA80-N6V	....
	½ to 1	Wet SFA60-M5V	....
	Over 1 to 3	Wet SFA54-L5V	....
	Over 3	Wet SFA46-K5V	....
	Under ½	Dry FA80-L6V	....
	½ to 1	Dry FA70-K8V	....
	Over 1 to 3	Dry FA60-J8V	FA60-H12VP
	Over 3	Dry FA46-J8V	FA54-H12VP

Variations from General Grinding Wheel Recommendations. — Recommendations for the selection of grinding wheels are usually based on average values with regard to both operational conditions and process objectives. In case of variations from such average values the composition of the grinding wheels must be adjusted for obtaining optimum results. While it is impossible to list and to appraise all possible variations and to define their effects on the selection of the best suited grinding wheels, some guidance can be obtained from experience. The following tabulation indicates the general directions in which the characteristics of the initially selected grinding wheel may have to be altered in order to approach optimum performance. Variations in a sense opposite to those mentioned will call for wheel characteristic changes in reverse.

Conditions or Objectives	Direction of Change
To increase cutting rate	Coarser grain, softer bond, higher porosity
To retain wheel size and/or form	Finer grain, harder bond
For small or narrow work surface	Finer grain, harder bond
For larger wheel diameter	Coarser grain
To improve finish or work	Finer grain, harder bond, or resilient bond
For increased work speed or feed rate	Harder bond
For increased wheel speed	Generally softer bond, except for high speed grinding which requires harder bond for added wheel strength
For interrupted or coarse work surface	Harder bond
For thin walled parts	Softer bond
To reduce load on the machine drive motor	Softer bond

Dressing and Truing Grinding Wheels. — The perfect grinding wheel operating under ideal conditions will be self sharpening; i.e., as the abrasive grains become dull, they will tend to fracture and be dislodged from the wheel by the grinding forces, thereby exposing new, sharp abrasive grains. While in precision machine grinding this ideal may be partially attained in some instances, it is almost never attained completely. Usually, the grinding wheel must be dressed and trued after mounting on the precision grinding machine spindle and periodically thereafter.

Dressing may be defined as any operation performed on the face of a grinding wheel that improves its cutting action. Truing is a dressing operation but is more precise, i.e., the face of the wheel may be made parallel to the spindle or made into a radius or special shape. Regularly applied truing is also needed for the accurate size control of the work, particularly in automatic grinding. The tools and processes generally used in grinding wheel dressing and truing are listed and described in Table 1.

Table 1. Tools and Methods for Grinding Wheel Dressing and Truing

Designation	Description	Application
Rotating Hand Dressers	Freely rotating discs, either star shaped with protruding points, or discs with corrugated or twisted perimeter, supported in a fork-type handle, the lugs of which can lean on the tool rest of the grinding machine.	Preferred for bench or floor type grinding machines, also for use on heavy portable grinders (snagging grinders) where free cutting properties of the grinding wheel are primarily sought and the accuracy of the trued profile is not critical.
Abrasive Sticks	Made of silicon carbide grains with a hard bond. Applied directly or supported in a handle. Less frequently abrasive sticks are also made of boron carbide.	Usually hand held and use limited to smaller size wheels. Because it also shears the grains of the grinding wheel, it is often used for roughing or pre-shaping, prior to final dressing with, e.g., a diamond.
Abrasive Wheels (Rolls)	Silicon carbide grains in a hard vitrified bond are cemented on ball bearing mounted spindles. Used either as hand tools with handles or rigidly held in a supporting member of the grinding machine. Generally freely rotating; also available with adjustable brake for diamond wheel dressing.	Preferred for large grinding wheels as a diamond saver, but also for the improved control of the dressed surface characteristics. When skewing the abrasive dresser wheel by a few degrees out of parallel with the grinding wheel axis, the basic crushing action is supplemented with wiping and shearing, thus producing the desired degree of wheel surface smoothness.
Single-point Diamonds	A diamond stone of selected size is mounted in a steel nib of cylindrical shape with or without head, dimensioned to fit the truing spindle of specific grinding machines. Proper orientation and retainment of the diamond point in the setting is an important requirement.	The most widely used tool for dressing and truing grinding wheels in precision grinding. Permits precisely controlled dressing action by regulating infeed and cross feed rate of the truing device. Most dependable in duplicating the movements of the truing spindle when latter is guided by cams or templates for accurate form truing.
Single-point Form Truing Diamonds	Selected diamonds having symmetrically located natural edges with precisely lapped diamond points, controlled cone angles and vertex radius, and the axis coinciding with that of the nib.	Truing operations requiring very accurately controlled, and often, also steeply inclined wheel profiles, such as in thread and gear grinding, where one or more diamond points participate in generating the resulting wheel periphery form, are dependent on specially designed and made truing diamonds and nibs.

Table 1 (*Continued*). Tools and Methods for Grinding Wheel Dressing and Truing

Designation	Description	Application
Cluster Type Diamond Dresser	Several, usually seven, smaller diamond stones are mounted in spaced relationship across the working surface of the nib. In some types of tools more than a single layer of such clusters is set at parallel levels in the matrix, the deeper positioned layer becoming active after the preceding layer has worn away.	Intended for straight face dressing and permits the utilization of smaller, less expensive diamond stones. In use, the holder is canted at a 3° to 10° angle, bringing two to five points into contact with the wheel. The multiple point contact permits faster cross feed rates during truing than used with single point diamonds for generating a specific degree of wheel face finish.
Impregnated Matrix Type Diamond Dressers	The operating surface consists of a layer of small, randomly distributed, yet rather uniformly spaced diamonds which are retained in a bond holding the points in an essentially common plane. Supplied either with straight or canted shaft, the latter being used to cancel the tilt of angular truing posts.	For the truing of wheel surfaces consisting of a single or several flat elements. The nib face should be held tangent to the grinding wheel periphery or parallel with a flat working surface. Offers economic advantages where technically applicable because of using less expensive diamond splinters presented in a manner permitting efficient utilization.
Form Generating Truing Devices	Swivelling diamond holder post with adjustable pivot location, arm length and swivel arc, mounted on angularly adjustable cross slides with controlled traverse movement, permit the generation of various straight and circular profile elements, kept in specific mutual locations.	Such devices are made in various degrees of complexity for the positionally controlled interrelation of several different profile elements. Limited to regular straight and circular sections, yet offers great flexibility of setup, very accurate adjustment and unique versatility for handling a large variety of frequently changing profiles.
Contour Duplicating Truing Devices	The form of a master, called cam or template, having the actual profile which is to be produced on the wheel, or its magnified version, is translated into the path of the diamond point by means of mechanical linkage, a fluid actuator or a pantograph device.	Preferred single-point truing method for profiles to be produced in quantities warranting the making of special profile bars or templates. Used also in small and medium volume production when the complexity of the profile to be produced, excludes alternate methods of form generation.

Table 1 (*Concluded*). **Tools and Methods for Grinding Wheel Dressing and Truing**

Designation	Description	Application
Grinding Wheel Contouring by Crush Truing	A hardened steel or carbide roll which is free to rotate and has the desired form of the workpiece, is fed gradually into the grinding wheel which runs at slow speed. The roll will, by crushing action produce its reverse form in the wheel. Crushing produces a free-cutting wheel face with sharp grains.	Requires grinding machines designed for crush truing, having stiff spindle bearings, rigid construction, slow wheel speed for truing, etc. Due to the cost of crush rolls and equipment the process is used for repetitive work only. It is one of the most efficient methods for precisely duplicating complex wheel profiles which are capable of grinding in the 8 mu in. AA range. Applicable for both surface and cylindrical grinding.
Rotating Diamond Roll Type Grinding Wheel Truing	Special rolls made to agree with specific profile specifications have their periphery coated with a large number of uniformly distributed diamonds, held in a matrix into which the individual stones are set by hand (for larger diamonds) or bonded by a plating process (for smaller elements).	The diamond rolls must be rotated by an air, hydraulic or electric motor at about one fourth of the grinding wheel surface speed and in opposite direction to the wheel rotation. While the initial costs are substantially higher than for single-point diamond truing, the savings in truing time warrants the method's application in large volume production of profile-ground components.
Diamond Dressing Blocks	Made as flat blocks for straight wheel surfaces, are also available for radius dressing and profile truing. The working surface consists of a layer of electroplated diamond grains, uniformly distributed and capable of truing even closely toleranced profiles.	For straight wheels it can reduce dressing time and offers easy installation on surface grinders, where the blocks mount on the magnetic plate. Recommended for small and medium volume production for truing intricate profiles on regular surface grinders, because the higher pressure developed in crush dressing is avoided.

Guidelines for Truing and Dressing With Single Point Diamonds. — The diamond nib should be canted at an angle of 10 to 15 degrees in the direction of the wheel rotation and also, if possible, by the same amount in the direction of the cross feed traverse during the truing (see diagram). The dragging effect resulting from this "angling", combined with the occasional rotation of the diamond nib in its holder, will prolong the diamond life by limiting the extent of wear facets and will also tend to produce a pyramid shape of the diamond tip. The diamond may also be set to contact the wheel by about ⅛ to ¼ inch below its centerline.

Depth of cut: This should not exceed 0.001 inch per pass for general work, and will have to be reduced to 0.0002 to 0.0004 inch per pass for wheels with fine grains used for precise finishing work.

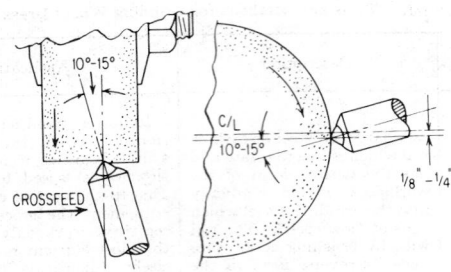

Diamond crossfeed rate: This may be varied to some extent depending on the required wheel surface: faster crossfeed for free cutting, and slower crossfeed for fine-finish producing characteristics. Such variations, however, must always stay within the limits set by the grain size of the wheel. Thus the advance rate of the truing diamond per wheel revolution should not exceed the diameter of a grain or be less than half of that rate. Consequently, the diamond crossfeed must be slower for a large wheel than for a smaller wheel having the same grain size number.

Typical crossfeed values for frequently used grain sizes are given in Table 2.

Table 2. Typical Diamond Truing Diamond Cross Feeds

Grain Size..........................	30	36	46	50
Cross Feed per Wheel Rev., inch....	.014–.024	.012–.019	.008–.014	.007–.012
Grain Size..........................	60	80	120	....
Cross Feed per Wheel Rev., inch....	.006–.010	.004–.007	.0025–.004	....

These values can then easily be converted into the more conveniently used inch-per-minute units, simply by multiplying them by the actual rpm of the grinding wheel. Example: For a 20-inch diameter wheel, Grain No. 46, running at 1200 rpm:

Crossfeed rate for roughing-cut truing — about 17 ipm
for finishing-cut truing — about 10 ipm

Coolant should be applied before the diamond is making contact with the wheel and must be continued in generous supply while truing.

The speed of the grinding wheel should be at the regular grinding rate, or not much lower. For that reason the feed wheels of centerless grinding machines have usually an additional speed rate higher than functionally needed, that speed being provided for wheel truing only.

The initial approach of the diamond to the wheel surface must be carried out carefully to prevent sudden contact with the diamond, resulting in penetration in excess of the selected depth of cut. It should be noted that the highest point of a worn wheel is often in its center portion and not at the edge from where the crossfeed of the diamond starts.

The general conditions of the truing device are important for best truing results and also for assuring extended diamond life. Rigid truing spindle, well-seated diamond nib and firmly set diamond point are mandatory. Sensitive infeed and smooth traverse movement at uniform speed must also be maintained.

Resetting of the diamond point. Never let the diamond point wear to a degree where the grinding wheel is contacting the steel nib. This can damage the setting of the diamond point and result in its loss. Expert resetting of a worn diamond can repeatedly add to its useful life, even when applied to lighter work because of reduced size.

Size Selection Guide for Single-Point Truing Diamonds. — There are no rigid rules for determining the proper size of diamond for any particular truing application because of the very large number of factors affecting that choice. Several of these factors are related to the condition, particularly the rigidity, of the grinding machine and truing device, as well as such characteristics of the diamond itself as purity, crystalline structure, etc. While these factors are difficult to evaluate in a generally applicable manner, the expected effects of several other conditions can be appraised and also considered in the selection of the proper diamond size.

The recommended sizes in Table 3 must be considered as informative only and as representing minimum values for generally favorable conditions. Factors calling for larger diamond sizes than listed are the following:

— Silicon carbide wheels (Table 3 refers to aluminum oxide wheels)
— Dry truing
— Grains sizes coarser than No. 46
— Bonds harder than M
— Wheel speed substantially higher than 6500 sfm.

It is advisable to consider any single or pair of these factors as justifying the selection of one size larger diamond. As an example: for using an SiC wheel, with grain size No. 36 and hardness P, select a diamond that is two sizes larger than that shown in Table 3 for the actually used wheel size.

Table 3. Recommended Minimum Sizes for Single-point Truing Diamonds

Diamond Size in Carats*	Index Number (Wheel Diam. × Width in Inches)	Examples of Max. Grinding Wheel Dimensions	
		Diameter	Width
0.25	3	4	0.75
0.35	6	6	1
0.50	10	8	1.25
0.60	15	10	1.50
0.75	21	12	1.75
1.00	30	12	2.50
1.25	48	14	3.50
1.50	65	16	4.00
1.75	80	20	4.00
2.00	100	20	5.00
2.50	150	24	6.00
3.00	200	24	8.00
3.50	260	30	8.00
4.00	350	36	10.00

* One carat equals 0.2 gram.

Single-point diamonds are available as loose stones, but are preferably procured from specialized manufacturers supplying the diamonds set into steel nibs. Expert setting, comprising both the optimum orientation of the stone and its firm retainment, are mandatory for assuring adequate diamond life and also satisfactory truing. Because the holding devices for truing diamonds are not yet standardized, the

required nib dimensions vary depending on the make and type of different grinding machines. Some nibs are made with angular heads, usually hexagonal, to permit occasional rotation of the nib either manually, with a wrench, or automatically.

Diamond Wheels

Diamond Wheels. — A diamond wheel is a special type of grinding wheel in which the abrasive elements are diamond grains held in a bond and applied to form a layer on the operating face of a non-abrasive core. Diamond wheels are used for grinding very hard or highly abrasive materials. Primary applications are the grinding of cemented carbides, such as the sharpening of carbide cutting tools; the grinding of glass, ceramics, asbestos and cement products; and the cutting and slicing of germanium and silicon.

Shapes of Diamond Wheels. — The industry-wide accepted Standard (ANSI B74.3-1974) specifies ten basic diamond wheel core shapes which are shown in Table 1 with the applicable designation symbols. The applied diamond abrasive layer may have different cross-sectional shapes. Those standardized are shown in Table 2. The third aspect which is standardized is the location of the diamond section on the wheel as shown by the diagrams in Table 3. Finally, modifications of the general core shape together with pertinent designation letters are given in Table 4.

The characteristics of the wheel shape listed in these four tables make up the components of the standard designation symbol for diamond wheel shapes. An example of that symbol with arbitrarily selected components is shown in Figure 1.

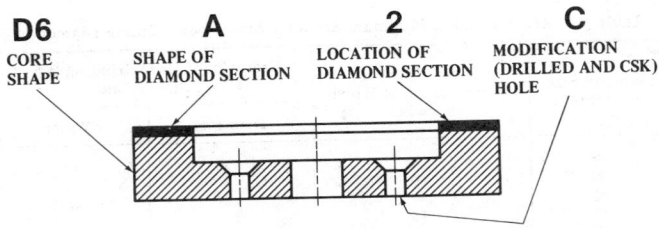

D6	**A**	**2**	**C**
CORE SHAPE	SHAPE OF DIAMOND SECTION	LOCATION OF DIAMOND SECTION	MODIFICATION (DRILLED AND CSK) HOLE

Fig. 1. A Typical Diamond Wheel Shape Designation Symbol

An explanation of these components is as follows:

Basic Core Shape: This portion of the symbol indicates the basic shape of the core on which the diamond abrasive section is mounted. The shape is actually designated by a number. The various core shapes and their designations are given in Table 1.

Diamond Cross-Section Shape: This, the second component, consisting of one or two letters, denotes the cross-sectional shape of the diamond abrasive section. The various shapes and their corresponding letter designations are given in Table 2.

Diamond Section Location: The third component of the symbol consists of a number which gives the location of the diamond section, i.e., periphery, side, corner, etc. An explanation of these numbers is shown in Table 3.

Modification: The fourth component of the symbol is a letter designating some modification, such as drilled and counterbored holes for mounting or special relieving of diamond section or core. This modification position of the symbol is used only when required. The modifications and their designations are given in Table 4.

Table 1. Diamond Wheel Core Shapes and Designations
(ANSI B74.3-1974)

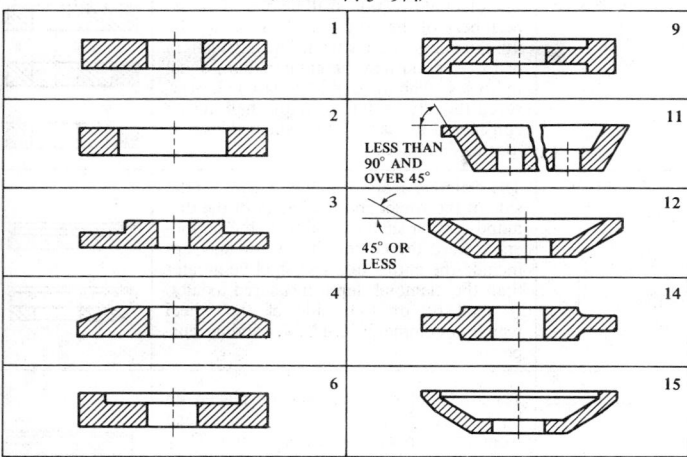

Table 2. Diamond Cross-sections and Designations
(ANSI B74.3-1974)

Table 3. Designations for Location of Diamond Section on Diamond Wheel
(ANSI B74.3-1974)

Designation No. and Location	Description	Illustration
1 — Periphery	The diamond section shall be placed on the periphery of the core and shall extend the full thickness of the wheel. The axial length of this section may be greater than, equal to, or less than the depth of diamond, measured radially. A hub or hubs shall not be considered as part of the wheel thickness for this definition.	
2 — Side	The diamond section shall be placed on the side of the wheel and the length of the diamond section shall extend from the periphery toward the center. It may or may not include the entire side and shall be greater than the diamond depth measured axially. It shall be on that side of the wheel which is commonly used for grinding purposes.	
3 — Both Sides	The diamond sections shall be placed on both sides of the wheel and shall extend from the periphery toward the center. They may or may not include the entire sides, and the radial length of the diamond section shall exceed the axial diamond depth.	
4 — Inside Bevel or Arc	This designation shall apply to the general wheel types 2, 6, 11, 12, and 15 and shall locate the diamond section on the side wall. This wall shall have an angle or arc extending from a higher point at the wheel periphery to a lower point toward the wheel center.	
5 — Outside Bevel or Arc	This designation shall apply to the general wheel types, 2, 6, 11, and 15 and shall locate the diamond section on the side wall. This wall shall have an angle or arc extending from a lower point at the wheel periphery to a higher point toward the wheel center.	
6 — Part of Periphery	The diamond section shall be placed on the periphery of the core but shall not extend the full thickness of the wheel and shall not reach to either side.	

Table 3 (*Concluded*). **Designations for Location of Diamond Section on Diamond Wheel (ANSI B74.3-1974)**

Designation No. and Location	Description	Illustration
7 — Part of Side	The diamond section shall be placed on the side of the core and shall not extend to the wheel periphery. It may or may not extend to the center.	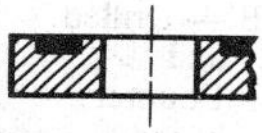
8 — Through-out	Designates wheels of solid diamond abrasive section without cores.	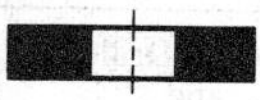
9 — Corner	Designates a location which would commonly be considered to be on the periphery except that the diamond section shall be on the corner but shall not extend to the other corner.	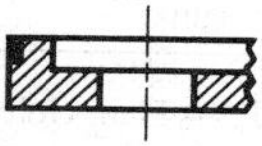
10 — Annular	Designates a location of the diamond abrasive section on the inner annular surface of the wheel.	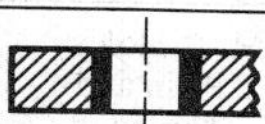

Composition of Diamond and Cubic Boron Nitride Wheels. — According to American National Standard ANSI B74.13-1982, a series of symbols is used to designate the composition of these wheels. An example is shown below.

Prefix	Abrasive	Grain Size	Grade	Concentration	Bond Type	Bond Modification	Depth of Abrasive	Manufacturer's Identification Symbol
M	D	120	R	100	B	56	1/8	

Fig. 2. Designation Symbols for Composition of Diamond and Cubic Boron Nitride Wheels

The meaning of each symbol is indicated by the following list:

1. *Prefix:* The prefix is a manufacturer's symbol indicating the exact kind of abrasive. Its use is optional.

2. *Abrasive Type:* The letter (B) is used for cubic boron nitride and (D) for diamond.

3. *Grain Size:* The grain sizes commonly used and varying from coarse to very fine are indicated by the following numbers: 8, 10, 12, 14, 16, 20, 24, 30, 36, 46, 54, 60, 70, 80, 90, 100, 120, 150, 180, and 220. The following additional sizes are used occasionally: 240, 280, 320, 400, 500, and 600. The wheel manufacturer may add to the regular grain number an additional symbol to indicate a special grain combination.

4. *Grade:* Grades are indicated by letters of the alphabet from A to Z in all bonds or processes. Wheel grades from A to Z range from soft to hard.

5. *Concentration:* The concentration symbol is a manufacturer's designation. It may be a number or a symbol.

Table 4. Designation Letters for Modifications of Diamond Wheels
(ANSI B74.3-1974)

Designation Letter*	Description	Illustration
B — Drilled and Counterbored	Holes drilled and counterbored in core.	
C — Drilled and Countersunk	Holes drilled and countersunk in core.	
H — Plain Hole	Straight hole drilled in core.	
M — Holes Plain and Threaded	Mixed holes, some plain, some threaded, are in core.	
P — Relieved One Side	Core relieved on one side of wheel. Thickness of core is less than wheel thickness.	
R — Relieved Two Sides	Core relieved on both sides of wheel. Thickness of core is less than wheel thickness.	
S — Segmented Diamond Section	Wheel has segmental diamond section mounted on core. (Clearance between segments has no bearing on definition.)	
SS — Segmental and Slotted	Wheel has separated segments mounted on a slotted core.	
T — Threaded Holes	Threaded holes are in core.	
Q — Diamond Inserted	Three surfaces of the diamond section are partially or completely enclosed by the core.	
V — Diamond Inverted	Any diamond cross section, which is mounted on the core so that the interior point of any angle, or the concave side of any arc, is exposed shall be considered inverted. *Exception:* Diamond cross section AH shall be placed on the core with the concave side of the arc exposed.	

* Y — Diamond Inserted and Inverted. See definitions for Q and V.

2500–3000 sfpm. However, this lower speed range can cause rapid wheel breakdown in finer grit wheels or in those with reduced diamond concentration.

Work Speeds: In diamond grinding, work rotation and table traverse are usually established by experience, adjusting these values to the selected infeed so as to avoid excessive wheel wear.

Infeed per Pass: Often referred to as downfeed and usually a function of the grit size of the wheel. The following are general values which may be increased for raising the productivity, or lowered to improve finish or to reduce wheel wear.

Wheel Grit Size Range	Infeed per Pass
100 to 120	0.001 inch
150 to 220	0.0005 inch
250 and finer	0.00025 inch

Grinding Wheel Safety

Safety in Operating Grinding Wheels. — Grinding wheels, although capable of exceptional cutting performance due to hardness and wear resistance, are prone to damage caused by improper handling and operation. Vitrified wheels, comprising the major part of grinding wheels used in industry, are held together by an inorganic bond which is actually a type of pottery product and therefore brittle and breakable. Although most of the organic bond types are somewhat more resistant to shocks, it must be realized that all grinding wheels are conglomerates of individual grains joined by a bond material whose strength is limited by the need of releasing the dull abrasive grains during use.

It must also be understood that during the grinding process very substantial forces act on the grinding wheel, including the centrifugal force due to rotation, the grinding forces resulting from the resistance of the work material, and shocks caused by sudden contact with the work. To be able to resist these forces, the grinding wheel must have a substantial minimum strength throughout that is well beyond that needed to hold the wheel together under static conditions.

Finally, a damaged grinding wheel can disintegrate during grinding, liberating dormant forces which normally are constrained by the resistance of the bond, thus presenting great hazards to both operator and equipment.

In order to avoid the breakage of the operating wheel and, should such a mishap occur, to prevent damage or injury, specific precautions have to be applied. These safeguards have been formulated into rules and regulations and are set forth in the American National Standard ANSI B7.1-1978, entitled the American National Standard Safety Requirements for the Use, Care, and Protection of Abrasive Wheels.

Handling, Storage and Inspection. — Grinding wheels should be hand carried, or transported, with proper support, by truck or conveyor. A grinding wheel must not be rolled around on its periphery.

The storage area, positioned not far from the location of the grinding machines, should be free from excessive temperature variations and humidity. Specially built racks are recommended on which the smaller or thin wheels are stacked lying on their sides and the larger wheels in an upright position on two-point cradle supports consisting of appropriately spaced wooden bars. Partitions should separate either the individual wheels, or a small group of identical wheels. Good accessibility to the stored wheels reduces the need of undesirable handling.

Inspection will primarily be directed at detecting visible damage, mostly originating from handling and shipping. Cracks which are not obvious can usually be detected by "ring testing"; this consists of suspending the wheel from its hole and tapping it

with a non-metallic implement. A clear metallic tone, a "ring" should be heard; a dead sound being indicative of a possible crack or cracks in the wheel.

Machine Conditions. — The general design of the grinding machines must ensure safe operation under normal conditions. The bearings and grinding wheel spindle must be dimensioned to withstand the expected forces and ample driving power should be provided to ensure maintenance of the rated spindle speed. For the protection of the operator, stationary machines used for dry grinding should have provision made for connection to an exhaust system and when used for off-hand grinding, a work support must be available.

Wheel guards are particularly important protection elements and their material specifications, wall thicknesses and construction principles should agree with the Standard's specifications. The exposure of the wheel should be just enough to avoid interference with the grinding operation. The need for access of the work to the grinding wheel will define the boundary of guard opening, particularly in the direction of the operator.

Grinding Wheel Mounting. — The mass and speed of the operating grinding wheel makes it particularly sensitive to imbalance. Vibrations which result from such conditions are harmful to the machine, particularly the spindle bearings and also affect the ground surface, i.e., wheel imbalance causes chatter marks and interferes with size control. While grinding wheels are shipped from the manufacturer's plant in a balanced condition, the retainment of the balanced state after the mounting of the wheel is quite uncertain. For that reason a balancing of the mounted wheel is indicated, and is particularly important for medium and large size wheels, as well as for producing acccurate and smooth surfaces. The most common way of balancing mounted wheels is by using balancing flanges with adjustable weights. The wheel and balancing flanges are mounted on a short balancing arbor, the two concentric and round stub ends of which are supported in a balancing stand. Such stands are of two types: (a) the parallel straight-edged which must be set up precisely level; (b) the disc type having two pairs of ball bearing mounted overlapping discs, which form a V for containing the arbor ends without hindering the free rotation of the wheel mounted on that arbor. The wheel will then only rotate when it is out of balance and its heavy spot is not in the lowest position. Rotating the wheel by hand to different positions will move the heavy spot, should such exist, from the bottom to a higher location where it can reveal its presence by causing the wheel to turn. Having detected the presence and location of the heavy spot, its effect can be cancelled by displacing the weights in the circular groove of the flange until a balanced condition has been accomplished.

Flanges are commonly used means for holding grinding wheels on the machine spindle. For that purpose, the wheel can either be mounted directly through its hole or by means of a sleeve which slips over a tapered section of the machine spindle. In either case, the flanges must be of equal diameter, usually not less than one-third of the new wheel's diameter. The purpose is to securely hold the wheel between the flanges, yet without interfering with the grinding operation even when the wheel becomes worn down to the point where it is ready to be discarded. Blotters or flange facings of compressible material should cover the entire contact area of the flanges.

One of the flanges is usually fixed while the other is loose and can be removed and adjusted along the machine spindle. The movable flange is held against the mounted grinding wheel by means of a nut engaging a threaded section of the machine spindle. The sense of that thread should be opposed to the direction of the wheel rotation, which is the direction in which the nut must be turned for removing it.

Safe Operating Speeds. — Safe grinding processes, while predicated on the proper use of the previously discussed equipment and procedures, are greatly dependent on the application of adequate operating speeds.

The Standard establishes maximum speeds at which grinding wheels can be operated, assigning the various types of wheels to several classification groups. Different values are listed according to bond type and also to wheel strength, distinguishing between low, medium and high strength wheels.

There is presented in the accompanying table, for the purpose of general information, an abbreviated version of the Standard's specification. However, for the actually permissible limits, the authoritative source is the manufacturer's tag on the wheel which, particularly for wheels of lower strength, might specify speeds below those of the table. All grinding wheels of 6 inches or greater diameter must be test run in the wheel manufacturer's plant at a speed which for all wheels having operating speeds in excess of 5000 sfpm is 1.5 times the maximum speed marked on the tag of the wheel.

The table shows the permissible wheel speeds in surface feet per minute (sfpm) units, whereas the tags on the grinding wheels state, for the convenience of the user, the maximum operating speed in revolutions per minute (rpm). The sfpm unit has the advantage of remaining valid for worn wheels too, whose rotational speed may be increased to the applicable sfpm value. The conversion from either to the other of these two kinds of units is a matter of simple calculation using the formulas:

$$sfpm = rpm \times \frac{D}{12} \times \pi$$

or

$$rpm = \frac{sfpm \times 12}{D \times \pi}$$

Where D = maximum diameter of the grinding wheel, in inches. Table 2 showing the conversion values from surface speed into rotational speed can be used for the direct reading of the rpm values corresponding to several different wheel diameters and surface speeds.

Special Speeds: Continuing progress in grinding methods has led to the recognition of certain advantages which can result from operating grinding wheels above, sometimes even higher than twice, the speeds considered earlier as the safe limits of grinding wheel operations. While advantages from the application of high speed grinding are limited to specific processes, the Standard admits, and offers code regulations for the use of wheels at special high speeds. These regulations define the structural requirements of the grinding machine and the responsibilities of the grinding wheel manufacturers, as well as of the users. High speed grinding should not be applied unless the machines, particularly guards, spindle assemblies and drive motors, are suitable for such methods. Also, appropriate grinding wheels expressly made for special high speeds must be used and, of course, the maximum operating speeds indicated on the wheel's tag must never be exceeded.

Portable Grinders. — The above discussed rules and regulations, devised primarily for stationary grinding machines are, in general, applicable for portable grinders too. In addition, the details of various other regulations, specially applicable to different types of portable grinders are discussed in the Standard, which should be consulted, particularly in the case of safety-wise critical uses of portable grinding machines.

Table 1. Maximum Peripheral Speeds for Grinding Wheels*
(Based on ANSI B7.1-1978)

Classifi-cation No.	Types of Wheels†	Maximum Operating Speeds, sfpm, Depending on Strength of Bond	
		Inorganic Bonds	Organic Bonds
1	Straight wheels — Type 1, except classifications 6, 7, 9, 10, 11, and 12 below Type 4** — Taper Side Wheels Types 5, 7, 20, 21, 22, 23, 24, 25, 26 Dish wheels — Type 12 Saucer wheels — Type 13 Cones and plugs — Types 16, 17, 18, 19	5,500 to 6,500	6,500 to 9,500
2	Cylinder wheels — Type 2 Segments	5,000 to 6,000	5,000 to 7,000
3	Cup shape tool grinding wheels — Types 6 and 11 (for fixed base machines)	4,500 to 6,000	6,000 to 8,500
4	Cup shape snagging wheels — Types 6 and 11 (for portable machines)	4,500 to 6,500	6,000 to 9,500
5	Abrasive discs	5,500 to 6,500	5,500 to 8,500
6	Reinforced wheels — except cutting-off wheels (depending on diameter and thickness)	. . .	9,500 to 14,200
7	Type 1 wheels for bench and pedestal grinders, Types 1 and 5 also in certain sizes for surface grinders	5,500 to 7,550	6,500 to 9,500
8	Diamond and cubic boron nitride wheels Metal bond Steel centered cutting off	to 6,500 to 12,000 to 16,000	to 9,500 . . . to 16,000
9	Cutting-off wheels — Larger than 16-inch diameter (incl. reinforced organic)	. . .	9,500 to 14,200
10	Cutting-off wheels — 16-inch diameter and smaller (incl. reinforced organic)	. . .	9,500 to 16,000
11	Thread and flute grinding wheels	8,000 to 12,000	8,000 to 12,000
12	Crankshaft and camshaft grinding wheels	5,500 to 8,500	6,500 to 9,500

* Values in this table are for general information only.   † See pages 1042–1048.
** Non-standard shape. For snagging wheels, 16 inches and larger — Type 1, internal wheels — Types 1 and 5, and mounted wheels, see ANSI B7.1-1978.   Under no conditions should a wheel be operated faster than the maximum operating speed established by the manufacturer.

Table 2. Revolutions per Minute for Various Grinding Speeds and Wheel Diameters

Wheel Diameter, Inch	Peripheral (Surface) Speed, Feet per Minute																Wheel Diameter, Inch
	16,000	14,000	12,000	10,000	9,500	9,000	8,500	8,000	7,500	7,000	6,500	6,000	5,500	5,000	4,500	4,000	
	Revolutions per Minute																
1	61,115	53,476	45,837	38,197	36,287	34,377	32,468	30,558	28,648	26,738	24,828	22,918	21,008	19,099	17,189	15,279	1
2	30,558	26,738	22,918	19,099	18,144	17,189	16,234	15,279	14,324	13,369	12,414	11,459	10,504	9,549	8,594	7,639	2
3	20,372	17,825	15,279	12,732	12,096	11,459	10,823	10,186	9,549	8,913	8,276	7,639	7,003	6,366	5,730	5,093	3
4	15,279	13,369	11,459	9,549	9,072	8,594	8,117	7,639	7,162	6,685	6,207	5,730	5,252	4,775	4,297	3,820	4
5	12,223	10,695	9,167	7,639	7,257	6,875	6,494	6,112	5,730	5,348	4,966	4,584	4,202	3,820	3,438	3,056	5
6	10,186	8,913	7,639	6,366	6,048	5,730	5,411	5,093	4,775	4,456	4,138	3,820	3,501	3,183	2,865	2,546	6
7	8,731	7,639	6,548	5,457	5,184	4,911	4,638	4,365	4,093	3,820	3,547	3,274	3,001	2,728	2,456	2,183	7
8	7,639	6,685	5,730	4,775	4,536	4,297	4,058	3,820	3,581	3,342	3,104	2,865	2,626	2,387	2,149	1,910	8
9	6,791	5,942	5,093	4,244	4,032	3,820	3,608	3,395	3,183	2,971	2,759	2,546	2,334	2,122	1,910	1,698	9
10	6,112	5,348	4,584	3,820	3,629	3,438	3,247	3,056	2,865	2,674	2,483	2,292	2,101	1,910	1,719	1,528	10
12	5,093	4,456	3,820	3,183	3,024	2,865	2,706	2,546	2,387	2,228	2,069	1,910	1,751	1,592	1,432	1,273	12
14	4,365	3,820	3,274	2,728	2,592	2,456	2,319	2,183	2,046	1,910	1,773	1,637	1,501	1,364	1,228	1,091	14
16	3,820	3,342	2,865	2,387	2,268	2,149	2,029	1,910	1,790	1,671	1,552	1,432	1,313	1,194	1,074	955	16
18	3,395	2,971	2,546	2,122	2,016	1,910	1,804	1,698	1,592	1,485	1,379	1,273	1,167	1,061	955	849	18
20	3,056	2,674	2,292	1,910	1,814	1,719	1,623	1,528	1,432	1,337	1,241	1,146	1,050	955	859	764	20
22	2,778	2,431	2,083	1,736	1,649	1,563	1,476	1,389	1,302	1,215	1,129	1,042	955	868	781	694	22
24	2,546	2,228	1,910	1,592	1,512	1,432	1,353	1,273	1,194	1,114	1,035	955	875	796	716	637	24
26	2,351	2,057	1,763	1,469	1,396	1,322	1,249	1,175	1,102	1,028	955	881	808	735	661	588	26
28	2,183	1,910	1,637	1,364	1,296	1,228	1,160	1,091	1,023	955	887	819	750	682	614	546	28
30	2,037	1,783	1,528	1,273	1,210	1,146	1,082	1,019	955	891	828	764	700	637	573	509	30
32	1,910	1,671	1,432	1,194	1,134	1,074	1,015	955	895	836	776	716	657	597	537	477	32
34	1,798	1,573	1,348	1,123	1,067	1,011	955	899	843	786	730	674	618	562	506	449	34
36	1,698	1,485	1,273	1,061	1,008	955	902	849	796	743	690	637	584	531	477	424	36
38	1,608	1,407	1,206	1,005	955	905	854	804	754	704	653	603	553	503	452	402	38
40	1,528	1,337	1,146	955	907	859	812	764	716	668	621	573	525	477	430	382	40
42	1,455	1,273	1,091	909	864	819	773	728	682	637	591	546	500	455	409	364	42
44	1,389	1,215	1,042	868	825	781	738	694	651	608	564	521	477	434	391	347	44
46	1,329	1,163	996	830	789	747	706	664	623	581	540	498	457	415	374	332	46
48	1,273	1,114	955	796	756	716	676	637	597	557	517	477	438	398	358	318	48
53	1,153	1,009	865	721	685	649	613	577	541	504	468	432	396	360	324	288	53
60	1,019	891	764	637	605	573	541	509	477	446	414	382	350	318	286	255	60
72	849	743	637	531	504	477	451	424	398	371	345	318	292	265	239	212	72

Cylindrical Grinding

Cylindrical grinding designates a general category of various grinding methods which have the common characteristic of rotating the workpiece around a fixed axis while grinding outside surface sections in controlled relation to that axis of rotation.

The form of the part or section being ground in this process is frequently cylindrical, hence the designation of the general category. However, the shape of the part may be tapered or of curvilinear profile; the position of the ground surface may also be perpendicular to the axis; and it is possible to grind concurrently several surface sections, adjacent or separated, of equal or different diameters, located in parallel or mutually inclined planes, etc., as long as the condition of a common axis of rotation is satisfied.

Size Range of Work Pieces and Machines: Cylindrical grinding is applied in the manufacture of miniature parts, such as instrument components and, at the opposite extreme, for grinding rolling mill rolls weighing several tons. Accordingly, there are cylindrical grinding machines of many different types, each adapted to a specific work size range. Machine capacities are usually expressed by such factors as maximum work diameter, work length and weight, complemented, of course, by many other significant data.

Plain, Universal and Limited-Purpose Cylindrical Grinding Machines. — The plain cylindrical grinding machine is considered the basic type of this general category, and is used for grinding parts with cylindrical or slightly tapered form.

The universal cylindrical grinder can be used, in addition to grinding the basic cylindrical forms, for the grinding of parts with steep tapers, of surfaces normal to the part axis, including the entire face of the workpiece, and also for internal grinding independently or in conjunction with the grinding of the part's outer surfaces. Such variety of part configurations requiring grinding is typical of work in the tool room, which constitutes the major area of application for universal cylindrical grinding machines.

Limited-purpose cylindrical grinders are needed for special work configurations and for high volume production, where productivity is more important than flexibility of adaptation. Examples of limited-purpose cylindrical grinding machines are crankshaft and camshaft grinders, polygonal grinding machines, roll grinders, etc.

Traverse or Plunge Grinding. — In traverse grinding the machine table carrying the work performs a reciprocating movement of specific travel length for transporting the rotating workpiece along the face of the grinding wheel. At each, or at alternate stroke ends, the wheel slide advances for the gradual feeding of the wheel into the work. The length of the surface which can be ground by this method is generally limited only by the stroke length of the machine table. In large roll grinders the relative movement between work and wheel is accomplished by the traverse of the wheel slide along a stationary machine table.

In plunge grinding the machine table, after having been set, is locked and, while the part is rotating, the wheel slide continually advances at a preset rate, until the finish size of the part is reached. The width of the grinding wheel is a limiting factor of the section length which can be ground in this process. Plunge grinding is required for profiled surfaces and for the simultaneous grinding of multiple surfaces of different diameters or located in different planes.

When the configuration of the part does not make the use of either method mandatory, the choice may be made on the basis of the following general considerations: traverse grinding usually produces a better finish, while the productivity of plunge grinding is generally higher.

Table 1. Work Holding Methods and Devices for Cylindrical Grinding

Designation	Description	Discussion
Centers, non-rotating ("dead"), with drive plate	Headstock with non-rotating spindle holds the center. Around the spindle an independently supported sleeve carries the drive plate for rotating the work. Tailstock for opposite center.	The simplest method of holding the work between two opposite centers is also the potentially most accurate, as long as correctly prepared and located centerholes are used in the work.
Centers, driving type	Work held between two centers obtains its rotation from the concurrently applied drive by the live headstock spindle and live tailstock spindle.	Eliminates the drawback of the common center-type grinding with driver plate which requires a dog attached to the workpiece. Driven spindles permit the grinding of the work up to both ends.
Chuck, geared or cam actuated.	Two, three or four jaws moved radially through mechanical elements, hand or power operated, exert concentrically acting clamping force.	Adaptable to workpieces of different configurations and within a generally wide capacity of the chuck. Flexible in uses which, however, do not include high precision work.
Chuck, diaphragm	Force applied by hand or power of a flexible diaphragm causes the attached jaws to deflect temporarily for accepting the work which is held when force is released.	Rapid action and flexible adaptation to different work configurations by means of special jaws offer varied uses for the grinding of disk shaped and similar parts.
Collets	Holding devices with externally or internally acting clamping force, easily adaptable to power actuation, assuring high centering accuracy.	Limited to parts with previously machined or ground holding surfaces, because of the small range of clamping movement of the collet jaws.
Face plate	Has four independently actuated jaws, any or several of which may be used, or entirely removed, using the base plate for supporting special clamps.	Used for holding bulky parts, or those of awkward shape, which are ground in small quantities not warranting special fixtures.
Magnetic plate	Flat plates, with pole distribution adapted to the work, are mounted on the spindle like chucks and may be used for work with locating face normal to the axis.	Applicable for light cuts such as are frequent in tool making, where the rapid clamping action, and easy access to both the O.D. and the exposed face, are sometimes of advantage.
Steady rests	Two basic types are used: (a) The two-jaw type supporting the work from the back (back rest), leaving access by the wheel; (b) The three-jaw type (center-rest).	A complementary work-holding device, used in conjunction with primary work holders, to provide additional support, particularly to long and/or slender parts.
Special fixtures	Single-purpose devices, designed for a particular workpiece, primarily for providing special locating elements.	Typical workpieces requiring special fixturing are, as examples: crankshafts where the holding is combined with balancing functions; internal gears located on the pitch circle of the teeth for O.D. grinding.

Work Holding on Cylindrical Grinding Machines. — The manner in which the work is located and held in the machine during the grinding process determines the configuration of the part which can be adapted for cylindrical grinding and affects the resulting accuracy of the ground surface. The method of work holding also affects the attainable production rate, because the mounting and dismounting of the part can represent a substantial portion of the total operational time.

Whatever method is used for holding the part on cylindrical types of grinding machines, two basic conditions must be satisfied: (1) The part should be located with respect to its correct axis of rotation, and (2) The work drive must cause the part to rotate, at a specific speed, around the established axis. The lengthwise location of the part, although controlled, is not too critical in traverse grinding; however, in plunge grinding, particularly when shoulder sections are also involved, it must be assured with great accuracy.

Table 1 presents a listing, with brief discussions, of work holding methods and devices which are the most frequently used in cylindrical grinding.

Table 2. Wheel Recommendations for Cylindrical Grinding

Material	Wheel Marking	Material	Wheel Marking
Aluminum	SFA46-18V	Forgings	A46-M5V
Armatures (laminated)	SFA100-18V	Gages (plug)	SFA80-K8V
Axles (auto & railway)	A54-M5V	General purpose grinding	SFA54-L5V
Brass	C36-KV	Glass	BFA220-011V
Bronze		Gun barrels	
Soft	C36-KV	Spotting and O.D.	BFA60-M5V
Hard	A46-M5V	Nitralloy	
Bushings (hardened steel)	BFA60-L5V	Before nitriding	A60-K5V
Bushings (cast iron)	C36-JV	After nitriding	
Cam lobes (cast alloy)		Commercial finish	SFA60-18V
Roughing	BFA54-N5V	High finish	C100-1V
Finishing	A70-P6B	Reflective finish	C500-19E
Cam lobes (hardened steel)		Pistons (aluminum)	SFA46-18V
Roughing	BFA54-L5V	(cast iron)	C36-KV
Finishing	BFA80-T8B	Plastics	C46-JV
Cast iron	C36-JV	Rubber	
Chromium plating		Soft	SFA20-K5B
Commercial finish	SFA60-J8V	Hard	C36-KB
High finish	A150-K5E	Spline shafts	SFA60-N5V
Reflective finish	C500-19E	Sprayed metal	C60-JV
Commutators (copper)	C60-M4E	Steel	
Crankshafts (airplane)		Soft	
Pins	BFA46-K5V	1″ dia. and smaller	SFA60-M5V
Bearings	A46-L5V	over 1″ dia.	SFA46-L5V
Crankshafts (automotive		Hardened	
pins and bearings)		1″ dia. and smaller	SFA80-L8V
Finishing	A54-N5V	over 1″ dia.	SFA60-K5V
Roughing & finishing	A54-O5V	300 series stainless	SFA46-K8V
Regrinding	A54-M5V	Stellite	BFA46-M5V
Regrinding, sprayed		Titanium	C60-JV
metal	C60-JV	Valve stems (automotive)	BFA54-N5V
Drills	BFA54-N5V	Valve tappets	BFA54-M5V

Note: Prefixes to the standard designation "A" of aluminum oxide indicate modified abrasives:

    BFA = Blended Friable (a blend of regular and friable)

    SFA = Semi-Friable

Selection of Grinding Wheels for Cylindrical Grinding. — For cylindrical grinding, as for grinding in general, the primary factor to be considered in wheel selection is the work material. Other factors are the amount of excess stock and its rate of removal (speeds and feeds), the desired accuracy and surface finish, the ratio of wheel and work diameter, wet or dry grinding, etc. In view of these many variables, it is not practical to set up a complete list of grinding wheel recommendations with general validity. Instead, examples of recommendations embracing a wide range of typical applications and assuming common practices are presented in Table 2. This is intended as a guide for the starting selection of grinding-wheel specifications which, in case of a not entirely satisfactory performance, can be refined subsequently. The content of the table is a version of the grinding-wheel recommendations for cylindrical grinding by the Norton Company using, however, non-proprietary designations for the abrasive types and bonds.

Operational Data for Cylindrical Grinding. — In cylindrical grinding, similarly to other metalcutting processes, the applied speed and feed rates must be adjusted to the operational conditions as well as to the objectives of the process. Grinding differs, however, from other types of metalcutting methods in regard to the cutting speed of the tool which, in grinding, is generally not a variable; it should be maintained at, or close to the optimum rate, commonly 6500 feet per minute peripheral speed.

In establishing the proper process values for grinding, of prime consideration are the work material, its condition (hardened or soft), and the type of operation (roughing or finishing). Other influencing factors are the characteristics of the grinding machine (stability, power), the specifications of the grinding wheel, the material allowance, the rigidity and balance of the workpiece, as well as several grinding process conditions, such as wet or dry grinding, the manner of wheel truing, etc.

Variables of the cylindrical grinding process, often referred to as *grinding data*, comprise the speed of work rotation (measured as the surface speed of the work); the infeed (in inches per pass for traverse grinding, or in inches per minute for plunge grinding); and, in the case of traverse grinding, the speed of the reciprocating table movement (expressed either in feet per minute, or as a fraction of the wheel width for each revolution of the work).

For the purpose of starting values in setting up a cylindrical grinding process, a brief listing of basic data for common cylindrical grinding conditions and involving frequently used materials, is presented in Table 3.

These data, which are, in general, considered conservative, are based on average operating conditions and may be modified subsequently,

— reducing the values in case of unsatisfactory quality of the grinding or the occurrence of failures;
— increasing the rates for raising the productivity of the process, particularly for rigid workpieces, substantial stock allowance, etc.

High Speed Cylindrical Grinding. — The maximum peripheral speed of the wheels in regular cylindrical grinding is generally 6500 feet per minute; the commonly used grinding wheels and machines are designed to operate efficiently at this speed. Recently, efforts were made to raise the productivity of different grinding methods, including cylindrical grinding, by increasing the peripheral speed of the grinding wheel to a substantially higher than traditional level, such as 12,000 feet per minute or more. Such methods are designated by the distinguishing term of high speed grinding.

Table 3.   Basic Process Data For Cylindrical Grinding

Work Material	Material Condition	Work Surface Speed, fpm	Infeed, Inch/Pass		Traverse for Each Work Revolution, In Fractions of The Wheel Width	
			Roughing	Finishing	Roughing	Finishing
TRAVERSE GRINDING						
Plain Carbon Steel	Annealed	100	.002	.0005	1/2	1/6
	Hardened	70	.002	.0003 to .0005	1/4	1/8
Alloy Steel	Annealed	100	.002	.0005	1/2	1/6
	Hardened	70	.002	.0002 to .0005	1/4	1/8
Tool Steel	Annealed	60	.002	.0005 max.	1/2	1/6
	Hardened	50	.002	.0001 to .0005	1/4	1/8
Copper Alloys	Annealed or Cold Drawn	100	.002	.0005 max.	1/3	1/6
Aluminum Alloys	Cold Drawn or Solution Treated	150	.002	.0005 max.	1/3	1/6

Work Material	Infeed Per Revolution of The Work, Inch	
	Roughing	Finishing
PLUNGE GRINDING		
Steel, soft	0.0005	0.0002
Plain carbon steel, hardened	0.0002	0.000050
Alloy and tool steel, hardened	0.0001	0.000025

For high speed grinding, special grinding machines have been built with high dynamic stiffness and static rigidity, equipped with powerful drive motors, extra-strong spindles and bearings, reinforced wheel guards, etc., and using grinding wheels expressly made and tested for operating at high peripheral speeds. The higher stock-removal rate accomplished by high speed grinding represents an advantage when the work configuration and material permit, and the removable stock allowance warrants its application.

CAUTION: High speed grinding must *not* be applied on standard types of equipment, such as general models of grinding machines and regular grinding wheels. Operating grinding wheels, even temporarily, at higher than approved speed constitutes a grave safety hazard.

Areas and Degrees of Automation in Cylindrical Grinding. — Power drive for the work rotation and for the reciprocating table traverse are fundamental machine movements which, once set for a certain rate, will function without requiring additional attention. Loading and removing the work, starting and stopping the main movements, and applying infeed by hand wheel, are carried out by the operator on cylindrical grinding machines in their basic degree of mechanization. Such equipment is still frequently used in tool room and jobbing type work.

More advanced levels of automation can be developed for cylindrical grinders and are being applied in different degrees, particularly in the following principal respects:

a. *Infeed*, in which different rates are provided for rapid approach, roughing and finishing, followed by a spark-out period, with presetting of the advance rates, the cutoff points, and the duration of time-related functions.

b. *Automatic cycling* actuated by a single lever to start work rotation, table reciprocation, grinding-fluid supply and infeed, followed at the end of the operation by wheel slide retraction, the successive stopping of the table movement, the work rotation and the fluid supply.

c. *Table traverse dwells* (tarry) in the extreme positions of the travel, over preset periods, to assure uniform exposure to the wheel contact of the entire work section.

d. *Mechanized work loading*, clamping and, after termination of the operation, unloading, combined with appropriate work-feeding devices such as indexing type drums.

e. *Size control* by in-process or post-process measurements. Signals originated by the gage will control the advance movement or cause automatic compensation of size variations by adjusting the cutoff points of the infeed.

f. *Automatic wheel dress-off* at preset frequency, combined with appropriate compensation in the infeed movement.

g. *Numerical control*, programmed on punched cards or tape, obviates the time-consuming setups for repetitive work which is carried out in small- or medium-size lots. As an application example: shafts with several sections of different lengths and diameters can be ground automatically in a single operation, grinding the sections in consecutive order to close dimensional limits, controlled by an in-process gage, which is also automatically set by means of the program card.

The choice of the grinding machine functions to be automated and the extent of automation will generally be guided by economic considerations, after a thorough review of the available standard and optional equipment.

Cylindrical Grinding Troubles and Their Correction. — Troubles that may be encountered in cylindrical grinding may be classified as work defects (chatter, checking, burning, scratching, and inaccuracies), improperly operating machines (jumpy in-feed or traverse), and wheel defects (too hard or soft action, loading, glazing, and breakage). The Landis Tool Company lists some of these troubles, their causes, and corrections as follows:

Chatter. — Sources of chatter include: (1) Faulty coolant, (2) wheel out of balance, (3) wheel out of round, (4) wheel too hard, (5) improper dressing, (6) faulty work support or rotation, (7) improper operation, (8) faulty traverse, (9) work vibration, (10) outside vibration transmitted to machine, (11) interference, (12) wheel base, and (13) headstock. Suggested procedures for correction of these troubles are:

(*1*) *Faulty coolant:* Clean tanks and lines. Replace dirty or heavy coolant with correct mixture.

(*2*) *Out-of-balance wheel:* Rebalance on mounting before and after dressing. Run wheel without coolant to remove excess water. Store a removed wheel on its side to keep retained water from causing a false heavy side. Tighten wheel mounting flanges. Make sure wheel center fits spindle.

(*3*) *Wheel out of round:* True before and after balancing. True sides to face.

(*4*) *Wheel too hard:* Use coarser grit, softer grade, more open bond. See "Wheel Defects."

(5) *Improper dressing:* Use sharp diamond and hold rigidly close to wheel. It must not overhang too far. Check diamond in mounting.

(6) *Faulty work support or rotation:* Use sufficient number of work rests and adjust them more carefully. Use proper angles in centers of work. Clean dirt from footstock spindle and be sure spindle is tight. Make certain that work centers fit properly in spindles.

(7) *Improper operation:* Reduce rate of wheel feed.

(8) *Faulty traverse:* See "Uneven Traverse or In-feed of Wheel Head."

(9) *Work vibration:* Reduce work speed. Check workpiece for balance.

(10) *Outside vibration transmitted to machine:* Check and make sure that machine is level and sitting solidly on foundation. Isolate machine or foundation.

(11) *Interference:* Check all guards for clearance.

(12) *Wheel base:* Check spindle bearing clearance. Use belts of equal lengths or uniform cross-section on motor drive. Check drive motor for unbalance. Check balance and fit of pulleys. Check wheel feed mechanism to see that all parts are tight.

(13) *Headstock:* Put belts of same length and cross-section on motor drive. Incorrect work speeds. Check drive motor for unbalance. Make certain that headstock spindle is not loose. Check work center fit in spindle. Check wear of face plate and jackshaft bearings.

Spirals on Work (traverse lines with same lead on work as rate of traverse). — Sources of spirals include: (1) Machine parts out of line, and (2) truing. Suggested procedures for correction of these troubles are:

(1) *Machine parts out of line:* Check wheel base, headstock, and footstock for proper alignment.

(2) *Truing:* Point truing tool down 3 degrees at the workwheel contact line. Make edges of wheel face round.

Check Marks on Work. — Sources of check marks include: (1) Improper operation, (2) improper heat treatment, (3) improper size control, (4) improper wheel, and (5) improper dressing. Suggested procedures for correction of these troubles are:

(1) *Improper operation:* Make wheel act softer. See "Wheel Defects." Do not force wheel into work. Use greater volume of coolant and a more even flow. Affirm the correct positioning of coolant nozzles to direct a copious flow of clean coolant at the proper location.

(2) *Improper heat treatment:* Take corrective measures in heat treating operations.

(3) *Improper size control:* Make sure that engineering establishes reasonable size limits. See that they are maintained.

(4) *Improper wheel:* Make wheel act softer. Use softer grade wheel. Review the grain size and type of abrasive. A finer grit or more friable abrasive or both may be called for.

(5) *Improper dressing:* Make sure you have a sharp, good quality diamond that is well set. Increase speed of the dressing cycle. Make sure diamond is not cracked.

Burning and Discoloration of Work. — Sources of burning and discoloration are: (1) Improper operation, and (2) improper wheel. Suggested procedures for correction of these troubles are:

(1) *Improper operation:* Decrease rate of in-feed. Don't stop work while in contact with wheel.

(2) *Improper wheel:* Use softer wheel or obtain softer effect. See "Wheel Defects." Use greater volume of coolant.

Isolated Deep Marks on Work. — Source of trouble is improper wheel. Use finer wheel and consider a change in abrasive type.

Fine Spiral or Thread on Work. — Sources of this trouble are: (1) Improper operation and (2) faulty wheel dressing. Suggested procedures for corrections of these troubles are:

(1) *Improper operation:* Reduce wheel pressure. Use more work rests. Reduce traverse with respect to work rotation. Use different traverse rates to break up pattern when making numerous passes. Keep edge of wheel from penetrating by dressing wheel face parallel to work.

(2) *Faulty wheel dressing:* Use slower or more even dressing traverse. Set dressing tool at least 3 degrees down and 30 degrees to the side from time to time. Tighten holder. Don't take too deep a cut. Round off wheel edges. Start dressing cut from wheel edge.

Narrow and Deep Regular Marks on Work. — Source of trouble is wheel too coarse. Use finer grain size.

Wide, Irregular Marks of Varying Depth on Work. — Source of trouble is wheel too soft. Use harder grade wheel. See "Wheel Defects."

Widely Spaced Spots on Work. — Source of trouble is oil spots or glazed areas on wheel face. Balance and true wheel. Keep oil from wheel face.

Irregular "Fish-tail" Marks of Various Lengths and Widths on Work. — Source of trouble is dirty coolant. Clean tank frequently. Use filter for fine finish grinding. Flush wheel guards after dressing or when changing to finer wheel.

Wavy Traverse Lines on Work. — Source of trouble is wheel edges. Round off. Check for loose thrust on spindle and correct if necessary.

Irregular Marks on Work. — Cause is loose dirt. Keep machine clean.

Deep, Irregular Marks on Work. — Source of trouble is loose wheel flanges. Tighten and make sure blotters are used.

Isolated Deep Marks on Work. — Source of trouble is: (1) Grains pull out, coolant too strong; (2) coarse grains or foreign matter in wheel face; and (3) improper dressing. Respective suggested procedures for corrections of these troubles are: (1) Decrease soda content in coolant mixture, (2) dress out, and (3) use sharper dressing tool. Brush wheel after dressing with stiff bristle brush.

Grain Marks on Work. — Source of trouble is: (1) Improper finishing cut; (2) grain sizes of roughing and finishing wheels differ too much; (3) dressing too coarse; and (4) wheel too coarse or too soft. Respective suggested procedures for corrections of these troubles are: (1) Start with high work and traverse speeds; finish with high work speed and slow traverse, letting wheel "spark-out" completely; (2) finish out better with roughing wheel or use finer roughing wheel; (3) use shallower and slower cut; and (4) use finer grain size or harder grade wheel.

Inaccuracies in Work. — Work out-of-round, out-of-parallel, or tapered. Source of trouble is: (1) Misalignment of machine parts, (2) work centers, (3) improper operation, (4) coolant, (5) wheel, (6) improper dressing, (7) spindle bearings, and (8) work. Suggested procedures for corrections of these troubles are:

(1) *Misalignment of machine parts:* Check headstock and tailstock for alignment and proper clamping.

(2) *Work centers:* Centers in work must be deep enough to clear center point. Keep work centers clean and lubricated. Check play of footstock spindle and see that footstock spindle is clean and tightly seated. Regrind work center if worn. Work centers must fit taper of work center holes. Footstock must be checked for proper tension.

(3) *Improper operation:* Don't let wheel traverse beyond end of work. Decrease wheel pressure so work won't spring. Use harder wheel or change feeds and speeds to make wheel act harder. Allow work to "spark-out." Decrease feed rate. Use proper number of work rests. Allow proper amount of tarry. Workpiece must be balanced if odd shape.

(4) *Coolant:* Use greater volume of coolant.

(5) *Wheel:* Rebalance wheel on mounting before and after truing.

(6) *Improper dressing:* Use same positions and machine conditions for dressing as in grinding.

(7) *Spindle bearings:* Check clearance.

(8) *Work:* Work must come to machine in reasonably accurate form.

Inaccurate Work Sizing (when wheel is fed to same position, it grinds one piece to correct size, another oversize, and still another undersize). — Sources of trouble are: (1) Improper work support or rotation, (2) wheel out of balance, (3) loaded wheel, (4) improper infeed, (5) improper traverse, (6) coolant, (7) misalignment, and (8) work. Suggested procedures for corrections of these troubles are:

(1) *Improper work support or rotation:* Keep work centers clean and lubricated. Regrind work center tips to proper angle. Be sure footstock spindle is tight. Use sufficient work rests properly spaced.

(2) *Wheel out of balance:* Balance wheel on mounting before and after truing.

(3) *Loaded wheel:* See "Wheel Defects."

(4) *Improper infeed:* Check forward stops of rapid feed and slow feed. When readjusting position of wheel base by means of the fine feed, move the wheel base back after making the adjustment and then bring it forward again to take up backlash and relieve strain in feed-up parts. Check wheel spindle bearings. Don't let excessive lubrication of wheel base slide cause "floating." Check and tighten wheel feed mechanism. Check parts for wear. Check pressure in hydraulic system. Set in-feed cushion properly. Check pistons to see that they are not sticking.

(5) *Improper traverse:* Check traverse hydraulic system and the operating pressure. Prevent excessive lubrication of carriage ways with resultant "floating" condition. Check to see if carriage traverse piston rods are binding. Carriage rack and driving gear must not bind. Change length of tarry period.

(6) *Coolant:* Use greater volume of clean coolant.

(7) *Misalignment:* Check level and alignment of machine.

(8) *Work:* Work pieces may vary too much in length permitting uneven center pressure.

Uneven Traverse or In-feed of Wheel Head. — Sources of uneven traverse or in-feed of wheel head are: (1) Carriage and wheel head, (2) hydraulic system, (3) interference, (4) unbalanced conditions, and (5) wheel out of balance. Suggested procedures for correction of these troubles are:

(1) *Carriage and wheel head:* Ways may be scored. Be sure to use recommended oil for both lubrication and hydraulic system. Make sure ways are not too smooth that they press out oil film. Check lubrication of ways. Check wheel feed mechanism, traverse gear and carriage rack clearance. Prevent binding of carriage traverse cylinder rods.

(2) *Hydraulic systems:* Remove air and check pressure of hydraulic oil. Check pistons and valves for oil leakage and for gumminess caused by incorrect oil. Check worn valves or pistons that permit leakage.

(3) *Interference:* Make sure guard strips do not interfere.

(4) *Unbalanced conditions:* Eliminate loose pulleys, unbalanced wheel drive motor, uneven belts, or high spindle keys.

(5) *Wheel out of balance:* Balance wheel on mounting before and after truing.

Wheel Defects. — When *wheel is acting too hard*, such defects as glazing, some loading, lack of cut, chatter, and burning of work result. Suggested procedures for correction of these faults are: (1) Increase work and traverse speeds as well as rate of in-feed; (2) decrease wheel speed, diameter, or width; (3) dress more sharply; (4) use thinner coolant; (5) don't tarry at end of traverse; (6) select softer wheel grade and coarser grain size; (7) avoid gummy coolant; and (8) on hardened work select finer grit, more fragile abrasive or both to get penetration. Use softer grade.

When *wheel is acting too soft*, such defects as wheel marks, tapered work, short wheel life, and not-holding-cut result. Suggested procedures for correction of these faults are: (1) Decrease work and traverse speeds as well as rate of in-feed; (2) increase wheel speed, diameter, or width; (3) dress with little in-feed and slow traverse; (4) user heavier coolants; (5) don't let wheel run off work at end of traverse; and (6) select harder wheel or less fragile grain or both.

Wheel Loading and Glazing. — Sources of the trouble of wheel loading or glazing are: (1) Incorrect wheel, (2) improper dress, (3) faulty operation, (4) faulty coolant, and (5) gummy coolant. Suggested procedures for correction of these faults are:

(*1*) *Incorrect wheel:* Use coarser grain size, more open bond, or softer grade.

(*2*) *Improper dressing:* Keep wheel sharp with sharp dresser, clean wheel after dressing, use faster dressing traverse, and deeper dressing cut.

(*3*) *Faulty operation:* Control speeds and feeds to soften action of wheel. Use less in-feed to prevent loading; more in-feed to stop glazing.

(*4*) *Faulty coolant:* Use more, cleaner and thinner coolant, and less oily coolant.

(*5*) *Gummy coolant:* To stop wheel glazing, increase soda content and avoid the use of soluble oils if water is hard. In using soluble oil coolant with hard water a suitable conditioner or "softener" should be added.

Wheel Breakage. — Suggested procedures for the correction of a radial break with three or more pieces are: (1) Reduce wheel speed to or below rated speed; (2) mount wheel properly, use blotters, tight arbors, even flange pressure and be sure to keep out dirt between flange and wheel; (3) use plenty of coolant to prevent over-heating; (4) use less in-feed; and (5) don't allow wheel to become jammed on work.

A radial break with two pieces may be caused by excessive side strain. To prevent an irregular wheel break, don't let wheel become jammed on work; don't allow striking of wheel; and never use wheels that have been damaged in handling. In general, do not use a wheel that is too tight on the arbor since the wheel is apt to break when started. Prevent excessive hammering action of wheel. Follow rules of the American National Standard Safety Requirements for the Use, Care, and Protection of Abrasive Wheels (ANSI B7.1-1978).

Centerless Grinding

In centerless grinding the work is supported on a work rest blade and is between the grinding wheel and a regulating wheel. The regulating wheel generally is a rubber bonded abrasive wheel. In the normal grinding position the grinding wheel forces the work downward against the work rest blade and also against the regulating wheel. The latter imparts a uniform rotation to the work giving it its same peripheral speed which is adjustable.

The higher the work center is placed above the line joining the centers of the grinding and regulating wheels the quicker the rounding action. Rounding action is also increased by a high work speed and a slow rate of traverse (if a through-feed operation). It is possible to have a higher work center when using softer wheels,

as their use gives decreased contact pressures and the tendency of the workpiece to lift off the work rest blade is lessened.

Long rods or bars are sometimes ground with their centers below the line-of-centers of the wheels to eliminate the whipping and chattering due to slight bends or kinks in the rods or bars, as they are held more firmly down on the blade by the wheels.

There are three general methods of centerless grinding which may be described as through-feed, in-feed, and end-feed methods.

Through-feed Method of Grinding. — The through-feed method is applied to straight cylindrical parts. The work is given an axial movement by the regulating wheel and passes between the grinding and regulating wheels from one side to the other. The rate of feed depends upon the diameter and speed of the regulating wheel and its inclination which is adjustable. It may be necessary to pass the work between the wheels more than once, the number of passes depending upon such factors as the amount of stock to be removed, the roundness and straightness of the unground work, and the limits of accuracy required.

The work rest fixture also contains adjustable guides on either side of the wheels that directs the work to and from the wheels in a straight line.

In-feed Method of Centerless Grinding. — When parts have shoulders, heads or some part larger than the ground diameter, the in-feed method usually is employed. This method is similar to "plungecut" form grinding on a center type of grinder. The length or sections to be ground in any one operation are limited by the width of the wheel. As there is no axial feeding movement, the regulating wheel is set with its axis approximately parallel to that of the grinding wheel, there being a slight inclination to keep the work tight against the end stop.

End-feed Method of Grinding. — The end-feed method is applied only to taper work. The grinding wheel, regulating wheel, and the work rest blade are set in a fixed relation to each other and the work is fed in from the front mechanically or manually to a fixed end stop. Either the grinding or regulating wheel, or both, are dressed to the proper taper.

Automatic Centerless Grinding. — The grinding of relatively small parts may be done automatically by equipping the machine with a magazine, gravity chute, or hopper feed, provided the shape of the part will permit using these feed mechanisms.

Internal Centerless Grinding. — Internal grinding machines based upon the centerless principle utilize the outside diameter of the work as a guide for grinding the bore which is concentric with the outer surface. In addition to straight and tapered bores, interrupted and "blind" holes can be ground by the centerless method. When two or more grinding operations such as roughing and finishing must be performed on the same part, the work can be rechucked in the same location as often as required.

Centerless Grinding Troubles. — A number of troubles and some corrective measures compiled by a manufacturer are listed here for the through-feed and in-feed methods of centerless grinding.

Chattermarks are caused by having the work center too high above the line joining the centers of the grinding and regulating wheels; using too hard or too fine a grinding wheel; using too steep an angle on the work support blade; using too thin a work support blade; "play" in the set-up due to loosely clamped members; having the grinding wheel fit loosely on the spindle; having vibration either transmitted to the

machine or caused by a defective drive in the machine; having the grinding wheel out-of-balance; using too heavy a stock removal; and having the grinding wheel or the regulating wheel spindles not properly adjusted.

Feed lines or spiral marks in through-feed grinding are caused by too sharp a corner on the exit side of the grinding wheel which may be alleviated by dressing the grinding wheel to a slight taper about ½ inch from the edge, dressing the edge to a slight radius, or swiveling the regulating wheel a bit.

Scored work is caused by burrs, abrasive grains, or removed material being imbedded in or fused to the work support blade. This condition may be alleviated by using a coolant with increased lubricating properties and if this does not help a softer grade wheel should be used.

Work not ground round may be due to the work center not being high enough above the line joining the centers of the grinding and regulating wheels. Placing the work center higher and using a softer grade wheel should help to alleviate this condition.

Work not ground straight in through-feed grinding may be due to an incorrect setting of the guides used in introducing and removing the work from the wheels, and the existence of convex or concave faces on the regulating wheel. For example if the work is tapered on the front end, the work guide on the entering side is deflected toward the regulating wheel. If tapered on the back end, then the work guide on the exit side is deflected toward the regulating wheel. If both ends are tapered, then both work guides are deflected toward the regulating wheel. The same barrel-shaped pieces are also obtained if the face of the regulating wheel is convex at the line of contact with the work. Conversely the work would be ground with hollow shapes if the work guides were deflected toward the grinding wheel or if the face of the regulating wheel were concave at the line of contact with the work. The use of a warped work rest blade may also result in the work not being ground straight and the blade should be removed and checked with a straight edge.

In in-feed grinding, in order to keep the wheel faces straight which will insure straightness of the cylindrical pieces being ground, the first item to be checked is the straightness and the angle of inclination of the work rest blade. If this is satisfactory then one of three corrective measures may be taken: the first might be to swivel the regulating wheel to compensate for the taper, the second might be to true the grinding wheel to that angle that will give a perfectly straight workpiece and the third might be to change the inclination of the regulating wheel (this is true only for correcting very slight tapers up to 0.0005 inch).

Difficulties in sizing the work in in-feed grinding are generally due to a worn in-feed mechanism and may be overcome by adjusting the in-feed nut.

Flat spots on the workpiece in in-feed grinding usually occur when grinding heavy work and generally when the stock removal is light. This condition is due to insufficient driving power between the work and the regulating wheel which may be alleviated by equipping the work rest with a roller that exerts a force against the workpiece; and by feeding the workpiece to the end stop using the upper slide.

Surface Grinding

The term surface grinding implies, in current technical usage, the grinding of surfaces which are essentially flat. Several methods of surface grinding, however, are adapted and used to produce surfaces characterized by parallel straight line elements in one direction, while normal to that direction the contour of the surface may consist of several straight line sections at different angles to each other (e.g., the guideways of a lathe bed); in other cases the contour may be curved or profiled (e.g., a thread cutting chaser).

Advantages of Surface Grinding. — Alternate methods for machining work surfaces similar to those produced by surface grinding are milling and, to a much more limited degree, planing. Surface grinding, however, has several advantages over alternate methods which are carried out with metal-cutting tools. Examples of such potential advantages are:

(1) Grinding is applicable to very hard and/or abrasive work materials, without significant effect on the efficiency of the stock removal.

(2) The desired form and dimensional accuracy of the work surface can be obtained to a much higher degree and in a more consistent manner.

(3) Surface textures of very high finish and—when the appropriate system is utilized—also with the required lay, are generally produced.

(4) Tooling for surface grinding is, as a rule, substantially less expensive, particularly for producing profiled surfaces, the shapes of which may be dressed into the wheel, often with simple devices, in processes which are much more economical than the making and the maintenance of form cutters.

(5) Fixturing for work holding is generally very simple in surface grinding, particularly when magnetic chucks are applicable, although the mechanical holding fixture can also be simpler, because of the lesser clamping force required than in milling or planing.

(6) Parallel surfaces on opposite sides of the work are produced accurately, either in consecutive operations using the first ground surface as a dependable reference plane, or, simultaneously, in double face grinding which usually operates without the need for holding the parts by clamping.

(7) Surface grinding is well adapted to process automation, particularly for size control, but also for mechanized work handling in the large volume production of a wide range of component parts.

Principal Systems of Surface Grinding. — Flat surfaces can be ground with different surface portions of the wheel, by different arrangements of the work and wheel, as well as by different interrelated movements. The various systems of surface grinding, with their respective capabilities, can best be reviewed by considering two major distinguishing characteristics:

(1) The operating surface of the grinding wheel, which may be the periphery or the face (the side);

(2) The movement of the work during the process, which may be traverse (generally reciprocating) or rotary (continuous), depending on the design of a particular category of surface grinders.

The accompanying table provides a concise review of the principal surface grinding systems, defined by the preceding characteristics. It should be noted that there are surface grinders built for specific applications, which do not fit exactly into any one of these major categories.

Selection of Grinding Wheels for Surface Grinding. — The most practical way to select a grinding wheel for surface grinding is to base the selection on the work material. Table 1a gives the grinding wheel recommendations for Types 1, 5, and 7 straight wheels used on reciprocating and rotary table surface grinders with horizontal spindles. Table 1b gives the grinding wheel recommendations for Type 2 cylinder wheels, Type 6 cup wheels, and segments used on vertical spindle surface grinders.

The wheel markings of the tables are those used by the Norton Company, complementing the basic standard markings with Norton symbols. The complementary symbols used in these tables, that is, those preceding the letter designating A (aluminum oxide) or C (silicon carbide) indicate the special type of basic abrasive which has the friability best suited for particular work materials. Those preceding A (aluminum oxide) are:

Table 1a. Grinding Wheel Recommendations for Surface Grinding—Using Straight Wheel Types 1, 5, and 7

Horizontal-spindle, reciprocating-table surface grinders		
Material	Wheels less than 16 inches in diameter	Wheels 16 inches in diameter and over
Cast iron	37C36-K8V or 23A46-I8VBE	23A36-I8VBE
Non-ferrous metals	37C36-K8V	37C36-K8V
Soft steel	23A46-J8VBE	23A36-J8VBE
Hardened steel— broad contact	32A46-H8VBE or 32A60-F12VBEP	32A36-H8VBE or 32A36-F12VBEP
Hardened steel— narrow contact or interrupted cut	32A46-I8VBE	32A36-J8VBE
General purpose wheel	23A46-H8VBE	23A36-I8VBE
Cemented carbides	Diamond wheels*	Diamond wheels*

Horizontal-spindle, rotary-table surface grinders	
Material	Wheels of any diameter
Cast iron	37C36-K8V or 23A46-I8VBE
Non-ferrous metals	37C36-K8V
Soft steel	23A46-J8VBE
Hardened steel—broad contact	32A46-I8VBE
Hardened steel—narrow contact or interrupted cut	32A46-J8VBE
General purpose wheel	23A46-I8VBE
Cemented carbides—roughing	Diamond wheels*

Courtesy of Norton Company

* General diamond wheel recommendations are listed in Table 5, page 1067.

57—a versatile abrasive suitable for grinding steel in either a hard or soft state,
38—the most friable abrasive,
32—the abrasive suited for tool steel grinding,
23—an abrasive with intermediate grinding action, and
19—the abrasive produced for less heat-sensitive steels.
Those preceding C (silicon carbide) are:
37—a general application abrasive, and
39—an abrasive for grinding hard cemented carbide.

Table 1b. Grinding Wheel Recommendations for Surface Grinding—Using Type 2 Cylinder Wheels, Type 6 Cup Wheels, and Segments

Material	Type 2 Cylinder Wheels	Type 6 Cup Wheels	Segments
High tensile cast iron and non-ferrous metals	37C24-HKV	37C24-HVK	37C24-HVK
Soft steel, malleable cast iron, steel castings, boiler plate	23A24-18VBE or 23A30-G12VBEP	23A24-I8VBE	23A24-I8VSM or 23A30-H12VSM
Hardened steel— broad contact	32A46-G8VBE or 32A36-E12VBEP	32A46-G8VBE or 32A60-E12VBEP	32A36-G8VBE or 32A46-E12VBEP
Hardened steel— narrow contact or interrupted cut	32A46-H8VBE	32A60-H8VBE	32A46-G8VBE or 32A60-G12VBEP
General purpose use	23A30-H8VBE or 23A30-E12VBEP	....	23A30-H8VSM or 23A30-G12VSM

Courtesy of Norton Company

Principal Systems of Surface Grinding — Diagrams

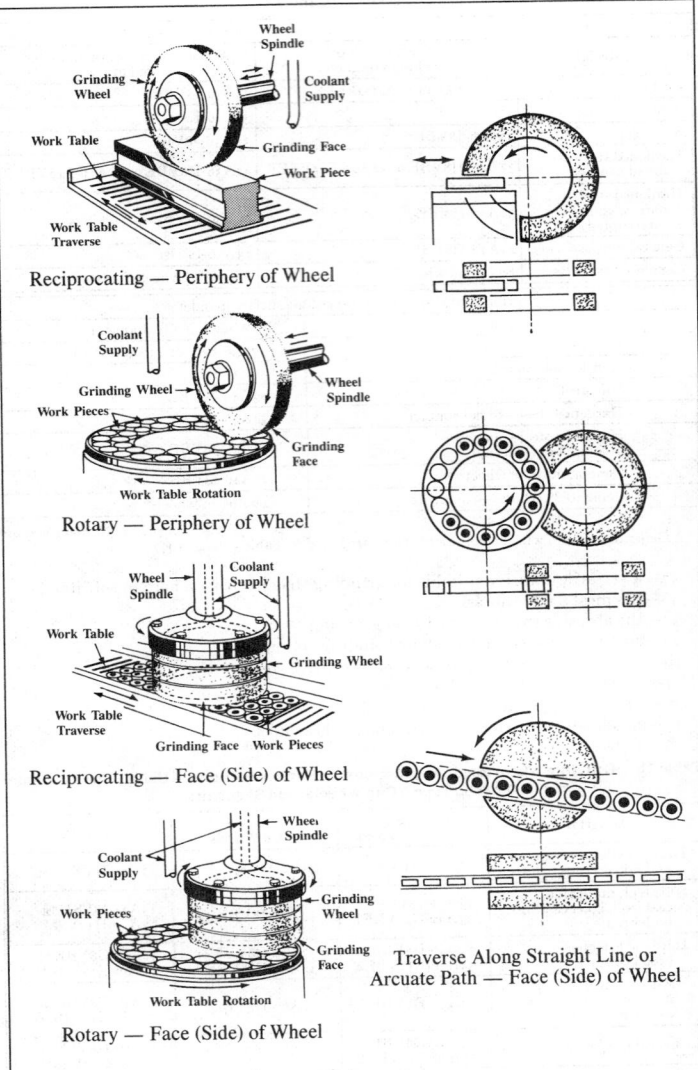

Reciprocating — Periphery of Wheel

Rotary — Periphery of Wheel

Reciprocating — Face (Side) of Wheel

Rotary — Face (Side) of Wheel

Traverse Along Straight Line or
Arcuate Path — Face (Side) of Wheel

Principal Systems of Surface Grinding—Principles of Operation

Effective Grinding Surface—Periphery of Wheel Movement of Work—Reciprocating
Work is mounted on the horizontal machine table which then traverses in a reciprocating movement at a speed generally selected from a steplessly variable range. The transverse movement, called cross feed of the table or of the wheel slide, operates at the end of the reciprocating stroke and assures the gradual exposure of the entire work surface which commonly exceeds the width of the wheel. The depth of the cut is controlled by the downfeed of the wheel, applied in increments at the reversal of the transverse movement.
Effective Grinding Surface—Periphery of Wheel Movement of Work—Rotary
Work is mounted, usually on the full diameter magnetic chuck of the circular machine table which rotates at a pre-set constant or automatically varying speed, the latter maintaining an approximately equal peripheral speed of the work surface area being ground. The wheelhead, installed on a cross slide, traverses over the table along a radial path, moving in alternating directions, toward and away from the center of the table. Infeed is by vertical movement of the saddle along the guideways of the vertical column, at the end of the radial wheelhead stroke. The saddle contains the guideways along which the wheelhead slide reciprocates.
Effective Grinding Surface—Face (Side) of Wheel Movement of Work—Reciprocating
Operation is similar to the reciprocating table type peripheral surface grinder, but grinding is with the face, usually with the rim of a cup shaped wheel, or a segmental wheel for large model machines. Capable of covering a much wider area of the work surface than the peripheral grinder, thus frequently no need for cross feed. Provides efficient stock removal, but is less adaptable than the reciprocating table type peripheral grinder.
Effective Grinding Surface—Face (Side) of Wheel Movement of Work—Rotary
The grinding wheel, usually of segmental type, is set in a position to cover either an annular area near the periphery of the table or, more commonly, to reach beyond the table center. A large circular magnetic chuck generally covers the entire table surface and facilitates the mounting of workpieces, even of fixtures, when needed. The uninterrupted passage of the work in contact with the large wheel face permits a very high rate of stock removal and the machine, with single or double wheelhead, can be adapted also to automatic operation with continuous part feed by mechanized work handling.
Effective Grinding Surface—Face (Side) of Wheel Movement of Work—Traverse Along Straight or Arcuate Path
Operates with practically the entire face of the wheel which is designated as an abrasive disc (hence "disc grinding") because of its narrow width in relation to the large diameter. Built either for one or, more frequently, for two discs operating with opposed faces for the simultaneous grinding of both sides of the workpiece. The parts pass between the operating faces of the wheel (a) pushed-in and retracted by the drawer-like movement of a feeding slide; (b) in an arcuate movement carried in the nests of a rotating feed wheel; (c) nearly diagonally advancing along a rail. Very well adapted to fully mechanized work handling.

The last letters (two or three) which may follow the bond designation V (vitrified) or B (resinoid) refer to: (1) bond modification, "BE" being especially suitable for surface grinding; (2) special structure, "P" type being distinctively porous; and (3) for segments made of 23A type abrasives, the term 12VSM implies porous structure, and the letter "P" is not needed.

Process Data for Surface Grinding. — In surface grinding, similarly to other metalcutting processes, the speed and feed rates that are applied must be adjusted to

Table 2.   Basic Process Data for Peripheral Surface Grinding on Reciprocating Table Surface Grinders

Work Material	Hardness	Material Condition	Wheel Speed, fpm	Table Speed, fpm	Downfeed, in. per pass		Cross Feed per Pass, fraction of wheel width
					Rough	Finish	
Plain carbon steel	52 R_c max.	Annealed, Cold drawn	5500 to 6500	50 to 100	0.003	0.0005 max.	1/4
	52 to 65 R_c	Carburized and/or quenched and tempered	5500 to 6500	50 to 100	0.003	0.0005 max.	1/10
Alloy steels	52 R_c max.	Annealed or quenched and tempered	5500 to 6500	50 to 100	0.003	0.001 max.	1/4
	52 to 65 R_c	Carburized and/or quenched and tempered	5500 to 6500	50 to 100	0.003	0.0005 max.	1/10
Tool steels	150 to 275 BHN	Annealed	5500 to 6500	50 to 100	0.002	0.0005 max.	1/5
	56 to 65 R_c	Quenched and tempered	5500 to 6500	50 to 100	0.002	0.0005 max.	1/10
Nitriding steels	200 to 350 BHN	Normalized, annealed	5500 to 6500	50 to 100	0.003	0.001 max.	1/4
	60 to 65 R_c	Nitrided	5500 to 6500	50 to 100	0.003	0.0005 max.	1/10
Cast steels	52 R_c max.	Normalized, annealed	5500 to 6500	50 to 100	0.003	0.001 max.	1/4
	Over 52 R_c	Carburized and/or quenched and tempered	5500 to 6500	50 to 100	0.003	0.0005 max.	1/10
Gray irons	52 R_c max.	As cast, annealed, and/or quenched and tempered	5000 to 6500	50 to 100	0.003	0.001 max.	1/3
Ductile irons	52 R_c max.	As cast, annealed or quenched and tempered	5500 to 6500	50 to 100	0.003	0.001 max.	1/5
Stainless steels, martensitic	135 to 235 BHN	Annealed or cold drawn	5500 to 6500	50 to 100	0.002	0.0005 max.	1/4
	Over 275 BHN	Quenched and tempered	5500 to 6500	50 to 100	0.001	0.0005 max.	1/8
Aluminum alloys	30 to 150 BHN	As cast, cold drawn or treated	5500 to 6500	50 to 100	0.003	0.001 max.	1/3

Table 3. Common Faults and Possible Causes in Surface Grinding

CAUSES	WORK DIMENSION — Work not flat	Work not parallel	Poor size holding	METALLURGICAL DEFECTS — Burnishing of work	Burning or checking	SURFACE QUALITY — Feed lines	Chatter marks	Scratches on surface	Poor finish	WHEEL CONDITION — Wheel loading	Wheel glazing	Rapid wheel wear	WORK RETAINMENT — Not firmly seated	Work sliding on chuck
WORK CONDITION														
Heat treat stresses	✓				✓									
Work too thin	✓	✓												
Work warped	✓												✓	
Abrupt section changes	✓													
GRINDING WHEEL														
Grit too fine				✓						✓	✓			
Grit too coarse								✓	✓					
Grade too hard				✓	✓					✓	✓			
Grade too soft		✓	✓									✓		
Wheel not balanced							✓							
Dense structure										✓	✓			
TOOLING AND COOLANT														
Improper coolant				✓	✓					✓				
Insufficient coolant				✓	✓					✓	✓			
Dirty coolant								✓		✓				
Diamond loose or chipped	✓							✓	✓					
Diamond dull		✓							✓	✓	✓			
No or poor magnetic force													✓	✓
Chuck surface worn or burred													✓	✓
MACHINE AND SETUP														
Chuck not aligned	✓	✓												
Vibrations in machine							✓							
Plane of movement out of parallel	✓	✓												
OPERATIONAL CONDITIONS														
Too low work speed										✓	✓			
Too light feed		✓									✓			
Too heavy cut	✓	✓			✓							✓		
Chuck retained swarf	✓												✓	✓
Chuck loading improper								✓						✓
Insufficient blocking of parts		✓	✓											
Wheel runs off the work	✓	✓												
Wheel dressing too fine									✓	✓	✓	✓		
Wheel edge not chamfered								✓						
Loose dirt under guard						✓		✓						

the operational conditions as well as to the objectives of the process. Grinding differs, however, from other types of metal cutting methods in regard to the cutting speed of the tool; the peripheral speed of the grinding wheel is maintained within a narrow range, generally 5500 to 6500 surface feet per minute. Speed ranges different from the common one are used in particular processes which require special wheels and equipment.

In establishing the proper process values for grinding, of prime consideration are the work material, its condition, and the type of operation (roughing or finishing). Table 2 gives basic process data for peripheral surface grinding on reciprocating table surface grinders. For different work materials and hardness ranges data are given regarding table speeds, downfeed (infeed) rates and cross feed, the latter as a function of the wheel width.

Common Faults and Possible Causes in Surface Grinding. — Approaching the ideal performance with regard to both the quality of the ground surface and the efficiency of surface grinding, requires the monitoring of the process and the correction of conditions adverse to the attainment of that goal.

Defective, or just not entirely satisfactory surface grinding may have any one or more of several causes. Exploring and determining the cause for eliminating its harmful effects is facilitated by knowing the possible sources of the experienced undesirable performance. Table 3, associating the common faults with their possible causes, is intended to aid in determining the actual cause, the correction of which should restore the desired performance level.

While the table lists the more common faults in surface grinding, and points out their frequent causes, other types of improper performance and/or other causes, in addition to those indicated, are not excluded.

Offhand Grinding

Offhand grinding consists of holding the wheel to the work or the work to the wheel and grinding to broad tolerances and includes such operations as certain types of tool sharpening, weld grinding, snagging castings and other rough grinding. Types of machines that are used for rough grinding in foundries are floor- and bench-stand machines. Wheels for these machines vary from 6 to 30 inches in diameter. Portable grinding machines (electric, flexible shaft, or air-driven) are used for cleaning and smoothing castings.

Many rough grinding operations on castings can be best done with shaped wheels, such as cup wheels (including plate mounted) or cone wheels, and it is advisable to have a good assortment of such wheels on hand to do the odd jobs the best way.

Floor- and Bench-Stand Grinding. — The most common method of rough grinding is on double-end floor and bench stands. In machine shops, welding shops, and automotive repair shops, these grinders are usually provided with a fairly coarse grit wheel on one end for miscellaneous rough grinding and a finer grit wheel on the other end for sharpening tools. The pressure exerted is a very important factor in selecting the proper grinding wheel. If grinding is to be done mostly on hard sharp fins, then durable, coarse and hard wheels are required, but if grinding is mostly on large gate and riser pads, then finer and softer wheels should be used for best cutting action.

Portable Grinding. — Portable grinding machines are usually classified as air grinders, flexible shaft grinders, and electric grinders. The electric grinders are of two types; namely, those driven by standard 60 cycle current and so-called high-cycle grinders. Portable grinders are used for grinding down and smoothing weld seams; cleaning metal before welding; grinding out imperfections, fins and parting lines in castings and smoothing castings; grinding punch press dies and patterns to proper size and shape; and grinding manganese steel castings.

Wheels used on portable grinders are of three bond types; namely, resinoid, rubber, and vitrified. By far the largest percentage is resinoid. Rubber bond is used for relatively thin wheels and where a good finish is required. Some of the smaller wheels such as cone and plug wheels are vitrified bonded.

Grit sizes most generally used in wheels from 4 to 8 inches in diameter are 16, 20, and 24. In the still smaller diameters, finer sizes are used, such as 30, 36, and 46.

The particular grit size to use depends chiefly on the kind of grinding to be done. If the work consists of sharp fins and the machine has ample power, a coarse grain size combined with a fairly hard grade should be used. If the job is more in the nature of smoothing or surfacing and a fairly good finish is required, then finer and softer wheels are called for.

Swing-Frame Grinding. — This type of grinding is employed where a considerable amount of material is to be removed as on snagging large castings. It may be possible to remove 10 times as much material from steel castings using swing-frame grinders as with portable grinders; and 3 times as much material as with high-speed floor-stand grinders.

The largest field of application for swing-frame machines is on castings which are too heavy to handle on a floor stand; but often it is found that comparatively large gates and risers on smaller castings can be ground more quickly with swing-frame grinders, even if fins and parting lines have to be ground on floor stands as a second operation.

In foundries, the swing-frame machines are usually suspended from a trolley on a jib that can be swung out of the way when placing the work on the floor with the help of an overhead crane. In steel mills when grinding billets, a number of swing-frame machines are usually suspended from trolleys on a line of beams which facilitate their use as required.

The grinding wheels used on swing-frame machines are made with coarser grit sizes and harder grades than wheels used on floor stands for the same work. The reason is that greater grinding pressures can be obtained on the swing-frame machines.

Mounted Wheels and Mounted Points. — These wheels and points are used in hard-to-get-at places and are available with a vitrified bond. The wheels are available with aluminum oxide or silicon carbide abrasive grains. The aluminum oxide wheels are used to grind tough and tempered die steels and the silicon carbide wheels, cast iron, chilled iron, bronze, and other non-ferrous metals.

The illustrations on pages 1106 and 1107 give the standard shapes of mounted wheels and points as published by the Grinding Wheel Institute. A note about the maximum operating speed for these wheels is given at the bottom of the first page of illustrations. Metric sizes are given on page 1105.

Abrasive Belt Grinding

Abrasive belts are used in the metalworking industry for removing stock, light cleaning up of metal surfaces, grinding welds, deburring, breaking and polishing hole edges, and finish grinding of sheet steel. The types of belts that are used may be coated with aluminum oxide (the most common coating) for stock removal and finishing of all alloy steels, high-carbon steel, and tough bronzes; and silicon carbide for use on hard, brittle, and low-tensile strength metals which would include aluminum and cast irons.

Table 1 is a guide to the selection of the proper abrasive belt, lubricant, and contact wheel. This table is entered on the basis of the material used and type of operation to be done and gives the abrasive belt specifications (type of bonding and

Table 1. Guide to the Selection and Application of Abrasive Belts

Material	Type of Operation	Abrasive Belt*	Grit	Belt Speed, fpm	Type of Grease Lubricant	Contact Wheel Type	Durometer Hardness
Hot- and Cold-Rolled Steel	Roughing	R/R Al_2O_3	24–60	4000–6500	Light-body or none	Cog-tooth, serrated rubber	70–90
	Polishing	R/G or R/R Al_2O_3	80–150	4500–7000	Light-body or none	Plain or serrated rubber, sectional or finger-type cloth wheel, free belt	20–60
	Fine Polishing	R/G or electro-coated Al_2O_3 cloth	180–500	4500–7000	Heavy or with abrasive compound	Smooth-faced rubber or cloth	20–40
Stainless Steel	Roughing	R/R Al_2O_3	50–80	3500–5000	Light-body or none	Cog-tooth, serrated rubber	70–90
	Polishing	R/G or R/R Al_2O_3	80–120	4000–5500	Light-body or none	Plain or serrated rubber, sectional or finger-type cloth wheel, free belt	30–60
	Fine Pol.	Closed-coat SiC	150–280	4500–5500	Heavy or oil mist	Smooth-faced rubber or cloth	20–40
Aluminum, Cast or Fabricated	Roughing	R/R SiC or Al_2O_3	24–80	5000–6500	Light	Cog-tooth, serrated rubber	70–90
	Polishing	R/G SiC or Al_2O_3	100–180	4500–6500	Light	Plain or serrated rubber, sectional or finger-type cloth wheel, free belt	30–50
	Fine Polishing	Closed-coat SiC or electro-coated Al_2O_3	220–320	4500–6500	Heavy or with abrasive compound	Plain faced rubber, finger-type cloth or free belt	20–50
Copper Alloys or Brass	Roughing	R/R SiC or Al_2O_3	36–80	2200–4500	Light-body	Cog-tooth, serrated rubber	70–90
	Polishing	Closed-coat SiC or electro-coated Al_2O_3 or R/G SiC or Al_2O_3	100–150	4000–6500	Light-body	Plain or serrated rubber, sectional or finger-type cloth wheel, free belt	30–50
	Fine Polishing	Closed-coat SiC or electro-coated Al_2O_3	180–320	4000–6500	Light or with abrasive compound	Same as for polishing	20–30
Non-ferrous Die-castings	Roughing	R/R SiC or Al_2O_3	24–80	4500–6500	Light-body	Hard wheel depending on application	50–70
	Polishing	R/G SiC or Al_2O_3	100–180	4500–6500	Light-body	Plain rubber, cloth or free belt	30–50
	Fine Polishing	Electro-coated Al_2O_3 or closed-coat SiC	220–320	4500–6500	Heavy or with abrasive compound	Plain or finger-type cloth wheel, or free belt	20–30
Cast Iron	Roughing	R/R Al_2O_3	24–60	2000–4000	None	Cog-tooth, serrated rubber	70–90
	Polishing	R/R Al_2O_3	80–150	4000–5500	None	Serrated rubber	30–70
	Fine Polishing	R/R Al_2O_3	120–240	4000–5500	Light-body	Smooth-faced rubber	30–40
Titanium	Roughing	R/R SiC or Al_2O_3	36–50	700–1500	Sulfur-chlorinated	Small-diameter, cog-tooth serrated rubber	70–80
	Polishing	R/R SiC	60–120	1200–2000	Light-body	Standard serrated rubber	50
	Fine Pol.	R/R SiC	120–240	1200–2000	Light-body	Smooth-faced rubber or cloth	20–40

* R/R indicates that both the making and sizing bond coats are resin. R/G indicates that the making coat is glue and the sizing coat is resin. The abbreviations Al_2O_3 for aluminum oxide and SiC for silicon carbide are used. Almost all R/R and R/G Al_2O_3 and SiC belts have a heavy-drill weight cloth backing. Most electro-coated Al_2O_3 and closed-coat SiC belts have a jeans weight cloth backing.

Table 2. Guide to the Selection and Application of Contact Wheels

Surface	Material	Hardness and Density	Purposes	Wheel Action	Comments
Cog-tooth	Rubber	70 to 90 durometer	Roughing	Fast cutting, allows long belt life.	For cutting down projections on castings and weld beads.
Standard serrated	Rubber	40 to 50 durometer, medium density	Roughing	Leaves rough- to medium-ground surface.	For smoothing projections and face defects.
X-shaped serrations	Rubber	20 to 50 durometer	Roughing and polishing	Flexibility of rubber allows entry into contours. Medium polishing, light removal.	Same as for standard serrated wheels but preferred for soft non-ferrous metals.
Plain face	Rubber	20 to 70 durometer	Roughing and polishing	Plain wheel face allows controlled penetration of abrasive grain. Softer wheels give better finishes.	For large or small flat faces.
Flat flexible	Compressed canvas	About nine densities from very hard to very soft	Roughing and polishing	Hard wheels can remove metal, but not as quickly as cog-tooth rubber wheels. Softer wheels polish well.	Good for medium-range grinding and polishing.
Flat flexible	Solid sectional canvas	Soft, medium, and hard	Polishing	Uniform polishing. Avoids abrasive pattern on work. Adjusts to contours. Can be preformed for contours.	A low-cost wheel with uniform density at the face. Handles all types of polishing.
Flat flexible	Buff section canvas	Soft	Contour polishing	For fine polishing and finishing.	Can be widened or narrowed by adding or removing sections. Low cost.
Flat flexible	Sponge rubber inserts	5 to durometer, soft	Polishing	Uniform polishing and finishing. Polishes and blends contours.	Has replaceable segments. Polishes and blends contours. Segments allow density changes.
Flexible	Fingers of canvas attached to hub	Soft	Polishing	Uniform polishing and finishing.	For polishing and finishing.
Flat flexible	Rubber segments	Varies in hardness	Roughing and polishing	Grinds or polishes depending on density and hardness of inserts.	For portable machines. Uses replaceable segments that save on wheel costs and allow density changes.
Flat flexible	Inflated rubber	Air pressure controls hardness	Roughing and polishing	Uniform finishing.	Adjusts to contours.

abrasive grain size and material), the range of speeds at which the belt may best be operated, the type of lubricant to use, and the type and hardness of the contact wheel to use. Table 2 serves as a guide in the selection of contact wheels. This table is entered on the basis of the type of contact wheel surface and the contact wheel material. The table gives the hardness and/or density, the type of abrasive belt grinding for which the contact wheel is intended, the character of the wheel action and such comments as the uses, and hints for best use. Both tables are intended only as guides for general shop practice; selections may be altered to suit individual requirements.

There are three types of abrasive belt grinding machines. One type employs a contact wheel behind the belt at the point of contact of the workpiece to the belt and facilitates a high rate of stock removal. Another type uses an accurate parallel ground platen over which the abrasive belt passes and facilitates the finishing of precision parts. A third type which has no platens or contact wheel is used for finishing parts having uneven surfaces or contours. In this type there is no support behind the belt at the point of contact of the belt with the workpiece. Some machines are so constructed that besides grinding against a platen or a contact wheel the workpiece may be moved and ground against an unsupported portion of the belt, thereby in effect making it a dual machine.

Although abrasive belts at the time of their introduction were used dry, since the advent of the improved waterproof abrasive belts, they have been used with coolants, oil-mists, and greases to aid the cutting action. The application of a coolant to the area of contact retards loading, resulting in a cool, free cutting action, a good finish and a long belt life.

Abrasive Cutting

Abrasive cut-off wheels are used for cutting steel, brass and aluminum bars and tubes of all shapes and hardnesses, ceramics, plastics, insulating materials, glass and cemented carbides. Originally a tool or stock room procedure, this method has developed into a high speed production operation. While the abrasive cut-off machine and cut-off wheel can be said to have revolutionized the practice of cutting-off materials, the metal saw continues to be the more economical method for cutting-off large cross-sections of certain materials. However, there are innumerable materials and shapes that can be cut with much greater speed and economy by the abrasive wheel method. On conventional chop-stroke abrasive cutting machines using 16-inch diameter wheels, 2-inch diameter bar stock is the maximum size that can be cut with satisfactory wheel efficiency, but bar stock up to 6 inches in diameter can be cut efficiently on oscillating-stroke machines. Tubing up to 3½ inches in diameter can also be cut efficiently.

Abrasive wheels are commonly available in four types of bonds: Resinoid, rubber, shellac and fiber or fabric reinforced. In general, resinoid bonded cut-off wheels are used for dry cutting where burrs and some burn are not objectionable and rubber bonded wheels are used for wet cutting where cuts are to be smooth, clean and free from burrs. Shellac bonded wheels have a soft, free cutting quality which makes them particularly useful in the tool room where tool steels are to be cut without discoloration. Fiber reinforced bonded wheels are able to withstand severe flexing and side pressures and fabric reinforced bonded wheels which are highly resistant to breakage caused by extreme side pressures, are fast cutting and have a low rate of wear.

The types of abrasives available in cut-off wheels are: Aluminum oxide, for cutting steel and most other metals; silicon carbide, for cutting non-metallic materials such as carbon, tile, slate, ceramics, etc.; and diamond, for cutting cemented carbides. The method of denoting abrasive type, grain size, grade, structure and

bond type by using a system of markings is the same as for grinding wheels (see page 1040). Maximum wheel speeds given in the American National Standard Safety Requirements for The Use, Care, and Protection of Abrasive Wheels (ANSI B7.1-1978) range from 9500 to 14,200 surface feet per minute for organic bonded cut-off wheels larger than 16 inches in diameter and from 9500 to 16,000 surface feet per minute for organic bonded cut-off wheels 16 inches in diameter and smaller. Maximum wheel speeds specified by the manufacturer should never be exceeded even though they may be lower than those given in the B7.1.

There are four basic types of abrasive cutting machines: Chop-stroke, oscillating stroke, horizontal stroke and work rotating. Each of these four types may be designed for dry cutting or for wet cutting (includes submerged cutting).

The accompanying table based upon information made available by The Carborundum Co. gives some of the probable causes of cutting off difficulties that might be experienced when using abrasive cut-off wheels.

Probable Causes of Cutting-Off Difficulties

Difficulty	Probable Cause
Angular Cuts and Wheel Breakage	(1) Inadequate clamping which allows movement of work while the wheel is in the cut. The work should be clamped on both sides of the cut. (2) Work vise higher on one side than the other causing wheel to be pinched. (3) Wheel vibration resulting from worn spindle bearings. (4) Too fast feeding into the cut when cutting wet.
Burning of Stock	(1) Insufficient power or drive allowing wheel to stall. (2) Cuts too heavy for grade of wheel being used. (3) Wheel fed through the work too slowly. This causes a heating up of the material being cut. This difficulty encountered chiefly in dry cutting.
Excessive Wheel Wear	(1) Too rapid cutting when cutting wet. (2) Grade of wheel too hard for work, resulting in excessive heating and burning out of bond. (3) Inadequate coolant supply in wet cutting. (4) Grade of wheel too soft for work. (5) Worn spindle bearings allowing wheel vibration.
Excessive Burring	(1) Feeding too slowly when cutting dry. (2) Grit size in wheel too coarse. (3) Grade of wheel too hard. (4) Wheel too thick for job.

Honing Process

The hone-abrading process for obtaining cylindrical forms with precise dimensions and surfaces can be applied to internal cylindrical surfaces with a wide range of diameters such as engine cylinders, bearing bores, pin holes, etc. and also to some external cylindrical surfaces. The process is used to: (1) eliminate inaccuracies resulting from previous operations by generating a true cylindrical form with respect to roundness and straightness within minimum dimensional limits; (2) generate final dimensional size accuracy within low tolerances, as may be required for interchangeability of parts; (3) provide rapid and economical stock removal consistent with accomplishment of the other results; and (4) generate surface finishes of a specified degree of surface smoothness with high surface quality.

Amount and Rate of Stock Removal. — Honing may be employed to increase bore diameters by as much as 0.100 inch or as little as 0.001 inch. The amount of stock removed by the honing process is entirely a question of processing economy. If other operations are performed before honing then the bulk of the stock should be taken off by the operation that can do it most economically. In large diameter bores that have been distorted in heat treating, it may be necessary to remove as much as 0.030 to 0.040 inch from the diameter to make the bore round and straight. For out-of-round or tapered bores, a good "rule of thumb" is to leave twice as much stock (on the diameter) for honing as there is error in the bore. Another general rule is: For bores over one inch in diameter, leave 0.001 to 0.0015 inch stock per inch of diameter. For example, 0.002 to 0.003 inch of stock is left in two-inch bores and 0.010 to 0.015 inch in ten-inch bores. Where parts are to be honed for finish only, the amount of metal to be left for removing tool marks may be as little as 0.0002 to 0.001 inch on the diameter.

In general, the honing process can be employed to remove stock from bore diameters at the rate of 0.009 to 0.012 inch per minute on cast-iron parts and from 0.005 to 0.008 inch per minute on steel parts having a hardness of 60 to 65 Rockwell C. These rates are based on parts having a length equal to three or four times the diameter. Stock has been removed from long parts such as gun barrels, at the rate of 65 cubic inches per hour. Recommended honing speeds for cast iron range from 110 to 200 surface feet per minute of rotation and from 50 to 110 lineal feet per minute of reciprocation. For steel, rotating surface speeds range from 50 to 110 feet per minute and reciprocation speeds from 20 to 90 lineal feet per minute. The exact rotation and reciprocation speeds to be used depend upon the size of the work, the amount and characteristics of the material to be removed and the quality of the finish desired. In general, the harder the material to be honed, the lower the speed. Interrupted bores are usually honed at faster speeds than plain bores.

Formula for Rotative Speeds. — Empirical formulas for determining rotative speeds for honing have been developed by the Micromatic Hone Corp. These formulas take into consideration the type of material being honed, its hardness and its surface characteristics; the abrasive area; and the type of surface pattern and degree of surface roughness desired. Because of the wide variations in material characteristics, abrasives available, and types of finishes specified, these formulas should be considered as a guide only in determining which of the available speeds (pulley or gear combinations) should be used for any particular application.

The formula for rotative speed, S, in surface feet per minute is:

$$S = \frac{K \times D}{W \times N}$$

The formula for rotative speed in revolutions per minute is:

$$\text{R.P.M.} = \frac{R}{W \times N}$$

where, K and R are factors taken from the table on the following page, D is the diameter of the bore in inches, W is the width of the abrasive stone or stock in inches, and N is the number of stones.

Although the actual speed of the abrasive is the resultant of both the rotative speed and the reciprocation speed, this latter quantity is seldom solved for or used. The reciprocation speed is not determined empirically but by testing under operating conditions. Changing the reciprocation speed affects the dressing action of the abrasive stones, therefore, the reciprocation speed is adjusted to provide for a desired surface finish which is usually a well lubricated bearing surface that will not scuff.

Table of Factors for Use in Rotative Speed Formulas

Character of Surface[1]	Material	Hardness[2]					
		Soft		Medium		Hard	
		Factors					
		K	R	K	R	K	R
Base Metal	Cast Iron	110	420	80	300	60	230
	Steel	80	300	60	230	50	190
Dressing Surface	Cast Iron	150	570	110	420	80	300
	Steel	110	420	80	300	60	230
Severe Dressing	Cast Iron	200	760	150	570	110	420
	Steel	150	570	110	420	80	300

[1] The character of the surface is classified according to its effect on the abrasive; *Base Metal* being a honed, ground or fine bored section that has little dressing action on the grit; *Dressing Surface* being a rough bored, reamed or broached surface or any surface broken by cross holes or ports; *Severe Dressing* being a surface interrupted by keyways, undercuts or burrs that dress the stones severely. If over half of the stock is to be removed after the surface is cleaned up, the speed should be computed using the *Base Metal* factors for K and R.

[2] Hardness designations of soft, medium and hard cover the following ranges on the Rockwell "C" hardness scale, respectively: 15 to 45, 45 to 60 and 60 to 70.

Possible Adjustments for Eliminating Undesirable Honing Conditions

Undesirable Condition	Adjustment Required to Correct Condition[1]								
	Abrasive[2]				Other				
	Friability	Grain Size	Hardness	Structure	Feed Pressure	Reciprocation	R.P.M.	Runout Time	Stroke Length
Abrasive Glazing	+	− −	− −	+	+ +	+ +	− −	−	0
Abrasive Loading	0	− −	−	−	+ +	+	− −	0	0
Too Rough Surface Finish	0	+ +	+ +	−	−	−	+ +	+	0
Too Smooth Surface Finish	0	− −	− −	+	+	+	− −	−	0
Poor Stone Life	−	+	+ +	−	−	−	+	0	0
Slow Stock Removal	+	− −	−	+	+ +	+ +	− −	0	0
Taper — Large at Ends	0	0	0	0	0	0	0	0	−
Taper — Small at Ends	0	0	0	0	0	0	0	0	+

Compiled by Micromatic Hone Corp.

[1] The + and + + symbols generally indicate that there should be an increase or addition while the − and − − symbols indicate that there should be a reduction or elimination. In each case, the double symbol indicates that the contemplated change would have the greatest effect. The 0 symbol means that a change would have no effect.

[2] For the abrasive adjustments the + and + + symbols indicate a more friable grain, a finer grain, a harder grade or a more open structure and the − and − − symbols just the reverse.

Abrasive Stones for Honing. — Honing stones consist of aluminum oxide, silicon carbide or diamond abrasive grits, held together in stick form by a vitrified clay, resinoid or metal bond. The grain and grade of abrasive to be used in any particular honing operation depend upon the quality of finish desired, the amount of stock to be removed, the material being honed and other factors. The following general rules may be followed in the application of abrasive for honing: (1) Silicon-carbide abrasive is commonly used for honing cast iron, while aluminum-oxide abrasive is generally used on steel; (2) The harder the material being honed, the softer the abrasive stick used; (3) A rapid reciprocating speed will tend to make the abrasive cut fast because the dressing action on the grits will be severe; and (4) To improve the finish, use a finer abrasive grit, incorporate more multi-direction action, allow more "run-out" time after honing to size, or increase the speed of rotation. Surface roughnesses ranging from less than 1 micro-inch r.m.s. to a relatively coarse roughness can be obtained by judicious choice of abrasive and honing time but the most common range is from 3 to 50 micro-inches r.m.s.

Adjustments for Eliminating Undesirable Honing Conditions. — The accompanying table indicates adjustments that may be made to correct certain undesirable conditions encountered in honing. Only one change should be made at a time and its effect noted before making other adjustments.

Tolerances. — For bore diameters above 4 inches the tolerance of honed surfaces with respect to roundness and straightness ranges from 0.0005 to 0.001 inch; for bore diameters from 1 to 4 inches, 0.0003 to 0.0005 inch; and for bore diameters below 1 inch, 0.00005 to 0.0003 inch.

Laps and Lapping

Material for Laps. — Laps are usually made of soft cast iron, copper, brass or lead. In general, the best material for laps to be used on very accurate work is soft, close-grained cast iron. If the grinding, prior to lapping, is of inferior quality, or an excessive allowance has been left for lapping, copper laps may be preferable. They can be charged more easily and cut more rapidly than cast iron, but do not produce as good a finish. Whatever material is used, the lap should be softer than the work, as, otherwise, the latter will become charged with the abrasive and cut the lap, the order of the operation being reversed. A common and inexpensive form of lap for holes is made of lead which is cast around a tapering steel arbor. The arbor usually has a groove or keyway extending lengthwise, into which the lead flows, thus forming a key that prevents the lap from turning. When the lap has worn slightly smaller than the hole and ceases to cut, the lead is expanded or stretched a little by the driving in of the arbor. When this expanding operation has been repeated two or three times, the lap usually must be trued or replaced with a new one, owing to distortion.

The tendency of lead laps to lose their form is an objectionable feature. They are, however, easily molded, inexpensive, and quickly charged with the cutting abrasive. A more elaborate form for holes is composed of a steel arbor and a split cast-iron or copper shell which is sometimes prevented from turning by a small dowel pin. The lap is split so that it can be expanded to accurately fit the hole being operated upon. For hardened work, some toolmakers prefer copper to either cast iron or lead. For holes varying from ¼ to ½ inch in diameter, copper or brass is sometimes used; cast iron is used for holes larger than ½ inch in diameter. The arbors for these laps should have a taper of about ¼ or ⅜ inch per foot. The length of the lap should be somewhat greater than the length of the hole, and the thickness of the shell or lap proper should be from ⅛ to ⅙ its diameter.

External laps are commonly made in the form of a ring, there being an outer ring or holder and an inner shell which forms the lap proper. This inner shell is made of cast iron, copper, brass or lead. Ordinarily the lap is split and screws are provided in the holder for adjustment. The length of an external lap should at least equal the diameter of the work, and might well be longer. Large ring laps usually have a handle for moving them across the work.

Laps for Flat Surfaces. — Laps for producing plane surfaces are made of cast iron. In order to secure accurate results, the lapping surface must be a true plane. A flat lap that is used for roughing or "blocking down" will cut better if the surface is scored by narrow grooves. These are usually located about ½ inch apart and extend both lengthwise and crosswise, thus forming a series of squares similar to those on a checker-board. An abrasive of No. 100 or 120 emery and lard oil can be used for charging the roughing lap. For finer work, a lap having an unscored surface is used, and the lap is charged with a finer abrasive. After a lap is charged, all loose abrasive should be washed off with gasoline, for fine work, and when lapping, the surface should be kept moist, preferably with kerosene. Gasoline will cause the lap to cut a little faster, but it evaporates so rapidly that the lap soon becomes dry and the surface caked and glossy in spots. Loose emery should not be applied while lapping, for if the lap is well charged with abrasive in the beginning, is kept well moistened and not crowded too hard, it will cut for a considerable time. The pressure upon the work should be just enough to insure constant contact. The lap can be made to cut only so fast, and if excessive pressure is applied it will become "stripped" in places. The causes of scratches are: Loose abrasive on the lap; too much pressure on the work, and poorly graded abrasive. To produce a perfectly smooth surface free from scratches, the lap should be charged with a very fine abrasive.

Grading Abrasives for Lapping. — For high-grade lapping, abrasives can be evenly graded as follows: A quantity of flour-emery or other abrasive is placed in a heavy cloth bag, which is gently tapped, causing very fine particles to be sifted through. When a sufficient quantity has been obtained in this way, it is placed in a dish of lard or sperm oil. The largest particles will then sink to the bottom and in about one hour the oil should be poured into another dish, care being taken not to disturb the sediment at the bottom. The oil is then allowed to stand for several hours, after which it is poured again, and so on, until the desired grade is obtained.

Charging Laps. — To charge a flat cast-iron lap, spread a very thin coating of the prepared abrasive over the surface and press the small cutting particles into the lap with a hard steel block. There should be as little rubbing as possible. When the entire surface is apparently charged, clean and examine for bright spots; if any are visible, continue charging until the entire surface has a uniform gray appearance. When the lap is once charged, it should be used without applying more abrasive until it ceases to cut. If a lap is over-charged and an excessive amount of abrasive is used, there is a rolling action between the work and lap which results in inaccuracy. The surface of a flat lap is usually finished true, prior to charging, by scraping and testing with a standard surface-plate, or by the well-known method of scraping-in three plates together, in order to secure a plane surface. In any case, the bearing marks or spots should be uniform and close together. These spots can be blended by covering the plates evenly with a fine abrasive and rubbing them together. While the plates are being ground in, they should be carefully tested and any high spots which may form should be reduced by rubbing them down with a smaller block.

To charge cylindrical laps for internal work, spread a thin coating of prepared abrasive over the surface of a hard steel block, preferably by rubbing lightly with a cast-iron or copper block; then insert an arbor through the lap and roll the latter over the steel block, pressing it down firmly to embed the abrasive into the surface of the lap. For external cylindrical laps, the inner surface can be charged by rolling-in the abrasive with a hard steel roller that is somewhat smaller in diameter than the lap. The taper cast-iron blocks which are sometimes used for lapping taper holes can also be charged by rolling-in the abrasive, as previously described; there is usually one roughing and one finishing lap, and when charging the former, it may be necessary to vary the charge in accordance with any error which might exist in the taper.

Rotary Diamond Lap. — This style of lap is used for accurately finishing very small holes, which, because of their size, cannot be ground. While the operation is referred to as lapping, it is, in reality, a grinding process, the lap being used the same as a grinding wheel. Laps employed for this work are made of mild steel, soft material being desirable because it can be charged readily. Charging is usually done by rolling the lap between two hardened steel plates. The diamond dust and a little oil is placed on the lower plate, and as the lap revolves, the diamond is forced into its surface. After charging, the lap should be washed in benzine. The rolling plates should also be cleaned before charging with dust of a finer grade. It is very important not to force the lap when in use, especially if it is a small size. The lap should just make contact with the high spots and gradually grind them off. If a diamond lap is lubricated with kerosene, it will cut freer and faster. These small laps are run at very high speeds, the rate depending upon the lap diameter. Soft work should never be ground with diamond dust because the dust will leave the lap and charge the work.

When using a diamond lap, it should be remembered that such a lap will not produce sparks like a regular grinding wheel; hence, it is easy to crowd the lap and "strip" some of the diamond dust. To prevent this, a sound intensifier or "harker" should be used. This is placed against some stationary part of the grinder spindle, and indicates when the lap touches the work, the sound produced by the slightest contact being intensified.

Grading Diamond Dust. — The grades of diamond dust used for charging laps are designated by numbers, the fineness of the dust increasing as the numbers increase. The diamond, after being crushed to powder in a mortar, is thoroughly mixed with high-grade olive oil. This mixture is allowed to stand five minutes and then the oil is poured into another receptacle. The coarse sediment which is left is removed and labeled No. 0, according to one system. The oil poured from No. 0 is again stirred and allowed to stand ten minutes, after which it is poured into another receptacle and the sediment remaining is labeled No. 1. This operation is repeated until practically all of the dust has been recovered from the oil, the time that the oil is allowed to stand being increased as shown by the following table. This is done in order to obtain the smaller particles that require a longer time for precipitation:

To obtain No. 1 — 10 minutes.	To obtain No. 4 — 2 hours.
To obtain No. 2 — 30 minutes.	To obtain No. 5 — 10 hours.
To obtain No. 3 — 1 hour.	To obtain No. 6 — until oil is clear.

The No. 0 or coarse diamond which is obtained from the first settling is usually washed in benzine, and re-crushed unless very coarse dust is required. This No. 0 grade is sometimes known as "ungraded" dust. In some places the time for settling, in order to obtain the various numbers, is greater than that given in the table.

Cutting Properties of Laps and Abrasives. — In order to determine the cutting properties of abrasives when used with different lapping materials and lubricants, a series of tests was conducted, the results of which were given in a paper by W. A. Knight and A. A. Case, presented before the American Society of Mechanical Engineers. In connection with these tests, a special machine was used, the construction being such that quantitative results could be obtained with various combinations of abrasive, lubricant, and lap material. These tests were confined to surface lapping.

It was not the intention to test a large variety of abrasives, three being selected as representative; namely, Naxos emery, carborundum, and alundum. Abrasive No. 150 was used in each case, and seven different lubricants, five different pressures, and three different lap materials were employed. The lubricants were lard oil, machine oil, kerosene, gasoline, turpentine, alcohol, and soda water.

These tests indicated throughout that there is, for each different combination of lap and lubricant, a definite size of grain that will give the maximum amount of cutting. With all the tests, except when using the two heavier lubricants, some reduction in the size of the grain below that used in the tests (No. 150) seemed necessary before the maximum rate of cutting was reached. This reduction, however, was continuous and soon passed below that which gave the maximum cutting rate.

Cutting Qualities with Different Laps. — The surfaces of the steel and cast-iron laps were finished by grinding. The hardness of the different laps, as determined by the scleroscope was, for cast-iron, 28; steel, 18; copper, 5. The total amount ground from the test-pieces with each of the three laps showed that, taking the whole number of tests as a standard, there is scarcely any difference between the steel and cast iron, but that copper has somewhat better cutting qualities, although, when comparing the laps on the basis of the highest and lowest values obtained with each lap, steel and cast iron are as good for all practical purposes as copper, when the proper abrasive and lubricant are used.

Wear of Laps. — The wear of laps depends upon the material from which they are made and the abrasive used. The wear on all laps was about twice as fast with carborundum as with emery, while with alundum the wear was about one and one-fourth times that with emery. On an average, the wear of the copper lap was about three times that of the cast-iron lap. This is not absolute wear, but wear in proportion to the amount ground from the test-pieces.

Lapping Abrasives. — As to the qualities of the three abrasives tested, it was found that carborundum usually began at a lower rate than the other abrasives, but, when once started, its rate was better maintained. The performance gave a curve that was more nearly a straight line. The charge or residue as the grinding proceeded remained cleaner and sharper and did not tend to become pasty or mucklike, as is so frequently the case with emery. When using a copper lap, carborundum shows but little gain over the cast-iron and steel laps, whereas, with emery and alundum, the gain is considerable.

Effect of Different Lapping Lubricants. — The action of the different lubricants, when tested, was found to depend upon the kind of abrasive and the lap material.

Lard and Machine Oil. — The test showed that lard oil, without exception, gave the higher rate of cutting, and that, in general, the initial rate of cutting is higher with the lighter lubricants, but falls off more rapidly as the test continues. The lowest results were obtained with machine oil, when using an emergy-charged, cast-iron lap. When using lard oil and a carborundum-charged steel lap, the highest results were obtained.

Gasoline and Kerosene. — On the cast-iron lap, gasoline was superior to any of the lubricants tested. Considering all three abrasives, the relative value of gasoline, when applied to the different laps, is as follows: Cast iron, 127; copper, 115; steel, 106. Kerosene, like gasoline, gives the best results on cast iron and the poorest on steel. The values obtained by carborundum were invariably higher than those obtained with emery, except when using gasoline and kerosene on a copper lap.

Turpentine and Alcohol. — Turpentine was found to do good work with carborundum on any lap. With emery, turpentine did fair work on the copper lap, but, with the emery on cast-iron and steel laps, it was distinctly inferior. Alcohol gives the lowest results with emery on the cast-iron and steel laps.

Soda Water. — Soda water gives medium results with almost any combination of lap and abrasives, the best work being on the copper lap and the poorest, on the steel lap. On the cast-iron lap, soda water is better than machine or lard oil, but not so good as gasoline or kerosene. Soda water when used with alundum on the copper lap, gave the highest results of any of the lubricants used with that particular combination.

Lapping Pressures. — Within the limits of the pressures used, that is, up to 25 pounds per square inch, the rate of cutting was found to be practically proportional to the pressure. The higher pressures of 20 and 25 pounds per square inch are not so effective on the copper lap as on the other materials.

Wet and Dry Lapping. — With the "wet method" of using a surface lap, there is a surplus of oil and abrasive on the surface of the lap. As the specimen being lapped is moved over it, there is more or less movement or shifting of the abrasive particles. With the "dry method," the lap is first charged by rubbing or rolling the abrasive into its surface. All surplus oil and abrasive are then washed off, leaving a clean surface, but one that has embedded uniformly over it small particles of the abrasive. It is then like the surface of a very fine oilstone and will cut away hardened steel that is rubbed over it. While this has been termed the dry method, in practice, the lap surface is kept moistened with kerosene or gasoline.

Experiments on dry lapping were carried out on the cast-iron, steel, and copper laps used in the previous tests, and also on one of tin made expressly for the purpose. Carborundum alone was used as the abrasive and a uniform pressure of 15 pounds per square inch was applied to the specimen throughout the tests. In dry lapping, much depends upon the manner of charging the lap. The rate of cutting decreased much more rapidly after the first 100 revolutions than with the wet method. Considering the amounts ground off during the first 100 revolutions, and the best result obtained with each lap taken as the basis of comparison, it was found that with a tin lap, charged by rolling No. 150 carborundum into the surface, the rate of cutting, when dry, approached that obtained with the wet method. With the other lap materials, the rate with the dry method was about one-half that of the wet method.

Summary of Lapping Tests. — The initial rate of cutting does not greatly differ for different abrasives. There is no advantage in using an abrasive coarser than No. 150. The rate of cutting is practically proportional to the pressure. The wear of the laps is in the following proportions: cast iron, 1.00; steel, 1.27; copper, 2.62. In general, copper and steel cut faster than cast iron, but, where permanence of form is a consideration, cast iron is the superior metal. Gasoline and kerosene are the best lubricants to use with a cast-iron lap. Machine and lard oil are the best lubricants to use with copper or steel laps. They are, however, least effective on a cast-iron lap. In general, wet lapping is from 1.2 to 6 times as fast as dry lapping, depending upon the material of the lap and the manner of charging.

Standard Shapes and Metric Sizes of Mounted Wheels and Points* (ANSI B74.2-1982)

Abrasive Shape No.†	Abrasive Shape Size		Abrasive Shape No.†	Abrasive Shape Size	
	Diameter	Thickness		Diameter	Thickness
A 1	20	65	A 24	6	20
A 3	22	70	A 25	25	...
A 4	30	30	A 26	16	...
A 5	20	28	A 31	35	26
A 11	21	45	A 32	25	20
A 12	18	30	A 34	38	10
A 13	25	25	A 35	25	10
A 14	18	22	A 36	40	10
A 15	6	25	A 37	30	6
A 21	25	25	A 38	25	25
A 23	20	25	A 39	20	20
B 41	16	16	B 97	3	10
B 42	13	20	B 101	16	18
B 43	6	8	B 103	16	5
B 44	5.6	10	B 104	8	10
B 51	11	20	B 111	11	18
B 52	10	20	B 112	10	13
B 53	8	16	B 121	13	...
B 61	20	8	B 122	10	...
B 62	13	10	B 123	5	...
B 71	16	3	B 124	3	...
B 81	20	5	B 131	13	13
B 91	13	16	B 132	10	13
B 92	6	6	B 133	10	10
B 96	3	6	B 135	6	13
W 144	3	6	W 196	16	26
W 145	3	10	W 197	16	50
W 146	3	13	W 200	20	3
W 152	5	6	W 201	20	6
W 153	5	10	W 202	20	10
W 154	5	13	W 203	20	13
W 158	6	3	W 204	20	20
W 160	6	6	W 205	20	25
W 162	6	10	W 207	20	40
W 163	6	13	W 208	20	50
W 164	6	20	W 215	25	3
W 174	10	6	W 216	25	6
W 175	10	10	W 217	25	10
W 176	10	13	W 218	25	13
W 177	10	20	W 220	25	25
W 178	10	25	W 221	25	40
W 179	10	30	W 222	25	50
W 181	13	1.5	W 225	30	6
W 182	13	3	W 226	30	10
W 183	13	6	W 228	30	20
W 184	13	10	W 230	30	30
W 185	13	13	W 232	30	50
W 186	13	20	W 235	40	6
W 187	13	25	W 236	40	13
W 188	13	40	W 237	40	25
W 189	13	50	W 238	40	40
W 195	16	20	W 242	50	25

*All dimensions are in millimeters.
† See shape diagrams on pages 1106 and 1107.

Standard Shapes and Inch Sizes of Mounted Wheels and Points
(ANSI B74.2-1982) — I

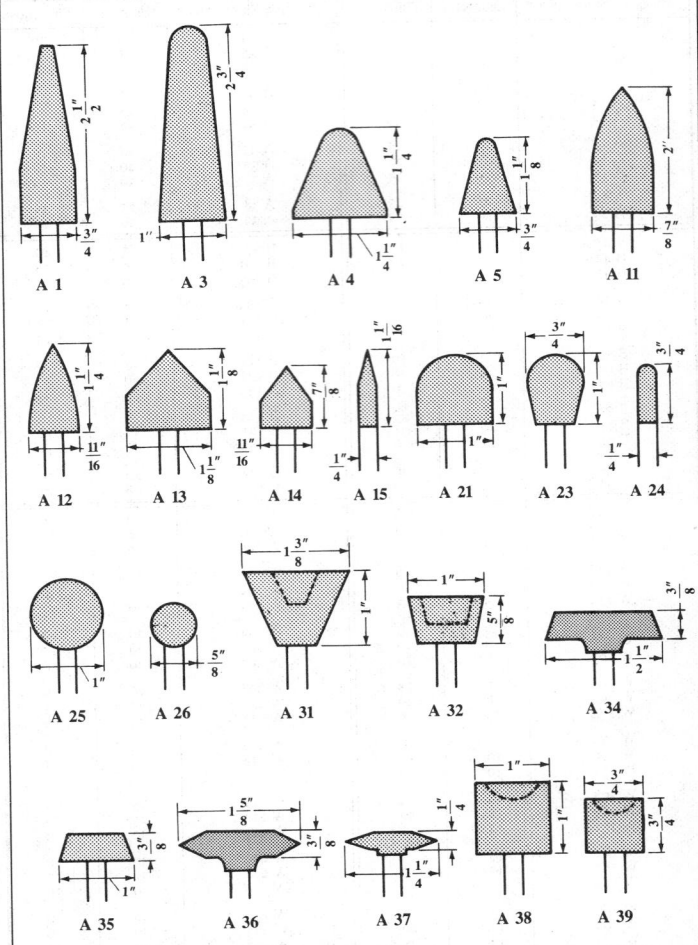

The maximum speeds of mounted vitrified wheels and points of average grade range from about 38,000 to 152,000 rpm for diameters of 1 inch down to ¼ inch. However, the safe operating speed usually is limited by the critical speed (speed at which vibration or whip tends to become excessive) which varies according to wheel or point dimensions, spindle diameter, and overhang.

Standard Shapes and Inch Sizes of Mounted Wheels and Points
(ANSI B74.2-1982) — 2

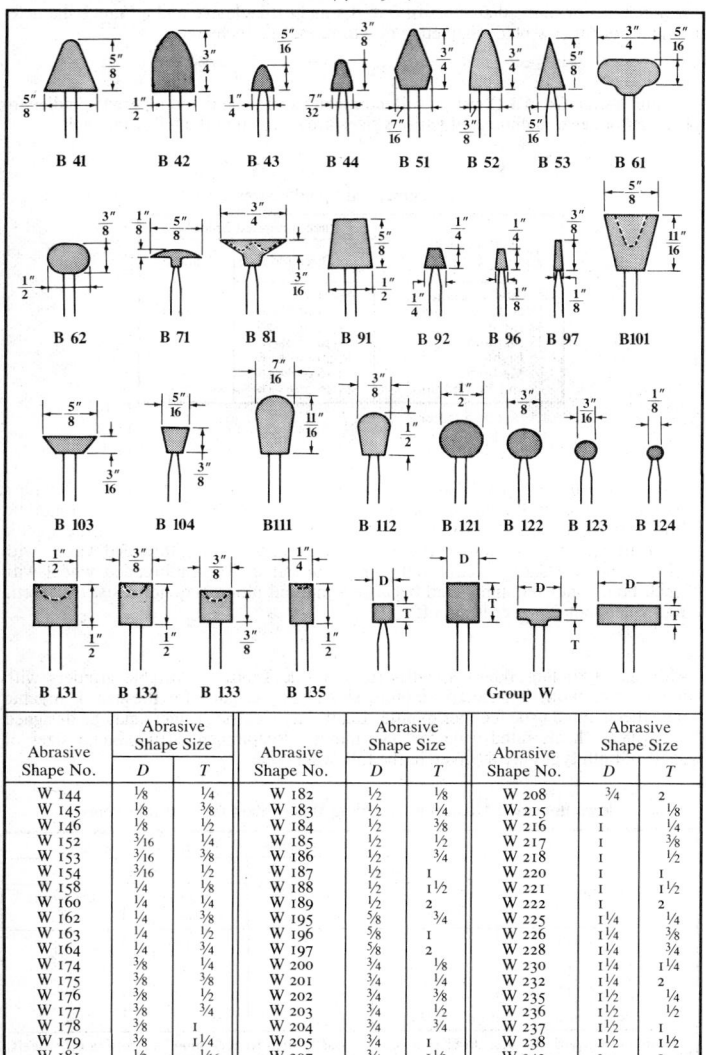

Abrasive Shape No.	Abrasive Shape Size		Abrasive Shape No.	Abrasive Shape Size		Abrasive Shape No.	Abrasive Shape Size	
	D	T		D	T		D	T
W 144	1/8	1/4	W 182	1/2	1/8	W 208	3/4	2
W 145	1/8	3/8	W 183	1/2	1/4	W 215	1	1/8
W 146	1/8	1/2	W 184	1/2	3/8	W 216	1	1/4
W 152	3/16	1/4	W 185	1/2	1/2	W 217	1	3/8
W 153	3/16	3/8	W 186	1/2	3/4	W 218	1	1/2
W 154	3/16	1/2	W 187	1/2	1	W 220	1	1
W 158	1/4	1/8	W 188	1/2	1 1/2	W 221	1	1 1/2
W 160	1/4	1/4	W 189	1/2	2	W 222	1	2
W 162	1/4	3/8	W 195	5/8	3/4	W 225	1 1/4	1/4
W 163	1/4	1/2	W 196	5/8	1	W 226	1 1/4	3/8
W 164	1/4	3/4	W 197	5/8	2	W 228	1 1/4	3/4
W 174	3/8	1/4	W 200	3/4	1/8	W 230	1 1/4	1 1/4
W 175	3/8	3/8	W 201	3/4	1/4	W 232	1 1/4	2
W 176	3/8	1/2	W 202	3/4	3/8	W 235	1 1/2	1/4
W 177	3/8	3/4	W 203	3/4	1/2	W 236	1 1/2	1/2
W 178	3/8	1	W 204	3/4	3/4	W 237	1 1/2	1
W 179	3/8	1 1/4	W 205	3/4	1	W 238	1 1/2	1 1/2
W 181	1/2	1/16	W 207	3/4	1 1/2	W 242	2	1

Circular Saw Arbors. — American Standard ASA B 5.38 "Driving and Spindle Ends for Portable and Air Electric Tools" calls for a round arbor of ⅝-inch diameter for nominal saw blade diameters of 6 to 8.5 inches, inclusive and a ¾-inch diameter round arbor for saw blade diameters of 9 to 12 inches, inclusive.

Spindles for Geared Chucks. — Recommended threaded and tapered spindles for portable tool geared chucks of various sizes are as given in the following table:

Recommended Spindle Sizes

Chuck Sizes, Inch	Recommended Spindles	
	Threaded	Taper*
³⁄₁₆ and ¼ Light	⅜-24	1
¼ and ⁵⁄₁₆ Medium	⅜-24 or ½-20	2 Short
⅜ Light	⅜-24 or ½-20	2
⅜ Medium	½-20 or ⅝-16	2
½ Light	½-20 or ⅝-16	33
½ Medium	⅝-16 or ¾-16	6
⅝ and ¾ Medium	⅝-16 or ¾-16	3

* Jacobs number.

Vertical and Angle Portable Tool Grinder Spindles. — The ⅝-11 spindle with a length of 1⅛ inches shown on page 1111 is designed to permit the use of a jam nut with threaded cup wheels. When a revolving guard is used, the length of the spindle is measured from the wheel bearing surface of the guard. For unthreaded wheels with a ⅞-inch hole, a safety sleeve nut is recommended. The unthreaded wheel with ⅝-inch hole is not recommended because a jam nut alone may not resist the inertia effect when motor power is cut off.

Straight Grinding Wheel Spindles for Portable Tools. — Portable grinders with pneumatic or induction electric motors should be designed for the use of organic bond wheels rated 9500 feet per minute. Light-duty electric grinders may be designed for vitrified wheels rated 6500 feet per minute. Recommended maximum sizes of wheels of both types are as given in the following table:

Recommended Maximum Grinding Wheel Sizes for Portable Tools

Spindle Size	Maximum Wheel Dimensions			
	9500 fpm		6500 fpm	
	Diameter D	Thickness T	Diameter D	Thickness T
⅜-24 × 1⅛	2½	½	4	½
½-13 × 1¾	4	¾	5	¾
⅝-11 × 2⅛	8	1	8	1
⅝-11 × 3⅛	6	2	...	...
⅝-11 × 3⅛	8	1½	...	...
¾-10 × 3¼	8	2	...	...

Minimum T with the first three spindles is about ⅛ inch to accommodate cutting off wheels. Flanges are assumed to be according to ANSI B7.1 and threads to ANSI B1.1.

American Standard Threaded and Tapered Spindles for Portable Air and Electric Tools (ASA B5.38-1958)

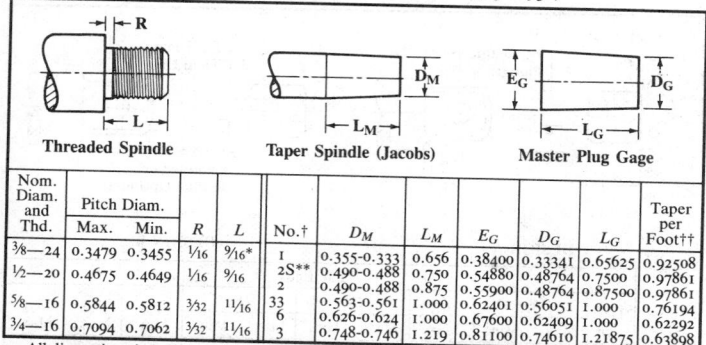

Threaded Spindle   **Taper Spindle (Jacobs)**   **Master Plug Gage**

Nom. Diam. and Thd.	Pitch Diam.		R	L	No.†	D_M	L_M	E_G	D_G	L_G	Taper per Foot††
	Max.	Min.									
³⁄₈—24	0.3479	0.3455	¹⁄₁₆	⁹⁄₁₆*	1	0.355-0.333	0.656	0.38400	0.33341	0.65625	0.92508
¹⁄₂—20	0.4675	0.4649	¹⁄₁₆	⁹⁄₁₆	2S**	0.490-0.488	0.750	0.54880	0.48764	0.7500	0.97861
					2	0.490-0.488	0.875	0.55900	0.48764	0.87500	0.97861
⁵⁄₈—16	0.5844	0.5812	³⁄₃₂	¹¹⁄₁₆	33	0.563-0.561	1.000	0.62401	0.56051	1.000	0.76194
³⁄₄—16	0.7094	0.7062	³⁄₃₂	¹¹⁄₁₆	6	0.626-0.624	1.000	0.67600	0.62409	1.000	0.62292
					3	0.748-0.746	1.219	0.81100	0.74610	1.21875	0.63898

All dimensions in inches.   Threads are per inch and right-hand.   * Also ⁷⁄₁₆ inch.   ** 2S stands for 2 Short.   † Jacobs taper number.   †† Calculated from E_G, D_G, L_G for the master plug gage. *Tolerances:* On R, plus or minus ¹⁄₆₄ inch; on L, plus 0.000, minus 0.030 inch.

American Standard Hexagonal Chucks for Portable Air and Electric Tools
(ASA B5.38-1958)

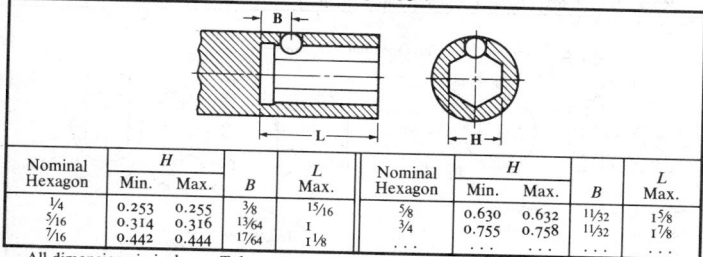

Nominal Hexagon	H		B	L Max.	Nominal Hexagon	H		B	L Max.
	Min.	Max.				Min.	Max.		
¹⁄₄	0.253	0.255	³⁄₈	¹⁵⁄₁₆	⁵⁄₈	0.630	0.632	¹¹⁄₃₂	1⁵⁄₈
⁵⁄₁₆	0.314	0.316	¹³⁄₆₄	1	³⁄₄	0.755	0.758	¹¹⁄₃₂	1⁷⁄₈
⁷⁄₁₆	0.442	0.444	¹⁷⁄₆₄	1¹⁄₈	...	...	...	...	...

All dimensions in inches.   Tolerance on B is plus or minus 0.005 inch.

American Standard Hexagon Shanks for Portable Air and Electric Tools
(ASA B5.38-1958)

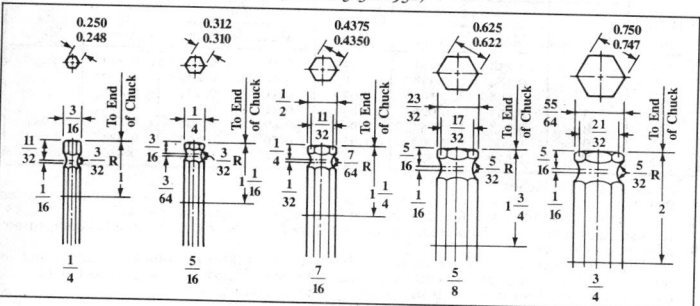

American Standard Square Drives for Portable Air and Electric Tools
(ASA B5.38-1958)

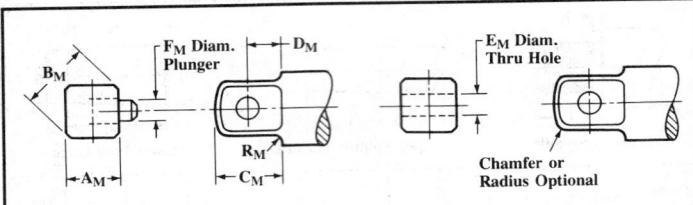

DESIGN A

DESIGN B

Drive Size	Design	A_M Max.	A_M Min.	B_M Max.	C_M Max.	C_M Min.	D_M Max.	D_M Min.	E_M Min.	F_M Max.	R_M Max.
1/4	A	0.252	0.247	0.330	0.312	0.265	0.165	0.153	...	0.078	0.015
3/8	A	0.377	0.372	0.500	0.438	0.406	0.227	0.215	...	0.156	0.031
1/2	A	0.502	0.497	0.665	0.625	0.531	0.321	0.309	...	0.187	0.031
5/8	A	0.627	0.622	0.834	0.656	0.594	0.321	0.309	...	0.187	0.047
3/4	B	0.752	0.747	1.000	0.938	0.750	0.415	0.403	0.216	...	0.047
1	B	1.002	0.997	1.340	1.125	1.000	0.602	0.590	0.234	...	0.063
1 1/2	B	1.503	1.498	1.968	1.625	1.562	0.653	0.641	0.310	...	0.094

Male End

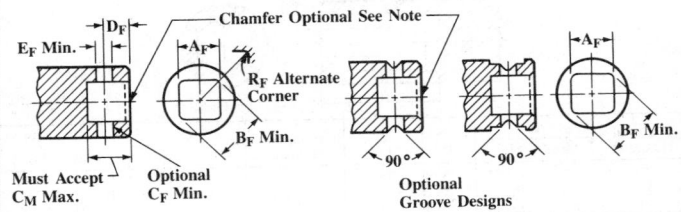

DESIGN A

DESIGN B

Drive Size	Design	A_F Max.	A_F Min.	B_F Min.	D_F Max.	D_F Min.	E_F Min.	R_F Max.
1/4	A	0.258	0.253	0.335	0.159	0.147	0.090	...
3/8	A	0.383	0.378	0.505	0.221	0.209	0.170	...
1/2	A	0.508	0.503	0.670	0.315	0.303	0.201	...
5/8	A	0.633	0.628	0.839	0.315	0.303	0.201	...
3/4	B	0.758	0.753	1.005	0.409	0.397	0.216	0.047
1	B	1.009	1.004	1.350	0.596	0.584	0.234	0.062
1 1/2	B	1.510	1.505	1.983	0.647	0.635	0.310	0.125

Female End

All dimensions in inches.

Incorporating fillet radius (R_M) at shoulder of male tang precludes use of minimum diameter cross-hole in socket (E_F), unless female drive end is chamfered (shown as optional).

If female drive end is not chamfered, socket cross-hole diameter (E_F) is increased to compensate for fillet radius R_M, max.

Minimum clearance across flats male to female is 0.001 inch through 3/4-inch size; 0.002 inch in 1- and 1 1/2-inch sizes. For impact wrenches A_M should be held as close to maximum as practical.

C_F, min. for both designs A and B should be equal to C_M, max.

American Standard Abrasion Tool Spindles for Portable Air and Electric Tools
(ASA B5.38-1958)

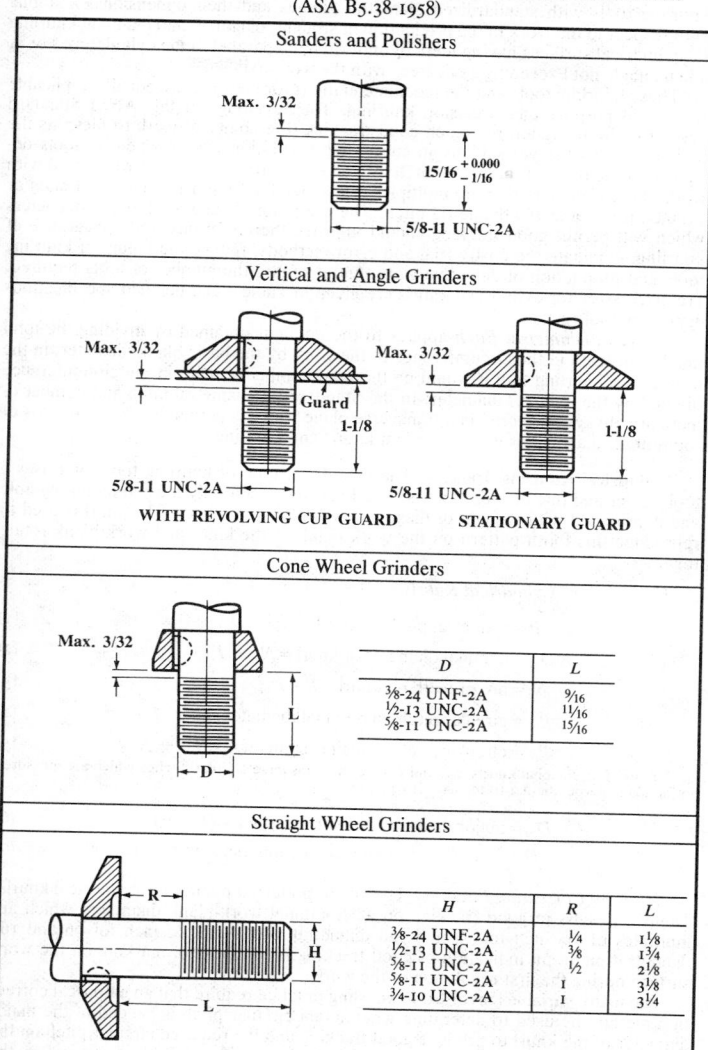

Sanders and Polishers

Max. 3/32

$15/16 \begin{smallmatrix} +0.000 \\ -1/16 \end{smallmatrix}$

5/8-11 UNC-2A

Vertical and Angle Grinders

Max. 3/32

Guard

1-1/8

5/8-11 UNC-2A

WITH REVOLVING CUP GUARD

Max. 3/32

1-1/8

5/8-11 UNC-2A

STATIONARY GUARD

Cone Wheel Grinders

Max. 3/32

L

D

D	L
3/8-24 UNF-2A	9/16
1/2-13 UNC-2A	11/16
5/8-11 UNC-2A	15/16

Straight Wheel Grinders

R

H

L

H	R	L
3/8-24 UNF-2A	1/4	1 1/8
1/2-13 UNC-2A	3/8	1 3/4
5/8-11 UNC-2A	1/2	2 1/8
5/8-11 UNC-2A	1	3 1/8
3/4-10 UNC-2A	1	3 1/4

All dimensions in inches.   Threads are right-hand.

ANSI Standard Knurls and Knurling. — The ANSI Standard B94.6-1984 covers knurling tools with standardized diametral pitches and their dimensional relations with respect to the work in the production of straight, diagonal, and diamond knurling on cylindrical surfaces having teeth of uniform pitch parallel to the cylinder axis or at a helix angle not exceeding 45 degrees with the work axis.

These knurling tools and the recommendations for their use are equally applicable to general purpose and precision knurling. The advantage of this ANSI Standard system is the provision by which good tracking (the ability of teeth to mesh as the tool penetrates the work blank in successive revolutions) is obtained by tools designed on the basis of diametral pitch instead of TPI (teeth per inch) when used with work blank diameters that are multiples of $1/64$ inch for 64 and 128 diametral pitch or $1/32$ inch for 96 and 160 diametral pitch. The use of knurls and work blank diameters which will permit good tracking should improve the uniformity and appearance of knurling, eliminate the costly trial and error methods, reduce the failure of knurling tools and production of defective work, and decrease the number of tools required. Preferred sizes for cylindrical knurls are given in Table 1 and detailed specifications appear in Table 2.

The term *Diametral Pitch* applies to the quotient obtained by dividing the total number of teeth in the circumference of the work by the basic blank diameter; in the case of the knurling tool it would be the total number of teeth in the circumference divided by the *nominal* diameter. In the Standard the diametral pitch and number of teeth are always measured in a transverse plane which is perpendicular to the axis of rotation for diagonal as well as straight knurls and knurling.

Cylindrical Knurling Tools. — The cylindrical type of knurling tool comprises a tool holder and one or more knurls. The knurl has a centrally located mounting hole and is provided with straight or diagonal teeth on its periphery. The knurl is used to reproduce this tooth pattern on the work blank as the knurl and work blank rotate together.

Formulas for Cylindrical Knurls:

$$P = \text{diametral pitch of knurl} = N_t \div D_{nt} \qquad (1)$$

$$D_{nt} = \text{nominal diameter of knurl} = N_t \div P \qquad (2)$$

$$N_t = \text{no. of teeth on knurl} = P \times D_{nt} \qquad (3)$$

$$*P_{nt} = \text{circular pitch on nominal diameter} = \pi \div P \qquad (4)$$

$$*P_{ot} = \text{circular pitch on major diameter} = \pi D_{ot} \div N_t \qquad (5)$$

* *Note:* For diagonal knurls, P_{nt} and P_{ot} are the transverse circular pitches which are measured in the plane perpendicular to the axis of rotation.

$$D_{ot} = \text{major diameter of knurl} = D_{nt} - (N_t Q \div \pi) \qquad (6)$$

$$Q = P_{nt} - P_{ot} = \text{tracking correction factor in Formula (6)} \qquad (7)$$

Tracking Correction Factor Q: Use of the preferred pitches for cylindrical knurls, Table 2, results in good tracking on all fractional work-blank diameters which are multiples of $1/64$ inch for 64 and 128 diametral pitch, and $1/32$ inch for 96 and 160 diametral pitch; an indication of good tracking is evenness of marking on the work surface during the first revolution of the work.

The many variables involved in knurling practice require that an empirical correction method be used to determine what actual circular pitch is needed at the major diameter of the knurl to produce good tracking and the required circular pitch on the workpiece. The empirical tracking correcton factor, Q, in Table 2 is used in the calculation of the major diameter of the knurl, Formula (6).

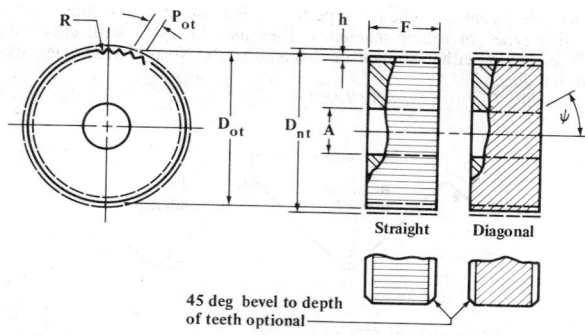

Cylindrical Knurl

45 deg bevel to depth
of teeth optional

Straight   Diagonal

Table 1. ANSI Standard Preferred Sizes for Cylindrical Type Knurls*
(ANSI B94.6-1984)

Nominal Outside Diameter D_{nt}	Width of Face F	Diameter of Hole A	Standard Diametral Pitches, P			
			64	96*	128	160
			Number of Teeth, N_t, for Standard Pitches			
1/2	3/16	3/16	32	48	64	80
5/8	1/4	1/4	40	60	80	100
3/4	3/8	1/4	48	72	96	120
7/8	3/8	1/4	56	84	112	140
Additional Sizes for Bench and Engine Lathe Tool Holders						
5/8	5/16	7/32	40	60	80	100
3/4	5/8	1/4	48	72	96	120
1	3/8	5/16	64	96	128	160

* The 96 diametral pitch knurl should be given preference in the interest of tool simplification. Dimensions D_{nt}, F, and A are in inches.

Table 2. ANSI Standard Specifications for Cylindrical Knurls with Straight or Diagonal Teeth* (ANSI B94.6-1984)

Diametral Pitch P	Nominal Diameter, D_{nt}					Tracking Correction Factor Q	Tooth Depth, h, +0.0015, −0.0000		Radius at Root R
	1/2	5/8	3/4	7/8	1		Straight	Diagonal	
	Major Diameter of Knurl, D_{ot}, +0.0000, −0.0015								
64	0.4932	0.6165	0.7398	0.8631	0.9864	0.0006676	0.024	0.021	0.0070
96	0.4960	0.6200	0.7440	0.8680	0.9920	0.0002618	0.016	0.014	0.0050 0.0060
128	0.4972	0.6215	0.7458	0.8701	0.9944	0.0001374	0.012	0.010	0.0040 0.0045
160	0.4976	0.6220	0.7464	0.8708	0.9952	0.00009425	0.009	0.008	0.0030 0.0040 0.0025

All dimensions except diametral pitch are in inches.
Approximate angle of space between sides of adjacent teeth for both straight and diagonal teeth is 80 degrees. The permissible eccentricity of teeth for all knurls is 0.002 inch maximum (total indicator reading).
Number of teeth in a knurl equals diametral pitch multiplied by nominal diameter.
* Diagonal teeth have 30-degree helix angle, ψ.

Flat Knurling Tools. — The flat type of tool is a knurling die, commonly used in reciprocating types of rolling machines. Dies may be made with either single or duplex faces having either straight or diagonal teeth. No preferred sizes are established for flat dies.

Flat Knurling Die with Straight Teeth:

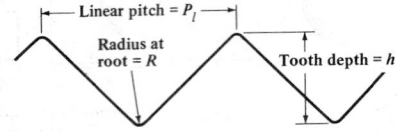

R = radius at root

P = diametral pitch = $N_w \div D_w$ $\hfill(8)$

D_w = work blank (pitch) diameter = $N_w \div P$ $\hfill(9)$

N_w = number of teeth on work = $P \times D_w$ $\hfill(10)$

h = tooth depth

Q = tracking correction factor (see Table 2)

P_l = linear pitch on die

= circular pitch on work pitch diameter = $P - Q$ $\hfill(11)$

Table 3. ANSI Standard Specifications for Flat Knurling Dies (ANSI B94.6-1984)

Diametral Pitch, P	Linear Pitch,* P_l	Tooth Depth, h		Radius at Root, R	Diametral Pitch, P	Linear Pitch,* P_l	Tooth Depth, h		Radius at Root, R
		Straight	Diagonal				Straight	Diagonal	
64	0.0484	0.024	0.021	0.0070	128	0.0244	0.012	0.010	0.0045
				0.0050					0.0030
96	0.0325	0.016	0.014	0.0060	160	0.0195	0.009	0.008	0.0040
				0.0040					0.0025

All dimensions except diametral pitch are in inches.

* The linear pitches are theoretical. The exact linear pitch produced by a flat knurling die may vary slightly from those shown depending upon the rolling condition and the material being rolled.

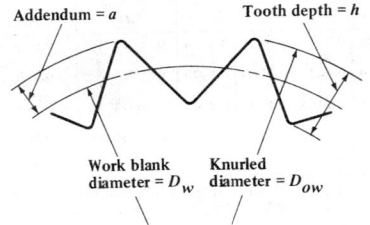

Teeth on Knurled Work

Formulas Applicable to Knurled Work. — The following formulas are applicable to knurled work with straight, diagonal, and diamond knurling.

Formulas for Straight or Diagonal Knurling with Straight or Diagonal Tooth Cylindrical Knurling Tools Set with Knurl Axis Parallel with Work Axis:

$$P = \text{diametral pitch} = N_w \div D_w \tag{12}$$

$$D_w = \text{work blank diameter} = N_w \div P \tag{13}$$

$$N_w = \text{no. of teeth on work} = P \times D_w \tag{14}$$

$$a = \text{``addendum'' of tooth on work} = (D_{ow} - D_w) \div 2 \tag{15}$$

$$h = \text{tooth depth (see Table 2)}$$

$$D_{ow} = \text{knurled diameter (outside diameter after knurling)} = D_w + 2a \tag{16}$$

Formulas for Diagonal and Diamond Knurling with Straight Tooth Knurling Tools Set at an Angle to the Work Axis:

If, ψ = angle between tool axis and work axis;
P = diametral pitch on tool;
P_ψ = diametral pitch produced on work blank (as measured in the transverse plane) by setting tool axis at an angle ψ with respect to work blank axis;
D_w = diameter of work blank; and
N_w = number of teeth produced on work blank (as measured in the transverse plane);

then, $$P_\psi = P \cos \psi \tag{17}$$

and, $$N_w = D_w P \cos \psi \tag{18}$$

For example, if 30 degree diagonal knurling were to be produced on 1-inch diameter stock with a 160 pitch straight knurl:

$$N_w = D_w P \cos 30° = 1.000 \times 160 \times 0.86603 = 138.56 \text{ teeth}$$

Good tracking is theoretically possible by changing the helix angle as follows to correspond to a whole number of teeth (138):

$$\cos \psi = N_w \div D_w P = 138 \div (1 \times 160) = 0.8625$$

$$\psi = 30\tfrac{1}{2} \text{ degrees, approximately}$$

Whenever it is more practical to machine the stock, good tracking can be obtained by reducing the work blank diameter as follows to correspond to a whole number of teeth (138):

$$D_w = \frac{N_w}{P \cos \psi} = \frac{138}{160 \times 0.866} = 0.996 \text{ inch}$$

Table 4. ANSI Standard Recommended Tolerances on Knurled Diameters
(ANSI B94.6-1984)

Toler-ance Class	Diametral Pitch							
	64	96	128	160	64	96	128	160
	Tolerance on Knurled Outside Diameter				Tolerance on Work-Blank Diameter Before Knurling			
I	+ 0.005 − 0.012	+ 0.004 − 0.010	+ 0.003 − 0.008	+ 0.002 − 0.006	±0.0015	±0.0010	±0.0007	±0.0005
II	+ 0.000 − 0.010	+ 0.000 − 0.009	+ 0.000 − 0.008	+ 0.000 − 0.006	±0.0015	±0.0010	±0.0007	±0.0005
III	+ 0.000 − 0.006	+ 0.000 − 0.005	+ 0.000 − 0.004	+ 0.000 − 0.003	+ 0.000 − 0.0015	+ 0.0000 − 0.0010	+ 0.000 − 0.0007	+ 0.0000 − 0.0005

Recommended Tolerances on Knurled Outside Diameters. — The recommended applications of the tolerance classes shown in Table 4 are as follows:

Class I: Tolerances in this classification may be applied to straight, diagonal and raised diamond knurling where the knurled outside diameter of the work need not be held to close dimensional tolerances. Such applications include knurling for decorative effect, grip on thumb screws, and inserts for moldings and castings.

Class II: Tolerances in this classification may be applied to straight knurling only and are recommended for applications requiring closer dimensional control of the knurled outside diameter than provided for by Class I tolerances.

Class III: Tolerances in this classification may be applied to straight knurling only and are recommended for applications requiring closest possible dimensional control of the knurled outside diameter. Such applications include knurling for close fits.

Note: The width of the knurling should not exceed the diameter of the blank, and knurling wider than the knurling tool cannot be produced unless the knurl starts at the end of the work.

Marking on Knurls and Dies. — Each knurl and die should be marked as follows: a. when straight to indicate its diametral pitch; b. when diagonal, to indicate its diametral pitch, helix angle, and hand of angle.

Concave Knurls. — The radius of a concave knurl should not be the same as the radius of the piece to be knurled. If the knurl and the work are of the same radius, the material compressed by the knurl will be forced down on the shoulder *D* and spoil the appearance of the work. A design of concave knurl is shown in the accompanying illustration, and all the important dimensions are designated by letters. To find these

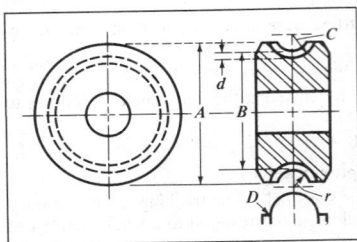

dimensions, the pitch of the knurl required must be known, and also, approximately, the throat diameter *B*. This diameter must suit the knurl holder used, and be such that the circumference contains an even number of teeth with the required pitch. When these dimensions have been decided upon, all the other unknown factors can be found by the following formulas: Let R = radius of piece to be knurled; r = radius of concave part of knurl; C = radius of cutter or hob for cutting the teeth in the knurl; B = diameter over concave part of knurl (throat diameter); A = outside diameter of knurl; d = depth of tooth in knurl; P = pitch of knurl (number of teeth per inch circumference); p = circular pitch of knurl; then $r = R + \frac{1}{2}d$; $C = r + d$; $A = B + 2r - (3d + 0.010$ inch); and $d = 0.5 \times p \times \cot \alpha/2$, where α is the included angle of the teeth.

As the depth of the tooth is usually very slight, the throat diameter *B* will be accurate enough for all practical purposes for calculating the pitch, and it is not necessary to take into consideration the pitch circle. For example, assume that the pitch of a knurl is 32, that the throat diameter *B* is 0.5561 inch, that the radius *R* of the piece to be knurled is ¹⁄₁₆ inch, and that the angle of the teeth is 90 degrees; find the dimensions of the knurl. Using the notation given:

$$p = \frac{1}{P} = \frac{1}{32} = 0.03125 \text{ inch}; \quad d = 0.5 \times 0.03125 \times \cot 45° = 0.0156 \text{ inch};$$

$$r = \frac{1}{16} + \frac{0.0156}{2} = 0.0703 \text{ inch}; \quad C = 0.0703 + 0.0156 = 0.0859 \text{ inch};$$

$$A = 0.5561 + 0.1406 - (0.0468 + 0.010) = 0.6399 \text{ inch}.$$

NUMERICAL CONTROL

Introduction. — The Electronic Industries Association (EIA) defines numerical control as "a system in which actions are controlled by the direct insertion of numerical data at some point." More specifically, numerical control, or NC as it will be termed here, involves machines controlled by electronic systems designed to accept numerical data and other instructions, usually in a coded form. These instructions may come directly from some source such as a punched tape, a floppy disc, directly from a computer, or from an operator. The latter three input methods are gaining popularity, but punched tape remains the most popular input medium.

Numerical Control Subject Index

The key to the success of numerical control lies in its flexibility. To machine a different part it is only necessary to "play" a different tape. NC machines are more productive than conventional equipment and consequently produce parts at less cost even when the higher investment is considered. NC machines also are more accurate and produce far less scrap than their conventional counterparts. By 1985, over 110,000 NC machine tools were operating in the United States. Over 80 per cent of the dollars being spent on the most common types of machine tools, namely, drilling, milling, boring, and turning machines, are going into NC equipment.

NC is a generic term for the whole field of numerical control and encompasses a complete field of endeavor. Sometimes CNC, which stands for Computer Numerical Control and applies only to the control system, is used erroneously as a replacement term for NC. Albeit a monumental development, use of the term CNC should be confined to installations where the older hardwire control systems have been replaced.

Metal cutting is the most popular application but NC is being applied successfully to other equipment including punch presses, EDM wire cutting machines, inspection machines, laser and other cutting and torching machines, tube bending machines, and sheet metal cutting and forming machines.

The items presented alphabetically in this section have been selected as being of particular interest to anyone working or planning to work with numerical control. The purpose of this section is not to provide a complete description of numerical control operation, for which a manual or text should be consulted, but to make available for quick reference information about topics on which readers may be unclear or may need to refresh their memory or understanding. The topics covered are listed above, and in the accompanying index.

Absolute Programming. — A part programming method which allows for definition of the coordinate locations of the points to which the cutter (or workpiece) is to move is called absolute programming. In Fig. 1, the X-Y coordinates of P1 are $X = 1.0$, $Y = 0.5$. The coordinates of P2 are $X = 2.0$, $Y = 1.1$. To indicate the movement from P1 to P2, the coordinates of P2 would be specified. Most CNC systems offer both absolute and incremental part programming. The choice is denoted by a G code; G90 for absolute programming and G91 for incremental programming. (See also *Incremental Programming*.)

Accuracy and Repeatability. — The National Machine Tool Builders Association* (NMTBA) defines accuracy as the signed (plus or minus) value of the difference

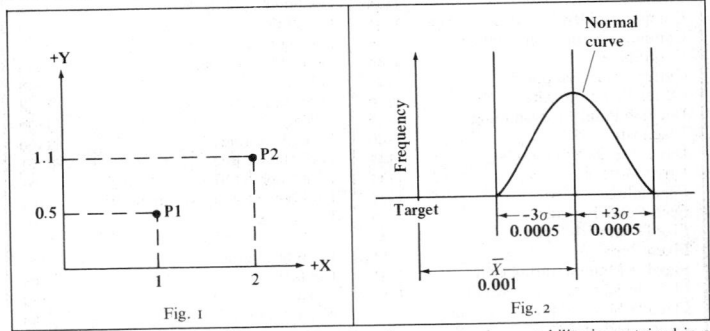

Fig. 1   Fig. 2

* An expanded and more detailed explanation of accuracy and repeatability is contained in a booklet published by the National Machine Tool Builders Association, 7901 Westpark Drive, McLean, VA 22102, phone (703) 893-2900

between the mean of the points arrived at and the given point (target), plus the value of the dispersion of the points arrived at that gives the largest absolute sum. It is not sufficient to note that accuracy and repeatability are based on the error of a single move to a programmed point, or even the average error of moves to a single point, but rather that the accuracy be based on the average error of moves to a point plus consideration of the spread of the points arrived at. The numerical measure of this spread is called the standard deviation, which is calculated by a statistical formula. Three times the standard deviation would account for over 99 per cent of the points in the spread. The accuracy therefore must be determined by the average error for the movements to the target plus three times the standard deviation of the spread, as illustrated in Fig. 2. Expressed mathematically:

$$A = \overline{X} \pm 3\sigma$$

where A = the accuracy

$\overline{X}$ = the average error from the target

σ = one standard deviation

The repeatability would be the spread, or $\pm 3\sigma$. The accuracy, as shown in Fig. 2, is +0.001 inch. In this instance, the repeatability would be ± 0.0005 inch. A rule of thumb is that the repeatability is generally one half the accuracy. If, for example, the positioning accuracy along an axis of a machine is given as plus or minus 0.001 inch, the repeatability would probably be plus or minus 0.0005 inch. It should be pointed out, however, that many of the newer CNC systems have the capability to compensate for certain errors in the machine tool. These errors can be corrected by features such as leadscrew compensation or axis alignment compensation. The CNC system could be preprogrammed to compensate, electronically, for the machine tool error. The term used for determining the error in the machine tool is deterministic metrology or mapping.

Adaptive Control. — Measuring performance of a process and then adjusting the process to obtain optimum performance is called adaptive control. In the machine tool field, adaptive control is a means of adjusting the feed and/or speed of the cutting tool, based on sensor feedback information, to maintain optimum cutting conditions. A typical arrangement is seen in Fig. 3. Adaptive control is used primarily for cutting higher strength materials such as titanium, although the concept is applicable to the cutting of any material. The cost of the sensors and software have restricted wider use of the feature.

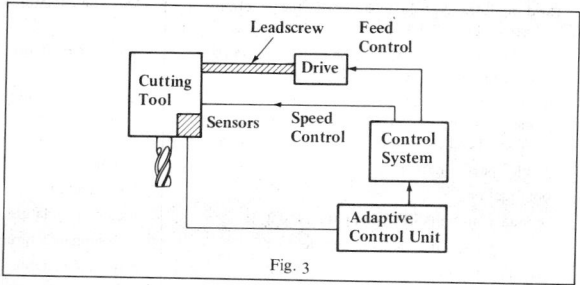

Fig. 3

Address. — In NC, an address refers to the letter that is the first notation of a word. For example, X is the letter address for the dimension word that requires a move in the direction of the X axis. As another example, F is the letter address for

the feed rate. The complete word consists of a letter plus numerical digits. Only two of the letters of the alphabet, H and L, are unassigned. The assigned addresses, as listed in the EIA Standard RS-274-D, are shown in the accompanying table.

Letter Addresses Used in Numerical Control

Letter Address	Description	Refers to
A	Angular dimension about the X axis. Measured in decimal parts of a degree	Axis nomenclature
B	Angular dimension about the Y axis. Measured in decimal parts of a degree	Axis nomenclature
C	Angular dimension about the Z axis. Measured in decimal parts of a degree	Axis nomenclature
D	Angular dimension about a special axis, or third feed function, or tool function for selection of tool compensation	Axis nomenclature
E	Angular dimension about a special axis or second feed function	Axis nomenclature
F	Feed word (code)	Feed words
G	Preparatory word (code)	Preparatory words
H	Unassigned	
I	Interpolation parameter or thread lead parallel to the X axis	Circular interpolation and threading
J	Interpolation parameter or thread lead parallel to the Y axis	Circular interpolation and threading
K	Interpolation parameter or thread lead parallel to the Z axis	Circular interpolation and threading
L	Unassigned	
M	Miscellaneous or auxiliary function	Miscellaneous functions
N	Sequence number	Sequence number
O	Sequence number for secondary head only	Sequence number
P	Third rapid-traverse dimension or tertiary motion dimension parallel to X	Axis nomenclature
Q	Second rapid-traverse dimension or tertiary motion dimension parallel to Y	Axis nomenclature
R	First rapid-traverse dimension or tertiary-motion dimension parallel to Z or radius for constant surface-speed calculation	Axis nomenclature
S	Spindle-speed function	Spindle speed
T	Tool function	Tool function
U	Secondary-motion dimension parallel to X	Axis nomenclature
V	Secondary-motion dimension parallel to Y	Axis nomenclature
W	Secondary-motion dimension parallel to Z	Axis nomenclature
X	Primary X-motion dimension	Axis nomenclature
Y	Primary Y-motion dimension	Axis nomenclature
Z	Primary Z-motion dimension	Axis nomenclature

APT. — APT stands for automatically programmed tool, and is a computer language designed for use with NC machine tools. The language is nonproprietary and in the public domain, and is the most advanced and most universal NC language, usable for parts of simple to exceptionally complex shape. Originally requiring a large mainframe computer, full APT can now be used on minicomputers, and two-dimensional systems, called ADAPT, have been developed for desk-top personal computers. APT has also been adopted as ANSI Standard X3.37, and by the International Organization for Standardization (ISO).

Owing to space limitations, the full capabilities of APT cannot be described here. What is described can be used as a ready-reference guide to the basic geometry patterns and motion statements of APT, which are suitable for programming the machining of the majority of cubic-type parts involving two-dimensional movements. Some of the three-dimensional geometry capabilities of APT and a description of its five-dimensional capability are also included. APT is a very dynamic program and is continually being updated. APT is even being used as a processor for part-programming graphic systems.

Many other computer-assisted part-programming languages are being offered; almost all of them are proprietary and require a fee for use. Most of these languages are two-dimensional and somewhat easier to use than APT. The selection of a computer-assisted part-programming language depends on the type and complexity of parts being machined more than on any other factor. APT has been selected for detailed description because it has the broadest range of capability, is nonproprietary, is probably the most utilized and is accepted worldwide.

As shown in Fig. 4, the APT system, within the computer, consists of five sections (0 through IV). The input program progresses through the first four sections and all four are controlled by a fifth section, 0. Section IV, the postprocessor, is the software package that is added to Sections II and III to customize the output and produce the necessary tape format (including the G words, M words, etc.), in which the coded instructions will be recognizable by the control system. Postprocessors are prepared, for a fee, by the machine tool builder, the control system builder, a third party, or, with the aid of a generalized instructional computer program, by the user. Postprocessors are not confined to APT and are used with most computerized part-programming languages.

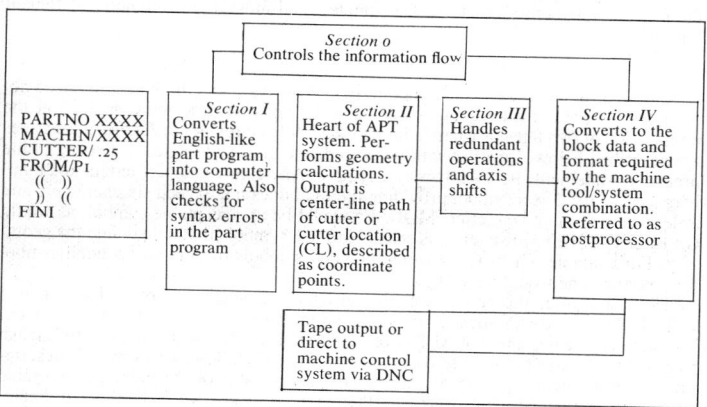

Fig. 4

APT Computational Statements. — Algebraic and trigonometric functions and computations can be performed with the APT system, as follows:

Arithmetic form	APT form	Arithmetic form	APT form
25×25	25*25	$\sqrt{25}$	SQRTF(25)
$25 \div 25$	25/25	$\sin \theta$	SINF(θ)
$25 + 25$	25 + 25	$\cos \theta$	COSF(θ)
$25 - 25$	25 − 25	$\tan \theta$	TANF(θ)
25^2	25**2	arctan .5000	ATANF(.5)
25^n	25**n		

Computations may be used in the APT system in two ways. One way is to let a factor equal the computation and then substitute the factor in a statement; the other is to put the computation directly into the statement. The following are a series of APT statements illustrating the first approach.

$$P1 = POINT/0,0,1$$
$$T = (25*2/3 + (3**2 - 1))$$
$$P2 = POINT/T,0,0$$

The second way would be as follows:

$$P1 = POINT/0,0,1$$
$$P2 = POINT/(25*2/3 + (3**2 - 1)),0,0$$

Note: The parentheses have been used as they would be in an algebraic formula so that the calculations will be carried out in proper sequence. The operations within the inner parentheses would be carried out first. It is important for the total number of left-hand parentheses to equal the total number of right-hand parentheses; otherwise the program will fail.

APT Geometry Statements. — Before movements around the geometry of a part can be described, the geometry must be defined. For example, in the statement GOTO/P1, the computer must know where P1 is located before the statement can be effective. P1 therefore must be described in a geometry statement, prior to its use in the motion statement GOTO/P1. The simplest and most direct geometry statement for a point is

$$P1 = POINT/X \text{ ordinate, Y ordinate, Z ordinate}$$

If the Z ordinate is zero and the point lies on the XY plane the Z location need not be noted. There are other ways of defining the position of a point, such as at the intersection of two lines or where a line is tangent to a circular arc. These alternatives are described below, together with ways to define lines and circles. Referring to the preceding statement, P1 is known as a symbol. Any combination of letters and numbers may be used as a symbol providing the total does not exceed six characters and at least one of them is a letter. MOUSE2 would be an acceptable symbol, as would CAT3 or FRISBE. However, it is sensible to use symbols that help define the geometry. For example, C1 or CIR3 would be good symbols for a circle. A good symbol for a vertical line would be VL5.

Next, and after the equal sign, the particular geometry is noted. Here, it is a POINT. This word is a vocabulary word and must be spelled exactly as prescribed. Throughout, the designers of APT have tried to use words that are as close to English as possible. A slash follows the vocabulary word and is followed by a specific description of the particular geometry, such as the coordinates of the point P1. A usable statement for P1 might appear as P1 = POINT/1,5,4. The 1 would be the X ordinate; the 5, the Y ordinate; and the 4, the Z ordinate.

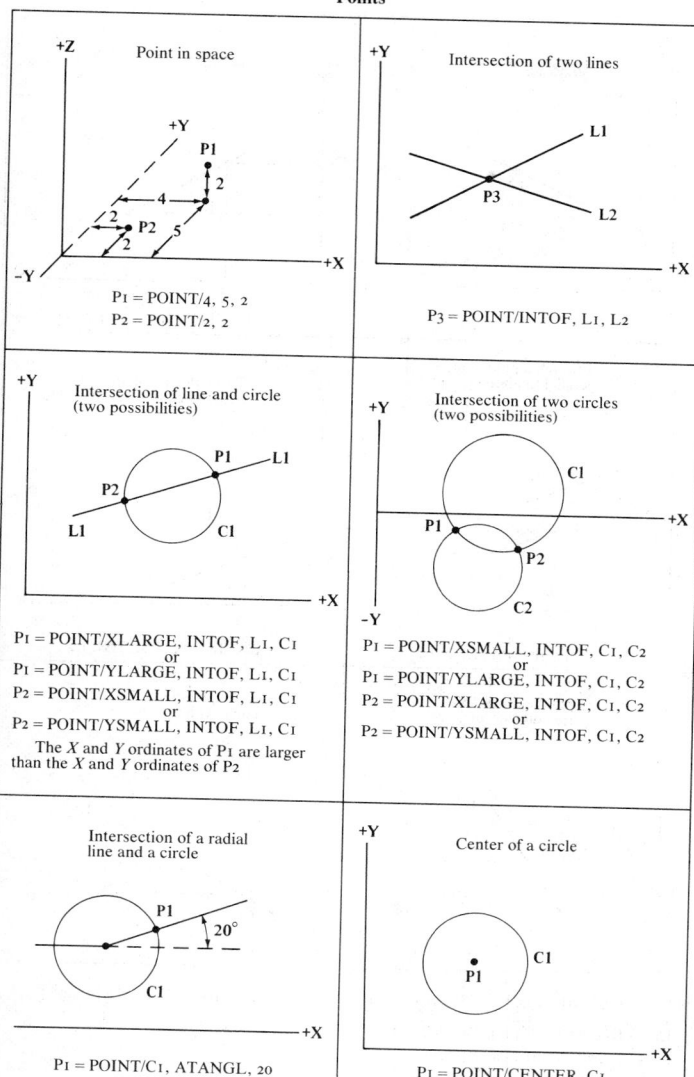

Points

Point in space

$P_1 = POINT/4, 5, 2$
$P_2 = POINT/2, 2$

Intersection of two lines

$P_3 = POINT/INTOF, L_1, L_2$

Intersection of line and circle
(two possibilities)

$P_1 = POINT/XLARGE, INTOF, L_1, C_1$
or
$P_1 = POINT/YLARGE, INTOF, L_1, C_1$
$P_2 = POINT/XSMALL, INTOF, L_1, C_1$
or
$P_2 = POINT/YSMALL, INTOF, L_1, C_1$
The X and Y ordinates of P_1 are larger than the X and Y ordinates of P_2

Intersection of two circles
(two possibilities)

$P_1 = POINT/XSMALL, INTOF, C_1, C_2$
or
$P_1 = POINT/YLARGE, INTOF, C_1, C_2$
$P_2 = POINT/XLARGE, INTOF, C_1, C_2$
or
$P_2 = POINT/YSMALL, INTOF, C_1, C_2$

Intersection of a radial
line and a circle

20°

$P_1 = POINT/C_1, ATANGL, 20$

Center of a circle

$P_1 = POINT/CENTER, C_1$

Lines

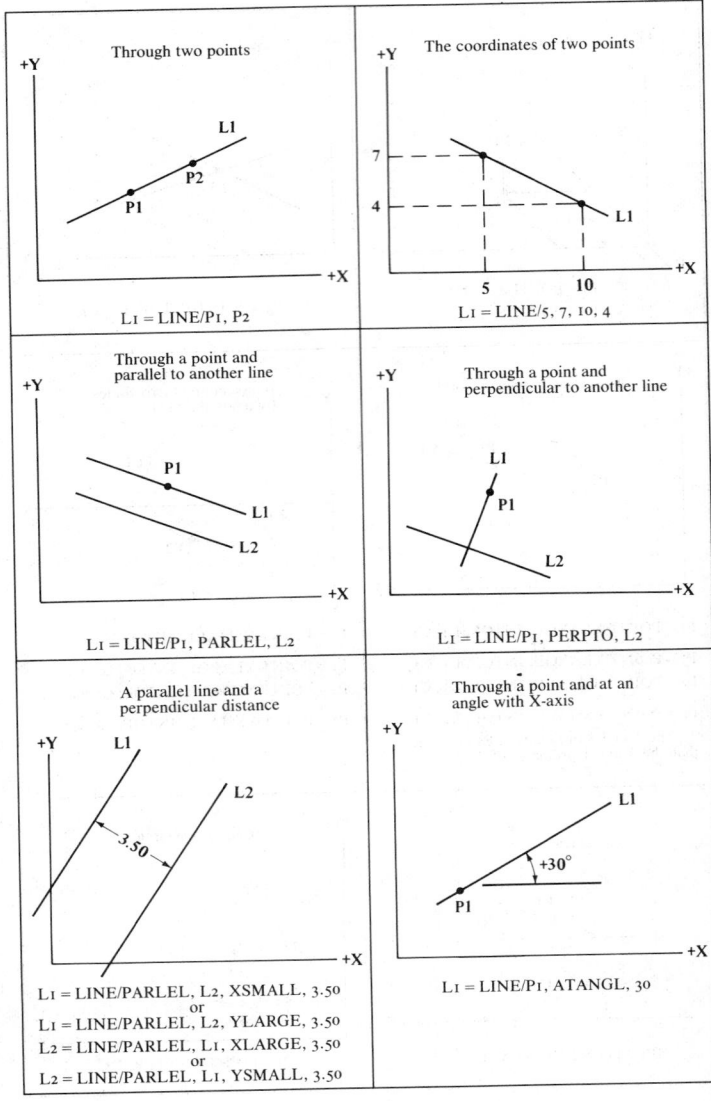

Through two points

L₁ = LINE/P₁, P₂

The coordinates of two points

L₁ = LINE/5, 7, 10, 4

Through a point and parallel to another line

L₁ = LINE/P₁, PARLEL, L₂

Through a point and perpendicular to another line

L₁ = LINE/P₁, PERPTO, L₂

A parallel line and a perpendicular distance

L₁ = LINE/PARLEL, L₂, XSMALL, 3.50
or
L₁ = LINE/PARLEL, L₂, YLARGE, 3.50
L₂ = LINE/PARLEL, L₁, XLARGE, 3.50
or
L₂ = LINE/PARLEL, L₁, YSMALL, 3.50

Through a point and at an angle with X-axis

L₁ = LINE/P₁, ATANGL, 30

Lines

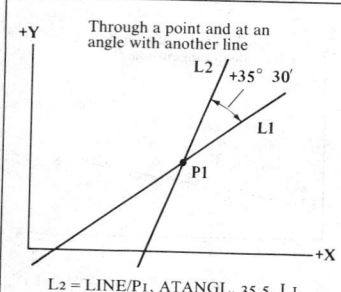

Through a point and at an angle with another line

L2 = LINE/P1, ATANGL, 35.5, L1

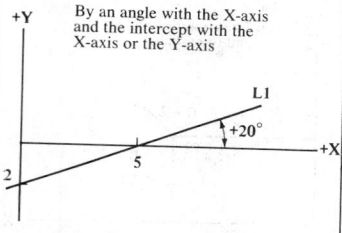

By an angle with the X-axis and the intercept with the X-axis or the Y-axis

L1 = LINE/ATANGL, 20, INTERC, XAXIS, 5
L1 = LINE/ATANGL, 20, INTERC, YAXIS, 2

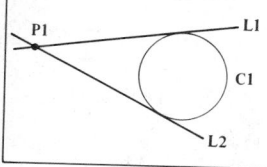

Through a point and tangent to a circle. Looking from the point towards the circle (two possibilities)

L1 = LINE/P1, LEFT, TANTO, C1
L2 = LINE/P1, RIGHT, TANTO, C1

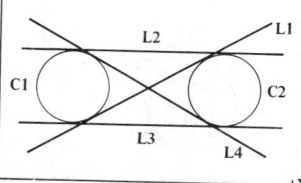

Tangent to two circles. Looking from the first circle noted in the statement towards the second circle noted (four possibilities)

L1 = LINE/RIGHT, TANTO, C2, LEFT, TANTO, C1
    or
L1 = LINE/RIGHT, TANTO, C1, LEFT, TANTO, C2

L2 = LINE/LEFT, TANTO, C1, LEFT, TANTO, C2
    or
L2 = LINE/RIGHT, TANTO, C2, RIGHT, TANTO, C1

L3 = LINE/RIGHT, TANTO, C1, RIGHT, TANTO, C2
    or
L3 = LINE/LEFT, TANTO, C2, LEFT, TANTO, C1

L4 = LINE/LEFT, TANTO, C2, RIGHT, TANTO, C1
    or
L4 = LINE/LEFT, TANTO, C1, RIGHT, TANTO, C2

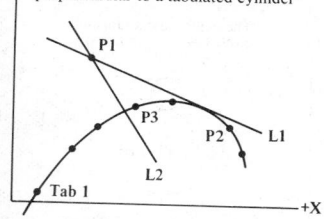

Through a point and tangent or perpendicular to a tabulated cylinder

L1 = LINE/P1, TANTO, TAB1, P3
L1 = LINE/P1, PERPTO, TAB1, P3

P2 and P3 are points close to the tangent points of L1 and the intersection point of L2, therefore cannot be end points of the tabulated cylinder

Circles

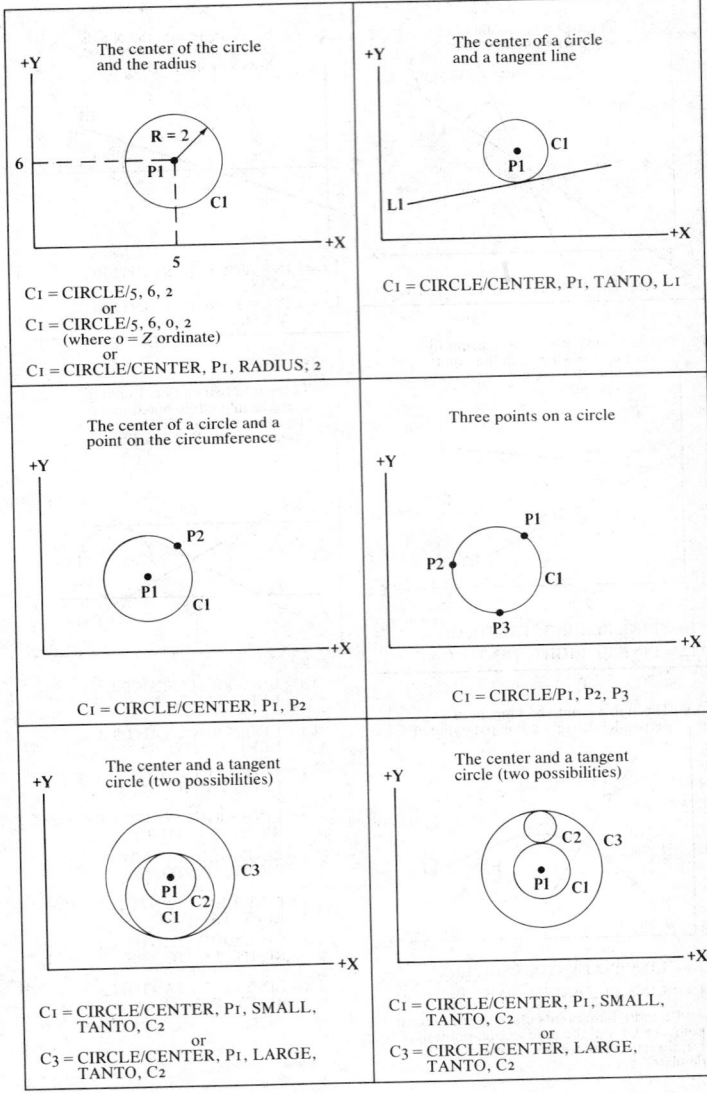

The center of the circle and the radius

C_I = CIRCLE/5, 6, 2
or
C_I = CIRCLE/5, 6, 0, 2
(where 0 = Z ordinate)
or
C_I = CIRCLE/CENTER, P_I, RADIUS, 2

The center of a circle and a tangent line

C_I = CIRCLE/CENTER, P_I, TANTO, L_I

The center of a circle and a point on the circumference

C_I = CIRCLE/CENTER, P_I, P_2

Three points on a circle

C_I = CIRCLE/P_I, P_2, P_3

The center and a tangent circle (two possibilities)

C_I = CIRCLE/CENTER, P_I, SMALL, TANTO, C_2
or
C_3 = CIRCLE/CENTER, P_I, LARGE, TANTO, C_2

The center and a tangent circle (two possibilities)

C_I = CIRCLE/CENTER, P_I, SMALL, TANTO, C_2
or
C_3 = CIRCLE/CENTER, LARGE, TANTO, C_2

Circles

Tangent to two intersecting lines, given the radius (four possibilities)	Tangent to a line and a point on the circumference, given the radius (two possibilities)
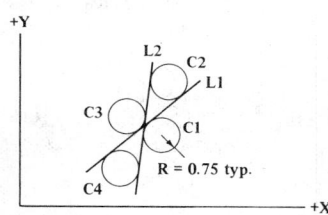	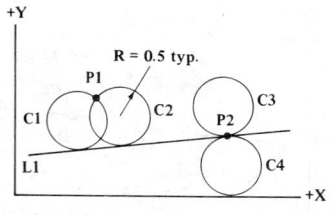
C1 = CIRCLE/XLARGE, L2, YSMALL, L1, RADIUS, .75 or C2 = CIRCLE/XLARGE, L2, YLARGE, L1, RADIUS, .75 or C3 = CIRCLE/XSMALL, L2, YLARGE, L1, RADIUS, .75 or C4 = CIRCLE/XSMALL, L2, YSMALL, L1, RADIUS, .75 The modifiers (XLARGE, etc.) are used to indicate which of the four circles is wanted	C1 = CIRCLE/TANTO, L1, XSMALL, P1, RADIUS, .5 or C2 = CIRCLE/TANTO, L1, XLARGE, P1, RADIUS, .5 if the point lies on the line, then: C3 = CIRCLE/TANTO, L1, YLARGE, P2, RADIUS, .5 or C4 = CIRCLE/TANTO, L1, YSMALL, P2, RADIUS, .5

Lines as calculated by the computer are infinitely long, and circles consist of 360 degrees. As the cutter is moved about the geometry under control of the motion statements, the lengths of the lines and the amounts of the arcs are "cut" to their proper size. [Some of the geometry statements shown in the accompanying illustrations for defining POINTS, LINES, CIRCLES, TABULATED CYLINDERS, CYLINDERS, CONES and SPHERES, in the APT language, may not be included in some two-dimensional (ADAPT) systems.]

APT Motion Statements. — APT is based on the concept that a milling cutter is guided by two surfaces when in a contouring mode. Examples of these surfaces are shown in Fig. 5, and they are called the "part" and the "drive" surfaces. Usually,

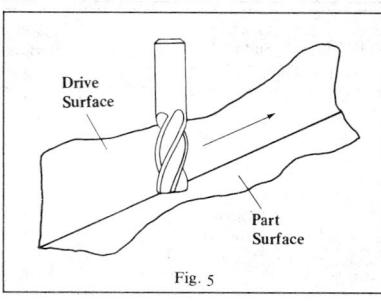

Fig. 5

Circles

Tangent to a line and a circle, given the radius (eight possibilities)	Tangent to two circles, given the radius (eight possibilities)

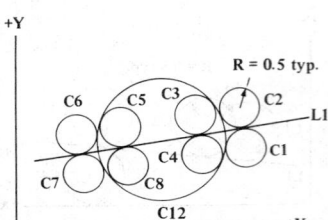 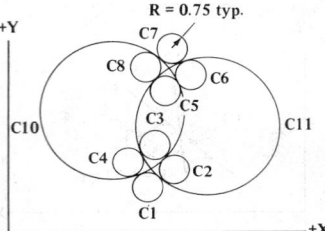

C_1 = CIRCLE/YSMALL, L_1, XLARGE, OUT, C_{12}, RADIUS, .5
C_2 = CIRCLE/YLARGE, L_1, XLARGE, OUT, C_{12}, RADIUS, .5
C_3 = CIRCLE/YLARGE, L_1, XLARGE, IN, C_{12}, RADIUS, .5
C_4 = CIRCLE/YSMALL, L_1, XLARGE, IN, C_{12}, RADIUS, .5
C_5 = CIRCLE/YLARGE, L_1, XSMALL, IN, C_{12}, RADIUS, .5
C_6 = CIRCLE/XLARGE, L_1, XSMALL, OUT, C_{12}, RADIUS, .5
C_7 = CIRCLE/YSMALL, L_1, XSMALL, OUT, C_{12}, RADIUS, .5
C_8 = CIRCLE/YSMALL, L_1, XSMALL, IN, C_{12}, RADIUS, .5

C_1 = CIRCLE/YSMALL, OUT, C_{10}, OUT, C_{11}, RADIUS, .75
C_2 = CIRCLE/YSMALL, OUT, C_{10}, IN, C_{11}, RADIUS, .75
C_3 = CIRCLE/YSMALL, IN, C_{10}, IN, C_{11}, RADIUS, .75
C_4 = CIRCLE/YSMALL, IN, C_{10}, OUT, C_{11}, RADIUS, .75
C_5 = CIRCLE/YLARGE, IN, C_{10}, IN, C_{11}, RADIUS, .75
C_6 = CIRCLE/YLARGE, OUT, C_{10}, IN, C_{11}, RADIUS, .75
C_7 = CIRCLE/YLARGE, OUT, C_{10}, OUT, C_{11}, RADIUS, .75
C_8 = CIRCLE/YLARGE, IN, C_{10}, OUT, C_{11}, RADIUS, .75

Recommendations:

1. Note which side of line circle is on (e.g., YSMALL, L_1)
2. Note whether the circle being defined is inside (IN), or outside (OUT), the known circle.
3. Of the two remaining circles, note whether the circle to be defined is XLARGE, XSMALL, or YLARGE or YSMALL, to arrive at the second modifier in the statement.

Recommendations

1. Apply IN, OUT modifiers
2. Apply XLARGE, etc., modifiers

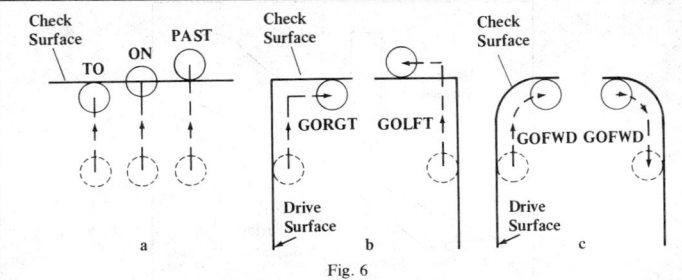

Fig. 6

Tabulated Cylinder

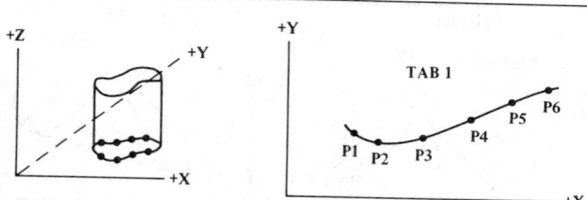

A tabulated cylinder is the line that is formed when an irregular cylinder intersects a plane. The plane intersected in the figure at the left is the *X-Y* plane.

A section of the line can be defined by a series of points on the line, as seen at the right. This line is called a TABCYL. The line must pass through all the points, therefore it is best not to use too many. The statement to the computer would read:

TAB1 = TABCYL/NOZ, SPLINE, P1, P2, P3, P4, P5, P6

or

TAB1 = TABCYL/NOZ, SPLINE, X., Y., X2, Y2, X3, Y3, X4, Y4, X5, Y5, X6, Y6
(where X and Y are the coordinates of the points)

Planes

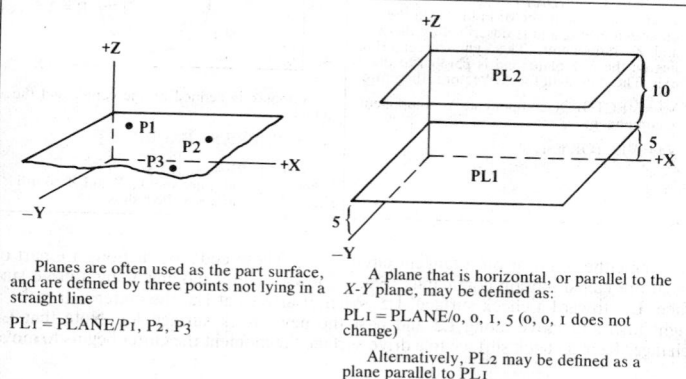

Planes are often used as the part surface, and are defined by three points not lying in a straight line

PL1 = PLANE/P1, P2, P3

A plane that is horizontal, or parallel to the *X-Y* plane, may be defined as:

PL1 = PLANE/0, 0, 1, 5 (0, 0, 1 does not change)

Alternatively, PL2 may be defined as a plane parallel to PL1

PL2 = PLANE/PARLEL, PL1, ZLARGE, 10

the part surface guides the bottom of the cutter and the drive surface guides the side of the cutter. These surfaces may or may not be actual surfaces on the part, and although they may be imaginary to the part programmer, they are very real to the computer. The cutter is either stopped or redirected by a third surface called a check surface. If one were to look directly down on these surfaces, they would appear as lines, as shown in Fig. 6.

When the cutter is moving towards the check surface, it may move to it, onto it, or past it, as illustrated in Fig. 6a. When the cutter meets the check surface, it may go right, denoted by the APT command GORGT, or go left, denoted by the command GOLFT, in Fig. 6b. Alternatively, the cutter may go forward, instructed by the command GOFWD, as in Fig. 6c. The command GOFWD is used when the cutter is

3-D Geometry

Cylinder

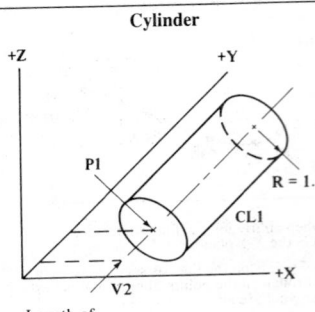

Length of vector = 1

A cylinder is defined by a vector, a point on the centerline, and the radius

CL1 = CYLNDR/P1, V2, 1.5

where V2 is a unit vector in line with the cylinder centerline, and is described by the X, Y and Z components. The cylinder centerline lies on the X-Y plane and is parallel to the Y axis. The statement for the vector is therefore:

V2 = VECTOR/X component, Y component, Z component

V2 = VECTOR/0, 1, 0

Cone

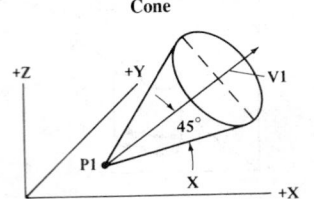

A cone is defined by its vertex, its axis as a unit vector, and the half angle. (refer to cylinder for an example of a vector statement)

CON1 = CONE/P1, V1, 45

Sphere

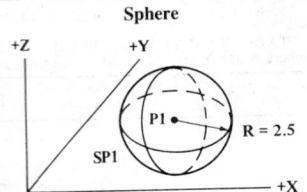

A sphere is defined by the center and the radius

SP1 = SPHERE/P1, RADIUS, 2.5
            or
SP1 = SPHERE/5, 5, 3, 2.5
(where 5, 5 and 3 are the X, Y and Z coordinates of P1, and 2.5 is the radius

moving either onto or off a tangent circular arc. These code instructions are part of what are called motion commands. Fig. 7 shows a cutter moving along a drive surface, L1, toward a check surface, L2. When it arrives at L2, the cutter will make a right turn and move along L2 and past the new check surface L3. Note that L2 changes from a check surface to a drive surface the moment the cutter begins to move

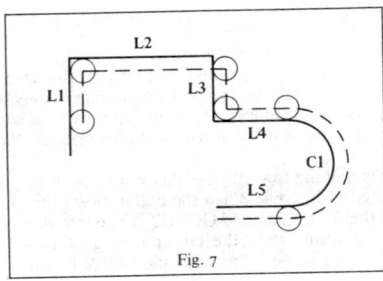

Fig. 7

along it. The APT motion statement for this move is:

GORGT/L2,PAST,L3

Still referring to Fig. 7, the cutter moves along L3 until it comes to L4. L3 now becomes the drive surface and L4 the check surface. The APT statement is:

GORGT/L3,TO,L4

The next statement is:

GOLFT/L4,TANTO,C1

Even though the cutter is moving to the right it makes a left turn if one is looking in the direction of travel of the cutter. In writing the motion statements, the part programmers must imagine they are steering the cutter. The drive surface now becomes L4 and the check surface, C1. The next statement will therefore be:

GOFWD/C1,TANTO,L5

This movement could continue indefinitely, with the cutter being guided by the drive, part, and check surfaces.

Start-up Statements. For the cutter to move along them, it must first be brought into contact with the three guiding surfaces by means of a start-up statement. There are three different start-up statements, depending on how many surfaces are involved. A three-surface start-up statement is one in which the cutter is moved to the drive, part, and check surfaces, as seen in Fig. 8a. A two-surface start-up is one in

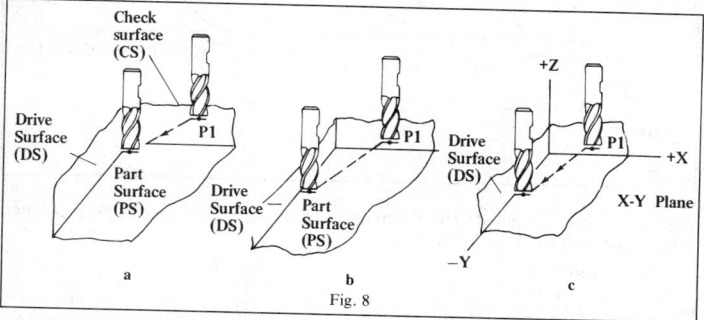

Fig. 8

which the cutter is moved to the drive and part surfaces, as in Fig. 8b. A one-surface start-up is one in which the cutter is moved to the drive surface and the *X-Y* plane, where $Z = 0$, as in Fig 8c. With the two- and one-surface start-up statements, the cutter moves in the most direct path, or perpendicular to the surfaces. Referring to Fig. 8a (three-surface start-up), the move is initiated from a point P1. The two statements that will move the cutter from P1 to the three surfaces are

FROM/P1
GO/TO,DS,TO,PS,TO,CS

DS is used as the symbol for the Drive Surface; PS as the symbol for the Part Surface; and CS as the symbol for the Check Surface. The surfaces must be denoted in this sequence. The drive surface is the surface that the cutter will move along after coming in contact with the three surfaces. The two statements applicable to the two-surface start-up (Fig. 8b) are:

FROM/P1
GO/TO,DS,TO,PS

The one-surface start-up (Fig. 8c) is:

FROM/P1
GO/TO,DS

Note that, in all three motion statements, the slash (/) mark lies between the GO and the TO. When the cutter is moving to a point rather than to surfaces, such as in a start-up statement, the statement is GOTO/ rather than GO/TO. A two-surface startup, Fig. 7, when completed, might appear as shown in Fig. 9, which includes the

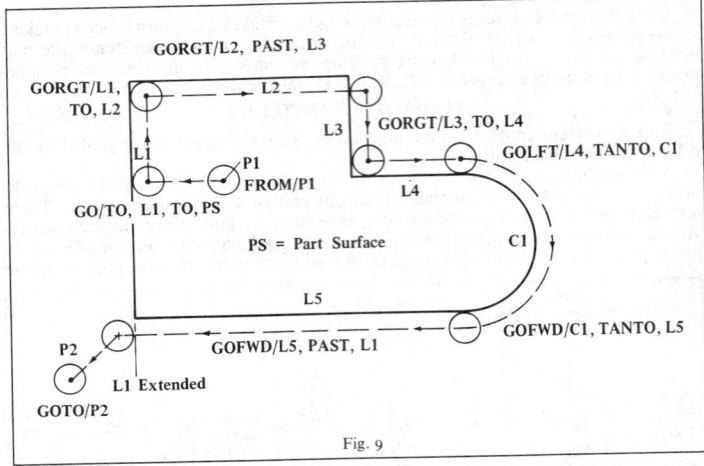

Fig. 9

motion statements needed. The motion statements, as they would appear in a part program are shown at the left, below:

FROM/P1	FROM/P1
GO/TO,L1,TO,PS	GOTO/P2
GORGT/L1,TO,L2	GOTO/P3
GORGT/L2,PAST,L3	GOTO/P4
GORGT/L3,TO,L4	GOTO/P5
GOLFT/L4,TANTO,C1	GOTO/P6
GOFWD/C1,TANTO,L5	GOTO/P7
GOFWD/L5,PAST,L1	
GOTO/P2	

GOTO statements can move the cutter throughout the range of the machine, as shown in Fig. 10. APT statements for such movements are shown at the right in the preceding example. The cutter may also be moved incrementally, as shown in Fig. 11. Here, the cutter is to move 2 inches in the $+X$ direction, 1 inch in the $+Y$ direction, and 1.5 inches in the $+Z$ direction. The incremental move statement (indicated by DLTA) is:

GODLTA/2,1,1.5

The first position after the slash is the X movement; the second the Y movement, and the third, the Z movement.

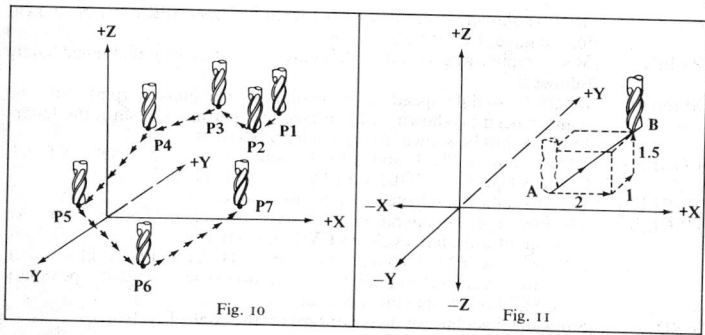

Fig. 10

Fig. 11

Five-Axis Machining: Machining on five axes is achieved by causing the APT program to generate automatically a unit vector that is normal to the surface being machined, as shown in Fig. 12. The vector would be described by its X, Y, and Z components. These components, along with the X, Y, and Z coordinate positions of the tool tip, are fed into the postprocessor, which determines the locations and angles for the machine tool head and/or table.

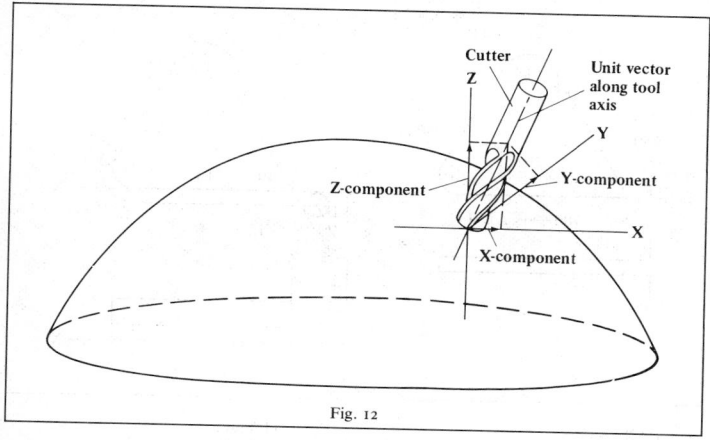

Fig. 12

APT Postprocessor Statements. — Statements that refer to the operation of the machine rather than to the geometry of the part or the motion of the cutter about the part, are called postprocessor statements. APT postprocessor statements have been standardized internationally. Some common statements and an explanation of their meanings follow:

MACHIN/   Specifies the postprocessor that is to be used. Every postprocessor has an identity code, and this code must follow the slash (/) mark. For example: MACHIN/LATH,82

FEDRAT/   Denotes the feed rate. If in inches per minute (ipm), only the number

need be shown. If in inches per revolution (ipr), IPR must be shown, for example: FEDRAT/.005,IPR

RAPID   Means rapid traverse and applies only to the statement that immediately follows it

SPINDL/   Refers to spindle speed. If in revolutions per minute (rpm), only the number need be shown. If in surface feet per minute (sfm), the letters SFM need to be shown, for example: SPINDL/100, SFM

COOLNT/   Means cutting fluid and can be subdivided into: COOLNT/ON, COOLNT/MIST, COOLNT/FLOOD, COOLNT/OFF

TURRET/   Used to call for a selected tool or turret position

CYCLE/   Specifies a cycle operation such as a drilling or boring cycle. An example of a drilling cycle is: CYCLE/DRILL,RAPTO,.45,FEDTO,0, IPR,.004 The next statement might be GOTO/P1 and the drill will then move to P1 and perform the cycle operation. The cycle will repeat until the CYCLE/OFF statement is read.

END   Stops the machine but does not turn off the control system

APT Example Program. — A dimensioned drawing of a part and a drawing with the symbols for the geometry elements are shown in Fig. 13. A complete APT program for this part, starting with the statement PARTNO 47F36542 and ending with FINI, is shown at the left below.

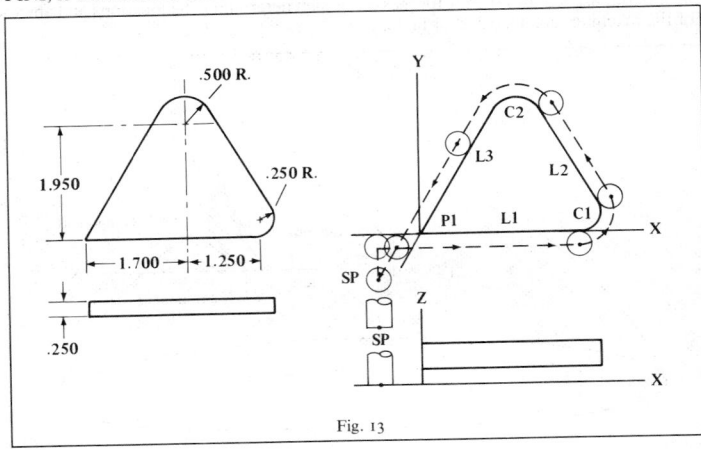

Fig. 13

(1) PARTNO	(1) PARTNO
(2) CUTTER/.25	(2) CUTTER/.25
(3) FEDRAT/5	(3) FEDRAT/5
(4) SP = POINT/ − .5, − .5, .75	(4) SP = POINT/ − .5, − .5, .75
(5) P1 = POINT/0, 0, 1	(5) P1 = POINT/0, 0, 1
(6) L1 = LINE/P1, ATANGL, 0	(6) L1 = LINE/P1, ATANGL, 0
(7) C1 = CIRCLE/(1.700 + 1.250), .250, .250	(7) C1 = CIRCLE/(1.700 + 1.250), .250, .250
(8) C2 = CIRCLE/1.700, 1.950, .5	(8) C2 = CIRCLE/1.700, 1.950, .5
(9) L2 = LINE/RIGHT, TANTO, C1, RIGHT, TANTO, C2	(9) L2 = LINE/RIGHT, TANTO, C1, RIGHT, TANTO, C2

(10) L3 = LINE/P1, LEFT, TANTO, C2
(11) FROM/SP
(12) GO/TO, L1
(13) GORGT/L1, TANTO, C1
(14) GOFWD/C1, TANTO, L2
(15) GOFWD/L2, TANTO, C2
(16) GOFWD/C2, TANTO, L3
(17) GOFWD/L3, PAST, L1
(18) GOTO/SP
(19) FINI

(10) L3 = LINE/P1, LEFT, TANTO, C2			
(11) FROM/SP			
(12) FROM	−.5000	−.5000	.7500
(13) GO/TO/, L1			
(14) GT	−.5000	−.1250	.0000
(15) GORGT/L1, TANTO, C1			
(16) GT	2.9500	−.1250	.0000
(17) GOFWD/C1, TANTO, L2			
(18) CIR	2.9500	.2500	.3750 CCLW
(19)	3.2763	.4348	.0000
(20) GOFWD/L2, TANTO, C2			
(21) GT	2.2439	2.2580	.0000
(22) GOFWD/C2, TANTO, L3			
(23) CIR	1.7000	1.9500	.6250 CCLW
(24)	1.1584	2.2619	.0000
(25) GOFWD/L3, PAST, L1			
(26) GT	−.2162	−.1250	.0000
(27) GOTO/SP			
(28) GT	−.5000	−.5000	.7500
(29) FINI			

The numbers at the left of the statements are for reference purposes only, and are not part of the program. The cutter is set initially at a point represented by the symbol SP, having coordinates $X = -0.5$, $Y = -0.5$, $Z = 0.75$, and moves to L1 (extended) with a one-surface start-up so that the bottom of the cutter rests on the X-Y plane. The cutter then moves counterclockwise around the part, past L1 (extended) and returns to SP. The coordinates of P1 are $X = 0$, $Y = 0$ and $Z = 1$.

Referring to the numbers at the left of the program:

(1) PARTNO must begin every program. Any identification can follow.
(2) The diameter of the cutter is specified. Here it is 0.25 inch.
(3) The feed rate is given as 5 inches per minute, which is contained in a postprocessor statement.
(4) — (10) Geometry statements
(11) — (18) Motion statements
(19) All APT programs end with FINI

A computer printout from Section II of the APT program is shown at the right above. This program was run on a desk-top personal computer. Lines (1) through (10) repeat the geometry statements from the original program. The motion statements are also repeated and below each motion statement are shown the X, Y, and Z coordinates of the end points of the center line (CL) movements for the cutter. Two lines of data follow those for the circular movements. For example, Line (18) which follows Line (17), GOFWD/C1,TANTO,L2, describes the X coordinate of the center of the arc, 2.9500, the Y coordinate of the center of the arc, 0.2500, and the radius of the arc required to be traversed by the cutter.

This radius is that of the arc shown on the part print, plus the radius of the cutter (0.2500 + 0.1250 = 0.3750). Line (18) also shows that the cutter is traveling in a counterclockwise (CCLW) motion. A circular motion is described in lines (22), (23), and (24). Finally, the cutter is directed to return to the starting point, SP, and this command is noted in line (27). The X, Y, and Z coordinates of SP are shown in line (28).

APT for Turning. — In its basic form, APT is not a good program for turning. Although APT is probably the most suitable program for three-, four-, and five-axis machining, it is awkward for the simple two-axis geometry required for lathe opera-

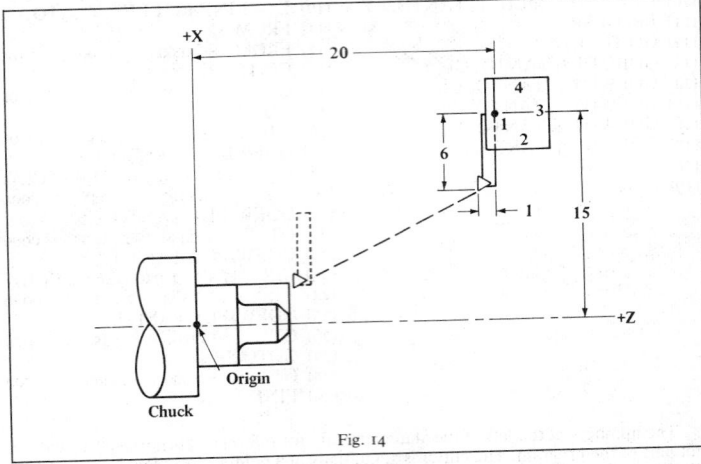

Fig. 14

tions. To overcome this problem, preprocessors have been developed especially for lathe part programming. The statements in the lathe program are automatically converted to basic APT statements in the computer and processed by the regular APT processor. An example of a lathe program, based on the APT processor and made available by the McDonnell Douglas Automation Co., is shown below. The numbers in parentheses are not part of the program, but are used only for reference. Fig. 14 shows the general setup for the part, and Fig. 15 shows an enlarged view of the part profile with dimensions expressed along what would be the X and Y axes on the part print.

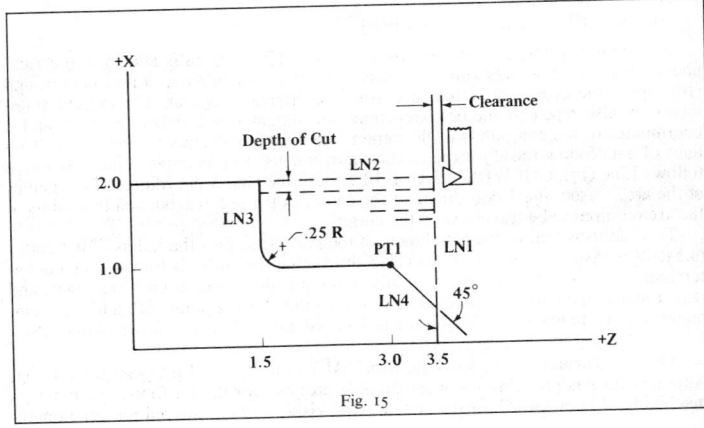

Fig. 15

(1) PARTNO LATHE EXAMPLE
(2) MACHIN/MODEL LATHE
(3)   T1 = TOOL/FACE, 1, XOFF, − 1, YOFF, − 6, RADIUS, .031
(4) BLANK1 = SHAPE/FACE, 3.5, TURN, 2
(5)   PART1 = SHAPE/FACE, 3.5, TAPER, 3.5, .5, ATANGL, − 45, TURN, 1,$
             FILLET, .25, FACE, 1.5, TURN, 2
(6)   FROM/(20 − 1), (15 − 6)
(7)   LATHE/ROUGH,BLANK1,PART1,STEP,.1,STOCK,.05,$
             SFM,300, IPR,.01,T1
(8)   LATHE/FINISH,PART1,SFM,400,IPR,.005,T1
(9)   END
(10)   FINI

Line (3) describes the tool. Here, the tool is located on face 1 of the turret and its tip is − 1 inch "off" (offset) in the X direction and − 6 inches "off" in the Y direction, when considering XY rather than XZ axes. The cutting tool tip radius is also noted in this statement. Line (4) describes the dimensions of the rough material, or blank. Lines parallel to the X axis are noted as FACE lines, and lines parallel to the Z axis are noted as TURN lines. The FACE line (LN1), is located 3.5 inches along the Z axis and parallel to the X axis. The TURN line (LN2) is located 2 inches above the Z axis and parallel to it. Note that in Figs. 14 and 15 the X axis is shown in a vertical position and the Z axis in a horizontal position. Line (5) describes the shape of the finished part. The term FILLET is used in this statement to describe a circle that is tangent to the line described by TURN,1 and the line that is described by FACE, 1.5. The $ sign means that the statement is continued on the next line. These geometry elements must be contiguous and must be described in sequence. Line (6) specifies the position of the tool tip at the start of the operation, relative to the point of origin. Line (7) describes the roughing operation and notes that the material to be roughed out lies between BLANK1 and PART1; that the STEP, or depth of roughing cuts, is to be 0.1 inch; that 0.05 inch is to be left for the finish cut; that the speed is to be 300 sfm and the feedrate is to be 0.01 ipr; and that the tool to be used is identified by the symbol T1. Line (8) describes the finish cut, which is to be along the contour described by PART1.

Axis Nomenclature. — To distinguish among the different motions, or axes, of a machine tool, a system of letter addresses has been developed. A letter is assigned, for example, to the table of the machine, another to the saddle, and still another to the spindle head. These letter addresses, or axis designations, are necessary for the electronic control system to assign movement instructions to the proper machine element. The assignment of these letter addresses has been standardized on a world-wide basis and is contained in three standards, all of which are in agreement. These standards are the EIA RS-267-B, issued by the Electronics Industries Association; the AIA NAS-938, issued by the Aerospace Industries Association; and the ISO/R 841, issued by the International Organization for Standardization.

The standards are based on a "right-hand rule," that describes the orientation of the motions as well as whether the motions are positive or negative. If a right hand is laid palm up on the table of a vertical milling machine as shown in Fig. 16, for example, the thumb will point in the positive X direction, the forefinger in the positive Y direction, and the erect middle finger in the positive Z direction, or up. The direction signs are based on the motion of the cutter relative to the workpiece. The movement of the table shown in Fig. 17 is therefore positive, even though the table is moving to the left, since the motion of the cutter relative to the workpiece is to the right, or in the positive direction. The motions are considered from the part programmers viewpoint, which assumes that the cutter always moves around the part, regard-

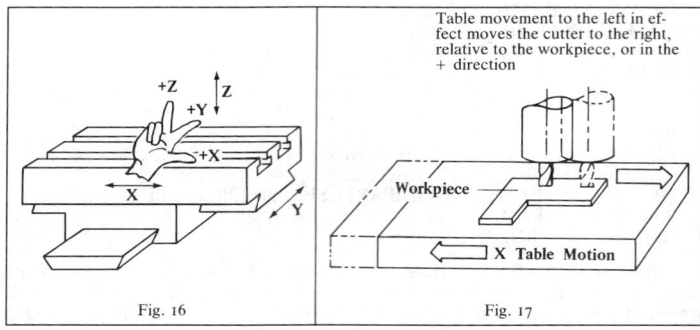

Fig. 16   Fig. 17

less of whether the cutter or the part moves. The right-hand rule also holds with a horizontal-spindle machine and a vertical table, or angle plate, as shown in Fig. 18. Here, spindle movement back and away from the angle plate, or workpiece, is a positive Z motion, and movement towards the angle plate is a negative Z motion.

Rotary motions also are governed by a right-hand rule, but the fingers are joined and the thumb is pointed in the positive direction of the axis. Fig. 19 shows the designations of the rotary motions about the three linear axes X, Y, and Z. Rotary motion about the X axis is designated as A; rotary motion about the Y axis is B; and rotary motion about the Z axis is C. The fingers point in the positive rotary directions. Movement of the rotary table around the Y axis shown in Fig. 19 is a B motion and is common with horizontal machining centers. Here, the view is from the spindle face, looking towards the rotary table. Referring, again, to linear motions, if the spindle is withdrawn axially from the work, the motion is a positive Z. A move towards the work is a negative Z.

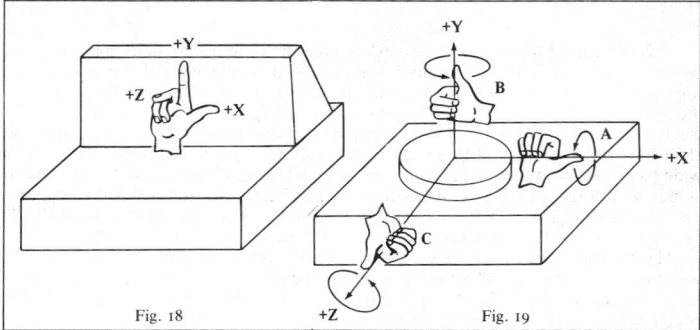

Fig. 18   Fig. 19

When a second linear motion is parallel to another linear motion, as with the horizontal boring mill seen in Fig. 20, the horizontal motion of the spindle, or quill, is designated as Z and a parallel motion of the angle plate is W. A movement parallel to the X axis is U and a movement parallel to the Y axis is V. Corresponding motions are summarized as follows:

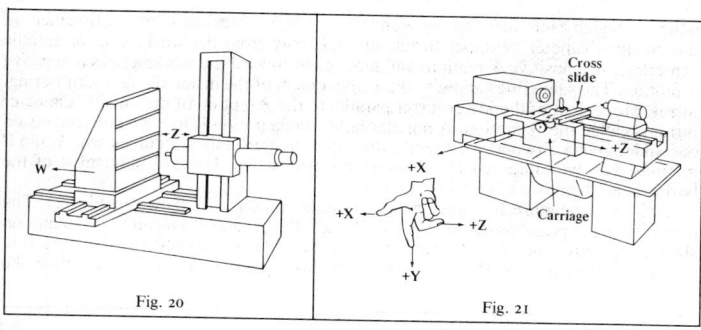

Fig. 20

Fig. 21

Linear	Rotary	Linear and Parallel
X	A	U
Y	B	V
Z	C	W

Axis designations for a lathe are shown in Fig. 21. Movement of the cross-slide away from the workpiece, or the centerline of the spindle, is noted as a plus X. Movement toward the workpiece is a minus X. The middle finger points in the positive Z direction; therefore, movement away from the headstock is positive and movement toward the headstock is negative. Generally, there is no Y movement.

The machine shown in Fig. 21 is of conventional design, but most NC lathes look more like that shown in Fig. 22. The same right-hand rule applies to this four-axis

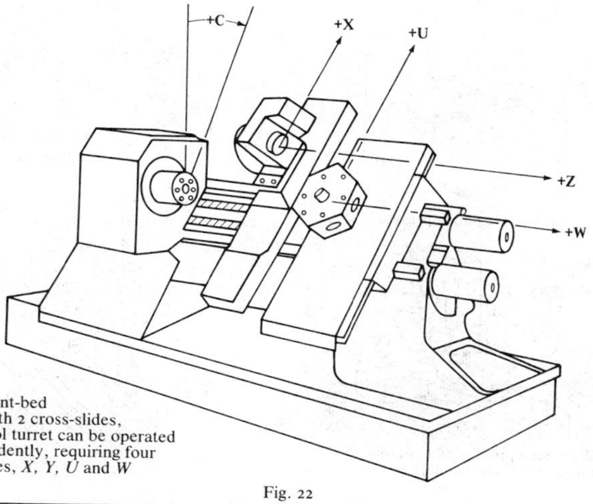

On a slant-bed lathe with 2 cross-slides, each tool turret can be operated independently, requiring four addresses, X, Y, U and W

Fig. 22

lathe, on which each turret moves along its own two independent axes. Movement of the outside-diameter or upper turret, up and away from the workpiece, or spindle centerline, is a positive X motion, and movement toward the workpiece is a negative X motion. The same rules apply to the U movement of the inside-diameter, or boring, turret. Movement of the lower turret parallel to the Z motion of the outside-diameter turret is called the W motion. A popular lathe configuration is to have both turrets on one slide, giving a two-axis system rather than the four-axis system shown. X and Z motions may be addressed for either of the two heads. Upward movement of the boring head therefore is a positive X motion.

Axis nomenclature for other machine configurations are shown in Fig. 23. The letters with the prime notation (e.g., X', Y', Z', W', A' and B') mean that the motion shown is positive, because the movement of the cutter with respect to the work is in a positive direction. In these instances, the workpiece is moving rather than the cutter.

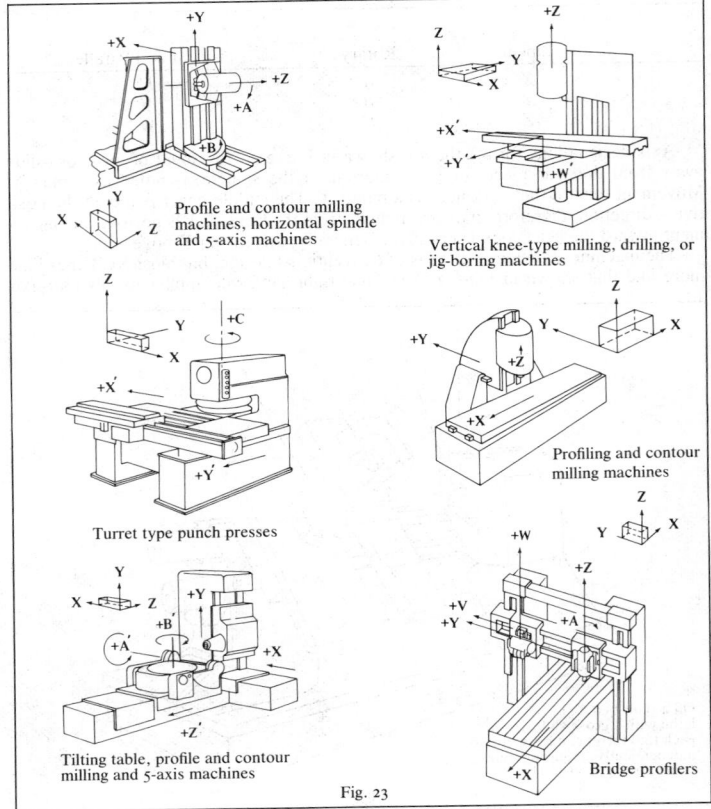

Profile and contour milling machines, horizontal spindle and 5-axis machines

Vertical knee-type milling, drilling, or jig-boring machines

Turret type punch presses

Profiling and contour milling machines

Tilting table, profile and contour milling and 5-axis machines

Bridge profilers

Fig. 23

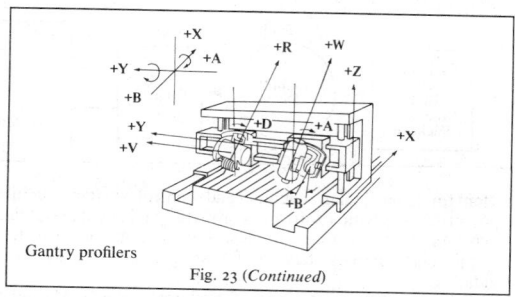

Gantry profilers

Fig. 23 (Continued)

Backlash. — The clearance, slack, or play between adjacent movable parts such as in a leadscrew and a nut, is called backlash. Backlash may be the result of wear, clearance, or both wear and clearance between mating parts.

Binary. — A system for describing numbers using combinations of only two digits, 0 and 1, is called a binary system. Most computers operate on a form of binary system that expresses numbers by combinations of "on" and "off" circuits, the digits 0 and 1 representing "off" and "on", respectively. The binary concept is based upon the placement of either a 0 or a 1 in a certain position to represent a number. Each position in the representation has either a 0 or a 1 placed there. If a 0 appears, the value of that position is 0; if a 1 appears, the value of that position is $2^{(n-1)}$, n being the number of the position beginning with 1 at the right and increasing by 1 for successive positions to the left, as shown below.

Position . . n		5	4	3	2	1
Position Value	$2^{(n-1)}$	2^4	2^3	2^2	2^1	2^0
Position Decimal Value	$2^{(n-1)}$	16	8	4	2	1
Example:		1	1	1	1	1 = 16 + 8 + 4 + 2 + 1 = 31
Example:			1	0	0	1 = 8 + 0 + 0 + 1 = 9

The binary concept is well suited for numerical control applications, since it allows a number to be expressed by either a "hole" or a combination of "holes" and "no holes" on a punched tape. (See *Binary Coded Decimal Format*.)

Binary Coded Decimal Format (BCD Format). — The BCD format is a system of representing numbers on tape for NC programming instructions. The system uses a combination of four binary bits ("holes" or "no holes") running across the tape to represent characters having a range of 1 through 9. Letters and punctuation also are expressed by combinations of binary bits. The BCD format used with NC instructions employs only the first four binary positions described under *Binary*. Zero (0) is expressed as a separate character on the tape with a hole in the sixth column for the EIA RS-244-B Standard, or holes in the fifth and sixth columns for the EIA RS-358-B Standard (also known as the ASCII Standard). These characters are arranged along the tape to form the required number. (See *Character* and *Character Code*.)

Binary Cutter Location. — BCL or binary cutter location is a part-programming and machine control system combination that eliminates the requirement for a postprocessor. In many installations, the CL data from a part program, which describes the centerline path of the cutter, is input to a postprocessor. The postprocessor resides in the computer (see *APT* and *Postprocessor*) along with the processor and is

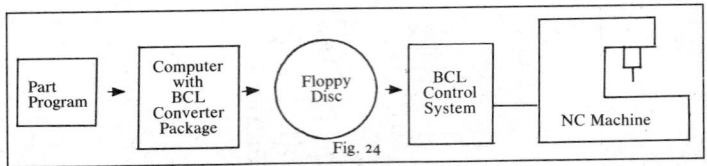

Fig. 24

specially written for a particular machine and control system combination. BCL replaces the postprocessor with a software converter package that outputs a standard (RS-494) format. As shown in Fig. 24, these binary-formatted instructions are recorded on a floppy disc whereby they can be input to a control system designed to accept BCL data.

In effect, the postprocessor statements, for example, the G, M, and other codes, are recognized by the control system software. The chief advantage of BCL is that the disc containing the machining instructions may be used in any machine (with a BCL control system) that has the capability for machining the part, without reprogramming. Substantial costs can be saved because a postprocessor is not needed. A conventional type of control system that accepts the more normal RS-294-D tape format is not suitable for BCL. However, some control systems builders include hardware packages to convert the BCL format to the RS-294-D format so that conventional control systems can be used.

Bit. — Short for binary digit, bit means the presence or absence of a signal. There is space for eight bits across an NC tape. The bit can be either "on" (a hole) or "off" (no hole). The combination of these bits forms a character that is a numeral, letter, punctuation, or carriage return. (See *Character*.)

Block. — A block is a *word* or group of words (see explanation of *Word*) considered as a unit and separated from other such units by an end-of-block (EOB) character. A block provides sufficient information for a complete command to the machine tool.

Block Delete. — When a block command is not needed for a specific version of a part, that command can be skipped over by programming a slash (/) before the block. The operator actuates a switch that causes the system to recognize the slash (or delete) code and skip that block.

Byte. — A sequence of eight binary digits (bits) operated upon as a unit and usually shorter than a word is called a byte. A word usually consists of 16 bits (2 bytes) or 32 bits (4 bytes). Most CNC systems utilize 16-bit words.

Central Processing Unit. — The central processing unit, or CPU, is the heart of the computer. The CPU handles the computations and the overall management and operation of the computer.

Character. — In an NC tape, a character is an arrangement of bits (holes and no holes) across a tape that represent a letter, numeral, punctuation, or command, such as a carriage return.

Character Code. — On an NC tape, the character code is the arrangement of a line of holes across the tape that represents a character. One such set of arrangements, known as the EIA code, was developed in the early days of NC by the Electronics Industries Association (EIA), as Standard RS-244-B. Another code, known as ASCII, for American (National) Standard Code for Information Inter-

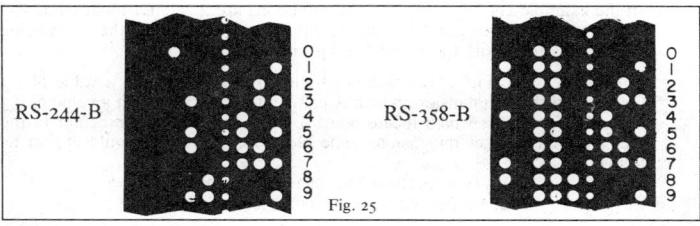

RS-244-B   RS-358-B

0
1
2
3
4
5
6
7
8
9

Fig. 25

change, initially was directed at computer communications. As NC became more computer-related, the ASCII code became more popular in the NC field and today it has all but supplanted the EIA RS-244-B code. The ASCII code also is an EIA Standard and is referred to as RS-358-B. Most control systems can accept either the EIA or ASCII code. The main difference between the codes is that the EIA code uses an odd number of holes across the tape (odd parity), and the ASCII code an even number of holes across the tape (even parity). Characters representing numbers used in the two codes are compared in Fig. 25.

Circular Interpolation. — A simplified means of programming circular arcs in one plane, using one block of data, is called circular interpolation. This procedure eliminates the need to break the arc into straight-line segments. Circular interpolation is usually handled in one plane, or two dimensions, although three-dimensional circular interpolation is described in the Standards. The plane to be used is selected by a G, or preparatory code. Referring to Fig. 26, G17 is used if the circle is to be formed in the X-Y plane, G18 if in the X-Z plane, and G19 if in the Y-Z plane. Often the control system is preset for the circular interpolation feature to operate in only one plane (e.g., the X-Y plane for milling machines or machining centers or the X-Z plane for lathes) and for these machines the G codes are not necessary.

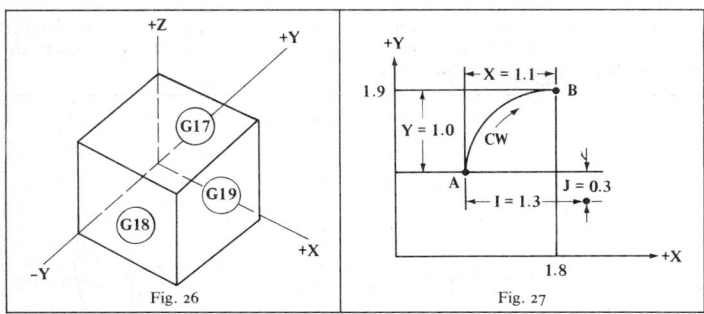

Fig. 26   Fig. 27

A circular arc may be described in several ways. Originally the RS-274 Standard specified that, with incremental programming, the block should contain:

1. A G code describing the direction of the arc; G02 for clockwise (CW), and G03 for counterclockwise (CCW).
2. Directions for the component movements around the arc parallel to the axes.

In the example shown in Fig. 27, the directions are $X = +1.1$ inches and $Y = +1.0$ inch. The signs are determined by the direction in which the arc is being generated. Here, both the X and Y are positive.

3. The I dimension (Fig. 27), which is parallel to the X axis with a value of 1.3 inches, and the J dimension, which is parallel to the Y axis with a value of 0.3 inch. These values, which locate point A with reference to the center of the arc, are called offset dimensions. The block for this work would appear as follows:

N0025 G02 X011000 Y010000 I013000 J003000
(The sequence number, N0025, is arbitrary)

The block would also contain the plane selection (i.e., G17, G18, or G19), if this selection is not preset in the system. Most of the newer control systems allow duplicate words in the same block but most of the older systems do not. In these older systems it is necessary to insert the plane selection code in a separate and prior block, for example, N0020G17.

Another stipulation in the Standard is that the arc is limited to one quadrant. Therefore, four blocks would be required to complete a circle. Four blocks would also be required to complete the arc shown in Fig. 28, which extends into all four quadrants.

When utilizing absolute programming the coordinates of the end point are described. Referring again to Fig. 27, the block, expressed in absolute coordinates, appears as:

N0055 G02 X018000 Y019000 I013000 J003000

Where the arc is continued from a previous block, the starting point for the arc in this block would be the end point of the previous block.

The Standard still contains the format discussed, but simpler alternatives have been developed. The latest version of the Standard (RS-274-D), allows multiple quadrant programming in one block, by inclusion of a G75 word. In the absolute-dimension mode (G90) the coordinates of the arc center are specified. In the incremental dimension mode (G91) the signed (plus or minus) incremental distances from the beginning point of the arc to the arc center are given. Most system builders have introduced some variations on this format. Referring to Fig. 29, one system builder utilizes the center and the end point of the arc when in an absolute mode and might describe the block, in going from A to B, as:

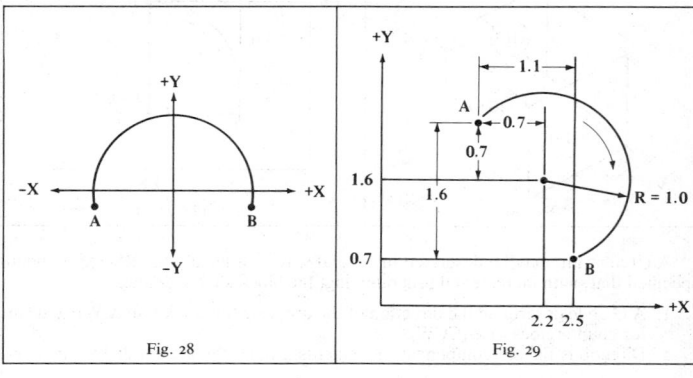

Fig. 28   Fig. 29

N0065 G75 G02 X2.5 Y0.7 I2.2 J1.6

The I and the J words are used to describe the coordinates of the arc center. Decimal point programming is also used here. A block for the same motion when programmed incrementally might appear as:

N0075 G75 G02 X1.1 Y − 1.6 I0.7 J0.7

This approach is more in conformance with the RS-274-D Standard in that the X and Y values describe the displacement between the starting and ending points (points A and B), and the I and J indicate the offsets of the starting point from the center.

Another and even more convenient way of formulating a circular motion block is to note the coordinates of the ending point and the radius of the arc. Referring to Fig. 29 and using absolute programming, the block might appear as:

N0085 G75 G02 X2.5 Y0.7 R10.0

The starting point is derived from the previous motion block. Multiquadrant circular interpolation is cancelled by a G74 code.

Closed Loop System. — Also referred to as a servo or feedback system, a closed-loop system is a control system that issues commands to the drive motors of an NC machine. The system then compares the results of these commands as measured by the movement or location of the machine component, such as the table or spindle-head. The feedback devices normally used for measuring movement or location of the component are called resolvers, encoders, Inductosyns or optical scales. The resolver, which is a rotary analog mechanism, is the least expensive, and has been the most popular since the first NC machines were developed. Resolvers are normally connected to the leadscrews of NC machines. Linear measurement is derived from monitoring the angle of rotation of the leadscrew and is quite accurate.

Encoders also are normally connected to the leadscrew of the NC machine, and measurements are in digital form. Pulses, or a binary code in digital form, are generated by rotation of the encoder, and represent turns or partial turns of the leadscrew. These pulses are well suited to the digital NC system, and encoders have therefore become very popular with such systems. Encoders generally are somewhat more expensive than resolvers.

The Inductosyn (a trade name of Farrand Controls, Inc.) also produces analog signals, but is attached to the slide or fixed part of a machine to measure the position of the table, spindle-head, or other component. The Inductosyn provides almost twice the measurement accuracy of the resolver, but is considerably more expensive, depending on the length of travel to be measured.

Optical scales generally produce information in digital form and, like the Inductosyn, are attached to the slide or fixed part of the machine. Optical scale measurements are more accurate than either resolvers or encoders and, because of their digital nature, are well suited to the digital computer in a CNC system. Like the Inductosyn, optical scales are more costly than either resolvers or encoders.

Computer-aided Part Programming. — As the name implies, computer-aided part programming uses a computer as an assistant in preparing the detailed instructions for operating an NC machine. Usually, the instructions are punched into a tape, but may be fed to the machine directly from the computer or transferred via a floppy disc. The instructions are the same as are produced by the manual part programming method. The program that is to be fed to the computer is called the source program and is usually in an Englishlike format. (See *APT*.)

Many of the newer systems use graphic images in which the part is "drawn" on a computer screen and the part programmer "moves" a cutter about the part to generate the part program, or the detailed block format instructions required by the

control system. Machining instructions, such as the speeds and feed rates, also are input via the keyboard. Some of the newer CNC systems have similar capabilities to a lesser degree. Using the computer as an assistant is faster and far more accurate than the manual part programming method.

Computer Numerical Control (also called CNC). — Applies to the use of a computer as the control system for an NC machine. From the mid-1950s until the mid-1970s practically all control systems were hardwired and designed for specific machine tools. These systems were replaced by minicomputers and, shortly thereafter, by microcomputers. Today, virtually all control systems being manufactured are of the softwire, or CNC, type. CNC systems are far superior to their predecessors. They are less expensive, more reliable from a maintenance standpoint, and have far more capability.

Coordinates. — A set of magnitudes by means of which the position of a point is determined from the zero points of the X, Y, and Z axes is termed the coordinates of that point. In NC, the term differs from ordinate, which is a measure of the distance from the zero point along any one of the axes.

Cutter Compensation. — Manual adjustment of the cutter center path to compensate for any variance between nominal and actual cutter radius is termed cutter compensation. The net effect is to move the path of the center of the cutter closer to, or away from, the edge of the workpiece, as shown in Fig. 30. The compensation may also be handled via a tool data table. (See *Tool Data Table.*) When cutter compensation is to be used it is necessary to include in the program a G41 code if the cutter is to be to the left of the part and a G42 code if to the right of the part, as shown in Fig. 30. A G40 code cancels cutter compensation. Cutter compensation with earlier hardwire systems was expensive, very limited, and usually held to $\pm$ 0.0999 inch. The range for cutter compensation with CNC control systems can go as high as $\pm$ 999.9999 inches, although adjustments of this magnitude are unlikely to be required.

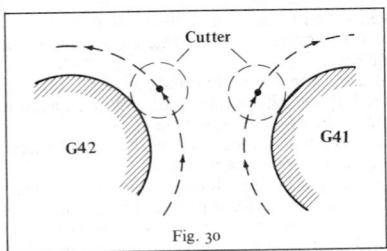

Fig. 30

Cutter Location Data (or CL Data). — Refer to the coordinates that describe the path of the center of the cutter.

Decimal Point Programming. — In decimal point programming, the decimal point is included in the dimension and feed rate words. This practice has become popular with users of CNC systems.

Diagnostics. — Diagnostics is a computer program, resident in the CNC system memory, designed to analyze malfunctions of the machine and/or control system and display information along with recommended corrective action, on the computer screen. Similar systems are used to analyze and display instructions for corrective action in connection with part programs.

Direct (or Distributed) NC. — Direct Numerical Control or Distributed Numerical Control (DNC) describe similar systems. Direct numerical control was the original usage, and was applied to the simultaneous operation of several machines by a single computer on a time-sharing basis. The computer fed blocks of data intermittently to the various machines as required. Generally, the computers were large mainframes and the control systems usually were hardwired. The instructions bypassed the tape reader and were fed directly to the control system. Distributed numerical control requires CNC systems, or storage facilities at the machine, that can store a complete program or large part thereof. The program is downloaded to the CNC system from the main computer, which can be as small as a desktop personal computer. The CNC system then operates alone to control machining of the part. Practically all DNC systems are now of the distributed numerical control type.

Drive Motors. — Four main types of motors are used for driving the leadscrews on NC machines, namely, hydraulic, stepping, DC brush, and brushless. Hydraulic motors offer compact power and are responsive, but their power supply units usually are noisy. Hydraulic systems are relatively expensive and sometimes leak so they are generally limited to use on very large NC machines. Stepping motors rotate a few degrees for each input pulse and do not require feedback, so are less expensive than other drive systems. These motors, however, are generally restricted in both power and speed compared to other drives. DC brush-type motors have been, by far, the most popular. They offer good ranges of power and speed, are reasonably priced and, except for the brushes, require little maintenance.

Brushless motors are relatively new to NC and offer at least one distinct advantage over the brush type in that there is no requirement for changing brushes. Besides not having brushes, the other main distinction of brushless motors is that the permanent magnetic field is in the rotor rather than in the stator, the rotor usually being made of a rare earth metal which is strongly magnetic. The price of the brushless motor is less than that of the brush type, but the power supply costs more, and the total package, including both motor and power supply, is therefore more expensive. This situation may change as technology improves the design of these systems.

Feedhold. — The term feedhold describes the ability of the control system to stop a machine anywhere in the cycle without losing tape data. The operation subsequently may be continued from where it was stopped.

Feed Rate Override. — Feed rate override refers to a control, usually a rotary dial, on the control system panel that allows the programmer or operator to override the programmed feed rate. With CNC systems, the range of override may extend from 0 to 150 per cent of the programmed feed rate. Hardwire systems are more restrictive and adjustment usually can not exceed 100 percent of the preset rate.

Fixed (Canned) Cycles. — Fixed (canned) cycles comprise sets of instructions providing for a preset sequence of events initiated by a single command or a block of data. Fixed cycles generally are offered by the builder of the control system or machine tool as part of the software package that accompanies the CNC system. Limited numbers of canned cycles began to appear on hardware control systems shortly before their demise. The canned cycles offered generally consist of the standard G codes covering drilling, boring, and tapping operations, plus options that have been developed by the system builder such as thread cutting and turning cycles. (See *Threadcutting* and *Turning Cycles.*) Some standard canned cycles included in RS-274-D are shown herewith. A block of data that might be used to generate the cycle functions is also shown above each illustration. Although the G codes for the functions are standardized, the other words in the block and the block format are not, and different control system builders have different arrangements. The blocks shown are reasonable examples of fixed cycles and do not represent those of any particular

system builder. The G81 block for a simple drilling cycle is:

N_____G81 X_____Y_____C_____D_____F_____EOB
N_____X_____Y_____EOB

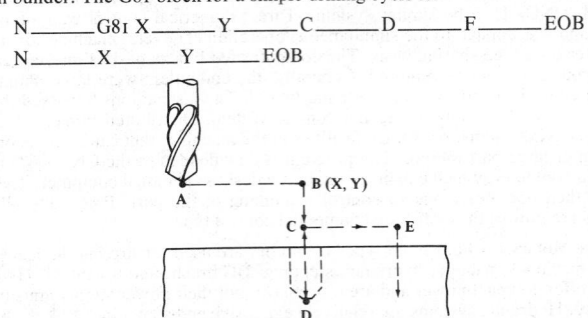

This G81 drilling cycle will move the drill point from position *A* to position *B* and then down to *C* at a rapid traverse rate; the drill point will next be fed from *C* to *D* at the programmed feed rate, then returned to *C* at the rapid traverse rate. If the cycle is to be repeated at a subsequent point, such as point *E* in the illustration, it is necessary only to give the required *X* and *Y* coordinates. This repetition capability is typical of canned cycles.

The G82 block for a spotfacing or drilling cycle with a dwell is:

N_____G82 X_____Y_____C_____D_____T_____F_____EOB

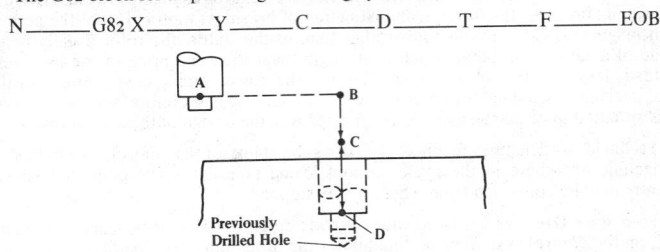

This G82 code produces a cycle that is very similar to the cycle of the G81 code except for the dwell period at point *D*. The dwell period allows the tool to smooth out the bottom of the counterbore or spotface. The time for the dwell, in seconds, is noted as a T word.

The G83 block for a peck-drilling cycle is:

N_____G83_____X_____Y_____C_____D_____K_____F_____EOB

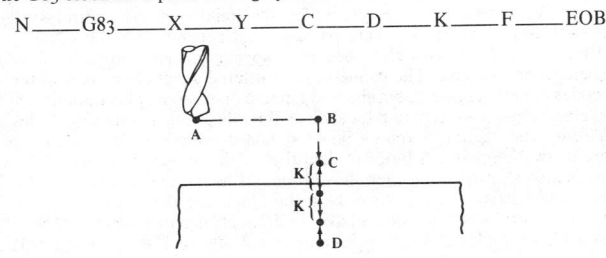

In the G83 peck-drilling cycle, the drill is moved from point *A* to point *B* and then to point *C* at the rapid-traverse rate; the drill is then fed the incremental distance *K*, followed by rapid return to *C*. Down feed again at the rapid traverse rate through the distance *K* is next, after which the drill is fed another distance *K*. The drill is then rapid-traversed back to *C*, followed by rapid traverse for a distance of *K* + *K*; down feed to *D* follows before the drill is rapid-traversed back to *C*, to end the cycle.

The G84 block for a tapping cycle is:

N_____G84_____X_____Y_____C_____D_____F_____EOB

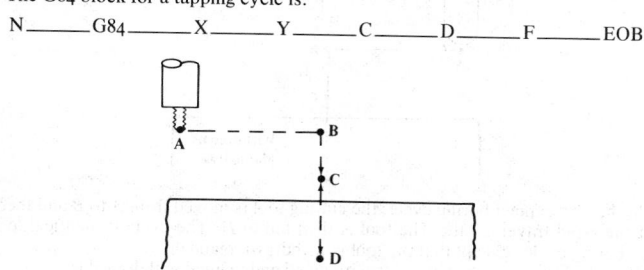

The G84 canned tapping cycle starts with the end of the tap being moved from point *A* to point *B* and then to point *C* at the rapid traverse rate. The tap is then fed to point *D*, reversed and moved back to point *C*.

The G85 block for a boring cycle with tool retraction at the feed rate is:

N_____G85_____X_____Y_____C_____D_____F_____EOB

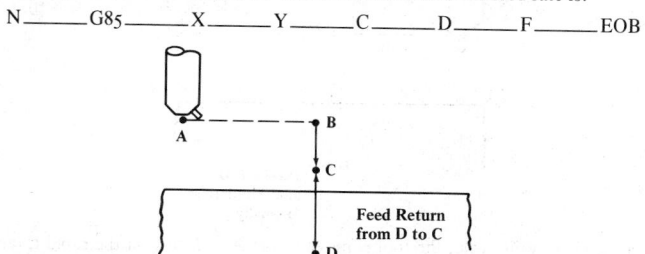

In the G85 boring cycle, the tool is moved from point *A* to point *B* and then to point *C* at the rapid traverse rate. The tool is next fed to point *D* and then, while still rotating, is moved back to point *C* at the same feed rate.

The G86 block for a boring cycle with rapid-traverse retraction is:

N_____G86 X_____Y_____C_____D_____F_____EOB

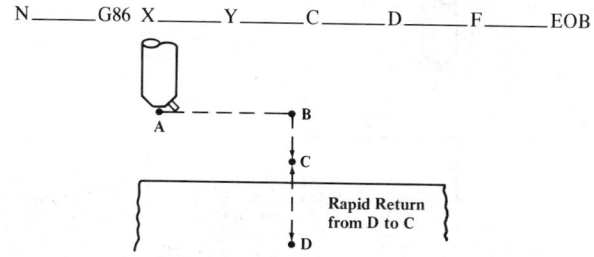

The G86 boring cycle is similar to the G85 cycle except that the tool is withdrawn at the rapid traverse rate.

The G87 block for a boring cycle with manual withdrawal of the tool is:

N_____G87 X_____Y_____C_____D_____F_____EOB

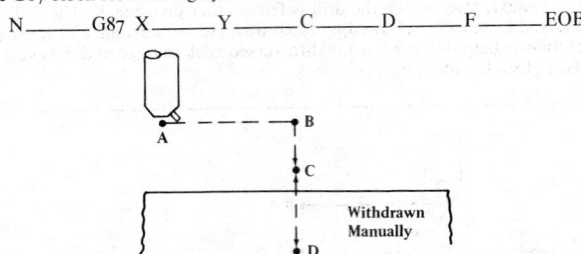

In the G87 canned boring cycle, the cutting tool is moved from A to B and then to C at the rapid traverse rate. The tool is then fed to D. The cycle is identical to the other boring cycles except that the tool is withdrawn manually.

The G88 block for a boring cycle with dwell and manual withdrawal is:

N_____G88 X_____Y_____C_____D_____T_____F_____EOB

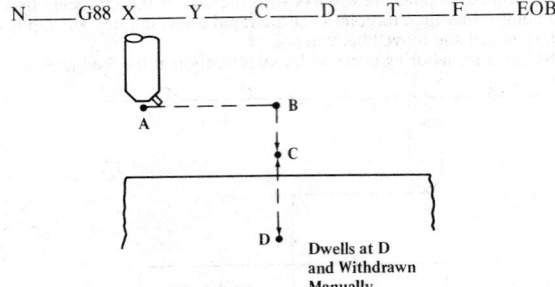

In the G88 dwell cycle, the tool is moved from A to B to C at the rapid traverse rate, then fed at the prescribed feed rate to D. The tool dwells at D then stops rotating and is withdrawn manually.

The G89 block for a boring cycle with dwell and withdrawal at the feed rate is:

N_____G89 X_____Y_____C_____D_____T_____F_____EOB

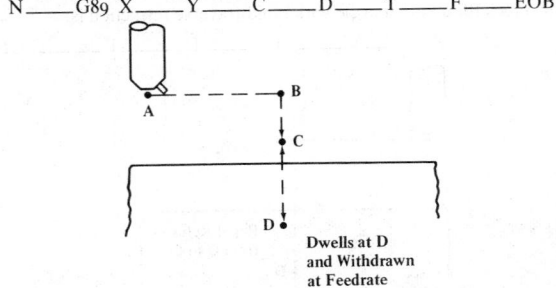

Tool movements in the G89 canned cycle are similar to those in the other boring cycles except that the cutter dwells at *D* and is then withdrawn, still rotating, at the preset feed rate. This cycle probably is the most-used of the canned boring cycles.

Fixed Zero. — Also referred to as machine zero, fixed zero is a point, physically located on the machine tool, where *X*, *Y*, and *Z* are zero. The arrangement is the opposite of a free-floating zero system. Fixed machine zero is especially popular with horizontal machining centers having rotary tables because the workpiece must be positioned accurately on the rotary table if it is to be machined on more than one side.

Flexible Manufacturing Cell. — A flexible manufacturing cell usually consists of two or three NC machines with some form of pallet-changing equipment or an industrial robot. Prismatic-type parts, such as would be processed on a machining center, are usually handled on pallets. Cylindrical parts, such as would be machined on an NC lathe, usually are handled with an overhead type of robot. The cell may be controlled by a computer, but is often run by programmable controllers. The systems can be operated without attendants but the mixture of parts usually must be less than with a flexible manufacturing system (FMS).

Flexible Manufacturing Module. — A flexible manufacturing module is defined as a single machining center (or turning center) with some type of automatic materials handling equipment such as multiple pallets for machining centers, or robots for manipulating cylindrical parts and chucks for turning centers.

Flexible Manufacturing System. — In the absence of a standard definition of a flexible manufacturing system (FMS) those proposed have varied widely. One definition is that an FMS consists of a number of NC work stations, connected by a materials-handling system, usually an automatic guided vehicle (AGV), and controlled by a supervisory computer. Such a system normally incorporates a washing unit and an inspection station (coordinate measuring machine), and can handle a mixture of parts for long periods without human attention. There is no agreed minimum for the number of work stations, or NC machines, but four has been common, as an economic number. Any less than four probably could not support the required ancillary equipment such as the computer and materials handling systems, the coordinate measuring machine and the washing station.

Format Detail. — A control system and machine tool are summarized very succinctly by the format detail. This NC shorthand is described in EIA standard RS-274-D, which gives the letter address words that are to be used to make up a block. Although it is not entirely comprehensive and falls short in describing all the features of modern CNC systems, the format detail does provide for defining the basic features of the control system and the type of machine tool to which it refers. For example, the format detail:

$$N4G2X + 24Y + 24Z + 24B24I24J24F31T4M2$$

specifies that the NC machine is a machining center (has *X*, *Y*, and *Z* axes and a tool changer with a four-digit tool selection code) (T4); the three linear axes are programmed with two digits before the decimal point and four after the decimal point (X24Y24Z24); probably has a horizontal spindle and rotary table (B24 = rotary motion about the *Y* axis); has circular interpolation (I24J24); has a feed rate range in which there are three digits before and one after the decimal point (F31); and can handle a four-digit sequence number (N4), two-digit G words (G2), and two-digit miscellaneous words (M2).

F-Word. — F-Word stands for feedrate word or feed rate, which can be expressed in inches or millimeters per minute (ipm or mmpm) or inches or millimeters per revolution (ipr or mmpr). A G94 word is used for ipm or mmpm and a G95 word

for ipr or mmpr. Some systems require an inverse time/feed command for certain types of cuts. The command value is calculated by dividing the feed rate by the distance moved in the block. The result is the reciprocal of time, i.e.:

$$\frac{\text{feed rate}}{\text{distance}} = \frac{\dfrac{\text{distance}}{\text{time}}}{\text{distance}} = \frac{\text{distance}}{\text{time}} \times \frac{1}{\text{distance}} = \frac{1}{\text{time}}$$

Free-floating Zero. — Provision for moving the cutter to any point within the range of the machine and reestablishing the origin, the point where X, Y, and Z equal 0, at that point, is called a free-floating zero. Many controls used on contour-milling machines provide this feature.

Hardwire Numerical Control. — As applied to NC, hardwire means a system in which the responses to data input, data-handling sequences and control functions are determined by fixed and committed circuit interconnections of discrete decision elements and storage devices. Changes in these functions can be made only by rewiring. In contrast, softwire NC, commonly known as CNC, or computer NC, permits changes to be made quickly and easily. Practically all the control systems being built today are softwire, or CNC systems.

Helical Interpolation. — Helical interpolation is used primarily for milling large threads and lubrication grooves, as shown in Fig. 31. Generally, circular interpolation (see *Circular Interpolation*) is used, together with the pitch and length of the thread. Cutting of the entire thread can usually be programmed in a single block.

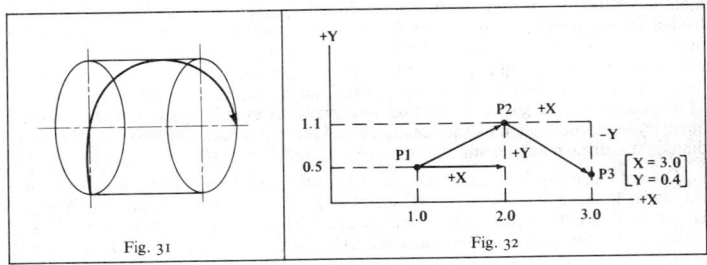

Fig. 31   Fig. 32

Incremental Programming. — The part-programming method in which the distance moved in the X, Y, and Z directions is specified, is known as incremental programming. In the example shown in Fig. 32, a move from P1 to P2 is written as $X + 1.0$, $Y + 0.6$. If there is no movement along the Z axis, Z is zero and normally is not noted. An X-Y incremental move from P2 to P3 in Fig. 1, is written as $X + 1.0$, $Y - 0.7$. Most CNC systems offer both absolute and incremental part programming. The choice is handled by a G word; G90 for absolute programming and G91 for incremental programming. (See also *Absolute Programming*.)

IGES. — IGES stands for Initial Graphics Exchange Specification, which is a means for exchanging or converting a computer graphics program for use in a different computer graphics system. The concept is shown diagrammatically in Fig. 33. Normally a program prepared on the computer graphics system supplied by company A would have to be rewritten before it would operate on the computer graphics system supplied by company B. However, with IGES, the program can be passed

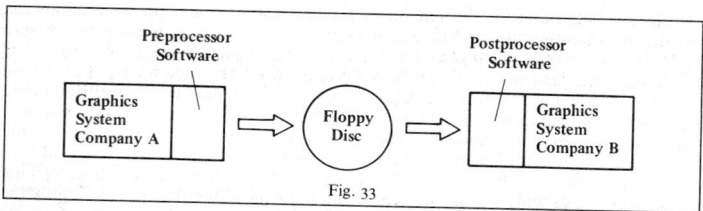

Fig. 33

through a software package called a preprocessor which will convert it into a standardized IGES format that can be stored on a magnetic disc. A postprocessor at company B is then used to convert the standard IGES format to that required for their graphics system. Both firms would be responsible for developing their own preprocessors and postprocessors, to suit their own machines and control systems. Almost all the major graphics-systems manufacturing companies today either have, or are developing, IGES preprocessor and postprocessor programs to convert software from one system to another.

Linear Interpolation. — The ability of the control system to guide the workpiece along a straight-line path at an angle to the slide movements is termed linear interpolation. Movements of the slides are controlled through simultaneous monitoring of pulses by the control system. For example, if monitoring of the pulses for the X axis of a milling machine is at the same rate as for the Y axis, the cutting tool will move at a 45-degree angle relative to the X axis. However, if the pulses are monitored at twice the rate for the X axis as for the Y axis, the angle that the line of travel will make with the X axis will be 26.57 degrees (tangent of 26.57 degrees = ½), as shown in Fig. 34. The data required are the distances traveled in the X and Y directions and from these data the control system will generate the straight line automatically. This monitoring concept also holds for linear motions along three axes. The required G-code for linear interpolation blocks is G01. The code is modal, which means that it will hold for succeeding blocks until it is changed.

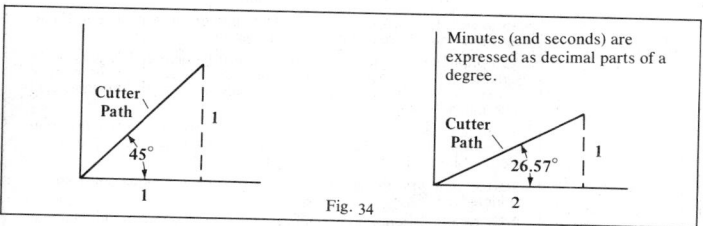

Fig. 34

Machining Center. — A machining center is an NC machine tool that can perform milling, drilling, boring, and usually tapping operations. Most machining centers also have automatic tool changing equipment.

Macro. — As with a parametric subroutine (see *Parametric Subroutine*), macro describes a type of program that can be recalled to allow insertion of finite values for letter variables that have been included. The difference between a macro and a parametric subroutine is that the term macro normally applies to the source program that is used with computer-assisted part programming and the parametric subroutine is a feature of the CNC system and can be input directly into that system.

Manual Data Input. — With manual data input (MDI) machining instructions can be inserted directly into the NC machine control system via push buttons, pressure pads, knobs, or other arrangements. MDI has been available since the earliest NC machines were designed. With hardwire systems, MDI provided for input of data in a format identical to that from the tape. However, the method was less efficient than tape for machining operations and was used primarily for setting up the NC machine. With their canned cycles and other computing capabilities, CNC systems have made the MDI concept more feasible and for some work MDI may be more practical than preparing a tape. The choice depends very much on the complexity of the machining work to be done, and, to a lesser degree, on the skill of the person who inputs the program.

Microcomputer. — A microcomputer is a class of computer having all major central processing functions on a single printed-circuit board constituting a stand-alone module. A microcomputer is constructed using a microprocessor as a basic element.

Microprocessor. — A microprocessor is a single integrated circuit and is the basic element of a central processing unit (CPU) used in a microcomputer. However, a microprocessor requires additional circuitry to make it suitable for use as a central processing unit. (See *Microcomputer*.)

Miscellaneous Functions. — Referred to also as auxiliary functions, miscellaneous functions constitute on-off commands. For example, these functions can be used to start or stop the machine, turn the coolant on or off, or clamp or unclamp a part in a fixture. The table lists and describes those miscellaneous functions (or words) that have been standardized under EIA Standard RS-274-D. The words marked unassigned are not assigned by the standard, but may be used by the builder of the system or the machine to address various NC functions that are not included in the standard.

Standard Miscellaneous Function Words from EIA Standard RS-274-D

Word (Code)	Explanation
M00	Automatically *stops* the machine. The operator must push a button to continue with the remainder of the program.
M01	An *optional stop* acted upon only when the operator has previously signaled for this command by pushing a button. When the control system senses the M01 code via the tape reader, the machine will automatically stop.
M02	At the completion of the machining operation, this *end of program* code stops the machine when all commands in the block are completed. May include rewinding of tape.
M03	Start *spindle rotation* in a *clockwise* direction — looking out from the spindle face.
M04	Start *spindle rotation* in a *counter-clockwise* direction — looking out from the spindle face.
M05	*Stop* the spindle in a normal and efficient manner.
M06	Command to *change a tool* (or tools) manually or automatically. Does not cover the selection of a tool as is possible with the T words.
M07	Code to turn a *coolant* on.
M08	Another code for turning on a coolant. May differ from M07 in that M07 may control *flood* coolant, and M08 *mist* coolant.
M09	Automatically shuts off the coolant.
M10 M11	M10 applies to automatic *clamping* of the machine slides, workpiece, fixture spindle, etc. M11 is an *unclamping* code.

Standard Miscellaneous Function Words from EIA Standard RS-274-D (*Cont'd*)

Word (Code)	Explanation
M12	An inhibiting code used for the synchronization of multiple sets of axes, such as a four-axis lathe having two independently-operated heads (turrets).
M13	Combines *clockwise spindle* motion and *coolant on* in the same command causing both to occur simultaneously.
M14	Combines *counter-clockwise* spindle motion and *coolant on* in the same command.
M15 M16	Rapid traverse or feed motion in either the + (M15) or − (M16) direction.
M17 M18	Unassigned.
M19	Oriented spindle stop. Causes the spindle to stop at a predetermined angular position.
M20 through M29	Unassigned.
M30	An *end of tape* command going slightly further than the M02 code in that this code *will* rewind the tape (assuming the control has this facility); also switch automatically to a second tape reader if incorporated in the control system.
M31	A command known as *interlock bypass* for temporarily circumventing a normally provided interlock.
M32 through M39	Unassigned.
M40 through M46	Used to signal gear changes if required at the machine; otherwise, unassigned.
M47	Continues program execution from the start of the program unless inhibited by an interlock signal.
M48	Cancel M49.
M49	Deactivates a manual spindle or feed override and returns the parameter to the programmed value.
M50 through M57	Unassigned.
M58	Cancel M59.
M59	Holds the rpm constant at the value in use when M59 is initiated.
M60 through M99	Unassigned.

Open-loop System. — A control system that issues commands to the drive motors of an NC machine and has no means of assessing the results of these commands is known as an open-loop system. In such a system, no provision is made for feedback of information concerning movement of the slide(s), or rotation of the leadscrew(s). Stepping motors are popular as drives for open-loop systems.

Parabolic Interpolation. — Simultaneous and coordinated control of two axes of motion such that the resulting cutter path describes part of a parabola is called parabolic interpolation. EIA Standard RS-274-D provides further details.

Parametric Subroutine. — Similar to a Subroutine (see *Subroutine*) except that a parametric subroutine permits letters or symbols to be inserted into the program in

place of numerical values. The parametric subroutine can then be called during part programming and values can be assigned to the letters or symbols. This facility is particularly helpful when dealing with families of parts.

Parity. — In NC, parity refers to one of a line of holes running across an NC tape that results in an odd or even number of holes representing a character. This extra hole provides a means to check that a correct character is being read. If the system is set to read an odd number of holes in the character, as with the EIA RS-244-B Standard, and a punching error had produced an even number of holes, the machine would stop. The opposite situation would occur with the ASCII RS-358-B standard, which will accept only an even number of holes in a character. (See *Character Code*.)

Part Program Edit. — Provision for changing or editing a part program at the machine with the aid of a display screen such as a CRT is called part program edit. Some control systems incorporate a tape punch with which a tape can be prepared from the edited program.

Part Zero. — Referred to also as workpiece zero, part zero is a reference point on the part drawing, established by the part programmer, where X, Y, and Z are zero.

Postprocessor. — A postprocessor is a set of computer instructions (also known as a software package) that tailors the cutter location (CL) data developed by a computer program to meet the requirements of a particular machine tool/system combination. (See *APT*.)

Preload Registers (G92 Word). — The G92 word is used to preload the registers in the control system with desired values. A common example is the loading of the axis position registers in the control system for a lathe. Fig. 35 shows a typical home position of the tool tip with respect to the zero point on the machine. The tool tip here is registered as being 15.0000 inches in the Z direction and 4.5000 inches in the X direction from machine zero. No movement of the tool is required. Although it will vary with different control system manufacturers, the block to accomplish the registration shown in Fig. 35 will be approximately: N0050 G92 X4.5 Z15.0.

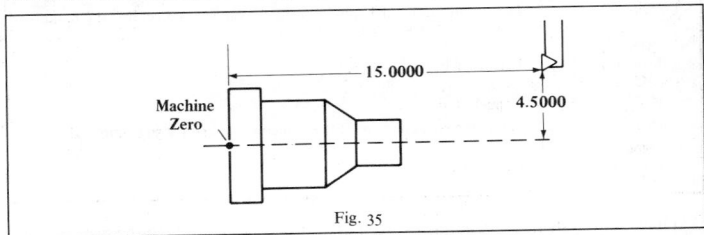

15.0000

Machine Zero

4.5000

Fig. 35

Preparatory Word. — A preparatory word (also referred to as a preparatory function or code) consists of the letter address G and two digits. The preparatory word is placed at the beginning of the block, normally following the sequence number. This word prepares the control system to accept the remainder of the block in a certain way. For example, G01 refers to linear interpolation and indicates that the words following in the block will move the cutter in a straight line. The code G02 indicates that the words following in the block will move the cutter in a clockwise circular path. The meanings of words in a block can be changed by a G word. For example, X is normally the address of a dimension word and describes the distance to be moved, or moved to, in the X direction.

The meaning of the X word may change completely if a G04 word precedes it in a block. In this case the G04 word, or code, means a dwell and the numbers in the X word give the time that the machine is to dwell, in seconds. Many G words have become part of the RS-274-D standard, but a lot of G words have not been standardized and remain unassigned. These G words are used by builders of control systems or machine tools for a variety of purposes, sometimes leading to confusion as to the meanings of the unassigned codes.

Some builders of systems or machine tools use some of the less popular assigned standardized codes for other than the meaning listed in the standards. The G codes are modal which means that they will hold for the succeeding blocks unless changed or cancelled. It is therefore not necessary to include the G code in succeeding blocks where it is to be applicable. Those G codes that have been standardized are shown below.

G Code Addresses

Code		Description	Code		Description
G00		Used to denote a rapid traverse movement, usually with point-to-point operations	G70		Programming in inches
			G71		Programming in millimeters
G01	D	Linear interpolation	G72		Circular interpolation — clockwise (three dimensional)
G02	D	Circular interpolation—clockwise movement			
G03	D	Circular interpolation—counterclockwise movement	G73		Circular interpolation — counterclockwise (three dimensional)
G04		Dwell—a calculated time delay usually less than a few seconds	G74	D	Cancel multiquadrant circular interpolation
			G75	D	Multiquadrant circular interpolation
G05		Unassigned	G76-G79		Unassigned
G06	D	Parabolic interpolation	G80	D	Cancel fixed (canned) cycle
G07		Unassigned	G81	D	Drill cycle
G08		Automatic acceleration	G82	D	Drill + dwell cycle
G09		Automatic deceleration	G83	D	Peck drill cycle
G10-G12		Unassigned	G84	D	Tapping cycle
G13-G16	D	Axis selection	G85	D	Boring cycle — rotating spindle retraction
G17	D	*XY* plane selection			
G18	D	*XZ* plane selection (described under circular interpolation)	G86	D	Boring cycle — rapid traverse retraction
G19	D	*YZ* plane selection	G87	D	Boring cycle — manual retraction
G20-G32		Unassigned	G88	D	Boring cycle — dwell and manual retraction
G33	D	Thread cutting-constant lead			
G34	D	Thread cutting-increasing lead	G89	D	Boring cycle — dwell and feed retraction
G35	D	Thread cutting-decreasing lead	G90	D	Absolute input programming
G36-G39		Unassigned	G91	D	Incremental input programming
G40	D	Cancel compensation/offset	G92	D	Preload of registers
G41	D	Cutter compensation — left	G93	D	Inverse time feed rate
G42		Cutter compensation — right	G94		Inches or millimeters per minute feed rate (programmed with F word)
G43	D	Cutter offset-inside corner	G95		Inches or millimeters per revolution (programmed with F word)
G44	D	Cutter offset-outside corner			
G45-G49		Unassigned	G96	D	Constant surface speed (programmed with F word)
G50-G59	D	Reserved for adaptive control			
G60-G69		Unassigned	G97		Revolutions per minute (programmed with S word)

D indicates that a detailed description appears under an alphabetic heading in this NC section.

Programmable Increment. — The smallest number that can be programmed into an NC machine tool is called the programmable increment. Most NC machines can be programmed to four decimal places in inches and three decimal places in millimeters.

Quadrant. — As shown in Fig. 36, a quadrant is defined as the fourth part of a graph in reference to the X and Y axes. The term is commonly used in NC part programming.

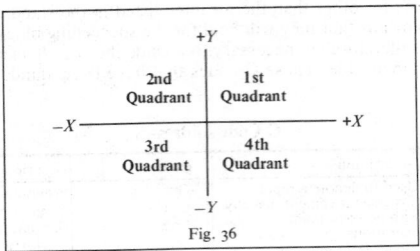

Fig. 36

Qualified Tooling. — Tooling and toolholders used for NC machining are usually made to more precise standards than ordinary tooling, and are then described as qualified to prescribed tolerances specified by the tool manufacturer. Some details of qualified tooling are given in the Handbook section on cutting tools (pages 727–731).

Resolution. — The smallest increment of distance that can be read and acted upon by an NC system is called its resolution. The resolution of the control system, however, may be smaller than the smallest increment the machine tool can respond to. For example, the resolution of a control system may be 0.00001 inch, but the smallest programmable increment may be 0.0001 inch.

RS-232-C Interface. — Standard ports, or interface connections, on control systems for the communication of data in serial form are made to the RS-232-C Standard. These standard ports accept multipin plugs attached to equipment used to input tape data and will accept input from distributed NC systems.

Sequence Number. — The sequence number is usually the first word in a block and it identifies a specific block in the part program. Usually each block has its own number. Most control systems offer sequence numbers running to four digits, or 9999, and the practice is to advance the sequence numbers in fives to leave spaces for additional blocks to be inserted if required. For example, the first block would be N0000; the next block N0005; the next N0010, and so on. The block that is being worked on by the machine is displayed on a digital readout so that the operator may know the precise operation being performed.

Software. — NC computer programs used to assist in part programming and in the operation of machine tools through CNC systems are generally known as software, as opposed to the machines and computer equipment, known as hardware.

Subroutine. — A subroutine is a set of instructions or blocks that can be inserted into a program and repeated whenever required. A typical subroutine for milling several identical rectangular patterns on a vertical milling machine, is shown in Fig. 37. There is no standard subroutine format, and, although typical, the example here

shown is not intended to conform to any manufacturer's program. The program for milling the three pockets shown in Fig. 37 might look as follows:

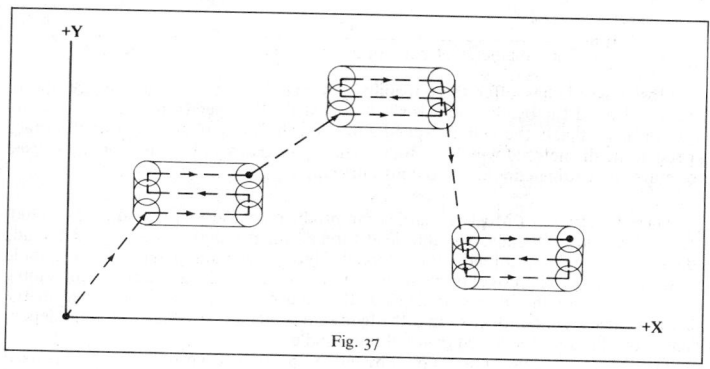

Fig. 37

N0010 G00 X.6 Y.85	Cutter is moved at a rapid traverse rate to a position over the corner of the first pocket to be cut.
N0020 M92	Tells the system that the subroutine is to start in the next block.
N0030 G01 Z − .25 F2.0	Cutter is moved axially into the workpiece 0.25 inch at 2.0 ipm.
N0040 X.8	Cutter is moved to the right 0.8 inch.
N0050 Y.2	Cutter is moved laterally up 0.2 inch.
N0060 X − .8	Cutter is moved to the left 0.8 inch.
N0070 Y.2	Cutter is moved laterally up 0.2 inch.
N0080 X.8	Cutter is moved to the right 0.8 inch.
N0090 G00 Z.25 M93	Cutter is moved axially out of pocket at rapid traverse rate. Last block of subroutine is signaled by word M93.
N0100 X.75 Y.5	Cutter is moved to bottom left hand corner of second pocket at rapid-traverse rate.
N0110 M94 N0030	Word M94 calls for repetition of subroutine which starts at sequence number N0030 and ends at sequence number N0090.
N0120 G00 X.2 Y − 1.3	After the second pocket is cut by repetition of sequence numbers N0030 through N0090, the cutter is moved to start the third pocket.
N0130 M94 N0030	Repetition of subroutine is called for by M94 word and third pocket is cut.

S-Word. — An S-word specifies the speed of rotation of the spindle in revolutions per minute (rpm). When speed is to be given in surface feet per minute (sfm), it is necessary to convert the value to rpm, which can be handled by the formula:

$$\text{rpm} = \frac{\text{sfm} \times 12}{\pi \times \text{diameter of part (in lathe) or milling cutter in inches}}$$

Most CNC lathes offer the capability for maintaining constant surface speed, which is called for by the G96 word, followed by the speed, in sfm. The control system then adjusts the spindle speed automatically to maintain a constant surface speed at the diameter at which the tool is cutting. A G97 word in the program is used to return the cutting conditions to constant rpm.

Thread Cutting. — Most NC lathes can produce a variety of thread types including constant lead threads, variable lead threads (increasing), variable lead threads (decreasing), multiple threads, taper threads, threads running parallel to the spindle axis, threads (spiral groove) perpendicular to the spindle axis, and threads containing a combination of the above. Instead of the feed rate, the lead is specified in the threading instruction block, so that the feed rate is made consistent with, and dependent upon, the selected speed (rpm) of the spindle.

The thread lead is generally noted by either an I or a K word. The I word is used if the thread is parallel to the X axis and the K word if the thread is parallel to the Z axis, the latter being by far the most common. The G word for a constant-lead thread is G33. The G word for an increasing variable lead is G34 and the G word for a decreasing variable lead thread is G35. Taper threads are obtained by noting the X and Z coordinates of the beginning and end points of the thread if the G90 code is in effect (absolute programming), or the incremental movement from the beginning point to the end point of the thread if the G91 code (incremental programming) is in effect.

Multiple threads are specified by a code in the block that spaces the starts of the threads equally around the cylinder being threaded. For example, if a triple thread is to be cut, the threads will start 120 degrees apart. Typical single-block thread cutting utilizing a plunge cut, is illustrated in Fig. 38, and shows two passes. The passes are

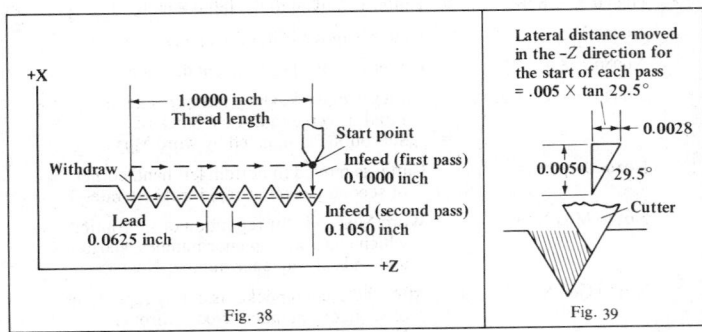

Fig. 38   Fig. 39

identical except for the distance of the plunge cut. Builders of control systems and machine tools use different code words for threading but those shown below can be considered typical. For clarity, both zeros and decimal points are shown.

N0001 G91	(Incremental programming)
N0002 G00 X − .1000	(Rapid traverse to depth)
N0003 G33 Z − 1.0000	(Produce a thread with a constant
K.0625	lead of 0.0625 inch)
N0004 G00 X.1000	(Withdraw at rapid traverse)
N0005 Z1.0000	(Move back to start point)

The only changes in the second pass are the depth of the plunge cut and the withdrawal. The blocks will appear as follows:

```
N0006 X − .1050
N0007 G33 Z − 1.0000 K.0625
N0008 G00 X.1050
N0009 Z1.0000
```

Compound thread cutting, rather than straight plunge thread cutting, is possible also, and is usually used on harder materials. As illustrated in Fig. 39, the starting point for the thread is moved laterally in the − Z direction by an amount equal to the depth of the cut times the tangent of an angle that is slightly less than 30 degrees. The program for the second pass for the example shown in Fig. 39, therefore, is as follows:

```
N0006 X − .1050 Z − .0028
N0007 G33 Z − 1.0000 K.0625
N0008 G00 X.1050
N0009 Z1.0000
```

Fixed (canned), one-block cycles also have been developed for CNC systems to produce the passes needed to complete a thread. These cycles may be offered by the builder of the control system or machine tool as standard or optional features. Subroutines also can generally be prepared by the user to accomplish the same purpose. (See *Subroutine*.) A one-block fixed threading cycle might look something like:

$$\text{N0048 G98 X} - .2000 \text{ Z} - 1.0000 \text{ D.0050 F.0010}$$

where: G98 = the preparatory code for the threading cycle
 X − .2000 = the total distance from the starting point to the bottom of the thread
 Z − 1.0000 = the length of the thread
 D.0050 = the depths of successive cuts
 F.0010 = the depth(s) of the finish cut(s)

Tool Data Table. — Information about the tools and the tool setups is input to the CNC system in the form of a tool data table. Details of specific tools are transferred from the table to the part program via the T word. The tool nose radius of a lathe tool, for example, is recorded in the tool data table so that the necessary tool path calculations can be made by the CNC system.

Tool Length Offset. — Changes to the programmed positions of cutting tool tip(s) can be made by tool length offset circuitry included in the control system. A dial or other means is provided, allowing the operator or part programmer to override the programmed axial, or Z axis, position, generally on milling, drilling and boring machines and machining centers. This feature is particularly helpful when setting the lengths of tools in their holders or setting a tool in a turret as shown in Fig. 40, because an exact setting is not necessary. The tool can be set to an approximate length and the discrepancy eliminated by the control system.

The amount of offset may be determined by noting the amount by which the cutter is moved manually to a fixed point on the fixture or on the part, from the programmed Z-axis location. For example, in Fig. 40 the programmed Z-axis motion results in the

cutter being moved to position *A*, whereas the required location for the tool is at *B*. Rather than resetting the tool or changing the part program, the tool length offset amount of 0.0500 inch is keyed into the control system. The 0.0500-inch amount is measured by moving the cutter tip manually to position *B* and reading the distance moved on the readout panel. Thereafter, every time that cutter is brought into the machining position, the programmed Z-axis location will be overridden by 0.0500 inch.

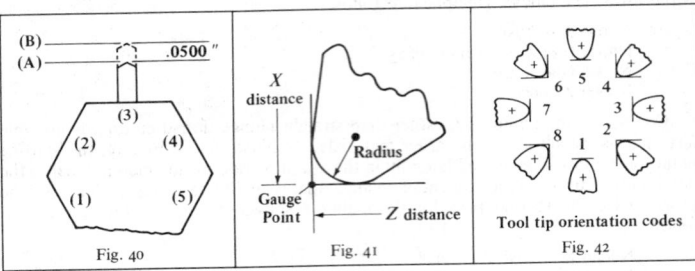

Fig. 40   Fig. 41   Tool tip orientation codes
                                   Fig. 42

Tool Nose Radius Compensation. — Compensation for variations in the tool nose radius, particularly on turning machines, allows the programmer to program the part geometry from the drawing and have the tool follow the correct path in spite of variations in the tool nose shape. Tooling and setup information entered into the tool data table is called upon by a T word. Typical of the data required (as shown in Fig. 41) is the nose radius of the cutter, the *X* and *Z* distances from the gauge point to some fixed reference point on the turret, and the orientation of the cutter (tool tip orientation code), as shown in Fig. 42.

Tool Offset. — Tool offset, also called cutter offset, is the amount of cutter adjustment in a direction parallel to the axes, and is provided almost universally on lathes. Adjustments can be made at the machine, overriding the tape dimension word instructions. For example, owing to wear, the tool tip in Fig. 43 is positioned a distance of 0.0065 inch from the location required for the work to be done. To compensate for this wear the operator (or part programmer), by means of the CNC control panel, adjusts the tool tip with reference to the *X* and *Z* axes, moving the tool closer to the work by 0.0065 inch throughout its traverse.

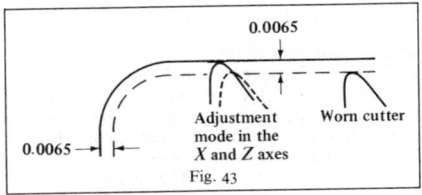

Fig. 43

Total Indicator Reading (TIR). — Total indicator reading is used as a measure of the range of machine tool error. TIR is particularly useful for describing the error in

a machine tool spindle, referred to as runout. As shown in Fig. 44, there are two types of runout: axial and radial. Axial runout refers to the wobble of a spindle and is measured at the spindle face. Radial runout is the range of movement of the spindle centerline and is measured on the side of the spindle or quill.

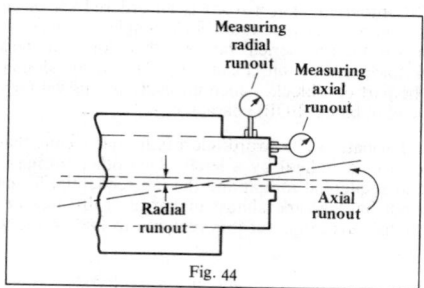

Fig. 44

Turning Cycles. — Canned turning cycles are available from most system builders and are designed to allow the programmer to describe a complete turning operation in one or a few blocks. There is no standard for this type of operation so a wide variety of programs has developed. Fig. 45 shows a hypothetical sequence in which the cutter is moved from the start point to depth for the first pass. If incremental programming is in effect, this distance is specified as D_1. The depths of the other cuts will also be programmed as D_2, D_3, and so on. The length of the cut will be set by the W word, and will remain the same with each pass. The preparatory word that calls for the roughing cycle is G77. The roughing feed rate is 0.03 ipr (inch per revolution), and the finishing feed rate (last pass) is 0.005 ipr. The block appears as follows:

$$N0054 \; G77 \; W = 3.1 \; D_1 = .4 \; D_2 = .3 \; D_3 = .3 \; D_4 = .1 \; F_1 = .03 \; F_2 = .005$$

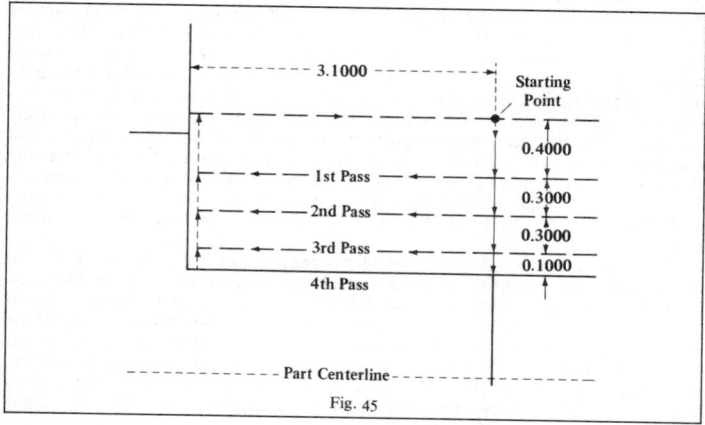

Fig. 45

T Word. — The T word calls out the tool that is to be selected on a machining center or lathe having an automatic tool changer or indexing turret. This word also specifies the proper turret face on a lathe. The word usually is accompanied by several numbers.

Word. — In NC programming, a word is an ordered set of characters used to cause a specific action of a machine tool. For example, dimension words include X, Y and Z, together with their numerical values. Other words having specific meanings include G (preparatory) and M (miscellaneous). The words alone cannot trigger an action, but must be part of a block, which normally begins with a sequence number and ends with an end of block (EOB) character.

Word Address Format. — The word address format means that each word in a block is addressed (or identified) by a letter. Two other formats used in the past comprised the fixed sequential format and the tab sequential format. However, the word address format has become almost universal. Tabs may be used with some word address systems for neatness if it is required to line up the words in columns when being printed.

Zero Suppression. — Zero suppression in a control system is an arrangement that allows zeros before the first significant figure to be dropped (leading zero suppression), or allows zeros after the last significant figure to be dropped (trailing zero suppression). An X axis movement of 05.3400, for example, could be expressed as 53400 (leading zero suppression) or 0543 (trailing zero suppression). This arrangement saves tape and computation time. With decimal-point programming, offered by many systems, the above number is expressed simply as 5.34.

Indexable Insert Holders for NC. — Indexable insert holders for numerical control lathes are usually made to more precise standards than ordinary holders. Where applicable, reference should be made to American National Standard B94.45M-1979, Precision Holders for Indexable Inserts. This standard covers the dimensional specifications, styles, and designations of precision holders for indexable inserts, which are defined as tool holders that locate the gage insert (a combination of shim and insert thicknesses) from the back or front and end surfaces to a specified dimension with a ± 0.003 inch (± 0.08 mm) tolerance. In NC programming, the programmed path is that followed by the center of the tool tip, which is the center of the point, or nose radius, of the insert. The surfaces produced are the result of the path of the nose and the major cutting edge, so it is necessary to compensate for the nose or point radius and the lead angle when writing the program. Table 1, from B94.45M, gives the compensating dimensions for different holder styles. The reference point is determined by the intersection of extensions from the major and minor cutting edges, which would be the location of the point of a sharp pointed tool. The distances from this point to the nose radius are $L1$ and $D1$; $L2$ and $D2$ are the distances from the sharp point to the center of the nose radius. Threading tools have sharp corners and do not require a radius compensation. Other dimensions of importance in programming threading tools also given in Table 2; the data were developed by Kennametal, Inc.

The C and F dimensions are tool holder dimensions other than the shank size. In all instances the C dimension is parallel to the length of the shank and the F dimension is parallel to the side dimension; actual dimensions must be obtained from the manufacturer. For all K style holders the C dimension is the distance from the end of the shank to the tangent point of the nose radius and the end cutting edge of the insert. For all other holders the C dimension is from the end of the shank to a tangent to the nose radius of the insert. The F dimension on all B, D, E, M, P, and V style holders is measured from the back side of the shank to the tangent point of the nose radius

and the side cutting edge of the insert. For all A, F, G, J, K, and L style holders the *F* dimension is the distance from the back side of the shank to the tangent of the nose radius of the insert. In all these designs, the nose radius is the standard radius corresponding to the inscribed circle as given in Table 2, on page 728.

Table 1. Insert Radius Compensation (ANSI B94.45M-1979)

Square Profile						

B Style*

		Turning 15° Lead Angle			
	Rad.	*L-1*	*L-2*	*D-1*	*D-2*
	1/64	.0035	.0191	.0009	.0110
	1/32	.0070	.0383	.0019	.0221
	3/64	.0105	.0574	.0028	.0331
	1/16	.0140	.0765	.0038	.0442

D Style*
Also Applies
to S Style

		Turning 45° Lead Angle			
	Rad.	*L-1*	*L-2*	*D-1*	*D-2*
	1/64	.0065	.0221	.0065	0
	1/32	.0129	.0442	.0129	0
	3/64	.0194	.0663	.0194	0
	1/16	.0259	.0884	.0259	0

K Style*

		Facing 15° Lead Angle			
	Rad.	*L-1*	*L-2*	*D-1*	*D-2*
	1/64	.0009	.0110	.0035	.0191
	1/32	.0019	.0221	.0070	.0383
	3/64	.0028	.0331	.0105	.0574
	1/16	.0038	.0442	.0140	.0765

Triangle Profile						

A Style*
Also Applies
to G Style

		Turning 0° Lead Angle			
	Rad.	*L-1*	*L-2*	*D-1*	*D-2*
	1/64	.0114	.0271	0	.0156
	1/32	.0229	.0541	0	.0312
	3/64	.0343	.0812	0	.0469
	1/16	.0458	.1082	0	.0625

B Style*

		Facing & Turning 15° Lead Angle			
	Rad.	*L-1*	*L-2*	*D-1*	*D-2*
	1/64	.0146	.0302	.0039	.0081
	1/32	.0291	.0604	.0078	.0162
	3/64	.0437	.0906	.0117	.0243
	1/16	.0582	.1207	.0156	.0324

All dimensions are given in inches.

* *L-1* and *D-1* over sharp point to nose radius; and *L-2* and *D-2* over sharp point to center of nose radius. The *D-1* dimension for the B, E, D, M, P, S, T, and V style tools are over the sharp point of insert to a sharp point at the intersection of a line on the lead angle on the cutting edge of the insert and the *C* dimension. The *L-1* dimensions on K style tools are over the sharp point of insert to sharp point intersection of lead angle and *F* dimensions.

Table 1 (*Continued*). **Insert Radius Compensation** (ANSI B94.45M-1979)

Triangle Profile		

E Style*
Also Applies
to T Style

	Turning 30° Lead Angle			
Rad.	L-1	L-2	D-1	D-2
1/64	.0156	.0312	.0090	0
1/32	.0312	.0625	.0180	0
3/64	.0469	.0938	.0271	0
1/16	.0625	.1250	.0361	0

F Style*
Also Applies
to C Style

	Facing 90° Lead Angle			
Rad.	L-1	L-2	D-1	D-2
1/64	0	.0156	.0114	.0271
1/32	0	.0312	.0229	.0541
3/64	0	.0469	.0343	.0812
1/16	0	.0625	.0458	.1082

J Style*

	Turning & Facing 3° Lead Angle			
Rad.	L-1	L-2	D-1	D-2
1/64	.0106	.0262	.0014	.0170
1/32	.0212	.0524	.0028	.0340
3/64	.0318	.0786	.0042	.0511
1/16	.0423	.1048	.0056	.0681

80° Diamond Profile		

G Style*

	Turning & Facing 0° Lead Angle			
Rad.	L-1	L-2	D-1	D-2
1/64	.0030	.0186	0	.0156
1/32	.0060	.0312	0	.0312
3/64	.0090	.0559	0	.0469
1/16	.0120	.0745	0	.0625

L Style*

	Turning & Facing 5° Reverse Lead Angle			
Rad.	L-1	L-2	D-1	D-2
1/64	.0016	.0172	.0016	.0172
1/32	.0031	.0344	.0031	.0344
3/64	.0047	.0516	.0047	.0516
1/16	.0062	.0688	.0062	.0688

F Style*

	Facing 0° Lead Angle			
Rad.	L-1	L-2	D-1	D-2
1/64	0	.0156	.0030	.0186
1/32	0	.0312	.0060	.0372
3/64	0	.0469	.0090	.0559
1/16	0	.0625	.0120	.0745

* See footnote page 1165.

Table 1 (*Concluded*). **Insert Radius Compensation** (ANSI B94.45M-1979)

80° Diamond Profile (cont.)						

B Style*

Turning 15° Lead Angle

Rad.	L-1	L-2	D-1	D-2
1/64	.0011	.0167	.0003	.0117
1/32	.0022	.0384	.0006	.0234
3/64	.0032	.0501	.0009	.0351
1/16	.0043	.0668	.0012	.0468

K Style*

Facing 15° Lead Angle

Rad.	L-1	L-2	D-1	D-2
1/64	.0003	.0117	.0011	.0167
1/32	.0006	.0234	.0022	.0334
3/64	.0009	.0351	.0032	.0501
1/16	.0012	.0468	.0043	.0668

M Style*

Turning 40° Lead Angle

Rad.	L-1	L-2	D-1	D-2
1/64	.0087	.0243	.0073	0
1/32	.0174	.0486	.0146	0
3/64	.0260	.0729	.0218	0
1/16	.0347	.0972	.0291	0

55° Profile						

J Style*

Profiling 3° Reverse Lead Angle

Rad.	L-1	L-2	D-1	D-2
1/64	.0135	.0292	.0015	.0172
1/32	.0271	.0583	.0031	.0343
3/64	.0406	.0875	.0046	.0519
1/16	.0541	.1166	.0062	.0687

35° Profile						

J Style* Negative rake holders have 6° back rake and 6° side rake

Profiling 3° Reverse Lead Angle

Rad.	L-1	L-2	D-1	D-2
1/64	.0330	.0487	.0026	.0182
1/32	.0661	.0973	.0051	.0364
3/64	.0991	.1460	.0077	.0546
1/16	.1322	.1947	.0103	.0728

L Style*

Profiling 5° Lead Angle

Rad.	L-1	L-2	D-1	D-2
1/64	.0324	.0480	.0042	.0198
1/32	.0648	.0360	.0086	.0398
3/64	.0971	.1440	.0128	.0597
1/16	.1205	.1920	.0170	.0795

* See footnote page 1165.

Table 2. Threading Tool Insert Radius Compensation for NC Programming

Insert Size	Threading					
	T	R	U	Y	X	Z
2	5/32 Wide	.040	.075	.040	.024	.140
3	3/16 Wide	.046	.098	.054	.031	.183
4	1/4 Wide	.053	.128	.054	.049	.239
5	3/8 Wide	.099	.190	. . .	. . .	. . .

Buttress Threading   29° Acme   60° V-Threading

NTB-B   NTB-A   NA   NTF   NT

All dimensions are given in inches.
Courtesy of Kennametal, Inc.

Machinery Noise. — Noise from machinery or other mechanical systems can be controlled to some degree in the design or development stage if quantified noise criteria are provided the designer. Manufacturers and consumers may also use the same information in deciding whether the noise generated by a particular machine will be acceptable for a specific purpose. Such criteria for noise may be classified into three types: (1) those relating to the degree of interference with speech communications; (2) those relating to physiological damage to humans, especially their hearing; and (3) those relating to psychological disturbances in people exposed to noise.

Sound Level Specifications: Noise criteria generally are specified in some system of units representing sound levels. One commonly used system specifies sound levels in units called decibels on the "A" scale, written dBA. The dBA scale designates a sound level system weighted to match human hearing responses to various frequencies and loudness. For example, to permit effective speech communication, typical criteria for indoor maximum noise levels are: meeting and conference rooms, 42 dBA; private offices and small meeting rooms, 38 to 47 dBA; supervisors' offices and reception rooms, 38 to 52 dBA; large offices and cafeterias, 42 to 52 dBA; laboratories, drafting rooms, and general office areas, 47 to 56 dBA; maintenance shops, computer rooms, and washrooms, 52 to 61 dBA; control and electrical equipment rooms, 56 to 66 dBA; and manufacturing areas and foremen's offices, 66 dBA. Similarly, there are standards and recommendations for daily permissible times of exposure at various steady sound levels to avoid hearing damage. For a working shift of 8 hours, a steady sound level of 90 dBA is the maximum generally permitted, with marked reduction in allowable exposure times for higher sound levels.*

Measuring Machinery Noise. — The noise level produced by a single machine can be measured by using a standard sound level meter of the handheld type set to the dBA scale. However, when other machines are running at the same time, or when there are other background noises, the noise of the machine cannot be measured

directly. In such cases, two measurements, taken as follows, can be used to calculate the noise level of the individual machine. The meter should be held at arm's length and well away from any bystanders to avoid possible significant error up to 5 dBA.

Step 1. At the point of interest, measure the total noise, T, in decibels; that is, measure the noise of the shop and the machine in question when all machines are running; Step 2. Turn off the machine in question and measure B, the remaining background noise level; Step 3. Calculate M, the noise of the machine alone, $M = 10\log_{10}[10^{(T/10)} - 10^{(B/10)}]$.

Example 1: With a machine running, the sound level meter reads 51 decibels as the total shop noise T; and with the machine shut off the meter reads 49 decibels as the remaining background noise B. What is the noise level M of the machine alone? $M = 10\log_{10}[10^{(51/10)} - 10^{(49/10)}] = 46.7$ decibels dBA.

Example 2: If in Example 1 the remaining background noise level B was 41 decibels instead of 49, what is the noise level of the machine alone? $M = 10\log_{10}[10^{(51/10)} - 10^{(41/10)}] = 50.5$ decibels dBA.

Note 1: From this example it is evident that when the background noise level B is approximately 10 or more decibels lower than the total noise level T measured at the machine in question, then the background noise does not contribute significantly to the sound level at the machine and, for practical purposes, $M = T$ and no calculation is required.

Note 2: In most industrial situations the sound level is not constant with time but rather it varies or fluctuates. The result is a continual movement of the needle of the sound-level meter. The combination of two sound levels is not a direct combination by addition, subtraction, or averaging of the two measured decibel levels; that is, the mean sound level in decibels is not the mean value of the fluctuating meter readings. This result is due to the nature of logarithmic quantities and their combination. An approximation used in practice to obtain a correct average of the fluctuations of the meter is to note the high and low readings and if the fluctuation is 6 decibels or less, the average of the two readings approximates the correct average. For fluctuations greater than 6 decibels, the average is taken as 3 decibels less than the larger value given by the meter. A simpler method than here described to obtain stable meter readings would make use of the "slow" response setting of the sound level meter as mentioned in the last paragraph of the preceding section "Machinery Noise." The "fast" and "slow" response positions on a typical meter are designed to minimize some of the needle fluctuations. The two positions on the meter determine the time response of the instrument; the "slow" position provides a longer time-average response, thereby reducing the needle fluctuations for rapidly varying sound levels. The "slow" response is required by OSHA for all steady-state (nonimpulsive) sound measurements.

Note 3: Where special measurements are needed or correction measures required, a noise control engineer should be consulted.**

---

* After April 1983, if employee noise exposures equal or exceed an 8-hour, time-weighted, average sound level of 85 dB, OSHA requires employers to administer an effective hearing conservation program.

** The material above is based upon information on pages 26, 27, and 113–139 of *Handbook of Industrial Noise Control*, L. L. Faulkner, Industrial Press Inc., New York.

PUNCHES, DIES, AND PRESS WORK

Clearance between Punches and Dies. — The amount of clearance between a punch and die for blanking and perforating is governed by the thickness and kind of stock to be operated upon. For thin material such as tin, for example, the punch should be a close sliding fit, as, otherwise, the punching will have ragged edges, but for heavier stock there should be some clearance. The clearance between the punch and die in cutting heavy material, lessens the danger of breaking the punch and reduces the pressure required for the punching operation.

Meaning of the Term "Clearance." — There is a difference of opinion among diemakers as to the method of designating clearance. The prevailing practice of fifteen firms specializing in die work is as follows: Ten of these firms define clearance as the space between the punch and die on *one side*, or one-half the difference between the punch and die sizes. The remaining five firms consider clearance as the total difference between the punch and die sizes; for example, if the die is round, clearance equals die diameter minus punch diameter. The advantage of designating clearance as the space on each side is particularly evident in the case of dies of irregular form or of angular shape. While the practice of designating clearance as the difference between the punch and die diameters, may be satisfactory in the case of round dies, it leads to confusion when the dies are of special unsymmetrical forms. The term "clearance" should not be used in specifications without indicating clearly just what it means. According to the practice of one manufacturer of dies, the term "cutting clearance" is used to indicate the space between the punch and die on each side, and the term "die clearance" refers to the angular clearance provided below the cutting edge so that the parts will clear as they fall through the die. The term "clearance" as here used means the space on one side only; hence, for round dies, clearance equals die radius minus punch radius.

Clearances Generally Allowed. — For brass and soft steel, most dies are given a clearance on one side equal to the stock thickness multiplied by 0.05 or 0.06; but one-half of this clearance is preferred for some classes of work, and a clearance equal to the stock thickness multiplied by 0.10 may give the cleanest fracture for certain other operations such, for example, as punching holes in ductile steel boiler plate.

Where Clearance is Applied. — Whether clearance is deducted from the diameter of the punch or added to the diameter of the die depends upon the nature of the work. If a blank of given size is required, the die is made to that size and the punch is made smaller. Inversely, when holes of a given size are required, the punch is made to the diameter wanted and the die is made larger. Therefore, for blanking to a given size, the clearance is deducted from the size of the punch, and for perforating, the clearance is added to the size of the die.

Effect of Clearance on Working Pressure. — Clearance not only affects the smoothness of the fracture, but also the pressure required for punching or blanking. This pressure is greatest when the punch diameter is small compared to the thickness of the stock. In one test, for example, a punching pressure of about 32,000 pounds was required to punch ¾-inch holes into ⁵⁄₁₆-inch mild steel plate when the clearance was about 10 per cent. With a clearance of about 4½ per cent, the pressure increased to 33,000 pounds and a clearance of 2¾ per cent resulted in a pressure of 34,500 pounds.

Soft ductile metal requires more clearance than hard metal, although it has been common practice to increase the clearance for the harder metals. In punching holes in fairly hard steel, a clean fracture was obtained with a clearance of only 0.03 times stock thickness.

Angular Clearance for Dies. — The amount of angular clearance ordinarily given a blanking die varies from one to two degrees, although dies that are to be used for producing a comparatively small number of blanks are sometimes given a clearance angle of four or five degrees to facilitate making the die quickly. When a large number of blanks are required, a clearance of about one degree is used. There are two methods of giving clearance to dies: In one case the clearance extends to the top face of the die; in the other, there is a space about ⅛ inch below the cutting edge which is left practically straight, or having a very small amount of clearance. For very soft metal, such as soft, thin brass, the first method is employed, but for harder material, such as hard brass, steel, etc., it is better to have a very shallow clearance for a short distance below the cutting edge. When a die is made in this way, thousands of blanks can be cut with little variation in their size, as grinding the die face will not enlarge the hole to any appreciable extent.

Lubricants for Press Work. — Dies are often run without lubrication, but they will last longer if oiled slightly. The oil is applied to the stock either from a saturated felt-roller, brush or pad, or by coating one sheet thickly and then feeding it through the rolls. By the latter method, the rolls are coated with sufficient lubricant for a number of sheets, and a very thin coat is applied to the material so that the work does not have to be cleaned, as is sometimes necessary when a felt-roller or pad is used. Lard or sperm oil is used when punching iron, steel or copper. For drawing steel, the following mixture is recommended: 25 per cent flaked graphite; 25 per cent beef tallow; and 50 per cent lard oil. This mixture should be heated and the work dipped into it. Oildag mixed with heavy grease is also used for steel, and a thin mixture of grease (preferably tallow) and white lead has proved satisfactory. The following compound is also used for drawing sheet steel of a mild grade: Mix one pound of white lead, one quart of fish oil, three ounces of black lead, and one pint of water. These ingredients should be boiled until thoroughly mixed. For drawing brass and copper, a solution obtained by dissolving soap in hot water is often used. (Ivory soap has given good results.) The quantity of soap to use depends upon the thickness of the metal, a thin solution being preferable for thin stock. For cutting aluminum, use kerosene, and for drawing aluminum, use kerosene or vaseline of a cheap grade. Lard oil is also applied to aluminum when drawing deep shells. Aluminum should never be worked without a lubricant. For many classes of die work, no lubricant is required, especially when the metal is of a "greasy" nature, like tin plate, for instance.

Annealing Drawn Shells. — When drawing steel, iron, brass or copper, annealing is necessary after two or three draws have been made, as the metal is hardened by the drawing process. For steel and brass, anneal between every other reduction, at least. Tin plate or stock that cannot be annealed without spoiling the finish must ordinarily be drawn to size in one or two operations. Aluminum can be drawn deeper and with less annealing than the other commercial metals, provided the proper grade is used. In case it is necessary to anneal aluminum, this can be done by heating it in a muffle furnace, care being taken to see that the temperature does not exceed 700 degrees F.

Drawing Brass. — When drawing brass shells or cup-shaped articles, it is usually possible to make the depth of the first draw equal to the diameter of the shell. By heating brass to a temperature just below what would show a dull red in a dark room, it is possible to draw difficult shapes, otherwise almost impossible, and to get shapes with square corners.

Drawing Rectangular Shapes. — When square or rectangular shapes are to be drawn, the radius of the corners should be as large as possible, because it is in the

corners that defects occur when drawing. Moreover, the smaller the radius, the less the depth which can be obtained in the first draw. The maximum depths which can be drawn with corners of a given radii are approximately as follows: With a radius of $\frac{3}{32}$ to $\frac{3}{16}$ inch, depth of draw, 1 inch; radius $\frac{3}{16}$ to $\frac{3}{8}$ inch, depth $1\frac{1}{2}$ inch; radius $\frac{3}{8}$ to $\frac{1}{2}$ inch, depth, 2 inches; radius $\frac{1}{2}$ to $\frac{3}{4}$ inch, depth, 3 inches. These figures are taken from actual practice and can doubtless be exceeded slightly when using extra good metal. If the box needs to be quite deep and the radius is quite small, two or more drawing operations will be necessary.

Speeds and Pressures for Presses. — The speeds for presses equipped with cutting dies depend largely upon the kind of material being worked, and its thickness. For punching and shearing ordinary metals not over $\frac{1}{4}$ inch thick, the speeds usually range between 50 and 200 strokes per minute, 100 strokes per minute being a fair average. For punching metal over $\frac{1}{4}$ inch thick, geared presses with speeds ranging from 25 to 75 strokes per minute are commonly employed.

The cutting pressures required depend upon the shearing strength of the material, and the actual area of the surface being severed. For round holes the pressure required equals the circumference of the hole $\times$ the thickness of the stock $\times$ the shearing strength. To allow for some excess pressure, the tensile strength may be substituted for the shearing strength; the tensile strength for these calculations may be roughly assumed as follows: Mild steel, 60,000 pounds per square inch; wrought iron, 50,000 pounds; bronze, 40,000 pounds; copper, 30,000 pounds; aluminum, 20,000 pounds; zinc, 10,000 pounds; tin and lead, 5,000 pounds.

Pressure required for Punching. — The following approximate rule may be used for rapidly finding the pressure in tons required for punching circular holes in sheet steel: Multiply the diameter of the hole in inches by the thickness of the sheet steel and multiply this product by 80. The result is the pressure in tons required. To find the pressure required for punching holes in brass, multiply the diameter of the hole by the thickness, and multiply this product by 65.

Example: — What pressure is required for punching a hole 2 inches in diameter through $\frac{1}{4}$-inch steel stock? According to the rule, $2 \times \frac{1}{4} \times 80 = 40$ tons.

If a hole is not circular, use as a factor, instead of the diameter of the hole, one-third of the total length of the outline of the hole to be punched.

Example: — What pressure is required for punching a 1 inch square hole in $\frac{1}{4}$ inch thick steel? According to the rule, the total length of the outline of the square is 4 inches. One-third of this is $1\frac{1}{3}$, and the pressure required is equal to $1\frac{1}{3} \times \frac{1}{4} \times 80 = 26\frac{2}{3}$ tons.

Example: — What pressure is required for punching a 1 by 2 inch rectangular hole in $\frac{1}{4}$ inch thick brass? According to the rule, the total length of the outline of the rectangle is 6 inches. One-third of this is 2, and the pressure required is equal to $2 \times \frac{1}{4} \times 65 = 32\frac{1}{2}$ tons.

Shut Height of Press. — The term "shut height" as applied to power presses, indicates the die space when the slide is at the bottom of its stroke and the slide connection has been adjusted upward as far as possible. The "shut height" is the distance from the lower face of the slide, either to the top of the bed or to the top of the bolster plate, there being two methods of determining it; hence, this term should always be accompanied by a definition explaining its meaning. According to one press manufacturer, the safest plan is to define "shut height" as the distance from the top of the bolster to the bottom of the slide, with the stroke down and the adjustment up, because most dies are mounted on bolster plates of standard thickness, and a misunderstanding which results in providing too much die space is less serious than having insufficient die space. It is believed that the expression

"shut height" was applied first to dies rather than to presses, the shut height of a die being the distance from the bottom of the lower section to the top of the upper section or punch, excluding the shank, and measured when the punch is in the lowest working position.

Diameters of Shell Blanks. — The diameters of blanks for drawing plain cylindrical shells can be obtained from the accompanying table, which gives a very close approximation for thin stock. The blank diameters given in this table are for sharp-cornered shells and are found by the following formula:

$$D = \sqrt{d^2 + 4\,dh}, \tag{1}$$

in which D = diameter of flat blank; d = diameter of finished shell; h = height of finished shell.

Example: — If the diameter of the finished shell is to be 1.5 inch, and the height, 2 inches, the trial diameter of the blank would be found as follows:

$$D = \sqrt{1.5^2 + 4 \times 1.5 \times 2} = \sqrt{14.25} = 3.78 \text{ inches.}$$

For a round-cornered cup, the following formula, in which r equals the radius of the corner, will give fairly accurate diameters, provided the radius does not exceed, say, ¼ the height of the shell:

$$D = \sqrt{d^2 + 4\,dh} - r. \tag{2}$$

These formulas are based on the assumption that the thickness of the drawn shell is the same as the original thickness of the stock, and that the blank is so proportioned that its area will equal the area of the drawn shell. This method of calculating the blank diameter is quite accurate for thin material, when there is only a slight reduction in the thickness of the metal incident to drawing; but when heavy stock is drawn and the thickness of the finished shell is much less than the original thickness of the stock, the blank diameter obtained from Formulas (1) or (2) will be too large, because when the stock is drawn thinner, there is an increase in area. When an appreciable reduction in thickness is to be made, the blank diameter can be obtained by first determining the "mean height" of the drawn shell by the following formula. This formula is only approximately correct, but will give results sufficiently accurate for most work:

$$M = \frac{ht}{T} \tag{3}$$

in which M = approximate mean height of drawn shell; h = height of drawn shell; t = thickness of shell; T = thickness of metal before drawing.

After determining the mean height, the blank diameter for the required shell diameter is obtained from the table previously referred to, the mean height being used instead of the actual height.

Example: — Suppose a shell 2 inches in diameter and 3¾ inches high is to be drawn, and that the original thickness of the stock is 0.050 inch, and thickness of drawn shell, 0.040 inch. To what diameter should the blank be cut? Using Formula (3) to obtain the mean height:

$$M = \frac{ht}{T} = \frac{3.75 \times 0.040}{0.050} = 3 \text{ inches.}$$

According to the table, the blank diameter for a shell 2 inches in diameter and 3 inches high is 5.29 inches. This formula is accurate enough for all practical purposes, unless the reduction in the thickness of the metal is greater than about one-fifth the original thickness. When there is considerable reduction, a blank calculated by this formula produces a shell that is too long. This, however, is an error in the right direction, as the edges of drawn shells are ordinarily trimmed.

If the shell has a rounded corner, the radius of the corner should be deducted from the figures given in the table. For example, if the shell referred to in the foregoing example had a corner of ¼-inch radius, the blank diameter would equal 5.29 — 0.25 = 5.04 inches.

Another formula which is sometimes used for obtaining blank diameters for shells, when there is a reduction in the thickness of the stock, is as follows:

$$D = \sqrt{a^2 + \left(a^2 - b^2\right)\frac{h}{t}} \qquad (4)$$

In this formula D = blank diameter; a = outside diameter; b = inside diameter; t = thickness of shell at bottom; h = depth of shell. This formula is based on the cubic contents of the drawn shell. It is assumed that the shells are cylindrical, and no allowance is made for a rounded corner at the bottom, or for trimming the shell after drawing. To allow for trimming, add the required amount to depth h. When a shell is of irregular cross-section, if its weight is known, the blank diameter can be determined by the following formula:

$$D = 1.1284 \sqrt{\frac{W}{wt}} \qquad (5)$$

in which D = blank diameter in inches; W = weight of shell; w = weight of metal per cubic inch; t = thickness of the shell.

In the construction of dies for producing shells, especially of irregular form, a common method of procedure is to make the drawing parts first. The actual blank diameter can then be determined by trial. One method is to cut a trial blank as near to size as can be estimated. The outline of this blank is then scribed on a flat sheet, after which the blank is drawn. If the finished shell shows that the blank is not of the right diameter, a new trial blank is cut either larger or smaller than the size indicated by the line previously scribed, this line acting as a guide. If a model shell is available, the blank diameter can also be determined as follows: First cut a blank somewhat large, and from the same material used for making the model; then, reduce the size of the blank until its weight equals the weight of the model.

Depth and Diameter Reductions of Drawn Shells. —The depth to which metal can be drawn in one operation depends upon the quality and kind of material, its thickness, the slant or angle of the dies, and the amount that the stock is thinned or "ironed" in drawing. A general rule for determining the depth to which cylindrical shells can be drawn in one operation is as follows: The depth or length of the first draw should never be greater than the diameter of the shell. If the shell is to have a flange at the top, it may not be practicable to draw as deeply as is indicated by this rule, unless the metal is extra good, because the stock is subjected to a higher tensile stress, owing to the larger blank which is necessary for forming the flange. According to another rule, the depth given the shell on the first draw should equal one-third the diameter of the blank. Ordinarily, it is possible to draw sheet steel of any thickness up to ¼ inch, so that the diameter of the first shell equals about six-tenths of the blank diameter. When drawing plain shells, the amount that the diameter is reduced for each draw must be governed by the quality of the metal and its susceptibility to drawing. The reduction for various thicknesses of metal is about as follows:

Approximate thickness of sheet steel	1/16	1/8	3/16	1/4	5/16
Possible reduction in diameter for each succeeding step, per cent	20	15	12	10	8

Diameters of Blanks for Drawn Shells

6	5½	5	4½	4	3¾	3½	3¼	3	2¾	2½	2¼	2	1¾	1½	1¼	1	¾	½	¼	Diam. of Shell
								Height of Shell												
2.46	2.36	2.25	2.14	2.01	1.95	1.89	1.82	1.75	1.68	1.60	1.52	1.44	1.35	1.25	1.14	1.03	0.90	0.75	0.56	¼
3.50	3.36	3.21	3.04	2.87	2.78	2.69	2.60	2.50	2.40	2.29	2.18	2.06	1.94	1.80	1.66	1.50	1.32	1.12	0.87	½
4.31	4.13	3.95	3.75	3.54	3.44	3.33	3.21	3.09	2.97	2.84	2.70	2.56	2.41	2.25	2.08	1.89	1.68	1.44	1.14	¾
5.00	4.80	4.58	4.36	4.12	4.00	3.87	3.74	3.61	3.46	3.32	3.16	3.00	2.83	2.65	2.45	2.24	2.00	1.73	1.41	1
5.62	5.39	5.15	4.91	4.64	4.51	4.37	4.22	4.07	3.91	3.75	3.58	3.40	3.21	3.01	2.79	2.56	2.30	2.01	1.68	1¼
6.18	5.94	5.68	5.41	5.12	4.98	4.82	4.66	4.50	4.33	4.15	3.97	3.78	3.57	3.36	3.12	2.87	2.60	2.29	1.94	1½
6.71	6.45	6.17	5.88	5.58	5.41	5.26	5.08	4.91	4.72	4.53	4.34	4.13	3.91	3.68	3.44	3.17	2.88	2.56	2.19	1¾
7.21	6.93	6.63	6.32	6.00	5.83	5.66	5.48	5.29	5.10	4.90	4.69	4.47	4.24	4.00	3.74	3.46	3.16	2.83	2.45	2
7.69	7.39	7.07	6.75	6.41	6.23	6.05	5.86	5.66	5.46	5.25	5.03	4.80	4.56	4.31	4.04	3.75	3.44	3.09	2.70	2¼
8.14	7.82	7.50	7.16	6.80	6.61	6.42	6.22	6.02	5.81	5.59	5.36	5.12	4.87	4.61	4.33	4.03	3.71	3.36	2.96	2½
8.58	8.25	7.91	7.55	7.18	6.99	6.79	6.58	6.37	6.15	5.92	5.68	5.44	5.18	4.91	4.62	4.31	3.98	3.61	3.21	2¾
9.00	8.66	8.31	7.94	7.55	7.35	7.14	6.93	6.71	6.48	6.25	6.00	5.74	5.48	5.20	4.90	4.58	4.24	3.87	3.46	3
9.41	9.06	8.69	8.31	7.91	7.70	7.49	7.27	7.04	6.80	6.56	6.31	6.04	5.77	5.48	5.18	4.85	4.51	4.13	3.71	3¼
9.81	9.45	9.07	8.67	8.26	8.05	7.83	7.60	7.36	7.12	6.87	6.61	6.34	6.06	5.77	5.45	5.12	4.77	4.39	3.97	3½
10.20	9.83	9.44	9.03	8.61	8.38	8.16	7.92	7.69	7.44	7.18	6.91	6.63	6.35	6.05	5.73	5.39	5.03	4.64	4.22	3¾
10.58	10.20	9.80	9.38	8.94	8.72	8.49	8.25	8.00	7.75	7.48	7.21	6.93	6.63	6.32	6.00	5.66	5.29	4.90	4.47	4
10.96	10.56	10.15	9.72	9.28	9.04	8.81	8.56	8.31	8.05	7.78	7.50	7.22	6.91	6.60	6.27	5.92	5.55	5.15	4.72	4¼
11.32	10.92	10.50	10.06	9.60	9.37	9.12	8.87	8.62	8.35	8.08	7.79	7.50	7.19	6.87	6.54	6.19	5.81	5.41	4.98	4½
11.69	11.27	10.84	10.40	9.93	9.69	9.44	9.18	8.92	8.65	8.37	8.08	7.78	7.47	7.15	6.80	6.45	6.07	5.66	5.22	4¾
12.04	11.62	11.18	10.72	10.25	10.00	9.75	9.49	9.22	8.94	8.66	8.37	8.06	7.75	7.42	7.07	6.71	6.32	5.92	5.48	5
12.39	11.96	11.51	11.05	10.56	10.31	10.05	9.79	9.52	9.24	8.95	8.65	8.34	8.02	7.68	7.33	6.97	6.58	6.17	5.73	5¼
12.74	12.30	11.84	11.37	10.87	10.62	10.36	10.10	9.81	9.53	9.23	8.93	8.62	8.29	7.95	7.60	7.23	6.84	6.42	5.98	5½
13.08	12.63	12.17	11.69	11.18	10.92	10.66	10.38	10.10	9.81	9.52	9.21	8.89	8.56	8.22	7.86	7.49	7.09	6.68	6.23	5¾
13.42	12.96	12.49	12.00	11.49	11.23	10.95	10.68	10.39	10.10	9.80	9.49	9.17	8.83	8.49	8.12	7.75	7.35	6.93	6.48	6

For example, if a shell made of $\frac{1}{16}$ inch stock is 3 inches in diameter after the first draw, it can be reduced 20 per cent on the next draw, and so on until the required diameter is obtained. These figures are based upon the assumption that the shell is annealed after the first drawing operation, and at least between every two of the following operations. Necking operations — that is, the drawing out of a short portion of the lower part of the cup into a long neck — may be done without such frequent annealings. In double-action presses, where the inside of the cup is supported by a bushing during drawing, the reductions possible may be increased to 30, 24, 18, 15 and 12 per cent, respectively. (The latter figures may also be used for brass in single-action presses.)

When a hole is to be pierced at the bottom of a cup and the remaining metal is to be drawn after the hole has been pierced or punched, always pierce from the opposite direction to that in which the stock is to be drawn after piercing. In extreme cases, it is necessary to machine the metal around the pierced hole in order to prevent the starting of cracks or flaws in the subsequent drawing operations.

The foregoing figures represent conservative practice and it is often possible to make greater reductions than are indicated by these figures, especially when using a good drawing metal. Taper shells require smaller reductions than cylindrical shells, because the metal tends to wrinkle if the shell to be drawn is much larger than the punch. The amount that the stock is "ironed" or thinned out while being drawn must also be considered, because a reduction in gage or thickness means greater pressure of the punch against the bottom of the shell; hence the amount that the shell diameter is reduced for each drawing operation must be lessened when much ironing is necessary. The extent to which a shell can be ironed in one drawing operation ranges between 0.002 and 0.004 inch per side, and should not exceed 0.001 inch on the final draw, if a good finish is required.

Allowances for Bending Sheet Metal. — In bending steel, brass, bronze or other metals, the problem is to find the length of straight stock required for each bend; then these lengths are added to the lengths of the straight sections to obtain the total length of the material before bending.

If L = length, in inches, of straight stock required before bending; T = thickness in inches; R = inside radius of bend in inches.

For 90-degree bends in soft brass and soft copper

$$\text{Table 1 or } L = (0.55 \times T) + (1.57 \times R) \qquad (1)$$

For 90-degree bends in half-hard copper and brass, soft steel, and aluminum

$$\text{Table 2 or } L = (0.64 \times T) + (1.57 \times R) \qquad (2)$$

For 90-degree bends in bronze, hard copper, cold-rolled steel and spring steel

$$\text{Table 3 or } L = (0.71 \times T) + (1.57 \times R) \qquad (3)$$

Angle of Bend Other Than 90 Degrees: For angles other than 90 degrees, find length L, using tables or formulas, and multiply L by angle of bend, in degrees, divided by 90, to find length of stock before bending. In using this rule, note that *angle of bend* is the angle through which the material has actually been bent; hence, it is not in all cases the angle as given on a drawing. To illustrate, in Fig. 1 (see diagram), the angle on the drawing is 60 degrees, but the angle of bend A is 120 degrees ($180 - 60 = 120$); in Fig. 2, the angle of bend A is 60 degrees; in Fig. 3, angle A is $90 - 30 = 60$ degrees. The Formulas (1), (2) and (3) are based upon extensive experiments of the Westinghouse Electric & Mfg. Co. They apply to parts bent with simple tools or on the bench, where limits of plus or minus $\frac{1}{64}$ inch are specified. If a part has two or more bends of the same radius, it is, of course, only necessary to obtain the length required for one of the bends and then multiply by the number of bends, thus obtaining the total allowance for the bent sections.

(Continued on page 1180)

Table 1. Lengths of Straight Stock Required for 90-Degree Bends in Soft Copper and Soft Brass

Radius R of Bend, Inches	Thickness T of Material, Inch												
	1/64	1/32	3/64	1/16	5/64	3/32	1/8	5/32	3/16	7/32	1/4	9/32	5/16
1/32	0.058	0.066	0.075	0.083	0.092	0.101	0.118	0.135	0.152	0.169	0.187	0.204	0.221
3/64	0.083	0.091	0.100	0.108	0.117	0.126	0.143	0.160	0.177	0.194	0.212	0.229	0.246
1/16	0.107	0.115	0.124	0.132	0.141	0.150	0.167	0.184	0.201	0.218	0.236	0.253	0.270
3/32	0.156	0.164	0.173	0.181	0.190	0.199	0.216	0.233	0.250	0.267	0.285	0.302	0.319
1/8	0.205	0.213	0.222	0.230	0.239	0.248	0.265	0.282	0.299	0.316	0.334	0.351	0.368
5/32	0.254	0.262	0.271	0.279	0.288	0.297	0.314	0.331	0.348	0.365	0.383	0.400	0.417
3/16	0.303	0.311	0.320	0.328	0.337	0.346	0.363	0.380	0.397	0.414	0.432	0.449	0.466
7/32	0.353	0.361	0.370	0.378	0.387	0.396	0.413	0.430	0.447	0.464	0.482	0.499	0.516
1/4	0.401	0.409	0.418	0.426	0.435	0.444	0.461	0.478	0.495	0.512	0.530	0.547	0.564
9/32	0.450	0.458	0.467	0.475	0.484	0.493	0.510	0.527	0.544	0.561	0.579	0.596	0.613
5/16	0.499	0.507	0.516	0.524	0.533	0.542	0.559	0.576	0.593	0.610	0.628	0.645	0.662
11/32	0.549	0.557	0.566	0.574	0.583	0.592	0.609	0.626	0.643	0.660	0.678	0.695	0.712
3/8	0.598	0.606	0.615	0.623	0.632	0.641	0.658	0.675	0.692	0.709	0.727	0.744	0.761
13/32	0.646	0.654	0.663	0.671	0.680	0.689	0.706	0.723	0.740	0.757	0.775	0.792	0.809
7/16	0.695	0.703	0.712	0.720	0.729	0.738	0.755	0.772	0.789	0.806	0.824	0.841	0.858
15/32	0.734	0.742	0.751	0.759	0.768	0.777	0.794	0.811	0.828	0.845	0.863	0.880	0.897
1/2	0.794	0.802	0.811	0.819	0.828	0.837	0.854	0.871	0.888	0.905	0.923	0.940	0.957
9/16	0.892	0.900	0.909	0.917	0.926	0.935	0.952	0.969	0.986	1.003	1.021	1.038	1.055
5/8	0.990	0.998	1.007	1.015	1.024	1.033	1.050	1.067	1.084	1.101	1.119	1.136	1.153
11/16	1.089	1.097	1.106	1.114	1.123	1.132	1.149	1.166	1.183	1.200	1.218	1.235	1.252
3/4	1.187	1.195	1.204	1.212	1.221	1.230	1.247	1.264	1.281	1.298	1.316	1.333	1.350
13/16	1.286	1.294	1.303	1.311	1.320	1.329	1.346	1.363	1.380	1.397	1.415	1.432	1.449
7/8	1.384	1.392	1.401	1.409	1.418	1.427	1.444	1.461	1.478	1.495	1.513	1.530	1.547
15/16	1.481	1.489	1.498	1.506	1.515	1.524	1.541	1.558	1.575	1.592	1.610	1.627	1.644
1	1.580	1.588	1.597	1.605	1.614	1.623	1.640	1.657	1.674	1.691	1.709	1.726	1.743
1 1/16	1.678	1.686	1.695	1.703	1.712	1.721	1.738	1.755	1.772	1.789	1.807	1.824	1.841
1 1/8	1.777	1.785	1.794	1.802	1.811	1.820	1.837	1.854	1.871	1.888	1.906	1.923	1.940
1 3/16	1.875	1.883	1.892	1.900	1.909	1.918	1.935	1.952	1.969	1.986	2.004	2.021	2.038
1 1/4	1.972	1.980	1.989	1.997	2.006	2.015	2.032	2.049	2.066	2.083	2.101	2.118	2.135

Table 2. Lengths of Straight Stock Required for 90-Degree Bends in Half-Hard Brass and Sheet Copper, Soft Steel and Aluminum

Radius R of Bend, Inches	Thickness T of Material, Inch												
	1/64	1/32	3/64	1/16	5/64	3/32	1/8	5/32	3/16	7/32	1/4	9/32	5/16
1/32	0.059	0.069	0.079	0.089	0.099	0.109	0.129	0.149	0.169	0.189	0.209	0.229	0.249
3/64	0.084	0.094	0.104	0.114	0.124	0.134	0.154	0.174	0.194	0.214	0.234	0.254	0.274
1/16	0.108	0.118	0.128	0.138	0.148	0.158	0.178	0.198	0.218	0.238	0.258	0.278	0.298
3/32	0.157	0.167	0.177	0.187	0.197	0.207	0.227	0.247	0.267	0.287	0.307	0.327	0.347
1/8	0.206	0.216	0.226	0.236	0.246	0.256	0.276	0.296	0.316	0.336	0.356	0.376	0.396
5/32	0.255	0.265	0.275	0.285	0.295	0.305	0.325	0.345	0.365	0.385	0.405	0.425	0.445
3/16	0.304	0.314	0.324	0.334	0.344	0.354	0.374	0.394	0.414	0.434	0.454	0.474	0.494
7/32	0.354	0.364	0.374	0.384	0.394	0.404	0.424	0.444	0.464	0.484	0.504	0.524	0.544
1/4	0.402	0.412	0.422	0.433	0.442	0.452	0.472	0.492	0.512	0.532	0.552	0.572	0.592
9/32	0.451	0.461	0.471	0.481	0.491	0.501	0.521	0.541	0.561	0.581	0.601	0.621	0.641
5/16	0.500	0.510	0.520	0.530	0.540	0.550	0.570	0.590	0.610	0.630	0.650	0.670	0.690
11/32	0.550	0.560	0.570	0.580	0.590	0.600	0.620	0.640	0.660	0.680	0.700	0.720	0.740
3/8	0.599	0.609	0.619	0.629	0.639	0.649	0.669	0.689	0.709	0.729	0.749	0.769	0.789
13/32	0.647	0.657	0.667	0.677	0.687	0.697	0.717	0.737	0.757	0.777	0.797	0.817	0.837
7/16	0.696	0.706	0.716	0.726	0.736	0.746	0.766	0.786	0.806	0.826	0.846	0.866	0.886
15/32	0.746	0.756	0.766	0.776	0.786	0.796	0.816	0.836	0.856	0.876	0.896	0.916	0.936
1/2	0.795	0.805	0.815	0.825	0.835	0.845	0.865	0.885	0.905	0.925	0.945	0.965	0.985
9/16	0.893	0.903	0.913	0.923	0.933	0.943	0.963	0.983	1.003	1.023	1.043	1.063	1.083
5/8	0.991	1.001	1.011	1.021	1.031	1.041	1.061	1.081	1.101	1.121	1.141	1.161	1.181
11/16	1.090	1.100	1.110	1.120	1.130	1.140	1.160	1.180	1.200	1.220	1.240	1.260	1.280
3/4	1.188	1.198	1.208	1.218	1.228	1.238	1.258	1.278	1.298	1.318	1.338	1.358	1.378
13/16	1.287	1.297	1.307	1.317	1.327	1.337	1.357	1.377	1.397	1.417	1.437	1.457	1.477
7/8	1.385	1.395	1.405	1.415	1.425	1.435	1.455	1.475	1.495	1.515	1.535	1.555	1.575
15/16	1.482	1.492	1.502	1.512	1.522	1.532	1.552	1.572	1.592	1.612	1.632	1.652	1.672
1	1.581	1.591	1.601	1.611	1.621	1.631	1.651	1.671	1.691	1.711	1.731	1.751	1.771
1 1/16	1.679	1.689	1.699	1.709	1.719	1.729	1.749	1.769	1.789	1.809	1.829	1.849	1.869
1 1/8	1.778	1.788	1.798	1.808	1.818	1.828	1.848	1.868	1.888	1.908	1.928	1.948	1.968
1 3/16	1.876	1.886	1.896	1.906	1.916	1.926	1.946	1.966	1.986	2.006	2.026	2.046	2.066
1 1/4	1.973	1.983	1.993	2.003	2.013	2.023	2.043	2.063	2.083	2.103	2.123	2.143	2.163

Table 3. Lengths of Straight Stock Required for 90-Degree Bends in Hard Copper, Bronze, Cold-Rolled Steel, and Spring Steel

Radius R of Bend, Inches	Thickness T of Material, Inch												
	1/64	1/32	3/64	1/16	5/64	3/32	1/8	5/32	3/16	7/32	1/4	9/32	5/16
1/32	0.060	0.071	0.082	0.093	0.104	0.116	0.138	0.160	0.182	0.204	0.227	0.249	0.271
3/64	0.085	0.096	0.107	0.118	0.129	0.141	0.163	0.185	0.207	0.229	0.252	0.274	0.296
1/16	0.109	0.120	0.131	0.142	0.153	0.165	0.187	0.209	0.231	0.253	0.276	0.298	0.320
3/32	0.158	0.169	0.180	0.191	0.202	0.214	0.236	0.258	0.280	0.302	0.325	0.347	0.369
1/8	0.207	0.218	0.229	0.240	0.251	0.263	0.285	0.307	0.329	0.351	0.374	0.396	0.418
5/32	0.256	0.267	0.278	0.289	0.300	0.312	0.334	0.356	0.378	0.400	0.423	0.445	0.467
3/16	0.305	0.316	0.327	0.338	0.349	0.361	0.383	0.405	0.427	0.449	0.472	0.494	0.516
7/32	0.355	0.365	0.377	0.388	0.399	0.411	0.433	0.455	0.477	0.499	0.522	0.544	0.566
1/4	0.403	0.414	0.425	0.436	0.447	0.459	0.481	0.503	0.525	0.547	0.570	0.592	0.614
9/32	0.452	0.463	0.474	0.485	0.496	0.508	0.530	0.552	0.574	0.596	0.619	0.641	0.663
5/16	0.501	0.512	0.523	0.534	0.545	0.557	0.579	0.601	0.623	0.645	0.668	0.690	0.712
11/32	0.551	0.562	0.573	0.584	0.595	0.607	0.629	0.651	0.673	0.695	0.718	0.740	0.762
3/8	0.600	0.611	0.622	0.633	0.644	0.656	0.678	0.700	0.722	0.744	0.767	0.789	0.811
13/32	0.648	0.659	0.670	0.681	0.692	0.704	0.726	0.748	0.770	0.792	0.815	0.837	0.859
7/16	0.697	0.708	0.719	0.730	0.741	0.753	0.775	0.797	0.819	0.841	0.864	0.886	0.908
15/32	0.736	0.747	0.758	0.769	0.780	0.792	0.814	0.836	0.858	0.880	0.903	0.925	0.947
1/2	0.796	0.807	0.818	0.829	0.840	0.852	0.874	0.896	0.918	0.940	0.963	0.985	1.007
9/16	0.894	0.905	0.916	0.927	0.938	0.950	0.972	0.994	1.016	1.038	1.061	1.083	1.105
5/8	0.999	1.003	1.014	1.025	1.036	1.048	1.070	1.092	1.114	1.136	1.159	1.181	1.203
11/16	1.091	1.102	1.113	1.124	1.135	1.147	1.169	1.191	1.213	1.235	1.258	1.280	1.302
3/4	1.189	1.200	1.211	1.222	1.233	1.245	1.267	1.289	1.311	1.333	1.356	1.378	1.400
13/16	1.288	1.299	1.310	1.321	1.332	1.344	1.366	1.388	1.410	1.432	1.455	1.477	1.499
7/8	1.386	1.397	1.408	1.419	1.430	1.442	1.464	1.486	1.508	1.530	1.553	1.575	1.597
15/16	1.483	1.494	1.505	1.516	1.527	1.539	1.561	1.583	1.605	1.627	1.650	1.672	1.694
1	1.582	1.593	1.604	1.615	1.626	1.638	1.660	1.682	1.704	1.726	1.749	1.771	1.793
1 1/16	1.680	1.691	1.702	1.713	1.724	1.736	1.758	1.780	1.802	1.824	1.847	1.869	1.891
1 1/8	1.779	1.790	1.801	1.812	1.823	1.835	1.857	1.879	1.901	1.923	1.946	1.968	1.990
1 3/16	1.877	1.888	1.899	1.910	1.921	1.933	1.955	1.977	1.999	2.021	2.044	2.066	2.088
1 1/4	1.974	1.985	1.996	2.007	2.018	2.030	2.052	2.074	2.096	2.118	2.141	2.163	2.185

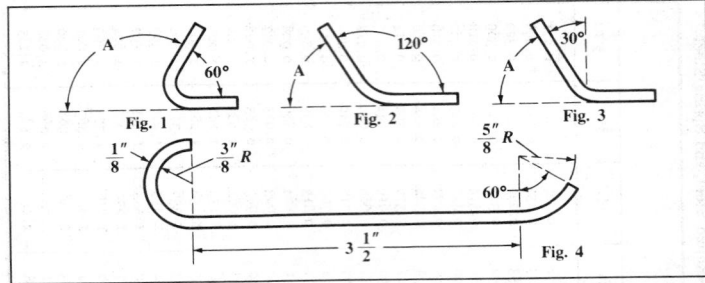

Fig. 1 Fig. 2 Fig. 3

Fig. 4

Example Showing Application of Formulas: Find the length before bending of the part illustrated by Fig. 4. Soft steel is to be used.

For bend at left-hand end (180-degree bend)

$$L = [(0.64 \times 0.125) + (1.57 \times 0.375)] \times \frac{180}{90} = 1.338$$

For bend at right-hand end (60-degree bend)

$$L = [(0.64 \times 0.125) + (1.57 \times 0.625)] \times \frac{60}{90} = 0.707$$

Total length before bending = 3.5 + 1.338 + 0.707 = 5.545 inches

Other Bending Allowance Formulas. — When bending sheet steel or brass, add from ⅓ to ½ of the thickness of the stock, for *each bend*, to the sum of the inside dimensions of the finished piece, to get the length of the straight blank. The harder the material the greater the allowance (⅓ of the thickness is added for soft stock and ½ of the thickness for hard material). The data given in the table, "Allowances for Bends in Sheet Metal," refer more particularly to the bending of sheet metal for counters, bank fittings and general office fixtures, for which purpose it is not absolutely essential to have the sections of the bends within very close limits. Absolutely accurate data for this work cannot be deduced, as the stock varies considerably as to hardness, etc. The figures given apply to sheet steel, aluminum, brass and bronze. Experience has demonstrated that for the semisquare corners, such as are formed in a V-die, the amount to be deducted from the sum of the outside bend dimensions, as shown in the accompanying illustration by the sum of the letters from *a* to *e*, is as follows: $X = 1.67\ BG$, where $X =$ the amount to be deducted; $B =$ the number of bends; and $G =$ the decimal equivalent of the gage. The values of X for different gages and numbers of bends are given in the table. Its application may be illustrated by an example: A strip having two bends is to have outside dimensions of 2, 1½ and 2 inches, and is made of stock 0.125 inch thick. The sum of the outside dimensions is thus 5½ inches, and from the table the amount to be deducted is found to be 0.416; hence the blank will be 5.5 − 0.416 = 5.084 inches long.

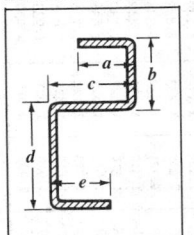

The lower part of the table applies to square bends which are either drawn through a block of steel made to the required shape, or else drawn through rollers in a drawbench. The pressure applied not only gives a much sharper corner, but it also elongates the material more than in the V-die process. In this case, the deduction is $X = 1.33\ BG$.

Allowances for Bends in Sheet Metal

Square Bends	Gage	Thickness, Inches	Amount to be Deducted from the Sum of the Outside Bend Dimensions, Inches			
			1 Bend	2 Bends	3 Bends	4 Bends
Formed in a Press by a V-die	18	0.0500	0.083	0.166	0.250	0.333
	16	0.0625	0.104	0.208	0.312	0.416
	14	0.0781	0.130	0.260	0.390	0.520
	13	0.0937	0.156	0.312	0.468	0.625
	12	0.1093	0.182	0.364	0.546	0.729
	11	0.1250	0.208	0.416	0.625	0.833
	10	0.1406	0.234	0.468	0.703	0.937
			5 Bends	6 Bends		7 Bends
	18	0.0500	0.416	0.500		0.583
	16	0.0625	0.520	0.625		0.729
	14	0.0781	0.651	0.781		0.911
	13	0.0937	0.781	0.937		1.093
	12	0.1093	0.911	1.093		1.276
	11	0.1250	1.041	1.250		1.458
	10	0.1406	1.171	1.406		1.643

Square Bends	Gage	Thickness, Inches	Amount to be Deducted from the Sum of the Outside Bend Dimensions, Inches			
			1 Bend	2 Bends	3 Bends	4 Bends
Rolled or Drawn in a Draw-bench	18	0.0500	0.066	0.133	0.200	0.266
	16	0.0625	0.083	0.166	0.250	0.333
	14	0.0781	0.104	0.208	0.312	0.416
	13	0.0937	0.125	0.250	0.375	0.500
	12	0.1093	0.145	0.291	0.437	0.583
	11	0.1250	0.166	0.333	0.500	0.666
	10	0.1406	0.187	0.375	0.562	0.750
			5 Bends	6 Bends		7 Bends
	18	0.0500	0.333	0.400		0.466
	16	0.0625	0.416	0.500		0.583
	14	0.0781	0.521	0.625		0.729
	13	0.0937	0.625	0.750		0.875
	12	0.1093	0.729	0.875		1.020
	11	0.1250	0.833	1.000		1.166
	10	0.1406	0.937	1.125		1.312

IRON AND STEEL CASTINGS

Cast irons and cast steels encompass a large family of ferrous alloys, which, as the name implies, are cast to shape rather than being formed by working in the solid state. In general, cast irons contain more than 2 per cent carbon and from 1 to 3 per cent silicon. Varying the balance between carbon and silicon, alloying with different elements, and changing melting, casting, and heat-treating practices can produce a broad range of properties. In most cases, the carbon exists in two forms: free carbon in the form of graphite and combined carbon in the form of iron carbide (cementite). Mechanical and physical properties depend strongly on the shape and distribution of the free graphite and the type of matrix surrounding the graphite particles.

The four basic types of cast iron are white iron, gray iron, malleable iron, and ductile iron. In addition to these basic types, there are other specific forms of cast iron to which special names have been applied, such as chilled iron, alloy iron, and compacted graphite cast iron.

Gray Cast Iron. — Gray cast iron may easily be cast into any desirable form and it may also be machined readily. It usually contains from 1.7 to 4.5 per cent carbon, and from 1 to 3 per cent silicon. The excess carbon is in the form of graphite flakes and these flakes impart to the material the dark-colored fracture which gives it its name. Gray iron castings are widely used for such applications as machine tools, automotive cylinder blocks, cast-iron pipe and fittings and agricultural implements.

The American National Standard Specifications for Gray Iron Castings — ANSI/ASTM A48-76 groups the castings into two categories. Gray iron castings in Classes 20A, 20B, 20C, 25A, 25B, 25C, 30A, 30B, 30C, 35A, 35B, and 35C are characterized by excellent machinability, high damping capacity, low modulus of elasticity, and comparative ease of manufacture. Castings in Classes 40B, 40C, 45B, 45C, 50B, 50C, 60B, and 60C are usually more difficult to machine, have lower damping capacity, a higher modulus of elasticity, and are more difficult to manufacture. The prefix number is indicative of the minimum tensile strength in pounds per square inch, i.e., 20 is 20,000 psi, 25 is 25,000 psi, 30 is 30,000 psi, etc.

High-strength iron castings produced by the Meehanite-controlled process may have various combinations of physical properties to meet different requirements. In addition to a number of general engineering types, there are heat-resisting, wear-resisting and corrosion-resisting Meehanite castings.

White Cast Iron. — When nearly all of the carbon in a casting is in the combined or cementite form, it is known as white cast iron. It is so named because it has a silvery-white fracture. White cast iron is very hard and also brittle; its ductility is practically zero. Castings of this material need particular attention with respect to design since sharp corners and thin sections result in material failures at the foundry. These castings are less resistant to impact loading than gray iron castings, but they have a compressive strength that is usually higher than 200,000 pounds per square inch as compared to 65,000 to 160,000 pounds per square inch for gray iron castings. Some white iron castings are used for applications that require maximum wear resistance but most of them are used in the production of malleable iron castings.

Chilled Cast Iron. — Many gray iron castings have wear-resisting surfaces of white cast iron. These surfaces are designated by the term "chilled cast iron" since they are produced in molds having metal chills for cooling the molten metal rapidly. This rapid cooling results in the formation of cementite and white cast iron.

Alloy Cast Iron. — This term designates castings containing alloying elements such as nickel, chromium, molybdenum, copper, and manganese in sufficient amounts to appreciably change the physical properties. These elements may be added either to increase the strength or to obtain special properties such as higher wear resistance, corrosion resistance, or heat resistance. Alloy cast irons are used extensively for

such parts as automotive cylinders, pistons, piston rings, crankcases, brake drums; for certain machine tool castings, for certain types of dies, for parts of crushing and grinding machinery, and for application where the casting must resist scaling at high temperatures. Machinable alloy cast irons having tensile strengths up to 70,000 pounds per square inch or even higher may be produced.

Malleable-iron Castings. — Malleable iron is produced by the annealing or graphitization of white iron castings. The graphitization in this case produces temper carbon which is graphite in the form of compact rounded aggregates. Malleable castings are used for many industrial applications where strength, ductility, machinability, and resistance to shock are important factors. In manufacturing these castings, the usual procedure is to first produce a hard, brittle white iron from a charge of pig iron and scrap. These hard white-iron castings are then placed in stationary batch-type furnaces or car-bottom furnaces and the graphitization (malleablizing) of the castings is accomplished by means of a suitable annealing heat treatment. During this annealing period the temperature is slowly (50 hours) increased to as much as 1650 or 1700 degrees F. after which time it is slowly (60 hours) cooled. The American National Standard Specifications for Malleable Iron Castings — ANSI/ASTM A47-77 specifies the following grades and their properties: No. 32520, having a minimum tensile strength of 50,000 pounds per square inch, a minimum yield strength of 32,500 psi., and a minimum elongation in 2 inches of 10 per cent; and No. 35018, having a minimum tensile strength of 53,000 psi., a minimum yield strength of 35,000 psi., and a minimum elongation in 2 inches of 18 per cent.

Cupola Malleable Iron. — Another method of producing malleable iron involves initially the use of a cupola or a cupola in conjunction with an air furnace. This type of malleble iron, called cupola malleable iron, exhibits good fluidity and will produce sound castings. It is used in the making of pipe fittings, valves, and similar parts and possesses the useful property of being well suited to galvanizing. The American National Standard Specifications for Cupola Malleable Iron — ANSI/ASTM 197-79 calls for a minimum tensile strength of 40,000 pounds per square inch; a minimum yield strength of 30,000 psi.; and a minimum elongation in 2 inches of 5 per cent.

Pearlitic Malleable Iron. — This type of malleable iron contains some combined carbon in various forms. It may be produced either by stopping the heat treatment of regular malleable iron during production before the combined carbon contained therein has all been transformed to graphite or by reheating regular malleable iron above the transformation range. Pearlitic malleable irons exhibit a wide range of properties and are used in place of steel castings or forgings or to replace malleable iron when a greater strength or wear resistance is required. Some forms are made rigid to resist deformation while others will undergo considerable deformation before breaking. This material has been used in axle housings, differential housings, camshafts, and crankshafts for automobiles; machine parts; ordnance equipment; and tools. Tension test requirements of pearlitic malleable iron castings called for in ASTM Specification A 220-79 are given in the accompanying table.

Tension Test Requirements of Pearlitic Malleable Iron Castings (ASTM A 220-79)

Casting Grade Numbers		40010	45008	45006	50005	60004	70003	80002	90001
Min. Tensile Strength	1000s Lbs. per Sq. In.	60	65	65	70	80	85	95	105
Min. Yield Strength		40	45	45	50	60	70	80	90
Min. Elong. in 2 In., Per Cent		10	8	6	5	4	3	2	1

Ductile Cast Iron. — A distinguishing feature of this widely used type of cast iron, also known as spheroidal graphite iron or nodular iron, is that the graphite is present in ball-like form instead of in flakes as in ordinary gray cast iron. The addition of small amounts of magnesium- or cerium-bearing alloys together with special processing produces this spheroidal graphite structure and results in a casting of high strength and appreciable ductility. Its toughness is intermediate between that of cast iron and steel, and its shock resistance is comparable to ordinary grades of mild carbon steel. Melting point and fluidity are similar to those of the high–carbon cast irons. It exhibits good pressure tightness under high stress and can be welded and brazed. It can be softened by annealing or hardened by normalizing and air cooling or oil quenching and drawing.

Five grades of this iron are specified in ASTM A 536-80 — Standard Specification for Ductile Iron Castings. The grades and their corresponding matrix microstructures and heat treatments are as follows: Grade 60-40-18, ferritic, may be annealed; Grade 65-45-12, mostly ferritic, as-cast or annealed; Grade 80-55-06, ferritic/pearlitic, as-cast; Grade 100-70-03, mostly pearlitic, may be normalized; Grade 120-90-02, martensitic, oil quenched and tempered. The grade nomenclature identifies the minimum tensile strength, 0.2 per cent yield strength, and per cent elongation in 2 inches. Thus, Grade 60-40-18 has a minimum tensile strength of 60,000 psi, a minimum 0.2 per cent yield strength of 40,000 psi, and minimum elongation in 2 inches of 18 per cent. Several other types are commercially available to meet specific needs. The common grades of ductile iron can also be specified by only Brinell hardness, although the appropriate microstructure for the indicated hardness is also a requirement. This method is used in SAE Specification J434C for automotive castings and similar applications. Other specifications not only specify tensile properties, but also have limitations in composition. Austenitic types with high nickel content, high corrosion resistance, and good strength at elevated temperatures, are specified in ASTM A439-80.

Ductile cast iron can be cast in molds containing metal chills if wear-resisting surfaces are desired. Hard carbide areas will form in a manner similar to the forming of areas of chilled cast iron in gray iron castings. Surface hardening by flame or induction methods is also feasible. Ductile cast iron can be machined with the same ease as gray cast iron. It finds use as crankshafts, pistons, and cylinder heads in the automotive industry; forging hammer anvils, cylinders, guides, and control levers in the heavy machinery field; and wrenches, clamp frames, face-plates, chuck bodies, and dies for forming metals in the tool and die field. The production of ductile iron castings involves complex metallurgy, the use of special melting stock, and close process control. The majority of applications of ductile iron have been made to utilize its excellent mechanical properties in combination with the castability, machinability, and corrosion resistance of gray iron.

Steel Castings. — Steel castings are especially adapted for machine parts that must withstand shocks or heavy loads. They are stronger than either wrought iron, cast iron, or malleable iron and are very tough. The steel used for making steel castings may be produced either by the open-hearth, electric arc, side-blow converter, or electric induction methods. The raw materials used are steel scrap, pig iron, and iron ore, the materials and their proportions varying according to the process and the type of furnace used. The open-hearth method is used when large tonnages are continually required while a small electric furnace might be used for steels of widely differing analyses, which are required in small lot production. The high frequency induction furnace is used for small quantity production of expensive steels of special composition such as high-alloy steels. Steel castings are used for such parts as hydroelectric turbine wheels, forging presses, gears, railroad car frames, valve bodies, pump casings, mining machinery, marine equipment, engine casings, etc.

Table 1. Mechanical Properties of Steel Castings

For general information only. Not for use as design or specification limit values.

Tensile Strength, Lbs. per Sq. In.	Yield Point, Lbs. per Sq. In.	Elongation in 2 In., Per Cent	Brinell Hardness Number	Type of Heat Treatment	Application Indicating Properties
Structural Grades of Carbon Steel Castings					
60,000	30,000	32	120	Annealed	Low electric resistivity. Desirable magnetic properties. Carburizing and case hardening grades. Weldability.
65,000	35,000	30	130	Normalized	Good weldability.
70,000	38,000	28	140	Normalized	Medium strength with good machinability and high ductility.
80,000	45,000	26	160	Normalized and tempered	High strength carbon steels with good machinability, toughness and good fatigue resistance.
85,000	50,000	24	175	Normalized and tempered	
100,000	70,000	20	200	Quenched and tempered	Wear resistance. Hardness.
Engineering Grades of Low Alloy Steel Castings					
70,000	45,000	26	150	Normalized and tempered	Good weldability. Medium strength with high toughness and good machinability. For high temperature service.
80,000	50,000	24	170	Normalized and tempered	
90,000	60,000	22	190	Normalized and tempered*	Certain steels of these classes have good high temperature properties and deep hardening properties. Toughness.
100,000	68,000	20	209	Normalized and tempered*	
110,000	85,000	20	235	Quenched and tempered	Impact resistance. Good low temperature properties for certain steels. Deep hardening. Good combination of strength and toughness.
120,000	95,000	16	245	Quenched and tempered	
150,000	125,000	12	300	Quenched and tempered	Deep hardening. High strength. Wear and fatigue resistance.
175,000	148,000	8	340	Quenched and tempered	High strength and hardness. Wear resistance. High fatigue resistance.
200,000	170,000	5	400	Quenched and tempered	

* Quench and temper heat treatments may also be employed for these classes.
The values listed above have been compiled by the Steel Founders' Society of America as those normally expected in the production of steel castings. The castings are classified according to tensile strength values which are given in the first column. Specifications covering steel castings are prepared by the American Society for Testing and Materials, the Association of American Railroads, the Society of Automotive Engineers, the United States Government (Federal and Military Specifications), etc. These specifications appear in publications issued by these organizations.

Steel castings can generally be made from any of the many types of carbon and alloy steels produced in wrought form and respond similarly to heat treatment; they also do not exhibit directionality effects that are typical of wrought steel. Steel castings are classified into two general groups: carbon steel and alloy steel.

Carbon Steel Castings. — Carbon steel castings may be designated as low-carbon, medium-carbon, and high-carbon. Low-carbon steel castings have a carbon content of less than 0.20 per cent (most are produced in the 0.16 to 0.19 per cent range). Other elements present are: manganese, 0.50 to 0.85 per cent; silicon, 0.25 to 0.70 per cent; phosphorus, 0.05 per cent max.; and sulfur, 0.06 per cent max. Their tensile strengths (annealed condition) range from 40,000 to 70,000 pounds per square inch. Medium-carbon steel castings have a carbon content of from 0.20 to 0.50 per cent. Other elements present are: manganese, 0.50 to 1.00 per cent; silicon, 0.20 to 0.80 per cent; phosphorus, 0.05 per cent max.; and sulfur, 0.06 per cent max. Their tensile strengths range from 65,000 to 105,000 pounds per square inch depending, in part, upon heat treatment. High-carbon steel castings have a carbon content of more than 0.50 per cent and also contain: manganese, 0.50 to 1.00 per cent; silicon, 0.20 to 0.70 per cent; and phosphorus and sulfur, 0.05 per cent max. each. Fully annealed high-carbon steel castings exhibit tensile strengths of from 95,000 to 125,000 pounds per square inch. See Table 1 for grades and properties of carbon steel castings.

Table 2. Nominal Chemical Composition and Mechanical Properties of Heat-Resistant Steel Castings (ASTM A297-81)

Grade	Nominal Chemical Composition, Per Cent†	Tensile Strength, min		0.2 Per Cent Yield Strength, min		Per Cent Elongation in 2 in., or 50 mm, min
		ksi	MPa	ksi	MPa	
HF	19 Chromium, 9 Nickel	70	485	35	240	25
HH	25 Chromium, 12 Nickel	75	515	35	240	10
HI	28 Chromium, 15 Nickel	70	485	35	240	10
HK	25 Chromium, 20 Nickel	65	450	35	240	10
HE	29 Chromium, 9 Nickel	85	585	40	275	9
HT	15 Chromium, 35 Nickel	65	450	...	...	4
HU	19 Chromium, 39 Nickel	65	450	...	...	4
HW	12 Chromium, 60 Nickel	60	415	...	...	...
HX	17 Chromium, 66 Nickel	60	415	...	...	...
HC	28 Chromium	55	380	...	...	...
HD	28 Chromium, 5 Nickel	75	515	35	240	8
HL	29 Chromium, 20 Nickel	65	450	35	240	10
HN	20 Chromium, 25 Nickel	63	435	...	...	8
HP	26 Chromium, 35 Nickel	62.5	430	34	235	4.5

† Remainder is iron.
ksi = kips per square inch = 1000s of pounds per square inch; MPa = megapascals.

Table 3. Nominal Chemical Composition and Mechanical Properties of Corrosion-Resistant Steel Castings (ASTM A743-81a)

Grade	Nominal Chemical Composition, Per Cent†	Tensile Strength, min ksi	MPa	0.2% Yield Strength, min ksi	MPa	Per Cent Elongation in 2 in., or 50 mm, min	Per Cent Reduction of Area, min
CF-8	9 Chromium, 9 Nickel	70*	485*	30*	205*	35	...
CG-12	22 Chromium, 12 Nickel	70	485	28	195	35	...
CF-20	19 Chromium, 9 Nickel	70	485	30	205	30	...
CF-8M	19 Chromium, 10 Nickel, with Molybdenum	70	485	30	205	30	...
CF-8C	19 Chromium, 10 Nickel with Niobium	70	485	30	205	30	...
CF-16 and CF-16Fa	19 Chromium, 9 Nickel, Free Machining	70	485	30	205	25	...
CH-20 and CH-10	25 Chromium, 12 Nickel	70	485	30	205	30	...
CK-20	25 Chromium, 20 Nickel	65	450	28	195	30	...
CE-30	29 Chromium, 9 Nickel	80	550	40	275	10	...
CA-15 and CA-15M	12 Chromium	90	620	65	450	18	30
CB-30	20 Chromium	65	450	30	205	...	...
CC-50	28 Chromium	55	380	...	...	...	...
CA-40	12 Chromium	100	690	70	485	15	25
CF-3	19 Chromium, 9 Nickel	70	485	30	205	35	...
CF-3M	19 Chromium, 10 Nickel, with Molybdenum	70	485	30	205	30	...
CG6MMN	Chromium-Nickel-Manganese-Molybdenum	75	515	35	240	30	...
CG-8M	19 Chromium, 11 Nickel, with Molybdenum	75	520	35	240	25	...
CN-7M	20 Chromium, 29 Nickel, with Copper and Molybdenum	62	425	25	170	35	...
CN-7MS	19 Chromium, 24 Nickel, with Copper and Molybdenum	70	485	30	205	35	...
CW-12M	Nickel, Molybdenum, Chromium	72	495	46	315	4.0	...
CY-40	Nickel, Chromium, Iron	70	485	28	195	30	...
CZ-100	Nickel Alloy	50	345	18	125	10	...
M-35-1	Nickel-Copper Alloy	65	450	25	170	25	...
M-35-2	Nickel-Copper Alloy	65	450	30	205	25	...
CA-6NM	12 Chromium, 4 Nickel	110	755	80	550	15	35
CD-4MCu	25 Chromium, 5 Nickel, 2 Molybdenum, 3 Copper	100	690	70	485	16	...
CA-6N	11 Chromium, 7 Nickel	140	965	135	930	15	50

* For low ferrite or non-magnetic castings of this grade, the following values shall apply: tensile strength, min, 65 ksi (450 MPa); yield point, min, 28 ksi (195 MPa). † Remainder is iron.

Alloy Steel Castings. — Alloy cast steels are those in which special alloying elements such as manganese, chromium, nickel, molybdenum, vanadium have been added in sufficient quantities to obtain or increase certain desirable properties. Alloy cast steels are comprised of two groups — the low-alloy steels with their alloy content totaling less than 8 per cent and the high-alloy steels with their alloy content totaling 8 per cent or more. The addition of these various alloying elements, in conjunction with suitable heat-treatments, makes it possible to secure steel castings having a wide range of properties. The three accompanying tables give information on these steels. The lower portion of Table 1 gives the engineering grades of low-alloy cast steels grouped according to tensile strengths and gives properties normally expected in the production of steel castings. Tables 2 and 3 give the standard designations and nominal chemical composition ranges of high-alloy castings which may be classified according to heat or corrosion resistance. The grades given in these tables are recognized in whole or in part by the Alloy Casting Institute (ACI), the American Society for Testing and Materials (ASTM), and the Society of Automotive Engineers (SAE).

Austenitic Manganese Cast Steel: Austenitic manganese cast steel is an important high-alloy cast steel which provides a high degree of shock and wear resistance. Its composition normally falls within the following ranges: carbon, 1.00 to 1.40 per cent; manganese, 10.00 to 14.00 per cent; silicon, 0.30 to 1.00 per cent; sulfur, 0.06 per cent max.; phosphorus, 0.10 per cent, max. In the as-cast condition, austenitic manganese steel is quite brittle. In order to strengthen and toughen the steel, it is heated to between 1830 and 1940 degrees F and quenched in cold water. Physical properties of quenched austenitic manganese steel that has been cast to size are as follows: tensile strength, 80,000 to 100,000 pounds per square inch; shear strength (single shear), 84,000 pounds per square inch; elongation in 2 inches, 15 to 35 per cent; reduction in area, 15 to 35 per cent; and Brinell hardness number, 180 to 220. When cold worked, the surface of such a casting increases to a Brinell hardness of from 450 to 550. In many cases the surfaces are cold worked to maximum hardness to assure immediate hardness in use. Heat-treated austenitic manganese steel is machined only with great difficulty since it hardens at and slightly ahead of the point of contact of the cutting tool. Grinding wheels mounted on specially adapted machines are used for boring, planing, keyway cutting, and similar operations on this steel. Where grinding cannot be employed and machining must be resorted to, high-speed tool steel or cemented carbide tools are used with heavy, rigid equipment and slow, steady operation. In any event, this procedure tends to be both tedious and expensive. Austenitic manganese cast steel can be arc-welded with manganese-nickel steel welding rods containing from 3 to 5 per cent nickel, 10 to 15 per cent manganese, and, usually, 0.60 to 0.80 per cent carbon.

Finishing Operations for Castings

Removal of Gates and Risers from Castings. — After the molten iron or steel has solidified and cooled, the castings are removed from their molds, either manually or by placing them on vibratory machines and shaking the sand loose from the castings. The gates and risers which are not broken off in the shake-out are removed by impact, sawing, shearing, or burning-off methods. In the impact method, a hammer is used to knock off the gates and risers. Where the possibility exists that the fracture would extend into the casting itself, the gates or risers are first notched to assure fracture in the proper place. Some risers have a necked down section at which the riser breaks off when struck. Sprue-cutter machines are also used to shear off gates. These machines facilitate the removal of a number of small castings from a central runner. Band saws, power saws, and abrasive cut-off wheel machines

are also used to remove gates and risers. The use of band saws permits following the contour of the casting when removing unwanted appendages. Abrasive cut-off wheels are used when the castings are too hard or difficult to saw. Oxy-acetylene cutting torches are used to cut off gates and risers and also to gouge out or remove surface defects on castings. These torches are used on steel castings where the gates and risers are of a relatively large size. Surface defects are subsequently repaired by conventional welding methods.

Any unwanted material in the form of fins, gate and riser pads which come above the casting surface, chaplets, parting-line flash, etc. is removed by chipping with pneumatic hammers, or by grinding with such equipment as floor or bench-stand grinders, portable grinders, and swing-frame grinders.

Blast Cleaning of Castings. — Blast cleaning of castings is performed to remove adhering sand, to remove cores, to improve the casting appearance, and to prepare the castings for their final finishing operation which includes painting, machining, or assembling. Scale produced as a result of heat-treating can also be removed. A variety of machines are used to handle all sizes of castings. The methods employed include blasting with sand, metal shot, or grit; and hydraulic cleaning or tumbling. In blasting, sharp sand, shot or grit is carried by a stream of compressed air or water or by centrifugal force (gained as a result of whirling in a rapidly rotating machine) and directed against the casting surface by means of nozzles. The operation is usually performed in cabinets or enclosed booths. In some set-ups the castings are placed on a revolving table and the nozzles which are either mechanically- or hand-held are played over all the casting surfaces. Tumbling machines are also employed for cleaning, the castings being placed in large revolving drums together with slugs, balls, pins, metal punchings or some abrasive, such as sandstone or granite chips, slag, silica, sand, or pumice. Quite frequently the tumbling and blasting methods are used together, the parts in this instance being tumbled and blasted simultaneously. Castings may also be cleaned by hydro-blasting. In this method the castings are placed in a water-tight room and a water and sand mixture under high pressure is played over the castings by means of nozzles. The action of the water and sand mixture cleans the castings very effectively.

Heat-treatment of Steel Castings. — Steel castings can be heat-treated to bring about diffusion of carbon or alloying elements, softening, hardening, stress-relieving, toughening, improved machinability, increased wear resistance, and removal of hydrogen entrapped at the surface of the casting. Heat treatment of steel castings of a given composition follows closely that of wrought steel of similar composition. For discussion of types of heat treatment refer to the Heat-treatment of Steel section of this Handbook.

Pattern Materials — Shrinkage, Draft and Finish Allowances

Woods for Patterns. — Woods commonly used for patterns are white pine, mahogany, cherry, maple, birch, white wood and fir. For most patterns, white pine is considered superior because it is easily worked, readily takes glue and varnish, and is fairly durable. For medium- and small-sized patterns, especially if they are to be extensively used, a harder wood is preferable. Mahogany is often used for patterns of this class, although many prefer cherry. As mahogany has a close grain, it is not as susceptible to atmospheric changes as a wood of coarser grain. Mahogany is superior in this respect to cherry, but is more expensive. In selecting cherry, never use young timber. Maple and birch are employed quite extensively, especially for turned parts, as they take a good finish. White wood is sometimes substituted for pine, but it is inferior to the latter in being more susceptible to atmospheric changes.

Selection of Wood. — It is very important to select wood for patterns that is well seasoned; that is, it should either be kiln dried or kept one or two years before using, the time depending upon the size of the lumber. During the seasoning or drying process the moisture leaves the wood cells and the wood shrinks, the shrinkage being almost entirely across the grain rather than in a lengthwise direction. Naturally, after this change takes place, the wood is less liable to warp, although it will absorb moisture in damp weather. Patterns also tend to absorb moisture from the damp sand of molds, and to minimize troubles from this source they are covered with varnish. Green or water-soaked lumber should not be put in a drying room, because the ends will dry out faster than the rest of the log, thus causing cracks. In a log there is what is called the "sap wood" and the "heart wood." The outer layers form the sap wood which is not as firm as the heart wood and is more likely to warp; hence, it should be avoided, if possible.

Pattern Varnish. — Patterns intended for repeated use are varnished to protect them against moisture, especially when in the damp molding sand. The varnish used should dry quickly to give a smooth surface that readily draws from the sand. Yellow shellac varnish is generally used. It is made by dissolving gum shellac in grain alcohol. Wood alcohol is sometimes substituted, but is inferior. The color of the varnish is commonly changed for covering core prints, in order to readily distinguish the prints from the body of the pattern. Black shellac varnish (which is the color generally used) is made by the addition of lamp black. This should be of good quality and free from grit. Red varnish can be made by adding Chinese vermilion. All coloring powders should be well pulverized. At least three coats of varnish should be applied to patterns, the surfaces being rubbed down with sand paper after applying the preliminary coats, in order to obtain a smooth surface.

Glue for Patterns. — There are many qualities of glue both in the liquid, sheet and pulverized form. Animal glue in the sheet or flake form is generally used for pattern work. As a rule, the best quality is of amber color and the flakes are rather thin. Where glue is used in small quantities, the pulverized form has the advantage of being quickly prepared. Freshly made glue is the strongest and, if of good quality, can be drawn out into thin threads. Whenever practicable, glued joints should be reinforced by nails or screws. A joint to be glued should be accurately fitted because glue does not get a grip unless the parts are in close contact.

Before applying glue, clean the surfaces of sand paper dust or other foreign material, so that the glue can enter the pores of the wood. This is very important. If the end grain must be glued, first apply a sizing coat to fill the openings among the fibers. When the sizing coat is dry, apply a second coat to the surface and unite. If the preliminary coat is not applied, the open end grain is liable to absorb the glue so rapidly as to weaken the joint. The hot glue should be thin enough to spread easily. It can be thicker, however, for pine than for wood of closer grain, like mahogany, because (aside from the quality of the glue) the holding or binding property depends upon the extent to which the glue enters the pores of the wood. All glued joints should be firmly pressed together with clamps immediately after applying the glue. The latter should be given plenty of time in which to set; ten or twelve hours in a dry place should be sufficient.

Shrinkage Allowance. — The shrinkage allowances ordinarily specified for patterns to compensate for the contraction of castings in cooling are as follows: cast iron, $3/32$ to $1/8$ inch per foot; common brass, $3/16$ inch per foot; yellow brass, $7/32$ inch per foot; bronze, $5/32$ inch per foot; aluminum, $1/8$ to $5/32$ inch per foot; magnesium, $1/8$ to $11/64$ inch per foot; steel casting, $3/16$ inch per foot. These shrinkage allowances are approximate values only since the exact allowance in any case de-

pends upon the size and shape of the casting and the resistance of the mold to the normal contraction of the casting during cooling. It is, therefore, possible that more than one shrinkage allowance will be required for different parts of the same pattern. Another factor that affects shrinkage allowance is the molding method, which may vary to such an extent from one foundry to another, that different shrinkage allowances for each would have to be used for the same pattern. For these reasons it is recommended that patterns be made at the foundry where the castings are to be produced to eliminate difficulties due to lack of accurate knowledge of shrinkage requirements.

An example of how casting shape can affect shrinkage allowance is given in the Steel Castings Handbook. In this example a straight round steel bar required a shrinkage allowance of approximately $\frac{9}{32}$ inch per foot. The same bar but with a large knob on each end required a shrinkage allowance of only $\frac{3}{16}$ inch per foot. A third steel bar with large flanges at each end required a shrinkage allowance of only $\frac{7}{64}$ inch per foot. This example would seem to indicate that the best practice in designing castings and making patterns is to obtain shrinkage values from the foundry that is to make the casting since there can be no fixed allowances.

Draft for Patterns. — The draft or the amount of taper given to patterns to facilitate withdrawing them from the mold, depends somewhat upon the size and shape of the pattern. A general rule is to taper each side $\frac{1}{8}$ inch for each foot of surface to be drawn. The average amount for small patterns is about $\frac{1}{16}$ inch per foot, although in some cases it can be less, but, as a rule, there should be at least $\frac{1}{32}$ inch draft. The draft slopes away from the pattern "face" which is usually uppermost in the mold when the pattern is drawn. Some patterns do not require draft because none of the surfaces are at right angles to the face. In some cases, very small patterns are made without draft.

Finish Allowance. — The amount added to a pattern to allow for machining the casting varies widely. It depends upon the method of machining, the size of the casting, and the importance of having a clean surface, free from flaws or defective spots. If castings are to be finished from the rough upon disk grinders, very little allowance is necessary; in fact, the molder may rap a pattern enough to allow for the finish. On small castings to be finished by milling or planing about $\frac{1}{8}$ inch is usually allowed for the machining operation, whereas large castings for engine beds, flywheels, pump cylinders, etc., often have an allowance of $\frac{3}{4}$ to 1 inch.

Standard Pattern Colors. — The color markings described in the following are recommended as standard for foundry patterns and core-boxes of wood construction. These standard colors have been accepted by the Bureau of Standards, the American Foundrymen's Association, the Steel Founders' Society of America, the Malleable Iron Research Institute, and by numerous other associations as well as prominent manufacturers.

(1) Surfaces to be left *unfinished* are to be painted *black*.
(2) Surfaces to be *machined* are to be painted *red*.
(3) *Seats of and for loose pieces* are to be marked by *red stripes on a yellow background*.
(4) *Core prints* and *seats for loose core prints* are to be painted *yellow*.
(5) *Stop-offs* are to be indicated by *diagonal black stripes on a yellow base*.
The colors may be obtained by mixing suitable inexpensive pigments with varnish or shellac to produce the type of coating desired.

Metal Patterns. — Metal patterns are especially adapted to molding machine practice, owing to their durability and superiority in retaining the required shape. The original master pattern is generally made of wood, the casting obtained from

the wood pattern being finished to make the metal pattern. The materials commonly used are brass, cast iron, aluminum and steel. Brass patterns should have a rather large percentage of tin, as this gives a good surface for the casting. Cast iron is generally used for patterns of large size, as it is cheaper than brass and more durable. Cast-iron patterns are largely used on molding machines. Aluminum patterns are light but they shrink considerably. White metal is sometimes used when it is necessary to avoid shrinkage. The gates for the mold may be cast or made of sheet brass. Some patterns are made of vulcanized rubber, especially for light match-board work.

Obtaining Weight of Casting from Pattern Weight. — To obtain the approximate weight of a casting, multiply the weight of the pattern by the factor given in the accompanying table. For example, if the weight of a white-pine pattern is 4 pounds, what is the weight of a solid cast-iron casting obtained from that pattern? Casting weight = $4 \times 16 = 64$ pounds. If the casting is cored, fill the core-boxes with dry sand, and multiply the weight of the sand by one of the following factors: For cast iron, 4; for brass, 4.65; for aluminum, 1.4. Then subtract the product of the sand weight and the factor just given from the weight of the solid casting, to obtain the weight of the cored casting. As the weight of

Factors for Obtaining Weight of Casting from Pattern Weight

Pattern Material	Factors				
	Cast Iron	Aluminum	Copper	Zinc	Brass, 70% Copper, 30% Zinc
White pine.............	16.00	5.70	19.60	15.00	19.00
Mahogany, Honduras	12.00	4.50	14.70	11.50	14.00
Cherry.................	10.50	3.80	13.00	10.00	12.50
Cast iron...............	1.00	0.35	1.22	0.95	1.17
Aluminum..............	2.85	1.00	3.44	2.70	3.30

wood varies considerably, the results obtained by the use of the table are only approximate, the factors being based on the average weight of the woods listed. For metal patterns, the results are more accurate.

Branch Pipes for Exhausting Shavings from Wood Working Machines. — The sizes of branch pipes, given in the accompanying table, are correct for pipes not exceeding twenty feet in length. Where branch pipes contain a number of elbows, or exceed twenty feet in length, the area should be proportionately increased. Where the work is light and the branch pipes short, smaller connections than those given can sometimes be used. The area of the main duct should be equal to, or slightly larger than, the sum of the areas of the connecting branches. This proportion should be carefully maintained. If the main pipe is too small, the suction will be impaired, and if it is too large, the velocity of the air may be reduced to such an extent that the material being exhausted will settle in the bottom of the pipe, thereby reducing the area. If the main pipe is unusually long (exceeding 100 feet) the area should be increased from 10 to 20 per cent. Avoid abrupt turns in the piping, and never enter branch pipes at right angles to the main pipe, but always connect them at an angle of from 30 to 45 degrees. Branch pipes should never enter the main at the bottom but always at the side, and two pipes should not enter directly opposite each other.

Branch Pipes for Exhausting Shavings from Wood Working Machines

Type of Machine	Diam. of Branch Pipe, Inches
Saws:	
Rip	
Cut-off	
Split — 18 inches diam. or less	4
Swing — 18 to 24 inches diam	5
Bracket	
Groove	
Heavy cut-off, 24 to 42 inches diameter..............	6
Band	4
Band resaw, ¾ to 1 inch	5
Band resaw, 1½ to 2½ ins.	6
Planer knives:	
Length of knife, 5 inches.....................	4
Length of knife, 10 inches....................	5
Length of knife, 14 inches....................	6
Length of knife, 24 inches....................	6
Length of knife, 30 inches....................	7
Matcher heads, each	5
Door tenoner.................................	5
Sash tenoner	5
Sticker machines, each head	4
Sand drum, 24 inches long	4
Sand drum, 30 inches long	5
Floor sweep-up	6
Heavy timber planer, each head	7
Diagonal planer for doors	7
Diagonal planer for doors, with sand drum	8

Speeds of Circular Saws for Wood

Size of Saw, Ins.	Rev. per Min.	Size of Saw, Ins.	Rev. per Min.	Size of Saw, Ins.	Rev. per Min.	Size of Saw, Ins.	Rev. per Min.	Size of Saw, Ins.	Rev. per Min.
8	4500	20	1800	32	1125	44	840	56	650
10	3600	22	1650	34	1050	46	800	58	625
12	3000	24	1500	36	1000	48	750	60	600
14	2600	26	1400	38	950	50	725	64	550
16	2200	28	1300	40	900	52	700	68	525
18	2000	30	1200	42	870	54	675	72	500

Die Casting

Die casting is a method of producing finished castings by forcing molten metal into a hard metal die, which is arranged to open after the metal has solidified so that the casting can be removed. The die casting process makes it possible to secure accuracy and uniformity in castings, and machining costs are either eliminated altogether or are greatly reduced. The greatest advantage of the die-casting process is that parts are accurately and often completely finished when taken from the die. When the dies are properly made, castings may be accurate within 0.001 inch or even less and a limit of 0.002 and 0.003 inch per inch of casting dimension can be maintained on many classes of work.

Die castings are used extensively in the manufacture of such products as cash registers, meters, time-controlling devices, small housings, washing machines, and parts for a great variety of mechanisms. Lugs and gear teeth are cast in place and both external and internal screw threads can be cast. Holes can be formed within about 0.001 inch of size and the most accurate bearings require only a finish-reaming operation. Figures and letters may be cast sunken or in relief on wheels for counting or printing devices, and with ingenious die designs, many shapes which formerly were believed too intricate for die casting are now produced successfully by this process.

As to the limitations of the die-casting process it may be mentioned that the cost of dies is high, and, therefore, die casting is economical only when large numbers of duplicate parts are required. The stronger and harder metals cannot be die cast, so that the process is not applicable for casting parts which must necessarily be made of iron or steel, although special alloys have been developed for die casting which have considerable tensile and compressive strength.

Many die castings are produced by the hot-chamber method in which the pressure chamber connected to the die cavity is immersed permanently in the molten metal and is automatically refilled through a hole that is uncovered as the (vertical) pressure plunger moves back after filling the die. This method can be used for alloys of low melting point and high fluidity such as zinc, lead, tin, and magnesium. Other alloys requiring higher pressures, such as brass, or which can attack and dissolve the ferrous pressure chamber material, such as aluminum, must use the slower cold-chamber method with a water-cooled (horizontal) pressure chamber outside the molten metal.

Porosity. — Molten metal injected into a die cavity displaces most of the air, but some of the air is trapped and is mixed with the metal. The high pressure applied to the metal squeezes the pores containing the air to very small size, but subsequent heating will soften the casting so that air in the surface pores can expand and cause blisters. Die castings are seldom solution heat treated or welded because of this blistering problem. The chilling effect of the comparatively cold die causes the outer layers of a die casting to be dense and relatively free of porosity. Vacuum die casting, in which the cavity atmosphere is evacuated before metal is injected, is sometimes used to reduce porosity. Another method is to displace the air by filling the cavity with oxygen just prior to injection. The oxygen is burned by the hot metal so that porosity does not occur.

When these special methods are not used, machining depths must be limited to 0.020–0.035 inch if pores are not to be exposed, but as-cast accuracy is usually good enough for only light finishing cuts to be needed. Special pore-sealing techniques must be used if pressure tightness is required.

Designing Die Castings. — Die castings are best designed with uniform wall thicknesses (to reduce cooling stresses) and cores of simple shapes (to facilitate extraction from the die). Heavy sections should be avoided or cored out to reduce metal con-

centrations that may attract trapped gases and cause porosity concentrations. Designs should aim at arranging for metal to travel through thick sections to reach thin ones if possible. Because of the high metal injection pressures, conventional sand cores cannot be used, so cored holes and apertures are made by metal cores that form part of the die. Small and slender cores are easily bent or broken, so should be avoided in favor of piercing or drilling operations on the finished castings. Ribbing adds strength to thin sections, and fillets should be used on all inside corners to avoid high stress concentrations in the castings. Sharp outside corners should be avoided. Draft allowances on a die casting are usually from 0.5 to 1.5 degrees per side to permit the castings to be pushed off cores or out of the cavity.

Alloys Used for Die Casting. — The alloys used in modern die casting practice are based on aluminum, zinc, and copper, with small numbers of castings also being made from magnesium-, tin-, and lead-based alloys.

Aluminum-base Alloys. — Aluminum-base die casting alloys are used more extensively than any other base metal alloy because of their superior strength combined with ease of castability. Shrinkage of aluminum on cooling is about 6.6 per cent. Casting temperatures are of the order of 1200 deg. F. Most aluminum die castings are produced in aluminum–silicon–copper alloys such as the Aluminum Association (AA) No. 380 (ASTM SC84A) (UNS A038000), containing silicon 7.5 to 9.5 and copper 3 to 4 per cent. Silicon increases fluidity for complete die filling, but reduces machinability, and copper adds hardness but reduces ductility in aluminum alloys. A less-used alloy having slightly greater fluidity is AA No. 384 (ASTM SC114A) (UNS A03840) containing silicon 10.5 to 12.0 and copper 3.0 to 4.5 per cent. For marine applications, AA 360 (ASTM 100A) (UNS A03600) containing silicon 9 to 10 and copper 0.6 per cent is recommended, the copper content being kept low to reduce susceptibility to corrosion in salt atmospheres. The tensile strengths of AA 380, 384, and 360 alloys are 47,000, 48,000, and 46,000 lb per square inch, respectively. Although 380, 384, and 360 are the most widely used die castable alloys, several other aluminum alloys are used for special applications. For instance, the AA 390 alloy, with its high silicon content (16 to 18 per cent), is used for internal combustion engine cylinder castings, to take advantage of the good wear resistance provided by the hard silicon grains. No. 390 alloy also contains 4 to 5 per cent copper, and has a hardness of 120 Brinell with low ductility, and a tensile strength of 41,000 lb per square inch.

Zinc-based Alloys. — In the molten state, zinc is extremely fluid and can therefore be cast into very intricate shapes. The metal also is plentiful and has good mechanical properties. Zinc die castings can be made to closer dimensional limits and with thinner walls than aluminum. A shrinkage allowance of 7.28 per cent must be made in dimensioning castings. The low casting temperatures (750–800 deg. F) and the hot-chamber process allow high production rates with simple automation. Zinc die castings can be produced with extremely smooth surfaces, lending themselves well to plating and other finishing methods. The established zinc alloys numbered 3, 5, and 7 [ASTM B86 (AG40A) (UNS Z33520), AG41A (UNS Z35531), and AG40B (UNS Z33522)] each contain 3.5 to 4.3 per cent of aluminum, which adds strength and hardness, plus carefully controlled amounts of other elements. Recent research has brought forward three new alloys of zinc containing 8, 12, and 27 per cent of aluminum, which confer tensile strength of 50,000–62,000 lb per square inch and hardness approaching that of cast iron (105–125 Brinell). These alloys can be used for gears and racks, for instance, and as housings for shafts that run directly in reamed or bored holes, with no need for bearing bushes.

Copper-base Alloys. — Brass alloys are used for plumbing, electrical, and marine components where resistance to corrosion must be combined with strength and wear resistance. With the development of the cold-chamber casting process, it became possible to make die castings from several standard alloys of copper and zinc such as yellow brass (ASTM B176-Z30A) (UNS C85800) containing copper 58, zinc 40, tin 1, and lead 1 per cent. Tin and lead are included to improve corrosion resistance and machinability, respectively, and this alloy has a tensile strength of 45,000 lb per square inch. Silicon brass (ASTM B176-ZS331A) (UNS C87800) with copper 65 and zinc 34 per cent, also contains 1 per cent silicon, giving it more fluidity for castability and with higher tensile strength (58,000 psi) and better resistance to corrosion. High silicon brass or tombasil (ASTM B176-ZS144A), containing copper 82, zinc 14, and silicon 4 per cent, has a tensile strength of 70,000 lb per square inch and good wear resistance, but at the expense of machinability.

Magnesium-base Alloys. — Light weight combined with strength and good mechanical properties are characteristics for which magnesium alloys are selected. Magnesium does not dissolve iron, so can be die cast by either the hot- or cold-chamber method, and its low specific heat allows the alloy to be melted at low cost in fuel. Special precautions must be taken in melting and casting magnesium because of its high affinity for oxygen, but the development of a new inert atmosphere known as SF^6 has simplified handling of the material. Magnesium is free machining but the large volumes of chips produced must be handled carefully to avoid a fire hazard. The most widely used alloy is the ASTM B93 (AZ 91B) (UNS 11912), which contains aluminum 8.3 to 9.7, and zinc 0.35 to 1.00 per cent, and has a tensile strength of 34,000 lb per square inch. The casting temperatures are about the same as for aluminum. The ASTM AZ 91A (UNS 11910) variant of this alloy can contain up to 0.35 per cent copper (compared with 0.1 per cent for AZ 91B) providing slightly more resistance to corrosion at a slightly higher cost.

Tin-base Alloys. — In this group tin is alloyed with copper, antimony, and lead. SAE Alloy No. 10 contains, as the principal ingredients, in percentages, tin, 90; copper, 4 to 5; antimony, 4 to 5; lead, maximum, 0.35. This high-quality babbitt mixture is used for main-shaft and connecting-rod bearings or bronze-backed bearings in the automotive and aircraft industries. SAE No. 110 contains tin, 87.75; antimony, 7.0 to 8.5; copper, maximum, 2.25 to 3.75 per cent and other constituents the same as No. 10. SAE No. 11, which contains a little more copper and antimony and about 4 per cent less tin than No. 10, is also used for bearings or other applications requiring a high-class tin-base alloy. These tin-base compositions are used chiefly for automotive bearings but they are also used for milking machines, soda fountains, syrup pumps, and similar apparatus requiring resistance against the action of acids, alkalies, and moisture.

Lead-base Alloys. — These alloys are employed usually where a cheap noncorrosive metal is needed and strength is relatively unimportant. Such alloys are used for parts of lead-acid batteries, for automobile wheel balancing weights, for parts which must withstand the action of strong mineral acids and for parts of X-ray apparatus. SAE Composition No. 13 contains (in percentages) lead, 86; antimony, 9.25 to 10.75; tin, 4.5 to 5.5 per cent. SAE Specification No. 14 contains less lead and more antimony and copper. The lead content is 76; antimony, 14 to 16; and tin, 9.25 to 10.75 per cent. These alloys, Nos. 13 and 14, are inexpensive owing to the high lead content and may be used for bearings which are large and subjected to light service.

Dies for Die-casting Machines. — Dies for die-casting machines are generally made of steel although cast iron and nonmetallic materials of a refractory nature have been

used, the latter being intended especially for bronze or brass castings, which, owing to their comparatively high melting temperatures, would damage ordinary steel dies. The steel most generally used is a low-carbon steel. Chromium-vanadium and tungsten steels are used for aluminum, magnesium and brass alloys, when dies must withstand relatively high temperatures.

Making die casting dies requires considerable skill and experience. Dies must be so designed that the metal will rapidly flow to all parts of the impression and at the same time allow the air to escape through shallow vent channels, 0.003 to 0.005 inch deep, cut into the parting of the die. To secure solid castings, the gates and vents must be located with reference to the particular shape to be cast. Shrinkage is another important feature, especially on accurate work. The amount usually varies from 0.002 to 0.007 inch per inch, but to determine the exact shrinkage allowance for an alloy containing three or four elements is difficult except by experiment.

Die Casting Bearing Metals in Place. — Practically all the metals that are suitable for bearings can be die cast in place. Automobile connecting-rods are an example of work to which this process has been applied successfully. After the bearings are cast in place, they are finished by boring or reaming. The best metals for the bearings, and those that also can be die cast most readily, are the babbitts containing about 85 per cent tin with the remainder copper and antimony. These metals should not contain over 9 per cent copper. The copper constitutes the hardening element in the bearing. A recommended composition for a high-class bearing metal is 85 per cent tin, 10 per cent antimony, and 5 per cent copper. The antimony may vary from 7 to 10 per cent and the copper from 5 to 8 per cent. To reduce costs, some bearing metals use lead instead of tin. One bearing alloy contains from 95 to 98 per cent lead. The die cast metal becomes harder upon seasoning a few days. In die casting bearings, the work is located from the bolt holes which are drilled previous to die casting. It is important that the bolt holes be drilled accurately with relation to the remainder of the machined surfaces.

Precision Investment Casting

Investment casting is a highly developed process that is capable of great casting accuracy and can form extremely intricate contours. The process may be utilized when metals are too hard to machine or otherwise fabricate; when it is the only practical method of producing a part; or when it is more economical than any other method of obtaining work of the quality required. Precision investment casting is especially applicable in producing either exterior or interior contours of intricate form with surfaces so located that they could not be machined readily if at all. The process provides efficient, accurate means of producing such parts as turbine blades, airplane or other parts made from alloys which have high melting points and must withstand exceptionally high temperatures, and many other products. The accuracy and finish of precision investment castings may either eliminate machining entirely or reduce it to a minimum. The quantity that may be produced economically may range from a few to thousands of duplicate parts.

Materials Which May Be Cast. — The precision investment process may be applied to a wide range of both ferrous and non-ferrous alloys. In industrial applications these include aluminum, and bronze alloys, stellite, Hastelloys, stainless and other alloy steels, and iron castings especially where thick and thin sections are encountered. In producing investment castings, it is possible to control the process in various ways so as to change the porosity or density of castings, obtain hardness variations in different sections, and vary the corrosion resistance and strength by special alloying.

General Procedure in Making Investment Castings. — Precision investment casting is similar in principle to the " lost wax " process which has long been used in manufacturing jewelry, ornamental pieces, dentures, inlays, and other items required in dentistry. When this process is employed, both the pattern and mold used in producing the casting are destroyed after each casting operation, but they may both be replaced readily. The " dispensable pattern " (or cluster of duplicate patterns) is first formed in a permanent mold or die and then is used to form the cavity in the mold or " investment " in which the casting (or castings) is made. The investment or casting mold consists of a refractory material contained within a reinforcing steel flask. The pattern is made of wax, plastic, or a mixture of the two. The material used is evacuated from the investment to form a cavity (without parting lines) for receiving the metal to be cast. Evacuation of the pattern (by the application of sufficient heat to melt and vaporize it) and the use of a master mold or die for reproducing it quickly and accurately in making duplicate castings, are distinguishing features of this casting process. Modern applications of the process include many developments such as variations in the preparation of molds, patterns, investments, etc., as well as in the casting procedure. Application of the process requires specialized knowledge and experience.

Master Mold for Making Dispensable Patterns. — Duplicate patterns for each casting operation are made by injecting the wax or other pattern material into the master mold or die which usually is made either of carbon steel or of a soft metal alloy. Rubber, alloy steels, and other materials are used in some cases. The mold cavity commonly is designed to form a cluster of patterns for multiple casting. The mold cavity is not, as a rule, an exact duplicate of the part to be cast because it is necessary to allow for shrinkage and perhaps to compensate for distortion which might affect the accuracy of the cast product. In producing master molds, there is considerable variation in practice. One general method is to form the cavity by machining; another is by pouring a molten alloy around a master pattern which usually is made of monel metal or of a high alloy stainless steel. Unless the cavity is machined, a master pattern is required. Sometimes, a sample of the product itself may be used as a master pattern, when, for example, a slight reduction in size due to shrinkage is not objectionable. The dispensable pattern material, which may consist of waxes, plastics, or a combination of these materials, is injected into the mold either by pressure, by gravity, or by the centrifugal method. The mold is made in sections to permit removal of the dispensable pattern. The mold while in use may be kept at the correct temperature either by electrical means, by using steam, or a water jacket.

Shrinkage Allowances for Patterns. — The shrinkage allowance varies considerably for different materials. In casting accurate parts, experimental preliminary casting operations may be necessary to determine the required shrinkage allowance and possible effects of distortion. Shrinkage allowances, in inches per inch, usually average about 0.022 for steel, 0.012 for gray iron, 0.016 for brass, 0.012 to 0.022 for bronze, 0.014 for aluminum and magnesium alloys. (See also page 1190.)

Investment Materials. — The investment materials which surround the dispensable pattern are made according to various formulas which may be simple or complex. The investment is in liquid form and the degree of liquidity depends chiefly upon the intricacy of the pattern forms. The investment must be fluid enough to enter such places as small holes, form sharp contours, threads, etc., and fill completely every part of the investment cavity. The accuracy of the product and its surface finish may be affected by the liquidity of the investment material. Porosity is another factor. If it is excessive, the finish will be marred but there

should be enough porosity to permit escape of air when the molten metal enters the cavity. The procedure in "investing" or applying the investment materials, varies according to these materials, the kind of metal to be cast and properties required in the casting. The hardening or setting up of the material may merely involve standing for a given time or hardening may be at a controlled temperature.

Evacuation of the Pattern. — In cases where heat is applied to control investment hardening or setting up, this heat may also result in evacuating most of the pattern material; however, complete evacuation is essential and methods are employed to remove all traces of the pattern. Evacuation may require closely regulated temperature and the use of boiling water, low pressure steam, or the application of a vacuum to the sprue bottom.

Casting Operations. — The temperature of the flask for casting may range all the way from a chilled condition up to 2000 degrees F. or higher, depending upon the metal to be cast, the size and shape of the casting or cluster, and the desired metallurgical conditions. Metals while being cast are nearly always subjected to centrifugal force or other pressure. The procedure is governed by the kind of alloy, the size of investment cavity, and its contours or shape.

Investment Removal. — The investments surrounding the casting or cluster is removed by destroying it. The investment may be soluble in water but those used for ferrous castings are broken by using pneumatic tools, hammers, or by shot or abrasive blasting and tumbling to remove all of the particles. Gates, sprues and runners may be removed from the castings by an abrasive cutting wheel or a band saw. The shape of the cluster and machinability of the material are factors governing the selection of the method.

Accuracy of Investment Castings. — The accuracy of precision investment castings may, in general, compare favorably with many machined parts. The over-all tolerance varies with the size of the work, the kind of metal, and the skill and experience back of the casting operations. Under normal conditions, the tolerances may vary from plus or minus 0.005 or 0.006 inch per inch, down to 0.0015 to 0.002 inch per inch, and even smaller tolerances are possible on very small dimensions. Where tolerances applying to a lengthwise dimension must be smaller than would be normal for the casting process, the casting gate may be at one end to permit controlling the length by a grinding operation when the gate is removed.

Casting Milling Cutters by Investment Method. — Possible applications of precision investment casting in tool manufacture and in other industrial applications, are indicated by its use in producing high-speed steel milling cutters of various forms and sizes. Thousands of these cutters have been precision cast in the Ford plant. Removal of the risers, sand-blasting to improve the appearance, and grinding the cutting edges are the only machining operations required. The bore is used as cast. Numerous tests have shown that the life of these cutters compares favorably with high-speed steel cutters made in the usual way.

Casting Weights and Sizes. — Investment castings may vary in weight from a fractional part of an ounce up to 25 pounds or more. Although the range of weights representing the practice of different firms specializing in investment casting may vary from about ½ pound up to 10 or 20 pounds, a practical limit of 3 or 4 pounds is common. The length of investment castings ordinarily does not exceed 5 or 6 inches but much longer parts may be cast. While it is possible to cast sections having a thickness of only a few thousandths of an inch, the preferable minimum thickness, as a general rule, is about 0.020 inch for alloys of high castability and 0.040 inch for alloys of low castability.

Extrusion of Metals

The Basic Process. — Extrusion is a metalworking process used to produce long, straight semifinished products such as bars, tubes, solid and hollow sections, wire, and strips by squeezing a solid slug of metal, either cast or wrought, from a closed container through a die. An analogy to the process is the dispensing of toothpaste from a collapsible tube.

During extrusion, compressive and shear, but no tensile, forces are developed in the stock, thus allowing the material to be heavily deformed without fracturing. The extrusion process can be performed at either room or high temperature, depending on the alloy and method. Cross sections of varying complexity can also be produced, depending on the materials and dies used.

In the specially constructed presses used for extrusion, the load is transmitted by a ram through an intermediate dummy block to the stock. The press container is usually fitted with a wear-resistant liner and is constructed to withstand high radial loads. The die stack consists of the die, die holder, and die backer, all of which are supported in the press end housing or platen, which resists the axial loads.

The following are characteristics of different extrusion methods and presses: (1) The movement of the extrusion relative to the ram. In "direct extrusion," the ram is advanced toward the die stack; in "indirect extrusion," the die moves down the container bore. (2) The position of the press axis, which is either horizontal or vertical. (3) The type of drive, which is either hydraulic or mechanical. (4) The method of load application, which is either conventional or hydrostatic.

In forming a hollow extrusion, such as a tube, a mandrel integral with the ram is pushed through the previously pierced raw billet.

Cold Extrusion: Cold extrusion has often been considered a separate process from hot extrusion; however, the only real difference is that cold or only slightly warm billets are used as starting stock. Cold extrusion is not limited to certain materials; the only limiting factor is the stresses in the tooling. In addition to the soft metals such as lead and tin, aluminum alloys, copper, zirconium, titanium, molybdenum, beryllium, vanadium, niobium, and steel can be extruded cold or at low deformation temperatures. Cold extrusion has many advantages, such as no oxidation or gas/metal reactions; high mechanical properties due to cold working if the heat of deformation does not initiate recrystallization; narrow tolerances; good surface finish if optimum lubrication is used; fast extrusion speeds can be used with alloys subject to hot shortness.

Examples of cold extruded parts are collapsible tubes, aluminum cans, fire extinguisher cases, shock absorber cylinders, automotive pistons, and gear blanks.

Hot Extrusion: Most hot extrusion is performed in horizontal hydraulic presses rated in size from 250 to 12,000 tons. The extrusions are long pieces of uniform cross sections, but complex cross sections are also produced. Most types of alloys can be hot extruded.

Owing to the temperatures and pressures encountered in hot extrusion, the major problems are the construction and the preservation of the equipment. The following are approximate temperature ranges used to extrude various types of alloys: magnesium, 650–850 degrees F.; aluminum, 650–900 degrees F.; copper, 1200–2000 degrees F.; steel, 2200–2400 degrees F.; titanium, 1300–2100 degrees F.; nickel 1900–2200 degrees F.; refractory alloys, up to 4000 degrees F. In addition, pressures range from as low as 5000 to over 100,000 psi. Therefore, lubrication and protection of the chamber, ram, and die are generally required. The use of oil and graphite mixtures is often sufficient at the lower temperatures; while at higher temperatures, glass powder, which becomes a molten lubricant, is used.

Extrusion Applications: The stress conditions in extrusion make it possible to work materials that are brittle and tend to crack when deformed by other primary metalworking processes. The most outstanding feature of the extrusion process, however, is its ability to produce a wide variety of cross-sectional configurations; shapes can be extruded that have complex, nonuniform, and nonsymmetrical sections that would be difficult or impossible to roll or forge. Extrusions in many instances can take the place of bulkier, more costly assemblies made by welding, bolting, or riveting. Many machining operations may also be reduced through the use of extruded sections. However, as extrusion temperatures increase, processing costs also increase, and the range of shapes and section sizes that can be obtained becomes narrower.

While many asymmetrical shapes are produced, symmetry is the most important factor in determining extrudability. Adjacent sections should be as nearly equal as possible to permit uniform metal flow through the die. The length of their protruding legs should not exceed 10 times their thickness.

The size and weight of extruded shapes are limited by the section configuration and properties of the material extruded. The maximum size that can be extruded on a press of a given capacity is determined by the "circumscribing circle," which is defined as the smallest diameter circle that will enclose the shape. This diameter controls the die size, which in turn is limited by the press size. For instance, the larger presses are generally capable of extruding aluminum shapes with a 25-in.-diam circumscribing circle and steel and titanium shapes with about 22-in.-diam circle.

The minimum cross-sectional area and minimum thickness that can be extruded on a given size press are dependent on the properties of the material, the extrusion ratio (ratio of the cross-sectional area of the billet to the extruded section), and the complexity of shape. As a rule thicker sections are required with increased section size.

The following table gives the approximate minimum cross section and minimum thickness of some commonly extruded metals.

Material	Minimum Cross Section (sq in.)	Minimum Thickness (in.)
Carbon and alloy steels	0.40	0.120
Stainless steels	0.45–0.70	0.120–0.187
Titanium	0.50	0.150
Aluminum	<0.40	0.040
Magnesium	<0.40	0.040

Extruded shapes minimize and sometimes eliminate the need for machining; however, they do not have the dimensional accuracy of machined parts. Smooth surfaces with finishes better than 30 μin. rms are attainable in magnesium and aluminum; an extruded finish of 125 μin. rms is generally obtained with most steels and titanium alloys. Minimum corner and fillet radii of $\frac{1}{64}$ in. are preferred for aluminum and magnesium alloys; while for steel, minimum corner radii of 0.030 in. and fillet radii of 0.125 in. are typical.

Extrusion of Tubes: In tube extrusion, the metal passes through a die, which determines its outer diameter, and around a central mandrel, which determines its inner diameter. Either solid or hollow billets may be used, with the solid billet being used most often. When a solid billet is extruded, the mandrel must pierce the billet by pushing axially through it before the metal can pass through the annular gap between the die and the mandrel. Special presses are used in tube extrusion to increase the output and improve the quality compared to what is obtained using ordinary extrusion presses. These special hydraulic presses independently control ram and mandrel positioning and movement.

Flame Spraying Process

In this process, the forerunner of which was called the metal spraying process, metals, alloys, ceramics, and cermets are deposited on metallic or other surfaces. The object may be to build up worn or undersize parts, provide wear-resisting or corrosion-resisting surfaces, correct defective castings, etc.

Different types of equipment are available that provide the means of depositing the coatings on the surfaces. In one, wire is fed automatically through the nozzle of the spray gun; then a combustible gas, oxygen and compressed air serve to melt and blow the atomized metal against the surface to be coated. The gas usually used is acetylene but other gases may be used. Any desired thickness of metal may be deposited and the metals include steels, ranging from low to high carbon content, various brass and bronze compositions, babbitt metal, tin, zinc, lead, nickel, copper, and aluminum. The movement of the spray gun, in covering a given surface, is controlled either mechanically or by hand. In enlarging worn or undersize shafts, spindles, etc., it is common practice to clamp the gun in a lathe toolholder and use the feed mechanism to traverse the gun at a uniform rate while the metal is being deposited upon the rotating workpiece. The spraying operation may be followed by machining or grinding to obtain a more precise dimension.

Some typical production applications using the wire process are the coating of automotive exhaust valves, refinishing of transfer ink rollers for the printing industry, and the rebuilding of worn truck clutch plates. Other production applications include the metallizing of glass meter box windows, the spraying of aluminum onto cloth gauze to produce electrolytic condenser plates, and the spraying of zinc or copper for coating ceramic insulators.

With another type of equipment, metal, refractory, and ceramic powder are used instead of wire. Ordinarily this equipment employs the use of two gases, oxygen and a fuel gas. The fuel gas is usually acetylene but in some instances hydrogen may be used. When hand-held, a small reservoir supplies the powder to the equipment but a larger reservoir is used for lathe-mounted equipment or for large-scale production work. The four basic types of coating powders used with this equipment are ceramics, oxidation-resistant metals and alloys, self-bonding alloys, and alloys for fused coatings. These powders are used to produce wear-resistant, corrosion-resistant, heat-resistant, and electrically conductive coatings.

Still other equipment employs the use of plasma flame with which vapors of materials are raised to a higher energy level than the ordinary gaseous state. Its use raises the temperature ceiling and provides a controlled atmosphere by permitting employment of an inert or chemically inactive gas so that chemical action, such as oxidation, during the heating and application of the spray material can be controlled. The temperatures that can be obtained with commercially available plasma equipment often exceed 30,000 degrees F, but for most plasma flame spray processes the temperature range of from 12,000 to 20,000 degrees F is optimum. Plasma flame spray materials include alumina, zirconia, tungsten, molybdenum, tantalum, copper, aluminum, carbides, and nickel-base alloys.

Regardless of the equipment used, what is important is the proper preparation of the surface that will receive the sprayed coating. Preparation activities include the degreasing or solvent cleaning of the surface, undercutting of the surface to provide room for the proper coating thickness, abrasive or grit blasting the substrate to provide a roughened surface, grooving (in the case of flat surfaces) or rough threading (in the case of cylindrical work) the surface to be coated, preheating the base metal. Methods of obtaining a bond between the sprayed material and the substrate are: heating the base, roughening the base, or spraying a "self-bonding" material onto a smooth surface; however, heating alone is seldom used in machine element work as the elevated temperatures required to obtain the proper bond causes problems of warpage and surface corrosion.

METAL JOINING, CUTTING, AND SURFACING

Metals may be joined without using fasteners by employing soldering, brazing, and welding. Soldering involves the use of a non-ferrous metal whose melting point is below that of the base metal and in all cases below 800 degrees F. Brazing entails the use of a non-ferrous filler metal with a melting point below that of the base metal but above 800 degrees F. In fusion welding, abutting metal surfaces are made molten, are joined in the molten state and then allowed to cool. The use of a filler metal and the application of pressure are considered to be optional in the practice of fusion welding.

Soldering

Soldering employs lead- or tin-base alloys with melting points below 800 degrees F and is commonly referred to as soft solders. Use of hard solders, silver solders and spelter solders which have silver, copper, or nickel bases and have melting points above 800 degrees F is known as brazing. Soldering is used to provide a convenient joint that does not require any great mechanical strength. It is used in a great many instances in combination with mechanical staking, crimping or folding; the solder being used only to seal against leakage or to assure electrical contact. The accompanying table gives some of the properties and uses of various solders that are generally available.

Forms Available. — Soft solders can be obtained in bar, cake, wire, pig, slab, ingot, ribbon, segment, powder, and foil-form for various uses to which they are put. In bar form they are commonly used for hand soldering. The pigs, ingots, and slabs are used in operations that employ melting kettles. The ribbon, segment, powder, and foil forms are used for special applications and the cake form is used for wiping. Wire forms are either solid or they contain acid or rosin cores for fluxing. These wire forms, both solid and core-containing, are used in hand and automatic machine applications. Prealloyed powders, suspended in a fluxing medium, are frequently applied by brush and upon heating, consistently wet the solderable surfaces to produce a satisfactory joint.

Fluxes for Soldering. — The surfaces of the metals being joined in the soldering operation must be clean in order to obtain an efficient joint. Fluxes clean the surfaces of the metal in the joint area by removing the oxide coating present, keep the area clean by preventing formation of oxide films, and lower the surface tension of the solder thereby increasing its wetting properties. Rosin, tallow, and stearin are mild fluxes which prevent oxidation but are not too effective in removing oxides present. Rosin is used for electrical applications since the residue is non-corrosive and non-conductive. Zinc chloride and ammonium chloride (sal ammoniac), used separately or in combination, are common fluxes that remove oxide films readily. The residue from these fluxes may in time cause trouble, due to their corrosive effects, if they are not removed or neutralized. Washing with water containing about 5 ounces of sodium citrate (for non-ferrous soldering) or 1 ounce of trisodium phosphate (for ferrous and non-ferrous soldering) per gallon followed by a clear water rinse or washing with commercial water-soluble detergents are methods of inactivating and removing this residue.

Methods of Application. — Solder is applied using a soldering iron, a torch, a solder bath, electric induction or resistance heating, a stream of hot neutral gas or by wiping. Clean surfaces which are hot enough to melt the solder being applied or accept molten solder are necessary to obtain a good clean bond. Parts being soldered should be free of oxides, dirt, oil, and scale. Scraping and the use of abrasives as well as fluxes are resorted to for preparing surfaces for soldering. The

Properties of Soft Solder Alloys (Appendix, ASTM:B 32-70)

Nominal Composition,* Per Cent				Specific Gravity†	Melting Ranges,‡ Degrees Fahrenheit		Uses
Sn	Pb	Sb	Ag		Solidus	Liquidus	
70	30	...	...	8.32	361	378	For coating metals.
63	37	...	...	8.40	361	361	As lowest melting solder for dip and hand soldering methods.
60	40	...	...	8.65	361	374	"Fine Solder." For general purposes, but particularly where the temperature requirements are critical.
50	50	...	...	8.85	361	421	For general purposes. Most popular of all.
45	55	...	...	8.97	361	441	For automobile radiator cores and roofing seams.
40	60	...	...	9.30	361	460	Wiping solder for joining lead pipes and cable sheaths. For automobile radiator cores and heating units.
35	65	...	...	9.50	361	477	General purpose and wiping solder.
30	70	...	...	9.70	361	491	For machine and torch soldering.
25	75	...	...	10.00	361	511	For machine and torch soldering.
20	80	...	...	10.20	361	531	For coating and joining metals. For filling dents or seams in automobile bodies.
15	85	...	...	10.50	440§	550	For coating and joining metals.
10	90	...	...	10.80	514§	570	For coating and joining metals.
5	95	...	...	11.30	518	594	For coating and joining metals.
40	58	2	...	9.23	365	448	Same uses as (50–50) tin-lead but not recommended for use on galvanized iron.
35	63.2	1.8	...	9.44	365	470	For wiping and all uses except on galvanized iron.
30	68.4	1.6	...	9.65	364	482	For torch soldering or machine soldering, except on galvanized iron.
25	73.7	1.3	...	9.96	364	504	For torch and maching soldering, except on galvanized iron.
20	79	1	...	10.17	363	517	For machine soldering and coating of metals, tipping, and like uses, but not recommended for use on galvanized iron.
95	...	5	...	7.25	452	464	For joints on copper in electrical, plumbing and heating work.
...	97.5	...	2.5	11.35	579	579	For use on copper, brass, and similar metals with torch heating. Not recommended in humid environments due to its known susceptibility to corrosion.
1	97.5	...	1.5	11.28	588	588	For use on copper, brass, and similar metals with torch heating.

* Abbreviations of alloying elements are as follows: Sn, tin; Pb, lead; Sb, antimony; and Ag, silver.

† The specific gravity multiplied by 0.0361 equals the density in pounds per cubic inch.

‡ The alloys are completely solid below the lower point given, designated "solidus," and completely liquid above the higher point given, designated "liquidus." In the range of temperatures between these two points the alloys are partly solid and partly liquid.

§ For some engineering design purposes, it is well to consider these alloys as having practically no mechanical strength above 360 degrees F.

procedures followed in soldering aluminum, magnesium and stainless steel differ somewhat from conventional soldering techniques and are indicated in the material which follows:

Soldering Aluminum: Two properties of aluminum which tend to make it more difficult to solder are its high thermal conductivity and the tenacity of its ever-present oxide film. Aluminum soldering is performed in a temperature range of from 550 to 770 degrees F, compared to 375 to 400 degrees F temperature range for ordinary metals, because of the metal's high thermal conductivity. Two methods can be used, one using flux and one using abrasion. The method employing flux is most widely used and is known as flow soldering. In this method flux dissolves the aluminum oxide and keeps it from re-forming. The flux should be fluid at soldering temperatures so that the solder can displace it in the joint. In the friction method the oxide film is mechanically abraded with a soldering iron, wire brush, or multi-toothed tool while being covered with molten solder. The molten solder keeps the oxygen in the atmosphere from reacting with the newly-exposed aluminum surface; thus wetting of the surface can take place.

The alloys that are used in soldering aluminum generally contain from 50 to 75 per cent tin with the remainder, zinc. The following aluminum alloys are listed in order of ease of soldering: commercial and high-purity aluminum, wrought alloys containing not more than 1 per cent manganese or magnesium, and finally the heat-treatable alloys which are the most difficult. Cast and forged aluminum parts are not generally soldered.

Soldering Magnesium: Magnesium is not ordinarily soldered to itself or other metals. Soldering is generally used for filling small surface defects, voids or dents in castings or sheets where the soldered area is not to be subjected to any load. Two solders can be used: one with a composition of 60 per cent cadmium, 30 per cent zinc, and 10 per cent tin has a melting point of 315 degrees F.; the other has a melting point of 500 degrees F. and has a nominal composition of 90 per cent cadmium and 10 per cent zinc.

The surfaces to be soldered are cleaned to a bright metallic luster by abrasive methods before soldering. The parts are preheated with a torch to the approximate melting temperature of the solder being used. The solder is applied and the surface under the molten solder is rubbed vigorously with a sharp pointed tool or wire brush. This action results in the wetting of the magnesium surface. To completely wet the surface, the solder is kept molten and the rubbing action continued. The use of flux is not recommended.

Soldering Stainless Steel: Stainless steel is somewhat more difficult to solder than other common metals. This is true because of a tightly adhering oxide film on the surface of the metal and because of its low thermal conductivity. The surface of the stainless steel must be thoroughly cleaned. This can be done by abrasion or by clean white pickling with acid. Muriatic (hydrochloric) acid saturated with zinc or combinations of this mixture and 25 per cent additional muriatic acid, or 10 per cent additional acetic acid, or 10 to 20 per cent additional water solution of orthophosphoric acid may all be used as fluxes for soldering stainless steel. Tin-lead solder can be used successfully. Because of the low thermal conductivity of stainless steel, a large soldering iron is needed to bring the surfaces to the proper temperature. The proper temperature is reached when the solder flows freely into the area of the joint. Removal of the corrosive flux is important in order to prevent joint failure. Soap and water or a suitable commercial detergent may be used to remove the flux residue.

Ultrasonic Fluxless Soldering. — This more recently introduced method of soldering makes use of ultrasonic vibrations which facilitates the penetration of surface films by the molten solder thus eliminating the need for flux. The equip-

ment offered by one manufacturer consists of an ultrasonic generator, ultrasonic soldering head which includes a transducer coupling, soldering tip, tip heater, and heating platen. Metals that can be soldered by this method include aluminum, copper, brass, silver, magnesium, germanium, and silicon.

Brazing

Brazing is a metal joining process which uses a non-ferrous filler metal with a melting point below that of the base metals but above 800 degrees F. The filler metal wets the base metal when molten in a manner similar to that of a solder and its base metal. There is a slight diffusion of the filler metal into the hot, solid base metal or a surface alloying of the base and filler metal. The molten metal flows between the close-fitting metals because of capillary forces.

Filler Metals for Brazing Applications. — Brazing filler metals have melting points that are lower than those of the base metals being joined and have the ability when molten to flow readily into closely fitted surfaces by capillary action. The commonly used brazing metals may be considered as grouped into the seven standard classifications shown in Table 1. These are aluminum-silicon; copper-phosphorus; silver; nickel; copper and copper-zinc; magnesium; and precious metals.

The solidus and liquidus are given in the table instead of the melting and flow points in order to avoid confusion. The solidus is the highest temperature at which the metal is completely solid or in other words the temperature above which the melting starts. The liquidus is the lowest temperature at which the metal is completely liquid, that is, the temperature below which the solidification starts.

Fluxes for Brazing. — In order to obtain a sound joint the surfaces in and adjacent to the joint must be free from dirt, oil, and oxides or other foreign matter at the time of brazing. Cleaning may be achieved by chemical or mechanical means. Some of the mechanical means employed are filing, grinding, scratch brushing and machining. The chemical means include the use of trisodium phosphate, carbon tetrachloride, and trichloroethylene for removing oils and greases.

Fluxes are used mainly to prevent the formation of oxides and to remove any oxides on the base and filler metals. They also promote free flow of the filler metal during the course of the brazing operation. They are made available in the following forms: powders, pastes or solutions, gases or vapors, and as coatings on the brazing rods. In the powder form a flux can be sprinkled along the joint, provided that the joint has been preheated sufficiently to permit the sprinkled flux to adhere and not be blown away by the torch flame during brazing. A thin paste or solution is easily applied and when spread on evenly, with no bare spots, gives a very satisfactory flux coating. Gases or vapors are used in controlled atmosphere furnace brazing where large amounts of assemblies are mass-brazed. Coatings on the brazing rods protect the filler metal from becoming oxidized and eliminate the need for dipping rods into the flux but it is recommended that flux be applied to the base metal since it may become oxidized in the heating operation. No matter which flux is used, however, it performs its task only if it is chemically active at the brazing temperature.

Chemical compounds incorporated into brazing fluxes include borates (sodium, potassium, lithium, etc.), fused borax, fluoborates (potassium, sodium, etc.), fluorides (sodium, potassium, lithium, etc.), chlorides (sodium, potassium, lithium), acids (boric, calcined boric acid), alkalies (potassium hydroxide, sodium hydroxide), wetting agents, and water (either as water of crystallization or as an addition for paste fluxes). The accompanying Table 2 provides a guide which will aid in the selection of brazing fluxes that are available commercially.

Methods of Steadying Work for Brazing. — Pieces to be joined by brazing after being properly jointed may be held in a stable position by means of clamping devices, spot welds, or mechanical means such as crimping, staking or spinning. When using clamping devices care must be taken to avoid the use of devices containing springs for applying pressure because springs tend to lose their properties under the influence of heat. Care must also be taken to be sure that the clamping devices are no larger than is necessary for strength considerations, since a large metal mass in contact with the base metal near the brazing area would tend to conduct heat away from the area too quickly and result in an inefficient braze. Thin sections that are to be brazed are frequently held together by spot welds. It must be remembered that these spot welds may interfere with the flow of the molten brazing alloy and appropriate steps taken to be sure that the alloy is placed where it can flow into all portions of the joint.

Methods of Supplying Heat for Brazing. — The methods of supplying heat for brazing form the basis of the classification of the different brazing methods and are given as follows:

Torch or Blowpipe Brazing: Air-gas, oxy-acetylene, air-acetylene and oxy-other fuel gas blowpipes are used to bring the areas of the joint and the filler material to the proper heat of brazing. The flames should generally be neutral or slightly reducing but in some instances some types of bronze welding required a slightly oxidizing flame.

Dip Brazing: Baths of molten alloy covered with flux or baths of molten salts are used. The parts, properly jigged, are dipped into the baths for brazing. In the case of the salt bath the filler metal has previously been inserted between the parts being joined or if in the form of wire has been wrapped around the area of the joint; flowing into the joint after immersion in the bath because of capillary action.

Furnace Brazing: Furnaces that are heated electrically or by gas or oil with auxiliary equipment that maintains a reducing or protective atmosphere and controlled temperatures therein are used for brazing a large number of units, usually without flux.

Resistance Brazing: Heat is supplied by means of hot or incandescent electrodes. The heat is produced by the resistance of the electrodes to the flow of electricity and the filler metal is frequently used as an insert between the parts being joined.

Induction Brazing: Parts to be joined are heated by being placed near a coil carrying an electric current. Eddy current losses of the induced electric current are dissipated in the form of heat. This method is both quick and clean.

Vacuum Furnace Brazing: Cold-wall vacuum furnaces, with electrical-resistance radiant heaters, and pumping systems capable of evacuating a conditioned chamber to moderate vacuum (about 0.01 micron) in five minutes are recommended for vacuum brazing. Metals commonly brazed in vacuum are the stainless steels, heat-resistant alloys, titanium, refractory metals, and aluminum. Fluxes and filler metals containing alloying elements with low boiling points or high vapor pressure are not used.

Brazing High-speed Steel Tips to Carbon Steel Shanks. — A method which is used in a large manufacturing plant is as follows: A seat is formed in the tool shank to receive the tip. A welding compound or flux is used in welding the tip to the shank. The flux is placed on the seat of the shank and the tip is then put on top of the flux in the desired position. The tool is placed in the preheating chamber of the furnace and heated to 1550–1600 degrees F., allowing sufficient time for complete penetration of the heat. The tool is then removed from the preheating chamber and the tip is pressed firmly to the seat of the shank to insure a close contact between the two pieces. Then the tool is placed in the main furnace chamber and heated rapidly to a temperature of approximately 2250 to 2400 de-

Table 1. Brazing Filler Metals [Based on Specification and Appendix of AWS (American Welding Society) A5.8-81]

AWS Classification[1]	Nominal Composition,[2] Per Cent						Temperature, Degrees F			Standard Form[3]	Uses
	Ag	Cu	Zn	Al	Ni	Other	Solidus	Liquidus	Brazing Range		
BAlSi-2	...	...	...	92.5	...	Si, 7.5	1070	1135	1110–1150	7	For joining the following aluminum alloys: 1060, EC, 1100, 3003, 3004, 5005, 5050, 6053, 6061, 6062, 6063, 6951 and cast alloys A612 and C612. All of these filler metals are suitable for furnace and dip brazing. BAlSi-3, -4 and -5 are suitable for torch brazing. Used with lap and tee joints rather than butt joints. Joint clearances run from .006 to .025 inch.
BAlSi-3	...	4	...	86	...	Si, 10	970	1085	1060–1120	2, 3, 5	
BAlSi-4	...	...	...	88	...	Si, 12	1070	1080	1080–1120	2, 3, 4, 5	
BAlSi-5	...	...	...	90	...	Si, 10	1070	1095	1090–1120	7	
BAlSi-6	...	...	...	90	...	Si, 7.5 Mg, 2.5	1038	1125	1110–1150	7	BAlSi-6 through -11 are vacuum brazing filler metals. Magnesium is present as an O_2 getter. When used in vacuum, solidus & liquidus temperatures are different from those shown.
BAlSi-7	...	...	...	88.5	...	Si, 10 Mg, 1.5	1038	1105	1090–1120	7	
BAlSi-8	...	...	...	86.5	...	Si, 12 Mg, 1.5	1038	1075	1080–1120	2, 7	
BAlSi-9	...	...	...	87	...	Si, 12 Mg, 0.3	1044	1080	1080–1120	7	
BAlSi-10	...	...	...	86.5	...	Si, 12 Mg, 2.5	1038	1086	1080–1120	2	
BAlSi-11	...	...	...	88.4	...	Si, 10 Mg, 1.5 Bi, 0.1	1038	1105	1090–1120	7	
BCuP-1	...	95	...	...	...	P, 5	1310	1695	1450–1700	1	For joining copper and its alloys with some limited use on silver, tungsten and molybdenum. Not for use on ferrous or nickel-base alloys. Are used for cupro-nickels but caution should be exercised when nickel content is greater than 30 per cent. Suitable for all brazing processes. Lap joints recommended but butt joints may be used. Clearances used range from .001 to .005 inch.
BCuP-2	...	93	...	...	...	P, 7	1310	1460	1350–1550	2, 3, 4	
BCuP-3	...	89	...	...	...	P, 6	1190	1485	1300–1500	2, 3, 4	
BCuP-4	5	87	...	...	...	P, 7	1190	1335	1300–1450	2, 3, 4	
BCuP-5	15	80	...	...	...	P, 5	1190	1475	1300–1500	1, 2, 3, 4	
BCuP-6	2	91	...	...	...	P, 7	1190	1450	1350–1500	2, 3, 4	
BCuP-7	5	88	...	...	...	P, 6.8	1190	1420	1300–1500	2, 3, 4	

For footnotes see end of table.

Table I (*Continued*). **Brazing Filler Metals [Based on Specification and Appendix of AWS (American Welding Society) A5.8-81]**

AWS Classification[1]	Nominal Composition,[2] Per Cent						Temperature, Degrees F			Standard Form[3]	Uses
	Ag	Cu	Zn	Al	Ni	Other	Solidus	Liquidus	Brazing Range		
BAg-1	45	15	16			Cd, 24	1125	1145	1145–1400	1, 2, 4	For joining most ferrous and nonferrous metals except aluminum and magnesium. These filler metals have good brazing properties and are suitable for preplacement in the joint or for manual feeding into the joint. All methods of heating may be used. Lap joints are generally used; however, butt joints may be used. Joint clearances of .002 to .005 inch are recommended. Flux is generally required.
BAg-1a	50	15.5	16.5			Cd, 18	1160	1175	1175–1400	1, 2, 4	
BAg-2	35	26	21			Cd, 18	1125	1295	1295–1550	1, 2, 4, 7	
BAg-2a	30	27	23			Cd, 20	1125	1310	1310–1550	1, 2, 4	
BAg-3	50	15.5	15.5		3	Cd, 16	1170	1270	1270–1500	1, 2, 4, 7	
BAg-4	40	30	28		2		1240	1435	1435–1650	1, 2	
BAg-5	45	30	25				1250	1370	1370–1550	1, 2	
BAg-6	50	34	16				1270	1425	1425–1600	1, 2	
BAg-7	56	22	17			Sn, 5	1145	1205	1205–1400	1, 2	
BAg-8	72	28					1435	1435	1435–1650	1, 2, 4	
BAg-8a	72	27.8				Li, 2	1410	1410	1410–1600	1, 2	
BAg-13	54	40	5		1		1325	1575	1575–1775	1, 2	
BAg-13a	56	42			2		1420	1640	1600–1800	1, 2	
BAg-18	60	30				Sn, 10	1115	1325	1325–1550	1, 2	
BAg-19	92.5	7.3				Li, 2	1435	1635	1610–1800	1, 2	
BAg-20	30	38	32				1250	1410	1410–1600	1, 2, 4	
BAg-21	63	28.5			2.5	Sn, 6	1275	1475	1475–1650	1, 2, 4	
BAg-22	49	16	23		4.5	Mn, 7.5	1260	1290	1290–1525	1, 2, 4, 7	
BAg-23	85					Mn, 15	1760	1780	1780–1900	1, 2, 4	
BAg-24	50	20	28		2		1220	1305	1305–1550	1, 2, 4	
BAg-25	20	40	35			Mn, 5	1360	1455	1455–1555	2, 4	
BAg-26	25	38	33		2	Mn, 2	1305	1475	1475–1600	1, 2	
BAg-27	25	35	26.5			Cd, 13.5	1125	1375	1375–1575	1, 2, 4, 7	
BAg-28	40	30	28			Sn, 2	1200	1310	1310–1550	1, 2, 4	

AWS Classification[1]	Nominal Composition,[2] Per Cent						Temperature, Degrees F			Standard Form[3]	Uses
	Ni	Cu	Cr	B	Si	Other	Solidus	Liquidus	Brazing Range		
BNi-1	74		14	3.5	4	Fe, 4.5	1790	1900	1950–2200	1, 2, 3, 4, 8	For brazing AISI 300 and 400 series stainless steels, and nickel- and cobalt-base alloys. Particularly suited to vacuum systems and vacuum tube applications because of their very low vapor pressure. The limiting element is chromium in those alloys in which it is employed. Special brazing procedures required with filler metal containing manganese.
BNi-2	82.5		7	3	4.5	Fe, 3	1780	1830	1850–2150	1, 2, 3, 4, 8	
BNi-3	91			3	4.5	Fe, 1.5	1800	1900	1850–2150	1, 2, 3, 4, 8	
BNi-4	93.5			1.5	3.5	Fe, 1.5	1800	1950	1850–2150	1, 2, 3, 4, 8	
BNi-5	71		19		10		1975	2075	2100–2200	1, 2, 3, 4, 8	
BNi-6	89					P, 11	1610	1610	1700–1875	1, 2, 3, 4, 8	
BNi-7	77		13			P, 10	1630	1630	1700–1900	1, 2, 3, 4, 8	
BNi-8	65.5	4.5			7	Mn, 23	1800	1850	1850–2000	1, 2, 3, 4, 8	

For footnotes see end of table.

Table I (Concluded). Brazing Filler Metals Based on Specification and Appendix of ASW (American Welding Society) A5.8-81]

AWS Classification[1]	Nominal Composition,[2] Per Cent						Temperature, Degrees F			Standard Form[3]	Uses
	Ni	Cu	Cr	B	Si	Other	Solidus	Liquidus	Brazing Range		
BCu-1	...	100	...	...	...	Ot, 1	1980	1980	2000–2100	1, 2	For joining various ferrous and nonferrous metals. They can also be used with various brazing processes. Avoid overheating the Cu-Zn alloys. Lap and butt joints are commonly used.
BCu-1a	...	99	...	...	...	O, 1.3.5	1980	1980	2000–2100	4	
BCu-2	...	86.5	...	...	...	O, 13.5	1980	1980	2000–2100	6	
RBCuZn-A	...	59	...	...	...	Zn, 41	1630	1650	1670–1750	1, 2, 3	
RBCuZn-C	...	58	...	...	0.1	Zn, 40 Fe, 0.7 Mn, 0.3 Sn, 1	1590	1630	1670–1750	2	
RBCuZn-D	10	48	...	...	0.2	Zn, 42	1690	1715	1720–1800	1, 2, 3	
BCuZn-E	...	50	...	...	...	Zn, 50	1595	1610	1610–1725	1, 2, 3, 4, 5	
BCuZn-F	...	50	...	...	...	Zn, 46.5 Sn, 3.5	1570	1580	1580–1700	1, 2, 3, 4, 5	
BCuZn-G	...	70	...	...	...	Zn, 30	1680	1750	1750–1850	1, 2, 3, 4, 5	
BCuZn-H	...	80	...	...	...	Zn, 20	1770	1830	1830–1950	1, 2, 3, 4, 5	
BMg-1	...	...	...	...	...	4	830	1100	1120–1160	2, 3	BMg-1 is used for joining AZ10A, K1A, and M1A magnesium-base metals.
BAu-1	...	63	...	...	...	Au, 37	1815	1860	1860–2000	1, 2, 4	For brazing of iron, nickel, and cobalt-base metals where resistance to oxidation or corrosion is required. Low rate of interaction with base metal facilitates use on thin base metals. Used with induction, furnace, or resistance heating in a reducing atmosphere or in a vacuum and with no flux. For other applications, a borax-boric acid flux is used.
BAu-2	...	20.5	...	...	...	Au, 79.5	1635	1635	1635–1850	1, 2, 4	
BAu-3	3	62.5	...	...	...	Au, 34.5	1785	1885	1885–1995	1, 2, 4	
BAu-4	18.5	...	...	...	...	Au, 81.5	1740	1740	1740–1840	1, 2, 4	
BAu-5	36	...	...	...	...	Au, 30 Pd, 34	2075	2130	2130–2250	1, 2, 4	
BAu-6	22	...	...	...	...	Au, 70 Pd, 8	1845	1915	1915–2050	1, 2, 4	
BCo-1	17	...	...	...	8	Cr, 19 W, 4 B, 0.8 C, 0.4 Co, 59	2050	2100	2100–2250	1, 3, 4, 8	Generally used for high temperature properties and compatability with cobalt-base metals.

[1] These classifications contain chemical symbols preceded by "B" which stands for brazing filler metal. [2] These are nominal compositions. Trace elements may be present in small amounts and are not shown. Abbreviations used are: Ag, silver; Cu, copper; Zn, zinc; Al, aluminum; Ni, nickel; Ot, other; Si, silicon; P, phosphorus; Cd, cadmium; Sn, tin; Li, lithium; Cr, chromium; B, boron; Fe, iron; O, oxygen; Mg, magnesium; W, tungsten; Pd, palladium; and Au, gold. [3] Numbers specify standard forms as follows: 1, strip; 2, wire; 3, rod; 4, powder; 5, paste; 6, sheet; 7, clad sheet or strip; and 8, transfer tape. [4] Al, 9; Zn, 2; Mg, 89.

Table 2.   Guide to Selection of Brazing Filler Metals and Fluxes

Base Metals Being Brazed	Filler Metals Recommended*	Flux				
		American Welding Society Brazing Flux Type No.	Effective Temperature Range, Degrees F.	Ingredients Contained Therein	Form Supplied In	Method of Use†
All brazeable aluminum alloys	BAlSi	1	700 to 1190	Chlorides Fluorides	Powder	1, 2 3, 4
All brazeable magnesium alloys	BMg	2	900 to 1200	Chlorides Fluorides	Powder	3, 4
Alloys such as aluminum-bronze; aluminum-brass containing additions of aluminum of 0.5 per cent or more	BCuZn BCuP	4§	1050 to 1800	Chlorides Fluorides Borates Wetting agent	Paste or Powder	1, 2, 3
Titanium and zirconium in base alloys	BAg	6	700 to 1600	Chlorides Fluorides Wetting agent	Paste or Powder	1, 2, 3
Any other brazeable alloys not listed above	All brazing filler metals except BAlSi and BMg	3	700 to 2000	Boric acid Borates Fluorides Fluoborates (must contain fluorine compound) Wetting agent	Paste, Powder, or Liquid	1, 2, 3
	All brazing filler metals except BAlSi, BMg, and BAg 1 through BAg 7	5	1000 to 2200	Borax Boric acid Borates Wetting agent No fluorine in any form	Paste, Powder, or Liquid	1, 2, 3

* Abbreviations used in this column are as follows: B, brazing filler metal; Al, aluminum; Si, silicon; Mg, magnesium; Cu, copper; Zn, zinc; P, phosphorus; and Ag, silver.
† Explanation of numbering system used is as follows: 1 — dry powder is sprinkled in joint region; 2 — heated metal filler rod is dipped into powder or paste; 3 — flux is mixed with alcohol, water, monochlorobenzene, etc. to form a paste or slurry; 4 — flux is used molten in a bath.
§ Types 1 and 3 fluxes, alone or in combination, may be used with some of these base metals also.

grees F., depending upon the kind of material used for the tip and its hardening requirements. The tool is next removed from the furnace and sufficient pressure applied to the tip to insure perfect cohesion. The press used for this purpose should be equipped with a pivoted pressure shoe and this shoe must be preheated to prevent cracking of the tip.

The hardening is accomplished at the same time as the tipping operation when tools are tipped with low-cobalt high-speed and high-cobalt high-speed steel. Tools that cannot be ground after hardening are often heated in barium chloride or some similar salt bath. After the pressing and welding operation, the tool is cooled to room temperature under an air blast or quenched in oil. It is advisable to maintain the oil quenching bath at a temperature of 150 to 200 degrees F. After the tipping operation, the tool should be reheated uniformly in an open furnace to a temperature of 1050 to 1150 degrees F. and allowed to cool in air. The hardened tools should have a minimum Rockwell C hardness of 63. Tips are cut from bar stock according to dimensions on standard detail drawings. The material generally used for the shanks of tools tipped with the cutting materials regularly employed contains 0.50 to 0.63 per cent carbon; 0.60 to 0.90 per cent manganese; 0.04 per cent phosphorus; and 0.15 per cent silicon.

Brazing Symbol Application. — According to American National Standard ANSI/ AWS 2.4-79 all symbols used for brazing are also used for welding with the exception of the symbol for a scarf joint (see groove symbols in Fig. 1). "Basic weld symbols" and the top figure of the illustration on the following page show typical applications of brazing symbols. The second, third, and fourth figures from the top show how joint clearances are indicated. If no special joint preparation is required, only the arrow is used with the brazing process indicated in the tail.

Brazing Carbide Tips to Steel Shanks. — Sintered carbide tips or blanks may be attached to steel shanks by brazing. Shanks usually are made of low-alloy steels having carbon contents ranging from 0.40 to 0.60 per cent. One prominent manufacturer of carbide tools recommends the following shank steels in the order listed: (1) High-silicon steel containing approximately 0.55 carbon, 0.85 manganese, 0.30 vanadium, 2.10 silicon, 0.25 chromium, and maximum sulfur and phosphorus contents of 0.025; (2) S.A.E. 2340 steel; (3) any low-alloy steel having 0.40 to 0.60 carbon.

Shank Preparation: The carbide tip usually is inserted into a milled recess or seat, but some prefer to omit the recess and braze the tip to the top of the shank. When a recess is used, the bottom should be flat to provide a firm even support for the tip. The corner radius of the seat should preferably be somewhat smaller than the radius on the tip to avoid contact and insure support along each side of the recess.

Cleaning: All surfaces to be brazed must be absolutely clean. Surfaces of the tip may be cleaned by grinding lightly or by sand-blasting. Cleaning with carbon tetrachloride is also recommended.

Brazing Materials and Equipment: The brazing metal may be copper, naval brass such as Tobin bronze, or silver solder. A flux such as borax is used to protect the clean surfaces and prevent oxidation. Heating may be done either in a furnace or by means of an oxy-acetylene torch or an oxy-hydrogen torch. Copper brazing usually is done in a furnace, although an oxy-hydrogen torch with excess hydrogen is sometimes used. An oxy-acetylene torch usually is employed for silver brazing or soldering.

Brazing Procedure: One method of brazing with a torch is to first place a thin sheet material, such as copper foil, around and beneath the carbide tip, the top of which is covered with flux. The flame is applied to the under side of the tool shank, and, when the materials melt, the tip is pressed firmly into its seat with

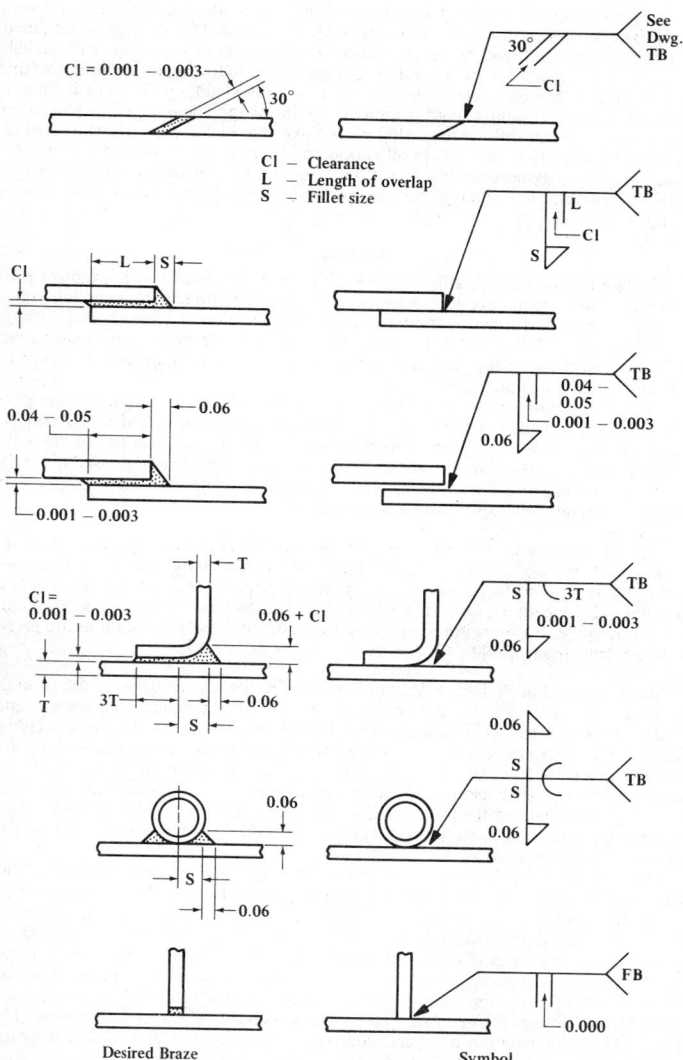

Typical Applications of Standard Brazing Symbols

tongs or with the end of a rod. If the brazing material is in the form of wire or rod, this may be used to coat or tin the surfaces of the recess after the flux melts and runs freely. The tip is then inserted, flux is applied to the top, and the heating is continued until the coatings melt and run freely. A firm which supplies carbide tips with nickel-coated surfaces ready for silver soldering suggests the following procedure: The tip, after coating with flux, is placed in the recess and the shank end is heated. Then a small piece of silver solder, having a melting point of 1325 degrees F., is placed on top of the tip. When this solder melts, it runs over the nickel-coated surfaces while the tip is held firmly in its seat. In all carbide tip brazing, the brazed tool should be cooled slowly to avoid cracking due to unequal contraction between the steel and carbide. To insure slow cooling, the tool may be buried in powdered charcoal, graphite, mica, or lime.

Welding

Welding or more specifically fusion welding may be defined as a group of processes in which metals are joined by bringing abutting surfaces to a molten state. Welding may be performed with or without the application of pressure and with or without the use of a filler metal. Heat may be provided by an electric arc, a gas flame, a chemical reaction, or the electrical resistance of the metals being joined to current passed through the joint.

The use of welding offers many advantages. It lends flexibility to machine designs. It facilitates lightweight construction and permits the use of standard rolled shapes. Standard shapes may be rolled or formed and joined with welding to cut costs by reducing material, machining, and finishing. As a joining process it is used not only for fabrication but also for the repair and maintenance of broken and worn parts. It requires relatively low capital investment costs.

Welding Processes. — Welding processes may be classified into the following broad categories: arc welding, gas welding, resistance welding, thermit welding, induction welding, and forge welding. Each broad category may be further modified into other classifications, i.e., shielded metal-arc, unshielded carbon-arc, spot, seam, oxy-acetylene, etc. Descriptions and uses according to these modifications are given in the accompanying Table 1.

Materials Used in Welds. — Materials used in welds are available in the form of electrodes or welding rods. Electrodes are used with arc-welding equipment and welding rods with gas welding equipment. Coiled wire for use in automatic welding machinery is also employed as a weld material. Electrodes and welding rods are available in a great many different chemical compositions which satisfy practically all the welding situations encountered. Although the electrodes are available as proprietary alloys a number of them have been classified jointly by the American Welding Society and American Society for Testing and Materials and assigned AWS-ASTM classification numbers or letters.

The AWS-ASTM classification numbers or letters indicate one or more of the following: the type of covering or coating on the electrode; the chemical composition of the filler metal, either in rod or electrode form, or as deposited; the minimum tensile and yield strengths and ductility of the filler metal as deposited; and a range of standard sizes (diameter of core) of a standard length or a range of standard lengths. Different coatings applied to electrodes provide shielding to protect the arc from the atmosphere, and determine operating characteristics of the electrode. Some coatings contain iron powder that deposits as filler metal with the core wire in the weld. The AWS-ASTM classification numbers, compositions, properties and related information are given in specifications published by the American Welding Society and American Society for Testing and Materials.

Table 1. Welding Processes — Designation, Description, and Use

Designation	Description	Uses
Induction Welding	Heat result of electric current induced in pieces to be joined. Workpieces are placed in or very close to inductor coil.	For welding steel pieces in thicknesses of from 0.010 inch to ⅛ inch and in diameters up to 3 inches by an economical mass-production method.
Arc Welding — Shielded Metal Electrode — Shielded Stud	Heat result of an electric arc between a metal stud and work. At proper temperature stud is brought against work under nominal pressure of stud welding gun. Shielding provided by ceramic ferrule surrounding the stud which reduces chance of oxidation.	For welding studs or fasteners onto various items such as decking, machine bases, and engine blocks. Used in the shipbuilding, railroad, building construction, equipment, and automotive industries. Both manual and machine welding performed.
Submerged Arc	Heat result of an electric arc or arcs maintained between bare or lightly coated electrodes and work. Shielding provided by decomposition of fusible granular material. Pressure not used. Filler metal obtained from electrode or supplementary welding rod.	Most commonly used for welding low-carbon and high-strength, low-alloy steels. Straight chromium and austenitic chromium-nickel steels are weldable in all thicknesses. Higher alloy air-hardening steels weldable using special techniques. Welding done automatically.
Inert Gas Metal-Arc	Heat result of an electric arc struck between an electrode (usually tungsten) and work. Shielding provided by inert gases (helium or argon). Use of pressure and filler metal optional.	For welding sheet steel, aluminum, and magnesium. Also for welding deoxidized copper, dissimilar metals, and for surfacing. Both manual and machine welding are performed.
Atomic Hydrogen	Heat result of an electric arc maintained between two electrodes in an atmosphere of hydrogen which provides shielding. Use of pressure and filler metal optional.	For welding practically all non-ferrous and ferrous metals and alloys. Flux required for aluminum, copper, copper alloys and some stainless steels. For repairing steel molds and dies. Production welding of tubing and fabrication of thin stock. Both manual and automatic welding are performed.
Shielded Metal-Arc	Heat result of electric arc struck between flux-coated electrode and base metal. Shielding provided by gases of decomposed flux coating. No pressure used. Filler metal supplied by consumable electrode.	For welding all ferrous metals particularly cast-iron, low-carbon, medium-carbon, high-carbon, low-alloy high-tensile, stainless, corrosion-resistant and copper bearing steels; and also the following non-ferrous metals: copper alloys, aluminum, nickel, nickel alloys, and bronze. These welding processes are used very widely, the main fields being structural, transportation, piping, appliance and machinery fabrication. Both manual and machine welding is performed employing these processes.

Table I (Continued). Welding Processes — Designation, Description, and Use

Designation		Description	Uses
Arc Welding— Unshielded Carbon Electrode	Carbon-Arc	Heat result of an electric arc struck between an electrode and work. Use of pressure and filler metal optional.	For welding steel, galvanized steel, copper, brass, bronze and aluminum. Both manual and automatic set-ups are used in these welding processes. All of these processes are also used for preheating small areas which are to be welded later, as well as for cutting metals. The twin-arc process may be used for brazing and soldering copper, galvanized and tinned parts both in light and heavy gages.
	Twin Carbon-Arc	Heat result of an electric arc maintained between two electrodes. Use of pressure and filler metal optional.	
Gas Welding	Air-Acetylene	Heat result of burning of mixture of air and acetylene. Lower flame temperatures achieved than with other gas welding processes. Pressure not used. Use of filler metal optional. Fluxes may be used.	For welding lead in thicknesses up to ¼ inch. Also used for torch brazing employing silver brazing alloys and for soft-soldering electrical connections and copper pipe joints. Manual equipment is used.
	Oxy-Acetylene	Heat result of burning of mixture of oxygen and acetylene. The use of pressure and filler metal optional. Fluxes may be used.	For welding many ferrous and non-ferrous metals. Heat obtained as high as original casting temperature. Metals of considerable thickness may be welded. Used in maintenance and repair. Both manual and machine equipment used.
	Oxy-Hydrogen	Heat result of burning of mixture of oxygen and hydrogen. The use of filler metal is optional. No pressure is used. Lower flame temperature than oxy-acetylene flame.	For low-temperature applications such as welding low-melting point metals, brazing and braze-welding. Aluminum, magnesium and lead are welded using this process. Manual equipment used.
	Gas Pressure	Heat result of burning of mixture of acetylene and oxygen. No filler metal used. Pressure used. Pressure may be applied while heating (closed joint method) or upsetting may be performed after heating (open-joint method).	For welding many steels of varying carbon contents including also low- and high-alloy steels. Non-ferrous alloys such as Monel, nickel-chromium, and copper-silicon alloys are welded by this process. Machine equipment used.
Thermit Welding	Pressure	Heat result of chemical reaction between a metal oxide (usually iron oxide) and aluminum both of which are in a fine-powder form. In the reaction, the aluminum displaces the metal in the oxide to form aluminum oxide. The metal liquefies and is used as a filler metal when a filler metal is used. Use of pressure optional. More non-pressure welding done.	For fabrication and repair of heavy sections of ferrous metals such as heavy machinery; stern frames of ships, sheet steel mill equipment; rail welding for railroads and mines; and locomotive repairs.
	Non-pressure		

Table 1 (*Continued*). **Welding Processes — Designation, Description, and Use**

Designation		Description	Uses
Resistance Welding	Spot	Heat result of electrical resistance offered by workpieces held under pressure between two electrodes. Weld formed immediately between workpieces.	For welding a great many different metals in a great range of different sizes and shapes. For welding landing gears, wing assemblies in the aircraft industry; bodies, cabs, seats in the automotive industry; trusses and partitions in the building construction industry; telephone equipment instruments in the communications industry; etc. All of these welding processes are performed by machine.
	Seam	Heat result of electrical resistance offered by workpieces held under pressure between two circular electrodes. Resulting weld is a continuous series of overlapping spot welds.	
	Projection	Heat result of electrical resistance offered by workpieces held under pressure between two electrodes. Welds are formed at projections, or other raised surfaces on workpieces specified by design.	
	Flash	Heat result of electrical resistance between two abutting surfaces. Pressure applied after heating which causes flashing, upsetting and an expulsion of metal from joint.	For welding sheets and other extended sections end-to-end. Used in high-production fabrication set-ups. Automotive and aircraft products, appliances and refrigerators are flash welded. Machine equipment used.
	Upset	Heat result of electrical resistance between two abutting surfaces. Pressure applied before and maintained during heating period.	For welding small ferrous and non-ferrous strips as well as the welding of longitudinal butt joints in tubing and pipe and transverse butt joints in heavy steel rings. Machine equipment used.
	Percussion	Heat result of an electric arc produced by a rapid discharge of stored electrical energy with pressure percussively applied during or immediately following the discharge.	For welding separate pieces of rod, tube or pipe to one another or to flat surfaces. For joining dissimilar metals not economically welded by flash welding method.
Forge Welding	Hammer	Heat result of external heating by furnace. Workpieces when at proper temperature are joined by blows of a hand or machine hammer.	Largely replaced by other welding methods but now used for some metal manufacturing and hardware applications.
	Die	Heat result of external heating by furnace. When proper temperature is reached, metals being joined are brought together under pressure with a bell or a mandrel and tube rolls.	Used in the fabrication of large diameter water transmission pipes and penstocks.
	Roll	Heat result of external heating by furnace. When proper temperature is reached, metals being joined are brought together under pressure between plate rolls.	For welding cladding metal to steel plate in the making of clad steels.

Table 2. Recommended Processes for Welding[1]

Metals and Alloys		Shielded metal-arc (coated electrode)	Submerged-arc	Atomic hydrogen	Inert-gas tungsten-arc	Inert-gas metal-arc	Flash welding	Seam welding	Spot welding	Gas welding (oxy-acetylene)	Brazing-furnace	Brazing-torch	Thermit
		Recommendation Based on Metal Used											
Steels	Low carbon; SAE 1010, 1020	R	R	S	S	S	R	R	R	R	R	S	S
	Medium carbon; SAE 1030, 1050	R	R	S	S	S	R	S	R	R	R	S	S
	Wrought alloy; SAE 4130, 4340	R	R	S	S	S	R	N	N	R	S	S	N
	Austenitic stainless; AISI 301, 309, 316	R	R	R	R	R	R	R	R	R	R,S	S	N
	Other stainless; AISI 405, 430	R	S	S	S	S	S	S	S	S	S	S	N
	High temperature alloys; 19-9DL, 16-25-6	R	S	S	S	S	S	S	R	S	N	N	N
	Cast iron, gray iron	S	N	N	S	N	N	D	D	R	N	R	S
Non-ferrous Alloys	Aluminum and its alloys	S	N	S	R	R	S	S	R	S	R	R	D
	Nickel and its alloys	R	S	S	R	R	S	S	R	S	S	R	N
	Copper and its alloys	N	N	N	R	R	S	N	S	S	S	R	N
	Magnesium and its alloys	D	D	N	R	S	N	N	S	N	N	N	D
	Silver	N	N	N	R	S	S	N	N	R	S	R	N
	Gold, platinum, iridium	N	N	R	R	S	S	N	S	R	S	R	N
	Titanium and its alloys	D	D	D	R	R	R	S	S	D	S²	S²	D
	Uranium, molybdenum, vanadium, zirconium, tungsten	D	D	N	R	S²	S	S	S	N	N	N	D
Joint Design[3]		Recommendation Based on Joint Used											
Butt Joint	Light section	S	S	R	R	N	N	D	D	R	N	S	D
	Heavy section	R	R	S	S	R	R	D	D	S	N	S	R
Lap Joint	Light section	R	S	S	R	N	D	R	R	R	R	R	D
	Heavy section	R	R	S	S	R	D	R	R	S	R	R	D
Fillet Joint	Light section	R	S	S	R	N	D	D	D	R	R	R	D
	Heavy section	R	R	S	S	R	N	D	D	S	R	R	D
Edge Joint	Light section	N	N	R	R	N	D	R	D	R	D	S	D
	Heavy section	R	S	S	S	S	D	R	D	S	D	S	D
Overlay Welding		R	R	R	R	R	D	D	D	R	N	S	N

[1] From data supplied by John J. Chyle, Director of Welding Research, A. O. Smith Corporation, Milwaukee, WI. Symbols used in the table have the following meanings: R — recommended, S — satisfactory, N — not recommended, D — does not apply.

[2] Questionable as to whether it can be satisfactorily welded.

[3] Light sections include sizes from 0.005 inch up to 0.078 inch; heavy sections include sizes 0.078 inch and larger.

Table 3. Resistance Welding — Ease of Weldability of Some Common Metals and Alloys[1]

Metals and Alloys	Aluminum	Brass	Copper	Galvanized Iron	Inconel	Lead	Magnesium	Molybdenum	Monel	Nichrome	Nickel	Nickel Silver	Phosphor Bronze	Stainless Steel	Steel	Tantalum	Terneplate	Tin Plate	Titanium	Tungsten	Zinc
Aluminum	1	4	5	5	6	5	3	6	6	6	6	6	6	5	5	5	6	5	5	6	5
Brass	4	1	3	5	5	5	5	5	5	5	3	3	2	5	5	4	4	4	5	5	4
Copper	5	3	4	5	5	5	5	5	4	4	3	3	5	5	4	5	5	4	5	5	5
Galvanized Iron	5	5	5	2	5	4	6	6	4	4	4	5	5	3	3	4	3	3	5	5	3
Inconel	6	5	5	5	1	5	5	3	2	3	2	2	5	2	2	3	4	4	4	3	5
Lead	5	5	5	4	5	3	6	6	5	5	5	5	5	5	5	5	5	5	5	5	3
Magnesium	3	5	5	6	5	6	1	3	6	6	6	6	6	5	5	5	6	5	6	5	5
Molybdenum	6	5	5	6	3	6	3	3	3	3	4	4	3	3	3	4	4	4	4	4	5
Monel	6	5	4	4	2	5	6	3	1	3	3	3	4	2	2	3	4	4	4	3	5
Nichrome	6	5	4	4	3	5	6	3	3	1	2	2	4	3	3	4	5	4	4	4	6
Nickel	6	3	4	4	2	5	6	3	3	2	1	2	3	3	3	4	5	4	4	4	6
Nickel Silver	6	3	3	5	2	5	6	4	3	2	2	1	2	4	4	4	4	4	5	4	6
Phosphor Bronze	6	2	3	5	5	5	6	4	4	4	3	2	2	5	5	4	4	4	5	5	4
Stainless Steel	5	5	5	3	2	5	5	3	2	3	3	4	5	1	1	2	3	3	3	3	5
Steel	5	5	5	3	2	5	5	3	2	3	3	4	5	1	1	3	3	3	3	3	5
Tantalum	5	4	4	4	3	5	5	3	3	4	4	4	4	2	1	2	5	4	4	5	5
Terneplate	6	4	5	3	4	5	6	4	5	5	4	4	3	3	5	2	3	5	5	3	3
Tin Plate	5	4	5	3	4	5	5	4	4	5	4	4	3	3	5	3	5	3	5	5	5
Titanium	5	5	4	5	4	5	5	4	4	4	4	5	3	3	4	5	5	5	1	5	5
Tungsten	6	5	5	5	3	5	6	4	3	3	4	4	5	3	3	4	5	5	5	2	5
Zinc	5	4	5	3	5	3	5	5	5	6	6	6	4	5	5	5	3	5	5	3	3

[1] *Key:* 1 — Welds excellently, 2 — Welds well, 3 — Welds fairly well, 4 — Welds poorly, 5 — Welds very poorly, 6 — Welding not practical. These data were made available by Sciaky Bros., Inc., Chicago, IL.

Electrodes contain fluxes in their coatings. Submerged arc fluxes are externally applied granular materials which completely cover and hide the arc. In gas welding, fluxes are used in the form of liquids, slurries or powders. Liquids or slurries may be used as a dip for welding rods or they may be used to paint the surface of the base metal. When powders are used the rod is heated and then dipped into the powder. The powder may also be sprinkled onto the base metal. Inert gases such as helium and argon are used to provide a protective shield for the deposition of the filler metal from the electrode.

Materials That Can Be Welded. — Almost all metallic materials can be welded, some more easily and by more processes than others. The accompanying Table 2 gives a list of common metals and alloys and their relative ease of weldability with respect to many of the available welding processes. It also lists different types of joints and gives the recommended and satisfactory welding methods that may be used in making these joints. In using the table as a guide to the selection of a possible welding process, it is suggested that the recommended welding processes prescribed by a particular type joint be determined and then compared with the welding processes prescribed as recommended or satisfactory for the metals or alloys to be used.

The accompanying Table 3 gives the ease of weldability of many similar and dissimilar metals and alloys employing the resistance welding processes. It must be remembered that these ratings are nominal and inclusive of the whole range of resistance welding processes, i.e., spot, seam, projection, etc., and are therefore just a general indication of weldability. Other factors which have to be considered and which may change the ease of weldability are thickness, plating, treatment (hot or cold), and the shape of the piece itself.

Typical Weldments. — Welded design is used for nearly every type of machine, ranging from steel mill equipment to jigs and fixtures. One advantage in designing for welding is that individual sections of a part can be considered as separate components which are then joined into one piece to form the entire machine part. Design is thereby simplified. Some of the elements which can be considered separately are shown in the following illustrations of examples of welded construction and rolled shapes available for use in weldments.

Checklist of Good Arc-Welding Practices

The checklist of arc-welding practices which follows has been prepared by The Lincoln Electric Company and has been designed to call attention to considerations that affect the quality and cost of arc welding. Anyone using the information in the checklist is less likely to run into difficulties when arc welding and especially those who are novices in the field.

Work. — *Material:* For easiest welding, use steel within the following preferred analysis range: carbon, 0.13 to 0.20 per cent; manganese, 0.40 to 0.60 per cent; silicon, 0.10 per cent max.; sulfur, 0.035 per cent max.; and phosphorus, 0.035 per cent max. Steels with a higher alloy content may require special treatment, such as a preheat, in order to produce sound welds.

Thickness: Exceptionally thin material (20 gage or thinner) and very heavy material (1 inch or thicker) may require special procedures or fixtures.

Accessibility: Joints must be accessible to the operator, if he is to weld them satisfactorily in a reasonable length of time.

Position: Material ³⁄₁₆ inch and thicker is welded fastest and best in the flat position, while lighter material is best welded in a 45-degree downhill position.

Examples of Welded Construction — 1

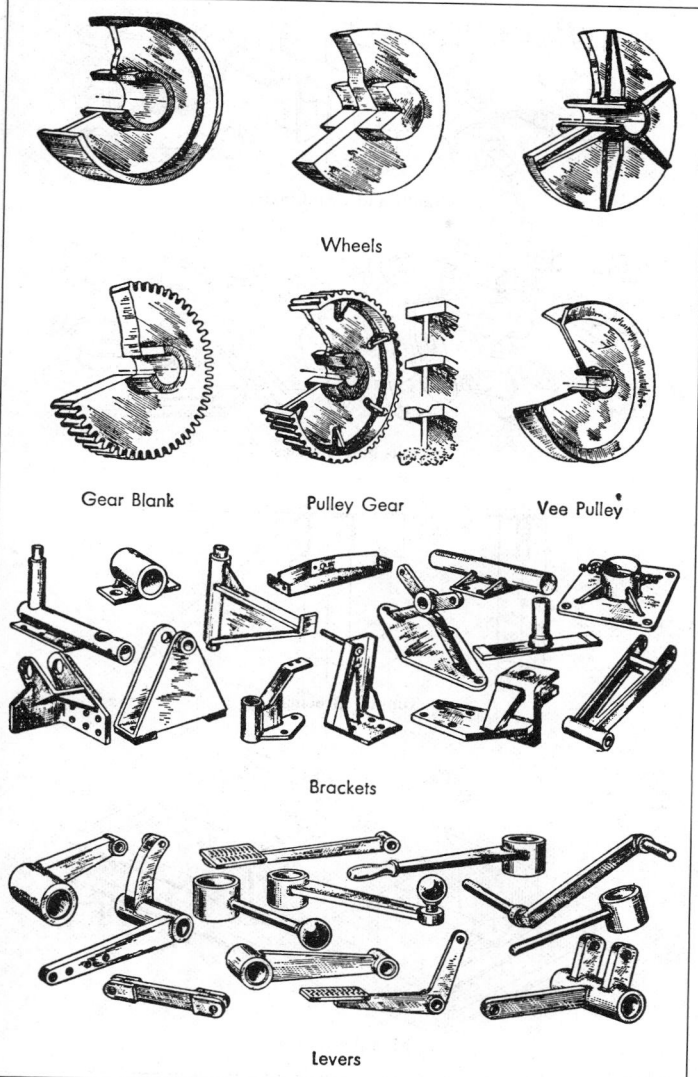

Wheels

Gear Blank   Pulley Gear   Vee Pulley

Brackets

Levers

Examples of Welded Construction — 2

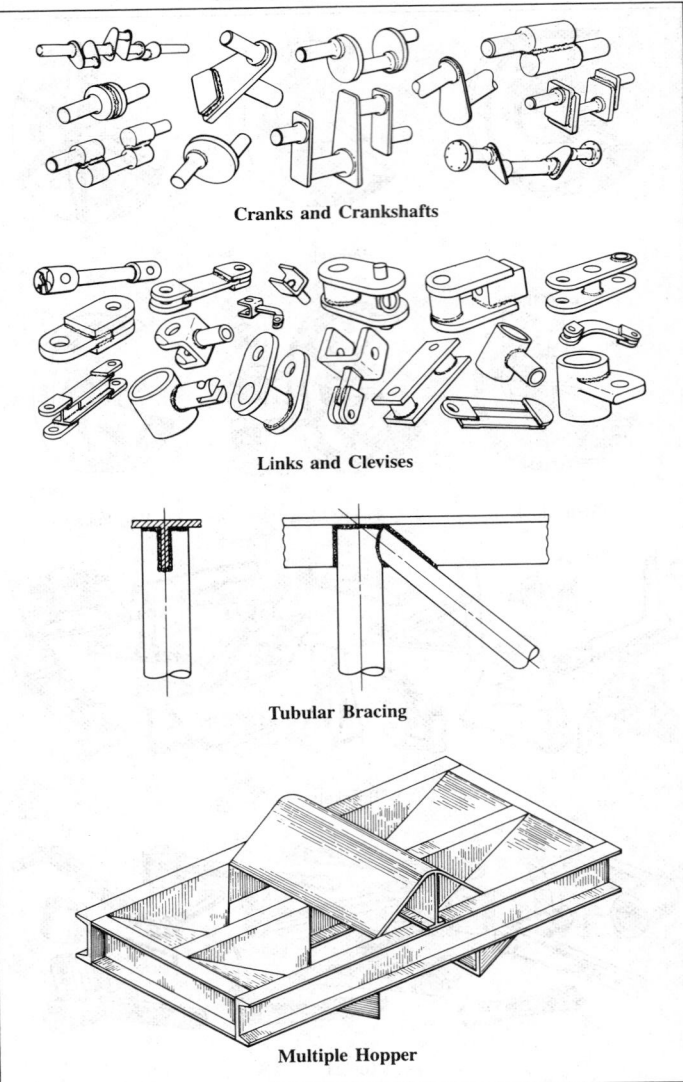

Cranks and Crankshafts

Links and Clevises

Tubular Bracing

Multiple Hopper

Examples of Welded Construction — 3

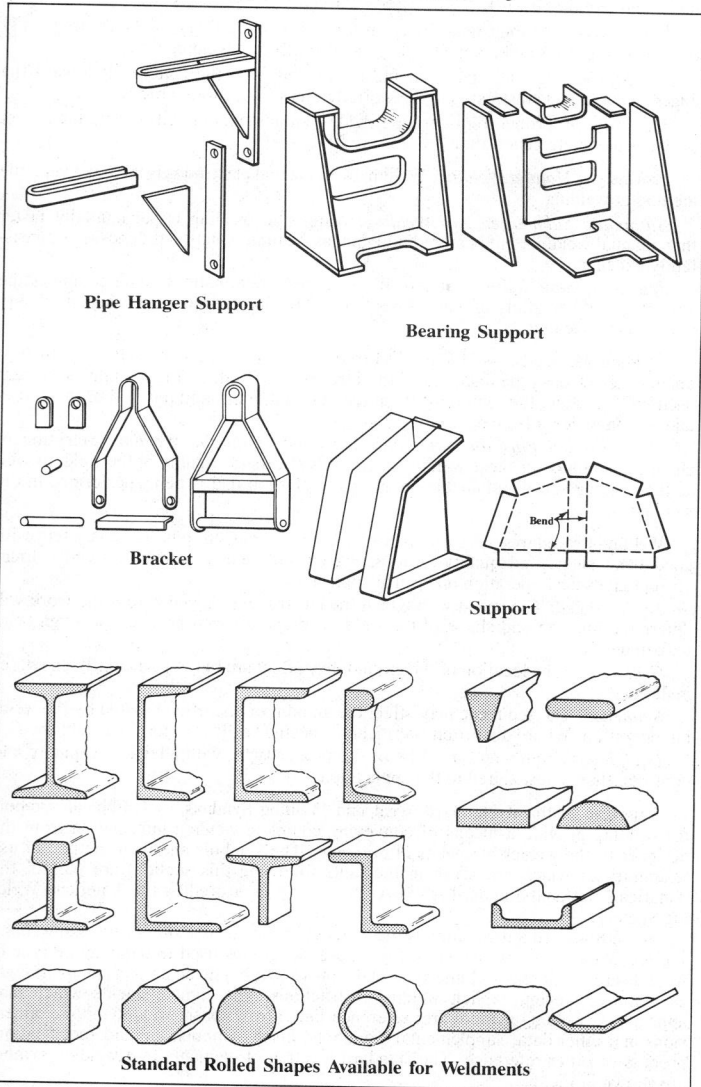

Pipe Hanger Support

Bearing Support

Bracket

Support

Bend

Standard Rolled Shapes Available for Weldments

Fixtures: Proper fixtures will reduce handling time, improve position, and maintain alignment of the work.

Cleanliness: The joint must be clean for best results. Dirty joints are frequently the cause of porous and cracked welds and also increase welding time.

Rigidity: Heavy, rigid parts should be welded from the center out toward the edges to avoid locked stresses in the work that may cause weld cracks.

Joint Design: Proper joint design provides a minimum amount of weld metal, yet provides adequate penetration.

Process. — *Manual Welding:* Welding with manual electrodes is the most versatile method of welding.

Automatic Submerged Arc Welding: Automatic welding is substantially faster than manual welding, but is more limited in application. Quality of deposits is consistently excellent.

Semi-Automatic Submerged Arc Welding: Semi-automatic welding combines the advantages of versatility of manual welding and the high speed and consistent quality of automatic welding.

Welding Materials. — *Manual Electrode:* For most efficient welding with best quality, select the type and size of the electrode, as well as the amount of current used on it, to meet the requirements of the material, joint, and position of the particular job on which it is being used.

Electrode and Flux for Automatic and Semi-Automatic Welding: Selection of electrode and flux for these welding operations affects the quality of the weld, as well as the welding speeds attainable on any particular job that is being performed in the welding shop.

Welding Procedures. — *Current:* Use the highest current possible consistent with good bead shape and quality. Alternating current reduces arc blow, while direct current gives best operation on critical applications.

Travel Speed: The speed with which the electrode is moved across the work will determine the size and shape of the weld. Too fast a speed produces a rough bead with undercut.

Grounding: The location of the ground may affect arc blow, particularly on lighter material.

Sequence: The sequence may affect the amount of distortion created by the welding operation. Where distortion is critical, a specific sequence must be established.

Equipment: Equipment must be of the proper type, with adequate capacity and controls, that is best suited to the application.

American National Standard Weld and Welding Symbols. — Graphical symbols for welding provide a means of conveying complete welding information from the designer to the welder by means of drawings. The symbols and their method of use (examples of which are given in the table following this section) are part of the American National Standard ANSI/AWS A2.4-79 sponsored by the American Welding Society.

In the Standard a distinction is made between the terms *weld symbol* and *welding symbol.* Weld symbols, shown in Fig. 1, are ideographs used to indicate the type of weld desired, whereas welding symbol denotes a symbol made up of as many as eight elements conveying explicit welding instructions. The eight elements which may appear in a welding symbol are: reference line, arrow, basic weld symbols, dimensions and other data, supplementary symbols, finish symbols, tail and specification, process or other reference. The standard location of elements of a welding symbol are shown in Fig. 2.

GROOVE WELD SYMBOLS

Square	‖ Scarf*	V	Bevel	U	J	Flare V	Flare-bevel

OTHER WELD SYMBOLS

Fillet	Plug or slot	Spot or projection	Seam	Back or backing	Surfacing	Flange	
						Edge	Corner

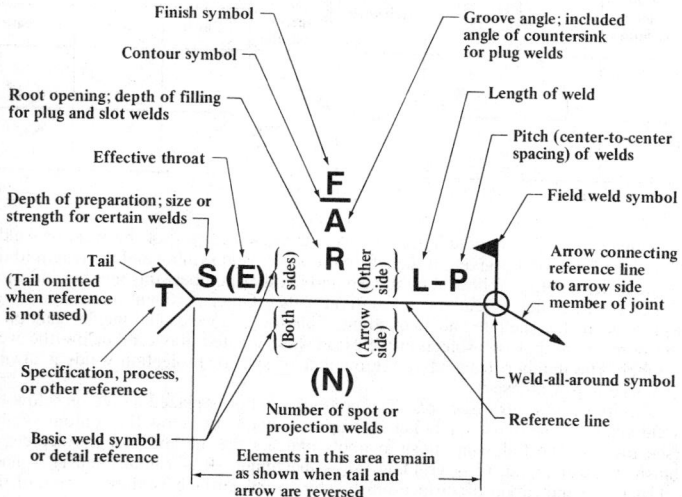

Fig. 1. Basic Weld Symbols†

Fig. 2. Standard Location of Elements of a Welding Symbol

Finish symbol

Contour symbol

Root opening; depth of filling for plug and slot welds

Effective throat

Depth of preparation; size or strength for certain welds

Tail (Tail omitted when reference is not used)

Specification, process, or other reference

Basic weld symbol or detail reference

Groove angle; included angle of countersink for plug welds

Length of weld

Pitch (center-to-center spacing) of welds

Field weld symbol

Arrow connecting reference line to arrow side member of joint

Weld-all-around symbol

Reference line

Number of spot or projection welds

Elements in this area remain as shown when tail and arrow are reversed

T S (E) (sides) R (Other side) L-P

(Both) (Arrow side)

(N)

*This scarf symbol used for brazing only (see page 1213).   † For examples of basic weld symbol applications see pages 1229 to 1233, inclusive.

Reference Line: This is the basis of the welding symbol. All other elements are oriented with respect to this line. The arrow is affixed to one end and a tail, when necessary, is affixed to the other.

Arrow: This connects the reference line to one side of the joint in the case of groove, fillet, flange, and flash or upset welding symbols. This side of the joint is known as the *arrow side* of the joint. The opposite side is known as the *other side* of the joint. In the case of plug, slot, projection, and seam welding symbols, the arrow connects the reference line to the outer surface of one of the members of the joint at the center line of the weld. In this case the member to which the arrow points is the *arrow side* member; the other member is the *other side* member. In the case of bevel and J-groove weld symbols, a two-directional arrow pointing toward a member indicates that the member is to be chamfered.

Basic Weld Symbols: These designate the type of welding to be performed. The basic symbols which are shown in Fig. 1 are placed approximately in the center of the reference line, either above or below it or on both sides of it as shown in Fig. 2. Welds on the arrow side of the joint are shown by placing the weld symbols on the side of the reference line towards the reader (lower side). Welds on the other side of the joint are shown by placing the weld symbols on the side of the reference line away from the reader (upper side).

Supplementary Symbols: These convey additional information relative to the extent of the welding, where the welding is to be performed, and the contour of the weld bead. The "weld-all-around" and "field" symbols are placed at the end of the reference line at the base of the arrow as shown in Fig. 2 and Fig. 3.

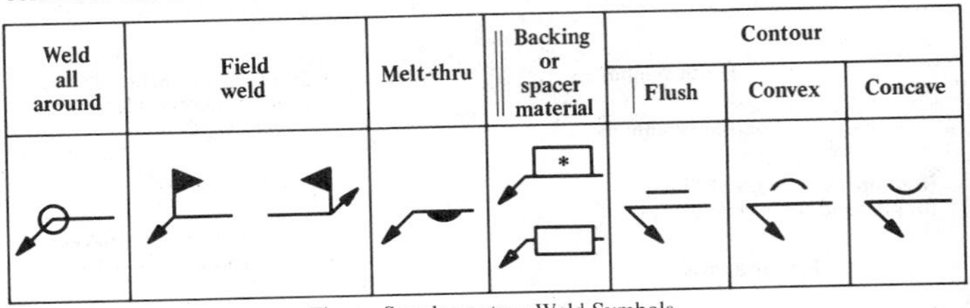

Weld all around	Field weld	Melt-thru	Backing or spacer material	Contour		
				Flush	Convex	Concave
			*			

Fig. 3. Supplementary Weld Symbols

Dimensions: These include the size, length, spacing, etc., of the weld or welds. The size of the weld is given to the left of the basic weld symbol and the length to the right. If the length is followed by a dash and another number, this number indicates the center-to-center spacing of intermittent welds. Other pertinent information such as groove angles, included angle of countersink for plug welds and the designation of the number of spot or projection welds are also located above or below the weld symbol. The number designating the number of spot or projection welds is always enclosed in parentheses.

Contour and Finish Symbols: The contour symbol is placed above or below the weld symbol. The finish symbol always appears above or below the contour symbol (see Fig. 2). The following finish symbols indicate the method, not the degrees of finish: C — chipping, G — grinding, M — machining, R — rolling, and H — hammering. For indication of surface finish refer to the Surface Texture section of this handbook.

Tail: The tail which appears on the end of the reference line opposite to the arrow end is used when a specification, process, or other reference is made in the welding symbol. When no specification, process, or other reference is used with a welding symbol, the tail may be omitted.

Melt-Thru Symbol: The melt-thru symbol is used only where 100 per cent joint or member penetration plus reinforcement is required.

Specification, Process, or Other Designation: These are placed in the tail of the welding symbol and are in accordance with the American National Standard. They do not have to be used if a note is placed on the drawing indicating that the welding is to be done to some specification or that instructions are given elsewhere as to the welding procedure to be used.

Letter Designations: American National Standard letter designations for welding and allied processes are shown in the table on page 1228.

Further Information: For complete information concerning welding specification by the use of standard symbols, reference should be made to American National Standard ANSI/AWS A2.4-79 which may be obtained from either the American National Standards Institute or the American Welding Society listed below.

Welding Codes, Rules, Regulations, and Specifications. — Codes recommending procedures for obtaining specified results in the welding of various structures have been established by societies, institutes, bureaus, and associations, as well as state and federal departments. The latest codes, rules, etc. may be obtained from these agencies, whose names and addresses are listed according to the fields in which their codes, rules, etc., are pertinent.

Pressure Vessels:
American Society of Mechanical Engineers, 345 E. 47th St., N.Y., N.Y. 10017.
American Petroleum Institute, 2101 L St., NW, Washington, D.C. 20037.
U.S. Government Printing Office, Washington, D.C. 20402.
American National Standards Institute, 1430 Broadway, N.Y., N.Y. 10018.
Insurance Services Office, 160 Water St., N.Y., N.Y. 10038.
Naval Ship Engineering Center, Dept. of the Navy, Hyattsville, Md. 20782.

Tanks:
American Welding Society, 550 Le Jeune Road, Miami, Florida 33135.

Piping:
American National Standards Institute, 1430 Broadway, N.Y., N.Y. 10018.
American Welding Society, 550 Le Jeune Road, Miami, Florida 33135.
Mechanical Contractors Association of America, 5530 Wisconsin Ave., Chevy Chase, Md. 20813.
Superintendent, State Division of Safety and Hygiene, Columbus, Ohio.

Structural and Bridges:
American Welding Society, 550 Le Jeune Road, Miami, Florida 33135.
Naval Facilities Engineering Command, 200 Stovall St., Alexandria, Va. 22332.
American Institute of Steel Construction, Inc., 400 North Michigan Ave., Chicago, Il. 60611.

Ships:
American Welding Society, 550 Le Jeune Road, Miami, Florida 33135.
Lloyd's Register of Shipping, 17 Battery Pl., N.Y., N.Y. 10004.
Naval Ship Engineering Center, Dept. of the Navy, Hyattsville, Md. 20782.
American Bureau of Shipping, 65 Broadway, N.Y., N.Y. 10006.

Aircraft Construction:
American Welding Society, 550 Le Jeune Road, Miami, Florida 33135.
Dept. of Transportation, Federal Aviation Administration, Washington, D.C. 20590.
AF/LGM, Dept. of the Air Force, Washington, D.C. 20330.

Electric Welding Machinery:
American National Standards Institute, 1430 Broadway, N.Y., N.Y. 10018.
National Electrical Manufacturers Assn., 2101 L St., NW, Washington, D.C. 20037.

American National Standard Letter Designations for Welding and Allied Processes
(ANSI/AWS A2.4-79)

Letter designation	Welding and allied processes	Letter designation	Welding and allied processes
AAC	air carbon arc cutting	IB	induction brazing
AAW	air acetylene welding	INS	iron soldering
ABD	adhesive bonding	IRB	infrared brazing
AB	arc brazing	IRS	infrared soldering
AC	arc cutting	IS	induction soldering
AHW	atomic hydrogen welding	IW	induction welding
AOC	oxygen arc cutting	LBC	laser beam cutting
AW	arc welding	LBW	laser beam welding
B	brazing	LOC	oxygen lance cutting
BB	block brazing	MAC	metal arc cutting
BMAW	bare metal arc welding	OAW	oxyacetylene welding
CAC	carbon arc cutting	OC	oxygen cutting
CAW	carbon arc welding	OFC	oxyfuel gas cutting
CAW-G	gas carbon arc welding	OFC-A	oxyacetylene cutting
CAW-S	shielded carbon arc welding	OFC-H	oxyhydrogen cutting
CAW-T	twin carbon arc welding	OFC-N	oxynatural gas cutting
CW	cold welding	OFC-P	oxypropane cutting
DB	dip brazing	OFW	oxyfuel gas welding
DFB	diffusion brazing	OHW	oxyhydrogen welding
DFW	diffusion welding	PAC	plasma arc cutting
DS	dip soldering	PAW	plasma arc welding
EASP	electric arc spraying	PEW	percussion welding
EBC	electron beam cutting	PGW	pressure gas welding
EBW	electron beam welding	POC	metal powder cutting
EGW	electrogas welding	PSP	plasma spraying
ESW	electroslag welding	RB	resistance brazing
EXW	explosion welding	RPW	projection welding
FB	furnace brazing	RS	resistance soldering
FCAW	flux cored arc welding	RSEW	resistance seam welding
FLB	flow brazing	RSW	resistance spot welding
FLOW	flow welding	ROW	roll welding
FLSP	flame spraying	RW	resistance welding
FOC	chemical flux cutting	S	soldering
FOW	forge welding	SAW	submerged arc welding
FRW	friction welding	SAW-S	series submerged arc welding
FS	furnace soldering	SMAC	shielded metal arc cutting
FW	flash welding	SMAW	shielded metal arc welding
GMAC	gas metal arc cutting	SSW	solid state welding
GMAW	gas metal arc welding	SW	stud arc welding
GMAW-P	gas metal arc welding — pulsed arc	TB	torch brazing
		TC	thermal cutting
GMAW-S	gas metal arc welding — short circuiting arc	TCAB	twin carbon arc brazing
		THSP	thermal spraying
GTAC	gas tungsten arc cutting	TS	torch soldering
GTAW	gas tungsten arc welding	TW	thermit welding
GTAW-P	gas tungsten arc welding — pulsed arc	USW	ultrasonic welding
		UW	upset welding
HFRW	high frequency resistance welding	WS	wave soldering
HPW	hot pressure welding		

Suffixes for Optional Use

| AU | automatic | ME | machine |
| MA | manual | SA | semiautomatic |

Obsolete or Seldom Used Processes

| DW | die welding | NTW | nonpressure thermit welding |
| HW | hammer welding | PTW | pressure thermit welding |

Application of American National Standard Welding Symbols — 1

Desired Weld	Symbol	Symbol Meaning
		Symbol indicates fillet weld on *arrow side* of the joint.
		Symbol indicates square-groove weld on *other side* of the joint.
		Symbol indicates bevel-groove weld on both sides of joint. Break in arrow indicates bevels on upper member of joint. Breaks in arrows are used on symbols designating bevel and J-groove welds.
		Symbol indicates plug weld on *arrow side* of joint.
		Symbol indicates resistance-seam weld. Weld symbol appears on both sides of reference line pointing up the fact that *arrow* and *other side* of joint references have no significance.
		Symbol indicates electron beam seam weld on *other side* of joint.

Application of American National Standard Welding Symbols — 2

Desired Weld	Symbol	Symbol Meaning
GROOVE WELD MADE BEFORE WELDING OTHER SIDE BACK WELD		Symbol indicates single-pass back weld.
		Symbol indicates a built-up surface ⅛ inch thick.
		Symbol indicates a bead-type back weld on the *other side* of joint, and a J-groove grooved horizontal member (shown by break in arrow) and fillet weld on *arrow side* of the joint.
		Symbol indicates two fillet welds, both with ½-inch leg dimensions.
		Symbol indicates a ½-inch fillet weld on *arrow side* of the joint and a ¼-inch fillet weld on *far side* of the joint.
	ORIENTATION SHOWN ON DRAWING	Symbol indicates a fillet weld on *arrow side* of joint with ¼- and ½-inch legs. Orientation of legs must be shown on drawing.

Application of American National Standard Welding Symbols — 3

Desired Weld	Symbol	Symbol Meaning
		Symbol indicates a 24-inch long fillet weld on the *arrow side* of the joint.
		Symbol indicates a series of intermittent fillet welds each 2 inches long and spaced 5 inches apart on centers directly opposite each other on both sides of the joint.
		Symbol indicates a series of intermittent fillet welds each 3 inches long and spaced 10 inches apart on centers. The centers of the welds on one side of the joint are displaced from those on the other.
		Symbol indicates a fillet weld around the perimeter of the member.
		Symbol indicates a ¼-inch V-groove weld with a ⅛-inch root penetration.
		Symbol indicates a ¼-inch bevel weld with a ⁵⁄₁₆-inch root penetration plus a subsequent ⅜-inch fillet weld.

Application of American National Standard Welding Symbols — 4

Desired Weld	Symbol	Symbol Meaning
		Symbol indicates a bevel weld with a root opening of $\frac{3}{16}$ inch.
		Symbol indicates a V-groove weld with a groove angle of 65 degrees on the *arrow side* and 90 degrees on the *other side*.
		Symbol indicates a flush surface with the reinforcement removed by chipping on the *other side* of the joint and a smooth grind on the *arrow side*. The symbols C and G should be the user's standard finish symbols.
		Symbol indicates a 2-inch U-groove weld with a 25-degree groove angle and no root opening for both sides of the joint.
		Symbol indicates plug welds of 1-inch diameter, a depth of filling of ½ inch and a 60-degree angle of countersink spaced 6 inches apart on centers.
		Symbol indicates all-around bevel and square-groove weld of these studs.

Application of American National Standard Welding Symbols — 5

Desired Weld	Symbol	Symbol Meaning
		Symbol indicates an electron beam seam weld with a minimum acceptable joint strength of 200 pounds per lineal inch.
		Symbol indicates four .10-inch diameter electron beam spot welds located at random.
		Symbol indicates a fillet weld on the *other side* of joint and a flare-bevel-groove weld and a fillet weld on the *arrow side* of the joint.
		Symbol indicates gas tungsten-arc seam weld on *arrow side* of joint.
		Symbol indicates edge-flange weld on *arrow side* of joint and flare-V-groove weld on *other side* of joint.
		Symbol indicates melt-thru weld. By convention this symbol is placed on the opposite side of the reference line from the corner-flange symbol.

Tool and Die Welding. — Arc welding is extensively used in the building and repairing of tools and dies. Inexpensive and truly functional tools and dies can be readily made using mild steel or medium carbon steel as the base material and surfacing the cutting or wearing surfaces with weld metal. Tool steel or high-speed steel can be quickly applied either in the fabrication or repair of dies and cutting tools with standard arc welding electrodes.

Nondestructive Testing

Nondestructive Testing Symbol Application. — The application of nondestructive testing symbols is also covered in American National Standard ANSI/AWS 2.4-79.
Basic Testing Symbols: These are shown in the following table.

American National Standard Basic Symbols for Nondestructive Testing
(ANSI/AWS 2.4-79)

Symbol	Type of Test	Symbol	Type of Test
AET	Acoustic Emission	PT	Penetrant
ET	Eddy Current	PRT	Proof
LT	Leak	RT	Radiographic
MT	Magnetic Particle	UT	Ultrasonic
NRT	Neutron Radiographic	VT	Visual

Testing Symbol Elements: The testing symbol consists of the following elements:
    Reference Line
        Arrow
        Basic Testing Symbol
        Test-all-around Symbol
        (N) Number of Tests
        Test in Field
    Tail
        Specification or other reference
The standard location of the testing symbol elements are shown in the following figure.

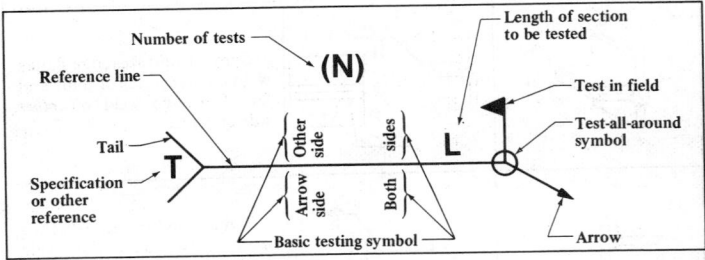

Locations of Testing Symbol Elements

The arrow connects the reference line to the part to be tested. The side of the part to which the arrow points is considered to be the *arrow side*. The side opposite the arrow side is considered to be the *other side*.

Location of Testing Symbol: Tests to be made on the arrow side of the part are indicated by the basic testing symbol on the side of the reference line toward the reader.

Tests to be made on the other side of the part are indicated by the basic testing symbol on the side of the reference line away from the reader.

To specify where only a certain length of a section is to be considered, the actual length or percentage of length to be tested is shown to the right of the basic test symbol. To specify the number of tests to be taken on a joint or part, the number of tests is shown in parentheses.

Tests to be made on both sides of the part are indicated by test symbols on both sides of the reference line. Where nondestructive symbols have no arrow or other significance, the testing symbols are centered in the reference line.

Combination of Symbols: Nondestructive basic testing symbols may be combined and nondestructive and welding symbols may be combined.

Direction of Radiation: When specified, the direction of radiation may be shown in conjunction with the radiographic or neutron radiographic basic testing symbols by means of a radiation symbol located on the drawing at the desired angle.

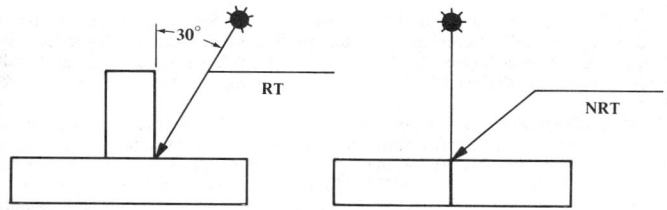

Tests Made All Around the Joint: To specify tests to be made all around a joint a circular test-all-around symbol is used.

Areas of Revolution: For nondestructive testing of areas of revolution, the area is indicated by the test-all-around symbol and appropriate dimensions.

Plane Areas: The area to be examined is enclosed by straight broken lines having a small circle around the angle apex at each change in direction.

Hard-Facing

Hard-facing is the process of welding onto parts or surfaces a coating, edge, or point of a metal highly capable of resisting one or more of the following: abrasion, corrosion, heat, and impact. The process can be applied equally well to new parts or old worn parts. The most common welding methods used to apply hard-facing materials include the oxy-acetylene gas, shielded-metal-arc, submerged arc, atomic hydrogen shielded arc, and inert-gas-shielded arc (consuming and nonconsuming electrode).

Hard-Facing Materials. — The first thing to be considered in the selection of a hard-facing material is to know what type of service the part in question is to undergo. Besides this, other considerations such as machinability, cost of hard-facing material, porosity of the deposit, appearance in use, and the ease of application may play a large part in the selection of a hard-facing material. Only generalized information can be given that can be used to guide in the selection of a material as the choice is dependent upon experience with a particular type of service. Generally the greater the hardness of the facing material the greater is its resistance to abrasion and shock or impact wear. Many hardenable materials may be used for hard-facing such as carbon steels, low-alloy steels, medium-alloy steels, and medium-high alloys but these are not outstanding. Some of the materials that might be considered more outstanding are high speed steel, austenitic manganese steel, austenitic high-chromium iron, cobalt-chromium alloy, copper-base alloy, and nickel-chromium-boron alloy.

High-Speed Steels. — These metals are available in the form of welding rods (RFe5) and electrodes (EFe5) for hard-facing where hardness is required at service temperatures up to 1100 degrees F. and where wear resistance and toughness are also required. Typical surfacing operations are done on cutting tools, shear blades, reamers, forming dies, shearing dies, guides, ingot tongs and broaches using these metals.

Hardness: They have a hardness of 55 to 60 on the Rockwell C scale in the as-welded condition and a hardness of 30 Rockwell C in the annealed condition. At a temperature of 1100 degrees F. the as-deposited hardness of 60 Rockwell C falls off very slowly to 47 Rockwell C. At about 1200 degrees F. the maximum Rockwell C hardness is 30.

Resistance Properties: As deposited the alloys can withstand only medium impact but when tempered the impact resistance is increased appreciably. Deposits of these alloys will oxidize readily because of their high molybdenum content but can withstand atmospheric corrosion. They do not withstand liquid corrosives.

Other Properties or Characteristics: The metals are well suited for metal-to-metal wear especially at elevated temperatures. They retain their hardness at elevated temperatures and can take a high polish. For machining, these alloys must first be annealed. Full hardness may be regained by a subsequent heat treatment of the metal.

Austenitic Manganese Steels. — These metals are available in the form of electrodes (EFeMn) for hard-facing when dealing with metal-to-metal wear and impact. Uses include facing rock-crushing equipment and railway frogs and crossings.

Hardness: Hardness of the as-deposited metals are 170 to 230 BHN but they can be work-hardened to 450 to 550 BHN very readily. For all practical purposes these metals have no hot hardness as they become brittle when reheated above 500 to 600 degrees F.

Resistance Properties: These metals are considered to be the outstanding engineering materials with respect to impact resistance. Their corrosion and oxidation resistance are similar to ordinary carbon steels. Their resistance to abrasion is only mediocre against hard abrasives like quartz.

Other Properties or Characteristics: The yield strength of the deposited metal in compression is low, but any compressive deformation rapidly raises it until plastic flow ceases. This property is an asset in impact wear situations. Machining is difficult with ordinary tools and equipment; finished surfaces for the most part are usually ground.

Austenitic High-Chromium Irons. — These metals are available in rod (RFeCr-A) and electrode (EFeCr-A) form for facing agricultural machinery parts, coke chutes, steel mill guides, sand blasting equipment and brick-making machinery.

Hardness: The as-welded deposit ranges in hardness from 51 to 62 Rockwell C. Under impact the deposit work hardens somewhat but the resulting deformation also leads to cracking and impact service is therefore avoided. Hot hardness decreases slowly at temperatures up to 800 and 900 degrees F. At 900 degrees F. the instantaneous hardness is 43 Rockwell C. In three minutes under load the hardness drops to 37 Rockwell C. At 1200 degrees F. the instantaneous hardness is 5 Rockwell C. The decrease in hardness during hot testing is practically recovered on cooling to ordinary temperatures.

Resistance Properties: Deposits will withstand only light impact without cracking. Dynamic compression stresses above 60,000 pounds per square inch should be avoided. These metals exhibit good oxidation resistance up to 1800 degrees F. and can be considered for hot wear applications where hot plasticity is not objectionable. They are not very resistant to corrosion from liquids and will rust in moist air but are more stable than ordinary iron and steel. Resistance to low stress scratching is outstanding and is related to the amount of hard carbides present. However, under high stress grinding abrasion, performance is only mediocre and they are not deemed advantageous for such service.

Other Properties or Characteristics: The deposited metals have a yield strength (0.1 per cent offset) of between 80,000 to 140,000 pounds per square inch in compression and an ultimate strength of from 150,000 to 280,000 pounds per square inch. Their tensile strength is low and therefore tension uses are avoided in design. These deposits are considered to be commercially unmachinable and are also very difficult to grind. When ground, a grinding wheel of aluminum oxide abrasive with a 24-grit size and a hard (Q) and medium spaced resinoid bond is recommended for off-hand high-speed work and a slightly softer (P) vitrified bond for off-hand low-speed work.

Cobalt-Base Alloys. — These metals are available in both rod (RCoCr) and electrode (ECoCr) form and are frequently used to surface the contact surfaces of exhaust valves in aircraft, truck, and bus engines. Other uses include parts such as valve trim in steam engines, and on pump shafts where conditions of corrosion and erosion are encountered. Several metals with a greater carbon content are available (CoCr-B, CoCr-C) which are used in applications requiring greater hardness and abrasion resistance but where impact resistance is not mandatory or expected to be a factor.

Hardness: Hardness ranges on the Rockwell C scale for gas welded deposits are as follows: CoCr-A, 38 to 47; CoCr-B, 45 to 49; and CoCr-C, 48 to 58. For arc-welded deposits hardness ranges (Rockwell C) as follows: CoCr-A, 23 to 47; CoCr-B, 34 to 47; and CoCr-C, 43 to 58. The values for the arc weld deposits depend for the most part on the base metal dilution. The greater the dilution the lower the hardness. Many surfacing alloys are softened permanently by heating to elevated

temperatures, however, these metals are an exception. They do exhibit lower hardness values when hot but return to their approximate original hardness values upon cooling. Elevated temperature strength and hardness are outstanding properties of this group. Their use at 1200 degrees F. and above is considered advantageous between 1000 and 1200 degrees F. their advantages are not definitely established and at temperatures below 1000 degrees F. other surfacing metals may prove better.

Resistance Properties: In the temperature range from 1000 to 1200 degrees F. weld deposits of these metals have a great resistance to creep. Tough martensitic steel deposits are considered superior to cobalt-base deposits in both flow resistance and toughness. The chromium in the deposited metal promotes the formation of a thin, tightly adherent scale which provides a scaling resistance to combustion products of internal combustion engines including deposits from leaded fuels. These metals are corrosion resistant in such media as air, food and certain acids. It would be well to conduct field tests to determine specific corrosion resistance for the instance in question.

Other Properties or Characteristics: They are able to take a high polish and have a low coefficient of friction and are therefore well suited for metal-to-metal wear. Machining of these deposits is difficult; the difficulty increases in proportion to the increase in carbon content. CoCr-A metals are preferably machined with sintered carbide tools. CoCr-C deposits are finished by grinding.

Copper-Base Alloys. — These metals are available in the rod (RCuAl-A2, RCuAl-B, RCuAl-C, RCuAl-D, RCuAl-E, RCuSi-A, RCuSn, RCuSn-D, RCuSn-E, and RCuZn-E and electrode (ECuAl-A2, ECuAl-B, ECuAl-C, ECuAl-D, ECuAl-E, ECuSi, ECuSn-A, ECuSn-C, ECuSn-E, and ECuZn-E) form and are used in depositing overlays and inlays for bearing, corrosion-resistant and wear-resistant surfaces. The CuAl-A2 rods and electrodes are used for surfacing bearing surfaces between the hardness ranges of 130 to 190 BHN as well as corrosion-resistant surfaces. The CuAl-B and CuAl-C rods and electrodes are used for surfacing bearing surfaces of hardness ranges 140 to 290 BHN. The CuAl-D and CuAl-E rods and electrodes are used to surface bearing and wear-resistant surfaces requiring the higher hardnesses of 230 to 390 BHN such as are on gears, cams, wear plates, and dies. The copper-tin (CuSn) metals are used where a lower hardness is required for surfacing, corrosion-resistant surfaces and sometimes for wear-resistant applications.

Hardness: Hardness of a deposit depends upon the welding process employed and the manner of depositing the metal. Deposits made by the inert-gas metal-arc process (both consumable and non-consumable electrode) will be higher in hardness than deposits made with the gas, metal-arc and carbon-arc processes since lower losses of aluminum, tin, silicon and zinc are achieved due to the better shielding from oxidation. Copper-base alloy metals are not recommended for use at elevated temperatures since their hardness and mechanical properties decrease consistently as the temperature goes above 400 degrees F.

Resistance Properties: The highest impact resistance of the copper-base alloy metals is exhibited by the CuAl-A2 deposit. As the aluminum content increases the impact resistance decreases markedly. CuSi weld deposits have good impact properties. CuSn metals as deposited have low impact resistance and CuZn-E deposits have a very low impact resistance. Deposits of the CuAl filler metals form a protective oxide coating upon exposure to the atmosphere. Oxidation resistance of CuSi deposits is fair and that of CuSn deposits are comparable to pure copper. These metals with the exception of the CuSn-E and CuZn-E alloys are widely used to resist many acids, mild alkalies and salt water. None of the copper-base alloy deposits are recommended for use where severe abrasion is encountered in service.

CuAl filler metals are used to overlay surfaces subjected to excessive wear from metal-to-metal contact such as gears, cams, sheaves, wear plates, and dies.

Other Properties or Characteristics: All copper-base alloy metals are used for overlays and inlays for bearing surfaces with the exception of the CuSi metals. Metals selected for bearing surfaces should have a Brinell hardness of 50 to 75 units below that of the mating metal surface. Slight porosity is generally acceptable in bearing service as a porous deposit is able to retain oil for lubricating purposes. CuAl deposits in compression have elastic limits ranging from 25,000 to 65,000 pounds per square inch and ultimate strengths of 120,000 to 171,000 pounds per square inch. The elastic limit and ultimate strength of CuSi deposits in compression are 22,000 pounds per square inch and 60,000 pounds per square inch, respectively. CuZn-E deposits in compression only have an elastic limit of about 5000 pounds per square inch and an ultimate strength of 20,000 pounds per square inch. All copper-base alloy deposits can be machined.

Nickel-Chromium-Boron Alloys. — These metals are available in both rod (RNiCr) and electrode (ENiCr) form and their deposits have good metal-to-metal wear resistance, good low-stress scratch abrasion resistance, corrosion resistance, and retention of hardness at elevated temperatures. These properties make their use suitable for seal rings, cement pump screws, valves, screw conveyors and cams. Three different formulations of these metals are recognized (NiCr-A, NiCr-B, and NiCr-C).

Hardness: Hardness of the deposited NiCr-A from rods range from 35 to 40 Rockwell C; of NiCr-B rods, 45 to 50 Rockwell C; of NiCr-C rods, 56 to 62 Rockwell C. Hardness of the deposited NiCr-A from electrodes ranges from 24 to 35 Rockwell C; of NiCr-B from electrodes, 30 to 45 Rockwell C; and of NiCr-C electrodes, 35 to 56. The lower hardness values and greater ranges of hardness values of the electrode deposits are attributed to the dilution of deposit and base metals. Hot Rockwell C hardness values of NiCr-A electrode deposits range from 30 to 19 in the temperature range from 600 to 1000 degrees F. from instantaneous loading to a 3-minute loading interval. NiCr-A rod deposits range from 34 to 24 in the same temperature range and under the same load conditions. Hot Rockwell C hardness values of NiCr-B electrode deposits range from 41 to 26 in the temperature range from 600 to 1000 degrees F. from instantaneous loading to a 3-minute loading interval. NiCr-B rod deposits range from 46 to 37 in the same temperature range and under the same load conditions. Hot Rockwell C hardness values of NiCr-C electrode deposits range from 49 to 31 in the temperature range from 600 to 1000 degrees F. from instantaneous loading to a 3-minute loading interval. NiCr-C rod deposits range from 55 to 40 in the same temperature range and under the same load conditions.

Resistance Properties: Deposits of these metal alloys will withstand light impact fairly well. When plastic deformation occurs cracks are more likely to appear in the NiCr-C deposit than in the NiCr-A and NiCr-B deposits. NiCr deposits are oxidation resistant up to 1800 degrees F. Their use above 1750 degrees F. is not recommended since fusion may begin near this temperature. NiCr deposits are completely resistant to atmospheric, steam, salt water, and salt spray corrosion and also resistant to the milder acids and many common corrosive chemicals. It would be well to conduct field tests when a corrosion application is contemplated. These metals are not recommended for high stress grinding abrasion. NiCr deposits have good metal-to-metal wear resistance, take a high polish under wearing conditions and are particularly resistant to galling. These properties are especially evident in the NiCr-C alloy.

Other Properties or Characteristics: In compression these alloys have an elastic limit of 42,000 pounds per square inch. Their yield strength in compression is

92,000 pounds per square inch (0.01 per cent offset), 150,000 pounds per square inch (0.10 per cent offset) and 210,000 pounds per square inch (0.20 per cent offset). Deposits of NiCr filler metals may be machined with tungsten carbide tools using slow speeds, light feeds and heavy tool shanks. They are also finished by grinding using a soft-to-medium vitrified silicon carbide wheel.

Chromium Plating. — Chromium plating is an electrolytic process of depositing chromium on metals either as a protection against corrosion or to increase the surface wearing qualities. The value of chromium-plating plug and ring gages has probably been more thoroughly demonstrated than any other one application of this treatment. Chromium-plated gages not only wear longer, but when worn, the chromium may be removed and the gage replated and reground to size.

In general, chromium-plated tools have operated well, giving greatly improved performance on nearly all classes of materials such as brass, bronze, copper, nickel, aluminum, cast iron, steel, plastics, asbestos compositions, and similar materials. Increased cutting life has been obtained with chromium-plated drills, taps, reamers, files, broaches, tool tips, saws, thread chasers, and the like. Dies for stamping, drawing, hot-forging, die-casting, and for molding plastic materials have shown greatly increased life after being plated with hard chromium.

Special care is essential in grinding and lapping tools preparatory to plating the cutting edges, because the chromium deposit is influenced materially by the grain structure and hardness of the base metal. The thickness of the plating may vary from 0.0001 to 0.001 or 0.002 inch, the thicker platings being used to build up undersize tools such as taps and reamers. Procedure followed by Westinghouse in the hard chromium-plating of tools, as well as parts salvaged by depositing chromium to increase diameters, is as follows: (1) Degrease with solvent; (2) mount the tools on racks; (3) clean in an anodic alkali bath held at a temperature of 82 degrees C. for from three to five minutes; (4) rinse in boiling water; (5) immerse in a 20 per cent hydro-chloric acid solution for two to three seconds; (6) rinse in cold water; (7) rinse in hot water; (8) etch in a reverse-current chromic acid bath for two to five minutes; (9) place work immediately in the chromium-plating bath; and (10) remove hydrogen embrittlement, if necessary, by immersing the plated tools for two hours in an oil bath maintained at 177 degrees C.

Chromium has a very low coefficient of friction. The static coefficient of friction for steel on chromium-plated steel is 0.17, and the sliding coefficient of friction is 0.16. This compares with static coefficient of friction for steel on steel of 0.30 and a sliding coefficient of friction of 0.20. The static coefficient of friction for steel on babbitt is 0.25, and the sliding coefficient of friction 0.20, whereas for chromium-plated steel on babbitt, the static coefficient of friction is 0.15, and the sliding coefficient of friction 0.13. These figures apply to highly polished bearing surfaces. Articles that are to be chromium-plated in order to resist frictional wear should be highly polished before plating so that full advantage can be taken of the low coefficient of friction that is characteristic of chromium. Chromium resists attack by almost all organic and inorganic compounds, except muriatic and sulphuric acids. The melting point of chromium is 2930 degrees F., and it remains bright up to 1200 degrees F. Above this temperature, it forms a light adherent oxide, which does not readily become detached. For this reason, chromium has been used successfully for protecting articles that must resist high temperatures, even above 2000 degrees F.

Cutting Metals with an Oxidizing Flame

The oxy-hydrogen and oxy-acetylene flames are especially adapted to cutting metals. When iron or steel is heated to a high temperature, it has a great affinity

for oxygen and readily combines with it to form different oxides, which causes the metal to be disintegrated and burned with great rapidity. The metal cutting or burning torch operates on this principle. A torch tip is designed to pre-heat the metal, which is then burned or oxidized by a jet of pure oxygen. The kerf or path left by the flame is suggestive of a saw cut when the cutting torch has been properly adjusted and used. The traversing motion of the torch along the work may be controlled either by hand or mechanically.

The Cutting Torch. — The ordinary cutting torch consists of a heating jet using oxygen and acetylene, oxygen and hydrogen, or, in fact, any other gas which, when combined with oxygen, will produce sufficient heat. By the use of this heating jet, the metal is first brought to a sufficiently high temperature, and an auxiliary jet of pure oxygen is then turned onto the red-hot metal, when the action just referred to takes place. Some cutting torches have a number of pre-heating flame ports surrounding the central oxygen port, so that a pre-heating flame will precede the oxygen regardless of the way in which the torch is moved. This arrangement has been used to advantage in mechanically guided torches. The rate of cutting varies with the thickness of the steel, the size of the tip and the oxygen pressure.

Adjustment and Use of Cutting Torch. — When using the cutting torch for the cutting of steel plate, the pre-heating flame first comes into contact with the edge of the plate and quickly raises it to a white hot temperature, and then the oxygen valve is opened by pulling a trigger on the torch and, as the pure oxygen comes into contact with the heated metal, the latter is burned or oxidized.

Metals that can be Cut. — Metals such as wrought iron and steels of comparatively low carbon content can be cut readily with the cutting torch. High carbon steels may be cut successfully if pre-heated to a temperature that depends somewhat on the carbon content. The higher the carbon content, the greater the degree of pre-heating. A black heat is sufficient for ordinary tool steel, but a low red may be required for some of the alloy tool steels. Brass and bronze plates have been cut by interposing them between steel plates.

Cutting Stainless Steel. — Stainless steel can be cut readily by the flux-injection method. The elements which give stainless steel their desirable properties produce oxides which reduce the operation to a slow melting away process when the conventional oxy-acetylene cutting equipment is used. By injecting a suitable flux directly into the stream of cutting oxygen before it enters the torch, the obstructing oxides are removed. A portable flux feeding unit is designed to inject a predetermined amount of the flux powder. The rate of flux flow is accurately regulated by a vibrator type of dispenser with rheostat control. The flux-injection method is applicable either to machine cutting or to a hand controlled torch. The operating procedure and speed of cutting are practically the same as in cutting mild steel.

Cutting Cast Iron. — The cutting of cast iron with the oxy-acetylene torch is practicable although it cannot be cut as readily as steel. The ease of cutting seems to depend largely on the physical character of the cast iron, very soft cast iron being more difficult to cut than harder varieties. The cost is much higher than that for cutting the same thickness of steel, because of the larger pre-heating flame necessary and the larger oxygen consumption. In spite of this, however, this method is economical in many cases. The slag from a cast iron cut contains considerable melted cast iron, while in the case of steel, the slag is practically free from particles of the metal. This indicates that cast iron cutting is partly a melting operation. Increased speed and decreased cost can often be obtained by feeding a steel rod, about ¼ inch in diameter, into the top of the cut, just beneath the torch tip. This

furnishes a large amount of slag which flows over the cut and increases the temperature of the cast iron. Special tips are used owing to the amount of heat and oxygen
required.

Mechanically Guided Torches. — Cutting torches used for cutting openings in plates or blocks or for cutting parts to some definite outline, often are
guided mechanically. Torches guided by pantograph mechanisms are especially
adapted for tracing the outline to be cut from a pattern or drawing. Other
designs are preferable for straight-line cutting and one type is designed for circular
cutting.

Cutting Steel Castings. — When cutting steel castings, care should be taken
to prevent burning pockets in the metal when the flame strikes a blow-hole. If
a blow-hole is penetrated, the molten oxide will splash into the cavity and the flame
will be diverted. The presence of the blow-hole is generally indicated by excessive
sparks. The operator should immediately move the torch back along the cut and
direct it at an angle so as to strike the metal beneath the blow-hole and burn it
away if possible beyond the cavity, when cutting in the normal position may be
resumed.

Thickness of Metal that can be Cut. — The maximum thickness of metal that
can be cut by these high-temperature flames depends largely upon the gases used
and the pressure of the oxygen, which may be as high as 150 pounds per square
inch; the thicker the metal the higher the pressure required. When using the
oxy-acetylene flame, it might be practicable to cut iron or steel up to 12 or 14
inches in thickness, whereas, the oxy-hydrogen flame has been used to cut steel
plates 24 inches thick. The oxy-hydrogen flame will cut thicker material principally because it is longer than the oxy-acetylene flame and can penetrate to the
full depth of the cut, thus keeping all the oxide in a molten condition so that it can
easily be blown out by the oxygen cutting jet. A mechanically guided torch will
cut thick material more satisfactorily than a hand-guided torch, because the flame
is directed straight into the cut and does not wobble, as it tends to do when the
torch is held by hand. With any flame, the cut is less accurate and the kerf wider,
as the thickness of the metal increases. When cutting light material, the kerf
might be $\frac{1}{16}$ inch wide, whereas, for heavy stock it might be $\frac{1}{4}$ or $\frac{3}{8}$ inch wide.

Arc Cutting of Metals

Arc Cutting. — According to the "Procedure Handbook of Arc-Welding Design
& Practice" published by The Lincoln Electric Co. a steel may be readily cut with
great accuracy by means of the oxy-acetylene torch. All metals, however, do not
cut as easily as steel. Cast iron, stainless steels, manganese steels and non-ferrous
materials are not as readily cut and shaped with the oxy-acetylene cutting process
because of their reluctance to oxidize. In these cases, arc cutting is often used to
good advantage.

The cutting of steel is a chemical action. The oxygen combines readily with
the iron to form iron oxide. In cast iron, this action is hindered by the presence of
carbon in graphite form. Thus, cast iron cannot be cut as readily as steel; higher
temperatures are necessary and cutting is slower. In steel, the action starts at
bright red heat, whereas, in cast iron, the temperature must be more near the melting
point in order to obtain a sufficient reaction.

Due to its very high temperature, the rate of cutting is usually fairly high. However, as the process is essentially one of melting without any great action tending to force the molten metal out of cut, some provision must be made for permitting the metal to flow readily away from the cut. This is usually done by starting at some point from which the molten metal may readily flow. This method is followed until the desired amount of metal has been melted away.

As an example, the general method is to apply the electric arc on the under side of the work, starting at a lower corner, working toward the center on the lower surface, and then up the side, repeating this action as many times as necessary. This will allow the molten metal to flow out of the cut.

A carbon electrode is generally used. Graphite electrodes are used to some extent because they permit use of higher currents. Shielded-arc type electrodes are also effective. In starting a cut, the arc is held at the point selected for the initial cut as, for example, a lower corner. When the metal begins to flow and run off, the arc is moved along at a rate to permit the metal to continuously flow out of the cut.

The width of the cut is dependent upon the ability of the operator to follow a straight line, the size electrode used, and the thickness of material. The width of the cut is greater on thick sections than on thin.

A relatively new development, the arc-air process, is also widely used for cutting and gouging. The process is essentially the melting of metal, any metal, with an electric arc and simultaneously mechanically removing the molten metal by means of a high velocity air jet, external and parallel to the electrode. This process is not dependent on oxidation and, for this reason, works on metals which do not readily oxidize, as well as those that do.

The equipment consists of a torch with a concentric cable which carries both air and current. The electrode is usually carbon graphite, though coated metal may also be used. An air line from an ordinary air compressor and the cable from a DC welding machine are both attached to the end of a concentric cable, which carries both the air and the current to the torch. The lever at the bottom of the torch controls air flow. The electrode is held in a rotating head which allows it to be set at any angle, but maintains the air stream always directed at the proper location.

An ordinary DC welding machine with reverse polarity is used. The current depends on the size of the electrode and varies from 70 amperes on $\frac{5}{32}$ inch electrodes to 600 amperes on $\frac{1}{2}$ inch electrodes. Although the higher the current density, the more efficient the process becomes.

The necessary air is obtained from an ordinary compressor. The torch is designed to operate at 90 to 100 psi pressure, which is the usual line pressure in most shops, and, since this is not a critical value, no regulator is needed.

The torch is used by holding the electrode at a leading angle and striking an arc between the electrode and the material to be cut. The air blast is directed immediately behind the point of arcing, and the electrode is pushed forward at a rapid rate with the air jet on continuously. The depth of the groove is determined by the angle of the electrode and the speed of travel.

Because of the small area, the metal being cut is instantly brought to the molten stage and speed of travel is very rapid; the surrounding metal does not reach a very high temperature; and there is little distortion or crack propagation during the operation.

The speed varies somewhat with individual shop conditions, such as operator technique and the current used. To give a general idea, a groove $\frac{3}{8}$ inch wide and $\frac{1}{4}$ inch deep is usually cut at about 3 fpm, or just about as fast as the operator can move.

FILES

Definitions of File Terms. — The following file terms apply to hand files but not to rotary files and burs.

Axis: Imaginary line extending the entire length of a file equidistant from faces and edges.

Back: The convex side of a file having the same or similar cross-section as a half-round file.

Bastard Cut: A grade of file coarseness between coarse and second cut of American pattern files and rasps.

Blank: A file in any process of manufacture before being cut.

Blunt: A file whose cross-sectional dimensions from point to tang remain unchanged.

Coarse Cut: The coarsest of all American pattern file and rasp cuts.

Coarseness: Term describing the relative number of teeth per unit length, the coarsest having the least number of file teeth per unit length; the smoothest, the most. American pattern files and rasps have four degrees of coarseness: coarse, bastard, second and smooth. Swiss pattern files usually have seven degrees of coarseness: oo, o, 1, 2, 3, 4, 6 (from coarsest to smoothest). Curved tooth files have three degrees of coarseness: standard, fine and smooth.

Curved Cut: File teeth which are made in curved contour across the file blank.

Cut: Term used to describe file teeth with respect to their coarseness or their character (single, double, rasp, curved, special).

Double Cut: A file tooth arrangement formed by two series of cuts, namely the overcut followed, at an angle, by the upcut.

Edge: Surface joining faces of a file. May have teeth or be smooth.

Face: Widest cutting surface or surfaces that are used for filing.

Heel or Shoulder: That portion of a file that abuts the tang.

Hopped: A term used among file makers to represent a very wide skip or spacing between file teeth.

Length: The distance from the heel to the point.

Overcut: The first series of teeth put on a double-cut file.

Point: The front end of a file; the end opposite the tang.

Rasp Cut: A file tooth arrangement of round-topped teeth, usually not connected, that are formed individually by means of a narrow, punch-like tool.

Re-cut: A worn-out file which has been re-cut and re-hardened after annealing and grinding off the old teeth.

Safe Edge: An edge of a file that is made smooth or uncut, so that it will not injure that portion or surface of the workpiece with which it may come in contact during filing.

Second Cut: A grade of file coarseness between bastard and smooth of American pattern files and rasps.

Set: To blunt the sharp edges or corners of file blanks before and after the overcut is made, in order to prevent weakness and breakage of the teeth along such edges or corners when the file is put to use.

Shoulder or Heel: See *Heel or Shoulder.*

Single Cut: A file tooth arrangement where the file teeth are composed of single unbroken rows of parallel teeth formed by a single series of cuts.

Smooth Cut: An American pattern file and rasp cut that is smoother than second cut.

Tang: The narrowed portion of a file which engages the handle.

Upcut: The series of teeth superimposed on the overcut, and at an angle to it, on a double-cut file.

File Characteristics. — Files are classified according to their shape or cross-section and according to the pitch or spacing of their teeth and the nature of the cut.

Cross-section and Outline: The cross-section may be quadrangular, circular, triangular, or some special shape. The outline or contour may be tapered or blunt. In the former, the point is more or less reduced in width and thickness by a gradually narrowing section that extends for one-half to two-thirds of the length. In the latter, the cross-section remains uniform from tang to point.

Cut: The character of the teeth is designated as single, double, rasp or curved. The *single cut file* (or *float* as the coarser cuts are sometimes called) has a single series of parallel teeth extending across the face of the file at an angle of from 45 to 85 degrees with the axis of the file. This angle depends upon the form of the file and the nature of the work for which it is intended. The single cut file is customarily used with a light pressure to produce a smooth finish. The *double cut file* has a multiplicity of small pointed teeth inclining toward the point of the file arranged in two series of diagonal rows that cross each other. For general work, the angle of the first series of rows is from 40 to 45 degrees and of the second from 70 to 80 degrees. For *double cut finishing files* the first series has an angle of about 30 degrees and the second, from 80 to 87 degrees. The second, or upcut, is almost always deeper than the first or overcut. Double cut files are usually employed, under heavier pressure, for fast metal removal and where a rougher finish is permissible. The *rasp* is formed by raising a series of individual rounded teeth from the surface of the file blank with a sharp, narrow, punch-like cutting tool and is used with a relatively heavy pressure on soft substances for fast removal of material. The curved tooth file has teeth that are in the form of parallel arcs extending across the face of the file, the middle portion of each arc being closest to the point of the file. The teeth are usually single cut and are relatively coarse. They may be formed by steel displacement but are more commonly formed by milling.

With reference to coarseness of cut the terms *coarse, bastard, second* and *smooth cuts* are used, the coarse or bastard files being used on the heavier classes of work and the second or smooth cut files for the finishing or more exacting work. These degrees of coarseness are only comparable when files of the same length are compared, as the number of teeth per inch of length decreases as the length of the file increases. The number of teeth per inch varies considerably for different sizes and shapes and for files of different makes. The coarseness range for the curved tooth files is given as standard, fine and smooth. In the case of Swiss pattern files, a series of numbers is used to designate coarseness instead of names; Nos. 00, 0, 1, 2, 3, 4 and 6 being the most common with No. 00 the coarsest and No. 6 the finest.

Classes of Files. — There are five main classes of files: mill or saw files; machinists' files; curved tooth files; Swiss pattern files; and rasps. The first two classes are commonly referred to as American pattern files.

Mill or Saw Files: These are used for sharpening mill or circular saws, large crosscut saws; for lathe work; for draw filing; for filing brass and bronze; and for smooth filing generally. *Cantsaw files* (1) have an obtuse isosceles triangular section, a blunt outline, are single cut and are used for sharpening saws having "M"-shaped teeth and teeth of less than 60-degree angle. *Crosscut files* (2) have a narrow triangular section with short side rounded, a blunt outline, are single cut and are used to sharpen crosscut saws. The rounded portion is used to deepen the gullets of saw teeth and the sides are used to sharpen the teeth themselves. *Double ender files* (3) have a triangular section, are tapered from the middle to both ends, are tangless, are single cut and are used reversibly for sharpening saws. The *mill file* (4), itself, is usually single cut, tapered in width, and often has two square cutting edges in addition to the cutting sides. Either or both edges may be rounded, however, for

filing the gullets between saw teeth. The *blunt mill file* has a uniform rectangular cross-section from tip to tang. The *triangular saw files* or *taper saw files* (5) have an equilateral triangular section, are tapered, are single cut and are used for filing saws with 60-degree angle teeth. They come in taper, slim taper, extra slim taper and double extra slim taper thicknesses. *Blunt triangular* and *blunt hand saw files* are without taper. *Web saw files* (6) have a diamond-shaped section, a blunt outline, are single cut and are used for sharpening pulpwood or web saws.

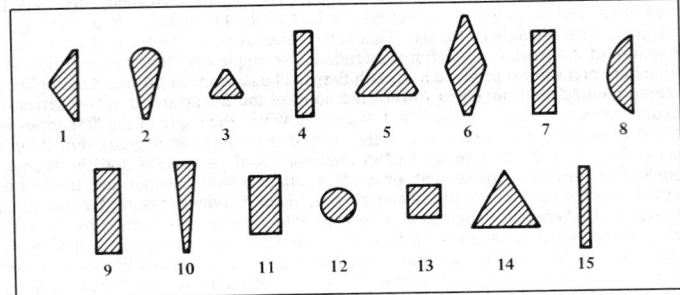

Machinists' Files: These files are used throughout industry where metal must be removed rapidly and finish is of secondary importance. Except for certain exceptions in the round and half-round shapes, all are double cut. *Flat files* (7) have a rectangular section, are tapered in width and thickness, are cut on both sides and edges and are used for general utility work. *Half round files* (8) have a circular segmental section, are tapered in width and thickness, have their flat side double cut, their rounded side mostly double but sometimes single cut, and are used to file rounded holes, concave corners, etc. in general filing work. *Hand files* (9) are similar to flat files but taper in thickness only. One edge is uncut or "safe." *Knife files* (10) have a "knife-blade" section, are tapered in width only, are double cut, and are used by tool and die makers on work having acute angles. *Machinist's General Purpose files* have a rectangular section, are tapered and have single cut teeth divided by angular serrations which produce short cutting edges. These edges help stock removal but still leave a smooth finish and are suitable for use on various materials including aluminum, bronze, cast iron, malleable iron, mild steels and annealed tool steels. *Pillar files* (11) are similar to hand files but are thicker and not as wide. *Round files* (12) have a circular section, are tapered, single cut, and are generally used to file circular openings or curved surfaces. *Square files* (13) have a square section, are tapered, and are used for filing slots, keyways and for general surface filing where a heavier section is preferred. *Three square files* (14) have an equilateral triangular section and are tapered on all sides. They are double cut and have sharp corners as contrasted with taper triangular files which are single cut and have somewhat rounded corners. They are used for filing accurate internal angles, for clearing out square corners, and for filing taps and cutters. *Warding files* (15) have a rectangular section, and taper in width to a narrow point. They are used for general narrow space filing. *Wood files* are made in the same sections as flat and half round files but with coarser teeth especially suited for working on wood.

Curved Tooth Files: Regular curved tooth files are made in both rigid and flexible forms. The rigid type has either a tang for a conventional handle or is made plain

with a hole at each end for mounting in a special holder. The flexible type is furnished for use in special holders only. The curved tooth files come in standard, fine and smooth cuts and in parallel flat, square, pillar, pillar narrow, half round and shell types. A special curved tooth file is available with teeth divided by long angular serrations. The teeth are cut in an "off center" arc. When moved across the work toward one edge of the file a fast cutting action is provided; when moved toward the other edge, a smoothing action; thus the file is made to serve a dual purpose.

Swiss Pattern Files: These are used by tool and die makers, model makers and delicate instrument parts finishers. They are made to closer tolerances than the conventional American pattern files although with similar cross-sections. The points of the Swiss pattern files are smaller, the tapers are longer and they are available in much finer cuts. They are primarily finishing tools for removing burrs left from previous finishing operations, truing up narrow grooves, notches and keyways, cleaning out corners and smoothing small parts. For very fine work, *round* and *square handled needle files,* available in numerous cross-sectional shapes in overall lengths from 4 to 7¾ inches, are used. Die sinkers use *die sinkers files* and *die sinkers rifflers.* The files, also made in many different cross-sectional shapes, are 3½ inches in length and are available in the cut Nos. 0, 1, 2 and 4. The rifflers are from 5½ to 6¾ inches long, have cutting surfaces on either end, and come in numerous cross-sectional shapes in cut Nos. 0, 2, 3, 4 and 6. These rifflers are used by die makers for getting into corners, crevices, holes and contours of intricate dies and molds. Used in the same fashion as die sinkers rifflers, *silversmiths rifflers,* that have a much heavier cross-section, are available in lengths from 6⅞ to 8 inches and in cuts Nos. 0, 1, 2, and 3. *Blunt machine files* in Cut Nos. 00, 0, and 2 for use in ordinary and bench filing machines are available in many different cross-sectional shapes, in lengths from 3 to 8 inches.

Rasps: Rasps are employed for work on relatively soft substances such as wood, leather, and lead where fast removal of material is required. They come in rectangular and half round cross-sections, the latter with and without a sharp edge.

Special Purpose Files: Falling under one of the preceding five classes of files, but modified to meet the requirements of some particular function, are a number of special purpose files. The *long angle lathe file* is used for filing work that is rotating in a lathe. The long tooth angle provides a clean shear, eliminates drag or tear and is self-clearing. This file has safe or uncut edges to protect shoulders of the work which are not to be filed. The *foundry file* has especially sturdy teeth with heavy set edges for the snagging of castings — the removing of fins, sprues, and other projections. The *die casting file* has extra strong teeth on corners and edges as well as sides for working on die castings of magnesium, zinc, or aluminum alloys. A special file for stainless steel is designed to stand up under the abrasive action of stainless steel alloys. *Aluminum rasps* and *files* are designed to eliminate clogging. A special tooth construction is used in one type of aluminum file which breaks up the filings, allows the file to clear itself and overcomes chatter. A *brass file* is designed so that with a little pressure the sharp, high-cut teeth bite deep while with less pressure, their short uncut angle produces a smoothing effect. The *lead float* has coarse, single cut teeth at almost right angles to the file axis. These shear away the metal under ordinary pressure and produce a smoothing effect under light pressure. The *shear tooth file* has a coarse single cut with a long angle for soft metals or alloys, plastics, hard rubber and wood. *Chain saw files* are designed to sharpen all types of chain saw teeth. These files come in round, rectangular, square and diamond-shaped sections. The round and square sectioned files have either double or single cut teeth, the rectangular files have single cut teeth and the diamond-shaped files have double cut teeth.

Filing Methods. — In *straight filing*, the file is pushed lengthwise, straight ahead, or slightly diagonally across the work. The amount of pressure to apply should be sufficient to enable the file to cut at all times. Too much pressure tends to clog the teeth and break them off. Too little pressure, especially on harder materials, tends to dull the teeth quickly. In *drawfiling*, the axis of the file is held at right angles to the direction of motion of the stroke. This method produces a somewhat finer finish than straight filing. In *lathe filing*, the file is stroked constantly across the revolving work and not held rigid or stationary. A slight gliding or lateral motion assists the file to clear itself of chips, and also avoids producing ridges or scores.

Rotary Files and Burs. — Rotary files and burs are used with power-operated tools, such as flexible- or stationary-shaft machines, drilling machines, lathes, and portable electric or pneumatic tools, for abrading or smoothing metals and other materials. Corners can be broken and chamfered, burrs and fins removed, holes and slots enlarged or elongated, and scale removed in die-sinking, metal pattern-making, mold finishing, toolmaking, and casting operations.

The difference between rotary files and rotary burs, as defined by most companies, is that the former have teeth cut by hand with hammer and chisel whereas the latter have teeth or flutes ground from the solid blank after hardening, or milled from the solid blank before hardening. (At least one company, however, prefers to differentiate the two by use and size: The larger-sized general purpose tools with ¼-inch shanks, whether hand cut or ground, are referred to as rotary files; the smaller-shanked — ⅛-inch — and correspondingly smaller-headed tools used by diesinkers and jewelers are referred to as burs.) Rotary files are made from high-speed steel and rotary burs from high-speed steel or cemented carbide in various cuts such as double extra coarse, extra coarse or rough, coarse or standard, medium, fine, and smooth. Standard shanks are ¼ inch in diameter.

Use of Rotary Files and Burs. — The choice between a rotary file and a bur depends on the type of job and the preference of the user. The rotary files, with their hand-cut interrupted teeth or flutes, are better suited for work on tough, dense metals, such as die steels, steel forgings, and electric and gas welds. Generally, the burs are more efficient on non-ferrous metals.

In using rotary files or burs, the tool should be moved at an even rate and pressure to avoid producing an uneven surface. The machine chuck should grip the rotary file or bur shank near the head or cut section for accurate control. For efficient operation, the tools should be kept sharp by regrinding.

The speed at which the tool should be operated varies with the skill and technique of the operator, the type of power used, the material being removed, the type of operation, and the size of the bur. Approximate speeds of medium-cut rotary files and burs for general applications are given in the accompanying table. There are many instances, however, where speeds much higher than those given in the table have been used successfully. The best method is to determine from experience what speeds will give the most satisfactory results.

It is advisable to start using a bur at a fairly low speed and increase the speed as the bur becomes broken in until the best speed is found for the particular operation being performed. For finishing operations, higher than normal speeds with decreased pressure provide a better finish, although possibly with some sacrifice in rate of stock removal. Excessive speeds sometimes make it difficult to control the bur, and should be avoided. Speeds slower than normal usually result in a poorer finish and tend to decrease the rate of stock removal, although the latter is dependent on such factors as the pressure applied, characteristics of the material being operated on, and the manner in which the bur is used.

Effectiveness of Rotary Files and Burs. — There is very little difference in the efficiency of rotary files or burs when used in electric tools and when used in air tools, provided the speeds have been reasonably well selected. Flexible-shaft and other machines used as a source of power for these tools have a limited number of speeds which govern the revolutions per minute at which the tools can be operated.

The carbide bur may be used on hard or soft materials with equally good results. The principle difference in construction of the carbide bur is that its teeth or flutes are provided with a negative rather than a radial rake. Carbide burs are relatively brittle, and must be treated more carefully than ordinary burs. They should be kept cutting freely, in order to prevent too much pressure, which might result in crumbling of the cutting edges.

At the same speeds, both high-speed steel and carbide burs remove approximately the same amount of metal. However, when carbide burs are used at their most efficient speeds, the rate of stock removal may be as much as four times that of ordinary burs. In certain cases, speeds much higher than those shown in the table can be used. It has been demonstrated that a carbide bur will last up to 100 times as long as a high-speed steel bur of corresponding size and shape.

Approximate Speeds of Rotary Files and Burs*

Tool Diam., Inches	Medium Cut, High-Speed Steel Bur or File		
	Mild Steel	Cast Iron	Bronze
	Speed, Revolutions per Minute		
1/8	4600	7000	15,000
1/4	3450	5250	11,250
3/8	2750	4200	9000
1/2	2300	3500	7500
5/8	2000	3100	6650
3/4	1900	2900	6200
7/8	1700	2600	5600
1	1600	2400	5150
1 1/8	1500	2300	4850
1 1/4	1400	2100	4500

Tool Diam., Inches	Medium Cut, High-Speed Steel Bur or File		Carbide Bur	
	Speed, Revolutions Per Minute		Medium Cut	Fine Cut
	Aluminum	Magnesium	Any Material	
1/8	20,000	30,000	45,000	30,000
1/4	15,000	22,500	30,000	20,000
3/8	12,000	18,000	24,000	16,000
1/2	10,000	15,000	20,000	13,350
5/8	8900	13,350	18,000	12,000
3/4	8300	12,400	16,000	10,650
7/8	7500	11,250	14,500	9650
1	6850	10,300	13,000	8650
1 1/8	6500	9750	...	...
1 1/4	6000	9000	...	...

* As recommended by the Nicholson File Company.

Power Brush Finishing

Power brush finishing is a production method of metal finishing that employs wire, elastomer bonded wire, or non-metallic (cord, natural fiber or synthetic) brushing wheels in automatic machines, semi-automatic machines and portable air tools to smooth or roughen surfaces, remove surface oxidation and weld scale or remove burrs.

Description of Brushes. — Brushes work in the following ways: the wire points of a brush can be considered to act as individual cutting tools so that the brush, in effect, is a multiple-tipped cutting tool. The fill material, as it is rotated, contacts the surface of the work and imparts an impact action which produces a coldworking effect. The type of finish produced depends upon the wheel material, wheel speed, and how the wheel is applied. Brushes differ in the following ways: (1) fill material (wire — carbon steel, stainless steel; synthetic; Tampico; and cord), (2) length of fill material (or trim), and (3) the density of the fill material.

To aid in wheel selection and use, the accompanying table made up from information supplied by The Osborn Manufacturing Company lists the characteristics and major uses of brushing wheels.

Use of Brushes. — The brushes should be located so as to bring the full face of the brush in contact with the work. Full face contact is necessary to avoid grooving the brush. Operations that are set up with the brush face not in full contact with the work require some provision for dressing the brush face. When the tips of a brush, used with full face contact, become dull during use with subsequent loss of working clearance, reconditioning and resharpening is necessary. This is accomplished simply and efficiently by alternately reversing the direction of rotation during use.

Deburring and Producing a Radius on the Tooth Profile of Gears. — The brush employed for deburring and producing a radius on the tooth profile of gears is a short trim, dense, wire-fill radial brush. The brush should be set up so as to brush across the edge as shown in Fig. 1A. Line contact brushing, as shown in Fig. 1B should be avoided because the brush face will wear non-uniformly; and the wire points, being flexible, tend to flare to the side, thus minimizing the effectiveness of the brushing operation. When brushing gears, the brushes are spaced and contact the tooth profile on the center line of the gear as shown in Fig. 2. This facilitates using brush reversal to maintain the wire brushing points at their maximum cutting efficiency.

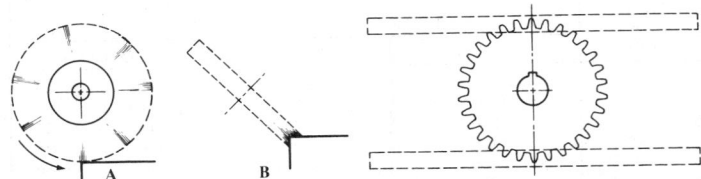

Fig. 1. (Left) — Methods of Brushing an Edge; (A) Correct, (B) Incorrect
Fig. 2. (Right) — Setup for Deburring Gears

The setup for brushing spline bores differs from brushing gears in that the brushes are located off-center, as illustrated in Fig. 3. When helical gears are brushed, it is

sometimes necessary to favor the acute side of the gear tooth to develop a generous radius prior to shaving. This can be accomplished by locating the brushes as shown in Fig. 4. Elastomer bonded wire-filled brushes are used for deburring fine pitch gears. These brushes remove the burrs without leaving any secondary roll. The use of bonded brushes is necessary when the gears are not shaved after hobbing or gear shaping.

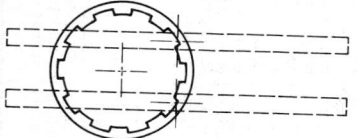

Fig. 3. — Setup for Brushing Broached Splines

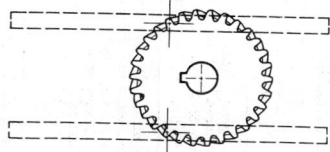

Fig. 4. — Setup for Finishing Helical Gears

Adjustments for Eliminating Undesirable Conditions in Power Brush Finishing

Undesirable Condition	Possible Adjustments for Eliminating Condition
Brush works too slowly	(1) Decrease trim length and increase fill density. (2) Increase filament diameter. (3) Increase surface speed by increasing R.P.M. or outside diameter.
Brush works too fast	(1) Reduce filament diameter. (2) Reduce surface speed by reducing R.P.M. or outside diameter. (3) Reduce fill density. (4) Increase trim length.
Action of brush peens burr to adjacent surface	(1) Decrease trim length and increase fill density. (2) If wire brush tests indicate metal too ductile (burr is peened rather than removed), change to nonmetallic brush such as a treated Tampico brush used with a burring compound.
Finer or smoother finish required	(1) Decrease trim length and increase fill density. (2) Decrease filament diameter. (3) Try treated Tampico or cord brushes with suitable compounds at recommended speeds. (4) Use auxiliary buffing compound with brush.
Finish too smooth and lustrous	(1) Increase trim length. (2) Reduce brush fill density. (3) Reduce surface speed. (4) Increase filament diameter.
Brushing action not sufficiently uniform	(1) Devise hand-held or mechanical fixture or machine which will avoid irregular off-hand manipulation. (2) Increase trim length and decrease fill density.

Characteristics and Applications of Brushes Used in Power Brush Finishing

Brush Type	Description	Operating Speed Range, sfpm	Uses	Remarks
Radial, short trim dense wire fill	Develops very little impact action but maximum cutting action.	6500	Removal of burrs from gear teeth and sprockets. Produces blends and radii at juncture of intersecting surfaces.	Brush should be set up so as to brush across any edge. Reversal of rotation needed to maintain maximum cutting efficiency of brush points.
Radial, medium to long trim twisted knot wire fill	Normally used singly and on portable tools. Brush is versatile and provides high impact action.	7500–9500 for high speeds. 1200 for slow speeds.	For cleaning welds in the automotive and pipeline industries. Also for cleaning surfaces prior to painting, stripping rubber flash from molded products and cleaning mesh-wire conveyor belts.	Surface speed plays an important role since at low speeds the brush is very flexible and at high speeds it is extremely hard and fast cutting.
Radial, medium to long trim crimped wire fill	With the 4- to 8-inch diameter brush, part is hand held. With the 10- to 15-inch diameter brush, part is held by machine.	4500–6000	Serves as utility tool on bench grinder for removing feather grinding burrs, machining burrs, and for cleaning and producing a satin or matte finish.	Good for hand held parts as brush is soft enough to conform to irregular surfaces and hard-to-reach areas. Smaller diameter brushes are not recommended for high-production operations.
Radial, sectional, non-metallic fill (treated and untreated Tampico or cord)	Provides means for improving finish or improving surface for plating. Works best with grease base deburring or buffing compound.	5500–6500 7500 for polishing	For producing radii and improving surface finish. Removes the sharp peaks that fixed abrasives leave on a surface so that surface will accept a uniform plating. Polishing marks and draw marks can be successfully blended.	Brush is selective to an edge which means that it removes metal from an edge but not from adjoining surfaces. It will produce a very uniform radius without peening or rolling any secondary metal.
Radial, wide-face, nonmetallic fill (natural fibers or synthetics)	Can be used with flow-through mounting which facilitates feeding of cold water and hot alkaline solutions through brush face to prevent buildup.	750–1200 for cleaning steel 600 when used with slurries	For cleaning steel. Used in electrolytic tinplate lines, continuous galvanizing and annealing lines, and cold reduction lines. Used to produce dull or matte-type finishes on stainless steel and synthetics.	Speeds above 3600 sfpm will not appreciably improve operation as brush wear will be excessive. Avoid excessive pressures. Ammeters should be installed in drive-motor circuit to indicate brushing pressure.

Characteristics and Applications of Brushes Used in Power Brush Finishing (*Concluded*)

Brush Type	Description	Operating Speed Range, sfpm	Uses	Remarks
Radial, wide face, metallic fill	This brush is made to customer's specifications. It is dynamically balanced at the speed at which it will operate.	2000-4000	Removes buildup of aluminum oxide from work rolls in aluminum mill. Removes lime or magnesium coatings from certain types of steel. Burnishes hot-dipped galvanized steel to produce a minimum spangled surface.	Each brush should have its own drive. An ammeter should be present in drive-motor circuit to measure brushing pressure. If strip is being brushed, a steel backup roll should be opposite the brush roll.
Radial, wide face, strip (interrupted brush face)	Performs cleaning operations that would cause a solid face brush to become loaded and unusable.	When cleaning conveyor belts, brush speed is 2 to 3 times that of conveyer belt.	Need for cleaning rubber and fabric conveyor belts of carry-back material which would normally foul snubber pulley and return idlers.	Designed for medium- to light-duty work. Brush face does not load.
Radial, Cup, Flared End, and Straight End, wire fill elastomer bonded	Extremely fast cutting with maximum operator safety. No loss of wire through fatigue. Always has uniform face.	3600-9000	For removing oxide weld scale, burrs, and insulation from wire.	Periodic reversing of brush direction will result in a brush life ten times greater than non-bonded wheels. Fast cutting action necessitates precise holding of part with respect to brush.
Cup, twisted knot wire fill	Fast cutting wheel used on portable tools to clean welds, scale, rust, and other oxides.	8000-10,000 4500-6500 for deburring and producing a radius around periphery of holes.	Used in shipyards and in structural steel industry. For cleaning outside diameter of pipe and removing burrs and producing radii on heat exchanger tube sheets and laminations for stator cores.	Fast acting brush cleans large areas economically. Setup time is short.
Radial, wire or treated Tampico or cord	For use with standard centerless grinders. Brush will not remove metal from a cylindrical surface. Parts must be ground to size before brushing.	....	For removing feather grinding burrs and improving surface finish. Parts of 25 microinches can be finished down to 15 to 10 microinches. Parts of 10 to 12 microinches can be finished down to 7 to 4 microinches.	Follows centerless grinding principles, except that accuracy in pressure and adjustment is not critical. A machine no longer acceptable for grinding can be used for brushing.

Polishing and Buffing

The terms "polishing" and "buffing" are sometimes applied to similar classes of work in different plants, but according to approved usage of the terms, there is the following distinction: Polishing is any operation performed with wheels having abrasive glued to the working surfaces, whereas, buffing is done with wheels having the abrasive applied loosely instead of imbedding it into glue; moreover, buffing is not so harsh an operation as ordinary polishing, and it is commonly utilized for obtaining very fine surfaces having a "grainless finish."

Polishing Wheels. — The principal materials from which polishing wheels are made are wood, leather, canvas, cotton cloth, felt, paper, walrus or sea-horse hide, sheepskin, impregnated rubber, canvas composition, and wool. Leather and canvas are the materials most commonly used in polishing wheel construction. Bull-neck leather wheels are made of oak tanned bullneck leather cut into disks of uniform thickness and cemented together. Wooden wheels covered with leather to which emery or some other abrasive is glued, are employed extensively for polishing flat surfaces, especially when good edges must be maintained. Canvas wheels are made in various ways; wheels having disks that are cemented together are very hard and used for rough, coarse work, whereas those having sewed disks are made of varying densities by sewing together a larger or smaller number of disks into sections and gluing them. Wheels in which the disks are held together by sewing and which are not stiffened by the use of glue, usually require metal side plates to support the canvas disks. Muslin wheels are made from sewed buffs glued together, but the outer edges of a wheel frequently are left open or free from glue to provide an open face of any desired depth. Wool felt wheels are flexible and resilient, and the density may be varied by sewing two or more disks together and then cementing these to form a wheel. Solid wheels made of Spanish or Mexican felt are quite popular for fine finishing but have little value as general utility wheels. Paper wheels are made from strawboard paper disks and are cemented together under pressure to form a very hard wheel for rough work. Softer wheels are similarly made from felt paper. Walrus leather or sea-horse hide may be used for fine polishing, but these wheels are expensive. The "compress" canvas wheel is commonly used in place of walrus wheels. This compress type of wheel has a cushion of polishing material formed by pieces of leather, canvas, felt, or whatever material is used, which are held in a crosswise radial position by two side plates attached to the wheel hub. This cushion of polishing material may be varied in density to suit the requirements; it may readily be shaped to conform to the curvature of the work and this shape can be maintained. Sheepskin polishing wheels and also paper wheels are used very little at the present time.

Polishing Operations and Abrasives. — Polishing operations on such parts as chisels, hammers, screwdrivers, wrenches, and other parts which are given a fine finish but are not plated, usually require four operations which are "roughing," "dry fining," "greasing" and "coloring." The roughing is frequently regarded as a solid grinding wheel job. Sometimes there are two steps to the greasing operation — rough and fine greasing. For some hardware, such as the cheaper screwdrivers, wrenches, etc., the operations of roughing and dry fining are considered sufficient. For knife blades and cutlery the roughing operation is performed with solid grinding wheels and the polishing is known as fine or blue glazing, but these terms are never used when referring to the polishing of hardware parts, plumbers' supplies, etc. A term used in finishing German silver, white metal, and similar materials is "sand-buffing," which, in distinction from the ordinary buffing operation that is used only to produce a very high finish, actually removes considerable metal, as in rough polishing or flexible grinding. For sand-buffing, rotten-stone and pumice are loosely applied.

Aluminum oxide abrasives are widely used for polishing high tensile strength metals such as carbon and alloy steels, tough iron, and non-ferrous alloys. Silicon carbide abrasives are recommended for hard, brittle substances such as grey iron, cemented carbide tools, and also materials of low tensile strength such as brass, aluminum and copper.

Buffing Wheels. — Buffing wheels, as defined by the Metal Finishers' Equipment Association, are wheels manufactured from disks (either whole or pieced) of bleached or unbleached cotton or woolen cloth, and they are used as the agent for carrying abrasive powders, such as tripoli, crocus, rouge, lime, etc., which are mixed with waxes or greases as a bond. There are two main classes of buffs known as the "pieced-sewed" buffs, which are made from various weaves and weights of cloths, and the "full disk" buffs which are made from the best sheeting and shirting. Bleached cloth is harder and stiffer than unbleached cloth, and is used for the faster cutting buffs. Coarsely woven unbleached cloth is recommended for highly colored work on soft metals, while the finer woven unbleached cloths are better adapted for the harder metals. A stiff buff when working at the usual speed is not suitable for "cutting down" soft metal or for use on light plated ware, but is used on the harder metals and for heavy nickel-plated articles.

Speed of Polishing Wheels. — The proper speed for polishing is governed to some extent by the nature of the work, but for ordinary operations the polishing wheel, according to one manufacturer, should have a peripheral speed of about 7500 feet per minute. If run at a lower rate of speed, the work tends to tear the polishing material from the wheel too readily, and the work is not as good in quality. Another manufacturer recommends the following speeds: Muslin, felt or sea-horse polishing wheels having wood or iron centers should be run at peripheral speeds varying from 3000 to 7000 feet per minute. It is rarely necessary to exceed 6000 feet per minute, and for most purposes 4000 feet per minute is sufficient. If the wheels are kept in good condition, in perfect balance, and are suitably mounted on substantial buffing lathes, they can safely be used for speeds within the limits given.

Grain Numbers of Emery. — The numbers commonly used in designating the different grains of emery, corundum and other abrasives are 10, 12, 14, 16, 18, 20, 24, 30, 36, 40, 46, 54, 60, 70, 80, 90, 100, 120, 150, 180 and 200, ranging from coarse to fine. These numbers represent the number of meshes per linear inch in the grading sieve. An abrasive finer than No. 200 is known as "flour" and the degree of fineness is designated by the letters CF, F, FF, FFF, FFFF and PCF or SF, ranging from coarse to fine. The methods of grading flour-emery adopted by different manufacturers do not exactly agree, the letters differing somewhat for the finer grades.

Grades of Emery Cloth. — The coarseness of emery cloth is indicated by letters and numbers corresponding to the grain number of loose emery. The letters and numbers for grits ranging from fine to coarse are as follows: FF, F, 120, 100, 90, 80, 70, 60, 54, 46, 40. For large work roughly filed, use coarse cloth such as Nos. 46 or 54, and then finer grades to obtain the required polish. If the work has been carefully filed, a good polish can be obtained with Nos. 60 and 90 cloth, and a brilliant polish by finishing with No. 120 and flour-emery.

Mixture for Cementing Emery Cloth to a Lapping Wheel. — Use 4½ pounds of rosin; 3 pounds of paraffine; 9 ounces of vaseline; melt the ingredients and mix them thoroughly. Heat the surface of the lapping wheel and spread on the mixture; then rub the emery cloth down so as to exclude all air from between the surface of the wheel and cloth. The surface of the lapping wheel should be clean before the cement is applied.

Etching and Etching Fluids

Etching Fluids for Different Metals. — A common method of etching names or simple designs upon steel is to apply a thin, even coating of beeswax or some similar substance which will resist acid; then mark the required lines or letters in the wax with a sharp-pointed scriber, thus exposing the steel (where the wax has been removed by the scriber point) to the action of an acid, which is finally applied. To apply a very thin coating of beeswax, place the latter in a silk cloth, warm the piece to be etched, and rub the pad over it. Regular coach varnish is also used instead of wax, as a "resist."

An etching fluid ordinarily used for carbon steel consists of nitric acid, 1 part; water, 4 parts. It may be necessary to vary the amount of water, as the exact proportion depends upon the carbon content and whether the steel is hard or soft. For hard steel, use nitric acid, 2 parts; acetic acid, 1 part. For high-speed steel, nickel or brass, use nitro-hydrochloric acid (nitric, 1 part; hydrochloric, 4 parts). For high-speed steel it is sometimes better to add a little more nitric acid. For etching bronze, use nitric acid, 100 parts; muriatic acid, 5 parts. For brass, nitric acid, 16 parts; water, 160 parts; dissolve 6 parts potassium chlorate in 100 parts of water; then mix the two solutions and apply.

A fluid which may be used either for producing a frosted effect or for deep etching (depending upon the time it is allowed to act) is composed of 1 ounce sulphate of copper (blue vitriol); ¼ ounce alum; ½ teaspoonful of salt; 1 gill of vinegar, and 20 drops of nitric acid. For aluminum, use a solution composed of alcohol, 4 ounces; acetic acid, 6 ounces; antimony chloride, 4 ounces; water, 40 ounces.

Various acid-resisting materials are used for covering the surfaces of steel rules, etc., prior to marking off the lines on a graduating machine. When the graduation lines are fine and very closely spaced, as on machinists' scales which are divided into hundredths or sixty-fourths, it is very important to use a thin resist that will cling to the metal and prevent any under-cutting of the acid; the resist should also enable fine lines to be drawn without tearing or crumbling as the tool passes through it. One resist that has been extensively used is composed of about 50 per cent of asphaltum, 25 per cent of beeswax, and, in addition, a small percentage of Burgundy pitch, black pitch, and turpentine. A thin covering of this resisting material is applied to the clean polished surface to be graduated and, after it is dry, the work is ready for the graduating machine. For some classes of work, paraffin is used for protecting the surface surrounding the graduation lines which are to be etched. The method of application consists in melting the paraffin and raising its temperature high enough so that it will flow freely; then the work is held at a slight angle and the paraffin is poured on its upper surface. The melted paraffin forms a thin protective coating.

Conversion Coatings and the Coloring of Metals

Conversion Coatings. — Conversion coatings are thin, adherent chemical compounds that are produced on metallic surfaces by chemical or electrochemical treatment. These coatings are insoluble, passive, and protective, and are divided into two basic systems: oxides or mixtures of oxides with other compounds, usually chromates or phosphates. Conversion coatings are used for corrosion protection, as an adherent paint base, and for decorative purposes because of their inherent color and because they can absorb dyes and colored sealants.

Conversion coatings are produced in three or four steps. First there is a pretreatment, which often involves mechanical surface preparation followed by degreasing and/or chemical or electrochemical cleaning or etching. Then thermal, chemical, or electrochemical surface conversion processes take place in acid or alkaline solutions applied by immersion, spraying, or brushing. A posttreatment follows, which includes rinsing and drying, and may also include sealing or dyeing. If coloring is the main purpose of the coating, then oiling, waxing, or lacquering may be required.

Passivation of Copper. — The blue-green patina that forms on copper alloys during atmospheric exposure is a passivated film; i.e., it prevents corrosion. This patina may be produced artificially or its growth may be accelerated by a solution of ammonium sulfate, 6 pounds; copper sulfate, 3 ounces; ammonia (technical grade, 0.90 specific gravity), 1.34 fluid ounces; and water, 6.5 gallons. This solution is applied as a fine spray to a chemically cleaned surface and is allowed to dry between each of five or six applications. In about 6 hours a patina somewhat bluer than natural begins to develop and continues after exposure to weathering.

Small copper parts can be coated with a passivated film by immersion in or brushing with a solution consisting of the following weight proportions: copper, 30; nitric acid, concentrated, 60; acetic acid (6%), 600; ammonium chloride, 11; and ammonium hydroxide (technical grade, 0.90 specific gravity), 20. To prepare the solution, the copper is dissolved in the nitric acid before the remaining chemicals are added, and the solution is allowed to stand for several days before use. A coating of linseed oil is applied to the treated parts.

Coloring of Copper Alloys. — Metals are colored to enhance their appearance, to produce an undercoat for an organic finish, or to reduce light reflection. Copper alloys can be treated to produce a variety of colors, with the final color depending on the base metal composition, the coloring solution's composition, the immersion time, and the operator's skill. Cleaning is an important part of the pretreatment; nitric and sulfuric acid solutions are used to remove oxides and to activate the surface.

The following solutions are used to color alloys that contain 85 per cent or more of copper. A dark red color is produced by immersing the parts in molten potassium nitrate, at 1200–1300°F, for up to 20 seconds, followed by a hot water quench. The parts must then be lacquered. A steel black color can be obtained by immersing the parts in a 180°F solution of arsenious oxide (white arsenic), 4 ounces; hydrochloric acid (1.16 specific gravity), 8 fluid ounces; and water, 1 gallon. The parts are immersed until a uniform color is obtained; they are scratch brushed while wet, and then dried and lacquered. A light brown color is obtained using a room-temperature solution of barium sufate, 0.5 ounce; ammonium carbonate, 0.25 ounce; and water, 1 gallon.

The following solutions are used to color alloys that contain less than 85 per cent copper. To color brass black, parts are placed in an oblique tumbling barrel made of stainless steel and covered with 3 to 5 gallons of water. Three ounces of copper sulfate and 6 ounces of sodium thiosulfate are dissolved in warm water and added to the barrel's contents. After tumbling for 15 to 30 minutes to obtain the finish, the solution is drained from the barrel, and the parts are washed thoroughly in clean water, dried in sawdust or air-blasted, and, if necessary, lacquered. To produce a blue-black color, the parts are immersed in a 130–175°F solution of copper carbonate, 1 pound; ammonium hydroxide (0.89 specific gravity), 1 quart; and water, 3 quarts. Excess copper carbonate should be present. The proper color is obtained in 1 minute. To color brass a hardware green, immerse the parts in a 160°F solution of ferric nitrate, 1 ounce; sodium thiosulfate, 6 ounces; and water, 1 gallon. To color brass a light brown, immerse the parts in a 195–212°F solution of potassium chlorate, 5.5 ounces; nickel sulfate, 2.75 ounces; copper sulfate, 24 ounces; and water, 1 gallon.

Posttreatment: The treated parts should be scratch brushed to remove any excess or loose deposits. A contrast of colors may be obtained by brushing with a slurry of fine pumice, hand rubbing with an abrasive paste, mass finishing, or buffing to remove the color from the highlights. In order to prolong the life of parts used for outdoor decorative purposes, a clear lacquer should be applied. Parts intended for indoor purposes are often used without additional protection.

Coloring of Iron and Steel. — Thin black oxide coatings are applied to steel by immersing the parts to be coated in a boiling solution of sodium hydroxide and

mixtures of nitrates and nitrites. These coatings serve as paint bases and, in some cases, as final finishes. When the coatings are impregnated with oil or wax, they furnish fairly good corrosion resistance. These finishes are relatively inexpensive compared to other coatings.

Phosphate Coatings: Phosphate coatings are applied to iron and steel parts by reacting them with a dilute solution of phosphoric acid and other chemicals. The surface of the metal is converted into an integral, mildly protective layer of insoluble crystalline phosphate. Small items are coated in tumbling barrels; large items are spray coated on conveyors.

The three types of phosphate coatings in general use are zinc, iron, and manganese. Zinc phosphate coatings vary from light to dark gray. The color depends on the carbon content and pretreatment of the steel's surface, as well as the composition of the solution. Zinc phosphate coatings are generally used as a base for paint or oil, as an aid in cold working, for increased wear resistance, or for rustproofing. Iron phosphate coatings were the first type to be used; they produce dark gray coatings and their chief application is as a paint base. Manganese phosphate coatings are usually dark gray; however, since they are used almost exclusively as an oil base, for break in and to prevent galling, they become black in appearance.

In general, stainless steels and certain alloy steels cannot be phosphated. Most cast irons and alloy steels accept coating with various degrees of difficulty depending on alloy content.

Anodizing Aluminum Alloys. — In the anodizing process, the aluminum object to be treated is immersed as the anode in an acid electrolyte, and a direct current is applied. Oxidation of the surface occurs, producing a greatly thickened, hard, porous film of aluminum oxide. The object is then immersed in boiling water to seal the porosity and render the film impermeable. Before sealing, the film can be colored by impregnation with dyes or pigments. Special electrolytes may also be used to produce colored anodic films directly in the anodizing bath. The anodic coatings are used primarily for corrosion protection and abrasion resistance, and as a paint base.

The three principal types of anodizing processes are: chromic, in which the active agent is chromic acid; sulfuric, in which the active agent is sulfuric acid; and hard anodizing, in which sulfuric acid is used by itself or with additives in a low-temperature electrolyte bath. Most of the anodic coatings range in thickness from 0.2 to 0.7 mil. The hard anodizing process can produce coatings up to 2 mils. The chromic acid coating is less brittle than the sulfuric, and, since the chromic electrolyte does not attack aluminum, it does not present a corrosion problem when it is trapped in crevices. The chromic coating is less resistant to abrasion than the sulfuric, but it cannot be used with alloys containing more than 5 per cent copper due to corrosion of the base metal.

Chemical Conversion Coatings for Aluminum: Chemical conversion coatings for aluminum alloys are adherent surface layers of low solubility oxide, phosphate, or chromate compounds produced by the reaction of the metal surface with suitable reagents. The conversion coatings are much thinner and softer than anodic coatings, but they are less expensive and serve as an excellent paint base.

Magnesium Alloys. — Chemical treatment of magnesium alloys is used to provide a paint base and to improve corrosion resistance. The popular conversion "dip" coatings are chrome pickle and dichromate treatments, and they are very thin. Anodic coatings are thicker and harder, and, after sealing, give the same protection against corrosion, although painting is still desirable.

Titanium Alloys. — Chemical conversion coatings are used on titanium alloys to improve lubricity by acting as a base for the retention of lubricants. The coatings are applied by immersion, spraying, or brushing. A popular coating bath is an aqueous solution of phosphates, fluorides, and hydrofluoric acid. The coating is comprised primarily of titanium and potassium fluorides and phosphates.

RIVETS AND RIVETED JOINTS

Classes and Types of Riveted Joints. — Riveted joints may be classified by application as: (1) pressure vessel; (2) structural; and (3) machine member. For information and data concerning joints for pressure vessels such as boilers, reference should be made to standard sources such as the ASME Boiler Code. The following sections will cover only structural and machine-member riveted joints.

Basically there are two kinds of riveted joints, the *lap-joint* and the *butt-joint*. In the ordinary *lap-joint* the plates overlap each other and are held together by one or more rows of rivets. In the *butt-joint* the plates being joined are in the same plane and are joined by means of a cover plate or butt strap which is riveted to both plates by one or more rows of rivets. The term *single riveting* means one row of rivets in a lap-joint or one row on each side of a butt-joint; *double riveting* means two rows of rivets in a lap-joint or two rows on each side of the joint in butt riveting. Joints are also triple and quadruple riveted. Lap-joints may also be made with inside or outside cover plates. Types of lap and butt joints are illustrated in the table on pages 1263 and 1264.

General Considerations of a Riveted Joint. — Factors to be considered in the design or specification of a riveted joint are: type of joint, spacing of rivets, type and size of rivet, type and size of hole, and rivet material.

Spacing of Rivets: The spacing between rivet centers is called *pitch* and between row center lines, *back pitch* or *transverse pitch*. The distance between centers of rivets nearest each other in adjacent rows is called *diagonal pitch*. The distance from the edge of the plate to the center line of the nearest row of rivets is called *margin*.

Examination of a riveted joint made up of several rows of rivets will reveal that after progressing along the joint a given distance the rivet pattern or arrangement is repeated. (For a butt joint, the length of a *repeating section* is usually equal to the *long pitch* or pitch of the rivets in the outer row, i.e. the row farthest from the edge of the joint.) For structural and machine-member joints the proper pitch may be determined by making the tensile strength of the plate over the length of the repeating section, i.e. distance between rivets in the outer row, equal to the total shear strength of the rivets in the repeating section. Minimum pitch and diagonal pitch are also governed by the clearance required for the hold-on (Dolly bar) and rivet set. Dimensions for different sizes of hold-ons and rivet sets are given in the table on page 1198.

When fastening thin plate it is particularly important to maintain accurate spacing to avoid buckling.

Size and Type of Rivets: The rivet diameter d commonly falls between $d = 1.2\sqrt{t}$ and $d = 1.4\sqrt{t}$, where t is the thickness of the plate. Dimensions for various types of American Standard large (½-inch diameter and up) rivets and small solid rivets are shown in tables which follow. It may be noted that countersunk heads are not as strong as other types.

Size and Type of Hole: Rivet holes may be punched, punched and reamed, or drilled. Rivet holes are usually made ¹⁄₁₆ inch larger in diameter than the nominal diameter of the rivet although in some classes of work in which the rivet is driven cold, as in automatic machine riveting, the holes are reamed to provide minimum clearance so that the rivet fills the hole completely.

When holes are punched in heavy steel plate, there may be considerable loss of strength unless the holes are reamed to remove the inferior metal immediately surrounding them. This results in the diameter of the punched hole being increased by from ¹⁄₁₆ to ⅛ inch. Annealing after punching tends to restore the strength of the plate in the vicinity of the holes.

Rivet Material: Rivets for structural and machine member purposes are usually made of wrought iron or soft steel but for aircraft and other applications where light weight or resistance to corrosion is important, copper, aluminum alloy, Monel, Inconel, etc., may be used as rivet material.

Failure of Riveted Joints. — Rivets may fail by:

1. Shearing through one cross-section (single shear)
2. Shearing through two cross-sections (double shear)
3. Crushing

Plates may fail by:

4. Shearing along two parallel lines extending from opposite sides of the rivet hole to the edge of the plate
5. Tearing along a single line from middle of rivet hole to edge of plate
6. Crushing
7. Tearing between adjacent rivets (tensile failure) in the same row or in adjacent rows.

Types 4 and 5 failures are caused by rivets being placed too close to the edge of the plate. These types of failure are avoided by placing the center of the rivet at a minimum of one and one-half times the rivet diameter away from the edge.

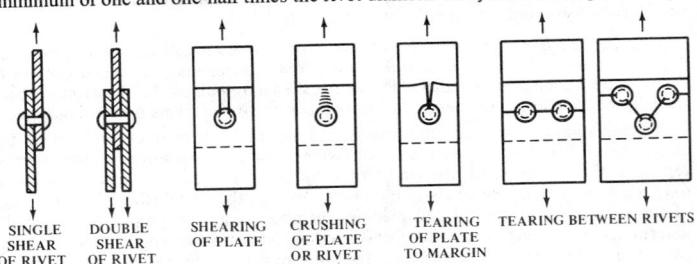

| SINGLE SHEAR OF RIVET | DOUBLE SHEAR OF RIVET | SHEARING OF PLATE | CRUSHING OF PLATE OR RIVET | TEARING OF PLATE TO MARGIN | TEARING BETWEEN RIVETS |

Types of Rivet and Plate Failure

Failure due to tearing on a diagonal between rivets in adjacent rows when the pitch is four times the rivet diameter or less is avoided by making the transverse pitch one and three-quarters times the rivet diameter.

Theoretical versus Actual Riveted Joint Failure. — If it is assumed that the rivets are placed the suggested distance from the edge of the plate and each row the suggested distance from another row, then the failure of a joint is most likely to occur as a result of shear failure of the rivets, bearing failure (crushing) of the plate or rivets, or tensile failure of the plate, alone or in combination depending on the make-up of the joints.

Joint failure in actuality is more complex than this. Rivets do not undergo pure shear especially in lap-joints where rivets are subjected to single shear. The rivet, in this instance, would be subject to a combination of tensile and shearing stresses and it would fail because of combined stresses, not a single stress. Furthermore, the shearing stress is usually considered to be distributed evenly over the cross-section which is also not the case.

Rivets that are usually driven hot contract upon cooling. This contraction in the length of the rivet draws the plates together and sets up a stress in the rivet estimated to be equal in magnitude to the yield point of the rivet steel. The contraction in the diameter of the rivet results in a little clearance between the rivet

and the hole in the plate. The tightness in the plates caused by the contraction in length of the rivet gives rise to a condition where quite a sizeable frictional force would have to be overcome before the plates would slip over one another and subject the rivets to a shearing force. It is European practice to design joints for resistance to this slipping. It has been found, however, that the strength-basis designs obtained in American and English practice are not very different from European designs.

Design of Riveted Joints. — In the design of riveted joints a simplified treatment is frequently used in which the following assumptions are made:

1. The load is carried equally by the rivets.
2. No combined stresses act on a rivet to cause failure.
3. The shearing stress in a rivet is uniform across the cross-section under question.
4. The load that would cause failure in single shear would have to be doubled to cause failure in double shear.
5. The bearing stress of rivet and plate is distributed equally over the projected area of the rivet.
6. The tensile stress is uniform in the section of metal between the rivets.

Allowable Stresses. — The design stresses for riveted joints are usually set by codes, practices, or specifications. The American Institute of Steel Construction issues specifications for the design, fabrication and erection of structural steel for buildings in which the allowable stress permitted in tension for structural steel and rivets is specified at 20,000 pounds per square inch, the allowable bearing stress for rivets is 40,000 psi in double shear and 32,000 psi in single shear, and the allowable shearing stress for rivets is 15,000 psi. The American Society of Mechanical Engineers in its Boiler Code lists the following ultimate stresses: tensile, 55,000 psi; shearing, 44,000 psi; compressive or bearing, 95,000 psi. The design stresses usually are one-fifth of these, i.e.: tensile, 11,000 psi; shearing, 8800 psi; compressive or bearing, 19,000 psi. In machine design work values close to these or somewhat lower are commonly used.

Analysis of Joint Strength. — The following examples and strength analyses of riveted joints are based upon the six assumptions previously outlined for a simplified treatment.

Example 1. Consider a 12-inch section of single-riveted lap-joint made up with plates of ¼-inch thickness and six rivets, ⅝ inch in diameter. Assume that rivet holes are ¹⁄₁₆ inch larger in diameter than the rivets. In this joint, the entire load is transmitted from one plate to the other by means of the rivets. Each plate and the six rivets carry the entire load. The safe tensile load L and the efficiency η may be determined in the following way: Design stresses of 8500 psi for shear, 20,000 psi for bearing, and 10,000 psi for tension are arbitrarily assigned and it is assumed that the rivets will not tear or shear through the plate to the edge of the joint.

(a) The safe tensile load L based on single shear of the rivets is equal to the number of rivets n times the cross-sectional area of one rivet A_r times the allowable shearing stress S_s or

$$L = n \times A_r \times S_s$$

$$L = 6 \times \frac{\pi}{4}(0.625)^2 \times 8500$$

$$L = 15,647 \text{ pounds}$$

(b) The safe tensile load L based on bearing stress is equal to the number of rivets n times the projected bearing area of the rivet A_b (diameter times thickness of

plate) times the allowable bearing stress S_c or $L = n \times A_b \times S_c = 6 \times (0.625 \times 0.25) \times 20,000 = 18,750$ pounds.

(c) The safe load L based on the tensile stress is equal to the net cross sectional area of the plate between rivet holes A_p times the allowable tensile stress S_t or $L = A_p \times S_t = 0.25[12 - 6(0.625 + 0.0625)] \times 10,000 = 19,688$ pounds.

The safe tensile load for the joint would be the least of the three loads just computed or 15,647 pounds and the efficiency η would be equal to this load divided by the tensile strength of the section of plate under consideration, if it were unperforated or $\eta = \dfrac{15,647}{12 \times 0.25 \times 10,000} \times 100 = 52.2$ per cent.

Example 2. Under consideration is a 12-inch section of double-riveted butt-joint with main plates ½ inch thick and two cover plates each ⁵⁄₁₆ inch thick. There are 3 rivets in the inner row and 2 on the outer and their diameters are ⅞ inch. Assume that the diameter of the rivet holes is ¹⁄₁₆ inch larger than that of the rivets. The rivets are so placed that the main plates will not tear diagonally from one rivet row to the others nor will they tear or fail in shear out to their edges. The safe tensile load L and the efficiency η may be determined in the following way: Design stresses for 8500 psi for shear, 20,000 psi for bearing, and 10,000 psi for tension are arbitrarily assigned.

(a) The safe tensile load L based on double shearing of the rivets is equal to the number of rivets n times the number of shearing planes per rivet times the cross sectional area of one rivet A_r times the allowable shearing stress S_s or $L = n \times 2 \times A_r \times S_s = 5 \times 2 \times \dfrac{\pi}{4}(0.875)^2 \times 8500 = 51,112$ pounds.

(b) The safe tensile load L based on bearing stress is equal to the number of rivets n times the projected bearing area of the rivet A_b (diameter times thickness of plate) times the allowable bearing stress S_c or $L = n \times A_b \times S_c = 5 \times (0.875 \times 0.5) \times 20,000 = 43,750$ pounds.

(Cover plates are not considered since their combined thickness is ¼ inch greater than the main plate thickness.)

(c) The safe tensile load L based on the tensile stress is equal to the net cross sectional area of the plate between the two rivets in the outer row A_p times the allowable tensile stress S_t or $L = A_p \times S_t = 0.5[12 - 2(0.875 + 0.0625)] \times 10,000 = 50,625$ pounds.

In completing the analysis, the sum of the load that would cause tearing between rivets in the three-hole section plus the load carried by the two rivets in the two-hole section is also investigated. The sum is necessary because if the joint is to fail, it must fail at both sections simultaneously. The least safe load that can be carried by the two rivets of the two-hole section is based on the bearing stress (see the foregoing calculations).

(1) The safe tensile load L based on the bearing strength of two rivets of the two-hole section is $L = n \times A_b \times S_c = 2 \times (0.875 \times 0.5) \times 20,000 = 17,500$ pounds.

(2) The safe tensile load L based on the tensile strength of the main plate between holes in the three-hole section is $L = A_p \times S_t = 0.5[12 - 3(0.875 + 0.0625)] \times 10,000 = 45,938$ pounds.

The total safe tensile load based on this combination is $17,500 + 45,938 = 63,438$ pounds which is greater than any of the other results obtained.

The safe tensile load for the joint would be the least of the loads just computed or 43,750 pounds and the efficiency η would be equal to this load divided by the tensile strength of the section of plate under consideration, if it were unperforated or

$$\eta = \frac{43,750}{0.5 \times 12 \times 10,000} \times 100 = 72.9 \text{ per cent.}$$

Analysis of Riveted Joints — 1

A riveted joint may fail by shearing through the rivets (single or double shear), crushing the rivets, tearing the plate between the rivets, crushing the plate or by a combination of two or more of the foregoing causes. Rivets placed too close to the edge of the plate may tear or shear the plate out to the edge but this type of failure is avoided by placing the center of the rivet 1.5 times the rivet diameter away from the edge.

The efficiency of a riveted joint is equal to the strength of the joint divided by the strength of the unriveted plate, expressed as a percentage.

In the following formulas, let,

d = diameter of rivets;	p = pitch of inner row of rivets;
D = diameter of holes;	P = pitch of outer row of rivets;
t = thickness of plate;	S_s = shear stress for rivets;
t_c = thickness of cover plates;	S_t = tensile stress for plates;
S_c = compressive or bearing stress for rivets or plates.	

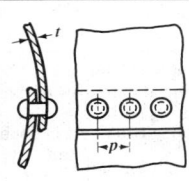

For Single-riveted Lap-joint

(1) Resistance to shearing one rivet $= \dfrac{\pi d^2}{4} S_s$

(2) Resistance to tearing plate between rivets
$= (p - D)t S_t$

(3) Resistance to crushing rivet or plate $= dt S_c$

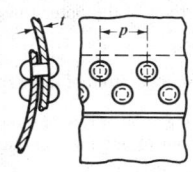

Double-riveted Lap-joint

(1) Resistance to shearing two rivets $= \dfrac{2\pi d^2}{4} S_s$

(2) Resistance to tearing between two rivets
$= (p - D)t S_t$

(3) Resistance to crushing in front of two rivets $= 2dt S_c$

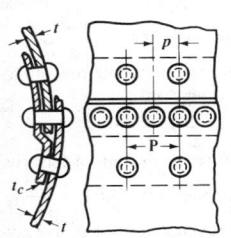

Single-riveted Lap-joint with Inside Cover Plate

(1) Resistance to tearing between outer row of rivets $= (P - D)t S_t$

(2) Resistance to tearing between inner row of rivets, and shearing outer row of rivets $=$
$(P - 2D)t S_t + \dfrac{\pi d^2}{4} S_s$

(3) Resistance to shearing three rivets $= \dfrac{3\pi d^2}{4} S_s$

(4) Resistance to crushing in front of three rivets $= 3td S_c$

(5) Resistance to tearing at inner row of rivets, and crushing in front of one rivet in outer row $= (P - 2D)t S_t + td S_c$

Analysis of Riveted Joints — 2

Double-riveted Lap-joint with Inside Cover Plate

(1) Resistance to tearing at outer row of rivets
$$= (P - D)tS_t$$

(2) Resistance to shearing four rivets $= \dfrac{4\pi d^2}{4}S_s$

(3) Resistance to tearing at inner row and shearing outer row of rivets
$$= (P - 1\tfrac{1}{2}D)\, tS_t + \frac{\pi d^2}{4}S_s$$

(4) Resistance to crushing in front of four rivets $= 4tdS_c$

(5) Resistance to tearing at inner row of rivets, and crushing in front of one rivet =
$$(P - 1\tfrac{1}{2}D)\, tS_t + tdS_c$$

Double-riveted Butt-joint

(1) Resistance to tearing at outer row of rivets
$$= (P - D)\, tS_t$$

(2) Resistance to shearing two rivets in double shear and one in single shear $= \dfrac{5\pi d^2}{4}S_s$

(3) Resistance to tearing at inner row of rivets and shearing one rivet of the outer row =
$$(P - 2D)\, tS_t + \frac{\pi d^2}{4}S_s$$

(4) Resistance to crushing in front of three rivets $= 3tdS_c$

(5) Resistance to tearing at inner row of rivets, and crushing in front of one rivet in outer row $= (P - 2D)\, tS_t + tdS_c$

Triple-riveted Butt-joint

(1) Resistance to tearing at outer row of rivets
$$= (P - D)\, tS_t$$

(2) Resistance to shearing four rivets in double shear and one in single shear $= \dfrac{9\pi d^2}{4}S_s$

(3) Resistance to tearing at middle row of rivets and shearing one rivet
$$= (P - 2D)\, tS_t + \frac{\pi d^2}{4}S_s$$

(4) Resistance to crushing in front of four rivets and shearing one rivet $= 4dtS_c + \dfrac{\pi d^2}{4}S_s$

(5) Resistance to crushing in front of five rivets $= 4dtS_c + dt_cS_c$

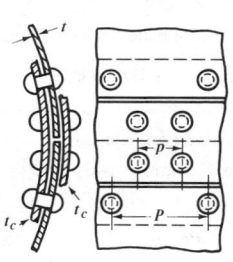

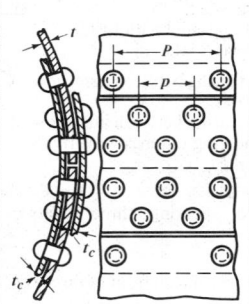

Dimensions are usually specified in inches and stresses in pounds per square inch. See page 1261 for a discussion of allowable stresses which may be used in calculating the strengths given by the formulas. The design stresses are usually set by codes, practices, or specifications.

Rivet Lengths for Forming Round and Countersunk Heads*

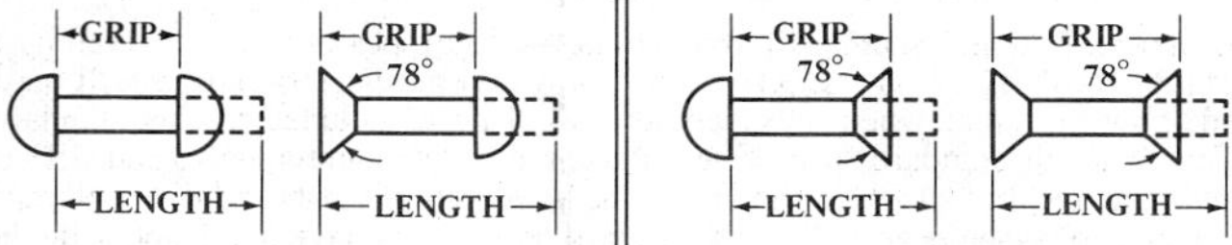

Grip in Inches	To Form Round Head						
	Diameter of Rivet in Inches						
	½	⅝	¾	⅞	1	1⅛	1¼
	Length of Rivet in Inches						
½	1⅝	1⅞	1⅞	2	2⅛	...	...
⅝	1¾	2	2	2⅛	2¼	...	...
¾	1⅞	2⅛	2⅛	2¼	2⅜	...	...
⅞	2	2¼	2¼	2⅜	2½	...	...
1	2¼	2⅜	2⅜	2½	2⅝	2¾	2⅞
1⅛	2⅜	2½	2½	2⅝	2¾	2⅞	3
1¼	2½	2⅝	2⅝	2¾	2⅞	3	3⅛
1⅜	2⅝	2¾	2¾	2⅞	3	3⅛	3¼
1½	2⅞	3	3	3⅛	3¼	3⅜	3½
1⅝	3	3⅛	3⅛	3¼	3⅜	3½	3½
1¾	3⅛	3¼	3¼	3½	3⅝	3¾	3¾
1⅞	3¼	3⅜	3⅜	3⅝	3¾	3⅞	3⅞
2	3½	3½	3⅝	3¾	3⅞	4	4
2⅛	3⅝	3⅝	3¾	3⅞	4	4⅛	4⅛
2¼	3¾	3⅞	3⅞	4	4⅛	4¼	4¼
2⅜	4	4	4	4⅛	4¼	4⅜	4⅜
2½	4⅛	4⅛	4⅛	4¼	4⅜	4½	4½
2⅝	4¼	4¼	4¼	4⅜	4½	4⅝	4⅝
2¾	4⅜	4⅜	4⅜	4½	4⅝	4¾	4¾
2⅞	4⅝	4⅝	4⅝	4⅝	4¾	4⅞	5
3	...	4¾	4¾	4⅞	5	5	5⅛
3⅛	...	4⅞	4⅞	5	5⅛	5¼	5¼
3¼	...	5	5	5⅛	5¼	5⅜	5⅜
3⅜	...	5⅛	5⅛	5¼	5⅜	5½	5½
3½	...	5⅜	5⅜	5⅜	5½	5⅝	5⅝
3⅝	...	5½	5½	5½	5⅝	5¾	5¾
3¾	...	5⅝	5⅝	5⅝	5¾	5⅞	5⅞
3⅞	...	5¾	5¾	5¾	5⅞	6	6
4	...	...	5⅞	6	6	6⅛	6¼
4⅛	...	...	6	6⅛	6¼	6⅜	6⅜
4¼	...	...	6⅛	6¼	6½	6½	6½
4⅜	...	...	6⅜	6½	6½	6⅝	6⅝
4½	...	...	6½	6⅝	6⅝	6¾	6¾
4⅝	...	...	6⅝	6¾	6¾	6¾	6⅞
4¾	...	...	6¾	6⅞	6⅞	7	7
4⅞	...	...	6⅞	7	7	7⅛	7⅛
5	...	...	...	7⅛	7⅛	7¼	7¼
5⅛	...	...	...	7¼	7¼	7⅜	7⅜
5¼	...	...	...	7⅜	7⅜	7½	7½
5⅜	...	...	...	7⅝	7⅝	7¾	7⅝
5½	...	...	...	7¾	7¾	7⅞	7⅞
5⅝	...	...	...	7⅞	7⅞	8	8
5¾	...	...	...	8	8	8⅛	8⅛
5⅞	...	...	...	8⅛	8⅛	8¼	8¼

Grip in Inches	To Form Countersunk Head						
	Diameter of Rivet in Inches						
	½	⅝	¾	⅞	1	1⅛	1¼
	Length of Rivet in Inches						
½	1	1	1⅛	1¼	1¼	...	...
⅝	1⅛	1¼	1¼	1⅜	1⅜	...	...
¾	1⅜	1⅜	1⅜	1½	1½	...	...
⅞	1½	1½	1½	1⅝	1⅝	...	...
1	1⅝	1⅝	1⅝	1¾	1¾	1⅞	1⅞
1⅛	1¾	1¾	1⅞	1⅞	1⅞	2	2
1¼	2	2	2	2	2	2⅛	2⅛
1⅜	2⅛	2⅛	2⅛	2¼	2¼	2⅜	2⅜
1½	2¼	2¼	2¼	2⅜	2⅜	2½	2½
1⅝	2⅜	2⅜	2⅜	2½	2½	2⅝	2⅝
1¾	2⅝	2⅝	2⅝	2⅝	2⅝	2¾	2¾
1⅞	2¾	2¾	2¾	2¾	2¾	2⅞	2⅞
2	2⅞	2⅞	2⅞	2⅞	2⅞	3	3
2⅛	3⅛	3	3	3	3	3⅛	3⅛
2¼	3¼	3⅛	3⅛	3⅛	3¼	3¼	3¼
2⅜	3⅜	3⅜	3⅜	3⅜	3⅜	3⅜	3⅜
2½	3½	3½	3½	3½	3½	3⅝	3⅝
2⅝	3¾	3⅝	3⅝	3⅝	3⅝	3¾	3¾
2¾	3⅞	3¾	3¾	3¾	3¾	3⅞	3⅞
2⅞	4	3⅞	3⅞	3⅞	3⅞	4	4
3	...	4⅛	4⅛	4⅛	4⅛	4⅛	4⅛
3⅛	...	4¼	4¼	4¼	4¼	4¼	4¼
3¼	...	4⅜	4⅜	4⅜	4⅜	4⅜	4⅜
3⅜	...	4½	4½	4½	4½	4½	4½
3½	...	4⅝	4⅝	4⅝	4⅝	4⅝	4⅝
3⅝	...	4¾	4¾	4¾	4¾	4⅞	4⅞
3¾	...	5	5	5	5	5	5
3⅞	...	5⅛	5⅛	5⅛	5⅛	5⅛	5⅛
4	...	...	5¼	5¼	5¼	5¼	5¼
4⅛	...	...	5⅜	5⅜	5⅜	5⅜	5⅜
4¼	...	...	5½	5½	5½	5½	5½
4⅜	...	...	5⅝	5⅝	5⅝	5⅝	5⅝
4½	...	...	5¾	5¾	5¾	5¾	5¾
4⅝	...	...	6	6	6	6	6
4¾	...	...	6⅛	6⅛	6⅛	6⅛	6⅛
4⅞	...	...	6¼	6¼	6¼	6¼	6¼
5	...	...	...	6⅜	6⅜	6⅜	6⅜
5⅛	...	...	...	6½	6½	6½	6½
5¼	...	...	...	6⅝	6⅝	6⅝	6⅝
5⅜	...	...	...	6¾	6¾	6¾	6¾
5½	...	...	...	6⅞	6⅞	6⅞	6⅞
5⅝	...	...	...	7	7	7	7
5¾	...	...	...	7¼	7¼	7¼	7¼
5⅞	...	...	...	7⅜	7⅜	7⅜	7⅜

* As given by the American Institute of Steel Construction. Values may vary from standard practice of individual fabricators and should be checked against fabricator's standard.

Standard Rivets

Standards for rivets published by the American National Standards Institute and the British Standards Institution are as follows:

American National Standard Large Rivets. — The types of rivets covered by this standard (ANSI B18.1.2-1972, R1977) are shown on pages 1267 and 1268. It may be noted, however, that when specified, the swell neck included in this standard is applicable to all standard large rivets except the flat countersunk head and oval countersunk head types. Also shown are the hold-on (dolly bar) and rivet set impression dimensions (see page 1269). All standard large rivets have fillets under the head not exceeding an 0.062-inch radius. The length tolerances for these rivets are given as follows: through 6 inches in length, ½- and ⅝-inch diameters, ±0.03 inch; ¾- and ⅞-inch diameters, ±0.06-inch; and 1- through 1¾-inch diameters, ±0.09 inch. For rivets over 6 inches in length, ½- and ⅝-inch diameters, ±0.06 inch; ¾- and ⅞-inch diameters, ±0.12 inch; and 1- through 1¾-inch diameters, ±0.19 inch. Steel and wrought iron rivet materials appear in ASTM Specifications A31, A131, A152 and A502.

American National Standard Small Solid Rivets. — The types of rivets covered by this standard (ANSI B18.1.1-1972, R1977) are shown on pages 1270 to 1272. In addition, the standard gives the dimensions of 60-degree flat countersunk head rivets used to assemble ledger plates and guards for mower cutter bars, but these are not shown. As the heads of standard rivets are not machined or trimmed, the circumference may be somewhat irregular and edges may be rounded or flat. Rivets other than countersunk types are furnished with a definite fillet under the head, whose radius should not exceed 10 per cent of the maximum shank diameter or 0.03 inch, whichever is the smaller. With regard to head dimensions, tolerances shown in the dimensional tables are applicable to rivets produced by the normal cold heading process. Unless otherwise specified, rivets should have plain sheared ends which should be at right angles within 2 degrees to the axis of the rivet and be reasonably flat. When so specified by user, rivets may have the standard header points shown on page 1270. Rivets may be made of ASTM Specification A31, Grade A steel; or may adhere to SAE Recommended Practice, Mechanical and Chemical Requirements for Nonthreaded Fasteners — SAE J430, Grade 0. When specified, rivets may be made of other materials.

ANSI B18.1.3M-1983, Metric Small Solid Rivets, provides data for small, solid rivets with flat, round, and flat countersunk heads in metric dimensions. The main series of rivets has body diameters, in millimeters, of 1.6, 2, 2.5, 3, 4, 5, 6, 8, 10, and 12. A secondary series (nonpreferred) consists of sizes 1, 1.2, 1.4, 3.5, 7, 9, and 11 millimeters.

British Standard Small Rivets for General Purposes. — Dimensions of small rivets for general purposes are given in British Standard 641:1951 and are shown in the table on page 1274. In addition, the standard lists the standard lengths of these rivets, gives the dimensions of washers to be used with countersunk head rivets (140°), indicates that the rivets may be made from mild steel, copper, brass and a range of aluminum alloys and pure aluminum specified in B.S. 1473, and gives the dimensions of Coopers' flat headed rivets ½ inch in diameter and below, in an appendix. In all types of rivets except those with countersunk heads there is a small radius or chamfer at the junction of the head and the shank.

British Standard Dimensions of Rivets (½ to 1¾ inch diameter). — The dimensions of rivets covered in B.S. 275:1927 are given on page 1273 and do not apply to boiler rivets. With regard to this standard the terms "nominal diameter" and "standard diameter" are synonymous. The term "tolerance" refers to the variation from the nominal diameter of the rivet and not to the difference between the diameter under the head and the diameter near the point.

American National Standard Large Rivets — 1 (ANSI B18.1.2-1972, R1977)

Button Head   High Button   Cone Head   Pan Head

Nom. Body Diam. D†	Head Diam. A		Height H		Head Diam. A		Height H	
	Mfd. Note 1	Driven Note 2	Mfd. Note 1	Driven Note 2	Mfd. Note 1	Driven Note 2	Mfd. Note 1	Driven Note 2
	Button Head				**High Button Head (Acorn)**			
1/2	0.875	0.922	0.375	0.344	0.781	0.875	0.500	0.375
5/8	1.094	1.141	0.469	0.438	0.969	1.062	0.594	0.453
3/4	1.312	1.375	0.562	0.516	1.156	1.250	0.688	0.531
7/8	1.531	1.594	0.656	0.609	1.344	1.438	0.781	0.609
1	1.750	1.828	0.750	0.688	1.531	1.625	0.875	0.688
1 1/8	1.969	2.062	0.844	0.781	1.719	1.812	0.969	0.766
1 1/4	2.188	2.281	0.938	0.859	1.906	2.000	1.062	0.844
1 3/8	2.406	2.516	1.031	0.953	2.094	2.188	1.156	0.938
1 1/2	2.625	2.734	1.125	1.031	2.281	2.375	1.250	1.000
1 5/8	2.844	2.969	1.219	1.125	2.469	2.562	1.344	1.094
1 3/4	3.062	3.203	1.312	1.203	2.656	2.750	1.438	1.172
	Cone Head				**Pan Head**			
1/2	0.875	0.922	0.438	0.406	0.800	0.844	0.350	0.328
5/8	1.094	1.141	0.547	0.516	1.000	1.047	0.438	0.406
3/4	1.312	1.375	0.656	0.625	1.200	1.266	0.525	0.484
7/8	1.531	1.594	0.766	0.719	1.400	1.469	0.612	0.578
1	1.750	1.828	0.875	0.828	1.600	1.687	0.700	0.656
1 1/8	1.969	2.063	0.984	0.938	1.800	1.891	0.788	0.734
1 1/4	2.188	2.281	1.094	1.031	2.000	2.094	0.875	0.812
1 3/8	2.406	2.516	1.203	1.141	2.200	2.312	0.962	0.906
1 1/2	2.625	2.734	1.312	1.250	2.400	2.516	1.050	0.984
1 5/8	2.844	2.969	1.422	1.344	2.600	2.734	1.138	1.062
1 3/4	3.062	3.203	1.531	1.453	2.800	2.938	1.225	1.141

All dimensions are given in inches.
† Tolerance for diameter of body is plus and minus from nominal and for 1/2-in. size equals + 0.020, − 0.022; for sizes 5/8 to 1-in., incl., equals + 0.030, − 0.025; for sizes 1 1/8 and 1 1/4-in. equals + 0.035, − 0.027; for sizes 1 3/8 and 1 1/2-in. equals + 0.040, − 0.030; for sizes 1 5/8 and 1 3/4-in. equals + 0.040, − 0.037.
Note 1. Basic dimensions of head as manufactured.
Note 2. Dimensions of manufactured head after driving and also of driven head.
Note 3. Slight flat permissible within the specified head-height tolerance.
The following formulas give the basic dimensions for manufactured shapes: *Button Head*, A = 1.750D; H = 0.750D; G = 0.885D. *High Button Head*, A = 1.500D; H = 0.750D + 0.125; F = 0.750D + 0.281; G = 0.750D − 0.281. *Cone Head*, A = 1.750D; B = 0.938D; H = 0.875D. *Pan Head*, A = 1.600D; B = 1.000D; H = 0.700D. Length L is measured parallel to the rivet axis, from the extreme end to the bearing surface plane for flat bearing surface head type rivets, or to the intersection of the head top surface with the head diameter for countersunk head type rivets.

American National Standard Large Rivets — 2 (ANSI B18.1.2-1972, R1977)

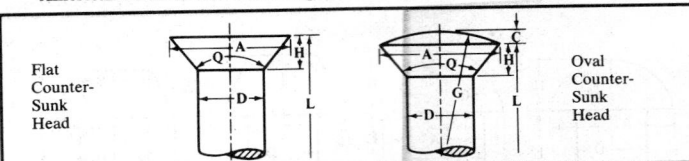

Flat Counter-Sunk Head

Oval Counter-Sunk Head

Flat and Oval Countersunk Head

Body Diameter D			Head Diam. A		Head Depth H	Oval Crown Height* C	Oval Crown Radius* G
Nominal*	Max.	Min.	Max.†	Min.‡	Ref.		
½ 0.500	0.520	0.478	0.936	0.872	0.260	0.095	1.125
⅝ 0.625	0.655	0.600	1.194	1.112	0.339	0.119	1.406
¾ 0.750	0.780	0.725	1.421	1.322	0.400	0.142	1.688
⅞ 0.875	0.905	0.850	1.647	1.532	0.460	0.166	1.969
1 1.000	1.030	0.975	1.873	1.745	0.520	0.190	2.250
1⅛ 1.125	1.160	1.098	2.114	1.973	0.589	0.214	2.531
1¼ 1.250	1.285	1.223	2.340	2.199	0.650	0.238	2.812
1⅜ 1.375	1.415	1.345	2.567	2.426	0.710	0.261	3.094
1½ 1.500	1.540	1.470	2.793	2.652	0.771	0.285	3.375
1⅝ 1.625	1.665	1.588	3.019	2.878	0.831	0.309	3.656
1¾ 1.750	1.790	1.713	3.262	3.121	0.901	0.332	3.938

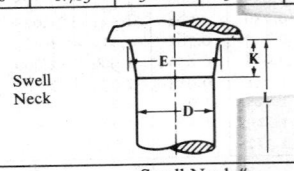

Swell Neck

Swell Neck#

Body Diameter D			Diameter Under Head E		Neck Length K*
Nominal*	Max.	Min.	Max. (Basic)	Min.	
½ 0.500	0.520	0.478	0.563	0.543	0.250
⅝ 0.625	0.655	0.600	0.688	0.658	0.312
¾ 0.750	0.780	0.725	0.813	0.783	0.375
⅞ 0.875	0.905	0.850	0.938	0.908	0.438
1 1.000	1.030	0.975	1.063	1.033	0.500
1⅛ 1.125	1.160	1.098	1.188	1.153	0.562
1¼ 1.250	1.285	1.223	1.313	1.278	0.625
1⅜ 1.375	1.415	1.345	1.438	1.398	0.688
1½ 1.500	1.540	1.470	1.563	1.523	0.750
1⅝ 1.625	1.665	1.588	1.688	1.648	0.812
1¾ 1.750	1.790	1.713	1.813	1.773	0.875

All dimensions are given in inches.
* Basic dimension as manufactured. For tolerances see table footnote on page 1267.
† Sharp edged head.
‡ Rounded or flat edged irregularly shaped head (heads are not machined or trimmed).
# The swell neck is applicable to all standard forms of large rivets except the flat countersunk and oval countersunk head types.

The following formulas give basic dimensions for manufactured shapes: *Flat Countersunk Head,* $A = 1.810D$; $H = 1.192$ (Max $A - D$)/2; included angle Q of head = 78 degrees. *Oval Countersunk Head,* $A = 1.810D$; $H = 1.192$ (Max $A - D$)/2; included angle of head = 78 degrees. *Swell Neck,* $E = D + 0.063$; $K = 0.500D$. Length L is measured parallel to the rivet axis, from the extreme end to the bearing surface plane for flat bearing surface head type rivets, or to the intersection of the head top surface with the head diameter for countersunk head type rivets.

American National Standard Dimensions for Hold-On (Dolly Bar) and Rivet Set Impression (ANSI B18.1.2-1972, R1977)

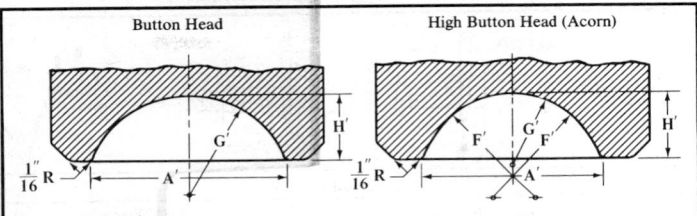

Rivet Body Diam.	Button Head			High Button Head			
	A'	H'	G'	A'	H'	F'	G'
½	0.906	0.312	0.484	0.859	0.344	0.562	0.375
⅝	1.125	0.406	0.594	1.047	0.422	0.672	0.453
¾	1.344	0.484	0.719	1.234	0.500	0.797	0.531
⅞	1.578	0.562	0.844	1.422	0.578	0.922	0.609
1	1.812	0.641	0.953	1.609	0.656	1.031	0.688
1⅛	2.031	0.719	1.078	1.797	0.719	1.156	0.766
1¼	2.250	0.797	1.188	1.984	0.797	1.266	0.844
1⅜	2.469	0.875	1.312	2.172	0.875	1.406	0.938
1½	2.703	0.953	1.438	2.344	0.953	1.500	1.000
1⅝	2.922	1.047	1.547	2.531	1.031	1.641	1.094
1¾	3.156	1.125	1.672	2.719	1.109	1.750	1.172

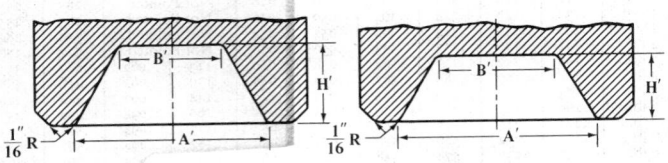

Rivet Body Diam.	Cone Head			Rivet Body Diam.	Pan Head		
	A'	B'	H'		A'	B'	H'
½	0.891	0.469	0.391	½	0.812	0.500	0.297
⅝	1.109	0.594	0.484	⅝	1.031	0.625	0.375
¾	1.328	0.703	0.578	¾	1.234	0.750	0.453
⅞	1.562	0.828	0.688	⅞	1.438	0.875	0.531
1	1.781	0.938	0.781	1	1.641	1.000	0.609
1⅛	2.000	1.063	0.875	1⅛	1.844	1.125	0.688
1¼	2.219	1.172	0.969	1¼	2.047	1.250	0.766
1⅜	2.453	1.297	1.078	1⅜	2.250	1.375	0.844
1½	2.672	1.406	1.172	1½	2.453	1.500	0.906
1⅝	2.891	1.531	1.266	1⅝	2.656	1.625	0.984
1¾	3.109	1.641	1.375	1¾	2.875	1.750	1.063

All dimensions are given in inches.

Table 1. American National Standard Small Solid Rivets[1]
(ANSI B18.1.1-1972, R1977 and Appendix)

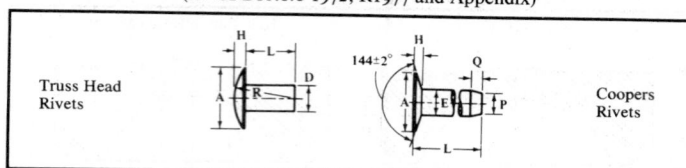

| Truss Head Rivets | | | | | | | | | | Coopers Rivets |

Truss Head Rivets[2]

Shank Diam.,[3] D		Head Dimensions						Shank Diam.,[3] D		Head Dimensions					
		Diam., A		Height, H		Rad. R				Diam., A		Height, H		Rad. R	
Nominal		Max.	Min.	Max.	Min.	Approx.		Nominal		Max.	Min.	Max.	Min.	Approx.	
3/32	.094	.226	.206	.038	.026	.239		9/32	.281	0.661	.631	.103	.085	0.706	
1/8	.125	.297	.277	.048	.036	.314		5/16	.312	0.732	.702	.113	.095	0.784	
5/32	.156	.368	.348	.059	.045	.392		11/32	.344	0.806	.776	.124	.104	0.862	
3/16	.188	.442	.422	.069	.055	.470		3/8	.375	0.878	.848	.135	.115	0.942	
7/32	.219	.515	.495	.080	.066	.555		13/32	.406	0.949	.919	.145	.123	1.028	
1/4	.250	.590	.560	.091	.075	.628		7/16	.438	1.020	.990	.157	.135	1.098	

Coopers Rivets

Size No.[4]	Shank Diameter, D		Head Diameter, A		Head Height, H		Point Dimensions		Length, L	
							Diam., P	Length, Q		
	Max.	Min.	Max.	Min.	Max.	Min.	Nom.	Nom.	Max.	Min.
1 lb.	.111	.105	.291	.271	.045	.031	Not Pointed		.249	.219
1¼ lb.	.122	.116	.324	.302	.050	.036	Not Pointed		.285	.255
1½ lb.	.132	.126	.324	.302	.050	.036	Not Pointed		.285	.255
1¾ lb.	.136	.130	.324	.302	.052	.034	Not Pointed		.318	.284
2 lb.	.142	.136	.355	.333	.056	.038	Not Pointed		.322	.288
3 lb.	.158	.152	.386	.364	.058	.040	.123	.062	.387	.353
4 lb.	.168	.159	.388	.362	.058	.040	.130	.062	.418	.388
5 lb.	.183	.174	.419	.393	.063	.045	.144	.062	.454	.420
6 lb.	.206	.197	.482	.456	.073	.051	.160	.094	.498	.457
7 lb.	.223	.214	.513	.487	.076	.054	.175	.094	.561	.523
8 lb.	.241	.232	.546	.516	.081	.059	.182	.094	.597	.559
9 lb.	.248	.239	.578	.548	.085	.063	.197	.094	.601	.563
10 lb.	.253	.244	.578	.548	.085	.063	.197	.094	.632	.594
12 lb.	.263	.251	.580	.546	.086	.060	.214	.094	.633	.575
14 lb.	.275	.263	.611	.577	.091	.065	.223	.094	.670	.612
16 lb.	.285	.273	.611	.577	.089	.063	.223	.094	.699	.641
18 lb.	.285	.273	.642	.608	.108	.082	.230	.125	.749	.691
20 lb.	.316	.304	.705	.671	.128	.102	.250	.125	.769	.711
3/8 in.	.380	.365	.800	.762	.136	.106	.312	.125	.840	.778

Note: When specified American National Standard Small Solid Rivets may be obtained with points. Point dimensions for belt and coopers rivets are given in the accompanying tables. Formulas for calculating point dimensions of other rivets are given with the diagram alongside.

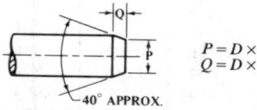

$$P = D \times 0.818$$
$$Q = D \times 0.25$$

[1] All dimensions in inches except where otherwise noted.   [2] Length tolerance of rivets is plus or minus .016 inch. Approximate proportions of rivets: $A = 2.300 \times D$, $H = 0.330 \times D$, $R = 2.512 \times D$.   [3] Tolerances on the nominal shank diameter in inches are given for the following body diameter ranges: 3/32 to 5/32, plus 0.002, minus 0.004; 3/16 to 1/4, plus 0.003, minus 0.006; 9/32 to 11/32, plus 0.004, minus 0.008; and 3/8 to 7/16, plus 0.005, minus 0.010.   [4] Size numbers in pounds refer to the approximate weight of 1000 rivets.

Table 2. American National Standard Small Solid Rivets[1]
(ANSI B18.1.1-1972, R1977)

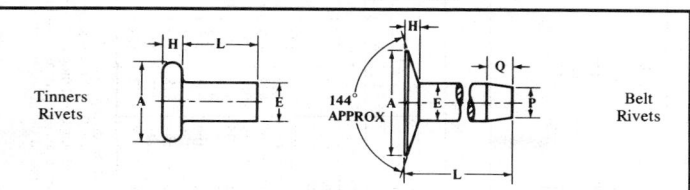

Tinners Rivets

Size No.[2]	Shank Diameter, E		Head Diam., A		Head Height, H		Length, L		
	Max.	Min.	Max.	Min.	Max.	Min.	Nom.	Max.	Min.
6 oz.	.081	.075	.213	.193	.028	.016	1/8	.135	.115
8 oz.	.091	.085	.225	.205	.036	.024	5/32	.166	.146
10 oz.	.097	.091	.250	.230	.037	.025	11/64	.182	.162
12 oz.	.107	.101	.265	.245	.037	.025	3/16	.198	.178
14 oz.	.111	.105	.275	.255	.038	.026	3/16	.198	.178
1 lb.	.113	.107	.285	.265	.040	.028	13/64	.213	.193
1¼ lb.	.122	.116	.295	.275	.045	.033	7/32	.229	.209
1½ lb.	.132	.126	.316	.294	.046	.034	15/64	.244	.224
1¾ lb.	.136	.130	.331	.309	.049	.035	1/4	.260	.240
2 lb.	.146	.140	.341	.319	.050	.036	17/64	.276	.256
2½ lb.	.150	.144	.311	.289	.069	.055	9/32	.291	.271
3 lb.	.163	.154	.329	.303	.073	.059	5/16	.323	.303
3½ lb.	.168	.159	.348	.322	.074	.060	21/64	.338	.318
4 lb.	.179	.170	.368	.342	.076	.062	11/32	.354	.334
5 lb.	.190	.181	.388	.362	.084	.070	3/8	.385	.365
6 lb.	.206	.197	.419	.393	.090	.076	25/64	.401	.381
7 lb.	.223	.214	.431	.405	.094	.080	13/32	.416	.396
8 lb.	.227	.218	.475	.445	.101	.085	7/16	.448	.428
9 lb.	.241	.232	.490	.460	.103	.087	29/64	.463	.443
10 lb.	.241	.232	.505	.475	.104	.088	15/32	.479	.459
12 lb.	.263	.251	.532	.498	.108	.090	1/2	.510	.490
14 lb.	.288	.276	.577	.543	.113	.095	33/64	.525	.505
16 lb.	.304	.292	.597	.563	.128	.110	17/32	.541	.521
18 lb.	.347	.335	.706	.668	.156	.136	19/32	.603	.583

Belt Rivets[3]

Size No.[4]	Shank Diameter, E		Head Diam., A		Head Height, H		Point Dimensions	
							Diam., P	Length, Q
	Max.	Min.	Max.	Min.	Max.	Min.	Nominal	Nominal
14	.085	.079	.260	.240	.042	.030	.065	.078
13	.097	.091	.322	.302	.051	.039	.073	.078
12	.111	.105	.353	.333	.054	.040	.083	.078
11	.122	.116	.383	.363	.059	.045	.097	.078
10	.136	.130	.417	.395	.065	.047	.109	.094
9	.150	.144	.448	.426	.069	.051	.122	.094
8	.167	.161	.481	.455	.072	.054	.135	.094
7	.183	.174	.513	.487	.075	.056	.151	.125
6	.206	.197	.606	.580	.090	.068	.165	.125
5	.223	.214	.700	.674	.105	.083	.185	.125
4	.241	.232	.921	.893	.138	.116	.204	.141

[1] All dimensions in inches.   [2] Size numbers refer to the approximate weight of 1000 rivets.
[3] Length tolerance on belt rivets is plus 0.031 inch, minus 0 inch.   [4] Size number refers to the Stub's iron wire gage number of the stock used in the shank of the rivet.

Note: American National Standard Small Solid Rivets may be obtained with or without points. Point proportions are given in the diagram in Table 1.

Table 3. American National Standard Small Solid Rivets¹ (ANSI B18.1.1-1972, R1977 and Appendix)

Shank Diameter			Flat Head				Flat Countersunk Head			Button Head					Pan Head						
										Head Dimensions											
D	E		Diam., A		Height, H		Diam., A Sharp		Height⁴ H	Diam., A		Height, H		Radius, R	Diam., A		Height, H			Radii	
Nominal	Max.	Min.	Max.	Min.	Max.	Min.	Max.²	Min.³	Ref.	Max.	Min.	Max.	Min.	Approx.	Max.	Min.	Max.	Min.	R1	R2	R3
1/16	.064	.059	.140	.120	.027	.017	.118	.110	.027	.122	.102	.052	.042	.055	.118	.098	.040	.030	.019	.052	.217
3/32	.096	.090	.200	.180	.038	.026	.176	.163	.040	.182	.162	.077	.065	.084	.173	.153	.060	.048	.030	.080	.326
1/8	.127	.121	.260	.240	.048	.036	.235	.217	.053	.235	.215	.100	.088	.111	.225	.205	.078	.066	.039	.106	.429
5/32	.158	.152	.323	.301	.059	.045	.293	.272	.066	.290	.268	.124	.110	.138	.279	.257	.096	.082	.049	.133	.535
3/16	.191	.182	.387	.361	.069	.055	.351	.326	.079	.348	.322	.147	.133	.166	.334	.308	.114	.100	.059	.159	.641
7/32	.222	.213	.453	.427	.080	.065	.413	.384	.094	.405	.379	.172	.158	.195	.391	.365	.133	.119	.069	.186	.754
1/4	.253	.244	.515	.485	.091	.075	.469	.437	.106	.460	.430	.196	.180	.221	.444	.414	.151	.135	.079	.213	.858
9/32	.285	.273	.579	.545	.103	.085	.528	.491	.119	.518	.484	.220	.202	.249	.499	.465	.170	.152	.088	.239	.963
5/16	.316	.304	.641	.607	.113	.095	.588	.547	.133	.572	.538	.243	.225	.276	.552	.518	.187	.169	.098	.266	1.070
11/32	.348	.336	.705	.667	.124	.104	.646	.602	.146	.630	.592	.267	.247	.304	.608	.570	.206	.186	.108	.292	1.176
3/8	.380	.365	.769	.731	.135	.115	.704	.656	.159	.684	.646	.291	.271	.332	.663	.625	.225	.205	.118	.319	1.286
13/32	.411	.396	.834	.796	.146	.124	.763	.710	.172	.743	.699	.316	.294	.358	.719	.675	.243	.221	.127	.345	1.392
7/16	.443	.428	.896	.852	.157	.135	.823	.765	.186	.798	.754	.339	.317	.387	.772	.728	.261	.239	.137	.372	1.500

¹ All dimensions in inches. Length tolerance of all rivets is plus or minus 0.016 inch. Approximate proportions of rivets: flat head, $A = 2.00 \times D$, $H = 0.33 D$; flat countersunk head, $A = 1.850 \times D$, $H = 0.425 \times D$; button head, $A = 1.750 \times D$, $H = 0.750 \times D$, $R = 0.885 \times D$; pan head, $A = 1.720 \times D$, $H = 0.570 \times D$, $R1 = 0.314 \times D$, $R2 = 0.850 \times D$, $R3 = 3.430 \times D$. ² Tabulated maximum values calculated on basic diameter of rivet and 92° included angle extended to a sharp edge. ³ Minimum of rounded or flat edged irregular shaped head. Rivet heads are not machined or trimmed and the circumference may be irregular and edges rounded or flat. ⁴ Given for reference purposes only. Variations in this dimension are controlled by the head and shank diameters and the included angle of the head.

Note: ANSI Small Solid Rivets may be obtained with or without points. Point proportions are given in the diagram in Table 1.

Head Dimensions and Diameters of British Standard Rivets (B.S. 275-1927)
(This standard does not apply to Boiler Rivets)

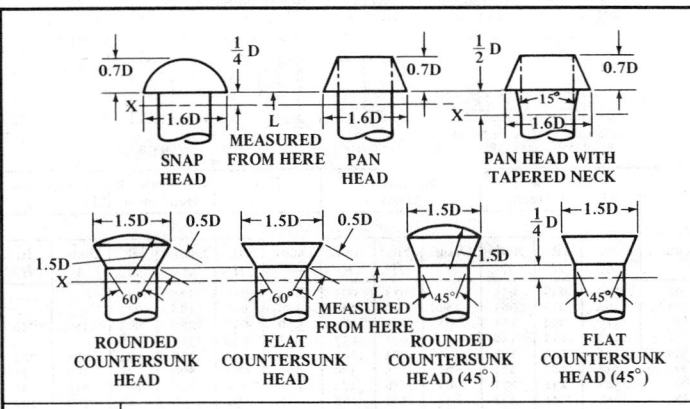

SNAP HEAD   PAN HEAD   PAN HEAD WITH TAPERED NECK

ROUNDED COUNTERSUNK HEAD   FLAT COUNTERSUNK HEAD   ROUNDED COUNTERSUNK HEAD (45°)   FLAT COUNTERSUNK HEAD (45°)

Nominal Rivet Diameter, D	Shank Diameter†				
	At Position X§		At Position Y§		At Position Z§
	Minimum	Maximum	Minimum	Maximum	Minimum
1/2	1/2	17/32	31/64	1/2	31/64
9/16*	9/16	19/32	35/64	9/16	35/64
5/8	5/8	21/32	39/64	5/8	39/64
11/16*	11/16	23/32	43/64	11/16	43/64
3/4	3/4	25/32	47/64	3/4	47/64
13/16*	13/16	27/32	51/64	13/16	51/64
7/8	7/8	29/32	55/64	7/8	55/64
15/16*	15/16	31/32	59/64	15/16	59/64
I	I	1 1/32	63/64	I	63/64
1 1/16*	1 1/16	1 3/32	1 3/64	1 1/16	1 3/64
1 1/8	1 1/8	1 5/32	1 7/64	1 1/8	1 7/64
1 3/16*	1 3/16	1 7/32	1 11/64	1 3/16	1 11/64
1 1/4	1 1/4	1 9/32	1 15/64	1 1/4	1 15/64
1 5/16*	1 5/16	1 11/32	1 19/64	1 5/16	1 19/64
1 3/8	1 3/8	1 13/32	1 23/64	1 3/8	1 23/64
1 7/16*	1 7/16	1 15/32	1 27/64	1 7/16	1 27/64
1 1/2	1 1/2	1 17/32	1 31/64	1 1/2	1 31/64
1 9/16*	1 9/16	1 19/32	1 35/64	1 9/16	1 35/64
1 5/8	1 5/8	1 21/32	1 39/64	1 5/8	1 39/64
1 11/16*	1 11/16	1 23/32	1 43/64	1 11/16	1 43/64
1 3/4	1 3/4	1 25/32	1 47/64	1 3/4	1 47/64

All dimensions that are tabulated are given in inches.
* At the recommendation of the British Standards Institution these sizes are to be dispensed with wherever possible.
† Tolerances of the rivet diameter are as follows: at position X, plus 1/32 inch, minus zero; at position Y, plus zero, minus 1/64 inch; at position Z, minus 1/64 inch but in no case shall the difference between the diameters at positions X and Y exceed 1/32 inch, nor shall the diameter of the shank between positions X and Y be less than the minimum diameter specified at position Y.
§ The location of positions Y and Z are as follows: Position Y is located 1/2D from the end of the rivet for rivet lengths 5 diameters long and under. For longer rivets, position Y is located 4 1/2D from the head of the rivet. Position Z (found only on rivets longer than 5D) is located 1/2D from the end of the rivet.

British Standard Small Rivets for General Purposes (B.S. 641:1951)

Head type figures with dimensional formulas:

- **SNAP (OR ROUND) HEAD:** A = 1.75D, H = 0.75D, R = 0.885D
- **MUSHROOM HEAD:** A = 2.25D, H = 0.5D, R = 1.516D
- **FLAT HEAD:** A = 2D, H = 0.25D
- **COUNTERSUNK HEAD (90°):** A = 2D, H = 0.5D
- **COUNTERSUNK HEAD (120°):** A = 2D, H = 0.29D

Nom. Diam. D	Snap (or Round) Head Diam. A	Ht. H	Rad. R	Mushroom Head Diam. A	Ht. H	Rad. R	Flat Head Diam. A	Ht. H	Countersunk Head (90°) Diam. A	Ht. H	Countersunk Head (120°) Diam. A	Ht. H
1/16	.109	.047	.055	.141	.031	.095	.125	.016	.125	.031	...	...
3/32	.164	.070	.083	.211	.047	.142	.188	.023	.188	.047	...	...
1/8	.219	.094	.111	.281	.063	.189	.250	.031	.250	.063	.250	.036
5/32	.273	.117	.138	.352	.078	.237	.313	.039	.313	.078	...	...
3/16	.328	.141	.166	.422	.094	.284	.375	.047	.375	.094	.375	.054
1/4	.438	.188	.221	.563	.125	.379	.500	.063	.500	.125	.500	.073
5/16	.547	.234	.277	.703	.156	.474	.625	.078	.625	.156	.625	.091
3/8	.656	.281	.332	.844	.188	.568	.750	.094	.750	.188	.750	.109
7/16	.766	.328	.387	.984	.219	.663	.875	.109	.875	.219	...	...

Head type figures with dimensional formulas:

- **PAN HEAD:** A = 1.6D, H = 0.7D
- **COUNTERSUNK HEAD (60°):** A = 1.75D, H = 0.65D
- **COUNTERSUNK HEAD (140°):** A = 2.75D, C = 0.4D, E = 0.79D
- **COUNTERSUNK HEAD REAPER:** A = 1.65D, H = 0.325D
- **SNAP (OR ROUND) HEAD REAPER:** A = 1.6D, H = 0.6D

Nominal Diameter D (Inch)	Gage No.*	Pan Head Diam. A	Ht. H	Countersunk Head (60°) Diam. A	Ht. H	Countersunk Head (140°) Diam. A	Ht. C	E	Countersunk Head Reaper Diam. A	Ht. H	Snap (or Round) Head Reaper Diam. A	Ht. H
.104	12	...	...	...	...	.286	.042	.082	...	...	...	...
.116	11	...	...	...	...	.319	.046	.092	...	...	...	...
.128	10	...	...	...	...	.352	.051	.101	...	...	...	...
.144	9	...	...	...	...	.396	.058	.114	...	...	...	...
.160	8	...	...	...	...	.440	.064	.126	...	...	...	...
.176	7	...	...	...	...	.484	.070	.139	...	...	...	...
3/16	...	.300	.131	.328	.122	...	...	...	...	...	...	...
.192	6	...	...	...	...	.528	.077	.152	.317	.062	.307	.115
.202	...	...	...	...	...	...	...	...	.333	.066	.323	.121
.212	5	...	...	...	...	.583	.085	.167	.350	.069	.339	.127
.232	4	...	...	...	...	.638	.093	.183	.383	.075	.371	.139
1/4	...	.400	.175	.438	.162	.688	.100	.198	...	...	...	...
.252	3	...	...	...	...	...	...	...	.416	.082	.403	.151
5/16	...	.500	.219	.547	.203	.859	.125	.247	...	...	...	...
3/8	...	.600	.263	.656	.244	1.031	.150	.296	...	...	...	...
7/16	...	.700	.306	.766	.284	...	...	...	...	...	...	...

All dimensions in inches unless specified otherwise.
* Gage numbers are British Standard Wire Gage (S.W.G.) numbers.

British Standard Rivets for General Engineering. — Dimensions in metric units of rivets for general engineering purposes are given in this British Standard, BS 4620:1970, which is based on ISO Recommendation ISO/R 1051. The snap head rivet dimensions of 14 millimeters and above are taken from the German Standard DIN 124, Round Head Rivets for Steel Structures. The shapes of heads have been restricted to those in common use in the United Kingdom. Table 2 shows the rivet dimensions. Table 1 shows a tentative range of preferred nominal lengths as given in an appendix to the Standard. It is stated that these lengths will be reviewed in the light of usage. The rivets are made by cold or hot forging methods from mild steel, copper, brass, pure aluminum, aluminum alloys, or other suitable metal. It is stated that the radius under the head of a rivet shall run smoothly into the face of the head and shank without step or discontinuity.

In this Standard, the following definitions apply: (1) *Nominal diameter.* The diameter of the shank. (2) *Nominal length of rivets other than countersunk or raised countersunk rivets.* The length from the underside of the head to the end of the shank. (3) *Nominal length of countersunk and raised countersunk rivets.* The distance from the periphery of the head to the end of the rivet measured parallel to the axis of the rivet. (4) *Manufactured head.* The head on the rivet as received from the manufacturer.

Table 1. Tentative Range of Lengths for Rivets (Appendix to B.S. 4620:1970)

Nom. Shank Diam.	Nominal Length																				
	3	4	5	6	8	10	12	14	16	(18)	20	(22)	25	(28)	30	(32)	35	(38)	40	45	..
1	X	X	X	X	X	X	X	X	X	X	..	X	..	..	..	..	..	..	..	..	..
1.2	X	X	X	X	X	X	X	X	X	X	..	X	..	..	..	..	..	..	..	..	..
1.6	X	X	X	X	X	X	X	X	X	X	X	..	X	X	..	..	..	..	..	..	..
2	X	X	X	X	X	X	X	X	X	X	..	X	X	X	..	..	..	..	..	..	..
2.5	X	X	X	X	X	X	X	X	X	X	X	..	X	X	..	..	..	..	..	..	..
3	..	X	X	X	X	X	X	X	X	X	X	X	X	X	..	..	..	..	..	..	..
(3.5)	..	..	..	..	..	..	..	..	..	..	..	..	..	..	..	..	..	..	..	..	..
4	..	..	..	X	X	X	X	X	X	X	..	X	X	..	..	..	X	X	..	..	..
5	..	..	X	X	X	X	X	X	X	X	X	X	..	X	X	..	..	..	..	..	..
6	..	..	X	X	X	X	X	X	X	X	X	X	X	X	..	X	X	..	..	X	..

Nom. Shank Diam.	Nominal Length																				
	10	12	14	16	(18)	20	(22)	25	(28)	30	(32)	35	(38)	40	45	50	55	60	65	70	75
(7)	..	..	..	..	..	..	..	..	..	..	..	..	..	..	..	..	..	..	..	..	..
8	X	X	X	X	X	X	X	X	X	X	..	X	X	X	X	..	..	..	..	..	..
10	..	..	X	X	X	X	X	X	X	X	..	X	X	X	X	..	..	..	..	..	..
12	..	..	..	..	..	..	..	X	..	X	..	X	..	X	X	..	..	..	..	..	X
(14)	..	..	..	..	..	..	..	..	..	..	..	..	..	..	..	..	..	..	..	..	..
16	..	..	..	..	..	..	..	..	..	..	..	..	..	X	X	X	X	X	X	..	X

Nom. Shank Diam.	Nominal Length																				
	45	50	55	60	65	70	75	80	85	90	(95)	100	(105)	110	(115)	120	(125)	130	140	150	160
(18)	..	..	..	..	..	..	..	..	..	..	..	..	..	..	..	..	..	..	..	..	..
20	X	..	..	..	X	..	X	..	X	..	..	..	..	..	..	..	..	..	..	..	..
(22)	..	..	..	..	..	..	..	..	..	..	..	..	..	..	..	..	..	..	..	..	..
24	..	..	..	X	..	X	..	X	X	..	X	..	..	..	..	..	..	..	..	..	..
(27)	..	..	..	..	..	..	..	..	..	..	..	..	..	..	..	..	..	..	..	..	..
30	..	..	..	..	..	..	..	X	X	..	..	X	..	..	X	..	..	X	..	..	..
(33)	..	..	..	..	..	..	..	..	..	..	..	..	..	..	..	..	..	..	..	..	..
36	..	..	..	..	..	..	..	..	..	..	X	..	..	X	..	..	X	..	X	X	X
(39)	..	..	..	..	..	..	..	..	..	..	X	..	..	X	..	..	X	..	X	X	X

All dimensions are in millimeters.

Note: Sizes and lengths shown in parenthesis are non-preferred and should be avoided if possible.

Table 2. British Standard Rivets for General Engineering Purposes (B.S. 4620:1970)

60° Csk. and Raised Csk. Head | 90° Csk. Head | Snap Head | Universal Head | Flat Head

*K = 0.43 d (for ref. only)   †K = 0.5 d (for ref. only)

Hot Forged Rivets

Nom. Shank Diam. d	Tol. on Diam. d	60° Csk. and Raised Csk. Head		Snap Head		Universal Head			
		Nom. Diam. D	Height of Raise W	Nom. Diam. D	Nom. Depth K	Nom. Diam. D	Nom. Depth K	Rad. R	Rad. r
(14)	±0.43	21	2.8	22	9	28	5.6	42	8.4
16		24	3.2	25	10	32	6.4	48	9.6
(18)		27	3.6	28	11.5	36	7.2	54	11
20	±0.52	30	4.0	32	13	40	8.0	60	12
(22)		33	4.4	36	14	44	8.8	66	13
24		36	4.8	40	16	48	9.6	72	14
(27)	±0.62	40	5.4	43	17	54	10.8	81	16
30		45	6.0	48	19	60	12.0	90	18
(33)		50	6.6	53	21	66	13.2	99	20
36		55	7.2	58	23	72	14.4	108	22
(39)		59	7.8	62	25	78	15.6	117	23

Cold Forged Rivets

Nom. Shank Diam. d	Tol. on Diam. d	90° Csk. Head	Snap Head		Universal Head				Flat Head	
		Nom. Diam. D	Nom. Diam. D	Nom. Depth K	Nom. Diam. D	Nom. Depth K	Rad. R	Rad. r	Nom. Diam. D	Nom. Depth K
1	±0.07	2	1.8	0.6	2	0.4	3.0	0.6	2	0.25
1.2		2.4	2.1	0.7	2.4	0.5	3.6	0.7	2.4	0.3
1.6		3.2	2.8	1.0	3.2	0.6	4.8	1.0	3.2	0.4
2		4	3.5	1.2	4	0.8	6.0	1.2	4	0.5
2.5		5	4.4	1.5	5	1.0	7.5	1.5	5	0.6
3		6	5.3	1.8	6	1.2	9.0	1.8	6	0.8
(3.5)	±0.09	7	6.1	2.1	7	1.4	10.5	2.1	7	0.9
4		8	7	2.4	8	1.6	12	2.4	8	1.0
5		10	8.8	3.0	10	2.0	15	3.0	10	1.3
6		12	10.5	3.6	12	2.4	18	3.6	12	1.5
(7)	±0.11	14	12.3	4.2	14	2.8	21	4.2	14	1.8
8		16	14	4.8	16	3.2	24	4.8	16	2
10		20	18	6.0	20	4.0	30	6	20	2.5
12	±0.14	24	21	7.2	24	4.8	36	7.2	...	...
(14)		...	25	8.4	28	5.6	42	8.4	...	...
16		...	28	9.6	32	6.4	48	9.6	...	...

All dimensions are in millimeters. Sizes shown in parentheses are non-preferred.

BOLTS, SCREWS, NUTS, AND WASHERS

Dimensions of bolts, screws, nuts, and washers used in machine construction are given here. For data on thread forms, see "Screw Thread Systems" section.

American Standard Square and Hexagon Bolts, Screws, and Nuts. — The 1941 American Standard ASA B18.2 covered head dimensions only. In 1952 and 1955 the Standard was revised to cover the entire product. Some bolt and nut classifications were simplified by elimination or consolidation in agreements reached with the British and Canadians. In 1965 ASA B18.2 was redesignated into two standards: B18.2.1 covering square and hexagon bolts and screws including hexagon cap screws and lag screws and B18.2.2 covering square and hexagon nuts. In B18.2.1-1965, hexagon head cap screws and finished hexagon bolts were consolidated into a single product; heavy semifinished hexagon bolts and heavy finished hexagon bolts were consolidated into a single product; regular semifinished hexagon bolts were eliminated; a new tolerance pattern for all bolts and screws and a positive identification procedure for determining whether an externally threaded product should be designated as a bolt or screw were established. Also included in this standard are heavy hexagon bolts and heavy hexagon structural bolts. In B18.2.2-1965, regular semifinished nuts were discontinued; regular hexagon and heavy hexagon nuts in sizes ¼ through 1 inch, finished hexagon nuts in sizes larger than 1½ inches, washer-faced semifinished style of finished nuts in sizes ⅝-inch and smaller and heavy series nuts in sizes ⁷⁄₁₆-inch and smaller were eliminated.

Further revisions and refinements including the addition of askew head bolts and hex head lag screws and the specifying of countersunk diameters for the various hex nuts and heavy hex nuts resulted in the issuance of the revised standards as ANSI B18.2.1-1972, revised in 1981 and ANSI B18.2.2-1972.

Unified Square and Hexagon Bolts, Screws, and Nuts. — In certain tables bold-faced items are shown. These are recognized in the Standard as "unified" dimensionally with British and Canadian standards. The others items in the same tables are based on formulas accepted and published by the British for sizes outside the range listed in their standards which, as a matter of information, are B.S. 1768:1963 for Precision (Normal Series) Unified Hexagon Bolts, Screws, Nuts (UNC and UNF Threads); B.S. 1769:1951 for Black (Heavy Series) Unified Hexagon Bolts, etc. and B.S. 2708:1956 for Black (Normal Series) Unified Square and Hexagon Bolts, etc. Tolerances applied to comparable dimensions of American and British Unified bolts and nuts may differ because of rounding off practices and other factors.

Differentiation between Bolt and Screw. — A bolt is an externally threaded fastener designed for insertion through holes in assembled parts, and is normally intended to be tightened or released by torquing a nut.

A screw is an externally threaded fastener capable of being inserted into holes in assembled parts, of mating with a preformed internal thread or forming its own thread and of being tightened or released by torquing the head.

An externally threaded fastener which is prevented from being turned during assembly, and which can be tightened or released only by torquing a nut is a *bolt*. (*Example:* round head bolts, track bolts, plow bolts.)

An externally threaded fastener which has a thread form which prohibits assembly with a nut having a straight thread of multiple pitch length is a *screw*. (*Example:* wood screws, tapping screws.)

An externally threaded fastener which must be assembled with a nut to perform its intended service is a *bolt*. (*Example:* heavy hex structural bolt.)

An externally threaded fastener which must be torqued by its head into a tapped or other performed hole to perform its intended service is a *screw*. (*Example:* square head set screw.)

Working Strength of Bolts. — When the nut on a bolt is tightened, an initial tensile load is placed on the bolt which must be taken into account in determining its safe working strength or external load-carrying capacity. The total load on the bolt theoretically varies from a maximum equal to the sum of the initial and external loads (when the bolt is absolutely rigid and the parts held together are elastic) to a minimum equal to either the initial or external loads, whichever is the greater (where the bolt is elastic and the parts held together are absolutely rigid). Since no material is absolutely rigid, in practice the total load values fall somewhere between these maximum and minimum limits, depending upon the relative elasticity of the bolt and joint members.

Some experiments made at Cornell University to determine the initial stress due to tightening nuts on bolts sufficiently to make a packed joint steam tight, showed that experienced machinists tighten nuts with a pull roughly proportional to the bolt diameter. It was also found that the stress due to nut tightening was often sufficient to break a ½-inch bolt, but not larger sizes, assuming that the nut is tightened by an experienced mechanic. It may be concluded, therefore, that bolts smaller than ⅝ inch should not be used for holding cylinder heads or other parts requiring a tight joint. As a result of these tests, the following empirical formula was established for the working strength of bolts used for packed joints or joints where the elasticity of a gasket is greater than the elasticity of the studs or bolts.

$$W = S_t(0.55\,d^2 - 0.25\,d)$$

In this formula, W = working strength of bolt or permissible load, in pounds after allowance is made for initial load due to tightening; S_t = allowable working stress in tension, pounds per square inch; and d = nominal outside diameter of stud or bolt, inches. A somewhat more convenient formula, and one which gives approximately the same results, is:

$$W = S_t(A - 0.25\,d)$$

In this formula, W, S_t and d are as previously given, and A = area at the root of the thread, square inches.

Example. — What is the working strength of a 1-inch bolt which is screwed up tightly in a packed joint when the allowable working stress is 10,000 psi?

$$W = 10{,}000(0.55 \times 1 - 0.25 \times 1) = 3000 \text{ pounds approx.}$$

Formulas for Stress Areas and Length of Engagement of Screw Threads. — The critical areas of stress of mating screw threads are: (1) The effective cross-sectional area, or tensile stress area, of the external thread; (2) the shear area of the external thread, which depends principally on the minor diameter of the tapped hole; and (3) the shear area of the internal thread, which depends principally on the major diameter of the external thread. The relation of these three stress areas to each other is an important factor in determining how a threaded connection will fail, whether by breakage in the threaded section of the screw (or bolt) or by stripping of either the external or internal thread.

If failure of a threaded assembly should occur, it is preferable for the screw to break rather than have either the external or internal thread strip. In other words, the length of engagement of mating threads should be sufficient to carry the full load necessary to break the screw without the threads stripping.

If mating internal and external threads are manufactured of materials having equal tensile strengths, then to prevent stripping of the external thread, the length of engagement should be not less than that given by Formula 1:

$$L_e = \frac{2 \times A_t}{3.1416\,K_n\text{max}\left[\dfrac{1}{2} + 0.57735 n(E_s\text{min} - K_n\text{max})\right]} \tag{1}$$

In this formula, the factor of 2 means that it is assumed that the area in shear of the screw must be twice the tensile stress area to develop the full strength of the screw (this value is slightly larger than required and thus provides a small factor of safety against stripping); L_e = length of engagement, in inches; n = number of threads per inch; $K_n max$ = maximum minor diameter of internal thread; $E_s min$ = minimum pitch diameter of external thread for the class of thread specified; and A_t = tensile stress area of screw thread given by Formula (2a) or (2b) or the thread tables on pages 1484 to 1493 for Unified threads which are based on Formula (2a).

For steels of up to 100,000 psi ultimate tensile strength,

$$A_t = 0.7854 \left(D - \frac{0.9743}{n} \right)^2 \tag{2a}$$

For steels of over 100,000 psi ultimate tensile strength,

$$A_t = 3.1416 \left(\frac{E_s min}{2} - \frac{0.16238}{n} \right)^2 \tag{2b}$$

In these formulas, D = basic major diameter of threads and the other symbols have the same meaning as before.

Stripping of Internal Thread: If the internal thread is made of material of lower strength than the external thread, stripping of the internal thread may take place before the screw breaks. To determine whether this condition exists it is necessary to calculate the factor J for the relative strength of the external and internal threads given by Formula (3):

$$J = \frac{A_s \times \text{tensile strength of external thread material}}{A_n \times \text{tensile strength of internal thread material}} \tag{3}$$

If J is less than or equal to 1, the length of engagement determined by Formula (1) is adequate to prevent stripping of the internal thread; if J is greater than 1, the required length of engagement Q to prevent stripping of the internal thread is obtained by multiplying the length of engagement L_e, Formula (1), by J:

$$Q = JL_e \tag{4}$$

In Formula (3), A_s and A_n are the shear areas of the external and internal threads, respectively, given by Formulas (5) and (6):

$$A_s = 3.1416 \, nL_e K_n max \left[\frac{1}{2n} + 0.57735(E_s min - K_n max) \right] \tag{5}$$

$$A_n = 3.1416 \, nL_e D_s min \left[\frac{1}{2n} + 0.57735(D_s min - E_n max) \right] \tag{6}$$

In these formulas, n = threads per inch; L_e = length of engagement from Formula (1); $K_n max$ = maximum minor diameter of internal thread; $E_s min$ = minimum pitch diameter of the external thread for the class of thread specified; $D_s min$ = minimum major diameter of the external thread; and $E_n max$ = maximum pitch diameter of internal thread.

Load to Break Threaded Portion of Screws and Bolts. — The direct tensile load P to break the threaded portion of a screw or bolt (assuming that no shearing or torsional stresses are acting) can be determined from the following formula:

$$P = SA_t$$

where, P = load in pounds to break screw; S = ultimate tensile strength of material of screw or bolt in pounds per square inch; and A_t = tensile stress area in square inches from Formulas (2a), (2b) or from the screw thread tables.

American National Standard Square and Hexagon Bolts and Nuts

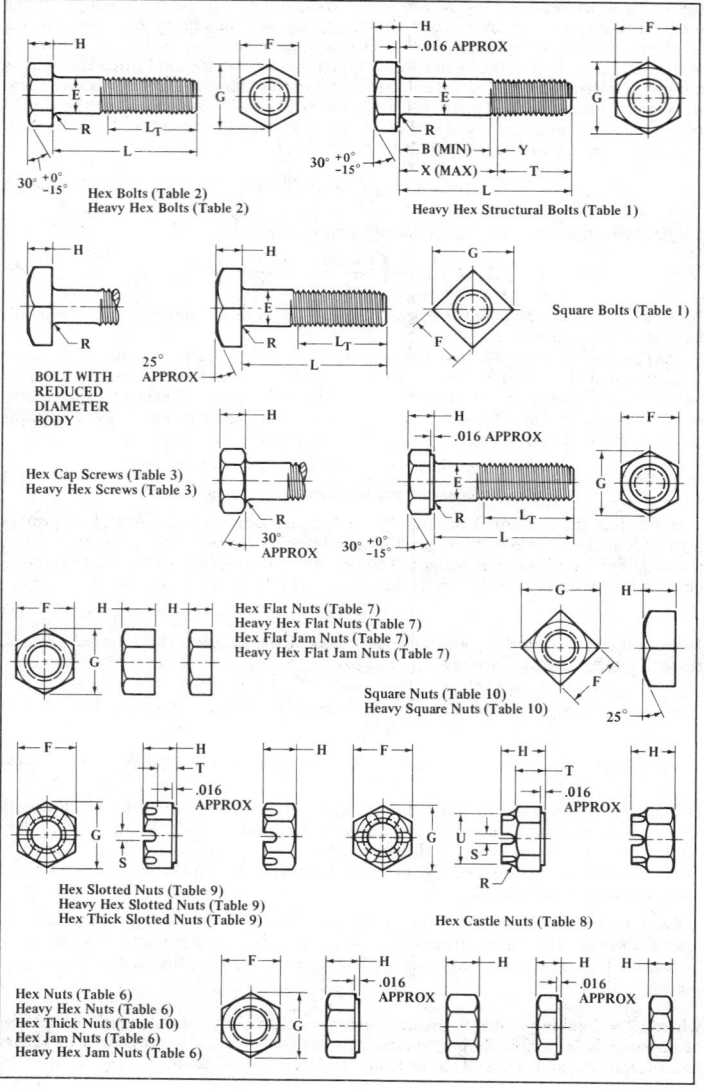

Square and Hex Bolts, Screws and Nuts. — The dimensions for square and hex bolts and screws given in the following tables have been taken from American National Standard ANSI B18.2.1-1981 and for nuts from American National Standard ANSI B18.2.2-1972. Reference should be made to these Standards for information or data not found in the following text and tables:

Designation: Bolts and screws should be designated by the following data in the sequence shown: nominal size (fractional and decimal equivalent); threads per inch (omit for lag screws); product length for bolts and screws (fractional or two-place decimal equivalent); product name; material, including specification, where necessary; and protective finish, if required. Examples: (1) 3/8-16 × 1½ Square Bolt, Steel, Zinc Plated; (2) ½-13 × 3 Hex Cap Screw, SAE Grade 8 Steel; and (3) .75 × 5.00 Hex Lag Screw, Steel. (4) ½-13 Square Nut, Steel, Zinc Plated; (5) ¾-16 Heavy Hex Nut, SAE Grade 5 Steel; and (6) 1.000-8 Hex Thick Slotted Nut, Corrosion-Resistant Steel.

Table 1. American National Standard and Unified Standard Square Bolts and American National Standard Heavy Hex Structural Bolts (ANSI B18.2.1-1981)

SQUARE BOLTS

Nominal Size* or Basic Product Diam.		Body Diam.† E	Width Across Flats F			Width Across Corners G		Height H			Thread Length§ L_T
		Max.	Basic	Max.	Min.	Max.	Min.	Basic	Max.	Min.	Basic
¼	0.2500	0.260	3/8	0.375	0.362	0.530	0.498	11/64	0.188	0.156	0.750
5/16	0.3125	0.324	½	0.500	0.484	0.707	0.665	13/64	0.220	0.186	0.875
3/8	0.3750	0.388	9/16	0.562	0.544	0.795	0.747	¼	0.268	0.232	1.000
7/16	0.4375	0.452	5/8	0.625	0.603	0.884	0.828	19/64	0.316	0.278	1.125
½	0.5000	0.515	¾	0.750	0.725	1.061	0.995	21/64	0.348	0.308	1.250
5/8	0.6250	0.642	15/16	0.938	0.906	1.326	1.244	27/64	0.444	0.400	1.500
¾	0.7500	0.768	1 1/8	1.125	1.088	1.591	1.494	½	0.524	0.476	1.750
7/8	0.8750	0.895	1 5/16	1.312	1.269	1.856	1.742	19/32	0.620	0.568	2.000
1	1.0000	1.022	1½	1.500	1.450	2.121	1.991	21/32	0.684	0.628	2.250
1 1/8	1.1250	1.149	1 11/16	1.688	1.631	2.386	2.239	¾	0.780	0.720	2.500
1¼	1.2500	1.277	1 7/8	1.875	1.812	2.652	2.489	27/32	0.876	0.812	2.750
1 3/8	1.3750	1.404	2 1/16	2.062	1.994	2.917	2.738	29/32	0.940	0.872	3.000
1½	1.5000	1.531	2¼	2.250	2.175	3.182	2.986	1	1.036	0.964	3.250

HEAVY HEX STRUCTURAL BOLTS

Nominal Size* or Basic Product Diam.		Body Diam. E		Width Across Flats F		Width Across Corners G		Height H		Radius of Fillet R		Thrd. Lgth. L_T	Transition Thrd. Y
		Max.	Min.	Max.	Min.	Max.	Min.	Max.	Min.	Max.	Min.	Basic	Max.
½	0.5000	0.515	0.482	0.875	0.850	1.010	0.969	0.323	0.302	0.031	0.009	1.00	0.19
5/8	0.6250	0.642	0.605	1.062	1.031	1.227	1.175	0.403	0.378	0.062	0.021	1.25	0.22
¾	0.7500	0.768	0.729	1.250	1.212	1.443	1.383	0.483	0.455	0.062	0.021	1.38	0.25
7/8	0.8750	0.895	0.852	1.438	1.394	1.660	1.589	0.563	0.531	0.062	0.031	1.50	0.28
1	1.0000	1.022	0.976	1.625	1.575	1.876	1.796	0.627	0.591	0.093	0.062	1.75	0.31
1 1/8	1.1250	1.149	1.098	1.812	1.756	2.093	2.002	0.718	0.658	0.093	0.062	2.00	0.34
1¼	1.2500	1.277	1.223	2.000	1.938	2.309	2.209	0.813	0.749	0.093	0.062	2.00	0.38
1 3/8	1.3750	1.404	1.345	2.188	2.119	2.526	2.416	0.878	0.810	0.093	0.062	2.25	0.44
1½	1.5000	1.531	1.470	2.375	2.300	2.742	2.622	0.974	0.902	0.093	0.062	2.25	0.44

All dimensions are in inches. See diagram on page 1280. **Bold type shows bolts unified dimensionally with British and Canadian Standards.** Threads, when rolled, shall be Unified Coarse, Fine, or 8-thread series (UNRC, UNRF, or 8 UNR Series), Class 2A. Threads produced by other methods may be Unified Coarse, Fine, or 8-thread series (UNC, UNF, or 8 UN Series), Class 2A. * Where specifying nominal size in decimals, zeros before the decimal point and in the fourth decimal place are omitted. † See *Body Diameter* footnote in Table 2. § Thread lengths, L_T, shown are for bolt lengths 6 inches and shorter. For longer bolt lengths add 0.250 inch to thread lengths shown.

Table 2. American National Standard and Unified Standard Hex and Heavy Hex Bolts (ANSI B18.2.1-1981)

Nominal Size* or Basic Diam.	Body Diam. E	Width Across Flats F			Width Across Corners G		Height H			Thread Length† L_T
	Max.	Basic	Max.	Min.	Max.	Min.	Basic	Max.	Min.	Basic
HEX BOLTS										
¼ 0.2500	0.260	7/16	0.438	0.425	0.505	0.484	11/64	0.188	0.150	0.750
5/16 0.3125	0.324	½	0.500	0.484	0.577	0.552	7/32	0.235	0.195	0.875
3/8 0.3750	0.388	9/16	0.562	0.544	0.650	0.620	¼	0.268	0.226	1.000
7/16 0.4375	0.452	5/8	0.625	0.603	0.722	0.687	19/64	0.316	0.272	1.125
½ 0.5000	0.515	¾	0.750	0.725	0.866	0.826	11/32	0.364	0.302	1.250
5/8 0.6250	0.642	15/16	0.938	0.906	1.083	1.033	27/64	0.444	0.378	1.500
¾ 0.7500	0.768	1⅛	1.125	1.088	1.299	1.240	½	0.524	0.455	1.750
7/8 0.8750	0.895	1 5/16	1.312	1.269	1.516	1.447	37/64	0.604	0.531	2.000
1 1.0000	1.022	1½	1.500	1.450	1.732	1.653	43/64	0.700	0.591	2.250
1⅛ 1.1250	1.149	1 11/16	1.688	1.631	1.949	1.859	¾	0.780	0.658	2.500
1¼ 1.2500	1.277	1 7/8	1.875	1.812	2.165	2.066	27/32	0.876	0.749	2.750
1⅜ 1.3750	1.404	2 1/16	2.062	1.994	2.382	2.273	29/32	0.940	0.810	3.000
1½ 1.5000	1.531	2¼	2.250	2.175	2.598	2.480	1	1.036	0.902	3.250
1¾ 1.7500	1.785	2 5/8	2.625	2.538	3.031	2.893	1 5/32	1.196	1.054	3.750
2 2.000	2.039	3	3.000	2.900	3.464	3.306	1 11/32	1.388	1.175	4.250
2¼ 2.2500	2.305	3⅜	3.375	3.262	3.897	3.719	1½	1.548	1.327	4.750
2½ 2.5000	2.559	3¾	3.750	3.625	4.330	4.133	1 21/32	1.708	1.479	5.250
2¾ 2.7500	2.827	4⅛	4.125	3.988	4.763	4.546	1 13/16	1.869	1.632	5.750
3 3.0000	3.081	4½	4.500	4.350	5.196	4.959	2	2.060	1.815	6.250
3¼ 3.2500	3.335	4 7/8	4.875	4.712	5.629	5.372	2 3/16	2.251	1.936	6.750
3½ 3.5000	3.589	5¼	5.250	5.075	6.062	5.786	2 5/16	2.380	2.057	7.250
3¾ 3.7500	3.858	5 5/8	5.625	5.437	6.495	6.198	2½	2.572	2.241	7.750
4 4.0000	4.111	6	6.000	5.800	6.928	6.612	2 11/16	2.764	2.424	8.250
HEAVY HEX BOLTS										
½ 0.5000	0.515	7/8	0.875	0.850	1.010	0.969	11/32	0.364	0.302	1.250
5/8 0.6250	0.642	1 1/16	1.062	1.031	1.227	1.175	27/64	0.444	0.378	1.500
¾ 0.7500	0.768	1¼	1.250	1.212	1.443	1.383	½	0.524	0.455	1.750
7/8 0.8750	0.895	1 7/16	1.438	1.394	1.660	1.589	37/64	0.604	0.531	2.000
1 1.0000	1.022	1 5/8	1.625	1.575	1.876	1.796	43/64	0.700	0.591	2.250
1⅛ 1.1250	1.149	1 13/16	1.812	1.756	2.093	2.002	¾	0.780	0.658	2.500
1¼ 1.2500	1.277	2	2.000	1.938	2.309	2.209	27/32	0.876	0.749	2.750
1⅜ 1.3750	1.404	2 3/16	2.188	2.119	2.526	2.416	29/32	0.940	0.810	3.000
1½ 1.5000	1.531	2⅜	2.375	2.300	2.742	2.622	1	1.036	0.902	3.250
1¾ 1.7500	1.785	2¾	2.750	2.662	3.175	3.035	1 5/32	1.196	1.054	3.750
2 2.0000	2.039	3⅛	3.125	3.025	3.608	3.449	1 11/32	1.388	1.175	4.250
2¼ 2.2500	2.305	3½	3.500	3.388	4.041	3.862	1½	1.548	1.327	4.750
2½ 2.5000	2.559	3 7/8	3.875	3.750	4.474	4.275	1 21/32	1.708	1.479	5.250
2¾ 2.7500	2.827	4¼	4.250	4.112	4.907	4.688	1 13/16	1.869	1.632	5.750
3 3.0000	3.081	4 5/8	4.625	4.475	5.340	5.102	2	2.060	1.815	6.250

All dimensions are in inches. See diagram on page 1280.
Bold type shows bolts unified dimensionally with British and Canadian Standards.
Threads: Threads, when rolled, are Unified Coarse, Fine, or 8-thread series (UNRC, UNRF, or 8 UNR Series), Class 2A. Threads produced by other methods may be Unified Coarse, Fine or 8-thread series (UNC, UNF, or 8 UN Series), Class 2A.
Body Diameter: Bolts may be obtained in "reduced diameter body." Where "reduced diameter body" is specified, the body diameter may be reduced to approximately the pitch diameter of the thread. A shoulder of full body diameter under the head may be supplied at the option of the manufacturer.
Material: Unless otherwise specified, chemical and mechanical properties of steel bolts conform to ASTM A307, Grade A. Other materials are as agreed upon by manufacturer and purchaser.
* *Nominal Size:* Where specifying nominal size in decimals, zeros preceding the decimal point and in the fourth decimal place are omitted.
† Thread lengths, L_T, shown are for bolt lengths 6 inches and shorter. For longer bolt lengths add 0.250 inch to thread lengths shown.

Table 3. American National Standard and Unified Standard Heavy Hex Screws and Hex Cap Screws (ANSI B18.2.1-1981)

Nominal Size* or Basic Diam.	Body Diam. E		Width Across Flats F			Width Across Corners G		Height H			Thread Length† L_T
	Max.	Min.	Basic	Max.	Min.	Max.	Min.	Basic	Max.	Min.	Basic
HEAVY HEX SCREWS											
½	**0.5000**	**0.5000**	0.482	⅞	0.875	0.850	1.010	0.969	⁵⁄₁₆	0.323	0.302	1.250
⅝	**0.6250**	**0.6250**	0.605	1¹⁄₁₆	1.062	1.031	1.227	1.175	²⁵⁄₆₄	0.403	0.378	1.500
¾	**0.7500**	**0.7500**	0.729	1¼	1.250	1.212	1.443	1.383	¹⁵⁄₃₂	0.483	0.455	1.750
⅞	**0.8750**	**0.8750**	0.852	1⁷⁄₁₆	1.438	1.394	1.660	1.589	³⁵⁄₆₄	0.563	0.531	2.000
1	**1.0000**	**1.0000**	0.976	1⅝	1.625	1.575	1.876	1.796	³⁹⁄₆₄	0.627	0.591	2.250
1⅛	**1.1250**	**1.1250**	1.098	1¹³⁄₁₆	1.812	1.756	2.093	2.002	¹¹⁄₁₆	0.718	0.658	2.500
1¼	**1.2500**	**1.2500**	1.223	2	2.000	1.938	2.309	2.209	²⁵⁄₃₂	0.813	0.749	2.750
1⅜	**1.3750**	**1.3750**	1.345	2³⁄₁₆	2.188	2.119	2.526	2.416	²⁷⁄₃₂	0.878	0.810	3.000
1½	**1.5000**	**1.5000**	1.470	2⅜	2.375	2.300	2.742	2.622	¹⁵⁄₁₆	0.974	0.902	3.250
1¾	**1.7500**	**1.7500**	1.716	2¾	2.750	2.662	3.175	3.035	1³⁄₃₂	1.134	1.054	3.750
2	**2.0000**	**2.0000**	1.964	3⅛	3.125	3.025	3.608	3.449	1⁷⁄₃₂	1.263	1.175	4.250
2¼	**2.2500**	**2.2500**	2.214	3½	3.500	3.388	4.041	3.862	1⅜	1.423	1.327	4.750
2½	**2.5000**	**2.5000**	2.461	3⅞	3.875	3.750	4.474	4.275	1¹⁷⁄₃₂	1.583	1.479	5.250
2¾	**2.7500**	**2.7500**	2.711	4¼	4.250	4.112	4.907	4.688	1¹¹⁄₁₆	1.744	1.632	5.750
3	**3.0000**	**3.0000**	2.961	4⅝	4.625	4.475	5.340	5.102	1⅞	1.935	1.815	6.250
HEX CAP SCREWS (Finished Hex Bolts)											
¼	**0.2500**	**0.2500**	**0.2450**	⁷⁄₁₆	0.438	0.428	0.505	0.488	⁵⁄₃₂	0.163	0.150	0.750
⁵⁄₁₆	**0.3125**	**0.3125**	**0.3065**	½	0.500	0.489	0.577	0.557	¹³⁄₆₄	0.211	0.195	0.875
⅜	**0.3750**	**0.3750**	**0.3690**	⁹⁄₁₆	0.562	0.551	0.650	0.628	¹⁵⁄₆₄	0.243	0.226	1.000
⁷⁄₁₆	**0.4375**	**0.4375**	**0.4305**	⅝	0.625	0.612	0.722	0.698	⁹⁄₃₂	0.291	0.272	1.125
½	**0.5000**	**0.5000**	**0.4930**	¾	0.750	0.736	0.866	0.840	⁵⁄₁₆	0.323	0.302	1.250
⁹⁄₁₆	**0.5625**	**0.5625**	**0.5545**	¹³⁄₁₆	0.812	0.798	0.938	0.910	²³⁄₆₄	0.371	0.348	1.375
⅝	**0.6250**	**0.6250**	**0.6170**	¹⁵⁄₁₆	0.938	0.922	1.083	1.051	²⁵⁄₆₄	0.403	0.378	1.500
¾	**0.7500**	**0.7500**	**0.7410**	1⅛	1.125	1.100	1.299	1.254	¹⁵⁄₃₂	0.483	0.455	1.750
⅞	**0.8750**	**0.8750**	**0.8660**	1⁵⁄₁₆	1.312	1.285	1.516	1.465	³⁵⁄₆₄	0.563	0.531	2.000
1	**1.0000**	**1.0000**	**0.9900**	1½	1.500	1.469	1.732	1.675	³⁹⁄₆₄	0.627	0.591	2.250
1⅛	**1.1250**	**1.1250**	**1.1140**	1¹¹⁄₁₆	1.688	1.631	1.949	1.859	¹¹⁄₁₆	0.718	0.658	2.500
1¼	**1.2500**	**1.2500**	**1.2390**	1⅞	1.875	1.812	2.165	2.066	²⁵⁄₃₂	0.813	0.749	2.750
1⅜	**1.3750**	**1.3750**	**1.3630**	2¹⁄₁₆	2.062	1.994	2.382	2.273	²⁷⁄₃₂	0.878	0.810	3.000
1½	**1.5000**	**1.5000**	**1.4880**	2¼	2.250	2.175	2.598	2.480	¹⁵⁄₁₆	0.974	0.902	3.250
1¾	**1.7500**	**1.7500**	**1.7380**	2⅝	2.625	2.538	3.031	2.893	1³⁄₃₂	1.134	1.054	3.750
2	**2.0000**	**2.0000**	**1.9880**	3	3.000	2.900	3.464	3.306	1⁷⁄₃₂	1.263	1.175	4.250
2¼	**2.2500**	**2.2500**	**2.2380**	3⅜	3.375	3.262	3.897	3.719	1⅜	1.423	1.327	4.750
2½	**2.5000**	**2.5000**	**2.4880**	3¾	3.750	3.625	4.330	4.133	1¹⁷⁄₃₂	1.583	1.479	5.250
2¾	**2.7500**	**2.7500**	**2.7380**	4⅛	4.125	3.988	4.763	4.546	1¹¹⁄₁₆	1.744	1.632	5.750
3	**3.0000**	**3.0000**	**2.9880**	4½	4.500	4.350	5.196	4.959	1⅞	1.935	1.815	6.250

All dimensions are in inches. See diagram on page 1280.

Unification: **Bold type indicates product features unified dimensionally with British and Canadian standards.** Unification of fine thread products is limited to sizes 1 inch and smaller.

Bearing Surface: Bearing surface is flat and washer faced. Diameter of bearing surface is equal to the maximum width across flats within a tolerance of minus 10 per cent.

Thread Series: Threads, when rolled, are Unified Coarse, Fine or 8-thread series (UNRC, UNRF, or 8 UNR Series), Class 2A. Threads produced by other methods shall preferably be UNRC, UNRF, or 8 UNR but, at manufacturer's option, may be Unified Coarse, Fine or 8-thread series (UNC, UNF or 8 UN Series), Class 2A.

Material: Chemical and mechanical properties of steel screws normally conform to Grades 2, 5, or 8 of SAE J429, ASTM A449 or ASTM A354 Grade BD. Where specified, screws may also be made from brass, bronze, corrosion-resisting steel, aluminum alloy or other materials.

* *Nominal Size:* Where specifying nominal size in decimals, zeros preceding the decimal and in the fourth decimal place are omitted.

† Thread lengths, L_T, shown are for bolt lengths 6 inches or shorter. For longer bolt lengths add 0.250 inch to thread lengths shown.

Table 4. American National Standard Square Lag Screws (ANSI B18.2.1-1981)

DETAIL OF THREAD — CONE POINT (60° APPROX), GIMLET POINT (60° APPROX), 25° APPROX

Nominal Size* or Basic Product Diam.	Body or Shoulder Diam. E		Width Across Flats F			Width Across Corners G		Height H			Shoulder Length S	Radius of Fillet R	Thread Dimensions				
	Max.	Min.	Basic	Max.	Min.	Max.	Min.	Basic	Max.	Min.	Min.	Max.	Thds. per Inch	Pitch P	Flat at Root B	Depth of Thd. T	Root Diam. D_1
No. 10 0.1900	0.199	0.178	9/32	0.281	0.271	0.398	0.372	1/8	0.140	0.110	0.094	0.03	11	0.091	0.039	0.035	0.120
1/4 0.2500	0.260	0.237	3/8	0.375	0.362	0.530	0.498	13/64	0.188	0.156	0.094	0.03	10	0.100	0.043	0.039	0.173
5/16 0.3125	0.324	0.298	1/2	0.500	0.484	0.707	0.665	13/64	0.220	0.186	0.125	0.03	9	0.111	0.048	0.043	0.227
3/8 0.3750	0.388	0.360	9/16	0.562	0.544	0.795	0.747	1/4	0.268	0.232	0.125	0.03	7	0.143	0.062	0.055	0.265
7/16 0.4375	0.452	0.421	5/8	0.625	0.603	0.884	0.828	19/64	0.316	0.278	0.156	0.03	7	0.143	0.062	0.055	0.328
1/2 0.5000	0.515	0.482	3/4	0.750	0.725	1.061	0.995	21/64	0.348	0.308	0.156	0.03	6	0.167	0.072	0.064	0.371
5/8 0.6250	0.642	0.605	15/16	0.938	0.906	1.326	1.244	27/64	0.444	0.400	0.312	0.06	5	0.200	0.086	0.077	0.471
3/4 0.7500	0.768	0.729	1 1/8	1.125	1.088	1.591	1.494	1/2	0.524	0.476	0.375	0.06	4 1/2	0.222	0.096	0.085	0.579
7/8 0.8750	0.895	0.852	1 5/16	1.312	1.269	1.856	1.742	19/32	0.620	0.568	0.375	0.06	4	0.250	0.108	0.096	0.683
1 1.0000	1.022	0.976	1 1/2	1.500	1.450	2.121	1.991	21/32	0.684	0.628	0.625	0.09	3 1/2	0.286	0.123	0.110	0.780
1 1/8 1.1250	1.149	1.098	1 11/16	1.688	1.631	2.386	2.239	3/4	0.780	0.720	0.625	0.09	3 1/4	0.308	0.133	0.119	0.887
1 1/4 1.2500	1.277	1.223	1 7/8	1.875	1.812	2.652	2.489	27/32	0.876	0.812	0.625	0.09	3 1/4	0.308	0.133	0.119	1.012

All dimensions in inches.   * When specifying decimal nominal size, zeros before decimal point and in fourth decimal place are omitted.

Minimum thread length is ½ length of screw plus ½ inch, or 6.00 inches, whichever is shorter. Screws too short for the formula thread length shall be threaded as close to the head as practicable.

Thread formulas: Pitch = 1 ÷ thds. per inch. Flat at root = 0.4305 × pitch. Depth of single thread = 0.385 × pitch.

Table 5. American National Standard Hex Lag Screws (ANSI B18.2.1-1981)

DETAIL OF THREAD

CONE POINT — 60° APPROX

GIMLET POINT — 60° APPROX

Nominal Size* or Basic Product Diam.	Body or Shoulder Diam. E		Width Across Flats F			Width Across Corners G		Height H			Shoulder Length S	Radius of Fillet R	Thread Dimensions				
	Max.	Min.	Basic	Max.	Min.	Max.	Min.	Basic	Max.	Min.	Min.	Max.	Thds. per Inch	Pitch P	Flat at Root B	Depth of Thd. T	Root Diam. D_1
No. 10 0.1900	0.199	0.178	9/32	0.281	0.271	0.323	0.309	1/8	0.140	0.110	0.094	0.03	11	0.091	0.039	0.035	0.120
1/4 0.2500	0.260	0.237	3/8	0.438	0.425	0.505	0.484	11/64	0.188	0.150	0.094	0.03	10	0.100	0.043	0.039	0.173
5/16 0.3125	0.324	0.298	1/2	0.500	0.484	0.577	0.552	7/32	0.235	0.195	0.125	0.03	9	0.111	0.048	0.043	0.227
3/8 0.3750	0.388	0.360	9/16	0.562	0.544	0.650	0.620	1/4	0.268	0.226	0.125	0.03	7	0.143	0.062	0.055	0.265
7/16 0.4375	0.452	0.421	5/8	0.625	0.603	0.722	0.687	19/64	0.316	0.272	0.156	0.03	7	0.143	0.062	0.055	0.328
1/2 0.5000	0.515	0.482	3/4	0.750	0.725	0.866	0.826	11/32	0.364	0.302	0.156	0.03	6	0.167	0.072	0.064	0.371
5/8 0.6250	0.642	0.605	15/16	0.938	0.906	1.083	1.033	27/64	0.444	0.378	0.312	0.06	5	0.200	0.086	0.077	0.471
3/4 0.7500	0.768	0.729	1 1/8	1.125	1.088	1.299	1.240	1/2	0.524	0.455	0.375	0.06	4 1/2	0.222	0.096	0.085	0.579
7/8 0.8750	0.895	0.852	1 5/16	1.312	1.269	1.516	1.447	37/64	0.604	0.531	0.375	0.06	4	0.250	0.108	0.096	0.683
1 1.0000	1.022	0.976	1 1/2	1.500	1.450	1.732	1.653	43/64	0.700	0.591	0.625	0.09	3 1/2	0.286	0.123	0.110	0.780
1 1/8 1.1250	1.149	1.098	1 11/16	1.688	1.631	1.949	1.859	3/4	0.780	0.658	0.625	0.09	3 3/4	0.308	0.133	0.119	0.887
1 1/4 1.2500	1.277	1.223	1 7/8	1.875	1.812	2.165	2.066	27/32	0.876	0.749	0.625	0.09	3 3/4	0.308	0.133	0.119	1.012

$30^{\circ} {}^{+0^{\circ}}_{-15^{\circ}}$

All dimensions in inches.
* When specifying decimal nominal size, zeros before decimal point and in fourth decimal place are omitted.
Minimum thread length is ½ length of screw plus 0.50 inch, or 6.00 inches, whichever is shorter. Screws too short for the formula thread length shall be threaded as close to the head as practicable.
Thread formulas: Pitch = 1 ÷ thds. per inch. Flat at root = 0.4305 × pitch. Depth of single thread = 0.385 × pitch.

ASTM and SAE Grade Markings for Steel Bolts and Screws
(ANSI B18.2.1-1981, Appendix III)

Grade Marking	Specification	Material
NO MARK	SAE — Grade 1	Low or Medium Carbon Steel
	ASTM — A307	Low Carbon Steel
	SAE — Grade 2	Low or Medium Carbon Steel
	SAE — Grade 5	Medium Carbon Steel, Quenched and Tempered
	ASTM — A 449	
	SAE — Grade 5.2	Low Carbon Martensite Steel, Quenched and Tempered
A 325	ASTM — A 325 Type 1	Medium Carbon Steel, Quenched and Tempered Radial dashes optional
A 325	ASTM — A 325 Type 2	Low Carbon Martensite Steel, Quenched and Tempered
A 325	ASTM — A 325 Type 3	Atmospheric Corrosion (Weathering) Steel, Quenched and Tempered
BC	ASTM — A 354 Grade BC	Alloy Steel, Quenched and Tempered
	SAE — Grade 7	Medium Carbon Alloy Steel, Quenched and Tempered, Roll Threaded After Heat Treatment
	SAE — Grade 8	Medium Carbon Alloy Steel, Quenched and Tempered
	ASTM — A 354 Grade BD	Alloy Steel, Quenched and Tempered
	SAE — Grade 8.2	Low Carbon Martensite Steel, Quenched and Tempered
A 490	ASTM — A 490 Type 1	Alloy Steel, Quenched and Tempered
A 490	ASTM — A 490 Type 3	Atmospheric Corrosion (Weathering) Steel, Quenched and Tempered

Table 6. American National Standard and Unified Standard Hex Nuts and Jam Nuts and Heavy Hex Nuts and Jam Nuts (ANSI B18.2.2-1972)

Nominal Size* or Basic Major Diam. of Thread		Width Across Flats F			Width Across Corners G		Thickness, Nuts H			Thickness, Jam Nuts H		
		Basic	Max.	Min.	Max.	Min.	Basic	Max.	Min.	Basic	Max.	Min.
Hex Nuts and Hex Jam Nuts												
1/4	0.2500	7/16	0.438	0.428	0.505	0.488	7/32	0.226	0.212	5/32	0.163	0.150
5/16	0.3125	1/2	0.500	0.489	0.577	0.557	17/64	0.273	0.258	3/16	0.195	0.180
3/8	0.3750	9/16	0.562	0.551	0.650	0.628	21/64	0.337	0.320	7/32	0.227	0.210
7/16	0.4375	11/16	0.688	0.675	0.794	0.768	3/8	0.385	0.365	1/4	0.260	0.240
1/2	0.5000	3/4	0.750	0.736	0.866	0.840	7/16	0.448	0.427	5/16	0.323	0.302
9/16	0.5625	7/8	0.875	0.861	1.010	0.982	31/64	0.496	0.473	5/16	0.324	0.301
5/8	0.6250	15/16	0.938	0.922	1.083	1.051	35/64	0.559	0.535	3/8	0.387	0.363
3/4	0.7500	1 1/8	1.125	1.088	1.299	1.240	41/64	0.665	0.617	27/64	0.446	0.398
7/8	0.8750	1 5/16	1.312	1.269	1.516	1.447	3/4	0.776	0.724	31/64	0.510	0.458
1	1.0000	1 1/2	1.500	1.450	1.732	1.653	55/64	0.887	0.831	35/64	0.575	0.519
1 1/8	1.1250	1 11/16	1.688	1.631	1.949	1.859	31/32	0.999	0.939	39/64	0.639	0.579
1 1/4	1.2500	1 7/8	1.875	1.812	2.165	2.066	1 1/16	1.094	1.030	23/32	0.751	0.687
1 3/8	1.3750	2 1/16	2.062	1.994	2.382	2.273	1 11/64	1.206	1.138	25/32	0.815	0.747
1 1/2	1.5000	2 1/4	2.250	2.175	2.598	2.480	1 9/32	1.317	1.245	27/32	0.880	0.808
Heavy Hex Nuts and Heavy Hex Jam Nuts												
1/4	0.2500	1/2	0.500	0.488	0.577	0.556	15/64	0.250	0.218	11/64	0.188	0.156
5/16	0.3125	9/16	0.562	0.546	0.650	0.622	19/64	0.314	0.280	13/64	0.220	0.186
3/8	0.3750	11/16	0.688	0.669	0.794	0.763	23/64	0.377	0.341	15/64	0.252	0.216
7/16	0.4375	3/4	0.750	0.728	0.866	0.830	27/64	0.441	0.403	17/64	0.285	0.247
1/2	0.5000	7/8	0.875	0.850	1.010	0.969	31/64	0.504	0.464	19/64	0.317	0.277
9/16	0.5625	15/16	0.938	0.909	1.083	1.037	35/64	0.568	0.526	21/64	0.349	0.307
5/8	0.6250	1 1/16	1.062	1.031	1.227	1.175	39/64	0.631	0.587	23/64	0.381	0.337
3/4	0.7500	1 1/4	1.250	1.212	1.443	1.382	47/64	0.758	0.710	27/64	0.446	0.398
7/8	0.8750	1 7/16	1.438	1.394	1.660	1.589	55/64	0.885	0.833	31/64	0.510	0.458
1	1.0000	1 5/8	1.625	1.575	1.876	1.796	63/64	1.012	0.956	35/64	0.575	0.519
1 1/8	1.1250	1 13/16	1.812	1.756	2.093	2.002	1 7/64	1.139	1.079	39/64	0.639	0.579
1 1/4	1.2590	2	2.000	1.938	2.309	2.209	1 7/32	1.251	1.187	23/32	0.751	0.687
1 3/8	1.3750	2 3/16	2.188	2.119	2.526	2.416	1 11/32	1.378	1.310	25/32	0.815	0.747
1 1/2	1.5000	2 3/8	2.375	2.300	2.742	2.622	1 15/32	1.505	1.433	27/32	0.880	0.808
1 5/8	1.6250	2 9/16	2.562	2.481	2.959	2.828	1 19/32	1.632	1.556	29/32	0.944	0.868
1 3/4	1.7500	2 3/4	2.750	2.662	3.175	3.035	1 23/32	1.759	1.679	31/32	1.009	0.929
1 7/8	1.8750	2 15/16	2.938	2.844	3.392	3.242	1 27/32	1.886	1.802	1 1/32	1.073	0.989
2	2.0000	3 1/8	3.125	3.025	3.608	3.449	1 31/32	2.013	1.925	1 3/32	1.138	1.050
2 1/4	2.2500	3 1/2	3.500	3.388	4.041	3.862	2 13/64	2.251	2.155	1 13/64	1.251	1.155
2 1/2	2.5000	3 7/8	3.875	3.750	4.474	4.275	2 29/64	2.505	2.401	1 29/64	1.505	1.401
2 3/4	2.7500	4 1/4	4.250	4.112	4.907	4.688	2 45/64	2.759	2.647	1 37/64	1.634	1.522
3	3.0000	4 5/8	4.625	4.475	5.340	5.102	2 61/64	3.013	2.893	1 45/64	1.763	1.643
3 1/4	3.2500	5	5.000	4.838	5.774	5.515	3 13/64	3.252	3.124	1 13/16	1.876	1.748
3 1/2	3.5000	5 3/8	5.375	5.200	6.207	5.928	3 7/16	3.506	3.370	1 15/16	2.006	1.870
3 3/4	3.7500	5 3/4	5.750	5.562	6.640	6.341	3 11/16	3.756	3.620	2 1/16	2.134	1.990
4	4.0000	6 1/8	6.125	5.925	7.073	6.755	3 15/16	4.014	3.862	2 3/16	2.264	2.112

All dimensions are in inches. See diagram on page 1280.

Bold type shows nuts unified dimensionally with British and Canadian Standards.

Threads are Unified Coarse-, Fine-, or 8-thread series (UNC, UNF or 8UN), Class 2B. Unification of fine-thread nuts is limited to sizes 1 inch and under.

* Where specifying nominal size in decimals, zeros before the decimal point and in the fourth decimal place are omitted.

Table 7. American National Standard and Unified Standard Hex Flat Nuts and Flat Jam Nuts and Heavy Hex Flat Nuts and Flat Jam Nuts (ANSI B18.2.2-1972)

Nominal Size* or Basic Major Diam. of Thread		Width Across Flats F			Width Across Corners G		Thickness, Flat Nuts H			Thickness, Flat Jam Nuts H		
		Basic	Max.	Min.	Max.	Min.	Basic	Max.	Min.	Basic	Max.	Min.
Hex Flat Nuts and Hex Flat Jam Nuts												
1⅛	1.1250	1¹¹⁄₁₆	1.688	1.631	1.949	1.859	1	1.030	0.970	⅝	0.655	0.595
1¼	1.2500	1⅞	1.875	1.812	2.165	2.066	1³⁄₃₂	1.126	1.062	¾	0.782	0.718
1⅜	1.3750	2¹⁄₁₆	2.062	1.994	2.382	2.273	1¹³⁄₆₄	1.237	1.169	¹³⁄₁₆	0.846	0.778
1½	1.5000	2¼	2.250	2.175	2.598	2.480	1⁵⁄₁₆	1.348	1.276	⅞	0.911	0.839
Heavy Hex Flat Nuts and Heavy Hex Flat Jam Nuts												
1⅛	1.1250	1¹³⁄₁₆	1.812	1.756	2.093	2.002	1⅛	1.155	1.079	⅝	0.655	0.579
1¼	1.2500	2	2.000	1.938	2.309	2.209	1¼	1.282	1.187	¾	0.782	0.687
1⅜	1.3750	2³⁄₁₆	2.188	2.119	2.526	2.416	1⅜	1.409	1.310	¹³⁄₁₆	0.846	0.747
1½	1.5000	2⅜	2.375	2.300	2.742	2.622	1½	1.536	1.433	⅞	0.911	0.808
1¾	1.7500	2¾	2.750	2.662	3.175	3.035	1¾	1.790	1.679	1	1.040	0.929
2	2.0000	3⅛	3.125	3.025	3.608	3.449	2	2.044	1.925	1⅛	1.169	1.050
2¼	2.2500	3½	3.500	3.388	4.041	3.862	2¼	2.298	2.155	1¼	1.298	1.155
2½	2.5000	3⅞	3.875	3.750	4.474	4.275	2½	2.552	2.401	1½	1.552	1.401
2¾	2.7500	4¼	4.250	4.112	4.907	4.688	2¾	2.806	2.647	1⅝	1.681	1.522
3	3.0000	4⅝	4.625	4.475	5.340	5.102	3	3.060	2.893	1¾	1.810	1.643
3¼	3.2500	5	5.000	4.838	5.774	5.515	3¼	3.314	3.124	1⅞	1.939	1.748
3½	3.5000	5⅜	5.375	5.200	6.207	5.928	3½	3.568	3.370	2	2.068	1.870
3¾	3.7500	5¾	5.750	5.562	6.640	6.341	3¾	3.822	3.616	2⅛	2.197	1.990
4	4.0000	6⅛	6.125	5.925	7.073	6.755	4	4.076	3.862	2¼	2.326	2.112

All dimensions are in inches. See diagram on page 1280.
Bold type indicates nuts unified dimensionally with British and Canadian Standards.
Threads are Unified Coarse-thread series (UNC), Class 2B.
* Where specifying nominal size in decimals, zeros before the decimal point and in the fourth decimal place are omitted.

Table 8. American National Standard Hex Castle Nuts (ANSI B18.2.2-1972)

Nominal Size* or Basic Major Diam. of Thread		Width Across Flats F		Width Across Corners G		Thickness H		Unslotted Thickness T		Width of Slot S		Radius of Fillet R	Diam. of Cylindrical Part U
		Max.	Min.	Max.	Min.	Max.	Min.	Max.	Min.	Max.	Min.	±.010	Min.
¼	0.2500	0.438	0.428	0.505	0.488	0.288	0.274	0.20	0.18	0.10	0.07	0.094	0.371
⁵⁄₁₆	0.3125	0.500	0.489	0.577	0.557	0.336	0.320	0.24	0.22	0.12	0.09	0.094	0.425
⅜	0.3750	0.562	0.551	0.650	0.628	0.415	0.398	0.29	0.27	0.15	0.12	0.094	0.478
⁷⁄₁₆	0.4375	0.688	0.675	0.794	0.768	0.463	0.444	0.31	0.29	0.15	0.12	0.094	0.582
½	0.5000	0.750	0.736	0.866	0.840	0.573	0.552	0.42	0.40	0.18	0.15	0.125	0.637
⁹⁄₁₆	0.5625	0.875	0.861	1.010	0.982	0.621	0.598	0.43	0.41	0.18	0.15	0.156	0.744
⅝	0.6250	0.938	0.922	1.083	1.051	0.731	0.706	0.51	0.49	0.24	0.18	0.156	0.797
¾	0.7500	1.125	1.088	1.299	1.240	0.827	0.798	0.57	0.55	0.24	0.18	0.156	0.941
⅞	0.8750	1.312	1.269	1.516	1.447	0.922	0.890	0.67	0.64	0.24	0.18	0.188	1.097
1	1.0000	1.500	1.450	1.732	1.653	1.018	0.982	0.73	0.70	0.30	0.24	0.188	1.254
1⅛	1.1250	1.688	1.631	1.949	1.859	1.176	1.136	0.83	0.80	0.33	0.24	0.250	1.411
1¼	1.2500	1.875	1.812	2.165	2.066	1.272	1.228	0.89	0.86	0.40	0.31	0.250	1.570
1⅜	1.3750	2.062	1.994	2.382	2.273	1.399	1.351	1.02	0.98	0.40	0.31	0.250	1.726
1½	1.5000	2.250	2.175	2.598	2.480	1.526	1.474	1.08	1.04	0.46	0.37	0.250	1.881

All dimensions are in inches. See diagram on page 1280.
Threads are Unified Coarse- or Fine-thread series (UNC or UNF), Class 2B.
* Where specifying nominal size in decimals, zeros before the decimal point and in the fourth decimal place are omitted.

Table 9. American National and Unified Standard Hex Slotted Nuts, Heavy Hex Slotted Nuts and Hex Thick Slotted Nuts (ANSI B18.2.2-1972)

Nominal Size* or Basic Major Diam. of Thread		Width Across Flats F			Width Across Corners G		Thickness H			Unslotted Thickness T		Width of Slot S	
		Basic	Max.	Min.	Max.	Min.	Basic	Max.	Min.	Max.	Min.	Max.	Min.
Hex Slotted Nuts													
1/4	0.2500	7/16 0.438	0.428	0.505	0.488	7/32 0.226	0.212	0.14	0.12	0.10	0.07		
5/16	0.3125	1/2 0.500	0.489	0.577	0.557	17/64 0.273	0.258	0.18	0.16	0.12	0.09		
3/8	0.3750	9/16 0.562	0.551	0.650	0.628	21/64 0.337	0.320	0.21	0.19	0.15	0.12		
7/16	0.4375	11/16 0.688	0.675	0.794	0.768	3/8 0.385	0.365	0.23	0.21	0.15	0.12		
1/2	0.5000	3/4 0.750	0.736	0.866	0.840	7/16 0.448	0.427	0.29	0.27	0.18	0.15		
9/16	0.5625	7/8 0.875	0.861	1.010	0.982	31/64 0.496	0.473	0.31	0.29	0.18	0.15		
5/8	0.6250	15/16 0.938	0.922	1.083	1.051	35/64 0.559	0.535	0.34	0.32	0.24	0.18		
3/4	0.7500	1 1/8 1.125	1.088	1.299	1.240	41/64 0.665	0.617	0.40	0.38	0.24	0.18		
7/8	0.8750	1 5/16 1.312	1.269	1.516	1.447	3/4 0.776	0.724	0.52	0.49	0.24	0.18		
1	1.0000	1 1/2 1.500	1.450	1.732	1.653	55/64 0.887	0.831	0.59	0.56	0.30	0.24		
1 1/8	1.1250	1 11/16 1.688	1.631	1.949	1.859	31/32 0.999	0.939	0.64	0.61	0.33	0.24		
1 1/4	1.2500	1 7/8 1.875	1.812	2.165	2.066	1 1/16 1.094	1.030	0.70	0.67	0.40	0.31		
1 3/8	1.3750	2 1/16 2.062	1.994	2.382	2.273	1 11/64 1.206	1.138	0.82	0.78	0.40	0.31		
1 1/2	1.5000	2 1/4 2.250	2.175	2.598	2.480	1 9/32 1.317	1.245	0.86	0.82	0.46	0.37		
Heavy Hex Slotted Nuts													
1/4	0.2500	1/2 0.500	0.488	0.577	0.556	1/4 0.250	0.218	0.15	0.13	0.10	0.07		
5/16	0.3125	9/16 0.562	0.546	0.650	0.622	19/64 0.314	0.280	0.21	0.19	0.12	0.09		
3/8	0.3750	11/16 0.688	0.669	0.794	0.763	23/64 0.377	0.341	0.24	0.22	0.15	0.12		
7/16	0.4375	3/4 0.750	0.728	0.866	0.829	27/64 0.441	0.403	0.28	0.26	0.15	0.12		
1/2	0.5000	7/8 0.875	0.850	1.010	0.969	31/64 0.504	0.464	0.34	0.32	0.18	0.15		
9/16	0.5625	15/16 0.938	0.909	1.083	1.037	35/64 0.568	0.526	0.37	0.35	0.18	0.15		
5/8	0.6250	1 1/16 1.062	1.031	1.227	1.175	39/64 0.631	0.587	0.40	0.38	0.24	0.18		
3/4	0.7500	1 1/4 1.250	1.212	1.443	1.382	47/64 0.758	0.710	0.49	0.47	0.24	0.18		
7/8	0.8750	1 7/16 1.438	1.394	1.660	1.589	55/64 0.885	0.833	0.62	0.59	0.24	0.18		
1	1.0000	1 5/8 1.625	1.575	1.876	1.796	63/64 1.012	0.956	0.72	0.69	0.30	0.24		
1 1/8	1.1250	1 13/16 1.812	1.756	2.093	2.002	1 7/64 1.139	1.079	0.78	0.75	0.33	0.24		
1 1/4	1.2500	2 2.000	1.938	2.309	2.209	1 7/32 1.251	1.187	0.86	0.83	0.40	0.31		
1 3/8	1.3750	2 3/16 2.188	2.119	2.526	2.416	1 11/32 1.378	1.310	0.99	0.95	0.40	0.31		
1 1/2	1.5000	2 3/8 2.375	2.300	2.742	2.622	1 15/32 1.505	1.433	1.05	1.01	0.46	0.37		
1 3/4	1.7500	2 3/4 2.750	2.662	3.175	3.035	1 23/32 1.759	1.679	1.24	1.20	0.52	0.43		
2	2.0000	3 1/8 3.125	3.025	3.608	3.449	1 31/32 2.013	1.925	1.43	1.38	0.52	0.43		
2 1/4	2.2500	3 1/2 3.500	3.388	4.041	3.862	2 13/64 2.251	2.155	1.67	1.62	0.52	0.43		
2 1/2	2.5000	3 7/8 3.875	3.750	4.474	4.275	2 29/64 2.505	2.401	1.79	1.74	0.64	0.55		
2 3/4	2.7500	4 1/4 4.250	4.112	4.907	4.688	2 45/64 2.759	2.647	2.05	1.99	0.64	0.55		
3	3.0000	4 5/8 4.625	4.475	5.340	5.102	2 61/64 3.013	2.893	2.23	2.17	0.71	0.62		
3 1/4	3.2500	5 5.000	4.838	5.774	5.515	3 3/16 3.252	3.124	2.47	2.41	0.71	0.62		
3 1/2	3.5000	5 3/8 5.375	5.200	6.207	5.928	3 7/16 3.506	3.370	2.72	2.65	0.71	0.62		
3 3/4	3.7500	5 3/4 5.750	5.562	6.640	6.341	3 11/16 3.760	3.616	2.97	2.90	0.71	0.62		
4	4.0000	6 1/8 6.125	5.925	7.073	6.755	3 15/16 4.013	3.862	3.22	3.15	0.71	0.62		
Hex Thick Slotted Nuts													
1/4	0.2500	7/16 0.438	0.428	0.505	0.488	9/32 0.288	0.274	0.20	0.18	0.10	0.07		
5/16	0.3125	1/2 0.500	0.489	0.577	0.557	21/64 0.336	0.320	0.24	0.22	0.12	0.09		
3/8	0.3750	9/16 0.562	0.551	0.650	0.628	13/32 0.415	0.398	0.29	0.27	0.15	0.12		
7/16	0.4375	11/16 0.688	0.675	0.794	0.768	29/64 0.463	0.444	0.31	0.29	0.15	0.12		
1/2	0.5000	3/4 0.750	0.736	0.866	0.840	9/16 0.573	0.552	0.42	0.40	0.18	0.15		
9/16	0.5625	7/8 0.875	0.861	1.010	0.982	39/64 0.621	0.598	0.43	0.41	0.18	0.15		
5/8	0.6250	15/16 0.938	0.922	1.083	1.051	23/32 0.731	0.706	0.51	0.49	0.24	0.18		
3/4	0.7500	1 1/8 1.125	1.088	1.299	1.240	13/16 0.827	0.798	0.57	0.55	0.24	0.18		
7/8	0.8750	1 5/16 1.312	1.269	1.516	1.447	29/32 0.922	0.890	0.67	0.64	0.24	0.18		
1	1.0000	1 1/2 1.500	1.450	1.732	1.653	1 1.018	0.982	0.73	0.70	0.30	0.24		
1 1/8	1.1250	1 11/16 1.688	1.631	1.949	1.859	1 5/32 1.176	1.136	0.83	0.80	0.33	0.24		
1 1/4	1.2500	1 7/8 1.875	1.812	2.165	2.066	1 1/4 1.272	1.228	0.89	0.86	0.40	0.31		
1 3/8	1.3750	2 1/16 2.062	1.994	2.382	2.273	1 3/8 1.399	1.351	1.02	0.98	0.40	0.31		
1 1/2	1.5000	2 1/4 2.250	2.175	2.598	2.480	1 1/2 1.474	1.474	1.08	1.04	0.46	0.37		

All dimensions are in inches. See diagram on page 1280. **Bold type indicates nuts unified dimensionally with British and Canadian Standards.** * See Table 10 footnote. Threads are Unified Coarse-, Fine-, or 8-thread series (UNC, UNF, or 8UN), Class 2B. Unification of fine-thread nuts is limited to sizes 1 inch and under.

Table 10. American National and Unified Standard Square Nuts and Heavy Square Nuts and American National Standard Hex Thick Nuts (ANSI B18.2.2-1972)

Nominal Size* or Basic Major Diam. of Thread		Width Across Flats F			Width Across Corners G		Thickness H		
		Basic	Max.	Min.	Max.	Min.	Basic	Max.	Min.
Square Nuts†									
¼	0.2500	⁷⁄₁₆	0.438	0.425	0.619	0.584	⁷⁄₃₂	0.235	0.203
⁵⁄₁₆	0.3125	⁹⁄₁₆	0.562	0.547	0.795	0.751	¹⁷⁄₆₄	0.283	0.249
⅜	0.3750	⅝	0.625	0.606	0.884	0.832	²¹⁄₆₄	0.346	0.310
⁷⁄₁₆	0.4375	¾	0.750	0.728	1.061	1.000	⅜	0.394	0.356
½	0.5000	¹³⁄₁₆	0.812	0.788	1.149	1.082	⁷⁄₁₆	0.458	0.418
⅝	0.6250	1	1.000	0.969	1.414	1.330	³⁵⁄₆₄	0.569	0.525
¾	0.7500	1⅛	1.125	1.088	1.591	1.494	²¹⁄₃₂	0.680	0.632
⅞	0.8750	1⁵⁄₁₆	1.312	1.269	1.856	1.742	⁴⁹⁄₆₄	0.792	0.740
1	1.0000	1½	1.500	1.450	2.121	1.991	⅞	0.903	0.847
1⅛	1.1250	1¹¹⁄₁₆	1.688	1.631	2.386	2.239	1	1.030	0.970
1¼	1.2500	1⅞	1.875	1.812	2.652	2.489	1³⁄₃₂	1.126	1.062
1⅜	1.3750	2¹⁄₁₆	2.062	1.994	2.917	2.738	1¹³⁄₆₄	1.237	1.169
1½	1.5000	2¼	2.250	2.175	3.182	2.986	1⁵⁄₁₆	1.348	1.276
Heavy Square Nuts†									
¼	0.2500	½	0.500	0.488	0.707	0.670	¼	0.266	0.218
⁵⁄₁₆	0.3125	⁹⁄₁₆	0.562	0.546	0.795	0.750	⁵⁄₁₆	0.330	0.280
⅜	0.3750	¹¹⁄₁₆	0.688	0.669	0.973	0.919	⅜	0.393	0.341
⁷⁄₁₆	0.4375	¾	0.750	0.728	1.060	1.000	⁷⁄₁₆	0.456	0.403
½	0.5000	⅞	0.875	0.850	1.237	1.167	½	0.520	0.464
⅝	0.6250	1¹⁄₁₆	1.062	1.031	1.503	1.416	⅝	0.647	0.587
¾	0.7500	1¼	1.250	1.212	1.768	1.665	¾	0.774	0.710
⅞	0.8750	1⁷⁄₁₆	1.438	1.394	2.033	1.914	⅞	0.901	0.833
1	1.0000	1⅝	1.625	1.575	2.298	2.162	1	1.028	0.956
1⅛	1.1250	1¹³⁄₁₆	1.812	1.756	2.563	2.411	1⅛	1.155	1.079
1¼	1.2500	2	2.000	1.938	2.828	2.661	1¼	1.282	1.187
1⅜	1.3750	2³⁄₁₆	2.188	2.119	3.094	2.909	1⅜	1.409	1.310
1½	1.5000	2⅜	2.375	2.300	3.359	3.158	1½	1.536	1.433
Hex Thick Nuts‡									
¼	0.2500	⁷⁄₁₆	0.438	0.428	0.505	0.488	⁹⁄₃₂	0.288	0.274
⁵⁄₁₆	0.3125	½	0.500	0.489	0.577	0.557	²¹⁄₆₄	0.336	0.320
⅜	0.3750	⁹⁄₁₆	0.562	0.551	0.650	0.628	¹³⁄₃₂	0.415	0.398
⁷⁄₁₆	0.4375	¹¹⁄₁₆	0.688	0.675	0.794	0.768	²⁹⁄₆₄	0.463	0.444
½	0.5000	¾	0.750	0.736	0.866	0.840	⁹⁄₁₆	0.573	0.552
⁹⁄₁₆	0.5625	⅞	0.875	0.861	1.010	0.982	³⁹⁄₆₄	0.621	0.598
⅝	0.6250	¹⁵⁄₁₆	0.938	0.922	1.083	1.051	²³⁄₃₂	0.731	0.706
¾	0.7500	1⅛	1.125	1.088	1.299	1.240	¹³⁄₁₆	0.827	0.798
⅞	0.8750	1⁵⁄₁₆	1.312	1.269	1.516	1.447	²⁹⁄₃₂	0.922	0.890
1	1.0000	1½	1.500	1.450	1.732	1.653	1	1.018	0.982
1⅛	1.1250	1¹¹⁄₁₆	1.688	1.631	1.949	1.859	1⁵⁄₃₂	1.176	1.136
1¼	1.2500	1⅞	1.875	1.812	2.165	2.066	1¼	1.272	1.228
1⅜	1.3750	2¹⁄₁₆	2.062	1.994	2.382	2.273	1⅜	1.399	1.351
1½	1.5000	2¼	2.250	2.175	2.598	2.480	1½	1.526	1.474

All dimensions are in inches. See diagram on page 1280.
Bold type indicates nuts unified dimensionally with British and Canadian Standards.
* Where specifying nominal size in decimals, zeros before the decimal point and in the fourth decimal place are omitted.
† Threads are Unified Coarse-thread series (UNC), Class 2B.
‡ Threads are Unified Coarse-, Fine-, or 8-thread series (UNC, UNF, or 8 UN), Class 2B.

Low and High Crown (Blind, Acorn) Nuts (SAE Recommended Practice J483a)

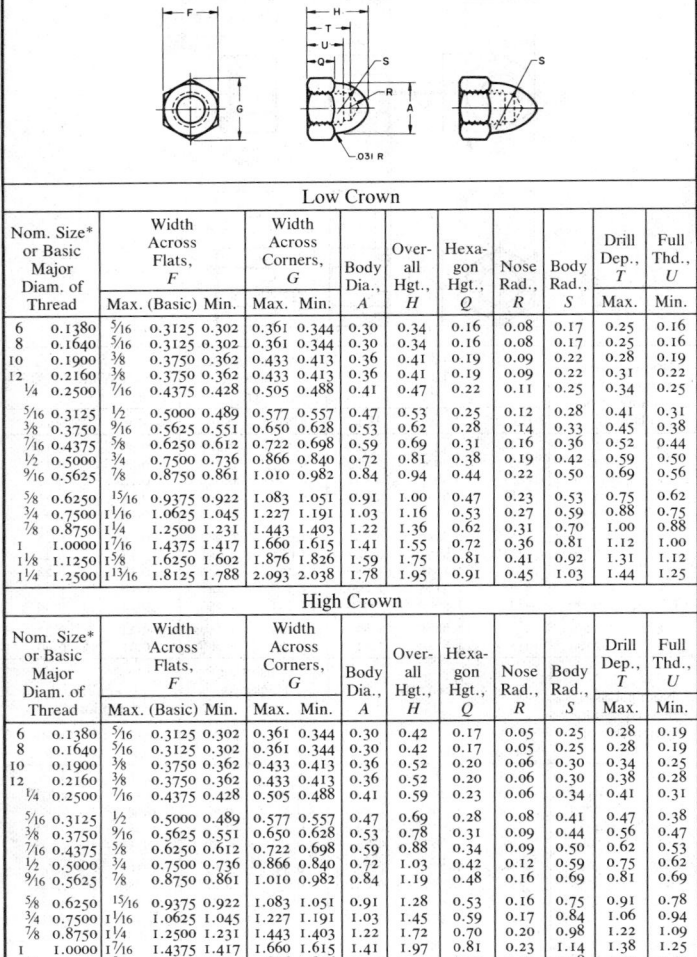

Low Crown

Nom. Size* or Basic Major Diam. of Thread	Width Across Flats, F		Width Across Corners, G		Body Dia., A	Over-all Hgt., H	Hexagon Hgt., Q	Nose Rad., R	Body Rad., S	Drill Dep., T Max.	Full Thd., U Min.
	Max. (Basic)	Min.	Max.	Min.							
6 0.1380	5/16 0.3125	0.302	0.361	0.344	0.30	0.34	0.16	0.08	0.17	0.25	0.16
8 0.1640	5/16 0.3125	0.302	0.361	0.344	0.30	0.34	0.16	0.08	0.17	0.25	0.16
10 0.1900	3/8 0.3750	0.362	0.433	0.413	0.36	0.41	0.19	0.09	0.22	0.28	0.19
12 0.2160	3/8 0.3750	0.362	0.433	0.413	0.36	0.41	0.19	0.09	0.22	0.31	0.22
1/4 0.2500	7/16 0.4375	0.428	0.505	0.488	0.41	0.47	0.22	0.11	0.25	0.34	0.25
5/16 0.3125	1/2 0.5000	0.489	0.577	0.557	0.47	0.53	0.25	0.12	0.28	0.41	0.31
3/8 0.3750	9/16 0.5625	0.551	0.650	0.628	0.53	0.62	0.28	0.14	0.33	0.45	0.38
7/16 0.4375	5/8 0.6250	0.612	0.722	0.698	0.59	0.69	0.31	0.16	0.36	0.52	0.44
1/2 0.5000	3/4 0.7500	0.736	0.866	0.840	0.72	0.81	0.38	0.19	0.42	0.59	0.50
9/16 0.5625	7/8 0.8750	0.861	1.010	0.982	0.84	0.94	0.44	0.22	0.50	0.69	0.56
5/8 0.6250	15/16 0.9375	0.922	1.083	1.051	0.91	1.00	0.47	0.23	0.53	0.75	0.62
3/4 0.7500	1 1/16 1.0625	1.045	1.227	1.191	1.03	1.16	0.53	0.27	0.59	0.88	0.75
7/8 0.8750	1 1/4 1.2500	1.231	1.443	1.403	1.22	1.36	0.62	0.31	0.70	1.00	0.88
1 1.0000	1 7/16 1.4375	1.417	1.660	1.615	1.41	1.55	0.72	0.36	0.81	1.12	1.00
1 1/8 1.1250	1 5/8 1.6250	1.602	1.876	1.826	1.59	1.75	0.81	0.41	0.92	1.31	1.12
1 1/4 1.2500	1 13/16 1.8125	1.788	2.093	2.038	1.78	1.95	0.91	0.45	1.03	1.44	1.25

High Crown

Nom. Size* or Basic Major Diam. of Thread	Width Across Flats, F		Width Across Corners, G		Body Dia., A	Over-all Hgt., H	Hexagon Hgt., Q	Nose Rad., R	Body Rad., S	Drill Dep., T Max.	Full Thd., U Min.
	Max. (Basic)	Min.	Max.	Min.							
6 0.1380	5/16 0.3125	0.302	0.361	0.344	0.30	0.42	0.17	0.05	0.25	0.28	0.19
8 0.1640	5/16 0.3125	0.302	0.361	0.344	0.30	0.42	0.17	0.05	0.25	0.28	0.19
10 0.1900	3/8 0.3750	0.362	0.433	0.413	0.36	0.52	0.20	0.06	0.30	0.34	0.25
12 0.2160	3/8 0.3750	0.362	0.433	0.413	0.36	0.52	0.20	0.06	0.30	0.38	0.28
1/4 0.2500	7/16 0.4375	0.428	0.505	0.488	0.41	0.59	0.23	0.06	0.34	0.41	0.31
5/16 0.3125	1/2 0.5000	0.489	0.577	0.557	0.47	0.69	0.28	0.08	0.41	0.47	0.38
3/8 0.3750	9/16 0.5625	0.551	0.650	0.628	0.53	0.78	0.31	0.09	0.44	0.56	0.47
7/16 0.4375	5/8 0.6250	0.612	0.722	0.698	0.59	0.88	0.34	0.09	0.50	0.62	0.53
1/2 0.5000	3/4 0.7500	0.736	0.866	0.840	0.72	1.03	0.42	0.12	0.59	0.75	0.62
9/16 0.5625	7/8 0.8750	0.861	1.010	0.982	0.84	1.19	0.48	0.16	0.69	0.81	0.69
5/8 0.6250	15/16 0.9375	0.922	1.083	1.051	0.91	1.28	0.53	0.16	0.75	0.91	0.78
3/4 0.7500	1 1/16 1.0625	1.045	1.227	1.191	1.03	1.45	0.59	0.17	0.84	1.06	0.94
7/8 0.8750	1 1/4 1.2500	1.231	1.443	1.403	1.22	1.72	0.70	0.20	0.98	1.22	1.09
1 1.0000	1 7/16 1.4375	1.417	1.660	1.615	1.41	1.97	0.81	0.23	1.14	1.38	1.25
1 1/8 1.1250	1 5/8 1.6250	1.602	1.876	1.826	1.59	2.22	0.92	0.27	1.28	1.59	1.41
1 1/4 1.2500	1 13/16 1.8125	1.788	2.093	2.038	1.78	2.47	1.03	0.28	1.44	1.75	1.56

All dimensions are in inches. Threads are Unified Standard Class 2B, UNC or UNF Series. * When specifying a nominal size in decimals, any zero in the fourth decimal place is omitted.

Hex High and Hex Slotted High Nuts (SAE Standard J482a)

Nominal Size* or Basic Major Diameter of Thread		Width Across Flats, F			Width Across Corners, G		Slot Width, S	
		Basic	Max.	Min.	Max.	Min.	Min.	Max.
1/4	0.2500	7/16	0.4375	0.428	0.505	0.488	0.07	0.10
5/16	0.3125	1/2	0.5000	0.489	0.577	0.557	0.09	0.12
3/8	0.3750	9/16	0.5625	0.551	0.650	0.628	0.12	0.15
7/16	0.4375	11/16	0.6875	0.675	0.794	0.768	0.12	0.15
1/2	0.5000	3/4	0.7500	0.736	0.866	0.840	0.15	0.18
9/16	0.5625	7/8	0.8750	0.861	1.010	0.982	0.15	0.18
5/8	0.6250	15/16	0.9375	0.922	1.083	1.051	0.18	0.24
3/4	0.7500	1 1/8	1.1250	1.088	1.299	1.240	0.18	0.24
7/8	0.8750	1 5/16	1.3125	1.269	1.516	1.447	0.18	0.24
1	1.0000	1 1/2	1.5000	1.450	1.732	1.653	0.24	0.30
1 1/8	1.1250	1 11/16	1.6875	1.631	1.949	1.859	0.24	0.33
1 1/4	1.2500	1 7/8	1.8750	1.812	2.165	2.066	0.31	0.40

Nominal Size* or Basic Major Diameter of Thread		Thickness, H			Unslotted Thickness, T		Counterbore (Optional)	
		Basic	Max.	Min.	Max.	Min.	Diam., A	Depth, D
1/4	0.2500	3/8	0.382	0.368	0.29	0.27	0.266	0.062
5/16	0.3125	29/64	0.461	0.445	0.37	0.35	0.328	0.078
3/8	0.3750	1/2	0.509	0.491	0.38	0.36	0.391	0.094
7/16	0.4375	39/64	0.619	0.599	0.46	0.44	0.453	0.109
1/2	0.5000	21/32	0.667	0.645	0.51	0.49	0.516	0.125
9/16	0.5625	49/64	0.778	0.754	0.59	0.57	0.591	0.141
5/8	0.6250	27/32	0.857	0.831	0.63	0.61	0.656	0.156
3/4	0.7500	1	1.015	0.985	0.76	0.73	0.781	0.188
7/8	0.8750	1 5/32	1.172	1.140	0.92	0.89	0.906	0.219
1	1.0000	1 5/16	1.330	1.292	1.05	1.01	1.031	0.250
1 1/8	1.1250	1 1/2	1.520	1.480	1.18	1.14	1.156	0.281
1 1/4	1.2500	1 11/16	1.710	1.666	1.34	1.29	1.281	0.312

All dimensions are in inches. Threads are Unified Standard Class 2B, UNC or UNF Series. * When specifying a nominal size in decimals, any zero in the fourth decimal place is omitted. *Reprinted with permission. Copyright © 1987, Society of Automotive Engineers, Inc. All rights reserved.*

American National Standard Round Head and Round Head Square Neck Bolts (ANSI B18.5-1978)

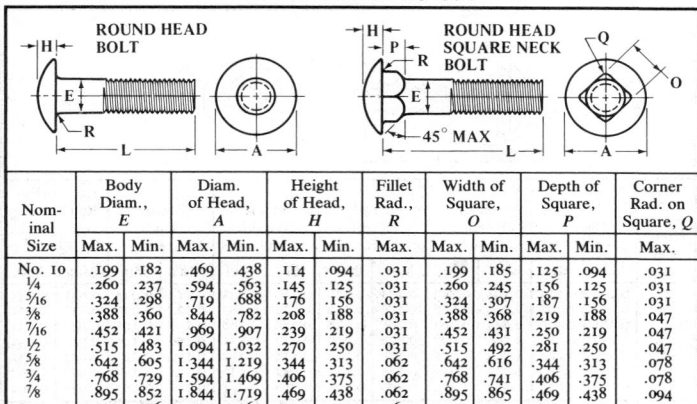

Nominal Size	Body Diam., E		Diam. of Head, A		Height of Head, H		Fillet Rad., R	Width of Square, O		Depth of Square, P		Corner Rad. on Square, Q
	Max.	Min.	Max.	Min.	Max.	Min.	Max.	Max.	Min.	Max.	Min.	Max.
No. 10	.199	.182	.469	.438	.114	.094	.031	.199	.185	.125	.094	.031
¼	.260	.237	.594	.563	.145	.125	.031	.260	.245	.156	.125	.031
5/16	.324	.298	.719	.688	.176	.156	.031	.324	.307	.187	.156	.031
3/8	.388	.360	.844	.782	.208	.188	.031	.388	.368	.219	.188	.047
7/16	.452	.421	.969	.907	.239	.219	.031	.452	.431	.250	.219	.047
½	.515	.483	1.094	1.032	.270	.250	.031	.515	.492	.281	.250	.047
5/8	.642	.605	1.344	1.219	.344	.313	.062	.642	.616	.344	.313	.078
¾	.768	.729	1.594	1.469	.406	.375	.062	.768	.741	.406	.375	.078
7/8	.895	.852	1.844	1.719	.469	.438	.062	.895	.865	.469	.438	.094
1	1.022	.976	2.094	1.969	.531	.500	.062	1.022	.990	.531	.500	.094

All dimensions are in inches unless otherwise specified.

Threads are Unified Standard, Class 2A, UNC Series, in accordance with ANSI B1.1. For threads with additive finish, the maximum diameters of Class 2A shall apply before plating or coating, whereas the basic diameters (Class 2A maximum diameters plus the allowance) shall apply to a bolt after plating or coating.

Bolts are designated in the sequence shown: nominal size (number, fraction or decimal equivalent); threads per inch; nominal length (fraction or decimal equivalent); product name; material; and protective finish, if required.

i.e.: ½–13 × 3 Round Head Square Neck Bolt, Steel
.375–16 × 2.50 Step Bolt, Steel, Zinc Plated

American National Standard T-Head Bolts (ANSI B18.5-1978)

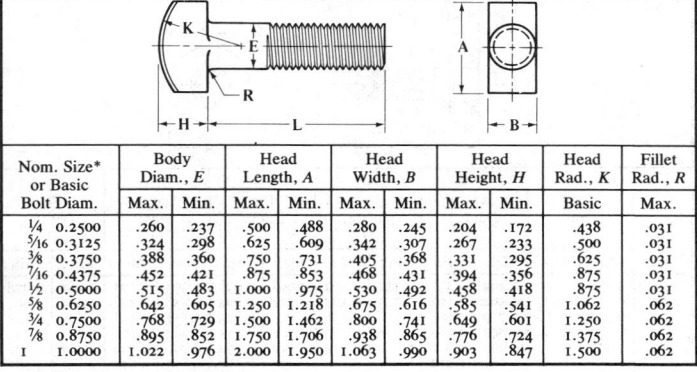

Nom. Size* or Basic Bolt Diam.		Body Diam., E		Head Length, A		Head Width, B		Head Height, H		Head Rad., K	Fillet Rad., R
		Max.	Min.	Max.	Min.	Max.	Min.	Max.	Min.	Basic	Max.
¼	0.2500	.260	.237	.500	.488	.280	.245	.204	.172	.438	.031
5/16	0.3125	.324	.298	.625	.609	.342	.307	.267	.233	.500	.031
3/8	0.3750	.388	.360	.750	.731	.405	.368	.331	.295	.625	.031
7/16	0.4375	.452	.421	.875	.853	.468	.431	.394	.356	.875	.031
½	0.5000	.515	.483	1.000	.975	.530	.492	.458	.418	.875	.031
5/8	0.6250	.642	.605	1.250	1.218	.675	.616	.585	.541	1.062	.062
¾	0.7500	.768	.729	1.500	1.462	.800	.741	.649	.601	1.250	.062
7/8	0.8750	.895	.852	1.750	1.706	.938	.865	.776	.724	1.375	.062
1	1.0000	1.022	.976	2.000	1.950	1.063	.990	.903	.847	1.500	.062

All dimensions are given in inches.

* Where specifying nominal size in decimals, zeros preceding the decimal point and in the fourth decimal place are omitted. For information as to threads and method of bolt designation, see footnotes to above table.

American National Standard Round Head Short Square Neck Bolts
(ANSI B18.5-1978)

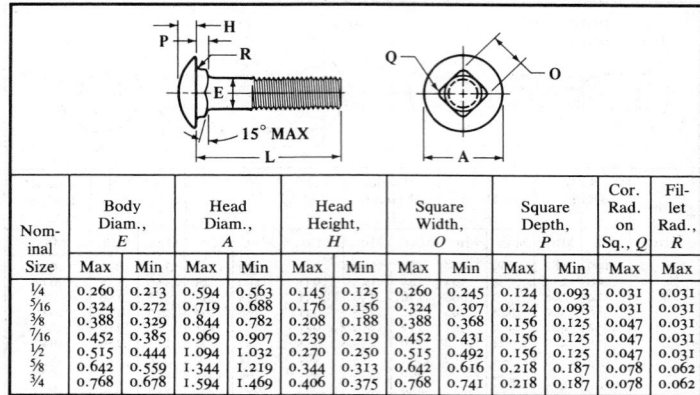

Nominal Size	Body Diam., E		Head Diam., A		Head Height, H		Square Width, O		Square Depth, P		Cor. Rad. on Sq., Q	Fillet Rad., R
	Max	Min	Max	Min	Max	Min	Max	Min	Max	Min	Max	Max
¼	0.260	0.213	0.594	0.563	0.145	0.125	0.260	0.245	0.124	0.093	0.031	0.031
5/16	0.324	0.272	0.719	0.688	0.176	0.156	0.324	0.307	0.124	0.093	0.031	0.031
3/8	0.388	0.329	0.844	0.782	0.208	0.188	0.388	0.368	0.156	0.125	0.047	0.031
7/16	0.452	0.385	0.969	0.907	0.239	0.219	0.452	0.431	0.156	0.125	0.047	0.031
½	0.515	0.444	1.094	1.032	0.270	0.250	0.515	0.492	0.156	0.125	0.047	0.031
5/8	0.642	0.559	1.344	1.219	0.344	0.313	0.642	0.616	0.218	0.187	0.078	0.062
¾	0.768	0.678	1.594	1.469	0.406	0.375	0.768	0.741	0.218	0.187	0.078	0.062

All dimensions are given in inches.

Threads are Unified Standard, Class 2A, UNC Series, in accordance with ANSI B1.1. For threads with additive finish, the maximum diameters of Class 2A apply before plating or coating, whereas the basic diameters (Class 2A maximum diameters plus the allowance) apply to a bolt after plating or coating.

Bolts are designated in the sequence shown: nominal size (number, fraction or decimal equivalent); threads per inch; nominal length (fraction or decimal equivalent); product name; material; and protective finish, if required.

           i.e., ½–13 × 3 Round Head Short Square Neck Bolt, Steel
                .375–16 × 2.50 Round Head Short Square Neck Bolt, Steel, Zinc Plated

American National Standard Round Head Fin Neck Bolts (ANSI B18.5-1978)

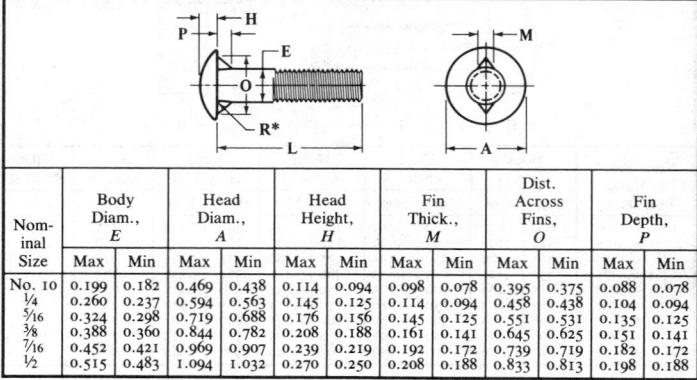

Nominal Size	Body Diam., E		Head Diam., A		Head Height, H		Fin Thick., M		Dist. Across Fins, O		Fin Depth, P	
	Max	Min	Max	Min	Max	Min	Max	Min	Max	Min	Max	Min
No. 10	0.199	0.182	0.469	0.438	0.114	0.094	0.098	0.078	0.395	0.375	0.088	0.078
¼	0.260	0.237	0.594	0.563	0.145	0.125	0.114	0.094	0.458	0.438	0.104	0.094
5/16	0.324	0.298	0.719	0.688	0.176	0.156	0.145	0.125	0.551	0.531	0.135	0.125
3/8	0.388	0.360	0.844	0.782	0.208	0.188	0.161	0.141	0.645	0.625	0.151	0.141
7/16	0.452	0.421	0.969	0.907	0.239	0.219	0.192	0.172	0.739	0.719	0.182	0.172
½	0.515	0.483	1.094	1.032	0.270	0.250	0.208	0.188	0.833	0.813	0.198	0.188

All dimensions are given in inches unless otherwise specified.
* Maximum fillet radius R is 0.031 inch for all sizes.
For information as to threads and method of bolt designation, see footnotes to above table.

American National Standard Round Head Ribbed Neck Bolts (ANSI B18.5-1978)

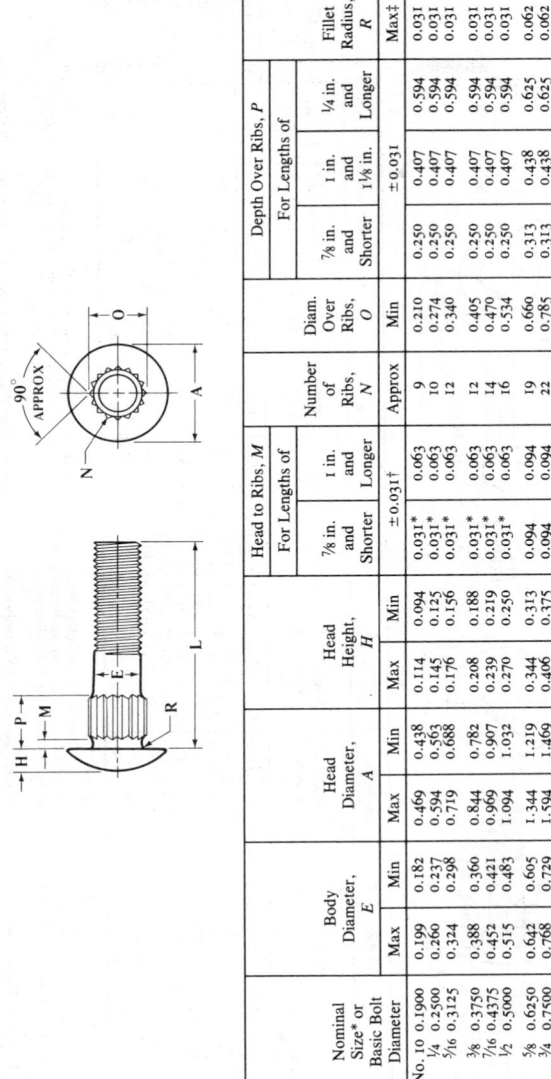

Nominal Size* or Basic Bolt Diameter	Body Diameter, E		Head Diameter, A		Head Height, H		Head to Ribs, M — For Lengths of		Number of Ribs, N	Diam. Over Ribs, O	Depth Over Ribs, P — For Lengths of			Fillet Radius, R
	Max	Min	Max	Min	Max	Min	⅞ in. and Shorter	1 in. and Longer	Approx	Min	⅞ in. and Shorter	1 in. and 1⅛ in.	¼ in. and Longer	Max‡
							±0.031†				±0.031			
No. 10 0.1900	0.199	0.182	0.469	0.438	0.114	0.094	0.031*	0.063	9	0.210	0.250	0.407	0.594	0.031
¼ 0.2500	0.260	0.237	0.594	0.563	0.145	0.125	0.031*	0.063	10	0.274	0.250	0.407	0.594	0.031
5⁄16 0.3125	0.324	0.298	0.719	0.688	0.176	0.156	0.031*	0.063	12	0.340	0.250	0.407	0.594	0.031
3⁄8 0.3750	0.388	0.360	0.844	0.782	0.208	0.188	0.031*	0.063	12	0.405	0.250	0.407	0.594	0.031
7⁄16 0.4375	0.452	0.421	0.969	0.907	0.239	0.219	0.031*	0.063	14	0.470	0.250	0.407	0.594	0.031
½ 0.5000	0.515	0.483	1.094	1.032	0.270	0.250	0.031*	0.063	16	0.534	0.250	0.407	0.594	0.031
5⁄8 0.6250	0.642	0.605	1.344	1.219	0.344	0.313	0.094	0.094	19	0.660	0.313	0.438	0.625	0.062
¾ 0.7500	0.768	0.729	1.594	1.469	0.406	0.375	0.094	0.094	22	0.785	0.313	0.438	0.625	0.062

All dimensions are given in inches unless otherwise specified.

* Where specifying nominal size in decimals, zeros preceding decimal and in the fourth decimal place shall be omitted.
† Tolerance on the No. 10 through ½ in. sizes for nominal lengths ⅞ in. and shorter shall be + 0.031 and − 0.000.
‡ The minimum radius is one half of the value shown.
For information as to threads and method of designating bolts, see following table.

American National Standard Step and 114 Degree Countersunk Square Neck Bolts (ANSI B18.5-1978)

STEP BOLT

114 DEGREE COUNTERSUNK SQUARE NECK BOLT

Nominal Size	Step & 114° Countersunk Bolts					Step Bolts							114° Countersunk Square Neck Bolts						
	Body Diam., E		Corner Rad. on Square, Q	Width of Square, O		Depth of Square, P		Diam. of Head, A		Height of Head, H		Fillet Radius, R	Depth of Square, P		Diam. of Head, A		Flat on Head, F	Height of Head, H	
	Max.	Min.	Max.	Max.	Min.	Max.	Min.	Max.	Min.	Max.	Min.	Max.	Max.	Min.	Max.	Min.	Min.	Max.	Min.
No. 10	.199	.182	.031	.199	.185	.125	.094	.656	.625	.114	.094	.031	.125	.094	.548	.500	.015	.131	.112
1/4	.260	.237	.031	.260	.245	.156	.125	.844	.813	.145	.125	.031	.156	.125	.682	.625	.018	.154	.135
5/16	.324	.298	.031	.324	.307	.187	.156	1.031	1.000	.176	.156	.031	.219	.188	.821	.750	.023	.184	.159
3/8	.388	.360	.047	.388	.368	.219	.188	1.219	1.188	.208	.188	.031	.250	.219	.960	.875	.027	.212	.183
7/16	.452	.421	.047	.452	.431	.250	.219	1.406	1.375	.239	.219	.031	.281	.250	1.093	1.000	.030	.235	.205
1/2	.515	.483	.047	.515	.492	.281	.250	1.594	1.563	.270	.250	.031	.312	.281	1.233	1.125	.035	.265	.229
5/8*	.642	.605	.078	.642	.616	…	…	…	…	…	…	…	.406	.375	1.495	1.375	.038	.316	.272
3/4*	.768	.729	.078	.768	.741	…	…	…	…	…	…	…	.500	.469	1.754	1.625	.041	.368	.314

All dimensions are in inches unless otherwise specified.

* These sizes pertain to 114 degree countersunk square neck bolts only. Dimensions given in last seven columns are for these bolts only.

Threads are Unified Standard, Class 2A, UNC Series, in accordance with ANSI B1.1. For threads with additive finish, the maximum diameters of Class 2A shall apply before plating or coating, whereas the basic diameters (Class 2A maximum diameters plus the allowance) shall apply to a bolt after plating or coating.

Bolts are designated in the sequence shown: nominal size (number, fraction or decimal equivalent); threads per inch; nominal length (fraction or decimal equivalent); product name; material; and protective finish, if required.

i.e.: 1/2–13 × 3 Round Head Square Neck Bolt, Steel

.375–16 × 2.50 Step Bolt, Steel, Zinc Plated

American National Standard Countersunk Bolts and Slotted Countersunk Bolts[1]
(ANSI B18.5-1978)

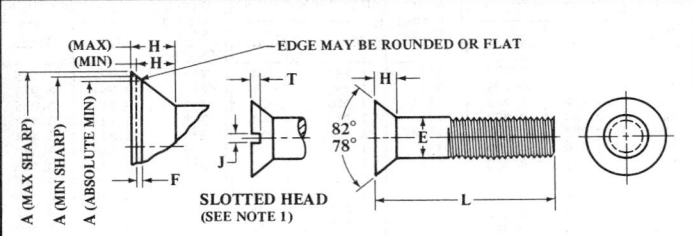

Nominal Size[2] or Basic Bolt Diameter	Body Diameter, E		Head Diameter, A			Flat on Min Diam. Head, F[5]
	Max	Min	Max Edge Sharp	Min Edge Sharp	Absolute Min Edge Rounded or Flat	Max
¼ 0.2500	0.260	0.237	0.493	0.477	0.445	0.018
⁵⁄₁₆ 0.3125	0.324	0.298	0.618	0.598	0.558	0.023
⅜ 0.3750	0.388	0.360	0.740	0.715	0.668	0.027
⁷⁄₁₆ 0.4375	0.452	0.421	0.803	0.778	0.726	0.030
½ 0.5000	0.515	0.483	0.935	0.905	0.845	0.035
⅝ 0.6250	0.642	0.605	1.169	1.132	1.066	0.038
¾ 0.7500	0.768	0.729	1.402	1.357	1.285	0.041
⅞ 0.8750	0.895	0.852	1.637	1.584	1.511	0.042
1 1.0000	1.022	0.976	1.869	1.810	1.735	0.043
1⅛ 1.1250	1.149	1.098	2.104	2.037	1.962	0.043
1¼ 1.2500	1.277	1.223	2.337	2.262	2.187	0.043
1⅜ 1.3750	1.404	1.345	2.571	2.489	2.414	0.043
1½ 1.5000	1.531	1.470	2.804	2.715	2.640	0.043

Nom. Size or Basic Bolt Diam.	Head Height, H		Slot Width, J		Slot Depth, T	
	Max[3]	Min[4]	Max	Min	Max	Min
¼ 0.2500	0.150	0.131	0.075	0.064	0.068	0.045
⁵⁄₁₆ 0.3125	0.189	0.164	0.084	0.072	0.086	0.057
⅜ 0.3750	0.225	0.196	0.094	0.081	0.103	0.068
⁷⁄₁₆ 0.4375	0.226	0.196	0.094	0.081	0.103	0.068
½ 0.5000	0.269	0.233	0.106	0.091	0.103	0.068
⅝ 0.6250	0.336	0.292	0.133	0.116	0.137	0.091
¾ 0.7500	0.403	0.349	0.149	0.131	0.171	0.115
⅞ 0.8750	0.470	0.408	0.167	0.147	0.206	0.138
1 1.0000	0.537	0.466	0.188	0.166	0.240	0.162
1⅛ 1.1250	0.604	0.525	0.196	0.178	0.257	0.173
1¼ 1.2500	0.671	0.582	0.211	0.193	0.291	0.197
1⅜ 1.3750	0.738	0.641	0.226	0.208	0.326	0.220
1½ 1.5000	0.805	0.698	0.258	0.240	0.360	0.244

All dimensions are given in inches.
[1] Heads are unslotted unless otherwise specified. For slot dimensions see Table 1 in Slotted Head Cap Screw section.
[2] Where specifying size in decimals, zeros preceding decimal and in fourth decimal place are omitted.
[3] Maximum head height calculated on maximum sharp head diameter, basic bolt diameter, and 78° head angle.
[4] Minimum head height calculated on minimum sharp head diameter, basic bolt diameter, and 82° head angle.
[5] Flat on minimum diameter head calculated on minimum sharp and absolute minimum head diameters and 82° head angle.
For thread information and method of bolt designation see footnotes to previous table.

Wrench Openings for Nuts (ANSI B18.2.2-1972 Appendix)

Max.* Width Across Flats of Nut	Wrench Opening†		Max.* Width Across Flats of Nut	Wrench Opening†		Max.* Width Across Flats of Nut	Wrench Opening†	
	Min.	Max.		Min.	Max.		Min.	Max.
5/32	0.158	0.163	1 1/8	1.132	1.142	2 13/16	2.827	2.845
3/16	0.190	0.195	1 1/4	1.257	1.267	2 15/16	2.954	2.973
7/32	0.220	0.225	1 5/16	1.320	1.331	3	3.016	3.035
1/4	0.252	0.257	1 3/8	1.383	1.394	3 1/8	3.142	3.162
9/32	0.283	0.288	1 7/16	1.446	1.457	3 3/8	3.393	3.414
5/16	0.316	0.322	1 1/2	1.508	1.520	3 1/2	3.518	3.540
11/32	0.347	0.353	1 5/8	1.634	1.646	3 3/4	3.770	3.793
3/8	0.378	0.384	1 11/16	1.696	1.708	3 7/8	3.895	3.918
7/16	0.440	0.446	1 13/16	1.822	1.835	4 1/8	4.147	4.172
1/2	0.504	0.510	1 7/8	1.885	1.898	4 1/4	4.272	4.297
9/16	0.566	0.573	2	2.011	2.025	4 1/2	4.524	4.550
5/8	0.629	0.636	2 1/16	2.074	2.088	4 5/8	4.649	4.676
11/16	0.692	0.699	2 3/16	2.200	2.215	4 7/8	4.900	4.928
3/4	0.755	0.763	2 1/4	2.262	2.277	5	5.026	5.055
13/16	0.818	0.826	2 3/8	2.388	2.404	5 1/4	5.277	5.307
7/8	0.880	0.888	2 7/16	2.450	2.466	5 3/8	5.403	5.434
15/16	0.944	0.953	2 9/16	2.576	2.593	5 5/8	5.654	5.686
1	1.006	1.015	2 5/8	2.639	2.656	5 3/4	5.780	5.813
1 1/16	1.068	1.077	2 3/4	2.766	2.783	6	6.031	6.065

All dimensions given in inches.   † Openings for 5/32 to 3/8 widths from old ASA B18.2-1960.
* Wrenches are marked with the "Nominal Size of Wrench" which is equal to the basic or maximum width across flats of the corresponding nut.
Minimum wrench opening equals (1.005W + 0.001).   Tolerance on wrench opening equals plus (0.005W + 0.004) from minimum, where W equals nominal size of wrench.

Wrench Clearance Dimensions. — Wrench clearances are given in Tables 1 and 2.   They are based on a wrench opening corresponding to the dimension across the flats of the fastener.   The listed values were obtained from a composite study of the alloy steel wrenches that are commercially available and military specifications.   They are suitable for general use as minimum requirements.

Table 1.   Wrench Clearances for Box Wrench — 12 Point
(From SAE Aeronautical Drafting Manual)

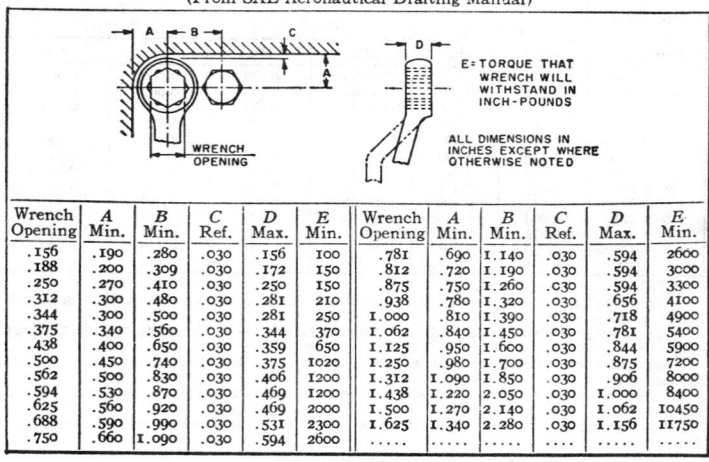

Wrench Opening	A Min.	B Min.	C Ref.	D Max.	E Min.	Wrench Opening	A Min.	B Min.	C Ref.	D Max.	E Min.
.156	.190	.280	.030	.156	100	.781	.690	1.140	.030	.594	2600
.188	.200	.309	.030	.172	150	.812	.720	1.190	.030	.594	3000
.250	.270	.410	.030	.250	150	.875	.750	1.260	.030	.594	3300
.312	.300	.480	.030	.281	210	.938	.780	1.320	.030	.656	4100
.344	.300	.560	.030	.281	250	1.000	.810	1.390	.030	.718	4900
.375	.340	.560	.030	.344	370	1.062	.840	1.450	.030	.781	5400
.438	.400	.650	.030	.359	650	1.125	.950	1.600	.030	.844	5900
.500	.450	.740	.030	.375	1020	1.250	.980	1.700	.030	.875	7200
.562	.500	.830	.030	.406	1200	1.312	1.090	1.850	.030	.906	8000
.594	.530	.870	.030	.469	1200	1.438	1.220	2.050	.030	1.000	8400
.625	.560	.920	.030	.469	2000	1.500	1.270	2.140	.030	1.062	10450
.688	.590	.990	.030	.531	2300	1.625	1.340	2.280	.030	1.156	11750
.750	.660	1.090	.030	.594	2600	......	...	...	...	...	...

Table 2. Wrench Clearances for Open End Engineers Wrench 15° and Socket Wrench (Regular Length)
(From SAE Aeronautical Drafting Manual; © Society of Automotive Engineers, Inc.)

H=THICKNESS OF WRENCH HEAD

J=TORQUE THAT WRENCH WILL WITHSTAND IN INCH-POUNDS

P=TORQUE THAT WRENCH WILL WITHSTAND IN INCH-POUNDS. *DOES NOT INCLUDE ALLOWANCE FOR TORQUE DEVICE

ALL DIMENSIONS IN INCHES EXCEPT WHERE OTHERWISE NOTED

OPEN END ENGINEERS WRENCH 15°

SQUARE DRIVE   SOCKET (REGULAR LENGTH)   WRENCH OPENING

Open End Engineers Wrench 15°

Wrench Opening	A Min.	B Max.	C Min.	D Min.	E Min.	F Max.	G Ref.	H Max.	J Min.	K Min.	L Ref.
.156	.220	.250	.390	.166	.250	.200	.030	.094	25	...	...
.188	.250	.286	.430	.190	.270	.230	.030	.172	40	.370	.030
.250	.280	.340	.530	.270	.310	.310	.030	.172	60	.470	.030
.312	.380	.470	.666	.280	.390	.390	.050	.203	125	.550	.030
.344	.420	.500	.750	.340	.450	.450	.050	.203	175	.580	.030
.375	.420	.500	.780	.360	.450	.520	.050	.219	175	.620	.030
.438	.470	.590	.890	.420	.520	.640	.050	.250	375	.750	.030
.500	.520	.640	1.000	.470	.580	.660	.050	.266	490	.810	.030
.562	.590	.770	1.130	.520	.666	.700	.050	.297	700	.870	.030
.594	.640	.830	1.210	.530	.700	.700	.050	.344	800	.920	.030
.625	.640	.830	1.230	.550	.700	.700	.060	.344	935	.950	.030
.688	.770	.920	1.470	.660	.880	.800	.060	.375	1250	1.030	.030
.750	.770	.920	1.510	.670	.880	.800	.060	.375	1500	1.120	.030
.781	.830	.950	1.550	.690	.900	.840	.060	.406	1615	1.150	.030
.812	.910	1.120	1.660	.720	.970	.860	.060	.438	1710	1.210	.030
.875	.910	1.120	1.810	.800	1.060	.910	.066	.438	2250	1.280	.030
.938	.970	1.150	1.850	.810	1.056	.950	.066	.438	2750	1.370	.030
1.000	1.050	1.230	2.000	.886	1.166	1.000	.080	.500	3250	1.470	.030
1.062	1.090	1.250	2.100	.970	1.200	1.060	.080	.500	3500	1.550	.030
1.125	1.140	1.370	2.210	1.000	1.270	1.230	.080	.500	4000	1.610	.030
1.250	1.270	1.420	2.440	1.080	1.390	1.310	.080	.562	5250	1.890	.030
1.312	1.390	1.690	2.630	1.250	1.520	1.340	.080	.562	6000	1.980	.030
1.438	1.470	1.720	2.800	1.250	1.590	1.340	.090	.641	7500	2.140	.030
1.500	1.470	1.720	2.840	1.270	1.590	1.450	.090	.641	8250	2.200	.030
1.625	1.560	1.880	3.100	1.386	1.750	1.560	.090	.641	9000	2.390	.030

Socket (Regular Length)

Wrench Opening	Q = .250			Q = .375			Q = .500			Q = .750		
	M Max.	N Max.	P Min.	M Max.	N Max.	P Min.	M Max.	N Max.	P Min.	M Max.	N Max.	P Min.
.188	...	.510	125	...	...	...	...	...	...	...	...	...
.250	...	.510	200	...	...	...	...	...	...	...	...	...
.312	1.000	.510	300	...	...	...	...	...	...	...	...	...
.344	1.000	.519	450	1.250	.690	250	...	...	...	...	...	...
.375	1.000	.580	550	1.250	.690	400	...	...	...	...	...	...
.438	1.000	.683	550	1.250	.690	675	...	...	...	...	...	...
.500	1.000	.692	600	1.250	.690	900	1.500	.880	1600	...	...	...
.562	...	...	...	1.250	.880	1250	1.500	.940	1700	...	...	...
.594	...	...	...	1.250	.932	1450	1.500	.940	2000	...	...	...
.625	...	...	...	1.250	.963	1600	1.500	.940	2700	...	...	...
.688	...	...	...	1.250	.995	1750	1.562	.970	3000	...	...	...
.750	...	...	...	1.250	1.058	2000	1.562	1.000	3000	...	...	...
.781	...	...	...	1.250	1.120	2000	1.562	1.065	4300	...	...	...
.812	...	...	...	1.250	1.126	2000	1.625	1.130	5000	...	...	...
.875	...	...	...	1.250	1.213	2000	1.625	1.130	5000	...	...	...
.938	...	...	...	...	...	...	1.750	1.222	5000	...	...	...
1.000	...	...	...	...	...	...	1.750	1.285	5000	...	...	...
1.062	...	...	...	...	...	...	1.750	1.410	5000	...	...	...
1.125	...	...	...	...	...	...	1.750	1.410	5000	...	...	...
1.250	...	...	...	...	...	...	1.844	1.505	5000	2.375	1.855	7250
1.312	...	...	...	...	...	...	1.938	1.507	5000	2.500	1.920	8000
1.438	...	...	...	...	...	...	2.000	1.723	5000	2.625	2.075	9550
1.500	...	...	...	...	...	...	...	...	...	2.625	2.170	10450
1.625	...	...	...	...	...	...	...	...	...	2.750	2.325	11750

Table 1A. American National Standard Type A Plain Washers — Preferred Sizes** (ANSI B18.22.1-1965, R1975)

Nominal Washer Size***		Series	Inside Diameter			Outside Diameter			Thickness		
			Basic	Tolerance		Basic	Tolerance		Basic	Max.	Min.
				Plus	Minus		Plus	Minus			
—	—		0.078	0.000	0.005	0.188	0.000	0.005	0.020	0.025	0.016
—	—		0.094	0.000	0.005	0.250	0.000	0.005	0.020	0.025	0.016
—	—		0.125	0.008	0.005	0.312	0.008	0.005	0.032	0.040	0.025
No. 6	0.138		0.156	0.008	0.005	0.375	0.015	0.005	0.049	0.065	0.036
No. 8	0.164		0.188	0.008	0.005	0.438	0.015	0.005	0.049	0.065	0.036
No. 10	0.190		0.219	0.008	0.005	0.500	0.015	0.005	0.049	0.065	0.036
3/16	0.188		0.250	0.015	0.005	0.562	0.015	0.005	0.049	0.065	0.036
No. 12	0.216		0.250	0.015	0.005	0.562	0.015	0.005	0.065	0.080	0.051
1/4	0.250	N	0.281	0.015	0.005	0.625	0.015	0.005	0.065	0.080	0.051
1/4	0.250	W	0.312	0.015	0.005	0.734*	0.015	0.007	0.065	0.080	0.051
5/16	0.312	N	0.344	0.015	0.005	0.688	0.015	0.007	0.065	0.080	0.051
5/16	0.312	W	0.375	0.015	0.005	0.875	0.030	0.007	0.083	0.104	0.064
3/8	0.375	N	0.406	0.015	0.005	0.812	0.015	0.007	0.065	0.080	0.051
3/8	0.375	W	0.438	0.015	0.005	1.000	0.030	0.007	0.083	0.104	0.064
7/16	0.438	N	0.469	0.015	0.005	0.922	0.015	0.007	0.065	0.080	0.051
7/16	0.438	W	0.500	0.015	0.005	1.250	0.030	0.007	0.083	0.104	0.064
1/2	0.500	N	0.531	0.015	0.005	1.062	0.030	0.007	0.095	0.121	0.074
1/2	0.500	W	0.562	0.015	0.005	1.375	0.030	0.007	0.109	0.132	0.086
9/16	0.562	N	0.594	0.015	0.005	1.156*	0.030	0.007	0.095	0.121	0.074
9/16	0.562	W	0.625	0.015	0.005	1.469*	0.030	0.007	0.109	0.132	0.086
5/8	0.625	N	0.656	0.030	0.007	1.312	0.030	0.007	0.095	0.121	0.074
5/8	0.625	W	0.688	0.030	0.007	1.750	0.030	0.007	0.134	0.160	0.108
3/4	0.750	N	0.812	0.030	0.007	1.469	0.030	0.007	0.134	0.160	0.108
3/4	0.750	W	0.812	0.030	0.007	2.000	0.030	0.007	0.148	0.177	0.122
7/8	0.875	N	0.938	0.030	0.007	1.750	0.030	0.007	0.134	0.160	0.108
7/8	0.875	W	0.938	0.030	0.007	2.250	0.030	0.007	0.165	0.192	0.136
1	1.000	N	1.062	0.030	0.007	2.000	0.030	0.007	0.134	0.160	0.108
1	1.000	W	1.062	0.030	0.007	2.500	0.030	0.007	0.165	0.192	0.136
1 1/8	1.125	N	1.250	0.030	0.007	2.250	0.030	0.007	0.134	0.160	0.108
1 1/8	1.125	W	1.250	0.030	0.007	2.750	0.030	0.007	0.165	0.192	0.136
1 1/4	1.250	N	1.375	0.030	0.007	2.500	0.030	0.007	0.165	0.192	0.136
1 1/4	1.250	W	1.375	0.030	0.007	3.000	0.030	0.007	0.165	0.192	0.136
1 3/8	1.375	N	1.500	0.030	0.007	2.750	0.030	0.007	0.165	0.192	0.136
1 3/8	1.375	W	1.500	0.045	0.010	3.250	0.045	0.010	0.180	0.213	0.153
1 1/2	1.500	N	1.625	0.030	0.007	3.000	0.030	0.007	0.165	0.192	0.136
1 1/2	1.500	W	1.625	0.045	0.010	3.500	0.045	0.010	0.180	0.213	0.153
1 5/8	1.625		1.750	0.045	0.010	3.750	0.045	0.010	0.180	0.213	0.153
1 3/4	1.750		1.875	0.045	0.010	4.000	0.045	0.010	0.180	0.213	0.153
1 7/8	1.875		2.000	0.045	0.010	4.250	0.045	0.010	0.180	0.213	0.153
2	2.000		2.125	0.045	0.010	4.500	0.045	0.010	0.180	0.213	0.153
2 1/4	2.250		2.375	0.045	0.010	4.750	0.045	0.010	0.220	0.248	0.193
2 1/2	2.500		2.625	0.045	0.010	5.000	0.045	0.010	0.238	0.280	0.210
2 3/4	2.750		2.875	0.065	0.010	5.250	0.065	0.010	0.259	0.310	0.228
3	3.000		3.125	0.065	0.010	5.500	0.065	0.010	0.284	0.327	0.249

All dimensions are in inches.

* The 0.734-inch, 1.156-inch, and 1.469-inch outside diameters avoid washers which could be used in coin operated devices.

** Preferred sizes are for the most part from series previously designated "Standard Plate" and "SAE." Where common sizes existed in the two series, the SAE size is designated "N" (narrow) and the Standard Plate "W" (wide). These sizes as well as all other sizes of Type A Plain Washers are to be ordered by ID, OD, and thickness dimensions.

*** Nominal washer sizes are intended for use with comparable nominal screw or bolt sizes.

Additional selected sizes of Type A Plain Washers are shown in Table 1B.

**Table 1B. American National Standard Type A Plain Washers —
Additional Selected Sizes (ANSI B18.22.1-1965, R1975)**

Inside Diameter			Outside Diameter			Thickness		
	Tolerance			Tolerance				
Basic	Plus	Minus	Basic	Plus	Minus	Basic	Max.	Min.
0.094	0.000	0.005 ·	0.219	0.000	0.005	0.020	0.025	0.016
0.125	0.000	0.005	0.250	0.000	0.005	0.022	0.028	0.017
0.156	0.008	0.005	0.312	0.008	0.005	0.035	0.048	0.027
0.172	0.008	0.005	0.406	0.015	0.005	0.049	0.065	0.036
0.188	0.008	0.005	0.375	0.015	0.005	0.049	0.065	0.036
0.203	0.008	0.005	0.469	0.015	0.005	0.049	0.065	0.036
0.219	0.008	0.005	0.438	0.015	0.005	0.049	0.065	0.036
0.234	0.008	0.005	0.531	0.015	0.005	0.049	0.065	0.036
0.250	0.015	0.005	0.500	0.015	0.005	0.049	0.065	0.036
0.266	0.015	0.005	0.625	0.015	0.005	0.049	0.065	0.036
0.312	0.015	0.005	0.875	0.015	0.007	0.065	0.080	0.051
0.375	0.015	0.005	0.734*	0.015	0.007	0.065	0.080	0.051
0.375	0.015	0.005	1.125	0.015	0.007	0.065	0.080	0.051
0.438	0.015	0.005	0.875	0.030	0.007	0.083	0.104	0.064
0.438	0.015	0.005	1.375	0.030	0.007	0.083	0.104	0.064
0.500	0.015	0.005	1.125	0.030	0.007	0.083	0.104	0.064
0.500	0.015	0.005	1.625	0.030	0.007	0.083	0.104	0.064
0.562	0.015	0.005	1.250	0.030	0.007	0.109	0.132	0.086
0.562	0.015	0.005	1.875	0.030	0.007	0.109	0.132	0.086
0.625	0.015	0.005	1.375	0.030	0.007	0.109	0.132	0.086
0.625	0.015	0.005	2.125	0.030	0.007	0.134	0.160	0.108
0.688	0.030	0.007	1.469*	0.030	0.007	0.134	0.160	0.108
0.688	0.030	0.007	2.375	0.030	0.007	0.165	0.192	0.136
0.812	0.030	0.007	1.750	0.030	0.007	0.148	0.177	0.122
0.812	0.030	0.007	2.875	0.030	0.007	0.165	0.192	0.136
0.938	0.030	0.007	2.000	0.030	0.007	0.165	0.192	0.136
0.938	0.030	0.007	3.375	0.045	0.010	0.180	0.213	0.153
1.062	0.030	0.007	2.250	0.030	0.007	0.165	0.192	0.136
1.062	0.045	0.010	3.875	0.045	0.010	0.238	0.280	0.210
1.250	0.030	0.007	2.500	0.030	0.007	0.165	0.192	0.136
1.375	0.030	0.007	2.750	0.030	0.007	0.165	0.192	0.136
1.500	0.045	0.010	3.000	0.045	0.010	0.180	0.213	0.153
1.625	0.045	0.010	3.250	0.045	0.010	0.180	0.213	0.153
1.688	0.045	0.010	3.500	0.045	0.010	0.180	0.213	0.153
1.812	0.045	0.010	3.750	0.045	0.010	0.180	0.213	0.153
1.938	0.045	0.010	4.000	0.045	0.010	0.180	0.213	0.153
2.062	0.045	0.010	4.250	0.045	0.010	0.180	0.213	0.153

All dimensions are in inches.
* The 0.734-inch and 1.469-inch outside diameters avoid washers which could be used in coin operated devices.
The above sizes are to be ordered by ID, OD, and thickness dimensions.
Preferred Sizes of Type A Plain Washers are shown in Table 1A.

ANSI Standard Plain Washers. — The Type A plain washers were originally developed in a light, medium, heavy and extra heavy series. These series have been discontinued and the washers are now designated by their nominal dimensions.

The Type B plain washers are available in a narrow, regular and wide series with proportions designed to distribute the load over larger areas of lower strength materials.

Plain washers are made of ferrous or non-ferrous metal, plastic or other material as specified. The tolerances indicated in the tables are intended for metal washers only.

Table 2. American National Standard Type B Plain Washers
(ANSI B18.22.1-1965, R1975)

Nominal Washer Size**		Series†	Inside Diameter			Outside Diameter			Thickness		
			Basic	Tolerance		Basic	Tolerance		Basic	Max.	Min.
				Plus	Minus		Plus	Minus			
No. 0	0.060	N	0.068	0.000	0.005	0.125	0.000	0.005	0.025	0.028	0.022
		R	0.068	0.000	0.005	0.188	0.000	0.005	0.025	0.028	0.022
		W	0.068	0.000	0.005	0.250	0.000	0.005	0.025	0.028	0.022
No. 1	0.073	N	0.084	0.000	0.005	0.156	0.000	0.005	0.025	0.028	0.022
		R	0.084	0.000	0.005	0.219	0.000	0.005	0.025	0.028	0.022
		W	0.084	0.000	0.005	0.281	0.000	0.005	0.032	0.036	0.028
No. 2	0.086	N	0.094	0.000	0.005	0.188	0.000	0.005	0.025	0.028	0.022
		R	0.094	0.000	0.005	0.250	0.000	0.005	0.032	0.036	0.028
		W	0.094	0.000	0.005	0.344	0.000	0.005	0.032	0.036	0.028
No. 3	0.099	N	0.109	0.000	0.005	0.219	0.000	0.005	0.025	0.028	0.022
		R	0.109	0.000	0.005	0.312	0.000	0.005	0.032	0.036	0.028
		W	0.109	0.008	0.005	0.406	0.008	0.005	0.040	0.045	0.036
No. 4	0.112	N	0.125	0.000	0.005	0.250	0.000	0.005	0.032	0.036	0.028
		R	0.125	0.008	0.005	0.375	0.008	0.005	0.040	0.045	0.036
		W	0.125	0.008	0.005	0.438	0.008	0.005	0.040	0.045	0.036
No. 5	0.125	N	0.141	0.000	0.005	0.281	0.000	0.005	0.032	0.036	0.028
		R	0.141	0.008	0.005	0.406	0.008	0.005	0.040	0.045	0.036
		W	0.141	0.008	0.005	0.500	0.008	0.005	0.040	0.045	0.036
No. 6	0.138	N	0.156	0.000	0.005	0.312	0.000	0.005	0.032	0.036	0.028
		R	0.156	0.008	0.005	0.438	0.008	0.005	0.040	0.045	0.036
		W	0.156	0.008	0.005	0.562	0.008	0.005	0.040	0.045	0.036
No. 8	0.164	N	0.188	0.008	0.005	0.375	0.008	0.005	0.040	0.045	0.036
		R	0.188	0.008	0.005	0.500	0.008	0.005	0.040	0.045	0.036
		W	0.188	0.008	0.005	0.625	0.015	0.005	0.063	0.071	0.056
No. 10	0.190	N	0.203	0.008	0.005	0.406	0.008	0.005	0.040	0.045	0.036
		R	0.203	0.008	0.005	0.562	0.008	0.005	0.040	0.045	0.036
		W	0.203	0.008	0.005	0.734*	0.015	0.007	0.063	0.071	0.056
No. 12	0.216	N	0.234	0.008	0.005	0.438	0.008	0.005	0.040	0.045	0.036
		R	0.234	0.008	0.005	0.625	0.015	0.005	0.063	0.071	0.056
		W	0.234	0.008	0.005	0.875	0.015	0.005	0.063	0.071	0.056
¼	0.250	N	0.281	0.015	0.005	0.500	0.015	0.005	0.063	0.071	0.056
		R	0.281	0.015	0.005	0.734*	0.015	0.007	0.063	0.071	0.056
		W	0.281	0.015	0.005	1.000	0.015	0.007	0.063	0.071	0.056
⁵⁄₁₆	0.312	N	0.344	0.015	0.005	0.625	0.015	0.005	0.063	0.071	0.056
		R	0.344	0.015	0.005	0.875	0.015	0.005	0.063	0.071	0.056
		W	0.344	0.015	0.005	1.125	0.015	0.007	0.063	0.071	0.056
⅜	0.375	N	0.406	0.015	0.005	0.734*	0.015	0.007	0.063	0.071	0.056
		R	0.406	0.015	0.005	1.000	0.015	0.007	0.063	0.071	0.056
		W	0.406	0.015	0.005	1.250	0.030	0.007	0.100	0.112	0.090
⁷⁄₁₆	0.438	N	0.469	0.015	0.005	0.875	0.015	0.007	0.063	0.071	0.056
		R	0.469	0.015	0.005	1.125	0.015	0.007	0.063	0.071	0.056
		W	0.469	0.015	0.005	1.469*	0.030	0.007	0.100	0.112	0.090

All dimensions are in inches.

* The 0.734-inch and 1.469-inch outside diameters avoid washers which could be used in coin operated devices.

** Nominal washer sizes are intended for use with comparable nominal screw or bolt sizes.

† N indicates Narrow; R, Regular; and W, Wide Series.

Inside and outside diameters shall be concentric within at least the inside diameter tolerance.

Washers shall be flat within 0.005 inch for basic outside diameters up to and including 0.875 inch, and within 0.010 inch for larger outside diameters.

Table 2 (*Concluded*). **American National Standard Type B Plain Washers**
(ANSI B18.22.1-1965, R1975).

Nominal Washer Size**		Series†	Inside Diameter			Outside Diameter			Thickness		
			Basic	Tolerance		Basic	Tolerance		Basic	Max.	Min.
				Plus	Minus		Plus	Minus			
½	0.500	N	0.531	0.015	0.005	1.000	0.015	0.007	0.063	0.071	0.056
		R	0.531	0.015	0.005	1.250	0.030	0.007	0.100	0.112	0.090
		W	0.531	0.015	0.005	1.750	0.030	0.007	0.100	0.112	0.090
9/16	0.562	N	0.594	0.015	0.005	1.125	0.015	0.007	0.063	0.071	0.056
		R	0.594	0.015	0.005	1.469*	0.030	0.007	0.100	0.112	0.090
		W	0.594	0.015	0.005	2.000	0.030	0.007	0.100	0.112	0.090
5/8	0.625	N	0.656	0.030	0.007	1.250	0.030	0.007	0.100	0.112	0.090
		R	0.656	0.030	0.007	1.750	0.030	0.007	0.100	0.112	0.090
		W	0.656	0.030	0.007	2.250	0.030	0.007	0.160	0.174	0.146
¾	0.750	N	0.812	0.030	0.007	1.375	0.030	0.007	0.100	0.112	0.090
		R	0.812	0.030	0.007	2.000	0.030	0.007	0.100	0.112	0.090
		W	0.812	0.030	0.007	2.500	0.030	0.007	0.160	0.174	0.146
7/8	0.875	N	0.938	0.030	0.007	1.469*	0.030	0.007	0.100	0.112	0.090
		R	0.938	0.030	0.007	2.250	0.030	0.007	0.160	0.174	0.146
		W	0.938	0.030	0.007	2.750	0.030	0.007	0.160	0.174	0.146
1	1.000	N	1.062	0.030	0.007	1.750	0.030	0.007	0.100	0.112	0.090
		R	1.062	0.030	0.007	2.500	0.030	0.007	0.160	0.174	0.146
		W	1.062	0.030	0.007	3.000	0.030	0.007	0.160	0.174	0.146
1⅛	1.125	N	1.188	0.030	0.007	2.000	0.030	0.007	0.100	0.112	0.090
		R	1.188	0.030	0.007	2.750	0.030	0.007	0.160	0.174	0.146
		W	1.188	0.030	0.007	3.250	0.030	0.007	0.160	0.174	0.146
1¼	1.250	N	1.312	0.030	0.007	2.250	0.030	0.007	0.160	0.174	0.146
		R	1.312	0.030	0.007	3.000	0.030	0.007	0.160	0.174	0.146
		W	1.312	0.045	0.010	3.500	0.045	0.010	0.250	0.266	0.234
1⅜	1.375	N	1.438	0.030	0.007	2.500	0.030	0.007	0.160	0.174	0.146
		R	1.438	0.030	0.007	3.250	0.030	0.007	0.160	0.174	0.146
		W	1.438	0.045	0.010	3.750	0.045	0.010	0.250	0.266	0.234
1½	1.500	N	1.562	0.030	0.007	2.750	0.030	0.007	0.160	0.174	0.146
		R	1.562	0.045	0.010	3.500	0.045	0.010	0.250	0.266	0.234
		W	1.562	0.045	0.010	4.000	0.045	0.010	0.250	0.266	0.234
1⅝	1.625	N	1.750	0.030	0.007	3.000	0.030	0.007	0.160	0.174	0.146
		R	1.750	0.045	0.010	3.750	0.045	0.010	0.250	0.266	0.234
		W	1.750	0.045	0.010	4.250	0.045	0.010	0.250	0.266	0.234
1¾	1.750	N	1.875	0.030	0.007	3.250	0.030	0.007	0.160	0.174	0.146
		R	1.875	0.045	0.010	4.000	0.045	0.010	0.250	0.266	0.234
		W	1.875	0.045	0.010	4.500	0.045	0.010	0.250	0.266	0.234
1⅞	1.875	N	2.000	0.045	0.010	3.500	0.045	0.010	0.250	0.266	0.234
		R	2.000	0.045	0.010	4.250	0.045	0.010	0.250	0.266	0.234
		W	2.000	0.045	0.010	4.750	0.045	0.010	0.250	0.266	0.234
2	2.000	N	2.125	0.045	0.010	3.750	0.045	0.010	0.250	0.266	0.234
		R	2.125	0.045	0.010	4.500	0.045	0.010	0.250	0.266	0.234
		W	2.125	0.045	0.010	5.000	0.045	0.010	0.250	0.266	0.234

All dimensions are in inches.

* The 1.469-inch outside diameter avoids washers which could be used in coin operated devices.

** Nominal washer sizes are intended for use with comparable nominal screw or bolt sizes.

† N indicates Narrow; R, Regular; and W, Wide Series.

Inside and outside diameters shall be concentric within at least the inside diameter tolerance.

Washers shall be flat within 0.005-inch for basic outside diameters up through 0.875-inch and within 0.010 inch for larger outside diameters.

For 2¼-, 2½-, 2¾-, and 3-inch sizes see ANSI B18.22.1-1965 (R1975).

Table 1. American National Standard Helical Spring Lock Washers (ANSI B18.21.1-1972)

Nominal Washer Size		Inside Diameter, A		Regular*			Heavy†			Extra Duty‡		
		Max.	Min.	O.D., B Max. ¶	Section Width, W	Section Thickness, T§	O.D., B Max. ¶	Section Width, W	Section Thickness, T§	O.D., B Max. ¶	Section Width, W	Section Thickness, T§
No. 2	0.086	0.094	0.088	0.172	0.035	0.020	0.182	0.040	0.025	0.208	0.053	0.027
No. 3	0.099	0.107	0.101	0.195	0.040	0.025	0.209	0.047	0.031	0.239	0.062	0.034
No. 4	0.112	0.120	0.114	0.209	0.040	0.025	0.223	0.047	0.031	0.253	0.062	0.034
No. 5	0.125	0.133	0.127	0.236	0.047	0.031	0.252	0.055	0.040	0.300	0.079	0.045
No. 6	0.138	0.148	0.141	0.250	0.047	0.031	0.266	0.055	0.040	0.314	0.079	0.045
No. 8	0.164	0.174	0.167	0.293	0.055	0.040	0.307	0.062	0.047	0.375	0.096	0.057
No. 10	0.190	0.200	0.193	0.334	0.062	0.047	0.350	0.070	0.056	0.434	0.112	0.068
No. 12	0.216	0.227	0.220	0.377	0.070	0.056	0.391	0.077	0.063	0.497	0.130	0.080
1/4	0.250	0.262	0.254	0.489	0.109	0.062	0.491	0.110	0.077	0.535	0.132	0.084
5/16	0.312	0.326	0.317	0.586	0.125	0.078	0.596	0.130	0.097	0.622	0.143	0.108
3/8	0.375	0.390	0.380	0.683	0.141	0.094	0.691	0.145	0.115	0.741	0.170	0.123
7/16	0.438	0.455	0.443	0.779	0.156	0.109	0.787	0.160	0.133	0.839	0.186	0.143
1/2	0.500	0.518	0.506	0.873	0.171	0.125	0.883	0.176	0.151	0.939	0.204	0.162
9/16	0.562	0.582	0.570	0.971	0.188	0.141	0.981	0.193	0.170	1.041	0.223	0.182
5/8	0.625	0.650	0.635	1.079	0.203	0.156	1.093	0.210	0.189	1.157	0.242	0.202
11/16	0.688	0.713	0.698	1.176	0.219	0.172	1.192	0.227	0.207	1.258	0.260	0.221
3/4	0.750	0.775	0.760	1.271	0.234	0.188	1.291	0.244	0.226	1.361	0.279	0.241
13/16	0.812	0.843	0.824	1.367	0.250	0.203	1.391	0.262	0.246	1.463	0.298	0.261
7/8	0.875	0.905	0.887	1.464	0.266	0.219	1.494	0.281	0.266	1.576	0.322	0.285
15/16	0.938	0.970	0.950	1.560	0.281	0.234	1.594	0.298	0.284	1.688	0.345	0.308
1	1.000	1.042	1.017	1.661	0.297	0.250	1.705	0.319	0.306	1.799	0.366	0.330
1 1/16	1.062	1.107	1.080	1.756	0.312	0.266	1.808	0.338	0.326	1.910	0.389	0.352
1 1/8	1.125	1.172	1.144	1.853	0.328	0.281	1.909	0.356	0.345	2.019	0.411	0.375
1 3/16	1.188	1.237	1.208	1.950	0.344	0.297	2.008	0.373	0.364	2.124	0.431	0.396
1 1/4	1.250	1.302	1.271	2.045	0.359	0.312	2.113	0.393	0.384	2.231	0.452	0.417
1 5/16	1.312	1.366	1.334	2.141	0.375	0.328	2.211	0.410	0.403	2.335	0.472	0.438
1 3/8	1.375	1.432	1.398	2.239	0.391	0.344	2.311	0.427	0.422	2.439	0.491	0.458
1 7/16	1.438	1.497	1.462	2.334	0.406	0.359	2.406	0.442	0.440	2.540	0.509	0.478
1 1/2	1.500	1.561	1.525	2.430	0.422	0.375	2.502	0.458	0.458	2.638	0.526	0.496

All dimensions are given in inches.   * Formerly designated Medium Helical Spring Lock Washers.   † Not recommended for new applications.   ‡ Formerly designated Extra Heavy Helical Spring Lock Washers.   ¶ The maximum outside diameters specified allow for the commercial tolerances on cold-drawn wire.   § T = mean section thickness = $(t_i + t_o) \div 2$.

**Table 2. American National Standard Hi-Collar* Helical Spring
Lock Washers (ANSI B18.21.1-1972)**

	Inside Diameter		Outside Diameter	Washer Section	
				Width	Thickness
Nominal Washer Size	Min.	Max.	Max.**	Min.	Min.
No. 4	0.112	0.114	0.120	0.173	0.022	0.022
No. 5	0.125	0.127	0.133	0.202	0.030	0.030
No. 6	0.138	0.141	0.148	0.216	0.030	0.030
No. 8	0.164	0.167	0.174	0.267	0.042	0.047
No. 10	0.190	0.193	0.200	0.294	0.042	0.047
¼	0.250	0.254	0.262	0.365	0.047	0.078
5/16	0.312	0.317	0.326	0.460	0.062	0.093
⅜	0.375	0.380	0.390	0.553	0.076	0.125
7/16	0.438	0.443	0.455	0.647	0.090	0.140
½	0.500	0.506	0.518	0.737	0.103	0.172
⅝	0.625	0.635	0.650	0.923	0.125	0.203
¾	0.750	0.760	0.775	1.111	0.154	0.218
⅞	0.875	0.887	0.905	1.296	0.182	0.234
1	1.000	1.017	1.042	1.483	0.208	0.250
1⅛	1.125	1.144	1.172	1.669	0.236	0.313
1¼	1.250	1.271	1.302	1.799	0.236	0.313
1⅜	1.375	1.398	1.432	2.041	0.292	0.375
1½	1.500	1.525	1.561	2.170	0.292	0.375
1¾	1.750	1.775	1.811	2.602	0.383	0.469
2	2.000	2.025	2.061	2.852	0.383	0.469
2¼	2.250	2.275	2.311	3.352	0.508	0.508
2½	2.500	2.525	2.561	3.602	0.508	0.508
2¾	2.750	2.775	2.811	4.102	0.633	0.633
3	3.000	3.025	3.061	4.352	0.633	0.633

* For use with 1960 Series Socket Head Cap Screws. See page 1389.
** The maximum outside diameters specified allow for the commercial tolerances on cold-drawn wire.

American National Standard Helical Spring and Tooth Lock Washers (ANSI B18.21.1-1972). — This standard covers helical spring lock washers of carbon steel; corrosion resistant steel, Types 302 and 305; aluminum-zinc alloy; phosphor-bronze; silicon-bronze; and K-Monel; in various series. It also covers tooth lock washers of carbon steel having internal teeth, external teeth, and both internal and external teeth, of two constructions, designated as Type A and Type B. These washers are intended for general industrial application.

Helical spring lock washers: They have the function of: (1) providing good bolt tension per unit of applied torque for tight assemblies; (2) providing hardened bearing surfaces to create uniform torque control; (3) providing uniform load distribution through controlled radii — section — cut-off; and (4) providing protection against looseness resulting from vibration and corrosion.

Nominal washer sizes are intended for use with comparable nominal screw or bolt sizes. These washers are designated by the following data in the sequence shown: Product name; nominal size (number, fraction or decimal equivalent); series; ma-

terial; and protective finish, if required. For example: Helical Spring Lock Washer, .375 Extra Duty, Steel, Phosphate Coated.

Carbon steel helical spring lock washers are available in four series: Regular, heavy, extra duty and hi-collar as given in Tables 1 and 2. All other helical spring lock washers made of other materials are available in the regular series as given in Table 1.

When carbon steel helical spring lock washers are to be hot-dipped galvanized for use with hot-dipped galvanized bolts or screws, they are to be coiled to limits 0.020 inch in excess of those specified in Tables 1 and 2 for minimum inside diameter and maximum outside diameter. Galvanizing on washers under ¼ inch nominal size is not recommended.

Tooth lock washers: They serve to lock fasteners, such as bolts and nuts, to the component parts of an assembly, or increase the friction between the fasteners and the assembly. These washers are designated in a manner similar to helical spring lock washers, and are available in carbon steel. Dimensions are given in Tables 3 and 4.

Table 3. American National Standard Internal-External Tooth Lock Washers
(ANSI B18.21.1-1972)

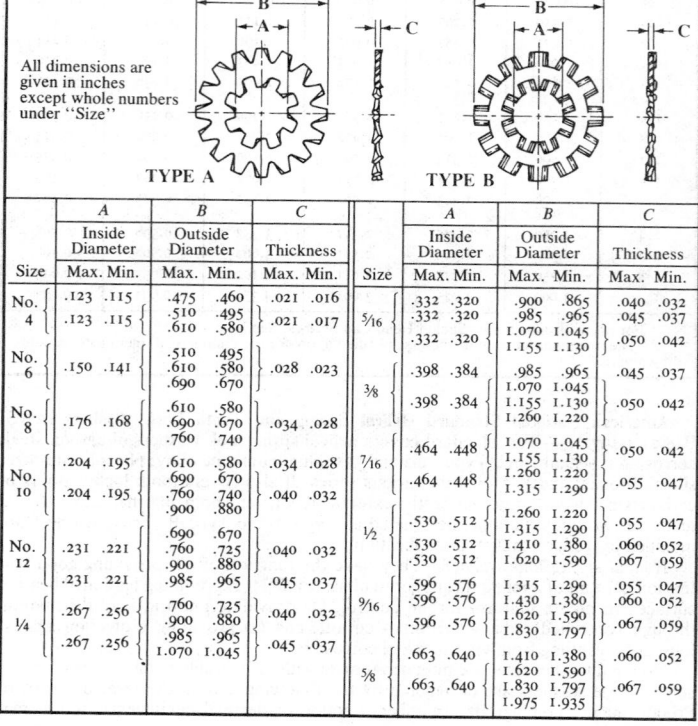

Size	A Inside Diameter Max.	A Inside Diameter Min.	B Outside Diameter Max.	B Outside Diameter Min.	C Thickness Max.	C Thickness Min.	Size	A Inside Diameter Max.	A Inside Diameter Min.	B Outside Diameter Max.	B Outside Diameter Min.	C Thickness Max.	C Thickness Min.
No. 4	.123	.115	.475	.460	.021	.016	5/16	.332	.320	.900	.865	.040	.032
	.123	.115	.510	.495	.021	.017		.332	.320	.985	.965	.045	.037
			.610	.580				.332	.320	1.070	1.045	.050	.042
										1.155	1.130		
No. 6	.150	.141	.510	.495			3/8	.398	.384	.985	.965	.045	.037
			.610	.580	.028	.023				1.070	1.045		
			.690	.670				.398	.384	1.155	1.130	.050	.042
										1.260	1.220		
No. 8	.176	.168	.610	.580			7/16	.464	.448	1.070	1.045	.050	.042
			.690	.670	.034	.028				1.155	1.130		
			.760	.740				.464	.448	1.260	1.220	.055	.047
										1.315	1.290		
No. 10	.204	.195	.610	.580	.034	.028	1/2	.530	.512	1.260	1.220	.055	.047
			.690	.670						1.315	1.290		
	.204	.195	.760	.740	.040	.032		.530	.512	1.410	1.380	.060	.052
			.900	.880				.530	.512	1.620	1.590	.067	.059
No. 12	.231	.221	.690	.670	.040	.032	9/16	.596	.576	1.315	1.290	.055	.047
			.760	.725				.596	.576	1.430	1.380	.060	.052
	.231	.221	.900	.880	.045	.037		.596	.576	1.620	1.590	.067	.059
			.985	.965						1.830	1.797		
¼	.267	.256	.760	.725	.040	.032	5/8	.663	.640	1.410	1.380	.060	.052
			.900	.880						1.620	1.590		
	.267	.256	.985	.965	.045	.037		.663	.640	1.830	1.797	.067	.059
			1.070	1.045						1.975	1.935		

Table 4. American National Standard Internal and External Tooth Lock Washers (ANSI B18.21.1-1972)

Diagrams: Internal Tooth — TYPE A, TYPE B; External Tooth — TYPE A, TYPE B; Countersunk External Tooth — TYPE A, TYPE B (80°–82°)

Internal Tooth Lock Washers

	Size	#2	#3	#4	#5	#6	#8	#10	#12	1/4	5/16	3/8	7/16	1/2	9/16	5/8	11/16	3/4	13/16	7/8	1	1 1/8	1 1/4
A	Max	0.095	0.109	0.123	0.136	0.150	0.176	0.204	0.231	0.267	0.332	0.398	0.464	0.530	0.596	0.663	0.728	0.795	0.861	0.927	1.060	1.192	1.325
	Min	0.089	0.102	0.115	0.129	0.141	0.168	0.195	0.221	0.256	0.320	0.384	0.448	0.512	0.576	0.640	0.704	0.769	0.832	0.894	1.019	1.144	1.275
B	Max	0.200	0.232	0.270	0.280	0.295	0.340	0.381	0.410	0.478	0.610	0.692	0.789	0.900	0.985	1.071	1.166	1.245	1.315	1.410	1.637	1.830	1.975
	Min	0.175	0.215	0.255	0.245	0.275	0.325	0.365	0.394	0.460	0.594	0.670	0.740	0.867	0.957	1.045	1.130	1.220	1.290	1.364	1.590	1.799	1.921
C	Max	0.015	0.019	0.019	0.021	0.021	0.023	0.025	0.025	0.028	0.034	0.040	0.040	0.045	0.045	0.050	0.050	0.055	0.055	0.060	0.067	0.067	0.067
	Min	0.010	0.012	0.015	0.017	0.017	0.018	0.020	0.020	0.023	0.028	0.032	0.032	0.037	0.037	0.042	0.042	0.047	0.047	0.052	0.059	0.059	0.059

External Tooth Lock Washers

	Size	#2	#3	#4	#5	#6	#8	#10	#12	1/4	5/16	3/8	7/16	1/2	9/16	5/8	11/16	3/4	13/16	7/8	1	1 1/8	1 1/4
A	Max	…	…	0.123	0.123	0.150	0.176	0.204	0.231	0.267	0.332	0.398	0.464	0.530	0.596	0.663	0.728	0.795	0.861	0.927	1.060		
	Min	…	…	0.115	0.115	0.141	0.168	0.195	0.221	0.256	0.320	0.384	0.448	0.513	0.576	0.641	0.704	0.768	0.833	0.897	1.025		
B	Max	…	…	0.260	0.285	0.320	0.381	0.410	0.475	0.510	0.610	0.694	0.760	0.900	0.985	1.070	1.155	1.260	1.315	1.410	1.620		
	Min	…	…	0.235	0.270	0.305	0.365	0.394	0.460	0.494	0.588	0.670	0.740	0.880	0.960	1.045	1.130	1.220	1.290	1.380	1.590		
C	Max	…	…	0.015	0.019	0.022	0.023	0.025	0.028	0.028	0.034	0.040	0.040	0.045	0.045	0.050	0.050	0.055	0.055	0.060	0.067		
	Min	…	…	0.012	0.015	0.016	0.018	0.020	0.023	0.023	0.028	0.032	0.032	0.037	0.037	0.042	0.042	0.047	0.047	0.052	0.052		

Heavy Internal Tooth Lock Washers

	Size	1/4	5/16	3/8	7/16	1/2	9/16	5/8	3/4	7/8
A	Max	0.267	0.332	0.398	0.464	0.530	0.596	0.663	0.795	0.927
	Min	0.256	0.320	0.384	0.448	0.512	0.576	0.640	0.768	0.894
B	Max	0.536	0.607	0.748	0.858	0.924	1.034	1.135	1.265	1.447
	Min	0.500	0.590	0.700	0.800	0.880	0.990	1.100	1.240	1.400
C	Max	0.045	0.050	0.050	0.067	0.067	0.067	0.067	0.084	0.084
	Min	0.035	0.040	0.042	0.050	0.050	0.055	0.059	0.070	0.075

Countersunk External Tooth Lock Washers*

	Size	#4	#6	#8	#10	#12	1/4	5/16	3/8	7/16	1/2
A	Max	0.123	0.150	0.177	0.205	0.231	0.267	0.287	0.333	0.398	0.463	0.529
	Min	0.113	0.140	0.167	0.195	0.220	0.255	0.273	0.318	0.381	0.448	0.512
C	Max	0.019	0.021	0.021	0.025	0.025	0.028	0.028	0.023	0.028	0.045	0.045
	Min	0.015	0.017	0.017	0.020	0.020	0.021	0.025	0.020	0.023	0.037	0.037
D	Max	0.065	0.092	0.099	0.128	0.118	0.113	0.137	0.165	0.192	0.255	0.304	0.294
	Min	0.050	0.082	0.083	0.118	0.083	0.088	0.105	0.128	0.147	0.165	0.242	0.260

All dimensions are given in inches. * Starting with #4, approx. O.D.'s are: 0.213, 0.289, 0.322, 0.354, 0.421, 0.454, 0.505, 0.599, 0.765, 0.867, and 0.976.

METRIC FASTENERS

A number of American National Standards covering metric bolts, screws, nuts, and washers have been established in cooperation with the Department of Defense in such a way that they could be used by the Government for procurement purposes. Extensive information concerning these metric fasteners is given in the following text and tables, but for additional manufacturing and acceptance specifications reference should be made to the respective Standards which may be obtained by nongovernmental agencies from the American National Standards Institute, 1430 Broadway, New York, N.Y. 10018. These Standards are:

ANSI B18.2.3.1M-1979 Metric Hex Cap Screws	Table 1
ANSI B18.2.3.2M-1979 Metric Formed Hex Screws	Table 2
ANSI B18.2.3.3M-1979 Metric Heavy Hex Screws	Table 3
ANSI B18.2.3.4M-1984 Metric Hex Flange Screws	Table 4
ANSI B18.2.3.5M-1979 Metric Hex Bolts	Table 6
ANSI B18.2.3.6M-1979 Metric Heavy Hex Bolts	Table 7
ANSI B18.2.3.7M-1979 Metric Heavy Hex Structural Bolts	Table 10
ANSI B18.2.3.8M-1981 Metric Hex Lag Screws	Table 12
ANSI B18.2.3.9M-1984 Metric Heavy Hex Flange Screws	Table 14
ANSI B18.5.2.2M-1982 Metric Round Head Square Neck Bolts	Table 15
ANSI B18.3.1M-1986 Socket Head Cap Screws (Metric Series)	Table 20
ANSI B18.2.4.1M-1979 Metric Hex Nuts, Style 1	Table 26
ANSI B18.2.4.2M-1979 Metric Hex Nuts, Style 2	Table 26
ANSI B18.2.4.3M-1979 Metric Slotted Hex Nuts	Table 27
ANSI B18.2.4.4M-1979 Metric Hex Flange Nuts	Table 28
ANSI B18.2.4.5M-1979 Metric Hex Jam Nuts	Table 29
ANSI B18.2.4.6M-1979 Metric Heavy Hex Nuts	Table 29
ANSI B18.16.3M-1982 Prevailing-Torque Metric Hex Nuts	Table 30
ANSI B18.16.3M-1982 Prevailing-Torque Metric Hex Flange Nuts	Table 31
ANSI B18.22M-1981 Metric Plain Washers	Table 32

Manufacturers should be consulted concerning which items and sizes are in stock production.

Comparison with ISO Standards. — American National Standards for Metric bolts, screws and nuts have been coordinated to the extent possible with the comparable ISO Standards or proposed Standards. The dimensional differences between the ANSI and the comparable ISO Standards or proposed Standards are few, relatively minor, and none will affect the functional interchangeability of bolts, screws, and nuts manufactured to the requirements of either.

Where no comparable ISO Standard had been developed, as was the case when the ANSI Standards for Metric Heavy Hex Screws, Metric Heavy Hex Bolts, and Metric Hex Lag Screws were adopted, nominal diameters, thread pitches, body diameters, widths across flats, head heights, thread lengths, thread dimensions, and nominal lengths are in accord with ISO Standards for related hex head screws and bolts. At the time of ANSI adoption (1982) there was no ISO Standard for round head square neck bolts.

The following functional characteristics of hex head screws and bolts are in agreement between the respective ANSI Standard and the comparable ISO Standard or proposed Standard: diameters and thread pitches, body diameters, widths across flats (see exception below), bearing surface diameters (except for metric hex bolts), flange diameters (for metric hex flange screws), head heights, thread lengths, thread dimensions, and nominal lengths.

Table 1. American National Standard Metric Hex Cap Screws
(ANSI B18.2.3.1M-1979)

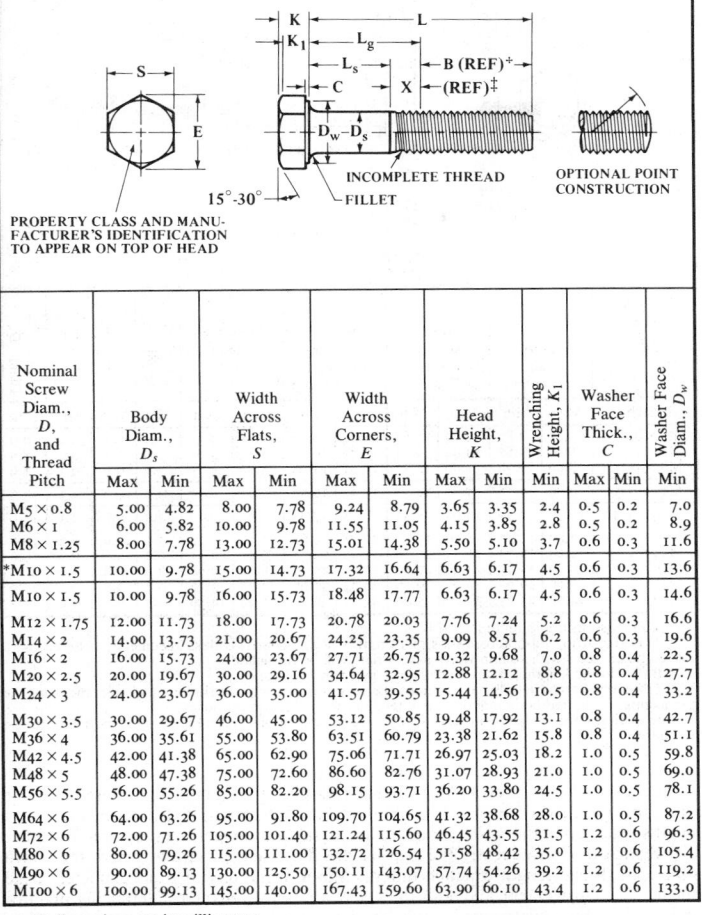

PROPERTY CLASS AND MANU-
FACTURER'S IDENTIFICATION
TO APPEAR ON TOP OF HEAD

INCOMPLETE THREAD

OPTIONAL POINT
CONSTRUCTION

Nominal Screw Diam., D, and Thread Pitch	Body Diam., D_s		Width Across Flats, S		Width Across Corners, E		Head Height, K		Wrenching Height, K_1	Washer Face Thick., C		Washer Face Diam., D_w
	Max	Min	Max	Min	Max	Min	Max	Min	Min	Max	Min	Min
M5 × 0.8	5.00	4.82	8.00	7.78	9.24	8.79	3.65	3.35	2.4	0.5	0.2	7.0
M6 × 1	6.00	5.82	10.00	9.78	11.55	11.05	4.15	3.85	2.8	0.5	0.2	8.9
M8 × 1.25	8.00	7.78	13.00	12.73	15.01	14.38	5.50	5.10	3.7	0.6	0.3	11.6
*M10 × 1.5	10.00	9.78	15.00	14.73	17.32	16.64	6.63	6.17	4.5	0.6	0.3	13.6
M10 × 1.5	10.00	9.78	16.00	15.73	18.48	17.77	6.63	6.17	4.5	0.6	0.3	14.6
M12 × 1.75	12.00	11.73	18.00	17.73	20.78	20.03	7.76	7.24	5.2	0.6	0.3	16.6
M14 × 2	14.00	13.73	21.00	20.67	24.25	23.35	9.09	8.51	6.2	0.6	0.3	19.6
M16 × 2	16.00	15.73	24.00	23.67	27.71	26.75	10.32	9.68	7.0	0.8	0.4	22.5
M20 × 2.5	20.00	19.67	30.00	29.16	34.64	32.95	12.88	12.12	8.8	0.8	0.4	27.7
M24 × 3	24.00	23.67	36.00	35.00	41.57	39.55	15.44	14.56	10.5	0.8	0.4	33.2
M30 × 3.5	30.00	29.67	46.00	45.00	53.12	50.85	19.48	17.92	13.1	0.8	0.4	42.7
M36 × 4	36.00	35.61	55.00	53.80	63.51	60.79	23.38	21.62	15.8	0.8	0.4	51.1
M42 × 4.5	42.00	41.38	65.00	62.90	75.06	71.71	26.97	25.03	18.2	1.0	0.5	59.8
M48 × 5	48.00	47.38	75.00	72.60	86.60	82.76	31.07	28.93	21.0	1.0	0.5	69.0
M56 × 5.5	56.00	55.26	85.00	82.20	98.15	93.71	36.20	33.80	24.5	1.0	0.5	78.1
M64 × 6	64.00	63.26	95.00	91.80	109.70	104.65	41.32	38.68	28.0	1.0	0.5	87.2
M72 × 6	72.00	71.26	105.00	101.40	121.24	115.60	46.45	43.55	31.5	1.2	0.6	96.3
M80 × 6	80.00	79.26	115.00	111.00	132.72	126.54	51.58	48.42	35.0	1.2	0.6	105.4
M90 × 6	90.00	89.13	130.00	125.50	150.11	143.07	57.74	54.26	39.2	1.2	0.6	119.2
M100 × 6	100.00	99.13	145.00	140.00	167.43	159.60	63.90	60.10	43.4	1.2	0.6	133.0

All dimensions are in millimeters.

* This size with width across flats of 15 mm is not standard. Unless specifically ordered, M10 hex cap screws with 16 mm width across flats will be furnished.

† Basic thread lengths, B, are the same as given in Table 6.

‡ Transition thread length, X, includes the length of incomplete threads and tolerances on grip gaging length and body length. It is intended for calculation purposes.

For additional manufacturing and acceptance specifications, reference should be made to the Standard.

Table 2. American National Standard Metric Formed Hex Screws
(ANSI B18.2.3.2M-1979)

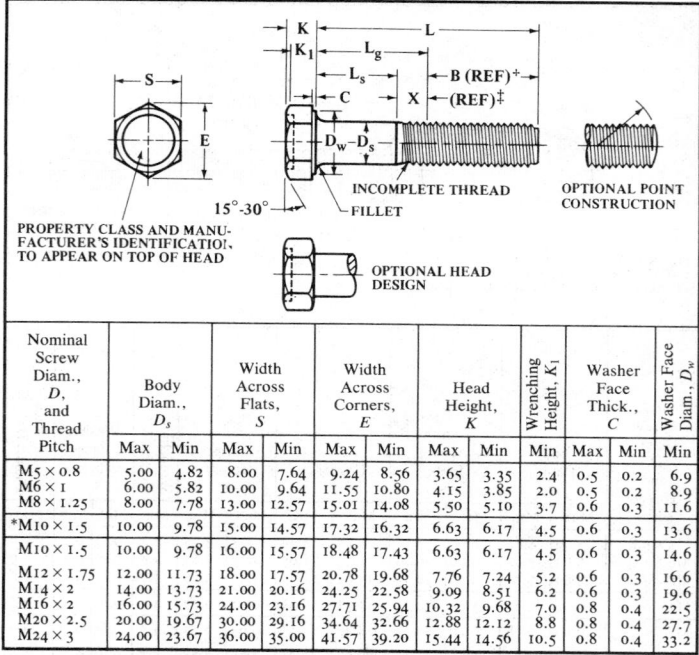

All dimensions are in millimeters.

Nominal Screw Diam., D, and Thread Pitch	Body Diam., D_s		Width Across Flats, S		Width Across Corners, E		Head Height, K		Wrenching Height, K_1	Washer Face Thick., C		Washer Face Diam., D_w
	Max	Min	Max	Min	Max	Min	Max	Min	Min	Max	Min	Min
M5 × 0.8	5.00	4.82	8.00	7.64	9.24	8.56	3.65	3.35	2.4	0.5	0.2	6.9
M6 × 1	6.00	5.82	10.00	9.64	11.55	10.80	4.15	3.85	2.0	0.5	0.2	8.9
M8 × 1.25	8.00	7.78	13.00	12.57	15.01	14.08	5.50	5.10	3.7	0.6	0.3	11.6
*M10 × 1.5	10.00	9.78	15.00	14.57	17.32	16.32	6.63	6.17	4.5	0.6	0.3	13.6
M10 × 1.5	10.00	9.78	16.00	15.57	18.48	17.43	6.63	6.17	4.5	0.6	0.3	14.6
M12 × 1.75	12.00	11.73	18.00	17.57	20.78	19.68	7.76	7.24	5.2	0.6	0.3	16.6
M14 × 2	14.00	13.73	21.00	20.16	24.25	22.58	9.09	8.51	6.2	0.6	0.3	19.6
M16 × 2	16.00	15.73	24.00	23.16	27.71	25.94	10.32	9.68	7.0	0.8	0.4	22.5
M20 × 2.5	20.00	19.67	30.00	29.16	34.64	32.66	12.88	12.12	8.8	0.8	0.4	27.7
M24 × 3	24.00	23.67	36.00	35.00	41.57	39.20	15.44	14.56	10.5	0.8	0.4	33.2

   * This size with width across flats of 15 mm is not standard. Unless specifically ordered, M10 formed hex screws with 16 mm width across flats will be furnished.
   † Basic thread lengths, B, are the same as given in Table 6.
   ‡ Transition thread length, X, includes the length of incomplete threads and tolerances on the grip gaging length and body length. It is intended for calculation purposes.
   For additional manufacturing and acceptance specifications, reference should be made to the Standard.

   Socket head cap screws (ANSI B18.3.1M-1982) are functionally interchangeable with screws which conform to ISO R861-1968 or ISO 4762-1977. However, the thread lengths specified in the ANSI Standard are equal to or longer than required by either ISO Standard. Consequently the grip lengths also vary on screws where the North American thread length practice differs. Minor variations in head diameter, head height, key engagement and wall thickness are due to diverse tolerancing practice and will be found documented in the ANSI Standard.

   One exception with respect to width across flats for metric hex cap screws, formed hex screws, and hex bolts is the M10 size. These are currently being produced in the United States with a width across flats of 15 mm. This size, however, is not an ISO Standard. Unless these M10 screws and bolts with 15 mm width across flats are specifically ordered, the M10 size with 16 mm across flats will be furnished.

Table 3. American National Standard Metric Heavy Hex Screws
(ANSI B18.2.3.3M-1979)

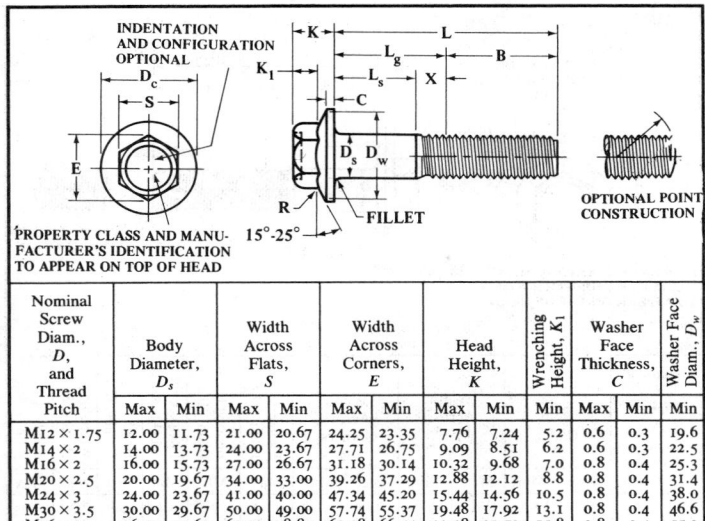

All dimensions are in millimeters.
Basic thread lengths, *B*, are the same as given in Table 6.
Transition thread length, *X*, includes the length of incomplete threads and tolerances on grip gaging length and body length. It is intended for calculation purposes.
For additional manufacturing and acceptance specifications, reference should be made to the Standard.

Nominal Screw Diam., *D*, and Thread Pitch	Body Diameter, D_s		Width Across Flats, *S*		Width Across Corners, *E*		Head Height, *K*		Wrenching Height, K_1	Washer Face Thickness, *C*		Washer Face Diam., D_w
	Max	Min	Max	Min	Max	Min	Max	Min	Min	Max	Min	Min
M12 × 1.75	12.00	11.73	21.00	20.67	24.25	23.35	7.76	7.24	5.2	0.6	0.3	19.6
M14 × 2	14.00	13.73	24.00	23.67	27.71	26.75	9.09	8.51	6.2	0.6	0.3	22.5
M16 × 2	16.00	15.73	27.00	26.67	31.18	30.14	10.32	9.68	7.0	0.8	0.4	25.3
M20 × 2.5	20.00	19.67	34.00	33.00	39.26	37.29	12.88	12.12	8.8	0.8	0.4	31.4
M24 × 3	24.00	23.67	41.00	40.00	47.34	45.20	15.44	14.56	10.5	0.8	0.4	38.0
M30 × 3.5	30.00	29.67	50.00	49.00	57.74	55.37	19.48	17.92	13.1	0.8	0.4	46.6
M36 × 4	36.00	35.61	60.00	58.80	69.28	66.44	23.38	21.72	15.8	0.8	0.4	55.9

ANSI letter symbols designating dimensional characteristics are in accord with those used in ISO Standards except capitals have been used for data processing convenience instead of the lower case letters used in the ISO Standards.

Metric Screw and Bolt Diameters. — Metric screws and bolts are furnished with full diameter body within the limits shown in the respective dimensional tables, or are threaded to the head (see Thread Length below) unless the purchaser specifies "reduced body diameter." Metric formed hex screws (Table 9), hex flange screws (Table 9), hex bolts (Table 9), heavy hex bolts (Table 9), hex lag screws (Table 12), heavy hex flange screws (Table 14), and round head square neck bolts (Table 18) may be obtained with reduced diameter body, if so specified; however, formed hex screws, hex flange screws, heavy hex flange screws, hex bolts, or heavy hex bolts with nominal lengths shorter than 4*D*, where *D* is the nominal diameter, are not recommended. Metric formed hex screws, hex flange screws, heavy hex flange screws, and hex lag screws with reduced body diameter will be furnished with a shoulder under the head. For metric hex bolts and heavy hex bolts this is optional with the manufacturer.

For bolts and lag screws there may be a reasonable swell, fin, or die seam on the body adjacent to the head not exceeding the nominal bolt diameter by: 0.50 mm for M5, 0.65 mm for M6, 0.75 mm for M8 through M14, 1.25 mm for M16, 1.50 mm for

Table 4. American National Standard Metric Hex Flange Screws
(ANSI B18.2.3.4M-1984)

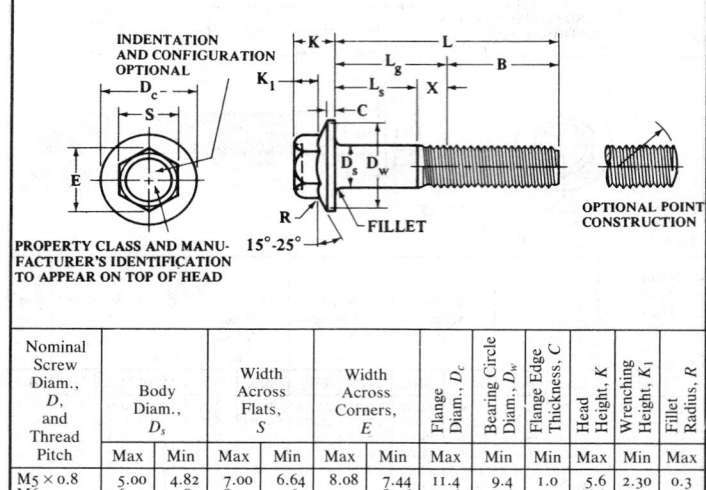

Nominal Screw Diam., D, and Thread Pitch	Body Diam., D_s		Width Across Flats, S		Width Across Corners, E		Flange Diam., D_c	Bearing Circle Diam., D_w	Flange Edge Thickness, C	Head Height, K		Wrenching Height, K_1	Fillet Radius, R
	Max	Min	Max	Min	Max	Min	Max	Min	Min	Max	Min	Min	Max
M5 × 0.8	5.00	4.82	7.00	6.64	8.08	7.44	11.4	9.4	1.0	5.6	2.30		0.3
M6 × 1	6.00	5.82	8.00	7.64	9.24	8.56	13.6	11.6	1.1	6.8	2.90		0.4
M8 × 1.25	8.00	7.78	10.00	9.64	11.55	10.80	17.0	14.9	1.2	8.5	3.80		0.5
M10 × 1.5	10.00	9.78	13.00	12.57	15.01	14.08	20.8	18.7	1.5	9.7	4.30		0.6
M12 × 1.75	12.00	11.73	15.00	14.57	17.32	16.32	24.7	22.0	1.8	11.9	5.40		0.7
M14 × 2	14.00	13.73	18.00	17.57	20.78	19.68	28.6	25.9	2.1	12.9	5.60		0.8
M16 × 2	16.00	15.73	21.00	20.48	24.25	22.94	32.8	30.1	2.4	15.1	6.70		1.0

All dimensions are in millimeters. Basic thread lengths, B, are the same as given in Table 6. Transition thread length, x, includes the length of incomplete threads and tolerances on grip gaging length and body length. This dimension is intended for calculation purposes only. For additional manufacturing and acceptance specifications, reference should be made to the Standard.

M20 through M30, 2.30 mm for M36 through M48, 3.00 mm for M56 through M72, and 4.80 mm for M80 through M100.

Metric Screw and Bolt Thread Lengths. — The length of thread on metric screws and bolts (except for metric lag screws) is controlled by the grip gaging length, L_g max. This is the distance measured parallel to the axis of the screw or bolt, from under the head bearing surface to the face of a noncounterbored or noncountersunk standard GO thread ring gage assembled by hand as far as the thread will permit. The maximum grip gaging length, as calculated and rounded to one decimal place, is equal to the nominal screw length, L, minus the basic thread length, B, or in the case of socket head cap screws, minus the minimum thread length L_T. B and L_T are reference dimensions intended for calculation purposes only and will be found in Tables 6 and 21, respectively.

The minimum thread length for hex lag screws is equal to one-half the nominal screw length plus 12 mm, or 150 mm, whichever is shorter. Screws too short for this formula to apply are threaded as close to the head as practicable.

Metric Screw and Bolt Diameter-Length Combinations. — For a given diameter, the recommended range of lengths of metric cap screws, formed hex screws, heavy hex

Table 5. Rec'd Diameter-Length Combinations for Metric Hex Cap Screws, Formed Hex Screws, Heavy Hex Screws, Hex Flange Screws, and Heavy Hex Flange Screws

Nominal Length*	Diameter—Pitch										
	M5 ×0.8	M6 ×1	M8 ×1.25	M10 ×1.5	M12 ×1.75	M14 ×2	M16 ×2	M20 ×2.5	M24 ×3	M30 ×3.5	M36 ×4
8	X	...	...	...	...	...	...	...	...	...	...
10	X	X	...	...	...	...	...	...	...	...	...
12	X	X	X	...	...	...	...	...	...	...	...
14	X	X	X	X†	...	...	...	...	...	...	...
16	X	X	X	X	X†	X†	...	...	...	...	...
20	X	X	X	X	X	X	...	...	...	...	...
25	X	X	X	X	X	X	X	...	...	...	...
30	X	X	X	X	X	X	X	X	...	...	...
35	X	X	X	X	X	X	X	X	X	...	...
40	X	X	X	X	X	X	X	X	X	X	...
45	X	X	X	X	X	X	X	X	X	X	...
50	X	X	X	X	X	X	X	X	X	X	X
(55)	...	X	X	X	X	X	X	X	X	X	X
60	...	X	X	X	X	X	X	X	X	X	X
(65)	...	...	X	X	X	X	X	X	X	X	X
70	...	...	X	X	X	X	X	X	X	X	X
(75)	...	...	X	X	X	X	X	X	X	X	X
80	...	...	X	X	X	X	X	X	X	X	X
(85)	...	...	...	X	X	X	X	X	X	X	X
90	...	...	...	X	X	X	X	X	X	X	X
100	...	...	...	X	X	X	X	X	X	X	X
110	...	...	...	...	X	X	X	X	X	X	X
120	...	...	...	...	X	X	X	X	X	X	X
130	...	...	...	...	...	X	X	X	X	X	X
140	...	...	...	...	...	X	X	X	X	X	X
150	...	...	...	...	...	...	X	X	X	X	X
160	...	...	...	...	...	...	X	X	X	X	X
(170)	...	...	...	...	...	...	...	X	X	X	X
180	...	...	...	...	...	...	...	X	X	X	X
(190)	...	...	...	...	...	...	...	X	X	X	X
200	...	...	...	...	...	...	...	X	X	X	X
220	...	...	...	...	...	...	...	...	X	X	X
240	...	...	...	...	...	...	...	...	X	X	X
260	...	...	...	...	...	...	...	...	...	X	X
280	...	...	...	...	...	...	...	...	...	X	X
300	...	...	...	...	...	...	...	...	...	X	X

All dimensions are in millimeters. * Lengths in parentheses are not recommended. Recommended lengths of formed hex screws, hex flange screws, and heavy hex flange screws do not extend above 150 mm. Recommended lengths of heavy hex screws do not extend below 20 mm. Standard sizes for government use. Recommended diameter-length combinations are indicated by the symbol X. Screws with lengths above heavy cross lines are threaded full length. † Does not apply to hex flange screws and heavy hex flange screws.

For available diameters of each type of screw, see respective dimensional table.

Table 6. American National Standard Metric Hex Bolts
(ANSI B18.2.3.5M-1979)

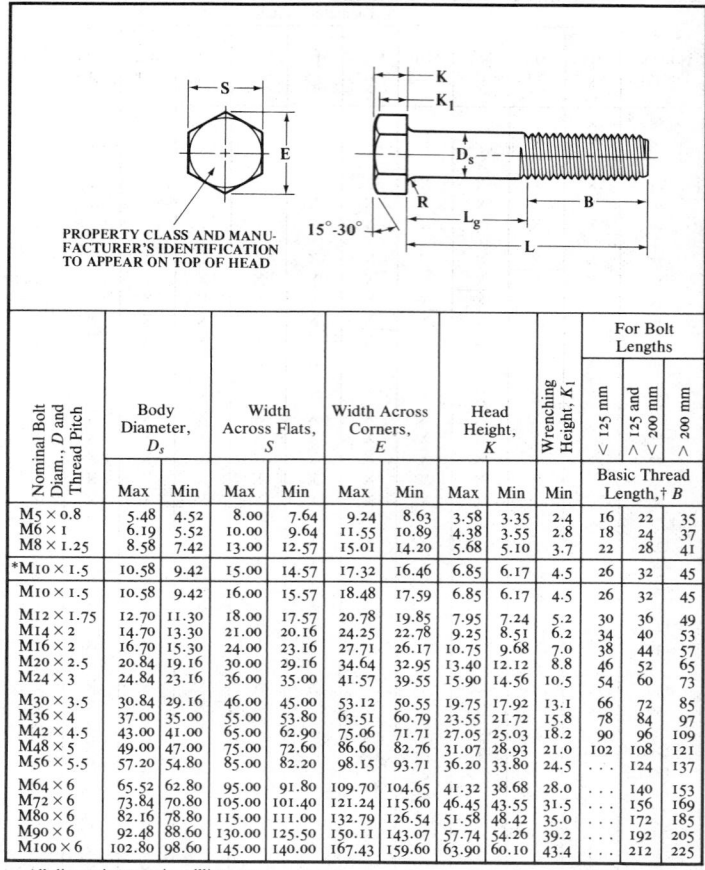

PROPERTY CLASS AND MANU-
FACTURER'S IDENTIFICATION
TO APPEAR ON TOP OF HEAD

Nominal Bolt Diam., D and Thread Pitch	Body Diameter, D_s		Width Across Flats, S		Width Across Corners, E		Head Height, K		Wrenching Height, K_1	For Bolt Lengths		
										< 125 mm	125 and 200 mm	> 200 mm
	Max	Min	Max	Min	Max	Min	Max	Min	Min	Basic Thread Length,† B		
M5 × 0.8	5.48	4.52	8.00	7.64	9.24	8.63	3.58	3.35	2.4	16	22	35
M6 × 1	6.19	5.52	10.00	9.64	11.55	10.89	4.38	3.55	2.8	18	24	37
M8 × 1.25	8.58	7.42	13.00	12.57	15.01	14.20	5.68	5.10	3.7	22	28	41
*M10 × 1.5	10.58	9.42	15.00	14.57	17.32	16.46	6.85	6.17	4.5	26	32	45
M10 × 1.5	10.58	9.42	16.00	15.57	18.48	17.59	6.85	6.17	4.5	26	32	45
M12 × 1.75	12.70	11.30	18.00	17.57	20.78	19.85	7.95	7.24	5.2	30	36	49
M14 × 2	14.70	13.30	21.00	20.16	24.25	22.78	9.25	8.51	6.2	34	40	53
M16 × 2	16.70	15.30	24.00	23.16	27.71	26.17	10.75	9.68	7.0	38	44	57
M20 × 2.5	20.84	19.16	30.00	29.16	34.64	32.95	13.40	12.12	8.8	46	52	65
M24 × 3	24.84	23.16	36.00	35.00	41.57	39.55	15.90	14.56	10.5	54	60	73
M30 × 3.5	30.84	29.16	46.00	45.00	53.12	50.55	19.75	17.92	13.1	66	72	85
M36 × 4	37.00	35.00	55.00	53.80	63.51	60.79	23.55	21.72	15.8	78	84	97
M42 × 4.5	43.00	41.00	65.00	62.90	75.06	71.71	27.05	25.03	18.2	90	96	109
M48 × 5	49.00	47.00	75.00	72.60	86.60	82.76	31.07	28.93	21.0	102	108	121
M56 × 5.5	57.20	54.80	85.00	82.20	98.15	93.71	36.20	33.80	24.5	...	124	137
M64 × 6	65.52	62.80	95.00	91.80	109.70	104.65	41.32	38.68	28.0	...	140	153
M72 × 6	73.84	70.80	105.00	101.40	121.24	115.60	46.45	43.55	31.5	...	156	169
M80 × 6	82.16	78.80	115.00	111.00	132.79	126.54	51.58	48.42	35.0	...	172	185
M90 × 6	92.48	88.60	130.00	125.50	150.11	143.07	57.74	54.26	39.2	...	192	205
M100 × 6	102.80	98.60	145.00	140.00	167.43	159.60	63.90	60.10	43.4	...	212	225

All dimensions are in millimeters.
  * This size with width across flats of 15 mm is not standard. Unless specifically ordered, M10 hex bolts with 16 mm width across flats will be furnished.
  † Basic thread length, B, is a reference dimension.
  For additional manufacturing and acceptance specifications, reference should be made to the Standard.

screws, hex flange screws, and heavy hex flange screws can be found in Table 5, for heavy hex structural bolts in Table 11, for hex lag screws in Table 13, for round head square neck bolts in Table 17, and for socket head cap screws in Table 24. No recommendations for diameter-length combinations are given in the Standards for hex bolts and heavy hex bolts.

Table 7. American National Standard Heavy Hex Bolts
(ANSI B18.2.3.6M-1979)

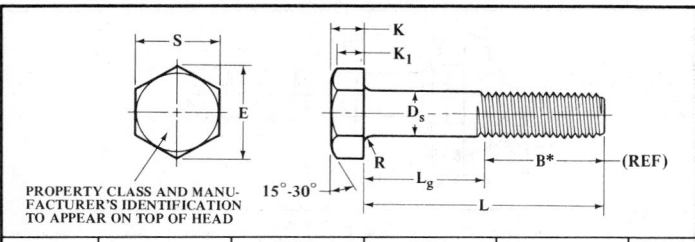

PROPERTY CLASS AND MANU-
FACTURER'S IDENTIFICATION
TO APPEAR ON TOP OF HEAD

Nominal Diam., D and Thread Pitch	Body Diameter, D_s		Width Across Flats, S		Width Across Corners, E		Head Height, K		Wrenching Height, K_1
	Max	Min	Max	Min	Max	Min	Max	Min	Min
M12 × 1.75	12.70	11.30	21.00	20.16	24.25	22.78	7.95	7.24	5.2
M14 × 2	14.70	13.30	24.00	23.16	27.71	26.17	9.25	8.51	6.2
M16 × 2	16.70	15.30	27.00	26.16	31.18	29.56	10.75	9.68	7.0
M20 × 2.5	20.84	19.16	34.00	33.00	39.26	37.29	13.40	12.12	8.8
M24 × 3	24.84	23.16	41.00	40.00	47.34	45.20	15.90	14.56	10.5
M30 × 3.5	30.84	29.16	50.00	49.00	57.74	55.37	19.75	17.92	13.1
M36 × 4	37.00	35.00	60.00	58.80	69.28	66.44	23.55	21.72	15.8

All dimensions are in millimeters.
* Basic thread lengths, B, are the same as given in Table 6.
For additional manufacturing and acceptance specifications, reference should be made to the Standard.

Table 8. Recommended Clearance Holes for Metric Hex Screws and Bolts*

Nominal Diam., D and Thread Pitch	Clearance Hole Diam., Basic, D_h			Nom. Diam., D and Thread Pitch	Clearance Hole Diam., Basic, D_h		
	Close	Normal, Preferred	Loose		Close	Normal, Preferred	Loose
M5 × 0.8	5.3	5.5	5.8	M30 × 3.5	31.0	33.0	35.0
M6 × 1	6.4	6.6	7.0	M36 × 4	37.0	39.0	42.0
M8 × 1.25	8.4	9.0	10.0	M42 × 4.5	43.0	45.0	48.0
M10 × 1.5	10.5	11.0	12.0	M48 × 5	50.0	52.0	56.0
M12 × 1.75	13.0	13.5	14.5	M56 × 5.5	58.0	62.0	66.0
M14 × 2	15.0	15.5	16.5	M64 × 6	66.0	70.0	74.0
M16 × 2	17.0	17.5	18.5	M72 × 6	74.0	78.0	82.0
M20 × 2.5	21.0	22.0	24.0	M80 × 6	82.0	86.0	91.0
M22 × 2.5†	23.0	24.0	26.0	M90 × 6	93.0	96.0	101.0
M24 × 3	25.0	26.0	28.0	M100 × 6	104.0	107.0	112.0
M27 × 3†	28.0	30.0	32.0	...	...	...	...

All dimensions are in millimeters.   * Does not apply to hex lag screws, hex socket head cap screws, or round head square neck bolts.   † Applies only to heavy hex structural bolts.

Normal Clearance: This is preferred for general purpose applications and should be specified unless special design considerations dictate the need for either a close or loose clearance hole.

Close Clearance: This should be specified only where conditions such as critical alignment of assembled parts, wall thickness or other limitations necessitate use of a minimum hole. When close clearance holes are specified, special provision (e.g. countersinking) must be made at the screw or bolt entry side to permit proper seating of the screw or bolt head.

Loose Clearance: This should be specified only for applications where maximum adjustment capability between components being assembled is necessary.

Recommended Tolerances: The clearance hole diameters given in this table are minimum size. Recommended tolerances are: for screw or bolt diameter M5, + 0.2 mm; for M6 through M16, + 0.3 mm; for M20 through M42, + 0.4 mm; for M48 through M72, + 0.5 mm; and for M80 through M100, + 0.6 mm.

Table 9. American National Standard Metric Hex Screws and Bolts — Reduced Body Diameters

Nominal Diam., D, and Thread Pitch	Shoulder Diameter,* D_s		Body Diameter, D_{si}		Shoulder Length,* L_{sh}		Nominal Diam., D, and Thread Pitch	Shoulder Diameter,* D_s		Body Diameter, D_{si}		Shoulder Length,* L_{sh}	
	Max	Min	Max	Min	Max	Min		Max	Min	Max	Min	Max	Min
Metric Formed Hex Screws (ANSI B18.2.3.2M-1979)													
M5 × 0.8	5.00	4.82	4.46	4.36	3.5	2.5	M14 × 2	14.00	13.73	12.77	12.50	8.0	7.0
M6 × 1	6.00	5.82	5.39	5.21	4.0	3.0	M16 × 2	16.00	15.73	14.77	14.50	9.0	8.0
M8 × 1.25	8.00	7.78	7.26	7.04	5.0	4.0	M20 × 2.5	20.00	19.67	18.49	18.16	11.0	10.0
M10 × 1.5	10.00	9.78	9.08	8.86	6.0	5.0	M24 × 3	24.00	23.67	22.13	21.80	13.0	12.0
M12 × 1.75	12.00	11.73	10.95	10.68	7.0	6.0	…	…	…	…	…	…	…
Metric Hex Flange Screws (ANSI B18.2.3.4M-1984)													
M5 × 0.8	5.00	4.82	4.46	4.36	3.5	2.5	M12 × 1.75	12.00	11.73	10.95	10.68	7.0	6.0
M6 × 1	6.00	5.82	5.39	5.21	4.0	3.0	M14 × 2	14.00	13.73	12.77	12.50	8.0	7.0
M8 × 1.25	8.00	7.78	7.26	7.04	5.0	4.0	M16 × 2	16.00	15.73	14.77	14.50	9.0	8.0
M10 × 1.5	10.00	9.78	9.08	8.86	6.0	5.0	…	…	…	…	…	…	…
Metric Hex Bolts (ANSI B18.2.3.5M-1979)													
M5 × 0.8	5.48	4.52	4.46	4.36	3.5	2.5	M14 × 2	14.70	13.30	12.77	12.50	8.0	7.0
M6 × 1	6.48	5.52	5.39	5.21	4.0	3.0	M16 × 2	16.70	15.30	14.77	14.50	9.0	8.0
M8 × 1.25	8.58	7.42	7.26	7.04	5.0	4.0	M20 × 2.5	20.84	19.16	18.49	18.16	11.0	10.0
M10 × 1.5	10.58	9.42	9.08	8.86	6.0	5.0	M24 × 3	24.84	23.16	22.13	21.80	13.0	12.0
M12 × 1.75	12.70	11.30	10.95	10.68	7.0	6.0	…	…	…	…	…	…	…
Metric Heavy Hex Bolts (ANSI B18.2.3.6M-1979)													
M12 × 1.75	12.70	11.30	10.95	10.68	7.0	6.0	M20 × 2.5	20.84	19.16	18.49	18.16	11.0	10.0
M14 × 2	14.70	13.30	12.77	12.50	8.0	7.0	M24 × 3	24.84	23.16	22.13	21.80	13.0	12.0
M16 × 2	16.70	15.30	14.77	14.50	9.0	8.0	…	…	…	…	…	…	…
Metric Heavy Hex Flange Screws (ANSI B18.2.3.9M-1984)													
M10 × 1.5	10.00	9.78	9.08	8.86	6.0	5.0	M16 × 2	16.00	15.73	14.77	14.50	9.0	8.0
M12 × 1.75	12.00	11.73	10.95	10.68	7.0	6.0	M20 × 2.5	20.00	19.67	18.49	18.16	11.0	10.0
M14 × 2	14.00	13.73	12.77	12.50	8.0	7.0	…	…	…	…	…	…	…

All dimensions are in millimeters.
* Shoulder is mandatory for formed hex screws, hex flange screws, and heavy hex flange screws. Shoulder is optional for hex bolts and heavy hex bolts.

Hex bolts in sizes M5 through M24 and heavy hex bolts in sizes M12 through M24 are standard only in lengths longer than 150 mm or 10D, whichever is shorter. When shorter lengths of these sizes are ordered, hex cap screws are normally supplied in place of hex bolts and heavy hex screws in place of heavy hex bolts. Hex bolts in sizes M30 and larger and heavy hex bolts in sizes M30 and M36 are standard in all lengths; however, at manufacturer's option, hex cap screws may be substituted for hex bolts and heavy hex screws for heavy hex bolts for any diameter-length combination.

(Continued on page 1318)

Table 10. American National Standard Metric Heavy Hex Structural Bolts (ANSI B18.2.3.7M-1979)

PROPERTY CLASS AND MANUFACTURER'S IDENTIFICATION TO APPEAR ON TOP OF HEAD

OPTIONAL POINT CONSTRUCTION

INCOMPLETE THREAD

FILLET

Nominal Bolt Diam. D, and Thread Pitch	Body Diameter, D_s		Width Across Flats, S		Width Across Corners, E		Head Height, K		Wrenching Height, K_1	Washer Face Diam., D_w	Washer Face Thickness, C		Thread Length, B†		Transition Thread Length, X*
													Bolt Lengths ≤ 100	Bolt Lengths > 100	
	Max	Min	Max	Min	Max	Min	Max	Min	Min	Min	Max	Min	Basic	Basic	Max
M16 × 2	16.70	15.30	27.00	26.16	31.18	29.56	10.75	9.25	6.5	24.9	0.8	0.4	31	38	6.0
M20 × 2.5	20.84	19.16	34.00	33.00	39.26	37.29	13.40	11.60	8.1	31.4	0.8	0.4	36	43	7.5
M22 × 2.5	22.84	21.16	36.00	35.00	41.57	39.55	14.90	13.10	9.2	33.3	0.8	0.4	38	45	7.5
M24 × 3	24.84	23.16	41.00	40.00	47.34	45.20	15.90	14.10	9.9	38.0	0.8	0.4	41	48	9.0
M27 × 3	27.84	26.16	46.00	45.00	53.12	50.85	17.90	16.10	11.3	42.8	0.8	0.4	44	51	9.0
M30 × 3.5	30.84	29.16	50.00	49.00	57.74	55.37	19.75	17.65	12.4	46.5	0.8	0.4	49	56	10.5
M36 × 4	37.00	35.00	60.00	58.80	69.28	66.44	23.55	21.45	15.0	55.9	0.8	0.4	56	63	12.0

All dimensions are in millimeters.

* Transition thread length, X, includes the length of incomplete threads and tolerances on grip gaging length and body length. It is intended for calculation purposes. † Basic thread length, B, is a reference dimension. ‡ $Z_{max} = 1.5 \times$ thread pitch and $Z_{min} = 0.5 \times$ thread pitch.

For additional manufacturing and acceptance specifications, reference should be made to the Standard.

Table 11. Recommended Diameter-Length Combinations for Metric Heavy Hex Structural Bolts

Nominal Length, L	Nominal Diameter and Thread Pitch						
	M16 × 2	M20 × 2.5	M22 × 2.5	M24 × 3	M27 × 3	M30 × 3.5	M36 × 4
45	X	...	...	...	...	...	...
50	X	X	...	...	...	...	...
55	X	X	X	...	...	...	...
60	X	X	X	X	...	...	...
65	X	X	X	X	X	...	...
70	X	X	X	X	X	X	...
75	X	X	X	X	X	X	...
80	X	X	X	X	X	X	X
85	X	X	X	X	X	X	X
90	X	X	X	X	X	X	X
95	X	X	X	X	X	X	X
100	X	X	X	X	X	X	X
110	X	X	X	X	X	X	X
120	X	X	X	X	X	X	X
130	X	X	X	X	X	X	X
140	X	X	X	X	X	X	X
150	X	X	X	X	X	X	X
160	X	X	X	X	X	X	X
170	X	X	X	X	X	X	X
180	X	X	X	X	X	X	X
190	X	X	X	X	X	X	X
200	X	X	X	X	X	X	X
210	X	X	X	X	X	X	X
220	X	X	X	X	X	X	X
230	X	X	X	X	X	X	X
240	X	X	X	X	X	X	X
250	X	X	X	X	X	X	X
260	X	X	X	X	X	X	X
270	X	X	X	X	X	X	X
280	X	X	X	X	X	X	X
290	X	X	X	X	X	X	X
300	X	X	X	X	X	X	X

All dimensions are in millimeters.
Recommended diameter-length combinations are indicated by the symbol X.
Bolts with lengths above the heavy cross lines are threaded full length.

Materials and Mechanical Properties. — Unless otherwise specified, steel metric screws and bolts, with the exception of heavy hex structural bolts, hex lag screws, and socket head cap screws, conform to the requirements specified in SAE J1199 or ASTM F568. Steel heavy hex structural bolts conform to ASTM A325M or ASTM A490M. Alloy steel socket head cap screws conform to ASTM 574M, property class 12.9, where the numeral 12 represents approximately one-hundredth of the minimum tensile strength in megapascals and the decimal .9 approximates the ratio of the minimum yield stress to the minimum tensile stress. This is in accord with ISO designation practice. Screws and bolts of other materials, and all materials for hex lag screws, have properties as agreed upon by the purchaser and the manufacturer.

Except for socket head cap screws, metric screws and bolts are furnished with a natural (as processed) finish, unplated or uncoated unless otherwise specified.

Table 12. American National Standard Metric Hex Lag Screws
(ANSI B18.2.3.8M-1981)

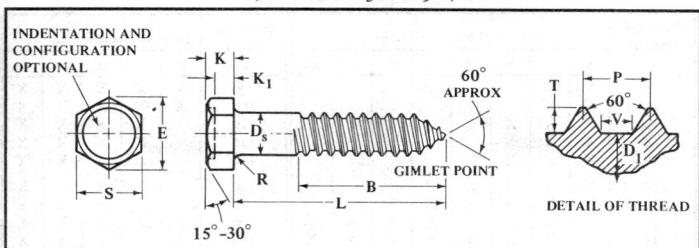

Nominal Screw Diam., D	Body Diameter, D_s		Width Across Flats, S		Width Across Corners, E		Head Height, K		Wrenching Height, K_1
	Max	Min	Max	Min	Max	Min	Max	Min	Min
5	5.48	4.52	8.00	7.64	9.24	8.63	3.9	3.1	2.4
6	6.48	5.52	10.00	9.64	11.55	10.89	4.4	3.6	2.8
8	8.58	7.42	13.00	12.57	15.01	14.20	5.7	4.9	3.7
10	10.58	9.42	16.00	15.57	18.48	17.59	6.9	5.9	4.5
12	12.70	11.30	18.00	17.57	20.78	19.85	8.0	7.0	5.2
16	16.70	15.30	24.00	23.16	27.71	26.17	10.8	9.3	7.0
20	20.84	19.16	30.00	29.16	34.64	32.95	13.4	11.6	8.8
24	24.84	23.16	36.00	35.00	41.57	39.55	15.9	14.1	10.5

Nominal Screw Diam., D	Thread Pitch, P	Flat at Root, V	Depth of Thread, T	Root Diam., D_1	Nominal Screw Diam., D	Thread Pitch, P	Flat at Root, V	Depth of Thread, T	Root Diam., D_1
	Thread Dimensions					Thread Dimensions			
5	2.3	1.0	0.9	3.2	12	4.2	1.8	1.6	8.7
6	2.5	1.1	1.0	4.0	16	5.1	2.2	2.0	12.0
8	2.8	1.2	1.1	5.8	20	5.6	2.4	2.2	15.6
10	3.6	1.6	1.4	7.2	24	7.3	3.1	2.8	18.1

REDUCED BODY DIAMETER

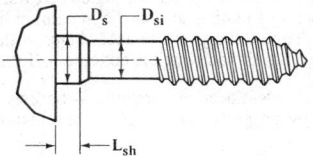

Nominal Screw Diam., D	Shoulder Diameter, D_s		Shoulder Length, L_{sh}		Nominal Screw Diam., D	Shoulder Diameter, D_s		Shoulder Length, L_{sh}	
	Max	Min	Max	Min		Max	Min	Max	Min
5	5.48	4.52	3.5	2.5	12	12.70	11.30	7.0	6.0
6	6.48	5.52	4.0	3.0	16	16.70	15.30	9.0	8.0
8	8.58	7.42	5.0	4.0	20	20.84	19.16	11.0	10.0
10	10.58	9.42	6.0	5.0	24	24.84	23.16	13.0	12.0

All dimensions are in millimeters. Reduced body diameter, D_{si}, is the blank diameter before rolling. Shoulder is mandatory when body diameter is reduced.

**Table 13. American National Standard Metric Hex Lag Screws —
Recommended Diameter-Length Combinations (ANSI B18.2.3.8M-1981)**

Nominal Length, L	Nominal Screw Diameter								Nominal Length, L	Nominal Screw Diameter				
	5	6	8	10	12	16	20	24		10	12	16	20	24
8	X	...	...	...	...	...	...	...	90	X	X	X	X	X
10	X	X	...	...	...	...	...	...	100	X	X	X	X	X
12	X	X	X	...	...	...	...	...	110	...	X	X	X	X
14	X	X	X	...	...	...	...	...	120	...	X	X	X	X
16	X	X	X	X	...	...	...	...	130	...	...	X	X	X
20	X	X	X	X	X	...	...	...	140	...	...	X	X	X
25	X	X	X	X	X	X	...	...	150	...	...	X	X	X
30	X	X	X	X	X	X	X	...	160	...	...	X	X	X
35	X	X	X	X	X	X	X	X	180	...	...	...	X	X
40	X	X	X	X	X	X	X	X	200	...	...	...	X	X
45	X	X	X	X	X	X	X	X	220	...	...	...	...	X
50	X	X	X	X	X	X	X	X	240	...	...	...	...	X
60	...	X	X	X	X	X	X	X	260	...	...	...	...	X
70	...	...	X	X	X	X	X	X	280	...	...	...	...	X
80	...	...	X	X	X	X	X	X	300	...	...	...	...	X

All dimensions are in millimeters.
Recommended diameter-length combinations are indicated by the symbol X.

Alloy steel socket head cap screws are furnished with an oiled black oxide coating (thermal or chemical) unless a protective plating or coating is specified by the purchaser.

Metric Screw and Bolt Thread Series. — Unless otherwise specified, metric screws and bolts, except for hex lag screws, are furnished with metric coarse threads conforming to the dimensions for general purpose threads given in ANSI B1.13M (see Metric Screw Threads in Index). Except for socket head cap screws, the tolerance class is 6g, which applies to plain finish (unplated or uncoated) screws or bolts and to plated or coated screws or bolts before plating or coating. For screws with additive finish, the 6g diameters may be exceeded by the amount of the allowance, i.e. the basic diameters apply to the screws or bolts after plating or coating. For socket head cap screws, the tolerance class is 4g6g, but for plated screws, the allowance g may be consumed by the thickness of plating so that the maximum limit of size after plating is tolerance class 4h6h. Thread limits are in accordance with ANSI B1.13M. Metric hex lag screws have a special thread which is covered in Table 12.

Metric Screw and Bolt Identification Symbols. — Screws and bolts are identified on the top of the head by property class symbols and manufacturer's identification symbol.

Metric Screw and Bolt Clearance Holes. — Clearance holes for screws and bolts with the exception of hex lag screws, socket head cap screws, and round head square neck bolts are given in Table 8. Clearance holes for round head square neck bolts are given in Table 18 and drill and counterbore sizes for socket head cap screws are given in Table 24.

Metric Screw and Bolt Designation. — Metric screws and bolts with the exception of socket head cap screws are designated by the following data, preferably in the sequence shown: product name, nominal diameter and thread pitch (except for hex

(Continued on page 1327)

Table 14. American National Standard Metric Heavy Hex Flange Screws
(ANSI B18.2.3.9M-1984)

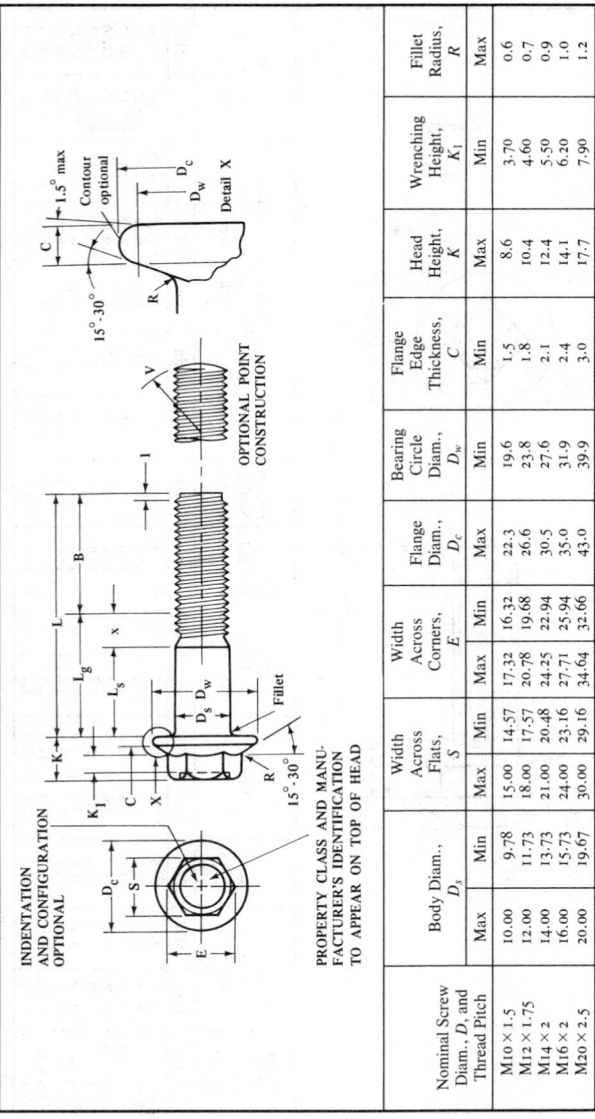

Nominal Screw Diam., D, and Thread Pitch	Body Diam., D_s		Width Across Flats, S		Width Across Corners, E		Flange Diam., D_c	Bearing Circle Diam., D_w	Flange Edge Thickness, C	Head Height, K	Wrenching Height, K_1	Fillet Radius, R
	Max	Min	Max	Min	Max	Min	Max	Min	Min	Max	Min	Max
M10 × 1.5	10.00	9.78	15.00	14.57	17.32	16.32	22.3	19.6	1.5	8.6	3.70	0.6
M12 × 1.75	12.00	11.73	18.00	17.57	20.78	19.68	26.6	23.8	1.8	10.4	4.60	0.7
M14 × 2	14.00	13.73	21.00	20.48	24.25	22.94	30.5	27.6	2.1	12.4	5.50	0.9
M16 × 2	16.00	15.73	24.00	23.16	27.71	25.94	35.0	31.9	2.4	14.1	6.20	1.0
M20 × 2.5	20.00	19.67	30.00	29.16	34.64	32.66	43.0	39.9	3.0	17.7	7.90	1.2

All dimensions are in millimeters. Basic thread lengths, B, are as given in Table 6. Transition thread length, x, includes the length of incomplete threads and tolerances on grip gaging length and body length. It is intended for calculation purposes. For additional manufacturing and acceptance specifications, reference should be made to the Standard.

Table 15. American National Standard Metric Round Head Square Neck Bolts (ANSI B18.5.2.2M-1982)

Nominal Bolt Diam., D and Thread Pitch	Diameter of Full Body, D_s		Head Radius, (R_k)	Head Height, K		Head Edge Thickness, C		Head Diam., D_c	Bearing Surface Diam., D_w	Square Depth, F		Square Corner Depth, F_1	Square Width Across Flats, V		Square Width Across Corners, E	
	Max	Min	Ref.	Max	Min	Max	Min	Max	Min	Max	Min	Min	Max	Min	Max	Min
M5 × 0.8	5.48	4.52	8.8	3.1	2.5	1.8	1.0	11.8	9.8	3.1	2.5	1.6	5.48	4.88	7.75	6.34
M6 × 1	6.48	5.52	10.7	3.6	3.0	1.9	1.1	14.2	12.2	3.6	3.0	1.9	6.48	5.88	9.16	7.64
M8 × 1.25	8.58	7.42	12.5	4.8	4.0	2.2	1.2	18.0	15.8	4.8	4.0	2.5	8.58	7.85	12.13	10.20
M10 × 1.5	10.58	9.42	15.5	5.8	5.0	2.5	1.5	22.3	19.6	5.8	5.0	3.2	10.58	9.85	14.96	12.80
M12 × 1.75	12.70	11.30	19.0	6.8	6.0	2.8	1.8	26.6	23.8	6.8	6.0	3.8	12.70	11.82	17.96	15.37
M14 × 2	14.70	13.30	21.9	7.9	7.0	3.3	2.1	30.5	27.6	7.9	7.0	4.4	14.70	13.82	20.79	17.97
M16 × 2	16.70	15.30	25.5	8.9	8.0	3.6	2.4	35.0	31.9	8.9	8.0	5.0	16.70	15.82	23.62	20.57
M20 × 2.5	20.84	19.16	31.9	10.9	10.0	4.2	3.0	43.0	39.9	10.9	10.0	6.3	20.84	19.79	29.47	25.73
M24 × 3	24.84	23.16	37.9	13.1	12.0	5.1	3.6	51.0	47.6	13.1	12.0	7.6	24.84	23.79	35.13	30.93

All dimensions are in millimeters.
† L_g is the grip gaging length which controls the length of thread B.
‡ B is the basic thread length and is a reference dimension (see Table 16).
For additional manufacturing and acceptance specifications, see Standard.

Table 16. Basic Thread Lengths for Metric Round Head Square Neck Bolts
(ANSI B18.5.2.2M-1982)

Nom. Bolt Diam., D and Thread Pitch	Bolt Length, L			Nom. Bolt Diam., D and Thread Pitch	Bolt Length, L		
	≤ 125	> 125 and ≤ 200	> 200		≤ 125	> 125 and ≤ 200	> 200
	Basic Thread Length, B				Basic Thread Length, B		
M5 × 0.8	16	22	35	M14 × 2	34	40	53
M6 × 1	18	24	37	M16 × 2	38	44	57
M8 × 1.25	22	28	41	M20 × 2.5	46	52	65
M10 × 1.5	26	32	45	M24 × 3	54	60	73
M12 × 1.75	30	36	49	...	...	...	...

All dimensions are in millimeters.
Basic thread length B is a reference dimension intended for calculation purposes only.

Table 17. Recommended Diameter-Length Combinations for Metric Round Head Square Neck Bolts

Nominal Length,* L	Nominal Diameter and Thread Pitch								
	M5 × 0.8	M6 × 1	M8 × 1.25	M10 × 1.5	M12 × 1.75	M14 × 2	M16 × 2	M20 × 2.5	M24 × 3
10	X	...	...	...	...	...	...	...	...
12	X	X	...	...	...	...	...	...	...
(14)	X	X	...	...	...	...	...	...	...
16	X	X	X	...	...	...	...	...	...
20	X	X	X	X	...	...	...	...	...
25	X	X	X	X	X	...	...	...	...
30	X	X	X	X	X	X	X	...	...
35	X	X	X	X	X	X	X	...	...
40	X	X	X	X	X	X	X	X	...
45	X	X	X	X	X	X	X	X	X
50	X	X	X	X	X	X	X	X	X
(55)	...	X	X	X	X	X	X	X	X
60	...	X	X	X	X	X	X	X	X
(65)	...	...	X	X	X	X	X	X	X
70	...	...	X	X	X	X	X	X	X
(75)	...	...	X	X	X	X	X	X	X
80	...	...	X	X	X	X	X	X	X
(85)	...	...	...	X	X	X	X	X	X
90	...	...	...	X	X	X	X	X	X
100	...	...	...	X	X	X	X	• X	X
110	...	...	...	...	X	X	X	X	X
120	...	...	...	...	X	X	X	X	X
130	...	...	...	...	...	X	X	X	X
140	...	...	...	...	...	X	X	X	X
150	...	...	...	...	...	...	X	X	X
160	...	...	...	...	...	...	X	X	X
(170)	...	...	...	...	...	...	...	X	X
180	...	...	...	...	...	...	...	X	X
(190)	...	...	...	...	...	...	...	X	X
200	...	...	...	...	...	...	...	X	X
220	...	...	...	...	...	...	...	...	X
240	...	...	...	...	...	...	...	...	X

All dimensions are in millimeters.   Recommended diameter-length combinations are indicated by the symbol X. Standard sizes for government use.
* Bolts with lengths above the heavy cross lines are threaded full length. Lengths in ( ) are not recommended.

**Table 18. American National Standard Metric Round Head Square Neck Bolts
Reduced Body Diameters** (ANSI B18.5.2.2M-1982)

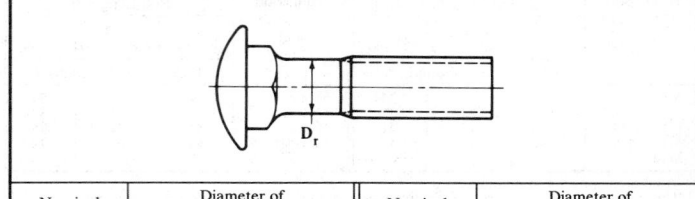

Nominal Bolt Diam., D and Thread Pitch	Diameter of Reduced Body D_r		Nominal Bolt Diam., D and Thread Pitch	Diameter of Reduced Body D_r	
	Max	Min		Max	Min
M5 × 0.8	5.00	4.36	M14 × 2	14.00	12.50
M6 × 1	6.00	5.21	M16 × 2	16.00	14.50
M8 × 1.25	8.00	7.04	M20 × 2.5	20.00	18.16
M10 × 1.5	10.00	8.86	M24 × 3	24.00	21.80
M12 × 1.75	12.00	10.68	. . .	. . .	. . .

All dimensions are in millimeters.

Table 19. Recommended Clearance Holes for Metric Round Head Square Neck Bolts

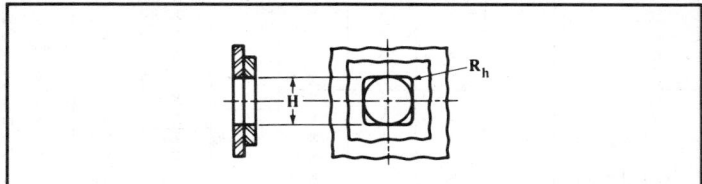

Nom. Bolt Diam., D and Thd. Pitch	Clearance			Corner Radius R_h	Nom. Bolt Diam., D and Thd. Pitch	Clearance			Corner Radius R_h
	Close*	Normal†	Loose‡			Close*	Normal†	Loose‡	
	Minimum Hole Diameter or Square Width, H					Minimum Hole Diameter or Square Width, H			
M5 × 0.8	5.5	. . .	5.8	0.2	M14 × 2	15.0	15.5	16.5	0.6
M6 × 1	6.6	. . .	7.0	0.3	M16 × 2	17.0	17.5	18.5	0.6
M8 × 1.25	. . .	9.0	10.0	0.4	M20 × 2.5	21.0	22.0	24.0	0.8
M10 × 1.5	. . .	11.0	12.0	0.4	M24 × 3	25.0	26.0	28.0	1.0
M12 × 1.75	13.0	13.5	14.5	0.6	. . .	. . .	. . .	. . .	. . .

All dimensions are in millimeters.

* *Close Clearance:* Close clearance should be specified only for square holes in very thin and/
or soft material, or for slots, or where conditions such as critical alignment of assembled parts,
wall thickness, or other limitations necessitate use of a minimal hole. Allowable swell or fins on
the bolt body and/or fins on the corners of the square neck may interfere with close clearance round
or square holes.

† *Normal Clearance:* Normal clearance hole sizes are preferred for general purpose applications
and should be specified unless special design considerations dictate the need for either a close or
loose clearance hole.

‡ *Loose Clearance:* Loose clearance hole sizes should be specified only for applications where
maximum adjustment capability between components being assembled is necessary. Loose clearance square hole or slots may not prevent bolt turning during wrenching.

Table 20. American National Standard Socket Head Cap Screws—Metric Series
(ANSI B18.3.1M-1986)

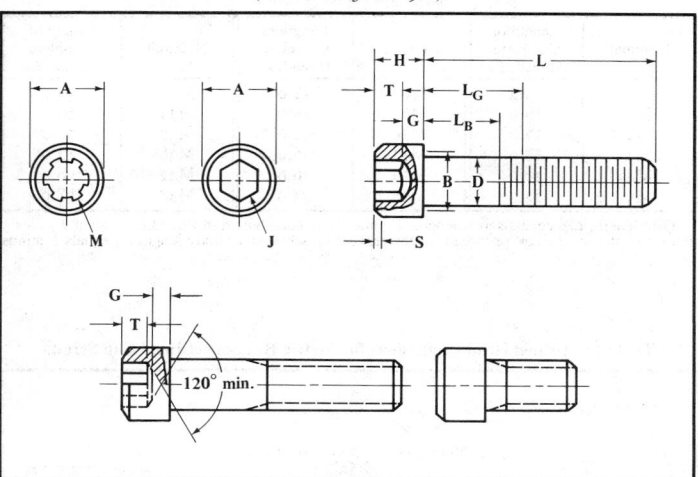

Nom. Size and Thread Pitch	Body Diameter, D		Head Diameter A		Head Height H		Chamfer or Radius S	Hexagon Socket Size‡ J	Spline Socket Size‡ M	Key Engagement T	Transition Diam. B†
	Max	Min	Max	Min	Max	Min	Max	Nom.	Nom.	Min	Max
M1.6 × 0.35	1.60	1.46	3.00	2.87	1.60	1.52	0.16	1.5	1.829	0.80	2.0
M2 × 0.4	2.00	1.86	3.80	3.65	2.00	1.91	0.20	1.5	1.829	1.00	2.6
M2.5 × 0.45	2.50	2.36	4.50	4.33	2.50	2.40	0.25	2.0	2.438	1.25	3.1
M3 × 0.5	3.00	2.86	5.50	5.32	3.00	2.89	0.30	2.5	2.819	1.50	3.6
M4 × 0.7	4.00	3.82	7.00	6.80	4.00	3.88	0.40	3.0	3.378	2.00	4.7
M5 × 0.8	5.00	4.82	8.50	8.27	5.00	4.86	0.50	4.0	4.648	2.50	5.7
M6 × 1	6.00	5.82	10.00	9.74	6.00	5.85	0.60	5.0	5.486	3.00	6.8
M8 × 1.25	8.00	7.78	13.00	12.70	8.00	7.83	0.80	6.0	7.391	4.00	9.2
M10 × 1.5	10.00	9.78	16.00	15.67	10.00	9.81	1.00	8.0	. . .	5.00	11.2
M12 × 1.75	12.00	11.73	18.00	17.63	12.00	11.79	1.20	10.0	. . .	6.00	14.2
M14 × 2*	14.00	13.73	21.00	20.60	14.00	13.77	1.40	12.0	. . .	7.00	16.2
M16 × 2	16.00	15.73	24.00	23.58	16.00	15.76	1.60	14.0	. . .	8.00	18.2
M20 × 2.5	20.00	19.67	30.00	29.53	20.00	19.73	2.00	17.0	. . .	10.00	22.4
M24 × 3	24.00	23.67	36.00	35.48	24.00	23.70	2.40	19.0	. . .	12.00	26.4
M30 × 3.5	30.00	29.67	45.00	44.42	30.00	29.67	3.00	22.0	. . .	15.00	33.4
M36 × 4	36.00	35.61	54.00	53.37	36.00	35.64	3.60	27.0	. . .	18.00	39.4
M42 × 4.5	42.00	41.61	63.00	62.31	42.00	41.61	4.20	32.0	. . .	21.00	45.6
M48 × 5	48.00	47.61	72.00	71.27	48.00	47.58	4.80	36.0	. . .	24.00	52.6

All dimensions are in millimeters
* The M14 × 2 size is not recommended for use in new designs.
† See Countersink footnote in Table 25. ‡ See also Table 23.
L_G is grip length and L_B is body length (see Table 21).
For length of complete thread, see Table 21.
For additional manufacturing and acceptance specifications, see Standard.

Table 21. Socket Head Cap Screws (Metric Series) — Length of Complete Thread
(ANSI B18.3.1M-1986)

Nominal Size	Length of Complete Thread, L_T	Nominal Size	Length of Complete Thread, L_T	Nominal Size	Length of Complete Thread, L_T
M1.6	15.2	M6	24.0	M20	52.0
M2	16.0	M8	28.0	M24	60.0
M2.5	17.0	M10	32.0	M30	72.0
M3	18.0	M12	36.0	M36	84.0
M4	20.0	M14	40.0	M42	96.0
M5	22.0	M16	44.0	M48	108.0

Grip length, L_G, equals screw length, L, minus L_T. Total length of thread L_{TT} equals L_T plus 5 times the pitch of the coarse thread for the respective screw size. Body length L_B equals L minus L_{TT}.

Table 22. Drilled Head Dimensions for Metric Hex Socket Head Cap Screws

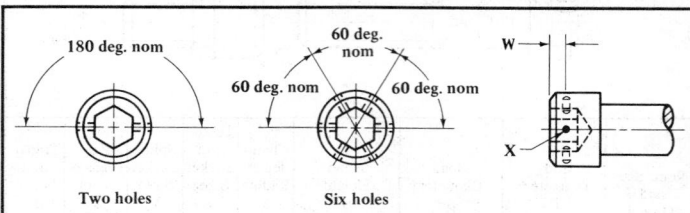

| | Two holes | | Six holes | | |

Nominal Size or Basic Screw Diameter	Hole Center Location, W		Drilled Hole Diameter, X		Hole Alignment Check Plug Diameter
	Max	Min	Max	Min	Basic
M3	1.20	0.80	0.95	0.80	0.75
M4	1.60	1.20	1.35	1.20	0.90
M5	2.00	1.50	1.35	1.20	0.90
M6	2.30	1.80	1.35	1.20	0.90
M8	2.70	2.20	1.35	1.20	0.90
M10	3.30	2.80	1.65	1.50	1.40
M12	4.00	3.50	1.65	1.50	1.40
M16	5.00	4.50	1.65	1.50	1.40
M20	6.30	5.80	2.15	2.00	1.80
M24	7.30	6.80	2.15	2.00	1.80
M30	9.00	8.50	2.15	2.00	1.80
M36	10.50	10.00	2.15	2.00	1.80

All dimensions are in millimeters.

Drilled head metric hexagon socket head cap screws normally are not available in screw sizes smaller than M3 nor larger than M36. The M3 and M4 nominal screw sizes have two drilled holes spaced 180 degrees apart. Nominal screw sizes M5 and larger have six drilled holes spaced 60 degrees apart unless the purchaser specifies two drilled holes. The positioning of holes on opposite sides of the socket should be such that the hole alignment check plug will pass completely through the head without any deflection. When so specified by the purchaser, the edges of holes on the outside surface of the head will be chamfered 45 degrees to a depth of 0.30 to 0.50 mm.

Table 23. American National Standard Hexagon and Spline Sockets for Socket Head Cap Screws — Metric Series (ANSI B18.3.1M-1986)

METRIC HEXAGON SOCKET   METRIC SPLINE SOCKET

Metric Hexagon Sockets

Nominal Hexagon Socket Size	Socket Width Across Flats, J		Socket Width Across Corners, C	Nominal Hexagon Socket Size	Socket Width Across Flats, J		Socket Width Across Corners, C
	Max	Min	Min		Max	Min	Min
1.5	1.545	1.520	1.73	12	12.146	12.032	13.80
2	2.045	2.020	2.30	14	14.159	14.032	16.09
2.5	2.560	2.520	2.87	17	17.216	17.050	19.56
3	3.071	3.020	3.44	19	19.243	19.065	21.87
4	4.084	4.020	4.58	22	22.319	22.065	25.31
5	5.084	5.020	5.72	24	24.319	24.065	27.60
6	6.095	6.020	6.86	27	27.319	27.065	31.04
8	8.115	8.025	9.15	32	32.461	32.080	36.80
10	10.127	10.025	11.50	36	36.461	36.080	41.38

Metric Spline Sockets†

Nominal Spline Socket Size	Socket Major Diameter, M		Socket Minor Diameter, N		Width of Tooth, P	
	Max	Min	Max	Min	Max	Min
1.829	1.8976	1.8542	1.6256	1.6002	0.4064	0.3810
2.438	2.4892	2.4638	2.0828	2.0320	0.5588	0.5334
2.819	2.9210	2.8702	2.4892	2.4384	0.6350	0.5842
3.378	3.4798	3.4290	2.9972	2.9464	0.7620	0.7112
4.648	4.7752	4.7244	4.1402	4.0894	0.9906	0.9398
5.486	5.6134	5.5626	4.8260	4.7752	1.2700	1.2192
7.391	7.5692	7.5184	6.4516	6.4008	1.7272	2.6764

All dimensions are in millimeters.

† The tabulated dimensions represent direct metric conversions of the equivalent inch size spline sockets shown in American National Standard Socket Cap, Shoulder and Set Screws — Inch Series (ANSI B18.3). Therefore, the spline keys and bits shown therein are applicable for wrenching the corresponding size metric spline sockets.

lag screws), nominal length, steel property class or material identification, and protective coating, if required.

Examples: Hex cap screw, M10 × 1.5 × 50, class 9.8, zinc plated
Heavy hex structural bolt, M24 × 3 × 80, ASTM A490M
Hex lag screw, 6 × 35, silicon bronze.

Socket head cap screws (metric series) are designated by the following data in the order shown: ANSI Standard number, nominal size, thread pitch, nominal screw length, name of product (may be abbreviated SHCS), material and property class (alloy steel screws are supplied to property class 12.9 as specified in ASTM A574M;

Table 24. Diameter-Length Combinations for Socket Head Cap Screws (Metric Series)

Nominal Length, L	Nominal Size													
	M1.6	M2	M2.5	M3	M4	M5	M6	M8	M10	M12	M14	M16	M20	M24
20	X	X												
25	X	X	X	X										
30	X	X	X	X	X	X								
35	⋯	X	X	X	X	X	X							
40	⋯	X	X	X	X	X	X							
45	⋯	⋯	X	X	X	X	X	X						
50	⋯	⋯	X	X	X	X	X	X	X					
55	⋯	⋯	⋯	X	X	X	X	X	X					
60	⋯	⋯	⋯	X	X	X	X	X	X	X				
65	⋯	⋯	⋯	X	X	X	X	X	X	X				
70	⋯	⋯	⋯	X	X	X	X	X	X	X	X			
80	⋯	⋯	⋯	⋯	X	X	X	X	X	X	X	X		
90	⋯	⋯	⋯	⋯	X	X	X	X	X	X	X	X	X	
100	⋯	⋯	⋯	⋯	⋯	X	X	X	X	X	X	X	X	X
110	⋯	⋯	⋯	⋯	⋯	X	X	X	X	X	X	X	X	X
120	⋯	⋯	⋯	⋯	⋯	⋯	X	X	X	X	X	X	X	X
130	⋯	⋯	⋯	⋯	⋯	⋯	⋯	X	X	X	X	X	X	X
140	⋯	⋯	⋯	⋯	⋯	⋯	⋯	X	X	X	X	X	X	X
150	⋯	⋯	⋯	⋯	⋯	⋯	⋯	X	X	X	X	X	X	X
160	⋯	⋯	⋯	⋯	⋯	⋯	⋯	X	X	X	X	X	X	X
180	⋯	⋯	⋯	⋯	⋯	⋯	⋯	⋯	X	X	X	X	X	X
200	⋯	⋯	⋯	⋯	⋯	⋯	⋯	⋯	⋯	X	X	X	X	X
220	⋯	⋯	⋯	⋯	⋯	⋯	⋯	⋯	⋯	⋯	X	X	X	X
240	⋯	⋯	⋯	⋯	⋯	⋯	⋯	⋯	⋯	⋯	⋯	X	X	X
260	⋯	⋯	⋯	⋯	⋯	⋯	⋯	⋯	⋯	⋯	⋯	⋯	X	X
300	⋯	⋯	⋯	⋯	⋯	⋯	⋯	⋯	⋯	⋯	⋯	⋯	⋯	X

All dimensions are in millimeters. Screws with lengths above heavy cross lines are threaded full length. Diameter-length combinations are indicated by the symbol X. Standard sizes for government use. In addition to the lengths shown, the following lengths are standard: 3, 4, 5, 6, 8, 10, 12, and 16 mm. No diameter-length combinations are given in the Standard for these lengths. Screws larger than M24 with lengths equal to or shorter than L_{TT} (see Table 21 footnote) are threaded full length.

Table 24. Drill and Counterbore Sizes for Metric Socket Head Cap Screws

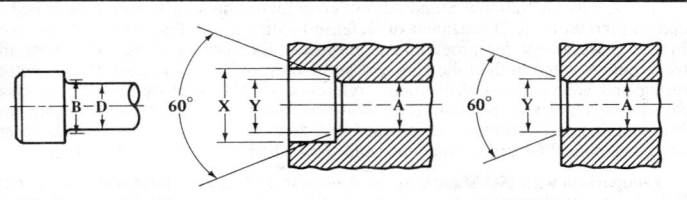

Nominal Size or Basic Screw Diameter	Nominal Drill Size, A		Counterbore Diameter, X	Countersink Diameter,‡ Y
	Close Fit*	Normal Fit†		
M1.6	1.80	1.95	3.50	2.0
M2	2.20	2.40	4.40	2.6
M2.5	2.70	3.00	5.40	3.1
M3	3.40	3.70	6.50	3.6
M4	4.40	4.80	8.25	4.7
M5	5.40	5.80	9.75	5.7
M6	6.40	6.80	11.25	6.8
M8	8.40	8.80	14.25	9.2
M10	10.50	10.80	17.25	11.2
M12	12.50	12.80	19.25	14.2
M14	14.50	14.75	22.25	16.2
M16	16.50	16.75	25.50	18.2
M20	20.50	20.75	31.50	22.4
M24	24.50	24.75	37.50	26.4
M30	30.75	31.75	47.50	33.4
M36	37.00	37.50	56.50	39.4
M42	43.00	44.00	66.00	45.6
M48	49.00	50.00	75.00	52.6

All dimensions are in millimeters.

* *Close Fit:* The close fit is normally limited to holes for those lengths of screws which are threaded to the head in assemblies where only one screw is to be used or where two or more screws are to be used and the mating holes are to be produced either at assembly or by matched and coordinated tooling.

† *Normal Fit:* The normal fit is intended for screws of relatively long length or for assemblies involving two or more screws where the mating holes are to be produced by conventional tolerancing methods. It provides for the maximum allowable eccentricity of the longest standard screws and for certain variations in the parts to be fastened, such as: deviations in hole straightness, angularity between the axis of the tapped hole and that of the hole for shank, differences in center distances of the mating holes, etc.

‡ *Countersink:* It is considered good practice to countersink or break the edges of holes which are smaller than *B* Max. (see Table 20) in parts having a hardness which approaches, equals, or exceeds the screw hardness. If such holes are not countersunk, the heads of screws may not seat properly or the sharp edges on holes may deform the fillets on screws, thereby making them susceptible to fatigue in applications involving dynamic loading. The countersink or corner relief, however, should not be larger than is necessary to ensure that the fillet on the screw is cleared. Normally, the diameter of countersink does not have to exceed *B* Max. Countersinks or corner reliefs in excess of this diameter reduce the effective bearing area and introduce the possibility of embedment where the parts to be fastened are softer than the screws or of brinnelling or flaring the heads of the screws where the parts to be fastened are harder than the screws.

corrosion-resistant steel screws are specified to the property class and material requirements in ASTM F837M), and protective finish, if required.

Examples: B18.3.1M—6 × 1 × 20 Hexagon Socket Head Cap Screw, Alloy Steel
B18.3.1M—10 × 1.5 × 40 SHCS, Alloy Steel Zinc Plated.

Metric Nuts

The American National Standards covering metric nuts have been established in cooperation with the Department of Defense in such a way that they could be used by the Government for procurement purposes. Extensive information concerning these nuts is given in the following text and tables, but for more complete manufacturing and acceptance specifications, reference should be made to the respective Standards, which may be obtained by non-governmental agencies from the American National Standards Institute, 1430 Broadway, New York, N.Y. 10018. Manufacturers should be consulted concerning items and sizes which are in stock production.

Comparison with ISO Standards. — American National Standards for metric nuts have been coordinated to the extent possible with comparable ISO Standards or proposed Standards, thus: ANSI B18.2.4.1M Metric Hex Nuts, Style 1 with ISO 4032; B18.2.4.2M Metric Hex Nuts, Style 2 with ISO 4033; B18.2.4.4M Metric Hex Flange Nuts with ISO 4161; B18.2.4.5M Metric Hex Jam Nuts with ISO 4035; and B18.2.4.3M Metric Slotted Hex Nuts, B18.2.4.6M Metric Heavy Hex Nuts in sizes M12 through M36, and B18.16.3M Prevailing-Torque Type Steel Metric Hex Nuts and Hex Flange Nuts with comparable draft ISO Standards. The dimensional differences between each ANSI Standard and the comparable ISO Standard or draft Standard are very few, relatively minor, and none will affect the interchangeability of nuts manufactured to the requirements of either.

At its meeting in Varna, May 1977, ISO/TC2 studied several technical reports analyzing design considerations influencing determination of the best series of widths across flats for hex bolts, screws, and nuts. A primary technical objective was to achieve a logical ratio between under head (nut) bearing surface area (which determines the magnitude of compressive stress on the bolted members) and the tensile stress area of the screw thread (which governs the clamping force that can be developed by tightening the fastener). The series of widths across flats in the ANSI Standards agree with those which were selected by ISO/TC2 to be ISO Standards.

One exception for width across flats of metric hex nuts, styles 1 and 2, metric slotted hex nuts, metric hex jam nuts, and prevailing-torque metric hex nuts is the M10 size. These nuts in M10 size are currently being produced in the United States with a width across flats of 15 mm. This width, however, is not an ISO Standard. Unless these M10 nuts with width across flats of 15 mm are specifically ordered, the M10 size with 16 mm width across flats will be furnished.

In ANSI Standards for metric nuts, letter symbols designating dimensional characteristics are in accord with those used in ISO Standards, except capitals have been used for data processing convenience instead of lower case letters used in ISO Standards.

Metric Nut Tops and Bearing Surfaces. — Metric hex nuts, styles 1 and 2, slotted hex nuts, and hex jam nuts are double chamfered in sizes M16 and smaller and in sizes M20 and larger may either be double chamfered or have a washer-faced bearing surface and a chamfered top at the option of the manufacturer. Metric heavy hex nuts are optional either way in all sizes. Metric hex flange nuts have a flange bearing surface and a chamfered top and prevailing-torque type metric hex nuts have a chamfered bearing surface. Prevailing-torque type metrix hex flange nuts have a flange bearing surface. All types of metric nuts have the tapped hole countersunk on the bearing face and metric slotted hex nuts, hex flange nuts, and prevailing-torque type hex nuts and hex flange nuts may be countersunk on the top face.

Materials and Mechanical Properties. — Nonheat-treated carbon steel metric hex nuts, style 1 and slotted hex nuts conform to material and property class requirements specified for property class 5 nuts; hex nuts, style 2 and hex flange nuts to property class 9 nuts; hex jam nuts to property class 04 nuts, and nonheat-treated carbon

Table 26. American National Standard Metric Hex Nuts, Styles 1 and 2
(ANSI B18.2.4.1M and B18.2.4.2M-1979)

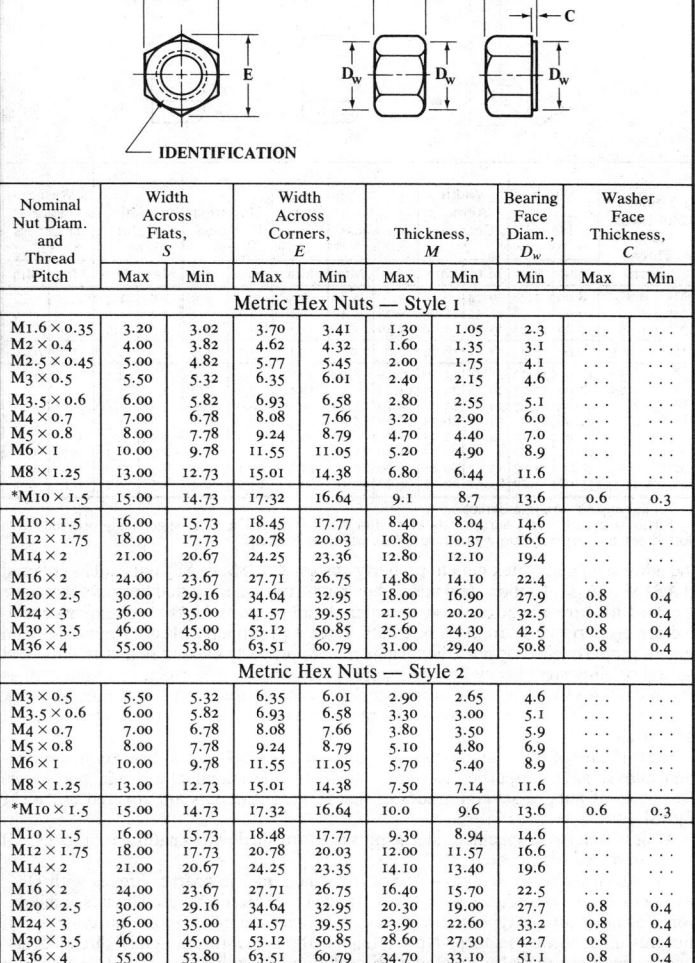

— IDENTIFICATION

Nominal Nut Diam. and Thread Pitch	Width Across Flats, S		Width Across Corners, E		Thickness, M		Bearing Face Diam., D_w	Washer Face Thickness, C	
	Max	Min	Max	Min	Max	Min	Min	Max	Min
Metric Hex Nuts — Style 1									
M1.6 × 0.35	3.20	3.02	3.70	3.41	1.30	1.05	2.3	...	...
M2 × 0.4	4.00	3.82	4.62	4.32	1.60	1.35	3.1	...	...
M2.5 × 0.45	5.00	4.82	5.77	5.45	2.00	1.75	4.1	...	...
M3 × 0.5	5.50	5.32	6.35	6.01	2.40	2.15	4.6	...	...
M3.5 × 0.6	6.00	5.82	6.93	6.58	2.80	2.55	5.1	...	...
M4 × 0.7	7.00	6.78	8.08	7.66	3.20	2.90	6.0	...	...
M5 × 0.8	8.00	7.78	9.24	8.79	4.70	4.40	7.0	...	...
M6 × 1	10.00	9.78	11.55	11.05	5.20	4.90	8.9	...	...
M8 × 1.25	13.00	12.73	15.01	14.38	6.80	6.44	11.6	...	...
*M10 × 1.5	15.00	14.73	17.32	16.64	9.1	8.7	13.6	0.6	0.3
M10 × 1.5	16.00	15.73	18.45	17.77	8.40	8.04	14.6	...	...
M12 × 1.75	18.00	17.73	20.78	20.03	10.80	10.37	16.6	...	...
M14 × 2	21.00	20.67	24.25	23.36	12.80	12.10	19.4	...	...
M16 × 2	24.00	23.67	27.71	26.75	14.80	14.10	22.4	...	...
M20 × 2.5	30.00	29.16	34.64	32.95	18.00	16.90	27.9	0.8	0.4
M24 × 3	36.00	35.00	41.57	39.55	21.50	20.20	32.5	0.8	0.4
M30 × 3.5	46.00	45.00	53.12	50.85	25.60	24.30	42.5	0.8	0.4
M36 × 4	55.00	53.80	63.51	60.79	31.00	29.40	50.8	0.8	0.4
Metric Hex Nuts — Style 2									
M3 × 0.5	5.50	5.32	6.35	6.01	2.90	2.65	4.6	...	...
M3.5 × 0.6	6.00	5.82	6.93	6.58	3.30	3.00	5.1	...	...
M4 × 0.7	7.00	6.78	8.08	7.66	3.80	3.50	5.9	...	...
M5 × 0.8	8.00	7.78	9.24	8.79	5.10	4.80	6.9	...	...
M6 × 1	10.00	9.78	11.55	11.05	5.70	5.40	8.9	...	...
M8 × 1.25	13.00	12.73	15.01	14.38	7.50	7.14	11.6	...	...
*M10 × 1.5	15.00	14.73	17.32	16.64	10.0	9.6	13.6	0.6	0.3
M10 × 1.5	16.00	15.73	18.48	17.77	9.30	8.94	14.6	...	...
M12 × 1.75	18.00	17.73	20.78	20.03	12.00	11.57	16.6	...	...
M14 × 2	21.00	20.67	24.25	23.35	14.10	13.40	19.6	...	...
M16 × 2	24.00	23.67	27.71	26.75	16.40	15.70	22.5	...	...
M20 × 2.5	30.00	29.16	34.64	32.95	20.30	19.00	27.7	0.8	0.4
M24 × 3	36.00	35.00	41.57	39.55	23.90	22.60	33.2	0.8	0.4
M30 × 3.5	46.00	45.00	53.12	50.85	28.60	27.30	42.7	0.8	0.4
M36 × 4	55.00	53.80	63.51	60.79	34.70	33.10	51.1	0.8	0.4

All dimensions are in millimeters.
* This size with width across flats of 15 mm is not standard. Unless specifically ordered, M10 hex nuts with 16 mm width across flats will be furnished.

Table 27. American National Standard Metric Slotted Hex Nuts
(ANSI B18.2.4.3M-1979)

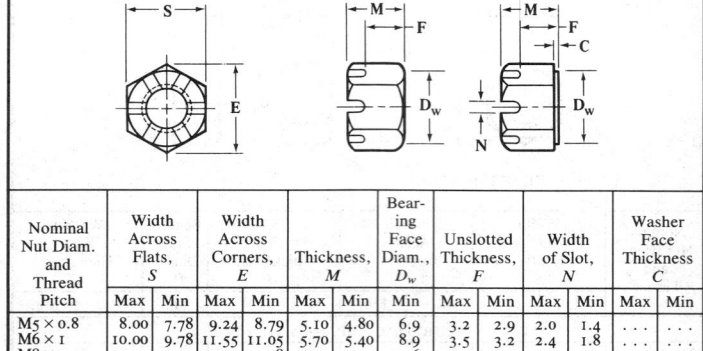

All dimensions are in millimeters.

Nominal Nut Diam. and Thread Pitch	Width Across Flats, S		Width Across Corners, E		Thickness, M		Bearing Face Diam., D_w	Unslotted Thickness, F		Width of Slot, N		Washer Face Thickness, C	
	Max	Min	Max	Min	Max	Min	Min	Max	Min	Max	Min	Max	Min
M5 × 0.8	8.00	7.78	9.24	8.79	5.10	4.80	6.9	3.2	2.9	2.0	1.4	...	...
M6 × 1	10.00	9.78	11.55	11.05	5.70	5.40	8.9	3.5	3.2	2.4	1.8	...	...
M8 × 1.25	13.00	12.73	15.01	14.38	7.50	7.14	11.6	4.4	4.1	2.9	2.3	...	...
*M10 × 1.5	15.00	14.73	17.32	16.64	10.0	9.6	13.6	5.7	5.4	3.4	2.8	0.6	0.3
M10 × 1.5	16.00	15.73	18.48	17.77	9.30	8.94	14.6	5.2	4.9	3.4	2.8	...	...
M12 × 1.75	18.00	17.73	20.78	20.03	12.00	11.57	16.6	7.3	6.9	4.0	3.2	...	...
M14 × 2	21.00	20.67	24.25	23.35	14.10	13.40	19.6	8.6	8.0	4.3	3.5	...	...
M16 × 2	24.00	23.67	27.71	26.75	16.40	15.70	22.5	9.9	9.3	5.3	4.5	...	...
M20 × 2.5	30.00	29.16	34.64	32.95	20.30	19.00	27.7	13.3	12.2	5.7	4.5	0.8	0.4
M24 × 3	36.00	35.00	41.57	39.55	23.90	22.60	33.2	15.4	14.3	6.7	5.5	0.8	0.4
M30 × 3.5	46.00	45.00	53.12	50.85	28.60	27.30	42.7	18.1	16.8	8.5	7.0	0.8	0.4
M36 × 4	55.00	53.80	63.51	60.79	34.70	33.10	51.1	23.7	22.4	8.5	7.0	0.8	0.4

* This size with width across flats of 15 mm is not standard. Unless specifically ordered, M10 slotted hex nuts with 16 mm width across flats will be furnished.

and alloy steel heavy hex nuts to property classes 5, 9, 8S, or 8S3 nuts; all as covered in ASTM A563M. Carbon steel metric hex nuts, style 1 and slotted hex nuts that have specified heat treatment conform to material and property class requirements specified for property class 10 nuts; hex nuts, style 2 to property class 12 nuts; hex jam nuts to property class 05 nuts; hex flange nuts to property classes 10 and 12 nuts; and carbon or alloy steel heavy hex nuts to property classes 10S, 10S3, or 12 nuts, all as covered in ASTM A563M. Carbon steel prevailing-torque type hex nuts and hex flange nuts conform to mechanical and property class requirements as given in ANSI B18.16.1M.

Metric nuts of other materials, such as stainless steel, brass, bronze, and aluminum alloys, have properties as agreed upon by the manufacturer and purchaser. Properties of nuts of several grades of non-ferrous materials are covered in ASTM F467M.

Unless otherwise specified, metric nuts are furnished with a natural (unprocessed) finish, unplated or uncoated.

Metric Nut Thread Series. — Metric nuts have metric coarse threads with class 6H tolerances in accordance with ANSI B1.13M (see Metric Screw Threads in index). For prevailing-torque type metric nuts this condition applies before introduction of the prevailing torque feature. Nuts intended for use with externally threaded fasteners which are plated or coated with a plating or coating thickness (e.g., hot dip galvanized) requiring overtapping of the nut thread to permit assembly, have overtapped threads in conformance with requirements specified in ASTM A563M.

Table 28. American National Standard Metric Hex Flange Nuts
(ANSI B18.2.4.4M-1979)

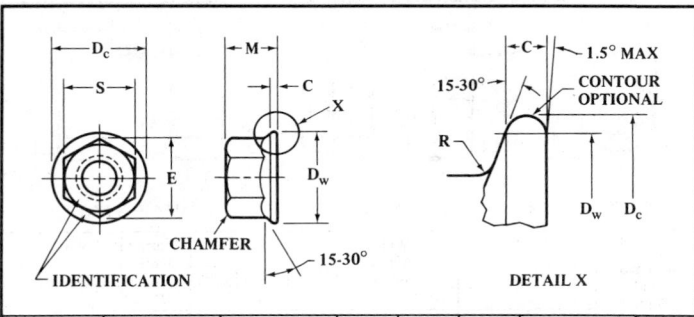

IDENTIFICATION

CHAMFER

15-30°

1.5° MAX

CONTOUR OPTIONAL

DETAIL X

Nominal Nut Diam. and Thread Pitch	Width Across Flats, S		Width Across Corners, E		Flange Diam., D_c	Bearing Circle Diam., D_w	Flange Edge Thickness, C	Thickness, M		Flange Top Fillet Radius, R
	Max	Min	Max	Min	Max	Min	Min	Max	Min	Max
M5 × 0.8	8.00	7.78	9.24	8.79	11.8	9.8	1.0	5.00	4.70	0.3
M6 × 1	10.00	9.78	11.55	11.05	14.2	12.2	1.1	6.00	5.70	0.4
M8 × 1.25	13.00	12.73	15.01	14.38	17.9	15.8	1.2	8.00	7.60	0.5
M10 × 1.5	15.00	14.73	17.32	16.64	21.8	19.6	1.5	10.00	9.60	0.6
M12 × 1.75	18.00	17.73	20.78	20.03	26.0	23.8	1.8	12.00	11.60	0.7
M14 × 2	21.00	20.67	24.25	23.35	29.9	27.6	2.1	14.00	13.30	0.9
M16 × 2	24.00	23.67	27.71	26.75	34.5	31.9	2.4	16.00	15.30	1.0
M20 × 2.5	30.00	29.16	34.64	32.95	42.8	39.9	3.0	20.00	18.90	1.2

All dimensions are in millimeters.

Types of Metric Prevailing-Torque Type Nuts. — There are three basic designs for prevailing-torque type nuts:

1. All-metal, one-piece construction nuts which derive their prevailing-torque characteristics from controlled distortion of the nut thread and/or body.

2. Metal nuts which derive their prevailing-torque characteristics from addition or fusion of a nonmetallic insert, plug. or patch in their threads.

3. Top insert, two-piece construction nuts which derive their prevailing-torque characteristics from an insert, usually a full ring of non-metallic material, located and retained in the nut at its top surface.

The first two designs are designated in Tables 30 and 31 as "all-metal" type and the third design as "top-insert" type.

Metric Nut Identification Symbols. — Carbon steel hex nuts, styles 1 and 2, hex flange nuts, and carbon and alloy steel heavy hex nuts are marked to identify the property class and manufacturer in accordance with requirements specified in ASTM A563M. The aforementioned nuts when made of other materials, as well as slotted hex nuts and hex jam nuts, are marked to identify the property class and manufacturer as agreed upon by manufacturer and purchaser. Carbon steel prevailing-torque type hex nuts and hex flange nuts are marked to identify property class and manufacturer as specified in ANSI B18.16.1M. Prevailing-torque

(Continued on page 1336)

Table 29. American National Standard Metric Hex Jam Nuts and Heavy Hex Nuts
(ANSI B18.2.4.5M and B18.2.4.6M-1979)

IDENTIFICATION   HEX JAM NUTS   HEAVY HEX NUTS

Nominal Nut Diam. and Thread Pitch	Width Across Flats, S		Width Across Corners, E		Thickness, M		Bearing Face Diam., D_w	Washer Face Thickness, C	
	Max	Min	Max	Min	Max	Min	Min	Max	Min
Metric Hex Jam Nuts									
M5 × 0.8	8.00	7.78	9.24	8.79	2.70	2.45	6.9	...	...
M6 × 1	10.00	9.78	11.55	11.05	3.20	2.90	8.9	...	...
M8 × 1.25	13.00	12.73	15.01	14.38	4.00	3.70	11.6	...	...
*M10 × 1.5	15.00	14.73	17.32	16.64	5.00	4.70	13.6	...	...
M10 × 1.5	16.00	15.73	18.48	17.77	5.00	4.70	14.6	...	...
M12 × 1.75	18.00	17.73	20.78	20.03	6.00	5.70	16.6	...	...
M14 × 2	21.00	20.67	24.25	23.35	7.00	6.42	19.6	...	...
M16 × 2	24.00	23.67	27.71	26.75	8.00	7.42	22.5	...	...
M20 × 2.5	30.00	29.16	34.64	32.95	10.00	9.10	27.7	0.8	0.4
M24 × 3	36.00	35.00	41.57	39.55	12.00	10.90	33.2	0.8	0.4
M30 × 3.5	46.00	45.00	53.12	50.85	15.00	13.90	42.7	0.8	0.4
M36 × 4	55.00	53.80	63.51	60.79	18.00	16.90	51.1	0.8	0.4
Metric Heavy Hex Nuts									
M12 × 1.75	21.00	20.16	24.25	22.78	12.3	11.9	19.2	0.8	0.4
M14 × 2	24.00	23.16	27.71	26.17	14.3	13.6	22.0	0.8	0.4
M16 × 2	27.00	26.16	31.18	29.56	17.1	16.4	24.9	0.8	0.4
M20 × 2.5	34.00	33.00	39.26	37.29	20.7	19.4	31.4	0.8	0.4
M22 × 2.5	36.00	35.00	41.57	39.55	23.6	22.3	33.3	0.8	0.4
M24 × 3	41.00	40.00	47.34	45.20	24.2	22.9	38.0	0.8	0.4
M27 × 3	46.00	45.00	53.12	50.85	27.6	26.3	42.8	0.8	0.4
M30 × 3.5	50.00	49.00	57.74	55.37	30.7	29.1	46.6	0.8	0.4
M36 × 4	60.00	58.80	69.28	66.44	36.6	35.0	55.9	0.8	0.4
M42 × 4.5	70.00	67.90	80.83	77.41	42.0	40.4	64.5	1.0	0.5
M48 × 5	80.00	77.60	92.38	88.46	48.0	46.4	73.7	1.0	0.5
M56 × 5.5	90.00	87.20	103.92	99.41	56.0	54.1	82.8	1.0	0.5
M64 × 6	100.00	96.80	115.47	110.35	64.0	62.1	92.0	1.0	0.5
M72 × 6	110.00	106.40	127.02	121.30	72.0	70.1	101.1	1.2	0.6
M80 × 6	120.00	116.00	138.56	132.24	80.0	78.1	110.2	1.2	0.6
M90 × 6	135.00	130.50	155.88	148.77	90.0	87.8	124.0	1.2	0.6
M100 × 6	150.00	145.00	173.21	165.30	100.0	97.8	137.8	1.2	0.6

All dimensions are in millimeters.
* This size with width across flats of 15 mm is not standard. Unless specifically ordered, M10 hex jam nuts with 16 mm width across flats will be furnished.

Table 30. American National Standard Prevailing-Torque Metric Hex Nuts — Property Classes 5, 9, and 10 (ANSI B18.16.3M-1982)

NOTE: SIZE, SHAPE AND LOCATION OF THE PREVAILING-TORQUE ELEMENT OPTIONAL

COUNTERSINK — 15 DEG/30 DEG

Nominal Nut Diam. and Thread Pitch	Width Across Flats, S		Width Across Corners, E		Thickness, M								Wrenching Height, M_1		Bearing Face Diam., D_w
					Property Classes 5 and 10 Nuts				Property Class 9 Nuts				Property Classes 5 and 10 Nuts	Property Class 9 Nuts	
					All Metal* Type		Top Insert Type		All Metal Type		Top Insert Type				
	Max	Min	Max	Min	Max	Min	Max	Min	Max	Min	Max	Min	Min	Min	Min
M3 × 0.5	5.50	5.32	6.35	6.01	3.10	2.65	4.50	3.90	3.10	2.65	4.50	3.90	1.4	1.4	4.6
M3.5 × 0.6	6.00	5.82	6.93	6.58	3.50	3.00	5.00	4.30	3.50	3.00	5.00	4.30	1.7	1.7	5.1
M4 × 0.7	7.00	6.78	8.08	7.66	4.00	3.50	6.00	5.30	4.00	3.50	6.00	5.30	1.9	1.9	5.9
M5 × 0.8	8.00	7.78	9.24	8.79	5.30	4.80	6.80	6.00	5.30	4.80	7.20	6.40	2.7	2.7	6.9
M6 × 1	10.00	9.78	11.55	11.05	5.90	5.40	8.00	7.20	6.70	5.40	8.50	7.70	3.0	3.0	8.9
M8 × 1.25	13.00	12.73	15.01	14.38	7.10	6.44	9.50	8.50	8.00	7.14	10.20	9.20	3.7	4.3	11.6
M10 × 1.5	15.00	14.73	17.32	16.64	9.70	8.70	12.50	11.50	11.20	9.60	13.50	12.50	5.6	6.2	13.6
M10 × 1.5	16.00	15.73	18.48	17.77	9.00	8.04	11.90	10.90	10.50	8.94	12.80	11.80	4.8	5.6	14.6
M12 × 1.75	18.00	17.73	20.78	20.03	11.60	10.37	14.90	13.90	13.30	11.57	16.10	15.10	6.7	7.7	16.6
M14 × 2	21.00	20.67	24.25	23.35	13.20	12.10	17.00	15.80	15.40	13.40	18.30	17.10	7.8	8.9	19.6
M16 × 2	24.00	23.67	27.71	26.75	15.20	14.10	19.10	17.90	17.90	15.70	20.70	19.50	9.1	10.5	22.5
M20 × 2.5	30.00	29.16	34.64	32.95	19.00	16.90	22.80	21.50	21.80	19.00	25.10	23.80	10.9	12.7	27.7
M24 × 3	36.00	35.00	41.57	39.55	23.00	20.20	27.10	25.60	26.40	22.60	29.50	28.00	13.0	15.1	33.2
M30 × 3.5	46.00	45.00	53.12	50.85	26.90	24.30	32.60	30.60	31.80	27.30	35.60	33.60	15.7	18.2	42.7
M36 × 4	55.00	53.80	63.51	60.79	32.50	29.40	38.90	36.90	38.50	33.10	42.60	40.60	19.0	22.1	51.1

All dimensions are in millimeters.
* Also includes metal nuts with non-metallic inserts, plugs, or patches in their threads.

**Table 31. American National Standard Prevailing-Torque
Metric Hex Flange Nuts (ANSI B18.16.3M-1982)**

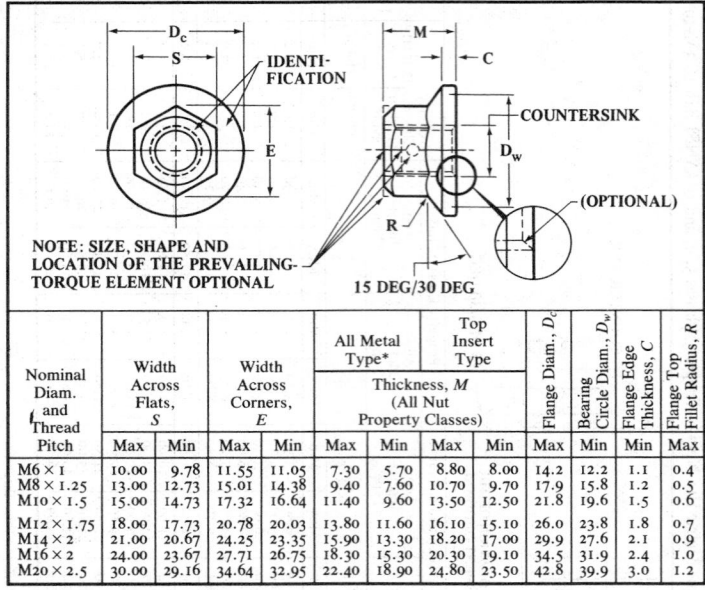

NOTE: SIZE, SHAPE AND
LOCATION OF THE PREVAILING-
TORQUE ELEMENT OPTIONAL

15 DEG/30 DEG

Nominal Diam. and Thread Pitch	Width Across Flats, S		Width Across Corners, E		All Metal Type*		Top Insert Type		Flange Diam., D_c		Bearing Circle Diam., D_w	Flange Edge Thickness, C		Flange Top Fillet Radius, R
					Thickness, M (All Nut Property Classes)									
	Max	Min	Max	Min	Max	Min	Max	Min	Max	Min	Min	Max	Min	Max
M6 × 1	10.00	9.78	11.55	11.05	7.30	5.70	8.80	8.00	14.2	12.2	1.1	0.4		
M8 × 1.25	13.00	12.73	15.01	14.38	9.40	7.60	10.70	9.70	17.9	15.8	1.2	0.5		
M10 × 1.5	15.00	14.73	17.32	16.64	11.40	9.60	13.50	12.50	21.8	19.6	1.5	0.6		
M12 × 1.75	18.00	17.73	20.78	20.03	13.80	11.60	16.10	15.10	26.0	23.8	1.8	0.7		
M14 × 2	21.00	20.67	24.25	23.35	15.90	13.30	18.20	17.00	29.9	27.6	2.1	0.9		
M16 × 2	24.00	23.67	27.71	26.75	18.30	15.30	20.30	19.10	34.5	31.9	2.4	1.0		
M20 × 2.5	30.00	29.16	34.64	32.95	22.40	18.90	24.80	23.50	42.8	39.9	3.0	1.2		

All dimensions are in millimeters.
* Also includes metal nuts with nonmetallic inserts, plugs, or patches in their threads.

type nuts of other materials are identified as agreed upon by the manufacturer and purchaser.

Metric Nut Designation: — Metric nuts are designated by the following data, preferably in the sequence shown: product name, nominal diameter and thread pitch, steel property class or material identification, and protective coating, if required. (Note: It is common practice in ISO Standards to omit thread pitch from the product designation when the nut threads are the metric coarse thread series, e.g., M10 stands for M10 × 1.5).

Examples: Hex nut, style 1, M10 × 1.5, ASTM A563M class 10, zinc plated
Heavy hex nut, M20 × 2.5, silicon bronze, ASTM F467, grade 651
Slotted hex nut, M20, ASTM A563M class 10.

Metric Washers

Metric Plain Washers. — American National Standard ANSI B18.22M-1981 covers general specifications and dimensions for flat, round-hole washers, both soft (as fabricated) and hardened, intended for use in general-purpose applications. Dimensions are given in the following table. Manufacturers should be consulted for current information on stock sizes.

Comparison with ISO Standards. — The washers covered by this ANSI Standard are nominally similar to those covered in various ISO documents. Outside diameters were selected, where possible, from ISO/TC2/WG6/N47 General Plan for Plain Washers for Metric Bolts, Screws, and Nuts. The thicknesses given in the ANSI Standard are similar to the nominal ISO thicknesses, however the tolerances differ. Inside diameters also differ.

Types of Metric Plain Washers. — Soft (as fabricated) washers are generally available in nominal sizes 1.6 mm through 36 mm in a variety of materials. They are normally used in low-strength applications to distribute bearing load, to provide a uniform bearing surface, and to prevent marring of the work surface.

Hardened steel washers are normally available in sizes 6 mm through 36 mm in the narrow and regular series. They are intended primarily for use in high-strength joints to minimize embedment, to provide a uniform bearing surface, and to bridge large clearance holes and slots.

Metric Plain Washer Materials and Finish. — Soft (as fabricated) washers are made of nonhardened steel unless otherwise specified by the purchaser. Hardened washers are made of through-hardened steel tempered to a hardness of 38 to 45 Rockwell C.

Unless otherwise specified, washers are furnished with a natural (as fabricated) finish, unplated or uncoated with a light film of oil or rust inhibitor.

Metric Plain Washer Designation. — When specifying metric plain washers, the designation should include the following data in the sequence shown: description, nominal size, series, material type, and finish, if required.

Example: Plain washer, 6 mm, narrow, soft, steel, zinc plated
Plain washer, 10 mm, regular, hardened steel.

Table 32. American National Standard Metric Plain Washers (ANSI B18.22.M-1981)

Nominal Washer Size*	Washer Series	Inside Diameter, A		Outside Diameter, B		Thickness, C	
		Max	Min	Max	Min	Max	Min
1.6	Narrow	2.09	1.95	4.00	3.70	0.70	0.50
	Regular	2.09	1.95	5.00	4.70	0.70	0.50
	Wide	2.09	1.95	6.00	5.70	0.90	0.60
2	Narrow	2.64	2.50	5.00	4.70	0.90	0.60
	Regular	2.64	2.50	6.00	5.70	0.90	0.60
	Wide	2.64	2.50	8.00	7.64	0.90	0.60
2.5	Narrow	3.14	3.00	6.00	5.70	0.90	0.60
	Regular	3.14	3.00	8.00	7.64	0.90	0.60
	Wide	3.14	3.00	10.00	9.64	1.20	0.80

All dimensions are in millimeters.   *See footnotes at end of table.

Table 32 (*Concluded*). **American National Standard Metric Plain Washers**
(ANSI B18.22M-1981)

Nominal Washer Size*	Washer Series	Inside Diameter, A		Outside Diameter, B		Thickness, C	
		Max	Min	Max	Min	Max	Min
3	Narrow	3.68	3.50	7.00	6.64	0.90	0.60
	Regular	3.68	3.50	10.00	9.64	1.20	0.80
	Wide	3.68	3.50	12.00	11.57	1.40	1.00
3.5	Narrow	4.18	4.00	9.00	8.64	1.20	0.80
	Regular	4.18	4.00	10.00	9.64	1.40	1.00
	Wide	4.18	4.00	15.00	14.57	1.75	1.20
4	Narrow	4.88	4.70	10.00	9.64	1.20	0.80
	Regular	4.88	4.70	12.00	11.57	1.40	1.00
	Wide	4.88	4.70	16.00	15.57	2.30	1.60
5	Narrow	5.78	5.50	11.00	10.57	1.40	1.00
	Regular	5.78	5.50	15.00	14.57	1.75	1.20
	Wide	5.78	5.50	20.00	19.48	2.30	1.60
6	Narrow	6.87	6.65	13.00	12.57	1.75	1.20
	Regular	6.87	6.65	18.80	18.37	1.75	1.20
	Wide	6.87	6.65	25.40	24.88	2.30	1.60
8	Narrow	9.12	8.90	18.80†	18.37†	2.30	1.60
	Regular	9.12	8.90	25.40†	24.48†	2.30	1.60
	Wide	9.12	8.90	32.00	31.38	2.80	2.00
10	Narrow	11.12	10.85	20.00	19.48	2.30	1.60
	Regular	11.12	10.85	28.00	27.48	2.80	2.00
	Wide	11.12	10.85	39.00	38.38	3.50	2.50
12	Narrow	13.57	13.30	25.40	24.88	2.80	2.00
	Regular	13.57	13.30	34.00	33.38	3.50	2.50
	Wide	13.57	13.30	44.00	43.38	3.50	2.50
14	Narrow	15.52	15.25	28.00	27.48	2.80	2.00
	Regular	15.52	15.25	39.00	38.38	3.50	2.50
	Wide	15.52	15.25	50.00	49.38	4.00	3.00
16	Narrow	17.52	17.25	32.00	31.38	3.50	2.50
	Regular	17.52	17.25	44.00	43.38	4.00	3.00
	Wide	17.52	17.25	56.00	54.80	4.60	3.50
20	Narrow	22.32	21.80	39.00	38.38	4.00	3.00
	Regular	22.32	21.80	50.00	49.38	4.60	3.50
	Wide	22.32	21.80	66.00	64.80	5.10	4.00
24	Narrow	26.12	25.60	44.00	43.38	4.60	3.50
	Regular	26.12	25.60	56.00	54.80	5.10	4.00
	Wide	26.12	25.60	72.00	70.80	5.60	4.50
30	Narrow	33.02	32.40	56.00	54.80	5.10	4.00
	Regular	33.02	32.40	72.00	70.80	5.60	4.50
	Wide	33.02	32.40	90.00	88.60	6.40	5.00
36	Narrow	38.92	38.30	66.00	64.80	5.60	4.50
	Regular	38.92	38.30	90.00	88.60	6.40	5.00
	Wide	38.92	38.30	110.00	108.60	8.50	7.00

All dimensions are in millimeters.

* Nominal washer sizes are intended for use with comparable screw and bolt sizes.

† The 18.80/18.37 and 25.40/24.48 mm outside diameters avoid washers which could be used in coin-operated devices.

BRITISH FASTENERS

British Standard Square and Hexagon Bolts, Screws and Nuts. — Important dimensions of precision hexagon bolts, screws and nuts (B.S.W. and B.S.F. threads) as covered by British Standard 1083:1965 are given in Tables 1 and 2. The use of fasteners in this standard will decrease as fasteners having Unified inch and ISO metric threads come into increasing use. Dimensions of Unified precision hexagon bolts, screws and nuts (UNC and UNF threads) are given in B.S. 1768:1963; of Unified black hexagon bolts, screws and nuts (UNC and UNF threads) in B.S. 1769:1951; and of Unified black square and hexagon bolts, screws and nuts (UNC and UNF threads) in B.S. 2708:1956. Unified nominal and basic dimensions in these British Standards are the same as the comparable dimensions in the American Standards, but the tolerances applied to these basic dimensions may differ because of rounding-off practices and other factors. For Unified dimensions of square and hexagon bolts and nuts as given in American National Standards B18.2.1-1981 and B18-2.2-1972, see Tables 1 to 3, 6, 7, 9 and 10 on Handbook pages 1281 to 1290. ISO precision hexagon bolts, screws and nuts are specified in the British Standard B.S. 3692:1967 (see Handbook pages 1343 to 1349), and ISO metric black hexagon bolts, screws and nuts are covered by British Standard B.S. 4190:1967.

British Standard Screwed Studs. — General purpose screwed studs are covered in British Standard 2693: Part 1:1956. The aim in this standard is to provide for a stud having tolerances which would not render it expensive to manufacture and which could be used in association with standard tapped holes for most purposes. Provision has been made for the use of both Unified Fine threads, Unified Coarse threads, British Standard Fine threads, and British Standard Whitworth threads as shown in the table on page 1350.

Designations: The *metal end* of the stud is the end which is screwed into the component. The *nut end* is the end of the screw of the stud which is not screwed into the component. The *plain portion* of the stud is the unthreaded length.

Recommended Fitting Practices for Metal End of Stud: It is recommended that holes tapped to Class 3B limits (see Handbook pages 1498 to 1520) in accordance with B.S. 1580 Unified screw threads or to Close Class limits in accordance with B.S. 84 "Screw Threads of Whitworth Form" as appropriate, be used in association with the metal end of the stud specified in this standard. Where fits are not critical, however, holes may be tapped to Class 2B limits (see tables on Handbook pages 1498 to 1520) in accordance with B.S. 1580 or Normal Class limits in accordance with B.S. 84.

It is recommended that the B.A. stud specified in this standard be associated with holes tapped to the limits specified for nuts in B.S. 93, 1919 edition. Where fits for these studs are not critical, holes may be tapped to limits specified for nuts in the current edition of B.S. 93.

In general, it will be found that the amount of oversize specified for the studs will produce a satisfactory fit in conjunction with the standard tapping as above. Even when interference is not present, locking will take place on the thread runout which has been carefully controlled for this purpose. Where it is considered essential to assure a true interference fit, higher grade studs should be used. It is recommended that standard studs be used even under special conditions where selective assembly may be necessary.

After several years of use of B.S. 2693: Part 1:1956, it was recognized that it would not meet the requirements of all stud users. The thread tolerances specified could result in clearance of interference fits and, in the former case, locking depended on the run-out threads. Thus, some users felt that true interference fits were essential for their needs. As a result, the British Standards Committee have incorporated the Class 5 interference fit threads specified in American Standard ASA B1.12 into their B.S. 2963: Part 2:1964, "Recommendations for High Grade Studs."

Britisch Standard Whitworth (B.S.W.) and Fine (B.S.F.) Precision Hexagon Bolts, Screws, and Nuts

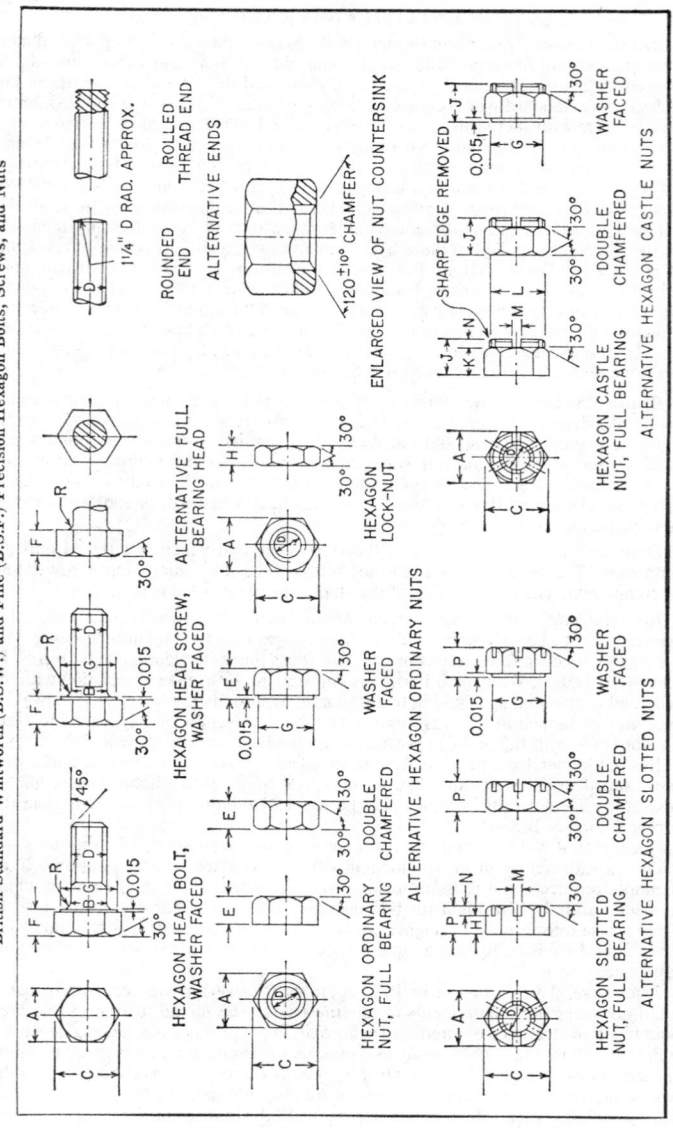

ROLLED THREAD END. APPROX.

1¼" D RAD. APPROX.

ROUNDED END

ALTERNATIVE ENDS

120 ±10° CHAMFER

ENLARGED VIEW OF NUT COUNTERSINK

WASHER FACED

DOUBLE CHAMFERED

SHARP EDGE REMOVED

HEXAGON CASTLE NUT, FULL BEARING

ALTERNATIVE HEXAGON CASTLE NUTS

ALTERNATIVE FULL BEARING HEAD

HEXAGON LOCK-NUT

HEXAGON HEAD SCREW, WASHER FACED

WASHER FACED

ALTERNATIVE HEXAGON ORDINARY NUTS

WASHER FACED

HEXAGON HEAD BOLT, WASHER FACED

DOUBLE CHAMFERED

DOUBLE CHAMFERED

HEXAGON ORDINARY NUT, FULL BEARING

HEXAGON SLOTTED NUT, FULL BEARING

ALTERNATIVE HEXAGON SLOTTED NUTS

For dimensions see Tables 1 and 2.

Table 1. British Standard Whitworth (B.S.W.) and Fine (B.S.F.) Precision Hexagon Bolts Screws, and Nuts
(B.S. 1083:1965)

Nominal Size D	Number of Threads per Inch		Bolts, Screws, and Nuts							Bolts and Screws				Nuts			
			Width			Diameter of Washer Face G		Radius Under Head R		Diameter of Unthreaded Portion of Shank B		Thickness Head F		Thickness			
			Across Flats A		Across Corners C									Ordinary E		Lock H	
	B.S.W.	B.S.F.	Max.	Min.†	Max.	Max.	Min.	Max.	Min.	Max.	Min.	Max.	Min.	Max.	Min.	Max.	Min.
1/4	20	26	0.445	0.438	0.51	0.428	0.418	0.025	0.015	0.2500	0.2465	0.176	0.166	0.200	0.190	0.185	0.180
5/16	18	22	0.525	0.518	0.61	0.508	0.498	0.025	0.015	0.3125	0.3090	0.218	0.208	0.250	0.240	0.210	0.200
3/8	16	20	0.600	0.592	0.69	0.582	0.572	0.025	0.015	0.3750	0.3715	0.260	0.250	0.312	0.302	0.260	0.250
7/16	14	18	0.710	0.702	0.82	0.690	0.680	0.025	0.015	0.4375	0.4335	0.302	0.292	0.375	0.365	0.275	0.265
1/2	12	16	0.820	0.812	0.95	0.800	0.790	0.025	0.015	0.5000	0.4960	0.343	0.333	0.437	0.427	0.300	0.290
9/16	12	16	0.920	0.912	1.06	0.900	0.890	0.045	0.020	0.5625	0.5585	0.375	0.365	0.500	0.490	0.333	0.323
5/8	11	14	1.010	1.000	1.17	0.985	0.975	0.045	0.020	0.6250	0.6190	0.417	0.407	0.562	0.552	0.375	0.365
3/4	10	12	1.200	1.190	1.39	1.175	1.165	0.045	0.020	0.7500	0.7440	0.500	0.480	0.687	0.677	0.458	0.448
7/8	9	11	1.300	1.288	1.50	1.273	1.263	0.065	0.040	0.8750	0.8670	0.583	0.563	0.750	0.740	0.500	0.490
1	8	10	1.480	1.468	1.71	1.453	1.443	0.095	0.060	1.0000	0.9920	0.666	0.636	0.875	0.865	0.583	0.573
1 1/8	7	9	1.670	1.640	1.93	1.620	1.610	0.095	0.060	1.1250	1.1170	0.750	0.710	1.000	0.990	0.666	0.656
1 1/4	7	9	1.860	1.815	2.15	1.795	1.785	0.095	0.060	1.2500	1.2420	0.830	0.790	1.125	1.105	0.750	0.730
1 3/8*	..	8	2.050	2.005	2.37	1.985	1.975	0.095	0.060	1.3750	1.3650	0.920	0.880	1.250	1.230	0.833	0.813
1 1/2	6	8	2.220	2.175	2.56	2.155	2.145	0.095	0.060	1.5000	1.4900	1.000	0.960	1.375	1.355	0.916	0.896
1 3/4	5	7	2.580	2.520	2.98	2.495	2.485	0.095	0.060	1.7500	1.7400	1.170	1.110	1.625	1.605	1.083	1.063
2	4.5	7	2.760	2.700	3.19	2.675	2.665	0.095	0.060	2.0000	1.9900	1.330	1.270	1.750	1.730	1.166	1.146

All dimensions in inches except where otherwise noted. * Not standard with B.S.W. thread. † When bolts from ¼ to 1 inch are hot forged, the tolerance on the width across flats shall be two and a half times the tolerance shown in the table and shall be unilaterally minus from maximum size. For dimensional notation, see diagram on page 1340.

Table 2. British Standard Whitworth (B.S.W.) and Fine (B.S.F.) Precision Hexagon Slotted and Castle Nuts
(B.S. 1083:1965)

Nominal Size D	Threads per Inch B.S.W.	Threads per Inch B.S.F.	Slotted Nuts Thickness P Max.	Min.	Slotted Nuts Lower Face to Bottom of Slot H Max.	Min.	Castle Nuts Total Thickness J Max.	Min.	Castle Nuts Lower Face to Bottom of Slot K Max.	Min.	Castellated Portion Diameter L Max.	Min.	Slots Width M Max.	Min.	Slots Depth N Approx.
1/4	20	26	0.200	0.190	0.170	0.160	0.290	0.280	0.200	0.190	0.430	0.425	0.100	0.090	0.090
5/16	18	22	0.250	0.240	0.190	0.180	0.340	0.330	0.250	0.240	0.510	0.500	0.100	0.090	0.090
3/8	16	20	0.312	0.302	0.222	0.212	0.402	0.392	0.312	0.302	0.585	0.575	0.100	0.090	0.090
7/16	14	18	0.375	0.365	0.235	0.225	0.515	0.505	0.375	0.365	0.695	0.685	0.135	0.125	0.140
1/2	12	16	0.437	0.427	0.297	0.287	0.577	0.567	0.437	0.427	0.805	0.795	0.135	0.125	0.140
9/16	12	16	0.500	0.490	0.313	0.303	0.687	0.677	0.500	0.490	0.905	0.895	0.175	0.165	0.187
5/8	11	14	0.562	0.552	0.375	0.365	0.749	0.739	0.562	0.552	0.995	0.985	0.175	0.165	0.187
3/4	10	12	0.687	0.677	0.453	0.443	0.921	0.911	0.687	0.677	1.185	1.105	0.218	0.208	0.234
7/8	9	11	0.750	0.740	0.516	0.506	0.984	0.974	0.750	0.740	1.285	1.265	0.218	0.208	0.234
1	8	10	0.875	0.865	0.595	0.585	1.155	1.145	0.875	0.865	1.465	1.445	0.260	0.250	0.280
1 1/8	7	9	1.000	0.990	0.720	0.710	1.280	1.270	1.000	0.990	1.655	1.635	0.260	0.250	0.280
1 1/4	7	9	1.125	1.105	0.797	0.777	1.453	1.433	1.125	1.105	1.845	1.825	0.300	0.290	0.328
1 3/8*	..	8	1.250	1.230	0.922	0.902	1.578	1.558	1.250	1.230	2.035	2.015	0.300	0.290	0.328
1 1/2	6	8	1.375	1.355	1.047	1.027	1.703	1.683	1.375	1.355	2.200	2.180	0.300	0.290	0.328
1 3/4	5	7	1.625	1.605	1.250	1.230	2.000	1.980	1.625	1.605	2.555	2.535	0.343	0.333	0.375
2	4.5	7	1.750	1.730	1.282	1.262	2.218	2.198	1.750	1.730	2.735	2.715	0.426	0.416	0.468

All dimensions in inches except where otherwise noted. * Not standard with B.S.W. thread. For widths across flats, widths across corners, and diameter of washer face see Table 1. For dimensional notation, see diagram on page 1340.

British Standard ISO Metric Precision Hexagon Bolts, Screws and Nuts. — This British Standard (BS 3692:1967) gives the general dimensions and tolerances of precision hexagon bolts, screws and nuts with ISO metric threads in diameters from 1.6 to 68 mm. It is based on the following ISO recommendations and draft recommendations: R 272, R 288, DR 911, DR 947, DR 950, DR 952 and DR 987. Mechanical properties are given only with respect to carbon or alloy steel bolts, screws and nuts, which are not to be used for special applications such as those requiring weldability, corrosion resistance or ability to withstand temperatures above 300°C or below −50°C. The dimensional requirements of this standard also apply to non-ferrous and stainless steel bolts, screws and nuts.

Finish: Finishes may be dull black which results from the heat-treating operation or may be bright finish, the result of bright drawing. Other finishes are possible by mutual agreement between purchaser and producer. It is recommended that reference be made to BS 3382 "Electroplated Coatings on Threaded Components," in this respect.

General Dimensions: The bolts, screws and nuts conform to the general dimensions given in Tables 1, 2, 3 and 4.

Nominal Lengths of Bolts and Screws: The nominal length of a bolt or screw is the distance from the underside of the head to the extreme end of the shank including any chamfer or radius. Standard nominal lengths and tolerances thereon are given in Table 5.

Bolt and Screw Ends: The ends of bolts and screws may be finished with either a 45-degree chamfer to a depth slightly exceeding the depth of thread or a radius approximately equal to $1\frac{1}{4}$ times the nominal diameter of the shank. With rolled threads, the lead formed at the end of the bolt by the thread rolling operation may be regarded as providing the necesssary chamfer to the end; the end being reasonably square with the center line of the shank.

Screw Thread Form: The form of thread and diameters and associated pitches of standard ISO metric bolts, screws and nuts are in accordance with BS 3643: Part 1, "Thread Data and Standard Thread Series." The screw threads are made to the tolerances for the medium class of fit ($6H/6g$) as specified in BS 3643: Part 2, "Limits and Tolerances for Coarse Pitch Series Threads."

Length of Thread on Bolts: The length of thread on bolts is the distance from the end of the bolt (including any chamfer or radius) to the leading face of a screw ring gage which has been screwed as far as possible onto the bolt by hand. Standard thread lengths of bolts are $2d + 6$ mm for a nominal length of bolt up to and including 125 mm, $2d + 12$ mm for a nominal bolt length over 125 mm up to and including 200 mm, and $2d + 25$ mm for a nominal bolt length over 200 mm. Bolts that are too short for minimum thread lengths are threaded as screws and designated as screws. The tolerance on bolt thread lengths are plus two pitches for all diameters.

Length of Thread on Screws: Screws are threaded to permit a screw ring gage being screwed by hand to within a distance from the underside of the head not exceeding two and a half times the pitch for diameters up to and including 52 mm and three and a half times the pitch for diameters over 52 mm.

Angularity and Eccentricity of Bolts, Screws and Nuts: The axis of the thread of the nut is square to the face of the nut subject to the "squareness tolerance" given in Table 3.

In gaging, the nut is screwed by hand onto a gage, having a truncated taper thread, until the thread of the nut is tight on the thread of the gage. A sleeve sliding on a parallel extension of the gage, which has a face of diameter equal to the minimum distance across the flats of the nut and exactly at 90 degrees to the axis of the gage, is brought into contact with the leading face of the nut. With

(*Continued on page* 1348)

Table 1. British Standard ISO Metric Precision Hexagon Bolts, Screws and Nuts
(BS 3692:1967)

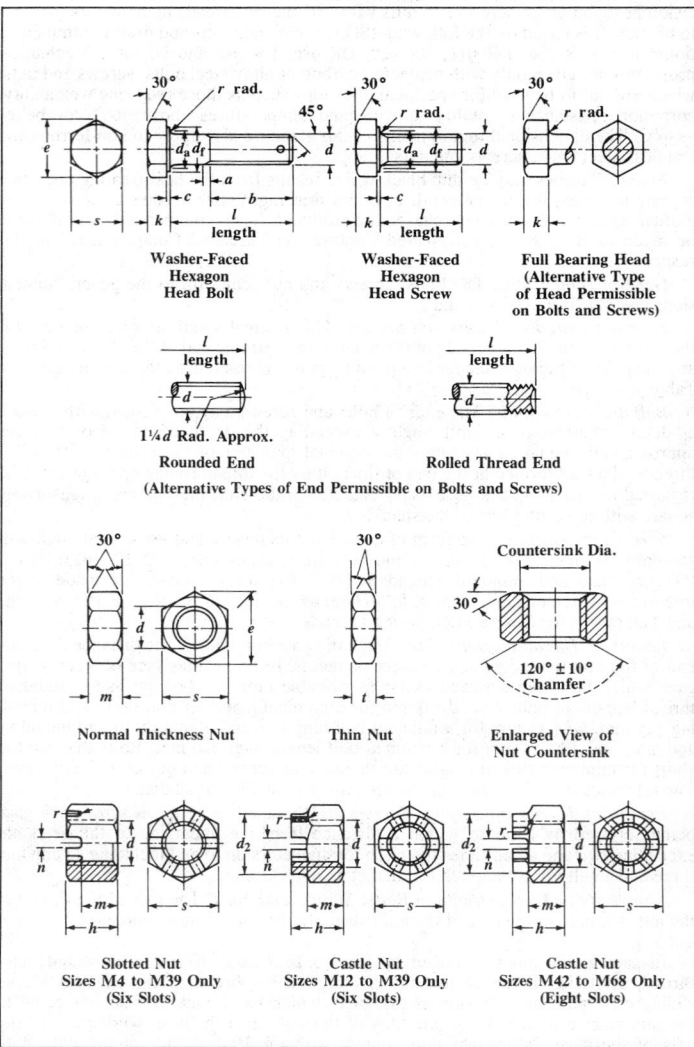

Washer-Faced Hexagon Head Bolt

Washer-Faced Hexagon Head Screw

Full Bearing Head (Alternative Type of Head Permissible on Bolts and Screws)

Rounded End

Rolled Thread End

(Alternative Types of End Permissible on Bolts and Screws)

Normal Thickness Nut

Thin Nut

Enlarged View of Nut Countersink

Slotted Nut
Sizes M4 to M39 Only
(Six Slots)

Castle Nut
Sizes M12 to M39 Only
(Six Slots)

Castle Nut
Sizes M42 to M68 Only
(Eight Slots)

For general dimensions see Tables 2, 3, 4 and 5.

Table 2. British Standard ISO Metric Precision Hexagon Bolts and Screws (BS 3692:1967)

Nom. Size and Thread Diam.* d	Pitch of Thread (Coarse Pitch Series)	Thread Run-out a Max.	Diam. of Unthreaded Shank d Max.	d Min.	Width Across Flats s Max.	s Min.	Width Across Corners e Max.	e Min.	Diam. of Washer Face d_w Max.	d_w Min.	Depth of Washer Face c	Transition Diam.† d_a Max.	Radius Under Head r Max.	r Min.	Height of Head k Max.	k Min.	Eccentricity of Head Max.	Eccentricity of Shank and Split Pin Hole to the Thread Max.
M1.6	0.35	0.8	1.6	1.46	3.2	3.08	3.7	3.48	...	...	...	2.0	0.2	0.1	1.225	0.975	0.18	0.14
M2	0.4	1.0	2.0	1.86	4.0	3.88	4.6	4.38	...	...	...	2.6	0.3	0.1	1.525	1.275	0.18	0.14
M2.5	0.45	1.0	2.5	2.36	5.0	4.88	5.8	5.51	...	...	...	3.1	0.3	0.1	1.825	1.575	0.18	0.14
M3	0.5	1.2	3.0	2.86	5.5	5.38	6.4	6.08	5.08	4.83	...	3.6	0.3	0.1	2.125	1.875	0.18	0.14
M4	0.7	1.6	4.0	3.82	7.0	6.85	8.1	7.74	6.55	6.30	...	4.7	0.35	0.2	2.925	2.675	0.22	0.18
M5	0.8	2.0	5.0	4.82	8.0	7.85	9.2	8.87	7.55	7.30	0.1	5.7	0.35	0.2	3.650	3.35	0.22	0.18
M6	1	2.5	6.0	5.82	10.0	9.78	11.5	11.05	9.48	9.23	0.1	6.8	0.4	0.25	4.15	3.85	0.22	0.18
M8	1.25	3.0	8.0	7.78	13.0	12.73	15.0	14.38	12.43	12.18	0.2	9.2	0.4	0.4	5.65	5.35	0.27	0.22
M10	1.5	3.5	10.0	9.78	17.0	16.73	19.6	18.90	16.43	16.18	0.3	11.2	0.6	0.4	7.18	6.82	0.27	0.22
M12	1.75	4.0	12.0	11.73	19.0	18.67	21.9	21.10	18.37	18.12	0.3	14.2	0.6	0.4	8.18	7.82	0.27	0.22
(M14)	2	5.0	14.0	13.73	22.0	21.67	25.4	24.49	21.37	21.12	0.4	16.2	1.1	0.6	9.18	8.82	0.33	0.27
M16	2	5.0	16.0	15.73	24.0	23.67	27.7	26.75	23.27	23.02	0.4	18.2	1.1	0.6	10.18	9.82	0.33	0.27
(M18)	2.5	6.0	18.0	17.73	27.0	26.67	31.2	30.14	26.27	26.02	0.4	20.2	1.1	0.6	12.215	11.785	0.33	0.27
M20	2.5	6.0	20.0	19.67	30.0	29.67	34.6	33.53	29.27	28.80	0.4	22.4	1.1	0.8	13.215	12.785	0.33	0.27
(M22)	2.5	6.0	22.0	21.67	32.0	31.61	36.9	35.72	31.21	30.74	0.4	24.4	1.2	0.8	14.215	13.785	0.39	0.33
M24	3	7.0	24.0	23.67	36.0	35.38	41.6	39.98	34.98	34.51	0.4	26.4	1.2	0.8	15.215	14.785	0.39	0.33
(M27)	3	7.0	27.0	26.67	41.0	40.38	47.3	45.63	39.98	39.36	0.5	30.4	1.2	0.8	17.215	16.785	0.39	0.33
M30	3.5	8.0	30.0	29.67	46.0	45.38	53.1	51.28	44.98	44.36	0.5	33.4	1.7	1.0	19.26	18.74	0.39	0.33
(M33)	3.5	8.0	33.0	32.61	50.0	49.38	57.7	55.80	48.98	48.36	0.5	36.4	1.7	1.0	21.26	20.74	0.39	0.33
M36	4	10.0	36.0	35.61	55.0	54.26	63.5	61.31	53.86	53.24	0.5	39.4	1.7	1.0	23.26	22.74	0.46	0.39
(M39)	4	10.0	39.0	38.61	60.0	59.26	69.3	66.96	58.86	58.24	0.6	42.4	1.7	1.0	25.26	24.74	0.46	0.39
M42	4.5	11.0	42.0	41.61	65.0	64.26	75.1	72.61	63.76	63.04	0.6	45.6	1.8	1.2	26.26	25.74	0.46	0.39
(M45)	4.5	11.0	45.0	44.61	70.0	69.26	80.8	78.26	68.76	68.04	0.6	48.6	1.8	1.2	28.26	27.74	0.46	0.39
M48	5	12.0	48.0	47.61	75.0	74.26	86.6	83.91	73.76	73.04	0.6	52.6	2.3	1.6	30.26	29.74	0.46	0.39
(M52)	5	19.0	52.0	51.54	80.0	79.26	92.4	89.56	...	...	...	56.6	3.5	2.0	33.31	32.69	0.46	0.39
M56	5.5	19.0	56.0	55.54	85.0	84.13	98.1	95.07	...	...	...	63.0	3.5	2.0	35.31	34.69	0.54	0.46
(M60)	5.5	19.0	60.0	59.54	90.0	89.13	103.9	100.72	...	...	...	67.0	3.5	2.0	38.31	37.69	0.54	0.46
M64	6	21.0	64.0	63.54	95.0	94.13	109.7	106.37	...	...	...	71.0	3.5	2.0	40.31	39.69	0.54	0.46
(M68)	6	21.0	68.0	67.54	100.0	99.13	115.5	112.02	...	...	...	75.0	3.5	2.0	43.31	42.69	0.54	0.46

All dimensions are in millimeters. For illustration of bolts and screws see Table 1.   * Sizes shown in parentheses are non-preferred.   † A true radius is not essential provided that the curve is smooth and lies wholly within the maximum radius, determined from the maximum transitional diameter, and the minimum radius specified.

Table 3. British Standard ISO Metric Precision Hexagon Nuts and Thin Nuts (BS 3692:1967)

Nominal Size and Thread Diameter* d	Pitch of Thread (Coarse Pitch Series)	Width Across Flats s Max.	Width Across Flats s Min.	Width Across Corners e Max.	Width Across Corners e Min.	Thickness of Normal Nut m Max.	Thickness of Normal Nut m Min.	Tolerance on Squareness of Thread to Face of Nut† Max.	Eccentricity of Hexagon Max.	Thickness of Thin Nut t Max.	Thickness of Thin Nut t Min.
M1.6	0.35	3.20	3.08	3.70	3.48	1.30	1.05	0.05	0.14	…	…
M2	0.4	4.00	3.88	4.60	4.38	1.60	1.35	0.06	0.14	…	…
M2.5	0.45	5.00	4.88	5.80	5.51	2.00	1.75	0.08	0.14	…	…
M3	0.5	5.50	5.38	6.40	6.08	2.40	2.15	0.09	0.14	…	…
M4	0.7	7.00	6.85	8.10	7.74	3.20	2.90	0.11	0.18	…	…
M5	0.8	8.00	7.85	9.20	8.87	4.00	3.70	0.13	0.18	…	…
M6	1	10.00	9.78	11.50	11.05	5.00	4.70	0.17	0.18	…	…
M8	1.25	13.00	12.73	15.00	14.38	6.50	6.14	0.22	0.22	5.0	4.70
M10	1.5	17.00	16.73	19.60	18.90	8.00	7.64	0.29	0.22	6.0	5.70
M12	1.75	19.00	18.67	21.90	21.10	10.00	9.64	0.32	0.27	7.0	6.64
(M14)	2	22.00	21.67	25.4	24.49	11.00	10.57	0.37	0.27	8.0	7.64
M16	2	24.00	23.67	27.7	26.75	13.00	12.57	0.41	0.27	8.0	7.64
(M18)	2.5	27.00	26.67	31.20	30.14	15.00	14.57	0.46	0.27	9.0	8.64
M20	2.5	30.00	29.67	34.60	33.53	16.00	15.57	0.51	0.33	9.0	8.64
(M22)	2.5	32.00	31.61	36.90	35.72	18.00	17.57	0.54	0.33	10.0	9.64
M24	3	36.00	35.38	41.60	39.98	19.00	18.48	0.61	0.33	10.0	9.64
(M27)	3	41.00	40.38	47.3	45.63	22.00	21.48	0.70	0.33	12.0	11.57
M30	3.5	46.00	45.38	53.1	51.28	24.00	23.48	0.78	0.33	12.0	11.57
(M33)	3.5	50.00	49.38	57.70	55.80	26.00	25.48	0.85	0.39	14.0	13.57
M36	4	55.00	54.26	63.50	61.31	29.00	28.48	0.94	0.39	14.0	13.57
(M39)	4	60.00	59.26	69.30	66.96	31.00	30.38	1.03	0.39	16.0	15.57
M42	4.5	65.00	64.26	75.10	72.61	34.00	33.38	1.11	0.39	16.0	15.57
(M45)	4.5	70.00	69.26	80.80	78.26	36.00	35.38	1.20	0.39	18.0	17.57
M48	5	75.00	74.26	86.60	83.91	38.00	37.38	1.29	0.39	18.0	17.57
(M52)	5	80.00	79.26	92.40	89.56	42.00	41.38	1.37	0.46	20.0	19.48
M56	5.5	85.00	84.13	98.10	95.07	45.00	44.38	1.46	0.46	…	…
(M60)	5.5	90.00	89.13	103.90	100.72	48.00	47.38	1.55	0.46	…	…
M64	6	95.00	94.13	109.70	106.37	51.00	50.26	1.63	0.46	…	…
M68	6	100.00	99.13	115.50	112.02	54.00	53.26	1.72	0.46	…	…

All dimensions are in millimeters. For illustration of hexagon nuts and thin nuts see Table 1.   * Sizes shown in parentheses are non-preferred.   † As measured with the nut squareness gage described in the text and illustrated in Appendix A of the Standard and a feeler gage.

Table 4. British Standard ISO Metric Precision Hexagon Slotted Nuts and Castle Nuts (BS 3692:1967)

Nominal Size and Thread Diameter* d	Width Across Flats s		Width Across Corners e		Diameter d_2		Thickness h		Lower Face of Nut to Bottom of Slot m		Width of Slot n		Radius $(0.25\,n)$ r	Eccentricity of the Slots
	Max.	Min.	Max.	Min.	Max.	Min.	Max.	Min.	Max.	Min.	Max.	Min.	Min.	Max.
M4	7.00	6.85	8.10	7.74	...	...	5	4.70	3.2	2.90	1.45	1.2	0.3	0.18
M5	8.00	7.85	9.20	8.87	...	...	6	5.70	4.0	3.70	1.65	1.4	0.35	0.18
M6	10.00	9.78	11.50	11.05	...	...	7.5	7.14	5	4.70	2.25	2	0.5	0.18
M8	13.00	12.73	15.00	14.38	...	...	9.5	9.14	6.5	6.14	2.75	2.5	0.625	0.22
M10	17.00	16.73	19.60	18.90	...	...	12	11.57	8	7.64	3.05	2.8	0.70	0.22
M12	19.00	18.67	21.90	21.10	17	16.57	15	14.57	10	9.64	3.80	3.5	0.875	0.27
(M14)	22.00	21.67	25.4	24.49	19	18.48	16	15.57	11	10.57	3.80	3.5	0.875	0.27
M16	24.00	23.67	27.7	26.75	22	21.48	19	18.48	13	12.57	4.80	4.5	1.125	0.27
(M18)	27.00	26.67	31.20	30.14	25	24.48	21	20.48	15	14.57	4.80	4.5	1.125	0.27
M20	30.00	29.67	34.60	33.53	28	27.48	22	21.48	16	15.57	4.80	4.5	1.125	0.33
(M22)	32.00	31.61	36.90	35.72	30	29.48	26	25.48	18	17.57	5.80	5.5	1.375	0.33
M24	36.00	35.38	41.60	39.98	34	33.38	27	26.48	19	18.48	5.80	5.5	1.375	0.33
(M27)	41.00	40.38	47.3	45.63	38	37.38	30	29.48	22	21.48	5.80	5.5	1.375	0.33
M30	46.00	45.38	53.1	51.28	42	41.38	33	32.38	24	23.48	7.36	7	1.75	0.33
(M33)	50.00	49.38	57.70	55.80	46	45.38	35	34.38	26	25.48	7.36	7	1.75	0.39
M36	55.00	54.26	63.50	61.31	50	49.38	38	37.38	29	28.48	7.36	7	1.75	0.39
(M39)	60.00	59.26	69.30	66.96	55	54.26	40	39.38	31	30.38	7.36	7	1.75	0.39
M42	65.00	64.26	75.00	72.61	58	57.26	46	45.38	34	33.38	9.36	9	2.25	0.39
(M45)	70.00	69.26	80.80	78.26	62	61.26	48	47.38	36	35.38	9.36	9	2.25	0.39
M48	75.00	74.26	86.60	83.91	65	64.26	50	49.38	38	37.38	9.36	9	2.25	0.39
(M52)	80.00	79.26	92.40	89.56	70	69.26	54	53.26	42	41.38	9.36	9	2.25	0.46
M56	85.00	84.13	98.10	95.07	75	74.26	57	56.26	45	44.38	9.36	9	2.25	0.46
(M60)	90.00	89.13	103.90	100.72	80	79.26	63	62.26	48	47.38	11.43	11	2.75	0.46
M64	95.00	94.13	109.70	106.37	85	84.13	66	65.26	51	50.26	11.43	11	2.75	0.46
M68	100.00	99.13	115.50	112.02	90	89.13	69	68.26	54	53.26	11.43	11	2.75	0.46

All dimensions are in millimeters. For illustration of hexagon slotted nuts and castle nuts see Table 1. * Sizes shown in parentheses are non-preferred.

Table 5. British Standard ISO Metric Bolt and Screw Nominal Lengths
(BS 3692:1967)

Nominal Length* l	Tolerance	Nominal Length* l	Tolerance	Nominal Length* l	Tolerance	Nominal Length* l	Tolerance
5	±0.24	30	±0.42	90	±0.70	200	±0.925
6	±0.24	(32)	±0.50	(95)	±0.70	220	±0.925
(7)	±0.29	35	±0.50	100	±0.70	240	±0.925
8	±0.29	(38)	±0.50	(105)	±0.70	260	±1.05
(9)	±0.29	40	±0.50	110	±0.70	280	±1.05
10	±0.29	45	±0.50	(115)	±0.70	300	±1.05
(11)	±0.35	50	±0.50	120	±0.70	325	±1.15
12	±0.35	55	±0.60	(125)	±0.80	350	±1.15
14	±0.35	60	±0.60	130	±0.80	375	±1.15
16	±0.35	65	±0.60	140	±0.80	400	±1.15
(18)	±0.35	70	±0.60	150	±0.80	425	±1.25
20	±0.42	75	±0.60	160	±0.80	450	±1.25
(22)	±0.42	80	±0.60	170	±0.80	475	±1.25
25	±0.42	85	±0.70	180	±0.80	500	±1.25
(28)	±0.42	...	...	190	±0.925	...	...

All dimensions are in millimeters.   * Nominal lengths shown in parentheses are non-preferred.

the sleeve in this position, it should not be possible for a feeler gage of thickness equal to the "squareness tolerance" to enter anywhere between the leading nut face and sleeve face.

The hexagon flats of bolts, screws and nuts are square to the bearing face, and the angularity of the head is within the limits of 90 degrees, plus or minus 1 degree. The eccentricity of the hexagon flats of nuts relative to the thread diameter should not exceed the values given in Table 3 and the eccentricity of the head relative to the width across flats and eccentricity between the shank and thread of bolts and screws should not exceed the values given in Table 2.

Chamfering, Washer Facing and Countersinking: Bolt and screw heads have a chamfer of approximately 30 degrees on their upper faces and, at the option of the manufacturer, a washer face or full bearing face on the underside. Nuts are countersunk at an included angle of 120 degrees plus or minus 10 degrees at both ends of the thread. The diameter of the countersink should not exceed the nominal major diameter of the thread plus 0.13 mm up to and including 12 mm diameter, and plus 0.25 mm above 12 mm diameter. This stipulation does not apply to slotted, castle or thin nuts.

Strength Grade Designation System for Steel Bolts and Screws: This Standard includes a strength grade designation system consisting of two figures. The first figure is one tenth of the minimum tensile strength in kgf/mm^2, and the second figure is one tenth of the ratio between the minimum yield stress (or stress at permanent set limit, $R_{0.2}$) and the minimum tensile strength, expressed as a percentage. For example with the strength designation grade 8.8, the first figure 8 represents V_{10} the minimum tensile strength of 80 kgf/mm^2 and the second figure 8 represents V_{10} the ratio

$$\frac{\text{stress at permanent set limit } R_{0.2}\%}{\text{minimum tensile strength}} = \frac{1}{10} \times \frac{64}{80} \times \frac{100}{1};$$

the numerical values of stress and strength being obtained from the accompanying table.

Strength Grade Designations of Steel Bolts and Screws

Strength Grade Designation	4.6	4.8	5.6	5.8	6.6	6.8	8.8	10.9	12.9	14.9
Tensile Strength (R_m), Min.	40	40	50	50	60	60	80	100	120	140
Yield Stress (R_e), Min.	24	32	30	40	36	48	...	...	...	...
Stress at Permanent Set Limit ($R_{0.2}$), Min.	...	...	...	...	...	...	64	90	108	126
All stress and strength values are in kgf/mm² units.										

Strength Grade Designation System for Steel Nuts: The strength grade designation system for steel nuts is a number which is one-tenth of the specified proof load stress in kgf/mm². The proof load stress corresponds to the minimum tensile strength of the highest grade of bolt or screw with which the nut can be used.

Strength Grade Designations of Steel Nuts

Strength Grade Designation	4	5	6	8	12	14
Proof Load Stress (kgf/mm²)	40	50	60	80	120	140

Recommended Bolt and Nut Combinations

Grade of Bolt	4.6	4.8	5.6	5.8	6.6	6.8	8.8	10.9	12.9	14.9
Recommended Grade of Nut	4	4	5	5	6	6	8	12	12	14
Note: Nuts of a higher strength grade may be substituted for nuts of a lower strength grade.										

Marking: The marking and identification requirements of this Standard are only mandatory for steel bolts, screws and nuts of 6 mm diameter and larger; manufactured to strength grade designations 8.8 (for bolts or screws) and 8 (for nuts) or higher. Bolts and screws are identified as ISO metric by either of the symbols "ISO M" or "M", embossed or indented on top of the head. Nuts may be indented or embossed by alternative methods depending on their method of manufacture.

Designation: Bolts 10 mm diameter, 50 mm long manufactured from steel of strength grade 8.8, would be designated:
"Bolts M10 × 50 to BS 3692 — 8.8."
Brass screws 8 mm diameter, 20 mm long would be designated:
"Brass screws M8 × 20 to BS 3692."
Nuts 12 mm diameter, manufactured from steel of strength grade 6, cadmium plated could be designated:
"Nuts M12 to BS 3692 — 6, plated to BS 3382: Part 1."

Miscellaneous Information: The Standard also gives mechanical properties of steel bolts, screws and nuts [i.e., tensile strengths; hardnesses (Brinell, Rockwell, Vickers); stresses (yield, proof load); etc.], material and manufacture of steel bolts, screws and nuts; and information on inspection and testing. Appendices to the Standard give information on gaging; chemical composition; testing of mechanical properties; examples of marking of bolts, screws and nuts; and a table of preferred standard sizes of bolts and screws, to name some.

British Standard General Purpose Studs (B.S. 2693: Part I:1956)

Limits for End Screwed into Component (All threads except B.A.)

Nom. Diam. D	Major Diam. Max.	Thds. per In.	Major Diam. Min.	Effective Diameter Max.	Min.	Minor Diameter Max.	Min.	Thds. per In.	Major Diam. Min.	Effective Diameter Max.	Min.	Minor Diam. Max.	Min.
UN TH'DS.			UNF THREADS					UNC THREADS					
1/4	.2500	28	.2435	.2294	.2265	.2088	.2037	20	.2419	.2201	.2172	.1913	.1849
5/16	.3125	24	.3053	.2883	.2852	.2643	.2586	18	.3038	.2793	.2762	.2472	.2402
3/8	.3750	24	.3678	.3510	.3478	.3270	.3211	16	.3656	.3375	.3343	.3014	.2936
7/16	.4375	20	.4294	.4084	.4050	.3796	.3729	14	.4272	.3945	.3911	.3533	.3447
1/2	.5000	20	.4919	.4712	.4675	.4424	.4356	13	.4891	.4537	.4500	.4093	.4000
9/16	.5625	18	.5538	.5302	.5264	.4981	.4907	12	.5511	.5122	.5084	.4641	.4542
5/8	.6250	18	.6163	.5929	.5889	.5608	.5533	11	.6129	.5700	.5660	.5175	.5069
3/4	.7500	16	.7406	.7137	.7094	.6776	.6693	10	.7371	.6893	.6850	.6316	.6200
7/8	.8750	14	.8647	.8332	.8286	.7920	.7828	9	.8611	.8074	.8028	.7433	.7306
1	1.0000	12	.9886	.9510	.9459	.9029	.8925	8	.9850	.9239	.9188	.8517	.8376
1 1/8	1.1250	12	1.1136	1.0762	1.0709	1.0281	1.0176	7	1.1086	1.0375	1.0322	.9550	.9393
1 1/4	1.2500	12	1.2386	1.2014	1.1959	1.1533	1.1427	7	1.2336	1.1627	1.1572	1.0802	1.0644
1 3/8	1.3750	12	1.3636	1.3265	1.3209	1.2784	1.2677	6	1.3568	1.2723	1.2667	1.1761	1.1581
1 1/2	1.5000	12	1.4886	1.4517	1.4459	1.4036	1.3928	6	1.4818	1.3975	1.3917	1.3013	1.2832
B.S.F. TH'DS.			B.S.F. THREADS					B.S.W. THREADS					
1/4	.2500	26	.2455	.2280	.2251	.2034	.1984	20	.2452	.2206	.2177	.1886	.1831
5/16	.3125	22	.3077	.2863	.2832	.2572	.2517	18	.3073	.2798	.2767	.2442	.2383
3/8	.3750	20	.3699	.3461	.3429	.3141	.3083	16	.3656	.3381	.3349	.2981	.2919
7/16	.4375	18	.4320	.4053	.4019	.3697	.3635	14	.4316	.3952	.3918	.3495	.3428
1/2	.5000	16	.4942	.4637	.4601	.4237	.4172	12	.4937	.4503	.4466	.3969	.3897
9/16	.5625	16	.5566	.5263	.5225	.4863	.4797	12	.5560	.5129	.5091	.4595	.4521
5/8	.6250	14	.6187	.5833	.5793	.5376	.5305	11	.6183	.5708	.5668	.5126	.5050
3/4	.7500	12	.7432	.7009	.6966	.6475	.6398	10	.7428	.6903	.6860	.6263	.6182
7/8	.8750	11	.8678	.8214	.8168	.7632	.7551	9	.8674	.8085	.8039	.7374	.7288
1	1.0000	10	.9924	.9411	.9360	.8771	.8686	8	.9920	.9251	.9200	.8451	.8360
1 1/8	1.1250	9	1.1171	1.0592	1.0539	.9881	.9792	7	1.1164	1.0388	1.0335	.9473	.9376
1 1/4	1.2500	9	1.2419	1.1844	1.1789	1.1133	1.1042	7	1.2413	1.1640	1.1585	1.0725	1.0627
1 3/8	1.3750	8	1.3665	1.3006	1.2950	1.2206	1.2110	. . .	. . .	. . .	. . .	. . .	. . .
1 1/2	1.5000	8	1.4913	1.4258	1.4200	1.3458	1.3360	6	1.4906	1.3991	1.3933	1.2924	1.2818

Limits for End Screwed into Component (B.A. Threads)[1]

Designation No.	Pitch	Major Diameter Max.	Min.	Effective Diameter Max.	Min.	Minor Diameter Max.	Min.
2	.8100 mm	4.700 mm	4.580 mm	4.275 mm	4.200 mm	3.790 mm	3.620 mm
	.03189 in.	.1850 in.	.1803 in.	.1683 in.	.1654 in.	.1492 in.	.1425 in.
4	.6600 mm	3.600 mm	3.500 mm	3.260 mm	3.190 mm	2.865 mm	2.720 mm
	.02598 in.	.1417 in.	.1378 in.	.1283 in.	.1256 in.	.1128 in.	.1071 in.

Minimum Nominal Lengths of Studs[2]

Nom. Stud Diam.	For Thread Length (Component End) of 1D	1.5D	Nom. Stud Diam.	For Thread Length (Component End) of 1D	1.5D	Nom. Stud Diam.	For Thread Length (Component End) of 1D	1.5D
1/4	7/8	1	9/16	2	2 3/8	1 1/8	4	4 5/8
5/16	1 1/8	1 3/8	5/8	2 1/4	2 5/8	1 1/4	4 3/4	5 1/2
3/8	1 3/8	1 5/8	3/4	2 5/8	3	1 3/8	5	5 3/4
7/16	1 5/8	1 7/8	7/8	3 1/8	3 5/8	1 1/2	5 1/4	6
1/2	1 3/4	2	1	3 1/2	4	. . .	. . .	. . .

All dimensions are in inches except where otherwise noted.
[1] Approximate inch equivalents are shown below the dimensions given in mm.
[2] The standard also gives preferred and standard lengths of studs:
  Preferred lengths of studs: 7/8, 1, 1 1/8, 1 1/4, 1 3/8, 1 1/2, 1 3/4, 2, 2 1/4, 2 1/2, 2 3/4, 3, 3 1/4, 3 1/2 and for lengths above 3 1/2 the preferred increment is 1/2.
  Standard lengths of studs: 7/8, 1, 1 1/8, 1 1/4, 1 3/8, 1 1/2, 1 5/8, 1 3/4, 1 7/8, 2, 2 1/8, 2 1/4, 2 3/8, 2 1/2, 2 5/8, 2 3/4, 2 7/8, 3, 3 1/8, 3 1/4, 3 3/8, 3 1/2 and for lengths above 3 1/2 the standard increment is 1/4.
  See page 1552 for interference-fit threads.

British Standard Single Coil Rectangular Section Spring Washers; Metric Series — Types B and BP (BS 4464: 1969)

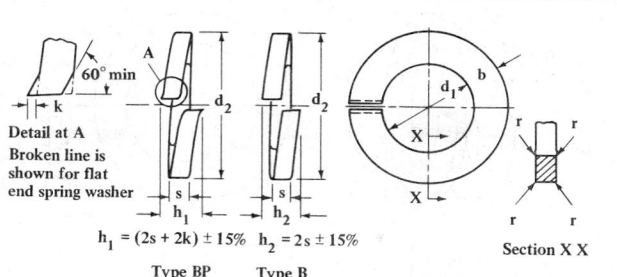

Detail at A
Broken line is
shown for flat
end spring washer

$h_1 = (2s + 2k) \pm 15\%$ $h_2 = 2s \pm 15\%$

Type BP   Type B

Section X X

Nom. Size & Thread Diam., d	Inside Diam., d_1		Width, b	Thickness, s	Outside Diam., d_2 Max	Radius, r Max	k (Type BP Only)
	Max	Min					
M1.6	1.9	1.7	0.7 ± 0.1	0.4 ± 0.1	3.5	0.15	...
M2	2.3	2.1	0.9 ± 0.1	0.5 ± 0.1	4.3	0.15	...
(M2.2)	2.5	2.3	1.0 ± 0.1	0.6 ± 0.1	4.7	0.2	...
M2.5	2.8	2.6	1.0 ± 0.1	0.6 ± 0.1	5.0	0.2	...
M3	3.3	3.1	1.3 ± 0.1	0.8 ± 0.1	6.1	0.25	...
(M3.5)	3.8	3.6	1.3 ± 0.1	0.8 ± 0.1	6.6	0.25	0.15
M4	4.35	4.1	1.5 ± 0.1	0.9 ± 0.1	7.55	0.3	0.15
M5	5.35	5.1	1.8 ± 0.1	1.2 ± 0.1	9.15	0.4	0.15
M6	6.4	6.1	2.5 ± 0.15	1.6 ± 0.1	11.7	0.5	0.2
M8	8.55	8.2	3 ± 0.15	2 ± 0.1	14.85	0.65	0.3
M10	10.6	10.2	3.5 ± 0.2	2.2 ± 0.15	18.0	0.7	0.3
M12	12.6	12.2	4 ± 0.2	2.5 ± 0.15	21.0	0.8	0.4
(M14)	14.7	14.2	4.5 ± 0.2	3 ± 0.15	24.1	1.0	0.4
M16	16.9	16.3	5 ± 0.2	3.5 ± 0.2	27.3	1.15	0.4
(M18)	19.0	18.3	5 ± 0.2	3.5 ± 0.2	29.4	1.15	0.4
M20	21.1	20.3	6 ± 0.2	4 ± 0.2	33.5	1.3	0.4
(M22)	23.3	22.4	6 ± 0.2	4 ± 0.2	35.7	1.3	0.4
M24	25.3	24.4	7 ± 0.25	5 ± 0.2	39.8	1.65	0.5
(M27)	28.5	27.5	7 ± 0.25	5 ± 0.2	43.0	1.65	0.5
M30	31.5	30.5	8 ± 0.25	6 ± 0.25	48.0	2.0	0.8
(M33)	34.6	33.5	10 ± 0.25	6 ± 0.25	55.1	2.0	0.8
M36	37.6	36.5	10 ± 0.25	6 ± 0.25	58.1	2.0	0.8
(M39)	40.8	39.6	10 ± 0.25	6 ± 0.25	61.3	2.0	0.8
M42	43.8	42.6	12 ± 0.25	7 ± 0.25	68.3	2.3	0.8
(M45)	46.8	45.6	12 ± 0.25	7 ± 0.25	71.3	2.3	0.8
M48	50.0	48.8	12 ± 0.25	7 ± 0.25	74.5	2.3	0.8
(M52)	54.1	52.8	14 ± 0.25	8 ± 0.25	82.6	2.65	1.0
M56	58.1	56.8	14 ± 0.25	8 ± 0.25	86.6	2.65	1.0
(M60)	62.3	60.9	14 ± 0.25	8 ± 0.25	90.8	2.65	1.0
M64	66.3	64.9	14 ± 0.25	8 ± 0.25	93.8	2.65	1.0
(M68)	70.5	69.0	14 ± 0.25	8 ± 0.25	99.0	2.65	1.0

All dimensions are given in millimeters.   Sizes shown in parentheses are non-preferred, and are not usually stock sizes.

British Standard Double Coil Rectangular Section Spring Washers; Metric Series — Type D (BS 4464: 1969)

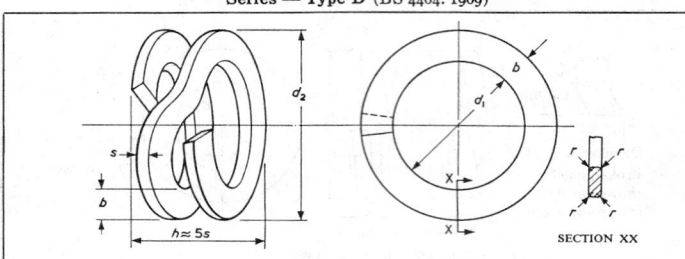

Nom. Size, d	Inside Diam., d_1		Width, b	Thick- ness, s	O.D., d_2 Max	Radius, r Max
	Max	Min				
M2	2.4	2.1	0.9 ± 0.1	0.5 ± 0.05	4.4	0.15
(M2.2)	2.6	2.3	1.0 ± 0.1	0.6 ± 0.05	4.8	0.2
M2.5	2.9	2.6	1.2 ± 0.1	0.7 ± 0.1	5.5	0.23
M3.0	3.6	3.3	1.2 ± 0.1	0.8 ± 0.1	6.2	0.25
(M3.5)	4.1	3.8	1.6 ± 0.1	0.8 ± 0.1	7.5	0.25
M4	4.6	4.3	1.6 ± 0.1	0.8 ± 0.1	8.0	0.25
M5	5.6	5.3	2 ± 0.1	0.9 ± 0.1	9.8	0.3
M6	6.6	6.3	3 ± 0.15	1 ± 0.1	12.9	0.33
M8	8.8	8.4	3 ± 0.15	1.2 ± 0.1	15.1	0.4
M10	10.8	10.4	3.5 ± 0.20	1.2 ± 0.1	18.2	0.4
M12	12.8	12.4	3.5 ± 0.2	1.6 ± 0.1	20.2	0.5
(M14)	15.0	14.5	5 ± 0.2	1.6 ± 0.1	25.4	0.5
M16	17.0	16.5	5 ± 0.2	2 ± 0.1	27.4	0.65
(M18)	19.0	18.5	5 ± 0.2	2 ± 0.1	29.4	0.65
M20	21.5	20.8	5 ± 0.2	2 ± 0.1	31.9	0.65
(M22)	23.5	22.8	6 ± 0.2	2.5 ± 0.15	35.9	0.8
M24	26.0	25.0	6.5 ± 0.2	3.25 ± 0.15	39.4	1.1
(M27)	29.5	28.0	7 ± 0.25	3.25 ± 0.15	44.0	1.1
M30	33.0	31.5	8 ± 0.25	3.25 ± 0.15	49.5	1.1
(M33)	36.0	34.5	8 ± 0.25	3.25 ± 0.15	52.5	1.1
M36	40.0	38.0	10 ± 0.25	3.25 ± 0.15	60.5	1.1
(M39)	43.0	41.0	10 ± 0.25	3.25 ± 0.15	63.5	1.1
M42	46.0	44.0	10 ± 0.25	4.5 ± 0.2	66.5	1.5
M48	52.0	50.0	10 ± 0.25	4.5 ± 0.2	72.5	1.5
M56	60.0	58.0	12 ± 0.25	4.5 ± 0.2	84.5	1.5
M64	70.0	67.0	12 ± 0.25	4.5 ± 0.2	94.5	1.5

All dimensions are given in millimeters.   Sizes shown in parentheses are non-preferred, and are not usually stock sizes.   The free height of double coil washers before compression is normally approximately five times the thickness but, if required, washers with other free heights may be obtained by arrangement with manufacturer.

British Standard Single Coil Square Section Spring Washers; Metric Series — Type A — 1 (BS 4464:1969)

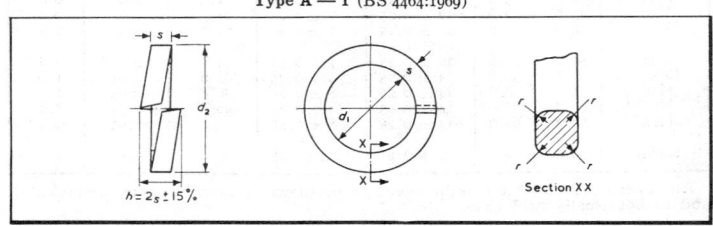

British Standard Single Coil Square Section Spring Washers; Metric Series —
Type A — 2 (BS 4464: 1969)

Nom. Size, d	Inside Diam., d_1		Thickness & Width, s	O.D., d_2 Max	Radius, r Max
	Max	Min			
M3	3.3	3.1	1 ± 0.1	5.5	0.3
(M3.5)	3.8	3.6	1 ± 0.1	6.0	0.3
M4	4.35	4.1	1.2 ± 0.1	6.95	0.4
M5	5.35	5.1	1.5 ± 0.1	8.55	0.5
M6	6.4	6.1	1.5 ± 0.1	9.6	0.5
M8	8.55	8.2	2 ± 0.1	12.75	0.65
M10	10.6	10.2	2.5 ± 0.15	15.9	0.8
M12	12.6	12.2	2.5 ± 0.15	17.9	0.8
(M14)	14.7	14.2	3 ± 0.2	21.1	1.0
M16	16.9	16.3	3.5 ± 0.2	24.3	1.15
(M18)	19.0	18.3	3.5 ± 0.2	26.4	1.15
M20	21.1	20.3	4.5 ± 0.2	30.5	1.5
(M22)	23.3	22.4	4.5 ± 0.2	32.7	1.5
M24	25.3	24.4	5 ± 0.2	35.7	1.65
(M27)	28.5	27.5	5 ± 0.2	38.9	1.65
M30	31.5	30.5	6 ± 0.2	43.9	2.0
(M33)	34.6	33.5	6 ± 0.2	47.0	2.0
M36	37.6	36.5	7 ± 0.25	52.1	2.3
(M39)	40.8	39.6	7 ± 0.25	55.3	2.3
M42	43.8	42.6	8 ± 0.25	60.3	2.65
(M45)	46.8	45.6	8 ± 0.25	63.3	2.65
M48	50.0	48.8	8 ± 0.25	66.5	2.65

All dimensions are given in millimeters. Sizes shown in parentheses are non-preferred and are not usually stock sizes.

British Standard for Metric Series Metal Washers. — BS 4320:1968 specifies bright and black metal washers for general engineering purposes.

Bright metal washers: These washers are made from either CS4 cold rolled strip steel (BS 1449:Part 3B) or from CZ 108 brass strip (BS 265), both in the hard condition. However, by mutual agreement between purchaser and supplier, washers may be made available with the material in any other condition, or they may be made from another material, or may be coated with a protective or decorative finish to some appropriate British Standard. Washers are reasonably flat and free from burrs and are normally supplied unchamfered. They may, however, have a 30-degree chamfer on one edge of the external diameter. These washers are made available in two size categories, normal and large diameter, and in two thicknesses, normal (Form A or C), and light (Form B or D). The thickness of a light range washer ranges from ½ to ⅔ the thickness of a normal range washer.

Black metal washers: These washers are made from mild steel, and can be supplied in three size categories designated normal, large, and extra large diameters. The normal diameter series are intended for bolts ranging from M5 to M68 (Form E washers); the large diameter series for bolts ranging from M8 to M39 (Form F washers) and the extra large series for bolts from M5 to M39 (Form G washers). A protective finish can be specified by the purchaser in accordance with any appropriate British Standard.

Washer designations: The Standard specifies the details that should be given when ordering or placing an inquiry for washers. They are the general description, namely, bright and black washers; the nominal size of the bolt or screw involved, for example, M5; the designated form, for example, Form A or Form E; the dimensions of any chamfer required on bright washers; the number of the Standard (BS 4320), and coating information if required, with the number of the appropriate British Standard and the coating thickness needed. As an example, in the use of this information, the designation for a chamfered, normal diameter series washer of normal range thickness to suit a 12-mm diameter bolt would be: Bright washers M12 (Form A) chamfered to BS 4320.

British Standard Bright Metal Washers — Metric Series (BS 4320:1968)

NORMAL DIAMETER SIZES

Nominal Size of Bolt or Screw	Inside Diameter			Outside Diameter			Thickness Form A (Normal Range)			Thickness Form B (Light Range)		
	Nom	Max	Min	Nom	Max	Min	Nom	Max	Min	Nom	Max	Min
M 1.0	1.1	1.25	1.1	2.5	2.5	2.3	0.3	0.4	0.2	....	....	....
M 1.2	1.3	1.45	1.3	3.0	3.0	2.8	0.3	0.4	0.2	....	....	....
(M 1.4)	1.5	1.65	1.5	3.0	3.0	2.8	0.3	0.4	0.2	....	....	....
M 1.6	1.7	1.85	1.7	4.0	4.0	3.7	0.3	0.4	0.2	....	....	....
M 2.0	2.2	2.35	2.2	5.0	5.0	4.7	0.3	0.4	0.2	....	....	....
(M 2.2)	2.4	2.55	2.4	5.0	5.0	4.7	0.5	0.6	0.4	....	....	....
M 2.5	2.7	2.85	2.7	6.5	6.5	6.2	0.5	0.6	0.4	....	....	....
M 3	3.2	3.4	3.2	7	7	6.7	0.5	0.6	0.4	....	....	....
(M 3.5)	3.7	3.9	3.7	7	7	6.7	0.5	0.6	0.4	....	....	....
M 4	4.3	4.5	4.3	9	9	8.7	0.8	0.9	0.7	....	....	....
(M 4.5)	4.8	5.0	4.8	9	9	8.7	0.8	0.9	0.7	....	....	....
M 5	5.3	5.5	5.3	10	10	9.7	1.0	1.1	0.9	....	....	....
M 6	6.4	6.7	6.4	12.5	12.5	12.1	1.6	1.8	1.4	0.8	0.9	0.7
(M 7)	7.4	7.7	7.4	14	14	13.6	1.6	1.8	1.4	0.8	0.9	0.7
M 8	8.4	8.7	8.4	17	17	16.6	1.6	1.8	1.4	1.0	1.1	0.9
M 10	10.5	10.9	10.5	21	21	20.5	2.0	2.2	1.8	1.25	1.45	1.05
M 12	13.0	13.4	13.0	24	24	23.5	2.5	2.7	2.3	1.6	1.80	1.40
(M 14)	15.0	15.4	15.0	28	28	27.5	2.5	2.7	2.3	1.6	1.8	1.4
M 16	17.0	17.4	17.0	30	30	29.5	3.0	3.3	2.7	2.0	2.2	1.8
(M 18)	19.0	19.5	19.0	34	34	33.2	3.0	3.3	2.7	2.0	2.2	1.8
M 20	21	21.5	21	37	37	36.2	3.0	3.3	2.7	2.0	2.2	1.8
(M 22)	23	23.5	23	39	39	38.2	3.0	3.3	2.7	2.0	2.2	1.8
M 24	25	25.5	25	44	44	43.2	4.0	4.3	3.7	2.5	2.7	2.3
(M 27)	28	28.5	28	50	50	49.2	4.0	4.3	3.7	2.5	2.7	2.3
M 30	31	31.6	31	56	56	55.0	4.0	4.3	3.7	2.5	2.7	2.3
(M 33)	34	34.6	34	60	60	59.0	5.0	5.6	4.4	3.0	3.3	2.7
M 36	37	37.6	37	66	66	65.0	5.0	5.6	4.4	3.0	3.3	2.7
(M 39)	40	40.6	40	72	72	71.0	6.0	6.6	5.4	3.0	3.3	2.7

LARGE DIAMETER SIZES

Nominal Size of Bolt or Screw	Inside Diameter			Outside Diameter			Thickness Form C (Normal Range)			Thickness Form D (Light Range)		
	Nom	Max	Min	Nom	Max	Min	Nom	Max	Min	Nom	Max	Min
M 4	4.3	4.5	4.3	10.0	10.0	9.7	0.8	0.9	0.7	....	....	....
M 5	5.3	5.5	5.3	12.5	12.5	12.1	1.0	1.1	0.9	....	....	....
M 6	6.4	6.7	6.4	14	14	13.6	1.6	1.8	1.4	0.8	0.9	0.7
M 8	8.4	8.7	8.4	21	21	20.5	1.6	1.8	1.4	1.0	1.1	0.9
M 10	10.5	10.9	10.5	24	24	23.5	2.0	2.2	1.8	1.25	1.45	1.05
M 12	13.0	13.4	13.0	28	28	27.5	2.5	2.7	2.3	1.6	1.8	1.4
(M 14)	15.0	15.4	15	30	30	29.5	2.5	2.7	2.3	1.6	1.8	1.4
M 16	17.0	17.4	17	34	34	33.2	3.0	3.3	2.7	2.0	2.2	1.8
(M 18)	19.0	19.5	19	37	37	36.2	3.0	3.3	2.7	2.0	2.2	1.8
M 20	21	21.5	21	39	39	38.2	3.0	3.3	2.7	2.0	2.2	1.8
(M 22)	23	23.5	23	44	44	43.2	3.0	3.3	2.7	2.0	2.2	1.8
M 24	25	25.5	25	50	50	49.2	4.0	4.3	3.7	2.5	2.7	2.3
(M 27)	28	28.5	28	56	56	55	4.0	4.3	3.7	2.5	2.7	2.3
M 30	31	31.6	31	60	60	59	4.0	4.3	3.7	2.5	2.7	2.3
(M 33)	34	34.6	34	66	66	65	5.0	5.6	4.4	3.0	3.3	2.7
M 36	37	37.6	37	72	72	71	5.0	5.6	4.4	3.0	3.3	2.7
(M 39)	40	40.6	40	77	77	76	6.0	6.6	5.4	3.0	3.3	2.7

All dimensions are given in millimeters.
Nominal bolt or screw sizes shown in parentheses are non-preferred.

British Standard Black Metal Washers — Metric Series (BS 4320:1968)

Nom Bolt or Screw Size	Inside Diameter			Outside Diameter			Thickness		
	Nom	Max	Min	Nom	Max	Min	Nom	Max	Min
NORMAL DIAMETER SIZES (Form E)									
M 5	5.5	5.8	5.5	10.0	10.0	9.2	1.0	1.2	0.8
M 6	6.6	7.0	6.6	12.5	12.5	11.7	1.6	1.9	1.3
(M 7)	7.6	8.0	7.6	14.0	14.0	13.2	1.6	1.9	1.3
M 8	9.0	9.4	9.0	17	17	16.2	1.6	1.9	1.3
M 10	11.0	11.5	11.0	21	21	20.2	2.0	2.3	1.7
M 12	14	14.5	14	24	24	23.2	2.5	2.8	2.2
(M 14)	16	16.5	16	28	28	27.2	2.5	2.8	2.2
M 16	18	18.5	18	30	30	29.2	3.0	3.6	2.4
(M 18)	20	20.6	20	34	34	32.8	3.0	3.6	2.4
M 20	22	22.6	22	37	37	35.8	3.0	3.6	2.4
(M 22)	24	24.6	24	39	39	37.8	3.0	3.6	2.4
M 24	26	26.6	26	44	44	42.8	4	4.6	3.4
(M 27)	30	30.6	30	50	50	48.8	4	4.6	3.4
M 30	33	33.8	33	56	56	54.5	4	4.6	3.4
(M 33)	36	36.8	36	60	60	58.5	5	6.0	4.0
M 36	39	39.8	39	66	66	64.5	5	6.0	4.0
(M 39)	42	42.8	42	72	72	70.5	6	7.0	5.0
(M 42)	45	45.8	45	78	78	76.5	7	8.2	5.8
(M 45)	48	48.8	48	85	85	83	7	8.2	5.8
M 48	52	53	52	92	92	90	8	9.2	6.8
(M 52)	56	57	56	98	98	96	8	9.2	6.8
M 56	62	63	62	105	105	103	9	10.2	7.8
(M 60)	66	67	66	110	110	108	9	10.2	7.8
M 64	70	71	70	115	115	113	9	10.2	7.8
(M 68)	74	75	74	120	120	118	10	11.2	8.8
LARGE DIAMETER SIZES (Form F)									
M 8	9	9.4	9.0	21	21	20.2	1.6	1.9	1.3
M 10	11	11.5	11	24	24	23.2	2	2.3	1.7
M 12	14	14.5	14	28	28	27.2	2.5	2.8	2.2
(M 14)	16	16.5	16	30	30	29.2	2.5	2.8	2.2
M 16	18	18.5	18	34	34	32.8	3	3.6	2.4
(M 18)	20	20.6	20	37	37	35.8	3	3.6	2.4
M 20	22	22.6	22	39	39	37.8	3	3.6	2.4
(M 22)	24	24.6	24	44	44	42.8	3	3.6	2.4
M 24	26	26.6	26	50	50	48.8	4	4.6	3.4
(M 27)	30	30.6	30	56	56	54.5	4	4.6	3.4
M 30	33	33.8	33	60	60	58.5	4	4.6	3.4
(M 33)	36	36.8	36	66	66	64.5	5	6.0	4
M 36	39	39.8	39	72	72	70.5	5	6.0	4
(M 39)	42	42.8	42	77	77	75.5	6	7	5
EXTRA LARGE DIAMETER SIZES (Form G)									
M 5	5.5	5.8	5.5	15	15	14.2	1.6	1.9	1.3
M 6	6.6	7.0	6.6	18	18	17.2	2	2.3	1.7
(M 7)	7.6	8.0	7.6	21	21	20.2	2	2.3	1.7
M 8	9	9.4	9.0	24	24	23.2	2	2.3	1.7
M 10	11	11.5	11.0	30	30	29.2	2.5	2.8	2.2
M 12	14	14.5	14.0	36	36	34.8	3	3.6	2.4
(M 14)	16	16.5	16.0	42	42	40.8	3	3.6	2.4
M 16	18	18.5	18	48	48	46.8	4	4.6	3.4
(M 18)	20	20.6	20	54	54	52.5	4	4.6	3.4
M 20	22	22.6	22	60	60	58.5	5	6.0	4
(M 22)	24	24.6	24	66	66	64.5	5	6.0	4
M 24	26	26.6	26	72	72	70.5	6	7	5
(M 27)	30	30.6	30	81	81	79	6	7	5
M 30	33	33.8	33	90	90	88	8	9.2	6.8
(M 33)	36	36.8	36	99	99	97	8	9.2	6.8
M 36	39	39.8	39	108	108	106	10	11.2	8.8
(M 39)	42	42.8	42	117	117	115	10	11.2	8.8

All dimensions are given in millimeters.
Nominal bolt or screw sizes shown in parentheses are non-preferred.

MACHINE SCREWS

American National Standard Machine Screws and Machine Screw Nuts. — This Standard (ANSI B18.6.3) covers both slotted and recessed head machine screws. Dimensions of various types of slotted machine screws, machine screw nuts, and header points are given in Tables 1 through 12. The Standard also covers flat trim head, oval trim head and drilled fillister head machine screws and gives cross recess dimensions and gaging dimensions for all types of machine screw heads. Information on metric machine screws (B18.6.7M) is given beginning on page 1366.

Threads: Except for sizes 0000, 000, and 00, machine screw threads may be either Unified Coarse (UNC) and Fine thread (UNF) Class 2A (see Unified Screw Threads starting on page 1479) or UNRC and UNRF Series, at option of manufacturer. Thread dimensions for sizes 0000, 000, and 00 are given in Table 7 on page 1361.

Threads for hexagon machine screw nuts may be either UNC or UNF, Class 2B, and for square machine screw nuts are UNC Class 2B.

Length of thread: Machine screws of sizes No. 5 and smaller with nominal lengths equal to 3 diameters and shorter have full form threads extending to within 1 pitch (thread) of the bearing surface of the head, or closer, if practicable. Nominal lengths greater than 3 diameters, up to and including 1⅛ inch, have full form threads extending to within two pitches (threads) of the bearing surface of the head, or closer, if practicable. Unless otherwise specified, screws of longer nominal length have a minimum length of full form thread of 1.00 inch.

Machine screws of sizes No. 6 and larger with nominal length equal to 3 diameters and shorter have full form threads extending to within 1 pitch (thread) of the bearing surface of the head, or closer, if practicable. Nominal lengths greater than 3 diameters, up to and including 2 inches, have full form threads extending to within 2 pitches (threads) of the bearing surface of the head, or closer, if practicable. Screws of longer nominal length, unless otherwise specified, have a minimum length of full form thread of 1.50 inches.

Table 1. American National Standard Square and Hexagon Machine Screw Nuts
(ANSI B18.6.3-1972, R1977)

Nom. Size	Basic Diam.	Basic F	Max. F	Min. F	Max. G	Min. G	Max. G_1	Min. G_1	Max. H	Min. H
0	.0600	⁵⁄₃₂	.156	.150	.221	.206	.180	.171	.050	.043
1	.0730	⁵⁄₃₂	.156	.150	.221	.206	.180	.171	.050	.043
2	.0860	³⁄₁₆	.188	.180	.265	.247	.217	.205	.066	.057
3	.0990	³⁄₁₆	.188	.180	.265	.247	.217	.205	.066	.057
4	.1120	¼	.250	.241	.354	.331	.289	.275	.098	.087
5	.1250	⁵⁄₁₆	.312	.302	.442	.415	.361	.344	.114	.102
6	.1380	⁵⁄₁₆	.312	.302	.442	.415	.361	.344	.114	.102
8	.1640	11⁄₃₂	.344	.332	.486	.456	.397	.378	.130	.117
10	.1900	⅜	.375	.362	.530	.497	.433	.413	.130	.117
12	.2160	⁷⁄₁₆	.438	.423	.619	.581	.505	.482	.161	.148
¼	.2500	⁷⁄₁₆	.438	.423	.619	.581	.505	.482	.193	.178
⁵⁄₁₆	.3125	⁹⁄₁₆	.562	.545	.795	.748	.650	.621	.225	.208
⅜	.3750	⅝	.625	.607	.884	.833	.722	.692	.257	.239

All dimensions in inches.   Hexagon machine screw nuts have tops flat and chamfered. Diameter of top circle should be the maximum width across flats within a tolerance of minus 15 per cent.   Bottoms are flat but may be chamfered if so specified.   Square machine screw nuts have tops and bottoms flat without chamfer.

Diameter of body: The diameter of machine screw bodies is not less than Class 2A thread minimum pitch diameter nor greater than the basic major diameter of the thread. Cross-recessed trim head machine screws not threaded to the head have an 0.062 in. minimum length shoulder under the head with diameter limits as specified in the dimensional tables in the standard.

Designation: Machine screws are designated by the following data in the sequence shown: Nominal size (number, fraction or decimal equivalent); threads per inch; nominal length (fraction or decimal equivalent); product name, including head type and driving provision; header point, if desired; material; and protective finish, if required. For example:

¼ - 20 x 1¼ Slotted Pan Head Machine Screw, Steel, Zinc Plated
6 - 32 x ¾ Type IA Cross Recessed Fillister Head Machine Screw, Brass.

Machine screw nuts are designated by the following data in sequence shown: Nominal size (number, fraction or decimal equivalent); threads per inch; product name; material; and protective finish, if required. For example:

10 - 24 Hexagon Machine Screw Nut, Steel, Zinc Plated
.138 - 32 Square Machine Screw Nut, Brass.

Table 2. American National Standard Slotted 100-Degree Flat Countersunk Head Machine Screws (ANSI B18.6.3-1972, R1977)

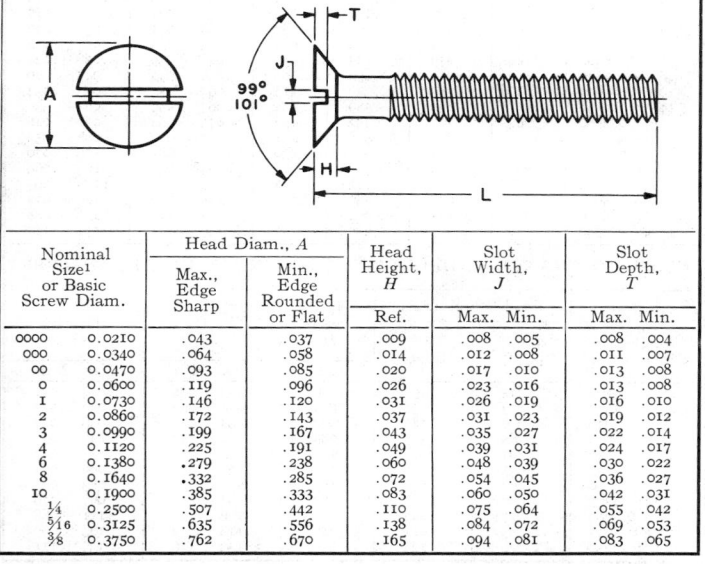

Nominal Size[1] or Basic Screw Diam.		Head Diam., A		Head Height, H	Slot Width, J		Slot Depth, T	
		Max., Edge Sharp	Min., Edge Rounded or Flat	Ref.	Max.	Min.	Max.	Min.
0000	0.0210	.043	.037	.009	.008	.005	.008	.004
000	0.0340	.064	.058	.014	.012	.008	.011	.007
00	0.0470	.093	.085	.020	.017	.010	.013	.008
0	0.0600	.119	.096	.026	.023	.016	.013	.008
1	0.0730	.146	.120	.031	.026	.019	.016	.010
2	0.0860	.172	.143	.037	.031	.023	.019	.012
3	0.0990	.199	.167	.043	.035	.027	.022	.014
4	0.1120	.225	.191	.049	.039	.031	.024	.017
6	0.1380	.279	.238	.060	.048	.039	.030	.022
8	0.1640	.332	.285	.072	.054	.045	.036	.027
10	0.1900	.385	.333	.083	.060	.050	.042	.031
¼	0.2500	.507	.442	.110	.075	.064	.055	.042
5⁄16	0.3125	.635	.556	.138	.084	.072	.069	.053
3⁄8	0.3750	.762	.670	.165	.094	.081	.083	.065

All dimensions are in inches.
[1] When specifying nominal size in decimals, zeros preceding the decimal point and in the fourth decimal place are omitted.

Table 3. American National Standard Slotted Flat Countersunk Head and Close Tolerance 100-Degree Flat Countersunk Head Machine Screws
(ANSI B18.6.3-1972, R1977)

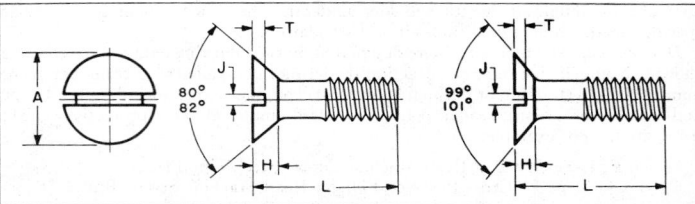

SLOTTED FLAT COUNTERSUNK HEAD TYPE

Nominal Size[1] or Basic Screw Diam.		Max., L[2]	Head Diam., A		Head Height, H	Slot Width, J		Slot Depth, T	
			Max., Edge Sharp	Min., Edge[3]	Ref.	Max.	Min.	Max.	Min.
0000	0.0210	....	.043	.037	.011	.008	.004	.007	.003
000	0.0340	....	.064	.058	.016	.011	.007	.009	.005
00	0.0470	....	.093	.085	.028	.017	.010	.014	.009
0	0.0600	⅛	.119	.099	.035	.023	.016	.015	.010
1	0.0730	⅛	.146	.123	.043	.026	.019	.019	.012
2	0.0860	⅛	.172	.147	.051	.031	.023	.023	.015
3	0.0990	⅛	.199	.171	.059	.035	.027	.027	.017
4	0.1120	³⁄₁₆	.225	.195	.067	.039	.031	.030	.020
5	0.1250	³⁄₁₆	.252	.220	.075	.043	.035	.034	.022
6	0.1380	³⁄₁₆	.279	.244	.083	.048	.039	.038	.024
8	0.1640	¼	.332	.292	.100	.054	.045	.045	.029
10	0.1900	⁵⁄₁₆	.385	.340	.116	.060	.050	.053	.034
12	0.2160	⅜	.438	.389	.132	.067	.056	.060	.039
¼	0.2500	⁷⁄₁₆	.507	.452	.153	.075	.064	.070	.046
⁵⁄₁₆	0.3125	½	.635	.568	.191	.084	.072	.088	.058
⅜	0.3750	⁹⁄₁₆	.762	.685	.230	.094	.081	.106	.070
⁷⁄₁₆	0.4375	⅝	.812	.723	.223	.094	.081	.103	.066
½	0.5000	¾	.875	.775	.223	.106	.091	.103	.065
⁹⁄₁₆	0.5625	...	1.000	.889	.260	.118	.102	.120	.077
⅝	0.6250	...	1.125	1.002	.298	.133	.116	.137	.088
¾	0.7500	...	1.375	1.230	.372	.149	.131	.171	.111

CLOSE TOLERANCE 100-DEGREE FLAT COUNTERSUNK HEAD TYPE

Nominal Size[1] or Basic Screw Diam.		Head Diameter, A		Head Height, H	Slot Width, J		Slot Depth, T	
		Max., Edge Sharp	Min., Edge[3]	Ref.	Max.	Min.	Max.	Min.
4	0.1120	.225	.191	.049	.039	.031	.024	.017
6	0.1380	.279	.238	.060	.048	.039	.030	..022
8	0.1640	.332	.285	.072	.054	.045	.036	.027
10	0.1900	.385	.333	.083	.060	.050	.042	.031
¼	0.2500	.507	.442	.110	.075	.064	.055	.042
⁵⁄₁₆	0.3125	.635	.556	.138	.084	.072	.069	.053
⅜	0.3750	.762	.670	.165	.094	.081	.083	.065
⁷⁄₁₆	0.4375	.890	.783	.193	.094	.081	.097	.076
½	0.5000	1.017	.897	.221	.106	.091	.111	.088
⁹⁄₁₆	0.5625	1.145	1.011	.249	.118	.102	.125	.099
⅝	0.6250	1.272	1.124	.276	.133	.116	.139	.111

All dimensions are in inches.   [1] When specifying nominal size in decimals, zeros preceding the decimal point and in the fourth decimal place are omitted.   [2] These lengths or shorter are undercut.   [3] May be rounded or flat.

Table 4. American National Standard Slotted Undercut Flat Countersunk Head and Plain and Slotted Hex Washer Head Machine Screws (ANSI B18.6.3-1972, R1977)

SLOTTED UNDERCUT FLAT COUNTERSUNK HEAD TYPE

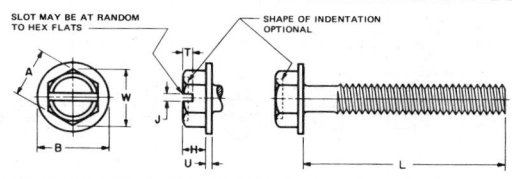

Nominal Size[1] or Basic Screw Diam.	Max., L[2]	Head Diam., A		Head Height, H		Slot Width, J		Slot Depth, T	
		Max., Edge Sharp	Min., Edge Rnded. or Flat	Max.	Min.	Max.	Min.	Max.	Min.
0	0.0600	1/8	.119	.099	.025	.018	.023	.016	.011	.007
1	0.0730	1/8	.146	.123	.031	.023	.026	.019	.014	.009
2	0.0860	1/8	.172	.147	.036	.028	.031	.023	.016	.011
3	0.0990	1/8	.199	.171	.042	.033	.035	.027	.019	.012
4	0.1120	3/16	.225	.195	.047	.038	.039	.031	.022	.014
5	0.1250	3/16	.252	.220	.053	.043	.043	.035	.024	.016
6	0.1380	3/16	.279	.244	.059	.048	.048	.039	.027	.017
8	0.1640	1/4	.332	.292	.070	.058	.054	.045	.032	.021
10	0.1900	5/16	.385	.340	.081	.068	.060	.050	.037	.024
12	0.2160	3/8	.438	.389	.092	.078	.067	.056	.043	.028
1/4	0.2500	7/16	.507	.452	.107	.092	.075	.064	.050	.032
5/16	0.3125	1/2	.635	.568	.134	.116	.084	.072	.062	.041
3/8	0.3750	9/16	.762	.685	.161	.140	.094	.081	.075	.049
7/16	0.4375	5/8	.812	.723	.156	.133	.094	.081	.072	.045
1/2	0.5000	3/4	.875	.775	.156	.130	.106	.091	.072	.046

PLAIN AND SLOTTED HEX WASHER HEAD TYPES

Nominal Size[1] or Basic Screw Diam.	Width Across Flats, A		Width Across Corn., W	Head Height, H		Washer Diam., B		Washer Thick., U		Slot[3] Width, J		Slot[3] Depth, T	
	Max.	Min.	Min.	Max.	Min.	Max.	Min.	Max.	Min.	Max.	Min.	Max.	Min.
2	0.0860	.125	.120	.134	.050	.040	.166	.154	.016	.010	....	....	....	...
3	0.0990	.125	.120	.134	.055	.044	.177	.163	.016	.010	....	....	....	...
4	0.1120	.188	.181	.202	.060	.049	.243	.225	.019	.011	.039	.031	.042	.025
5	0.1250	.188	.181	.202	.070	.058	.260	.240	.025	.015	.043	.035	.049	.030
6	0.1380	.250	.244	.272	.093	.080	.328	.302	.025	.015	.048	.039	.053	.033
8	0.1640	.250	.244	.272	.110	.096	.348	.322	.031	.019	.054	.045	.074	.052
10	0.1900	.312	.305	.340	.120	.105	.414	.384	.031	.019	.060	.050	.080	.057
12	0.2160	.312	.305	.340	.155	.139	.432	.398	.039	.022	.067	.056	.103	.077
1/4	0.2500	.375	.367	.409	.190	.172	.520	.480	.050	.030	.075	.064	.111	.083
5/16	0.3125	.500	.489	.545	.230	.208	.676	.624	.055	.035	.084	.072	.134	.100
3/8	0.3750	.562	.551	.614	.295	.270	.780	.720	.063	.037	.094	.081	.168	.131

All dimensions are in inches.
[1] When specifying nominal size in decimals, zeros preceding the decimal point and in the fourth decimal place are omitted.
[2] These lengths or shorter are undercut.
[3] Unless otherwise specified, hexagon washer head machine screws are not slotted.

Table 5. **American National Standard Slotted Truss Head and Plain and Slotted Hexagon Head Machine Screws** (ANSI B18.6.3-1972, R1977)

SLOTTED TRUSS HEAD TYPE

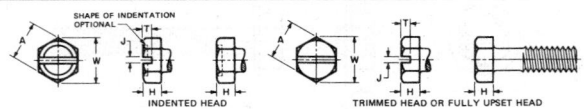

Nominal Size[1] or Basic Screw Diam.		Head Diam., A		Head Height, H		Head Radius, R	Slot Width, J		Slot Depth, T	
		Max.	Min.	Max.	Min.	Max.	Max.	Min.	Max.	Min.
0000	0.0210	.049	.043	.014	.010	.032	.009	.005	.009	.005
000	0.0340	.077	.071	.022	.018	.051	.013	.009	.013	.009
00	0.0470	.106	.098	.030	.024	.070	.017	.010	.018	.012
0	0.0600	.131	.119	.037	.029	.087	.023	.016	.022	.014
1	0.0730	.164	.149	.045	.037	.107	.026	.019	.027	.018
2	0.0860	.194	.180	.053	.044	.129	.031	.023	.031	.022
3	0.0990	.226	.211	.061	.051	.151	.035	.027	.036	.026
4	0.1120	.257	.241	.069	.059	.169	.039	.031	.040	.030
5	0.1250	.289	.272	.078	.066	.191	.043	.035	.045	.034
6	0.1380	.321	.303	.086	.074	.211	.048	.039	.050	.037
8	0.1640	.384	.364	.102	.088	.254	.054	.045	.058	.045
10	0.1900	.448	.425	.118	.103	.283	.060	.050	.068	.053
12	0.2160	.511	.487	.134	.118	.336	.067	.056	.077	.061
¼	0.2500	.573	.546	.150	.133	.375	.075	.064	.087	.070
⁵⁄₁₆	0.3125	.698	.666	.183	.162	.457	.084	.072	.106	.085
⅜	0.3750	.823	.787	.215	.191	.538	.094	.081	.124	.100
⁷⁄₁₆	0.4375	.948	.907	.248	.221	.619	.094	.081	.142	.116
½	0.5000	1.073	1.028	.280	.250	.701	.106	.091	.161	.131
⁹⁄₁₆	0.5625	1.198	1.149	.312	.279	.783	.118	.102	.179	.146
⅝	0.6250	1.323	1.269	.345	.309	.863	.133	.116	.196	.162
¾	0.7500	1.573	1.511	.410	.368	1.024	.149	.131	.234	.182

PLAIN AND SLOTTED HEXAGON HEAD TYPES

Nominal Size[1] or Basic Screw Diam.		Regular Head			Large Head			Head Height, H		Slot[2] Width, J		Slot[2] Depth, T	
		Width Across Flats, A		Across Corn., W	Width Across Flats, A		Across Corn., W						
		Max.	Min.	Min.	Max.	Min.	Min.	Max.	Min.	Max.	Min.	Max.	Min.
1	0.0730	.125	.120	.134	....	....	....	.044	.036	....	....	....	....
2	0.0860	.125	.120	.134	....	....	....	.050	.040	....	....	....	....
3	0.0990	.188	.181	.202	....	....	....	.055	.044	....	....	....	....
4	0.1120	.188	.181	.202	.219	.213	.238	.060	.049	.039	.031	.036	.025
5	0.1250	.188	.181	.202	.250	.244	.272	.070	.058	.043	.035	.042	.030
6	0.1380	.250	.244	.272	....	....	....	.093	.080	.048	.039	.046	.033
8	0.1640	.250	.244	.272	.312	.305	.340	.110	.096	.054	.045	.066	.052
10	0.1900	.312	.305	.340	....	....	....	.120	.105	.060	.050	.072	.057
12	0.2160	.312	.305	.340	.375	.367	.409	.155	.139	.067	.056	.093	.077
¼	0.2500	.375	.367	.409	.438	.428	.477	.190	.172	.075	.064	.101	.083
⁵⁄₁₆	0.3125	.500	.489	.545	....	....	....	.230	.208	.084	.072	.122	.100
⅜	0.3750	.562	.551	.614	....	....	....	.295	.270	.094	.081	.156	.131

All dimensions are in inches. [1] Where specifying nominal size in decimals, zeros preceding decimal points and in the fourth decimal place are omitted. [2] Unless otherwise specified, hexagon head machine screws are not slotted.

Table 6.   American National Standard Slotted Pan Head Machine Screws
(ANSI B18.6.3-1972, R1977)

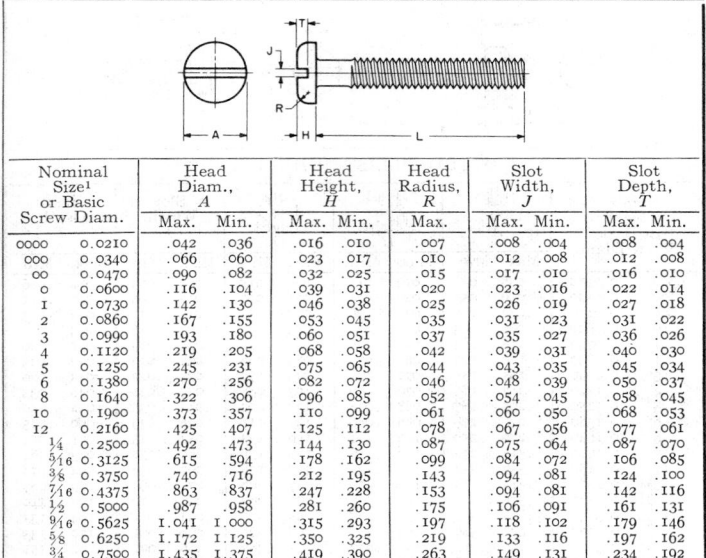

Nominal Size[1] or Basic Screw Diam.		Head Diam., A		Head Height, H		Head Radius, R	Slot Width, J		Slot Depth, T	
		Max.	Min.	Max.	Min.	Max.	Max.	Min.	Max.	Min.
0000	0.0210	.042	.036	.016	.010	.007	.008	.004	.008	.004
000	0.0340	.066	.060	.023	.017	.010	.012	.008	.012	.008
00	0.0470	.090	.082	.032	.025	.015	.017	.010	.016	.010
0	0.0600	.116	.104	.039	.031	.020	.023	.016	.022	.014
1	0.0730	.142	.130	.046	.038	.025	.026	.019	.027	.018
2	0.0860	.167	.155	.053	.045	.035	.031	.023	.031	.022
3	0.0990	.193	.180	.060	.051	.037	.035	.027	.036	.026
4	0.1120	.219	.205	.068	.058	.042	.039	.031	.040	.030
5	0.1250	.245	.231	.075	.065	.044	.043	.035	.045	.034
6	0.1380	.270	.256	.082	.072	.046	.048	.039	.050	.037
8	0.1640	.322	.306	.096	.085	.052	.054	.045	.058	.045
10	0.1900	.373	.357	.110	.099	.061	.060	.050	.068	.053
12	0.2160	.425	.407	.125	.112	.078	.067	.056	.077	.061
¼	0.2500	.492	.473	.144	.130	.087	.075	.064	.087	.070
⁵⁄₁₆	0.3125	.615	.594	.178	.162	.099	.084	.072	.106	.085
⅜	0.3750	.740	.716	.212	.195	.143	.094	.081	.124	.100
⁷⁄₁₆	0.4375	.863	.837	.247	.228	.153	.094	.081	.142	.116
½	0.5000	.987	.958	.281	.260	.175	.106	.091	.161	.131
⁹⁄₁₆	0.5625	1.041	1.000	.315	.293	.197	.118	.102	.179	.146
⅝	0.6250	1.172	1.125	.350	.325	.219	.133	.116	.197	.162
¾	0.7500	1.435	1.375	.419	.390	.263	.149	.131	.234	.192

All dimensions are in inches.
[1] Where specifying nominal size in decimals, zeros preceding decimal and in the fourth decimal place are omitted.

Table 7.   Nos. 0000, 000 and 00 Threads (ANSI B18.6.3, R1977 Appendix)

Nominal Size[1] and Threads Per Inch	Series Designat.	Class	External[2]						Internal[3]			
			Major Diameter		Pitch Diameter		Minor Diam.	Class	Pitch Diameter		Major Diam.	
			Max.	Min.	Max.	Min.	Tol.			Min.	Max.	Tol.	Min.
0000 – 160 or 0.0210 – 160	NS	2	.0210	.0195	.0169	.0158	.0011	.0128	2	.0169	.0181	.0012	.0210
000 – 120 or 0.0340 – 120	NS	2	.0340	.0325	.0286	.0272	.0014	.0232	2	.0286	.0300	.0014	.0340
00 – 90 or 0.0470 – 90	NS	2	.0470	.0450	.0398	.0382	.0016	.0326	2	.0398	.0414	.0016	.0470
00 – 96 or 0.0470 – 96	NS	2	.0470	.0450	.0402	.0386	.0016	.0334	2	.0402	.0418	.0016	.0470

All dimensions are in inches.
[1] Where specifying nominal size in decimals, zeros preceding decimal and in the fourth decimal place are omitted.
[2] There is no allowance provided on the external threads.
[3] The minor diameter limits for internal threads are not specified, they being determined by the amount of thread engagement necessary to satisfy the strength requirements and tapping performance in the intended application.

Table 8. American National Standard Slotted Fillister and Slotted Drilled Fillister Head Machine Screws (ANSI B18.6.3-1972, R1977)

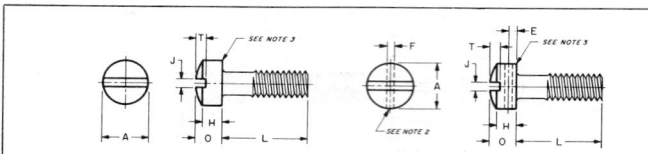

SLOTTED FILLISTER HEAD TYPE

Nominal Size[1] or Basic Screw Diam.		Head Diam., A		Head Side Height, H		Total Head Height, O		Slot Width, J		Slot Depth, T	
		Max.	Min.	Max.	Min.	Max.	Min.	Max.	Min.	Max.	Min.
0000	0.0210	.038	.032	.019	.011	.025	.015	.008	.004	.012	.006
000	0.0340	.059	.053	.029	.021	.035	.027	.012	.006	.017	.011
00	0.0470	.082	.072	.037	.028	.047	.039	.017	.010	.022	.015
0	0.0600	.096	.083	.043	.038	.055	.047	.023	.016	.025	.015
1	0.0730	.118	.104	.053	.045	.066	.058	.026	.019	.031	.020
2	0.0860	.140	.124	.062	.053	.083	.066	.031	.023	.037	.025
3	0.0990	.161	.145	.070	.061	.095	.077	.035	.027	.043	.030
4	0.1120	.183	.166	.079	.069	.107	.088	.039	.031	.048	.035
5	0.1250	.205	.187	.088	.078	.120	.100	.043	.035	.054	.040
6	0.1380	.226	.208	.096	.086	.132	.111	.048	.039	.060	.045
8	0.1640	.270	.250	.113	.102	.156	.133	.054	.045	.071	.054
10	0.1900	.313	.292	.130	.118	.180	.156	.060	.050	.083	.064
12	0.2160	.357	.334	.148	.134	.205	.178	.067	.056	.094	.074
1/4	0.2500	.414	.389	.170	.155	.237	.207	.075	.064	.109	.087
5/16	0.3125	.518	.490	.211	.194	.295	.262	.084	.072	.137	.110
3/8	0.3750	.622	.590	.253	.233	.355	.315	.094	.081	.164	.133
7/16	0.4375	.625	.589	.265	.242	.368	.321	.094	.081	.170	.135
1/2	0.5000	.750	.710	.297	.273	.412	.362	.106	.091	.190	.151
9/16	0.5625	.812	.768	.336	.308	.466	.410	.118	.102	.214	.172
5/8	0.6250	.875	.827	.375	.345	.521	.461	.133	.116	.240	.193
3/4	0.7500	1.000	.945	.441	.406	.612	.542	.149	.131	.281	.226

SLOTTED DRILLED FILLISTER HEAD TYPE

Nominal Size[1] or Basic Screw Diam.		Head Diam., A		Head Side Height, H		Total Head Height, O		Slot Width, J		Slot Depth, T		Drilled Hole Locat., E	Drilled Hole Diam., F
		Max.	Min.	Max.	Min.	Max.	Min.	Max.	Min.	Max.	Min.	Basic	Basic
2	0.0860	.140	.124	.062	.055	.083	.070	.031	.023	.030	.022	.026	.031
3	0.0990	.161	.145	.070	.064	.095	.082	.035	.027	.034	.026	.030	.037
4	0.1120	.183	.166	.079	.072	.107	.094	.039	.031	.038	.030	.035	.037
5	0.1250	.205	.187	.088	.081	.120	.106	.043	.035	.042	.033	.038	.046
6	0.1380	.226	.208	.096	.089	.132	.118	.048	.039	.045	.035	.043	.046
8	0.1640	.270	.250	.113	.106	.156	.141	.054	.045	.065	.054	.043	.046
10	0.1900	.313	.292	.130	.123	.180	.165	.060	.050	.075	.064	.043	.046
12	0.2160	.357	.334	.148	.139	.205	.188	.067	.056	.087	.074	.053	.046
1/4	0.2500	.414	.389	.170	.161	.237	.219	.075	.064	.102	.087	.062	.062
5/16	0.3125	.518	.490	.211	.201	.295	.276	.084	.072	.130	.110	.078	.070
3/8	0.3750	.622	.590	.253	.242	.355	.333	.094	.081	.154	.134	.094	.070

All dimensions are in inches.

[1] Where specifying nominal size in decimals, zeros preceding decimal points and in the fourth decimal place are omitted.

[2] Drilled hole shall be approximately perpendicular to the axis of slot and may be permitted to break through bottom of the slot. Edges of the hole shall be free from burrs.

[3] A slight rounding of the edges at periphery of head is permissible provided the diameter of the bearing circle is equal to no less than 90 per cent of the specified minimum head diameter.

Table 9. American National Standard Slotted Oval Countersunk Head Machine Screws (ANSI B18.6.3-1972, R1977)

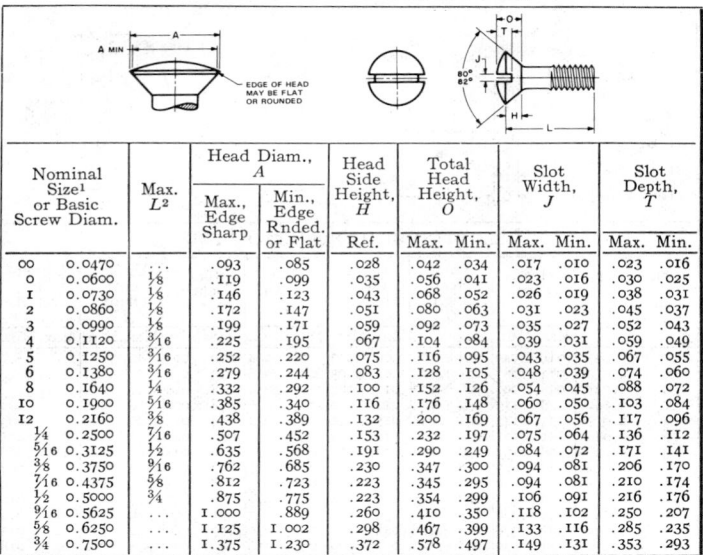

Nominal Size[1] or Basic Screw Diam.		Max. L[2]	Head Diam., A		Head Side Height, H	Total Head Height, O		Slot Width, J		Slot Depth, T	
			Max., Edge Sharp	Min., Edge Rnded. or Flat	Ref.	Max.	Min.	Max.	Min.	Max.	Min.
00	0.0470	...	.093	.085	.028	.042	.034	.017	.010	.023	.016
0	0.0600	⅛	.119	.099	.035	.056	.041	.023	.016	.030	.025
1	0.0730	⅛	.146	.123	.043	.068	.052	.026	.019	.038	.031
2	0.0860	⅛	.172	.147	.051	.080	.063	.031	.023	.045	.037
3	0.0990	⅛	.199	.171	.059	.092	.073	.035	.027	.052	.043
4	0.1120	3/16	.225	.195	.067	.104	.084	.039	.031	.059	.049
5	0.1250	3/16	.252	.220	.075	.116	.095	.043	.035	.067	.055
6	0.1380	3/16	.279	.244	.083	.128	.105	.048	.039	.074	.060
8	0.1640	¼	.332	.292	.100	.152	.126	.054	.045	.088	.072
10	0.1900	5/16	.385	.340	.116	.176	.148	.060	.050	.103	.084
12	0.2160	⅜	.438	.389	.132	.200	.169	.067	.056	.117	.096
¼	0.2500	7/16	.507	.452	.153	.232	.197	.075	.064	.136	.112
5/16	0.3125	½	.635	.568	.191	.290	.249	.084	.072	.171	.141
⅜	0.3750	9/16	.762	.685	.230	.347	.300	.094	.081	.206	.170
7/16	0.4375	⅝	.812	.723	.223	.345	.295	.094	.081	.210	.174
½	0.5000	¾	.875	.775	.223	.354	.299	.106	.091	.216	.176
9/16	0.5625	...	1.000	.889	.260	.410	.350	.118	.102	.250	.207
⅝	0.6250	...	1.125	1.002	.298	.467	.399	.133	.116	.285	.235
¾	0.7500	...	1.375	1.230	.372	.578	.497	.149	.131	.353	.293

All dimensions are in inches. [1] Where specifying nominal size in decimals, zeros preceding decimal points and in the fourth decimal place are omitted. [2] These lengths or shorter are undercut.

Table 10. American National Standard Header Points for Machine Screws before Threading (ANSI B18.6.3-1972, R1977)

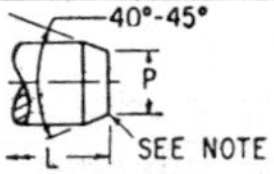

Nom. Size	Threads per Inch	Max. P	Min. P	Max. L
2	56	0.057	0.050	½
	64	0.060	0.053	½
4	40	0.074	0.065	½
	48	0.079	0.070	½
5	40	0.086	0.076	½
	44	0.088	0.079	½
6	32	0.090	0.080	¾
	40	0.098	0.087	¾
8	32	0.114	0.102	1
	36	0.118	0.106	1

Nom. Size	Threads per Inch	Max. P	Min. P	Max. L
10	24	0.125	0.112	1¼
	32	0.138	0.124	1¼
12	24	0.149	0.134	1⅜
	28	0.156	0.141	1⅜
¼	20	0.170	0.153	1½
	28	0.187	0.169	1½
5/16	18	0.221	0.200	1½
	24	0.237	0.215	1½
⅜	16	0.270	0.244	1½
	24	0.295	0.267	1½
7/16	14	0.316	0.287	1½
	20	0.342	0.310	1½
½	13	0.367	0.333	1½
	20	0.399	0.362	1½

All dimensions in inches. Edges of point may be rounded and end of point need not be flat nor perpendicular to shank. Machine screws normally have plain sheared ends but when specified may have header points as shown above.

Table 11. American National Standard Slotted Binding Head and Slotted Undercut Oval Countersunk Head Machine Screws (ANSI B18.6.3-1972, R1977)

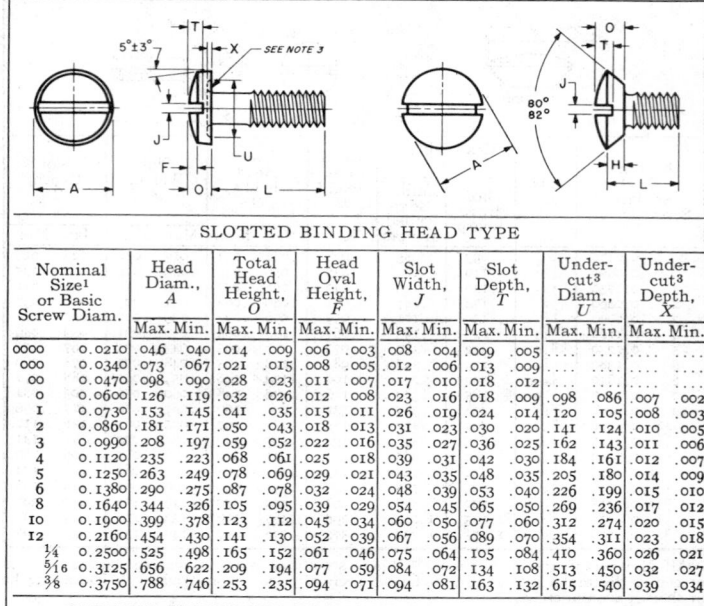

SLOTTED BINDING HEAD TYPE

Nominal Size[1] or Basic Screw Diam.		Head Diam., A		Total Head Height, O		Head Oval Height, F		Slot Width, J		Slot Depth, T		Undercut[3] Diam., U		Undercut[3] Depth, X	
		Max.	Min.	Max.	Min.	Max.	Min.	Max.	Min.	Max.	Min.	Max.	Min.	Max.	Min.
0000	0.0210	.046	.040	.014	.009	.006	.003	.008	.004	.009	.005	....	....	....	....
000	0.0340	.073	.067	.021	.015	.008	.005	.012	.006	.013	.009	....	....	....	....
00	0.0470	.098	.090	.028	.023	.011	.007	.017	.010	.018	.012	....	....	....	....
0	0.0600	.126	.119	.032	.026	.012	.008	.023	.016	.018	.009	.098	.086	.007	.002
1	0.0730	.153	.145	.041	.035	.015	.011	.026	.019	.024	.014	.120	.105	.008	.003
2	0.0860	.181	.171	.050	.043	.018	.013	.031	.023	.030	.020	.141	.124	.010	.005
3	0.0990	.208	.197	.059	.052	.022	.016	.035	.027	.036	.025	.162	.143	.011	.006
4	0.1120	.235	.223	.068	.061	.025	.018	.039	.031	.042	.030	.184	.161	.012	.007
5	0.1250	.263	.249	.078	.069	.029	.021	.043	.035	.048	.035	.205	.180	.014	.009
6	0.1380	.290	.275	.087	.078	.032	.024	.048	.039	.053	.040	.226	.199	.015	.010
8	0.1640	.344	.326	.105	.095	.039	.029	.054	.045	.065	.050	.269	.236	.017	.012
10	0.1900	.399	.378	.123	.112	.045	.034	.060	.050	.077	.060	.312	.274	.020	.015
12	0.2160	.454	.430	.141	.130	.052	.039	.067	.056	.089	.070	.354	.311	.023	.018
¼	0.2500	.525	.498	.165	.152	.061	.046	.075	.064	.105	.084	.410	.360	.026	.021
⁵⁄₁₆	0.3125	.656	.622	.209	.194	.077	.059	.084	.072	.134	.108	.513	.450	.032	.027
⅜	0.3750	.788	.746	.253	.235	.094	.071	.094	.081	.163	.132	.615	.540	.039	.034

SLOTTED UNDERCUT OVAL COUNTERSUNK HEAD TYPES

Nominal Size[1] or Basic Screw Diam.		Max. L[2]	Head Diam., A		Head Side Height, H	Total Head Height, O		Slot Width, J		Slot Depth, T	
			Max., Edge Sharp	Min., Edge Rnded. or Flat	Ref.	Max.	Min.	Max.	Min.	Max.	Min.
0	0.0600	⅛	.119	.099	.025	.046	.033	.023	.016	.028	.022
1	0.0730	⅛	.146	.123	.031	.056	.042	.026	.019	.034	.027
2	0.0860	⅛	.172	.147	.036	.065	.050	.031	.023	.040	.033
3	0.0990	⅛	.199	.171	.042	.075	.059	.035	.027	.047	.038
4	0.1120	³⁄₁₆	.225	.195	.047	.084	.067	.039	.031	.053	.043
5	0.1250	³⁄₁₆	.252	.220	.053	.094	.076	.043	.035	.059	.048
6	0.1380	³⁄₁₆	.279	.244	.059	.104	.084	.048	.039	.065	.053
8	0.1640	¼	.332	.292	.070	.123	.101	.054	.045	.078	.064
10	0.1900	⁵⁄₁₆	.385	.340	.081	.142	.118	.060	.050	.090	.074
12	0.2160	⅜	.438	.389	.092	.161	.135	.067	.056	.103	.085
¼	0.2500	⁷⁄₁₆	.507	.452	.107	.186	.158	.075	.064	.119	.098
⁵⁄₁₆	0.3125	½	.635	.568	.134	.232	.198	.084	.072	.149	.124
⅜	0.3750	⁹⁄₁₆	.762	.685	.161	.278	.239	.094	.081	.179	.149
⁷⁄₁₆	0.4375	⅝	.812	.723	.156	.279	.239	.094	.081	.184	.154
½	0.5000	¾	.875	.775	.156	.288	.244	.106	.091	.204	.169

All dimensions are in inches.   [1] Where specifying nominal size in decimals, zeros preceding decimal points and in the fourth decimal place are omitted.   [2] These lengths or shorter are undercut.   [3] Unless otherwise specified, slotted binding head machine screws are not undercut.

Table 12.   Slotted Round Head Machine Screws
(ANSI B18.6.3-1972, R1977 Appendix)*

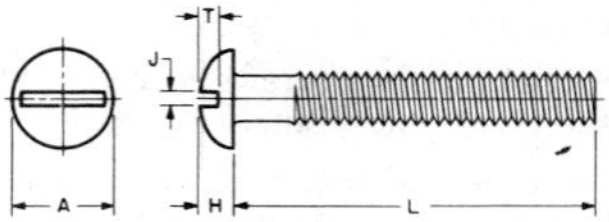

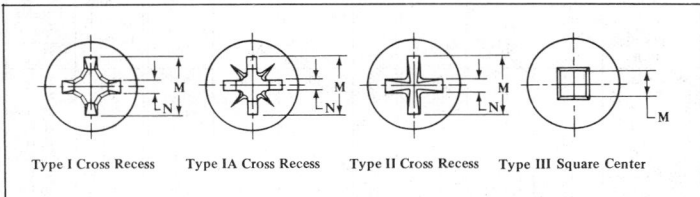

Nominal Size[1] or Basic Screw Diam.		Head Diameter, A		Head Height, H		Slot Width, J		Slot Depth, T	
		Max.	Min.	Max.	Min.	Max.	Min.	Max.	Min.
0000	0.0210	.041	.035	.022	.016	.008	.004	.017	.013
000	0.0340	.062	.056	.031	.025	.012	.008	.018	.012
00	0.0470	.089	.080	.045	.036	.017	.010	.026	.018
0	0.0600	.113	.099	.053	.043	.023	.016	.039	.029
1	0.0730	.138	.122	.061	.051	.026	.019	.044	.033
2	0.0860	.162	.146	.069	.059	.031	.023	.048	.037
3	0.0990	.187	.169	.078	.067	.035	.027	.053	.040
4	0.1120	.211	.193	.086	.075	.039	.031	.058	.044
5	0.1250	.236	.217	.095	.083	.043	.035	.063	.047
6	0.1380	.260	.240	.103	.091	.048	.039	.068	.051
8	0.1640	.309	.287	.120	.107	.054	.045	.077	.058
10	0.1900	.359	.334	.137	.123	.060	.050	.087	.065
12	0.2160	.408	.382	.153	.139	.067	.056	.096	.073
¼	0.2500	.472	.443	.175	.160	.075	.064	.109	.082
5⁄16	0.3125	.590	.557	.216	.198	.084	.072	.132	.099
⅜	0.3750	.708	.670	.256	.237	.094	.081	.155	.117
7⁄16	0.4375	.750	.707	.328	.307	.094	.081	.196	.148
½	0.5000	.813	.766	.355	.332	.106	.091	.211	.159
9⁄16	0.5625	.938	.887	.410	.385	.118	.102	.242	.183
⅝	0.6250	1.000	.944	.438	.411	.133	.116	.258	.195
¾	0.7500	1.250	1.185	.547	.516	.149	.131	.320	.242

All dimensions are in inches.
* Not recommended, use Pan Head machine screws.
[1] Where specifying nominal size in decimals, zeros preceding decimal point and in the fourth decimal place are omitted.

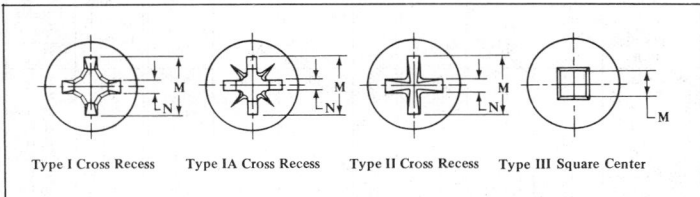

| Type I Cross Recess | Type IA Cross Recess | Type II Cross Recess | Type III Square Center |

American National Standard Cross Recesses for
Machine Screws and Metric Machine Screws

Machine Screw Cross Recesses. — Four cross recesses, Types I, IA, II, and III, may be used in lieu of slots in machine screw heads. Dimensions for recess diameter M, width N, and depth T (not shown above) together with recess penetration gaging depths are given in American National Standard ANSI B18.6.3-1972, R1977, for machine screws, and in ANSI B18.6.7M-1985 for metric machine screws.

American National Standard Metric Machine Screws. — This Standard (B18.6.7M) covers metric flat and oval countersunk and slotted and recessed pan head machine screws and metric hex head and hex flange head machine screws. Dimensions are given in Tables 2 through 4 and 6.

Threads: Threads for metric machine screws are coarse M profile threads, as given in ANSI B1.13M (see page 1521), unless otherwise specified.

Length of Thread: The lengths of threads on metric machine screws are given in Table 1 for the applicable screw type, size, and length.

Diameter of Body: The body diameters of metric machine screws are within the limits specified in the dimensional tables (Tables 2 through 4 and 6).

Table 1. American National Standard Thread Lengths for Metric Machine Screws (ANSI B18.6.7M-1985)

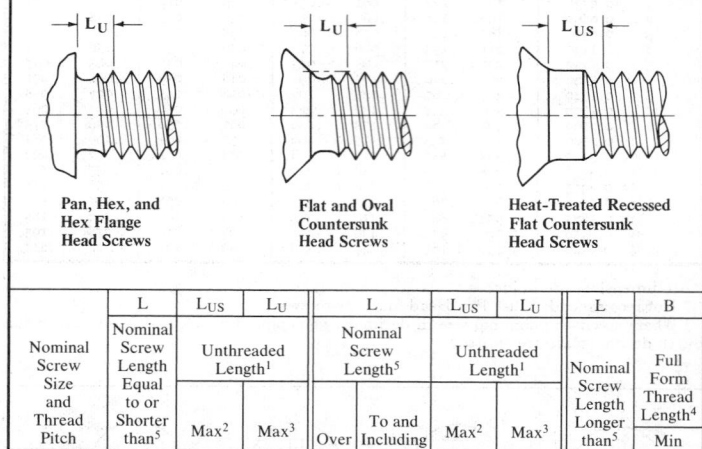

Pan, Hex, and Hex Flange Head Screws

Flat and Oval Countersunk Head Screws

Heat-Treated Recessed Flat Countersunk Head Screws

Nominal Screw Size and Thread Pitch	L — Nominal Screw Length Equal to or Shorter than[5]	L_US — Unthreaded Length[1] Max[2]	L_U — Unthreaded Length[1] Max[3]	L — Nominal Screw Length[5] Over	L — Nominal Screw Length[5] To and Including	L_US — Unthreaded Length[1] Max[2]	L_U — Unthreaded Length[1] Max[3]	L — Nominal Screw Length Longer than[5]	B — Full Form Thread Length[4] Min
M2 × 0.4	6	1.0	0.4	6	30	1.0	0.8	30	25.0
M2.5 × 0.45	8	1.1	0.5	8	30	1.1	0.9	30	25.0
M3 × 0.5	9	1.2	0.5	9	30	1.2	1.0	30	25.0
M3.5 × 0.6	10	1.5	0.6	10	50	1.5	1.2	50	38.0
M4 × 0.7	12	1.8	0.7	12	50	1.8	1.4	50	38.0
M5 × 0.8	15	2.0	0.8	15	50	2.0	1.6	50	38.0
M6 × 1	18	2.5	1.0	18	50	2.5	2.0	50	38.0
M8 × 1.25	24	3.1	1.2	24	50	3.1	2.5	50	38.0
M10 × 1.5	30	3.8	1.5	30	50	3.8	3.0	50	38.0
M12 × 1.75	36	4.4	1.8	36	50	4.4	3.5	50	38.0

All dimensions in millimeters.

[1] Unthreaded lengths L_U and L_{US} represent the distance, measured parallel to the axis of screw, from the underside of the head to the face of a nonchamfered or noncounterbored standard GO thread ring gage assembled by hand as far as the thread will permit.

[2] The L_{US} values apply only to heat treated recessed flat countersunk head screws.

[3] The L_U values apply to all screws except heat treated recessed flat countersunk head screws.

[4] Refer to the illustrations for respective screw head styles.

[5] The length tolerances for metric machine screws are: up to 3 mm, incl., ± 0.2 mm; over 3 to 10 mm, incl., ± 0.3 mm; over 10 to 16 mm, incl., ± 0.4 mm; over 16 to 50 mm, incl., ± 0.5 mm; over 50 mm, ± 1.0 mm.

Table 2. American National Standard Slotted and Cross and Square Recessed Flat Countersunk Head Metric Machine Screws (ANSI B18.6.7M-1985)

Slotted and Style A[1] for Cross and Square Recessed Heads

Style B[1] for Cross and Square Recessed Heads

Nominal Screw Size and Thread Pitch	Slotted and Style A D_S Body Diameter Max	Min	Style B D_{SH} Body and Shoulder Diameter Max	Shoulder Diameter Min	D_S Body Diameter Min	L_{SH}[1] Shoulder Length Max	Min	D_K Head Diameter Theoretical Sharp Max	Min	Actual Min	K Head Height Max Ref	R Underhead Fillet Radius Max	Min	N Slot Width Max	Min	T Slot Depth Max	Min
M2 × 0.4[2]	2.00	1.65	2.00	1.86	1.65	0.50	0.30	4.4	4.1	3.5	1.2	0.8	0.4	0.7	0.5	0.6	0.4
M2.5 × 0.45	2.50	2.12	2.50	2.36	2.12	0.55	0.35	5.5	5.1	4.4	1.5	1.0	0.5	0.8	0.6	0.7	0.5
M3 × 0.5	3.00	2.58	3.00	2.86	2.58	0.60	0.40	6.3	5.9	5.2	1.7	1.2	0.6	1.0	0.8	0.9	0.6
M3.5 × 0.6	3.50	3.00	3.50	3.32	3.00	0.70	0.50	8.2	7.7	6.9	2.3	1.4	0.7	1.2	1.0	1.2	0.9
M4 × 0.7	4.00	3.43	4.00	3.82	3.43	0.80	0.60	9.4	8.9	8.0	2.7	1.6	0.8	1.5	1.2	1.3	1.0
M5 × 0.8	5.00	4.36	5.00	4.82	4.36	0.90	0.70	10.4	9.8	8.9	2.7	2.0	1.0	1.5	1.2	1.4	1.1
M6 × 1	6.00	5.21	6.00	5.82	5.21	1.10	0.90	12.6	11.9	10.9	3.3	2.4	1.2	1.9	1.6	1.6	1.2
M8 × 1.25	8.00	7.04	8.00	7.78	7.04	1.40	1.10	17.3	16.5	15.4	4.6	3.2	1.6	2.3	2.0	2.3	1.8
M10 × 1.5	10.00	8.86	10.00	9.78	8.86	1.70	1.30	20.0	19.2	17.8	5.0	4.0	2.0	2.8	2.5	2.6	2.0

All dimensions in millimeters. [1] All recessed head heat-treated steel screws of property class 9.8 or higher strength have the Style B head form. Recessed head screws other than those specifically designated to be Style B have the Style A head form. The underhead shoulder on the Style B head form is mandatory and all other head dimensions are common to both the Style A and Style B head forms. [2] This size is not specified for Type III square recessed flat countersunk heads; Type II cross recess is not specified for any size.

Table 3. American National Standard Slotted and Cross and Square Recessed Oval Countersunk Head Metric Machine Screws (ANSI B18.6.7M-1985)

Nominal Screw Size and Thread Pitch	D_S Body Diameter		D_K Head Diameter			K Head Side Height	F Raised Head Height	R_F Head Top Radius	R Underhead Fillet Radius		N Slot Width		T Slot Depth	
			Theoretical Sharp		Actual									
	Max	Min	Max	Min	Min	Max Ref	Max	Approx	Max	Min	Max	Min	Max	Min
M2 × 0.4[1]	2.00	1.65	4.4	4.1	3.5	1.2	0.5	5.0	0.8	0.4	0.7	0.5	1.0	0.8
M2.5 × 0.45	2.50	2.12	5.5	5.1	4.4	1.5	0.6	6.6	1.0	0.5	0.8	0.6	1.2	1.0
M3 × 0.5	3.00	2.58	6.3	5.9	5.2	1.7	0.7	7.4	1.2	0.6	1.0	0.8	1.5	1.2
M3.5 × 0.6	3.50	3.00	8.2	7.7	6.9	2.3	0.8	10.9	1.4	0.7	1.2	1.0	1.7	1.4
M4 × 0.7	4.00	3.43	9.4	8.9	8.0	2.7	1.0	11.6	1.6	0.8	1.5	1.2	1.9	1.6
M5 × 0.8	5.00	4.36	10.4	9.8	8.9	2.7	1.2	11.9	2.0	1.0	1.5	1.2	2.4	2.0
M6 × 1	6.00	5.21	12.6	11.9	10.9	3.3	1.4	14.9	2.4	1.2	1.9	1.6	2.8	2.4
M8 × 1.25	8.00	7.04	17.3	16.5	15.4	4.6	2.0	19.7	3.2	1.6	2.3	2.0	3.7	3.2
M10 × 1.5	10.00	8.86	20.0	19.2	17.8	5.0	2.3	22.9	4.0	2.0	2.8	2.5	4.4	3.8

All dimensions are in millimeters.
[1] This size is not specified for Type III square recessed oval countersunk heads; Type II cross recess is not specified for any size.

Table 4. American National Standard Slotted and Cross and Square Recessed Pan Head Metric Machine Screws (ANSI B18.6.7M-1985)

SEE TABLE 1
SEE TABLE 7

Nominal Screw Size and Thread Pitch	D_S Body Diameter		D_K Head Diameter		Slotted			Cross and Square Recess			D_A Underhead Fillet		N Slot Width		T Slot Depth	W Unslotted Head Thickness
					K Head Height		R_1 Head Radius	K Head Height		R_1 Head Radius	Transition Dia	Radius				
	Max	Min	Max	Min	Max	Min	Max	Max	Min	Ref	Max	Min	Max	Min	Min	Min
M2 × 0.4¹	2.00	1.65	4.0	3.7	1.3	1.1	0.8	1.6	1.4	3.2	2.6	0.1	0.7	0.5	0.5	0.4
M2.5 × 0.45	2.50	2.12	5.0	4.7	1.5	1.3	1.0	2.1	1.9	4.0	3.1	0.1	0.8	0.6	0.6	0.5
M3 × 0.5	3.00	2.58	5.6	5.3	1.8	1.6	1.2	2.4	2.2	5.0	3.6	0.1	1.0	0.8	0.7	0.7
M3.5 × 0.6	3.50	3.00	7.0	6.6	2.1	1.9	1.4	2.6	2.3	6.0	4.1	0.1	1.2	1.0	0.8	0.8
M4 × 0.7	4.00	3.43	8.0	7.6	2.4	2.2	1.6	3.1	2.8	6.5	4.7	0.2	1.5	1.2	1.0	0.9
M5 × 0.8	5.00	4.36	9.5	9.1	3.0	2.7	2.0	3.7	3.4	8.0	5.7	0.2	1.5	1.2	1.2	1.2
M6 × 1	6.00	5.21	12.0	11.5	3.6	3.3	2.5	4.6	4.3	10.0	6.8	0.3	1.9	1.6	1.4	1.4
M8 × 1.25	8.00	7.04	16.0	15.5	4.8	4.5	3.2	6.0	5.6	13.0	9.2	0.4	2.3	2.0	1.9	1.9
M10 × 1.5	10.00	8.86	20.0	19.4	6.0	5.7	4.0	7.5	7.1	16.0	11.2	0.4	2.8	2.5	2.4	2.4

All dimensions in millimeters. ¹ This size not specified for Type III square recessed pan heads; Type II cross recess is not specified for any size.

Table 5. American National Standard Header Points for Metric Machine Screws before Threading (ANSI B18.6.7M-1985)

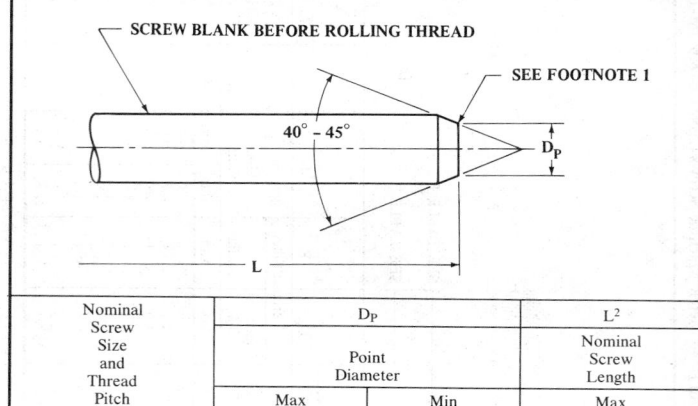

Nominal Screw Size and Thread Pitch	D_P		L^2
	Point Diameter		Nominal Screw Length
	Max	Min	Max
M2 × 0.4	1.33	1.21	13
M2.5 × 0.45	1.73	1.57	13
M3 × 0.5	2.12	1.93	16
M3.5 × 0.6	2.46	2.24	20
M4 × 0.7	2.80	2.55	25
M5 × 0.8	3.60	3.28	30
M6 × 1	4.25	3.85	40
M8 × 1.25	5.82	5.30	40
M10 × 1.5	7.36	6.71	40
M12 × 1.75	8.90	8.11	45

All dimensions in millimeters.

[1] The edge of the point may be rounded and the end of point need not be flat nor perpendicular to the axis of screw shank.

[2] Header points apply to these nominal lengths or shorter. The pointing of longer lengths may require machining to the dimensions specified.

Designation: Metric machine screws are designated by the following data in the sequence shown: Nominal size and thread pitch; nominal length; product name, including head type and driving provision; header point if desired; material (including property class, if steel); and protective finish, if required. For example:

M8 × 1.25 × 30 Slotted Pan Head Machine Screw, Class 4.8 Steel, Zinc Plated

M3.5 × 0.6 × 20 Type IA Cross Recessed Oval Countersunk Head Machine Screw, Header Point, Brass

It is common ISO practice to omit the thread pitch from the product size designation when screw threads are the metric coarse thread series, e.g., M10 stands for M10 × 1.5.

Table 6. American National Standard Hex and Hex Flange Head Metric Machine Screws (ANSI B18.6.7M-1985)

Hex Head

Shape of Indentation Optional — See Footnote 1 — Indented Head [3] — Trimmed or Fully Upset Head [3] — See Table 1 — See Table 7

Nominal Screw Size and Thread Pitch	D_S Body Diameter		S^2 Hex Width Across Flats		E^2 Hex Width Across Corners	K Head Height		D_A Underhead Fillet Transition Dia	R Underhead Fillet Radius
	Max	Min	Max	Min	Min	Max	Min	Max	Min
M2 × 0.4	2.00	1.65	3.20	3.02	3.38	1.6	1.3	2.6	0.1
M2.5 × 0.45	2.50	2.12	4.00	3.82	4.28	2.1	1.8	3.1	0.1
M3 × 0.5	3.00	2.58	5.00	4.82	5.40	2.3	2.0	3.6	0.1
M3.5 × 0.6	3.50	3.00	5.50	5.32	5.96	2.6	2.3	4.1	0.1
M4 × 0.7	4.00	3.43	7.00	6.78	7.59	3.0	2.6	4.7	0.2
M5 × 0.8	5.00	4.36	8.00	7.78	8.71	3.8	3.3	5.7	0.2
M6 × 1	6.00	5.21	10.00	9.78	10.95	4.7	4.1	6.8	0.3
M8 × 1.25	8.00	7.04	13.00	12.73	14.26	6.0	5.2	9.2	0.4
M10 × 1.5	10.00	8.86	16.00	15.73	17.62	7.5	6.5	11.2	0.4
M12 × 1.75	12.00	10.68	18.00	17.73	19.86	9.0	7.8	13.2	0.4
M10 × 1.5[5]	10.00	8.86	15.00	14.73	16.50	7.5	6.5	11.2	0.4

See footnotes at end of table.

Table 6 (*Concluded*). American National Standard Hex and Hex Flange Head Metric Machine Screws (ANSI B18.6.7M-1985)

Hex Flange Head

Nominal Screw Size and Thread Pitch	D_S Body Diameter		S^2 Hex Width Across Flats		E^2 Hex Width Across Corners	D_C Flange Diameter		K Overall Head Height	K_1 Hex Height	C^5 Flange Edge Thickness	R_1 Flange Top Fillet Radius	D_A Underhead Fillet Transition Dia	R Underhead Fillet Radius
	Max	Min	Max	Min	Min	Max	Min	Max	Min	Min	Max	Max	Min
M2 × 0.4	2.00	1.65	3.00	2.84	3.16	4.5	4.1	2.2	1.3	0.3	0.1	2.6	0.1
M2.5 × 0.45	2.50	2.12	3.20	3.04	3.39	5.4	5.0	2.7	1.6	0.3	0.2	3.1	0.1
M3 × 0.5	3.00	2.58	4.00	3.84	4.27	6.4	5.9	3.2	1.9	0.4	0.2	3.6	0.1
M3.5 × 0.6	3.50	3.00	5.00	4.82	5.36	7.5	6.9	3.8	2.4	0.5	0.2	4.1	0.1
M4 × 0.7	4.00	3.43	5.50	5.32	5.92	8.5	7.8	4.3	2.8	0.6	0.2	4.7	0.2
M5 × 0.8	5.00	4.36	7.00	6.78	7.55	10.6	9.8	5.4	3.5	0.7	0.3	5.7	0.2
M6 × 1	6.00	5.21	8.00	7.78	8.66	12.8	11.8	6.7	4.2	1.0	0.4	6.8	0.3
M8 × 1.25	8.00	7.04	10.00	9.78	10.89	16.8	15.5	8.6	5.6	1.2	0.5	9.2	0.4
M10 × 1.5	10.00	8.86	13.00	12.72	14.16	21.0	19.3	10.7	7.0	1.4	0.6	11.2	0.4
M12 × 1.75	12.00	10.68	15.00	14.72	16.38	24.8	23.3	13.7	8.4	1.8	0.7	13.2	0.4

All dimensions in millimeters. [1] A slight rounding of all edges of the hexagon surfaces of indented hex heads is permissible provided the diameter of the bearing circle is not less than the equivalent of 90 per cent of the specified minimum width across flats dimension. [2] Dimensions across flats and across corners of the head are measured at the point of maximum metal. Taper of sides of head (angle between one side and the axis) shall not exceed 2° or 0.10 mm, whichever is greater, the specified width across flats being the large dimension. [3] Heads may be indented, trimmed, or fully upset at the option of the manufacturer. [4] The M10 size screws having heads with 15 mm width across flats are not ISO Standard. Unless M10 size screws with 15 mm width across flats are specifically ordered, M10 size screws with 16 mm width across flats shall be furnished. [5] The contour of the edge at periphery of flange is optional provided the minimum flange thickness is maintained at the minimum flange diameter. The top surface of flange may be straight or slightly rounded (convex) upward.

Table 7. Recommended Nominal Screw Lengths for Metric Machine Screws

Nominal Screw Length	Nominal Screw Size									
	M2	M2.5	M3	M3.5	M4	M5	M6	M8	M10	M12
2.5	PH									
3	A	PH								
4	A	A	PH							
5	A	A	A	PH	PH					
6	A	A	A	A	A	PH				
8	A	A	A	A	A	A	A			
10	A	A	A	A	A	A	A	A		
13	A	A	A	A	A	A	A	A	A	
16	A	A	A	A	A	A	A	A	A	H
20	A	A	A	A	A	A	A	A	A	H
25		A	A	A	A	A	A	A	A	H
30			A	A	A	A	A	A	A	H
35				A	A	A	A	A	A	H
40					A	A	A	A	A	H
45						A	A	A	A	H
50						A	A	A	A	H
55							A	A	A	H
60							A	A	A	H
65								A	A	H
70								A	A	H
80								A	A	H
90									A	H

All dimensions in millimeters.

[1] The nominal screw lengths included between the heavy lines are recommended for the respective screw sizes and screw head styles as designated by the symbols.

A — Signifies screws of all head styles covered in this standard.

P — Signifies pan head screws.

H — Signifies hex and hex flange head screws.

Table 8. Clearance Holes for Metric Machine Screws
(ANSI B18.6.7M-1985, Appendix)

Nominal Screw Size	Basic Clearance Hole Diameter[1]		
	Close Clearance[2]	Normal Clearance (Preferred)[2]	Loose Clearance[2]
M2	2.40	2.40	2.60
M2.5	2.70	2.90	3.10
M3	3.20	3.40	3.60
M3.5	3.70	3.90	4.20
M4	4.30	4.50	4.80
M5	5.30	5.50	5.80
M6	6.40	6.60	7.00
M8	8.40	9.00	10.00
M10	10.50	11.00	12.00
M12	13.00	13.50	14.50

All dimensions in millimeters.

[1] The values given in this table are minimum limits. The recommended plus tolerances are as follows: for clearance hole diameters over 1.70 to and including 5.80 mm, plus 0.12, 0.20, and 0.30 mm for close, normal, and loose clearances, respectively; for clearance hole diameters over 5.80 to 14.50 mm, plus 0.18, 0.30, and 0.45 mm for close, normal, and loose clearances, respectively.

[2] Normal clearance hole sizes are preferred. Close clearance hole sizes are for situations such as critical alignment of assembled components, wall thickness, or other limitations which necessitate the use of a minimal hole. Countersinking or counterboring at the fastener entry side may be necessary for the proper seating of the head. Loose clearance hole sizes are for applications where maximum adjustment capability between the components being assembled is necessary.

British Machine Screws. — Currently (1982) covering fasteners of this type are British Standards B.S. 57:1951, "B.A. Screws, Bolts and Nuts"; B.S. 450:1958, "Machine Screws and Machine Screw Nuts (B.S.W. and B.S.F. Threads)"; B.S. 1981:1953, "Unified Machine Screws and Machine Screw Nuts"; B.S. 2827:1957, "Machine Screw Nuts, Pressed Type (B.A. and Whitworth Form Threads)"; B.S. 3155:1960, "American Machine Screws and Nuts in Sizes Below ¼ inch Diameter"; and B.S. 4183: 1967, "Machine Screws and Machine Screw Nuts, Metric Series." At a conference organized by the British Standards Institution in 1965 at which the major sectors of British industry were represented, a policy statement was approved which urged British firms to regard the traditional screw thread systems — Whitworth, B.A. and B.S.F. — as obsolescent, and to make the internationally-agreed ISO metric thread their first choice (with ISO Unified thread as second choice) for all future designs. It is recognized that some sections of British industry already using ISO inch (Unified) screw threads may find it necessary, for various reasons, to continue with their use for some time; Whitworth and B.A. threads should, however, be superseded by ISO metric threads in preference to an intermediate change to ISO inch threads. This means that eventually fasteners covered by B.S. 57, B.S. 450 and B.S. 2827 would be superseded and replaced by fasteners specified by B.S. 4183.

British Standard Whitworth (B.S.W.) and Fine (B.S.F.) Machine Screws. — British Standard B.S. 450:1958 covers machine screws and nuts with British Standard Whitworth and British Standard Fine threads. It covers all of the various heads in common use in both slotted and recessed forms. Head shapes are shown on page 1384 and dimensions on page 1386. It is intended that this standard will eventually be superseded and replaced by B.S. 4183, "Machine Screws and Machine Screw Nuts, Metric Series."

British Standard Machine Screws and Machine Screw Nuts, Metric Series. — British Standard B.S. 4183:1967 gives dimensions and tolerances for: countersunk head, raised countersunk head, and cheese head slotted head screws in a diameter range from M1 (1 mm) to M20 (20 mm); pan head slotted head screws in a diameter range from M2.5 (2.5 mm) to M10 (10 mm); countersunk head and raised countersunk head recessed head screws in a diameter range from M2.5 (2.5 mm) to M12 (12 mm); pan head recessed head screws in a diameter range from M2.5 (2.5 mm) to M10 (10 mm); and square and hexagon machine screw nuts in a diameter range from M1.6 (1.6 mm) to M10 (10 mm). Mechanical properties are also specified for steel, brass and aluminum alloy machine screws and machine screw nuts in this standard.

Material: The materials from which the screws and nuts are manufactured have a tensile strength not less than the following: steel, 40 kgf/mm² (392 N/mm²); brass, 32 kgf/mm² (314 N/mm²); and aluminum alloy, 32 kgf/mm² (314 N/mm²). The unit, kgf/mm² is in accordance with ISO DR 911 and the unit in parentheses has the relationship, 1 kgf = 9.80665 Newtons. These minimum strengths are applicable to the finished products. Steel machine screws conform to the requirements for strength grade designation 4.8. The strength grade designation system for machine screws consists of two figures, the first is ⅒ of the minimum tensile strength in kgf/mm², the second is ⅒ of the ratio between the yield stress and the minimum tensile strength expressed as a percentage: ⅒ minimum tensile strength of 40 kgf/mm² gives the symbol "4"; ⅒ ratio $\dfrac{\text{yield stress}}{\text{minimum tensile strength}}$ % = ⅒ × 32/40 × 100/1 = "8"; giving the strength grade designation "4.8." Multiplication of these two figures gives the minimum yield stress in kgf/mm².

Coating of Screws and Nuts: It is recommended that the coating comply with the appropriate part of B.S. 3382 "Electroplated Coatings on Threaded Components."

Screw Threads: Screw threads are ISO metric coarse pitch series threads in accordance with B.S. 3643, "ISO Metric Screw Threads," Part 1, "Thread Data and Standard Thread Series." The external threads used for screws conform to tolerance Class 6g limits (medium fit) as given in B.S. 3643, "ISO Metric Screw Threads," Part 2, "Limits and Tolerances for Coarse Pitch Series Threads." The internal threads used for nuts conform to tolerance Class 6H limits (medium fit) as given in B.S. 3643: Part 2.

Nominal Length of Screws: For countersunk head screws the nominal length is the distance from the upper surface of the head to the extreme end of the shank, including any chamfer, radius, or cone point. For raised countersunk head screws the nominal length is the distance from the upper surface of the head (excluding the raised portion) to the extreme end of the shank, including any chamfer, radius, or cone point. For pan and cheese head the nominal length is the distance from the underside of the head to the extreme end of the shank, including any chamfer, radius, or cone point. Standard nominal lengths and tolerances are given in Table 5.

Length of Thread on Screws: The length of thread is the distance from the end of the screw (including any chamfer, radius, or cone point) to the leading face of a nut without countersink which has been screwed as far as possible onto the screw by hand. The minimum thread length is as given below:

Nominal Thread Diam. d†	M1	M1.2	(M1.4)	M1.6	M2	(M2.2)	M2.5	M3	(M3.5)	M4
Thread Length b (Min.)	*	*	*	15	16	17	18	19	20	22

Nominal Thread Diam. d†	(M4.5)	M5	M6	M8	M10	M12	(M14)	M16	(M18)	M20
Thread Length b (Min.)	24	25	28	34	40	46	52	58	64	70

All dimensions are in millimeters.
† Items shown in parentheses are non-preferred.
* Threaded up to the head.

Screws of nominal thread diameter M1, M1.2 and M1.4 and screws of larger diameters which are too short for the above thread lengths are threaded as far as possible up to the head. In these the length of unthreaded shank under the head does not exceed 1½ pitches for lengths up to twice the diameter and 2 pitches for longer lengths, and is defined as the distance from the leading face of a nut which has been screwed as far as possible onto the screw by hand to: (1) the junction of the basic major diameter and the countersunk portion of the head on countersunk and raised countersunk heads; (2) the underside of the head on other types of heads.

Diameter of Unthreaded Shank on Screws: The diameter of the unthreaded portion of the shank on screws is not greater than the basic major diameter of the screw head and not less than the minimum effective diameter of the screw thread. The diameter of the unthreaded portion of shank is closely associated with the method of manu-

facture; it will generally be nearer the major diameter of the thread for turned screws and nearer the effective diameter for those produced by cold heading.

Radius Under the Head of Screws: The radius under the head of pan and cheese head screws runs smoothly into the face of the head and shank without any step or discontinuity. A true radius is not essential providing that the curve is smooth and lies wholly within the maximum radius. Any radius under the head of countersunk head screws runs smoothly into the conical bearing surface of the head and the shank without any step or discontinuity. The radius values given in Tables 1 and 2 are regarded as the maximum where the shank diameter is equal to the major diameter of the thread and minimum where the shank diameter is approximately equal to the effective diameter of the thread.

Ends of Screws: When screws are made with rolled threads the "lead" formed by the thread rolling operation is normally regarded as providing the necessary chamfer and no other machining is necessary. The ends of screws with cut threads are normally finished with a chamfer conforming to the dimension in Fig. 1. At the option of the manufacturer, the ends of screws smaller than M6 (6 mm diameter) may be finished with a radius approximately equal to $1\frac{1}{4}$ times the nominal diameter of the shank. When cone point ends are required they should have the dimensions given in Fig. 1.

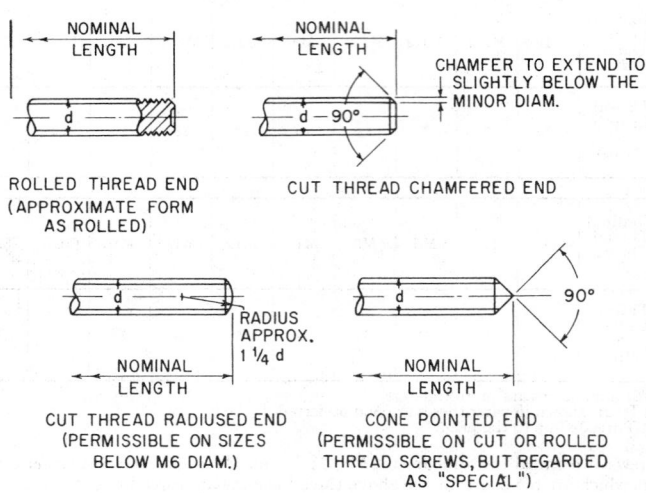

ROLLED THREAD END
(APPROXIMATE FORM AS ROLLED)

CUT THREAD CHAMFERED END

CUT THREAD RADIUSED END
(PERMISSIBLE ON SIZES BELOW M6 DIAM.)

CONE POINTED END
(PERMISSIBLE ON CUT OR ROLLED THREAD SCREWS, BUT REGARDED AS "SPECIAL")

Fig. 1. Alternative Types of End Permissible on Screws

Dimensions of 90-Degree Countersunk Head Screws: One of the appendices to this British Standard states that countersunk head screws should fit into the countersunk hole with as great a degree of flushness as possible. To achieve this, it is necessary for the dimensions of both the head of the screw and the countersunk hole to be controlled within prescribed limits. The maximum or design size of the head is controlled by a theoretical diameter to a sharp corner and the minimum head

Illustrative matter continued on page 1377 and text on page 1383.

British Standard Machine Screws and Machine Screw Nuts — Metric Series

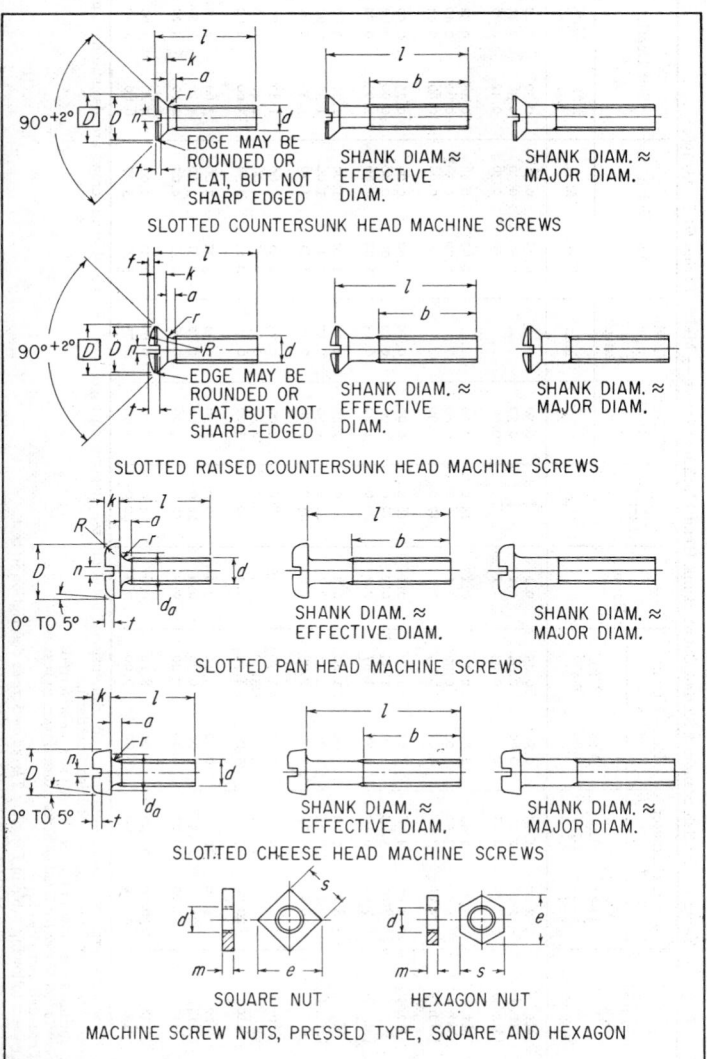

90°+2°   EDGE MAY BE ROUNDED OR FLAT, BUT NOT SHARP EDGED

SHANK DIAM. ≈ EFFECTIVE DIAM.   SHANK DIAM. ≈ MAJOR DIAM.

SLOTTED COUNTERSUNK HEAD MACHINE SCREWS

90°+2°   EDGE MAY BE ROUNDED OR FLAT, BUT NOT SHARP-EDGED

SHANK DIAM. ≈ EFFECTIVE DIAM.   SHANK DIAM. ≈ MAJOR DIAM.

SLOTTED RAISED COUNTERSUNK HEAD MACHINE SCREWS

0° TO 5°

SHANK DIAM. ≈ EFFECTIVE DIAM.   SHANK DIAM. ≈ MAJOR DIAM.

SLOTTED PAN HEAD MACHINE SCREWS

0° TO 5°

SHANK DIAM. ≈ EFFECTIVE DIAM.   SHANK DIAM. ≈ MAJOR DIAM.

SLOTTED CHEESE HEAD MACHINE SCREWS

SQUARE NUT   HEXAGON NUT

MACHINE SCREW NUTS, PRESSED TYPE, SQUARE AND HEXAGON

For dimensions see Tables 1 through 5.

Table 1. British Standard Slotted Countersunk Head Machine Screws — Metric Series (B.S. 4183:1967)

Nominal Size d*	Head Diameter D		Head Height k		Radius r†	Thread Length b	Thread Run-out a	Flushness Tolerance¶	Slot Width n		Slot Depth t	
	Max. (Theor. Sharp) $2d$	Min. $1.75d$	Max. $0.5d$	Min. $0.45d$		Min.	Max. 2β§	Max.	Max.	Min.	Max. $0.3d$	Min. $0.2d$
M1	2.00	1.75	0.50	0.45	0.1	‡‡‡‡‡‡	0.50	...	0.45	0.31	0.30	0.20
M1.2	2.40	2.10	0.60	0.54	0.1		0.50	...	0.50	0.36	0.36	0.24
(M1.4)	2.80	2.45	0.70	0.63	0.1		0.60	...	0.50	0.36	0.42	0.28
M1.6	3.20	2.80	0.80	0.72	0.1	15.0	0.70	...	0.60	0.46	0.48	0.32
M2.0	4.00	3.50	1.00	0.90	0.1	16.0	0.80	...	0.70	0.56	0.60	0.40
(M2.2)	4.40	3.85	1.10	0.99	0.1	17.0	0.90	...	0.80	0.66	0.66	0.44
M2.5	5.00	4.38	1.25	1.12	0.1	18.0	0.90	0.10	0.80	0.66	0.75	0.50
M3	6.00	5.25	1.50	1.35	0.1	19.0	1.00	0.12	1.00	0.86	0.90	0.60
(M3.5)	7.00	6.10	1.75	1.57	0.2	20.0	1.20	0.13	1.00	0.86	1.05	0.70
M4	8.00	7.00	2.00	1.80	0.2	22.0	1.40	0.15	1.20	1.06	1.20	0.80
(M4.5)	9.00	7.85	2.25	2.03	0.2	24.0	1.50	0.17	1.20	1.06	1.35	0.90
M5	10.00	8.75	2.50	2.25	0.2	25.0	1.60	0.19	1.51	1.26	1.50	1.00
M6	12.00	10.50	3.00	2.70	0.25	28.0	2.00	0.23	1.91	1.66	1.80	1.20
M8	16.00	14.00	4.00	3.60	0.4	34.0	2.50	0.29	2.31	2.06	2.40	1.60
M10	20.00	17.50	5.00	4.50	0.4	40.0	3.00	0.37	2.81	2.56	3.00	2.00
M12	24.00	21.00	6.00	5.40	0.6	46.0	3.50	0.44	3.31	3.06	3.60	2.40
(M14)	28.00	24.50	7.00	6.30	0.6	52.0	4.00	0.52	3.31	3.06	4.20	2.80
M16	32.00	28.00	8.00	7.20	0.6	58.0	4.00	0.60	4.37	4.07	4.80	3.20
(M18)	36.00	31.50	9.00	8.10	0.6	64.0	5.00	0.67	4.37	4.07	5.40	3.60
M20	40.00	35.00	10.00	9.00	0.8	70.0	5.00	0.75	5.37	5.07	6.00	4.00

All dimensions are given in millimeters. For dimensional notation, see diagram on page 1377. Recessed head screws are also standard and are available. For dimensions see British Standard.

* Nominal sizes shown in parentheses are non-preferred.

† See *Radius Under the Head of Screws* description in text.

‡ Threaded up to head.

§ See text following table in *Length of Thread on Screws* description in text.

¶ See *Dimensions of 90-Degree Countersunk Head of Screws* description in text.

Table 2. British Standard Slotted Raised Countersunk Head Machine Screws — Metric Series (B.S. 4183:1967)

Nominal Size d*	Head Diameter D		Head Height k		Radius Under Head r†	Thread Length b	Thread Run-out a	Height of Raised Portion f	Head Radius R	Slot Width n		Slot Depth t	
	Max. (Theor. Sharp) 2d	Min. 1.75d	Max. 0.5d	Min. 0.45d		Min.	Max. 2⅛δ	Nom. 0.25d	Nom.	Max.	Min.	Max. 0.5d	Min. 0.4d
M1	2.00	1.75	0.50	0.45	0.1	‡	0.50	0.25	2.0	0.45	0.31	0.50	0.40
M1.2	2.40	2.10	0.60	0.54	0.1	‡	0.50	0.30	2.5	0.50	0.36	0.60	0.48
(M1.4)	2.80	2.45	0.70	0.63	0.1	‡	0.60	0.35	2.5	0.50	0.36	0.70	0.56
M1.6	3.20	2.80	0.80	0.72	0.1	15.0	0.70	0.40	3.0	0.60	0.46	0.80	0.64
M2.0	4.00	3.50	1.00	0.90	0.1	16.0	0.80	0.50	4.0	0.70	0.56	1.00	0.80
(M2.2)	4.40	3.85	1.10	0.99	0.1	17.0	0.90	0.55	4.0	0.80	0.66	1.10	0.88
M2.5	5.00	4.38	1.25	1.12	0.1	18.0	0.90	0.60	5.0	0.80	0.66	1.25	1.00
M3	6.00	5.25	1.50	1.35	0.1	19.0	1.00	0.75	6.0	1.00	0.86	1.50	1.20
(M3.5)	7.00	6.10	1.75	1.57	0.2	20.0	1.20	0.90	6.0	1.00	0.86	1.75	1.40
M4	8.00	7.00	2.00	1.80	0.2	22.0	1.40	1.00	8.0	1.20	1.06	2.00	1.60
(M4.5)	9.00	7.85	2.25	2.03	0.2	24.0	1.50	1.10	8.0	1.20	1.06	2.25	1.80
M5	10.00	8.75	2.50	2.25	0.2	25.0	1.60	1.25	10.0	1.51	1.26	2.50	2.00
M6	12.00	10.50	3.00	2.70	0.25	28.0	2.00	1.50	12.0	1.91	1.66	3.00	2.40
M8	16.00	14.00	4.00	3.60	0.4	34.0	2.50	2.00	16.0	2.31	2.06	4.00	3.20
M10	20.00	17.50	5.00	4.50	0.4	40.0	3.00	2.50	20.0	2.81	2.56	5.00	4.00
M12	24.00	21.00	6.00	5.40	0.6	46.0	3.50	3.00	25.0	3.31	3.06	6.00	4.80
(M14)	28.00	24.50	7.00	6.30	0.6	52.0	4.00	3.50	25.0	3.37	3.06	7.00	5.60
M16	32.00	28.00	8.00	7.20	0.6	58.0	4.00	4.00	32.0	4.37	4.07	8.00	6.40
(M18)	36.00	31.50	9.00	8.10	0.6	64.0	5.00	4.50	32.0	4.37	4.07	9.00	7.20
M20	40.00	35.00	10.00	9.00	0.8	70.0	5.00	5.00	40.0	5.37	5.07	10.00	8.00

All dimensions are given in millimeters. For dimensional notation see diagram on page 1377. Recessed head screws are also standard and available. For dimensions see British Standard.

* Nominal sizes shown in parentheses are non-preferred.
† See *Radius Under the Head of Screws* description in text.
‡ Threaded up to head.
§ See text following table in *Length of Thread on Screws* description in text.

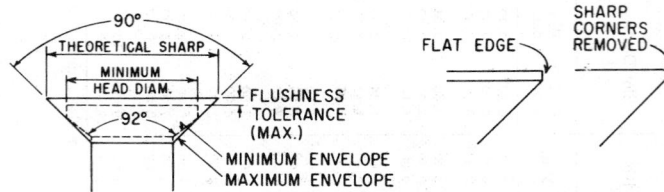

Fig. 2. Head Configuration   Fig. 3. Edge Configuration

Table 3. British Standard Slotted Pan Head Machine Screws —
Metric Series (B.S. 4183:1967)

Nominal Size d^*	Head Diameter D		Head Height k		Head Radius R	Radius Under Head r	Transition Diameter d_a
	Max. $2d$	Min.	Max. $0.6d$	Min.	Max. $0.4d$	Min.	Max.
M2.5	5.00	4.70	1.50	1.36	1.00	0.10	3.10
M3	6.00	5.70	1.80	1.66	1.20	0.10	3.60
(M3.5)	7.00	6.64	2.10	1.96	1.40	0.20	4.30
M4	8.00	7.64	2.40	2.26	1.60	0.20	4.70
(M4.5)	9.00	8.64	2.70	2.56	1.80	0.20	5.20
M5	10.00	9.64	3.00	2.86	2.00	0.20	5.70
M6	12.00	11.57	3.60	3.42	2.50	0.25	6.80
M8	16.00	15.57	4.80	4.62	3.20	0.40	9.20
M10	20.00	19.48	6.00	5.82	4.00	0.40	11.20

Nominal Size d^*	Thread Length b	Thread Run-out a	Slot Width n		Slot Depth t	
	Min.	Max. $2p\dagger$	Max.	Min.	Max. $0.6k$	Min. $0.4k$
M2.5	18.00	0.90	0.80	0.66	0.90	0.60
M3	19.00	1.00	1.00	0.86	1.08	0.72
(M3.5)	20.00	1.20	1.00	0.86	1.26	0.84
M4	22.00	1.40	1.20	1.06	1.44	0.96
(M4.5)	24.00	1.50	1.20	1.06	1.62	1.08
M5	25.00	1.60	1.51	1.26	1.80	1.20
M6	28.00	2.00	1.91	1.66	2.16	1.44
M8	34.00	2.50	2.31	2.06	2.88	1.92
M10	40.00	3.00	2.81	2.56	3.60	2.40

All dimensions are given in millimeters. For dimensional notation, see diagram on page 1377. Recessed head screws are also standard and available. For dimensions see British Standard.
* Nominal sizes shown in parentheses are non-preferred.
† See text following table in *Length of Thread on Screws* description in text.

Table 4. British Standard Slotted Cheese Head Machine Screws — Metric Series (B.S. 4183:1967)

Nominal Size d*	Head Diameter D		Head Height k		Radius r†	Transition Diameter d_a	Thread Length b	Thread Run-out a	Slot Width n		Slot Depth t	
	Max.	Min.	Max.	Min.	Min.	Max.	Min.	Max.§	Max.	Min.	Max.	Min.
M1	2.00	1.75	0.70	0.56	0.10	1.30	†	0.50	0.45	0.31	0.44	0.30
M1.2	2.30	2.05	0.80	0.66	0.10	1.50	†	0.50	0.50	0.36	0.49	0.35
(M1.4)	2.60	2.35	0.90	0.76	0.10	1.70	†	0.60	0.50	0.36	0.60	0.40
M1.6	3.00	2.75	1.00	0.86	0.10	2.00	15.00	0.70	0.60	0.46	0.65	0.45
M2	3.80	3.50	1.30	1.16	0.10	2.60	16.00	0.80	0.70	0.56	0.85	0.60
(M2.2)	4.00	3.70	1.50	1.36	0.10	2.80	17.00	0.90	0.80	0.66	1.00	0.70
M2.5	4.50	4.20	1.60	1.46	0.10	3.10	18.00	0.90	0.80	0.66	1.00	0.70
M3	5.50	5.20	2.00	1.86	0.10	3.60	19.00	1.00	1.00	0.86	1.30	0.90
(M3.5)	6.00	5.70	2.40	2.26	0.10	4.10	20.00	1.20	1.00	0.86	1.40	1.00
M4	7.00	6.64	2.60	2.46	0.20	4.70	22.00	1.40	1.20	1.06	1.60	1.20
(M4.5)	8.00	7.64	3.10	2.92	0.20	5.20	24.00	1.50	1.20	1.06	1.80	1.40
M5	8.50	8.14	3.30	3.12	0.20	5.70	25.00	1.60	1.51	1.26	2.00	1.50
M6	10.00	9.64	3.90	3.72	0.25	6.80	28.00	2.00	1.91	1.66	2.30	1.80
M8	13.00	12.57	5.00	4.82	0.40	9.20	34.00	2.50	2.31	2.06	2.80	2.30
M10	16.00	15.57	6.00	5.82	0.40	11.20	40.00	3.00	2.81	2.56	3.20	2.70
M12	18.00	17.57	7.00	6.78	0.60	14.20	46.00	3.50	3.31	3.06	3.80	3.20
(M14)	21.00	20.48	8.00	7.78	0.60	16.20	52.00	4.00	3.31	3.06	4.20	3.60
M16	24.00	23.48	9.00	8.78	0.60	18.20	58.00	4.00	4.37	4.07	4.60	4.00
(M18)	27.00	26.48	10.00	9.78	0.60	20.20	64.00	5.00	4.37	4.07	5.10	4.50
M20	30.00	29.48	11.00	10.73	0.80	22.40	70.00	5.00	5.27	5.07	5.60	5.00

All dimensions are given in millimeters. For dimensional notation, see diagram on page 1377.

* Nominal sizes shown in parentheses are non-preferred.

† Threaded up to head.

§ See text following table in *Length of Thread on Screws* description in text.

Table 5. British Standard Machine Screws and Nuts — Metric Series (B.S. 4183:1967)

CONCENTRICITY TOLERANCES

COUNTERSUNK & RAISED COUNTERSUNK HEADS — IT 13, IT 13, HEAD TO SHANK, SLOT TO HEAD

PAN & CHEESE HEADS — SLOT TO HEAD, HEAD TO HEAD

Nominal Size d^*	Head to Shank and Slot to Head (IT 13) Csk., Raised Csk. and Pan Heads	Cheese Heads
M1	0.14	0.14
M1.2	0.14	0.14
(M1.4)	0.14	0.14
M1.6	0.14	0.18
M2	0.18	0.18
(M2.2)	0.18	0.18
M2.5	0.18	0.18
M3	0.18	0.18
(M3.5)	0.22	0.18
M4	0.22	0.22
(M4.5)	0.22	0.22
M5	0.22	0.22
M6	0.27	0.27
M8	0.27	0.27
M10	0.33	0.27
M12	0.33	0.33
(M14)	0.33	0.33
M16	0.39	0.33
(M18)	0.39	0.33
M20	0.39	0.33

NOMINAL LENGTHS AND TOLERANCES ON LENGTH FOR MACHINE SCREWS

Nominal Length*	Tolerance	Nominal Length*	Tolerance
1.5	±0.12	45	±0.50
2	±0.12	50	±0.50
2.5	±0.20	55	±0.60
3	±0.20	60	±0.60
4	±0.24	65	±0.60
5	±0.24	70	±0.60
6	±0.24	75	±0.60
(7)	±0.29	80	±0.60
8	±0.29	85	±0.70
(9)	±0.29	90	±0.70
10	±0.29	(95)	±0.70
(11)	±0.35	100	±0.70
12	±0.35	(105)	±0.70
14	±0.35	110	±0.70
16	±0.35	(115)	±0.70
(18)	±0.42	120	±0.70
20	±0.42	(125)	±0.80
(22)	±0.42	130	±0.80
25	±0.42	140	±0.80
(28)	±0.42	150	±0.80
30	±0.42	160	±0.80
(32)	±0.50	170	±0.80
35	±0.50	180	±0.80
(38)	±0.50	190	±0.925
40	±0.50	200	±0.925

DIMENSIONS OF MACHINE SCREW NUTS, PRESSED TYPE, SQUARE AND HEXAGON

Nominal Size d^*	Width Across Flats s Max.	Width Across Flats s Min.	Width Across Corners e Square
M1.6	3.2	3.02	4.5
M2	4.0	3.82	5.7
(M2.2)	4.5	4.32	6.4
M2.5	5.0	4.82	7.1
M3	5.5	5.32	7.8
(M3.5)	6.0	5.82	8.5
M4	7.0	6.78	9.9
M5	8.0	7.78	11.3
M6	10.0	9.78	14.1
M8	13.0	12.73	18.4
M10	17.0	16.73	24.0

Nominal Size d^*	Width Across Corners e Hexagon	Thickness m Max.	Thickness m Min.
M1.6	3.7	1.0	0.75
M2	4.6	1.2	0.95
(M2.2)	5.2	1.2	0.95
M2.5	5.8	1.6	1.35
M3	6.4	1.6	1.35
(M3.5)	6.9	2.0	1.75
M4	8.1	2.0	1.75
M5	9.2	2.5	2.25
M6	11.5	2.75	2.75
M8	15.0	4.0	3.70
M10	19.6	5.0	4.70

All dimensions are given in millimeters. For dimensional notation see diagram on page 1377.
* Nominal sizes and lengths shown in parentheses are non-preferred.

angle of 90 degrees. The minimum head size is controlled by a minimum head diameter, the maximum head angle of 92 degrees and a flushness tolerance (see Fig. 2, page 1380). The edge of the head may be flat or rounded, as shown in Fig. 3 on page 1380.

General Dimensions: The general dimensions and tolerances for screws and nuts are given in the accompanying tables. Although slotted screw dimensions are given, recessed head screws are also standard and available. Dimensions of recessed head screws are given in B.S. 4183:1967.

British Unified Machine Screws and Nuts. — British Standard B.S. 1981:1953 covers certain types of machine screws and machine screw nuts for which agreement has been reached with the United States and Canada as to general dimensions for interchangeability. These types are: countersunk, raised-countersunk, pan, and raised-cheese head screws with slotted or recessed heads; small hexagon head screws; and precision and pressed nuts. All have Unified threads. Head shapes are shown on page 1384 and dimensions are given on page 1385.

Identification: As revised by Amendment No. 1 in February 1955, this standard now requires that the above-mentioned screws and nuts which conform to this standard should have a distinguishing feature applied to identify them as Unified. All *recessed head screws* are to be identified as Unified by a groove in the form of four arcs of a circle in the upper surface of the head. All *hexagon head screws* are to be identified as Unified by: (1) a circular recess in the upper surface of the head; or

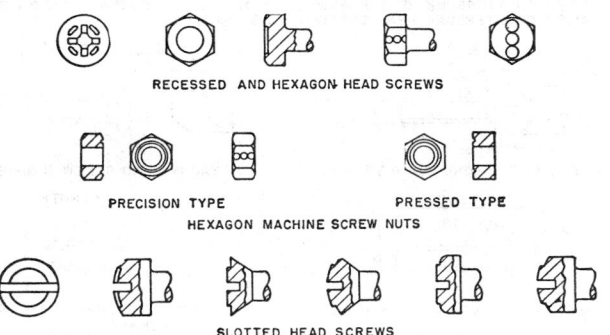

RECESSED AND HEXAGON HEAD SCREWS

PRECISION TYPE   PRESSED TYPE

HEXAGON MACHINE SCREW NUTS

SLOTTED HEAD SCREWS

Identification Markings for British Standard Unified Machine Screws

(2) a continuous line of circles indented on one or more of the flats of the hexagon and parallel to the screw axis; or (3) at least two contiguous circles indented on the upper surface of the head. All *machine screw nuts* of the pressed type shall be identified as Unified by means of the application of a groove indented in one face of the nut approximately midway between the major diameter of the thread and flats of the square or hexagon. *Slotted head screws* shall be identified as Unified either by a circular recess or by a circular platform or raised portion on the upper surface of the head. *Machine screw nuts* of the *precision type* shall be identified as Unified by either a groove indented on one face of the front approximately midway between the major diameter of the thread and the flats of the hexagon or a continuous line of circles indented on one or more of the flats of the hexagon and parallel to the nut axis.

British Standard Machine Screws and Nuts (B.S. 450:1958 and B.S. 1981:1953)

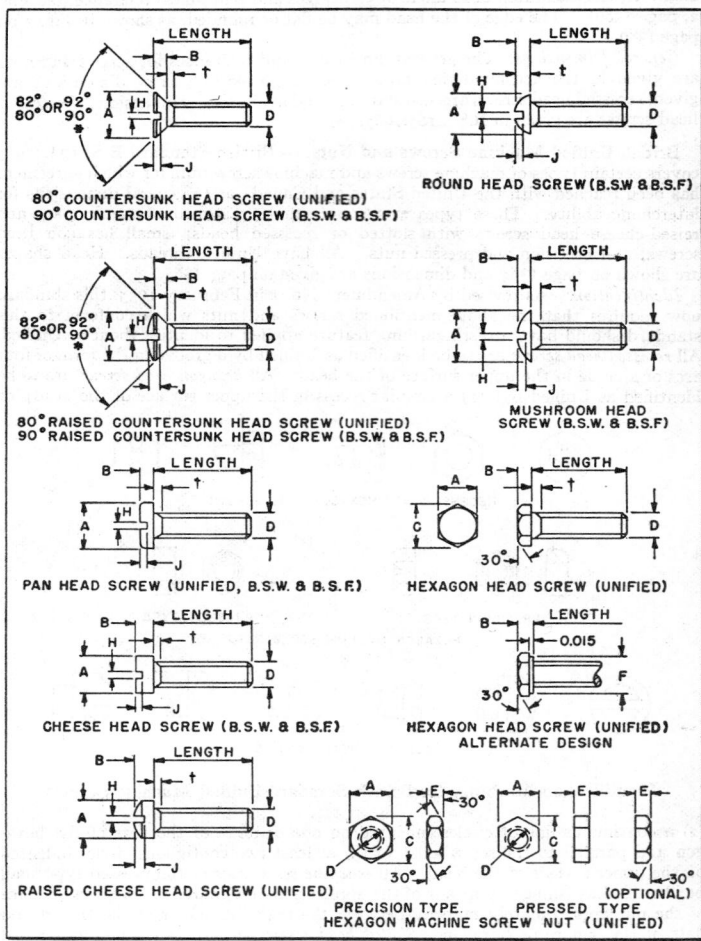

* Countersinks to suit the screws should have a maximum angle of 80° (Unified) or 90° (B.S.F. and B.S.W.) with a negative tolerance.

† Unified countersunk and raised countersunk head screws 2 inches long and under are threaded right up to the head. Other Unified, B.S.W. and B.S.F. machine screws 2 inches long and under have an unthreaded shank equal to twice the pitch. All Unified, B.S.W. and B.S.F. machine screws longer than 2 inches have a minimum thread length of 1¾ inches.

British Standard Unified Machine Screws and Machine Screw Nuts (B.S. 1981:1953)[1]

Nom. Size of Screw	Basic Diam. D	Threads per Inch		Diam. of Head A		Depth of Head B		Width of Slot H		Depth of Slot J
		UNC	UNF	Max.	Min.	Max.	Min.	Max.	Min.	
80° COUNTERSUNK HEAD SCREWS[2,3]										
4	.112	40	..	.211	.194	.067	....	.039	.031	.025
6	.138	32	..	.260	.242	.083	....	.048	.039	.031
8	.164	32	..	.310	.291	.100	....	.054	.045	.037
10	.190	24*	32	.359	.339	.116	....	.060	.050	.044
¼	.250	20	28	.473	.450	.153	....	.075	.064	.058
⁵⁄₁₆	.3125	18	24	.593	.565	.191	....	.084	.072	.073
⅜	.375	16	24	.712	.681	.230	....	.094	.081	.086
⁷⁄₁₆	.4375	14	20	.753	.719	.223	....	.094	.081	.086
½	.500	13	20	.808	.770	.223	....	.106	.091	.086
⅝	.625	11	18	1.041	.996	.298	....	.133	.116	.113
¾	.750	10	16	1.275	1.223	.372	....	.149	.131	.141
PAN HEAD SCREWS[3]										
4	.112	40	..	.219	.205	.068	.058	.039	.031	.036
6	.138	32	..	.270	.256	.082	.072	.048	.039	.044
8	.164	32	..	.322	.306	.096	.085	.054	.045	.051
10	.190	24*	32	.373	.357	.110	.099	.060	.050	.059
¼	.250	20	28	.492	.473†	.144	.130	.075	.064	.079
⁵⁄₁₆	.3125	18	24	.615	.594	.178	.162	.084	.072	.101
⅜	.375	16	24	.740	.716	.212	.195	.094	.081	.122
⁷⁄₁₆	.4375	14	20	.863	.838	.247	.227	.094	.081	.133
½	.500	13	20	.987	.958	.281	.260	.106	.091	.152
⅝	.625	11	18	1.125	1.090	.350	.325	.133	.116	.189
¾	.750	10	16	1.250	1.209	.419	.390	.149	.131	.226
RAISED CHEESE-HEAD SCREWS[3]										
4	.112	40	..	.183	.166	.107	.088	.039	.031	.042
6	.138	32	..	.226	.208	.132	.111	.048	.039	.053
8	.164	32	..	.270	.250	.156	.133	.054	.045	.063
10	.190	24*	32	.313	.292	.180	.156	.060	.050	.074
¼	.250	20	28	.414	.389	.237	.207	.075	.064	.098
⁵⁄₁₆	.3125	18	24	.518	.490	.295	.262	.084	.072	.124
⅜	.375	16	24	.622	.590	.355	.315	.094	.081	.149
⁷⁄₁₆	.4375	14	20	.625	.589	.368	.321	.094	.081	.153
½	.500	13	20	.750	.710	.412	.362	.106	.091	.171
⅝	.625	11	18	.875	.827	.521	.461	.133	.116	.217
¾	.750	10	16	1.000	.945	.612	.542	.149	.131	.254

Nom. Size	Basic Diam. D	Threads per Inch		Width Across			H'd Depth B Nut Thick. E		Wash. Face Diam. F	
				Flats A		Corners C				
		UNC	UNF	Max.	Min.	Max.	Max.	Min.	Max.	Min.
HEXAGON HEAD SCREWS										
4	.112	40	..	.1875	.1835	.216	.060	.055	.183	.173
6	.138	32	..	.2500	.2450	.289	.080	.074	.245	.235
8	.164	32	..	.2500	.2450	.289	.110	.104	.245	.235
10	.190	24*	32	.3125	.3075	.361	.120	.113	.307	.297
HEXAGON MACHINE SCREW NUTS — PRECISION TYPE										
4	.112	40	..	.1875	.1835	.216	.098	.087	....	....
6	.138	32	..	.2500	.2450	.289	.114	.102	....	....
8	.164	32	..	.3125	.3075	.361	.130	.117	....	....
10	.190	24*	32	.3125	.3075	.361	.130	.117	....	....
HEXAGON MACHINE SCREW NUTS — PRESSED TYPE										
4	.112	40	..	.2500	.2410	.289	.087	.077	....	....
6	.138	32	..	.3125	.3020	.361	.114	.102	....	....
8	.164	32	..	.3438	.3320	.397	.130	.117	....	....
10	.190	24*	32	.3750	.3620	.433	.130	.117	....	....
¼	.250	20	28	.4375	.4230	.505	.193	.178	....	....
⁵⁄₁₆	.3125	18	24	.5625	.5450	.649	.225	.208	....	....
⅜	.375	16	24	.6250	.6070	.722	.257	.239	....	....

All dimensions in inches. [1] See page 1384, for a pictorial representation and letter dimensions. [2] All dimensions, except J, given for the No. 4 to ⅜-inch sizes, incl., also apply to all the 80° Raised Countersunk Head Screws given in the Standard. [3] Also available with recessed heads. * Non-preferred. † By arrangement may also be .468.

British Standard Whitworth (B.S.W.) and Fine (B.S.F.) Machine Screws (B.S. 450:1958)[1]

Nom. Size of Screw	Basic Diam. D	Threads per Inch		Diam. of Head A		Depth of Head B		Width of Slot H		Depth of Slot J
		B.S.W.	B.S.F.	Max.	Min.	Max.	Min.	Max.	Min.	
90° COUNTERSUNK HEAD SCREWS[2,3]										
1/8	.1250	40	..	.219	.201	.056	....	.039	.032	.027
3/16	.1875	24	32*	.328	.307	.084	....	.050	.042	.041
7/32	.2188	..	28*	.383	.360	.098	....	.055	.046	.048
1/4	.2500	20	26	.438	.412	.113	....	.061	.051	.055
5/16	.3125	18	22	.547	.518	.141	....	.071	.061	.069
3/8	.3750	16	20	.656	.624	.169	....	.082	.072	.083
7/16	.4375	14	18	.766	.729	.197	....	.093	.082	.097
1/2	.5000	12	16	.875	.835	.225	....	.104	.092	.111
9/16	.5625	12*	16*	.984	.941	.253	....	.115	.103	.125
5/8	.6250	11	14	1.094	1.046	.281	....	.126	.113	.138
3/4	.7500	10	12	1.312	1.257	.338	....	.148	.134	.166
ROUND HEAD SCREWS[3]										
1/8	.1250	40	..	.219	.206	.087	.082	.039	.032	.048
3/16	.1875	24	32*	.328	.312†	.131	.124	.050	.042	.072
7/32	.2188	..	28*	.383	.365	.153	.145	.055	.046	.084
1/4	.2500	20	26	.438	.417	.175	.165	.061	.051	.096
5/16	.3125	18	22	.547	.524	.219	.207	.071	.061	.120
3/8	.3750	16	20	.656	.629	.262	.249	.082	.072	.144
7/16	.4375	14	18	.766	.735	.306	.291	.093	.082	.168
1/2	.5000	12	16	.875	.840	.350	.333	.104	.092	.192
9/16	.5625	12*	16*	.984	.946	.394	.375	.115	.103	.217
5/8	.6250	11	14	1.094	1.051	.437	.417	.126	.113	.240
3/4	.7500	10	12	1.312	1.262	.525	.500	.148	.134	.288
PAN HEAD SCREWS[3]										
1/8	.1250	40	..	.245	.231	.075	.065	.039	.032	.040
3/16	.1875	24	32*	.373	.375	.110	.099	.050	.042	.061
7/32	.2188	..	28*	.425	.407	.125	.112	.055	.046	.069
1/4	.2500	20	26	.492	.473§	.144	.130	.061	.051	.078
5/16	.3125	18	22	.615	.594	.178	.162	.071	.061	.095
3/8	.3750	16	20	.740	.716	.212	.195	.082	.072	.112
7/16	.4375	14	18	.863	.838	.247	.227	.093	.082	.129
1/2	.5000	12	16	.987	.958	.281	.260	.104	.092	.145
9/16	.5625	12*	16*	1.031	.999	.315	.293	.115	.103	.162
5/8	.6250	11	14	1.125	1.090	.350	.325	.126	.113	.179
3/4	.7500	10	12	1.250	1.209	.419	.390	.148	.134	.213
CHEESE HEAD SCREWS[3]										
1/8	.1250	40	..	.188	.180	.087	.082	.039	.032	.039
3/16	.1875	24	32*	.281	.270	.131	.124	.050	.042	.059
7/32	.2188	..	28*	.328	.315	.153	.145	.055	.046	.069
1/4	.2500	20	26	.375	.360	.175	.165	.061	.051	.079
5/16	.3125	18	22	.469	.450	.219	.207	.071	.061	.098
3/8	.3750	16	20	.562	.540	.262	.249	.082	.072	.118
7/16	.4375	14	18	.656	.630	.306	.291	.093	.082	.138
1/2	.5000	12	16	.750	.720	.350	.333	.104	.092	.157
9/16	.5625	12*	16*	.844	.810	.394	.375	.115	.103	.177
5/8	.6250	11	14	.938	.900	.437	.417	.126	.113	.197
3/4	.7500	10	12	1.125	1.080	.525	.500	.148	.134	.236
MUSHROOM HEAD SCREWS[3]										
1/8	.1250	40	..	.289	.272	.078	.066	.043	.035	.040
3/16	.1875	24	32*	.448	.425	.118	.103	.060	.050	.061
1/4	.2500	20	26	.573	.546	.150	.133	.075	.064	.079
5/16	.3125	18	22	.698	.666	.183	.162	.084	.072	.096
3/8	.3750	16	20	.823	.787	.215	.191	.094	.081	.112

All dimensions in inches. [1] See diagram on page 1384 for a pictorial representation of screws and letter dimensions. [2] All dimensions, except J, given for the 1/8-through 3/8-inch sizes also apply to all the 90° Raised Countersunk Head Screw dimensions given in the Standard. [3] These screws are also available with recessed heads; dimensions of recess are not given here but may be found in the Standard. * Non-preferred size; avoid use whenever possible. † By arrangement this may also be .309. § By arrangement this may also be .468.

Slotted Head Cap Screws. — American National Standard ANSI B18.6.2-1972, R1977 is intended to cover the complete general and dimensional data for the various styles of slotted head cap screws as well as square head and slotted headless set screws (see page 1394). Reference should be made to this Standard for information or data not found in the following text or tables.

Length of Thread: The length of complete (full form) thread on cap screws is equal to twice the basic screw diameter plus 0.250 in. with a plus tolerance of 0.188 in. or an amount equal to 2½ times the pitch of the thread, whichever is greater. Cap screws of lengths too short to accommodate the minimum thread length have full form threads extending to within a distance equal to 2½ pitches (threads) of the head.

Designation: Slotted head cap screws are designated by the following data in the sequence shown: Nominal size (fraction or decimal equivalent); threads per inch; screw length (fraction or decimal equivalent); product name; material; and protective finish, if required. Examples: ½-13 × 3 Slotted Round Head Cap Screw, SAE Grade 2 Steel, Zinc Plated. .750-16 × 2.25 Slotted Flat Countersunk Head Cap Screw, Corrosion Resistant Steel.

Table 1. American National Standard Slotted Flat Countersunk Head Cap Screws (ANSI B18.6.2-1972, R1977)

Nominal Size* or Basic Screw Diam.	Body Diam., *E*		Head Dia., *A*			Head Hgt., *H*	Slot Width, *J*		Slot Depth, *T*		Filet Rad., *U*
			Edge Sharp	Edge Rnd'd. or Flat							
	Max.	Min.	Max.	Min.		Ref.	Max.	Min.	Max.	Min.	Max.
¼ 0.2500	.2500	.2450	.500	.452		.140	.075	.064	.068	.045	.100
5⁄16 0.3125	.3125	.3070	.625	.567		.177	.084	.072	.086	.057	.125
3⁄8 0.3750	.3750	.3690	.750	.682		.210	.094	.081	.103	.068	.150
7⁄16 0.4375	.4375	.4310	.812	.736		.210	.094	.081	.103	.068	.175
½ 0.5000	.5000	.4930	.875	.791		.210	.106	.091	.103	.068	.200
9⁄16 0.5625	.5625	.5550	1.000	.906		.244	.118	.102	.120	.080	.225
5⁄8 0.6250	.6250	.6170	1.125	1.020		.281	.133	.116	.137	.091	.250
¾ 0.7500	.7500	.7420	1.375	1.251		.352	.149	.131	.171	.115	.300
7⁄8 0.8750	.8750	.8660	1.625	1.480		.423	.167	.147	.206	.138	.350
1 1.0000	1.000	.9900	1.875	1.711		.494	.188	.166	.240	.162	.400
1⅛ 1.1250	1.1250	1.1140	2.062	1.880		.529	.196	.178	.257	.173	.450
1¼ 1.2500	1.2500	1.2390	2.312	2.110		.600	.211	.193	.291	.197	.500
1⅜ 1.3750	1.3750	1.3630	2.562	2.340		.665	.226	.208	.326	.220	.550
1½ 1.5000	1.5000	1.4880	2.812	2.570		.742	.258	.240	.360	.244	.600

All dimensions are in inches.
Threads: Threads are Unified Standard Class 2A; UNC, UNF and 8 UN Series or UNRC, UNRF, and 8 UNR Series.
* When specifying a nominal size in decimals, the zero preceding the decimal point is omitted as is any zero in the fourth decimal place.

Table 2. American National Standard Slotted Round Head Cap Screws
(ANSI B18.6.2-1972, R1977)

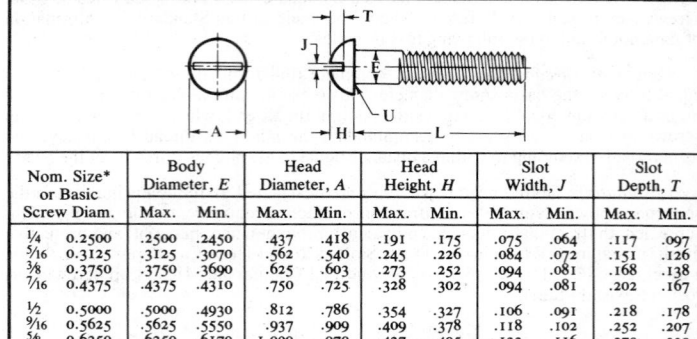

Nom. Size* or Basic Screw Diam.		Body Diameter, E		Head Diameter, A		Head Height, H		Slot Width, J		Slot Depth, T	
		Max.	Min.	Max.	Min.	Max.	Min.	Max.	Min.	Max.	Min.
¼	0.2500	.2500	.2450	.437	.418	.191	.175	.075	.064	.117	.097
⁵⁄₁₆	0.3125	.3125	.3070	.562	.540	.245	.226	.084	.072	.151	.126
⅜	0.3750	.3750	.3690	.625	.603	.273	.252	.094	.081	.168	.138
⁷⁄₁₆	0.4375	.4375	.4310	.750	.725	.328	.302	.094	.081	.202	.167
½	0.5000	.5000	.4930	.812	.786	.354	.327	.106	.091	.218	.178
⁹⁄₁₆	0.5625	.5625	.5550	.937	.909	.409	.378	.118	.102	.252	.207
⅝	0.6250	.6250	.6170	1.000	.970	.437	.405	.133	.116	.270	.220
¾	0.7500	.7500	.7420	1.250	1.215	.546	.507	.149	.131	.338	.278

All dimensions are in inches.
Fillet Radius, U: For fillet radius see footnote to table below.
Threads: Threads are Unified Standard Class 2A; UNC, UNF and 8 UN Series or UNRC, UNRF and 8 UNR Series.
  * When specifying a nominal size in decimals, the zero preceding the decmial point is omitted as is any zero in the fourth decimal place.

Table 3. American National Standard Slotted Fillister Head Cap Screws
(ANSI B18.6.2-1972, R1977)

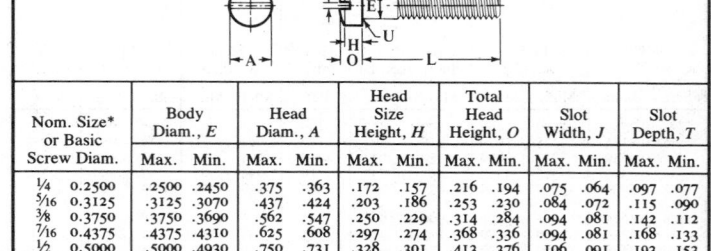

Nom. Size* or Basic Screw Diam.		Body Diam., E		Head Diam., A		Head Size Height, H		Total Head Height, O		Slot Width, J		Slot Depth, T	
		Max.	Min.	Max.	Min.	Max.	Min.	Max.	Min.	Max.	Min.	Max.	Min.
¼	0.2500	.2500	.2450	.375	.363	.172	.157	.216	.194	.075	.064	.097	.077
⁵⁄₁₆	0.3125	.3125	.3070	.437	.424	.203	.186	.253	.230	.084	.072	.115	.090
⅜	0.3750	.3750	.3690	.562	.547	.250	.229	.314	.284	.094	.081	.142	.112
⁷⁄₁₆	0.4375	.4375	.4310	.625	.608	.297	.274	.368	.336	.094	.081	.168	.133
½	0.5000	.5000	.4930	.750	.731	.328	.301	.413	.376	.106	.091	.193	.153
⁹⁄₁₆	0.5625	.5625	.5550	.812	.792	.375	.346	.467	.427	.118	.102	.213	.168
⅝	0.6250	.6250	.6170	.875	.853	.422	.391	.521	.478	.133	.116	.239	.189
¾	0.7500	.7500	.7420	1.000	.976	.500	.466	.612	.566	.149	.131	.283	.223
⅞	0.8750	.8750	.8660	1.125	1.098	.594	.556	.720	.668	.167	.147	.334	.264
I	1.0000	1.0000	.9900	1.312	1.282	.656	.612	.803	.743	.188	.166	.371	.291

All dimensions are in inches.
Fillet Radius, U: The fillet radius is as follows: For screw sizes ¼ to ⅜ incl., .031 max. and .016 min.; ⁷⁄₁₆ to ⁹⁄₁₆, incl., .047 max., .016 min.; and for ⅝ to 1, incl., .062 max., .031 min.
Threads: Threads are Unified Standard Class 2A; UNC, UNF and 8 UN Series or UNRC, UNRF and 8 UNR Series.
  * When specifying nominal size in decimals, the zero preceding the decmial point is omitted as is any zero in the fourth decimal place.

Table 1. American National Standard Hexagon and Spline Socket Head Cap Screws (1960 Series) (ANSI B18.3-1986)

Nominal Size	Body Diameter Max	Body Diameter Min	Head Diameter Max	Head Diameter Min	Head Height Max	Head Height Min	Spline Socket Size Nom	Hex. Socket Size Nom	Fillet Ext. Max	Key Engagement*
	D		*A*		*H*		*M*	*J*	*F*	*T*
0	0.0600	0.0568	0.096	0.091	0.060	0.057	0.060	0.050	0.007	0.025
1	0.0730	0.0695	0.118	0.112	0.073	0.070	0.072	1/16 0.062	0.007	0.031
2	0.0860	0.0822	0.140	0.134	0.086	0.083	0.096	5/64 0.078	0.008	0.038
3	0.0990	0.0949	0.161	0.154	0.099	0.095	0.096	5/64 0.078	0.008	0.044
4	0.1120	0.1075	0.183	0.176	0.112	0.108	0.111	3/32 0.094	0.009	0.051
5	0.1250	0.1202	0.205	0.198	0.125	0.121	0.111	3/32 0.094	0.010	0.057
6	0.1380	0.1329	0.226	0.218	0.138	0.134	0.133	7/64 0.109	0.010	0.064
8	0.1640	0.1585	0.270	0.262	0.164	0.159	0.168	9/64 0.141	0.012	0.077
10	0.1900	0.1840	0.312	0.303	0.190	0.185	0.183	5/32 0.156	0.014	0.090
1/4	0.2500	0.2435	0.375	0.365	0.250	0.244	0.216	3/16 0.188	0.014	0.120
5/16	0.3125	0.3053	0.469	0.457	0.312	0.306	0.291	1/4 0.250	0.017	0.151
3/8	0.3750	0.3678	0.562	0.550	0.375	0.368	0.372	5/16 0.312	0.020	0.182
7/16	0.4375	0.4294	0.656	0.642	0.438	0.430	0.454	3/8 0.375	0.023	0.213
1/2	0.5000	0.4919	0.750	0.735	0.500	0.492	0.454	3/8 0.375	0.026	0.245
5/8	0.6250	0.6163	0.938	0.921	0.625	0.616	0.595	1/2 0.500	0.032	0.307
3/4	0.7500	0.7406	1.125	1.107	0.750	0.740	0.620	5/8 0.625	0.039	0.370
7/8	0.8750	0.8647	1.312	1.293	0.875	0.864	0.698	3/4 0.750	0.044	0.432
1	1.0000	0.9886	1.500	1.479	1.000	0.988	0.790	3/4 0.750	0.050	0.495
1 1/8	1.1250	1.1086	1.688	1.665	1.125	1.111	. . .	7/8 0.875	0.055	0.557
1 1/4	1.2500	1.2336	1.875	1.852	1.250	1.236	. . .	7/8 0.875	0.060	0.620
1 3/8	1.3750	1.3568	2.062	2.038	1.375	1.360	. . .	1 1.000	0.065	0.682
1 1/2	1.5000	1.4818	2.250	2.224	1.500	1.485	. . .	1 1.000	0.070	0.745
1 3/4	1.7500	1.7295	2.625	2.597	1.750	1.734	. . .	1 1/4 1.250	0.080	0.870
2	2.0000	1.9780	3.000	2.970	2.000	1.983	. . .	1 1/2 1.500	0.090	0.995
2 1/4	2.2500	2.2280	3.375	3.344	2.250	2.232	. . .	1 3/4 1.750	0.100	1.120
2 1/2	2.5000	2.4762	3.750	3.717	2.500	2.481	. . .	1 3/4 1.750	0.110	1.245
2 3/4	2.7500	2.7262	4.125	4.090	2.750	2.730	. . .	2 2.000	0.120	1.370
3	3.0000	2.9762	4.500	4.464	3.000	2.979	. . .	2 1/4 2.250	0.130	1.495
3 1/4	3.2500	3.2262	4.875	4.837	3.250	3.228	. . .	2 1/4 2.250	0.140	1.620
3 1/2	3.5000	3.4762	5.250	5.211	3.500	3.478	. . .	2 3/4 2.750	0.150	1.745
3 3/4	3.7500	3.7262	5.625	5.584	3.750	3.727	. . .	2 3/4 2.750	0.160	1.870
4	4.0000	3.9762	6.000	5.958	4.000	3.976	. . .	3 3.000	0.170	1.995

* Key engagement depths are minimum.

All dimensions in inches. The body length L_B of the screw is the length of the unthreaded cylindrical portion of the shank. The length of thread, L_T, is the distance from the extreme point to the last complete (full form) thread. Standard length increments for screw diameters up to 1 inch are 1/16 inch for lengths 1/8 through 1/4 inch, 1/8 inch for lengths 1/4 through 1 inch, 1/4 inch for lengths 1 through 3 1/2 inches, 1/2 inch for lengths 3 1/2 through 7 inches, 1 inch for lengths 7 through 10 inches and for diameters over 1 inch are 1/2 inch for lengths 1 through 7 inches, 1 inch for lengths 7 through 10 inches and 2 inches for lengths over 10 inches. Heads may be plain or knurled, and chamfered to an angle E of 30 to 45 degrees with the surface of the flat. The thread conforms to the Unified Standard with radius root, Class 3A, UNRC and UNRF for screw sizes No. 0 through 1 inch inclusive, Class 2A, UNRC and UNRF for over 1 inch through 1 1/2 inches inclusive, and Class 2A UNRC for sizes larger than 1 1/2 inches. Socket dimensions are given in Table 9. For manufacturing details not shown, including materials, see American National Standard ANSI B18.3-1986.

Table 2. Drill and Counterbore Sizes For Socket Head Cap Screws (1960 Series)

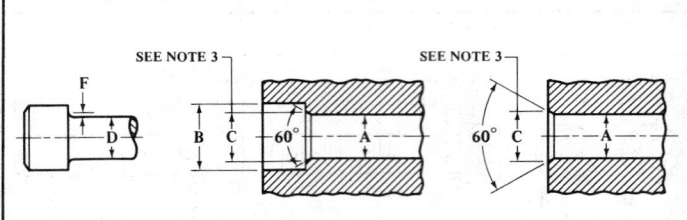

Nominal Size or Basic Screw Diameter		Nominal Drill Size				Counterbore Diameter	Countersink Diameter[3]
		Close Fit[1]		Normal Fit[2]			
		Number or Fractional Size	Decimal Size	Number or Fractional Size	Decimal Size		
			A			*B*	*C*
0	0.0600	51	0.067	49	0.073	1/8	0.074
1	0.0730	46	0.081	43	0.089	5/32	0.087
2	0.0860	3/32	0.094	36	0.106	3/16	0.102
3	0.0990	36	0.106	31	0.120	7/32	0.115
4	0.1120	1/8	0.125	29	0.136	7/32	0.130
5	0.1250	9/64	0.141	23	0.154	1/4	0.145
6	0.1380	23	0.154	18	0.170	9/32	0.158
8	0.1640	15	0.180	10	0.194	5/16	0.188
10	0.1900	5	0.206	2	0.221	3/8	0.218
1/4	0.2500	17/64	0.266	9/32	0.281	7/16	0.278
5/16	0.3125	21/64	0.328	11/32	0.344	17/32	0.346
3/8	0.3750	25/64	0.391	13/32	0.406	5/8	0.415
7/16	0.4375	29/64	0.453	15/32	0.469	23/32	0.483
1/2	0.5000	33/64	0.516	17/32	0.531	13/16	0.552
5/8	0.6250	41/64	0.641	21/32	0.656	1	0.689
3/4	0.7500	49/64	0.766	25/32	0.781	1 3/16	0.828
7/8	0.8750	57/64	0.891	29/32	0.906	1 3/8	0.963
1	1.0000	1 1/64	1.016	1 1/32	1.031	1 5/8	1.100
1 1/4	1.2500	1 9/32	1.281	1 5/16	1.312	2	1.370
1 1/2	1.5000	1 17/32	1.531	1 9/16	1.562	2 3/8	1.640
1 3/4	1.7500	1 25/32	1.781	1 13/16	1.812	2 3/4	1.910
2	2.0000	2 1/32	2.031	2 1/16	2.062	3 1/8	2.180

All dimensions in inches.

[1] *Close Fit:* The close fit is normally limited to holes for those lengths of screws which are threaded to the head in assemblies where only one screw is to be used or where two or more screws are to be used and the mating holes are to be produced either at assembly or by matched and coordinated tooling.

[2] *Normal Fit:* The normal fit is intended for screws of relatively long length or for assemblies involving two or more screws where the mating holes are to be produced by conventional toleranc-ing methods. It provides for the maximum allowable eccentricity of the longest standard screws and for certain variations in the parts to be fastened, such as: deviations in hole straightness, angularity between the axis of the tapped hole and that of the hole for the shank, differences in center distances of the mating holes, etc.

[3] *Countersink:* It is considered good practice to countersink or break the edges of holes which are smaller than (*D* Max + 2*F* Max) in parts having a hardness which approaches, equals or exceeds the screw hardness. If such holes are not countersunk, the heads of screws may not seat properly or the sharp edges on holes may deform the fillets on screws thereby making them susceptible to fatigue in applications involving dynamic loading. The countersink or corner relief, however, should not be larger than is necessary to insure that the fillet on the screw is cleared.

Source: Appendix to American National Standard ANSI B18.3-1986.

Table 3. American National Standard Hexagon and Spline Socket Flat Countersunk Head Cap Screws (ANSI B18.3-1986)

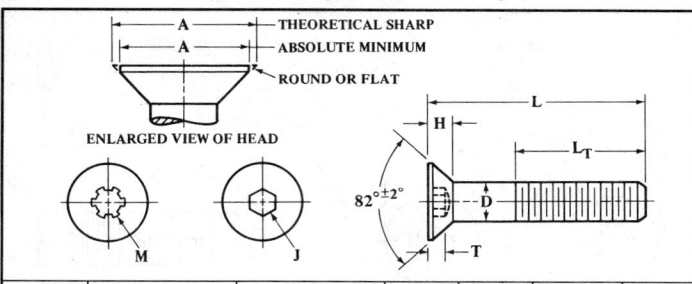

Nominal Size	Body Diam. Max.	Body Diam. Min.	Head Diameter Theoretical Sharp Max.	Head Diameter Abs. Min.	Head Height Reference	Spline Socket Size	Hexagon Socket Size Nom.	Key Engagement Min.
	D		A		H	M	J	T
0	0.0600	0.0568	0.138	0.117	0.044	0.048	0.035	0.025
1	0.0730	0.0695	0.168	0.143	0.054	0.060	0.050	0.031
2	0.0860	0.0822	0.197	0.168	0.064	0.060	0.050	0.038
3	0.0990	0.0949	0.226	0.193	0.073	0.072	1/16	0.044
4	0.1120	0.1075	0.255	0.218	0.083	0.072	1/16	0.055
5	0.1250	0.1202	0.281	0.240	0.090	0.096	5/64	0.061
6	0.1380	0.1329	0.307	0.263	0.097	0.096	5/64	0.066
8	0.1640	0.1585	0.359	0.311	0.112	0.111	3/32	0.076
10	0.1900	0.1840	0.411	0.359	0.127	0.145	1/8	0.087
1/4	0.2500	0.2435	0.531	0.480	0.161	0.183	5/32	0.111
5/16	0.3125	0.3053	0.656	0.600	0.198	0.216	3/16	0.135
3/8	0.3750	0.3678	0.781	0.720	0.234	0.251	7/32	0.159
7/16	0.4375	0.4294	0.844	0.781	0.234	0.291	1/4	0.159
1/2	0.5000	0.4919	0.938	0.872	0.251	0.372	5/16	0.172
5/8	0.6250	0.6163	1.188	1.112	0.324	0.454	3/8	0.220
3/4	0.7500	0.7406	1.438	1.355	0.396	0.454	1/2	0.220
7/8	0.8750	0.8647	1.688	1.604	0.468	. . .	9/16	0.248
1	1.0000	0.9886	1.938	1.841	0.540	. . .	5/8	0.297
1 1/8	1.1250	1.1086	2.188	2.079	0.611	. . .	3/4	0.325
1 1/4	1.2500	1.2336	2.438	2.316	0.683	. . .	7/8	0.358
1 3/8	1.3750	1.3568	2.688	2.553	0.755	. . .	7/8	0.402
1 1/2	1.5000	1.4818	2.938	2.791	0.827	. . .	1	0.435

All dimensions in inches.

The body of the screw is the unthreaded cylindrical portion of the shank where not threaded to the head; the shank being the portion of the screw from the point of juncture of the conical bearing surface and the body to the flat of the point. The length of thread L_T is the distance measured from the extreme point to the last complete (full form) thread.

Standard length increments of No. 0 through 1-inch sizes are as follows: 1/16 inch for nominal screw lengths of 1/8 through 1/4 inch; 1/8 inch for lengths of 1/4 through 1 inch; 1/4 inch for lengths of 1 inch through 3½ inches; ½ inch for lengths of 3½ through 7 inches; and 1 inch for lengths of 7 through 10 inches. For screw sizes over 1 inch, length increments are: ½ inch for nominal screw lengths of 1 inch through 7 inches; 1 inch for lengths of 7 through 10 inches; and 2 inches for lengths over 10 inches.

Threads shall be Unified external threads with radius root; Class 3A UNRC and UNRF series for sizes No. 0 through 1 inch and Class 2A UNRC and UNRF series for sizes over 1 inch to 1½ inches, incl.

For manufacturing details not shown, including materials, see American National Standard ANSI B18.3-1986. Socket dimensions are given in Table 9.

Table 4. American National Standard Hexagon Socket and Spline Socket Button Head Cap Screws* (ANSI B18.3-1986)

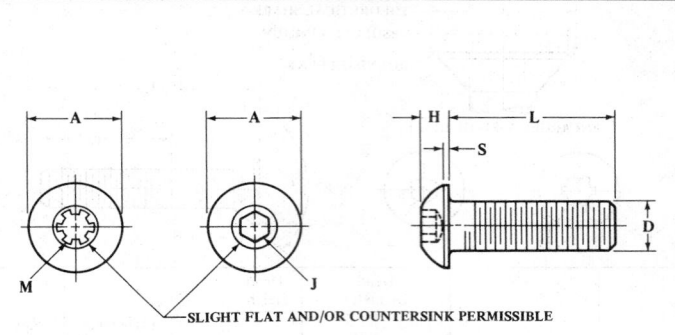

SLIGHT FLAT AND/OR COUNTERSINK PERMISSIBLE

Nominal Size	Screw Diameter Basic D	Head Diameter Max. A	Head Diameter Min. A	Head Height Max. H	Head Height Min. H	Head Side Height Ref. S	Spline Socket Size† Nom. M	Hexagon Socket Size† Nom. J	Standard Length Max. L
0	0.0600	0.114	0.104	0.032	0.026	0.010	0.048	0.035	½
1	0.0730	0.139	0.129	0.039	0.033	0.010	0.060	0.050	½
2	0.0860	0.164	0.154	0.046	0.038	0.010	0.060	0.050	½
3	0.0990	0.188	0.176	0.052	0.044	0.010	0.072	1/16	½
4	0.1120	0.213	0.201	0.059	0.051	0.015	0.072	1/16	½
5	0.1250	0.238	0.226	0.066	0.058	0.015	0.096	5/64	½
6	0.1380	0.262	0.250	0.073	0.063	0.015	0.096	5/64	5/8
8	0.1640	0.312	0.298	0.087	0.077	0.015	0.111	3/32	3/4
10	0.1900	0.361	0.347	0.101	0.091	0.020	0.145	1/8	1
¼	0.2500	0.437	0.419	0.132	0.122	0.031	0.183	5/32	1
5/16	0.3125	0.547	0.527	0.166	0.152	0.031	0.216	3/16	1
3/8	0.3750	0.656	0.636	0.199	0.185	0.031	0.251	7/32	1¼
½	0.5000	0.875	0.851	0.265	0.245	0.046	0.372	5/16	2
5/8	0.6250	1.000	0.970	0.331	0.311	0.062	0.454	3/8	2

All dimensions in inches.

* These cap screws have been designed and recommended for light fastening applications. They are not suggested for use in critical high-strength applications where socket head cap screws should normally be used.

Standard length increments for socket button head cap screws are as follows: 1/16 inch for nominal screw lengths of ⅛ through ¼ inch, ⅛ inch for nominal screw lengths of ¼ through 1 inch, and ¼ inch for nominal screw lengths of 1 inch through 2 inches. Tolerances on lengths are −0.03 inch for lengths up to 1 inch inclusive. For lengths from 1 through 2 inches, inclusive, length tolerances are −0.04 inch.

The thread conforms to the Unified standard, Class 3A, with radius root, UNRC and UNRF.

† Socket dimensions are given in Table 9.

To prevent interference, American National Standard ANSI B18.3.4M-1986 gives metric dimensional and general requirements for a lower head profile hexagon socket button head cap screw. Because of its design, wrenchability and other design factors are reduced; therefore, B18.3.4M should be reviewed carefully. Available only in metric sizes and with metric threads.

For manufacturing details, including materials, not shown, see American National Standard ANSI B18.3-1986.

Table 5. American National Standard Hexagon Socket Head Shoulder Screws
(ANSI B18.3-1986)

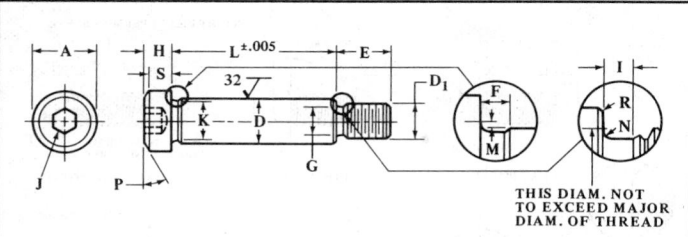

Nomi-nal Size	Shoulder Diameter		Head Diameter		Head Height		Head Side Height	Nominal Thread Size	Thread Length
	Max.	Min.	Max.	Min.	Max.	Min.	Min.		
	D		A		H		S	D_1	E
¼	0.2480	0.2460	0.375	0.357	0.188	0.177	0.157	10–24	0.375
⁵⁄₁₆	0.3105	0.3085	0.438	0.419	0.219	0.209	0.183	¼–20	0.438
⅜	0.3730	0.3710	0.562	0.543	0.250	0.240	0.209	⁵⁄₁₆–18	0.500
½	0.4980	0.4960	0.750	0.729	0.312	0.302	0.262	⅜–16	0.625
⅝	0.6230	0.6210	0.875	0.853	0.375	0.365	0.315	½–13	0.750
¾	0.7480	0.7460	1.000	0.977	0.500	0.490	0.421	⅝–11	0.875
1	0.9980	0.9960	1.312	1.287	0.625	0.610	0.527	¾–10	1.000
1¼	1.2480	1.2460	1.750	1.723	0.750	0.735	0.633	⅞–9	1.125
1½	1.4980	1.4960	2.125	2.095	1.000	0.980	0.842	1⅛–7	1.500
1¾	1.7480	1.7460	2.375	2.345	1.125	1.105	0.948	1¼–7	1.750
2	1.9980	1.9960	2.750	2.720	1.250	1.230	1.054	1½–6	2.000

Nomi-nal Size	Thread Neck Diameter		Thread Neck Width	Shoulder Neck Diam.	Shoulder Neck Width	Thread Neck Fillet		Head Fillet Extension Above D	Hexagon Socket Size
	Max.	Min.	Max.	Min.	Max.	Max.	Min.	Max.	Nom.
	G		I	K	F	N		M	J
¼	0.142	0.133	0.083	0.227	0.093	0.023	0.017	0.014	⅛
⁵⁄₁₆	0.193	0.182	0.100	0.289	0.093	0.028	0.022	0.017	⁵⁄₃₂
⅜	0.249	0.237	0.111	0.352	0.093	0.031	0.025	0.020	³⁄₁₆
½	0.304	0.291	0.125	0.477	0.093	0.035	0.029	0.026	¼
⅝	0.414	0.397	0.154	0.602	0.093	0.042	0.036	0.032	⁵⁄₁₆
¾	0.521	0.502	0.182	0.727	0.093	0.051	0.045	0.039	⅜
1	0.638	0.616	0.200	0.977	0.125	0.055	0.049	0.050	½
1¼	0.750	0.726	0.222	1.227	0.125	0.062	0.056	0.060	⅝
1½	0.964	0.934	0.286	1.478	0.125	0.072	0.066	0.070	⅞
1¾	1.089	1.059	0.286	1.728	0.125	0.072	0.066	0.080	1
2	1.307	1.277	0.333	1.978	0.125	0.102	0.096	0.090	1¼

All dimensions are in inches. The shoulder is the enlarged, unthreaded portion of the screw. Standard length increments for shoulder screws are: ⅛ inch for nominal screw lengths of ¼ through ¾ inch; ¼ inch for lengths above ¾ through 5 inches; and ½ inch for lengths over 5 inches. The thread conforms to the Unified Standard Class 3A, UNC. Hexagon socket sizes for the respective shoulder screw sizes are the same as for set screws of the same nominal size (see Table 7) except for shoulder screw size 1 inch, socket size is ½ inch, for screw size 1½ inches, socket size is ⅞ inch and for screw size 2 inches, socket size is 1¼ inches. For manufacturing details not shown, including materials, see American National Standard ANSI B18.3-1986.

Table 6. American National Standard Slotted Headless Set Screws
(ANSI B18.6.2-1972, R1977)

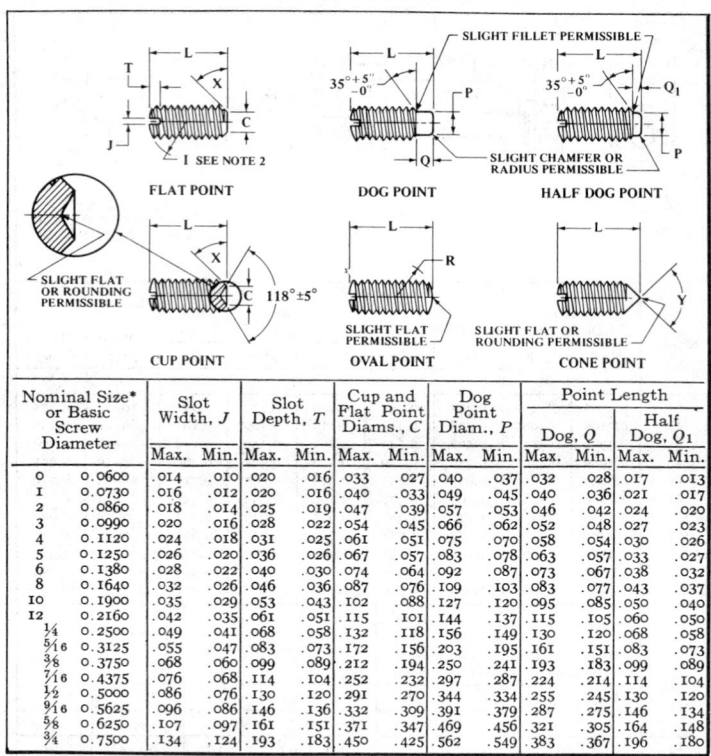

FLAT POINT   DOG POINT   HALF DOG POINT

CUP POINT   OVAL POINT   CONE POINT

Nominal Size* or Basic Screw Diameter		Slot Width, J		Slot Depth, T		Cup and Flat Point Diams., C		Dog Point Diam., P		Point Length			
										Dog, Q		Half Dog, Q_1	
		Max.	Min.	Max.	Min.	Max.	Min.	Max.	Min.	Max.	Min.	Max.	Min.
0	0.0600	.014	.010	.020	.016	.033	.027	.040	.037	.032	.028	.017	.013
1	0.0730	.016	.012	.020	.016	.040	.033	.049	.045	.040	.036	.021	.017
2	0.0860	.018	.014	.025	.019	.047	.039	.057	.053	.046	.042	.024	.020
3	0.0990	.020	.016	.028	.022	.054	.045	.066	.062	.052	.048	.027	.023
4	0.1120	.024	.018	.031	.025	.061	.051	.075	.070	.058	.054	.030	.026
5	0.1250	.026	.020	.036	.026	.067	.057	.083	.078	.063	.057	.033	.027
6	0.1380	.028	.022	.040	.030	.074	.064	.092	.087	.073	.067	.038	.032
8	0.1640	.032	.026	.046	.036	.087	.076	.109	.103	.083	.077	.043	.037
10	0.1900	.035	.029	.053	.043	.102	.088	.127	.120	.095	.085	.050	.040
12	0.2160	.042	.035	.061	.051	.115	.101	.144	.137	.115	.105	.060	.050
¼	0.2500	.049	.041	.068	.058	.132	.118	.156	.149	.130	.120	.068	.058
5⁄16	0.3125	.055	.047	.083	.073	.172	.156	.203	.195	.161	.151	.083	.073
3⁄8	0.3750	.068	.060	.099	.089	.212	.194	.250	.241	.193	.183	.099	.089
7⁄16	0.4375	.076	.068	.114	.104	.252	.232	.297	.287	.224	.214	.114	.104
½	0.5000	.086	.076	.130	.120	.291	.270	.344	.334	.255	.245	.130	.120
9⁄16	0.5625	.096	.086	.146	.136	.332	.309	.391	.379	.287	.275	.146	.134
5⁄8	0.6250	.107	.097	.161	.151	.371	.347	.469	.456	.321	.305	.164	.148
¾	0.7500	.134	.124	.193	.183	.450	.425	.562	.549	.383	.367	.196	.180

All dimensions are in inches.

Crown Radius, I: The crown radius has the same value as the basic screw diameter to three decimal places.

Oval Point Radius, R: Values of the oval point radius according to nominal screw size are: For a screw size of 0, a radius of .045; 1, .055; 2, .064; 3, .074; 4, .084; 5, .094; 6, .104; 8, .123; 10, .142; 12, .162; ¼, .188; 5⁄16, .234; 3⁄8, .281; 7⁄16, .328; ½, .375; 9⁄16, .422; 5⁄8, .469; and for ¾, .562.

Cone Point Angle, Y: The cone point angle is 90° ± 2° for the following nominal lengths, or longer, shown according to screw size: For nominal size 0, a length of 5⁄64; 1, 3⁄32; 2, 7⁄64; 3, ⅛; 4, 5⁄32; 5, 3⁄16; 6, 3⁄16; 8, ¼; 10, ¼; 12, 5⁄16; ¼, 5⁄16; 5⁄16, 3⁄8; 3⁄8, 7⁄16; 7⁄16, ½; ½, 9⁄16; 9⁄16, 5⁄8; 5⁄8, ¾; and for ¾, ⅞. For shorter screws, the cone point angle is 118° ± 2°.

Point Angle X: The point angle is 45°, + 5°, − 0°, for screws of nominal lengths, or longer, as given just above for cone point angle, and 30°, min. for shorter screws.

Threads: Threads are Unified Standard Class 2A; UNC and UNF Series or UNRC and UNRF Series.

* When specifying a nominal size in decimals a zero preceding the decimal point or any zero in the fourth decimal place is omitted.

Table 7. American National Standard Hexagon and Spline Socket Set Screws
(ANSI B18.3-1986)

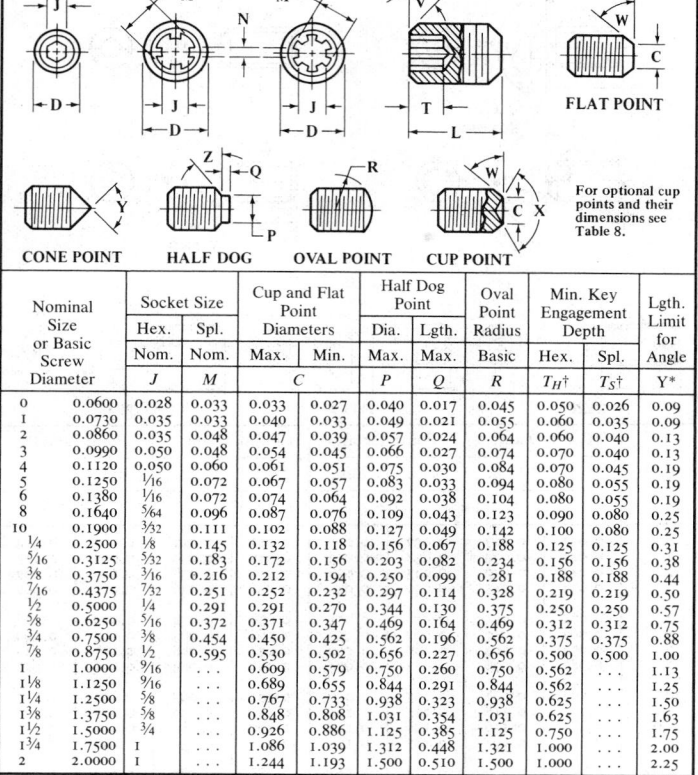

CONE POINT   HALF DOG   OVAL POINT   CUP POINT

FLAT POINT

For optional cup points and their dimensions see Table 8.

Nominal Size or Basic Screw Diameter		Socket Size		Cup and Flat Point Diameters		Half Dog Point		Oval Point Radius	Min. Key Engagement Depth		Lgth. Limit for Angle
		Hex.	Spl.			Dia.	Lgth.		Hex.	Spl.	
		Nom.	Nom.	Max.	Min.	Max.	Max.	Basic			
		J	M	C		P	Q	R	T_H†	T_S†	Y*
0	0.0600	0.028	0.033	0.033	0.027	0.040	0.017	0.045	0.050	0.026	0.09
1	0.0730	0.035	0.033	0.040	0.033	0.049	0.021	0.055	0.060	0.035	0.09
2	0.0860	0.035	0.048	0.047	0.039	0.057	0.024	0.064	0.060	0.040	0.13
3	0.0990	0.050	0.048	0.054	0.045	0.066	0.027	0.074	0.070	0.040	0.13
4	0.1120	0.050	0.060	0.061	0.051	0.075	0.030	0.084	0.070	0.045	0.19
5	0.1250	1/16	0.072	0.067	0.057	0.083	0.033	0.094	0.080	0.055	0.19
6	0.1380	1/16	0.072	0.074	0.064	0.092	0.038	0.104	0.080	0.055	0.19
8	0.1640	5/64	0.096	0.087	0.076	0.109	0.043	0.123	0.090	0.080	0.25
10	0.1900	3/32	0.111	0.102	0.088	0.127	0.049	0.142	0.100	0.080	0.25
1/4	0.2500	1/8	0.145	0.132	0.118	0.156	0.067	0.188	0.125	0.125	0.31
5/16	0.3125	5/32	0.183	0.172	0.156	0.203	0.082	0.234	0.156	0.156	0.38
3/8	0.3750	3/16	0.216	0.212	0.194	0.250	0.099	0.281	0.188	0.188	0.44
7/16	0.4375	7/32	0.251	0.252	0.232	0.297	0.114	0.328	0.219	0.219	0.50
1/2	0.5000	1/4	0.291	0.291	0.270	0.344	0.130	0.375	0.250	0.250	0.57
5/8	0.6250	5/16	0.372	0.371	0.347	0.469	0.164	0.469	0.312	0.312	0.75
3/4	0.7500	3/8	0.454	0.450	0.425	0.562	0.196	0.562	0.375	0.375	0.88
7/8	0.8750	1/2	0.595	0.530	0.502	0.656	0.227	0.656	0.500	0.500	1.00
1	1.0000	9/16	. . .	0.609	0.579	0.750	0.260	0.750	0.562	. . .	1.13
1 1/8	1.1250	9/16	. . .	0.689	0.655	0.844	0.291	0.844	0.562	. . .	1.25
1 1/4	1.2500	5/8	. . .	0.767	0.733	0.938	0.323	0.938	0.625	. . .	1.50
1 3/8	1.3750	5/8	. . .	0.848	0.808	1.031	0.354	1.031	0.625	. . .	1.63
1 1/2	1.5000	3/4	. . .	0.926	0.886	1.125	0.385	1.125	0.750	. . .	1.75
1 3/4	1.7500	1	. . .	1.086	1.039	1.312	0.448	1.321	1.000	. . .	2.00
2	2.0000	1	. . .	1.244	1.193	1.500	0.510	1.500	1.000	. . .	2.25

All dimensions are in inches. The thread conforms to the Unified Standard, Class 3A, UNC and UNF series. The socket depth T is included in the Standard and some are shown here. The nominal length L of all socket type set screws is the total or overall length. For nominal screw lengths of 1/16 through 3/16 inch (0 through 3 sizes incl.) the standard length increment is 0.06 inch; for lengths 1/8 through 1 inch the increment is 1/8 inch; for lengths 1 through 2 inches the increment is 1/4 inch; for lengths 2 through 6 inches the increment is 1/2 inch; for lengths 6 inches and longer the increment is 1 inch. Socket dimensions are given in Table 9.

* Cone point angle Y is 90 degrees plus or minus 2 degrees for these nominal lengths or longer and 118 degrees plus or minus 2 degrees for shorter nominal lengths.

† Reference should be made to the Standard for shortest optimum nominal lengths to which the minimum key engagement depths T_H and T_S apply.

Length Tolerance: The allowable tolerance on length L for all set screws of the socket type is ± 0.01 inch for set screws up to 5/8 inch long; ± 0.02 inch for screws over 5/8 to 2 inches long; ± 0.03 inch for screws over 2 to 6 inches long and ± 0.06 inch for screws over 6 inches long. Socket dimensions are given in Table 9.

For manufacturing details, including materials, not shown, see American National Standard ANSI B18.3-1986.

Table 8. American National Standard Hexagon and Spline Socket Set Screw Optional Cup Points (ANSI B18.3-1986)

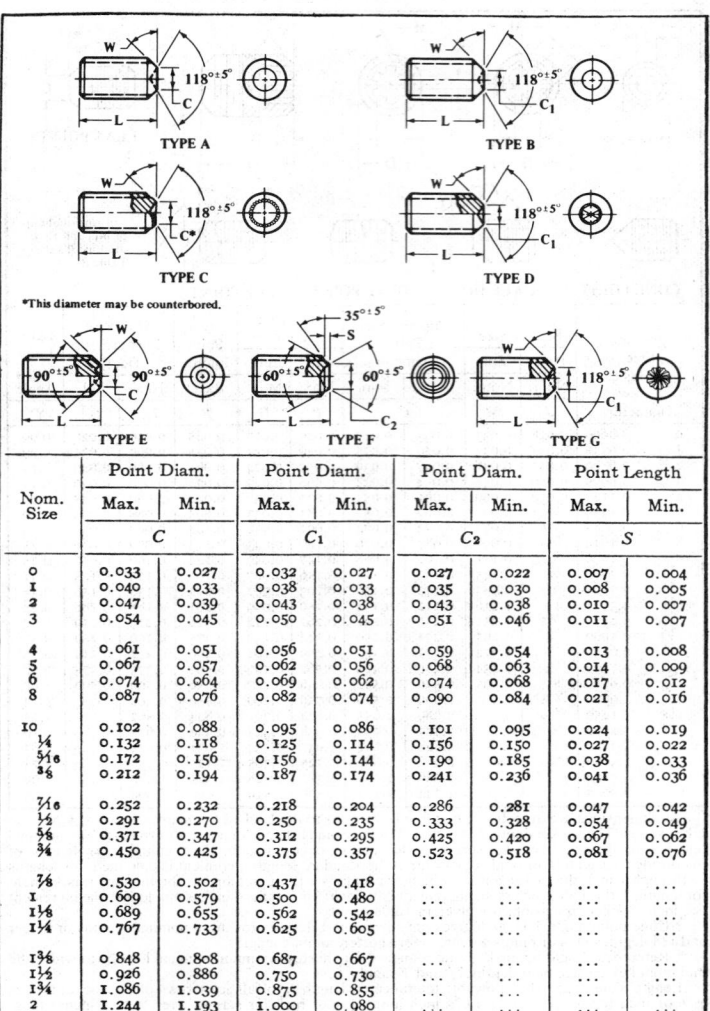

*This diameter may be counterbored.

All dimensions are in inches.

Nom. Size	Point Diam.		Point Diam.		Point Diam.		Point Length	
	Max.	Min.	Max.	Min.	Max.	Min.	Max.	Min.
	C		C_1		C_2		S	
0	0.033	0.027	0.032	0.027	0.027	0.022	0.007	0.004
1	0.040	0.033	0.038	0.033	0.035	0.030	0.008	0.005
2	0.047	0.039	0.043	0.038	0.043	0.038	0.010	0.007
3	0.054	0.045	0.050	0.045	0.051	0.046	0.011	0.007
4	0.061	0.051	0.056	0.051	0.059	0.054	0.013	0.008
5	0.067	0.057	0.062	0.056	0.068	0.063	0.014	0.009
6	0.074	0.064	0.069	0.062	0.074	0.068	0.017	0.012
8	0.087	0.076	0.082	0.074	0.090	0.084	0.021	0.016
10	0.102	0.088	0.095	0.086	0.101	0.095	0.024	0.019
1/4	0.132	0.118	0.125	0.114	0.156	0.150	0.027	0.022
5/16	0.172	0.156	0.156	0.144	0.190	0.185	0.038	0.033
3/8	0.212	0.194	0.187	0.174	0.241	0.236	0.041	0.036
7/16	0.252	0.232	0.218	0.204	0.286	0.281	0.047	0.042
1/2	0.291	0.270	0.250	0.235	0.333	0.328	0.054	0.049
5/8	0.371	0.347	0.312	0.295	0.425	0.420	0.067	0.062
3/4	0.450	0.425	0.375	0.357	0.523	0.518	0.081	0.076
7/8	0.530	0.502	0.437	0.418	...	...	...	...
1	0.609	0.579	0.500	0.480	...	...	...	...
1 1/8	0.689	0.655	0.562	0.542	...	...	...	...
1 1/4	0.767	0.733	0.625	0.605	...	...	...	...
1 3/8	0.848	0.808	0.687	0.667	...	...	...	...
1 1/2	0.926	0.886	0.750	0.730	...	...	...	...
1 3/4	1.086	1.039	0.875	0.855	...	...	...	...
2	1.244	1.193	1.000	0.980	...	...	...	...

All dimensions are in inches.
The cup point types shown are those available from various manufacturers.

Table 9. American National Standard Hexagon and Spline Sockets (ANSI B18.3-1986)

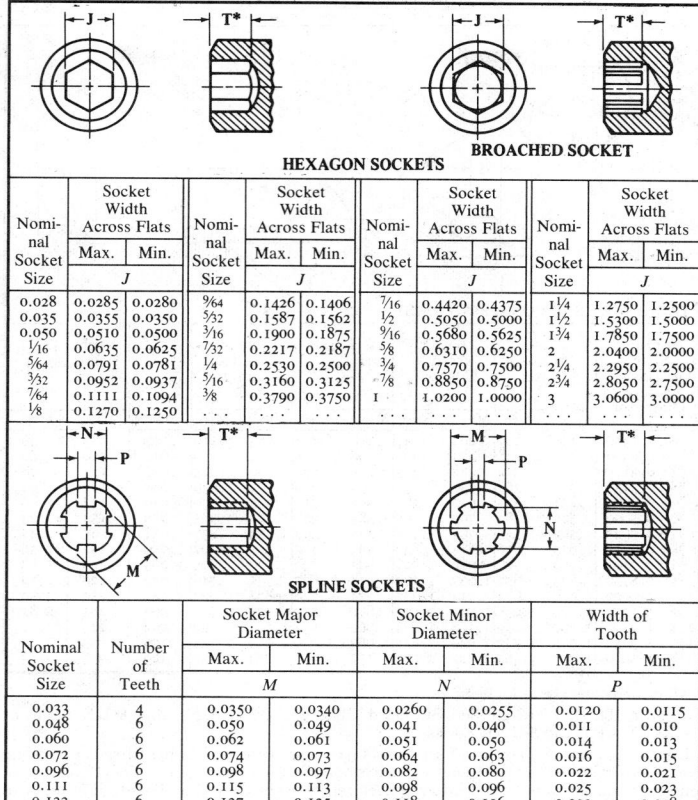

HEXAGON SOCKETS

BROACHED SOCKET

Nominal Socket Size	Socket Width Across Flats		Nominal Socket Size	Socket Width Across Flats		Nominal Socket Size	Socket Width Across Flats		Nominal Socket Size	Socket Width Across Flats	
	Max.	Min.		Max.	Min.		Max.	Min.		Max.	Min.
	J			J			J			J	
0.028	0.0285	0.0280	9/64	0.1426	0.1406	7/16	0.4420	0.4375	1¼	1.2750	1.2500
0.035	0.0355	0.0350	5/32	0.1587	0.1562	½	0.5050	0.5000	1½	1.5300	1.5000
0.050	0.0510	0.0500	3/16	0.1900	0.1875	9/16	0.5680	0.5625	1¾	1.7850	1.7500
1/16	0.0635	0.0625	7/32	0.2217	0.2187	5/8	0.6310	0.6250	2	2.0400	2.0000
5/64	0.0791	0.0781	¼	0.2530	0.2500	¾	0.7570	0.7500	2¼	2.2950	2.2500
3/32	0.0952	0.0937	5/16	0.3160	0.3125	7/8	0.8850	0.8750	2¾	2.8050	2.7500
7/64	0.1111	0.1094	3/8	0.3790	0.3750	1	1.0200	1.0000	3	3.0600	3.0000
1/8	0.1270	0.1250	…	…	…						

SPLINE SOCKETS

Nominal Socket Size	Number of Teeth	Socket Major Diameter		Socket Minor Diameter		Width of Tooth	
		Max.	Min.	Max.	Min.	Max.	Min.
		M		N		P	
0.033	4	0.0350	0.0340	0.0260	0.0255	0.0120	0.0115
0.048	6	0.050	0.049	0.041	0.040	0.011	0.010
0.060	6	0.062	0.061	0.051	0.050	0.014	0.013
0.072	6	0.074	0.073	0.064	0.063	0.016	0.015
0.096	6	0.098	0.097	0.082	0.080	0.022	0.021
0.111	6	0.115	0.113	0.098	0.096	0.025	0.023
0.133	6	0.137	0.135	0.118	0.116	0.030	0.028
0.145	6	0.149	0.147	0.128	0.126	0.032	0.030
0.168	6	0.173	0.171	0.150	0.147	0.036	0.033
0.183	6	0.188	0.186	0.163	0.161	0.039	0.037
0.216	6	0.221	0.219	0.190	0.188	0.050	0.048
0.251	6	0.256	0.254	0.221	0.219	0.060	0.058
0.291	6	0.298	0.296	0.254	0.252	0.068	0.066
0.372	6	0.380	0.377	0.319	0.316	0.092	0.089
0.454	6	0.463	0.460	0.386	0.383	0.112	0.109
0.595	6	0.604	0.601	0.509	0.506	0.138	0.134
0.620	6	0.631	0.627	0.535	0.531	0.149	0.145
0.698	6	0.709	0.705	0.604	0.600	0.168	0.164
0.790	6	0.801	0.797	0.685	0.681	0.189	0.185

All dimensions are in inches.
* Socket depths, T, for various screw types are given in the standard but are not shown here.
Where sockets are chamfered, the depth of chamfer shall not exceed 10 per cent of the nominal socket size for sizes up to and including ¹⁄₁₆ inch for hexagon sockets and 0.060 for spline sockets, and 7.5 per cent for larger sizes.

Table 10. American National Standard Square Head Set Screws
(ANSI B18.6.2-1972, R1977)

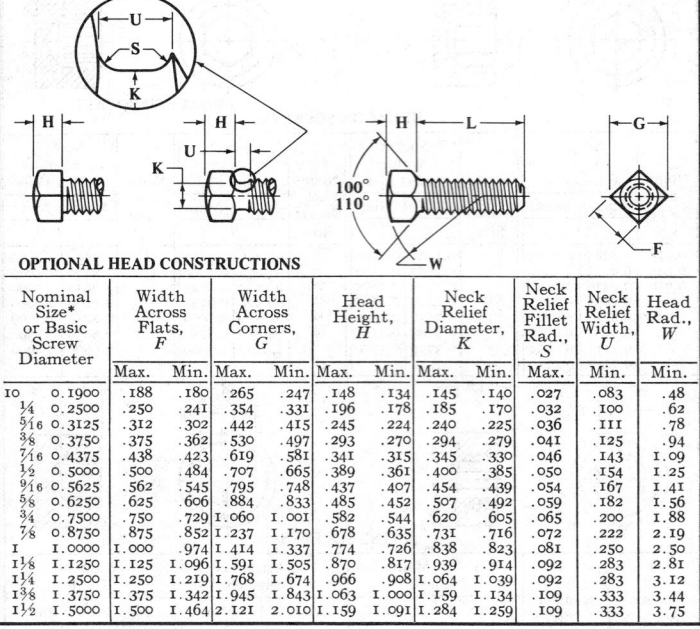

OPTIONAL HEAD CONSTRUCTIONS

Nominal Size* or Basic Screw Diameter		Width Across Flats, F		Width Across Corners, G		Head Height, H		Neck Relief Diameter, K		Neck Relief Fillet Rad., S	Neck Relief Width, U	Head Rad., W
		Max.	Min.	Max.	Min.	Max.	Min.	Max.	Min.	Max.	Min.	Min.
10	0.1900	.188	.180	.265	.247	.148	.134	.145	.140	.027	.083	.48
¼	0.2500	.250	.241	.354	.331	.196	.178	.185	.170	.032	.100	.62
⁵⁄₁₆	0.3125	.312	.302	.442	.415	.245	.224	.240	.225	.036	.111	.78
⅜	0.3750	.375	.362	.530	.497	.293	.270	.294	.279	.041	.125	.94
⁷⁄₁₆	0.4375	.438	.423	.619	.581	.341	.315	.345	.330	.046	.143	1.09
½	0.5000	.500	.484	.707	.665	.389	.361	.400	.385	.050	.154	1.25
⁹⁄₁₆	0.5625	.562	.545	.795	.748	.437	.407	.454	.439	.054	.167	1.41
⅝	0.6250	.625	.606	.884	.833	.485	.452	.507	.492	.059	.182	1.56
¾	0.7500	.750	.729	1.060	1.001	.582	.544	.620	.605	.065	.200	1.88
⅞	0.8750	.875	.852	1.237	1.170	.678	.635	.731	.716	.072	.222	2.19
1	1.0000	1.000	.974	1.414	1.337	.774	.726	.838	.823	.081	.250	2.50
1⅛	1.1250	1.125	1.096	1.591	1.505	.870	.817	.939	.914	.092	.283	2.81
1¼	1.2500	1.250	1.219	1.768	1.674	.966	.908	1.064	1.039	.092	.283	3.12
1⅜	1.3750	1.375	1.342	1.945	1.843	1.063	1.000	1.159	1.134	.109	.333	3.44
1½	1.5000	1.500	1.464	2.121	2.010	1.159	1.091	1.284	1.259	.109	.333	3.75

All dimensions are in inches.

Threads: Threads are Unified Standard Class 2A; UNC, UNF and 8 UN Series or UNRC, UNRF and 8 UNR Series.

Length of Thread: Square head set screws have complete (full form) threads extending over that portion of the screw length which is not affected by the point. For the respective constructions, threads extend into the neck relief, to the conical underside of head, or to within one thread (as measured with a thread ring gage) from the flat underside of the head. Threads through angular or crowned portions of points have fully formed roots with partial crests.

Point Types: Unless otherwise specified, square head set screws are supplied with cup points. Cup points as furnished by some manufacturers may be externally or internally knurled. Where so specified by the purchaser, screws have cone, dog, half-dog, flat or oval points as given on the following page.

Designation: Square head set screws are designated by the following data in the sequence shown: Nominal size (number, fraction or decimal equivalent); threads per inch; screw length (fraction or decimal equivalent); product name; point style; material; and protective finish, if required.   Examples: ¼ - 20 x ¾ Square Head Set Screw, Flat Point, Steel, Cadmium Plated. .500 - 13 x 1.25 Square Head Set Screw, Cone Point, Corrosion Resistant Steel.

* When specifying a nominal size in decimals, the zero preceding the decimal point is omitted as is any zero in the fourth decimal place.

Table 10 (*Concluded*). **American National Standard Square Head Set Screws** (ANSI B18.6.2-1972, R1977)

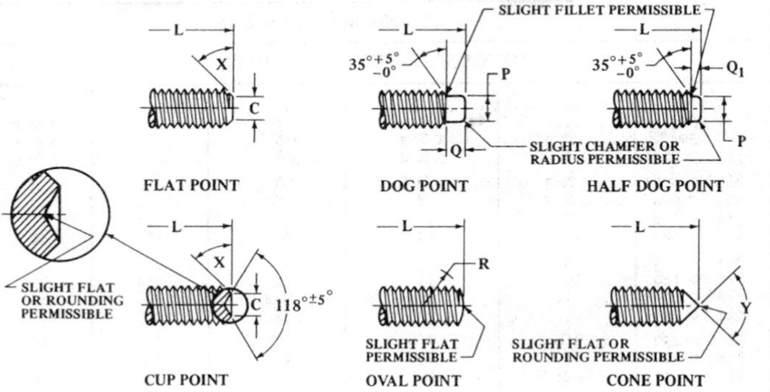

Nominal Size* or Basic Screw Diameter		Cup and Flat Point Diams., C		Dog and Half Dog Point Diams., P		Point Length: Dog, Q		Point Length: Half Dog, Q_1		Oval Point Rad., R
		Max.	Min.	Max.	Min.	Max.	Min.	Max.	Min.	+.031 −.000
10	0.1900	.102	.088	.127	.120	.095	.085	.050	.040	.142
¼	0.2500	.132	.118	.156	.149	.130	.120	.068	.058	.188
5/16	0.3125	.172	.156	.203	.195	.161	.151	.083	.073	.234
3/8	0.3750	.212	.194	.250	.241	.193	.183	.099	.089	.281
7/16	0.4375	.252	.232	.297	.287	.224	.214	.114	.104	.328
½	0.5000	.291	.270	.344	.334	.255	.245	.130	.120	.375
9/16	0.5625	.332	.309	.391	.379	.287	.275	.146	.134	.422
5/8	0.6250	.371	.347	.469	.456	.321	.305	.164	.148	.469
¾	0.7500	.450	.425	.562	.549	.383	.367	.196	.180	.562
7/8	0.8750	.530	.502	.656	.642	.446	.430	.227	.211	.656
1	1.0000	.609	.579	.750	.734	.510	.490	.260	.240	.750
1⅛	1.1250	.689	.655	.844	.826	.572	.552	.291	.271	.844
1¼	1.2500	.767	.733	.938	.920	.635	.615	.323	.303	.938
1⅜	1.3750	.848	.808	1.031	1.011	.698	.678	.354	.334	1.031
1½	1.5000	.926	.886	1.125	1.105	.760	.740	.385	.365	1.125

All dimensions are in inches.

Cone Point Angle, Y: For the following nominal lengths, or longer, shown according to nominal size, the cone point angle is 90° ± 2°: For size No. 10, ¼; ¼, 5/16; 5/16, 3/8; 3/8, 7/16; 7/16, ½; ½, 9/16; 9/16, 5/8; 5/8, ¾; ¾, 7/8; 7/8, 1; 1, 1⅛; 1⅛, 1¼; 1¼, 1½; 1⅜, 1⅝; and for 1½, 1¾. For shorter screws the cone point angle is 118° ± 2°.

Point Angle, X: The point angle is 45°, + 5°, − 0° for screws of the nominal lengths, or longer, given just above for cone point angle, and 30° min. for shorter lengths.

* When specifying a nominal size in decimals the zero preceding the decimal point is omitted as is any zero in the fourth decimal place.

Table 11. Applicability of Hexagon and Spline Keys and Bits

Nominal Key or Bit Size	Cap Screws 1960 Series	Flat Countersunk Head Cap Screws	Button Head Cap Screws	Shoulder Screws	Set Screws
			Nominal Screw Sizes		
HEXAGON KEYS AND BITS					
0.028	...	...	...	...	o
0.035	...	o	o	...	1 & 2
0.050	o	1 & 2	1 & 2	...	3 & 4
1/16 0.062	1	3 & 4	3 & 4	...	5 & 6
5/64 0.078	2 & 3	5 & 6	5 & 6	...	8
3/32 0.094	4 & 5	8	8	...	10
7/64 0.109	6	...	...	...	...
1/8 0.125	...	10	10	1/4	1/4
9/64 0.141	8	...	...	...	...
5/32 0.156	10	1/4	1/4	5/16	5/16
3/16 0.188	1/4	5/16	5/16	3/8	3/8
7/32 0.219	...	3/8	3/8	...	7/16
1/4 0.250	5/16	7/16	...	1/2	1/2
5/16 0.312	3/8	1/2	1/2	5/8	5/8
3/8 0.375	7/16 & 1/2	5/8	5/8	3/4	3/4
7/16 0.438	...	...	...	...	...
1/2 0.500	5/8	3/4	...	...	7/8
9/16 0.562	...	7/8	...	1	1 & 1 1/8
5/8 0.625	3/4	1	...	1 1/4	1 1/4 & 1 3/8
3/4 0.750	7/8 & 1	1 1/8	...	...	1 1/2
7/8 0.875	1 1/8 & 1 1/4	1 1/4 & 1 3/8	...	1 1/2	...
1 1.000	1 3/8 & 1 1/2	1 1/2	...	1 3/4	1 3/4 & 2
1 1/4 1.250	1 3/4	...	...	2	...
1 1/2 1.500	2	...	...	...	...
1 3/4 1.750	2 1/4 & 2 1/2	...	...	...	...
2 2.000	2 3/4	...	...	...	...
2 1/4 2.250	3 & 3 1/4	...	...	...	...
2 3/4 2.750	3 1/2 & 3 3/4	...	...	...	...
3 3.000	4	...	...	...	...
SPLINE KEYS AND BITS					
0.033	...	...	...	...	o & 1
0.048	...	o	o	...	2 & 3
0.060	o	1 & 2	1 & 2	...	4
0.072	1	3 & 4	3 & 4	...	5 & 6
0.096	2 & 3	5 & 6	5 & 6	...	8
0.111	4 & 5	8	8	...	10
0.133	6	...	...	...	...
0.145	...	10	10	...	1/4
0.168	8	...	...	...	...
0.183	10	1/4	1/4	...	5/16
0.216	1/4	5/16	5/16	...	3/8
0.251	...	3/8	3/8	...	7/16
0.291	5/16	7/16	...	...	1/2
0.372	3/8	1/2	...	1/2	5/8
0.454	7/16 & 1/2	5/8 & 3/4	...	5/8	3/4
0.595	5/8	...	...	...	7/8
0.620	3/4	...	...	...	...
0.698	7/8	...	...	...	...
0.790	1	...	...	...	...

Source: Appendix to American National Standard ANSI B18.3-1986.

British Standard Hexagon Socket Screws — Metric Series. — The first five parts of British Standard BS 4168: 1981 provide specifications for hexagon socket head cap screws and hexagon socket set screws.

Hexagon Socket Head Cap Screws: The dimensional data in Table 1 are based upon BS 4168: Part 1: 1981. These screws are available in stainless steel and alloy steel, the latter having class 12.9 properties as specified in BS 6104:Part 1. When ordering these screws, the designation "Hexagon socket head cap screw BS 4168 M5 $\times$ 20–12.9" would mean, as an example, a cap screw having a thread size of d = M5, nominal length l = 20 mm, and property class 12.9. Alloy steel cap screws are furnished with a black oxide finish (thermal or chemical); stainless steel cap screws with a plain finish. Combinations of thread size, nominal length, and length of thread are shown in Table 2; the screw threads in these combinations are in the ISO metric coarse pitch series specified in BS 3643 with tolerances in the 5g6g class. (See Metric Screw Threads in Index.)

Hexagon Socket Set Screws: Part 2 of BS 4168:1981 specifies requirements for hexagon socket set screws with flat point having ISO metric threads, and diameters from 1.6 mm up to and including 24 mm. The dimensions of these set screws along with those of cone-point, dog-point, and cup-point set screws in accord, respectively, with Parts 3, 4, and 5 of the Standard are given in Table 3 and the accompanying illustration. All of these set screws are available in either steel processed to mechanical properties class 45H (BS 6104:Part 3); or stainless steel processed to mechanical properties described in BS 6105. Steel set screws are furnished with black oxide (thermal or chemical) finish; stainless steel set screws are furnished plain. The tolerances applied to the threads of these set screws are for ISO product grade A, based on ISO 4759/1-1978 "Tolerances for fasteners — Part 1: Bolts, screws, and nuts with thread diameters greater than or equal to 1.6 mm and less than or equal to 150 mm and product grades A, B, and C."

Hexagon socket set screws are designated by the type, the thread size, nominal length, and property class. As an example, for a flat-point set screw of thread size d = M6, nominal length l = 12 mm, and property class 45H:

Hexagon socket set screw flat point BS 4168 M6 $\times$ 12–45H

British Standard Hexagon Socket Countersunk and Button Head Screws — Metric Series. — British Standard BS 4168:1967 provides a metric series of hexagon socket countersunk and button head screws. The dimensions of these screws are given in Table 5. The revision of this Standard will constitute Parts 6 and 8 of BS 4168.

British Standards for Mechanical Properties of Fasteners. — BS 6104: Part 1:1981 specifies mechanical properties for bolts, screws, and studs with nominal diameters up to and including 39 mm of any triangular ISO thread and made of carbon or alloy steel. It does not apply to set screws and similar threaded fasteners. Part 2 of this Standard specifies the mechanical properties of set screws and similar fasteners, not under tensile stress, in the range from M1.6 up to and including M39 and made of carbon or alloy steel.

BS 6105:1981 provides specifications for bolts, screws, studs, and nuts made from austenitic, ferritic, and martensitic grades of corrosion-resistant steels. This Standard applies only to fastener components after completion of manufacture with nominal diameters from M1.6 up to and including M39. These Standards are not described further here. Copies may be obtained from the British Standards Institution, 2 Park Street, London W1A 2BS and also from the American National Standards Institute, 1430 Broadway, New York, N.Y. 10018.

Table 1. British Standard Hexagon Socket Head Cap Screws—Metric Series (BS 4168: Part 1: 1981)

Nominal Size,[1] d	Body Diameter, D		Head Diameter, A			Head Height, H		Hexagon Socket Size, J[4]	Key Engagement, K	Wall Thickness, W	Fillet	
	Max	Min	Max[2]	Max[3]	Min	Max	Min	Nom.	Min	Min	Rad., F Min	Diam., d_a Max
M1.6	1.6	1.46	3	3.14	2.86	1.6	1.46	1.5	0.7	0.55	0.1	2
M2	2	1.86	3.8	3.98	3.62	2	1.86	1.5	1	0.55	0.1	2.6
M2.5	2.5	2.36	4.5	4.68	4.32	2.5	2.36	2	1.1	0.85	0.1	3.1
M3	3	2.86	5.5	5.68	5.32	3	2.86	2.5	1.3	1.15	0.1	3.6
M4	4	3.82	7	7.22	6.78	4	3.82	3	2	1.4	0.2	4.7
M5	5	4.82	8.5	8.72	8.28	5	4.82	4	2.5	1.9	0.2	5.7
M6	6	5.82	10	10.22	9.78	6	5.70	5	3	2.3	0.25	6.8
M8	8	7.78	13	13.27	12.73	8	7.64	6	4	3.3	0.4	9.2
M10	10	9.78	16	16.27	15.73	10	9.64	8	5	4	0.4	11.2
M12	12	11.73	18	18.27	17.73	12	11.57	10	6	4.8	0.6	14.2
(M14)	14	13.73	21	21.33	20.67	14	13.57	12	7	5.8	0.6	16.2
M16	16	15.73	24	24.33	23.67	16	15.57	14	8	6.8	0.6	18.2
M20	20	19.67	30	30.33	29.67	20	19.48	17	10	8.6	0.8	22.4
M24	24	23.67	36	36.39	35.61	24	23.48	19	12	10.4	0.8	26.4
M30	30	29.67	45	45.39	44.61	30	29.48	22	15.5	13.1	1	33.4
M36	36	35.61	54	54.46	53.54	36	35.38	27	19	15.3	1	39.4

All dimensions are given in millimeters.   [1] The size shown in ( ) is non-preferred.   [2] For plain heads.   [3] For knurled heads.   [4] See Table 2 for min/max.

Table 2. British Standard Hexagon Socket Screws — Metric Series
(BS 4168: Part 1: 1981)

Dimensions of Hexagon Sockets		

Nominal Socket Size	Socket Width Across Flats, J		Nominal Socket Size	Socket Width Across Flats, J	
	Max.	Min.		Max.	Min.
1.5	1.545	1.52	6	6.095	6.02
2.0	2.045	2.02	8	8.115	8.025
2.5	2.56	2.52	10	10.115	10.025
3	3.08	3.02	12	12.142	12.032
4	4.095	4.02	14	14.142	14.032
5	5.095	5.02	17	17.23	17.05
...	...	...	19	19.275	19.065

Association of Nominal and Thread Lengths for Each Thread Size	

All dimensions are in millimeters.

The popular lengths are those between the stepped solid lines. Lengths above the dashed lines are threaded to the head within 3 pitch lengths ($3P$). Lengths below the dashed lines have values of L_g and L_s (see Table 1) given by the formulas: $L_g$max. = L nom. − b ref., and $L_s$min. = $L_g$max. − $5P$.

Table 3. British Standard Hexagon Socket Set Screws — Metric Series (BS 4168: Parts 2, 3, 4, and 5: 1981)

Nom. Size, d	Pitch, P	Socket Size, s	Depth of Key Engagement, t†		Range of Popular Lengths				Length of Dog on Dog Point Screws*					End Diameters			
					Flat Point, l	Cone Point, l	Dog Point, l	Cup Point, l	Short Dog, z		Long Dog, z			Flat Point, d_z	Cone Point, d_t	Dog Point, d_p	Cup Point, d_z
		nom	min	min	l	l	l	l	min	max	min	max	*	max	max	max	max
M1.6	0.35	0.7	1.5	0.7	2–8	2–8	2–8	2–8	0.4	0.65	0.8	1.05	2.5	0.8	0	0.8	0.8
M2	0.4	0.9	1.7	0.8	2–10	2–10	2.5–10	2–10	0.5	0.75	1.0	1.25	3.0	1.0	0	1.0	1.0
M2.5	0.45	1.3	2.0	1.2	2–12	2.5–12	3–12	2.5–12	0.63	0.88	1.25	1.5	4	1.5	0	1.5	1.2
M3	0.5	1.5	2.0	1.2	2–16	2.5–16	4–16	2.5–16	0.75	1.0	1.5	1.75	5	2.0	0	2.0	1.4
M4	0.7	2.0	2.5	1.5	2.5–20	3–20	5–20	3–20	1.0	1.25	2.0	2.25	6	2.5	0	2.5	2.0
M5	0.8	2.5	3.0	2.0	3–25	4–25	6–25	4–25	1.25	1.5	2.5	2.75	6	3.5	0	3.5	2.5
M6	1.0	3.0	3.5	2.0	4–30	5–30	8–30	5–30	1.5	1.75	3.0	3.25	8	4.0	1.5	4.0	3.0
M8	1.25	4.0	5.0	3.0	5–40	6–40	8–40	6–40	2.0	2.25	4.0	4.3	10	5.5	2.0	5.5	5.0
M10	1.5	5.0	6.0	4.0	6–50	8–50	10–50	8–50	2.5	2.75	5.0	5.3	12	7.0	2.5	7.0	6.0
M12	1.75	6.0	8.0	4.8	8–60	10–60	12–60	10–60	3.0	3.25	6.0	6.3	16	8.5	3.0	8.5	8.0
M16	2.0	8.0	10.0	6.4	10–60	12–60	16–60	12–60	4.0	4.3	8.0	8.36	20	12.0	4.0	12.0	10.0
M20	2.5	10.0	12.0	8.0	12–60	16–60	20–60	16–60	5.0	5.3	10.0	10.36	25	15.0	5.0	15.0	14.0
M24	3.0	12.0	15.0	10.0	16–60	20–60	25–60	20–60	6.0	6.3	12.0	12.43	30	18.0	6.0	18.0	16.0

All dimensions are in millimeters. For dimensional notation, see diagram, page 1405.

* A dog point set screw having a nominal length equal to or less than the length shown in the (*) column of the table is supplied with length z shown in the short dog column. For set screws of lengths greater than shown in the (*) column, z for long dogs applies.

† The smaller of the two t min. values applies to certain short-length set screws. These short-length screws are those whose length is approximately equal to the diameter of the screw. The larger t min. values apply to longer-length screws.

British Standard Hexagon Socket Set Screws — Metric Series
(BS 4168: Parts 2, 3, 4, and 5: 1981)

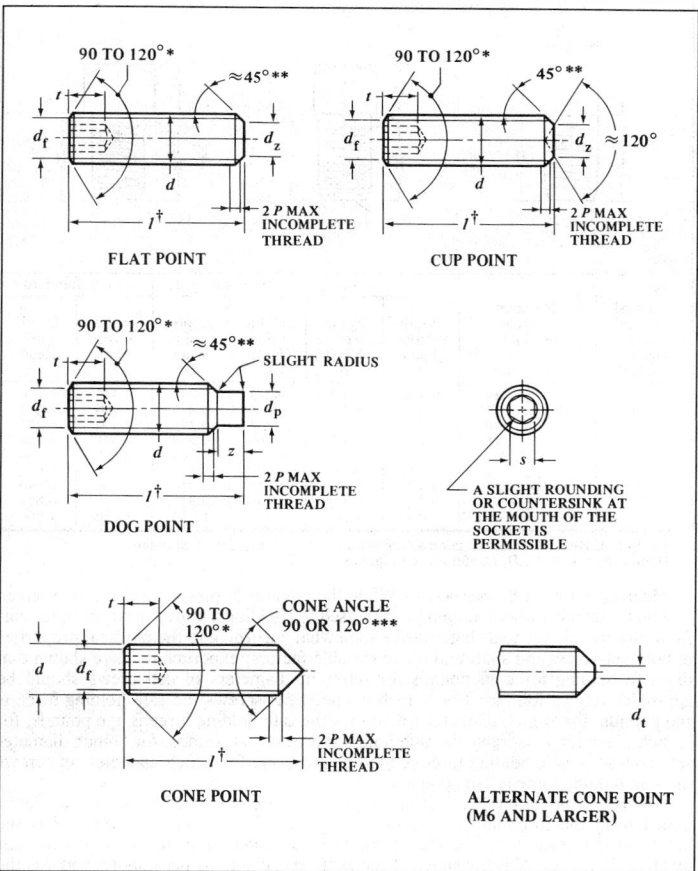

FLAT POINT

CUP POINT

DOG POINT

A SLIGHT ROUNDING OR COUNTERSINK AT THE MOUTH OF THE SOCKET IS PERMISSIBLE

CONE POINT

ALTERNATE CONE POINT (M6 AND LARGER)

* The 120° angle is mandatory for short-length screws shown in the Standard. Short-length screws are those whose length is, approximately, equal to the diameter of the screw.

** The 45° angle applies only to that portion of the point below the root diameter, d_f, of the thread.

*** The cone angle applies only to the portion of the point below the root diameter, d_f, of the thread and shall be 120° for certain short lengths listed in the Standard. All other lengths have a 90° cone angle.

† The popular length ranges of these set screws are listed in Table 3. These lengths have been selected from the following nominal lengths: 2, 2.5, 3, 4, 5, 6, 8, 10, 12, 16, 20, 25, 30, 35, 40, 45, 50, 55, and 60 millimeters.

Table 4. British Standard Whitworth (B.S.W.) and British Standard Fine (B.S.F.) Bright Square Head Set-Screws (With Flat Chamfered Ends)

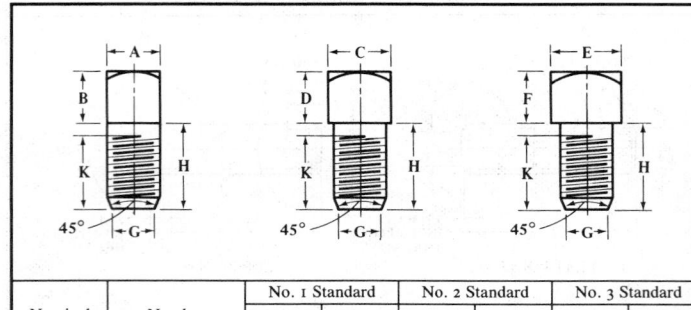

Nominal Size and Max. Diam., Inches	Number of Threads per Inch		No. 1 Standard		No. 2 Standard		No. 3 Standard	
	B.S.W.	B.S.F.	Width Across Flats A	Depth of Head B	Width Across Flats C	Depth of Head D	Width Across Flats E	Depth of Head F
¼	20	26	0.250	0.250	0.313	0.250	0.375	0.250
5/16	18	22	0.313	0.313	0.375	0.313	0.438	0.313
⅜	16	20	0.375	0.375	0.438	0.375	0.500	0.375
7/16	14	18	0.438	0.438	0.500	0.438	0.625	0.438
½	12	16	0.500	0.500	0.563	0.500	0.750	0.500
⅝	11	14	0.625	0.625	0.750	0.625	0.875	0.625
¾	10	12	0.750	0.750	0.875	0.750	1.000	0.750
⅞	9	11	0.875	0.875	1.000	0.875	1.125	0.875
1	8	10	1.000	1.000	1.125	1.000	1.250	1.000

* Depth of Head B, D and F same as for Width Across Flats, No. 1 Standard.
Dimensions A, B, C, D, E, and F are in inches.

Holding Power of Set-screws. — While the amount of power a set-screw of given size will transmit without slipping (when used for holding a pulley, gear, or other part from turning relative to a shaft) varies somewhat according to the physical properties of both set-screw and shaft and other variable factors, experiments have shown that the safe holding force in pounds for different diameters of set-screws should be approximately as follows: For ¼-inch diameter set-screws the safe holding force is 100 pounds, for ⅜-inch diameter set-screws the safe holding force is 250 pounds, for ½-inch diameter set-screws the safe holding force is 500 pounds, for ¾-inch diameter set-screws the safe holding force is 1300 pounds, and for 1-inch diameter set-screws the safe holding force is 2500 pounds.

The power or torque that can be safely transmitted by a set-screw may be determined from the formulas, $P = (DNd^{2.3}) \div 50$; or $T = 1250Dd^{2.3}$ in which P is the horsepower transmitted; T is the torque in inch-pounds transmitted; D is the shaft diameter in inches; N is the speed of the shaft in revolutions per minute; and d is the diameter of the set-screw in inches.

Example: — How many ½-inch diameter set-screws would be required to transmit 3 horsepower at a shaft speed of 1000 rpm if the shaft diameter is 1 inch?

Using the first formula given above, the power transmitted by a single ½-inch diameter set-screw is determined: $P = [1 \times 1000 \times (\frac{1}{2})^{2.3}] \div 50 = 4.1$ hp. Therefore a single ½-inch diameter set-screw is sufficient.

Example: — In the previous example, how many ⅜-inch diameter set-screws would be required? $P = [1 \times 1000 \times (\frac{3}{8})^{2.3}] \div 50 = 2.1$ hp. Therefore two ⅜-inch diameter set-screws are required.

Table 5. British Standard Hexagon Socket Countersunk and Button Head Screws — Metric Series (B.S. 4168:1967)

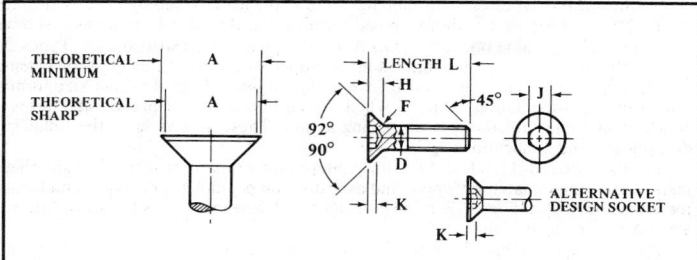

COUNTERSUNK HEAD SCREWS

Nom. Size*	Body Diameter, D		Head Diameter, A		Head Height, H		Hexagon Socket Size, J	Key Engagement, K	Fillet Radius, F
	Max.	Min.	Theor. Sharp Max.	Absolute Min.	Ref.	Flushness Tolerance	Nom.	Min.	Max.
M3	3.00	2.86	6.72	5.82	1.86	0.20	2.00	1.05	0.40
M4	4.00	3.82	8.96	7.78	2.48	0.20	2.50	1.49	0.40
M5	5.00	4.82	11.20	9.78	3.10	0.20	3.00	1.86	0.40
M6	6.00	5.82	13.44	11.73	3.72	0.20	4.00	2.16	0.60
M8	8.00	7.78	17.92	15.73	4.96	0.24	5.00	2.85	0.70
M10	10.00	9.78	22.40	19.67	6.20	0.30	6.00	3.60	0.80
M12	12.00	11.73	26.88	23.67	7.44	0.36	8.00	4.35	1.10
(M14)	14.00	13.73	30.24	26.67	8.12	0.40	10.00	4.65	1.10
M16	16.00	15.73	33.60	29.67	8.80	0.45	10.00	4.89	1.10
(M18)	18.00	17.73	36.96	32.61	9.48	0.50	12.00	5.25	1.10
M20	20.00	19.67	40.32	35.61	10.16	0.54	12.00	5.45	1.10

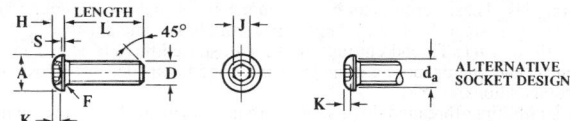

BUTTON HEAD SCREWS

Nom. Size, D	Head Diameter, A		Head Height, H		Head Side Height, S	Hexagon Socket Size, J	Key Engagement, K	Fillet Radius	
	Max.	Min.	Max.	Min.	Ref.	Nom.	Min.	F Min.	d_a Max.
M3	5.50	5.32	1.60	1.40	0.38	2.00	1.04	0.10	3.60
M4	7.50	7.28	2.10	1.85	0.38	2.50	1.30	0.20	4.70
M5	9.50	9.28	2.70	2.45	0.50	3.00	1.56	0.20	5.70
M6	10.50	10.23	3.20	2.95	0.80	4.00	2.08	0.25	6.80
M8	14.00	13.73	4.30	3.95	0.80	5.00	2.60	0.40	9.20
M10	18.00	17.73	5.30	4.95	0.80	6.00	3.12	0.40	11.20
M12	21.00	20.67	6.40	5.90	0.80	8.00	4.16	0.60	14.20

All dimensions are given in millimeters.
* Sizes shown in parentheses are non-preferred.

SELF-THREADING SCREWS

ANSI Standard Sheet Metal, Self-Tapping and Metallic Drive Screws. — Table 1 shows the various types of "self-tapping" screw threads covered by the ANSI Standard ANSI B18.6.4-1981. (Metric thread forming and thread cutting tapping screws are discussed beginning on page 1424.) ANSI designations are also shown. Types A, AB, B, BP and C when turned into a hole of proper size form a thread by a displacing action. Types D, F, G, T, BF and BT when turned into a hole of proper size form a thread by a cutting action. Type U when driven into a hole of proper size forms a series of multiple threads by a displacing action. These screws have the following descriptions and applications:

Type A: Spaced-thread screw with gimlet point primarily for use in light sheet metal, resin-impregnated plywood, and asbestos compositions. This type is no longer recommended. Use Type AB in new designs and whenever possible substitute for Type A in existing designs.

Type AB: Spaced-thread screw with same pitches as Type B but with gimlet point, primarily for similar uses as for Type A.

Type B: Spaced-thread screw with a blunt point with pitches generally somewhat finer than Type A. Used for thin metal, non-ferrous castings, plastics, resin-impregnated plywood, and asbestos compositions.

Type BP: Spaced-thread screw, the same as Type B but having a conical point extending beyond incomplete entering threads. Used for piercing fabrics or in assemblies where holes are misaligned.

Type C: Screws having machine screw diameter-pitch combinations with threads approximately Unified Form and with blunt tapered points. Used where the use of a machine screw thread is preferable to the use of the spaced-thread types of thread forming screws. Also useful when chips from machine screw thread-cutting screws are objectionable. In view of the declining use of Type C screws, which in general require high driving torques, in favor of more efficient designs of thread tapping screws, they are not recommended for new designs.

Types D, F, G, and T: Thread-cutting screws with threads approximating machine screw threads, with blunt point, and with tapered entering threads having one or more cutting edges and chip cavities. The tapered threads of the Type F may be complete or incomplete at the producer's option; all other types have incomplete tapered threads. These screws can be used in materials such as aluminum, zinc, and lead die-castings; steel sheets and shapes; cast iron; brass; and plastics.

Types BF and BT: Thread-cutting screws with spaced threads as in Type B, with blunt points, and one or more cutting grooves. Used in plastics, asbestos, and other similar compositions.

Type U: Multiple-threaded drive screw with large helix angle, having a pilot point, for use in metal and plastics. This screw is forced into the work by pressure and is intended for making permanent fastenings.

ANSI Standard Head Types for Tapping and Metallic Drive Screws. — Many of the head types used with "self-tapping" screw threads are similar to the head types of American National Standard machine screws shown in the section with that heading.

Round Head: The round head has a semi-elliptical top surface and a flat bearing surface. Because of the superior slot driving characteristics of pan head screws over round head screws, and the overlap in dimensions of cross recessed pan heads and round heads, it is recommended that pan head screws be used in new designs and wherever possible substituted in existing designs.

Flat Countersunk Head: The flat countersunk head has a flat top surface and a conical bearing surface with a head angle for one design of approximately 82 degrees and for another design of approximately 100 degrees. Because of its limited usage and in the interest of curtailing product varieties, the 100-degree flat countersunk head is considered non-preferred.

Oval Countersunk Head: The oval countersunk head has a rounded top surface and a conical bearing surface with a head angle of approximately 82 degrees.

Undercut Flat and Oval Countersunk Heads: For short lengths, 82-degree flat and oval countersunk head tapping screws have heads undercut to 70 per cent of normal side height to afford greater length of thread on the screws.

Flat and Oval Countersunk Trim Heads: Flat and oval countersunk trim heads are similar to the 82-degree flat and oval countersunk heads except that the size of head for a given size screw is one (large trim head) or two (small trim head) sizes smaller than the regular flat and oval countersunk head size. Oval countersunk trim heads have a definite radius where the curved top surface meets the conical bearing surface. Trim heads are furnished only in cross recessed types.

Pan Head: The slotted pan head has a flat top surface rounded into cylindrical sides and a flat bearing surface. The recessed pan head has a rounded top and a flat bearing surface. This head type is now preferred to the round head.

Fillister Head: The fillister head has a rounded top surface, cylindrical sides, and a flat bearing surface.

Hex Head: The hex head has a flat or indented top surface, six flat sides, and a flat bearing surface. Because the slotted hex head requires a secondary operation in manufacture which often results in burrs at the extremity of the slot which interfere with socket wrench engagement and the wrenching capability of the hex far exceeds that of the slot, it is not recommended for new designs.

Hex Washer Head: The hex washer head has an indented top surface and six flat sides formed integrally with a flat washer which projects beyond the sides and provides a flat bearing surface. Because the slotted hex washer head requires a secondary operation in manufacture which often results in burrs at the extremity of the slot which often interferes with socket wrench engagement and since the wrenching capability of the hex far exceeds that of the slot in the indented head, it is not recommended for new designs.

Truss Head: The truss head has a low rounded top surface with a flat bearing surface, the diameter of which for a given screw size is larger than the diameter of the corresponding round head. In the interest of product simplification and recognizing that the truss head is an inherently weak design, it is not recommended for new designs.

Method of Designation. — Tapping screws are designated by the following data in the sequence shown: Nominal size (number, fraction or decimal equivalent); threads per inch; nominal length (fraction or decimal equivalent); point type; product name, including head type and driving provision; material; and protective finish, if required. Examples:

> $1/4$–$14 \times 1\frac{1}{2}$ Type AB Slotted Pan Head Tapping Screw, Steel, Nickel Plated
> 6–$32 \times 3/4$ Type T, Type 1A Cross Recessed Pan Head Tapping Screw, Corrosion Resistant Steel
> .375–16 × 1.50 Type D, Washer Head Tapping Screw, Steel

Metallic Drive Screws: Type U metallic drive screws are designated by the following data in the sequence shown: Nominal size (number, fraction, or decimal equivalent); nominal length (fraction or decimal equivalent); product name, including head type; material; and protective finish, if required. Examples:

> $10 \times 5/16$ Round Head Metallic Drive Screw, Steel
> .312 × .50 Round Head Metallic Drive Screw, Steel, Zinc Plated

Table 1. ANSI Standard Threads and Points for Thread Forming Self-Tapping Screws (ANSI B18.6.4-1981)

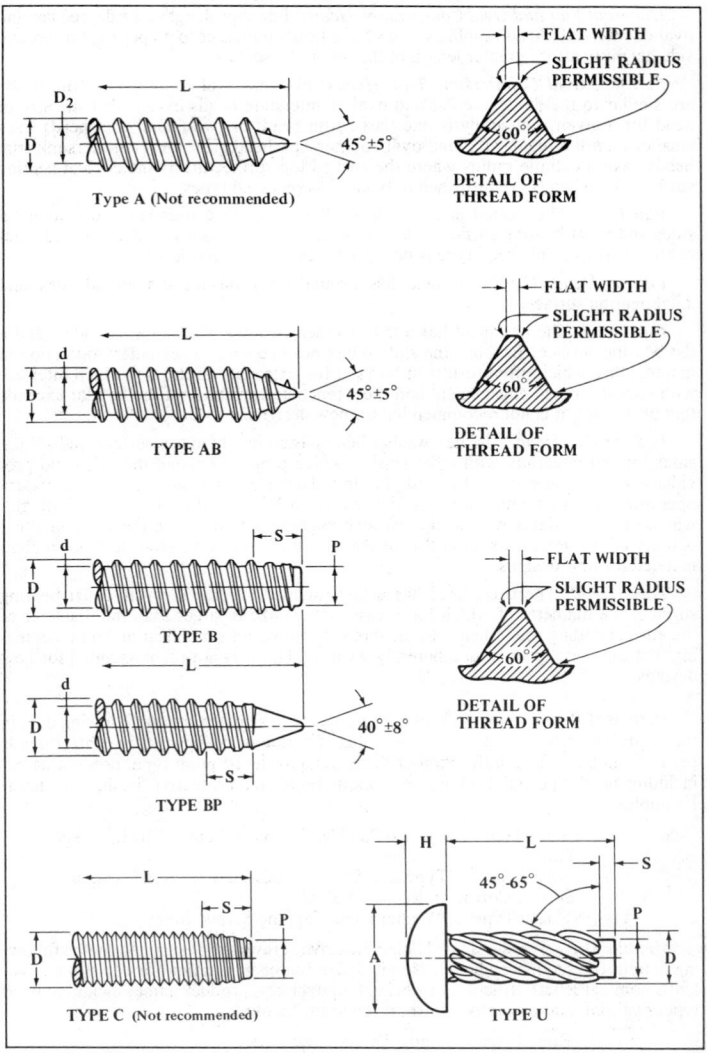

See Tables 4, 5, and 6 for thread data.

Table 2. ANSI Standard Threads and Points for Thread Cutting Self-Tapping Screws (ANSI B18.6.4-1981)

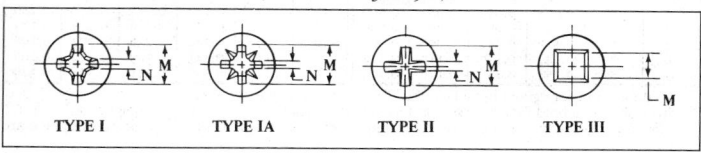

TYPE BF

TYPE BT

FLAT WIDTH
SLIGHT RADIUS PERMISSIBLE
60°

DETAIL OF THREAD FORM

TYPE D

TYPE F

TYPE G

TYPE T

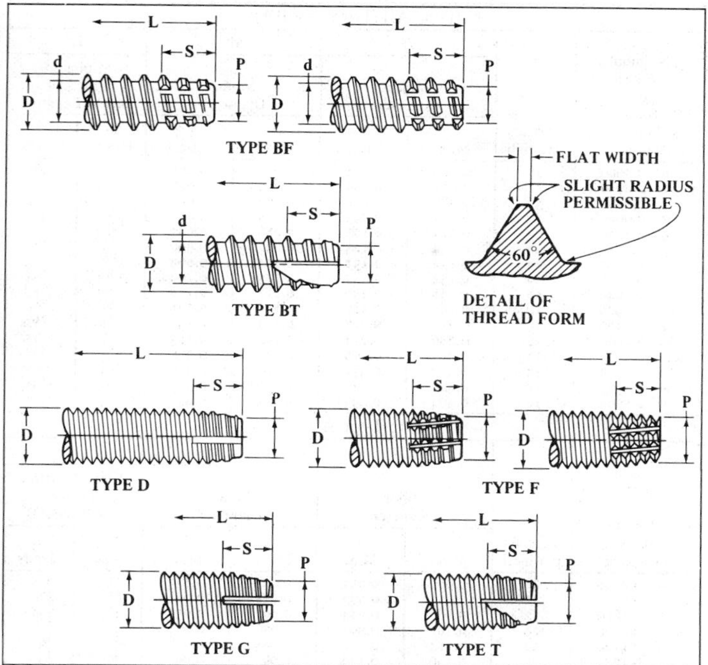

See Tables 5 and 7 for thread data.

Cross Recesses. — Type I cross recess has a large center opening, tapered wings, and blunt bottom, with all edges relieved or rounded. Type IA cross recess has a large center opening, wide straight wings, and blunt bottom, with all edges relieved or rounded. Type II consists of two intersecting slots with parallel sides converging to a slightly truncated apex at the bottom of the recess. Type III has a square center opening, slightly tapered side walls, and a conical bottom, with top edges relieved or rounded.

Table 3. ANSI Standard Cross Recesses for Self-Tapping Screws (ANSI B18.6.4-1981) and Metric Thread Forming and Thread Cutting Tapping Screws (ANSI B18.6.5M-1986)

TYPE I

TYPE IA

TYPE II

TYPE III

Table 4. ANSI Standard Thread and Point Dimensions for Types AB, A and U Thread Forming Tapping Screws (ANSI B18.6.4-1981)

Type AB (Formerly BA)

Nominal Size or Basic Screw Diameter	Threads per inch	D Major Diameter Max.	D Major Diameter Min.	d Minor Diameter Max.	d Minor Diameter Min.	L Minimum Practical Screw Lengths 90° Heads	L Minimum Practical Screw Lengths Csk. Heads
0 0.0600	48	0.060	0.054	0.036	0.033	1/8	5/32
1 0.0730	42	0.075	0.069	0.049	0.046	5/32	3/16
2 0.0860	**32**	**0.088**	**0.082**	**0.064**	**0.060**	3/16	7/32
3 0.0990	28	0.101	0.095	0.075	0.071	3/16	1/4
4 0.1120	**24**	**0.114**	**0.108**	**0.086**	**0.082**	7/32	9/32
5 0.1250	20	0.130	0.123	0.094	0.090	1/4	5/16
6 0.1380	**20**	**0.139**	**0.132**	**0.104**	**0.099**	9/32	11/32
7 0.1510	19	0.154	0.147	0.115	0.109	5/16	3/8
8 0.1640	**18**	**0.166**	**0.159**	**0.122**	**0.116**	5/16	3/8
10 0.1900	**16**	**0.189**	**0.182**	**0.141**	**0.135**	3/8	7/16
12 0.2160	14	0.215	0.208	0.164	0.157	7/16	21/32
1/4 0.2500	**14**	**0.246**	**0.237**	**0.192**	**0.185**	1/2	19/32
5/16 0.3125	12	0.315	0.306	0.244	0.236	5/8	3/4
3/8 0.3750	12	0.380	0.371	0.309	0.299	3/4	29/32
7/16 0.4375	10	0.440	0.429	0.359	0.349	7/8	1 1/32
1/2 0.5000	10	0.504	0.493	0.423	0.413	1	1 5/32

Type A

Nominal Size* or Basic Screw Diameter	Threads per inch	D Major Diameter Max.	D Major Diameter Min.	d Minor Diameter Max.	d Minor Diameter Min.	L These Lengths or Shorter — Use Type AB 90° Heads	L These Lengths or Shorter — Use Type AB Csk. Heads
0 0.0600	40	0.060	0.057	0.042	0.039	1/8	3/16
1 0.0730	32	0.075	0.072	0.051	0.048	1/8	3/16
2 0.0860	32	0.088	0.084	0.061	0.056	5/32	3/16
3 0.0990	28	0.101	0.097	0.076	0.071	3/16	7/32
4 0.1120	24	0.114	0.110	0.083	0.078	3/16	1/4
5 0.1250	20	0.130	0.126	0.095	0.090	3/16	1/4
6 0.1380	18	0.141	0.136	0.102	0.096	1/4	5/16
7 0.1510	16	0.158	0.152	0.114	0.108	5/16	3/8
8 0.1640	15	0.168	0.162	0.123	0.116	3/8	7/16
10 0.1900	12	0.194	0.188	0.133	0.126	3/8	1/2
12 0.2160	11	0.221	0.215	0.162	0.155	7/16	9/16
14 0.2420	10	0.254	0.248	0.185	0.178	1/2	5/8
16 0.2680	10	0.280	0.274	0.197	0.189	9/16	3/4
18 0.2940	9	0.306	0.300	0.217	0.209	5/8	13/16
20 0.3200	9	0.333	0.327	0.234	0.226	11/16	13/16
24 0.3720	9	0.390	0.383	0.291	0.282	3/4	1

Type U Metallic Drive Screws

Nom. Size	No. of Starts	Out. Dia. Max.	Out. Dia. Min.	Pilot Dia. Max.	Pilot Dia. Min.	Nom. Size	No. of Starts	Out. Dia. Max.	Out. Dia. Min.	Pilot Dia. Max.	Pilot Dia. Min.
00	6	0.060	0.057	0.049	0.046	8	8	0.167	0.162	0.136	0.132
0	6	0.075	0.072	0.063	0.060	10	8	0.182	0.177	0.150	0.146
2	8	0.100	0.097	0.083	0.080	12	8	0.212	0.206	0.177	0.173
4	7	0.116	0.112	0.096	0.092	14	9	0.242	0.236	0.202	0.198
6	7	0.140	0.136	0.116	0.112	5/16	11	0.315	0.309	0.272	0.267
7	8	0.154	0.150	0.126	0.122	3/8	12	0.378	0.371	0.334	0.329

All dimensions are given in inches. See Table 1 for thread diagrams.

Sizes shown in bold face type are preferred. Type A screws no longer recommended.

* Where specifying nominal size in decimals, zeros preceding decimal and in fourth place are omitted.

Table 5. ANSI Standard Thread and Point Dimensions for Types B and BP Thread Forming and Types BF and BT Thread Cutting Tapping Screws (ANSI B18.6.4-1981)

Nominal Size[1] or Basic Screw Diameter	Thds per Inch[2]	D Major Diameter		d Minor Diameter		P Point Diameter[3]		S Point Taper Length[4]		L Minimum Practical Nominal Screw Lengths			
										Type B		Type BP	
		Max	Min	Max	Min	Max	Min	Max	Min	90° Heads	Csk Heads	90° Heads	Csk Heads
0 0.0600	48	0.060	0.054	0.036	0.033	0.031	0.027	0.042	0.031	1/8	1/8	5/32	3/16
1 0.0730	42	0.075	0.069	0.049	0.046	0.044	0.040	0.048	0.036	1/8	5/32	3/16	7/32
2 0.0860	32	0.088	0.082	0.064	0.060	0.058	0.054	0.062	0.047	5/32	3/16	1/4	9/32
3 0.0990	28	0.101	0.095	0.075	0.071	0.068	0.063	0.071	0.054	3/16	7/32	9/32	5/16
4 0.1120	24	0.114	0.108	0.086	0.082	0.079	0.074	0.083	0.063	3/16	1/4	5/16	11/32
5 0.1250	20	0.130	0.123	0.094	0.090	0.087	0.082	0.100	0.075	7/32	9/32	11/32	13/32
6 0.1380	20	0.139	0.132	0.104	0.099	0.095	0.089	0.100	0.075	1/4	9/32	3/8	7/16
7 0.1510	19	0.154	0.147	0.115	0.109	0.105	0.099	0.105	0.079	1/4	5/16	13/32	15/32
8 0.1640	18	0.166	0.159	0.122	0.116	0.112	0.106	0.111	0.083	9/32	11/32	7/16	1/2
10 0.1900	16	0.189	0.182	0.141	0.135	0.130	0.123	0.125	0.094	5/16	3/8	1/2	19/32
12 0.2160	14	0.215	0.208	0.164	0.157	0.152	0.145	0.143	0.107	11/32	7/16	9/16	21/32
1/4 0.2500	14	0.246	0.237	0.192	0.185	0.179	0.171	0.143	0.107	3/8	1/2	21/32	3/4
5/16 0.3125	12	0.315	0.306	0.244	0.236	0.230	0.222	0.167	0.125	15/32	19/32	27/32	31/32
3/8 0.3750	12	0.380	0.371	0.309	0.299	0.293	0.285	0.167	0.125	17/32	11/16	15/16	1 1/8
7/16 0.4375	10	0.440	0.429	0.359	0.349	0.343	0.335	0.200	0.150	5/8	25/32	1 1/8	1 1/4
1/2 0.5000	10	0.504	0.493	0.423	0.413	0.407	0.399	0.200	0.150	11/16	27/32	1 1/4	1 13/32

THREAD CUTTING TYPES BF AND BT[4]

Nominal Size[1] or Basic Screw Diameter	Thds per Inch[2]	D Major Diameter		d Minor Diameter		P Point Diameter[3]		S Point Taper Length[4]		L Minimum Practical Nominal Screw Lengths	
		Max	Min	Max	Min	Max	Min	Max	Min	90° Heads	Csk Heads
0 0.0600	48	0.060	0.054	0.036	0.033	0.031	0.027	0.042	0.031	1/8	1/8
1 0.0730	42	0.075	0.069	0.049	0.046	0.044	0.040	0.048	0.036	1/8	5/32
2 0.0860	32	0.088	0.082	0.064	0.060	0.058	0.054	0.062	0.047	5/32	3/16
3 0.0990	28	0.101	0.095	0.075	0.071	0.068	0.063	0.071	0.054	3/16	7/32
4 0.1120	24	0.114	0.108	0.086	0.082	0.079	0.074	0.083	0.063	3/16	1/4
5 0.1250	20	0.130	0.123	0.094	0.090	0.087	0.082	0.100	0.075	7/32	9/32
6 0.1380	20	0.139	0.132	0.104	0.099	0.095	0.089	0.100	0.075	1/4	9/32
7 0.1510	19	0.154	0.147	0.115	0.109	0.105	0.099	0.105	0.079	1/4	5/16
8 0.1640	18	0.166	0.159	0.122	0.116	0.112	0.106	0.111	0.083	9/32	11/32
10 0.1900	16	0.189	0.182	0.141	0.135	0.130	0.123	0.125	0.094	5/16	3/8
12 0.2160	14	0.215	0.208	0.164	0.157	0.152	0.145	0.143	0.107	11/32	7/16
1/4 0.2500	14	0.246	0.237	0.192	0.185	0.179	0.171	0.143	0.107	3/8	1/2
5/16 0.3125	12	0.315	0.306	0.244	0.236	0.230	0.222	0.167	0.125	15/32	19/32
3/8 0.3750	12	0.380	0.371	0.309	0.299	0.293	0.285	0.167	0.125	17/32	11/16
7/16 0.4375	10	0.440	0.429	0.359	0.349	0.343	0.335	0.200	0.150	5/8	25/32
1/2 0.5000	10	0.504	0.493	0.423	0.413	0.407	0.399	0.200	0.150	11/16	27/32

All dimensions are given in inches. See Tables 1 and 2 for thread diagrams.

[1] Where specifying nominal size in decimals, zeros preceding decimal and in the fourth decimal place shall be omitted.

[2] The width of flat at crest of thread shall not exceed 0.004 inch for sizes up to No. 8, inclusive, and 0.006 inch for larger sizes.

[3] Point diameters specified apply to screw threads before roll threading.

[4] Points of screws are tapered and fluted or slotted. The flute on Type BT screws has an included angle of 90 to 95 degrees and the thread cutting edge is located above the axis of the screw. Flutes and slots extend through first full form thread beyond taper except for Type BF screw on which tapered threads may be complete at manufacturer's option and flutes may be one pitch short of first full form thread.

Table 6. Thread and Point Dimensions for Type C Thread Forming Tapping Screws (ANSI B18.6.4-1981, Appendix)

Nominal Size[1] or Basic Screw Diameter	Threads per inch	D Major Diameter		P Point Diameter[2]		S Point Taper Length[3]				Determinant Lengths for Point Taper		L Minimum Practical Nominal Screw Lengths	
						For Short Screws		For Long Screws					
		Max	Min	Max	Min	Max	Min	Max	Min	90° Heads	Csk Heads	90° Heads	Csk Heads
2 0.0860	56	0.0860	0.0813	0.068	0.061	0.062	0.045	0.080	0.062	5/32	3/16	5/32	3/16
2 0.0860	64	0.0860	0.0816	0.070	0.064	0.055	0.039	0.070	0.055	1/8	3/16	1/8	5/32
3 0.0990	48	0.0990	0.0938	0.078	0.070	0.073	0.052	0.094	0.073	3/16	7/32	5/32	7/32
3 0.0990	56	0.0990	0.0942	0.081	0.074	0.062	0.045	0.080	0.062	3/32	3/16	5/32	3/16
4 0.1120	40	0.1120	0.1061	0.087	0.078	0.088	0.062	0.112	0.088	7/32	1/4	3/16	1/4
4 0.1120	48	0.1120	0.1068	0.091	0.083	0.073	0.052	0.094	0.073	3/16	9/32	3/16	7/32
5 0.1250	40	0.1250	0.1191	0.100	0.091	0.088	0.062	0.112	0.088	7/32	1/4	3/16	1/4
5 0.1250	44	0.1250	0.1195	0.102	0.094	0.080	0.057	0.102	0.080	1/4	5/16	3/16	1/4
6 0.1380	32	0.1380	0.1312	0.107	0.096	0.109	0.078	0.141	0.109	1/4	9/32	3/16	5/16
6 0.1380	40	0.1380	0.1321	0.113	0.104	0.088	0.062	0.112	0.088	7/32	11/32	3/16	1/4
8 0.1640	32	0.1640	0.1571	0.132	0.122	0.109	0.078	0.141	0.109	1/4	5/16	1/4	5/16
8 0.1640	36	0.1640	0.1577	0.136	0.126	0.097	0.069	0.125	0.097	7/32	11/32	7/32	13/32
10 0.1900	24	0.1900	0.1818	0.148	0.135	0.146	0.104	0.188	0.146	5/16	7/16	1/4	5/16
10 0.1900	32	0.1900	0.1831	0.158	0.148	0.109	0.078	0.141	0.109	1/4	11/32	1/4	13/32
12 0.2160	24	0.2160	0.2078	0.174	0.161	0.146	0.104	0.188	0.146	5/16	7/16	5/16	3/8
12 0.2160	28	0.2160	0.2085	0.180	0.168	0.125	0.089	0.161	0.125	9/32	13/32	3/8	1/2
1/4 0.2500	20	0.2500	0.2408	0.200	0.184	0.175	0.125	0.225	0.175	9/32	17/32	9/32	9/16
1/4 0.2500	28	0.2500	0.2425	0.214	0.202	0.125	0.089	0.161	0.125	5/16	13/32	7/16	15/32
5/16 0.3125	18	0.3125	0.3026	0.257	0.239	0.194	0.139	0.250	0.194	7/16	19/32	5/16	5/8
5/16 0.3125	24	0.3125	0.3042	0.271	0.257	0.146	0.104	0.188	0.146	1/2	5/8	9/16	1/2
3/8 0.3750	16	0.3750	0.3643	0.312	0.293	0.219	0.156	0.281	0.219	11/32	1/2	3/8	23/32
3/8 0.3750	24	0.3750	0.3667	0.333	0.319	0.146	0.104	0.188	0.146	1/2	11/16	19/32	17/32
7/16 0.4375	14	0.4375	0.4258	0.366	0.344	0.250	0.179	0.321	0.250	19/32	3/4	3/8	3/4
7/16 0.4375	20	0.4375	0.4281	0.387	0.371	0.175	0.125	0.225	0.175	13/32	25/32	19/32	17/32
1/2 0.5000	13	0.5000	0.4876	0.423	0.399	0.269	0.192	0.346	0.269	5/8	9/16	5/8	3/4
1/2 0.5000	20	0.5000	0.4906	0.450	0.433	0.175	0.125	0.225	0.175	13/32	9/16	13/32	17/32

All dimensions in inches. See Table 1 for thread diagrams. Type C is not recommended for new designs. Tapered threads shall have unfinished crests. [1] Where specifying nominal size in decimals, zeros preceding decimal and in the fourth decimal place shall be omitted. [2] The tabulated values apply to screw blanks before roll threading. [3] Screws of these nominal lengths and shorter shall have point taper length specified above for short screws. Longer lengths shall have point taper length specified for long screws.

Table 7. ANSI Standard Thread and Point Dimensions for Types D*, F, G, and T† Thread Cutting Tapping Screws
(ANSI B18.6.4-1981)

Nominal Size or Basic Screw Diameter	Threads per inch	D Major Diameter Max	Min	P Point Diameter[2] Max	Min	S Point Taper Length[3] For Short Screws Max	Min	For Long Screws Max	Min	L Determinant Lengths for Point Taper[3] 90° Heads	Csk Heads	Minimum Practical Nominal Screw Lengths 90° Heads	Csk Heads
2 0.0860	56	0.0860	0.0813	0.068	0.061	0.062	0.045	0.080	0.062	5/32	3/16	5/32	3/16
2 0.0860	64	0.0860	0.0816	0.070	0.064	0.055	0.039	0.070	0.055	1/8	3/16	1/8	5/32
3 0.0990	48	0.0990	0.0938	0.078	0.070	0.073	0.052	0.094	0.073	3/16	7/32	7/32	7/32
3 0.0990	56	0.0990	0.0942	0.081	0.074	0.062	0.045	0.080	0.062	5/32	3/16	5/32	3/16
4 0.1120	40	0.1120	0.1061	0.087	0.078	0.088	0.062	0.112	0.088	7/32	1/4	3/16	1/4
4 0.1120	48	0.1120	0.1068	0.091	0.083	0.073	0.052	0.094	0.073	3/16	7/32	5/32	3/16
5 0.1250	40	0.1250	0.1191	0.100	0.091	0.088	0.062	0.112	0.088	7/32	9/32	3/16	5/32
5 0.1250	44	0.1250	0.1195	0.102	0.094	0.080	0.057	0.102	0.080	3/16	1/4	3/16	1/4
6 0.1380	32	0.1380	0.1312	0.107	0.096	0.109	0.078	0.141	0.109	1/4	9/32	1/4	1/4
6 0.1380	40	0.1380	0.1321	0.113	0.104	0.088	0.062	0.112	0.088	7/32	1/4	3/16	3/16
8 0.1640	32	0.1640	0.1571	0.132	0.122	0.109	0.078	0.141	0.109	1/4	9/32	1/4	5/16
8 0.1640	36	0.1640	0.1577	0.136	0.126	0.097	0.069	0.125	0.097	7/32	5/16	7/32	1/4
10 0.1900	24	0.1900	0.1818	0.148	0.135	0.146	0.104	0.188	0.146	5/16	11/32	5/16	5/16
10 0.1900	32	0.1900	0.1831	0.158	0.148	0.109	0.078	0.141	0.109	1/4	7/16	1/4	1/4
12 0.2160	24	0.2160	0.2078	0.174	0.161	0.146	0.104	0.188	0.146	1/4	7/16	5/16	5/16
12 0.2160	28	0.2160	0.2085	0.180	0.168	0.125	0.089	0.161	0.125	5/16	13/32	9/32	13/32
1/4 0.2500	20	0.2500	0.2408	0.200	0.184	0.175	0.125	0.225	0.175	13/32	17/32	3/8	3/8
1/4 0.2500	28	0.2500	0.2425	0.214	0.202	0.125	0.089	0.161	0.125	5/16	13/32	9/32	3/8
5/16 0.3125	18	0.3125	0.3026	0.257	0.239	0.194	0.139	0.250	0.194	15/32	19/32	7/16	9/16
5/16 0.3125	24	0.3125	0.3042	0.271	0.257	0.146	0.104	0.188	0.146	11/32	15/32	5/16	15/32
3/8 0.3750	16	0.3750	0.3643	0.312	0.293	0.219	0.156	0.281	0.219	1/2	19/32	15/32	5/8
3/8 0.3750	24	0.3750	0.3667	0.366	0.319	0.146	0.104	0.188	0.146	11/32	11/16	5/16	1/2
7/16 0.4375	14	0.4375	0.4258	0.387	0.344	0.250	0.179	0.321	0.250	19/32	1/2	9/16	23/32
7/16 0.4375	20	0.4375	0.4281	0.387	0.371	0.175	0.125	0.225	0.175	13/32	3/4	3/8	9/16
1/2 0.5000	13	0.5000	0.4876	0.423	0.399	0.269	0.192	0.346	0.269	5/8	25/32	19/32	3/4
1/2 0.5000	20	0.5000	0.4906	0.450	0.433	0.175	0.125	0.225	0.175	13/32	9/16	3/8	17/32

All dimensions are given in inches. See Table 2 for thread diagrams. * Otherwise designated "Type 1." † Otherwise designated "Type 23." [1] Where specifying nominal size in decimals, zeros preceding decimal and in the fourth decimal place shall be omitted. [2] The tabulated values apply to screw blanks before roll threading. [3] Screws of these nominal lengths and shorter shall have point taper length specified above for short screws. Longer lengths shall have point taper length specified for long screws.

Table 8. Approximate Hole Sizes for Type A Steel Thread Forming Screws*

In Steel, Stainless Steel, Monel Metal, Brass, and Aluminum Sheet Metal

Screw Size	Metal Thick-ness	Hole Size Pierced or Extruded	Hole Size Drilled or Clean Punched	Drill Size	Screw Size	Metal Thick-ness	Hole Size Pierced or Extruded	Hole Size Drilled or Clean Punched	Drill Size
4	.015	...	.086	44	8	.024	.136	.125	⅛
	.018	...	.086	44		.030	.136	.125	⅛
	.024	.098	.094	42		.036	.136	.125	⅛
	.030	.098	.094	42		.048	.136	.128	30
	.036	.098	.098	40	10	.018	...	.136	29
6	.015	...	.104	37		.024	.157	.136	29
	.018	...	.104	37		.030	.157	.136	29
	.024	.111	.104	37		.036	.157	.136	29
	.030	.111	.104	37		.048	.157	.149	25
	.036	.111	.106	36	12	.024	...	.161	20
7	.015	...	.116	32		.030	.185	.161	20
	.018	...	.116	32		.036	.185	.161	20
	.024	.120	.116	32		.048	.185	.161	20
	.030	.120	.116	32	14	.024	...	.185	13
	.036	.120	.116	32		.030	.209	.189	12
	.048	.120	.120	31		.036	.209	.191	11
8	.018	...	.125	⅛		.048	.209	.196	9

Screw Size	In Plywood (Resin Impregnated) Hole Size	Drill Size	Min. Mat'l Thick-ness	Penetration in Blind Holes Min.	Penetration in Blind Holes Max.	Screw Size	In Asbestos Compositions Hole Size	Drill Size	Min. Mat'l Thick-ness	Penetration in Blind Holes Min.	Penetration in Blind Holes Max.
4	.098	40	.188	.250	.750	4	.094	42	.188	.250	.750
6	.110	35	.188	.250	.750	6	.106	36	.188	.250	.750
7	.128	30	.250	.312	.750	7	.125	⅛	.250	.312	.750
8	.140	28	.250	.312	.750	8	.136	29	.250	.312	.750
10	.170	18	.312	.375	1.000	10	.161	20	.312	.375	1.000
12	.189	12	.312	.375	1.000	12	.185	13	.312	.375	1.000
14	.228	1	.438	.500	1.000	14	.213	3	.438	.500	1.000

See footnote at bottom of Table 9. * Type A is not recommended, use Type AB.

Table 9. Approximate Hole Sizes for Type C Steel Thread Forming Screws*

In Sheet Steel

Screw Size	Metal Thick-ness	Hole Size	Drill Size	Screw Size	Metal Thick-ness	Hole Size	Drill Size	Screw Size	Metal Thick-ness	Hole Size	Drill Size
4–40	.037	.094	42	10–24	.037	.154	23	¼–20	.037	.221	2
	.048	.094	42		.048	.161	20		.048	.221	2
	.062	.096	41		.062	.166	19		.062	.228	1
	.075	.100	39		.075	.170	18		.075	.234	A
	.105	.102	38		.105	.173	17		.105	.234	A
	.134	.102	38		.134	.177	16		.134	.236	6mm
6–32	.037	.113	33	10–32	.037	.170	18	¼–28	.037	.224	5.7mm
	.048	.116	32		.048	.170	18		.048	.228	1
	.062	.116	32		.062	.170	18		.062	.232	5.9mm
	.075	.122	3.1mm		.075	.173	17		.075	.234	A
	.105	.125	⅛		.105	.177	16		.105	.238	B
	.134	.125	⅛		.134	.177	16		.134	.238	B
8–32	.037	.136	29	12–24	.037	.189	12	5⁄16–18	.037	.290	L
	.048	.144	27		.048	.194	10		.048	.290	L
	.062	.144	27		.062	.194	10		.062	.290	L
	.075	.147	26		.075	.199	8		.075	.295	M
	.105	.150	25		.105	.199	8		.105	.295	M
	.134	.150	25		.134	.199	8		.134	.295	M

All dimensions in inches except drill sizes. It may be necessary to vary the hole size to suit a particular application. * Type C is not recommended for new designs.

Table 10. Approximate Pierced or Extruded Hole Sizes for Types AB, B and BP Steel Thread Forming Screws*

Screw Size	Metal Thickness	Pierced or Extruded Hole Size	Screw Size	Metal Thickness	Pierced or Extruded Hole Size	Screw Size	Metal Thickness	Pierced or Extruded Hole Size
In Steel, Stainless Steel, Monel Metal, and Brass Sheet Metal								
4	.015	.086	7	.024	.120	10	.030	.157
	.018	.086		.030	.120		.036	.157
	.024	.098		.036	.120		.048	.157
	.030	.098		.048	.120	12	.024	.185
	.036	.098	8	.018	.136		.030	.185
6	.015	.111		.024	.136		.036	.185
	.018	.111		.030	.136		.048	.185
	.024	.111		.036	.136	1/4	.030	.209
	.030	.111		.048	.136		.036	.209
	.036	.111	10	.018	.157		.048	.209
7	.018	.120		.024	.157	...	...	...
In Aluminum Alloy Sheet Metal								
4	.024	.086	6	.048	.111	8	.036	.136
	.030	.086	7	.024	.120		.048	.136
	.036	.086		.030	.120	10	.024	.157
	.048	.086		.036	.120		.030	.157
6	.024	.111		.048	.120		.036	.157
	.030	.111	8	.024	.136		.048	.157
	.036	.111		.030	.136	...	...	...

See footnotes at bottom of page.

Table 11. Drilled Hole Sizes for Types AB, B and BP Steel Thread Forming Screws*

Screw Size	Hole Size*	Drill Size	Min. Mat'l Thickness	Penetration in Blind Holes Min.	Penetration in Blind Holes Max.	Screw Size	Hole Size*	Drill Size	Min. Mat'l Thickness	Penetration in Blind Holes Min.	Penetration in Blind Holes Max.
In Plywood (Resin Impregnated)						In Asbestos Compositions					
2	.073	49	.125	.188	.500	2	.076	48	.125	.188	.500
4	.100	39	.188	.250	.625	4	.101	38	.188	.250	.625
6	.125	1/8	.188	.250	.625	6	.120	31	.188	.250	.625
7	.136	29	.188	.250	.750	7	.136	29	.250	.312	.750
8	.144	27	.188	.250	.750	8	.147	26	.312	.375	.750
10	.173	17	.250	.312	1.000	10	.166	19	.312	.375	1.000
12	.194	10	.312	.375	1.000	12	.196	9	.312	.375	1.000
1/4	.228	1	.312	.375	1.000	1/4	.228	1	.438	.500	1.000
In Aluminum, Magnesium, Zinc, Brass, and Bronze Castings†						In Phenol Formaldehyde Plastics†					
2	.078	47	...	.125	...	2	.078	47	...	.188	...
4	.104	37	...	.188	...	4	.100	39	...	.250	...
6	.128	30	...	.250	...	6	.128	30	...	.250	...
7	.144	27	...	.250	...	7	.136	29	...	.250	...
8	.152	24	...	.250	...	8	.150	25	...	.312	...
10	.177	16	...	.250	...	10	.177	16	...	.312	...
12	.199	8	...	.281	...	12	.199	8	...	.375	...
1/4	.234	15/64	...	.312	...	1/4	.234	15/64	...	.375	...
In Cellulose Acetate and Nitrate, and Acrylic and Styrene Resins†											
2	.078	47	...	.188	...	8	.144	27	...	.312	...
4	.094	42	...	.250	...	10	.170	18	...	.312	...
6	.120	31	...	.250	...	12	.191	11	...	.375	...
7	.128	30	...	.250	...	1/4	.221	2	...	.375	...

All dimensions are given in inches except whole number screw and drill sizes.
* Since conditions differ widely, it may be necessary to vary the hole size to suit a particular application. † Data below apply to Types B and BP only.

Table 12. Approximate Drilled or Clean-Punched Hole Sizes for Types AB, B and BP Steel Thread Forming Screws*

In Steel, Stainless Steel, Monel Metal, and Brass Sheet Metal

Screw Size	Metal Thickness	Hole Size	Drill Size	Screw Size	Metal Thickness	Hole Size	Drill Size	Screw Size	Metal Thickness	Hole Size	Drill Size
2	.015	.064	52	7	.018	.116	32	10	.125	.170	18
	.018	.064	52		.024	.116	32		.135	.170	18
	.024	.067	51		.030	.116	32		.164	.173	17
	.030	.070	50		.036	.116	32	12	.024	.166	19
	.036	.073	49		.048	.120	31		.030	.166	19
	.048	.073	49		.060	.128	30		.036	.166	19
	.060	.076	48		.075	.136	29		.048	.170	18
4	.015	.086	44		.105	.140	28		.060	.177	16
	.018	.086	44	8	.024	.125	1/8		.075	.182	14
	.024	.089	43		.030	.125	1/8		.105	.185	13
	.030	.094	42		.036	.125	1/8		.125	.196	9
	.036	.094	42		.048	.128	30		.135	.196	9
	.048	.096	41		.060	.136	29		.164	.201	7
	.060	.100	39		.075	.140	28	1/4	.030	.194†	10†
	.075	.102	38		.105	.150	25		.036	.194†	10†
6	.015	.104	37		.125	.150	25		.048	.194†	10†
	.018	.104	37		.135	.152	24		.060	.199†	8†
	.024	.106	36	10	.024	.144	27		.075	.204†	6†
	.030	.106	36		.030	.144	27		.105	.209	4
	.036	.110	35		.036	.147	26		.125	.228	1
	.048	.111	34		.048	.152†	24†		.135	.228	1
	.060	.116	32		.060	.152†	24†		.164	.234	15/64
	.075	.120	31		.075	.157	22		.187	.234	15/64
	.105	.128	30		.105	.161	20		.194	.234	15/64

In Aluminum Alloy Sheet Metal

Screw Size	Metal Thickness	Hole Size	Drill Size	Screw Size	Metal Thickness	Hole Size	Drill Size	Screw Size	Metal Thickness	Hole Size	Drill Size
2	.024	.064	52	7	.060	.120	31	10	.164	.159	21
	.030	.064	52		.075	.128	30		.200 to .375	.166	19
	.036	.064	52		.105	.136	29	12	.048	.161	20
	.048	.067	51		.128 to .250	.136	29		.060	.166	19
	.060	.070	50	8	.030	.116	32		.075	.173	17
4	.030	.086	44		.036	.120	31		.105	.180	15
	.036	.086	44		.048	.128	30		.125	.182	14
	.048	.086	44		.060	.136	29		.135	.182	14
	.060	.089	43		.075	.140	28		.164	.189	12
	.075	.089	43		.105	.147	26		.200 to .375	.196	9
	.105	.094	42		.125	.147	26	1/4	.060	.199	8
6	.030	.104	37		.135	.149	25		.075	.201	7
	.036	.104	37		.162 to .375	.152	24		.105	.204	6
	.048	.104	37	10	.036	.144	27		.125	.209	4
	.060	.106	36		.048	.144	27		.135	.209	4
	.075	.110	35		.060	.144	27		.164	.213	3
	.105	.111	34		.075	.147	26		.187	.213	3
	.128 to .250	.120	31		.105	.147	26		.194	.221	2
7	.030	.113	33		.125	.154	23		.200 to .375	.228	1
	.036	.113	33		.135	.154	23				
	.048	.116	32								

All dimensions are given in inches except whole number screw and drill sizes.

* Since conditions differ widely, it may be necessary to vary the hole size to suit a particular application. Hole sizes for metal thicknesses above .075 inch are for Types B and BP only.

† For Types B and BP only; for Type AB see concluded Table 12 following.

Table 12 *(Concluded)*. **Supplementary Data for Type AB Thread Forming Screws in Steel, Stainless Steel, Monel Metal, and Brass Sheet Metals**

Screw Size	Thickness	Hole Size	Drill Size	Screw Size	Thickness	Hole Size	Drill Size	Screw Size	Thickness	Hole Size	Drill Size
10	0.018	0.144	27	¼	0.018	0.196	9	¼	0.048	0.205	5
10	0.048	0.149	25	¼	0.024	0.196	9	¼	0.060	0.228	1
10	0.060	0.154	23	¼	0.030	0.196	9	¼	0.075	0.232	5.9 mm
...	...	...	...	¼	0.036	0.196	9	...	...	...	...

All dimensions are given in inches except numbered screw and drill sizes. See also footnotes on previous page.

Table 13. Approximate Hole Sizes for Types D, F, G, and T Steel Thread Cutting Screws in Sheet Metals

Screw Size	Thickness	Steel		Aluminum Alloy		Screw Size	Thickness	Steel		Aluminum Alloy	
		Hole Size	Drill Size	Hole Size	Drill Size			Hole Size	Drill Size	Hole Size	Drill Size
2-56	0.050	0.073	49	0.070	50	8-32	0.187	0.150	25	0.147	26
	0.060	0.073	49	0.073	49		0.250	0.150	25	0.150	25
	0.083	0.073	49	0.073	49		0.312	0.150	25	0.150	25
	0.109	0.073	49	0.073	49	10-24	0.050	0.152	24	0.150	25
	0.125	0.076	48	0.073	49		0.060	0.154	23	0.152	24
	0.140	0.076	48	0.073	49		0.083	0.161	20	0.154	23
3-48	0.050	0.081	46	0.078	5/64		0.109	0.161	20	0.157	22
	0.060	0.081	46	0.081	46		0.125	0.166	19	0.159	21
	0.083	0.082	45	0.082	45		0.140	0.170	18	0.161	20
	0.109	0.086	44	0.082	45		0.187	0.173	17	0.166	19
	0.125	0.086	44	0.082	45		0.250	0.173	17	0.172	11/64
	0.140	0.086	44	0.086	44		0.312	0.173	17	0.173	17
	0.187	0.089	43	0.086	44		0.375	0.173	17	0.173	17
4-40	0.050	0.089	43	0.089	43	10-32	0.050	0.159	21	0.161	20
	0.060	0.089	43	0.089	43		0.060	0.166	19	0.161	20
	0.083	0.094	42	0.089	43		0.083	0.166	19	0.161	20
	0.109	0.096	41	0.094	42		0.109	0.170	18	0.166	19
	0.125	0.098	40	0.094	42		0.125	0.170	18	0.166	19
	0.140	0.098	40	0.094	3/32		0.140	0.170	18	0.166	19
	0.187	0.102	38	0.098	40		0.187	0.177	16	0.172	11/64
5-40	0.050	0.106	36	0.102	38		0.250	0.177	16	0.177	16
	0.060	0.106	36	0.102	38		0.312	0.177	16	0.177	16
	0.083	0.106	36	0.104	37		0.375	0.177	16	0.177	16
	0.109	0.106	36	0.104	37	12-24	0.060	0.180	15	0.177	16
	0,125	0.109	7/64	0.106	36		0.083	0.182	14	0.180	15
	0.140	0.110	35	0.106	36		0.109	0.188	3/16	0.182	14
	0.187	0.116	32	0.110	35		0.125	0.191	11	0.185	13
	0.250	0.116	32	0.113	33		0.140	0.191	11	0.188	3/16
6-32	0.050	0.110	35	0.109	7/64		0.187	0.199	8	0.191	11
	0.060	0.113	33	0.109	7/64		0.250	0.199	8	0.199	8
	0.083	0.116	32	0.111	34		0.312	0.199	8	0.199	8
	0.109	0.116	32	0.113	33		0.375	0.199	8	0.199	8
	0.125	0.116	32	0.116	32		0.500	0.199	8	0.199	8
	0.140	0.120	31	0.116	32	¼-20	0.083	0.213	3	0.206	5
	0.187	0.125	1/8	0.120	31		0.109	0.219	7/32	0.209	4
	0.250	0.125	1/8	0.125	1/8		0.125	0.221	2	0.213	3
8-32	0.050	0.136	29	0.136	29		0.140	0.221	2	0.213	3
	0.060	0.140	28	0.136	29		0.187	0.228	1	0.221	2
	0.083	0.140	28	0.136	29		0.250	0.228	1	0.228	1
	0.109	0.144	27	0.140	28		0.312	0.228	1	0.228	1
	0.125	0.144	27	0.140	28		0.375	0.228	1	0.228	1
	0.140	0.147	26	0.144	27		0.500	0.228	1	0.228	1

For footnotes see concluded Table 13 on following page.

Table 13 (*Concluded*). Approximate Hole Sizes for Types D, F, G, and T Steel Thread Cutting Screws in Sheet Metals

Screw Size	Thickness	Steel Hole Size	Steel Drill Size	Aluminum Alloy Hole Size	Aluminum Alloy Drill Size	Screw Size	Thickness	Steel Hole Size	Steel Drill Size	Aluminum Alloy Hole Size	Aluminum Alloy Drill Size
1/4-28	0.083	0.221	2	0.219	7/32	5/16-24	0.187	0.295	M	0.290	L
	0.109	0.228	1	0.221	2		0.250	0.295	M	0.295	M
	0.125	0.228	1	0.221	2		0.312	0.295	M	0.295	M
	0.140	0.234	A	0.221	2		0.375	0.295	M	0.295	M
	0.187	0.234	15/64	0.228	1		0.500	0.295	M	0.295	M
	0.250	0.234	15/64	0.234	15/64	3/8-16	0.125	0.339	R	0.328	21/64
	0.312	0.234	15/64	0.234	15/64		0.140	0.339	R	0.332	Q
	0.375	0.234	15/64	0.234	15/64		0.187	0.348	S	0.339	R
	0.500	0.234	15/64	0.234	15/64		0.250	0.358	T	0.348	S
5/16-18	0.109	0.277	J	0.266	H		0.312	0.358	T	0.348	S
	0.125	0.277	J	0.272	I		0.375	0.358	T	0.348	S
	0.140	0.281	9/32	0.272	I		0.500	0.358	T	0.348	S
	0.187	0.290	L	0.281	K	3/8-24	0.125	0.348	S	0.344	11/32
	0.250	0.290	L	0.290	L		0.140	0.348	S	0.344	11/32
	0.312	0.290	L	0.290	L		0.187	0.358	T	0.348	S
	0.375	0.290	L	0.290	L		0.250	0.358	T	0.358	T
	0.500	0.290	L	0.290	L		0.312	0.358	T	0.358	T
5/16-24	0.109	0.290	L	0.281	K		0.375	0.358	T	0.358	T
	0.125	0.290	L	0.281	9/32		0.500	0.358	T	0.358	T
	0.140	0.290	L	0.281	9/32		...	...	...	...	...

All dimensions are given in inches except numbered drill and screw sizes. It may be necessary to vary the hole size to suit a particular application.

Table 14. Approximate Hole Sizes for Types D, F, G, and T Steel Thread Cutting Screws in Cast Metals and Plastics

Screw Size	Thickness	Cast Iron Hole Size	Cast Iron Drill Size	Zinc and Aluminum* Hole Size	Zinc and Aluminum* Drill Size	Screw Size	Thickness	Cast Iron Hole Size	Cast Iron Drill Size	Zinc and Aluminum* Hole Size	Zinc and Aluminum* Drill Size
2-56	0.050	0.076	48	0.073	49	5-40	0.083	0.113	33	0.106	36
	0.060	0.076	48	0.073	49		0.109	0.113	33	0.110	35
	0.083	0.076	48	0.076	48		0.125	0.116	32	0.110	35
	0.109	0.078	5/64	0.076	48		0.140	0.116	32	0.110	35
	0.125	0.078	5/64	0.076	48		0.187	0.116	32	0.111	34
	0.140	0.078	5/64	0.076	48		0.250	0.116	32	0.113	33
3-48	0.050	0.089	43	0.082	45	6-32	0.050	0.120	31	0.116	32
	0.060	0.089	43	0.082	45		0.060	0.120	31	0.120	31
	0.083	0.089	43	0.082	45		0.083	0.125	1/8	0.120	31
	0.109	0.089	43	0.086	44		0.109	0.125	1/8	0.120	31
	0.125	0.089	43	0.089	43		0.125	0.125	1/8	0.120	31
	0.140	0.094	42	0.089	43		0.140	0.125	1/8	0.120	31
	0.187	0.094	42	0.089	43		0.187	0.128	30	0.120	31
4-40	0.050	0.100	39	0.090	41		0.250	0.128	30	0.120	31
	0.060	0.100	39	0.096	41	8-32	0.050	0.147	26	0.144	27
	0.083	0.102	38	0.096	41		0.060	0.150	25	0.144	27
	0.109	0.102	38	0.096	41		0.083	0.150	25	0.144	27
	0.125	0.102	38	0.100	39		0.109	0.150	25	0.144	27
	0.140	0.102	38	0.100	39		0.125	0.150	25	0.147	26
	0.187	0.104	37	0.100	39		0.140	0.150	25	0.147	26
5-40	0.050	0.111	34	0.106	36		0.187	0.154	23	0.147	26
	0.060	0.111	34	0.106	36		0.250	0.154	23	0.150	25
							0.312	0.154	23	0.150	25

For footnotes see Table 13 above.
* Die Castings

Table 14 (*Concluded*). Approximate Hole Sizes for Types D, F, G, and T Steel Thread Cutting Screws in Cast Metals and Plastics

Screw Size	Thickness	Cast Iron Hole Size	Cast Iron Drill Size	Zinc and Aluminum* Hole Size	Zinc and Aluminum* Drill Size	Screw Size	Thickness	Cast Iron Hole Size	Cast Iron Drill Size	Zinc and Aluminum* Hole Size	Zinc and Aluminum* Drill Size
10-24	0.050	0.170	18	0.161	20	1/4-28	0.083	0.234	A	0.228	I
	0.060	0.170	18	0.166	19		0.109	0.234	15/64	0.228	I
	0.083	0.172	11/64	0.166	19		0.125	0.234	15/64	0.228	I
	0.109	0.173	17	0.166	19		0.140	0.234	15/64	0.228	I
	0.125	0.173	17	0.166	19		0.187	0.238	B	0.228	I
	0.140	0.173	17	0.166	19		0.250	0.238	B	0.234	A
	0.187	0.177	16	0.170	18		0.312	0.238	B	0.234	A
	0.250	0.177	16	0.170	18		0.375	0.238	B	0.234	15/64
	0.312	0.177	16	0.172	11/64		0.500	0.238	B	0.234	15/64
	0.375	0.177	16	0.172	11/64	5/16-18	0.109	0.290	L	0.277	J
10-32	0.050	0.173	17	0.170	18		0.125	0.290	L	0.281	K
	0.060	0.173	17	0.170	18		0.140	0.290	L	0.281	K
	0.083	0.177	16	0.172	11/64		0.187	0.295	M	0.281	9/32
	0.109	0.177	16	0.172	11/64		0.250	0.295	M	0.281	9/32
	0.125	0.177	16	0.172	11/64		0.312	0.295	M	0.290	L
	0.140	0.177	16	0.172	11/64		0.375	0.295	M	0.290	L
	0.187	0.180	15	0.172	11/64		0.500	0.295	M	0.290	L
	0.250	0.180	15	0.173	17	5/16-24	0.109	0.295	M	0.290	L
	0.312	0.180	15	0.173	17		0.125	0.295	M	0.290	L
	0.375	0.180	15	0.177	16		0.140	0.295	M	0.290	L
12-24	0.060	0.196	9	0.189	12		0.187	0.302	N	0.290	L
	0.083	0.199	8	0.191	11		0.250	0.302	N	0.290	L
	0.109	0.199	8	0.191	11		0.312	0.302	N	0.295	M
	0.125	0.199	8	0.191	11		0.375	0.302	N	0.295	M
	0.140	0.199	8	0.194	10		0.500	0.302	N	0.295	M
	0.187	0.203	13/64	0.194	10	3/8-16	0.125	0.348	S	0.339	R
	0.250	0.204	6	0.196	9		0.140	0.348	S	0.339	R
	0.312	0.204	6	0.196	9		0.187	0.348	S	0.339	R
	0.375	0.204	6	0.199	8		0.250	0.348	S	0.344	11/32
	0.500	0.204	6	0.199	8		0.312	0.348	S	0.344	11/32
1/4-20	0.083	0.228	1	0.219	7/32		0.375	0.348	S	0.348	S
	0.109	0.228	1	0.219	7/32		0.500	0.348	S	0.348	S
	0.125	0.228	1	0.221	2	3/8-24	0.125	0.358	T	0.348	S
	0.140	0.228	1	0.221	2		0.140	0.358	T	0.348	S
	0.187	0.234	15/64	0.221	2		0.187	0.358	T	0.348	S
	0.250	0.234	15/64	0.228	1		0.250	0.358	T	0.358	T
	0.312	0.234	15/64	0.228	1		0.312	0.358	T	0.358	T
	0.375	0.234	15/64	0.228	1		0.375	0.358	T	0.358	T
	0.500	0.234	15/64	0.228	1		0.500	0.358	T	0.358	T

Screw Size	Phenol Formaldehyde† Hole Size	Phenol Formaldehyde† Drill Size	Depth of Penetration Min	Depth of Penetration Max	Cellulose Acetate, Cellulose Nitrate, Acrylic Resin, and Styrene Resin† Hole Size	Drill Size	Depth of Penetration Min	Depth of Penetration Max
2-56	0.078	5/64	0.219	0.375	0.076	48	0.219	0.375
3-48	0.089	43	0.219	0.375	0.086	44	0.219	0.375
4-40	0.098	40	0.250	0.312	0.093	42	0.250	0.312
5-40	0.113	33	0.250	0.438	0.110	35	0.250	0.438
6-32	0.116	32	0.250	0.312	0.116	32	0.250	0.312
8-32	0.144	27	0.312	0.500	0.144	27	0.312	0.500
10-24	0.161	20	0.375	0.500	0.161	20	0.375	0.500
10-32	0.166	19	0.375	0.500	0.166	19	0.375	0.500
1/4-20	0.228	1	0.375	0.625	0.228	1	0.375	1.000

For footnotes see Table 13.   * Die Castings   † Plastics

Table 15. Approximate Hole Sizes for Types BF and BT Steel Thread Cutting Screws in Cast Metals

			In Die Cast Zinc and Aluminum				
Screw Size	Thickness	Hole Size	Drill Size	Screw Size	Thickness	Hole Size	Drill Size
2	0.060	0.073	49	10	0.125	0.166	19
	0.083	0.073	49		0.140	0.166	19
	0.109	0.076	48		0.188	0.166	19
	0.125	0.076	48		0.250	0.170	18
	0.140	0.076	48		0.312	0.172	11/64
					0.375	0.172	11/64
3	0.060	0.086	44	12	0.125	0.191	11
	0.083	0.086	44		0.140	0.191	11
	0.109	0.086	44		0.188	0.191	11
	0.125	0.086	44		0.250	0.196	9
	0.140	0.089	43		0.312	0.196	9
	0.188	0.089	43		0.375	0.196	9
4	0.109	0.098	40	1/4	0.125	0.221	2
	0.125	0.100	39		0.140	0.221	2
	0.140	0.100	39		0.188	0.221	2
	0.188	0.100	39		0.250	0.228	1
	0.250	0.102	38		0.312	0.228	1
					0.375	0.228	1
5	0.109	0.111	34	5/16	0.125	0.281	K
	0.125	0.111	34		0.140	0.281	K
	0.140	0.113	33		0.188	0.281	K
	0.188	0.113	33		0.250	0.281	K
	0.250	0.116	32		0.312	0.290	L
					0.375	0.290	L
6	0.125	0.120	31	3/8	0.125	0.344	11/32
	0.140	0.120	31		0.140	0.344	11/32
	0.188	0.120	31		0.188	0.344	11/32
	0.250	0.125	1/8		0.250	0.344	11/32
	0.312	0.125	1/8		0.312	0.348	S
8	0.125	0.149	25		0.375	0.348	S
	0.140	0.149	25		...	...	...
	0.188	0.149	25				
	0.250	0.152	24				
	0.312	0.152	24				

All dimensions are given in inches except numbered drill and screw sizes. It may be necessary to vary the hole size to suit a particular application.

Table 16. Approximate Hole Size for Types BF and BT Steel Thread Cutting Screws in Plastics

Screw Size	Phenol Formaldehyde				Cellulose Acetate, Cellulose Nitrate, Acrylic Resin and Styrene Resin			
	Hole Size	Drill Size	Depth of Penetration		Hole Size	Drill Size	Depth of Penetration	
			Min	Max			Min	Max
2	0.078	5/64	0.094	0.250	0.076	48	0.094	0.250
3	0.089	43	0.125	0.312	0.089	43	0.125	0.312
4	0.104	37	0.125	0.312	0.100	39	0.125	0.312
5	0.116	32	0.188	0.375	0.113	33	0.188	0.375
6	0.125	1/8	0.188	0.375	0.120	31	0.188	0.375
8	0.147	26	0.250	0.500	0.144	27	0.250	0.500
10	0.170	18	0.312	0.625	0.166	19	0.312	0.625
12	0.194	10	0.375	0.625	0.189	12	0.375	0.625
1/4	0.228	1	0.375	0.750	0.221	2	0.375	0.750

For footnotes see above table.

Table 17. Approximate Hole Sizes for Type U Hardened Steel Metallic Drive Screws

In Ferrous and Non-Ferrous Castings, Sheet Metals, Plastics, Plywood (Resin-Impregnated) and Fiber								
Screw Size	Hole Size	Drill Size	Screw Size	Hole Size	Drill Size	Screw Size	Hole Size	Drill Size
00	.052	55	6	.120	31	12	.191	11
0	.067	51	7	.136	29	14	.221	2
2	.086	44	8	.144	27	5/16	.295	M
4	.104	37	10	.161	20	3/8	.358	T

All dimensions are given in inches except whole number screw and drill sizes and letter drill sizes.

Table 18. ANSI Standard Torsional Strength Requirements for Tapping Screws
(ANSI B18.6.4-1981)

Nom. Screw Size	Type A	Types AB, B, BF, BP, and BT	Types C, D, F, G, and T		Nom. Screw Size	Type A	Types AB, B, BF, BP, and BT	Types C, D, F, G, and T	
			Coarse Thread	Fine Thread				Coarse Thread	Fine Thread
2	4	4	5	6	1/4	...	142	140	179
3	9	9	9	10	16	152	...	...	...
4	12	13	13	15	18	196	...	...	...
5	18	18	18	20	5/16	...	290	306	370
6	24	24	23	27	20	250	...	...	...
7	30	30	...	...	24	492	...	...	...
8	39	39	42	47	3/8	...	590	560	710
10	48	56	56	74	7/16	...	620	700	820
12	83	88	93	108	1/2	...	1020	1075	1285
14	125	...	...	...					

Torsional strength data are in pound-inches.

Self-tapping Thread Inserts. —Self-tapping screw thread inserts are essentially hard bushings with internal and external threads. The internal threads conform to Unified and American standard classes 2B and 3B, depending on the type of insert used. The external thread has cutting edges on the end that provide the self-tapping feature. These inserts may be used in magnesium, aluminum, cast iron, zinc, plastics, and other materials. Self-tapping inserts are made of case-hardened carbon steel, stainless steel, and brass, the brass type being designed specifically for installation in wood.

Screw Thread Inserts. — Screw thread inserts are helically formed coils of diamond-shaped stainless steel or phosphor bronze wire that screw into a threaded hole to form a mating internal thread for a screw or stud. These inserts provide a convenient means of repairing stripped-out threads and are also used to provide stronger threads in soft materials such as aluminum, zinc die castings, wood, magnesium, etc. than can be obtained by direct tapping of the base metal involved.

According to the Heli-Coil Corp., conventional design practice in specifying boss diameters or edge distances can usually be applied since the major diameter of a hole tapped to receive a thread insert is not much larger than the major diameter of thread the insert provides.

Screw thread inserts are available in thread sizes from 4-40 to 1½-6 inch National and Unified Coarse Thread Series and in 6-40 to 1½-12 sizes in the fine-thread series. When used in conjunction with appropriate taps and gages, screw thread inserts will meet requirements of 2, 2B, 3 and 3B thread classes.

ANSI Standard Metric Thread Forming and Thread Cutting Tapping Screws. — Table 1 shows the various types of metric thread forming and thread cutting screw threads covered by ANSI Standard ANSI B18.6.5M-1986. The designations of the American National Standards Institute are shown.

Thread Forming Tapping Screws: These are generally for application in materials where large internal stresses are permissible or desirable, to increase resistance to loosening. These screws have the following descriptions and applications:

Type AB: Spaced thread screw with gimlet point primarily intended for use in thin metal, resin impregnated plywood, and asbestos compositions.

Type B: Spaced thread screw with a blunt point that has tapered entering threads with unfinished crests and same pitches as Type AB. Used for thin metal, nonferrous castings, resin impregnated plywood, certain resilient plastics, and asbestos compositions.

Thread Cutting Tapping Screws: These are generally for application in materials where disruptive internal stresses are undesirable or where excessive driving torques are encountered with thread forming tapping screws. These screws have the following descriptions and applications:

Types BF and BT: Spaced threads with blunt point and tapered entering threads having unfinished crests, as on Type B, with one or more cutting edges or chip cavities, intended for use in plastics, asbestos compositions, and other similar materials.

Types D, F, and T: Tapping screws with threads of machine screw diameter-pitch combinations (metric coarse thread series) approximating a 60 degree basic thread form (not necessarily conforming to any standard thread profile) with a blunt point and tapered entering threads with unfinished crests and having one or more cutting edges and chip cavities, intended for use in materials such as aluminum, zinc, and lead die castings; steel sheets and shapes; cast iron; brass; and plastics.

ANSI Standard Head Types for Metric Thread Forming and Cutting Tapping Screws. — The head types covered by ANSI B18.6.5M-1986 include those commonly recognized as being applicable to metric tapping screws and are described as follows:

Flat Countersunk Head: The flat countersunk head has a flat top surface and a conical bearing surface with a head angle of 90 to 92 degrees.

Oval Countersunk Head: The oval countersunk head has a rounded top surface and a conical bearing surface with a head angle of 90 to 92 degrees.

Pan Head: The slotted pan head has a flat top surface rounding into cylindrical sides and a flat bearing surface. The recessed pan head has a rounded top surface blending into cylindrical sides and a flat bearing surface.

Hex Head: The hex head has a flat or indented top surface, six flat sides, and a flat bearing surface.

Hex Flange Head: The hex flange head has a flat or indented top surface and six flat sides formed integrally with a frustroconical or slightly rounded (convex) flange which projects beyond the sides and provides a flat bearing surface.

Method of Designation. — Metric tapping screws are designated with the following data, preferably in the sequence shown: Nominal size; thread pitch; nominal length; thread and point type; product name, including head style and driving provision; material; and protective finish, if required.

Examples:

6.3 × 1.8 × 30 Type AB, Slotted Pan Head Tapping Screw, Steel, Zinc Plated

6 × 1 × 20 Type T, Type 1A Cross Recessed Pan Head Tapping Screw, Corrosion Resistant Steel

4.2 × 1.4 × 13 Type BF, Type 1 Cross Recessed Oval Countersunk Head Tapping Screw, Steel, Chromium Plated

10 × 1.5 × 40 Type D, Hex Flange Head Tapping Screw, Steel

Table 1. ANSI Standard Threads and Points for Metric Thread Forming and Thread Cutting Tapping Screws (ANSI B18.6.5M-1986)

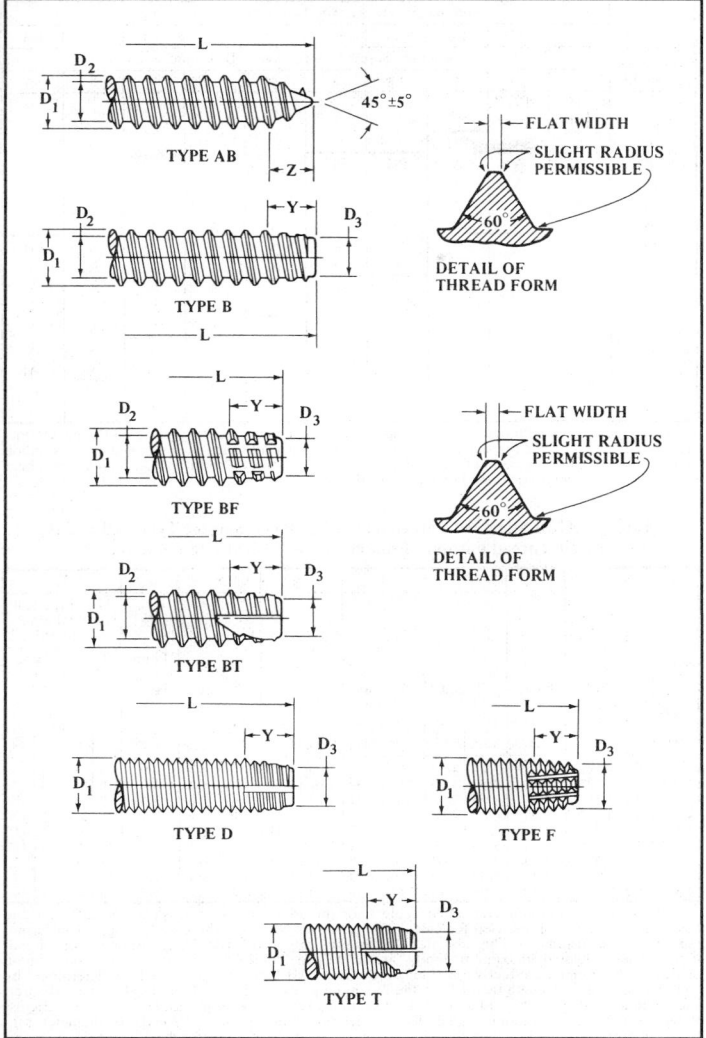

See Tables 3 and 4 for thread data.

Table 2. Recommended Nominal Screw Lengths for Metric Tapping Screws
(ANSI B18.6.5M-1986)

Nominal Screw Length[1]	Nominal Screw Size for Types AB, B, BF, and BT									
	2.2	—	2.9	3.5	4.2	4.8	5.5	6.3	8	9.5
	Nominal Screw Size for Types D, F, and T									
	2	2.5	3	3.5	4	5	—	6	8	10
4	PH	PH								
5	PH	PH								
6	A	A	PH							
8	A	A	A	PH	PH					
10	A	A	A	A	A	PH				
13	A	A	A	A	A	A	PH	PH		
16		A	A	A	A	A	A	A	PH	
20			A	A	A	A	A	A	A	PH
25				A	A	A	A	A	A	A
30						A	A	A	A	A
35							A	A	A	A
40								A	A	A
45									A	A
50									A	A
55										A
60										A

All dimensions in millimeters. [1] The nominal screw lengths included between the heavy lines are recommended for the respective screw sizes and screw head styles as designated by the symbols below: A — Signifies screws of all head styles covered in this standard. P — Signifies pan head screws. H — Signifies hex and hex flange head screws.

Table 3. ANSI Standard Thread and Point Dimensions for Types AB and B Metric Thread Forming Tapping Screws (ANSI B18.6.5M-1986)

Nominal Screw Size and Thread Pitch[6]	Basic Screw Diameter Ref[1]	Basic Thread Pitch Ref[1]	D_1 Thread Major Diameter Max	Min	D_2 Thread Minor Diameter Max	Min	D_3 Point Diameter[2] Max	Min	Y Point Taper Length Type B[3] Max	Min	Z Point Length Factor Type AB Ref[4]	L Min. Practical Nominal Screw Length[5] Type AB Note[7]	Type AB Note[8]	Type B Note[7]	Type B Note[8]
2.2 × 0.8	2.184	0.79	2.24	2.10	1.63	1.52	1.47	1.37	1.6	1.2	2.0	4	6	4	5
2.9 × 1	2.845	1.06	2.90	2.76	2.18	2.08	2.01	1.88	2.1	1.6	2.6	6	7	5	7
3.5 × 1.3	3.505	1.27	3.53	3.35	2.64	2.51	2.41	2.26	2.5	1.9	3.2	7	9	6	8
4.2 × 1.4	4.166	1.41	4.22	4.04	3.10	2.95	2.84	2.69	2.8	2.1	3.7	8	10	7	10
4.8 × 1.6	4.826	1.59	4.80	4.62	3.58	3.43	3.30	3.12	3.2	2.4	4.3	9	12	8	11
5.5 × 1.8	5.486	1.81	5.46	5.28	4.17	3.99	3.86	3.68	3.6	2.7	5.0	11	14	9	12
6.3 × 1.8	6.350	1.81	6.25	6.03	4.88	4.70	4.55	4.34	3.6	2.7	6.0	12	16	10	13
8 × 2.1	7.938	2.12	8.00	7.78	6.20	5.99	5.84	5.64	4.2	3.2	7.5	16	20	12	17
9.5 × 2.1	9.525	2.12	9.65	9.43	7.85	7.59	7.44	7.24	4.2	3.2	8.0	19	24	14	19

All dimensions in millimeters. See Table 1 for thread diagrams. [1] Basic screw diameter and basic thread pitch shall be used for calculation purposes wherever these factors appear in formulations for dimensions. [2] The tabulated values shall apply to screw blanks prior to roll threading. [3] The tabulated maximum limits are equal to approximately two times the thread pitch. [4] The minimum effective grip length on Type AB tapping screws shall be determined by subtracting the point length factor from the minimum screw length. [5] Lengths shown are theoretical minimums and are intended to assist the user in the selection of appropriate short screw lengths. Refer to Table 2 for recommended diameter-length combinations. [6] The body diameter (unthreaded portion) is not less than the minimum minor diameter nor greater than the maximum major diameter of the thread. [7] Pan, hex, and hex flange heads. [8] Flat and oval countersunk heads.

Table 4. ANSI Standard Thread and Point Dimensions for Types BF, BT, D, F, and T Metric Thread Cutting Tapping Screws (ANSI B18.6.5M-1986)

Types BF and BT												
Nominal Screw Size and Thread Pitch	Basic Screw Diameter Ref[1]	Basic Thread Pitch Ref[1]	D_1 Thread Major Diameter		D_2 Thread Minor Diameter		D_3 Point Diameter[2]		Y Point Taper Length[3]		L Minimal Practical Nominal Screw Length[4]	
			Max	Min	Max	Min	Max	Min	Max	Min	Pan, Hex and Hex Flange Heads	Flat and Oval Csunk Heads
2.2 × 0.8	2.184	0.79	2.24	2.10	1.63	1.52	1.47	1.37	1.6	1.2	4	5
2.9 × 1	2.845	1.06	2.90	2.76	2.18	2.08	2.01	1.88	2.1	1.6	5	7
3.5 × 1.3	3.505	1.27	3.53	3.35	2.64	2.51	2.41	2.26	2.5	1.9	6	8
4.2 × 1.4	4.166	1.41	4.22	4.04	3.10	2.95	2.84	2.69	2.8	2.1	7	10
4.8 × 1.6	4.826	1.59	4.80	4.62	3.58	3.43	3.30	3.12	3.2	2.4	8	11
5.5 × 1.8	5.486	1.81	5.46	5.28	4.17	3.99	3.86	3.68	3.6	2.7	9	12
6.3 × 1.8	6.350	1.81	6.25	6.03	4.88	4.70	4.55	4.34	3.6	2.7	10	13
8 × 2.1	7.938	2.12	8.00	7.78	6.20	5.99	5.84	5.64	4.2	3.2	12	17
9.5 × 2.1	9.525	2.12	9.65	9.43	7.85	7.59	7.44	7.24	4.2	3.2	14	19

Types D, F, and T											
Nominal Screw Size and Thread Pitch	D_1 Thread Major Diameter		D_3 Point Diameter[2]		D_S Body Diameter[5]	Y Point Taper Length				L Minimum Practical Nominal Screw Length[4]	
						For Short Screws		For Long Screws[6]			
	Max	Min	Max	Min	Min	Max	Min	Max	Min	Pan, Hex and Hex Flange Heads	Flat and Oval Csunk Heads
2 × 0.4	2.00	1.88	1.45	1.39	1.65	1.4	1.0	1.8	1.4	4	5
2.5 × 0.45	2.50	2.37	1.88	1.82	2.12	1.6	1.1	2.0	1.6	4	6
3 × 0.5	3.00	2.87	2.32	2.26	2.58	1.8	1.3	2.3	1.8	5	6
3.5 × 0.6	3.50	3.35	2.68	2.60	3.00	2.1	1.5	2.7	2.1	5	8
4 × 0.7	4.00	3.83	3.07	2.97	3.43	2.5	1.8	3.2	2.5	6	9
5 × 0.8	5.00	4.82	3.94	3.84	4.36	2.8	2.0	3.6	2.8	7	10
6 × 1	6.00	5.79	4.69	4.55	5.21	3.5	2.5	4.5	3.5	9	12
8 × 1.25	8.00	7.76	6.40	6.24	7.04	4.4	3.1	5.6	4.4	11	16
10 × 1.5	10.00	9.73	8.08	7.88	8.86	5.3	3.8	6.8	5.3	13	18

All dimensions in millimeters. See Table 1 for thread diagrams.
[1] Basic screw diameter and basic thread pitch are used for calculation purposes whenever these factors appear in formulations for dimensions.
[2] The tabulated values apply to screw blanks prior to roll threading.
[3] The tabulated maximum limits are equal to approximately two times the thread pitch.
[4] Lengths shown are theoretical minimums and are intended to assist in the selection of appropriate short screw lengths. See Table 2 for recommended length-diameter combinations. For Types D, F, and T, shorter screws are available with the point length reduced to the limits tabulated for short screws.
[5] Minimum limits for body diameter (unthreaded portion) are tabulated for convenient reference. For Types BF and BT, the body diameter is not less than the minimum minor diameter nor greater than the maximum major diameter of the thread.
[6] Long screws are screws of nominal lengths equal to or longer than those listed under L.

Material and Heat Treatment. — Tapping screws are normally fabricated from carbon steel and are usually processed to meet the performance and test requirements outlined in the standard, B18.6.5M. Tapping screws may also be made from corrosion resistant steel, monel, brass, and aluminum alloys. The materials, properties, and performance characteristics applicable to such screws should be mutually agreed upon between the manufacturer and the purchaser.

Table 5. Clearance Holes for Metric Tapping Screws
(ANSI B18.6.5M-1986, Appendix)

Nominal Screw Size and Thread Pitch	Basic Clearance Hole Diameter[1]			Nominal Screw Size and Thread Pitch	Basic Clearance Hole Diameter[1]		
	Close Clearance[2]	Normal Clearance (Preferred)[2]	Loose Clearance[2]		Close Clearance[2]	Normal Clearance (Preferred)[2]	Loose Clearance[2]
Types AB, B, BF, and BT				Types D, F, and T			
2.2 × 0.8	2.40	2.60	2.80	2 × 0.4	2.20	2.40	2.60
2.9 × 1	3.10	3.30	3.50	2.5 × 0.45	2.70	2.90	3.10
3.5 × 1.3	3.70	3.90	4.20	3 × 0.5	3.20	3.40	3.60
4.2 × 1.4	4.50	4.70	5.00	3.5 × 0.6	3.70	3.90	4.20
4.8 × 1.6	5.10	5.30	5.60	4 × 0.7	4.30	4.50	4.80
5.5 × 1.8	5.90	6.10	6.50	5 × 0.8	5.30	5.50	5.80
6.3 × 1.8	6.70	6.90	7.30	6 × 1	6.40	6.60	7.00
8 × 2.1	8.40	9.00	10.00	8 × 1.25	8.40	9.00	10.00
9.5 × 2.1	10.00	10.50	11.50	10 × 1.5	10.50	11.00	12.00

All dimensions in millimeters.
[1] The values given in this table are minimum limits. The recommended plus tolerances are as follows: for clearance hole diameters over 1.70 to and including 5.80 mm, plus 0.12, 0.20, and 0.30 mm for close, normal, and loose clearances, respectively; over 5.80 to and including 14.50 mm, plus 0.18, 0.30, and 0.45 mm for close, normal, and loose clearances, respectively.
[2] Normal clearance hole sizes are preferred. Close clearance hole sizes are for situations such as critical alignment of assembled components, wall thickness, or other limitations that necessitate the use of a minimal hole. Countersinking or counterboring at the fastener entry side may be necessary for the proper seating of the head. Loose clearance hole sizes are for applications where maximum adjustment capability between the components being assembled is necessary.

Table 6. Approximate Drilled or Clean-Punched Hole Sizes for Steel Type AB Metric Thread Forming Tapping Screws in Sheet Metal

Nominal Screw Size and Thread Pitch	Metal Thickness	Hole Size	Drill Size[1]	Nominal Screw Size and Thread Pitch	Metal Thickness	Hole Size	Drill Size[1]	Nominal Screw Size and Thread Pitch	Metal Thickness	Hole Size	Drill Size[1]
In Steel, Stainless Steel, Monel, and Brass Sheet Metal											
2.2 × 0.8	0.38	1.63	52	3.5 × 1.3	0.61	2.69	36	4.8 × 1.6	1.22	3.78	25
	0.46	1.63	52		0.76	2.69	36		1.52	3.91	23
	0.61	1.70	51		0.91	2.79	35		1.90	3.99	22
	0.76	1.78	50		1.22	2.82	34	5.5 × 1.8	0.46	...	...
	0.91	1.85	49		1.52	2.95	32		0.61	4.22	19
	1.22	1.85	49		1.90	3.05	31		0.76	4.22	19
	1.52	1.93	48	4.2 × 1.4	0.46	...	...		0.91	4.22	19
2.9 × 1	0.38	2.18	44		0.61	3.18	...		1.22	4.32	18
	0.46	2.18	44		0.76	3.18	...		1.52	4.50	16
	0.61	2.26	43		0.91	3.18	...		1.90	4.62	14
	0.76	2.39	42		1.22	3.25	30	6.3 × 1.8	0.46	4.98	9
	0.91	2.39	42		1.52	3.45	29		0.61	4.98	9
	1.22	2.44	41		1.90	3.56	28		0.76	4.98	9
	1.52	2.54	39	4.8 × 1.6	0.46	3.66	27		0.91	4.98	9
	1.90	2.59	38		0.61	3.66	27		1.22	5.21	W
3.5 × 1.3	0.38	2.64	37		0.76	3.66	27		1.52	5.79	1
	0.46	2.64	37		0.91	3.73	26		1.90	5.89	...

All dimensions in millimeters except drill sizes.
[1] Customary drill size references have been retained where the metric hole diameters are direct conversions of their decimal inch equivalents.

Table 6 (*Concluded*). **Approximate Drilled or Clean-Punched Hole Sizes for Steel Type AB Metric Thread Forming Tapping Screws in Sheet Metal**

Nominal Screw Size and Thread Pitch	Metal Thickness	Hole Size	Drill Size[1]	Nominal Screw Size and Thread Pitch	Metal Thickness	Hole Size	Drill Size[1]	Nominal Screw Size and Thread Pitch	Metal Thickness	Hole Size	Drill Size[1]
In Aluminum Alloy Sheet Metal											
2.2 × 0.8	0.38	...	...	3.5 × 1.3	0.61	...	...	4.8 × 1.6	1.22	3.66	27
	0.46	...	...		0.76	2.64	37		1.52	3.66	27
	0.61	1.63	52		0.91	2.64	37		1.90	3.73	26
	0.76	1.63	52		1.22	2.64	37	5.5 × 1.8	0.46	...	...
	0.91	1.63	52		1.52	2.69	36		0.61	...	...
	1.22	1.70	51		1.90	2.79	35		0.76	...	...
	1.52	1.78	50	4.2 × 1.4	0.46	...	...		0.91	...	...
2.9 × 1	0.38	...	...		0.61	...	...		1.22	4.09	20
	0.46	...	...		0.76	2.95	32		1.52	4.22	19
	0.61	...	...		0.91	3.05	31		1.90	4.39	17
	0.76	2.18	44		1.22	3.25	30	6.3 × 1.8	0.46	...	...
	0.91	2.18	44		1.52	3.45	29		0.61	...	...
	1.22	2.18	44		1.90	3.56	28		0.76	...	...
	1.52	2.26	43	4.8 × 1.6	0.46	...	...		0.91	...	...
	1.90	2.26	43		0.61	...	...		1.22	...	...
3.5 × 1.3	0.38	...	...		0.76	...	...		1.52	5.05	8
	0.46	...	...		0.91	3.66	27		1.90	5.11	7

All dimensions in millimeters except drill sizes.
[1] Customary drill size references have been retained where the metric hole diameters are direct conversions of their decimal inch equivalents.

Approximate Installation Hole Sizes for Metric Tapping Screws. — The approximate hole sizes given in Tables 6 through 10 provide general guidance in selecting holes for installing the respective types of metric thread forming and thread cutting tapping screws in various commonly used materials. Types AB, B, BF, and BT metric tapping screws are covered in these tables; hole sizes for Types D, F, and T metric thread cutting tapping screws are still under development.

Table 7. Approximate Hole Sizes for Steel Type AB Metric Thread Forming Tapping Screws in Plywoods and Asbestos

Nominal Screw Size and Thread Pitch	Hole Size	Drill Size[1]	Min Mat'l Thickness	Penetration in Blind Holes Min	Penetration in Blind Holes Max	Hole Size	Drill Size[1]	Min Mat'l Thickness	Penetration in Blind Holes Min	Penetration in Blind Holes Max
In Plywood (Resin Impregnated)						In Asbestos Compositions				
2.2 × 0.8	1.85	49	3.18	4.78	12.70	1.93	48	3.18	4.78	12.70
2.9 × 1	2.54	39	4.78	6.35	15.88	2.57	38	4.78	6.35	15.88
3.5 × 1.3	3.18	...	4.78	6.35	15.88	3.05	31	4.78	6.35	15.88
4.2 × 1.4	3.66	27	4.78	6.35	19.05	3.73	26	7.92	9.52	19.05
4.8 × 1.6	4.39	17	6.35	7.92	25.40	4.22	19	7.92	9.52	25.40
5.5 × 1.8	4.93	10	7.92	9.52	25.40	4.98	9	7.92	9.52	25.40
6.3 × 1.8	5.79	5	7.92	9.52	25.40	5.79	1	11.13	12.70	25.40

All dimensions in millimeters except drill sizes.
[1] Customary drill size references have been retained where the metric hole diameters are direct conversions of their decimal inch equivalents.

Table 8. Approximate Pierced or Extruded Hole Sizes for Steel Types AB and B Metric Thread Forming Tapping Screws

Nominal Screw Size and Thread Pitch	Metal Thickness	Hole Size	Nominal Screw Size and Thread Pitch	Metal Thickness	Hole Size	Nominal Screw Size and Thread Pitch	Metal Thickness	Hole Size
In Steel, Stainless Steel, Monel, and Brass Sheet Metal								
2.9 × 1	0.38	2.18	4.2 × 1.4	0.46	3.45	5.5 × 1.8	0.61	4.70
	0.46	2.18		0.61	3.45		0.76	4.70
	0.61	2.49		0.76	3.45		0.91	4.70
	0.76	2.49		0.91	3.45		1.22	4.70
	0.91	2.49		1.22	3.45		...	...
3.5 × 1.3	0.38	2.82	4.8 × 1.6	0.46	3.99	6.3 × 1.8	0.76	5.31
	0.46	2.82		0.61	3.99		0.91	5.31
	0.61	2.82		0.76	3.99		1.22	5.31
	0.76	2.82		0.91	3.99		...	...
	0.91	2.82		1.22	3.99		...	...
In Aluminum Alloy								
2.9 × 1	0.61	2.18	3.5 × 1.3	0.91	2.82	4.8 × 1.6	0.61	3.99
	0.76	2.18		1.22	2.82		0.76	3.99
	0.91	2.18	4.2 × 1.4	0.61	3.45		0.91	3.99
	1.22	2.18		0.76	3.45		1.22	3.99
3.5 × 1.3	0.61	2.82		0.91	3.45			
	0.76	2.82		1.22	3.45			

All dimensions in millimeters.

Table 9. Approximate Drilled or Clean-Punched Hole Sizes for Steel Type B Metric Thread Forming Tapping Screws in Sheet Metal and Cast Metals

Nominal Screw Size and Thread Pitch	Metal Thickness	Hole Size	Drill Size[1]	Nominal Screw Size and Thread Pitch	Metal Thickness	Hole Size	Drill Size[1]	Nominal Screw Size and Thread Pitch	Metal Thickness	Hole Size	Drill Size[1]
In Steel, Stainless Steel, Monel, and Brass Sheet Metal											
2.2 × 0.8	0.38	1.63	52	3.5 × 1.3	1.90	3.05	31	5.5 × 1.8	0.61	4.22	19
	0.46	1.63	52		2.67	3.25	30		0.76	4.22	19
	0.61	1.70	51	4.2 × 1.4	0.61	3.18	...		0.91	4.22	19
	0.76	1.78	50		0.76	3.18	...		1.22	4.32	18
	0.91	1.85	49		0.91	3.18	...		1.52	4.50	16
	1.22	1.85	49		1.22	3.25	30		1.90	4.62	14
	1.52	1.93	48		1.52	3.45	29		2.67	4.70	13
2.9 × 1	0.38	2.18	44		1.90	3.56	28		3.18	4.98	9
	0.46	2.18	44		2.67	3.81	25		3.43	4.98	9
	0.61	2.26	43		3.18	3.81	25		4.17	5.11	7
	0.76	2.39	42		3.43	3.86	24	6.3 × 1.8	0.76	4.93	10
	0.91	2.39	42	4.8 × 1.6	0.61	3.66	27		0.91	4.93	10
	1.22	2.44	41		0.76	3.66	27		1.22	4.93	10
	1.52	2.54	39		0.91	3.73	26		1.52	5.05	8
	1.90	2.59	38		1.22	3.86	24		1.90	5.18	6
3.5 × 1.3	0.38	2.64	37		1.52	3.86	24		2.67	5.31	4
	0.46	2.64	37		1.90	3.99	22		3.18	5.79	1
	0.61	2.69	36		2.67	4.09	20		3.43	5.79	1
	0.76	2.69	36		3.18	4.32	18		4.17	5.94	...
	0.91	2.79	35		3.43	4.32	18		4.75	5.94	...
	1.22	2.82	34		4.17	4.39	17		4.93	5.94	...
	1.52	2.95	32								

All dimensions in millimeters except drill sizes.   [1] Customary drill size references have been retained where the metric hole diameters are direct conversions of their decimal inch equivalents.

Table 9 *(Concluded)*. **Approximate Drilled or Clean-Punched Hole Sizes for Steel Type B Metric Thread Forming Tapping Screws in Sheet Metal and Cast Metals**

Nominal Screw Size and Thread Pitch	Metal Thickness	Hole Size	Drill Size[1]	Nominal Screw Size and Thread Pitch	Metal Thickness	Hole Size	Drill Size[1]	Nominal Screw Size and Thread Pitch	Metal Thickness	Hole Size	Drill Size[1]
In Aluminum Alloy Sheet Metal											
2.2 × 0.8	0.61	1.63	52	4.2 × 1.4	0.76	2.95	32	5.5 × 1.8	1.22	4.09	20
	0.76	1.63	52		0.91	3.05	31		1.52	4.22	19
	0.91	1.63	52		1.22	3.25	30		1.90	4.39	17
	1.22	1.70	51		1.52	3.45	29		2.67	4.57	15
	1.52	1.78	50		1.90	3.56	28		3.18	4.62	14
2.9 × 1	0.76	2.18	44		2.67	3.73	26		3.43	4.62	14
	0.91	2.18	44		3.18	3.73	26		4.17	4.80	12
	1.22	2.18	44		3.43	3.78	25		5.08 to 9.52	4.98	9
	1.52	2.26	43		4.11 to 9.52	3.86	24	6.3 × 1.8	1.52	5.05	8
	1.90	2.26	43	4.8 × 1.6	0.91	3.66	27		1.90	5.11	7
	2.67	2.39	42		1.22	3.66	27		2.67	5.18	6
3.5 × 1.3	0.76	2.64	37		1.52	3.66	27		3.18	5.31	4
	0.91	2.64	37		1.90	3.73	26		3.43	5.31	4
	1.22	2.64	37		2.67	3.73	26		4.17	5.41	3
	1.52	2.69	36		3.18	3.91	23		4.75	5.41	3
	1.90	2.79	35		3.43	3.91	23		4.93	5.61	2
	2.67	2.82	34		4.17	4.04	21		5.08 to 9.52	5.79	1
	3.25 to 6.25	3.05	31		5.08 to 9.52	4.22	19				

In Aluminum, Magnesium, Zinc, Brass, and Bronze Cast Metals

Nominal Screw Size and Thread Pitch	Hole Size	Drill Size[1]	Min Penetration in Blind Holes	Nominal Screw Size and Thread Pitch	Hole Size	Drill Size[1]	Min Penetration in Blind Holes
2.2 × 0.8	1.98	47	3.18	4.8 × 1.6	4.50	16	6.35
2.9 × 1	2.64	37	4.78	5.5 × 1.8	5.05	8	7.14
3.5 × 1.3	3.25	30	6.35	6.3 × 1.8	5.94	4	7.92
4.2 × 1.4	3.86	24	6.35	...	...	...	...

All dimensions in millimeters, except drill sizes.
[1] Customary drill size references have been retained where the metric hole diameters are direct conversions of their decimal inch equivalents.

Table 10. Approximate Hole Sizes for Steel Type B Metric Thread Forming Tapping Screws in Plywoods, Asbestos, and Plastics

Nominal Screw Size and Thread Pitch	Hole Size	Drill Size[1]	Min Mat'l Thickness	Penetration in Blind Holes Min	Penetration in Blind Holes Max	Nominal Screw Size and Thread Pitch	Hole Size	Drill Size[1]	Min Mat'l Thickness	Penetration in Blind Holes Min	Penetration in Blind Holes Max
In Plywood (Resin Impregnated)											
2.2 × 0.8	1.85	49	3.18	4.78	12.70	4.8 × 1.6	4.39	17	6.35	7.92	25.40
2.9 × 1	2.54	39	4.78	6.35	15.88	5.5 × 1.8	4.93	10	7.92	9.52	25.40
3.5 × 1.3	3.18	...	4.78	6.35	15.88	6.3 × 1.8	5.79	1	7.92	9.52	25.40
4.2 × 1.4	3.66	27	4.78	6.35	19.05	...	...	...	...	...	...

See footnotes at end of table.

Table 10 (*Concluded*). **Approximate Hole Sizes for Steel Type B Metric Thread Forming Tapping Screws in Plywoods, Asbestos, and Plastics**

Nominal Screw Size and Thread Pitch	Hole Size	Drill Size[1]	Min Mat'l Thickness	Penetration in Blind Holes	
				Min	Max
In Asbestos Compositions					
2.2 × 0.8	1.93	48	3.18	4.78	12.70
2.9 × 1	2.57	38	4.78	6.35	15.88
3.5 × 1.3	3.05	31	4.78	6.35	15.88
4.2 × 1.4	3.73	26	7.92	9.52	19.05
4.8 × 1.6	4.22	19	7.92	9.52	25.40
5.5 × 1.8	4.98	9	7.92	9.52	25.40
6.3 × 1.8	5.79	1	11.13	12.70	25.40

Nominal Screw Size and Thread Pitch	Hole Size	Drill Size[1]	Min Penetration in Blind Holes	Hole Size	Drill Size[1]	Min Penetration in Blind Holes
In Phenol Formaldehyde				In Cellulose Acetate & Nitrate, Acrylic and Styrene Resins		
2.2 × 0.8	1.98	47	4.78	1.98	47	4.78
2.9 × 1	2.54	39	6.35	2.39	42	6.35
3.5 × 1.3	3.25	30	6.35	3.05	32	6.35
4.2 × 1.4	3.81	25	7.92	3.66	27	7.92
4.8 × 1.6	4.50	16	7.92	4.32	18	7.92
5.5 × 1.8	5.05	8	9.52	4.85	11	9.52
6.3 × 1.8	5.94	. . .	9.52	5.61	2	9.52

All dimensions in millimeters except drill sizes.
[1] Customary drill size references have been retained where the metric hole diameters are direct conversions of their decimal inch equivalents.

Table 11. Approximate Hole Sizes for Steel Types BF and BT Metric Thread Cutting Tapping Screws for Cast Metals and Plastics

Nominal Screw Size and Thread Pitch	Material Thickness	Hole Size	Drill Size[1]	Nominal Screw Size and Thread Pitch	Material Thickness	Hole Size	Drill Size[1]
In Die Cast Zinc and Aluminum							
2.2 × 0.8	1.52	1.85	49	3.5 × 1.3	3.18	3.05	31
	2.11	1.85	49		3.56	3.05	31
	2.77	1.93	48		4.78	3.05	31
	3.18	1.93	48		6.35	3.18	. . .
	3.56	1.93	48		7.92	3.18	. . .
2.9 × 1	2.77	2.49	40	4.2 × 1.4	3.18	3.78	25
	3.18	2.54	39		3.56	3.78	25
	3.56	2.54	39		4.78	3.78	25
	4.78	2.54	39		6.35	3.86	24
	6.35	2.59	38		7.92	3.86	24

All dimensions in millimeters except drill sizes.
[1] Customary drill size references have been retained where the metric hole diameters are direct conversions of their decimal inch equivalents.

Table 11 (*Concluded*). **Approximate Hole Sizes for Steel Types BF and BT Metric Thread Cutting Tapping Screws for Cast Metals and Plastics**

Nominal Screw Size and Thread Pitch	Material Thickness	Hole Size	Drill Size[1]	Nominal Screw Size and Thread Pitch	Material Thickness	Hole Size	Drill Size[1]
In Die Cast Zinc and Aluminum							
4.8 × 1.6	3.18	4.22	19	6.3 × 1.8	6.35	5.79	I
	3.56	4.22	19		7.92	5.79	I
	4.78	4.22	19		9.52	5.79	I
	6.35	4.32	18	8 × 2.1	3.18	7.14	K
	7.92	4.37	...		3.56	7.14	K
	9.52	4.37	...		4.78	7.14	K
5.5 × 1.8	3.18	4.85	11		6.35	7.14	K
	3.56	4.85	11		7.92	7.37	L
	4.78	4.85	11		9.52	7.37	L
	6.35	4.98	9	9.5 × 2.1	3.18	8.74	...
	7.92	4.98	9		3.56	8.74	...
	9.52	4.98	9		4.78	8.74	...
6.3 × 1.8	3.18	5.61	2		6.35	8.74	...
	3.56	5.61	2		7.92	8.84	S
	4.78	5.61	2		9.52	8.84	S

Nominal Screw Size and Thread Pitch	Hole Size	Drill Size[1]	Depth of Penetration	
			Min	Max
In Phenol Formaldehyde				
2.2 × 0.8	1.98	...	2.39	6.35
2.9 × 1	2.64	37	3.18	7.92
3.5 × 1.3	3.18	...	4.78	9.52
4.2 × 1.4	3.73	26	6.35	12.70
4.8 × 1.6	4.32	18	7.92	15.88
5.5 × 1.8	4.93	10	9.52	15.88
6.3 × 1.8	5.79	I	9.52	19.05
In Cellulose Acetate and Nitrate, Acrylic and Styrene Resins				
2.2 × 0.8	1.93	48	2.39	6.35
2.9 × 1	2.54	39	3.18	7.92
3.5 × 1.3	3.05	31	4.78	9.52
4.2 × 1.4	3.66	27	6.35	12.70
4.8 × 1.6	4.22	19	7.92	15.88
5.5 × 1.8	4.80	12	9.52	15.88
6.3 × 1.8	5.61	2	9.52	19.05

All dimensions in millimeters except drill sizes.
[1] Customary drill size references have been retained where the metric hole sizes are direct conversions of their decimal inch equivalents.

The finish (plating or coating) on metric tapping screws and the material composition and hardness of the mating component are factors that affect assembly torques in individual applications. Although the recommended installation hole sizes given in Tables 6 through 10 were based on the use of plain unfinished carbon steel metric tapping screws, experience has shown that the specified holes are also suitable for screws having most types of commercial finishes. However, owing to various finishes providing different degrees of lubricity, some adjustment of installation torques may be necessary to suit individual applications. Also, where exceptionally heavy finishes are involved or screws are to be assembled into materials of higher hardness, some deviation from the specified hole sizes may be required to provide optimum assembly. The necessity and extent of such deviations can best be determined by experiment in the particular assembly environment.

Table 1. American National Standard T-Slots (ANSI B5.1M-1985)

BASIC DIMENSIONS — A₁, D₁, B₁, C₁

T-SLOTS — W₁, U₁, R₁, 45°

SUGGESTED APPROXIMATE DIMENSIONS FOR ROUNDING OR BREAKING OF CORNERS

Nominal T-Bolt Size		Width of Throat A_1		Headspace Dimensions								Depth of Throat D_1				Rounding or Breaking of Corners					
				Width B_1				Depth C_1								inch			mm		
				inch		mm		inch		mm		inch		mm		R_1 max	W_1 max	U_1 max	R_1 max	W_1 max	U_1 max
inch	mm	inch	mm	min	max	min	max	min	max	min	max	min	max	min	max						
	4		5			10	11			3	3.5			4.5	7				0.5	0.8	0.8
	5		6			11	12.5			5	6			5	8				0.5	0.8	0.8
0.250	6	0.282	8	0.500	0.562	14.5	16	0.203	0.234	6	8	0.125	0.375	7	11	0.02	0.02		0.5	0.8	0.8
0.312	8	0.344	10	0.594	0.656	16	18	0.234	0.266	7	8	0.156	0.438	9	14	0.02	0.03	0.03	0.5	0.8	0.8
0.375	10	0.438	12	0.719	0.781	19	21	0.297	0.328	7	9	0.219	0.562	11	17	0.02	0.03	0.03	0.5	0.8	0.8
0.500	12	0.562	14	0.906	0.969	23	25	0.359	0.391	9	11	0.312	0.688	16	19	0.02	0.03	0.03	0.8	0.8	1.3
0.625	16	0.688	18	1.188	1.250	30	32	0.453	0.484	12	14	0.438	0.875	20	24	0.03	0.03	0.05	0.8	0.8	1.3
0.750	20	0.812	22	1.375	1.469	37	40	0.594	0.625	16	18	0.562	1.062	26	29	0.03	0.03	0.05	0.8	1.5	1.3
1.000	24	1.062	28	1.750	1.844	46	50	0.781	0.828	20	22	0.750	1.250	33	36	0.03	0.06	0.05	0.8	1.5	1.3
1.250	30	1.312	36	2.125	2.219	56	60	1.031	1.094	25	28	1.000	1.562	39	46	0.03	0.06		1.5	2.5	2
1.500	36	1.562	42	2.562	2.656	68	72	1.281	1.344	32	35	1.250	1.938	44	53		0.06		1.5	2.5	2
	42		48			80	85			36	40			50	59						
	48		54			90	95			40	44				66						

* Throat dimensions are basic. When slots are intended to be used for holding only, tolerances can be 0.0 + 0.010 inch or H12 Metric (ISO/R286); when intended for location, tolerance can be 0.0 + 0.001 inch or H8 Metric. B5.1M.

† Width of tongue (tenon) to be used with the above T-Slots will be found in the complete standard. B5.1M.

§ Corners of T-Slots may be square or may be rounded or broken to the indicated maximum dimensions at the manufacturer's option. B5.1M.

For the dimensions of tongue seats, inserted tongues, and solid tongues refer to the complete standard, B5.1M.

Table 2. American National Standard T-Bolts (ANSI B5.1M-1985)

Nominal T-Bolt Size and Thread A_2*§		Bolt Head Dimensions										Rounding of Corners†			
		Width Across Flats B_2				Width Across Corners		Height C_2				R_2		W_2	
		inch		mm		inch	mm	inch		mm		inch	mm	inch	mm
inch UNC-2A	metric ISO**	max	min	max	min	max	max	max	min	max	min	max	max	max	max
0.250–20	M4	0.469	0.438	9	8.5	0.663	12.7	0.156	0.141	2.5	2.1		0.3		0.5
0.312–18	M5	0.562	0.531	10	9.5	0.796	14.1	0.188	0.172	4	3.6		0.3	0.03	0.5
0.375–16	M6	0.688	0.656	13	12	0.972	18.4	0.250	0.234	6	5.6	0.02	0.5	0.03	0.8
0.500–13	M8	0.875	0.844	15	14	1.238	21.2	0.312	0.297	6	5.6	0.02	0.5	0.03	0.8
0.625–11	M10	1.125	1.094	18	17	1.591	25.5	0.406	0.391	7	6.6	0.02	0.5	0.06	0.8
0.750–10	M12	1.312	1.281	22	21	1.856	31.1	0.531	0.500	8	7.6	0.03	0.8	0.06	1.5
1.000–8	M16	1.688	1.656	28	27	2.387	39.6	0.688	0.656	10	9.6	0.03	0.8	0.06	1.5
1.250–7	M20	2.062	2.031	34	33	2.917	48.1	0.938	0.906	14	13.2	0.03	0.8	0.06	1.5
1.500–6	M24	2.500	2.469	43	42	3.536	60.8	1.188	1.156	18	17.2	0.03	0.8	0.06	1.5
	M30			53	52		75			23	22.2		0.8		1.5
	M36			64	63		90.5			28	27.2		0.8		1.5
	M42			75	74		106.1			32	30.5		1		2
	M48			85	84		120.2			36	34.5		1		2

* For inch tolerances for thread diameters of bolts or studs and for threads see page 1498.
† Corners of T-bolts may be square or may be rounded or broken to the indicated maximum dimensions at the manufacturer's option.
§ T-slots to be used with these bolts will be found in Table 1.
** Metric thread grade and tolerance position is 5g 6g (see page 1529).

Table 3. American National Standard T-Nuts (ANSI B5.1M-1985)

T-NUTS

(Diagram showing T-nut cross-sections with dimensions labeled: E_3, L_3, W_3, A_3, B_3, K_3, C_3, R_3.)

Nominal T-Bolt Size§		Width of Tongue A_3				Tap for Stud† E_3		Width of Nut B_3				Height of Nut C_3				Total Thickness Including Tongue* K_3		Length of Nut* L_3		Rounding of Corners			
		inch		mm		inch		inch		mm		inch		mm		inch	mm	inch	mm	R_3		W_3	
inch	mm	max	min	max	min	UNC-3B	ISO** mm	max	min	max	min	max	min	max	min					inch max	mm max	inch max	mm max
	4																						
	5																						
0.250	6					0.250-20	M6																
0.312	8	0.330	0.320	8.7	8.5	0.312-18	M8	0.562	0.531	15	14	0.188	0.172	6	5.6	0.281	9	0.562	18	0.02	0.5	0.03	0.8
0.375	10	0.418	0.408	11	10.75	0.375-16	M10	0.688	0.656	18	17	0.250	0.234	7	6.6	0.375	10.5	0.688	20	0.02	0.5	0.03	0.8
0.500	12	0.543	0.533	13.5	13.25	0.500-13	M12	0.875	0.844	22	21	0.312	0.297	8	7.6	0.531	12	0.875	23	0.02	0.5	0.06	1.5
0.625	16	0.668	0.658	17.25	17	0.625-11	M16	1.125	1.094	28	27	0.406	0.391	10	9.6	0.781	15	1.125	27	0.03	0.8	0.06	1.5
0.750	20	0.783	0.773	20.5	20.25	0.750-10	M20	1.312	1.281	34	33	0.531	0.500	14	13.2		21	1.312	35	0.03	0.8	0.06	1.5
1.000	24	1.033	1.018	26.5	26	1.000-8	M24	1.688	1.656	43	42	0.688	0.656	18	17.2	1.000	27	1.688	46	0.03	0.8	0.06	1.5
1.250	30	1.273	1.258	33	32.5	1.250-7	M30	2.062	2.031	53	52	0.938	0.906	23	22.2	1.312	34	2.062	53	0.03	0.8	0.06	1.5
1.500	36	1.523	1.508	39.25	38.75		M36	2.500	2.469	64	63	1.188	1.156	28	27.2	1.625	42	2.500	65	0.03	0.8	0.06	1.5
	42			46.75	46.25		M42			75	74			32	30.5		48		75		1		2
	48			52.5	51.75					85	84			36	34.5		54		85		1		2

* There are no tolerances given for ''Total Thickness'' or ''Nut Length'' as they need not be held to close limits.
† For tolerances of inch threads see page 1498.
§ T-slot dimensions to fit the above nuts will be found in Table 1.
** Metric tapped thread grade and tolerance position is 5H (see page 1529).

American National Standard Cotter Pins (ANSI B18.8.1-1972, R1977)

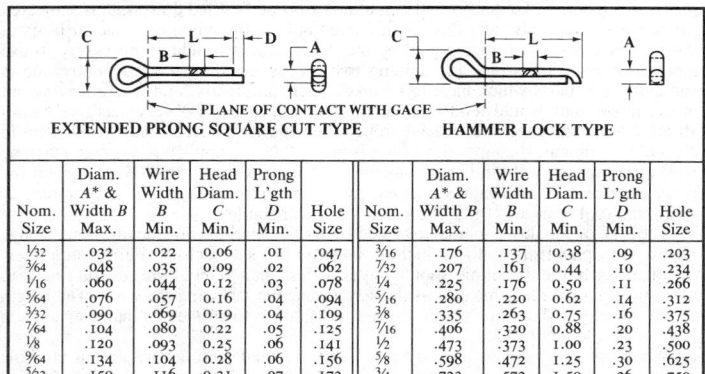

EXTENDED PRONG SQUARE CUT TYPE   **HAMMER LOCK TYPE**

Nom. Size	Diam. A^* & Width B Max.	Wire Width B Min.	Head Diam. C Min.	Prong L'gth D Min.	Hole Size	Nom. Size	Diam. A^* & Width B Max.	Wire Width B Min.	Head Diam. C Min.	Prong L'gth D Min.	Hole Size
1/32	.032	.022	0.06	.01	.047	3/16	.176	.137	0.38	.09	.203
3/64	.048	.035	0.09	.02	.062	7/32	.207	.161	0.44	.10	.234
1/16	.060	.044	0.12	.03	.078	1/4	.225	.176	0.50	.11	.266
5/64	.076	.057	0.16	.04	.094	5/16	.280	.220	0.62	.14	.312
3/32	.090	.069	0.19	.04	.109	3/8	.335	.263	0.75	.16	.375
7/64	.104	.080	0.22	.05	.125	7/16	.406	.320	0.88	.20	.438
1/8	.120	.093	0.25	.06	.141	1/2	.473	.373	1.00	.23	.500
9/64	.134	.104	0.28	.06	.156	5/8	.598	.472	1.25	.30	.625
5/32	.156	.116	0.31	.07	.172	3/4	.723	.572	1.50	.36	.750

All dimensions are given in inches. * Tolerances are: − .004 inch for the 1/32- to 3/16-inch sizes, incl.; − .005 inch for the 7/32- to 5/16-inch sizes, incl.; − .006 inch for the 3/8- to 1/2- inch sizes, incl.; and − .008 inch for the 5/8- and 3/4-inch sizes.

American National Standard Clevis Pins (ANSI B18.8.1-1972, R1977)

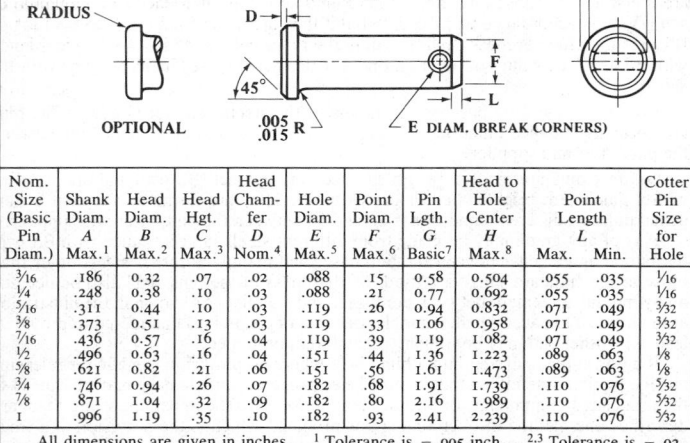

Nom. Size (Basic Pin Diam.)	Shank Diam. A Max.[1]	Head Diam. B Max.[2]	Head Hgt. C Max.[3]	Head Chamfer D Nom.[4]	Hole Diam. E Max.[5]	Point Diam. F Max.[6]	Pin Lgth. G Basic[7]	Head to Hole Center H Max.[8]	Point Length L Max.	Point Length L Min.	Cotter Pin Size for Hole
3/16	.186	0.32	.07	.02	.088	.15	0.58	0.504	.055	.035	1/16
1/4	.248	0.38	.10	.03	.088	.21	0.77	0.692	.055	.035	1/16
5/16	.311	0.44	.10	.03	.119	.26	0.94	0.832	.071	.049	3/32
3/8	.373	0.51	.13	.03	.119	.33	1.06	0.958	.071	.049	3/32
7/16	.436	0.57	.16	.04	.119	.39	1.19	1.082	.071	.049	3/32
1/2	.496	0.63	.16	.04	.151	.44	1.36	1.223	.089	.063	1/8
5/8	.621	0.82	.21	.06	.151	.56	1.61	1.473	.089	.063	1/8
3/4	.746	0.94	.26	.07	.182	.68	1.91	1.739	.110	.076	5/32
7/8	.871	1.04	.32	.09	.182	.80	2.16	1.989	.110	.076	5/32
1	.996	1.19	.35	.10	.182	.93	2.41	2.239	.110	.076	5/32

All dimensions are given in inches. [1] Tolerance is − .005 inch. [2,3] Tolerance is − .02 inch. [4] Tolerance is ± .01 inch. [5] Tolerance is − .015 inch. [6] Tolerance is − .01 inch. [7] Lengths tabulated are intended for use with standard clevises, without spacers. When required, it is recommended that other pin lengths be limited wherever possible to nominal lengths in .06-inch increments. [8] Tolerance is − .020 inch.

Dowel-Pins. — Dowel-pins are used either to retain parts in a fixed position or to preserve alignment. Under normal conditions a properly fitted dowel-pin is subjected to shearing strain only, and this strain occurs only at the junction of the surfaces of the two parts which are being held by the dowel-pin. It is seldom necessary to use more than two dowel-pins for holding two pieces together and frequently one is sufficient. For parts which have to be taken apart frequently, and where driving out of the dowel-pins would tend to wear the holes, and also for very accurately constructed tools and gages which have to be taken apart, or which require to be kept in absolute alignment, the taper dowel-pin is preferable. As applied to average machine work, the taper dowel-pin is most commonly used but the straight type is given the preference on tool and gage work, except where extreme accuracy is required, or where the tool or gage is to be subjected to rough handling.

The size of the dowel-pin is governed by its application. For locating nests, gage plates, etc., pins from ⅛ to ³⁄₁₆ inch in diameter are satisfactory. For locating dies, the diameter of the dowel-pin should never be less than ¼ inch; the general rule is to use dowel-pins of the same size as the screws used in fastening the work. The length of the dowel-pin should be about one and one-half to two times its diameter in each plate or part to be doweled.

When hardened cylindrical dowel-pins are inserted in soft parts, ream the hole about 0.001 inch smaller than the dowel-pin. If the doweled parts are hardened, grind (or lap) the hole 0.0002 to 0.0003 inch under size. The hole should be ground or lapped straight or without taper or "bell-mouth."

British Standard for Metric Series Dowel Pins. — Steel parallel dowel pins specified in British Standard 1804:Part 2:1968 are divided into three grades which provide different degrees of pin accuracy:

Grade 1 is a precision ground pin made from En 32A or En 32B low carbon steel (BS 970) or from high carbon steel to BS 1407 or BS 1423. Pins below 4 mm diameter are unhardened. Those of 4 mm diameter and above are hardened to a minimum of 750 HV 30 in accordance with BS 427, but if they are made from steels to BS 1407 or BS 1423 then the hardness shall be within the range 600 to 700 HV 30, in accordance with BS 427. The values of other hardness scales may be used in accordance with BS 860.

Grade 2 is a ground pin made from any of the steels used for Grade 1. The pins are normally supplied unhardened, unless a different condition is agreed on between the purchaser and supplier.

Grade 3 pins are made from En 1A free cutting steel (BS 970) and are supplied with a machined, bright rolled or drawn finish. They are normally supplied unhardened unless a different condition is agreed on between the purchaser and supplier.

Pins of any grade may be made from different steels in accordance with BS 970, by mutual agreement between the purchaser and manufacturer. If steels other than those in the standard range are used, the hardness of the pins shall also be decided on by mutual agreement between purchaser and supplier. As shown in the illustration at the head of the accompanying table, one end of each pin is chamfered to provide a lead. The other end may be similarly chamfered, or domed.

If a dowel pin is driven into a blind hole where no provision is made for releasing air, the worker assembling the pin may be endangered, and damage may be caused to the associated component, or stresses may be set up. The appendix of the Standard describes one method of overcoming this problem, which is the provision of a small flat surface along the length of a pin to permit the release of air.

For purposes of marking, the Standard states that each package or lot of dowel pins shall bear the manufacturer's name or trademark, the BS number, and the grade of pin.

British Standard Parallel Steel Dowel Pins — Metric Series (BS 1804: Part 2: 1968)

Limits of Tolerance on Diameter

Nom. Diam., mm		Grade*		
		1	2	3
		Tolerance Zone		
		m5	h7	h11
Over	To & Incl.	Limits of Tolerance, 0.001 mm		
—	3	+7 / +2	0 / −12†	0 / −60
3	6	+9 / +4	0 / −12	0 / −75
6	10	+12 / +6	0 / −15	0 / −90
10	14	+15 / +7	0 / −18	0 / −110
14	18	+15 / +7	0 / −18	0 / −110
18	24	+17 / +8	0 / −21	0 / −130
24	30	+17 / +8	0 / −21	0 / −130

* The limits of tolerance for grades 1 and 2 dowel pins have been chosen to provide satisfactory assembly when used in standard reamed holes (H7 and H8 tolerance zones). If the assembly is not satisfactory, BS 1916: Part 1, Limits and Fits for Engineering should be consulted, and a different class of fit chosen.

† This tolerance is larger than that given in BS 1916, and has been included because the use of a closer tolerance would involve precision grinding by the manufacturer, which is uneconomic for a grade 2 dowel pin.

The tolerance limits on the overall length of all grades of dowel pin up to and including 50 mm long are + 0.5, − 0.0 mm, and for pins over 50 mm long, are + 0.8, − 0.0 mm. The Standard specifies that the roughness of the cylindrical surface of grades 1 and 2 dowel pins, when assessed in accordance with BS 1134, shall not be greater than 0.4 μm CLA (16 CLA).

Standard Sizes

Nominal Diameter D, mm	1	1.5	2	2.5	3	4	5	6	8	10	12	16	20	25
Chamfer a max, mm	0.3	0.3	0.3	0.4	0.45	0.6	0.75	0.9	1.2	1.5	1.8	2.5	3	4
Nom. Length L, mm														
4	o	o												
6	o	o	o											
8	o	o	o	o										
10		o	o	o	o									
12		o	o	o	o	o								
16			o	o	o	o	o	o						
20				o	o	o	o	o	o					
25					o	o	o	o	o	o				
30						o	o	o	o	o	o			
35						o	o	o	o	o	o			
40							o	o	o	o	o	o		
45								o	o	o	o	o		
50									o	o	o	o	o	
60									o	o	o	o	o	o
70										o	o	o	o	o
80											o	o	o	o
90												o	o	o
100												o	o	o
110													o	o
120													o	o

(Diagram: dowel pin with overall length L, diameter D, chamfer a, and a chamfer angle of 20°–40°.)

American National Standard Hardened Ground Machine Dowel Pins. — Hardened ground machine dowel pins are furnished in two diameter series: Standard Series having basic diameters 0.0002 inch over the nominal diameter, intended for initial installations; and Oversize Series having basic diameters 0.001 inch over the nominal diameter, intended for replacement use.

Preferred Lengths and Sizes: The preferred lengths and sizes in which these pins are normally available are given in Table 1. Other sizes and lengths are produced as required by the purchaser.

Effective Length: The effective length, L_e, must not be less than 75 per cent of the overall length of the pin.

Shear Strength: These pins must have a single shear strength of 130,000 psi minimum and be capable of withstanding the minimum double shear loads given in Table 1 when tested in accordance with the procedure outlined in ANSI B18.8.2-1978.

Designation: These pins are designated by the following data in the sequence shown: Product name (noun first), including pin series, nominal pin diameter (fraction or decimal equivalent), length (fraction or decimal equivalent), material, and protective finish, if required.

Examples: Pins, Hardened Ground Machine Dowel — Standard Series, 3/8 × 1 1/2, Steel, Phosphate Coated.

Pins, Hardened Ground Machine Dowel — Oversize Series, .625 × 2.500, Steel

Installation Precaution: Pins should not be installed by striking or hammering and when installing with a press, a shield should be used and safety glasses worn.

American National Standard Hardened Ground Production Dowel Pins. — Hardened ground production dowel pins have basic diameters which are 0.0002 inch over the nominal pin diameter.

Preferred Lengths and Sizes: The preferred lengths and sizes in which these pins are available are given in Table 2. Other sizes and lengths are produced as required by the purchaser.

Shear Strength: These pins must have a single shear strength of 102,000 psi minimum and be capable of withstanding the double shear loads given in Table 2 when tested in accordance with the procedure outlined in ANSI B18.8.2-1978.

Ductility: These standard pins are sufficiently ductile to withstand being pressed into holes 0.0005 inch smaller than the nominal pin diameter in hardened steel without cracking or shattering.

Designation: These pins are designated by the following data in the sequence shown: Product name (noun first), nominal pin diameter (fraction or decimal equivalent), length (fraction or decimal equivalent), material, and protective finish, if required.

Examples: Pins, Hardened Ground Production Dowel, 1/8 × 3/4, Steel, Phosphate Coated

Pins, Hardened Ground Production Dowel, .375 × 1.500, Steel

American National Standard Unhardened Ground Dowel Pins. — Unhardened ground dowel pins are normally produced by grinding the outside diameter of commercial wire or rod material to size. Consequently, the maximum diameters of the pins, as specified, in Table 3 are below the minimum commercial stock sizes by graduated amounts from 0.0005 inch on the 1/16-inch nominal pin size to 0.0028 inch on the 1-inch nominal pin size.

Table 9. American National Standard Hardened Ground Machine Dowel Pins (ANSI B18.8.2-1978)

Nominal Size[1] or Nominal Pin Diameter	Pin Diameter, A						Point Diameter, B		Crown Height or Radius, C		Range of Preferred Lengths,[2] L	Double Shear Load, Min, lb for Carbon or Alloy Steel	Suggested Hole Diameter[3]	
	Standard Series Pins			Oversize Series Pins										
	Basic	Max	Min	Basic	Max	Min	Max	Min	Max	Min			Max	Min
1/16 0.0625	0.0627	0.0628	0.0626	0.0635	0.0636	0.0634	0.058	0.048	0.020	0.008	3/16-3/4	800	0.0625	0.0620
*5/64 0.0781	0.0783	0.0784	0.0782	0.0791	0.0792	0.0790	0.074	0.064	0.026	0.010	···	1,240	0.0781	0.0776
3/32 0.0938	0.0940	0.0941	0.0939	0.0948	0.0949	0.0947	0.089	0.079	0.031	0.012	5/16-1	1,800	0.0937	0.0932
1/8 0.1250	0.1252	0.1253	0.1251	0.1260	0.1261	0.1259	0.120	0.110	0.041	0.016	3/8-2	3,200	0.1250	0.1245
*5/32 0.1562	0.1564	0.1565	0.1563	0.1572	0.1573	0.1571	0.150	0.140	0.052	0.020	···	5,000	0.1562	0.1557
3/16 0.1875	0.1877	0.1878	0.1876	0.1885	0.1886	0.1884	0.180	0.170	0.062	0.023	1/2-2	7,200	0.1875	0.1870
1/4 0.2500	0.2502	0.2503	0.2501	0.2510	0.2511	0.2509	0.240	0.230	0.083	0.031	1/2-2 1/2	12,800	0.2500	0.2495
5/16 0.3125	0.3127	0.3128	0.3126	0.3135	0.3136	0.3134	0.302	0.290	0.104	0.039	1/2-2 1/2	20,000	0.3125	0.3120
3/8 0.3750	0.3752	0.3753	0.3751	0.3760	0.3761	0.3759	0.365	0.350	0.125	0.047	1/2-3	28,700	0.3750	0.3745
7/16 0.4375	0.4377	0.4378	0.4376	0.4385	0.4386	0.4384	0.424	0.409	0.146	0.055	7/8-3	39,100	0.4375	0.4370
1/2 0.5000	0.5002	0.5003	0.5001	0.5010	0.5011	0.5009	0.486	0.471	0.167	0.063	3/4,1-4	51,000	0.5000	0.4995
5/8 0.6250	0.6252	0.6253	0.6251	0.6260	0.6261	0.6259	0.611	0.595	0.208	0.078	1 1/2-5	79,800	0.6250	0.6245
3/4 0.7500	0.7502	0.7503	0.7501	0.7510	0.7511	0.7509	0.735	0.715	0.250	0.094	1 1/2-6	114,000	0.7500	0.7495
7/8 0.8750	0.8752	0.8753	0.8751	0.8760	0.8761	0.8759	0.860	0.840	0.293	0.109	2,2 1/2-6	156,000	0.8750	0.8745
1 1.0000	1.0002	1.0003	1.0001	1.0010	1.0011	1.0009	0.980	0.960	0.333	0.125	2,2 1/2-5,6	204,000	1.0000	0.9995

All dimensions are in inches.

* Nonpreferred sizes, not recommended for use in new designs.

[1] Where specifying nominal size as basic diameter, zeros preceding decimal and in the fourth decimal place are omitted.

[2] Lengths increase in 1/16-inch steps up to 3/8 inch, in 1/8-inch steps from 3/8 inch to 1 inch, in 1/4-inch steps from 1 inch to 2 1/2 inches, and in 1/2-inch steps above 2 1/2 inches. Tolerance on length is ±0.010 inch.

[3] These hole sizes have been commonly used for press fitting Standard Series machine dowel pins into materials such as mild steels and cast iron. In soft materials such as aluminum or zinc die castings, hole size limits are usually decreased by 0.0005 inch to increase the press fit.

Table 2. American National Standard Hardened Ground Production Dowel Pins
(ANSI B18.8.2-1978)

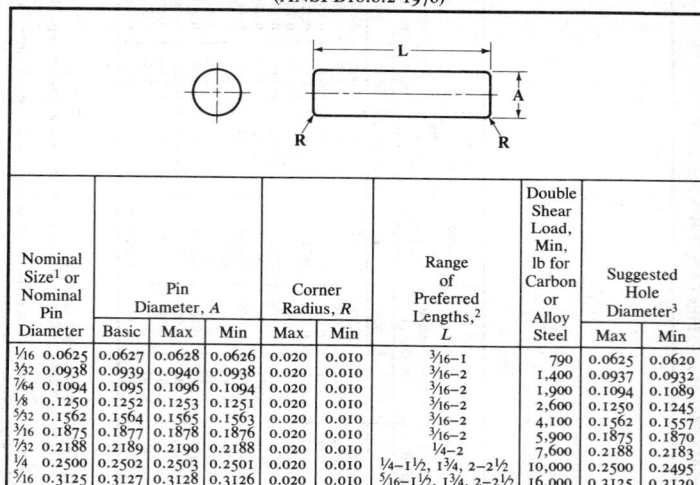

Nominal Size[1] or Nominal Pin Diameter	Pin Diameter, A			Corner Radius, R		Range of Preferred Lengths,[2] L	Double Shear Load, Min, lb for Carbon or Alloy Steel	Suggested Hole Diameter[3]	
	Basic	Max	Min	Max	Min			Max	Min
1/16 0.0625	0.0627	0.0628	0.0626	0.020	0.010	3/16–1	790	0.0625	0.0620
3/32 0.0938	0.0939	0.0940	0.0938	0.020	0.010	3/16–2	1,400	0.0937	0.0932
7/64 0.1094	0.1095	0.1096	0.1094	0.020	0.010	3/16–2	1,900	0.1094	0.1089
1/8 0.1250	0.1252	0.1253	0.1251	0.020	0.010	3/16–2	2,600	0.1250	0.1245
5/32 0.1562	0.1564	0.1565	0.1563	0.020	0.010	3/16–2	4,100	0.1562	0.1557
3/16 0.1875	0.1877	0.1878	0.1876	0.020	0.010	3/16–2	5,900	0.1875	0.1870
7/32 0.2188	0.2189	0.2190	0.2188	0.020	0.010	1/4–2	7,600	0.2188	0.2183
1/4 0.2500	0.2502	0.2503	0.2501	0.020	0.010	1/4–1½, 1¾, 2–2½	10,000	0.2500	0.2495
5/16 0.3125	0.3127	0.3128	0.3126	0.020	0.010	5/16–1½, 1¾, 2–2½	16,000	0.3125	0.3120
3/8 0.3750	0.3752	0.3753	0.3751	0.020	0.010	3/8–1½, 1¾, 2–3	23,000	0.3750	0.3745

All dimensions are in inches.
[1] Where specifying nominal pin size in decimals, zeros preceding decimal and in the fourth decimal place are omitted.
[2] Lengths increase in 1/16-inch steps up to 1 inch, in 1/8-inch steps from 1 inch to 2 inches and then are 2¼, 2½, and 3 inches.
[3] These hole sizes have been commonly used for press fitting production dowel pins into materials such as mild steels and cast iron. In soft materials such as aluminum or zinc die castings, hole size limits are usually decreased by 0.0005 inch to increase the press fit.

Preferred Lengths and Sizes: The preferred lengths and sizes in which unhardened ground pins are normally available are given in Table 3. Other sizes and lengths are produced as required by the purchaser.

Shear Strength: These pins must have a single shear strength of 64,000 psi minimum for pins made from steel and 40,000 psi minimum for pins made from brass and must be capable of withstanding the minimum double shear loads given in Table 3 when tested in accordance with the procedure outlined in ANSI B18.8.2-1978.

Designation: These pins are designated by the following data in the order shown: Product name (noun first), nominal pin diameter (fraction or decimal equivalent), length (fraction or decimal equivalent), material, and protective finish, if required.

Examples: Pins, Unhardened Ground Dowel, 1/8 × 3/4, Steel
        Pins, Unhardened Ground Dowel, .025 × 2.500, Steel, Zinc Plated

American National Standard Straight Pins. — The diameter of both chamfered and square end straight pins is that of the commercial wire or rod from which the pins are made. The tolerances shown in Table 4 are applicable to carbon steel and some deviations in the diameter limits may be necessary for pins made from other materials.

Length Increments: Lengths are as specified by the purchaser, however it is recommended the nominal pin lengths be limited to increments of not less than 0.062 inch.

Table 3. American National Standard Unhardened Ground Dowel Pins
(ANSI B18.8.2-1978)

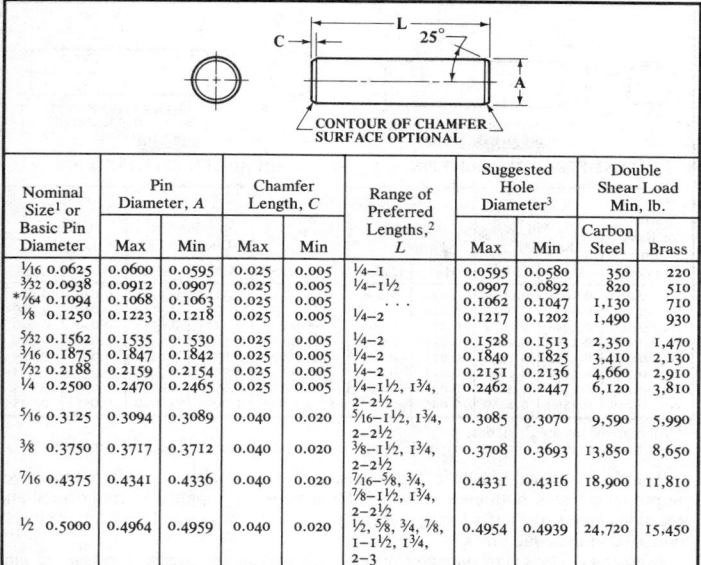

CONTOUR OF CHAMFER
SURFACE OPTIONAL

Nominal Size[1] or Basic Pin Diameter	Pin Diameter, A		Chamfer Length, C		Range of Preferred Lengths,[2] L	Suggested Hole Diameter[3]		Double Shear Load Min. lb.	
	Max	Min	Max	Min		Max	Min	Carbon Steel	Brass
1/16 0.0625	0.0600	0.0595	0.025	0.005	1/4–1	0.0595	0.0580	350	220
3/32 0.0938	0.0912	0.0907	0.025	0.005	1/4–1 1/2	0.0907	0.0892	820	510
*7/64 0.1094	0.1068	0.1063	0.025	0.005	. . .	0.1062	0.1047	1,130	710
1/8 0.1250	0.1223	0.1218	0.025	0.005	1/4–2	0.1217	0.1202	1,490	930
5/32 0.1562	0.1535	0.1530	0.025	0.005	1/4–2	0.1528	0.1513	2,350	1,470
3/16 0.1875	0.1847	0.1842	0.025	0.005	1/4–2	0.1840	0.1825	3,410	2,130
7/32 0.2188	0.2159	0.2154	0.025	0.005	1/4–2	0.2151	0.2136	4,660	2,910
1/4 0.2500	0.2470	0.2465	0.025	0.005	1/4–1 1/2, 1 3/4, 2–2 1/2	0.2462	0.2447	6,120	3,810
5/16 0.3125	0.3094	0.3089	0.040	0.020	5/16–1 1/2, 1 3/4, 2–2 1/2	0.3085	0.3070	9,590	5,990
3/8 0.3750	0.3717	0.3712	0.040	0.020	3/8–1 1/2, 1 3/4, 2–2 1/2	0.3708	0.3693	13,850	8,650
7/16 0.4375	0.4341	0.4336	0.040	0.020	7/16–5/8, 3/4, 7/8–1 1/2, 1 3/4, 2–2 1/2	0.4331	0.4316	18,900	11,810
1/2 0.5000	0.4964	0.4959	0.040	0.020	1/2, 5/8, 3/4, 7/8, 1–1 1/2, 1 3/4, 2–3	0.4954	0.4939	24,720	15,450
5/8 0.6250	0.6211	0.6206	0.055	0.035	5/8, 3/4, 7/8, 1–1 1/2, 1 3/4, 2, 2 1/2–4	0.6200	0.6185	38,710	24,190
3/4 0.7500	0.7458	0.7453	0.055	0.035	3/4, 7/8, 1, 1 1/4, 1 1/2, 1 3/4, 2, 2 1/2–4	0.7446	0.7431	55,840	34,900
7/8 0.8750	0.8705	0.8700	0.070	0.050	7/8, 1, 1 1/4, 1 1/2, 1 3/4, 2, 2 1/2–4	0.8692	0.8677	76,090	47,550
1 1.0000	0.9952	0.9947	0.070	0.050	1, 1 1/4, 1 1/2, 1 3/4, 2, 2 1/2–4	0.9938	0.9923	99,460	62,160

All dimensions are in inches.   * Nonpreferred size, not recommended for use in new designs.
[1] Where specifying pin size in decimals, zeros preceding decimal and in the fourth decimal place are omitted.
[2] Lengths increase in 1/16-inch increments from 1/4 to 1 inch, in 1/8-inch increments from 1 inch to 2 inches, and in 1/4-inch increments from 2 to 2 1/2 inches, and in 1/2-inch increments from 2 1/2 to 4 inches.
[3] These hole sizes have been found to be satisfactory for press fitting pins into mild steel and cast and malleable irons. In soft materials such as aluminum alloys or zinc die castings, hole size limits are usually decreased by 0.0005 inch to increase the press fit.

Material: Straight pins are normally made from cold drawn steel wire or rod having a maximum carbon content of 0.28 per cent. Where required, pins may also be made from corrosion resistant steel, brass, or other metals.

Designation: Straight pins are designated by the following data, in the sequence shown: Product name (noun first), nominal size (fraction or decimal equivalent), material, and protective finish, if required.

Examples: Pin, Chamfered Straight, 1/8 × 1.500, Steel
Pin, Square End Straight, .250 × 2.250, Steel, Zinc Plated

Table 4. American National Standard Chamfered and Square End Straight Pins
(ANSI B18.8.2-1978)

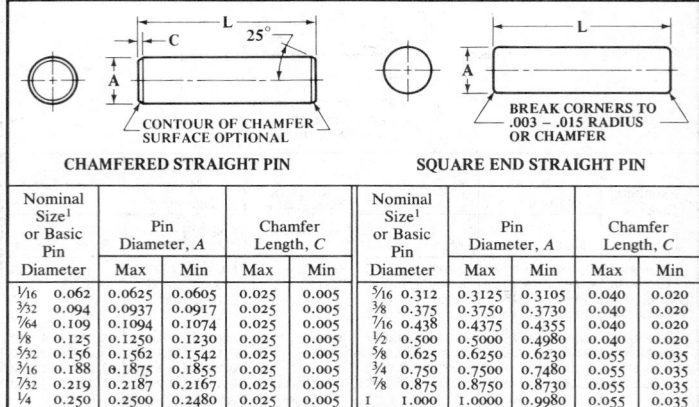

CONTOUR OF CHAMFER
SURFACE OPTIONAL

CHAMFERED STRAIGHT PIN

BREAK CORNERS TO
.003 – .015 RADIUS
OR CHAMFER

SQUARE END STRAIGHT PIN

Nominal Size[1] or Basic Pin Diameter	Pin Diameter, A		Chamfer Length, C		Nominal Size[1] or Basic Pin Diameter	Pin Diameter, A		Chamfer Length, C	
	Max	Min	Max	Min		Max	Min	Max	Min
1/16 0.062	0.0625	0.0605	0.025	0.005	5/16 0.312	0.3125	0.3105	0.040	0.020
3/32 0.094	0.0937	0.0917	0.025	0.005	3/8 0.375	0.3750	0.3730	0.040	0.020
7/64 0.109	0.1094	0.1074	0.025	0.005	7/16 0.438	0.4375	0.4355	0.040	0.020
1/8 0.125	0.1250	0.1230	0.025	0.005	1/2 0.500	0.5000	0.4980	0.040	0.020
5/32 0.156	0.1562	0.1542	0.025	0.005	5/8 0.625	0.6250	0.6230	0.055	0.035
3/16 0.188	0.1875	0.1855	0.025	0.005	3/4 0.750	0.7500	0.7480	0.055	0.035
7/32 0.219	0.2187	0.2167	0.025	0.005	7/8 0.875	0.8750	0.8730	0.055	0.035
1/4 0.250	0.2500	0.2480	0.025	0.005	1 1.000	1.0000	0.9980	0.055	0.035

All dimensions are in inches.
[1] Where specifying nominal size in decimals, zeros preceding decimal point are omitted.

American National Standard Taper Pins. — Taper pins have a uniform taper over the pin length with both ends crowned. Most sizes are supplied in commercial and precision classes, the latter having generally tighter tolerances and being more closely controlled in manufacture.

Diameters: The major diameter of both commercial and precision classes of pins is the diameter of the large end and is the basis for pin size. The diameter at the small end is computed by multiplying the nominal length of the pin by the factor 0.02083 and subtracting the result from the basic pin diameter. See also Table 6.

Taper: The taper on commercial class pins is 0.250 ± 0.006 inch per foot and on the precision class pins is 0.250 ± 0.004 inch per foot of length.

Materials: Unless otherwise specified, taper pins are made from AISI 1211 steel or cold drawn AISI 1212 or 1213 steel or equivalents, and no mechanical properties apply.

Hole Sizes: Under most circumstances, holes for taper pins require taper reaming. Sizes and lengths of taper pins for which standard reamers are available are given in Table 5. Drilling specifications for taper pins are given below.

Designation: Taper pins are designated by the following data in the sequence shown: Product name (noun first), class, size number (or decimal equivalent), length (fraction or three-place decimal equivalent), material, and protective finish, if required.

Examples: Pin, Taper (Commercial Class) No. 0 × 3/4, Steel

Pin, Taper (Precision Class) .219 × 1.750, Steel, Zinc Plated

Drilling Specifications for Taper Pins. — When helically fluted taper pin reamers are used, the diameter of the through hole drilled prior to reaming is equal to the diameter at the small end of the taper pin. (See Table 6.) However, when straight fluted taper reamers are to be used, it may be necessary, in the case of long pins, to step drill the hole before reaming, the number and sizes of the drills to be used depending on the depth of the hole (pin length).

Table 5. American National Standard Taper Pins (ANSI B18.8.2-1978)

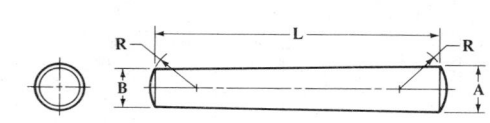

Pin Size Number and Basic Pin Diameter[1]	Major Diameter (Large End), A				End Crown Radius, R		Range of Lengths,[2] L	
	Commercial Class		Precision Class					
	Max	Min	Max	Min	Max	Min	Stand. Reamer Avail.[3]	Other
7/0 0.0625	0.0638	0.0618	0.0635	0.0625	0.072	0.052	...	¼–1
6/0 0.0780	0.0793	0.0773	0.0790	0.0780	0.088	0.068	...	¼–1½
5/0 0.0940	0.0953	0.0933	0.0950	0.0940	0.104	0.084	¼–1	1¼, 1½
4/0 0.1090	0.1103	0.1083	0.1100	0.1090	0.119	0.099	¼–1	1¼–2
3/0 0.1250	0.1263	0.1243	0.1260	0.1250	0.135	0.115	¼–1	1¼–2
2/0 0.1410	0.1423	0.1403	0.1420	0.1410	0.151	0.131	½–1¼	1½–2½
0 0.1560	0.1573	0.1553	0.1570	0.1560	0.166	0.146	½–1¼	1½–3
1 0.1720	0.1733	0.1713	0.1730	0.1720	0.182	0.162	¾–1¼	1½–3
2 0.1930	0.1943	0.1923	0.1940	0.1930	0.203	0.183	¾–1½	1¾–3
3 0.2190	0.2203	0.2183	0.2200	0.2190	0.229	0.209	¾–1¾	2–4
4 0.2500	0.2513	0.2493	0.2510	0.2500	0.260	0.240	¾–2	2¼–4
5 0.2890	0.2903	0.2883	0.2900	0.2890	0.299	0.279	1–2½	2¾–6
6 0.3410	0.3423	0.3403	0.3420	0.3410	0.351	0.331	1¼–3	3¼–6
7 0.4090	0.4103	0.4083	0.4100	0.4090	0.419	0.399	1¼–3¾	4–8
8 0.4920	0.4933	0.4913	0.4930	0.4920	0.502	0.482	1¼–4½	4¾–8
9 0.5910	0.5923	0.5903	0.5920	0.5910	0.601	0.581	1¼–5¼	5½–8
10 0.7060	0.7073	0.7053	0.7070	0.7060	0.716	0.696	1½–6	6¼–8
11 0.8600	0.8613	0.8593	...	...	0.870	0.850	...	2–8
12 1.0320	1.0333	1.0313	...	...	1.042	1.022	...	2–9
13 1.2410	1.2423	1.2403	...	...	1.251	1.231	...	3–11
14 1.5210	1.5223	1.5203	...	...	1.531	1.511	...	3–13

All dimensions are in inches.
For nominal diameters, B, see Table 6.
[1] When specifying nominal pin size in decimals, zeros preceding the decimal and in the fourth decimal place are omitted.
[2] Lengths increase in ⅛-inch steps up to 1 inch and in ¼-inch steps above 1 inch.
[3] Standard reamers are available for pin lengths in this column.

To determine the number and sizes of step drills required: (1) find the length of pin to be used at the top of the chart on page 1446 and follow this length down to the intersection with that heavy line which represents the size of taper pin (see taper pin numbers at the right-hand end of each heavy line). (2) If the length of pin falls between the first and second dots, counting from the left, only one drill is required. Its size is indicated by following the nearest horizontal line from the point of intersection (of the pin length) on the heavy line over to the drill diameter values at the left. (3) If the intersection of pin length comes between the second and third dots, then two drills are required. In this case the smaller has a size corresponding to the intersection of the pin length and heavy line and the larger is the corresponding drill diameter for the intersection of one-half this length with the heavy line. (4) Should the pin length fall between the third and fourth dots, then three drills are required. The smallest will have a diameter corresponding to the intersection of the total pin length with the heavy line, the next in size will have a diameter corresponding to the intersection of two-thirds of this length with the heavy line and the largest will have a diameter corresponding to the intersection of one-third of this length with the heavy line. Where the intersection falls between two drill sizes, use the smaller.

Chart to Facilitate Selection of Number and Sizes of Drills for Step-Drilling Prior to Taper Reaming

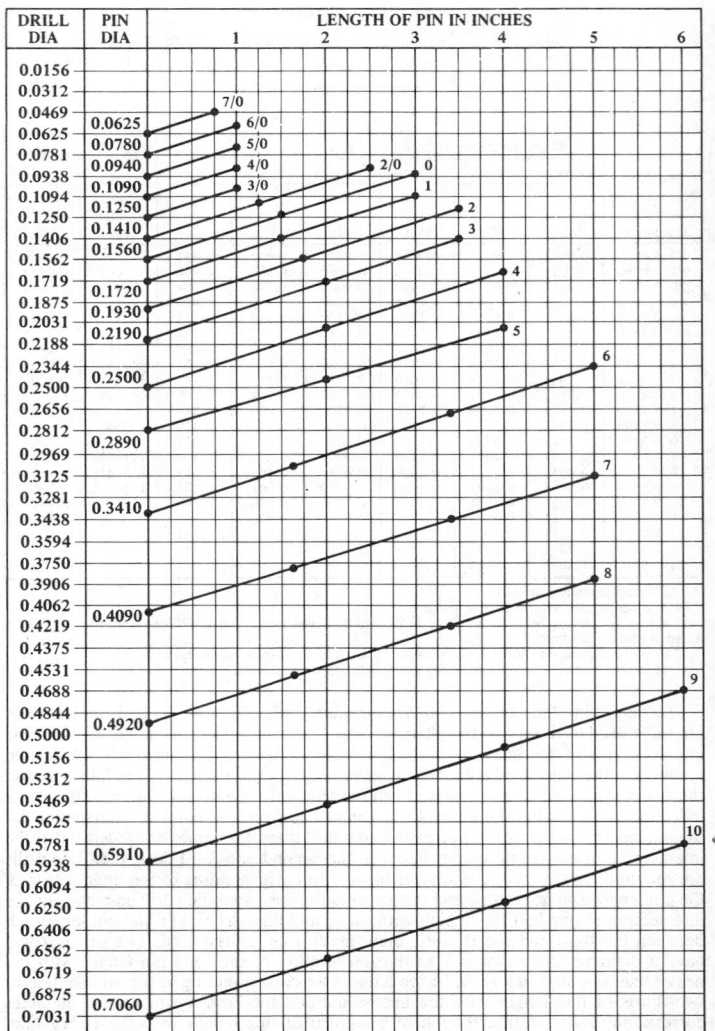

Table 6. Nominal Diameter at Small Ends of Standard Taper Pins

Pin Length in Inches	Pin Number and Small End Diameter for Given Length										
	0	1	2	3	4	5	6	7	8	9	10
¾	0.140	0.156	0.177	0.203	0.235	0.273	0.325	0.393	0.476	0.575	0.690
1	0.135	0.151	0.172	0.198	0.230	0.268	0.320	0.388	0.471	0.570	0.685
1¼	0.130	0.146	0.167	0.192	0.224	0.263	0.315	0.382	0.466	0.565	0.680
1½	0.125	0.141	0.162	0.187	0.219	0.258	0.310	0.377	0.460	0.560	0.675
1¾	0.120	0.136	0.157	0.182	0.214	0.252	0.305	0.372	0.455	0.554	0.669
2	0.114	0.130	0.151	0.177	0.209	0.247	0.299	0.367	0.450	0.549	0.664
2¼	0.109	0.125	0.146	0.172	0.204	0.242	0.294	0.362	0.445	0.544	0.659
2½	0.104	0.120	0.141	0.166	0.198	0.237	0.289	0.356	0.440	0.539	0.654
2¾	0.099	0.115	0.136	0.161	0.193	0.232	0.284	0.351	0.434	0.534	0.649
3	0.094	0.110	0.131	0.156	0.188	0.227	0.279	0.346	0.429	0.528	0.643
3¼	…	…	…	0.151	0.182	0.221	0.273	0.340	0.424	0.523	0.638
3½	…	…	…	0.146	0.177	0.216	0.268	0.335	0.419	0.518	0.633
3¾	…	…	…	0.141	0.172	0.211	0.263	0.330	0.414	0.513	0.628
4	…	…	…	0.136	0.167	0.206	0.258	0.326	0.409	0.508	0.623
4¼	…	…	…	0.131	0.162	0.201	0.253	0.321	0.403	0.502	0.617
4½	…	…	…	0.125	0.156	0.195	0.247	0.315	0.398	0.497	0.612
5	…	…	…	…	0.146	0.185	0.237	0.305	0.389	0.487	0.602
5½	…	…	…	…	…	…	…	0.294	0.377	0.476	0.591
6	…	…	…	…	…	…	…	0.284	0.367	0.466	0.581

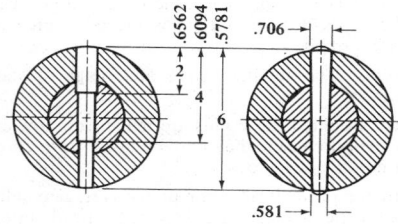

Examples: For a No. 10 taper pin 6-inches long, three drills would be used, of the sizes and for the depths shown in the accompanying diagram.

For a No. 10 taper pin 3-inches long, two drills would be used since the 3-inch length falls between the second and third dots. The first or through drill will be 0.6406 inch and the second drill, 0.6719 inch for a depth of 1½ inches.

American National Standard Grooved Pins. — These pins have three equally spaced longitudinal grooves and an expanded diameter over the crests of the ridges formed by the material displaced when the grooves are produced. The grooves are aligned with the axis of the pins. There are seven types of grooved pins as shown in the illustration on page 1449.

Standard Sizes and Lengths: The standard sizes and lengths in which grooved pins are normally available are given in Table 7.

Materials: Grooved pins are normally made from cold drawn low carbon steel wire or rod. Where additional performance is required, carbon steel pins may be supplied surface hardened and heat treated to a hardness consistent with the performance requirements. Pins may also be made from alloy steel, corrosion resistant steel, brass, Monel and other non-ferrous metals having chemical properties as agreed upon between manufacturer and purchaser.

Performance Requirements: Grooved pins are required to withstand the minimum double shear loads which are given in Table 7 for the respective materials shown, when tested in accordance with the Double Shear Testing of Pins as set forth in ANSI B18.8.2-1978.

Holes Sizes: To obtain maximum product retention under average conditions, it is recommended that holes for the installation of grooved pins be held as close as possible to the limits shown in Table 7. The minimum limits correspond to the drill size which is the same as the basic pin diameter. The maximum limits are generally suitable for length-diameter ratios of not less than 4 to 1 or greater than 10 to 1. For smaller length-to-diameter ratios, the hole should be held closer to the minimum limits where retention is critical. Conversely for larger ratios where retention requirements are not essential, it may be desirable to increase the hole diameters beyond the maximum limits shown.

Designation: Grooved pins are designated by the following data in the sequence shown: Product name (noun first) including type designation, nominal size (number, fraction or decimal equivalent), length (fraction or decimal equivalent), material, including specification or heat treatment where necessary, protective finish, if required.

Examples: Pin, Type A Grooved, ³/₃₂ × ¾, Steel, Zinc Plated

Pin, Type F Grooved, .250 × 1.500, Corrosion Resistant Steel

American National Standard Grooved T-Head Cotter Pins and Round Head Grooved Drive Studs. — The cotter pins have a T-head and the studs a round head. Both pins and studs have three equally spaced longitudinal grooves and an expanded diameter over the crests of the raised ridges formed by the material displaced when the grooves are formed.

Standard Sizes and Lengths: The standard sizes and range of standard lengths are given in Tables 8 and 9.

Material: Unless otherwise specified these pins are made from low carbon steel. Where so indicated by the purchaser, they may be made from corrosion resistant steel, brass or other non-ferrous alloys.

Hole Sizes: To obtain optimum product retention under average conditions, it is recommended that holes for the installation of grooved T-head cotter pins and grooved drive studs be held as close as possible to the limits tabulated in the tables. The minimum limits given correspond to the drill size which is equivalent to the basic shank diameter. The maximum limits shown are generally suitable for length-diameter ratios of not less than 4 to 1 and not greater than 10 to 1. For smaller length-to-diameter ratios, the holes should be held closer to minimum limits where retention is critical. Conversely, for larger length-to-diameter ratios or where retention requirements are not essential, it may be desirable to increase the hole diameter beyond the maximum limits shown.

Designation: Grooved T-head cotter pins and round head grooved drive studs are designated by the following data, in the order shown: Product name (noun first), nominal size (number, fraction or decimal equivalent), length (fraction or decimal equivalent) material including specification or heat treatment where necessary, and protective finish, if required.

Examples: Pin, Grooved T-Head Cotter, ¼ × 1¼, Steel, Zinc Plated

Drive Stud, Round Head Grooved, No. 10 × ½, Corrosion Resistant Steel

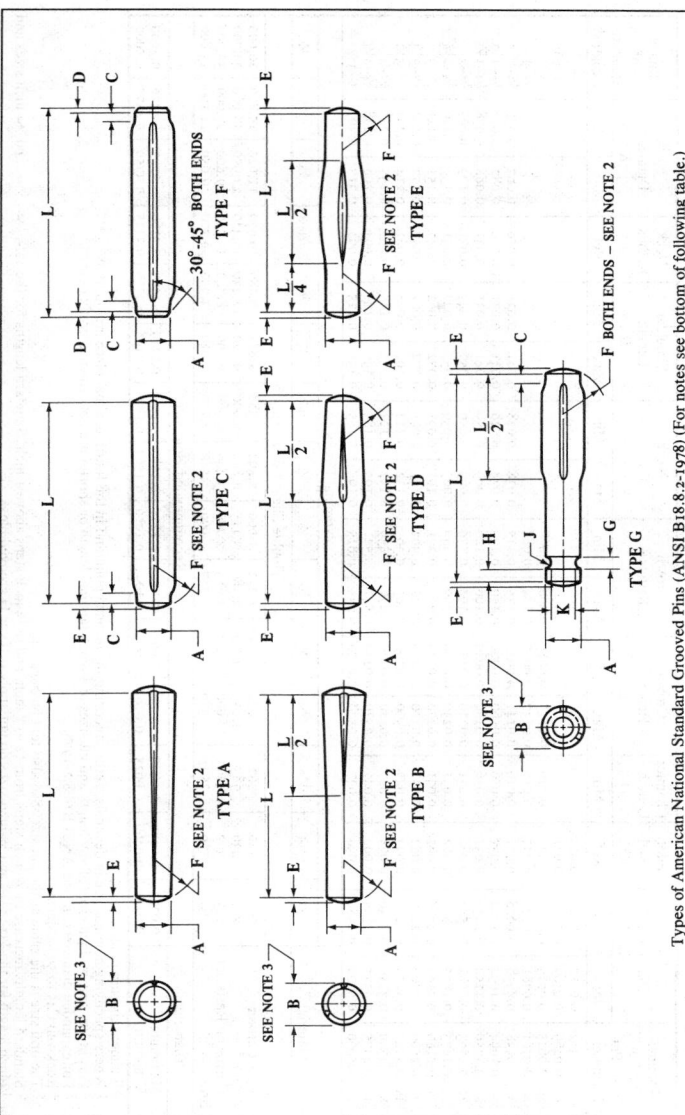

Types of American National Standard Grooved Pins (ANSI B18.8.2-1978) (For notes see bottom of following table.)

Table 7. American National Standard Grooved Pins (ANSI B18.8.2-1978)

Nominal Size[1] or Basic Pin Diameter	Pin Diameter,[3] A Max	Pin Diameter,[3] A Min	Pilot Length, C Ref	Chamfer Length,[2] D Min	Crown Height,[2] E Max	Crown Height,[2] E Min	Crown Radius,[2] F Max	Crown Radius,[2] F Min	Neck Width, G Max	Neck Width, G Min	Shoulder Length, H Max	Shoulder Length, H Min	Neck Radius, J Ref	Neck Diameter, K Max	Neck Diameter, K Min	Range of Standard Lengths[6]
1/32	0.0312	0.0302	0.015	…	…	…	…	…	…	…	…	…	…	…	…	1/8–1/2
3/64	0.0469	0.0459	0.031	…	…	…	…	…	…	…	…	…	…	…	…	1/8–5/8
1/16	0.0625	0.0615	0.031	0.016	0.0115	0.0015	0.088	0.068	…	…	…	…	…	…	…	1/8–1
5/64	0.0781	0.0771	0.031	0.016	0.0137	0.0037	0.104	0.084	…	…	…	…	…	…	…	1/8–1
3/32	0.0938	0.0928	0.031	0.016	0.0141	0.0041	0.135	0.115	0.038	0.028	0.041	0.031	0.016	0.067	0.057	1/4–1
7/64	0.1094	0.1074	0.031	0.016	0.0160	0.0060	0.150	0.130	0.038	0.028	0.041	0.031	0.016	0.082	0.072	1/4–1 1/4
1/8	0.1250	0.1230	0.031	0.016	0.0180	0.0080	0.166	0.146	0.069	0.059	0.041	0.031	0.031	0.088	0.078	1/4–1 1/4
5/32	0.1563	0.1543	0.062	0.031	0.0220	0.0120	0.198	0.178	0.069	0.059	0.057	0.047	0.031	0.109	0.099	1/4–1 1/2
3/16	0.1875	0.1855	0.062	0.031	0.0230	0.0130	0.260	0.240	0.069	0.059	0.057	0.047	0.031	0.130	0.120	3/8–2
7/32	0.2188	0.2168	0.062	0.031	0.0270	0.0170	0.291	0.271	0.101	0.091	0.057	0.047	0.031	0.151	0.141	3/8–2 1/4
1/4	0.2500	0.2480	0.062	0.031	0.0310	0.0210	0.322	0.302	0.101	0.091	0.072	0.062	0.047	0.172	0.162	3/8–2 3/4
5/16	0.3125	0.3105	0.094	0.047	0.0390	0.0290	0.385	0.365	0.132	0.122	0.104	0.094	0.062	0.214	0.204	1/2–3 1/2
3/8	0.3750	0.3730	0.094	0.047	0.0440	0.0340	0.479	0.459	0.132	0.122	0.135	0.125	0.062	0.255	0.245	3/4–4 1/4
7/16	0.4375	0.4355	0.094	0.047	0.0520	0.0420	0.541	0.521	0.195	0.185	0.135	0.125	0.094	0.298	0.288	7/8–4 1/2
1/2	0.5000	0.4980	0.094	0.047	0.0570	0.0470	0.635	0.615	0.195	0.185	0.135	0.125	0.094	0.317	0.307	1–4 1/2

Nominal Pin Size

Material	1/32	3/64	1/16	5/64	3/32	7/64	1/8	5/32	3/16	1/4	5/16	3/8	7/16	1/2
Double Shear Load, Min, lb														
Steels — Low Carbon	104	220	402	624	896	1,222	1,600	2,494	3,588	6,380	9,970	11,620	15,820	20,600
Steels — Alloy[4]	202	430	785	1,215	1,750	2,380	3,115	4,860	6,990	12,430	19,420	27,950	38,060	49,700
Corrosion Resistant	143	300	540	860	1,240	1,685	2,200	3,440	4,960	8,840	13,750	19,800	27,000	35,200
Brass	64	136	250	386	555	757	990	1,540	2,220	3,950	6,170	9,050	12,100	15,800
Recommended Hole Sizes[5]														
Diameter — Maximum	0.0324	0.0482	0.0640	0.0798	0.0956	0.1113	0.1271	0.1587	0.1903	0.2534	0.3166	0.3797	0.4428	0.5060
Diameter — Minimum	0.0312	0.0469	0.0625	0.0781	0.0938	0.1094	0.1250	0.1563	0.1875	0.2500	0.3125	0.3750	0.4375	0.5000

All dimensions are in inches.

[1] Where specifying nominal size in decimals, zeros preceding the decimal point and in the fourth decimal place are omitted.
[2] Pins in 1/32- and 3/64-inch sizes of any length and all sizes of 1/4-inch nominal length or shorter are not crowned or chamfered.
[3] For expanded diameters, B, see ANSI B18.8.2-1978.
[4] Rockwell C45 to 50 hardness.
[5] The drill size is the same as the pin size. See also text on page 1448.
[6] Standard lengths increase in 1/8-inch steps from 1/8 to 1 inch, and in 1/4-inch steps above 1 inch. Standard lengths for the 1/32-, 3/64-, and 1/16-inch sizes and the 1/4-inch length for the 3/32-, 7/64-, and 1/8-inch sizes do not apply to Type G grooved pins.

Table 8. American National Standard Grooved T-Head Cotter Pins
(ANSI B.18.8.2-1978)

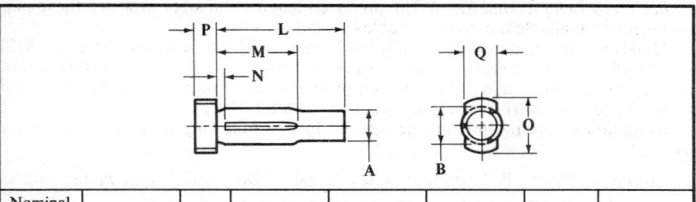

Nominal Size[1] or Basic Shank Diameter	Shank Diameter, A		Lgth, N	Head Diameter, O		Head Height, P		Head Width, Q		Range of Standard Lengths, L	Recommended Hole Size	
	Max	Min	Max	Max	Min	Max	Min	Max	Min		Max	Min
5⁄32 0.156	0.154	0.150	0.08	0.26	0.24	0.11	0.09	0.18	0.15	¾–1⅛	0.161	0.156
3⁄16 0.187	0.186	0.182	0.09	0.30	0.28	0.13	0.11	0.22	0.18	¾–1¼	0.193	0.187
¼ 0.250	0.248	0.244	0.12	0.40	0.38	0.17	0.15	0.28	0.24	1–1½	0.257	0.250
5⁄16 0.312	0.310	0.305	0.16	0.51	0.48	0.21	0.19	0.34	0.30	1⅛–2	0.319	0.312
23⁄64 0.359	0.358	0.353	0.18	0.57	0.54	0.24	0.22	0.38	0.35	1¼–2	0.366	0.359
½ 0.500	0.498	0.493	0.25	0.79	0.76	0.32	0.30	0.54	0.49	2–3	0.508	0.500

All dimensions are in inches.

When specifying nominal size in decimals, zeros preceding decimal point and in the fourth decimal place are omitted.

For expanded diameter, B, dimensions, see ANSI B18.8.2-1978.

Lengths increase in ⅛-inch steps from ¾ to 1¼ inch and in ¼-inch steps above 1¼ inches. For groove length, M, dimensions see ANSI B18.8.2-1978.

Table 9. American National Standard Round Head Grooved Drive Studs
(ANSI B18.8.2-1978)

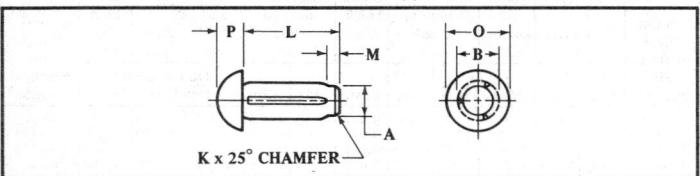

K x 25° CHAMFER

Stud Size Number and Basic Shank Diameter[1]	Shank Diameter, A		Head Diameter, O		Head Height, P		Range of Standard Lengths, L	Recommended Hole Size		Drill Size
	Max	Min	Max	Min	Max	Min		Max	Min	
0 0.067	0.067	0.065	0.130	0.120	0.050	0.040	⅛–¼	0.0686	0.0670	51
2 0.086	0.086	0.084	0.162	0.146	0.070	0.059	⅛–¼	0.0877	0.0860	44
4 0.104	0.104	0.102	0.211	0.193	0.086	0.075	3⁄16–⅜	0.1059	0.1040	37
6 0.120	0.120	0.118	0.260	0.240	0.103	0.091	¼–⅜	0.1220	0.1200	31
7 0.136	0.136	0.134	0.309	0.287	0.119	0.107	5⁄16–½	0.1382	0.1360	29
8 0.144	0.144	0.142	0.309	0.287	0.119	0.107	⅜–⅝	0.1463	0.1440	27
10 0.161	0.161	0.159	0.359	0.334	0.136	0.124	⅜–⅝	0.1636	0.1610	20
12 0.196	0.196	0.194	0.408	0.382	0.152	0.140	½–¾	0.1990	0.1960	9
14 0.221	0.221	0.219	0.457	0.429	0.169	0.156	½–¾	0.2240	0.2210	2
16 0.250	0.250	0.248	0.472	0.443	0.174	0.161	½	0.2534	0.2500	¼

All dimensions are in inches.

Where specifying nominal size in decimals, zeros preceding decimal point and in the fourth decimal place are omitted.

Lengths increase in 1⁄16-inch steps from ⅛ to ⅜ inch and in ⅛-inch steps above ⅜ inch.

For pilot length, M, and expanded diameter, B, dimensions see ANSI B18.8.2-1978.

American National Standard Spring Pins. — These pins are made in two types: one type has a slot throughout its length; the other is shaped in the form of a coil.

Preferred Lengths and Sizes: The preferred lengths and sizes in which these pins are normally available are given in Tables 10 and 11.

Materials: Spring pins are normally made from AISI 1070-1095 carbon steel, AISI 6150 H alloy steel, AISI Types 410 through 420 and 302 corrosion resistant steels, and beryllium copper alloy, heat treated or cold worked to attain the hardness and performance characteristics set forth in ANSI B18.8.2-1978.

Designation: Spring pins are designated by the following data in the sequence shown:

Examples: Pin, Coiled Spring, ¼ × 1¼, Standard Duty, Steel, Zinc Plated

Pin, Slotted Spring, ½ × 3, Steel, Phosphate Coated

Table 10. American National Standard Slotted Type Spring Pins (ANSI B18.8.2-1978)

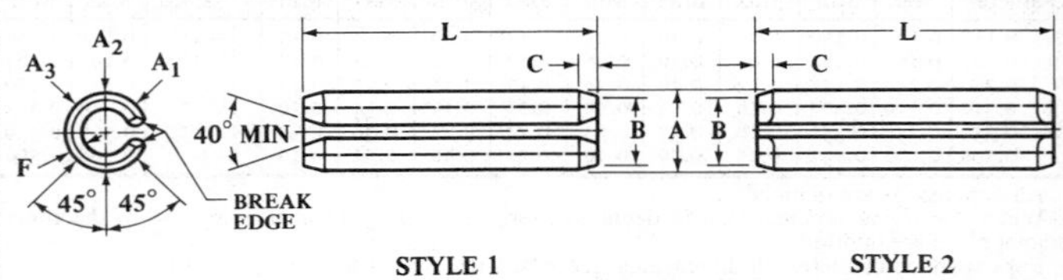

STYLE 1　　　　　　　　　　　STYLE 2

Nominal Size[1] or Basic Pin Diameter	Average Pin Diameter, A		Chamfer Diam., B	Chamfer Length, C		Stock Thickness, F	Recommended Hole Size		AISI 1070-1095 and AISI 420	AISI 302	Beryllium Copper	Range of Practical Lengths[3]
	Max	Min	Max	Max	Min	Basic	Max	Min	Double Shear Load, Min, lb			
1/16 0.062	0.069	0.066	0.059	0.028	0.007	0.012	0.065	0.062	425	350	270	3/16-1
5/64 0.078	0.086	0.083	0.075	0.032	0.008	0.018	0.081	0.078	650	550	400	3/16-1½
3/32 0.094	0.103	0.099	0.091	0.038	0.008	0.022	0.097	0.094	1,000	800	660	3/16-1½
1/8 0.125	0.135	0.131	0.122	0.044	0.008	0.028	0.129	0.125	2,100	1,500	1,200	5/16-2
9/64 0.141	0.149	0.145	0.137	0.044	0.008	0.028	0.144	0.140	2,200	1,600	1,400	3/8-2
5/32 0.156	0.167	0.162	0.151	0.048	0.010	0.032	0.160	0.156	3,000	2,000	1,800	7/16-2½
3/16 0.188	0.199	0.194	0.182	0.055	0.011	0.040	0.192	0.187	4,400	2,800	2,600	½-2½
7/32 0.219	0.232	0.226	0.214	0.065	0.011	0.048	0.224	0.219	5,700	3,550	3,700	½-3
1/4 0.250	0.264	0.258	0.245	0.065	0.012	0.048	0.256	0.250	7,700	4,600	4,500	½-3½
5/16 0.312	0.328	0.321	0.306	0.080	0.014	0.062	0.318	0.312	11,500	7,095	6,800	¾-4
3/8 0.375	0.392	0.385	0.368	0.095	0.016	0.077	0.382	0.375	17,600	10,000	10,100	¾, 7/8, 1, 1¼, 1½, 1¾, 2-4
7/16 0.438	0.456	0.448	0.430	0.095	0.017	0.077	0.445	0.437	20,000	12,000	12,200	1, 1¼, 1½, 1¾, 2-4
1/2 0.500	0.521	0.513	0.485	0.110	0.025	0.094	0.510	0.500	25,800	15,500	16,800	1¼, 1½, 1¾, 2-4
5/8 0.625	0.650	0.640	0.608	0.125	0.030	0.125	0.636	0.625	46,000[2]	18,800	. . .	2-6
3/4 0.750	0.780	0.769	0.730	0.150	0.030	0.150	0.764	0.750	66,000[2]	23,200	. . .	. . .

All dimensions are in inches.

[1] Where specifying nominal size in decimals, zeros preceding decimal point are omitted.

[2] Sizes 5/8 and 3/4 inch are produced from AISI 6150 alloy steel, not AISI 1070-1095 carbon steel.

[3] Length increments are 1/16 inch from 1/8 to 1 inch; 1/8 inch from 1 inch to 2 inches; and 1/4 inch from 2 inches to 6 inches.

Table 11. American National Standard Coiled Type Spring Pins (ANSI B18.8.2-1978)

SWAGED CHAMFER BOTH ENDS, CONTOUR OF CHAMFER OPTIONAL

BREAK EDGE

Nominal Size[1] or Basic Pin Diameter	Pin Diameter, A						Chamfer		Recommended Hole Size		Double Shear Load, Min, lb					
	Standard Duty		Heavy Duty		Light Duty		Dia., B	Length, C			Standard Duty		Heavy Duty		Light Duty	
	Max	Min	Max	Min	Max	Min	Max	Ref	Max	Min	AISI 1070-1095 and AISI 420	AISI 302	AISI 1070-1095 and AISI 420	AISI 302	AISI 1070-1095 and AISI 420	AISI 302
1/32 0.031	0.035	0.033	…	…	…	…	0.029	0.024	0.032	0.031	75[2]	60	…	…	…	…
0.039	0.044	0.041	…	…	…	…	0.037	0.024	0.040	0.039	120[2]	100	…	…	…	…
3/64 0.047	0.052	0.049	…	…	…	…	0.045	0.024	0.048	0.046	170[2]	140	…	…	…	…
0.052	0.057	0.054	…	…	…	…	0.050	0.024	0.053	0.051	230[2]	190	…	…	…	…
1/16 0.062	0.072	0.067	0.070	0.066	0.073	0.067	0.059	0.028	0.065	0.061	300	250	450	350	…	135
5/64 0.078	0.088	0.083	0.086	0.082	0.089	0.083	0.075	0.032	0.081	0.077	475	400	700	550	…	225
3/32 0.094	0.105	0.099	0.103	0.098	0.106	0.099	0.091	0.038	0.097	0.093	700	550	1,000	800	375	300
7/64 0.109	0.120	0.114	0.118	0.113	0.121	0.114	0.106	0.038	0.112	0.108	950	750	1,400	1,125	525	425
1/8 0.125	0.138	0.131	0.136	0.130	0.139	0.131	0.121	0.044	0.129	0.124	1,250	1,000	2,100	1,700	675	550
5/32 0.156	0.171	0.163	0.168	0.161	0.172	0.163	0.152	0.048	0.160	0.155	1,925	1,550	3,000	2,400	1,100	875
3/16 0.188	0.205	0.196	0.202	0.194	0.207	0.196	0.182	0.055	0.192	0.185	2,800	2,250	4,400	3,500	1,500	1,200
7/32 0.219	0.238	0.228	0.235	0.226	0.240	0.228	0.214	0.065	0.224	0.217	3,800	3,000	5,700	4,600	2,100	1,700
1/4 0.250	0.271	0.260	0.268	0.258	0.273	0.260	0.243	0.065	0.256	0.247	5,000	4,000	7,700	6,200	2,700	2,200
5/16 0.312	0.337	0.324	0.334	0.322	0.339	0.324	0.304	0.080	0.319	0.308	7,700	6,200	11,500	9,200	4,440	3,500
3/8 0.375	0.403	0.388	0.400	0.386	0.405	0.388	0.366	0.095	0.383	0.370	11,200	9,000	17,600	14,000	6,000	5,000
7/16 0.438	0.469	0.452	0.466	0.450	0.471	0.452	0.427	0.095	0.446	0.431	15,200	13,000	22,500	18,000	8,400	6,400
1/2 0.500	0.535	0.516	0.532	0.514	0.537	0.516	0.488	0.110	0.510	0.493	21,000[3]	16,000	30,000	24,000	11,000	8,800
5/8 0.625	0.661	0.642	0.658	0.640	…	…	0.613	0.125	0.635	0.618	31,000[3]	25,000	46,000[3]	37,000	…	…
3/4 0.750	0.787	0.768	0.784	0.766	…	…	0.738	0.150	0.760	0.743	45,000[3]	36,000	66,000[3]	53,000	…	…

All dimensions are in inches. [1] Where specifying nominal size in decimals, zeros preceding decimal are omitted. [2] Sizes 1/32 inch through 0.052 inch are not available in AISI 1070-1095 carbon steel. [3] Sizes 3/8 inch and larger are produced from AISI 6150 alloy steel, not AISI 1070-1095 carbon steel. Practical lengths, L, for sizes 1/32 through 0.052 inch are 1/8 through 3/4 inches, and for the 7/64-inch size, 1/4 through 1 3/4 inches. For lengths of other sizes see Table 10.

Table 1. American National Standard Metric Tapered Retaining Rings — Basic External Series — 3AM1 (ANSI B27.7-1977)

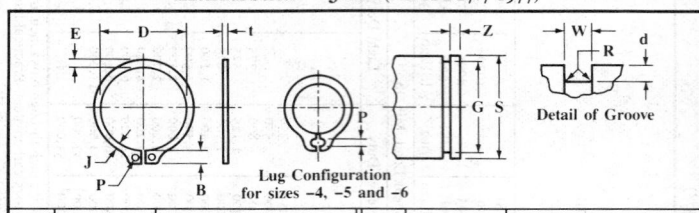

Detail of Groove

Lug Configuration for sizes −4, −5 and −6

	Ring		Groove					Ring		Groove			
Shaft Diam.	Free Diam.	Thickness	Diam.	Width	Depth	Edge Margin	Shaft Diam.	Free Diam.	Thickness	Diam.	Width	Depth	Edge Margin
S	D	t	G	W	d ref	Z min	S	D	t	G	W	d ref	Z min
4	3.60	0.25	3.80	0.32	0.1	0.3	36	33.25	1.3	33.85	1.4	1.06	3.2
5	4.55	0.4	4.75	0.5	0.13	0.4	38	35.20	1.3	35.8	1.4	1.10	3.3
6	5.45	0.4	5.70	0.5	0.15	0.5	40	36.75	1.6	37.7	1.75	1.15	3.4
7	6.35	0.6	6.60	0.7	0.20	0.6	42	38.80	1.6	39.6	1.75	1.20	3.6
8	7.15	0.6	7.50	0.7	0.25	0.8	43	39.85	1.6	40.5	1.75	1.25	3.8
9	8.15	0.6	8.45	0.7	0.28	0.8	45	41.60	1.6	42.4	1.75	1.30	3.9
10	9.00	0.6	9.40	0.7	0.30	0.9	46	42.55	1.6	43.3	1.75	1.35	4.0
11	10.00	0.6	10.35	0.7	0.33	1.0	48	44.40	1.6	45.2	1.75	1.40	4.2
12	10.85	0.6	11.35	0.7	0.33	1.0	50	46.20	1.6	47.2	1.75	1.40	4.2
13	11.90	0.9	12.30	1.0	0.35	1.0	52	48.40	2.0	49.1	2.15	1.45	4.3
14	12.90	0.9	13.25	1.0	0.38	1.2	54	49.9	2.0	51.0	2.15	1.50	4.5
15	13.80	0.9	14.15	1.0	0.43	1.3	55	50.6	2.0	51.8	2.15	1.60	4.8
16	14.70	0.9	15.10	1.0	0.45	1.4	57	52.9	2.0	53.8	2.15	1.60	4.8
17	15.75	0.9	16.10	1.0	0.45	1.4	58	53.6	2.0	54.7	2.15	1.65	4.9
18	16.65	1.1	17.00	1.2	0.50	1.5	60	55.8	2.0	56.7	2.15	1.65	4.9
19	17.60	1.1	17.95	1.2	0.53	1.6	62	57.3	2.0	58.6	2.15	1.70	5.1
20	18.35	1.1	18.85	1.2	0.58	1.7	65	60.4	2.0	61.6	2.15	1.70	5.1
21	19.40	1.1	19.80	1.2	0.60	1.8	68	63.1	2.0	64.5	2.15	1.75	5.3
22	20.30	1.1	20.70	1.2	0.65	1.9	70	64.6	2.4	66.4	2.55	1.80	5.4
23	21.25	1.1	21.65	1.2	0.67	2.0	72	66.6	2.4	68.3	2.55	1.85	5.5
24	22.20	1.1	22.60	1.2	0.70	2.1	75	69.0	2.4	71.2	2.55	1.90	5.7
25	23.10	1.1	23.50	1.2	0.75	2.3	78	72.0	2.4	74.0	2.55	2.00	6.0
26	24.05	1.1	24.50	1.2	0.75	2.3	80	74.2	2.4	75.9	2.55	2.05	6.1
27	24.95	1.3	25.45	1.4	0.78	2.3	82	76.4	2.4	77.8	2.55	2.10	6.3
28	25.80	1.3	26.40	1.4	0.80	2.4	85	78.5	2.4	80.6	2.55	2.20	6.6
30	27.90	1.3	28.35	1.4	0.83	2.5	88	81.4	2.8	83.5	2.95	2.25	6.7
32	29.60	1.3	30.20	1.4	0.90	2.7	90	83.2	2.8	85.4	2.95	2.30	6.9
34	31.40	1.3	32.00	1.4	1.00	3.0	95	88.1	2.8	90.2	2.95	2.40	7.2
35	32.30	1.3	32.90	1.4	1.05	3.1	100	92.5	2.8	95.0	2.95	2.50	7.5

All dimensions are in millimeters. Sizes −4, −5, and −6 are available in beryllium copper only.

These rings are designated by series symbol and shaft diameter, thus: for a 4 mm diameter shaft, 3AM1–4; for a 20 mm diameter shaft, 3AM1–20; etc.

Ring Free Diameter Tolerances: For ring sizes −4 through −6, +0.05, −0.10 mm; for sizes −7 through −12, +0.05, −0.15 mm; for sizes −13 through −26, +0.15, −0.25 mm; for sizes −27 through −38, +0.25, −0.40 mm; for sizes −40 through −50, +0.35, −0.50 mm; for sizes −52 through −62, +0.35, −0.65 mm; and for sizes −65 through −100, +0.50, −0.75 mm.

Groove Diameter Tolerances: For ring sizes −4 through −6, −0.08 mm; for sizes −7 through −10, −0.10 mm; for sizes −11 through −15, −0.12 mm; for sizes −16 through −26, −0.15 mm; for sizes −27 through −36, −0.020 mm; for sizes −38 through −55, −0.30 mm; and for sizes −57 through −100, −0.40 mm.

Groove Diameter F.I.M. (full indicator movement) or maximum allowable deviation of concentricity between groove and shaft: For ring sizes −4 through −6, 0.03 mm; for ring sizes −7 through −12, 0.05 mm; for sizes −13 through −28, 0.10 mm; for sizes −30 through −55, 0.15 mm; and for sizes −57 through −100, 0.20 mm.

Groove Width Tolerances: For ring size −4, +0.05 mm; for sizes −5 and −6, +0.10 mm, for sizes −7 through −38, +0.15 mm; and for sizes −40 through −100, +0.20 mm.

Groove Maximum Bottom Radii, R: For ring sizes −4 through −6, none; for sizes −7 through −18, 0.1 mm; for sizes −19 through −30, 0.2 mm; for sizes −32 through −50, 0.3 mm; and for sizes −52 through −100, 0.4 mm. For manufacturing details not shown, including materials, see ANSI B27.7-1977.

Table 2. American National Standard Metric Tapered Retaining Rings — Basic Internal Series — 3BM1 (ANSI B27.7-1977)

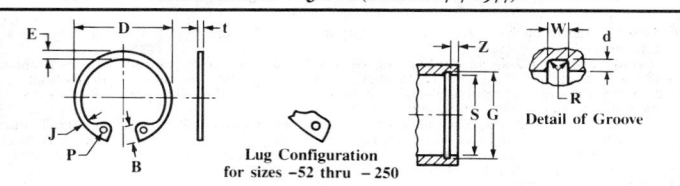

Lug Configuration
for sizes −52 thru −250

Detail of Groove

	Ring		Groove					Ring		Groove			
Shaft Diam.	Free Diam.	Thickness	Diam.	Width	Depth	Edge Margin	Shaft Diam.	Free Diam.	Thickness	Diam.	Width	Depth	Edge Margin
S	D	t	G	W	d ref	Z min	S	D	t	G	W	d ref	Z min
8	8.80	0.4	8.40	0.5	0.2	0.6	65	72.2	2.4	69.0	2.55	2.00	6.0
9	10.00	0.6	9.45	0.7	0.23	0.7	68	75.7	2.4	72.2	2.55	2.10	6.3
10	11.10	0.6	10.50	0.7	0.25	0.8	70	77.5	2.4	74.4	2.55	2.20	6.6
11	12.20	0.6	11.60	0.7	0.3	0.9	72	79.6	2.4	76.5	2.55	2.25	6.7
12	13.30	0.6	12.65	0.7	0.33	1.0	75	83.3	2.4	79.7	2.55	2.35	7.1
13	14.25	0.9	13.70	1.0	0.35	1.1	78	86.8	2.8	82.8	2.95	2.40	7.2
14	15.45	0.9	14.80	1.0	0.40	1.2	80	89.1	2.8	85.0	2.95	2.50	7.5
15	16.60	0.9	15.85	1.0	0.43	1.3	82	91.1	2.8	87.2	2.95	2.60	7.8
16	17.70	0.9	16.90	1.0	0.45	1.4	85	94.4	2.8	90.4	2.95	2.70	8.1
17	18.90	0.9	18.00	1.0	0.50	1.5	88	97.9	2.8	93.6	2.95	2.80	8.4
18	20.05	0.9	19.05	1.0	0.53	1.6	90	100.0	2.80	95.7	2.95	2.85	8.6
19	21.10	0.9	20.10	1.0	0.55	1.7	92	102.2	2.8	97.8	2.95	2.90	8.7
20	22.25	0.9	21.15	1.0	0.57	1.7	95	105.6	2.8	101.0	2.95	3.00	9.0
21	23.30	0.9	22.20	1.0	0.60	1.8	98	109.0	2.8	104.2	2.95	3.10	9.3
22	24.40	1.1	23.30	1.2	0.65	1.9	100	110.7	2.8	106.3	2.95	3.15	9.5
23	25.45	1.1	24.35	1.2	0.67	2.0	102	112.4	2.8	108.4	2.95	3.20	9.6
24	26.55	1.1	25.4	1.2	0.70	2.1	105	115.8	2.8	111.5	2.95	3.25	9.8
25	27.75	1.1	26.6	1.2	0.80	2.4	108	119.2	2.8	114.6	2.95	3.30	9.9
26	28.85	1.1	27.7	1.2	0.85	2.6	110	120.8	2.8	116.7	2.95	3.35	10.1
27	29.95	1.3	28.8	1.4	0.90	2.7	115	126.0	2.8	121.9	2.95	3.45	10.4
28	31.10	1.3	29.8	1.4	0.90	2.7	120	132.4	2.8	127.0	2.95	3.50	10.5
30	33.40	1.3	31.9	1.4	0.90	2.9	125	137.1	2.8	132.1	2.95	3.55	10.7
32	35.35	1.3	33.9	1.4	0.95	2.9	130	142.5	2.8	137.2	2.95	3.60	10.8
34	37.75	1.3	36.1	1.4	1.05	3.2	135	148.5	3.2	142.3	3.40	3.65	11.0
35	38.75	1.3	37.2	1.4	1.10	3.3	140	154.1	3.2	147.4	3.40	3.70	11.1
36	40.00	1.3	38.3	1.4	1.15	3.5	145	159.5	3.2	152.5	3.40	3.75	11.3
37	41.05	1.3	39.3	1.4	1.15	3.5	150	164.5	3.2	157.6	3.40	3.80	11.4
38	42.15	1.3	40.4	1.4	1.20	3.6	155	168.8	3.2	162.7	3.40	3.85	11.6
40	44.25	1.6	42.4	1.75	1.20	3.6	160	175.1	4.0	167.8	4.25	3.90	11.7
42	46.60	1.6	44.5	1.75	1.25	3.7	165	180.3	4.0	172.9	4.25	3.95	11.9
45	49.95	1.6	47.6	1.75	1.30	3.9	170	185.6	4.0	178.0	4.25	4.00	12.0
46	51.05	1.6	48.7	1.75	1.35	4.0	175	191.3	4.0	183.2	4.25	4.10	12.3
47	52.15	1.6	49.8	1.75	1.40	4.2	180	196.6	4.0	188.4	4.25	4.20	12.6
48	53.30	1.6	50.9	1.75	1.45	4.3	185	202.7	4.8	193.6	5.10	4.30	12.9
50	55.35	1.6	53.1	1.75	1.55	4.6	190	207.7	4.8	198.8	5.10	4.40	13.2
52	57.90	2.0	55.3	2.15	1.65	5.0	200	217.8	4.8	209.0	5.10	4.50	13.5
55	61.10	2.0	58.4	2.15	1.70	5.1	210	230.3	4.8	219.4	5.10	4.70	14.1
57	63.25	2.0	60.5	2.15	1.75	5.3	220	240.5	4.8	230.0	5.10	5.00	15.0
58	64.4	2.0	61.6	2.15	1.80	5.4	230	251.4	4.8	240.6	5.10	5.30	15.9
60	66.8	2.0	63.8	2.15	1.90	5.7	240	262.3	4.8	251.0	5.10	5.50	16.5
62	68.6	2.0	65.8	2.15	1.90	5.7	250	273.3	4.8	261.4	5.10	5.70	17.1
63	69.9	2.0	66.9	2.15	1.95	5.9	...	...	...	...	...	...	...

All dimensions are in millimeters.

These rings are designated by series symbol and shaft diameter, thus: for a 9 mm diameter shaft, 3BM1-9; for a 22 mm diameter shaft, 3BM1-22; etc.

Ring Free Diameter Tolerances: For ring sizes −8 through −20, +0.25, −0.13 mm; for sizes −21 through −26, +0.40, −0.25 mm; for sizes −27 through −38, +0.65, −0.50 mm; for sizes −40 through −50, +0.90, −0.65 mm; for sizes −52 through −75, +1.00, −0.75 mm; for sizes −78 through −92, +1.40, −1.40 mm; for sizes −95 through −155, +1.65, −1.65 mm;

(Continued on next page)

Table 2 (*Concluded*). American National Standard Metric Tapered Retaining Rings — Basic Internal Series — 3BMI (ANSI B27.7-1977)

for sizes − 160 through − 180, + 2.05, − 2.05 mm; and for sizes − 185 through − 250, + 2.30, − 2.30 mm.

Groove Diameter Tolerances: For ring sizes − 8 and − 9, + 0.06 mm; for sizes − 10 through − 18, + 0.10 mm; for sizes − 19 through − 28, + 0.15 mm; for sizes − 30 through − 50, + 0.20 mm; for sizes − 52 through − 98, + 0.30; for sizes − 100 through − 160, + 0.40 mm; and for sizes − 165 through − 250, + 0.50 mm.

Groove Diameter F.I.M. (full indicator movement) or maximum allowable deviation of concentricity between groove and shaft: For ring sizes − 8 through − 10, 0.03 mm; for sizes − 11 through − 15, 0.05 mm; for sizes − 16 through − 25, 0.10 mm; for sizes − 26 through − 45, 0.15 mm; for sizes − 46 through − 80, 0.20 mm; for sizes − 82 through − 150, 0.25 mm; and for sizes − 155 through − 250, 0.30 mm.

Groove Width Tolerances: For ring size − 8, + 0.10 mm; for sizes − 9 through − 38, + 0.15 mm; for sizes − 40 through − 130, + 0.20 mm; and for sizes − 135 through − 250, + 0.25 mm.

Groove Maximum Bottom Radii: For ring sizes − 8 through − 17, 0.1 mm; for sizes − 18 through − 30, 0.2 mm; for sizes − 32 through − 55, 0.3 mm; and for sizes − 56 through − 250, 0.4 mm.

For manufacturing details not shown, including materials, see ANSI B27.7-1977.

Table 3. American National Standard Metric Reduced Cross Section Retaining Rings — E Ring External Series — 3CMI (ANSI B27.7-1977)

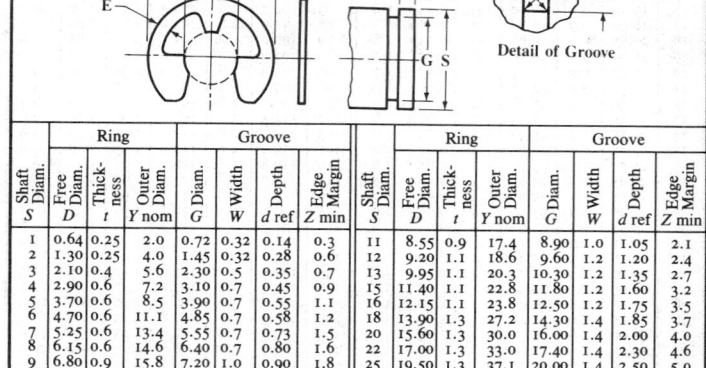

Detail of Groove

Shaft Diam. S	Ring Free Diam. D	Ring Thickness t	Ring Outer Diam. Y nom	Groove Diam. G	Groove Width W	Groove Depth d ref	Groove Edge Margin Z min	Shaft Diam. S	Ring Free Diam. D	Ring Thickness t	Ring Outer Diam. Y nom	Groove Diam. G	Groove Width W	Groove Depth d ref	Groove Edge Margin Z min
1	0.64	0.25	2.0	0.72	0.32	0.14	0.3	11	8.55	0.9	17.4	8.90	1.0	1.05	2.1
2	1.30	0.25	4.0	1.45	0.32	0.28	0.6	12	9.20	1.1	18.6	9.60	1.2	1.20	2.4
3	2.10	0.4	5.6	2.30	0.5	0.35	0.7	13	9.95	1.1	20.3	10.30	1.2	1.35	2.7
4	2.90	0.6	7.2	3.10	0.7	0.45	0.9	15	11.40	1.1	22.8	11.80	1.2	1.60	3.2
5	3.70	0.6	8.5	3.90	0.7	0.55	1.1	16	12.15	1.1	23.8	12.50	1.2	1.75	3.5
6	4.70	0.6	11.1	4.85	0.7	0.58	1.2	18	13.90	1.3	27.2	14.30	1.4	1.85	3.7
7	5.25	0.6	13.4	5.55	0.7	0.73	1.5	20	15.60	1.3	30.0	16.00	1.4	2.00	4.0
8	6.15	0.6	14.6	6.40	0.7	0.80	1.6	22	17.00	1.3	33.0	17.40	1.4	2.30	4.6
9	6.80	0.9	15.8	7.20	1.0	0.90	1.8	25	19.50	1.3	37.1	20.00	1.4	2.50	5.0
10	7.60	0.9	16.8	8.00	1.0	1.00	2.0								

All dimensions are in millimeters. Size − 1 is available in beryllium copper only.

These rings are designated by series symbol and shaft diameter, thus: for a 2 mm diameter shaft, 3CM1-2; for a 13 mm shaft, 3CM1-13; etc.

Ring Free Diameter Tolerances: For sizes − 1 through − 7, + 0.03, − 0.08 mm; for sizes − 8 through − 13, + 0.05, − 0.10 mm; for sizes − 15 through − 25, + 0.10, − 0.15 mm.

Groove Diameter Tolerances: For ring sizes − 1 and − 2, − 0.05 mm; for sizes − 3 through − 6, − 0.08; for sizes − 7 through − 11, − 0.10 mm; for sizes − 12 through − 18, − 0.15 mm; and for sizes − 20 through − 25, − 0.20 mm.

Groove Diameter F.I.M. (Full Indicator Movement) or maximum allowable deviation of concentricity between groove and shaft: For ring sizes − 1 through − 3, 0.04 mm; for − 4 through − 6, 0.05 mm; for − 7 through − 10, 0.08 mm; for − 11 through − 25, 0.10 mm.

Groove Width Tolerances: For ring sizes − 1 and − 2, + 0.05 mm; for size − 3, + 0.10 mm; and for sizes − 4 through − 25, + 0.15 mm.

Groove Maximum Bottom Radii: For ring sizes − 1 and − 2, 0.05 mm; for − 3 through − 7, 0.15 mm; for − 8 through − 13, 0.25 mm; and for − 15 through − 25, 0.4 mm.

For manufacturing details not shown, including materials. See ANSI B27.7-1977.

Retaining Rings. — The preceding Table 1 covers Type 3AM1 tapered external retaining rings which are spread over a shaft by means of pliers or a special tool and allowed to relax and seat in a circumferential groove, thereby providing an external protruding shoulder which can be used for locating and retaining a part on a shaft.

Table 2 covers Type 3BM1 tapered internal retaining rings which are compressed into a housing by means of pliers or special tool and allowed to relax and seat in a circumferential groove, thereby providing an internal protruding shoulder which can be used for locating and retaining a part contained inside the housing.

Table 3 covers Type 3CM1 reduced cross section external retaining rings which contain 3 prongs connected by a reduced section bridge to provide greater resilience during installation. They are installed radially, usually by means of an applicator, and provide a high shoulder for abutment by a retained part.

Table 4. American National Standard Metric Basic External
Series 3AM1 Retaining Rings — Checking and Performance Data

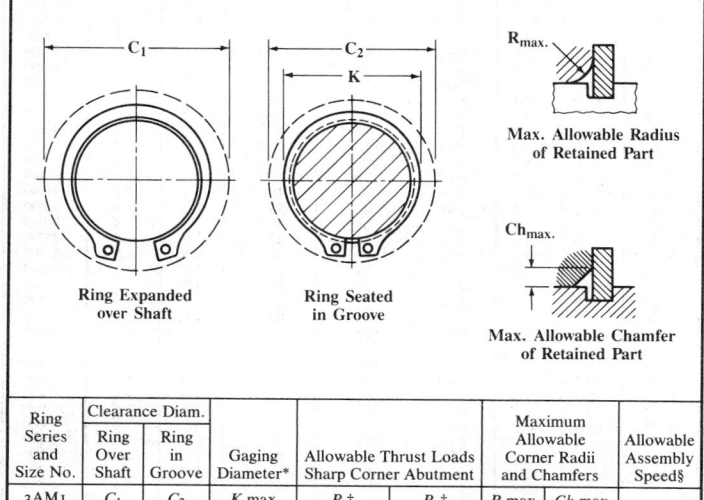

Ring Series and Size No.	Clearance Diam.		Gaging Diameter*	Allowable Thrust Loads Sharp Corner Abutment		Maximum Allowable Corner Radii and Chamfers		Allowable Assembly Speed§
	Ring Over Shaft	Ring in Groove						
3AM1	C_1	C_2	K max	P_r†	P_g‡	R max	Ch max	...
No.	mm	mm	mm	kN	kN	mm	mm	rpm
−4*	7.0	6.8	4.90	0.6	0.2	0.35	0.25	70 000
−5*	8.2	7.9	5.85	1.1	0.3	0.35	0.25	70 000
−6*	9.1	8.8	6.95	1.4	0.4	0.35	0.25	70 000
−7	12.3	11.8	8.05	2.6	0.7	0.45	0.3	60 000
−8	13.6	13.0	9.15	3.1	1.0	0.5	0.35	55 000
−9	14.5	13.8	10.35	3.5	1.2	0.6	0.35	48 000
−10	15.5	14.7	11.50	3.9	1.5	0.7	0.4	42 000
−11	16.4	15.6	12.60	4.3	1.8	0.75	0.45	38 000
−12	17.4	16.6	13.80	4.7	2.0	0.8	0.45	34 000
−13	19.7	18.8	15.05	7.5	2.2	0.8	0.5	31 000
−14	20.7	19.7	15.60	8.1	2.6	0.9	0.5	28 000
−15	21.7	20.6	17.20	8.7	3.2	1.0	0.6	27 000
−16	22.7	21.6	18.35	9.3	3.5	1.1	0.6	25 000
−17	23.7	22.6	19.35	9.9	4.0	1.1	0.6	24 000

See footnotes at end of table.   *(Table concluded on next page.)*

Table 4 (*Concluded*). **American National Standard Metric Basic External Series 3AM1 Retaining Rings — Checking and Performance Data**

Ring Series and Size No.	Clearance Diam. Ring Over Shaft	Clearance Diam. Ring in Groove	Gaging Diameter*	Allowable Thrust Loads Sharp Corner Abutment		Maximum Allowable Corner Radii and Chamfers		Allowable Assembly Speed§
3AM1	C_1	C_2	K max	P_r†	P_g‡	R max	Ch max	...
No.	mm	mm	mm	kN	kN	mm	mm	rpm
− 18	26.2	25.0	20.60	16.0	4.4	1.2	0.7	23 000
− 19	27.2	25.9	21.70	16.9	4.9	1.2	0.7	21 500
− 20	28.2	26.8	22.65	17.8	5.7	1.2	0.7	20 000
− 21	29.2	27.7	23.80	18.6	6.2	1.3	0.7	19 000
− 22	30.3	28.7	24.90	19.6	7.0	1.3	0.8	18 500
− 23	31.3	29.6	26.00	20.5	7.6	1.3	0.8	18 000
− 24	34.1	32.4	27.15	21.4	8.2	1.4	0.8	17 500
− 25	35.1	33.3	28.10	22.3	9.2	1.4	0.8	17 000
− 26	36.0	34.2	29.25	23.2	9.6	1.5	0.9	16 500
− 27	37.8	35.9	30.35	28.4	10.3	1.5	0.9	16 300
− 28	38.8	36.9	31.45	28.4	11.0	1.6	1.0	15 800
− 30	40.8	38.8	33.6	31.6	12.3	1.6	1.0	15 000
− 32	42.8	40.7	35.9	33.6	14.1	1.7	1.0	14 800
− 34	44.9	42.5	37.9	36	16.7	1.7	1.1	14 000
− 35	45.9	43.4	39.0	37	18.1	1.8	1.1	13 500
− 36	48.6	46.1	40.2	38	18.9	1.9	1.2	13 300
− 38	50.6	48.0	42.5	40	20.5	2.0	1.2	12 700
− 40	54.0	51.3	44.5	52	22.6	2.1	1.2	12 000
− 42	56.0	53.2	46.9	54	24.8	2.2	1.3	11 000
− 43	57.0	54.0	47.9	55	26.4	2.3	1.4	10 800
− 45	59.0	55.9	50.0	58	28.8	2.3	1.4	10 000
− 46	60.0	56.8	50.9	59	30.4	2.4	1.4	9 500
− 48	62.4	59.1	53.0	62	33	2.4	1.4	8 800
− 50	64.4	61.1	55.2	64	35	2.4	1.4	8 000
− 52	67.6	64.1	57.4	84	37	2.5	1.5	7 700
− 54	69.6	66.1	59.5	87	40	2.5	1.5	7 500
− 55	70.6	66.9	60.4	89	44	2.5	1.5	7 400
− 57	72.6	68.9	62.7	91	45	2.6	1.5	7 200
− 58	73.6	69.8	63.6	93	46	2.6	1.6	7 100
− 60	75.6	71.8	65.8	97	49	2.6	1.6	7 000
− 62	77.6	73.6	67.9	100	52	2.7	1.6	6 900
− 65	80.6	76.6	71.2	105	54	2.8	1.7	6 700
− 68	83.6	79.5	74.5	110	58	2.9	1.7	6 500
− 70	88.1	83.9	76.4	136	62	2.9	1.7	6 400
− 72	90.1	85.8	78.5	140	65	2.9	1.7	6 200
− 75	93.1	88.7	81.7	147	69	3.0	1.8	5 900
− 78	95.4	92.1	84.6	151	76	3.0	1.8	5 600
− 80	97.9	93.1	87.0	155	80	3.1	1.9	5 400
− 82	100.0	95.1	89.0	159	84	3.2	1.9	5 200
− 85	103.0	97.9	92.1	165	91	3.2	1.9	5 000
− 88	107.0	100.8	95.1	199	97	3.2	1.9	4 800
− 90	109.0	103.6	97.1	204	101	3.2	1.9	4 500
− 95	114.0	108.6	102.7	215	112	3.4	2.1	4 350
− 100	119.5	113.7	108.0	227	123	3.5	2.1	4 150

* For checking when ring is seated in groove.

† These values apply to rings made from SAE 1060-1090 steels and PH 15-7 Mo stainless steel used on shafts hardened to R_c 50 minimum, with the exception of sizes − 4, − 5, and − 6 which are supplied in beryllium copper only. Values for other sizes made from beryllium copper can be calculated by multiplying the listed values by 0.75. The values listed include a safety factor of 4.

‡ These values are for all standard rings used on low carbon steel shafts. They include a safety factor of 2.

§ These values have been calculated for steel rings.

Maximum allowable assembly loads with R max or Ch max are: For rings sizes − 4, 0.2 kN; for sizes − 5 and − 6, 0.5 kN; for sizes − 7 through − 12, 2.1 kN; for sizes − 13 through − 17, 4.0 kN; for sizes − 18 through − 26, 6.0 kN; for sizes − 27 through − 38, 8.6 kN; for sizes − 40 through − 50, 13.2 kN; for sizes − 52 through − 68, 22.0 kN; for sizes − 70 through − 85, 32 kN; and for sizes − 88 through − 100, 47 kN.

Source: Appendix to American National Standard ANSI B27.7-1977.

RETAINING RINGS 1459

Table 5. American National Standard Metric Basic Internal Series 3BM1 Retaining Rings — Checking and Performance Data

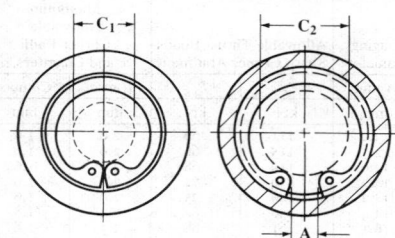

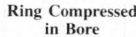

Ring Compressed in Bore

Ring Seated in Groove

R max.

Max. Allowable Radius of Retained Part

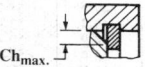

Ch max.

Max. Allowable Chamfer of Retained Part

Ring Series and Size No.	Clearance Diam.		Gaging Diameter*	Allowable Thrust Loads Sharp Corner Abutment		Maximum Allowable Corner Radii and Chamfers	
	Ring in Bore	Ring in Groove					
3BM1	C_1	C_2	A min	P_r†	P_g‡	R max	Ch max
No.	mm	mm	mm	kN	kN	mm	mm
− 8	4.4	4.8	1.40	2.4	1.0	0.4	0.3
− 9	4.6	5.0	1.50	4.4	1.2	0.5	0.35
− 10	5.5	6.0	1.85	4.9	1.5	0.5	0.35
− 11	5.7	6.3	1.95	5.4	2.0	0.6	0.4
− 12	6.7	7.3	2.25	5.8	2.4	0.6	0.4
− 13	6.8	7.5	2.35	8.9	2.6	0.7	0.5
− 14	6.9	7.7	2.65	9.7	3.2	0.7	0.5
− 15	7.9	8.7	2.80	10.4	3.7	0.7	0.5
− 16	8.8	9.7	2.80	11.0	4.2	0.7	0.5
− 17	9.8	10.8	3.35	11.7	4.9	0.75	0.6
− 18	10.3	11.3	3.40	12.3	5.5	0.75	0.6
− 19	11.4	12.5	3.40	13.1	6.0	0.8	0.65
− 20	11.6	12.7	3.8	13.7	6.6	0.9	0.7
− 21	12.6	13.8	4.2	14.5	7.3	0.9	0.7
− 22	13.5	14.8	4.3	22.5	8.3	0.9	0.7
− 23	14.5	15.9	4.9	23.5	8.9	1.0	0.8
− 24	15.5	16.9	5.2	24.8	9.7	1.0	0.8
− 25	16.5	18.1	6.0	25.7	11.6	1.0	0.8
− 26	17.5	19.2	5.7	26.8	12.7	1.2	1.0
− 27	17.4	19.2	5.9	33	14.0	1.2	1.0
− 28	18.2	20.0	6.0	34	14.6	1.2	1.0
− 30	20.0	21.9	6.0	37	16.5	1.2	1.0
− 32	22.0	23.9	7.3	39	17.6	1.2	1.0
− 34	24.0	26.1	7.6	42	20.6	1.2	1.0
− 35	25.0	27.2	8.0	43	22.3	1.2	1.0
− 36	26.0	28.3	8.3	44	23.9	1.2	1.0
− 37	27.0	29.3	8.4	45	24.6	1.2	1.0
− 38	28.0	30.4	8.6	46	26.4	1.2	1.0
− 40	29.2	31.6	9.7	62	27.7	1.7	1.3
− 42	29.7	32.2	9.0	65	30.2	1.7	1.3
− 45	32.3	34.9	9.6	69	33.8	1.7	1.3
− 46	33.3	36.0	9.7	71	36	1.7	1.3
− 47	34.3	37.1	10.0	72	38	1.7	1.3
− 48	35.0	37.9	10.5	74	40	1.7	1.3
− 50	36.9	40.0	12.1	77	45	1.7	1.3
− 52	38.6	41.9	11.7	99	50	2.0	1.6
− 55	40.8	44.2	11.9	105	54	2.0	1.6
− 57	42.2	45.7	12.5	109	58	2.0	1.6

See footnotes at end of table.

(*Table concluded on next page*)

Table 5 (*Concluded*). **American National Standard Metric Basic Internal Series 3BM1 Retaining Rings — Checking and Performance Data**

Ring Series and Size No.	Clearance Diam.		Gaging Diameter*	Allowable Thrust Loads Sharp Corner Abutment		Maximum Allowable Corner Radii and Chamfers	
	Ring in Bore	Ring in Groove					
3BM1	C_1	C_2	A min	P_r†	P_g‡	R max	Ch max
No.	mm	mm	mm	kN	kN	mm	mm
−58	43.2	46.8	13.0	111	60	2.0	1.6
−60	45.5	49.3	12.7	115	66	2.0	1.6
−62	47.0	50.8	14.0	119	68	2.0	1.6
−63	47.8	51.7	14.2	120	71	2.0	1.6
−65	49.4	53.4	14.2	149	75	2.0	1.6
−68	52.0	56.2	14.4	156	82	2.3	1.8
−70	53.8	58.2	16.1	161	88	2.3	1.8
−72	55.9	60.4	17.4	166	93	2.3	1.8
−75	58.2	62.9	16.8	172	101	2.3	1.8
−78	61.2	66.0	17.6	209	108	2.5	2.0
−80	63.0	68.0	17.2	215	115	2.5	2.0
−82	63.5	68.7	18.8	220	122	2.6	2.1
−85	66.8	72.2	19.1	228	131	2.6	2.1
−88	69.6	75.2	20.4	236	141	2.8	2.2
−90	71.6	77.3	21.4	241	147	2.8	2.2
−92	73.6	79.4	22.2	247	153	2.9	2.4
−95	76.7	82.7	22.6	255	164	3.0	2.5
−98	78.3	84.5	22.6	263	174	3.0	2.5
−100	80.3	86.6	24.1	269	181	3.1	2.5
−102	82.2	88.6	25.5	273	187	3.2	2.6
−105	85.1	91.6	26.0	281	196	3.3	2.6
−108	88.1	94.7	26.4	290	205	3.5	2.7
−110	88.4	95.1	27.5	295	212	3.6	2.8
−115	93.2	100.1	29.4	309	227	3.7	2.9
−120	98.2	105.2	27.2	321	241	3.9	3.1
−125	103.1	110.2	30.3	335	255	4.0	3.2
−130	108.0	115.2	31.0	349	269	4.0	3.2
−135	110.4	117.7	30.4	415	283	4.3	3.4
−140	115.3	122.7	30.4	429	298	4.3	3.4
−145	120.4	127.9	31.6	444	313	4.3	3.4
−150	125.3	132.9	33.5	460	327	4.3	3.4
−155	130.4	138.1	37.0	475	343	4.3	3.4
−160	133.8	141.6	35.0	613	359	4.5	3.6
−165	138.7	146.6	33.1	632	374	4.6	3.7
−170	143.6	151.6	38.2	651	390	4.6	3.7
−175	146.0	154.2	37.7	670	403	4.8	3.8
−180	151.4	159.8	39.0	690	434	5.0	4.0
−185	154.7	163.3	37.3	851	457	5.1	4.1
−190	159.5	168.3	35.0	873	480	5.3	4.3
−200	169.2	178.2	43.9	919	517	5.4	4.3
−210	177.5	186.9	40.6	965	566	5.8	4.6
−220	184.1	194.1	38.3	1000	608	6.1	4.9
−230	194.0	204.6	49.0	1060	686	6.3	5.1
−240	200.4	211.4	45.4	1090	725	6.6	5.3
−250	210.0	221.4	53.0	1150	808	6.7	5.4

* For checking when ring is seated in groove.

† These values apply to rings made from SAE 1060-1090 steels and PH 15-7 Mo stainless steel used in bores hardened to R_c 50 minimum. Values for rings made from beryllium copper can be calculated by multiplying the listed values by 0.75. The values listed include a safety factor of 4.

‡ These values are for standard rings used in low carbon steel bores. They include a safety factor of 2.

Maximum allowable assembly loads for R max or Ch max are: for ring size −8, 0.8 kN; for sizes −9 through −12, 2.0 kN; for sizes −13 through −22, 4.0 kN; for sizes −22 through −26, 7.4 kN; for sizes −27 through −38, 10.8 kN; for sizes −40 through −50, 17.4 kN; for sizes −52 through −63, 27.4 kN; for size −65, 42.0 kN; for sizes −68 through −72, 39 kN; for sizes −75 through −130, 54 kN; for sizes −135 through −155, 67 kN; for sizes −160 through −180, 102 kN; and for sizes −185 through −250, 151 kN.

Source: Appendix to American National Standard ANSI B27.7-1977.

Table 6. American National Standard Metric E-Type External Series 3CM1 Retaining Rings — Checking and Performance Data

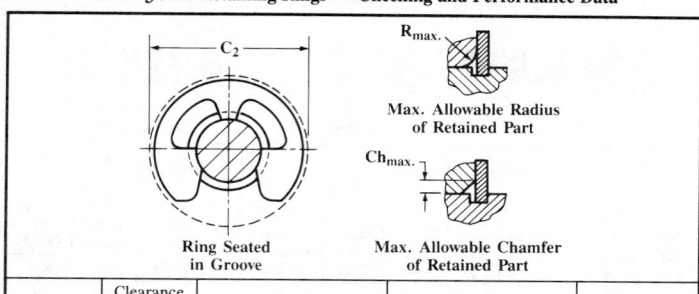

Ring Seated in Groove

Max. Allowable Radius of Retained Part

Max. Allowable Chamfer of Retained Part

Ring Series and Size No.	Clearance Diameter Ring in Groove	Allowable Thrust Loads Sharp Corner Abutment		Maximum Allowable Corner Radii and Chamfers		Allowable Assembly Speed§
3CM1	C_2	P_r†	P_g‡	R max	Ch max	...
No.	mm	kN	kN	mm	mm	rpm
− 1	2.2	0.06	0.02	0.4	0.25	40 000
− 2	4.3	0.13	0.09	0.8	0.5	40 000
− 3	6.0	0.3	0.17	1.1	0.7	34 000
− 4	7.6	0.7	0.3	1.6	1.2	31 000
− 5	8.9	0.9	0.4	1.6	1.2	27 000
− 6	11.5	1.1	0.6	1.6	1.2	25 000
− 7	14.0	1.2	0.8	1.6	1.2	23 000
− 8	15.1	1.4	1.0	1.7	1.3	21 500
− 9	16.5	3.0	1.3	1.7	1.3	19 500
− 10	17.5	3.4	1.6	1.7	1.3	18 000
− 11	18.0	3.7	1.9	1.7	1.3	16 500
− 12	19.3	4.9	2.3	1.9	1.4	15 000
− 13	21.0	5.4	2.9	2.0	1.5	13 000
− 15	23.5	6.2	4.0	2.0	1.5	11 500
− 16	24.5	6.6	4.5	2.0	1.5	10 000
− 18	27.9	8.7	5.4	2.1	1.6	9 000
− 20	30.7	9.8	6.5	2.2	1.7	8 000
− 22	33.7	10.8	8.1	2.2	1.7	7 000
− 25	37.9	12.2	10.1	2.4	1.9	5 000

† These values apply to rings made from SAE 1060-1090 steels and PH 15-7 Mo stainless steel used on shafts hardened to R_c 50 minimum, with the exception of size − 1 which is supplied in beryllium copper only. Values for other sizes made from beryllium copper can be calculated by multiplying the listed values by 0.75. The values listed include a safety factor of 4.

‡ These values apply to all standard rings used on low carbon steel shafts. They include a safety factor of 2.

§ These values have been calculated for steel rings.

Maximum allowable assembly loads with R max or Ch max are as follows:

Ring Size No.	Maximum Allowable Load, kN	Ring Size No.	Maximum Allowable Load, kN	Ring Size No.	Maximum Allowable Load, kN
− 1	0.06	− 8	1.4	− 16	6.6
− 2	0.13	− 9	3.0	− 18	8.7
− 3	0.3	− 10	3.4	− 20	9.8
− 4	0.7	− 11	3.7	− 22	10.8
− 5	0.9	− 12	4.9	− 25	12.2
− 6	1.1	− 13	5.4	...	...
− 7	1.2	− 15	6.2	...	...

Source: Appendix to American National Standard ANSI B27.7-1977.

American National Standard Flat, Pan, and Oval Head Wood Screws
(ANSI B18.6.1-1981)

FLAT HEAD   PAN HEAD   OVAL HEAD

		D†	J		A		B		P	H
					Head Diameter					
Nominal Size	Threads per Inch	Basic Diam. of Screw	Width of Slot		Max., Sharp Edge	Min., Edge Rounded or Flat	Head Diameter		Head Radius	Height of Head‡
			Max.	Min.			Max.	Min.	Max.	Ref.
0	32	.060	.023	.016	.119	.099	.116	.104	.020	.035
1	28	.073	.026	.019	.146	.123	.142	.130	.025	.043
2	26	.086	.031	.023	.172	.147	.167	.155	.035	.051
3	24	.099	.035	.027	.199	.171	.193	.180	.037	.059
4	22	.112	.039	.031	.225	.195	.219	.205	.042	.067
5	20	.125	.043	.035	.252	.220	.245	.231	.044	.075
6	18	.138	.048	.039	.279	.244	.270	.256	.046	.083
7	16	.151	.048	.039	.305	.268	.296	.281	.049	.091
8	15	.164	.054	.045	.332	.292	.322	.306	.052	.100
9	14	.177	.054	.045	.358	.316	.348	.331	.056	.108
10	13	.190	.060	.050	.385	.340	.373	.357	.061	.116
12	11	.216	.067	.056	.438	.389	.425	.407	.078	.132
14	10	.242	.075	.064	.507	.452	.492	.473	.087	.153
16	9	.268	.075	.064	.544	.485	.528	.508	.094	.164
18	8	.294	.084	.072	.635	.568	.615	.594	.099	.191
20	8	.320	.084	.072	.650	.582	.631	.608	.121	.196
24	7	.372	.094	.081	.762	.685	.740	.716	.143	.230

		O		K		T		U		V	
Nominal Size	Threads per Inch	Tot. Hgt. of Head		Height of Head		Depth of Slot		Depth of Slot		Depth of Slot	
		Max.	Min.	Max.	Min.	Max.	Min.	Max.	Min.	Max.	Min.
0	32	.056	.041	.039	.031	.015	.010	.022	.014	.030	.025
1	28	.068	.052	.046	.038	.019	.012	.027	.018	.038	.031
2	26	.080	.063	.053	.045	.023	.015	.031	.022	.045	.037
3	24	.092	.073	.060	.051	.027	.017	.035	.027	.052	.043
4	22	.104	.084	.068	.058	.030	.020	.040	.030	.059	.049
5	20	.116	.095	.075	.065	.034	.022	.045	.034	.067	.055
6	18	.128	.105	.082	.072	.038	.024	.050	.037	.074	.060
7	16	.140	.116	.089	.079	.041	.027	.054	.041	.081	.066
8	15	.152	.126	.096	.085	.045	.029	.058	.045	.088	.072
9	14	.164	.137	.103	.092	.049	.032	.063	.049	.095	.078
10	13	.176	.148	.110	.099	.053	.034	.068	.053	.103	.084
12	11	.200	.169	.125	.112	.060	.039	.077	.061	.117	.096
14	10	.232	.197	.144	.130	.070	.046	.087	.070	.136	.112
16	9	.248	.212	.153	.139	.075	.049	.093	.074	.146	.120
18	8	.290	.249	.178	.162	.083	.054	.106	.085	.171	.141
20	8	.296	.254	.182	.166	.090	.059	.108	.087	.175	.144
24	7	.347	.300	.212	.195	.106	.070	.124	.100	.204	.168

All dimensions in inches. The edges of flat and oval head screws may be flat or rounded. Wood screws are also available with Types I, IA, and II recessed heads. Consult the standard for recessed head dimensions.

* The length of the thread, L_T, on wood screws having cut threads shall be equivalent to approximately two-thirds of the nominal length of the screw. For rolled threads, L_T shall be at least four times the basic screw diameter or two-thirds of the nominal screw length, whichever is greater. Screws of nominal lengths that are too short to accommodate the minimum thread length shall have threads extending as close to the underside of the head as practicable.

† *Diameter Tolerance:* Equals + .004 in. and − .007 in. for cut threads. For rolled thread body diameter tolerances, see ANSI 18.6.1-1981.

Standard Wire Nails and Spikes
(Size, Length and Approximate Number to Pound)

Size of Nail	Length, Inches	Common Wire Nails and Brads		Flooring Brads		Fence Nails		Casing, Smooth and Barbed Box		Finishing Nails	
		Gage	No. to Lb.	Gage	No. to Lb.	Gage	No. to Lb.	Gage	No. to Lb.	Gage	No. to Lb.
2 d	1	15	876	...	...	...	...	15½	1010	16½	1351
3 d	1¼	14	568	...	...	...	...	14½	635	15½	807
4 d	1½	12½	316	...	...	...	...	14	473	15	584
5 d	1¾	12½	271	...	...	10	142	14	406	15	500
6 d	2	11½	181	11	157	10	124	12½	236	13½	309
7 d	2¼	11½	161	11	139	9	92	12½	210	13	238
8 d	2½	10¼	106	10	99	9	82	11½	145	12½	189
9 d	2¾	10¼	96	10	90	8	62	11½	132	12½	172
10 d	3	9	69	9	69	7	50	10½	94	11½	121
12 d	3¼	9	64	8	54	6	40	10½	87	11½	113
16 d	3½	8	49	7	43	5	30	10	71	11	90
20 d	4	6	31	6	31	4	23	9	52	10	62
30 d	4½	5	24	...	...	...	...	9	46	...	...
40 d	5	4	18	...	...	...	...	8	35	...	...
50 d	5½	3	16	...	...	...	...	...	...	...	...
60 d	6	2	11	...	...	...	...	...	...	...	...

Size and Length		Hinge Nails, Heavy		Hinge Nails, Light		Clinch Nails		Barbed Car Nails, Heavy		Barbed Car Nails, Light	
2 d	1	...	...	...	...	14	710	...	...	...	...
3 d	1¼	...	...	...	...	13	429	...	...	...	...
4 d	1½	3	50	6	82	12	274	10	165	12	274
5 d	1¾	3	38	6	62	12	235	9	118	10	142
6 d	2	3	30	6	50	11	157	9	103	10	124
7 d	2¼	00	12	3	25	11	139	8	76	9	92
8 d	2½	00	11	3	23	10	99	8	69	9	82
9 d	2¾	00	10	3	22	10	90	7	54	8	62
10 d	3	00	9	3	19	9	69	7	50	8	57
12 d	3¼	...	...	...	...	9	62	6	42	7	50
16 d	3½	...	...	...	...	8	49	6	35	7	43
20 d	4	...	...	...	...	7	37	5	26	6	31
30 d	4½	...	...	...	...	...	...	5	24	6	28
40 d	5	...	...	...	...	...	...	4	18	5	21
50 d	5½	...	...	...	...	...	...	3	15	4	17
60 d	6	...	...	...	...	...	...	3	13	4	15

Size and Length		Boat Nails, Heavy		Boat Nails, Light		Slating Nails	
2 d	1	...	...	...	...	12	411
3 d	1¼	...	...	...	...	10½	225
4 d	1½	¼	44	3/16	82	10½	187
5 d	1¾	...	...	...	...	10	142
6 d	2	¼	32	3/16	62	9	103
7 d	2¼	...	...	...	...	...	...
8 d	2½	¼	26	3/16	50	...	...
9 d	2¾	...	...	...	...	...	...
10 d	3	3/8	14	¼	22	...	...
12 d	3¼	3/8	13	¼	20	...	...
16 d	3½	3/8	12	¼	18	...	...
20 d	4	3/8	10	¼	16		
30 d	4½	...	...	...	...		
40 d	5	...	...				
50 d	5½						
60 d	6						

Spikes

Size and Length		Gage	No. to Lb.
10 d	3	6	41
12 d	3¼	6	38
16 d	3½	5	30
20 d	4	4	23
30 d	4½	3	17
40 d	5	2	13
50 d	5½	1	10
60 d	6	1	8
...	7	0	7
...	8	00	6
...	9	00	5
...	10	3/8	4
...	12	3/8	3

Wing Nuts, Wing Screws and Thumb Screws

Wing Nuts. — A wing nut is a nut having wings designed for manual turning without driver or wrench. As covered by ANSI Standard B18.17-1968 (R1975) wing nuts are classified first, by type on the basis of the method of manufacture; and second, by style on the basis of design characteristics. They consist of:

Type A: Type A wing nuts are cold forged or cold formed solid nuts having wings of moderate height. In some sizes they are produced in regular, light, and heavy series to best suit the requirements of specific applications. Dimensions are given in Table 1.

Type B: Type B wing nuts are hot forged solid nuts available in two wing styles: Style 1, having wings of moderate height; and Style 2, having high wings. Dimensions are given in Table 2.

Type C: Type C wing nuts are die cast solid nuts and are available in three wing styles: Style 1, having wings of moderate height; Style 2, having low wings; and Style 3, having high wings. In some sizes, the Style 1 nuts are produced in regular, light, and heavy series to best suit the requirements of specific applications. Dimensions are given in Table 3.

Type D: Type D wing nuts are stamped sheet metal nuts and are available in three styles: Style 1, having wings of moderate height; Style 2, having low wings; and Style 3, having wings of moderate height and a larger bearing surface. In some sizes, Styles 2 and 3 are produced in regular, light, and heavy series to best suit the requirements of specific applications. Dimensions are given in Table 4.

Specification of Wing Nuts. — When specifying wing nuts, the following data should be included in the designation and should appear in the following sequence: nominal size (number, fraction or decimal equivalent), threads per inch, type, style and/or series, material, and finish.

Examples:

    10-32 Type A Wing Nut, Regular Series, Steel, Zinc Plated.

    .250-20 Type C Wing Nut, Style 1, Zinc Alloy, Plain.

Threads for Wing Nuts. — Threads are in conformance with the ANSI Standard Unified Thread, Class 2B for all types of wing nuts except type D which have a modified Class 2B thread. Because of the method of manufacture, the minor diameter of the thread in type D nuts may be somewhat larger than the Unified Thread Class 2B maximum but shall in no case exceed the minimum pitch diameter.

Materials and Finish for Wing Nuts. — Types A, B, and D wing nuts are normally supplied as specified by the user in carbon steel, brass or corrosion resistant steel of good quality and adaptable to the manufacturing process. Type C wing nuts are made from die cast zinc alloy. Unless otherwise specified, wing nuts are supplied with a plain (unplated or uncoated) finish.

Wing Screws. — A wing screw is a screw having a wing-shaped head designed for manual turning without a driver or wrench. As covered by ANSI Standard B18.17-1968 (R1975) wing screws are classified first, by type on the basis of the method of manufacture, and second, by style on the basis of design characteristics. They consist of the following:

Type A: Type A wing screws are of two-piece construction having cold formed or cold forged wing portions of moderate height. In some sizes they are produced in regular, light, and heavy series to best suit the requirements of specific applications. Dimensions are given in Table 5.

Type B: Type B wing screws are of hot forged one-piece construction available

Table 1. American National Standard Type A Wing Nuts (ANSI B18.17-1968, R1975)

Nominal Size or Basic Major Diameter of Thread*	Thds. per Inch	Series†	Nut Blank Size (Ref)	A Wing Spread Max	A Min	B Wing Height Max	B Min	C Wing Thick. Max	C Min	D Between Wings Max	D Min	E Boss Diam. Max	E Min	G Boss Height Max	G Min
3 (0.0990)	48, 56	Hvy.	AA	0.72	0.59	0.41	0.28	0.11	0.07	0.21	0.17	0.33	0.29	0.14	0.10
4 (0.1120)	40, 38	Hvy.	AA	0.72	0.59	0.41	0.28	0.11	0.07	0.21	0.17	0.33	0.29	0.14	0.10
5 (0.1250)	40, 44	Lgt.	AA	0.72	0.59	0.41	0.28	0.11	0.07	0.21	0.17	0.33	0.29	0.14	0.10
		Hvy.	A	0.91	0.78	0.47	0.34	0.14	0.10	0.27	0.22	0.43	0.39	0.18	0.14
6 (0.1380)	32, 40	Lgt.	AA	0.72	0.59	0.41	0.28	0.11	0.07	0.21	0.17	0.33	0.29	0.14	0.10
		Hvy.	A	0.91	0.78	0.47	0.34	0.14	0.10	0.27	0.22	0.43	0.39	0.18	0.14
8 (0.1640)	32, 36	Lgt.	A	0.91	0.78	0.47	0.34	0.14	0.10	0.27	0.22	0.43	0.39	0.18	0.14
		Hvy.	B	1.10	0.97	0.57	0.43	0.18	0.14	0.33	0.26	0.50	0.45	0.22	0.17
10 (0.1900)	24, 32	Lgt.	A	0.91	0.78	0.47	0.34	0.14	0.10	0.27	0.22	0.43	0.39	0.18	0.14
		Hvy.	B	1.10	0.97	0.57	0.43	0.18	0.14	0.33	0.26	0.50	0.45	0.22	0.17
12 (0.2160)	24, 28	Lgt.	B	1.10	0.97	0.57	0.43	0.18	0.14	0.33	0.26	0.50	0.45	0.22	0.17
		Hvy.	C	1.25	1.12	0.66	0.53	0.21	0.17	0.39	0.32	0.58	0.51	0.25	0.20
¼ (0.2500)	20, 28	Lgt.	B	1.10	0.97	0.57	0.43	0.18	0.14	0.39	0.26	0.50	0.45	0.22	0.17
		Reg.	C	1.25	1.12	0.66	0.53	0.21	0.17	0.39	0.32	0.58	0.51	0.25	0.20
		Hvy.	D	1.44	1.31	0.79	0.65	0.24	0.20	0.48	0.42	0.70	0.64	0.30	0.26
5⁄16 (0.3125)	18, 24	Lgt.	C	1.25	1.12	0.66	0.53	0.21	0.17	0.39	0.32	0.58	0.51	0.25	0.20
		Reg.	D	1.44	1.31	0.79	0.65	0.24	0.20	0.48	0.42	0.70	0.64	0.30	0.26
		Hvy.	E	1.94	1.81	1.00	0.87	0.33	0.26	0.65	0.54	0.93	0.86	0.39	0.35
⅜ (0.3750)	16, 24	Lgt.	D	1.44	1.31	0.79	0.65	0.24	0.20	0.48	0.42	0.70	0.64	0.30	0.26
		Reg.	E	1.94	1.81	1.00	0.87	0.33	0.26	0.65	0.54	0.93	0.86	0.39	0.35
7⁄16 (0.4375)	14, 20	Lgt.	E	1.94	1.81	1.00	0.87	0.33	0.26	0.65	0.54	0.93	0.86	0.39	0.35
		Hvy.	F	2.76	2.62	1.44	1.31	0.40	0.34	0.90	0.80	1.19	1.13	0.55	0.51
½ (0.5000)	13, 20	Lgt.	E	1.94	1.81	1.00	0.87	0.33	0.26	0.65	0.54	0.93	0.86	0.39	0.35
		Hvy.	F	2.76	2.62	1.44	1.31	0.40	0.34	0.90	0.80	1.19	1.13	0.55	0.51
9⁄16 (0.5625)	12, 18	Hvy.	F	2.76	2.62	1.44	1.31	0.40	0.34	0.90	0.80	1.19	1.13	0.55	0.51
⅝ (0.6250)	11, 18	Hvy.	F	2.76	2.62	1.44	1.31	0.40	0.34	0.90	0.80	1.19	1.13	0.55	0.51
¾ (0.7500)	10, 16	Hvy.	F	2.76	2.62	1.44	1.31	0.40	0.34	0.90	0.80	1.19	1.13	0.55	0.51

All dimensions in inches.
* Where specifying nominal size in decimals, zeros in the fourth decimal place are omitted.
† Lgt. = Light; Hvy. = Heavy; Reg. = Regular. Sizes shown in **bold face** are preferred.

in two wing styles: Style 1, having wings of moderate height; and Style 2, having high wings. Dimensions are given in Table 5.

Type C: Type C wing screws are available in two styles: Style 1, of a one-piece die cast construction having wings of moderate height; and Style 2, of a two-piece construction having a die cast wing portion of moderate height. Dimensions are given in Table 6.

Table 2. American National Standard Type B Wing Nuts (ANSI B18.17-1968, R1975)

STYLE 1   STYLE 2

Nominal Size or Basic Major Diameter of Thread*	Thds. per Inch	A Wing Spread		B Wing Height		C Wing Thick.		D Between Wings		E Boss Diam.		G Boss Height	
		Max	Min	Max	Min	Max	Min	Max	Min	Max	Min	Max	Min
Type B, Style 1													
5 (0.1250)	40	0.78	0.72	0.36	0.30	0.13	0.10	0.28	0.22	0.31	0.28	0.22	0.16
10 (0.1900)	24	0.97	0.91	0.45	0.39	0.15	0.12	0.34	0.28	0.39	0.36	0.28	0.22
¼ (0.2500)	20	1.16	1.09	0.56	0.50	0.17	0.14	0.41	0.34	0.47	0.44	0.34	0.28
5⁄16 (0.3125)	18	1.44	1.38	0.67	0.61	0.18	0.15	0.50	0.44	0.55	0.52	0.41	0.34
3⁄8 (0.3750)	16	1.72	1.66	0.80	0.73	0.20	0.17	0.59	0.53	0.63	0.60	0.47	0.41
7⁄16 (0.4375)	14	2.00	1.94	0.91	0.84	0.21	0.18	0.69	0.62	0.71	0.68	0.53	0.47
½ (0.5000)	13	2.31	2.22	1.06	0.94	0.23	0.20	0.78	0.69	0.79	0.76	0.62	0.50
9⁄16 (0.5625)	12	2.59	2.47	1.17	1.05	0.25	0.21	0.88	0.78	0.88	0.84	0.69	0.56
5⁄8 (0.6250)	11	2.84	2.72	1.31	1.19	0.27	0.23	0.94	0.84	0.96	0.92	0.75	0.62
¾ (0.7500)	10	3.31	3.19	1.52	1.39	0.29	0.25	1.10	1.00	1.12	1.08	0.88	0.75
Type B, Style 2													
5 (0.1250)	40	0.81	0.75	0.62	0.56	0.12	0.09	0.28	0.22	0.31	0.28	0.22	0.16
10 (0.1900)	24	1.01	0.95	0.78	0.72	0.14	0.11	0.35	0.29	0.39	0.36	0.28	0.22
¼ (0.2500)	20	1.22	1.16	0.94	0.88	0.16	0.13	0.41	0.35	0.47	0.44	0.34	0.28
5⁄16 (0.3125)	18	1.43	1.37	1.09	1.03	0.17	0.14	0.48	0.42	0.55	0.52	0.41	0.34
3⁄8 (0.3750)	16	1.63	1.57	1.25	1.19	0.18	0.15	0.55	0.49	0.63	0.60	0.47	0.41
7⁄16 (0.4375)	14	1.90	1.84	1.42	1.36	0.19	0.16	0.62	0.56	0.71	0.68	0.53	0.47
½ (0.5000)	13	2.13	2.04	1.58	1.45	0.20	0.17	0.69	0.60	0.79	0.76	0.62	0.50
9⁄16 (0.5625)	12	2.40	2.28	1.75	1.62	0.22	0.18	0.76	0.67	0.88	0.84	0.69	0.56
5⁄8 (0.6250)	11	2.60	2.48	1.91	1.78	0.23	0.19	0.83	0.74	0.96	0.92	0.75	0.62
¾ (0.7500)	10	3.02	2.90	2.22	2.09	0.24	0.20	0.97	0.88	1.12	1.08	0.88	0.75

All dimensions in inches.
* Where specifying nominal size in decimals, zeros in the fourth decimal place are omitted.

Type D: Type D wing screws are of two-piece welded construction having stamped sheet metal wing portions of moderate height. Dimensions are given in Table 6.

Thumb Screws. — A thumb screw is a screw having a flattened head designed for manual turning without a driver or wrench. As covered by ANSI Standard B18.17-1968 (R1975) thumb screws are classified by type on the basis of design characteristics. They consist of the following:

(Continued on page 1469)

Table 3. American National Standard Type C Wing Nuts (ANSI B18.17-1968, R1975)

STYLE 1 STYLE 2 STYLE 3

Nominal Size or Basic Major Diameter of Thread*	Thds. per Inch	Series	Nut Blank Size (Ref)	A Wing Spread Max	A Min	B Wing Height Max	B Min	C Wing Thick. Max	C Min	D Between Wings Max	D Min	E Boss Diam. Max	E Min	F Boss Diam. Max	F Min	G Boss Height Max	G Min
Type C, Style 1																	
4 (0.1120)	40	Reg.	AA	0.66	0.64	0.36	0.35	0.11	0.09	0.18	0.16	0.27	0.25	0.32	0.30	0.16	0.14
5 (0.1250)	40	Reg.	AA	0.66	0.64	0.36	0.35	0.11	0.09	0.18	0.16	0.27	0.25	0.32	0.30	0.16	0.14
6 (0.1380)	**32**	**Reg.**	**AA**	**0.66**	**0.64**	**0.36**	**0.35**	**0.11**	**0.09**	**0.18**	**0.16**	**0.27**	**0.25**	**0.32**	**0.30**	**0.16**	**0.14**
		Hvy.	A	0.85	0.83	0.43	0.42	0.14	0.12	0.29	0.27	0.38	0.36	0.41	0.40	0.20	0.18
8 (0.1640)	32	Reg.	A	0.85	0.83	0.43	0.42	0.14	0.12	0.29	0.27	0.38	0.36	0.41	0.40	0.20	0.18
10 (0.1900)	24, 32	Reg.	A	0.85	0.83	0.43	0.42	0.14	0.12	0.29	0.27	0.38	0.36	0.41	0.40	0.20	0.18
12 (0.2160)	**24**	**Reg.**	**A**	**0.85**	**0.83**	**0.43**	**0.42**	**0.14**	**0.12**	**0.29**	**0.27**	**0.38**	**0.36**	**0.41**	**0.40**	**0.20**	**0.18**
		Hvy.	B	1.08	1.05	0.57	0.53	0.16	0.14	0.32	0.30	0.44	0.42	0.48	0.46	0.23	0.21
¼ (0.2500)	20, 28	Reg.	B	1.08	1.05	0.57	0.53	0.16	0.14	0.32	0.30	0.44	0.42	0.48	0.46	0.23	0.21
5/16 (0.3125)	18, 24	Reg.	C	1.23	1.20	0.64	0.62	0.20	0.18	0.39	0.35	0.50	0.49	0.57	0.55	0.26	0.24
3/8 (0.3750)	16, 24	Reg.	D	1.45	1.42	0.74	0.72	0.23	0.21	0.46	0.42	0.62	0.60	0.69	0.67	0.29	0.27
7/16 (0.4375)	**14, 20**	**Reg.**	**E**	**1.89**	**1.86**	**0.91**	**0.90**	**0.29**	**0.28**	**0.67**	**0.65**	**0.75**	**0.73**	**0.83**	**0.82**	**0.38**	**0.37**
		Hvy.	EH	1.89	1.86	0.93	0.91	0.34	0.33	0.63	0.62	0.81	0.79	0.89	0.87	0.42	0.40
½ (0.5000)	**13, 20**	**Reg.**	**E**	**1.89**	**1.86**	**0.91**	**0.90**	**0.29**	**0.28**	**0.67**	**0.65**	**0.75**	**0.73**	**0.83**	**0.82**	**0.38**	**0.37**
		Hvy.	EH	1.89	1.86	0.93	0.91	0.34	0.33	0.63	0.62	0.81	0.79	0.89	0.87	0.42	0.40
Type C, Style 2																	
5 (0.1250)	40	...	...	0.82	0.80	0.25	0.23	0.09	0.08	0.21	0.19	0.26	0.24	...	...	0.17	0.15
6 (0.1380)	32	...	...	0.82	0.80	0.25	0.23	0.09	0.08	0.21	0.19	0.26	0.24	...	...	0.17	0.15
8 (0.1640)	32	...	...	1.01	0.99	0.28	0.27	0.11	0.09	0.29	0.28	0.36	0.34	...	...	0.19	0.18
10 (0.1900)	24, 32	...	...	1.01	0.99	0.28	0.27	0.11	0.09	0.29	0.28	0.36	0.34	...	...	0.19	0.18
12 (0.2160)	24	...	...	1.20	1.18	0.32	0.31	0.12	0.11	0.38	0.37	0.44	0.43	...	...	0.22	0.20
¼ (0.2500)	20	...	...	1.20	1.18	0.32	0.31	0.12	0.11	0.38	0.37	0.44	0.43	...	...	0.22	0.20
5/16 (0.3125)	18	...	...	1.51	1.49	0.36	0.35	0.14	0.12	0.44	0.43	0.51	0.49	...	...	0.24	0.23
3/8 (0.3750)	16	...	...	1.89	1.86	0.58	0.55	0.20	0.17	0.44	0.43	0.63	0.62	...	...	0.37	0.35
Type C, Style 3																	
5 (0.1250)	40	...	...	0.92	0.89	0.70	0.67	0.16	0.15	0.26	0.24	0.38	0.36	...	...	0.25	0.24
6 (0.1380)	32	...	...	0.92	0.89	0.70	0.67	0.16	0.15	0.26	0.24	0.38	0.36	...	...	0.25	0.24
8 (0.1640)	32	...	...	0.92	0.89	0.70	0.67	0.16	0.15	0.26	0.24	0.38	0.36	...	...	0.25	0.24
10 (0.1900)	24, 32	...	...	1.14	1.12	0.85	0.83	0.19	0.17	0.32	0.30	0.44	0.42	...	...	0.29	0.27
12 (0.2160)	24	...	...	1.14	1.12	0.85	0.83	0.19	0.17	0.32	0.30	0.44	0.42	...	...	0.29	0.27
¼ (0.2500)	20	...	...	1.14	1.12	0.85	0.83	0.19	0.17	0.32	0.30	0.44	0.42	...	...	0.29	0.27
5/16 (0.3125)	18	...	...	1.29	1.27	1.04	1.02	0.23	0.22	0.39	0.36	0.50	0.49	...	...	0.35	0.34
3/8 (0.3750)	16	...	...	1.51	1.49	1.20	1.18	0.25	0.25	0.45	0.42	0.62	0.60	...	...	0.43	0.42

All dimensions in inches. Sizes shown in **bold face** are preferred.
* Where specifying nominal size in decimals, zeros in the fourth decimal place are omitted.

Table 4. American National Standard Type D Wing Nuts (ANSI B18.17-1968, R1975)

STYLE 1   STYLE 2 (LOW WING)   STYLE 3 (LARGE BASE)

Nominal Size or Basic Major Diameter of Thread*	Thds. per Inch	Series†	A Wing Spread		B Wing Height		C Wing Thick.		D Between Wings	E Boss Diam.		G Boss Hgt.	H Wall Hgt.	T Stock Thick.	
			Max	Min	Max	Min	Max	Min	Min	Max	Min	Min	Min	Max	Min
Type D, Style 1															
8 (0.1640)	32, 36	...	0.78	0.72	0.40	0.34	0.18	0.14	0.25	0.41	0.35	0.08	0.12	0.04	0.03
10 (0.1900)	24, 32	...	0.91	0.85	0.47	0.41	0.21	0.17	0.34	0.53	0.47	0.10	0.12	0.04	0.03
12 (0.2160)	24, 28	...	1.09	1.03	0.47	0.41	0.21	0.17	0.34	0.53	0.47	0.10	0.12	0.05	0.04
¼ (0.2500)	20, 28	...	1.11	1.05	0.50	0.44	0.25	0.21	0.34	0.62	0.56	0.11	0.12	0.05	0.04
⁵⁄₁₆ (0.3125)	18, 24	...	1.30	1.24	0.59	0.53	0.30	0.26	0.46	0.73	0.67	0.14	0.18	0.06	0.05
⅜ (0.3750)	16, 24	...	1.41	1.34	0.67	0.61	0.34	0.30	0.69	0.83	0.77	0.16	0.18	0.06	0.05
Type D, Style 2															
5 (0.1250)	40	Reg.	1.03	0.97	0.25	0.19	0.19	0.13	0.30	0.40	0.34	0.07	0.09	0.04	0.03
6 (0.1380)	32	Reg.	1.03	0.97	0.25	0.19	0.19	0.13	0.30	0.40	0.34	0.08	0.09	0.04	0.03
8 (0.1640)	32	Reg.	1.03	0.97	0.25	0.19	0.19	0.13	0.30	0.40	0.34	0.08	0.09	0.04	0.03
10 (0.1900)	24, 32	Reg.	1.40	1.34	0.34	0.28	0.25	0.18	0.32	0.53	0.47	0.09	0.16	0.05	0.04
		Hvy.	1.21	1.16	0.28	0.26	0.31	0.25	0.60	0.61	0.55	0.09	0.13	0.05	0.04
12 (0.2160)	24	Reg.	1.21	1.16	0.28	0.26	0.31	0.25	0.60	0.61	0.55	0.11	0.13	0.05	0.04
¼ (0.2500)	20	Reg.	1.21	1.16	0.28	0.26	0.31	0.25	0.60	0.61	0.55	0.11	0.13	0.05	0.04
Type D, Style 3															
10 (0.1900)	24, 32	Lgt.	1.31	1.25	0.48	0.42	0.29	0.23	0.47	0.65	0.59	0.08	0.12	0.04	0.03
		Reg.	1.40	1.34	0.53	0.47	0.25	0.19	0.50	0.75	0.69	0.08	0.14	0.04	0.03
12 (0.2160)	24	Reg.	1.28	1.22	0.40	0.34	0.23	0.17	0.59	0.73	0.67	0.11	0.12	0.04	0.03
¼ (0.2500)	20	Lgt.	1.28	1.22	0.40	0.34	0.23	0.17	0.59	0.73	0.67	0.11	0.12	0.04	0.03
		Reg.	1.78	1.72	0.66	0.60	0.31	0.25	0.70	1.03	0.97	0.14	0.17	0.06	0.04
		Hvy.	1.47	1.40	0.50	0.44	0.37	0.31	0.66	1.03	0.97	0.14	0.14	0.08	0.06
⁵⁄₁₆ (0.3125)	18	Reg.	1.78	1.72	0.66	0.60	0.31	0.25	0.70	1.03	0.97	0.14	0.17	0.06	0.04
		Hvy.	1.47	1.40	0.50	0.44	0.37	0.31	0.66	1.03	0.97	0.14	0.14	0.08	0.06

All dimensions in inches.
* Where specifying nominal size in decimals, zeros in the fourth decimal place are omitted.
† Lgt. = Light; Hvy. = Heavy; Reg. = Regular.

Type A: Type A thumb screws are forged one-piece screws having a shoulder under the head and are available in two series: regular and heavy. Dimensions are given in Table 7.

Type B: Type B thumb screws are forged one-piece screws without a shoulder and are available in two series: regular and heavy. Dimensions are given in Table 7.

Wing Screw and Thumb Screw Designation. — When specifying wing and thumb screws, the following data should be included in the designation and should appear in the following sequence: nominal size (number, fraction or decimal equivalent), threads per inch, length (fractions or decimal equivalents), type, style and/or series, point (if other than plain point), materials, and finish.

Examples:

> 10–32 × 1¼, Thumb Screw, Type A, Regular, Steel, Zinc Plated.
> .375–16 × 2.00, Wing Screw, Type B, Style 2, Steel, Cadmium Plated.
> .250–20 × 1.50, Wing Screw, Type C, Style 2, Zinc Alloy Wings, Steel Shank, Brass Plated.

Lengths of Wing and Thumb Screws. — The length of wing or thumb screws is measured parallel to the axis of the screw from the intersection of the head or shoulder with the shank to the extreme point of the screw. Standard length increments are as follows: For sizes No. 4 through ¼ inch and for nominal lengths of 0.25 to 0.75 inch, 0.12-inch increments; from 0.75- to 1.50-inch lengths, 0.25-inch increments; and for 1.50- to 3.00-inch lengths, 0.50-inch increments. For sizes 5⁄16 through ½ inch and for 0.50- to 1.50-inch lengths, 0.25-inch increments; for 1.50- to 3.00-inch lengths, 0.50-inch increments; and for 3.00- to 4.00-inch lengths, 1.00-inch increments.

Threads for Wing Screws and Thumb Screws. — Threads for all types of wing screws and thumb screws are in conformance with ANSI Standard Unified Thread, Class 2A. For threads with an additive finish the Class 2A maximum diameters apply to an unplated screw or to a screw before plating, whereas the basic diameters (Class 2A maximum diameters plus the allowance) apply to a screw after plating. All types of wing and thumb screws should have complete (full form) threads extending as close to the head or shoulder as practicable.

Points for Wing and Thumb Screws. — Wing and thumb screws are normally supplied with plain points (sheared ends). Where so specified, these screws may be obtained with cone, cup, dog, flat or oval points as shown in Table 8.

Materials for Wing and Thumb Screws. — Type A wing screws are normally supplied in carbon steel with the shank portion case hardened. When so specified, they also may be made from corrosion resistant steel, brass or other materials as agreed upon by the manufacturer and user.

Type B wing screws are normally made from carbon steel but also may be made from corrosion resistant steel, brass or other materials.

Type C, Style 1, wing screws are supplied only in die cast zinc alloy. Type C, Style 2, wing screws have the wing portion made from die cast zinc alloy with the shank portion normally made from carbon steel. Where so specified, the shank portion may be made from corrosion resistant steel, brass or other materials as agreed upon by the manufacturer and user.

(Continued on page 1473)

Table 5. American National Standard Types A and B Wing Screws
(ANSI B18.17-1968, R1975)

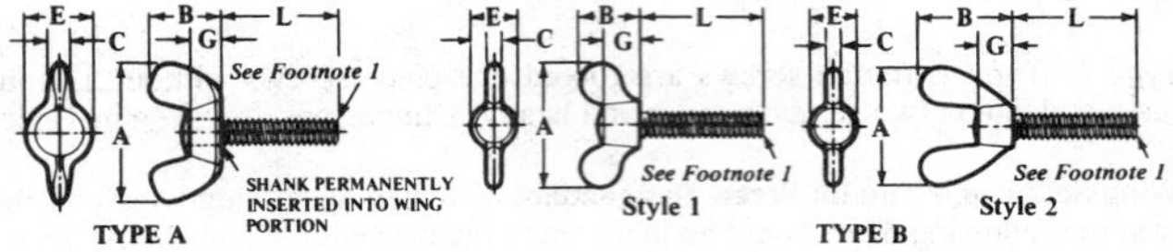

Nominal Size or Basic Screw Diameter*	Thds. per Inch	Series†	Head Blank size (Ref)	A Wing Spread Max	A Wing Spread Min	B Wing Height Max	B Wing Height Min	C Wing Thick. Max	C Wing Thick. Min	E Boss Diam. Max	E Boss Diam. Min	G Boss Height Max	G Boss Height Min	L Practical Screw Lengths Max	L Practical Screw Lengths Min
Type A															
4 (0.1120)	40	Hvy.	AA	0.72	0.59	0.41	0.28	0.11	0.07	0.33	0.29	0.14	0.10	0.75	0.25
6 (0.1380)	32	Lgt.	AA	0.72	0.59	0.41	0.28	0.11	0.07	0.33	0.29	0.14	0.10	0.75	0.25
		Hvy.	**A**	**0.91**	**0.78**	**0.47**	**0.34**	**0.14**	**0.10**	**0.43**	**0.39**	**0.18**	**0.14**		
8 (0.1640)	32	**Lgt.**	**A**	**0.91**	**0.78**	**0.47**	**0.34**	**0.14**	**0.10**	**0.43**	**0.39**	**0.18**	**0.14**	0.75	0.38
		Hvy.	B	1.10	0.97	0.57	0.43	0.18	0.14	0.50	0.45	0.22	0.17		
10 (0.1900)	24, 32	**Lgt.**	**A**	**0.91**	**0.78**	**0.47**	**0.34**	**0.14**	**0.10**	**0.43**	**0.39**	**0.18**	**0.14**	1.00	0.38
		Hvy.	B	1.10	0.97	0.57	0.43	0.18	0.14	0.50	0.45	0.22	0.17		
12 (0.2160)	24	**Lgt.**	**B**	**1.10**	**0.97**	**0.57**	**0.43**	**0.18**	**0.14**	**0.50**	**0.45**	**0.22**	**0.17**	1.00	0.38
		Hvy.	C	1.25	1.12	0.66	0.53	0.21	0.17	0.58	0.51	0.25	0.20		
¼ (0.2500)	20	**Lgt.**	**B**	**1.10**	**0.97**	**0.57**	**0.43**	**0.18**	**0.14**	**0.50**	**0.45**	**0.22**	**0.17**	1.50	0.50
		Reg.	C	1.25	1.12	0.66	0.53	0.21	0.17	0.58	0.51	0.25	0.20		
		Hvy.	D	1.44	1.31	0.79	0.65	0.24	0.20	0.70	0.64	0.30	0.26		
⁵⁄₁₆ (0.3125)	18	**Lgt.**	**C**	**1.25**	**1.12**	**0.66**	**0.53**	**0.21**	**0.17**	**0.58**	**0.51**	**0.25**	**0.20**	1.50	0.50
		Reg.	D	1.44	1.31	0.79	0.65	0.24	0.20	0.70	0.64	0.30	0.26		
		Hvy.	E	1.94	1.81	1.00	0.87	0.33	0.26	0.93	0.86	0.39	0.35		
⅜ (0.3750)	16	**Lgt.**	**D**	**1.44**	**1.31**	**0.79**	**0.65**	**0.24**	**0.20**	**0.70**	**0.64**	**0.30**	**0.26**	2.00	0.75
		Reg.	E	1.94	1.81	1.00	0.87	0.33	0.26	0.93	0.86	0.39	0.35		
		Hvy.	F	2.76	2.62	1.44	1.31	0.40	0.34	1.19	1.13	0.55	0.51		
⁷⁄₁₆ (0.4375)	14	**Lgt.**	**E**	**1.94**	**1.81**	**1.00**	**0.87**	**0.33**	**0.26**	**0.93**	**0.86**	**0.39**	**0.35**	4.00	1.00
		Hvy.	F	2.76	2.62	1.44	1.31	0.40	0.34	1.19	1.13	0.55	0.51		
½ (0.5000)	13	**Lgt.**	**E**	**1.94**	**1.81**	**1.00**	**0.87**	**0.33**	**0.26**	**0.93**	**0.86**	**0.39**	**0.35**	4.00	1.00
		Hvy.	F	2.76	2.62	1.44	1.31	0.40	0.34	1.19	1.13	0.55	0.51		
⅝ (0.6250)	11	Hvy.	F	2.76	2.62	1.44	1.31	0.40	0.34	1.19	1.13	0.55	0.51	4.00	1.25
Type B, Style 1															
10 (0.1900)	24	...	...	0.97	0.91	0.45	0.39	0.15	0.12	0.39	0.36	0.28	0.22	2.00	0.50
¼ (0.2500)	20	...	...	1.16	1.09	0.56	0.50	0.17	0.14	0.47	0.44	0.34	0.28	3.00	0.50
⁵⁄₁₆ (0.3125)	18	...	...	1.44	1.38	0.67	0.61	0.18	0.15	0.55	0.52	0.41	0.34	3.00	0.50
⅜ (0.3750)	16	...	...	1.72	1.66	0.80	0.73	0.20	0.17	0.63	0.60	0.47	0.41	4.00	0.50
⁷⁄₁₆ (0.4375)	14	...	...	2.00	1.94	0.91	0.84	0.21	0.18	0.71	0.68	0.53	0.47	3.00	1.00
½ (0.5000)	13	...	...	2.31	2.22	1.06	0.94	0.23	0.20	0.79	0.76	0.62	0.50	3.00	1.00
⅝ (0.6250)	11	...	...	2.84	2.72	1.31	1.19	0.27	0.23	0.96	0.92	0.75	0.62	2.50	1.00
Type B, Style 2															
10 (0.1900)	24	...	...	1.01	0.95	0.78	0.72	0.14	0.11	0.39	0.36	0.28	0.22	1.25	0.50
¼ (0.2500)	20	...	...	1.22	1.16	0.94	0.88	0.16	0.13	0.47	0.44	0.34	0.28	2.00	0.50
⁵⁄₁₆ (0.3125)	18	...	...	1.43	1.37	1.09	1.03	0.17	0.14	0.55	0.52	0.41	0.34	2.00	0.50
⅜ (0.3750)	16	...	...	1.63	1.57	1.25	1.19	0.18	0.15	0.63	0.60	0.47	0.41	2.00	0.50

All dimensions in inches. Sizes shown in **bold face** are preferred.
[1] Plain point, unless alternate point from styles shown in Table 8 is specified by user.
* Where specifying nominal size in decimals, zeros in the fourth decimal place are omitted.
† Hvy. = Heavy; Lgt. = Light; Reg. = Regular.

Table 6. American National Standard Types C and D Wing Screws
(ANSI B18.17-1968, R1975)

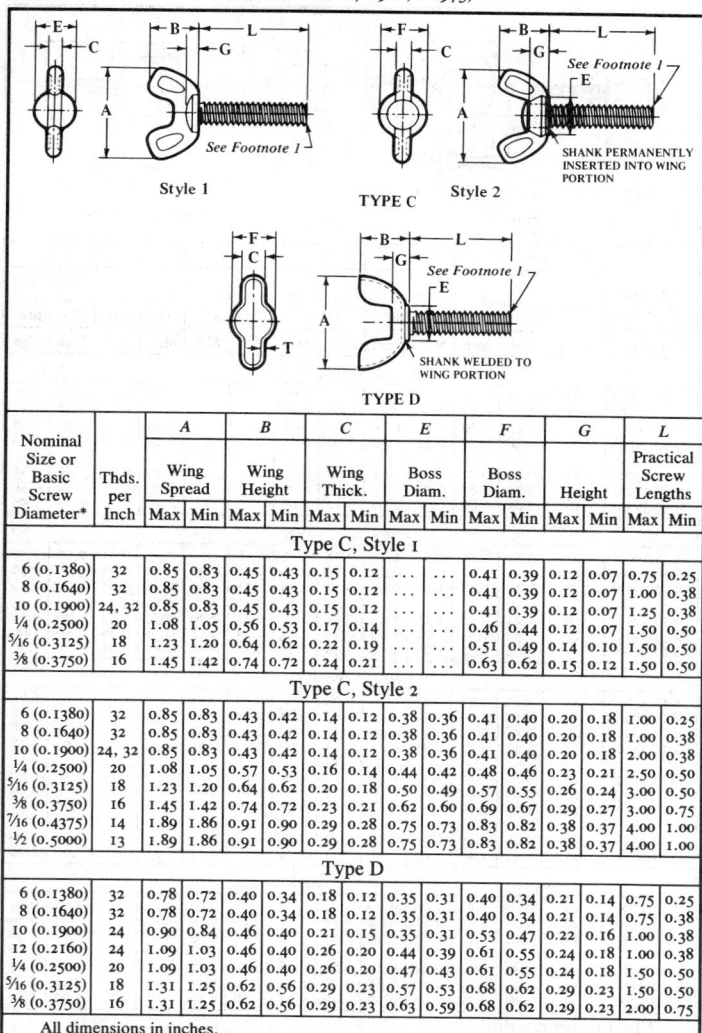

Nominal Size or Basic Screw Diameter*	Thds. per Inch	A Wing Spread		B Wing Height		C Wing Thick.		E Boss Diam.		F Boss Diam.		G Height		L Practical Screw Lengths	
		Max	Min	Max	Min	Max	Min	Max	Min	Max	Min	Max	Min	Max	Min
Type C, Style 1															
6 (0.1380)	32	0.85	0.83	0.45	0.43	0.15	0.12	...	...	0.41	0.39	0.12	0.07	0.75	0.25
8 (0.1640)	32	0.85	0.83	0.45	0.43	0.15	0.12	...	...	0.41	0.39	0.12	0.07	1.00	0.38
10 (0.1900)	24, 32	0.85	0.83	0.45	0.43	0.15	0.12	...	...	0.41	0.39	0.12	0.07	1.25	0.38
¼ (0.2500)	20	1.08	1.05	0.56	0.53	0.17	0.14	...	...	0.46	0.44	0.12	0.07	1.50	0.50
⁵⁄₁₆ (0.3125)	18	1.23	1.20	0.64	0.62	0.22	0.19	...	...	0.51	0.49	0.14	0.10	1.50	0.50
⅜ (0.3750)	16	1.45	1.42	0.74	0.72	0.24	0.21	...	...	0.63	0.62	0.15	0.12	1.50	0.50
Type C, Style 2															
6 (0.1380)	32	0.85	0.83	0.43	0.42	0.14	0.12	0.38	0.36	0.41	0.40	0.20	0.18	1.00	0.25
8 (0.1640)	32	0.85	0.83	0.43	0.42	0.14	0.12	0.38	0.36	0.41	0.40	0.20	0.18	1.00	0.38
10 (0.1900)	24, 32	0.85	0.83	0.43	0.42	0.14	0.12	0.38	0.36	0.41	0.40	0.20	0.18	2.00	0.38
¼ (0.2500)	20	1.08	1.05	0.57	0.53	0.16	0.14	0.44	0.42	0.48	0.46	0.23	0.21	2.50	0.50
⁵⁄₁₆ (0.3125)	18	1.23	1.20	0.64	0.62	0.20	0.18	0.50	0.49	0.57	0.55	0.26	0.24	3.00	0.50
⅜ (0.3750)	16	1.45	1.42	0.74	0.72	0.23	0.21	0.62	0.60	0.69	0.67	0.29	0.27	3.00	0.75
⁷⁄₁₆ (0.4375)	14	1.89	1.86	0.91	0.90	0.29	0.28	0.75	0.73	0.83	0.82	0.38	0.37	4.00	1.00
½ (0.5000)	13	1.89	1.86	0.91	0.90	0.29	0.28	0.75	0.73	0.83	0.82	0.38	0.37	4.00	1.00
Type D															
6 (0.1380)	32	0.78	0.72	0.40	0.34	0.18	0.12	0.35	0.31	0.40	0.34	0.21	0.14	0.75	0.25
8 (0.1640)	32	0.78	0.72	0.40	0.34	0.18	0.12	0.35	0.31	0.40	0.34	0.21	0.14	0.75	0.38
10 (0.1900)	24	0.90	0.84	0.46	0.40	0.21	0.15	0.35	0.31	0.53	0.47	0.22	0.16	1.00	0.38
12 (0.2160)	24	1.09	1.03	0.46	0.40	0.26	0.20	0.44	0.39	0.61	0.55	0.24	0.18	1.00	0.38
¼ (0.2500)	20	1.09	1.03	0.46	0.40	0.26	0.20	0.47	0.43	0.61	0.55	0.24	0.18	1.50	0.50
⁵⁄₁₆ (0.3125)	18	1.31	1.25	0.62	0.56	0.29	0.23	0.57	0.53	0.68	0.62	0.29	0.23	1.50	0.50
⅜ (0.3750)	16	1.31	1.25	0.62	0.56	0.29	0.23	0.63	0.59	0.68	0.62	0.29	0.23	2.00	0.75

All dimensions in inches.
[1] Plain point, unless alternate point from styles shown in Table 8 is specified by user.
* Where specifying nominal size in decimals, zeros in the fourth decimal place are omitted.

Table 7. American National Standard Types A and B Thumb Screws
(ANSI B18.17-1968, R1975)

Nominal Size or Basic Screw Diameter*	Thds. per Inch	A Head Width		B Head Height		C Head Thick.		C' Head Thick.		E Shoulder Diameter		L Practical Screw Lengths	
		Max	Min	Max	Min	Max	Min	Max	Min	Max	Min	Max	Min
Type A, Regular													
6 (0.1380)	32	0.31	0.29	0.33	0.31	0.05	0.04	...	...	0.25	0.23	0.75	0.25
8 (0.1640)	32	0.36	0.34	0.38	0.36	0.06	0.05	...	...	0.31	0.29	0.75	0.38
10 (0.1900)	24, 32	0.42	0.40	0.48	0.46	0.06	0.05	...	...	0.35	0.32	1.00	0.38
12 (0.2160)	24	0.48	0.46	0.54	0.52	0.06	0.05	...	...	0.40	0.38	1.00	0.38
1/4 (0.2500)	20	0.55	0.52	0.64	0.61	0.07	0.05	...	...	0.47	0.44	1.50	0.50
5/16 (0.3125)	18	0.70	0.67	0.78	0.75	0.09	0.07	...	...	0.59	0.56	1.50	0.50
3/8 (0.3750)	16	0.83	0.80	0.95	0.92	0.11	0.09	...	...	0.76	0.71	2.00	0.75
Type A, Heavy													
10 (0.1900)	24	0.89	0.83	0.84	0.72	0.18	0.16	0.10	0.08	0.33	0.31	2.00	0.50
1/4 (0.2500)	20	1.05	0.99	0.94	0.81	0.24	0.22	0.10	0.08	0.40	0.38	3.00	0.50
5/16 (0.3125)	18	1.21	1.15	1.00	0.88	0.27	0.25	0.11	0.09	0.46	0.44	4.00	0.50
3/8 (0.3750)	16	1.41	1.34	1.16	1.03	0.30	0.28	0.11	0.09	0.55	0.53	4.00	0.50
7/16 (0.4375)	14	1.59	1.53	1.22	1.09	0.36	0.34	0.13	0.11	0.71	0.69	2.50	1.00
1/2 (0.5000)	13	1.81	1.72	1.28	1.16	0.40	0.38	0.14	0.12	0.83	0.81	3.00	1.00
Type B, Regular													
6 (0.1380)	32	0.45	0.43	0.28	0.26	0.08	0.06	0.03	0.02	...	...	1.00	0.25
8 (0.1640)	32	0.51	0.49	0.32	0.30	0.09	0.07	0.04	0.02	...	...	1.00	0.38
10 (0.1900)	24, 32	0.58	0.54	0.39	0.36	0.10	0.08	0.05	0.03	...	...	2.00	0.38
12 (0.2160)	24	0.71	0.67	0.45	0.43	0.11	0.09	0.05	0.03	...	...	2.00	0.38
1/4 (0.2500)	20	0.83	0.80	0.52	0.48	0.16	0.14	0.06	0.03	...	...	2.50	0.50
5/16 (0.3125)	18	0.96	0.91	0.64	0.60	0.17	0.14	0.09	0.06	...	...	3.00	0.50
3/8 (0.3750)	16	1.09	1.03	0.71	0.67	0.22	0.18	0.11	0.08	...	...	3.00	0.75
7/16 (0.4375)	14	1.40	1.35	0.96	0.91	0.27	0.24	0.14	0.11	...	...	4.00	1.00
1/2 (0.5000)	13	1.54	1.46	1.09	1.03	0.33	0.29	0.15	0.11	...	...	4.00	1.00
Type B, Heavy													
10 (0.1900)	24	0.89	0.83	0.78	0.66	0.18	0.16	0.08	0.06	...	...	2.00	0.50
1/4 (0.2500)	20	1.05	0.99	0.81	0.72	0.24	0.22	0.11	0.09	...	...	3.00	0.50
5/16 (0.3125)	18	1.21	1.15	0.88	0.78	0.27	0.25	0.11	0.09	...	...	4.00	0.50
3/8 (0.3750)	16	1.41	1.34	0.94	0.84	0.30	0.28	0.14	0.12	...	...	4.00	0.50
7/16 (0.4375)	14	1.59	1.53	1.00	0.91	0.36	0.34	0.14	0.12	...	...	3.00	1.00
1/2 (0.5000)	13	1.81	1.72	1.09	0.97	0.40	0.38	0.18	0.16	...	...	3.00	1.00

All dimensions in inches.
[1] Plain point, unless alternate point from styles shown in Table 8 is specified by user.
* Where specifying nominal size in decimals, zeroes in fourth decimal place are omitted.

**Table 8. American National Standard Alternate Points for Wing and
Thumb Screws (ANSI B18.17-1968, R1975)**

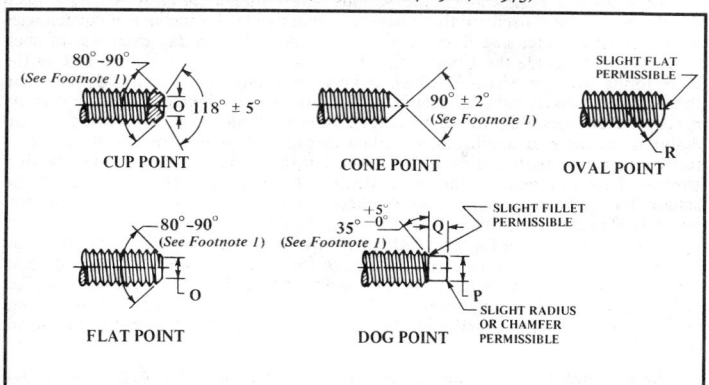

Nominal Size or Basic Screw Diameter*	*O* Cup and Flat Point Diameter		*P* Dog Point† Diameter		*Q* Dog Point† Length		*R* Oval Point Radius	
	Max	Min	Max	Min	Max	Min	Max	Min
4 (0.1120)	0.061	0.051	0.075	0.070	0.061	0.051	0.099	0.084
6 (0.1380)	0.074	0.064	0.092	0.087	0.075	0.065	0.140	0.109
8 (0.1640)	0.087	0.076	0.109	0.103	0.085	0.075	0.156	0.125
10 (0.1900)	0.102	0.088	0.127	0.120	0.095	0.085	0.172	0.141
12 (0.2160)	0.115	0.101	0.144	0.137	0.115	0.105	0.188	0.156
¼ (0.2500)	0.132	0.118	0.156	0.149	0.130	0.120	0.219	0.188
⁵⁄₁₆ (0.3125)	0.172	0.156	0.203	0.195	0.161	0.151	0.256	0.234
³⁄₈ (0.3750)	0.212	0.194	0.250	0.241	0.193	0.183	0.312	0.281
⁷⁄₁₆ (0.4375)	0.252	0.232	0.297	0.287	0.224	0.214	0.359	0.328
½ (0.5000)	0.291	0.270	0.344	0.334	0.255	0.245	0.406	0.375
⁵⁄₈ (0.6250)	0.371	0.347	0.469	0.456	0.321	0.305	0.500	0.469

All dimensions in inches.
[1] The external point angles specified shall apply to those portions of the angles which lie below the thread root diameter, it being recognized the angle within the thread profile may be varied due to the manufacturing processes.
*Where specifying nominal size in decimals, zeros in the fourth decimal place are omitted.
†The axis of dog points shall not be eccentric with the axis of the screw by more than 3 per cent of the basic screw diameter or 0.005 in., whichever is the smaller.

Type D wing screws are normally supplied in carbon steel but also may be made from corrosion resistant steel, brass or other materials.

Thumb screws of all types are normally made from a good commercial quality of carbon steel having a maximum ultimate tensile strength of 48,000 psi. Where so specified, carbon steel thumb screws are case hardened. They are also made from corrosion resistant steel, brass, and other materials as agreed upon by the manufacturer and user.

Unless otherwise specified, wing screws and thumb screws are supplied with a plain (unplated or uncoated) finish.

SCREW THREAD SYSTEMS

Screw Thread Forms. — Of the various screw thread forms which have been developed, the most used are those having symmetrical sides inclined at equal angles with a vertical center line through the thread apex. Present-day examples of such threads would include the Unified, the Whitworth and the Acme forms. One of the early forms was the Sharp V which is now used only occasionally. Symmetrical threads are relatively easy to manufacture and inspect and hence are widely used on mass-produced general-purpose threaded fasteners of all types. In addition to general-purpose fastener applications, certain threads are used to repeatedly move or translate machine parts against heavy loads. For these so-called translation threads a stronger form is required. The most widely used translation thread forms are the square, the Acme, and the buttress. Of these, the square thread is the most efficient, but it is also the most difficult to cut owing to its parallel sides and it cannot be adjusted to compensate for wear. Although less efficient, the Acme form of thread has none of the disadvantages of the square form and has the advantage of being somewhat stronger. The buttress form is used for translation of loads in one direction only because of its non-symmetrical form and combines the high efficiency and strength of the square thread with the ease of cutting and adjustment of the Acme thread.

Sharp V-thread. — The sides of the thread form an angle of 60 degrees with each other. The top and bottom of the thread are, theoretically, sharp, but in practice it is necessary to make the thread with a slight flat. There is no standard adopted for this flat, but it is usually made about one-twenty-fifth of the pitch. If p = pitch of thread, and d = depth of thread, then:

$$d = p \times \cos 30 \text{ deg.} = 0.866\, p = \frac{0.866}{\text{no. of threads per inch}}$$

Some modified V-threads, for locomotive boiler taps particularly, have a depth of $0.8 \times$ pitch.

American National and Unified Screw Thread Forms. — The American National form (formerly known as the United States Standard) was used for many years for most screws, bolts, and miscellaneous threaded products produced in the United States. The American National Standard for Unified Screw Threads now in use includes certain modifications of the former standard as is explained on page 1479. The Basic Profile is shown below and is identical for both UN and UNR screw threads. In this figure H is the height of a sharp V-thread.

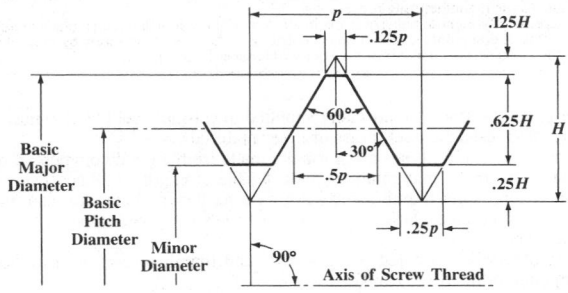

Definitions of Screw Threads. — The following definitions are based on American National Standard ANSI B1.7M-1984, "Nomenclature, Definitions, and Letter Symbols for Screw Threads," and refer to both straight and taper threads.

Actual Size: An actual size is a measured size.

Allowance: An allowance is the prescribed difference between the design (maximum material) size and the basic size. It is numerically equal to the absolute value of the ISO term *fundamental deviation*.

Axis of Thread: Thread axis is coincident with the axis of its pitch cylinder or cone.

Basic Profile of Thread: The basic profile of a thread is the cyclical outline, in an axial plane, of the permanently established boundary between the provinces of the external and internal threads. All deviations are with respect to this boundary.

Basic Size: The basic size is that size from which the limits of size are derived by the application of allowances and tolerances.

Bilateral Tolerance: This is a tolerance in which variation is permitted in both directions from the specified dimension.

Black Crest Thread: This is a thread whose crest displays an unfinished cast, rolled, or forged surface.

Blunt Start Thread: "Blunt start" designates the removal of the incomplete thread at the starting end of the thread. This is a feature of threaded parts that are repeatedly assembled by hand, such as hose couplings and thread plug gages, to prevent cutting of hands and crossing of threads. It was formerly known as a Higbee cut.

Chamfer: This is a conical surface at the starting end of a thread.

Class of Thread: The class of a thread is an alphanumerical designation to indicate the standard grade of tolerance and allowance specified for a thread.

Clearance Fit: This is a fit having limits of size so prescribed that a clearance always results when mating parts are assembled at their maximum material condition.

Complete Thread: The complete thread is that thread whose profile lies within the size limits. (See also *Effective Thread* and *Length of Complete Thread.*) *Note:* Formerly in pipe thread terminology this was referred to as "the perfect thread" but that term is no longer considered desirable.

Crest: This is that surface of a thread which joins the flanks of the thread and is farthest from the cylinder or cone from which the thread projects.

Crest Truncation: This is the radial distance between the sharp crest (crest apex) and the cylinder or cone that would bound the crest.

Depth of Thread Engagement: The depth (or height) of thread engagement between two coaxially assembled mating threads is the radial distance by which their thread forms overlap each other.

Design Size: This is the basic size with allowance applied, from which the limits of size are derived by the application of a tolerance. If there is no allowance, the design size is the same as the basic size.

Deviation: Deviation is a variation from an established dimension, position, standard, or value. In ISO usage, it is the algebraic difference between a size (actual, maximum, or minimum) and the corresponding basic size. The term deviation does not necessarily indicate an error. (See also *Error.*)

Deviation, Fundamental (ISO term): For standard threads, the fundamental deviation is the upper or lower deviation closer to the basic size. It is the upper deviation *es* for an external thread and the lower deviation *EI* for an internal thread. (See also *Allowance* and *Tolerance Position.*)

Deviation, Lower (ISO term): The algebraic difference between the minimum limit of size and the basic size. It is designated *EI* for internal and *ei* for external thread diameters.

Deviation, Upper (ISO term): The algebraic difference between the maximum limit of size and the basic size. It is designated *ES* for internal and *es* for external thread diameters.

Dimension: This is a numerical value expressed in appropriate units of measure and indicated on drawings along with lines, symbols, and notes to define the geometrical characteristic of an object.

Effective Size: See *Pitch Diameter, Functional Diameter.*

Effective Thread: The effective (or useful) thread includes the complete thread, and those portions of the incomplete thread which are fully formed at the root but not at the crest (in taper pipe threads this includes the so-called black crest threads); thus excluding the vanish thread.

Error: This is the algebraic difference between an observed or measured value beyond tolerance limits, and the specified value.

External Thread: This is a thread on a cylindrical or conical external surface.

Fit: Fit is the relationship resulting from the designed difference, before assembly, between the sizes of two mating parts which are to be assembled.

Flank: The flank of a thread is either surface connecting the crest with the root. The flank surface intersection with an axial plane is theoretically a straight line.

Flank Angle: The flank angles are the angles between the individual flanks and the perpendicular to the axis of the thread, measured in an axial plane. A flank angle of a symmetrical thread is commonly termed the half-angle of thread.

Flank Diametral Displacement: In a boundary profile defined system, flank diametral displacement is twice the radial distance between the straight thread flank segments of the maximum and minimum boundary profiles. The value of flank diametral displacement is equal to pitch diameter tolerance in a pitch line reference thread system.

Height of Thread: The height (or depth) of thread is the distance, measured radially, between the major and minor cylinders or cones, respectively.

Helix Angle: On a straight thread, the helix angle is the angle made by the helix of the thread and its relation to the thread axis. On a taper thread, the helix angle at a given axial position is the angle made by the conical spiral of the thread with the axis of the thread. The helix angle is the complement of the lead angle. (See also page 1717 for diagram.)

Higbee Cut: See *Blunt Start Thread.*

Imperfect Thread: See *Incomplete Thread.*

Included Angle: See *Thread Angle.*

Incomplete Thread: This is a threaded profile having either crests or roots or both, not fully formed, resulting from their intersection with the cylindrical or end surface of the work or the vanish cone. It may occur at either end of the thread.

Interference Fit: This is a fit having limits of size so prescribed that an interference always results when mating parts are assembled.

Internal Thread: This is a thread on a cylindrical or conical internal surface.

Lead: Lead is the axial distance between two consecutive points of intersection of a helix by a line parallel to the axis of the cylinder on which it lies, i.e., the axial movement of a threaded part rotated one turn in its mating thread.

Lead Angle: On a straight thread, the lead angle is the angle made by the helix of the thread at the pitch line with a plane perpendicular to the axis. On a taper thread, the lead angle at a given axial position is the angle made by the conical spiral of the thread with the perpendicular to the axis at the pitch line.

Lead Thread: This is that portion of the incomplete thread that is fully formed at the root but not fully formed at the crest which occurs at the entering end of either an external or internal thread.

Left-hand Thread: A thread is a left-hand thread if, when viewed axially, it winds in a counterclockwise and receding direction. Left-hand threads are designated LH.

Length of Complete Thread: This is the axial length of a thread section having full form at both crest and root but also including a maximum of two pitches at the start of the thread which may have a chamfer or incomplete crests.

Length of Thread Engagement: The length of thread engagement of two mating threads is the axial distance over which the two threads, each having full form at both crest and root, are designed to contact. (See also *Length of Complete Thread*.)

Limits of Size: These are the applicable maximum and minimum sizes.

Major Clearance: This is the radial distance between the root of the internal thread and the crest of the external thread of the coaxially assembled designed forms of mating threads.

Major Cone: This is the imaginary cone that would bound the crests of an external taper thread or the roots of an internal taper thread.

Major Cylinder: This is the imaginary cylinder that would bound the crests of an external straight thread or the roots of an internal straight thread.

Major Diameter: On a straight thread the major diameter is that of the major cylinder. On a taper thread the major diameter at a given position on the thread axis is that of the major cone at that position. (See also *Major Cylinder* and *Major Cone*.).

Maximum Material Condition (MMC): This is the condition where a feature of size contains the maximum amount of material within the stated limits of size. For example, minimum internal thread size or maximum external thread size.

Minimum Material Condition (Least Material Condition (LMC)): This is the condition where a feature of size contains the least amount of material within the stated limits of size. For example, maximum internal thread size or minimum external thread size.

Minor Clearance: This is the radial distance between the crest of the internal thread and the root of the external thread of the coaxially assembled design forms of mating threads.

Minor Cone: This is the imaginary cone that would bound the roots of an external taper thread or the crests of an internal taper thread.

Minor Cylinder: This is the imaginary cylinder that would bound the roots of an external straight thread or the crests of an internal straight thread.

Minor Diameter: On a straight thread the minor diameter is that of the minor cylinder. On a taper thread the minor diameter at a given position on the thread axis is that of the minor cone at that position. (See also *Minor Cylinder* and *Minor Cone*.)

Multiple-Start Thread: This is a thread in which the lead is an integral multiple, other than one, of the pitch.

Nominal Size: This is the designation used for the purpose of general identification.

Parallel Thread: See *Screw Thread*.

Partial Thread: See *Vanish Thread*.

Pitch: The pitch of a thread having uniform spacing is the distance measured parallel with its axis between corresponding points on adjacent thread forms in the same axial plane and on the same side of the axis. Pitch is equal to the lead divided by the number of thread starts.

Pitch Cone: The pitch cone is an imaginary cone of such apex angle and location of its vertex and axis that its surface would pass through a taper thread in such a manner as to make the widths of the thread ridge and the thread groove equal. It is, therefore, located equidistantly between the sharp major and minor cones of a given thread form. On a theoretically perfect taper thread, these widths are equal to one-half the basic pitch. (See also *Axis of Thread* and *Pitch Diameter*.)

Pitch Cylinder: The pitch cylinder is an imaginary cylinder of such diameter and location of its axis that its surface would pass through a straight thread in such a manner as to make the widths of the thread ridge and groove equal. It is, therefore, located equidistantly between the sharp major and minor cylinders of a given thread form. On a theoretically perfect thread these widths are equal to one-half the basic pitch. (See also *Axis of Thread* and *Pitch Diameter*.)

Pitch Diameter: On a straight thread the pitch diameter is the diameter of the pitch cylinder. On a taper thread the pitch diameter at a given position on the thread

axis is the diameter of the pitch cone at that position. *Note:* When the crest of a thread is truncated beyond the pitch line, the pitch diameter and pitch cylinder or pitch cone would be based on a theoretical extension of the thread flanks.

Pitch Diameter, Functional Diameter: The functional diameter is the pitch diameter of an enveloping thread with perfect pitch, lead, and flank angles and having a specified length of engagement. It includes the cumulative effect of variations in lead (pitch), flank angle, taper, straightness, and roundness. Variations at the thread crest and root are excluded. Other, nonpreferred terms are *virtual diameter, effective size, virtual effective diameter,* and *thread assembly diameter.*

Pitch Line: This is the generator of the cylinder or cone specified in *Pitch Cylinder* and *Pitch Cone.*

Right-hand Thread: A thread is a right-hand thread if, when viewed axially, it winds in a clockwise and receding direction. A thread is considered to be right-hand unless specifically indicated otherwise.

Root: A root is that surface of the thread which joins the flanks of adjacent thread forms and is immediately adjacent to the cylinder or cone from which the thread projects.

Root Truncation: This is the radial distance between the sharp root (root apex) and the cylinder or cone that would bound the root.

Runout: As applied to screw threads, unless otherwise specified, runout refers to circular runout of major and minor cylinders with respect to the pitch cylinder. Circular runout, in accordance with ANSI Y14.5M, controls cumulative variations of circularity and coaxiality. Runout includes variations due to eccentricity and out-of-roundness. The amount of runout is usually expressed in terms of full indicator movement (FIM).

Screw Thread: A screw thread is a continuous and projecting helical ridge usually of uniform section on a cylindrical or conical surface.

Sharp Crest (Crest Apex): This is the apex formed by the intersection of the flanks of a thread when extended, if necessary, beyond the crest.

Sharp Root (Root Apex): This is the apex formed by the intersection of the adjacent flanks of adjacent threads when extended, if necessary, beyond the root.

Standoff: This is the axial distance between specified reference points on external and internal taper thread members or gages, when assembled with a specified torque or under other specified conditions.

Straight Thread: A straight thread is a screw thread projecting from a cylindrical surface.

Taper Thread: A taper thread is a screw thread projecting from a conical surface.

Tensile Stress Area: The tensile stress area is an arbitrarily selected area for computing the tensile strength of an externally threaded fastener so that the fastener strength is consistent with the basic material strength of the fastener. It is typically defined as a function of pitch diameter and/or minor diameter to calculate a circular cross section of the fastener correcting for the notch and helix effects of the threads.

Thread: A thread is a portion of a screw thread encompassed by one pitch. On a single-start thread it is equal to one turn. (See also *Threads per Inch* and *Turns per Inch.*)

Thread Runout: See *Vanish Thread.*

Thread Series: Thread Series are groups of diameter/pitch combinations distinguished from each other by the number of threads per inch applied to specific diameters.

Thread Shear Area: The thread shear area is the total ridge cross-sectional area intersected by a specified cylinder with diameter and length equal to the mating thread engagement. Usually the cylinder diameter for external thread shearing is the minor diameter of the internal thread and for internal thread shearing it is the major diameter of the external thread.

Threads per Inch: The number of threads per inch is the reciprocal of the axial pitch in inches.

Tolerance: The total amount by which a specific dimension is permitted to vary. The tolerance is the difference between the maximum and minimum limits.

Tolerance Class (metric): The tolerance class (metric) is the combination of a tolerance position with a tolerance grade. It specifies the allowance (fundamental deviation), pitch diameter tolerance (flank diametral displacement), and the crest diameter tolerance.

Tolerance Grade (metric): The tolerance grade (metric) is a numerical symbol that designates the tolerances of crest diameters and pitch diameters applied to the design profiles.

Tolerance Limit: This is the variation, positive or negative, by which a size is permitted to depart from the design size.

Tolerance Position (metric): The tolerance position (metric) is a letter symbol that designates the position of the tolerance zone in relation to the basic size. This position provides the allowance (fundamental deviation).

Total Thread: This includes the complete and all of the incomplete thread, thus including the vanish thread and the lead thread.

Transition Fit: This is a fit having limits of size so prescribed that either a clearance or an interference may result when mating parts are assembled.

Turns per Inch: The number of turns per inch is the reciprocal of the lead in inches.

Unilateral Tolerance: A tolerance in which variation is permitted in one direction from the specified dimension.

Vanish Thread (Partial Thread, Washout Thread, or Thread Runout): This is that portion of the incomplete thread which is not fully formed at the root or at crest and root. It is produced by the chamfer at the starting end of the thread forming tool.

Virtual Diameter: See *Pitch Diameter, Functional Diameter.*

Washout Thread: See *Vanish Thread.*

American Standard for Unified Screw Threads. — American Standard B1.1-1949 was the first American standard to cover those Unified Thread Series agreed upon by the United Kingdom, Canada, and the United States to obtain screw thread interchangeability among these three nations. These Unified threads are now the basic American standard for fastening types of screw threads. In relation to previous American practice, Unified threads have substantially the same thread form and are mechanically interchangeable with the former American National threads of the same diameter and pitch. The principal differences between the two systems lie in: (1) application of allowances; (2) variation of tolerances with size; (3) difference in amount of pitch diameter tolerance on external and internal threads; and (4) differences in thread designation.

In the Unified system an allowance is provided on both the Classes 1A and 2A external threads whereas in the American National system only the Class 1 external thread has an allowance. Also, in the Unified system the pitch diameter tolerance of an internal thread is 30 per cent greater than that of the external thread, whereas they are equal in the American National system.

Revised Standard: The revised screw thread standard ANSI B1.1-1982 is much the same as that of ANSI B1.1-1974. The definition of screw thread acceptability criteria in reference to B1.1-1960 has been deleted. Acceptability criteria are described in ANSI B1.3M-1986, Screw Thread Gaging Systems for Dimensional Acceptability, Inch or Metric Screw Threads (UN, UNR, UNJ, M, and MJ).

Where the letters U, A or B do not appear in the thread designations, the threads conform to the outdated American National screw threads.

Advantages of Unified Threads: The Unified standard is designed to correct certain production difficulties resulting from the former standard. Often, under the old

system, the tolerances of the product were practically absorbed by the combined tool and gage tolerances, leaving little for a working tolerance in manufacture. Somewhat greater tolerances are now provided for nut threads. As contrasted with the old "classes of fit" 1, 2, and 3, for each of which the pitch diameter tolerance on the external and internal threads were equal, the Classes 1B, 2B, and 3B (internal) threads in the new standard have, respectively, a 30 per cent larger pitch diameter tolerance than the 1A, 2A, and 3A (external) threads. Relatively more tolerance is provided for fine threads than for coarse threads of the same pitch. In cases where previous tolerances were more liberal than required, they were reduced.

Thread Form. — The Design Profiles for Unified screw threads, shown on page 1481, define the maximum material condition for external and internal threads with no allowance and are derived from the Basic Profile, shown on page 1474.

UN External Screw Threads: A flat root contour is specified, but it is necessary to provide for some threading tool crest wear, hence a rounded root contour cleared beyond the 0.25p flat width of the Basic Profile is optional.

UNR External Screw Threads: In order to reduce the rate of threading tool crest wear and to improve fatigue strength of a flat root thread, the Design Profile of the UNR thread has a smooth, continuous, non-reversing contour with a radius of curvature not less than 0.108p at any point and blends tangentially into the flanks and any straight segment. At the maximum material condition, the point of tangency is specified to be at a distance not less than 0.625H (where H is the height of a sharp V-thread) below the basic major diameter.

UN and UNR External Screw Threads: The Design Profiles of both UN and UNR external screw threads have flat crests. However, in practice, product threads are produced with partially or completely rounded crests. A rounded crest tangent at 0.125p flat is shown as an option on page 1481.

UN Internal Screw Thread: In practice it is necessary to provide for some threading tool crest wear, therefore the root of the Design Profile is rounded and cleared beyond the 0.125p flat width of the Basic Profile.

There is no internal UNR screw thread.

Thread Series. — Thread series are groups of diameter-pitch combinations distinguished from each other by the number of threads per inch applied to a specific diameter. The various diameter-pitch combinations of eleven standard series are shown in Table 2. The limits of size of threads in the eleven standard series together with certain selected combinations of diameter and pitch, as well as the symbols for designating the various threads, are given in Table 4.

Coarse-Thread Series: This series, UNC/UNRC, is the one most commonly used in the bulk production of bolts, screws, nuts and other general engineering applications. It is also used for threading into lower tensile strength materials such as cast iron, mild steel and softer materials (bronze, brass, aluminum, magnesium and plastics) to obtain the optimum resistance to stripping of the internal thread. It is applicable for rapid assembly or disassembly, or if corrosion or slight damage is possible.

Fine-Thread Series: This series, UNF/UNRF, is suitable for the production of bolts, screws, and nuts and for other applications where the Coarse series is not applicable. External threads of this series have greater tensile stress area than comparable sizes of the Coarse series. The Fine series is suitable when the resistance to stripping of both external and mating internal threads equals or exceeds the tensile load carrying capacity of the externally threaded member (see p. 1278). It is also used where the length of engagement is short, where a smaller lead angle is desired, where the wall thickness demands a fine pitch, or where finer adjustment is needed.

Extra-Fine-Thread Series: This series, UNEF/UNREF, is applicable where even finer pitches of threads are desirable, as for short lengths of engagement and for thin-walled tubes, nuts, ferrules, or couplings. It is also generally applicable under the conditions stated above for the fine threads.

American National Standard Unified Internal and External Screw Thread Design Profiles (Maximum Material Condition)

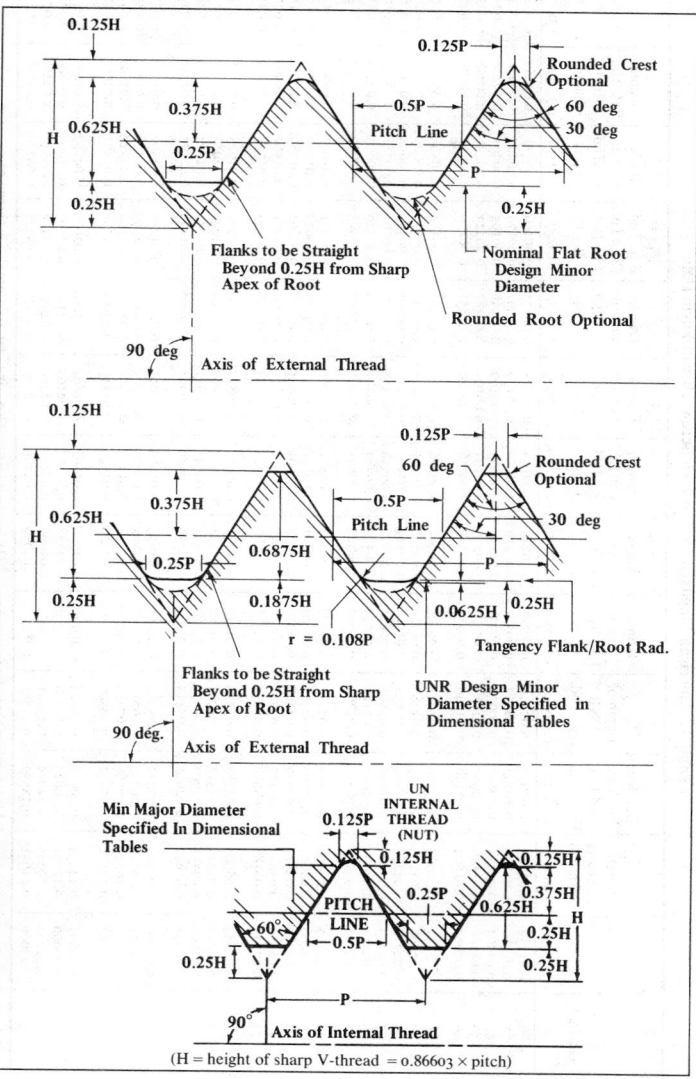

(H = height of sharp V-thread = 0.86603 × pitch)

Table 1. American Standard Unified Inch Screw Thread Form Data

Threads per Inch, n	Pitch, p	Depth of Sharp V-Thread	Depth of Int. Thd. and UN Ext. Thd.*	Depth of UNR Ext. Thd.	Truncation of Ext. Thd. Root	Truncation of UNR Ext. Thd. Root†	Truncation of Ext. Thd. Crest	Truncation of Int. Thd. Root	Truncation of Int. Thd. Crest	Flat at Ext. Thd. Crest and Root	Basic Flat at Int. Thd. Crest‡	Maximum Ext. Thd. Root Radius	Addendum of Ext. Thd.
		0.86603p	0.54127p	0.59539p	0.21651p	0.16238p	0.10825p	0.10825p	0.2165p	0.125p	0.25p	0.14434p	0.32476p
80	0.01250	0.01083	0.00677	0.00744	0.00271	0.00203	0.00135	0.00135	0.00271	0.00156	0.00312	0.00180	0.00406
72	0.01389	0.01203	0.00752	0.00827	0.00301	0.00226	0.00150	0.00150	0.00301	0.00174	0.00347	0.00200	0.00451
64	0.01563	0.01353	0.00846	0.00930	0.00338	0.00254	0.00169	0.00169	0.00338	0.00195	0.00391	0.00226	0.00507
56	0.01786	0.01546	0.00967	0.01063	0.00387	0.00290	0.00193	0.00193	0.00387	0.00223	0.00446	0.00258	0.00580
48	0.02083	0.01804	0.01128	0.01240	0.00451	0.00338	0.00226	0.00226	0.00451	0.00260	0.00521	0.00301	0.00677
44	0.02273	0.01968	0.01230	0.01353	0.00492	0.00369	0.00246	0.00246	0.00492	0.00284	0.00568	0.00328	0.00738
40	0.02500	0.02165	0.01353	0.01488	0.00541	0.00406	0.00271	0.00271	0.00541	0.00312	0.00625	0.00361	0.00812
36	0.02778	0.02406	0.01504	0.01654	0.00601	0.00451	0.00301	0.00301	0.00601	0.00347	0.00694	0.00401	0.00902
32	0.03125	0.02706	0.01691	0.01861	0.00677	0.00507	0.00338	0.00338	0.00677	0.00391	0.00781	0.00451	0.01015
28	0.03571	0.03093	0.01933	0.02126	0.00773	0.00580	0.00387	0.00387	0.00773	0.00446	0.00893	0.00515	0.01160
27	0.03704	0.03208	0.02005	0.02205	0.00802	0.00601	0.00401	0.00401	0.00802	0.00463	0.00926	0.00535	0.01203
24	0.04167	0.03608	0.02255	0.02481	0.00902	0.00677	0.00451	0.00451	0.00902	0.00521	0.01042	0.00601	0.01353
20	0.05000	0.04330	0.02706	0.02977	0.01083	0.00812	0.00541	0.00541	0.01083	0.00625	0.01250	0.00722	0.01624
18	0.05556	0.04811	0.03007	0.03308	0.01203	0.00902	0.00601	0.00601	0.01203	0.00694	0.01389	0.00802	0.01804
16	0.06250	0.05413	0.03383	0.03721	0.01353	0.01015	0.00677	0.00677	0.01353	0.00781	0.01562	0.00902	0.02030
14	0.07143	0.06186	0.03866	0.04253	0.01546	0.01160	0.00773	0.00773	0.01546	0.00893	0.01786	0.01031	0.02320
13	0.07692	0.06662	0.04164	0.04580	0.01655	0.01249	0.00833	0.00833	0.01665	0.00962	0.01923	0.01110	0.02498
12	0.08333	0.07217	0.04511	0.04962	0.01804	0.01353	0.00902	0.00902	0.01804	0.01042	0.02083	0.01203	0.02706
11½	0.08696	0.07531	0.04707	0.05177	0.01883	0.01412	0.00941	0.00941	0.01883	0.01087	0.02174	0.01255	0.02824
11	0.09091	0.07873	0.04921	0.05413	0.01968	0.01476	0.00984	0.00984	0.01968	0.01136	0.02273	0.01312	0.02952
10	0.10000	0.08660	0.05413	0.05954	0.02165	0.01624	0.01083	0.01083	0.02165	0.01250	0.02500	0.01443	0.03248
9	0.11111	0.09623	0.06014	0.06615	0.02406	0.01804	0.01203	0.01203	0.02406	0.01389	0.02778	0.01604	0.03608
8	0.12500	0.10825	0.06766	0.07442	0.02706	0.02030	0.01353	0.01353	0.02706	0.01562	0.03125	0.01804	0.04059
7	0.14286	0.12372	0.07732	0.08506	0.03093	0.02320	0.01546	0.01546	0.03093	0.01786	0.03571	0.02062	0.04639
6	0.16667	0.14434	0.09021	0.09923	0.03608	0.02706	0.01804	0.01804	0.03608	0.02083	0.04167	0.02406	0.05413
5	0.20000	0.17321	0.10825	0.11908	0.04330	0.03248	0.02165	0.02165	0.04330	0.02500	0.05000	0.02887	0.06495
4½	0.22222	0.19245	0.12028	0.13231	0.04811	0.03608	0.02406	0.02406	0.04811	0.02778	0.05556	0.03208	0.07217
4	0.25000	0.21651	0.13532	0.14885	0.05413	0.04059	0.02706	0.02706	0.05413	0.03125	0.06250	0.03608	0.08119

All dimensions are in inches.   * Also depth of thread engagement.   † Design profile.   ‡ Also basic flat at external UN thread root.

Table 2. Diameter-Pitch Combinations for Standard Series of Threads (UN/UNR)*

Sizes† No. or Inches	Basic Major Diam. Inches	Coarse UNC	Fine^a UNF	Extra-fine^b UNEF	4UN	6UN	8UN	12UN	16UN	20UN	28UN	32UN
		Series with Graded Pitches			Threads per Inch — Series with Uniform (Constant) Pitches							
0	0.0600	...	80		Series designation shown indicates the UN thread form; however, the UNR thread form may be specified by substituting UNR in place of UN in all designations for external threads.							
(1)	0.0730	64	72									
2	0.0860	56	64									
(3)	0.0990	48	56									
4	0.1120	40	48									
5	0.1250	40	44	...	...	...	...	...	...	...	...	...
6	0.1380	32	40	...	...	...	...	...	...	...	...	UNC
8	0.1640	32	36	...	...	...	...	...	...	...	...	UNC
10	0.1900	24	32	...	...	...	...	...	...	...	...	UNF
(12)	0.2160	24	28	32	...	...	...	...	...	...	UNF	UNEF
¼	0.2500	20	28	32	...	...	...	...	...	UNC	UNF	UNEF
5/16	0.3125	18	24	32	...	...	...	...	...	20	28	UNEF
3/8	0.3750	16	24	32	...	...	...	...	UNC	20	28	UNEF
7/16	0.4375	14	20	28	...	...	...	...	16	UNF	UNEF	32
½	0.5000	13	20	28	...	...	...	...	16	UNF	UNEF	32
9/16	0.5625	12	18	24	...	...	...	UNC	16	20	28	32
5/8	0.6250	11	18	24	...	...	...	12	16	20	28	32
(11/16)	0.6875	...	...	24	...	...	...	12	16	20	28	32
¾	0.7500	10	16	20	...	...	...	12	UNF	UNEF	28	32
(13/16)	0.8125	...	...	20	...	...	...	12	16	UNEF	28	32
7/8	0.8750	9	14	20	...	...	...	12	16	UNEF	28	32
(15/16)	0.9375	...	...	20	...	...	...	12	16	UNEF	28	32
1	1.0000	8	12	20	...	...	UNC	UNF	16	UNEF	28	32
(1 1/16)	1.0625	...	...	18	...	...	8	12	16	20	28	...
1 1/8	1.1250	7	12	18	...	...	8	UNF	16	20	28	...
(1 3/16)	1.1875	...	...	18	...	...	8	12	16	20	28	...
1 ¼	1.2500	7	12	18	...	...	8	UNF	16	20	28	...
(1 5/16)	1.3125	...	...	18	...	...	8	12	16	20	28	...
1 3/8	1.3750	6	12	18	...	UNC	8	UNF	16	20	28	...
(1 7/16)	1.4375	...	...	18	...	6	8	12	16	20	28	...
1 ½	1.5000	6	12	18	...	UNC	8	UNF	16	20	28	...
(1 9/16)	1.5625	...	...	18	...	6	8	12	16	20	...	...
1 5/8	1.6250	...	...	18	...	6	8	12	16	20	...	...
(1 11/16)	1.6875	...	...	18	...	6	8	12	16	20	...	...
1 ¾	1.7500	5	...	...	...	6	8	12	16	20	...	...
(1 13/16)	1.8125	...	...	...	...	6	8	12	16	20	...	...
1 7/8	1.8750	...	...	...	...	6	8	12	16	20	...	...
(1 15/16)	1.9375	...	...	...	...	6	8	12	16	20	...	...
2	2.0000	4½	...	...	...	6	8	12	16	20	...	...
(2 1/8)	2.1250	...	...	...	...	6	8	12	16	20	...	...
2 ¼	2.2500	4½	...	...	...	6	8	12	16	20	...	...
(2 3/8)	2.3750	...	...	...	...	6	8	12	16	20	...	...
2 ½	2.5000	4	...	...	UNC	6	8	12	16	20	...	...
(2 5/8)	2.6250	...	...	...	4	6	8	12	16	20	...	...
2 ¾	2.7500	4	...	...	UNC	6	8	12	16	20	...	...
(2 7/8)	2.8750	...	...	...	4	6	8	12	16	20	...	...
3	3.0000	4	...	...	UNC	6	8	12	16	20	...	...
(3 1/8)	3.1250	...	...	...	4	6	8	12	16	...	...	...
3 ¼	3.2500	4	...	...	UNC	6	8	12	16	...	...	...
(3 3/8)	3.3750	...	...	...	4	6	8	12	16	...	...	...
3 ½	3.5000	4	...	...	UNC	6	8	12	16	...	...	...
(3 5/8)	3.6250	...	...	...	4	6	8	12	16	...	...	...
3 ¾	3.7500	4	...	...	UNC	6	8	12	16	...	...	...
(3 7/8)	3.8750	...	...	...	4	6	8	12	16	...	...	...
4	4.0000	4	...	...	UNC	6	8	12	16	...	...	...

* For UNR thread form substitute UNR for UN for external threads only.

† Sizes shown in parentheses are secondary sizes. Primary sizes of 4¼, 4½, 4¾, 5, 5¼, 5½, 5¾ and 6 inches also are in the 4, 6, 8, 12, and 16 thread series; secondary sizes of 4⅛, 4⅜, 4⅝, 4⅞, 5⅛, 5⅜, 5⅝, and 5⅞ also are in the 4, 6, 8, 12, and 16 thread series.

^a For diameters over 1½ inches, use 12-thread series.

^b For diameters over 1 11/16 inches, use 16-thread series.

Table 3a. Coarse-Thread Series, UNC and UNRC — Basic Dimensions

Sizes No. or Inches	Basic Major Diam., D Inches	Thds. per Inch, n	Basic Pitch Diam.,[a] E Inches	Minor Diameter Ext. Thds.,[c] K_s (Ref.) Inches	Minor Diameter Int. Thds.,[d] K_n Inches	Lead Angle λ at Basic P.D. Deg. Min.	Area of Minor Diam. at D-$2h_b$ Sq. In.	Tensile Stress Area[b] Sq. In.
1 (.073)*	0.0730	64	0.0629	0.0544	0.0561	4 31	0.00218	0.00263
2 (.086)	0.0860	56	0.0744	0.0648	0.0667	4 22	0.00310	0.00370
3 (.099)*	0.0990	48	0.0855	0.0741	0.0764	4 26	0.00406	0.00487
4 (.112)	0.1120	40	0.0958	0.0822	0.0849	4 45	0.00496	0.00604
5 (.125)	0.1250	40	0.1088	0.0952	0.0979	4 11	0.00672	0.00796
6 (.138)	0.1380	32	0.1177	0.1008	0.1042	4 50	0.00745	0.00909
8 (.164)	0.1640	32	0.1437	0.1268	0.1302	3 58	0.01196	0.0140
10 (.190)	0.1900	24	0.1629	0.1404	0.1449	4 39	0.01450	0.0175
12 (.216)*	0.2160	24	0.1889	0.1664	0.1709	4 1	0.0206	0.0242
¼	0.2500	20	0.2175	0.1905	0.1959	4 11	0.0269	0.0318
⁵⁄₁₆	0.3125	18	0.2764	0.2464	0.2524	3 40	0.0454	0.0524
⅜	0.3750	16	0.3344	0.3005	0.3073	3 24	0.0678	0.0775
⁷⁄₁₆	0.4375	14	0.3911	0.3525	0.3602	3 20	0.0933	0.1063
½	0.5000	13	0.4500	0.4084	0.4167	3 7	0.1257	0.1419
⁹⁄₁₆	0.5625	12	0.5084	0.4633	0.4723	2 59	0.162	0.182
⅝	0.6250	11	0.5660	0.5168	0.5266	2 56	0.202	0.226
¾	0.7500	10	0.6850	0.6309	0.6417	2 40	0.302	0.334
⅞	0.8750	9	0.8028	0.7427	0.7547	2 31	0.419	0.462
1	1.0000	8	0.9188	0.8512	0.8647	2 29	0.551	0.606
1⅛	1.1250	7	1.0322	0.9549	0.9704	2 31	0.693	0.763
1¼	1.2500	7	1.1572	1.0799	1.0954	2 15	0.890	0.969
1⅜	1.3750	6	1.2667	1.1766	1.1946	2 24	1.054	1.155
1½	1.5000	6	1.3917	1.3016	1.3196	2 11	1.294	1.405
1¾	1.7500	5	1.6201	1.5119	1.5335	2 15	1.74	1.90
2	2.0000	4½	1.8557	1.7353	1.7594	2 11	2.30	2.50
2¼	2.2500	4½	2.1057	1.9853	2.0094	1 55	3.02	3.25
2½	2.5000	4	2.3376	2.2023	2.2294	1 57	3.72	4.00
2¾	2.7500	4	2.5876	2.4523	2.4794	1 46	4.62	4.93
3	3.0000	4	2.8376	2.7023	2.7294	1 36	5.62	5.97
3¼	3.2500	4	3.0876	2.9523	2.9794	1 29	6.72	7.10
3½	3.5000	4	3.3376	3.2023	3.2294	1 22	7.92	8.33
3¾	3.7500	4	3.5876	3.4523	3.4794	1 16	9.21	9.66
4	4.0000	4	3.8376	3.7023	3.7294	1 11	10.61	11.08

* Secondary sizes.
[a] British: Effective Diameter.
[b] See formula, page 1279.
[c] Design form for UNR threads. (See figure on page 1481.)
[d] Basic minor diameter.

Constant Pitch Series: The various constant-pitch series, UN, with 4, 6, 8, 12, 16, 20, 28 and 32 threads per inch, given in Table 4, offer a comprehensive range of diameter-pitch combinations for those purposes where the threads in the Coarse, Fine, and Extra-Fine series do not meet the particular requirements of the design.

When selecting threads from these constant-pitch series, preference should be given wherever possible to those tabulated in the 8-, 12-, or 16-thread series.

8-Thread Series: The 8-thread series (8UN) is a uniform-pitch series for large diameters. Although originally intended for high-pressure-joint bolts and nuts, it is now widely used as a substitute for the Coarse-Thread Series for diameters larger than 1 inch.

12-Thread Series: The 12-thread series (12UN) is a uniform pitch series for large diameters requiring threads of medium-fine pitch. Although originally intended for boiler practice, it is now used as a continuation of the Fine-Thread Series for diameters larger than 1½ inches.

Table 3b. Fine-Thread Series, UNF and UNRF — Basic Dimensions

Sizes No. or Inches	Basic Major Diam., D Inches	Thds. per Inch, n	Basic Pitch Diam.,[a] E Inches	Minor Diameter Ext. Thds.,[c] K_s (Ref.) Inches	Minor Diameter Int. Thds.,[d] K_n Inches	Lead Angle λ at Basic P.D. Deg.	Min.	Area of Minor Diam. at $D-2h_b$ Sq. In.	Tensile Stress Area[b] Sq. In.
0 (.060)	0.0600	80	0.0519	0.0451	0.0465	4	23	0.00151	0.00180
1 (.073)*	0.0730	72	0.0640	0.0565	0.0580	3	57	0.00237	0.00278
2 (.086)	0.0860	64	0.0759	0.0674	0.0691	3	45	0.00339	0.00394
3 (.099)*	0.0990	56	0.0874	0.0778	0.0797	3	43	0.00451	0.00523
4 (.112)	0.1120	48	0.0985	0.0871	0.0894	3	51	0.00566	0.00661
5 (.125)	0.1250	44	0.1102	0.0979	0.1004	3	45	0.00716	0.00830
6 (.138)	0.1380	40	0.1218	0.1082	0.1109	3	44	0.00874	0.01015
8 (.164)	0.1640	36	0.1460	0.1309	0.1339	3	28	0.01285	0.01474
10 (.190)	0.1900	32	0.1697	0.1528	0.1562	3	21	0.0175	0.0200
12 (.216)*	0.2160	28	0.1928	0.1734	0.1773	3	22	0.0226	0.0258
¼	0.2500	28	0.2268	0.2074	0.2113	2	52	0.0326	0.0364
5/16	0.3125	24	0.2854	0.2629	0.2674	2	40	0.0524	0.0580
3/8	0.3750	24	0.3479	0.3254	0.3299	2	11	0.0809	0.0878
7/16	0.4375	20	0.4050	0.3780	0.3834	2	15	0.1090	0.1187
½	0.5000	20	0.4675	0.4405	0.4459	1	57	0.1486	0.1599
9/16	0.5625	18	0.5264	0.4964	0.5024	1	55	0.189	0.203
5/8	0.6250	18	0.5889	0.5589	0.5649	1	43	0.240	0.256
¾	0.7500	16	0.7094	0.6763	0.6823	1	36	0.351	0.373
7/8	0.8750	14	0.8286	0.7900	0.7977	1	34	0.480	0.509
1	1.0000	12	0.9459	0.9001	0.9098	1	36	0.625	0.663
1⅛	1.1250	12	1.0709	1.0258	1.0348	1	25	0.812	0.856
1¼	1.2500	12	1.1959	1.1508	1.1598	1	16	1.024	1.073
1⅜	1.3750	12	1.3209	1.2758	1.2848	1	9	1.260	1.315
1½	1.5000	12	1.4459	1.4008	1.4098	1	3	1.521	1.581

* Secondary sizes.
[a] British: Effective Diameter.
[b] See formula, page 1279.
[c] Design form for UNR threads. (See figure on page 1481.)
[d] Basic minor diameter.

16-Thread Size: The 16-thread series (16UN) is a uniform pitch series for large diameters requiring fine-pitch threads. It is suitable for adjusting collars and retaining nuts, and also serves as a continuation of the Extra-fine Thread Series for diameters larger than 1 11/16 inches.

4-, 6-, 20-, 28-, and 32-Thread Series: These thread series have been used more or less widely in industry for various applications where the Standard Coarse, Fine or Extra-fine Series were not as applicable. They are now recognized as Standard Unified Thread Series in a specified selection of diameters for each pitch (see Table 2). Whenever a thread in a constant-pitch series also appears in the UNC, UNF, or UNEF series, the symbols and tolerances for limits of size of UNC, UNF, or UNEF series are applicable, as will be seen in Tables 2 and 4.

Fine Threads for Thin-Wall Tubing: Dimensions for a 27-thread series, ranging from ¼- to 1-inch nominal size, also are included in Table 4. These threads are recommended for general use on thin-wall tubing. The minimum length of complete thread is one-third of the basic major diameter plus 5 threads (+ 0.185 in.).

Selected Combinations: Thread data are tabulated in Table 4 for certain additional selected special combinations of diameter and pitch, with pitch diameter tolerances based on a length of thread engagement of 9 times the pitch. The pitch diameter limits are applicable to a length of engagement of from 5 to 15 times the pitch. (This should not be confused with the lengths of thread on mating parts, as they may exceed the length of engagement by a considerable amount.) Thread symbols are UNS and UNRS.

Table 3c. Extra-Fine-Thread Series, UNEF and UNREF — Basic Dimensions

Sizes No. or Inches	Basic Major Diam., D	Thds. per Inch, n	Basic Pitch Diam.,[a] E	Minor Diameter		Lead Angle λ at Basic P.D.		Area of Minor Diam. at $D-2h_b$	Tensile Stress Area[b]
				Ext. Thds.,[c] K_s (Ref.)	Int. Thds.,[d] K_n				
	Inches	n	Inches	Inches	Inches	Deg.	Min.	Sq. In.	Sq. In.
12 (.216)*	0.2160	32	0.1957	0.1788	0.1822	2	55	0.0242	0.0270
1/4	0.2500	32	0.2297	0.2128	0.2162	2	29	0.0344	0.0379
5/16	0.3125	32	0.2922	0.2753	0.2787	1	57	0.0581	0.0625
3/8	0.3750	32	0.3547	0.3378	0.3412	1	36	0.0878	0.0932
7/16	0.4375	28	0.4143	0.3949	0.3988	1	34	0.1201	0.1274
1/2	0.5000	28	0.4768	0.4574	0.4613	1	22	0.162	0.170
9/16	0.5625	24	0.5354	0.5129	0.5174	1	25	0.203	0.214
5/8	0.6250	24	0.5979	0.5754	0.5799	1	16	0.256	0.268
11/16*	0.6875	24	0.6604	0.6379	0.6424	1	9	0.315	0.329
3/4	0.7500	20	0.7175	0.6905	0.6959	1	16	0.369	0.386
13/16*	0.8125	20	0.7800	0.7530	0.7584	1	10	0.439	0.458
7/8	0.8750	20	0.8425	0.8155	0.8209	1	5	0.515	0.536
15/16*	0.9375	20	0.9050	0.8780	0.8834	1	0	0.598	0.620
1	1.0000	20	0.9675	0.9405	0.9459	0	57	0.687	0.711
1 1/16*	1.0625	18	1.0264	0.9964	1.0024	0	59	0.770	0.799
1 1/8	1.1250	18	1.0889	1.0589	1.0649	0	56	0.871	0.901
1 3/16*	1.1875	18	1.1514	1.1214	1.1274	0	53	0.977	1.009
1 1/4	1.2500	18	1.2139	1.1839	1.1899	0	50	1.090	1.123
1 5/16*	1.3125	18	1.2764	1.2464	1.2524	0	48	1.208	1.244
1 3/8	1.3750	18	1.3389	1.3089	1.3149	0	45	1.333	1.370
1 7/16*	1.4375	18	1.4014	1.3714	1.3774	0	43	1.464	1.503
1 1/2	1.5000	18	1.4639	1.4339	1.4399	0	42	1.60	1.64
1 9/16*	1.5625	18	1.5264	1.4964	1.5024	0	40	1.74	1.79
1 5/8	1.6250	18	1.5889	1.5589	1.5649	0	38	1.89	1.94
1 11/16*	1.6875	18	1.6514	1.6214	1.6274	0	37	2.05	2.10

* Secondary sizes.
[a] British: Effective Diameter.
[b] See formula, page 1279.
[c] Design form for UNR threads. (See figure on page 1481.)
[d] Basic minor diameter.

Other Threads of Special Diameters, Pitches, and Lengths of Engagement: Thread data for special combinations of diameter, pitch, and length of engagement not included in selected combinations are also given in the Standard but are not given here. Also, when design considerations require non-standard pitches or extreme conditions of engagement not covered by the tables, the allowance and tolerances should be derived from the formulas in the Standard. The thread symbol for such special threads is UNS.

Thread Classes. — Thread classes are distinguished from each other by the amounts of tolerance and allowance. Classes identified by a numeral followed by the letters A and B are derived from certain Unified formulas (not shown here) in which the pitch diameter tolerances are based on increments of the basic major (nominal) diameter, the pitch, and the length of engagement. These formulas and the class identification or symbols apply to all of the Unified threads.

Classes 1A, 2A, and 3A apply to external threads only, and Classes 1B, 2B, and 3B apply to internal threads only. The disposition of the tolerances, allowances, and crest clearances for the various classes is illustrated on pages 1494 and 1495.

Classes 2A and 2B: Classes 2A and 2B are the most commonly used for general applications, including production of bolts, screws, nuts, and similar fasteners.

Table 3d. 4-Thread Series, 4UN and 4UNR — Basic Dimensions

| Sizes | | Basic Major Diam., D | Basic Pitch Diam.,[a] E | Minor Diameter | | Lead Angle λ at Basic P.D. | | Area of Minor Diam. at $D-2h_b$ | Tensile Stress Area[b] |
| Primary | Secondary | | | Ext. Thds.,[c] K_s (Ref.) | Int. Thds.,[d] K_n | Deg. | Min. | | |
Inches	Inches	Inches	Inches	Inches	Inches	Deg.	Min.	Sq. In.	Sq. In.
2½†		2.5000	2.3376	2.2023	2.2294	1	57	3.72	4.00
	2⅝	2.6250	2.4626	2.3273	2.3544	1	51	4.16	4.45
2¾†		2.7500	2.5876	2.4523	2.4794	1	46	4.62	4.93
	2⅞	2.8750	2.7126	2.5773	2.6044	1	41	5.11	5.44
3†		3.0000	2.8376	2.7023	2.7294	1	36	5.62	5.97
	3⅛	3.1250	2.9626	2.8273	2.8544	1	32	6.16	6.52
3¼†		3.2500	3.0876	2.9523	2.9794	1	29	6.72	7.10
	3⅜	3.3750	3.2126	3.0773	3.1044	1	25	7.31	7.70
3½†		3.5000	3.3376	3.2023	3.2294	1	22	7.92	8.33
	3⅝	3.6250	3.4626	3.3273	3.3544	1	19	8.55	9.00
3¾†		3.7500	3.5876	3.4523	3.4794	1	16	9.21	9.66
	3⅞	3.8750	3.7126	3.5773	3.6044	1	14	9.90	10.36
4†		4.0000	3.8376	3.7023	3.7294	1	11	10.61	11.08
	4⅛	4.1250	3.9626	3.8273	3.8544	1	9	11.34	11.83
4¼		4.2500	4.0876	3.9523	3.9794	1	7	12.10	12.61
	4⅜	4.3750	4.2126	4.0773	4.1044	1	5	12.88	13.41
4½		4.5000	4.3376	4.2023	4.2294	1	3	13.69	14.23
	4⅝	4.6250	4.4626	4.3273	4.3544	1	1	14.52	15.1
4¾		4.7500	4.5876	4.4523	4.4794	1	0	15.4	15.9
	4⅞	4.8750	4.7126	4.5773	4.6044	0	58	16.3	16.8
5		5.0000	4.8376	4.7023	4.7294	0	57	17.2	17.8
	5⅛	5.1250	4.9626	4.8273	4.8544	0	55	18.1	18.7
5¼		5.2500	5.0876	4.9523	4.9794	0	54	19.1	19.7
	5⅜	5.3750	5.2126	5.0773	5.1044	0	52	20.0	20.7
5½		5.5000	5.3376	5.2023	5.2294	0	51	21.0	21.7
	5⅝	5.6250	5.4626	5.3273	5.3544	0	50	22.1	22.7
5¾		5.7500	5.5876	5.4523	5.4794	0	49	23.1	23.8
	5⅞	5.8750	5.7126	5.5773	5.6044	0	48	24.2	24.9
6		6.0000	5.8376	5.7023	5.7294	0	47	25.3	26.0

† These are standard sizes of the UNC series.
[a] British: Effective Diameter.
[b] See formula, page 1279.
[c] Design form for UNR threads. (See figure on page 1481.)
[d] Basic minor diameter.

The maximum diameters of Class 2A (external) uncoated threads are less than basic by the amount of the allowance. The allowance minimizes galling and seizing in high-cycle wrench assembly, or it can be used to accommodate plated finishes or other coating. However, for threads with additive finish, the maximum diameters of Class 2A may be exceeded by the amount of the allowance; for example, the 2A maximum diameters apply to an unplated part or to a part before plating whereas the 2A maximum diameter plus allowance) apply to a part after plating. The minimum diameters of Class 2B (internal) threads, whether or not plated or coated, are basic, affording no allowance or clearance in assembly at maximum metal limits.

Class 2AG: Certain applications require an allowance for rapid assembly to permit application of the proper lubricant or for residual growth due to high-temperature expansion. In these applications, when the thread is coated and the 2A allowance is not permitted to be consumed by such coating, the thread class symbol is qualified by G following the class symbol.

(*Continued on page* 1492)

Table 3e. 6-Thread Series, 6UN and 6UNR — Basic Dimensions

Sizes		Basic Major Diam., D	Basic Pitch Diam.,[a] E	Minor Diameter		Lead Angle λ at Basic P.D.		Area of Minor Diam. at $D\text{-}2h_b$	Tensile Stress Area[b]
Primary	Secondary			Ext. Thds.,[c] K_s (Ref.)	Int. Thds.,[d] K_n	Deg.	Min.		
Inches	Inches	Inches	Inches	Inches	Inches			Sq. In.	Sq. In.
1³⁄₈†		1.3750	1.2667	1.1766	1.1946	2	24	1.054	1.155
	1⁷⁄₁₆	1.4375	1.3292	1.2391	1.2571	2	17	1.171	1.277
1¹⁄₂†		1.5000	1.3917	1.3016	1.3196	2	11	1.294	1.405
	1⁹⁄₁₆	1.5625	1.4542	1.3641	1.3821	2	5	1.423	1.54
1⁵⁄₈		1.6250	1.5167	1.4271	1.4446	2	0	1.56	1.68
	1¹¹⁄₁₆	1.6875	1.5792	1.4891	1.5071	1	55	1.70	1.83
1³⁄₄		1.7500	1.6417	1.5516	1.5696	1	51	1.85	1.98
	1¹³⁄₁₆	1.8125	1.7042	1.6141	1.6321	1	47	2.00	2.14
1⁷⁄₈		1.8750	1.7667	1.6766	1.6946	1	43	2.16	2.30
	1¹⁵⁄₁₆	1.9375	1.8292	1.7391	1.7571	1	40	2.33	2.47
2		2.0000	1.8917	1.8016	1.8196	1	36	2.50	2.65
	2¹⁄₈	2.1250	2.0167	1.9266	1.9446	1	30	2.86	3.03
2¹⁄₄		2.2500	2.1417	2.0516	2.0696	1	25	3.25	3.42
	2³⁄₈	2.3750	2.2667	2.1766	2.1946	1	20	3.66	3.85
2¹⁄₂		2.5000	2.3917	2.3016	2.3196	1	16	4.10	4.29
	2⁵⁄₈	2.6250	2.5167	2.4266	2.4446	1	12	4.56	4.76
2³⁄₄		2.7500	2.6417	2.5516	2.5696	1	9	5.04	5.26
	2⁷⁄₈	2.8750	2.7667	2.6766	2.6946	1	6	5.55	5.78
3		3.0000	2.8917	2.8016	2.8196	1	3	6.09	6.33
	3¹⁄₈	3.1250	3.0167	2.9266	2.9446	1	0	6.64	6.89
3¹⁄₄		3.2500	3.1417	3.0516	3.0696	0	58	7.23	7.49
	3³⁄₈	3.3750	3.2667	3.1766	3.1946	0	56	7.84	8.11
3¹⁄₂		3.5000	3.3917	3.3016	3.3196	0	54	8.47	8.75
	3⁵⁄₈	3.6250	3.5167	3.4266	3.4446	0	52	9.12	9.42
3³⁄₄		3.7500	3.6417	3.5516	3.5696	0	50	9.81	10.11
	3⁷⁄₈	3.8750	3.7667	3.6766	3.6946	0	48	10.51	10.83
4		4.0000	3.8917	3.8016	3.8196	0	47	11.24	11.57
	4¹⁄₈	4.1250	4.0167	3.9266	3.9446	0	45	12.00	12.33
4¹⁄₄		4.2500	4.1417	4.0516	4.0696	0	44	12.78	13.12
	4³⁄₈	4.3750	4.2667	4.1766	4.1946	0	43	13.58	13.94
4¹⁄₂		4.5000	4.3917	4.3016	4.3196	0	42	14.41	14.78
	4⁵⁄₈	4.6250	4.5167	4.4266	4.4446	0	40	15.3	15.6
4³⁄₄		4.7500	4.6417	4.5516	4.5696	0	39	16.1	16.5
	4⁷⁄₈	4.8750	4.7667	4.6766	4.6946	0	38	17.0	17.5
5		5.0000	4.8917	4.8016	4.8196	0	37	18.0	18.4
	5¹⁄₈	5.1250	5.0167	4.9266	4.9446	0	36	18.9	19.3
5¹⁄₄		5.2500	5.1417	5.0516	5.0696	0	35	19.9	20.3
	5³⁄₈	5.3750	5.2667	5.1766	5.1946	0	35	20.9	21.3
5¹⁄₂		5.5000	5.3917	5.3016	5.3196	0	34	21.9	22.4
	5⁵⁄₈	5.6250	5.5167	5.4266	5.4446	0	33	23.0	23.4
5³⁄₄		5.7500	5.6417	5.5516	5.5696	0	32	24.0	24.5
	5⁷⁄₈	5.8750	5.7667	5.6766	5.6946	0	32	25.1	25.6
6		6.0000	5.8917	5.8016	5.8196	0	31	26.3	26.8

† These are standard sizes of the UNC Series.
[a] British: Effective Diameter.
[b] See formula, page 1279.
[c] Design form for UNR threads. (See figure on page 1481.)
[d] Basic minor diameter.

Table 3f. 8-Thread Series, 8UN and 8UNR — Basic Dimensions

Sizes		Basic Major Diam., D	Basic Pitch Diam.,[a] E	Minor Diameter		Lead Angle λ at Basic P.D.		Area of Minor Diam. at D-2h_b	Tensile Stress Area[b]
Primary	Secondary			Ext. Thds.,[c] K_s (Ref.)	Int. Thds.,[d] K_n	Deg.	Min.		
Inches	Inches	Inches	Inches	Inches	Inches	Deg.	Min.	Sq. In.	Sq. In.
1[†]		1.0000	0.9188	0.8512	0.8647	2	29	0.551	0.606
	1 1/16	1.0625	0.9813	0.9137	0.9272	2	19	0.636	0.695
1 1/8		1.1250	1.0438	0.9792	0.9897	2	11	0.728	0.790
	1 3/16	1.1875	1.1063	1.0387	1.0522	2	4	0.825	0.892
1 1/4		1.2500	1.1688	1.1012	1.1147	1	57	0.929	1.000
	1 5/16	1.3125	1.2313	1.1637	1.1772	1	51	1.039	1.114
1 3/8		1.3750	1.2938	1.2262	1.2397	1	46	1.155	1.233
	1 7/16	1.4375	1.3563	1.2887	1.3022	1	41	1.277	1.360
1 1/2		1.5000	1.4188	1.3512	1.3647	1	36	1.405	1.492
	1 9/16	1.5625	1.4813	1.4137	1.4272	1	32	1.54	1.63
1 5/8		1.6250	1.5438	1.4806	1.4897	1	29	1.68	1.78
	1 11/16	1.6875	1.6063	1.5387	1.5522	1	25	1.83	1.93
1 3/4		1.7500	1.6688	1.6012	1.6147	1	22	1.98	2.08
	1 13/16	1.8125	1.7313	1.6637	1.6772	1	19	2.14	2.25
1 7/8		1.8750	1.7938	1.7262	1.7397	1	16	2.30	2.41
	1 15/16	1.9375	1.8563	1.7887	1.8022	1	14	2.47	2.59
2		2.0000	1.9188	1.8512	1.8647	1	11	2.65	2.77
	2 1/8	2.1250	2.0438	1.9762	1.9897	1	7	3.03	3.15
2 1/4		2.2500	2.1688	2.1012	2.1147	1	3	3.42	3.56
	2 3/8	2.3750	2.2938	2.2262	2.2397	1	0	3.85	3.99
2 1/2		2.5000	2.4188	2.3512	2.3647	0	57	4.29	4.44
	2 5/8	2.6250	2.5438	2.4762	2.4897	0	54	4.76	4.92
2 3/4		2.7500	2.6688	2.6012	2.6147	0	51	5.26	5.43
	2 7/8	2.8750	2.7938	2.7262	2.7397	0	49	5.78	5.95
3		3.0000	2.9188	2.8512	2.8647	0	47	6.32	6.51
	3 1/8	3.1250	3.0438	2.9762	2.9897	0	45	6.89	7.08
3 1/4		3.2500	3.1688	3.1012	3.1147	0	43	7.49	7.69
	3 3/8	3.3750	3.2938	3.2262	3.2397	0	42	8.11	8.31
3 1/2		3.5000	3.4188	3.3512	3.3647	0	40	8.75	8.96
	3 5/8	3.6250	3.5438	3.4762	3.4897	0	39	9.42	9.64
3 3/4		3.7500	3.6688	3.6012	3.6147	0	37	10.11	10.34
	3 7/8	3.8750	3.7938	3.7262	3.7397	0	36	10.83	11.06
4		4.0000	3.9188	3.8512	3.8647	0	35	11.57	11.81
	4 1/8	4.1250	4.0438	3.9762	3.9897	0	34	12.34	12.59
4 1/4		4.2500	4.1688	4.1012	4.1147	0	33	13.12	13.38
	4 3/8	4.3750	4.2938	4.2262	4.2397	0	32	13.94	14.21
4 1/2		4.5000	4.4188	4.3512	4.3647	0	31	14.78	15.1
	4 5/8	4.6250	4.5438	4.4762	4.4897	0	30	15.6	15.9
4 3/4		4.7500	4.6688	4.6012	4.6147	0	29	16.5	16.8
	4 7/8	4.8750	4.7938	4.7262	4.7397	0	29	17.4	17.7
5		5.0000	4.9188	4.8512	4.8647	0	28	18.4	18.7
	5 1/8	5.1250	5.0438	4.9762	4.9897	0	27	19.3	19.7
5 1/4		5.2500	5.1688	5.1012	5.1147	0	26	20.3	20.7
	5 3/8	5.3750	5.2938	5.2262	5.2397	0	26	21.3	21.7
5 1/2		5.5000	5.4188	5.3512	5.3647	0	25	22.4	22.7
	5 5/8	5.6250	5.5438	5.4762	5.4897	0	25	23.4	23.8
5 3/4		5.7500	5.6688	5.6012	5.6147	0	24	24.5	24.9
	5 7/8	5.8750	5.7938	5.7262	5.7397	0	24	25.6	26.0
6		6.0000	5.9188	5.8512	5.8647	0	23	26.8	27.1

† This is a standard size of the UNC Series.
[a] British: Effective Diameter.
[b] See formula, page 1279.
[c] Design form for UNR threads. See figure on page 1481.)
[d] Basic minor diameter.

Table 3g. 12-Thread Series, 12UN and 12UNR — Basic Dimensions

Sizes Primary	Sizes Secondary	Basic Major Diam., D	Basic Pitch Diam.,[a] E	Minor Diameter Ext. Thds.,[c] K_s (Ref.)	Minor Diameter Int. Thds.,[d] K_n	Lead Angle λ at Basic P.D. Deg.	Lead Angle λ at Basic P.D. Min.	Area of Minor Diam. at $D-2h_b$	Tensile Stress Area[b]
Inches	Inches	Inches	Inches	Inches	Inches	Deg.	Min.	Sq. In.	Sq. In.
9/16†		0.5625	0.5084	0.4633	0.4723	2	59	0.162	0.182
5/8		0.6250	0.5709	0.5258	0.5348	2	40	0.210	0.232
	11/16	0.6875	0.6334	0.5883	0.5973	2	24	0.264	0.289
3/4		0.7500	0.6959	0.6508	0.6598	2	11	0.323	0.351
	13/16	0.8125	0.7584	0.7133	0.7223	2	0	0.390	0.420
7/8		0.8750	0.8209	0.7758	0.7848	1	51	0.462	0.495
	15/16	0.9375	0.8834	0.8383	0.8473	1	43	0.540	0.576
1†		1.0000	0.9459	0.9008	0.9098	1	36	0.625	0.663
	1 1/16	1.0625	1.0084	0.9633	0.9723	1	30	0.715	0.756
1 1/8†		1.1250	1.0709	1.0258	1.0348	1	25	0.812	0.856
	1 3/16	1.1875	1.1334	1.0883	1.0973	1	20	0.915	0.961
1 1/4†		1.2500	1.1959	1.1508	1.1598	1	16	1.024	1.073
	1 5/16	1.3125	1.2584	1.2133	1.2223	1	12	1.139	1.191
1 3/8†		1.3750	1.3209	1.2758	1.2848	1	9	1.260	1.315
	1 7/16	1.4375	1.3834	1.3383	1.3473	1	6	1.388	1.445
1 1/2†		1.5000	1.4459	1.4008	1.4098	1	3	1.52	1.58
	1 9/16	1.5625	1.5084	1.4633	1.4723	1	0	1.66	1.72
1 5/8		1.6250	1.5709	1.5258	1.5348	0	58	1.81	1.87
	1 11/16	1.6875	1.6334	1.5883	1.5973	0	56	1.96	2.03
1 3/4		1.7500	1.6959	1.6508	1.6598	0	54	2.12	2.19
	1 13/16	1.8125	1.7584	1.7133	1.7223	0	52	2.28	2.35
1 7/8		1.8750	1.8209	1.7758	1.7848	0	50	2.45	2.53
	1 15/16	1.9375	1.8834	1.8383	1.8473	0	48	2.63	2.71
2		2.0000	1.9459	1.9008	1.9098	0	47	2.81	2.89
	2 1/8	2.1250	2.0709	2.0258	2.0348	0	44	3.19	3.28
2 1/4		2.2500	2.1959	2.1508	2.1598	0	42	3.60	3.69
	2 3/8	2.3750	2.3209	2.2758	2.2848	0	39	4.04	4.13
2 1/2		2.5000	2.4459	2.4008	2.4098	0	37	4.49	4.60
	2 5/8	2.6250	2.5709	2.5258	2.5348	0	35	4.97	5.08
2 3/4		2.7500	2.6959	2.6508	2.6598	0	34	5.48	5.59
	2 7/8	2.8750	2.8209	2.7758	2.7848	0	32	6.01	6.13
3		3.0000	2.9459	2.9008	2.9098	0	31	6.57	6.69
	3 1/8	3.1250	3.0709	3.0258	3.0348	0	30	7.15	7.28
3 1/4		3.2500	3.1959	3.1508	3.1598	0	29	7.75	7.89
	3 3/8	3.3750	3.3209	3.2758	3.2848	0	27	8.38	8.52
3 1/2		3.5000	3.4459	3.4008	3.4098	0	26	9.03	9.18
	3 5/8	3.6250	3.5709	3.5258	3.5348	0	26	9.71	9.86
3 3/4		3.7500	3.6959	3.6508	3.6598	0	25	10.42	10.57
	3 7/8	3.8750	3.8209	3.7758	3.7848	0	24	11.14	11.30
4		4.0000	3.9459	3.9008	3.9098	0	23	11.90	12.06
	4 1/8	4.1250	4.0709	4.0258	4.0348	0	22	12.67	12.84
4 1/4		4.2500	4.1959	4.1508	4.1598	0	22	13.47	13.65
	4 3/8	4.3750	4.3209	4.2758	4.2848	0	21	14.30	14.48
4 1/2		4.5000	4.4459	4.4008	4.4098	0	21	15.1	15.3
	4 5/8	4.6250	4.5709	4.5258	4.5348	0	20	16.0	16.2
4 3/4		4.7500	4.6959	4.6508	4.6598	0	19	16.9	17.1
	4 7/8	4.8750	4.8209	4.7758	4.7848	0	19	17.8	18.0
5		5.0000	4.9459	4.9008	4.9098	0	18	18.8	19.0
	5 1/8	5.1250	5.0709	5.0258	5.0348	0	18	19.8	20.0
5 1/4		5.2500	5.1959	5.1508	5.1598	0	18	20.8	21.0
	5 3/8	5.3750	5.3209	5.2758	5.2848	0	17	21.8	22.0
5 1/2		5.5000	5.4459	5.4008	5.4098	0	17	22.8	23.1
	5 5/8	5.6250	5.5709	5.5258	5.5348	0	16	23.9	24.1
5 3/4		5.7500	5.6959	5.6508	5.6598	0	16	25.0	25.2
	5 7/8	5.8750	5.8209	5.7758	5.7848	0	16	26.1	26.4
6		6.0000	5.9459	5.9008	5.9098	0	15	27.3	27.5

† These are standard sizes of the UNC or UNF Series.
[a] British: Effective Diameter.
[b] See formula, page 1279.
[c] Design form for UNR threads. (See figure on page 1481.)
[d] Basic minor diameter.

Table 3h. 16-Thread Series, 16UN and 16UNR — Basic Dimensions

Sizes		Basic Major Diam., D	Basic Pitch Diam.,a E	Minor Diameter		Lead Angle λ at Basic P.D.		Area of Minor Diam. at D-2hb	Tensile Stress Areab
Primary	Secondary			Ext. Thds.,c Ks (Ref.)	Int. Thds.,d Kn	Deg.	Min.		
Inches	Inches	Inches	Inches	Inches	Inches			Sq. In.	Sq. In.
3/8†		0.3750	0.3344	0.3005	0.3073	3	24	0.0678	0.0775
7/16		0.4375	0.3969	0.3630	0.3698	2	52	0.0997	0.1114
1/2		0.5000	0.4594	0.4255	0.4323	2	29	0.1378	0.151
9/16		0.5625	0.5219	0.4880	0.4948	2	11	0.182	0.198
5/8		0.6250	0.5844	0.5505	0.5573	1	57	0.232	0.250
	11/16	0.6875	0.6469	0.6130	0.6198	1	46	0.289	0.308
3/4†		0.7500	0.7094	0.6755	0.6823	1	36	0.351	0.373
	13/16	0.8125	0.7719	0.7380	0.7448	1	29	0.420	0.444
7/8		0.8750	0.8344	0.8005	0.8073	1	22	0.495	0.521
	15/16	0.9375	0.8969	0.8630	0.8698	1	16	0.576	0.604
1		1.0000	0.9594	0.9255	0.9323	1	11	0.663	0.693
	1 1/16	1.0625	1.0219	0.9880	0.9948	1	7	0.756	0.788
1 1/8		1.1250	1.0844	1.0505	1.0573	1	3	0.856	0.889
	1 3/16	1.1875	1.1469	1.1130	1.1198	1	0	0.961	0.997
1 1/4		1.2500	1.2094	1.1755	1.1823	0	57	1.073	1.111
	1 5/16	1.3125	1.2719	1.2380	1.2448	0	54	1.191	1.230
1 3/8		1.3750	1.3344	1.3005	1.3073	0	51	1.315	1.356
	1 7/16	1.4375	1.3969	1.3630	1.3698	0	49	1.445	1.488
1 1/2		1.5000	1.4594	1.4255	1.4323	0	47	1.58	1.63
	1 9/16	1.5625	1.5219	1.4880	1.4948	0	45	1.72	1.77
1 5/8		1.6250	1.5844	1.5505	1.5573	0	43	1.87	1.92
	1 11/16	1.6875	1.6469	1.6130	1.6198	0	42	2.03	2.08
1 3/4		1.7500	1.7094	1.6755	1.6823	0	40	2.19	2.24
	1 13/16	1.8125	1.7719	1.7380	1.7448	0	39	2.35	2.41
1 7/8		1.8750	1.8344	1.8005	1.8073	0	37	2.53	2.58
	1 15/16	1.9375	1.8969	1.8630	1.8698	0	36	2.71	2.77
2		2.0000	1.9594	1.9255	1.9323	0	35	2.89	2.95
	2 1/8	2.1250	2.0844	2.0505	2.0573	0	33	3.28	3.35
2 1/4		2.2500	2.2094	2.1755	2.1823	0	31	3.69	3.76
	2 3/8	2.3750	2.3344	2.3005	2.3073	0	29	4.13	4.21
2 1/2		2.5000	2.4594	2.4255	2.4323	0	28	4.60	4.67
	2 5/8	2.6250	2.5844	2.5505	2.5573	0	26	5.08	5.16
2 3/4		2.7500	2.7094	2.6755	2.6823	0	25	5.59	5.68
	2 7/8	2.8750	2.8344	2.8005	2.8073	0	24	6.13	6.22
3		3.0000	2.9594	2.9255	2.9323	0	23	6.69	6.78
	3 1/8	3.1250	3.0844	3.0505	3.0573	0	22	7.28	7.37
3 1/4		3.2500	3.2094	3.1755	3.1823	0	21	7.89	7.99
	3 3/8	3.3750	3.3344	3.3005	3.3073	0	21	8.52	8.63
3 1/2		3.5000	3.4594	3.4255	3.4323	0	20	9.18	9.29
	3 5/8	3.6250	3.5844	3.5505	3.5573	0	19	9.86	9.98
3 3/4		3.7500	3.7094	3.6755	3.6823	0	18	10.57	10.69
	3 7/8	3.8750	3.8344	3.8005	3.8073	0	18	11.30	11.43
4		4.0000	3.9594	3.9255	3.9323	0	17	12.06	12.19
	4 1/8	4.1250	4.0844	4.0505	4.0573	0	17	12.84	12.97
4 1/4		4.2500	4.2094	4.1755	4.1823	0	16	13.65	13.78
	4 3/8	4.3750	4.3344	4.3005	4.3073	0	16	14.48	14.62
4 1/2		4.5000	4.4594	4.4255	4.4323	0	15	15.34	15.5
	4 5/8	4.6250	4.5844	4.5505	4.5573	0	15	16.2	16.4
4 3/4		4.7500	4.7094	4.6755	4.6823	0	15	17.1	17.3
	4 7/8	4.8750	4.8344	4.8005	4.8073	0	14	18.0	18.2
5		5.0000	4.9594	4.9255	4.9323	0	14	19.0	19.2
	5 1/8	5.1250	5.0844	5.0505	5.0573	0	13	20.0	20.1
5 1/4		5.2500	5.2094	5.1755	5.1823	0	13	21.0	21.1
	5 3/8	5.3750	5.3344	5.3005	5.3073	0	13	22.0	22.2
5 1/2		5.5000	5.4594	5.4255	5.4323	0	13	23.1	23.2
	5 5/8	5.6250	5.5844	5.5505	5.5573	0	12	24.1	24.3
5 3/4		5.7500	5.7094	5.6755	5.6823	0	12	25.2	25.4
	5 7/8	5.8750	5.8344	5.8005	5.8073	0	12	26.4	26.5
6		6.0000	5.9594	5.9255	5.9323	0	11	27.5	27.7

†These are standard sizes of the UNC or UNF Series. aBritish: Effective Diameter. bSee formula, page 1279. c Design form for UNR threads. (See figure on page 1481.) d Basic minor diameter.

Table 3i. 20-Thread Series, 20UN and 20UNR — Basic Dimensions

Sizes		Basic Major Diam., D	Basic Pitch Diam.,[a] E	Minor Diameter		Lead Angle λ at Basic P.D.		Area of Minor Diam. at D-2h[b]	Tensile Stress Area[b]
Primary	Secondary			Ext. Thds.,[c] K_s (Ref.)	Int. Thds.,[d] K_n	Deg.	Min.		
Inches	Inches	Inches	Inches	Inches	Inches	Deg.	Min.	Sq. In.	Sq. In.
1/4†		0.2500	0.2175	0.1905	0.1959	4	11	0.0269	0.0318
5/16		0.3125	0.2800	0.2530	0.2584	3	15	0.0481	0.0547
3/8		0.3750	0.3425	0.3155	0.3209	2	40	0.0755	0.0836
7/16†		0.4375	0.4050	0.3780	0.3834	2	15	0.1090	0.1187
1/2†		0.5000	0.4675	0.4405	0.4459	1	57	0.1486	0.160
9/16		0.5625	0.5300	0.5030	0.5084	1	43	0.194	0.207
5/8		0.6250	0.5925	0.5655	0.5709	1	32	0.246	0.261
	11/16	0.6875	0.6550	0.6280	0.6334	1	24	0.304	0.320
3/4†		0.7500	0.7175	0.6905	0.6959	1	16	0.369	0.386
	13/16	0.8125	0.7800	0.7530	0.7584	1	10	0.439	0.458
7/8†		0.8750	0.8425	0.8155	0.8209	1	5	0.515	0.536
	15/16†	0.9375	0.9050	0.8780	0.8834	1	0	0.598	0.620
1†		1.0000	0.9675	0.9405	0.9459	0	57	0.687	0.711
	1 1/16	1.0625	1.0300	1.0030	1.0084	0	53	0.782	0.807
1 1/8		1.1250	1.0925	1.0655	1.0709	0	50	0.882	0.910
	1 3/16	1.1875	1.1550	1.1280	1.1334	0	47	0.990	1.018
1 1/4		1.2500	1.2175	1.1905	1.1959	0	45	1.103	1.133
	1 5/16	1.3125	1.2800	1.2530	1.2584	0	43	1.222	1.254
1 3/8		1.3750	1.3425	1.3155	1.3209	0	41	1.348	1.382
	1 7/16	1.4375	1.4050	1.3780	1.3834	0	39	1.479	1.51
1 1/2		1.5000	1.4675	1.4405	1.4459	0	37	1.62	1.65
	1 9/16	1.5625	1.5300	1.5030	1.5084	0	36	1.76	1.80
1 5/8		1.6250	1.5925	1.5655	1.5709	0	34	1.91	1.95
	1 11/16	1.6875	1.6550	1.6280	1.6334	0	33	2.07	2.11
1 3/4		1.7500	1.7175	1.6905	1.6959	0	32	2.23	2.27
	1 13/16	1.8125	1.7800	1.7530	1.7584	0	31	2.40	2.44
1 7/8		1.8750	1.8425	1.8155	1.8209	0	30	2.57	2.62
	1 15/16	1.9375	1.9050	1.8780	1.8834	0	29	2.75	2.80
2		2.0000	1.9675	1.9405	1.9459	0	28	2.94	2.99
	2 1/8	2.1250	2.0925	2.0655	2.0709	0	26	3.33	3.39
2 1/4		2.2500	2.2175	2.1905	2.1959	0	25	3.75	3.81
	2 3/8	2.3750	2.3425	2.3155	2.3209	0	23	4.19	4.25
2 1/2		2.5000	2.4675	2.4405	2.4459	0	22	4.66	4.72
	2 5/8	2.6250	2.5925	2.5655	2.5709	0	21	5.15	5.21
2 3/4		2.7500	2.7175	2.6905	2.6959	0	20	5.66	5.73
	2 7/8	2.8750	2.8425	2.8155	2.8209	0	19	6.20	6.27
3		3.0000	2.9675	2.9405	2.9459	0	18	6.77	6.84

† These are standard sizes of the UNC, UNF, or UNEF Series.
[a] British: Effective Diameter.
[b] See formula, page 1279.
[c] Design form for UNR threads. (See figure on page 1481.)
[d] Basic minor diameter.

Classes 3A and 3B: Classes 3A and 3B may be used if closer tolerances are desired than those provided by Classes 2A and 2B. The maximum diameters of Class 3A (external) threads and the minimum diameters of Class 3B (internal) threads, whether or not plated or coated, are basic, affording no allowance or clearance for assembly of maximum metal components.

Classes 1A and 1B: Classes 1A and 1B threads replaced American National Class 1. These classes are intended for ordnance and other special uses. They are used on threaded components where quick and easy assembly is necessary and where a liberal

Table 3j. 28-Thread Series, 28UN and 28UNR — Basic Dimensions

Sizes		Basic Major Diam., D	Basic Pitch Diam.,[a] E	Minor Diameter		Lead Angle λ at Basic P.D.		Area of Minor Diam. at $D-2h_b$	Tensile Stress Area[b]
Primary	Secondary			Ext. Thds.,[c] K_s (Ref.)	Int. Thds.,[d] K_n	Deg.	Min.		
Inches	Inches	Inches	Inches	Inches	Inches	Deg.	Min.	Sq. In.	Sq. In.
	12 (.216)†	0.2160	0.1928	0.1734	0.1773	3	22	0.0226	0.0258
1/4†		0.2500	0.2268	0.2074	0.2113	2	52	0.0326	0.0364
5/16		0.3125	0.2893	0.2699	0.2738	2	15	0.0556	0.0606
3/8		0.3750	0.3518	0.3324	0.3363	1	51	0.0848	0.0909
7/16†		0.4375	0.4143	0.3949	0.3988	1	34	0.1201	0.1274
1/2†		0.5000	0.4768	0.4574	0.4613	1	22	0.162	0.170
9/16		0.5625	0.5393	0.5199	0.5238	1	12	0.209	0.219
5/8		0.6250	0.6018	0.5824	0.5863	1	5	0.263	0.274
	11/16	0.6875	0.6643	0.6449	0.6488	0	59	0.323	0.335
3/4		0.7500	0.7268	0.7074	0.7113	0	54	0.389	0.402
	13/16	0.8125	0.7893	0.7699	0.7738	0	50	0.461	0.475
7/8		0.8750	0.8518	0.8324	0.8363	0	46	0.539	0.554
	15/16	0.9375	0.9143	0.8949	0.8988	0	43	0.624	0.640
1		1.0000	0.9768	0.9574	0.9613	0	40	0.714	0.732
	1 1/16	1.0625	1.0393	1.0199	1.0238	0	38	0.811	0.830
1 1/8		1.1250	1.1018	1.0824	1.0863	0	35	0.914	0.933
	1 3/16	1.1875	1.1643	1.1449	1.1488	0	34	1.023	1.044
1 1/4		1.2500	1.2268	1.2074	1.2113	0	32	1.138	1.160
	1 5/16	1.3125	1.2893	1.2699	1.2738	0	30	1.259	1.282
1 3/8		1.3750	1.3518	1.3324	1.3363	0	29	1.386	1.411
	1 7/16	1.4375	1.4143	1.3949	1.3988	0	28	1.52	1.55
1 1/2		1.5000	1.4768	1.4574	1.4613	0	26	1.66	1.69

† These are standard sizes of the UNF or UNEF Series.
[a] British: Effective Diameter.
[b] See formula, page 1279.
[c] Design form for UNR threads. (See figure on page 1481.)
[d] Basic minor diameter.

Table 3k. 32-Thread Series, 32UN and 32UNR — Basic Dimensions

Sizes		Basic Major Diam., D	Basic Pitch Diam.,[a] E	Minor Diameter		Lead Angle λ at Basic P.D.		Area of Minor Diam. at $D-2h_b$	Tensile Stress Area[b]
Primary	Secondary			Ext. Thds.,[c] K_s (Ref.)	Int. Thds.,[d] K_n	Deg.	Min.		
Inches	Inches	Inches	Inches	Inches	Inches	Deg.	Min.	Sq. In.	Sq. In.
6 (.138)†		0.1380	0.1177	0.1008	0.1042	4	50	0.00745	0.00909
8 (.164)†		0.1640	0.1437	0.1268	0.1302	3	58	0.01196	0.0140
10 (.190)†		0.1900	0.1697	0.1528	0.1562	3	21	0.01750	0.0200
	12 (.216)	0.2160	0.1957	0.1788	0.1822	2	55	0.0242	0.0270
1/4†		0.2500	0.2297	0.2128	0.2162	2	29	0.0344	0.0379
5/16†		0.3125	0.2922	0.2753	0.2787	1	57	0.0581	0.0625
3/8†		0.3750	0.3547	0.3378	0.3412	1	36	0.0878	0.0932
7/16		0.4375	0.4172	0.4003	0.4037	1	22	0.1237	0.1301
1/2		0.5000	0.4797	0.4628	0.4662	1	11	0.166	0.173
9/16		0.5625	0.5422	0.5253	0.5287	1	3	0.214	0.222
5/8		0.6250	0.6047	0.5878	0.5912	0	57	0.268	0.278
	11/16	0.6875	0.6672	0.6503	0.6537	0	51	0.329	0.339
3/4		0.7500	0.7297	0.7128	0.7162	0	47	0.395	0.407
	13/16	0.8125	0.7922	0.7753	0.7787	0	43	0.468	0.480
7/8		0.8750	0.8547	0.8378	0.8412	0	40	0.547	0.560
	15/16	0.9375	0.9172	0.9003	0.9037	0	37	0.632	0.646
1		1.0000	0.9797	0.9628	0.9662	0	35	0.723	0.738

† These are standard sizes of the UNC, UNF, or UNEF Series.
[a] British: Effective Diameter.
[b] See formula, page 1279.
[c] Design form for UNR threads. (See figure on page 1481.)
[d] Basic minor diameter.

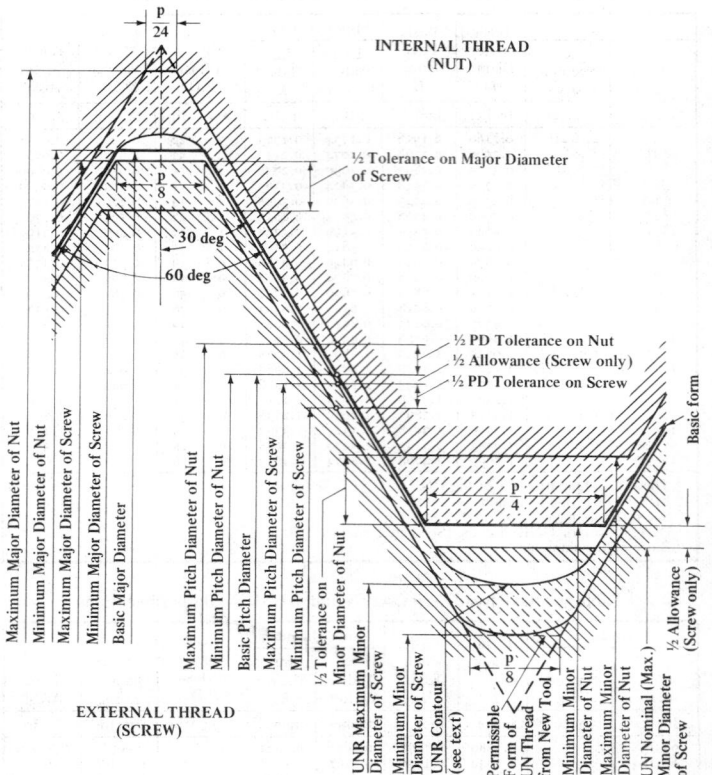

Limits of Size Showing Tolerances, Allowances (Neutral Space), and Crest Clearances for Unified Classes 1A, 2A, 1B, and 2B

allowance is required to permit ready assembly, even with slightly bruised or dirty threads.

Maximum diameters of Class 1A (external) threads are less than basic by the amount of the same allowance as applied to Class 2A. For the intended applications in American practice the allowance is not available for plating or coating. Where the thread is plated or coated, special provisions are necessary. The minimum diameters of Class 1B (internal) threads, whether or not plated or coated, are basic, affording no allowance or clearance for assembly with maximum metal external thread components having maximum diameters which are basic.

Coated 60-deg. Threads. — Although the Standard does not make recommendations for thicknesses of, or specify limits for coatings, it does outline certain princi-

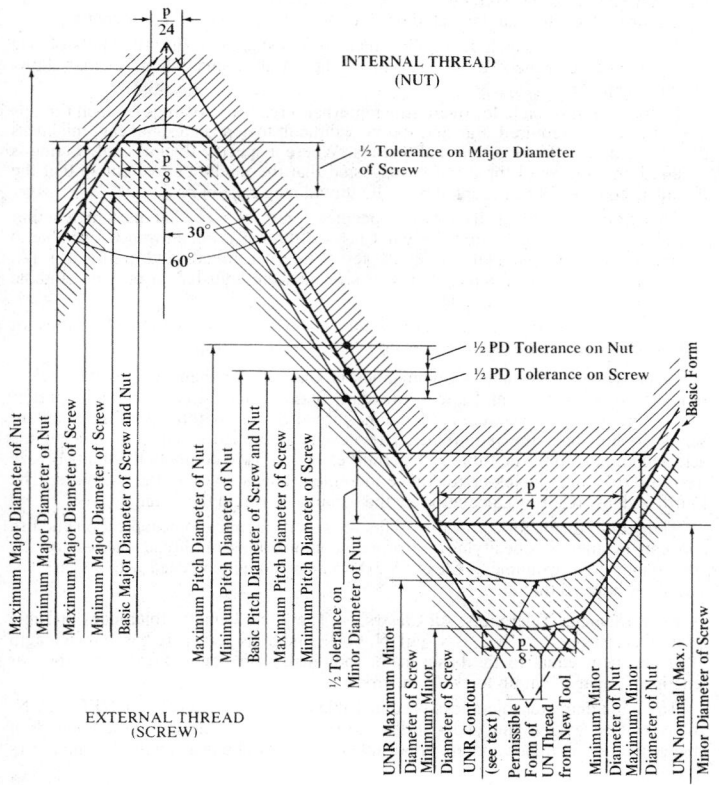

Limits of Size Showing Tolerances and Crest Clearances for Unified Classes 3A and 3B and American National Classes 2 and 3

ples that will aid mechanical interchangeability if followed whenever conditions permit.

To keep finished threads within the limits of size established in the Standard, external threads should not exceed basic size after plating and internal threads should not be below basic size after plating. This recommendation does not apply to threads coated by certain commonly used processes such as hot-dip galvanizing where it may not be required to maintain these limits.

Class 2A provides both a tolerance and an allowance. Many thread requirements call for coatings such as those deposited by electro-plating processes and, in general, the 2A allowance provides adequate undercut for such coatings. There may be variations in thickness and symmetry of coating resulting from commercial processes but after plating the threads should be accepted by a basic Class 3A size GO gage and a

Class 2A gage as a NOT-GO gage. Class 1A provides an allowance which is maintained for both coated and uncoated product, i.e., it is not available for coating.

Class 3A does not include an allowance so it is suggested that the limits of size before plating be reduced by the amount of the 2A allowance whenever that allowance is adequate.

No provision is made for overcutting internal threads as coatings on such threads are not generally required. Further, it is very difficult to deposit a significant thickness of coating on the flanks of internal threads. Where a specific thickness of coating is required on an internal thread, it is suggested that the thread be overcut so that the thread as coated will be accepted by a GO thread plug gage of basic size.

This Standard (ANSI B1.1-1982) specifies limits of size that pertain whether threads are coated or uncoated. Only in Class 2A threads is an allowance available to accommodate coatings. Thus, in all classes of internal threads and in all Class 1A, 2AG, and 3A external threads, limits of size must be adjusted to provide suitable provision for the desired coating.

For further information concerning dimensional accommodation of coating or plating for 60-degree threads, see Section 7, ANSI B1.1a-1974.

Screw Thread Selection — Combination of Classes. — Whenever possible, selection should be made from Table 2, Standard Series Unified Screw Threads, preference being given to the Coarse- and Fine-thread Series. If threads in the standard series do not meet the requirements of design, reference should be made to the selected combinations in Table 4. The third expedient is to compute the limits of size from the tolerance tables or tolerance increment tables given in the Standard. The fourth and last resort is calculation by the formulas given in the Standard.

The requirements for screw thread fits for specific applications depend on end use and can be met by specifying the proper combinations of thread classes for the components. For example, a Class 2A external thread may be used with a Class 1B, 2B, or 3B internal thread.

Pitch Diameter Tolerances, All Classes. — The pitch diameter tolerances in Table 4 for all classes of the UNC, UNF, 4UN, 6UN, and 8UN series are based on a length of engagement equal to the basic major (nominal) diameter and are applicable for lengths of engagement up to 1½ diameters.

The pitch diameter tolerances used in Table 4 for all classes of the UNEF, 12UN, 16UN, 20UN, 28UN, and 32UN series and the UNS series are based on a length of engagement of 9 threads and are applicable for lengths of engagement of from 5 to 15 threads.

Screw Thread Designation. — The basic method of designating a screw thread is used where the standard tolerances or limits of size based on the standard length of engagement are applicable. The designation specifies in sequence the nominal size, number of threads per inch, thread series symbol, thread class symbol, and, finally, gaging system number per ANSI B1.3. The nominal size is the basic major diameter and is specified as the fractional diameter, screw number, or their decimal equivalent. Where decimal equivalents are used for size call-out they shall be interpreted as being nominal size designations only and shall have no dimensional significance beyond the fractional size or number designation. The symbol LH is placed after the thread class symbol to indicate a left-hand thread:

Examples: ¼–20 UNC–2A (21) or 0.250–20 UNC–2A (21)
                10–32 UNF–2A (22) or 0.190–32 UNF–2A (22)
                ⁷⁄₁₆–20 UNRF–2A (23) or 0.4375–20 UNRF–2A (23)
                2–12 UN–2A (21) or 2.000–12 UN–2A (21)
                ¼–20 UNC–3A–LH (21) or 0.250–20 UNC–3A–LH (21)

For uncoated standard series threads these designations may optionally be supplemented by the addition of the pitch diameter limits of size.

Example: ¼–20 UNC–2A (21)
PD 0.2164–0.2127 (Optional for uncoated threads)

Designating Coated Threads. — For coated (or plated) Class 2A external threads, the basic (max) major and basic (max) pitch diameters are given followed by the words AFTER COATING. The major and pitch diameter limits of size before coating are also given followed by the words BEFORE COATING.

Example: ¾–10 UNC–2A (21)
*Major dia 0.7500 max
PD 0.6850 max   } AFTER COATING

†Major dia 0.7482–0.7353
PD 0.6832–0.6773   } BEFORE COATING

Certain applications require an allowance for rapid assembly, to permit application of a proper lubricant, or for residual growth due to high temperature expansion. In such applications where the thread is to be coated and the 2A allowance is not permitted to be consumed by such coating, the thread class symbol is qualified by the addition of the letter G (symbol for allowance) following the class symbol, and the maximum major and maximum pitch diameters are reduced below basic size by the amount of the 2A allowance and followed by the words AFTER COATING. This insures that the allowance is maintained. The major and pitch diameter limits of size before coating are also given followed by SPL and BEFORE COATING. For information concerning the designating of this and other special coating conditions reference should be made to American National Standard ANSI B1.1-1982.

Designating UNS Threads. — UNS screw threads which have special combinations of diameter and pitch with tolerance to Unified formulation have the basic form designation set out first followed always by the limits of size.

Designating Multiple Start Threads. — If a screw thread is of multiple start it is designated by specifying in sequence the nominal size, pitch (in decimals or threads per inch) and lead (in decimals or fractions).

Other Special Designations. — For other special designations including threads with modified limits of size or with special lengths of engagement, reference should be made to American National Standard ANSI B1.1-1982.

Hole Sizes for Tapping. — Hole size limits for tapping Classes 1B, 2B, and 3B threads of various lengths of engagement are given in the Tapping Section.

Internal Thread Minor Diameter Tolerances. — Internal thread minor diameter tolerances in Table 4 are based on a length of engagement equal to the nominal diameter. For general applications these tolerances are suitable for lengths of engagement up to 1½ diameters. However, some thread applications have lengths of engagement which are greater than 1½ diameters or less than the nominal diameter. For such applications it may be advantageous to increase or decrease the tolerance, respectively, as explained in the Tapping Section.

*Major and PD values are equal to basic and correspond to those in Table 4 for Class 3A.
†Major and PD limits are those in Table 4 for Class 2A.

Table 4. Standard Series and Selected Combinations† — Unified Screw Threads

Nominal Size, Threads per Inch, and Series Designation^e	Class	Allowance	External^c Major Diameter Max^b	Min	Min^d	Pitch Diameter Max^b	Min	UNR Minor Diam.^a Max (Ref.)	Class	Internal^c Minor Diameter Min	Max	Pitch Diameter Min	Max	Major Diameter Min
0–80 UNF	2A	0.0005	0.0595	0.0563	—	0.0514	0.0496	0.0446	2B	0.0465	0.0514	0.0519	0.0542	0.0600
	3A	0.0000	0.0600	0.0568	—	0.0519	0.0506	0.0451	3B	0.0465	0.0514	0.0519	0.0536	0.0600
1–64 UNC	2A	0.0006	0.0724	0.0686	—	0.0623	0.0603	0.0538	2B	0.0561	0.0623	0.0629	0.0655	0.0730
	3A	0.0000	0.0730	0.0692	—	0.0629	0.0614	0.0544	3B	0.0561	0.0623	0.0629	0.0648	0.0730
1–72 UNF	2A	0.0006	0.0724	0.0689	—	0.0634	0.0615	0.0559	2B	0.0580	0.0635	0.0640	0.0665	0.0730
	3A	0.0000	0.0730	0.0695	—	0.0640	0.0626	0.0565	3B	0.0580	0.0635	0.0640	0.0659	0.0730
2–56 UNC	2A	0.0006	0.0854	0.0813	—	0.0738	0.0717	0.0642	2B	0.0667	0.0737	0.0744	0.0772	0.0860
	3A	0.0000	0.0860	0.0819	—	0.0744	0.0728	0.0648	3B	0.0667	0.0737	0.0744	0.0765	0.0860
2–64 UNF	2A	0.0006	0.0854	0.0816	—	0.0753	0.0733	0.0668	2B	0.0691	0.0753	0.0759	0.0786	0.0860
	3A	0.0000	0.0860	0.0822	—	0.0759	0.0744	0.0674	3B	0.0691	0.0753	0.0759	0.0779	0.0860
3–48 UNC	2A	0.0007	0.0983	0.0938	—	0.0848	0.0825	0.0734	2B	0.0764	0.0845	0.0855	0.0885	0.0990
	3A	0.0000	0.0990	0.0945	—	0.0855	0.0838	0.0741	3B	0.0764	0.0845	0.0855	0.0877	0.0990
3–56 UNF	2A	0.0007	0.0983	0.0942	—	0.0867	0.0845	0.0771	2B	0.0797	0.0865	0.0874	0.0902	0.0990
	3A	0.0000	0.0990	0.0949	—	0.0874	0.0858	0.0778	3B	0.0797	0.0865	0.0874	0.0895	0.0990
4–40 UNC	2A	0.0008	0.1112	0.1061	—	0.0950	0.0925	0.0814	2B	0.0849	0.0939	0.0958	0.0991	0.1120
	3A	0.0000	0.1120	0.1069	—	0.0958	0.0939	0.0822	3B	0.0849	0.0939	0.0958	0.0982	0.1120
4–48 UNF	2A	0.0007	0.1113	0.1068	—	0.0978	0.0954	0.0864	2B	0.0894	0.0968	0.0985	0.1016	0.1120
	3A	0.0000	0.1120	0.1075	—	0.0985	0.0967	0.0871	3B	0.0894	0.0968	0.0985	0.1008	0.1120
5–40 UNC	2A	0.0008	0.1242	0.1191	—	0.1080	0.1054	0.0944	2B	0.0979	0.1062	0.1088	0.1121	0.1250
	3A	0.0000	0.1250	0.1199	—	0.1088	0.1069	0.0952	3B	0.0979	0.1062	0.1088	0.1113	0.1250
5–44 UNF	2A	0.0007	0.1243	0.1195	—	0.1095	0.1070	0.0972	2B	0.1004	0.1079	0.1102	0.1134	0.1250
	3A	0.0000	0.1250	0.1202	—	0.1102	0.1083	0.0979	3B	0.1004	0.1079	0.1102	0.1126	0.1250
6–32 UNC	2A	0.0008	0.1372	0.1312	—	0.1169	0.1141	0.1000	2B	0.104	0.114	0.1177	0.1214	0.1380
	3A	0.0000	0.1380	0.1320	—	0.1177	0.1156	0.1008	3B	0.1040	0.1140	0.1177	0.1204	0.1380
6–40 UNF	2A	0.0008	0.1372	0.1321	—	0.1210	0.1184	0.1074	2B	0.111	0.119	0.1218	0.1252	0.1380
	3A	0.0000	0.1380	0.1329	—	0.1218	0.1198	0.1082	3B	0.1110	0.1186	0.1218	0.1243	0.1380
8–32 UNC	2A	0.0009	0.1631	0.1571	—	0.1428	0.1399	0.1259	2B	0.130	0.139	0.1437	0.1475	0.1640
	3A	0.0000	0.1640	0.1580	—	0.1437	0.1415	0.1268	3B	0.1300	0.1389	0.1437	0.1465	0.1640

Size	Class	Allowance	Major Dia Max	Major Dia Min	Minor Dia (Ref)	Pitch Dia Max	Pitch Dia Min	Minor Dia Max	Class	Minor Dia Min	Minor Dia Max	Pitch Dia Min	Pitch Dia Max	Major Dia Min
8–36 UNF	2A	0.0008	0.1632	0.1577	—	0.1452	0.1424	0.1301	2B	0.134	0.142	0.1460	0.1496	0.1640
	3A	0.0000	0.1640	0.1585	—	0.1460	0.1439	0.1309	3B	0.1340	0.1416	0.1460	0.1487	0.1640
10–24 UNC	2A	0.0010	0.1890	0.1818	—	0.1619	0.1586	0.1394	2B	0.145	0.156	0.1629	0.1672	0.1900
	3A	0.0000	0.1900	0.1828	—	0.1629	0.1604	0.1404	3B	0.1450	0.1555	0.1629	0.1661	0.1900
10–28 UNS	2A	0.0010	0.1890	0.1825	—	0.1658	0.1625	0.1464	2B	0.151	0.160	0.1668	0.1711	0.1900
10–32 UNF	2A	0.0009	0.1891	0.1831	—	0.1688	0.1658	0.1519	2B	0.156	0.164	0.1697	0.1736	0.1900
	3A	0.0000	0.1900	0.1840	—	0.1697	0.1674	0.1528	3B	0.1560	0.1641	0.1697	0.1726	0.1900
10–36 UNS	2A	0.0009	0.1891	0.1836	—	0.1711	0.1681	0.1560	2B	0.160	0.166	0.1720	0.1759	0.1900
10–40 UNS	2A	0.0009	0.1891	0.1840	—	0.1729	0.1700	0.1592	2B	0.163	0.169	0.1738	0.1775	0.1900
10–48 UNS	2A	0.0008	0.1892	0.1847	—	0.1757	0.1731	0.1644	2B	0.167	0.172	0.1765	0.1799	0.1900
10–56 UNS	2A	0.0007	0.1893	0.1852	—	0.1777	0.1752	0.1681	2B	0.171	0.175	0.1784	0.1816	0.1900
12–24 UNC	2A	0.0010	0.2150	0.2078	—	0.1879	0.1845	0.1654	2B	0.171	0.181	0.1889	0.1933	0.2160
	3A	0.0000	0.2160	0.2088	—	0.1889	0.1863	0.1664	3B	0.1710	0.1807	0.1889	0.1922	0.2160
12–28 UNF	2A	0.0010	0.2150	0.2085	—	0.1918	0.1886	0.1724	2B	0.177	0.186	0.1928	0.1970	0.2160
	3A	0.0000	0.2160	0.2095	—	0.1928	0.1904	0.1734	3B	0.1770	0.1857	0.1928	0.1959	0.2160
12–32 UNEF	2A	0.0009	0.2151	0.2091	—	0.1948	0.1917	0.1779	2B	0.182	0.190	0.1957	0.1998	0.2160
	3A	0.0000	0.2160	0.2100	—	0.1957	0.1933	0.1788	3B	0.1820	0.1895	0.1957	0.1988	0.2160
12–36 UNS	2A	0.0009	0.2151	0.2096	—	0.1971	0.1941	0.1821	2B	0.186	0.192	0.1980	0.2019	0.2160
12–40 UNS	2A	0.0009	0.2151	0.2100	—	0.1989	0.1960	0.1835	2B	0.189	0.195	0.1998	0.2035	0.2160
12–48 UNS	2A	0.0008	0.2152	0.2107	—	0.2017	0.1991	0.1904	2B	0.193	0.198	0.2025	0.2059	0.2160
12–56 UNS	2A	0.0007	0.2153	0.2112	—	0.2037	0.2012	0.1941	2B	0.197	0.201	0.2044	0.2076	0.2160
1/4–20 UNC	1A	0.0011	0.2489	0.2367	—	0.2164	0.2108	0.1894	1B	0.196	0.207	0.2175	0.2248	0.2500
	2A	0.0011	0.2489	0.2408	—	0.2164	0.2127	0.1894	2B	0.196	0.207	0.2175	0.2224	0.2500
	3A	0.0000	0.2500	0.2419	0.2367	0.2175	0.2147	0.1905	3B	0.1960	0.2067	0.2175	0.2211	0.2500
1/4–24 UNS	2A	0.0011	0.2489	0.2417	—	0.2218	0.2181	0.1993	2B	0.205	0.215	0.2229	0.2277	0.2500
1/4–27 UNS	2A	0.0010	0.2490	0.2423	—	0.2249	0.2214	0.2049	2B	0.210	0.219	0.2259	0.2304	0.2500
1/4–28 UNF	1A	0.0010	0.2490	0.2392	—	0.2258	0.2208	0.2064	1B	0.211	0.220	0.2268	0.2333	0.2500
	2A	0.0010	0.2490	0.2425	—	0.2258	0.2225	0.2064	2B	0.211	0.220	0.2268	0.2311	0.2500
	3A	0.0000	0.2500	0.2435	—	0.2268	0.2243	0.2074	3B	0.2110	0.2190	0.2268	0.2300	0.2500
1/4–32 UNEF	2A	0.0010	0.2490	0.2430	—	0.2287	0.2255	0.2118	2B	0.216	0.224	0.2297	0.2339	0.2500
	3A	0.0000	0.2500	0.2440	—	0.2297	0.2273	0.2128	3B	0.2160	0.2229	0.2297	0.2328	0.2500

All dimensions in inches.   † Use UNS threads only if Standard Series do not meet requirements (see p. 1479). See footnotes *a*, *b*, *c*, *d* and *e* at end of table.

Table 4 (*Continued*). Standard Series and Selected Combinations† — Unified Screw Threads

| Nominal Size, Threads per Inch, and Series Designation^c | External^c | | | | | | | | Internal^c | | | | | |
| | Class | Allowance | Major Diameter | | | Pitch Diameter | | UNR Minor Diam.,^a Max (Ref.) | Class | Minor Diameter | | Pitch Diameter | | Major Diameter Min |
			Max^b	Min	Min^d	Max^b	Min			Min	Max	Min	Max	
¼–36 UNS	2A	0.0009	0.2491	0.2436	—	0.2311	0.2280	0.2161	2B	0.220	0.226	0.2320	0.2360	0.2500
¼–40 UNS	2A	0.0009	0.2491	0.2440	—	0.2329	0.2300	0.2193	2B	0.223	0.229	0.2338	0.2376	0.2500
¼–48 UNS	2A	0.0008	0.2492	0.2447	—	0.2357	0.2330	0.2243	2B	0.227	0.232	0.2365	0.2401	0.2500
¼–56 UNS	2A	0.0008	0.2492	0.2451	—	0.2376	0.2350	0.2280	2B	0.231	0.235	0.2384	0.2417	0.2500
⁵⁄₁₆–18 UNC	1A	0.0012	0.3113	0.2982	0.2982	0.2752	0.2691	0.2452	1B	0.252	0.265	0.2764	0.2843	0.3125
	2A	0.0012	0.3113	0.3026	—	0.2752	0.2712	0.2452	2B	0.252	0.265	0.2764	0.2817	0.3125
	3A	0.0000	0.3125	0.3038	—	0.2764	0.2734	0.2464	3B	0.2520	0.2630	0.2764	0.2803	0.3125
⁵⁄₁₆–20 UN	2A	0.0012	0.3113	0.3032	—	0.2788	0.2748	0.2518	2B	0.258	0.270	0.2800	0.2852	0.3125
	3A	0.0000	0.3125	0.3044	—	0.2800	0.2770	0.2530	3B	0.2580	0.2680	0.2800	0.2839	0.3125
⁵⁄₁₆–24 UNF	1A	0.0011	0.3114	0.3006	—	0.2843	0.2788	0.2618	1B	0.267	0.277	0.2854	0.2925	0.3125
	2A	0.0011	0.3114	0.3042	—	0.2843	0.2806	0.2618	2B	0.267	0.277	0.2854	0.2902	0.3125
	3A	0.0000	0.3125	0.3053	—	0.2854	0.2827	0.2629	3B	0.2670	0.2754	0.2854	0.2890	0.3125
⁵⁄₁₆–27 UNS	2A	0.0010	0.3115	0.3048	—	0.2874	0.2839	0.2674	2B	0.272	0.281	0.2884	0.2929	0.3125
⁵⁄₁₆–28 UN	2A	0.0010	0.3115	0.3050	—	0.2883	0.2849	0.2689	2B	0.274	0.282	0.2893	0.2937	0.3125
	3A	0.0000	0.3125	0.3060	—	0.2893	0.2867	0.2699	3B	0.2740	0.2807	0.2893	0.2926	0.3125
⁵⁄₁₆–32 UNEF	2A	0.0010	0.3115	0.3055	—	0.2912	0.2880	0.2743	2B	0.279	0.286	0.2922	0.2964	0.3125
	3A	0.0000	0.3125	0.3065	—	0.2922	0.2898	0.2753	3B	0.2790	0.2847	0.2922	0.2953	0.3125
⁵⁄₁₆–36 UNS	2A	0.0009	0.3116	0.3061	—	0.2936	0.2905	0.2785	2B	0.282	0.289	0.2945	0.2985	0.3125
⁵⁄₁₆–40 UNS	2A	0.0009	0.3116	0.3065	—	0.2954	0.2925	0.2818	2B	0.285	0.291	0.2963	0.3001	0.3125
⁵⁄₁₆–48 UNS	2A	0.0008	0.3117	0.3072	—	0.2982	0.2955	0.2869	2B	0.290	0.295	0.2990	0.3026	0.3125
³⁄₈–16 UNC	1A	0.0013	0.3737	0.3595	0.3595	0.3331	0.3266	0.2992	1B	0.307	0.321	0.3344	0.3429	0.3750
	2A	0.0013	0.3737	0.3643	—	0.3331	0.3287	0.2992	2B	0.307	0.321	0.3344	0.3401	0.3750
	3A	0.0000	0.3750	0.3656	—	0.3344	0.3311	0.3005	3B	0.3070	0.3182	0.3344	0.3387	0.3750
³⁄₈–18 UNS	2A	0.0013	0.3737	0.3650	—	0.3376	0.3333	0.3076	2B	0.315	0.328	0.3389	0.3445	0.3750
³⁄₈–20 UN	2A	0.0012	0.3738	0.3657	—	0.3413	0.3372	0.3143	2B	0.321	0.332	0.3425	0.3479	0.3750
	3A	0.0000	0.3750	0.3669	—	0.3425	0.3394	0.3155	3B	0.3210	0.3397	0.3425	0.3465	0.3750
³⁄₈–24 UNF	1A	0.0011	0.3739	0.3631	—	0.3468	0.3411	0.3243	1B	0.330	0.340	0.3479	0.3553	0.3750
	2A	0.0011	0.3739	0.3667	—	0.3468	0.3430	0.3243	2B	0.330	0.340	0.3479	0.3528	0.3750

3/8–24 UNF	3A	0.0000	0.3750	0.3678	—	0.3479	0.3450	0.3254	3B	0.3300	0.3372	0.3479	0.3516	0.3750
3/8–27 UNS	2A	0.0011	0.3739	0.3672	—	0.3498	0.3462	0.3298	2B	0.335	0.344	0.3509	0.3556	0.3750
3/8–28 UN	2A	0.0011	0.3739	0.3674	—	0.3507	0.3471	0.3313	2B	0.336	0.345	0.3518	0.3564	0.3750
	3A	0.0000	0.3750	0.3685	—	0.3518	0.3491	0.3324	3B	0.3360	0.3426	0.3518	0.3553	0.3750
3/8–32 UNEF	2A	0.0010	0.3740	0.3680	—	0.3537	0.3503	0.3368	2B	0.341	0.349	0.3547	0.3591	0.3750
	3A	0.0000	0.3750	0.3690	—	0.3547	0.3522	0.3378	3B	0.3410	0.3409	0.3547	0.3580	0.3750
3/8–36 UNS	2A	0.0010	0.3740	0.3685	—	0.3560	0.3528	0.3409	2B	0.345	0.352	0.3570	0.3612	0.3750
3/8–40 UNS	2A	0.0009	0.3741	0.3690	—	0.3579	0.3548	0.3443	2B	0.348	0.354	0.3588	0.3628	0.3750
0.390–27 UNS	2A	0.0011	0.3889	0.3822	—	0.3648	0.3612	0.3448	2B	0.350	0.359	0.3659	0.3706	0.3900
7/16–14 UNC	1A	0.0014	0.4361	0.4206	0.4206	0.3897	0.3826	0.3511	1B	0.360	0.376	0.3911	0.4003	0.4375
	2A	0.0014	0.4361	0.4258	—	0.3897	0.3850	0.3511	2B	0.360	0.376	0.3911	0.3972	0.4375
	3A	0.0000	0.4375	0.4272	—	0.3911	0.3876	0.3525	3B	0.3600	0.3717	0.3911	0.3957	0.4375
7/16–16 UN	2A	0.0014	0.4361	0.4267	—	0.3955	0.3909	0.3616	2B	0.370	0.384	0.3969	0.4028	0.4375
	3A	0.0000	0.4375	0.4281	—	0.3969	0.3935	0.3630	3B	0.3700	0.3800	0.3969	0.4014	0.4375
7/16–18 UNS	2A	0.0013	0.4362	0.4275	—	0.4001	0.3958	0.3701	2B	0.377	0.390	0.4014	0.4070	0.4375
7/16–20 UNF	1A	0.0013	0.4362	0.4240	—	0.4037	0.3975	0.3767	1B	0.383	0.395	0.4050	0.4131	0.4375
	2A	0.0013	0.4362	0.4281	—	0.4037	0.3995	0.3767	2B	0.383	0.395	0.4050	0.4104	0.4375
	3A	0.0000	0.4375	0.4294	—	0.4050	0.4019	0.3780	3B	0.3830	0.3916	0.4050	0.4091	0.4375
7/16–24 UNS	2A	0.0011	0.4364	0.4292	—	0.4093	0.4055	0.3868	2B	0.392	0.402	0.4104	0.4153	0.4375
7/16–27 UNS	2A	0.0011	0.4364	0.4297	—	0.4123	0.4087	0.3923	2B	0.397	0.406	0.4134	0.4181	0.4375
7/16–28 UNEF	2A	0.0011	0.4364	0.4299	—	0.4132	0.4096	0.3938	2B	0.399	0.407	0.4143	0.4189	0.4375
	3A	0.0000	0.4375	0.4310	—	0.4143	0.4116	0.3949	3B	0.3990	0.4051	0.4143	0.4178	0.4375
7/16–32 UN	2A	0.0010	0.4365	0.4305	—	0.4162	0.4128	0.3993	2B	0.404	0.411	0.4172	0.4216	0.4375
	3A	0.0000	0.4375	0.4315	—	0.4172	0.4147	0.4003	3B	0.4040	0.4094	0.4172	0.4205	0.4375
1/2–12 UNS	2A	0.0016	0.4984	0.4870	—	0.4443	0.4389	0.3992	2B	0.410	0.428	0.4459	0.4529	0.5000
	3A	0.0000	0.5000	0.4886	—	0.4459	0.4419	0.4008	3B	0.4100	0.4223	0.4459	0.4511	0.5000
1/2–13 UNC	1A	0.0015	0.4985	0.4822	0.4822	0.4485	0.4411	0.4069	1B	0.417	0.434	0.4500	0.4597	0.5000
	2A	0.0015	0.4985	0.4876	—	0.4485	0.4435	0.4069	2B	0.417	0.434	0.4500	0.4565	0.5000
	3A	0.0000	0.5000	0.4891	—	0.4500	0.4463	0.4084	3B	0.4170	0.4284	0.4500	0.4548	0.5000
1/2–14 UNS	2A	0.0015	0.4985	0.4882	—	0.4521	0.4471	0.4135	2B	0.423	0.438	0.4536	0.4601	0.5000

All dimensions in inches. † Use UNS threads only if Standard Series do not meet requirements (see p. 1479). See footnotes *a*, *b*, *c*, *d*, and *e* at end of table.

Table 4 (*Continued*). Standard Series and Selected Combinations† — Unified Screw Threads

Nominal Size, Threads per Inch, and Series Designation^c	Class	External^c							Internal^c					
		Allowance	Major Diameter Max^b	Major Diameter Min	Major Diameter Min^d	Pitch Diameter Max^b	Pitch Diameter Min	UNR Minor Diam.^a Max (Ref.)	Class	Minor Diameter Min	Minor Diameter Max	Pitch Diameter Min	Pitch Diameter Max	Major Diameter Min
1/2–16 UN	2A	0.0014	0.4986	0.4892	—	0.4580	0.4533	0.4241	2B	0.432	0.446	0.4594	0.4655	0.5000
	3A	0.0000	0.5000	0.4906	—	0.4594	0.4559	0.4255	3B	0.4320	0.4419	0.4594	0.4640	0.5000
1/2–18 UNS	2A	0.0013	0.4987	0.4900	—	0.4626	0.4582	0.4326	2B	0.440	0.453	0.4639	0.4697	0.5000
1/2–20 UNF	1A	0.0013	0.4987	0.4865	—	0.4662	0.4598	0.4392	1B	0.446	0.457	0.4675	0.4759	0.5000
	2A	0.0013	0.4987	0.4906	—	0.4662	0.4619	0.4392	2B	0.446	0.457	0.4675	0.4731	0.5000
	3A	0.0000	0.5000	0.4919	—	0.4675	0.4643	0.4405	3B	0.4460	0.4537	0.4675	0.4717	0.5000
1/2–24 UNS	2A	0.0012	0.4988	0.4916	—	0.4717	0.4678	0.4492	2B	0.455	0.465	0.4729	0.4780	0.5000
1/2–27 UNS	2A	0.0011	0.4989	0.4922	—	0.4748	0.4711	0.4548	2B	0.460	0.469	0.4759	0.4807	0.5000
1/2–28 UNEF	2A	0.0011	0.4989	0.4924	—	0.4757	0.4720	0.4563	2B	0.461	0.470	0.4768	0.4816	0.5000
	3A	0.0000	0.5000	0.4935	—	0.4768	0.4740	0.4574	3B	0.4610	0.4076	0.4768	0.4804	0.5000
1/2–32 UN	2A	0.0010	0.4990	0.4930	—	0.4787	0.4752	0.4618	2B	0.466	0.474	0.4797	0.4842	0.5000
	3A	0.0000	0.5000	0.4940	—	0.4797	0.4771	0.4628	3B	0.4660	0.4719	0.4797	0.4831	0.5000
9/16–12 UNC	1A	0.0016	0.5609	0.5437	0.5437	0.5068	0.4990	0.4617	1B	0.472	0.490	0.5084	0.5186	0.5625
	2A	0.0016	0.5609	0.5495	—	0.5068	0.5016	0.4617	2B	0.472	0.490	0.5084	0.5152	0.5625
	3A	0.0000	0.5625	0.5511	—	0.5084	0.5045	0.4633	3B	0.4720	0.4843	0.5084	0.5135	0.5625
9/16–14 UNS	2A	0.0015	0.5610	0.5507	—	0.5146	0.5096	0.4760	2B	0.485	0.501	0.5161	0.5226	0.5625
9/16–16 UN	2A	0.0014	0.5611	0.5517	—	0.5205	0.5158	0.4866	2B	0.495	0.509	0.5219	0.5280	0.5625
	3A	0.0000	0.5625	0.5531	—	0.5219	0.5184	0.4880	3B	0.4950	0.5040	0.5219	0.5265	0.5625
9/16–18 UNF	1A	0.0014	0.5611	0.5480	—	0.5250	0.5182	0.4950	1B	0.502	0.515	0.5264	0.5353	0.5625
	2A	0.0014	0.5611	0.5524	—	0.5250	0.5205	0.4950	2B	0.502	0.515	0.5264	0.5323	0.5625
	3A	0.0000	0.5625	0.5538	—	0.5264	0.5230	0.4964	3B	0.5020	0.5106	0.5264	0.5308	0.5625
9/16–20 UN	2A	0.0013	0.5612	0.5531	—	0.5287	0.5245	0.5017	2B	0.508	0.520	0.5300	0.5355	0.5625
	3A	0.0000	0.5625	0.5544	—	0.5300	0.5268	0.5030	3B	0.5080	0.5162	0.5300	0.5341	0.5625
9/16–24 UNEF	2A	0.0012	0.5613	0.5541	—	0.5342	0.5303	0.5117	2B	0.517	0.527	0.5354	0.5405	0.5625
	3A	0.0000	0.5625	0.5553	—	0.5354	0.5325	0.5129	3B	0.5170	0.5244	0.5354	0.5392	0.5625
9/16–27 UNS	2A	0.0011	0.5614	0.5547	—	0.5373	0.5336	0.5173	2B	0.522	0.531	0.5384	0.5432	0.5625
9/16–28 UN	2A	0.0011	0.5614	0.5549	—	0.5382	0.5345	0.5188	2B	0.524	0.532	0.5393	0.5441	0.5625
	3A	0.0000	0.5625	0.5560	—	0.5393	0.5365	0.5199	3B	0.5240	0.5301	0.5393	0.5429	0.5625
9/16–32 UN	2A	0.0010	0.5615	0.5555	—	0.5412	0.5377	0.5243	2B	0.529	0.536	0.5422	0.5467	0.5625
	3A	0.0000	0.5625	0.5565	—	0.5422	0.5396	0.5253	3B	0.5290	0.5344	0.5422	0.5456	0.5625

Nominal Size, Threads per Inch, and Series	Class	Allowance	Ext. Max Major	Ext. Min Major	Ext. Minor (UNR) Max	Ext. Max Pitch	Ext. Min Pitch	Ext. Min Minor	Class	Int. Min Minor	Int. Max Minor	Int. Min Pitch	Int. Max Pitch	Int. Min Major
5/8–11 UNC	1A	0.0016	0.6234	0.6052	—	0.5644	0.5561	0.5152	1B	0.527	0.546	0.5660	0.5767	0.6250
	2A	0.0016	0.6234	0.6113	—	0.5644	0.5589	0.5152	2B	0.527	0.546	0.5660	0.5732	0.6250
	3A	0.0000	0.6250	0.6129	—	0.5660	0.5619	0.5168	3B	0.5270	0.5391	0.5660	0.5714	0.6250
5/8–12 UN	2A	0.0016	0.6234	0.6120	—	0.5693	0.5639	0.5242	2B	0.535	0.553	0.5709	0.5780	0.6250
	3A	0.0000	0.6250	0.6136	—	0.5709	0.5668	0.5258	3B	0.5350	0.5463	0.5709	0.5762	0.6250
5/8–14 UNS	2A	0.0015	0.6235	0.6132	—	0.5771	0.5720	0.5385	2B	0.548	0.564	0.5786	0.5852	0.6250
5/8–16 UN	2A	0.0014	0.6236	0.6142	—	0.5830	0.5782	0.5491	2B	0.557	0.571	0.5844	0.5906	0.6250
	3A	0.0000	0.6250	0.6156	—	0.5844	0.5808	0.5505	3B	0.5570	0.5662	0.5844	0.5890	0.6250
5/8–18 UNF	1A	0.0014	0.6236	0.6105	—	0.5875	0.5805	0.5575	1B	0.565	0.578	0.5889	0.5980	0.6250
	2A	0.0014	0.6236	0.6149	—	0.5875	0.5828	0.5575	2B	0.565	0.578	0.5889	0.5949	0.6250
	3A	0.0000	0.6250	0.6163	—	0.5889	0.5854	0.5589	3B	0.5650	0.5730	0.5889	0.5934	0.6250
5/8–20 UN	2A	0.0013	0.6237	0.6156	—	0.5912	0.5869	0.5642	2B	0.571	0.582	0.5925	0.5981	0.6250
	3A	0.0000	0.6250	0.6169	—	0.5925	0.5893	0.5655	3B	0.5710	0.5787	0.5925	0.5967	0.6250
5/8–24 UNEF	2A	0.0012	0.6238	0.6166	—	0.5967	0.5927	0.5742	2B	0.580	0.590	0.5979	0.6031	0.6250
	3A	0.0000	0.6250	0.6178	—	0.5979	0.5949	0.5754	3B	0.5800	0.5869	0.5979	0.6018	0.6250
5/8–27 UNS	2A	0.0011	0.6239	0.6172	—	0.5998	0.5960	0.5798	2B	0.585	0.594	0.6009	0.6059	0.6250
5/8–28 UN	2A	0.0011	0.6239	0.6174	—	0.6007	0.5969	0.5813	2B	0.586	0.595	0.6018	0.6067	0.6250
	3A	0.0000	0.6250	0.6185	—	0.6018	0.5990	0.5824	3B	0.5860	0.5926	0.6018	0.6055	0.6250
5/8–32 UN	2A	0.0011	0.6239	0.6179	—	0.6036	0.6000	0.5867	2B	0.591	0.599	0.6047	0.6093	0.6250
	3A	0.0000	0.6250	0.6190	—	0.6047	0.6020	0.5878	3B	0.5910	0.5969	0.6047	0.6082	0.6250
11/16–12 UN	2A	0.0016	0.6859	0.6745	—	0.6318	0.6264	0.5867	2B	0.597	0.615	0.6334	0.6405	0.6875
	3A	0.0000	0.6875	0.6761	—	0.6334	0.6293	0.5883	3B	0.5970	0.6085	0.6334	0.6387	0.6875
11/16–16 UN	2A	0.0014	0.6861	0.6767	—	0.6455	0.6407	0.6116	2B	0.620	0.634	0.6469	0.6531	0.6875
	3A	0.0000	0.6875	0.6781	—	0.6469	0.6433	0.6130	3B	0.6200	0.6284	0.6469	0.6515	0.6875
11/16–20 UN	2A	0.0013	0.6862	0.6781	—	0.6537	0.6494	0.6267	2B	0.633	0.645	0.6550	0.6606	0.6875
	3A	0.0000	0.6875	0.6794	—	0.6550	0.6518	0.6280	3B	0.6330	0.6412	0.6550	0.6592	0.6875
11/16–24 UNEF	2A	0.0012	0.6863	0.6791	—	0.6592	0.6552	0.6367	2B	0.642	0.652	0.6604	0.6656	0.6875
	3A	0.0000	0.6875	0.6803	—	0.6604	0.6574	0.6379	3B	0.6420	0.6494	0.6604	0.6643	0.6875
11/16–28 UN	2A	0.0011	0.6864	0.6799	—	0.6632	0.6594	0.6438	2B	0.649	0.657	0.6643	0.6692	0.6875
	3A	0.0000	0.6875	0.6810	—	0.6643	0.6615	0.6449	3B	0.6490	0.6551	0.6643	0.6680	0.6875
11/16–32 UN	2A	0.0011	0.6864	0.6804	—	0.6661	0.6625	0.6492	2B	0.654	0.661	0.6672	0.6718	0.6875
	3A	0.0000	0.6875	0.6815	—	0.6672	0.6645	0.6503	3B	0.6540	0.6594	0.6672	0.6707	0.6875

All dimensions in inches.   † Use UNS threads only if Standard Series do not meet requirements (see p. 1479). See footnotes a, b, c, d, and e at end of table.

Table 4 (Continued). Standard Series and Selected Combinations‡ — Unified Screw Threads

Nominal Size, Threads per Inch, and Series Designation^e	Class	Allowance	Major Diameter Max^b	Major Diameter Min	Major Diameter Min^d	Pitch Diameter Max^b	Pitch Diameter Min	UNR Minor Diam.^a Max (Ref.)	Class	Minor Diameter Min	Minor Diameter Max	Pitch Diameter Min	Pitch Diameter Max	Major Diameter Min
			External^c						Internal^c					
¾–10 UNC	1A	0.0018	0.7482	0.7288	—	0.6832	0.6744	0.6291	1B	0.642	0.663	0.6850	0.6965	0.7500
	2A	0.0018	0.7482	0.7353	0.7288	0.6832	0.6773	0.6291	2B	0.642	0.663	0.6850	0.6927	0.7500
	3A	0.0000	0.7500	0.7371	—	0.6850	0.6806	0.6309	3B	0.6420	0.6545	0.6850	0.6907	0.7500
¾–12 UN	2A	0.0017	0.7483	0.7369	—	0.6942	0.6887	0.6491	2B	0.660	0.678	0.6959	0.7031	0.7500
	3A	0.0000	0.7500	0.7386	—	0.6959	0.6918	0.6508	3B	0.6600	0.6707	0.6959	0.7013	0.7500
¾–14 UNS	2A	0.0015	0.7485	0.7382	—	0.7021	0.6970	0.6635	2B	0.673	0.688	0.7036	0.7103	0.7500
¾–16 UNF	1A	0.0015	0.7485	0.7343	—	0.7079	0.7004	0.6740	1B	0.682	0.696	0.7094	0.7192	0.7500
	2A	0.0015	0.7485	0.7391	—	0.7079	0.7029	0.6740	2B	0.682	0.696	0.7094	0.7159	0.7500
	3A	0.0000	0.7500	0.7406	—	0.7094	0.7056	0.6755	3B	0.6820	0.6908	0.7094	0.7143	0.7500
¾–18 UNS	2A	0.0014	0.7486	0.7399	—	0.7125	0.7079	0.6825	2B	0.690	0.703	0.7139	0.7199	0.7500
¾–20 UNEF	2A	0.0013	0.7487	0.7406	—	0.7162	0.7118	0.6892	2B	0.696	0.707	0.7175	0.7232	0.7500
	3A	0.0000	0.7500	0.7419	—	0.7175	0.7142	0.6905	3B	0.6960	0.7037	0.7175	0.7218	0.7500
¾–24 UNS	2A	0.0012	0.7488	0.7416	—	0.7217	0.7176	0.6992	2B	0.705	0.715	0.7229	0.7282	0.7500
¾–27 UNS	2A	0.0012	0.7488	0.7421	—	0.7247	0.7208	0.7047	2B	0.710	0.719	0.7259	0.7310	0.7500
¾–28 UN	2A	0.0012	0.7488	0.7423	—	0.7256	0.7218	0.7062	2B	0.711	0.720	0.7268	0.7318	0.7500
	3A	0.0000	0.7500	0.7435	—	0.7268	0.7239	0.7074	3B	0.7110	0.7176	0.7268	0.7305	0.7500
¾–32 UN	2A	0.0011	0.7489	0.7429	—	0.7286	0.7250	0.7117	2B	0.716	0.724	0.7297	0.7344	0.7500
	3A	0.0000	0.7500	0.7440	—	0.7297	0.7270	0.7128	3B	0.7160	0.7219	0.7297	0.7333	0.7500
13/16–12 UN	2A	0.0017	0.8108	0.7994	—	0.7567	0.7512	0.7116	2B	0.722	0.740	0.7584	0.7656	0.8125
	3A	0.0000	0.8125	0.8011	—	0.7584	0.7543	0.7133	3B	0.7220	0.7329	0.7584	0.7638	0.8125
13/16–16 UN	2A	0.0015	0.8110	0.8016	—	0.7704	0.7655	0.7365	2B	0.745	0.759	0.7719	0.7782	0.8125
	3A	0.0000	0.8125	0.8031	—	0.7719	0.7683	0.7380	3B	0.7450	0.7533	0.7719	0.7766	0.8125
13/16–20 UNEF	2A	0.0013	0.8112	0.8031	—	0.7787	0.7743	0.7517	2B	0.758	0.770	0.7800	0.7857	0.8125
	3A	0.0000	0.8125	0.8044	—	0.7800	0.7767	0.7530	3B	0.7580	0.7662	0.7800	0.7843	0.8125
13/16–28 UN	2A	0.0012	0.8113	0.8048	—	0.7881	0.7843	0.7687	2B	0.774	0.782	0.7893	0.7943	0.8125
	3A	0.0000	0.8125	0.8060	—	0.7893	0.7864	0.7699	3B	0.7740	0.7801	0.7893	0.7930	0.8125
13/16–32 UN	2A	0.0011	0.8114	0.8054	—	0.7911	0.7875	0.7742	2B	0.779	0.786	0.7922	0.7969	0.8125
	3A	0.0000	0.8125	0.8065	—	0.7922	0.7895	0.7753	3B	0.7790	0.7844	0.7922	0.7958	0.8125

All dimensions in inches.

Nominal Size, Threads per Inch	Class	Allowance	Ext. Major Dia. Max	Ext. Major Dia. Min	(UNR)	Ext. Pitch Dia. Max	Ext. Pitch Dia. Min	Ext. Minor Dia.	Class	Int. Minor Dia. Min	Int. Minor Dia. Max	Int. Pitch Dia. Min	Int. Pitch Dia. Max	Int. Major Dia. Min
7/8–9 UNC	1A	0.0019	0.8731	0.8523	—	0.8009	0.7914	0.7408	1B	0.755	0.778	0.8028	0.8151	0.8750
	2A	0.0019	0.8731	0.8592	0.8523	0.8009	0.7946	0.7408	2B	0.755	0.778	0.8028	0.8110	0.8750
	3A	0.0000	0.8750	0.8611	—	0.8028	0.7981	0.7427	3B	0.7550	0.7681	0.8028	0.8089	0.8750
7/8–10 UNS	2A	0.0018	0.8732	0.8603	—	0.8082	0.8022	0.7542	2B	0.767	0.788	0.8100	0.8178	0.8750
7/8–12 UN	2A	0.0017	0.8733	0.8619	—	0.8192	0.8137	0.7741	2B	0.785	0.803	0.8209	0.8281	0.8750
	3A	0.0000	0.8750	0.8636	—	0.8209	0.8168	0.7758	3B	0.7850	0.7948	0.8209	0.8263	0.8750
7/8–14 UNF	1A	0.0016	0.8734	0.8579	—	0.8270	0.8189	0.7884	1B	0.798	0.814	0.8286	0.8392	0.8750
	2A	0.0016	0.8734	0.8631	—	0.8270	0.8216	0.7884	2B	0.798	0.814	0.8286	0.8356	0.8750
	3A	0.0000	0.8750	0.8647	—	0.8286	0.8245	0.7900	3B	0.7980	0.8068	0.8286	0.8339	0.8750
7/8–16 UN	2A	0.0015	0.8735	0.8641	—	0.8329	0.8280	0.8005	2B	0.807	0.821	0.8344	0.8407	0.8750
	3A	0.0000	0.8750	0.8656	—	0.8344	0.8308	0.8075	3B	0.8070	0.8158	0.8344	0.8391	0.8750
7/8–18 UNS	2A	0.0014	0.8736	0.8649	—	0.8375	0.8329	—	2B	0.815	0.828	0.8389	0.8449	0.8750
7/8–20 UNEF	2A	0.0013	0.8737	0.8656	—	0.8412	0.8368	0.8142	2B	0.821	0.832	0.8425	0.8482	0.8750
	3A	0.0000	0.8750	0.8669	—	0.8425	0.8392	0.8155	3B	0.8210	0.8287	0.8425	0.8468	0.8750
7/8–24 UNS	2A	0.0012	0.8738	0.8666	—	0.8467	0.8426	0.8242	2B	0.830	0.840	0.8479	0.8532	0.8750
7/8–27 UNS	2A	0.0012	0.8738	0.8671	—	0.8497	0.8458	0.8297	2B	0.835	0.844	0.8509	0.8560	0.8750
7/8–28 UN	2A	0.0012	0.8738	0.8673	—	0.8506	0.8468	0.8312	2B	0.836	0.845	0.8518	0.8568	0.8750
	3A	0.0000	0.8750	0.8685	—	0.8518	0.8489	0.8334	3B	0.8360	0.8426	0.8518	0.8555	0.8750
7/8–32 UN	2A	0.0011	0.8739	0.8679	—	0.8536	0.8500	0.8367	2B	0.841	0.849	0.8547	0.8594	0.8750
	3A	0.0000	0.8750	0.8690	—	0.8547	0.8520	0.8378	3B	0.8410	0.8469	0.8547	0.8583	0.8750
15/16–12 UN	2A	0.0017	0.9358	0.9244	—	0.8817	0.8760	0.8366	2B	0.847	0.865	0.8834	0.8908	0.9375
	3A	0.0000	0.9375	0.9261	—	0.8834	0.8793	0.8383	3B	0.8470	0.8575	0.8834	0.8889	0.9375
15/16–16 UN	2A	0.0015	0.9360	0.9266	—	0.8954	0.8904	0.8615	2B	0.870	0.884	0.8969	0.9034	0.9375
	3A	0.0000	0.9375	0.9281	—	0.8969	0.8932	0.8630	3B	0.8700	0.8783	0.8969	0.9018	0.9375
15/16–20 UNEF	2A	0.0014	0.9361	0.9280	—	0.9036	0.8991	0.8766	2B	0.883	0.895	0.9050	0.9109	0.9375
	3A	0.0000	0.9375	0.9294	—	0.9050	0.9016	0.8780	3B	0.8830	0.8912	0.9050	0.9094	0.9375
15/16–28 UN	2A	0.0012	0.9363	0.9298	—	0.9131	0.9091	0.8937	2B	0.899	0.907	0.9143	0.9195	0.9375
	3A	0.0000	0.9375	0.9310	—	0.9143	0.9113	0.8949	3B	0.8990	0.9051	0.9143	0.9182	0.9375
15/16–32 UN	2A	0.0011	0.9364	0.9304	—	0.9161	0.9123	0.8992	2B	0.904	0.911	0.9172	0.9221	0.9375
	3A	0.0000	0.9375	0.9315	—	0.9172	0.9144	0.9003	3B	0.9040	0.9094	0.9172	0.9209	0.9375
1–8 UNC	1A	0.0020	0.9980	0.9755	—	0.9168	0.9067	0.8492	1B	0.865	0.890	0.9188	0.9320	1.0000
	2A	0.0020	0.9980	0.9830	0.9755	0.9168	0.9100	0.8492	2B	0.865	0.890	0.9188	0.9276	1.0000
	3A	0.0000	1.0000	0.9850	—	0.9188	0.9137	0.8512	3B	0.8650	0.8797	0.9188	0.9254	1.0000

† Use UNR threads only if Standard Series do not meet requirements (see p. 1479). See footnotes a, b, c, d, and e at end of table.

Table 4 (*Continued*). Standard Series and Selected Combinations† — Unified Screw Threads

Nominal Size, Threads per Inch, and Series Designation^c	Class	External^c							Internal^c					
		Allowance	Major Diameter			Pitch Diameter		UNR Minor Diam.^a	Class	Minor Diameter		Pitch Diameter		Major Diameter
			Max^b	Min	Min^d	Max^b	Min	Max (Ref.)		Min	Max	Min	Max	Min
1–10 UNS	2A	0.0018	0.9982	0.9853	—	0.9332	0.9270	0.8792	2B	0.892	0.913	0.9350	0.9430	1.0000
1–12 UNF	1A	0.0018	0.9982	0.9810	—	0.9441	0.9353	0.8990	1B	0.910	0.928	0.9459	0.9573	1.0000
	2A	0.0018	0.9982	0.9868	—	0.9441	0.9382	0.8990	2B	0.910	0.928	0.9459	0.9535	1.0000
	3A	0.0000	1.0000	0.9886	—	0.9459	0.9415	0.9008	3B	0.9100	0.9198	0.9459	0.9516	1.0000
1–14 UNS^f	1A	0.0017	0.9983	0.9828	—	0.9519	0.9435	0.9132	1B	0.923	0.938	0.9536	0.9645	1.0000
	2A	0.0017	0.9983	0.9880	—	0.9519	0.9463	0.9132	2B	0.923	0.938	0.9536	0.9609	1.0000
	3A	0.0000	1.0000	0.9897	—	0.9536	0.9494	0.9149	3B	0.9230	0.9315	0.9536	0.9590	1.0000
1–16 UN	2A	0.0015	0.9985	0.9891	—	0.9579	0.9529	0.9240	2B	0.932	0.946	0.9594	0.9659	1.0000
	3A	0.0000	1.0000	0.9906	—	0.9594	0.9557	0.9255	3B	0.9320	0.9408	0.9594	0.9643	1.0000
1–18 UNS	2A	0.0014	0.9986	0.9899	—	0.9625	0.9578	0.9325	2B	0.940	0.953	0.9639	0.9701	1.0000
1–20 UNEF	2A	0.0014	0.9986	0.9905	—	0.9661	0.9616	0.9391	2B	0.946	0.957	0.9675	0.9734	1.0000
	3A	0.0000	1.0000	0.9919	—	0.9675	0.9641	0.9405	3B	0.9460	0.9537	0.9675	0.9719	1.0000
1–24 UNS	2A	0.0013	0.9987	0.9915	—	0.9716	0.9674	0.9491	2B	0.955	0.965	0.9729	0.9784	1.0000
1–27 UNS	2A	0.0012	0.9988	0.9921	—	0.9747	0.9707	0.9547	2B	0.960	0.969	0.9759	0.9811	1.0000
1–28 UN	2A	0.0012	0.9988	0.9923	—	0.9756	0.9716	0.9562	2B	0.961	0.970	0.9768	0.9820	1.0000
	3A	0.0000	1.0000	0.9935	—	0.9768	0.9738	0.9574	3B	0.9610	0.9676	0.9768	0.9807	1.0000
1–32 UN	2A	0.0011	0.9989	0.9929	—	0.9786	0.9748	0.9617	2B	0.966	0.974	0.9797	0.9846	1.0000
	3A	0.0000	1.0000	0.9940	—	0.9797	0.9769	0.9628	3B	0.9660	0.9719	0.9797	0.9834	1.0000
1 1/16–8 UN	2A	0.0020	1.0605	1.0455	—	0.9793	0.9725	0.9117	2B	0.927	0.952	0.9813	0.9902	1.0625
	3A	0.0000	1.0625	1.0475	—	0.9813	0.9762	0.9137	3B	0.9270	0.9422	0.9813	0.9880	1.0625
1 1/16–12 UN	2A	0.0017	1.0608	1.0494	—	1.0067	1.0010	0.9616	2B	0.972	0.990	1.0084	1.0158	1.0625
	3A	0.0000	1.0625	1.0511	—	1.0084	1.0042	0.9633	3B	0.9720	0.9823	1.0084	1.0139	1.0625
1 1/16–16 UN	2A	0.0015	1.0610	1.0516	—	1.0204	1.0154	0.9865	2B	0.995	1.009	1.0219	1.0284	1.0625
	3A	0.0000	1.0625	1.0531	—	1.0219	1.0182	0.9880	3B	0.9950	1.0033	1.0219	1.0268	1.0625
1 1/16–18 UNEF	2A	0.0014	1.0611	1.0524	—	1.0250	1.0203	0.9950	2B	1.002	1.015	1.0264	1.0326	1.0625
	3A	0.0000	1.0625	1.0538	—	1.0264	1.0228	0.9964	3B	1.0020	1.0105	1.0264	1.0310	1.0625
1 1/16–20 UN	2A	0.0014	1.0611	1.0530	—	1.0286	1.0241	1.0016	2B	1.008	1.020	1.0300	1.0359	1.0625
	3A	0.0000	1.0625	1.0544	—	1.0300	1.0266	1.0030	3B	1.0080	1.0162	1.0300	1.0344	1.0625

Nominal Size, Threads per Inch	Allowance	Class	Major Dia Max	Major Dia Min	†	Pitch Dia Max	Pitch Dia Min	Minor Dia Max	Class	Minor Dia Min	Minor Dia Max	Pitch Dia Min	Pitch Dia Max	Major Dia Min
1 1/16–28 UN	0.0012	2A	1.0613	1.0548	—	1.0381	1.0341	1.0187	2B	1.024	1.032	1.0393	1.0445	1.0625
	0.0000	3A	1.0625	1.0560	—	1.0393	1.0363	1.0199	3B	1.0240	1.0301	1.0393	1.0432	1.0625
1 1/8–7 UNC	0.0022	1A	1.1228	1.0982	—	1.0300	1.0191	0.9527	1B	0.970	0.998	1.0322	1.0463	1.1250
	0.0022	2A	1.1228	1.1064	—	1.0300	1.0228	0.9527	2B	0.970	0.998	1.0322	1.0416	1.1250
	0.0000	3A	1.1250	1.1086	1.0982	1.0322	1.0268	0.9549	3B	0.9700	0.9875	1.0322	1.0393	1.1250
1 1/8–8 UN	0.0021	2A	1.1229	1.1079	—	1.0417	1.0348	0.9741	2B	0.990	1.015	1.0438	1.0528	1.1250
	0.0000	3A	1.1250	1.1100	1.1004	1.0438	1.0386	0.9762	3B	0.9900	1.0047	1.0438	1.0595	1.1250
1 1/8–10 UNS	0.0018	2A	1.1232	1.1103	—	1.0582	1.0520	1.0042	2B	1.017	1.038	1.0600	1.0680	1.1250
1 1/8–12 UNF	0.0018	1A	1.1232	1.1060	—	1.0691	1.0601	1.0240	1B	1.035	1.053	1.0709	1.0826	1.1250
	0.0018	2A	1.1232	1.1118	—	1.0691	1.0631	1.0240	2B	1.035	1.053	1.0709	1.0787	1.1250
	0.0000	3A	1.1250	1.1136	—	1.0709	1.0664	1.0258	3B	1.0350	1.0448	1.0709	1.0768	1.1250
1 1/8–14 UNS	0.0016	2A	1.1234	1.1131	—	1.0770	1.0717	1.0384	2B	1.048	1.064	1.0786	1.0855	1.1250
1 1/8–16 UN	0.0015	2A	1.1235	1.1141	—	1.0829	1.0779	1.0490	2B	1.057	1.071	1.0844	1.0999	1.1250
	0.0000	3A	1.1250	1.1156	—	1.0844	1.0807	1.0505	3B	1.0570	1.0658	1.0844	1.0893	1.1250
1 1/8–18 UNEF	0.0014	2A	1.1236	1.1149	—	1.0875	1.0828	1.0575	2B	1.065	1.078	1.0889	1.0951	1.1250
	0.0000	3A	1.1250	1.1163	—	1.0889	1.0853	1.0589	3B	1.0650	1.0730	1.0889	1.0935	1.1250
1 1/8–20 UN	0.0014	2A	1.1236	1.1155	—	1.0911	1.0866	1.0641	2B	1.071	1.082	1.0925	1.0984	1.1250
	0.0000	3A	1.1250	1.1169	—	1.0925	1.0891	1.0655	3B	1.0710	1.0787	1.0925	1.0969	1.1250
1 1/8–24 UNS	0.0013	2A	1.1237	1.1165	—	1.0966	1.0924	1.0742	2B	1.080	1.090	1.0979	1.1034	1.1250
1 1/8–28 UN	0.0012	2A	1.1238	1.1173	—	1.1006	1.0966	1.0812	2B	1.086	1.095	1.1018	1.1070	1.1250
	0.0000	3A	1.1250	1.1185	—	1.1018	1.0988	1.0824	3B	1.0860	1.0926	1.1018	1.1057	1.1250
1 3/16–8 UN	0.0021	2A	1.1854	1.1704	—	1.1042	1.0972	1.0366	2B	1.052	1.077	1.1063	1.1154	1.1875
	0.0000	3A	1.1875	1.1725	—	1.1063	1.1011	1.0387	3B	1.0520	1.0672	1.1063	1.1131	1.1875
1 3/16–12 UN	0.0017	2A	1.1858	1.1744	—	1.1317	1.1259	1.0866	2B	1.097	1.115	1.1334	1.1409	1.1875
	0.0000	3A	1.1875	1.1761	—	1.1334	1.1291	1.0883	3B	1.0970	1.1073	1.1334	1.1390	1.1875
1 3/16–16 UN	0.0015	2A	1.1860	1.1766	—	1.1454	1.1403	1.1115	2B	1.120	1.134	1.1469	1.1535	1.1875
	0.0000	3A	1.1875	1.1781	—	1.1469	1.1431	1.1130	3B	1.1200	1.1283	1.1469	1.1519	1.1875
1 3/16–18 UNEF	0.0015	2A	1.1860	1.1773	—	1.1499	1.1450	1.1199	2B	1.127	1.140	1.1514	1.1577	1.1875
	0.0000	3A	1.1875	1.1788	—	1.1514	1.1478	1.1214	3B	1.1270	1.1355	1.1514	1.1561	1.1875
1 3/16–20 UN	0.0014	2A	1.1861	1.1780	—	1.1536	1.1489	1.1266	2B	1.133	1.145	1.1550	1.1611	1.1875
	0.0000	3A	1.1875	1.1794	—	1.1550	1.1515	1.1280	3B	1.1330	1.1412	1.1550	1.1595	1.1875
1 3/16–28 UN	0.0012	2A	1.1863	1.1798	—	1.1631	1.1590	1.1437	2B	1.149	1.157	1.1643	1.1696	1.1875
	0.0000	3A	1.1875	1.1810	—	1.1643	1.1612	1.1449	3B	1.1490	1.1551	1.1643	1.1683	1.1875

All dimensions in inches. † Use UNS threads only if Standard Series do not meet requirements (see p. 1479). See footnotes a, b, c, d, e, and f at end of table.

Table 4 (*Continued*). Standard Series and Selected Combinations† — Unified Screw Threads

Nominal Size, Threads per Inch, and Series Designation[e]	External[c]								Internal[c]					
			Major Diameter			Pitch Diameter		UNR Minor Diam.[a]		Minor Diameter		Pitch Diameter		Major Diameter
	Class	Allowance	Max[b]	Min	Min[d]	Max[b]	Min	Max (Ref.)	Class	Min	Max	Min	Max	Min
1¼–7 UNC	1A	0.0022	1.2478	1.2232	—	1.1550	1.1439	1.0777	1B	1.095	1.123	1.1572	1.1716	1.2500
	2A	0.0022	1.2478	1.2314	1.2232	1.1550	1.1476	1.0777	2B	1.095	1.123	1.1572	1.1668	1.2500
	3A	0.0000	1.2500	1.2336	1.2254	1.1572	1.1517	1.0799	3B	1.0950	1.1125	1.1572	1.1644	1.2500
1¼–8 UN	2A	0.0021	1.2479	1.2329	—	1.1667	1.1597	1.0991	2B	1.115	1.140	1.1688	1.1780	1.2500
	3A	0.0000	1.2500	1.2350	—	1.1688	1.1635	1.1012	3B	1.1150	1.1297	1.1688	1.1757	1.2500
1¼–10 UNS	2A	0.0019	1.2481	1.2352	—	1.1831	1.1768	1.1291	2B	1.142	1.163	1.1850	1.1932	1.2500
1¼–12 UNF	1A	0.0018	1.2482	1.2310	—	1.1941	1.1849	1.1490	1B	1.160	1.178	1.1959	1.2079	1.2500
	2A	0.0018	1.2482	1.2368	—	1.1941	1.1879	1.1490	2B	1.160	1.178	1.1959	1.2039	1.2500
	3A	0.0000	1.2500	1.2386	—	1.1959	1.1913	1.1508	3B	1.1600	1.1698	1.1959	1.2019	1.2500
1¼–14 UNS	2A	0.0016	1.2484	1.2381	—	1.2020	1.1966	1.1634	2B	1.173	1.188	1.2036	1.2106	1.2500
1¼–16 UN	2A	0.0015	1.2485	1.2391	—	1.2079	1.2028	1.1740	2B	1.182	1.196	1.2094	1.2160	1.2500
	3A	0.0000	1.2500	1.2406	—	1.2094	1.2056	1.1755	3B	1.1820	1.1908	1.2094	1.2144	1.2500
1¼–18 UNEF	2A	0.0015	1.2485	1.2398	—	1.2124	1.2075	1.1824	2B	1.190	1.203	1.2139	1.2202	1.2500
	3A	0.0000	1.2500	1.2413	—	1.2139	1.2103	1.1839	3B	1.1900	1.1980	1.2139	1.2186	1.2500
1¼–20 UN	2A	0.0014	1.2486	1.2405	—	1.2161	1.2114	1.1891	2B	1.196	1.207	1.2175	1.2236	1.2500
	3A	0.0000	1.2500	1.2419	—	1.2175	1.2140	1.1905	3B	1.1960	1.2037	1.2175	1.2220	1.2500
1¼–24 UNS	2A	0.0013	1.2487	1.2415	—	1.2216	1.2173	1.1991	2B	1.205	1.215	1.2229	1.2285	1.2500
1¼–28 UN	2A	0.0012	1.2488	1.2423	—	1.2256	1.2215	1.2062	2B	1.211	1.220	1.2268	1.2321	1.2500
	3A	0.0000	1.2500	1.2435	—	1.2268	1.2237	1.2074	3B	1.2110	1.2176	1.2268	1.2308	1.2500
1⁵⁄₁₆–8 UN	2A	0.0021	1.3104	1.2954	—	1.2292	1.2221	1.1616	2B	1.177	1.202	1.2313	1.2405	1.3125
	3A	0.0000	1.3125	1.2975	—	1.2313	1.2260	1.1637	3B	1.1770	1.1922	1.2313	1.2382	1.3125
1⁵⁄₁₆–12 UN	2A	0.0017	1.3108	1.2994	—	1.2567	1.2509	1.2116	2B	1.222	1.240	1.2584	1.2659	1.3125
	3A	0.0000	1.3125	1.3011	—	1.2584	1.2541	1.2133	3B	1.2220	1.2323	1.2584	1.2640	1.3125
1⁵⁄₁₆–16 UN	2A	0.0015	1.3110	1.3016	—	1.2704	1.2653	1.2365	2B	1.245	1.259	1.2719	1.2785	1.3125
	3A	0.0000	1.3125	1.3031	—	1.2719	1.2681	1.2380	3B	1.2450	1.2533	1.2719	1.2769	1.3125
1⁵⁄₁₆–18 UNEF	2A	0.0015	1.3110	1.3023	—	1.2749	1.2700	1.2449	2B	1.252	1.265	1.2764	1.2827	1.3125
	3A	0.0000	1.3125	1.3038	—	1.2764	1.2728	1.2464	3B	1.2520	1.2605	1.2764	1.2811	1.3125
1⁵⁄₁₆–20 UN	2A	0.0014	1.3111	1.3030	—	1.2786	1.2739	1.2516	2B	1.258	1.270	1.2800	1.2861	1.3125
	3A	0.0000	1.3125	1.3044	—	1.2800	1.2765	1.2530	3B	1.2580	1.2662	1.2800	1.2845	1.3125

Note: The column headings for this table are continued from the preceding page. The left-hand group of data is for external threads (Classes 1A, 2A, 3A); the right-hand group is for internal threads (Classes 1B, 2B, 3B). Columns, left to right: Identification; Class; Allowance; Major Diameter Max; Major Diameter Min; (reference); Pitch Diameter Max; Pitch Diameter Min; UNR Minor Dia Max; Class; Minor Diameter Min; Minor Diameter Max; Pitch Diameter Min; Pitch Diameter Max; Major Diameter Min.

Identification	Class	Allow.	Maj. Max	Maj. Min		PD Max	PD Min	UNR Minor Max	Class	Minor Min	Minor Max	PD Min	PD Max	Maj. Min
1 5/16–28 UN	2A	0.0012	1.3113	1.3048	—	1.2881	1.2840	1.2687	2B	1.274	1.282	1.2893	1.2946	1.3125
	3A	0.0000	1.3125	1.3060	—	1.2893	1.2862	1.2699	3B	1.2740	1.2801	1.2893	1.2933	1.3125
1 3/8–6 UNC	1A	0.0024	1.3726	1.3453	—	1.2643	1.2523	1.1742	1B	1.195	1.225	1.2667	1.2822	1.3750
	2A	0.0024	1.3726	1.3544	—	1.2643	1.2563	1.1742	2B	1.195	1.225	1.2667	1.2771	1.3750
	3A	0.0000	1.3750	1.3568	1.3453	1.2667	1.2607	1.1766	3B	1.1950	1.2146	1.2667	1.2745	1.3750
1 3/8–8 UN	2A	0.0022	1.3728	1.3578	—	1.2916	1.2844	1.2240	2B	1.240	1.265	1.2938	1.3031	1.3750
	3A	0.0000	1.3750	1.3600	1.3503	1.2938	1.2884	1.2262	3B	1.2400	1.2547	1.2938	1.3008	1.3750
1 3/8–10 UNS	2A	0.0019	1.3731	1.3602	—	1.3081	1.3018	1.2541	2B	1.267	1.288	1.3100	1.3182	1.3750
1 3/8–12 UNF	1A	0.0019	1.3731	1.3559	—	1.3190	1.3096	1.2739	1B	1.285	1.303	1.3209	1.3332	1.3750
	2A	0.0019	1.3731	1.3617	—	1.3190	1.3127	1.2739	2B	1.285	1.303	1.3209	1.3291	1.3750
	3A	0.0000	1.3750	1.3636	—	1.3209	1.3162	1.2758	3B	1.2850	1.2948	1.3209	1.3270	1.3750
1 3/8–14 UNS	2A	0.0016	1.3734	1.3631	—	1.3270	1.3216	1.2884	2B	1.298	1.314	1.3286	1.3356	1.3750
1 3/8–16 UN	2A	0.0015	1.3735	1.3641	—	1.3329	1.3278	1.2990	2B	1.307	1.321	1.3344	1.3410	1.3750
	3A	0.0000	1.3750	1.3656	—	1.3344	1.3306	1.3005	3B	1.3070	1.3158	1.3344	1.3394	1.3750
1 3/8–18 UNEF	2A	0.0015	1.3735	1.3648	—	1.3374	1.3325	1.3074	2B	1.315	1.328	1.3389	1.3452	1.3750
	3A	0.0000	1.3750	1.3663	—	1.3389	1.3353	1.3089	3B	1.3150	1.3230	1.3389	1.3436	1.3750
1 3/8–20 UN	2A	0.0014	1.3736	1.3655	—	1.3411	1.3364	1.3141	2B	1.321	1.332	1.3445	1.3486	1.3750
	3A	0.0000	1.3750	1.3669	—	1.3445	1.3390	1.3155	3B	1.3210	1.3287	1.3445	1.3470	1.3750
1 3/8–24 UNS	2A	0.0013	1.3737	1.3665	—	1.3466	1.3423	1.3241	2B	1.330	1.340	1.3479	1.3535	1.3750
1 3/8–28 UN	2A	0.0012	1.3738	1.3673	—	1.3506	1.3465	1.3312	2B	1.336	1.345	1.3518	1.3571	1.3750
	3A	0.0000	1.3750	1.3685	—	1.3518	1.3487	1.3324	3B	1.3360	1.3426	1.3518	1.3558	1.3750
1 7/16–6 UN	2A	0.0024	1.4351	1.4169	—	1.3268	1.3188	1.2367	2B	1.257	1.288	1.3292	1.3396	1.4375
	3A	0.0000	1.4375	1.4193	—	1.3292	1.3232	1.2391	3B	1.2570	1.2771	1.3292	1.3370	1.4375
1 7/16–8 UN	2A	0.0022	1.4353	1.4203	—	1.3541	1.3469	1.2865	2B	1.302	1.327	1.3563	1.3657	1.4375
	3A	0.0000	1.4375	1.4225	—	1.3563	1.3509	1.2887	3B	1.3020	1.3172	1.3563	1.3634	1.4375
1 7/16–12 UN	2A	0.0018	1.4357	1.4243	—	1.3816	1.3757	1.3365	2B	1.347	1.365	1.3834	1.3910	1.4375
	3A	0.0000	1.4375	1.4261	—	1.3834	1.3790	1.3383	3B	1.3470	1.3573	1.3834	1.3891	1.4375
1 7/16–16 UN	2A	0.0016	1.4359	1.4265	—	1.3953	1.3901	1.3614	2B	1.370	1.384	1.3969	1.4037	1.4375
	3A	0.0000	1.4375	1.4281	—	1.3969	1.3930	1.3630	3B	1.3700	1.3783	1.3969	1.4020	1.4375
1 7/16–18 UNEF	2A	0.0015	1.4360	1.4273	—	1.3999	1.3949	1.3699	2B	1.377	1.390	1.4014	1.4079	1.4375
	3A	0.0000	1.4375	1.4288	—	1.4014	1.3977	1.3714	3B	1.3770	1.3855	1.4014	1.4062	1.4375
1 7/16–20 UN	2A	0.0014	1.4361	1.4280	—	1.4036	1.3988	1.3766	2B	1.383	1.395	1.4050	1.4112	1.4375
	3A	0.0000	1.4375	1.4294	—	1.4050	1.4014	1.3780	3B	1.3830	1.3912	1.4050	1.4096	1.4375
1 7/16–28 UN	2A	0.0013	1.4362	1.4297	—	1.4130	1.4088	1.3936	2B	1.399	1.407	1.4143	1.4198	1.4375
	3A	0.0000	1.4375	1.4310	—	1.4143	1.4112	1.3949	3B	1.3990	1.4051	1.4143	1.4184	1.4375

All dimensions in inches. † Use UNS threads only if Standard Series do not meet requirements (see p. 1479). See footnotes a, b, c, d, and e at end of table.

Table 4 (*Continued*). Standard Series and Selected Combinations† — Unified Screw Threads

Nominal Size, Threads per Inch, and Series Designation[e]	Class	External[c]							Class	Internal[c]				
		Allowance	Major Diameter			Pitch Diameter		UNR Minor Diam.[a] Max (Ref.)		Minor Diameter		Pitch Diameter		Major Diameter
			Max[b]	Min	Min[d]	Max[b]	Min			Min	Max	Min	Max	Min
1½–6 UNC	1A	0.0024	1.4976	1.4703	—	1.3893	1.3772	1.2992	1B	1.320	1.350	1.3917	1.4075	1.5000
	2A	0.0024	1.4976	1.4794	1.4703	1.3893	1.3812	1.2992	2B	1.320	1.350	1.3917	1.4022	1.5000
	3A	0.0000	1.5000	1.4818		1.3917	1.3856	1.3016	3B	1.3200	1.3396	1.3917	1.3996	1.5000
1½–8 UN	2A	0.0022	1.4978	1.4828	1.4753	1.4166	1.4093	1.3490	2B	1.365	1.390	1.4188	1.4283	1.5000
	3A	0.0000	1.5000	1.4850		1.4188	1.4133	1.3512	3B	1.3650	1.3797	1.4188	1.4259	1.5000
1½–10 UNS	2A	0.0019	1.4981	1.4852		1.4331	1.4267	1.3791	2B	1.392	1.413	1.4350	1.4433	1.5000
1½–12 UNF	1A	0.0019	1.4981	1.4809		1.4440	1.4344	1.3989	1B	1.410	1.428	1.4459	1.4584	1.5000
	2A	0.0019	1.4981	1.4867		1.4440	1.4376	1.3989	2B	1.410	1.428	1.4459	1.4542	1.5000
	3A	0.0000	1.5000	1.4886		1.4459	1.4411	1.4008	3B	1.4100	1.4198	1.4459	1.4522	1.5000
1½–14 UNS	2A	0.0017	1.4983	1.4880		1.4519	1.4464	1.4133	2B	1.423	1.438	1.4536	1.4608	1.5000
1½–16 UN	2A	0.0016	1.4984	1.4890		1.4578	1.4526	1.4239	2B	1.432	1.446	1.4594	1.4662	1.5000
	3A	0.0000	1.5000	1.4906		1.4594	1.4555	1.4255	3B	1.4320	1.4408	1.4594	1.4645	1.5000
1½–18 UNEF	2A	0.0015	1.4985	1.4898		1.4624	1.4574	1.4324	2B	1.440	1.452	1.4639	1.4704	1.5000
	3A	0.0000	1.5000	1.4913		1.4639	1.4602	1.4339	3B	1.4400	1.4480	1.4639	1.4687	1.5000
1½–20 UN	2A	0.0014	1.4986	1.4905		1.4661	1.4613	1.4391	2B	1.446	1.457	1.4675	1.4737	1.5000
	3A	0.0000	1.5000	1.4919		1.4675	1.4639	1.4405	3B	1.4460	1.4537	1.4675	1.4721	1.5000
1½–24 UNS	2A	0.0013	1.4987	1.4915		1.4716	1.4672	1.4491	2B	1.455	1.465	1.4729	1.4787	1.5000
1½–28 UN	2A	0.0013	1.4987	1.4922		1.4755	1.4713	1.4561	2B	1.461	1.470	1.4768	1.4823	1.5000
	3A	0.0000	1.5000	1.4935		1.4768	1.4737	1.4574	3B	1.4610	1.4676	1.4768	1.4809	1.5000
1⁹⁄₁₆–6 UN	2A	0.0024	1.5601	1.5419	—	1.4518	1.4436	1.3617	2B	1.382	1.413	1.4542	1.4648	1.5625
	3A	0.0000	1.5625	1.5443		1.4542	1.4481	1.3641	3B	1.3820	1.4021	1.4542	1.4622	1.5625
1⁹⁄₁₆–8 UN	2A	0.0022	1.5603	1.5453		1.4791	1.4717	1.4115	2B	1.427	1.452	1.4813	1.4909	1.5625
	3A	0.0000	1.5625	1.5475		1.4813	1.4758	1.4137	3B	1.4270	1.4422	1.4813	1.4885	1.5625
1⁹⁄₁₆–12 UN	2A	0.0018	1.5607	1.5493		1.5066	1.5007	1.4615	2B	1.472	1.490	1.5084	1.5160	1.5625
	3A	0.0000	1.5625	1.5511		1.5084	1.5040	1.4633	3B	1.4720	1.4823	1.5084	1.5141	1.5625
1⁹⁄₁₆–16 UN	2A	0.0016	1.5609	1.5515		1.5203	1.5151	1.4864	2B	1.495	1.509	1.5219	1.5287	1.5625
	3A	0.0000	1.5625	1.5531		1.5219	1.5180	1.4880	3B	1.4950	1.5033	1.5219	1.5270	1.5625
1⁹⁄₁₆–18 UNEF	2A	0.0015	1.5610	1.5523		1.5249	1.5199	1.4949	2B	1.502	1.515	1.5264	1.5329	1.5625
	3A	0.0000	1.5625	1.5538		1.5264	1.5227	1.4964	3B	1.5020	1.5105	1.5264	1.5312	1.5625

Identification	Class	Allowance	Major Dia Max	Major Dia Min	UNR Minor Dia Max	Pitch Dia Max	Pitch Dia Min	Minor Dia	Class	Minor Dia Min	Minor Dia Max	Pitch Dia Min	Pitch Dia Max	Major Dia Min
1⁹⁄₁₆–20 UN	2A	0.0014	1.5611	1.5530	—	1.5286	1.5238	1.5016	2B	1.508	1.520	1.5300	1.5362	1.5625
	3A	0.0000	1.5625	1.5544	—	1.5300	1.5264	1.5030	3B	1.5080	1.5162	1.5300	1.5346	1.5625
1⁵⁄₈–6 UN	2A	0.0025	1.6225	1.6043	—	1.5142	1.5060	1.4246	2B	1.445	1.475	1.5167	1.5274	1.6250
	3A	0.0000	1.6250	1.6068	—	1.5167	1.5105	1.4271	3B	1.4450	1.4646	1.5167	1.5247	1.6250
1⁵⁄₈–8 UN	2A	0.0022	1.6228	1.6078	—	1.5416	1.5342	1.4784	2B	1.490	1.515	1.5438	1.5535	1.6250
	3A	0.0000	1.6250	1.6100	1.6003	1.5438	1.5382	1.4806	3B	1.4900	1.5047	1.5438	1.5510	1.6250
1⁵⁄₈–10 UNS	2A	0.0019	1.6231	1.6102	—	1.5581	1.5517	1.5041	2B	1.517	1.538	1.5600	1.5683	1.6250
1⁵⁄₈–12 UN	2A	0.0018	1.6232	1.6118	—	1.5691	1.5632	1.5240	2B	1.535	1.553	1.5709	1.5785	1.6250
	3A	0.0000	1.6250	1.6136	—	1.5709	1.5665	1.5258	3B	1.5350	1.5448	1.5709	1.5766	1.6250
1⁵⁄₈–14 UNS	2A	0.0017	1.6233	1.6130	—	1.5769	1.5714	1.5383	2B	1.548	1.564	1.5786	1.5858	1.6250
1⁵⁄₈–16 UN	2A	0.0016	1.6234	1.6140	—	1.5828	1.5776	1.5489	2B	1.557	1.571	1.5844	1.5912	1.6250
	3A	0.0000	1.6250	1.6156	—	1.5844	1.5805	1.5505	3B	1.5570	1.5658	1.5844	1.5895	1.6250
1⁵⁄₈–18 UNEF	2A	0.0015	1.6235	1.6148	—	1.5874	1.5824	1.5574	2B	1.565	1.578	1.5889	1.5954	1.6250
	3A	0.0000	1.6250	1.6163	—	1.5889	1.5852	1.5589	3B	1.5650	1.5730	1.5889	1.5937	1.6250
1⁵⁄₈–20 UN	2A	0.0014	1.6236	1.6155	—	1.5911	1.5863	1.5641	2B	1.571	1.582	1.5925	1.5987	1.6250
	3A	0.0000	1.6250	1.6169	—	1.5925	1.5889	1.5655	3B	1.5710	1.5787	1.5925	1.5971	1.6250
1⁵⁄₈–24 UNS	2A	0.0013	1.6237	1.6165	—	1.5966	1.5922	1.5741	2B	1.580	1.590	1.5979	1.6037	1.6250
1¹¹⁄₁₆–6 UN	2A	0.0025	1.6850	1.6668	—	1.5767	1.5684	1.4866	2B	1.507	1.538	1.5792	1.5900	1.6875
	3A	0.0000	1.6875	1.6693	—	1.5792	1.5730	1.4891	3B	1.5070	1.5271	1.5792	1.5873	1.6875
1¹¹⁄₁₆–8 UN	2A	0.0022	1.6853	1.6703	—	1.6041	1.5966	1.5365	2B	1.552	1.577	1.6063	1.6160	1.6875
	3A	0.0000	1.6875	1.6725	—	1.6063	1.6007	1.5387	3B	1.5520	1.5672	1.6063	1.6136	1.6875
1¹¹⁄₁₆–12 UN	2A	0.0018	1.6857	1.6743	—	1.6316	1.6256	1.5865	2B	1.597	1.615	1.6334	1.6412	1.6875
	3A	0.0000	1.6875	1.6761	—	1.6334	1.6289	1.5883	3B	1.5970	1.6073	1.6334	1.6392	1.6875
1¹¹⁄₁₆–16 UN	2A	0.0016	1.6859	1.6765	—	1.6453	1.6400	1.6114	2B	1.620	1.634	1.6469	1.6538	1.6875
	3A	0.0000	1.6875	1.6781	—	1.6469	1.6429	1.6130	3B	1.6200	1.6283	1.6469	1.6521	1.6875
1¹¹⁄₁₆–18 UNEF	2A	0.0015	1.6860	1.6773	—	1.6499	1.6448	1.6199	2B	1.627	1.640	1.6514	1.6580	1.6875
	3A	0.0000	1.6875	1.6788	—	1.6514	1.6476	1.6214	3B	1.6270	1.6355	1.6514	1.6563	1.6875
1¹¹⁄₁₆–20 UN	2A	0.0015	1.6860	1.6779	—	1.6535	1.6487	1.6265	2B	1.633	1.645	1.6550	1.6613	1.6875
	3A	0.0000	1.6875	1.6794	—	1.6550	1.6514	1.6280	3B	1.6330	1.6412	1.6550	1.6597	1.6875
1³⁄₄–5 UNC	1A	0.0027	1.7473	1.7165	—	1.6174	1.6040	1.5092	1B	1.534	1.568	1.6201	1.6375	1.7500
	2A	0.0027	1.7473	1.7268	—	1.6174	1.6085	1.5092	2B	1.534	1.568	1.6201	1.6317	1.7500
	3A	0.0000	1.7500	1.7295	1.7165	1.6201	1.6134	1.5119	3B	1.5340	1.5575	1.6201	1.6288	1.7500

All dimensions in inches. † Use UNS threads only if Standard Series do not meet requirements (see p. 1479). See footnotes a, b, c, d, and e at end of table.

Table 4 (Continued). Standard Series and Selected Combinations† — Unified Screw Threads

| Nominal Size, Threads per Inch, and Series Designation[c] | Class | Allowance | External[c] | | | | | | Internal[c] | | | | | |
| | | | Major Diameter | | | Pitch Diameter | | UNR Minor Diam.[a] Max (Ref.) | Class | Minor Diameter | | Pitch Diameter | | Major Diameter |
			Max[b]	Min	Min[d]	Max[b]	Min			Min	Max	Min	Max	Min
1¾-6 UN	2A	0.0025	1.7475	1.7293	—	1.6392	1.6309	1.5491	2B	1.570	1.600	1.6417	1.6525	1.7500
	3A	0.0000	1.7500	1.7318	—	1.6417	1.6354	1.5516	3B	1.5700	1.5896	1.6417	1.6498	1.7500
1¾-8 UN	2A	0.0023	1.7477	1.7327	1.7252	1.6665	1.6590	1.5989	2B	1.615	1.640	1.6688	1.6786	1.7500
	3A	0.0000	1.7500	1.7350	—	1.6688	1.6632	1.6012	3B	1.6150	1.6297	1.6688	1.6762	1.7500
1¾-10 UNS	2A	0.0019	1.7481	1.7352	—	1.6831	1.6766	1.6291	2B	1.642	1.663	1.6850	1.6934	1.7500
1¾-12 UN	2A	0.0018	1.7482	1.7368	—	1.6941	1.6881	1.6490	2B	1.660	1.678	1.6959	1.7037	1.7500
	3A	0.0000	1.7500	1.7386	—	1.6959	1.6914	1.6508	3B	1.6600	1.6698	1.6959	1.7017	1.7500
1¾-14 UNS	2A	0.0017	1.7483	1.7380	—	1.7019	1.6963	1.6632	2B	1.673	1.688	1.7036	1.7109	1.7500
1¾-16 UN	2A	0.0016	1.7484	1.7390	—	1.7078	1.7025	1.6739	2B	1.682	1.696	1.7094	1.7163	1.7500
	3A	0.0000	1.7500	1.7406	—	1.7094	1.7054	1.6755	3B	1.6820	1.6908	1.7094	1.7146	1.7500
1¾-18 UNS	2A	0.0015	1.7485	1.7398	—	1.7124	1.7073	1.6824	2B	1.690	1.703	1.7139	1.7205	1.7500
1¾-20 UN	2A	0.0015	1.7485	1.7404	—	1.7160	1.7112	1.6800	2B	1.696	1.707	1.7175	1.7238	1.7500
	3A	0.0000	1.7500	1.7419	—	1.7175	1.7139	1.6905	3B	1.6960	1.7037	1.7175	1.7222	1.7500
1¹³⁄₁₆-6 UN	2A	0.0025	1.8100	1.7918	—	1.7017	1.6933	1.6116	2B	1.632	1.663	1.7042	1.7151	1.8125
	3A	0.0000	1.8125	1.7943	—	1.7042	1.6979	1.6141	3B	1.6320	1.6521	1.7042	1.7124	1.8125
1¹³⁄₁₆-8 UN	2A	0.0023	1.8102	1.7952	—	1.7290	1.7214	1.6614	2B	1.677	1.702	1.7313	1.7412	1.8125
	3A	0.0000	1.8125	1.7975	—	1.7313	1.7256	1.6637	3B	1.6770	1.6922	1.7313	1.7387	1.8125
1¹³⁄₁₆-12 UN	2A	0.0018	1.8107	1.7993	—	1.7566	1.7506	1.7115	2B	1.722	1.740	1.7584	1.7662	1.8125
	3A	0.0000	1.8125	1.8011	—	1.7584	1.7539	1.7133	3B	1.7220	1.7323	1.7584	1.7642	1.8125
1¹³⁄₁₆-16 UN	2A	0.0016	1.8109	1.8015	—	1.7703	1.7650	1.7364	2B	1.745	1.759	1.7719	1.7788	1.8125
	3A	0.0000	1.8125	1.8031	—	1.7719	1.7679	1.7380	3B	1.7450	1.7533	1.7719	1.7771	1.8125
1¹³⁄₁₆-20 UN	2A	0.0015	1.8110	1.8029	—	1.7785	1.7737	1.7515	2B	1.758	1.770	1.7800	1.7863	1.8125
	3A	0.0000	1.8125	1.8044	—	1.7800	1.7764	1.7530	3B	1.7580	1.7662	1.7800	1.7847	1.8125
1⅞-6 UN	2A	0.0025	1.8725	1.8543	—	1.7642	1.7558	1.6741	2B	1.695	1.725	1.7667	1.7777	1.8750
	3A	0.0000	1.8750	1.8568	—	1.7667	1.7604	1.6766	3B	1.6950	1.7146	1.7667	1.7749	1.8750
1⅞-8 UN	2A	0.0023	1.8727	1.8577	1.8502	1.7915	1.7838	1.7239	2B	1.740	1.765	1.7938	1.8038	1.8750
	3A	0.0000	1.8750	1.8600	—	1.7938	1.7881	1.7262	3B	1.7400	1.7547	1.7938	1.8013	1.8750
1⅞-10 UNS	2A	0.0019	1.8731	1.8602	—	1.8081	1.8016	1.7541	2B	1.767	1.788	1.8100	1.8184	1.8750

Standard Series and Selected Combinations — Unified Screw Threads

Nominal Size, Threads per Inch, Series	Class	Allowance	Major Dia Max	Major Dia Min	UNR Minor Dia Max (Ref.)	Pitch Dia Max	Pitch Dia Min	Minor Dia Max (Ref.)	Class	Minor Dia Min	Minor Dia Max	Pitch Dia Min	Pitch Dia Max	Major Dia Min
1⅞–12 UN	2A	0.0018	1.8732	1.8618	—	1.8191	1.8131	1.7740	2B	1.785	1.803	1.8209	1.8287	1.8750
	3A	0.0000	1.8750	1.8636	—	1.8209	1.8164	1.7758	3B	1.7850	1.7948	1.8209	1.8267	1.8750
1⅞–14 UNS	2A	0.0017	1.8733	1.8630	—	1.8269	1.8213	1.7883	2B	1.798	1.814	1.8286	1.8359	1.8750
1⅞–16 UN	2A	0.0016	1.8734	1.8640	—	1.8328	1.8275	1.7989	2B	1.807	1.821	1.8344	1.8413	1.8750
	3A	0.0000	1.8750	1.8656	—	1.8344	1.8304	1.8005	3B	1.8070	1.8158	1.8344	1.8396	1.8750
1⅞–18 UNS	2A	0.0015	1.8735	1.8648	—	1.8374	1.8323	1.8074	2B	1.815	1.828	1.8389	1.8455	1.8750
1⅞–20 UN	2A	0.0015	1.8735	1.8654	—	1.8410	1.8362	1.8140	2B	1.821	1.832	1.8425	1.8488	1.8750
	3A	0.0000	1.8750	1.8669	—	1.8425	1.8389	1.8155	3B	1.8210	1.8287	1.8425	1.8472	1.8750
1 15/16–6 UN	2A	0.0026	1.9349	1.9167	—	1.8266	1.8181	1.7305	2B	1.757	1.788	1.8292	1.8403	1.9375
	3A	0.0000	1.9375	1.9193	—	1.8292	1.8228	1.7391	3B	1.7570	1.7771	1.8292	1.8375	1.9375
1 15/16–8 UN	2A	0.0023	1.9352	1.9202	—	1.8540	1.8463	1.7864	2B	1.802	1.827	1.8563	1.8663	1.9375
	3A	0.0000	1.9375	1.9225	—	1.8563	1.8505	1.7887	3B	1.8020	1.8172	1.8563	1.8638	1.9375
1 15/16–12 UN	2A	0.0018	1.9357	1.9243	—	1.8816	1.8755	1.8365	2B	1.847	1.865	1.8834	1.8913	1.9375
	3A	0.0000	1.9375	1.9261	—	1.8834	1.8789	1.8383	3B	1.8470	1.8573	1.8834	1.8893	1.9375
1 15/16–16 UN	2A	0.0016	1.9359	1.9265	—	1.8953	1.8899	1.8614	2B	1.870	1.884	1.8969	1.9039	1.9375
	3A	0.0000	1.9375	1.9281	—	1.8969	1.8929	1.8630	3B	1.8700	1.8783	1.8969	1.9021	1.9375
1 15/16–20 UN	2A	0.0015	1.9360	1.9279	—	1.9035	1.8986	1.8765	2B	1.883	1.895	1.9050	1.9114	1.9375
	3A	0.0000	1.9375	1.9294	—	1.9050	1.9013	1.8780	3B	1.8830	1.8912	1.9050	1.9098	1.9375
2–4½ UNC	1A	0.0029	1.9971	1.9641	—	1.8528	1.8385	1.7334	1B	1.759	1.795	1.8557	1.8743	2.0000
	2A	0.0029	1.9971	1.9751	—	1.8528	1.8433	1.7334	2B	1.759	1.795	1.8557	1.8681	2.0000
	3A	0.0000	2.0000	1.9780	—	1.8557	1.8486	1.7353	3B	1.7590	1.7861	1.8557	1.8650	2.0000
2–6 UN	2A	0.0026	1.9974	1.9792	—	1.8891	1.8805	1.7990	2B	1.820	1.850	1.8917	1.9028	2.0000
	3A	0.0000	2.0000	1.9818	—	1.8917	1.8853	1.8016	3B	1.8200	1.8396	1.8917	1.9000	2.0000
2–8 UN	2A	0.0023	1.9977	1.9827	—	1.9165	1.9087	1.8489	2B	1.865	1.890	1.9188	1.9289	2.0000
	3A	0.0000	2.0000	1.9850	—	1.9188	1.9130	1.8512	3B	1.8650	1.8797	1.9188	1.9264	2.0000
2–10 UNS	2A	0.0020	1.9980	1.9851	—	1.9330	1.9265	1.8790	2B	1.892	1.913	1.9350	1.9435	2.0000
2–12 UN	2A	0.0018	1.9982	1.9868	—	1.9441	1.9380	1.8990	2B	1.910	1.928	1.9459	1.9538	2.0000
	3A	0.0000	2.0000	1.9886	—	1.9459	1.9414	1.9008	3B	1.9100	1.9198	1.9459	1.9518	2.0000
2–14 UNS	2A	0.0017	1.9983	1.9880	—	1.9519	1.9462	1.9133	2B	1.923	1.938	1.9536	1.9610	2.0000
2–16 UN	2A	0.0016	1.9984	1.9890	—	1.9578	1.9524	1.9239	2B	1.932	1.946	1.9594	1.9664	2.0000
	3A	0.0000	2.0000	1.9906	—	1.9594	1.9554	1.9255	3B	1.9320	1.9408	1.9594	1.9646	2.0000
2–18 UNS	2A	0.0015	1.9985	1.9898	—	1.9624	1.9573	1.9334	2B	1.940	1.953	1.9639	1.9706	2.0000

See footnotes a, b, c, d, and e at end of table.

All dimensions in inches.   † Use UNS threads only if Standard Series do not meet requirements (see p. 1479).

Table 4 (Continued). Standard Series and Selected Combinations† — Unified Screw Threads

Columns 2–9 are External[c]; columns 10–15 are Internal[c].

Nominal Size, Threads per Inch, and Series Designation[e]	Class	Allowance	Major Dia Max[b]	Major Dia Min	Major Dia Min[d]	Pitch Dia Max[b]	Pitch Dia Min	UNR Minor Diam Max (Ref.)[a]	Class	Minor Dia Min	Minor Dia Max	Pitch Dia Min	Pitch Dia Max	Major Dia Min
2–20 UN	2A	0.0015	1.9985	1.9904	—	1.9660	1.9611	1.9390	2B	1.946	1.957	1.9675	1.9739	2.0000
	3A	0.0000	2.0000	1.9919	—	1.9675	1.9638	1.9405	3B	1.9460	1.9537	1.9675	1.9723	2.0000
2¹/₁₆–16 UNS	2A	0.0016	2.0609	2.0515	—	2.0203	2.0149	1.9864	2B	1.995	2.009	2.0219	2.0289	2.0625
	3A	0.0000	2.0625	2.0531	—	2.0219	2.0179	1.9880	3B	1.9950	2.0033	2.0219	2.0271	2.0625
2¹/₈–6 UN	2A	0.0026	2.1224	2.1042	—	2.0141	2.0054	1.9240	2B	1.945	1.975	2.0167	2.0280	2.1250
	3A	0.0000	2.1250	2.1068	—	2.0167	2.0102	1.9266	3B	1.9450	1.9646	2.0167	2.0251	2.1250
2¹/₈–8 UN	2A	0.0024	2.1226	2.1076	2.1001	2.0414	2.0335	1.9738	2B	1.990	2.015	2.0438	2.0540	2.1250
	3A	0.0000	2.1250	2.1100	—	2.0438	2.0379	1.9762	3B	1.9900	2.0047	2.0438	2.0515	2.1250
2¹/₈–12 UN	2A	0.0018	2.1232	2.1118	—	2.0691	2.0630	2.0240	2B	2.035	2.053	2.0709	2.0788	2.1250
	3A	0.0000	2.1250	2.1136	—	2.0709	2.0664	2.0258	3B	2.0350	2.0448	2.0709	2.0768	2.1250
2¹/₈–16 UN	2A	0.0016	2.1234	2.1140	—	2.0828	2.0774	2.0489	2B	2.057	2.071	2.0844	2.0914	2.1250
	3A	0.0000	2.1250	2.1156	—	2.0844	2.0803	2.0505	3B	2.0570	2.0658	2.0844	2.0896	2.1250
2¹/₈–20 UN	2A	0.0015	2.1235	2.1154	—	2.0910	2.0861	2.0640	2B	2.071	2.082	2.0925	2.0989	2.1250
	3A	0.0000	2.1250	2.1169	—	2.0925	2.0888	2.0655	3B	2.0710	2.0787	2.0925	2.0973	2.1250
2³/₁₆–16 UNS	2A	0.0016	2.1859	2.1765	—	2.1453	2.1399	2.1114	2B	2.120	2.134	2.1469	2.1539	2.1875
	3A	0.0000	2.1875	2.1781	—	2.1469	2.1428	2.1130	3B	2.1200	2.1283	2.1469	2.1521	2.1875
2¹/₄–4¹/₂ UNC	1A	0.0029	2.2471	2.2141	—	2.1028	2.0882	1.9824	1B	2.009	2.045	2.1057	2.1247	2.2500
	2A	0.0029	2.2471	2.2251	2.2141	2.1028	2.0931	1.9824	2B	2.009	2.045	2.1057	2.1183	2.2500
	3A	0.0000	2.2500	2.2280	—	2.1057	2.0984	1.9853	3B	2.0090	2.0361	2.1057	2.1152	2.2500
2¹/₄–6 UN	2A	0.0026	2.2474	2.2292	2.2251	2.1391	2.1303	2.0490	2B	2.070	2.100	2.1417	2.1531	2.2500
	3A	0.0000	2.2500	2.2318	—	2.1417	2.1351	2.0516	3B	2.0700	2.0896	2.1417	2.1502	2.2500
2¹/₄–8 UN	2A	0.0024	2.2476	2.2326	—	2.1664	2.1584	2.0988	2B	2.115	2.140	2.1688	2.1792	2.2500
	3A	0.0000	2.2500	2.2350	—	2.1688	2.1628	2.1012	3B	2.1150	2.1297	2.1688	2.1766	2.2500
2¹/₄–10 UNS	2A	0.0020	2.2480	2.2351	—	2.1830	2.1765	2.1290	2B	2.142	2.163	2.1830	2.1935	2.2500
2¹/₄–12 UN	2A	0.0018	2.2482	2.2368	—	2.1941	2.1880	2.1490	2B	2.160	2.178	2.1959	2.2038	2.2500
	3A	0.0000	2.2500	2.2386	—	2.1959	2.1914	2.1508	3B	2.1600	2.1698	2.1959	2.2018	2.2500
2¹/₄–14 UNS	2A	0.0017	2.2483	2.2380	—	2.2019	2.1962	2.1633	2B	2.173	2.188	2.2036	2.2110	2.2500
2¹/₄–16 UN	2A	0.0016	2.2484	2.2390	—	2.2078	2.2024	2.1739	2B	2.182	2.196	2.2094	2.2164	2.2500
	3A	0.0000	2.2500	2.2406	—	2.2094	2.2053	2.1755	3B	2.1820	2.1908	2.2094	2.2146	2.2500

Standard Series and Selected Combinations — Unified Screw Threads (continued)

Identification	Class	Allowance	External — Major Dia Max	Major Dia Min	UNR Minor Dia Max	Pitch Dia Max	Pitch Dia Min	Minor Dia	Class	Internal — Minor Dia Min	Minor Dia Max	Pitch Dia Min	Pitch Dia Max	Major Dia Min
2¼–18 UNS	2A	0.0015	2.2485	2.2398	—	2.2124	2.2073	2.1824	2B	2.190	2.203	2.2139	2.2206	2.2500
	3A	0.0000	2.2500	2.2404	—	2.2139	2.2111	2.1890	3B	2.1900	2.200	2.2139	2.2189	2.2500
2¼–20 UN	2A	0.0015	2.2485	2.2419	—	2.2160	2.2137	2.1895	2B	2.196	2.207	2.2175	2.2239	2.2500
	3A	0.0000	2.2500	2.2437	—	2.2175	2.2149	2.1921	3B	2.1960	2.2037	2.2175	2.2223	2.2500
2 5/16–16 UNS	2A	0.0017	2.3108	2.3014	—	2.2702	2.2647	2.2363	2B	2.245	2.259	2.2719	2.2791	2.3125
	3A	0.0000	2.3125	2.3031	—	2.2719	2.2678	2.2380	3B	2.2450	2.2533	2.2719	2.2773	2.3125
2⅜–6 UN	2A	0.0027	2.3723	2.3541	—	2.2640	2.2551	2.1739	2B	2.195	2.226	2.2667	2.2782	2.3750
	3A	0.0000	2.3750	2.3568	—	2.2667	2.2601	2.1766	3B	2.1950	2.2146	2.2667	2.2753	2.3750
2⅜–8 UN	2A	0.0024	2.3726	2.3576	—	2.2914	2.2833	2.2238	2B	2.240	2.265	2.2938	2.3043	2.3750
	3A	0.0000	2.3750	2.3600	—	2.2938	2.2878	2.2262	3B	2.2400	2.2547	2.2938	2.3017	2.3750
2⅜–12 UN	2A	0.0019	2.3731	2.3617	—	2.3190	2.3128	2.2739	2B	2.285	2.303	2.3209	2.3290	2.3750
	3A	0.0000	2.3750	2.3636	—	2.3209	2.3163	2.2758	3B	2.2850	2.2948	2.3209	2.3269	2.3750
2⅜–16 UN	2A	0.0017	2.3733	2.3639	—	2.3327	2.3272	2.2988	2B	2.307	2.321	2.3344	2.3416	2.3750
	3A	0.0000	2.3750	2.3656	—	2.3344	2.3303	2.3005	3B	2.3070	2.3158	2.3344	2.3398	2.3750
2⅜–20 UN	2A	0.0015	2.3735	2.3654	—	2.3410	2.3359	2.3140	2B	2.321	2.332	2.3425	2.3491	2.3750
	3A	0.0000	2.3750	2.3669	—	2.3425	2.3387	2.3155	3B	2.3210	2.3287	2.3425	2.3475	2.3750
2 7/16–16 UNS	2A	0.0017	2.4358	2.4264	—	2.3952	2.3897	2.3613	2B	2.370	2.384	2.3969	2.4041	2.4375
	3A	0.0000	2.4375	2.4281	—	2.3969	2.3928	2.3630	3B	2.3700	2.3783	2.3969	2.4023	2.4375
2½–4 UNC	1A	0.0031	2.4969	2.4612	—	2.3345	2.3190	2.1992	1B	2.229	2.267	2.3376	2.3578	2.5000
	2A	0.0031	2.4969	2.4731	2.4612	2.3345	2.3241	2.1992	2B	2.229	2.267	2.3376	2.3511	2.5000
	3A	0.0000	2.5000	2.4762	—	2.3376	2.3298	2.2023	3B	2.2294	2.2594	2.3376	2.3477	2.5000
2½–6 UN	2A	0.0027	2.4973	2.4791	—	2.3890	2.3800	2.2989	2B	2.320	2.350	2.3917	2.4033	2.5000
	3A	0.0000	2.5000	2.4818	—	2.3917	2.3850	2.3016	3B	2.3200	2.3396	2.3917	2.4004	2.5000
2½–8 UN	2A	0.0024	2.4976	2.4826	2.4751	2.4164	2.4082	2.3488	2B	2.365	2.390	2.4188	2.4294	2.5000
	3A	0.0000	2.5000	2.4850	—	2.4188	2.4127	2.3512	3B	2.3650	2.3797	2.4188	2.4268	2.5000
2½–10 UNS	2A	0.0020	2.4980	2.4851	—	2.4330	2.4263	2.3790	2B	2.392	2.413	2.4350	2.4437	2.5000
2½–12 UN	2A	0.0019	2.4981	2.4867	—	2.4440	2.4378	2.3989	2B	2.410	2.428	2.4459	2.4540	2.5000
	3A	0.0000	2.5000	2.4886	—	2.4459	2.4413	2.4008	3B	2.4100	2.4198	2.4459	2.4519	2.5000
2½–14 UNS	2A	0.0017	2.4983	2.4880	—	2.4519	2.4461	2.4133	2B	2.423	2.438	2.4536	2.4612	2.5000
2½–16 UN	2A	0.0017	2.4983	2.4889	—	2.4577	2.4522	2.4238	2B	2.432	2.446	2.4594	2.4666	2.5000
	3A	0.0000	2.5000	2.4906	—	2.4594	2.4553	2.4255	3B	2.4320	2.4408	2.4594	2.4648	2.5000
2½–18 UNS	2A	0.0016	2.4984	2.4897	—	2.4623	2.4570	2.4323	2B	2.440	2.453	2.4639	2.4708	2.5000
2½–20 UN	2A	0.0015	2.4985	2.4904	—	2.4660	2.4609	2.4390	2B	2.446	2.457	2.4675	2.4741	2.5000
	3A	0.0000	2.5000	2.4919	—	2.4675	2.4637	2.4405	3B	2.4460	2.4537	2.4675	2.4725	2.5000

All dimensions in inches. † Use UNS threads only if Standard Series do not meet requirements (see p. 1479). See footnotes *a*, *b*, *c*, *d*, and *e* at end of table.

Table 4 (*Continued*). **Standard Series and Selected Combinations† — Unified Screw Threads**

Nominal Size, Threads per Inch, and Series Designation[e]	Class	Allowance	External[c] Major Diameter Max[b]	Major Diameter Min	Major Diameter Min[d]	Pitch Diameter Max[b]	Pitch Diameter Min	UNR Minor Diam.[a] Max (Ref.)	Class	Internal[c] Minor Diameter Min	Minor Diameter Max	Pitch Diameter Min	Pitch Diameter Max	Major Diameter Min
2⅝–6 UN	2A	0.0027	2.6223	2.6041	—	2.5140	2.5050	2.4239	2B	2.445	2.475	2.5167	2.5285	2.6250
	3A	0.0000	2.6250	2.6068	—	2.5167	2.5099	2.4266	3B	2.4450	2.4646	2.5167	2.5255	2.6250
2⅝–8 UN	2A	0.0025	2.6225	2.6075	—	2.5413	2.5331	2.4737	2B	2.490	2.515	2.5438	2.5545	2.6250
	3A	0.0000	2.6250	2.6100	—	2.5438	2.5376	2.4762	3B	2.4900	2.5047	2.5438	2.5518	2.6250
2⅝–12 UN	2A	0.0019	2.6231	2.6117	—	2.5690	2.5628	2.5239	2B	2.535	2.553	2.5709	2.5790	2.6250
	3A	0.0000	2.6250	2.6136	—	2.5709	2.5663	2.5258	3B	2.5350	2.5448	2.5709	2.5769	2.6250
2⅝–16 UN	2A	0.0017	2.6233	2.6139	—	2.5827	2.5772	2.5488	2B	2.557	2.571	2.5844	2.5916	2.6250
	3A	0.0000	2.6250	2.6156	—	2.5844	2.5803	2.5505	3B	2.5570	2.5658	2.5844	2.5898	2.6250
2⅝–20 UN	2A	0.0015	2.6235	2.6154	—	2.5910	2.5859	2.5640	2B	2.571	2.582	2.5925	2.5991	2.6250
	3A	0.0000	2.6250	2.6169	—	2.5925	2.5887	2.5655	3B	2.5710	2.5787	2.5925	2.5975	2.6250
2¾–4 UNC	1A	0.0032	2.7468	2.7111	—	2.5844	2.5686	2.4491	1B	2.479	2.517	2.5876	2.6082	2.7500
	2A	0.0032	2.7468	2.7230	—	2.5844	2.5739	2.4491	2B	2.479	2.517	2.5876	2.6013	2.7500
	3A	0.0000	2.7500	2.7262	2.7111	2.5876	2.5797	2.4523	3B	2.4790	2.5094	2.5876	2.5979	2.7500
2¾–6 UN	2A	0.0027	2.7473	2.7291	—	2.6390	2.6299	2.5489	2B	2.570	2.600	2.6417	2.6536	2.7500
	3A	0.0000	2.7500	2.7318	—	2.6417	2.6349	2.5516	3B	2.5700	2.5896	2.6417	2.6506	2.7500
2¾–8 UN	2A	0.0025	2.7475	2.7325	—	2.6663	2.6580	2.5987	2B	2.615	2.640	2.6688	2.6796	2.7500
	3A	0.0000	2.7500	2.7350	2.7250	2.6688	2.6625	2.6012	3B	2.6150	2.6297	2.6688	2.6769	2.7500
2¾–10 UNS	2A	0.0020	2.7480	2.7351	—	2.6830	2.6763	2.6290	2B	2.642	2.663	2.6850	2.6937	2.7500
2¾–12 UN	2A	0.0019	2.7481	2.7367	—	2.6940	2.6878	2.6489	2B	2.660	2.678	2.6959	2.7040	2.7500
	3A	0.0000	2.7500	2.7386	—	2.6959	2.6913	2.6508	3B	2.6600	2.6698	2.6959	2.7019	2.7500
2¾–14 UNS	2A	0.0017	2.7483	2.7380	—	2.7019	2.6961	2.6633	2B	2.673	2.688	2.7036	2.7112	2.7500
2¾–16 UN	2A	0.0017	2.7483	2.7389	—	2.7077	2.7022	2.6738	2B	2.682	2.696	2.7094	2.7166	2.7500
	3A	0.0000	2.7500	2.7406	—	2.7094	2.7053	2.6755	3B	2.6820	2.6908	2.7094	2.7148	2.7500
2¾–18 UNS	2A	0.0016	2.7484	2.7397	—	2.7123	2.7070	2.6823	2B	2.690	2.703	2.7139	2.7208	2.7500
2¾–20 UN	2A	0.0015	2.7485	2.7404	—	2.7160	2.7109	2.6890	2B	2.696	2.707	2.7175	2.7241	2.7500
	3A	0.0000	2.7500	2.7419	—	2.7175	2.7137	2.6905	3B	2.6960	2.7037	2.7175	2.7225	2.7500
2⅞–6 UN	2A	0.0028	2.8722	2.8540	—	2.7639	2.7547	2.6738	2B	2.695	2.725	2.7667	2.7787	2.8750
	3A	0.0000	2.8750	2.8568	—	2.7667	2.7598	2.6766	3B	2.6950	2.7146	2.7667	2.7757	2.8750

Nominal Size, Threads per Inch, and Series Designation	Class	Allowance	Major Dia Max	Major Dia Min	UNR Minor Dia Max (Ref)	Pitch Dia Max	Pitch Dia Min	Minor Dia Max	Class	Minor Dia Min	Minor Dia Max	Pitch Dia Min	Pitch Dia Max	Major Dia Min
2⅞-8 UN	2A	0.0025	2.8725	2.8575	—	2.7913	2.7829	2.7237	2B	2.740	2.765	2.7938	2.8048	2.8750
	3A	0.0000	2.8750	2.8600	—	2.7938	2.7875	2.7262	3B	2.7400	2.7547	2.7938	2.8020	2.8750
2⅞-12 UN	2A	0.0019	2.8731	2.8617	—	2.8190	2.8127	2.7739	2B	2.785	2.803	2.8209	2.8291	2.8750
	3A	0.0000	2.8750	2.8636	—	2.8209	2.8162	2.7758	3B	2.7850	2.7948	2.8209	2.8271	2.8750
2⅞-16 UN	2A	0.0017	2.8733	2.8639	—	2.8327	2.8271	2.7988	2B	2.807	2.821	2.8344	2.8417	2.8750
	3A	0.0000	2.8750	2.8656	—	2.8344	2.8302	2.8005	3B	2.8070	2.8158	2.8344	2.8399	2.8750
2⅞-20 UN	2A	0.0016	2.8734	2.8653	—	2.8409	2.8357	2.8139	2B	2.821	2.832	2.8425	2.8493	2.8750
	3A	0.0000	2.8750	2.8669	—	2.8425	2.8386	2.8155	3B	2.8210	2.8287	2.8425	2.8476	2.8750
3-4 UNC	1A	0.0032	2.9968	2.9611	2.9611	2.8344	2.8183	2.6991	1B	2.729	2.767	2.8376	2.885	3.0000
	2A	0.0032	2.9968	2.9730	—	2.8344	2.8237	2.6991	2B	2.729	2.767	2.8376	2.8515	3.0000
	3A	0.0000	3.0000	2.9762	—	2.8376	2.8296	2.7023	3B	2.7290	2.7594	2.8376	2.8480	3.0000
3-6 UN	2A	0.0028	2.9972	2.9790	—	2.8889	2.8796	2.7988	2B	2.820	2.850	2.8917	2.9038	3.0000
	3A	0.0000	3.0000	2.9818	—	2.8917	2.8847	2.8016	3B	2.8200	2.8396	2.8917	2.9008	3.0000
3-8 UN	2A	0.0026	2.9974	2.9824	2.9749	2.9162	2.9077	2.8486	2B	2.865	2.890	2.9188	2.9299	3.0000
	3A	0.0000	3.0000	2.9850	—	2.9188	2.9124	2.8512	3B	2.8650	2.8797	2.9188	2.9271	3.0000
3-10 UNS	2A	0.0020	2.9980	2.9851	—	2.9330	2.9262	2.8790	2B	2.892	2.913	2.9350	2.9439	3.0000
3-12 UN	2A	0.0019	2.9981	2.9867	—	2.9440	2.9377	2.8989	2B	2.910	2.928	2.9459	2.9541	3.0000
	3A	0.0000	3.0000	2.9886	—	2.9459	2.9412	2.9008	3B	2.9100	2.9198	2.9459	2.9521	3.0000
3-14 UNS	2A	0.0018	2.9982	2.9879	—	2.9518	2.9459	2.9132	2B	2.923	2.938	2.9536	2.9613	3.0000
3-16 UN	2A	0.0017	2.9983	2.9889	—	2.9577	2.9521	2.9238	2B	2.932	2.946	2.9594	2.9667	3.0000
	3A	0.0000	3.0000	2.9906	—	2.9594	2.9552	2.9255	3B	2.9320	2.9408	2.9594	2.9649	3.0000
3-18 UNS	2A	0.0016	2.9984	2.9897	—	2.9623	2.9559	2.9323	2B	2.940	2.953	2.9639	2.9709	3.0000
3-20 UN	2A	0.0016	2.9984	2.9903	—	2.9659	2.9607	2.9389	2B	2.946	2.957	2.9675	2.9743	3.0000
	3A	0.0000	3.0000	2.9919	—	2.9675	2.9636	2.9405	3B	2.9460	2.9537	2.9675	2.9726	3.0000
3⅛-6 UN	2A	0.0028	3.1222	3.1040	—	3.0139	3.0045	2.9238	2B	2.945	2.975	3.0167	3.0289	3.1250
	3A	0.0000	3.1250	3.1068	—	3.0167	3.0097	2.9266	3B	2.9450	2.9646	3.0167	3.0259	3.1250
3⅛-8 UN	2A	0.0026	3.1224	3.1074	—	3.0412	3.0326	2.9736	2B	2.990	3.015	3.0438	3.0550	3.1250
	3A	0.0000	3.1250	3.1100	—	3.0438	3.0374	2.9762	3B	2.9900	3.0047	3.0438	3.0522	3.1250
3⅛-12 UN	2A	0.0019	3.1231	3.1117	—	3.0690	3.0627	3.0239	2B	3.035	3.053	3.0709	3.0791	3.1250
	3A	0.0000	3.1250	3.1136	—	3.0709	3.0662	3.0258	3B	3.0350	3.0448	3.0709	3.0771	3.1250
3⅛-16 UN	2A	0.0017	3.1233	3.1139	—	3.0827	3.0771	3.0488	2B	3.057	3.071	3.0844	3.0917	3.1250
	3A	0.0000	3.1250	3.1156	—	3.0844	3.0802	3.0505	3B	3.0570	3.0658	3.0844	3.0899	3.1250

All dimensions in inches. † Use UNS threads only if Standard Series do not meet requirements (see p. 1479). See footnotes *a*, *b*, *c*, *d*, and *e* at end of table.

Table 4 (*Continued*). Standard Series and Selected Combinations† — Unified Screw Threads

Nominal Size, Threads per Inch, and Series Designation^c	Class	External^c							Internal^c					
		Allow-ance	Major Diameter			Pitch Diameter		UNR Minor Diam.^a Max (Ref.)	Class	Minor Diameter		Pitch Diameter		Major Diameter
			Max^b	Min	Min^d	Max^b	Min			Min	Max	Min	Max	Min
3¼-4 UNC	1A	0.0033	3.2467	3.2110	3.2110	3.0843	3.0680	2.9490	1B	2.979	3.017	3.0876	3.1088	3.2500
	2A	0.0033	3.2467	3.2229		3.0843	3.0734	2.9490	2B	2.979	3.017	3.0876	3.1017	3.2500
	3A	0.0000	3.2500	3.2262		3.0876	3.0794	2.9523	3B	2.9790	3.0094	3.0876	3.0982	3.2500
3¼-6 UN	2A	0.0028	3.2472	3.2290		3.1389	3.1294	3.0488	2B	3.070	3.100	3.1417	3.1540	3.2500
	3A	0.0000	3.2500	3.2318		3.1417	3.1346	3.0516	3B	3.0700	3.0896	3.1417	3.1509	3.2500
3¼-8 UN	2A	0.0026	3.2474	3.2334		3.1662	3.1575	3.0986	2B	3.115	3.140	3.1688	3.1801	3.2500
	3A	0.0000	3.2500	3.2350		3.1688	3.1623	3.1012	3B	3.1150	3.1297	3.1688	3.1773	3.2500
3¼-10 UNS	2A	0.0020	3.2480	3.2351	3.2249	3.1830	3.1762	3.1290	2B	3.142	3.163	3.1850	3.1939	3.2500
3¼-12 UN	2A	0.0019	3.2481	3.2367		3.1940	3.1877	3.1489	2B	3.160	3.178	3.1959	3.2041	3.2500
	3A	0.0000	3.2500	3.2386		3.1959	3.1912	3.1508	3B	3.1600	3.1698	3.1959	3.2021	3.2500
3¼-14 UNS	2A	0.0018	3.2482	3.2379		3.2018	3.1959	3.1632	2B	3.173	3.188	3.2036	3.2113	3.2500
3¼-16 UN	2A	0.0017	3.2483	3.2389		3.2077	3.2021	3.1738	2B	3.182	3.196	3.2094	3.2167	3.2500
	3A	0.0000	3.2500	3.2406		3.2094	3.2052	3.1755	3B	3.1820	3.1908	3.2094	3.2149	3.2500
3¼-18 UNS	2A	0.0016	3.2484	3.2397		3.2123	3.2069	3.1823	2B	3.190	3.203	3.2139	3.2209	3.2500
3⅜-6 UN	2A	0.0029	3.3721	3.3539		3.2638	3.2543	3.1737	2B	3.195	3.225	3.2667	3.2791	3.3750
	3A	0.0000	3.3750	3.3568		3.2667	3.2595	3.1766	3B	3.1950	3.2146	3.2667	3.2760	3.3750
3⅜-8 UN	2A	0.0026	3.3724	3.3574		3.2912	3.2824	3.2236	2B	3.240	3.265	3.2938	3.3052	3.3750
	3A	0.0000	3.3750	3.3600		3.2938	3.2872	3.2262	3B	3.2400	3.2547	3.2938	3.3023	3.3750
3⅜-12 UN	2A	0.0019	3.3731	3.3617		3.3190	3.3126	3.2739	2B	3.285	3.303	3.3209	3.3293	3.3750
	3A	0.0000	3.3750	3.3636		3.3209	3.3161	3.2758	3B	3.2850	3.2948	3.3209	3.3272	3.3750
3⅜-16 UN	2A	0.0017	3.3733	3.3639		3.3327	3.3269	3.2988	2B	3.307	3.321	3.3344	3.3419	3.3750
	3A	0.0000	3.3750	3.3656		3.3344	3.3301	3.3005	3B	3.3070	3.3158	3.3344	3.3400	3.3750
3½-4 UNC	1A	0.0033	3.4967	3.4610	3.4610	3.3343	3.3177	3.1990	1B	3.229	3.267	3.3376	3.3591	3.5000
	2A	0.0033	3.4967	3.4729		3.3343	3.3233	3.1990	2B	3.229	3.267	3.3376	3.3519	3.5000
	3A	0.0000	3.5000	3.4762		3.3376	3.3293	3.2023	3B	3.2290	3.2594	3.3376	3.3484	3.5000
3½-6 UN	2A	0.0029	3.4971	3.4789		3.3888	3.3792	3.2987	2B	3.320	3.350	3.3917	3.4042	3.5000
	3A	0.0000	3.5000	3.4818		3.3917	3.3845	3.3016	3B	3.3200	3.3396	3.3917	3.4011	3.5000
3½-8 UN	2A	0.0026	3.4974	3.4824		3.4162	3.4074	3.3486	2B	3.365	3.390	3.4188	3.4303	3.5000
	3A	0.0000	3.5000	3.4850		3.4188	3.4122	3.3512	3B	3.3650	3.3797	3.4188	3.4274	3.5000

Identification	Class	Allowance	Major Dia Max	Major Dia Min	UNR Minor	Pitch Dia Max	Pitch Dia Min	Minor Dia Max	Class	Minor Dia Min	Minor Dia Max	Pitch Dia Min	Pitch Dia Max	Major Dia Min
3½–10 UNS	2A	0.0021	3.4979	3.4850	—	3.4329	3.4260	3.3789	2B	3.392	3.413	3.4350	3.4440	3.5000
3½–12 UN	2A	0.0019	3.4981	3.4867	—	3.4440	3.4376	3.3989	2B	3.410	3.428	3.4459	3.4543	3.5000
	3A	0.0000	3.5000	3.4886	—	3.4459	3.4411	3.4008	3B	3.4100	3.4198	3.4459	3.4522	3.5000
3½–14 UNS	2A	0.0018	3.4982	3.4879	—	3.4518	3.4457	3.4132	2B	3.423	3.438	3.4536	3.4615	3.5000
3½–16 UN	2A	0.0017	3.4983	3.4889	—	3.4577	3.4519	3.4238	2B	3.432	3.446	3.4594	3.4669	3.5000
	3A	0.0000	3.5000	3.4906	—	3.4594	3.4551	3.4255	3B	3.4320	3.4408	3.4594	3.4650	3.5000
3½–18 UNS	2A	0.0017	3.4983	3.4896	—	3.4622	3.4567	3.4322	2B	3.440	3.453	3.4639	3.4711	3.5000
3⅝–6 UN	2A	0.0029	3.6221	3.6039	—	3.5138	3.5041	3.4237	2B	3.445	3.475	3.5167	3.5293	3.6250
	3A	0.0000	3.6250	3.6068	—	3.5167	3.5094	3.4266	3B	3.4450	3.4646	3.5167	3.5262	3.6250
3⅝–8 UN	2A	0.0027	3.6223	3.6073	—	3.5411	3.5322	3.4735	2B	3.490	3.515	3.5438	3.5554	3.6250
	3A	0.0000	3.6250	3.6100	—	3.5438	3.5371	3.4762	3B	3.4900	3.5047	3.5438	3.5525	3.6250
3⅝–12 UN	2A	0.0019	3.6231	3.6117	—	3.5690	3.5626	3.5239	2B	3.535	3.553	3.5709	3.5793	3.6250
	3A	0.0000	3.6250	3.6136	—	3.5709	3.5661	3.5258	3B	3.5350	3.5448	3.5709	3.5772	3.6250
3⅝–16 UN	2A	0.0017	3.6233	3.6139	—	3.5827	3.5769	3.5488	2B	3.557	3.571	3.5844	3.5919	3.6250
	3A	0.0000	3.6250	3.6156	—	3.5844	3.5801	3.5505	3B	3.5570	3.5658	3.5844	3.5900	3.6250
3¾–4 UNC	1A	0.0034	3.7466	3.7109	—	3.5842	3.5674	3.4489	1B	3.479	3.517	3.5876	3.6094	3.7500
	2A	0.0034	3.7466	3.7228	—	3.5842	3.5730	3.4489	2B	3.479	3.517	3.5876	3.6021	3.7500
	3A	0.0000	3.7500	3.7262	3.7109	3.5876	3.5792	3.4523	3B	3.4790	3.5094	3.5876	3.5985	3.7500
3¾–6 UN	2A	0.0029	3.7471	3.7289	—	3.6388	3.6290	3.5487	2B	3.615	3.640	3.6417	3.6544	3.7500
	3A	0.0000	3.7500	3.7318	—	3.6417	3.6344	3.5516	3B	3.6150	3.6297	3.6417	3.6512	3.7500
3¾–8 UN	2A	0.0027	3.7473	3.7323	—	3.6661	3.6571	3.5985	2B	3.660	3.678	3.6688	3.6805	3.7500
	3A	0.0000	3.7500	3.7350	3.7248	3.6688	3.6621	3.6012	3B	3.6600	3.6698	3.6688	3.6776	3.7500
3¾–10 UNS	2A	0.0021	3.7479	3.7350	—	3.6829	3.6760	3.6289	2B	3.642	3.663	3.6850	3.6940	3.7500
3¾–12 UN	2A	0.0019	3.7481	3.7367	—	3.6940	3.6876	3.6489	2B	3.660	3.678	3.6959	3.7043	3.7500
	3A	0.0000	3.7500	3.7386	—	3.6959	3.6911	3.6508	3B	3.6600	3.6698	3.6959	3.7022	3.7500
3¾–14 UNS	2A	0.0018	3.7482	3.7379	—	3.7018	3.6957	3.6632	2B	3.673	3.688	3.7036	3.7115	3.7500
3¾–16 UN	2A	0.0017	3.7483	3.7389	—	3.7077	3.7019	3.6738	2B	3.682	3.696	3.7094	3.7169	3.7500
	3A	0.0000	3.7500	3.7406	—	3.7094	3.7051	3.6755	3B	3.6820	3.6908	3.7094	3.7150	3.7500
3¾–18 UNS	2A	0.0017	3.7483	3.7396	—	3.7122	3.7067	3.6822	2B	3.690	3.703	3.7139	3.7211	3.7500
3⅞–6 UN	2A	0.0030	3.8720	3.8538	—	3.7637	3.7538	3.6736	2B	3.695	3.725	3.7667	3.7795	3.8750
	3A	0.0000	3.8750	3.8568	—	3.7667	3.7593	3.6766	3B	3.6950	3.7146	3.7667	3.7763	3.8750

All dimensions in inches.

† Use UNS threads only if Standard Series do not meet requirements (see p. 1479). See footnotes a, b, c, d, and e at end of table.

Table 4 (Concluded). Standard Series and Selected Combinations† — Unified Screw Threads

Nominal Size, Threads per Inch, and Series Designation^e	External^c								Internal^c					
	Class	Allow- ance	Major Diameter			Pitch Diameter		UNR Minor Diam.^a Max (Ref.)	Class	Minor Diameter		Pitch Diameter		Major Diameter Min
			Max^b	Min	Min^d	Max^b	Min			Min	Max	Min	Max	
3⁷/₈-8 UN	2A	0.0027	3.8723	3.8573	—	3.7911	3.7820	3.7235	2B	3.740	3.765	3.7938	3.8056	3.8750
	3A	0.0000	3.8750	3.8600	—	3.7938	3.7870	3.7262	3B	3.7400	3.7547	3.7938	3.8026	3.8750
3⁷/₈-12 UN	2A	0.0020	3.8730	3.8616	—	3.8189	3.8124	3.7738	2B	3.785	3.803	3.8209	3.8294	3.8750
	3A	0.0000	3.8750	3.8636	—	3.8209	3.8160	3.7758	3B	3.7850	3.7948	3.8209	3.8273	3.8750
3⁷/₈-16 UN	2A	0.0018	3.8732	3.8638	—	3.8326	3.8267	3.7987	2B	3.807	3.821	3.8344	3.8420	3.8750
	3A	0.0000	3.8750	3.8656	—	3.8344	3.8300	3.8005	3B	3.8070	3.8158	3.8344	3.8401	3.8750
4-4 UNC	1A	0.0034	3.9966	3.9609	—	3.9342	3.9172	3.6989	1B	3.729	3.767	3.8376	3.8597	4.0000
	2A	0.0034	3.9966	3.9728	3.9609	3.9342	3.9229	3.6989	2B	3.729	3.767	3.8376	3.8523	4.0000
	3A	0.0000	4.0000	3.9762	—	3.9376	3.9291	3.7023	3B	3.7290	3.7594	3.8376	3.8487	4.0000
4-6 UN	2A	0.0030	3.9970	3.9788	—	3.8887	3.8788	3.7986	2B	3.820	3.850	3.8917	3.9046	4.0000
	3A	0.0000	4.0000	3.9818	—	3.8917	3.8843	3.8016	3B	3.8200	3.8396	3.8917	3.9014	4.0000
4-8 UN	2A	0.0027	3.9973	3.9823	3.9748	3.9161	3.9070	3.8485	2B	3.865	3.890	3.9188	3.9307	4.0000
	3A	0.0000	4.0000	3.9850	—	3.9188	3.9120	3.8512	3B	3.8650	3.8797	3.9188	3.9277	4.0000
4-10 UNS	2A	0.0021	3.9979	3.9850	—	3.9329	3.9259	3.8768	2B	3.892	3.913	3.9350	3.9441	4.0000
4-12 UN	2A	0.0020	3.9980	3.9866	—	3.9439	3.9374	3.8988	2B	3.910	3.928	3.9459	3.9544	4.0000
	3A	0.0000	4.0000	3.9886	—	3.9459	3.9410	3.9008	3B	3.9100	3.9198	3.9459	3.9523	4.0000
4-14 UNS	2A	0.0018	3.9982	3.9879	—	3.9518	3.9456	3.9132	2B	3.923	3.938	3.9536	3.9616	4.0000
4-16 UN	2A	0.0018	3.9982	3.9888	—	3.9576	3.9517	3.9237	2B	3.932	3.946	3.9594	3.9670	4.0000
	3A	0.0000	4.0000	3.9906	—	3.9594	3.9550	3.9255	3B	3.9330	3.9408	3.9594	3.9651	4.0000

All dimensions in inches. † Use UNS threads only if Standard Series do not meet requirements (see p. 1479). For sizes above 4 inches see ANSI B1.1-1982.
a UN series external thread maximum minor diameter is basic for Class 3A and basic minus allowance for Classes 1A and 2A.
b For Class 2A threads having an additive finish the maximum is increased, by the allowance, to the basic size, the value being the same as for Class 3A.
c Regarding combinations of thread classes, see text on p. 1479.
d For unfinished hot-rolled material not including standard fasteners with rolled threads.
e Use UNR designation instead of UN wherever UNR thread form is desired for external use.
f Formerly NF. Tolerances and allowances are based on one diameter length of engagement.

American National Standard Metric Screw Threads — M Profile. — American National Standard ANSI B1.13M-1983 describes a system of metric threads for general fastening purposes in mechanisms and structures. The standard is in basic agreement with ISO screw standards and resolutions, as of the date of publication, and features detailed information for diameter-pitch combinations selected as to preferred standard sizes. This Standard contains general metric standards for a 60-degree symmetrical screw thread with a basic ISO 68 designated profile.

Application Comparison with Inch Threads. — The metric M profile threads of tolerance class 6H/6g (see page 1528) are intended for metric applications where the inch class 2A/2B have been used. At the minimum material limits, the 6H/6g results in a looser fit than the 2A/2B. Tabular data are also provided for a tighter tolerance fit external thread of class 4g6g which is approximately equivalent to the inch class 3A but with an allowance applied. It may be noted that a 4H5H/4h6h fit is approximately equivalent to class 3A/3B fit in the inch system.

Interchangeability with Other System Threads. — Threads produced to this Standard (ANSI B1.13M) are fully interchangeable with threads conforming to other National Standards which are based on ISO 68 basic profile and ISO 965/1 tolerance practices.

Threads produced to this Standard should be mechanically interchangeable with those produced to ANSI B1.18M-1982, "Metric Screw Threads for Commercial Mechanical Fasteners — Boundary Profile Defined," of the same size and tolerance class. However, there is a possibility that some parts may be accepted by conventional gages used for threads made to ANSI B1.13M and rejected by the Double-NOT-GO gages required for threads made to ANSI B1.18M.

Threads produced in accordance with M profile and MJ profile (ANSI B1.21M) design data will assemble with each other. However, external MJ threads will encounter interference on the root radii with internal M thread crests when both threads are at maximum material condition.

Definitions. — The following definitions apply to metric screw threads — M profile.

Allowance: The minimum nominal clearance between a prescribed dimension and its basic dimension. Allowance is not an ISO metric screw thread term but it is numerically equal to the absolute value of the ISO term *fundamental deviation*.

Basic Thread Profile: The cyclical outline in an axial plane of the permanently established boundary between the provinces of the external and internal threads. All deviations are with respect to this boundary. (See Figs. 1 and 5.)

Bolt Thread (External Thread): The term used in ISO metric thread standards to describe all external threads. All symbols associated with external threads are designated with lower case letters. This Standard uses the term external threads in accordance with United States practice.

Clearance: The difference between the size of the internal thread and the size of the external thread when the latter is smaller.

Crest Diameter: The major diameter of an external thread and the minor diameter of an internal thread.

Design Profiles: The maximum material profiles permitted for external and internal threads for a specified tolerance class. (See Figs. 2 and 3.)

Deviation: An ISO term for the algebraic difference between a given size (actual, measured, maximum, minimum, etc.) and the corresponding basic size. The term deviation does not necessarily indicate an error.

Fit: The relationship existing between two corresponding external and internal threads with respect to the amount of clearance or interference which is present when they are assembled.

Fundamental Deviation: For Standard threads, the deviation (upper or lower) closer to the basic size. It is the upper deviation, *es*, for an external thread and the lower deviation, *EI*, for an internal thread. (See Fig. 5.)

Limiting Profiles: The limiting M profile for internal threads is shown in Fig. 6. The limiting M profile for external threads is shown in Fig. 7.

Lower Deviation: The algebraic difference between the minimum limit of size and the corresponding basic size.

Nut Thread (Internal Thread): A term used in ISO metric thread standards to describe all internal threads. All symbols associated with internal threads are designated with upper case letters. This Standard uses the term *internal thread* in accordance with United States practice.

Tolerance: The total amount of variation permitted for the size of a dimension. It is the difference between the maximum limit of size and the minimum limit of size (i.e., the algebraic difference between the upper deviation and the lower deviation). The tolerance is an absolute value without sign. Tolerance for threads is applied to the design size in the direction of the minimum material. On external threads the tolerance is applied negatively. On internal threads the tolerance is applied positively.

Tolerance Class: The combination of a tolerance position with a tolerance grade. It specifies the allowance (fundamental deviation) and tolerance for the pitch and major diameters of external threads and pitch and minor diameters of internal threads.

Tolerance Grade: A numerical symbol that designates the tolerances of crest diameters and pitch diameters applied to the design profiles.

Tolerance Position: A letter symbol that designates the position of the tolerance zone in relation to the basic size. This position provides the allowance (fundamental deviation).

Upper Deviation: The algebraic difference between the maximum limit of size and the corresponding basic size.

General Symbols. — The general symbols used to describe the metric screw thread forms are shown in Table 1.

Basic M Profile. — The basic M thread profile also known as ISO 68 basic profile for metric screw threads is shown in Fig. 1 with associated dimensions listed in Table 2.

Design M Profile for Internal Thread. — The design M profile for the internal thread at maximum material condition is the basic ISO 68 profile. It is shown in Fig. 2 with associated thread data listed in Table 2.

Design M Profile for External Thread. — The design M profile for the external thread at the no allowance maximum material condition is the basic ISO 68 profile except where a rounded root is required. For the standard $0.125P$ minimum radius, the ISO 68 profile is modified at the root with a $0.17783H$ truncation blending into two arcs with radii of $0.125P$ tangent to the thread flanks as shown in Fig. 3 with associated thread data in Table 2.

M Crest and Root Form. — The form of crest at the major diameter of the external thread is flat, permitting corner rounding. The external thread is truncated $0.125H$ from a sharp crest. The form of the crest at the minor diameter of the internal thread is flat. It is truncated $0.25H$ from a sharp crest.

The crest and root tolerance zones at the major and minor diameters will permit rounded crest and root forms in both external and internal threads.

The root profile of the external thread must lie within the "section lined" tolerance zone shown in Fig. 4. For the rounded root thread, the root profile must lie within the "section lined" rounded root tolerance zone shown in Fig. 4. The profile must be a continuous smoothly blended non-reversing curve, no part of which has a radius of less than $0.125P$, and which is tangential to the thread flank. The profile may comprise tangent flank arcs that are joined by a tangential flat at the root.

(*Continued on page* 1525)

Table 1. American National Standard Symbols for Metric Threads
(ANSI B1.13M-1983)

Symbol	Explanation
D	Major Diameter Internal Thread
D_1	Minor Diameter Internal Thread
D_2	Pitch Diameter Internal Thread
d	Major Diameter External Thread
d_1	Minor Diameter External Thread
d_2	Pitch Diameter External Thread
d_3	Rounded Form Minor Diameter External Thread
P	Pitch
r	External Thread Root Radius
T	Tolerance
T_{D1}, T_{D2}	Tolerances for D_1, D_2
T_d, T_{d2}	Tolerances for d, d_2
ES	Upper Deviation, Internal Thread [Equals the Allowance (Fundamental Deviation) Plus the Tolerance]. See Fig. 5.
EI	Lower Deviation, Internal Thread Allowance (Fundamental Deviation). See Fig. 5.
G, H	Letter Designations for Tolerance Positions for Lower Deviation, Internal Thread
g, h	Letter Designations for Tolerance Positions for Upper Deviation, External Thread
es	Upper Deviation, External Thread Allowance (Fundamental Deviation). See Fig. 5. In the ISO system es is always negative for an allowance fit or zero for no allowance.
ei	Lower Deviation, External Thread [Equals the Allowance (Fundamental Deviation) Plus the Tolerance]. See Fig. 5. In the ISO system ei is always negative for an allowance fit.
H	Height of Fundamental Triangle
LE	Length of Engagement
LH	Left Hand Thread

Table 2. American National Standard Metric Thread — M Profile Data (ANSI B1.13M-1979)

Pitch P	Truncation of Internal Thread Root and External Thread Crest $\frac{H}{8}$ 0.108253P	Addendum of Internal Thread and Truncation of Internal Thread $\frac{H}{4}$ 0.216506P	Dedendum of Internal Thread and Addendum of External Thread $\frac{3H}{8}$ 0.324760P	Difference[1] $\frac{H}{2}$ 0.433013P	Height of Internal Thread and Depth of Thread Engagement $\frac{5H}{8}$ 0.541266P	Difference[2] 0.711325H 0.616025P	Twice the External Thread Addendum $\frac{3H}{4}$ 0.649519P	Difference[3] $\frac{11H}{12}$ 0.793857P	Height of Sharp V-Thread 0.8660254P	Double Height of Internal Thread $\frac{5H}{4}$ 1.082532P
0.2	0.02165	0.04330	0.06495	0.08660	0.10825	0.12321	0.12990	0.15877	0.17321	0.21651
0.25	0.02706	0.05413	0.08119	0.10825	0.13532	0.15401	0.16238	0.19846	0.21651	0.27063
0.3	0.03248	0.06495	0.09743	0.12990	0.16238	0.18481	0.19486	0.23816	0.25981	0.32476
0.35	0.03789	0.07578	0.11367	0.15155	0.18944	0.21561	0.22733	0.27785	0.30311	0.37889
0.4	0.04330	0.08660	0.12990	0.17321	0.21651	0.24541	0.25981	0.31754	0.34641	0.43301
0.45	0.04871	0.09743	0.14614	0.19486	0.24357	0.27721	0.29228	0.35724	0.38971	0.48714
0.5	0.05413	0.10825	0.16238	0.21651	0.27063	0.30801	0.32476	0.39693	0.43301	0.54127
0.6	0.06495	0.12990	0.19486	0.25981	0.32476	0.36962	0.38971	0.47631	0.51962	0.64952
0.7	0.07578	0.15155	0.22733	0.30311	0.37889	0.43122	0.45466	0.55570	0.60622	0.75777
0.75	0.08119	0.16238	0.24357	0.32476	0.40595	0.46202	0.48714	0.59539	0.64952	0.81190
0.8	0.08660	0.17321	0.25981	0.34641	0.43301	0.49282	0.51962	0.63509	0.69282	0.86603
1	0.10825	0.21651	0.32476	0.43301	0.54127	0.61603	0.64952	0.79386	0.86603	1.08253
1.25	0.13532	0.27063	0.40595	0.54127	0.67658	0.77003	0.81190	0.99232	1.08253	1.35316
1.5	0.16238	0.32476	0.48714	0.64952	0.81190	0.92404	0.97428	1.19079	1.29904	1.62380
1.75	0.18944	0.37889	0.56853	0.75777	0.94722	1.07804	1.13666	1.38925	1.51554	1.89443
2	0.21651	0.43301	0.64952	0.86603	1.08253	1.23205	1.29904	1.58771	1.73205	2.16506
2.5	0.27063	0.54127	0.81190	1.08253	1.35316	1.54006	1.62380	1.98464	2.16506	2.70633
3	0.32476	0.64952	0.97428	1.29904	1.62380	1.84808	1.94856	2.38157	2.59808	3.24760
3.5	0.37889	0.75777	1.13666	1.51554	1.89443	2.15609	2.27332	2.77850	3.03109	3.78886
4	0.43301	0.86603	1.29904	1.73205	2.16506	2.46410	2.59808	3.17543	3.46410	4.33013
4.5	0.48714	0.97428	1.46142	1.94856	2.43570	2.77211	2.92284	3.57235	3.89711	4.87139
5	0.54127	1.08253	1.62380	2.16506	2.70633	3.08013	3.24760	3.96928	4.33013	5.41266
5.5	0.59539	1.19078	1.78618	2.38157	2.97696	3.38814	3.57235	4.36621	4.76314	5.95392
6	0.64952	1.29904	1.94856	2.59808	3.24760	3.69615	3.89711	4.76314	5.19615	6.49519
8	0.86603	1.73205	2.59808	3.46410	4.33013	4.92820	5.19615	6.35085	6.92820	8.66025

All dimensions are in millimeters.
[1] Difference between max theoretical pitch diameter and max minor diameter of external thread and between min theoretical pitch diameter and min minor diameter of internal thread.
[2] Difference between min theoretical pitch diameter and min design minor diameter of external thread for 0.125P root radius.
[3] Difference between max major diameter and max theoretical pitch diameter of internal thread.

The root profile of the internal thread must not be smaller than the basic profile. The maximum major diameter must not be sharp.

Standard M Profile Screw Thread Series. — The standard metric screw thread series for general purpose equipment's threaded components design and mechanical fasteners is a *coarse thread* series. Their diameter/pitch combinations are shown in Table 3. These diameter/pitch combinations are the preferred sizes and should be the first choice as applicable. Additional *fine pitch* diameter/pitch combinations are shown in Table 4.

Limits and Fits for Metric Screw Threads — M Profile. — The International (ISO) metric tolerance system is based on a system of limits and fits. The limits of the tolerances on the mating parts together with their allowances (fundamental deviations) determine the fit of the assembly. For simplicity the system is described for cylindrical parts (see page 659) but in this Standard it is applied to screw threads. Holes are equivalent to internal threads and shafts to external threads.

Basic Size: This is the zero line or surface at assembly where the interface of the two mating parts have a common reference.*

Upper Deviation: This is the algebraic difference between the maximum limit of size and the basic size. It is designated by the French term "écart supérieur" (*ES* for internal and *es* for external threads).

Lower Deviation: This is the algebraic difference between the minimum limit of size and the basic size. It is designated by the French term "écart inférieur" (*EI* for internal and *ei* for external threads).

Fundamental Deviations (Allowances): These are the deviations which closest to the basic size. In the accompanying figure they would be *EI* and *es*.

Fits: Fits are determined by the fundamental deviations assigned to the mating parts and may be positive or negative. The selected fits can be clearance, transition, or interference. To illustrate the fits schematically, a zero line is drawn to represent the basic size as shown in Fig. 5. By convention, the external thread lies below the zero line and the internal thread lies above it (except for interference fits). This makes the fundamental deviation negative for the external thread and equal to its upper deviation (*es*). The fundamental deviation is positive for the internal thread and equal to its lower deviation (*EI*).

* Basic," when used to identify a particular dimension in this Standard, such as basic major diameter, refers to the h/H tolerance position (zero fundamental deviation) value.

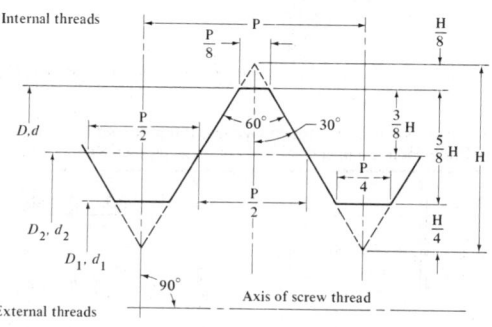

$$H = \frac{\sqrt{3}}{2} \times P = 0.866\ 025P$$

$0.125H = 0.108\ 253P \qquad 0.250H = 0.216\ 506P \qquad 0.375H = 0.324\ 760P \qquad 0.625H = 0.541\ 266P$

Fig. 1. Basic M Thread Profile (ISO 68 Basic Profile)

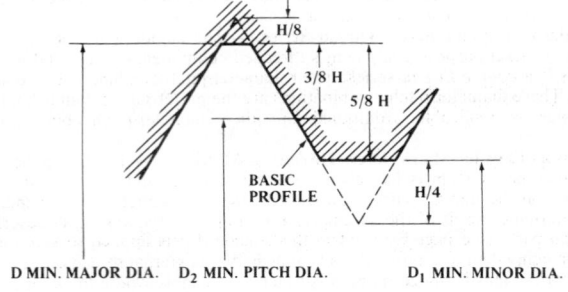

D MIN. MAJOR DIA.　　D₂ MIN. PITCH DIA.　　D₁ MIN. MINOR DIA.

Fig. 2. Internal Thread Design M Profile with No Allowance (Fundamental Deviation)
(Maximum Material Condition). For Dimensions see Table 2

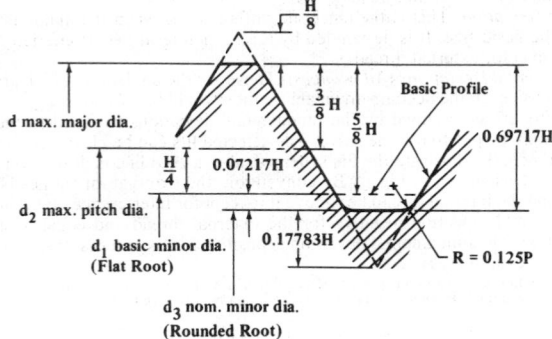

Fig. 3. External Thread Design M Profile with No Allowance (Fundamental Deviation)
(Flanks at Maximum Material Condition). For Dimensions see Table 2

**Table 3. American National Standard General Purpose and Mechanical Fastener
Coarse Pitch Metric Thread — M Profile Series (ANSI B1.13M-1983)**

Nom. Size	Pitch	Nom. Size	Pitch	Nom. Size	Pitch	Nom. Size	Pitch
1.6	0.35	6	1	22	2.5*	56	5.5
2	0.4	8	1.25	24	3	64	6
2.5	0.45	10	1.5	27	3*	72	6
3	0.5	12	1.75	30	3.5	80	6
3.5	0.6	14	2	36	4	90	6
4	0.7	16	2	42	4.5	100	6
5	0.8	20	2.5	48	5	. . .	. . .

All dimensions are in millimeters.

* For high strength structural steel fasteners only.

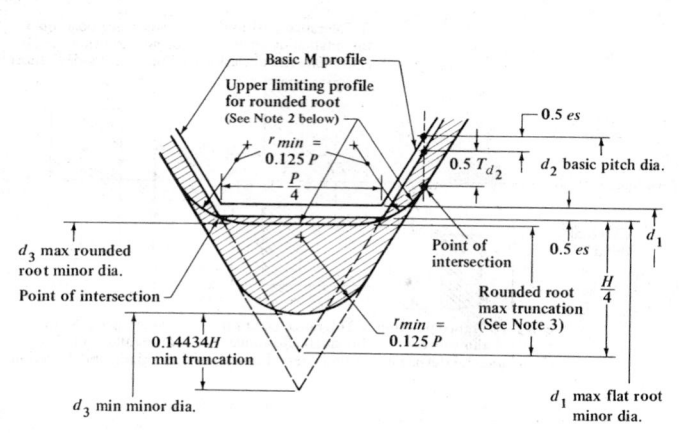

Notes: 1. "Section lined" portions identify tolerance zone and unshaded portions identify allowance (fundamental deviation).

2. The upper limiting profile for rounded root is not a design profile; rather it indicates the limiting acceptable condition for the rounded root which will pass a GO thread gage.

3. Max truncation $= \dfrac{H}{4} - r_{min}\left(1 - \cos\left[60° - \arccos\left(1 - \dfrac{T_{d2}}{4r_{min}}\right)\right]\right)$

where: H = Height of fundamental triangle
r_{min} = Minimum external thread root radius
T_{d2} = Tolerance on pitch diameter of external thread

Fig. 4. M Profile, External Thread Root, Upper and Lower Limiting Profiles for $r_{min} = 0.125\,P$ and for Flat Root (Shown for Tolerance Position g)

Table 4. American National Standard Fine Pitch Metric Thread — M Profile Series
(ANSI B1.13M-1983)

Nom. Size	Pitch			Nom. Size	Pitch		Nom. Size	Pitch		Nom. Size	Pitch
8	1		. . .	27	. . .	2	56	. . .	2	105	2
10	0.75		1.25	30	1.5	2	60	1.5	. . .	110	2
12	1	1.5*	1.25	33	. . .	2	64	. . .	2	120	2
14	. . .		1.5	35	1.5	. . .	65	1.5	. . .	130	2
15	1		. . .	36	. . .	2	70	1.5	. . .	140	2
16	. . .		1.5	39	. . .	2	72	. . .	2	150	2
17	1		. . .	40	1.5	. . .	75	1.5	. . .	160	3
18	. . .		1.5	42	. . .	2	80	1.5	2	170	3
20	1		1.5	45	1.5	. . .	85	. . .	2	180	3
22	. . .		1.5	48	. . .	2	90	. . .	2	190	3
24	. . .		2	50	1.5	. . .	95	. . .	2	200	3
25	1.5		. . .	55	1.5	. . .	100	. . .	2		

All dimensions are in millimeters.   * Only for wheel studs and nuts.

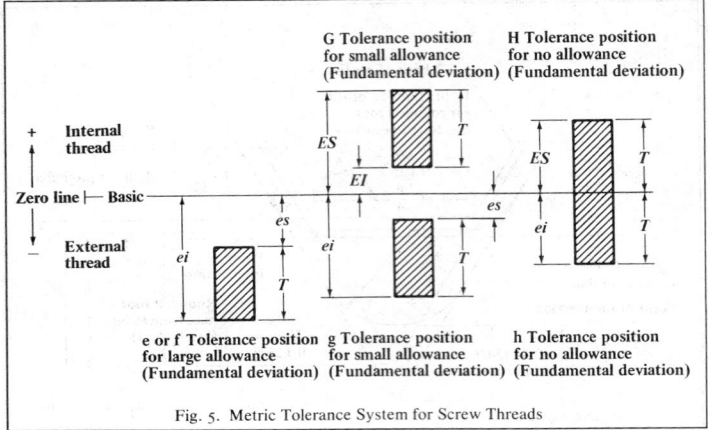

Fig. 5. Metric Tolerance System for Screw Threads

Tolerance: The tolerance is defined by a series of numerical grades. Each grade provides numerical values for the various nominal sizes corresponding to the standard tolerance for that grade.

In the schematic diagram the tolerance for the external thread is shown as negative. Thus the tolerance plus the fit define the lower deviation (ei). The tolerance for the mating internal thread is shown as positive. Thus the tolerance plus the fit defines the upper deviation (ES).

Tolerance Grade: This is indicated by a number. The system provides for a series of tolerance grades for each of the four screw thread parameters: minor diameter, internal thread, D_1; major diameter, external thread, d; pitch diameter, internal thread, D_2; and pitch diameter, external thread, d_2. The tolerance grades for this Standard (ANSI B1.13M) were selected from those given in ISO 965/1.

Dimension	Tolerance Grades	Table
D_1	4, 5, <u>6</u>, 7, 8	7
d	4, <u>6</u>, 8	10
D_2	4, 5, <u>6</u>, 7, 8	8
d_2	3, <u>4</u>, 5, <u>6</u>, 7, 8, 9	9

Note: The underlined tolerance grades are used with normal length of thread engagement.

Tolerance Position: This position is the allowance (fundamental deviation) and is indicated by a letter. A capital letter is used for internal threads and a lower case letter for external threads. The system provides a series of tolerance positions for internal and external threads. The underlined letters are used in this Standard:

Internal threads	G, <u>H</u>	Table 5
External threads	e, f, <u>g</u>, h	Table 5

Table 5. American National Standard Allowance (Fundamental Deviation) for Internal and External Metric Threads (ISO 965/1) (ANSI B1.13M-1983)

Pitch P	Internal Thread D_2, D_1 G EI	H EI	External Thread d, d_2 e es	f es	g es	h es
0.2	+ 0.017	0	. . .	. . .	− 0.017	0
0.25	+ 0.018	0	. . .	. . .	− 0.018	0
0.3	+ 0.018	0	. . .	. . .	− 0.018	0
0.35	+ 0.019	0	. . .	− 0.034	− 0.019	0
0.4	+ 0.019	0	. . .	− 0.034	− 0.019	0
0.45	+ 0.020	0	. . .	− 0.035	− 0.020	0
0.5	+ 0.020	0	− 0.050	− 0.036	− 0.020	0
0.6	+ 0.021	0	− 0.053	− 0.036	− 0.021	0
0.7	+ 0.022	0	− 0.056	− 0.038	− 0.022	0
0.75	+ 0.022	0	− 0.056	− 0.038	− 0.022	0
0.8	+ 0.024	0	− 0.060	− 0.038	− 0.024	0
1	+ 0.026	0	− 0.060	− 0.040	− 0.026	0
1.25	+ 0.028	0	− 0.063	− 0.042	− 0.028	0
1.5	+ 0.032	0	− 0.067	− 0.045	− 0.032	0
1.75	+ 0.034	0	− 0.071	− 0.048	− 0.034	0
2	+ 0.038	0	− 0.071	− 0.052	− 0.038	0
2.5	+ 0.042	0	− 0.080	− 0.058	− 0.042	0
3	+ 0.048	0	− 0.085	− 0.063	− 0.048	0
3.5	+ 0.053	0	− 0.090	− 0.070	− 0.053	0
4	+ 0.060	0	− 0.095	− 0.075	− 0.060	0
4.5	+ 0.063	0	− 0.100	− 0.080	− 0.063	0
5	+ 0.071	0	− 0.106	− 0.085	− 0.071	0
5.5	+ 0.075	0	− 0.112	− 0.090	− 0.075	0
6	+ 0.080	0	− 0.118	− 0.095	− 0.080	0

All dimensions are in millimeters.
* Allowance is the absolute value of fundamental deviation.

Designations of Tolerance Grade, Tolerance Position, and Tolerance Class: The tolerance grade is given first followed by the tolerance position, thus: 4g or 5H. To designate the tolerance class the grade and position of the pitch diameter is shown first followed by that for the major diameter in the case of the external thread or that for the minor diameter in the case of the internal thread, thus 4g6g for an external thread and 5H6H for an internal thread. If the two grades and positions are identical, it is not necessary to repeat the symbols, thus 4g, alone, stands for 4g4g and 5H, alone, stands for 5H5H.

Lead and Flank Angle Tolerances: For acceptance of lead and flank angles of product screw threads reference should be made to Section 10 of ANSI B1.13M-1983.

Short and Long Lengths of Thread Engagement when Gaged with Normal Length Contacts: For short lengths of thread engagement, LE, reduce the pitch diameter tolerance of the external thread by one tolerance grade number. For long lengths of thread engagement, LE, increase the allowance (fundamental deviation) at the pitch diameter of the external thread. Examples of tolerance classes required for normal, short, and long gage length contacts are given in the following table.

For lengths of thread engagement classified as normal, short, and long, see Table 6.

Normal LE	Short LE	Long LE
6g	5g6g	6e6g
4g6g	3g6g	4e6g
6h*	5h6h	6g6h
4h6h*	3h6h	4g6h

* Applies to maximum material functional size (GO thread gage) for plated 6g and 4g6g class threads, respectively.

Coated or Plated Threads: Coating is one or more applications of additive material to the threads, including dry-film lubricants, but excluding soft or liquid lubricants that are readily displaced in assembly or gaging. Plating is included as coating in the Standard. Unless otherwise specified, size limits for standard external tolerance classes 6g and 4g6g apply prior to coating. The external thread allowance may thus be used to accommodate the coating thickness on coated parts, provided that the maximum coating thickness is no more than one-quarter of the allowance. Thus, the thread after coating is subject to acceptance using a basic (tolerance position h) size GO thread gage and tolerance position g thread gage for either minimum material, LO, or NOT-GO. Where the external thread has no allowance or the allowance must be maintained after coating, and for standard internal threads, sufficient allowance must be provided prior to coating to insure that finished product threads do not exceed the maximum material limits specified. For thread classes with tolerance position H or h, coating allowances in accordance with Table 5 for position G or g, respectively, should be applied wherever possible.

Dimensional Effect of Coating. — On a cylindrical surface, the effect of coating is to change the diameter by twice the coating thickness. On a 60-degree thread, however, since the coating thickness is measured perpendicular to the thread surface while the pitch diameter is measured perpendicular to the thread axis, the effect of a uniformly coated flank on the pitch diameter is to change it by four times the thickness of the coating on the flank.

External Thread with No Allowance for Coating: To determine gaging limits before coating for a uniformly coated thread, decrease: (a) maximum pitch diameter by four times maximum coating thickness, (b) minimum pitch diameter by four times minimum coating thickness, (c) maximum major diameter by two times maximum coating thickness, and (d) minimum major diameter by two times minimum coating thickness.

External Thread with Only Nominal or Minimum Thickness Coating: If no coating thickness tolerance is given, it is recommended that a tolerance of plus 50 per cent of the nominal or minimum thickness be assumed. Then, to determine before coating gaging limits for a uniformly coated thread, decrease: (a) maximum pitch diameter by six times coating thickness, (b) minimum pitch diameter by four times coating thickness, (c) maximum major diameter by three times coating thickness, and (d) minimum major diameter by two times coating thickness.

Adjusted Size Limits: It should be noted that the before coating material limit tolerances are less than the tolerance after coating. This is because the coating tolerance consumes some of the product tolerance. In cases there may be insufficient pitch diameter tolerance available in the before coating condition so that additional adjustments and controls will be necessary.

Strength: On small threads (5 mm and smaller) there is a possibility that coating thickness adjustments will cause base material minimum material conditions which may significantly affect strength of externally threaded parts. Limitations on coating thickness or part redesign may then be necessary.

Internal Threads: Standard internal threads provide no allowance for coating thickness. To determine before coating, gaging limits for a uniformly coated thread,

increase: (a) minimum pitch diameter by four times maximum coating thickness, if specified, or by six times minimum or nominal coating thickness when a tolerance is not specified, (b) maximum pitch diameter by four times minimum or nominal coating thickness, (c) minimum minor diameter by two times maximum coating thickness, if specified, or by three times minimum or nominal coating thickness, and (d) maximum minor diameter by two times minimum or nominal coating thickness.

Other Considerations. — It is essential to review all possibilities adequately and consider limitations in the threading and coating production processes before finally deciding on the coating process and the allowance required to accommodate the coating. A no-allowance thread after coating must not transgress the basic profile and is, therefore, subject to acceptance using a basic (tolerance position H/h) size GO thread gage.

Formulas for M Profile Screw Thread Limiting Dimensions. — The limiting dimensions for M profile screw threads are calculated from the following formulas.

Internal Threads:

Min major dia. = basic major dia. + EI (Table 5)

Min pitch dia. = basic major dia. − $0.649519P$ (Table 2) + EI for D_2 (Table 5)

Max pitch dia. = min pitch dia. + T_{D2} (Table 8)

Max major dia. = max pitch dia. + $0.793857P$ (Table 2)

Min minor dia. = min major dia. − $1.082532P$ (Table 2)

Max minor dia. = min minor dia. + T_{D1} (Table 7)

External Threads:

Max major dia. = basic major dia. − es (Table 5) (Note that es is an absolute value.)

Min major dia. = max major dia. − T_d (Table 10)

Max pitch dia. = basic major dia. − $0.649519P$ (Table 2) − es for d_2 (Table 5)

Min pitch dia. = max pitch dia. − T_{d2} (Table 9)

Max flat form minor dia. = max pitch dia. − $0.433013P$ (Table 2)

Max rounded root minor dia. = max pitch dia. − $2 \times$ max trunc. (See Fig. 4)

Min rounded root minor dia. = min pitch dia. − $0.616025P$ (Table 2)

Min root radius = $0.125P$

Tolerance Grade Comparisons. — The approximate ratios of the tolerance grades shown in Tables 7, 8, 9, and 10 in terms of Grade 6 are as follows:

Minor Diameter Tolerance of Internal Thread, T_{D1} (Table 7): Grade 4 is $0.63\ T_{D1}$ (6); Grade 5 is $0.8\ T_{D1}$ (6); Grade 7 is $1.25\ T_{D1}$ (6); and Grade 8 is $1.6\ T_{D1}$ (6).

Pitch Diameter Tolerance of Internal Thread, T_{D2} (Table 8): Grade 4 is $0.85T_{d2}$ (6); Grade 5 is $1.06\ T_{d2}$ (6); Grade 6 is $1.32\ T_{d2}$ (6); Grade 7 is $1.7\ T_{d2}$ (6); and Grade 8 is $2.12\ T_{d2}$ (6). It should be noted that these ratios are in terms of the Grade 6 pitch diameter tolerance for the external thread.

Major Diameter Tolerance of External Thread, T_d (Table 10): Grade 4 is $0.63\ T_d$ (6); and Grade 8 is $1.6\ T_d$ (6).

Pitch Diameter Tolerance of External Thread, T_{d2} (Table 9): Grade 3 is $0.5\ T_{d2}$ (6); Grade 4 is $0.63\ T_{d2}$ (6); Grade 5 is $0.8\ T_{d2}$ (6); Grade 7 is $1.25\ T_{d2}$ (6); Grade 8 is $1.6\ T_{d2}$ (6); and Grade 9 is $2\ T_{d2}$ (6).

Standard M Profile Screw Threads, Limits of Size. — The limiting M profile for internal threads is shown in Fig. 6 with associated dimensions for standard sizes in Table 11. The limiting M profiles for external threads are shown in Fig. 7 with associated dimensions for standard sizes in Table 12.

If the required values are not listed in these tables, they may be calculated using the data in Tables 2, 5, 6, 7, 8, 9, and 10 together with the preceding formulas. If the required data are not included in any of the tables listed above, reference should be made to Sections 6 and 9.3 of ANSI B1.13M which gives design formulas.

Table 6. American National Standard Length of Metric Thread Engagement
(ISO 965/1) (ANSI B1.13M-1983)

Basic Major Diameter d		Pitch P	Length of Thread Engagement			
			l_S	l_N		l_L
Over	Up to and incl.		Up to and incl.	Over	Up to and incl.	Over
1.5	2.8	0.2	0.5	0.5	1.5	1.5
		0.25	0.6	0.6	1.9	1.9
		0.35	0.8	0.8	2.6	2.6
		0.4	1	1	3	3
		0.45	1.3	1.3	3.8	3.8
2.8	5.6	0.35	1	1	3	3
		0.5	1.5	1.5	4.5	4.5
		0.6	1.7	1.7	5	5
		0.7	2	2	6	6
		0.75	2.2	2.2	6.7	6.7
		0.8	2.5	2.5	7.5	7.5
5.6	11.2	0.75	2.4	2.4	7.1	7.1
		1	3	3	9	9
		1.25	4	4	12	12
		1.5	5	5	15	15
11.2	22.4	1	3.8	3.8	11	11
		1.25	4.5	4.5	13	13
		1.5	5.6	5.6	16	16
		1.75	6	6	18	18
		2	8	8	24	24
		2.5	10	10	30	30
22.4	45	1	4	4	12	12
		1.5	6.3	6.3	19	19
		2	8.5	8.5	25	25
		3	12	12	36	36
		3.5	15	15	45	45
		4	18	18	53	53
		4.5	21	21	63	63
45	90	1.5	7.5	7.5	22	22
		2	9.5	9.5	28	28
		3	15	15	45	45
		4	19	19	56	56
		5	24	24	71	71
		5.5	28	28	85	85
		6	32	32	95	95
90	180	2	12	12	36	36
		3	18	18	53	53
		4	24	24	71	71
		6	36	36	106	106
180	355	3	20	20	60	60
		4	26	26	80	80
		6	40	40	118	118

All dimensions are in millimeters.

Metric Screw Thread Designations. — Metric screw threads are identified by the letter (M) for the thread form profile, followed by the nominal diameter size and the pitch expressed in millimeters, separated by the sign (×) and followed by the tolerance class separated by a dash (−) from the pitch.

The simplified international practice for designating coarse pitch M profile metric screw threads is to leave off the pitch. Thus a M14 × 2 thread is designated just M14. However, to prevent misunderstanding, it is mandatory to use the value for pitch in all designations.

Thread acceptability gaging system requirements of ANSI B1.3M may be added to the thread size designation as noted in the examples (numbers in parentheses) or as specified in pertinent documentation, such as the drawing or procurement document.

Unless otherwise specified in the designation, the screw thread is right hand.

Examples: External thread of M profile, right hand: M6 × 1 − 4g6g (22)

Internal thread of M profile, right hand: M6 × 1 − 5H6H (21)

Designation of Left Hand Thread: When a left hand thread is specified, the tolerance class designation is followed by a dash and LH.

Example: M6 × 1 − 5H6H − LH (23)

(Continued on page 1537)

Table 7. American National Standard Minor Diameter Tolerances of Internal Metric Threads T_{DI} **(ISO 965/1) (ANSI B1.13M-1983)**

Pitch P	Tolerance Grade				
	4	5	6	7	8
0.2	0.038	. . .	. . .	. . .	. . .
0.25	0.045	0.056	. . .	. . .	. . .
0.3	0.053	0.067	0.085	. . .	. . .
0.35	0.063	0.080	0.100	. . .	. . .
0.4	0.071	0.090	0.112	. . .	. . .
0.45	0.080	0.100	0.125	. . .	. . .
0.5	0.090	0.112	0.140	0.180	. . .
0.6	0.100	0.125	0.160	0.200	. . .
0.7	0.112	0.140	0.180	0.224	. . .
0.75	0.118	0.150	0.190	0.236	. . .
0.8	0.125	0.160	0.200	0.250	0.315
1	0.150	0.190	0.236	0.300	0.375
1.25	0.170	0.212	0.265	0.335	0.425
1.5	0.190	0.236	0.300	0.375	0.475
1.75	0.212	0.265	0.335	0.425	0.530
2	0.236	0.300	0.375	0.475	0.600
2.5	0.280	0.355	0.450	0.560	0.710
3	0.315	0.400	0.500	0.630	0.800
3.5	0.355	0.450	0.560	0.710	0.900
4	0.375	0.475	0.600	0.750	0.950
4.5	0.425	0.530	0.670	0.850	1.060
5	0.450	0.560	0.710	0.900	1.120
5.5	0.475	0.600	0.750	0.950	1.180
6	0.500	0.630	0.800	1.000	1.250

All dimensions are in millimeters.

Table 8. American National Standard Pitch Diameter Tolerances of Internal Metric Threads, T_{D2} (ISO 965/1) (ANSI B1.13M-1983)

Basic Major Diameter, D		Pitch P	Tolerance Grade				
Over	Up to and incl.		4	5	6	7	8
1.5	2.8	0.2	0.042	. . .	. . .	. . .	. . .
		0.25	0.048	0.060	. . .	. . .	. . .
		0.35	0.053	0.067	0.085	. . .	. . .
		0.4	0.056	0.071	0.090	. . .	. . .
		0.45	0.060	0.075	0.095	. . .	. . .
2.8	5.6	0.35	0.056	0.071	0.090	. . .	. . .
		0.5	0.063	0.080	0.100	0.125	. . .
		0.6	0.071	0.090	0.112	0.140	. . .
		0.7	0.075	0.095	0.118	0.150	. . .
		0.75	0.075	0.095	0.118	0.150	. . .
		0.8	0.080	0.100	0.125	0.160	0.200
5.6	11.2	0.75	0.085	0.106	0.132	0.170	. . .
		1	0.095	0.118	0.150	0.190	0.236
		1.25	0.100	0.125	0.160	0.200	0.250
		1.5	0.112	0.140	0.180	0.224	0.280
11.2	22.4	1	0.100	0.125	0.160	0.200	0.250
		1.25	0.112	0.140	0.180	0.224	0.280
		1.5	0.118	0.150	0.190	0.236	0.300
		1.75	0.125	0.160	0.200	0.250	0.315
		2	0.132	0.170	0.212	0.265	0.335
		2.5	0.140	0.180	0.224	0.280	0.355
22.4	45	1	0.106	0.132	0.170	0.212	. . .
		1.5	0.125	0.160	0.200	0.250	0.315
		2	0.140	0.180	0.224	0.280	0.355
		3	0.170	0.212	0.265	0.335	0.425
		3.5	0.180	0.224	0.280	0.355	0.450
		4	0.190	0.236	0.300	0.375	0.475
		4.5	0.200	0.250	0.315	0.400	0.500
45	90	1.5	0.132	0.170	0.212	0.265	0.335
		2	0.150	0.190	0.236	0.300	0.375
		3	0.180	0.224	0.280	0.355	0.450
		4	0.200	0.250	0.315	0.400	0.500
		5	0.212	0.265	0.335	0.425	0.530
		5.5	0.224	0.280	0.355	0.450	0.560
		6	0.236	0.300	0.375	0.475	0.600
90	180	2	0.160	0.200	0.250	0.315	0.400
		3	0.190	0.236	0.300	0.375	0.475
		4	0.212	0.265	0.335	0.425	0.530
		6	0.250	0.315	0.400	0.500	0.630
180	355	3	0.212	0.265	0.335	0.425	0.530
		4	0.236	0.300	0.375	0.475	0.600
		6	0.265	0.335	0.425	0.530	0.670

All dimensions are in millimeters.

Table 9. American National Standard Pitch Diameter Tolerances of External Metric Thread, T_{d2} (ISO 965/1) (ANSI B1.13M-1983)

Basic Major Diameter, d		Pitch P	Tolerance Grade						
Over	Up to and incl.		3	4	5	6	7	8	9
1.5	2.8	0.2	0.025	0.032	0.040	0.050	...	...	...
		0.25	0.028	0.036	0.045	0.056	...	...	...
		0.35	0.032	0.040	0.050	0.063	0.080	...	...
		0.4	0.034	0.042	0.053	0.067	0.085	...	...
		0.45	0.036	0.045	0.056	0.071	0.090	...	...
2.8	5.6	0.35	0.034	0.042	0.053	0.067	0.085	...	...
		0.5	0.038	0.048	0.060	0.075	0.095	...	...
		0.6	0.042	0.053	0.067	0.085	0.106	...	...
		0.7	0.045	0.056	0.071	0.090	0.112	...	...
		0.75	0.045	0.056	0.071	0.090	0.112	...	...
		0.8	0.048	0.060	0.075	0.095	0.118	0.150	0.190
5.6	11.2	0.75	0.050	0.063	0.080	0.100	0.125	...	...
		1	0.056	0.071	0.090	0.112	0.140	0.180	0.224
		1.25	0.060	0.075	0.095	0.118	0.150	0.190	0.236
		1.5	0.067	0.085	0.106	0.132	0.170	0.212	0.265
11.2	22.4	1	0.060	0.075	0.095	0.118	0.150	0.190	0.236
		1.25	0.067	0.085	0.106	0.132	0.170	0.212	0.265
		1.5	0.071	0.090	0.112	0.140	0.180	0.224	0.280
		1.75	0.075	0.095	0.118	0.150	0.190	0.236	0.300
		2	0.080	0.100	0.125	0.160	0.200	0.250	0.315
		2.5	0.085	0.106	0.132	0.170	0.212	0.265	0.335
22.4	45	1	0.063	0.080	0.100	0.125	0.160	0.200	0.250
		1.5	0.075	0.095	0.118	0.150	0.190	0.236	0.300
		2	0.085	0.106	0.132	0.170	0.212	0.265	0.335
		3	0.100	0.125	0.160	0.200	0.250	0.315	0.400
		3.5	0.106	0.132	0.170	0.212	0.265	0.335	0.425
		4	0.112	0.140	0.180	0.224	0.280	0.355	0.450
		4.5	0.118	0.150	0.190	0.236	0.300	0.375	0.475
45	90	1.5	0.080	0.100	0.125	0.160	0.200	0.250	0.315
		2	0.090	0.112	0.140	0.180	0.224	0.280	0.355
		3	0.106	0.132	0.170	0.212	0.265	0.335	0.425
		4	0.118	0.150	0.190	0.236	0.300	0.375	0.475
		5	0.125	0.160	0.200	0.250	0.315	0.400	0.500
		5.5	0.132	0.170	0.212	0.265	0.335	0.425	0.530
		6	0.140	0.180	0.224	0.280	0.355	0.450	0.560
90	180	2	0.095	0.118	0.150	0.190	0.236	0.300	0.375
		3	0.112	0.140	0.180	0.224	0.280	0.355	0.450
		4	0.125	0.160	0.200	0.250	0.315	0.400	0.500
		6	0.150	0.190	0.236	0.300	0.375	0.475	0.600
180	355	3	0.125	0.160	0.200	0.250	0.315	0.400	0.500
		4	0.140	0.180	0.224	0.280	0.355	0.450	0.560
		6	0.160	0.200	0.250	0.315	0.400	0.500	0.630

All dimensions are in millimeters.

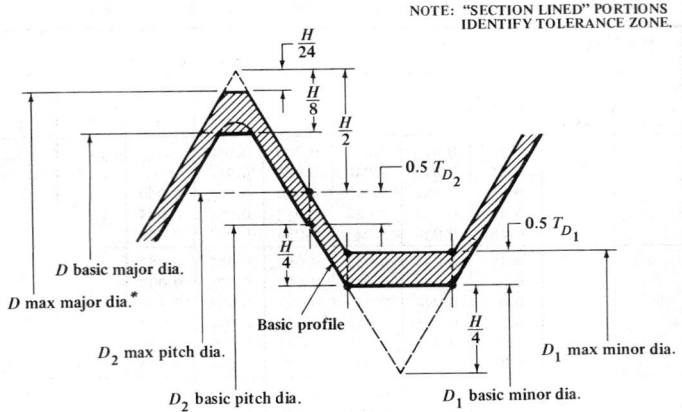

NOTE: "SECTION LINED" PORTIONS IDENTIFY TOLERANCE ZONE.

*This dimension is used in the design of tools, etc. In dimensioning internal threads it is not normally specified. Generally, major diameter acceptance is based on maximum material condition gaging.

Fig. 6. Internal Thread — Limiting M Profile, Tolerance Position H

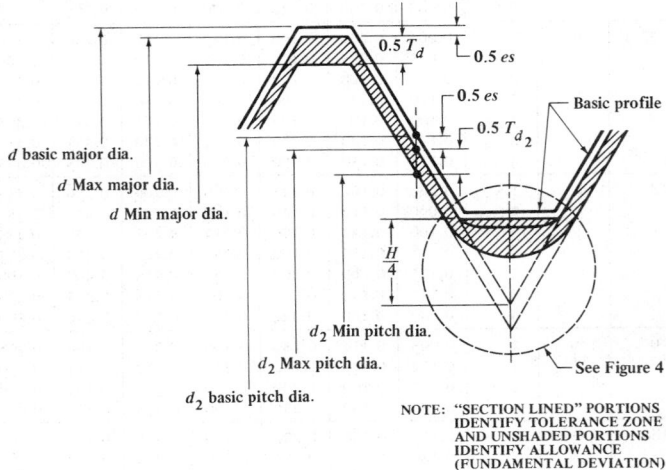

NOTE: "SECTION LINED" PORTIONS IDENTIFY TOLERANCE ZONE AND UNSHADED PORTIONS IDENTIFY ALLOWANCE (FUNDAMENTAL DEVIATION)

Fig. 7. External Thread — Limiting M Profile, Tolerance Position g

Table 10. American National Standard Major Diameter Tolerances of External Metric Threads, T_d (ISO 965/1) (ANSI B1.13M-1983)

Pitch P	Tolerance Grade			Pitch P	Tolerance Grade		
	4	6	8		4	6	8
0.2	0.036	0.056	. . .	1.25	0.132	0.212	0.335
0.25	0.042	0.067	. . .	1.5	0.150	0.236	0.375
0.3	0.048	0.075	. . .	1.75	0.170	0.265	0.425
0.35	0.053	0.085	. . .	2	0.180	0.280	0.450
0.4	0.060	0.095	. . .	2.5	0.212	0.335	0.530
0.45	0.063	0.100	. . .	3	0.236	0.375	0.600
0.5	0.067	0.106	. . .	3.5	0.265	0.425	0.670
0.6	0.080	0.125	. . .	4	0.300	0.475	0.750
0.7	0.090	0.140	. . .	4.5	0.315	0.500	0.800
0.75	0.090	0.140	. . .	5	0.335	0.530	0.850
0.8	0.095	0.150	0.236	5.5	0.355	0.560	0.900
1	0.112	0.180	0.280	6	0.375	0.600	0.950

All dimensions are in millimeters.

Designation for Identical Tolerance Classes: If the two tolerance class designations for a thread are identical, it is not necessary to repeat the symbols.

Example: M6 × 1 – 6H (21)

Designation Using All Capital Letters: When computer and teletype thread designations use all capital letters, the external or internal thread may need further identification. Thus the tolerance class is followed by the abbreviations EXT or INT in capital letters.

Examples: M6 × 1 – 4G6G EXT;   M6 × 1 – 6H INT

Designation for Thread Fit: A fit between mating threads is indicated by the internal thread tolerance class followed by the external thread tolerance class and separated by a slash.

Examples: M6 × 1 – 6H/6g;   M6 × 1 – 6H/4g6g

Designation for Rounded Root External Thread: The M profile with a minimum root radius of 0.125P on the external thread is desirable for all threads but is mandatory for threaded mechanical fasteners of ISO 898/I property class 8.8 (minimum tensile strength 800 MPa) and stronger. No special designation is required for these threads. Other parts requiring a 0.125P root radius must have that radius specified.

When a special rounded root is required, its external thread designation is suffixed by the minimum root radius value in millimeters and the letter R.

Example: M42 × 4.5 – 6g – 0.63R

Designation of Threads Having Modified Crests: Where the limits of size of the major diameter of an external thread or the minor diameter of an internal thread are modified, the thread designation is suffixed by the letters MOD followed by the modified diameter limits.

Examples: External thread M profile, major diameter reduced 0.075 mm.

M6 × 1 – 4h6h MOD

Major dia = 5.745 – 5.925 MOD

Internal thread M profile, minor diameter increased 0.075 mm.

M6 × 1 – 4H5H MOD

Minor dia = 5.101 – 5.291 MOD

(Continued on page 1544)

Table 11. Internal Metric Thread — M Profile Limiting Dimensions
(ANSI B1.13M-1983)

Basic Thread Designation	Toler. Class	Minor Diameter D_1		Pitch Diameter D_2			Major Diameter D	
		Min	Max	Min	Max	Tol	Min	Max†
M1.6 × 0.35	6H	1.221	1.321	1.373	1.458	0.085	1.600	1.736
M2 × 0.4	6H	1.567	1.679	1.740	1.830	0.090	2.000	2.148
M2.5 × 0.45	6H	2.013	2.138	2.208	2.303	0.095	2.500	2.660
M3 × 0.5	6H	2.459	2.599	2.675	2.775	0.100	3.000	3.172
M3.5 × 0.6	6H	2.850	3.010	3.110	3.222	0.112	3.500	3.699
M4 × 0.7	6H	3.242	3.422	3.545	3.663	0.118	4.000	4.219
M5 × 0.8	6H	4.134	4.334	4.480	4.605	0.125	5.000	5.240
M6 × 1	6H	4.917	5.153	5.350	5.500	0.150	6.000	6.294
M8 × 1.25	6H	6.647	6.912	7.188	7.348	0.160	8.000	8.340
M8 × 1	6H	6.917	7.153	7.350	7.500	0.150	8.000	8.294
M10 × 1.5	6H	8.376	8.676	9.026	9.206	0.180	10.000	10.396
M10 × 1.25	6H	8.647	8.912	9.188	9.348	0.160	10.000	10.340
M10 × 0.75	6H	9.188	9.378	9.513	9.645	0.132	10.000	10.240
M12 × 1.75	6H	10.106	10.441	10.863	11.063	0.200	12.000	12.453
M12 × 1.5	6H	10.376	10.676	11.026	11.216	0.190	12.000	12.406
M12 × 1.25	6H	10.647	10.912	11.188	11.368	0.180	12.000	12.360
M12 × 1	6H	10.917	11.153	11.350	11.510	0.160	12.000	12.304
M14 × 2	6H	11.835	12.210	12.701	12.913	0.212	14.000	14.501
M14 × 1.5	6H	12.376	12.676	13.026	13.216	0.190	14.000	14.406
M15 × 1	6H	13.917	14.153	14.350	14.510	0.160	15.000	15.304
M16 × 2	6H	13.835	14.210	14.701	14.913	0.212	16.000	16.501
M16 × 1.5	6H	14.376	14.676	15.026	15.216	0.190	16.000	16.406
M17 × 1	6H	15.917	16.153	16.350	16.510	0.160	17.000	17.304
M18 × 1.5	6H	16.376	16.676	17.026	17.216	0.190	18.000	18.406
M20 × 2.5	6H	17.294	17.744	18.376	18.600	0.224	20.000	20.585
M20 × 1.5	6H	18.376	18.676	19.026	19.216	0.190	20.000	20.406
M20 × 1	6H	18.917	19.153	19.350	19.510	0.160	20.000	20.304
M22 × 2.5	6H	19.294	19.744	20.376	20.600	0.224	22.000	22.585
M22 × 1.5	6H	20.376	20.676	21.026	21.216	0.190	22.000	22.406
M24 × 3	6H	20.752	21.252	22.051	22.316	0.265	24.000	24.698
M24 × 2	6H	21.835	22.210	22.701	22.925	0.224	24.000	24.513
M25 × 1.5	6H	23.376	23.676	24.026	24.226	0.200	25.000	25.416
M27 × 3	6H	23.752	24.252	25.051	25.316	0.265	27.000	27.698
M27 × 2	6H	24.835	25.210	25.701	25.925	0.224	27.000	27.513
M30 × 3.5	6H	26.211	26.771	27.727	28.007	0.280	30.000	30.785
M30 × 2	6H	27.835	28.210	28.701	28.925	0.224	30.000	30.513
M30 × 1.5	6H	28.376	28.676	29.026	29.226	0.200	30.000	30.416
M33 × 2	6H	30.835	31.210	31.701	31.925	0.224	33.000	33.513
M35 × 1.5	6H	33.376	33.676	34.026	34.226	0.200	35.000	35.416
M36 × 4	6H	31.670	32.270	33.402	33.702	0.300	36.000	36.877

All dimensions are in millimeters.
† This reference dimension is used in design of tools, etc., and in dimensioning internal threads is not normally specified. Generally, major diameter acceptance is based upon maximum material condition gaging.

Table 11 *(Concluded).* **Internal Metric Thread — M Profile Limiting Dimensions**
(ANSI B1.13M-1983)

Basic Thread Designation	Toler. Class	Minor Diameter D_1		Pitch Diameter D_2			Major Diameter D	
		Min	Max	Min	Max	Tol	Min	Max†
M36 × 2	6H	33.835	34.210	34.701	34.925	0.224	36.000	36.513
M39 × 2	6H	36.835	37.210	37.701	37.925	0.224	39.000	39.513
M40 × 1.5	6H	38.376	38.676	39.026	39.226	0.200	40.000	40.416
M42 × 4.5	6H	37.129	37.799	39.077	39.392	0.315	42.000	42.965
M42 × 2	6H	39.835	40.210	40.701	40.925	0.224	42.000	42.513
M45 × 1.5	6H	43.376	43.676	44.026	44.226	0.200	45.000	45.416
M48 × 5	6H	42.587	43.297	44.752	45.087	0.335	48.000	49.057
M48 × 2	6H	45.835	46.210	46.701	46.937	0.236	48.000	48.525
M50 × 1.5	6H	48.376	48.676	49.026	49.238	0.212	50.000	50.428
M55 × 1.5	6H	53.376	53.676	54.026	54.238	0.212	55.000	55.428
M56 × 5.5	6H	50.046	50.796	52.428	52.783	0.355	56.000	57.149
M56 × 2	6H	53.835	54.210	54.701	54.937	0.236	56.000	56.525
M60 × 1.5	6H	58.376	58.676	59.026	59.238	0.212	60.000	60.428
M64 × 6	6H	57.505	58.305	60.103	60.478	0.375	64.000	65.241
M64 × 2	6H	61.835	62.210	62.701	62.937	0.236	64.000	64.525
M65 × 1.5	6H	63.376	63.676	64.026	64.238	0.212	65.000	65.428
M70 × 1.5	6H	68.376	68.676	69.026	69.238	0.212	70.000	70.428
M72 × 6	6H	65.505	66.305	68.103	68.478	0.375	72.000	73.241
M72 × 2	6H	69.835	70.210	70.701	70.937	0.236	72.000	72.525
M75 × 1.5	6H	73.376	73.676	74.026	74.238	0.212	75.000	75.428
M80 × 6	6H	73.505	74.305	76.103	76.478	0.375	80.000	81.241
M80 × 2	6H	77.835	78.210	78.701	78.937	0.236	80.000	80.525
M80 × 1.5	6H	78.376	78.676	79.026	79.238	0.212	80.000	80.428
M85 × 2	6H	82.835	83.210	83.701	83.937	0.236	85.000	85.525
M90 × 6	6H	83.505	84.305	86.103	86.478	0.375	90.000	91.241
M90 × 2	6H	87.835	88.210	88.701	88.937	0.236	90.000	90.525
M95 × 2	6H	92.835	93.210	93.701	93.951	0.250	95.000	95.539
M100 × 6	6H	93.505	94.305	96.103	96.503	0.400	100.000	101.266
M100 × 2	6H	97.835	98.210	98.701	98.951	0.250	100.000	100.539
M105 × 2	6H	102.835	103.210	103.701	103.951	0.250	105.000	105.539
M110 × 2	6H	107.835	108.210	108.701	108.951	0.250	110.000	110.539
M120 × 2	6H	117.835	118.210	118.701	118.951	0.250	120.000	120.539
M130 × 2	6H	127.835	128.210	128.701	128.951	0.250	130.000	130.539
M140 × 2	6H	137.835	138.210	138.701	138.951	0.250	140.000	140.539
M150 × 2	6H	147.835	148.210	148.701	148.951	0.250	150.000	150.539
M160 × 3	6H	156.752	157.252	158.051	158.351	0.300	160.000	160.733
M170 × 3	6H	166.752	167.252	168.051	168.351	0.300	170.000	170.733
M180 × 3	6H	176.752	177.252	178.051	178.351	0.300	180.000	180.733
M190 × 3	6H	186.752	187.252	188.051	188.386	0.335	190.000	190.768
M200 × 3	6H	196.752	197.252	198.051	198.386	0.335	200.000	200.768

All dimensions are in millimeters.
† This reference dimension is used in design of tools, etc., and is not normally specified.
Generally, major diameter acceptance is based upon maximum material condition gaging.

Table 12. External Metric Thread — M Profile Limiting Dimensions
(ANSI B1.13M-1983)

Basic Thread Desig.	Toler. Class	Allow. es**	Major Diam.* d		Pitch Diam.* d2			Minor Diam., d1*	Minor Diam., d3†
			Max	Min	Max	Min	Tol.	Max	Min
M1.6 × 0.35	6g	0.019	1.581	1.496	1.354	1.291	0.063	1.202	1.075
M1.6 × 0.35	4g6g	0.019	1.581	1.496	1.354	1.314	0.040	1.202	1.098
M2 × 0.4	6g	0.019	1.981	1.886	1.721	1.654	0.067	1.548	1.408
M2 × 0.4	4g6g	0.019	1.981	1.886	1.721	1.679	0.042	1.548	1.433
M2.5 × 0.45	6g	0.020	2.480	2.380	2.188	2.117	0.071	1.993	1.840
M2.5 × 0.45	4g6g	0.020	2.480	2.380	2.188	2.143	0.045	1.993	1.866
M3 × 0.5	6g	0.020	2.980	2.874	2.655	2.580	0.075	2.439	2.272
M3 × 0.5	4g6g	0.020	2.980	2.874	2.655	2.607	0.048	2.439	2.299
M3.5 × 0.6	6g	0.021	3.479	3.354	3.089	3.004	0.085	2.829	2.635
M3.5 × 0.6	4g6g	0.021	3.479	3.354	3.089	3.036	0.053	2.829	2.667
M4 × 0.7	6g	0.022	3.978	3.838	3.523	3.433	0.090	3.220	3.002
M4 × 0.7	4g6g	0.022	3.978	3.838	3.523	3.467	0.056	3.220	3.036
M5 × 0.8	6g	0.024	4.976	4.826	4.456	4.361	0.095	4.110	3.869
M5 × 0.8	4g6g	0.024	4.976	4.826	4.456	4.396	0.060	4.110	3.904
M6 × 1	6g	0.026	5.974	5.794	5.324	5.212	0.112	4.891	4.596
M6 × 1	4g6g	0.026	5.974	5.794	5.324	5.253	0.071	4.891	4.637
M8 × 1.25	6g	0.028	7.972	7.760	7.160	7.042	0.118	6.619	6.272
M8 × 1.25	4g6g	0.028	7.972	7.760	7.160	7.085	0.075	6.619	6.315
M8 × 1	6g	0.026	7.974	7.794	7.324	7.212	0.112	6.891	6.596
M8 × 1	4g6g	0.026	7.974	7.794	7.324	7.253	0.071	6.891	6.637
M10 × 1.5	6g	0.032	9.968	9.732	8.994	8.862	0.132	8.344	7.938
M10 × 1.5	4g6g	0.032	9.968	9.732	8.994	8.909	0.085	8.344	7.985
M10 × 1.25	6g	0.028	9.972	9.760	9.160	9.042	0.118	8.619	8.272
M10 × 1.25	4g6g	0.028	9.972	9.760	9.160	9.085	0.075	8.619	8.315
M10 × 0.75	6g	0.022	9.978	9.838	9.491	9.391	0.100	9.166	8.929
M10 × 0.75	4g6g	0.022	9.978	9.838	9.491	9.428	0.063	9.166	8.966
M12 × 1.75	6g	0.034	11.966	11.701	10.829	10.679	0.150	10.072	9.601
M12 × 1.75	4g6g	0.034	11.966	11.701	10.829	10.734	0.095	10.072	9.656
M12 × 1.5	6g	0.032	11.968	11.732	10.994	10.854	0.140	10.344	9.930
M12 × 1.25	6g	0.028	11.972	11.760	11.160	11.028	0.132	10.619	10.258
M12 × 1.25	4g6g	0.028	11.972	11.760	11.160	11.075	0.085	10.619	10.305
M12 × 1	6g	0.026	11.974	11.794	11.324	11.206	0.118	10.891	10.590
M12 × 1	4g6g	0.026	11.974	11.794	11.324	11.249	0.075	10.891	10.633
M14 × 2	6g	0.038	13.962	13.682	12.663	12.503	0.160	11.797	11.271
M14 × 2	4g6g	0.038	13.962	13.682	12.663	12.563	0.100	11.797	11.331
M14 × 1.5	6g	0.032	13.968	13.732	12.994	12.854	0.140	12.344	11.930
M14 × 1.5	4g6g	0.032	13.968	13.732	12.994	12.904	0.090	12.344	11.980
M15 × 1	6g	0.026	14.974	14.794	14.324	14.206	0.118	13.891	13.590
M15 × 1	4g6g	0.026	14.974	14.794	14.324	14.249	0.075	13.891	13.633
M16 × 2	6g	0.038	15.962	15.682	14.663	14.503	0.160	13.797	13.271
M16 × 2	4g6g	0.038	15.962	15.682	14.663	14.563	0.100	13.797	13.331

All dimensions are in millimeters. For footnotes see end of table.

Table 12 (*Continued*). **External Metric Thread — M Profile Limiting Dimensions**
(ANSI B1.13M-1983)

Basic Thread Desig.	Toler. Class	Allow. es**	Major Diam.* d		Pitch Diam.* d₂			Minor Diam., d₁*	Minor Diam., d₃†
			Max	Min	Max	Min	Tol.	Max	Min
M16 × 1.5	6g	0.032	15.968	15.732	14.994	14.854	0.140	14.344	13.930
M16 × 1.5	4g6g	0.032	15.968	15.732	14.994	14.904	0.090	14.344	13.980
M17 × 1	6g	0.026	16.974	16.794	16.324	16.206	0.118	15.891	15.590
M17 × 1	4g6g	0.026	16.974	16.794	16.324	16.249	0.075	15.891	15.633
M18 × 1.5	6g	0.032	17.968	17.732	16.994	16.854	0.140	16.344	15.930
M18 × 1.5	4g6g	0.032	17.968	17.732	16.994	16.904	0.090	16.344	15.980
M20 × 2.5	6g	0.042	19.958	19.623	18.334	18.164	0.170	17.252	16.624
M20 × 2.5	4g6g	0.042	19.958	19.623	18.334	18.228	0.106	17.252	16.688
M20 × 1.5	6g	0.032	19.968	19.732	18.994	18.854	0.140	18.344	17.930
M20 × 1.5	4g6g	0.032	19.968	19.732	18.994	18.904	0.090	18.344	17.980
M20 × 1	6g	0.026	19.974	19.794	19.324	19.206	0.118	18.891	18.590
M20 × 1	4g6g	0.026	19.974	19.794	19.324	19.249	0.075	18.891	18.633
M22 × 2.5	6g	0.042	21.958	21.623	20.334	20.164	0.170	19.252	18.624
M22 × 1.5	6g	0.032	21.968	21.732	20.994	20.854	0.140	20.344	19.930
M22 × 1.5	4g6g	0.032	21.968	21.732	20.994	20.904	0.090	20.344	19.980
M24 × 3	6g	0.048	23.952	23.577	22.003	21.803	0.200	20.704	19.955
M24 × 3	4g6g	0.048	23.952	23.577	22.003	21.878	0.125	20.704	20.030
M24 × 2	6g	0.038	23.962	23.682	22.663	22.493	0.170	21.797	21.261
M24 × 2	4g6g	0.038	23.962	23.682	22.663	22.557	0.106	21.797	21.325
M25 × 1.5	6g	0.032	24.968	24.732	23.994	23.844	0.150	23.344	22.920
M25 × 1.5	4g6g	0.032	24.968	24.732	23.994	23.899	0.095	23.344	22.975
M27 × 3	6g	0.048	26.952	26.577	25.003	24.803	0.200	23.704	22.955
M27 × 2	6g	0.038	26.962	26.682	25.663	25.493	0.170	24.797	24.261
M27 × 2	4g6g	0.038	29.962	26.682	25.663	25.557	0.106	24.797	24.325
M30 × 3.5	6g	0.053	29.947	29.522	27.674	27.462	0.212	26.158	25.306
M30 × 3.5	4g6g	0.053	29.947	29.522	27.674	27.542	0.132	26.158	25.386
M30 × 2	6g	0.038	29.962	29.682	28.663	28.493	0.170	27.797	27.261
M30 × 2	4g6g	0.038	29.962	29.682	28.663	28.557	0.106	27.797	27.325
M30 × 1.5	6g	0.032	29.968	29.732	28.994	28.844	0.150	28.344	27.920
M30 × 1.5	4g6g	0.032	29.968	29.732	28.994	28.899	0.095	28.344	27.975
M33 × 2	6g	0.038	32.962	32.682	31.663	31.493	0.170	30.797	30.261
M33 × 2	4g6g	0.038	32.962	32.682	31.663	31.557	0.106	30.797	30.325
M35 × 1.5	6g	0.032	34.968	34.732	33.994	33.844	0.150	33.344	33.920
M36 × 4	6g	0.060	35.940	35.465	33.342	33.118	0.224	31.610	30.654
M36 × 4	4g6g	0.060	35.940	35.465	33.342	33.202	0.140	31.610	30.738
M36 × 2	6g	0.038	35.962	35.682	34.663	34.493	0.170	33.797	33.261
M36 × 2	4g6g	0.038	35.962	35.682	34.663	34.557	0.106	33.797	33.325
M39 × 2	6g	0.038	38.962	38.682	37.663	37.493	0.170	36.797	36.261
M39 × 2	4g6g	0.038	38.962	38.682	37.663	37.557	0.106	36.797	36.325
M40 × 1.5	6g	0.032	39.968	39.732	38.994	38.844	0.150	38.344	37.920
M40 × 1.5	4g6g	0.032	39.968	39.732	38.994	38.899	0.095	38.344	37.975

All dimensions are in millimeters.
For footnotes see end of table.

Table 12 (*Continued*). **External Metric Thread — M Profile Limiting Dimensions**
(ANSI B1.13M-1983)

Basic Thd. Desig.	Toler. Class	Allow. *es***	Major Diam.* d		Pitch Diam.* d_2			Minor Diam., d_1*	Minor Diam., d_3†
			Max	Min	Max	Min	Tol.	Max	Min
M42 × 4.5	6g	0.063	41.937	41.437	39.014	38.778	0.236	37.066	36.006
M42 × 4.5	4g6g	0.063	41.937	41.437	39.014	38.864	0.150	37.066	36.092
M42 × 2	6g	0.038	41.962	41.682	40.663	40.493	0.170	39.797	39.261
M42 × 2	4g6g	0.038	41.962	41.682	40.663	40.557	0.106	39.797	39.325
M45 × 1.5	6g	0.032	44.968	44.732	43.994	43.844	0.150	43.344	42.920
M45 × 1.5	4g6g	0.032	44.968	44.732	43.994	43.899	0.095	43.344	42.975
M48 × 5	6g	0.071	47.929	47.399	44.681	44.431	0.250	42.516	41.351
M48 × 5	4g6g	0.071	47.929	47.399	44.681	44.521	0.160	42.516	41.441
M48 × 2	6g	0.038	47.962	47.682	46.663	46.483	0.180	45.797	45.251
M48 × 2	4g6g	0.038	47.962	47.682	46.663	46.551	0.112	45.797	45.319
M50 × 1.5	6g	0.032	49.968	49.732	48.994	48.834	0.160	48.344	47.910
M50 × 1.5	4g6g	0.032	49.968	49.732	48.994	48.894	0.100	48.344	47.970
M55 × 1.5	6g	0.032	54.968	54.732	53.994	53.834	0.160	53.344	52.910
M55 × 1.5	4g6g	0.032	54.968	54.732	53.994	53.894	0.100	53.344	52.970
M56 × 5.5	6g	0.075	55.925	55.365	52.353	52.088	0.265	49.971	48.700
M56 × 5.5	4g6g	0.075	55.925	55.365	52.353	52.183	0.170	49.971	48.795
M56 × 2	6g	0.038	55.962	55.682	54.663	54.483	0.180	53.797	53.251
M56 × 2	4g6g	0.038	55.962	55.682	54.663	54.551	0.112	53.797	53.319
M60 × 1.5	6g	0.032	59.968	59.732	58.994	58.834	0.160	58.344	57.910
M60 × 1.5	4g6g	0.032	59.968	59.732	58.994	58.894	0.100	58.344	57.970
M64 × 6	6g	0.080	63.920	63.320	60.023	59.743	0.280	57.425	56.047
M64 × 6	4g6g	0.080	63.920	63.320	60.023	59.843	0.180	57.425	56.147
M64 × 2	6g	0.038	63.962	63.682	62.663	62.483	0.180	61.797	61.251
M64 × 2	4g6g	0.038	63.962	63.682	62.663	62.551	0.112	61.797	61.319
M65 × 1.5	6g	0.032	64.968	64.732	63.994	63.834	0.160	63.344	62.910
M65 × 1.5	4g6g	0.032	64.968	64.732	63.994	63.894	0.100	63.344	62.970
M70 × 1.5	6g	0.032	69.968	69.732	68.994	68.834	0.160	68.344	67.910
M70 × 1.5	4g6g	0.032	69.968	69.732	68.994	68.894	0.100	68.344	67.970
M72 × 6	6g	0.080	71.920	71.320	68.023	67.743	0.280	65.425	64.047
M72 × 6	4g6g	0.080	71.920	71.320	68.023	67.843	0.180	65.425	64.147
M72 × 2	6g	0.038	71.962	71.682	70.663	70.483	0.180	69.797	69.251
M72 × 2	4g6g	0.038	71.962	71.682	70.663	70.551	0.112	69.797	69.319
M75 × 1.5	6g	0.032	74.968	74.732	73.994	73.834	0.160	73.344	72.910
M75 × 1.5	4g6g	0.032	74.968	74.732	73.994	73.894	0.100	73.344	72.970
M80 × 6	6g	0.080	79.920	79.320	76.023	75.743	0.280	73.425	72.047
M80 × 6	4g6g	0.080	79.920	79.320	76.023	75.843	0.180	73.425	72.147
M80 × 2	6g	0.038	79.962	79.682	78.663	78.483	0.180	77.797	77.251
M80 × 2	4g6g	0.038	79.962	79.682	78.663	78.551	0.112	77.797	77.319
M80 × 1.5	6g	0.032	79.968	79.732	78.994	78.834	0.160	78.344	77.910
M80 × 1.5	4g6g	0.032	79.968	79.732	78.994	78.894	0.100	78.344	77.970
M85 × 2	6g	0.038	84.962	84.682	83.663	83.483	0.180	82.797	82.251
M85 × 2	4g6g	0.038	84.962	84.682	83.663	83.551	0.112	82.797	82.319

All dimensions are in millimeters.
For footnotes see end of table.

Table 12 (*Concluded*). **External Metric Thread — M Profile Limiting Dimensions**
(ANSI B1.13M-1983)

Basic Thread Desig.	Toler. Class	Allow. es**	Major Diam.* d		Pitch Diam.* d2			Minor Diam., d1*	Minor Diam., d3†
			Max	Min	Max	Min	Tol.	Max	Min
M90 × 6	6g	0.080	89.920	89.320	86.023	85.743	0.280	83.425	82.047
M90 × 6	4g6g	0.080	89.920	89.320	86.023	85.843	0.180	83.425	82.147
M90 × 2	6g	0.038	89.962	89.682	88.663	88.483	0.180	87.797	87.251
M90 × 2	4g6g	0.038	89.962	89.682	88.663	88.551	0.112	87.797	87.319
M95 × 2	6g	0.038	94.962	94.682	93.663	93.473	0.190	92.797	92.241
M95 × 2	4g6g	0.038	94.962	94.682	93.663	93.545	0.118	92.797	92.313
M100 × 6	6g	0.080	99.920	99.320	96.023	95.723	0.300	93.425	92.027
M100 × 6	4g6g	0.080	99.920	99.320	96.023	95.833	0.190	93.425	92.137
M100 × 2	6g	0.038	99.962	99.682	98.663	98.473	0.190	97.797	97.241
M100 × 2	4g6g	0.038	99.962	99.682	98.663	98.545	0.118	97.797	97.313
M105 × 2	6g	0.038	104.962	104.682	103.663	103.473	0.190	102.797	102.241
M105 × 2	4g6g	0.038	104.962	104.682	103.663	103.545	0.118	102.797	102.313
M110 × 2	6g	0.038	109.962	109.682	108.663	108.473	0.190	107.797	107.241
M110 × 2	4g6g	0.038	109.962	109.682	108.663	108.545	0.118	107.797	107.313
M120 × 2	6g	0.038	119.962	119.682	118.663	118.473	0.190	117.797	117.241
M120 × 2	4g6g	0.038	119.962	119.682	118.663	118.545	0.118	117.797	117.313
M130 × 2	6g	0.038	129.962	129.682	128.663	128.473	0.190	127.797	127.241
M130 × 2	4g6g	0.038	129.962	129.682	128.663	128.545	0.118	127.797	127.313
M140 × 2	6g	0.038	139.962	139.682	138.663	138.473	0.190	137.797	137.241
M140 × 2	4g6g	0.038	139.962	139.682	138.663	138.545	0.118	137.797	137.313
M150 × 2	6g	0.038	149.962	149.682	148.663	148.473	0.190	147.797	147.241
M150 × 2	4g6g	0.038	149.962	149.682	148.663	148.545	0.118	147.797	147.313
M160 × 3	6g	0.048	159.952	159.577	158.003	157.779	0.224	156.704	155.931
M160 × 3	4g6g	0.048	159.952	159.577	158.003	157.863	0.140	156.704	156.015
M170 × 3	6g	0.048	169.952	169.577	168.003	167.779	0.224	166.704	165.931
M170 × 3	4g6g	0.048	169.952	169.577	168.003	167.863	0.140	166.704	166.015
M180 × 3	6g	0.048	179.952	179.577	178.003	177.779	0.224	176.704	175.931
M180 × 3	4g6g	0.048	179.952	179.577	178.003	177.863	0.140	176.704	176.015
M190 × 3	6g	0.048	189.952	189.577	188.003	187.753	0.250	186.704	185.905
M190 × 3	4g6g	0.048	189.952	189.577	188.003	187.843	0.160	186.704	185.995
M200 × 3	6g	0.048	199.952	199.577	198.003	197.753	0.250	196.704	195.905
M200 × 3	4g6g	0.048	199.952	199.577	198.003	197.843	0.160	196.704	195.995

All dimensions are in millimeters.

* (Flat form) For screw threads at maximum limits of tolerance position *h*, add the absolute value *es* to the maximum diameters required. For maximum major diameter this value is the basic thread size listed in Table 11 as Minimum Major Diameter (D_{min}); for maximum pitch diameter this value is the same as listed in Table 11 as Minimum Pitch Diameter ($D_{2\ min}$); and for maximum minor diameter this value is the same as listed in Table 11 as Minimum Minor Diameter ($D_{1\ min}$).

** *es* is an absolute value.

† (Rounded form) This reference dimension is used in the design of tools, etc. In dimensioning external threads it is not normally specified. Generally minor diameter acceptance is based upon maximum material condition gaging.

Designation of Special Threads: Special diameter-pitch threads developed in accordance with this Standard (ANSI B1.13M) are identified by the letters SPL following the tolerance class. The limits of size for the major diameter, pitch diameter, and minor diameter are specified below this designation.

Examples: External thread.

$$M6.5 \times 1 - 4h6h - SPL (22)$$
Major dia = 6.320 − 6.500
Pitch dia = 5.779 − 5.850
Minor dia = 5.163 − 5.386

Internal thread.

$$M6.5 \times 1 - 4H5H - SPL (23)$$
Major dia = 6.500 min
Pitch dia = 5.850 − 5.945
Minor dia = 5.417 − 5.607

Designation of Multiple Start Threads: When a thread is required with a multiple start, it is designated by specifying sequentially: M for metric thread, nominal diameter size, × L for lead, lead value, dash, P for pitch, pitch value, dash, tolerance class, parenthesis, script number of starts, and the word starts, close parenthesis.

Examples: M16 × L4 − P2 − 4h6h (TWO STARTS)

M14 × L6 − P2 − 6H (THREE STARTS)

Designation of Coated or Plated Threads: In designating coated or plated M threads the tolerance class should be specified as after coating or after plating. If no designation of after coating or after plating is specified, the tolerance class applies before coating or plating in accordance with ISO practice. After plating, the thread must not transgress the maximum material limits for the tolerance position H/h.

Examples: M6 × 1 − 6h AFTER COATING or AFTER PLATING

M6 × 1 − 6g AFTER COATING or AFTER PLATING

Where the tolerance position G/g is insufficient relief for the application to hold the threads within product limits, the coating or plating allowance may be specified as the maximum and minimum limits of size for minor and pitch diameters of internal threads or major and pitch diameters for external threads before coating or plating.

Example: Allowance on external thread M profile based on 0.010 mm min coating thickness.

$$M6 \times 1 - 4h6h - AFTER COATING$$

BEFORE COATING
Major dia = 5.780 − 5.940
Pitch dia = 5.239 − 5.290

Metric Screw Threads — MJ Profile. — The MJ screw thread is intended for aerospace metric threaded parts and for other highly stressed applications requiring high fatigue strength, or for "no allowance" applications. The MJ profile thread is a hard metric version similar to the UNJ inch, MIL-S-8879, which has a $0.15P$ to $0.18P$ controlled root radius in the external thread and the internal thread minor diameter truncated to accommodate the external thread maximum root radius.

The American National Standard ANSI B1.21M-1978 establishes the basic triangular profile for the MJ form of thread; gives a system of designations; lists the standard series of diameter-pitch combinations for diameters from 1.6 to 200 mm; and specifies limiting dimensions and tolerances.

Diameter-Pitch Combinations: This Standard includes a selected series of diameter-pitch combinations of threads taken from International Standard ISO 261 plus some additional sizes in the constant pitch series. It also includes the standard series of diameter-pitch combinations for aerospace screws, bolts, and nuts as shown below.

American National Standard Thread Series for Aerospace Screws, Bolts, and Nuts
(ANSI B1.21M-1978)

Nom. Size	Pitch	Nom. Size	Pitch	Nom. Size	Pitch	Nom. Size	Pitch
1.6	0.35	5	0.8	14	1.5	27	2
2	0.4	6	1	16	1.5	30	2
2.5	0.45	7	1	18	1.5	33	2
3	0.5	8	1	20	1.5	36	2
3.5	0.6	10	1.25	22	1.5	39	2
4	0.7	12	1.25	24	2	...	...

All dimensions are in millimeters.

Tolerances: One tolerance class, 4h6h, is specified in this Standard for all sizes of external threads after processing, including coating or electroplating. The tolerance position h provides no allowance. The pitch diameter tolerance is grade 4 and the major diameter tolerance is grade 6. For coated or plated external threads having pitches of 2 mm or smaller, the tolerance class before processing is applied is 4g6g. The tolerance position g provides an allowance for coating or plating only. For pitches larger than 2 mm, special allowances are provided.

For internal threads, after all processing including coating or plating has been completed, tolerance class 4H6H is specified for sizes 1 through 5 mm and 4H5H for sizes 6 mm and larger. The tolerance position H provides no allowance. The pitch diameter tolerance is grade 4 for all sizes and the minor diameter tolerance is grade 6 for the 5 mm size and smaller and grade 5 for the 6 mm size and larger. For coated or plated internal threads having pitches 2 mm or smaller, the tolerance class is 4G6G for sizes 1 through 5 mm and 4G5G for sizes 6 mm and larger. The tolerance position G provides an allowance for coating or plating.

The above class tolerances are positive for internal threads and negative for external threads, that is, in the direction of minimum material.

Symbols: Standard symbols appearing in the following diagrams are:
D = Basic major diameter of internal thread
D_2 = Basic pitch diameter of internal thread
D_1 = Basic minor diameter of internal thread
d = Basic major diameter of external thread
d_2 = Basic pitch diameter of external thread
d_1 = Basic minor diameter of internal thread
d_3 = Diameter to bottom of external thread root radius
H = Height of fundamental triangle P = Pitch

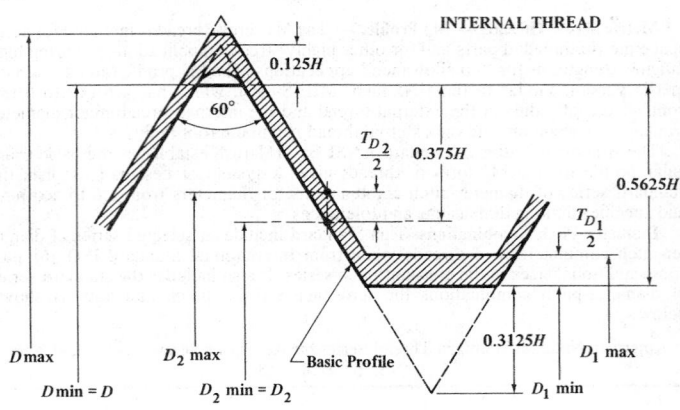

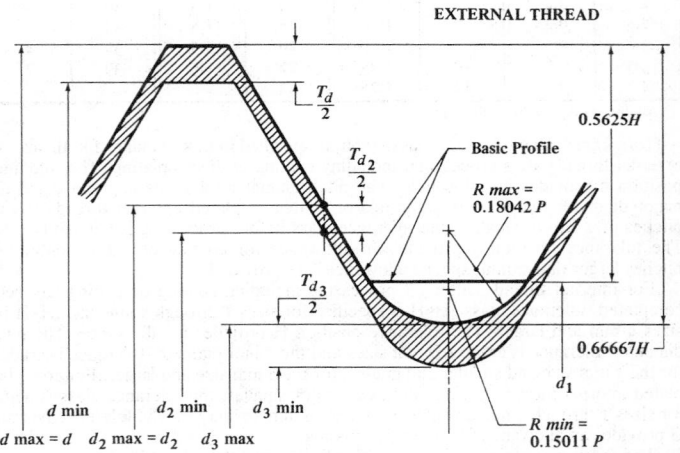

Internal MJ Thread Basic and Design Profiles (Above) and External MJ Thread Basic and Design Profiles (Below) Showing Tolerance Zones

Basic Designations: The aerospace metric screw thread is designated by the letters "MJ" to identify the metric J thread form, followed by the nominal size and pitch in millimeters (separated by the sign "×") and followed by the tolerance class (separated by a dash from the pitch). Unless otherwise specified in the designation, the thread helix is right hand. Example: MJ6 × 1 – 4h6h

For further details concerning limiting dimensions, allowances for coating and plating, modified and special threads, etc., reference should be made to the Standard.

American Standard for Unified Miniature Screw Threads. — This American Standard (B1.10-1958) introduces a new thread series to be known as Unified Miniature Screw Threads and intended for general purpose fastening screws and similar uses in watches, instruments, and miniature mechanisms. Use of this series is recommended on all new products in place of the many improvised and unsystematized sizes now in existence which have never achieved broad acceptance nor recognition by standardization bodies. The series covers a diameter range from 0.30 to 1.40 millimeters (0.0118 to 0.0551 inch) and thus supplements the Unified and American thread series which begins at 0.060 inch (number 0 of the machine screw series). It comprises a total of fourteen sizes which, together with their respective pitches, are those endorsed by the American-British-Canadian Conference of April 1955 as the basis for a Unified standard among the inch-using countries, and coincide with the corresponding range of sizes in ISO (International Organization for Standardization) Recommendation No. 68. Additionally, it utilizes thread forms which are compatible in all significant respects with both the Unified and ISO basic thread profiles. Thus, threads in this series are interchangeable with the corresponding sizes in both the American-British-Canadian and ISO standardization programs.

Basic Form of Thread: The basic profile by which the design forms of the threads covered by this standard are governed is shown in Table 1. The thread angle is 60 degrees and except for basic height and depth of engagement which are $0.52p$, instead of $0.54127p$, the basic profile for this thread standard is identical with the Unified and American basic thread form. The selection of 0.52 as the exact value of the coefficient for the height of this basic form is based on practical manufacturing considerations and a plan evolved to simplify calculations and achieve more precise agreement between the metric and inch dimensional tables.

Products made to this standard will be interchangeable with products made to other standards which allow a maximum depth of engagement (or combined addendum height) of $0.54127p$. The resulting difference is negligible (only 0.00025 inch for the coarsest pitch) and is completely offset by practical considerations in tapping, since internal thread heights exceeding $0.52p$ are avoided in these (Unified Miniature) small thread sizes in order to escape excessive tap breakage.

Design Forms of Threads: The design (maximum material) forms of the external and internal threads are shown in Table 2. These forms are derived from the basic profile shown in Table 1 by the application of clearances for the crests of the addenda at the roots of the mating dedendum forms. Basic and design form dimensions are given in Table 3.

Nominal Sizes: The thread sizes comprising this series and their respective pitches are shown in the first two columns of Table 5. The fourteen sizes shown in Table 5 have been systematically distributed to provide a uniformly proportioned selection over the entire range. They are separated alternately into two categories: The sizes shown in bold type are selections made in the interest of simplification and are those to which it is recommended that usage be confined wherever the circumstances of design permit. Where these sizes do not meet requirements the intermediate sizes shown in light type are available.

Limits of Size: Formulas used to determine limits of size are given in Table 4; the limits of size are given in Table 5. The diagram on page 1551 illustrates the limits of size and Table 6 gives values for the minimum flat at the root of the external thread shown on the diagram.

Classes of Threads: The standard establishes one class of thread with zero allowance on all diameters. When coatings of a measurable thickness are required, they should be included within the maximum material limits of the threads since these limits apply to both coated and uncoated threads.

Hole Sizes for Tapping: Suggested hole sizes are given in the Tapping Section.

Table 1. Unified Miniature Screw Threads — Basic Thread Form

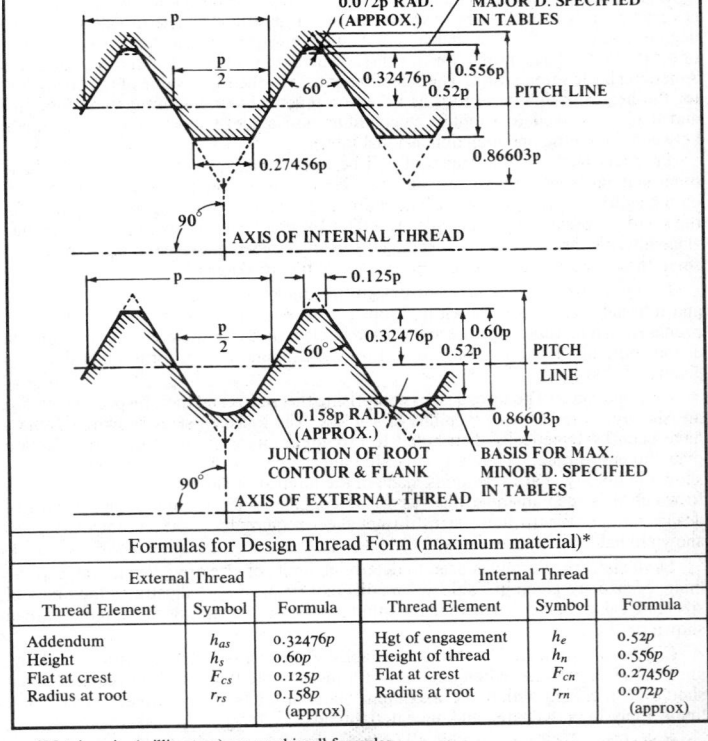

	Formulas for Basic Thread Form*		
	Thread Element	Symbol	Formula
	Angle of thread	2α	$60°$
	Half angle of thread	α	$30°$
	Pitch of thread	p	
	No. of threads per inch	n	$25.4/p$
	Height of sharp V thread	H	$0.86603p$
	Addendum of basic thread	h_{ab}	$0.32476p$
	Height of basic thread	h_b	$0.52p$

*Metric units (millimeters) are used in all formulas.

Table 2. Unified Miniature Screw Threads — Design Thread Form

Formulas for Design Thread Form (maximum material)*						
External Thread				Internal Thread		
Thread Element	Symbol	Formula		Thread Element	Symbol	Formula
Addendum	h_{as}	$0.32476p$		Hgt of engagement	h_e	$0.52p$
Height	h_s	$0.60p$		Height of thread	h_n	$0.556p$
Flat at crest	F_{cs}	$0.125p$		Flat at crest	F_{cn}	$0.27456p$
Radius at root	r_{rs}	$0.158p$ (approx)		Radius at root	r_{rn}	$0.072p$ (approx)

*Metric units (millimeters) are used in all formulas.

Table 3. Unified Miniature Screw Threads — Basic and Design Form Dimensions

Threads per inch $n*$	Pitch p	Height of Sharp V $H = 0.86603p$	Height $h_b = 0.52p$	Addendum $h_{ab} = h_{as} = 0.32476p$	Height $h_s = 0.60p$	Flat at Crest $F_{cs} = 0.125p$	Radius at Root $r_{rs} = 0.158p$	Height $h_n = 0.556p$	Flat at Crest $F_{cn} = 0.27456p$	Radius at Root $r_{rn} = 0.072p$
					Millimeter Dimensions					
...	.080	.0693	.0416	.0260	.048	.0100	.0126	.0445	.0220	.0058
...	.090	.0779	.0468	.0292	.054	.0112	.0142	.0500	.0247	.0065
...	.100	.0866	.0520	.0325	.060	.0125	.0158	.0556	.0275	.0072
...	.125	.1083	.0650	.0406	.075	.0156	.0198	.0695	.0343	.0090
...	.150	.1299	.0780	.0487	.090	.0188	.0237	.0834	.0412	.0108
...	.175	.1516	.0910	.0568	.105	.0219	.0277	.0973	.0480	.0126
...	.200	.1732	.1040	.0650	.120	.0250	.0316	.1112	.0549	.0144
...	.225	.1949	.1170	.0731	.135	.0281	.0356	.1251	.0618	.0162
...	.250	.2165	.1300	.0812	.150	.0312	.0395	.1390	.0686	.0180
...	.300	.2598	.1560	.0974	.180	.0375	.0474	.1668	.0824	.0216
					Inch Dimensions					
317½	.003150	.00273	.00164	.00102	.00189	.00039	.00050	.00175	.00086	.00023
282 2/9	.003543	.00307	.00184	.00115	.00213	.00044	.00056	.00197	.00097	.00026
254	.003937	.00341	.00205	.00128	.00236	.00049	.00062	.00219	.00108	.00028
203⅛	.004921	.00426	.00256	.00160	.00295	.00062	.00078	.00274	.00135	.00035
169⅓	.005906	.00511	.00307	.00192	.00354	.00074	.00093	.00328	.00162	.00043
145 1/7	.006890	.00597	.00358	.00224	.00413	.00086	.00109	.00383	.00189	.00050
127	.007874	.00682	.00409	.00256	.00472	.00098	.00124	.00438	.00216	.00057
112 8/9	.008858	.00767	.00461	.00288	.00531	.00111	.00140	.00493	.00243	.00064
101⅗	.009843	.00852	.00512	.00320	.00591	.00123	.00156	.00547	.00270	.00071
84⅔	.011811	.01023	.00614	.00384	.00709	.00148	.00187	.00657	.00324	.00085

*In Tables 5 and 6 these values are shown rounded to the nearest whole number.

Table 4. Unified Miniature Screw Threads — Formulas for Basic and Design Dimensions and Tolerances

Formulas for Basic Dimensions
D = Basic Major Diameter and Nominal Size in millimeters; p = Pitch in millimeters; E = Basic Pitch Diameter in millimeters = $D - 0.64952p$; and K = Basic Minor Diameter in millimeters = $D - 1.04p$

Formulas for Design Dimensions (Maximum Material)	
External Thread	Internal Thread
D_s = Major Diameter = D E_s = Pitch Diameter = E K_s = Minor Diameter = $D - 1.20p$	D_n = Major Diameter = $D + 0.072p$ E_n = Pitch Diameter = E K_n = Minor Diameter = K

Formulas for Tolerances on Design Dimensions‡	
External Thread (−)	Internal Thread (+)
Major Diameter Tol., $0.12p + 0.006$ Pitch Diameter Tol., $0.08p + 0.008$ *Minor Diameter Tol., $0.16p + 0.008$	†Major Diameter Tol., $0.168p + 0.008$ Pitch Diameter Tol., $0.08p + 0.008$ Minor Diameter Tol., $0.32p + 0.012$

Metric units (millimeters) apply in all formulas. Inch tolerances are not derived by direct conversion of the metric values. They are the differences between the rounded off limits of size in inch units.

* This tolerance establishes the minimum limit of the minor diameter of the external thread. In practice, this limit is applied to the threading tool and only gaged on the product in confirming new tools. Values for this tolerance are, therefore, not given in Table 5.

† This tolerance establishes the maximum limit of the major diameter of the internal thread. In practice, this limit is applied to the threading tool (tap) and not gaged on the product. Values for this tolerance are, therefore, not given in Table 5.

‡ These tolerances are based on lengths of engagement of ⅔ D to 1½D.

Table 5. Unified Miniature Screw Threads — Limits of Size and Tolerances

Dimensions in mm

Size Designation[a]	Pitch (mm)	External Threads Major Diam. Max*	External Threads Major Diam. Min	External Threads Pitch Diam. Max*	External Threads Pitch Diam. Min	External Threads Minor Diam. Max[b]	External Threads Minor Diam. Min[c]	Internal Threads Minor Diam. Min*	Internal Threads Minor Diam. Max	Internal Threads Pitch Diam. Min*	Internal Threads Pitch Diam. Max	Internal Threads Major Diam. Min[d]	Internal Threads Major Diam. Max[c]	Lead Angle deg	Lead Angle min	Sectional Area at Minor Diam. (sq mm)
0.30 UNM	**0.080**	**0.300**	**0.284**	**0.248**	**0.234**	**0.204**	**0.183**	**0.217**	**0.254**	**0.248**	**0.262**	**0.306**	**0.327**	**5**	**52**	**0.0307**
0.35 UNM	0.090	0.350	0.333	0.292	0.277	0.242	0.220	0.256	0.297	0.292	0.307	0.356	0.380	5	37	0.0433
0.40 UNM	**0.100**	**0.400**	**0.382**	**0.335**	**0.319**	**0.280**	**0.256**	**0.296**	**0.340**	**0.335**	**0.351**	**0.407**	**0.432**	**5**	**26**	**0.0581**
0.45 UNM	0.100	0.450	0.432	0.385	0.369	0.330	0.306	0.346	0.390	0.385	0.401	0.457	0.482	5	44	0.0814
0.50 UNM	**0.125**	**0.500**	**0.479**	**0.419**	**0.401**	**0.350**	**0.322**	**0.370**	**0.422**	**0.419**	**0.437**	**0.509**	**0.538**	**4**	**26**	**0.0908**
0.55 UNM	0.125	0.550	0.529	0.469	0.451	0.400	0.372	0.420	0.472	0.469	0.487	0.559	0.588	5	26	0.1195
0.60 UNM	**0.150**	**0.600**	**0.576**	**0.503**	**0.483**	**0.420**	**0.388**	**0.444**	**0.504**	**0.503**	**0.523**	**0.611**	**0.644**	**5**	**26**	**0.1307**
0.70 UNM	0.175	0.700	0.673	0.586	0.564	0.490	0.454	0.518	0.586	0.586	0.608	0.713	0.750	5	26	0.1780
0.80 UNM	**0.200**	**0.800**	**0.770**	**0.670**	**0.646**	**0.560**	**0.520**	**0.592**	**0.668**	**0.670**	**0.694**	**0.814**	**0.856**	**5**	**26**	**0.232**
0.90 UNM	0.225	0.900	0.867	0.754	0.728	0.630	0.586	0.666	0.750	0.754	0.780	0.916	0.962	5	26	0.294
1.00 UNM	**0.250**	**1.000**	**0.964**	**0.838**	**0.810**	**0.700**	**0.652**	**0.740**	**0.832**	**0.838**	**0.866**	**1.018**	**1.068**	**4**	**51**	**0.363**
1.10 UNM	0.250	1.100	1.064	0.938	0.910	0.800	0.752	0.840	0.932	0.938	0.966	1.118	1.168	4	51	0.478
1.20 UNM	**0.250**	**1.200**	**1.164**	**1.038**	**1.010**	**0.900**	**0.852**	**0.940**	**1.032**	**1.038**	**1.066**	**1.218**	**1.268**	**4**	**23**	**0.608**
1.40 UNM	0.300	1.400	1.358	1.205	1.173	1.040	0.984	1.088	1.196	1.205	1.237	1.422	1.480	4	32	0.811

Dimensions in inch

Size Designation[a]	Thds. per in.	External Threads Major Diam. Max*	External Threads Major Diam. Min	External Threads Pitch Diam. Max*	External Threads Pitch Diam. Min	External Threads Minor Diam. Max[b]	External Threads Minor Diam. Min[c]	Internal Threads Minor Diam. Min*	Internal Threads Minor Diam. Max	Internal Threads Pitch Diam. Min*	Internal Threads Pitch Diam. Max	Internal Threads Major Diam. Min[d]	Internal Threads Major Diam. Max[c]	Lead Angle deg	Lead Angle min	Sectional Area at Minor Diam. (sq in)
0.30 UNM	**318**	**0.0118**	**0.0112**	**0.0098**	**0.0092**	**0.0080**	**0.0072**	**0.0085**	**0.0100**	**0.0098**	**0.0104**	**0.0120**	**0.0129**	**5**	**52**	**0.0000475**
0.35 UNM	282	0.0138	0.0131	0.0115	0.0109	0.0095	0.0086	0.0101	0.0117	0.0115	0.0121	0.0140	0.0149	5	37	0.0000671
0.40 UNM	**254**	**0.0157**	**0.0150**	**0.0132**	**0.0126**	**0.0110**	**0.0101**	**0.0117**	**0.0134**	**0.0132**	**0.0138**	**0.0160**	**0.0170**	**5**	**26**	**0.0000901**
0.45 UNM	254	0.0177	0.0170	0.0152	0.0145	0.0130	0.0120	0.0136	0.0154	0.0152	0.0158	0.0180	0.0190	5	44	0.0001262
0.50 UNM	**203**	**0.0197**	**0.0189**	**0.0165**	**0.0158**	**0.0138**	**0.0127**	**0.0146**	**0.0166**	**0.0165**	**0.0172**	**0.0200**	**0.0212**	**4**	**26**	**0.0001407**
0.55 UNM	203	0.0217	0.0208	0.0185	0.0177	0.0157	0.0146	0.0165	0.0186	0.0185	0.0192	0.0220	0.0231	5	26	0.0001852
0.60 UNM	**169**	**0.0236**	**0.0227**	**0.0198**	**0.0190**	**0.0165**	**0.0153**	**0.0175**	**0.0198**	**0.0198**	**0.0206**	**0.0240**	**0.0254**	**5**	**26**	**0.000203**
0.70 UNM	145	0.0276	0.0265	0.0231	0.0222	0.0193	0.0179	0.0204	0.0231	0.0231	0.0240	0.0281	0.0295	5	26	0.000276
0.80 UNM	**127**	**0.0315**	**0.0303**	**0.0264**	**0.0254**	**0.0220**	**0.0205**	**0.0233**	**0.0263**	**0.0264**	**0.0273**	**0.0321**	**0.0337**	**5**	**26**	**0.000360**
0.90 UNM	113	0.0354	0.0341	0.0297	0.0287	0.0248	0.0231	0.0262	0.0295	0.0297	0.0307	0.0361	0.0379	5	26	0.000456
1.00 UNM	**102**	**0.0394**	**0.0380**	**0.0330**	**0.0319**	**0.0276**	**0.0257**	**0.0291**	**0.0327**	**0.0330**	**0.0341**	**0.0401**	**0.0420**	**4**	**51**	**0.000563**
1.10 UNM	102	0.0433	0.0419	0.0369	0.0358	0.0315	0.0296	0.0331	0.0367	0.0369	0.0380	0.0440	0.0460	4	51	0.000741
1.20 UNM	**102**	**0.0472**	**0.0458**	**0.0409**	**0.0397**	**0.0354**	**0.0335**	**0.0370**	**0.0406**	**0.0409**	**0.0420**	**0.0480**	**0.0499**	**4**	**23**	**0.000943**
1.40 UNM	85	0.0551	0.0535	0.0474	0.0462	0.0409	0.0387	0.0428	0.0471	0.0474	0.0487	0.0560	0.0583	4	32	0.001257

* This is also the basic dimension.   [a] Sizes shown in bold type are preferred.   [b] This limit, in conjunction with root form shown in Table 2, is advocated for use when optical projection methods of gaging are employed. For mechanical gaging the minimum minor diameter of the internal thread is applied.   [c] This limit is provided for reference only. In practice, the form of the threading tool is relied upon for this limit.   [d] This limit is provided for reference only, and is not gaged. For gaging, the maximum major diameter of the external thread is applied.

Table 6. Unified Miniature Screw Threads — Minimum Root Flats for External Threads

Pitch	No. of Threads	Thread Height for Min. Flat at Root 0.64p		Minimum Flat at Root $F_{rs} = 0.136p$	
mm	Per Inch	mm	Inch	mm	Inch
0.080	318	0.0512	0.00202	0.0109	0.00043
0.090	282	0.0576	0.00227	0.0122	0.00048
0.100	254	0.0640	0.00252	0.0136	0.00054
0.125	203	0.0800	0.00315	0.0170	0.00067
0.150	169	0.0960	0.00378	0.0204	0.00080
0.175	145	0.1120	0.00441	0.0238	0.00094
0.200	127	0.1280	0.00504	0.0272	0.00107
0.225	113	0.1440	0.00567	0.0306	0.00120
0.250	102	0.1600	0.00630	0.0340	0.00134
0.300	85	0.1920	0.00756	0.0408	0.00161

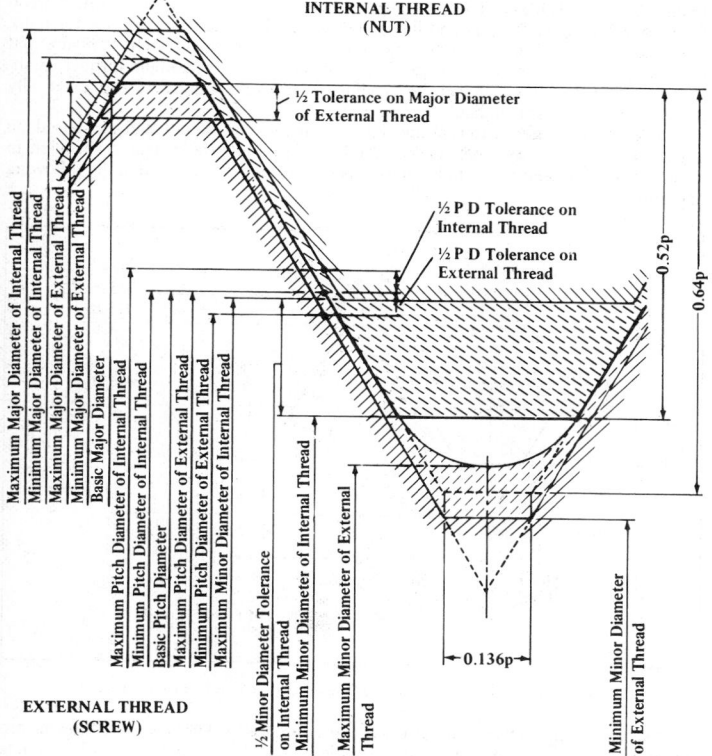

Limits of Size Showing Tolerances and Crest Clearances for UNM Threads

Interference-Fit Threads.—Interference-fit threads are threads in which the externally threaded member is larger than the internally threaded member when both members are in the free state and which, when assembled, become the same size and develop a holding torque through elastic compression, plastic movement of material, or both. By custom, these threads are designated Class 5.

The data in Tables 1, 2, and 3, which is based on ten years of research, testing and field study, represents the first attempt to establish an American standard for interference fit threads that would overcome the difficulties experienced with previous interference fit recommendations such as are given in Federal Screw Thread Handbook H28. These data were adopted as American Standard ASA B1.12-1963. Subsequently the standard was revised and issued as American National Standard ANSI B1.12-1972.

The data in Tables 1, 2, and 3 provide dimensions for external and internal interference-fit (Class 5) threads of modified American National form in the Coarse Thread series, sizes ¼ inch to 1½ inches. It is intended that interference-fit threads conforming with this standard will provide adequate torque conditions which fall within the limits shown in Table 3. These torque limits are from the Federal Screw Thread Handbook H28-1957 Part III and have been generally accepted in service for more than twenty years. The minimum torques are intended to be sufficient to insure that externally threaded members will not loosen in service; the maximum torques establish a ceiling below which seizing, galling or torsional failure of the externally threaded components is unlikely.

Tables 1 and 2 give external and internal thread dimensions and are based on engagement lengths, external thread lengths, and tapping hole depths specified in Table 3 and in compliance with the design and application data given in the following paragraphs.

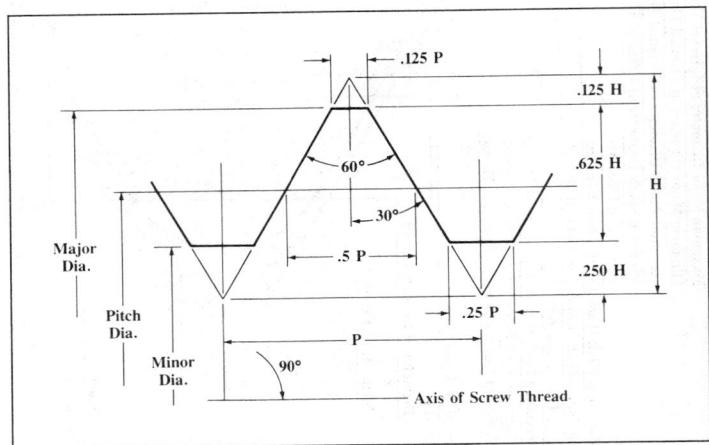

Basic Profile of American National Standard Class 5 Interference Fit Thread

Design and Application Data for Class 5 Interference-Fit Threads. — Following are conditions of usage and inspection on which satisfactory application of products made to dimensions in Tables 1, 2, and 3 are based.

Thread Designations: The following thread designations provide a means of distinguishing the American Standard Class 5 Threads from the tentative Class 5 and alternate Class 5 threads, specified in Handbook H28. It also distinguishes between external and internal American Standard Class 5 Threads.

Class 5 External Threads are designated as follows:

NC5 HF — For driving in hard ferrous material of hardness over 160 BHN.

NC5 CSF — For driving in copper alloy and soft ferrous material of 160 BHN or less.

NC5 ONF — For driving in other non-ferrous material (non-ferrous materials other than copper alloys), any hardness.

Class 5 Internal Threads are designated as follows:

NC5 IF — Entire ferrous material range.

NC5 INF — Entire non-ferrous material range.

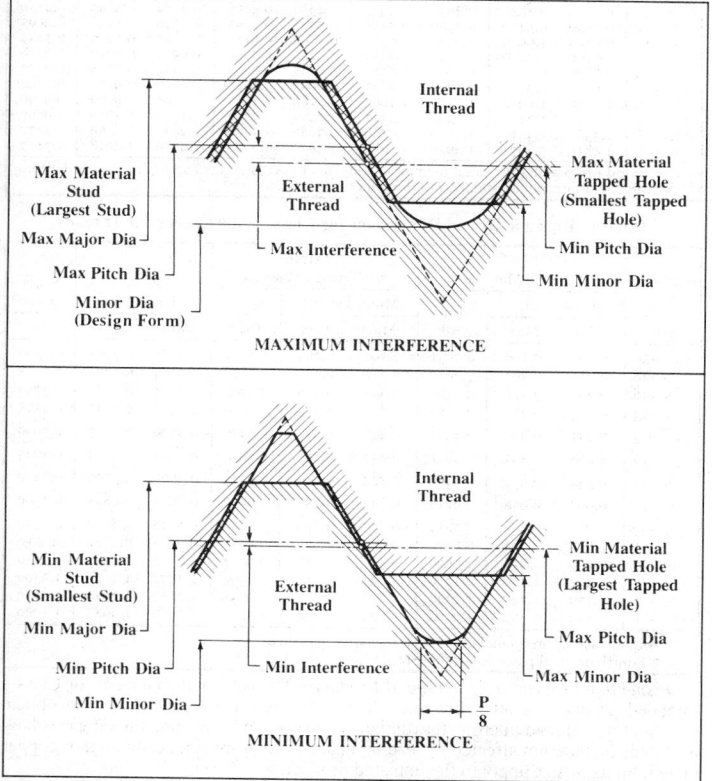

Maximum and Minimum Material Limits for Class 5 Interference-Fit Threads

Table 1. External Thread Dimensions for Class 5 Interference-Fit Threads *

| | Major Diameter, Inches | | | | | | Pitch Diameter, Inches | | Minor Diameter, Inches |
| | NC5HF for driving in ferrous material with hardness greater than 160 BHN $L_e = 1\tfrac{1}{4}$ Diam. | | NC5CSF for driving in brass and ferrous material with hardness equal to or less than 160 BHN $L_e = 1\tfrac{1}{4}$ Diam. | | NC5ONF for driving in nonferrous except brass (any hardness) $L_e = 2\tfrac{1}{2}$ Diam. | | | | |
Size	Max	Min	Max	Min	Max	Min	Max	Min	Max
¼ –20	0.2470	0.2408	0.2470	0.2408	0.2470	0.2408	0.2230	0.2204	0.1932
⁵⁄₁₆–18	0.3080	0.3020	0.3090	0.3030	0.3090	0.3030	0.2829	0.2799	0.2508
⅜ –16	0.3690	0.3626	0.3710	0.3646	0.3710	0.3646	0.3414	0.3382	0.3053
⁷⁄₁₆–14	0.4305	0.4233	0.4330	0.4258	0.4330	0.4258	0.3991	0.3955	0.3579
½ –13	0.4920	0.4846	0.4950	0.4876	0.4950	0.4876	0.4584	0.4547	0.4140
⁹⁄₁₆–12	0.5540	0.5460	0.5580	0.5495	0.5580	0.5495	0.5176	0.5136	0.4695
⅝ –11	0.6140	0.6056	0.6195	0.6111	0.6195	0.6111	0.5758	0.5716	0.5233
¾ –10	0.7360	0.7270	0.7440	0.7350	0.7440	0.7350	0.6955	0.6910	0.6378
⅞ – 9	0.8600	0.8502	0.8685	0.8587	0.8685	0.8587	0.8144	0.8095	0.7503
1 – 8	0.9835	0.9727	0.9935	0.9827	0.9935	0.9827	0.9316	0.9262	0.8594
1⅛ – 7	1.1070	1.0952	1.1180	1.1062	1.1180	1.1062	1.0465	1.0406	0.9640
1¼ – 7	1.232	1.220	1.2430	1.2312	1.2430	1.2312	1.1715	1.1656	1.0890
1⅜ – 6	1.356	1.341	1.3680	1.3538	1.3680	1.3538	1.2839	1.2768	1.1877
1½ – 6	1.481	1.467	1.4930	1.4788	1.4930	1.4788	1.4089	1.4018	1.3127

* Based on external threaded members being steel ASTM A-325 (SAE Grade 5) or better. For rolled, cut or ground threads. L_e = length of engagement.

Table 2. Internal Thread Dimensions for Class 5 Interference-Fit Threads

| | NC5IF Ferrous Material | | | NC5INF Nonferrous Material | | | Pitch Diameter | | Major Diam. |
| | Minor Diam.* | | Tap Drill | Minor Diam.* | | Tap Drill | | | |
Size	Min	Max		Min	Max		Min	Max	Min
¼ –20	0.196	0.206	0.2031	0.196	0.206	0.2031	0.2175	0.2201	0.2500
⁵⁄₁₆–18	0.252	0.263	0.2610	0.252	0.263	0.2610	0.2764	0.2794	0.3125
⅜ –16	0.307	0.318	0.3160	0.307	0.318	0.3160	0.3344	0.3376	0.3750
⁷⁄₁₆–14	0.374	0.381	0.3750	0.360	0.372	0.3680	0.3911	0.3947	0.4375
½ –13	0.431	0.440	0.4331	0.417	0.429	0.4219	0.4500	0.4537	0.5000
⁹⁄₁₆–12	0.488	0.497	0.4921	0.472	0.485	0.4844	0.5084	0.5124	0.5625
⅝ –11	0.544	0.554	0.5469	0.527	0.540	0.5313	0.5660	0.5702	0.6250
¾ –10	0.667	0.678	0.6719	0.642	0.655	0.6496	0.6850	0.6895	0.7500
⅞ – 9	0.777	0.789	0.7812	0.755	0.769	0.7656	0.8028	0.8077	0.8750
1 – 8	0.890	0.904	0.8906	0.865	0.880	0.8750	0.9188	0.9242	1.000
1⅛ – 7	1.000	1.015	1.0000	0.970	0.986	0.9844	1.0322	1.0381	1.1250
1¼ – 7	1.125	1.140	1.1250	1.095	1.111	1.1094	1.1572	1.1631	1.2500
1⅜ – 6	1.229	1.247	1.2344	1.195	1.213	1.2031	1.2667	1.2738	1.3750
1½ – 6	1.354	1.372	1.3594	1.320	1.338	1.3281	1.3917	1.3988	1.5000

All dimensions are in inches, unless otherwise specified.
* Fourth decimal place is 0 for all sizes.

Inspection of Externally Threaded Products: The controlling element for Class 5 threaded products is pitch diameter. This element can be checked by an optical comparator, a thread micrometer, thread snap gages, or indicating thread gages having anvils that are not affected by lead or angle. If studs are zinc, cadmium, or copper plated, limits of size apply to the unplated product.

Points of externally threaded components should be chamfered or otherwise reduced to a diameter below the minimum minor diameter of the thread. The

Table 3. Torques, Interferences, and Engagement Lengths for Class 5 Interference-Fit Threads

| Size | Inter-ference on Pitch Diameter | | Engagement Lengths, External Thread Lengths and Tapped Hole Depths* | | | | | | Approx. Torque at Full Engagement of 1-¼D in Ferrous Material | |
| | | | In Brass and Ferrous | | | In Nonferrous Except Brass | | | | |
	Max	Min	L_e	T_s	T_h min	L_e	T_s	T_h min	Max, Ft-lbs	Min, Ft-lbs
¼ –20	.0055	.0003	0.312	0.375 + .125 – 0	0.375	0.625	0.688 + .125 – 0	¹¹⁄₁₆	12	3
⁵⁄₁₆–18	.0065	.0005	0.391	0.469 + .139 – 0	0.469	0.781	0.859 + .139 – 0	⁵⁵⁄₆₄	19	6
⅜ –16	.0070	.0006	0.469	0.562 + .156 – 0	0.562	0.938	1.031 + .156 – 0	1¹⁄₃₂	35	10
⁷⁄₁₆–14	.0080	.0008	0.547	0.656 + .179 – 0	0.656	1.094	1.203 + .179 – 0	1¹³⁄₆₄	45	15
½ –13	.0084	.0010	0.625	0.750 + .192 – 0	0.750	1.250	1.375 + .192 – 0	1⅜	75	20
⁹⁄₁₆–12	.0092	.0012	0.703	0.844 + .208 – 0	0.844	1.406	1.547 + .208 – 0	1³⁵⁄₆₄	90	30
⅝ –11	.0098	.0014	0.781	0.938 + .227 – 0	0.938	1.562	1.719 + .227 – 0	1²³⁄₃₂	120	37
¾ –10	.0105	.0015	0.938	1.125 + .250 – 0	1.125	1.875	2.062 + .250 – 0	2¹⁄₁₆	190	60
⅞ – 9	.0116	.0018	1.094	1.312 + .278 – 0	1.312	2.188	2.406 + .278 – 0	2¹³⁄₃₂	250	90
1 – 8	.0128	.0020	1.250	1.500 + .312 – 0	1.500	2.500	2.750 + .312 – 0	2¾	400	125
1⅛ – 7	.0143	.0025	1.406	1.688 + .357 – 0	1.688	2.812	3.094 + .357 – 0	3³⁄₃₂	470	155
1¼ – 7	.0143	.0025	1.562	1.875 + .357 – 0	1.875	3.125	3.438 + .357 – 0	3⁷⁄₁₆	580	210
1⅜ – 6	.0172	.0030	1.719	2.062 + .419 – 0	2.062	3.438	3.781 + .419 – 0	3²⁵⁄₃₂	705	250
1½ – 6	.0172	.0030	1.875	2.250 + .419 – 0	2.250	3.750	4.125 + .419 – 0	4⅛	840	325

All dimensions are in inches. * L_e = Length of engagement. T_s = External thread length. T_h = Depth of full form thread in hole.

threads should be free from excessive nicks, burrs, chips, grit or other extraneous material before driving.

Materials for Externally Threaded Products: The length of engagement, depth of thread engagement and pitch diameter in Tables 1, 2, and 3 are designed to produce adequate torque conditions when heat-treated medium-carbon steel studs, ASTM A-325 (SAE Grade 5) or better, are used. In many applications, case-carburized and unheat-treated medium-carbon steel products of SAE Grade 4, are satisfactory. SAE Grades 1, 2, 8 and 8.1 may be desirable under certain conditions. This standard is not intended to cover the use of products made of stainless steel, silicon bronze, brass or similar materials. When such materials are used, the tabulated dimensions will probably require adjustment based on pilot experimental work with the materials involved.

Holes: GO plain plug and GO thread plug gages should be inserted to full depth in order to detect the effect of excessive drill or tap wear at the bottom of the hole. NOT–GO thread plug gages should enter not more than 1½ threads. Holes must be clean from grit, chips, oil or other extraneous material prior to gaging and before driving studs or screws. Holes should be countersunk to a diameter greater than the major diameter to facilitate starting of the externally threaded product and to prevent raising a lip around the hole after driving.

Lead and Angle Variations: Angle and lead errors are not normally objectionable since they contribute to interference which is the purpose of the Class 5 thread. Experience may dictate the need for imposing some limits under certain conditions.

Lubrication: For driving in ferrous material, a good lubricant sealer should be used, particularly in the hole. A non-carbonizing type of lubricant (such as a rubber-in-water dispersion) is suggested. The lubricant must be applied to the hole and it may be applied to the male member. In applying it to the hole, care must be taken so that an excess amount of lubricant will not cause the male member to be impeded by hydraulic pressure in a blind hole. Where sealing is involved, the lubricant selected should be insoluble in the medium being sealed.

For driving, in nonferrous material, lubrication may not be needed. Recent British research recommends the use of medium gear oil for driving in aluminum. American research has observed that the minor diameter of lubricated tapped holes in nonferrous materials may tend to close in, that is, be reduced in driving; whereas with an unlubricated hole the minor diameter may tend to open up in some cases.

Driving Speed: This standard makes no recommendation for driving speed. Some opinion has been advanced that careful selection and control of driving speed is desirable to obtain optimum results with various combinations of surface hardness and roughness. Experience with threads made to this standard may indicate what limitations should be placed on driving speeds.

Relation of Driving Torque to Length of Engagement: Torques increase directly as the length of engagement. American research indicates that this increase is proportionately more rapid as size increases.

Breakloose Torques after Reapplication: The standard does not establish recommended reapplication breakloose torques in cases where repeated usage is involved.

Assembly Torques for Reapplication: The standard does not establish assembly torques for reapplication.

Bottoming and Shouldering of Studs. — Among the conclusions drawn from stud research is the fact that studs should be driven to a predetermined depth. "Bottoming" or "Shouldering" should be avoided. 'Bottoming," which is engagement of the threads of the stud with the imperfect threads at the bottom of a shallow drilled and tapped hole causes the stud to stop suddenly, thus inviting failure in torsional shear. "Shouldering," which is the practice of driving the stud until the thread runout engages with the top threads of the hole, creates radial compressive stresses and upward bulging of the material at the top of the hole. This results in erratic variations in free stud length after driving.

Extension of the Standard. — By using the new principles upon which this standard is based, thread sizes may be extended downward. However, adequate data are not now available to permit setting a standard. American research indicates that on smaller sizes the main reliance for producing adequate breakloose torque should be placed on pitch diameter interference and not on increasing the length of engagement.

Although there is some current usage of interference fits on large size threads, adequate data is not now developed to permit setting a standard on larger sizes.

Use of the coarse thread series is urged unless requirements for strength of the male members make a finer pitch necessary. No research data is now available to enable the setting of a trial standard for fine thread products. Indications are, however, that the product of the ratio:

$$\frac{\text{Class 2A UNF PD tolerance}}{\text{Class 2A UNC PD tolerance}}$$

times the coarse thread dimensions given in Tables 1 and 2 will probably work for:

a. Major diameter tolerance, external threads
b. Pitch diameter tolerance, external threads
c. Pitch diameter tolerance, internal threads
d. Minor diameter tolerance, internal threads
e. Mimimum interference

Similarly, the principles observed in setting the pitch diameter and major diameter limits on the fine series Class 5 external threads above may be followed in deriving the pitch diameter and minor diameter of the fine series internal threads.

American National Standard Acme Screw Threads. — This American National Standard ANSI B1.5-1977 is a revision of American Standard ASA B1.5-1973 and provides for two general applications of Acme threads, namely, General Purpose and Centralizing.

The limits and tolerances in this standard relate to single-start Acme threads, and may be used, if considered suitable, for multi-start Acme threads, which provide fast relative traversing motion when this is necessary. For information on additional allowances for multi-start Acme threads, see later section on page 1558.

General Purpose Acme Threads. — Three classes of General Purpose threads, 2G, 3G, and 4G, are provided in the standard, each having clearance on all diameters for free movement, and may be used in assemblies with the internal thread rigidly fixed and movement of the external thread in a direction perpendicular to its axis limited by its bearing or bearings. It is suggested that external and internal threads of the same class be used together for general purpose assemblies, Class 2G being the preferred choice. If less backlash or end play is desired, Classes 3G and 4G are provided.

Where minimal backlash or end play is required, Class 5G is provided. Assemblies of internal and external class 5G threads normally require some fitting-up for satisfactory results. External threads of any class may be assembled with internal threads of any class to provide other degrees of backlash or end play.

Thread Form: The accompanying figure shows the thread form of these General Purpose threads, and the formulas accompanying the figure determine their basic dimensions. Table 1 gives the basic dimensions for the most generally used pitches.

Angle of Thread: The angle between the sides of the thread, measured in an axial plane, is 29 degrees. The line bisecting this 29-degree angle shall be perpendicular to the axis of the screw thread.

Thread Series: A series of diameters and associated pitches is recommended in the Standard as preferred. These diameters and pitches have been chosen to meet present needs with the fewest number of items in order to reduce to a minimum the inventory of both tools and gages. This series of diameters and associated pitches is given in Table 3.

Chamfers and Fillets: General Purpose external threads may have the crest corner chamfered to an angle of 45 degrees with the axis to a maximum width of $p/15$, where p is the pitch. This corresponds to a maximum depth of chamfer flat of $0.0945p$.

Basic Diameters: The maximum major diameter of the external thread is basic and is the nominal major diameter for all classes. The minimum pitch diameter of the internal thread is basic and is equal to the basic major diameter minus the basic height of the thread, h. The basic minor diameter is the minimum minor diameter of the internal thread. It is equal to the basic major diameter minus twice the basic thread height, $2h$.

Length of Engagement: The tolerances specified in this standard are applicable to lengths of engagement not exceeding twice the nominal major diameter.

Major and Minor Diameter Allowances: A minimum diametral clearance is provided at the minor diameter of all external threads by establishing the maximum minor diameter 0.020 inch below the basic minor diameter of the nut for pitches of 10 threads per inch and coarser, and 0.010 inch for finer pitches. A minimum diametral clearance at the major diameter is obtained by establishing the minimum major diameter of the internal thread 0.020 inch above the basic major diameter of the screw for pitches of 10 threads per inch and coarser, and 0.010 inch for finer pitches.

Major and Minor Diameter Tolerances: The tolerance on the external thread major diameter is $0.05p$, where p is the pitch, with a minimum of 0.005 inch. The tolerance on the internal thread major diameter is 0.020 inch for 10 threads per inch and coarser and 0.010 for finer pitches. The tolerance on the external thread minor diameter is $1.5 \times$ pitch diameter tolerance. The tolerance on the internal thread minor

American National Standard General Purpose Acme and Centralizing Acme Screw Thread Form (ANSI B1.5-1977), and Stub Acme Screw Thread Form (ANSI B1.8-1977)

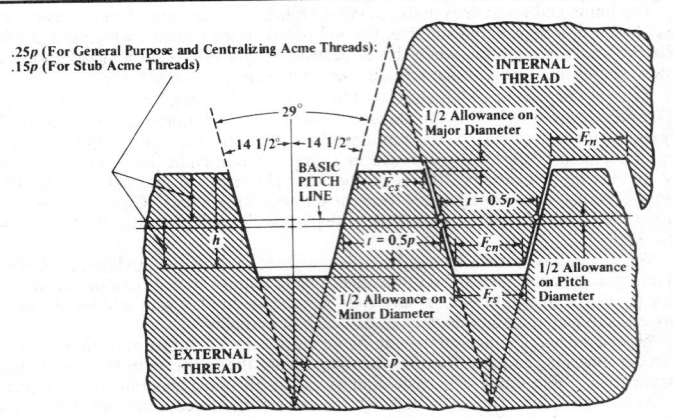

Formulas for Finding Basic Dimensions of General Purpose Acme, Centralizing Acme and Stub Acme Screw Threads

General Purpose and Centralizing Acme Threads	Stub Acme Threads
Pitch = $p = 1 \div$ No. threads per inch, n	Pitch = $p = 1 \div$ No. threads per inch, n
Basic thread height $h = 0.5p$	Basic thread height $h = 0.3p$
Basic thread thickness $t = 0.5p$	Basic thread thickness $t = 0.5p$
Basic flat at crest $F_{cn} = 0.3707p$ (internal thread)	Basic flat at crest $F_{cn} = 0.4224p$ (internal thread)
Basic flat at crest $F_{cs} = 0.3707p - 0.259 \times$ (P.D. allowance on ext. thd.)	Basic flat at crest $F_{cs} = 0.4224p$ (external thread)
$F_{rn} = 0.3707p - 0.259 \times$ (major diam. allowance on internal thread)	$F_{rn} = 0.4224p - 0.259 \times$ (major diam. allowance on internal thread)
$F_{rs} = 0.3707p - 0.259 \times$ (minor diam. allowance on ext. thread – pitch diam. allowance on ext. thread)	$F_{rs} = 0.4224p - 0.259 \times$ (minor diam. allowance on ext. thread – pitch diam. allowance on ext. thread)

Stress Area of General Purpose Acme Threads. — For computing the tensile strength of the thread section, the minimum stress area based on the mean of the minimum pitch diameter E_s and the minimum minor diameter K_s of the external thread is used:

$$\text{Stress Area} = 3.1416 \left(\frac{E_s + K_s}{4} \right)^2,$$

where E_s and K_s may be computed by Formulas 4 and 6, Table 2a or taken from Table 2b.

Shear Area of General Purpose Acme Threads. For computing the shear area per inch length of engagement of the external thread, the maximum minor diameter of the internal thread K_n, and the minimum pitch diameter of the external thread E_s, Table 2b or Formulas 12 and 4, Table 2a, are used:

$$\text{Shear Area} = 3.1416 K_n [0.5 + n \tan 14\tfrac{1}{2}° (E_s - K_n)]$$

diameter is 0.05p with a minimum of 0.005 inch.

Pitch Diameter Allowances and Tolerances: Allowances on the pitch diameter of General Purpose Acme threads are given in Table 4. Pitch diameter tolerances are given in Table 5. The ratios of the pitch diameter tolerances of Classes 2G, 3G, 4G, and 5G General Purpose threads are 3.0, 1.4, 1, and 0.8, respectively.

An increase of 10 per cent in the allowance is recommended for each inch, or fraction thereof, that the length of engagement exceeds two diameters.

Application of Tolerances: The tolerances specified are such as to assure interchangeability and maintain a high grade of product. The tolerances on diameters of the internal thread are plus, being applied from minimum sizes to above the minimum sizes. The tolerances on diameters of the external thread are minus, being applied from the maximum sizes to below the maximum sizes. The pitch diameter (or thread thickness) tolerances for an external or internal thread of a given class are the same. The thread thickness tolerance is 0.259 times the pitch diameter tolerance.

Limiting Dimensions: Limiting dimensions of General Purpose Acme screw threads in the recommended series are given in Table 2b. These are based on the formulas in Table 2a.

For combinations of pitch and diameter other than those in the recommended series, the formulas in Table 2a and the data in Tables 4 and 5 make it possible to readily determine the limiting dimensions required.

A diagram showing the disposition of allowances, tolerances, and crest clearances for General Purpose Acme threads appears on page 1566.

Acme Thread Abbreviations. — The following abbreviations are recommended for use on drawings and in specifications, and on tools and gages:

Acme = Acme threads
G = General Purpose
C = Centralizing
p = pitch
L = lead
LH = left hand

Designation of General Purpose Acme Threads. — The following examples listed below are given here to show how these Acme threads are designated on drawings and tools:

1¾–4 Acme–2G indicates a General Purpose Class 2G Acme thread of 1¾-inch major diameter, 4 threads per inch, single thread, right hand. The same thread, but left hand, is designated 1¾–4 Acme–2G–LH.

2⅞–0.4p–0.8L–Acme–3G indicates a General Purpose Class 3G Acme thread of 2⅞-inch major diameter, pitch 0.4 inch, lead 0.8 inch, double thread, right hand.

Multiple Start Acme Threads. — The tabulated diameter-pitch data with allowances and tolerances relate to single-start threads. These data, as tabulated, may be and often are used for two-start Class 2G threads but this usage generally requires reduction of the full working tolerances to provide a greater allowance or clearance zone between the mating threads to assure satisfactory assembly.

When the class of thread requires smaller working tolerances than the 2G class or when threads with 3, 4, or more starts are required, some additional allowances and/or increased tolerances may be needed to insure adequate working tolerances and satisfactory assembly of mating parts.

It is suggested that the allowances shown in Table 4 be used for all external threads and that allowances be applied to internal threads in the following ratios: for two-start threads, 50 per cent of the allowances shown in the third, fourth, and fifth columns of

(Continued on page 1564)

**Table 1. American National Standard General Purpose Acme Screw
Thread Form — Basic Dimensions* (ANSI B1.5-1977)**

					Width of Flat	
Thds. per Inch n	Pitch, $p = 1/n$	Height of Thread (Basic), $p/2$	Total Height of Thread, $p/2 + \frac{1}{2}$ allowance†	Thread Thickness (Basic), $p/2$	Crest of Internal Thread (Basic), $0.3707p$	Root of Internal Thread, $0.3707p -$ $0.259 \times$ allowance†
16	0.06250	0.03125	0.0362	0.03125	0.0232	0.0206
14	0.07143	0.03571	0.0407	0.03571	0.0265	0.0239
12	0.08333	0.04167	0.0467	0.04167	0.0309	0.0283
10	0.10000	0.05000	0.0600	0.05000	0.0371	0.0319
8	0.12500	0.06250	0.0725	0.06250	0.0463	0.0411
6	0.16667	0.08333	0.0933	0.08333	0.0618	0.0566
5	0.20000	0.10000	0.1100	0.10000	0.0741	0.0689
4	0.25000	0.12500	0.1350	0.12500	0.0927	0.0875
3	0.33333	0.16667	0.1767	0.16667	0.1236	0.1184
2½	0.40000	0.20000	0.2100	0.20000	0.1483	0.1431
2	0.50000	0.25000	0.2600	0.25000	0.1853	0.1802
1½	0.66667	0.33333	0.3433	0.33333	0.2471	0.2419
1⅓	0.75000	0.37500	0.3850	0.37500	0.2780	0.2728
1	1.00000	0.50000	0.5100	0.50000	0.3707	0.3655

* All dimensions are in inches.
† Allowance is 0.020 inch for 10 threads per inch and coarser, and 0.010 inch for finer threads.

**Table 2a. American National Standard General Purpose Acme Single-Start
Screw Threads — Formulas for Determining Diameters (ANSI B1.5-1977)**

	D = Basic Major Diameter and Nominal Size, in Inches.
	p = Pitch = 1 ÷ Number of Threads per Inch.
	E = Basic Pitch Diameter = $D - 0.5p$
	K = Basic Minor Diameter = $D - p$

No.	External Threads (Screws)
1	Major Diam., Max. = D
2	Major Diam., Min. = D minus 0.05p* but not less than 0.005.
3	Pitch Diam., Max. = E minus allowance from Table 4.
4	Pitch Diam., Min. = Pitch Diam., Max. (Formula 3) minus tolerance from Table 5.
5	Minor Diam., Max. = K minus 0.020 for 10 threads per inch and coarser and 0.010 for finer pitches.
6	Minor Diam., Min. = Minor Diam., Max. (Formula 5) minus 1.5 × pitch diameter tolerance from Table 5.

	Internal Threads (Nuts)
7	Major Diam., Min. = D plus 0.020 for 10 threads per inch and coarser and 0.010 for finer pitches.
8	Major Diam., Max. = Major Diam., Min. (Formula 7) plus 0.020 for 10 threads per inch and coarser and 0.010 for finer pitches.
9	Pitch Diam., Min. = E
10	Pitch Diam., Max. = Pitch Diam., Min. (Formula 9) plus tolerance from Table 5.
11	Minor Diam., Min. = K
12	Minor Diam., Max. = Minor Diam., Min. (Formula 11) plus 0.05p* but not less than 0.005.

* If p is between two recommended pitches listed in Table 3, use the coarser of the two pitches in this formula instead of the actual value of p.

Table 2b. Limiting Dimensions of American National Standard General Purpose Acme Single-Start Screw Threads

Limiting Diameters		1/4	5/16	3/8	7/16	1/2	5/8	3/4	7/8	1	1 1/8	1 1/4	1 3/8
Nominal Diameter, D — Threads per Inch*		16	14	12	12	10	8	6	6	5	5	5	4
External Threads													
Classes 2G, 3G, 4G, and 5G, Major Diameter	Max (D)	0.2500	0.3125	0.3750	0.4375	0.5000	0.6250	0.7500	0.8750	1.0000	1.1250	1.2500	1.3750
	Min	0.2450	0.3075	0.3700	0.4325	0.4950	0.6188	0.7417	0.8667	0.9900	1.1150	1.2400	1.3025
Classes 2G, 3G, 4G, and 5G, Minor Diameter	Max	0.1775	0.2311	0.2817	0.3442	0.3800	0.4800	0.5633	0.6883	0.7800	0.9050	1.0300	1.1050
Class 2G, Minor Diameter	Min	0.1618	0.2140	0.2632	0.3253	0.3594	0.4569	0.5372	0.6615	0.7599	0.8753	0.9998	1.0720
Class 3G, Minor Diameter	Min	0.1702	0.2231	0.2730	0.3354	0.3704	0.4692	0.5511	0.6758	0.7664	0.8912	1.0159	1.0896
Class 4G, Minor Diameter	Min	0.1722	0.2254	0.2755	0.3379	0.3731	0.4723	0.5546	0.6794	0.7703	0.8951	1.0199	1.0940
Class 5G, Minor Diameter	Min	0.1733	0.2266	0.2767	0.3391	0.3745	0.4738	0.5563	0.6811	0.7722	0.8971	1.0219	1.0962
Class 2G, Pitch Diameter	Max	0.2148	0.2728	0.3284	0.3909	0.4443	0.5562	0.6598	0.7842	0.8920	1.0165	1.1411	1.2406
	Min	0.2043	0.2614	0.3161	0.3783	0.4306	0.5408	0.6424	0.7663	0.8726	0.9967	1.1210	1.2188
Class 3G, Pitch Diameter	Max	0.2158	0.2738	0.3296	0.3921	0.4458	0.5578	0.6615	0.7861	0.8940	1.0186	1.1433	1.2430
	Min	0.2109	0.2685	0.3238	0.3862	0.4394	0.5506	0.6534	0.7778	0.8849	1.0094	1.1339	1.2327
Class 4G, Pitch Diameter	Max	0.2168	0.2748	0.3309	0.3934	0.4472	0.5593	0.6632	0.7880	0.8960	1.0208	1.1455	1.2453
	Min	0.2133	0.2710	0.3268	0.3892	0.4426	0.5542	0.6574	0.7820	0.8895	1.0142	1.1388	1.2380
Class 5G, Pitch Diameter	Max	0.2188	0.2768	0.3333	0.3958	0.4500	0.5625	0.6667	0.7917	0.9000	1.0250	1.1500	1.2500
	Min	0.2160	0.2738	0.3300	0.3924	0.4463	0.5584	0.6620	0.7869	0.8948	1.0197	1.1446	1.2441
Internal Threads													
Classes 2G, 3G, 4G, and 5G, Minor Diameter	Min	0.2600	0.3225	0.3850	0.4475	0.5200	0.6450	0.7700	0.8950	1.0200	1.1450	1.2700	1.3950
	Max	0.2700	0.3325	0.3950	0.4575	0.5400	0.6650	0.7900	0.9150	1.0400	1.1650	1.2900	1.4150
Classes 2G, 3G, 4G, and 5G, Major Diameter	Min	0.1875	0.2411	0.2917	0.3542	0.4000	0.5000	0.5833	0.7083	0.8000	0.9250	1.0500	1.1250
	Max	0.1925	0.2461	0.2967	0.3592	0.4050	0.5062	0.5916	0.7166	0.8100	0.9350	1.0600	1.1375
Class 2G, Pitch Diameter	Min	0.2188	0.2768	0.3333	0.3958	0.4500	0.5625	0.6667	0.7917	0.9000	1.0250	1.1500	1.2500
	Max	0.2293	0.2882	0.3456	0.4084	0.4617	0.5779	0.6841	0.8096	0.9194	1.0448	1.1701	1.2720
Class 3G, Pitch Diameter	Min	0.2188	0.2768	0.3333	0.3958	0.4500	0.5625	0.6667	0.7917	0.9000	1.0250	1.1500	1.2500
	Max	0.2237	0.2821	0.3391	0.4017	0.4564	0.5697	0.6748	0.8000	0.9091	1.0342	1.1594	1.2603
Class 4G, Pitch Diameter	Min	0.2188	0.2768	0.3333	0.3958	0.4500	0.5625	0.6667	0.7917	0.9000	1.0250	1.1500	1.2500
	Max	0.2223	0.2806	0.3374	0.4000	0.4546	0.5676	0.6725	0.7977	0.9065	1.0316	1.1567	1.2573
Class 5G, Pitch Diameter	Min	0.2188	0.2768	0.3333	0.3958	0.4500	0.5625	0.6667	0.7917	0.9000	1.0250	1.1500	1.2500
	Max	0.2216	0.2798	0.3366	0.3992	0.4537	0.5666	0.6714	0.7965	0.9052	1.0303	1.1554	1.2559

* All other dimensions are given in inches.

Table 2b (Concluded). Limiting Dimensions of American National Standard General Purpose Acme Single-Start Screw Threads

Limiting Diameters		Nominal Diameter, D										
		1½	1¾	2	2¼	2½	2¾	3	3½	4	4½	5
	Threads per Inch* →	4	4	4	3	3	3	2	2	2	2	2
External Threads												
Classes 2G, 3G, 4G, and 5G, Major Diameter	Max (D)	1.5000	1.7500	2.0000	2.2500	2.5000	2.7500	3.0000	3.5000	4.0000	4.5000	5.0000
	Min	1.4875	1.7375	1.9875	2.2333	2.4833	2.7333	2.9750	3.4750	3.9750	4.4750	4.9750
Classes 2G, 3G, 4G, and 5G, Minor Diameter	Max	1.2300	1.4800	1.7300	1.8967	2.1467	2.3967	2.4800	2.9800	3.4800	3.9800	4.4800
Class 2G, Minor Diameter	Min	1.1965	1.4456	1.6948	1.8572	2.1065	2.3558	2.4326	2.9314	3.4302	3.9291	4.4281
Class 3G, Minor Diameter	Min	1.2144	1.4640	1.7135	1.8783	2.1279	2.3776	2.4579	2.9574	3.4568	3.9553	4.4557
Class 4G, Minor Diameter	Min	1.2189	1.4686	1.7183	1.8835	2.1333	2.3831	2.4642	2.9638	3.4634	3.9631	4.4627
Class 5G, Minor Diameter	Min	1.2210	1.4708	1.7206	1.8862	2.1360	2.3858	2.4674	2.9669	3.4666	3.9663	4.4662
Class 2G, Pitch Diameter	Max	1.3652	1.6145	1.8637	2.0713	2.3207	2.5700	2.7360	3.2350	3.7340	4.2330	4.7319
	Min	1.3429	1.5916	1.8402	2.0450	2.2939	2.5427	2.7044	3.2026	3.7008	4.1991	4.6973
Class 3G, Pitch Diameter	Max	1.3677	1.6171	1.8665	2.0743	2.3238	2.5734	2.7395	3.2388	3.7380	4.2373	4.7364
	Min	1.3573	1.6064	1.8555	2.0620	2.3113	2.5607	2.7248	3.2237	3.7225	4.2215	4.7202
Class 4G, Pitch Diameter	Max	1.3701	1.6198	1.8693	2.0773	2.3270	2.5767	2.7430	3.2425	3.7420	4.2415	4.7409
	Min	1.3627	1.6122	1.8615	2.0685	2.3181	2.5676	2.7325	3.2317	3.7309	4.2302	4.7294
Class 5G, Pitch Diameter	Max	1.3750	1.6250	1.8750	2.0833	2.3333	2.5833	2.7500	3.2500	3.7500	4.2500	4.7500
	Min	1.3690	1.6189	1.8687	2.0763	2.3262	2.5760	2.7416	3.2413	3.7411	4.2409	4.7408
Internal Threads												
Classes 2G, 3G, 4G, and 5G, Major Diameter	Min	1.5200	1.7700	2.0200	2.2700	2.5200	2.7700	3.0200	3.5200	4.0200	4.5200	5.0200
	Max	1.5400	1.7900	2.0400	2.2900	2.5400	2.7900	3.0400	3.5400	4.0400	4.5400	5.0400
Classes 2G, 3G, 4G, and 5G, Minor Diameter	Min	1.2500	1.5000	1.7500	1.9167	2.1667	2.4167	2.5000	3.0000	3.5000	4.0000	4.5000
	Max	1.2625	1.5125	1.7625	1.9334	2.1834	2.4334	2.5250	3.0250	3.5250	4.0250	4.5250
Class 2G, Pitch Diameter	Min	1.3750	1.6250	1.8750	2.0833	2.3333	2.5833	2.7500	3.2500	3.7500	4.2500	4.7500
	Max	1.3973	1.6479	1.8985	2.1096	2.3601	2.6106	2.7816	3.2824	3.7832	4.2839	4.7846
Class 3G, Pitch Diameter	Min	1.3750	1.6250	1.8750	2.0833	2.3333	2.5833	2.7500	3.2500	3.7500	4.2500	4.7500
	Max	1.3854	1.6357	1.8860	2.0956	2.3458	2.5960	2.7647	3.2651	3.7655	4.2658	4.7662
Class 4G, Pitch Diameter	Min	1.3750	1.6250	1.8750	2.0833	2.3333	2.5833	2.7500	3.2500	3.7500	4.2500	4.7500
	Max	1.3824	1.6326	1.8828	2.0921	2.3422	2.5924	2.7605	3.2608	3.7611	4.2613	4.7615
Class 5G, Pitch Diameter	Min	1.3750	1.6250	1.8750	2.0833	2.3333	2.5833	2.7500	3.2500	3.7500	4.2500	4.7500
	Max	1.3810	1.6311	1.8813	2.0903	2.3404	2.5906	2.7584	3.2587	3.7589	4.2591	4.7592

* All other dimensions are given in inches.

Table 3. General Purpose Acme Single-Start Screw Thread Data (ANSI B1.5-1977)

Nominal Sizes (All Classes)	Threads per Inch,* n	Major Diameter, D	Pitch Diameter, E = D − h	Minor Diameter, K = D − 2h	Pitch, p	Thickness at Pitch Line, t = p/2	Basic Height of Thread, h = p/2	Basic Width of Flat, F = 0.3707p	Lead Angle Deg	Lead Angle Min	Shear Area† Class 3G	Stress Area‡ Class 3G
1/4	16	0.2500	0.2188	0.1875	0.06250	0.03125	0.03125	0.0232	5	12	0.350	0.0285
5/16	14	0.3125	0.2768	0.2411	0.07143	0.03571	0.03571	0.0265	4	42	0.451	0.0474
3/8	12	0.3750	0.3333	0.2917	0.08333	0.04167	0.04167	0.0309	4	33	0.545	0.0699
3/8	10	0.3750	0.3250	0.2750	0.10000	0.05000	0.05000	0.0371	⋯	⋯	⋯	⋯
7/16	12	0.4375	0.3958	0.3542	0.08333	0.04167	0.04167	0.0309	3	50	0.660	0.1022
7/16	10	0.4375	0.3875	0.3375	0.10000	0.05000	0.05000	0.0371	⋯	⋯	⋯	⋯
1/2	10	0.5000	0.4500	0.4000	0.10000	0.05000	0.05000	0.0371	4	3	0.749	0.1287
5/8	8	0.6250	0.5625	0.5000	0.12500	0.06250	0.06250	0.0463	4	3	0.941	0.2043
3/4	6	0.7500	0.6667	0.5833	0.16667	0.08333	0.08333	0.0618	4	33	1.108	0.2848
7/8	6	0.8750	0.7917	0.7083	0.16667	0.08333	0.08333	0.0618	3	50	1.339	0.4150
1	5	1.0000	0.9000	0.8000	0.20000	0.10000	0.10000	0.0741	4	3	1.519	0.5354
1 1/8	5	1.1250	1.0250	0.9250	0.20000	0.10000	0.10000	0.0741	3	33	1.751	0.709
1 1/4	5	1.2500	1.1500	1.0500	0.20000	0.10000	0.10000	0.0741	3	10	1.983	0.907
1 3/8	4	1.3750	1.2500	1.1250	0.25000	0.12500	0.12500	0.0927	3	39	2.139	1.059
1 1/2	4	1.5000	1.3750	1.2500	0.25000	0.12500	0.12500	0.0927	3	19	2.372	1.298
1 3/4	4	1.7500	1.6250	1.5000	0.25000	0.12500	0.12500	0.0927	2	48	2.837	1.851
2	4	2.0000	1.8750	1.7500	0.25000	0.12500	0.12500	0.0927	2	26	3.301	2.501
2 1/4	3	2.2500	2.0833	1.9167	0.33333	0.16667	0.16667	0.1236	2	55	3.643	3.049
2 1/2	3	2.5000	2.3333	2.1667	0.33333	0.16667	0.16667	0.1236	2	36	4.110	3.870
2 3/4	3	2.7500	2.5833	2.4167	0.33333	0.16667	0.16667	0.1236	2	21	4.577	4.788
3	2	3.0000	2.7500	2.5000	0.50000	0.25000	0.25000	0.1853	3	19	4.786	5.27
3 1/2	2	3.5000	3.2500	3.0000	0.50000	0.25000	0.25000	0.1853	2	48	5.73	7.50
4	2	4.0000	3.7500	3.5000	0.50000	0.25000	0.25000	0.1853	2	26	6.67	10.12
4 1/2	2	4.5000	4.2500	4.0000	0.50000	0.25000	0.25000	0.1853	2	9	7.60	13.13
5	2	5.0000	4.7500	4.5000	0.50000	0.25000	0.25000	0.1853	1	55	8.54	16.53

Column groups: *Identification* (Nominal Sizes, Threads per Inch). *Basic Diameters* — Classes 2G, 3G, 4G & 5G (Major, Pitch, Minor Diameter). *Thread Data* — Pitch; Thickness at Pitch Line; Basic Height of Thread; Basic Width of Flat; Lead Angle at Basic Pitch Diameter* (Classes 2G, 3G, 4G & 5G), λ (Deg, Min); Shear Area† Class 3G; Stress Area‡ Class 3G.

* All other dimensions are given in inches.

† Per inch length of engagement of the external thread in line with the minor diameter crests of the internal thread. Figures given are the minimum shear area based on max K_n and min E_s.

‡ Figures given are the minimum stress area based on the mean of the minimum minor and pitch diameters of the external thread. See formulas for shear area and stress area on page 1559.

**Table 4. American National Standard General Purpose Acme Single-Start
Screw Threads — Pitch Diameter Allowances* (ANSI B1.5-1977)**

Nominal Size Range**		Allowances on External Threads†		
Above	To and Including	Class 2G, $0.008 \sqrt{D}$	Class 3G, $0.006 \sqrt{D}$	Class 4G, $0.004 \sqrt{D}$
0	3/16	0.0024	0.0018	0.0012
3/16	5/16	0.0040	0.0030	0.0020
5/16	7/16	0.0049	0.0037	0.0024
7/16	9/16	0.0057	0.0042	0.0028
9/16	11/16	0.0063	0.0047	0.0032
11/16	13/16	0.0069	0.0052	0.0035
13/16	15/16	0.0075	0.0056	0.0037
15/16	1 1/16	0.0080	0.0060	0.0040
1 1/16	1 3/16	0.0085	0.0064	0.0042
1 3/16	1 5/16	0.0089	0.0067	0.0045
1 5/16	1 7/16	0.0094	0.0070	0.0047
1 7/16	1 9/16	0.0098	0.0073	0.0049
1 9/16	1 7/8	0.0105	0.0079	0.0052
1 7/8	2 1/8	0.0113	0.0085	0.0057
2 1/8	2 3/8	0.0120	0.0090	0.0060
2 3/8	2 5/8	0.0126	0.0095	0.0063
2 5/8	2 7/8	0.0133	0.0099	0.0066
2 7/8	3 1/4	0.0140	0.0105	0.0070
3 1/4	3 3/4	0.0150	0.0112	0.0075
3 3/4	4 1/4	0.0160	0.0120	0.0080
4 1/4	4 3/4	0.0170	0.0127	0.0085
4 3/4	5 1/2	0.0181	0.0136	0.0091

All dimensions in inches. Class 5G has no allowance on pitch diameter of either external or internal threads.

* Allowances for Class 2G threads in column 3 also apply to American National Standard Stub Acme threads (ANSI B1.8-1977).

** The values in columns 3 to 5 are to be used for any size within the range shown in columns 1 and 2. These values are calculated from the mean of the range.

It is recommended that the sizes given in Table 3 be used whenever possible.

† An increase of 10 per cent in the allowance is recommended for each inch, or fraction thereof, that the length of engagement exceeds two diameters.

Table 4; for three-start threads, 75 per cent of these allowances; and for four-start threads, 100 per cent of these same values.

These values will provide for a 1/4"–16 Acme 2G thread size, 0.002, 0.003, and 0.004 inch additional clearance for 2-, 3-, and 4-start threads, respectively. For a 5"–2 Acme 3G thread size the additional clearances would be 0.0068, 0.0102, and 0.0136 inch, respectively. GO thread plug gages and taps would be increased by these same values. To maintain the same working tolerances on multi-start threads, the pitch diameter of the NOT GO thread plug gage would also be increased by these same values.

For multi-start threads with more than four starts, it is believed that the 100 per cent allowance provided by the above procedures would be adequate as index spacing variables would generally be no greater than on a four-start thread.

In general, for multi-start threads of Classes 2G, 3G, and 4G the percentages would be applied, usually, to allowances for the same class, respectively. However, in cases where exceptionally good control over lead, angle, and spacing variables would produce close to theoretical values in the product, it is conceivable that these percentages could be applied to Class 3G or Class 4G allowances used on Class 2G internally threaded product. Also these percentages could be applied to Class 4G allowances used on Class 3G internally threaded product. It is not advocated that any change be made in externally threaded products.

Designations for gages or tools for internal threads could cover allowance requirements as follows:

GO and NOT GO thread plug gages for: 2 7/8–0.4p–0.8L Acme 2G with 4G internal thread allowance.

Table 5. American National Standard General Purpose Acme Single-Start Screw Threads — Pitch Diameter Tolerances[1,2] (ANSI B1.5-1977)

(For any particular size of thread, the pitch diameter tolerance is obtained by adding the *diameter increment* from the upper half of the table to the *pitch increment* from the lower half of the table. *Example:* A ¼–16 Acme–2G thread has a pitch diameter tolerance of 0.00300 + 0.00750 = 0.0105 inch.)

	Class of Thread					Class of Thread			
	2G	3G	4G	5G		2G	3G	4G	5G
	Diameter Increment					Diameter Increment			
Nom. Dia.,[3] D	$.006\sqrt{D}$	$.0028\sqrt{D}$	$.002\sqrt{D}$	$.0016\sqrt{D}$	Nom. Dia.,[3] D	$.006\sqrt{D}$	$.0028\sqrt{D}$	$.002\sqrt{D}$	$.0016\sqrt{D}$
¼	.00300	.00140	.00100	.00080	1½	.00735	.00343	.00245	.00196
5/16	.00335	.00157	.00112	.00089	1¾	.00794	.00370	.00265	.00212
⅜	.00367	.00171	.00122	.00098	2	.00849	.00396	.00283	.00226
7/16	.00397	.00185	.00132	.00106	2¼	.00900	.00420	.00300	.00240
½	.00424	.00198	.00141	.00113	2½	.00949	.00443	.00316	.00253
⅝	.00474	.00221	.00158	.00126	2¾	.00995	.00464	.00332	.00265
¾	.00520	.00242	.00173	.00139	3	.01039	.00485	.00346	.00277
⅞	.00561	.00262	.00187	.00150	3½	.01122	.00524	.00374	.00299
1	.00600	.00280	.00200	.00160	4	.01200	.00560	.00400	.00320
1⅛	.00636	.00297	.00212	.00170	4½	.01273	.00594	.00424	.00339
1¼	.00671	.00313	.00224	.00179	5	.01342	.00626	.00447	.00358
1⅜	.00704	.00328	.00235	.00188	…	…	…	…	…

	Class of Thread					Class of Thread			
	2G	3G	4G	5G		2G	3G	4G	5G
	Pitch Increment					Pitch Increment			
Thds. per Inch, n	$.030\sqrt{1/n}$	$.014\sqrt{1/n}$	$.010\sqrt{1/n}$	$.008\sqrt{1/n}$	Thds. per Inch, n	$.030\sqrt{1/n}$	$.014\sqrt{1/n}$	$.010\sqrt{1/n}$	$.008\sqrt{1/n}$
16	.00750	.00350	.00250	.00200	4	.01500	.00700	.00500	.00400
14	.00802	.00374	.00267	.00214	3	.01732	.00808	.00577	.00462
12	.00866	.00404	.00289	.00231	2½	.01897	.00885	.00632	.00506
10	.00949	.00443	.00316	.00253	2	.02121	.00990	.00707	.00566
8	.01061	.00495	.00354	.00283	1½	.02449	.01143	.00816	.00653
6	.01225	.00572	.00408	.00327	1⅓	.02598	.01212	.00866	.00693
5	.01342	.00626	.00447	.00358	1	.03000	.01400	.01000	.00800

All dimensions are given in inches.

[1] The equivalent tolerance on thread thickness is 0.259 times the pitch diameter tolerance.

[2] Columns for the 2G Class of thread in this table also apply to American National Standard Stub Acme threads, ANSI B1.8-1977.

[3] For a nominal diameter between any two tabulated nominal diameters, use the diameter increment for the larger of the two tabulated nominal diameters.

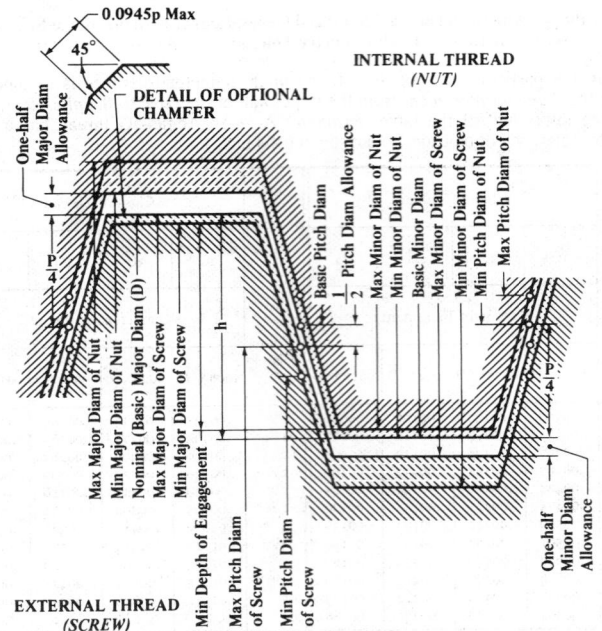

Disposition of Allowances, Tolerances, and Crest Clearances for General
Purpose Single-start Acme Threads (All Classes)

Centralizing Acme Threads. — The five classes of Centralizing Acme threads in
American National Standard ANSI B1.5-1977, designated as 2C, 3C, 4C, 5C and 6C,
have limited clearance at the major diameters of internal and external threads so that
a bearing at the major diameters maintains approximate alignment of the thread axis
and prevents wedging on the flanks of the thread. An alternative series having centralizing control on the *minor* diameter is described on page 1583. For any combination of the five classes of threads covered in this standard some end play or backlash
will result.

Application: These five classes together with the accompanying specifications are
for the purpose of ensuring the interchangeable manufacture of Centralizing Acme
threaded parts. Each user is free to select the classes best adapted to his particular
needs. It is suggested that external and internal threads of the same class be used
together for centralizing assemblies, Class 2C providing the maximum end play or
backlash. If less backlash or end play is desired, Classes 3C, 4C, 5C, and 6C are
provided. The requirement for a centralizing fit is that the sum of the major diameter
tolerance plus the major diameter allowance on the internal thread, and the major
diameter tolerance on the external thread shall equal or be less than the pitch diameter
allowance on the external thread. A Class 2C external thread, which has a larger pitch
diameter allowance than either a Class 3C or 4C, can be used interchangeably with a
Class 2C, 3C, or 4C internal thread and fulfill this requirement. Similarly, a Class 3C
external thread can be used interchangeably with a Class 3C or 4C internal thread,
but only a Class 4C internal thread can be used with a Class 4C external thread.

Table 6. American National Standard Centralizing Acme Screw Thread Form — Basic Dimensions* (ANSI B1.5-1977)

| Thds per Inch, n | Pitch, p | Height of Thread (Basic), $h = p/2$ | Total Height of Thread (All External Threads) $h_s = h + \frac{1}{2}$ allowance† | Thread Thickness (Basic), $t = p/2$ | 45-Deg Chamfer Crest of External Threads | | Max Fillet Radius, Root of Tapped Hole, 0.06p | Fillet Radius at Minor Diameter of Screws | |
					Min Depth, 0.05p	Min Width of Chamfer Flat, 0.0707p		Min (Classes 5 and 6 Only), 0.07p	Max (All), 0.10p
16	0.06250	0.03125	0.0362	0.03125	0.0031	0.0044	0.0040	0.0044	0.0062
14	0.07143	0.03571	0.0407	0.03571	0.0036	0.0050	0.0040	0.0050	0.0071
12	0.08333	0.04167	0.0467	0.04167	0.0042	0.0060	0.0050	0.0058	0.0083
10	0.10000	0.05000	0.0600	0.05000	0.0050	0.0070	0.0060	0.0070	0.0100
8	0.12500	0.06250	0.0725	0.06250	0.0062	0.0090	0.0075	0.0088	0.0125
6	0.16667	0.08333	0.0933	0.08333	0.0083	0.0120	0.0100	0.0117	0.0167
5	0.20000	0.10000	0.1100	0.10000	0.0100	0.0140	0.0120	0.0140	0.0200
4	0.25000	0.12500	0.1350	0.12500	0.0125	0.0180	0.0150	0.0175	0.0250
3	0.33333	0.16667	0.1767	0.16667	0.0167	0.0240	0.0200	0.0233	0.0333
2½	0.40000	0.20000	0.2100	0.20000	0.0200	0.0280	0.0240	0.0280	0.0400
2	0.50000	0.25000	0.2600	0.25000	0.0250	0.0350	0.0300	0.0350	0.0500
1½	0.66667	0.33333	0.3433	0.33333	0.0330	0.0470	0.0400	0.0467	0.0667
1⅓	0.75000	0.37500	0.3850	0.37500	0.0380	0.0530	0.0450	0.0525	0.0750
1	1.00000	0.50000	0.5100	0.50000	0.0500	0.0710	0.0600	0.0700	0.1000

* All dimensions in inches. See diagram on page 1559.
† Allowance is 0.020 inch for 10 or less threads per inch and 0.010 inch for more than 10 threads per inch.

Classes 5C and 6C external and internal threads can be used interchangeably. The average backlash for any cross combination will be between the backlash when both members are Class 5C and when both members are Class 6C.

Thread Form: The thread form is the same as the General Purpose Acme Thread and is shown in the figure on page 1559. The formulas accompanying the figure determine the basic dimensions which are given in Table 6 for the most generally used pitches.

Angle of Thread: The angle between the sides of the thread measured in an axial plane is 29 degrees. The line bisecting this 29-degree angle shall be perpendicular to the axis of the thread.

Thread Series: A series of diameters and pitches is recommended in the Standard as preferred. These diameters and pitches have been chosen to meet present needs with the fewest number of items in order to reduce to a minimum the inventory of both tools and gages. This series of diameters and associated pitches is given in Table 8.

Chamfers and Fillets: External threads have the crest corners chamfered at an angle of 45 degrees with the axis to a minimum depth of $p/20$ and a maximum depth of $p/15$. This corresponds to a minimum width of chamfer flat of $0.0707p$ and a maximum width of $0.0945p$ (see Table 6, columns 6 and 7).

External threads for Classes 2C, 3C, and 4C may have a fillet at the minor diameter not greater than $0.1p$ and for Classes 5C and 6C the minimum fillet is $0.07p$ and the maximum fillet $0.1p$.

Basic Diameters: The maximum major diameter of the external thread is basic and is the nominal major diameter for all classes, except Classes 5C and 6C. The maximum major diameter of Classes 5C and 6C external threads is the major diameter B established by subtracting $0.025\sqrt{D}$ from the nominal diameter D. The minimum pitch diameter of the internal thread is basic for all classes and is equal to the basic major diameter D for Class 2C, 3C, and 4C and the major diameter B for Classes 5C

(Continued on page 1574)

Table 7a. American National Standard Centralizing Acme Single-Start Screw Threads — Formulas for Determining Diameters (ANSI B1.5-1977)

	D = Nominal Size or Diameter in Inches p = Pitch = 1 ÷ Number of Threads per Inch
No.	Classes 2C, 3C, and 4C External Threads (Screws)
1	Major Diam., Max = D (Basic)
2	Major Diam., Min = D minus tolerance from Table 11, cols. 7, 8, or 10.
3	Pitch Diam., Max = Int. Pitch Diam., Min (Formula 9) minus allowance from Table 9, cols. 3, 4, or 5.
4	Pitch Diam., Min = Ext. Pitch Diam., Max (Formula 3) minus tolerance from Table 10.
5	Minor Diam., Max = D minus p minus allowance from Table 11, col. 3.
6	Minor Diam., Min = Ext. Minor Diam., Max (Formula 5) minus 1.5 × Pitch Diam. tolerance from Table 10.
	Classes 2C, 3C, and 4C Internal Threads (Nuts)
7	Major Diam., Min = D plus allowance from Table 11, col. 4.
8	Major Diam., Max = Int. Major Diam., Min (Formula 7) plus tolerance from Table 11, cols. 7, 9, or 11.
9	Pitch Diam., Min = D minus $p/2$. (Basic)
10	Pitch Diam., Max = Int. Pitch Diam., Min (Formula 9) plus tolerance from Table 10.
11	Minor Diam., Min = D minus $0.9p$.
12	Minor Diam., Max = Int. Minor Diam., Min (Formula 11) plus tolerance from Table 11, col. 6.
	Classes 5C and 6C External Threads (Screws)
13	Major Diam., Max = D minus $0.025\sqrt{D}$ (Basic)
14	Major Diam., Min = Ext. Major Diam., Max (Formula 13) minus tolerance from Table 11, cols. 8 or 10.
15	Pitch Diam., Max = Int. Pitch Diam., Min (Formula 21) minus allowance from Table 9, cols. 3 or 4.
16	Pitch Diam., Min = Ext. Pitch Diam., Max (Formula 15) minus tolerance from Table 10.
17	Minor Diam., Max = Ext. Major Diam., Max (Formula 13) minus p minus allowance from Table 11, col. 3.
18	Minor Diam., Min = Ext. Minor Diam., Max (Formula 17) minus 1.5 × Pitch Diam. tolerance from Table 10.
	Classes 5C and 6C Internal Threads (Nuts)
19	Major Diam., Min = Ext. Major Diam., Max (Formula 13) plus allowance from Table 11, col. 4.
20	Major Diam., Max = Int. Major Diam., Min (Formula 19) plus tolerance from Table 11, cols. 9 or 11.
21	Pitch Diam., Min = Ext. Major Diam., Max (Formula 13) minus $p/2$ (Basic)
22	Pitch Diam., Max = Int. Pitch Diam., Min (Formula 21) plus tolerance from Table 10.
23	Minor Diam., Min = Ext. Major Diam., Max (Formula 13) minus $0.9p$.
24	Minor Diam., Max = Int. Minor Diam., Min (Formula 23) plus tolerance from Table 11, col. 6.

Table 7b. Limiting Dimensions of American National Standard Centralizing Acme Single-Start Screw Threads, Classes 2C, 3C, and 4C (ANSI B1.5-1977)

Limiting Diameters		1/2	5/8	3/4	7/8	1	1 1/8	1 1/4	1 3/8	1 1/2
Threads per Inch*		10	8	6	6	5	5	5	4	4
External Threads										
Classes 2C, 3C, and 4C, Major Diameter	Max	0.5000	0.6250	0.7500	0.8750	1.0000	1.1250	1.2500	1.3750	1.5000
Class 2C, Major Diameter	Min	0.4975	0.6222	0.7470	0.8717	0.9965	1.1213	1.2461	1.3709	1.4957
Class 3C, Major Diameter	Min	0.4989	0.6238	0.7487	0.8736	0.9985	1.1234	1.2483	1.3732	1.4982
Class 4C, Major Diameter	Min	0.4993	0.6242	0.7491	0.8741	0.9990	1.1239	1.2489	1.3738	1.4988
Classes 2C, 3C, and 4C, Minor Diameter	Max	0.3800	0.4800	0.5633	0.6883	0.7800	0.9050	1.0300	1.1050	1.2300
Class 2C, Minor Diameter	Min	0.3594	0.4570	0.5371	0.6615	0.7509	0.8753	0.9998	1.0719	1.1965
Class 3C, Minor Diameter	Min	0.3704	0.4693	0.5511	0.6758	0.7664	0.8912	1.0159	1.0896	1.2144
Class 4C, Minor Diameter	Min	0.3731	0.4723	0.5546	0.6794	0.7703	0.8951	1.0199	1.0940	1.2188
Class 2C, Pitch Diameter	Max	0.4443	0.5562	0.6598	0.7842	0.8920	1.0165	1.1411	1.2406	1.3652
Class 2C, Pitch Diameter	Min	0.4306	0.5408	0.6424	0.7663	0.8726	0.9967	1.1210	1.2186	1.3429
Class 3C, Pitch Diameter	Max	0.4458	0.5578	0.6615	0.7861	0.8940	1.0186	1.1433	1.2433	1.3677
Class 3C, Pitch Diameter	Min	0.4394	0.5506	0.6534	0.7778	0.8849	1.0094	1.1339	1.2330	1.3573
Class 4C, Pitch Diameter	Max	0.4472	0.5593	0.6632	0.7880	0.8960	1.0208	1.1455	1.2453	1.3701
Class 4C, Pitch Diameter	Min	0.4426	0.5542	0.6574	0.7820	0.8895	1.0142	1.1388	1.2380	1.3627
Internal Threads										
Classes 2C, 3C, and 4C, Major Diameter	Min	0.5007	0.6258	0.7509	0.8759	1.0010	1.1261	1.2511	1.3762	1.5012
Classes 2C and 3C, Major Diameter	Max	0.5032	0.6286	0.7539	0.8792	1.0045	1.1298	1.2550	1.3803	1.5055
Class 4C, Major Diameter	Max	0.5021	0.6274	0.7526	0.8778	1.0030	1.1282	1.2533	1.3785	1.5036
Classes 2C, 3C, and 4C, Minor Diameter	Min	0.4100	0.5125	0.6000	0.7250	0.8200	0.9450	1.0700	1.1500	1.2750
Classes 2C, 3C, and 4C, Minor Diameter	Max	0.4150	0.5187	0.6083	0.7333	0.8300	0.9550	1.0800	1.1625	1.2875
Class 2C, Pitch Diameter	Min	0.4500	0.5625	0.6667	0.7917	0.9000	1.0250	1.1500	1.2500	1.3750
Class 2C, Pitch Diameter	Max	0.4637	0.5779	0.6841	0.8096	0.9194	1.0448	1.1701	1.2720	1.3973
Class 3C, Pitch Diameter	Min	0.4500	0.5625	0.6667	0.7917	0.9000	1.0250	1.1500	1.2500	1.3750
Class 3C, Pitch Diameter	Max	0.4564	0.5697	0.6748	0.8000	0.9091	1.0342	1.1594	1.2603	1.3854
Class 4C, Pitch Diameter	Min	0.4500	0.5625	0.6667	0.7917	0.9000	1.0250	1.1500	1.2500	1.3750
Class 4C, Pitch Diameter	Max	0.4646	0.5676	0.6725	0.7977	0.9065	1.0316	1.1567	1.2573	1.3824

Nominal Diameter, D

* All other dimensions are in inches. The selection of threads per inch is arbitrary and for the purpose of establishing a standard.

Table 7b (Concluded). Limiting Dimensions of Centralizing Acme Single-Start Screw Threads, Classes 2C, 3C, and 4C (ANSI B1.5-1977)

Limiting Diameters		Nominal Diameter, D									
		5	4½	4	3½	3	2¾	2½	2¼	2	1¾
		Threads per Inch*									
		2	2	2	2	2	3	3	3	4	4
External Threads											
Classes 2C, 3C, and 4C, Major Diameter	Max	5.0000	4.5000	4.0000	3.5000	3.0000	2.7500	2.5000	2.2500	2.0000	1.7500
Class 2C, Major Diameter	Min	4.9922	4.4926	3.9930	3.4935	2.9939	2.7442	2.4945	2.2448	1.9951	1.7454
Class 3C, Major Diameter	Min	4.9966	4.4968	3.9970	3.4972	2.9974	2.7475	2.4976	2.2478	1.9979	1.7480
Class 4C, Major Diameter	Min	4.9978	4.4979	3.9980	3.4981	2.9983	2.7483	2.4984	2.2485	1.9986	1.7487
Classes 2C, 3C, and 4C, Minor Diameter	Max	4.4800	3.9800	3.4800	2.9800	2.4800	2.3967	2.1467	1.8967	1.7300	1.4800
Class 2C, Minor Diameter	Min	4.4281	3.9291	3.4302	2.9314	2.4326	2.3558	2.1065	1.8572	1.6948	1.4456
Class 3C, Minor Diameter	Min	4.4558	3.9563	3.4568	2.9573	2.4579	2.3776	2.1279	1.8783	1.7136	1.4640
Class 4C, Minor Diameter	Min	4.4627	3.9630	3.4614	2.9618	2.4642	2.3831	2.1333	1.8835	1.7183	1.4685
Class 2C, Pitch Diameter	Max	4.7319	4.2330	3.7340	3.2350	2.7360	2.5700	2.3207	2.0713	1.8637	1.6145
	Min	4.6973	4.1991	3.7008	3.2026	2.7044	2.5427	2.2939	2.0450	1.8402	1.5916
Class 3C, Pitch Diameter	Max	4.7364	4.2373	3.7380	3.2388	2.7395	2.5734	2.3238	2.0743	1.8665	1.6171
	Min	4.7202	4.2215	3.7225	3.2237	2.7248	2.5607	2.3113	2.0620	1.8555	1.6064
Class 4C, Pitch Diameter	Max	4.7409	4.2415	3.7420	3.2425	2.7430	2.5767	2.3270	2.0773	1.8693	1.6198
	Min	4.7294	4.2302	3.7309	3.2317	2.7325	2.5676	2.3181	2.0685	1.8615	1.6122
Internal Threads											
Classes 2C, 3C, and 4C, Major Diameter	Min	5.0022	4.5021	4.0020	3.5019	3.0017	2.7517	2.5016	2.2515	2.0014	1.7513
Classes 2C and 3C, Major Diameter	Max	5.0100	4.5095	4.0090	3.5084	3.0078	2.7575	2.5071	2.2567	2.0063	1.7559
Class 4C, Major Diameter	Max	5.0067	4.5063	4.0060	3.5056	3.0052	2.7550	2.5048	2.2545	2.0042	1.7539
Classes 2C, 3C, and 4C, Minor Diameter	Max	4.5500	4.0500	3.5500	3.0500	2.5500	2.4500	2.2000	1.9500	1.7750	1.5250
	Min	4.5750	4.0750	3.5750	3.0750	2.5750	2.4667	2.2167	1.9667	1.7875	1.5375
Class 2C, Pitch Diameter	Min	4.7500	4.2500	3.7500	3.2500	2.7500	2.5833	2.3333	2.0833	1.8750	1.6250
	Max	4.7846	4.2839	3.7832	3.2824	2.7816	2.6106	2.3601	2.1096	1.8985	1.6479
Class 3C, Pitch Diameter	Min	4.7500	4.2500	3.7500	3.2500	2.7500	2.5833	2.3333	2.0833	1.8750	1.6250
	Max	4.7662	4.2658	3.7655	3.2651	2.7647	2.5960	2.3458	2.0956	1.8860	1.6357
Class 4C, Pitch Diameter	Min	4.7500	4.2500	3.7500	3.2500	2.7500	2.5833	2.3333	2.0833	1.8750	1.6250
	Max	4.7615	4.2613	3.7611	3.2608	2.7605	2.5924	2.3422	2.0921	1.8828	1.6326

* All other dimensions are given in inches. The selection of threads per inch is arbitrary and for the purpose of establishing a standard.

Table 7c. Limiting Dimensions of American National Standard Centralizing Acme Single-Start Screw Threads, Classes 5C and 6C (ANSI B1.5-1977)

Limiting Diameters and Tolerances		1/2	5/8	3/4	7/8	1	1 1/8	1 1/4	1 3/8	1 1/2
		Nominal Diameter, D								
		10	8	6	6	5	5	5	4	4
						Threads per Inch*				
External Threads										
Classes 5C and 6C, Major Diameter	Max	0.4823	0.6052	0.7283	0.8516	0.9750	1.0985	1.2220	1.3457	1.4694
Class 5C, Major Diameter	Min	0.4812	0.6040	0.7270	0.8502	0.9735	1.0969	1.2203	1.3439	1.4676
Class 6C, Major Diameter	Min	0.4816	0.6044	0.7274	0.8507	0.9740	1.0974	1.2209	1.3445	1.4682
Classes 5C and 6C, Minor Diameter	Max	0.3623	0.4602	0.5417	0.6649	0.7550	0.8785	1.0020	1.0757	1.1994
Class 5C, Minor Diameter	Min	0.3527	0.4495	0.5295	0.6524	0.7414	0.8647	0.9879	1.0603	1.1838
Class 6C, Minor Diameter	Min	0.3554	0.4525	0.5330	0.6560	0.7453	0.8686	0.9919	1.0647	1.1882
Class 5C, Pitch Diameter	Max	0.4266	0.5364	0.6381	0.7608	0.8670	0.9900	1.1131	1.2113	1.3346
	Min	0.4202	0.5292	0.6300	0.7525	0.8579	0.9808	1.1037	1.2010	1.3242
Class 6C, Pitch Diameter	Max	0.4281	0.5380	0.6398	0.7627	0.8690	0.9921	1.1153	1.2137	1.3371
	Min	0.4235	0.5329	0.6340	0.7567	0.8625	0.9855	1.1086	1.2064	1.3397
Internal Threads										
Classes 5C and 6C, Major Diameter	Min	0.4830	0.6060	0.7292	0.8525	0.9760	1.0996	1.2231	1.3469	1.4706
Class 5C, Major Diameter	Max	0.4855	0.6088	0.7322	0.8558	0.9795	1.1033	1.2270	1.3510	1.4749
Class 6C, Major Diameter	Max	0.4844	0.6076	0.7309	0.8544	0.9780	1.1017	1.2253	1.3492	1.4730
Classes 5C and 6C, Minor Diameter	Min	0.3923	0.4927	0.5783	0.7016	0.7950	0.9185	1.0420	1.1207	1.2444
	Max	0.3973	0.4989	0.5866	0.7099	0.8050	0.9285	1.0520	1.1332	1.2569
Class 5C, Pitch Diameter	Min	0.4323	0.5427	0.6450	0.7683	0.8750	0.9985	1.1220	1.2207	1.3444
	Max	0.4387	0.5499	0.6531	0.7766	0.8841	1.0077	1.1314	1.2310	1.3548
Class 6C, Pitch Diameter	Min	0.4323	0.5427	0.6450	0.7683	0.8750	0.9985	1.1220	1.2207	1.3444
	Max	0.4369	0.5478	0.6508	0.7743	0.8815	1.0051	1.1287	1.2280	1.3518

* All other dimensions are given in inches. The selection of threads per inch is arbitrary and for the purpose of establishing a standard.

Table 7c (*Concluded*). **Limiting Dimensions of Centralizing Acme Single-Start Screw Threads, Classes 5C and 6C (ANSI B1.5-1977)**

Limiting Diameters		Nominal Diameter, D									
		1¾	2	2¼	2½	2¾	3	3½	4	4½	5
		Threads per Inch*									
		4	4	3	3	3	2	2	2	2	2
External Threads											
Classes 5C and 6C, Major Diameter	Max	1.7169	1.9646	2.2125	2.4605	2.7085	2.9567	3.4532	3.9500	4.4470	4.9441
Class 5C, Major Diameter	Min	1.7149	1.9625	2.2103	2.4581	2.7060	2.9541	3.4504	3.9470	4.4438	4.9407
Class 6C, Major Diameter	Min	1.7156	1.9632	2.2110	2.4589	2.7068	2.9550	3.4513	3.9480	4.4449	4.9419
Classes 5C and 6C, Minor Diameter	Max	1.4469	1.6946	1.8592	2.1072	2.3552	2.4367	2.9332	3.4300	3.9270	4.4241
Class 5C, Minor Diameter	Min	1.4308	1.6782	1.8408	2.0884	2.3361	2.4146	2.9106	3.4068	3.9033	4.3999
Class 6C, Minor Diameter	Min	1.4354	1.6829	1.8460	2.0938	2.3416	2.4209	2.9170	3.4134	3.9101	4.4068
Class 5C, Pitch Diameter	Max	1.5814	1.8283	2.0338	2.2812	2.5285	2.6927	3.1882	3.6840	4.1800	4.6760
	Min	1.5707	1.8173	2.0215	2.2687	2.5158	2.6780	3.1731	3.6685	4.1642	4.6598
Class 6C, Pitch Diameter	Max	1.5840	1.8311	2.0368	2.2843	2.5319	2.6962	3.1920	3.6880	4.1843	4.6805
	Min	1.5764	1.8233	2.0280	2.2754	2.5228	2.6857	3.1812	3.6769	4.1730	4.6690
Internal Threads											
Classes 5C and 6C, Major Diameter	Min	1.7182	1.9660	2.2140	2.4621	2.7102	2.9584	3.4551	3.9520	4.4491	4.9463
Class 5C, Major Diameter	Max	1.7228	1.9709	2.2192	2.4676	2.7160	2.9645	3.4616	3.9590	4.4565	4.9541
Class 6C, Major Diameter	Max	1.7208	1.9688	2.2170	2.4653	2.7135	2.9619	3.4588	3.9560	4.4533	4.9508
Classes 5C and 6C, Minor Diameter	Min	1.4919	1.7396	1.9125	2.1605	2.4085	2.5067	3.0032	3.5000	3.9970	4.4941
Class 5C, Minor Diameter	Max	1.5044	1.7521	1.9292	2.1772	2.4252	2.5317	3.0282	3.5250	4.0020	4.5191
Class 5C, Pitch Diameter	Min	1.5919	1.8396	2.0458	2.2938	2.5418	2.7067	3.2032	3.7000	4.1970	4.6941
	Max	1.6026	1.8506	2.0581	2.3063	2.5545	2.7214	3.2183	3.7155	4.2128	4.7103
Class 6C, Pitch Diameter	Min	1.5919	1.8396	2.0458	2.2938	2.5418	2.7067	3.2032	3.7000	4.1970	4.6941
	Max	1.5995	1.8474	2.0546	2.3027	2.5509	2.7172	3.2140	3.7111	4.2083	4.7056

* All other dimensions are given in inches. The selection of threads per inch is arbitrary and for the purpose of establishing a standard.

Table 8. American National Standard Centralizing Acme Single-Start Screw Thread Data (ANSI B1.5-1977)

Nominal Sizes (All Classes)	Threads per Inch, n	Basic Major Diameter, D	Pitch Diameter, E=(D−h)	Minor Diameter, K=(D−2h)	Major Diameter, B=(D−0.025√D)	Pitch Diameter, E=(B−h)	Minor Diameter, K=(B−2h)	Pitch, p	Thickness at Pitch Line, t=p/2	Basic Height of Thread, h=p/2	Basic Width of Flat, F=0.3707p	Lead Angle 2C,3C,4C λ Deg	Lead Angle 2C,3C,4C λ Min	Lead Angle 5C,6C λ Deg	Lead Angle 5C,6C λ Min
1/4	16	0.2500	0.2188	0.1875	....	....	....	0.06250	0.03125	0.03125	0.0232	5	12	....	....
5/16	14	0.3125	0.2768	0.2411	....	....	....	0.07143	0.03571	0.03571	0.0265	4	42	....	....
3/8	12	0.3750	0.3333	0.2917	....	....	....	0.08333	0.04167	0.04167	0.0309	4	33	....	....
7/16	12	0.4375	0.3958	0.3542	....	....	....	0.08333	0.04167	0.04167	0.0309	3	50	....	....
1/2	10	0.5000	0.4500	0.4000	0.4823	0.4323	0.3823	0.10000	0.05000	0.05000	0.0371	4	3	4	13
5/8	8	0.6250	0.5625	0.5000	0.6052	0.5427	0.4802	0.12500	0.06250	0.06250	0.0463	4	3	4	12
3/4	6	0.7500	0.6667	0.5833	0.7284	0.6451	0.5617	0.16667	0.08333	0.08333	0.0618	4	33	4	42
7/8	6	0.8750	0.7917	0.7083	0.8516	0.7683	0.6849	0.16667	0.08333	0.08333	0.0618	3	50	3	57
1	5	1.0000	0.9000	0.8000	0.9750	0.8750	0.7750	0.20000	0.10000	0.10000	0.0741	4	3	4	10
1 1/8	5	1.1250	1.0250	0.9250	1.0985	0.9985	0.8985	0.20000	0.10000	0.10000	0.0741	3	33	3	39
1 1/4	5	1.2500	1.1500	1.0500	1.2220	1.1220	1.0220	0.20000	0.10000	0.10000	0.0741	3	10	3	15
1 3/8	4	1.3750	1.2500	1.1250	1.3457	1.2207	1.0957	0.25000	0.12500	0.12500	0.0927	3	39	3	44
1 1/2	4	1.5000	1.3750	1.2500	1.4694	1.3444	1.2194	0.25000	0.12500	0.12500	0.0927	3	19	3	23
1 3/4	4	1.7500	1.6250	1.5000	1.7169	1.5919	1.4669	0.25000	0.12500	0.12500	0.0927	2	48	2	52
2	4	2.0000	1.8750	1.7500	1.9646	1.8396	1.7146	0.25000	0.12500	0.12500	0.0927	2	26	2	29
2 1/4	3	2.2500	2.0833	1.9167	2.2125	2.0458	1.8792	0.33333	0.16667	0.16667	0.1236	2	55	2	58
2 1/2	3	2.5000	2.3333	2.1667	2.4605	2.2938	2.1272	0.33333	0.16667	0.16667	0.1236	2	36	2	39
2 3/4	3	2.7500	2.5833	2.4167	2.7085	2.5418	2.3752	0.33333	0.16667	0.16667	0.1236	2	21	2	23
3	2	3.0000	2.7500	2.5000	2.9567	2.7067	2.4567	0.50000	0.25000	0.25000	0.1853	3	19	3	22
3 1/2	2	3.5000	3.2500	3.0000	3.4532	3.2032	2.9532	0.50000	0.25000	0.25000	0.1853	2	48	2	51
4	2	4.0000	3.7500	3.5000	3.9500	3.7000	3.4500	0.50000	0.25000	0.25000	0.1853	2	26	2	28
4 1/2	2	4.5000	4.2500	4.0000	4.4470	4.1970	3.9470	0.50000	0.25000	0.25000	0.1853	2	9	2	10
5	2	5.0000	4.7500	4.5000	4.9441	4.6941	4.4441	0.50000	0.25000	0.25000	0.1853	1	55	1	56

* All other dimensions are given in inches.

Table 9. American National Standard Centralizing Acme Single-Start Screw Threads — Pitch Diameter Allowances (ANSI B1.5-1977)

Nominal Size Range*		Allowances on External Threads†		
		Centralizing		
Above	To and Including	Classes 2C and 5C, $0.008\sqrt{D}$	Classes 3C and 6C, $0.006\sqrt{D}$	Class 4C, $0.004\sqrt{D}$
0	³⁄₁₆	0.0024	0.0018	0.0012
³⁄₁₆	⁵⁄₁₆	0.0040	0.0030	0.0020
⁵⁄₁₆	⁷⁄₁₆	0.0049	0.0037	0.0024
⁷⁄₁₆	⁹⁄₁₆	0.0057	0.0042	0.0028
⁹⁄₁₆	¹¹⁄₁₆	0.0063	0.0047	0.0032
¹¹⁄₁₆	¹³⁄₁₆	0.0069	0.0052	0.0035
¹³⁄₁₆	¹⁵⁄₁₆	0.0075	0.0056	0.0037
¹⁵⁄₁₆	1¹⁄₁₆	0.0080	0.0060	0.0040
1¹⁄₁₆	1³⁄₁₆	0.0085	0.0064	0.0042
1³⁄₁₆	1⁵⁄₁₆	0.0089	0.0067	0.0045
1⁵⁄₁₆	1⁷⁄₁₆	0.0094	0.0070	0.0047
1⁷⁄₁₆	1⁹⁄₁₆	0.0098	0.0073	0.0049
1⁹⁄₁₆	1⁷⁄₈	0.0105	0.0079	0.0052
1⁷⁄₈	2¹⁄₈	0.0113	0.0085	0.0057
2¹⁄₈	2³⁄₈	0.0120	0.0090	0.0060
2³⁄₈	2⁵⁄₈	0.0126	0.0095	0.0063
2⁵⁄₈	2⁷⁄₈	0.0133	0.0099	0.0066
2⁷⁄₈	3¹⁄₄	0.0140	0.0105	0.0070
3¹⁄₄	3³⁄₄	0.0150	0.0112	0.0075
3³⁄₄	4¹⁄₄	0.0160	0.0120	0.0080
4¹⁄₄	4³⁄₄	0.0170	0.0127	0.0085
4³⁄₄	5¹⁄₂	0.0181	0.0136	0.0091

All dimensions are given in inches.
* The values in cols. 3 to 5 are to be used for any size within the range shown in cols. 1 and 2. These values are calculated from the mean of the range.
It is recommended that the sizes given in Table 8 be used whenever possible.
† An increase of 10 per cent in the allowance is recommended for each inch, or fraction thereof, that the length of engagement exceeds two diameters.

and 6C, minus the basic height of thread, h. The basic minor diameter is equal to the basic major diameter D for Classes 2C, 3C, and 4C, and the major diameter B for Classes 5C and 6C minus twice the basic thread height, $2h$. The minimum minor diameter of the internal thread for all classes is $0.1p$ above basic.

Length of Engagement: The tolerances specified in this Standard are applicable to lengths of engagement not exceeding twice the nominal major diameter.

Major and Minor Diameter Allowances: A minimum diametral clearance is provided at the minor diameter of all external threads by establishing the maximum minor diameter 0.020 inch below the basic minor diameter for 10 threads per inch and coarser, and 0.010 inch for finer pitches and by establishing the minimum minor diameter of the internal thread $0.1p$ greater than the basic minor diameter.

A minimum diametral clearance at the major diameter is obtained by establishing the minimum major diameter of the internal thread $0.001\sqrt{D}$ above the basic major diameter. These allowances are shown in Table 11.

Major and Minor Diameter Tolerances: The tolerances on the major and minor diameters of the external and internal threads are listed in Table 11 and are based upon the formulas given in the column headings.

Pitch Diameter Allowances: Allowances applied to the pitch diameter of the external thread for all classes are given in Table 9. For Classes 5C and 6C, the pitch diameter allowances are equal to those of Classes 3C and 4C, respectively, plus an additional amount of $0.002\sqrt{D}$ which is required to maintain the centralizing fit and minimum end play of $0.0005\sqrt{D}$ for Classes 5C and 6C.

(*Continued on page* 1578)

Table 10. American National Standard Centralizing Acme Single-Start Screw Threads — Pitch Diameter Tolerances[1] (ANSI B1.5-1977)

(For any particular size of thread, the pitch diameter tolerance is obtained by adding the *diameter increment* from the upper half of the table to the *pitch increment* from the lower half of the table. *Example:* A ¼–16 Acme–2C thread has a pitch diameter tolerance of 0.00300 + 0.00750 = 0.0105 inch.)

	Class of Thread				Class of Thread		
	2C	3C and 5C	4C and 6C		2C	3C and 5C	4C and 6C
	Diameter Increment				Diameter Increment		
Nom. Dia.,[2] D	$.006\sqrt{D}$	$.0028\sqrt{D}$	$.002\sqrt{D}$	Nom. Dia.,[2] D	$.006\sqrt{D}$	$.0028\sqrt{D}$	$.002\sqrt{D}$
¼	.00300	.00140	.00100	1½	.00735	.00343	.00245
5/16	.00335	.00157	.00112	1¾	.00794	.00370	.00265
3/8	.00367	.00171	.00122	2	.00849	.00396	.00283
7/16	.00397	.00185	.00132	2¼	.00900	.00420	.00300
½	.00424	.00198	.00141	2½	.00949	.00443	.00316
5/8	.00474	.00221	.00158	2¾	.00995	.00464	.00332
¾	.00520	.00242	.00173	3	.01039	.00485	.00346
7/8	.00561	.00262	.00187	3½	.01122	.00524	.00374
1	.00600	.00280	.00200	4	.01200	.00560	.00400
1 1/8	.00636	.00297	.00212	4½	.01273	.00594	.00424
1¼	.00671	.00313	.00224	5	.01342	.00626	.00447
1 3/8	.00704	.00328	.00235	. . .	. . .	. . .	. . .

	Class of Thread				Class of Thread		
	2C	3C and 5C	4C and 6C		2C	3C and 5C	4C and 6C
	Pitch Increment				Pitch Increment		
Thds. per Inch, n	$.030\sqrt{1/n}$	$.014\sqrt{1/n}$	$.010\sqrt{1/n}$	Thds. per Inch, n	$.030\sqrt{1/n}$	$.014\sqrt{1/n}$	$.010\sqrt{1/n}$
16	.00750	.00350	.00250	4	.01500	.00700	.00500
14	.00802	.00374	.00267	3	.01732	.00808	.00577
12	.00866	.00404	.00289	2½	.01897	.00885	.00632
10	.00949	.00443	.00316	2	.02121	.00990	.00707
8	.01061	.00495	.00354	1½	.02449	.01143	.00816
6	.01225	.00572	.00408	1⅓	.02598	.01212	.00866
5	.01342	.00626	.00447	1	.03000	.01400	.01000

All dimensions are given in inches.
[1] The equivalent tolerance on thread thickness is 0.259 times the pitch diameter tolerance.
[2] For a nominal diameter between any two tabulated nominal diameters, use the diameter increment for the larger of the two tabulated nominal diameters.

Table 11. American National Standard Centralizing Acme Single-Start Screw Threads — Tolerances and Allowances for Major and Minor Diameters (ANSI B1.5-1977)

Size (Nom.)	Thds* per Inch	Allowance From Basic Major and Minor Diameters (All Classes)			Toler. on Minor Diam,*** All Internal Threads, (Plus 0.05p)	Tolerance on Major Diameter Plus on Internal, Minus on External Threads				
		Minor Diam,† All External Threads (Minus)	Internal Thread			Class 2C External and Internal Threads, 0.0035√D	Classes 3C and 5C		Classes 4C and 6C	
			Major Diam,‡ (Plus 0.0010√D)	Minor Diam,† (Plus 0.1p)			External Thread, 0.0015√D	Internal Thread, 0.0035√D	External Thread, 0.0010√D	Internal Thread, 0.0020√D
¼	16	0.010	...	...	0.0050	...	...	...	...	...
5/16	14	0.010	...	...	0.0050	...	...	...	...	...
⅜	12	0.010	...	...	0.0050	...	...	...	...	...
7/16	12	0.010	...	...	0.0050	...	...	...	...	...
½	10	0.020	0.0007	0.0100	0.0050	0.0025	0.0011	0.0025	0.0007	0.0014
⅝	8	0.020	0.0008	0.0125	0.0062	0.0028	0.0012	0.0028	0.0008	0.0016
¾	6	0.020	0.0009	0.0167	0.0083	0.0030	0.0013	0.0030	0.0009	0.0017
⅞	6	0.020	0.0009	0.0167	0.0083	0.0033	0.0014	0.0033	0.0009	0.0019
1	5	0.020	0.0010	0.0200	0.0100	0.0035	0.0015	0.0035	0.0010	0.0020
1⅛	5	0.020	0.0011	0.0200	0.0100	0.0037	0.0016	0.0037	0.0011	0.0021
1¼	5	0.020	0.0011	0.0200	0.0100	0.0039	0.0017	0.0039	0.0011	0.0022
1⅜	4	0.020	0.0012	0.0250	0.0125	0.0041	0.0018	0.0041	0.0012	0.0023
1½	4	0.020	0.0012	0.0250	0.0125	0.0043	0.0018	0.0043	0.0012	0.0024
1¾	4	0.020	0.0013	0.0250	0.0125	0.0046	0.0020	0.0046	0.0013	0.0026
2	4	0.020	0.0014	0.0250	0.0125	0.0049	0.0021	0.0049	0.0014	0.0028
2¼	3	0.020	0.0015	0.0333	0.0167	0.0052	0.0022	0.0052	0.0015	0.0030
2½	3	0.020	0.0016	0.0333	0.0167	0.0055	0.0024	0.0055	0.0016	0.0032
2¾	2	0.020	0.0017	0.0500	0.0250	0.0058	0.0025	0.0058	0.0017	0.0033
3	2	0.020	0.0017	0.0500	0.0250	0.0061	0.0026	0.0061	0.0017	0.0035
3½	2	0.020	0.0019	0.0500	0.0250	0.0065	0.0028	0.0065	0.0019	0.0037
4	2	0.020	0.0020	0.0500	0.0250	0.0070	0.0030	0.0070	0.0020	0.0040
4½	2	0.020	0.0021	0.0500	0.0250	0.0074	0.0032	0.0074	0.0021	0.0042
5	2	0.020	0.0022	0.0500	0.0250	0.0078	0.0034	0.0078	0.0022	0.0045

* All other dimensions are given in inches. Intermediate pitches take the values of the next coarser pitch listed.
** Values for intermediate diameters should be calculated from the formulas in column headings, but ordinarily may be interpolated.
*** To avoid a complicated formula and still provide an adequate tolerance, the pitch factor is used as a basis, with the minimum tolerance set at 0.005 in.
† The minimum clearance at the minor diameter between the internal and external thread is 1.5 × pitch diameter tolerance.
‡ The minimum clearance at the major diameter between the internal and external thread is equal to col. 4.
Tolerance on minor diameter of all external threads is 1.5 × pitch diameter tolerance.

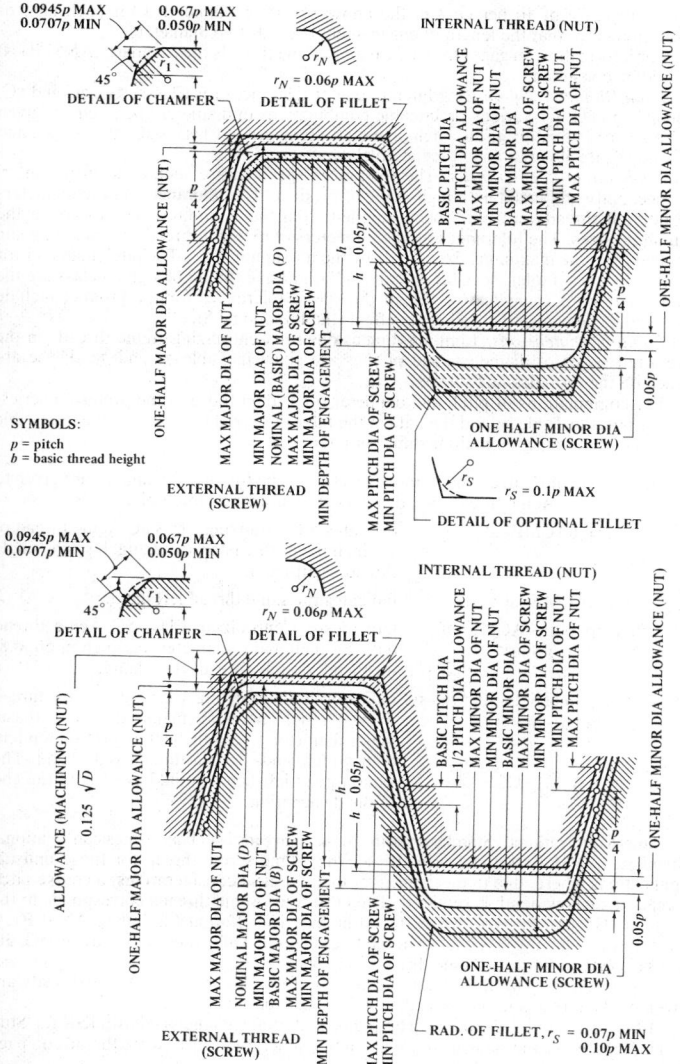

Disposition of Allowances, Tolerances, and Crest Clearances for Centralizing Single-Start Acme Threads — Classes 2C, 3C, and 4C (Upper Diagram) and Classes 5C and 6C (Lower Diagram)

An increase of 10 per cent in the allowance is recommended for each inch or fraction thereof that the length of engagement exceeds two diameters.

For information on gages for Centralizing Acme threads the Standard (ANSI B1.5) should be consulted.

Pitch Diameter Tolerances: Pitch diameter tolerances for Classes 2C, 3C and 5C, and 4C and 6C, for various practicable combinations of diameter and pitch are given in Table 10. The ratios of the pitch diameter tolerances of Classes 2C, 3C and 5C, and 4C and 6C are 3.0, 1.4, and 1, respectively.

Application of Tolerances: The tolerances specified are such as to insure interchangeability and maintain a high grade of product. The tolerances on the diameters of internal threads are plus, being applied from the minimum sizes to above the minimum sizes. The tolerances on the diameters of external threads are minus, being applied from the maximum sizes to below the maximum sizes. The pitch diameter (or thread thickness) tolerances for an external or internal thread of a given class are the same. The pitch diameter (or thread thickness) tolerances for the product include lead and angle errors. This gives, in effect, the Functional Size.

Limiting Dimensions: Limiting dimensions for Centralizing Acme threads in the preferred series of diameters and pitches are given in Table 7b and 7c. These are based on the formulas in Table 7a.

For combinations of pitch and diameter other than those in the preferred series, the formulas in Table 7a and the data in the tables referred to therein make it possible to readily determine the limiting dimension required.

Designation of Centralizing Acme Threads. — The following examples are given to show how these Acme threads are designated on drawings and tools:

1¾–6ACME–4C	indicates a Centralizing Class 4C Acme thread of 1¾-inch major diameter, 0.1667-inch pitch, single thread, right-hand.
1¾–6ACME–4C–LH	indicates the same thread left-hand.
2⅞–0.4p–0.8L–ACME–3C (Two Start)	indicates a Centralizing Class 3C Acme thread with 2⅞-inch major diameter, 0.4-inch pitch, 0.8-inch lead, double thread, right-hand.
2½–0.3333p–0.6667L–ACME–5C (Two Start)	indicates a Centralizing Class 5C Acme thread with 2½-inch nominal major diameter (basic major diameter 2.4605 inches), 0.3333-inch pitch, 0.6667-inch lead, double thread, right-hand. The same thread left-hand would have LH at the end of the designation.

American National Standard Stub Acme Threads. — This American National Standard (ANSI B1.8-1977) provides a Stub Acme screw thread for those unusual applications where, due to mechanical or metallurgical considerations, a coarse-pitch thread of shallow depth is required. The fit of Stub Acme threads corresponds to the Class 2G General Purpose Acme thread in American National Standard ANSI B1.5-1977. For a fit having less backlash, the tolerances and allowances for Classes 3G, 4G and 5G General Purpose Acme threads may be used.

Thread Form: The thread form and basic formulas for Stub Acme threads are given on page 1559 and the basic dimensions in Table 12.

Allowances and Tolerances: The major and minor diameter allowances for Stub Acme threads are the same as those given for General Purpose Acme threads on page 1557.

Pitch diameter allowances for Stub Acme threads are the same as for Class 2G

(Continued on page 1582)

**Table 12. American National Standard Stub Acme Screw Thread Form —
Basic Dimensions (ANSI B1.8-1977)**

					Width of Flat	
Thds. per Inch* n	Pitch, $p = 1/n$	Height of Thread (Basic), $0.3p$	Total Height of Thread, $0.3p + \frac{1}{2}$ allowance†	Thread Thickness (Basic), $p/2$	Crest of Internal Thread (Basic), $0.4224p$	Root of Internal Thread, $0.4224p - 0.259 \times$ allowance†
16	0.06250	0.01875	0.0238	0.03125	0.0264	0.0238
14	0.07143	0.02143	0.0264	0.03571	0.0302	0.0276
12	0.08333	0.02500	0.0300	0.04167	0.0352	0.0326
10	0.10000	0.03000	0.0400	0.05000	0.0422	0.0370
9	0.11111	0.03333	0.0433	0.05556	0.0469	0.0417
8	0.12500	0.03750	0.0475	0.06250	0.0528	0.0476
7	0.14286	0.04285	0.0529	0.07143	0.0603	0.0551
6	0.16667	0.05000	0.0600	0.08333	0.0704	0.0652
5	0.20000	0.06000	0.0700	0.10000	0.0845	0.0793
4	0.25000	0.07500	0.0850	0.12500	0.1056	0.1004
3½	0.28571	0.08571	0.0957	0.14286	0.1207	0.1155
3	0.33333	0.10000	0.1100	0.16667	0.1408	0.1356
2½	0.40000	0.12000	0.1300	0.20000	0.1690	0.1638
2	0.50000	0.15000	0.1600	0.25000	0.2112	0.2060
1½	0.66667	0.20000	0.2100	0.33333	0.2816	0.2764
1⅓	0.75000	0.22500	0.2350	0.37500	0.3168	0.3116
1	1.00000	0.30000	0.3100	0.50000	0.4224	0.4172

* All other dimensions in inches. See diagram, page 1559.
† Allowance is 0.020 inch for 10 or less threads per inch and 0.010 inch for more than 10 threads per inch.

**Table 13a. American National Standard Stub Acme Single-Start Screw Threads —
Formulas for Determining Diameters (ANSI B1.8-1977)**

D = Basic Major Diameter and Nominal Size in Inches
p = Pitch = 1 ÷ Number of Threads per Inch
E = Basic Pitch Diameter = $D - 0.3p$
K = Basic Minor Diameter = $D - 0.6p$

No.	External Threads (Screws)
1	Major Diam., Max = D
2	Major Diam., Min. = D *minus* $0.05p$.
3	Pitch Diam., Max. = E *minus* allowance from column 3, Table 4.
4	Pitch Diam., Min. = Pitch Diam., Max. (Formula 3) *minus* Class 2G tolerance from Table 5.
5	Minor Diam., Max. = K *minus* 0.020 for 10 threads per inch and coarser and 0.010 for finer pitches.
6	Minor Diam., Min. = Minor Diam., Max. (Formula 5) *minus* Class 2G pitch diameter tolerance from Table 5.

No.	Internal Threads (Nuts)
7	Major Diam., Min. = D *plus* 0.020 for 10 threads per inch and coarser and 0.010 for finer pitches.
8	Major Diam., Max. = Major Diam., Min. (Formula 7) *plus* Class 2G pitch diameter tolerance from Table 5.
9	Pitch Diam., Min. = E
10	Pitch Diam., Max. = Pitch Diam., Min. (Formula 9) *plus* Class 2G tolerance from Table 5.
11	Minor Diam., Min. = K
12	Minor Diam., Max = Minor Diam., Min. (Formula 11) *plus* $0.05p$.

Table 13b. Limiting Dimensions for American National Standard Stub Acme Single-Start Screw Threads (ANSI B1.8-1977)

			1/4	5/16	3/8	7/16	1/2	5/8	3/4	7/8	1	1 1/8	1 1/4	1 3/8
		Nominal Diameter, D												
		Threads per Inch*	16	14	12	12	10	8	6	6	5	5	5	4
Limiting Diameters		**External Threads**												
Major Diam.	Max (D)		0.2500	0.3125	0.3750	0.4375	0.5000	0.6250	0.7500	0.8750	1.0000	1.1250	1.2500	1.3750
	Min		0.2469	0.3089	0.3708	0.4333	0.4950	0.6188	0.7417	0.8667	0.9900	1.1150	1.2400	1.3625
Pitch Diam.	Max		0.2272	0.2871	0.3451	0.4076	0.4643	0.5812	0.6931	0.8175	0.9320	1.0565	1.1811	1.2906
	Min		0.2167	0.2757	0.3328	0.3950	0.4506	0.5658	0.6757	0.7996	0.9126	1.0367	1.1610	1.2686
Minor Diam.	Max		0.2024	0.2597	0.3150	0.3775	0.4200	0.5300	0.6300	0.7550	0.8600	0.9850	1.1100	1.2050
	Min		0.1919	0.2483	0.3027	0.3649	0.4063	0.5146	0.6126	0.7371	0.8406	0.9652	1.0899	1.1830
		Internal Threads												
Major Diam.	Min		0.2600	0.3225	0.3850	0.4475	0.5200	0.6450	0.7700	0.8950	1.0200	1.1450	1.2700	1.3950
	Max		0.2705	0.3339	0.3973	0.4601	0.5337	0.6604	0.7874	0.9129	1.0394	1.1648	1.2901	1.4170
Pitch Diam.	Min		0.2312	0.2911	0.3500	0.4125	0.4700	0.5875	0.7000	0.8250	0.9400	1.0650	1.1900	1.3000
	Max		0.2417	0.3025	0.3623	0.4251	0.4837	0.6029	0.7174	0.8429	0.9594	1.0848	1.2101	1.3220
Minor Diam.	Min		0.2125	0.2696	0.3250	0.3875	0.4400	0.5500	0.6500	0.7750	0.8800	1.0050	1.1300	1.2250
	Max		0.2156	0.2732	0.3292	0.3917	0.4450	0.5562	0.6583	0.7833	0.8900	1.0150	1.1400	1.2375

* All other dimensions are given in inches.

Table 13b. (Concluded). Limiting Dimensions for American National Standard Stub Acme Single-Start Screw Threads (ANSI B1.8-1977)

Limiting Diameters		1½	1¾	2	2¼	2½	2¾	3	3½	4	4½	5
Nominal Diameter, D		1½	1¾	2	2¼	2½	2¾	3	3½	4	4½	5
Threads per Inch*		4	4	4	3	3	3	2	2	2	2	2
External Threads												
Major Diam.	Max (D)	1.5000	1.7500	2.0000	2.2500	2.5000	2.7500	3.0000	3.5000	4.0000	4.5000	5.0000
	Min	1.4875	1.7375	1.9875	2.2333	2.4833	2.7333	2.9750	3.4750	3.9750	4.4750	4.9750
Pitch Diam.	Max	1.4152	1.6645	1.9137	2.1380	2.3874	2.6367	2.8360	3.3350	3.8340	4.3330	4.8319
	Min	1.3929	1.6416	1.8902	2.1117	2.3606	2.6094	2.8044	3.3026	3.8008	4.2991	4.7973
Minor Diam.	Max	1.3300	1.5800	1.8300	2.0300	2.2800	2.5300	2.6800	3.1800	3.6800	4.1800	4.6800
	Min	1.3077	1.5571	1.8065	2.0037	2.2532	2.5027	2.6484	3.1476	3.6468	4.1461	4.6454
Internal Threads												
Major Diam.	Min	1.5200	1.7700	2.0200	2.2700	2.5200	2.7700	3.0200	3.5200	4.0200	4.5200	5.0200
	Max	1.5423	1.7929	2.0435	2.2963	2.5468	2.7973	3.0516	3.5524	4.0532	4.5539	5.0546
Pitch Diam.	Min	1.4250	1.6750	1.9250	2.1500	2.4000	2.6500	2.8500	3.3500	3.8500	4.3500	4.8500
	Max	1.4473	1.6979	1.9485	2.1763	2.4268	2.6773	2.8816	3.3824	3.8832	4.3839	4.8846
Minor Diam.	Min	1.3500	1.6000	1.8500	2.0500	2.3000	2.5500	2.7000	3.2000	3.7000	4.2000	4.7000
	Max	1.3625	1.6125	1.8625	2.0667	2.3167	2.5667	2.7250	3.2250	3.7250	4.2250	4.7250

* All other dimensions are given in inches.

General Purpose Acme threads and are given in column 3 of Table 4. Pitch diameter tolerances for Stub Acme threads are the same as for Class 2G General Purpose Acme threads and are given in columns 2 and 7 of Table 5.

Limiting Dimensions: Limiting dimensions of American Standard Stub Acme threads may be determined by using the formulas given in Table 13a, or directly from Table 13b. The diagram below shows the limits of size for Stub Acme threads.

Thread Series: The same preferred series of diameters and pitches for General Purpose Acme threads (Table 3) are recommended for Stub Acme threads.

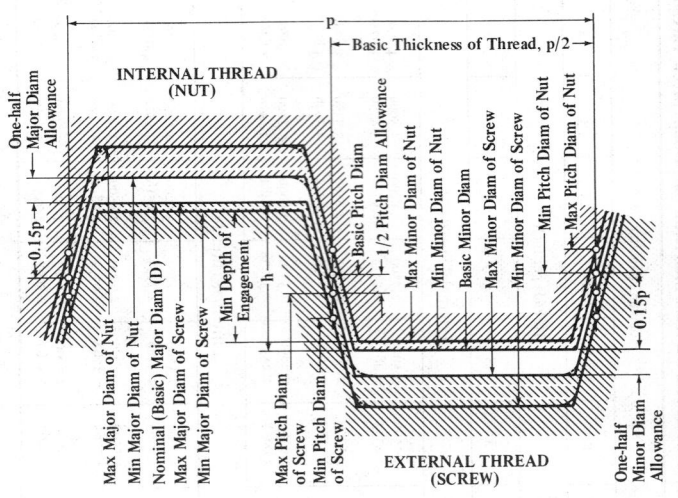

Limits of Size, Allowances, Tolerances, and Crest
Clearances for American National Standard Stub Acme Threads

Stub Acme Thread Designations. — The method of designation for Standard Stub Acme threads is illustrated in the following examples: ½–20 Stub Acme indicates a ½-inch major diameter, 20 threads per inch, right hand, single thread, Standard Stub Acme thread. The designation ½–20 Stub Acme–LH indicates the same thread except that it is left hand.

Alternative Stub Acme Threads. — Since one Stub Acme thread form may not meet the requirements of all applications, basic data for two of the other commonly used forms is included in the appendix of the American Standard for Stub Acme Threads. These so-called Modified Form 1 and Modified Form 2 threads utilize the same tolerances and allowances as Standard Stub Acme threads and have the same major diameter and basic thread thickness at the pitchline (0.5p). The basic height of Form 1 threads, h, is 0.375p; for Form 2 it is 0.250p. The basic width of flat at the crest of the internal thread is 0.4030p for Form 1 and 0.4353p for Form 2.

The pitch diameter and minor diameter for Form 1 threads will be smaller than similar values for the Standard Stub Acme Form and for Form 2 they will be larger owing to the differences in basic thread height h. Therefore, in calculating the dimensions of Form 1 and Form 2 threads using Formulas 1 through 12 in Table 13a, it is

only necessary to substitute the following values in applying the formulas: For Form 1, $E = D - 0.375p$, $K = D - 0.75p$; for Form 2, $E = D - 0.25p$, $K = D - 0.5p$.

Thread Designation: These threads are designated in the same manner as Standard Stub Acme threads except for the insertion of either M1 or M2 after "Acme." Thus, ½–20 Stub Acme M1 for a Form 1 thread; and ½–20 Stub Acme M2 for a Form 2 thread.

Acme Centralizing Threads — Alternative Series with Minor Diameter Centralizing Control. — When Acme centralizing threads are produced in single units or in very small quantities (and principally in sizes larger than the range of commercial taps and dies) where the manufacturing process employs cutting tools (such as lathe cutting), it may be economically advantageous and therefore desirable to have the centralizing control of the mating threads located at the *minor diameters*.

Particularly under the above-mentioned type of manufacturing, the advantages cited for minor diameter centralizing control over centralizing control at the major diameters of the mating threads are:

(1) Greater ease and faster checking of machined thread dimensions. It is much easier to measure the minor diameter (root) of the external thread and the mating minor diameter (crest or bore) of the internal thread than it is to determine the major diameter (root) of the internal thread and the major diameter (crest or turn) of the external thread;

(2) better manufacturing control of the machined size due to greater ease of checking;

(3) lower manufacturing costs.

60-Degree Stub Thread. — Former American Standard B1.3-1941 included a 60-degree stub thread for use where design or operating conditions could be better satisfied by the use of this thread, or other modified threads, than by Acme threads. Data for 60-Degree Stub thread form are given in the accompanying diagram.

60-Degree Stub Thread

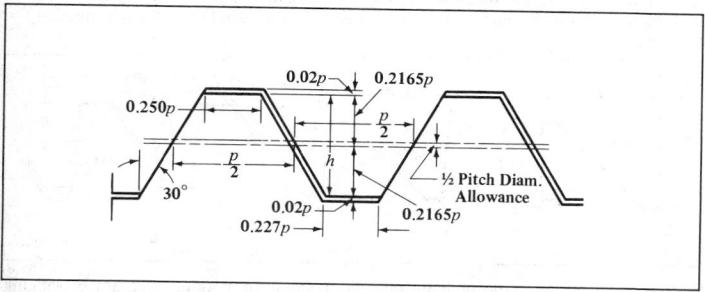

A clearance of at least $0.02 \times$ pitch is added to depth h to produce extra depth, thus avoiding interference with threads of mating part at minor or major diameters.

Basic thread thickness at pitch line $= 0.5 \times$ pitch p; basic depth $h = 0.433 \times$ pitch; basic width of flat at crest $= 0.25 \times$ pitch; width of flat at root of screw thread $= 0.227 \times$ pitch; basic pitch diameter $=$ basic major diameter $- 0.433 \times$ pitch; basic minor diameter $=$ basic major diameter $- 0.866 \times$ pitch.

10-Degree Modified Square Thread. — The included angle between the sides of the thread is 10 degrees (see accompanying diagram). The angle of 10 degrees results in a thread which is the practical equivalent of a "square thread," and yet is capable of economical production. Multiple thread milling cutters and ground thread taps should not be specified for modified square threads of the larger lead angles without consulting the cutting tool manufacturer.

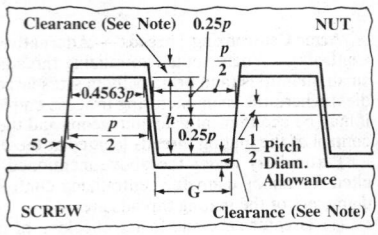

Formulas: In the following formulas, D = basic major diameter; E = basic pitch diameter; K = basic minor diameter; p = pitch; h = basic depth of thread on screw depth when there is no clearance between root of screw and crest of thread on nut; t = basic thickness of thread at pitch line; F = basic width of flat at crest of screw thread; G = basic width of flat at root of screw thread; C = clearance between root of screw and crest of thread on nut: $E = D - 0.5p$; $K = D - p$; $h = 0.5p$ (see Note); $t = 0.5p$; $F = 0.4563p$; $G = 0.4563p - (0.17 \times C)$. *Note:* A clearance should be added to depth h to avoid interference with threads of mating parts at minor or major diameters.

Threads of Buttress Form. — The buttress form of thread has certain advantages in applications involving exceptionally high stresses along the thread axis in one direction only. The contacting flank of the thread, which takes the thrust, is referred to as the *pressure flank* and is so nearly perpendicular to the thread axis that the radial component of the thrust is reduced to a minimum. Because of the small radial thrust, this form of thread is particularly applicable where tubular members are screwed together, as in the case of breech mechanisms of large guns and airplane propeller hubs.

Diagram A shows a common form. The front or load-resisting face is perpendicular to the axis of the screw and the thread angle is 45 degrees. According to one rule, the pitch $P = 2 \times$ screw diameter $\div$ 15. The thread depth d may equal $\frac{3}{4} \times$ pitch, making the flat $f = \frac{1}{8} \times$ pitch. Sometimes depth d is reduced to $\frac{2}{3} \times$ pitch, making $f = \frac{1}{6} \times$ pitch.

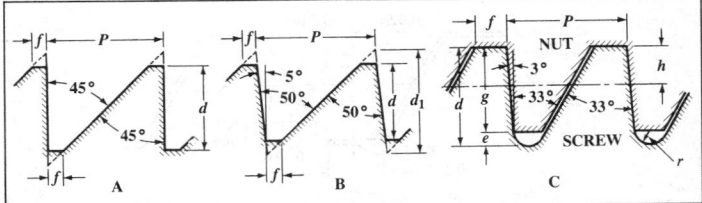

The load-resisting side or flank may be inclined an amount (diagram B) ranging usually from 1 to 5 degrees to avoid cutter interference in milling the thread. With an angle of 5 degrees and an included thread angle of 50 degrees, if the width of the flat f at both crest and root equals $\frac{1}{8} \times$ pitch, then the thread depth equals $0.69 \times$ pitch or $\frac{3}{4} d_1$.

The saw-tooth form of thread illustrated by diagram C is known in Germany as the "Sägengewinde" and in Italy as the "Fillettatura a dente di Sega." Pitches are standardized from 2 millimeters up to 48 millimeters in the German and Italian

specifications. The front face inclines 3 degrees from the perpendicular and the included angle is 33 degrees.

The thread depth d for the screw = 0.86777 × pitch P. The thread depth g for the nut = 0.75 × pitch. Dimension $h = 0.341 \times P$. The width f of flat at the crest of the thread on the screw = 0.26384 × pitch. Radius r at the root = 0.12427 × pitch. The clearance space $e = 0.11777 \times$ pitch.

American National Standard Buttress Inch Screw Threads. — The buttress form of thread has certain advantages in applications involving exceptionally high stresses along the thread axis in one direction only. As the thrust side (load flank) of the standard buttress thread is made very nearly perpendicular to the thread axis, the radial component of the thrust is reduced to a minimum. On account of the small radial thrust, the buttress form of thread is particularly applicable when tubular members are screwed together. Examples of actual applications are the breech assemblies of large guns, airplane propeller hubs, and columns for hydraulic presses.

**Table 1. American National Standard Buttress Inch Screw Threads —
Basic Dimensions (ANSI B1.9-1973)**

Thds.* per Inch	Pitch, p	Basic Height of Thread, $h = 0.6p$	Height of Sharp-V Thread, $H = 0.89064p$	Crest Truncation, $f = 0.14532p$	Height of Thread, h_s or $h_n = 0.66271p$	Max. Root Truncation,† $s = 0.0826p$	Max. Root Radius,‡ $r = 0.0714p$	Width of Flat at Crest, $F = 0.16316p$
20	0.0500	0.0300	0.0445	0.0073	0.0331	0.0041	0.0036	0.0082
16	0.0625	0.0375	0.0557	0.0091	0.0414	0.0052	0.0045	0.0102
12	0.0833	0.0500	0.0742	0.0121	0.0552	0.0069	0.0059	0.0136
10	0.1000	0.0600	0.0891	0.0145	0.0663	0.0083	0.0071	0.0163
8	0.1250	0.0750	0.1113	0.0182	0.0828	0.0103	0.0089	0.0204
6	0.1667	0.1000	0.1484	0.0242	0.1105	0.0138	0.0119	0.0271
5	0.2000	0.1200	0.1781	0.0291	0.1325	0.0165	0.0143	0.0326
4	0.2500	0.1500	0.2227	0.0363	0.1657	0.0207	0.0179	0.0408
3	0.3333	0.2000	0.2969	0.0484	0.2209	0.0275	0.0238	0.0543
2½	0.4000	0.2400	0.3563	0.0581	0.2651	0.0330	0.0286	0.0653
2	0.5000	0.3000	0.4453	0.0727	0.3314	0.0413	0.0357	0.0816
1½	0.6667	0.4000	0.5938	0.0969	0.4418	0.0551	0.0476	0.1088
1¼	0.8000	0.4800	0.7125	0.1163	0.5302	0.0661	0.0572	0.1305
1	1.0000	0.6000	0.8906	0.1453	0.6627	0.0826	0.0714	0.1632

* All other dimensions are in inches.
† Minimum root truncation is one-half of maximum.
‡ Minimum root radius is one-half of maximum.

**Table 2. American National Standard Diameter — Pitch Combinations
for 7°/45° Buttress Threads (ANSI B1.9-1973)**

Preferred Nominal Major Diameters, Inches	Threads per Inch*	Preferred Nominal Major Diameters, Inches	Threads per Inch*
0.5, 0.625, 0.75	(20, 16, 12)	4.5, 5, 5.5, 6	12, 10, 8, (6, 5, 4), 3
0.875, 1.0	(16, 12, 10)	7, 8, 9, 10	10, 8, 6, (5, 4, 3), 2.5, 2
1.25, 1.375, 1.5	16, (12, 10, 8), 6	11, 12, 14, 16	10, 8, 6, 5, (4, 3, 2.5), 2, 1.5, 1.25
1.75, 2, 2.25, 2.5	16, 12, (10, 8, 6), 5, 4	18, 20, 22, 24	8, 6, 5, 4, (3, 2.5, 2), 1.5, 1.25, 1
2.75, 3, 3.5, 4	16, 12, 10, (8, 6, 5), 4		

* Preferred pitches are in parentheses.

In selecting the form of thread recommended as standard (ANSI B1.9-1973), man-ufacture by milling, grinding, rolling, or other suitable means, has been taken into consideration. All dimensions are in inches.

Form of Thread: The form of the buttress thread is shown in the accompanying figure and has the following characteristics:

a. A load flank angle, measured in an axial plane, of 7 degrees from the normal to the axis. (*Continued on page* 1588.)

Form of American National Standard 7°/45° Buttress Thread with 0.6p Basic Height of Thread Engagement

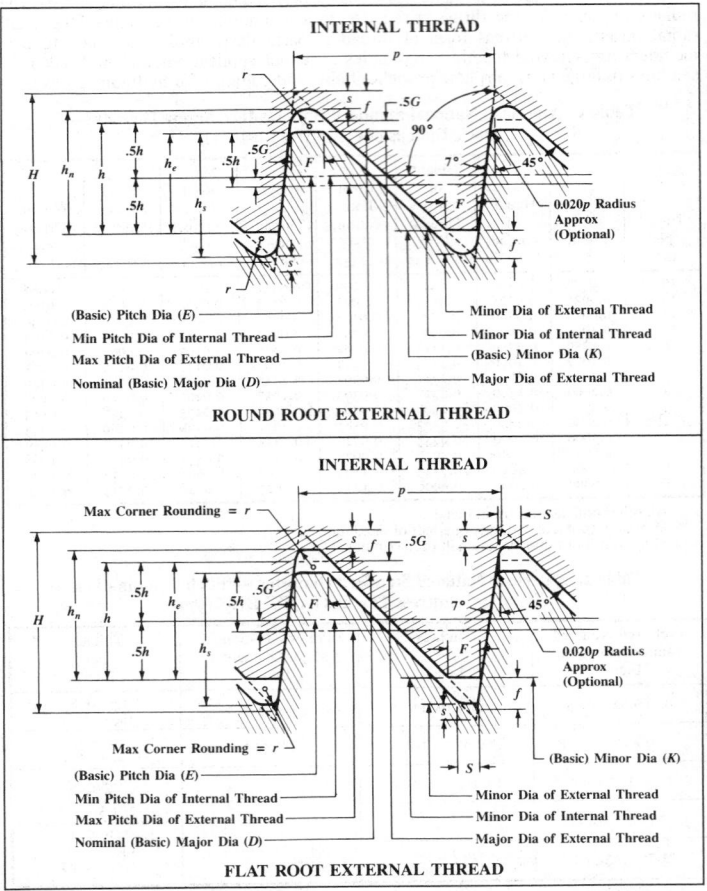

ROUND ROOT EXTERNAL THREAD

FLAT ROOT EXTERNAL THREAD

Heavy Line Indicates Basic Form

Table 3. **American National Standard Buttress Inch Screw Thread Symbols and Formulas** (ANSI B1.9-1973)

Thread Element	Max. Material (Basic)	Min. Material
Pitch	p	
Height of sharp-V thread	$H = 0.89064p$	
Basic height of thread engagement	$h = 0.6p$	
Root radius (theoretical) (see footnote [a])	$r = 0.07141p$	Min. $r = 0.0357p$
Root truncation	$s = 0.0826p$	Min. $s = 0.5$; Max. $s = 0.0413p$
Root truncation for flat root form	$s = 0.0826p$	Min. $s = 0.5$; Max. $s = 0.0413p$
Flat width for flat root form	$S = 0.0928p$	Min. $S = 0.0464p$
Allowance	G (see text)	
Height of thread engagement	$h_e = h - 0.5G$	Min. $h_e =$ Max. $h_e -$ [0.5 tol. on major diam. external thread $+ 0.5$ tol. on minor diam. internal thread].
Crest truncation	$f = 0.14532p$	
Crest width	$F = 0.16316p$	
Major diameter	D	
Major diameter of internal thread	$D_n = D + 0.12542p$	Max. $D_n =$ Max. pitch diam. of internal thread $+ 0.80803p$
Major diameter of external thread	$D_s = D - G$	Min. $D_s = D - G - D$ tol.
Pitch diameter	E	
Pitch diameter of internal thread (see footnote [b])	$E_n = D - h$	Max. $E_n = D - h + PD$ tol.
Pitch diameter of external thread (see footnote [c])	$E_s = D - h - G$	Min. $E_s = D - h - G - PD$ tol.
Minor diameter	K	
Minor diameter of external thread	$K_s = D - 1.32542p - G$	Min. $K_s =$ Min. pitch diam. of external thread $- 0.80803p$
Minor diameter of internal thread	$K_n = D - 2h$	Min. $K_n = D - 2h + K$ tol.
Height of thread of internal thread	$h_n = 0.66271p$	
Height of thread of external thread	$h_s = 0.66271p$	
Pitch diameter increment for lead	ΔE_l	
Pitch diameter increment for 45° clearance flank angle	ΔE_{α_1}	
Pitch diameter increment for 7° load flank angle	ΔE_{α_2}	
Length of engagement	L_e	

[a] Unless the flat root form is specified, the rounded root form of the external and internal thread shall be a continuous, smoothly blended curve within the zone defined by $0.07141p$ maximum to $0.0357p$ minimum radius. The resulting curve shall have no reversals and sudden angular variations, and shall be tangent to the flanks of the thread. There is, in practice, almost no chance that the rounded thread form will be achieved strictly as basically specified, that is, as a true radius.

[b] The pitch diameter X tolerances for GO and NOT GO threaded plug gages are applied to the internal product limits for E_n and Max. E_n.

[c] The pitch diameter W tolerances for GO and NOT GO threaded setting plug gages are applied to the external product limits for E_s and Min. E_s.

b. A clearance flank angle, measured in an axial plane, of 45 degrees from the normal to the axis.

c. Equal truncations at the crests of the external and internal threads such that the basic height of thread engagement (assuming no allowance) is equal to 0.6 of the pitch.

d. Equal radii, at the roots of the external and internal basic thread forms tangential to the load flank and the clearance flank. (There is, in practice, almost no chance that the thread forms will be achieved strictly as basically specified, that is, as true radii.) When specified, equal flat roots of the external and internal thread may be supplied.

Buttress Thread Tolerances. — Tolerances from basic size on external threads are applied in a minus direction and on internal threads in a plus direction.

Pitch Diameter Tolerances: The following formula is used for determining the pitch diameter product tolerance for Class 2 (standard grade) external or internal threads:

$$\text{PD tolerance} = 0.002 \sqrt[3]{D} + 0.00278 \sqrt{L_e} + 0.00854 \sqrt{p}$$

where D = basic major diameter of external thread (assuming no allowance)
L_e = length of engagement
p = pitch of thread

When the length of engagement is taken as $10p$, the formula reduces to:

$$0.002 \sqrt[3]{D} + 0.0173 \sqrt{p}$$

It is to be noted that this formula relates specifically to Class 2 (standard grade) PD tolerances. Class 3 (precision grade) PD tolerances are two-thirds of Class 2 PD tolerances. Pitch diameter tolerances based on this latter formula, for various diameter pitch combinations, are given in Tables 4 and 5.

Functional Size: Deviations in lead and flank angle of product threads increase the functional size of an external thread and decrease the functional size of an internal thread by the cumulative effect of the diameter equivalents of these deviations. The functional size of all buttress product threads shall not exceed the maximum-material limit.

Tolerances on Major Diameter of External Thread and Minor Diameter of Internal Thread: Unless otherwise specified, these tolerances should be the same as the pitch diameter tolerance for the class used.

Tolerances on Minor Diameter of External Thread and Major Diameter of Internal Thread: It will be sufficient in most instances to state only the maximum minor diameter of the external thread and the minimum major diameter of the internal thread without any tolerance. However, the root truncation from a sharp V should not be greater than $0.0826p$ nor less than $0.0413p$.

Lead and Flank Angle Deviations for Class 2: The deviations in lead and flank angles may consume the entire tolerance zone between maximum and minimum material product limits given in Table 4.

Diameter Equivalents for Variations in Lead and Flank Angles for Class 3: The combined diameter equivalents of variations in lead (including helix deviations), and flank angle for Class 3, shall not exceed 50 percent of the pitch diameter tolerances given in Table 5.

Tolerances on Taper and Roundness: There are no requirements for taper and roundness for Class 2 buttress screw threads.

The major and minor diameter of Class 3 buttress thread shall not taper or be out-of-round to the extent that specified limits for major and minor diameter are

Table 4. American National Standard Class 2 (Standard Grade) Tolerances for Buttress Inch Screw Threads (ANSI B1.9-1973)

Thds. per Inch	Pitch,* p Inch	Basic Major Diameter, Inch									Pitch† Increment, 0.0173$\sqrt{p}$ Inch
		From 0.5 thru 0.7	Over 0.7 thru 1.0	Over 1.0 thru 1.5	Over 1.5 thru 2.5	Over 2.5 thru 4	Over 4 thru 6	Over 6 thru 10	Over 10 thru 16	Over 16 thru 24	
		Tolerance on Major Diameter of External Thread, Pitch Diameter of External and Internal Threads, and Minor Diameter of Internal Thread, Inch									
20	0.0500	.0056	….	….	….	….	….	….	….	….	.00387
16	0.0625	.0060	.0062	.0065	.0068	.0073	….	….	….	….	.00432
12	0.0833	.0067	.0069	.0071	.0075	.0080	.0084	….	….	….	.00499
10	0.1000	….	.0074	.0076	.0080	.0084	.0089	.0095	.0102	….	.00547
8	0.1250	….	….	.0083	.0086	.0091	.0095	.0101	.0108	.0115	.00612
6	0.1667	….	….	.0092	.0096	.0100	.0105	.0111	.0118	.0125	.00706
5	0.2000	….	….	….	.0103	.0107	.0112	.0117	.0124	.0132	.00774
4	0.2500	….	….	….	.0112	.0116	.0121	.0127	.0134	.0141	.00865
3	0.3333	….	….	….	….	.0134	.0140	.0147	.0154	….	.00999
2.5	0.4000	….	….	….	….	….	.0149	.0156	.0164	….	.01094
2.0	0.5000	….	….	….	….	….	.0162	.0169	.0177	….	.01223
1.5	0.6667	….	….	….	….	….	….	.0188	.0196	….	.01413
1.25	0.8000	….	….	….	….	….	….	.0202	.0209	….	.01547
1.0	1.0000	….	….	….	….	….	….	….	.0227	….	.01730
Diameter Increment,‡ 0.002$\sqrt[3]{D}$		.00169	.00189	.00215	.00252	.00296	.00342	.00400	.00470	.00543	

* For threads with pitches not shown in this table, pitch increment to be used in tolerance formula is to be determined by use of formula P.D. Tolerance = $0.002\sqrt[3]{D} + 0.00278\sqrt{L_e} + 0.00854\sqrt{p}$ where: D = basic major diameter of external thread (assuming no allowance), L_e = length of engagement, and p = pitch of thread.

† When the length of engagement is taken as $10p$, the formula reduces to: $0.002\sqrt[3]{D} + 0.0173\sqrt{p}$.

‡ Diameter D, used in diameter increment formula, is based on the average of the range.

Table 5. American National Standard Class 3 (Precision Grade) Tolerances for Buttress Inch Screw Threads (ANSI B1.9-1973)

Threads per Inch	Pitch, p Inch	Basic Major Diameter, Inch								
		From 0.5 thru 0.7	Over 0.7 thru 1.0	Over 1.0 thru 1.5	Over 1.5 thru 2.5	Over 2.5 thru 4	Over 4 thru 6	Over 6 thru 10	Over 10 thru 16	Over 16 thru 24
		Tolerance on Major Diameter of External Thread, Pitch Diameter of External and Internal Threads, and Minor Diameter of Internal Thread, Inch								
20	0.0500	.0037	….	….	….	….	….	….	….	….
16	0.0625	.0040	.0042	.0043	.0046	.0049	….	….	….	….
12	0.0833	.0044	.0046	.0048	.0050	.0053	.0056	….	….	….
10	0.1000	….	.0049	.0051	.0053	.0056	.0059	.0063	.0068	….
8	0.1250	….	….	.0055	.0058	.0061	.0064	.0067	.0072	.0077
6	0.1667	….	….	.0061	.0064	.0067	.0070	.0074	.0078	.0083
5	0.2000	….	….	….	.0068	.0071	.0074	.0078	.0083	.0088
4	0.2500	….	….	….	.0074	.0077	.0080	.0084	.0089	.0094
3	0.3333	….	….	….	….	.0089	.0093	.0098	.0103	….
2.5	0.4000	….	….	….	….	….	.0100	.0104	.0109	….
2.0	0.5000	….	….	….	….	….	.0108	.0113	.0118	….
1.5	0.6667	….	….	….	….	….	….	.0126	.0130	….
1.25	0.8000	….	….	….	….	….	….	.0135	.0139	….
1.0	1.0000	….	….	….	….	….	….	….	.0152	….

Table 6. American National Standard External Thread Allowances for Classes 2 and 3 Buttress Inch Screw Threads (ANSI B1.9-1973)

Threads per Inch	Pitch, p Inch	Basic Major Diameter, Inch								
		From 0.5 thru 0.7	Over 0.7 thru 1.0	Over 1.0 thru 1.5	Over 1.5 thru 2.5	Over 2.5 thru 4	Over 4 thru 6	Over 6 thru 10	Over 10 thru 16	Over 16 thru 24
		Allowance on Major, Minor and Pitch Diameters of External Thread, Inch								
20	0.0500	.0037	....	....	....	....	....	....	....	....
16	0.0625	.0040	.0042	.0043	.0046	.0049	....	....	....	....
12	0.0833	.0044	.0046	.0048	.0050	.0053	.0056	....	....	....
10	0.1000	....	.0049	.0051	.0053	.0056	.0059	.0063	.0068	....
8	0.1250	....	....	.0055	.0058	.0061	.0064	.0067	.0072	.0077
6	0.1667	....	....	.0061	.0064	.0067	.0070	.0074	.0078	.0083
5	0.2000	....	....	....	.0068	.0071	.0074	.0078	.0083	.0088
4	0.2500	....	....	....	....	.0074	.0077	.0080	.0084	.0089	.0094
3	0.3333	....	....	....	....	....	.0089	.0093	.0098	.0103
2.5	0.4000	....	....	....	....	....	....	.0100	.0104	.0109
2.0	0.5000	....	....	....	....	....	....	.0108	.0113	.0118
1.5	0.6667	....	....	....	....	....	....	....	.0126	.0130
1.25	0.8000	....	....	....	....	....	....	....	.0135	.0139
1.0	1.0000	....	....	....	....	....	....	....	....	.0152

exceeded. The taper and out-of-roundness of the pitch diameter for Class 3 buttress threads shall not exceed 50 percent of the pitch diameter tolerances.

Allowances for Easy Assembly. — An allowance (clearance) should be provided on all external threads to secure easy assembly of parts. The amount of the allowance is deducted from the nominal major, pitch and minor diameters of the external thread in order to determine the maximum material condition of the external thread.

The minimum internal thread is basic.

The amount of the allowance is the same for both classes and is equal to the Class 3 pitch diameter tolerance as calculated by the formulas previously given. The allowances for various diameter-pitch combinations are given in Table 6.

Example Showing Dimensions for a Typical Buttress Thread. — The dimensions for a 2-inch diameter, 4 threads per inch, Class 2 buttress thread with flank angles of 7 degrees and 45 degrees are:

h = Basic thread height = 0.1500 (Table 1)
$h_s = h_n$ = Height of thread in external and internal threads = 0.1657 (Table 1)
G = Pitch diameter allowance on external thread = 0.0074 (Table 6)
Tolerance on PD of external and internal threads = 0.0112 (Table 4)
Tolerance on major diameter of external thread and minor diameter of internal thread = 0.0112 (Table 4)

Internal Thread
Basic Major Diameter = D = 2.0000
Min Major Diameter = $D - 2h + 2h_n$ = 2.0314 (see Table 1)
Min Pitch Diameter = $D - h$ = 1.8500 (see Table 1)
Max Pitch Diameter = $D - h + PD$ Tolerance = 1.8612 (see Table 4)
Min Minor Diameter = $D - 2h$ = 1.7000 (see Table 1)
Max Minor Diameter = $D - 2h$ + Minor Diameter Tolerance
= 1.7112 (see Table 4)

External Thread

Max Major Diameter $= D - G = 1.9926$ (see Table 6)
Min Major Diameter $= D - G -$ Major Diameter Tolerance
$= 1.9814$ (see Tables 4 and 6)
Max Pitch Diameter $= D - h - G = 1.8426$ (see Tables 1 and 6)
Min Pitch Diameter $= D - h - G - PD$ Tolerance $= 1.8314$ (see Table 4)
Max Minor Diameter $= D - G - 2h_s = 1.6612$ (see Tables 1 and 6)

Buttress Thread Designations. — When only the designation, BUTT is used, the thread is "pull" type buttress (external thread pulls) with the clearance flank leading and the pressure flank 7° following. When the designation, PUSH–BUTT is used, the thread is a push type buttress (external thread pushes) with the load flank 7° leading and the 45° clearance flank following. Whenever possible this description should be confirmed by a simplified view showing thread angles on the drawing of the product that has the buttress thread.

Standard Buttress Threads: A buttress thread is considered to be standard when: (a) opposite flank angles are 7° and 45°; (b) basic thread height is $0.6p$; (c) tolerances and allowances are as shown in Tables 4 through 6; and (d) length of engagement is $10p$ or less.

Thread Designation Abbreviations: In thread designations on drawings, tools, gages, and in specifications, the following abbreviations and letters are to be used:

BUTT for buttress thread, pull type
PUSH-BUTT for buttress thread, push type
LH for left-hand thread (Absence of LH indicates that the thread is a right-hand thread.)
P for pitch
L for lead
A for external thread
B for internal thread

Note: Absence of A or B after thread class indicates that designation covers both the external and internal threads.

Le for length of thread engagement
SPL for special
FL for flat root thread
E for pitch diameter
TPI for threads per inch
THD for thread

Designation Sequence for Buttress Inch Screw Threads. — When designating single-start standard buttress threads the nominal size is given first, the threads per inch next, then PUSH if the internal member is to push, but nothing if it is to pull, then the class of thread (2 or 3), then whether external (A) or internal (B), then LH if left-hand, but nothing if right-hand, and finally FL if a flat root thread, but nothing if a radiused root thread; thus, 2.5-8 BUTT-2A indicates a 2.5 inch, 8 threads per inch buttress thread, Class 2 external, right-hand, internal member to pull, with radiused root of thread. The designation 2.5-8 PUSH-BUTT-2A-LH-FL signifies a 2.5 inch size, 8 threads per inch buttress thread with internal member to push, Class 2 external, left-hand, and flat root.

A multiple-start standard buttress thread is similarly designated but the pitch is given instead of the threads per inch, this is followed by the lead and the number of starts is indicated in parentheses after the class of thread. Thus, 10-0.25P-0.5L — BUTT-3B (2 start) indicates a 10-inch thread with 4 threads per inch, 0.5 inch lead, buttress form with internal member to pull, Class 3 internal, 2 starts, with radiused root of thread.

British Standard Pipe Threads for Non-pressure-tight Joints. — The threads in BS 2779:1973 — "Specifications for Pipe Threads where Pressure-tight Joints are not Made on the Threads" are Whitworth form parallel fastening threads that are generally used for fastening purposes such as the mechanical assembly of component parts of fittings, cocks and valves. They are not suitable where pressure-tight joints are made on the threads.

The crests of the basic Whitworth thread form may be truncated to certain limits of size given in the Standard except on internal threads, when they are likely to be assembled with external threads conforming to the requirements of BS 21 British Standard Pipe Threads for Pressure-tight Joints (see page 1621).

For external threads two classes of tolerance are provided and for internal, one class. The two classes of tolerance for external threads are Class A and Class B. For economy of manufacture the class B fit should be chosen whenever possible. The class A is reserved for those applications where the closer tolerance is essential. Class A tolerance is an entirely negative value, equivalent to the internal thread tolerance. Class B tolerance is an entirely negative value twice that of class A tolerance. Tables showing limits and dimensions are given in the Standard.

The thread series specified in this Standard shall be designated by the letter "G". A typical reference on a drawing might be "G½", for internal thread; "G ½ A", for external thread, class A; and "G ½ B", for external thread, class B. Where no class reference is stated for external threads, that of class B will be assumed. The designation of truncated threads shall have the addition of the letter "T" to the designation, i.e., G ½ T and G ½ BT.

British Standard Pipe Threads (Non-pressure-tight Joints) — Metric and Inch Basic Sizes* (BS 2779:1973)

Nominal Size, Inches	Threads per Inch†	Depth of Thread	Major Diameter	Pitch Diameter	Minor Diameter	Nominal Size, Inches	Threads per Inch†	Depth of Thread	Major Diameter	Pitch Diameter	Minor Diameter
1/16	28	0.581 / 0.0229	7.723 / 0.3041	7.142 / 0.2812	6.561 / 0.2583	1¾	11	1.479 / 0.0582	53.746 / 2.1160	52.267 / 2.0578	50.788 / 1.9996
1/8	28	0.581 / 0.0229	9.728 / 0.3830	9.147 / 0.3601	8.566 / 0.3372	2	11	1.479 / 0.0582	59.614 / 2.3470	58.135 / 2.2888	56.656 / 2.2306
1/4	19	0.856 / 0.0337	13.157 / 0.5180	12.301 / 0.4843	11.445 / 0.4506	2¼	11	1.479 / 0.0582	65.710 / 2.5870	64.231 / 2.5288	62.752 / 2.4706
3/8	19	0.856 / 0.0337	16.662 / 0.6560	15.806 / 0.6223	14.950 / 0.5886	2½	11	1.479 / 0.0582	75.184 / 2.9600	73.705 / 2.9018	72.226 / 2.8436
1/2	14	1.162 / 0.0457	20.955 / 0.8250	19.793 / 0.7793	18.631 / 0.7336	2¾	11	1.479 / 0.0582	81.534 / 3.2100	80.055 / 3.1518	78.576 / 3.0936
5/8	14	1.162 / 0.0457	22.911 / 0.9020	21.749 / 0.8563	20.587 / 0.8106	3	11	1.479 / 0.0582	87.884 / 3.4600	86.405 / 3.4018	84.926 / 3.3436
3/4	14	1.162 / 0.0457	26.441 / 1.0410	25.279 / 0.9953	24.117 / 0.9496	3½	11	1.479 / 0.0582	100.330 / 3.9500	98.851 / 3.8918	97.372 / 3.8336
7/8	14	1.162 / 0.0457	30.201 / 1.1890	29.039 / 1.1433	27.877 / 1.0976	4	11	1.479 / 0.0582	113.030 / 4.4500	111.551 / 4.3918	110.072 / 4.3336
1	11	1.479 / 0.0582	33.249 / 1.3090	31.770 / 1.2508	30.291 / 1.1926	4½	11	1.479 / 0.0582	125.730 / 4.9500	124.251 / 4.8918	122.772 / 4.8336
1⅛	11	1.479 / 0.0582	37.897 / 1.4920	36.418 / 1.4338	34.939 / 1.3756	5	11	1.479 / 0.0582	138.430 / 5.4500	136.951 / 5.3918	135.472 / 5.3336
1¼	11	1.479 / 0.0582	41.910 / 1.6500	40.431 / 1.5918	38.952 / 1.5336	5½	11	1.479 / 0.0582	151.130 / 5.9500	149.651 / 5.8918	148.172 / 5.8336
1½	11	1.479 / 0.0582	47.803 / 1.8820	46.324 / 1.8238	44.845 / 1.7656	6	11	1.479 / 0.0582	163.830 / 6.4500	162.351 / 6.3918	160.872 / 6.3336

* Each basic metric dimension is given in roman figures (nominal sizes excepted) and each basic inch dimension is shown in italics directly beneath it.

† The thread pitches in millimeters are as follows: 0.907 for 28 threads per inch, 1.337 for 19 threads per inch, 1.814 for 14 threads per inch, and 2.309 for 11 threads per inch.

British Standard Unified Screw Threads of UNJ Basic Profile. — This British Standard (B.S. 4084:1978) arises from a request originating from within the British aircraft industry and is based upon specifications for Unified screw threads and American military standard MIL-S-8879.

These UNJ threads, having an enlarged root radius, were introduced for applications requiring high fatigue strength where working stress levels are high, in order to minimize size and weight, as in aircraft engines, airframes, missiles, space vehicles and similar designs where size and weight are critical. To meet these requirements, the root radius of external Unified threads is controlled between appreciably enlarged limits, the minor diameter of the mating internal threads being appropriately increased to insure the necessary clearance. The requirement for high strength is further met by restricting the tolerances for UNJ threads to the highest classes, Classes 3A and 3B, of Unified screw threads.

The standard, not described further here, contains both a coarse and a fine pitch series of threads.

British Standard ISO Metric Screw Threads. — BS 3643: Part 1: 1981 provides principles and basic data for ISO metric screw threads. It covers single-start, parallel screw threads of from 1 to 300 millimeters in diameter. Part 2 of the Standard gives the specifications for selected limits of size.

Basic Profile: The ISO basic profile for triangular screw threads is shown in Fig. 1 and basic dimensions of this profile are given in Table 1.

Tolerance System: The tolerance system defines *tolerance classes* in terms of a combination of a *tolerance grade* (figure) and a *tolerance position* (letter). The tolerance position is defined by the distance between the basic size and the nearest end of the tolerance zone, this distance being known as the *fundamental deviation*, EI, in the case of internal threads, and es in the case of external threads. These tolerance positions with respect to the basic size (zero line) are shown in Fig. 2 and fundamental deviations for nut and bolt threads are given in Table 5.

Tolerance Grades: Tolerance grades specified in the Standard for each of the four main screw thread diameters are as follows:

Minor diameter of nut threads (D_1): tolerance grades 4, 5, 6, 7, and 8.
Major diameter of bolt threads (d): tolerance grades 4, 6, and 8.
Pitch diameter of nut threads (D_2): tolerance grades 4, 5, 6, 7, and 8.
Pitch diameter of bolt threads (d_2): tolerance grades 3, 4, 5, 6, 7, 8, and 9.

Tolerance Positions: Tolerance positions are G and H for nut threads and e, f, g, and h for bolt threads. The relationship of these tolerance position identifying letters to the amount of fundamental deviation is shown in Table 5.

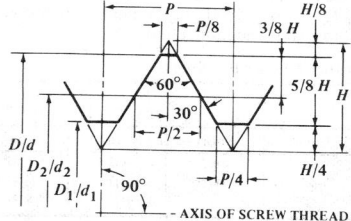

D = maj. diam. of internal thread;
d = maj. diam. of external thread;
D_2 = pitch diam. of internal thread;
d_2 = pitch diam. of external thread;
D_1 = minor diam. of internal thread;
d_1 = minor diam. of external thread;
P = pitch;
H = height of fundamental triangle.

Fig. 1. Basic Profile of ISO Metric Thread

Table 1. British Standard ISO Metric Screw Threads — Basic Profile Dimensions
(BS 3643: 1981)

Pitch P	H 0.86603P	$\frac{5}{8}H$ 0.54127P	$\frac{3}{8}H$ 0.32476P	$H/4$ 0.21651P	$H/8$ 0.10825P
0.2	0.173 205	0.108 253	0.064 952	0.043 301	0.021 651
0.25	0.216 506	0.135 316	0.081 190	0.054 127	0.027 063
0.3	0.259 808	0.162 380	0.097 428	0.064 952	0.032 476
0.35	0.303 109	0.189 443	0.113 666	0.075 777	0.037 889
0.4	0.346 410	0.216 506	0.129 904	0.086 603	0.043 301
0.45	0.389 711	0.243 570	0.146 142	0.097 428	0.048 714
0.5	0.433 013	0.270 633	0.162 380	0.108 253	0.054 127
0.6	0.519 615	0.324 760	0.194 856	0.129 904	0.064 952
0.7	0.606 218	0.378 886	0.227 332	0.151 554	0.075 777
0.75	0.649 519	0.405 949	0.243 570	0.162 380	0.081 190
0.8	0.692 820	0.433 013	0.259 808	0.173 205	0.086 603
1	0.866 025	0.541 266	0.324 760	0.216 506	0.108 253
1.25	1.082 532	0.676 582	0.405 949	0.270 633	0.135 316
1.5	1.299 038	0.811 899	0.487 139	0.324 760	0.162 380
1.75	1.515 544	0.947 215	0.568 329	0.378 886	0.189 443
2	1.732 051	1.082 532	0.649 519	0.433 013	0.216 506
2.5	2.165 063	1.353 165	0.811 899	0.541 266	0.270 633
3	2.598 076	1.623 798	0.974 279	0.649 519	0.324 760
3.5	3.031 089	1.894 431	1.136 658	0.757 772	0.378 886
4	3.464 102	2.165 063	1.299 038	0.866 025	0.433 013
4.5	3.897 114	2.435 696	1.461 418	0.974 279	0.487 139
5	4.330 127	2.706 329	1.623 798	1.082 532	0.541 266
5.5	4.763 140	2.976 962	1.786 177	1.190 785	0.595 392
6	5.196 152	3.247 595	1.948 557	1.299 038	0.649 519
8*	6.928 203	4.330 127	2.598 076	1.732 051	0.866 025

All dimensions are given in millimeters.
* This pitch is not used in any of the ISO metric standard series.

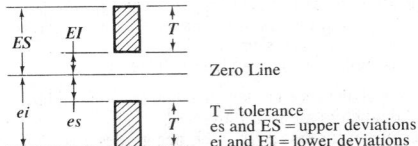

Zero Line

T = tolerance
es and ES = upper deviations
ei and EI = lower deviations

Fig. 2. Tolerance Positions with Respect to Zero Line (Basic Size)

Tolerance Classes: To reduce the number of gages and tools, the Standard specifies that the tolerance positions and classes shall be chosen from those listed in Tables 2 and 3 for short, normal, and long lengths of thread engagement. The following rules apply for the choice of tolerance quality: *Fine:* for precision threads when little variation of fit character is needed; *Medium:* for general use; and *Coarse:* for cases where manufacturing difficulties can arise as, for example, when threading hot-rolled bars and long blind holes. If the actual length of thread engagement is unknown, as in the manufacturing of standard bolts, normal is recommended.

Table 2. Tolerance Classes for Nuts

Tolerance Quality	Tolerance Position G			Tolerance Position H		
	Short	Normal	Long	Short	Normal	Long
Fine	. . .	. . .	. . .	4H[2]	5H[2]	6H[2]
Medium	5G[3]	6G[3]	7G[3]	5H[1]	6H[1,4]	7H[1]
Coarse	. . .	7G[3]	8G[3]	. . .	7H[2]	8H[2]

[1] First choice.   [2] Second choice.   [3] Third choice; these are to be avoided.   [4] For commercial nut and bolt threads.   See Table 4 for short, normal, and long categories.

Table 3. Tolerance Classes for Bolts

Tolerance Quality	Tolerance Position e			Tolerance Position f			Tolerance Position g			Tolerance Position h		
	Short	Normal	Long	Short	Normal	Long	Short	Normal	Long	Short	Normal	Long
Fine	...	...	...	...	...	...	...	...	...	3h4h[3]	4h[1]	5h4h[3]
Medium	...	6e[1]	...	...	6f[1]	...	5g6g[3]	6g[1,4]	7g6g[3]	5h6h[3]	6h[2]	7h6h[3]
Coarse	...	7e6e[3]	...	...	...	...	...	8g[2]	9g8g[3]	...	...	...

See footnotes to Table 2, and see Table 4 for short, normal, and long categories.

Note: Any of the recommended tolerance classes for nuts can be combined with any of the recommended tolerance classes for bolts with the exception of sizes M1.4 and smaller for which the combination 5H/6h or finer shall be chosen. However, to guarantee a sufficient overlap, the finished components should preferably be made to form the fits H/g, H/h, or G/h.

Table 4. Lengths of Thread Engagements for Short, Normal, and Long Categories

Basic Major Diameter d		Pitch P	Short	Normal		Long
Over	Up to and Incl.		Up to and Incl.	Over	Up to and Incl.	Over
0.99	1.4	0.2	0.5	0.5	1.4	1.4
		0.25	0.6	0.6	1.7	1.7
		0.3	0.7	0.7	2	2
1.4	2.8	0.2	0.5	0.5	1.5	1.5
		0.25	0.6	0.6	1.9	1.9
		0.35	0.8	0.8	2.6	2.6
		0.4	1	1	3	3
		0.45	1.3	1.3	3.8	3.8
2.8	5.6	0.35	1	1	3	3
		0.5	1.5	1.5	4.5	4.5
		0.6	1.7	1.7	5	5
		0.7	2	2	6	6
		0.75	2.2	2.2	6.7	6.7
		0.8	2.5	2.5	7.5	7.5
5.6	11.2	0.75	2.4	2.4	7.1	7.1
		1	3	3	9	9
		1.25	4	4	12	12
		1.5	5	5	15	15
11.2	22.4	1	3.8	3.8	11	11
		1.25	4.5	4.5	13	13
		1.5	5.6	5.6	16	16
		1.75	6	6	18	18
		2	8	8	24	24
		2.5	10	10	30	30
22.4	45	1	4	4	12	12
		1.5	6.3	6.3	19	19
		2	8.5	8.5	25	25
		3	12	12	36	36
		3.5	15	15	45	45
		4	18	18	53	53
		4.5	21	21	63	63
45	90	1.5	7.5	7.5	22	22
		2	9.5	9.5	28	28
		3	15	15	45	45
		4	19	19	56	56
		5	24	24	71	71
		5.5	28	28	85	85
		6	32	32	95	95
90	180	2	12	12	36	36
		3	18	18	53	53
		4	24	24	71	71
		6	36	36	106	106
180	300	3	20	20	60	60
		4	26	26	80	80
		6	40	40	118	118

All dimensions are given in millimeters.

Design Profiles: The design profiles for ISO metric internal and external screw threads are shown in Fig. 3. These represent the profiles of the threads at their maximum metal condition. It may be noted that the root of each thread is deepened so as to clear the basic flat crest of the other thread. The contact between the thread is thus confined to their sloping flanks. However, for nut threads as well as bolt threads, the actual root contours shall not at any point violate the basic profile.

Designation: Screw threads complying with the requirements of the Standard shall be designated by the letter M followed by values of the nominal diameter and of the pitch, expressed in millimeters, and separated by the sign ×. Example: M6 × 0.75. The absence of the indication of pitch means that a coarse pitch is specified.

The complete designation of a screw thread consists of a designation for the thread system and size, and a designation for the crest diameter tolerance. Each class designation consists of: a figure indicating the tolerance grade; and a letter indicating the tolerance position, capital for nuts, lower case for bolts. If the two class designations for a thread are the same (one for the pitch diameter and one for the crest diameter), it is not necessary to repeat the symbols. As examples, a bolt thread designated M10-6g signifies a thread of 10 mm nominal diameter in the Coarse Thread Series having a tolerance class 6g for both pitch and major diameters. A designation M10 × 1-5g6g signifies a bolt thread of 10 mm nominal diameter having a pitch of 1 mm, a tolerance class 5g for pitch diameter, and a tolerance class 6g for major diameter. A designation M10-6H signifies a nut thread of 10 mm diameter in the Coarse Thread Series having a tolerance class 6H for both pitch and minor diameters.

A fit between mating parts is indicated by the nut thread tolerance class followed by the bolt thread tolerance class separated by an oblique stroke. *Examples:* M6-6H/6g and M20 × 2-6H/5g6g. For coated threads, the tolerances apply to the parts before coating, unless otherwise specified. After coating, the actual thread profile shall not at any point exceed the maximum material limits for either tolerance position H or h.

NUT (INTERNAL THREAD)

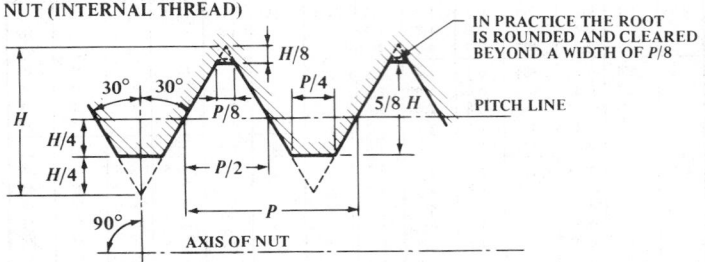

BOLT (EXTERNAL THREAD)

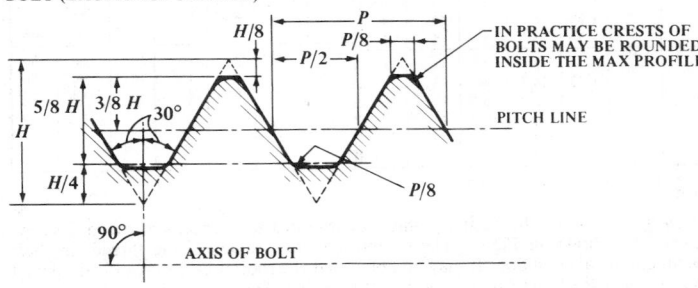

Fig. 3. Maximum Material Profiles for Internal and External Threads

Fundamental Deviation Formulas: The formulas used to calculate the fundamental deviations in Table 5 are:

$$EI_G = +(15 + 11P);$$
$$EI_H = 0;$$
$$es_e = -(50 + 11P) \text{ except for threads with } P \leq 0.45 \text{ mm};$$
$$es_f = -(30 + 11P);$$
$$es_g = -(15 + 11P);$$
$$\text{and } es_h = 0.$$

In these formulas, EI and es are expressed in micrometers and P is in millimeters.

Crest Diameter Tolerance Formulas: The tolerances for the major diameter of bolt threads (T_d), grade 6, in Table 7, were calculated from the formula:

$$T_d (6) = 180 \sqrt[3]{P^2} - \frac{3.15}{\sqrt{P}}$$

In this formula, $T_d (6)$ is in micrometers and P is in millimeters. For tolerance grades 4 and 8: $T_d (4) = 0.63 \, T_d (6)$ and $T_d (8) = 1.6 \, T_d (6)$, respectively.

The tolerances for the minor diameter of nut threads (T_{D1}), grade 6, in Table 7, were calculated as follows:

For pitches 0.2 to 0.8 mm, $T_{D1} (6) = 433P - 190P^{1.22}$.

For pitches 1 mm and coarser, $T_{D1} (6) = 230P^{0.7}$.

In these formulas, $T_{D1} (6)$ is in micrometers and P is in millimeters. For tolerance grades 4, 5, 7, and 8: $T_{D1} (4) = 0.63 \, T_{D1} (6)$; $T_{D1} (5) = 0.8 \, T_{D1} (6)$; $T_{D1} (7) = 1.25 \, T_{D1} (6)$; and $T_{D1} (8) = 1.6 \, T_{D1} (6)$, respectively.

Table 5. Fundamental Deviations for Nut Threads and Bolt Threads

Pitch P	Nut Thread D_2, D_1		Bolt Thread d, d_2				Pitch P	Nut Thread D_2, D_1		Bolt Thread d, d_2			
	G	H	e	f	g	h		G	H	e	f	g	h
	EI	EI	es	es	es	es		EI	EI	es	es	es	es
mm	μm	μm	μm	μm	μm	μm	mm	μm	μm	μm	μm	μm	μm
0.2	+17	0	...	...	−17	0	1.25	+28	0	−63	−42	−28	0
0.25	+18	0	...	...	−18	0	1.5	+32	0	−67	−45	−32	0
0.3	+18	0	...	...	−18	0	1.75	+34	0	−71	−48	−34	0
0.35	+19	0	...	−34	−19	0	2	+38	0	−71	−52	−38	0
0.4	+19	0	...	−34	−19	0	2.5	+42	0	−80	−58	−42	0
0.45	+20	0	...	−35	−20	0	3	+48	0	−85	−63	−48	0
0.5	+20	0	−50	−36	−20	0	3.5	+53	0	−90	−70	−53	0
0.6	+21	0	−53	−36	−21	0	4	+60	0	−95	−75	−60	0
0.7	+22	0	−56	−38	−22	0	4.5	+63	0	−100	−80	−63	0
0.75	+22	0	−56	−38	−22	0	5	+71	0	−106	−85	−71	0
0.8	+24	0	−60	−38	−24	0	5.5	+75	0	−112	−90	−75	0
1	+26	0	−60	−40	−26	0	6	+80	0	−118	−95	−80	0

See Figs. 1 and 2 for meaning of symbols.

Table 6. British Standard ISO Metric Screw Threads — Diameter/Pitch Combinations
(BS 3643: Part 1: 1981)

Nominal Diameter			Coarse Pitch	Fine Pitch	Constant Pitch
1st	2nd	3rd			
1	...	...	0.25	0.2	...
...	1.1	...	0.25	0.2	...
1.2	...	...	0.25	0.2	...
...	1.4	...	0.3	0.2	...
1.6	...	...	0.35	0.2	...
...	1.8	...	0.35	0.2	...
2.0	...	...	0.4	0.25	...
...	2.2	...	0.45	0.25	...
2.5	...	...	0.45	0.35	...
3	...	...	0.5	0.35	...
...	3.5	...	0.6	0.35	...
4	...	...	0.7	0.5	...
...	4.5	...	0.75	0.5	...
5	...	...	0.8	0.5	...
...	...	5.5	...	(0.5)	...
6	...	...	1	0.75	...
...	7	...	1	0.75	...
8	...	...	1.25	1	0.75
...	...	9	1.25	1	1,0.75
10	...	...	1.5	1.25	1,0.75
...	...	11	1.5	1.5	1,0.75
12	...	...	1.75	1.25	1.5,1
...	14	...	2	1.5	1.25*,1
...	...	15	...	...	1.5,1
16	...	...	2	1.5	1
...	...	17	...	...	1.5,1
...	18	...	2.5	1.5	2,1
20	...	...	2.5	1.5	2,1
...	22	...	2.5	1.5	2,1
24	...	...	3	2	1.5,1
...	...	25	...	...	2,1.5,1
...	...	26	...	...	1.5
...	27	...	3	2	1.5,1
...	...	28	...	...	2,1.5,1
30	...	...	3.5	2	(3),1.5,1
...	...	32	...	...	2,1.5
...	33	...	3.5	2	(3),1.5
...	...	35**	...	...	1.5
36	...	...	4	...	3,2,1.5
...	...	38	...	...	1.5
...	39	...	4	...	3,2,1.5
...	...	40	...	...	3,2,1.5
42	45	...	4.5	...	4,3,2,1.5
48	...	...	5	...	4,3,2,1.5
...	...	50	...	...	3,2,1.5
...	52	...	5	...	4,3,2,1.5
...	...	55	...	...	4,3,2,1.5
56	...	...	5.5	...	4,3,2,1.5
...	...	58	...	...	4,3,2,1.5
...	60	...	5.5	...	4,3,2,1.5
...	...	62	...	...	4,3,2,1.5
64	...	...	6	...	4,3,2,1.5
...	68	...	6	...	4,3,2,1.5

Nominal Diameter			Constant Pitch
1st	2nd	3rd	
...	...	70	6,4,3,2,1.5
72	...	...	6,4,3,2,1.5
...	...	75	4,3,2,1.5
...	76	...	6,4,3,2,1.5
...	...	78	2
80	...	...	6,4,3,2,1.5
...	...	82	2
...	85	...	6,4,3,2
90	...	...	6,4,3,2
...	95	...	6,4,3,2
100	...	...	6,4,3,2
...	105	...	6,4,3,2
110	...	...	6,4,3,2
...	115	...	6,4,3,2
...	120	...	6,4,3,2
125	...	...	6,4,3,2
...	130	...	6,4,3,2
...	...	135	6,4,3,2
140	...	...	6,4,3,2
...	...	145	6,4,3,2
...	150	...	6,4,3,2
...	...	155	6,4,3
160	...	...	6,4,3
...	...	165	6,4,3
...	170	...	6,4,3
...	...	175	6,4,3
180	...	...	6,4,3
...	...	185	6,4,3
...	190	...	6,4,3
...	...	195	6,4,3
200	...	...	6,4,3
...	...	205	6,4,3
...	210	...	6,4,3
...	...	215	6,4,3
220	...	...	6,4,3
...	...	225	6,4,3
...	...	230	6,4,3
...	...	235	6,4,3
...	240	...	6,4,3
...	...	245	6,4,3
250	...	...	6,4,3
...	...	255	6,4
...	260	...	6,4
...	...	265	6,4
...	...	270	6,4
...	...	275	6,4
280	...	...	6,4
...	...	285	6,4
...	...	290	6,4
...	...	295	6,4
...	300	...	6,4
...	...	...	...
...	...	...	...

All dimensions are given in millimeters.
* Only for spark plugs for engines.
** Only for locking nuts for bearings.
Pitches in parentheses ( ) are to be avoided as far as possible.

Diameter/Pitch Combinations: Part 1 of BS 3643 provides a choice of diameter/pitch combinations shown here in Table 6. The use of first-choice items is preferred, but if necessary, second, then third choice combinations may be selected. If pitches finer than those given in Table 6 are necessary, only the following pitches should be used: 3, 2, 1.5, 1, 0.75, 0.5, 0.35, 0.25, and 0.2 mm. When selecting such pitches it should be noted that there is increasing difficulty in meeting tolerance requirements as the diameter is increased for a given pitch. It is suggested that diameters greater than the following should not be used with the pitches indicated:

Pitch, mm	0.5	0.75	1	1.5	2	3
Maximum Diameter, mm	22	33	80	150	200	300

In cases where it is necessary to use a thread with a pitch larger than 6 mm, in the diameter range of 150 to 300 mm, the 8 mm pitch should be used.

Limits and Tolerances for Finished Uncoated Threads: Part 2 of BS 3643 specifies the fundamental deviations, tolerances, and limits of size for the tolerance classes 4H, 5H, 6H, and 7H for internal threads (nuts) and 4h, 6g, and 8g for external threads (bolts) for coarse-pitch series within the range of 1 to 68 mm; fine-pitch series within the range of 1 to 33 mm; and constant pitch series within the range of 8 to 300 mm diameter.

The data in Table 7 provide the first, second, and third choice combinations shown in Table 6 except that constant-pitch series threads are omitted. For diameters larger than shown in Table 7, and for constant-pitch series data, refer to the Standard.

Table 7. British Standard ISO Metric Screw Threads: Limits and Tolerances for Finished Uncoated Threads for Normal Lengths of Engagement
(BS 3643: Part 2: 1981)

Nominal Diameter**	Pitch Coarse	Pitch Fine	Tol. Class	External: Fund dev.	External Major Diam. Max	External Major Diam. Tol (−)	External Pitch Diam. Max	External Pitch Diam. Tol (−)	External Minor Diam. Min	Tol. Class	Internal Major Diam. Min	Internal Pitch Diam. Max	Internal Pitch Diam. Tol (−)	Internal Minor Diam. Max	Internal Minor Diam. Tol (−)
1		0.2	4h	0	1.000	0.036	0.870	0.030	0.717	4H	1.000	0.910	0.040	0.821	0.038
			6g	0.017	0.983	0.056	0.853	0.048	0.682						
	0.25		4h	0	1.000	0.042	0.838	0.034	0.649	4H	1.000	0.883	0.045	0.774	0.045
			6g	0.018	0.982	0.067	0.820	0.053	0.613	5H	1.000	0.894	0.056	0.785	0.056
1.1		0.2	4h	0	1.100	0.036	0.970	0.030	0.817	4H	1.100	1.010	0.040	0.921	0.038
			6g	0.017	1.083	0.056	0.953	0.048	0.782						
	0.25		4h	0	1.100	0.042	0.938	0.034	0.750	4H	1.100	0.983	0.045	0.874	0.045
			6g	0.018	1.082	0.067	0.920	0.053	0.713	5H	1.100	0.994	0.056	0.885	0.056
1.2		0.2	4h	0	1.200	0.036	1.070	0.030	0.917	4H	1.200	1.110	0.040	1.021	0.038
			6g	0.017	1.183	0.056	1.053	0.048	0.882						
	0.25		4h	0	1.200	0.042	1.038	0.034	0.850	4H	1.200	1.083	0.045	0.974	0.045
			6g	0.018	1.182	0.067	1.020	0.053	0.813	5H	1.200	1.094	0.056	0.985	0.056
1.4		0.2	4h	0	1.400	0.036	1.270	0.030	1.117	4H	1.400	1.310	0.040	1.221	0.038
			6g	0.017	1.383	0.056	1.253	0.048	1.082						
	0.3		4h	0	1.400	0.048	1.205	0.036	0.984	4H	1.400	1.253	0.048	1.128	0.053
			6g	0.018	1.382	0.075	1.187	0.056	0.946	5H	1.400	1.265	0.060	1.142	0.067
										6H	1.400	1.280	0.075	1.160	0.085
1.6		0.2	4h	0	1.600	0.036	1.470	0.032	1.315	4H	1.600	1.512	0.042	1.421	0.038
			6g	0.017	1.583	0.056	1.453	0.050	1.280						
	0.35		4h	0	1.600	0.053	1.373	0.040	1.117	4H	1.600	1.426	0.053	1.284	0.063
			6g	0.019	1.581	0.085	1.354	0.063	1.075	5H	1.600	1.440	0.067	1.301	0.080
										6H	1.600	1.458	0.085	1.321	0.100

All dimensions are in millimeters.   See footnotes at end of table.

Table 7 (*Continued*). **British Standard ISO Metric Screw Threads: Limits and Tolerances for Finished Uncoated Threads for Normal Lengths of Engagement**
(BS 3643: Part 2: 1981)

Nominal Diameter**	Coarse	Fine	Tol. Class	Fund dev.	Major Diam. Max	Major Diam. Tol (−)	Pitch Diam. Max	Pitch Diam. Tol (−)	Minor Diam. Min	Tol. Class	Major Diam. Min	Pitch Diam. Max	Pitch Diam. Tol (−)	Minor Diam. Max	Minor Diam. Tol (−)
		0.2	4h	0	1.800	0.036	1.670	0.032	1.515	4H	1.800	1.712	0.042	1.621	0.038
			6g	0.017	1.783	0.056	1.653	0.050	1.480						
1.8	0.35		4h	0	1.800	0.053	1.573	0.040	1.317	4H	1.800	1.626	0.053	1.484	0.063
			6g	0.019	1.781	0.085	1.554	0.063	1.275	5H	1.800	1.640	0.067	1.501	0.080
										6H	1.800	1.658	0.085	1.521	0.100
		0.25	4h	0	2.000	0.042	1.838	0.036	1.648	4H	2.000	1.886	0.048	1.774	0.045
			6g	0.018	1.982	0.067	1.820	0.056	1.610	5H	2.000	1.898	0.060	1.785	0.056
2	0.4		4h	0	2.000	0.060	1.740	0.042	1.452	4H	2.000	1.796	0.056	1.638	0.071
			6g	0.019	1.981	0.095	1.721	0.067	1.408	5H	2.000	1.811	0.071	1.657	0.090
										6H	2.000	1.830	0.090	1.679	0.112
		0.25	4h	0	2.200	0.042	2.038	0.036	1.848	4H	2.200	2.086	0.048	1.974	0.045
			6g	0.018	2.182	0.067	2.020	0.056	1.810	5H	2.200	2.098	0.060	1.985	0.056
2.2	0.45		4h	0	2.200	0.063	1.908	0.045	1.585	4H	2.200	1.968	0.060	1.793	0.080
			6g	0.020	2.180	0.100	1.888	0.071	1.539	5H	2.200	1.983	0.075	1.813	0.100
										6H	2.200	2.003	0.095	1.838	0.125
		0.35	4h	0	2.500	0.053	2.273	0.040	2.017	4H	2.500	2.326	0.053	2.184	0.063
			6g	0.019	2.481	0.085	2.254	0.063	1.975	5H	2.500	2.340	0.067	2.201	0.080
2.5										6H	2.500	2.358	0.085	2.221	0.100
	0.45		4h	0	2.500	0.063	2.208	0.045	1.885	4H	2.500	2.268	0.060	2.093	0.080
			6g	0.020	2.480	0.100	2.188	0.071	1.839	5H	2.500	2.283	0.075	2.113	0.100
										6H	2.500	2.303	0.095	2.138	0.125
		0.35	4h	0	3.000	0.053	2.773	0.042	2.515	4H	3.000	2.829	0.056	2.684	0.063
			6g	0.019	2.981	0.085	2.754	0.067	2.471	5H	3.000	2.844	0.071	2.701	0.080
3										6H	3.000	2.863	0.090	2.721	0.100
	0.5		4h	0	3.000	0.067	2.675	0.048	2.319	5H	3.000	2.755	0.080	2.571	0.112
			6g	0.020	2.980	0.106	2.655	0.075	2.272	6H	3.000	2.775	0.100	2.599	0.140
										7H	3.000	2.800	0.125	2.639	0.180
		0.35	4h	0	3.500	0.053	3.273	0.042	3.015	4H	3.500	3.329	0.056	3.184	0.063
			6g	0.019	3.481	0.085	3.254	0.067	2.971	5H	3.500	3.344	0.071	3.201	0.080
3.5										6H	3.500	3.363	0.090	3.221	0.100
	0.6		4h	0	3.500	0.080	3.110	0.053	2.688	5H	3.500	3.200	0.090	2.975	0.125
			6g	0.021	3.479	0.125	3.089	0.085	2.635	6H	3.500	3.222	0.112	3.010	0.160
										7H	3.500	3.250	0.140	3.050	0.200
		0.5	4h	0	4.000	0.067	3.675	0.048	3.319	5H	4.000	3.755	0.080	3.571	0.112
			6g	0.020	3.980	0.106	3.655	0.075	3.272	6H	4.000	3.775	0.100	3.599	0.140
4										7H	4.000	3.800	0.125	3.639	0.180
	0.7		4h	0	4.000	0.067	3.545	0.056	3.058	5H	4.000	3.640	0.095	3.382	0.140
			6g	0.022	3.978	0.140	3.523	0.090	3.002	6H	4.000	3.663	0.118	3.422	0.180
										7H	4.000	3.695	0.150	3.466	0.224
		0.5	4h	0	4.500	0.067	4.175	0.048	3.819	5H	4.500	4.255	0.080	4.071	0.112
			6g	0.020	4.480	0.106	4.155	0.075	3.772	6H	4.500	4.275	0.100	4.099	0.140
4.5										7H	4.500	4.300	0.125	4.139	0.180
	0.75		4h	0	4.500	0.090	4.013	0.056	3.495	5H	4.500	4.108	0.095	3.838	0.150
			6g	0.022	4.478	0.140	3.991	0.090	3.439	6H	4.500	4.131	0.118	3.878	0.190
										7H	4.500	4.163	0.150	3.924	0.236
		0.5	4h	0	5.000	0.067	4.675	0.048	4.319	5H	5.000	4.755	0.080	4.571	0.112
			6g	0.020	4.980	0.106	4.655	0.075	4.272	6H	5.000	4.775	0.100	4.599	0.140
5										7H	5.000	4.800	0.125	4.639	0.180
	0.8		4h	0	5.000	0.095	4.480	0.060	3.927	5H	5.000	4.580	0.100	4.294	0.160
			6g	0.024	4.976	0.150	4.456	0.095	3.868	6H	5.000	4.605	0.125	4.334	0.200
										7H	5.000	4.640	0.160	4.384	0.250

All dimensions are in millimeters.   See footnotes at end of table.

Table 7 (Continued). British Standard ISO Metric Screw Threads: Limits and Tolerances for Finished Uncoated Threads for Normal Lengths of Engagement
(BS 3643: Part 2: 1981)

Nominal Diameter**	Coarse	Fine	Tol. Class	Fund dev.	Major Max	Major Tol (−)	Pitch Max	Pitch Tol (−)	Minor Min	Tol. Class	Major Min	Pitch Max	Pitch Tol (−)	Minor Max	Minor Tol (−)
5.5		0.5	4h	0	5.500	0.067	5.175	0.048	4.819	5H	5.500	5.255	0.080	5.071	0.112
			6g	0.020	5.480	0.106	5.155	0.075	4.772	6H	5.500	5.275	0.100	5.099	0.140
										7H	5.500	5.300	0.125	5.139	0.180
6		0.75	4h	0	6.000	0.090	5.513	0.063	4.988	5H	6.000	5.619	0.106	5.338	0.150
			6g	0.022	5.978	0.140	5.491	0.100	4.929	6H	6.000	5.645	0.132	5.378	0.190
										7H	6.000	5.683	0.170	5.424	0.236
	1		4h	0	6.000	0.112	5.350	0.071	4.663	5H	6.000	5.468	0.118	5.107	0.190
			6g	0.026	5.974	0.180	5.324	0.112	4.597	6H	6.000	5.500	0.150	5.153	0.236
			8g	0.026	5.974	0.280	5.324	0.180	4.528	7H	6.000	5.540	0.190	5.217	0.300
7		0.75	4h	0	7.000	0.090	6.513	0.063	5.988	5H	7.000	6.619	0.106	6.338	0.150
			6g	0.022	6.978	0.140	6.491	0.100	5.929	6H	7.000	6.645	0.132	6.378	0.190
										7H	7.000	6.683	0.170	6.424	0.236
	1		4h	0	7.000	0.112	6.350	0.071	5.663	5H	7.000	6.468	0.118	6.107	0.190
			6g	0.026	6.974	0.180	6.324	0.112	5.596	6H	7.000	6.500	0.150	6.153	0.236
			8g	0.026	6.974	0.280	6.324	0.180	5.528	7H	7.000	6.540	0.190	6.217	0.300
8		1	4h	0	8.000	0.112	7.350	0.071	6.663	5H	8.000	7.468	0.118	7.107	0.190
			6g	0.026	7.974	0.180	7.324	0.112	6.596	6H	8.000	7.500	0.150	7.153	0.236
			8g	0.026	7.974	0.280	7.324	0.180	6.528	7H	8.000	7.540	0.190	7.217	0.300
	1.25		4h	0	8.000	0.132	7.188	0.075	6.343	5H	8.000	7.313	0.125	6.859	0.212
			6g	0.028	7.972	0.212	7.160	0.118	6.272	6H	8.000	7.348	0.160	6.912	0.265
			8g	0.028	7.972	0.335	7.160	0.190	6.200	7H	8.000	7.388	0.200	6.982	0.335
9	1.25		4h	0	9.000	0.132	8.188	0.075	7.343	5H	9.000	8.313	0.125	7.859	0.212
			6g	0.028	8.972	0.212	8.160	0.118	7.272	6H	9.000	8.348	0.160	7.912	0.265
			8g	0.028	8.972	0.335	8.160	0.190	7.200	7H	9.000	8.388	0.200	7.982	0.335
10		1.25	4h	0	10.000	0.132	9.188	0.075	8.343	5H	10.000	9.313	0.125	8.859	0.212
			6g	0.028	9.972	0.212	9.160	0.118	8.272	6H	10.000	9.348	0.160	8.912	0.265
			8g	0.028	9.972	0.335	9.160	0.190	8.200	7H	10.000	9.388	0.200	8.982	0.335
	1.5		4h	0	10.000	0.150	9.026	0.085	8.018	5H	10.000	9.166	0.140	8.612	0.236
			6g	0.032	9.968	0.236	8.994	0.132	7.938	6H	10.000	9.206	0.180	8.676	0.300
			8g	0.032	9.968	0.375	8.994	0.212	7.858	7H	10.000	9.250	0.224	8.751	0.375
11	1.5		4h	0	11.000	0.150	10.026	0.085	9.018	5H	11.000	10.166	0.140	9.612	0.236
			6g	0.032	10.968	0.236	9.994	0.132	8.938	6H	11.000	10.206	0.180	9.676	0.300
			8g	0.032	10.968	0.375	9.994	0.212	8.858	7H	11.000	10.250	0.224	9.751	0.375
12		1.25	4h	0	12.000	0.132	11.188	0.085	10.333	5H	12.000	11.328	0.140	10.859	0.212
			6g	0.028	11.972	0.212	11.160	0.132	10.257	6H	12.000	11.368	0.180	10.912	0.265
			8g	0.028	11.972	0.335	11.160	0.212	10.177	7H	12.000	11.412	0.224	10.982	0.335
	1.75		4h	0	12.000	0.170	10.863	0.095	9.692	5H	12.000	11.023	0.160	10.371	0.265
			6g	0.034	11.966	0.265	10.829	0.150	9.602	6H	12.000	11.063	0.200	10.441	0.335
			8g	0.034	11.966	0.425	10.829	0.236	9.516	7H	12.000	11.113	0.250	10.531	0.425
14		1.5	4h	0	14.000	0.150	13.026	0.090	12.012	5H	14.000	13.176	0.150	12.612	0.236
			6g	0.032	13.968	0.236	12.994	0.140	11.930	6H	14.000	13.216	0.190	12.676	0.300
			8g	0.032	13.968	0.375	12.994	0.224	11.846	7H	14.000	13.262	0.236	12.751	0.375
	2		4h	0	14.000	0.180	12.701	0.100	11.369	5H	14.000	12.871	0.170	12.135	0.300
			6g	0.038	13.962	0.280	12.663	0.160	11.271	6H	14.000	12.913	0.212	12.210	0.375
			8g	0.038	13.962	0.450	12.663	0.250	11.181	7H	14.000	12.966	0.265	12.310	0.475
16		1.5	4h	0	16.000	0.150	15.026	0.090	14.012	5H	16.000	15.176	0.150	14.612	0.236
			6g	0.032	15.968	0.236	14.994	0.140	13.930	6H	16.000	15.216	0.190	14.676	0.300
			8g	0.032	15.968	0.375	14.994	0.224	13.846	7H	16.000	15.262	0.236	14.751	0.375
	2		4h	0	16.000	0.180	14.701	0.100	13.369	5H	16.000	14.871	0.170	14.135	0.300
			6g	0.038	15.962	0.280	14.663	0.160	13.271	6H	16.000	14.913	0.212	14.210	0.375
			8g	0.038	15.962	0.450	14.663	0.250	13.181	7H	16.000	14.966	0.265	14.310	0.475

The table header groups: Pitch (Coarse, Fine); External Threads (Bolts): Tol. Class, Fund dev., Major Diam. (Max / Tol −), Pitch Diam. (Max / Tol −), Minor Diam. (Min); Internal Threads (Nuts)*: Tol. Class, Major Diam. (Min), Pitch Diam. (Max / Tol −), Minor Diam. (Max / Tol −).

All dimensions are in millimeters. See footnotes at end of table.

Table 7 (*Concluded*). British Standard ISO Metric Screw Threads: Limits and Tolerances for Finished Uncoated Threads for Normal Lengths of Engagement
(BS 3643: Part 2: 1981)

Nominal Diameter**	Pitch Coarse	Pitch Fine	Tol. Class	Fund dev.	Major Diam. Max	Major Diam. Tol (−)	Pitch Diam. Max	Pitch Diam. Tol (−)	Minor Diam. Min	Tol. Class	Major Diam. Min	Pitch Diam. Max	Pitch Diam. Tol (−)	Minor Diam. Max	Minor Diam. Tol (−)
					External Threads (Bolts)						Internal Threads (Nuts)*				
18		1.5	4h	0	18.000	0.150	17.026	0.090	16.012	5H	18.000	17.176	0.150	16.612	0.236
			6g	0.032	17.968	0.236	16.994	0.140	15.930	6H	18.000	17.216	0.190	16.676	0.300
			8g	0.032	17.968	0.375	16.994	0.224	15.846	7H	18.000	17.262	0.236	16.751	0.375
	2.5		4h	0	18.000	0.212	16.376	0.106	14.730	5H	18.000	16.556	0.180	15.649	0.355
			6g	0.042	17.958	0.335	16.334	0.170	14.624	6H	18.000	16.600	0.224	15.744	0.450
			8g	0.042	17.958	0.530	16.334	0.265	14.529	7H	18.000	16.656	0.280	15.854	0.560
20		1.5	4h	0	20.000	0.150	19.026	0.090	18.012	5H	20.000	19.176	0.150	18.612	0.236
			6g	0.032	19.968	0.236	18.994	0.140	17.930	6H	20.000	19.216	0.190	18.676	0.300
			8g	0.032	19.968	0.375	18.994	0.224	17.846	7H	20.000	19.262	0.236	18.751	0.375
	2.5		4h	0	20.000	0.212	18.376	0.106	16.730	5H	20.000	18.556	0.180	17.649	0.355
			6g	0.042	19.958	0.335	18.334	0.170	16.624	6H	20.000	18.600	0.224	17.744	0.450
			8g	0.042	19.958	0.530	18.334	0.265	16.529	7H	20.000	18.656	0.280	17.854	0.560
22		1.5	4h	0	22.000	0.150	21.026	0.090	20.012	5H	22.000	21.176	0.150	20.612	0.236
			6g	0.032	21.968	0.236	20.994	0.140	19.930	6H	22.000	21.216	0.190	20.676	0.300
			8g	0.032	21.968	0.375	20.994	0.224	19.846	7H	22.000	21.262	0.236	20.751	0.375
	2.5		4h	0	22.000	0.212	20.376	0.106	18.730	5H	22.000	20.556	0.180	19.649	0.335
			6g	0.042	21.958	0.335	20.334	0.170	18.624	6H	22.000	20.600	0.224	19.744	0.450
			8g	0.042	21.958	0.530	20.334	0.265	18.529	7H	22.000	20.656	0.280	19.854	0.560
24		2	4h	0	24.000	0.180	22.701	0.106	21.363	5H	24.000	22.881	0.180	22.135	0.300
			6g	0.038	23.962	0.280	22.663	0.170	21.261	6H	24.000	22.925	0.224	22.210	0.375
			8g	0.038	23.962	0.450	22.663	0.265	21.166	7H	24.000	22.981	0.280	22.310	0.475
	3		4h	0	24.000	0.236	22.051	0.125	20.078	5H	24.000	22.263	0.212	21.152	0.400
			6g	0.048	23.952	0.375	22.003	0.200	19.955	6H	24.000	22.316	0.265	21.252	0.500
			8g	0.048	23.952	0.600	22.003	0.315	19.840	7H	24.000	22.386	0.335	21.382	0.630
27		2	4h	0	27.000	0.180	25.701	0.106	24.363	5H	27.000	25.881	0.180	25.135	0.300
			6g	0.038	26.962	0.280	25.663	0.170	24.261	6H	27.000	25.925	0.224	25.210	0.375
			8g	0.038	26.962	0.450	25.663	0.265	24.166	7H	27.000	25.981	0.280	25.310	0.475
	3		4h	0	27.000	0.236	25.051	0.125	23.078	5H	27.000	25.263	0.212	24.152	0.400
			6g	0.048	26.952	0.375	25.003	0.200	22.955	6H	27.000	25.316	0.265	24.252	0.500
			8g	0.048	26.952	0.600	25.003	0.315	22.840	7H	27.000	25.386	0.335	24.382	0.630
30		2	4h	0	30.000	0.180	28.701	0.106	27.363	5H	30.000	28.881	0.180	28.135	0.300
			6g	0.038	29.962	0.280	28.663	0.170	27.261	6H	30.000	28.925	0.224	28.210	0.375
			8g	0.038	29.962	0.450	28.663	0.265	27.166	7H	30.000	28.981	0.280	28.310	0.475
	3.5		4h	0	30.000	0.265	27.727	0.132	25.438	5H	30.000	27.951	0.224	26.661	0.450
			6g	0.053	29.947	0.425	27.674	0.212	25.305	6H	30.000	28.007	0.280	26.771	0.560
			8g	0.053	29.947	0.670	27.674	0.335	25.182	7H	30.000	28.082	0.355	26.921	0.710
33		2	4h	0	33.000	0.180	31.701	0.106	30.363	5H	33.000	31.881	0.180	31.135	0.300
			6g	0.038	32.962	0.280	31.663	0.170	30.261	6H	33.000	31.925	0.224	31.210	0.375
			8g	0.038	32.962	0.450	31.663	0.265	30.166	7H	33.000	31.981	0.280	31.310	0.475
	3.5		4h	0	33.000	0.265	30.727	0.132	28.438	5H	33.000	30.951	0.224	29.661	0.450
			6g	0.053	32.947	0.425	30.674	0.212	28.305	6H	33.000	31.007	0.280	29.771	0.560
			8g	0.053	32.947	0.670	30.674	0.335	28.182	7H	33.000	31.082	0.355	29.921	0.710
36	4		4h	0	36.000	0.300	33.402	0.140	30.798	5H	36.000	33.638	0.236	32.145	0.475
			6g	0.060	35.940	0.475	33.342	0.224	30.654	6H	36.000	33.702	0.300	32.270	0.600
			8g	0.060	35.940	0.750	33.342	0.355	30.523	7H	36.000	33.777	0.375	32.420	0.750
39	4		4h	0	39.000	0.300	36.402	0.140	33.798	5H	39.000	36.638	0.236	35.145	0.475
			6g	0.060	38.940	0.475	36.342	0.224	33.654	6H	39.000	36.702	0.300	35.270	0.600
			8g	0.060	38.940	0.750	36.342	0.355	33.523	7H	39.000	36.777	0.375	35.420	0.750

All dimensions are in millimeters.

* The fundamental deviation for internal threads (nuts) is zero for threads listed in this table.

** This table provides coarse- and fine-pitch series data for threads listed in Table 6 for first, second, and third choices. For constant-pitch series and for larger sizes than are shown, refer to the Standard.

British Standard Buttress Threads (B.S. 1657:1950). — Specifications for buttress threads in this standard are similar to those in the American Standard except: (1) A basic depth of thread of 0.4*p* is used instead of 0.6*p*; (2) Sizes below 1 inch are not included; (3) Tolerances on major and minor diameters are the same as the pitch diameter tolerances, whereas in the American Standard separate tolerances are provided; however, provision is made for smaller major and minor diameter tolerances when crest surfaces of screws or nuts are used as datum surfaces, or when the resulting reduction in depth of engagement must be limited; and (4) Certain combinations of large diameters with fine pitches are provided that are not encouraged in the American Standard.

Löwenherz Thread. — The Löwenherz thread has flats at the top and bottom the same as the U.S. standard form, but the angle is 53 degrees 8 minutes. The depth equals 0.75 × the pitch, and the width of the flats at the top and bottom is equal to 0.125 × the pitch. This screw thread is based on the metric system and is used for measuring instruments, especially in Germany.

Löwenherz Thread

Diameter		Pitch, Milli- meters	Approxi- mate No. of Threads per Inch	Diameter		Pitch, Milli- meters	Approxi- mate No. of Threads per Inch
Milli- meters	Inches			Milli- meters	Inches		
1.0	0.0394	0.25	101.6	9.0	0.3543	1.30	19.5
1.2	0.0472	0.25	101.6	10.0	0.3937	1.40	18.1
1.4	0.0551	0.30	84.7	12.0	0.4724	1.60	15.9
1.7	0.0669	0.35	72.6	14.0	0.5512	1.80	14.1
2.0	0.0787	0.40	63.5	16.0	0.6299	2.00	12.7
2.3	0.0905	0.40	63.5	18.0	0.7087	2.20	11.5
2.6	0.1024	0.45	56.4	20.0	0.7874	2.40	10.6
3.0	0.1181	0.50	50.8	22.0	0.8661	2.80	9.1
3.5	0.1378	0.60	42.3	24.0	0.9450	2.80	9.1
4.0	0.1575	0.70	36.3	26.0	1.0236	3.20	7.9
4.5	0.1772	0.75	33.9	28.0	1.1024	3.20	7.9
5.0	0.1968	0.80	31.7	30.0	1.1811	3.60	7.1
5.5	0.2165	0.90	28.2	32.0	1.2599	3.60	7.1
6.0	0.2362	1.00	25.4	36.0	1.4173	4.00	6.4
7.0	0.2756	1.10	23.1	40.0	1.5748	4.40	5.7
8.0	0.3150	1.20	21.1	. . .	. . .	. . .	. . .

International Metric Thread System. — The Système Internationale (S.I.) Thread was adopted at the International Congress for the standardization of screw threads held in Zurich in 1898. The thread form is similar to the American standard (formerly U.S. Standard), excepting the depth which is greater. There is a clearance between the root and mating crest fixed at a maximum of $\frac{1}{16}$ the height of the fundamental triangle or 0.054 × pitch. A rounded root profile is recommended. This system formed the basis of the normal metric series of many European countries.

Depth $d = 0.7035\ P$ max; $0.6855\ P$ min.

Flat $f = 0.125\ P$

Radius $r = 0.0633\ P$ max.; $0.054\ P$ min.

Tap drill diam. = major diam. − pitch

S.A.E. Standard Threads for Spark Plugs

Size Nom. × Pitch	Major Diameter		Pitch Diameter		Minor Diameter	
	Max.	Min.	Max.	Min.	Max.	Min.
Spark Plug Threads, mm (inches)						
M18 × 1.5	17.955 (0.7069)	17.803 (0.7009)	16.980 (0.6685)	16.853 (0.6635)	16.053 (0.6320)	...
M14 × 1.25	13.868 (0.5460)	13.741 (0.5410)	13.104 (0.5159)	12.997 (0.5117)	12.339 (0.4858)	...
M12 × 1.25	11.862 (0.4670)	11.735 (0.4620)	11.100 (0.4370)	10.998 (0.4330)	10.211 (0.4020)	...
M10 × 1.0	9.974 (0.3927)	9.794 (0.3856)	9.324 (0.3671)	9.212 (0.3627)	8.747 (0.3444)	...
Tapped Hole Threads, mm (inches)						
M18 × 1.5	...	18.039 (0.7102)	17.153 (0.6753)	17.026 (0.6703)	16.426 (0.6467)	16.266 (0.6404)
M14 × 1.25	...	14.034 (0.5525)	13.297 (0.5235)	13.188 (0.5192)	12.692 (0.4997)	12.499 (0.4921)
M12 × 1.25	...	11.935 (0.4699)	11.242 (0.4426)	11.138 (0.4385)	10.559 (0.4157)	10.366 (0.4081)
M10 × 1.0	...	10.000 (0.3937)	9.500 (0.3740)	9.350 (0.3681)	9.153 (0.3604)	9.020 (0.3551)

In order to keep the wear on the threading tools within permissible limits, the threads in the spark plug GO (ring) gage shall be truncated to the maximum minor diameter of the spark plug, and in the tapped hole GO (plug) gage to the minimum major diameter of the tapped hole. The plain plug gage for checking the minor diameter of the tapped hole shall be the minimum specified. The thread form is that of the ISO metric (see page 1593).

Reprinted with permission © 1987 Society of Automotive Engineers, Inc.

British Standard for Spark Plugs (BS 45:1972). — This revised British Standard refers solely to spark plugs used in automobiles and industrial spark ignition internal combustion engines. The basic thread form is that of the ISO metric (see page 1593). In assigning tolerances to the threads of the spark plug and the tapped holes, full consideration has been given to the desirability of achieving the closest possible measure of interchangeability between British spark plugs and engines, and those made to the standards of other ISO Member Bodies.

Basic Thread Dimensions for Spark Plug and Tapped Hole in Cylinder Head

Nom. Size	Pitch	Thread	Major Diam.		Pitch Diam.		Minor Diam.	
			Max.	Min.	Max.	Min.	Max.	Min.
14	1.25	Plug	13.937	13.725	13.125	12.993	12.402	12.181
14	1.25	Hole	*	14.00	13.368	13.188	12.912	12.647
18	1.5	Plug	17.933	17.697	16.959	16.819	16.092	15.845
18	1.5	Hole	*	18.00	17.216	17.026	16.676	16.376

All dimensions are given in millimeters.

* Not specified.

The tolerance grades for finished spark plugs and corresponding tapped holes in the cylinder head are: for 14 mm size, 6e for spark plugs and 6H for tapped holes which gives a minimum clearance of 0.063 mm; and for 18 mm size, 6e for spark plugs and 6H for tapped holes which gives a minimum clearance of 0.067 mm.

These minimum clearances are intended to prevent the possibility of seizure, as a result of combustion deposits on the bare threads, when removing the spark plugs and applies to both ferrous and non-ferrous materials. These clearances are also intended to enable spark plugs with threads in accordance with this standard to be fitted into existing tapped holes.

ANSI Standard Hose Coupling Screw Threads. — Threads for hose couplings, valves, and all other fittings used in direct connection with hose intended for domestic, industrial and general service in sizes ½, ⅝, ¾, 1, 1¼, 1½, 2, 2½, 3, 3½, and 4 inches are covered by American National Standard ANSI B2.4-1966 (R1974). These threads are designated as follows:

NH — Standard hose coupling threads of full form as produced by cutting or rolling.

NHR — Standard hose coupling threads for garden hose applications where the design utilizes thin walled material which is formed to the desired thread.

NPSH — Standard straight hose coupling thread series in sizes ½ to 4 inches for joining to American National Standard taper pipe threads using a gasket to seal the joint.

Thread dimensions are given in Table 1 and thread lengths in Table 2.

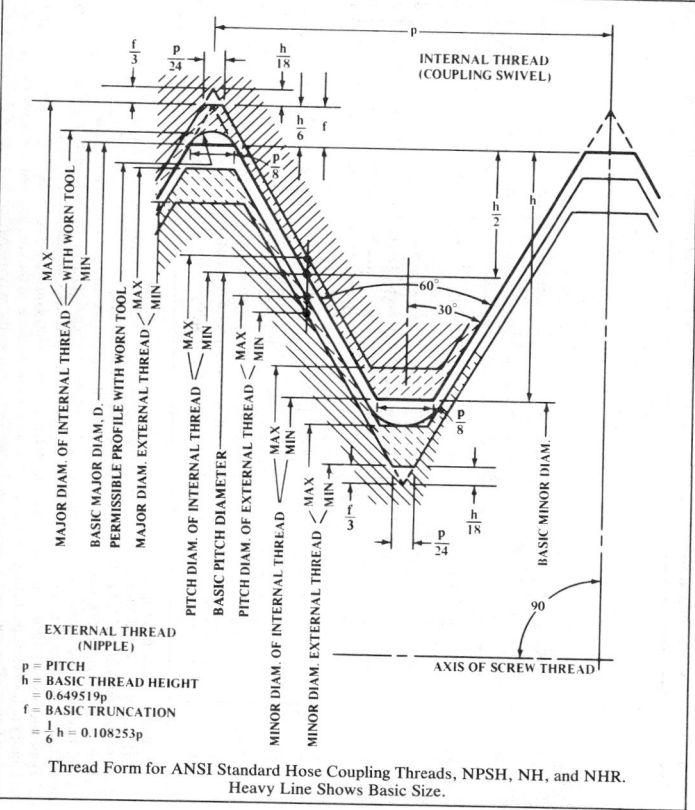

EXTERNAL THREAD (NIPPLE)

p = PITCH
h = BASIC THREAD HEIGHT
  = 0.649519p
f = BASIC TRUNCATION
  = ⅙ h = 0.108253p

Thread Form for ANSI Standard Hose Coupling Threads, NPSH, NH, and NHR.
Heavy Line Shows Basic Size.

Table 1. ANSI Standard Hose Coupling Threads for NPSH, NH and NHR Nipples and Coupling Swivels* (ANSI B2.4-1966, R1974)

Nom. Size of Hose	Thds. per Inch	Thread Designation	Pitch	Basic Height of Thread	Nipple (External) Thread					Coupling (Internal) Thread				
					Major Diam.		Pitch Diam.		Minor Diam.	Minor Diam.		Pitch Diam.		Major Diam.
					Max.	Min.	Max.	Min.	Max.	Min.	Max.	Min.	Max.	Min.
½, ⅝, ¾	11.5	.75-11.5NH	.08696	.05648	1.0625	1.0455	1.0060	0.9975	0.9495	0.9595	0.9765	1.0160	1.0245	1.0725
½, ⅝, ¾	11.5	.75-11.5NHR	.08696	.05648	1.0520	1.0350	1.0100	0.9930	0.9495	0.9720	0.9930	1.0160	1.0280	1.0680
½	14	.5-14NPSH	.07143	.04639	0.8248	0.8108	0.7784	0.7714	0.7320	0.7395	0.7535	0.7859	0.7929	0.8323
¾	14	.75-14NPSH	.07143	.04639	1.0353	1.0213	0.9889	0.9819	0.9425	0.9500	0.9640	0.9964	1.0034	1.0428
1	11.5	1-11.5NPSH	.08696	.05648	1.2951	1.2781	1.2386	1.2301	1.1821	1.1921	1.2091	1.2486	1.2571	1.3051
1¼	11.5	1.25-11.5NPSH	.08696	.05648	1.6399	1.6229	1.5834	1.5749	1.5629	1.5369	1.5539	1.5934	1.6019	1.6499
1½	11.5	1.5-11.5 NPSH	.08696	.05648	1.8788	1.8618	1.8223	1.8138	1.7658	1.7758	1.7928	1.8323	1.8408	1.8888
2	11.5	2-11.5NPSH	.08696	.05648	2.3528	2.3358	2.2963	2.2878	2.2398	2.2498	2.2668	2.3063	2.3148	2.3628
2½	8	2.5-8NPSH	.12500	.08119	2.8434	2.8212	2.7622	2.7511	2.6810	2.6930	2.7152	2.7742	2.7853	2.8554
3	8	3-8NPSH	.12500	.08119	3.4697	3.4475	3.3885	3.3774	3.3073	3.3193	3.3415	3.4005	3.4116	3.4817
3½	8	3.5-8NPSH	.12500	.08119	3.9700	3.9478	3.8888	3.8777	3.8076	3.8196	3.8418	3.9008	3.9119	3.9820
4	8	4-8NPSH	.12500	.08119	4.4683	4.4461	4.3871	4.3760	4.3059	4.3179	4.3401	4.3991	4.4102	4.4803
4	6	4-6NH (SPL)	.16667	.10825	4.9082	4.8722	4.7999	4.7819	4.6916	4.7117	4.7477	4.8200	4.8380	4.9283

All dimensions are given in inches.

*NH and NHR threads are used for garden hose applications. NPSH threads are used for steam, air and all other hose connections to be made up with standard pipe threads. NH (SPL) threads are used for marine applications.

Dimensions given for the maximum minor diameter of the nipple are figured to the intersection of the worn tool arc with a centerline through crest and root. The minimum minor diameter of the nipple shall be that corresponding to a flat at the minor diameter of the minimum nipple equal to $\frac{1}{24}p$, and may be determined by subtracting $0.7939p$ from the minimum pitch diameter of the nipple. (See diagram on p. 1605.)

Dimensions given for the minimum major diameter of the coupling correspond to the basic flat, $\frac{1}{8}p$, and the profile at the major diameter produced by a worn tool must not fall below the basic outline. The maximum major diameter of the coupling shall be that corresponding to a flat at the major diameter of the maximum coupling equal to $\frac{1}{24}p$ and may be determined by adding $0.7939p$ to the maximum pitch diameter of the coupling. (See diagram on p. 1605.)

Table 2. ANSI Standard Hose Coupling Screw Thread Lengths
(ANSI B2.4-1966, R1974)

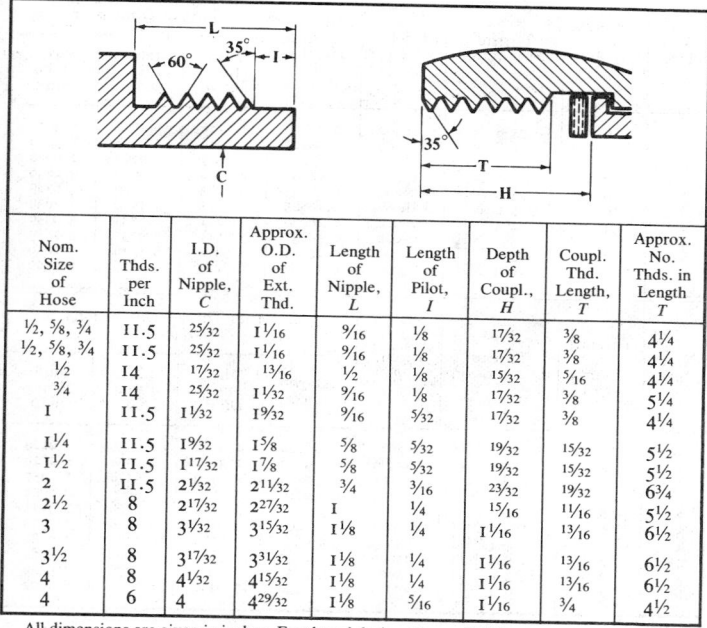

Nom. Size of Hose	Thds. per Inch	I.D. of Nipple, C	Approx. O.D. of Ext. Thd.	Length of Nipple, L	Length of Pilot, I	Depth of Coupl., H	Coupl. Thd. Length, T	Approx. No. Thds. in Length T
½, ⅝, ¾	11.5	25/32	1 1/16	9/16	⅛	17/32	⅜	4¼
½, ⅝, ¾	11.5	25/32	1 1/16	9/16	⅛	17/32	⅜	4¼
½	14	17/32	13/16	½	⅛	15/32	5/16	4¼
¾	14	25/32	1 1/32	9/16	⅛	17/32	⅜	5¼
1	11.5	1 1/32	1 9/32	9/16	5/32	17/32	⅜	4¼
1¼	11.5	1 9/32	1⅝	⅝	5/32	19/32	15/32	5½
1½	11.5	1 17/32	1⅞	⅝	5/32	19/32	15/32	5½
2	11.5	2 1/32	2 11/32	¾	3/16	23/32	19/32	6¾
2½	8	2 17/32	2 27/32	1	¼	15/16	11/16	5½
3	8	3 1/32	3 15/32	1⅛	¼	1 1/16	13/16	6½
3½	8	3 17/32	3 31/32	1⅛	¼	1 1/16	13/16	6½
4	8	4 1/32	4 15/32	1⅛	¼	1 1/16	13/16	6½
4	6	4	4 29/32	1⅛	5/16	1 1/16	¾	4½

All dimensions are given in inches. For thread designation see Table 1.

American National Fire Hose Connection Screw Thread. — This thread is specified in the National Fire Protection Association's Standard NFPA No. 194-1974. It covers the dimensions for screw thread connections for fire hose couplings, suction hose couplings, relay supply hose couplings, fire pump suctions, discharge valves, fire hydrants, nozzles, adaptors, reducers, caps, plugs, wyes, siamese connections, standpipe connections, and sprinkler connections.

Form of thread: The basic form of thread is as shown on page 1605. It has an included angle of 60 degrees and is truncated top and bottom. The flat at the root and crest of the basic thread form is equal to ⅛ (0.125) times the pitch in inches. The height of the thread is equal to 0.649519 times the pitch. The outer ends of both external and internal threads are terminated by the blunt start or "Higbee Cut" on full thread to avoid crossing and mutilation of thread.

Thread Designation: The thread is designated by specifying in sequence the nominal size of the connection, number of threads per inch followed by the thread symbol NH. Thus, .75-8NH indicates a nominal size connection of 0.75 inch diameter with 8 threads per inch.

Basic Dimensions: The basic dimensions of the thread are as given in Table 1.

Thread Limits of Size: Limits of size for NH external threads are given in Table 2. Limits of size for NH internal threads are given in Table 3.

Table 1. Basic Dimensions of NH Threads (NFPA 194-1974)

Nom. Size	Threads per Inch, (tpi)	Thread Designation	Pitch, p	Basic Thread Height, h	Minimum Internal Thread Dimensions		
					Min. Minor Diam.	Basic Pitch Diam.	Basic Major Diam.
¾	8	.75-8 NH	0.12500	0.08119	1.2246	1.3058	1.3870
1	8	1-8 NH	0.12500	0.08119	1.2246	1.3058	1.3870
1½	9	1.5-9 NH	0.11111	0.07217	1.8577	1.9298	2.0020
2½	7.5	2.5-7.5 NH	0.13333	0.08660	2.9104	2.9970	3.0836
3	6	3-6 NH	0.16667	0.10825	3.4223	3.5306	3.6389
3½	6	3.5-6 NH	0.16667	0.10825	4.0473	4.1556	4.2639
4	4	4-4 NH	0.25000	0.16238	4.7111	4.8735	5.0359
4½	4	4.5-4 NH	0.25000	0.16238	5.4611	5.6235	5.7859
5	4	5-4 NH	0.25000	0.16238	5.9602	6.1226	6.2850
6	4	6-4 NH	0.25000	0.16238	6.7252	6.8876	7.0500

Nom. Size	Threads per Inch, (tpi)	Thread Designation	Pitch, p	External Thread Dimensions (Nipple)			
				Allowance	Max. Major Diam.	Max. Pitch Diam.	Max. Minor Diam.
¾	8	.75-8 NH	0.12500	0.0120	1.3750	1.2938	1.2126
1	8	1-8 NH	0.12500	0.0120	1.3750	1.2938	1.2126
1½	9	1.5-9 NH	0.11111	0.0120	1.9900	1.9178	1.8457
2½	7.5	2.5-7.5 NH	0.13333	0.0150	3.0686	2.9820	2.8954
3	6	3-6 NH	0.16667	0.0150	3.6239	3.5156	3.4073
3½	6	3.5-6 NH	0.16667	0.0200	4.2439	4.1356	4.0273
4	4	4-4 NH	0.25000	0.0250	5.0109	4.8485	4.6861
4½	4	4.5-4 NH	0.25000	0.0250	5.7609	5.5985	5.4361
5	4	5-4 NH	0.25000	0.0250	6.2600	6.0976	5.9352
6	4	6-4 NH	0.25000	0.0250	7.0250	6.8626	6.7002

All dimensions are in inches.

Tolerances: The pitch diameter tolerances for a mating external and internal thread are the same. Pitch diameter tolerances include lead and half-angle deviations. Lead deviations consuming one-half of the pitch diameter tolerance are 0.0032 inch for ¾, 1 and 1½-inch sizes; 0.0046 inch for 2½-inch size; 0.0052 inch for 3 and 3½-inch sizes; and 0.0072 inch for 4, 4½, 5, and 6-inch sizes. Half-angle deviations consuming one-half of the pitch diameter tolerance are 1 degree, 42 minutes for ¾, and 1-inch sizes; 1 degree, 54 minutes for 1½-inch size; 2 degrees, 17 minutes for 2½-inch size; 2 degrees, 4 minutes for 3 and 3½-inch sizes; and 1 degree, 55 minutes for 4, 4½, 5, and 6-inch sizes.

Tolerances for the external threads are:

$$\text{Major diameter tolerance} = 2 \times \text{Pitch diameter tolerance}$$
$$\text{Minor diameter tolerance} = \text{Pitch diameter tolerance} + 2h/9$$

The minimum minor diameter of the external thread is such as to result in a flat equal to one-third of the $p/8$ basic flat, or $p/24$, at the root when the pitch diameter of the external thread is at its minimum value. The maximum minor diameter is basic, but may be such as results from the use of a worn or rounded threading tool. This is the maximum minor diameter shown in the figure on page 1605 and is the diameter upon which the minor diameter tolerance formula shown above is based.

Tolerances for the internal threads are:

$$\text{Minor diameter tolerance} = 2 \times \text{Pitch diameter tolerance}$$

The minimum minor diameter of the internal thread is such as to result in a basic flat, $p/8$, at the crest when the pitch diameter of the thread is at its minimum value.

$$\text{Major diameter tolerance} = \text{Pitch diameter tolerance} - 2h/9$$

Gages and Gaging: Full information on gage dimensions and the use of gages in checking the NH thread are given in NFPA Standard No. 194-1974, published by the National Fire Protection Association, 470 Atlantic Avenue, Boston, Mass. 02210.

The information and data taken from this standard are reproduced with the permission of the Association.

Table 2.   Limits of Size and Tolerances for NH External Threads, (Nipples)
(NFPA 194-1974)

Nom. Size	Threads per Inch, (tpi)	External Thread (Nipple)							
		Major Diameter			Pitch Diameter			Minor* Diam.
		Max.	Min.	Toler.	Max.	Min.	Toler.	Max.
¾	8	1.3750	1.3528	0.0222	1.2938	1.2827	0.0111	1.2126
1	8	1.3750	1.3528	0.0222	1.2938	1.2827	0.0111	1.2126
1½	9	1.9900	1.9678	0.0222	1.9178	1.9067	0.0111	1.8457
2½	7.5	3.0686	3.0366	0.0320	2.9820	2.9660	0.0160	2.8954
3	6	3.6239	3.5879	0.0360	3.5156	3.4976	0.0180	3.4073
3½	6	4.2439	4.2079	0.0360	4.1356	4.1176	0.0180	4.0273
4	4	5.0109	4.9609	0.0500	4.8485	4.8235	0.0250	4.6861
4½	4	5.7609	5.7109	0.0500	5.5985	5.5735	0.0250	5.4361
5	4	6.2600	6.2100	0.0500	6.0976	6.0726	0.0250	5.9352
6	4	7.0250	6.9750	0.0500	6.8626	6.8376	0.0250	6.7002

All dimensions are in inches.
* Dimensions given for the maximum minor diameter of the nipple are figured to the intersection of the worn tool arc with a centerline through crest and root.   The minimum minor diameter of the nipple shall be that corresponding to a flat at the minor diameter of the minimum nipple equal to $p/24$ and may be determined by subtracting $11h/9$ (or $0.7939p$) from the minimum pitch diameter of the nipple.

Table 3.   Limits of Size and Tolerances for NH Internal Threads, (Couplings)
(NFPA 194-1974)

Nom. Size	Threads per Inch, (tpi)	Internal Thread (Coupling)							
		Minor Diameter			Pitch Diameter			Major* Diam.
		Min.	Max.	Toler.	Min.	Max.	Toler.	Min.
¾	8	1.2246	1.2468	0.0222	1.3058	1.3169	0.0111	1.3870
1	8	1.2246	1.2468	0.0222	1.3058	1.3169	0.0111	1.3870
1½	9	1.8577	1.8799	0.0222	1.9298	1.9409	0.0111	2.0020
2½	7.5	2.9104	2.9424	0.0320	2.9970	3.0130	0.0160	3.0836
3	6	3.4223	3.4583	0.0360	3.5306	3.5486	0.0180	3.6389
3½	6	4.0473	4.0833	0.0360	4.1556	4.1736	0.0180	4.2639
4	4	4.7111	4.7611	0.0500	4.8735	4.8985	0.0250	5.0359
4½	4	5.4611	5.5111	0.0500	5.6235	5.6485	0.0250	5.7859
5	4	5.9602	6.0102	0.0500	6.1226	6.1476	0.0250	6.2850
6	4	6.7252	6.7752	0.0500	6.8876	6.9126	0.0250	7.0500

All dimensions are in inches.
* Dimensions for the minimum major diameter of the coupling correspond to the basic flat ($p/8$), and the profile at the major diameter produced by a worn tool must not fall below the basic outline.   The maximum major diameter of the coupling shall be that corresponding to a flat at the major diameter of the maximum coupling equal to $p/24$ and may be determined by adding $11h/9$ (or $0.7939p$) to the maximum pitch diameter of the coupling.

Rolled Threads for Screw Shells of Electric Sockets and Lamp Bases — American Standard

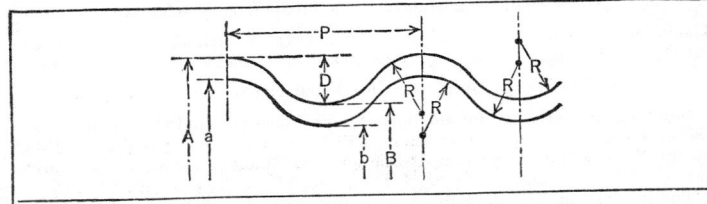

Male or Base Screw Shells Before Assembly

Size	Threads per Inch	Pitch P	Depth of Thread D	Radius Crest Root R	Major Diam.		Minor Diam.	
					Max. A	Min. a	Max. B	Min. b
Miniature	14	0.07143	0.020	0.0210	0.375	0.370	0.335	0.330
Candelabra	10	0.10000	0.025	0.0312	0.465	0.460	0.415	0.410
Intermediate	9	0.11111	0.027	0.0353	0.651	0.645	0.597	0.591
Medium	7	0.14286	0.033	0.0470	1.037	1.031	0.971	0.965
Mogul	4	0.25000	0.050	0.0906	1.555	1.545	1.455	1.445

Socket Screw Shells Before Assembly

Size	Threads per Inch	Pitch P	Depth of Thread D	Radius Crest Root R	Major Diam.		Minor Diam.	
					Max. A	Min. a	Max. B	Min. b
Miniature	14	0.07143	0.020	0.0210	0.3835	0.3775	0.3435	0.3375
Candelabra	10	0.10000	0.025	0.0312	0.476	0.470	0.426	0.420
Intermediate	9	0.11111	0.027	0.0353	0.664	0.657	0.610	0.603
Medium	7	0.14286	0.033	0.0470	1.053	1.045	0.987	0.979
Mogul	4	0.25000	0.050	0.0906	1.577	1.565	1.477	1.465

All dimensions given in inches.

Base Screw Shell Gage Tolerances: Threaded ring gages — "Go," Max. thread size to minus 0.0003 inch; "Not Go," Min. thread size to plus 0.0003 inch. Plain ring gages — "Go," Max. thread O.D. to minus 0.0002 inch; "Not Go", Min. thread O.D. to plus 0.0002. inch.

Socket Screw Shell Gages: Threaded plug gages — "Go," Min. thread size to plus 0.0003 inch; "Not Go," Max. thread size to minus 0.0003 inch. Plain plug gages — "Go," Min. minor diam. to plus 0.0002 inch; "Not Go," Max. minor diam. to minus 0.0002 inch.

Check Gages for Base Screw Shell Gages: Threaded plugs for checking threaded ring gages — "Go," Max. thread size to minus 0.0003 inch; "Not Go," Min. thread size to plus 0.0003 inch.

Instrument Makers' System. — The standard screw system of the Royal Microscopical Society of London, England, also known as the "Society Thread," is employed for microscope objectives and the nose pieces of the microscope into which these objectives screw. The form of the thread is the standard Whitworth form. The number of threads per inch is 36. The dimensions are as follows:

Male thread, outside diam.,	max. 0.7982 inch,	min. 0.7952 inch;	
root diam.,	max. 0.7626 inch,	min. 0.7596 inch;	
Female thread, root of thread,	max. 0.7674 inch,	min. 0.7644 inch;	
top of thread,	max. 0.8030 inch,	min. 0.8000 inch.	

Pipe Threads. — The types of threads used on pipe and pipe fittings may be classed according to their intended use: (1) threads which when assembled with a sealer will produce a pressure-tight joint; (2) threads which when assembled without a sealer will produce a pressure-tight joint; (3) threads which provide free- and loose-fitting mechanical joints without pressure tightness; and (4) threads which produce rigid mechanical joints without pressure tightness. American National Standard pipe threads described in the following paragraphs provide taper and straight pipe threads for use in various combinations and with certain modifications to meet these specific needs.

American National Standard Taper Pipe Threads. — The basic dimensions of the ANSI Standard taper pipe thread are given in Table 3.

Form of Thread: The angle between the sides of the thread is 60 degrees when measured in an axial plane, and the line bisecting this angle is perpendicular to the axis. The depth of the truncated thread is based on factors entering into the manufacture of cutting tools and the making of tight joints and is given by the formulas in Table 3 or the data in Table 1 obtained from these formulas. While the standard shows flat surfaces at the crest and root of the thread, some rounding may occur in commercial practice, and it is intended that the pipe threads of product shall be acceptable when crest and root of the tools or chasers lie within the limits shown in Table 1.

Table 1.   Limits on Crest and Root of American National Standard External and Internal Taper Pipe Threads, NPT (ANSI B2.1-1968)

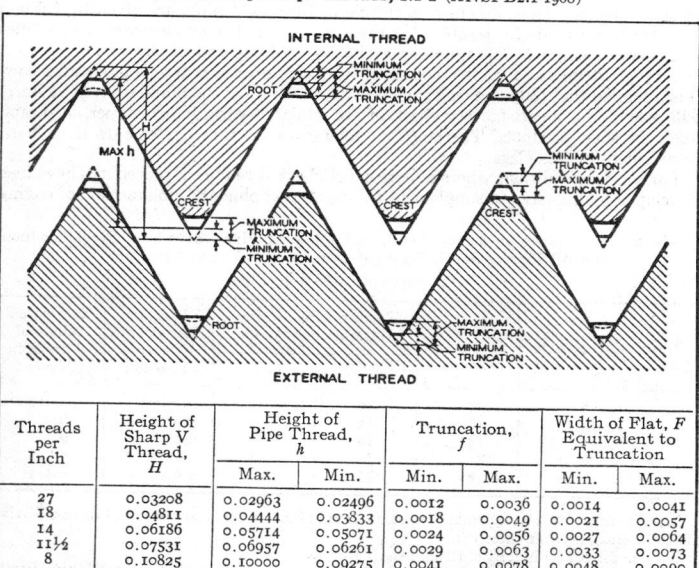

Threads per Inch	Height of Sharp V Thread, H	Height of Pipe Thread, h		Truncation, f		Width of Flat, F Equivalent to Truncation	
		Max.	Min.	Min.	Max.	Min.	Max.
27	0.03208	0.02963	0.02496	0.0012	0.0036	0.0014	0.0041
18	0.04811	0.04444	0.03833	0.0018	0.0049	0.0021	0.0057
14	0.06186	0.05714	0.05071	0.0024	0.0056	0.0027	0.0064
11½	0.07531	0.06957	0.06261	0.0029	0.0063	0.0033	0.0073
8	0.10825	0.10000	0.09275	0.0041	0.0078	0.0048	0.0090

All dimensions are in inches and are given to four or five decimal places only to avoid errors in computations, not to indicate required precision.

Pitch Diameter Formulas: In the following formulas, which apply to the ANSI Standard taper pipe thread, E_0 = pitch diameter at end of pipe; E_1 = pitch diameter at the large end of the internal thread and also at the gaging notch; D = outside diameter of pipe; L_1 = length of hand-tight or normal engagement between external and internal threads; L_2 = basic length of effective external taper thread; p = pitch = 1 ÷ number of threads per inch.

$$E_0 = D - (0.05D + 1.1)p$$

$$E_1 = E_0 + 0.0625L_1$$

Thread Length: The formula for L_2 determines the length of the effective thread and includes approximately two usable threads which are slightly imperfect at the crest. The normal length of engagement L_1 between external and internal taper threads, when assembled by hand, is controlled by the use of the gages.

$$L_2 = (0.80D + 6.8)p$$

Taper: The taper of the thread is 1 in 16 or 0.75 inch per foot measured on the diameter and along the axis. The corresponding half-angle of taper or angle with the center line is one degree, 47 minutes.

Tolerances on Thread Elements. — The maximum allowable variation in the commercial product (manufacturing tolerance) is one turn large or small from the basic dimensions.

The permissible variations in thread elements on steel products and all pipe made of steel, wrought iron, or brass, exclusive of butt-weld pipe, are given in Table 2. This table is a guide for establishing the limits of the thread elements of taps, dies, and thread chasers. These limits may be required on product threads.

On pipe fittings and valves (not steel) for steam pressures 300 pounds and below, it is intended that plug and ring gage practice as set up in the Standard (ANSI B2.1) will provide for a satisfactory check of accumulated variations of taper, lead, and angle in such product. Therefore no tolerances on thread elements have been established for this class.

For service conditions where a more exact check is required, procedures have been developed by industry to supplement the regulation plug and ring method of gaging.

Table 2. Tolerances on Taper, Lead, and Angle of Pipe Threads of Steel Products and All Pipe of Steel, Wrought-Iron, or Brass (ANSI B2.1-1968)

(Exclusive of Butt-Weld Pipe)

Nominal Pipe Size	Threads per Inch	Taper on Pitch Line, Inches per Foot		Lead in Length of Effective Threads	60 Degree Angle of Threads, Degrees
		Max.	Min.		
1/16, 1/8	27	7/8	1 1/16	±0.003	±2 1/2
1/4, 3/8	18	7/8	1 1/16	±0.003	±2
1/2, 3/4	14	7/8	1 1/16	±0.003(a)	±2
1, 1 1/4, 1 1/2, 2	11 1/2	7/8	1 1/16	±0.003(a)	±1 1/2
2 1/2 and larger	8	7/8	1 1/16	±0.003(a)	±1 1/2

(a) The tolerance on lead shall be ±0.003 in. per inch on any size threaded to an effective thread length greater than 1 in.

For tolerances on height of thread see Table 1.

The limits specified in this table are intended to serve as a guide for establishing limits of the thread elements of taps, dies, and thread chasers. These limits may be required on product threads.

Table 3. Basic Dimensions, American National Standard Taper Pipe Threads,[1] NPT
(ANSI B2.1-1968)

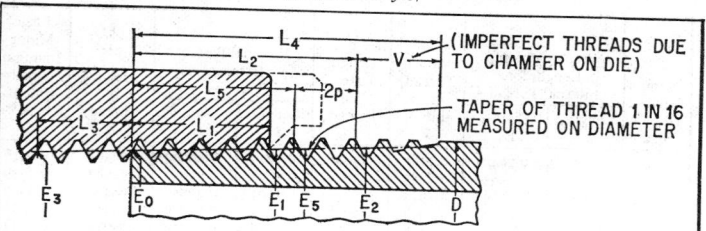

For all dimensions see corresponding reference letters in table.

Angle between sides of thread is 60 degrees.   Taper of thread, on diameter, is ¾ inch per foot.   Angle of taper with centerline is 1°47'.

The basic maximum thread height, h, of the truncated thread is 0.8 × pitch of thread.   The crest and root are truncated a minimum of 0.033 × pitch for all pitches.   For maximum depth of truncation see Table 1.

Nominal Pipe Size	Outside Diam. of Pipe, D	Threads per Inch, n	Pitch of Thread, p	Pitch Diameter at Beginning of External Thread, E_0	Handtight Engagement		Effective Thread, External	
					Length,[2] L_1 In.	Diam.,[3] E_1	Length,[4] L_2 In.	Diam., E_2
1⁄16	0.3125	27	0.03704	0.27118	0.160	0.28118	0.2611	0.28750
1⁄8	0.405	27	0.03704	0.36351	0.1615	0.37360	0.2639	0.38000
1⁄4	0.540	18	0.05556	0.47739	0.2278	0.49163	0.4018	0.50250
3⁄8	0.675	18	0.05556	0.61201	0.240	0.62701	0.4078	0.63750
1⁄2	0.840	14	0.07143	0.75843	0.320	0.77843	0.5337	0.79179
3⁄4	1.050	14	0.07143	0.96768	0.339	0.98887	0.5457	1.00179
1	1.315	11½	0.08696	1.21363	0.400	1.23863	0.6828	1.25630
1¼	1.660	11½	0.08696	1.55713	0.420	1.58338	0.7068	1.60130
1½	1.900	11½	0.08696	1.79609	0.420	1.82234	0.7235	1.84130
2	2.375	11½	0.08696	2.26902	0.436	2.29627	0.7565	2.31630
2½	2.875	8	0.12500	2.71953	0.682	2.76216	1.1375	2.79062
3	3.500	8	0.12500	3.34062	0.766	3.38850	1.2000	3.41562
3½	4.000	8	0.12500	3.83750	0.821	3.88881	1.2500	3.91562
4	4.500	8	0.12500	4.33438	0.844	4.38712	1.3000	4.41562
5	5.563	8	0.12500	5.39073	0.937	5.44929	1.4063	5.47862
6	6.625	8	0.12500	6.44609	0.958	6.50597	1.5125	6.54062
8	8.625	8	0.12500	8.43359	1.063	8.50003	1.7125	8.54062
10	10.750	8	0.12500	10.54531	1.210	10.62094	1.9250	10.66562
12	12.750	8	0.12500	12.53281	1.360	12.61781	2.1250	12.66562
14 OD	14.000	8	0.12500	13.77500	1.562	13.87262	2.2500	13.91562
16 OD	16.000	8	0.12500	15.76250	1.812	15.87575	2.4500	15.91562
18 OD	18.000	8	0.12500	17.75000	2.000	17.87500	2.6500	17.91562
20 OD	20.000	8	0.12500	19.73750	2.125	19.87031	2.8500	19.91562
24 OD	24.000	8	0.12500	23.71250	2.375	23.86094	3.2500	23.91562

All dimensions given in inches.
[1] The basic dimensions of the ANSI Standard Taper Pipe Thread are given in inches to four or five decimal places.   While this implies a greater degree of precision than is ordinarily attained, these dimensions are the basis of gage dimensions and are so expressed for the purpose of eliminating errors in computations.
[2] Also length of thin ring gage and length from gaging notch to small end of plug gage.
[3] Also pitch diameter at gaging notch (handtight plane).
[4] Also length of plug gage.

Table 3. (*Concluded*). Basic Dimensions, American National Standard Taper Pipe Threads, NPT (ANSI B2.1-1968)

| Nominal Pipe Size | Wrench Makeup Length for Internal Thread | | Vanish Thread, (3.47 thds.), V | Overall Length External Thread, L_4 | Nominal Perfect External Threads[5] | | Height of Thread, h | Basic Minor Diam. at Small End of Pipe,[6] K_0 |
	Length,[7] L_3	Diam., E_3			Length, L_5	Diam., E_5		
1/16	0.1111	0.26424	0.1285	0.3896	0.1870	0.28287	0.02963	0.2416
1/8	0.1111	0.35656	0.1285	0.3924	0.1898	0.37537	0.02963	0.3339
1/4	0.1667	0.46697	0.1928	0.5946	0.2907	0.49556	0.04444	0.4329
3/8	0.1667	0.60160	0.1928	0.6006	0.2967	0.63056	0.04444	0.5676
1/2	0.2143	0.74504	0.2478	0.7815	0.3909	0.78286	0.05714	0.7013
3/4	0.2143	0.95429	0.2478	0.7935	0.4029	0.99286	0.05714	0.9105
1	0.2609	1.19733	0.3017	0.9845	0.5089	1.24543	0.06957	1.1441
1 1/4	0.2609	1.54083	0.3017	1.0085	0.5329	1.59043	0.06957	1.4876
1 1/2	0.2609	1.77978	0.3017	1.0252	0.5496	1.83043	0.06957	1.7265
2	0.2609	2.25272	0.3017	1.0582	0.5826	2.30543	0.06957	2.1995
2 1/2	0.2500[8]	2.70391	0.4337	1.5712	0.8875	2.77500	0.100000	2.6195
3	0.2500[8]	3.32500	0.4337	1.6337	0.9500	3.40000	0.100000	3.2406
3 1/2	0.2500	3.82188	0.4337	1.6837	1.0000	3.90000	0.100000	3.7375
4	0.2500	4.31875	0.4337	1.7337	1.0500	4.40000	0.100000	4.2344
5	0.2500	5.37511	0.4337	1.8400	1.1563	5.46300	0.100000	5.2907
6	0.2500	6.43047	0.4337	1.9462	1.2625	6.52500	0.100000	6.3461
8	0.2500	8.41797	0.4337	2.1462	1.4625	8.52500	0.100000	8.3336
10	0.2500	10.52969	0.4337	2.3587	1.6750	10.65000	0.100000	10.4453
12	0.2500	12.51719	0.4337	2.5587	1.8750	12.65000	0.100000	12.4328
14 OD	0.2500	13.75938	0.4337	2.6837	2.0000	13.90000	0.100000	13.6750
16 OD	0.2500	15.74688	0.4337	2.8837	2.2000	15.90000	0.100000	15.6625
18 OD	0.2500	17.73438	0.4337	3.0837	2.4000	17.90000	0.100000	17.6500
20 OD	0.2500	19.72188	0.4337	3.2837	2.6000	19.90000	0.100000	19.6375
24 OD	0.2500	23.69688	0.4337	3.6837	3.0000	23.90000	0.100000	23.6125

[5] The length L_5 from the end of the pipe determines the plane beyond which the thread form is imperfect at the crest. The next two threads are perfect at the root. At this plane the cone formed by the crests of the thread intersects the cylinder forming the external surface of the pipe. $L_5 = L_2 - 2p$.
[6] Given as information for use in selecting tap drills.
[7] Three threads for 2-inch size and smaller; two threads for larger sizes.
[8] Military Specification MIL — P — 7105 gives the wrench makeup as three threads for 3 in. and smaller. The E_3 dimensions are then as follows: Size 2 1/2 in., 2.69609 and size 3 in., 3.31719. Increase in diameter per thread is equal to $0.0625/n$.

Engagement between External and Internal Taper Threads. — The normal length of engagement-between external and internal taper threads when screwed together handtight is shown as L_1 in Table 3. This length is controlled by the construction and use of the pipe thread gages. It is recognized that in special applications, such as flanges for high-pressure work, longer thread engagement is used, in which case the pitch diameter E_1 (Table 3) is maintained and the pitch diameter E_0 at the end of the pipe is proportionately smaller.

Railing Joint Taper Pipe Threads, NPTR. — Railing joints require a rigid mechanical thread joint with external and internal taper threads. The external thread is basically the same as the ANSI Standard Taper Pipe Thread, except that sizes 1/2 through 2 inches are shortened by 3 threads and sizes 2 1/2 through 4 inches are shortened by 4 threads to permit the use of the larger end of the pipe thread. A recess in the fitting covers the last scratch or imperfect threads on the pipe.

Table 4. American National Standard Internal Threads in Pipe Couplings, NPSC for Pressuretight Joints with Lubricant or Sealer (ANSI B2.1-1968)

Nom. Pipe Size	Thds. per Inch	Minor[2] Diam. Min.	Pitch Diameter[1] Min.	Pitch Diameter[1] Max.	Nom. Pipe Size	Thds. per Inch	Minor[2] Diam. Min.	Pitch Diameter[1] Min.	Pitch Diameter[1] Max.
1/8	27	0.342	0.3701	0.3771	1½	11½	1.745	1.8142	1.8305
1/4	18	0.440	0.4864	0.4968	2	11½	2.219	2.2881	2.3044
3/8	18	0.577	0.6218	0.6322	2½	8	2.650	2.7504	2.7739
1/2	14	0.715	0.7717	0.7851	3	8	3.277	3.3768	3.4002
3/4	14	0.925	0.9822	0.9956	3½	8	3.777	3.8771	3.9005
1	11½	1.161	1.2305	1.2468	4	8	4.275	4.3754	4.3988
1¼	11½	1.506	1.5752	1.5915	...	...	...	...	...

[1] The actual pitch diameter of the straight tapped hole will be slightly smaller than the value given when gaged with a taper plug gage as called for in ANSI B2.1.

[2] As the ANSI Standard Pipe Thread form is maintained, the major and minor diameters of the internal thread vary with the pitch diameter. All dimensions are given in inches.

Straight Pipe Threads in Pipe Couplings, NPSC. — Threads in pipe couplings made in accordance with the ANSI B2.1 specifications are straight (parallel) threads of the same thread form as the ANSI Standard Taper Pipe Thread. They are used to form pressuretight joints when assembled with an ANSI Standard external taper pipe thread and made up with lubricant or sealant. These joints are recommended for comparatively low pressures only.

Straight Pipe Threads for Mechanical Joints, NPSM, NPSL, and NPSH. — While external and internal taper pipe threads are recommended for pipe joints in practically every service, there are mechanical joints where straight threads are used to advantage. Three types covered by ANSI B2.1 are:

Free-Fitting Mechanical Joints for Fixtures (External and Internal), NPSM: Standard iron, steel, and brass pipe are often used for special applications where there are no internal pressures. Where straight thread joints are required for mechanical assemblies, straight pipe threads are often found more suitable or convenient. Dimensions of these threads are given in Table 5.

Loose-Fitting Mechanical Joints With Locknuts (External and Internal), NPSL: This thread is designed to produce a pipe thread having the largest diameter that it is possible to cut on standard pipe. The dimensions of these threads are given in Table 5. It will be noted that the maximum major diameter of the external thread is slightly greater than the nominal outside diameter of the pipe. The normal manufacturer's variation in pipe diameter provides for this increase.

Loose-Fitting Mechanical Joints for Hose Couplings (External and Internal), NPSH: Hose coupling joints are ordinarily made with straight internal and external loose-fitting threads. There are several standards of hose threads having various diameters and pitches. One of these is based on the ANSI Standard pipe thread and by the use of this thread series, it is possible to join small hose couplings in sizes ½ to 4 inches, inclusive, to ends of standard pipe having ANSI Standard External Pipe Threads, using a gasket to seal the joints. For the hose coupling thread dimensions see pages 1605 to 1609.

Thread Designation and Notation. — American National Standard Pipe Threads are designated by specifying in sequence the nominal size, number of threads per inch, and the symbols for the thread series and form, as: ⅜ — 18 NPT. The symbol designations are as follows: NPT — American National Standard Taper Pipe Thread; NPTR — American National Standard Taper Pipe Thread for Railing Joints; NPSC — American National Standard Straight Pipe Thread for Couplings; NPSM — American National Standard Straight Pipe Thread for Free-fitting Mechanical Joints; NPSL — American National Standard Straight Pipe Thread for Loose-fitting Mechanical Joints with Locknuts; and NPSH — American National Standard Straight Pipe Thread for Hose Couplings.

Table 5. American National Standard Straight Pipe Threads for Mechanical Joints, NPSM and NPSL (ANSI B2.1-1968)

Nominal Pipe Size	Threads per Inch	Allowance	External Thread				Internal Thread			
			Major Diameter		Pitch Diameter		Minor Diameter		Pitch Diameter	
			Max.[2]	Min.	Max.	Min.	Min.[2]	Max.	Min.[1]	Max.
Free-fitting Mechanical Joints for Fixtures — NPSM										
1/8	27	0.0011	0.397	0.390	0.3725	0.3689	0.358	0.364	0.3736	0.3783
1/4	18	0.0013	0.526	0.517	0.4903	0.4859	0.468	0.481	0.4916	0.4974
3/8	18	0.0014	0.662	0.653	0.6256	0.6211	0.603	0.612	0.6270	0.6329
1/2	14	0.0015	0.823	0.813	0.7769	0.7718	0.747	0.759	0.7784	0.7851
3/4	14	0.0016	1.034	1.024	0.9873	0.9820	0.958	0.970	0.9889	0.9958
1	11½	0.0017	1.293	1.281	1.2369	1.2311	1.201	1.211	1.2386	1.2462
1¼	11½	0.0018	1.638	1.626	1.5816	1.5756	1.546	1.555	1.5834	1.5912
1½	11½	0.0018	1.877	1.865	1.8205	1.8144	1.785	1.794	1.8223	1.8302
2	11½	0.0019	2.351	2.339	2.2944	2.2882	2.259	2.268	2.2963	2.3044
2½	8	0.0022	2.841	2.826	2.7600	2.7526	2.708	2.727	2.7622	2.7720
3	8	0.0023	3.467	3.452	3.3862	3.3786	3.334	3.353	3.3885	3.3984
3½	8	0.0023	3.968	3.953	3.8865	3.8788	3.835	3.848	3.8888	3.8988
4	8	0.0023	4.466	4.451	4.3848	4.3771	4.333	4.346	4.3871	4.3971
5	8	0.0024	5.528	5.513	5.4469	5.4390	5.395	5.408	5.4493	5.4598
6	8	0.0024	6.585	6.570	6.5036	6.4955	6.452	6.464	6.5060	6.5165
Loose-fitting Mechanical Joints for Locknut Connections — NPSL										
1/8	27	...	0.409	...	0.3840	0.3805	0.362	...	0.3863	0.3898
1/4	18	...	0.541	...	0.5038	0.4986	0.470	...	0.5073	0.5125
3/8	18	...	0.678	...	0.6409	0.6357	0.607	...	0.6444	0.6496
1/2	14	...	0.844	...	0.7963	0.7896	0.753	...	0.8008	0.8075
3/4	14	...	1.054	...	1.0067	1.0000	0.964	...	1.0112	1.0179
1	11½	...	1.318	...	1.2604	1.2523	1.208	...	1.2658	1.2739
1¼	11½	...	1.663	...	1.6051	1.5970	1.553	...	1.6106	1.6187
1½	11½	...	1.902	...	1.8441	1.8360	1.792	...	1.8495	1.8576
2	11½	...	2.376	...	2.3180	2.3099	2.265	...	2.3234	2.3315
2½	8	...	2.877	...	2.7934	2.7817	2.718	...	2.8012	2.8129
3	8	...	3.503	...	3.4198	3.4081	3.344	...	3.4276	3.4393
3½	8	...	4.003	...	3.9201	3.9084	3.845	...	3.9279	3.9396
4	8	...	4.502	...	4.4184	4.4067	4.343	...	4.4262	4.4379
5	8	...	5.564	...	5.4805	5.4688	5.405	...	5.4884	5.5001
6	8	...	6.620	...	6.5372	6.5255	6.462	...	6.5450	6.5567
8	8	...	8.615	...	8.5313	8.5196	8.456	...	8.5391	8.5508
10	8	...	10.735	...	10.6522	10.6405	10.577	...	10.6600	10.6717
12	8	...	12.732	...	12.6491	12.6374	12.574	...	12.6569	12.6686

All dimensions are given in inches.

Notes for Free-fitting Fixture Threads:

[1] This is the same as the pitch diameter at end of internal thread, E_1 Basic. (See Table 3.)

The minor diameters of external threads and major diameters of internal threads are those as produced by commercial straight pipe dies and commercial ground straight pipe taps.

The major diameter of the external thread has been calculated on the basis of a truncation of $0.10825p$, and the minor diameter of the internal thread has been calculated on the basis of a truncation of $0.21651p$, to provide no interference at crest and root when product is gaged with gages made in accordance with the Standard.

Notes for Loose-fitting Locknut Threads:

[2] As the ANSI Standard Straight Pipe Thread form of thread is maintained, the major and the minor diameters of the internal thread and the minor diameter of the external thread vary with the pitch diameter. The major diameter of the external thread is usually determined by the diameter of the pipe. These theoretical diameters result from adding the depth of the truncated thread ($0.666025 \times p$) to the maximum pitch diameters, and it should be understood that commercial pipe will not always have these maximum major diameters.

The locknut thread is established on the basis of retaining the greatest possible amount of metal thickness between the bottom of the thread and the inside of the pipe.

In order that a locknut may fit loosely on the externally threaded part, an allowance equal to the "increase in pitch diameter per turn" is provided, with a tolerance of 1½ turns for both external and internal threads.

American National Standard Dryseal Pipe Threads for Pressure-Tight Joints. — Dryseal pipe threads are based on the USA (American) pipe thread; however, they differ in that they are designed to seal pressure-tight joints without the necessity of using sealing compounds. To accomplish this, some modification of thread form and greater accuracy in manufacture is required. The roots of both the external and internal threads are truncated slightly more than the crests, i.e., roots have wider flats than crests so that metal-to-metal contact occurs at the crests and roots coincident with, or prior to, flank contact. Thus, as the threads are assembled by wrenching, the roots of the threads crush the sharper crests of the mating threads. This sealing action at both major and minor diameters tends to prevent spiral leakage and makes the joints pressure-tight without the necessity of using sealing compounds, provided that the threads are in accordance with standard specifications and tolerances and are not damaged by galling in assembly. The control of crest and root truncation is simplified by the use of properly designed threading tools. Also, it is desirable that both external and internal threads have full thread height for the length of hand engagement. Where not functionally objectionable, the use of a compatible lubricant or sealant is permissible to minimize the possibility of galling. This is desirable in assembling Dryseal pipe threads in refrigeration and other systems to effect a pressure-tight seal. The crest and root of Dryseal pipe threads may be slightly rounded, but are acceptable if they lie within the truncation limits given in Table 6.

Table 6. American National Standard Dryseal Pipe Threads — Limits on Crest and Root Truncation (ANSI B1.20.3-1976)

Threads Per Inch	Height of Sharp V Thread (H)	Truncation							
		Minimum				Maximum			
		At Crest		At Root		At Crest		At Root	
		Formula	Inch	Formula	Inch	Formula	Inch	Formula	Inch
27	0.03208	$0.047p$	0.0017	$0.094p$	0.0035	$0.094p$	0.0035	$0.140p$	0.0052
18	0.04811	$0.047p$	0.0026	$0.078p$	0.0043	$0.078p$	0.0043	$0.109p$	0.0061
14	0.06180	$0.036p$	0.0026	$0.060p$	0.0043	$0.060p$	0.0043	$0.085p$	0.0061
11½	0.07531	$0.040p$	0.0035	$0.060p$	0.0052	$0.060p$	0.0052	$0.090p$	0.0078
8	0.10825	$0.042p$	0.0052	$0.055p$	0.0069	$0.055p$	0.0069	$0.076p$	0.0095

All dimensions are given in inches. In the formulas, p = pitch.

Types of Dryseal Pipe Thread. — American National Standard ANSI B1.20.3-1976 covers four types of standard Dryseal pipe threads:

NPTF — Dryseal USA (American) Standard Taper Pipe Thread

PTF-SAE SHORT — Dryseal SAE Short Taper Pipe Thread

NPSF — Dryseal USA (American) Standard Fuel Internal Straight Pipe Thread

NPSI — Dryseal USA (American) Standard Intermediate Internal Straight Pipe Thread

NPTF Threads: This type applies to both external and internal threads and is suitable for pipe joints in practically every type of service. Of all Dryseal pipe threads, NPTF external and internal threads mated are generally conceded to be superior for strength and seal since they have the longest length of thread and, theoretically, interference (sealing) occurs at every engaged thread root and crest. Use of tapered internal threads, such as NPTF or PTF-SAE SHORT in hard or brittle materials having thin sections will minimize the possibility of fracture.

There are two classes of NTPF threads. Class 1 threads are made to interfere (seal) at root and crest when mated, but inspection of crest and root truncation is not required. Consequently, Class 1 threads are intended for applications where close control of tooling is required for conformance of truncation or where sealing is accomplished by means of a sealant applied to the threads.

Class 2 threads are theoretically identical to those made to Class 1, however, inspection of root and crest truncation is required. Consequently, where a sealant is not used, there is more assurance of a pressure-tight seal for Class 2 threads than for Class 1 threads.

PTF-SAE SHORT Threads: External threads of this type conform in all respects with NPTF threads except that the thread length has been shortened by eliminating one thread from the small (entering) end. These threads are designed for applications where clearance is not sufficient for the full length of the NPTF threads or for economy of material where the full thread length is not necessary.

Internal threads of this type conform in all respects with NPTF threads, except that the thread length has been shortened by eliminating one thread from the large (entry) end. These threads are designed for thin materials where thickness is not sufficient for the full thread length of the NPTF threads or for economy in tapping where the full thread length is not necessary.

Pressure-tight joints without the use of lubricant or sealer can best be ensured where mating components are both threaded with NPTF threads. This should be considered before specifying PTF-SAE SHORT external or internal threads.

NPSF Threads: Threads of this type are straight (cylindrical) instead of tapered and are internal only. They are more economical to produce than tapered internal threads, but when assembled do not offer as strong a guarantee of sealing since root and crest interference will not occur for all threads. NPSF threads are generally used with soft or ductile materials which will tend to adjust at assembly to the taper of external threads, but may be used in hard or brittle materials where the section is thick.

NPSI Threads: Threads of this type are straight (cylindrical) instead of tapered, are internal only and are slightly larger in diameter than NPSF threads but have the same tolerance and thread length. They are more economical to produce than tapered

Table 7. Recommended Limitation of Assembly among the Various Types of Dryseal Threads[1]

External Dryseal Thread		For Assembly with Internal Dryseal Thread	
Type	Description	Type	Description
1	NPTF (tapered), ext thd	1 2[2,4] 3[2,5] 4[2,5,6]	NPTF (tapered), int thd PTF-SAE SHORT (tapered), int thd NPSF (straight), int thd NPSI (straight), int thd
2[2,3]	PTF-SAE SHORT (tapered) ext thd	4 1	NPSI (straight), int thd NPTF (tapered), int thd

[1] An assembly with straight internal pipe threads and taper external pipe threads is frequently more advantageous than an all taper thread assembly, particularly in automotive and other allied industries where economy and rapid production are major considerations. Dryseal threads are not used in assemblies in which both components have straight pipe threads.

[2] Pressure-tight joints without the use of a sealant can best be ensured where both components are threaded with NPTF (full length threads), since theoretically interference (sealing) occurs at all threads, but there are two less threads engaged than for NPTF assemblies. When straight internal threads are used, there is interference only at one thread depending on ductility of materials.

[3] PTF-SAE SHORT external threads are primarily intended for assembly with type 4-NPSI internal threads but can also be used with type 1-NPTF internal threads. They are not designed for, and at extreme tolerance limits may not assemble with, type 2-PTF-SAE SHORT internal threads or type 3-NPSF internal threads.

[4] PTF-SAE SHORT internal threads are primarily intended for assembly with type 1-NPTF external threads. They are not designed for, and at extreme tolerance limits may not assemble with, type 2-PTF-SAE SHORT external threads.

[5] There is no external straight Dryseal thread.

[6] NPSI internal threads are primarily intended for assembly with type 2-PTF-SAE SHORT external threads but will also assemble with full length type 1 NPTF external threads.

Table 8. Suggested Tap Drill Sizes for Internal Dryseal Pipe Threads

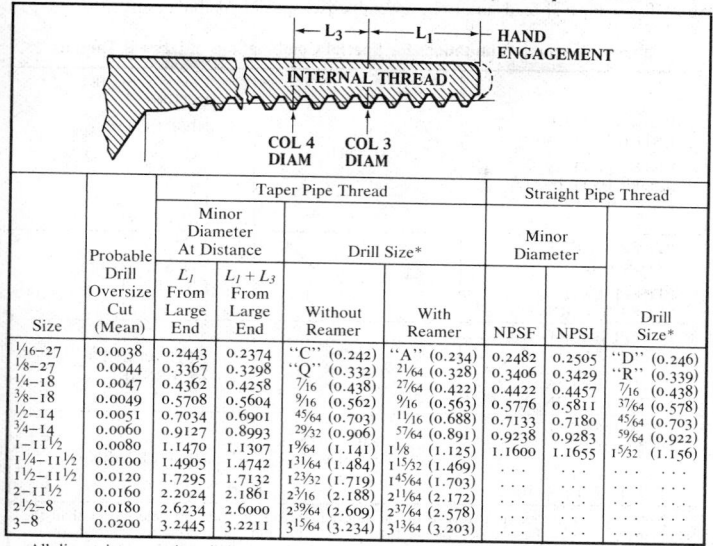

Size	Probable Drill Oversize Cut (Mean)	Taper Pipe Thread				Straight Pipe Thread		
		Minor Diameter At Distance		Drill Size*		Minor Diameter		Drill Size*
		L_1 From Large End	$L_1 + L_3$ From Large End	Without Reamer	With Reamer	NPSF	NPSI	
1/16–27	0.0038	0.2443	0.2374	"C" (0.242)	"A" (0.234)	0.2482	0.2505	"D" (0.246)
1/8–27	0.0044	0.3367	0.3298	"Q" (0.332)	21/64 (0.328)	0.3406	0.3429	"R" (0.339)
1/4–18	0.0047	0.4362	0.4258	7/16 (0.438)	27/64 (0.422)	0.4422	0.4457	7/16 (0.438)
3/8–18	0.0049	0.5708	0.5604	9/16 (0.562)	9/16 (0.563)	0.5776	0.5811	37/64 (0.578)
1/2–14	0.0051	0.7034	0.6901	45/64 (0.703)	11/16 (0.688)	0.7133	0.7180	45/64 (0.703)
3/4–14	0.0060	0.9127	0.8993	29/32 (0.906)	57/64 (0.891)	0.9238	0.9283	59/64 (0.922)
1–11 1/2	0.0080	1.1470	1.1307	1 9/64 (1.141)	1 1/8 (1.125)	1.1600	1.1655	1 5/32 (1.156)
1 1/4–11 1/2	0.0100	1.4905	1.4742	1 31/64 (1.484)	1 15/32 (1.469)	…	…	…
1 1/2–11 1/2	0.0120	1.7295	1.7132	1 23/32 (1.719)	1 45/64 (1.703)	…	…	…
2–11 1/2	0.0160	2.2024	2.1861	2 3/16 (2.188)	2 11/64 (2.172)	…	…	…
2 1/2–8	0.0180	2.6234	2.6000	2 39/64 (2.609)	2 37/64 (2.578)	…	…	…
3–8	0.0200	3.2445	3.2211	3 15/64 (3.234)	3 13/64 (3.203)	…	…	…

All dimensions are given in inches.
* Some drill sizes listed may not be standard drills.

threads and may be used in hard or brittle materials where the section is thick or where there is little expansion at assembly with external taper threads. As with NPSF threads, NPSI threads when assembled do not offer as strong a guarantee of sealing as do tapered internal threads.

For more complete specifications for production and acceptance of Dryseal pipe threads, see ANSI B1.20.3 (Inch) and B1.20.4 (Metric Translation), and for gaging and inspection, see ANSI B1.20.5 (Inch) and B1.20.6M (Metric Translation).

Designation of Dryseal Pipe Threads: The standard Dryseal pipe threads are designated by specifying in sequence nominal size, thread series symbol, and class:

Examples: 1/8–27 NPTF-1
　　　　　　1/8–27 PTF-SAE SHORT
　　　　　　3/8–18 NPTF-1 AFTER PLATING

Special Dryseal Threads. — Where design limitations, economy of material, permanent installation, or other limiting conditions prevail, consideration may be given to using a special Dryseal thread series.

Dryseal Special Short Taper Pipe Thread, PTF-SPL SHORT: Threads of this series conform in all respects to PTF-SAE SHORT threads except that the full thread length has been further shortened by eliminating one thread at the small end of internal threads or one thread at the large end of external threads.

Dryseal Special Extra Short Taper Pipe Thread, PTF-SPL EXTRA SHORT: Threads of this series conform in all respects to PTF-SAE SHORT threads except that the full thread length has been further shortened by eliminating two threads at the small end of internal threads or two threads at the large end of external threads.

Limitations of Assembly: Table 9 applies where Dryseal Special Short or Extra Short Taper Pipe Threads are to be assembled as special combinations.

Table 9. Assembly Limitations for Special Combinations of Dryseal Threads

PTF SPL SHORT EXTERNAL PTF SPL EXTRA SHORT EXTERNAL	May Assemble with*	PTF-SAE SHORT INTERNAL NPSF INTERNAL PTF SPL SHORT INTERNAL PTF SPL EXTRA SHORT INTERNAL
PTF SPL SHORT INTERNAL PTF SPL EXTRA SHORT INTERNAL	May Assemble with*	PTF-SAE SHORT EXTERNAL
PTF SPL SHORT EXTERNAL PTF SPL EXTRA SHORT EXTERNAL	May Assemble with‡	NPTF or NPSI INTERNAL
PTF SPL SHORT INTERNAL PTF SPL EXTRA SHORT INTERNAL	May Assemble with‡	NPTF EXTERNAL

* Only when the external thread or the internal thread or both are held closer than the standard tolerance, the external thread toward the minimum and the internal thread toward the maximum pitch diameter to provide a minimum of one turn hand engagement. At extreme tolerance limits the shortened full-thread lengths reduce hand engagement and the threads may not start to assemble.

‡ Only when the internal thread or the external thread or both are held closer than the standard tolerance, the internal thread toward the minimum and the external thread toward the maximum pitch diameter to provide a minimum of two turns for wrench make-up and sealing. At extreme tolerance limits the shortened full-thread lengths reduce wrench make-up and the threads may not seal.

Dryseal Fine Taper Thread Series, F-PTF: The need for finer pitches for nominal pipe sizes has brought into use applications of 27 threads per inch to ¼- and ⅜-inch pipe sizes. There may be other needs that require finer pitches for larger pipe sizes. It is recommended that the existing threads per inch be applied to the next larger pipe size for a fine thread series, thus: ¼-27, ⅜-27, ½-18, ¾-18, 1-14, 1¼-14, 1½-14, and 2-14. This series applies to external and internal threads of full length and is suitable for applications where threads finer than NPTF are required.

Dryseal Special Diameter-Pitch Combination Series, SPL-PTF: Other applications of diameter-pitch combinations have come into use where taper pipe threads are applied to nominal size thin wall tubing. These combinations are: ½-27, ⅝-27, ¾-27, ⅞-27, and 1-27. This series applies to external and internal threads of full length and is applicable to thin wall nominal diameter outside tubing.

Designation of Special Dryseal Pipe Threads: The designations used for these special dryseal pipe threads are as follows:

    ⅛-27 PTF-SPL SHORT
    ⅛-27 PTF-SPL EXTRA SHORT
    ½-27 SPL PTF, OD 0.500

Note that in the last designation the OD of tubing is given.

British Standard Pipe Threads for Pressure-tight Joints. — The threads in BS 21:1973 — "Specification for Pipe Threads where Pressure-tight Joints are Made on the Threads" are based on the Whitworth thread form and are specified as: (1) *Jointing threads.* These relate to pipe threads for joints made pressure-tight by the mating of the threads; they include taper external threads for assembly with either taper or parallel internal threads (parallel external pipe threads are not suitable as jointing threads). (2) *Longscrew threads.* These relate to parallel external pipe threads used for longscrews (connectors) specified in BS 1387 where a pressure-tight joint is achieved by the compression of a soft material onto the surface of the external thread by tightening a back nut against a socket.

British Standard External and Internal Pipe Threads (Pressure-tight Joints) — Metric and Inch Dimensions and Limits of Size* (BS 21:1973)

Nominal Size	No. of Threads per Inch†	Basic Diameters at Gage Plane			Gage Length		Number of Useful Threads on Pipe for Basic Gage Length‡	Tol., + and −, Gage Plane to Face of Int. Taper Thread	Tol., + and −, on Diameter of Parallel Int. Threads
		Major	Pitch	Minor	Basic	Tolerance (+ and −)			
1/16	28	7.723 / 0.304	7.142 / 0.2812	6.561 / 0.2583	(4⅜) / 4.0	(1) / 0.9	(7⅛) / 6.5	(1¼) / 1.1	0.071 / 0.0028
1/8	28	9.728 / 0.383	9.147 / 0.3601	8.566 / 0.3372	(4⅜) / 4.0	(1) / 0.9	(7⅛) / 6.5	(1¼) / 1.1	0.071 / 0.0028
1/4	19	13.157 / 0.518	12.301 / 0.4843	11.445 / 0.4506	(4½) / 6.0	(1) / 1.3	(7¼) / 9.7	(1¼) / 1.7	0.104 / 0.0041
3/8	19	16.662 / 0.656	15.806 / 0.6223	14.950 / 0.5886	(4¾) / 6.4	(1) / 1.3	(7½) / 10.1	(1¼) / 1.7	0.104 / 0.0041
1/2	14	20.955 / 0.825	19.793 / 0.7793	18.631 / 0.7336	(4½) / 8.2	(1) / 1.8	(7¼) / 13.2	(1¼) / 2.3	0.142 / 0.0056
3/4	14	26.441 / 1.041	25.279 / 0.9953	24.117 / 0.9496	(5¼) / 9.5	(1) / 1.8	(8) / 14.5	(1¼) / 2.3	0.142 / 0.0056
1	11	33.249 / 1.309	31.770 / 1.2508	30.291 / 1.1926	(4½) / 10.4	(1) / 2.3	(7¼) / 16.8	(1¼) / 2.9	0.180 / 0.0071
1¼	11	41.910 / 1.650	40.431 / 1.5918	38.952 / 1.5336	(5½) / 12.7	(1) / 2.3	(8¼) / 19.1	(1¼) / 2.9	0.180 / 0.0071
1½	11	47.803 / 1.882	46.324 / 1.8238	44.845 / 1.7656	(5½) / 12.7	(1) / 2.3	(8¼) / 19.1	(1¼) / 2.9	0.180 / 0.0071
2	11	59.614 / 2.347	58.135 / 2.2888	56.656 / 2.2306	(6⅞) / 15.9	(1) / 2.3	(10⅛) / 23.4	(1¼) / 2.9	0.180 / 0.0071
2½	11	75.184 / 2.960	73.705 / 2.9018	72.226 / 2.8436	(7⁹⁄₁₆) / 17.5	(1½) / 3.5	(11⁹⁄₁₆) / 26.7	(1½) / 3.5	0.216 / 0.0085
3	11	87.884 / 3.460	86.405 / 3.4018	84.926 / 3.3436	(8¹⁵⁄₁₆) / 20.6	(1½) / 3.5	(12¹⁵⁄₁₆) / 29.8	(1½) / 3.5	0.216 / 0.0085
4	11	113.030 / 4.450	111.551 / 4.3918	110.072 / 4.3336	(11) / 25.4	(1½) / 3.5	(15½) / 35.8	(1½) / 3.5	0.216 / 0.0085
5	11	138.430 / 5.450	136.951 / 5.3918	135.472 / 5.3336	(12⅜) / 28.6	(1½) / 3.5	(17⅜) / 40.1	(1½) / 3.5	0.216 / 0.0085
6	11	163.830 / 6.450	162.351 / 6.3918	160.872 / 6.3336	(12⅜) / 28.6	(1½) / 3.5	(17⅜) / 40.1	(1½) / 3.5	0.216 / 0.0085

* Each basic metric dimension is given in roman figures (nominal sizes excepted) and each basic inch dimension is shown in italics directly beneath it. Figures in ( ) are numbers of turns of thread with metric linear equivalents given beneath. Taper of taper thread is 1 in 16 on diameter. †In the Standard (BS 21:1973) the thread pitches in millimeters are as follows: 0.907 for 28 threads per inch, 1.337 for 19 threads per inch, 1.814 for 14 threads per inch, and 2.309 for 11 threads per inch. ‡This is the minimum number of useful threads on the pipe for the basic gage length; for the maximum and minimum gage lengths, the minimum numbers of useful threads are, respectively, greater and less by the amount of tolerance in the column to the left. The design of internally threaded parts shall make allowance for receiving pipe ends of up to the minimum number of useful threads corresponding to the maximum gage length; the minimum number of useful *internal* threads shall be no less than 80 per cent of the minimum number of useful external threads for the minimum gage length.

Measuring Screw Threads

Pitch and Lead of Screw Threads. — The *pitch* of a screw thread is the distance from the center of one thread to the center of the next thread. This applies no matter whether the screw has a single, double, triple or quadruple thread. The *lead* of a screw thread is the distance the nut will move forward on the screw if it is turned around one full revolution. In a single-threaded screw, the pitch and lead are equal, because the nut would move forward the distance from one thread to the next, if turned around once. In a double-threaded screw, the nut will move forward two threads, or twice the pitch, so that in this case the lead equals twice the pitch. In a triple-threaded screw, the lead equals three times the pitch, and so on.

The word "pitch" is often, although improperly, used to denote the *number of threads per inch*. Screws are spoken of as having a 12-pitch thread, when twelve threads per inch is what is really meant. The number of threads per inch equals 1 divided by the pitch, or expressed as a formula:

$$\text{Number of threads per inch} = \frac{1}{\text{pitch}}$$

The pitch of a screw equals 1 divided by the number of threads per inch, or:

$$\text{Pitch} = \frac{1}{\text{number of threads per inch}}$$

If the number of threads per inch equals 16, the pitch = $\frac{1}{16}$. If the pitch equals 0.05, the number of threads per inch is $1 \div 0.05 = 20$. If the pitch is $\frac{2}{5}$ inch, the number of threads per inch equals $1 \div \frac{2}{5} = 2\frac{1}{2}$.

Confusion is often caused by the indefinite designation of multiple-thread screws (double, triple, quadruple, etc.). The expression, "four threads per inch, triple," for example, is not to be recommended. It means that the screw is cut with four triple threads or with twelve threads per inch, if the threads are counted by placing a scale alongside the screw. To cut this screw, the lathe would be geared to cut four threads per inch, but they would be cut only to the depth required for twelve threads per inch. The best expression, when a multiple-thread is to be cut, is to say, in this case, "¼ inch lead, $\frac{1}{12}$ inch pitch, triple thread." For single-threaded screws, only the number of threads per inch and the form of the thread are specified. The word "single" is not required.

Measuring Screw Thread Pitch Diameters by Thread Micrometers. — As the pitch or angle diameter of a tap or screw is the most important dimension, it is necessary that the pitch diameter of screw threads be measured, in addition to the outside diameter. One method of measuring in the angle of a thread is by means of a special screw thread micrometer, as shown in the accompanying engraving, Fig. 1. The fixed anvil is W-shaped to engage two thread flanks, and the movable point is cone-shaped so as to enable it to enter the space between two threads, and at the same time be at liberty to revolve. The contact points are on the sides of the thread, as they necessarily must be in order that the

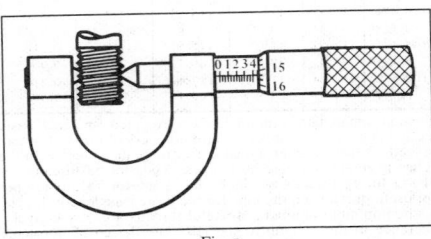

Fig. 1

pitch diameter may be determined. The cone-shaped point of the measuring screw is slightly rounded so that it will not bear in the bottom of the thread. There is also sufficient clearance at the bottom of the V-shaped anvil to prevent it from bearing on the top of the thread. The movable point is adapted to measuring all pitches, but the fixed anvil is limited in its capacity. To cover the whole range of pitches, from the finest to the coarsest, a number of fixed anvils are therefore required.

To find the theoretical pitch diameter, which is measured by the micrometer, subtract twice the addendum of the thread from the standard outside diameter. The addendum of the thread for the American and other standard threads is given in the section on screw thread systems.

Ball-point Micrometers. — If standard plug gages are available, it is not necessary to actually measure the pitch diameter, but merely to compare it with the standard gage. In this case, a ball-point micrometer, as shown in Fig. 2, may be employed. Two types of ball-point micrometers are ordinarily used. One is simply a regular plain micrometer with ball points made to slip over both measuring points. (See *B*, Fig. 2.) This makes a kind of combination plain and ball-point micrometer, the ball points being easily removed. These ball points, however, do not fit solidly on their seats, even if they are split, as shown, and are apt to cause errors in measurements. The best, and, in the long run, the cheapest, method is to use a regular micrometer arranged as shown at *A*. Drill and ream out both the end of the measuring screw or spindle and the anvil, and fit ball points into them as shown. Care should be taken to have the ball point in the spindle run true. The holes in the micrometer spindle and anvil and the shanks on the points are tapered to insure a good fit. The hole *H* in spindle *G* is provided so that the ball point can be easily driven out when a change for a larger or smaller size of ball point is required.

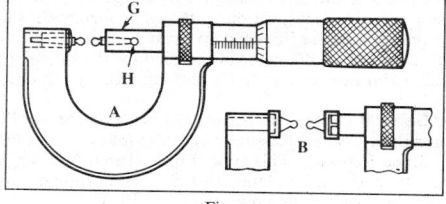

Fig. 2

A ball-point micrometer may be used for comparing the *angle* of a screw thread, with that of a gage. This can be done by using different sizes of ball points, comparing the size first near the root of the thread, then (using a larger ball point) at about the point of the pitch diameter, and finally near the top of the thread (using in the latter case, of course, a much larger ball point). If the gage and thread measurements are the same at each of the three points referred to, this indicates that the thread angle is correct.

Measuring Screw Threads by Three-wire Method. — The *effective* or *pitch diameter* of a screw thread may be measured very accurately by means of some form of micrometer and three wires of equal diameter. This method is extensively used in checking the accuracy of threaded plug gages and other precision screw threads. Two of the wires are placed in contact with the thread on one side and the third wire in a position diametrically opposite as illustrated by the diagram, (see table "Formulas for Checking Pitch Diameters of Screw Threads") and the dimension over the wires is determined by means of a micrometer. An ordinary micrometer is commonly used but some form of "floating micrometer" is preferable, especially for measuring thread gages and other precision work. The floating micrometer is mounted upon a compound slide so that it can move freely in directions parallel or at right angles to the axis of the screw, which is held in a horizontal position between adjustable centers. With this arrangement the microm-

eter is held constantly at right angles to the axis of the screw so that only one wire on each side may be used instead of having two on one side and one on the other, as is necessary when using an ordinary micrometer. The accuracy of the pitch diameter may be determined provided the correct micrometer reading for wires of a given size is known.

Classes of Formulas for Three-wire Measurement. — Various formulas have been established for checking the pitch diameters of screw threads by measurement over wires of known size. These formulas differ in regard to their simplicity or complexity and resulting accuracy. They also differ in that some show what measurement M over the wires should be to obtain a given pitch diameter E, whereas others show the value of the pitch diameter E for a given measurement M.

Formulas for Finding Measurement M: In using a formula for finding the value of measurement M, the required pitch diameter E is inserted in the formula. Then, in cutting or grinding a screw thread, the *actual* measurement M is made to conform to the *calculated* value of M. Formulas for finding measurement M may be modified so that the basic major or outside diameter is inserted in the formula instead of the pitch diameter; however, the pitch diameter type of formula is preferable because this is a more important dimension than the major diameter.

Formulas for Finding Pitch Diameters E: Some formulas are arranged to show the value of the pitch diameter E when measurement M is known. Thus the value of M is first determined by actual measurement and then it is inserted in the formula for finding the corresponding pitch diameter E. This type of formula is useful for determining the pitch diameter of an existing thread gage or other screw thread in connection with inspection work. The formula for finding measurement M is more convenient to use in the shop or tool-room in cutting or grinding new threads, because the pitch diameter is specified on the drawing and the problem is to find the value of measurement M for obtaining this pitch diameter.

General Classes of Screw Thread Profiles. — Thread profiles may be divided into three general classes or types as follows:

Screw Helicoid: This type is represented by a screw thread having a straight-line profile in the axial plane. Such a screw thread may be cut in a lathe by using a straight-sided single-point tool, provided the top surface lies in the axial plane.

Involute Helicoid: This type is represented either by a screw thread or a helical gear tooth having an involute profile in a plane perpendicular to the axis. A rolled screw thread, theoretically at least, is an exact involute helicoid.

Intermediate Profiles: An intermediate profile which lies somewhere between the screw helicoid and the involute helicoid will be formed on a screw thread either by milling or grinding with a straight-sided wheel set in alignment with the thread groove. The resulting form will approach closely the involute helicoid form. In milling or grinding a thread, the included cutter or wheel angle may either equal the standard thread angle (which is always measured in the axial plane) or the cutter or wheel angle may be reduced to approximate, at least, the thread angle in the normal plane. These variations in practice all affect the three-wire measurement.

Accuracy of Formulas for Checking Pitch Diameters by Three-wire Method. — The exact measurement M for a given pitch diameter depends upon the lead angle, the thread angle, and the profile or cross-sectional shape of the thread. As pointed out in the preceding paragraph, the profile depends upon the method of cutting or forming the thread. In the case of a milled or ground thread, the profile is affected not only by the cutter or wheel angle but also by the diameter of the cutter or wheel; hence, because of these variations, an absolutely exact and reasonably simple general formula for measurement M cannot be established; however, if the lead angle

is low, as in the case of a standard single-thread screw, and especially if the thread angle is high like a 60-degree thread, simple formulas which are not arranged to compensate for the lead angle are used ordinarily and meet most practical requirements, particularly in measuring 60-degree threads. If lead angles are large enough to decidedly affect the result, as in the case of most multiple threads (especially Acme or 29-degree worm threads), a formula should be used which compensates for the lead angle sufficiently to obtain the necessary accuracy.

The formulas which follow include (1) a very simple type in which the effect of the lead angle on measurement M is entirely ignored. This simple formula usually is applicable to the measurement of 60-degree single-thread screws, except possibly when gage-making accuracy is required; (2) formulas which do include the effect of the lead angle but, nevertheless, are approximations and not always suitable for the higher lead angles when extreme accuracy is required; (3) formulas for the higher lead angles and the most precise classes of work.

Where approximate formulas are applied consistently in the measurement of both thread plug gages and the threaded "setting plugs" for ring gages, interchangeability might be secured assuming that such approximate formulas were universally employed.

Wire Sizes for Checking Pitch Diameters of Screw Threads. — In checking screw threads by the 3-wire method, the general practice is to use measuring wires of the so-called "best size." The "best size" wire is one which contacts at the

Diameters of Wires for Measuring American Standard and British Standard Whitworth Screw Threads

Threads per Inch	Pitch, Inch	Wire Diameters for American Standard Threads			Wire Diameters for Whitworth Standard Threads		
		Max.	Min.	Pitch-line Contact	Max.	Min.	Pitch-line Contact
4	0.2500	0.2250	0.1400	0.1443	0.1900	0.1350	0.1409
4½	0.2222	0.2000	0.1244	0.1283	0.1689	0.1200	0.1253
5	0.2000	0.1800	0.1120	0.1155	0.1520	0.1080	0.1127
5½	0.1818	0.1636	0.1018	0.1050	0.1382	0.0982	0.1025
6	0.1667	0.1500	0.0933	0.0962	0.1267	0.0900	0.0939
7	0.1428	0.1286	0.0800	0.0825	0.1086	0.0771	0.0805
8	0.1250	0.1125	0.0700	0.0722	0.0950	0.0675	0.0705
9	0.1111	0.1000	0.0622	0.0641	0.0844	0.0600	0.0626
10	0.1000	0.0900	0.0560	0.0577	0.0760	0.0540	0.0564
11	0.0909	0.0818	0.0509	0.0525	0.0691	0.0491	0.0512
12	0.0833	0.0750	0.0467	0.0481	0.0633	0.0450	0.0470
13	0.0769	0.0692	0.0431	0.0444	0.0585	0.0415	0.0434
14	0.0714	0.0643	0.0400	0.0412	0.0543	0.0386	0.0403
16	0.0625	0.0562	0.0350	0.0361	0.0475	0.0337	0.0352
18	0.0555	0.0500	0.0311	0.0321	0.0422	0.0300	0.0313
20	0.0500	0.0450	0.0280	0.0289	0.0380	0.0270	0.0282
22	0.0454	0.0409	0.0254	0.0262	0.0345	0.0245	0.0256
24	0.0417	0.0375	0.0233	0.0240	0.0317	0.0225	0.0235
28	0.0357	0.0321	0.0200	0.0206	0.0271	0.0193	0.0201
32	0.0312	0.0281	0.0175	0.0180	0.0237	0.0169	0.0176
36	0.0278	0.0250	0.0156	0.0160	0.0211	0.0150	0.0156
40	0.0250	0.0225	0.0140	0.0144	0.0190	0.0135	0.0141

pitch line or mid-slope of the thread because then the measurement of the pitch diameter is least affected by an error in the thread angle. In the following formula for determining approximately the "best size" wire or the diameter for pitch-line contact, A = one-half included angle of thread in axial plane.

$$\text{Best size wire} = \frac{0.5 \text{ pitch}}{\cos A} = 0.5 \text{ pitch} \times \sec A$$

For 60-degree threads this formula reduces to

$$\text{Best size wire} = 0.57735 \times \text{pitch}$$

These formulas are based upon a thread groove of zero lead angle because ordinary variations in the lead angle have little effect on the wire diameter and it is desirable to use one wire size for a given pitch regardless of the lead angle. A theoretically correct solution for finding the *exact* size for pitch-line contact, involves the use of cumbersome indeterminate equations with solution by successive trials. The accompanying table gives the wire sizes for both American Standard (formerly U. S. Standard) and the Whitworth Standard Threads. The following formulas for determining wire diameters do not give the extreme theoretical limits but the smallest and largest sizes which are practicable. The diameters in the table are based upon these approximate formulas.

American Standard
$$\begin{cases} \text{Smallest wire diameter} = 0.56 \times \text{pitch} \\ \text{Largest wire diameter} = 0.90 \times \text{pitch} \\ \text{Diameter for pitch-line contact} = 0.57735 \times \text{pitch} \end{cases}$$

Whitworth
$$\begin{cases} \text{Smallest wire diameter} = 0.54 \times \text{pitch} \\ \text{Largest wire diameter} = 0.76 \times \text{pitch} \\ \text{Diameter for pitch-line contact} = 0.56368 \times \text{pitch} \end{cases}$$

Measuring Wire Accuracy. — A set of three measuring wires should have the same diameter within 0.00002 inch. In order to measure the pitch diameter of a screw-thread gage to an accuracy of 0.0001 inch by means of wires, it is necessary to know the wire diameters to 0.00002 inch. If the diameters of the wires are known only to an accuracy of 0.0001 inch, an accuracy better than 0.0003 inch in the measurement of pitch diameter cannot be expected. The wires should be accurately finished hardened steel cylinders of the maximum possible hardness without being brittle. The hardness should not be less than that corresponding to a Knoop indentation number of 630. A wire of this hardness can be cut with a file only with difficulty. The surface should not be rougher than the equivalent of one measuring 3 microinches deviation from a true cylindrical surface.

Measuring or Contact Pressure. — In measuring screw threads or screw thread gages by the 3-wire method, variations in contact pressure will result in different readings. The effect of a variation in contact pressure in measuring threads of fine pitches is indicated by the difference in readings obtained with pressures of 2 pounds and 5 pounds in checking a thread plug gage having 24 threads per inch. The reading over the wires with 5 pounds pressure was 0.00013 inch less than with 2 pounds pressure. For pitches finer than 20 threads per inch a pressure of 16 ounces is recommended by the National Bureau of Standards. For pitches of 20 threads per inch and coarser, a pressure of 2½ pounds is recommended.

In the case of Acme threads, the wire presses against the sides of the thread with a pressure of approximately twice that of the measuring instrument. To limit the tendency of the wires to wedge in between the sides of an Acme thread, it is recommended that pitch diameter measurements on 8 threads per inch and finer be made at 1 pound, and at 2½ pounds for pitches coarser than 8 threads per inch.

Notation Used in Formulas for Checking Pitch Diameters by Three-wire Method

M = dimension over wires

D = basic major or outside diameter

E = pitch diameter (basic, maximum, or minimum) for which M is required, or pitch diameter corresponding to measurement M

T = 0.5 pitch P = width of thread in axial plane at diameter E

T_a = arc thickness on pitch cylinder in plane perpendicular to axis

P = pitch = 1 ÷ number of threads per inch

W = wire or pin diameter

A = one-half included thread angle in the axial plane

A_n = one-half included thread angle in the normal plane or in plane perpendicular to sides of thread = one-half included angle of cutter when thread is milled (tan A_n = tan A × cos B). (Note: Included angle of milling cutter or grinding wheel may equal the nominal included angle of thread, or cutter angle may be reduced to whatever normal angle is required to make the thread angle standard in the axial plane. In either case, A_n = one-half cutter angle.)

S = number of "starts" or threads on a multiple-threaded worm or screw (2 for double thread, 3 for triple thread, etc.)

B = lead angle at pitch diameter = helix angle of thread as measured from a plane perpendicular to the axis. Tan $B = L ÷ 3.1416E$

L = lead of thread = pitch P × number of threads S

H = helix angle at pitch diam. and measured from axis = 90° − B or tan H = cot B

H_b = helix angle at R_b measured from axis

R_b = radius required in Formulas (4e) and (4)

F = angle required in Formulas (4b), (4d) and (4e)

G = angle required in Formula (4)

Approximate Three-wire Formulas which do not Compensate for Lead Angle. — A general formula in which the effect of lead angle is ignored is as follows (see accompanying notation used in formulas):

$$M = E − T \cot A + W (1 + \operatorname{cosec} A) \qquad (1)$$

This formula can be simplified for any given thread angle and pitch. To illustrate, since $T = 0.5\ P$, $M = E − 0.5\ P \cot 30° + W\ (1 + 2)$, for a 60-degree thread, such as the American Standard,

$$M = E − 0.86603\ P + 3\ W$$

The accompanying table contains these simplified formulas for different standard threads. Two formulas are given in each case. The upper one is used when the measurement over wires, M, is known and the corresponding pitch diameter, E, is required; the lower formula gives the measurement M for a specified value of pitch diameter. These formulas are sufficiently accurate for practically all checking of standard 60-degree single-thread screws because of the low lead angles which vary from 1° 11' to 4° 31' in the American Standard Coarse-thread Series.

Bureau of Standards General Formula. — The Formula (2) which follows compensates quite largely for the effect of the lead angle. It is from the National Bureau of Standards Handbook H 28 (1944). The formula, however, as here given has been arranged for finding the value of M (instead of E).

$$M = E − T \cot A + W (1 + \operatorname{cosec} A + 0.5 \tan^2 B \cos A \cot A) \qquad (2)$$

The Bureau of Standards uses Formula (2) in preference to Formula (1) when the value of 0.5 $W \tan^2 B \cos A \cot A$ exceeds 0.00015, as in the case of the larger lead angles. If this test is applied to American Standard 60-degree threads it will show that Formula (1) is generally applicable; but in case of 29-degree Acme or worm threads, Formula (2) (or some other which includes effect of lead angle) should be employed.

Formulas for Checking Pitch Diameters of Screw Threads

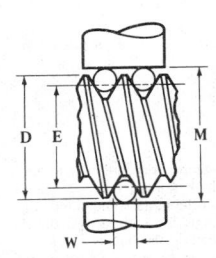

The formulas below do not compensate for the effect of the lead angle upon measurement M, but they are sufficiently accurate for checking standard single-thread screws unless exceptional accuracy is required. See accompanying information on effect of lead angle; also matter relating to measuring wire sizes, accuracy required for such wires, and contact or measuring pressure. The approximate best wire size for pitch-line contact may be obtained by the formula

$$W = 0.5 \times \text{pitch} \times \sec \tfrac{1}{2} \text{ included thread angle}$$

For 60-degree threads, $W = 0.57735 \times \text{pitch}$.

Form of Thread	Formulas for determining measurement M corresponding to correct pitch diameter and the pitch diameter E corresponding to a given measurement over wires.*
American National Standard Unified	When measurement M is known. $$E = M + 0.86603P - 3W$$ When pitch diameter E is used in formula. $$M = E - 0.86603P + 3W$$ The American Standard formerly was known as U.S. Standard.
British Standard Whitworth	When measurement M is known. $$E = M + 0.9605P - 3.1657W$$ When pitch diameter E is used in formula. $$M = E - 0.9605P + 3.1657W$$
British Association Standard	When measurement M is known. $$E = M + 1.1363P - 3.4829W$$ When pitch diameter E is used in formula. $$M = E - 1.1363P + 3.4829W$$
Lowenherz Thread	When measurement M is known. $$E = M + P - 3.2359W$$ When pitch diameter E is used in formula. $$M = E - P + 3.2359W$$
Sharp V-Thread	When measurement M is known. $$E = M + 0.86603P - 3W$$ When pitch diameter E is used in formula. $$M = E - 0.86603P + 3W$$
International Standard	Use the formula given above for the American National Standard Unified Thread.
Pipe Thread	See accompanying paragraph on Measuring Taper Screw Threads by Three-wire Method.
Acme and Worm Threads	See Buckingham Formulas; also Three-wire Method for Checking Thickness of Acme Threads.
Buttress Form of Thread	Different forms of buttress threads are used. See paragraph on wire method applied to buttress threads.

* The wires must be lapped to a uniform diameter and it is very important to insert in the rule or formula the wire diameter as determined by precise means of measurement. Any error will be multiplied. See paragraph on Wire Sizes for Checking Pitch Diameters.

Values of Constants Used in Formulas for Measuring Pitch Diameters of Screws by the Three-wire System

No. of Threads per Inch	American Standard Unified and Sharp V-Thread 0.86603P	Whitworth Thread 0.9605P	No. of Threads per Inch	American Standard Unified and Sharp V-Thread 0.86603P	Whitworth Thread 0.9605P
2¼	0.38490	0.42689	18	0.04811	0.05336
2⅜	0.36464	0.40442	20	0.04330	0.04803
2½	0.34641	0.38420	22	0.03936	0.04366
2⅝	0.32992	0.36590	24	0.03608	0.04002
2¾	0.31492	0.34927	26	0.03331	0.03694
2⅞	0.30123	0.33409	28	0.03093	0.03430
3	0.28868	0.32017	30	0.02887	0.03202
3¼	0.26647	0.29554	32	0.02706	0.03002
3½	0.24744	0.27443	34	0.02547	0.02825
4	0.21651	0.24013	36	0.02406	0.02668
4½	0.19245	0.21344	38	0.02279	0.02528
5	0.17321	0.19210	40	0.02165	0.02401
5½	0.15746	0.17464	42	0.02062	0.02287
6	0.14434	0.16008	44	0.01968	0.02183
7	0.12372	0.13721	46	0.01883	0.02088
8	0.10825	0.12006	48	0.01804	0.02001
9	0.09623	0.10672	50	0.01732	0.01921
10	0.08660	0.09605	52	0.01665	0.01847
11	0.07873	0.08732	56	0.01546	0.01715
12	0.07217	0.08004	60	0.01443	0.01601
13	0.06662	0.07388	64	0.01353	0.01501
14	0.06186	0.06861	68	0.01274	0.01412
15	0.05774	0.06403	72	0.01203	0.01334
16	0.05413	0.06003	80	0.01083	0.01201

Constants Used for Measuring Pitch Diameters of Metric Screws by the Three-wire System

Pitch in mm	0.86603P in Inches	W in Inches	Pitch in mm	0.86603P in Inches	W in Inches	Pitch in mm	0.86603P in Inches	W in Inches
0.2	0.00682	0.00455	0.75	0.02557	0.01705	3.5	0.11933	0.07956
0.25	0.00852	0.00568	0.8	0.02728	0.01818	4	0.13638	0.09092
0.3	0.01023	0.00682	1	0.03410	0.02273	4.5	0.15343	0.10229
0.35	0.01193	0.00796	1.25	0.04262	0.02841	5	0.17048	0.11365
0.4	0.01364	0.00909	1.5	0.05114	0.03410	5.5	0.18753	0.12502
0.45	0.01534	0.01023	1.75	0.05967	0.03978	6	0.20457	0.13638
0.5	0.01705	0.01137	2	0.06819	0.04546	8	0.30686	0.18184
0.6	0.02046	0.01364	2.5	0.08524	0.05683	...	...	...
0.7	0.02387	0.01591	3	0.10229	0.06819	...	...	...

This table may be used for American National Standard Metric Threads. The formulas for American Standard Unified Threads on page 1628 are used. In the table above, the values of 0.86603P and W are in inches so that the values for E and M calculated from the formulas on page 1628 are also in inches.

Why Small Thread Angle Affects Accuracy of Three-wire Measurement. — In measuring or checking Acme threads, or any others having a comparatively small thread angle A, it is particularly important to use a formula which compensates largely, if not entirely, for the effect of the lead angle, especially in all gage and precision work. The effect of the lead angle on the position of the wires and upon the resulting measurement M is much greater in the case of a 29-degree thread than in the case of a higher thread angle such, for example, as a 60-degree thread. This is because the cotangent of the thread angle increases as this angle becomes smaller; consequently, the reduction in the width of the thread groove in the normal plane due to the lead angle causes a wire of given size to rest higher in the groove of a thread having a small thread angle A (like a 29-degree thread) than in the groove with a larger angle (like a 60-degree American Standard).

Acme Threads: Three-wire measurements of high accuracy require the use of Formula 4. For most cases, however, Formula 2 or 3 gives satisfactory results. The table on page 1635 lists suitable wire sizes for use in Formulas 2 and 4.

Dimensions Over Wires of Given Diameter for Checking Screw Threads of American National form (U. S. Standard) and the V-form

Diam. of Screw	No. of Threads per Inch	Diam. of Wire used	Dimension over Wires, V-Thread	Dimension over Wires, U. S. Thread	Diam. of Screw	No. of Threads per Inch	Diam. of Wire used	Dimension over Wires, V-Thread	Dimension over Wires, U. S. Thread
1/4	18	0.035	0.2588	0.2708	7/8	8	0.090	0.9285	0.9556
1/4	20	0.035	0.2684	0.2792	7/8	9	0.090	0.9525	0.9766
1/4	22	0.035	0.2763	0.2861	7/8	10	0.090	0.9718	0.9935
1/4	24	0.035	0.2828	0.2919	15/16	8	0.090	0.9910	1.0181
5/16	18	0.035	0.3213	0.3333	15/16	9	0.090	1.0150	1.0391
5/16	20	0.035	0.3309	0.3417	1	8	0.090	1.0535	1.0806
5/16	22	0.035	0.3388	0.3486	1	9	0.090	1.0775	1.1016
5/16	24	0.035	0.3453	0.3544	1 1/8	7	0.090	1.1476	1.1785
3/8	16	0.040	0.3867	0.4003	1 1/4	7	0.090	1.2726	1.3035
3/8	18	0.040	0.3988	0.4108	1 3/8	6	0.150	1.5363	1.5724
3/8	20	0.040	0.4084	0.4192	1 1/2	6	0.150	1.6613	1.6974
7/16	14	0.050	0.4638	0.4793	1 5/8	5 1/2	0.150	1.7601	1.7995
7/16	16	0.050	0.4792	0.4928	1 3/4	5	0.150	1.8536	1.8969
1/2	12	0.050	0.5057	0.5237	1 7/8	5	0.150	1.9786	2.0219
1/2	13	0.050	0.5168	0.5334	2	4 1/2	0.150	2.0651	2.1132
1/2	14	0.050	0.5263	0.5418	2 1/4	4 1/2	0.150	2.3151	2.3632
9/16	12	0.050	0.5682	0.5862	2 1/2	4	0.150	2.5170	2.5711
9/16	14	0.050	0.5888	0.6043	2 3/4	4	0.150	2.7670	2.8211
5/8	10	0.070	0.6618	0.6835	3	3 1/2	0.200	3.1051	3.1670
5/8	11	0.070	0.6775	0.6972	3 1/4	3 1/2	0.200	3.3551	3.4170
5/8	12	0.070	0.6907	0.7087	3 1/2	3 1/4	0.250	3.7171	3.7837
1 1/16	10	0.070	0.7243	0.7460	3 3/4	3	0.250	3.9226	3.9948
1 1/16	11	0.070	0.7400	0.7597	4	3	0.250	4.1726	4.2448
3/4	10	0.070	0.7868	0.8085	4 1/4	2 7/8	0.250	4.3975	4.4729
3/4	11	0.070	0.8025	0.8222	4 1/2	2 3/4	0.250	4.6202	4.6989
3/4	12	0.070	0.8157	0.8337	4 3/4	2 5/8	0.250	4.8402	4.9227
13/16	9	0.070	0.8300	0.8541	5	2 1/2	0.250	5.0572	5.1438
13/16	10	0.070	0.8493	0.8710	....	....	.....	.....	.....

Table for Measuring Whitworth Standard Threads by the Three-wire Method

Diam. of Thread	No. of Threads per Inch	Diam. of Wire Used	Diam. Measured over Wires	Diam. of Thread	No. of Threads per Inch	Diam. of Wire Used	Diam. Measured over Wires
1/8	40	0.018	0.1420	2 1/4	4	0.150	2.3247
3/16	24	0.030	0.2158	2 3/8	4	0.150	2.4497
1/4	20	0.035	0.2808	2 1/2	4	0.150	2.5747
5/16	18	0.040	0.3502	2 5/8	4	0.150	2.6997
3/8	16	0.040	0.4015	2 3/4	3 1/2	0.200	2.9257
7/16	14	0.050	0.4815	2 7/8	3 1/2	0.200	3.0507
1/2	12	0.050	0.5249	3	3 1/2	0.200	3.1757
9/16	12	0.050	0.5874	3 1/8	3 1/2	0.200	3.3007
5/8	11	0.070	0.7011	3 1/4	3 1/4	0.200	3.3905
11/16	11	0.070	0.7636	3 3/8	3 1/4	0.200	3.5155
3/4	10	0.070	0.8115	3 1/2	3 1/4	0.200	3.6405
13/16	10	0.070	0.8740	3 5/8	3 1/4	0.200	3.7655
7/8	9	0.070	0.9187	3 3/4	3	0.200	3.8495
15/16	9	0.070	0.9812	3 7/8	3	0.200	3.9745
1	8	0.090	1.0848	4	3	0.200	4.0995
1 1/16	8	0.090	1.1473	4 1/8	3	0.200	4.2245
1 1/8	7	0.090	1.1812	4 1/4	2 7/8	0.250	4.4846
1 3/16	7	0.090	1.2437	4 3/8	2 7/8	0.250	4.6096
1 1/4	7	0.090	1.3062	4 1/2	2 7/8	0.250	4.7346
1 5/16	7	0.090	1.3687	4 5/8	2 7/8	0.250	4.8596
1 3/8	6	0.120	1.4881	4 3/4	2 3/4	0.250	4.9593
1 7/16	6	0.120	1.5506	4 7/8	2 3/4	0.250	5.0843
1 1/2	6	0.120	1.6131	5	2 3/4	0.250	5.2093
1 9/16	6	0.120	1.6756	5 1/8	2 3/4	0.250	5.3343
1 5/8	5	0.120	1.6847	5 1/4	2 5/8	0.250	5.4316
1 11/16	5	0.120	1.7472	5 3/8	2 5/8	0.250	5.5566
1 3/4	5	0.120	1.8097	5 1/2	2 5/8	0.250	5.6816
1 13/16	5	0.120	1.8722	5 5/8	2 5/8	0.250	5.8066
1 7/8	4 1/2	0.150	1.9942	5 3/4	2 1/2	0.250	5.9011
1 15/16	4 1/2	0.150	2.0567	5 7/8	2 1/2	0.250	6.0261
2	4 1/2	0.150	2.1192	6	2 1/2	0.250	6.1511
2 1/8	4 1/2	0.150	2.2442	...	...	...	...

All dimensions are given in inches.

Buckingham Simplified Formula which Includes Effect of Lead Angle. — The Formula 3 which follows gives very accurate results for the lower lead angles in determining measurement M. However, if extreme accuracy is essential, it may be advisable to use the involute helicoid formulas as explained later.

$$M = E + W(1 + \sin A_n) \text{ where } W = \frac{T \times \cos B}{\cos A_n} \qquad (3 \text{ and } 3a)$$

Theoretically correct equations for determining measurement M are complex and cumbersome to apply. Formula 3 combines simplicity with a degree of accuracy which meets all but the most exacting requirements, particularly for lead angles below 8 or 10 degrees and the higher thread angles. However, the wire diameter used in Formula (3) must conform to that obtained by Formula (3a) to permit a direct solution or one not involving indeterminate equations and successive trials.

Application of Buckingham Formula: In the application of Formula (3) to screw threads or to worms, there are two general cases to be considered.

Case 1: The screw thread or worm is to be milled with a cutter having an included angle equal to the nominal or standard thread angle which is assumed to be the angle in the axial plane. For example, a 60-degree cutter is to be used for milling a thread. In this case, the thread angle in the plane of the axis will exceed 60 degrees by an amount increasing with the lead angle. This variation from the standard angle may be of little or no practical importance if the lead angle is small or if the mating nut (or teeth in the case of worm gearing) is formed to suit the thread as milled.

Case 2: The screw thread or worm is to be milled with a cutter reduced to whatever normal angle is equivalent to the standard thread angle in the axial plane. For example, a 29-degree Acme thread is to be milled with a cutter having some angle smaller than 29 degrees (the reduction increasing with the lead angle) to make the thread angle standard in the plane of the axis. Theoretically, the milling cutter angle should always be corrected to suit the normal angle; but if the lead angle is small, such correction may be unnecessary.

If the thread is cut in a lathe to the standard angle as measured in the axial plane, Case 2 applies in determining the pin size W and the over-all measurement M.

In solving all problems under Case 1, the angle A_n used in Formulas (3) and (3a) equals one-half the included angle of the milling cutter.

When Case 2 applies, the angle A_n for milled threads also equals one-half the included angle of the cutter, but the cutter angle is reduced and is determined as follows:

$$\tan A_n = \tan A \times \cos B$$

The included angle of the cutter or the normal included angle of the thread groove $= 2 A_n$. Examples 1 and 2 which follow illustrate Cases 1 and 2.

Example 1 (Case 1): The example which follows illustrates the case of an Acme screw thread which is milled with a cutter having an included angle of 29 degrees; consequently, the angle of the thread is over 29 degrees in the axial section.

The outside or major diameter is 3 inches; the pitch, ½ inch; the lead, 1 inch; the number of threads or "starts," 2. Find pin size W and measurement M.

Pitch diameter $E = 2.75$; $T = 0.25$; $L = 1.0$; $A_n = 14.50°$; $\tan A_n = 0.258618$; $\sin A_n = 0.25038$; $\cos A_n = 0.968148$.

$$\tan B = \frac{1.0}{3.1416 \times 2.75} = 0.115749; \quad B = 6.6025°$$

$$W = \frac{0.25 \times 0.993368}{0.968148} = 0.25651 \text{ inch}$$

$$M = 2.75 + 0.25651 \times (1 + 0.25038) = 3.0707 \text{ inches.}$$

Note: This value of M is only 0.0001 inch larger than that obtained by using the very accurate involute helicoid formula (4) referred to on the following page.

Example 2 (Case 2): A triple-threaded worm has a pitch diameter of 2.481 inches; pitch of 1.5 inches; lead of 4.5 inches; lead angle of 30 degrees and nominal thread angle of 60 degrees in axial plane. Milling cutter angle is to be reduced. $T = 0.75$ inch; $\cos B = 0.866025$; $\tan A = 0.57735$. Again use Formula (3) to see if it is applicable in this case.

$\tan A_n = \tan A \times \cos B = 0.57735 \times 0.866025 = 0.5000$; hence $A_n = 26.565°$ making the included cutter angle 53.13°. $\cos A_n = 0.89443$; $\sin A_n = 0.44721$.

$$W = \frac{0.75 \times 0.866025}{0.89443} = 0.72618 \text{ inch.}$$

$$M = 2.481 + 0.72618 \times (1 + 0.44721) = 3.532 \text{ inches.}$$

Note: If the value of measurement M is determined by using the following

Formula (4) it will be found that $M = 3.515 +$ inches; hence the error equals $3.532 - 3.515 = 0.017$ inch approximately, which indicates that Formula (3) is not accurate enough in this case. The application of this simpler Formula (3) will depend upon the lead angle and thread angle (as previously explained) and also upon the class of work.

Buckingham Exact Involute Helicoid Formula Applied to Screw Threads. — When extreme accuracy is required in finding measurement M for obtaining a given pitch diameter, the equations which follow, although somewhat cumbersome to apply, have the merit of providing a direct and very accurate solution; consequently, they are preferable to the indeterminate equations and successive trial solutions heretofore employed when extreme precision is required. These equations are exact for involute helical gears and, consequently, give theoretically correct results when applied to a screw thread of the involute helicoidal form; they also give very close approximations for threads having intermediate profiles.

Helical Gear Equation Applied to Screw Thread Measurement: In applying the helical gear equations to a screw thread, use either the axial or normal thread angle and the lead angle of the helix. In order to keep the solution on a practical basis, either thread angle A or A_n, as the case may be, is assumed to equal the cutter angle of a milled thread. Actually, the profile of a milled thread will have some curvature in both axial and normal sections; hence angles A and A_n represent the angular approximations of these slightly curved profiles. The equations which follow give the values needed to solve the screw thread problem as a helical gear problem.

$$M = \frac{2 R_b}{\cos G} + W \qquad (4)$$

$$\tan F = \frac{\tan A}{\tan B} = \frac{\tan A_n}{\sin B}; \qquad R_b = \frac{E}{2}\cos F \qquad \text{(4a and 4b)}$$

$$T_a = \frac{T}{\tan B}; \qquad \tan H_b = \cos F \tan H \qquad \text{(4c and 4d)}$$

$$\operatorname{inv} G = \frac{T_a}{E} + \operatorname{inv} F + \frac{W}{2 R_b \cos H_b} - \frac{\pi}{S} \qquad (4e)$$

A table of involute functions is required (see Handbook pages 82 to 126).

Example 3: To illustrate the application of Formula (4) and the supplementary formulas, assume that the number of starts $S = 6$; pitch diameter $E = 0.6250$; normal thread angle $A_n = 20°$; lead of thread $L = 0.864$ inch; $T = 0.072$; $W = 0.07013$ inch.

$$\tan B = \frac{L}{\pi E} = \frac{0.864}{1.9635} = 0.44003; \quad B = 23.751°$$

Helix angle $H = 90° - 23.751° = 66.249°$

$$\tan F = \frac{\tan A_n}{\sin B} = \frac{0.36397}{0.40276} = 0.90369; \quad F = 42.104°$$

$$R_b = \frac{E}{2}\cos F = \frac{0.6250}{2} \times 0.74193 = 0.23185$$

$$T_a = \frac{T}{\tan B} = \frac{0.072}{0.44003} = 0.16362$$

$$\tan H_b = \cos F \tan H = 0.74193 \times 2.27257 = 1.68609; \quad H_b = 59.328°$$

The involute function of G is found next by Formula (4e).

$$\text{inv } G = \frac{0.16362}{0.625} + 0.16884 + \frac{0.07013}{2 \times 0.23185 \times 0.51012} - \frac{3.1416}{6} = 0.20351$$

A table of involute functions shows that $44° \ 21'$ or $44.350°$ is the angular equivalent of 0.20351; hence $G = 44.350°$.

$$M = \frac{2 \ R_b}{\cos G} + W = \frac{2 \times 0.23185}{0.71508} + 0.07013 = 0.71859 \text{ inch}$$

Accuracy of Formulas (3) and (4) Compared. — With the involute helicoid Formula (4) any wire size which makes contact with the flanks of the thread may be used; however, in the preceding example, the wire diameter W was obtained by Formula (3a) in order to compare Formula (4) with (3). If Example (3) is solved by Formula (3), $M = 0.71912$; hence the difference between the values of M obtained with Formulas (3) and (4) equals $0.71912 - 0.71859 = 0.00053$ inch. The included thread angle in this case is 40 degrees. If Formulas (3) and (4) are applied to a 29-degree thread, the difference in measurements M or the error resulting from the use of Formula (3) will be larger. For example, in case of an Acme thread having a lead angle of about 34 degrees, the difference in values of M obtained by the two formulas equals 0.0008 inch.

Three-wire Measurement of Acme and Stub Acme Thread Pitch Diameter. — For single- and multiple-start Acme and Stub Acme threads having lead angles of less than 5 degrees, the approximate three-wire formula given on page 1627 and the best wire size taken from the table on page 1635 may be used.

Multiple-start Acme and Stub Acme threads commonly have a lead angle of greater than 5 degrees. For these, a direct determination of the actual pitch diameter is obtained by using the formula: $E = M - (C + c)$ in conjunction with the table on pages 1636 and 1637. To enter the table, the lead angle B of the thread to be measured must be known. It is found by the formula: $\tan B = L \div 3.1416 E_1$ where L is the lead of the thread and E_1 is the nominal pitch diameter. The best wire size is now found by taking the value of w_1 as given in the table for lead angle B, with interpolation, and dividing it by the number of threads per inch. The value of $(C + c)_1$ given in the table for lead angle B is also divided by the number of threads per inch to get $(C + c)$. Using the best size wires, the actual measurement over wires M is made and the actual pitch diameter E found by using the formula: $E = M - (C + c)$.

Example: For a 5 tpi, 4-start Acme thread with a $13.952°$ lead angle, using three 0.10024-inch wires, $M = 1.1498$ inches, hence $E = 1.1498 - 0.1248 = 1.0250$ inches.

Under certain conditions, a wire may contact one thread flank at two points, in which case it is advisable to substitute balls of the same diameter as the wires.

Checking Thickness of Acme Screw Threads. — In some instances it may be preferable to check the thread thickness instead of the pitch diameter, especially if there is a thread thickness tolerance.

A direct method, applicable to the larger pitches, is to use a vernier gear-tooth caliper for measuring the thickness in the *normal* plane of the thread. This measurement, for an American Standard General Purpose Acme thread, should be made at a distance below the *basic* outside diameter equal to $p/4$. The thickness at this basic pitch-line depth in the axial plane should be $p/2 - 0.259 \times$ the pitch diameter allowance from Table 4 on page 1564 with a tolerance of *minus* $0.259 \times$ the pitch diameter tolerance from Table 5 on page 1565. The thickness in the normal plane or plane of measurement is equal to the thickness in the axial plane multiplied by the cosine of the helix angle. The helix angle may be determined from the formula:

tangent of helix angle = lead of thread $\div$ ($3.1416 \times$ pitch diameter)

Three-Wire Method for Checking Thickness of Acme Threads. — The application of the 3-wire method of checking the thickness of an Acme screw thread

Wire Sizes for Three-Wire Measurement of Acme Threads with Lead Angles of
Less than 5 Degrees

Threads per Inch	Best Size	Max.	Min.	Threads per Inch	Best Size	Max.	Min.
1	0.51645	0.65001	0.48726	5	0.10329	0.13000	0.09745
1⅛	0.38734	0.48751	0.36545	6	0.08608	0.10834	0.08121
1½	0.34430	0.43334	0.32484	8	0.06456	0.08125	0.06091
2	0.25822	0.32501	0.24363	10	0.05164	0.06500	0.04873
2½	0.20658	0.26001	0.19491	12	0.04304	0.05417	0.04061
3	0.17215	0.21667	0.16242	14	0.03689	0.04643	0.03480
4	0.12911	0.16250	0.12182	16	0.03228	0.04063	0.03045

Wire sizes are based upon zero helix angle. Best size = 0.51645 × pitch; maximum size = 0.650013 × pitch; minimum size = 0.487263 × pitch.

is included in Report of the National Screw Thread Commission. In applying the 3-wire method for checking thread thickness, the procedure is the same as in checking pitch diameter although a different formula is required. Assume that D = basic major diameter of screw; M = measurement over wires; W = diameter of wires; S = tangent of helix angle at pitch line; P = pitch; T = thread thickness at depth equal to $0.25P$.

$$T = 1.12931 \times P + 0.25862 \times (M - D) - W \times (1.29152 + 0.48407 S^2)$$

This formula transposed to show the correct measurement M equivalent to a given required thread thickness is as follows:

$$M = D + \frac{W \times (1.29152 + 0.48407 \times S^2) + T - 1.12931 \times P}{0.25862}$$

Example: An Acme General Purpose thread, Class 2G, has a 5-inch basic major diameter, 0.5-inch pitch, and 1-inch lead (double thread). Assume the wire size is 0.258 inch. Determine measurement M for a thread thickness T at the basic pitch line of 0.2454 inch. (This is the maximum thickness at the basic pitch line and equals the basic thickness, 0.5P, − 0.259 × allowance from Table 4, page 1306.)

$$M = 5 + \frac{0.258 \times (1.29152 + 0.48407 \times 0.06701^2) + 0.2454 - 1.12931 \times 0.5}{0.25862}$$

$$= 5.056 \text{ inches}$$

Testing Angle of Thread by Three-wire Method. — The error in the angle of a thread may be determined by using sets of wires of two diameters, the measurement over the two sets of wires being followed by calculations to determine the amount of error, assuming that the angle cannot be tested by comparison with a standard plug gage which is known to be correct. The diameter of the small wires for the American standard thread is usually about 0.6 times the pitch and the diameter of the large wires, about 0.9 times the pitch. The total difference between the measurements over the large and small sets of wires is first determined. If the thread is an American standard or any other form having an included angle of 60 degrees, the difference between the two measurements should equal three times the difference between the diameters of the wires used. Thus, if the wires are 0.116 and 0.076 inch in diameter, respectively, the difference equals 0.116 − 0.076 = 0.040 inch. Therefore the difference between the micrometer readings for a standard angle of 60 degrees equals 3 × 0.040 = 0.120 inch in this case. If the

Best Wire Diameters and Constants for Three-wire Measurement of Acme and Stub Acme Threads with Large Lead Angles — 1-inch Axial Pitch*

Lead angle, B, deg.	1-start threads		2-start threads	
	w_1	$(C+c)_1$	w_1	$(C+c)_1$
5.0	0.51450	0.64311	0.51443	0.64290
5.1	0.51442	0.64301	0.51435	0.64279
5.2	0.51435	0.64291	0.51427	0.64268
5.3	0.51427	0.64282	0.51418	0.64256
5.4	0.51419	0.64272	0.51410	0.64245
5.5	0.51411	0.64261	0.51401	0.64233
5.6	0.51403	0.64251	0.51393	0.64221
5.7	0.51395	0.64240	0.51384	0.64209
5.8	0.51386	0.64229	0.51375	0.64196
5.9	0.51377	0.64218	0.51366	0.64184
6.0	0.51368	0.64207	0.51356	0.64171
6.1	0.51359	0.64195	0.51346	0.64157
6.2	0.51350	0.64184	0.51336	0.64144
6.3	0.51340	0.64172	0.41327	0.64131
6.4	0.51330	0.64160	0.51317	0.64117
6.5	0.51320	0.64147	0.51306	0.64103
6.6	0.51310	0.64134	0.51296	0.64089
6.7	0.51300	0.64122	0.51285	0.64075
6.8	0.51290	0.64110	0.51275	0.64061
6.9	0.51280	0.64097	0.51264	0.64046
7.0	0.51270	0.64085	0.51254	0.64032
7.1	0.51259	0.64072	0.51243	0.64017
7.2	0.51249	0.64060	0.51232	0.64002
7.3	0.51238	0.64047	0.51221	0.63987
7.4	0.51227	0.64034	0.51209	0.63972
7.5	0.51217	0.64021	0.51198	0.63957
7.6	0.51206	0.64008	0.51186	0.63941
7.7	0.51196	0.63996	0.51174	0.63925
7.8	0.51186	0.63983	0.51162	0.63909
7.9	0.51175	0.63970	0.51150	0.63892
8.0	0.51164	0.63957	0.51138	0.63876
8.1	0.51153	0.63944	0.51125	0.63859
8.2	0.51142	0.63930	0.51113	0.63843
8.3	0.51130	0.63916	0.51101	0.63827
8.4	0.51118	0.63902	0.51088	0.63810
8.5	0.51105	0.63887	0.51075	0.63793
8.6	0.51093	0.63873	0.51062	0.63775
8.7	0.51081	0.63859	0.51049	0.63758
8.8	0.51069	0.63845	0.51035	0.63740
8.9	0.51057	0.63831	0.51022	0.63722
9.0	0.51044	0.63817	0.51008	0.63704
9.1	0.51032	0.63802	0.50993	0.63685
9.2	0.51019	0.63788	0.50979	0.63667
9.3	0.51006	0.63774	0.50965	0.63649
9.4	0.50993	0.63759	0.50951	0.63630
9.5	0.50981	0.63744	0.50937	0.63612
9.6	0.50968	0.63730	0.50922	0.63593
9.7	0.50955	0.63715	0.50908	0.63574
9.8	0.50941	0.63700	0.50893	0.63555
9.9	0.50927	0.63685	0.50879	0.63537
10.0	0.50913	0.63670	0.50864	0.63518

Lead angle, B, deg.	2-start threads		3-start threads	
	w_1	$(C+c)_1$	w_1	$(C+c)_1$
10.0	0.50864	0.63518	0.50847	0.63463
10.1	0.50849	0.63498	0.50831	0.63442
10.2	0.50834	0.63478	0.50815	0.63420
10.3	0.50818	0.63457	0.50800	0.63399
10.4	0.50802	0.63436	0.50784	0.63378
10.5	0.40786	0.63416	0.50768	0.63356
10.6	0.50771	0.63395	0.50751	0.63333
10.7	0.50755	0.63375	0.50735	0.63311
10.8	0.50739	0.53354	0.50718	0.63288
10.9	0.50723	0.63333	0.50701	0.63265
11.0	0.50707	0.63313	0.50684	0.63242
11.1	0.50691	0.63292	0.50667	0.63219
11.2	0.50674	0.63271	0.50649	0.63195
11.3	0.50658	0.63250	0.50632	0.63172
11.4	0.50641	0.63228	0.50615	0.63149
11.5	0.50623	0.63206	0.50597	0.63126
11.6	0.50606	0.63184	0.50579	0.63102
11.7	0.50589	0.63162	0.50561	0.63078
11.8	0.50571	0.63140	0.50544	0.63055
11.9	0.50553	0.63117	0.50526	0.63031
12.0	0.50535	0.63095	0.50507	0.63006
12.1	0.50517	0.63072	0.50488	0.62981
12.2	0.50500	0.63050	0.50470	0.62956
12.3	0.50482	0.63027	0.50451	0.62931
12.4	0.50464	0.63004	0.50432	0.62906
12.5	0.50445	0.62981	0.50413	0.62881
12.6	0.50427	0.62958	0.50394	0.62856
12.7	0.50408	0.62934	0.50375	0.62830
12.8	0.50389	0.62911	0.50356	0.62805
12.9	0.50371	0.62888	0.50336	0.62779
13.0	0.50352	0.62865		
13.1	0.50333	0.62841		
13.2	0.50313	0.62817		
13.3	0.50293	0.62792		
13.4	0.50274	0.62778		
13.5	0.50254	0.62743		
13.6	0.50234	0.62718	For these	
13.7	0.50215	0.62694	3-start	
13.8	0.50195	0.62670	thread	
13.9	0.50175	0.62645	values	
14.0	0.50155	0.62621	see	
14.1	0.50135	0.62596	table	
14.2	0.50115	0.62571	on	
14.3	0.50094	0.62546	following	
14.4	0.50073	0.62520	page.	
14.5	0.50051	0.62494		
14.6	0.50030	0.62468		
14.7	0.50009	0.62442		
14.8	0.49988	0.62417		
14.9	0.49966	0.62391		
15.0	0.49945	0.62365		

All dimensions are in inches.

Courtesy of Van Keuren Co.

* Values given for w_1 and $(C+c)_1$ in table are for 1-inch pitch axial threads. For other pitches, divide table values by number of threads per inch.

Best Wire Diameters and Constants for Three-wire Measurement of Acme and Stub Acme Threads with Large Lead Angles — 1-inch Axial Pitch*

Lead angle, B, deg.	3-start threads		4-start threads		Lead angle, B, deg.	3-start threads		4-start threads	
	w_1	$(C+c)_1$	w_1	$(C+c)_1$		w_1	$(C+c)_1$	w_1	$(C+c)_1$
13.0	0.50316	0.62752	0.50297	0.62694	18.0	0.49154	0.61250	0.49109	0.61109
13.1	0.50295	0.62725	0.50277	0.62667	18.1	0.49127	0.61216	0.49082	0.61073
13.2	0.50275	0.62699	0.50256	0.62639	18.2	0.49101	0.61182	0.49054	0.61037
13.3	0.50255	0.62672	0.50235	0.62611	18.3	0.49074	0.61148	0.49027	0.61001
13.4	0.50235	0.62646	0.50215	0.62583	18.4	0.49047	0.61114	0.48999	0.60964
13.5	0.50214	0.62619	0.50194	0.62555	18.5	0.49020	0.61080	0.48971	0.69928
13.6	0.50194	0.62592	0.50173	0.62526	18.6	0.48992	0.61045	0.48943	0.60981
13.7	0.50173	0.62564	0.50152	0.62498	18.7	0.48965	0.61011	0.48915	0.60854
13.8	0.50152	0.62537	0.50131	0.62469	18.8	0.48938	0.60976	0.48887	0.60817
13.9	0.50131	0.62509	0.50109	0.62440	18.9	0.48910	0.60941	0.48859	0.60780
14.0	0.50110	0.62481	0.50087	0.62411	19.0	0.48882	0.60906	0.48830	0.60742
14.1	0.50089	0.62453	0.50065	0.62381	19.1	0.48854	0.60871	0.48800	0.60704
14.2	0.50068	0.62425	0.50043	0.62351	19.2	0.48825	0.60835	0.48771	0.60666
14.3	0.50046	0.62397	0.50021	0.62321	19.3	0.48797	0.60799	0.48742	0.60628
14.4	0.50024	0.62368	0.49999	0.62291	19.4	0.48769	0.60764	0.48713	0.60590
14.5	0.50003	0.62340	0.49977	0.62262	19.5	0.48741	0.60729	0.48684	0.60552
14.6	0.49981	0.62312	0.49955	0.62232	19.6	0.48712	0.60693	0.48655	0.60514
14.7	0.49959	0.62883	0.49932	0.62202	19.7	0.48638	0.60657	0.48625	0.60475
14.8	0.49936	0.62253	0.49910	0.62172	19.8	0.48655	0.60621	0.48596	0.60437
14.9	0.49914	0.62224	0.49887	0.62141	19.9	0.48626	0.60585	0.48566	0.60398
15.0	0.49891	0.62195	0.49864	0.62110	20.0	0.48597	0.60549	0.48536	0.60359
15.1	0.49869	0.62166	0.49842	0.62080	20.1	...	...	0.48506	0.60320
15.2	0.49846	0.62137	0.49819	0.62049	20.2	...	...	0.48476	0.60281
15.3	0.49824	0.62108	0.49795	0.62017	20.3	...	...	0.48445	0.60241
15.4	0.42801	0.62078	0.49771	0.61985	20.4	...	...	0.48415	0.60202
15.5	0.49778	0.62048	0.49747	0.61953	20.5	...	...	0.48384	0.60162
15.6	0.49754	0.62017	0.49723	0.61921	20.6	...	...	0.48354	0.60123
15.7	0.49731	0.61987	0.49699	0.61889	20.7	...	...	0.48323	0.60083
15.8	0.49707	0.61956	0.49675	0.61857	20.8	...	...	0.48292	0.60042
15.9	0.49683	0.61926	0.49651	0.61825	20.9	...	...	0.48261	0.60002
16.0	0.49659	0.61895	0.49627	0.61793	21.0	...	...	0.48230	0.59961
16.1	0.49635	0.61864	0.49602	0.61760	21.1	...	...	0.48198	0.49920
16.2	0.49611	0.61833	0.49577	0.61727	21.2	...	...	0.48166	0.59879
16.3	0.49586	0.61801	0.49552	0.61694	21.3	...	...	0.48134	0.59838
16.4	0.49562	0.61770	0.49527	0.61661	21.4	...	...	0.48103	0.59797
16.5	0.49537	0.61738	0.49502	0.61628	21.5	...	...	0.48071	0.59756
16.6	0.49512	0.61706	0.49476	0.61594	21.6	...	...	0.48040	0.59715
16.7	0.49488	0.61675	0.49451	0.61560	21.7	...	...	0.48008	0.59674
16.8	0.40463	0.61643	0.49425	0.61526	21.8	...	...	0.47975	0.59632
16.9	0.49438	0.61611	0.49400	0.61492	21.9	...	...	0.47943	0.59590
17.0	0.49414	0.61580	0.49375	0.61458	22.0	...	...	0.47910	0.59548
17.1	0.49389	0.61548	0.49349	0.61424	22.1	...	...	0.47878	0.59507
17.2	0.49363	0.61515	0.49322	0.61389	22.2	...	...	0.47845	0.59465
17.3	0.49337	0.61482	0.49296	0.61354	22.3	...	...	0.47812	0.59422
17.4	0.49311	0.61449	0.49269	0.61319	22.4	...	...	0.47778	0.59379
17.5	0.49285	0.61416	0.49243	0.61284	22.5	...	...	0.47745	0.59336
17.6	0.49259	0.61383	0.49217	0.61250	22.6	...	...	0.47711	0.52993
17.7	0.49233	0.61350	0.49191	0.61215	22.7	...	...	0.47677	0.59250
17.8	0.49206	0.61316	0.49164	0.61180	22.8	...	...	0.47643	0.59207
17.9	0.49180	0.61283	0.49137	0.61144	22.9	...	...	0.47610	0.59164
...	...	...	...	...	23.0	...	...	0.47577	0.59121

All dimensions are in inches. *Courtesy of Van Keuren Co.*

* Values given for w_1 and $(C + c)_1$ in table are for 1-inch pitch axial pitch. For other pitches, divide table values by number of threads per inch.

angle is incorrect, the amount of error may be determined by the following formula, which applies to any thread regardless of angle:

$$\text{Sin } a = \frac{A}{B - A}$$

In this formula,

A = difference in diameters of the large and small wires used;

B = total difference between the measurements over the large and small wires;

a = one-half the included thread angle.

Example: The diameter of the large wires used for testing the angle of a thread is 0.116 inch and of the small wires 0.076 inch. The measurement over the two sets of wires shows a total difference of 0.122 inch instead of the correct difference, 0.120 inch, for a standard angle of 60 degrees when using the sizes of wires mentioned. Therefore the amount of error is determined as follows:

$$\text{Sin } a = \frac{0.040}{0.122 - 0.040} = \frac{0.040}{0.082} = 0.4878$$

By referring to a table of sines it will be seen that this value (0.4878) is the sine of 29 degrees 12 minutes, approximately. Therefore the angle of the thread is 58 degrees, 24 minutes or 1 degree 36 minutes less than the standard angle.

Measuring Taper Screw Threads by 3-Wire Method. — When the 3-wire method is used in measuring a taper screw thread the measurement is along a line that is not perpendicular to the axis of the screw thread, the inclination from the perpendicular equalling one-half the included angle of the taper. The formula which follows compensates for this inclination resulting from contact of the measuring instrument surfaces, with two wires on one side and one on the other. The taper thread is measured over the wires in the usual manner excepting that the single wire must be located in the thread at a point where the effective diameter is to be checked (as described more fully later). The formula shows the dimension equivalent to the correct pitch diameter at this given point. The general formula for taper screw threads follows:

M = measurement over the 3 wires; E = pitch diameter; a = one-half the angle of the thread; N = number of threads per inch; W = diameter of wires; b = one-half the angle of taper

$$M = \frac{E - \dfrac{\cot a}{2\,N} + W\,(1 + \text{cosec } a)}{\sec b}$$

This formula is not theoretically correct but it is, however, accurate for screw threads having tapers of ¾ inch per foot or less. This general formula can be simplified for a given thread angle and taper. The simplified formula following (in which P = pitch) is for an American Standard Pipe Thread:

$$M = \frac{E - (0.86603 \times P) + 3 \times W}{1.00049}$$

Standard pitch diameters for pipe threads will be found in the table "American Standard Pipe Thread." The location of this pitch diameter or distance from the end of the pipe is also shown by the table. In using the formula for finding dimen-

sion M over the wires, the single wire is placed in whatever part of the thread groove locates it at the point where the pitch diameter is to be checked. The wire must be accurately located at this point. The other wires are then placed on each side of that thread which is diametrically opposite the single wire. If the pipe thread is straight or without taper,

$$M = E - (0.86603 \times P) + 3 \times W$$

Application of Formula to Pipe Thread: To illustrate the use of the formula for taper pipe threads, assume that dimension M is required for an American Standard 3-inch pipe thread gage. The table " American Standard Pipe Thread " shows that the 3-inch size has 8 threads per inch or a pitch of 0.125 inch and a pitch diameter at the gaging notch of 3.3885 inches. Assume that the wire diameter is 0.07217 inch: Then when the pitch diameter is correct

$$M = \frac{3.3885 - (0.86603 \times 0.125) + 3 \times 0.07217}{1.00049} = 3.495 \text{ inch}$$

Pitch Diameter Equivalent to a Given Measurement Over the Wires: The formula following may be used to check the pitch diameter at any point along a tapering thread when measurement M over wires of a given diameter is known. In this formula E = the effective or pitch diameter at the position occupied by the single wire. The formula is not theoretically correct but gives very accurate results when applied to tapers of ¾ inch per foot or less.

$$E = 1.00049 \times M + (0.86603 \times P) - 3 \times W$$

Example: Measurement M = 3.495 inches at the gaging notch of a 3-inch pipe thread and the wire diameter = 0.07217 inch. Then:

$$E = 1.00049 \times 3.495 + (0.86603 \times 0.125) - 3 \times 0.07217 = 3.3885 \text{ inches}$$

Pitch Diameter at Any Point Along Taper Screw Thread: When the pitch diameter in any position along a tapering thread is known, the pitch diameter at any other position may be determined as follows:

Multiply the distance (measured along the axis) between the location of the known pitch diameter and the location of the required pitch diameter, by the taper per inch or by 0.0625 for American Standard Pipe Threads. Add this product to the known diameter, if the required diameter is at a larger part of the taper, or subtract if the required diameter is smaller.

Example: The pitch diameter of a 3-inch American Standard Pipe Thread is 3.3885 at the gaging notch. Determine the pitch diameter at the small end. The American Standard Pipe Thread Table shows that the distance between the gaging notch and the small end of a 3-inch pipe is 0.77 inch. Hence the pitch diameter at the small end = 3.3885 − (0.77 × 0.0625) = 3.3404 inches.

Three-Wire Method Applied to Buttress Threads. — The angles of buttress threads vary somewhat, especially on the front or load-resisting side. Formula (1) which follows may be applied to any angles required. In this formula M = measurement over wires when *pitch diameter* E is correct; A = included angle of thread and thread groove; a = angle of front face or load-resisting side, measured from a line perpendicular to screw thread axis; P = pitch of thread; W = wire diameter.

$$M = E - \left[\frac{P}{\tan a + \tan (A - a)} \right] + W \left[1 + \cos \left(\frac{A}{2} - a \right) \times \csc \frac{A}{2} \right] \quad (1)$$

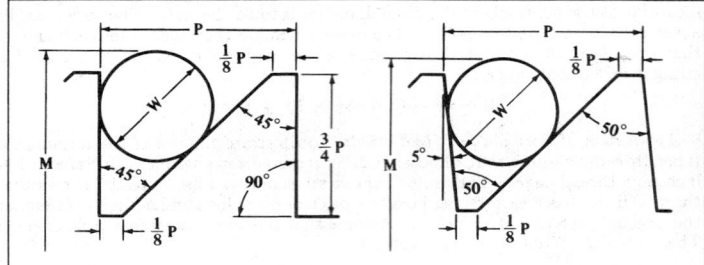

For given angles A and a, this general formula may be simplified as shown by Formulas (3) and (4). These simplified formulas contain constants with values depending upon angles A and a.

Wire Diameter: The wire diameter for obtaining pitch-line contact at the back of a buttress thread may be determined by the following general formula (2):

$$W = P\left(\frac{\cos a}{1 + \cos A}\right) \tag{2}$$

45-Degree Buttress Thread: The buttress thread shown by the diagram at the left, has a front or load-resisting side that is perpendicular to the axis of the screw. Measurement M equivalent to a correct pitch diameter E may be determined by formula (3):

$$M = E - P + (W \times 3.4142) \tag{3}$$

Wire diameter W for pitch-line contact at back of thread = 0.586 × pitch.

50-Degree Buttress Thread with Front-face Inclination of 5 Degrees: This buttress thread form is illustrated by the diagram at the right. Measurement M equivalent to the correct pitch diameter E may be determined by formula (4):

$$M = E - (P \times 0.91955) + (W \times 3.2235) \tag{4}$$

Wire diameter W for pitch-line contact at back of thread = 0.606 × pitch. If the width of flat at crest and root = ⅛ × pitch, depth = 0.69 × pitch.

American National Standard Buttress Threads (ANSI B1.9-1973): This buttress screw thread has an included thread angle of 52 degrees and a front face inclination of 7 degrees. Measurements M equivalent to a pitch diameter E may be determined by formula (5):

$$M = E - 0.89064P + 3.15689W + c \tag{5}$$

The wire angle correction factor c is less than 0.0004 inch for recommended combinations of thread diameters and pitches and may be neglected. Use of wire diameter $W = 0.54147P$ is recommended.

Measurement of Pitch Diameter of Thread Ring Gages. — The application of direct methods of measurement to determine the pitch diameter of thread ring gages presents serious difficulties, particularly in securing proper contact pressure when a high degree of precision is required. The usual practice is to fit the ring gage to a master setting plug. When the thread ring gage is of correct lead, angle, and thread form, within close limits, this method is quite satisfactory and represents standard American practice. It is the only method available for small sizes of threads. For the larger sizes, various more or less satisfactory methods have been devised, but none of these have found wide application.

Screw Thread Gage Classification. — Screw thread gages are classified by their degree of accuracy, that is, by the amount of tolerance afforded the gage manufacturer and the wear allowance, if any. There are also three classifications according to use: (1) Working gages for controlling production; (2) inspection gages for rejection or acceptance of the finished product; (3) reference gages for determining the accuracy of the working and inspection gages.

American National Standard for Gages and Gaging for Unified Inch Screw Threads (ANSI B1.2-1983). — This standard covers gaging methods for conformance of Unified Screw Threads and provides the essential specifications for applicable gages required for unified inch screw threads.

The standard includes the following gages for *Product Internal Thread:*

GO Working Thread Plug Gage for inspecting the maximum-material GO functional limit.

NOT GO (HI) Thread Plug Gage for inspecting the NOT GO (HI) functional diameter limit.

Thread Snap Gage — GO Segments or Rolls for inspecting the maximum-material GO functional limit.

Thread Snap Gage — NOT GO (HI) Segments or Rolls for inspecting the NOT GO (HI) functional diameter limit.

Thread Snap Gages — Minimum Material: Pitch Diameter Cone Type and Vee and Thread Groove Diameter Type for inspecting the minimum-material limit pitch diameter.

Thread-Setting Solid Ring Gage for setting internal thread indicating and snap gages.

Plain Plug, Snap, and Indicating Gages for checking the minor diameter of internal threads.

Snap and Indicating Gages for checking the major diameter of internal threads.

Functional Indicating Thread Gage for inspecting the maximum-material GO functional limit and size and the NOT GO (HI) functional diameter limit and size.

Minimum-Material Indicating Thread Gage for inspecting the minimum-material limit and size.

Indicating Runout Thread Gage for inspecting runout of the minor diameter to pitch diameter.

In addition to these gages for product internal threads, the Standard also covers differential gaging and such instruments as pitch micrometers, thread-measuring balls, optical comparator and toolmaker's microscope, profile tracing instrument, surface roughness measuring instrument, and roundness measuring equipment.

The Standard includes the following gages for *Product External Thread:*

GO Working Thread Ring Gage for inspecting the maximum-material GO functional limit.

NOT GO (LO) Thread Ring Gage for inspecting the NOT GO (LO) functional diameter limit.

Thread Snap Gage — GO Segments or Rolls for inspecting the maximum-material GO functional limit.

Thread Snap Gage — NOT GO (LO) Segments or Rolls for inspecting the NOT GO (LO) functional diameter limit.

Thread Snap Gages — Cone and Vee Type and Minimum Material Thread Groove Diameter Type for inspecting the minimum-material pitch diameter limit.

Plain Ring and Snap Gages for checking the major diameter.

Snap Gage for checking the minor diameter.

Functional Indicating Thread Gage for inspecting the maximum-material GO functional limit and size and the NOT GO (LO) functional diameter limit and size.

Minimum-Material Indicating Thread Gage for inspecting the minimum-material limit and size.

Indicating Runout Gage for inspecting the runout of the major diameter to the pitch diameter.

W Tolerance Thread-Setting Plug Gage for setting adjustable thread ring gages, checking solid thread ring gages, setting thread snap limit gages, and setting indicating thread gages.

Plain Check Plug Gage for Thread Ring Gage for verifying the minor diameter limits of thread ring gages after the thread rings have been properly set with the applicable thread-setting plug gages.

Indicating Plain Diameter Gage for checking the major diameter.

Indicating Gage for checking the minor diameter.

In addition to these gages for product external threads, the Standard also covers differential gaging and such instruments as thread micrometers, thread-measuring wires, optical comparator and toolmaker's microscope, profile tracing instrument, electromechanical lead tester, helical path attachment used with GO type thread indicating gage, helical path analyzer, surface roughness measuring equipment, and roundness measuring equipment.

The standard lists the following for use of Threaded and Plain Gages for verification of product internal threads:

Tolerance. Unless otherwise specified all thread gages which directly check the product thread shall be X tolerance for all classes.

GO Thread Plug Gages. GO thread plug gages must enter and pass through the full threaded length of the product freely. The GO thread plug gage is a cumulative check of all thread elements except the minor diameter.

NOT GO (HI) Thread Plug Gages. NOT GO (HI) thread plug gages when applied to the product internal thread may engage only the end threads (which may not be representative of the complete thread). Entering threads on product are incomplete and permit gage to start. Starting threads on NOT GO (HI) plugs are subject to greater wear than the remaining threads. Such wear in combination with the incomplete product threads permits further entry of the gage. NOT GO (HI) functional diameter is acceptable when the NOT GO (HI) thread plug gage applied to the product internal thread does not enter more than three complete turns. The gage should not be forced. Special requirements such as exceptionally thin or ductile material, small number of threads, etc., may necessitate modification of this practice.

GO and NOT GO Plain Plug Gages for Minor Diameter of Product Internal Thread. (Recommended in Class Z tolerance.) GO plain plug gages must completely enter and pass through the length of the product without force. NOT GO cylindrical plug gage must not enter.

The standard lists the following for use of Threaded and Plain Ring, Snap and Indicating Thread Gages for verification of product external threads:

GO Thread Ring Gages. Adjustable GO thread ring gages must be set to the applicable W tolerance setting plugs to assure they are within specified limits. The product thread must freely enter the GO thread ring gage for the entire length of the threaded portion. The GO thread ring gage is a cumulative check of all thread elements except the major diameter.

NOT GO (LO) Thread Ring Gages. NOT GO (LO) thread ring gages must be set to the applicable W tolerance setting plugs to assure that they are within specified limits. NOT GO (LO) thread ring gages when applied to the product external thread may engage only the end threads (which may not be representative of the complete product thread).

Starting threads on NOT GO (LO) rings are subject to greater wear than the remaining threads. Such wear in combination with the incomplete threads at the end of the product thread permit further entry in the gage. NOT GO (LO) functional diameter is acceptable when the NOT GO (LO) thread ring gage applied to the product external thread does not pass over the thread more than three complete

turns. The gage should not be forced. Special requirements such as exceptionally thin or ductile material, small number of threads, etc., may necessitate modification of this practice.

GO and NOT GO Plain Ring and Snap Gages for Checking Major Diameter of Product External Thread. The GO gage must completely receive or pass over the major diameter of the product external thread to ensure that the major diameter does not exceed the maximum-material-limit. The NOT GO gage must not pass over the major diameter of the product external thread to ensure that the major diameter is not less than the minimum-material-limit.

Limitations concerning the use of gages are given in the standard as follows:

Product threads accepted by a gage of one type may be verified by other types. It is possible, however, that parts which are near either rejection limit may be accepted by one type and rejected by another. Also, it is possible for two individual limit gages of the same type to be at the opposite extremes of the gage tolerances permitted, and borderline product threads accepted by one gage could be rejected by another. For these reasons, a product screw thread is considered acceptable when it passes a test by any of the permissible gages in ANSI B1.3 for the gaging system

(Continued on page 1645)

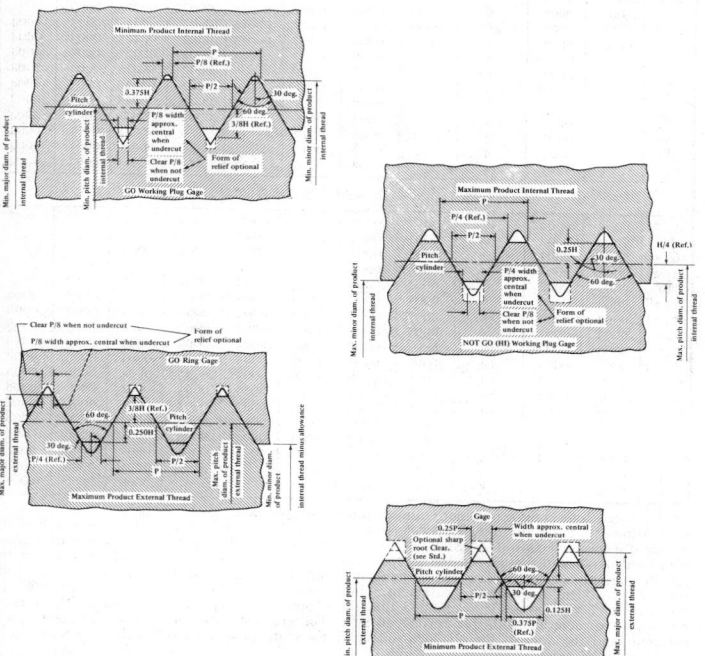

Fig. 1. Thread Forms of Gages for Product Internal and External Threads

Table 1. American National Standard Tolerance for GO, HI, and LO Thread Gages for Unified Inch Screw Threads (ANSI B1.2.1983)

Thds. per Inch	Tolerance on Lead*		Tol. on Thread	Tol. on Major and Minor Diams.‡			Tolerance on Pitch Diameter‡				
	To & incl. ½ in. Diam.	Above ½ in. Diam.	Half-angle (±), minutes	To & incl. ½ in. Diam.	Above ½ to 4 in. Diam.	Above 4 in. Diam.	To & incl. ½ in. Diam.	Above ½ to 1½ in. Diam.	Above 1½ to 4 in. Diam.	Above 4 to 8 in. Diam.	Above 8 to 12 in. Diam.†
W GAGES											
80, 72	.0001	.00015	20	.0003	.0003	...	.0001	.00015	...	...	...
64	.0001	.00015	20	.0003	.0004	...	.0001	.00015	...	...	...
56	.0001	.00015	20	.0003	.0004	...	.0001	.00015	.0002	...	...
48	.0001	.00015	18	.0003	.0004	...	.0001	.00015	.0002	...	...
44, 40	.0001	.00015	15	.0003	.0004	...	.0001	.00015	.0002	...	...
36	.0001	.00015	12	.0003	.0004	...	.0001	.00015	.0002	...	...
32	.0001	.00015	12	.0003	.0005	.0007	.0001	.00015	.0002	.00025	.0003
28, 27	.00015	.00015	8	.0005	.0005	.0007	.0001	.00015	.0002	.00025	.0003
24, 20	.00015	.00015	8	.0005	.0005	.0007	.0001	.00015	.0002	.00025	.0003
18	.00015	.00015	8	.0005	.0005	.0007	.0001	.00015	.0002	.00025	.0003
16	.00015	.00015	8	.0006	.0006	.0009	.0001	.0002	.00025	.0003	.0004
14, 13	.0002	.0002	6	.0006	.0006	.0009	.00015	.0002	.00025	.0003	.0004
12	.0002	.0002	6	.0006	.0006	.0009	.00015	.0002	.00025	.0003	.0004
11½	.0002	.0002	6	.0006	.0006	.0009	.00015	.0002	.00025	.0003	.0004
11	.0002	.0002	6	.0006	.0006	.0009	.00015	.0002	.00025	.0003	.0004
10	...	.00025	6	...	.0006	.0009	...	.0002	.00025	.0003	.0004
9	...	.00025	6	...	.0007	.0011	...	.0002	.00025	.0003	.0004
8	...	.00025	5	...	.0007	.0011	...	.0002	.00025	.0003	.0004
7	...	.0003	5	...	.0007	.0011	...	.0002	.00025	.0003	.0004
6	...	.0003	5	...	.0008	.0013	...	.0002	.00025	.0003	.0004
5	...	.0003	4	...	.0008	.0013	...	...	.00025	.0003	.0004
4½	...	.0003	4	...	.0008	.0013	...	...	.00025	.0003	.0004
4	...	.0003	4	...	.0009	.0015	...	...	.00025	.0003	.0004
X GAGES											
80, 72	.0002	.0002	30	.0003	.0003	...	.0002	.0002	...	...	...
64	.0002	.0002	30	.0004	.0004	...	.0002	.0002	...	...	...
56, 48	.0002	.0002	30	.0004	.0004	...	.0002	.0002	.0003	...	...
44, 40	.0002	.0002	20	.0004	.0004	...	.0002	.0002	.0003	...	...
36	.0002	.0002	20	.0004	.0004	...	.0002	.0002	.0003	...	...
32, 28	.0003	.0003	15	.0005	.0005	.0007	.0003	.0003	.0004	.0005	.0006
27, 24	.0003	.0003	15	.0005	.0005	.0007	.0003	.0003	.0004	.0005	.0006
20	.0003	.0003	15	.0005	.0005	.0007	.0003	.0003	.0004	.0005	.0006
18	.0003	.0003	10	.0005	.0005	.0007	.0003	.0003	.0004	.0005	.0008
16, 14	.0003	.0003	10	.0006	.0006	.0009	.0003	.0003	.0004	.0006	.0008
13, 12	.0003	.0003	10	.0006	.0006	.0009	.0003	.0003	.0004	.0006	.0008
11½	.0003	.0003	10	.0006	.0006	.0009	.0003	.0003	.0004	.0006	.0008
11, 10	.0003	.0003	10	.0006	.0006	.0009	.0003	.0003	.0004	.0006	.0008
9	.0003	.0003	10	.0007	.0007	.0011	.0003	.0003	.0004	.0006	.0008
8, 7	.0004	.0004	5	.0007	.0007	.0011	.0004	.0004	.0005	.0006	.0008
6	.0004	.0004	5	.0008	.0008	.0013	.0004	.0004	.0005	.0006	.0008
5, 4½	.0004	.0004	5	.0008	.0008	.0013	...	...	.0005	.0006	.0008
4	.0004	.0004	5	.0009	.0009	.0015	...	...	.0005	.0006	.0008

All dimensions are given in inches unless otherwise specified.

* Allowable variation in lead between any two threads not farther apart than the length of the standard gage as shown in ANSI B47.1. The tolerance on lead establishes the width of a zone, measured parallel to the axis of the thread, within which the actual helical path must lie for the specified length of the thread. Measurements are taken from a fixed reference point, located at the start of the first full thread, to a sufficient number of positions along the entire helix to detect all types of lead variations. The amounts that these positions vary from their basic (theoretical) positions are recorded with due respect to sign. The greatest variation in each direction (±) is selected, and the sum of their values, disregarding sign, must not exceed the tolerance limits specified for W gages.

† Above 12 in. the tolerance is directly proportional to the tolerance given in this column below, in the ratio of the diameter to 12 in.

‡ Tolerances apply to designated size of thread. The application of the tolerances is specified in the Standard.

**Table 2. Formulas for Limits of American National Standard Gages
for Unified Inch Screw Threads* (ANSI B1.2-1983) — 1**

No.	THREAD GAGES FOR EXTERNAL THREADS
1	GO Pitch Diameter = Maximum pitch diameter of external thread. Gage tolerance is *minus*.
2	GO Minor Diameter = Maximum pitch diameter of external thread minus $H/2$. Gage tolerance is *minus*.
3	NOT GO (LO) Pitch Diameter (for plus tolerance gage) = Minimum pitch diameter of external thread. Gage tolerance is *plus*.
4	NOT GO (LO) Minor Diameter = Minimum pitch diameter of external thread minus $H/4$. Gage tolerance is *plus*.

	PLAIN GAGES FOR MAJOR DIAMETER OF EXTERNAL THREADS
5	GO = Maximum major diameter of external thread. Gage tolerance is *minus*.
6	NOT GO = Minimum major diameter of external thread. Gage tolerance is *plus*.

	THREAD GAGES FOR INTERNAL THREADS
7	GO Major Diameter = Minimum major diameter of internal thread. Gage tolerance is *plus*.
8	GO Pitch Diameter = Minimum pitch diameter of internal thread. Gage tolerance is *plus*.
9	NOT GO (HI) Major Diameter = Maximum pitch diameter of internal thread plus $H/2$. Gage tolerance is *minus*.
10	NOT GO (HI) Pitch Diameter = Maximum pitch diameter of internal thread. Gage tolerance is *minus*.

	PLAIN GAGES FOR MINOR DIAMETER OF INTERNAL THREADS
11	GO = Minimum minor diameter of internal thread. Gage tolerance is *plus*.
12	NOT GO = Maximum minor diameter of internal thread. Gage tolerance is *minus*.

	FULL FORM AND TRUNCATED SETTING PLUGS
13	GO Major Diameter (Truncated Portion) = Maximum major diameter of external thread (= minimum major diameter of full portion of GO setting plug) minus $(0.060 \sqrt[3]{p^2} + 0.017p)$. Gage tolerance is *minus*.
14	GO Major Diameter (Full Portion) = Maximum major diameter of external thread. Gage tolerance is *plus*.
15	GO Pitch Diameter = Maximum pitch diameter of external thread. Gage tolerance is *minus*.
16	†NOT GO (LO) Major Diameter (Truncated Portion) = Minimum pitch diameter of external thread plus $H/2$. Gage tolerance is *minus*.

	FULL FORM AND TRUNCATED SETTING PLUGS
17	NOT GO (LO) Major Diameter (Full Portion) = Maximum major diameter of external thread provided major diameter crest width shall not be less than 0.001 in. (0.0009 in. truncation). Apply W tolerance *plus* for maximum size except that for 0.001 in. crest width apply tolerance *minus*. For the 0.001 in. crest width, major diameter is equal to maximum major diameter of external thread plus $0.216506p$ minus the sum of external thread pitch diameter tolerance and 0.0017 in.
18	NOT GO (LO) Pitch Diameter = Minimum pitch diameter of external thread. Gage tolerance is *plus*.

	SOLID THREAD-SETTING RINGS FOR SNAP AND INDICATING GAGES
19	‡GO Pitch Diameter = Minimum pitch diameter of internal thread. W gage tolerance is *plus*.
20	GO Minor Diameter = Minimum minor diameter of internal thread. W gage tolerance is *minus*.
21	‡NOT GO (HI) Pitch Diameter = Maximum pitch diameter of internal thread. W gage tolerance is *minus*.
22	NOT GO (HI) Minor Diameter = Maximum minor diameter of internal thread. W gage tolerance is *minus*.

* See data in Screw Thread Systems section for symbols and dimensions of Unified Screw Threads.   † Truncated portion is required when optional sharp root profile is used.   ‡ Tolerances greater than W tolerance for pitch diameter are acceptable when internal indicating or snap gage can accommodate a greater tolerance and when agreed upon by supplier and user.

specified, provided the gages being used are within the tolerances specified in ANSI B1.2.

Table 3. American National Standard Tolerances for Plain Cylindrical Gages
(ANSI B1.2-1983)

Size Range		Tolerance Class*				
Above	To and Including	XX	X	Y	Z	ZZ
		Tolerance				
0.020	0.825	.00002	.00004	.00007	.00010	.00020
0.825	1.510	.00003	.00006	.00009	.00012	.00024
1.510	2.510	.00004	.00008	.00012	.00016	.00032
2.510	4.510	.00005	.00010	.00015	.00020	.00040
4.510	6.510	.000065	.00013	.00019	.00025	.00050
6.510	9.010	.00008	.00016	.00024	.00032	.00064
9.010	12.010	.00010	.00020	.00030	.00040	.00080

All dimensions are given in inches.
* Tolerances apply to actual diameter of plug or ring. Apply tolerances as specified in the Standard. Symbols XX, X, Y, Z, and ZZ are standard gage tolerance classes.

Gaging large product external and internal threads equal to above 6.25-inch nominal size with plain and threaded plug and ring gages presents problems for technical and economic reasons. In these instances, verification may be based on use of modified snap or indicating gages or measurement of thread elements. Various types of gages or measuring devices in addition to those defined in the Standard are available and acceptable when properly correlated to this Standard. Producer and user should agree on the method and equipment used.

Thread Forms of Gages. — Thread forms of gages for product internal and external threads are given in Fig. 1. The Standard (ANSI B1.2-1983) also gives illustrations of the thread forms of truncated thread setting plug gages, the thread forms of full-form thread setting plug gages, the thread forms of solid thread setting ring gages, and an illustration that shows the chip groove and removal of partial thread.

Thread Gage Tolerances. — Gage tolerances of thread plug and ring gages, thread setting plugs, and setting rings for Unified screw threads, designated as W and X tolerances, are given in Table 1. W tolerances represent the highest commercial grade of accuracy and workmanship, and are specified for thread setting gages; X tolerances are larger than W tolerances and are used for product inspection gages. Tolerances for plain gages are given in Table 3.

Determining Size of Gages. — The three-wire method of determining pitch diameter size of plug gages is recommended for gages covered by American National Standard B1.2. This method is described in Appendix B of the 1983 issue of that Standard.

Size limit adjustments of thread ring and external thread snap gages are determined by their fit on their respective calibrated setting plugs. Indicating gages and thread gages for product external threads are controlled by reference to appropriate calibrated setting plugs.

Size limit adjustments of internal thread snap gages are determined by their fit on their respective calibrated setting rings. Indicating gages and other adjustable thread gages for product internal threads are controlled by reference to appropriate calibrated setting rings or by direct measuring methods.

Interpretation of Tolerances. — Tolerances on lead, half-angle, and pitch diameter are variations which may be taken independently for each of these elements and may be taken to the extent allowed by respective tabulated dimensional limits. The tabulated tolerance on any one element must not be exceeded, even though variations in the other two elements are smaller than the respective tabulated tolerances.

Direction of Tolerance on Gages. — At the maximum-material limit (GO), the dimensions of all gages used for final conformance gaging are to be within limits of size of the product thread. At the functional diameter limit, using NOT GO (HI and LO)

Table 4. Constants for Computing Thread Gage Dimensions (ANSI B1.2-1983)

Threads per Inch	Pitch, p	$.060\sqrt[3]{p^2} + .017p$	$.05p$	$.087p$	Height of Sharp V-Thread, $H = .866025p$	$H/2 = .43301p$	$H/4 = .216506p$
80	.012500	.0034	.00063	.00109	.010825	.00541	.00271
72	.013889	.0037	.00069	.00122	.012028	.00601	.00301
64	.015625	.0040	.00078	.00136	.013532	.00677	.00338
56	.017857	.0044	.00089	.00155	.015465	.00773	.00387
48	.020833	.0049	.00104	.00181	.018042	.00902	.00451
44	.022727	.0052	.00114	.00198	.019682	.00984	.00492
40	.025000	.0056	.00125	.00218	.021651	.01083	.00541
36	.027778	.0060	.00139	.00242	.024056	.01203	.00601
32	.031250	.0065	.00156	.00272	.027063	.01353	.00677
28	.035714	.0071	.00179	.00311	.030929	.01546	.00773
27	.037037	.0073	.00185	.00322	.032075	.01604	.00802
24	.041667	.0079	.00208	.00361	.036084	.01804	.00902
20	.050000	.0090	.00250	.00435	.043301	.02165	.01083
18	.055556	.0097	.00278	.00483	.048113	.02406	.01203
16	.062500	.0105	.00313	.00544	0.54127	.02706	.01353
14	.071429	.0115	.00357	.00621	.061859	.03093	.01546
13	.076923	.0122	.00385	.00669	.066617	.03331	.01665
12	.083333	.0129	.00417	.00725	.072169	.03608	.01804
11½	.086957	.0133	.00435	.00757	.075307	.03765	.01883
11	.090909	.0137	.00451	.00791	.078730	.03936	.01968
10	.100000	.0146	.00500	.00870	.086603	.04330	.02165
9	.111111	.0158	.00556	.00967	.096225	.04811	.02406
8	.125000	.0171	.00625	.01088	.108253	.05413	.02706
7	.142857	.0188	.00714	.01243	.123718	.06186	.03093
6	.166667	.0210	.00833	.01450	.144338	.07217	.03608
5	.200000	.0239	.01000	.01740	.173205	.08660	.04330
4½	.222222	.0258	.01111	.01933	.192450	.09623	.04811
4	.250000	.0281	.01250	.02175	.216506	.10825	.05413

All dimensions are given in inches unless otherwise specified.

thread gages, the standard practice is to have the gage tolerance within the limits of size of the product thread.

Formulas for Limits of Gages. — Formulas for limits of American National Standard Gages for Unified screw threads are given in Table 2. Some constants which are required to determine gage dimensions are tabulated in Table 4.

Standard Temperature. — A temperature of 68°F (20°C) is the standard temperature used internationally for linear measurements. According to the Standard, nominal dimensions of gages and product as specified and actual dimensions as measured should be within specified limits at this temperature. For screw thread gaging, the acceptable tolerance on the standard temperature is ± 2°F (± 1°C).

Since product threads are frequently checked at temperatures that are not controlled, it is desirable that the coefficient of the thermal expansion of the gages be the same as that of the product on which they are used. Inasmuch as the majority of threaded product consists of iron or steel, and thread gages are made of hardened steel, this condition is usually fulfilled without special attention, provided thread gages and product have stabilized to the same temperature. When the materials of the product thread and the gage are dissimilar, the differing thermal coefficients can cause serious complications and must be taken into account, unless both product and gage at the time of gaging are at a temperature of:

(a) 68°F ± 4°F (20°C ± 2°C) for 1 in. and smaller;

(b) 68°F ± 2°F (20°C ± 1°C) for sizes above 1 in. to 3 in.;

(c) 68°F ± 1°F (20°C ± 0.5°C) for sizes above 3 in. to 6 in.

TAPPING AND THREAD CUTTING

Selection of Taps. — For most applications, a standard tap supplied by the manufacturer can be used, but some jobs may require special taps. A variety of standard taps can be obtained. In addition to specifying the size of the tap it is necessary to be able to select the one most suitable for the application at hand. The elements of standard taps that are varied are: the number of flutes; the type of flute, whether straight, spiral pointed, or spiral fluted; the chamfer length; the relief of the land, if any; the tool steel used to make the tap; and the surface treatment of the tap. Details regarding the nomenclature of tap elements are given on pages 863–896 along with a listing of the standard sizes available.

Factors to consider in selecting a tap include: the method of tapping, by hand or by machine; the material to be tapped and its heat treatment; the length of thread, or depth of the tapped hole; the required tolerance or class of fit; the production requirement and the type of machine to be used. The diameter of the hole must also be considered, although this is usually a matter of design and the specification of the tap drill size.

Method of Tapping: The term *hand tap* is used for both hand and machine taps, and almost all taps can be applied to the hand or machine method. While any tap can be used for hand tapping, those having a concentric land without the relief are preferable. In hand tapping the tool is reversed periodically to break the chip, and the heel of the land of a tap with a concentric land (without relief) will cut the chip off cleanly or any portion of it that is attached to the work, whereas a tap with an eccentric or con-eccentric relief may leave a small burr that becomes wedged between the relieved portion of the land and the work. This creates a pressure towards the cutting face of the tap that may cause it to chip; it tends to roughen the threads in the hole, and it increases the overall torque required to turn the tool. When tapping by machine, however, the tap is usually turned only in one direction until the operation is complete, and an eccentric or con-eccentric relief is often an advantage.

Chamfer Length: Three types of hand taps, used both for hand and machine tapping, are available, and they are distinguished from each other by the length of chamfer. *Taper taps* have a chamfer angle of 4 degrees which reduces the height about 8-10 teeth; *plug taps* have a 9 degree chamfer angle with 3-5 threads reduced in height; and *bottoming taps* have a 30 degree chamfer angle with 1½ threads reduced in height. Since the teeth that are reduced in height do practically all of the cutting, the chip load or chip thickness per tooth will be least for a taper tap, greater for a plug tap, and greatest for a bottoming tap.

For most through hole tapping applications it is necessary to use only a plug type tap, which is also most suitable for blind holes in which the tap drill hole is deeper than the required thread. If the tap must bottom in a blind hole, the hole is usually threaded first with a plug tap and then finished with a bottoming tap to catch the last threads in the bottom of the hole. Taper taps are used on hard materials having a high tensile strength where the chip load per tooth must be kept to a minimum. However, taper taps should not be used on materials that have a strong tendency to work harden, such as the austenitic stainless steels.

Spiral Point Taps: Spiral point taps offer a special advantage when machine tapping through holes in ductile materials because they are designed to handle the long continuous chips that form and which otherwise cause a disposal problem. An angular gash is ground at the point or end of the tap along the face of the chamfered threads or lead teeth of the tap. This gash forms a left-hand helix in the flutes adjacent to the lead teeth which causes the chips to flow ahead of the tap and through the hole. The gash is usually formed to produce a rake angle on the cutting face that increases progressively towards the end of the tool. Since the

flutes are used primarily to provide a passage for the cutting fluid, they are usually made narrower and shallower thereby strengthening the tool. For tapping thin workpieces short fluted spiral taps are recommended. They have a spiral point gash along the cutting teeth; the remainder of the threaded portion of the tap has no flute. Most spiral pointed taps are of plug type; however, spiral point bottoming taps are also made.

Spiral Fluted Taps: Spiral fluted taps have a helical flute; the helix angle of the flute may be between 15 and 52 degrees and the hand of the helix is the same as that of the threads on the tap. The spiral flute and the axial rake that it forms on the cutting face of the tap combine to induce the chips to flow backward along the helix and out of the hole. Thus, they are ideally suited for tapping blind holes and they are available as plug and bottoming types. A higher helix angle should be specified for tapping very ductile materials; when tapping harder materials chipping at the cutting edge may result and the helix angle must be reduced. For tapping very high strength materials serial taps are recommended having a small helix angle.

Holes having a pronounced interruption such as a groove or a keyway can be tapped with spiral fluted taps on which the helix has the opposite hand as the threads. The land bridges the interruption and allows the tap to cut relatively smoothly.

Serial Taps and Close Tolerance Threads: For tapping holes to close tolerances a set of serial taps is used. They are usually available in sets of three: the No. 1 tap is undersize and is the first rougher; the No. 2 tap is of intermediate size and is the second rougher; and the No. 3 tap is used for finishing. The different taps are identified by one, two, and three annular grooves in the shank adjacent to the square. For some applications involving finer pitches only two serial taps are required. Sets are also used to tap hard or tough materials having a high tensile strength, deep blind holes in normal materials, and large coarse threads. A set of more than three taps is sometimes required to produce threads of coarse pitch. Threads to some commercial tolerances, such as American Standard Unified 2B, or ISO Metric 6H, can be produced in one cut using a ground tap; sometimes even closer tolerances can be produced with a single tap. Ground taps are recommended for all close tolerance tapping operations. For much ordinary work, cut taps are satisfactory and more economical than ground taps.

Tap Steels: Most taps are made from high speed steel, although carbon tool steel taps are available and in many instances are economical to use. The type of tool steel used is determined by the tap manufacturer and is usually satisfactory when correctly applied except in a few exceptional cases. Typical grades of high speed steel used to make taps are M-1, M-7, and M-10. Carbon tool steel taps are satisfactory where the operating temperature of the tap is low and where a high resistance to abrasion is not required.

Surface Treatment: The life of high speed steel taps can sometimes be increased significantly by treating the surface of the tap. A very common treatment is oxide coating, which forms a thin non-metallic oxide coating on the tap that has lubricity and is somewhat porous to absorb and retain oil. This coating reduces the friction between the tap and the work and it makes the surface virtually impervious to rust. It does not increase the hardness of the surface but it significantly reduces or prevents entirely galling, or the tendency of the work material to weld or stick to the cutting edge and to other areas on the tap with which it is in contact. For this reason oxide coated taps are recommended for metals that tend to gall and stick such as non-free cutting low carbon steels and soft copper. It is also useful for tapping other steels having higher strength properties.

Nitriding provides a very hard and wear resistant case on high speed steel. Nitrided taps are especially recommended for tapping plastics; they have also been used successfully on a variety of other materials including high strength high alloy

steels. However, some caution must be used in specifying nitrided taps because the nitride case is very brittle and may have a tendency to chip.

Chrome plating has been used to increase the wear resistance of taps but its application has been limited because of the high cost and the danger of hydrogen embrittlement which can cause cracks to form in the tool. A flash plate of about .0001 to .0005 in. in thickness is applied to the tap. Chrome-plated taps have been used successfully to tap a variety of ferrous and nonferrous materials including plastics, hard rubber, mild steel, and tool steel. Other surface treatments that have been used successfully to a limited extent are vapor blasting and liquid honing.

Rake Angle: For the majority of applications in both ferrous and nonferrous materials the rake angle machined on the tap by the manufacturer is satisfactory. This angle is approximately 5 to 7 degrees. In some instances it may be desirable to alter the rake angle of the tap to obtain beneficial results and Table 1 provides a guide that can be used. In selecting a rake angle from this table, consideration must be given to the size of the tap and the strength of the land.

Table 1. Tap Rake Angles for Tapping Different Materials

Material	Rake Angle, Degrees	Material	Rake Angle, Degrees
Cast Iron	0–3	Aluminum	10–20
Malleable Iron	5–8	Brass	2–5
Steel		Naval Brass	5–8
AISI 1100 Series	9–12	Phosphor Bronze	9–12
Low Carbon (up	9–12	Tobin Bronze	5–8
to .25 per cent)		Manganese Bronze	9–12
Medium Carbon, Annealed	7–10	Magnesium	10–20
(.30 to .60 per cent)		Monel	9–12
Heat Treated, 225–283	0–8	Copper	15–18
Brinell. (.30 to .60 per cent)		Zinc Die Castings	12–15
High Carbon and	0–5	Plastic	
High Speed		Thermoplastic	5–8
Stainless	10–15	Thermosetting	0–3
Titanium	6–10	Hard Rubber	0–3

Cutting Speed. — The cutting speed for machine tapping is treated in detail on page 969. It suffices to say here that many variables must be considered in selecting this cutting speed and any tabulation may have to be modified greatly. Where cutting speeds are mentioned in the following section, they are intended only to provide a guideline to show the possible range of speeds that could be used.

Tapping Specific Materials. — The work material has a great influence on the ease with which a hole can be tapped. For production work, in many instances, modified taps are recommended; however, for toolroom or short batch work, standard hand taps can be used on most jobs, providing reasonable care is taken when tapping. The following concerns the tapping of metallic materials; information on the tapping of plastics is given on page 1036.

Low Carbon Steel (Less than 0.15% C): These steels are very soft and ductile resulting in a tendency for the work material to tear and to weld to the tap. They produce a continuous chip that is difficult to break and spiral pointed taps are recommended for tapping through holes; for blind holes a spiral fluted tap is recommended. To prevent galling and welding, a liberal application of a sulfur base or other suitable cutting fluid is essential and the selection of an oxide coated tap is very helpful.

Low Carbon Steels (0.15 to 0.30% C): The additional carbon in these steels is beneficial as it reduces the tendency to tear and to weld; their machinability is further improved by cold drawing. These steels present no serious problems in

tapping provided a suitable cutting fluid is used. An oxide coated tap is recommended, particularly in the lower carbon range.

Medium Carbon Steels (0.30 to 0.60 % C): These steels can be tapped without too much difficulty, although a lower cutting speed must be used in machine tapping. The cutting speed is dependent on the carbon content and the heat treatment. Steels that have a higher carbon content must be tapped more slowly, especially if the heat treatment has produced a pearlitic microstructure. The cutting speed and ease of tapping is significantly improved by heat treating to produce a spheroidized microstructure. A suitable cutting fluid must be used.

High Carbon Steels (More than 0.6 % C): Usually these materials are tapped in the annealed or normalized condition although sometimes tapping is done after hardening and tempering to a hardness below 55 Rc. Recommendations for tapping after hardening and tempering are given under High Tensile Strength Steels. In the annealed and normalized condition these steels have a higher strength and are more abrasive than steels with a lower carbon content; thus, they are more difficult to tap. The microstructure resulting from the heat treatment has a significant effect on the ease of tapping and the tap life, a spherodite structure being better in this respect than a pearlitic structure. The rake angle of the tap should not exceed 5 degrees and for the harder materials a concentric tap is recommended. The cutting speed is considerably lower for these steels and an activated sulfur-chlorinated cutting fluid is recommended.

Alloy Steels: This classification includes a wide variety of steels, each of which may be heat treated to have a wide range of properties. When annealed and normalized they are similar to medium to high carbon steels and usually can be tapped without difficulty, although for some alloy steels a lower tapping speed may be required. Standard taps can be used and for machine tapping a con-eccentric relief may be helpful. A suitable cutting fluid must be used in all cases.

High Tensile Strength Steels: Any steel that must be tapped after being heat treated to a hardness range of 40-55 Rc is included in this classification. Low tap life and excessive tap breakage are characteristics of tapping these materials; those that have a high chromium content are particularly troublesome. Best results are obtained with taps that have concentric lands, a rake angle that is at or near zero degrees, and 6 to 8 chamfered threads on the end to reduce the chip load per tooth. The chamfer relief should be kept to a minimum. The load on the tap should be kept to a minimum by every possible means, including using the largest possible tap drill size; keeping the hole depth to a minimum; avoidance of bottoming holes; and, in the larger sizes, using fine instead of coarse pitches. Oxide coated taps are recommended although in some cases a nitrided tap can be used to reduce tap wear. An active sulfur-chlorinated oil is recommended as a cutting fluid and the tapping speed should not exceed about 10 feet per minute.

Stainless Steels: Ferritic and martensitic type stainless steels are somewhat like alloy steels that have a high chromium content, and they can be tapped in a similar manner, although a slightly slower cutting speed may have to be used. Standard rake angle oxide coated taps are recommended and a cutting fluid containing molybdenum disulphide is helpful to reduce the friction in tapping. Austenitic stainless steels are very difficult to tap because of their high resistance to cutting and their great tendency to work harden. A work-hardened layer is formed by a cutting edge of the tap and the depth of this layer depends on the severity of the cut and the sharpness of the tool. The next cutting edge must penetrate below the work-hardened layer, if it is to be able to cut. Therefore, the tap must be kept sharp and each succeeding cutting edge on the tool must penetrate below the work-hardened layer formed by the preceding cutting edge. For this reason, a taper tap should not be used, but rather a plug tap having 3-5 chamfered threads. To reduce the rubbing of the lands, an eccentric or con-eccentric relieved land should be used

and a 10-15 degree rake angle is recommended. A tough continuous chip is formed that is difficult to break. To control this chip, spiral pointed taps are recommended for through holes and low-helix angle spiral fluted taps for blind holes. An oxide coating on the tap is very helpful and a sulfur-chlorinated mineral lard oil is recommended, although heavy duty soluble oils have also been used successfully.

Free Cutting Steels: There is a large number of free-cutting steels, including free cutting stainless steels, which are also called free-machining steels. Sulfur, lead, or phosphorus are added to these steels to improve their machinability. In all cases, the free-machining steels are easier to tap than their counterparts that do not have the free-machining additives. Tool life is usually increased and a somewhat higher cutting speed can be used. The type of tap recommended depends on the particular type of free machining steel and the nature of the tapping operation; in most cases a standard tap can be used.

High Temperature Alloys: These are cobalt or nickel base nonferrous alloys that cut like austenitic stainless steel, but are often even more difficult to machine. The recommendations given for austenitic stainless steel also apply to tapping these alloys but the rake angle should be 0 to 10 degrees to strengthen the cutting edge. For most applications a nitrided tap or one made from M41, M42, M43, or M44 steel is recommended. The tapping speed is usually in the range of 5 to 10 feet per minute.

Titanium and Titanium Alloys: Titanium and its alloys have a low specific heat and a pronounced tendency to weld on to the tool material; therefore, oxide coated taps are recommended to minimize galling and welding. The rake angle of the tap should be from 6 to 10 degrees. To minimize the contact between the work and the tap an eccentric or con-eccentric relief land should be used. In some cases taps having interrupted threads are helpful. Pure titanium is comparatively easy to tap but the alloys are very difficult. The cutting speed depends on the composition of the alloy and may vary from 40 to 10 feet per minute. Special cutting oils are recommended for tapping titanium.

Gray Cast Iron: The microstructure of gray cast iron can vary, even within a single casting, and compositions are used that vary in tensile strength from about 20,000 to 60,000 psi (160 to 250 Bhn). It is seen then that this is not a single material, although in general it is not difficult to tap. The cutting speed may vary from 90 feet per minute for the softer grades to 30 feet per minute for the harder grades. The chip is discontinuous and straight fluted taps should be used for all applications. Oxide coated taps are helpful and in most cases gray cast iron can be tapped dry, although water soluble oils and chemical emulsions are sometimes used.

Malleable Cast Iron: Commercial malleable cast irons are also available having a rather wide range of properties, although within a single casting they tend to be quite uniform. They are relatively easy to tap and standard taps can be used. The cutting speed for ferritic cast irons is 60-90 feet per minute, for pearlitic malleable irons 40-50 feet per minute, and for martensitic malleable irons 30-35 feet per minute. A soluble oil cutting fluid is recommended except for martensitic malleable iron where a sulfur base oil may work better.

Ductile or Nodular Cast Iron: Several classes of nodular iron are used having a tensile strength varying from 60,000 to 120,000 psi. Moreover, the microstructure in a single casting and in castings produced at different times vary rather widely. The chips are easily controlled but have some tendency to weld to the face and flanks of cutting tools. For this reason oxide coated taps are recommended. The cutting speed may vary from 15 fpm for the harder martensitic ductile irons to 60 fpm for the softer ferritic grades. A suitable cutting fluid should be used.

Aluminum: Aluminum and aluminum alloys are relatively soft materials that have little resistance to cutting. The danger in tapping these alloys is that the tap will ream the hole instead of cutting threads, or that it will cut a thread eccentric

to the hole. For these reasons, extra care must be taken when aligning the tap and starting the thread. For production tapping a spiral pointed tap is recommended for through holes and a spiral fluted tap for blind holes; preferably these taps should have a 10 to 15 degree rake angle. A lead screw tapping machine is helpful in cutting accurate threads. A heavy duty soluble oil or a light base mineral oil should be used as a cutting fluid.

Copper Alloys: Most copper alloys are not difficult to tap, except beryllium copper and a few other hard alloys. Pure copper offers some difficulty because of its ductility and the ductile continuous chip formed, which can be difficult to control. However, with reasonable care and the use of medium heavy duty mineral lard oil it can be tapped successfully. Red brass, yellow brass, and similar alloys containing not more than 35 per cent zinc produce a continuous chip. While straight fluted taps can be used for hand tapping these alloys, machine tapping should be done with spiral pointed or spiral fluted taps for through and blind holes respectively. Naval brass, leaded brass, and cast brasses produce a discontinuous chip and a straight fluted tap can be used for machine tapping. These alloys do exhibit a tendency to close in on the tap and sometimes an interrupted thread tap is used to reduce the resulting jamming effect. Beryllium copper and the silicon bronzes are the strongest of the copper alloys. Their strength combined with their ability to work harden can cause difficulties in tapping. For these alloys plug type taps should be used and the taps should be kept as sharp as possible. A medium or heavy duty water soluble oil is recommended as a cutting fluid.

Diameter of Tap Drill. — Tapping troubles are sometimes caused by tap drills that are too small in diameter. The tap drill should not be smaller than is necessary to give the required strength to the thread as even a very small decrease in the diameter of the drill will increase the torque required and the possibility of broken taps. Tests have shown that any increase in the percentage of full thread over 60 per cent does not significantly increase the strength of the thread. In many cases a 55 to 60 per cent thread is satisfactory, although 75 per cent threads are commonly used to provide an extra measure of safety. The present thread specifications do not always allow the use of the smaller thread depths. In all cases the specification given on a part drawing must be adhered to and these may require smaller minor diameters than might otherwise be recommended.

The depth of the thread in the tapped hole is dependent on the length of thread engagement and on the material from which it is made. In general, when the engagement length is more than one and one-half times the nominal diameter a 50 or 55 per cent thread is satisfactory. Soft ductile materials may permit use of a slightly larger tapping hole than brittle materials such as gray cast iron.

It must be remembered that a twist drill is a roughing tool that may be expected to drill slightly oversize and that some variations in the size of the tapping holes are to be expected. When a closer control of the hole size is required it must be reamed. Reaming is recommended for the larger thread diameters and for some fine pitch threads.

For threads of Unified form (see American Standard Unified Threads, page 1474) the selection of tap drills is covered in the following section, Factors Influencing Minor Diameter Tolerances of Tapped Holes and the hole size limits are given in Table 2. Tables 3 and 4 give tap drill sizes for American National Form threads based on 75 per cent of full thread depth. In the case of smaller-size threads the use of slightly larger drills, if permissible, will reduce tap breakage. The selection of tap drills for these threads also may be based on the hole size limits given in Table 2 for unified threads which take into account lengths of engagement.

The size of the tap drill hole for any desired percentage of full thread depth can be calculated by the formulas below. In these formulas the Per Cent Full Thread is

Table 2. Recommended Hole Size Limits Before Tapping Unified Threads

Length of Engagement (D = Nominal Size of Thread) — Recommended Hole Size Limits

Thread Size	Classes 1B and 2B — To and Including 1/3D Min*	Max	Above 1/3D to 2/3D Min	Max	Above 2/3D to 1½D Min	Maxt	Above 1½D to 3D Min	Max	Class 3B — To and Including 1/3D Min*	Max	Above 1/3D to 2/3D Min	Max	Above 2/3D to 1½D Min	Maxt	Above 1½D to 3D Min	Max
0–80	0.0465	0.0500	0.0479	0.0514	0.0479	0.0514	0.0479	0.0514	0.0465	0.0500	0.0479	0.0514	0.0479	0.0514	0.0479	0.0514
1–64	0.0561	0.0599	0.0585	0.0623	0.0585	0.0623	0.0585	0.0623	0.0561	0.0599	0.0585	0.0623	0.0585	0.0623	0.0585	0.0623
1–72	0.0580	0.0613	0.0596	0.0629	0.0602	0.0635	0.0602	0.0635	0.0580	0.0613	0.0596	0.0629	0.0602	0.0635	0.0602	0.0635
2–56	0.0667	0.0705	0.0686	0.0724	0.0699	0.0737	0.0699	0.0737	0.0667	0.0705	0.0686	0.0724	0.0699	0.0737	0.0699	0.0737
2–64	0.0691	0.0724	0.0707	0.0740	0.0720	0.0753	0.0720	0.0753	0.0691	0.0724	0.0707	0.0740	0.0720	0.0753	0.0720	0.0753
3–48	0.0764	0.0804	0.0785	0.0825	0.0805	0.0845	0.0806	0.0846	0.0764	0.0804	0.0785	0.0825	0.0805	0.0845	0.0806	0.0846
3–56	0.0797	0.0831	0.0814	0.0848	0.0831	0.0865	0.0833	0.0867	0.0797	0.0831	0.0814	0.0848	0.0831	0.0865	0.0833	0.0867
4–40	0.0849	0.0894	0.0871	0.0916	0.0894	0.0939	0.0902	0.0947	0.0849	0.0894	0.0871	0.0916	0.0894	0.0939	0.0902	0.0947
4–48	0.0894	0.0931	0.0912	0.0949	0.0931	0.0968	0.0939	0.0976	0.0894	0.0931	0.0912	0.0949	0.0931	0.0968	0.0939	0.0976
5–40	0.0979	0.1020	0.1000	0.1041	0.1021	0.1062	0.1036	0.1077	0.0979	0.1020	0.1000	0.1041	0.1021	0.1062	0.1036	0.1077
5–44	0.1004	0.1042	0.1023	0.1060	0.1042	0.1079	0.1060	0.1097	0.1004	0.1042	0.1023	0.1066	0.1042	0.1079	0.1060	0.1097
6–32	0.104	0.109	0.106	0.112	0.109	0.114	0.112	0.117	0.1040	0.1091	0.1066	0.1115	0.1091	0.1140	0.1115	0.1164
6–40	0.111	0.115	0.113	0.117	0.115	0.119	0.117	0.121	0.1110	0.1148	0.1128	0.1167	0.1147	0.1186	0.1166	0.1205
8–32	0.130	0.134	0.132	0.137	0.134	0.139	0.137	0.141	0.1300	0.1345	0.1324	0.1367	0.1346	0.1389	0.1367	0.1410
8–36	0.134	0.138	0.136	0.140	0.138	0.142	0.140	0.144	0.1340	0.1377	0.1359	0.1397	0.1378	0.1416	0.1397	0.1435
10–24	0.145	0.150	0.148	0.154	0.150	0.156	0.152	0.159	0.1450	0.1502	0.1475	0.1528	0.1502	0.1555	0.1528	0.1581
10–32	0.156	0.160	0.158	0.162	0.160	0.164	0.162	0.166	0.1560	0.1601	0.1581	0.1621	0.1601	0.1641	0.1621	0.1661
12–24	0.171	0.176	0.174	0.179	0.176	0.181	0.178	0.184	0.1710	0.1758	0.1733	0.1782	0.1758	0.1807	0.1782	0.1831
12–28	0.177	0.182	0.179	0.184	0.182	0.186	0.184	0.188	0.1770	0.1815	0.1794	0.1836	0.1815	0.1857	0.1836	0.1878
12–32	0.182	0.186	0.184	0.188	0.186	0.190	0.188	0.192	0.1820	0.1858	0.1837	0.1877	0.1855	0.1895	0.1873	0.1913
1/4–20	0.196	0.202	0.199	0.204	0.202	0.207	0.204	0.210	0.1960	0.2013	0.1986	0.2040	0.2013	0.2067	0.2040	0.2094
1/4–28	0.211	0.216	0.213	0.218	0.216	0.220	0.218	0.222	0.2110	0.2152	0.2172	0.2212	0.2150	0.2190	0.2169	0.2209
1/4–32	0.216	0.220	0.218	0.222	0.220	0.224	0.222	0.226	0.2160	0.2196	0.2199	0.2243	0.2214	0.2258	0.2206	0.2246
5/16–18	0.252	0.259	0.255	0.262	0.259	0.265	0.262	0.268	0.2520	0.2577	0.2551	0.2604	0.2577	0.2630	0.2604	0.2657
5/16–24	0.267	0.272	0.270	0.275	0.272	0.277	0.275	0.280	0.2670	0.2714	0.2694	0.2734	0.2714	0.2754	0.2734	0.2774
5/16–32	0.279	0.283	0.281	0.285	0.283	0.286	0.285	0.289	0.2790	0.2817	0.2792	0.2832	0.2807	0.2847	0.2822	0.2862
3/8–16	0.307	0.314	0.310	0.316	0.313	0.319	0.315	0.321	0.3070	0.3127	0.3101	0.3155	0.3128	0.3182	0.3155	0.3209
3/8–24	0.330	0.335	0.333	0.338	0.335	0.340	0.338	0.343	0.3300	0.3354	0.3314	0.3354	0.3332	0.3372	0.3351	0.3391
3/8–32	0.341	0.345	0.343	0.347	0.345	0.349	0.347	0.351	0.3410	0.3441	0.3415	0.3455	0.3429	0.3469	0.3444	0.3484

36-36	0.345	0.349	0.346	0.350	0.347	0.352	0.349	0.350	0.3450	0.3488	0.3449	0.3488	0.3461	0.3501	0.3474	0.3514
7/16-14	0.360	0.368	0.364	0.372	0.368	0.376	0.372	0.380	0.3600	0.3660	0.3630	0.3688	0.3659	0.3717	0.3688	0.3746
7/16-20	0.383	0.389	0.386	0.391	0.389	0.395	0.391	0.397	0.3830	0.3875	0.3855	0.3896	0.3875	0.3916	0.3896	0.3937
7/16-28	0.399	0.403	0.401	0.406	0.403	0.407	0.406	0.410	0.3990	0.4020	0.3995	0.4035	0.4011	0.4051	0.4017	0.4067
1/2-12	0.417	0.426	0.421	0.430	0.414	0.433	0.424	0.438	0.4170	0.4225	0.4196	0.4254	0.4160	0.4284	0.4255	0.4313
1/2-13	0.410	0.414	0.414	0.424	0.428	0.434	0.452	0.457	0.4100	0.4161	0.4129	0.4192	0.4226	0.4255	0.4192	0.4255
1/2-20	0.446	0.452	0.449	0.454	0.452	0.457	0.454	0.460	0.4460	0.4517	0.4477	0.4517	0.4497	0.4537	0.4516	0.4556
1/2-28	0.461	0.476	0.463	0.468	0.470	0.472	0.468	0.472	0.4610	0.4645	0.4620	0.4660	0.4636	0.4676	0.4652	0.4692
9/16-12	0.472	0.476	0.476	0.486	0.476	0.490	0.486	0.495	0.4720	0.4783	0.4753	0.4813	0.4783	0.4843	0.4813	0.4873
9/16-18	0.502	0.509	0.505	0.512	0.509	0.515	0.512	0.518	0.5020	0.5065	0.5045	0.5086	0.5065	0.5086	0.5086	0.5127
9/16-24	0.517	0.522	0.520	0.525	0.522	0.527	0.525	0.530	0.5170	0.5209	0.5186	0.5226	0.5204	0.5244	0.5221	0.5261
9/16-28	0.524	0.528	0.526	0.531	0.528	0.532	0.531	0.535	0.5240	0.5270	0.5245	0.5285	0.5261	0.5301	0.5277	0.5317
5/8-11	0.527	0.536	0.532	0.541	0.536	0.546	0.541	0.551	0.5270	0.5328	0.5298	0.5360	0.5329	0.5391	0.5360	0.5422
5/8-18	0.535	0.544	0.540	0.549	0.544	0.553	0.549	0.558	0.5350	0.5406	0.5377	0.5435	0.5405	0.5463	0.5434	0.5492
5/8-24	0.565	0.572	0.568	0.575	0.572	0.578	0.575	0.581	0.5650	0.5690	0.5670	0.5711	0.5690	0.5730	0.5711	0.5752
5/8-28	0.580	0.585	0.583	0.588	0.585	0.590	0.588	0.593	0.5800	0.5834	0.5811	0.5851	0.5829	0.5869	0.5846	0.5886
11/16-24	0.586	0.591	0.588	0.593	0.591	0.595	0.593	0.597	0.5800	0.5895	0.5870	0.5910	0.5886	0.5926	0.5902	0.5942
11/16-28	0.597	0.606	0.602	0.611	0.606	0.615	0.611	0.620	0.5970	0.6029	0.6001	0.6057	0.6029	0.6085	0.6057	0.6113
3/4-10	0.642	0.647	0.645	0.650	0.647	0.652	0.650	0.655	0.6420	0.6459	0.6436	0.6476	0.6454	0.6494	0.6471	0.6511
3/4-16	0.660	0.653	0.647	0.658	0.653	0.663	0.658	0.668	0.6600	0.6481	0.6449	0.6513	0.6481	0.6545	0.6513	0.6577
3/4-20	0.682	0.689	0.686	0.674	0.669	0.678	0.674	0.683	0.6820	0.6652	0.6626	0.6680	0.6653	0.6707	0.6680	0.6734
3/4-28	0.696	0.702	0.699	0.702	0.689	0.696	0.693	0.700	0.6900	0.6866	0.6844	0.6887	0.6865	0.6908	0.6886	0.6929
13/16-16	0.711	0.716	0.713	0.718	0.702	0.707	0.704	0.710	0.6960	0.6998	0.6977	0.7017	0.6997	0.7037	0.7016	0.7056
13/16-20	0.722	0.731	0.727	0.736	0.716	0.720	0.718	0.722	0.7110	0.7145	0.7120	0.7160	0.7136	0.7176	0.7152	0.7192
7/8-9	0.745	0.752	0.749	0.756	0.731	0.740	0.736	0.745	0.7220	0.7276	0.7259	0.7303	0.7276	0.7329	0.7303	0.7356
7/8-12	0.758	0.764	0.761	0.766	0.752	0.759	0.756	0.763	0.7450	0.7491	0.7459	0.7512	0.7490	0.7533	0.7511	0.7554
7/8-16	0.765	0.767	0.765	0.773	0.764	0.770	0.766	0.772	0.7580	0.7623	0.7602	0.7642	0.7622	0.7662	0.7641	0.7681
7/8-28	0.785	0.794	0.790	0.799	0.767	0.778	0.773	0.785	0.7550	0.7614	0.7580	0.7647	0.7614	0.7681	0.7647	0.7714
15/16-12	0.798	0.806	0.802	0.810	0.794	0.803	0.799	0.808	0.7850	0.7900	0.7874	0.7926	0.7900	0.7952	0.7926	0.7978
15/16-16	0.807	0.814	0.811	0.818	0.806	0.814	0.810	0.818	0.7980	0.8022	0.8000	0.8045	0.8023	0.8068	0.8045	0.8090
1-8	0.821	0.827	0.824	0.829	0.814	0.821	0.818	0.825	0.8070	0.8116	0.8094	0.8137	0.8115	0.8158	0.8136	0.8179
1-12	0.836	0.840	0.838	0.843	0.827	0.832	0.829	0.835	0.8210	0.8248	0.8227	0.8267	0.8247	0.8287	0.8266	0.8306
1-14	0.847	0.856	0.852	0.861	0.840	0.845	0.843	0.847	0.8360	0.8395	0.8370	0.8410	0.8386	0.8426	0.8402	0.8442
1-16	0.865	0.870	0.874	0.881	0.856	0.865	0.861	0.870	0.8470	0.8524	0.8499	0.8550	0.8524	0.8575	0.8550	0.8601
1-20	0.883	0.889	0.886	0.891	0.877	0.884	0.881	0.888	0.8700	0.8741	0.8719	0.8759	0.8740	0.8783	0.8761	0.8804
1-28	0.910	0.919	0.915	0.924	0.889	0.896	0.891	0.897	0.8830	0.8873	0.8852	0.8892	0.8872	0.8912	0.8890	0.8931

*† See footnotes at end of table.

Table 2 (Continued). Recommended Hole Size Limits Before Tapping Unified Threads

Thread Size	Classes 1B and 2B								Class 3B							
	To and including ⅓D		Above ⅓D to ⅔D		Above ⅔D to 1½D		Above 1½D to 3D		To and Including ⅓D		Above ⅓D to ⅔D		Above ⅔D to 1½D		Above 1½D to 3D	
	Min*	Max	Min	Max	Min	Maxt	Min	Max	Min*	Max	Min	Max	Min	Maxt	Min	Max
$1\frac{1}{16}$-12	0.972	0.981	0.977	0.986	0.981	0.990	0.986	0.995	0.9720	0.9773	0.9748	0.9798	0.9773	0.9823	0.9798	0.9848
$1\frac{1}{16}$-16	0.995	1.002	0.999	1.005	1.002	1.009	1.005	1.013	0.9950	0.9991	0.9969	1.0012	0.9990	1.0033	1.0011	1.0054
$1\frac{1}{16}$-18	1.002	1.009	1.005	1.012	1.009	1.015	1.012	1.018	1.0020	1.0065	1.0044	1.0085	1.0064	1.0105	1.0085	1.0126
$1\frac{1}{8}$-7	0.970	0.984	0.977	0.991	0.984	0.998	0.991	1.005	0.9700	0.9790	0.9747	0.9833	0.9789	0.9875	0.9832	0.9918
$1\frac{1}{8}$-12	0.990	1.003	0.996	1.009	1.003	1.015	1.009	1.021	0.9900	0.9972	0.9934	1.0009	0.9972	1.0047	1.0010	1.0085
$1\frac{1}{8}$-18	1.035	1.044	1.040	1.049	1.044	1.053	1.049	1.058	1.0350	1.0398	1.0373	1.0423	1.0398	1.0448	1.0423	1.0473
$1\frac{3}{16}$-12	1.057	1.064	1.061	1.068	1.064	1.071	1.068	1.075	1.0570	1.0616	1.0594	1.0637	1.0615	1.0658	1.0636	1.0679
$1\frac{3}{16}$-18	1.065	1.072	1.068	1.075	1.072	1.078	1.075	1.081	1.0650	1.0690	1.0669	1.0710	1.0689	1.0730	1.0710	1.0751
$1\frac{3}{16}$-28	1.071	1.077	1.074	1.079	1.077	1.082	1.081	1.085	1.0710	1.0748	1.0727	1.0767	1.0747	1.0787	1.0766	1.0806
$1\frac{1}{4}$-7	1.086	1.091	1.088	1.093	1.091	1.095	1.093	1.097	1.0860	1.0895	1.0870	1.0910	1.0886	1.0926	1.0902	1.0942
$1\frac{1}{4}$-8	1.097	1.106	1.102	1.111	1.106	1.115	1.111	1.120	1.0970	1.1023	1.0998	1.1048	1.1023	1.1073	1.1048	1.1098
$1\frac{1}{4}$-12	1.120	1.127	1.124	1.131	1.127	1.134	1.131	1.138	1.1200	1.1241	1.1219	1.1262	1.1240	1.1283	1.1261	1.1304
$1\frac{1}{4}$-16	1.127	1.134	1.130	1.137	1.134	1.140	1.137	1.143	1.1270	1.1315	1.1294	1.1335	1.1314	1.1355	1.1335	1.1376
$1\frac{1}{4}$-18	1.095	1.109	1.102	1.116	1.109	1.123	1.116	1.130	1.0959	1.1040	1.0997	1.1083	1.1039	1.1125	1.1082	1.1168
$1\frac{1}{4}$-20	1.115	1.128	1.121	1.134	1.128	1.140	1.134	1.146	1.1150	1.1222	1.1184	1.1259	1.1222	1.1297	1.1260	1.1335
$1\frac{5}{16}$-12	1.160	1.169	1.165	1.174	1.169	1.178	1.174	1.183	1.1600	1.1648	1.1623	1.1673	1.1648	1.1698	1.1673	1.1723
$1\frac{5}{16}$-16	1.182	1.189	1.186	1.193	1.189	1.196	1.193	1.200	1.1820	1.1866	1.1844	1.1887	1.1865	1.1968	1.1886	1.1929
$1\frac{5}{16}$-18	1.190	1.197	1.193	1.200	1.197	1.203	1.200	1.206	1.1900	1.1940	1.1919	1.1960	1.1939	1.1980	1.1960	1.2001
$1\frac{3}{8}$-6	1.196	1.202	1.199	1.207	1.203	1.207	1.206	1.210	1.1960	1.1998	1.1977	1.2017	1.1997	1.2037	1.2016	1.2056
$1\frac{3}{8}$-8	1.222	1.231	1.227	1.236	1.231	1.240	1.236	1.245	1.2220	1.2273	1.2248	1.2298	1.2273	1.2323	1.2298	1.2348
$1\frac{3}{8}$-12	1.245	1.252	1.249	1.256	1.252	1.259	1.256	1.263	1.2450	1.2491	1.2469	1.2512	1.2490	1.2533	1.2511	1.2554
$1\frac{3}{8}$-16	1.252	1.259	1.256	1.262	1.259	1.265	1.262	1.268	1.2520	1.2565	1.2544	1.2585	1.2564	1.2605	1.2585	1.2626
$1\frac{3}{8}$-18	1.195	1.203	1.203	1.221	1.210	1.225	1.221	1.239	1.1950	1.2046	1.1996	1.2096	1.2046	1.2146	1.2096	1.2196
$1\frac{7}{16}$-6	1.240	1.246	1.246	1.259	1.253	1.265	1.259	1.271	1.2400	1.2472	1.2434	1.2559	1.2472	1.2547	1.2510	1.2585
$1\frac{7}{16}$-12	1.285	1.290	1.290	1.299	1.294	1.303	1.299	1.308	1.2850	1.2898	1.2873	1.2923	1.2898	1.2948	1.2923	1.2973
$1\frac{7}{16}$-16	1.307	1.311	1.311	1.318	1.314	1.321	1.318	1.325	1.3070	1.3116	1.3094	1.3137	1.3115	1.3158	1.3136	1.3179
$1\frac{7}{16}$-18	1.315	1.318	1.318	1.325	1.322	1.328	1.325	1.331	1.3150	1.3190	1.3169	1.3210	1.3189	1.3230	1.3210	1.3251
$1\frac{1}{2}$-6	1.347	1.350	1.350	1.361	1.354	1.365	1.361	1.370	1.3470	1.3523	1.3498	1.3548	1.3523	1.3573	1.3548	1.3598
$1\frac{1}{2}$-8	1.370	1.374	1.374	1.381	1.377	1.384	1.381	1.388	1.3700	1.3741	1.3719	1.3762	1.3740	1.3783	1.3761	1.3804
$1\frac{1}{2}$-12	1.320	1.328	1.328	1.346	1.335	1.350	1.346	1.364	1.3200	1.3296	1.3246	1.3340	1.3296	1.3396	1.3346	1.3446
$1\frac{1}{2}$-16	1.365	1.371	1.371	1.384	1.378	1.390	1.384	1.396	1.3650	1.3722	1.3684	1.3759	1.3722	1.3797	1.3760	1.3835

Length of Engagement (D = Nominal Size of Thread)

Recommended Hole Size Limits

1½-12	1.410	1.419	1.415	1.424	1.419	1.428	1.424	1.433	1.4100	1.4148	1.4123	1.4173	1.4148	1.4198	1.4173	1.4223
1½-16	1.432	1.439	1.436	1.443	1.439	1.446	1.443	1.450	1.4320	1.4366	1.4344	1.4387	1.4365	1.4408	1.4386	1.4429
1½-18	1.440	1.446	1.443	1.450	1.446	1.452	1.450	1.456	1.4400	1.4440	1.4419	1.4460	1.4439	1.4480	1.4460	1.4501
1½-20	1.446	1.452	1.449	1.454	1.452	1.457	1.454	1.460	1.4460	1.4498	1.4477	1.4517	1.4497	1.4537	1.4516	1.4556
1 9/16-16	1.495	1.502	1.499	1.506	1.502	1.509	1.506	1.513	1.4950	1.4991	1.4969	1.5012	1.4990	1.5033	1.5011	1.5054
1 9/16-18	1.502	1.509	1.505	1.512	1.509	1.515	1.512	1.518	1.5020	1.5065	1.5044	1.5085	1.5064	1.5105	1.5085	1.5126
1⅝-8	1.490	1.498	1.494	1.509	1.498	1.515	1.509	1.521	1.4900	1.4972	1.4934	1.5009	1.4972	1.5047	1.5010	1.5085
1⅝-12	1.535	1.544	1.540	1.549	1.544	1.553	1.549	1.558	1.5350	1.5398	1.5373	1.5423	1.5398	1.5448	1.5423	1.5473
1⅝-16	1.557	1.564	1.561	1.568	1.564	1.571	1.568	1.575	1.5570	1.5616	1.5594	1.5637	1.5615	1.5658	1.5636	1.5679
1⅝-18	1.565	1.572	1.568	1.575	1.572	1.578	1.575	1.581	1.5650	1.5690	1.5669	1.5710	1.5689	1.5730	1.5710	1.5751
1 11/16-16	1.620	1.627	1.624	1.631	1.627	1.634	1.631	1.638	1.6200	1.6241	1.6219	1.6262	1.6240	1.6283	1.6261	1.6304
1 11/16-18	1.627	1.634	1.630	1.637	1.634	1.640	1.637	1.643	1.6270	1.6315	1.6294	1.6335	1.6314	1.6355	1.6335	1.6376
1¾-5	1.534	1.551	1.543	1.560	1.551	1.568	1.560	1.577	1.5340	1.5455	1.5395	1.5515	1.5455	1.5575	1.5515	1.5635
1¾-8	1.615	1.628	1.621	1.634	1.628	1.640	1.634	1.646	1.6150	1.6222	1.6184	1.6259	1.6222	1.6297	1.6260	1.6335
1¾-12	1.660	1.669	1.665	1.674	1.669	1.678	1.674	1.683	1.6600	1.6648	1.6623	1.6673	1.6648	1.6698	1.6673	1.6723
1¾-16	1.682	1.689	1.686	1.693	1.689	1.696	1.693	1.700	1.6820	1.6866	1.6844	1.6887	1.6865	1.6908	1.6886	1.6929
1¾-20	1.696	1.702	1.699	1.704	1.702	1.707	1.704	1.710	1.6960	1.6998	1.6977	1.7017	1.6997	1.7037	1.7016	1.7056
1 13/16-16	1.745	1.752	1.749	1.756	1.752	1.759	1.756	1.763	1.7450	1.7491	1.7469	1.7512	1.7490	1.7533	1.7511	1.7554
1⅞-8	1.740	1.752	1.746	1.759	1.752	1.765	1.759	1.771	1.7400	1.7472	1.7434	1.7509	1.7472	1.7547	1.7510	1.7585
1⅞-12	1.785	1.794	1.790	1.799	1.794	1.803	1.799	1.808	1.7850	1.7898	1.7873	1.7923	1.7898	1.7948	1.7923	1.7973
1⅞-16	1.807	1.814	1.810	1.818	1.814	1.821	1.818	1.825	1.8070	1.8116	1.8094	1.8137	1.8115	1.8158	1.8136	1.1879
1 15/16-16	1.870	1.877	1.874	1.881	1.877	1.884	1.881	1.888	1.8700	1.8741	1.8719	1.8762	1.8740	1.8783	1.8761	1.8804
2-4½	1.759	1.777	1.768	1.786	1.777	1.795	1.786	1.804	1.7590	1.7727	1.7661	1.7794	1.7728	1.7861	1.7794	1.7927
2-8	1.865	1.878	1.871	1.884	1.878	1.890	1.884	1.896	1.8650	1.8722	1.8684	1.8759	1.8722	1.8797	1.8760	1.8835
2-12	1.910	1.919	1.915	1.924	1.919	1.928	1.924	1.933	1.9100	1.9148	1.9123	1.9173	1.9148	1.9198	1.9173	1.9223
2-16	1.932	1.939	1.936	1.943	1.939	1.946	1.943	1.950	1.9320	1.9366	1.9344	1.9387	1.9365	1.9408	1.9386	1.9429
2-20	1.946	1.952	1.949	1.954	1.952	1.957	1.954	1.960	1.9460	1.9498	1.9477	1.9517	1.9497	1.9537	1.9516	1.9556
2 1/16-16	1.995	2.002	2.000	2.006	2.002	2.009	2.006	2.012	1.9950	1.9991	1.9969	2.0012	1.9990	2.0033	2.0011	2.0054
2⅛-8	1.990	2.003	1.996	2.009	2.003	2.015	2.009	2.021	1.9900	1.9972	1.9934	2.0009	1.9972	2.0047	2.0010	2.0085
2⅛-12	2.035	2.044	2.040	2.049	2.044	2.053	2.049	2.058	2.0350	2.0398	2.0373	2.0423	2.0398	2.0448	2.0423	2.0473
2⅛-16	2.057	2.064	2.061	2.068	2.064	2.071	2.068	2.075	2.0570	2.0616	2.0594	2.0637	2.0615	2.0658	2.0636	2.0679
2 3/16-16	2.120	2.127	2.124	2.131	2.127	2.134	2.131	2.138	2.1200	2.1241	2.1219	2.1262	2.1240	2.1283	2.1261	2.1304
2¼-4½	2.009	2.027	2.018	2.036	2.027	2.045	2.036	2.054	2.0090	2.0227	2.0161	2.0294	2.0228	2.0361	2.0294	2.0427
2¼-8	2.115	2.128	2.121	2.134	2.128	2.140	2.134	2.146	2.1150	2.1222	2.1184	2.1259	2.1222	2.1297	2.1260	2.1335
2¼-12	2.160	2.169	2.165	2.174	2.169	2.178	2.174	2.182	2.1600	2.1648	2.1623	2.1673	2.1648	2.1698	2.1673	2.1723
2¼-16	2.182	2.189	2.186	2.193	2.189	2.196	2.193	2.200	2.1820	2.1866	2.1844	2.1887	2.1865	2.1908	2.1886	2.1929
2¼-20	2.196	2.202	2.199	2.204	2.202	2.207	2.204	2.210	2.1960	2.1998	2.1977	2.2017	2.1997	2.2037	2.2016	2.2056
2 5/16-16	2.245	2.252	2.249	2.256	2.252	2.259	2.256	2.263	2.2450	2.2491	2.2469	2.2512	2.2490	2.2533	2.2511	2.2554
2⅜-12	2.285	2.294	2.290	2.299	2.294	2.303	2.299	2.308	2.2850	2.2898	2.2873	2.2923	2.2898	2.2948	2.2923	2.2973
2⅜-16	2.307	2.314	2.311	2.318	2.314	2.321	2.318	2.325	2.3070	2.3116	2.3094	2.3137	2.3115	2.3158	2.3136	2.3179
2 7/16-16	2.370	2.377	2.374	2.381	2.377	2.384	2.381	2.388	2.3700	2.3741	2.3719	2.3762	2.3740	2.3783	2.3761	2.3804
2½-4	2.229	2.248	2.238	2.258	2.248	2.267	2.258	2.277	2.2290	2.2444	2.2369	2.2519	2.2444	2.2594	2.2519	2.2669

*† See footnotes at end of table.

Table 2 (*Continued*). **Recommended Hole Size Limits Before Tapping Unified Threads**

Thread Size	Classes 1B and 2B								Class 3B							
	Length of Engagement (D = Nominal Size of Thread)															
	To and Including ⅓D		Above ⅓D to ⅔D		Above ⅔D to 1½D		Above 1½D to 3D		To and Including ⅓D		Above ⅓D to ⅔D		Above ⅔D to 1½D		Above 1½D to 3D	
	Recommended Hole Size Limits															
	Min*	Max	Min	Max	Min	Max†	Min	Max	Min*	Max	Min	Max	Min	Max†	Min	Max
2½–8	2.365	2.378	2.371	2.384	2.378	2.390	2.384	2.396	2.3650	2.3722	2.3684	2.3759	2.3722	2.3797	2.3760	2.3835
2½–12	2.410	2.419	2.415	2.424	2.419	2.428	2.424	2.433	2.4100	2.4148	2.4123	2.4173	2.4148	2.4198	2.4173	2.4223
2½–16	2.432	2.439	2.436	2.443	2.439	2.446	2.443	2.450	2.4320	2.4366	2.4344	2.4387	2.4365	2.4408	2.4386	2.4429
2½–20	2.446	2.452	2.449	2.454	2.452	2.457	2.454	2.460	2.4460	2.4498	2.4478	2.4517	2.4497	2.4537	2.4516	2.4556
2⅝–12	2.535	2.544	2.540	2.549	2.544	2.553	2.549	2.558	2.5350	2.5398	2.5373	2.5423	2.5398	2.5448	2.5423	2.5473
2⅝–16	2.557	2.564	2.561	2.568	2.564	2.571	2.568	2.575	2.5570	2.5616	2.5594	2.5637	2.5615	2.5658	2.5636	2.5679
2¾–4	2.479	2.498	2.489	2.508	2.498	2.517	2.508	2.527	2.4790	2.4944	2.4869	2.5019	2.4944	2.5094	2.5019	2.5169
2¾–8	2.615	2.628	2.621	2.634	2.628	2.640	2.634	2.644	2.6150	2.6222	2.6184	2.6259	2.6222	2.6297	2.6260	2.6335
2¾–12	2.660	2.669	2.665	2.674	2.669	2.678	2.674	2.683	2.6600	2.6648	2.6623	2.6673	2.6648	2.6698	2.6673	2.6723
2¾–16	2.682	2.689	2.686	2.693	2.689	2.696	2.693	2.700	2.6820	2.6866	2.6844	2.6887	2.6865	2.6908	2.6886	2.6929
2⅞–12	2.785	2.794	2.790	2.809	2.794	2.803	2.809	2.808	2.7850	2.7898	2.7873	2.7923	2.7898	2.7948	2.7923	2.7973
2⅞–16	2.807	2.814	2.811	2.818	2.814	2.821	2.818	2.825	2.8070	2.8116	2.8094	2.8137	2.8115	2.8158	2.8136	2.8179
3–4	2.729	2.748	2.739	2.758	2.748	2.767	2.758	2.777	2.7290	2.7444	2.7369	2.7519	2.7444	2.7594	2.7519	2.7669
3–8	2.865	2.878	2.871	2.884	2.878	2.890	2.884	2.896	2.8650	2.8722	2.8684	2.8759	2.8722	2.8797	2.8760	2.8835
3–12	2.910	2.919	2.915	2.924	2.919	2.928	2.924	2.933	2.9100	2.9148	2.9123	2.9173	2.9148	2.9198	2.9173	2.9223
3–16	2.932	2.939	2.936	2.943	2.939	2.946	2.943	2.950	2.9320	2.9366	2.9344	2.9387	2.9365	2.9408	2.9386	2.9429
3⅛–12	3.035	3.044	3.040	3.049	3.044	3.053	3.049	3.058	3.0350	3.0398	3.0373	3.0423	3.0398	3.0448	3.0423	3.0473
3⅛–16	3.057	3.064	3.061	3.068	3.064	3.071	3.068	3.075	3.0570	3.0616	3.0594	3.0637	3.0615	3.0658	3.0636	3.0679
3¼–4	2.979	2.998	2.989	3.008	2.998	3.017	3.008	3.027	2.9790	2.9944	2.9869	3.0019	2.9944	3.0094	3.0019	3.0169
3¼–8	3.115	3.128	3.121	3.134	3.128	3.140	3.134	3.146	3.1150	3.1222	3.1184	3.1259	3.1222	3.1297	3.1260	3.1335
3¼–12	3.160	3.169	3.165	3.174	3.169	3.178	3.174	3.183	3.1600	3.1648	3.1623	3.1673	3.1648	3.1698	3.1673	3.1723
3¼–16	3.182	3.189	3.186	3.193	3.189	3.196	3.193	3.200	3.1820	3.1866	3.1844	3.1887	3.1865	3.1908	3.1886	3.1929
3⅜–12	3.285	3.294	3.290	3.299	3.294	3.303	3.299	3.299	3.2850	3.2898	3.2873	3.2923	3.2898	3.2948	3.2923	3.2973
3⅜–16	3.307	3.314	3.311	3.318	3.314	3.321	3.317	3.325	3.3070	3.3116	3.3094	3.3137	3.3115	3.3158	3.3136	3.3179
3½–4	3.229	3.248	3.239	3.258	3.248	3.267	3.258	3.277	3.2290	3.2444	3.2369	3.2519	3.2444	3.2594	3.2519	3.2669
3½–8	3.365	3.378	3.371	2.384	3.378	3.390	3.384	3.396	3.3650	3.3722	3.3684	3.3759	3.3722	3.3797	3.3760	3.3835
3½–12	3.410	3.419	3.415	3.424	3.419	3.428	3.424	3.433	3.4100	3.4148	3.4123	3.4173	3.4148	3.4198	3.4173	3.4223
3½–16	3.432	3.439	3.436	3.443	3.439	3.446	3.443	3.450	3.4320	3.4366	3.4344	3.4387	3.4365	3.4408	3.4386	3.4429
3⅝–12	3.535	3.544	3.544	3.549	3.544	3.553	3.549	3.553	3.5350	3.5398	3.5373	3.5423	3.5398	3.5448	3.5423	3.5473
3⅝–16	3.557	3.564	3.561	3.568	3.567	3.571	3.568	3.575	3.5570	3.5616	3.5594	3.5637	3.5615	3.5658	3.5636	3.5679
3¾–4	3.479	3.498	3.489	3.508	3.498	3.517	3.508	3.527	3.4790	3.4944	3.4869	3.5019	3.4944	3.5094	3.5019	3.5169
3¾–8	3.615	3.628	3.615	3.634	3.628	3.640	3.634	3.646	3.6150	3.6222	3.6184	3.6259	3.6222	3.6297	3.6260	3.6335

All dimensions are in inches.

For basis of recommended hole size limits see accompanying text.

As an aid in selecting suitable drills, see the listing of American Standard drill sizes in the twist drill section. For amount of expected drill over-size, see page 846.

* This is the minimum minor diameter specified in the thread tables, page 1498.

† This is the maximum minor diameter specified in the thread tables, page 1498.

Size	1	2	3	4	5	6	7	8	9	10	11	12	13	14	15	16
3¾-12	3.6723	3.6673	3.6698	3.6648	3.6673	3.6623	3.6648	3.6600	3.683	3.674	3.678	3.669	3.674	3.665	3.669	3.660
3¾-16	3.6929	3.6886	3.6908	3.6865	3.6887	3.6844	3.6866	3.6820	3.700	3.693	3.696	3.689	3.693	3.686	3.689	3.682
3⅞-12	3.7973	3.7923	3.7948	3.7898	3.7923	3.7873	3.7898	3.7850	3.808	3.799	3.803	3.794	3.799	3.790	3.794	3.785
3⅞-16	3.8179	3.8136	3.8158	3.8115	3.8137	3.8094	3.8116	3.8070	3.825	3.818	3.821	3.814	3.818	3.811	3.814	3.807
4-4	3.7669	3.7519	3.7594	3.7444	3.7519	3.7369	3.7444	3.7290	3.777	3.758	3.767	3.748	3.758	3.739	3.748	3.729
4-8	3.8835	3.8760	3.8797	3.8722	3.8759	3.8684	3.8722	3.8650	3.896	3.884	3.890	3.878	3.884	3.871	3.878	3.865
4-12	3.9223	3.9173	3.9198	3.9148	3.9173	3.9123	3.9148	3.9100	3.933	3.924	3.928	3.919	3.924	3.915	3.919	3.910
4-16	3.9429	3.9386	3.9408	3.9365	3.9387	3.9344	3.9366	3.9320	3.950	3.943	3.946	3.939	3.943	3.936	3.939	3.932
4¼-6	4.0169	4.0019	4.0094	3.9944	4.0019	3.9869	3.9944	3.9790	4.027	4.008	4.017	3.998	4.008	3.989	3.998	3.979
4¼-8	4.1335	4.1260	4.1297	4.1222	4.1259	4.1184	4.1222	4.1150	4.146	4.134	4.140	4.128	4.134	4.121	4.128	4.115
4¼-12	4.1723	4.1673	4.1698	4.1648	4.1673	4.1623	4.1648	4.1600	4.183	4.174	4.178	4.169	4.174	4.165	4.169	4.160
4¼-16	4.1929	4.1886	4.1908	4.1865	4.1887	4.1844	4.1866	4.1820	4.200	4.193	4.196	4.189	4.193	4.186	4.189	4.182
4½-6	4.2669	4.2519	4.2594	4.2444	4.2519	4.2369	4.2444	4.2290	4.277	4.258	4.267	4.248	4.258	4.239	4.248	4.229
4½-8	4.3835	4.3760	4.3797	4.3722	4.3759	4.3684	4.3722	4.3650	4.396	4.384	4.390	4.378	4.384	4.371	4.378	4.365
4½-12	4.4223	4.4173	4.4198	4.4148	4.4173	4.4123	4.4148	4.4100	4.433	4.424	4.428	4.419	4.424	4.415	4.419	4.410
4½-16	4.4429	4.4386	4.4408	4.4365	4.4387	4.4344	4.4366	4.4320	4.450	4.443	4.446	4.439	4.443	4.436	4.439	4.432
4¾-6	4.5169	4.5019	4.5094	4.4944	4.5019	4.4869	4.4944	4.4790	4.527	4.508	4.517	4.498	4.508	4.489	4.498	4.479
4¾-8	4.6335	4.6260	4.6297	4.6222	4.6259	4.6184	4.6222	4.6150	4.646	4.634	4.640	4.628	4.634	4.621	4.628	4.615
4¾-12	4.6723	4.6673	4.6698	4.6648	4.6673	4.6623	4.6648	4.6600	4.683	4.674	4.678	4.669	4.674	4.665	4.669	4.660
4¾-16	4.6929	4.6886	4.6908	4.6865	4.6887	4.6844	4.6866	4.6820	4.700	4.693	4.696	4.689	4.693	4.686	4.689	4.682
5-8	4.8835	4.8760	4.8797	4.8722	4.8759	4.8684	4.8722	4.8650	4.896	4.884	4.890	4.878	4.884	4.871	4.878	4.865
5-12	4.9223	4.9173	4.9198	4.9148	4.9173	4.9123	4.9148	4.9100	4.933	4.924	4.928	4.919	4.924	4.915	4.919	4.910
5-16	4.9429	4.9386	4.9408	4.9365	4.9387	4.9344	4.9366	4.9320	4.950	4.943	4.946	4.939	4.943	4.936	4.939	4.932
5¼-6	5.0169	5.0019	5.0094	4.9944	5.0019	4.9869	4.9944	4.9790	5.027	5.008	5.017	4.998	5.008	4.989	4.998	4.979
5¼-8	5.1335	5.1260	5.1297	5.1222	5.1259	5.1184	5.1222	5.1150	5.146	5.134	5.140	5.128	5.134	5.121	5.128	5.115
5¼-12	5.1723	5.1673	5.1698	5.1648	5.1673	5.1623	5.1648	5.1600	5.183	5.174	5.178	5.169	5.174	5.165	5.169	5.160
5¼-16	5.1929	5.1886	5.1908	5.1865	5.1887	5.1844	5.1866	5.1820	5.200	5.193	5.196	5.189	5.193	5.186	5.189	5.182
5½-8	5.3835	5.3760	5.3797	5.3722	5.3759	5.3684	5.3722	5.3650	5.396	5.384	5.390	5.378	5.384	5.371	5.378	5.365
5½-12	5.4223	5.4173	5.4198	5.4148	5.4173	5.4123	5.4148	5.4100	5.433	5.424	5.428	5.419	5.424	5.415	5.419	5.410
5½-16	5.4429	5.4386	5.4408	5.4365	5.4387	5.4344	5.4366	5.4320	5.450	5.443	5.446	5.439	5.443	5.436	5.439	5.432
5¾-8	5.6335	5.6260	5.6297	5.6222	5.6259	5.6184	5.6222	5.6150	5.646	5.634	5.640	5.628	5.634	5.621	5.628	5.615
5¾-12	5.6723	5.6673	5.6698	5.6648	5.6673	5.6623	5.6648	5.6600	5.683	5.674	5.678	5.669	5.674	5.665	5.669	5.660
5¾-16	5.6929	5.6886	5.6908	5.6865	5.6887	5.6844	5.6866	5.6820	5.700	5.693	5.696	5.689	5.693	5.686	5.689	5.682
6-8	5.8835	5.8760	5.8797	5.8722	5.8759	5.8684	5.8722	5.8650	5.896	5.884	5.890	5.878	5.884	5.871	5.878	5.865
6-12	5.9223	5.9173	5.9198	5.9148	5.9173	5.9123	5.9148	5.9100	5.933	5.924	5.928	5.919	5.924	5.915	5.919	5.910
6-16	5.9429	5.9386	5.9408	5.9365	5.9387	5.9344	5.9366	5.9320	5.950	5.943	5.946	5.939	5.943	5.936	5.939	5.932

Table 3. Tap Drill Sizes for Threads of American National Form

Screw Thread Outside Diam. Pitch	Root Diam.	Commercial Tap Drills* Size or Number	Decimal Equiv.	Screw Thread Outside Diam. Pitch	Root Diam.	Commercial Tap Drills* Size or Number	Decimal Equiv.
1/16-64	0.0422	3/64	0.0469	27	0.4519	15/32	0.4687
72	0.0445	3/64	0.0469	9/16-12	0.4542	31/64	0.4844
5/64-60	0.0563	1/16	0.0625	18	0.4903	33/64	0.5156
72	0.0601	52	0.0635	27	0.5144	17/32	0.5312
3/32-48	0.0667	49	0.0730	5/8-11	0.5069	17/32	0.5312
50	0.0678	49	0.0730	12	0.5168	35/64	0.5469
7/64-48	0.0823	43	0.0890	18	0.5528	37/64	0.5781
1/8-32	0.0844	3/32	0.0937	27	0.5769	19/32	0.5937
40	0.0925	38	0.1015	11/16-11	0.5694	19/32	0.5937
9/64-40	0.1081	32	0.1160	16	0.6063	5/8	0.6250
5/32-32	0.1157	1/8	0.1250	3/4-10	0.6201	21/32	0.6562
36	0.1202	30	0.1285	12	0.6418	43/64	0.6719
11/64-32	0.1313	9/64	0.1406	16	0.6688	11/16	0.6875
3/16-24	0.1334	26	0.1470	27	0.7019	23/32	0.7187
32	0.1469	22	0.1570	13/16-10	0.6826	23/32	0.7187
13/64-24	0.1490	20	0.1610	7/8-9	0.7307	49/64	0.7656
7/32-24	0.1646	16	0.1770	12	0.7668	51/64	0.7969
32	0.1782	12	0.1890	14	0.7822	13/16	0.8125
15/64-24	0.1806	10	0.1935	18	0.8028	53/64	0.8281
1/4-20	0.1850	7	0.2010	27	0.8269	27/32	0.8437
24	0.1959	4	0.2090	15/16-9	0.7932	53/64	0.8281
27	0.2019	3	0.2130	1-8	0.8376	7/8	0.8750
28	0.2036	3	0.2130	12	0.8918	59/64	0.9219
32	0.2094	7/32	0.2187	14	0.9072	15/16	0.9375
5/16-18	0.2403	F	0.2570	27	0.9519	31/32	0.9687
20	0.2476	17/64	0.2656	1 1/8-7	0.9394	63/64	0.9844
24	0.2584	I	0.2720	12	1.0168	1 3/64	1.0469
27	0.2644	J	0.2770	1 1/4-7	1.0644	1 7/64	1.1094
32	0.2719	9/32	0.2812	12	1.1418	1 11/64	1.1719
3/8-16	0.2938	5/16	0.3125	1 3/8-6	1.1585	1 7/32	1.2187
20	0.3100	21/64	0.3281	12	1.2668	1 19/64	1.2969
24	0.3209	Q	0.3320	1 1/2-6	1.2835	1 11/32	1.3437
27	0.3269	R	0.3390	12	1.3918	1 27/64	1.4219
7/16-14	0.3447	U	0.3680	1 5/8-5 1/2	1.3888	1 29/64	1.4531
20	0.3726	25/64	0.3906	1 3/4-5	1.4902	1 9/16	1.5625
24	0.3834	X	0.3970	1 7/8-5	1.6152	1 11/16	1.6875
27	0.3894	Y	0.4040	2 -4 1/2	1.7113	1 25/32	1.7812
1/2-12	0.3918	27/64	0.4219	2 1/8-4 1/2	1.8363	1 29/32	1.9062
13	0.4001	27/64	0.4219	2 1/4-4 1/2	1.9613	2 1/32	2.0312
20	0.4351	29/64	0.4531	2 3/8-4	2.0502	2 1/8	2.1250
24	0.4459	29/64	0.4531	2 1/2-4	2.1752	2 1/4	2.2500

* These tap drill diameters allow approximately 75 per cent of a full thread. For small thread sizes the use of larger drills will reduce tap breakage.

Table 4.   Tap Drills and Clearance Drills for Machine Screws with American National Thread Form

Size of Screw		No. of Threads per Inch	Tap Drills		Clearance Hole Drills			
					Close Fit		Free Fit	
No. or Diam.	Decimal Equiv.		Drill Size	Decimal Equiv.	Drill Size	Decimal Equiv.	Drill Size	Decimal Equiv.
0	.060	80	³⁄₆₄	.0469	52	.0635	50	.0700
1	.073	64 72	53 53	.0595 .0595	48	.0760	46	.0810
2	.086	56 64	50 50	.0700 .0700	43	.0890	41	.0960
3	.099	48 56	47 45	.0785 .0820	37	.1040	35	.1100
4	.112	36* 40 48	44 43 42	.0860 .0890 .0935	32	.1160	30	.1285
5	.125	40 44	38 37	.1015 .1040	30	.1285	29	.1360
6	.138	32 40	36 33	.1065 .1130	27	.1440	25	.1495
8	.164	32 36	29 29	.1360 .1360	18	.1695	16	.1770
10	.190	24 32	25 21	.1495 .1590	9	.1960	7	.2010
12	.216	24 28	16 14	.1770 .1820	2	.2210	1	.2280
14	.242	20* 24*	10 7	.1935 .2010	D	.2460	F	.2570
¼	.250	20 28	7 3	.2010 .2130	F	.2570	H	.2660
⁵⁄₁₆	.3125	18 24	F I	.2570 .2720	P	.3230	Q	.3320
⅜	.375	16 24	⁵⁄₁₆ Q	.3125 .3320	W	.3860	X	.3970
⁷⁄₁₆	.4375	14 20	U ²⁵⁄₆₄	.3680 .3906	²⁹⁄₆₄	.4531	¹⁵⁄₃₂	.4687
½	.500	13 20	²⁷⁄₆₄ ²⁹⁄₆₄	.4219 .4531	³³⁄₆₄	.5156	¹⁷⁄₃₂	.5312

* Screws marked with asterisk (*) are not in the American Standard but are from the former A.S.M.E. Standard.

expressed as a decimal; e.g., 75% is expressed as .75. The tap drill size is the size nearest to the calculated hole size.

For American Unified Thread form:

$$\text{Hole Size} = \text{Basic Major Diameter} - \frac{1.08253 \times \text{Per Cent Full Thread}}{\text{Number of Threads per Inch}}$$

For American National Thread form:

$$\text{Hole Size} = \text{Basic Major Diameter} - \frac{1.29904 \times \text{Per Cent Full Thread}}{\text{Number of Threads per Inch}}$$

For ISO Metric threads (all dimensions in millimeters):

$$\text{Hole Size} = \text{Basic Major Diameter} - (1.08253 \times \text{Pitch} \times \text{Per Cent Full Thread})$$

Factors Influencing Minor Diameter Tolerances of Tapped Holes. — As stated in the Unified screw thread standard, the principal practical factors which govern minor diameter tolerances of internal threads are tapping difficulties, particularly tap breakage in the small sizes, availability of standard drill sizes in the medium and large sizes, and depth (radial) of engagement. Depth of engagement is related to the stripping strength of the thread assembly, and thus also, to the length of engagement. It also has an influence on the tendency toward disengagement of the threads on one side when assembly is eccentric. The amount of possible eccentricity is one-half of the sum of the pitch diameter allowance and tolerances on both mating threads. For a given pitch, or height of thread, this sum increases with the diameter, and accordingly this factor would require a decrease in minor diameter tolerance with increase in diameter. However, such decrease in tolerance would often require the use of special drill sizes; therefore, to facilitate the use of standard drill sizes, for any given pitch the minor diameter tolerance for Unified thread classes 1B and 2B threads of ¼ inch diameter and larger is constant, in accordance with a formula given in the American Standard for Unified Screw Threads.

Effect of Length of Engagement on Minor Diameter Tolerances: There may be applications where the lengths of engagement of mating threads is relatively short or the combination of materials used for mating threads are such that the maximum minor diameter tolerance given in the Standard (based on a length of engagement equal to the nominal diameter) may not provide the desired strength of the fastening. Experience has shown that for lengths of engagement less than ⅔ D (the minimum thickness of standard nuts) the minor diameter tolerance may be reduced without causing tapping difficulties. In other applications the length of engagement of mating threads may be long because of design considerations or the combination of materials used for mating threads. As the threads engaged increase in number, a shallower depth of engagement may be permitted and still develop stripping strength greater than the external thread breaking strength. In these cases the maximum tolerance given in the Standard should be increased to reduce the possibility of tapping difficulties. The following paragraphs indicate how the afore-mentioned considerations were taken into account in determining the minor diameter limits for various lengths of engagement given in Table 2.

Recommended Hole Sizes before Tapping. — Recommended hole size limits before threading to provide for optimum strength of fastenings and tapping conditions are shown in Table 2 for classes 1B, 2B, and 3B. The hole size limits before threading, and the tolerances between them, are derived from the minimum and maximum minor diameters of the internal thread given in the dimensional tables for Unified threads in the screw thread section using the following rules:

1. For lengths of engagement in the range to and including ⅓D, where D equals nominal diameter, the minimum hole size will be equal to the minimum minor diameter of the internal thread and the maximum hole size will be larger by one-half the minor diameter tolerance.

2. For the range from ⅓D to ⅔D, the minimum and maximum hole sizes will each be one quarter of the minor diameter tolerance larger than the corresponding limits for the length of engagement to and including ⅓D.

3. For the range from ⅔D to 1½D the minimum hole size will be larger than the minimum minor diameter of the internal thread by one-half the minor diameter tolerance and the maximum hole size will be equal to the maximum minor diameter.

4. For the range from 1½D to 3D the minimum and maximum hole sizes will each be one-quarter of the minor diameter tolerance of the internal thread larger than the corresponding limits for the ⅔D to 1½D length of engagement.

From the foregoing it will be seen that the difference between limits in each range is the same and equal to one-half of the minor diameter tolerance given in the Unified screw thread dimensional tables. This is a general rule, except that the minimum differences for sizes below ¼ inch are equal to the minor diameter tolerances calculated on the basis of lengths of engagement to and including ⅓D. Also, for lengths of engagement greater than ⅓D and for sizes ¼ inch and larger the values are adjusted so that the difference between limits is never less than 0.004 inch.

For diameter-pitch combinations other than those given in Table 2, the foregoing rules should be applied to the tolerances given in the dimensional tables in the screw thread section or the tolerances derived from the formulas given in the Standard to determine the hole size limits.

Selection of Tap Drills: In selecting standard drills to produce holes within the limits given in Table 2 it should be recognized that drills have a tendency to cut oversize. The material on page 846 may be used as a guide to the expected amount of oversize.

Hole Sizes for Tapping Unified Miniature Screw Threads. — Table 5 indicates the hole size limits recommended for tapping. These limits are derived from the internal thread minor diameter limits given in the American Standard for Unified Miniature Screw Threads (ASA B1.10-1958) and are disposed so as to provide the optimum conditions for tapping. The maximum limits are based on providing a functionally adequate fastening for the most common applications, where the material of the externally threaded member is of a strength essentially equal to or greater than that of its mating part. In applications where, because of considerations other than the fastening, the screw is made of an appreciably weaker material, the use of smaller hole sizes is usually necessary to extend thread engagement to a greater depth on the external thread. Recommended minimum hole sizes are greater than the minimum limits of the minor diameters to allow for the spin-up developed in tapping.

In selecting drills to produce holes within the limits given in Table 5 it should be recognized that drills have a tendency to cut oversize. The material on page 846 may be used as a guide to the expected amount of oversize.

British Standard Tapping Drill Sizes for Screw and Pipe Threads. — British Standard BS 1157:1975 provides recommendations for tapping drill sizes for use with fluted taps for various ISO metric, Unified, British Standard fine, British Association, and British Standard Whitworth screw threads as well as British Standard parallel and taper pipe threads.

In the accompanying Table 6, recommended and alternative drill sizes are given for producing holes for ISO metric coarse pitch series threads. These coarse pitch threads are suitable for the large majority of general-purpose applications, and the limits and tolerances for internal coarse threads are given on pages 1599 to

Table 5. Unified Miniature Screw Threads — Recommended Hole Size Limits Before Tapping*

Thread Size		Internal Threads		Lengths of Engagement					
		Minor Diameter Limits		To and including ⅔D		Above ⅔D to 1½D		Above 1½D to 3D	
				Recommended Hole Size Limits					
Designation	Pitch	Min	Max	Min	Max	Min	Max	Min	Max
	mm	mm	mm	mm	mm	mm	mm	mm	mm
0.30 UNM	**0.080**	**0.217**	**0.254**	**0.226**	**0.240**	**0.236**	**0.254**	**0.245**	**0.264**
0.35 UNM	0.090	0.256	0.297	0.267	0.282	0.277	0.297	0.287	0.307
0.40 UNM	**0.100**	**0.296**	**0.340**	**0.307**	**0.324**	**0.318**	**0.340**	**0.329**	**0.351**
0.45 UNM	0.100	0.346	0.390	0.357	0.374	0.368	0.390	0.379	0.401
0.50 UNM	**0.125**	**0.370**	**0.422**	**0.383**	**0.402**	**0.396**	**0.422**	**0.409**	**0.435**
0.55 UNM	0.125	0.420	0.472	0.433	0.452	0.446	0.472	0.459	0.485
0.60 UNM	**0.150**	**0.444**	**0.504**	**0.459**	**0.482**	**0.474**	**0.504**	**0.489**	**0.519**
0.70 UNM	0.175	0.518	0.586	0.535	0.560	0.552	0.586	0.569	0.603
0.80 UNM	**0.200**	**0.592**	**0.668**	**0.611**	**0.640**	**0.630**	**0.668**	**0.649**	**0.687**
0.90 UNM	0.225	0.666	0.750	0.687	0.718	0.708	0.750	0.729	0.771
1.00 UNM	**0.250**	**0.740**	**0.832**	**0.763**	**0.798**	**0.786**	**0.832**	**0.809**	**0.855**
1.10 UNM	0.250	0.840	0.932	0.863	0.898	0.886	0.932	0.909	0.955
1.20 UNM	**0.250**	**0.940**	**1.032**	**0.963**	**0.998**	**0.986**	**1.032**	**1.009**	**1.055**
1.40 UNM	0.300	1.088	1.196	1.115	1.156	1.142	1.196	1.169	1.223

Designation	Thds. per in.	inch	inch	inch	inch	inch	inch	inch	inch
0.30 UNM	318	**0.0085**	**0.0100**	**0.0089**	**0.0095**	**0.0093**	**0.0100**	**0.0096**	**0.0104**
0.35 UNM	282	0.0101	0.0117	0.0105	0.0111	0.0109	0.0117	0.0113	0.0121
0.40 UNM	254	**0.0117**	**0.0134**	**0.0121**	**0.0127**	**0.0125**	**0.0134**	**0.0130**	**0.0138**
0.45 UNM	254	0.0136	0.0154	0.0141	0.0147	0.0145	0.0154	0.0149	0.0158
0.50 UNM	203	**0.0146**	**0.0166**	**0.0150**	**0.0158**	**0.0156**	**0.0166**	**0.0161**	**0.0171**
0.55 UNM	203	0.0165	0.0186	0.0170	0.0178	0.0176	0.0186	0.0181	0.0191
0.60 UNM	169	**0.0175**	**0.0198**	**0.0181**	**0.0190**	**0.0187**	**0.0198**	**0.0193**	**0.0204**
0.70 UNM	145	0.0204	0.0231	0.0211	0.0221	0.0217	0.0231	0.0224	0.0237
0.80 UNM	127	**0.0233**	**0.0263**	**0.0241**	**0.0252**	**0.0248**	**0.0263**	**0.0256**	**0.0270**
0.90 UNM	113	0.0262	0.0295	0.0270	0.0283	0.0279	0.0295	0.0287	0.0304
1.00 UNM	102	**0.0291**	**0.0327**	**0.0300**	**0.0314**	**0.0309**	**0.0327**	**0.0319**	**0.0337**
1.10 UNM	102	0.0331	0.0367	0.0340	0.0354	0.0349	0.0367	0.0358	0.0376
1.20 UNM	102	**0.0370**	**0.0406**	**0.0379**	**0.0393**	**0.0388**	**0.0406**	**0.0397**	**0.0415**
1.40 UNM	85	0.0428	0.0471	0.0439	0.0455	0.0450	0.0471	0.0460	0.0481

* As an aid in selecting suitable drills, see the listing of American Standard drill sizes in the twist drill section. Thread sizes in heavy type are preferred sizes.

1602. It should be noted that Table 6 is for fluted taps only since a fluteless tap will require for the same screw thread a different size of twist drill than will a fluted tap. When tapped, holes produced with drills of the recommended sizes provide for a theoretical radial engagement with the external thread of about 81 per cent in most cases. Holes produced with drills of the alternative sizes provide for a theoretical radial engagement with the external thread of about 70 to 75 per cent. In some cases, as indicated in Table 6, the alternative drill sizes are suitable only for medium (6H) or for free (7H) thread tolerance classes.

When relatively soft material is being tapped, there is a tendency for the metal to be squeezed down towards the root of the tap thread, and in such instances, the minor diameter of the tapped hole may become smaller than the diameter of the drill employed. Users may wish to choose different tapping drill sizes to overcome this problem or for special purposes, and reference can be made to the pages mentioned above to obtain the minor diameter limits for internal pitch series threads.

Reference should be made to this standard (BS 1157:1975) for recommended tapping hole sizes for other types of British Standard screw threads and pipe threads.

Table 6. British Standard Tapping Drill Sizes for ISO Metric Coarse Pitch Series Threads (BS 1157:1975)

Nom. Size and Thread Diam.	Standard Drill Sizes§				Nom. Size and Thread Diam.	Standard Drill Sizes§			
	Recommended		Alternative			Recommended		Alternative	
	Size	Theoretical Radial Engagement with Ext. Thread (Per Cent)	Size	Theoretical Radial Engagement with Ext. Thread (Per Cent)		Size	Theoretical Radial Engagement with Ext. Thread (Per Cent)	Size	Theoretical Radial Engagement with Ext. Thread (Per Cent)
M 1	0.75	81.5	0.78	71.7	M 12	10.20	83.7	10.40	74.5*
M 1.1	0.85	81.5	0.88	71.7	M 14	12.00	81.5	12.20	73.4*
M 1.2	0.95	81.5	0.98	71.7	M 16	14.00	81.5	14.25	71.3†
M 1.4	1.10	81.5	1.15	67.9	M 18	15.50	81.5	15.75	73.4†
M 1.6	1.25	81.5	1.30	69.9	M 20	17.50	81.5	17.75	73.4†
M 1.8	1.45	81.5	1.50	69.9	M 22	19.50	81.5	19.75	73.4†
M 2	1.60	81.5	1.65	71.3	M 24	21.00	81.5	21.25	74.7*
M 2.2	1.75	81.5	1.80	72.5	M 27	24.00	81.5	24.25	74.7*
M 2.5	2.05	81.5	2.10	72.5	M 30	26.50	81.5	26.75	75.7*
M 3	2.50	81.5	2.55	73.4	M 33	29.50	81.5	29.75	75.7*
M 3.5	2.90	81.5	2.95	74.7	M 36	32.00	81.5	...	...
M 4	3.30	81.5	3.40	69.9*	M 39	35.00	81.5	...	...
M 4.5	3.70	86.8	3.80	76.1	M 42	37.50	81.5	...	...
M 5	4.20	81.5	4.30	71.3*	M 45	40.50	81.5	...	...
M 6	5.00	81.5	5.10	73.4	M 48	43.00	81.5	...	...
M 7	6.00	81.5	6.10	73.4	M 52	47.00	81.5	...	...
M 8	6.80	78.5	6.90	71.7*	M 56	50.50	81.5	...	...
M 9	7.80	78.5	7.90	71.7*	M 60	54.50	81.5	...	...
M 10	8.50	81.5	8.60	76.1	M 64	58.00	81.5	...	...
M 11	9.50	81.5	9.60	76.1	M 68	62.00	81.5	...	...

Drill sizes are given in millimeters.
§ These tapping drill sizes are for fluted taps only.
* For tolerance class 6H and 7H threads only.
† For tolerance class 7H threads only.

British Standard Clearance Holes for Metric Bolts and Screws. — The dimensions of the clearance holes specified in this British Standard (BS 4186:1967) have been chosen in such a way as to require the use of the minimum number of drills. The recommendations cover three series of clearance holes, namely close fit (H 12), medium fit (H 13), and free fit (H 14) and are suitable for use with bolts and screws specified in the following metric British Standards: BS 3692, ISO metric precision hexagon bolts, screws, and nuts; BS 4168, Hexagon socket screws and wrench keys; BS 4183, Machine screws and machine screw nuts; and BS 4190, ISO metric black hexagon bolts, screws, and nuts. The sizes are in accordance with those given in ISO Recommendation R273, and the range has been extended up to 150 millimeters diameter in accordance with an addendum to that recommendation. The selection of clearance holes sizes to suit particular design requirements can of course be dependent upon many variable factors. It is however felt that the medium fit series should suit the majority of general purpose applications. In the Standard, limiting dimensions are given in a table which is included for reference purposes only, for use in instances where it may be desirable to specify tolerances.

To avoid any risk of interference with the radius under the head of bolts and screws, it is necessary to countersink slightly all recommended clearance holes in the close and medium fit series. Dimensional details for the radius under the head of fasteners made according to BS 3692 are given on page 1345; those for fasteners to BS 4168 are given on page 1402; those to BS 4183 are given on pages 1378 to 1381.

Table 7. British Standard Metric Bolt and Screw Clearance Holes
(BS 4186: 1967)

Nominal Thread Diameter	Clearance Hole Sizes			Nominal Thread Diameter	Clearance Hole Sizes		
	Close Fit Series	Medium Fit Series	Free Fit Series		Close Fit Series	Medium Fit Series	Free Fit Series
1.6	1.7	1.8	2.0	52.0	54.0	56.0	62.0
2.0	2.2	2.4	2.6	56.0	58.0	62.0	66.0
2.5	2.7	2.9	3.1	60.0	62.0	66.0	70.0
3.0	3.2	3.4	3.6	64.0	66.0	70.0	74.0
4.0	4.3	4.5	4.8	68.0	70.0	74.0	78.0
5.0	5.3	5.5	5.8	72.0	74.0	78.0	82.0
6.0	6.4	6.6	7.0	76.0	78.0	82.0	86.0
7.0	7.4	7.6	8.0	80.0	82.0	86.0	91.0
8.0	8.4	9.0	10.0	85.0	87.0	91.0	96.0
10.0	10.5	11.0	12.0	90.0	93.0	96.0	101.0
12.0	13.0	14.0	15.0	95.0	98.0	101.0	107.0
14.0	15.0	16.0	17.0	100.0	104.0	107.0	112.0
16.0	17.0	18.0	19.0	105.0	109.0	112.0	117.0
18.0	19.0	20.0	21.0	110.0	114.0	117.0	122.0
20.0	21.0	22.0	24.0	115.0	119.0	122.0	127.0
22.0	23.0	24.0	26.0	120.0	124.0	127.0	132.0
24.0	25.0	26.0	28.0	125.0	129.0	132.0	137.0
27.0	28.0	30.0	32.0	130.0	134.0	137.0	144.0
30.0	31.0	33.0	35.0	140.0	144.0	147.0	155.0
33.0	34.0	36.0	38.0	150.0	155.0	158.0	165.0
36.0	37.0	39.0	42.0	...	...	...	...
39.0	40.0	42.0	45.0	...	...	...	...
42.0	43.0	45.0	48.0	...	...	...	...
45.0	46.0	48.0	52.0	...	...	...	...
48.0	50.0	52.0	56.0	...	...	...	...

All dimensions are given in millimeters.

Cold Form Tapping. — Cold form taps do not have cutting edges or conventional flutes; the threads on the tap form the threads in the hole by displacing the metal in an extrusion or swaging process. The threads thus produced are stronger than conventionally cut threads because the grains in the metal are unbroken and the displaced metal is work hardened. The surface of the thread is burnished and has an excellent finish. Although chip problems are eliminated, cold form tapping does displace the metal surrounding the hole and countersinking or chamfering before tapping is recommended. Cold form tapping is not recommended if the wall thickness of the hole is less than two-thirds of the nominal diameter of the thread. If possible, blind holes should be drilled deep enough to permit a cold form tap having a four thread lead to be used as this will require less torque, produce less burr surrounding the hole, and give a greater tool life.

The operation requires 0 to 50 per cent more torque than conventional tapping, and the cold form tap will pick up its own lead when entering the hole; thus, conventional tapping machines and tapping heads can be used. Another advantage is the better tool life obtained. The best results are obtained by using a good lubricating oil instead of a conventional cutting oil.

Table 8. Theoretical and Tap Drill or Core Hole Sizes for Cold Form Tapping Unified Threads

Tap Size	Threads Per Inch	Percentage of Full Thread								
		75			65			55		
		Theor. Hole Size	Nearest Drill Size	Dec. Equiv.	Theor. Hole Size	Nearest Drill Size	Dec. Equiv.	Theor. Hole Size	Nearest Drill Size	Dec. Equiv.
0	80	.0536	1.35 mm	.0531	.0545	....	....	.0554	54	.055
1	64	.0650	1.65 mm	.0650	.0661	....	....	.0672	51	.0670
	72	.0659	1.65 mm	.0650	.0669	1.7 mm	.0669	.0679	51	.0670
2	56	.0769	1.95 mm	.0768	.0781	5⁄64	.0781	.0794	2.0 mm	.0787
	64	.0780	5⁄64	.0781	.0791	2.0 mm	.0787	.0802	....	....
3	48	.0884	2.25 mm	.0886	.0898	43	.089	.0913	2.3 mm	.0906
	56	.0889	43	.089	.0911	2.3 mm	.0906	.0924	2.35 mm	.0925
4	40	.0993	2.5 mm	.0984	.1010	39	.0995	.1028	2.6 mm	.1024
	48	.0104	38	.1015	.1028	2.6 mm	.1024	.1043	37	.1040
5	40	.1123	34	.1110	.1140	33	.113	.1158	32	.1160
	44	.1134	33	.113	.1150	2.9 mm	.1142	.1166	32	....
6	32	.1221	3.1 mm	.1220	.1243	....	....	.1264	3.2 mm	.1260
	40	.1253	⅛	.1250	.1270	3.2 mm	.1260	.1288	30	.1285
8	32	.1481	3.75 mm	.1476	.1503	25	.1495	.1524	24	.1520
	36	.1498	25	.1495	.1518	24	.1520	.1537	3.9 mm	.1535
10	24	.1688	....	....	.1717	11⁄64	.1719	.1746	17	.1730
	32	.1741	17	.1730	.1763	....	....	.1784	4.5 mm	.1772
12	24	.1948	10	.1935	.1977	5.0 mm	.1968	.2006	5.1 mm	.2008
	28	.1978	5.0 mm	.1968	.2003	8	.1990	.2028	....	....
¼	20	.2245	5.7 mm	.2244	.2280	1	.2280	.2315	....	....
	28	.2318	....	....	.2343	A	.2340	.2368	6.0 mm	.2362
5⁄16	18	.2842	7.2 mm	.2835	.2879	7.3 mm	.2874	.2917	7.4 mm	.2913
	24	.2912	7.4 mm	.2913	.2941	M	.2950	.2969	19⁄64	.2969
⅜	16	.3431	11⁄32	.3437	.3474	S	.3480	.3516	....	....
	24	.3537	9.0 mm	.3543	.3566	....	....	.3594	23⁄64	.3594
7⁄16	14	.4011	Z	....	.4059	13⁄32	.4062	.4108	....	....
	20	.4120	Z	.413	.4154	....	....	.4188	....	....
½	13	.4608	....	....	.4660	....	....	.4712	12 mm	.4724
	20	.4745	....	....	.4779	....	....	.4813	....	....
9⁄16	12	.5200	....	....	.5257	....	....	.5313	17⁄32	.5312
	18	.5342	13.5 mm	.5315	.5380	....	....	.5417	....	....
⅝	11	.5787	37⁄64	.5781	.5848	....	....	.5910	15 mm	.5906
	18	.5976	19⁄32	.5937	.6004	....	....	.6042	....	....
¾	10	.6990	....	....	.7058	45⁄64	.7031	.7126	....	....
	16	.7181	23⁄32	.7187	.7224	....	....	.7266	....	....

The method can be applied only to relatively ductile metals, such as low carbon steel, leaded steels, austenitic stainless steels, wrought aluminum, low silicon aluminum die casting alloys, zinc die casting alloys, magnesium, copper, and ductile copper alloys. A higher than normal tapping speed can be used, sometimes by as much as 100 per cent.

Conventional tap drill sizes should not be used for cold form tapping because the metal is displaced to form the thread. Since the cold formed thread is stronger than the conventionally tapped thread, the thread height can be reduced to 60 per cent without a loss of strength; however, the use of a 65 per cent thread is strongly recommended. The following formula is used to calculate the theoretical hole size for cold form tapping:

$$\text{Theoretical Hole Size} = \text{Basic Tap O.D.} - \frac{.0068 \times \text{Per Cent of Full Thread}}{\text{Threads per inch}}$$

The theoretical hole size and the tap drill sizes for American Unified threads are given in Table 8, and Table 9 lists drills for ISO Metric threads. Sharp drills should be used to prevent cold working the walls of the hole, especially on metals that are prone to work hardening. Such damage may cause the torque to increase, possibly stopping the machine or breaking the tap. On materials that can be die cast, cold form tapping can be done in cored holes provided the correct core pin size is used. Since the core pins are slightly tapered, the theoretical hole size should be at the point on the pin where this point is equal to one-half of the required engagement length of the thread in the hole. The core pins should be designed to form a chamfer on the hole to accept the vertical extrusion.

Table 9. Tap Drill or Core Hole Sizes† for Cold Form Tapping ISO Threads

Nominal Size of Tap	Pitch	Recommended Tap Drill Size	Nominal Size of Tap	Pitch	Recommended Tap Drill Size
1.6 mm	0.35 mm	1.45 mm	4.0 mm	0.70 mm	3.7 mm
1.8 mm	0.35 mm	1.65 mm	4.5 mm	0.75 mm	4.2 mm*
2.0 mm	0.40 mm	1.8 mm	5.0 mm	0.80 mm	4.6 mm
2.2 mm	0.45 mm	2.0 mm	6.0 mm	1.00 mm	5.6 mm*
2.5 mm	0.45 mm	2.3 mm	7.0 mm	1.00 mm	6.5 mm
3.0 mm	0.50 mm	2.8 mm*	8.0 mm	1.25 mm	7.4 mm
3.5 mm	0.60 mm	3.2 mm	10.0 mm	1.50 mm	9.3 mm

† The sizes are calculated to provide 60 to 75 per cent of full thread.
* These diameters are the nearest stocked drill sizes and not the theoretical hole size, and may not produce 60 to 75 per cent full thread.

Removing a Broken Tap.—Broken taps can be removed by electrodischarge machining (EDM), and this method is recommended when available. When an EDM machine is not available, broken taps may be removed by using a tap extractor, which has fingers that enter the flutes of the tap; the tap is backed out of the hole by turning the extractor with a wrench. Sometimes the injection of a small amount of a proprietary solvent into the hole will be helpful. A solvent can be made by diluting about one part nitric acid with five parts water. The action of the proprietary solvent or the diluted nitric acid on the steel loosens the tap so that it can be removed with pliers or with a tap extractor. The hole should be washed out afterwards so that the acid will not continue to work on the part.

Another method is to add, by electric arc welding, additional metal to the shank of the broken tap, above the level of the hole. Care must be taken to prevent depositing metal on the threads in the tapped hole. After the shank has been built up, the head of a bolt or a nut is welded to it and then the tap may be backed out.

Tap Drills for Pipe Taps*

Size of Tap	Drills for Briggs Pipe Taps	Drills for Whitworth Pipe Taps	Size of Tap	Drills for Briggs Pipe Taps	Drills for Whitworth Pipe Taps	Size of Tap	Drills for Briggs Pipe Taps	Drills for Whitworth Pipe Taps
⅛	11/32	5/16	1¼	1½	1 15/32	3¼	...	3½
¼	7/16	27/64	1½	1 23/32	1 25/32	3½	3¾	3¾
⅜	19/32	9/16	1¾	...	1 15/16	3¾	...	4
½	23/32	11/16	2	2 3/16	2 5/32	4	4¼	4¼
⅝	...	25/32	2¼	...	2 13/32	4½	4¾	4¾
¾	15/16	29/32	2½	2⅝	2 25/32	5	5 5/16	5¼
⅞	...	1 1/16	2¾	...	3 1/32	5½	...	5¾
1	1 5/32	1⅛	3	3¼	3 9/32	6	6⅜	6¼

All dimensions are given in inches.
* To secure the best results, the hole should be reamed before tapping with a reamer having a taper of ¾ inch per foot.

Power for Pipe Taps. — The power required for driving pipe taps is given in the following table, which includes nominal pipe tap sizes from 2 to 8 inches.

The holes to be tapped were reamed with standard pipe tap reamers before tapping. The horsepower recorded was read off just before the tap was reversed. The

Power Required for Pipe Taps

Nominal Tap Size	Rev. per Min.	Net H.P.	Thickness of Metal	Nominal Tap Size	Rev. per Min.	Net H.P.	Thickness of Metal
2	40	4.24	1⅛	3½	25.6	7.20	1¾
2½	40	5.15	1⅛	4	18	6.60	2
*2½	38.5	9.14	1⅛	5	18	7.70	2
3	40	5.75	1⅛	6	17.8	8.80	2
*3	38.5	9.70	1⅛	8	14	7.96	2½

Tap size and metal thickness are given in inches.
* Tapping steel casting; other tests in cast iron.

table gives the net horsepower, deductions being made for the power required to run the machine without a load. The material tapped was cast iron, except in two instances, where steel casting was tapped. It will be seen that nearly double the power is required for tapping steel casting. The power varies, of course, with the conditions. More power than that indicated in the table will be required if the cast iron is of a harder quality or if the taps are not properly relieved. The taps used in these experiments were of the inserted-blade type, the blades being made of high-speed steel.

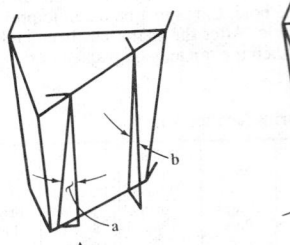 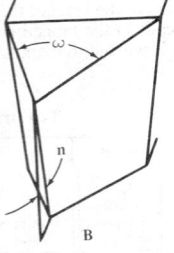

Two similar diagrams showing relationships of
various relief angles of thread cutting tools

Relief Angles for Single-Point Thread Cutting Tools. — The surface finish on threads cut with single-point thread cutting tools is influenced by the relief angles on the tools. The leading and trailing cutting edges that form the sides of the thread, and the cutting edge at the nose of the tool must all be provided with an adequate amount of relief. Moreover, it is recommended that the effective relief angle, a_e, for all of these cutting edges be made equal, although the practice in some shops is to use slightly less relief at the trailing cutting edge. While too much relief may weaken the cutting edge, causing it to chip, an inadequate amount of relief will result in rough threads and in a shortened tool life. Other factors that influence the finish produced on threads include the following: the work material; the cutting speed; the cutting fluid used; the method used to cut the thread; and, the condition of the cutting edge.

Relief angles on single-point thread cutting tools are often specified on the basis of experience. While this method may give satisfactory results in many instances, better results can usually be obtained by calculating these angles, using the formulas provided further on. When special high helix angle threads are to be cut, the magnitude of the relief angles should always be calculated. These calculations are based on the effective relief angle, a_e; this is the angle between the flank of the tool and the sloping sides of the thread, measured in a direction parallel to the axis of the thread. Recommended values of this angle are 8 to 14 degrees for high speed steel tools, and 5 to 10 degrees for cemented carbide tools. The larger values are recommended for cutting threads on soft and gummy materials, and the smaller values are for the harder materials, which inherently take a better surface finish. Harder materials also require more support below the cutting edges, which is provided by using a smaller relief angle. These values are recommended for the relief angle below the cutting edge at the nose without any further modification. The angles below the leading and trailing side cutting edges are modified, using the formulas provided. The angles b and b' are the relief angles actually ground on the tool below the leading and trailing side cutting edges respectively; they are measured perpendicular to the side cutting edges. When designing or grinding the thread cutting tool, it is sometimes helpful to know the magnitude of the angle, n, for which a formula is provided. This angle would occur only in the event that the tool were ground to a sharp point. It is the angle of the edge formed by the intersection of the flank surfaces.

$$\tan \phi = \frac{\text{lead of thread}}{\pi K} \qquad\qquad \tan \phi' = \frac{\text{lead of thread}}{\pi D}$$

$$a = a_e + \phi$$
$$a' = a_e - \phi'$$

$$\tan b = \tan a \cos (\tfrac{1}{2}\omega)$$

$$\tan b' = \tan a' \cos (\tfrac{1}{2}\omega)$$

$$\tan n = \frac{\tan a - \tan a'}{2 \tan (\tfrac{1}{2}\omega)}$$

Where: ϕ = Helix angle of thread at minor diameter
ϕ' = Helix angle of thread at major diameter
K = Minor diameter of thread
D = Major diameter of thread
a = Side relief angle parallel to thread axis at leading edge of tool
a' = Side relief angle parallel to thread axis at trailing edge of tool
a_e = Effective relief angle
b = Side relief angle perpendicular to leading edge of tool
b' = Side relief angle perpendicular to trailing edge of tool
ω = Included angle of thread cutting tool
n = Nose angle resulting from intersection of flank surfaces

Example: Calculate the relief angles and the nose angle, n, for a single-point thread cutting tool which is to be used to cut a 1-inch diameter, 5 threads-per-inch, double Acme thread. The lead of this thread is $2 \times .200 = .400$ inch. The included angle, ω, of this thread is 29 degrees, the minor diameter, K, is .780 inch, and the effective relief angle, a_e, below all cutting edges, is to be 10 degrees.

$$\tan \phi = \frac{\text{Lead of thread}}{\pi K} = \frac{.400}{\pi \times .780}$$

$$\phi = 9.27° \ (9°16')$$

$$\tan \phi' = \frac{\text{Lead of thread}}{\pi D} = \frac{.400}{\pi \times 1.000}$$

$$\phi' = 7.26° \ (7°15')$$

$$a = a_e + \phi = 10° + 9.27° = 19.27°$$

$$a' = a_e - \phi' = 10° - 7.26° = 2.74°$$

$$\tan b = \tan a \cos \tfrac{1}{2}\omega = \tan 19.27 \cos 14.5$$

$$b = 18.70° \ (18°42')$$

$$\tan b' = \tan a' \cos \tfrac{1}{2}\omega = \tan 2.74 \cos 14.5$$

$$b' = 2.65° \ (2°39')$$

$$\tan n = \frac{\tan a - \tan a'}{2 \tan (\tfrac{1}{2}\omega)} = \frac{\tan 19.27 - \tan 2.74}{2 \tan 14.5}$$

$$n = 30.26° \ (30°16')$$

Lathe Change Gears

Change Gears for Thread Cutting. — To determine the change gears to use for cutting a thread of given pitch, first find what number of threads per inch will be cut when gears of the same size are placed on the lead-screw and spindle stud, either by actual trial or by referring to the index plate; then multiply this number, called the "lathe screw constant," by some trial number to obtain the number of teeth in the gear for the spindle stud, and multiply the threads per inch to be cut by the *same* trial number to obtain the number of teeth in the gear for the lead-screw. Expressing this rule as a formula:

$$\frac{\text{Trial number} \times \text{lathe screw constant}}{\text{Trial number} \times \text{threads per inch to be cut}} = \frac{\text{teeth in gear on spindle stud}}{\text{teeth in gear on lead-screw}}$$

For example, suppose the available change gears supplied with the lathe have 24, 28, 32, 36 teeth, etc., the number increasing by four up to one hundred, and that 10 threads per inch are to be cut in a lathe having a lathe screw constant of 6; then, if the screw constant is written as the numerator, and the number of threads per inch to be cut, as the denominator of a fraction, and both numerator and denominator are multiplied by some trial number, say 4, it is found that gears having 24 and 40 teeth can be used. Thus:

$$\frac{6}{10} = \frac{6 \times 4}{10 \times 4} = \frac{24}{40}$$

The 24-tooth gear goes on the spindle stud and the 40-tooth gear on the lead-screw.

The lathe screw constant is, of course, equal to the number of threads per inch on the lead-screw, provided the spindle stud and spindle are geared in the ratio of 1 to 1, which, however, is not always the case.

Compound Gearing. — To find the change gears used in compound gearing, place the screw constant as the numerator, and the number of threads per inch to be cut as the denominator of a fraction; resolve both numerator and denominator into two factors each, and multiply each "pair" of factors by the same number, until values are obtained representing suitable numbers of teeth for the change gears. (One factor in the numerator and one in the denominator make a "pair" of factors.)

Example: — $1\frac{3}{4}$ threads per inch are to be cut in a lathe having a screw constant of 8; the available gears have 24, 28, 32, 36, 40 teeth, etc., increasing by four up to one hundred. Following the rule:

$$\frac{8}{1\frac{3}{4}} = \frac{2 \times 4}{1 \times 1\frac{3}{4}} = \frac{(2 \times 36) \times (4 \times 16)}{(1 \times 36) \times (1\frac{3}{4} \times 16)} = \frac{72 \times 64}{36 \times 28}$$

The gears having 72 and 64 teeth are the *driving* gears and those with 36 and 28 teeth are the *driven* gears.

Fractional Threads. — Sometimes the lead of a thread is given as a fraction of an inch instead of stating the number of threads per inch. For example, a thread may be required to be cut, having $\frac{3}{8}$ inch lead. The expression "$\frac{3}{8}$ inch lead" should first be transformed to "number of threads per inch." The number of threads per inch (the thread being single) equals:

$$\frac{1}{\frac{3}{8}} = 1 \div \frac{3}{8} = \frac{8}{3} = 2\frac{2}{3}$$

To find the change gears to cut $2\frac{2}{3}$ threads per inch in a lathe having a screw

constant 8 and change gears running from 24 to 100 teeth, increasing by 4, proceed as below:

$$\frac{8}{2\frac{2}{3}} = \frac{2 \times 4}{1 \times 2\frac{2}{3}} = \frac{(2 \times 36) \times (4 \times 24)}{(1 \times 36) \times (2\frac{2}{3} \times 24)} = \frac{72 \times 96}{36 \times 64}$$

Change Gears for Metric Pitches. — When screws are cut in accordance with the metric system, it is the usual practice to give the lead of the thread in millimeters, instead of the number of threads per unit of measurement. To find the change gears for cutting metric threads, when using a lathe having an English lead-screw, first determine the number of threads per inch corresponding to the given lead in millimeters. Suppose a thread of 3 millimeters lead is to be cut in a lathe having an English lead-screw and a screw constant of 6. As there are 25.4 millimeters per inch, the number of threads per inch will equal 25.4 ÷ 3. Place the screw constant as the numerator, and the number of threads per inch to be cut as the denominator:

$$\frac{6}{\frac{25.4}{3}} = 6 \div \frac{25.4}{3} = \frac{6 \times 3}{25.4}$$

The numerator and denominator of this fractional expression of the change gear ratio is next multiplied by some trial number to determine the size of the gears. The first whole number by which 25.4 can be multiplied so as to get a whole number as the result is 5. Thus, 25.4 × 5 = 127. Hence, one gear having 127 teeth is always used when cutting metric threads with an English lead-screw. The other gear required in this case has 90 teeth. Thus:

$$\frac{6 \times 3 \times 5}{25.4 \times 5} = \frac{90}{127}$$

Therefore, the following rule can be used to find the change gears for cutting metric pitches with an English lead-screw:

Rule: Place the lathe screw constant multiplied by the lead of the required thread in millimeters multiplied by 5, as the numerator of the fraction, and 127 as the denominator. The product of the numbers in the numerator equals the number of teeth for the spindle-stud gear, and 127 is the number of teeth for the lead-screw gear.

If the lathe has a metric pitch lead-screw, and a screw having a given number of threads per inch is to be cut, first find the "metric screw constant" of the lathe or the lead of thread in millimeters that would be cut with change gears of equal size on the lead-screw and spindle stud; then the method of determining the change gears is simply the reverse of the one already explained for cutting a metric thread with an English lead-screw.

Rule: To find the change gears for cutting English threads with a metric lead-screw, place 127 in the numerator and the threads per inch to be cut, multiplied by the metric screw constant multiplied by 5, in the denominator; 127 is the number of teeth on the spindle-stud gear and the product of the numbers in the denominator equals the number of teeth in the lead-screw gear.

Threads per Inch Obtained with a Given Combination of Gears. — To determine the number of threads per inch that will be obtained with a given combination of gearing, multiply the lathe screw constant by the number of teeth in the *driven* gear (or by the product of the numbers of teeth in both driven gears of compound gearing), and divide the product thus obtained by the number of teeth in the *driving* gear (or by the product of the two driving gears of a compound train). The quotient equals the number of threads per inch.

Change Gears for Fractional Ratios. — When gear ratios cannot be expressed exactly in whole numbers which are within the range of ordinary gearing, the combination of gearing required for the fractional ratio may be determined quite easily, in some cases, by the "cancellation method." To illustrate this method, assume that the speeds of two gears are to be in the ratio of 3.423 to 1. The number 3.423 is first changed to $\frac{3423}{1000}$ to clear it of decimals. Then, in order to secure a fraction that can be reduced, 3423 is changed to 3420;

$$\frac{3420}{1000} = \frac{342}{100} = \frac{3 \times 2 \times 57}{2 \times 50} = \frac{3 \times 57}{1 \times 50}$$

Then, multiplying $\frac{3}{1}$ by some trial number, say, 24, the following gear combination is obtained:

$$\frac{72}{24} \times \frac{57}{50} = \frac{4104}{1200} = \frac{3.42}{1}$$

As the desired ratio is 3.423 to 1, there is an error of 0.003. When the ratios are comparatively simple, the cancellation method is not difficult and is frequently used; but by the logarithmic method to be described, more accurate results are possible in most cases.

Modifying the Quick Change Gearbox Output. — On most modern lathes the gear train connecting the headstock spindle with the lead screw contains a quick change gearbox. Instead of using different change gears, it is only necessary to position the handles of the gearbox to adjust the speed ratio between the spindle and the lead screw in preparation for cutting a thread. There are, however, occasions when a thread must be cut for which there is no quick change gearbox setting. In this event, it is necessary to modify the normal, or standard, gear ratio between the spindle and the gearbox by installing modifying change gears to replace the standard gears normally used. Metric and other odd pitch threads can be cut on lathes that have an inch thread lead screw and a quick change gearbox having only settings for inch threads by using modifying change gears in the gear train. Likewise, inch threads and other odd pitch threads can be cut on metric lead screw lathes having a gearbox on which only metric thread setting can be made. Modifying change gears also can be used for cutting odd pitch threads on lathes having a quick change gearbox that has both inch and metric thread settings.

The sizes of the modifying change gears can be calculated by formulas, which are given further on; they depend on the thread to be cut and on the setting of the quick change gearbox. Many different sets of gears can be found for each thread to be cut. It is recommended that in each case several calculations be made in order to find the set of gears that is most suitable for installation on the lathe. The modifying change gear formulas that follow are based on the type of lead screw; i.e., whether the lead screw has inch or metric threads.

Metric Threads on Inch Lead Screw Lathes: A 127-tooth translating gear must be used in the modifying change gear train in order to be able to cut metric threads on inch lead screw lathes. The formula for calculating the modifying change gears is:

$$\frac{5 \times \text{gearbox setting in thds/in.} \times \text{pitch in mm to be cut}}{127} = \frac{\text{driving gears}}{\text{driven gears}}$$

The numerator and denominator of this formula are multiplied by equal numbers, called trial numbers, to find the gears. If suitable gears cannot be found with one set, then another set of equal trial numbers is used. (Since these numbers are equal, such as 15/15 or 24/24, they are equal to the number one when thought of as a fraction; their inclusion has the effect of multiplying the formula by one, which does not change its value.) It is necessary to select the gearbox setting in threads per inch which must be

used to cut the metric thread when using the gears calculated by the formula. One method is to select a quick change gearbox setting that is close to the actual number of metric threads in a one-inch length, called the equivalent threads per inch, which can be calculated by the following formula: Equivalent thds/in. = 25.4 ÷ pitch in mm to be cut.

Example: Select the quick change gearbox setting and calculate the modifying change gears required to set up a lathe having an inch-thread lead screw in order to cut an M12 × 1.75 metric thread.

$$\text{Equivalent thds/in.} = \frac{25.4}{\text{pitch in mm to be cut}} = \frac{25.4}{1.75} = 14.5 \ (\text{Use 14 thds/in.})$$

$$\frac{5 \times \text{gearbox setting in thds/in.} \times \text{pitch in mm to be cut}}{127} = \frac{5 \times 14 \times 1.75}{127}$$

$$= \frac{(24) \times 5 \times 14 \times 1.75}{(24) \times 127} = \frac{(5 \times 14) \times (24 \times 1.75)}{24 \times 127}$$

$$= \frac{70 \times 42}{24 \times 127} = \frac{\text{driving gears}}{\text{driven gears}}$$

Odd Inch Pitch Threads: The calculation of the modifying change gears used for cutting odd pitch threads that are specified by their pitch in inches involves the sizes of the standard gears, which can be found by counting their teeth. Standard gears are those used to enable the lathe to cut the thread for which the gearbox setting is made; they are the gears that are normally used. The threads on worms used with worm gears are among the odd pitch threads that can be cut by this method. As before, it is usually advisable to calculate the actual number of threads per inch of the odd pitch thread and to select a gearbox setting that is close to this value. The following formula is used to calculate the modifying change gears to cut odd inch pitch threads:

$$\frac{\text{standard driving gear} \times \text{pitch to be cut in inches} \times \text{gearbox setting in thds/in.}}{\text{standard driven gear}}$$

$$= \frac{\text{driving gears}}{\text{driven gears}}$$

Example: Select the quick change gearbox setting and calculate the modifying change gears required to cut a thread having a pitch equal to .195 inch. The standard driving and driven gears both have 48 teeth. To find equivalent threads per inch:

$$\text{thds/in.} = \frac{1}{\text{pitch}} = \frac{1}{.195} = 5.13 \quad (\text{Use 5 thds/in.})$$

$$\frac{\text{standard driving gear} \times \text{pitch to be cut in inches} \times \text{gearbox setting in thds/in.}}{\text{standard driven gear}}$$

$$= \frac{48 \times .195 \times 5}{48} = \frac{(1000) \times .195 \times 5}{(1000)} = \frac{195 \times 5}{500 \times 2} = \frac{39 \times 5}{100 \times 2} = \frac{39 \times 5 \times (8)}{50 \times 2 \times 2 \times (8)}$$

$$= \frac{39 \times 40}{50 \times 32} = \frac{\text{driving gears}}{\text{driven gears}}$$

It will be noted that in the second step above 1000/1000 has been substituted for 48/48. This substitution does not change the ratio. The reason why this substitution was made

is that $1000 \times .195 = 195$, a whole number. Actually 200/200 might have been substituted since $200 \times .195 = 39$, also a whole number.

The procedure for calculating the modifying gears using the following formulas is the same as illustrated by the two previous examples.

Odd Threads per Inch on Inch Lead Screw Lathes:

$$\frac{\text{standard driving gear} \times \text{gearbox setting in thds/in.}}{\text{standard driven gear} \times \text{thds/in. to be cut}} = \frac{\text{driving gears}}{\text{driven gears}}$$

Inch Threads on Metric Lead Screw Lathes:

$$\frac{127}{5 \times \text{gearbox setting in mm pitch} \times \text{thds/in. to be cut}} = \frac{\text{driving gears}}{\text{driven gears}}$$

Odd Metric Pitch Threads on Metric Lead Screw Lathes:

$$\frac{\text{standard driving gear} \times \text{mm pitch to be cut}}{\text{standard driven gear} \times \text{gearbox setting in mm pitch}} = \frac{\text{driving gears}}{\text{driven gears}}$$

Tabulated Logarithms of Change-gear Ratios. — Change-gear problems can be solved readily by the use of the accompanying tables which contain the six-place logarithms of the ratios of all gear combinations between 15 and 120 teeth, inclusive, excepting the 1 to 1 ratios. To illustrate how these logarithms of ratios were obtained, take as an example gears having 72 and 41 teeth, respectively; the ratio equals 72 divided by 41, and to divide by means of logarithms, the logarithm of one number is subtracted from the logarithm of the other, thus,

$$\log 72 = 1.857\underline{333} \text{ (from a calculator)}$$
$$\log 41 = 1.612\underline{784} \text{ (from a calculator)}$$
$$\text{ratio log} = 0.244549$$

The logarithms for ratios of pairs of gears having between 16 and 120 teeth have been arranged in numerical order in the tables. In a number of cases, more than one combination gives the same logarithm, so that the different pairs of gears that equal the logarithm have been repeated. In some simple cases, only the ratio has been given in order to shorten the table; for instance, all the gear combinations that equal a 2 to 1 ratio have been omitted and only the ratio is given.

There are nearly 5000 different ratios represented in the gear tables between the extremes $1.0084 +$ to 1 (120 : 119) and $7\frac{1}{2}$ to 1 (120 : 16). As the sum of any two two-gear logarithms equals a four-gear logarithm, the tables represent over 12,000,000 four-gear combinations; and by using three pairs of gears in a train, there are over 20,000,000,000 six-gear combinations available.

Driving and Driven Gears. — Insofar as the use of the gear logarithm tables is concerned, it is immaterial which is the driver and which is the driven gear, and by comparing the gears selected with the ratio, no confusion should result. For example, gears for the ratio 7.32 : 4.17 are selected in the same manner as gears for the ratio 4.17 : 7.32, each ratio being the reciprocal of the other so that the first ratio can be obtained from the second, and vice versa, by merely inverting the ratio. Thus, the logarithm of the smaller number is *always* subtracted from the logarithm of the larger to get the logarithm of the ratio, or its reciprocal, as the case may be, since the table is set up to correspond with this order.

The speeds of the driving and driven gears in a four-gear train are related to the numbers of teeth as follows:

$$\frac{\text{Speed of Driving Shaft}}{\text{Speed of Driven Shaft}} = \frac{\text{Teeth in 1st Driven Gear} \times \text{Teeth in 2nd Driven Gear}}{\text{Teeth in 1st Driving Gear} \times \text{Teeth in 2nd Driving Gear}}$$

Solving Two-gear Change-gear Problems by Use of Logarithms. — Suppose that two gears having the ratio 3.423:1 are desired. Log 3.423 = 0.534407. From the table, the logarithm nearest to this is log 89:26 = 0.534417; therefore, the gears having 89 and 26 teeth are the nearest to the ratio 3.423 to 1, and as 89 ÷ 26 = 3.423077, the ratio error is 0.000077.

When solving gear problems, the ratio should be reduced to terms of 1. For example, what two gears will have teeth numbers in the ratio of 7.182 to 3.902? $\frac{7.182}{3.902} = \frac{1.84059}{1}$; the log of 1.84059 is 0.264958. From the table, 81 : 44 = log 0.265032. As 81 ÷ 44 = 1.84091, the error is 0.00032.

A more rapid solution of the same problem makes use of logarithms:

$$\log 7.182 = 0.856245 \text{ (from a calculator)}$$
$$\log 3.902 = \underline{0.591287} \text{ (from a calculator)}$$
$$\log \text{ of ratio} = 0.264958 \text{ is closest to log 81:44 from table.}$$

Finding Four-gear Ratios. — When four gears must be used, the gear logarithms make it possible to obtain results of high accuracy quickly and with minimum effort. For example, suppose it is desired to find four gears that will have numbers of teeth in a ratio of 2.105399 to 1. Log 2.105399 = 0.323334. To keep the ratio about equal in each pair of gears, select from the table that set of gears the logarithm of which is equal to about one-half the ratio logarithm, as log 57:37 = 0.187673. By subtracting this from the log of 2.105399, 0.323334, the other logarithm is found to be 0.135661.

From the table, log 41:30 = 0.135663. The result obtained is: $\frac{57}{37} \times \frac{41}{30} = \frac{2337}{1110} =$ 2.105405. The error in the ratio is 0.000006.

In case no combination can be found that nearly equals the logarithm of the ratio, a suitable four-gear combination may be found by a slightly different procedure. For example, what gears will have teeth in the ratio of 595 to 594? As before, take the logarithms from a six-place table, or a calculator,

$$\log 595 = 2.774517$$
$$\log 594 = \underline{2.773786}$$
$$\log \text{ of ratio} = 0.000731$$

Next, from the table of logarithms for gear ratios, select any ratio, say log 72:70 = 0.012235, and add the logarithm of the ratio 595:594, or 0.000731; the sum is 0.012966. Select the logarithm nearest the sum from the table; this is found to be log 68:66 = 0.012965. Now by inverting the first pair selected, 72:70, and multiplying this inverted pair by the ratio just determined, 68:66, the desired ratio will be obtained. Thus, inverting 72:70 gives 70:72. Multiplying by 68:66 gives $\frac{70}{72} \times \frac{68}{66} = \frac{4760}{4752} = \frac{595}{594}$.

Lathe Change-gears. — For calculating the change-gears to cut any pitch on a lathe, the "constant" of the machine must be known. For any lathe, $C:L$ = driver: driven gear, in which C = constant of machine and L = threads per inch.

For example, what change-gears are required to cut 1.7345 threads per inch on a lathe having a constant of 4?

$$C : L = 4 : 1.7345$$
$$\log 4 = 0.602060$$
$$\log 1.7345 = \underline{0.239174}$$
$$\text{ratio log} = 0.362886$$

From the table, $\log 113 : 49 = \underline{0.362882}$

$$\text{error in log of ratio} = 0.000004$$

Therefore, the driver has 113 teeth, and the driven gear, 49 teeth.

Relieving Helical Fluted Hobs. — The problem of relieving hobs that have been fluted at right angles to the thread is an example of the special application of the gear logarithms to difficult problems. The usual method is to alter the angle of the helical flutes to agree with previously calculated change-gears. The ratio between the hob and relieving attachment cam is expressed by the following terms:

$$\frac{N}{\cos^2 \alpha} : C = \text{drivers : driven gears}$$

$$\text{Tan } \alpha = \frac{P}{H_c}$$

in which

N = number of flutes in hob;
α = helix angle of thread from plane perpendicular to axis;
C = constant of relieving attachment;
P = axial lead of hob;
H_c = hob pitch circumference, or 3.1416 × pitch diameter.

The constant of a relieving attachment can be found on its index-plate, and is determined by the number of flutes that require equal gears on the change-gear studs. This will vary with different makes of lathes.

Example. — What four change-gears must be used to relieve a helical fluted worm-gear hob, 24 diametral pitch, sextuple thread, 13 degrees, 41 minutes helix angle of thread, with eleven helical flutes, assuming that a relieving attachment having a constant of 4 is to be used?

$$\text{Cos 13 degrees, 41 minutes} = 0.97162$$
$$0.97162^2 = 0.944045$$

$$\frac{N \div \cos^2 \alpha}{C} = \frac{N}{C \times \cos^2 \alpha} ; \quad \frac{11}{4 \times 0.944045} = \frac{11}{3.776}$$

$$\log 11 = 1.04139$$
$$\log 3.776 = \underline{0.57703}$$
$$\log \text{ratio} = 0.46436$$

From tables, $\log 67 : 39 = \underline{0.23501}$

Subtracting from log ratio = 0.22935
From table, $\log 78 : 46 = 0.22933$

Therefore, the gears are $\dfrac{67}{39} \times \dfrac{78}{46} = \dfrac{\text{drivers}}{\text{driven}}$

The ratio of these gears equals 2.913 which is the ratio represented by 11 ÷ 3.776. In relieving hobs for spur gears, the *normal* pitch (or lead of a single-threaded hob) should equal the circular pitch of the gear, and the *axial* pitch = normal pitch ÷ cos α. Sine α = normal pitch ÷ H_c.

Logarithms of Gear Ratios from 1.0084+ to 7.5

Numbers of Teeth	Logarithm of Ratio	Numbers of Teeth	Logarithm of Ratio	Numbers of Teeth	Logarithm of Ratio	Numbers of Teeth	Logarithm of Ratio
120 : 16	0.875061	99 : 16	0.791515	101 : 18	0.749049	115 : 22	0.718275
119 : 16	.871427	105 : 17	.790740	112 : 20	.748188	94 : 18	.717855
118 : 16	.867762	111 : 18	.790051	95 : 17	.747275	120 : 23	.717453
117 : 16	.864065	98 : 16	.787106	106 : 19	.746552	99 : 19	.716882
116 : 16	.860338	104 : 17	.786584	117 : 21	.745967	109 : 21	.715207
115 : 16	.856578	110 : 18	.786120	89 : 16	.745270	83 : 16	.714958
114 : 16	.852785	116 : 19	.785704	100 : 18	.744728	114 : 22	.714482
113 : 16	.848958	97 : 16	.782652	111 : 20	.744293	88 : 17	.714034
120 : 17	0.848732	103 : 17	.782388	94 : 17	0.742680	119 : 23	0.713819
119 : 17	.845098	109 : 18	.782154	105 : 19	.742436	93 : 18	.713210
112 : 16	.845098	115 : 19	.781944	116 : 21	.742239	98 : 19	.712473
118 : 17	.841433	114 : 19	.778151	110 : 20	.740363	103 : 20	.711807
111 : 16	.841203	108 : 18	.778151	99 : 18	.740363	108 : 21	.711205
117 : 17	.837737	102 : 17	.778151	88 : 16	.740363	113 : 22	.710656
110 : 16	.837272	96 : 16	.778151	115 : 21	.738479	118 : 23	.710154
116 : 17	.834009	119 : 20	.774517	104 : 19	.738280	82 : 16	.709694
109 : 16	0.833307	113 : 19	0.774325	93 : 17	0.738034	87 : 17	0.709070
115 : 17	.830249	107 : 18	.774111	120 : 22	.736759	92 : 18	.708515
108 : 16	.829304	101 : 17	.773873	109 : 20	.736397	97 : 19	.708018
114 : 17	.826456	95 : 16	.773604	98 : 18	.735954	102 : 20	.707570
107 : 16	.825264	112 : 19	.770464	87 : 16	.735400	107 : 21	.707165
120 : 18	.823909	106 : 18	.770033	114 : 21	.734686	112 : 22	.706795
113 : 17	.822629	100 : 17	.769551	103 : 19	.734084	117 : 23	.706458
106 : 16	.821186	94 : 16	.769008	92 : 17	.733339	81 : 16	.704365
119 : 18	0.820275	117 : 20	0.767156	119 : 22	0.733124	86 : 17	0.704050
112 : 17	.818769	111 : 19	.766570	108 : 20	.732394	91 : 18	.703769
105 : 16	.817069	105 : 18	.765917	97 : 18	.731499	96 : 19	.703518
118 : 18	.816609	99 : 17	.765186	113 : 21	.730859	101 : 20	.703291
111 : 17	.814874	93 : 16	.764363	86 : 16	.730379	106 : 21	.703087
117 : 18	.812913	116 : 20	.763429	102 : 19	.729847	111 : 22	.702900
104 : 16	.812913	110 : 19	.762639	118 : 22	.729459	116 : 23	.702730
110 : 17	.810944	104 : 18	.761761	91 : 17	.728593	120 : 24	.698970
116 : 18	0.809186	98 : 17	0.760777	107 : 20	0.728354	115 : 23	0.698970
103 : 16	.808717	115 : 20	.759668	112 : 21	.726999	110 : 22	.698970
109 : 17	.806978	92 : 16	.759668	96 : 18	.726999	105 : 21	.698970
115 : 18	.805425	109 : 19	.758673	117 : 22	.725763	100 : 20	.698970
102 : 16	.804480	103 : 18	.757565	101 : 19	.725578	95 : 19	.698970
108 : 17	.802975	120 : 21	.756962	85 : 16	.725299	90 : 18	.698970
114 : 18	.801632	114 : 20	.755875	106 : 20	.724276	85 : 17	.698970
120 : 19	.800428	91 : 16	.754921	90 : 17	.723794	80 : 16	.698970
101 : 16	0.800201	108 : 19	0.754670	111 : 21	0.723104	119 : 24	0.695336
107 : 17	.798936	119 : 21	.753328	95 : 18	.722451	114 : 23	.695171
113 : 18	.797806	102 : 18	.753328	116 : 22	.722035	109 : 22	.695004
119 : 19	.796793	113 : 20	.752048	100 : 19	.721246	104 : 21	.694814
100 : 16	.795880	96 : 17	.751822	105 : 20	.720159	99 : 20	.694605
106 : 17	.794857	107 : 19	.750630	84 : 16	.720159	94 : 19	.694374
112 : 18	.793946	90 : 16	.750123	110 : 21	.719173	89 : 18	0.694118
118 : 19	.793128	118 : 21	.749663	89 : 17	.718941	84 : 17	.693830

Logarithms of Gear Ratios from 1.0084+ to 7.5

Numbers of Teeth	Logarithm of Ratio	Numbers of Teeth	Logarithm of Ratio	Numbers of Teeth	Logarithm of Ratio	Numbers of Teeth	Logarithm of Ratio
79 : 16	0.693507	103 : 22	0.670415	76 : 17	0.650365	90 : 21	0.632023
118 : 24	.691671	117 : 25	.670246	116 : 26	.649485	107 : 25	.631444
113 : 23	.691351	112 : 24	.669007	107 : 24	.649173	77 : 18	.631218
108 : 22	.691001	98 : 21	.669007	98 : 22	.648803	94 : 22	.630705
103 : 21	.690618	84 : 18	.669007	89 : 20	.648360	111 : 26	.630350
98 : 20	.690196	107 : 23	.667656	80 : 18	.647817	81 : 19	.629731
93 : 19	.689729	93 : 20	.667453	120 : 27	.647817	98 : 23	.629498
88 : 18	.689210	79 : 17	.667178	111 : 25	.647383	115 : 27	.629334
83 : 17	0.688629	116 : 25	0.666518	71 : 16	0.647138	119 : 28	0.628389
117 : 24	.687975	102 : 22	.666178	102 : 23	.646872	85 : 20	.628389
78 : 16	.687975	88 : 19	.665730	93 : 21	.646264	68 : 16	.628389
112 : 23	.687490	111 : 24	.665112	115 : 26	.645725	106 : 25	.627366
107 : 22	.686961	74 : 16	.665112	84 : 19	.645526	89 : 21	.627171
102 : 21	.686381	97 : 21	.664552	106 : 24	.645095	72 : 17	.626884
97 : 20	.685742	120 : 26	.664208	75 : 17	.644612	110 : 26	.626419
92 : 19	.685034	83 : 18	.663806	97 : 22	.644349	93 : 22	.626060
116 : 24	0.684247	106 : 23	0.663578	119 : 27	0.644183	114 : 27	0.625541
87 : 18	.684247	115 : 25	.662758	110 : 25	.643453	76 : 18	.625541
111 : 23	.683595	92 : 20	.662758	88 : 20	.643453	97 : 23	.625044
82 : 17	.683365	101 : 22	.661899	101 : 23	.642594	118 : 28	.624724
106 : 22	.682883	78 : 17	.661645	79 : 18	.642355	80 : 19	.624336
77 : 16	.682371	110 : 24	.661182	114 : 26	.641932	101 : 24	.624110
101 : 21	.682102	87 : 19	.660766	92 : 21	.641569	105 : 25	.623249
120 : 25	.681241	119 : 26	.660574	105 : 24	.640978	84 : 20	.623249
96 : 20	0.681241	96 : 21	0.660052	70 : 16	0.640978	109 : 26	0.622453
115 : 24	.680487	105 : 23	.659462	118 : 27	.640518	88 : 21	.622263
91 : 19	.680288	73 : 16	.659203	83 : 19	.640324	67 : 16	.621955
110 : 23	.679665	114 : 25	.658965	96 : 22	.639849	113 : 27	.621715
86 : 18	.679226	82 : 18	.658541	109 : 25	.639487	92 : 22	.621365
105 : 22	.678767	91 : 20	.658011	74 : 17	.638783	117 : 28	.621028
81 : 17	.678036	100 : 22	.657577	87 : 20	.638489	71 : 17	.620809
100 : 21	.677781	109 : 24	.657215	100 : 23	.638272	96 : 23	.620543
119 : 25	0.677607	118 : 26	0.656909	113 : 26	0.638105	100 : 24	0.619789
114 : 24	.676694	77 : 17	.656042	117 : 27	.636822	75 : 18	.619789
95 : 20	.676694	86 : 19	.655745	104 : 24	.636822	104 : 25	.619093
76 : 16	.676694	95 : 21	.655504	91 : 21	.636822	79 : 19	.618874
109 : 23	.675699	104 : 23	.655306	78 : 18	.636822	108 : 26	.618451
90 : 19	.675489	113 : 25	.655138	108 : 25	.635484	83 : 20	.618048
104 : 22	.674611	117 : 26	.653213	95 : 22	.635301	112 : 27	.617854
85 : 18	.674146	108 : 24	.653213	82 : 19	.635060	116 : 28	.617300
118 : 25	0.673942	99 : 22	0.653213	69 : 16	0.634729	87 : 21	0.617300
99 : 21	.673416	90 : 20	.653213	112 : 26	.634245	120 : 29	.616783
113 : 24	.672867	81 : 18	.653213	99 : 23	.633907	91 : 22	.616619
80 : 17	.672641	72 : 16	.653213	86 : 20	.633469	95 : 23	.615996
94 : 20	.672098	112 : 25	.651278	116 : 27	.633094	99 : 24	.615424
108 : 23	.671696	103 : 23	.651109	73 : 17	.632874	66 : 16	.615424
75 : 16	.670941	94 : 21	.650909	103 : 24	.632626	103 : 25	.614897
89 : 19	.670636	85 : 19	.650665	120 : 28	.632023	70 : 17	.614649

Logarithms of Gear Ratios from 1.0084+ to 7.5

Numbers of Teeth	Logarithm of Ratio	Numbers of Teeth	Logarithm of Ratio	Numbers of Teeth	Logarithm of Ratio	Numbers of Teeth	Logarithm of Ratio
107 : 26	0.614411	67 : 17	0.595626	99 : 26	0.580662	81 : 22	0.566062
111 : 27	.613959	63 : 16	.595221	118 : 31	.580520	92 : 25	.565848
74 : 18	.613959	118 : 30	.594761	114 : 30	.579784	103 : 28	.565679
115 : 28	.613540	114 : 29	.594507	95 : 25	.579784	114 : 31	.565543
78 : 19	.613341	110 : 28	.594235	76 : 20	.579784	88 : 24	.564271
119 : 29	.613149	106 : 27	.593942	110 : 29	.578995	110 : 30	.564271
82 : 20	.612784	102 : 26	.593627	91 : 24	.578830	99 : 27	.564271
86 : 21	.612279	98 : 25	.593286	72 : 19	.578579	77 : 21	.564271
90 : 22	0.611820	94 : 24	0.592917	106 : 28	0.578148	66 : 18	0.564271
94 : 23	.611400	90 : 23	.592515	87 : 23	.577792	117 : 32	.563036
98 : 24	.611015	86 : 22	.592076	102 : 27	.577236	106 : 29	.562908
102 : 25	.610660	82 : 21	.591595	68 : 18	.577236	95 : 26	.562750
106 : 26	.610333	117 : 30	.591065	117 : 31	.576824	84 : 23	.562552
110 : 27	.610029	78 : 20	.591065	83 : 22	.576655	73 : 20	.562293
114 : 28	.609747	113 : 29	.590680	98 : 26	.576253	62 : 17	.561943
118 : 29	.609484	74 : 19	.590478	113 : 30	.575957	113 : 31	.561717
65 : 16	0.608793	109 : 28	0.590269	64 : 17	0.575731	102 : 28	0.561442
69 : 17	.608400	105 : 27	.589826	79 : 21	.575408	91 : 25	.561101
73 : 18	.608050	70 : 18	.589826	94 : 25	.575188	120 : 33	.560667
77 : 19	.607737	101 : 26	.589348	109 : 29	.575029	80 : 22	.560667
81 : 20	.607455	66 : 17	.589105	120 : 32	.574031	109 : 30	.560305
85 : 21	.607200	97 : 25	.588832	105 : 28	.574031	69 : 19	.560096
89 : 22	.606967	62 : 16	.588272	90 : 24	.574031	98 : 27	.559862
93 : 23	.606755	93 : 24	.588272	75 : 20	.574031	116 : 32	.559308
97 : 24	0.606561	120 : 31	0.587820	60 : 16	.574031	87 : 24	0.559308
101 : 25	.606381	89 : 23	.587662	116 : 31	.573096	105 : 29	.558791
105 : 26	.606216	116 : 30	.587337	86 : 23	.572770	76 : 21	.558594
109 : 27	.606063	85 : 22	.586996	71 : 19	.572505	94 : 26	.558155
113 : 28	.605920	112 : 29	.586820	112 : 30	.572097	112 : 31	.557856
117 : 29	.605788	108 : 28	.586266	97 : 26	.571798	65 : 18	.557641
64 : 16		81 : 21	.586266	82 : 22	.571391	83 : 23	.557350
or		104 : 27	.585670	108 : 29	.571026	101 : 28	.557163
any	0.602060	77 : 20	0.585461	67 : 18	0.570802	119 : 33	0.557033
4 to 1		100 : 26	.585027	93 : 25	.570543	108 : 30	.556303
ratio		73 : 19	.584569	119 : 32	.570397	90 : 25	.556303
119 : 30	.598426	96 : 25	.584331	104 : 28	.569875	72 : 20	.556303
115 : 29	.598300	119 : 31	.584185	78 : 21	.569875	115 : 32	.555548
111 : 28	.598165	115 : 30	.583577	115 : 31	.569336	97 : 27	.555408
107 : 27	.598020	92 : 24	.583577	89 : 24	.569179	79 : 22	.555204
103 : 26	.597864	69 : 18	.583577	63 : 17	.568892	61 : 17	.554881
99 : 25	0.597695	111 : 29	0.582925	100 : 27	0.568636	104 : 29	0.554635
95 : 24	.597512	88 : 23	.582755	111 : 30	.568202	86 : 24	.554287
91 : 23	.597314	65 : 17	.582465	74 : 20	.568202	111 : 31	.553961
87 : 22	.597007	107 : 28	.582226	85 : 23	.567691	68 : 19	.553755
83 : 21	.596859	84 : 22	.581857	96 : 26	.567298	93 : 26	.553510
79 : 20	.596597	103 : 27	.581473	107 : 29	.566986	118 : 33	.553368
75 : 19	.596308	61 : 16	.581210	118 : 32	.566732	100 : 28	.552842
71 : 18	.595986	80 : 21	.580871	70 : 19	.566344	75 : 21	.552842

Logarithms of Gear Ratios from 1.0084+ to 7.5

Numbers of Teeth	Logarithm of Ratio	Numbers of Teeth	Logarithm of Ratio	Numbers of Teeth	Logarithm of Ratio	Numbers of Teeth	Logarithm of Ratio
107 : 30	0.552263	76 : 22	0.538391	104 : 31	0.525672	101 : 31	0.512960
82 : 23	.552086	107 : 31	.538022	114 : 34	.525426	114 : 35	.512837
114 : 32	.551755	69 : 20	.537819	67 : 20	.525045	117 : 36	.511883
89 : 25	.551450	100 : 29	.537602	77 : 23	.524763	104 : 32	.511883
96 : 27	.550907	93 : 27	.537119	87 : 26	.524546	91 : 28	.511883
64 : 18	.550907	62 : 18	.537119	97 : 29	.524374	65 : 20	.511883
103 : 29	.550439	117 : 34	.536707	107 : 32	.524234	120 : 37	.510980
71 : 20	.550228	86 : 25	.536559	117 : 35	.524118	107 : 33	.510870
110 : 31	0.550031	110 : 32	0.536243	120 : 36	0.522879	94 : 29	0.510730
117 : 33	.549672	55 : 16	.536243	110 : 33	.522879	81 : 25	.510545
78 : 22	.549672	79 : 23	.535899	100 : 30	.522879	68 : 21	.510290
85 : 24	.549208	103 : 30	.535716	90 : 27	.522879	110 : 34	.509914
92 : 26	.548815	120 : 35	.535113	80 : 24	.522879	97 : 30	.509650
99 : 28	.548477	96 : 28	.535113	70 : 21	.522879	84 : 26	.509306
106 : 30	.548185	72 : 21	.535113	60 : 18	.522879	113 : 35	.509010
113 : 32	.547928	113 : 33	.534565	113 : 34	.521600	71 : 22	.508836
120 : 34	0.547702	89 : 26	0.534417	103 : 31	0.521476	100 : 31	0.508638
67 : 19	.547321	65 : 19	.534160	93 : 28	.521325	116 : 36	.508156
74 : 21	.547012	106 : 31	.533944	83 : 25	.521138	87 : 27	.508156
81 : 23	.546757	82 : 24	.533603	73 : 22	.520900	103 : 32	.507687
88 : 25	.546543	99 : 29	.533237	63 : 19	.520587	74 : 23	.507504
95 : 27	.546360	116 : 34	.532980	116 : 35	.520390	119 : 37	.507345
102 : 29	.546202	75 : 22	.532639	106 : 32	.520156	90 : 28	.507085
109 : 31	.546065	92 : 27	.532424	96 : 29	.519873	106 : 33	.506792
119 : 34	0.544068	109 : 32	0.532277	86 : 26	0.519525	61 : 19	0.506576
112 : 32	.544068	119 : 35	.531479	119 : 36	.519245	77 : 24	.506280
105 : 30	.544068	102 : 30	.531479	76 : 23	.519086	93 : 29	.506085
98 : 28	.544068	85 : 25	.531479	109 : 33	.518913	109 : 34	.505948
91 : 26	.544068	68 : 20	.531479	99 : 30	.518514	112 : 35	.505150
84 : 24	.544068	112 : 33	.530704	89 : 27	.518026	96 : 30	.505150
77 : 22	.544068	95 : 28	.530566	112 : 34	.517739	80 : 25	.505150
70 : 20	.544068	78 : 23	.530367	79 : 24	.517416	64 : 20	.505150
63 : 18	0.544068	61 : 18	0.530057	102 : 31	0.517239	115 : 36	0.504395
56 : 16	.544068	105 : 31	.529828	115 : 35	.516630	99 : 31	.504274
115 : 33	.542184	88 : 26	.529509	69 : 21	.516630	83 : 26	.504105
108 : 31	.542062	115 : 34	.529219	92 : 28	.516629	67 : 21	.503856
101 : 29	.541923	71 : 21	.529039	105 : 32	.516039	118 : 37	.503680
87 : 25	.541579	98 : 29	.528828	82 : 25	.515874	102 : 32	.503450
80 : 23	.541362	108 : 32	.528274	118 : 36	.515580	86 : 27	.503135
73 : 21	.541104	81 : 24	.528274	95 : 29	.515326	105 : 33	.502675
66 : 19	0.540790	118 : 35	0.527814	108 : 33	0.514910	70 : 22	0.502675
118 : 34	.540403	91 : 27	.527678	72 : 22	.514910	89 : 28	.502232
111 : 32	.540173	64 : 19	.527426	85 : 26	.514446	108 : 34	.501945
104 : 30	.539912	101 : 30	.527200	98 : 30	.514105	73 : 23	.501595
97 : 28	.539614	111 : 33	.526809	111 : 34	.513844	92 : 29	.501390
90 : 26	.539269	74 : 22	.526809	62 : 19	.513638	111 : 35	.501255
83 : 24	.538867	84 : 25	.526339	75 : 23	.513334	95 : 30	.500602
114 : 33	.538391	94 : 28	.525970	88 : 27	.513119	76 : 24	.500602

Logarithms of Gear Ratios from 1.0084+ to 7.5

Numbers of Teeth	Logarithm of Ratio	Numbers of Teeth	Logarithm of Ratio	Numbers of Teeth	Logarithm of Ratio	Numbers of Teeth	Logarithm of Ratio
117 : 37	0.499984	120 : 39	0.488117	92 : 31	0.472426	81 : 28	0.461327
98 : 31	.499864	80 : 26	.488117	89 : 30	.472269	107 : 37	.461182
79 : 25	.499687	83 : 27	.487714	86 : 29	.472101	104 : 36	.460731
120 : 38	.499398	86 : 28	.487341	83 : 28	.471920	78 : 27	.460731
101 : 32	.499171	89 : 29	.486992	80 : 27	.471726	101 : 35	.460253
82 : 26	.498841	92 : 30	.486667	77 : 26	.471517	75 : 26	.460088
104 : 33	.498519	95 : 31	.486362	74 : 25	.471291	98 : 34	.459747
63 : 20	.498311	98 : 32	.486076	71 : 24	.471047	72 : 25	.459393
85 : 27	0.498055	101 : 33	0.485808	68 : 23	0.470781	95 : 33	0.459210
107 : 34	.497905	104 : 34	.485554	65 : 22	.470491	118 : 41	.459098
110 : 35	.497325	107 : 35	.485316	62 : 21	.470172	115 : 40	.458638
88 : 28	.497325	110 : 36	.485090	118 : 40	.469822	92 : 32	.458638
113 : 36	.496776	113 : 37	.484877	115 : 39	.469633	69 : 24	.458638
91 : 29	.496643	116 : 38	.484674	112 : 38	.469434	112 : 39	.458153
69 : 22	.496426	119 : 39	.484482	109 : 37	.469225	89 : 31	.458028
116 : 37	.496256	61 : 20	.484300	106 : 36	.469003	66 : 23	.457816
94 : 30	0.496007	64 : 21	0.483961	103 : 35	0.468769	109 : 38	0.457643
119 : 38	.495763	67 : 22	.483652	100 : 34	.468521	86 : 30	.457377
72 : 23	.495605	70 : 23	.483370	97 : 33	.468258	106 : 37	.457104
97 : 31	.495410	73 : 24	.483112	94 : 32	.467978	63 : 22	.456918
100 : 32	.494850	76 : 25	.482874	91 : 31	.467680	83 : 29	.456680
75 : 24	.494850	79 : 26	.482654	88 : 30	.467361	103 : 36	.456535
103 : 33	.494323	82 : 27	.482450	85 : 29	.467021	120 : 42	.455932
78 : 25	.494155	85 : 28	.482261	82 : 28	.466656	100 : 35	.455932
106 : 34	.493827	88 : 29	0.482085	120 : 41	0.466397	80 : 28	0.455932
81 : 26	.493511	91 : 30	.481920	79 : 27	.466263	60 : 21	.455932
109 : 35	.493359	94 : 31	.481766	117 : 40	.466126	117 : 41	.455402
112 : 36	.492916	97 : 32	.481622	114 : 39	.465840	97 : 34	.455293
84 : 27	.492916	100 : 33	.481486	76 : 26	.465840	77 : 27	.455127
115 : 37	.492496	103 : 34	.481358	111 : 38	.465539	114 : 40	.454845
87 : 28	.492361	106 : 35	.481237	73 : 25	.465383	94 : 33	.454614
118 : 38	.492098	109 : 36	.481124	108 : 37	.465222	111 : 39	.454258
90 : 29	0.491845	112 : 37	0.481016	70 : 24	0.464887	74 : 26	0.454258
93 : 30	.491362	115 : 38	.480914	102 : 35	.464532	91 : 32	.453891
62 : 20	.491362	118 : 39	.480817	67 : 23	.464347	71 : 25	.453318
96 : 31	.490910	48 : 16		99 : 34	.464156	88 : 31	.453121
65 : 21	.490694	or any	.477121	96 : 33	.463757	105 : 37	.452988
99 : 32	.490485	3 to 1		64 : 22	.463757	119 : 42	.452298
102 : 33	.490086	ratio		93 : 32	.463333	102 : 36	.452298
68 : 22	.490086			61 : 21	.463111	85 : 30	.452298
105 : 34	0.489710	119 : 40	.473487	90 : 31	0.462881	68 : 24	0.452298
71 : 23	.489531	116 : 39	0.473393	119 : 41	.462763	116 : 41	.451674
108 : 35	.489356	113 : 38	.473295	116 : 40	.462398	99 : 35	.451567
111 : 36	.489021	110 : 37	.473191	87 : 30	.462398	82 : 29	.451416
74 : 24	.489021	107 : 36	.473081	58 : 20	.462398	65 : 23	.451186
114 : 37	.488703	104 : 35	.472965	113 : 39	.462013	113 : 40	.451018
77 : 25	.488551	101 : 34	.472843	84 : 29	.461881	96 : 34	.450792
117 : 38	.488402	98 : 33	.472712	110 : 38	.461609	79 : 28	.450469
		95 : 32	.472574				

Logarithms of Gear Ratios from 1.0084+ to 7.5

Numbers of Teeth	Logarithm of Ratio	Numbers of Teeth	Logarithm of Ratio	Numbers of Teeth	Logarithm of Ratio	Numbers of Teeth	Logarithm of Ratio
110 : 39	0.450328	77 : 28	0.439333	94 : 35	0.429060	84 : 32	0.419129
93 : 33	.449969	66 : 24	.439333	102 : 38	.428817	118 : 45	.418670
62 : 22	.449969	55 : 20	.439333	110 : 41	.428609	97 : 37	.418570
107 : 38	.449600	118 : 43	.438414	59 : 22	.428429	76 : 29	.418416
76 : 27	.449450	107 : 39	.438319	67 : 25	.428135	55 : 21	.418143
90 : 32	.449092	96 : 35	.438203	75 : 28	.427903	89 : 34	.417911
104 : 37	.448832	85 : 31	.438057	83 : 31	.427716	102 : 39	.417536
118 : 42	.448633	74 : 27	.437868	91 : 34	.427563	68 : 26	.417536
73 : 26	0.448350	63 : 23	.437613	99 : 37	0.427434	115 : 44	0.417245
87 : 31	.448158	115 : 42	0.437449	107 : 40	.427324	81 : 31	.417123
101 : 36	.448019	104 : 38	.437250	115 : 43	.427229	94 : 36	.416825
115 : 41	.447914	93 : 34	.437004	120 : 45	.425969	107 : 41	.416600
98 : 35	.447158	82 : 30	.436693	112 : 42	.425969	60 : 23	.416424
84 : 30	.447158	112 : 41	.436434	104 : 39	.425969	73 : 28	.416165
112 : 40	.447158	71 : 26	.436285	96 : 36	.425969	86 : 33	.415985
70 : 25	.447158	101 : 37	.436120	88 : 33	.425969	99 : 38	.415852
109 : 39	0.446362	90 : 33	.435729	80 : 30	0.425969	112 : 43	0.415750
95 : 34	.446245	60 : 22	0.435729	72 : 27	.425969	117 : 45	.414973
81 : 29	.446087	109 : 40	.435367	64 : 24	.425969	91 : 35	.414973
67 : 24	.445864	79 : 29	.435229	56 : 21	.425969	78 : 30	.414973
120 : 43	.445713	98 : 36	.434924	117 : 44	.424733	52 : 20	.414973
106 : 38	.445522	117 : 43	.434717	109 : 41	.424643	109 : 42	.414177
92 : 33	.445274	68 : 25	.434569	101 : 38	.424538	96 : 37	.414070
117 : 42	.444937	87 : 32	.434369	93 : 35	.424415	83 : 32	.413928
78 : 28	0.444937	106 : 39	.434241	85 : 32	0.424269	70 : 27	0.413734
103 : 37	.444636	114 : 42	0.433656	77 : 29	.424093	57 : 22	.413452
64 : 23	.444452	95 : 35	.433656	69 : 26	.423876	101 : 39	.413257
89 : 32	.444240	76 : 28	.433656	61 : 23	.423602	88 : 34	.413004
114 : 41	.444121	57 : 21	.433656	114 : 43	.423436	119 : 46	.412789
100 : 36	.443698	103 : 38	.433054	106 : 40	.423246	75 : 29	.412663
75 : 27	.443698	84 : 31	.432918	98 : 37	.423024	106 : 41	.412522
111 : 40	.443263	65 : 24	.432702	90 : 34	.422764	62 : 24	.412180
86 : 31	0.443137	111 : 41	.432539	82 : 31	0.422452	93 : 36	0.412180
61 : 22	.442907	92 : 34	0.432309	119 : 45	.422335	111 : 43	.411855
97 : 35	.442704	119 : 44	.432094	111 : 42	.422074	80 : 31	.411728
72 : 26	.442359	73 : 27	.431959	74 : 28	.422074	98 : 38	.411443
108 : 39	.442359	100 : 37	.431798	103 : 39	.421773	116 : 45	.411246
119 : 43	.442079	108 : 40	.431364	95 : 36	.421421	67 : 26	.411102
83 : 30	.441957	81 : 30	.431364	87 : 33	.421005	85 : 33	.410905
94 : 34	.441649	54 : 20	.431364	58 : 22	.421005	103 : 40	.410777
105 : 38	0.441406	116 : 43	.430990	108 : 41	0.420640	90 : 35	0.410175
58 : 21	.441209	89 : 33	0.430876	79 : 30	.420506	72 : 28	.410175
69 : 25	.440909	62 : 23	.430664	100 : 38	.420216	54 : 21	.410175
80 : 29	.440692	97 : 36	.430469	71 : 27	.419895	113 : 44	.409626
91 : 33	.440528	105 : 39	.430125	92 : 35	.419720	95 : 37	.409522
102 : 37	.440399	70 : 26	.430125	113 : 43	.419610	77 : 30	.409369
113 : 41	.440295	113 : 42	.429829	105 : 40	.419130	59 : 23	.409124
99 : 36	.439333	78 : 29	.429697	63 : 24	.419130	100 : 39	.408935
88 : 32	.439333	86 : 32	.429349	. . . . . .	. . . . . .	. . . . . .	. . . . . .

Logarithms of Gear Ratios from 1.0084+ to 7.5

Numbers of Teeth	Logarithm of Ratio	Numbers of Teeth	Logarithm of Ratio	Numbers of Teeth	Logarithm of Ratio	Numbers of Teeth	Logarithm of Ratio
82 : 32	0.408664	40 : 16		110 : 45	0.388180	48 : 20	0.380211
105 : 41	.408405	or any	0.397940	88 : 36	.388180	115 : 48	.379457
64 : 25	.408240	2½ to 1		66 : 27	.388180	103 : 43	.379369
87 : 34	.408040	ratio		105 : 43	.387721	91 : 38	.379258
110 : 43	.407924	117 : 47	.396088	83 : 34	.387599	79 : 33	.379113
115 : 45	.407485	112 : 45	.396006	61 : 25	.387390	67 : 28	.378916
92 : 36	.407485	107 : 43	.395915	100 : 41	.387216	55 : 23	.378635
69 : 27	.407485	102 : 41	.395816	117 : 48	.386945	98 : 41	.378442
120 : 47	0.407083	97 : 39	0.395707	78 : 32	0.386945	86 : 36	0.378196
97 : 38	.406988	92 : 37	.395586	95 : 39	.386659	117 : 49	.377990
74 : 29	.406834	87 : 35	.395451	112 : 46	.386460	74 : 31	.377870
51 : 20	.406540	82 : 33	.395300	56 : 23	.386460	105 : 44	.377737
79 : 31	.406265	77 : 31	.395129	73 : 30	.386202	93 : 39	.377418
107 : 42	.406135	72 : 29	.394935	90 : 37	.386041	62 : 26	.377418
84 : 33	.405765	67 : 27	.394711	107 : 44	.385931	112 : 47	.377120
56 : 22	.405765	62 : 25	.394452	102 : 42	.385351	81 : 34	.377006
117 : 46	0.405428	119 : 48	0.394306	119 : 49	0.385351	100 : 42	0.376751
89 : 35	.405322	57 : 23	.394147	85 : 35	.385351	50 : 21	.376751
61 : 24	.405119	109 : 44	.393974	68 : 28	.385351	119 : 50	.376577
94 : 37	.404926	52 : 21	.393784	51 : 21	.385351	69 : 29	.376451
99 : 39	.404571	99 : 40	.393575	114 : 47	.384807	88 : 37	.376281
66 : 26	.404571	94 : 38	.393344	97 : 40	.384712	107 : 45	.376171
104 : 41	.404249	89 : 36	.393088	80 : 33	.384576	95 : 40	.375664
71 : 28	.404100	84 : 34	.392800	63 : 26	.384367	76 : 32	.375664
109 : 43	0.403958	79 : 32	0.392477	109 : 45	0.384214	57 : 24	0.375664
114 : 45	.403692	116 : 47	.392360	92 : 38	.384004	102 : 43	.375132
76 : 30	.403692	111 : 45	.392111	75 : 31	.383700	83 : 35	.375010
119 : 47	.403449	74 : 30	.392111	104 : 43	.383565	64 : 27	.374816
81 : 32	.403335	106 : 43	.391837	87 : 36	.383217	109 : 46	.374669
86 : 34	.403020	69 : 28	.391692	58 : 24	.383217	90 : 38	.374459
91 : 36	.402739	101 : 41	.391538	99 : 41	.382851	116 : 49	.374262
96 : 38	.402488	96 : 39	.391207	70 : 29	.382700	71 : 30	.374137
101 : 40	0.402261	64 : 26	0.391207	111 : 46	0.382565	97 : 41	0.373988
53 : 21	.402057	91 : 37	.390840	82 : 34	.382335	78 : 33	.373581
111 : 44	.401870	59 : 24	.390641	94 : 39	.382063	52 : 22	.373581
58 : 23	.401700	86 : 35	.390431	53 : 22	.381853	111 : 47	.373225
63 : 25	.401401	113 : 46	.390321	118 : 49	.381686	85 : 36	.373116
68 : 27	.401145	108 : 44	.389971	65 : 27	.381550	59 : 25	.372912
73 : 29	.400925	81 : 33	.389971	77 : 32	.381341	92 : 39	.372723
78 : 31	.400733	54 : 22	.389971	89 : 37	.381188	66 : 28	.372386
83 : 33	0.400564	103 : 42	0.389588	101 : 42	0.381072	99 : 42	0.372386
88 : 35	.400415	76 : 31	.389452	113 : 47	.380981	106 : 45	.372093
93 : 37	.400281	98 : 40	.389166	120 : 50	.380211	73 : 31	.371961
98 : 39	.400162	49 : 20	.389166	108 : 45	.380211	113 : 48	.371837
103 : 41	.400053	120 : 49	.388985	96 : 40	.380211	120 : 51	.371611
108 : 43	.399955	71 : 29	.388860	84 : 35	.380211	80 : 34	.371611
113 : 45	.399866	93 : 38	.388699	72 : 30	.380211	87 : 37	.371318
118 : 47	.399784	115 : 47	.388600	60 : 25	.380211	94 : 40	.371068

Logarithms of Gear Ratios from 1.0084+ to 7.5

Numbers of Teeth	Logarithm of Ratio	Numbers of Teeth	Logarithm of Ratio	Numbers of Teeth	Logarithm of Ratio	Numbers of Teeth	Logarithm of Ratio
47 : 20	0.371068	115 : 50	0.361728	70 : 31	0.353736	62 : 28	0.345234
101 : 43	.370853	92 : 40	.361728	79 : 35	.353559	104 : 47	.344935
54 : 23	.370666	69 : 30	.361728	88 : 39	.353418	73 : 33	.344809
115 : 49	.370502	46 : 20	.361728	97 : 43	.353303	115 : 52	.344695
61 : 26	.370357	108 : 47	.361326	106 : 47	.353208	84 : 38	.344496
68 : 29	.370111	85 : 37	.361217	115 : 51	.353128	95 : 43	.344255
75 : 32	.369911	62 : 27	.361028	117 : 52	.352183	53 : 24	.344065
82 : 35	.369746	101 : 44	.360869	108 : 48	.352183	117 : 53	.343910
89 : 38	0.369606	117 : 51	0.360616	99 : 44	0.352183	64 : 29	0.343782
96 : 41	.369487	78 : 34	.360616	90 : 40	.352183	75 : 34	.343582
103 : 44	.369385	94 : 41	.360344	81 : 36	.352183	86 : 39	.343434
110 : 47	.369295	55 : 24	.360152	72 : 32	.352183	97 : 44	.343319
117 : 50	.369216	71 : 31	.359897	63 : 28	.352183	108 : 49	.343228
119 : 51	.367977	87 : 38	.359736	45 : 20	.352183	119 : 54	.343153
105 : 45	.367977	103 : 45	.359625	119 : 53	.351271	99 : 45	.342423
98 : 42	.367977	119 : 52	.359544	110 : 49	.351197	88 : 40	.342423
91 : 39	0.367977	112 : 49	0.359022	101 : 45	0.351109	77 : 35	0.342423
84 : 36	.367977	96 : 42	.359022	92 : 41	.351004	66 : 30	.342423
77 : 33	.367977	80 : 35	.359022	83 : 37	.350876	55 : 25	.342423
70 : 30	.367977	64 : 28	.359022	74 : 33	.350718	112 : 51	.341648
63 : 27	.367977	48 : 21	.359022	65 : 29	.350515	101 : 46	.341564
56 : 24	.367977	105 : 46	.358432	56 : 25	.350248	90 : 41	.341459
49 : 21	.367977	89 : 39	.358325	103 : 46	.350079	79 : 36	.341325
114 : 49	.366709	73 : 32	.358173	94 : 42	.349879	57 : 26	.340902
107 : 46	0.366626	57 : 25	0.357935	85 : 38	0.349635	103 : 47	0.340739
100 : 43	.366532	98 : 43	.357758	114 : 51	.349335	92 : 42	.340539
93 : 40	.366423	82 : 36	.357511	76 : 34	.349335	46 : 21	.340539
86 : 37	.366297	107 : 47	.357286	105 : 47	.349091	81 : 37	.340283
79 : 34	.366148	66 : 29	.357146	67 : 30	.348954	116 : 53	.340182
72 : 31	.365971	91 : 40	.356981	96 : 43	.348803	105 : 48	.339948
65 : 28	.365755	116 : 51	.356888	87 : 39	.348455	70 : 32	.339948
58 : 25	.365488	75 : 33	.356547	58 : 26	.348455	94 : 43	.339659
109 : 47	0.365329	50 : 22	0.356547	107 : 48	0.348143	118 : 54	0.339488
102 : 44	.365148	100 : 44	.356547	78 : 35	.348027	59 : 27	.339488
95 : 41	.364940	109 : 48	.356185	98 : 44	.347773	83 : 38	.339295
88 : 38	.364700	84 : 37	.356078	49 : 22	.347773	107 : 49	.339188
81 : 35	.364417	59 : 26	.355879	118 : 53	.347606	120 : 55	.338819
118 : 51	.364312	93 : 41	.355699	69 : 31	.347487	96 : 44	.338819
111 : 48	.364082	102 : 45	.355388	89 : 40	.347330	72 : 33	.338819
74 : 32	.364082	68 : 30	.355388	109 : 49	.347230	48 : 22	.338819
104 : 45	0.363821	111 : 49	0.355127	100 : 45	0.346788	109 : 50	0.338457
67 : 29	.363677	77 : 34	.355012	80 : 36	.346788	85 : 39	.338354
90 : 39	.363178	120 : 53	.354905	60 : 27	.346788	61 : 28	.338172
60 : 26	.363178	86 : 38	.354715	111 : 50	.346353	98 : 45	.338014
113 : 49	.362882	95 : 42	.354474	91 : 41	.346258	111 : 51	.337753
53 : 23	.362548	52 : 23	.354276	51 : 23	.345842	74 : 34	.337753
76 : 33	.362300	113 : 50	.354108	113 : 51	.345508	87 : 40	.337459
99 : 43	.362167	61 : 27	.353966	93 : 42	.345234	100 : 46	.337242

Logarithms of Gear Ratios from 1.0084+ to 7.5

Numbers of Teeth	Logarithm of Ratio	Numbers of Teeth	Logarithm of Ratio	Numbers of Teeth	Logarithm of Ratio	Numbers of Teeth	Logarithm of Ratio
113 : 52	0.337075	111 : 52	0.329320	88 : 42	0.321233	107 : 52	0.313381
63 : 29	.336943	96 : 45	.329059	44 : 21	.321233	72 : 35	.313265
76 : 35	.336746	64 : 30	.329059	111 : 53	.321047	109 : 53	.313151
89 : 41	.336606	113 : 53	.328803	67 : 32	.320925	74 : 36	.312929
102 : 47	.336502	81 : 38	.328701	90 : 43	.320774	111 : 54	.312929
115 : 53	.336422	98 : 46	.328468	113 : 54	.320685	113 : 55	.312716
117 : 54	.335792	115 : 54	.328304	115 : 55	.320335	76 : 37	.312612
91 : 42	.335792	66 : 31	.328182	92 : 44	.320335	115 : 56	.312510
78 : 36	0.335792	83 : 39	0.328014	69 : 33	0.320335	117 : 57	0.312311
65 : 30	.335792	100 : 47	.327902	46 : 22	.320335	78 : 38	.312311
52 : 24	.335792	117 : 55	.327823	117 : 56	.319998	119 : 58	.312119
119 : 55	.335184	119 : 56	.327359	71 : 34	.319779	80 : 39	.312025
106 : 49	.335110	85 : 40	.327359	119 : 57	.319672	82 : 40	.311754
93 : 43	.335014	68 : 32	.327359	96 : 46	.319513	41 : 20	.311754
80 : 37	.334888	51 : 24	.327359	73 : 35	.319255	84 : 41	.311495
67 : 31	.334713	104 : 49	.326837	98 : 47	.319128	86 : 42	.311249
54 : 25	0.334454	87 : 41	0.326735	100 : 48	0.318759	43 : 21	0.311249
95 : 44	.334271	70 : 33	.326584	75 : 36	.318759	88 : 43	.311014
82 : 38	.334030	89 : 42	.326141	102 : 49	.318404	90 : 44	.310790
110 : 51	.333823	108 : 51	.325854	77 : 37	.318289	45 : 22	.310790
69 : 32	.333699	72 : 34	.325854	52 : 25	.318063	92 : 45	.310575
97 : 45	.333559	91 : 43	.325573	79 : 38	.317844	94 : 46	.310370
84 : 39	.333215	55 : 26	.325389	106 : 51	.317736	96 : 47	.310173
56 : 26	.333215	74 : 35	.325164	81 : 39	.317420	98 : 48	.309985
99 : 46	0.332877	93 : 44	0.325030	110 : 53	0.317117	100 : 49	0.309804
71 : 33	.332744	112 : 53	.324942	83 : 40	.317018	102 : 50	.309630
114 : 53	.332629	95 : 45	.324511	112 : 54	.316824	51 : 25	.309630
86 : 40	.332439	76 : 36	.324511	85 : 41	.316635	104 : 51	.309463
43 : 20	.332439	57 : 27	.324511	114 : 55	.316542	53 : 26	.309303
101 : 47	.332224	116 : 55	.324095	87 : 42	.316270	108 : 53	.309148
58 : 27	.332064	97 : 46	.324014	58 : 28	.316270	55 : 27	.308999
73 : 34	.331844	78 : 37	.323893	118 : 57	.316007	112 : 55	.308855
88 : 41	0.331699	59 : 28	0.323694	89 : 43	0.315922	57 : 28	0.308717
103 : 48	.331596	99 : 47	.323537	120 : 58	.315753	116 : 57	.308583
118 : 55	.331519	120 : 57	.323306	60 : 29	.315753	59 : 29	.308454
105 : 49	.330993	80 : 38	.323306	91 : 44	.315589	118 : 58	.308454
90 : 42	.330993	101 : 48	.323080	93 : 45	.315270	120 : 59	.308329
75 : 35	.330993	61 : 29	.322932	62 : 30	.315270	61 : 30	.308209
60 : 28	.330993	82 : 39	.322749	95 : 46	.314966	63 : 31	.307979
45 : 21	.330993	103 : 49	.322641	64 : 31	.314818	65 : 32	.307763
107 : 50	0.330414	105 : 50	0.322219	97 : 47	0.314674	67 : 33	0.307561
92 : 43	.330319	84 : 40	.322219	99 : 48	.314394	69 : 34	.307370
77 : 36	.330188	63 : 30	.322219	66 : 32	.314394	71 : 35	.307190
62 : 29	.329994	42 : 20	.322219	101 : 49	.314125	73 : 36	.307020
109 : 51	.329855	107 : 51	.321814	68 : 33	.313995	75 : 37	.306860
94 : 44	.329675	86 : 41	.321715	103 : 50	.313867	77 : 38	.306707
47 : 22	.329675	65 : 31	.321552	105 : 51	.313619	79 : 39	.306563
79 : 37	.329425	109 : 52	.321423	70 : 34	.313619	81 : 40	.306425

Logarithms of Gear Ratios from 1.0084+ to 7.5

Numbers of Teeth	Logarithm of Ratio	Numbers of Teeth	Logarithm of Ratio	Numbers of Teeth	Logarithm of Ratio	Numbers of Teeth	Logarithm of Ratio
83 : 41	0.306294	71 : 36	0.294956	93 : 48	0.287242	99 : 52	0.279632
85 : 42	.306170	69 : 35	.294781	62 : 32	.287242	59 : 31	.279490
87 : 43	.306051	67 : 34	.294596	91 : 47	.286944	78 : 41	.279311
89 : 44	.305937	65 : 33	.294400	120 : 62	.286790	97 : 51	.279202
91 : 45	.305829	63 : 32	.294191	89 : 46	.286632	116 : 61	.279128
93 : 46	.305725	61 : 31	.293968	118 : 61	.286552	95 : 50	.278754
95 : 47	.305626	120 : 61	.293851	116 : 60	.286307	76 : 40	.278754
97 : 48	.305531	59 : 30	.293731	58 : 30	.286307	57 : 30	.278754
99 : 49	0.305439	116 : 59	0.293606	114 : 59	0.286053	38 : 20	0.278754
101 : 50	.305351	57 : 29	.293477	85 : 44	.285966	112 : 59	.278366
103 : 51	.305267	112 : 57	.293343	56 : 29	.285790	93 : 49	.278287
105 : 52	.305186	55 : 28	.293205	83 : 43	.285610	74 : 39	.278167
107 : 53	.305108	108 : 55	.293061	110 : 57	.285518	55 : 29	.277965
109 : 54	.305033	53 : 27	.292912	81 : 42	.285236	91 : 48	.277800
111 : 55	.304960	104 : 53	.292757	54 : 28	.285235	108 : 57	.277549
113 : 56	.304890	51 : 26	.292597	106 : 55	.284943	72 : 38	.277549
115 : 57	0.304823	100 : 51	0.292430	79 : 41	0.284843	89 : 47	0.277292
117 : 58	.304758	98 : 50	.292256	52 : 27	.284640	53 : 28	.277118
119 : 59	.304695	96 : 49	.292075	77 : 40	.284431	87 : 46	.276762
32 : 16		94 : 48	.291887	102 : 53	.284324	104 : 55	.276671
or		92 : 47	.291690	100 : 52	.283997	119 : 63	.276207
any	.301030	90 : 46	.291485	75 : 39	.283997	85 : 45	.276207
2 to 1		88 : 45	.291270	98 : 51	.283656	68 : 36	.276206
ratio		86 : 44	.291046	73 : 38	.283540	117 : 62	.275794
119 : 60	0.297396	84 : 43	0.290811	96 : 50	0.283301	100 : 53	0.275724
117 : 59	.297334	82 : 42	.290565	119 : 62	.283155	83 : 44	.275625
115 : 58	.297270	41 : 21	.290565	71 : 37	.283057	66 : 35	.275476
113 : 57	.297204	80 : 41	.290306	94 : 49	.282932	115 : 61	.275368
111 : 56	.297135	119 : 61	.290217	117 : 61	.282856	98 : 52	.275223
109 : 55	.297064	78 : 40	.290035	92 : 48	.282547	81 : 43	.275017
107 : 54	.296990	39 : 20	.290035	69 : 36	.282547	113 : 60	.274927
105 : 53	.296913	115 : 59	.289846	113 : 59	.282226	96 : 51	.274701
103 : 52	0.296833	76 : 39	0.289749	90 : 47	0.282145	64 : 34	0.274701
101 : 51	.296751	113 : 58	.289650	67 : 35	.282007	111 : 59	.274471
99 : 50	.296665	111 : 57	.289448	111 : 58	.281895	79 : 42	.274378
97 : 49	.296576	74 : 38	.289448	88 : 46	.281725	94 : 50	.274158
95 : 48	.296482	109 : 56	.289239	109 : 57	.281552	109 : 58	.273999
93 : 47	.296385	72 : 37	.289131	65 : 34	.281435	62 : 33	.273878
91 : 46	.296284	107 : 55	.289021	86 : 45	.281286	77 : 41	.273707
89 : 45	.296178	70 : 36	.288796	107 : 56	.281196	92 : 49	.273592
87 : 44	0.296067	103 : 53	0.288561	84 : 44	0.280827	107 : 57	0.273509
85 : 43	.295950	68 : 35	.288441	63 : 33	.280827	120 : 64	.273001
83 : 42	.295829	101 : 52	.288318	103 : 54	.280443	105 : 56	.273001
81 : 41	.295701	66 : 34	.288065	82 : 43	.280345	90 : 48	.273001
79 : 40	.295567	99 : 51	.288065	61 : 32	.280180	75 : 40	.273001
77 : 39	.295426	97 : 50	.287802	101 : 53	.280046	60 : 32	.273001
75 : 38	.295278	64 : 33	.287666	80 : 42	.279841	118 : 63	.272542
73 : 37	.295121	95 : 49	.287528	40 : 21	.279841	103 : 55	.272475

Logarithms of Gear Ratios from 1.0084+ to 7.5

Numbers of Teeth	Logarithm of Ratio	Numbers of Teeth	Logarithm of Ratio	Numbers of Teeth	Logarithm of Ratio	Numbers of Teeth	Logarithm of Ratio
88 : 47	0.272385	94 : 51	0.265558	87 : 48	0.258278	116 : 65	0.251545
73 : 39	.272258	70 : 38	.265314	58 : 32	.258278	91 : 51	.251471
58 : 31	.272066	116 : 63	.265118	96 : 53	.257995	66 : 37	.251342
101 : 54	.271928	81 : 44	.265032	67 : 37	.257873	107 : 60	.251233
86 : 46	.271741	92 : 50	.264818	105 : 58	.257761	82 : 46	.251056
114 : 61	.271575	103 : 56	.264649	38 : 21	.257564	98 : 55	.250863
99 : 53	.271360	68 : 37	.264307	76 : 42	.257564	57 : 32	.250725
112 : 60	.271067	79 : 43	.264159	85 : 47	.257321	73 : 41	.250539
84 : 45	0.271067	90 : 49	0.264046	94 : 52	0.257125	89 : 50	0.250420
56 : 30	.271067	101 : 55	.263959	103 : 57	.256962	105 : 59	.250337
97 : 52	.270768	112 : 61	.263888	112 : 62	.256826	112 : 63	.249878
69 : 37	.270647	110 : 60	.263241	56 : 31	.256826	96 : 54	.249878
110 : 59	.270541	99 : 54	.263241	65 : 36	.256611	80 : 45	.249878
82 : 44	.270361	88 : 48	.263241	74 : 41	.256448	64 : 36	.249878
41 : 22	.270361	77 : 42	.263241	83 : 46	.256320	119 : 67	.249472
95 : 51	.270153	66 : 36	.263241	92 : 51	.256218	103 : 58	.249409
54 : 29	0.269996	55 : 30	0.263241	101 : 56	0.256133	87 : 49	0.249323
67 : 36	.269772	119 : 65	.262634	110 : 61	.256063	71 : 40	.249198
80 : 43	.269622	108 : 59	.262572	119 : 66	.256003	110 : 62	.249001
93 : 50	.269513	97 : 53	.262496	99 : 55	.255273	55 : 31	.249001
106 : 57	.269431	86 : 47	.262401	90 : 50	.255273	94 : 53	.248852
119 : 64	.269367	75 : 41	.262277	81 : 45	.255273	117 : 66	.248642
117 : 63	.268845	64 : 35	.262112	72 : 40	.255273	78 : 44	.248642
91 : 49	.268845	53 : 29	.261878	63 : 35	.255273	101 : 57	.248447
78 : 42	0.268845	95 : 52	0.261720	54 : 30	0.255273	62 : 35	0.248324
65 : 35	.268845	84 : 46	.261522	115 : 64	.254518	85 : 48	.248178
52 : 28	.268845	115 : 63	.261357	106 : 59	.254454	108 : 61	.248094
39 : 21	.268845	73 : 40	.261262	97 : 54	.254378	115 : 65	.247785
115 : 62	.268306	104 : 57	.261158	88 : 49	.254287	92 : 52	.247785
102 : 55	.268238	62 : 34	.260913	79 : 44	.254174	69 : 39	.247785
89 : 48	.268149	93 : 51	.260912	70 : 39	.254033	99 : 56	.247447
76 : 41	.268030	113 : 62	.260687	61 : 34	.253851	76 : 43	.247345
63 : 34	0.267862	82 : 45	0.260601	113 : 63	0.253738	106 : 60	0.247155
113 : 61	.267749	51 : 28	.260412	52 : 29	.253605	53 : 30	.247155
100 : 54	.267606	71 : 39	.260194	95 : 53	.253448	83 : 47	.246980
87 : 47	.267421	91 : 50	.260071	86 : 48	.253257	113 : 64	.246898
111 : 60	.267172	111 : 61	.259993	120 : 67	.253106	120 : 68	.246672
74 : 40	.267172	100 : 55	.259637	77 : 43	.253022	90 : 51	.246672
37 : 20	.267172	80 : 44	.259637	111 : 62	.252931	60 : 34	.246672
98 : 53	.266950	40 : 22	.259637	102 : 57	.252725	97 : 55	.246409
61 : 33	0.266816	109 : 60	0.259275	68 : 38	0.252725	67 : 38	0.246291
85 : 46	.266661	89 : 49	.259194	93 : 52	.252480	104 : 59	.246181
109 : 59	.266575	69 : 38	.259066	118 : 66	.252338	111 : 63	.245982
96 : 52	.266268	118 : 65	.258969	59 : 33	.252338	74 : 42	.245982
72 : 39	.266268	98 : 54	.258832	84 : 47	.252181	37 : 21	.245982
107 : 58	.265956	78 : 43	.258626	109 : 61	.252097	118 : 67	.245807
118 : 64	.265702	107 : 59	.258532	100 : 56	.251812	81 : 46	.245727
59 : 32	.265702	116 : 64	.258278	75 : 42	.251812	88 : 50	.245513

Logarithms of Gear Ratios from 1.0084+ to 7.5

Numbers of Teeth	Logarithm of Ratio	Numbers of Teeth	Logarithm of Ratio	Numbers of Teeth	Logarithm of Ratio	Numbers of Teeth	Logarithm of Ratio
95 : 54	0.245330	52 : 30	0.238882	111 : 65	0.232410	111 : 66	0.225779
51 : 29	.245172	97 : 56	.238584	70 : 41	.232314	74 : 44	.225779
109 : 62	.245035	71 : 41	.238474	87 : 51	.231949	116 : 69	.225609
58 : 33	.244914	116 : 67	.238383	58 : 34	.231949	79 : 47	.225529
65 : 37	.244712	90 : 52	.238239	104 : 61	.231704	84 : 50	.225309
72 : 41	.244549	109 : 63	.238086	75 : 44	.231608	89 : 53	.225114
79 : 45	.244415	64 : 37	.237978	92 : 54	.231394	94 : 56	.224940
86 : 49	.244302	83 : 48	.237837	109 : 64	.231247	99 : 59	.224783
93 : 53	0.244207	102 : 59	0.237748	63 : 37	0.231139	52 : 31	0.224642
100 : 57	.244125	95 : 55	.237361	80 : 47	.230992	109 : 65	.224513
107 : 61	.244054	76 : 44	.237361	97 : 57	.230897	114 : 68	.224396
114 : 65	.243992	57 : 33	.237361	114 : 67	.230830	57 : 34	.224396
119 : 68	.243038	38 : 22	.237361	85 : 50	.230449	119 : 71	.224289
112 : 64	.243038	107 : 62	.236992	68 : 40	.230449	62 : 37	.224190
98 : 56	.243038	88 : 51	.236913	51 : 30	.230449	67 : 40	.224015
91 : 52	.243038	69 : 40	.236789	107 : 63	.230043	72 : 43	.223864
84 : 48	0.243038	119 : 69	0.236698	90 : 53	0.229967	77 : 46	0.223733
77 : 44	.243038	100 : 58	.236572	73 : 43	.229854	82 : 49	.223618
70 : 40	.243038	50 : 29	.236572	112 : 66	.229674	87 : 52	.223516
63 : 36	.243038	81 : 47	.236387	56 : 33	.229674	92 : 55	.223425
35 : 20	.243038	112 : 65	.236305	95 : 56	.229536	97 : 58	.223344
117 : 67	.242111	62 : 36	.236089	117 : 69	.229337	102 : 61	.223270
110 : 63	.242052	93 : 54	.236089	78 : 46	.229337	107 : 64	.223204
103 : 59	.241985	105 : 61	.235860	100 : 59	.229148	112 : 67	.223143
96 : 55	0.241909	74 : 43	0.235763	61 : 36	0.229027	117 : 70	0.223088
89 : 51	.241820	117 : 68	.235677	83 : 49	.228882	100 : 60	.221849
82 : 47	.241716	86 : 50	.235529	105 : 62	.228798	95 : 57	.221849
75 : 43	.241593	98 : 57	.235351	110 : 65	.228479	90 : 54	.221849
68 : 39	.241444	110 : 64	.235213	88 : 52	.228479	85 : 51	.221849
61 : 35	.241262	55 : 32	.235213	66 : 39	.228479	80 : 48	.221849
115 : 66	.241154	67 : 39	.235010	115 : 68	.228187	75 : 45	.221849
108 : 62	.241032	79 : 46	.234869	93 : 55	.228120	70 : 42	.221849
54 : 31	0.241032	91 : 53	0.234766	71 : 42	0.228009	65 : 39	0.221849
101 : 58	.240893	103 : 60	.234686	120 : 71	.227923	60 : 36	.221849
94 : 54	.240734	115 : 67	.234623	98 : 58	.227798	55 : 33	.221849
87 : 50	.240549	96 : 56	.234083	76 : 45	.227601	50 : 30	.221849
120 : 69	.240332	84 : 49	.234083	103 : 61	.227507	118 : 71	.220624
80 : 46	.240332	72 : 42	.234083	81 : 48	.227244	113 : 68	.220570
113 : 65	.240165	60 : 35	.234083	54 : 32	.227244	108 : 65	.220510
73 : 42	.240074	113 : 66	.233535	113 : 67	.227004	103 : 62	.220446
106 : 61	0.239976	101 : 59	0.233469	86 : 51	0.226928	98 : 59	0.220374
99 : 57	.239760	89 : 52	.233388	59 : 35	.226784	93 : 56	.220295
66 : 38	.239760	77 : 45	.233278	91 : 54	.226648	88 : 53	.220207
92 : 53	.239512	65 : 38	.233130	64 : 38	.226396	83 : 50	.220109
59 : 34	.239373	118 : 69	.233033	96 : 57	.226396	78 : 47	.219997
85 : 49	.239223	53 : 31	.232914	101 : 60	.226170	73 : 44	.219870
111 : 64	.239143	94 : 55	.232765	69 : 41	.226065	68 : 41	.219725
78 : 45	.238882	82 : 48	.232573	106 : 63	.225965	63 : 38	.219557

Logarithms of Gear Ratios from 1.0084+ to 7.5

Numbers of Teeth	Logarithm of Ratio	Numbers of Teeth	Logarithm of Ratio	Numbers of Teeth	Logarithm of Ratio	Numbers of Teeth	Logarithm of Ratio
58 : 35	0.219360	116 : 71	0.213200	29 : 18	0.207126	81 : 51	0.200915
111 : 67	.219248	98 : 60	.213075	95 : 59	.206872	54 : 34	.200915
53 : 32	.219126	49 : 30	.213075	66 : 41	.206760	100 : 63	.200660
101 : 61	.218992	80 : 49	.212894	103 : 64	.206657	73 : 46	.200565
96 : 58	.218843	111 : 68	.212814	111 : 69	.206474	119 : 75	.200486
91 : 55	.218679	93 : 57	.212608	74 : 46	.206474	92 : 58	.200360
86 : 52	.218495	62 : 38	.212608	119 : 74	.206315	111 : 70	.200225
81 : 49	.218289	106 : 65	.212393	82 : 51	.206244	65 : 41	.200130
119 : 72	0.218215	75 : 46	0.212304	90 : 56	0.206055	84 : 53	0.200003
76 : 46	.218056	119 : 73	.212224	98 : 61	.205896	103 : 65	.199924
109 : 66	.217883	88 : 54	.212089	53 : 33	.205762	114 : 72	.199572
71 : 43	.217790	101 : 62	.211930	114 : 71	.205647	95 : 60	.199572
104 : 63	.217693	57 : 35	.211807	61 : 38	.205546	76 : 48	.199572
66 : 40	.217484	70 : 43	.211630	69 : 43	.205380	57 : 36	.199572
94 : 57	.217253	96 : 59	.211419	77 : 48	.205250	106 : 67	.199231
61 : 37	.217128	109 : 67	.211352	85 : 53	.205143	87 : 55	.199157
89 : 54	0.216996	117 : 72	0.210853	93 : 58	0.205055	68 : 43	0.199040
117 : 71	.216928	91 : 56	.210853	101 : 63	.204981	117 : 74	.198954
112 : 68	.216709	78 : 48	.210853	109 : 68	.204918	98 : 62	.198834
84 : 51	.216709	65 : 40	.210853	117 : 73	.204863	49 : 31	.198834
56 : 34	.216709	112 : 69	.210369	96 : 60	.204120	79 : 50	.198657
107 : 65	.216470	99 : 61	.210305	88 : 55	.204120	109 : 69	.198577
79 : 48	.216386	86 : 53	.210223	80 : 50	.204120	90 : 57	.198368
102 : 62	.216209	73 : 45	.210110	72 : 45	.204120	60 : 38	.198368
51 : 31	0.216209	120 : 74	0.209950	64 : 40	0.204120	101 : 64	0.198141
74 : 45	.216019	60 : 37	.209950	56 : 35	.204120	71 : 45	.198046
97 : 59	.215920	107 : 66	.209840	32 : 20	.204120	112 : 71	.197960
120 : 73	.215858	94 : 58	.209700	115 : 72	.203365	82 : 52	.197811
115 : 70	.215600	81 : 50	.209515	107 : 67	.203309	93 : 59	.197631
92 : 56	.215600	115 : 71	.209440	99 : 62	.203244	52 : 33	.197489
69 : 42	.215600	68 : 42	.209260	91 : 57	.203167	115 : 73	.197375
110 : 67	.215318	34 : 21	.209260	83 : 52	.203075	63 : 40	.197281
87 : 53	0.215243	89 : 55	0.209027	75 : 47	0.202963	74 : 47	0.197134
64 : 39	.215115	55 : 34	.208884	67 : 42	.202826	85 : 54	.197025
105 : 64	.215009	76 : 47	.208716	59 : 37	.202650	96 : 61	.196941
82 : 50	.214844	97 : 60	.208620	110 : 69	.202544	107 : 68	.196875
100 : 61	.214670	118 : 73	.208559	51 : 32	.202420	118 : 75	.196821
118 : 72	.214550	84 : 52	.208276	94 : 59	.202276	110 : 70	.196295
59 : 36	.214550	63 : 39	.208276	86 : 54	.202105	99 : 63	.196295
77 : 47	.214393	113 : 70	.207980	78 : 49	.201899	88 : 56	.196295
95 : 58	0.214296	92 : 57	0.207913	113 : 71	0.201820	77 : 49	0.196295
113 : 69	.214229	71 : 44	.207806	105 : 66	.201645	66 : 42	.196295
90 : 55	.213880	100 : 62	.207608	70 : 44	.201645	55 : 35	.196295
72 : 44	.213880	50 : 31	.207608	97 : 61	.201442	113 : 72	.195746
54 : 33	.213880	79 : 49	.207431	62 : 39	.201327	91 : 58	.195613
36 : 22	.213880	108 : 67	.207349	89 : 56	.201202	102 : 65	.195687
103 : 63	.213497	87 : 54	.207126	116 : 73	.201135	69 : 44	.195396
67 : 41	.213291	58 : 36	.207126	108 : 68	.200915	58 : 37	.195226

Logarithms of Gear Ratios from 1.0084+ to 7.5

Numbers of Teeth	Logarithm of Ratio	Numbers of Teeth	Logarithm of Ratio	Numbers of Teeth	Logarithm of Ratio	Numbers of Teeth	Logarithm of Ratio
105 : 67	0.195115	99 : 64	0.189455	110 : 72	0.184060	71 : 47	0.179160
94 : 60	.194977	116 : 75	.189397	55 : 36	.184060	74 : 49	.179036
47 : 30	.194977	119 : 77	.189056	84 : 55	.183917	77 : 51	.178921
83 : 53	.194802	102 : 66	.189056	113 : 74	.183847	80 : 53	.178814
119 : 76	.194733	85 : 55	.189056	116 : 76	.183644	83 : 55	.178715
108 : 69	.194575	68 : 44	.189056	87 : 57	.183644	86 : 57	.178624
97 : 62	.194380	51 : 33	.189056	58 : 38	.183644	89 : 59	.178538
61 : 39	.194265	34 : 22	.189056	119 : 78	.183452	92 : 61	.178458
86 : 55	0.194136	105 : 68	0.188680	90 : 59	0.183391	95 : 63	0.178383
111 : 71	.194065	88 : 57	.188608	61 : 40	.183270	98 : 65	.178312
100 : 64	.193820	71 : 46	.188501	93 : 61	.183153	101 : 67	.178247
75 : 48	.193820	54 : 35	.188326	96 : 63	.182931	104 : 69	.178184
50 : 32	.193820	91 : 59	.188190	64 : 42	.182931	107 : 71	.178126
114 : 73	.193582	111 : 72	.187991	32 : 21	.182931	110 : 73	.178070
89 : 57	.193515	74 : 48	.187991	99 : 65	.182722	113 : 75	.178017
64 : 41	.193396	94 : 61	.187798	67 : 44	.182622	116 : 77	.177967
103 : 66	0.193293	57 : 37	0.187673	102 : 67	0.182525	119 : 79	0.177920
117 : 75	.193125	77 : 50	.187521	105 : 69	.182340	24 : 16	
78 : 50	.193125	97 : 63	.187431	70 : 46	.182340	or	
92 : 59	.192936	117 : 76	.187372	108 : 71	.182166	any	
53 : 34	.192797	100 : 65	.187087	73 : 48	.182082	3 to 2	
120 : 77	.192691	80 : 52	.187087	111 : 73	.182000	ratio	.176091
67 : 43	.192606	60 : 39	.187087	114 : 75	.181844	118 : 79	.174255
81 : 52	.192482	103 : 67	.186762	76 : 50	.181844	115 : 77	.174207
95 : 61	0.192394	83 : 54	0.186684	117 : 77	0.181695	112 : 75	0.174157
109 : 70	.192329	63 : 41	.186557	79 : 52	.181624	109 : 73	.174104
112 : 72	.191886	106 : 69	.186457	120 : 79	.181554	106 : 71	.174048
98 : 63	.191886	86 : 56	.186311	82 : 54	.181420	103 : 69	.173988
84 : 54	.191886	109 : 71	.186168	85 : 56	.181231	100 : 67	.173925
70 : 45	.191886	66 : 43	.186075	88 : 58	.181055	97 : 65	.173858
56 : 36	.191886	89 : 58	.185962	91 : 60	.180890	94 : 63	.173787
115 : 74	.191466	112 : 73	.185895	94 : 62	.180736	91 : 61	.173712
101 : 65	0.191408	115 : 75	0.185637	47 : 31	0.180736	88 : 59	0.173631
87 : 56	.191331	92 : 60	.185637	97 : 64	.180592	85 : 57	.173544
73 : 47	.191225	69 : 45	.185637	100 : 66	.180456	82 : 55	.173451
59 : 38	.191068	46 : 30	.185637	50 : 33	.180456	79 : 53	.173351
104 : 67	.190959	118 : 77	.185391	103 : 68	.180328	76 : 51	.173243
90 : 58	.190815	95 : 62	.185332	106 : 70	.180208	73 : 49	.173127
76 : 49	.190618	72 : 47	.185235	53 : 35	.180208	70 : 47	.173000
107 : 69	.190535	98 : 64	.185046	109 : 72	.180094	67 : 45	.172862
93 : 60	0.190332	49 : 32	0.185046	112 : 74	0.179986	64 : 43	0.172712
62 : 40	.190332	75 : 49	.184865	56 : 37	.179986	61 : 41	.172546
110 : 71	.190134	101 : 66	.184778	115 : 76	.179884	119 : 80	.172457
79 : 51	.190057	104 : 68	.184524	118 : 78	.179787	58 : 39	.172363
96 : 62	.189880	78 : 51	.184524	59 : 39	.179787	113 : 76	.172265
113 : 73	.189756	52 : 34	.184524	62 : 41	.179608	55 : 37	.172161
65 : 42	.189664	107 : 70	.184286	65 : 43	.179445	107 : 72	.172051
82 : 53	.189538	81 : 53	.184209	68 : 45	.179296	104 : 70	.171935

Logarithms of Gear Ratios from 1.0084+ to 7.5

Numbers of Teeth	Logarithm of Ratio	Numbers of Teeth	Logarithm of Ratio	Numbers of Teeth	Logarithm of Ratio	Numbers of Teeth	Logarithm of Ratio
52 : 35	0.171935	47 : 32	0.166948	77 : 53	0.162215	102 : 71	0.157342
101 : 68	.171813	116 : 79	.166831	61 : 42	.162081	79 : 55	.157264
98 : 66	.171682	69 : 47	.166751	106 : 73	.161983	112 : 78	.157123
49 : 33	.171682	91 : 62	.166650	90 : 62	.161851	56 : 39	.157123
95 : 64	.171544	113 : 77	.166588	45 : 31	.161851	89 : 62	.156998
92 : 62	.171396	110 : 75	.166331	119 : 82	.161733	99 : 69	.156786
46 : 31	.171396	88 : 60	.166331	74 : 51	.161662	66 : 46	.156786
89 : 60	.171239	66 : 45	.166331	103 : 71	.161579	109 : 76	.156613
86 : 58	0.171071	44 : 30	0.166331	116 : 80	0.161368	76 : 53	0.156538
83 : 56	.170890	107 : 73	.166061	87 : 60	.161368	119 : 83	.156469
120 : 81	.170696	85 : 58	.165991	58 : 40	.161368	86 : 60	.156347
80 : 54	.170696	63 : 43	.165872	100 : 69	.161151	43 : 30	.156347
117 : 79	.170559	104 : 71	.165775	71 : 49	.161062	96 : 67	.156196
77 : 52	.170487	82 : 56	.165626	113 : 78	.160984	106 : 74	.156074
114 : 77	.170414	101 : 69	.165472	84 : 58	.160851	53 : 37	.156074
111 : 75	.170262	120 : 82	.165367	97 : 67	.160697	116 : 81	.155973
74 : 50	0.170262	60 : 41	0.165367	110 : 76	0.160579	63 : 44	0.155888
108 : 73	.170101	79 : 54	.165233	55 : 38	.160579	73 : 51	.155753
71 : 48	.170017	98 : 67	.165151	68 : 47	.160411	93 : 65	.155570
105 : 71	.169931	117 : 80	.165096	81 : 56	.160297	103 : 72	.155505
68 : 46	.169751	114 : 78	.164810	94 : 65	.160215	113 : 79	.155541
99 : 67	.169560	95 : 65	.164810	107 : 74	.160152	120 : 84	.154902
65 : 44	.169461	76 : 52	.164810	120 : 83	.160103	110 : 77	.154902
96 : 65	.169358	57 : 39	.164810	117 : 81	.159701	100 : 70	.154902
93 : 63	0.169142	111 : 76	0.164509	104 : 72	0.159701	90 : 63	0.154902
62 : 42	.169142	92 : 63	.164447	91 : 63	.159701	80 : 56	.154902
31 : 21	.169142	73 : 50	.164353	78 : 54	.159701	70 : 49	.154902
90 : 61	.168913	108 : 74	.164192	65 : 45	.159701	60 : 42	.154902
59 : 40	.168792	54 : 37	.164192	52 : 36	.159701	30 : 21	.154902
118 : 80	.168792	89 : 61	.164060	114 : 79	.159278	117 : 82	.154372
87 : 59	.168667	105 : 72	.163857	101 : 70	.159223	107 : 75	.154323
115 : 78	.168603	70 : 48	.163857	88 : 61	.159153	97 : 68	.154263
112 : 76	0.168404	86 : 59	0.163647	75 : 52	0.159058	87 : 61	0.154190
84 : 57	.168404	102 : 70	.163502	62 : 43	.158923	77 : 54	.154097
56 : 38	.168404	51 : 35	.163502	111 : 77	.158832	67 : 47	.153977
109 : 74	.168196	118 : 81	.163397	98 : 68	.158717	114 : 80	.153815
81 : 55	.168122	67 : 46	.163317	49 : 34	.158717	57 : 40	.153815
106 : 72	.167973	83 : 57	.163203	85 : 59	.158567	104 : 73	.153710
53 : 36	.167973	99 : 68	.163106	108 : 75	.158363	94 : 66	.153584
78 : 53	.167819	115 : 79	.163071	72 : 50	.158363	47 : 33	.153584
103 : 70	0.167739	112 : 77	0.162727	95 : 66	0.158180	84 : 59	0.153427
100 : 68	.167491	96 : 66	.162727	59 : 41	.158068	111 : 78	.153228
75 : 51	.167491	80 : 55	.162727	82 : 57	.157939	74 : 52	.153228
50 : 34	.167491	64 : 44	.162727	105 : 73	.157866	101 : 71	.153063
97 : 66	.167228	48 : 33	.162727	115 : 80	.157608	64 : 45	.152968
72 : 49	.167136	32 : 22	.162727	92 : 64	.157608	91 : 64	.152861
119 : 81	.167062	109 : 75	.162365	69 : 48	.157608	118 : 83	.152804
94 : 64	.166948	93 : 64	.162303	46 : 32	.157608	108 : 76	.152610

Logarithms of Gear Ratios from 1.0084+ to 7.5

Numbers of Teeth	Logarithm of Ratio	Numbers of Teeth	Logarithm of Ratio	Numbers of Teeth	Logarithm of Ratio	Numbers of Teeth	Logarithm of Ratio
81 : 57	0.152610	45 : 32	0.148063	114 : 82	0.143091	106 : 77	0.138815
54 : 38	.152610	97 : 69	.147923	57 : 41	.143091	117 : 85	.138767
98 : 69	.152377	104 : 74	.147802	82 : 59	.142962	110 : 80	.138303
71 : 50	.152288	52 : 37	.147802	107 : 77	.142893	99 : 72	.138303
115 : 81	.152213	111 : 79	.147696	100 : 72	.142668	88 : 64	.138303
88 : 62	.152091	118 : 84	.147603	75 : 54	.142668	77 : 56	.138303
44 : 31	.152091	66 : 47	.147446	50 : 36	.142668	66 : 48	.138303
105 : 74	.151958	73 : 52	.147320	118 : 85	.142463	55 : 40	.138303
61 : 43	0.151861	80 : 57	0.147215	93 : 67	0.142408	44 : 32	0.138303
78 : 55	.151732	87 : 62	.147128	68 : 49	.142313	114 : 83	.137827
95 : 67	.151649	94 : 67	.147053	111 : 80	.142233	103 : 75	.137776
112 : 79	.151591	101 : 72	.146989	86 : 62	.142107	92 : 67	.137713
119 : 84	.151268	108 : 77	.146933	43 : 31	.142107	81 : 59	.137633
102 : 72	.151268	115 : 82	.146884	104 : 75	.141972	70 : 51	.137528
85 : 60	.151268	98 : 70	.146128	61 : 44	.141877	118 : 86	.137384
68 : 48	.151268	91 : 65	.146128	79 : 57	.141752	59 : 43	.137384
51 : 36	0.151268	84 : 60	0.146128	97 : 70	0.141674	107 : 78	0.137289
109 : 77	.150936	77 : 55	.146128	115 : 83	.141620	96 : 70	.137173
92 : 65	.150874	70 : 50	.146128	108 : 78	.141329	48 : 35	.137173
75 : 53	.150785	63 : 45	.146128	90 : 65	.141329	85 : 62	.137027
116 : 82	.150644	56 : 40	.146128	72 : 52	.141329	111 : 81	.136838
58 : 41	.150644	49 : 35	.146128	54 : 39	.141329	74 : 54	.136838
99 : 70	.150537	42 : 30	.146128	119 : 86	.141049	100 : 73	.136677
82 : 58	.150386	35 : 25	.146128	101 : 73	.140999	63 : 46	.136583
106 : 75	0.150245	116 : 83	0.145380	83 : 60	0.140927	89 : 65	0.136477
65 : 46	.150156	109 : 78	.145332	65 : 47	.140816	115 : 84	.136419
89 : 63	.150050	102 : 73	.145277	112 : 81	.140733	104 : 76	.136220
113 : 80	.149988	88 : 63	.145142	94 : 68	.140619	78 : 57	.136220
120 : 85	.149762	81 : 58	.145057	47 : 34	.140619	52 : 38	.136220
96 : 68	.149762	74 : 53	.144956	76 : 55	.140451	119 : 87	.136028
72 : 51	.149762	67 : 48	.144834	105 : 76	.140376	93 : 68	.135974
48 : 34	.149762	120 : 86	.144683	116 : 84	.140179	67 : 49	.135879
103 : 73	0.149514	60 : 43	0.144683	87 : 63	0.140179	108 : 79	0.135797
79 : 56	.149439	113 : 81	.144593	58 : 42	.140179	82 : 60	.135663
110 : 78	.149298	106 : 76	.144492	29 : 21	.140179	41 : 30	.135663
55 : 39	.149298	53 : 38	.144492	98 : 71	.139988	97 : 71	.135513
86 : 61	.149169	99 : 71	.144377	69 : 50	.139879	112 : 82	.135404
117 : 83	.149108	92 : 66	.144244	109 : 79	.139799	56 : 41	.135404
93 : 66	.148939	46 : 33	.144244	120 : 87	.139662	71 : 52	.135255
62 : 44	.148939	85 : 61	.144089	80 : 58	.139662	86 : 63	.135158
31 : 22	0.148939	117 : 84	0.143907	91 : 66	0.139498	101 : 74	0.135090
100 : 71	.148742	78 : 56	.143907	102 : 74	.139369	116 : 85	.135039
69 : 49	.148653	110 : 79	.143766	51 : 37	.139369	105 : 77	.134699
107 : 76	.148570	71 : 51	.143688	113 : 82	.139265	90 : 66	.134699
114 : 81	.148420	103 : 74	.143606	62 : 45	.139179	75 : 55	.134699
76 : 54	.148420	96 : 69	.143422	73 : 53	.139047	60 : 44	.134699
83 : 59	.148226	64 : 46	.143422	84 : 61	.138950	45 : 33	.134699
90 : 64	.148063	89 : 64	.143210	95 : 69	.138875	30 : 22	.134699

Logarithms of Gear Ratios from 1.0084+ to 7.5

Numbers of Teeth	Logarithm of Ratio	Numbers of Teeth	Logarithm of Ratio	Numbers of Teeth	Logarithm of Ratio	Numbers of Teeth	Logarithm of Ratio
109 : 80	0.134337	58 : 43	0.129960	104 : 78	0.124939	95 : 72	0.120391
94 : 69	.134279	89 : 66	.129846	88 : 66	.124939	62 : 47	.120294
79 : 58	.134199	120 : 89	.129791	84 : 63	.124939	91 : 69	.120192
64 : 47	.134082	93 : 69	.129634	64 : 48	.124939	120 : 91	.120140
113 : 83	.134000	62 : 46	.129634	60 : 45	.124939	116 : 88	.119975
98 : 72	.133894	97 : 72	.129439	44 : 33	.124939	87 : 66	.119975
49 : 36	.133894	66 : 49	.129348	40 : 30	.124939	58 : 44	.119975
83 : 61	.133748	101 : 75	.129260	28 : 21	.124939	29 : 22	.119975
117 : 86	0.133687	105 : 78	0.129095	117 : 88	0.123703	112 : 85	0.119799
102 : 75	.133539	70 : 52	.129095	113 : 85	.123660	83 : 63	.119738
68 : 50	.133539	109 : 81	.128942	109 : 82	.123613	108 : 82	.119610
87 : 64	.133339	74 : 55	.128869	105 : 79	.123562	54 : 41	.119610
106 : 78	.133211	113 : 84	.128799	101 : 76	.123508	79 : 60	.119476
72 : 53	.133057	117 : 87	.128667	97 : 73	.123449	104 : 79	.119406
91 : 67	.132967	78 : 58	.128667	93 : 70	.123385	100 : 76	.119186
110 : 81	.132908	82 : 61	.128484	89 : 67	.123315	75 : 57	.119186
114 : 84	0.132626	86 : 64	0.128319	85 : 64	0.123238	50 : 38	0.119186
95 : 70	.132626	43 : 32	.128319	81 : 61	.123155	96 : 73	.118948
76 : 56	.132626	90 : 67	.128168	77 : 58	.123063	71 : 54	.118865
57 : 42	.132626	94 : 70	.128030	73 : 55	.122960	117 : 89	.118796
118 : 87	.132363	47 : 35	.128030	69 : 52	.122846	92 : 70	.118690
99 : 73	.132312	98 : 73	.127903	65 : 49	.122717	46 : 35	.118690
80 : 59	.132238	102 : 76	.127787	61 : 46	.122572	113 : 86	.118580
61 : 45	.132117	51 : 38	.127787	118 : 89	.122492	67 : 51	.118505
103 : 76	0.132024	106 : 79	0.127679	114 : 86	0.122406	88 : 67	0.118408
84 : 62	.131888	110 : 82	.127579	57 : 43	.122406	109 : 83	.118348
42 : 31	.131888	55 : 41	.127579	110 : 83	.122315	84 : 64	.118099
107 : 79	.131757	114 : 85	.127486	53 : 40	.122216	63 : 48	.118099
65 : 48	.131672	118 : 88	.127399	106 : 80	.122216	42 : 32	.118090
88 : 65	.131570	59 : 44	.127399	102 : 77	.122110	105 : 80	.118090
111 : 82	.131509	63 : 47	.127243	98 : 74	.121994	101 : 77	.117831
115 : 85	.131279	67 : 50	.127105	49 : 37	.121994	80 : 61	.117760
92 : 68	0.131279	71 : 53	0.126982	94 : 71	0.121870	118 : 90	0.117640
69 : 51	.131279	75 : 56	.126873	90 : 68	.121734	59 : 45	.117640
46 : 34	.131279	79 : 59	.126775	45 : 34	.121734	97 : 74	.117540
119 : 88	.131064	83 : 62	.126686	86 : 65	.121585	114 : 87	.117386
96 : 71	.131013	87 : 65	.126606	82 : 62	.121422	76 : 58	.117386
73 : 54	.130929	91 : 68	.126533	41 : 31	.121422	93 : 71	.117225
100 : 74	.130768	95 : 71	.126465	119 : 90	.121305	110 : 84	.117113
50 : 37	.130768	99 : 74	.126404	78 : 59	.121243	55 : 42	.117113
77 : 57	0.130616	103 : 77	0.126347	115 : 87	0.121179	72 : 55	0.116970
104 : 77	.130543	107 : 80	.126294	111 : 84	.121044	89 : 68	.116881
81 : 60	.130334	111 : 83	.126245	74 : 56	.121044	106 : 81	.116821
54 : 40	.130334	115 : 86	.126199	107 : 81	.120899	119 : 91	.116506
27 : 20	.130334	119 : 89	.126157	70 : 53	.120822	102 : 78	.116506
112 : 83	.130140	120 : 90	.124939	103 : 78	.120743	85 : 65	.116506
85 : 63	.130078	112 : 84	.124939	99 : 75	.120574	68 : 52	.116506
116 : 86	.129960	108 : 81	.124939	66 : 50	.120574	51 : 39	.116506

Logarithms of Gear Ratios from 1.0084+ to 7.5

Numbers of Teeth	Logarithm of Ratio	Numbers of Teeth	Logarithm of Ratio	Numbers of Teeth	Logarithm of Ratio	Numbers of Teeth	Logarithm of Ratio
34 : 26	0.116506	101 : 78	0.112227	118 : 92	0.108094	75 : 59	0.104209
115 : 88	.116215	110 : 85	.111974	59 : 46	.108094	61 : 48	.104089
98 : 75	.116165	88 : 68	.111974	109 : 85	.108008	108 : 85	.104005
81 : 62	.116093	44 : 34	.111974	100 : 78	.107905	94 : 74	.103896
64 : 49	.115984	66 : 51	.111974	50 : 39	.107905	47 : 37	.103896
111 : 85	.115904	119 : 92	.111759	91 : 71	.107783	80 : 63	.103750
94 : 72	.115795	97 : 75	.111710	82 : 64	.107634	113 : 89	.103688
47 : 36	.115795	75 : 58	.111633	41 : 32	.107634	99 : 78	.103541
77 : 59	0.115639	106 : 82	0.111492	114 : 89	0.107515	66 : 52	0.103541
107 : 82	.115570	53 : 41	.111492	73 : 57	.107448	118 : 93	.103399
120 : 92	.115393	84 : 65	.111366	105 : 82	.107375	85 : 67	.103344
90 : 69	.115393	115 : 89	.111308	96 : 75	.107210	104 : 82	.103219
60 : 46	.115393	93 : 72	.111151	64 : 50	.107210	52 : 41	.103219
30 : 23	.115393	62 : 48	.111151	119 : 93	.107064	90 : 71	.102984
103 : 79	.115210	102 : 79	.110973	87 : 68	.107010	109 : 86	.102928
73 : 56	.115135	71 : 55	.110896	110 : 86	.106894	95 : 75	.102662
116 : 89	0.115068	111 : 86	0.110825	55 : 43	0.106894	76 : 60	0.102662
86 : 66	.114955	120 : 93	.110698	78 : 61	.106765	57 : 45	.102662
43 : 33	.114955	80 : 62	.110698	101 : 79	.106694	38 : 30	.102662
99 : 76	.114822	40 : 31	.110698	115 : 90	.106455	119 : 94	.102419
112 : 86	.114720	89 : 69	.110541	92 : 72	.106455	100 : 79	.102373
56 : 43	.114720	98 : 76	.110413	69 : 54	.106455	81 : 64	.102305
69 : 53	.114573	49 : 38	.110413	46 : 36	.106455	62 : 49	.102196
82 : 63	.114473	107 : 83	.110306	106 : 83	.106228	105 : 83	.102111
95 : 73	0.114401	116 : 90	0.110216	83 : 65	0.106165	86 : 68	0.101990
108 : 83	.114346	58 : 45	.110216	120 : 94	.106053	43 : 34	.101990
117 : 90	.113943	67 : 52	.110072	60 : 47	.106053	110 : 87	.101873
104 : 80	.113943	76 : 59	.109962	97 : 76	.105958	67 : 53	.101799
91 : 70	.113943	85 : 66	.109875	111 : 87	.105804	91 : 72	.101709
78 : 60	.113943	94 : 73	.109805	74 : 58	.105804	115 : 91	.101656
65 : 50	.113943	103 : 80	.109747	88 : 69	.105634	120 : 95	.101458
52 : 40	.113943	112 : 87	.109699	102 : 80	.105510	96 : 76	.101458
39 : 30	0.113943	117 : 91	0.109145	51 : 40	0.105510	72 : 57	0.101458
26 : 20	.113943	108 : 84	.109145	116 : 91	.105417	48 : 38	.101458
113 : 87	.113559	99 : 77	.109145	65 : 51	.105343	101 : 80	.101231
100 : 77	.113509	90 : 70	.109145	79 : 62	.105235	77 : 61	.101161
87 : 67	.113445	81 : 63	.109145	93 : 73	.105160	106 : 84	.101027
74 : 57	.113357	72 : 56	.109145	107 : 84	.105105	53 : 42	.101027
61 : 47	.113232	63 : 49	.109145	112 : 88	.104735	82 : 65	.100901
109 : 84	.113147	54 : 42	.109145	98 : 77	.104735	111 : 88	.100840
96 : 74	0.113040	45 : 35	0.109145	84 : 66	0.104735	116 : 92	0.100670
48 : 37	.113040	27 : 21	.109145	70 : 55	.104735	87 : 69	.100670
83 : 64	.112898	113 : 88	.108596	56 : 44	.104735	58 : 46	.100670
118 : 91	.112841	104 : 81	.108548	42 : 33	.104735	29 : 23	.100670
70 : 54	.112704	95 : 74	.108492	28 : 22	.104735	92 : 73	.100465
92 : 71	.112530	86 : 67	.108424	117 : 92	.104398	63 : 50	.100371
57 : 44	.112422	77 : 60	.108339	103 : 81	.104352	97 : 77	.100281
79 : 61	.112297	68 : 53	.108233	89 : 70	.104292	102 : 81	.100115

Logarithms of Gear Ratios from 1.0084+ to 7.5

Numbers of Teeth	Logarithm of Ratio	Numbers of Teeth	Logarithm of Ratio	Numbers of Teeth	Logarithm of Ratio	Numbers of Teeth	Logarithm of Ratio
68 : 54	0.100115	117 : 94	0.095058	74 : 60	0.091080	88 : 72	0.087150
34 : 27	.100115	56 : 45	.094976	37 : 30	.091080	77 : 63	.087150
107 : 85	.099964	107 : 86	.094885	90 : 73	.090920	66 : 54	.087150
73 : 58	.099894	102 : 82	.094786	106 : 86	.090807	55 : 45	.087150
112 : 89	.099828	51 : 41	.094786	53 : 43	.090807	44 : 36	.087150
78 : 62	.099703	97 : 78	.094677	69 : 56	.090661	116 : 95	.086734
39 : 31	.099703	92 : 74	.094556	85 : 69	.090570	105 : 86	.086691
83 : 66	.099534	46 : 37	.094556	101 : 82	.090508	94 : 77	.086637
88 : 70	0.099385	87 : 70	0.094421	117 : 95	0.090462	83 : 68	0.086569
44 : 35	.099385	82 : 66	.094270	112 : 91	.090177	72 : 59	.086481
93 : 74	.099251	41 : 33	.094270	96 : 78	.090177	61 : 50	.086360
98 : 78	.099132	118 : 95	.094158	80 : 65	.090177	111 : 91	.086282
49 : 39	.099132	77 : 62	.094090	64 : 52	.090177	100 : 82	.086186
103 : 82	.099023	113 : 91	.094037	48 : 39	.090177	50 : 41	.086186
108 : 86	.098925	108 : 87	.093905	107 : 87	.089865	89 : 73	.086067
54 : 43	.098925	72 : 58	.093905	91 : 74	.089810	117 : 96	.085915
113 : 90	0.098836	103 : 83	0.093759	75 : 61	0.089732	78 : 64	0.085915
118 : 94	.098754	67 : 54	.093681	118 : 96	.089611	39 : 32	.085915
59 : 47	.098754	98 : 79	.093599	59 : 48	.089611	106 : 87	.085787
64 : 51	.098610	93 : 75	.093422	102 : 83	.089522	67 : 55	.085712
69 : 55	.098486	62 : 50	.093422	86 : 70	.089401	95 : 78	.085629
74 : 59	.098380	119 : 96	.093276	43 : 35	.089401	112 : 92	.085430
79 : 63	.098287	88 : 71	.093224	113 : 92	.089291	84 : 69	.085430
84 : 67	.098205	114 : 92	.093117	70 : 57	.089223	56 : 46	.085430
89 : 71	0.098132	57 : 46	0.093117	97 : 79	0.089145	28 : 23	0.085430
94 : 75	.098067	83 : 67	.093003	108 : 88	.088941	101 : 83	.085243
99 : 79	.098008	109 : 88	.092944	81 : 66	.088941	118 : 97	.085110
104 : 83	.097955	104 : 84	.092754	54 : 44	.088941	90 : 74	.085011
109 : 87	.097907	78 : 63	.092754	27 : 22	.088941	45 : 37	.085011
114 : 91	.097864	52 : 42	.092754	119 : 97	.088775	107 : 88	.084901
119 : 95	.097823	26 : 21	.092754	65 : 53	.088638	62 : 51	.084822
20 : 16		99 : 80	.092545	103 : 84	.088558	79 : 65	.084714
or		73 : 59	0.092471	114 : 93	0.088422	96 : 79	0.084644
any	0.096910	120 : 97	.092410	76 : 62	.088422	113 : 93	.084596
5 to 4		94 : 76	.092314	38 : 31	.088422	119 : 98	.084321
ratio		47 : 38	.092314	87 : 71	.088261	102 : 84	.084321
116 : 93	.095975	115 : 93	.092215	98 : 80	.088136	85 : 70	.084321
111 : 89	.095933	68 : 55	.092146	49 : 40	.088136	68 : 56	.084321
106 : 85	.095887	89 : 72	.092058	109 : 89	.088037	51 : 42	.084321
101 : 81	.095836	110 : 89	.092003	120 : 98	.087955	108 : 89	.084034
96 : 77	0.095781	105 : 85	0.091770	60 : 49	0.087955	91 : 75	0.083980
91 : 73	.095719	84 : 68	.091770	71 : 58	.087830	74 : 61	.083902
86 : 69	.095649	63 : 51	.091770	82 : 67	.087739	114 : 94	.083777
81 : 65	.095572	42 : 34	.091770	93 : 76	.087669	57 : 47	.083777
76 : 61	.095484	116 : 94	.091330	104 : 85	.087614	97 : 80	.083682
71 : 57	.095383	58 : 47	.091330	115 : 94	.087570	120 : 99	.083546
66 : 53	.095268	95 : 77	0.091233	110 : 90	.087150	80 : 66	.083546
61 : 49	.095134	111 : 90	.091080	99 : 81	.087150	40 : 33	.083546

Logarithms of Gear Ratios from 1.0084+ to 7.5

Numbers of Teeth	Logarithm of Ratio	Numbers of Teeth	Logarithm of Ratio	Numbers of Teeth	Logarithm of Ratio	Numbers of Teeth	Logarithm of Ratio
103 : 85	0.083418	115 : 96	0.078427	114 : 96	0.074634	73 : 62	0.070931
63 : 52	.083337	109 : 91	.078385	95 : 80	.074634	93 : 79	.070856
86 : 71	.083240	103 : 86	.078393	76 : 64	.074634	113 : 96	.070808
109 : 90	.083184	97 : 81	.078287	57 : 48	.074634	100 : 85	.070581
115 : 95	.082974	91 : 76	.078228	38 : 32	.074634	80 : 68	.070581
92 : 76	.082974	85 : 71	.078161	108 : 91	.074382	60 : 51	.070581
69 : 57	.082974	79 : 66	.078083	89 : 75	.074329	40 : 34	.070581
46 : 38	.082974	73 : 61	.077993	70 : 59	.074246	107 : 91	.070342
98 : 81	0.082741	67 : 56	0.077887	102 : 86	0.074102	87 : 74	0.070288
75 : 62	.082670	61 : 51	.077760	51 : 43	.074102	67 : 57	.070200
52 : 43	.082535	116 : 97	.077686	83 : 70	.073980	114 : 97	.070133
81 : 67	.082410	55 : 46	.077605	115 : 97	.073927	94 : 80	.070038
110 : 91	.082351	104 : 87	.077514	96 : 81	.073786	47 : 40	.070038
116 : 96	.082187	98 : 82	.077412	64 : 54	.073786	74 : 63	.069891
87 : 72	.082187	49 : 41	.077412	109 : 92	.073639	101 : 86	.069823
58 : 48	.082187	92 : 77	.077297	77 : 65	.073577	81 : 69	.069636
29 : 24	0.082187	86 : 72	0.077166	90 : 76	0.073429	54 : 46	0.069636
93 : 77	.081992	43 : 36	.077166	45 : 38	.073429	115 : 98	.069472
64 : 53	.081904	80 : 67	.077015	103 : 87	.073318	88 : 75	.069421
99 : 82	.081821	117 : 98	.076960	116 : 98	.073232	61 : 52	.069327
105 : 87	.081670	111 : 93	.076840	58 : 49	.073232	95 : 81	.069239
70 : 58	.081670	74 : 62	.076840	71 : 60	.073107	68 : 58	.069081
76 : 63	.081473	37 : 31	.076840	84 : 71	.073021	109 : 93	.068944
117 : 97	.081414	105 : 88	.076707	97 : 82	.072958	75 : 64	.068881
82 : 68	0.081305	68 : 57	0.076634	110 : 93	0.072910	116 : 99	0.068823
41 : 34	.081305	99 : 83	.076557	117 : 99	.072551	82 : 70	.068716
88 : 73	.081160	93 : 78	.076388	104 : 88	.072551	41 : 35	.068716
94 : 78	.081033	62 : 52	.076388	91 : 77	.072551	89 : 76	.068576
47 : 39	.081033	118 : 99	.076247	78 : 66	.072551	96 : 82	.068457
100 : 83	.080922	87 : 73	.076196	65 : 55	.072551	48 : 41	.068457
106 : 88	.080823	112 : 94	.076090	39 : 33	.072551	103 : 88	.068355
53 : 44	.080823	56 : 47	.076090	52 : 44	.072551	55 : 47	.068265
118 : 98	0.080656	81 : 68	0.075976	111 : 94	0.072195	117 : 100	0.068186
59 : 49	.080656	106 : 89	.075916	98 : 83	.072148	62 : 53	.068116
65 : 54	.080520	100 : 84	.075721	85 : 72	.072086	69 : 59	.067997
71 : 59	.080406	75 : 63	.075721	72 : 61	.072003	76 : 65	.067900
77 : 64	.080311	50 : 42	.075721	118 : 100	.071882	83 : 71	.067820
83 : 69	.080229	25 : 21	.075721	59 : 50	.071882	90 : 77	.067752
89 : 74	.080158	119 : 100	0.075547	105 : 89	.071797	97 : 83	.067694
95 : 79	.080097	94 : 79	.075501	92 : 78	.071693	104 : 89	.067643
107 : 89	0.079994	69 : 58	0.075421	46 : 39	.071693	111 : 95	0.067599
113 : 94	.079951	113 : 95	.075355	79 : 67	0.071552	98 : 84	.066947
119 : 99	.079912	88 : 74	.075251	112 : 95	.071494	91 : 78	.066947
30 : 25		44 : 37	.075251	99 : 84	.071456	84 : 72	.066947
or		107 : 90	.075141	66 : 56	.071356	77 : 66	.066947
any	.079181	63 : 53	.075065	86 : 73	.071176	70 : 60	.066947
6 to 5		82 : 69	.074965	106 : 90	.071063	63 : 54	.066947
ratio		101 : 85	.074903	53 : 45	.071063	56 : 48	.066947

Logarithms of Gear Ratios from 1.0084+ to 7.5

Numbers of Teeth	Logarithm of Ratio	Numbers of Teeth	Logarithm of Ratio	Numbers of Teeth	Logarithm of Ratio	Numbers of Teeth	Logarithm of Ratio
49 : 42	0.066947	75 : 65	0.062148	113 : 99	0.057443	69 : 61	0.053519
42 : 36	.066947	60 : 52	.062148	105 : 92	.057402	95 : 84	.053444
35 : 30	.066947	45 : 39	.062148	97 : 85	.057353	78 : 69	.053246
113 : 97	.066307	113 : 98	.061852	89 : 78	.057295	52 : 46	.053246
106 : 91	.066265	98 : 85	.061807	81 : 71	.057227	113 : 100	.053078
99 : 85	.066216	83 : 72	.061746	73 : 64	.057143	87 : 77	.053029
92 : 79	.066161	68 : 59	.061657	65 : 57	.057039	61 : 54	.052936
85 : 73	.066096	53 : 46	.061518	57 : 50	.056905	96 : 85	.052852
78 : 67	0.066020	91 : 79	0.061414	106 : 93	0.056823	70 : 62	0.052706
71 : 61	.065929	76 : 66	.061270	98 : 86	.056728	35 : 31	.052706
64 : 55	.065817	38 : 33	.061270	49 : 43	.056728	79 : 70	.052529
57 : 49	.065679	99 : 86	.061137	90 : 79	.056615	88 : 78	.052388
107 : 92	.065596	61 : 53	.061054	82 : 72	.056481	44 : 39	.052388
100 : 86	.065502	84 : 73	.060956	41 : 36	.056481	97 : 86	.052273
50 : 43	.065502	107 : 93	.060901	74 : 65	.056318	53 : 47	.052178
93 : 80	.065393	92 : 80	.060698	107 : 94	.056256	62 : 55	.052029
86 : 74	0.065267	69 : 60	0.060698	99 : 87	0.056116	71 : 63	0.051918
43 : 37	.065267	46 : 40	.060698	66 : 58	.056116	80 : 71	.051832
79 : 68	.065118	100 : 87	.060481	91 : 80	.055951	89 : 79	.051763
115 : 99	.065063	77 : 67	.060416	58 : 51	.055858	98 : 87	.051707
72 : 62	.064941	54 : 47	.060296	83 : 73	.055755	107 : 95	.051660
36 : 31	.064941	85 : 74	.060187	108 : 95	.055700	99 : 88	.051153
101 : 87	.064802	93 : 81	.059998	100 : 88	.055517	90 : 80	.051153
65 : 56	.064725	62 : 54	.059998	75 : 66	.055517	81 : 72	.051153
94 : 81	0.064643	101 : 88	0.059839	50 : 44	0.055517	72 : 64	0.051153
87 : 75	.064458	70 : 61	.059768	92 : 81	.055303	63 : 56	.051153
109 : 94	.064299	109 : 95	.059703	67 : 59	.055223	54 : 48	.051153
80 : 69	.064241	78 : 68	.059586	109 : 96	.055155	45 : 40	.051153
102 : 88	.064118	39 : 34	.059586	84 : 74	.055048	36 : 32	.051153
51 : 44	.064118	86 : 75	.059437	42 : 37	.055048	109 : 97	.050655
73 : 63	.063982	47 : 41	.059314	101 : 89	.054931	100 : 89	.050610
95 : 82	.063910	94 : 82	.059314	59 : 52	.054849	91 : 81	.050556
88 : 76	0.063669	102 : 89	0.059210	76 : 67	0.054739	82 : 73	0.050491
66 : 57	.063669	55 : 48	.059122	93 : 82	.054669	73 : 65	.050410
44 : 38	.063669	63 : 55	.058978	110 : 97	.054621	64 : 57	.050305
103 : 89	.063447	71 : 62	.058867	102 : 90	.054358	55 : 49	.050167
81 : 70	.063387	79 : 69	.058778	85 : 75	.054358	101 : 90	.050079
59 : 51	.063282	87 : 76	.058706	68 : 60	.054358	92 : 82	.049974
96 : 83	.063193	95 : 83	.058646	51 : 45	.054358	46 : 41	.049974
74 : 64	.063052	103 : 90	.058595	34 : 30	.054358	83 : 74	.049846
37 : 32	0.063052	96 : 84	0.057992	111 : 98	0.054097	74 : 66	0.049688
89 : 77	.062899	88 : 77	.057992	94 : 83	.054050	37 : 33	.049688
52 : 45	.062791	80 : 70	.057992	77 : 68	.053982	102 : 91	.049558
67 : 58	.062647	72 : 63	.057992	60 : 53	.053875	65 : 58	.049485
82 : 71	.062556	64 : 56	.057992	103 : 91	.053796	93 : 83	.049405
97 : 84	.062493	56 : 49	.057992	86 : 76	.053685	56 : 50	.049218
112 : 97	.062446	48 : 42	.057992	43 : 38	.053685	84 : 75	.049218
90 : 78	.062148	40 : 35	.057992	112 : 99	.053583	103 : 92	.049049

Logarithms of Gear Ratios from 1.0084+ to 7.5

Numbers of Teeth	Logarithm of Ratio	Numbers of Teeth	Logarithm of Ratio	Numbers of Teeth	Logarithm of Ratio	Numbers of Teeth	Logarithm of Ratio
75 : 67	0.048987	41 : 37	0.044582	57 : 52	0.039872	38 : 35	0.035716
94 : 84	.048849	72 : 65	.044419	80 : 73	.039767	89 : 82	.035576
47 : 42	.048849	103 : 93	.044354	103 : 94	.039709	51 : 47	.035472
66 : 59	.048692	93 : 84	.044204	92 : 84	.039509	64 : 59	.035328
85 : 76	.048605	62 : 56	.044204	69 : 63	.039509	77 : 71	.035232
104 : 93	.048550	83 : 75	.044017	46 : 42	.039509	90 : 83	.035164
95 : 85	.048305	52 : 47	.043905	104 : 95	.039310	103 : 95	.035114
76 : 68	.048305	73 : 66	.043779	81 : 74	.039253	91 : 84	.034762
57 : 51	0.048305	94 : 85	0.043709	58 : 53	0.039152	78 : 72	0.034762
38 : 34	.048305	84 : 76	.043466	93 : 85	.039064	65 : 60	.034762
105 : 94	.048061	63 : 57	.043466	105 : 96	.038918	52 : 48	.034762
86 : 77	.048008	42 : 38	.043466	70 : 64	.038918	39 : 36	.034762
67 : 60	.047924	95 : 86	.043225	35 : 32	.038918	105 : 97	.034418
96 : 86	.047773	74 : 67	.043157	82 : 75	.038753	92 : 85	.034369
48 : 43	.047773	53 : 48	.043035	94 : 86	.038629	79 : 73	.034304
77 : 69	.047642	85 : 77	.042928	47 : 43	.038629	66 : 61	.034214
106 : 95	0.047582	96 : 87	0.042752	106 : 97	0.038534	53 : 49	0.034080
87 : 78	.047425	64 : 58	.042752	59 : 54	.038458	93 : 86	.033984
58 : 52	.047425	107 : 97	.042612	71 : 65	.038345	80 : 74	.033858
29 : 26	.047425	75 : 68	.042552	83 : 76	.038265	40 : 37	.033858
97 : 87	.047252	86 : 78	.042404	95 : 87	.038204	107 : 99	.033749
39 : 35	.046997	43 : 39	.042404	107 : 98	.038158	67 : 62	.033683
88 : 79	.046856	97 : 88	.042289	96 : 88	.037789	94 : 87	.033609
98 : 88	.046743	54 : 49	.042198	84 : 77	0.037789	81 : 75	.033424
49 : 44	0.046743	65 : 59	0.042061	72 : 66	.037789	54 : 50	0.033424
108 : 97	.046652	76 : 69	.041965	60 : 55	.037789	95 : 88	.033241
59 : 53	.046576	87 : 79	.041892	48 : 44	.037789	68 : 63	.033168
69 : 62	.046457	98 : 89	.041836	36 : 33	.037789	82 : 76	.033000
79 : 71	.046369	109 : 99	.041791	109 : 100	.037427	41 : 38	.033000
89 : 80	.046300	99 : 90	.041393	97 : 89	.037382	96 : 89	.032881
99 : 89	.046245	88 : 80	.041393	85 : 78	.037324	55 : 51	.032793
109 : 98	.046200	77 : 70	.041393	73 : 67	.037248	69 : 64	.032669
100 : 90	0.045758	66 : 60	0.041393	61 : 56	0.037142	83 : 77	0.032587
90 : 81	.045758	55 : 50	.041393	98 : 90	.036984	97 : 90	.032529
80 : 72	.045758	44 : 40	.041393	49 : 45	.036984	98 : 91	.032185
70 : 63	.045758	33 : 30	.041393	86 : 79	.036871	84 : 78	.032185
60 : 54	.045758	100 : 91	.040959	74 : 68	.036723	70 : 65	.032185
50 : 45	.045758	89 : 81	.040905	37 : 34	.036723	56 : 52	.032185
40 : 36	.045758	78 : 71	.040836	99 : 91	.036594	42 : 39	.032185
111 : 100	.045323	67 : 61	.040745	62 : 57	.036517	99 : 92	.031847
101 : 91	0.045280	56 : 51	0.040618	87 : 80	0.036429	85 : 79	0.031792
91 : 82	.045228	101 : 92	.040534	100 : 92	.036212	71 : 66	.031714
81 : 73	.045162	90 : 82	.040429	75 : 69	.036212	57 : 53	.031599
71 : 64	.045078	45 : 41	.040429	50 : 46	.036212	100 : 93	.031517
61 : 55	.044967	79 : 72	.040295	88 : 81	.035998	86 : 80	.031409
51 : 46	.044812	68 : 62	.040117	63 : 58	.035913	43 : 40	.031409
92 : 83	.044710	34 : 31	.040117	101 : 93	.035839	72 : 67	.031258
82 : 74	.044582	91 : 83	.039963	76 : 70	.035716	101 : 94	.031194

Logarithms of Gear Ratios from 1.0084+ to 7.5

Numbers of Teeth	Logarithm of Ratio	Numbers of Teeth	Logarithm of Ratio	Numbers of Teeth	Logarithm of Ratio	Numbers of Teeth	Logarithm of Ratio
87 : 81	0.031034	50 : 47	0.026872	80 : 76	0.022276	100 : 96	0.017729
58 : 54	.031034	67 : 63	.026734	60 : 57	.022276	75 : 72	.017729
102 : 95	.030877	84 : 79	.026652	40 : 38	.022276	50 : 48	.017729
73 : 68	.030814	101 : 95	.026598	20 : 19	.022276	101 : 97	.017550
88 : 82	.030669	85 : 80	.026329	101 : 96	.022050	76 : 73	.017491
44 : 41	.030669	68 : 64	.026329	81 : 77	.021994	51 : 49	.017374
103 : 96	.030566	51 : 48	.026329	61 : 58	.021902	77 : 74	.017259
59 : 55	.030489	34 : 32	.026329	103 : 98	.021611	103 : 99	.017202
74 : 69	0.030383	103 : 97	0.026066	62 : 59	0.021540	78 : 75	0.017033
89 : 83	.030312	86 : 81	.026014	83 : 79	.021451	52 : 50	.017033
104 : 97	.030262	69 : 65	.025936	104 : 99	.021398	79 : 76	.016814
105 : 98	.029963	52 : 49	.025807	84 : 80	.021189	53 : 51	.016706
90 : 84	.029963	87 : 82	.025705	63 : 60	.021189	80 : 77	.016599
75 : 70	.029963	70 : 66	.025554	42 : 40	.021189	81 : 78	.016391
60 : 56	.029963	35 : 33	.025554	85 : 81	.020934	54 : 52	.016391
45 : 42	.029963	88 : 83	.025405	64 : 61	.020850	82 : 79	.016187
106 : 99	0.029671	53 : 50	0.025306	86 : 82	0.020685	55 : 53	0.016087
91 : 85	.029623	71 : 67	.025184	43 : 41	.020685	83 : 80	.015988
76 : 71	.029555	89 : 84	.025111	65 : 62	.020522	84 : 81	.015794
61 : 57	.029455	90 : 85	.024824	87 : 83	.020441	56 : 54	.015794
107 : 100	.029384	72 : 68	.024824	88 : 84	.020203	85 : 82	.015605
92 : 86	.029289	54 : 51	.024824	66 : 63	.020203	57 : 55	.015512
46 : 43	.029289	36 : 34	.024824	44 : 42	.020203	86 : 83	.015420
77 : 72	.029158	91 : 86	.024543	89 : 85	.019971	87 : 84	.015240
93 : 87	0.028964	73 : 69	0.024474	67 : 64	0.019895	58 : 56	0.015240
62 : 58	.028964	55 : 52	.024359	90 : 86	.019744	88 : 85	.015064
78 : 73	.028772	92 : 87	.024269	45 : 43	.019744	59 : 57	.014977
47 : 44	.028645	74 : 70	.024134	68 : 65	.019596	89 : 86	.014892
94 : 88	.028645	37 : 35	.024134	91 : 87	.019522	90 : 87	.014723
63 : 59	.028489	93 : 88	.024000	92 : 88	.019305	60 : 58	.014723
79 : 74	.028395	56 : 53	.023912	69 : 66	.019305	91 : 88	.014559
95 : 89	.028334	75 : 71	.023803	46 : 44	.019305	61 : 59	.014478
96 : 90	0.028029	94 : 89	0.023738	93 : 89	0.019093	92 : 89	0.014398
80 : 75	.028029	95 : 90	.023481	70 : 67	.019023	93 : 90	.014240
64 : 60	.028029	76 : 72	.023481	94 : 90	.018885	62 : 60	.014240
48 : 45	.028029	57 : 54	.023481	47 : 45	.018885	31 : 30	.014240
32 : 30	.028029	38 : 36	.023481	71 : 68	.018749	94 : 91	.014087
97 : 91	.027730	96 : 91	.023230	95 : 91	.018682	63 : 61	.014011
81 : 76	.027671	77 : 73	.023168	96 : 92	.018483	95 : 92	.013936
65 : 61	.027584	58 : 55	.023065	72 : 69	.018483	96 : 93	.013788
98 : 92	0.027438	97 : 92	0.022984	48 : 46	0.018483	64 : 62	0.013788
49 : 46	.027438	78 : 74	.022863	24 : 23	.018483	32 : 31	.013788
82 : 77	.027323	39 : 37	.022863	97 : 93	.018289	97 : 94	.013644
99 : 93	.027152	98 : 93	.022743	73 : 70	.018225	65 : 63	.013573
66 : 62	.027152	59 : 56	.022664	98 : 94	.018098	98 : 95	.013503
33 : 31	.027152	79 : 75	.022566	49 : 47	.018098	99 : 96	.013364
83 : 78	.026984	99 : 94	.022507	74 : 71	.017973	66 : 64	.013364
100 : 94	.026872	100 : 95	.022276	99 : 95	.017912	33 : 32	.013364

Logarithms of Gear Ratios from 1.0084+ to 7.5

Numbers of Teeth	Logarithm of Ratio	Numbers of Teeth	Logarithm of Ratio	Numbers of Teeth	Logarithm of Ratio	Numbers of Teeth	Logarithm of Ratio
100 : 97	0.013228	87 : 85	0.010100	61 : 60	0.007179	93 : 92	0.004695
67 : 65	.013161	88 : 86	.009984	62 : 61	.007062	94 : 93	.004645
101 : 98	.013095	44 : 43	.009984	63 : 62	.006949	95 : 94	.004596
68 : 66	.012965	89 : 87	.009871	64 : 63	.006840	96 : 95	.004548
34 : 33	.012965	90 : 88	.009760	65 : 64	.006733	97 : 96	.004501
103 : 100	.012837	45 : 44	.009760	66 : 65	.006631	98 : 97	.004454
69 : 67	.012774	91 : 89	.009651	67 : 66	.006531	99 : 98	.004409
70 : 68	.012589	92 : 90	.009545	68 : 67	.006434	100 : 99	.004365
35 : 34	0.012589	46 : 45	0.009545	69 : 68	0.006340	101 : 100	0.004321
72 : 70	.012235	93 : 91	.009442	70 : 69	.006249	102 : 101	.004279
36 : 35	.012235	94 : 92	.009340	71 : 70	.006160	103 : 102	.004237
73 : 71	.012065	47 : 46	.009340	72 : 71	.006074	104 : 103	.004196
74 : 72	.011899	95 : 93	.009241	73 : 72	.005990	105 : 104	.004156
37 : 36	.011899	96 : 94	.009143	74 : 73	.005909	106 : 105	.004117
75 : 73	.011738	48 : 47	.009143	75 : 74	.005830	107 : 106	.004078
76 : 74	.011582	97 : 95	.009048	76 : 75	.005752	108 : 107	.004040
38 : 37	0.011582	98 : 96	0.008955	77 : 76	0.005677	109 : 108	0.004002
77 : 75	.011429	49 : 48	.008955	78 : 77	.005604	110 : 109	.003967
78 : 76	.011281	99 : 97	.008864	79 : 78	.005533	111 : 110	.003930
39 : 38	.011281	100 : 98	.008774	80 : 79	.005463	112 : 111	.003895
79 : 77	.011136	50 : 49	.008774	81 : 80	.005395	113 : 112	.003860
80 : 78	.010995	101 : 99	.008686	82 : 81	.005329	114 : 113	.003827
40 : 39	.010995	51 : 50	.008600	83 : 82	.005264	115 : 114	.003793
81 : 79	.010858	52 : 51	.008433	84 : 83	.005201	116 : 115	.003760
82 : 80	0.010724	53 : 52	0.008273	85 : 84	0.005140	117 : 116	0.003728
41 : 40	.010724	54 : 53	.008118	86 : 85	.005080	118 : 117	.003696
83 : 81	.010593	55 : 54	.007969	87 : 86	.005021	119 : 118	.003665
84 : 82	.010465	56 : 55	.007825	88 : 87	.004963	120 : 119	.003634
42 : 41	.010465	57 : 56	.007687	89 : 88	.004907	......	......
85 : 83	.010341	58 : 57	.007553	90 : 89	.004853	......	......
86 : 84	.010219	59 : 58	.007424	91 : 90	.004799	......	......
43 : 42	.010219	60 : 59	.007299	92 : 91	.004746	......	......

Example: A driven shaft is to rotate 6.9078 revolutions while the driving shaft rotates 1.3961 revolutions. Determine sizes of driving and driven gears.

$$\frac{\text{Driving gear size}}{\text{Driven gear size}} = \frac{\text{Driven gear speed}}{\text{Driving gear speed}} = \frac{6.9078}{1.3961}$$

Log 6.9078 = 0.8393398

Log 1.3961 = 0.1449165

$$0.6944233 = \log \frac{\text{Driving gear size}}{\text{Driven gear size}}$$

The nearest logarithm in table is 0.694374 which is the log for ratio 94/19. The driving gear is to have 94 teeth and the driven gear 19 teeth.

Actual revolutions of driven shaft (while driving shaft makes 1.3961 turns) = 1.3961 × 94/19 = 6.9070 or a difference from the desired rotation of 0.0008 revolution. Should a more exact solution be required, some form of compound gearing might be used. (See also examples preceding table.)

THREAD ROLLING

Screw threads may be formed by rolling either by using some type of thread-rolling machine or by equipping an automatic screw machine or turret lathe with a suitable threading roll. If a thread-rolling machine is used, the unthreaded screw, bolt or other "blank," is placed (either automatically or by hand) between dies having thread-shaped ridges which sink into the blank, and by displacing the metal, form a thread of the required shape and pitch. The thread-rolling process is applied where bolts, screws, studs, threaded rods, etc., are required in large quantities. Screw threads that are within the range of the rolling process may be produced more rapidly by this method than in any other way. The rolled thread, due to the cold-working action of the dies, is 10 to 20 per cent stronger than a cut or ground thread, and the increase may be much higher when tested for fatigue resistance. Another advantage of the rolling process is that no stock is wasted in forming the thread, and the surface of a rolled thread is harder than that of a cut thread, thus increasing wear resistance.

Thread-Rolling Machine of Flat-die Type. — One type of machine which is used extensively is equipped with a pair of flat or straight dies One die is stationary and the other has a reciprocating movement when the machine is in use. The ridges on these dies, which form the screw thread, incline at an angle equal to the helix angle of the thread. In making dies for precision thread rolling, the threads may be formed either by milling and grinding after heat-treatment, or by grinding "from the solid" after heat-treating. A vitrified wheel is used. The thread is formed in one passage of the work, which is inserted at one end of the dies, either by hand or automatically, and then rolls between the die faces until it is ejected at the opposite end. The relation between the position of the dies and a screw thread being rolled is such that the top of the thread-shaped ridge of one die, at the point of contact with the screw thread, is directly opposite the bottom of the thread groove in the other die at the point of contact. Some form of mechanism insures starting the blank at the right time and also square with the dies.

Thread-Rolling Machine of Cylindrical Die Type. — With machines of this type, the blank is threaded while being rolled between two or three cylindrical dies (depending upon type of machine) which are pressed into the blank at a rate of penetration adjusted to the hardness of the material, or wall thickness in the case of threading operations on tubing or hollow parts. The dies have ground or ground and lapped threads and a pitch diameter that is a multiple of the pitch diameter of the thread to be rolled. As the dies are much larger in diameter than the work, a multiple thread is required to obtain the same lead angle as that of the work. The thread may be formed in one die revolution or even less, or several revolutions may be required (as in rolling hard materials) to obtain a gradual rate of penetration equivalent to that obtained with flat or straight dies if extended to a length of possibly 15 or 20 feet. Provisions for accurately adjusting or matching the thread rolls to bring them into proper alignment with each other, are important features of these machines.

Two-Roll Type of Machine: With a two-roll type of machine, the work is rotated between two horizontal power-driven threading rolls and is supported by a hardened rest bar on the lower side. One roll is fed inward by hydraulic pressure to a depth that is governed automatically.

Three-Roll Type of Machine: With this machine the blank to be threaded is held in a "floating position" while being rolled between three cylindrical dies which, through toggle arms, are moved inward at a predetermined rate of penetration until the required pitch diameter is obtained. The die movement is governed by a cam driven through change gears selected to give the required cycle of squeeze, dwell and release.

Rate of Production. — Production rates in thread rolling depend upon the type of machine, the size of both machine and work, and whether the parts to be threaded are inserted by hand or automatically. A reciprocating flat die type of machine, applied to ordinary steels, may thread 30 or 40 parts per minute in diameters ranging from about ⅝ to 1⅛ inch, and 150 to 175 per minute in machine screw sizes from No. 10 (.190) to No. 6 (.138). In the case of heat-treated alloy steels in the usual hardness range of 26 to 32 Rockwell C, the production may be 30 or 40 per minute or less. With a cylindrical die type of machine, which is designed primarily for precision work and hard metals, 10 to 30 parts per minute are common production rates, the amount depending upon the hardness of material and allowable rate of die penetration per work revolution. These production rates are intended as a general guide only. The diameters of rolled threads usually range from the smallest machine screw sizes up to 1 or 1½ inches, depending upon the type and size of machine.

Precision Thread Rolling. — Both flat and cylindrical dies are used in aeronautical and other plants for precision work. With accurate dies and blank diameters held to close limits, it is practicable to produce rolled threads for American Standard Class 3 and Class 4 fits. The blank sizing may be by centerless grinding or by means of a die in conjunction with the heading operations. The blank should be round, and, as a general rule, the diameter tolerance should not exceed ½ to ⅔ the pitch diameter tolerance. The blank diameter should range from the correct size (which is close to the pitch diameter, but should be determined by actual trial), down to the allowable minimum, the tolerance being minus to insure a correct pitch diameter, even though the major diameter may vary slightly. Precision thread rolling has become an important method of threading alloy steel studs and other threaded parts, especially in aeronautical work where precision and high-fatigue resistance are required. Micrometer screws are also an outstanding example of precision thread rolling. This process has also been applied in tap making, although it is the general practice to finish rolled taps by grinding when the Class 3 and Class 4 fits are required.

Steels for Thread Rolling. — Steels vary from soft low-carbon types for ordinary screws and bolts, to nickel, nickel-chromium and molybdenum steels for aircraft studs, bolts, etc., or for any work requiring exceptional strength and fatigue resistance. Typical SAE alloy steels are No. 2330, 3135, 3140, 4027, 4042, 4640 and 6160. The hardness of these steels after heat-treatment usually ranges from 26 to 32 Rockwell C, with tensile strengths varying from 130,000 to 150,000 pounds per square inch. While harder materials might be rolled, grinding is more practicable when the hardness exceeds 40 Rockwell C. Thread rolling is applicable not only to a wide range of steels but for non-ferrous materials, especially if there is difficulty in cutting due to "tearing" the threads.

Diameter of Blank for Thread Rolling. — The diameter of the screw blank or cylindrical part upon which a thread is to be rolled should be less than the outside screw diameter by an amount that will just compensate for the metal that is displaced and raised above the original surface by the rolling process. The increase in diameter is approximately equal to the depth of one thread. While there are rules and formulas for determining blank diameters, it may be necessary to make slight changes in the calculated size in order to secure a well-formed thread. The blank diameter should be verified by trial, especially when rolling accurate screw threads. Some stock offers greater resistance to displacement than other stock, owing to the greater hardness or tenacity of the metal. The following figures may prove useful in establishing trial sizes. The blank diameters for screws varying from ¼ to ½ are from 0.002 to 0.0025 inch larger than the pitch diameter, and for screws varying from ½ to 1 inch or larger, the blank diameters are from 0.0025 to

0.003 inch larger than the pitch diameter. Blanks which are slightly less than the pitch diameter are intended for bolts, screws, etc., which are to have a comparatively free fit. Blanks for this class of work may vary from 0.002 to 0.003 inch less than the pitch diameter for screw thread sizes varying from ¼ to ½ inch, and from 0.003 to 0.005 inch less than the pitch diameter for sizes above ½ inch. If the screw threads are smaller than ¼ inch, the blanks are usually from 0.001 to 0.0015 inch less than the pitch diameter for ordinary grades of work.

Thread Rolling in Automatic Screw Machines. — Screw threads are sometimes rolled in automatic screw machines and turret lathes when the thread is behind a shoulder so that it cannot be cut with a die. In such cases, the advantage of rolling the thread is that a second operation is avoided. A circular roll is used for rolling threads in screw machines. The roll may be presented to the work either in a tangential direction or radially, either method producing a satisfactory thread. In the former case, the roll gradually comes into contact with the periphery of the work and completes the thread as it passes across the surface to be threaded. When the roll is held in a radial position, it is simply forced against one side until a complete thread is formed. The method of applying the roll may depend upon the relation between the threading operation and other machining operations. Thread rolling in automatic screw machines is generally applied only to brass and other relatively soft metals, owing to the difficulty of rolling threads in steel. Thread rolls made of chrome-nickel steel containing from 0.15 to 0.20 per cent of carbon have given fairly good results, however, when applied to steel. A 3 per cent nickel steel containing about 0.12 per cent carbon has also proved satisfactory for threading brass.

Factors Governing the Diameter of Thread Rolling. — The threading roll used in screw machines may be about the same diameter as the screw thread, but for sizes smaller than, say, ¾ inch, the roll diameter is some multiple of the thread diameter minus a slight amount to obtain a better rolling action. When the diameters of the thread and roll are practically the same, a single-threaded roll is used to form a single thread on the screw. If the diameter of the roll is made double that of the screw, in order to avoid using a small roll, then the roll must have a double thread. If the thread roll is three times the size of the screw thread, a triple thread is used, and so on. These multiple threads are necessary when the roll diameter is some multiple of the work, in order to obtain corresponding helix angles on the roll and work.

Diameter of Threading Roll. — The pitch diameter of a threading roll having a single thread is slightly less than the pitch diameter of the screw thread to be rolled, and in the case of multiple-thread rolls, the pitch diameter is not an exact multiple of the screw thread pitch diameter but is also reduced somewhat. The amount of reduction recommended by one screw machine manufacturer is given by the formula shown at the end of this paragraph. A description of the terms used in the formula is given as follows: D = pitch diameter of threading roll, d = pitch diameter of screw thread, N = number of single threads or "starts" on the roll (this number is selected with reference to diameter of roll desired), T = single depth of thread:

$$D = N\left(d - \frac{T}{2}\right) - T$$

Example: Find, by using above formula, the pitch diameter of a double-thread roll for rolling a ½-inch American standard screw thread. Pitch diameter d = 0.4500 inch and thread depth T = 0.0499 inch.

$$D = 2\left(0.4500 - \frac{0.0499}{2}\right) - 0.0499 = 0.8001 \text{ inch}$$

Kind of Thread on Roll and Its Shape. — The thread (or threads) on the roll should be left hand for rolling a right-hand thread, and *vice versa*. The roll should be wide enough to overlap the part to be threaded, provided there are clearance spaces at the ends, which should be formed if possible. The thread on the roll should be sharp on top for rolling an American (National) standard form of thread, so that less pressure will be required to displace the metal when rolling the thread. The bottom of the thread groove on the roll may also be left sharp or it may have a flat. If the bottom is sharp, the roll is sunk only far enough into the blank to form a thread having a flat top, assuming that the thread is the American form. The number of threads on the roll (whether double, triple, quadruple, etc.) is selected, as a rule, so that the diameter of the thread roll will be somewhere between 1¼ and 2¼ inches. In making a thread roll, the ends are beveled at an angle of 45 degrees, to prevent the threads on the ends of the roll from chipping. Precautions should be taken in hardening, because if the sharp edges are burnt, the roll will be useless. Thread rolls, as a rule, are lapped after hardening. This is done by holding them on an arbor in the lathe and using emery and oil on a piece of hard wood. A thread roll, to give good results, should fit closely in the holder. If the roll is made to fit loosely, it will mar the threads.

Application of Thread Roll. — The shape of the work, and the character of the operations necessary to produce it, govern, to a large extent, the method employed in applying the thread roll. Some of the points to consider are as follows: 1. Diameter of the part to be threaded. 2. Location of the part to be threaded. 3. Length of the part to be threaded. 4. Relation that the thread rolling operation bears to the other operations. 5. Shape of the part to be threaded, whether straight, tapered or otherwise. 6. Method of applying the support. When the diameter to be rolled is much smaller than the diameter of the shoulder preceding it, a cross-slide knurl-holder should be used. If the part to be threaded is not behind a shoulder, a holder on the swing principle should be used. When the work is long (greater in length than two-and-one-half times its diameter) a swing roll-holder should be employed, carrying a support. When the work can be cut off after the thread is rolled, a cross-slide roll-holder should be used. The method of applying the support to the work also governs to some extent the method of applying the thread roll. When no other tool is working at the same time as the thread roll, and when there is freedom from chips, the roll can be held more rigidly by passing it under instead of over the work. When passing the roll over the work, it has a tendency to raise the cross-slide. Where the part to be threaded is tapered, the roll can best be presented to the work by holding it in a cross-slide roll-holder.

Speeds and Feeds for Thread Rolling. — When the thread roll is made from high-carbon steel and used on brass, a surface speed as high as 200 feet per minute can be used. Better results, however, are obtained by using a lower speed than this. When the roll is held in a holder attached to the cross-slide, and is presented either tangentially or radially to the work, a considerably higher speed can be used than if it is held in a swing tool. This is due to the lack of rigidity in a holder of the swing type. The feeds to be used when a cross-slide roll-holder is used are given in the upper half of the table "Feeds for Thread Rolling"; the lower half of the table gives the feeds for thread rolling with swing tools. These feeds are applicable for rolling threads without a support, when the root diameter of the blank is not less than five times the double depth of the thread. When the root diameter is less than this, a support should be used. A support should also be used when the width of the roll is more than two-and-one-half times the smallest diameter of the piece to be rolled, irrespective of the pitch of the thread. When the smallest diameter of the piece to be rolled is much less than the root diameter of the thread, the smallest diameter should be taken as the deciding factor for the feed to be used.

Feeds for Thread Rolling

Root Diam. of Blank	Number of Threads per Inch													
	72	64	56	48	44	40	36	32	28	24	22	20	18	14
	Cross-slide Holders — Feed per Revolution in Inches													
1/8	0.0045	0.0040	0.0035	0.0030	0.0025	0.0020	0.0015	0.0010	.....	.....	.....	.....	.....	.....
3/16	0.0050	0.0045	0.0040	0.0035	0.0030	0.0025	0.0020	0.0015	0.0005	.....	.....	.....	.....	.....
1/4	0.0055	0.0050	0.0045	0.0040	0.0035	0.0030	0.0025	0.0020	0.0010	0.0005	0.0005	.....	.....	.....
5/16	0.0060	0.0055	0.0050	0.0045	0.0040	0.0035	0.0030	0.0025	0.0015	0.0010	0.0010	0.0005	0.0005	.....
3/8	0.0065	0.0060	0.0055	0.0050	0.0045	0.0040	0.0035	0.0030	0.0020	0.0015	0.0015	0.0010	0.0010	0.0005
7/16	0.0070	0.0065	0.0060	0.0055	0.0050	0.0045	0.0040	0.0035	0.0025	0.0020	0.0020	0.0015	0.0015	0.0010
1/2	0.0075	0.0070	0.0065	0.0060	0.0055	0.0050	0.0045	0.0040	0.0030	0.0025	0.0025	0.0020	0.0020	0.0015
5/8	0.0080	0.0075	0.0070	0.0065	0.0060	0.0055	0.0050	0.0045	0.0035	0.0030	0.0030	0.0025	0.0025	0.0020
3/4	0.0085	0.0080	0.0075	0.0070	0.0065	0.0060	0.0055	0.0050	0.0040	0.0035	0.0035	0.0030	0.0030	0.0025
7/8	0.0090	0.0085	0.0080	0.0075	0.0070	0.0065	0.0060	0.0055	0.0045	0.0040	0.0040	0.0035	0.0035	0.0030
1	0.0095	0.0090	0.0085	0.0080	0.0075	0.0070	0.0065	0.0060	0.0050	0.0045	0.0045	0.0040	0.0040	0.0035
Root Diam.	Swing Holders — Feed per Revolution in Inches													
1/8	0.0025	0.0020	0.0015	0.0010	0.0005	.....	.....	.....	.....	.....	.....	.....	.....	.....
3/16	0.0028	0.0025	0.0020	0.0015	0.0008	0.0005	.....	.....	.....	.....	.....	.....	.....	.....
1/4	0.0030	0.0030	0.0025	0.0020	0.0010	0.0010	0.0005	0.0005	0.0005	.....	.....	.....	.....	.....
5/16	0.0035	0.0035	0.0030	0.0025	0.0015	0.0015	0.0010	0.0010	0.0010	0.0005	.....	.....	.....	.....
3/8	0.0040	0.0040	0.0035	0.0030	0.0020	0.0020	0.0015	0.0015	0.0015	0.0010	0.0005	0.0005	0.0005	.....
7/16	0.0045	0.0045	0.0040	0.0035	0.0030	0.0025	0.0020	0.0020	0.0020	0.0015	0.0010	0.0010	0.0010	.....
1/2	0.0048	0.0048	0.0045	0.0040	0.0035	0.0030	0.0025	0.0025	0.0025	0.0020	0.0015	0.0015	0.0015	0.0005
5/8	0.0050	0.0050	0.0048	0.0043	0.0040	0.0035	0.0030	0.0030	0.0028	0.0025	0.0020	0.0020	0.0018	0.0010
3/4	0.0055	0.0052	0.0050	0.0045	0.0043	0.0040	0.0035	0.0035	0.0030	0.0028	0.0025	0.0022	0.0020	0.0013
7/8	0.0058	0.0055	0.0052	0.0048	0.0045	0.0043	0.0040	0.0038	0.0032	0.0030	0.0028	0.0025	0.0022	0.0015
1	0.0060	0.0058	0.0054	0.0050	0.0048	0.0047	0.0043	0.0040	0.0035	0.0032	0.0030	0.0028	0.0025	0.0018

THREAD GRINDING

Thread grinding is employed for precision tool and gage work and also in producing certain classes of threaded parts. Thread grinding may be utilized (1) because of the accuracy and finish obtained, (2) hardness of material to be threaded, (3) economy in grinding certain classes of screw threads when using modern machines, wheels, and thread-grinding oils. In some cases pre-cut threads are finished by grinding; but usually, threads are ground "from the solid," being formed entirely by the grinding process. Examples of work include thread gages and taps of steel and tungsten carbide, hobs, worms, lead-screws, adjusting or traversing screws, alloy steel studs, etc. Grinding is applied to external, internal, straight, and tapering threads, and to various thread forms.

Accuracy Obtainable by Thread Grinding. — With single-edge or single-ribbed wheels it is possible to grind threads on gages to a degree of accuracy that requires but very little lapping to produce a so-called "master" thread gage. As far as lead is concerned, some thread grinding machine manufacturers guarantee to hold the lead within 0.0001 inch per inch of thread; and while it is not guaranteed that a higher degree of accuracy for lead is obtainable, it is known that threads have been ground to closer tolerances than this on the lead. Pitch diameter accuracies for either Class 3 or Class 4 fits are obtainable according to the grinding method used; with single-edge wheels, the thread angle can be ground to an accuracy of within two or three minutes in half the angle.

Wheels for Thread Grinding. — The wheels used for steel have an aluminous abrasive and, ordinarily, either a resinoid bond or a vitrified bond. The general rule is to use resinoid wheels when extreme tolerances are not required, and it is desirable to form the thread with a minimum number of passes, as in grinding threaded machine parts, such as studs, adjusting screws which are not calibrated, and for some classes of taps. *Resinoid wheels,* as a rule, will hold a fine edge longer than a vitrified wheel but they are more flexible and, consequently, less suitable for accurate work, especially when there is lateral grinding pressure that causes wheel deflection. *Vitrified wheels* are utilized for obtaining extreme accuracy in thread form and lead because they are very rigid and not easily deflected by side pressure in grinding. This rigidity is especially important in grinding pre-cut threads on such work as gages, taps and lead-screws. The progressive lead errors in long lead-screws, for example, might cause an increasing lateral pressure that would deflect a resinoid wheel. Vitrified wheels are also recommended for internal grinding.

Diamond Wheels: Diamond wheels set in a rubber or plastic bond are also used for thread grinding, especially for grinding threads in carbide materials and in other hardened alloys. Thread grinding is now being done successfully on a commercial basis on both taps and gages made from carbides. Gear hobs made from carbides have also been tested with successful results. Diamond wheels are dressed by means of silicon-carbide grinding wheels which travel past the diamond-wheel thread form at the angle required for the flanks of the thread to be ground. The action of the dressing wheels is, perhaps, best described as a "scrubbing" of the bond which holds the diamond grits. Obviously, the silicon-carbide wheels do not dress the diamonds, but they loosen the bond until the diamonds not wanted drop out.

Thread Grinding with Single-Edge Wheel. — With this type of wheel, the edge is trued to the cross-sectional shape of the thread groove. The wheel, when new, may have a diameter of 18 or 20 inches and, when grinding a thread, the wheel is inclined to align it with the thread groove. On some machines, lead variations are obtained by means of change-gears which transmit motion from the work-driving spindle to the lead-screw. Other machines are so designed that a lead-

screw is selected to suit the lead of thread to be ground and transmits motion directly to the work-driving spindle.

Wheels with Edges for Roughing and Finishing. — The "three-ribbed" type of wheel has a roughing edge or rib which removes about two-thirds of the metal. This is followed by an intermediate rib which leaves about 0.005 inch for the third or finishing rib. The accuracy obtained with this triple-edge type compares with

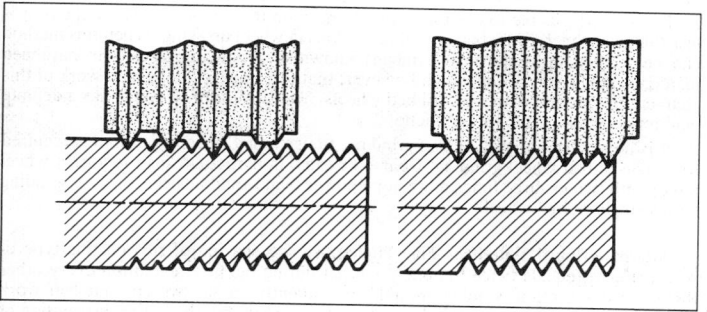

Fig. 1. Wheel with Edges for Roughing and Finishing

Fig. 2. Multi-ribbed Type of Thread-grinding Wheel

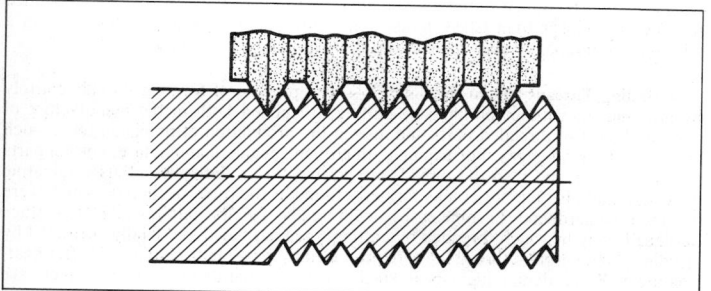

Fig. 3. Alternate-ribbed Wheel for Grinding the Finer Pitches

that of a single-edge wheel, which means that it may be used for the greatest accuracy obtainable in thread grinding. When the accuracy required makes it necessary, this wheel can be inclined to the helix angle of the thread, the same as is the single-edge wheel.

The three-ribbed wheel is recommended not only for precision work but for grinding threads which are too long for the multi-ribbed wheel referred to later. It is also well adapted to tap grinding, because it is possible to dress a portion of the wheel adjacent to the finish rib for the purpose of grinding the outside diameter of the thread, as indicated in Fig. 1. Furthermore, the wheel can be dressed for grinding or relieving both crests and flanks at the same time.

Multi-ribbed Wheels. — This type of wheel is employed when rapid production is more important than extreme accuracy, which means that it is intended

primarily for the grinding of duplicate parts in manufacturing. A wheel 1¼ to 2 inches wide has formed upon its face a series of annular thread-shaped ridges (see Fig. 2); hence, if the length of the thread is not greater than the wheel width, a thread may be ground in one work revolution plus about one-half revolution for feeding in and withdrawing the wheel. The principle of operation is the same as that of thread milling with a multiple type cutter. This type of wheel is not inclined to the lead angle. To obtain a Class 3 fit, the lead angle should not exceed 4 degrees.

It is not practicable to use this form of wheel on thread pitches where the root is less than 0.007 inch wide, because of difficulties in wheel dressing. When this method can be applied, it is the fastest means known of producing threads in hardened materials. It is not recommended, however, that thread gages, taps, and work of this character be ground with multi-ribbed wheels. The single-ribbed wheel has a definite field for accurate, small-lot production.

It is necessary, in multi-ribbed grinding, to use more horsepower than is required for single-ribbed wheel grinding. Coarse threads, in particular, may require a wheel motor with two or three times more horsepower than would be necessary for grinding with a single-ribbed wheel.

Alternate-ribbed Wheel for Fine Pitches. — The spacing of ribs on this type of wheel (Fig. 3) equals twice the pitch, so that during the first revolution every other thread groove section is being ground; consequently, about two and one-half work revolutions are required for grinding a complete thread, but the better distribution of cooling oil and resulting increase in work speeds makes this wheel very efficient. This alternate-type of wheel is adapted for grinding threads of fine pitch. Since these wheels cannot be tipped to the helix angle of the thread, they are not recommended for anything closer than Class 3 fits. The "three-ribbed" wheels referred to in a previous paragraph are also made in the alternate type for the finer pitches.

Grinding Threads "from the Solid." — The process of forming threads entirely by grinding, or without preliminary cutting, is applied both in the manufacture of certain classes of threaded parts and also in the production of precision tools, such as taps and thread gages. For example, in airplane engine manufacture, certain parts are heat-treated and then the threads are ground "from the solid," thus eliminating distortion and also minute cracks formerly found at the roots of threads which were cut and then hardened. In some cases steel threads of coarse pitch, which are surface hardened, may be rough threaded by cutting, then hardened and finally corrected by grinding. Many ground thread taps are produced by grinding from the solid after heat-treatment. By hardening high-speed steel taps before the thread is formed, there are no narrow or delicate crests to interfere with the application of the high temperature required for uniform hardness and the best steel structure.

Number of Wheel Passes. — The number of cuts or passes for grinding from the solid depends upon the type of wheel and accuracy required. In general, threads of 12 or 14 per inch and finer may be ground in one pass of a single-edge wheel unless the "unwrapped" thread length is much greater than normal. Unwrapped length = pitch circumference × total number of thread turns, approximately. For example, a thread gage 1¼ inches long with 24 threads per inch would have an unwrapped length equal to 30 × pitch circumference. (If more convenient, outside circumference may be used instead of pitch circumference.) Assume that there are 6 or 7 feet of unwrapped length on a screw thread having 12 threads per inch. In this case, one pass might be sufficient for a Class 3 fit, whereas two passes might be recommended for a Class 4 fit. When two passes are required, too deep a roughing cut may break down the narrow edge of the wheel. To prevent this, try a roughing

cut depth equal to about two-thirds the total thread depth, thus leaving one-third for the finishing cut.

Wheel and Work Rotation. — When a screw thread, on the side being ground, is moving *upward* or *against* the grinding wheel rotation, less heat is generated and the grinding operation is more efficient than when wheel and work are moving in the same direction on the grinding side; however, to avoid running a machine idle during its return stroke, many screw threads are ground during both the forward and return traversing movements, by reversing the work rotation at the end of the forward stroke. For this reason, thread grinders generally are equipped so that both forward and return work speeds may be changed; they may also be designed to accelerate the return movement when grinding in one direction only.

Wheel Speeds. — Wheel speeds should always be limited to the maximum specified on the wheel by the manufacturer. According to the American National Standard Safety Code, resinoid and vitrified wheels are limited to 12,000 surface feet per minute; however, according to Norton Co., the most efficient speeds are from 9,000 to 10,500 for resinoid wheels and 7,500 to 9,500 for vitrified wheels. Only tested wheels recommended by the wheel manufacturer should be used. After a suitable surface speed has been established, it should be maintained by increasing the rpm of the wheel, as the latter is reduced in diameter by wear.

Since thread grinding wheels work close to the limit of their stock-removing capacity, some adjustment of the wheel or work speed may be required to get the best results. If the wheel speed is too slow for a given job and excessive heat is generated, try an increase in speed, assuming that such increase is within the safety limits. If the wheel is too soft and the edge wears excessively, again an increase in wheel speed will give the effect of a harder wheel and result in better form-retaining qualities.

Work Speeds. — The work speed usually ranges from 3 to 10 feet per minute. In grinding with a comparatively heavy feed, and a mininum number of passes, the speed may not exceed 2½ or 3 feet per minute. If very light feeds are employed as in grinding hardened high-speed steel, the work speed may be much higher than 3 feet per minute and should be determined by test. If excessive heat is generated by removing stock too rapidly, a work speed reduction is one remedy. If a wheel is working below its normal capacity, an increase in work speed would prevent dulling of the grains and reduce the tendency to heat or "burn" the work. An increase in work speed and reduction in feed may also be employed to prevent burning while grinding hardened steel.

Truing Grinding Wheels. — Thread grinding wheels are trued both to maintain the required thread form and also an efficient grinding surface. Thread grinders ordinarily are equipped with precision truing devices which function automatically. One type automatically dresses the wheel and also compensates for the slight amount removed in dressing, thus automatically maintaining size control of the work. While truing the wheel, a small amount of grinding oil should be used to reduce diamond wear. Light truing cuts are advisable, especially in truing resinoid wheels which may be deflected by excessive truing pressure. A master former for controlling the path followed by the truing diamond may require a modified profile to prevent distortion of the thread form, especially when the lead angles are comparatively large. Such modification usually is not required for 60-degree threads when the pitches for a given diameter are standard because then the resulting lead angles are less than 4½ degrees. In grinding Acme threads or 29-degree worm threads having lead angles greater than 4 or 5 degrees, modified formers may be required to prevent a bulge in the thread profile. The highest point of this bulge is approximately at the pitch line. A bulge of about 0.001 inch may be within allowable

limits on some commercial worms but precision worms for gear hobbers, etc., require straight flanks in the axial plane.

Crushing Method: Thread grinding wheels are also dressed or formed by the crushing method, which is used in connection with some types of thread grinding machines. When this method is used, the annular ridge or ridges on the wheel are formed by a hardened steel cylindrical dresser or crusher. The crusher has a series of smooth annular ridges which are shaped and spaced like the thread that is to be ground. During the wheel dressing operation, the crusher is positively driven instead of the grinding wheel, and the ridges on the wheel face are formed by the rotating crusher being forced inward.

Wheel Hardness or Grade. — Wheel hardness or grade selection is based upon a compromise between efficient cutting and durability of the grinding edge. Grade selection depends on the bond and the character of the work. The following general recommendations are based upon Norton grading.

Vitrified wheels usually range from J to M, and resinoid wheels from R to U. For heat-treated screws or studs and the Unified Standard Thread, try the following. For 8 to 12 threads per inch, grade S resinoid wheel; for 14 to 20 threads per inch, grade T resinoid; for 24 threads per inch and finer, grades T or U resinoid. For high-speed steel taps 4 to 12 threads per inch, grade J vitrified or S resinoid; 14 to 20 threads per inch, grade K vitrified or T resinoid; 24 to 36 threads per inch, grade M vitrified or T resinoid.

Grain Size. — A thread grinding wheel usually operates close to its maximum stock-removing capacity, and the narrow edge which forms the root of the thread is the most vulnerable part. In grain selection, the general rule is to use the coarsest grained wheel that will hold its form while grinding a reasonable amount of work. Pitch of thread and quality of finish are two governing factors. Thus, to obtain an exceptionally fine finish, the grain size might be smaller than is needed to retain the edge profile. The usual grain sizes range from 120 to 150. For heat-treated screws and studs with Unified Standard Threads, 100 to 180 is the usual range. For precision screw threads of very fine pitch, the grain size may range from 220 to 320. For high-speed steel taps, the usual range is from 150 to 180 for Unified Standard Threads, and from 80 to 150 for pre-cut Acme threads.

Thread Grinding by Centerless Method. — Screw threads may be ground from the solid by the centerless method. A centerless thread grinder is similar in its operating principle to a centerless grinder designed for general work, in that it has a grinding wheel, a regulating or feed wheel (with speed adjustments), and a work-rest. Adjustments are provided to accommodate work of different sizes and for varying the rates of feed. The grinding wheel is a multi-ribbed type, being a series of annular ridges across the face. These ridges conform in pitch and profile with the thread to be ground. The grinding wheel is inclined to suit the helix or lead angle of the thread. In grinding threads on such work as socket type set-screws, the blanks are fed automatically and passed between the grinding and regulating wheels in a continuous stream. To illustrate production possibilities, hardened socket set-screws of ¼-20 size may be ground from the solid at the rate of 60 to 70 per minute and with the wheel operating continuously for 8 hours without redressing. The lead errors of centerless ground screw threads may be limited to 0.0005 inch per inch or even less by reducing the production rate. The pitch diameter tolerances are within 0.0002 to 0.0003 inch of the basic size. The grain size for the wheel is selected with reference to the pitch of the thread, the following sizes being recommended: For 11 to 13 threads per inch, 150; for 16 threads per inch, 180; for 18 to 20 threads per inch, 220; for 24 to 28 threads per inch, 320; for 40 threads per inch, 400.

THREAD MILLING

Single-cutter Method: Whenever a single cutter is used, the axis of the cutter is inclined an amount equal to the lead angle of the screw thread, in order to locate the cutter in line with the thread groove at the point where the cutting action takes place. Tangent of lead angle = lead of screw thread ÷ pitch circumference of screw.

The helical thread groove is generated in practically the same way as when a lathe is used. The single cutter process is especially applicable to the milling of large screw threads of coarse pitch, and either single or multiple threads.

The cutter should revolve as fast as possible without dulling the cutting edges excessively, in order to mill a smooth thread and prevent the unevenness that would result with a slow-moving cutter, on account of the tooth spaces. As the cutter rotates, the part on which a thread is to be milled is also revolved, but at a very slow rate (a few inches per minute), since this rotation of the work is practically a feeding movement. The cutter is ordinarily set to the full depth of the thread groove and finishes a single thread in one passage, although deep threads of coarse pitch may require two or even three cuts. For fine pitches and short threads, the multiple-cutter method (described in the next paragraph) usually is preferable, because it is more rapid. The milling of taper screw threads may be done on a single-cutter type of machine by traversing the cutter laterally as it feeds along in a lengthwise direction, the same as when using a taper attachment on a lathe.

Multiple-cutter Method: The multiple cutter for thread milling is practically a series of single cutters, although formed of one solid piece of steel, at least so far as the cutter proper is concerned. The rows of teeth do not lie in a helical path, like the teeth of a hob or tap, but they are annular or without lead. If the cutter had helical teeth the same as a gear hob, it would have to be geared to revolve in a certain fixed ratio with the screw being milled, but a cutter having annular teeth may rotate at any desired cutting speed, while the screw blank is rotated slowly to provide a suitable rate of feed. (The multiple cutters used for thread milling are frequently called "hobs," but the term hob should be applied only to cutters having a helical row of teeth like a gear-cutting hob.)

The object in using a multiple cutter instead of a single cutter is to finish a screw thread complete in approximately one revolution of the work, a slight amount of over-travel being allowed to insure milling the thread to the full depth where the end of cut joins the starting point. The cutter which is at least one or two threads or pitches wider than the thread to be milled, is fed in to the full thread depth and then either the cutter or screw blank is moved in a lengthwise direction a distance equal to the lead of the thread during one revolution of the work.

The multiple cutter is used for milling comparatively short threads and usually medium or fine pitches. The accompanying illustration shows typical examples of external and internal work for which the multiple-cutter type of thread milling has proved very efficient, although its usefulness is not confined to shoulder work and "blind" holes.

In using multiple cutters either for internal or external thread milling, the axis of the cutter is set parallel with the axis of the work, instead of inclining the cutter to suit the lead angle of the thread, as when using a single cutter. Theoretically, this is not the correct position for a cutter, since each cutting edge is revolving in a plane at right angles to the screw's axis while milling a thread groove of helical form. However, as a general rule, interference between the cutter and the thread, does not result a decided change in the standard thread form. Usually the defect is very slight and may be disregarded except when milling threads which incline considerably relative to the axis like a thread of multiple form and large lead angle. Multiple cutters are suitable for external threads having lead angles under 3½ degrees and for internal threads having lead angles under 2½ degrees. Threads

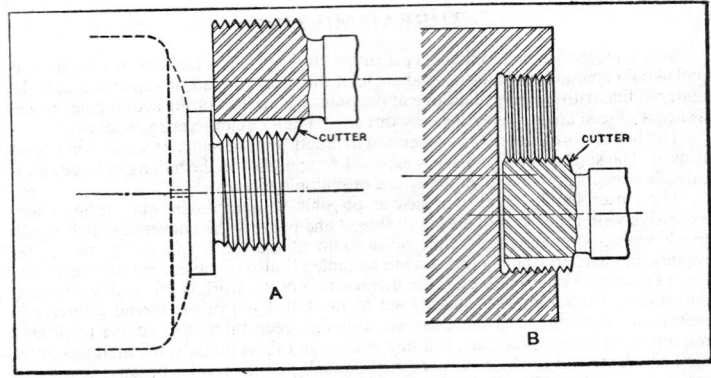

Examples of External and Internal Thread Milling with
a Multiple Type of Cutter

which have steeper sides or smaller included angles than the American Standard
or Whitworth forms should ordinarily be milled with a single cutter, assuming that
the milling process is preferable to other methods. For instance, in milling an Acme
thread which has an included angle between the sides of 29 degrees, there might be
considerable interference if a multiple cutter were used, unless the screw thread
diameter were large enough in proportion to the pitch to prevent such interference.
If an attempt were made to mill a square thread with a multiple cutter, the results
would be unsatisfactory owing to the interference.

Interference between the cutter and work is more pronounced when milling internal threads, because the cutter does not clear itself so well. Experiments have
shown that multiple cutters for internal work should preferably not exceed one-
third the diameter of the hole to be threaded. A cutter that is one-quarter the
diameter of the thread will do very satisfactory work. It is preferable to use as
small a cutter as practicable, either for internal or external work, not only to avoid
interference, but to reduce the strain on the driving mechanism. Some thread milling cutters, known as "topping cutters," are made for milling the outside diameter
of the thread as well as the angular sides and root.

Planetary Method: The planetary method of thread milling is similar in principle
to planetary milling. The part to be threaded is held stationary and the thread
milling cutter, while revolving about its own axis, is given a planetary movement
around the work in order to mill the thread in one planetary revolution. The machine spindle and the cutter which is held by it is moved longitudinally for thread
milling, an amount equal to the thread lead during one planetary revolution. This
operation is applicable to both internal and external threads. For the latter operation, the thread milling cutter surrounds the work. This thread milling is frequently accompanied by milling operations on other adjoining surfaces. For example, a planetary type of machine may be used for milling a screw thread and a
concentric cylindrical surface simultaneously. When the milling operation begins,
the eccentrically mounted cutter-spindle feeds the cutter into the right depth and
then the planetary movement begins, thus milling the thread and the cylindrical
surface. Thin sharp starting edges are eliminated on threads milled by the planetary method and the thread begins with a smooth gradual approach. One design of

machine will mill internal and external threads simultaneously. These threads may be of the same hand or one may be right hand and the other left hand. The threads may also be either of the same pitch or of a different pitch, and either straight or tapered.

Classes of Work for Thread Milling Machines. — Thread milling machines are used in preference to lathes or taps and dies for certain threading operations. There are four general reasons why a thread milling machine may be preferred: (1) Because the pitch of the thread is too coarse for cutting with a die; (2) because the milling process is more efficient than using a single-point tool in a lathe; (3) to secure a smoother and more accurate thread than would be obtained with a tap or die; (4) because the thread is so located relative to a shoulder or other surface that the milling method is superior, if not the only practicable way. A thread milling machine having a single cutter is especially adapted for coarse pitches, multiple-threaded screws, or any form or size of thread requiring the removal of a relatively large amount of metal, particularly if the pitch of the thread is large in proportion to the screw diameter, since the torsional strain due to the milling process is relatively small. While thread milling has little, if any, advantage over the lathe in regard to accuracy of lead, it gives a higher rate of production, and a thread is usually finished by means of a single passage of the cutter. The multiple-cutter type of thread milling machine frequently comes into competition with dies and taps, and especially self-opening dies and collapsing taps. The use of a multiple cutter is desirable when a thread must be cut close to a shoulder or to the bottom of a shallow recess, although the usefulness of the multiple cutter is not confined to shoulder work and "blind" holes.

Maximum Pitches of Die-cut Threads. — Dies of special design could be constructed for practically any pitch, if the screw blank were strong enough to resist the cutting strains and the size and cost of the die were immaterial; but, as a general rule, when the pitch is coarser than four or five threads per inch, the difficulty of cutting threads with dies increases rapidly, although in a few cases some dies are used successfully on screw threads having two or three threads per inch or less. Much depends upon the design of the die, the finish or smoothness required, and the relation between the pitch of the thread and the diameter of the screw. When the screw diameter is relatively small in proportion to the pitch, there may be considerable distortion due to the twisting strains set up when the thread is being cut. If the number of threads per inch is only one or two less than the standard number for a given diameter, a screw blank ordinarily will be strong enough to permit the use of a die.

Changing Pitch of Screw Thread Slightly. — A very slight change in the pitch of a screw thread may be necessary as, for example, when the pitch of a tap is increased a small amount to compensate for shrinkage in hardening. One method of obtaining slight variations in pitch is by means of a taper attachment. This attachment is set at an angle and the work is located at the same angle by adjusting the tailstock center. The result is that the tool follows an angular path relative to the movement of the carriage and, consequently, the pitch of the thread is increased slightly, the amount depending upon the angle to which the work and taper attachment are set. The cosine of this angle, for obtaining a given increase in pitch, equals the standard pitch (which would be obtained with the lathe used in the regular way) divided by the increased pitch necessary to compensate for shrinkage.

Example: — If the pitch of a ¾-inch American standard screw is to be increased from 0.100 to 0.1005, the cosine of the angle to which the taper attachment and work should be set is found as follows:

$$\text{Cosine of required angle} = \frac{0.100}{0.1005} = 0.9950$$

which is the cosine of 5 degrees 45 minutes, nearly.

CHANGE GEARS FOR HELICAL MILLING

Lead of a Milling Machine. — If gears with an equal number of teeth are placed on the table feed-screw and the worm-gear stud, then the *lead of the milling machine* is the distance the table will travel while the index spindle makes one complete revolution. This distance is a constant used in figuring the change gears.

The lead of a helix or "spiral" is the distance, measured along the axis of the work, in which the helix makes one full turn around the work. The lead of the milling machine may, therefore, also be expressed as the lead of the helix that will be cut when gears with an equal number of teeth are placed on the feed-screw and the worm-gear stud, and an idler of suitable size is interposed between the gears.

Rule: To find the lead of a milling machine, place equal gears on the worm-gear stud and on the feed-screw, and multiply the number of revolutions made by the feed-screw to produce one revolution of the index head spindle, by the lead of the thread on the feed-screw. Expressing the rule given as a formula:

$$\frac{\text{lead of milling}}{\text{machine}} = \frac{\text{rev. of feed-screw for one}}{\text{revolution of index spindle}} \times \frac{\text{lead of}}{\text{feed-screw.}}$$
$$\text{with equal gears}$$

Assume that it is necessary to make 40 revolutions of the feed-screw to turn the index head spindle one complete revolution, when the gears are equal, and that the lead of the thread on the feed-screw of the miling machine is ¼ inch; then the lead of the machine equals 40 × ¼ inch = 10 inches.

Change Gears for Helical Milling. — To find the change gears to be used in the compound train of gears for helical milling, place the lead of the helix to be cut in the numerator and the lead of the milling machine in the denominator of a fraction; divide numerator and denominator into two factors each; and multiply each "pair" of factors by the *same* number until suitable numbers of teeth for the change gears are obtained. (One factor in the numerator and one in the denominator are considered as one "pair" in this calculation.)

Example: Assume that the lead of a machine is 10 inches, and that a helix having a 48-inch lead is to be cut. Following the method explained:

$$\frac{48}{10} = \frac{6 \times 8}{2 \times 5} = \frac{(6 \times 12) \times (8 \times 8)}{(2 \times 12) \times (5 \times 8)} = \frac{72 \times 64}{24 \times 40}$$

The gear having 72 teeth is placed on the worm-gear stud and meshes with the 24-tooth gear on the intermediate stud. On the same intermediate stud is then placed the gear having 64 teeth, which is driven by the gear having 40 teeth placed on the feed-screw. This makes the gears having 72 and 64 teeth the driven gears, and the gears having 24 and 40 teeth the driving gears. In general, for compound gearing, the following formula may be used:

$$\frac{\text{lead of helix to be cut}}{\text{lead of machine}} = \frac{\text{product of driven gears}}{\text{product of driving gears}}$$

Short-lead Milling. — If the lead to be milled is exceptionally short, the drive may be direct from the table feed-screw to the dividing head spindle to avoid excessive load on feed-screw and change-gears. If the table feed-screw has 4 threads per inch (usual standard), then

$$\text{Change-gear ratio} = \frac{\text{Lead to be milled}}{0.25} = \frac{\text{Driven gears}}{\text{Driving gears}}$$

For indexing, the number of teeth on the spindle change-gear should be some multiple of the number of divisions required, to permit indexing by disengaging and turning the gear.

Helix. — A helix is a curve generated by a point moving about a cylindrical surface (real or imaginary) at a constant rate in the direction of the cylinder's axis. The curvature of a screw thread is one common example of a helical curve.

Lead of Helix: The lead of a helix is the distance that it advances in an axial direction, in one complete turn about the cylindrical surface. To illustrate, the lead of a screw thread equals the distance that a thread advances in one turn; it also equals the distance that a nut would advance in one turn.

Development of Helix: If one turn of a helical curve were unrolled onto a plane surface (as shown by diagram), the helix would become a straight line forming the hypotenuse of a right angle triangle. The length of one side of this triangle would equal the circumference of the cylinder with which the helix coincides, and the length of the other side of the triangle would equal the lead of the helix.

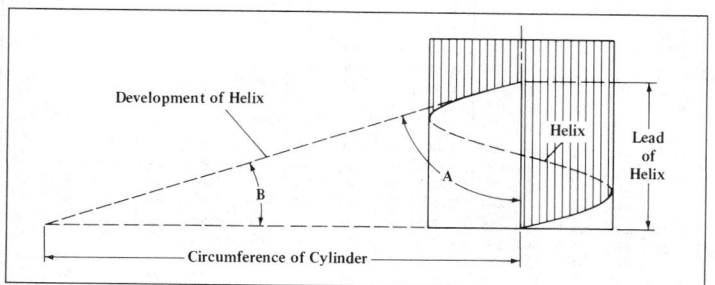

Helix Angles. — The triangular development of a helix has one angle *A* subtended by the circumference of the cylinder, and another angle *B* subtended by the lead of the helix. The term "helix angle" applies to angle *A*. For example, the helix angle of a helical gear, according to the general usage of the term, is always angle *A*, because this is the angle used in helical gear-designing formulas. Helix angle *A* would also be applied in milling the helical teeth of cutters, reamers, etc. Angle *A* of a gear or cutter tooth is a measure of its inclination relative to the axis of the gear or cutter.

Lead Angle: Angle *B* is applied to screw threads and worm threads and is referred to as the lead angle of the screw thread or worm. This angle *B* is a measure of the inclination of a screw thread from a plane that is perpendicular to the screw thread axis. Angle *B* is called the "lead angle" because it is subtended by the lead of the thread, and to distinguish it from the term "helix angle" as applied to helical gears.

Finding Helix Angle of Helical Gear: A helical gear tooth has an infinite number of helix angles, but the angle at the pitch diameter or mid-working depth is the one required in gear designing and gear cutting. This angle *A*, relative to the axis of the gear, is found as follows:

$$\tan \text{ helix angle} = \frac{3.1416 \times \text{pitch diameter of gear}}{\text{Lead of gear tooth}}$$

Finding Lead Angle of Screw Thread: The lead or helix angle at the pitch diameter of a screw thread usually is required when, for example, a thread milling cutter must be aligned with the thread. This angle measured from a plane perpendicular to the screw thread axis, is found as follows:

$$\tan \text{ lead angle} = \frac{\text{Lead of screw thread}}{3.1416 \times \text{pitch diameter of screw thread}}$$

Change Gears for Different Leads — 0.670 Inch to 2.658 Inches

Lead in Inches	Driven (Gear on Worm)	Driver (First Gear on Stud)	Driven (Second Gear on Stud)	Driver (Gear on Screw)	Lead in Inches	Driven (Gear on Worm)	Driver (First Gear on Stud)	Driven (Second Gear on Stud)	Driver (Gear on Screw)	Lead in Inches	Driven (Gear on Worm)	Driver (First Gear on Stud)	Driven (Second Gear on Stud)	Driver (Gear on Screw)
0.670	24	86	24	100	1.711	28	72	44	100	2.182	24	44	40	100
0.781	24	86	28	100	1.714	24	56	40	100	2.188	24	48	28	64
0.800	24	72	24	100	1.744	24	64	40	86	2.193	24	56	44	86
0.893	24	86	32	100	1.745	24	44	32	100	2.200	24	48	44	100
0.930	24	72	24	86	1.750	28	64	40	100	2.222	24	48	32	72
1.029	24	56	24	100	1.776	24	44	28	86	2.233	40	86	48	100
1.042	28	86	32	100	1.778	32	72	40	100	2.238	28	64	44	86
1.047	24	64	24	86	1.786	24	86	64	100	2.240	28	40	32	100
1.050	24	64	28	100	1.800	24	64	48	100	2.250	24	40	24	64
1.067	24	72	32	100	1.809	28	72	40	86	2.274	32	72	44	86
1.085	24	72	28	86	1.818	24	44	24	72	2.286	32	56	40	100
1.116	24	86	40	100	1.823	28	86	56	100	2.292	24	64	44	72
1.196	24	56	24	86	1.860	28	56	32	86	2.326	32	64	40	86
1.200	24	48	24	100	1.861	24	72	48	86	2.333	28	48	40	100
1.221	24	64	28	86	1.867	28	48	24	100	2.338	24	44	24	56
1.228	24	86	44	100	1.875	24	48	24	64	2.344	28	86	72	100
1.240	24	72	32	86	1.886	24	56	44	100	2.368	28	44	32	86
1.250	24	64	24	72	1.905	24	56	32	72	2.381	32	86	64	100
1.302	28	86	40	100	1.919	24	64	44	86	2.386	24	44	28	64
1.309	24	44	24	100	1.920	24	40	32	100	2.392	24	56	48	86
1.333	24	72	40	100	1.925	28	64	44	100	2.400	28	56	48	100
1.340	24	86	48	100	1.944	24	48	28	72	2.424	24	44	32	72
1.371	24	56	32	100	1.954	24	40	28	86	2.431	28	64	40	72
1.395	24	48	24	86	1.956	32	72	44	100	2.442	24	32	28	86
1.400	24	48	28	100	1.990	28	72	44	86	2.445	40	72	44	100
1.429	24	56	24	72	1.993	24	56	40	86	2.450	28	64	56	100
1.440	24	40	24	100	2.000	24	40	24	72	2.456	24	86	48	100
1.458	24	64	28	72	2.009	24	86	72	100	2.481	32	72	48	86
1.467	24	72	44	100	2.030	24	44	32	86	2.489	32	72	56	100
1.488	32	86	40	100	2.035	28	64	40	86	2.500	24	48	28	56
1.500	24	64	40	100	2.036	28	44	32	100	2.514	32	56	44	100
1.522	24	44	24	86	2.045	24	44	24	64	2.532	28	72	56	86
1.550	24	72	40	86	2.047	40	86	44	100	2.537	24	44	40	86
1.563	24	86	56	100	2.057	24	28	24	100	2.546	28	44	40	100
1.595	24	56	32	86	2.067	32	72	40	86	2.558	32	64	44	86
1 600	24	48	32	100	2.083	24	64	40	72	2.567	28	48	44	100
1.607	24	56	24	64	2.084	28	86	64	100	2.571	24	40	24	56
1.628	24	48	28	86	2.093	24	64	48	86	2.593	28	48	32	72
1.637	32	86	44	100	2.100	24	64	56	100	2.605	28	40	32	86
1.650	24	64	44	100	2.121	24	44	28	72	2.618	24	44	48	100
1.667	24	56	28	72	2.133	24	72	64	100	2.619	24	56	44	72
1.674	24	40	24	86	2.143	24	56	32	64	2.625	24	40	28	64
1.680	24	40	28	100	2.171	24	72	56	86	2.640	24	40	44	100
1.706	24	72	44	86	2.178	28	72	56	100	2.658	32	56	40	86

Change Gears for Different Leads — 2.667 Inches to 4.040 Inches

Lead in Inches	Driven Gear on Worm	Driver First Gear on Stud	Driven Second Gear on Stud	Driver Gear on Screw	Lead in Inches	Driven Gear on Worm	Driver First Gear on Stud	Driven Second Gear on Stud	Driver Gear on Screw	Lead in Inches	Driven Gear on Worm	Driver First Gear on Stud	Driven Second Gear on Stud	Driver Gear on Screw
2.667	40	72	48	100	3.140	24	86	72	64	3.588	72	56	24	86
2.674	28	64	44	72	3.143	40	56	44	100	3.600	72	48	24	100
2.678	24	56	40	64	3.150	28	100	72	64	3.618	56	72	40	86
2.679	32	86	72	100	3.175	32	56	40	72	3.636	24	44	32	48
2.700	24	64	72	100	3.182	28	44	32	64	3.637	48	44	24	72
2.713	28	48	40	86	3.189	32	56	48	86	3.646	40	48	28	64
2.727	24	44	32	64	3.190	24	86	64	56	3.655	40	56	44	86
2.743	24	56	64	100	3.198	40	64	44	86	3.657	64	56	32	100
2.750	40	64	44	100	3.200	28	100	64	56	3.663	72	64	28	86
2.778	32	64	40	72	3.214	24	56	48	64	3.667	40	48	44	100
2.791	28	56	48	86	3.225	24	100	86	64	3.673	24	28	24	56
2.800	24	24	28	100	3.241	28	48	40	72	3.684	44	86	72	100
2.812	24	32	24	64	3.256	24	24	28	86	3.686	86	56	24	100
2.828	28	44	32	72	3.267	28	48	56	100	3.704	32	48	40	72
2.843	40	72	44	86	3.273	24	40	24	44	3.721	24	24	32	86
2.845	32	72	64	100	3.275	44	86	64	100	3.733	48	72	56	100
2.849	28	64	56	86	3.281	24	32	28	64	3.750	24	32	24	48
2.857	24	48	32	56	3.300	44	64	48	100	3.763	86	64	28	100
2.865	44	86	56	100	3.308	32	72	64	86	3.771	44	56	48	100
2.867	86	72	24	100	3.333	32	64	48	72	3.772	24	28	44	100
2.880	24	40	48	100	3.345	28	100	86	72	3.799	56	48	28	86
2.894	28	72	64	86	3.349	40	86	72	100	3.809	24	28	32	72
2.909	32	44	40	100	3.360	56	40	24	100	3.810	64	56	24	72
2.917	24	64	56	72	3.383	32	44	40	86	3.818	24	40	28	44
2.924	32	56	44	86	3.403	28	64	56	72	3.819	40	64	44	72
2.933	44	72	48	100	3.409	24	44	40	64	3.822	86	72	32	100
2.934	32	48	44	100	3.411	32	48	44	86	3.837	24	32	24	86
2.946	24	56	44	64	3.422	44	72	56	100	3.840	64	40	24	100
2.960	28	44	40	86	3.428	24	40	32	56	3.850	44	64	56	100
2.977	40	86	64	100	3.429	40	28	24	100	3.876	24	72	100	86
2.984	28	48	44	86	3.438	24	48	44	64	3.889	32	64	56	72
3.000	24	40	28	56	3.488	40	64	48	86	3.896	24	44	40	56
3.030	24	44	40	72	3.491	64	44	24	100	3.907	56	40	24	86
3.044	24	44	48	86	3.492	32	56	44	72	3.911	44	72	64	100
3.055	28	44	48	100	3.500	40	64	56	100	3.920	28	40	56	100
3.056	32	64	44	72	3.520	32	40	44	100	3.927	72	44	24	100
3.070	24	40	44	86	3.535	28	44	40	72	3.929	32	56	44	64
3.080	28	40	44	100	3.552	56	44	24	86	3.977	28	44	40	64
3.086	24	56	72	100	3.556	40	72	64	100	3.979	44	72	56	86
3.101	40	72	48	86	3.564	56	44	28	100	3.987	24	28	40	86
3.111	28	40	32	72	3.565	28	44	44	72	4.000	24	40	32	48
3.117	24	44	32	56	3.571	24	48	40	56	4.011	28	48	44	64
3.125	28	56	40	64	3.572	48	86	64	100	4.019	72	86	48	100
3.126	48	86	56	100	3.582	44	40	28	86	4.040	32	44	40	72

Change Gears for Different Leads — 4.059 Inches to 5.568 Inches

Lead in Inches	Driven — Gear on Worm	Driver — First Gear on Stud	Driven — Second Gear on Stud	Driver — Gear on Screw	Lead in Inches	Driven — Gear on Worm	Driver — First Gear on Stud	Driven — Second Gear on Stud	Driver — Gear on Screw	Lead in Inches	Driven — Gear on Worm	Driver — First Gear on Stud	Driven — Second Gear on Stud	Driver — Gear on Screw
4.059	32	44	48	86	4.567	72	44	24	86	5.105	28	48	56	64
4.060	64	44	24	86	4.572	40	56	64	100	5.116	44	24	24	86
4.070	28	32	40	86	4.582	72	44	28	100	5.119	86	56	24	72
4.073	64	44	28	100	4.583	44	64	48	72	5.120	64	40	32	100
4.074	32	48	44	72	4.584	32	48	44	64	5.133	56	48	44	100
4.091	24	44	48	64	4.651	40	24	24	86	5.134	44	24	28	100
4.093	32	40	44	86	4.655	64	44	32	100	5.142	72	56	40	100
4.114	48	28	24	100	4.667	28	40	32	48	5.143	24	28	24	40
4.125	24	40	44	64	4.675	24	28	24	44	5.156	44	32	24	64
4.135	40	72	64	86	4.687	40	32	24	64	5.160	86	40	24	100
4.144	56	44	28	86	4.688	56	86	72	100	5.168	100	72	32	86
4.167	28	48	40	56	4.691	86	44	24	100	5.185	28	24	32	72
4.186	72	64	32	86	4.714	44	40	24	56	5.186	64	48	28	72
4.200	48	64	56	100	4.736	64	44	28	86	5.195	32	44	40	56
4.242	28	44	32	48	4.762	40	28	24	72	5.209	100	64	24	72
4.253	64	56	32	86	4.773	24	32	28	44	5.210	64	40	28	86
4.264	40	48	44	86	4.778	86	72	40	100	5.226	86	64	28	72
4.267	64	48	32	100	4.784	72	56	32	86	5.233	72	64	40	86
4.278	28	40	44	72	4.785	48	28	24	86	5.236	72	44	32	100
4.286	24	28	24	48	4.800	48	24	24	100	5.238	44	28	24	72
4.300	86	56	28	100	4.813	44	40	28	64	5.250	24	32	28	40
4.320	72	40	24	100	4.821	72	56	24	64	5.256	86	72	44	100
4.341	48	72	56	86	4.849	32	44	48	72	5.280	48	40	44	100
4.342	64	48	28	86	4.861	40	32	28	72	5.303	28	44	40	48
4.361	100	64	24	86	4.884	48	64	56	86	5.316	40	28	32	86
4.363	24	40	32	44	4.889	32	40	44	72	5.328	72	44	28	86
4.364	40	44	48	100	4.898	24	28	32	56	5.333	40	24	32	100
4.365	40	56	44	72	4.900	56	32	28	100	5.347	44	64	56	72
4.375	24	24	28	64	4.911	40	56	44	64	5.348	44	32	28	72
4.386	24	28	44	86	4.914	86	56	32	100	5.357	44	28	24	64
4.400	24	24	44	100	4.950	56	44	28	72	5.358	64	86	72	100
4.444	64	56	28	72	4.961	64	48	32	86	5.375	86	64	40	100
4.465	64	40	24	86	4.978	56	72	64	100	5.400	72	32	24	100
4.466	48	40	32	86	4.984	100	56	24	86	5.413	64	44	32	86
4.477	44	32	28	86	5.000	24	28	28	56	5.426	40	24	28	86
4.479	86	64	24	72	5.017	86	48	28	100	5.427	40	48	56	86
4.480	56	40	32	100	5.023	72	40	24	86	5.444	56	40	28	72
4.500	72	64	40	100	5.029	28	32	100	100	5.455	48	44	28	56
4.522	100	72	28	86	5.040	72	40	28	100	5.469	40	32	24	64
4.537	56	48	28	72	5.074	40	44	48	86	5.473	86	44	28	100
4.545	24	44	40	48	5.080	64	56	32	72	5.486	64	28	24	100
4.546	28	44	40	56	5.088	100	64	28	86	5.500	44	40	24	48
4.548	44	72	64	86	5.091	56	44	40	100	5.556	40	24	24	72
4.558	56	40	28	86	5.093	40	48	44	72	5.568	56	44	28	64

Change Gears for Different Leads — 5.581 Inches to 7.500 Inches

Lead in Inches	Driven: Gear on Worm	Driver: First Gear on Stud	Driven: Second Gear on Stud	Driver: Gear on Screw	Lead in Inches	Driven: Gear on Worm	Driver: First Gear on Stud	Driven: Second Gear on Stud	Driver: Gear on Screw	Lead in Inches	Driven: Gear on Worm	Driver: First Gear on Stud	Driven: Second Gear on Stud	Driver: Gear on Screw
5.581	64	32	24	86	6.172	72	28	24	100	6.825	86	56	32	72
5.582	48	24	24	86	6.202	40	24	32	86	6.857	32	28	24	40
5.600	56	24	24	100	6.222	64	40	28	72	6.875	44	24	24	64
5.625	48	32	24	64	6.234	32	28	24	44	6.880	86	40	32	100
5.657	56	44	32	72	6.250	24	24	40	64	6.944	100	48	24	72
5.698	56	32	28	86	6.255	56	44	32	100	6.945	100	56	28	72
5.714	48	28	24	72	6.279	72	64	48	86	6.968	86	48	28	72
5.730	40	48	44	64	6.286	44	40	32	56	6.977	48	32	40	86
5.733	86	48	32	100	6.300	72	32	28	100	6.982	64	44	48	100
5.756	72	64	44	86	6.343	100	44	24	86	6.984	44	28	32	72
5.759	86	56	24	64	6.350	40	28	32	72	7.000	28	24	24	40
5.760	72	40	32	100	6.364	56	44	24	48	7.013	72	44	24	56
5.788	64	72	56	86	6.379	64	28	24	86	7.040	64	40	44	100
5.814	100	64	32	86	6.396	44	32	40	86	7.071	56	44	40	72
5.818	64	44	40	100	6.400	64	24	24	100	7.104	56	44	48	86
5.833	28	24	24	48	6.417	44	40	28	48	7.106	100	72	44	86
5.847	64	56	44	86	6.429	24	28	24	32	7.111	64	40	32	72
5.848	44	28	32	86	6.450	86	64	48	64	7.130	44	24	28	72
5.861	72	40	28	86	6.460	100	72	40	86	7.143	40	28	32	64
5.867	44	24	32	100	6.465	64	44	32	72	7.159	72	44	28	64
5.893	44	32	24	56	6.482	56	48	40	72	7.163	56	40	44	86
5.912	86	64	44	100	6.512	56	24	24	86	7.167	86	40	24	72
5.920	56	44	40	86	6.515	86	44	24	72	7.176	72	28	24	86
5.926	64	48	32	72	6.534	56	24	28	100	7.200	72	24	24	100
5.952	100	56	24	72	6.545	44	40	24	44	7.268	100	64	40	86
5.954	64	40	32	86	6.548	44	48	40	56	7.272	64	44	28	56
5.969	44	24	28	86	6.563	56	32	24	64	7.273	32	24	24	44
5.972	86	48	24	72	6.578	72	56	44	86	7.292	56	48	40	64
5.980	72	56	40	86	6.600	48	32	44	100	7.310	44	28	40	86
6.000	48	40	28	56	6.645	100	56	32	86	7.314	64	28	32	100
6.016	44	32	28	64	6.667	64	48	28	56	7.326	72	32	28	86
6.020	86	40	28	100	6.689	86	72	56	100	7.330	86	44	24	64
6.061	40	44	32	48	6.697	100	56	24	64	7.333	44	24	40	100
6.077	100	64	28	72	6.698	72	40	32	86	7.334	44	40	32	48
6.089	72	44	32	86	6.719	86	48	24	64	7.347	48	28	24	56
6.109	56	44	48	100	6.720	56	40	48	100	7.371	86	56	48	100
6.112	24	24	44	72	6.735	44	28	24	56	7.372	86	28	24	100
6.122	40	28	24	56	6.750	72	40	24	64	7.400	100	44	28	86
6.125	56	40	28	64	6.757	86	56	44	100	7.408	40	24	32	72
6.137	72	44	24	64	6.766	64	44	40	86	7.424	56	44	28	48
6.140	48	40	44	86	6.784	100	48	28	86	7.442	64	24	24	86
6.143	86	56	40	100	6.806	56	32	28	72	7.465	86	64	40	72
6.160	56	40	44	100	6.818	40	32	24	44	7.467	64	24	28	100
6.171	72	56	48	100	6.822	44	24	32	86	7.500	48	24	24	64

Change Gears for Different Leads — 7.525 Inches to 9.598 Inches

Lead in Inches	Driven (Gear on Worm)	Driver (First Gear on Stud)	Driven (Second Gear on Stud)	Driver (Gear on Screw)	Lead in Inches	Driven (Gear on Worm)	Driver (First Gear on Stud)	Driven (Second Gear on Stud)	Driver (Gear on Screw)	Lead in Inches	Driven (Gear on Worm)	Driver (First Gear on Stud)	Driven (Second Gear on Stud)	Driver (Gear on Screw)
7.525	86	32	28	100	8.140	56	32	40	86	8.800	48	24	44	100
7.543	48	28	44	100	8.145	64	44	56	100	8.838	100	44	28	72
7.576	100	44	24	72	8.148	64	48	44	72	8.839	72	56	44	64
7.597	56	24	28	86	8.149	44	24	32	72	8.909	56	40	28	44
7.601	86	44	28	72	8.163	40	28	32	56	8.929	100	48	24	56
7.611	72	44	40	86	8.167	56	40	28	48	8.930	64	40	48	86
7.619	64	48	32	56	8.182	48	32	24	44	8.953	56	32	44	86
7.620	64	28	24	72	8.186	64	40	44	86	8.959	86	48	28	56
7.636	56	40	24	44	8.212	86	64	44	72	8.960	64	40	56	100
7.639	44	32	40	72	8.229	72	28	32	100	8.980	44	28	32	56
7.644	86	72	64	100	8.250	44	32	24	40	9.000	88	32	24	40
7.657	56	32	28	64	8.306	100	56	40	86	9.044	100	72	56	86
7.674	72	48	44	86	8.312	64	44	32	56	9.074	56	24	28	72
7.675	48	32	44	86	8.333	40	24	24	48	9.091	40	24	24	44
7.679	86	48	24	56	8.334	40	24	28	56	9.115	100	48	28	64
7.680	64	40	48	100	8.361	86	40	28	72	9.134	72	44	48	86
7.700	56	32	44	100	8.372	72	24	24	86	9.137	100	56	44	86
7.714	72	40	24	56	8.377	86	44	24	56	9.143	56	40	32	56
7.752	100	48	32	86	8.400	72	24	28	100	9.164	72	44	56	100
7.778	32	24	28	48	8.437	72	32	24	64	9.167	44	24	24	48
7.792	40	28	24	44	8.457	100	44	32	86	9.210	72	40	44	86
7.813	100	48	24	64	8.484	32	24	28	44	9.214	86	40	24	56
7.815	56	40	48	86	8.485	64	44	28	48	9.260	100	48	32	72
7.818	86	44	40	100	8.485	56	44	32	48	9.302	48	24	40	86
7.838	86	48	28	64	8.506	64	28	32	86	9.303	56	28	40	86
7.855	72	44	48	100	8.523	100	44	24	64	9.333	64	40	28	48
7.857	44	24	24	56	8.527	44	24	40	86	9.334	32	24	28	40
7.872	44	28	32	64	8.532	86	56	40	72	9.351	48	28	24	44
7.875	72	40	28	64	8.534	64	24	32	100	9.375	48	32	40	64
7.883	86	48	44	100	8.552	86	44	28	64	9.382	86	44	48	100
7.920	72	40	44	100	8.556	56	40	44	72	9.385	86	56	56	72
7.936	100	56	32	72	8.572	64	32	24	56	9.406	86	40	28	64
7.954	40	32	28	44	8.572	48	24	24	56	9.428	44	28	24	40
7.955	56	44	40	64	8.594	44	32	40	64	9.429	48	40	44	56
7.963	86	48	32	72	8.600	86	24	24	100	9.460	86	40	44	100
7.974	48	28	40	86	8.640	72	40	48	100	9.472	64	44	56	86
7.994	100	64	44	86	8.681	100	64	40	72	9.524	40	28	32	48
8.000	64	32	40	100	8.682	64	24	28	86	9.545	72	44	28	48
8.021	44	32	28	48	8.687	86	44	32	72	9.546	56	32	24	44
8.035	72	56	40	64	8.721	100	32	24	86	9.547	56	44	48	64
8.063	86	40	24	64	8.727	48	40	32	44	9.549	100	64	44	72
8.081	64	44	40	72	8.730	44	28	40	72	9.556	86	40	32	72
8.102	100	48	28	72	8.750	28	24	24	32	9.569	72	28	32	86
8.119	64	44	48	86	8.772	48	28	44	86	9.598	86	56	40	64

Change Gears for Different Leads — 9.600 Inches to 12.375 Inches

Lead in Inches	Driven — Gear on Worm	Driver — First Gear on Stud	Driven — Second Gear on Stud	Driver — Gear on Screw
9.600	72	24	32	100
9.625	44	32	28	40
9.643	72	32	24	56
9.675	86	64	72	100
9.690	100	48	40	86
9.697	64	48	32	44
9.723	40	24	28	48
9.741	100	44	24	56
9.768	72	48	56	86
9.773	86	44	24	48
9.778	64	40	44	72
9.796	64	28	24	56
9.818	72	40	24	44
9.822	44	32	40	56
9.828	86	28	32	100
9.844	72	32	28	64
9.900	72	32	44	100
9.921	100	56	40	72
9.923	64	24	32	86
9.943	100	44	28	64
9.954	86	48	40	72
9.967	100	56	48	86
9.968	100	28	24	86
10.000	56	28	24	48
10.033	86	24	28	100
10.046	72	40	48	86
10.057	64	28	44	100
10.078	86	32	24	64
10.080	72	40	56	100
10.101	100	44	32	72
10.159	64	28	32	72
10.175	100	32	28	86
10.182	64	40	28	44
10.186	44	24	40	72
10.209	56	24	28	64
10.228	72	44	40	64
10.233	48	24	44	86
10.238	86	28	24	72
10.267	56	24	44	100
10.286	48	28	24	40
10.312	48	32	44	64
10.313	72	48	44	64
10.320	86	40	48	100
10.336	100	72	64	86
10.370	64	24	28	72
10.371	64	48	56	72
10.390	40	28	32	44
10.417	100	32	24	72
10.419	64	40	56	86
10.451	86	32	28	72
10.467	72	32	40	86
10.473	72	44	64	100
10.476	44	24	32	56
10.477	48	28	44	72
10.500	56	32	24	40
10.558	86	56	44	64
10.571	100	44	40	86
10.606	56	44	40	48
10.631	64	28	40	86
10.655	72	44	56	86
10.659	100	48	44	86
10.667	64	40	48	72
10.694	44	24	28	48
10.713	40	28	24	32
10.714	48	32	40	56
10.750	86	40	24	48
10.800	72	32	48	100
10.853	56	24	40	86
10.859	86	44	40	72
10.909	72	44	32	48
10.913	100	56	44	72
10.937	56	32	40	64
10.945	86	44	56	100
10.949	86	48	44	72
10.972	64	28	48	72
11.000	44	24	24	40
11.021	72	28	24	56
11.057	86	56	72	100
11.111	40	24	32	48
11.137	56	32	28	44
11.160	100	56	40	64
11.163	72	24	32	86
11.169	86	44	32	56
11.198	86	48	40	64
11.200	56	24	48	100
11.225	44	28	40	56
11.250	72	24	24	64
11.313	64	44	56	72
11.314	72	28	44	100
11.363	100	44	24	48
11.401	86	44	28	48
11.429	32	24	24	28
11.454	72	40	28	44
11.459	44	24	40	64
11.467	86	24	32	100
11.512	72	32	44	86
11.518	86	28	24	64
11.520	72	40	64	100
11.574	100	48	40	72
11.629	100	24	24	86
11.638	64	40	32	44
11.667	56	24	24	48
11.688	72	44	40	56
11.695	64	24	48	86
11.719	100	32	24	64
11.721	72	40	56	86
11.728	86	40	24	44
11.733	64	24	44	100
11.757	86	32	28	64
11.785	72	48	44	56
11.786	44	28	24	32
11.825	86	32	44	100
11.905	100	28	24	72
11.938	56	24	44	86
11.944	86	24	24	72
11.960	72	28	40	86
12.000	48	24	24	40
12.031	56	32	44	64
12.040	86	40	56	100
12.121	40	24	32	44
12.153	100	32	28	72
12.178	72	44	64	86
12.216	86	44	40	64
12.222	44	24	32	48
12.245	48	28	40	56
12.250	56	32	28	40
12.272	72	32	24	44
12.277	100	56	44	64
12.286	86	28	40	100
12.318	86	48	44	64
12.343	72	28	48	100
12.375	72	40	44	64

Change Gears for Different Leads — 12.403 Inches to 16.000 Inches

Lead in Inches	Driven (Gear on Worm)	Driver (First Gear on Stud)	Driven (Second Gear on Stud)	Driver (Gear on Screw)	Lead in Inches	Driven (Gear on Worm)	Driver (First Gear on Stud)	Driven (Second Gear on Stud)	Driver (Gear on Screw)	Lead in Inches	Driven (Gear on Worm)	Driver (First Gear on Stud)	Driven (Second Gear on Stud)	Driver (Gear on Screw)
12.403	64	24	40	86	13.438	86	24	24	64	14.668	44	24	32	40
12.444	64	40	56	72	13.469	48	28	44	56	14.694	72	28	32	56
12.468	64	28	24	44	13.500	72	32	24	40	14.743	86	28	48	100
12.500	40	24	24	32	13.514	86	28	44	100	14.780	86	40	44	64
12.542	86	40	28	48	13.566	100	24	28	86	14.800	100	44	56	86
12.508	86	44	64	100	13.611	56	24	28	48	14.815	64	24	40	72
12.558	72	32	48	86	13.636	48	32	40	44	14.849	56	24	28	44
12.571	64	40	44	56	13.643	64	24	44	86	14.880	100	48	40	56
12.572	44	28	32	40	13.650	86	28	32	72	14.884	64	28	56	86
12.600	72	32	56	100	13.672	100	32	28	64	14.931	86	32	40	72
12.627	100	44	40	72	13.682	86	40	28	44	14.933	64	24	56	100
12.686	100	44	48	86	13.713	64	40	48	56	14.950	100	56	72	86
12.698	64	28	40	72	13.715	64	28	24	40	15.000	48	24	24	32
12.727	64	32	28	44	13.750	44	24	24	32	15.050	86	32	56	100
12.728	56	24	24	44	13.760	86	40	64	100	15.150	100	44	32	48
12.732	100	48	44	72	13.889	100	24	24	72	15.151	100	44	48	72
12.758	64	28	48	86	13.933	86	48	56	72	15.202	86	44	56	72
12.791	100	40	44	86	13.935	86	24	28	72	15.238	64	28	48	72
12.798	86	48	40	56	13.953	72	24	40	86	15.239	64	28	32	48
12.800	64	28	56	100	13.960	86	44	40	56	15.272	56	40	48	44
12.834	56	40	44	48	13.968	64	28	44	72	15.278	44	24	40	48
12.857	72	28	32	64	14.000	56	24	24	40	15.279	100	40	44	72
12.858	48	28	24	32	14.025	72	44	48	56	15.306	100	28	24	56
12.900	86	32	48	100	14.026	72	28	24	44	15.349	72	24	44	86
12.963	56	24	40	72	14.063	72	32	40	64	15.357	86	28	24	48
12.987	100	44	32	56	14.071	86	44	72	100	15.429	72	40	48	56
13.020	100	48	40	64	14.078	86	48	44	56	15.469	72	32	44	64
13.024	56	24	48	86	14.142	72	40	44	56	15.480	86	40	72	100
13.030	86	44	32	48	14.204	100	44	40	64	15.504	100	48	64	86
13.062	64	28	32	56	14.260	56	24	44	72	15.556	64	32	56	72
13.082	100	64	72	86	14.286	40	24	24	28	15.584	48	28	40	44
13.090	72	40	32	44	14.318	72	32	28	44	15.625	100	24	24	64
13.096	44	28	40	48	14.319	72	44	56	64	15.636	86	40	32	44
13.125	72	32	28	48	14.322	100	48	44	64	15.677	86	32	28	48
13.139	86	40	44	72	14.333	86	40	32	48	15.714	44	24	24	28
13.157	72	28	44	86	14.352	72	28	48	86	15.750	72	32	28	40
13.163	86	28	24	56	14.400	72	24	48	100	15.767	86	24	44	100
13.200	72	24	44	100	14.536	100	32	40	86	15.873	100	56	64	72
13.258	100	44	28	48	14.545	64	24	24	44	15.874	100	28	32	72
13.289	100	28	32	86	14.583	56	32	40	48	15.909	100	40	28	44
13.333	64	24	24	48	14.584	40	24	28	32	15.925	86	48	64	72
13.393	100	56	48	64	14.651	72	32	56	86	15.926	86	24	32	72
13.396	72	40	64	86	14.659	86	44	48	64	15.989	100	32	44	86
13.437	86	32	28	56	14.667	64	40	44	48	16.000	64	24	24	40

Change Gears for Different Leads — 16.042 Inches to 21.39 Inches

Lead in Inches	Driven — Gear on Worm	Driver — First Gear on Stud	Driven — Second Gear on Stud	Driver — Gear on Screw	Lead in Inches	Driven — Gear on Worm	Driver — First Gear on Stud	Driven — Second Gear on Stud	Driver — Gear on Screw	Lead in Inches	Driven — Gear on Worm	Driver — First Gear on Stud	Driven — Second Gear on Stud	Driver — Gear on Screw
16.042	56	24	44	64	17.442	100	32	48	86	19.350	86	32	72	100
16.043	44	24	28	32	17.454	64	40	48	44	19.380	100	24	40	86
16.071	72	32	40	56	17.500	56	24	24	32	19.394	64	24	32	44
16.125	86	32	24	40	17.550	86	28	32	56	19.444	40	24	28	24
16.204	100	24	28	72	17.677	100	44	56	72	19.480	100	28	24	44
16.233	100	44	40	56	17.679	72	32	44	56	19.531	100	32	40	64
16.280	100	40	56	86	17.778	64	24	32	48	19.535	72	24	56	86
16.288	86	44	40	48	17.858	100	24	24	56	19.545	86	24	24	44
16.296	64	24	44	72	17.917	86	24	32	64	19.590	64	28	48	56
16.327	64	28	40	56	17.918	86	24	24	48	19.635	72	40	48	44
16.333	56	24	28	40	17.959	64	28	44	56	19.642	100	40	44	56
16.364	72	24	24	44	18.000	72	24	24	40	19.643	44	28	40	32
16.370	100	48	44	56	18.181	56	28	40	44	19.656	86	28	64	100
16.423	86	32	44	72	18.182	48	24	40	44	19.687	72	32	56	64
16.456	72	28	64	100	18.229	100	32	28	48	19.710	86	40	44	48
16.500	72	40	44	48	18.273	100	28	44	86	19.840	100	28	40	72
16.612	100	28	40	86	18.285	64	28	32	40	19.886	100	44	56	64
16.623	64	28	32	44	18.333	56	28	44	48	19.887	100	32	28	44
16.667	56	28	40	48	18.367	72	28	40	56	19.908	86	24	40	72
16.722	86	40	56	72	18.428	86	28	24	40	19.934	100	28	48	86
16.744	72	24	48	86	18.476	86	32	44	64	20.00	72	24	32	48
16.752	86	44	48	56	18.519	100	24	32	72	20.07	86	24	56	100
16.753	86	28	24	44	18.605	100	40	64	86	20.09	100	56	72	64
16.797	86	32	40	44	18.663	100	64	86	72	20.16	86	48	72	64
16.800	72	24	56	100	18.667	64	24	28	40	20.20	100	44	64	72
16.875	72	32	48	64	18.700	72	44	64	56	20.35	100	32	56	86
16.892	86	40	44	56	18.750	100	32	24	40	20.36	64	40	56	44
16.914	100	44	64	86	18.750	72	32	40	48	20.41	100	28	32	56
16.969	64	44	56	48	18.770	86	28	44	72	20.42	56	24	28	32
16.970	64	24	28	44	18.812	86	32	28	40	20.45	72	32	40	44
17.045	100	32	24	44	18.858	48	28	44	40	20.48	86	48	64	56
17.046	100	44	48	64	18.939	100	44	40	48	20.57	72	40	64	56
17.062	86	28	40	72	19.029	100	44	72	86	20.63	72	32	44	48
17.101	86	44	56	64	19.048	40	24	32	28	20.74	64	24	56	72
17.102	86	32	28	44	19.090	56	32	48	44	20.78	64	28	40	44
17.141	64	32	48	56	19.091	72	24	28	44	20.83	100	32	48	72
17.143	64	28	24	32	19.096	100	32	44	72	20.90	86	32	56	72
17.144	48	24	24	28	19.111	86	40	64	72	20.93	100	40	72	86
17.188	100	40	44	64	19.136	72	28	64	86	20.95	100	28	44	48
17.200	86	32	64	100	19.197	86	32	40	56	21.00	56	32	48	40
17.275	86	56	72	64	19.200	72	24	64	100	21.12	86	32	44	56
17.361	100	32	40	72	19.250	56	32	44	40	21.32	100	24	44	86
17.364	64	24	56	86	19.285	72	32	48	56	21.33	100	56	86	72
17.373	86	44	64	72	19.286	72	28	24	32	21.39	44	24	28	24

Change Gears for Different Leads — 21.43 Inches to 32.09 Inches

Lead in Inches	Driven — Gear on Worm	Driver — First Gear on Stud	Driven — Second Gear on Stud	Driver — Gear on Screw	Lead in Inches	Driven — Gear on Worm	Driver — First Gear on Stud	Driven — Second Gear on Stud	Driver — Gear on Screw	Lead in Inches	Driven — Gear on Worm	Driver — First Gear on Stud	Driven — Second Gear on Stud	Driver — Gear on Screw
21.43	100	40	48	56	24.88	100	72	86	48	28.05	72	28	48	44
21.48	100	32	44	64	24.93	64	28	48	44	28.06	100	28	44	56
21.50	86	24	24	40	25.00	72	24	40	48	28.13	100	40	72	64
21.82	72	44	64	48	25.08	86	24	28	40	28.15	86	28	44	48
21.88	100	40	56	64	25.09	86	40	56	64	28.29	72	28	44	40
21.90	86	24	44	72	25.13	86	44	72	56	28.41	100	32	40	44
21.94	86	28	40	56	25.14	86	28	44	40	28.57	100	56	64	40
21.99	86	44	72	64	25.45	64	44	56	32	28.64	72	44	56	32
22.00	64	32	44	40	25.46	100	24	44	72	28.65	100	32	44	48
22.04	72	28	48	56	25.51	100	28	40	56	28.67	86	40	64	48
22.11	86	28	72	100	25.57	100	64	72	44	29.09	64	24	48	44
22.22	100	40	64	72	25.60	86	28	40	48	29.17	100	40	56	48
22.34	86	44	64	56	25.67	86	44	44	40	29.22	100	56	72	44
22.40	86	32	40	48	25.71	72	24	48	56	29.32	86	48	72	44
22.50	72	24	48	64	25.72	72	24	24	28	29.34	64	24	44	40
22.73	100	24	24	44	25.80	86	24	72	100	29.39	72	28	64	56
22.80	86	48	56	44	25.97	100	44	64	56	29.56	86	32	44	40
22.86	64	24	24	28	26.04	100	32	40	48	29.76	100	28	40	48
22.91	72	44	56	40	26.06	86	44	64	48	29.86	100	40	86	72
22.92	100	40	44	48	26.16	100	32	72	86	29.90	100	28	72	86
22.93	86	24	64	100	26.18	72	40	64	44	30.00	56	28	48	32
23.04	86	56	72	48	26.19	44	24	40	28	30.23	86	32	72	64
23.14	100	24	40	72	26.25	72	32	56	48	30.30	100	48	64	44
23.26	100	32	64	86	26.33	86	28	48	56	30.48	64	24	32	28
23.33	64	32	56	48	26.52	100	44	56	48	30.54	100	44	86	64
23.38	72	28	40	44	26.58	100	28	64	86	30.56	64	24	40	24
23.44	100	48	72	64	26.67	64	28	56	44	30.61	100	28	48	56
23.45	86	40	48	44	26.79	100	48	72	56	30.71	86	24	48	56
23.52	86	32	56	64	26.88	86	28	56	64	30.72	86	24	24	28
23.57	72	28	44	48	27.00	72	32	48	40	30.86	72	28	48	40
23.81	100	48	64	56	27.13	100	24	56	86	31.01	100	24	64	86
23.89	86	32	64	72	27.15	100	44	86	72	31.11	64	24	56	48
24.00	64	40	72	48	27.22	56	24	28	24	31.25	100	28	56	64
24.13	86	28	44	56	27.27	100	40	48	44	31.27	86	40	64	44
24.19	86	40	72	64	27.30	86	28	64	72	31.35	86	32	56	48
24.24	64	24	40	44	27.34	100	32	56	64	31.36	86	24	28	32
24.31	100	32	56	72	27.36	86	40	56	40	31.43	64	28	44	32
24.43	86	32	40	44	27.43	64	28	48	40	31.50	72	32	56	40
24.44	44	24	32	24	27.50	86	32	44	28	31.75	100	72	64	28
24.54	72	32	48	44	27.64	86	40	72	56	31.82	100	44	56	40
24.55	100	32	44	56	27.78	100	32	64	72	31.85	86	24	64	72
24.57	86	40	64	56	27.87	86	24	56	72	31.99	100	56	86	48
24.64	86	24	44	64	27.92	86	28	40	44	32.00	64	28	56	40
24.75	72	32	44	40	28.00	100	64	86	48	32.09	56	24	44	32

Change Gears for Different Leads — 32.14 Inches to 60.00 Inches

Lead in Inches	Driven — Gear on Worm	Driver — First Gear on Stud	Driven — Second Gear on Stud	Driver — Gear on Screw	Lead in Inches	Driven — Gear on Worm	Driver — First Gear on Stud	Driven — Second Gear on Stud	Driver — Gear on Screw	Lead in Inches	Driven — Gear on Worm	Driver — First Gear on Stud	Driven — Second Gear on Stud	Driver — Gear on Screw
32.14	100	56	72	40	38.20	100	24	44	48	46.07	86	28	72	48
32.25	86	48	72	40	38.39	100	40	86	56	46.67	64	24	56	32
32.41	100	24	56	72	38.57	72	28	48	32	46.88	100	32	72	48
32.47	100	28	40	48	38.89	56	24	40	24	47.15	72	24	44	28
32.58	86	24	40	44	38.96	100	28	48	44	47.62	100	28	64	48
32.73	72	32	64	44	39.09	86	32	64	44	47.78	86	24	64	48
32.74	100	28	44	48	39.29	100	28	44	40	47.99	100	32	86	56
32.85	86	24	44	48	39.42	86	24	44	40	48.00	72	24	64	40
33.00	72	24	44	40	39.49	86	28	72	56	48.38	86	32	72	40
33.33	100	24	32	40	39.77	100	32	56	44	48.61	100	24	56	48
33.51	86	28	48	44	40.00	72	24	64	44	48.86	100	40	86	44
33.59	100	64	86	40	40.18	100	32	72	56	48.89	64	24	44	24
33.79	86	28	44	40	40.31	86	32	72	48	49.11	100	28	44	32
33.94	64	24	56	44	40.72	100	44	86	48	49.14	86	28	64	40
34.09	100	48	72	44	40.82	100	28	64	56	49.27	86	24	44	32
34.20	86	44	56	32	40.91	100	40	72	44	49.77	100	24	86	72
34.29	72	48	64	28	40.95	86	28	64	48	50.00	100	28	56	40
34.38	100	32	44	40	40.96	86	24	32	28	50.17	86	24	56	40
34.55	86	32	72	56	41.14	72	28	64	40	50.26	86	28	72	44
34.72	100	24	40	48	41.25	72	24	44	32	51.14	100	32	72	44
34.88	100	24	72	86	41.67	100	32	64	48	51.19	86	24	40	28
34.90	100	56	86	44	41.81	86	24	56	48	51.43	72	28	64	32
35.00	72	24	56	48	41.91	64	24	44	28	51.95	100	28	64	44
35.10	86	28	64	56	41.99	100	32	86	64	52.12	86	24	64	44
35.16	100	32	72	64	42.00	72	24	56	40	52.50	72	24	56	32
35.18	86	44	72	40	42.23	86	28	44	32	53.03	100	24	56	44
35.36	72	32	44	28	42.66	100	28	86	72	53.33	64	24	56	28
35.56	64	24	32	24	42.78	56	24	44	24	53.57	100	28	72	48
35.71	100	32	64	56	42.86	100	28	48	40	53.75	86	24	48	32
35.72	100	24	24	28	43.00	86	32	64	40	54.85	100	28	86	56
35.83	86	32	64	48	43.64	72	24	64	44	55.00	72	24	44	24
36.00	72	32	64	40	43.75	100	32	56	40	55.28	86	28	72	40
36.36	100	44	64	40	43.98	86	32	72	44	55.56	100	24	32	24
36.46	100	48	56	32	44.44	64	24	40	24	55.99	100	24	86	64
36.67	48	24	44	24	44.64	100	28	40	32	56.25	100	32	72	40
36.86	86	28	48	40	44.68	86	28	64	44	56.31	86	24	44	28
37.04	100	24	64	72	44.79	100	40	86	48	57.14	100	28	64	40
37.33	100	32	86	72	45.00	72	28	56	32	57.30	100	24	44	32
37.40	72	28	64	44	45.45	100	32	64	44	57.33	86	24	64	40
37.50	100	48	72	40	45.46	100	28	56	44	58.33	100	24	56	40
37.63	86	32	56	40	45.61	86	24	56	44	58.44	100	28	72	44
37.88	100	24	40	44	45.72	64	24	48	28	58.64	86	24	72	44
38.10	64	24	40	28	45.84	100	24	44	40	59.53	100	24	40	28
38.18	72	24	56	44	45.92	100	28	72	56	60.00	72	24	64	32

Lead of Helix for Given Helix Angle Relative to Axis, When Diameter = 1

Deg.	o'	6'	12'	18'	24'	30'	36'	42'	48'	54'	60'
o	Infin.	1800.001	899.997	599.994	449.993	359.992	299.990	257.130	224.986	199.983	179.982
1	179.982	163.616	149.978	138.438	128.545	119.973	112.471	105.851	99.967	94.702	89.964
2	89.964	85.676	81.778	78.219	74.956	71.954	69.183	66.617	64.235	62.016	59.945
3	59.945	58.008	56.191	54.485	52.879	51.365	49.934	48.581	47.299	46.082	44.927
4	44.927	43.827	42.780	41.782	40.829	39.918	39.046	38.212	37.412	36.645	35.909
5	35.909	35.201	34.520	33.866	33.235	32.627	32.040	31.475	30.928	30.400	29.890
6	29.890	29.397	28.919	28.456	28.008	27.573	27.152	26.743	26.346	25.961	25.586
7	25.586	25.222	24.868	24.524	24.189	23.863	23.545	23.236	22.934	22.640	22.354
8	22.354	22.074	21.801	21.535	21.275	21.021	20.773	20.530	20.293	20.062	19.835
9	19.835	19.614	19.397	19.185	18.977	18.773	18.574	18.379	18.188	18.000	17.817
10	17.817	17.637	17.460	17.287	17.117	16.950	16.787	16.626	16.469	16.314	16.162
11	16.162	16.013	15.866	15.722	15.581	15.441	15.305	15.170	15.038	14.908	14.780
12	14.780	14.654	14.530	14.409	14.289	14.171	14.055	13.940	13.828	13.717	13.608
13	13.608	13.500	13.394	13.290	13.187	13.086	12.986	12.887	12.790	12.695	12.600
14	12.600	12.507	12.415	12.325	12.237	12.148	12.061	11.975	11.890	11.807	11.725
15	11.725	11.643	11.563	11.484	11.405	11.328	11.252	11.177	11.102	11.029	10.956
16	10.956	10.884	10.813	10.743	10.674	10.606	10.538	10.471	10.405	10.340	10.276
17	10.276	10.212	10.149	10.086	10.025	9.964	9.904	9.844	9.785	9.727	9.669
18	9.669	9.612	9.555	9.499	9.444	9.389	9.335	9.281	9.228	9.176	9.124
19	9.124	9.072	9.021	8.971	8.921	8.872	8.823	8.774	8.726	8.679	8.631
20	8.631	8.585	8.539	8.493	8.447	8.403	8.358	8.314	8.270	8.227	8.184
21	8.184	8.142	8.099	8.058	8.016	7.975	7.935	7.894	7.855	7.815	7.776
22	7.776	7.737	7.698	7.660	7.622	7.584	7.547	7.510	7.474	7.437	7.401
23	7.401	7.365	7.330	7.295	7.260	7.225	7.191	7.157	7.123	7.089	7.056
24	7.056	7.023	6.990	6.958	6.926	6.894	6.862	6.830	6.799	6.768	6.737
25	6.737	6.707	6.676	6.646	6.617	6.586	6.557	6.528	6.499	6.470	6.441
26	6.441	6.413	6.385	6.357	6.329	6.300	6.274	6.246	6.219	6.192	6.166
27	6.166	6.139	6.113	6.087	6.061	6.035	6.009	5.984	5.959	5.933	5.908
28	5.908	5.884	5.859	5.835	5.810	5.786	5.762	5.738	5.715	5.691	5.668
29	5.668	5.644	5.621	5.598	5.575	5.553	5.530	5.508	5.486	5.463	5.441

Lead of Helix for Given Helix Angle Relative to Axis, When Diameter = 1

Deg.	0'	6'	12'	18'	24'	30'	36'	42'	48'	54'	60'
30	5.441	5.420	5.398	5.376	5.355	5.333	5.312	5.291	5.270	5.249	5.228
31	5.228	5.208	5.187	5.167	5.147	5.127	5.107	5.087	5.067	5.047	5.028
32	5.028	5.008	4.989	4.969	4.950	4.931	4.912	4.894	4.875	4.856	4.838
33	4.838	4.819	4.801	4.783	4.764	4.746	4.728	4.711	4.693	4.675	4.658
34	4.658	4.640	4.623	4.605	4.588	4.571	4.554	4.537	4.520	4.503	4.487
35	4.487	4.470	4.453	4.437	4.421	4.404	4.388	4.372	4.356	4.340	4.324
36	4.324	4.308	4.292	4.277	4.261	4.246	4.230	4.215	4.199	4.184	4.169
37	4.169	4.154	4.139	4.124	4.109	4.094	4.079	4.065	4.050	4.036	4.021
38	4.021	4.007	3.992	3.978	3.964	3.950	3.935	3.921	3.907	3.893	3.880
39	3.880	3.866	3.852	3.838	3.825	3.811	3.798	3.784	3.771	3.757	3.744
40	3.744	3.731	3.718	3.704	3.691	3.678	3.665	3.652	3.640	3.627	3.614
41	3.614	3.601	3.589	3.576	3.563	3.551	3.538	3.526	3.514	3.501	3.489
42	3.489	3.477	3.465	3.453	3.440	3.428	3.416	3.405	3.393	3.381	3.369
43	3.369	3.358	3.346	3.334	3.322	3.311	3.299	3.287	3.276	3.265	3.253
44	3.253	3.242	3.231	3.219	3.208	3.197	3.186	3.175	3.164	3.153	3.142
45	3.142	3.131	3.120	3.109	3.098	3.087	3.076	3.066	3.055	3.044	3.034
46	3.034	3.023	3.013	3.002	2.992	2.981	2.971	2.960	2.950	2.940	2.930
47	2.930	2.919	2.909	2.899	2.889	2.879	2.869	2.859	2.849	2.839	2.829
48	2.829	2.819	2.809	2.799	2.789	2.779	2.770	2.760	2.750	2.741	2.731
49	2.731	2.721	2.712	2.702	2.693	2.683	2.674	2.664	2.655	2.645	2.636
50	2.636	2.627	2.617	2.608	2.599	2.590	2.581	2.571	2.562	2.553	2.544
51	2.544	2.535	2.526	2.517	2.508	2.499	2.490	2.481	2.472	2.463	2.454
52	2.454	2.446	2.437	2.428	2.419	2.411	2.402	2.393	2.385	2.376	2.367
53	2.367	2.359	2.350	2.342	2.333	2.325	2.316	2.308	2.299	2.291	2.282
54	2.282	2.274	2.266	2.257	2.249	2.241	2.233	2.224	2.216	2.208	2.200
55	2.200	2.192	2.183	2.175	2.167	2.159	2.151	2.143	2.135	2.127	2.119
56	2.119	2.111	2.103	2.095	2.087	2.079	2.072	2.064	2.056	2.048	2.040
57	2.040	2.032	2.025	2.017	2.009	2.001	1.994	1.986	1.978	1.971	1.963
58	1.963	1.955	1.948	1.940	1.933	1.925	1.918	1.910	1.903	1.895	1.888
59	1.888	1.880	1.873	1.865	1.858	1.851	1.843	1.836	1.828	1.821	1.814

Lead of Helix for Given Helix Angle Relative to Axis, When Diameter = 1

Deg.	0'	6'	12'	18'	24'	30'	36'	42'	48'	54'	60'
60	1.814	1.806	1.799	1.792	1.785	1.777	1.770	1.763	1.756	1.749	1.741
61	1.741	1.734	1.727	1.720	1.713	1.706	1.699	1.692	1.685	1.677	1.670
62	1.670	1.663	1.656	1.649	1.642	1.635	1.628	1.621	1.615	1.608	1.601
63	1.601	1.594	1.587	1.580	1.573	1.566	1.559	1.553	1.546	1.539	1.532
64	1.532	1.525	1.519	1.512	1.505	1.498	1.492	1.485	1.478	1.472	1.465
65	1.465	1.458	1.452	1.445	1.438	1.432	1.425	1.418	1.412	1.405	1.399
66	1.399	1.392	1.386	1.379	1.372	1.366	1.359	1.353	1.346	1.340	1.334
67	1.334	1.327	1.321	1.314	1.308	1.301	1.295	1.288	1.282	1.276	1.269
68	1.269	1.263	1.257	1.250	1.244	1.237	1.231	1.225	1.219	1.212	1.206
69	1.206	1.200	1.193	1.187	1.181	1.175	1.168	1.162	1.156	1.150	1.143
70	1.143	1.137	1.131	1.125	1.119	1.112	1.106	1.100	1.094	1.088	1.082
71	1.082	1.076	1.069	1.063	1.057	1.051	1.045	1.039	1.033	1.027	1.021
72	1.021	1.015	1.009	1.003	0.997	0.991	0.985	0.978	0.972	0.966	0.960
73	0.960	0.954	0.948	0.943	0.937	0.931	0.925	0.919	0.913	0.907	0.901
74	0.901	0.895	0.889	0.883	0.877	0.871	0.865	0.859	0.854	0.848	0.842
75	0.842	0.836	0.830	0.824	0.818	0.812	0.807	0.801	0.795	0.789	0.783
76	0.783	0.777	0.772	0.766	0.760	0.754	0.748	0.743	0.737	0.731	0.725
77	0.725	0.720	0.714	0.708	0.702	0.696	0.691	0.685	0.679	0.673	0.668
78	0.668	0.662	0.656	0.651	0.645	0.639	0.633	0.628	0.622	0.616	0.611
79	0.611	0.605	0.599	0.594	0.588	0.582	0.577	0.571	0.565	0.560	0.554
80	0.554	0.548	0.543	0.537	0.531	0.526	0.520	0.514	0.509	0.503	0.498
81	0.498	0.492	0.486	0.481	0.475	0.469	0.464	0.458	0.453	0.447	0.441
82	0.441	0.436	0.430	0.425	0.419	0.414	0.408	0.402	0.397	0.391	0.386
83	0.386	0.380	0.375	0.369	0.363	0.358	0.352	0.347	0.341	0.336	0.330
84	0.330	0.325	0.319	0.314	0.308	0.302	0.297	0.291	0.286	0.280	0.275
85	0.275	0.269	0.264	0.258	0.253	0.247	0.242	0.236	0.231	0.225	0.220
86	0.220	0.214	0.209	0.203	0.198	0.192	0.187	0.181	0.176	0.170	0.165
87	0.165	0.159	0.154	0.148	0.143	0.137	0.132	0.126	0.121	0.115	0.110
88	0.110	0.104	0.099	0.093	0.088	0.082	0.077	0.071	0.066	0.060	0.055
89	0.055	0.049	0.044	0.038	0.033	0.027	0.022	0.016	0.011	0.005	0.000

Leads, Change Gears and Angles for Helical Milling

Lead of Helix, Inches	Change Gears: Gear on Worm	Change Gears: First Gear on Stud	Change Gears: Second Gear on Stud	Change Gears: Gear on Screw	Diameter of Work, Inches: 1/8	1/4	3/8	1/2	5/8	3/4	7/8	1	1¼	1½
					Approximate Angles for Milling Machine Table									
0.67	24	86	24	100	30¼	...	...	...	...	...	...	...	...	...
0.78	24	86	28	100	26	44½	...	...	...	...	...	...	...	...
0.89	24	86	32	100	23½	41	...	...	...	...	...	...	...	...
1.12	24	86	40	100	19	34½	...	...	...	...	...	...	...	...
1.34	24	86	48	100	16	30¼	41½	...	...	...	...	...	...	...
1.46	24	64	28	72	14¾	28	38½	...	...	...	...	...	...	...
1.56	24	86	56	100	13¾	26½	37	...	...	...	...	...	...	...
1.67	24	64	32	72	12¾	25	34¾	43¼	...	...	...	...	...	...
1.94	32	64	28	72	11¼	21¾	31	39	45	...	...	...	...	...
2.08	24	64	40	72	10¼	20½	29½	37	43¼	...	...	...	...	...
2.22	32	56	28	72	9¾	19¼	27½	35	41¼	...	...	...	...	...
2.50	24	64	48	72	8¾	17	25	32	38	43¼	...	...	...	...
2.78	40	56	28	72	8	15½	23	29½	35¼	40½	44¾	...	...	...
2.92	24	64	56	72	7½	15	21¾	28¼	34	39	43¼	...	...	...
3.24	40	48	28	72	6¾	13¼	19¾	25¾	31¼	36	40½	44¼	...	...
3.70	40	48	32	72	6	11¾	17½	23	28	32½	36½	40½	...	...
3.89	56	48	24	72	5½	11¼	16¾	22	26¾	31¼	35¼	39	...	...
4.17	40	72	48	64	5¼	10½	15¾	20½	25¼	29½	33½	37	43¼	...
4.46	48	40	32	86	4¾	9¾	14¾	19¼	23¾	27¾	31½	35	41½	...
4.86	40	64	56	72	4½	9	13½	17¾	22	25¾	29½	33	39	44¼
5.33	48	40	32	72	4	8¼	12¼	16½	20¼	23¾	27¼	30½	36½	41½
5.44	56	40	28	72	4	8	12	16	20	23½	26¾	30	36	41
6.12	56	40	28	64	3½	7¼	11	14½	17¾	21	24¼	27	33	37¾
6.22	56	40	32	72	3½	7	10¾	14¼	17½	20¾	23¾	26¾	32½	37¼
6.48	56	48	40	72	3¼	6¾	10¼	13½	16¾	20	23	25¾	31½	36¼
6.67	64	48	28	56	3¼	6½	10	13¼	16½	19½	22½	25¼	30¾	35¼
7.29	56	48	40	64	3	6¼	9¼	12¼	15	18	20½	23½	28½	33
7.41	64	48	40	72	3	6	9	12	14¾	17¾	20¼	22¾	28¼	32½
7.62	64	48	32	56	2¾	5¾	8¾	11½	14½	17¼	19¾	22¼	27½	32
8.33	48	32	40	72	2½	5¼	8	10½	13¼	15¾	18¼	20½	25½	29½
8.95	86	48	28	56	2½	5	7½	10	12½	14¾	17	19¼	24	28
9.33	56	40	48	72	2¼	4¾	7¼	9½	11¾	14	16¼	18½	23	27
9.52	64	48	40	56	2¼	4½	7	9¼	11½	13¾	16	18¼	22½	26½
10.29	72	40	32	56	2	4¼	6½	8¾	10¾	12¾	15	17¼	21	24¾
10.37	64	48	56	72	2	4¼	6½	8½	10½	12¾	14¾	17	20¾	24½
10.50	48	40	56	64	2	4¼	6¼	8½	10½	12½	14½	16¾	20½	24¼
10.67	64	40	48	72	2	4	6¼	8¼	10¼	12¼	14¼	16½	20¼	24
10.94	56	32	40	64	2	4	6	8¼	10¼	12	14	16¼	20	23½
11.11	64	32	40	72	2	4	6	8	10	11¾	13¾	16	19¾	23
11.66	56	32	48	72	1¾	3¾	5¾	7½	9½	11¼	13¼	15¼	18¾	22
12.00	72	40	32	48	1¾	3¾	5½	7¼	9¼	11	12¾	15	18¼	21½
13.12	56	32	48	64	1½	3½	5¼	6¾	8½	10¼	11¾	13½	16¾	20
13.33	56	28	48	72	1½	3¼	5	6½	8¼	10	11½	13¼	16½	19½
13.71	64	40	48	56	1½	3¼	4¾	6½	8	9¾	11¼	13	16	19
15.24	64	28	48	72	1½	3	4½	5¾	7¼	8¾	10¼	11¾	14½	17¼
15.56	64	32	56	72	1¼	2¾	4¼	5¾	7¼	8¾	10	11½	14¼	17
15.75	56	64	72	40	1¼	2¾	4¼	5½	7	8½	9¾	11¼	14	16¾
16.87	72	32	48	64	1¼	2½	4	5¼	6¾	7¾	9¼	10½	13¼	15¾
17.14	64	32	48	56	1¼	2½	4	5¼	6½	7¾	9	10¼	13	15½
18.75	72	32	40	48	1	2¼	3½	4¾	6	7¼	8¼	9½	12	14¼
19.29	72	32	48	56	1	2¼	3½	4½	5¾	7	8	9¼	11½	13¾
19.59	64	28	48	56	1	2¼	3¼	4½	5¾	6¾	8	9¼	11½	13½
19.69	72	32	56	64	1	2¼	3¼	4½	5¾	6¾	8	9	11½	13½
21.43	72	24	40	56	1	2	3¼	4¼	5¼	6¼	7½	8½	10½	12½
22.50	72	28	56	64	1	2	3	4	5	6	7	8	10	12
23.33	64	32	56	48	1	2	3	4	5	5¾	6¾	7¾	9¾	11½
26.25	72	24	56	64	1	1¾	2¾	3½	4¼	5	6	7	8½	10¼
26.67	64	28	56	48	¾	1¾	2¾	3½	4¼	5	6	6¾	8½	10
28.00	64	32	56	40	¾	1¾	2½	3¼	4	4¾	5¾	6½	8	9½
30.86	72	28	48	40	¾	1½	2¼	3	3¾	4½	5	5¾	7¼	8¾

Leads, Change Gears and Angles for Helical Milling

Lead of Helix, Inches	Change Gears				Diameter of Work, Inches									
	Gear on Worm	First Gear on Stud	Second Gear on Stud	Gear on Screw	1¾	2	2¼	2½	2¾	3	3¼	3½	3¾	4
					Approximate Angles for Milling Machine Table									
6.12	56	40	28	64	42	...	...	...	...	...	...	...	...	...
6.22	56	40	32	72	41½	...	...	...	...	...	...	...	...	...
6.48	56	48	40	72	40¼	44¼	...	...	...	...	...	...	...	...
6.67	64	48	28	56	39½	43½	...	...	...	...	...	...	...	...
7.29	56	48	40	64	37	41	44¼	...	...	...	...	...	...	...
7.41	64	48	40	72	36½	40¼	43¾	...	...	...	...	...	...	...
7.62	64	48	32	56	36	39½	43	...	...	...	...	...	...	...
8.33	48	32	40	72	33½	37	40½	43½	...	...	...	...	...	...
8.95	86	48	28	56	31¾	35¼	38½	41¼	44	...	...	...	...	...
9.33	56	40	48	72	30½	34	37¼	40¼	43	...	...	...	...	...
9.52	64	48	40	56	30	33½	36½	39½	42¼	45	...	...	...	...
10.29	72	40	32	56	28¼	31½	34½	37½	40	42½	45	...	...	...
10.37	64	48	56	72	28	31¼	34¼	37¼	39¾	42¼	44¾	...	...	...
10.50	48	40	56	64	27¾	31	34	36¾	39½	42	44¼	...	...	...
10.67	64	40	48	72	27¼	30½	33½	36½	39	41½	43¾	...	...	...
10.94	56	32	40	64	26¾	30	33	35¾	38¼	40¾	43	...	...	...
11.11	64	32	40	72	26½	29½	32½	35¼	38	40¼	42½	44¾	...	...
11.66	56	32	48	72	25¼	28½	31¼	34	36½	39	41¼	43½	...	...
12.00	72	40	32	48	24¾	27¾	30½	33¼	35¾	38	40¼	42½	44¾	...
13.12	56	32	48	64	22¾	25¾	28¼	31	33¼	35¾	37¾	40	42	43¾
13.33	56	28	48	72	22½	25½	28	30½	33	35¼	37½	39½	41½	43¼
13.71	64	40	48	56	22	24¾	27¼	30	32¼	34½	36½	38¾	40¾	42½
15.24	64	28	48	72	20	22½	25	27¼	29½	31¾	34	35¾	37¾	39½
15.56	64	32	56	72	19½	22	24½	27	29	31¼	33¼	35¼	37	39
15.75	56	64	72	40	19¼	21¾	24¼	26½	28¾	31	33	35	36¾	38½
16.87	72	32	48	64	18¼	20½	22¾	25	27	29¼	31¼	33¼	35	36½
17.14	64	32	48	56	17¾	20¼	22¼	24¾	26¾	29	30¾	32¾	34½	36
18.75	72	32	40	48	16¼	18½	20¾	22¾	25	26¾	28½	30¼	32	33¾
19.29	72	32	48	56	16	18¼	20¼	22¼	24	26	28	29¾	31½	33
19.59	64	28	48	56	15¾	18	20	22	23¾	25¾	27½	29¼	31	32¾
19.69	72	32	56	64	15¾	17¾	20	21¾	23¾	25½	27½	29¼	31	32½
21.43	72	24	40	56	14½	16½	18½	20¼	22	23¾	25½	27¼	29	30¼
22.50	72	28	56	64	13¾	15¾	17½	19¼	21	22¾	24½	26	27¾	29¼
23.33	64	32	56	48	13¼	15¼	17	18¾	20¼	22	23½	25¼	27	28¼
26.25	72	24	56	64	12	13½	15	16¾	18¼	19¾	21¼	22¾	24¼	25½
26.67	64	28	56	48	11¾	13¼	14¾	16½	18	19½	21	22¼	23¾	25¼
28.00	64	32	56	40	11¼	12¾	14¼	15¾	17¼	18¾	20	21½	22¾	24
30.86	72	28	48	40	10	11½	13	14¼	15½	17	18½	19½	21	22
31.50	72	32	56	40	10	11¼	12¾	14	15¼	16½	18	19¼	20½	21¾
36.00	72	32	64	40	8¾	10	11	12¼	13½	14¾	16	17	18¼	19¼
41.14	72	28	64	40	7¾	8¾	9¾	10¾	11¾	13	14	15	16	17
45.00	72	28	56	32	7	8	9	10	11	11¾	12¾	13¾	14¾	15½
48.00	72	24	64	40	6½	7½	8½	9¼	10¼	11¼	12	13	13¾	14½
51.43	72	28	64	32	6	7	7¾	8¾	9½	10½	11¼	12	12¾	13¾
60.00	72	24	64	32	5¼	6	6¾	7½	8¼	9	9½	10¼	11	11¾
68.57	72	24	64	28	4¼	5¼	5¾	6½	7¼	8	8½	9	9¾	10¼

Helix Angle for Given Lead and Diameter. — The table on this and the preceding page gives helix angles (relative to axis) equivalent to a range of leads and diameters. The expression "Diameter of Work" at the top of the table might mean pitch diameter or outside diameter, depending upon the class of work. Assume, for example, that a plain milling cutter 4 inches in diameter is to have helical teeth and a helix angle of about 25 degrees is desired. The table shows that this

angle will be obtained approximately by using change-gears which will give a lead of 26.67 inches. As the outside diameter of the cutter is 4 inches, the helix angle of 25¼ degrees is at the tops of the teeth. The angles listed for different diameters are used in setting the table of a milling machine. In milling a right-hand helix (or cutter teeth which turn to the right as seen from the end of the cutter), swivel the right-hand end of the machine table toward the rear, and, inversely, for a left-hand helix, swivel the left-hand end of the table toward the rear. The angles in the table are based upon the following formula:

$$\text{Cot helix angle relative to axis} = \frac{\text{Lead of helix}}{3.1416 \times \text{diameter}}$$

Lead of Helix for Given Angle. — The lead of a helix or "spiral" for given angles measured with the axis of the work is given in the table, pages 1728–1730 for a diameter of 1. For other diameters, lead equals the value found in the table multiplied by the given diameter. Suppose the angle is 55 degrees, and the diameter 5 inches, what would be the lead? By referring to the table (Part 2), it is found that the lead for a diameter of 1 and an angle of 55 degrees 0 minutes equals 2.200. Multiply this value by 5; 5 × 2.200 = 11 inches, which is the required lead. If the lead and diameter are given, and the angle is wanted, divide the given lead by the given diameter, thus obtaining the lead for a diameter equal to 1; then find the angle corresponding to this lead in the table. If the lead and angle are given, and the diameter is wanted, divide the lead by the value in the table for the angle.

SIMPLE, COMPOUND, DIFFERENTIAL, AND BLOCK INDEXING

Simple Indexing. — A general rule for determining the number of turns the crank of a dividing head must make, to obtain a given number of divisions, is as follows: Divide the number of turns required for one revolution of the dividing-head spindle by the number of divisions into which the periphery of the work is to be divided.

Example: — If 40 turns of the index crank are required for one revolution of the spindle, and 12 divisions are required, the number of turns of the index crank for each indexing would equal 40 ÷ 12 = 3⅓ turns.

Compound Indexing. — This method is sometimes used to obtain divisions which are beyond the range of those secured by the simple method. The crank is first turned a definite amount in the regular way, and then the index plate is also turned either in the same or opposite direction, in order to locate the index crank in the proper position. Thus, there are two separate movements which are, in reality, two simple indexing operations. The following rule is for determining what circles of holes can be used for indexing by the compound method.

Rule: Resolve into its factors the number of divisions required; then choose at random two circles of holes, subtract one from the other, and factor the difference; place the two sets of factors thus obtained above a horizontal line. Next factor the number of turns of the crank required for one revolution of the spindle, and also the number of holes in each of the chosen circles; place the three sets of factors thus obtained below the horizontal line. If all the factors *above* the line can be canceled by those below, the two circles chosen will give the required number of divisions; if not, other circles must be chosen and another trial made.

Example: — Assume that 69 divisions are required, and that circles having 33 and 23 holes are chosen for the first trial. Then, by applying the foregoing rule, it is found that all the factors above the line cancel:

$$\frac{3 \times 23 \times 2 \times 5}{2 \times 2 \times 2 \times 5 \times 3 \times 11 \times 23} = \frac{1}{2 \times 2 \times 11}$$

Compound Indexing*

No. of Divisions	Indexing Movements	No. of Times Around	No. of Divisions	Indexing Movements	No. of Times Around	No. of Divisions	Indexing Movements	No. of Times Around
51	$8\frac{41}{47} - \frac{12}{49}$	11	133	$3\frac{23}{29} - \frac{16}{33}$	11	198*	$\frac{3}{27} + \frac{3}{33}$	...
53	$6\frac{43}{47} - \frac{6}{49}$	9	134	$3\frac{27}{47} + \frac{15}{49}$	13	199	$2\frac{12}{41} - \frac{7}{49}$	11
57	$4\frac{40}{47} + \frac{3}{49}$	7	137	$3\frac{17}{43} - \frac{9}{49}$	11	201	$2\frac{18}{47} + \frac{10}{49}$	13
59	$7\frac{10}{47} + \frac{12}{49}$	11	138*	$\frac{11}{33} - \frac{1}{23}$	...	202	$3\frac{10}{41} + \frac{6}{49}$	17
61	$3\frac{42}{47} + \frac{2}{49}$	6	139	$2\frac{25}{37} + \frac{24}{49}$	11	203	$1\frac{23}{39} + \frac{9}{49}$	9
63	$4\frac{19}{29} + \frac{14}{33}$	8	141	$1\frac{32}{39} + \frac{22}{49}$	8	204	$2\frac{20}{41} + \frac{3}{49}$	13
67	$2\frac{27}{41} + \frac{16}{49}$	5	142	$4\frac{1}{47} + \frac{10}{49}$	15	206	$2\frac{34}{39} + \frac{2}{49}$	15
69*	$\frac{21}{23} - \frac{11}{33}$	...	143	$1\frac{36}{47} - \frac{18}{49}$	5	207	$3\frac{8}{41} - \frac{24}{49}$	14
71	$3\frac{34}{41} - \frac{22}{49}$	6	146	$2\frac{3}{37} - \frac{8}{49}$	7	208	$1\frac{19}{47} + \frac{16}{49}$	9
73	$6\frac{28}{47} - \frac{1}{49}$	12	147*	$\frac{13}{39} - \frac{3}{49}$	...	209	$\frac{8}{49} + \frac{9}{41}$	2
77*	$\frac{9}{21} + \frac{3}{33}$	...	149	$3\frac{5}{43} - \frac{8}{49}$	11	211	$1\frac{28}{39} + \frac{18}{49}$	11
79	$2\frac{42}{43} + \frac{3}{49}$	6	151	$1\frac{42}{43} - \frac{6}{49}$	7	212	$3\frac{4}{47} + \frac{6}{49}$	17
81	$5\frac{5}{41} - \frac{9}{49}$	10	153	$2\frac{45}{47} - \frac{4}{49}$	11	213	$1\frac{18}{39} + \frac{2}{49}$	8
83	$3\frac{45}{47} - \frac{5}{49}$	8	154*	$\frac{8}{21} - \frac{4}{33}$	...	214	$3\frac{9}{47} - \frac{19}{49}$	15
87*	$\frac{23}{29} - \frac{11}{33}$	...	157	$2\frac{23}{31} + \frac{2}{33}$	11	217	$2\frac{3}{43} + \frac{16}{49}$	13
89	$3\frac{28}{39} - \frac{6}{49}$	8	158	$5\frac{5}{43} - \frac{15}{49}$	19	218	$1\frac{22}{47} - \frac{9}{49}$	7
91*	$\frac{6}{39} + \frac{14}{49}$	...	159	$2\frac{7}{37} + \frac{16}{49}$	10	219	$3\frac{29}{43} - \frac{10}{49}$	19
93*	$\frac{3}{31} + \frac{11}{33}$	...	161	$2\frac{10}{39} - \frac{1}{49}$	9	221	$1\frac{5}{47} - \frac{1}{49}$	6
96*	$\frac{3}{18} + \frac{5}{20}$	...	162	$1\frac{30}{39} - \frac{2}{49}$	7	222	$2\frac{8}{43} - \frac{10}{49}$	11
97	$4\frac{27}{41} - \frac{6}{49}$	11	163	$3\frac{7}{37} - \frac{24}{49}$	11	223	$2\frac{26}{43} + \frac{13}{49}$	16
99*	$\frac{15}{27} - \frac{5}{33}$	...	166	$1\frac{19}{43} + \frac{12}{49}$	7	224	$2\frac{6}{23} + \frac{2}{33}$	13
101	$4\frac{32}{43} - \frac{19}{49}$	11	167	$2\frac{1}{29} + \frac{4}{33}$	9	225*	$\frac{5}{18} - \frac{2}{20}$	...
102	$4\frac{17}{43} - \frac{4}{49}$	11	169	$1\frac{32}{37} + \frac{13}{49}$	9	226	$1\frac{38}{39} + \frac{16}{49}$	13
103	$1\frac{8}{43} + \frac{18}{49}$	4	171	$1\frac{29}{47} + \frac{1}{49}$	7	227	$3\frac{3}{43} + \frac{5}{49}$	18
106	$2\frac{38}{41} + \frac{23}{49}$	9	173	$1\frac{7}{43} + \frac{11}{49}$	6	228	$2\frac{8}{41} - \frac{13}{49}$	11
107	$2\frac{21}{31} - \frac{2}{33}$	7	174*	$\frac{11}{33} - \frac{3}{29}$	...	229	$2\frac{19}{41} - \frac{18}{49}$	12
109	$2\frac{19}{39} + \frac{4}{49}$	7	175	$1\frac{4}{31} + \frac{8}{33}$	6	231*	$\frac{3}{21} + \frac{1}{33}$	...
111	$3\frac{29}{47} + \frac{17}{49}$	11	176	$1\frac{14}{43} + \frac{13}{49}$	7	233	$1\frac{36}{47} + \frac{6}{49}$	11
112	$4\frac{10}{31} - \frac{13}{33}$	11	177	$2\frac{19}{47} + \frac{4}{49}$	11	234	$2\frac{21}{29} + \frac{6}{33}$	17
113	$3\frac{26}{47} - \frac{18}{49}$	9	178	$3\frac{28}{47} + \frac{11}{49}$	17	236	$2\frac{30}{43} + \frac{9}{49}$	17
114	$1\frac{35}{37} + \frac{25}{49}$	7	179	$2\frac{34}{47} - \frac{13}{49}$	11	237	$2\frac{12}{47} - \frac{3}{49}$	13
117	$7\frac{1}{47} - \frac{9}{49}$	20	181	$2\frac{8}{43} + \frac{12}{49}$	11	238	$2\frac{3}{31} + \frac{14}{33}$	15
118	$1\frac{8}{39} + \frac{24}{49}$	5	182*	$\frac{3}{39} + \frac{7}{49}$	...	239	$1\frac{23}{43} + \frac{15}{49}$	11
119	$3\frac{4}{23} - \frac{16}{33}$	8	183	$1\frac{24}{41} + \frac{8}{49}$	8	241	$1\frac{1}{41} + \frac{23}{49}$	9
121	$1\frac{14}{47} - \frac{15}{49}$	3	186*	$\frac{17}{31} - \frac{11}{33}$	...	242	$2\frac{23}{41} - \frac{4}{49}$	15
122	$3\frac{41}{43} - \frac{17}{49}$	11	187	$1\frac{20}{47} + \frac{14}{49}$	8	243	$1\frac{29}{41} - \frac{3}{49}$	10
123	$1\frac{12}{43} + \frac{17}{49}$	5	189	$2\frac{26}{41} - \frac{15}{49}$	11	244	$2\frac{15}{31} + \frac{10}{33}$	17
125	$2\frac{33}{41} - \frac{12}{49}$	8	191	$1\frac{38}{47} + \frac{14}{49}$	10	246	$1\frac{9}{43} - \frac{16}{49}$	5
126	$3\frac{16}{19} - \frac{7}{20}$	11	192	$2\frac{22}{41} - \frac{12}{49}$	11	247	$2\frac{15}{43} - \frac{4}{49}$	14
127	$2\frac{23}{39} + \frac{12}{49}$	9	193	$1\frac{5}{37} - \frac{15}{49}$	4	249	$3\frac{4}{43} - \frac{2}{49}$	19
129	$5\frac{24}{41} + \frac{15}{49}$	19	194	$2\frac{22}{37} - \frac{16}{49}$	11	250	$2\frac{9}{37} - \frac{8}{49}$	13
131	$2\frac{40}{43} + \frac{21}{49}$	11	197	$1\frac{39}{43} + \frac{16}{49}$	11	...	...	...

* The indexing movements are exact for the divisions marked with an asterisk (*); the errors of the other divisions are so slight as to be negligible for all ordinary classes of work, such as gear-cutting, etc.

This shows that these circles can be used. The factors 2, 2 and 11 remain uncanceled below the line. The amount the crank and index plate must be moved in their respective circles is next determined by multiplying together all these uncanceled factors. Thus $2 \times 2 \times 11 = 44$. This means that we can index $\frac{1}{88}$ revolution by turning the crank forward 44 holes in the 23-hole circle, and the index plate backward 44 holes in the 33-hole circle. The movement could also be forward 44 holes in the 33-hole circle and backward 44 holes in the 23-hole circle, without affecting the result. The movements obtained by the foregoing rule are expressed in compound indexing tables in the form of fractions, as, for example: $+ \frac{44}{23} - \frac{44}{33}$. The numerators represent the number of holes indexed and the denominators the circles used; the + and − signs show that the movements of the crank and index plate are opposite in direction. These fractions can often be reduced and simplified, so that it will not be necessary to move so many holes, by adding some number to them algebraically. The number is chosen by trial, and its sign should be opposite that of the fraction to which it is added. Suppose, for example, a fraction is added representing one complete turn, to each of the fractions referred to; then there will be a movement of 21 holes in the 23-hole circle, and a movement of 11 holes in the opposite direction, in the 33-hole circle.

Differential Indexing. — This method is the same, in principle, as compound indexing, but differs from the latter in that the index plate is rotated by suitable gearing which connects it to the spiral-head spindle. This rotation or differential motion of the index plate takes place when the crank is turned, the plate moving either in the same direction as the crank or opposite to it, as may be required. The result is that the *actual* movement of the crank, at every indexing, is either greater or less than its movement with relation to the index plate. The differential method makes it possible to obtain almost any division, by using only one circle of holes for that division and turning the index crank in one direction, the same as for plain indexing. The gears to use for moving the index plate the required amount (when gears are required) are shown by the tables, "Simple and Differential Indexing." This table shows what divisions can be obtained by plain indexing, and also when it is necessary to use gears and the differential system. For example, if 50 divisions are required, the 20-hole index circle is used and the crank is moved 16 holes, but no gears are required. For 51 divisions, a 24-tooth gear is placed on the worm-shaft and a 48-tooth gear is mounted on the spindle. These two gears are connected by two idler gears having 24 and 44 teeth, respectively. To illustrate the principle of differential indexing, suppose a dividing head is to be geared for 271 divisions. The table calls for a gear on the worm-shaft having 56 teeth; a spindle gear with 72 teeth; and a 24-toothed idler which serves to rotate the index plate in the same direction as the crank. The sector should be set for giving the crank a movement of 3 holes in the 21-hole circle. If the spindle and index plate were not connected through gearing, 280 divisions would be obtained by successively moving the crank 3 holes in the 21-hole circle, but the gears cause the index plate to turn in the same direction as the crank at such a rate that, when 271 indexings have been made, the work is turned one complete revolution; therefore, we have 271 divisions instead of 280, the number being reduced because the total movement of the crank, for each indexing, is equal to its movement relative to the index plate, *plus* the movement of the plate itself when (as in this case) the crank and plate rotate in the same direction. If they were rotated in opposite directions, the crank would have a total movement equal to the amount it turned relative to the plate, *minus* the plate's movement. Sometimes it is necessary to use compound gearing, in order to move the index plate the required amount for each turn of the crank. The differential method cannot be used in connection with helical or spiral milling, because the spiral head is then geared to the lead-screw of the machine.

To Find Ratio of Gearing for Differential Indexing. — To find the gearing ratio for differential indexing, first select some approximate number A of divisions either greater or less than the required number N. To illustrate, if the required number N is 67, the approximate number A might be 70; then if 40 turns of the index crank are required for 1 revolution of the spindle,

$$\text{Gearing ratio } R = (A - N) \times \frac{40}{A}$$

If the approximate number A is less than N, the formula is the same as above except that $A - N$ is replaced by $N - A$.

Example: Find the gearing ratio and indexing movement for 67 divisions. If $A = 70$,

$$\text{Gearing ratio} = (70 - 67) \times \frac{40}{70} = \frac{12}{7} = \frac{\text{Gear on spindle (driver)}}{\text{Gear on worm (driven)}}$$

The fraction $\frac{12}{7}$ is raised to obtain a numerator and denominator equivalent to available gears. For example, $\frac{12}{7} = \frac{48}{28}$.

Various combinations of gearing and index circles are possible for a given number of divisions. The index movements and gear combinations in the accompanying table apply to a given series of index circles and gear-tooth numbers. The approximate number A upon which any combination is based may be determined by dividing 40 by the fraction representing the indexing movement. For example, the approximate number used for 109 divisions equals $40 \div \frac{6}{18}$ or $40 \times \frac{16}{18} = 106\frac{2}{3}$. If this approximate number is inserted in the preceding formula, it will be found that the gear ratio is $\frac{7}{8}$ as shown in the table.

Second Method of Determining Gear Ratio. — In illustrating a somewhat different method of determining the gear ratio, 67 divisions will again be used. If 70 is selected as the approximate number, then $\frac{40}{70} = \frac{4}{7}$ or $\frac{12}{21}$ turn of the index crank will be required. If the crank is indexed four-sevenths of a turn sixty-seven times, it will make $\frac{4}{7} \times 67 = 38\frac{2}{7}$ revolutions. This is $1\frac{5}{7}$ turns less than the forty required for one revolution of the work (indicating that the gearing should be arranged to rotate the index plate in the same direction as the index crank to increase the indexing movement); hence the gear ratio is $1\frac{5}{7} = \frac{12}{7}$.

To Find the Indexing Movement. — The indexing movement is represented by the fraction $\frac{40}{A}$. For example, if 70 is the approximate number A used in calculating the gear ratio for 67 divisions, then, to find the required movement of the index crank, reduce $\frac{40}{70}$ to any fraction of equal value and having as denominator any number equal to the number of holes available in an index circle. To illustrate,

$$\frac{40}{70} = \frac{4}{7} = \frac{12}{21} = \frac{\text{number of holes indexed}}{\text{number of holes in index circle}}$$

Use of Idler Gears. — In differential indexing, idler gears are used (1) to rotate the index plate in the same direction as the index crank, thus *increasing* the actual indexing movement, or (2) to rotate the index plate in the opposite direction, thus *reducing* the actual indexing movement.

Case 1: If the approximate number A is *greater* than the actual number of divisions N, simple gearing will require one idler, and compound gearing no idler. Index plate and crank rotate in the same direction.

Case 2: If the approximate number A is *less* than the actual number of divisions N, simple gearing requires two idlers, and compound gearing one idler. Index plate and crank rotate in opposite directions.

When Compound Gearing Is Required. — In some cases, as will be noted by referring to the table, it is necessary to use a train of four gears in order to obtain the required ratio with gear-tooth numbers in the available series.

Example: Find the gear combination and indexing movement for 99 divisions, assuming that an approximate number A of 100 is used.

$$\text{Ratio} = (100 - 99) \times \frac{40}{100} = \frac{4}{10} = \frac{4 \times 1}{5 \times 2} = \frac{32}{40} \times \frac{28}{56}$$

These final numbers conform to available gear sizes. The gears having 32 and 28 teeth are the drivers (gear on spindle and first gear on stud), and gears having 40 and 56 teeth are driven (second gear on stud and gear on worm). The indexing movement is represented by the fraction $\frac{40}{100}$ which is reduced to $\frac{8}{20}$, the 20-hole index circle being used in this case.

Example: Determine the gear combination to use for indexing 53 divisions. If 56 is used as an approximate number (possibly after one or more trial solutions to find an approximate number and resulting gear ratio coinciding with available gears).

$$\text{Gearing ratio} = (56 - 53) \times \frac{40}{56} = \frac{15}{7} = \frac{3 \times 5}{1 \times 7} = \frac{72 \times 40}{24 \times 56}$$

The tooth numbers above the line represent *gear on spindle* and *first gear on stud.* The numbers below the line represent *second gear on stud* and *gear on worm.*

$$\text{Indexing movement} = \frac{40}{56} = \frac{5}{7} = \frac{5 \times 7}{7 \times 7} = \frac{35 \text{ holes}}{49\text{-hole circle}}$$

In setting sector arms, do not count the hole containing the index crank pin.

To Check the Number of Divisions Obtained with a Given Gear Ratio and Index Movement. — Invert the fraction representing the indexing movement and let C equal this inverted fraction. R = gearing ratio.

Case 1: If simple gearing is used with one idler or compound gearing with no idler,

$$\text{Number of divisions } N = 40\,C - RC$$

Case 2: If simple gearing is used with two idlers or compound gearing with one idler,

$$\text{Number of divisions } N = 40\,C + RC$$

Example: The gear ratio is $\frac{12}{7}$; there is simple gearing and one idler (Case 1), and the indexing movement is $\frac{12}{21}$, making the inverted fraction $C = \frac{21}{12}$; find the number of divisions N

$$N = (40 \times \tfrac{21}{12}) - (\tfrac{12}{7} \times \tfrac{21}{12}) = 70 - \tfrac{21}{7} = 67$$

Example: The gear ratio is $\frac{7}{8}$; two idlers are used with simple gearing (Case 2) and the indexing movement is 6 holes in the 16-hole circle. Then

$$N = (40 \times \tfrac{16}{6}) + (\tfrac{7}{8} \times \tfrac{16}{6}) = 109$$

Simple and Differential Indexing — Brown & Sharpe Milling Machines

GEAR ON SPINDLE 64 T.

IDLER 24 T.

NO. 1 HOLE

NO. 2 HOLE

1ST GEAR ON STUD 56 T.

GEAR ON WORM 40 T.

2D GEAR ON STUD 32 T.

GEARED FOR 107 DIVISIONS

A

B

Note: Graduations in table indicate setting for sector arms when index crank moves through arc A, except figures marked *, when crank moves through arc B.

Number of Divisions	Index Circle	Number of Turns of Crank	Graduation on Sector	Number of Divisions	Index Circle	Number of Turns of Crank	Graduation on Sector	Gear on Worm	No. 1 Hole		Idlers		
									First Gear on Stud	Second Gear on Stud	Gear on Spindle	No. 1 Hole	No. 2 Hole†
2	Any	20	....	33	33	$1^{7}/_{33}$	41		*Differential Indexing*				
3	39	$13^{13}/_{39}$	65	34	17	$1^{8}/_{17}$	33		Certain divisions such as 51, 53, 57, etc. require the use of differential indexing. In differential indexing, change gears are used to transmit motion from main spindle of dividing head to index plate which turns (either in the same direction as the index crank, or in the opposite direction) whatever amount is required to obtain the correct indexing movement.				
4	Any	10	....	35	49	$1^{7}/_{49}$	26						
5	Any	8	....	36	27	$1^{3}/_{27}$	21						
6	39	$6^{26}/_{39}$	132	37	37	$1^{3}/_{37}$	15						
7	49	$5^{35}/_{49}$	140	38	19	$1^{1}/_{19}$	9						
8	Any	5	....	39	39	$1^{1}/_{39}$	3						
9	27	$4^{12}/_{27}$	88	40	Any	1	...						
10	Any	4	....	41	41	$4^{0}/_{41}$	3*						
11	33	$3^{21}/_{33}$	126	42	21	$2^{0}/_{21}$	9*	The numbers in the columns below represent tooth numbers of the necessary change gears. Where no numbers are shown, simple indexing, which does not require change gears, is used.					
12	39	$3^{13}/_{39}$	65	43	43	$4^{0}/_{43}$	12*						
13	39	$3^{3}/_{39}$	14	44	33	$3^{0}/_{33}$	17*						
14	49	$2^{42}/_{49}$	169	45	27	$2^{4}/_{27}$	21*						
15	39	$2^{26}/_{39}$	132	46	23	$2^{0}/_{23}$	172						
16	20	$2^{10}/_{20}$	98	47	47	$4^{0}/_{47}$	168						
17	17	$2^{6}/_{17}$	69	48	18	$1^{5}/_{18}$	165						
18	27	$2^{6}/_{27}$	43	49	49	$4^{0}/_{49}$	161						
19	19	$2^{2}/_{19}$	19	50	20	$1^{6}/_{20}$	158		Differential Gears				
20	Any	2	....	51	17	$1^{4}/_{17}$	33*	24	...	...	48	24	44
21	21	$1^{19}/_{21}$	18*	52	39	$3^{0}/_{39}$	152	...	...	...	...	...	...
22	33	$1^{27}/_{33}$	161	53	49	$3^{5}/_{49}$	140	56	40	24	72	...	...
23	23	$1^{17}/_{23}$	147	54	27	$2^{0}/_{27}$	147	...	...	...	...	...	...
24	39	$1^{26}/_{39}$	132	55	33	$2^{4}/_{33}$	144	...	...	...	...	...	...
25	20	$1^{12}/_{20}$	118	56	49	$3^{5}/_{49}$	140	...	...	...	...	...	...
26	39	$1^{21}/_{39}$	106	57	21	$1^{5}/_{21}$	142	56	...	...	40	24	44
27	27	$1^{13}/_{27}$	95	58	29	$2^{0}/_{29}$	136	...	...	...	...	...	...
28	49	$1^{21}/_{49}$	83	59	39	$2^{6}/_{39}$	132	48	...	...	32	44	...
29	29	$1^{11}/_{29}$	75	60	39	$2^{6}/_{39}$	132	...	...	...	...	...	...
30	39	$1^{13}/_{39}$	65	61	39	$2^{6}/_{39}$	132	48	...	...	32	24	44
31	31	$1^{9}/_{31}$	56	62	31	$2^{0}/_{31}$	127	...	...	...	...	...	...
32	20	$1^{12}/_{20}$	48	63	39	$2^{6}/_{39}$	132	24	...	...	48	24	44

† On Nos. 1, 1½ and 2 machines, No. 2 hole is in machine table. On Nos. 3 and 4 machines, No. 2 hole is in head.

Simple and Differential Indexing

No. of Divisions	Index Circle	No. of Turns of Crank	Graduation on Sector	Gear on Worm	No. 1 Hole First Gear on Stud	No. 1 Hole Second Gear on Stud	Gear on Spindle	Idlers No. 1 Hole	Idlers No. 2 Hole
64	16	1 9/16	123	....	....	....	....	....	....
65	39	24/39	121	....	....	....	....	....	....
66	33	20/33	120	....	....	....	....	....	....
67	21	12 12/21	113	28	....	....	48	44	....
68	17	10/17	116	....	....	....	....	....	....
69	20	12 12/20	118	40	....	....	56	24	44
70	49	28/49	112	....	....	....	....	....	....
71	18	10/18	109	72	....	....	40	24	....
72	27	15/27	110	....	....	....	....	....	....
73	21	12 12/21	113	28	....	....	48	24	44
74	37	20/37	107	....	....	....	....	....	....
75	15	8/15	105	....	....	....	....	....	....
76	19	10/19	103	....	....	....	....	....	....
77	20	10 10/20	98	32	....	....	48	44	....
78	39	20/39	101	....	....	....	....	....	....
79	20	10/20	98	48	....	....	24	44	....
80	20	10/20	98	....	....	....	....	....	....
81	20	10/20	98	48	....	....	24	24	44
82	41	20/41	96	....	....	....	....	....	....
83	20	10/20	98	32	....	....	48	24	44
84	21	10/21	94	....	....	....	....	....	....
85	17	8/17	92	....	....	....	....	....	....
86	43	20/43	91	....	....	....	....	....	....
87	15	7/15	92	40	....	....	24	24	44
88	33	15/33	89	....	....	....	....	....	....
89	18	8/18	87	72	....	....	32	44	....
90	27	12 12/27	88	....	....	....	....	....	....
91	39	18/39	91	24	....	....	48	24	44
92	23	10/23	86	....	....	....	....	....	....
93	18	8/18	87	24	....	....	32	24	44
94	47	20/47	83	....	....	....	....	....	....
95	19	8/19	82	....	....	....	....	....	....
96	21	9/21	85	28	....	....	32	24	44
97	20	8/20	78	40	....	....	48	44	....
98	49	20/49	79	....	....	....	....	....	....
99	20	8/20	78	56	28	40	32	....	....
100	20	8/20	78	....	....	....	....	....	....
101	20	8/20	78	72	24	40	48	....	24
102	20	8/20	78	40	....	....	32	24	44
103	20	8/20	78	40	....	....	48	24	44
104	39	15/39	75	....	....	....	....	....	....
105	21	8/21	75	....	....	....	....	....	....
106	43	16/43	73	86	24	24	48	....	....

Simple and Differential Indexing

No. of Divisions	Index Circle	No. of Turns of Crank	Graduation on Sector	Gear on Worm	No. 1 Hole First Gear on Stud	No. 1 Hole Second Gear on Stud	Gear on Spindle	Idlers No. 1 Hole	Idlers No. 2 Hole
107	20	8/20	78	40	56	32	64	....	24
108	27	10/27	73	....	....	....	....	....	....
109	16	9/16	73	32	....	....	28	24	44
110	33	12/33	71	....	....	....	....	....	....
111	39	18/39	65	24	....	....	72	32	....
112	39	18/39	65	24	....	....	64	44	....
113	39	18/39	65	24	....	....	56	44	....
114	39	18/39	65	24	....	....	48	44	....
115	23	8/23	68	....	....	....	....	....	....
116	29	10/29	68	....	....	....	....	....	....
117	39	18/39	65	24	....	....	24	56	....
118	39	18/39	65	48	....	....	32	44	....
119	39	18/39	65	72	....	....	24	44	....
120	39	18/39	65	....	....	....	....	....	....
121	39	18/39	65	72	....	....	24	24	44
122	39	18/39	65	48	....	....	32	24	44
123	39	18/39	65	24	....	....	24	24	44
124	31	10/31	63	....	....	....	....	....	....
125	39	13/39	65	24	....	....	40	24	44
126	39	18/39	65	24	....	....	48	24	44
127	39	18/39	65	24	....	....	56	24	44
128	16	5/16	61	....	....	....	....	....	....
129	39	18/39	65	24	....	....	72	24	44
130	39	12/39	60	....	....	....	....	....	....
131	20	9/20	58	40	....	....	28	44	....
132	33	10/33	59	....	....	....	....	....	....
133	21	6/21	56	24	....	....	48	44	....
134	21	6/21	56	28	....	....	48	44	....
135	27	8/27	58	....	....	....	....	....	....
136	17	5/17	57	....	....	....	....	....	....
137	21	6/21	56	28	....	....	24	56	....
138	21	6/21	56	56	....	....	32	44	....
139	21	6/21	56	56	32	48	24	....	....
140	49	14/49	55	....	....	....	....	....	....
141	18	5/18	54	48	....	....	40	44	....
142	21	6/21	56	56	....	....	32	24	44
143	21	6/21	56	28	....	....	24	24	44
144	18	5/18	54	....	....	....	....	....	....
145	29	8/29	54	....	....	....	....	....	....
146	21	6/21	56	28	....	....	48	24	44
147	21	6/21	56	24	....	....	48	24	44
148	37	10/37	53	....	....	....	....	....	....
149	21	6/21	56	28	....	....	72	24	44

Simple and Differential Indexing

No of Divisions	Index Circle	No. of Turns of Crank	Graduation on Sector	Gear on Worm	No. 1 Hole		Gear on Spindle	Idlers	
					First Gear on Stud	Second Gear on Stud		No. 1 Hole	No. 2 Hole
150	15	4/15	52	....	....	....	....	....	....
151	20	5/20	48	32	....	....	72	44	....
152	19	5/19	51	....	....	....	....	....	....
153	20	5/20	48	32	....	....	56	44	....
154	20	5/20	48	32	....	....	48	44	....
155	31	8/31	50	....	....	....	....	....	....
156	39	10/39	50	....	....	....	....	....	....
157	20	5/20	48	32	....	....	24	56	....
158	20	5/20	48	48	....	....	24	44	....
159	20	5/20	48	64	32	56	28	....	....
160	20	5/20	48	....	....	....	....	....	....
161	20	5/20	48	64	32	56	28	....	24
162	20	5/20	48	48	....	....	24	24	44
163	20	5/20	48	32	....	....	24	24	44
164	41	10/41	47	....	....	....	....	....	....
165	33	8/33	47	....	....	....	....	....	....
166	20	5/20	48	32	....	....	48	24	44
167	20	5/20	48	32	....	....	56	24	44
168	21	5/21	47	....	....	....	....	....	....
169	20	5/20	48	32	....	....	72	24	44
170	17	4/17	45	....	....	....	....	....	....
171	21	5/21	47	56	....	....	40	24	44
172	43	10/43	44	....	....	....	....	....	....
173	18	4/18	43	72	56	32	64	....	....
174	18	4/18	43	24	....	....	32	56	....
175	18	4/18	43	72	40	32	64	....	....
176	18	4/18	43	72	24	24	64	....	....
177	18	4/18	43	72	....	....	48	24	....
178	18	4/18	43	72	....	....	32	44	....
179	18	4/18	43	72	24	48	32	....	....
180	18	4/18	43	....	....	....	....	....	....
181	18	4/18	43	72	24	48	32	....	24
182	18	4/18	43	72	....	....	32	24	44
183	18	4/18	43	48	....	....	32	24	44
184	23	5/23	42	....	....	....	....	....	....
185	37	8/37	42	....	....	....	....	....	....
186	18	4/18	43	48	....	....	64	24	44
187	18	4/18	43	72	48	24	56	....	24
188	47	10/47	40	....	....	....	....	....	....
189	18	4/18	43	32	....	....	64	24	44
190	19	4/19	40	....	....	....	....	....	....
191	20	5/20	38	40	....	....	72	24	....
192	20	5/20	38	40	....	....	64	44	....

Simple and Differential Indexing

No. of Divisions	Index Circle	No. of Turns of Crank	Graduation on Sector	Gear on Worm	No. 1 Hole		Gear on Spindle	Idlers	
					First Gear on Stud	Second Gear on Stud		No. 1 Hole	No. 2 Hole
193	20	4/20	38	40	....	....	56	44	....
194	20	4/20	38	40	....	....	48	44	....
195	39	8/39	39	....	....	....	....	....	....
196	49	10/49	38	....	....	....	....	....	....
197	20	4/20	38	40	....	....	24	56	....
198	20	4/20	38	56	28	40	32	....	....
199	20	4/20	38	100	40	64	32	....	....
200	20	4/20	38	....	....	....	....	....	....
201	20	4/20	38	72	24	40	24	....	24
202	20	4/20	38	72	24	40	48	....	24
203	20	4/20	38	40	....	....	24	24	44
204	20	4/20	38	40	....	....	32	24	44
205	41	8/41	37	....	....	....	....	....	....
206	20	4/20	38	40	....	....	48	24	44
207	20	4/20	38	40	....	....	56	24	44
208	20	4/20	38	40	....	....	64	24	44
209	20	4/20	38	40	....	....	72	24	44
210	21	4/21	37	....	....	....	....	....	....
211	16	3/16	36	64	....	....	28	44	....
212	43	8/43	35	86	24	24	48	....	....
213	27	5/27	36	72	....	....	40	44	....
214	20	4/20	38	40	56	32	64	....	24
215	43	8/43	35	....	....	....	....	....	....
216	27	5/27	36	....	....	....	....	....	....
217	21	4/21	37	48	....	....	64	24	44
218	16	3/16	36	64	....	....	56	24	44
219	21	4/21	37	28	....	....	48	24	44
220	33	9/33	35	....	....	....	....	....	....
221	17	8/17	33	24	....	....	24	56	....
222	18	3/18	32	24	....	....	72	44	....
223	43	8/43	35	86	48	24	64	....	24
224	18	3/18	32	24	....	....	64	44	....
225	27	5/27	36	24	....	....	40	24	44
226	18	3/18	32	24	....	....	56	44	....
227	49	8/49	30	56	64	28	72	....	....
228	18	3/18	32	24	....	....	48	44	....
229	18	3/18	32	24	....	....	44	48	....
230	23	4/23	34	....	....	....	....	....	....
231	18	3/18	32	32	....	....	48	44	....
232	29	5/29	33	....	....	....	....	....	....
233	18	3/18	32	48	....	....	56	44	....
234	18	3/18	32	24	....	....	24	56	....
235	47	8/47	32	....	....	....	....	....	....

Simple and Differential Indexing

No. of Divisions	Index Circle	No. of Turns of Crank	Graduation on Sector	Gear on Worm	No. 1 Hole		Gear on Spindle	Idlers	
					First Gear on Stud	Second Gear on Stud		No. 1 Hole	No. 2 Hole
236	18	8/18	32	48	....	....	32	44	....
237	18	8/18	32	48	....	....	24	44	....
238	18	8/18	32	72	....	....	24	44	....
239	18	8/18	32	72	24	64	32	....	....
240	18	8/18	32	....	....	....	....	....	....
241	18	8/18	32	72	24	64	32	....	24
242	18	8/18	32	72	....	....	24	24	44
243	18	8/18	32	64	....	....	32	24	44
244	18	8/18	32	48	....	....	32	24	44
245	49	8/49	30	....	....	....	....	....	....
246	18	8/18	32	24	....	....	24	24	44
247	18	8/18	32	48	....	....	56	24	44
248	31	5/31	31	....	....	....	....	....	....
249	18	8/18	32	32	....	....	48	24	44
250	18	8/18	32	24	....	....	40	24	44
251	18	8/18	32	48	44	32	64	....	24
252	18	8/18	32	24	....	....	48	24	44
253	33	5/33	29	24	....	....	40	56	....
254	18	8/18	32	24	....	....	56	24	44
255	18	8/18	32	48	40	24	72	....	24
256	18	8/18	32	24	....	....	64	24	44
257	49	8/49	30	56	48	28	64	....	24
258	43	7/43	31	32	....	....	64	24	44
259	21	3/21	28	24	....	....	72	44	....
260	39	6/39	29	....	....	....	....	....	....
261	29	4/29	26	48	64	24	72	....	....
262	20	3/20	28	40	....	....	28	44	....
263	49	8/49	30	56	64	28	72	....	24
264	33	5/33	29	....	....	....	....	....	....
265	21	3/21	28	56	40	24	72	....	....
266	21	3/21	28	32	....	....	64	44	....
267	27	4/27	28	72	....	....	32	44	....
268	21	3/21	28	28	....	....	48	44	....
269	20	3/20	28	64	32	40	28	....	24
270	27	4/27	28	....	....	....	....	....	....
271	21	3/21	28	56	24	24	72	....	....
272	21	3/21	28	56	....	....	64	24	....
273	21	3/21	28	24	....	....	24	56	....
274	21	5/21	28	56	....	....	48	44	....
275	21	5/21	28	56	....	....	40	44	....
276	21	3/21	28	56	....	....	32	44	....
277	21	3/21	28	56	....	....	24	44	....
278	21	3/21	28	56	32	48	24	....	....

Simple and Differential Indexing

No. of Divisions	Index Circle	No. of Turns of Crank	Graduation on Sector	Gear on Worm	No. 1 Hole First Gear on Stud	No. 1 Hole Second Gear on Stud	Gear on Spindle	Idlers No. 1 Hole	Idlers No. 2 Hole
279	27	4/27	28	24	....	....	32	24	44
280	49	7/49	26	....	....	....	....	....	....
281	21	8/21	28	72	24	56	24	....	24
282	43	6/43	26	86	24	24	56	....	....
283	21	8/21	28	56	....	....	24	24	44
284	21	8/21	28	56	....	....	32	24	44
285	21	8/21	28	56	....	....	40	24	44
286	21	8/21	28	56	....	....	48	24	44
287	21	8/21	28	24	....	....	24	24	44
288	21	8/21	28	28	....	....	32	24	44
289	21	8/21	28	56	24	24	72	....	24
290	29	4/29	26	....	....	....	....	....	....
291	15	2/15	25	40	....	....	48	44	....
292	21	8/21	28	28	....	....	48	24	44
293	15	2/15	25	48	32	40	56	....	....
294	21	8/21	28	24	....	....	48	24	44
295	15	2/15	25	48	....	....	32	44	....
296	37	5/37	26	....	....	....	....	....	....
297	33	4/33	23	28	48	24	56	....	....
298	21	8/21	28	28	....	....	72	24	44
299	23	3/23	25	24	....	....	24	56	....
300	15	2/15	25	....	....	....	....	....	....
301	43	6/43	26	24	....	....	48	24	44
302	16	2/16	24	32	....	....	72	24	....
303	15	2/15	25	72	24	40	48	....	24
304	16	2/16	24	24	....	....	48	44	....
305	15	2/15	25	48	....	....	32	24	44
306	15	2/15	25	40	....	....	32	24	44
307	15	2/15	25	72	48	40	56	....	24
308	16	2/16	24	32	....	....	48	44	....
309	15	2/15	25	40	....	....	48	24	44
310	31	4/31	24	....	....	....	....	....	....
311	16	2/16	24	64	24	24	72	....	....
312	39	5/39	24	....	....	....	....	....	....
313	16	2/16	24	32	....	....	28	56	....
314	16	2/16	24	32	....	....	24	56	....
315	16	2/16	24	64	....	....	40	24	....
316	16	2/16	24	64	....	....	32	44	....
317	16	2/16	24	64	....	....	24	44	....
318	16	2/16	24	56	28	48	24	....	....
319	29	4/29	26	48	64	24	72	....	24
320	16	2/16	24	....	....	....	....	....	....
321	16	2/16	24	72	24	64	24	....	24

Simple and Differential Indexing

No. of Divisions	Index Circle	No. of Turns of Crank	Graduation on Sector	Gear on Worm	No. 1 Hole		Gear on Spindle	Idlers	
					First Gear on Stud	Second Gear on Stud		No. 1 Hole	No. 2 Hole
322	23	9/23	25	32	....	....	64	24	44
323	16	2/16	24	64	....	....	24	24	44
324	16	2/16	24	64	....	....	32	24	44
325	16	2/16	24	64	....	....	40	24	44
326	16	2/16	24	32	....	....	24	24	44
327	16	2/16	24	32	....	....	28	24	44
328	41	5/41	23	....	....	....	....	....	....
329	16	2/16	24	64	24	24	72	....	24
330	33	4/33	23	....	....	....	....	....	....
331	16	2/16	24	64	44	24	48	....	24
332	16	2/16	24	32	....	....	48	24	44
333	18	2/18	21	24	....	....	72	44	....
334	16	2/16	24	32	....	....	56	24	44
335	33	4/33	23	72	48	44	40	....	24
336	16	2/16	24	32	....	....	64	24	44
337	43	5/43	21	86	40	32	56	....	....
338	16	2/16	24	32	....	....	72	24	44
339	18	2/18	21	24	....	....	56	44	....
340	17	2/17	22	....	....	....	....	....	....
341	43	5/43	21	86	24	32	40	....	....
342	18	2/18	21	32	....	....	64	44	....
343	15	3/15	25	40	64	24	86	....	24
344	43	5/43	21	....	....	....	....	....	....
345	18	2/18	21	24	....	....	40	56	....
346	18	2/18	21	72	56	32	64	....	....
347	43	5/43	21	86	24	32	40	....	24
348	18	2/18	21	24	....	....	32	56	....
349	18	2/18	21	72	44	24	48	....	....
350	18	2/18	21	72	40	32	64	....	....
351	18	2/18	21	24	....	....	24	56	....
352	18	2/18	21	72	24	24	64	....	....
353	18	2/18	21	72	24	24	56	....	....
354	18	2/18	21	72	....	....	48	24	....
355	18	2/18	21	72	....	....	40	24	....
356	18	2/18	21	72	....	....	32	24	....
357	18	2/18	21	72	....	....	24	44	....
358	18	2/18	21	72	32	48	24	....	....
359	43	5/43	21	86	48	32	100	....	24
360	18	2/18	21	....	....	....	....	....	....
361	19	2/19	19	32	....	....	64	44	....
362	18	2/18	21	72	28	56	32	....	24
363	18	2/18	21	72	....	....	24	24	44
364	18	2/18	21	72	....	....	32	24	44

Indexing Movements for Standard Index Plate — Cincinnati Milling Machine

The standard index plate indexes all numbers up to and including 60; all even numbers and those divisible by 5 up to 120; and all divisions listed below up to 400. This plate is drilled on both sides, and has holes as follows:

First side: 24, 25, 28, 30, 34, 37, 38, 39, 41, 42, 43.
Second side: 46, 47, 49, 51, 53, 54, 57, 58, 59, 62, 66.

No. of Divisions	Circle	Turns	Holes	No. of Divisions	Circle	Holes	No. of Divisions	Circle	Holes	No. of Divisions	Circle	Holes
2	Any	20	...	44	66	60	104	39	15	205	41	8
3	24	13	8	45	54	48	105	42	16	210	42	8
4	Any	10	...	46	46	40	106	53	20	212	53	10
5	Any	8	...	47	47	40	108	54	20	215	43	8
6	24	6	16	48	24	20	110	66	24	216	54	10
7	28	5	20	49	49	40	112	28	10	220	66	12
8	Any	5	...	50	25	20	114	57	20	224	28	5
9	54	4	24	51	51	40	115	46	16	228	57	10
10	Any	4	...	52	39	30	116	58	20	230	46	8
11	66	3	42	53	53	40	118	59	20	232	58	10
12	24	3	8	54	54	40	120	66	22	235	47	8
13	39	3	3	55	66	48	124	62	20	236	59	10
14	49	2	42	56	28	20	125	25	8	240	66	11
15	24	2	16	57	57	40	130	39	12	245	49	8
16	24	2	12	58	58	40	132	66	20	248	62	10
17	34	2	12	59	59	40	135	54	16	250	25	4
18	54	2	12	60	42	28	136	34	10	255	51	8
19	38	2	4	62	62	40	140	28	8	260	39	6
20	Any	2	...	64	24	15	144	54	15	264	66	10
21	42	1	38	65	39	24	145	58	16	270	54	8
22	66	1	54	66	66	40	148	37	10	272	34	5
23	46	1	34	68	34	20	150	30	8	280	28	4
24	24	1	16	70	28	16	152	38	10	290	58	8
25	25	1	15	72	54	30	155	62	16	296	37	5
26	39	1	21	74	37	20	156	39	10	300	30	4
27	54	1	26	75	30	16	160	28	7	304	38	5
28	42	1	18	76	38	20	164	41	10	310	62	8
29	58	1	22	78	39	20	165	66	16	312	39	5
30	24	1	8	80	34	17	168	42	10	320	24	3
31	62	1	18	82	41	20	170	34	8	328	41	5
32	28	1	7	84	42	20	172	43	10	330	66	8
33	66	1	14	85	34	16	176	66	15	336	42	5
34	34	1	6	86	43	20	180	54	12	340	34	4
35	28	1	4	88	66	30	184	46	10	344	43	5
36	54	1	6	90	54	24	185	37	8	360	54	6
37	37	1	3	92	46	20	188	47	10	368	46	5
38	38	1	2	94	47	20	190	38	8	370	37	4
39	39	1	1	95	38	16	192	24	5	376	47	5
40	Any	1	...	96	24	10	195	39	8	380	38	4
41	41	...	40	98	49	20	196	49	10	390	39	4
42	42	...	40	100	25	10	200	30	6	392	49	5
43	43	...	40	102	51	20	204	51	10	400	30	3

Indexing Movements for High Numbers — Cincinnati Milling Machine

This set of 3 index plates indexes all numbers up to and including 200; all even numbers and those divisible by 5 up to and including 400. The plates are drilled on each side, making six sides *A*, *B*, *C*, *D*, *E* and *F*.

Example: — It is required to index 35 divisions. The preferred side is *F*, since this requires the least number of holes; but should one of plates *D*, *A* or *E* be in place, either can be used, thus avoiding the changing of plates.

No. of Divisions	Side	Circle	Turns	Holes	No. of Divisions	Side	Circle	Turns	Holes	No. of Divisions	Side	Circle	Turns	Holes
2	Any	Any	20	....	15	C	93	2	62	28	D	77	1	33
3	A	30	13	10	15	F	159	2	106	28	A	91	1	39
3	B	36	13	12	16	E	26	2	13	29	E	87	1	33
3	E	42	13	14	16	F	28	2	14	30	A	30	1	10
3	C	93	13	31	16	A	30	2	15	30	B	36	1	12
3	F	159	13	53	16	D	32	2	16	30	E	42	1	14
4	Any	Any	10	....	16	C	34	2	17	30	C	93	1	31
5	Any	Any	8	....	16	B	36	2	18	30	F	159	1	53
6	A	30	6	20	17	C	34	2	12	31	C	93	1	27
6	B	36	6	24	17	E	119	2	42	32	F	28	1	7
6	E	42	6	28	17	C	153	2	54	32	D	32	1	8
6	C	93	6	62	17	F	187	2	66	32	B	36	1	9
6	F	159	6	106	18	B	36	2	8	32	A	48	1	12
7	F	28	5	20	18	A	99	2	22	33	A	99	1	21
7	E	42	5	30	18	C	153	2	34	34	C	34	1	6
7	D	77	5	55	19	F	38	2	4	34	E	119	1	21
7	A	91	5	65	19	E	133	2	14	34	F	187	1	33
8	Any	Any	5	....	19	A	171	2	18	35	F	28	1	4
9	B	36	4	16	20	Any	Any	2	....	35	D	77	1	11
9	A	99	4	44	21	E	42	1	38	35	A	91	1	13
9	C	153	4	68	21	A	147	1	133	35	E	119	1	17
10	Any	Any	4	....	22	D	44	1	36	36	B	36	1	4
11	D	44	3	28	22	A	99	1	81	36	A	99	1	11
11	A	99	3	63	22	F	143	1	117	36	C	153	1	17
11	F	143	3	91	23	C	46	1	34	37	B	111	1	9
12	A	30	3	10	23	A	69	1	51	38	F	38	1	2
12	B	36	3	12	23	E	161	1	119	38	E	133	1	7
12	E	42	3	14	24	A	30	1	20	38	A	171	1	9
12	C	93	3	31	24	B	36	1	24	39	A	117	1	3
12	F	159	3	53	24	E	42	1	28	40	Any	Any	1	....
13	E	26	3	2	24	C	93	1	62	41	C	123	....	120
13	A	91	3	7	24	F	159	1	106	42	E	42	....	40
13	F	143	3	11	25	A	30	1	18	42	A	147	....	140
13	B	169	3	13	25	E	175	1	105	43	A	129	....	120
14	F	28	2	24	26	F	26	1	14	44	D	44	....	40
14	E	42	2	36	26	A	91	1	49	44	A	99	....	90
14	D	77	2	66	26	B	169	1	91	44	F	143	....	130
14	A	91	2	78	27	B	81	1	39	45	B	36	....	32
15	A	30	2	20	27	A	189	1	91	45	A	99	....	88
15	B	36	2	24	28	F	28	1	12	45	C	153	....	136
15	E	42	2	28	28	E	42	1	18	46	C	46	....	40

Indexing Movements for High Numbers — Cincinnati Milling Machine

No. of Divisions	Side	Circle	Holes	No. of Divisions	Side	Circle	Holes	No. of Divisions	Side	Circle	Holes
46	A	69	60	70	E	119	68	96	B	36	15
46	E	161	140	71	F	71	40	96	A	48	20
47	B	141	120	72	B	36	20	97	B	97	40
48	A	30	25	72	A	117	65	98	A	147	60
48	B	36	30	72	C	153	85	99	A	99	40
49	A	147	120	73	E	73	40	100	A	30	12
50	A	30	24	74	B	111	60	100	E	175	70
50	E	175	140	75	A	30	16	101	F	101	40
51	C	153	120	76	F	38	20	102	C	153	60
52	E	26	20	76	E	133	70	103	E	103	40
52	A	91	70	76	A	171	90	104	E	26	10
52	F	143	110	77	D	77	40	104	A	91	35
52	B	169	130	78	A	117	60	104	F	143	55
53	F	159	120	79	C	79	40	104	B	169	65
54	B	81	60	80	E	26	13	105	E	42	16
54	A	189	140	80	F	28	14	105	A	147	56
55	D	44	32	80	A	30	15	106	F	159	60
55	F	143	104	80	D	32	16	107	D	107	40
56	F	28	20	80	C	34	17	108	B	81	30
56	E	42	30	80	B	36	18	108	A	189	70
56	D	77	55	80	E	42	21	109	C	109	40
56	A	91	65	81	B	81	40	110	D	44	16
57	A	171	120	82	C	123	60	110	A	99	36
58	E	87	60	83	F	83	40	110	F	143	52
59	A	177	120	84	E	42	20	111	B	111	40
60	A	30	20	84	A	147	70	112	F	28	10
60	B	36	24	85	C	34	16	112	E	42	15
60	E	42	28	85	E	119	56	113	F	113	40
60	F	159	106	85	F	187	88	114	A	171	60
61	B	183	120	86	A	129	60	115	C	46	16
62	C	93	60	87	E	87	40	115	A	69	24
63	A	189	120	88	D	44	20	115	E	161	56
64	D	32	20	88	A	99	45	116	E	87	30
64	A	48	30	88	F	143	65	117	A	117	40
65	E	26	16	89	D	89	40	118	A	177	60
65	A	91	56	90	B	36	16	119	E	119	40
65	F	143	88	90	A	99	44	120	A	30	10
65	B	169	104	90	C	153	68	120	B	36	12
66	A	99	60	91	A	91	40	120	E	42	14
67	B	67	40	92	C	46	20	120	C	93	31
68	C	34	20	92	A	69	30	120	F	159	53
68	E	119	70	92	E	161	70	121	D	121	40
68	F	187	110	93	C	93	40	122	B	183	60
69	A	69	40	94	B	141	60	123	C	123	40
70	F	28	16	95	F	38	16	124	C	93	30
70	D	42	24	95	E	133	56	125	E	175	56
70	A	91	52	95	A	171	72	126	A	189	60

Indexing Movements for High Numbers — Cincinnati Milling Machine

No. of Divisions	Side	Circle	Holes	No. of Divisions	Side	Circle	Holes	No. of Divisions	Side	Circle	Holes
127	B	127	40	160	A	48	12	198	A	99	20
128	D	32	10	161	E	161	40	199	B	199	40
128	A	48	15	162	B	81	20	200	A	30	6
129	A	129	40	163	D	163	40	200	E	175	35
130	E	26	8	164	C	123	30	202	F	101	20
130	A	91	28	165	A	99	24	204	C	153	30
130	F	143	44	166	F	83	20	205	C	123	24
130	B	169	52	167	C	167	40	206	E	103	20
131	F	131	40	168	E	42	10	208	E	26	5
132	A	99	30	168	A	147	35	210	E	42	8
133	E	133	40	169	B	169	40	210	A	147	28
134	B	67	20	170	C	34	8	212	F	159	30
135	B	81	24	170	E	119	28	214	D	107	20
135	A	189	56	170	F	187	44	215	A	129	24
136	C	34	10	171	A	171	40	216	B	81	15
136	E	119	35	172	A	129	30	216	A	189	35
137	D	137	40	173	F	173	40	218	C	109	20
138	A	69	20	174	E	87	20	220	D	44	8
139	C	139	40	175	E	175	40	220	A	99	18
140	F	28	8	176	D	44	10	220	F	143	26
140	E	42	12	177	A	177	40	222	B	111	20
140	D	77	22	178	D	89	20	224	F	28	5
140	A	91	26	179	D	179	40	226	F	113	20
141	B	141	40	180	B	36	8	228	A	171	30
142	F	71	20	180	A	99	22	230	C	46	8
143	F	143	40	180	C	153	34	230	A	69	12
144	B	36	10	181	C	181	40	230	E	161	28
145	E	87	24	182	A	91	20	232	E	87	15
146	E	73	20	183	B	183	40	234	A	117	20
147	A	147	40	184	C	46	10	235	B	141	24
148	B	111	30	184	A	69	15	236	A	177	30
149	E	149	40	184	E	161	35	238	E	119	20
150	A	30	8	185	B	111	24	240	A	30	5
151	D	151	40	186	C	93	20	240	B	36	6
152	F	38	10	187	F	187	40	240	E	42	7
152	E	133	35	188	B	141	30	240	A	48	8
152	A	171	45	189	A	189	40	242	D	121	20
153	C	153	40	190	F	38	8	244	B	183	30
154	D	77	20	190	E	133	28	245	A	147	24
155	C	93	24	190	A	171	36	246	C	123	20
156	A	117	30	191	E	191	40	248	C	93	15
157	B	157	40	192	A	48	10	250	E	175	28
158	C	79	20	193	D	193	40	252	A	189	30
159	F	159	40	194	B	97	20	254	B	127	20
160	F	28	7	195	A	117	24	255	C	153	24
160	D	32	8	196	A	147	30	256	D	32	5
160	B	36	9	197	C	197	40	258	A	129	20

Indexing Movements for High Numbers — Cincinnati Milling Machine

No. of Divisions	Side	Circle	Holes	No. of Divisions	Side	Circle	Holes	No. of Divisions	Side	Circle	Holes
260	E	26	4	304	F	38	5	354	A	177	20
260	A	91	14	305	B	183	24	355	F	71	8
260	F	143	22	306	C	153	20	356	D	89	10
260	B	169	26	308	D	77	10	358	D	179	20
262	F	131	20	310	C	93	12	360	B	36	4
264	A	99	15	312	A	117	15	360	A	99	11
265	F	159	24	314	B	157	20	360	C	153	17
266	E	133	20	315	A	189	24	362	C	181	20
268	B	67	10	316	C	79	10	364	A	91	10
270	B	81	12	318	F	159	20	365	E	73	8
270	A	189	28	320	D	32	4	366	B	183	20
272	C	34	5	320	A	48	6	368	C	46	5
274	D	137	20	322	E	161	20	370	B	111	12
276	A	69	10	324	B	81	10	372	C	93	10
278	C	139	20	326	D	163	20	374	F	187	20
280	F	28	4	328	C	123	15	376	B	141	15
280	E	42	6	330	A	99	12	378	A	189	20
280	D	77	11	332	F	83	10	380	F	38	4
280	A	91	13	334	C	167	20	380	E	133	14
282	B	141	20	335	B	67	8	380	A	171	18
284	F	71	10	336	E	42	5	382	E	191	20
285	A	171	24	338	B	169	20	384	A	48	5
286	F	143	20	340	C	34	4	385	D	77	8
288	B	36	5	340	E	119	14	386	D	193	20
290	E	87	12	340	F	187	22	388	B	97	10
292	E	73	10	342	A	171	20	390	A	117	12
294	A	147	20	344	A	129	15	392	A	147	15
295	A	177	24	345	A	69	8	394	C	197	20
296	B	111	15	346	F	173	20	395	C	79	8
298	E	149	20	348	E	87	10	396	A	99	10
300	A	30	4	350	E	175	20	398	B	199	20
302	D	151	20	352	D	44	5	400	A	30	3

Angular Indexing. — With the ordinary indexing head, in which 40 turns of the index crank are required for one revolution of the work, one turn of the index crank equals 9 degrees. Hence, when one complete turn of the index crank equals 9 degrees, two holes in the 18-hole circle, or 3 holes in the 27-hole circle, must correspond to one degree. The first principle or rule for indexing for angles is therefore that two holes in the 18-hole circle or 3 holes in the 27-hole circle equals a movement of one degree of the index head spindle and the work.

Assume that an indexing movement of 35 degrees is required. One complete turn of the index crank equals 9 degrees; therefore, first divide the number of degrees for which to index, by 9, in order to find how many complete turns the index crank should make. The number of degrees left to turn after having completed the full turns are indexed by taking two holes in the 18-hole circle for each degree. In

this case, $\frac{35}{9} = 3\frac{8}{9}$, which indicates that the index crank must be turned three full revolutions, and then 8 degrees more are indexed by moving 16 holes in the 18-hole circle.

To index for 11½ degrees, for example, first turn the index crank one revolution, this being a 9-degree movement. Then to index 2½ degrees, move the index crank 5 holes in the 18-hole circle (4 holes for the two whole degrees and one hole for the ½ degree equals the total movement of 5 holes).

Below is shown how this calculation may be carried out to plainly indicate the movement required for this angle:

11½ deg. = 9 deg. + 2 deg. + ½ deg.

1 turn + 4 holes + 1 hole in the 18-hole circle.

Should it be required to index only ⅓ degree, this may be done by using the 27-hole circle. In this circle a three-hole movement equals one degree, and a one-hole movement in that circle thus equals ⅓ degree, or 20 minutes. Assume that it is required to index the work through an angle of 48 degrees 40 minutes. Below is plainly shown how this calculation may be carried out:

48 deg. 40 min. = 45 deg. + 3 deg. + 40 min.

5 turns + 9 holes + 2 holes in the 27-hole circle.

Angular Values of One-Hole Moves — B. & S. Index Plates

15-hole circle = 36 minutes	29-hole circle = 18.621 minutes
16-hole circle = 33.750 minutes	31-hole circle = 17.419 minutes
17-hole circle = 31.765 minutes	33-hole circle = 16.364 minutes
18-hole circle = 30 minutes	37-hole circle = 14.595 minutes
19-hole circle = 28.421 minutes	39-hole circle = 13.846 minutes
20-hole circle = 27 minutes	41-hole circle = 13.171 minutes
21-hole circle = 25.714 minutes	43-hole circle = 12.558 minutes
23-hole circle = 23.478 minutes	47-hole circle = 11.489 minutes
27-hole circle = 20 minutes	49-hole circle = 11.020 minutes

Approximate Indexing for Angles.— The following general rule for *approximate* indexing of small angles is applicable to any index head requiring 40 revolutions of the index crank for one revolution of the work.

Rule: Divide 540 by the total number of minutes to be indexed. If the quotient is approximately equal to the number of holes in any index circle available, the angular movement is obtained by moving the crank one hole in this index circle; but if the quotient is not approximately equal, multiply it by any trial number which will give a product equal to the number of holes in an available index circle and move the index crank as many holes as are indicated by the trial number. (If the quotient of 540 divided by the total number of minutes is greater than the number of holes in any of the index circles, it is not possible to obtain the required movement for the angle by simple indexing.)

Example: — Assume that it is required to index to an angle of 2 degrees 46 minutes. Changing this to minutes gives a total of 166 minutes. Dividing 540 by 166 we have 540 ÷ 166 = 3.253. This quotient is next multiplied by some trial number to obtain a product which equals the number of holes in an available index circle. Multiplying by 12, we have 3.253 × 12 = 39.036. Therefore, for indexing 2 degrees 46 minutes, the 39-hole circle can be used and the index crank would be moved 12 holes.

Tables for Angular Indexing. — The table, "Angular Indexing," gives the number of turns of the index crank for indexing various angles. In the column headed, "Turns of Index Crank," the whole number (where given) indicates the number of full revolutions; the numerator of the fraction, the number of holes additional; and the denominator, the number of holes in the index circle to be used. The angular movement obtained for a movement of one hole, in various index plates is given in the table, "Angular Values of One-Hole Moves."

Angular Indexing

Angle in Degs.	Turns of Index Crank	Angle in Degs.	Turns of Index Crank	Angle in Degs.	Turns of Index Crank	Angle in Degs.	Turns of Index Crank	Angle in Degs.	Turns of Index Crank
1	2/18	10	1 2/18	19	2 2/18	28	3 2/18	37	4 2/18
1⅓	4/27	10⅓	1 4/27	19⅓	2 4/27	28⅓	3 4/27	37⅓	4 4/27
1½	3/18	10½	1 3/18	19½	2 3/18	28½	3 3/18	37½	4 3/18
1⅔	5/27	10⅔	1 5/27	19⅔	2 5/27	28⅔	3 5/27	37⅔	4 5/27
2	4/18	11	1 4/18	20	2 4/18	29	3 4/18	38	4 4/18
2⅓	7/27	11⅓	1 7/27	20⅓	2 7/27	29⅓	3 7/27	38⅓	4 7/27
2½	5/18	11½	1 5/18	20½	2 5/18	29½	3 5/18	38½	4 5/18
2⅔	8/27	11⅔	1 8/27	20⅔	2 8/27	29⅔	3 8/27	38⅔	4 8/27
3	6/18	12	1 6/18	21	2 6/18	30	3 6/18	39	4 6/18
3⅓	10/27	12⅓	1 10/27	21⅓	2 10/27	30⅓	3 10/27	39⅓	4 10/27
3½	7/18	12½	1 7/18	21½	2 7/18	30½	3 7/18	39½	4 7/18
3⅔	11/27	12⅔	1 11/27	21⅔	2 11/27	30⅔	3 11/27	39⅔	4 11/27
4	8/18	13	1 8/18	22	2 8/18	31	3 8/18	40	4 8/18
4⅓	13/27	13⅓	1 13/27	22⅓	2 13/27	31⅓	3 13/27	40⅓	4 13/27
4½	9/18	13½	1 9/18	22½	2 9/18	31½	3 9/18	40½	4 9/18
4⅔	14/27	13⅔	1 14/27	22⅔	2 14/27	31⅔	3 14/27	40⅔	4 14/27
5	10/18	14	1 10/18	23	2 10/18	32	3 10/18	41	4 10/18
5⅓	16/27	14⅓	1 16/27	23⅓	2 16/27	32⅓	3 16/27	41⅓	4 16/27
5½	11/18	14½	1 11/18	23½	2 11/18	32½	3 11/18	41½	4 11/18
5⅔	17/27	14⅔	1 17/27	23⅔	2 17/27	32⅔	3 17/27	41⅔	4 17/27
6	12/18	15	1 12/18	24	2 12/18	33	3 12/18	42	4 12/18
6⅓	19/27	15⅓	1 19/27	24⅓	2 19/27	33⅓	3 19/27	42⅓	4 19/27
6½	13/18	15½	1 13/18	24½	2 13/18	33½	3 13/18	42½	4 13/18
6⅔	20/27	15⅔	1 20/27	24⅔	2 20/27	33⅔	3 20/27	42⅔	4 20/27
7	14/18	16	1 14/18	25	2 14/18	34	3 14/18	43	4 14/18
7⅓	22/27	16⅓	1 22/27	25⅓	2 22/27	34⅓	3 22/27	43⅓	4 22/27
7½	15/18	16½	1 15/18	25½	2 15/18	34½	3 15/18	43½	4 15/18
7⅔	23/27	16⅔	1 23/27	25⅔	2 23/27	34⅔	3 23/27	43⅔	4 23/27
8	16/18	17	1 16/18	26	2 16/18	35	3 16/18	44	4 16/18
8⅓	25/27	17⅓	1 25/27	26⅓	2 25/27	35⅓	3 25/27	44⅓	4 25/27
8½	17/18	17½	1 17/18	26½	2 17/18	35½	3 17/18	44½	4 17/18
8⅔	26/27	17⅔	1 26/27	26⅔	2 26/27	35⅔	3 26/27	44⅔	4 26/27
9	1	18	2	27	3	36	4	45	5
9⅓	1 1/27	18⅓	2 1/27	27⅓	3 1/27	36⅓	4 1/27	45⅓	5 1/27
9½	1 1/18	18½	2 1/18	27½	3 1/18	36½	4 1/18	45½	5 1/18
9⅔	1 2/27	18⅔	2 2/27	27⅔	3 2/27	36⅔	4 2/27	45⅔	5 2/27

Accurate Angular Indexing Movements — 1 *

Fractional Indexing Movement	B. & S., Becker, Hendey, K. & T. and Rockford	Cincinnati and LeBlond *	Fractional Indexing Movement	B. & S., Becker, Hendey, K. & T. and Rockford	Cincinnati and LeBlond *	Fractional Indexing Movement	B. & S., Becker, Hendey, K. & T. and Rockford	Cincinnati and LeBlond *
0.0152	.....	1/66	0.0541	2/37	2/37	0.1000	2/20	3/30
0.0161	.....	1/62	0.0556	1/18	3/54	0.1017	.....	6/59
0.0169	.....	1/59	0.0566	.....	3/53	0.1020	5/49	5/49
0.0172	.....	1/58	0.0588	1/17	2/34	0.1026	4/39	4/39
0.0175	.....	1/57	0.0588	.....	3/51	0.1034	3/29	6/58
0.0185	.....	1/54	0.0606	2/33	4/66	0.1053	2/19	4/38
0.0189	.....	1/53	0.0612	3/49	3/49	0.1053	.....	6/57
0.0196	.....	1/51	0.0625	1/16	.....	0.1061	.....	7/66
0.0204	1/49	1/49	0.0638	3/47	3/47	0.1064	5/47	5/47
0.0213	1/47	1/47	0.0645	2/31	4/62	0.1071	.....	3/28
0.0217	.....	1/46	0.0652	.....	3/46	0.1081	4/37	4/37
0.0233	1/43	1/43	0.0667	.....	2/30	0.1087	.....	5/46
0.0238	.....	1/42	0.0678	.....	4/59	0.1111	.....	2/18
0.0244	1/41	1/41	0.0690	2/29	4/58	0.1111	3/27	6/54
0.0256	1/39	1/39	0.0698	3/43	3/43	0.1129	.....	7/62
0.0263	.....	1/38	0.0702	.....	4/57	0.1132	.....	6/53
0.0270	1/37	1/37	0.0714	.....	2/28	0.1163	5/43	5/43
0.0294	.....	1/34	0.0714	.....	3/42	0.1176	2/17	4/34
0.0303	1/33	2/66	0.0732	3/41	3/41	0.1176	.....	6/51
0.0323	1/31	2/62	0.0741	2/27	4/54	0.1186	.....	7/59
0.0333	.....	1/30	0.0755	.....	4/53	0.1190	.....	5/42
0.0338	.....	2/59	0.0758	.....	5/66	0.1200	.....	3/25
0.0345	1/29	2/58	0.0769	3/39	3/39	0.1207	.....	7/58
0.0351	.....	2/57	0.0784	.....	4/51	0.1212	4/33	8/66
0.0357	.....	1/28	0.0789	.....	3/38	0.1220	5/41	5/41
0.0370	1/27	2/54	0.0800	.....	2/25	0.1224	6/49	6/49
0.0377	.....	2/53	0.0806	.....	5/62	0.1228	.....	7/57
0.0392	.....	2/51	0.0811	3/37	3/37	0.1250	2/16	3/24
0.0400	.....	1/25	0.0816	4/49	4/49	0.1277	6/47	6/47
0.0408	2/49	2/49	0.0833	.....	2/24	0.1282	5/39	5/39
0.0417	.....	1/24	0.0847	.....	5/59	0.1290	4/31	8/62
0.0426	2/47	2/47	0.0851	4/47	4/47	0.1296	.....	7/54
0.0435	1/23	2/46	0.0862	.....	5/58	0.1304	3/23	6/46
0.0454	.....	3/66	0.0870	2/23	4/46	0.1316	.....	5/38
0.0465	2/43	2/43	0.0877	.....	5/57	0.1321	.....	7/53
0.0476	1/21	2/42	0.0882	.....	3/34	0.1333	2/15	4/30
0.0484	.....	3/62	0.0909	3/33	6/66	0.1351	5/37	5/37
0.0488	2/41	2/41	0.0926	.....	5/54	0.1356	.....	8/59
0.0500	1/20	.....	0.0930	4/43	4/43	0.1364	.....	9/66
0.0508	.....	3/59	0.0943	.....	5/53	0.1372	.....	7/51
0.0513	2/39	2/39	0.0952	2/21	4/42	0.1379	4/29	8/58
0.0517	.....	3/58	0.0968	3/31	6/62	0.1395	6/43	6/43
0.0526	1/19	2/38	0.0976	4/41	4/41	0.1404	.....	8/57
0.0526	.....	3/57	0.0980	.....	5/51	0.1429	.....	4/28

* See explanatory note below Table 8.

Accurate Angular Indexing Movements — 2 *

Fractional Indexing Movement	B. & S., Becker, Hendey, K. & T. and Rockford	Cincinnati and LeBlond *	Fractional Indexing Movement	B. & S., Becker, Hendey, K. & T. and Rockford	Cincinnati and LeBlond *	Fractional Indexing Movement	B. & S., Becker, Hendey, K. & T. and Rockford	Cincinnati and LeBlond *
0.1429	3/21	6/42	0.1864		11/59	0.2308	9/39	9/39
0.1429	7/49	7/49	0.1875	3/16		0.2326	10/43	10/43
0.1452		9/62	0.1887		10/53	0.2333		7/30
0.1463	6/41	6/41	0.1892	7/37	7/37	0.2340	11/47	11/47
0.1471		5/34	0.1897		11/58	0.2353	4/17	8/34
0.1481	4/27	8/54	0.1905	4/21	8/42	0.2353		12/51
0.1489	7/47	7/47	0.1915	9/47	9/47	0.2368		9/38
0.1500	3/20		0.1930		11/57	0.2373		14/59
0.1509		8/53	0.1935	6/31	12/62	0.2381	5/21	10/42
0.1515	5/33	10/66	0.1951	8/41	8/41	0.2391		11/46
0.1522		7/46	0.1957		9/46	0.2400		6/25
0.1525		9/59	0.1961		10/51	0.2407		13/54
0.1538	6/39	6/39	0.1970		13/66	0.2414	7/29	14/58
0.1552		9/58	0.2000	3/15	5/25	0.2419		15/62
0.1569		8/51	0.2000	4/20	6/30	0.2424	8/33	16/66
0.1579	3/19	6/38	0.2034		12/59	0.2432	9/37	9/37
0.1579		9/57	0.2037		11/54	0.2439	10/41	10/41
0.1600		4/25	0.2041	10/49	10/49	0.2449	12/49	12/49
0.1613	5/31	10/62	0.2051	8/39	8/39	0.2453		13/53
0.1622	6/37	6/37	0.2059		7/34	0.2456		14/57
0.1628	7/43	7/43	0.2069	6/29	12/58	0.2500	4/16	6/24
0.1633	8/49	8/49	0.2075		11/53	0.2500	5/20	7/28
0.1667	3/18	11/66	0.2083		5/24	0.2542		15/59
0.1667		9/54	0.2093	9/43	9/43	0.2549		13/51
0.1667		7/42	0.2097		13/62	0.2553	12/47	12/47
0.1667		5/30	0.2105	4/19	8/38	0.2558	11/43	11/43
0.1667		4/24	0.2105		12/57	0.2564	10/39	10/39
0.1695		10/59	0.2121	7/33	14/66	0.2576		17/66
0.1698		9/53	0.2128	10/47	10/47	0.2581	8/31	16/62
0.1702	8/47	8/47	0.2143		6/28	0.2586		15/58
0.1707	7/41	7/41	0.2143		9/42	0.2593	7/27	14/54
0.1724	5/29	10/58	0.2157		11/51	0.2609	6/23	12/46
0.1739	4/23	8/46	0.2162	8/37	8/37	0.2619		11/42
0.1754		10/57	0.2174	5/23	10/46	0.2632	5/19	10/38
0.1765	3/17	6/34	0.2195	9/41	9/41	0.2632		15/57
0.1765		9/51	0.2203		13/59	0.2642		14/53
0.1774		11/62	0.2222	4/18		0.2647		9/34
0.1786		5/28	0.2222	6/27	12/54	0.2653	13/49	13/49
0.1795	7/39	7/39	0.2241		13/58	0.2667	4/15	8/30
0.1818	6/33	12/66	0.2245	11/49	11/49	0.2683	11/41	11/41
0.1839	9/49	9/49	0.2258	7/31	14/62	0.2703	10/37	10/37
0.1842		7/38	0.2264		12/53	0.2712		16/59
0.1852	5/27	10/54	0.2273		15/66	0.2727	9/33	18/66
0.1860	8/43	8/43	0.2281		13/57	0.2742		17/62

* See explanatory note below Table 8.

Accurate Angular Indexing Movements — 3 *

Fractional Indexing Movement	B. & S., Becker, Hendey, K. & T. and Rockford	Cincinnati and LeBlond *	Fractional Indexing Movement	B. & S., Becker, Hendey, K. & T. and Rockford	Cincinnati and LeBlond *	Fractional Indexing Movement	B. & S., Becker, Hendey, K. & T. and Rockford	Cincinnati and LeBlond *
0.2745	.....	14/51	0.3191	15/47	15/47	0.3617	17/47	17/47
0.2759	8/29	16/58	0.3200	.....	8/25	0.3621	.....	21/58
0.2766	13/47	13/47	0.3208	.....	17/53	0.3636	12/33	24/66
0.2778	5/18	15/54	0.3214	.....	9/28	0.3659	15/41	15/41
0.2791	12/43	12/43	0.3220	.....	19/59	0.3667	.....	11/30
0.2800	.....	7/25	0.3226	10/31	20/62	0.3673	18/49	18/49
0.2807	.....	16/57	0.3235	.....	11/34	0.3684	7/19	14/38
0.2821	11/39	11/39	0.3243	12/37	12/37	0.3684	.....	21/57
0.2826	.....	13/46	0.3256	14/43	14/43	0.3696	.....	17/46
0.2830	.....	15/53	0.3261	.....	15/46	0.3704	10/27	20/54
0.2857	.....	8/28	0.3265	16/49	16/49	0.3710	.....	23/62
0.2857	14/49	14/49	0.3276	.....	19/58	0.3721	16/43	16/43
0.2857	6/21	12/42	0.3333	6/18	8/24	0.3725	.....	19/51
0.2879	.....	19/66	0.3333	5/15	10/30	0.3729	.....	22/59
0.2881	.....	17/59	0.3333	13/39	13/39	0.3750	6/16	9/24
0.2895	.....	11/38	0.3333	7/21	14/42	0.3774	.....	20/53
0.2903	9/31	18/62	0.3333	.....	17/51	0.3784	14/37	14/37
0.2917	.....	7/24	0.3333	9/27	18/54	0.3788	.....	25/66
0.2927	12/41	12/41	0.3333	11/33	22/66	0.3793	11/29	22/58
0.2931	.....	17/58	0.3387	.....	21/62	0.3810	8/21	16/42
0.2941	.....	15/51	0.3390	.....	20/59	0.3824	.....	13/34
0.2941	5/17	10/34	0.3396	.....	18/53	0.3830	18/47	18/47
0.2963	8/27	16/54	0.3404	16/47	16/47	0.3846	15/39	15/39
0.2973	11/37	11/37	0.3415	14/41	14/41	0.3860	.....	22/57
0.2979	14/47	14/47	0.3421	.....	13/38	0.3871	12/31	24/62
0.2982	.....	17/57	0.3448	10/29	20/58	0.3878	19/49	19/49
0.3000	6/20	9/30	0.3469	17/49	17/49	0.3889	7/18	21/54
0.3019	.....	16/53	0.3478	8/23	16/46	0.3898	.....	23/59
0.3023	13/43	13/43	0.3485	.....	23/66	0.3902	16/41	16/41
0.3030	10/33	20/66	0.3488	15/43	15/43	0.3913	9/23	18/46
0.3043	7/23	14/46	0.3500	7/20	.....	0.3922	.....	20/51
0.3051	.....	18/59	0.3509	.....	20/57	0.3929	.....	11/28
0.3061	15/49	15/49	0.3514	13/37	13/37	0.3939	13/33	26/66
0.3065	.....	19/62	0.3519	.....	19/54	0.3947	.....	15/38
0.3077	12/39	12/39	0.3529	6/17	12/34	0.3953	17/43	17/43
0.3095	.....	13/42	0.3529	.....	18/51	0.3962	.....	21/53
0.3103	9/29	18/58	0.3548	11/31	22/62	0.3966	.....	23/58
0.3125	5/16	.....	0.3559	.....	21/59	0.4000	6/15	10/25
0.3137	.....	16/51	0.3571	.....	10/28	0.4000	8/20	12/30
0.3148	.....	17/54	0.3571	.....	15/42	0.4032	.....	25/62
0.3158	6/19	12/38	0.3585	.....	19/53	0.4035	.....	23/57
0.3158	.....	18/57	0.3590	14/39	14/39	0.4043	19/47	19/47
0.3171	13/41	13/41	0.3600	.....	9/25	0.4048	.....	17/42
0.3182	.....	21/66				0.4054	15/37	15/37

* See explanatory note below Table 8.

Accurate Angular Indexing Movements — 4°

Fractional Indexing Movement	B. & S., Becker, Hendey, K. & T. and Rockford	Cincinnati and LeBlond *	Fractional Indexing Movement	B. & S., Becker, Hendey, K. & T. and Rockford	Cincinnati and LeBlond *	Fractional Indexing Movement	B. & S., Becker, Hendey, K. & T. and Rockford	Cincinnati and LeBlond *
0.4068	.....	$\frac{24}{59}$	0.4490	$\frac{22}{49}$	$\frac{22}{49}$	0.4912	.....	$\frac{28}{57}$
0.4074	$\frac{11}{27}$	$\frac{22}{54}$	0.4500	$\frac{9}{20}$	.....	0.4915	.....	$\frac{29}{59}$
0.4082	$\frac{20}{49}$	$\frac{20}{49}$	0.4510	.....	$\frac{23}{51}$	0.5000	$\frac{8}{16}$	$\frac{12}{24}$
0.4091	.....	$\frac{27}{66}$	0.4516	$\frac{14}{31}$	$\frac{28}{62}$	0.5000	$\frac{9}{18}$	$\frac{14}{28}$
0.4103	$\frac{16}{39}$	$\frac{16}{39}$	0.4524	.....	$\frac{19}{42}$	0.5000	$\frac{10}{20}$	$\frac{15}{30}$
0.4118	$\frac{7}{17}$	$\frac{14}{34}$	0.4528	.....	$\frac{24}{53}$	0.5000	.....	$\frac{17}{34}$
0.4118	.....	$\frac{21}{51}$	0.4545	$\frac{15}{33}$	$\frac{30}{66}$	0.5000	.....	$\frac{19}{38}$
0.4130	.....	$\frac{19}{46}$	0.4561	.....	$\frac{26}{57}$	0.5000	.....	$\frac{21}{42}$
0.4138	$\frac{12}{29}$	$\frac{24}{58}$	0.4565	.....	$\frac{21}{46}$	0.5000	.....	$\frac{23}{46}$
0.4146	$\frac{17}{41}$	$\frac{17}{41}$	0.4576	.....	$\frac{27}{59}$	0.5000	.....	$\frac{27}{54}$
0.4151	.....	$\frac{22}{53}$	0.4583	.....	$\frac{11}{24}$	0.5000	.....	$\frac{29}{58}$
0.4167	.....	$\frac{10}{24}$	0.4595	$\frac{17}{37}$	$\frac{17}{37}$	0.5000	.....	$\frac{31}{62}$
0.4186	$\frac{18}{43}$	$\frac{18}{43}$	0.4615	$\frac{18}{39}$	$\frac{18}{39}$	0.5000	.....	$\frac{33}{66}$
0.4194	$\frac{13}{31}$	$\frac{26}{62}$	0.4630	.....	$\frac{25}{54}$	0.5085	.....	$\frac{30}{59}$
0.4211	$\frac{8}{19}$	$\frac{16}{38}$	0.4634	$\frac{19}{41}$	$\frac{19}{41}$	0.5088	.....	$\frac{29}{57}$
0.4211	.....	$\frac{24}{57}$	0.4643	.....	$\frac{13}{28}$	0.5094	.....	$\frac{27}{53}$
0.4237	.....	$\frac{25}{59}$	0.4651	$\frac{20}{43}$	$\frac{20}{43}$	0.5098	.....	$\frac{26}{51}$
0.4242	$\frac{14}{33}$	$\frac{28}{66}$	0.4655	.....	$\frac{27}{58}$	0.5102	$\frac{25}{49}$	$\frac{25}{49}$
0.4255	$\frac{20}{47}$	$\frac{20}{47}$	0.4667	$\frac{7}{15}$	$\frac{14}{30}$	0.5106	$\frac{24}{47}$	$\frac{24}{47}$
0.4259	.....	$\frac{23}{54}$	0.4677	.....	$\frac{29}{62}$	0.5116	$\frac{22}{43}$	$\frac{22}{43}$
0.4286	.....	$\frac{12}{28}$	0.4681	$\frac{22}{47}$	$\frac{22}{47}$	0.5122	$\frac{21}{41}$	$\frac{21}{41}$
0.4286	$\frac{9}{21}$	$\frac{18}{42}$	0.4694	$\frac{23}{49}$	$\frac{23}{49}$	0.5128	$\frac{20}{39}$	$\frac{20}{39}$
0.4286	$\frac{21}{49}$	$\frac{21}{49}$	0.4697	.....	$\frac{31}{66}$	0.5135	$\frac{19}{37}$	$\frac{19}{37}$
0.4310	.....	$\frac{25}{58}$	0.4706	$\frac{8}{17}$	$\frac{16}{34}$	0.5152	$\frac{17}{33}$	$\frac{34}{66}$
0.4314	.....	$\frac{22}{51}$	0.4706	.....	$\frac{24}{51}$	0.5161	$\frac{16}{31}$	$\frac{32}{62}$
0.4324	$\frac{16}{37}$	$\frac{16}{37}$	0.4717	.....	$\frac{25}{53}$	0.5172	$\frac{15}{29}$	$\frac{30}{58}$
0.4333	.....	$\frac{13}{30}$	0.4737	$\frac{9}{19}$	$\frac{27}{57}$	0.5185	$\frac{14}{27}$	$\frac{28}{54}$
0.4340	.....	$\frac{23}{53}$	0.4746	.....	$\frac{28}{59}$	0.5200	.....	$\frac{13}{25}$
0.4348	$\frac{10}{23}$	$\frac{20}{46}$	0.4762	$\frac{10}{21}$	$\frac{20}{42}$	0.5217	$\frac{12}{23}$	$\frac{24}{46}$
0.4355	.....	$\frac{27}{62}$	0.4783	$\frac{11}{23}$	$\frac{22}{46}$	0.5238	$\frac{11}{21}$	$\frac{22}{42}$
0.4359	$\frac{17}{39}$	$\frac{17}{39}$	0.4800	.....	$\frac{12}{25}$	0.5254	.....	$\frac{31}{59}$
0.4375	$\frac{7}{16}$	.....	0.4814	.....	$\frac{26}{54}$	0.5263	$\frac{10}{19}$	$\frac{20}{38}$
0.4386	.....	$\frac{25}{57}$	0.4815	$\frac{13}{27}$	.....	0.5263	.....	$\frac{30}{57}$
0.4390	$\frac{18}{41}$	$\frac{18}{41}$	0.4828	$\frac{14}{29}$	$\frac{28}{58}$	0.5283	.....	$\frac{28}{53}$
0.4394	.....	$\frac{29}{66}$	0.4839	$\frac{15}{31}$	$\frac{30}{62}$	0.5294	$\frac{9}{17}$	$\frac{18}{34}$
0.4400	.....	$\frac{11}{25}$	0.4848	$\frac{16}{33}$	$\frac{32}{66}$	0.5294	.....	$\frac{27}{51}$
0.4407	.....	$\frac{26}{59}$	0.4865	$\frac{18}{37}$	$\frac{18}{37}$	0.5303	.....	$\frac{35}{66}$
0.4412	.....	$\frac{15}{34}$	0.4872	$\frac{19}{39}$	$\frac{19}{39}$	0.5306	$\frac{26}{49}$	$\frac{26}{49}$
0.4419	$\frac{19}{43}$	$\frac{19}{43}$	0.4878	$\frac{20}{41}$	$\frac{20}{41}$	0.5319	$\frac{25}{47}$	$\frac{25}{47}$
0.4444	$\frac{8}{18}$	.....	0.4884	$\frac{21}{43}$	$\frac{21}{43}$	0.5323	.....	$\frac{33}{62}$
0.4444	$\frac{12}{27}$	$\frac{24}{54}$	0.4894	$\frac{23}{47}$	$\frac{23}{47}$	0.5333	$\frac{8}{15}$	$\frac{16}{30}$
0.4468	$\frac{21}{47}$	$\frac{21}{47}$	0.4898	$\frac{24}{49}$	$\frac{24}{49}$	0.5345	.....	$\frac{31}{58}$
0.4474	.....	$\frac{17}{38}$	0.4902	.....	$\frac{25}{51}$	0.5349	$\frac{23}{43}$	$\frac{23}{43}$
0.4483	$\frac{13}{29}$	$\frac{26}{58}$	0.4906	.....	$\frac{26}{53}$	0.5357	.....	$\frac{15}{28}$

* See explanatory note below Table 8.

Accurate Angular Indexing Movements — 5 *

Fractional Indexing Movement	B. & S., Becker, Hendey, K. & T. and Rockford	Cincinnati and LeBlond *	Fractional Indexing Movement	B. & S., Becker, Hendey, K. & T. and Rockford	Cincinnati and LeBlond *	Fractional Indexing Movement	B. & S., Becker, Hendey, K. & T. and Rockford	Cincinnati and LeBlond *
0.5366	$22\tfrac{2}{41}$	$22\tfrac{2}{41}$	0.5789	.....	$33\tfrac{5}{57}$	0.6250	$10\tfrac{9}{16}$	$15\tfrac{5}{24}$
0.5370	.....	$29\tfrac{5}{44}$	0.5806	$18\tfrac{5}{31}$	$36\tfrac{5}{62}$	0.6271	.....	$37\tfrac{5}{59}$
0.5385	$21\tfrac{5}{39}$	$21\tfrac{5}{39}$	0.5814	.....	$25\tfrac{5}{43}$	0.6275	.....	$32\tfrac{5}{51}$
0.5405	$29\tfrac{2}{37}$	$29\tfrac{2}{37}$	0.5833	.....	$14\tfrac{5}{24}$	0.6279	$27\tfrac{4}{43}$	$27\tfrac{4}{43}$
0.5417	.....	$13\tfrac{5}{24}$	0.5849	.....	$31\tfrac{5}{53}$	0.6290	.....	$39\tfrac{5}{62}$
0.5424	.....	$32\tfrac{5}{59}$	0.5854	$24\tfrac{5}{41}$	$24\tfrac{5}{41}$	0.6296	$17\tfrac{5}{27}$	$34\tfrac{5}{54}$
0.5435	.....	$25\tfrac{5}{46}$	0.5862	$17\tfrac{5}{29}$	$34\tfrac{5}{58}$	0.6304	.....	$29\tfrac{5}{46}$
0.5439	.....	$31\tfrac{5}{57}$	0.5870	.....	$27\tfrac{5}{46}$	0.6316	$12\tfrac{5}{19}$	$24\tfrac{5}{38}$
0.5455	$18\tfrac{5}{33}$	$39\tfrac{6}{66}$	0.5882	$19\tfrac{1}{17}$	$20\tfrac{5}{34}$	0.6316	.....	$39\tfrac{5}{67}$
0.5472	.....	$29\tfrac{5}{53}$	0.5882	.....	$39\tfrac{5}{61}$	0.6327	$31\tfrac{4}{49}$	$31\tfrac{4}{49}$
0.5476	.....	$23\tfrac{5}{42}$	0.5897	$23\tfrac{5}{39}$	$23\tfrac{5}{39}$	0.6333	.....	$19\tfrac{5}{30}$
0.5484	$17\tfrac{5}{31}$	$34\tfrac{5}{62}$	0.5909	.....	$39\tfrac{5}{66}$	0.6341	$26\tfrac{5}{41}$	$26\tfrac{5}{41}$
0.5490	.....	$28\tfrac{5}{51}$	0.5918	$29\tfrac{4}{49}$	$29\tfrac{4}{49}$	0.6364	$21\tfrac{5}{33}$	$42\tfrac{5}{66}$
0.5500	$11\tfrac{5}{20}$	.....	0.5926	$16\tfrac{5}{27}$	$32\tfrac{5}{54}$	0.6379	.....	$37\tfrac{5}{58}$
0.5510	$27\tfrac{4}{49}$	$27\tfrac{4}{49}$	0.5932	.....	$35\tfrac{5}{59}$	0.6383	$30\tfrac{4}{47}$	$30\tfrac{4}{47}$
0.5517	$16\tfrac{5}{29}$	$32\tfrac{5}{58}$	0.5946	$22\tfrac{5}{37}$	$22\tfrac{5}{37}$	0.6400	.....	$16\tfrac{5}{25}$
0.5526	.....	$21\tfrac{5}{38}$	0.5952	.....	$25\tfrac{5}{42}$	0.6410	$25\tfrac{4}{39}$	$25\tfrac{5}{39}$
0.5532	$26\tfrac{4}{47}$	$26\tfrac{4}{47}$	0.5957	$28\tfrac{5}{47}$	$28\tfrac{5}{47}$	0.6415	.....	$34\tfrac{5}{53}$
0.5556	$10\tfrac{1}{18}$	.....	0.5965	.....	$34\tfrac{5}{57}$	0.6429	.....	$18\tfrac{5}{28}$
0.5556	$15\tfrac{5}{27}$	$30\tfrac{5}{54}$	0.5968	.....	$37\tfrac{5}{62}$	0.6429	.....	$27\tfrac{5}{42}$
0.5581	$24\tfrac{5}{43}$	$24\tfrac{5}{43}$	0.6000	$9\tfrac{5}{15}$	$15\tfrac{5}{25}$	0.6441	.....	$38\tfrac{5}{59}$
0.5588	.....	$19\tfrac{5}{34}$	0.6000	$12\tfrac{5}{20}$	$18\tfrac{5}{30}$	0.6452	$20\tfrac{5}{31}$	$40\tfrac{5}{62}$
0.5593	.....	$33\tfrac{5}{59}$	0.6034	.....	$35\tfrac{5}{58}$	0.6471	$11\tfrac{1}{17}$	$22\tfrac{5}{34}$
0.5600	.....	$14\tfrac{5}{25}$	0.6038	.....	$32\tfrac{5}{53}$	0.6471	.....	$33\tfrac{5}{51}$
0.5606	.....	$37\tfrac{5}{66}$	0.6047	$26\tfrac{4}{43}$	$26\tfrac{4}{43}$	0.6481	.....	$35\tfrac{5}{54}$
0.5610	$23\tfrac{5}{41}$	$23\tfrac{5}{41}$	0.6053	.....	$23\tfrac{5}{38}$	0.6486	$24\tfrac{5}{37}$	$24\tfrac{5}{37}$
0.5614	.....	$32\tfrac{5}{57}$	0.6061	$20\tfrac{5}{33}$	$40\tfrac{5}{66}$	0.6491	.....	$37\tfrac{5}{57}$
0.5625	$9\tfrac{5}{16}$	.....	0.6071	$17\tfrac{5}{28}$	.....	0.6500	$13\tfrac{5}{20}$	.....
0.5641	$22\tfrac{5}{39}$	$22\tfrac{5}{39}$	0.6078	.....	$31\tfrac{5}{51}$	0.6512	$28\tfrac{4}{43}$	$28\tfrac{4}{43}$
0.5645	.....	$35\tfrac{5}{62}$	0.6087	$14\tfrac{5}{23}$	$28\tfrac{5}{46}$	0.6515	.....	$43\tfrac{5}{66}$
0.5652	$13\tfrac{5}{23}$	$29\tfrac{5}{46}$	0.6098	$25\tfrac{4}{41}$	$25\tfrac{5}{41}$	0.6522	$15\tfrac{5}{23}$	$30\tfrac{5}{46}$
0.5660	.....	$39\tfrac{5}{63}$	0.6102	.....	$36\tfrac{5}{59}$	0.6531	$32\tfrac{4}{49}$	$32\tfrac{4}{49}$
0.5667	.....	$17\tfrac{5}{30}$	0.6111	$11\tfrac{1}{18}$	$33\tfrac{5}{54}$	0.6552	$19\tfrac{5}{29}$	$38\tfrac{5}{58}$
0.5676	$21\tfrac{5}{37}$	$21\tfrac{5}{37}$	0.6122	$30\tfrac{4}{49}$	$30\tfrac{4}{49}$	0.6579	.....	$25\tfrac{5}{38}$
0.5686	.....	$29\tfrac{5}{51}$	0.6129	$19\tfrac{5}{31}$	$38\tfrac{5}{62}$	0.6585	$27\tfrac{5}{41}$	$27\tfrac{5}{41}$
0.5690	.....	$33\tfrac{5}{58}$	0.6140	.....	$35\tfrac{5}{57}$	0.6596	$31\tfrac{4}{47}$	$31\tfrac{4}{47}$
0.5714	.....	$19\tfrac{5}{28}$	0.6154	.....	$24\tfrac{5}{39}$	0.6604	.....	$35\tfrac{5}{53}$
0.5714	$13\tfrac{5}{21}$	$24\tfrac{5}{42}$	0.6170	$29\tfrac{4}{47}$	$29\tfrac{4}{47}$	0.6610	.....	$39\tfrac{5}{59}$
0.5714	$28\tfrac{4}{49}$	$28\tfrac{4}{49}$	0.6176	.....	$21\tfrac{5}{34}$	0.6613	.....	$41\tfrac{5}{62}$
0.5741	.....	$31\tfrac{5}{54}$	0.6190	$13\tfrac{5}{21}$	$26\tfrac{5}{42}$	0.6667	$12\tfrac{5}{18}$	$16\tfrac{5}{24}$
0.5745	$27\tfrac{4}{47}$	$27\tfrac{4}{47}$	0.6207	$18\tfrac{5}{29}$	$36\tfrac{5}{58}$	0.6667	$10\tfrac{5}{15}$	$20\tfrac{5}{30}$
0.5758	$19\tfrac{5}{33}$	$38\tfrac{5}{66}$	0.6212	.....	$41\tfrac{5}{66}$	0.6667	$26\tfrac{5}{39}$	$26\tfrac{5}{39}$
0.5763	.....	$34\tfrac{5}{59}$	0.6216	$23\tfrac{5}{37}$	$23\tfrac{5}{37}$	0.6667	$14\tfrac{5}{21}$	$28\tfrac{5}{42}$
0.5789	$11\tfrac{1}{19}$	$22\tfrac{5}{38}$	0.6226	.....	$33\tfrac{5}{53}$	0.6667	.....	$34\tfrac{5}{51}$

* See explanatory note below Table 8.

Accurate Angular Indexing Movements — 6 *

Fractional Indexing Movement	B. & S., Becker, Hendey, K. & T. and Rockford	Cincinnati and LeBlond *	Fractional Indexing Movement	B. & S., Becker, Hendey, K. & T. and Rockford	Cincinnati and LeBlond *	Fractional Indexing Movement	B. & S., Becker, Hendey, K. & T. and Rockford	Cincinnati and LeBlond *
0.6667	$\tfrac{18}{27}$	$\tfrac{36}{54}$	0.7119	.....	$\tfrac{42}{59}$	0.7576	$\tfrac{25}{33}$	$\tfrac{50}{66}$
0.6667	.....	$\tfrac{38}{57}$	0.7121	.....	$\tfrac{47}{66}$	0.7581	.....	$\tfrac{47}{62}$
0.6667	$\tfrac{22}{33}$	$\tfrac{44}{66}$	0.7143	.....	$\tfrac{20}{28}$	0.7586	$\tfrac{22}{29}$	$\tfrac{44}{58}$
0.6724	.....	$\tfrac{39}{58}$	0.7143	$\tfrac{15}{21}$	$\tfrac{30}{42}$	0.7593	.....	$\tfrac{41}{54}$
0.6735	$\tfrac{33}{49}$	$\tfrac{33}{49}$	0.7143	$\tfrac{35}{49}$	$\tfrac{35}{49}$	0.7600	.....	$\tfrac{19}{25}$
0.6739	.....	$\tfrac{31}{46}$	0.7170	.....	$\tfrac{38}{53}$	0.7609	.....	$\tfrac{35}{46}$
0.6744	$\tfrac{29}{43}$	$\tfrac{29}{43}$	0.7174	.....	$\tfrac{33}{46}$	0.7619	$\tfrac{16}{21}$	$\tfrac{32}{42}$
0.6757	$\tfrac{25}{37}$	$\tfrac{25}{37}$	0.7179	$\tfrac{28}{39}$	$\tfrac{28}{39}$	0.7627	.....	$\tfrac{45}{59}$
0.6765	.....	$\tfrac{23}{34}$	0.7193	.....	$\tfrac{41}{57}$	0.7632	.....	$\tfrac{29}{38}$
0.6774	$\tfrac{21}{31}$	$\tfrac{42}{62}$	0.7200	.....	$\tfrac{18}{25}$	0.7647	$\tfrac{13}{17}$	$\tfrac{26}{34}$
0.6780	.....	$\tfrac{40}{59}$	0.7209	$\tfrac{31}{43}$	$\tfrac{31}{43}$	0.7647	.....	$\tfrac{39}{51}$
0.6786	.....	$\tfrac{19}{28}$	0.7222	$\tfrac{13}{18}$	$\tfrac{39}{54}$	0.7660	$\tfrac{36}{47}$	$\tfrac{36}{47}$
0.6792	.....	$\tfrac{36}{53}$	0.7234	$\tfrac{34}{47}$	$\tfrac{34}{47}$	0.7667	.....	$\tfrac{23}{30}$
0.6800	.....	$\tfrac{17}{25}$	0.7241	$\tfrac{21}{29}$	$\tfrac{42}{58}$	0.7674	$\tfrac{33}{43}$	$\tfrac{33}{43}$
0.6809	$\tfrac{32}{47}$	$\tfrac{32}{47}$	0.7255	.....	$\tfrac{37}{51}$	0.7692	$\tfrac{30}{39}$	$\tfrac{30}{39}$
0.6818	.....	$\tfrac{45}{66}$	0.7258	.....	$\tfrac{45}{62}$	0.7719	.....	$\tfrac{44}{57}$
0.6829	$\tfrac{28}{41}$	$\tfrac{28}{41}$	0.7273	$\tfrac{24}{33}$	$\tfrac{48}{66}$	0.7727	.....	$\tfrac{51}{66}$
0.6842	$\tfrac{13}{19}$	$\tfrac{26}{38}$	0.7288	.....	$\tfrac{43}{59}$	0.7736	.....	$\tfrac{41}{53}$
0.6842	.....	$\tfrac{39}{57}$	0.7297	$\tfrac{27}{37}$	$\tfrac{27}{37}$	0.7742	$\tfrac{24}{31}$	$\tfrac{48}{62}$
0.6852	.....	$\tfrac{37}{54}$	0.7317	$\tfrac{30}{41}$	$\tfrac{30}{41}$	0.7755	$\tfrac{38}{49}$	$\tfrac{38}{49}$
0.6863	.....	$\tfrac{35}{51}$	0.7333	$\tfrac{11}{15}$	$\tfrac{22}{30}$	0.7759	.....	$\tfrac{45}{58}$
0.6875	$\tfrac{11}{16}$	.....	0.7347	.....	$\tfrac{36}{49}$	0.7778	$\tfrac{14}{18}$	.....
0.6897	$\tfrac{20}{29}$	$\tfrac{40}{58}$	0.7353	.....	$\tfrac{25}{34}$	0.7778	$\tfrac{21}{27}$	$\tfrac{42}{54}$
0.6905	.....	$\tfrac{29}{42}$	0.7358	.....	$\tfrac{39}{53}$	0.7797	.....	$\tfrac{46}{59}$
0.6923	$\tfrac{27}{39}$	$\tfrac{27}{39}$	0.7368	$\tfrac{14}{19}$	$\tfrac{28}{38}$	0.7805	$\tfrac{32}{41}$	$\tfrac{32}{41}$
0.6935	.....	$\tfrac{43}{62}$	0.7368	.....	$\tfrac{42}{57}$	0.7826	$\tfrac{18}{23}$	$\tfrac{36}{46}$
0.6939	$\tfrac{34}{49}$	$\tfrac{34}{49}$	0.7381	.....	$\tfrac{31}{42}$	0.7838	$\tfrac{29}{37}$	$\tfrac{29}{37}$
0.6949	.....	$\tfrac{41}{59}$	0.7391	$\tfrac{17}{23}$	$\tfrac{34}{46}$	0.7843	.....	$\tfrac{40}{51}$
0.6957	$\tfrac{16}{23}$	$\tfrac{32}{46}$	0.7407	$\tfrac{20}{27}$	$\tfrac{40}{54}$	0.7857	$\tfrac{22}{28}$	$\tfrac{22}{28}$
0.6970	$\tfrac{23}{33}$	$\tfrac{46}{66}$	0.7414	.....	$\tfrac{43}{58}$	0.7857	.....	$\tfrac{33}{42}$
0.6977	$\tfrac{30}{43}$	$\tfrac{30}{43}$	0.7419	$\tfrac{23}{31}$	$\tfrac{46}{62}$	0.7872	$\tfrac{37}{47}$	$\tfrac{37}{47}$
0.6981	.....	$\tfrac{37}{53}$	0.7424	.....	$\tfrac{49}{66}$	0.7879	$\tfrac{26}{33}$	$\tfrac{52}{66}$
0.7000	$\tfrac{14}{20}$	$\tfrac{21}{30}$	0.7436	$\tfrac{29}{39}$	$\tfrac{29}{39}$	0.7895	$\tfrac{15}{19}$	$\tfrac{30}{38}$
0.7018	.....	$\tfrac{40}{57}$	0.7442	$\tfrac{32}{43}$	$\tfrac{32}{43}$	0.7895	.....	$\tfrac{45}{57}$
0.7021	$\tfrac{33}{47}$	$\tfrac{33}{47}$	0.7447	$\tfrac{35}{47}$	$\tfrac{35}{47}$	0.7903	.....	$\tfrac{49}{62}$
0.7027	$\tfrac{26}{37}$	$\tfrac{26}{37}$	0.7451	.....	$\tfrac{38}{51}$	0.7907	$\tfrac{34}{43}$	$\tfrac{34}{43}$
0.7037	$\tfrac{19}{27}$	$\tfrac{38}{54}$	0.7458	.....	$\tfrac{44}{59}$	0.7917	.....	$\tfrac{19}{24}$
0.7059	$\tfrac{12}{17}$	$\tfrac{24}{34}$	0.7500	$\tfrac{12}{16}$	$\tfrac{18}{24}$	0.7925	.....	$\tfrac{42}{53}$
0.7059	.....	$\tfrac{36}{51}$	0.7500	$\tfrac{15}{20}$	$\tfrac{21}{28}$	0.7931	$\tfrac{23}{29}$	$\tfrac{46}{58}$
0.7069	.....	$\tfrac{41}{58}$	0.7544	.....	$\tfrac{43}{57}$	0.7941	.....	$\tfrac{27}{34}$
0.7073	$\tfrac{29}{41}$	$\tfrac{29}{41}$	0.7547	.....	$\tfrac{40}{53}$	0.7949	$\tfrac{31}{39}$	$\tfrac{31}{39}$
0.7083	.....	$\tfrac{17}{24}$	0.7551	$\tfrac{37}{49}$	$\tfrac{37}{49}$	0.7959	$\tfrac{39}{49}$	$\tfrac{39}{49}$
0.7097	$\tfrac{22}{31}$	$\tfrac{44}{62}$	0.7561	$\tfrac{31}{41}$	$\tfrac{31}{41}$	0.7963	.....	$\tfrac{43}{54}$
0.7105	.....	$\tfrac{27}{38}$	0.7568	$\tfrac{28}{37}$	$\tfrac{28}{37}$	0.7966	.....	$\tfrac{47}{59}$

* See explanatory note below Table 8.

Accurate Angular Indexing Movements — 7 *

Fractional Indexing Movement	B. & S., Becker, Hendey, K. & T. and Rockford	Cincinnati and LeBlond *	Fractional Indexing Movement	B. & S., Becker, Hendey, K. & T. and Rockford	Cincinnati and LeBlond *	Fractional Indexing Movement	B. & S., Becker, Hendey, K. & T. and Rockford	Cincinnati and LeBlond *
0.8000	$\frac{16}{20}$	$\frac{20}{25}$	0.8431	.....	$\frac{43}{51}$	0.8871	.....	$\frac{55}{62}$
0.8000	$\frac{12}{15}$	$\frac{24}{30}$	0.8448	.....	$\frac{49}{58}$	0.8889	$\frac{16}{18}$	.....
0.8030	.....	$\frac{53}{66}$	0.8462	$\frac{33}{39}$	$\frac{33}{39}$	0.8889	$\frac{24}{27}$	$\frac{48}{54}$
0.8039	.....	$\frac{41}{51}$	0.8475	.....	$\frac{50}{59}$	0.8913	.....	$\frac{41}{46}$
0.8043	.....	$\frac{37}{46}$	0.8478	.....	$\frac{39}{46}$	0.8919	$\frac{33}{37}$	$\frac{33}{37}$
0.8049	$\frac{33}{41}$	$\frac{33}{41}$	0.8485	$\frac{28}{33}$	$\frac{56}{66}$	0.8929	.....	$\frac{25}{28}$
0.8065	$\frac{25}{31}$	$\frac{50}{62}$	0.8491	.....	$\frac{45}{53}$	0.8936	$\frac{42}{47}$	$\frac{42}{47}$
0.8070	.....	$\frac{46}{57}$	0.8500	$\frac{17}{20}$	.....	0.8939	.....	$\frac{59}{66}$
0.8085	$\frac{38}{47}$	$\frac{38}{47}$	0.8511	$\frac{40}{47}$	$\frac{40}{47}$	0.8947	$\frac{17}{19}$	$\frac{34}{38}$
0.8095	$\frac{17}{21}$	$\frac{34}{42}$	0.8519	$\frac{23}{27}$	$\frac{46}{54}$	0.8947	.....	$\frac{51}{57}$
0.8103	.....	$\frac{47}{58}$	0.8529	.....	$\frac{29}{34}$	0.8966	$\frac{26}{29}$	$\frac{52}{58}$
0.8108	$\frac{30}{37}$	$\frac{30}{37}$	0.8537	$\frac{35}{41}$	$\frac{35}{41}$	0.8974	$\frac{35}{39}$	$\frac{35}{39}$
0.8113	.....	$\frac{43}{53}$	0.8548	.....	$\frac{53}{62}$	0.8980	$\frac{44}{49}$	$\frac{44}{49}$
0.8125	$\frac{13}{16}$	.....	0.8571	.....	$\frac{24}{28}$	0.8983	.....	$\frac{53}{59}$
0.8136	.....	$\frac{48}{59}$	0.8571	$\frac{18}{21}$	$\frac{36}{42}$	0.9000	$\frac{18}{20}$	$\frac{27}{30}$
0.8140	.....	$\frac{35}{43}$	0.8571	$\frac{42}{49}$	$\frac{42}{49}$	0.9020	.....	$\frac{46}{51}$
0.8148	.....	$\frac{44}{54}$	0.8596	.....	$\frac{49}{57}$	0.9024	$\frac{37}{41}$	$\frac{37}{41}$
0.8158	.....	$\frac{31}{38}$	0.8605	$\frac{37}{43}$	$\frac{37}{43}$	0.9032	$\frac{28}{31}$	$\frac{56}{62}$
0.8163	$\frac{40}{49}$	$\frac{40}{49}$	0.8621	$\frac{25}{29}$	$\frac{50}{58}$	0.9048	$\frac{19}{21}$	$\frac{38}{42}$
0.8182	$\frac{27}{33}$	$\frac{54}{66}$	0.8627	.....	$\frac{44}{51}$	0.9057	.....	$\frac{48}{53}$
0.8205	$\frac{32}{39}$	$\frac{32}{39}$	0.8636	.....	$\frac{57}{66}$	0.9070	$\frac{39}{43}$	$\frac{39}{43}$
0.8214	.....	$\frac{23}{28}$	0.8644	.....	$\frac{51}{59}$	0.9074	.....	$\frac{49}{54}$
0.8226	.....	$\frac{51}{62}$	0.8649	$\frac{32}{37}$	$\frac{32}{37}$	0.9090	$\frac{30}{33}$	$\frac{60}{66}$
0.8235	$\frac{14}{17}$	$\frac{28}{34}$	0.8667	$\frac{13}{15}$	$\frac{26}{30}$	0.9118	.....	$\frac{31}{34}$
0.8235	.....	$\frac{42}{51}$	0.8679	.....	$\frac{46}{53}$	0.9123	.....	$\frac{52}{57}$
0.8246	.....	$\frac{47}{57}$	0.8684	.....	$\frac{33}{38}$	0.9130	$\frac{21}{23}$	$\frac{42}{46}$
0.8261	$\frac{19}{23}$	$\frac{38}{46}$	0.8696	$\frac{20}{23}$	$\frac{40}{46}$	0.9138	.....	$\frac{53}{58}$
0.8276	$\frac{24}{29}$	$\frac{48}{58}$	0.8704	.....	$\frac{47}{54}$	0.9149	$\frac{43}{47}$	$\frac{43}{47}$
0.8293	$\frac{34}{41}$	$\frac{34}{41}$	0.8710	$\frac{27}{31}$	$\frac{54}{62}$	0.9153	.....	$\frac{54}{59}$
0.8298	$\frac{39}{47}$	$\frac{39}{47}$	0.8718	$\frac{34}{39}$	$\frac{34}{39}$	0.9167	.....	$\frac{22}{24}$
0.8302	.....	$\frac{44}{53}$	0.8723	$\frac{41}{47}$	$\frac{41}{47}$	0.9184	$\frac{45}{49}$	$\frac{45}{49}$
0.8305	.....	$\frac{49}{59}$	0.8750	$\frac{14}{16}$	$\frac{21}{24}$	0.9189	$\frac{34}{37}$	$\frac{34}{37}$
0.8333	$\frac{15}{18}$	$\frac{20}{24}$	0.8772	.....	$\frac{50}{57}$	0.9194	.....	$\frac{57}{62}$
0.8333	.....	$\frac{25}{30}$	0.8776	$\frac{43}{49}$	$\frac{43}{49}$	0.9200	.....	$\frac{23}{25}$
0.8333	.....	$\frac{35}{42}$	0.8780	$\frac{36}{41}$	$\frac{36}{41}$	0.9211	.....	$\frac{35}{38}$
0.8333	.....	$\frac{45}{54}$	0.8788	$\frac{29}{33}$	$\frac{58}{66}$	0.9216	.....	$\frac{47}{51}$
0.8333	.....	$\frac{55}{66}$	0.8793	.....	$\frac{51}{58}$	0.9231	$\frac{36}{39}$	$\frac{36}{39}$
0.8367	$\frac{41}{49}$	$\frac{41}{49}$	0.8800	.....	$\frac{22}{25}$	0.9242	.....	$\frac{61}{66}$
0.8372	$\frac{36}{43}$	$\frac{36}{43}$	0.8810	.....	$\frac{37}{42}$	0.9245	.....	$\frac{49}{53}$
0.8378	$\frac{31}{37}$	$\frac{31}{37}$	0.8814	.....	$\frac{52}{59}$	0.9259	$\frac{25}{27}$	$\frac{50}{54}$
0.8387	$\frac{26}{31}$	$\frac{52}{62}$	0.8824	$\frac{15}{17}$	$\frac{30}{34}$	0.9268	$\frac{38}{41}$	$\frac{38}{41}$
0.8400	.....	$\frac{21}{25}$	0.8824	.....	$\frac{45}{51}$	0.9286	.....	$\frac{26}{28}$
0.8421	.....	$\frac{32}{38}$	0.8837	$\frac{38}{43}$	$\frac{38}{43}$	0.9286	.....	$\frac{39}{42}$
0.8421	$\frac{16}{19}$	$\frac{48}{57}$	0.8868	.....	$\frac{47}{53}$	0.9298	.....	$\frac{53}{57}$

* See explanatory note below Table 8.

Accurate Angular Indexing Movements — 8 *

Fractional Indexing Movement	B. & S., Becker, Hendey, K. & T. and Rockford	Cincinnati and LeBlond *	Fractional Indexing Movement	B. & S., Becker, Hendey, K. & T. and Rockford	Cincinnati and LeBlond *	Fractional Indexing Movement	B. & S., Becker, Hendey, K. & T. and Rockford	Cincinnati and LeBlond *
0.9302	$\frac{40}{43}$	$\frac{40}{43}$	0.9500	$\frac{19}{20}$	....	0.9697	$\frac{32}{33}$	$\frac{64}{66}$
0.9310	$\frac{27}{29}$	$\frac{54}{58}$	0.9512	$\frac{39}{41}$	$\frac{39}{41}$	0.9706	....	$\frac{33}{34}$
0.9322	....	$\frac{55}{59}$	0.9516	....	$\frac{59}{62}$	0.9730	$\frac{36}{37}$	$\frac{36}{37}$
0.9333	$\frac{14}{15}$	$\frac{28}{30}$	0.9524	$\frac{20}{21}$	$\frac{40}{42}$	0.9737	....	$\frac{37}{38}$
0.9348	....	$\frac{43}{46}$	0.9535	$\frac{41}{43}$	$\frac{41}{43}$	0.9744	$\frac{38}{39}$	$\frac{38}{39}$
0.9355	$\frac{29}{31}$	$\frac{58}{62}$	0.9545	....	$\frac{63}{66}$	0.9756	$\frac{40}{41}$	$\frac{40}{41}$
0.9362	$\frac{44}{47}$	$\frac{44}{47}$	0.9565	$\frac{22}{23}$	$\frac{44}{46}$	0.9762	....	$\frac{41}{42}$
0.9375	$\frac{15}{16}$	....	0.9574	$\frac{45}{47}$	$\frac{45}{47}$	0.9767	$\frac{42}{43}$	$\frac{42}{43}$
0.9388	$\frac{46}{49}$	$\frac{46}{49}$	0.9583	....	$\frac{23}{24}$	0.9783	....	$\frac{45}{46}$
0.9394	$\frac{31}{33}$	$\frac{62}{66}$	0.9592	$\frac{47}{49}$	$\frac{47}{49}$	0.9787	$\frac{46}{47}$	$\frac{46}{47}$
0.9412	$\frac{16}{17}$	$\frac{32}{34}$	0.9600	....	$\frac{24}{25}$	0.9796	$\frac{48}{49}$	$\frac{48}{49}$
0.9412	....	$\frac{48}{51}$	0.9608	....	$\frac{49}{51}$	0.9804	....	$\frac{50}{51}$
0.9434	....	$\frac{50}{53}$	0.9623	....	$\frac{51}{53}$	0.9811	....	$\frac{52}{53}$
0.9444	$\frac{17}{18}$	$\frac{51}{54}$	0.9630	$\frac{26}{27}$	$\frac{52}{54}$	0.9815	....	$\frac{53}{54}$
0.9459	$\frac{35}{37}$	$\frac{35}{37}$	0.9643	....	$\frac{27}{28}$	0.9825	....	$\frac{56}{57}$
0.9474	$\frac{18}{19}$	$\frac{36}{38}$	0.9649	....	$\frac{55}{57}$	0.9828	....	$\frac{57}{58}$
0.9474	....	$\frac{54}{57}$	0.9655	$\frac{28}{29}$	$\frac{56}{58}$	0.9831	....	$\frac{58}{59}$
0.9483	....	$\frac{55}{58}$	0.9661	....	$\frac{57}{59}$	0.9839	....	$\frac{61}{62}$
0.9487	$\frac{37}{39}$	$\frac{37}{39}$	0.9667	....	$\frac{29}{30}$	0.9848	....	$\frac{65}{66}$
0.9492	....	$\frac{56}{59}$	0.9677	$\frac{30}{31}$	$\frac{60}{62}$	......	....	....

* The foregoing tables may be used when indexing for angles in degrees, minutes or seconds. The tables are used as follows: Reduce the angle to seconds and divide the value thus obtained by 32,400. The quotient gives the number of complete turns and decimal fraction of a turn required. Then find the decimal (or nearest decimal) to this decimal fraction in the tables. Opposite this decimal will be found the fractional number indicating the indexing movement.

Example: — Assume that an angle of 10 degrees, 32 minutes, 12 seconds is to be indexed. Then, 10° 32′ 12″ = 37,932 seconds and 37.932 ÷ 32,400 = 1.1707; therefore this indexing can be made by one complete turn and 0.1707 part of a turn. The second table shows that 0.1707 part of a turn is obtained by moving 7 holes in the 41-hole circle.

Example: — Two slots are to be milled in the edge of a disk and the angle between their center-lines is 58 degrees 51 minutes and 53 seconds. Determine the indexing movement.

The angle 58° 51′ 53″ reduced to seconds = 211,913 seconds and 211,913 ÷ 32,400 = 6.5405; therefore this indexing movement requires six complete turns and 0.5405 part of a turn. The fifth table shows that 0.5405 part of a turn is obtained by moving 20 holes in the 37-hole circle.

The number of holes in the index circles of the indexing-heads made by the Brown & Sharpe Mfg. Co., Becker Milling Machine Co., Hendey Machine Co., Kearney & Trecker Co., and the Rockford Milling Machine Co. are the same. The index circles of the Cincinnati Milling Machine Co. differ from these; hence, a separate column is given in the table for the "Cincinnati" index-head. The R. K. LeBlond Machine Tool Co.'s dividing head has the same index circles as that of the Cincinnati Milling Machine Co., except that the former does not have the 24-, 25-, 28-, and 30-hole circles, but has, instead, 36-, 48-, and 56-hole circles. The movements in the 24- and 28-hole circles of the Cincinnati index-head may be made on the LeBlond index-head by taking double the number of holes in the 48-hole and 56-hole circles, respectively. In this way, the table can be used for practically all movements with LeBlond milling machines.

Block or Multiple Indexing for Gear Cutting

Teeth to be Cut	Number Indexed at Once	First Driver	First Follower	Second Driver	Second Follower	Turns of Locking Disk	Teeth to be Cut	Number Indexed at Once	First Driver	First Follower	Second Driver	Second Follower	Turns of Locking Disk
25	4	100	50	72	30	4	77	4	100	70	96	44	2
26	3	100	50	90	52	4	78	5	100	30	90	78	2
27	2	100	50	60	54	4	80	3	100	50	90	80	2
28	3	100	50	90	56	4	81	7	100	30	84	52	2
29	3	100	50	90	58	4	82	5	100	30	90	82	2
30	7	100	30	84	40	4	84	5	100	30	90	84	2
31	3	100	50	90	62	4	85	4	100	50	96	68	2
32	3	100	50	90	64	4	86	5	100	30	90	86	2
33	4	100	50	80	44	4	87	7	100	30	84	58	2
34	3	100	50	90	68	4	88	5	100	30	90	88	2
35	4	100	50	96	56	4	90	7	100	30	70	50	2
36	5	100	48	80	40	4	91	3	100	70	72	52	2
37	5	100	30	90	74	4	92	5	100	30	90	92	2
38	5	100	30	90	76	4	93	7	100	30	84	62	2
39	5	100	30	90	78	4	94	5	100	30	90	94	2
40	3	100	50	90	80	4	95	4	100	50	96	76	2
41	5	100	30	90	82	4	96	5	100	30	90	96	2
42	5	100	30	90	84	4	98	5	100	30	90	98	2
43	5	100	30	90	86	4	99	10	100	30	80	44	2
44	5	100	30	90	88	4	100	7	100	50	84	40	2
45	7	100	50	70	30	4	102	5	100	30	60	68	2
46	5	100	30	90	92	4	104	5	100	60	90	52	2
47	5	100	30	90	94	4	105	4	100	70	96	60	2
48	5	100	30	90	96	4	108	7	100	30	70	60	2
49	5	100	30	90	98	4	110	7	100	50	84	44	2
50	7	100	50	84	40	4	111	5	100	74	80	40	2
51	4	100	30	96	68	2	112	5	100	60	90	56	2
52	5	100	30	90	52	2	114	7	100	30	84	76	2
54	5	100	30	90	54	2	115	8	100	50	96	46	2
55	4	100	50	96	44	2	116	5	100	60	90	58	2
56	5	100	30	90	56	2	117	8	100	30	96	78	2
57	4	100	30	96	76	2	119	3	100	70	72	68	2
58	5	100	30	90	58	2	120	7	100	50	70	40	2
60	7	100	30	84	40	2	121	4	60	66	96	44	2
62	5	100	30	90	62	2	123	7	100	30	84	82	2
63	5	100	30	80	56	2	124	5	100	60	90	62	2
64	5	100	30	90	64	2	125	7	100	50	84	50	2
65	4	100	50	96	52	2	126	5	100	50	50	42	2
66	5	100	44	80	40	2	128	5	100	60	90	64	2
67	5	100	30	90	67	2	129	7	100	30	84	86	2
68	5	100	30	90	68	2	130	7	100	50	84	52	2
69	5	100	46	80	40	2	132	5	100	88	80	40	2
70	3	100	50	90	70	2	133	4	100	70	96	76	2
72	5	100	30	90	72	2	134	5	100	60	90	67	2
74	5	100	30	90	74	2	135	7	100	50	84	54	2
75	7	100	30	84	50	2	136	5	100	60	90	68	2
76	5	100	30	90	76	2	138	5	100	92	80	40	2

Block or Multiple Indexing for Gear Cutting

Teeth to be Cut	Number Indexed at Once	First Driver	First Follower	Second Driver	Second Follower	Turns of Locking Disk	Teeth to be Cut	Number Indexed at Once	First Driver	First Follower	Second Driver	Second Follower	Turns of Locking Disk
140	3	50	50	90	70	2	170	7	100	50	84	68	2
141	5	100	94	80	40	2	171	5	70	42	80	76	2
143	6	90	66	96	52	2	172	5	100	60	90	86	2
144	5	100	60	90	72	2	174	7	100	60	84	58	2
145	6	100	50	72	58	2	175	8	100	50	96	70	2
147	5	100	98	80	40	2	176	5	100	60	90	88	2
148	5	100	60	90	74	2	180	7	100	60	70	50	2
150	7	100	60	84	50	2	182	9	90	56	96	52	2
152	5	100	60	90	76	2	184	5	100	60	90	92	2
153	5	100	68	80	60	2	185	6	100	50	72	74	2
154	5	100	56	72	66	2	186	7	100	60	84	62	2
155	6	100	50	72	62	2	187	5	100	44	48	68	2
156	5	100	60	90	78	2	188	5	100	60	90	94	2
160	7	100	50	84	64	2	189	5	100	60	80	84	2
161	5	100	70	60	46	2	190	7	100	50	84	76	2
162	7	100	60	84	52	2	192	5	100	60	90	96	2
164	5	100	60	90	82	2	195	7	100	60	84	78	2
165	7	100	50	84	66	2	196	5	100	60	90	98	2
168	5	100	60	90	84	2	198	7	100	50	70	66	2
169	6	96	52	90	78	2	200	7	60	60	84	40	2

Block or Multiple Indexing for Gear Cutting. — With the block system of indexing, a number of teeth are indexed at one time, instead of cutting the teeth consecutively, and the gear is revolved several times before the teeth are all finished. For example, when cutting a gear having 25 teeth, the indexing mechanism is geared to index four teeth at once (see table) and the first time around, six widely separated tooth spaces are cut. The second time around, the cutter is one tooth behind the spaces previously milled. On the third indexing, the cutter has dropped back another tooth, thus finishing the gear (in this case) by indexing it around four times. The various combinations of change gears to use for block or multiple indexing are given in the accompanying table. The advantage claimed for block indexing is that the heat generated by the cutter (especially when cutting cast-iron gears of coarse pitch) is distributed more evenly about the rim and dissipated to a greater extent, thus avoiding distortion due to local heating and permitting higher speeds and feeds. The table given is intended for use with Brown & Sharpe automatic gear-cutting machines, but the gears for any other machine equipped with a similar indexing mechanism can be calculated. Assume, for example, that a gear cutter requires the following change gears for indexing a certain number of teeth: Driving gears having 20 and 30 teeth, respectively, and driven gears having 50 and 60 teeth. Then if it is desired to cut, say, every fifth tooth, multiply the fractions $\frac{20}{60}$ and $\frac{30}{50}$ by 5. Then, $\frac{20}{60} \times \frac{30}{50} \times \frac{5}{1} = \frac{1}{1}$. In this particular instance, then, the blank could be divided so that every fifth space would be cut, by using gears of equal size. The number of teeth in the gear and the number of teeth indexed in each block, must not have a common factor.

Indexing Movements for 60-Tooth Worm-Wheel Dividing Head

Divisions	Index Circle	No. of Turns	No. of Holes	Divisions	Index Circle	No. of Turns	No. of Holes	Divisions	Index Circle	No. of Holes	Divisions	Index Circle	No. of Holes
2	Any	30	..	50	60	1	12	98	49	30	146	73	30
3	Any	20	..	51	17	1	3	99	33	20	147	49	20
4	Any	15	..	52	26	1	4	100	60	36	148	37	15
5	Any	12	..	53	53	1	7	101	101	60	149	149	60
6	Any	10	..	54	27	1	3	102	17	10	150	60	24
7	21	8	12	55	33	1	3	103	103	60	151	151	60
8	26	7	13	56	28	1	2	104	26	15	152	76	30
9	21	6	14	57	19	1	1	105	21	12	153	51	20
10	Any	6	..	58	29	1	1	106	53	30	154	77	30
11	33	5	15	59	59	1	1	107	107	60	155	31	12
12	Any	5	..	60	Any	1	..	108	27	15	156	26	10
13	26	4	16	61	61	..	60	109	109	60	157	157	60
14	21	4	6	62	31	..	30	110	33	18	158	79	30
15	Any	4	..	63	21	..	20	111	37	20	159	53	20
16	28	3	21	64	32	..	30	112	28	15	160	32	12
17	17	3	9	65	26	..	24	113	113	60	161	161	60
18	21	3	7	66	33	..	30	114	19	10	162	27	10
19	19	3	3	67	67	..	60	115	23	12	163	163	60
20	Any	3	..	68	17	..	15	116	29	15	164	41	15
21	21	2	18	69	23	..	20	117	39	20	165	33	12
22	33	2	24	70	21	..	18	118	59	30	166	83	30
23	23	2	14	71	71	..	60	119	119	60	167	167	60
24	26	2	13	72	60	..	50	120	26	13	168	28	10
25	60	2	24	73	73	..	60	121	121	60	169	169	60
26	26	2	8	74	37	..	30	122	61	30	170	17	6
27	27	2	6	75	60	..	48	123	41	20	171	57	20
28	21	2	3	76	19	..	15	124	31	15	172	43	15
29	29	2	2	77	77	..	60	125	100	48	173	173	60
30	Any	2	..	78	26	..	20	126	21	10	174	29	10
31	31	1	29	79	79	..	60	127	127	60	175	35	12
32	32	1	28	80	28	..	21	128	32	15	176	44	15
33	33	1	27	81	27	..	20	129	43	20	177	59	20
34	17	1	13	82	41	..	30	130	26	12	178	89	30
35	21	1	15	83	83	..	60	131	131	60	179	179	60
36	21	1	14	84	21	..	15	132	33	15	180	21	7
37	37	1	23	85	17	..	12	133	133	60	181	181	60
38	19	1	11	86	43	..	30	134	67	30	182	91	30
39	26	1	14	87	29	..	20	135	27	12	183	61	20
40	26	1	13	88	44	..	30	136	68	30	184	46	15
41	41	1	19	89	89	..	60	137	137	60	185	37	12
42	21	1	9	90	21	..	14	138	23	10	186	31	10
43	43	1	17	91	91	..	60	139	139	60	187	187	60
44	33	1	12	92	23	..	15	140	21	9	188	47	15
45	21	1	7	93	31	..	20	141	47	20	189	63	20
46	23	1	7	94	47	..	30	142	71	30	190	19	6
47	47	1	13	95	19	..	12	143	143	60	191	191	60
48	28	1	7	96	32	..	20	144	60	25	192	32	12
49	49	1	11	97	97	..	60	145	29	12	193	193	60

Indexing for Rack Cutting. — When racks are cut on a milling machine, there are two general methods of indexing. One is by using the graduated dial on the feed-screw and the other is by using an indexing attachment. The accompanying table shows the indexing movements when the first method is employed. This table applies to milling machines having feed-screws with the usual lead of ¼ inch and 250 dial graduations each equivalent to 0.001 inch of table movement.

$$\text{Actual rotation of feed-screw} = \frac{\text{Linear pitch of rack}}{\text{Lead of feed-screw}}$$

Multiply *decimal* part of turn (obtained by above formula) by 250, to obtain dial reading for fractional part of indexing movement, assuming that dial has 250 graduations.

Indexing Movements for Cutting Rack Teeth on Milling Machine

These movements are for table feed-screws having the usual lead of ¼ inch

Pitch of Rack Tooth		Indexing, Movement		Pitch of Rack Teeth		Indexing, Movement	
Diametral Pitch	Linear or Circular	No. of Whole Turns	No. of .001 Inch Divisions	Diametral Pitch	Linear or Circular	No. of Whole Turns	No. of .001 Inch Divisions
2	1.5708	6	70.8	12	0.2618	1	11.8
2¼	1.3963	5	146.3	13	0.2417	0	241.7
2½	1.2566	5	6.6	14	0.2244	0	224.4
2¾	1.1424	4	142.4	15	0.2094	0	208.4
3	1.0472	4	47.2	16	0.1963	0	196.3
3½	0.8976	3	147.6	17	0.1848	0	184.8
4	0.7854	3	35.4	18	0.1745	0	174.8
5	0.6283	2	128.3	19	0.1653	0	165.3
6	0.5236	2	23.6	20	0.1571	0	157.1
7	0.4488	1	198.8	22	0.1428	0	142.8
8	0.3927	1	142.7	24	0.1309	0	130.9
9	0.3491	1	99.1	26	0.1208	0	120.8
10	0.3142	1	64.2	28	0.1122	0	112.2
11	0.2856	1	35.6	30	0.1047	0	104.7

Note: The linear pitch of the rack equals the circular pitch of gear or pinion which is to mesh with the rack. The table gives both standard diametral pitches and their equivalent linear or circular pitches.

Example: Find indexing movement for cutting rack to mesh with a pinion of 10 diametral pitch.

Indexing movement equals 1 whole turn of feed-screw plus 64.2 thousandths or divisions on feed-screw dial. The feed-screw may be turned this fractional amount by setting dial back to its zero position for each indexing (without backward movement of feed-screw), or, if preferred, 64.2 (in this example) may be added to each successive dial position as shown below.

Dial reading for second position = 64.2 × 2 = 128.4 (complete movement = 1 turn + 64.2 additional divisions by turning feed-screw until dial reading is 128.4).

Third dial position = 64.2 × 3 = 192.6 (complete movement = 1 turn + 64.2 additional divisions by turning until dial reading is 192.6).

Fourth position = 64.2 × 4 − 250 = 6.8 (1 turn + 64.2 additional divisions by turning feed-screw until dial reading is 6.8 divisions past the zero mark); or, to simplify operation, set dial back to zero for fourth indexing (without moving feed-screw) and then repeat settings for the three previous indexings or whatever number can be made before making a complete turn of the dial.

GEARS AND GEARING

Definitions of Gear Terms. — The terms which follow are commonly applied to various classes of gearing.

Addendum: Height of tooth above pitch circle or the radial distance between the pitch circle and the top of the tooth (see illustration).

Approach Ratio: The ratio of the arc of approach to the arc of action.

Arc of Action: Arc of the pitch circle through which a tooth travels from the first point of contact with the mating tooth to the point where contact ceases.

Arc of Approach: Arc of the pitch circle through which a tooth travels from the first point of contact with the mating tooth to the pitch point.

Arc of Recess: Arc of the pitch circle through which a tooth travels from its contact with the mating tooth at the pitch point to the point where its contact ceases.

Axial Plane: In a pair of gears it is the plane that contains the two axes. In a single gear, it may be any plane containing the axis and a given point.

Backlash: The amount by which the width of a tooth space exceeds the thickness of the engaging tooth on the pitch circles. As actually indicated by measuring devices, backlash may be determined variously in the transverse, normal or axial planes, and either in the direction of the pitch circles or on the line of action. Such measurements should be converted to corresponding values on transverse pitch circles for general comparison.

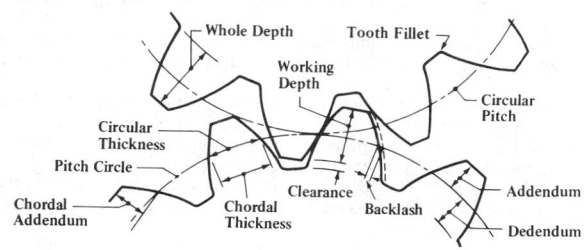

Gear Tooth Parts

Base Circle: The circle from which an involute tooth curve is generated or developed.

Base Helix Angle: The angle, at the base cylinder of an involute gear, that the tooth makes with the gear axis.

Base Pitch: In an involute gear it is the pitch on the base circle or along the line of action. Corresponding sides of involute teeth are parallel curves, and the base pitch is the constant and fundamental distance between them along a common normal in a plane of rotation. The *Normal Base Pitch* is the base pitch in the normal plane, and the *Axial Base Pitch* is the base pitch in the axial plane.

Center Distance: The distance between the parallel axes of spur gears and parallel helical gears, or between the crossed axes of crossed helical gears and worm gears. Also, it is the distance between the centers of the pitch circles.

Central Plane: In a worm gear this is the plane perpendicular to the gear axis and contains the common perpendicular of the gear and worm axes. In the usual case with the axes at right angles, it contains the worm axis.

Chordal Addendum: The height from the top of the tooth to the chord subtending the circular-thickness arc.

Chordal Thickness: Length of the chord subtended by the circular thickness arc the dimension obtained when a gear-tooth caliper is used to measure the thickness t the pitch circle).

Circular Pitch: Length of the arc of the pitch circle between the centers or other corresponding points of adjacent teeth (see illustration). *Normal Circular Pitch* is the circular pitch in the normal plane.

Circular Thickness: The length of arc between the two sides of a gear tooth, on the pitch circle unless otherwise specified. *Normal Circular Thickness* is the circular thickness in the normal plane.

Clearance: The amount by which the dedendum in a given gear exceeds the addendum of its mating gear. It is also the radial distance between the top of a tooth and the bottom of the mating tooth space.

Contact Diameter: The smallest diameter on a gear tooth with which the mating gear makes contact.

Contact Ratio: The ratio of the arc of action to the circular pitch. It is sometimes thought of as the average number of teeth in contact. For involute gears, the contact ratio is obtained most directly as the ratio of the length of action to the base pitch.

Contact Stress: The maximum compressive stress within the contact area between mating gear tooth profiles. It is also called Hertz stress.

Cycloid: The curve formed by the path of a point on a circle as it rolls along a straight line. When this circle rolls along the outer side of another circle, the curve is called an *Epicycloid*; when it rolls along the inner side of another circle it is called a *Hypocycloid*. These curves are used in defining the former American Standard composite tooth form.

Dedendum: The depth of tooth space below the pitch circle or the radial dimension between the pitch circle and the bottom of the tooth space (see illustration).

Diametral Pitch: The ratio of the number of teeth to the number of inches of pitch diameter — equals number of gear teeth to each inch of pitch diameter. *Normal Diametral Pitch* is the diametral pitch as calculated in the normal plane and is equal to the diametral pitch divided by the cosine of the helix angle.

Effective Face Width: That portion of the face width that actually comes into contact with mating teeth, as occasionally one member of a pair of gears may have a greater face width than the other.

Efficiency: The actual torque ratio of a gear set divided by its gear ratio.

External Gear: A gear with teeth on the outer cylindrical surface.

Face of Tooth: That surface of the tooth which is between the pitch circle and the top of the tooth.

Fillet Curve: The concave portion of the tooth profile where it joins the bottom of the tooth space. The approximate radius of this curve is called the *Fillet Radius*.

Fillet Stress: The maximum tensile stress in the gear tooth fillet.

Flank of Tooth: That surface which is between the pitch circle and the bottom land. The flank includes the fillet.

Helical Overlap: The effective face width of a helical gear divided by the gear axial pitch; also called the *Face Overlap*.

Helix Angle: The angle that a helical gear tooth makes with the gear axis at the pitch circle unless otherwise specified.

Hertz Stress: See Contact Stress.

Highest Point of Single Tooth Contact: The largest diameter on a spur gear at which a single tooth is in contact with the mating gear. Often referred to as HPSTC.

Internal Diameter: The diameter of a circle coinciding with the tops of the teeth of an internal gear.

Internal Gear: A gear with teeth on the inner cylindrical surface.

Involute: The curve formed by the path of a point on a straight line, called the generatrix, as it rolls along a convex base curve. (The base curve is usually a circle. This curve is generally used as the profile of gear teeth.

Land: The *Top Land* is the top surface of a tooth, and the *Bottom Land* is the surface of the gear between the fillets of adjacent teeth.

Lead: The distance a helical gear or worm would thread along its axis in one revolution if it were free to move axially.

Length of Action: The distance on an involute line of action through which the point of contact moves during the action of the tooth profile.

Line of Action: The path of contact in involute gears. It is the straight line passing through the pitch point and tangent to the base circles.

Lowest Point of Single Tooth Contact: The smallest diameter on a spur gear at which a single tooth of one gear is in contact with its mating gear. Often referred to as LPSTC. Gear set contact stress is determined with a load placed at this point on the pinion.

Module: Ratio of the pitch diameter to the number of teeth. Ordinarily, module is understood to mean ratio of pitch diameter in *millimeters* to the number of teeth. The *English Module* is a ratio of the pitch diameter in inches to the number of teeth.

Normal Plane: A plane normal to the tooth surfaces at a point of contact, and perpendicular to the pitch plane.

Pitch: The distance between similar, equally-spaced tooth surfaces, in a given direction and along a given curve or line. The single word "pitch" without qualification has been used to designate circular pitch, axial pitch, and diametral pitch, but such confusing usage should be avoided.

Pitch Circle: A circle the radius of which is equal to the distance from the gear axis to the pitch point (see illustration).

Pitch Diameter: The diameter of the pitch circle. In parallel shaft gears the pitch diameters can be determined directly from the center distance and the numbers of teeth by proportionality. *Operating Pitch Diameter* is the pitch diameter at which the gears operate. *Generating Pitch Diameter* is the pitch diameter at which the gear is generated. In a bevel gear the pitch diameter is understood to be at the outer ends of the teeth unless otherwise specified. (See also reference to standard pitch diameter under *Pressure Angle*.)

Pitch Plane: In a pair of gears it is the plane perpendicular to the axial plane and tangent to the pitch surfaces. In a single gear it may be any plane tangent to its pitch surface.

Pitch Point: This is the point of tangency of two pitch circles (or of a pitch circle and a pitch line) and is on the line of centers. The pitch point of a tooth profile is at its intersection with the pitch circle.

Plane of Rotation: Any plane perpendicular to a gear axis.

Pressure Angle: The angle between a tooth profile and a radial line at its pitch point. In involute teeth, pressure angle is often described as the angle between the line of action and the line tangent to the pitch circle. *Standard Pressure Angles* are established in connection with standard gear-tooth proportions. A given pair of involute profiles will transmit smooth motion at the same velocity ratio even when the center distance is changed. Changes in center distance, however, in gear design and gear manufacturing operations, are accompanied by changes in pitch diameter, pitch, and pressure angle. Different values of pitch diameter and pressure angle therefore may occur in the same gear under different conditions. Usually in gear design, and unless otherwise specified, the pressure angle is the *standard pressure angle* at the *standard pitch diameter,* and is standard for the hob or cutter used to generate the teeth. The *Operating Pressure Angle* is determined by the center distance at which a pair of gears operates. The *Generating Pressure Angle* is the angle at the pitch diameter in effect when the gear is generated. Other pressure angles may be considered in gear calculations. In gear cutting tools and cutters, the pressure angle indicates the direction of the cutting edge as referred to some principal direction. In oblique teeth, that is helical, spiral, etc., the pressure angle

may be specified in the *transverse, normal,* or *axial* plane. For a spur gear or a straight bevel gear, in which only one direction of cross-section needs to be considered, the general term pressure angle may be used without qualification to indicate transverse pressure angle. In spiral bevel gears, unless otherwise specified, pressure angle means normal pressure angle at the mean cone distance.

Principal Reference Planes: These are a pitch plane, axial plane, and transverse plane, all intersecting at a point and mutually perpendicular.

Rack: A gear with teeth spaced along a straight line, and suitable for straightline motion. A *Basic Rack* is one that is adopted as the basis of a system of interchangeable gears. Standard gear-tooth proportions are often illustrated on an outline of the basic rack (see diagrams on page 1773). A *Generating Rack* is a rack outline used to indicate tooth details and dimensions for the design of a required generating tool, such as a hob or gear-shaper cutter.

Ratio of Gearing: Ratio of the numbers of teeth on mating gears. Ordinarily the ratio is found by dividing the number of teeth on the larger gear by the number of teeth on the smaller gear or pinion. For example, if the ratio is 2 or "2 to 1," this usually means that the smaller gear or pinion makes two revolutions to one revolution of the larger mating gear.

Roll Angle: The angle subtended at the center of a base circle from the origin of an involute to the point of tangency of the generatrix from any point on the same involute. The radian measure of this angle is the tangent of the pressure angle of the point on the involute.

Root Circle: A circle coinciding with or tangent to the bottoms of the tooth spaces.

Root Diameter: Diameter of the root circle.

Tangent Plane: A plane tangent to the tooth surfaces at a point or line of contact.

Tip Relief: An arbitrary modification of a tooth profile whereby a small amount of material is removed near the tip of the gear tooth.

Total Face Width: The actual width dimension of a gear blank. It may exceed the effective face width, as in the case of double-helical gears where the total face width includes any distance separating the right-hand and left-hand helical teeth.

Transverse Plane: A plane perpendicular to the axial plane and to the pitch plane. In gears with parallel axes, the transverse plane and the plane of rotation coincide.

Trochoid: The curve formed by the path of a point on the extension of a radius of a circle as it rolls along a curve or line. It is also the curve formed by the path of a point on a perpendicular to a straight line as the straight line rolls along the convex side of a base curve. By the first definition the trochoid is derived from the cycloid; by the second definition it is derived from the involute.

True Involute Form Diameter: The smallest diameter on the tooth at which the involute exists. Usually this is the point of tangency of the involute tooth profile and the fillet curve. This is usually referred to as the *TIF diameter.*

Undercut: A condition in generated gear teeth when any part of the fillet curve lies inside of a line drawn tangent to the working profile at its lowest point. Undercut may be deliberately introduced to facilitate finishing operations, as in preshaving.

Whole Depth: The total depth of a tooth space, equal to addendum plus dedendum, also equal to working depth plus clearance.

Working Depth: The depth of engagement of two gears, that is, the sum of their addendums. The standard working distance is the depth to which a tooth extends into the tooth space of a mating gear when the center distance is standard.

Definitions of gear terms are given in AGMA Standards 112.05, 115.01, and 116.01 entitled "Terms, Definitions, Symbols and Abbreviations," "Reference Information — Basic Gear Geometry," and "Glossary — Terms Used in Gearing," respectively; obtainable from American Gear Manufacturers Assn., 1500 King. St., Arlington, Va. 22314.

Gear Teeth of Different Diametral Pitch, Full Size

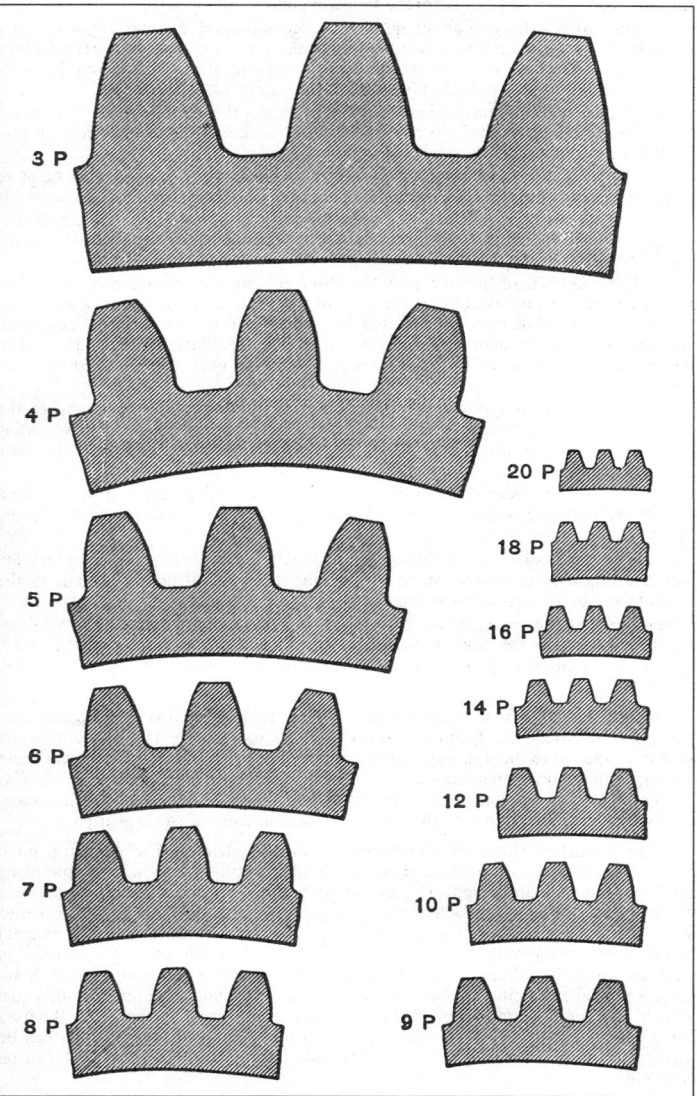

Properties of the Involute Curve. — The involute curve is used almost exclusively for gear-tooth profiles, because of the following important properties.

1. The form or shape of an involute curve depends upon the diameter of the base circle from which it is derived. (If a taut line were unwound from the circumference of a circle — the *base circle* of the involute — the end of that line or any point on the unwound portion, would describe an involute curve.)

2. If a gear tooth of involute curvature acts against the involute tooth of a mating gear while rotating at a uniform rate, the angular motion of the driven gear will also be uniform, even though the center-to-center distance is varied.

3. The relative rate of motion between driving and driven gears having involute tooth curves is established by the diameters of their base circles.

4. Contact between intermeshing involute teeth on a driving and driven gear is along a straight line that is tangent to the two base circles of these gears. This is the *line of action*.

5. The point where the line of action intersects the common center-line of the mating involute gears, establishes the radii of the pitch circles of these gears; hence true pitch circle diameters are affected by a change in the center distance. (Pitch diameters obtained by dividing the number of teeth by the diametral pitch apply when the center distance equals the total number of teeth on both gears divided by twice the diametral pitch.)

6. The pitch diameters of mating involute gears are directly proportional to the diameters of their respective base circles; thus, if the base circle of one mating gear is three times as large as the other, the pitch circle diameters will be in the same ratio.

7. The angle between the line of action and a line perpendicular to the common center-line of mating gears, is the *pressure angle;* hence the pressure angle is affected by any change in the center distance.

8. When an involute curve acts against a straight line (as in the case of an involute pinion acting against straight-sided rack teeth), the straight line is tangent to the involute and perpendicular to its line of action.

9. The pressure angle, in the case of an involute pinion acting against straight-sided rack teeth, is the angle between the line of action and the line of the rack's motion. If the involute pinion rotates at a uniform rate, movement of the rack will also be uniform.

Diametral and Circular Pitch Systems. — Gear tooth system standards are established by specifying the tooth proportions of the basic rack. The diametral pitch system is applied to most of the gearing produced in the United States. If gear teeth are larger than about one diametral pitch, it is common practice to use the circular pitch system. The circular pitch system is also applied to cast gearing and it is commonly used in connection with the design and manufacture of worm gearing.

Pitch Diameters Obtained with Diametral Pitch System. — The diametral pitch system is arranged to provide a series of standard tooth sizes, the principle being similar to the standardization of screw thread pitches. Inasmuch as there must be a whole number of teeth on each gear, the increase in pitch diameter per tooth varies according to the pitch. For example, the pitch diameter of a gear having, say, 20 teeth of 4 diametral pitch, will be 5 inches; 21 teeth, 5¼ inches; and so on, the increase in diameter for each additional tooth being equal to ¼ inch for 4 diametral pitch. Similarly, for 2 diametral pitch the variations for successive numbers of teeth would equal ½ inch, and for 10 diametral pitch the variations would equal ¹⁄₁₀ inch, etc. Where a given center distance must be maintained and no standard diametral pitch can be used, gears should be designed to the Variable Center Distance System shown on page 1782.

Table 1. Formulas for Dimensions of Standard Spur Gears

Notation		
ϕ = Pressure Angle		D_O = Outside Diameter
a = Addendum		D_R = Root Diameter
a_G = Addendum of Gear		F = Face Width
a_P = Addendum of Pinion		h_k = Working Depth of Tooth
b = Dedendum		h_t = Whole Depth of Tooth
c = Clearance		m_G = Gear Ratio
C = Center Distance		N = Number of Teeth
D = Pitch Diameter		N_G = Number of Teeth in Gear
D_G = Pitch Diameter of Gear		N_P = Number of Teeth in Pinion
D_P = Pitch Diameter of Pinion		p = Circular Pitch
D_B = Base Circle Diameter		P = Diametral Pitch

No.	To Find	Formula	No.	To Find	Formula
		General Formulas			
1	Base Circle Diameter	$D_B = D \cos \phi$	6a	Number of Teeth	$N = P \times D$
2a	Circular Pitch	$p = \dfrac{3.1416 D}{N}$	6b	Number of Teeth	$N = \dfrac{3.1416 D}{p}$
2b	Circular Pitch	$p = \dfrac{3.1416}{P}$	7a	Outside Diameter (Full-depth Teeth)	$D_O = \dfrac{N+2}{P}$
3a	Center Distance	$C = \dfrac{N_P(m_G + 1)}{2P}$	7b	Outside Diameter (Full-depth Teeth)	$D_O = \dfrac{(N+2)p}{3.1416}$
3b	Center Distance	$C = \dfrac{D_P + D_G}{2}$	8a	Outside Diameter (Amer. Stnd. Stub Teeth)	$D_O = \dfrac{N+1.6}{P}$
3c	Center Distance	$C = \dfrac{N_G + N_P}{2P}$	8b	Outside Diameter (Amer. Stnd. Stub Teeth)	$D_O = \dfrac{(N+1.6)p}{3.1416}$
3d	Center Distance	$C = \dfrac{(N_G + N_P)p}{6.2832}$	9	Outside Diameter	$D_O = D + 2a$
4a	Diametral Pitch	$P = \dfrac{3.1416}{p}$	10a	Pitch Diameter	$D = \dfrac{N}{P}$
4b	Diametral Pitch	$P = \dfrac{N}{D}$	10b	Pitch Diameter	$D = \dfrac{Np}{3.1416}$
4c	Diametral Pitch	$P = \dfrac{N_P(m_G + 1)}{2C}$	11	Root Diameter*	$D_R = D - 2b$
5	Gear Ratio	$m_G = \dfrac{N_G}{N_P}$	12	Whole Depth	$a + b$
			13	Working Depth	$a_G + a_P$

* See also formulas in table on pages 1772 and 1775.

American National Standard Coarse Pitch Spur Gear Tooth Forms. — The American National Standard (ANSI B6.1-1968, R1974) provides tooth proportion information on two involute spur gear forms. These two forms are identical except that one has a pressure angle of 20 degrees and a minimum allowable tooth number of 18 while the other has a pressure angle of 25 degrees and a minimum allowable tooth number of 12. (For pinions with fewer teeth, see tables of tooth proportions for long addendum pinions and their mating short addendum gears on pages 1797 to 1799.) A gear tooth standard is established by specifying the tooth proportions of the basic rack. Gears made to this standard will thus be conjugate with the specified rack and with each other. The basic rack forms for the 20-degree and 25-degree standard are shown on the following page; basic formulas for these proportions as a function of the gear diametral pitch and also of the circular pitch are given in Table 2. Tooth parts data are given in Table 3.

Table 2. Formulas for Tooth Parts — American National Standard Coarse Pitch Spur Gear Tooth Forms (ANSI B6.1-1968, R1974)

To Find	Diametral Pitch, P, Known	Circular Pitch, p, Known
20-DEGREE INVOLUTE FULL-DEPTH TEETH		
25-DEGREE INVOLUTE FULL-DEPTH TEETH		
Addendum	$a = 1.000 \div P$	$a = 0.3183 \times p$
Dedendum (Preferred) (Shaved or Ground Teeth)*	$b = 1.250 \div P$ $b = 1.350 \div P$	$b = 0.3979 \times p$ $b = 0.4297 \times p$
Working Depth	$h_k = 2.000 \div P$	$h_k = 0.6366 \times p$
Whole Depth (Preferred) (Shaved or Ground Teeth)	$h_t = 2.250 \div P$ $h_t = 2.350 \div P$	$h_t = 0.7162 \times p$ $h_t = 0.7480 \times p$
Clearance (Preferred)† (Shaved or Ground Teeth)	$c = 0.250 \div P$ $c = 0.350 \div P$	$c = 0.0796 \times p$ $c = 0.1114 \times p$
Fillet Radius (Rack)‡	$r_f = 0.300 \div P$	$r_f = 0.0955 \times p$
Pitch Diameter	$D = N \div P$	$D = 0.3183 \times Np$
Outside Diameter	$D_O = (N + 2) \div P$	$D_O = 0.3183 \times (N + 2)p$
Root Diameter (Preferred) (Shaved or Ground Teeth)	$D_R = (N - 2.5) \div P$ $D_R = (N - 2.7) \div P$	$D_R = 0.3183 \times (N - 2.5)p$ $D_R = 0.3183 \times (N - 2.7)p$
Circular Thickness — Basic	$t = 1.5708 \div P$	$t = p \div 2$

* When gears are preshave cut on a gear shaper the dedendum will usually need to be increased to $1.40/P$ to allow for the higher fillet trochoid produced by the shaper cutter. This is of particular importance on gears of few teeth or if the gear blank configuration requires the use of a small diameter shaper cutter, in which case the dedendum may need to be increased to as much as $1.45/P$. This should be avoided on highly loaded gears where the consequently reduced J factor will increase gear tooth stress excessively.

† A minimum clearance of $0.157/P$ may be used for the basic 20-degree and 25-degree pressure angle rack in the case of shallow root sections and use of existing hobs or cutters. However, whenever less than standard clearance is used, the location of the TIF diameter should be determined by the method shown on page 1805. The TIF diameter must be less than the Contact Diameter determined by the method shown on page 1803.

‡ The fillet radius of the basic rack should not exceed $0.235/P$ for a 20-degree pressure angle rack or $0.270/P$ for a 25-degree pressure angle rack for a clearance of $0.157/P$. The basic rack fillet radius must be *reduced* for teeth with a 25-degree pressure angle having a clearance in excess of $0.250/P$.

American National Standard and Former American Standard Gear Tooth Forms (ANSI B6.1-1968, R1974 and ASA B6.1-1932)

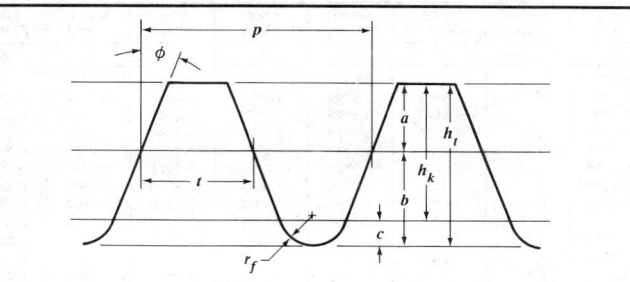

a = addendum h_k = working depth r_f = fillet radius of basic rack
b = dedendum h_t = whole depth   t = circular tooth thickness — basic
c = clearance   p = circular pitch ϕ = pressure angle

Basic Rack of the 20-Degree and 25-Degree Full-Depth Involute Systems

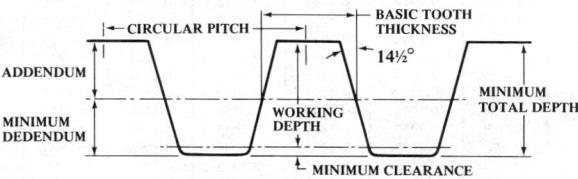

Basic Rack of the 14½-Degree Full-Depth Involute System

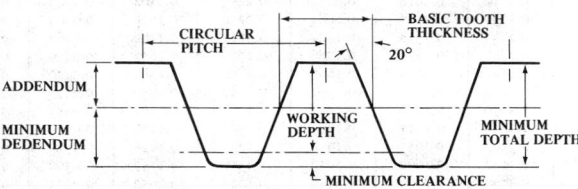

Basic Rack of the 20-Degree Stub Involute System

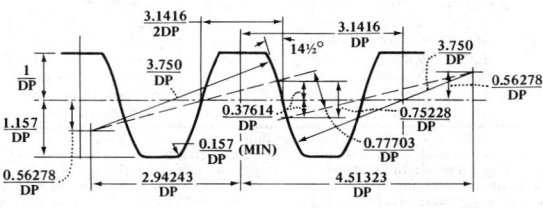

Approximation of Basic Rack for the 14½-Degree Composite System

Table 3. Gear Tooth Parts for American National Standard Coarse Pitch 20- and 25-Degree Pressure Angle Gears

Diam. Pitch	Circ. Pitch	Stand. Addend.*	Stand. Dedend.	Spec. Dedend.†	Min. Dedend.	Stand. F. Rad.	Min. F. Rad.
P	p	a	b	b	b	r_f	r_f
0.3142	10.	3.1831	3.9789	4.2972	3.6828	0.9549	0.4997
0.3307	9.5	3.0239	3.7799	4.0823	3.4987	0.9072	0.4748
0.3491	9.	2.8648	3.5810	3.8675	3.3146	0.8594	0.4498
0.3696	8.5	2.7056	3.3820	3.6526	3.1304	0.8117	0.4248
0.3927	8.	2.5465	3.1831	3.4377	2.9463	0.7639	0.3998
0.4189	7.5	2.3873	2.9842	3.2229	2.7621	0.7162	0.3748
0.4488	7.	2.2282	2.7852	3.0080	2.5780	0.6685	0.3498
0.4833	6.5	2.0690	2.5863	2.7932	2.3938	0.6207	0.3248
0.5236	6.	1.9099	2.3873	2.5783	2.2097	0.5730	0.2998
0.5712	5.5	1.7507	2.1884	2.3635	2.0256	0.5252	0.2749
0.6283	5.	1.5915	1.9894	2.1486	1.8414	0.4775	0.2499
0.6981	4.5	1.4324	1.7905	1.9337	1.6573	0.4297	0.2249
0.7854	4.	1.2732	1.5915	1.7189	1.4731	0.3820	0.1999
0.8976	3.5	1.1141	1.3926	1.5040	1.2890	0.3342	0.1749
1.	3.1416	1.0000	1.2500	1.3500	1.1570	0.3000	0.1570
1.25	2.5133	0.8000	1.0000	1.0800	0.9256	0.2400	0.1256
1.5	2.0944	0.6667	0.8333	0.9000	0.7713	0.2000	0.1047
1.75	1.7952	0.5714	0.7143	0.7714	0.6611	0.1714	0.0897
2.	1.5708	0.5000	0.6250	0.6750	0.5785	0.1500	0.0785
2.25	1.3963	0.4444	0.5556	0.6000	0.5142	0.1333	0.0698
2.5	1.2566	0.4000	0.5000	0.5400	0.4628	0.1200	0.0628
2.75	1.1424	0.3636	0.4545	0.4909	0.4207	0.1091	0.0571
3.	1.0472	0.3333	0.4167	0.4500	0.3857	0.1000	0.0523
3.25	0.9666	0.3077	0.3846	0.4154	0.3560	0.0923	0.0483
3.5	0.8976	0.2857	0.3571	0.3857	0.3306	0.0857	0.0449
3.75	0.8378	0.2667	0.3333	0.3600	0.3085	0.0800	0.0419
4.	0.7854	0.2500	0.3125	0.3375	0.2893	0.0750	0.0392
4.5	0.6981	0.2222	0.2778	0.3000	0.2571	0.0667	0.0349
5.	0.6283	0.2000	0.2500	0.2700	0.2314	0.0600	0.0314
5.5	0.5712	0.1818	0.2273	0.2455	0.2104	0.0545	0.0285
6.	0.5236	0.1667	0.2083	0.2250	0.1928	0.0500	0.0262
6.5	0.4833	0.1538	0.1923	0.2077	0.1780	0.0462	0.0242
7.	0.4488	0.1429	0.1786	0.1929	0.1653	0.0429	0.0224
7.5	0.4189	0.1333	0.1667	0.1800	0.1543	0.0400	0.0209
8.	0.3927	0.1250	0.1563	0.1687	0.1446	0.0375	0.0196
8.5	0.3696	0.1176	0.1471	0.1588	0.1361	0.0353	0.0185
9.	0.3491	0.1111	0.1389	0.1500	0.1286	0.0333	0.0174
9.5	0.3307	0.1053	0.1316	0.1421	0.1218	0.0316	0.0165
10.	0.3142	0.1000	0.1250	0.1350	0.1157	0.0300	0.0157
11.	0.2856	0.0909	0.1136	0.1227	0.1052	0.0273	0.0143
12.	0.2618	0.0833	0.1042	0.1125	0.0964	0.0250	0.0131
13.	0.2417	0.0769	0.0962	0.1038	0.0890	0.0231	0.0121
14.	0.2244	0.0714	0.0893	0.0964	0.0826	0.0214	0.0112
15.	0.2094	0.0667	0.0833	0.0900	0.0771	0.0200	0.0105
16.	0.1963	0.0625	0.0781	0.0844	0.0723	0.0188	0.0098
17.	0.1848	0.0588	0.0735	0.0794	0.0681	0.0176	0.0092
18.	0.1745	0.0556	0.0694	0.0750	0.0643	0.0167	0.0087
19.	0.1653	0.0526	0.0658	0.0711	0.0609	0.0158	0.0083
20.	0.1571	0.0500	0.0625	0.0675	0.0579	0.0150	0.0079

The working depth is equal to twice the addendum.

The whole depth is equal to the addendum plus the dedendum.

* When using equal addendums on pinion and gear the minimum number of teeth on the pinion is 18 and the minimum total number of teeth in the pair is 36 for 20-degree full depth involute tooth form and 12 and 24, respectively, for 25-degree full depth tooth form.

† The dedendum in this column is used when the gear tooth is shaved. It allows for the higher fillet cut by a protuberance hob.

In recent years the established standard of almost universal use is the ANSI 20-degree standard spur gear form. It provides a gear with good strength and without fillet undercut in pinions of as few as eighteen teeth. Some more recent applications have required a tooth form of even greater strength and fewer teeth than eighteen. This requirement has stimulated the establishment of the ANSI 25-degree standard. This 25-degree form will give greater tooth strength than the 20-degree standard, will provide pinions of as few as twelve teeth without fillet undercut and will provide a lower contact compressive stress for greater gear set surface durability.

Table 4. Tooth Proportions for Fine-Pitch Involute Spur and Helical Gears of 14½-, 20-, and 25-Degree Pressure Angle — American National Standard (ANSI B6.7-1977)

Item	Spur	Helical
Addendum, a	$\dfrac{1.000}{P}$	$\dfrac{1.000}{P_n}$
Dedendum, b	$\dfrac{1.200}{P} + 0.002$ (min.)	$\dfrac{1.200}{P_n} + 0.002$ (min.)
Working Depth, h_k	$\dfrac{2.000}{P}$	$\dfrac{2.000}{P_n}$
Whole Depth, h_t	$\dfrac{2.200}{P} + 0.002$ (min.)	$\dfrac{2.200}{P_n} + 0.002$ (min.)
Clearance, c (Standard)	$\dfrac{0.200}{P} + 0.002$ (min.)	$\dfrac{0.200}{P_n} + 0.002$ (min.)
(Shaved or Ground Teeth)	$\dfrac{0.350}{P} + 0.002$ (min.)	$\dfrac{0.350}{P_n} + 0.002$ (min.)
Tooth Thickness, t At Pitch Diameter	$t = \dfrac{1.5708}{P}$	$t_n = \dfrac{1.5708}{P_n}$
Circular Pitch, p	$p = \dfrac{\pi D}{N}$ or $\dfrac{\pi d}{n}$ or $\dfrac{\pi}{P}$	$p_n = \dfrac{\pi}{P_n}$
Pitch Diameter Pinion, d	$\dfrac{n}{P}$	$\dfrac{n}{P_n \cos \psi}$
Gear, D	$\dfrac{N}{P}$	$\dfrac{N}{P_n \cos \psi}$
Outside Diameter Pinion, d_o	$\dfrac{n+2}{P}$	$\dfrac{1}{P_n}\left(\dfrac{n}{\cos \psi} + 2\right)$
Gear, D_o	$\dfrac{N+2}{P}$	$\dfrac{1}{P_n}\left(\dfrac{N}{\cos \psi} + 2\right)$
Center Distance, C	$\dfrac{N+n}{2P}$	$\dfrac{N+n}{2P_n \cos \psi}$

All dimensions are in inches.

P = Transverse Diametral Pitch
P_n = Normal Diametral Pitch
t_n = Normal Tooth Thickness at Pitch Diameter
p_n = Normal Circular Pitch

ψ = Helix Angle
n = Number of pinion teeth
N = Number of gear teeth

American National Standard Tooth Proportions for Fine-Pitch Involute Spur and Helical Gears. — The proportions of spur gears in this Standard (ANSI B6.7-1977) follow closely ANSI B6.1-1968, R1974, "Tooth Proportions for Coarse-Pitch Involute Spur Gears." The main difference between fine-pitch and coarse-pitch gears is the greater clearance specified for fine-pitch gears. The increased clearance provides for any foreign material that may tend to accumulate at the bottoms of the teeth and also the relatively larger fillet radius resulting from proportionately greater wear on the tips of fine-pitch cutting tools.

Pressure Angle: The standard pressure angle for fine-pitch gears is 20 degrees and is recommended for most applications. For helical gears this pressure angle applies in the *normal* plane. In certain cases, notably sintered or molded gears, or in gearing where greatest strength and wear resistance are desired, a 25-degree pressure angle may be required. However, pressure angles greater than 20 degrees tend to require use of generating tools having very narrow point widths, and higher pressure angles require closer control of center distance when backlash requirements are critical.

In those cases where consideration of angular position or backlash is critical and both pinion and gear contain relatively large numbers of teeth, a 14½-degree pressure angle may be desirable. In general, pressure angles less than 20 degrees require greater amounts of tooth modification to avoid undercutting problems and are limited to larger total numbers of teeth in pinion and gear when operating at a standard center distance. Information Sheet B in the Standard provides tooth proportions for both 14½- and 25-degree pressure angle fine-pitch gears. Table 4 provides tooth proportions for fine-pitch spur and helical gears with 14½-, 20-, and 25-degree pressure angles, and Table 5 provides tooth parts.

Diametral Pitches: Diametral pitches preferred are: 20, 24, 32, 40, 48, 64, 72, 80, 96, and 120.

Other American Spur Gear Standards. — An appended information sheet in the American National Standard (ANSI B6.1-1968, R1974) provides tooth proportion information for three spur gear forms with the notice that they are "not recommended for new designs." These forms are therefore considered to be obsolescent but the information is given on their proportions because they have been used widely in the past. These forms are the 14½-degree full depth form, the 20-degree stub involute form and the 14½-degree composite form which were covered in the former American Standard (ASA B6.1-1932). The basic rack for the 14½-degree

Table 5. American National Standard Fine Pitch Standard Gear Tooth Parts — 14½-, 20-, and 25-Degree Pressure Angles

Diametral Pitch	Circular Pitch	Circular Thickness	Standard Addend.	Standard Dedend.	Special Dedend.*
P	p	t	a	b	b
20	.1571	.0785	.0500	.0620	.0695
24	.1309	.0654	.0417	.0520	.0582
32	.0982	.0491	.0313	.0395	.0442
40	.0785	.0393	.0250	.0320	.0358
48	.0654	.0327	.0208	.0270	.0301
64	.0491	.0245	.0156	.0208	.0231
72	.0436	.0218	.0139	.0187	.0208
80	.0393	.0196	.0125	.0170	.0189
96	.0327	.0164	.0104	.0145	.0161
120	.0262	.0131	.0083	.0120	.0132

* Based upon clearance for shaved or ground teeth.

The working depth is equal to twice the addendum. The whole depth is equal to the addendum plus the dedendum. For minimum number of teeth see page 1797.

Table 6. Formulas for Tooth Parts — Former American Standard Spur Gear Tooth Forms (ASA B6.1-1932)

To Find	Diametral Pitch, P Known	Circular Pitch, p Known
14½-DEGREE INVOLUTE FULL-DEPTH TEETH		
14½-DEGREE COMPOSITE FULL-DEPTH TEETH		
Addendum	$a = 1.000 \div P$	$a = 0.3183 \times p$
Minimum Dedendum	$b = 1.157 \div P$	$b = 0.3683 \times p$
Working Depth	$h_k = 2.000 \div P$	$h_k = 0.6366 \times p$
Minimum Whole Depth	$h_t = 2.157 \div P$	$h_t = 0.6866 \times p$
Basic Tooth Thickness on Pitch Line	$t = 1.5708 \div P$	$t = 0.500 \times p$
Minimum Clearance	$c = 0.157 \div P$	$c = 0.050 \times p$
20-DEGREE INVOLUTE STUB TEETH		
Addendum	$a = 0.800 \div P$	$a = 0.2546 \times p$
Minimum Dedendum	$b = 1.000 \div P$	$b = 0.3183 \times p$
Working Depth	$h_k = 1.600 \div P$	$h_k = 0.5092 \times p$
Minimum Whole Depth	$h_t = 1.800 \div P$	$h_t = 0.5729 \times p$
Basic Tooth Thickness on Pitch Line	$t = 1.5708 \div P$	$t = 0.500 \times p$
Minimum Clearance	$c = 0.200 \div P$	$c = 0.0637 \times p$

Note: Radius of fillet equals $1\frac{1}{3} \times$ clearance for 14½-degree full-depth teeth and $1\frac{1}{2} \times$ clearance for 20-degree full-depth teeth.

Note: A suitable working tolerance should be considered in connection with all minimum recommendations.

full depth form is shown on page 1773; basic formulas for these proportions are given in Table 6. Tooth parts data are given in Tables 7 to 9, inclusive.

Fellows Stub Tooth. — The system of stub gear teeth introduced by the Fellows Gear Shaper Co. is based upon the use of two diametral pitches. One diametral pitch, say, 8, is used as the basis for obtaining the dimensions for the addendum and dedendum, while another diametral pitch, say, 6, is used for obtaining the dimensions of the thickness of the tooth, the number of teeth, and the pitch diameter. Teeth made according to this system are designated as ⁶⁄₈ pitch, ¹²⁄₁₄ pitch, etc., the numerator in this fraction indicating the pitch determining the thickness of the tooth, and the denominator, the pitch determining the depth of the tooth. The clearance is made greater than in the ordinary gear-tooth system and equals $0.25 \div$ denominator of the diametral pitch. The pressure angle is 20 degrees.

This type of stub gear tooth is now used infrequently. For information as to the tooth part dimensions see 18th and earlier editions of Machinery's Handbook.

Table 7. Gear Tooth Parts — Former American Standard 14½-degree Involute Full-depth and 14½-degree Composite Systems* (ASA B6.1-1932)
(Diametral Pitch Gears)

Diametral Pitch	Equivalent Circular Pitch	Standard Circular Thickness	Standard Addendum	Working Depth of Tooth	Standard Dedendum	Whole Depth of Tooth
P	p	t	a	h_k	b	h_t
½	6.2832	3.1416	2.0000	4.0000	2.3142	4.3142
¾	4.1888	2.0944	1.3333	2.6666	1.5428	2.8761
1	3.1416	1.5708	1.0000	2.0000	1.1571	2.1571
1¼	2.5133	1.2566	0.8000	1.6000	0.9257	1.7257
1½	2.0944	1.0472	0.6666	1.3333	0.7714	1.4381
1¾	1.7952	0.8976	0.5714	1.1429	0.6612	1.2326
2	1.5708	0.7854	0.5000	1.0000	0.5785	1.0785
2¼	1.3963	0.6981	0.4444	0.8888	0.5143	0.9587
2½	1.2566	0.6283	0.4000	0.8000	0.4628	0.8628
2¾	1.1424	0.5712	0.3636	0.7273	0.4208	0.7844
3	1.0472	0.5236	0.3333	0.6666	0.3857	0.7190
3½	0.8976	0.4488	0.2857	0.5714	0.3306	0.6163
4	0.7854	0.3927	0.2500	0.5000	0.2893	0.5393
5	0.6283	0.3142	0.2000	0.4000	0.2314	0.4314
6	0.5236	0.2618	0.1666	0.3333	0.1928	0.3595
7	0.4488	0.2244	0.1429	0.2857	0.1653	0.3081
8	0.3927	0.1963	0.1250	0.2500	0.1446	0.2696
9	0.3491	0.1745	0.1111	0.2222	0.1286	0.2397
10	0.3142	0.1571	0.1000	0.2000	0.1157	0.2157
11	0.2856	0.1428	0.0909	0.1818	0.1052	0.1961
12	0.2618	0.1309	0.0833	0.1666	0.0964	0.1798
13	0.2417	0.1208	0.0769	0.1538	0.0890	0.1659
14	0.2244	0.1122	0.0714	0.1429	0.0826	0.1541
15	0.2094	0.1047	0.0666	0.1333	0.0771	0.1438
16	0.1963	0.0982	0.0625	0.1250	0.0723	0.1348
17	0.1848	0.0924	0.0588	0.1176	0.0681	0.1269
18	0.1745	0.0873	0.0555	0.1111	0.0643	0.1198
19	0.1653	0.0827	0.0526	0.1053	0.0609	0.1135
20	0.1571	0.0785	0.0500	0.1000	0.0579	0.1079
22	0.1428	0.0714	0.0455	0.0909	0.0526	0.0980
24	0.1309	0.0654	0.0417	0.0833	0.0482	0.0898
26	0.1208	0.0604	0.0385	0.0769	0.0445	0.0829
28	0.1122	0.0561	0.0357	0.0714	0.0413	0.0770
30	0.1047	0.0524	0.0333	0.0666	0.0386	0.0719
32	0.0982	0.0491	0.0312	0.0625	0.0362	0.0674
34	0.0924	0.0462	0.0294	0.0588	0.0340	0.0634
36	0.0873	0.0436	0.0278	0.0555	0.0321	0.0599
38	0.0827	0.0413	0.0263	0.0526	0.0304	0.0568
40	0.0785	0.0393	0.0250	0.0500	0.0289	0.0539
42	0.0748	0.0374	0.0238	0.0476	0.0275	0.0514
44	0.0714	0.0357	0.0227	0.0455	0.0263	0.0490
46	0.0683	0.0341	0.0217	0.0435	0.0252	0.0469
48	0.0654	0.0327	0.0208	0.0417	0.0241	0.0449
50	0.0628	0.0314	0.0200	0.0400	0.0231	0.0431

* For 20-degree full-depth teeth of 20 diametral pitch and finer see table on page 1776.

Table 8. Gear Tooth Parts — Former American Standard 14½-degree Involute Full-depth and 14½-degree Composite Systems* (ASA B6.1-1932)
(Circular Pitch Gears)

Circular Pitch	Equivalent Diametral Pitch	Standard Circular Thickness	Standard Addendum	Working Depth of Tooth	Standard Dedendum	Whole Depth of Tooth
p	P	t	a	h_k	b	h_t
4	0.7854	2.0000	1.2732	2.5464	1.4732	2.7464
3½	0.8976	1.7500	1.1140	2.2281	1.2890	2.4031
3	1.0472	1.5000	0.9549	1.9098	1.1049	2.0598
2¾	1.1424	1.3750	0.8753	1.7506	1.0128	1.8881
2½	1.2566	1.2500	0.7957	1.5915	0.9207	1.7165
2¼	1.3963	1.1250	0.7162	1.4323	0.8287	1.5448
2	1.5708	1.0000	0.6366	1.2732	0.7366	1.3732
1⅞	1.6755	0.9375	0.5968	1.1937	0.6906	1.2874
1¾	1.7952	0.8750	0.5570	1.1141	0.6445	1.2016
1⅝	1.9333	0.8125	0.5173	1.0345	0.5985	1.1158
1½	2.0944	0.7500	0.4775	0.9549	0.5525	1.0299
1⁷⁄₁₆	2.1855	0.7187	0.4576	0.9151	0.5294	0.9870
1⅜	2.2848	0.6875	0.4377	0.8754	0.5064	0.9441
1⁵⁄₁₆	2.3936	0.6562	0.4178	0.8356	0.4834	0.9012
1¼	2.5133	0.6250	0.3979	0.7958	0.4604	0.8583
1³⁄₁₆	2.6456	0.5937	0.3780	0.7560	0.4374	0.8154
1⅛	2.7925	0.5625	0.3581	0.7162	0.4143	0.7724
1¹⁄₁₆	2.9568	0.5312	0.3382	0.6764	0.3913	0.7295
1	3.1416	0.5000	0.3183	0.6366	0.3683	0.6866
¹⁵⁄₁₆	3.3510	0.4687	0.2984	0.5968	0.3453	0.6437
⅞	3.5904	0.4375	0.2785	0.5570	0.3223	0.6007
¹³⁄₁₆	3.8666	0.4062	0.2586	0.5173	0.2993	0.5579
¾	4.1888	0.3750	0.2387	0.4775	0.2762	0.5150
¹¹⁄₁₆	4.5696	0.3437	0.2189	0.4377	0.2532	0.4720
⅔	4.7124	0.3333	0.2122	0.4244	0.2455	0.4577
⅝	5.0265	0.3125	0.1989	0.3979	0.2301	0.4291
⁹⁄₁₆	5.5851	0.2812	0.1790	0.3581	0.2071	0.3862
½	6.2832	0.2500	0.1592	0.3183	0.1842	0.3433
⁷⁄₁₆	7.1808	0.2187	0.1393	0.2785	0.1611	0.3003
⅖	7.8540	0.2000	0.1273	0.2546	0.1473	0.2746
⅜	8.3776	0.1875	0.1194	0.2387	0.1381	0.2575
⅓	9.4248	0.1666	0.1061	0.2122	0.1228	0.2289
⁵⁄₁₆	10.0531	0.1562	0.0995	0.1989	0.1151	0.2146
²⁄₇	10.9956	0.1429	0.0909	0.1819	0.1052	0.1962
¼	12.5664	0.1250	0.0796	0.1591	0.0921	0.1716
²⁄₉	14.1372	0.1111	0.0707	0.1415	0.0818	0.1526
⅕	15.7080	0.1000	0.0637	0.1273	0.0737	0.1373
³⁄₁₆	16.7552	0.0937	0.0597	0.1194	0.0690	0.1287
⅙	18.8496	0.0833	0.0531	0.1061	0.0614	0.1144
⅐	21.9911	0.0714	0.0455	0.0910	0.0526	0.0981
⅛	25.1327	0.0625	0.0398	0.0796	0.0460	0.0858
⅑	28.2743	0.0555	0.0354	0.0707	0.0409	0.0763
¹⁄₁₀	31.4159	0.0500	0.0318	0.0637	0.0368	0.0687
¹⁄₁₆	50.2655	0.0312	0.0199	0.0398	0.0230	0.0429

* For 20-degree full-depth teeth of 20 diametral pitch and finer see table on page 1776.

Table 9. Gear Tooth Parts — Former American Standard 20-degree Involute Stub Tooth System (ASA B6.1-1932)

Diametral Pitch	Circular Pitch	Standard Circular Thickness	Standard Addendum	Working Depth of Tooth	Standard Dedendum	Whole Depth of Tooth
P	p	t	a	h_k	b	h_t
½	6.2832	3.1416	1.6000	3.2000	2.0000	3.6000
¾	4.1888	2.0944	1.0667	2.1334	1.3333	2.4000
1	3.1416	1.5708	0.8000	1.6000	1.0000	1.8000
1¼	2.5133	1.2566	0.6400	1.2800	0.8000	1.4400
1½	2.0944	1.0472	0.5333	1.0666	0.6667	1.2000
1¾	1.7952	0.8976	0.4571	0.9142	0.5714	1.0285
2	1.5708	0.7854	0.4000	0.8000	0.5000	0.9000
2¼	1.3963	0.6981	0.3556	0.7112	0.4444	0.8000
2½	1.2566	0.6283	0.3200	0.6400	0.4000	0.7200
2¾	1.1424	0.5712	0.2909	0.5818	0.3636	0.6545
3	1.0472	0.5236	0.2667	0.5334	0.3333	0.6000
3½	0.8976	0.4488	0.2286	0.4572	0.2857	0.5143
4	0.7854	0.3927	0.2000	0.4000	0.2500	0.4500
5	0.6283	0.3142	0.1600	0.3200	0.2000	0.3600
6	0.5236	0.2618	0.1333	0.2666	0.1667	0.3000
7	0.4488	0.2244	0.1143	0.2286	0.1428	0.2571
8	0.3927	0.1963	0.1000	0.2000	0.1250	0.2250
9	0.3491	0.1745	0.0889	0.1778	0.1111	0.2000
10	0.3142	0.1571	0.0800	0.1600	0.1000	0.1800
11	0.2856	0.1428	0.0727	0.1454	0.0909	0.1636
12	0.2618	0.1309	0.0667	0.1334	0.0833	0.1500
13	0.2417	0.1208	0.0615	0.1230	0.0769	0.1384
14	0.2244	0.1122	0.0571	0.1142	0.0714	0.1285
15	0.2094	0.1047	0.0533	0.1066	0.0667	0.1200
16	0.1963	0.0982	0.0500	0.1000	0.0625	0.1125
17	0.1848	0.0924	0.0470	0.0940	0.0588	0.1058
18	0.1745	0.0873	0.0444	0.0888	0.0556	0.1000
19	0.1653	0.0827	0.0421	0.0842	0.0526	0.0947
20	0.1571	0.0785	0.0400	0.0800	0.0500	0.0900
22	0.1428	0.0714	0.0364	0.0728	0.0454	0.0818
24	0.1309	0.0654	0.0333	0.0666	0.0417	0.0750
26	0.1208	0.0604	0.0308	0.0616	0.0384	0.0692
28	0.1122	0.0561	0.0286	0.0572	0.0357	0.0643
30	0.1047	0.0524	0.0267	0.0534	0.0333	0.0600
32	0.0982	0.0491	0.0250	0.0500	0.0312	0.0562
34	0.0924	0.0462	0.0236	0.0472	0.0294	0.0530
36	0.0873	0.0436	0.0222	0.0444	0.0278	0.0500
38	0.0827	0.0413	0.0210	0.0420	0.0263	0.0473
40	0.0785	0.0393	0.0200	0.0400	0.0250	0.0450

**Table 10. Formulas for Outside and Root Diameters of Spur Gears that are
Finish-hobbed, Shaped, or Pre-shaved**

Notation

D	= Pitch Diameter	a	= Standard Addendum
D_O	= Outside Diameter	b	= Standard Minimum Dedendum
D_R	= Root Diameter	b_s	= Standard Dedendum
P	= Diametral Pitch	b_{ps}	= Dedendum for Pre-shaving

$14\frac{1}{2}$-, 20-, AND 25-DEGREE INVOLUTE FULL-DEPTH TEETH
($19P$ and coarser)*

$$D_O = D + 2a = \frac{N}{P} + \left(2 \times \frac{1}{P}\right)$$

$$D_R = D - 2b = \frac{N}{P} - \left(2 \times \frac{1.157}{P}\right) \quad \text{(Hobbed)}[1]$$

$$D_R = D - 2b_s = \frac{N}{P} - \left(2 \times \frac{1.25}{P}\right) \quad \text{(Shaped)}[2]$$

$$D_R = D - 2b_{ps} = \frac{N}{P} - \left(2 \times \frac{1.35}{P}\right) \quad \text{(Pre-shaved)}[3]$$

$$D_R = D - 2b_{ps} = \frac{N}{P} - \left(2 \times \frac{1.40}{P}\right) \quad \text{(Pre-shaved)}[4]$$

20-DEGREE INVOLUTE FINE-PITCH FULL-DEPTH TEETH ($20P$ and finer)

$$D_O = D + 2a = \frac{N}{P} + \left(2 \times \frac{1}{P}\right)$$

$$D_R = D - 2b = \frac{N}{P} - 2\left(\frac{1.2}{P} + 0.002\right) \quad \text{(Hobbed or Shaped)}[5]$$

$$D_R = D - 2b_{ps} = \frac{N}{P} - 2\left(\frac{1.35}{P} + 0.002\right) \quad \text{(Pre-shaved)}[6]$$

20-DEGREE INVOLUTE STUB TEETH*

$$D_O = D + 2a = \frac{N}{P} + \left(2 \times \frac{0.8}{P}\right)$$

$$D_R = D - 2b = \frac{N}{P} - \left(2 \times \frac{1}{P}\right) \quad \text{(Hobbed)}$$

$$D_R = D - 2b_{ps} = \frac{N}{P} - \left(2 \times \frac{1.35}{P}\right) \quad \text{(Pre-shaved)}$$

* $14\frac{1}{2}$-degree full-depth and 20-degree stub teeth are not recommended for new designs.

[1] According to ANSI B6.1-1968 a minimum clearance of $0.157/P$ may be used for the basic 20-degree and 25-degree pressure angle rack in the case of shallow root sections and the use of existing hobs and cutters.

[2] According to ANSI B6.1-1968 the preferred clearance is $0.250/P$.

[3] According to ANSI B6.1-1968 the clearance for teeth which are shaved or ground is $0.350/P$.

[4] When gears are preshave cut on a gear shaper the dedendum will usually need to be increased to $1.40/P$ to allow for the higher fillet trochoid produced by the shaper cutter; this is of particular importance on gears of few teeth or if the gear blank configuration requires the use of a small diameter shaper cutter, in which case the dedendum may need to be increased to as much as $1.45/P$. This should be avoided on highly loaded gears where the consequently reduced J factor will increase gear tooth stress excessively.

[5] According to ANSI B6.7-1967 the standard clearance is $0.200/P + 0.002$ (min.).

[6] According to ANSI B6.7-1967 the clearance for shaved or ground teeth is $0.350/P + 0.002$ (min.).

Basic Gear Dimensions. — The basic dimensions for all involute spur gears may be obtained using the formulas shown in Table 1. This table is used in conjunction with Table 3 to obtain dimensions for coarse pitch gears and Table 5 to obtain dimensions for fine pitch standard spur gears. To obtain the dimensions of gears that are specified at a standard circular pitch, the equivalent diametral pitch is first calculated by using the formula in Table 1. If the required number of teeth in the pinion (N_P) is less than the minimum specified in either Table 3 or Table 5, whichever is applicable, the gears must be proportioned by the long and short addendum method shown on page 1797.

Gear Set Center Distance. — For any set of gears that operate on parallel shafts, the overall size of the gear set is essentially determined by the center distance. This center distance C is a function of the pinion pitch diameter D_P and the gear ratio m_G by this relation:

$$C = \frac{D_P (m_G + 1)}{2}$$

On page 1770 it is indicated that the pitch diameter will change in steps; using the above relation it is seen that the center distance will also change in steps. The following Table 11, "Gear Set Center Distance" is given to assist in preliminary gear set sizing. It gives center distances to three significant figures for pinion pitch diameters ranging from one-half to 24 inches and for a gear set ratio of from 1.25 to 10. This table must not be used to obtain the final center distance; for this determination the appropriate formula in Table 1 must be used. This table may also be used to estimate the required size of the space within a housing that is needed for the gear set. The length of this space is:

$$2(C + a)$$

where a = the gear addendum. The width of this space is the large gear outside diameter which may be estimated as:

$$D_P m_G + 2a$$

Gears for Given Center Distance and Ratio. — When it is necessary to use a pair of gears of given ratio at a specified center distance C_1, it may be found that no gears of standard diametral pitch will satisfy the center distance requirement. In this case, gears of standard diametral pitch P may be redesigned to operate at other than their standard pitch diameter D and standard pressure angle ϕ. The diametral pitch P_1 at which these gears will operate is:

$$P_1 = \frac{N_P + N_G}{2C_1} \tag{1}$$

where N_P = number of teeth in pinion
N_G = number of teeth in gear

and their operating pressure angle ϕ_1 is:

$$\phi_1 = \text{arc cos} \left(\frac{P_1}{P} \cos \phi \right) \tag{2}$$

Thus the pair of gears are cut to a diametral pitch P and a pressure angle ϕ, however, they operate as standard gears of diametral pitch P_1 and pressure angle ϕ_1.

Table 11. Gear Set Center Distance

Pitch Dia.	Ratio									
	1.25	1.5	1.75	2.	2.25	2.5	2.75	3.	3.5	4.
	Center Distance, Inches									
0.5	0.563	0.625	0.688	0.75	0.813	0.875	0.938	1.	1.13	1.25
0.6	0.675	0.75	0.825	0.9	0.975	1.05	1.12	1.2	1.35	1.5
0.7	0.787	0.875	0.962	1.05	1.14	1.22	1.31	1.4	1.57	1.75
0.8	0.9	1.	1.1	1.2	1.3	1.4	1.5	1.6	1.8	2.
0.9	1.01	1.12	1.24	1.35	1.46	1.57	1.69	1.8	2.02	2.25
1.	1.12	1.25	1.37	1.5	1.62	1.75	1.87	2.	2.25	2.5
1.1	1.24	1.37	1.51	1.65	1.79	1.92	2.06	2.2	2.47	2.75
1.2	1.35	1.6	1.65	1.8	1.95	2.1	2.25	2.4	2.7	3.
1.3	1.46	1.62	1.79	1.95	2.11	2.27	2.44	2.6	2.92	3.25
1.4	1.57	1.75	1.92	2.1	2.27	2.45	2.62	2.8	3.15	3.5
1.5	1.69	1.87	2.06	2.25	2.44	2.62	2.81	3.	3.37	3.75
1.6	1.8	2.	2.2	2.4	2.6	2.8	3.	3.2	3.6	4.
1.7	1.91	2.12	2.34	2.55	2.76	2.97	3.19	3.4	3.82	4.25
1.8	2.02	2.25	2.47	2.7	2.92	3.15	3.37	3.6	4.05	4.5
1.9	2.14	2.37	2.61	2.85	3.09	3.32	3.56	3.8	4.27	4.75
2.	2.25	2.5	2.75	3.	3.25	3.5	3.75	4.	4.5	5.
2.25	2.53	2.81	3.09	3.38	3.66	3.94	4.22	4.5	5.06	5.63
2.5	2.81	3.13	3.44	3.75	4.06	4.38	4.69	5.	5.63	6.25
2.75	3.09	3.44	3.78	4.13	4.47	4.81	5.16	5.5	6.19	6.88
3.	3.38	3.75	4.13	4.5	4.88	5.25	5.63	6.	6.75	7.5
3.25	3.66	4.06	4.47	4.88	5.28	5.69	6.09	6.5	7.31	8.13
3.5	3.94	4.38	4.81	5.25	5.69	6.13	6.56	7.	7.88	8.75
3.75	4.22	4.69	5.16	5.63	6.09	6.56	7.03	7.5	8.44	9.38
4.	4.5	5.	5.5	6.	6.5	7.	7.5	8.	9.	10.
4.25	4.78	5.31	5.84	6.38	6.91	7.44	7.97	8.5	9.56	10.6
4.5	5.06	5.63	6.19	6.75	7.31	7.88	8.44	9.	10.1	11.3
4.75	5.34	5.94	6.53	7.13	7.72	8.31	8.91	9.5	10.7	11.9
5.	5.63	6.25	6.88	7.5	8.13	8.75	9.38	10.	11.3	12.5
5.25	5.91	6.56	7.22	7.88	8.53	9.19	9.84	10.5	11.8	13.1
5.5	6.19	6.88	7.56	8.25	8.94	9.63	10.3	11.	12.4	13.8
5.75	6.47	7.19	7.91	8.63	9.34	10.1	10.8	11.5	12.9	14.4
6.	6.75	7.5	8.25	9.	9.75	10.5	11.3	12.	13.5	15.
6.5	7.31	8.13	8.94	9.75	10.6	11.4	12.2	13.	14.6	16.3
7.	7.88	8.75	9.63	10.5	11.4	12.3	13.1	14.	15.8	17.5
7.5	8.44	9.38	10.3	11.3	12.2	13.1	14.1	15.	16.9	18.8
8.	9.	10.	11.	12.	13.	14.	15.	16.	18.	20.
8.5	9.56	10.6	11.7	12.8	13.8	14.9	15.9	17.	19.1	21.3
9.	10.1	11.3	12.4	13.5	14.6	15.8	16.9	18.	20.3	22.5
9.5	10.7	11.9	13.1	14.3	15.4	16.6	17.8	19.	21.4	23.8
10.	11.3	12.5	13.8	15.	16.3	17.5	18.8	20.	22.5	25.
10.5	11.8	13.1	14.4	15.8	17.1	18.4	19.7	21.	23.6	26.3
11.	12.4	13.8	15.1	16.5	17.9	19.3	20.6	22.	24.8	27.5
11.5	12.9	14.4	15.8	17.3	18.7	20.1	21.6	23.	25.9	28.8
12.	13.5	15.	16.5	18.	19.5	21.	22.5	24.	27.	30.
13.	14.6	16.3	17.9	19.5	21.1	22.8	24.4	26.	29.3	32.5
14.	15.8	17.5	19.3	21.	22.8	24.5	26.3	28.	31.5	35.
15.	16.9	18.8	20.6	22.5	24.4	26.3	28.1	30.	33.8	37.5
16.	18.	20.	22.	24.	26.	28.	30.	32.	36.	40.
17.	19.1	21.3	23.4	25.5	27.6	29.8	31.9	34.	38.3	42.5
18.	20.3	22.5	24.8	27.	29.3	31.5	33.8	36.	40.5	45.
19.	21.4	23.8	26.1	28.5	30.9	33.3	35.6	38.	42.8	47.5
20.	22.5	25.	27.5	30.	32.5	35.	37.5	40.	45.	50.
21.	23.6	26.3	28.9	31.5	34.1	36.8	39.4	42.	47.3	52.5
22.	24.8	27.5	30.3	33.	35.8	38.5	41.3	44.	49.5	55.
23.	25.9	28.8	31.6	34.5	37.4	40.3	43.1	46.	51.8	57.5
24.	27.	30.	33.	36.	39.	42.	45.	48.	54.	60.

Table 11 (*Continued*). Gear Set Center Distance

Pitch Dia.	Ratio									
	4.5	5.	5.5	6.	6.5	7.	7.5	8.	9.	10.
	Center Distance, Inches									
0.5	1.38	1.5	1.63	1.75	1.88	2.	2.13	2.25	2.5	2.75
0.6	1.65	1.8	1.95	2.1	2.25	2.4	2.55	2.7	3.	3.3
0.7	1.92	2.1	2.27	2.45	2.62	2.8	2.97	3.15	3.5	3.85
0.8	2.2	2.4	2.6	2.8	3.	3.2	3.4	3.6	4.	4.4
0.9	2.47	2.7	2.92	3.15	3.37	3.6	3.82	4.05	4.5	4.95
1.	2.75	3.	3.25	3.5	3.75	4.	4.25	4.5	5.	5.5
1.1	3.02	3.3	3.57	3.85	4.12	4.4	4.67	4.95	5.5	6.05
1.2	3.3	3.6	3.9	4.2	4.5	4.8	5.1	5.4	6.	6.6
1.3	3.57	3.9	4.22	4.55	4.87	5.2	5.52	5.85	6.5	7.15
1.4	3.85	4.2	4.55	4.9	5.25	5.6	5.95	6.3	7.	7.7
1.5	4.12	4.5	4.87	5.25	5.62	6.	6.37	6.75	7.5	8.25
1.6	4.4	4.8	5.2	5.6	6.	6.4	6.8	7.2	8.	8.8
1.7	4.67	5.1	5.52	5.95	6.37	6.8	7.22	7.65	8.5	9.35
1.8	4.95	5.4	5.85	6.3	6.75	7.2	7.65	8.1	9.	9.9
1.9	5.22	5.7	6.17	6.65	7.12	7.6	8.07	8.55	9.5	10.4
2.	5.5	6.	6.5	7.	7.5	8.	8.5	9.	10.	11.
2.25	6.19	6.75	7.31	7.88	8.44	9.	9.56	10.1	11.3	12.4
2.5	6.88	7.5	8.13	8.75	9.38	10.	10.6	11.3	12.5	13.8
2.75	7.56	8.25	8.94	9.63	10.3	11.	11.7	12.4	13.8	15.1
3.	8.25	9.	9.75	10.5	11.3	12.	12.8	13.5	15.	16.5
3.25	8.94	9.75	10.6	11.4	12.2	13.	13.8	14.6	16.3	17.9
3.5	9.63	10.5	11.4	12.3	13.1	14.	14.9	15.8	17.5	19.3
3.75	10.3	11.3	12.2	13.1	14.1	15.	15.9	16.9	18.8	20.6
4.	11.	12.	13.	14.	15.	16.	17.	18.	20.	22.
4.25	11.7	12.8	13.8	14.9	15.9	17.	18.1	19.1	21.3	23.4
4.5	12.4	13.5	14.6	15.8	16.9	18.	19.1	20.3	22.5	24.8
4.75	13.1	14.3	15.4	16.6	17.8	19.	20.2	21.4	23.8	26.1
5.	13.8	15.	16.3	17.5	18.8	20.	21.3	22.5	25.	27.5
5.25	14.4	15.8	17.1	18.4	19.7	21.	22.3	23.6	26.3	28.9
5.5	15.1	16.5	17.9	19.3	20.6	22.	23.4	24.8	27.5	30.3
5.75	15.8	17.3	18.7	20.1	21.6	23.	24.4	25.9	28.8	31.6
6.	16.5	18.	19.5	21.	22.5	24.	25.5	27.	30.	33.
6.5	17.9	19.5	21.1	22.8	24.4	26.	27.6	29.3	32.5	35.8
7.	19.3	21.	22.8	24.5	26.3	28.	29.8	31.5	35.	38.5
7.5	20.6	22.5	24.4	26.3	28.1	30.	31.9	33.8	37.5	41.3
8.	22.	24.	26.	28.	30.	32.	34.	36.	40.	44.
8.5	23.4	25.5	27.6	29.8	31.9	34.	36.1	38.3	42.5	46.8
9.	24.8	27.	29.3	31.5	33.8	36.	38.3	40.5	45.	49.5
9.5	26.1	28.5	30.9	33.3	35.6	38.	40.4	42.8	47.5	52.3
10.	27.5	30.	32.5	35.	37.5	40.	42.5	45.	50.	55.
10.5	28.9	31.5	34.1	36.8	39.4	42.	44.6	47.3	52.5	57.8
11.	30.3	33.	35.8	38.5	41.3	44.	46.8	49.5	55.	60.5
11.5	31.6	34.5	37.4	40.3	43.1	46.	48.9	51.8	57.5	63.3
12.	33.	36.	39.	42.	45.	48.	51.	54.	60.	66.
13.	35.8	39.	42.3	45.5	48.8	52.	55.3	58.5	65.	71.5
14.	38.5	42.	45.5	49.	52.5	56.	59.5	63.	70.	77.
15.	41.3	45.	48.8	52.5	56.3	60.	63.8	67.5	75.	82.5
16.	44.	48.	52.	56.	60.	64.	68.	72.	80.	88.
17.	46.8	51.	55.3	59.5	63.8	68.	72.3	76.5	85.	93.5
18.	49.5	54.	58.5	63.	67.5	72.	76.5	81.	90.	99.
19.	52.3	57.	61.8	66.5	71.3	76.	80.8	85.5	95.	105.
20.	55.	60.	65.	70.	75.	80.	85.	90.	100.	110.
21.	57.8	63.	68.3	73.5	78.8	84.	89.3	94.5	105.	116.
22.	60.5	66.	71.5	77.	82.5	88.	93.5	99.	110.	121.
23.	63.3	69.	74.8	80.5	86.3	92.	97.8	104.	115.	127.
24.	66.	72.	78.	84.	90.	96.	102.	108.	120.	132.

The pitch P and pressure angle ϕ should be chosen so that ϕ_1 lies between about 18 and 25 degrees.

The operating pitch diameters of the pinion D_{P_1} and of the gear D_{G_1} are:

$$D_{P_1} = \frac{N_P}{P_1} \text{ and } D_{G_1} = \frac{N_G}{P_1} \qquad \text{(3a) and (3b)}$$

The base diameters of the pinion D_{PB_1} and of the gear D_{GB_1} are:

$$D_{PB_1} = D_{P_1} \cos\phi_1 \text{ and } D_{GB_1} = D_{G_1} \cos\phi_1 \qquad \text{(4a) and (4b)}$$

The basic tooth thickness, t_1, at the operating pitch diameter for both pinion and gear is:

$$t_1 = \frac{1.5708}{P_1} \qquad (5)$$

The root diameters of the pinion D_{PR_1} and gear D_{GR_1} and the corresponding outside diameters D_{PO_1} and D_{GO_1} are not standard because each gear is to be cut with a cutter that is not standard for the operating pitch diameters D_{P_1} and D_{G_1}.

The root diameters are:

$$D_{PR_1} = \frac{N_P}{P} - 2b_{P_1} \text{ and } D_{GR_1} = \frac{N_G}{P} - 2b_{G_1} \qquad \text{(6a) and (6b)}$$

where:

$$b_{P_1} = b_c - \left(\frac{t_{P_2} - \frac{1.5708}{P}}{2 \tan\phi} \right) \qquad (7a)$$

and

$$b_{G_1} = b_c - \frac{t_{G_2} - \frac{1.5708}{P}}{2 \tan\phi} \qquad (7b)$$

where b_c is the hob or cutter addendum for the pinion and gear.

The tooth thicknesses of the pinion t_{P_2} and the gear t_{G_2} are:

$$t_{P_2} = \frac{N_P}{P} \left(\frac{1.5708}{N_P} + \text{inv}\,\phi_1 - \text{inv}\,\phi \right) \qquad (8a)$$

$$t_{G_2} = \frac{N_G}{P} \left(\frac{1.5708}{N_G} + \text{inv}\,\phi_1 - \text{inv}\,\phi \right) \qquad (8b)$$

The outside diameter of the pinion D_{PO} and the gear D_{GO} are:

$$D_{PO} = 2 \times C_1 - D_{GR_1} - 2 \left(b_c - \frac{1}{P} \right) \qquad (9a)$$

and

$$D_{GO} = 2 \times C_1 - D_{PR_1} - 2 \left(b_c - \frac{1}{P} \right) \qquad (9b)$$

Example: Design gears of 8 diametral pitch, 20-degree pressure angle and 28 and 88 teeth to operate at 7.50-inch center distance. The gears are to be cut with a hob of 0.169-inch addendum.

$$P_1 = \frac{28 + 88}{2 \times 7.50} = 7.7333 \qquad (1)$$

$$\phi_1 = \text{arc cos}\left(\frac{7.7333}{8} \times 0.93969\right) = 24.719° \tag{2}$$

$$D_{P_1} = \frac{28}{7.7333} = 3.6207 \text{ in.} \tag{3a}$$

and $$D_{G_1} = \frac{88}{7.7333} = 11.3794 \text{ in.} \tag{3b}$$

$$D_{PB_1} = 3.6207 \times 0.90837 = 3.2889 \text{ in.} \tag{4a}$$

and $$D_{GB_1} = 11.3794 \times 0.90837 = 10.3367 \text{ in.} \tag{4b}$$

$$t_1 = \frac{1.5708}{7.7333} = 0.20312 \text{ in.} \tag{5}$$

$$D_{PR_1} = \frac{28}{8} - 2 \times 0.1016 = 3.2968 \text{ in.} \tag{6a}$$

and $$D_{GR_1} = \frac{88}{8} - 2 \times (-0.0428) = 11.0856 \text{ in.} \tag{6b}$$

$$b_{P_1} = 0.169 - \left(\frac{0.2454 - \dfrac{1.5708}{8}}{2 \times 0.36397}\right) = 0.1016 \text{ in.} \tag{7a}$$

$$b_{G_1} = 0.169 - \left(\frac{0.3505 - \dfrac{1.5708}{8}}{2 \times 0.36397}\right) = -0.0428 \text{ in.} \tag{7b}$$

$$t_{P_2} = \frac{28}{8}\left(\frac{1.5708}{28} + 0.028922 - 0.014904\right) = 0.2454 \text{ in.} \tag{8a}$$

$$t_{G_2} = \frac{88}{8}\left(\frac{1.5708}{88} + 0.028922 - 0.014904\right) = 0.3505 \text{ in.} \tag{8b}$$

$$D_{PO_1} = 2 \times 7.50 - 11.0856 - 2\left(0.169 - \frac{1}{8}\right) = 3.8264 \text{ in.} \tag{9a}$$

$$D_{GO_1} = 2 \times 7.50 - 3.2968 - 2\left(0.169 - \frac{1}{8}\right) = 11.6152 \text{ in.} \tag{9b}$$

Tooth Thickness Allowance for Shaving. — Proper stock allowance is important for good results in shaving operations. If too much stock is left for shaving, the life of the shaving tool is reduced and, in addition, shaving time is increased. The following figures represent the amount of stock to be left on the teeth for removal by shaving under average conditions: For diametral pitches of 2 to 4, a thickness of 0.003 to 0.004 inch (one-half on each side of the tooth); for 5 to 6 diametral pitch, 0.0025 to 0.0035 inch; for 7 to 10 diametral pitch, 0.002 to 0.003 inch; for 11 to 14 diametral pitch, 0.0015 to 0.0020 inch; for 16 to 18 diametral pitch, 0.001 to 0.002 inch; for 20 to 48 diametral pitch, 0.0005 to 0.0015 inch; and for 52 to 72 diametral pitch, 0.0003 to 0.0007 inch.

The thickness of the gear teeth may be measured in several ways to determine the amount of stock left on the sides of the teeth to be removed by shaving. If it

is necessary to measure the tooth thickness during the pre-shaving operation while the gear is in the gear shaper or hobbing machine, a gear tooth caliper or pins would be employed. Caliper methods of measuring gear teeth are explained in detail on pages 1790 and 1791 for measurements over single teeth and on pages 1976 to 1978 for measurements over two or more teeth.

When the pre-shaved gear can be removed from the machine for checking, the center distance method may be employed. In this method the pre-shaved gear is meshed without backlash with a gear of standard tooth thickness and the increase in center distance is noted. The amount of total tooth thickness over standard that is left on the pre-shaved gear can then be determined by the formula: $t_2 = 2 \tan \phi \times d$, where: $t_2 =$ amount that total thickness of the tooth exceeds the standard thickness, $\phi =$ pressure angle, and $d =$ amount that the center distance between the two gears exceeds the standard center distance.

Circular Pitch for Given Center Distance and Ratio. — When it is necessary to use a pair of gears of given ratio at a specified center distance it may be found that no gears of standard diametral pitch will satisfy the center distance requirement. Hence, circular pitch gears may be selected. To find the required circular pitch p, when the center distance C and total number of teeth N in both gears are known, use the following formula:

$$p = \frac{C \times 6.2832}{N}$$

Example: A pair of gears having a ratio of 3 is to be used at a center distance of 10.230 inches. If one gear has 60 teeth and the other 20, what must be their circular pitch?

$$p = \frac{10.230 \times 6.2832}{60 + 20} = 0.8035 \text{ inch}$$

Circular Thickness of Tooth when Outside Diameter is Standard. — For a full-depth or stub tooth gear of standard outside diameter, the tooth thickness on the pitch circle (circular thickness or arc thickness) is found by the following formula:

$$t = \frac{1.5708}{P}$$

where $t =$ circular thickness and $P =$ diametral pitch. In the case of Fellows stub tooth gears the diametral pitch used is the numerator of the pitch fraction (for example, 6 if the pitch is 6/8).

Example 1: Find the tooth thickness on the pitch circle of a 14½-degree full-depth tooth of 12 diametral pitch.

$$t = \frac{1.5708}{12} = 0.1309 \text{ inch}$$

Example 2: Find the tooth thickness on the tooth circle of a 20-degree full-depth involute tooth having a diametral pitch of 5.

$$t = \frac{1.5708}{5} = 0.31416, \text{ say } 0.3142 \text{ inch}$$

The tooth thickness on the pitch circle can be determined very accurately by means of measurement over wires which are located in tooth spaces that are diametrically opposite or as nearly diametrically opposite as possible. Where

measurement over wires is not feasible, the circular or arc tooth thickness can be used in determining the chordal thickness which is the dimension measured with a gear tooth caliper.

Circular Thickness of Tooth when Outside Diameter has been Enlarged. — When the outside diameter of a small pinion is not standard but is enlarged to avoid undercut and to improve tooth action, the teeth are located farther out radially relative to the standard pitch diameter and consequently the circular arc thickness at the standard pitch diameter is increased. To find this increased arc thickness the following formula is used, where t = tooth thickness; e = amount outside diameter is increased over standard; ϕ = pressure angle; and p = circular pitch at the standard pitch diameter.

$$t = \frac{p}{2} + e \tan \phi$$

Example: The outside diameter of a pinion having 10 teeth of 5 diametral pitch and a pressure angle of 14½ degrees is to be increased by 0.2746 inch. The circular pitch equivalent to 5 diametral pitch is 0.6283 inch. Find the arc tooth thickness at the standard pitch diameter.

$$t = \frac{0.6283}{2} + (0.2746 \times \tan 14\tfrac{1}{2}°)$$

$$t = 0.3142 + (0.2746 \times 0.25862) = 0.3852 \text{ inch}$$

Circular Thickness of Tooth when Outside Diameter has been Reduced. — If the outside diameter of a gear is reduced, as is frequently done to maintain the standard center distance when the outside diameter of the mating pinion is increased, the circular thickness of the gear teeth at the standard pitch diameter will be reduced. This decreased circular thickness can be found by the following formula where t = circular thickness at the standard pitch diameter; e = amount outside diameter is reduced under standard; ϕ = pressure angle; and p = circular pitch.

$$t = \frac{p}{2} - e \tan \phi$$

Example: The outside diameter of a gear having a pressure angle of 14½ degrees is to be reduced by 0.2746 inch or an amount equal to the increase in diameter of its mating pinion. The circular pitch is 0.6283 inch. Determine the circular tooth thickness at the standard pitch diameter.

$$t = \frac{0.6283}{2} - (0.2746 \times \tan 14\tfrac{1}{2}°)$$

$$t = 0.3142 - (0.2746 \times 0.25862) = 0.2432 \text{ inch}$$

Chordal Thickness of Tooth when Outside Diameter is Standard. — To find the chordal or straight line thickness of a gear tooth the following formula can be used where t_c = chordal thickness; D = pitch diameter; and N = number of teeth.

$$t_c = D \sin\left(\frac{90°}{N}\right)$$

Example: A pinion has 15 teeth of 3 diametral pitch; the pitch diameter is equal to $15 \div 3$ or 5 inches. Find the chordal thickness at the standard pitch diameter.

$$t_c = 5 \ \sin\left(\frac{90°}{15}\right) = 5 \sin 6°$$

$$t_c = 5 \times 0.10453 = 0.5226 \text{ inch}$$

Chordal Thickness of Tooth when Outside Diameter is Special. — When the outside diameter is larger or smaller than standard the chordal thickness at the standard pitch diameter is found by the following formula where t_c = chordal thickness at the standard pitch diameter D; t = circular thickness at the standard pitch diameter of the enlarged pinion or reduced gear being measured.

$$t_c = t - \frac{t^3}{6 \times D^2}$$

Example 1: The outside diameter of a pinion having 10 teeth of 5 diametral pitch has been *enlarged* by 0.2746 inch. This enlargement has increased the circular tooth thickness at the standard pitch diameter (as determined by the formula previously given) to 0.3852 inch. Find the equivalent chordal thickness.

$$t_c = 0.3852 - \frac{(0.385)^3}{6 \times (2)^2} = 0.3852 - 0.0024 = 0.3828 \text{ inch.}$$

(The error introduced by rounding the circular thickness to three significant figures before cubing it only affects the fifth decimal place in the result.)

Example 2: A gear having 30 teeth is to mesh with the pinion in Example 1 and is *reduced* so that the circular tooth thickness at the standard pitch diameter is 0.2432 inch. Find the equivalent chordal thickness.

$$t_c = 0.2432 - \frac{(0.243)^3}{6 \times (6)^2} = 0.2432 - 0.00007 = 0.2431 \text{ inch.}$$

Chordal Addendum. — In measuring the chordal thickness, the vertical scale of a gear tooth caliper is set to the chordal or "corrected" addendum to locate the caliper jaws at the pitch line (see illustration on page 1791). The simplified formula which follows may be used in determining the chordal addendum either when the addendum is standard for full-depth or stub teeth or when the addendum is either longer or shorter than standard as in case of an enlarged pinion or a gear which is to mesh with an enlarged pinion and has a reduced addendum to maintain the standard center distance. If a_c = chordal addendum; a = addendum; and t = circular thickness of tooth at pitch diameter D; then,

$$a_c = a + \frac{t^2}{4D}$$

Example 1: The outside diameter of an 8 diametral pitch 14-tooth pinion with 20-degree full-depth teeth is to be increased by using an enlarged addendum of $1.234 \div 8 = 0.1542$ inch (see Table 1 on page 1797). The basic tooth thickness of the enlarged pinion is $1.741 \div 8 = 0.2176$ inch. What is the chordal addendum?

$$\text{Chordal addendum} = 0.1542 + \frac{0.2176^2}{4(14 \div 8)}$$

$$= 0.1610 \text{ inch.}$$

Example 2: The outside diameter of a 14½-degree pinion having 12 teeth of 2 diametral pitch is to be enlarged 0.624 inch to avoid undercut (see Table 2, on page 652), thus increasing the addendum from 0.5000 to 0.8120 inch and the arc thickness at the pitch line from 0.7854 to 0.9467 inch. Then,

$$\text{Chordal addendum of pinion} = 0.8120 + \frac{0.9467^2}{4 \times (12 \div 2)} = 0.8493 \text{ inch}$$

Example 3: The outside diameter of the mating gear for the pinion in Example 2 is to be reduced 0.624 inch. The gear has 60 teeth and the addendum is reduced from 0.5000 to 0.1881 inch (to maintain the standard center distance), thus reducing the arc thickness to 0.6240 inch. Then,

$$\text{Chordal addendum of gear} = 0.1881 + \frac{0.6240^2}{4 \times (60 \div 2)} = 0.1913 \text{ inch}$$

When a gear addendum is reduced as much as the mating pinion addendum is enlarged, the minimum number of gear teeth required to prevent undercutting depends upon the enlargement of the mating pinion. To illustrate, if a 14½-degree pinion with 13 teeth is enlarged 1.185 inches, then the reduced mating gear should have a minimum of 51 teeth to avoid undercut (see Table 2, page 1798).

Table for Chordal Thicknesses and Chordal Addenda of Full-depth Teeth. — The table on page 1792 gives values for chordal thickness and chordal addendum of full-depth spur gear teeth of 1 diametral pitch and from 10 to 156 teeth for gears of standard outside diameter. For any other diametral pitch the values are to be divided by the required pitch.

Helical Gears: In applying this table to helical gears, especially when the number of teeth is small and the helix angle large, the equivalent number of teeth N_e for entering the table is found by the formula, $N_e = N \div \cos^3 \psi$, where N is the actual number of teeth in the helical gear and ψ is the helix angle. The values obtained from the table should be divided by the *normal* diametral pitch of the helical gear to get the normal chordal thickness and the normal chordal addendum.

Example: Find the normal chordal thickness and the normal chordal addendum of a helical gear having 54 teeth of 6 normal diametral pitch and a helix angle of 45 degrees.

$$N_e = \frac{54}{\cos^3 45°} = \frac{54}{(0.70711)^3} = 153 \text{ teeth}$$

$$\text{Normal chordal thickness} = \frac{1.57077}{6} = 0.26180 \text{ inch}$$

$$\text{Normal chordal addendum} = \frac{1.00405}{6} = 0.16734 \text{ inch}$$

Tables for Chordal Thicknesses and Chordal Addenda of Milled, Full-depth Teeth. — Two convenient tables for checking gears with milled, full-depth teeth are given on pages 1795 and 1796. The first shows chordal thicknesses and chordal addenda for the lowest number of teeth cut by gear cutters Nos. 1 through 8, and for the commonly used diametral pitches. The second gives similar data for commonly used circular pitches. In each case the data shown are accurate for the number of gear teeth indicated, but are approximate for other numbers of teeth within the range of the cutter under which they appear in the table. For the higher diametral pitches and lower circular pitches, the error introduced by using the data for any tooth number within the range of the cutter under which it appears is comparatively small. The chordal thicknesses and chordal addenda for gear cutters Nos. 1 through 8 of the more commonly used diametral and circular pitches can be obtained from the table and formulas on pages 1795 and 1796.

Caliper Measurement of Gear Tooth. — In cutting gear teeth, the general practice is to adjust the cutter or hob until it grazes the outside diameter of the blank; the cutter is then sunk to the total depth of the tooth space plus whatever slight additional amount may be required to provide the necessary play or backlash

between the teeth. (For recommendations concerning backlash and excess depth of cut required, see pages 1829 to 1832.) If the outside diameter of the gear blank is correct, the tooth thickness should also be correct after the cutter has been sunk to the depth required for a given pitch and backlash. However, it is advisable to check

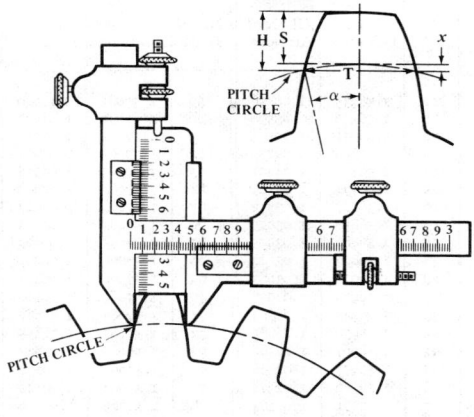

the tooth thickness by measuring it, and the vernier gear-tooth caliper (see accompanying illustration) is commonly used in measuring the thickness.

The vertical scale of this caliper is set so that when it rests upon the top of the tooth as shown, the lower ends of the caliper jaws will be at the height of the pitch circle; the horizontal scale then shows the chordal thickness of the tooth at this point. If the gear is being cut on a milling machine or with the type of gear-cutting machine employing a formed milling cutter, the tooth thickness is checked by first taking a trial cut for a short distance at one side of the blank; then the gear blank is indexed for the next space and another cut is taken far enough to mill the full outline of the tooth. The tooth thickness is then measured.

Before the gear-tooth caliper can be used, it is necessary to determine the correct chordal thickness and also the chordal addendum (or "corrected addendum" as it is sometimes called). The vertical scale is set to the chordal addendum, thus locating

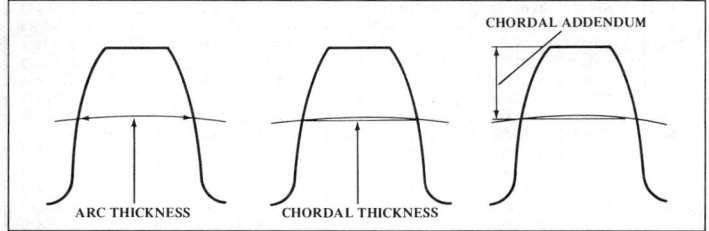

ARC THICKNESS   CHORDAL THICKNESS   CHORDAL ADDENDUM

the ends of the jaws at the height of the pitch circle. The rules or formulas to use in determining the chordal thickness and chordal addendum will depend upon the outside diameter of the gear; for example, if the outside diameter of a small pinion is enlarged to avoid undercut and improve the tooth action, this must be taken into account in figuring the chordal thickness and chordal addendum as shown by the accompanying rules. The detail of a gear tooth included with the gear-tooth caliper illustration, represents the chordal thickness T, the addendum S, and the chordal addendum H. For caliper measurements over two or more teeth see pages 1976 to 1978.

Chordal Thicknesses and Chordal Addenda of Full-depth Gear Teeth

This table is for spur gears of one diametral pitch. For any other diametral pitch, divide the given value by the required pitch. Table gives the chordal thickness and chordal addendum at the pitch circle when addendum is standard for full-depth teeth. Table is applicable to helical gears as explained on page 1790.

No. of Teeth	Chordal Thickness	Chordal Addend.	No. of Teeth	Chordal Thickness	Chordal Addend.	No. of Teeth	Chordal Thickness	Chordal Addend.
10	1.56435	1.06156	59	1.57061	1.01046	108	1.57074	1.00570
11	1.56546	1.05598	60	1.57062	1.01029	109	1.57075	1.00565
12	1.56631	1.05133	61	1.57062	1.01011	110	1.57075	1.00560
13	1.56698	1.04739	62	1.57063	1.00994	111	1.57075	1.00556
14	1.56752	1.04401	63	1.57063	1.00978	112	1.57075	1.00551
15	1.56794	1.04109	64	1.57064	1.00963	113	1.57075	1.00546
16	1.56827	1.03852	65	1.57064	1.00947	114	1.57075	1.00541
17	1.56856	1.03625	66	1.57065	1.00933	115	1.57075	1.00537
18	1.56880	1.03425	67	1.57065	1.00920	116	1.57075	1.00533
19	1.56899	1.03244	68	1.57066	1.00907	117	1.57075	1.00529
20	1.56918	1.03083	69	1.57066	1.00893	118	1.57075	1.00524
21	1.56933	1.02936	70	1.57067	1.00880	119	1.57075	1.00519
22	1.56948	1.02803	71	1.57067	1.00867	120	1.57075	1.00515
23	1.56956	1.02681	72	1.57067	1.00855	121	1.57075	1.00511
24	1.56967	1.02569	73	1.57068	1.00843	122	1.57075	1.00507
25	1.56977	1.02466	74	1.57068	1.00832	123	1.57076	1.00503
26	1.56986	1.02371	75	1.57068	1.00821	124	1.57076	1.00499
27	1.56991	1.02284	76	1.57069	1.00810	125	1.57076	1.00495
28	1.56998	1.02202	77	1.57069	1.00799	126	1.57076	1.00491
29	1.57003	1.02127	78	1.57069	1.00789	127	1.57076	1.00487
30	1.57008	1.02055	79	1.57069	1.00780	128	1.57076	1.00483
31	1.57012	1.01990	80	1.57070	1.00772	129	1.57076	1.00479
32	1.57016	1.01926	81	1.57070	1.00762	130	1.57076	1.00475
33	1.57019	1.01869	82	1.57070	1.00752	131	1.57076	1.00472
34	1.57021	1.01813	83	1.57070	1.00743	132	1.57076	1.00469
35	1.57025	1.01762	84	1.57071	1.00734	133	1.57076	1.00466
36	1.57028	1.01714	85	1.57071	1.00725	134	1.57076	1.00462
37	1.57032	1.01667	86	1.57071	1.00716	135	1.57076	1.00457
38	1.57035	1.01623	87	1.57071	1.00708	136	1.57076	1.00454
39	1.57037	1.01582	88	1.57071	1.00700	137	1.57076	1.00451
40	1.57039	1.01542	89	1.57072	1.00693	138	1.57076	1.00447
41	1.57041	1.01504	90	1.57072	1.00686	139	1.57076	1.00444
42	1.57043	1.01471	91	1.57072	1.00679	140	1.57076	1.00441
43	1.57045	1.01434	92	1.57072	1.00672	141	1.57076	1.00439
44	1.57047	1.01404	93	1.57072	1.00665	142	1.57076	1.00435
45	1.57048	1.01370	94	1.57072	1.00658	143	1.57076	1.00432
46	1.57050	1.01341	95	1.57073	1.00651	144	1.57076	1.00429
47	1.57051	1.01311	96	1.57073	1.00644	145	1.57077	1.00425
48	1.57052	1.01285	97	1.57073	1.00637	146	1.57077	1.00422
49	1.57053	1.01258	98	1.57073	1.00630	147	1.57077	1.00419
50	1.57054	1.01233	99	1.57073	1.00623	148	1.57077	1.00416
51	1.57055	1.01209	100	1.57073	1.00617	149	1.57077	1.00413
52	1.57056	1.01187	101	1.57074	1.00611	150	1.57077	1.00411
53	1.57057	1.01165	102	1.57074	1.00605	151	1.57077	1.00409
54	1.57058	1.01143	103	1.57074	1.00599	152	1.57077	1.00407
55	1.57058	1.01121	104	1.57074	1.00593	153	1.57077	1.00405
56	1.57059	1.01102	105	1.57074	1.00587	154	1.57077	1.00402
57	1.57060	1.01083	106	1.57074	1.00581	155	1.57077	1.00400
58	1.57061	1.01064	107	1.57074	1.00575	156	1.57077	1.00397

Circular Pitch Gears — Pitch Diameters, Outside Diameters, and Root Diameters

For any particular circular pitch and number of teeth, use the table as shown in the example to find the pitch diameter, outside diameter, and root diameter. *Example:* Pitch diameter for 57 teeth of 6-inch circular pitch = 10 × pitch diameter given under factor for 5 teeth plus pitch diameter given under factor for 7 teeth. (10 × 9.5493) + 13.3690 = 108.862 inches.

Outside diameter of gear equals pitch diameter plus outside diameter factor from next-to-last column in table = 108.862 + 3.8197 = 112.682 inches.

Root diameter of gear equals pitch diameter minus root diameter factor from last column in table = 108.862 − 4.4194 = 104.443 inches.

Circular Pitch in Inches	Factor for Number of Teeth									Outside Diam. Factor	Root Diameter Factor
	1	2	3	4	5	6	7	8	9		
	Pitch Diameter Corresponding to Factor for Number of Teeth										
6	1.9099	3.8197	5.7296	7.6394	9.5493	11.4591	13.3690	15.2788	17.1887	3.8197	4.4194
5½	1.7507	3.5014	5.2521	7.0028	8.7535	10.5042	12.2549	14.0056	15.7563	3.5014	4.0511
5	1.5915	3.1831	4.7746	6.3662	7.9577	9.5493	11.1408	12.7324	14.3239	3.1831	3.6828
4½	1.4324	2.8648	4.2972	5.7296	7.1620	8.5943	10.0267	11.4591	12.8915	2.8648	3.3146
4	1.2732	2.5465	3.8197	5.0929	6.3662	7.6394	8.9127	10.1859	11.4591	2.5465	2.9463
3½	1.1141	2.2282	3.3422	4.4563	5.5704	6.6845	7.7986	8.9127	10.0267	2.2282	2.5780
3	0.9549	1.9099	2.8648	3.8197	4.7746	5.7296	6.6845	7.6394	8.5943	1.9099	2.2097
2½	0.7958	1.5915	2.3873	3.1831	3.9789	4.7746	5.5704	6.3662	7.1620	1.5915	1.8414
2	0.6366	1.2732	1.9099	2.5465	3.1831	3.8197	4.4563	5.0929	5.7296	1.2732	1.4731
1⅞	0.5968	1.1937	1.7905	2.3873	2.9841	3.5810	4.1778	4.7746	5.3715	1.1937	1.3811
1¾	0.5570	1.1141	1.6711	2.2282	2.7852	3.3422	3.8993	4.4563	5.0134	1.1141	1.2890
1⅝	0.5173	1.0345	1.5518	2.0690	2.5863	3.1035	3.6208	4.1380	4.6553	1.0345	1.1969
1½	0.4775	0.9549	1.4324	1.9099	2.3873	2.8648	3.3422	3.8197	4.2972	0.9549	1.1049
1⁷⁄₁₆	0.4576	0.9151	1.3727	1.8303	2.2878	2.7454	3.2030	3.6606	4.1181	0.9151	1.0588
1⅜	0.4377	0.8754	1.3130	1.7507	2.1884	2.6261	3.0637	3.5014	3.9391	0.8754	1.0128
1⁵⁄₁₆	0.4178	0.8356	1.2533	1.6711	2.0889	2.5067	2.9245	3.3422	3.7600	0.8356	0.9667
1¼	0.3979	0.7958	1.1937	1.5915	1.9894	2.3873	2.7852	3.1831	3.5810	0.7958	0.9207
1³⁄₁₆	0.3780	0.7560	1.1340	1.5120	1.8900	2.2680	2.6459	3.0239	3.4019	0.7560	0.8747
1⅛	0.3581	0.7162	1.0743	1.4324	1.7905	2.1486	2.5067	2.8648	3.2229	0.7162	0.8286
1¹⁄₁₆	0.3382	0.6764	1.0146	1.3528	1.6910	2.0292	2.3674	2.7056	3.0438	0.6764	0.7826
1	0.3183	0.6366	0.9549	1.2732	1.5915	1.9099	2.2282	2.5465	2.8648	0.6366	0.7366
15⁄16	0.2984	0.5968	0.8952	1.1937	1.4921	1.7905	2.0889	2.3873	2.6857	0.5968	0.6905
⅞	0.2785	0.5570	0.8356	1.1141	1.3926	1.6711	1.9496	2.2282	2.5067	0.5570	0.6445
13⁄16	0.2586	0.5173	0.7759	1.0345	1.2931	1.5518	1.8104	2.0690	2.3276	0.5173	0.5985
¾	0.2387	0.4775	0.7162	0.9549	1.1937	1.4324	1.6711	1.9099	2.1486	0.4775	0.5524
11⁄16	0.2188	0.4377	0.6565	0.8754	1.0942	1.3130	1.5319	1.7507	1.9695	0.4377	0.5064
⅔	0.2122	0.4244	0.6366	0.8488	1.0610	1.2732	1.4854	1.6977	1.9099	0.4244	0.4910
⅝	0.1989	0.3979	0.5968	0.7958	0.9947	1.1937	1.3926	1.5915	1.7905	0.3979	0.4604
9⁄16	0.1790	0.3581	0.5371	0.7162	0.8952	1.0743	1.2533	1.4324	1.6114	0.3581	0.4143
½	0.1592	0.3183	0.4775	0.6366	0.7958	0.9549	1.1141	1.2732	1.4324	0.3183	0.3683
7⁄16	0.1393	0.2785	0.4178	0.5570	0.6963	0.8356	0.9748	1.1141	1.2533	0.2785	0.3222
⅜	0.1194	0.2387	0.3581	0.4775	0.5968	0.7162	0.8356	0.9549	1.0743	0.2387	0.2762
⅓	0.1061	0.2122	0.3183	0.4244	0.5305	0.6366	0.7427	0.8488	0.9549	0.2122	0.2455
5⁄16	0.0995	0.1989	0.2984	0.3979	0.4974	0.5968	0.6963	0.7958	0.8952	0.1989	0.2302
¼	0.0796	0.1592	0.2387	0.3183	0.3979	0.4775	0.5570	0.6366	0.7162	0.1592	0.1841
3⁄16	0.0597	0.1194	0.1790	0.2387	0.2984	0.3581	0.4178	0.4775	0.5371	0.1194	0.1381
⅛	0.0398	0.0796	0.1194	0.1592	0.1989	0.2387	0.2785	0.3183	0.3581	0.0796	0.0921
1⁄16	0.0199	0.0398	0.0597	0.0796	0.0995	0.1194	0.1393	0.1592	0.1790	0.0398	0.0460

Selection of Involute Gear Milling Cutter for a Given Diametral Pitch and Number of Teeth. — When gear teeth are cut by using formed milling cutters, the cutter must be selected to suit both the pitch and the number of teeth, because the shapes of the tooth spaces vary according to the number of teeth. For instance, the tooth spaces of a small pinion are not of the same shape as the spaces of a large gear of equal pitch. Theoretically, there should be a different formed cutter for every tooth number, but such refinement is unnecessary in practice. The involute formed cutters commonly used are made in series of eight cutters for each diametral pitch (see accompanying table). The shape of each cutter in this series is correct for a certain number of teeth only, but it can be used for other numbers within the limits given. For instance, a No. 6 cutter may be used for gears having from 17 to 20 teeth, but the tooth outline is correct only for 17 teeth or the lowest number in the range, which is also true of the other cutters listed. When this cutter is used for a gear having, say, 19 teeth, too much material is removed from the upper surfaces of the teeth, although the gear meets ordinary requirements. When greater accuracy of tooth shape is desired to ensure smoother or quieter operation, an intermediate series of cutters having half-numbers may be used provided the number of gear teeth is between the number listed for the regular cutters (see table).

Involute gear milling cutters are designed to cut a composite tooth form, the center portion being a true involute while the top and bottom portions are cycloidal. This composite form is necessary to prevent tooth interference when milled mating gears are meshed with each other. Because of their composite form, milled gears will not mate satisfactorily enough for high grade work with those of generated, full-involute form. Composite form hobs are available, however, which will produce generated gears that mesh with those cut by gear milling cutters.

Metric Module Gear Cutters: The accompanying table for selecting the cutter number to be used to cut a given number of teeth may be used also to select metric module gear cutters except that the numbers are designated in reverse order. For example, cutter No. 1, in the metric module system, is used for 12–13 teeth, cutter No. 2 for 14–16 teeth, etc.

Series of Involute, Finishing Gear Milling Cutters for Each Pitch*

Number of Cutter	Will cut Gears from	Number of Cutter	Will cut Gears from
1	135 teeth to a rack	5	21 to 25 teeth
2	55 to 134 teeth	6	17 to 20 teeth
3	35 to 54 teeth	7	14 to 16 teeth
4	26 to 34 teeth	8	12 to 13 teeth

The regular cutters listed above are used ordinarily. The cutters listed below (an intermediate series having half numbers) may be used when greater accuracy of tooth shape is essential in cases where the number of teeth is between the numbers for which the regular cutters are intended.

Number of Cutter	Will cut Gears from	Number of Cutter	Will cut Gears from
1½	80 to 134 teeth	5½	19 to 20 teeth
2½	42 to 54 teeth	6½	15 to 16 teeth
3½	30 to 34 teeth	7½	13 teeth
4½	23 to 25 teeth	…	…

* Roughing cutters are made with No. 1 form only. Dimensions of roughing and finishing cutters are given on page 780. Dimensions of cutters for bevel gears are given on page 781.

Chordal Thicknesses and Chordal Addenda of Milled, Full-depth Gear Teeth and of Gear Milling Cutters

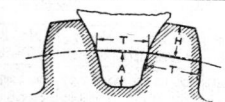

T = chordal thickness of gear tooth and cutter tooth at pitch line;
H = chordal addendum for full-depth gear tooth;
A = chordal addendum of cutter = $(2.157 \div$ diametral pitch$) - H = (0.6866 \times$ circular pitch$) - H$.

Diametral Pitch	Dimension	No. 1 135 Teeth	No. 2 55 Teeth	No. 3 35 Teeth	No. 4 26 Teeth	No. 5 21 Teeth	No. 6 17 Teeth	No. 7 14 Teeth	No. 8 12 Teeth
1	T	1.5707	1.5706	1.5702	1.5698	1.5694	1.5686	1.5675	1.5663
	H	1.0047	1.0112	1.0176	1.0237	1.0294	1.0362	1.0440	1.0514
1½	T	1.0471	1.0470	1.0468	1.0465	1.0462	1.0457	1.0450	1.0442
	H	0.6698	0.6741	0.6784	0.6824	0.6862	0.6908	0.6960	0.7009
2	T	0.7853	0.7853	0.7851	0.7849	0.7847	0.7843	0.7837	0.7831
	H	0.5023	0.5056	0.5088	0.5118	0.5147	0.5181	0.5220	0.5257
2½	T	0.6283	0.6282	0.6281	0.6279	0.6277	0.6274	0.6270	0.6265
	H	0.4018	0.4044	0.4070	0.4094	0.4117	0.4144	0.4176	0.4205
3	T	0.5235	0.5235	0.5234	0.5232	0.5231	0.5228	0.5225	0.5221
	H	0.3349	0.3370	0.3392	0.3412	0.3431	0.3454	0.3480	0.3504
3½	T	0.4487	0.4487	0.4486	0.4485	0.4484	0.4481	0.4478	0.4475
	H	0.2870	0.2889	0.2907	0.2919	0.2935	0.2954	0.2977	0.3004
4	T	0.3926	0.3926	0.3926	0.3924	0.3923	0.3921	0.3919	0.3915
	H	0.2511	0.2528	0.2544	0.2559	0.2573	0.2590	0.2610	0.2628
5	T	0.3141	0.3141	0.3140	0.3139	0.3138	0.3137	0.3135	0.3132
	H	0.2009	0.2022	0.2035	0.2047	0.2058	0.2072	0.2088	0.2102
6	T	0.2618	0.2617	0.2617	0.2616	0.2615	0.2614	0.2612	0.2610
	H	0.1674	0.1685	0.1696	0.1706	0.1715	0.1727	0.1740	0.1752
7	T	0.2244	0.2243	0.2243	0.2242	0.2242	0.2240	0.2239	0.2237
	H	0.1435	0.1444	0.1453	0.1462	0.1470	0.1480	0.1491	0.1502
8	T	0.1963	0.1963	0.1962	0.1962	0.1961	0.1960	0.1959	0.1958
	H	0.1255	0.1264	0.1272	0.1279	0.1286	0.1295	0.1305	0.1314
9	T	0.1745	0.1745	0.1744	0.1744	0.1743	0.1743	0.1741	0.1740
	H	0.1116	0.1123	0.1130	0.1137	0.1143	0.1151	0.1160	0.1168
10	T	0.1570	0.1570	0.1570	0.1569	0.1569	0.1568	0.1567	0.1566
	H	0.1004	0.1011	0.1017	0.1023	0.1029	0.1036	0.1044	0.1051
11	T	0.1428	0.1428	0.1427	0.1427	0.1426	0.1426	0.1425	0.1424
	H	0.0913	0.0919	0.0925	0.0930	0.0935	0.0942	0.0949	0.0955
12	T	0.1309	0.1309	0.1308	0.1308	0.1308	0.1307	0.1306	0.1305
	H	0.0837	0.0842	0.0848	0.0853	0.0857	0.0863	0.0870	0.0876
14	T	0.1122	0.1122	0.1121	0.1121	0.1121	0.1120	0.1119	0.1118
	H	0.0717	0.0722	0.0726	0.0731	0.0735	0.0740	0.0745	0.0751
16	T	0.0981	0.0981	0.0981	0.0981	0.0980	0.0980	0.0979	0.0979
	H	0.0628	0.0632	0.0636	0.0639	0.0643	0.0647	0.0652	0.0657
18	T	0.0872	0.0872	0.0872	0.0872	0.0872	0.0871	0.0870	0.0870
	H	0.0558	0.0561	0.0565	0.0568	0.0571	0.0575	0.0580	0.0584
20	T	0.0785	0.0785	0.0785	0.0785	0.0784	0.0784	0.0783	0.0783
	H	0.0502	0.0505	0.0508	0.0511	0.0514	0.0518	0.0522	0.0525

Chordal Thicknesses and Chordal Addenda of Milled, Full-depth Gear Teeth and of Gear Milling Cutters

Circular Pitch	Dimension	Number of Gear Cutter, and Corresponding Number of Teeth							
		No. 1 135 Teeth	No. 2 55 Teeth	No. 3 35 Teeth	No. 4 26 Teeth	No. 5 21 Teeth	No. 6 17 Teeth	No. 7 14 Teeth	No. 8 12 Teeth
¼	T	0.1250	0.1250	0.1249	0.1249	0.1249	0.1248	0.1247	0.1246
	H	0.0799	0.0804	0.0809	0.0814	0.0819	0.0824	0.0830	0.0836
⁵⁄₁₆	T	0.1562	0.1562	0.1562	0.1561	0.1561	0.1560	0.1559	0.1558
	H	0.0999	0.1006	0.1012	0.1018	0.1023	0.1030	0.1038	0.1045
³⁄₈	T	0.1875	0.1875	0.1874	0.1873	0.1873	0.1872	0.1871	0.1870
	H	0.1199	0.1207	0.1214	0.1221	0.1228	0.1236	0.1245	0.1254
⁷⁄₁₆	T	0.2187	0.2187	0.2186	0.2186	0.2185	0.2184	0.2183	0.2181
	H	0.1399	0.1408	0.1416	0.1425	0.1433	0.1443	0.1453	0.1464
½	T	0.2500	0.2500	0.2499	0.2498	0.2498	0.2496	0.2495	0.2493
	H	0.1599	0.1609	0.1619	0.1629	0.1638	0.1649	0.1661	0.1673
⁹⁄₁₆	T	0.2812	0.2812	0.2811	0.2810	0.2810	0.2808	0.2806	0.2804
	H	0.1799	0.1810	0.1821	0.1832	0.1842	0.1855	0.1868	0.1882
⁵⁄₈	T	0.3125	0.3125	0.3123	0.3123	0.3122	0.3120	0.3118	0.3116
	H	0.1998	0.2012	0.2023	0.2036	0.2047	0.2061	0.2076	0.2091
¹¹⁄₁₆	T	0.3437	0.3437	0.3436	0.3435	0.3434	0.3432	0.3430	0.3427
	H	0.2198	0.2213	0.2226	0.2239	0.2252	0.2267	0.2283	0.2300
¾	T	0.3750	0.3750	0.3748	0.3747	0.3747	0.3744	0.3742	0.3740
	H	0.2398	0.2414	0.2428	0.2443	0.2457	0.2473	0.2491	0.2509
¹³⁄₁₆	T	0.4062	0.4062	0.4060	0.4059	0.4059	0.4056	0.4054	0.4050
	H	0.2598	0.2615	0.2631	0.2647	0.2661	0.2679	0.2699	0.2718
⁷⁄₈	T	0.4375	0.4375	0.4373	0.4372	0.4371	0.4368	0.4366	0.4362
	H	0.2798	0.2816	0.2833	0.2850	0.2866	0.2885	0.2906	0.2927
¹⁵⁄₁₆	T	0.4687	0.4687	0.4685	0.4684	0.4683	0.4680	0.4678	0.4674
	H	0.2998	0.3018	0.3035	0.3054	0.3071	0.3092	0.3114	0.3137
1	T	0.5000	0.5000	0.4998	0.4997	0.4996	0.4993	0.4990	0.4986
	H	0.3198	0.3219	0.3238	0.3258	0.3276	0.3298	0.3322	0.3346
1⅛	T	0.5625	0.5625	0.5623	0.5621	0.5620	0.5617	0.5613	0.5610
	H	0.3597	0.3621	0.3642	0.3665	0.3685	0.3710	0.3737	0.3764
1¼	T	0.6250	0.6250	0.6247	0.6246	0.6245	0.6241	0.6237	0.6232
	H	0.3997	0.4023	0.4047	0.4072	0.4095	0.4122	0.4152	0.4182
1⅜	T	0.6875	0.6875	0.6872	0.6870	0.6869	0.6865	0.6861	0.6856
	H	0.4397	0.4426	0.4452	0.4479	0.4504	0.4534	0.4567	0.4600
1½	T	0.7500	0.7500	0.7497	0.7495	0.7494	0.7489	0.7485	0.7480
	H	0.4797	0.4828	0.4857	0.4887	0.4914	0.4947	0.4983	0.5019
1¾	T	0.8750	0.8750	0.8746	0.8744	0.8743	0.8737	0.8732	0.8726
	H	0.5596	0.5633	0.5666	0.5701	0.5733	0.5771	0.5813	0.5855
2	T	1.0000	1.0000	0.9996	0.9994	0.9992	0.9986	0.9980	0.9972
	H	0.6396	0.6438	0.6476	0.6516	0.6552	0.6596	0.6644	0.6692
2¼	T	1.1250	1.1250	1.1246	1.1242	1.1240	1.1234	1.1226	1.1220
	H	0.7195	0.7242	0.7285	0.7330	0.7371	0.7420	0.7474	0.7528
2½	T	1.2500	1.2500	1.2494	1.2492	1.2490	1.2482	1.2474	1.2464
	H	0.7995	0.8047	0.8095	0.8145	0.8190	0.8245	0.8305	0.8365
3	T	1.5000	1.5000	1.4994	1.4990	1.4990	1.4978	1.4970	1.4960
	H	0.9594	0.9657	0.9714	0.9774	0.9828	0.9894	0.9966	1.0038

Increasing Pinion Diameter to Avoid Undercut or Interference. — On coarse-pitch pinions with small numbers of teeth (10 to 17 for 20 degree- and 10 and 11 for 25-degree pressure angle involute tooth forms) undercutting of the tooth profile or fillet interference with the tip of the mating gear can be avoided by making certain changes from the standard tooth proportions that are specified in Table 3 on page 622. These changes consist essentially in increasing the addendum and hence the outside diameter of the pinion and decreasing the addendum and hence the outside diameter of the mating gear. These changes in outside diameters of pinion and gear do not change the velocity ratio or the procedures in cutting the teeth on a hobbing machine or generating type of shaper or planer.

Data in Table 1 which follows are taken from ANSI Standard B6.1-1968, reaffirmed 1974, and show for 20-degree and 25-degree full-depth standard tooth forms, respectively, the addendums and tooth thicknesses for long addendum pinions and their mating short addendum gears when the number of teeth in the pinion is as given. Similar data for former standard 14½-degree full-depth teeth (20 diametral pitch and coarser) are given in Table 2.

Table 1. Addendums and Tooth Thicknesses for Coarse-Pitch Long-Addendum Pinions and their Mating Short-Addendum Gears — 20- and 25-degree Pressure Angles* (ANSI-B6.1-1968, R 1974)

Number of Teeth in Pinion	Addendum		Basic Tooth Thickness		Number of Teeth in Gear
	Pinion	Gear	Pinion	Gear	
N_P	a_P	a_G	t_P	t_G	N_G (min)
20-DEGREE INVOLUTE FULL DEPTH TOOTH FORM (Less than 20 Diametral Pitch)					
10	1.468	.532	1.912	1.230	25
11	1.409	.591	1.868	1.273	24
12	1.351	.649	1.826	1.315	23
13	1.292	.708	1.783	1.358	22
14	1.234	.766	1.741	1.400	21
15	1.175	.825	1.698	1.443	20
16	1.117	.883	1.656	1.486	19
17	1.058	.942	1.613	1.529	18
25-DEGREE INVOLUTE FULL DEPTH TOOTH FORM (Less than 20 Diametral Pitch)					
10	1.184	.816	1.742	1.399	15
11	1.095	.905	1.659	1.482	14

* All values are for 1 diametral pitch. For any other sizes of teeth all linear dimensions should be divided by the diametral pitch. Basic tooth thicknesses do not include an allowance for backlash.

Example: A 14-tooth, 20-degree pressure angle pinion of 6 diametral pitch is to be enlarged. What will be the outside diameters of the pinion and a 60-tooth mating gear? If the mating gear is to have the minimum number of teeth to avoid undercut, what will be its outside diameter?

$$D_o \text{ (pinion)} = \frac{N_P}{P} + 2a = \frac{14}{6} + 2\left(\frac{1.234}{6}\right) = 2.745 \text{ inches}$$

$$D_o \text{ (gear)} = \frac{N_G}{P} + 2a = \frac{60}{6} + 2\left(\frac{0.766}{6}\right) = 10.255 \text{ inches}$$

For a mating gear with minimum number of teeth to avoid undercut:

$$D_o \text{ (gear)} = \frac{N_G}{P} + 2a = \frac{21}{6} + 2\left(\frac{0.766}{6}\right) = 3.755 \text{ inches}$$

**Table 2. Enlarged Pinion and Reduced Gear Dimensions to Avoid Interference —
Coarse Pitch 14½-degree Involute Full Depth Teeth**

Number of Pinion Teeth	Changes in Pinion and Gear Diameters	Circular Tooth Thickness		Min. No. of Teeth in Mating Gear	
		Pinion	Mating Gear	To Avoid Undercut	For Full Invo- lute Action
10	1.3731	1.9259	1.2157	54	27
11	1.3104	1.9097	1.2319	53	27
12	1.2477	1.8935	1.2481	52	28
13	1.1850	1.8773	1.2643	51	28
14	1.1223	1.8611	1.2805	50	28
15	1.0597	1.8449	1.2967	49	28
16	0.9970	1.8286	1.3130	48	28
17	0.9343	1.8124	1.3292	47	28
18	0.8716	1.7962	1.3454	46	28
19	0.8089	1.7800	1.3616	45	28
20	0.7462	1.7638	1.3778	44	28
21	0.6835	1.7476	1.3940	43	28
22	0.6208	1.7314	1.4102	42	27
23	0.5581	1.7151	1.4265	41	27
24	0.4954	1.6989	1.4427	40	27
25	0.4328	1.6827	1.4589	39	26
26	0.3701	1.6665	1.4751	38	26
27	0.3074	1.6503	1.4913	37	26
28	0.2447	1.6341	1.5075	36	25
29	0.1820	1.6179	1.5237	35	25
30	0.1193	1.6017	1.5399	34	24
31	0.0566	1.5854	1.5562	33	24

All dimensions are given in inches and are for 1 diametral pitch. For other pitches divide tabular values by desired diametral pitch.

Add to the standard outside diameter of the pinion the amount given in the second column of the table divided by the desired diametral pitch, and (to maintain standard center distance) subtract the same amount from the outside diameter of the mating gear. Long addendum pinions will mesh with standard gears, but the center distance will be greater than standard.

Enlarged Fine-Pitch Pinions. — American Standard ANSI B6.7-1977, Information Sheet A provides a different system for 20-degree pressure angle pinion enlargement than is used for coarse-pitch gears. Pinions with 11 through 23 teeth (9 through 14 teeth for 25-degree pressure angle) are enlarged so that a standard tooth thickness rack with addendum $1.05/P$ will start contact 5° of roll above the base circle radius. The use of $1.05/P$ for the addendum allows for center distance variation and eccentricity of the mating gear outside diameter; the 5° roll angle avoids the fabrication of the involute in the troublesome area near the base circle.

Pinions with less than 11 teeth (9 teeth for 25-degree pressure angle) are enlarged to the extent that the highest point of undercut coincides with the start of contact with the standard rack described previously. The height of undercut considered is that produced by a sharp-cornered 120 pitch hob. Pinions with less than 13 teeth (11 teeth for 25-degree pressure angle) are truncated to provide a top land of $0.275/P$. Data for enlarged pinions may be found in Tables 3a, 3b, 3c, and 3d.

Minimum Number of Teeth to Avoid Undercutting by Hob. — The data in the above tables give tooth proportions for low numbers of teeth to avoid interference between the gear tooth tip and the pinion tooth flank. Consideration must also be given to possible undercutting of the pinion tooth flank by the hob used to cut the pinion. The minimum number of teeth N_{min} of standard proportion that may be cut without undercut is: $N_{min} = 2P \csc^2 \phi \, [a_H - r_t \, (1 - \sin \phi)]$ where: a_H = cutter addendum; r_t = radius at cutter tip or corners; ϕ = cutter pressure angle; and P = diametral pitch.

(Continued on page 1802)

Table 3a. Increase in Dedendum, Δ, for 20-, and 25-Degree Pressure Angle Fine-Pitch Enlarged Pinions and Reduced Gears (ANSI B6.7-1977)

Diametral Pitch, P	Δ	Diametral Pitch, P	Δ	Diametral Pitch, P	Δ	Diametral Pitch, P	Δ	Diametral Pitch, P	Δ
20	0.0000	32	0.0007	48	0.0012	72	0.0015	96	0.0016
24	0.0004	40	0.0010	64	0.0015	80	0.0015	120	0.0017

Δ = increase in standard dedendum to provide increased clearance. See footnote to Table 3d.

Table 3b. Dimensions Required when Using Enlarged, Fine-pitch, 14½-Degree Pressure Angle Pinions (ANSI B6.7-1977, Information Sheet B)

			Standard Center-distance System (Long and Short Addendum)				Enlarged Center-distance System		
Enlarged Pinion			Reduced Mating Gear						
No. of Teeth n	Outside Diameter	Cir. Tooth Thickness at Standard Pitch Diam.	Decrease in Standard Outside Diam.[1]	Cir. Tooth Thickness at Standard Pitch Diam.	Recommended Minimum No. of Teeth N	Contact Ratio, n Mating with N	Enlarged Pinion Mating with St'd. Gear	Two Equal Enlarged Mating Pinions[2]	Contact Ratio of Two Equal Enlarged Mating Pinions
							Increase over St'd. Center Distance		
10	13.3731	1.9259	1.3731	1.2157	54	1.831	0.6866	1.3732	1.053
11	14.3104	1.9097	1.3104	1.2319	53	1.847	0.6552	1.3104	1.088
12	15.2477	1.8935	1.2477	1.2481	52	1.860	0.6239	1.2477	1.121
13	16.1850	1.8773	1.1850	1.2643	51	1.873	0.5925	1.1850	1.154
14	17.1223	1.8611	1.1223	1.2805	50	1.885	0.5612	1.2223	1.186
15	18.0597	1.8448	1.0597	1.2967	49	1.896	0.5299	1.0597	1.217
16	18.9970	1.8286	0.9970	1.3130	48	1.906	0.4985	0.9970	1.248
17	19.9343	1.8124	0.9343	1.3292	47	1.914	0.4672	0.9343	1.278
18	20.8716	1.7962	0.8716	1.3454	46	1.922	0.4358	0.8716	1.307
19	21.8089	1.7800	0.8089	1.3616	45	1.929	0.4045	0.8089	1.336
20	22.7462	1.7638	0.7462	1.3778	44	1.936	0.3731	0.7462	1.364
21	23.6835	1.7476	0.6835	1.3940	43	1.942	0.3418	0.6835	1.392
22	24.6208	1.7314	0.6208	1.4102	42	1.948	0.3104	0.6208	1.419
23	25.5581	1.7151	0.5581	1.4265	41	1.952	0.2791	0.5581	1.446
24	26.4954	1.6989	0.4954	1.4427	40	1.956	0.2477	0.4954	1.472
25	27.4328	1.6827	0.4328	1.4589	39	1.960	0.2164	0.4328	1.498
26	28.3701	1.6665	0.3701	1.4751	38	1.963	0.1851	0.3701	1.524
27	29.3074	1.6503	0.3074	1.4913	37	1.965	0.1537	0.3074	1.549
28	30.2447	1.6341	0.2448	1.5075	36	1.967	0.1224	0.2448	1.573
29	31.1820	1.6179	0.1820	1.5237	35	1.969	0.0910	0.1820	1.598
30	32.1193	1.6017	0.1193	1.5399	34	1.970	0.0597	0.1193	1.622
31	33.0566	1.5854	0.0566	1.5562	33	1.971	0.0283	0.0566	1.646

All dimensions are given in inches and are for 1 diametral pitch. For other pitches divide tabulated dimensions by the diametral pitch.

[1] To maintain standard center distance when using an enlarged pinion, the mating gear diameter must be decreased by the amount of the pinion enlargement.

[2] If enlarged mating pinions are of unequal size, the center distance is increased by an amount equal to one-half the sum of their increase over standard outside diameters. Data in this column are not given in the standard.

Table 3c. Tooth Proportions Recommended for Enlarging Fine-Pitch Pinions of 20-Degree Pressure Angle — 20 Diametral Pitch and Finer (ANSI B6.7-1977)

	Enlarged Pinion Dimensions				Enlarged C.D. System Pinion Mating with Standard Gear		Standard Center Distance System (Long and Short Addendums) Reduced Gear Dimensions				
Number of Teeth,* n	Outside Diameter,* D_{oP}	Addendum, a_P	Basic Tooth Thickness, t_P	Dedendum Based on 20 Pitch,† b_P	Contact Ratio Two Equal Pinions	Contact Ratio with a 24-Tooth Gear	Addendum, a_G	Basic Tooth Thickness, t_G	Dedendum Based on 20 Pitch,† b_G	Recommended Minimum No. of Teeth, N	Contact Ratio n Mating with N
7	10.0102	1.5051	2.14114	0.4565	0.697	1.003	0.2165	1.00045	2.0235	42	1.079
8	11.0250	1.5125	2.09854	0.5150	0.792	1.075	0.2750	1.04305	1.9650	40	1.162
9	12.0305	1.5152	2.05594	0.5735	0.893	1.152	0.3335	1.08565	1.9065	39	1.251
10	13.0279	1.5140	2.01355	0.6321	0.982	1.211	0.3921	1.12824	1.8479	38	1.312
11	14.0304	1.5152	1.97937	0.6787	1.068	1.268	0.4387	1.16222	1.8013	37	1.371
12	15.0296	1.5148	1.94703	0.7232	1.151	1.322	0.4832	1.19456	1.7568	36	1.427
13	15.9448	1.4724	1.91469	0.7676	1.193	1.353	0.5276	1.22690	1.7124	35	1.457
14	16.8560	1.4280	1.88235	0.8120	1.232	1.381	0.5720	1.25924	1.6680	34	1.483
15	17.7671	1.3836	1.85001	0.8564	1.270	1.408	0.6164	1.29158	1.6236	33	1.507
16	18.6782	1.3391	1.81766	0.9009	1.323	1.434	0.6609	1.32393	1.5791	32	1.528
17	19.5894	1.2947	1.78532	0.9453	1.347	1.458	0.7053	1.35627	1.5347	31	1.546
18	20.5006	1.2503	1.75298	0.9897	1.385	1.482	0.7497	1.38861	1.4903	30	1.561
19	21.4116	1.2058	1.72064	1.0342	1.423	1.505	0.7942	1.42095	1.4458	29	1.574
20	22.3228	1.1614	1.68839	1.0786	1.461	1.527	0.8386	1.45320	1.4014	28	1.584
21	23.2340	1.1170	1.65595	1.1230	1.498	1.548	0.8830	1.48564	1.3570	27	1.592
22	24.1450	1.0725	1.62361	1.1675	1.536	1.568	0.9275	1.51798	1.3125	26	1.598
23	25.0561	1.0281	1.59127	1.2119	1.574	1.588	0.9719	1.55032	1.2681	25	1.601
24	26.0000	1.0000	1.57080	1.2400	1.602	1.602	1.0000	1.57080	1.2400	24	1.602

For footnotes * and †, see bottom of Table 3d.   All dimensions are given in inches.

Table 3d. Tooth Proportions Recommended for Enlarging Fine-Pitch Pinions of 25-Degree Pressure Angle — 20 Diametral Pitch and Finer (ANSI B6.7-1977, Information Sheet B)

Number of Teeth,* n	Enlarged Pinion Dimensions				Enlarged C.D. System Pinion Mating with Standard Gear		Standard Center Distance System (Long and Short Addendums) Reduced Gear Dimensions				
	Outside Diameter, D_{oP}	Addendum, a_P	Basic Tooth Thickness, t_P	Dedendum Based on 20 Pitch,† b_P	Contact Ratio Two Equal Pinions	Contact Ratio with 15-Tooth Gear	Addendum, a_G	Basic Tooth Thickness, t_G	Dedendum Based on 20 Pitch,† b_G	Recommended Minimum No. of Teeth N	Contact Ratio, n Mating with N
6	8.7645	1.3822	2.18362	0.5829	0.696	0.954	0.3429	0.95797	1.8971	24	1.030
7	9.7253	1.3626	2.10029	0.6722	0.800	1.026	0.4322	1.04130	1.8078	23	1.108
8	10.6735	1.3368	2.01701	0.7616	0.904	1.094	0.5216	1.12459	1.7184	22	1.177
9	11.6203	1.3102	1.94110	0.8427	1.003	1.156	0.6029	1.20048	1.6371	20	1.234
10	12.5691	1.2846	1.87345	0.9155	1.095	1.211	0.6755	1.26814	1.5645	19	1.282
11	13.5939	1.2520	1.80579	0.9880	1.183	1.261	0.7480	1.33581	1.4920	18	1.322
12	14.3588	1.1794	1.73813	1.0606	1.231	1.290	0.8206	1.40346	1.4194	17	1.337
13	15.2138	1.1069	1.67047	1.1331	1.279	1.317	0.8931	1.47112	1.3469	16	1.347
14	16.0686	1.0343	1.60281	1.2057	1.328	1.343	0.9657	1.53878	1.2743	15	1.352
15	17.0000	1.0000	1.57030	1.2400	1.358	1.358	1.0000	1.57080	1.2400	15	1.358

All dimensions are given in inches.

All values are for 1 diametral pitch. For any other sizes of teeth, all linear dimensions should be divided by the diametral pitch.

* Caution should be exercised in the use of pinions above the horizontal lines. They should be checked for suitability, particularly in the areas of contact ratio (less than 1.2 is not recommended), center distance, clearance, and tooth strength.

† The actual dedendum is calculated by dividing the values in this column by the desired diametral pitch and then adding to the result an amount Δ found in Table 3a. As an example, a 20-degree pressure angle 7-tooth pinion meshing with a 42-tooth gear would have, for 24 diametral pitch, a dedendum of 0.4565 ÷ 24 + 0.0004 = 0.0194. The 42-tooth gear would have a dedendum of 2.0235 ÷ 24 + 0.0004 = 0.0847 inch.

Note: The tables in the Standard also specify Form Diameter, Roll Angle to Form Diameter, and Top Land. These are not shown here. The top land is in no case less than 0.275/P. The form diameters and the roll angles to form diameter shown in the Standard are the values which should be met with a standard hob when generating the tooth thicknesses shown in the tables. These form diameters provide more than enough length of involute profile for any mating gear smaller than a rack. However, since these form diameters are based on gear tooth generation using standard hobs, they should impose little or no hardship on manufacture except in cases of the most critical quality levels. In such cases, form diameter specifications and master gear design should be based upon actual mating conditions.

Gear to Mesh with Enlarged Pinion. — Data in the fifth column of Table 2 show minimum number of teeth in a mating gear which can be cut with hob or rack type cutter without undercut, when outside diameter of gear has been reduced an amount equal to the pinion enlargement to retain the standard center distance. To calculate N for the gear, insert addendum a of enlarged mating pinion in the formula $N = 2a \times \csc^2\phi$.

Example: A gear is to mesh with a 24-tooth pinion of 1 diametral pitch which has been enlarged 0.4954 inch, as shown by the table. The pressure angle is 14½ degrees. Find minimum number of teeth N for reduced gear.

Pinion addendum = $1 + (0.4954 \div 2) = 1.2477$; hence

$$N = 2 \times 1.2477 \times 15.95 = 39.8 \text{ (use 40)}$$

In the case of fine pitch gears with reduced outside diameters, the recommended minimum numbers of teeth given in Tables 3b, 3c, and 3d, are somewhat more than the minimum numbers required to prevent undercutting and are based upon studies made by the American Gear Manufacturers Association.

Standard Center-distance System for Enlarged Pinions. — In this system, sometimes referred to as "long and short addendums," the center distance is made standard for the numbers of teeth in pinion and gear. The outside diameter of the gear is decreased by the same amount that the outside of the pinion is enlarged. The advantages of this system are: (1) No change in center distance or ratio is required; (2) The operating pressure angle remains standard; and (3) A slightly greater contact ratio is obtained than when the center distance is increased. The disadvantages are (1) The gears as well as the pinion must be changed from standard dimensions; (2) Pinions having fewer than the minimum number of teeth to avoid undercut cannot be satisfactorily meshed together; and (3) In most cases where gear trains include idler gears, the standard center-distance system cannot be used.

Enlarged Center-distance System for Enlarged Pinions. — If an enlarged pinion is meshed with another enlarged pinion or with a gear of standard outside diameter, the center distance must be increased. For fine-pitch gears, it is usually satisfactory to increase the center distance by an amount equal to one-half of the enlargements (see eighth column of Table 3b). This is an approximation as theoretically there is a slight increase in backlash. The advantages of this system are: (1) Only the pinions need be changed from the standard dimensions; (2) Pinions having fewer than 18 teeth may engage other pinions in this range; (3) The pinion tooth, which is the weaker member, is made stronger by the enlargement; and (4) The tooth contact stress, which controls gear durability, is lowered by being moved away from the pinion base circle. The disadvantages are: (1) Center distances must be enlarged over the standard; (2) The operating pressure angle increases slightly with different combinations of pinions and gears, which is usually not important; and (3) The contact ratio is slightly smaller than that obtained with the standard center-distance system. This consideration is of minor importance as in the worst case the loss is approximately only 6 per cent.

Enlarged Pinions Meshing without Backlash: When two enlarged pinions are to mesh without backlash, their center distance will be greater than the standard and less than that for the enlarged center-distance system. This center distance may be calculated by the formulas given in the following section.

Center Distance at Which Modified Mating Spur Gears Will Mesh with No Backlash. — When the tooth thickness of one or both of a pair of mating spur gears has been increased or decreased from the standard value ($\pi \div 2P$), the center distance at which they will mesh tightly (without backlash) may be calculated from the following formulas:

$$\text{inv } \phi_1 = \text{inv } \phi + \frac{P(t + T) - \pi}{n + N}$$

$$C = \frac{n + N}{2P}$$

$$C_1 = \frac{\cos \phi}{\cos \phi_1} \times C$$

In these formulas, P = diametral pitch; n = number of teeth in pinion; N = number of teeth in gear; t and T are the actual tooth thicknesses of the pinion and gear, respectively, on their standard pitch circles; inv ϕ = involute function of standard pressure angle of gears; C = standard center distance for the gears; C_1 = center distance at which the gears mesh without backlash; and inv ϕ_1 = involute function of operating pressure angle when gears are meshed tightly at center distance C_1.

Example: Calculate the center distance for no backlash when an enlarged 10-tooth pinion of 100 diametral pitch and 20-degree pressure angle is meshed with a standard 30-tooth gear, the circular tooth thickness of the pinion and gear, respectively, being 0.01873 and 0.015708 inch.

$$\text{inv } \phi_1 = \text{inv } 20° + \frac{100(0.01873 + 0.015708) - \pi}{(10 + 30)}$$

From the table of involute functions, inv 20-degrees = 0.014904. Therefore,

$$\text{inv } \phi_1 = 0.014904 + \frac{0.34438 - 0.31416}{4} = 0.022459$$

$$\phi_1 = 22°49' \text{ from page } 109$$

$$C = \frac{n + N}{2P} = \frac{10 + 30}{2 \times 100} = 0.2000 \text{ inch}$$

$$C_1 = \frac{\cos 20°}{\cos 22°49'} \times 0.2000 = \frac{0.93969}{0.92175} \times 0.2000 = 0.2039 \text{ inch}$$

Contact Diameter. — For two meshing gears it is important to know the contact diameter of each. A first gear with number of teeth, n, and outside diameter, d_o, meshes at a standard center distance with a second gear with number of teeth, N, and outside diameter, D_o; both gears have a diametral pitch, P, and pressure angle, ϕ. a, A, b, and B are unnamed angles used only in the calculations. The contact diameter, d_c, is found by a three-step calculation that can be done by hand using a trigonometric table and a logarithmic table or a desk calculator. Slide rule calculation is not recommended because it is not accurate enough to give good results. The three-step formulas to find the contact diameter, d_c, of the first gear are:

$$\cos A = \frac{N \cos \phi}{D_o \times P} \qquad (1)$$

$$\tan b = \tan \phi - \frac{N}{n} (\tan A - \tan \phi) \qquad (2)$$

$$d_c = \frac{n \cos \phi}{P \cos b} \qquad (3)$$

Similarly the three-step formulas to find the contact diameter, D_c, of the second gear are:

$$\cos a = \frac{n \cos \phi}{d_o \times P} \tag{4}$$

$$\tan B = \tan \phi - \frac{n}{N}(\tan a - \tan \phi) \tag{5}$$

$$D_c = \frac{N \cos \phi}{P \cos B} \tag{6}$$

Contact Ratio. — The contact ratio of a pair of mating spur gears must be well over 1.0 to assure a smooth transfer of load from one pair of teeth to the next pair as the two gears rotate under load. Because of a reduction in contact ratio due to such factors as tooth deflection, tooth spacing errors, tooth tip breakage, and outside diameter and center distance tolerances, the contact ratio of gears for power transmission as a general rule should not be less than about 1.4. A contact ratio of as low as 1.15 may be used in extreme cases, provided the tolerance effects mentioned above are accounted for in the calculation. The formula for determining the contact ratio, m_f, using the nomenclature in the previous section is:

$$m_f = \frac{N}{6.28318}(\tan A - \tan B) \tag{7a}$$

or

$$m_f = \frac{n}{6.28318}(\tan a - \tan b) \tag{7b}$$

Both formulas should give the same answer. It is good practice to use both formulas as a check on the previous calculations.

Lowest Point of Single Tooth Contact. — This diameter on the pinion (sometimes referred to as LPSTC) is used to find the maximum contact compressive stress (sometimes called the Hertz Stress) of a pair of mating spur gears. The two-step formulas for determining this pinion diameter, d_L, using the same nomenclature as in the previous sections with c and C as unnamed angles used only in the calculations are:

$$\tan c = \tan a - \frac{6.28318}{n} \tag{8}$$

$$d_L = \frac{n \cos \phi}{P \cos c} \tag{9}$$

In some cases it is necessary to have a plot of the compressive stress over the whole cycle of contact; in this case the LPSTC for the gear is required also. The similar two-step formulas for this gear diameter are:

$$\tan C = \tan A - \frac{6.28318}{N} \tag{10}$$

$$D_L = \frac{N \cos \phi}{P \cos C} \tag{11}$$

Maximum Hob Tip Radius. — The standard gear tooth proportions given by the formulas in Table 2 on page 1772 provide a specified size for the rack fillet radius in the general form of (a constant) × (pitch). For any given standard this constant may vary up to a maximum which it is geometrically impossible to exceed; this maximum constant, $r_c(\max)$, is found by the formula:

$$r_c(\max) = \frac{0.785398 \cos \phi - b \sin \phi}{1 - \sin \phi} \tag{12}$$

where b is the similar constant in the specified formula for the gear dedendum. The hob tip radius of any standard hob to finish cut any standard gear may vary from zero up to this limiting value.

Undercut Limit for Hobbed Involute Gears. — It is well to avoid designing and specifying gears that will have a hobbed trochoidal fillet that undercuts the involute gear tooth profile. This should be avoided because it may cause the involute profile to be cut away up to a point above the required contact diameter with the mating gear so that involute action is lost and the contact ratio reduced to a level that may be too low for proper conjugate action. An undercut fillet will also weaken the beam strength and thus raise the fillet tensile stress of the gear tooth. To assure that the hobbed gear tooth will not have an undercut fillet, the following formula must be satisfied:

$$\frac{b - r_c}{\sin \phi} + r_c \leq 0.5n \sin \phi \tag{13}$$

where b is the dedendum constant; r_c is the hob or rack tip radius constant; n is the number of teeth in the gear; and ϕ is the gear and hob pressure angle. If the gear is not standard or the hob does not roll at the gear pitch diameter, this formula can not be applied and the determination of the expected existence of undercut becomes a considerably more complicated procedure.

Highest Point of Single Tooth Contact. — This diameter is used to place the maximum operating load for the determination of the gear tooth fillet stress. The two-step formulas for determining this diameter, d_H, of the pinion using the same nomenclature as in the previous sections with d and D as unnamed angles used only in the calculations are:

$$\tan d = \tan b + \frac{6.28318}{n} \tag{14}$$

$$d_H = \frac{n \cos \phi}{P \cos d} \tag{15}$$

Similarly for the gear: —

$$\tan D = \tan B + \frac{6.28318}{N} \tag{16}$$

$$D_H = \frac{N \cos \phi}{P \cos D} \tag{17}$$

True Involute Form Diameter. — The point on the gear tooth at which the fillet and the involute profile are tangent to each other should be determined to assure that it lies at a smaller diameter than the required contact diameter with the mating

gear. If the TIF diameter is larger than the contact diameter, then fillet interference will occur with severe damage to the gear tooth profile and rough action of the gear set. This two-step calculation is made by using the following two formulas with e and E as unnamed angles used only in the calculations:

$$\tan e = \tan \phi - \frac{4}{n}\left(\frac{b - r_c}{\sin 2 \phi} + \frac{r_c}{2 \cos \phi}\right) \qquad (18)$$

$$d_{TIF} = \frac{n \cos \phi}{P \cos e} \qquad (19)$$

As in the previous sections, ϕ is the pressure angle of the gear; n is the number of teeth in the pinion; b is the dedendum constant, r_c is the rack or hob tip radius constant, P is the gear diametral pitch and d_{TIF} is the true involute form diameter.

Similarly, for the mating gear:

$$\tan E = \tan \phi - \frac{4}{N}\left(\frac{b - r_c}{\sin 2 \phi} + \frac{r_c}{2 \cos \phi}\right) \qquad (20)$$

$$D_{TIF} = \frac{N \cos \phi}{P \cos E} \qquad (21)$$

Where N is the number of teeth in this mating gear and D_{TIF} is the true involute form diameter.

Profile Checker Settings. — The actual tooth profile tolerance will need to be determined on high performance gears that operate either at high unit loads or at high pitch-line velocity. This is done on an involute checker, a machine which requires two settings, the gear base radius and the roll angle in degrees to significant points on the involute. From the smallest diameter outward these significant points are: TIF, Contact Diameter, LPSTC, Pitch Diameter, HPSTC, and Outside Diameter.

The base radius is:

$$R_b = \frac{N \cos \phi}{2P} \qquad (22)$$

The roll angle, in degrees, at any point is equal to the tangent of the pressure angle at that point multiplied by 57.2958. The following table shows the tangents to be used at each significant diameter.

Significant Point on Tooth Profile	Pinion	Gear	For Computation
TIF	$\tan e$	$\tan E$	(See Formulas 18 & 20)
Contact Diam.	$\tan b$	$\tan B$	(See Formulas 2 & 5)
LPSTC	$\tan c$	$\tan C$	(See Formulas 8 & 10)
Pitch Diam.	$\tan \phi$	$\tan \phi$	(ϕ = Pressure angle)
HPSTC	$\tan d$	$\tan D$	(See Formulas 14 & 16)
Outside Diam.	$\tan a$	$\tan A$	(See Formulas 4 & 1)

Example: Find the significant diameters, contact ratio and hob tip radius for a 10-diametral pitch, 23-tooth, 20-degree pressure angle pinion of 2.5-inch outside diameter if it is to mesh with a 31-tooth gear of 3.3-inch outside diameter.

Thus: $n = 23$ $N = 31$
$d_O = 2.5$ $D_O = 3.3$
$P = 10$ $\phi = 20°$

1. Pinion contact diameter d_c:

$$\cos A = \frac{31 \times 0.93969}{3.3 \times 10} \tag{1}$$

$$= 0.88274 \qquad A = 28°1'30''$$

$$\tan b = 0.36397 - \frac{31}{23} \ (0.53227 - 0.36397) \tag{2}$$

$$= 0.13713 \qquad b = 7°48'26''$$

$$d_c = \frac{23 \times 0.93969}{10 \times 0.99073} \tag{3}$$

$$= 2.1815 \text{ inches}$$

2. Gear contact diameter, D_c

$$\cos a = \frac{23 \times 0.93963}{2.5 \times 10} \tag{4}$$

$$= 0.86452 \qquad a = 30°10'20''$$

$$\tan B = 0.36397 - \frac{23}{31} \ (0.58136 - 0.36937) \tag{5}$$

$$= 0.20267 \qquad B = 11°27'26''$$

$$D_c = \frac{31 \times 0.93969}{10 \times 0.98000} \tag{6}$$

$$= 2.9725 \text{ inches}$$

3. Contact ratio, m_f

$$m_f = \frac{31}{6.28318} \ (0.53227 - 0.20267) \tag{7a}$$

$$= 1.626$$

$$m_f = \frac{23}{6.28318} \ (0.58136 - 0.13713) \tag{7b}$$

$$= 1.626$$

4. Pinion LPSTC, d_L

$$\tan c = 0.58136 - \frac{6.28318}{23} \tag{8}$$

$$= 0.30818 \qquad c = 17°7'41''$$

$$d_L = \frac{23 \times 0.93969}{10 \times 0.95565} \tag{9}$$

$$= 2.2616 \text{ inches}$$

5. Gear LPSTC, D_L

$$\tan C = 0.53227 - \frac{6.28318}{31} \tag{10}$$

$$= 0.32959 \qquad C = 18°14'30''$$

$$D_L = \frac{31 \times 0.93969}{10 \times 0.94974} \tag{11}$$

$$= 3.0672 \text{ inches}$$

6. Maximum permissible hob tip radius, r_c (max). The dedendum factor is 1.25.

$$r_c \text{ (max)} = \frac{0.785398 \times 0.93969 - 1.25 \times 0.34202}{1 - 0.34202} \tag{12}$$

$$= 0.4719 \text{ inch.}$$

7. If the hob tip radius r_c is 0.30, determine if the pinion involute is undercut.

$$\frac{1.25 - 0.30}{0.34202} + 0.30 \leqq 0.5 \times 23 \times 0.34202 \tag{13}$$

$$3.0776 < 3.9332$$

therefore there is no involute undercut.

8. Pinion HPSTC, D_H

$$\tan d = 0.13713 + \frac{6.28318}{23} \tag{14}$$

$$= 0.41031 \qquad d = 22°18'32''$$

$$d_H = \frac{23 \times 0.93969}{10 \times 0.92515} \tag{15}$$

$$= 2.3362 \text{ inches}$$

9. Gear HPSTC, D_H

$$\tan D = 0.20267 + \frac{6.28318}{31} \tag{16}$$

$$= 0.40535 \qquad D = 22°3'55''$$

$$D_H = \frac{31 \times 0.93969}{10 \times 0.92676} \tag{17}$$

$$= 3.1433 \text{ inches}$$

10. Pinion TIF diameter, d_{TIF}

$$\tan e = 0.36397 - \frac{4}{23}\left(\frac{1.25 - 0.30}{0.64279} + \frac{0.30}{2 \times 0.93969}\right) \tag{18}$$

$$= 0.07917 \qquad e = 4°31'36''$$

$$d_{TIF} = \frac{23 \times 0.93969}{10 \times 0.99688} \tag{19}$$

$$= 2.1681 \text{ inches}$$

11. Gear TIF diameter D_{TIF}

$$\tan E = 0.36397 - \frac{4}{31}\left(\frac{1.25 - 0.30}{0.64279} + \frac{0.30}{2 \times 0.93969}\right) \tag{20}$$

$$= 0.15267 \qquad E = 8°40'50''$$

$$D_{TIF} = \frac{31 \times 0.93969}{10 \times 0.98855} = 2.9468 \text{ inches} \tag{21}$$

Gear Blanks for Fine-pitch Gears. — The accuracy to which gears can be produced is considerably affected by the design of the gear blank and the accuracy to which the various surfaces of the blank are machined. The following recommendations should not be regarded as inflexible rules, but rather as minimum average requirements for gear-blank quality compatible with the expected quality class of the finished gear.

Design of Gear Blanks: The accuracy to which gears can be produced is affected by the design of the blank, so the following points of design should be noted: (1) Gears designed with a hole should have the hole large enough that the blank can be adequately supported during machining of the teeth and yet not so large as to cause distortion; (2) Face widths should be wide enough, in proportion to outside diameters, to avoid springing and to permit obtaining flatness in important surfaces; (3) Short bore lengths should be avoided wherever possible. It is feasible, however, to machine relatively thin blanks in stacks, provided the surfaces are flat and parallel to each other; (4) Where gear blanks with hubs are to be designed, attention should be given to the wall sections of the hubs. Too thin a section will not permit proper clamping of the blank during machining operations and may also affect proper mounting of the gear; and (5) Where pinions or gears integral with their shafts are to be designed, deflection of the shaft can be minimized by having the shaft length and shaft diameter well proportioned to the gear or pinion diameter. The foregoing general principles may also be useful when applied to blanks for coarser pitch gears.

Specifying Spur and Helical Gear Data on Drawings. — The data that may be shown on drawings of spur and helical gears falls into three groups: The first group consists of data basic to the design of the gear; the second group consists of data used in manufacturing and inspection; and the third group consists of engineering reference data. The accompanying table may be used as a checklist for the various data which may be placed on gear drawings and the sequence in which they should appear.

Explanation of Terms Used in Gear Specifications:

1. Number of teeth is the number of teeth in 360 deg of gear circumference. In a sector gear, both the actual number of teeth in the sector and the theoretical number of teeth in 360 deg should be given.

2. Diametral pitch is the ratio of the number of teeth in the gear to the number of inches in the standard pitch diameter. It is used in this standard as a nominal specification of tooth size.

2a. Normal diametral pitch is the diametral pitch in the normal plane.

2b. Transverse diametral pitch is the diametral pitch in the transverse plane.

3. Pressure angle is the angle between the gear tooth profile and a radial line at the pitch point. It is used in this standard to specify the pressure angle of the basic rack used in defining the gear tooth profile.

3a. Normal pressure angle is the pressure angle in the normal plane.

3b. Transverse pressure angle is the pressure angle in the transverse plane.

4. Helix angle is the angle between the pitch helix and an element of the pitch cylinder, unless otherwise specified.

4a. Hand of helix is the direction in which the teeth twist as they recede from an observer along the axis. A right hand helix twists clockwise and a left hand helix twists counterclockwise.

5. Standard pitch diameter is the diameter of the pitch circle. It equals the number of teeth divided by the transverse diametral pitch.

6. Tooth form may be specified as standard addendum, long addendum, short

addendum, modified involute or special. If a modified involute or special tooth form is required, a detailed view should be shown on the drawing. If a special tooth form is specified, roll angles must be supplied (see page 1806).

7. Addendum is the radial distance between the standard pitch circle and the outside circle. The actual value depends on the specification of outside diameter.

8. Whole depth is the total radial depth of the tooth space. The actual value is dependent on the specification of outside diameter and root diameter.

9. Maximum calculated circular thickness on the standard pitch circle is the tooth thickness which will provide the desired minimum backlash when the gear is assembled in mesh with its mate on minimum center distance. Control may best be exerted by testing in tight mesh with a master which integrates all errors in the several teeth in mesh through the arc of action as explained on page 1835. This value is independent of the effect of runout.

9a. Maximum calculated *normal* circular thickness is the circular tooth thickness in the normal plane which satisfies requirements explained in (9).

10. Gear testing radius is the distance from its axis of rotation to the standard pitch line of a standard master when in intimate contact under recommended pressure on a variable-center-distance running gage. Maximum testing radius should be calculated to provide the maximum circular tooth thickness specified in (9) when checked as explained on page 1835. This value is affected by the runout of the gear. Tolerance on testing radius must be equal to or greater than the total composite error permitted by the quality class specified in (11).

11. Quality class is specified for convenience when talking or writing about the accuracy of the gear. These classes are explained on page 1980.

12. Maximum total composite error, and 13. Maximum tooth-to-tooth composite error. Actual tolerance values (12 and 13) permitted by the quality class (11) are specified in inches to provide machine operator or inspector with tolerances required to inspect the gear.

14. Testing pressure recommendations are given on page 1835. Incorrect testing pressure will result in incorrect measurement of testing radius.

15. Master specifications by tool or code number may be required to call for the use of a special master gear when tooth thickness deviates excessively from standard.

16. Measurement over two 0.xxxx diameter pins may be specified to assist the manufacturing department in determining size at machine for setup only.

17. Outside diameter is usually shown on the drawing of the gear together with other blank dimensions so that it will not be necessary for machine operators to search gear tooth data for this dimension. Since outside diameter is also frequently used in the manufacture and inspection of the teeth, it may be included in the data block with other tooth specifications if preferred. To permit use of topping hobs for cutting gears on which the tooth thickness has been modified from standard, the outside diameter should be related to the specified gear testing radius (10).

18. Maximum root diameter is specified to assure adequate clearance for the outside diameter of the mating gear. This dimension is usually considered acceptable if the gear is checked with a master and meets specifications (10) through (13).

19. Active profile diameter of a gear is the smallest diameter at which the mating gear tooth profile can make contact. Because of difficulties involved in checking, this specification is not recommended for gears finer than 48 pitch.

20. Surface roughness on active profile surfaces may be specified in microinches to be checked by instrument up to about 32 pitch, or by visual comparison in the finer pitch ranges. It is difficult to determine accurately the surface roughness of fine pitch gears. For many commercial applications surface roughness may be considered

acceptable on gears which meet the maximum tooth-to-tooth-error specification (13).

21. Mating gear part number may be shown as a convenient reference. If the gear is used in several applications, all mating gears may be listed but usual practice is to record this information in a reference file.

22. Number of teeth in mating gear, and 23. Minimum operating center distance. This information is often specified to eliminate the necessity of getting prints of the mating gear and assemblies for checking the design specifications, interference, backlash, determination of master gear specification, and acceptance or rejection of gears made out of tolerance.

Data for Spur and Helical Gear Drawings

Type of Data	Min. Spur Gear Data	Min. Helical Gear Data	Add'l Optional Data	Item Number*	Data*
Basic Specifications	X	X		1	Number of teeth
	X			2	Diametral pitch
		X		2a	Normal diametral pitch
			X	2b	Transverse diametral pitch
	X			3	Pressure angle
		X		3a	Normal pressure angle
			X	3b	Transverse pressure angle
		X		4	Helix angle
		X		4a	Hand of helix
	X	X		5	Standard pitch diameter
	X	X		6	Tooth form
			X	7	Addendum
			X	8	Whole depth
	X			9	Max. calc. circular thickness on std. pitch circle
		X		9a	Max. calc. normal circular thickness on std. pitch circle
Manufacturing and Inspection			X	10	Roll angles
	X	X		11	A.G.M.A. quality class
	X	X		12	Max. total composite error
	X	X		13	Max. tooth-to-tooth composite error
			X	14	Testing pressure (Ounces)
	X	X		15	Master specification
			X	16	Meas. over two .xxxx dia. pins (For setup only)
	X	X		17	Outside diameter (Preferably shown on drawing of gear)
			X	18	Max. root diameter
			X	19	Active profile diameter
			X	20	Surface roughness of active profile
Engineering Reference			X	21	Mating gear part number
			X	22	Number of teeth in mating gear
			X	23	Minimum operating center distance

* An item-by-item explanation of the terms used in this table is given beginning on page 1809.

Power Transmitting Capacity of Spur Gears. — If a set of spur gears are made, installed and lubricated properly they normally may be subject to three primary modes of failure:

1. *Pitting:* This is a surface durability or fatigue failure mode that may be anticipated in the gear design by a determination of the gear set contact compressive stress and will most usually occur at a point just below the pitch surface on the driving pinion.

2. *Tooth Breakage:* This is usually a tensile fatigue failure at the weakest section of the gear tooth when considered as a cantilever beam. The weakest point is normally the tensile side of the gear tooth fillet and it may be anticipated in the gear design by determining the stress at this weakest section of the gear tooth.

3. *Tooth Scoring:* This is a scuffing or welding type of failure. It is not a fatigue failure but rather a failure of the lubricant in the presence of high tooth sliding velocity under high load. Well proportioned commercial gears with a pitch line velocity of less than 7,000 feet per minute will normally not score if they have a reasonably good surface finish and are properly lubricated. If scoring does occur or if it is suspected to be critical in a new high speed design, the scoring temperature index should be determined by the method shown in American Gear Manufacturers Standard AGMA 217.01 or by some similar method.

Surface Durability Stress. — The following method of determining the surface durability stress, s_c, in pounds per square inch, of a spur gear set is derived from American Gear Manufacturers Standard AGMA 210.02. The surface compressive stress in a gear set is a maximum at the lowest point of single tooth contact (LPSTC) of the pinion.

It is determined by the following formula:

$$s_c = 3239 \, C \sqrt{\frac{T_P}{F d_P^2 I}} \tag{1}$$

$$\text{where: } C = \frac{C_p C_t C_r}{C_l C_h} \sqrt{\frac{C_o C_s C_m C_f}{C_v}} \tag{1a}$$

$$\text{and } I = \frac{\cos^2 \phi}{2\left(\dfrac{1}{\tan c} + \dfrac{N_P}{N_G \tan D}\right)} \tag{2}$$

(For standard tooth forms the values of the geometry factor I are tabulated in Table 5 on page 1815; the interpolation procedure for finding I for values of gear ratio m_g and of number of teeth in pinion that are not tabulated in Table 5, is shown on page 1817 in the section Spur Gear Sizing.)

where d_P = pinion pitch diameter in inches

ϕ = pressure angle in degrees

F = face width in inches

N_G = number of teeth in gear

T_P = pinion max. operating torque in lbs.-in.

C_p = material factor — equals 1.0 for steel pinion and steel gear; for steel and iron use 0.87; and for iron and iron use 0.78

C_t = temperature factor — equals 1.0 except when gears operate near the tempering temperature of the gear material

C_r = reliability factor — normally equals 1.0; for very high reliability use 1.25

C_l = life factor — equals 1.0 for 10^7 cycles or more; 1.3 for 10^5 cycles; and 1.5 for 10^4 cycles

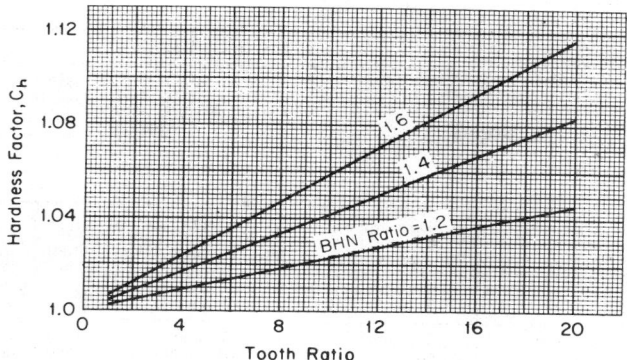

C_h = hardness factor — for a Brinell hardness ratio of pinion to gear of 1.2 or less, use 1.0, otherwise use value taken from curve in chart shown above.

C_o = overload factor — use value from Table 1.

C_s = Size factor, equals 1 provided gear metallurgy and case depth are appropriate.

C_f = surface factor — equals 1.0 for normal surface finish and condition.

C_v = dynamic factor — equals 1.0 for precision gears. For commercial gears use value from Table 2.

C_m = Load distribution factor. For normal commercial gearing when the face width to pinion pitch diameter ratio is not greater than 2, see Table 3.

tan c = as determined by formula on page 1804

tan D = as determined by formula on page 1805

Surface durability stress, s_c, must not be greater than the limiting values of allowable contact stress, s_{ac}, as given in Table 4.

Fillet Tensile Stress. — The following method of determining the fillet tensile stress of a spur gear is taken from American Gear Manufacturers Standard AGMA 220.02. The fillet tensile stress of a gear is normally determined with the full maximum operating load applied at the highest point of single tooth contact (HPSTC) diameter; however, if tooth spacing errors are significantly high compared with the expected full load deflection of the gear tooth mesh, the maximum operating load should be applied at the gear outside diameter for this stress determination. For all normal cases using standard gear tooth proportions the fillet tensile stress is highest on the pinion, this, therefore, is the only one that need be calculated. However, it can be seen that the comparable stress on the mating gear is inversely proportional to the value of the J factor of the two gears, all other parameters being equal. The fillet tensile stress is determined by the formula:

$$s_t = \frac{2K \times T_1 \times n}{F \times d_1^2 \times J} \text{ , where: } \quad K = \frac{K_o \times K_s \times K_m \times K_r \times K_t}{K_v \times K_l} \qquad (3), (4)$$

For definitions of symbols in formulas (3) and (4) see page 1815.

Table 1. Overload Factor, C_o

Power Source Shock*	Driven Load Shock		
	None	Moderate	Heavy
	Overload Factor, C_o		
None	1.00	1.25	1.75
Light	1.25	1.50	2.00
Medium	1.50	1.75	2.25

Note: For speed increaser drives add $0.01\sqrt{(N/n)}$ to the tabular value.
*For example: No shock — electric motor or turbine drive; Light shock — 4 or more cylinder internal combustion engine drive; and Medium Shock — 1 or 2 cylinder internal combustion engine drive.

Table 2. Dynamic Factor C_v

Pitch Line Velocity, fpm	Dynamic Factor, C_v	Pitch Line Velocity, fpm	Dynamic Factor, C_v
Up to 100	0.9	701 to 1500	0.6
101 to 300	0.8	1501 to 3500	0.5
301 to 700	0.7	3501 to 7000	0.4

Table 3. Load Distribution Factor, C_m

Face Width, In.	C_m and K_m	Face Width, In.	C_m and K_m	Face Width, In.	C_m and K_m
Up to 2	1.3	6 to 9	1.5	12 to 16	1.7
2 to 6	1.4	9 to 12	1.6	...	...

Table 4. Allowable Contact Stress Number — s_{ac}

Material	Surface Hardness, Minimum	s_{ac}	Material	Surface Hardness, Minimum	s_{ac}
	Through Hardened		Cast Iron		
	180 Bhn	85- 95,000	AGMA Grade 20	—	50-60,000
			AGMA Grade 30	175 Bhn	65-75,000
	240 Bhn	105-115,000	AGMA Grade 40	200 Bhn	75-85,000
	300 Bhn	120-135,000	Nodular Iron		90-100% of the s_{ac} value of steel with the same hardness
	360 Bhn	145-160,000	Annealed	165 Bhn	
			Normalized	210 Bhn	
	440 Bhn	170-190,000	Oil Quench and Temper	255 Bhn	
Steel	Case Carburized				
	55 R_c	180-200,000	Bronze	Tensile Strength psi (Min.)	s_{ac}
	60 R_c	200-225,000	Tin Bronze AGMA 2C (10-12% Tin)	40,000	30,000
	Flame or Induction Hardened		Aluminum Bronze ASTM B 148-52 (Alloy 9C-H.T.)	90,000	65,000
	50 R_c	170-190,000			

Table 5. *J* Tooth Form Factor and *I* Geometry Factor for
American National Standard Involute Gears

Gear Ratio	No. of Pinion Teeth	20-degree Pressure Angle			25-degree Pressure Angle		
		I Factor	*J* Factor, Ded. = 1.25*	*J* Factor, Ded. = 1.35*	*I* Factor	*J* Factor, Ded. = 1.25*	*J* Factor, Ded. = 1.35†
1	17	.0755	.320	.302	.0900	.383	.353
1	20	.0774	.343	.326	.0919	.409	.379
1	25	.0789	.372	.356	.0936	.441	.410
1	30	.0795	.395	.379	.0944	.465	.435
1	40	.0800	.428	.411	.0951	.499	.468
1	60	.0802	.467	.449	.0955	.539	.505
1	100	.0803	.505	.487	.0957	.575	.540
1	200	.0803	.540	.521	.0957	.607	.570
3	17	.0983	.340	.319	.1173	.403	.369
3	20	.1037	.365	.344	.1227	.430	.396
3	25	.1089	.396	.376	.1283	.463	.429
3	30	.1121	.419	.399	.1316	.487	.452
3	40	.1154	.452	.432	.1354	.519	.485
3	60	.1180	.490	.469	.1387	.557	.520
3	100	.1195	.524	.503	.1410	.589	.553
3	200	.1202	.552	.531	.1424	.615	.576
10	17	.1128	.351	.328	.1347	.413	.377
10	20	.1204	.377	.354	.1423	.440	.404
10	25	.1282	.409	.387	.1502	.474	.437
10	30	.1328	.432	.410	.1553	.498	.461
10	40	.1379	.465	.443	.1611	.530	.492
10	60	.1420	.500	.478	.1662	.564	.526
10	100	.1444	.531	.509	.1698	.594	.555
10	200	.1456	.557	.535	.1721	.618	.579

* Rack fillet radius = .30/*P*.   † Rack fillet radius = .21/*P*.

K_o = Overload factor — see Table 1 (same as for C_o).

K_s = Size factor — equals 1.0 provided metallurgy and case depth are appropriate.

K_m = Load distribution factor — see Table 3 (same as for C_m).

K_r = Reliability factor — normally equals 1.0; for very high reliability use 1.50.

K_t = Temperature factor — equals 1.0 except when gears are operated near the tempering temperature of the gear material.

K_v = Dynamic factor — equals 1.0 for precision gears. For commercial gears use value from Table 2 (same as for C_v).

K_l = Life factor — equals 1.0 for 10^7 cycles or more; for 10^6 cycles use 1.1.

J = Tooth form factor — for standard gears use value from Table 5. (*J* tooth form factors for values of gear ratio m_g and numbers of teeth in pinion that are not tabulated in Table 5 must be determined by interpolation using the method shown on page 1818 in the section Spur Gear Sizing.)

T_1 = Pinion maximum operating torque in pound-inches.

n = Number of teeth in the pinion.

F = Gear face width in inches.

d_1 = Pinion pitch diameter.

Example: Find the surface durability stress and the fillet tensile stress of a 23-tooth, 20-degree AISI standard pinion with a $1.25/P$ dedendum and a diametral pitch of 10 driving a 31-tooth gear. The face width is 1.00 inch; the pinion torque is 500 pounds-inch; the pinion speed is 3600 rpm; and the pinion pitch line velocity is 2000 feet per minute. The gears are made of case carburized steel.

$$n = 23 \qquad\qquad N = 31$$
$$d = 2.3 \qquad\qquad P = 10$$
$$d_o = \frac{n+2}{10} = 2.5 \qquad T_1 = 500$$
$$DF = 1.00 \qquad\qquad V_1 = 2{,}000$$

From the formulas for significant diameters beginning on page 1803:

$$\cos a = \frac{n \cos \phi}{d_o P} = \frac{23 \times 0.93969}{2.5 \times 10} = 0.86451 \qquad (4)$$

$$a = 30°10'24''$$

$$\tan c = \tan a - \frac{6.28318}{n} = 0.58139 - \frac{6.28318}{23} = 0.30821 \qquad (8)$$

$$\tan B = \tan \phi - \frac{n}{N} (\tan a - \tan \phi) \qquad (5)$$

$$= 0.36397 - \frac{23}{31} (0.58139 - 0.36397) = 0.20266$$

$$\tan D = \tan B + \frac{6.28318}{N} = 0.20266 + \frac{6.28318}{31} = 0.40534 \qquad (16)$$

Surface Durability Stress:

Based on these data, the following surface durability factors apply: $C_p = 1$, $C_t = 1$, $C_r = 1$, $C_l = 1$, $C_h = 1$, $C_o = 1$. $C_s = 1$, $C_v = 0.55$, $C_m = 1.3$, $C_f = 1$, and

$$I = \frac{0.93969^2}{2 \left(\frac{1}{0.30821} + \frac{23}{31 \times 0.40534} \right)} = 0.087 \qquad (2)$$

$$C = \frac{1 \times 1 \times 1}{1 \times 1} \sqrt{\frac{1 \times 1 \times 1.3 \times 1}{0.55}} = 1.54 \qquad (1a)$$

$$s_c = 3239 \times 1.54 \sqrt{\frac{500}{1 \times (2.3)^2 \times 0.087}} = 164{,}000 \text{ psi} \qquad (1)$$

Fillet Tensile Stress:

Based on the above given data, the following fillet tensile stress factors apply: $K_o = 1$, $K_s = 1$, $K_m = 1.3$, $K_r = 1$, $K_t = 1$, $K_v = 0.55$, and $K_l = 1$.

Using the formula for fillet tensile stress:

For the 23-tooth pinion, $J = 0.369$.

$$K = \frac{1 \times 1 \times 1.3 \times 1 \times 1}{0.55 \times 1} = 2.36 \qquad (4)$$

$$s_t = \frac{2 \times 2.36 \times 500 \times 23}{1 \times 5.3 \times 0.369} = 27,800 \text{ psi.} \qquad (3)$$

The actual tensile stress s_t must not be greater than the allowable tensile stress, s_{at}, as given in Table 6.

Table 6. Limiting Values of Fillet Tensile Stress, s_{at}

Material and Brinell Hardness	Allowable Stress, psi	Material and Brinell Hardness	Allowable Stress, psi
Cast Iron — 175	8,500	Steel — 300	36,000
Cast Iron — 200	13,000	Steel — 350	39,000
Steel, Unhardened	22,000	Steel — 400	42,000
Steel — 200	27,000	Steel — 450	44,000
Steel — 250	32,000	Steel — Carburized	55,000 to 65,000

Optimum Spur Gear Sizing for Standard Tooth Form. — The formulas shown in the previous two sections for the determination of surface durability stress s_c and fillet tensile stress s_t may be inverted in such a manner that they can be used to determine the optimum gear size and number of teeth of standard tooth form to be used to satisfy required operating load, stresses, gear ratio, and tooth form.

Data that must be given or assumed are:

1. *Gear tooth form:* (a) pressure angle, 20° or 25°; (b) dedendum factor, 1.25 or 1.35; and (c) diametral pitch.

2. *Operating conditions:* (a) maximum pinion torque, inch pounds; (b) gear load factor; and (c) gear reduction ratio.

3. *Gear material hardness:* Brinell hardness number. Material must be steel.

The data to be computed are: (a) gear tooth fillet tensile stress; (b) gear surface durability stress; (c) number of teeth in pinion and gear; and (d) gear face width.

Using these data, complete gear dimensions may be obtained using the formulas in Table 1 on page 1771.

The first step in this sizing procedure is to obtain a first estimate of the number of teeth N_P in the pinion using the following formula:

$$\log N_P = \frac{1}{1-q} \log \left(\frac{73000 \times K_m}{6H_b} \right) \qquad (1)$$

where: H_b = Gear material hardness, Bhn
K_m = Material factor — 0.6 for through-hardening steels
0.9 for case-hardening steels.

$$q = \frac{5}{90} + \frac{1}{7 \sqrt[3]{m_g^2}} \qquad (2)$$

where: m_g = gear reduction ratio; must be greater than 1.

This pinion tooth number N_P and the given gear ratio m_g are used to enter Table 5 on page 1815 for the specified tooth form to get values for the I and J factors. In most cases the specified gear ratio and the estimated number of teeth are not tabulated values, thus the I and J factors must be found by interpolation. The interpolation factor i to be used is found by the formula:

$$\text{Factor } i = \frac{h}{m} \left(\frac{m-l}{h-l} \right) \qquad (3)$$

where: l = tabulated input next lower than the specified value
m = specified input value
h = tabulated input next higher than the specified value

Two interpolation factors are usually required, one is a gear ratio interpolation factor and the other is a pinion tooth number interpolation factor. In later calculations for the I and J factors care must be taken not to confuse one with the other. In the event that the required gear ratio is a tabulated value, no gear ratio interpolation factor is needed; likewise if the estimated pinion tooth number is a tabulated value, no tooth number interpolation factor is needed.

Both the I and J factors are determined in the same manner; the procedure will be explained in terms of I. The interpolation formulas are:

$$I = I_l + i(I_h - I_l) \quad \text{and} \quad J = J_l + i(J_h - J_l) \qquad \text{(4a) and (4b)}$$

where: I_l = the tabulated I for the input next lower than the specified value
I_h = the tabulated I for the input next higher than the specified value

Each formula is used three times; first, at the tabulated gear ratio lower than the specified value to obtain the I factor for the estimated number of teeth N_P at this tabulated value of gear ratio and second, at the tabulated gear ratio higher than the specified value. Both of these calculations use the tooth number interpolation factor i_N. The third interpolation calculation uses the two values of I calculated above to determine the I factor for the specified tooth number at the specified gear ratio; this calculation uses the gear ratio interpolation factor i_m.

The allowable fillet tensile stress may be obtained by dividing the appropriate value from Table 6 on page 1817 by the gear load factor K (see page 1813); or the following formula may be used:

$$\log (s_t) = \frac{3}{4} \log (H_b) - 0.3010 - \log (K) \qquad (5)$$

where: s_t = Fillet stress in kpsi
H_b = Gear material hardness in Brinell hardness number
K = Load factor (see formula 4, page 1813)
(Logarithms are to the base 10)

Note that the stress must be given in thousands of pounds per square inch (kpsi). If the hardness is given in terms of the Rockwell C scale as is often done with case-hardening materials, conversion to Bhn may be done with the conversion table given on page 559; the required hardness is given in this table in the column labeled 'Tungsten Carbide Ball.' For case-hardening materials, the hardness of the case is to be used in this formula.

To satisfy the requirements of this sizing method, the allowable surface durability stress s_c must be obtained from the following formula:

$$s_c = \sqrt{\frac{s_t \times H_b}{K_m}} \qquad (6)$$

where: s_c = Durability stress in kpsi
s_t = Fillet tensile stress in kpsi
H_b = Material hardness in Bhn
K_m = Material factor, 0.6 for through-hardening steels, 0.9 for case-hardening steels.

The stress given by this formula is almost the same as that given in Table 4 on page 1814, divided by the square root of K.

All data needed to size the gear set are now available. To determine the gear size, the following formula is used:

$$FD_P{}^2 = \frac{10.58T_P}{Is_c{}^2} \qquad (7)$$

where: F = Effective face width of gear and pinion in inches
D_P = Pinion pitch diameter in inches
s_c = Surface durability stress in kpsi
I = Gear geometry factor
T_P = Pinion maximum operating torque in inch-pounds

and the number of teeth in the pinion N_P is determined by:

$$N_P = \frac{s_t(FD_P^2)J_P}{2T_P} \times 10^3 \tag{8}$$

where: s_t = Fillet tensile stress in kpsi
(FD_P^2) = is as determined above
J_P = pinion tooth form factor

This formula will most probably give a result that is not an integer; if so, the actual pinion tooth number must be the next lower integer. The number of teeth N_G in the mating gear is:

$$N_G = N_P m_g \text{ to the nearest integer.} \tag{9}$$

where: m_g = the gear ratio

It is good practice to back check to determine the actual ratio by:

$$m_g = \frac{N_G}{N_P} \tag{10}$$

If this results in an unsatisfactory deviation from the specified gear ratio, the number of teeth in the pinion may be reduced by one or two and a new number is determined for the number of teeth in the mating gear. If this renumbering process is continued excessively, a revised I factor and a revised FD_P^2 will need to be calculated. As the tooth numbers are reduced, the fillet tensile stress is also reduced, thus a redetermination of s_t by the formula on page 1813 need not be made unless the actual reduced value of this stress is needed for some purpose.

If the diametral pitch P has been specified, the pinion pitch diameter D_P may be determined by:

$$D_P = \frac{N_P}{P} \tag{11}$$

If the diametral pitch is not specified, some other criterion is needed to obtain the pinion pitch diameter. This criterion may be the required center distance C, in which case the pinion pitch diameter is:

$$D_P = \frac{2C}{m_g + 1} \tag{12}$$

where: m_g = the actual gear ratio.

On the other hand it might be that the gear face width F_G is specified by some ancillary requirement, in which case the pinion pitch diameter is determined by:

$$D_P = \sqrt{\frac{(F_G D_P^2)}{F_G}} \tag{13}$$

Note that although D_P is to be determined, $(F_G D_P^2)$ has already been calculated.

Or, as is the case with many aerospace applications, it may be required that the smallest practical diameter shall be used; in this case the face width should be no more than 0.7 of the pinion pitch diameter unless special consideration is

given to the load patterning of the gear teeth; where the pinion pitch diameter is then determined by:

$$D_P = \sqrt[3]{\frac{(FD_P{}^2)}{0.70}} \qquad (14)$$

The diametral pitch may now be determined by:

$$P = \frac{N_P}{D_P} \qquad (15)$$

and the pitch diameter of the mating gear is:

$$D_G = \frac{N_G}{P} \qquad (16)$$

If the pinion pitch diameter is directly specified because of needed clearance over a bearing, spline, seal, key, or other associated element, then the face width may be directly determined by:

$$F_G = \frac{(FD_P{}^2)}{D_P{}^2} \qquad (17)$$

All data are now at hand for the completion of the required data by use of the formulas in Table 1 (page 1771).

A numerical example of this method will illustrate its use; particular attention should be given to the method of interpolation of the tabular values of I and J in Table 5 on page 1815.

Spur Gear Sizing Example: The following example shows how complete gear dimensions may be calculated using the formulas and procedure just outlined.

Given data:

> Pressure angle = 20°
> Dedendum factor K_o = 1.25
> Rack tooth fillet radius r_f = 0.30 in.
> Gear reduction ratio m_g = 2.25
> Pinion torque T_P = 1300 in.-lbs.
> Load factor K = 1.4
> Material hardness H_b = 340 Bhn
> Material factor K_m = 0.60
> Diametral pitch P = 16

Ratio factor:

$$q = \frac{5}{90} + \frac{1}{7\sqrt[3]{m_g{}^2}} \qquad (2)$$

$$q = \frac{5}{90} + \frac{1}{7\sqrt[3]{2.25^2}} = 0.139$$

First estimate of number of teeth in pinion:

$$\log N_P = \frac{1}{1-q} \log\left(\frac{12.17\, K_m \times 10^3}{H_b}\right) \qquad (1)$$

$$= \frac{1}{1-0.139} \log\left(\frac{12.17 \times 0.60 \times 10^3}{340}\right)$$

$$= 1.5470$$

$$N_P = 35.2, \text{ nearest integer } 35$$

Interpolation factor for Table 5 (p. 1815):

$$i = \frac{h}{m}\left(\frac{m-l}{h-l}\right) \tag{3}$$

For tooth number interpolation: $l = 30$, $m = 35$ and $h = 40$

$$i_N = \frac{40}{35}\left(\frac{35-30}{40-30}\right) = 0.5714$$

For gear ratio interpolation: $l = 1$, $m = 2.25$ and $h = 3$

$$i_m = \frac{3}{2.25}\left(\frac{2.25-1}{3-1}\right) = 0.8333$$

Determination of geometry factor I:

$$I = I_l + i_N(I_h - I_l) \tag{4a}$$

For $m_g = 1$ and $N_P = 35$, $I_h = 0.0800$, $I_l = 0.0795$ and $i_N = 0.5714$
$$I_1 = 0.0795 + 0.5714\,(0.0800 - 0.0795)$$
$$= 0.0798$$

For $m_g = 3$ and $N_P = 35$, $I_h = 0.1154$, $I_l = 0.1121$ and $i_N = 0.5714$
$$I_2 = 0.1121 + 0.5714\,(0.1154 - 0.1121)$$
$$= 0.1140$$

$$I = I_l + i_m\,(I_h - I_l) \tag{4a}$$

For $m_g = 2.25$ and $N_P = 35$, $I_l = 0.0798$, $I_h = 0.1140$ and $i_m = 0.8333$
$$I = 0.0798 + 0.8333\,(0.1140 - 0.0798)$$
$$= 0.1083$$

Determination of form factor J:

$$J = J_l + i_N\,(J_h - J_l) \tag{4b}$$

For $m_g = 1$ and $N_P = 35$, $J_l = 0.395$, $J_h = 0.428$ and $i_N = 0.5714$
$$J_1 = 0.395 + 0.5714\,(0.428 - 0.395)$$
$$= 0.414$$

For $m_g = 3$ and $N_P = 35$, $J_l = 0.419$, $J_h = 0.452$ and $i_N = 0.5714$
$$J_2 = 0.419 + 0.5714\,(0.452 - 0.419)$$
$$= 0.438$$

$$J = J_l + i_m\,(J_h - J_l) \tag{4b}$$

For $m_g = 2.25$ and $N_P = 35$, $J_l = 0.414$, $J_h = 0.438$ and $i_m = 0.8333$
$$J = 0.414 + 0.8333\,(0.438 - 0.414)$$
$$= 0.434$$

Determination of allowable stresses s_t and s_c

$$\log s_t = \frac{3}{4}\log H_b - 0.301 - \log K \tag{5}$$

$$= \frac{3}{4}\log 340 - 0.301 - \log 1.4$$

$$= 1.4515$$

$$s_t = 28.28, \text{ say } 28 \text{ kpsi}$$

$$s_c = \sqrt{\frac{s_t H_b}{K_m}} \tag{6}$$

$$= \sqrt{\frac{28 \times 340}{0.6}} = \sqrt{15{,}870}$$

$$s_c = 126 \text{ kpsi}$$

Determination of Gear Size and Tooth Numbers:

$$FD_P{}^2 = \frac{10.58 T_P}{I s_c{}^2} \tag{7}$$

$$= \frac{10.58 \times 1300}{0.1083 \times 15{,}870}$$

$$= 8.00 \text{ in.}^3$$

$$N_P = \frac{s_t (FD_P{}^2) J_P \times 10^3}{2 T_P} \tag{8}$$

$$= \frac{28 \times 8.00 \times 0.434 \times 1000}{2 \times 1300}$$

$$= 37.4, \text{ say, } 37 \text{ teeth}$$

$$N_G = N_P m_g \tag{9}$$

$$= 37 \times 2.25 = 83.25, \text{ say, } 83 \text{ teeth}$$

Checking on actual ratio:

$$m_g = \frac{N_G}{N_P} \tag{10}$$

$$m_g = \frac{83}{37} = 2.24$$

Determination of pinion and gear pitch diameters:

$$D_P = \frac{N_P}{P} \tag{11}$$

$$= \frac{37}{16} = 2.3125 \text{ in., diameter of pinion}$$

$$D_G = \frac{N_G}{P} \tag{11}$$

$$= \frac{83}{16} = 5.1875 \text{ in., diameter of gear}$$

Determination of face width:

$$F = \frac{FD_P{}^2}{D_P{}^2} \tag{17}$$

$$= \frac{8.00}{2.3125^2} = 1.50 \text{ in. wide}$$

The designer will need to use judgement in any particular case. This should not be difficult as long as it is remembered that the method described here will determine the maximum tooth numbers and the minimum size (FD^2) that must be used to satisfy the specified input criteria. Production cost is usually a function of the amount of metal removal; thus fewer teeth mean increased production cost. An increase in size (FD^2) means an increase in gear weight and, in most cases, an increase in gear-box or transmission weight; this, in turn, will mean an increase in production cost.

British Standard Horsepower Formulas. — The horsepower formulas which follow are included in the revised British standard specifications No. 436 — 1940, amended in 1941, 1943, and 1956, for machine cut spur gears and also for helical gears used in driving parallel shafts. Two formulas are given. One indicates the horsepower with reference to tooth strength; the other, the horsepower as limited by tooth wear. In deciding upon the horsepower capacity formula, the Committee gave special consideration to the question whether the speed factor should be based on revolutions per minute or pitch-line speed in feet per minute, and the former was adopted. The power capacity of both gear and pinion should be checked (1) for tooth wear, and (2) for tooth strength. The smallest of the four power ratings thus obtained should be used.

In the following formulas S_c = surface stress factor (Table 1); X_c = speed factor for wear (Table 2); Z = zone factor (see chart, page 1826); F = face width, inches; N = revolutions per minute; T = number of teeth; K = pitch factor = $P^{0.8}$; P = diametral pitch; S_b = bending stress (Table 1); X_b = speed factor for strength (Table 3); Y = strength factor (see chart, page 1827). The wear and strength formulas follow:

$$\text{Horsepower for wear} = \frac{S_c X_c Z F N T}{126,000 K P}$$

$$\text{Horsepower for strength} = \frac{S_b X_b Y F N T}{126,000 P^2}$$

Example: Find the allowable horsepower for spur gears. The pinion and gear speeds are 500 and 100 revolutions per minute; continuous operation 12 hours per day; diametral pitch, 3; face width, 4 inches; pressure angle 20 degrees; pinion, 20 teeth; pinion material, 0.40% carbon steel normalized; gear, 100 teeth; gear material, cast iron of ordinary grade.

Horsepower formulas for wear and for strength are applied to both gear and pinion as shown below and the smallest horsepower rating (approximately 40 in this case) is used.

$$\text{Pinion H.P. for wear} = \frac{1600 \times 0.305 \times 2.20 \times 4 \times 500 \times 20}{126,000 \times 2.40 \times 3} = 47$$

$$\text{Gear H.P. for wear} = \frac{1000 \times 0.410 \times 2.20 \times 4 \times 100 \times 100}{126,000 \times 2.40 \times 3} = 40$$

$$\text{Pinion H.P. for strength} = \frac{19,000 \times 0.305 \times 0.72 \times 4 \times 500 \times 20}{126,000 \times 3 \times 3} = 147$$

$$\text{Gear H.P. for strength} = \frac{5800 \times 0.410 \times 0.61 \times 4 \times 100 \times 100}{126,000 \times 3 \times 3} = 51$$

Hence, the power rating would be 40 horsepower, assuming sufficient gear usage to warrant a wear rating instead of a tooth breakage rating.

Table 1. Basic Surface and Bending-Stress Factors of Spur and Helical Gears

Factors for use in British Standard Horsepower Formula

Type of Material (Numbers in Parentheses Indicate Footnotes)	Minimum Tensile Strength Tons per Sq. In.	Minimum Brinell Hardness Number	Surface Stress Factor S_c	Bending Stress Lb. per Sq. In. S_b
Fabric	..	..	560	4,500
Cast Iron, Ordinary Grade	12	165	1,000	5,800
" ", Medium Grade	16	210	1,350	7,600
" ", High Grade, as Cast	22	220	1,450	10,400
Castings, Malleable	20	140	850	11,000
Phosphor Bronze, Sand Cast	12	69	700	7,000
" ", Chill Cast	15	82	850	8,500
" ", Centrifugally Cast	17	90	1,000	10,000
Cast Steel, 0.35% to 0.45% Carbon	35	145	1,400	19,000
" ", 0.50% to 0.55% Carbon (1)	38	160 (3)	3,100	13,000
Forged Carbon Steel, 0.15% Carbon (2)	32	140 (4)	9,000	28,000
" " ", 0.40% Carbon (6)	35	145	1,400	17,000
" " ", 0.40% Carbon (7)	35	145	1,600	19,000
" " ", 0.40% Carbon (1)	35	145 (8)	2,800	12,000
" " ", 0.40% Carbon (9)	40	175	1,800	20,000
" " ", 0.40% Carbon(10)	40	175	2,000	22,000
" " ", 0.55% Carbon (6)	45	200	2,000	21,600
" " ", 0.55% Carbon (7)	45	200	2,300	24,000
" " ", 0.55% Carbon (1)	45	200 (11)	4,000	15,000
Nickel Steel, 1% nickel (12)	40	175	2,000	22,000
" ", 3% nickel (12)	45	200	2,300	24,000
" ", 3% nickel (2)	45	200 (13)	10,200	40,000
" ", 3½% nickel (1)	55	250 (14)	5,100	18,500
" ", 3½% nickel (12)	55	250	3,000	30,000
" ", 3½% nickel (2)	45	200 (15)	10,200	40,000
" ", 5% nickel (2)	55	250 (16)	11,200	47,000
Nickel-chromium, 1½% Ni, 1% Cr(17)	55	250	3,000	30,000
" ", 1½% Ni, 1% Cr (1)	55	250 (13)	5,100	18,500
" ", 1½% (17)	100	440	5,500	40,000
" ", 3½% (17)	55	250	3,000	30,000
" ", 3½% (1)	55	250 (14)	5,100	18,500
" ", 3½% (2)	55	250 (16)	11,200	47,000
Carbon-chromium, 0.55% carbon (18)	55	250	3,000	30,000
" ", 0.55% carbon (18)	65	290	3,500	36,000
" ", 0.55% carbon (1)	55	250 (14)	5,100	18,500

(1) Surface Hardened; (2) Casehardened; (3) Core, 160; case, 530; (4) Hardness of core; (5) Core, 140; case, 640; (6) normalized; for sections thicker than 5 inches; (7) normalized; for sections less than 5 inches thick; (8) Core, 145; case, 460; (9) Heat-treated; for sections thicker than 5 inches; (10) Heat-treated; for sections less 5 inches thick; (11) Core, 200; case, 520; (12) Heat-treated; (13) Core; (14) Core, 250; case, 500; (15) Core, 200; case, 620; (16) Core, 250; case, 600; (17) Oil hardened and tempered to strength given in second col.; (18) Heat-treated to strength given in second column.

Table 2. Speed Factors X_c for Wear

Rev. per Minute	Running Time — Hours per Day							
	1	2	4	6	8	12	18	24
	Speed Factors X_c for Wear							
100	0.935	0.735	0.585	0.515	0.470	0.410	0.350	0.320
150	0.865	0.685	0.540	0.475	0.435	0.370	0.330	0.300
200	0.825	0.650	0.520	0.460	0.415	0.360	0.310	0.280
300	0.775	0.615	0.485	0.425	0.380	0.330	0.290	0.270
400	0.730	0.580	0.460	0.400	0.360	0.320	0.270	0.250
500	0.700	0.550	0.440	0.380	0.350	0.305	0.260	0.240
600	0.680	0.530	0.425	0.370	0.340	0.290	0.250	0.230
800	0.635	0.500	0.400	0.350	0.320	0.270	0.240	0.220
1,000	0.610	0.480	0.380	0.335	0.305	0.260	0.230	0.210
1,500	0.550	0.440	0.345	0.310	0.275	0.240	0.210	0.190
2,000	0.520	0.415	0.325	0.290	0.260	0.220	0.200	0.180
2,500	0.480	0.380	0.305	0.265	0.240	0.210	0.185	0.165
3,000	0.450	0.355	0.280	0.250	0.225	0.195	0.170	0.155
4,000	0.415	0.325	0.260	0.225	0.207	0.180	0.155	0.145
5,000	0.380	0.305	0.240	0.210	0.190	0.165	0.155	0.132
6,000	0.355	0.285	0.225	0.200	0.180	0.155	0.145	0.125
7,000	0.340	0.270	0.215	0.190	0.170	0.150	0.135	0.118
8,000	0.325	0.260	0.205	0.180	0.165	0.142	0.130	0.113
9,000	0.315	0.250	0.200	0.175	0.157	0.135	0.125	0.108
10,000	0.305	0.240	0.190	0.165	0.152	0.130	0.115	0.105

Table 3. Speed Factors X_b for Strength

Running Time, Hours per Day	Revolutions per Minute									
	100	150	200	300	400	500	600	800	1000	1500
	Speed Factors X_b for Strength									
1	0.600	0.550	0.525	0.445	0.435	0.420	0.415	0.410	0.385	0.350
3	0.510	0.435	0.425	0.410	0.400	0.380	0.370	0.345	0.330	0.300
6	0.430	0.415	0.405	0.380	0.360	0.345	0.330	0.310	0.295	0.275
12	0.410	0.380	0.360	0.340	0.320	0.310	0.300	0.285	0.270	0.245
24	0.375	0.350	0.330	0.310	0.295	0.285	0.275	0.255	0.245	0.225

Running Time, Hours per Day	Revolutions per Minute									
	2000	2500	3000	4000	5000	6000	7000	8000	9000	10,000
	Speed Factors X_b for Strength									
1	0.325	0.305	0.285	0.260	0.240	0.225	0.215	0.208	0.200	0.192
3	0.285	0.260	0.245	0.225	0.208	0.195	0.185	0.178	0.170	0.165
6	0.255	0.235	0.220	0.200	0.185	0.175	0.165	0.160	0.153	0.148
12	0.230	0.215	0.200	0.182	0.168	0.158	0.150	0.145	0.140	0.135
24	0.210	0.195	0.180	0.165	0.152	0.143	0.138	0.130	0.126	0.120

Table 4. Pitch Factors K

Diametral Pitch	Factor K	Diametral Pitch	Factor K	Diametral Pitch	Factor K	Diametral Pitch	Factor K
1	1.00	2¼	1.90	4	3.05	9	5.80
1¼	1.20	2½	2.10	5	3.65	10	6.40
1½	1.40	2¾	2.25	6	4.25	12	7.40
1¾	1.55	3	2.40	7	4.80	14	8.30
2	1.75	3½	2.70	8	5.40	16	9.25

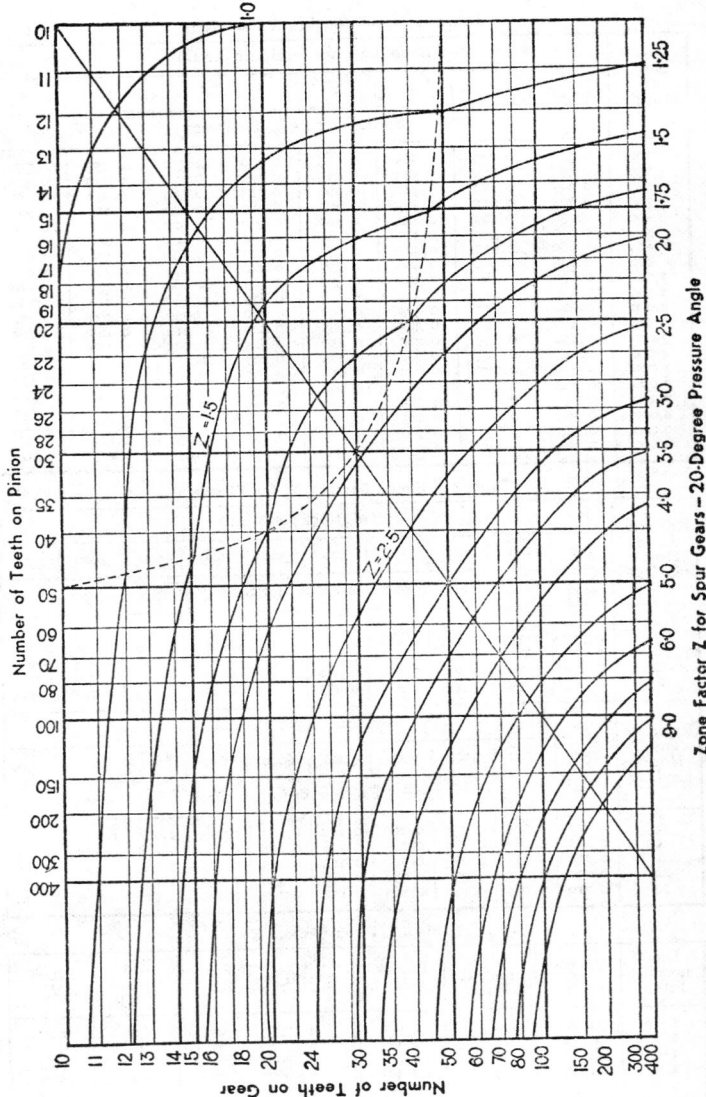

Number of Teeth on Pinion

Number of Teeth on Gear

Zone Factor Z for Spur Gears—20-Degree Pressure Angle

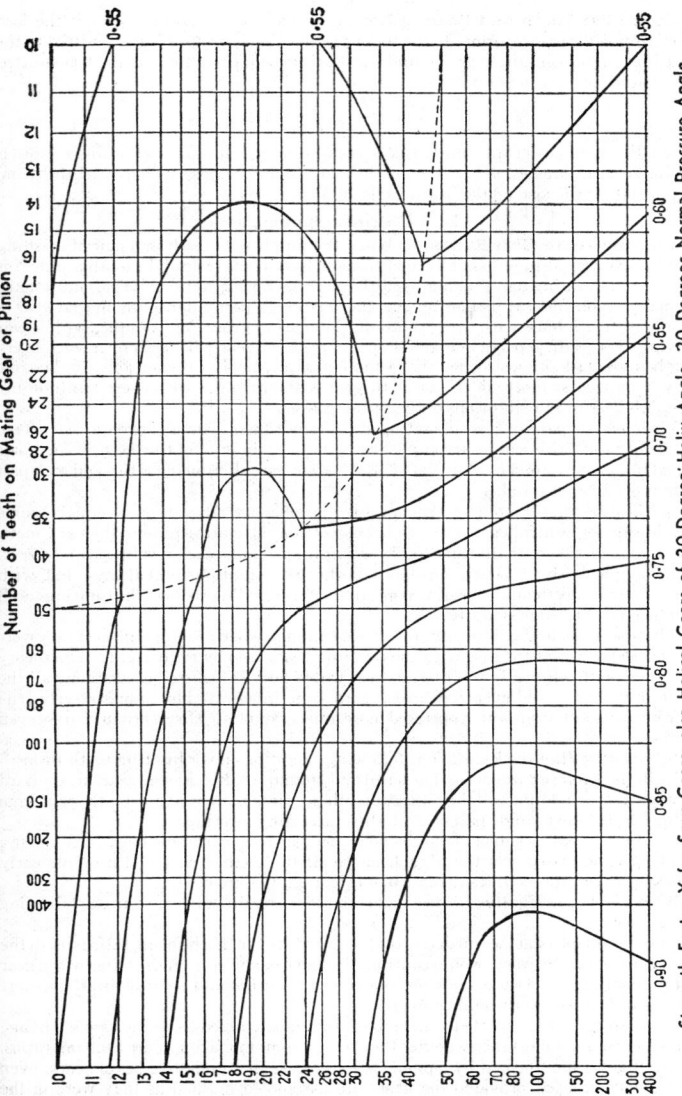

Spur Gear Tooth and Bearing Loads. — The tooth load on a gear is the load exerted in a direction normal to the tooth profile. The same load is transmitted to the bearings supporting the gear. To find the bearing load, first calculate the transmitted tangential load, L, in pounds:

$$L = \frac{P \times 63{,}000}{R \times r}$$

where P = horsepower transmitted; R = revolutions per minute of gear; and r = pitch radius of gear, inches. Next, the tooth load, L_T in pounds, which is of the same magnitude as the load on the bearings, is found:

$$L_T = L \div \cos \text{ pressure angle}$$

Hunting Tooth Gear Ratios. — When the numbers of teeth in a pair of meshing gears are such that they do not have a common divisor, the ratio of the tooth numbers is said to be a "hunting tooth" or non-factorizing ratio. Each gear in such a pair must make as many revolutions as its mating gear has teeth for a particular tooth on that gear to "hunt" or mesh once with every tooth on the mating gear. For example, with a pair having $^{73}/_{24}$ teeth, a particular tooth on the 73-tooth gear will mesh once with every tooth on the 24-tooth gear when the 73-tooth gear has made 24 revolutions. In the same way, a particular tooth on the 24-tooth gear will mesh once with every tooth on the 73-tooth gear in 73 revolutions of the 24-tooth gear.

When several pairs of gears operating at the same center distance are required to have hunting ratios, this can be accomplished by having the sum of the teeth in each pair equal to a prime number. Thus $^{77}/_{20}$, $^{76}/_{21}$, $^{75}/_{19}$, etc., are all hunting ratios and each pair sum is 97, a prime number.

Origin of Hunting Tooth Use: Whenever the number of teeth on a gear is equal to some whole number multiplied by the number of teeth on the mating pinion, then any tooth on the gear meshes with the same tooth on the pinion during each revolution of the gear. For example, with a $^{72}/_{24}$ pair, one tooth on the 24-tooth pinion will always mesh with the same three teeth on the gear. Changing the ratio to $^{73}/_{24}$ produces the hunting tooth condition described previously.

The practice of using "hunting" ratios was common when gears of high accuracy were not available as was the case with the cast iron gears used by millwrights. The theory was that, with a hunting tooth, gear wear would be evenly distributed among all the teeth and they would eventually wear into some indefinite but comparatively true (conjugate) tooth form which operated more quietly and lasted longer than gears of even ratios.

When to Use Hunting Ratios: The following conditions favor hunting tooth ratios:

1. Both gears are hardened or hardened and ground. In this case the gears are too hard to wear in much, but use of the hunting tooth prevents high dynamic tooth loads from being confined to a single pair of teeth thus increasing gear life.

2. The gear pair is subject to high cyclic loads such as in a crankshaft. In this case a hunting ratio distributes the high loads to all the teeth evenly and prevents early breakage of certain teeth resulting when integral ratios are used.

When Not to Use Hunting Ratios: Hunting ratios should not be used in the following situations:

1. For multiple-start worm drives that require long life. If a hunting ratio is used, the hard material of the worm would continually shave particles from the softer worm gear throughout the life of the drive whereas an integral ratio would bed itself in after a short time, and further wear would be small.

2. When gears are repeatedly assembled and disassembled as in the case of change gears; it is of no importance whether they have a common factor or are hunting ratios.

3. If gears are finished by lapping as in the case of automotive bevel gears, even ratios may be used provided the gears are assembled meshed as they were in the lapping process. If this is not practical, then a hunting ratio is more satisfactory.

Backlash in Gears. — In general, backlash in gears is play between mating teeth. For purposes of measurement and calculation, backlash is defined as the amount by which a tooth space exceeds the thickness of an engaging tooth. It does not include the effect of center-distance changes of the mountings and variations in bearings. When not otherwise specified, numerical values of backlash are understood to be given on the pitch circles.

The general purpose of backlash is to prevent gears from jamming together and making contact on both sides of their teeth simultaneously. Lack of backlash may cause noise, overloading, overheating of the gears and bearings, and even seizing and failure.

Excessive backlash is objectionable, particularly if the drive is frequently reversing, or if there is an overrunning load as in cam drives. On the other hand, specification of an unnecessarily small amount of backlash allowance will increase the cost of gears, because errors in runout, pitch, profile, and mounting must be held correspondingly smaller. Backlash in no way affects involute action and usually is not detrimental to proper gear action.

Determining Proper Amount of Backlash. — In specifying proper backlash and tolerances for a pair of gears, the most important factor is probably the maximum permissible amount of runout in both gear and pinion (or worm). Next are the allowable errors in profile, pitch, tooth thickness, and helix angle. It can be seen that the backlash between a pair of gears will vary as successive teeth make contact because of the effect of composite tooth errors, particularly runout, and errors in the gear mounting centers and bearings.

Other important considerations are speed and space for lubricant film. Slow-moving gears, in general, require the least backlash. Fast-moving fine-pitch gears are usually lubricated with relatively light oil, but if there is insufficient clearance for an oil film, and particularly if oil trapped at the root of the teeth cannot escape, heat and excessive tooth loading will occur.

Heat is a factor because gears may operate warmer than, and therefore expand more than, the housings. The heat may result from oil churning or from frictional losses between the teeth, at bearings or oil seals, or from external causes. Moreover, for the same temperature rise the material of the gears — for example, bronze and aluminum — may expand more than the material of the housings, usually steel or cast iron.

The higher the helix angle or spiral angle, the more transverse backlash is required for a given normal backlash. The transverse backlash is equal to the normal backlash divided by the cosine of the helix angle.

In designs employing normal pressure angles higher than 20 degrees, special consideration must be given to backlash, since more backlash is required on the pitch circles to obtain a given amount of backlash in a direction normal to the tooth profiles.

Errors in boring the gear housings, both in center distance and alignment, are of extreme importance in determining allowance to obtain the backlash desired. The same is true in the mounting of the gears, which is affected by the type and adjustment of bearings, and similar factors. Other influences in backlash specification are heat-treatment subsequent to cutting the teeth, lapping operations, the possible necessity for recutting for any reason, and reduction of tooth thickness through normal wear.

Minimum backlash is necessary for timing, indexing, gun-sighting, and certain instrument gear trains. If the operating speed is very low and the necessary precautions are taken in the manufacture of such gear trains, the backlash may be held to extremely small limits. However, the specification of "zero backlash," so commonly stipulated for gears of this nature, usually involves special and expensive techniques, and is difficult to obtain.

Recommended Backlash. — In the following tables American Gear Manufacturers Association recommendations for backlash ranges for various kinds of gears are given.* For purposes of measurement and calculation, backlash is defined as the amount by which a tooth space exceeds the thickness of an engaging tooth. When not otherwise specified, numerical values of backlash are understood to be measured at the tightest point of mesh on the pitch circle in a direction normal to the tooth surface when the gears are mounted in their specified position.

Coarse-Pitch Gears: Table 1 gives the recommended backlash range for coarse-pitch spur, helical and herringbone gearing. Because backlash for helical and herringbone gears is more conveniently measured in the normal plane, Table 1 has been prepared to show backlash in the normal plane for coarse-pitch helical and herringbone gearing and in the transverse plane for spur gears. To obtain backlash in the transverse plane for helical and herringbone gears, divide the normal plane backlash in Table 1 by the cosine of the helix angle.

The backlash tolerances given in this table contain allowances for gear expansion due to a differential in the operating temperature of the gearing and their supporting structure. The values may be used where the operating temperature is up to 70 degrees F. higher than the ambient temperature. These backlash values will provide proper running clearances for most of the applications listed in Table 9, beginning on page 2000.

The following important factors must be considered in establishing backlash tolerances: (a) Center distance tolerance, (b) Parallelism of gear axes, (c) Side runout or wobble, (d) Tooth thickness tolerance, (e) Pitch line runout tolerance, (f) Profile tolerance, (g) Pitch tolerance, (h) Lead tolerance, (i) Types of bearings and subsequent wear, (j) Deflection under load, (k) Gear tooth wear, (l) Pitch line velocity, (m) Lubrication requirements, (n) Thermal expansion of gears and housing.

A tight mesh may result in objectionable gear sound, increased power losses, overheating, rupture of the lubricant film, overloaded bearings and premature gear failure. However, it is recognized that there are some gearing applications where a tight mesh (zero backlash) may be required.

Specifying unnecessarily close backlash tolerances will increase the cost of the gearing. It is obvious from the above summary that the desired amount of backlash is difficult to evaluate. It is, therefore, recommended that when a designer, user or purchaser includes a reference to backlash in a gearing specification and drawing, consultation be arranged with the manufacturer.

Bevel and Hypoid Gears: Table 2 gives similar backlash range values for bevel and hypoid gears. These are values based upon average conditions for general purpose gearing, but may require modification to meet specific needs.

Backlash on bevel and hypoid gears can be controlled to some extent by axial adjustment of the gears during assembly. However, due to the fact that actual adjustment of a bevel or hypoid gear in its mounting will alter the amount of backlash, it is imperative that the amount of backlash cut into the gears during manufacture is not excessive. Bevel and hypoid gears must always be capable of operation without interference when adjusted for zero backlash. This requirement is imposed by the fact that a failure of the axial thrust bearing might permit the gears to operate under this condition. Therefore, bevel and hypoid gears should never be designed to operate with normal backlash in excess of $0.080/P$ where P is diametral pitch.

Fine-Pitch Gears: Table 3 gives similar backlash range values for fine-pitch spur, helical and herringbone gearing.

* Extracted from Gear Classification Manual, AGMA 390.03 with permission of the publisher, the American Gear Manufacturers Association, 1500 King St., Arlington, Va. 22314.

Table 1. AGMA Recommended Backlash Range for Coarse-Pitch Spur, Helical and Herringbone Gearing

Center Distance (Inches)	Normal Diametral Pitches				
	0.5–1.99	2–3.49	3.5–5.99	6–9.99	10–19.99
	Backlash, Normal Plane, Inches*				
Up to 5					.005–.015
Over 5 to 10				.010–.020	.010–.020
Over 10 to 20			.020–.030	.015–.025	.010–.020
Over 20 to 30		.030–.040	.025–.030	.020–.030	
Over 30 to 40	.040–.060	.035–.045	.030–.040	.025–.035	
Over 40 to 50	.050–.070	.040–.055	.035–.050	.030–.040	
Over 50 to 80	.060–.080	.045–.065	.040–.060		
Over 80 to 100	.070–.095	.050–.080			
Over 100 to 120	.080–.110				

* Suggested backlash, on nominal centers, measured after rotating to the point of closest engagement. For helical and herringbone gears, divide above values by the cosine of the helix angle to obtain the transverse backlash.

The above backlash tolerances contain allowance for gear expansion due to differential in the operating temperature of the gearing and their supporting structure. The values may be used where the operating temperatures are up to 70 deg F higher than the ambient temperature.

For most of the gearing applications listed in Table 9, page 2000 the recommended backlash ranges will provide proper running clearance between the engaging teeth of mating gears. Deviation below the minimum or above the maximum values shown, which do not affect operational use of the gearing, should not be cause for rejection.

Definite backlash tolerances on coarse-pitch gearing are to be considered binding on the gear manufacturer only when agreed upon in writing.

Some applications may require less backlash than shown in the above table. In such cases the amount and tolerance should be by agreement between the manufacturer and purchaser.

Table 2. AGMA Recommended Backlash Range for Bevel and Hypoid Gears*

Diametral Pitch	Normal Backlash, Inch		Diametral Pitch	Normal Backlash, Inch	
	Quality Numbers 7 through 13	Quality Numbers 3 through 6		Quality Numbers 7 through 13	Quality Numbers 3 through 6
1.00 to 1.25	0.020–0.030	0.045–0.065	5.00 to 6.00	0.005–0.007	0.006–0.013
1.25 to 1.50	0.018–0.026	0.035–0.055	6.00 to 8.00	0.004–0.006	0.005–0.010
1.50 to 1.75	0.016–0.022	0.025–0.045	8.00 to 10.00	0.003–0.005	0.004–0.008
1.75 to 2.00	0.014–0.018	0.020–0.040	10.00 to 16.00	0.002–0.004	0.003–0.005
2.00 to 2.50	0.012–0.016	0.020–0.030	16.00 to 20.00	0.001–0.003	0.002–0.004
2.50 to 3.00	0.010–0.013	0.015–0.025	20 to 50	0.000–0.002	0.000–0.002
3.00 to 3.50	0.008–0.011	0.012–0.022	50 to 80	0.000–0.001	0.000–0.001
3.50 to 4.00	0.007–0.009	0.010–0.020	80 and finer	0.000–0.0007	0.000–0.0007
4.00 to 5.00	0.006–0.008	0.008–0.016	. . .	. . .	. . .

* Measured at tightest point of mesh

Table 3. AGMA Backlash Allowance and Tolerance for Fine-Pitch Spur, Helical and Herringbone Gearing

Backlash Designation	Normal Diametral Pitch Range	Tooth Thinning to Obtain Backlash (Note 1)		Resulting Approximate Backlash (per Mesh) Normal Plane (Note 2), Inch
		Allowance, per Gear, Inch	Tolerance, per Gear, Inch	
A	20 thru 45	.002	0 to .002	.004 to .008
	46 thru 70	.0015	0 to .002	.003 to .007
	71 thru 90	.001	0 to .00175	.002 to .0055
	91 thru 200	.00075	0 to .00075	.0015 to .003
B	20 thru 60	.001	0 to .001	.002 to .004
	61 thru 120	.00075	0 to .00075	.0015 to .003
	121 thru 200	.0005	0 to .0005	.001 to .002
C	20 thru 60	.0005	0 to .0005	.001 to .002
	61 thru 120	.00035	0 to .0004	.0007 to .0015
	121 thru 200	.0002	0 to .0003	.0004 to .001
D	20 thru 60	.00025	0 to .00025	.0005 to .001
	61 thru 120	.0002	0 to .0002	.0004 to .0008
	121 thru 200	.0001	0 to .0001	.0002 to .0004
E	20 thru 60	Zero (Note 3)	0 to .00025	0 to .0005
	61 thru 120		0 to .0002	0 to .0004
	121 thru 200		0 to .0001	0 to .0002

Backlash in gears is the play between mating tooth surfaces. For purposes of measurement and calculation, backlash is defined as the amount by which a tooth space exceeds the thickness of an engaging tooth. When not otherwise specified, numerical values of backlash are understood to be measured at the tightest point of mesh on the pitch circle in a direction normal to the tooth surface when the gears are mounted in their specified position.

Allowance is the basic amount that a tooth is thinned from basic calculated circular tooth thickness to obtain the required backlash class.

Tolerance is the total permissible variation in the circular thickness of the teeth.

Note 1: These dimensions are shown primarily for the benefit of the gear manufacturer and represent the amount that the thickness of teeth should be reduced in the pinion and gear below the standard calculated value, to provide for backlash in the mesh. In some cases, particularly with pinions involving small numbers of teeth, it may be desirable to provide for total backlash by thinning the teeth in the gear member only by twice the allowance value shown in column (3). In this case both members will have the tolerance shown in column (4).

In some cases, particularly in meshes with a small total number of teeth, backlash may be achieved by an increase in basic center distance. In such cases, neither member is reduced by the allowance shown in column (3).

Note 2: These dimensions indicate the approximate backlash that will occur in a mesh in which each of the mating pairs of gears have the teeth thinned by the amount referred to in Note 1, and are meshed on theoretical centers.

Note 3: Backlash in gear sets can also be achieved by increasing the center distance above nominal and using the teeth at standard tooth thickness. Class E backlash designation infers gear sets operating under these conditions.

Providing Backlash. — In order to obtain the amount of backlash desired, it is necessary to decrease tooth thicknesses. However, because of manufacturing and assembling inaccuracies not only in the gears but also in other parts, the allowances made on tooth thickness almost always must exceed the desired amount of backlash. Since the amounts of these allowances depend on the closeness of control exercised on all manufacturing operations, no general recommendations for them can be given.

It is customary to make half of the allowance for backlash on the tooth thickness of each gear of a pair, although there are exceptions. For example, on pinions having very low numbers of teeth it is desirable to provide all of the allowance on the mating gear, so as not to weaken the pinion teeth. In worm gearing, ordinary practice is to provide all of the allowance on the worm which is usually made of a material stronger than that of the worm gear.

In some instances the backlash allowance is provided in the cutter, and the cutter is then operated at the standard tooth depth. In still other cases, backlash is obtained by setting the distance between two tools for cutting the two sides of the teeth, as in straight bevel gears, or by taking side cuts, or by changing the center distance between the gears in their mountings. In spur and helical gearing, backlash allowance is usually obtained by sinking the cutter deeper into the blank than the standard depth. The accompanying table gives the excess depth of cut for various pressure angles.

Excess Depth of Cut E to Provide Backlash Allowance

Distribution of Backlash	Pressure Angle ϕ, Degrees				
	$14\frac{1}{2}$	$17\frac{1}{2}$	20	25	30
Excess Depth of Cut E to Obtain Circular Backlash B*					
All on One Gear	$1.93B$	$1.59B$	$1.37B$	$1.07B$	$0.87B$
One-half on Each Gear	$0.97B$	$0.79B$	$0.69B$	$0.54B$	$0.43B$
Excess Depth of Cut E to Obtain Backlash B_b Normal to Tooth Profile†					
All on One Gear	$2.00B_b$	$1.66B_b$	$1.46B_b$	$1.18B_b$	$1.99B_b$
One-half on Each Gear	$1.00B_b$	$0.83B_b$	$0.73B_b$	$0.59B_b$	$0.50B_b$

* Circular backlash is the amount by which the width of a tooth space is greater than the thickness of the engaging tooth on the pitch circles. As described on pages 1829 and 1834, this is what is meant by backlash unless otherwise specified.
† Backlash measured normal to the tooth profile by inserting a feeler gage between meshing teeth; to convert to circular backlash, $B = B_b \div \cos \phi$.

Control of Backlash Allowances in Production. — Measurement of the tooth thickness of gears is perhaps the simplest way of controlling backlash allowances in production. There are several ways in which this may be done including: (1) chordal thickness measurements as described on page 1790; (2) caliper measurements over two or more teeth as described on page 1976; and (3) measurements over wires. In this last method, first the theoretical measurement over wires when the backlash allowance is zero is determined by the method described on page 1960; then the amount this measurement must be reduced to obtain a desired backlash allowance is taken from the table on page 1970.

It should be understood, as explained in the section "Measurement of Backlash," that merely making tooth thickness allowances will not guarantee the amount of backlash in the ready-to-run assembly of two or more gears. Manufacturing limitations will introduce such gear errors as runout, pitch error, profile error, and lead error, and gear-housing errors in both center distance and alignment. All of these make the backlash of the assembled gears different from that indicated by tooth thickness measurements on the individual gears.

Measurement of Backlash. — Backlash is commonly measured by holding one gear of a pair stationary and rocking the other back and forth. The movement is registered by a dial indicator having its pointer or finger in a plane of rotation at or

near the pitch diameter and in a direction parallel to a tangent to the pitch circle of the moving gear. If the direction of measurement is normal to the teeth, or other than as specified above, it is recommended that readings be converted to the plane of rotation and in a tangent direction at or near the pitch diameter, for purposes of standardization and comparison.

In spur gears, parallel helical gears, and bevel gears, it is immaterial whether the pinion or gear is held stationary for the test. In crossed helical and hypoid gears, readings may vary according to which member is stationary; hence it is customary to hold the pinion stationary and measure on the gear.

In some instances backlash is measured by thickness gages or feelers. A similar method utilizes lead wire inserted between the teeth as they pass through mesh. In both cases it is likewise recommended that readings be converted to the plane of rotation and in a tangent direction at or near the pitch diameter, taking into account the normal pressure angle, and the helix angle or spiral angle of the teeth.

Sometimes backlash in parallel helical or herringbone gears is checked by holding the gear stationary, and axially moving the pinion back and forth, readings being taken on the face or shaft of the pinion, and converted to the plane of rotation by calculation. Another method consists of meshing a pair of gears tightly together on centers and observing the variation from the specified center distance. Such readings should also be converted to the plane of rotation and in a tangent direction at or near the pitch diameter for the reasons previously given.

Measurements of backlash may vary in the same pair of gears, depending on accuracy of manufacturing and assembling. Incorrect tooth profiles will cause a change of backlash at different phases of the tooth action. Eccentricity may cause a substantial difference between maximum and minimum backlash at different positions around the gears. In stating amounts of backlash, it should always be remembered that merely making allowances on tooth thickness does not guarantee the minimum amount of backlash that will exist in actual gears when assembled.

Fine-pitch Gears: The measurement of backlash of fine-pitch gears, when assembled, cannot be made in the same manner and by the same techniques employed for gears of coarser pitches. In the very fine pitches it is virtually impossible to use indicating devices for measuring backlash. Sometimes a toolmaker's microscope is used for this purpose to good advantage on very small mechanisms.

Another means of measuring backlash in fine-pitch gears is to attach a beam to one of the shafts and measure the angular displacement in inches when one member is held stationary. The ratio of the length of the beam to the nominal pitch radius of the gear or pinion to which the beam is attached gives the approximate ratio of indicator reading to circular backlash. Because of the limited means of measuring backlash between a pair of fine-pitch gears, gear centers and tooth thickness of the gears when cut must be held to very close limits. Tooth thickness of fine-pitch spur and helical gears can best be checked on a variable center distance fixture using a master gear. When checked in this manner, tooth thickness change = 2 × center distance change × tangent of transverse pressure angle, approximately.

Control of Backlash in Assemblies. — Provision is often made for adjusting one gear relative to the other, thereby affording complete control over backlash at initial assembly and throughout the life of the gears. Such practice is most common in bevel gearing. It is fairly common in spur and helical gearing when the application permits slight changes between shaft centers. It is practical in worm gearing only for single thread worms with low lead angles. Otherwise faulty contact results.

Another method of controlling backlash quite common in bevel gears and less common in spur and helical gears is to match the high and low spots of the runout gears of one to one ratio and mark the engaging teeth at the point where the runout of one gear cancels the runout of the mating gear.

Angular Backlash in Gears. — When the backlash on the pitch circles of a meshing pair of gears is known, the angular backlash or angular play corresponding to this backlash may be computed from the following formulas.

$$\theta_D = \frac{6875B}{D} \text{ minutes}; \qquad \theta_d = \frac{6875B}{d} \text{ minutes}$$

In these formulas, B = backlash between gears, in inches; D = pitch diameter of larger gear, in inches; d = pitch diameter of smaller gear in inches; θ_D = angular backlash or angular movement of larger gear in minutes when smaller gear is held fixed and larger gear rocked back and forth; and θ_d = angular backlash or angular movement of smaller gear in minutes when larger gear is held fixed and smaller gear rocked back and forth.

Inspection of Gears. — Perhaps the most widely used method of determining relative accuracy in a gear is to rotate the gear through at least one complete revolution in intimate contact with a master gear of known accuracy. The gear to be tested and the master gear are mounted on a variable-center-distance fixture and the resulting radial displacements or changes in center distance during rotation of the gear are measured by a suitable device. Except for the effect of backlash, this so-called "composite check" approximates the action of the gear under operating conditions and gives the combined effect of the following errors: Runout; pitch error; tooth-thickness variation; profile error; and lateral runout (sometimes called wobble).

Tooth-to-tooth Composite Error (Table 5. page 1990) is the error which shows up as flicker on the indicator of a variable-center-distance fixture as the gear being tested is rotated from tooth to tooth in intimate contact with the master gear. Such flicker shows the combined or composite effect of circular pitch error, tooth-thickness variation, and profile error.

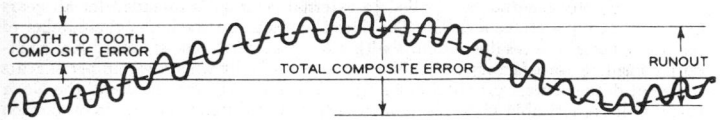

Chart Showing Nature of Composite Errors

Total Composite Error (Table 4) is made up of runout, wobble, and the tooth-to-tooth composite error; it is the total center distance displacement read on the indicating device of the testing fixture as shown in the accompanying diagram.

Pressure for Composite Checking of Fine-pitch Gears. — In using a variable-center-distance fixture, excessive pressure on fine-pitch gears of narrow face width will result in incorrect readings due to deflection of the teeth. Based on tests, the following checking pressures are recommended for gears of 0.100-inch face width: 20 to 29 diametral pitch, 28 ounces; 30 to 39 pitch, 24 ounces; 40 to 49 pitch, 20 ounces; 50 to 59 pitch, 16 ounces; 60 to 79 pitch, 12 ounces; 80 to 99 pitch, 8 ounces; 100 to 149 pitch, 4 ounces; and 150 and finer pitches, 2 ounces, minimum. These recommended checking pressures are based on the use of antifriction mountings for the movable head of the checking fixture and include the pressure of the indicating device. For face widths less than 0.100-inch, the recommended pressures should be reduced proportionately; for larger widths no increase is necessary although this may be done safely in the proper proportion.

INTERNAL GEARING

Internal Spur Gears. — An internal gear may be proportioned like a standard spur gear turned "outside in" or with addendum and dedendum in reverse positions; however, to avoid interference or improve the tooth form and action, the internal diameter of the gear should be increased and the outside diameter of the mating pinion is also made larger than the size based upon standard or conventional tooth proportions. The extent of these enlargements will be illustrated by means of examples given in connection with the Rules for Internal Gears. The 20-degree involute full-depth tooth form is recommended for internal gears; the 20-degree stub tooth and the 14½-degree full-depth tooth are also used.

Methods of Cutting Internal Gears. — Internal spur gears are cut by methods similar in principle to those employed for external spur gears. They may be cut by one of the following methods: (1) By a generating process, as when using a Fellows gear shaper; (2) by using a formed cutter and milling the teeth; (3) by planing, using a machine of the templet or form-copying type (especially applicable to gears of large pitch); and (4) by using a formed tool which reproduces its shape and is given a planing action either on a slotting or a planing type of machine. Internal gears frequently have a web at one side which limits the amount of clearance space at the ends of the teeth. Such gears may be cut readily on a gear shaper. The most practical method of cutting very large internal gears is on a planer of the form-copying type. A regular spur gear planer is equipped with a special tool-holder for locating the tool in the position required for cutting internal teeth.

Formed Cutters for Internal Gears. — When formed cutters are used, a special cutter usually is desirable, because the tooth spaces of an internal gear are not the same shape as the tooth spaces of external gearing having the same pitch and number of teeth. This is due to the fact that an internal gear is a spur gear "turned outside in." According to one rule, the standard No. 1 cutter for external gearing may be used for internal gears of 4 diametral pitch and finer, when there are sixty teeth or more. This No. 1 cutter, as applied to external gearing, is intended for all gears having from 135 teeth to a rack. The finer the pitch and the larger the number of teeth, the better the results obtained with a No. 1 cutter. The standard No. 1 cutter is considered satisfactory for jobbing work, and usually when the number of gears to be cut does not warrant obtaining a special cutter, although the use of the No. 1 cutter is not practicable when the number of teeth in the pinion is large in proportion to the number of teeth in the internal gear.

Arc Thickness of Internal Gear Tooth. — *Rule:* If internal diameter of internal gear is enlarged as determined by Rules 1 and 2 for Internal Diameters (see Rules for Internal Gears), the arc tooth thickness at the pitch circle equals 1.3888 divided by the diametral pitch, assuming a pressure angle of 20 degrees.

Arc Thickness of Pinion Tooth. — *Rule:* If pinion for internal gear is larger than conventional size (see Outside Diameter of Pinion for Internal Gear, under Rules for Internal Gears), then the arc tooth thickness on pitch circle equals 1.7528 divided by the diametral pitch, assuming a pressure angle of 20 degrees.

Note: For chordal thickness and chordal addendum, see rules and formulas for spur gears.

Relative Sizes of Internal Gear and Pinion. — If a pinion is too large or too near the size of its mating internal gear, serious interference or modification of the tooth shape may occur.

Rule: For internal gears having a 20-degree pressure angle and full-depth teeth, the difference between number of teeth in gear and pinion should not be less than 12. For teeth of stub form, the smallest difference should be 7 or 8 teeth. For a pressure angle of 14½ degrees, the difference in tooth numbers should not be less than 15.

Rules for Internal Gears — 20-degree Full-depth Teeth

To Find	Rule
Pitch Diameter	*Rule:* To find the pitch diameter of an internal gear, divide number of internal gear teeth by the diametral pitch. The pitch diameter of mating pinion also equals number of pinion teeth divided by diametral pitch, the same as for external spur gears.
Internal Diameter (Enlarged to avoid Interference)	*Rule 1:* For internal gears to mesh with pinions having 16 teeth or more, subtract 1.2 from the number of teeth and divide remainder by diametral pitch. *Example:* An internal gear has 72 teeth of 6 diametral pitch and the mating pinion has 18 teeth; then $$\text{Internal diameter} = \frac{72 - 1.2}{6} = 11.8 \text{ inches}$$ *Rule 2:* If circular pitch is used, subtract 1.2 from the number of internal gear teeth, multiply remainder by the circular pitch, and divide the product by 3.1416.
Internal Diameter (Based upon Spur Gear Reversed)	*Rule:* If the internal gear is to be designed to conform to a spur gear turned outside in, then subtract 2 from the number of teeth and divide remainder by the diametral pitch to find internal diameter. *Example:* (Same as Example above.) $$\text{Internal diameter} = \frac{72 - 2}{6} = 11.666 \text{ inches}$$
Outside Diameter of Pinion for Internal Gear	*Note:* If the internal gearing is to be proportioned like standard spur gearing, use the rule or formula previously given for spur gears in determining the outside diameter. The rule and formula following apply to a pinion that is enlarged and intended to mesh with an internal gear enlarged as determined by the preceding Rules 1 and 2 above. *Rule:* For pinions having 16 teeth or more, add 2.5 to the number of pinion teeth and divide by the diametral pitch. *Example 1:* A pinion for driving an internal gear is to have 18 teeth (full depth) of 6 diametral pitch; then $$\text{Outside diameter} = \frac{18 + 2.5}{6} = 3.416 \text{ inches}$$ Using the rule for external spur gears, the outside diameter = 3.333 inches.
Center Distance	*Rule:* Subtract the number of pinion teeth from the number of internal gear teeth and divide remainder by two times the diametral pitch.
Tooth Thickness	See paragraphs, Arc Thickness of Internal Gear Tooth and Arc Thickness of Pinion Tooth.

Bevel Gearing

Types of Bevel Gears. — Bevel gears are conical gears, that is, gears in the shape of cones, and are used to connect shafts having intersecting axes. Hypoid gears are similar in general form to bevel gears, but operate on axes that are offset. With few exceptions, most bevel gears may be classified as being either of the straight-tooth type or of the curved tooth type. The latter type includes spiral bevels, Zerol bevels, and hypoid gears. The following is a brief description of the distinguishing characteristics of the different types of bevel gears.

Straight Bevel Gears: The teeth of this most commonly used type of bevel gear are straight but their sides are tapered so that they would intersect the axis at a common point called the pitch cone apex if extended inward. The face cone elements of most straight bevel gears, however, are now made parallel to the root cone elements of the mating gear to obtain uniform clearance along the length of the teeth. The face cone elements of such gears would, therefore, intersect the axis at a point inside the pitch cone. Straight bevel gears are the easiest to calculate and are economical to produce.

Straight bevel gear teeth may be generated for full length contact or for localized contact. The latter are slightly convex in a lengthwise direction so that some adjustment of the gears during assembly is possible and small displacements due to load deflections can occur without undesirable load concentration on the ends of the teeth. This slight lengthwise rounding of the tooth sides need not be computed in the design but is taken care of automatically in the cutting operation on the newer types of bevel gear generators.

Zerol* Bevel Gears: The teeth of Zerol bevel gears are curved but lie in the same general direction as the teeth of straight bevel gears. They may be thought of as spiral bevel gears of zero spiral angle and are manufactured on the same machines as spiral bevel gears. The face cone elements of Zerol bevel gears do not pass through the pitch cone apex but instead are approximately parallel to the root cone elements of the mating gear to provide uniform tooth clearance. The root cone elements also do not pass through the pitch cone apex because of the manner in which these gears are cut. Zerol bevel gears are used in place of straight bevel gears when generating equipment of the spiral-type but not the straight-type is available, and may be used when hardened bevel gears of high accuracy (produced by grinding) are required.

Spiral Bevel Gears: Spiral bevel gears have curved oblique teeth on which contact begins gradually and continues smoothly from end to end. They mesh with a rolling contact similar to straight bevel gears. As a result of their overlapping tooth action, however, spiral bevel gears will transmit motion more smoothly than straight bevel or Zerol bevel gears, reducing noise and vibration which become especially noticeable at high speeds.

One of the advantages associated with spiral bevel gears is the complete control of the localized tooth contact. By making a slight change in the radii of curvature of the mating tooth surfaces, the amount of surface over which tooth contact takes place can be changed to suit the specific requirements of each job. Localized tooth contact promotes smooth, quiet running spiral bevel gears, and permits some mounting deflections without concentrating the load dangerously near either end of the tooth. Permissible deflections established by experience are given under the heading "Mountings for Bevel Gears."

Because their tooth surfaces can be ground, spiral bevel gears have a definite advantage in applications requiring hardened gears of high accuracy. The bottom of the tooth spaces and the tooth profiles may be ground simultaneously, resulting in a smooth blending of the tooth profile, the tooth fillet, and the bottom of the tooth

* Registered in U.S. Patent Office.

space. This is important from a strength standpoint because it eliminates cutter marks and other surface interruptions which frequently result in stress concentrations.

Hypoid Gears: Hypoid gears resemble, in general appearance, spiral bevel gears, except that the axis of the pinion is offset relative to the gear axis. If there is sufficient offset, the shafts may pass one another thus permitting the use of a compact straddle mounting on the gear and pinion. Whereas a spiral bevel pinion has equal pressure angles and symmetrical profile curvatures on both sides of the teeth, a hypoid pinion properly conjugate to a mating gear having equal pressure angles on both sides of the teeth must have non-symmetrical profile curvatures for proper tooth action. In addition, in order to obtain equal arcs of motion for both sides of the teeth, it is necessary to use unequal pressure angles on hypoid pinions. Hypoid gears are usually designed so that the pinion has a larger spiral angle than the gear. The advantage of such design is that the pinion diameter is increased and is stronger than a corresponding spiral bevel pinion. This diameter increment permits the use of comparatively high ratios without the pinion becoming too small to allow a bore or shank of adequate size. The sliding action along the lengthwise direction of their teeth in hypoid gears is a function of the difference in the spiral angles on the gear and pinion. This sliding effect makes such gears even smoother running than spiral bevel gears. Grinding of hypoid gears can be accomplished on the same machines used for grinding spiral bevel and Zerol bevel gears.

Application of Bevel and Hypoid Gears. — Bevel and hypoid gears may be used to transmit power between shafts at practically any angle and speed. The particular type of gearing best suited for a specific job, however, depends on the mountings and the operating conditions.

Straight and Zerol Bevel Gears: For peripheral speeds up to 1000 feet per minute, where maximum smoothness and quietness are not the primary consideration, straight and Zerol bevel gears are recommended. For such applications, plain bearings may be used for radial and axial loads, although the use of anti-friction bearings is always preferable. Plain bearings permit a more compact and less expensive design; one reason why straight and Zerol bevel gears are much used in differentials. This type of bevel gearing is the simplest to calculate and set up for cutting, and is ideal for small lots where fixed charges must be kept to a minimum. Zerol bevel gears are recommended for use in place of straight bevel gears: (1) Where hardened gears of high accuracy are required, since Zerol gears may be ground; and (2) when only spiral-type equipment is available for cutting bevel gears.

Spiral Bevel and Hypoid Gears: Spiral bevel and hypoid gears are recommended for applications where peripheral speeds exceed 1000 feet per minute or 1000 revolutions per minute, whichever occurs first. In many instances they may be used to advantage at lower speeds, particularly where extreme smoothness and quietness are desired. For peripheral speeds above 8000 feet per minute, ground gears should be used.

For large reduction ratios the use of spiral and hypoid gears will reduce the overall size of the installation since the continuous pitch line contact of these gears makes it practical to obtain smooth performance with a smaller number of teeth in the pinion than is possible with straight or Zerol bevel gears.

Hypoid gears are recommended for industrial applications: (1) When maximum smoothness of operation is desired; (2) for high reduction ratios where compactness of design, smoothness of operation, and maximum pinion strength are important; and (3) for non-intersecting shafts.

Bevel and hypoid gears may be used for both speed-reducing and speed-increasing drives. In speed-increasing drives, however, the ratio should be kept as low as

possible and the pinion mounted on anti-friction bearings; otherwise bearing friction will cause the drive to lock.

Notes on the Design of Bevel Gear Blanks. — The quality of any finished gear is dependent, to a large degree, on the design and accuracy of the gear blank. A number of factors which affect manufacturing economy as well as performance must be considered.

A gear blank should be designed to avoid localized stresses and serious deflections within itself. Sufficient thickness of metal should be provided under the roots of gear teeth to give them proper support. As a general rule, the amount of metal under the root should equal the whole depth of the tooth; this metal depth should be maintained under the small ends of the teeth as well as under the middle. On webless type ring gears, the minimum stock between the root line and the bottom of tap drill holes should be one-third the tooth depth. For heavily loaded gears, a preliminary analysis of the direction and magnitude of the forces is helpful in the design of both the gear and its mounting. Rigidity is also necessary for proper chucking when cutting the teeth. For this reason, bores, hubs, and other locating surfaces must be in proper proportion to the diameter and pitch of the gear. Small bores, thin webs, or any condition which necessitates excessive overhang in cutting should be avoided.

Other factors to be considered are the ease of machining and, in the case of gears that are to be hardened, proper design to insure the best hardening conditions. It is desirable to provide a locating surface of generous size on the back of gears. This surface should be machined or ground square with the bore and is used both for locating the gear axially in assembly and for holding it when the teeth are cut. The front clamping surface must, of course, be flat and parallel to the back surface. In connection with cutting the teeth on Zerol bevel, spiral bevel, and hypoid gears, clearance must be provided for face-mill type cutters; front and rear hubs should not intersect the extended root line of the gear or they will interfere with the path of the cutter. In addition, there must be enough room in the front of the gear for the clamp nut which holds the gear on the arbor, or in the chuck, while cutting the teeth. The same considerations must be given to straight bevel gears which are to be generated using a circular-type cutter instead of reciprocating tools.

Mountings for Bevel Gears. — Rigid mountings should be provided for bevel gears to keep the displacements of the gears under operating loads within recommended limits. In order to align gears properly, care should be exercised to insure accurately machined mountings, properly fitted keys, and couplings that run true and square.

As a result of deflection tests on gears and their mountings, and having observed these same units in service, the Gleason Works recommends that the following allowable deflections be used for gears from 6 to 15 inches in diameter: (1) Neither the pinion nor the gear should lift or depress more than 0.003 inch at the center of the face width; (2) the pinion should not yield axially more than 0.003 inch in either direction; and (3) the gear should not yield axially more than 0.003 inch in either direction on 1 to 1 ratio gears (miter gears), or near miters, or more than 0.010 inch away from the pinion on higher ratios.

When deflections exceed these limits, additional problems are involved in obtaining satisfactory gears. It becomes necessary to narrow and shorten the tooth contacts to suit the more flexible mounting. This decreases the bearing area, raises the unit tooth pressure, and reduces the number of teeth in contact, which results in increased noise and the danger of surface failure as well as tooth breakage.

Spiral bevel and hypoid gears should, in general, be mounted on anti-friction bearings in an oil-tight case. While designs for a given set of conditions may use

plain bearings for radial and thrust loads, maintaining gears in satisfactory alignment is usually more easily accomplished with ball or roller bearings.

Bearing Spacing and Shaft Stiffness: Bearing spacing and shaft stiffness are extremely important if gear deflections are to be minimized. For both straddle mounted and overhung mounted gears the spread between bearings should never be less than 70 per cent of the pitch diameter of the gear. On overhung mounted gears the spread should be at least 2½ times the overhang and, in addition, the shaft diameter should be equal to or preferably greater than the overhang to provide sufficient shaft stiffness. When two spiral bevel or hypoid gears are mounted on the same shaft, the axial thrust should be taken at one place only and near the gear where the greater thrust is developed. Provision should be made for adjusting both the gear and pinion axially in assembly. Details on how this may be accomplished are given in the Gleason Works booklet, "Assembling Bevel Gears."

Cutting Bevel Gear Teeth. — A correctly formed bevel gear tooth has the same sectional shape throughout its length, but on a uniformly diminishing scale from the large to the small end. The only way to obtain this correct form is by using a generating type of bevel gear cutting machine. This accounts, in part, for the extensive use of generating type gear cutting equipment in the production of bevel gears.

Bevel gears too large to be cut by generating equipment (100 inches or over in diameter) may be produced by a form-copying type of gear planer. With this method, a template or former is used to mechanically guide a single cutting tool in the proper path to cut the profile of the teeth. Since the tooth profile produced by this method is dependent on the contour of the template used, it is possible to produce tooth profiles to suit a variety of requirements.

Although generating methods are to be preferred, there are still some cases where straight bevel gears are produced by milling. Milled gears cannot be produced with the accuracy of generated gears and generally are not suitable for use in high-speed applications or where angular motion must be transmitted with a high degree of accuracy. Milled gears are used chiefly as replacement gears in certain applications, and gears which are subsequently to be finished on generating type equipment are sometimes roughed out by milling. Formulas and methods used for the cutting of bevel gears are given in the latter part of this section.

In producing gears by generating methods, the tooth curvature is generated from a straight-sided cutter or tool having an angle equal to the required pressure angle. This tool represents the side of a crown gear tooth. The teeth of a true involute crown gear, however, have sides which are very slightly curved. If the curvature of the cutting tool conforms to that of the involute crown gear, an involute form of bevel gear tooth will be obtained. The use of a straight-sided tool is more practical and results in a very slight change of tooth shape to what is known as the "octoid" form. Both the octoid and involute forms of bevel gear tooth give theoretically correct action.

Bevel gear teeth, like those for spur gears, differ as to pressure angle and tooth proportions. The whole depth and the addendum at the large end of the tooth may be the same as for a spur gear of equal pitch. Most bevel gears, however, both of the straight tooth and spiral-bevel types, have lengthened pinion addendums and shortened gear addendums as in the case of some spur gears, the amount of departure from equal addendums varying with the ratio of gearing. Long addendums on the pinion are used principally to avoid undercut and to increase tooth strength. In addition, where long and short addendums are used, the tooth thickness of the gear is decreased and that of the pinion increased to provide a better balance of strength. See the Gleason Works System for straight and spiral bevel gears and also the British Standard.

Nomenclature for Bevel Gears. — The accompanying diagram illustrates various angles and dimensions referred to in describing bevel gears. In connection with the face angles shown in the diagram, it should be noted that the face cones are made parallel to the root cones of the mating gears to provide uniform clearance along the length of the teeth.

Bevel Gear Nomenclature

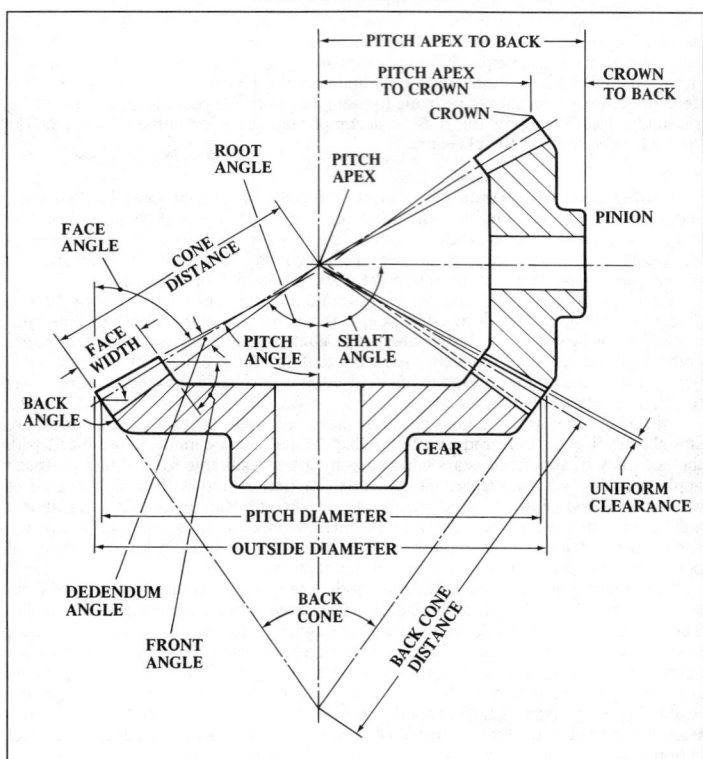

American Standard Straight Bevel Gears. — American Standard ANSI/AGMA 208.03-1979 presents a universal system for straight bevel gear teeth and provides recommended tooth proportions for generated bevel gears in general use. The system includes machined, molded, forged, and formed gears in cases where the profile curve resembles the generated curve. Straight bevel gear teeth of special design to comply with particular production processes and specific applications are not covered.

Gleason System Straight Bevel Gears. — Bevel gears made to the Gleason System for straight bevel gears meet the requirements of the universal system for straight bevel gears covered in ANSI/AGMA 208.03.

Pressure Angle: The basic pressure angle is 20 degrees. However, for those who wish to use a 14½-degree pressure angle, the following can be cut without undercut: 29 or more teeth in the pinion; and pinion and gear combinations of 28/29 teeth and higher; 27/32 and higher; 26/35 and higher; 25/42 and higher; and 24/59 and higher. For a 25-degree pressure angle, the following can be cut without undercut: 13 or more teeth in the pinion.

For the basic pressure angle of 20 degrees, the numbers of teeth for no undercut are: 16 or more teeth in the pinion; and combinations of 15/17 teeth and higher; 14/20 and higher; and 13/31 and higher.

Addendums: Except when the numbers of teeth are equal, the pinion has a long addendum and the gear a short addendum. The amount of departure from equal addendums varies with the ratio. Long addendums are used on the pinion mainly to avoid undercut and to increase tooth strength.

Face Angles: The face cone of a blank is made parallel to the root cone of the mating gear. This gives constant clearance along the tooth and allows the use of larger edge radii on generating tools without fillet interference at the small end of the tooth, thus increasing tooth strength.

Tooth Thicknesses: The tooth thicknesses of gear and pinion are proportioned to obtain approximately equal strength according to the present method of tooth design. In determining tooth thicknesses to balance the strength of gear and pinion, the load is taken in the position where it is carried entirely on one tooth.

Diametral Pitches: Practical values of diametral pitches range from 1 to 64.

Formulas for Gleason System Straight Bevel Gears (Revised 1980). — The data in Table 2 are used with the formulas in Table 1 to compute the dimensions of Gleason System straight bevel gears with axes at right angles. The following explanatory notes apply to the numbered items in Table 1.

Number of Teeth (Items 1 and 2): The Gleason System covers 20-degree pressure angle straight bevel gears for all combinations in which the pinion has 16 or more teeth and combinations in which: (a) the pinion has 15 and the gear 17 or more teeth; (b) the pinion has 14 and the gear 20 or more teeth; and (c) the pinion has 13 and the gear 31 or more teeth.

Face Width (Item 4): The face width should not exceed one-third of the cone distance (Item 11) or 10 inches ÷ diametral pitch, whichever is smaller. In best design practice the ratio of face width to cone distance will be from 0.25 to 0.3. Increasing the face width over recommended proportions adds strength and durability, but at a rapidly diminishing rate. Extra face width results in manufacturing difficulties by requiring tools of less point width and decreases possible fillet radii. It may even increase the danger of breakage and wear if the load becomes concentrated on the small ends of the teeth.

Pressure Angle (Item 5): The standard pressure angle is 20 degrees. However, if a lower pressure angle is preferred, 14½ degrees may be used without undercut for: combinations with 29 or more teeth in the pinion; 28 teeth in the pinion and 29 or more teeth in the gear; 27 teeth in the pinion and 32 or more in the gear; 26 teeth in the pinion and 35 or more in the gear; 25 teeth in the pinion and 42 or more in the gear; and 24 teeth in the pinion and 59 or more in the gear. The 20-degree pressure angle should be used on all aircraft applications where strength is of prime importance, and on all instrument gears which are to run with "zero" backlash.

Whole Depth (Item 8): Unless smooth bottoms and fillets are required, it is recommended that gears of 10 diametral pitch and coarser be roughed 0.005 inch deeper than the calculated whole depth so that the finishing tools will not cut on their ends. The calculated whole depth is, of course, used for the finishing operation.

Dedendum (Item 14): The value computed in this item is used in subsequent calculations. The actual dedendum, however, will be greater by 0.002 inch.

Backlash (Item 22): The table on page 1831 gives recommended backlash values when the gear and pinion are assembled and ready to run. Because of manufacturing tolerances and changes caused by heat treatment, it may be necessary in cutting the teeth to allow for more than one-half the tabular value in computing chordal thickness (Item 23) in order to obtain the desired backlash in assembly.

Tooth Angle (Item 25): The tooth angle is a machine setting in Gleason two-tool straight bevel gear generators. It is the angle at the machine center between the line to the center of the tooth on the root circle and the line to the point of the tool.

Limit Point Width (Item 26): The limit point width is the width of the point of a straight-sided V tool of given pressure angle which will touch both sides and the bottom of a finished tooth space at the small end. The point width of the tools used must be less than the limit point width.

Tool Advance: The tool advance is a machine setting to extend the tool and cut deeper. Its purpose is to increase the clearance along the full length of the tooth. The 0.002 inch in the formula for clearance is obtained in this way.

Formulas for Gleason System Angular Straight Bevel Gears. — Angular bevel gears are those whose shaft angle is other than 90 degrees. The formulas in Table 1 are not directly applicable to angular gears but must be modified as explained in the summary of procedure outlined below in the order of calculation; the item numbers referred to are those in Table 1:

Items 7, 8, and 9: Formulas same as given.

Item 10, Pitch Angles: There are two cases which must be distinguished.

Case I — Shaft Angle (Σ) less than 90 degrees, for which

$$\tan \gamma = \frac{\sin \Sigma}{\dfrac{N}{n} + \cos \Sigma} \; ; \quad \Gamma = \Sigma - \gamma$$

Case II — Shaft Angle (Σ) greater than 90 degrees, for which

$$\tan \gamma = \frac{\sin (180 - \Sigma)}{\dfrac{N}{n} - \cos (180 - \Sigma)} \; ; \quad \Gamma = \Sigma - \gamma$$

A gear pitch angle (Γ) greater than 90 degrees indicates an *internal* gear and calculations should be referred to the manufacturer of the generating equipment used to determine whether the gear can be cut.

In either Case I or II,

$$\frac{\sin \gamma}{\sin \Gamma} = \frac{n}{N}$$

Items 11 and 12: Formulas same as given.

Items 13 and 14: Formulas same as given. However, determination of the tooth proportions requires the use of the equivalent 90-degree bevel gear ratio computed from the formula

$$\text{Equivalent 90-degree ratio } m_{90} = \sqrt{\frac{N \cos \gamma}{n \cos \Gamma}}$$

This equivalent ratio is used as the ratio (N/n) when determining the gear addendum.

Items 15, 16, 17, 18, and 19: Formulas same as given.

Item 20, Pitch Apex to Crown: The following formulas must be used:

$$x_O = A_O \cos \gamma - a_P \sin \gamma$$
$$X_O = A_O \cos \Gamma - a_G \sin \Gamma$$

Table 1. Formulas for Gleason System 20-degree Straight Bevel Gears — 90-degree Shaft Angle

Given					
1	Number of Pinion Teeth, n		4	Face Width, F	
2	Number of Gear Teeth, N		5	Pressure Angle, $\phi = 20°$	
3	Diametral Pitch, P		6	Shaft Angle, $\Sigma = 90°$	

		To Find	
		Formula	
No.	Item	PINION	GEAR
7	Working Depth	$h_k = \dfrac{2.000}{P}$	Same as pinion
8	Whole Depth	$h_t = \dfrac{2.188}{P} + 0.002$	Same as pinion
9	Pitch Diameter	$d = \dfrac{n}{P}$	$D = \dfrac{N}{P}$
10	Pitch Angle	$\gamma = \tan^{-1}\dfrac{n}{N}$	$\Gamma = 90° - \gamma$
11	Cone Distance	$A_O = \dfrac{D}{2 \sin \Gamma}$	Same as pinion
12	Circular Pitch	$p = \dfrac{3.1416}{P}$	Same as pinion
13	Addendum	$a_P = h_k - a_G$	$a_G = \dfrac{0.540}{P} + \dfrac{0.460}{P(N/n)^2}$
14	Dedendum (See Note 1)	$b_P = \dfrac{2.188}{P} - a_P$	$b_G = \dfrac{2.188}{P} - a_G$
15	Clearance	$c = h_t - h_k$	Same as pinion
16	Dedendum Angle	$\delta_P = \tan^{-1}\dfrac{b_P}{A_O}$	$\delta_G = \tan^{-1}\dfrac{b_G}{A_O}$
17	Face Angle of Blank	$\gamma_O = \gamma + \delta_G$	$\Gamma_O = \Gamma + \delta_P$
18	Root Angle	$\gamma_R = \gamma - \delta_P$	$\Gamma_R = \Gamma - \delta_G$
19	Outside Diameter	$d_O = d + 2a_P \cos \gamma$	$D_O = D + 2a_G \cos \Gamma$
20	Pitch Apex to Crown	$x_O = \dfrac{D}{2} - a_P \sin \gamma$	$X_O = \dfrac{d}{2} - a_G \sin \Gamma$
21	Circular Thickness	$t = p - T$	$T = \dfrac{p}{2} - (a_P - a_G) \tan \phi - \dfrac{K}{P}$ $K = (\text{Chart } 1)$
22	Backlash	$B = (\text{See Table on page 1831})$	
23	Chordal Thickness	$t_C = t - \dfrac{t^3}{6d^2} - \dfrac{B}{2}$	$T_C = T - \dfrac{T^3}{6D^2} - \dfrac{B}{2}$
24	Chordal Addendum	$a_{CP} = a_P + \dfrac{t^2 \cos \gamma}{4d}$	$a_{CG} = a_G + \dfrac{T^2 \cos \Gamma}{4D}$
25	Tooth Angle	$\dfrac{3438}{A_O}\left(\dfrac{t}{2} + b_P \tan \phi\right)$ minutes	$\dfrac{3438}{A_O}\left(\dfrac{T}{2} + b_G \tan \phi\right)$ minutes
26	Limit Point Width	$\dfrac{A_O - F}{A_O}(T - 2b_P \tan \phi) - 0.0015$	$\dfrac{A_O - F}{A_O}(t - 2b_G \tan \phi) - 0.0015$

All linear dimensions are in inches.
Note 1: The actual dedendum will be 0.002-inch greater than calculated due to tool advance.

Item 5, Pressure Angle: The minimum pressure angle necessary to avoid undercut is computed from the formula:

$$\cos \phi = \sqrt{\frac{\cos \delta \sin (\gamma - \delta)}{\sin \gamma}}$$

The selected pressure angle should be equal to or greater than this calculated value.

Items 21, 22, 23, 24, 25 and 26: Formulas same as given.

Ratio of Generator Roll Gears: Angular gears require special ratio of roll gears for cutting on Gleason generators. The decimal ratio for the NC/75 ratio gears is found by the formula

$$\text{Decimal ratio of gears} = \frac{A_O P}{37.5}$$

The work roll is then found in the usual manner using the decimal ratio.

Chart 1. Circular Thickness Factor K for Straight Bevel Gears with 20-degree Pressure Angle — Gleason System, 1980

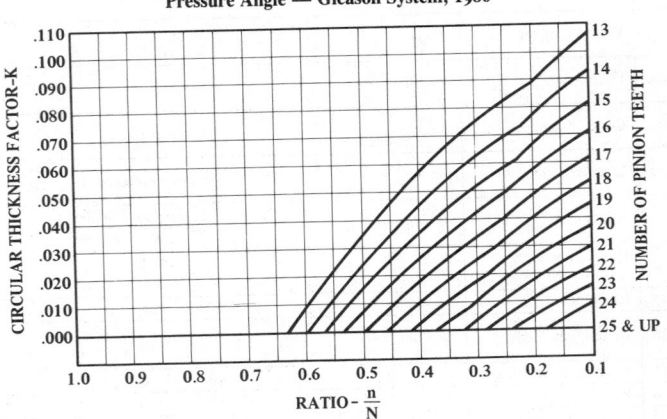

Zerol* Bevel Gears. — As in the case of straight bevel gears, Zerol gears may be mounted on plain journal bearings to obtain both a smaller and a less-costly assembly. As with spiral bevel gears, their tooth form can easily be manufactured to have a localized tooth contact that will allow for assembly position tolerance and operating deflection without causing end contact. These teeth may also be ground if high accuracy is required for reduced noise, or for high pitchline velocity. The normal limit is a pitchline velocity of 1,000 feet per minute; however, applications have been successful, under special conditions, at speeds up to 15,000 feet per minute. Zerol gears have the same radial and axial loading as straight bevel gears of the same size and proportions. Therefore, they may be used to supplant a straight bevel gear set where gear performance requires improvement, with no need to also change the gear mounting bearings.

Zerol bevel gears of 20-degree pressure angle should have a pinion with no less than 17 teeth; if a lesser tooth number must be used, the pressure angle may be increased to 22½ or 25 degrees to avoid undercut or involute interference. In this case the tooth thickness and the dedendum angles must be modified as indicated in AGMA 202.03, or Gleason publication, "Zerol Bevel Gear System."

* Registered in U.S. Patent Office.

Zerol bevel gears of 20-degree pressure angle and 90-degree shaft angle may be proportioned by the use of Table 1, page 1845, except that Item 16, dedendum angles (δ_P) and (δ_G) must be reduced by:

$$\Delta\delta = \frac{1}{2PA_O}\left(6668 - \frac{300}{F}\sqrt{d \sin\Gamma} - 14P\right) \text{ minutes}$$

Gleason System Spiral Bevel Gears. — The 1982 revision of the Gleason System provides a basis for designing spiral bevel gears to meet practical operating requirements. The system also provides tooth proportions for spiral bevel gears with fewer than 12 teeth, namely, 6, 7, 8, 9, 10, and 11 teeth for ratios above 3 to 1. The basic pressure angle of Gleason System spiral bevel gears is 20 degrees which is in line with the general trend in gear standards for all types of gears.

The System has been designed for a spiral angle of 35 degrees. If smaller spiral angles are used, undercut may occur and the contact ratio reduced.

There are many spiral-bevel-gear applications and cutting methods which require special tooth designs, and for these the standard system does not apply. Typical examples would be: (1) automotive axle drives; (2) gears and pinions of 12 diametral pitch and finer when cut by the duplex spread-blade methods, which is the usual case; (3) gear cut spread blade and pinion cut single-side, with a spiral angle less than 20 degrees; (4) ratios with fewer than 17 teeth in the pinion, 16/18 teeth and higher, 15/19 and higher, 14/20 and higher, 13/22 and higher, and 12/26 and higher; and (5) large spiral bevel gears cut on the planing-type generators where the spiral angle should not exceed 30 degrees.

The formulas and data that follow apply to spiral bevel gears of 12 diametral pitch and coarser with generated teeth; axes at right angles; and with 12 or more teeth in the pinion. Formulas for use when the axes are not at right angles are also provided.

Working Depth: The working depth is 1.700 inches ÷ diametral pitch.

Clearance: The clearance is 0.188 inches ÷ diametral pitch.

Whole Depth: The whole depth is usually determined as the working depth plus the clearance. It is common practice, however, with gears of 10 diametral pitch and coarser, to rough cut 0.005 inch deeper than the calculated depth in order to avoid having the finishing blades cut on their ends. The calculated whole depth, however, is used in the calculation of tooth proportions and for the finishing operation.

Pressure Angle: The basic pressure angle is 20 degrees and permits the following ratios to be cut without undercut: Ratios with 17 or more teeth in the pinion; 16/18 and higher; 15/19 and higher; 14/20 and higher; 13/22 and higher; and 12/26 and higher. However, for those who wish to use a lower pressure angle, the following ratios for 14½-degree and 16-degree pressure angles can be cut without undercut.

14½-degree Pressure Angle: Ratios with 28 or more teeth in the pinion; 27/29 and higher; 26/30 and higher; 25/32 and higher; 24/33 and higher; 23/36 and higher; 22/40 and higher; 21/42 and higher; 20/50 and higher; 19/70 and higher.

16-degree Pressure Angle: Ratios with 24 or more teeth in the pinion; 23/25 and higher; 22/26 and higher; 21/27 and higher; 20/29 and higher; 19/31 and higher; 18/36 and higher; 17/45 and higher; and 16/59 and higher.

Addendums: Except where the numbers of teeth are equal, the pinion has a long addendum and the gear a short addendum, the amount of departure from equal addendums varying with the ratio. Long addendums are used on the pinion primarily to avoid undercut and to increase tooth strength.

Face Angles: The face cone of a blank is made parallel to the root cone of the mating gear and therefore intersects the gear axis at a point inside the pitch apex. This gives constant clearance along the tooth and allows the use of larger edge radii on generating tools without fillet interference at the small end, thus increasing tooth strength.

Tooth Thickness: The tooth thicknesss are proportioned so that the stresses in

the gear and pinion will be approximately equal with a left-hand pinion driving clockwise or a right-hand pinion driving counterclockwise. This will give a satisfactory balance of life for gears operating below the endurance limit. If the gears are to operate above the endurance limit, special proportions will be required. Also, on reversible drives which must be designed for optimum load capacity, special proportions will be required. The method of determining the balance of strength for these special cases may be found in the Gleason publication "Strength of Bevel and Hypoid Gears." Determination of tooth thickness measurements will vary with different methods of cutting and will not be given here.

Spiral Angle: The system has been designed for a spiral angle of 35 degrees. For smaller spiral angles, undercut may occur and the contact ratio reduced.

Face Width: The recommended maximum face width is 0.3 times the cone distance or 10 inches ÷ diametral pitch, whichever is smaller.

Backlash: Recommended backlash values for general purpose bevel gearing are given in the table on page 1831.

Standard Cutter Edge Radius: On cutters for general use, the standard edge radius is made as large as practical for the various jobs for which the cutter may be suitable. Where the maximum fillet radius is required in gears for a specific case, as in aircraft gears, a special determination of the edge radius must be made.

Formulas for Gleason System Spiral Bevel Gears. — The formulas in Table 2 may be used to compute the dimensions of Gleason System spiral bevel gears. These formulas apply only to gears with axes at right angles. For other than 90-degree axes, the formulas in the following paragraphs apply.

Formulas for Gleason System Angular Spiral Bevel Gears. — Angular spiral bevel gears are those whose shaft angle is other than 90 degrees. The formulas in Table 2 are not directly applicable to angular spiral bevel gears but must be modified as explained in the summary of procedure outlined below in the order of calculation; item numbers referred to are those in Table 2.

Items 7, 8, and 9: Formulas same as given.

Item 10, Pitch Angles: There are two cases which must be distinguished.

Case I — Shaft Angle (Σ) less than 90 degrees, for which

$$\tan \gamma = \frac{\sin \Sigma}{\frac{N}{n} + \cos \Sigma} \; ; \quad \Gamma = \Sigma - \gamma$$

Case II — Shaft Angle (Σ) greater than 90 degrees, for which

$$\tan \gamma = \frac{\sin (180 - \Sigma)}{\frac{N}{n} - \cos (180 - \Sigma)} \; ; \quad \Gamma = \Sigma - \gamma$$

A gear pitch angle (Γ) greater than 90 degrees indicates an *internal* gear and calculations should be referred to the manufacturer of the generating equipment used to determine whether the gear can be cut.

In either Case I or II, $\dfrac{\sin \gamma}{\sin \Gamma} = \dfrac{n}{N}$

Items 11 and 12: Formulas same as given.

Items 13 and 14: Formulas same as given. However, determination of the tooth proportions requires the use of a modified bevel gear ratio computed from the formula:

$$\text{Equivalent 90° ratio, } m_{90} = \sqrt{\frac{N \cos \gamma}{n \cos \Gamma}}$$

Table 2. Formulas for Gleason System 20-degree Pressure Angle, Spiral Bevel Gears — 90-degree Shaft Angle

	Given			
I	Number of Pinion Teeth, n		4	Face Width, F
2	Number of Gear Teeth, N		5	Pressure Angle, $\phi = 20°$
3	Diametral Pitch, P		6	Shaft Angle, $\Sigma = 90°$

	To Find		
		Formula	
No.	Item	PINION	GEAR
7	Working Depth	$h_k = \dfrac{1.700}{P}$	Same as pinion
8	Whole Depth	$h_t = \dfrac{1.888}{P}$	Same as pinion
9	Pitch Diameter	$d = \dfrac{n}{P}$	$D = \dfrac{N}{P}$
10	Pitch Angle	$\gamma = \tan^{-1} \dfrac{n}{N}$	$\Gamma = 90 - \gamma$
11	Cone Distance	$A_O = \dfrac{D}{2 \sin \Gamma}$	Same as pinion
12	Circular Pitch	$p = \dfrac{3.1416}{P}$	Same as pinion
13	Addendum	$a_P = h_k - a_G$	$a_G = \dfrac{1}{P} \left[0.46 + \dfrac{0.39}{(N/n)^2} \right]$
14	Dedendum	$b_P = h_t - a_P$	$b_G = h_t - a_G$
15	Clearance	$c = h_t - h_k$	Same as pinion
16	Dedendum Angle	$\delta_P = \tan^{-1} \dfrac{b_P}{A_O}$	$\delta_G = \tan^{-1} \dfrac{b_G}{A_O}$
17	Face Angle of Blank	$\gamma_O = \gamma + \delta_G$	$\Gamma_O = \Gamma + \delta_P$
18	Root Angle	$\gamma_R = \gamma - \delta_P$	$\Gamma_R = \Gamma - \delta_G$
19	Outside Diameter	$d_O = d + 2a_P \cos \gamma$	$D_O = D + 2a_G \cos \Gamma$
20	Pitch Apex to Crown	$x_O = \dfrac{D}{2} - a_P \sin \gamma$	$X_O = \dfrac{d}{2} - a_G \sin \Gamma$
21	Circular Thickness	$t = p - T$	$T = \dfrac{(1.5708 - K)}{P} - \dfrac{\tan \phi}{\cos \psi}(a_P - a_G)$ For K, see Chart 1, p. 1851
22	Backlash*	$B =$ (See table on page 1831)	

* When the gear is cut spread-blade, all the backlash is taken from the pinion thickness. When both members are cut single-side, each thickness is reduced by half the backlash.

This equivalent ratio is used when determining the gear addendum.

Items 15, 16, 17, 18, and 19: Formulas same as given.

Item 20, Pitch Apex to Crown:

$$x_O = A_O \cos \gamma - a_P \sin \gamma$$

$$X_O = A_O \cos \Gamma - a_G \sin \Gamma$$

Item 5, Pressure Angle: The minimum pressure angle necessary to avoid undercut can be obtained from Chart 2. The point of intersection on the graph of the dedendum angle and the pitch angle must be on or below the line representing the selected pressure angle if undercut is to be avoided.

Item 21: Formulas same as given. However, the value of K from Chart 1 must be determined by using the reciprocal, $1/m_{90}$, of the equivalent 90-degree bevel gear ratio m_{90} which was computed for Item 13, and the number of teeth in the equivalent 90-degree bevel pinion, n_{90}, which is computed as follows:

$$n_{90} = \frac{n \sin (\tan^{-1} m_{90})}{\cos \gamma}$$

Item 22: Same formulas as given.

Bevel Gear Mounting. — To assure successful operation, bevel gears must be rigidly mounted within the housing such that the deflection of one gear relative to the other does not exceed 0.006 inch either axially or tangential to the pitch diameters. Gear mountings that are more flexible will require special treatment in manufacture that will provide a smaller tooth contact area whose position is carefully controlled near the toe end of the face. This will raise the tooth contact stress over the expected value and increase the operating noise level. Satisfactory mounting stiffness is most easily obtained by using large, long-life bearings broadly spaced, and with a toe end bearing on the pinion if at all possible. It is good practice to have the bearing spacing on both gear and pinion at least equal to three-quarters of the gear pitch diameter if at all possible; on high ratio gear sets this will not always be practical on the pinion.

Provision shall be made for axially shimming the position of both gears at the time of assembly. The mounting distance of the gear is usually the Pitch Apex to Back (p. 1842) and should be stamped or etched on every bevel gear. At assembly, shims are used to obtain this distance exactly. The gears, when so mounted, should have the proper tooth bearing pattern and backlash. These should be checked; the pattern is checked by bluing the tooth and turning it over under light load while noting the pattern transfer; the backlash is checked with feeler gages.

More complete data are given in "Installation of Bevel Gears" and "Bevel and Hypoid Gear Design" both published by Gleason Works, Rochester, New York.

Spiral Bevel Gears with Spiral Angle other than 35 Degrees. — In general, the same tooth proportions used for Gleason System spiral bevel gears may be used for spiral bevel gears with spiral angle of from, say, 20 to 45 degrees. Below 20-degree spiral angle, however, special proportions may be required. In the description of the Gleason System for spiral bevel gears, the reason for stating that the system is based on a 35-degree spiral angle is that the tooth-thickness balance for equal stress in gear and pinion was determined using this spiral angle. This spiral angle is most commonly selected for spiral bevel gears. In general, the higher the spiral angle the greater the face contact ratio becomes and the smoother and quieter the gears will be. The choice of spiral angle has no effect on efficiency, but does affect the thrust loads produced by the gears.

Chart 1. Circular Thickness Factors *K* for Gleason System Spiral Bevel Gears*

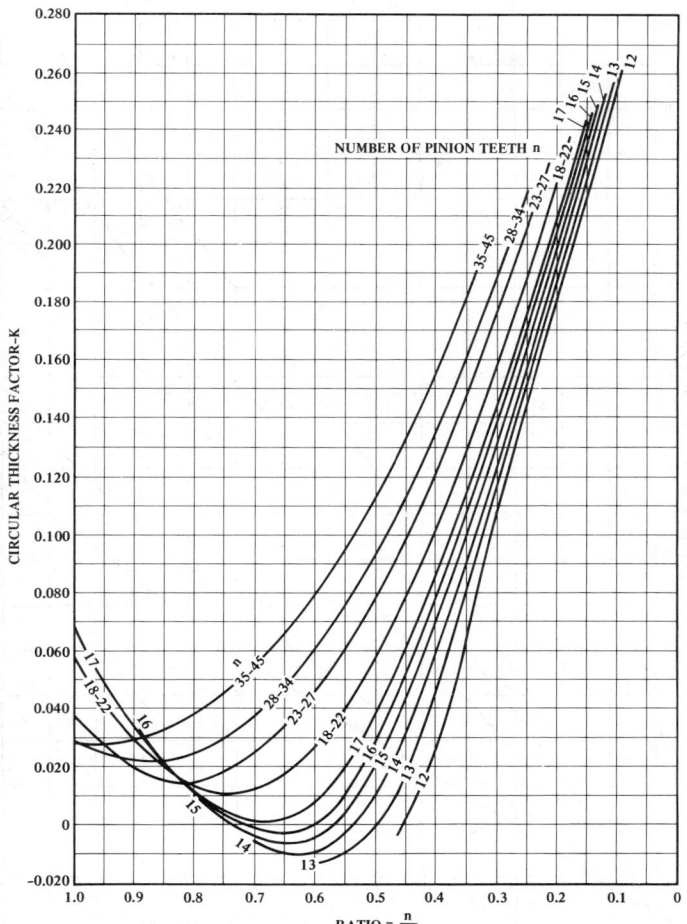

* These circular thickness factors are for spiral bevel gears with 20-degree pressure angle and 35-degree spiral angle for combinations with a left-hand pinion driving clockwise, or a right-hand pinion driving counter-clockwise. (See "Hand of Spiral, Hand of Rotation, and Spiral Angle" for definitions.) When used in the circular thickness formula given in Table 2, the tooth thickness of gear and pinion will be proportioned to provide a satisfactory balance of life for gears operating below the endurance limit. For gears operating above the endurance limit, special proportions will be required.

Chart 2.   Relation between Pressure Angle, Dedendum Angle, and Pitch Angle for No Undercut in Spiral Bevel Gears with 35-degree Spiral Angle*

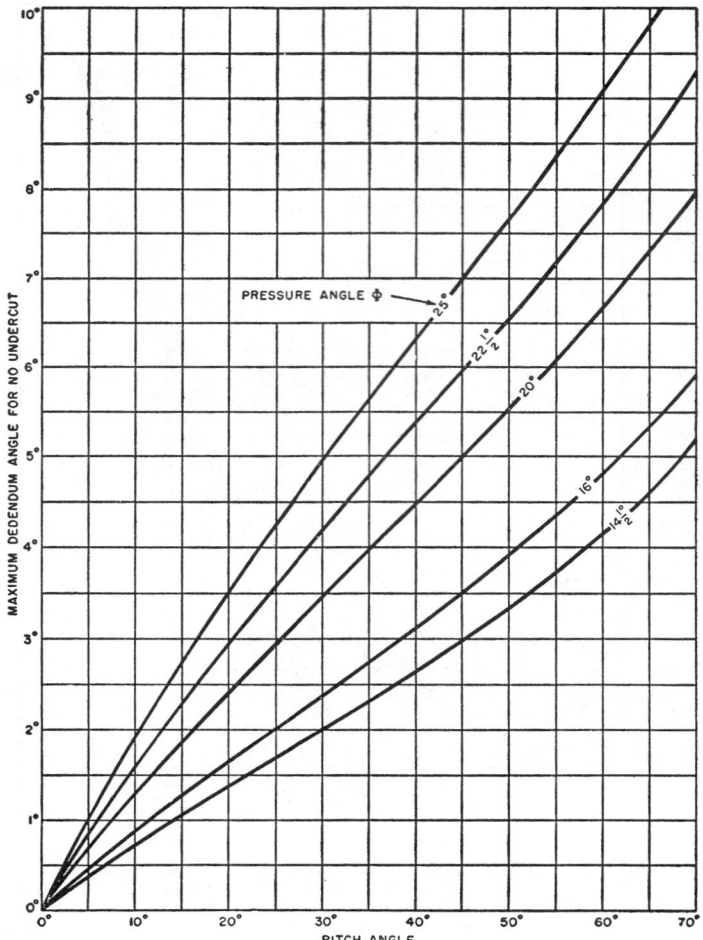

* To determine whether or not a 35-degree spiral angle bevel gear will be undercut, first, locate the point of intersection of the pitch angle and the dedendum angle. If this point of intersection is on or below the curve representing the selected pressure angle ϕ, then no undercutting will take place. If, however, the intersection is above the curve, undercutting will occur and, therefore, a higher pressure angle should be selected or other changes made, such as using larger tooth numbers in the gear and pinion.

Hand of Spiral, Direction of Rotation, and Spiral Angle. — The hand of spiral on spiral bevel gears is indicated by the direction in which the teeth curve away from the axis. Left-hand teeth curve away from the axis in a counter-clockwise direction when an observer looks at the face of the gear; right-hand teeth curve away in the clockwise direction. The hand of spiral of one member is always the opposite of that of its mate, and it is customary to specify the hand of spiral of the pinion in identifying the combination.

The *direction of rotation* of a bevel gear is determined as clockwise, or counter-clockwise, by viewing the gear from the back.

Effect of Hand of Spiral on Thrust Loads: A right-hand spiral pinion driving clockwise tends to move toward the cone center while a left-hand pinion tends to move away because of the oblique direction of the curved teeth. If there is any end play in the pinion shaft, the movement of a right-hand spiral pinion driving clockwise will take up the backlash under heavy load and the teeth of gear and pinion may wedge together; while a left-hand spiral pinion under the same conditions would back away and merely introduce additional backlash between the teeth. When the ratio, pressure angle, and spiral angle are such that it is possible, the hand of spiral should be selected to give an axial thrust that tends to move both the gear and pinion out of mesh; otherwise the hand of spiral should be selected to give an axial thrust that tends to move the pinion out of mesh. Often the mounting conditions will dictate the hand of spiral to be used; in a reversible drive, there is no choice unless the pair performs a heavier duty in one direction a greater part of the time. (See thrust formulas on page 1854.)

Selection of Spiral Angle: The spiral angle should be selected, if possible, to give a face contact ratio of at least 1.25 to assure smooth tooth action and quiet gears. On straight or Zerol bevel gears the spiral angle will, of course, be zero. The face contact ratio for various spiral angles and diametral pitches may be determined from the chart on page 1860. Before finally deciding on the spiral angle and hand of spiral, the loads should be computed to determine whether adequate bearings can be provided.

Bearing Loads Produced by Bevel Gears. — The normal load on the tooth surfaces of bevel gears may be resolved into three components: (1) One in a direction parallel to the axis of the gear (axial or thrust component); (2) one perpendicular to the axis of the gear (radial or separating component); and (3) one tangent to the mean pitch radius of the gear (tangential or driving component). The direction and magnitude of the normal load on the tooth surfaces, and hence of the three components, will depend on the gear ratio, hand of spiral, direction of rotation, and whether the gear is the driving or driven member.

Axial Thrust Load Produced on the Bearings: The axial thrust load produced on the bearings of a bevel gear is equal to the axial component of the normal load on the tooth surfaces. Its value may be computed by the formulas given in the table on page 1854.

Radial Load Produced on the Bearings: Unlike the axial thrust load, which is the direct result of the axial component of the normal tooth load, the radial load on the bearings results from the *moments* produced by each of the three components of the normal tooth load.

To determine the radial load on the bearings resulting from the moments produced by the components, first, compute the magnitudes and directions of these components of the normal tooth load using the formulas in the table on page 1854. The values thus obtained, together with the dimensions of the bearing mountings, are then used to determine the radial bearing loads by applying the principle of moments. It should be noted that in computing the moment due to the axial thrust component, this component is considered to act at the mean pitch radius of

Formulas for Tangential, Axial, and Separating Components of the Normal Tooth Load for Straight and Spiral Bevel Gears

W_t = tangential load on driving gear at its mean pitch diameter, in pounds; also equal to the tangential load on the driven gear at its mean pitch diameter;

W_x = axial thrust load, in pounds;

W_s = separating component in pounds;

d_m = mean pitch diameter of driving member in inches = $d - F \sin \gamma_d$,

where d = pitch diameter of driving member, F = face width, and
γ_d = pitch angle;

n = speed of driving member in rpm;

P = horsepower transmitted;

ϕ = normal pressure angle;

ψ = spiral angle;

γ_d = pitch angle of driving member;

γ_D = pitch angle of driven member.

Component of Normal Tooth Load to be Computed	Hand of Spiral and Direction of Rotation of Driving Member	
	Right-hand Spiral and Clockwise Rotation, or Left-hand Spiral and Counter-clockwise Rotation	Right-hand Spiral and Counter-clockwise Rotation, or Left-hand Spiral and Clockwise Rotation
Tangential, W_t Driving Member and Driven Member	$W_t = \dfrac{126{,}050\,P}{n d_m}$	
*Axial, W_x Driving Member	$W_x = \dfrac{W_t}{\cos \psi}(\tan \phi \sin \gamma_d - \sin \psi \cos \gamma_d)$	$W_x = \dfrac{W_t}{\cos \psi}(\tan \phi \sin \gamma_d + \sin \psi \cos \gamma_d)$
Driven Member	$W_x = \dfrac{W_t}{\cos \psi}(\tan \phi \sin \gamma_D + \sin \psi \cos \gamma_D)$	$W_x = \dfrac{W_t}{\cos \psi}(\tan \phi \sin \gamma_D - \sin \psi \cos \gamma_D)$
†Separating, W_s Driving Member	$W_s = \dfrac{W_t}{\cos \psi}(\tan \phi \cos \gamma_d + \sin \psi \sin \gamma_d)$	$W_s = \dfrac{W_t}{\cos \psi}(\tan \phi \cos \gamma_d - \sin \psi \sin \gamma_d)$
Driven Member	$W_s = \dfrac{W_t}{\cos \psi}(\tan \phi \cos \gamma_D - \sin \psi \sin \gamma_D)$	$W_s = \dfrac{W_t}{\cos \psi}(\tan \phi \cos \gamma_D + \sin \psi \sin \gamma_D)$

* If the computed value of W_x is positive, then the thrust is *away* from the cone center. A negative value indicates the thrust is *toward* the cone center.

† If the computed value of W_s is positive, then the force is *away* from the mating member (separating force). A negative value indicates the *force* is *toward* the mating member (attracting force).

the gear or pinion, that is, at the pitch radius of the mid-face. The mean pitch radius may be computed by subtracting from the pitch radius one-half the product of the face width and the sine of the pitch angle.

Design of Bevel Gears for Durability and Strength — The design or selection of a pair of bevel gears to meet a given set of operating conditions is accomplished by a four-step sequence of calculations: (1) Determine the equivalent service horsepower or design load; (2) Select approximate gear sizes by means of a series of design charts; (3) Check the surface durability of the gears selected; and (4) Check the strength of the gears selected. The following information for determining gear ratings is based on methods used by the Gleason Works for general industrial work.

Equivalent Service Horsepower (Design Load). — The equivalent service horsepower or design load of a pair of bevel gears is the load for which the gears must be designed. It is computed as follows:

$$\text{Design Load } P_2 = C_s P_1 \quad \text{or,} \quad P_2 = \frac{P_M}{2}, \quad \text{whichever is greater} \qquad (1) \text{ and } (2)$$

where P_1 = normal operating load in horsepower; P_M = momentary peak load in horsepower; C_s = service factor from Table 1 on page 1856.

If the value of P_M to be used in formula (2) is unknown, but the momentary peak pinion torque T_M in inch-pounds is known, then P_M should be computed from the following formula in which n is the pinion speed in revolutions per minute:

$$P_M = \frac{nT_M}{63,025} \qquad (3)$$

Example: An electric motor is to deliver 65 horsepower at 1800 rpm to a centrifugal pump which is to run continuously at 600 rpm. The drive is to be through a pair of oil-hardened steel spiral bevel gears at 90-degree shaft angle. The starting torque will be 4000 inch-pounds.

From Formula (3),

$$P_M = \frac{1800 \times 4000}{63,025} = 114 \text{ horsepower}$$

From Formula (1),

$$P_2 = 1.0 \times 65 = 65 \text{ horsepower}$$

From Formula (2),

$$P_2 = \frac{114}{2} = 57 \text{ horsepower}$$

Therefore the value $P_2 = 65$ horsepower is used as the design load in subsequent calculations of the gear sizes.

Approximate Pinion and Gear Sizes for Design Load. — The approximate pinion size necessary to transmit design load P_2 can be determined from Chart 1 on page 1858 by using the gear ratio, and the horsepower to be transmitted per 100 rpm of the pinion computed from the following formula:

$$\text{Horsepower per 100 rpm of pinion } P_{100} = \frac{100 \, P_2}{nC_M} \qquad (4)$$

where C_M = material factor from Table 2; P_2 = design load from Formula (1) or (2); and n = speed of pinion in rpm. In the previous example, $P_2 = 65$ horsepower; $n = 1800$ rpm; and C_M for an oil-hardened steel gear and pinion from Table 2 is 0.65. Thus, using Formula (4),

$$P_{100} = \frac{100 \times 65}{1800 \times 0.65} = 5.6 \text{ horsepower}$$

and from Chart 1, for a 3:1 ratio, the minimum pinion pitch diameter corresponding to $P_{100} = 5.6$ horsepower is 4.1 inches. The approximate pitch diameter of the gear is equal to the approximate pitch diameter of the pinion multiplied by the gear ratio and in this case is $4.1 \times 3 = 12.3$ inches.

Number of Teeth: The relation between pinion size and number of pinion teeth to give a well-balanced design is shown on Chart 2A for spiral bevel gears, and on Chart 2B for straight bevel and Zerol bevel gears. In the previous example the pinion pitch diameter was determined to be approximately 4.1 inches. For a 3:1 gear ratio the number of teeth in the pinion, from Chart 2A, is found to be 15. The number of teeth in the gear is equal to the number of pinion teeth multiplied by the gear ratio and in this case is $15 \times 3 = 45$ teeth. (If hardened gears are to be lapped it is advisable, if possible, to alter the ratio slightly so that the numbers of teeth in gear and pinion do not have a common factor to obtain better results in the lapping process.)

Diametral Pitch: The diametral pitch is obtained by dividing the number of teeth in the pinion by the approximate pinion pitch diameter. For the previous example,

$$\text{Diametral Pitch} = \frac{15}{4.1} = 3.7$$

Although it is not necessary to round 3.7 to an integral value, a diametral pitch of

Table 1. Service Factors C_s for Bevel Gear Drives*

Character of Power Source	Character of Load on Driven Machine		
	Uniform	Moderate Shock	Heavy Shock
Uniform	1.00	1.25	1.75
Light Shock	1.10	1.35	1.80
Medium Shock	1.25	1.50	1.85

* Service factors given are for speed-decreasing drives; for speed-increasing drives, add 0.15 to these factors. Extreme repetitive shock and other applications where exceedingly high energy loads must be absorbed require special consideration and are not covered by the service factors in this table.

4 will be used for convenience in subsequent calculations in this problem. Based on this diametral pitch of 4, the pinion and gear pitch diameters are recomputed:

$$\text{Pitch Diameter of Pinion} = \frac{\text{Number of Pinion Teeth}}{\text{Diametral Pitch}} = \frac{15}{4} = 3.750 \text{ inches}$$

$$\text{Pitch Diameter of Gear} = \frac{\text{Number of Gear Teeth}}{\text{Diametral Pitch}} = \frac{45}{4} = 11.250 \text{ inches}$$

Face Width: The face width of the pair of bevel gears can now be determined from the following formula:

$$F = 0.15d\sqrt{1 + \left(\frac{N}{n}\right)^2} \tag{5}$$

where F = face width in inches; d = pitch diameter of pinion in inches; N = number of teeth in gear; and n = number of teeth in pinion. The face width determined by this formula will be equal to 0.3 times the cone distance, which is good design practice. In the previous example, $d = 3.750$ inches, $N = 45$, and $n = 15$. Therefore,

$$F = 0.15 \times 3.750\sqrt{1 + \left(\frac{45}{15}\right)^2} = 1.78, \quad \text{say,} \quad 1\tfrac{3}{4} \text{ inches}$$

Spiral Angle and Hand of Spiral: For spiral bevel gears the spiral angle should be sufficient to give a face contact ratio of at least 1.25 to assure smooth tooth action and quiet gears. Design Chart 3 may be used as a guide to assist in the selection of the spiral angle. (See also "Hand of Spiral, Hand of Rotation, and Spiral Angle.") Before making a final decision on the spiral angle, the bearing loads should be computed to determine whether adequate bearings can be provided. (See "Bearing Loads Produced by Bevel Gears.")

In the previous example, the diametral pitch was 4 and the face width of the pinion 1¾ inches. The product of the diametral pitch and the pinion face width, $4 \times 1\tfrac{3}{4} = 7$, is used to enter Chart 3 to find a spiral angle sufficient to give at least 1.25 face contact ratio. From the chart it is found that spiral angles above 25 degrees will have face contact ratios above 1.25. Therefore a spiral angle of, say, 35 degrees may be selected, provided the gear mountings are rigid and the bearings large enough to absorb the radial and thrust loads associated with this spiral angle. Calculations of the thrust loads may indicate that the spiral angle should not be larger than, say, 30 degrees.

Summary: Thus far, the following approximate sizes of gear and pinion have been determined by means of relatively simple formulas and design charts: Teeth in pinion, 15; teeth in gear, 45; face width, 1.75 inches; spiral angle, 35 degrees; pinion pitch diameter, 3.75 inches; and gear pitch diameter, 11.25 inches. The next step is to check this approximate design using the surface durability equation in the following paragraph.

Table 2. Material Factors C_M for Surface Durability of Bevel Gears

Gear			Pinion			Surface Durability Factor C_M
Material	Hardness		Material	Hardness		
	Brinell	Rockwell "C"		Brinell	Rockwell "C"	
Cast Iron	...	...	Cast Iron	...	...	0.30
Cast Iron	...	...	Annealed Steel	160–200	...	0.30
Cast Iron	...	...	Flame-hardened Steel	...	50*	0.40
Cast Iron	...	...	Casehardened Steel	...	55*	0.40
Heat-treated Steel	210–245	...	Heat-treated Steel	245–280	...	0.35
Heat-treated Steel	250–300	...	Casehardened Steel	...	55*	0.50
Oil-hardened Steel	...	...	Oil-hardened Steel	...	...	0.65
Flame-hardened Steel	...	50*	Flame-hardened Steel	...	50*	1.00
Flame-hardened Steel	...	50*	Casehardened Steel	...	55*	1.00
Casehardened Steel	...	55*	Casehardened Steel	...	55*	1.00

*Minimum values.

Surface Durability of Bevel Gears. — Except for cast iron gears, flame-hardened gears, or gears of brittle materials, surface durability is usually the determining factor in tooth failure. Bevel gears are, therefore, usually designed on a surface durability basis and then checked for strength.

The method previously described is a simplified approach of a more detailed method for determining bevel gear sizes for a given application. The gear sizes thus obtained should, therefore, be regarded as tentative until a further check is made using the following formula that takes into account a number of important factors not included in the initial determination.

$$P = FC_M C_s K_3 L_T \qquad (6)$$

where P = maximum equivalent service horsepower. This should be equal or greater than the design horsepower P_2 determined by Formula (1) or (2), whichever is greater

F = face width in inches. (For hypoids, use face width of gear.)

C_M = material factor for surface durability from Table 2.

C_s = factor for contact ratio = $\sqrt{0.4(m_F + m_P)}$, where m_F = face contact ratio from Chart 3. If m_F is greater than 2, use $m_F = 2$ in this formula. For straight bevel and Zerol bevel gears, $m_F = 0$. For hypoids use the average of gear and pinion spiral angle when obtaining m_F from Chart 3. m_P = profile contact ratio from Table 3.

K_3 = combined factor for pinion diameter, velocity, and allowable unit load. Values are obtained from Chart 4. For hypoids, the pitch diameter of the corresponding bevel pinion is used to enter the chart.

L_T = lubrication factor = 1 for bevel gears lubricated with mineral oil, or hypoid gear lubricated with extreme pressure lubricant. Use a value of 0.6 for hypoid gears lubricated with mineral oil.

An example using Formula (6) is given on page 1859.

Chart 1.   Horsepower per 100 RPM of the Pinion for Bevel Gears Operating at 90-degree Shaft Angle*

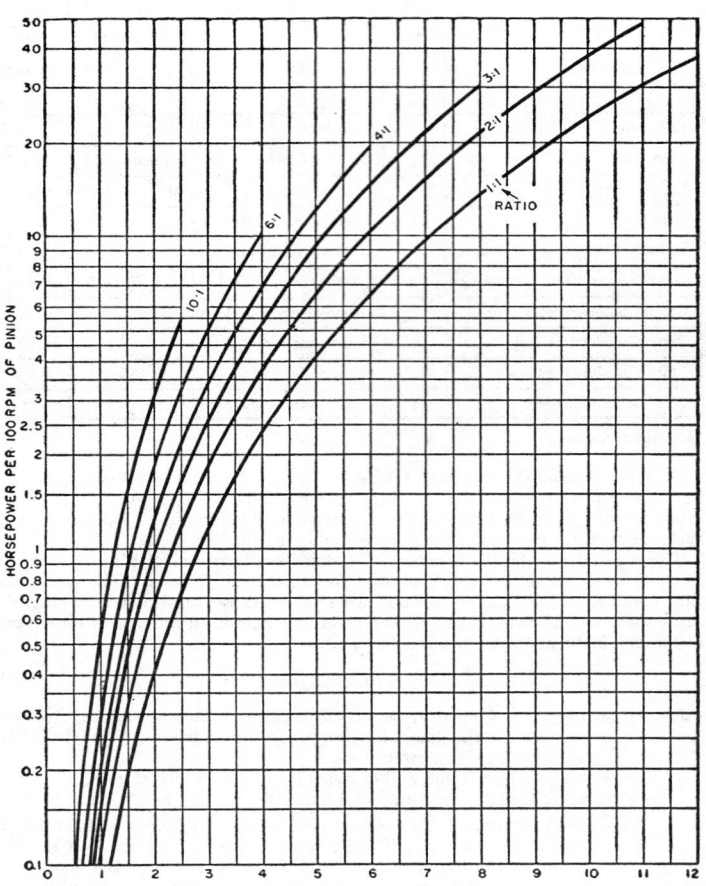

* This design chart is based on gears having a face width equal to three-tenths of the cone distance and a pitch line velocity of 900 feet per minute for straight and Zerol bevel gears and 2500 feet per minute for spiral bevel gears. If this chart is used for selecting an approximate pinion size for gears operating at lower speeds than indicated, the size obtained may be somewhat larger than actually needed.

For ratios in between those shown on the curves, interpolation may be employed to select the approximate pinion size required.

Chart 2A. Approximate Number of Teeth in Spiral Bevel Pinions of 35-degree Spiral Angle for Well-balanced Design

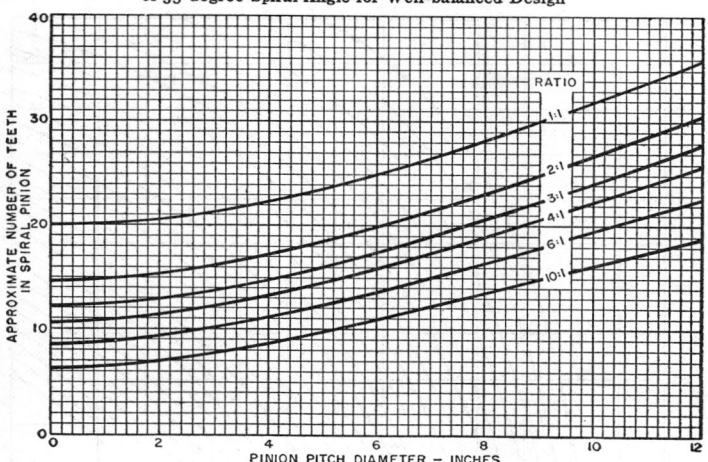

Chart 2B. Approximate Number of Teeth in Straight and Zerol Bevel Pinions for Well-balanced Design

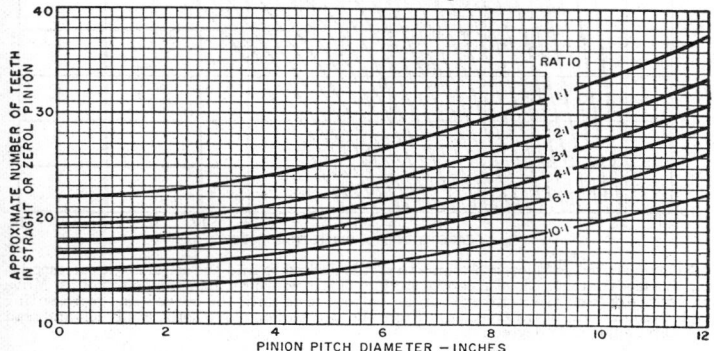

Example: Using the pinion and gear sizes previously determined, check the maximum equivalent service horsepower P that the gears can carry according to Formula (6)

$$F = 1.75 \text{ inches}$$
$$C_M = 0.65 \text{ (Table 2)}$$
$$m_F = 1.85 \text{ for 35-degree spiral angle (Chart 3)}$$
$$m_P = 1.22 \text{ for 15 teeth (Table 3)}$$
$$C_s = \sqrt{0.4(1.85 + 1.22)} = 1.11$$

Chart 3. Spiral Angles Corresponding to Various Face Contact Ratios and Products of Face Width by Diametral Pitch*

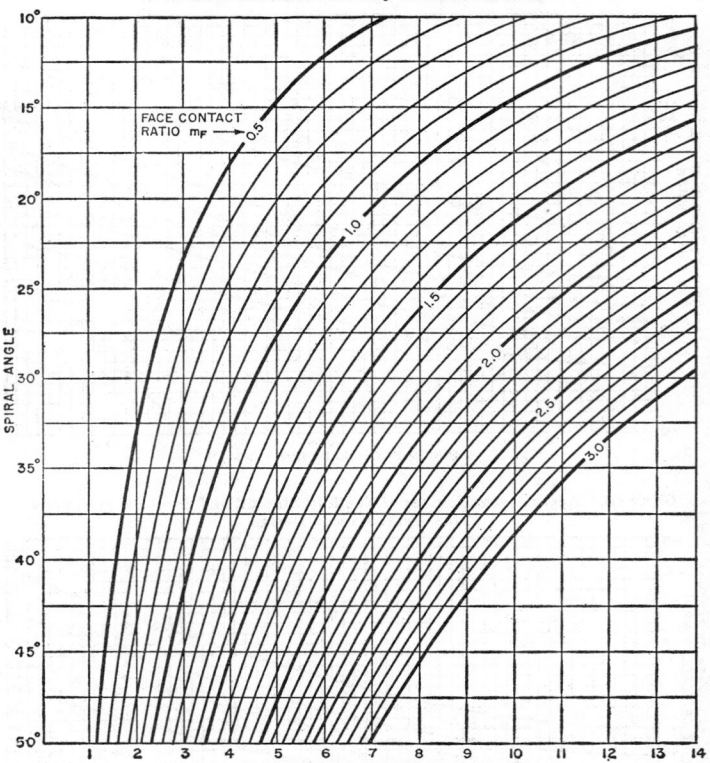

* Chart is used as follows: Assume that it is desired to find the spiral angle corresponding to a face contact ratio of 1.5 and a product of face width by diametral pitch of 8. Move up vertically from the number 8 on the bottom scale until the curved line representing 1.5 face contact ratio is intersected. From the intersection move horizontally left and read the spiral angle of 27 degrees.

$$\text{Pinion rpm} = 1800$$
$$\text{Pinion pitch diameter} = 3.750 \text{ inches}$$
$$K_3 = 50 \text{ (Chart 4)}$$
$$P = 1.75 \times 0.65 \times 1.11 \times 50 \times 1 = 63 \text{ horsepower maximum}$$

The gears selected by means of the approximate method are, therefore, too small to carry the design load of 65 horsepower. In this case, therefore, it will be necessary to change the proportions somewhat to increase the capacity of the gears by 2 horsepower. One way to accomplish this is to increase the number of teeth

Chart 4.   Combined Durability Factor K_3 for Bevel Gears*

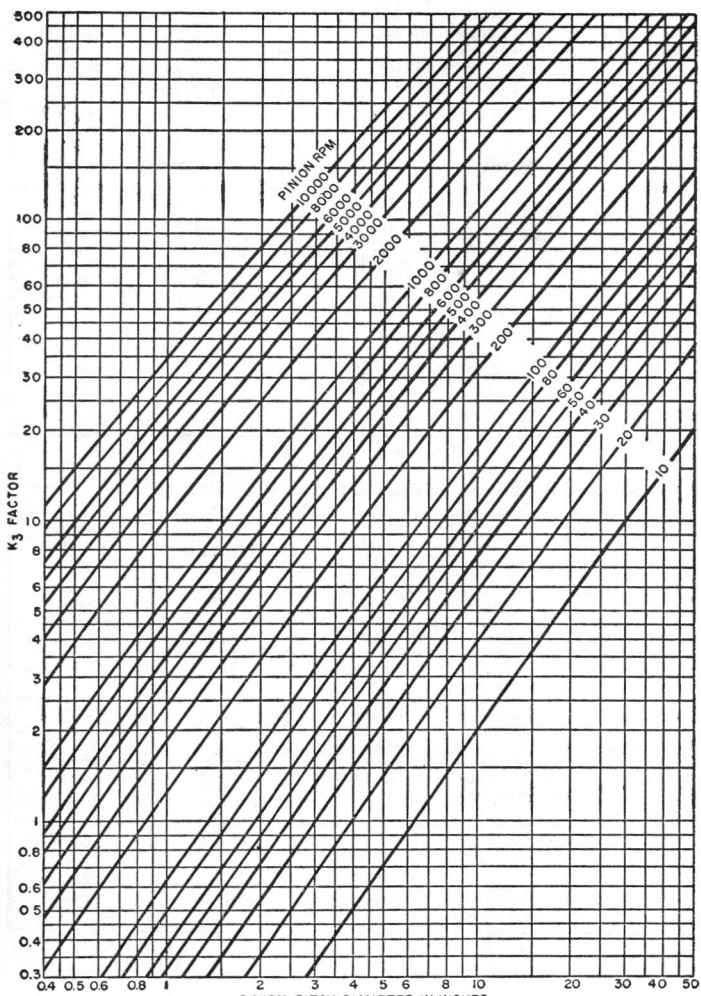

* This combined durability factor, which is used in the surface durability equation on page 1857, is based on the pinion pitch diameter, velocity, and allowable unit load.

Table 3. Profile Contact Ratios m_p for 20-degree Pressure Angle Bevel and Hypoid Gears

Number of Teeth in Pinion	Profile Contact Ratio m_p		Number of Teeth in Pinion	Profile Contact Ratio m_p	
	Spiral Bevel and Hypoid Gears	Straight and Zerol Bevel Gears		Spiral Bevel and Hypoid Gears	Straight and Zerol Bevel Gears
6	0.87	....	21	1.26	1.57
7	0.95	....	22	1.27	1.58
8	1.02	....	23, 24	1.28	1.59
9	1.07	....	25	1.29	1.60
10	1.11	....	26	1.29	1.61
11	1.15	....	27, 28	1.30	1.62
12	1.19	....	29	1.30	1.63
13	1.20	1.48	30, 31	1.31	1.64
14	1.21	1.49	32, 33	1.32	1.65
15	1.22	1.51	34, 35	1.32	1.66
16	1.23	1.52	36, 37	1.33	1.67
17	1.24	1.53	38, 39, 40	1.34	1.68
18	1.25	1.54			
19	1.25	1.55			
20	1.26	1.56			

Table 4. Allowable Bending Stress s' for Various Materials Used in Bevel Gears

Material	Condition of Material	Heat Treatment	Brinell Hardness Number	Allowable Bending Stress s'
Cast Iron	Ordinary	As Cast	....	4,600
Cast Iron	High grade	As Cast	200–300	7,000
Steel	Forged or Equivalent	Normalized	140–180	11,000
Steel	Forged or Equivalent	Hardened and Tempered	180–220	13,500
Steel	Forged or Equivalent	Hardened and Tempered	300–350	19,000
Steel	Forged or Equivalent	Flame or Induction Hardened*	450–550	13,500*
Steel	Forged or Equivalent	Carburized	575–625	30,000

* Unhardened root fillets.

in the pinion slightly. This will increase the pitch diameter of the pinion and, in proper proportion, the face width recomputed by Formula (5). Another way would be to use a coarser pitch, say, 3.75, since in bevel gear practice the generating tools are not limited to integral pitches.

Chart 5A. Geometry Factors J for Straight Bevel Gears with 20-degree
Pressure Angle and 90-degree Shaft Angle

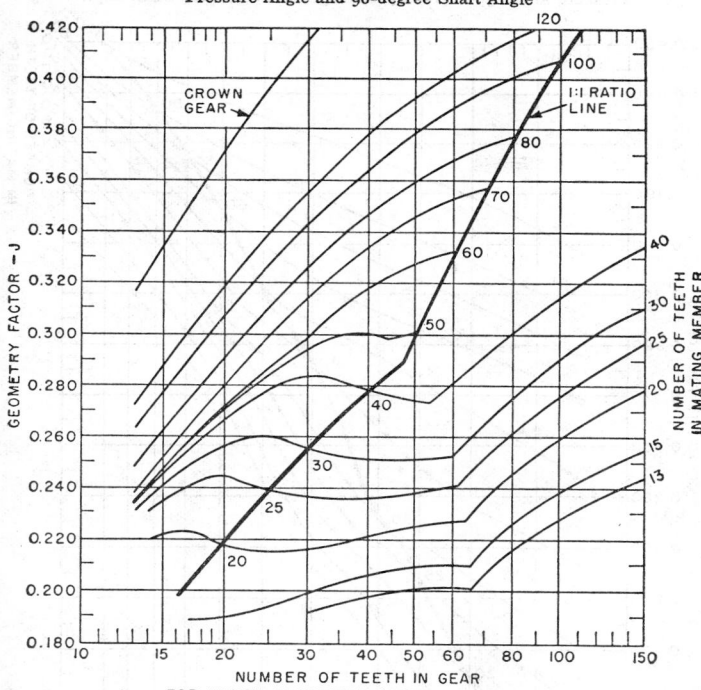

NUMBER OF TEETH IN GEAR
FOR WHICH GEOMETRY FACTOR IS DESIRED

Strength of Bevel Gears. — Although surface durability is usually the determining factor in gear tooth failure, there are cases where failure takes place by tooth breakage. This may be particularly true where loads are high or where brittle materials are used for the gears.

The following formula may be used to check the power which a pair of bevel gears may transmit without failing by tooth breakage:

$$P_3 = \frac{W_t' n d L_f}{126,000} \text{ horsepower} \qquad (7)$$

where P_3 = maximum horsepower. This must be equal to or greater than the permissible horsepower determined by surface durability requirements, using Formula (6) on page 1857.

$W_t' = \dfrac{s'FJ}{K_m P^{0.75}}$ = maximum allowable tangential load in pounds

s' = maximum allowable bending stress in pounds per square inch from Table 4

F = face width in inches (assumed equal on both members)

Chart 5B.   Geometry Factors J for Spiral Bevel Gears of 20-degree
Pressure Angle, 35-degree Spiral Angle, and 90-degree Shaft Angle

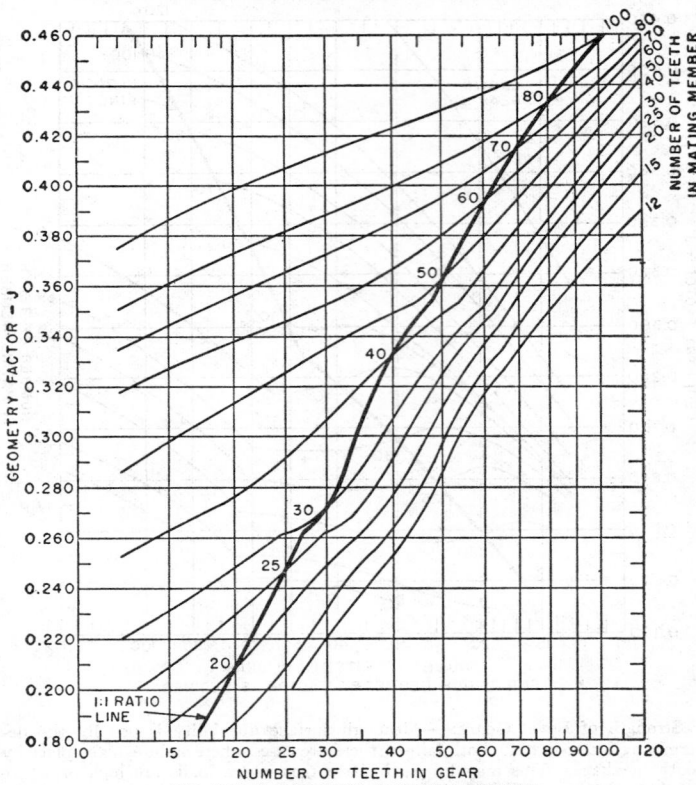

J = geometry factor incorporating stress concentration factor, effective
face width, load location, load distribution, and inertia factor.
Charts 5A and 5B give values for gear and pinion for straight bevel
gears and spiral bevel gears of 35-degree spiral angle with 20-degree
pressure angle, 90-degree shaft angle, and a tool edge radius of 0.240
inch ÷ diametral pitch.   Values of J should be reduced by 11 per cent
if tools with standard edge radius of approximately 0.120 inch ÷
diametral pitch are employed.

Because the geometry factor J will be different for the gear and the
mating pinion of straight bevel gears, the smaller of these J values
should be used in computing W_t'.   For spiral bevel gears, both J
values will be the same.

K_m = mounting factor. In general, use a value of 1.0 when both members are straddle mounted and a value of 1.1 when only one member is straddle mounted. Somewhat higher values may be required if mountings are poor.

P = diametral pitch at large end

n = pinion rpm

d = pinion pitch diameter in inches at large end

L_f = life factor. The life factor should be taken as 1.0 if stresses are to be kept below the endurance limit of the material used, that is, if the gears are to have indefinite life. Gears designed for heavy loading, and hence for shorter life, may employ life factors up to 4.6. Such life factors are used only for special applications where indefinite life is not required. In these applications the use of large life factors permits the use of smaller, less expensive gears. Life factors corresponding to various numbers of cycles of repeated loading required for failure are as follows: $L_f = 1.45$ for 1,000,000 cycles; $L_f = 2.0$ for 100,000 cycles; $L_f = 3.0$ for 10,000 cycles; and $L_f = 4.6$ for 1000 cycles.

Example: Determine whether or not the pinion and gear previously designed will fail by tooth breakage, assuming that these gears are to have an indefinite life. The data previously determined, are: Diametral pitch $P = 4$; face width $F = 1\frac{3}{4}$ inches; pinion speed $n = 1800$ rpm; pinion pitch diameter $d = 3.750$ inches; number of teeth in pinion = 15; and number of teeth in gear = 45.

Assuming that the gear and pinion were hardened and tempered to a Brinell hardness number of 300, the allowable bending stress s' from Table 4 is 19,000 psi; the geometry factor J from Chart 5B is 0.280 for both pinion and gear; the mounting factor, assuming only one member of the pair to be straddle mounted, is 1.1; and the life factor L_f, for indefinite life, is 1.0.

$$W_t' = \frac{sFJ}{K_m P^{0.75}} = \frac{19,000 \times 1.75 \times 0.280}{1.1 \times 4^{0.75}} = 3000 \text{ pounds}$$

$$P_3 = \frac{W_t' n d L_f}{126,000} = \frac{3000 \times 1800 \times 3.750 \times 1.0}{126,000} = 161 \text{ horsepower}$$

Since this computed value of permissible horsepower based on tooth breakage exceeds the permissible horsepower based on surface durability (63 horsepower), surface durability will be the determining factor in tooth failure for this pair of gears.

Materials for Bevel Gears. — The material most commonly used in the manufacture of bevel gears is steel. In particular, gears that are heavily loaded are usually made of a grade of steel suitable for case hardening after the teeth are cut. The table on page 1866 lists those steels most frequently specified for bevel gears, and may be used as a guide in selecting the type of steel suitable for a particular application. Gears of small size are usually made from bar stock, while larger gears are machined from forgings. For very large gears, for which forgings cannot be obtained, alloy steel castings are used.

Forgings: Where forgings are to be used, any of the commonly accepted forging methods in which the grain flow in the original steel billet follows the direction of the gear axis may be employed. Correct grain flow minimizes hardening distortion.

Cast Iron: Bevel gears may be made of high quality cast iron if the loads are relatively light and the size and weight of the gears are not of prime importance.

Other Gear Materials: Bevel gears that are lightly loaded and subject to corrosion may be made of brass, bronze, duralumin, or stainless steel. Gears made from plastic or phenolic materials are satisfactory only under very light loads.

Representative Steels Used for Bevel Gear Applications*

SAE or AISI No.	Type of Steel	Preliminary Heat Treatment	Brinell Hardness Number	ASTM Grain Size	Remarks
CARBURIZING STEELS					
1024	Manganese	Normalize			Low Alloy — oil quench limited to thin sections
2512	Nickel Alloy	Normalize — Anneal	163–228	5–8	Aircraft quality
3310 3312X	Nickel-Chromium (Krupp) Nickel-Chromium	Normalize, then heat to 1450°F, cool in furnace. Reheat to 1170°F — cool in air	163–228	5–8	Used for maximum resistance to wear and fatigue
4028	Molybdenum	Normalize	163–217		Low Alloy
4615 4620	Nickel-Molybdenum Nickel-Molybdenum	Normalize — 1700°F–1750°F	163–217	5–8	Good machining qualities. Well adapted to direct quench — gives tough core with minimum distortion
4815 4820	Nickel-Molybdenum Nickel-Molybdenum	Normalize	163–241	5–8	For aircraft and heavily loaded service
5120	Chromium	Normalize	163–217	5–8	
8615 8620 8715 8720	Chromium-Nickel-Molybdenum	Normalize — cool at hammer	163–217	5–8	Used as an alternate for 4620
OIL HARDENING AND FLAME HARDENING STEELS					
1141	Sulphurized free cutting carbon steel	Normalize Heat treated	179–228 255–269	5 or Coarser	Free cutting steel used for unhardened gears, oil treated gears, and for gears to be surface hardened where stresses are low
4140 4640	Chromium-Molybdenum Nickel-Molybdenum	For oil hardening: Normalize — Anneal For surface hardening: Normalize, reheat, quench and draw	179–212 235–269 269–302 302–341		Used for heat-treated, oil hardened, and surface hardened gears. Machining qualities of 4640 are superior to 4140, and it is the preferred steel for flame hardening
6145	Chromium-Vanadium	Normalize — reheat, quench, and draw	235–269 269–302 302–341		Fair machining qualities. Used for surface hardened gears when 4640 is not available
8640 8739	Chromium-Nickel-Molybdenum	Same as for 4640			Used as an alternate for 4640
NITRIDING STEELS					
Nitralloy H & G	Special Alloy	Anneal	163–192		Normal hardness range for cutting is 20–28 Rockwell C

* Any other steels equivalent to those listed in the table may also be used.

Formulas for Dimensions of Milled Bevel Gears. — Most bevel gears, as explained on page 1841, are produced by generating methods. Even so, there are applications for which it may be desired to cut a pair of mating bevel gears by using rotary formed milling cutters. Examples of such cases include replacement gears for certain types of equipment and gears for use in experimental developments.

The tooth proportions of milled bevel gears differ in some respects from those of generated gears, the principal difference being that for milled bevel gears the tooth thicknesses of pinion and gear are made equal, and the addendum and dedendum of the pinion are respectively the same as those of the gear. The formulas in the accompanying table may be used to calculate the dimensions of milled bevel gears with shafts at a right angle, an acute angle, and an obtuse angle.

In the accompanying diagram and list of notation, the various terms and symbols applied to milled bevel gears are indicated.

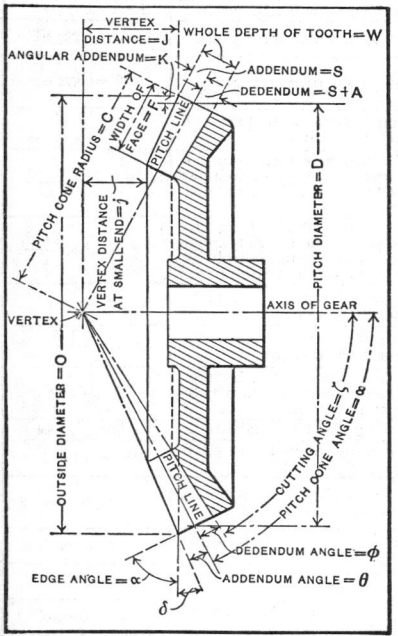

N = number of teeth;
P = diametral pitch;
p = circular pitch;
α = pitch cone angle and edge angle;
Σ = angle between shafts;
D = pitch diameter;
S = addendum;
$S+A$ = dedendum (A = clearance);
W = whole depth of tooth;
T = thickness of tooth at pitch line;
C = pitch cone radius;
F = width of face;
s = addendum at small end of tooth;
t = thickness of tooth at pitch line at small end;
θ = addendum angle;
ϕ = dedendum angle;
γ = face angle = pitch cone angle + addendum angle;
δ = angle of compound rest;
ζ = cutting angle;
K = angular addendum;
O = outside diameter;
J = vertex distance;
j = vertex distance at small end;
N' = number of teeth for which to select cutter.

The Brown & Sharpe Mfg. Co. recommends that the formulas for milled bevel gears be modified to make the clearance at the bottom of the teeth uniform instead of tapering toward the vertex. If this recommendation is followed, then the cutting angle (root angle) should be determined by subtracting the *addendum* angle from the pitch cone angle instead of subtracting the dedendum angle as in the formula given in the table.

Rules and Formulas for Calculating Dimensions of Milled Bevel Gears

To Find	Rule	Formula
Pitch Cone Angle of Pinion	Divide the sine of the shaft angle by the sum of the cosine of the shaft angle and the quotient obtained by dividing the number of teeth in the gear by the number of teeth in the pinion; this gives the tangent. *Note:* For shaft angles greater than 90° the cosine is negative.	$\tan \alpha_P = \dfrac{\sin \Sigma}{\dfrac{N_G}{N_P} + \cos \Sigma}$ For 90° shaft angle, $\tan \alpha_P = \dfrac{N_P}{N_G}$
Pitch Cone Angle of Gear	Subtract the pitch cone angle of the pinion from the shaft angle.	$\alpha_G = \Sigma - \alpha_P$
Pitch Diameter	Divide the number of teeth by the diametral pitch.	$D = N \div P$
Addendum	Divide 1 by the diametral pitch.	$S = 1 \div P$
Dedendum	Divide 1.157 by the diametral pitch.	$S + A = 1.157 \div P$
Whole Depth of Tooth	Divide 2.157 by the diametral pitch.	$W = 2.157 \div P$
Thickness of Tooth at Pitch Line	Divide 1.571 by the diametral pitch.	$T = 1.571 \div P$
Pitch Cone Radius	Divide the pitch diameter by twice the sine of the pitch cone angle.	$C = \dfrac{D}{2 \times \sin \alpha}$
Addendum of Small End of Tooth	Subtract the width of face from the pitch cone radius, divide the remainder by the pitch cone radius and multiply by the addendum.	$s = S \times \dfrac{C - F}{C}$
Thickness of Tooth at Pitch Line at Small End	Subtract the width of face from the pitch cone radius, divide the remainder by the pitch cone radius and multiply by the thickness of the tooth at pitch line.	$t = T \times \dfrac{C - F}{C}$
Addendum Angle	Divide the addendum by the pitch cone radius to get the tangent.	$\tan \theta = \dfrac{S}{C}$
Dedendum Angle	Divide the dedendum by the pitch cone radius to get the tangent.	$\tan \phi = \dfrac{S + A}{C}$
Face Width (Max.)	Divide the pitch cone radius by 3 or divide 8 by the diametral pitch, whichever gives the smaller value.	$F = \dfrac{C}{3}$ or $F = \dfrac{8}{P}$
Circular Pitch	Divide 3.1416 by the diametral pitch.	$p = 3.1416 \div P$
Face Angle	Add the addendum angle to the pitch cone angle.	$\gamma = \alpha + \theta$
Compound Rest Angle for Turning Blank	Subtract both the pitch cone angle and the addendum angle from 90 degrees.	$\delta = 90° - \alpha - \theta$
Cutting Angle	Subtract the dedendum angle from the pitch cone angle.	$\zeta = \alpha - \phi$
Angular Addendum	Multiply the addendum by the cosine of the pitch cone angle.	$K = S \times \cos \alpha$
Outside Diameter	Add twice the angular addendum to the pitch diameter.	$O = D + 2K$
Vertex or Apex Distance	Multiply one-half the outside diameter by the cotangent of the face angle.	$J = \dfrac{O}{2} \times \cot \gamma$
Vertex Distance at Small End of Tooth	Subtract the width of face from the pitch cone radius; divide the remainder by the pitch cone radius and multiply by the apex distance.	$j = J \times \dfrac{C - F}{C}$
Number of Teeth for which to Select Cutter	Divide the number of teeth by the cosine of the pitch cone angle.	$N' = \dfrac{N}{\cos \alpha}$

(Rows from Addendum through Circular Pitch are bracketed at the left: "These dimensions are the same for both gear and pinion.")

Numbers of Formed Cutters Used to Mill Teeth in Mating Bevel Gear and Pinion with Shafts at Right Angles

(Number of cutter for gear given first, followed by number for pinion. See text, page 1871.)

Number of Teeth in Gear	Number of Teeth in Pinion																
	12	13	14	15	16	17	18	19	20	21	22	23	24	25	26	27	28
12	7-7	…	…	…	…	…	…	…	…	…	…	…	…	…	…	…	…
13	6-7	6-6	…	…	…	…	…	…	…	…	…	…	…	…	…	…	…
14	5-7	6-6	6-6	…	…	…	…	…	…	…	…	…	…	…	…	…	…
15	5-7	5-6	5-6	5-5	…	…	…	…	…	…	…	…	…	…	…	…	…
16	4-7	5-7	5-6	5-6	5-5	…	…	…	…	…	…	…	…	…	…	…	…
17	4-7	4-7	4-6	5-6	5-5	5-5	…	…	…	…	…	…	…	…	…	…	…
18	4-7	4-7	4-6	4-6	4-5	4-5	5-5	…	…	…	…	…	…	…	…	…	…
19	3-7	4-7	4-6	4-6	4-5	4-5	4-4	…	…	…	…	…	…	…	…	…	…
20	3-7	3-7	4-6	4-6	4-6	4-5	4-5	4-4	4-4	…	…	…	…	…	…	…	…
21	3-8	3-7	3-7	3-6	4-6	4-5	4-5	4-5	4-4	4-4	…	…	…	…	…	…	…
22	3-8	3-7	3-7	3-6	3-6	3-5	4-5	4-5	4-4	4-4	4-4	…	…	…	…	…	…
23	3-8	3-7	3-7	3-6	3-6	3-5	3-5	3-5	4-4	4-4	4-4	4-4	…	…	…	…	…
24	3-8	3-7	3-7	3-6	3-6	3-6	3-5	3-5	3-4	3-4	3-4	4-4	4-4	…	…	…	…
25	2-8	2-7	3-7	3-6	3-6	3-6	3-5	3-5	3-5	3-4	3-4	3-4	4-4	3-3	…	…	…
26	2-8	2-7	3-7	3-6	3-6	3-6	3-5	3-5	3-5	3-4	3-4	3-4	3-4	3-3	3-3	…	…
27	2-8	2-7	2-7	2-6	3-6	3-6	3-5	3-5	3-5	3-4	3-4	3-4	3-4	3-4	3-3	3-3	…
28	2-8	2-7	2-7	2-6	2-6	3-6	3-5	3-5	3-5	3-4	3-4	3-4	3-4	3-3	3-3	3-3	3-3
29	2-8	2-7	2-7	2-7	2-6	2-6	3-5	3-5	3-5	3-4	3-4	3-4	3-4	3-3	3-3	3-3	3-3
30	2-8	2-7	2-7	2-7	2-6	2-6	2-5	2-5	3-5	3-5	3-4	3-4	3-4	3-4	3-4	3-3	3-3
31	2-8	2-7	2-7	2-7	2-6	2-6	2-6	2-5	2-5	3-5	3-4	3-4	3-4	3-4	3-4	3-3	3-3
32	2-8	2-7	2-7	2-7	2-6	2-6	2-6	2-5	2-5	2-5	2-4	3-4	3-4	3-4	3-4	3-4	3-3
33	2-8	2-8	2-7	2-7	2-6	2-6	2-6	2-5	2-5	2-5	2-4	2-4	3-4	3-4	3-4	3-4	3-3
34	2-8	2-8	2-7	2-7	2-7	2-6	2-6	2-5	2-5	2-5	2-4	2-4	2-4	3-4	3-4	3-4	3-3
35	2-8	2-8	2-7	2-7	2-7	2-6	2-6	2-5	2-5	2-5	2-4	2-4	2-4	2-4	2-4	2-4	2-3
36	2-8	2-8	2-7	2-7	2-6	2-6	2-6	2-5	2-5	2-5	2-5	2-4	2-4	2-4	2-4	2-4	2-3
37	2-8	2-8	2-7	2-7	2-6	2-6	2-6	2-5	2-5	2-5	2-5	2-4	2-4	2-4	2-4	2-4	2-3
38	2-8	2-8	2-7	2-7	2-6	2-6	2-6	2-5	2-5	2-5	2-5	2-4	2-4	2-4	2-4	2-4	2-4
39	2-8	2-8	2-7	2-7	2-6	2-6	2-6	2-5	2-5	2-5	2-5	2-4	2-4	2-4	2-4	2-4	2-4
40	1-8	2-8	2-7	2-7	2-6	2-6	2-6	2-5	2-5	2-5	2-5	2-4	2-4	2-4	2-4	2-4	2-4
41	1-8	1-8	2-7	2-7	2-6	2-6	2-6	2-6	2-5	2-5	2-5	2-4	2-4	2-4	2-4	2-4	2-4
42	1-8	1-8	2-7	2-7	2-6	2-6	2-6	2-6	2-5	2-5	2-5	2-5	2-4	2-4	2-4	2-4	2-4
43	1-8	1-8	1-7	2-7	2-6	2-6	2-6	2-6	2-5	2-5	2-5	2-5	2-4	2-4	2-4	2-4	2-4
44	1-8	1-8	1-7	1-7	2-6	2-6	2-6	2-6	2-5	2-5	2-5	2-5	2-4	2-4	2-4	2-4	2-4
45	1-8	1-8	1-7	1-7	1-6	2-6	2-6	2-6	2-5	2-5	2-5	2-5	2-4	2-4	2-4	2-4	2-4
46	1-8	1-8	1-7	1-7	1-7	2-6	2-6	2-6	2-5	2-5	2-5	2-5	2-4	2-4	2-4	2-4	2-4
47	1-8	1-8	1-7	1-7	1-7	1-6	2-6	2-6	2-5	2-5	2-5	2-5	2-4	2-4	2-4	2-4	2-4
48	1-8	1-8	1-7	1-7	1-7	1-6	1-6	2-6	2-5	2-5	2-5	2-4	2-4	2-4	2-4	2-4	2-4
49	1-8	1-8	1-7	1-7	1-7	1-6	1-6	1-6	2-5	2-5	2-5	2-4	2-4	2-4	2-4	2-4	2-4
50	1-8	1-8	1-7	1-7	1-7	1-6	1-6	1-6	2-5	2-5	2-5	2-5	2-4	2-4	2-4	2-4	2-4
51	1-8	1-8	1-7	1-7	1-7	1-6	1-6	1-6	1-5	2-5	2-5	2-5	2-4	2-4	2-4	2-4	2-4
52	1-8	1-8	1-7	1-7	1-7	1-6	1-6	1-6	1-5	1-5	2-5	2-5	2-4	2-4	2-4	2-4	2-4
53	1-8	1-8	1-7	1-7	1-7	1-6	1-6	1-6	1-5	1-5	1-5	2-5	2-4	2-4	2-4	2-4	2-4
54	1-8	1-8	1-7	1-7	1-7	1-6	1-6	1-6	1-5	1-5	1-5	1-5	2-4	2-4	2-4	2-4	2-4
55	1-8	1-8	1-7	1-7	1-7	1-6	1-6	1-6	1-5	1-5	1-5	1-5	1-4	2-4	2-4	2-4	2-4

Numbers of Formed Cutters Used to Mill Teeth in Mating Bevel Gear and Pinion with Shafts at Right Angles (*Continued*)

(Number of cutter for gear given first, followed by number for pinion. See text, page 1871.)

Number of Teeth in Gear	Number of Teeth in Pinion																
	12	13	14	15	16	17	18	19	20	21	22	23	24	25	26	27	28
56	1-8	1-8	1-7	1-7	1-6	1-6	1-6	1-6	1-5	1-5	1-5	1-5	1-4	1-4	2-4	2-4	2-4
57	1-8	1-8	1-7	1-7	1-6	1-6	1-6	1-6	1-5	1-5	1-5	1-5	1-4	1-4	1-4	2-4	2-4
58	1-8	1-8	1-7	1-7	1-6	1-6	1-6	1-6	1-5	1-5	1-5	1-5	1-4	1-4	1-4	1-4	2-4
59	1-8	1-8	1-7	1-7	1-6	1-6	1-6	1-6	1-5	1-5	1-5	1-5	1-5	1-4	1-4	1-4	1-4
60	1-8	1-8	1-7	1-7	1-6	1-6	1-6	1-6	1-5	1-5	1-5	1-5	1-5	1-4	1-4	1-4	1-4
61	1-8	1-8	1-7	1-7	1-6	1-6	1-6	1-6	1-5	1-5	1-5	1-5	1-5	1-4	1-4	1-4	1-4
62	1-8	1-8	1-7	1-7	1-6	1-6	1-6	1-6	1-5	1-5	1-5	1-5	1-5	1-4	1-4	1-4	1-4
63	1-8	1-8	1-7	1-7	1-6	1-6	1-6	1-6	1-5	1-5	1-5	1-5	1-5	1-4	1-4	1-4	1-4
64	1-8	1-8	1-7	1-7	1-6	1-6	1-6	1-6	1-6	1-5	1-5	1-5	1-5	1-4	1-4	1-4	1-4
65	1-8	1-8	1-7	1-7	1-7	1-6	1-6	1-6	1-6	1-5	1-5	1-5	1-5	1-4	1-4	1-4	1-4
66	1-8	1-8	1-7	1-7	1-7	1-6	1-6	1-6	1-6	1-5	1-5	1-5	1-5	1-4	1-4	1-4	1-4
67	1-8	1-8	1-7	1-7	1-7	1-6	1-6	1-6	1-6	1-5	1-5	1-5	1-5	1-4	1-4	1-4	1-4
68	1-8	1-8	1-7	1-7	1-7	1-6	1-6	1-6	1-6	1-5	1-5	1-5	1-5	1-4	1-4	1-4	1-4
69	1-8	1-8	1-7	1-7	1-7	1-6	1-6	1-6	1-6	1-5	1-5	1-5	1-5	1-4	1-4	1-4	1-4
70	1-8	1-8	1-7	1-7	1-7	1-6	1-6	1-6	1-6	1-5	1-5	1-5	1-5	1-4	1-4	1-4	1-4
71	1-8	1-8	1-7	1-7	1-7	1-6	1-6	1-6	1-6	1-5	1-5	1-5	1-5	1-4	1-4	1-4	1-4
72	1-8	1-8	1-7	1-7	1-7	1-6	1-6	1-6	1-6	1-5	1-5	1-5	1-5	1-4	1-4	1-4	1-4
73	1-8	1-8	1-7	1-7	1-7	1-6	1-6	1-6	1-6	1-5	1-5	1-5	1-5	1-4	1-4	1-4	1-4
74	1-8	1-8	1-7	1-7	1-7	1-6	1-6	1-6	1-6	1-5	1-5	1-5	1-5	1-4	1-4	1-4	1-4
75	1-8	1-8	1-7	1-7	1-7	1-6	1-6	1-6	1-6	1-5	1-5	1-5	1-5	1-4	1-4	1-4	1-4
76	1-8	1-8	1-7	1-7	1-7	1-6	1-6	1-6	1-6	1-5	1-5	1-5	1-5	1-4	1-4	1-4	1-4
77	1-8	1-8	1-7	1-7	1-7	1-6	1-6	1-6	1-6	1-5	1-5	1-5	1-5	1-4	1-4	1-4	1-4
78	1-8	1-8	1-7	1-7	1-7	1-6	1-6	1-6	1-6	1-5	1-5	1-5	1-5	1-4	1-4	1-4	1-4
79	1-8	1-8	1-7	1-7	1-7	1-6	1-6	1-6	1-6	1-5	1-5	1-5	1-5	1-4	1-4	1-4	1-4
80	1-8	1-8	1-7	1-7	1-7	1-6	1-6	1-6	1-6	1-5	1-5	1-5	1-5	1-4	1-4	1-4	1-4
81	1-8	1-8	1-7	1-7	1-7	1-6	1-6	1-6	1-6	1-5	1-5	1-5	1-5	1-4	1-4	1-4	1-4
82	1-8	1-8	1-7	1-7	1-7	1-6	1-6	1-6	1-6	1-5	1-5	1-5	1-5	1-4	1-4	1-4	1-4
83	1-8	1-8	1-7	1-7	1-7	1-6	1-6	1-6	1-6	1-5	1-5	1-5	1-5	1-4	1-4	1-4	1-4
84	1-8	1-8	1-7	1-7	1-7	1-6	1-6	1-6	1-6	1-5	1-5	1-5	1-5	1-4	1-4	1-4	1-4
85	1-8	1-8	1-7	1-7	1-7	1-6	1-6	1-6	1-6	1-5	1-5	1-5	1-5	1-4	1-4	1-4	1-4
86	1-8	1-8	1-7	1-7	1-7	1-6	1-6	1-6	1-6	1-5	1-5	1-5	1-5	1-4	1-4	1-4	1-4
87	1-8	1-8	1-7	1-7	1-7	1-6	1-6	1-6	1-6	1-5	1-5	1-5	1-5	1-4	1-4	1-4	1-4
88	1-8	1-8	1-7	1-7	1-7	1-6	1-6	1-6	1-6	1-5	1-5	1-5	1-5	1-4	1-4	1-4	1-4
89	1-8	1-8	1-7	1-7	1-7	1-6	1-6	1-6	1-6	1-5	1-5	1-5	1-5	1-4	1-4	1-4	1-4
90	1-8	1-8	1-7	1-7	1-7	1-6	1-6	1-6	1-6	1-5	1-5	1-5	1-5	1-4	1-4	1-4	1-4
91	1-8	1-8	1-7	1-7	1-7	1-6	1-6	1-6	1-6	1-5	1-5	1-5	1-5	1-4	1-4	1-4	1-4
92	1-8	1-8	1-7	1-7	1-7	1-6	1-6	1-6	1-6	1-5	1-5	1-5	1-5	1-4	1-4	1-4	1-4
93	1-8	1-8	1-7	1-7	1-7	1-6	1-6	1-6	1-6	1-5	1-5	1-5	1-5	1-4	1-4	1-4	1-4
94	1-8	1-8	1-7	1-7	1-7	1-6	1-6	1-6	1-6	1-5	1-5	1-5	1-5	1-4	1-4	1-4	1-4
95	1-8	1-8	1-7	1-7	1-7	1-6	1-6	1-6	1-6	1-5	1-5	1-5	1-5	1-4	1-4	1-4	1-4
96	1-8	1-8	1-7	1-7	1-7	1-6	1-6	1-6	1-6	1-5	1-5	1-5	1-5	1-4	1-4	1-4	1-4
97	1-8	1-8	1-7	1-7	1-7	1-6	1-6	1-6	1-6	1-5	1-5	1-5	1-5	1-4	1-4	1-4	1-4
98	1-8	1-8	1-7	1-7	1-7	1-6	1-6	1-6	1-6	1-5	1-5	1-5	1-5	1-4	1-4	1-4	1-4
99	1-8	1-8	1-7	1-7	1-7	1-6	1-6	1-6	1-6	1-5	1-5	1-5	1-5	1-4	1-4	1-4	1-4
100	1-8	1-8	1-7	1-7	1-7	1-6	1-6	1-6	1-6	1-5	1-5	1-5	1-5	1-4	1-4	1-4	1-4

Selecting Formed Cutters for Milling Bevel Gears. — For milling 14½-degree pressure angle bevel gears, the standard cutter series furnished by manufacturers of formed milling cutters is commonly used. There are 8 cutters in the series for each diametral pitch to cover the full range from a 12-tooth pinion to a crown gear. The difference between formed cutters used for milling spur gears and those used for bevel gears is that bevel gear cutters are thinner, since they must pass through the narrow tooth space at the small end of the bevel gear; otherwise the shape of the cutter and hence, the cutter number, are the same.

To select the proper number of cutter to be used when a bevel gear is to be milled, it is necessary, first, to compute what is called the " Number of Teeth, N' for which to Select Cutter." This number of teeth can then be used to select the proper number of bevel gear cutter from the spur gear milling cutter table on page 1794. The value of N' may be computed using the last formula on page 1868.

Example 1: What numbers of cutters are required for a pair of bevel gears of 4 diametral pitch and 70 degree shaft angle if the gear has 50 teeth and the pinion 20 teeth?

The pitch cone angle of the pinion is determined by using the first formula on page 1868:

$$\tan \alpha_P = \frac{\sin \Sigma}{\dfrac{N_G}{N_P} + \cos \Sigma} = \frac{\sin 70°}{\dfrac{50}{20} + \cos 70°} = 0.33064; \quad \alpha_P = 18°18'$$

The pitch cone angle of the gear is determined from the second formula on page 1868:

$$\alpha_G = \Sigma - \alpha_P = 70° - 18°18' = 51°42'$$

The numbers of teeth N' for which to select the cutters for the gear and pinion may now be determined from the last formula on page 1868:

$$N' \text{ for the pinion} = \frac{N_P}{\cos \alpha_P} = \frac{20}{\cos 18°18'} = 21.1, \text{ say, } 21 \text{ teeth}$$

$$N' \text{ for the gear} = \frac{N_G}{\cos \alpha_G} = \frac{50}{\cos 51°42'} = 80.7, \text{ say, } 81 \text{ teeth}$$

From the table on page 1794 the numbers of the cutters for pinion and gear are found to be, respectively, 5 and 2.

Example 2: Required the cutters for a pair of bevel gears where the gear has 24 teeth and the pinion 12 teeth. The shaft angle is 90 degrees. As in the first example, the formulas given on page 1868 will be used.

$$\tan \alpha_P = N_P \div N_G = 12 \div 24 = 0.5000; \text{ and } \alpha_P = 26°34'$$

$$\alpha_G = \Sigma - \alpha_P = 90° - 26°34' = 63°26'$$

$$N' \text{ for pinion} = 12 \div \cos 26°34' = 13.4, \text{ say, } 13 \text{ teeth}$$

$$N' \text{ for gear} = 24 \div \cos 63°26' = 53.6, \text{ say, } 54 \text{ teeth}$$

And from the table on page 1794 the cutters for pinion and gear are found to be, respectively, 8 and 3.

Use of Table for Selecting Formed Cutters for Milling Bevel Gears. — The table beginning on page 1869 gives the numbers of cutters to use for milling various numbers of teeth in the gear and pinion. The table applies only to bevel gears with axes at right angles. Thus, in Example 2 given above, the numbers of the cutters could have been obtained directly by entering the table with the actual numbers of teeth in the gear, 24, and the pinion, 12.

Offset of Cutter for Milling Bevel Gears. — When milling bevel gears with a rotary formed cutter, it is necessary to take two cuts through each tooth space with the gear blank slightly off center, first on one side and then on the other, to obtain a tooth of approximately the correct form. The gear blank is also rotated proportionately to obtain the proper tooth thickness at the large and small ends. The amount that the gear blank or cutter should be offset from the central position can be determined quite accurately by the use of the table " Factors for Obtaining Offset for Milling Bevel Gears," in conjunction with the following rule: Find the factor in the table corresponding to the number of cutter used and to the ratio of the pitch cone radius to the face width; then divide this factor by the diametral pitch and subtract the result from half the thickness of the cutter at the pitch line.

Factors for Obtaining Offset for Milling Bevel Gears

No. of Cutter	Ratio of Pitch Cone Radius to Width of Face $\left(\dfrac{C}{F}\right)$												
	$\dfrac{3}{1}$	$\dfrac{3\frac{1}{4}}{1}$	$\dfrac{3\frac{1}{2}}{1}$	$\dfrac{3\frac{3}{4}}{1}$	$\dfrac{4}{1}$	$\dfrac{4\frac{1}{4}}{1}$	$\dfrac{4\frac{1}{2}}{1}$	$\dfrac{4\frac{3}{4}}{1}$	$\dfrac{5}{1}$	$\dfrac{5\frac{1}{2}}{1}$	$\dfrac{6}{1}$	$\dfrac{7}{1}$	$\dfrac{8}{1}$
1	0.254	0.254	0.255	0.256	0.257	0.257	0.257	0.258	0.258	0.259	0.260	0.262	0.264
2	0.266	0.268	0.271	0.272	0.273	0.274	0.274	0.275	0.277	0.279	0.280	0.283	0.284
3	0.266	0.268	0.271	0.273	0.275	0.278	0.280	0.282	0.283	0.286	0.287	0.290	0.292
4	0.275	0.280	0.285	0.287	0.291	0.293	0.296	0.298	0.298	0.302	0.305	0.308	0.311
5	0.280	0.285	0.290	0.293	0.295	0.296	0.298	0.300	0.302	0.307	0.309	0.313	0.315
6	0.311	0.318	0.323	0.328	0.330	0.334	0.337	0.340	0.343	0.348	0.352	0.356	0.362
7	0.289	0.298	0.308	0.316	0.324	0.329	0.334	0.338	0.343	0.350	0.360	0.370	0.376
8	0.275	0.286	0.296	0.309	0.319	0.331	0.338	0.344	0.352	0.361	0.368	0.380	0.386

Note. — For obtaining offset by above table, use formula:

$$\text{Offset} = \frac{T}{2} - \frac{\text{factor from table}}{P}$$

P = diametral pitch of gear to be cut;

T = thickness of cutter used, measured at pitch line.

To illustrate, what would be the amount of offset for a bevel gear having 24 teeth, 6 diametral pitch, 30-degree pitch cone angle and 1¼-inch face or tooth length? In order to obtain a factor from the table, the ratio of the pitch cone radius to the face width must be determined. The pitch cone radius equals the pitch diameter divided by twice the sine of the pitch cone angle = 4 ÷ (2 × 0.5) = 4 inches. As the face width is 1.25, the ratio is 4 ÷ 1.25 or about 3¼ to 1. The factor in the table for this ratio is 0.280 with a No. 4 cutter, which would be the cutter number for this particular gear. The thickness of the cutter at the pitch line is measured by using a vernier gear tooth caliper. The depth $S + A$ (see following illustration; S = addendum; A = clearance) at which to take the measurement equals 1.157 divided by the diametral·pitch; thus, 1.157 ÷ 6 = 0.1928 inch. The cutter thickness at this depth will vary with different cutters and even with the same cutter as it is ground away, because formed bevel gear cutters are commonly provided with side relief. Assuming that the thickness is 0.1745 inch, and substituting the values in the formula given, we have:

$$\text{Offset} = \frac{0.1745}{2} - \frac{0.280}{6} = 0.0406 \text{ inch}$$

Adjusting the Gear Blank for Milling. — After the offset is determined, the blank is adjusted laterally this amount, and the tooth spaces are milled around the blank. After having milled one side of each tooth to the proper dimensions, the blank is set over in the opposite direction the same amount from a position central with the cutter, and is rotated to line up the cutter with a tooth space at the small end. A trial cut is then taken, which will leave the tooth being milled a little too thick, provided the cutter is thin enough — as it should be — to pass

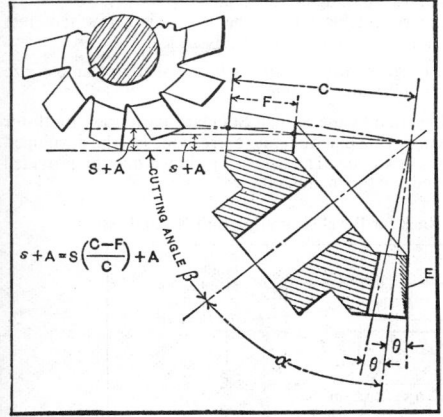

through the small end of the tooth space of the finished gear. This trial tooth is made the proper thickness by rotating the blank toward the cutter. To test the amount of offset measure the tooth thickness (with a vernier caliper) at the large and small ends. The caliper should be set so that the addendum at the small end is in proper proportion to the addendum at the large end; that is, in the ratio, $\dfrac{C-F}{C}$ (see illustration). In taking these measurements, if the thicknesses at both ends (which should be in this same ratio) are too great, rotate the tooth toward the

cutter and take trial cuts until the proper thickness at either the large or small end is obtained. If the large end of the tooth is the right thickness and the small end too thick, the blank was offset too much; inversely, if the small end is correct and the large end too thick, the blank was not set enough off center, and, in either case, its position should be changed accordingly. The formula and table previously referred to will enable a properly turned blank to be set accurate enough for general work. The dividing head should be set to the cutting angle β (see illustration), which is found by subtracting the addendum angle θ from the pitch cone angle α. After cutting a bevel gear by the method described, the sides of the teeth at the small end should be filed as indicated by the shade lines at E; that is, by filing off a triangular area from the point of the tooth at the large end to the point at the small end, thence down to the pitch line and back diagonally to a point at the large end.

Circular Thickness, Chordal Thickness, and Chordal Addendum of Milled Bevel Gear Teeth. — In the formulas that follow, T = circular tooth thickness on pitch circle at large end of tooth; t = circular thickness at small end; T_c and t_c = chordal thickness at large and small end respectively; S_c and s_c = chordal addendum at large and small end respectively; D = pitch diameter at large end; and $C, F, P, S, s,$ and α are as defined on page 1867.

$$T = \frac{1.5708}{P}; \qquad t = \frac{T(C-F)}{C}$$

$$T_c = T - \frac{T^3}{6\,D^2}; \qquad t_c = t - \frac{t^3}{6\,(D - 2\,F \sin \alpha)^2}$$

$$S_c = S + \frac{T^2 \cos \alpha}{4\,D}; \qquad s_c = s + \frac{t^2 \cos \alpha}{4\,(D - 2\,F \sin \alpha)}$$

British Standard for Bevel Gears (B.S. 545: 1949). — This British Standard for machine cut bevel gearing applies to all bevel gears connecting intersecting shafts and includes either straight or curved teeth having a normal pressure angle of 20 degrees at the pitch cone. These gears are divided into three classes:

Class A — Precision gears recommended for peripheral speeds above 2000 feet per minute when transmitting normal loads.

Class B — High-class cut gears suitable for peripheral speeds below 3000 feet per minute when transmitting normal loads.

Class C — Commercial cut gears suitable for peripheral speeds below 1200 feet per minute when transmitting normal loads.

This standard assumes that the gears will be supported on shafts of ample size provided with suitable, effectively-lubricated bearings.

Form of Tooth: The normal section of the British Standard basic rack tooth for bevel gears is shown in the accompanying diagram, and corresponds to the developed section of the crown gear on the back cone. The tip easing shown may be provided at the root of the tooth, if preferred, by the manufacturer.

British Standard Basic Rack for Bevel Gears (for Unit Normal Pitch)

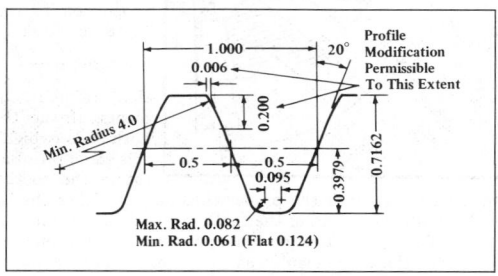

Cutting Tool: A cutting tool of form counterpart to a single flank and root fillet of the basic rack may be used to generate single flanks and root fillets of bevel gear teeth of British Standard form. In general, neither the amount of tip easing, nor the fillet radius on the bevel gear tooth will be a constant multiple of the pitch at all distances from the apex.

Face Width: The face width should not normally exceed one-third of the pitch cone distance or three times the pitch.

Tooth Spiral: The form of tooth spiral is left to the manufacturer as it is dependent on the type of cutter and method of cutting employed. For this reason no formula is given for the spiral angle at different positions along the face, although reference is made both to the spiral angle at midface and to that at the pitch circle. The following notation will be used:

A = gear addendum	p = circular pitch = $\pi \div P$
a = pinion addendum	p_n = normal pitch = $p \cos \sigma_2$
C = cone distance = $D \div 2 \sin \gamma_\omega$	T = number of gear teeth
D = gear pitch diameter	t = number of pinion teeth
d = pinion pitch diameter	σ_1 = spiral angle at midface

k_p = pinion addendum coefficient

k_w = gear addendum coefficient

P = diametral pitch

P_n = normal diametral pitch = $P \sec \sigma_2$

σ_2 = spiral angle at pitch circle

γ_p = pitch angle of pinion

γ_w = pitch angle of gear

ψ_n = normal pressure angle

Q = interference factor corresponding to various values of σ_1. The following values are taken from a chart in the Standard: $0°$, 34.2; $5°$, 34; $10°$, 33.3; $20°$, 30.5; $30°$, 26.5; $40°$, 21.7; and $50°$, 16.5.

Avoiding Combinations that Give Interference: Where the number of teeth in the two mating gears is small, that is, when $(t \sec \gamma_p + T \sec \gamma_w)$ is less than Q, interference or undercutting will be present. Use of such gears should be avoided. For straight bevel gears $(t \sec \gamma_p + T \sec \gamma_w)$ should not be less than 34.2, and for bevel gears with 30-degree spiral angle, not less than 26.5.

Addendum: The addendum shall be determined by the following formulas:

$$\text{Pinion addendum } a = \frac{p_n}{\pi}(1 + k_p) = \frac{1 + k_p}{P_n}$$

$$\text{Gear addendum } A = \frac{p_n}{\pi}(1 + k_w) = \frac{1 + k_w}{P_n}$$

where k_p and k_w, the addendum coefficients, are determined by the formulas,

$$k_p = 0.4\left(1 - \frac{t \sec \gamma_p}{T \sec \gamma_w}\right)$$

or

$$k_p = \left(1 - \frac{2t \sec \gamma_p}{Q}\right), \text{ whichever is greater.}$$

Normally,

$$k_w = -k_p.$$

Dedendum: The dedendum of the tooth is made equal to the difference between the whole depth of the tooth and the addendum as determined above.

Reduced Addendum and Dedendum: If t is less than p_n ($11.63 \div \sec^2 \sigma_2 \sec \gamma_p$), addendum and dedendum should each be reduced by p_n ($0.4 - 0.0344t \sec^2 \sigma_2 \sec \gamma_p$) unless this quantity exceeds $0.065 p_n$, in which case the gears are outside the scope of the Standard.

Caliper Settings: Caliper settings shall be obtained as follows:

$$\text{Let } S = \frac{p}{2} + 2 \tan \psi_n \left(A \sec \sigma_2 - \frac{p}{\pi}\right)$$

$$\text{and } s = p - S$$

Then,

$$\text{Caliper thickness setting for gear, } G = \frac{S}{\sec\left(\sigma_2 + \frac{S}{2C}\right)}$$

$$\text{Caliper thickness setting for pinion, } g = \frac{s}{\sec\left(\sigma_2 + \frac{s}{2C}\right)}$$

$$\text{Caliper height setting for gear, } H = A + \frac{G_2}{4D \sec^2 \sigma_2 \sec \gamma_w}$$

$$\text{Caliper height setting for pinion, } h = a + \frac{g^2}{4d \sec^2 \sigma_2 \sec \gamma_p}$$

Worm Gearing

Worm Gearing. — Worm gearing may be divided into two general classes, fine-pitch worm gearing, and coarse-pitch worm gearing. Fine-pitch worm gearing is segregated from coarse-pitch worm gearing for the following reasons:

1. Fine-pitch worms and wormgears are used largely to transmit motion rather than power. Tooth strength except at the coarser end of the fine-pitch range is seldom an important factor; durability and accuracy, as they affect the transmission of uniform angular motion, are of greater importance.

2. Housing constructions and lubricating methods are, in general, quite different for fine-pitch worm gearing.

3. Because fine-pitch worms and wormgears are so small, profile deviations and tooth bearings cannot be measured with the same accuracy as can those of coarse pitches.

4. Equipment generally available for cutting fine-pitch wormgears has restrictions which limit the diameter, the lead range, the degree of accuracy attainable, and the kind of tooth bearing obtainable.

5. Special consideration must be given to top lands in fine-pitch hardened worms and wormgear-cutting tools.

6. Interchangeability and high production are important factors in fine-pitch worm gearing; individual matching of the worm to the gear, as often practiced with coarse-pitch precision worms, is impractical in the case of fine-pitch worm drives.

American Standard Design for Fine-pitch Worm Gearing (ANSI B6.9-1977). — This standard is intended as a design procedure for fine-pitch worms and wormgears having axes at right angles. It covers cylindrical worms with helical threads, and wormgears hobbed for fully conjugate tooth surfaces. It does not cover helical gears used as wormgears.

Hobs: The hob for producing the gear is a duplicate of the mating worm with regard to tooth profile, number of threads, and lead. The hob differs from the worm principally in that the outside diameter of the hob is larger to allow for resharpening and to provide bottom clearance in the wormgear.

Pitches: Eight standard axial pitches have been established to provide adequate coverage of the pitch range normally required: 0.030, 0.040, 0.050, 0.065, 0.080, 0.100, 0.130, and 0.160 inch.

Axial pitch is used as a basis for this design standard because: (1) Axial pitch establishes lead which is a basic dimension in the production and inspection of worms; (2) the axial pitch of the worm is equal to the circular pitch of the gear in the central plane; (3) only one set of change gears or one master lead cam is required for a given lead, regardless of lead angle, on commonly-used worm-producing equipment.

Lead Angles: Fifteen standard lead angles have been established to provide adequate coverage: 0.5, 1, 1.5, 2, 3, 4, 5, 7, 9, 11, 14, 17, 21, 25, and 30 degrees.

This series of lead angles has been standardized to: (1) Minimize tooling; (2) permit obtaining geometric similarity between worms of different axial pitch by keeping the same lead angle; and (3) take into account the production distribution found in fine-pitch worm gearing applications. For example, most fine-pitch worms have either one or two threads. This requires smaller increments at the low end of the lead angle series. For the less frequently used thread numbers, proportionately greater increments at the high end of the lead angle series are sufficient.

Pressure Angle of Worm: A pressure angle of 20 degrees has been selected as standard for cutters and grinding wheels used to produce worms within the scope of this Standard because it avoids objectionable undercutting regardless of lead angle.

Although the pressure angle of the cutter or grinding wheel used to produce the worm is 20 degrees, the normal pressure angle produced in the worm will actually be slightly greater, and will vary with the worm diameter, lead angle, and diameter of cutter or grinding wheel. A method for calculating the pressure angle change is

Table 1. Formulas for Proportions of American Standard Fine-pitch Worms and Wormgears (ANSI B6.9-1977)

LETTER SYMBOLS

P = Circular pitch of wormgear
= axial pitch of the worm, P_x, in the central plane

P_x = Axial pitch of worm

P_n = Normal circular pitch of worm and wormgear = $P_x \cos \lambda = P \cos \psi$

λ = Lead angle of worm

ψ = Helix angle of wormgear

n = Number of threads in worm

N = Number of teeth in wormgear = nm_G

m_G = Ratio of gearing = $N \div n$

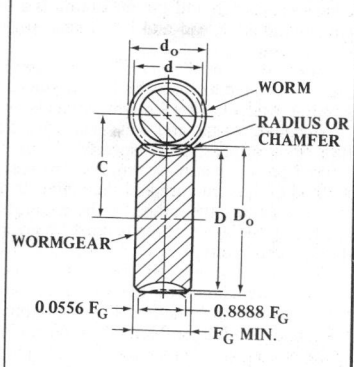

Item	Formula	Item	Formula
WORM DIMENSIONS		**WORMGEAR DIMENSIONS****	
Lead	$l = nP_x$	Pitch Diameter	$D = NP \div \pi = NP_x \div \pi$
Pitch Diameter	$d = l \div (\pi \tan \lambda)$	Outside Diameter	$D_o = 2C - d + 2a$
Outside Diameter	$d_o = d + 2a$	Face Width	$F_{Gmin} = 1.125 \times$
Safe Minimum Length of Threaded Portion of Worm*	$F_w = \sqrt{D_o^2 - D^2}$		$\sqrt{(d_o + 2c)^2 - (d_o - 4a)^2}$
DIMENSIONS FOR BOTH WORM AND WORMGEAR			
Addendum	$a = 0.3183P_n$	Tooth thickness	$t_n = 0.5P_n$
Whole Depth	$h_t = 0.7003P_n + 0.002$	Approximate normal pressure angle†	$\phi_n = 20$ degrees
Working Depth	$h_k = 0.6366P_n$		
Clearance	$c = h_t - h_k$	Center distance	$C = 0.5 (d + D)$

All dimensions in inches unless otherwise indicated.

* This formula allows a sufficient length for fine-pitch worms.

** Current practice for fine-pitch worm gearing does not require the use of throated blanks. This results in the much simpler blank shown in the diagram which is quite similar to that for a spur or helical gear. The slight loss in contact resulting from the use of non-throated blanks has little effect on the load-carrying capacity of fine-pitch worm gears.

It is sometimes desirable to use topping hobs for producing wormgears in which the size relation between the outside and pitch diameters must be closely controlled. In such cases the blank is made slightly larger than D_o by an amount (usually from 0.010 to 0.020) depending on the pitch. Topped wormgears will appear to have a small throat which is the result of the hobbing operation. For all intents and purposes, this throating is negligible and a blank so made is not to be considered as being a throated blank.

† As stated in the text on page 1876, the actual pressure angle will be slightly greater due to the manufacturing process.

given under the heading "Effect of Production Method on Worm Profile and Pressure Angle."

Pitch Diameter Range of Worms: The minimum recommended worm pitch diameter is 0.250 inch and the maximum is 2.000 inches. Pitch diameters for all possible combinations of lead and lead angle, together with the number of threads for each lead, are given in Table 2.

Tooth Form of Worm and Wormgear: The shape of the worm thread in the normal plane is defined as that which is produced by a symmetrical double-conical cutter or grinding wheel having straight elements and an included angle of 40 degrees.

Because worms and wormgears are closely related to their method of manufacture, it is impossible to specify clearly the tooth form of the wormgear without referring to the mating worm. For this reason, worm specifications should include the method of manufacture and the diameter of cutter or grinding wheel used. Similarly, for determining the shape of the generating tool, information about the method of producing the worm threads must be given to the manufacturer if the tools are to be designed correctly.

The worm profile will be a curve that departs from a straight line by varying amounts, depending on the worm diameter, lead angle, and the cutter or grinding wheel diameter. A method for calculating this deviation is given under the heading "Effect of Production Method on Worm Profile and Pressure Angle." The tooth form of the wormgear is understood to be made fully conjugate to the mating worm thread.

Proportions of Fine-pitch Worms and Wormgears. — Hardened worms and cutting tools for wormgears should have adequate top lands. To automatically provide sufficient top lands, regardless of lead angle or axial pitch, the addendum and whole depth proportions of fine-pitch worm gearing are based on the normal circular pitch. Tooth proportions based on normal pitch for all combinations of standard axial pitches and lead angles are given in Table 3. Formulas for the proportions of worms and wormgears are given in Table 1.

Example 1. Determine the design of a worm and wormgear for a center distance of approximately 3 inches if the ratio is to be 10 to 1; axial pitch, 0.1600 inch; and lead angle, 30 degrees.

From Table 2 it can be determined that there are eight possible worm diameters that will satisfy the given conditions of lead angle and pitch. These worms have from 3 to 10 threads.

To satisfy the 3-inch center distance requirement it is now necessary to determine which of these eight worms, together with its mating wormgear, will come closest to making up this center distance. One way of doing this is as follows:

First use the formula given below to obtain the approximate number of threads necessary. Then from the eight possible worms in Table 2, choose the one whose number of threads is nearest this approximate value:

Approximate number of threads needed for required center distance =

$$\frac{2\pi \times \text{required center distance}}{P_x\,(\cot \lambda + m_G)}$$

$$\text{Approximate number of threads} = \frac{2 \times 3.1416 \times 3}{0.1600\,(1.7320 + 10)} = 10.04 \text{ threads}$$

Of the eight possible worms in Table 2, the one having a number of threads nearest this value is the 10-thread worm with a pitch diameter of 0.8821 inch. Since the ratio of gearing is given as 10, N may now be computed as follows:

$$N = 10 \times 10 = 100 \text{ teeth} \qquad \text{(from Table 1)}$$

Table 2. Pitch Diameters of Fine-pitch Worms for American Standard Combinations of Lead and Lead Angle (ANSI B6.9-1977)

Lead in Inches, l	Number of Threads, n	Lead Angle λ in Degrees							
		0.5	1.0	1.5	2.0	3.0	4.0	5.0	7.0
		Pitch Diameter d in Inches							
0.030	1	1.0937	0.5472	0.3647	0.2735	...	...	...	...
0.040	1	1.4583	0.7297	0.4863	0.3646	0.2429	...	...	...
0.050	1	1.8228	0.9121	0.6079	0.4558	0.3037	0.2276	...	...
0.060	2	2.1874	1.0945	0.7295	0.5469	0.3644	0.2731	...	...
0.065	1	...	1.1857	0.7903	0.5925	0.3948	0.2959	0.2365	...
0.080	1,2	...	1.4593	0.9726	0.7293	0.4859	0.3641	0.2911	...
0.090	3	...	1.6417	1.0942	0.8204	0.5466	0.4097	0.3274	0.2333
0.100	1,2	...	1.8242	1.2158	0.9116	0.6073	0.4552	0.3638	0.2592
0.120	3,4	...	2.1890	1.4590	1.0939	0.7288	0.5462	0.4366	0.3111
0.130	1,2	...	...	1.5805	1.1851	0.7896	0.5917	0.4730	0.3370
0.150	3,5	...	...	1.8237	1.3674	0.9110	0.6828	0.5457	0.3889
0.160	1,2,4	...	...	1.9453	1.4585	0.9718	0.7283	0.5821	0.4148
0.180	6	...	...	2.1884	1.6408	1.0932	0.8193	0.6549	0.4667
0.195	3	...	...	...	1.7776	1.1843	0.8876	0.7095	0.5055
0.200	2,4,5	...	...	...	1.8232	1.2147	0.9104	0.7276	0.5185
0.210	7	...	...	...	1.9143	1.2754	0.9559	0.7640	0.5444
0.240	3,6,8	...	...	...	2.1878	1.4576	1.0924	0.8732	0.6222
0.250	5					1.5184	1.1380	0.9096	0.6481
0.260	2,4					1.5791	1.1835	0.9459	0.6741
0.270	9					1.6398	1.2290	0.9823	0.7000
0.280	7					1.7006	1.2745	1.0187	0.7259
0.300	3,6,10					1.8220	1.3656	1.0915	0.7778
0.320	2,4,8					1.9435	1.4566	1.1642	0.8296
0.325	5					1.9739	1.4794	1.1824	0.8426
0.350	7					2.1257	1.5932	1.2734	0.9074
0.360	9					...	1.6387	1.3098	0.9333
0.390	3,6					...	1.7752	1.4189	1.0111
0.400	4,5,8,10					...	1.8207	1.4553	1.0370
0.450	9					...	2.0483	1.6372	1.1666
0.455	7					...	...	1.6554	1.1796
0.480	3,6					...	...	1.7463	1.2444
0.500	5,10					...	...	1.8191	1.2963
0.520	4,8					...	...	1.8919	1.3481
0.560	7					...	...	2.0374	1.4518
0.585	9					...	...	...	1.5166
0.600	6					...	...	...	1.5555
0.640	4,8					...	...	...	1.6592
0.650	5,10					...	...	...	1.6852
0.700	7					...	...	...	1.8148
0.720	9					...	...	...	1.8666
0.780	6					...	...	...	2.0222

For each lead shown in the first column, the numbers of threads given in the second column are those corresponding to standard axial pitches.

The standard axial pitch for each pitch diameter shown in the body of the table is obtained by dividing the lead given in column 1 by the number of threads given in column 2. Thus, where more than one number of threads is given in column 2, there will be more than one standard axial pitch for the corresponding pitch diameter.

Example: For a lead angle of 2.0 degrees and a lead of 0.240 inch, the corresponding pitch diameter is given in the table as 2.1878 inches. Since there are three values of n given in column 2, namely, 3, 6, and 8, there will also be three standard axial pitches available for this pitch diameter. These pitches are: $0.240 \div 3 = 0.080$; $0.240 \div 6 = 0.040$; and $0.240 \div 8 = 0.030$.

Table 2 (*Concluded*). **Pitch Diameters of Fine-pitch Worms for American Standard Combinations of Lead and Lead Angle** (ANSI B6.9-1977)

Lead in Inches, l	Number of Threads, n	Lead Angle λ in Degrees						
		9.0	11.0	14.0	17.0	21.0	25.0	30.0
		Pitch Diameter d in Inches						
0.120	3,4	0.2412	...	...	...	...	...	...
0.130	1,2	0.2613	...	...	...	...	...	...
0.150	3,5	0.3015	0.2456	...	...	...	...	...
0.160	1,2,4	0.3216	0.2620	...	...	...	...	...
0.180	6	0.3618	0.2948	...	...	...	...	...
0.195	3	0.3919	0.3193	0.2490	...	...	...	...
0.200	2,4,5	0.4020	0.3275	0.2553	...	...	...	...
0.210	7	0.4221	0.3439	0.2681	...	...	...	...
0.240	3,6,8	0.4823	0.3930	0.3064	0.2499	...	...	...
0.250	5	0.5024	0.4094	0.3192	0.2603	...	...	...
0.260	2,4	0.5225	0.4258	0.3319	0.2707	...	...	...
0.270	9	0.5426	0.4421	0.3447	0.2811	...	...	...
0.280	7	0.5627	0.4585	0.3575	0.2915	...	...	...
0.300	3,6,10	0.6029	0.4913	0.3830	0.3123	0.2488	...	...
0.320	2,4,8	0.6431	0.5240	0.4085	0.3332	0.2654	...	...
0.325	5	0.6532	0.5322	0.4149	0.3384	0.2695	...	...
0.350	7	0.7034	0.5731	0.4468	0.3644	0.2902	...	...
0.360	9	0.7235	0.5895	0.4596	0.3748	0.2985	0.2457	...
0.390	3,6	0.7838	0.6387	0.4979	0.4060	0.3234	0.2662	...
0.400	4,5,8,10	0.8039	0.6550	0.5107	0.4165	0.3317	0.2730	...
0.450	9	0.9044	0.7369	0.5745	0.4685	0.3732	0.3072	0.2481
0.455	7	0.9144	0.7451	0.5809	0.4737	0.3773	0.3106	0.2509
0.480	3,6	0.9647	0.7860	0.6128	0.4998	0.3980	0.3277	0.2646
0.500	5,10	1.0049	0.8188	0.6383	0.5206	0.4146	0.3413	0.2757
0.520	4,8	1.0451	0.8515	0.6639	0.5414	0.4312	0.3550	0.2867
0.560	7	1.1255	0.9170	0.7149	0.5830	0.4644	0.3823	0.3087
0.585	9	1.1757	0.9580	0.7469	0.6091	0.4851	0.3993	0.3225
0.600	6	1.2059	0.9825	0.7660	0.6247	0.4975	0.4096	0.3308
0.640	4,8	1.2863	1.0480	0.8171	0.6663	0.5307	0.4369	0.3529
0.650	5,10	1.3064	1.0644	0.8298	0.6767	0.5390	0.4437	0.3584
0.700	7	1.4068	1.1463	0.8937	0.7288	0.5805	0.4778	0.3859
0.720	9	1.4470	1.1791	0.9192	0.7496	0.5971	0.4915	0.3970
0.780	6	1.5676	1.2773	0.9958	0.8121	0.6468	0.5324	0.4300
0.800	5,8,10	1.6078	1.3101	1.0213	0.8329	0.6634	0.5461	0.4411
0.900	9	1.8088	1.4738	1.1490	0.9370	0.7463	0.6144	0.4962
0.910	7	1.8289	1.4902	1.1618	0.9474	0.7546	0.6212	0.5017
0.960	6	1.9294	1.5721	1.2256	0.9995	0.7961	0.6553	0.5293
1.000	10	2.0098	1.6376	1.2767	1.0412	0.8292	0.6826	0.5513
1.040	8	...	1.7031	1.3277	1.0828	0.8624	0.7099	0.5734
1.120	7	...	1.8341	1.4299	1.1661	0.9287	0.7645	0.6175
1.170	9	...	1.9160	1.4937	1.2181	0.9720	0.7987	0.6451
1.280	8	...	2.0961	1.6341	1.3327	1.0614	0.8738	0.7057
1.300	10	...	...	1.6597	1.3535	1.0780	0.8874	0.7167
1.440	9	...	...	1.8384	1.4993	1.1941	0.9830	0.7939
1.600	10	...	...	2.0427	1.6658	1.3268	1.0922	0.8821

Other worm and wormgear dimensions may now be calculated using the formulas given in Table 1 or may be taken from the data presented in Tables 2 and 3.

$l = 1.600$ inches	(from Table 2)
$d = 0.8821$ inch	(from Table 2)
$D = 100 \times 0.1600 \div 3.1416 = 5.0930$ inches	(from Table 1)
$C = 0.5(0.8821 + 5.0930) = 2.9876$ inches	(from Table 1)
$P_n = 0.1386$ inch	(from Table 3)
$a = 0.0441$ inch	(from Table 3)
$h_t = 0.0990$ inch	(from Table 3)
$h_k = 0.6366 \times 0.1386 = 0.0882$ inch	(from Table 1)
$c = 0.0990 - 0.0882 = 0.0108$ inch	(from Table 1)
$t_n = 0.5 \times 0.1386 = 0.0693$ inch	(from Table 1)
$d_o = 0.8821 + (2 \times 0.0441) = 0.9703$ inch	(from Table 1)
$D_o = (2 \times 2.9876) - 0.8821 + (2 \times 0.0441) = 5.1813$	(from Table 1)

$$F_G = 1.125\sqrt{(0.9703 + 2 \times 0.0108)^2 \; - \; (0.9703 - 4 \times 0.0441)^2} = 0.6689 \text{ inch}$$
(from Table 1)

$$F_W = \sqrt{5.1813^2 - 5.0930^2} = 0.9525 \text{ inch}$$
(from Table 1)

Example 2: Determine the design of a worm and wormgear for a center distance of approximately 0.550 inch if the ratio is to be 50 to 1 and the axial pitch is to be 0.050 inch.

Assume that $n = 1$ (since most fine-pitch worms have either one or two threads). The lead of the worm will then be $nP_x = 1 \times 0.050 = 0.050$ inch. From Table 2 it can be determined that there are six possible lead angles and corresponding worm diameters that will satisfy this lead. The approximate lead angle required to meet the conditions of the example can be computed from the following formula:

$$\text{Cotangent of approx. lead angle} = \frac{2\pi \times \text{approximate center distance required}}{\text{assumed number of threads} \times \text{axial pitch}} - m_G$$

Using letter symbols, this formula becomes:

$$\cot \lambda = \frac{2\pi \times C}{n \times P_x} - m_G = \frac{2 \times 3.1416 \times 0.550}{1 \times 0.050} - 50 = 19.1152$$

or

$$\lambda = 2° \ 59'$$

Of the six possible worms in Table 2, the one with the 3-degree lead angle is closest to the calculated 2° 59' lead angle. This worm, which has a pitch diameter of 0.3037 inch, is therefore selected.

The remaining worm and wormgear dimensions may now be determined from the data in Tables 2 and 3 and by computation using the formulas given in Table 1.

$N = 50 \times 1 = 50$ teeth	(from Table 1)
$d = 0.3037$ inch	(from Table 2)
$D = 50 \times 0.050 \div 3.1416 = 0.7958$ inch	(from Table 1)
$C = 0.5(0.3037 + 0.7958) = 0.5498$ inch	(from Table 1)
$P_n = 0.0499$ inch	(from Table 3)
$a = 0.0159$ inch	(from Table 3)
$h_t = 0.0370$ inch	(from Table 3)
$h_k = 0.6366 \times 0.0499 = 0.0318$ inch	(from Table 1)
$c = 0.0370 - 0.0318 = 0.0052$ inch	(from Table 1)

Table 3. Tooth Proportions of American Standard Fine-pitch Worms and Wormgears (ANSI B6.9-1977)

Standard Axial Pitch in Inches, P_x	Tooth Parts*	Lead Angle λ in Degrees														
		0.5	1	1.5	2	3	4	5	7	9	11	14	17	21	25	30
		Dimensions of Tooth Parts in Inches†														
0.030	a	.0095	.0095	.0095	.0095	.0095	.0095	.0095	.0095	.0094	.0094	.0093	.0091	.0089	…	…
	h_t	.0229	.0229	.0229	.0229	.0229	.0229	.0229	.0229	.0227	.0227	.0225	.0220	.0216	…	…
	P_n	.0300	.0300	.0300	.0300	.0300	.0299	.0299	.0298	.0296	.0294	.0291	.0287	.0280	…	…
0.040	a	.0127	.0127	.0127	.0127	.0127	.0127	.0127	.0126	.0126	.0125	.0124	.0122	.0119	.0115	…
	h_t	.0299	.0299	.0299	.0400	.0399	.0399	.0399	.0397	.0397	.0395	.0293	.0288	.0282	.0273	…
	P_n	.0400	.0400	.0400	.0400	.0399	.0399	.0398	.0397	.0395	.0393	.0388	.0383	.0373	.0363	…
0.050	a	.0159	.0159	.0159	.0159	.0159	.0159	.0159	.0158	.0157	.0156	.0154	.0152	.0149	.0144	.0138
	h_t	.0370	.0370	.0370	.0370	.0370	.0370	.0370	.0368	.0365	.0363	.0359	.0354	.0348	.0337	.0324
	P_n	.0500	.0500	.0500	.0500	.0499	.0499	.0498	.0496	.0494	.0491	.0485	.0478	.0467	.0453	.0433
0.065	a	.0207	.0207	.0207	.0207	.0207	.0206	.0206	.0205	.0204	.0203	.0201	.0198	.0193	.0188	.0179
	h_t	.0475	.0475	.0475	.0475	.0475	.0473	.0473	.0471	.0469	.0467	.0462	.0456	.0445	.0434	.0414
	P_n	.0650	.0650	.0650	.0650	.0649	.0648	.0648	.0645	.0642	.0638	.0631	.0622	.0607	.0589	.0563
0.080	a	…	.0255	.0255	.0254	.0254	.0254	.0254	.0253	.0252	.0250	.0247	.0244	.0238	.0231	.0221
	h_t	…	.0581	.0581	.0579	.0579	.0579	.0579	.0577	.0574	.0570	.0563	.0557	.0544	.0528	.0506
	P_n	…	.0800	.0800	.0800	.0799	.0798	.0797	.0794	.0790	.0785	.0776	.0765	.0747	.0725	.0693
0.100	a	…	.0318	.0318	.0318	.0318	.0318	.0317	.0316	.0314	.0312	.0309	.0304	.0297	.0288	.0276
	h_t	…	.0720	.0720	.0720	.0720	.0720	.0717	.0716	.0711	.0706	.0700	.0689	.0673	.0654	.0627
	P_n	…	.1000	.1000	.0999	.0999	.0998	.0996	.0993	.0988	.0982	.0970	.0956	.0934	.0906	.0866
0.130	a	…	…	.0414	.0414	.0413	.0413	.0412	.0411	.0409	.0406	.0402	.0396	.0386	.0375	.0358
	h_t	…	…	.0931	.0931	.0929	.0929	.0926	.0924	.0920	.0913	.0904	.0891	.0869	.0845	.0808
	P_n	…	…	.1300	.1299	.1298	.1297	.1295	.1290	.1284	.1276	.1261	.1243	.1214	.1178	.1126
0.160	a	…	…	.0509	.0509	.0509	.0508	.0507	.0506	.0503	.0500	.0494	.0487	.0475	.0462	.0441
	h_t	…	…	.1140	.1140	.1140	.1138	.1135	.1133	.1127	.1120	.1107	.1091	.1065	.1036	.0990
	P_n	…	…	.1599	.1599	.1598	.1596	.1594	.1588	.1580	.1571	.1552	.1530	.1494	.1450	.1386

* a = addendum; h_t = whole depth; and P_n = normal circular pitch.
† Tooth proportions are based on the formulas given in Table 1.

$t_n = 0.5 \times 0.0499 = 0.0250$ inch   (from Table 1)
$d_o = 0.3037 + (2 \times 0.0159) = 0.3355$ inch   (from Table 1)
$D_o = (2 \times 0.5498) - 0.3037 + (2 \times 0.0159) = 0.8277$ inch   (from Table 1)
$F_{Gmin} = 1.125\sqrt{(0.3355 + 2 \times 0.0052)^2 - (0.3355 - 4 \times 0.0159)^2}$
$= 0.2405$ inch   (from Table 1)
$F_W = \sqrt{0.8277^2 - 0.7958^2} = 0.2276$ inch   (from Table 1)

Effect of Production Method on Worm Profile and Pressure Angle. — In worm gearing, tooth bearing is usually used as the means of judging tooth profile accuracy since direct profile measurements on fine-pitch worms or wormgears is not practical. According to AGMA 370.01, Design Manual for Fine-Pitch Gearing, a minimum of 50 per cent initial area of contact is suitable for most fine-pitch worm gearing, although in some cases, such as when the load fluctuates widely, a more restricted initial area of contact may be desirable.

Except where single-pointed lathe tools, end mills, or cutters of special shape are used in the manufacture of worms, the pressure angle and profile produced by the cutter are different from those of the cutter itself. The amounts of these differences depend on several factors, namely, diameter and lead angle of the worm, thickness and depth of the worm thread, and diameter of the cutter or grinding wheel. The accompanying diagram shows the curvature and pressure angle effects produced in the worm by cutters and grinding wheels, and how the amount of variation in worm profile and pressure angle is influenced by the diameter of the cutting tool used.

Effect of Diameter of Cutting Tool on Profile and Pressure Angle of Worms

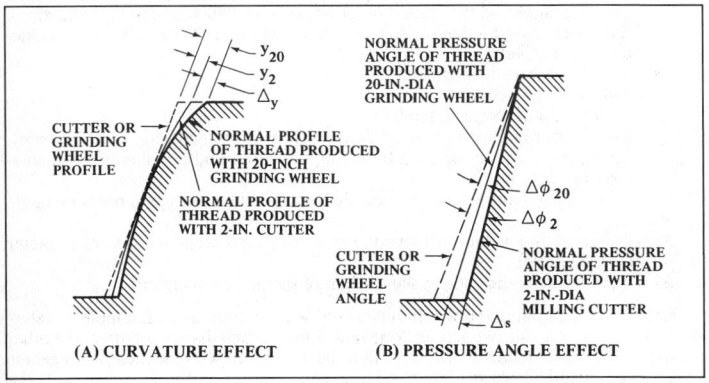

(A) CURVATURE EFFECT   (B) PRESSURE ANGLE EFFECT

Calculating Worm Deviations and Pressure Angle Changes: Included in American Standard ANSI B6.9-1977 is an extensive tabulation of profile deviations and pressure angle changes produced by cutters and grinding wheels of 2-inch and 20-inch diameters. These diameters represent the limits of the range commonly used, and the data given are useful in specifying worm profile tolerances. The data also aid in the selection of the method to be used in producing the worm and in specifying the hobs for wormgears. The formulas used to compute the data in the Standard are given here in slightly modified form, and may be used to calculate the profile

deviations and pressure angle changes produced in the worm by cutters or grinding wheels.

$$\rho_{ni} = \frac{r \sin \phi_n}{\sin^2 \lambda} \text{ inches} \tag{1}$$

$$\rho_n = \rho_{ni} + \frac{r\rho_{ni}}{R \cos^2 \lambda} \text{ inches} \tag{2}$$

$$\Delta\phi = \frac{5400r \sin^3 \lambda}{n(R \cos^2 \lambda + r)} \text{ minutes} \tag{3}$$

$$q = a \sec \phi_n \text{ inches} \tag{4}$$

$$y = \frac{q^2}{2\rho_n} \text{ inches} \tag{5}$$

$$s = 0.000582 \, q\Delta\phi \text{ inches} \tag{6}$$

$$\Delta y = y_w - y_c \text{ inches} \tag{7}$$

$$\Delta s = s_c - s_w \text{ inches} \tag{8}$$

In these formulas,

ρ_{ni} = radius of curvature of normal thread profile for involute thread;

r = pitch radius of worm;

ϕ_n = normal pressure angle of cutter or grinding wheel;

λ = lead angle of worm;

ρ_n = radius of curvature of normal thread profile;

R = radius of cutter or grinding wheel;

$\Delta\phi$ = difference between the normal pressure angle of the thread and the normal pressure angle of the cutter or grinding wheel in minutes (see diagram). Subscripts c and w are used to denote the cutter and grinding wheel diameters, respectively;

n = number of threads in worm;

a = addendum of worm;

q = slant height of worm addendum;

y = amount normal worm profile departs from a straight side (see diagram). Subscripts c and w are used to denote the cutter and grinding wheel diameters, respectively;

s = effect along slant height of worm thread caused by change in pressure angle $\Delta\phi$;

Δy = difference in y values of two cutters or grinding wheels of different diameter (see diagram);

Δs = effect of $\Delta\phi_c - \Delta\phi_w$ along slant height of thread (see diagram).

Example 3: Assuming the worm dimensions are the same as in Example 1, determine the corrections for two worms, one milled by a 2-inch diameter cutter, the other ground by a 20-inch diameter wheel, both to be assembled with identical wormgears.

To make identical worms when using a 2-inch cutter and a 20-inch wheel, the pressure angle of either the cutter or the wheel must be corrected by an amount corresponding to Δs and the profile of the cutter or wheel must be a curve which departs from a straight line by an amount Δy. The calculations are as follows:

For the 2-inch diameter cutter, using Formulas (1) to (6),

$$\rho_{ni} = \frac{0.4410 \times 0.3420}{0.5000^2} = 0.6033 \text{ inch} \tag{1}$$

$$\rho_n = 0.6033 + \frac{0.4410 \times 0.6033}{1 \times 0.8660^2} = 0.9581 \text{ inch} \tag{2}$$

$$\Delta\phi_c = \frac{5400 \times 0.4410 \times 0.5000^3}{10(1 \times 0.8660^2 + 0.4410)} = 24.99 \text{ minutes} \tag{3}$$

$$q = 0.0441 \times 1.0642 = 0.0469 \text{ inch} \tag{4}$$

$$y_c = \frac{0.0469^2}{2 \times 0.9581} = 0.00115 \text{ inch} \tag{5}$$

$$s_c = 0.000582 \times 0.0469 \times 24.99 = 0.000682 \text{ inch} \tag{6}$$

For the 20-inch diameter wheel, using Formulas (1) to (6)

$$\rho_{ni} = \frac{0.4410 \times 0.3420}{0.5000^2} = 0.6033 \text{ inch} \tag{1}$$

$$\rho_n = 0.6033 + \frac{0.4410 \times 0.6033}{10 \times 0.8660^2} = 0.6387 \text{ inch} \tag{2}$$

$$\Delta\phi_w = \frac{5400 \times 0.4410 \times 0.5000^3}{10(10 \times 0.8660^2 + 0.4410)} = 3.749 \text{ minutes} \tag{3}$$

$$q = 0.0441 \times 1.0642 = 0.0469 \text{ inch} \tag{4}$$

$$y_w = \frac{0.0469^2}{2 \times 0.6387} = 0.00172 \text{ inch} \tag{5}$$

$$s_w = 0.000582 \times 0.0469 \times 3.749 = 0.000102 \text{ inch} \tag{6}$$

Applying formulas (7) and (8):

$$\Delta y = 0.00172 - 0.00115 = 0.00057 \text{ inch} \tag{7}$$

$$\Delta s = 0.000682 - 0.000102 = 0.000580 \text{ inch} \tag{8}$$

Therefore the pressure angle of either the cutter or the wheel must be corrected by an amount corresponding to a Δs of 0.000580 inch and the profile of the cutter or wheel must be a curve which departs from a straight line by 0.00057 inch.

Industrial Worm Gearing. — The primary considerations in industrial worm gearing are usually: (1) To transmit power efficiently; (2) to transmit power at a considerable reduction in velocity; and (3) to provide a considerable "mechanical advantage" when a given applied force must overcome a comparatively high resisting force. Worm gearing for use in such applications is usually of relatively coarse pitch. The notation below is used in the formulas on the following pages.

a = addendum, worm thread
A = addendum, wormgear tooth
B = dedendum, wormgear tooth
b = dedendum, worm thread
C = center distance (Fig. 1, p. 1889)
c = clearance
D = pitch diameter of wormgear
d = pitch diameter of worm
d_o = outside diameter of worm
D_o = outside or over-all diameter of wormgear
D_t = throat diameter of wormgear
E = efficiency of worm gearing, per cent
F = nominal face width of wormgear rim
F_e = effective face width (Fig. 1, p. 1889)
f = coefficient of friction
G = length of worm threaded section
H = horsepower rating

L = lead of worm thread = pitch × number of threads or "starts"
L_a = lead angle of worm = helix angle measured from a plane perpendicular to worm axis
M = torque applied to wormgear, pound inches
m = module = 0.3183 × axial pitch
N = revolutions per minute of wormgear
n = revolutions per minute of worm
P = axial pitch of worm and circular pitch of wormgear
P_n = normal pitch of worm
Q = arc length of wormgear tooth measured along root
R = ratio of worm gearing = No. of wormgear teeth ÷ No. of worm threads.

S_c = surface stress factor (Table 4)
S_b = bending stress factors, lbs. per sq. in. (Table 4)
T = number of teeth on wormgear
t = number of threads or "starts" on worm — 2 for double thread, 3 for triple thread, 4 for quadruple thread, etc.
U = radius of wormgear throat (Fig. 1)

V = rubbing speed of worm in feet per minute
W = whole tooth depth (worm and wormgear)
X_{cp} and X_{cw} = speed factor when load rating is limited by *wear* (Fig. 2)
X_{bp} and X_{bw} = speed factor when load rating is limited by *strength* (Table 5)
ϕ = angle of friction (tan ϕ = coefficient of friction)

Materials for Worm Gearing — Worm gearing, especially for power transmission, should have steel worms and phosphor bronze wormgears. This combination is used extensively. The worms should be hardened and ground to obtain accuracy and a smooth finish.

The phosphor bronze wormgears should contain from 10 to 12 per cent of tin. The S.A.E. phosphor gear bronze (No. 65) contains 88–90% copper, 10–12% tin, 0.50% lead, 0.50% zinc (but with a maximum total lead, zinc and nickel content of 1.0 per cent), phosphorous 0.10–0.30%, aluminum 0.005%. The S.A.E. nickel phosphor gear bronze (No. 65 + Ni) contains 87% copper, 11% tin, 2% nickel and 0.2% phosphorous. Table 4 shows the British Standard wormgear and worm materials.

Single-thread Worms. — The ratio of the worm speed to the wormgear speed may range from 1.5 or even less up to 100 or more. Worm gearing having high ratios are not very efficient as transmitters of power; nevertheless high as well as low ratios often are required. Since the ratio equals the number of wormgear teeth divided by the number of threads or "starts" on the worm, single-thread worms are used to obtain a high ratio. As a general rule, a ratio of 50 is about the maximum recommended for a single worm and wormgear combination, although ratios up to 100 or higher are possible. When a high ratio is required, it may be preferable to use, in combination, two sets of worm gearing of the multi-thread type in preference to one set of the single-thread type in order to obtain the same total reduction and a higher combined efficiency.

Single-thread worms are comparatively inefficient because of the effect of the low lead angle (see Efficiency of Worm Gearing); consequently, single-thread worms are not used when the primary purpose is to transmit power as efficiently as possible but they may be employed either when a large speed reduction with one set of gearing is necessary, or possibly as a means of adjustment, especially if "mechanical advantage" or self-locking are important factors.

Multi-thread Worms. — When worm gearing is designed primarily for transmitting power efficiently, the lead angle of the worm should be as high as is consistent with other requirements and preferably between, say, 25 or 30 and 45 degrees, as indicated by Tables 1 and 2. This means that the worm must be multi-threaded. To obtain a given ratio, some number of wormgear teeth divided by some number of worm theads must equal the ratio. Thus, if the ratio is 6, combinations such as the following might be used:

$$\frac{24}{4}, \frac{30}{5}, \frac{36}{6}, \frac{42}{7}$$

The numerators represent the number of wormgear teeth and the denominators, the number of worm threads or "starts." The number of wormgear teeth may

Rules and Formulas for Worm Gearing — 1

No.	To Find	Rule	Formula
1	Addendum	Addendum may be affected by lead angle. See paragraph, Addendum and Dedendum.	
2	Center Distance	Add pitch diameter of wormgear to pitch diameter of worm, and divide sum by 2.	$C = \dfrac{(D+d)}{2}$
3		Divide number of worm threads by tangent lead angle, add number of wormgear teeth and multiply sum by quotient obtained by dividing pitch by 6.2832.	$C = \dfrac{P}{6.2832}\left(\dfrac{t}{\tan L_a} + T\right)$
4	Dedendum	Dedendum may be affected by lead angle. See paragraph, Addendum and Dedendum.	
5	Clearance	British Standard — multiply cosine lead angle by 0.2 times module.	$c = 0.2\ m \cos L_a$
6	Face Width, Wormgear	For single and double thread worms, multiply pitch by 2.38 and add 0.25. (Shell type worm.)	$F = 2.38P + 0.25$
7		For triple and quadruple thread worms, multiply pitch by 2.15 and add 0.2. (Shell type.)	$F = 2.15P + 0.2$
8		When worm threads are integral with shaft, face width of wormgear may equal $C^{0.875}$ divided by 3.	$F = \dfrac{C^{0.875}}{3}$
9	Lead of Worm Thread	Multiply pitch by number of worm threads or "starts."	$L = tP$
10		Multiply pitch circumference of worm by tangent of lead angle.	$L = \pi d \times \tan L_a$
11		Divide pitch circumference of wormgear by ratio.	$L = \pi D \div R$
12	Lead Angle, Worm	Divide lead by pitch circumference of worm; quotient is tangent of lead angle.	$\tan L_a = \dfrac{L}{3.1416d}$
13	Outside Diam., Worm	Add to pitch diameter twice the addendum. See paragraph, Pitch Diameter of Worm; also Addendum and Dedendum.	$d_o = d + 2a$
14	Outside Diam., Wormgear	For outside or over-all diameter of wormgear, see paragraph, Outside Diameter of Wormgear.	
15	Pitch of Worm and Wormgear	Divide lead by number of threads or "starts" on worm = axial pitch of worm and circular pitch of wormgear.	$P = \dfrac{L}{t}$
16		Subtract the worm pitch diameter from twice the center distance. Multiply by 3.1416 and divide by number of wormgear teeth.	$P = \dfrac{(2C - d) \times 3.1416}{T}$

Rules and Formulas for Worm Gearing — 2

No.	To Find	Rule	Formula
17	Pitch of Worm, Normal	Multiply axial pitch by cosine of lead angle to find normal pitch.	$P_n = P \times \cos L_a$
18		Subtract pitch diameter of worm-gear from twice the center distance.	$d = 2C - D$
19	Pitch Diam., Worm	Subtract twice the addendum from outside diameter. See Addendum and Dedendum.	$d = d_o - 2a$
20		Multiply lead by cotangent lead angle and divide product by 3.1416.	$d = \dfrac{L \times \cot L_a}{3.1416}$
21	Pitch Diam., Wormgear	Subtract pitch diameter of worm from twice the center distance.	$D = 2C - d$
22		Multiply number of wormgear teeth by axial pitch of worm and divide product by 3.1416.	$D = \dfrac{TP}{3.1416}$
23	Radius of Rim Corner, Wormgear	Multiply pitch by 0.25	$\text{Rad.} = 0.25P$
24		British Standard: Radius = 0.5 × module.	$\text{Rad.} = 0.5m$
25	Ratio	Divide number of wormgear teeth by number of worm threads.	$R = T \div t$
26	Rubbing Speed, Ft. per Minute	Divide wormgear pitch diameter by ratio; square quotient and add to square of worm pitch diameter; multiply square root of this sum by 0.262 × R.P.M. of worm.	$V = 0.262n\sqrt{d^2 + \left(\dfrac{D}{R}\right)^2}$
27		Multiply 0.262 × pitch diameter of worm by R.P.M. of worm; then multiply product by secant of lead angle.	$V = 0.262dn \times \sec L_a$
28	Throat Diam., Wormgear	Add twice the addendum to pitch diameter — see paragraph, Addendum and Dedendum.	$D_t = D + 2A$
29	Throat Radius, Wormgear	Subtract twice worm addendum from outside radius of worm.	$U = \dfrac{d_o}{2} - 2a$
30	Tooth Depth	Whole depth equals addendum + dedendum. See paragraph, Addendum and Dedendum.	$W = a + b$ or $A + B$
31	Worm Thread Length	Multiply the number of wormgear teeth by 0.02, add 4.5 and multiply sum by pitch.	$G = P(4.5 + 0.02T)$
32		British Standard — Subtract square of wormgear pitch diameter from square of outside diameter and extract square root of remainder.	$G = \sqrt{D^2_o - D^2}$

not be an exact multiple of the number of threads on a multi-thread worm (as explained later) in order to obtain a "hunting tooth" action.

Number of Threads or "Starts" on Worm: The number of threads on the worm ordinarily varies from one to six or eight, depending upon the ratio of the gearing. As the ratio is increased, the number of worm threads is reduced, as a general rule. In some cases, however, the higher of two ratios may also have a larger number of threads. For example, a ratio of 6⅕ would have 5 threads whereas a ratio of 6⅚ would have 6 threads. Whenever the ratio is fractional, the number of threads on the worm equals the denominator of the fractional part of the ratio.

Ratio for Obtaining "Hunting Tooth" Action. — In designing wormgears having multi-thread worms, it is common practice to select a number of wormgear teeth that is not an exact multiple of the number of worm threads. To illustrate, if the desired ratio is about 5 or 6, the actual ratio might be 5⅙, 5⅚, 5²⁄₇, 6⅕, etc., so that combinations such as ³¹⁄₆, ³⁵⁄₆, ³⁷⁄₇ or ³¹⁄₅ would be obtained. Since the number of wormgear teeth and number of worm threads do not have a common divisor, the threads of the worm will mesh with all of the wormgear teeth in succession, thus obtaining a "hunting tooth" or self-indexing action. This progressive change will also occur during the wormgear hobbing operation, and its primary purpose is to produce more accurate wormgears by uniformly distributing among all of the teeth, any slight errors which might exist in the hob teeth. Another object is to improve the "running-in" action between the hardened and ground worm and the phosphor bronze wormgear, but in order to obtain this advantage, the threads on the worm must be accurately or uniformly spaced by precise indexing. With a "hunting tooth ratio," if the thread spacing of a multi-thread worm is inaccurate, load distribution on the threads will be unequal and some threads might not even make contact with the wormgear teeth. For this reason, if the indexing is inaccurate, it is preferable to avoid a hunting tooth ratio, but in that case, if the gearing is disassembled, the same worm and wormgear teeth should be mated when reassembled.

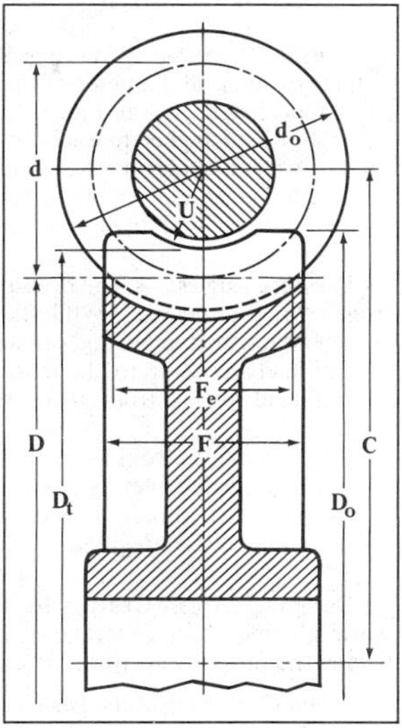

Fig. 1

Pitch Diameter of Worm. — The worm must be strong enough to transmit its maximum load without excessive deflection but the diameter should be as small as is consistent with the necessary strength in order to minimize the rubbing speed. It is impracticable to give a rule or formula that is generally applicable, but the following empirical rules are based upon actual practice and may prove useful as a general guide. They apply to casehardened alloy steel worms which are integral with the shaft.

For ratios of 5, 6, or 7, the pitch diameter ranges approximately from 0.38 *C* when center distance *C* is 4 inches to 0.33 *C* when *C* is 20 inches.

For ratios of 8, 9, 10, the pitch diameter ranges approximately from 0.38 *C* when center distance *C* is 4 inches to 0.25 *C* when *C* is 30 inches.

For ratios of 10 to 20, the pitch diameter ranges approximately from 0.37 C when center distance C is 4 inches to 0.24 C when C is 30 inches.

For ratios of 20 to 40, the pitch diameter ranges approximately from 0.36 C when center distance C is 4 inches to 0.23 C when C is 30 inches.

According to another empirical formula pitch diameter $d = C^{0.875} \div 2.2$.

Addendum and Dedendum. — The following A.G.M.A. formulas are applicable to industrial worm gearing. For single- and double-thread worms, addendum $a = 0.318\ P$ and whole depth $W = 0.686\ P$; for triple and quadruple threads, addendum $a = 0.286\ P$ and whole depth $W = 0.623\ P$.

According to the British standard, a = module $m = 0.3183\ P$; $b = m(2.2 \cos L_a - 1)$; $A = m(2 \cos L_a - 1)$; $B = m(1 + 0.2 \cos L_a)$.

Outside Diameter of Wormgear. — Practice varies somewhat in determining the outside or over-all diameter of the wormgear, as indicated by the following formulas. For usual rim shape, see Fig. 1.

  1. For lead angles up to about 15 or 20 degrees, $D_o = D + (3 \times 0.3183\ P)$
  2. For lead angles over 20 degrees, $D_o = D + (3 \times 0.3183\ P \times \cos L_a)$
  3. For single and double thread, $D_o = D_t + 0.4775\ P$
  4. For triple and quadruple thread, $D_o = D_t + 0.3183\ P$

Pressure Angles. — The pressure angle (one-half the included thread angle) ranges from $14\frac{1}{2}$ to 30 degrees. While the practice varies somewhat, the following relationship between lead angle and pressure angle may be used as a general guide.

For lead angles up to about 10 or 12 degrees, pressure angle = $14\frac{1}{2}$ degrees.

For lead angles from 10 or 12 to about 20 or 25 degrees, pressure angle = 20 degrees.

For lead angles from 25 to about 35 degrees, pressure angle = 25 degrees.

For lead angles over 35 degrees, pressure angle = 30 degrees.

In the British Standard specifications, the recommended thread form has a normal pressure angle of 20 degrees.

Designing Worm Gearing Relative to Center Distance and Ratio. — In designing worm gearing, three general cases or types of problems may be encountered in establishing the proportions of the worm and wormgear.

When Center Distance is Fixed and Ratio may be Varied: The ratio in this case is nominal and may be varied somewhat to meet other conditions. Assume that the required center distance is 6 inches, the desired ratio is about 7, and the pitch of the worm and wormgears is to be approximately 1 inch. Combinations of wormgears and worms such as the following might be used in this case:

$$\frac{28}{4}, \ \frac{35}{5}, \ \frac{42}{6}, \ \frac{56}{8}, \ \text{etc.}$$

Suppose we select the $\frac{28}{4}$ combination for trial but change the number of worm-gear teeth from 28 to 29 to obtain a self-indexing or "hunting tooth" action. The ratio now equals $\frac{29}{4}$ or 7.25. Then, for trial purposes

Pitch diameter D of wormgear $= \dfrac{T \times P}{\pi} = \dfrac{29 \times 1}{3.1416} = 9.231$ inches

Pitch diameter d of worm $= 2C - D = 2 \times 6 - 9.231 = 2.769$ inches

Assume that experience, tests, or calculations show that a worm of smaller diameter will have the necessary bending and torsional strength and that a pitch of 1.0625 will be satisfactory. Then the pitch diameter of the worm will be decreased to 2.192 inches and the pitch diameter of the wormgear will be increased to 9.808 inches. A check of the lead-angle will show that it equals 31°41′ which is conducive to high efficiency.

When Ratio is Fixed and Center Distance may be Varied: Assume that the required ratio is 7¼ and that the center distance may be any value consistent with approved designing practice. This ratio may be obtained with a number of different worm and wormgear sizes. For example, in a series of commercial wormgears, the following combinations are employed for gearing having a ratio of 7¼ with center distances varying from 4 to 8.25 inches. The number of worm threads is 4 and the number of teeth on the wormgear is 29 in all cases.

When $C = 4$ inches, $d = 1.654$; $D = 6.346$; $P = 0.6875$; $L_a = 27°54′$
When $C = 5$ inches, $d = 1.923$; $D = 8.077$; $P = 0.875$; $L_a = 30°5′$
When $C = 6$ inches, $d = 2.192$; $D = 9.808$; $P = 1.0625$; $L_a = 31°41′$
When $C = 7$ inches, $d = 2.461$; $D = 11.539$; $P = 1.25$; $L_a = 32°53′$
When $C = 8.25$ inches, $d = 2.942$; $D = 13.558$; $P = 1.4687$, $L_a = 32°27′$

The horsepower rating increases considerably as the proportions of the worm gearing increase; hence if the gears are intended primarily for power transmission, the general proportions must be selected with reference to the power-transmitting capacity, and, usually the smallest and most compact design that will give satisfactory performance should be selected. The power capacity of the transmission, however, does not depend solely upon the proportions of the worm and wormgear. For example, the quality and viscosity of the lubricant is an important factor. The load transmitting capacity of the lubricant may also be increased decidedly when excessive temperature rises are prevented by special means such as forced air cooling. (See "Water and Forced-Air Cooling.")

When Both Ratio and Center Distance are Fixed: When both ratio and center distance are fixed, the problem usually is to obtain the *best proportions* of worm and wormgear conforming to these fixed values.

Example: The required ratio is 6 (6 to 1) and the center distance is fixed at 3.600 inches. Assume that experience or tests show that an axial pitch of 0.50 inch will meet strength requirements. If normal pitch P_n is given, change to axial pitch ($P = P_n \div \cos L_a$). With a ratio of 6, some of the combinations for trial are:

$$\frac{30}{5}; \frac{36}{6}; \frac{42}{7}.$$

Trial calculations will show that the 36/6 combination gives the best proportions of worm and wheel for the center distance and pitch specified. Thus

$$D = \frac{TP}{\pi} = \frac{36 \times 0.5}{3.1416} = 5.729; \quad d = 2C - D = 2 \times 3.6 - 5.729 = 1.471$$

The lead angle is about 33 degrees. The effect of lead angle on efficiency is dealt with in a following paragraph. The total obtained by adding the number of wormgear teeth to the number of worm threads, equals $36 + 6 = 42$ (a total of 40 is a desirable minimum). With the 42/7 combination of the same pitch, the worm would be too small (0.516 inch); and with the 30/5 combination it would be too large (2.426 inches). The present trend in gear designing practice is to use finer pitches than in the past. In the case of worm gearing, the pitch may, in certain instances,

be changed somewhat either to permit cutting with available equipment or to improve the proportions of worm and wheel.

When Ratio, Pitch and Lead Angle are Fixed. Assume that $R = 10$, axial pitch $P = 0.16$ inch, $L_a = 30$ degrees and $C = 3$ inches, approximately.

The first step is to determine for the given ratio, pitch and lead angle, the number of worm threads t which will give a center distance nearest 3 inches.

$$t = \frac{C \times 2\pi}{P \times (\cot L_a + R)} = \frac{3 \times 6.2832}{0.16 \times (1.7320 + 10)} = 10.04$$

The whole number nearest 10.04, or 10, is the required number of worm threads; hence number of teeth on wormgear equals $R \times 10 = 100$

$$d = (L \cot L_a) \div \pi = (10 \times 0.16 \times 1.732) \div \pi = 0.8821 \text{ inches}$$

$$D = (TP) \div \pi = (100 \times 0.16) \div \pi = 5.0929 \text{ inches}$$

$$C = (D + d) \div 2 = (5.0929 + 0.8821) \div 2 = 2.9875 \text{ inches}$$

Efficiency of Worm Gearing. — The efficiency at a given speed, depends upon the worm lead angle, the workmanship, the lubrication, and the general design of the transmission. When worm gearing consists of a hardened and ground worm running with an accurately hobbed wormgear properly lubricated, the efficiency depends chiefly upon the lead angle and coefficient of friction between the worm and wormgear. In the lower range of lead angles, the efficiency increases considerably as the lead angle increases, as shown by Tables 1 and 2. This increase in efficiency remains practically constant for lead angles between 30 and 45 degrees. Several formulas for obtaining efficiency percentage follow:

With worm driving:

$$E = 100 - \frac{R}{2} \text{ (empirical rule)}; \quad E = \frac{100 \times \tan L_a}{\tan (L_a + \phi)}; \quad E = \frac{100 \times L}{L + f\pi d}$$

With wormgear driving:

$$E = 100 - 2R \text{ (empirical rule)}; \quad E = \frac{100 \times \tan (L_a - \phi)}{\tan L_a}$$

Table 1. Efficiency of Worm Gearing for Different Lead Angles and Frictional Coefficients

Coefficient of Friction	Lead Angle of Worm								
	5 Deg.	10 Deg.	15 Deg.	20 Deg.	25 Deg.	30 Deg.	35 Deg.	40 Deg.	45 Deg.
0.01	89.7	94.5	96.1	97.0	97.4	97.7	97.9	98.0	98.0
0.02	81.3	89.5	92.6	94.1	95.0	95.5	95.9	96.0	96.1
0.03	74.3	85.0	89.2	91.4	92.7	93.4	93.9	94.1	94.2
0.04	68.4	80.9	86.1	88.8	90.4	91.4	92.0	92.2	92.3
0.05	63.4	77.2	83.1	86.3	88.2	89.4	90.1	90.4	90.5
0.06	59.0	73.8	80.4	84.0	86.1	87.5	88.2	88.6	88.7
0.07	55.2	70.7	77.8	81.7	84.1	85.6	86.4	86.9	86.9
0.08	51.9	67.8	75.4	79.6	82.2	83.8	84.7	85.2	85.2
0.09	48.9	65.2	73.1	77.6	80.3	82.0	83.0	83.5	83.5
0.10	46.3	62.7	70.9	75.6	78.5	80.3	81.4	81.9	81.8

The efficiencies obtained by these formulas and other modifications of them differ somewhat and do not take into account bearing and oil-churning losses. The efficiency may be improved somewhat after the "running in" period.

Self-locking or Irreversible Worm Gearing. — Neglecting friction in the bearings, worm gearing is irreversible when the efficiency is zero or negative, the lead angle being equal to or less than the angle ϕ of friction (tan ϕ = coefficient of friction). When worm gearing is self-locking or irreversible, this means that the worm-

Table 2. Efficiency of Worm Gearing for Different Lead Angles and Pitch-line Velocities

Velocity at Pitch Line, Feet per Minute	Lead or Helix Angle, Degrees					
	5	10	20	30	40	45
	Efficiency, Per Cent					
5	40	56	69	76	79	80
10	47	62	74	79	82	82
20	52	67	78	83	85	86
30	56	71	81	85	87	87
40	60	74	83	87	88	88
50	63	76	85	88	89	89
75	67	80	87	90	90	90
100	70	82	88	91	91	91
150	74	84	90	92	92	92
200	76	85	91	92	92	92

Table 3. Coefficients of Friction (f) for Worm Gearing*

Rubbing Speed, Ft. per Min.	Coefficients of Friction	Rubbing Speed, Ft. per Min.	Coefficients of Friction	Rubbing Speed, Ft. per Min.	Coefficients of Friction	Rubbing Speed, Ft. per Min.	Coefficients of Friction
30	0.073	180	0.045	550	0.028	1600	0.0175
40	0.070	190	0.044	600	0.027	1700	0.0170
50	0.066	200	0.043	650	0.026	1800	0.0165
60	0.062	225	0.041	700	0.026	1900	0.0165
70	0.060	250	0.040	750	0.025	2000	0.0160
80	0.058	275	0.038	800	0.024	2100	0.0160
90	0.056	300	0.036	850	0.023	2200	0.0155
100	0.054	325	0.035	900	0.023	2300	0.0150
110	0.052	350	0.034	950	0.022	2400	0.0150
120	0.051	375	0.033	1000	0.022	2500	0.0150
130	0.050	400	0.033	1100	0.021	2600	0.0145
140	0.049	425	0.032	1200	0.020	2700	0.0145
150	0.048	450	0.031	1300	0.019	2800	0.0140
160	0.047	475	0.030	1400	0.019	2900	0.0140
170	0.046	500	0.030	1500	0.018	3000	0.0140

* These values for different rubbing speeds are based upon the use of phosphor bronze worm-gears with case-hardened ground and polished steel worms lubricated with mineral oil.

gear cannot drive the worm. Since the angle of friction changes rapidly with the rubbing speed, and the static angle of friction may be reduced by external vibration, it is usually impracticable to design irreversible worm gearing with any security. If irreversibility is desired, it is recommended that some form of brake be employed.

Worm Gearing Operating Temperatures. — The load capacity of a worm gearing lubricant at operating temperature is an important factor in establishing the continuous power-transmitting capacity of the gearing. If the churning or turbulence of the oil generates excessive heat, the viscosity of the lubricant may be reduced below its load-supporting capacity. The temperature measured in the oil sump should not, as a rule, exceed 180 to 200 degrees F. or rise more than 120 to 140 degrees F. above a surrounding air temperature of 60 degrees F. In rear axle motor vehicle transmissions, the maximum operating temperature may be somewhat higher than the figures given and usually is limited to about 220 degrees F.

Thermal Rating. — In some cases, especially when the worm speed is comparatively high, the horsepower capacity of worm gearing should be based upon its thermal rating instead of the mechanical rating. To illustrate, worm gearing may have a thermal rating of, say, 60 H.P., and mechanical ratings which are considerably higher than 60 for the higher speed ranges. This means that the gearing is capable of transmitting more than 60 H.P. so far as wear and strength are concerned but not without overheating; hence, in this case a rating of 60 should be considered maximum. Of course, if the power to be transmitted is less than the thermal rating for a given ratio, then the thermal rating may be ignored.

Water and Forced-Air Cooling. — One method of increasing the thermal rating of a speed-reducing unit of the worm gearing type, is by installing a water-cooling coil through which water is circulated to prevent an excessive rise of the oil temperature. According to one manufacturer, the thermal rating may be increased as much as 35 per cent in this manner. Much larger increases have been obtained by means of a forced air cooling system incorporated in the design of the speed-reducing unit. A fan which is mounted on the worm shaft draws air through a double walled housing, thus maintaining a comparatively low oil bath temperature. A fan cooling system makes it possible to transmit a given amount of power through a worm-gearing unit that is much smaller than one not equipped with a fan.

Horsepower Rating for Worm Gearing. — According to the British Standard for worm gearing, B.S. 721:1963, the permissible torque for a pair of worm gears is limited either by consideration of surface stress (conveniently referred to as "wear") or of bending stress (referred to as "strength"), in both the worm threads and wormgear teeth. Consequently, the determination of the load capacity of a pair of gears involves four calculations concerned with wear and strength of worm and wormgear, the permissible torque on the wormgear being the least of the four values.

In the formulas that follow, the permissible horsepower or wormgear torque obtained is the loading to which the gears may be subjected for a total running time of 26,000 hours. For any other life basis, the permissible horsepower or wormgear torque calculated from the formulas must be multiplied by

$$\left(\frac{27,000}{1,000 + U_{ec}}\right)^{1/3} \text{ for wear}$$

$$\text{and} \left(\frac{26,200}{200 + U_{eb}}\right)^{1/7} \text{ for strength}$$

Table 4. Bending Stress and Surface Stress Factors for Wormgears

Example: Wormgear is Phosphor Bronze, Centrifugally Cast (Material in Classification A.) Worm is Nickel Casehardening Steel (In Classification E.)

Find Factor S_c for wormgear under worm material E (upper section of table) and opposite "Phosphor Bronze, Centrifugally Cast." (S_c equals 2200.)

Find Factor S_c for worm under wormgear material A (lower section of table) and opposite "Nickel Casehardening Steel." (S_c equals 7700.)

Find Factor S_b for wormgear in column "Bending Stress Factor" and opposite "Phosphor Bronze, Centrifugally Cast." (S_b equals 10,000.)

Find Factor S_b for worm in column "Bending Stress Factor" and opposite "Nickel Casehardening Steel." (S_b equals 47,000.)

Wormgear Materials		Bending Stress Factor S_b Pounds per Sq. In.	Find surface stress factor S_c for wormgear, under worm material classification letter				
			A	B	C	D	E
A	Phosphor Bronze, Sand Cast	7,200	. . .	660*	660	770	1500
	Phosphor Bronze, Chill Cast	9,100	. . .	900*	900	1000	1800
	Phosphor Bronze, Centrifugally Cast	10,000	. . .	1200*	1200	1300	2200
B	Cast Iron (Gray)	5,800	900*	600*	600†	600†	750†

Worm Materials		Bending Stress Factor S_b Pounds per Sq. In.	Find surface stress factor S_c for worm, under wormgear material classification letter				
			A	B	C	D	E
C	0.4 Per Cent Carbon Steel, Normalized	20,000	1550	1000†	. . .	. . .	. . .
D	0.55 Per Cent Carbon Steel, Normalized	25,000	2200	1200†	. . .	. . .	. . .
E	Carbon Casehardening Steel	40,000	7000	4400†	. . .	. . .	2200†
	Nickel and Nickel-Molybdenum Casehardening Steel	47,000	7700	4400†	. . .	. . .	2200†
	Nickel-Chromium and Nickel-Chromium-Molybdenum Casehardening Steel	50,000	8800	4400†	. . .	. . .	2200†

* Maximum permissible rubbing speeds, 500 feet per minute.
† Should not be used except for hand operated gearing.

where U_{ec} and U_{eb} are the total running times for wear and strength respectively corresponding to the expected life.

Permissible Load for Wear: For the normal rating (26,000 hours) the permissible torque on the wormgear, M_w, in pound-inches, is limited by wear to the lower of the values

$$X_{cp}S_{cp}ZD_f^{1.8}m$$

$$X_{cw}S_{cw}ZD_f^{1.8}m$$

where X_{cp} is the speed factor for wear, worm; S_{cp} is the surface stress factor, worm (Table 4); X_{cw} is the speed factor for wear, wormgear; S_{cw} is the surface stress factor, wormgear (Table 4); D_f is the wormgear reference circle diameter, in inches; and m is the axial module in inches. The zone factor Z may be taken from Fig. 3. The speed factors for wear, X_{cp} and X_{cw} may be determined from Fig. 2 by entering the chart with the reference circle diameter of the worm, and that of the wormgear, together with the rubbing speed V_S in feet per minute calculated from the formula

$$V_s = 0.262\, d_f n \sec \lambda$$

where d_f is the worm reference circle diameter in inches; n is the worm speed in rpm; and λ is the lead angle of the worm.

Permissible Load for Strength: The permissible torque on the wormgear, M_w, in pound-inches, is limited by strength to the lower of the values

$$1.8\, X_{bp}S_{bp}ml_rD_f \cos \lambda$$

$$1.8\, X_{bw}S_{bw}ml_rD_f \cos \lambda$$

In these formulas, X_{bp} and X_{bw}, the speed factors for strength of the worm and wormgear, respectively, are obtained from Table 5 based upon rotational speed; the stress factors S_{bp} and S_{bw} for the worm and wormgear, respectively, are obtained from Table 4; and l_r, the length of root of the wormgear teeth, is obtained from the formula

$$l_r = (d_a + 2c) \sin^{-1} \left[\frac{F_e}{d_a + 2c} \right] \text{ where the angle is in radians.}$$

in which d_a is the worm tip diameter, c is the clearance, and F_e is the wormgear effective face width.

Table 5. Speed Factors X_{bp} and X_{bw} for Worm Gearing (Strength)

R.P.M.	Speed Factor X_{bp} or X_{bw}	R.P.M.	Speed Factor X_{bp} or X_{bw}	R.P.M.	Speed Factor X_{bp} or X_{bw}	R.P.M.	Speed Factor X_{bp} or X_{bw}
10	0.560	90	0.420	500	0.310	3000	0.200
15	0.540	100	0.415	600	0.300	3500	0.190
20	0.520	150	0.385	700	0.290	4000	0.180
30	0.500	200	0.365	800	0.280	4500	0.175
40	0.480	250	0.350	900	0.270	5000	0.170
50	0.460	300	0.340	1000	0.260	6000	0.160
60	0.450	350	0.335	1500	0.240	7000	0.150
70	0.440	400	0.330	2000	0.225	8000	0.140
80	0.430	450	0.320	2500	0.210	9000	0.135

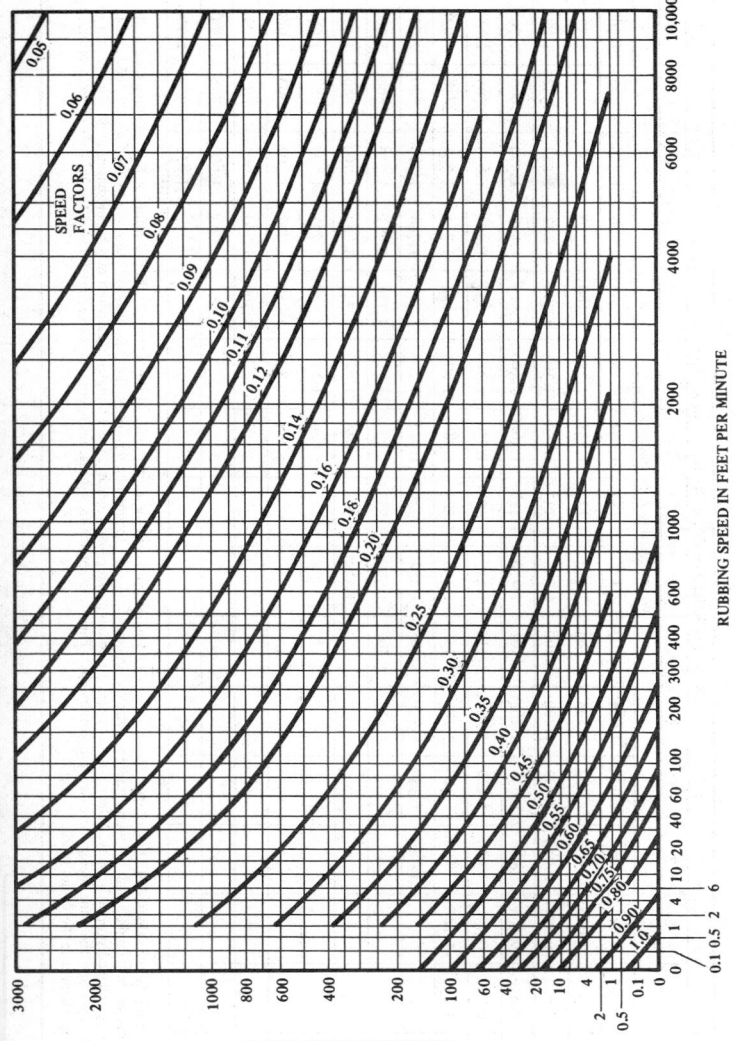

Fig. 2. Speed Factors X_{cp} and X_{cw} when Load Rating is Limited by Wear

No. of Threads, t	Ratio q of Reference Circle Diameter d_f to Module m														
	6	6.5	7	7.5	8	8.5	9	9.5	10	11	12	13	14	17	20
	Worm Gearing Zone Factor Z														
1	1.045	1.048	1.052	1.065	1.084	1.107	1.128	1.137	1.143	1.160	1.202	1.260	1.318	1.402	1.508
2	0.991	1.028	1.055	1.099	1.144	1.183	1.214	1.223	1.231	1.250	1.280	1.320	1.360	1.447	1.575
3	0.822	0.890	0.989	1.109	1.209	1.260	1.305	1.333	1.350	1.365	1.393	1.422	1.442	1.532	1.674
4	0.826	0.883	0.981	1.098	1.204	1.301	1.380	1.428	1.460	1.490	1.515	1.545	1.570	1.666	1.798
5	0.947	0.991	1.050	1.122	1.216	1.315	1.417	1.490	1.550	1.610	1.632	1.652	1.675	1.765	1.886
6	1.132	1.145	1.172	1.220	1.287	1.350	1.438	1.521	1.588	1.675	1.694	1.714	1.733	1.818	1.928
7			1.316	1.340	1.370	1.405	1.452	1.540	1.614	1.704	1.725	1.740	1.760	1.846	1.950
8					1.437	1.462	1.500	1.557	1.623	1.715	1.738	1.753	1.778	1.868	1.960
9							1.573	1.604	1.648	1.720	1.743	1.767	1.790	1.880	1.970
10									1.680	1.728	1.748	1.773	1.798	1.888	1.980
11										1.732	1.753	1.777	1.802	1.892	1.987
12											1.760	1.780	1.806	1.895	1.992
13												1.784	1.806	1.898	1.998
14													1.811	1.900	2.000

The zone factor values Z are based upon $F_e = 2m \sqrt{q+1}$ symmetrical about the center plane of the wormgear. In this formula, F_e = effective face width of wormgear; m = module; and q = worm reference circle diameter divided by the module m. For smaller face widths than given by the formula, the value of Z must be reduced proportionately. When it is necessary to obtain greater load capacity, the wormgear face width may be increased up to a maximum of $2.3m \sqrt{q+1}$ and the zone factor increased proportionately. The table applies to wormgears having 30 teeth; variations in the number of teeth produce negligible changes in the values of Z.

Fig. 3. Worm Gearing Zone Factors Z

Horsepower: The normal horsepower rating of the wormgear is given by

$$\frac{M_w N}{63,000} \text{ horsepower}$$

where M_w is the smallest of the four torque values calculated previously; and N is the speed of the wormgear in rpm.

Methods of Machining Worm Threads. — In producing worm threads, the method employed may depend upon quantity required, number of threads on worm or its lead angle, and equipment available. Methods of cutting threads on worms of cylindrical form are described on page 1900.

Table 6. Change Gears for Cutting Diametral Pitch Worms*

Diametral Pitch to be Cut	Width of Tool Point	Threads per Inch on Lead Screw							
		2	3	4	5	6	7	8	10
2	0.487	22/7	33/7	44/7	55/7	66/7	77/7	88/7	110/7
2¼	0.433	176/63	88/21	352/63	440/63	176/21	88/9	704/63	880/63
2½	0.390	88/35	132/35	176/35	44/7	264/35	308/35	352/35	440/35
2¾	0.354	16/7	24/7	32/7	40/7	48/7	56/7	64/7	72/7
3	0.325	44/21	22/7	88/21	110/21	44/7	22/3	176/21	220/21
3½	0.278	88/49	132/49	176/49	220/49	264/49	44/7	352/49	440/49
4	0.243	11/7	33/14	22/7	55/14	33/7	11/2	44/7	55/7
4½	0.217	88/63	44/21	176/63	220/63	88/21	44/9	352/63	440/63
5	0.195	44/35	66/35	88/35	22/7	132/35	22/5	176/35	44/7
6	0.162	22/21	11/7	44/21	55/21	22/7	11/3	88/21	110/21
7	0.139	44/49	66/49	88/49	110/49	132/49	22/7	176/49	220/49
8	0.122	11/14	33/28	11/7	55/28	33/14	11/4	22/7	55/14
9	0.108	44/63	22/21	88/63	110/63	44/21	22/9	176/63	220/63
10	0.097	22/35	33/35	44/35	11/7	66/35	11/5	88/35	22/7
11	0.088	4/7	6/7	8/7	10/7	12/7	14/7	16/7	20/7
12	0.081	11/21	11/14	22/21	55/42	11/7	11/6	44/21	55/21
14	0.069	22/49	33/49	44/49	55/49	66/49	11/7	88/49	110/49
16	0.061	11/28	33/56	22/28	55/56	33/28	77/56	11/7	55/28
18	0.054	22/63	11/21	44/63	55/63	22/21	11/9	88/63	110/63
20	0.049	11/35	33/70	22/35	11/14	33/35	77/70	44/35	11/7
22	0.044	2/7	3/7	4/7	5/7	6/7	7/7	8/7	10/7
24	0.040	11/42	33/84	11/21	55/84	33/42	77/84	22/21	55/42
26	0.037	22/91	33/91	44/91	55/91	66/91	77/91	88/91	110/91
28	0.035	11/49	33/98	22/49	55/98	33/49	11/14	44/49	55/49
30	0.032	22/105	11/35	44/105	11/21	22/35	77/105	88/105	22/21
32	0.030	11/56	33/112	11/28	55/112	33/56	77/112	11/14	55/56
40	0.024	11/70	33/140	11/35	11/28	33/70	77/140	22/35	11/14
48	0.020	11/84	33/168	11/42	55/168	33/84	77/168	11/21	55/84

* The ratio of change gears for cutting diametral pitch worms is as 22 times the threads per inch on lead-screw is to 7 times the diametral pitch to be cut. Thus,

$$\frac{22 \times \text{Threads per Inch}}{7 \times \text{Diametral Pitch}} = \text{Ratio of Change Gears}$$

1. *Milling with a Disk-shaped Cutter.* The cutter has straight sides or edges and it is inclined an amount equal to the lead angle to locate the cutting side in alignment with the thread groove. There is a traversing movement per work revolution equal to the required lead. If the worm has two or more threads or starts, these are, of course, milled one at a time and should be uniformly spaced by indexing. *Precise indexing is very important.* A worm- or thread-milling machine is used.

2. *Hobbing.* A regular gear-hobbing machine is very efficient, especially for multi-thread worms, because all of the threads are finished simultaneously instead of taking separate cuts and indexing. The hobbing machine is geared with reference to the number of threads on the worm, the procedure being similar to that followed in cutting a helical gear.

3. *Generating by using a Worm Thread Generator.* The machine is equipped with a helical type gear shaper cutter, the axis of which is at right angles to the axis of the worm. The cutter generates the thread or threads as it rolls in mesh with the rotating work.

4. *Cutting in a Lathe.* One method is to locate the top cutting face of the tool normal or perpendicular to the worm thread. A second method is to locate the top cutting face in the same plane as the axis of the worm. With this method, cutting difficulties are encountered when the lead angles are comparatively large, due to the negative rake on the following side of the tool. As a general rule, the first method is preferable when a lathe must be used.

Grinding Worm Threads and Hobs. — A common method of producing worm threads is by milling with a disk-shaped cutter and then grinding after the hardening operation. The milling cutter has straight edges and the grinding wheel straight sides, because of the practical advantage in producing and maintaining these straight edges and sides.

In finishing hardened steel worms by grinding, the straight-sided grinding wheel produces a thread having convex sides and this curvature should be duplicated on the wormgear hob. This can be accomplished by finishing both worm and hob with grinding wheels which are maintained within diameter limits established by experience. In the case of multi-thread worms, accuracy of the indexing is essential when grinding both the worm and hob and accurate uniform lead is also important.

Double-enveloping Worm Gearing. — Contact between the worm and wormgear of the conventional type of worm gearing is theoretically a line contact; however, due to deflection of the materials under load, the line is increased to a narrow band or contact zone. In attempting to produce a double-enveloping type of worm gearing (with the worm curved longitudinally to fit the curvature of the gear as shown by illustration), the problem primarily was that of generating the worm and wormgear in such a manner as to obtain *area contact* between the engaging teeth. A practical method of obtaining such contact was developed by Samuel I. Cone at the Norfolk Navy Yard, and this is known as "Cone-Drive" worm gearing. The Cone generating method makes it possible to cut the worm and wormgear without any interference which

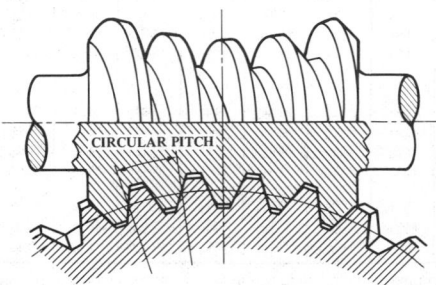

CIRCULAR PITCH

Table 1. AGMA Input Mechanical Horsepower Ratings of Cone-Drive Worm Gearing*

2-Inch Center Distance

Ratio	100	300	720	870	1150	1750
5:1	0.40	1.04	2.18	2.51	3.02	3.81
10:1	0.25	0.66	1.40	1.62	1.98	2.52
15:1	0.18	0.47	0.99	1.15	1.40	1.79
20:1	0.13	0.36	0.76	0.88	1.07	1.38
25:1	0.11	0.29	0.61	0.71	0.87	1.11
30:1	0.09	0.24	0.51	0.59	0.73	0.93
40:1	0.07	0.18	0.38	0.45	0.55	0.70
50:1	0.05	0.15	0.31	0.36	0.44	0.56

2.5-Inch Center Distance

Ratio	100	300	720	870	1150	1750
5:1	0.78	2.04	4.13	4.68	5.48	6.87
10:1	0.49	1.30	2.67	3.05	3.62	4.54
15:1	0.35	0.91	1.89	2.16	2.57	3.22
20:1	0.27	0.70	1.44	1.65	1.97	2.48
25:1	0.21	0.56	1.16	1.33	1.59	2.00
30:1	0.18	0.47	0.98	1.12	1.33	1.68
40:1	0.13	0.35	0.73	0.84	1.00	1.26
50:1	0.11	0.28	0.59	0.68	0.81	1.01

3-Inch Center Distance

Ratio	100	300	720	870	1150	1750
5:1	1.38	3.60	6.99	7.79	9.06	11.30
10:1	0.88	2.31	4.65	5.26	6.16	7.72
15:1	0.62	1.62	3.29	3.74	4.38	5.49
20:1	0.47	1.24	2.52	2.87	3.37	4.22
25:1	0.38	1.00	2.04	2.31	2.72	3.41
30:1	0.32	0.84	1.71	1.94	2.28	2.86
40:1	0.24	0.63	1.28	1.46	1.72	2.15
50:1	0.19	0.51	1.03	1.17	1.38	1.73
60:1	0.16	0.42	0.86	0.98	1.15	1.44

3.5-Inch Center Distance

Ratio	100	300	720	870	1150	1750
5:1	2.55	6.60	12.30	13.70	15.90	19.70
10:1	1.63	4.24	8.27	9.21	10.70	13.40
15:1	1.14	2.99	5.85	6.54	7.62	9.53
20:1	0.88	2.28	4.49	5.03	5.86	7.33
25:1	0.71	1.84	3.62	4.06	4.73	5.92
30:1	0.59	1.54	3.04	3.40	3.96	4.96
40:1	0.44	1.16	2.29	2.56	2.99	3.74
50:1	0.36	0.93	1.83	2.06	2.40	3.00
60:1	0.30	0.78	1.53	1.72	2.00	2.51

4-Inch Center Distance

Ratio	100	300	720	870	1150	1750
5:1	3.66	9.40	16.90	18.70	21.70	26.70
10:1	2.35	6.09	11.50	12.70	14.80	18.40
15:1	1.65	4.29	8.15	9.06	10.50	13.10
20:1	1.26	3.28	6.26	6.96	8.09	10.10
25:1	1.02	2.65	5.05	5.62	6.53	8.13
30:1	0.85	2.22	4.24	4.71	5.48	6.82
40:1	0.64	1.67	3.19	3.55	4.12	5.14
50:1	0.51	1.34	2.56	2.85	3.31	4.12
60:1	0.43	1.12	2.14	2.38	2.76	3.44

5-Inch Center Distance

Ratio	100	300	720	870	1150	1750
5:1	7.21	18.20	31.00	34.30	39.60	47.40
10:1	4.63	11.80	21.10	23.40	27.20	33.30
15:1	3.25	8.36	15.00	16.70	19.30	23.80
20:1	2.49	6.40	11.60	12.80	14.90	18.30
25:1	2.01	5.16	9.34	10.30	12.00	14.80
30:1	1.68	4.32	7.83	8.67	10.10	12.40
40:1	1.27	3.25	5.90	6.53	7.58	9.34
50:1	1.02	2.61	4.73	5.24	6.09	7.50

5-Inch Center Distance (Continued)

Ratio	100	300	720	870	1150	1750
60:1	0.85	2.18	3.95	4.37	5.08	6.26
70:1	0.73	1.87	3.39	3.76	4.36	5.38

6-Inch Center Distance

Ratio	100	300	720	870	1150	1750
5:1	11.10	27.30	45.30	50.10	57.50	66.80
10:1	7.08	17.80	30.40	33.70	38.80	46.40
15:1	4.98	12.60	21.60	23.90	27.60	33.20
20:1	3.81	9.64	16.60	18.40	21.20	25.60
25:1	3.07	7.78	13.40	14.90	17.20	20.70
30:1	2.57	6.52	11.30	12.50	14.40	17.40
40:1	1.94	4.90	8.47	9.38	10.80	13.10
50:1	1.55	3.93	6.80	7.53	8.70	10.50
60:1	1.30	3.28	5.68	6.29	7.26	8.79
70:1	1.11	2.82	4.87	5.40	6.23	7.55

7-Inch Center Distance

Ratio	100	300	720	870	1150	1750
5:1	17.50	41.60	67.30	73.90	83.60	96.40
10:1	11.20	27.70	46.20	51.20	58.70	68.40
15:1	7.88	19.60	32.90	36.50	41.90	49.20
20:1	6.03	15.00	25.30	28.00	32.20	37.90
25:1	4.86	12.20	20.50	22.60	26.00	30.70
30:1	4.07	10.20	17.20	19.00	21.80	25.80
40:1	3.06	7.66	12.90	14.30	16.40	19.40
50:1	2.46	6.15	10.40	11.50	13.20	15.60
60:1	2.05	5.13	8.66	9.58	11.00	13.00
70:1	1.76	4.41	7.43	8.23	9.46	11.20

8-Inch Center Distance

Ratio	100	300	720	870	1150	1750
5:1	25.90	59.60	95.20	104.00	116.00	134.00
10:1	16.70	40.90	67.40	74.40	85.10	98.70
15:1	11.80	29.00	48.20	53.20	61.20	71.00
20:1	9.00	22.20	37.00	40.90	47.00	54.70
25:1	7.26	18.00	29.90	33.10	38.00	44.30
30:1	6.08	15.10	25.10	27.80	31.90	37.20
40:1	4.58	11.30	18.90	20.90	24.00	28.00
50:1	3.67	9.10	15.20	16.80	19.30	22.50
60:1	3.07	7.59	12.70	14.00	16.10	18.80
70:1	2.63	6.52	10.90	12.00	13.80	16.10

10-Inch Center Distance

Ratio	100	300	720	870	1150	1750
5:1	48.50	105.00	164.00	178.00	194.00	226.00
10:1	31.40	73.20	117.00	129.00	144.00	166.00
15:1	22.10	52.10	83.70	91.90	104.00	119.00
20:1	16.90	40.00	64.40	70.70	79.80	91.80
25:1	13.60	32.30	52.10	57.20	64.60	74.30
30:1	11.40	27.10	43.70	48.00	54.20	62.40
40:1	8.60	20.40	32.90	36.10	40.90	47.00
50:1	6.90	16.40	26.40	29.00	32.80	37.80
60:1	5.76	13.70	22.10	24.20	27.40	31.60
70:1	4.94	11.70	18.90	20.80	23.50	27.10

12-Inch Center Distance

Ratio	100	300	720	870	1150	1750
5:1	81.30	167.00	257.00	271.00	300.00	...
10:1	53.20	118.00	186.00	202.00	221.00	...
15:1	37.50	83.90	133.00	145.00	159.00	...
20:1	28.70	64.40	102.00	111.00	122.00	...
25:1	23.10	52.10	82.70	90.10	99.10	...
30:1	19.40	43.70	69.40	75.70	83.20	...
40:1	14.60	32.90	52.20	57.00	62.70	...
50:1	11.70	26.40	41.90	45.80	50.40	...
60:1	9.77	22.10	35.00	38.20	42.10	...
70:1	8.39	18.90	30.10	32.80	36.10	...

* Horsepower ratings are for Class 1 service, using splash lubrication, except for that shown in italics, for which force feed lubrication is required. Other ratios and center distances are available.
Source: Ex-Cell-O Corp., Cone Drive Operations.

would alter the required tooth form. The larger tooth bearing area and multiple tooth contact obtained with this type of worm gearing, increases the load-carrying or horse-power capacity so that as compared with a conventional worm drive a double-enveloping worm drive may be considerably smaller in size. Table 1, which is intended as a general guide, gives input horsepower ratings for Cone-Drive worm gearing for various center distances of from 2 to 12 inches. These ratings are based on AGMA specifications 341, 441, and 641. They allow for starting and momentary peak over-loads of up to 300 per cent of the values shown in Table 1 using a service factor of 1. Factors for various types of service are given in Table 2. To obtain the mechanical horsepower rating required, multiply the appropriate rating given in Table 1 by the service factor taken from Table 2.

Table 2. Service Factors for Cone-Drive Worm Gearing

Hours/Day	Uniform Motion	Moderate Shock	Heavy Shock	Extreme Shock
0.5	0.6	0.8	0.9	1.1
1	0.7	0.9	1	1.2
2	0.9	1	1.2	1.3
10	1	1.2	1.3	1.5
24	1.2	1.3	1.5	1.75

Thermal Horsepower Rating: When the operation is to be continuous, considera-tion must be given to the possibility of overheating. For this possibility, the thermal horsepower rating given in Ex-Cell-O Corporation, Cone Drive Operations catalog must be checked. The thermal rating defines the maximum horsepower which can be transmitted continuously (30 minutes or longer). This is based on an oil sump temper-ature rise of 100 deg. F above the ambient temperature. This rise must not exceed 200 deg. F. If the thermal rating is lower than the mechanical rating, the unit must be selected on the basis of the thermal rating.

Type of Drive Connection: If either input or output shaft is connected to driver or driven mechanism by other than a direct shaft coupling, the overhead load require-ment (chain pull) must be calculated by dividing the torque demand by the pitch radius of the sprocket, sheave, spur gear, or helical gear used. The result is multiplied by the overhung load factor which is: for a chain sprocket, 1.00; for spur or helical gearing, 1.25; for a V-belt sheave, 1.50; and for a flat belt sheave, 2.50. As modified by the applicable overhung load factor, this load may not exceed the overhung load rating given in the company catalog.

Locking Considerations: It is a common misconception that all worm gears are self-locking or non-overhauling. Actually, wormgear ratios up to 15 to 1 will overhaul quite freely. Ratios from 20:1 to 40:1 can generally be considered as overhauling with difficulty, particularly from rest. Ratios above 40:1 may or may not overhaul depend-ing on the loading, lubrication, and amount of vibration present. Therefore it is not acceptable to rely on a wormgear to prevent movement in a system. Whenever a load must be stopped or held in place, a positive mechanical device must be incorporated into the system to prevent rotation of the gearset.

Backdriving or Overhauling: Applications such as wheel drives that require a brake on the motor or input shaft to decelerate a high inertial load require special attention to brake selection. Wherever possible, these applications should utilize freely overhauling ratios (15:1 or less). If higher ratios are used with a brake, the gearset can, under certain conditions, lockup during deceleration and impose severe shock loading on the reducer and driven equipment.

Stairstepping: Self-locking ratios (generally 40:1 and higher) are susceptible to the phenomenon of "stairstepping" when backdriving or overhauling. This erratic rota-tion of the gearset occasionally occurs when the gearset is backdriven at worm

speeds less than the theoretical lockup speed of the gearset and can be amplified by the rest of the drive train creating a very undesirable operating condition. "Stairstepping" can occur on drives where there is a high inertial load at the output shaft.

Backlash: Defined as the amount of movement at the pitch line of the gear with the worm locked and the gearset on exact center distance, backlash normally ranges from 0.003 to 0.008 inch for a 2-inch center distance set up to 0.012 to 0.020 inch for a 12-inch center distance set. When the gearset is assembled into a machine or reducer, the assembled backlash may fall outside these limits depending upon worm and gear bearing looseness and the actual center distance used.

Lubrication: Lubricating oils for use in double-enveloping worm drive units should be well refined petroleum oils of high quality. They should not be corrosive to gears or bearings and they must be neutral in reaction and free from grit or abrasives. They should have good defoaming properties and good resistance to oxidation. For worm gears, add up to 3 to 10 per cent of acidless tallow or similar animal fat.

The oil bath temperature should not exceed 200 degrees F. Where worm speed exceeds 3600 revolutions per minute, or 2000 feet per minute rubbing speed, a force-feed lubrication may be required. Auxiliary cooling by forced air, water coils in sump, or an oil heat exchanger may be provided in a unit where mechanical horsepower rating is in excess of the thermal rating, if full advantage of mechanical capacity is to be realized. The rubbing speed (V), in feet per minute, may be calculated from the formula: $V = 0.262 \times$ worm throat diameter in inches $\times$ worm RPM $\div \cos$ lead angle.

Wormgear Hobs. — An ideal hob would have exactly the same pitch diameter and lead angle as the worm; repeated sharpening, however, would reduce the hob size because of the form-relieved teeth. Hence, the general practice is to make hobs (especially the radial or in-feed type) "over-size" to provide a grinding allowance and increase the hob life. An over-size hob has a larger pitch diameter and smaller lead angle than the worm, but repeated sharpenings gradually reduce these differences. To compensate for the smaller lead angle of an over-size hob, the hob axis may be set 90-degrees relative to the wormgear axis plus the difference between the *lead* angle of the worm at the pitch line, and the *lead* angle of the over-size hob at its pitch line. This angular adjustment is in the direction required to increase the inclination of the wormgear teeth so that the axis of the assembled worm will be 90 degrees from the wormgear axis. ("Lead angle" is measured from a plane perpendicular to worm or hob axis.)

Hob Diameter Formulas: If D = pitch diameter of worm; D_h = pitch diameter of hob; A = addendum of worm and wormgear; C = clearance between worm and wormgear; and S = increase in hob diameter or "over-size" allowance for sharpening.

Outside diameter O of hob = $D + 2A + 2C + S$
Root diameter of hob = $D - 2A$
Pitch diameter D_h of hob = $O - (2A + 2C)$

Sharpening Allowance: Hobs for ordinary commercial work are given the following sharpening allowance, according to the recommended practice of the AGMA: In this formula, h = helix angle of hob at outside diameter measured from axis; H = helix angle of hob at pitch diameter measured from axis.

$$\text{Sharpening allowance} = 0.075 \times \text{normal pitch} \times \left[\frac{16 - (h - H)}{16} \right] + 0.010$$

Number of Flutes or Gashes in Hobs. — For finding the approximate number of flutes in a hob, the following rule may be used: Multiply the diameter of the hob by 3, and divide this product by twice the linear pitch. This rule gives suitable results for hobs for general purposes. Certain modifications, however, are necessary as explained in the following paragraph.

It is important that the number of flutes or gashes in hobs bear a certain relation to the number of threads in the hob and the number of teeth in the wormgear to be hobbed. In the first place, avoid having a common factor between the number of threads in the hob and the number of flutes; that is, if the worm is double-threaded, the number of gashes should be, say, 7 or 9, rather than 8. If it is triple-threaded, the number of gashes should be 7 or 11, rather than 6 or 9. The second requirement is to avoid having a common factor between the number of threads in the hob and the number of teeth in the wormgear. For example, if the number of teeth in the wheel is 28, it would be best to have the hob triple-threaded, as 3 is not a factor of 28. Again, if there were to be 36 threads in the wormgear, it would be preferable to have 5 threads in the hob.

The cutter used in gashing hobs should be from ⅛ to ¼ inch thick at the periphery, according to the pitch of the thread of the hob. The width of the gash at the periphery of the hob should be about 0.4 times the pitch of the flutes. The cutter should be sunk into the hob blank so that it reaches from 3⁄16 to ¼ inch below the root of the thread.

Helical Fluted Hobs. — Hobs are generally fluted parallel with the axis, but it is obvious that the cutting action will be better if they are fluted on a helix at right angles with the thread helix. The difficulty of relieving the teeth with the ordinary backing-off attachment is the cause for using a flute parallel with the axis. Flutes cut at right angles to the direction of the thread can, however, also be relieved, if the angle of the flutes is slightly modified. In order to relieve hobs with a regular relieving attachment, it is necessary that the number of teeth in one revolution along the thread helix be such that the relieving attachment can be geared to suit it. The following method makes it possible to select an angle of flute that will make the flute come *approximately* at right angles to the thread, and at the same time the angle is so selected that the relieving attachment can be properly geared for relieving the hob.

Let

C = pitch circumference;

T = developed length of thread in one turn;

N = number of teeth in one turn along thread helix;

F = number of flutes;

α = angle of thread helix.

Then (see illustration on the following page):

$C \div F$ = length of each small division on pitch circumference;

$(C \div F) \times \cos \alpha$ = length of division on developed thread;

$C \div \cos \alpha = T$

Hence

$$\frac{T}{(C \div F)\cos \alpha} = N = \frac{F}{\cos^2 \alpha}$$

Now, if

$\alpha = 30$ degrees, $N = 1\tfrac{1}{3}F$;

$\alpha = 45$ degrees, $N = 2F$;

$\alpha = 60$ degrees, $N = 4F$.

In most cases, however, such simple relations are not obtained. Suppose for example that $F = 7$, and $\alpha = 35$ degrees. Then $N = 10.432$, and no gears could be selected that would relieve this hob. By a very slight change in the helix angle of the flute, however, we can change N to 10 or 10½; in either case we can find suitable gears for the relieving attachment.

The rule for finding the modified helical lead of the flute is: Multiply the lead of the hob by F, and divide the product by the difference between the desired values of N and F.

Hence, the lead of flute required to make $N = 10$ is:

Lead of hob $\times$ (7 ÷ 3).

To make $N = 10\frac{1}{2}$, we have:

Lead of flute = lead of hob $\times$ (7 ÷ 3.5).

From this the angle of the flute can easily be found.

That the rule given is correct will be understood from the following consideration. Change the angle of the flute helix β so that AG contains the required number of parts N desired. Then EG contains $N - F$ parts. But $\cot \beta = BD \div ED$ and by the law of similar triangles,

$$BD = \frac{F}{N} \times BG, \text{ and } ED = \frac{N-F}{N} C$$

The lead of the helix of the flute, however, is $C \times \cot \beta$.

Hence, the required lead of the helix of the flute:

$$C \times \cot \beta = \frac{F}{N-F} L$$

This formula makes it possible always to flute hobs so that they can be conveniently relieved, and at the same time have the flutes at approximately right angles to the thread.

Helical Gearing

Basic Rules and Formulas for Helical Gear Calculations. — The ten rules and formulas in the accompanying table may be called the basic rules for helical gear calculations. The following definitions should be clearly understood in order to avoid misunderstandings. The *center angle* of a pair of helical gears is the angle made by the two center lines or axes of the gears. The *tooth angle* is the angle which the direction of the tooth makes with the axis of the gear. The *normal diametral pitch* is the diametral pitch of the cutter used for cutting the teeth in helical gears. In the formulas in the table of "Basic Rules and Formulas for Helical Gear Calculations" the following notation is used:

P_n = normal diametral pitch (pitch of cutter);

D = pitch diameter;

N = number of teeth;

α = helix angle;

γ = center angle, or angle between shafts;

C = center distance;

N' = number of teeth for which to select a formed cutter;

L = lead of tooth helix;

S = addendum;

W = whole depth of tooth;

T_n = normal tooth thickness at pitch line;

O = outside diameter.

The rules and formulas are given in the same order as they would ordinarily be used by the designer when calculating a pair of helical gears. The formulas, how-

Basic Rules and Formulas for Helical Gear Calculations

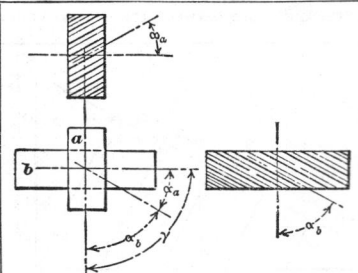

In the formulas, N, α, etc., are the numbers of teeth, helix angle, etc., for *either* gear or pinion; the notations N_a, N_b, α_a, α_b, etc., refer to the teeth or angles in the pinion or gear, respectively, in a pair of gears a and b.

No.	To Find	Rule	Formula
1	Relation between Shaft and Tooth Angles.	See rules at bottom of page 1909.	
2	Pitch Diameter.	Divide the number of teeth by the product of the normal pitch and the cosine of the tooth angle.	$D = \dfrac{N}{P_n \cos \alpha}$
3	Center Distance.	Add together the pitch diameters of the two gears and divide by 2.	$C = \dfrac{D_a + D_b}{2}$
4	Checking Calculations in (2) and (3); for use when angle between shafts is 90 degrees.	To prove the calculations for pitch diameters and center distance, multiply the number of teeth in the first gear by the tangent of the tooth angle of that gear, and add the number of teeth in the second gear to the product; the sum should equal twice the product of the center distance multiplied by the normal diametral pitch, multiplied by the sine of the tooth angle of the first gear.	$N_b + (N_a \times \tan \alpha_a) = 2\,CP_n \times \sin \alpha_a$
5	Number of Teeth for which to Select Formed Cutter.	Follow procedure outlined under heading "Selecting Cutter for Milling Helical Gears," page 1927.	
6	Lead of Tooth Helix.	Multiply the pitch diameter by 3.1416 times the cotangent of the tooth angle.	$L = \pi D \times \cot \alpha$
7	Addendum.	Divide 1 by the normal diametral pitch.	$S = \dfrac{1}{P_n}$
8	Whole Depth of Tooth.	Divide 2.157 by the normal diametral pitch.*	$W = \dfrac{2.157}{P_n}$
9	Normal Tooth Thickness at Pitch Line.	Divide 1.571 by the normal diametral pitch.	$T_n = \dfrac{1.571}{P_n}$
10	Outside Diameter.	Add twice the addendum to the pitch diameter.	$O = D + 2S$

* For hobbed 20° pressure angle gears of 20 diametral pitch and finer, $W = (2.200 \div P_n) + 0.002$.

ever, cannot be directly applied to all cases of helical gear problems, and a complete set of formulas for each of the seventeen different cases which are frequently met with is, therefore, given in the following, together with an example for each case. These seventeen cases are:

1. Shafts parallel, ratio 1 ("1 to 1"), center distance approximate.
2. Shafts parallel, ratio 1, center distance exact.
3. Shafts parallel, ratio other than 1, center distance approximate.
4. Shafts parallel, ratio other than 1, center distance exact.
5. Shafts at right angles, ratio 1, center distance approximate.
6. Shafts at right angles, ratio 1, center distance exact.
7. Shafts at right angles, ratio other than 1, center distance approximate.
8A. Shafts at right angles, ratio other than 1, center distance exact.
8B. Shafts at right angles, any ratio, helix angle for minimum center distance.
9. Shafts at 45-degree angle, ratio 1, center distance approximate.
10. Shafts at 45-degree angle, ratio 1, center distance exact.
11. Shafts at 45-degree angle, ratio other than 1, center distance approximate.
12. Shafts at 45-degree angle, ratio other than 1, center distance exact.
13. Shafts at any angle, ratio 1, center distance approximate.
14. Shafts at any angle, ratio 1, center distance exact.
15. Shafts at any angle, ratio other than 1, center distance approximate.
16. Shafts at any angle, ratio other than 1, center distance exact.

Pitch of Cutter for Helical Gears — The thickness of the cutter at the pitch line for cutting helical gears should equal one-half the normal circular pitch n (see illustration). If a cutter were used having a thickness, at the pitch line, equal to one-half the circular pitch P, as for spur gearing, the spaces between the teeth would be cut too wide, thus producing thin teeth. The normal pitch varies with the angle α of the helix; hence, the helix angle must be considered when selecting a cutter. The cutter should be of the same pitch as the *normal* diametral pitch of the gear. This normal pitch is found by dividing the transverse diametral pitch by the cosine of the helix angle. To illustrate, if the pitch diameter of a helical gear is 6.718 and there are 38 teeth having a helix angle of 45 degrees, the transverse diametral pitch equals 38 ÷ 6.718 = 5.656; then, the normal diametral pitch equals 5.656 divided by the cosine of 45 degrees or 5.656 ÷ 0.707 = 8. A cutter, then, of 8 diametral pitch is the one to use for this particular gear. This same result could

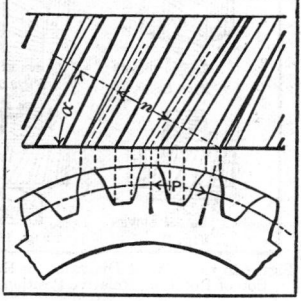

also be obtained as follows: If the circular pitch P is 0.5554, the normal circular pitch n can be found by multiplying the circular pitch P by the cosine of the helix angle. For example, 0.5554 × 0.707 = 0.3927. The normal diametral pitch is then found by dividing 3.1416 by the normal circular pitch. Thus 3.1416 ÷ 0.3927 = 8, which is the diametral pitch of the cutter.

Helical gears should preferably be cut on some type of gear-cutting machine such as a "hobber," or on a shaper or planer of the generating type. Milling machines are used in some shops, especially when a gear-cutting machine is not available. The pitch of the formed cutter used in milling a helical gear, must not only conform to the normal diametral pitch of the gear, but the cutter number must also be determined. See page 1927."Selecting Cutter for Milling Helical Gears."

Procedure in Calculating Helical Gears. — One of the first steps necessary in helical gear design is to determine the direction of the thrust, if the thrust is to be taken in one direction only. When the direction of the thrust has been determined and the relative position of the driver and driven gear is known, the direction of helix (right- or left-hand) may be found. The thrust diagrams, Figs. 1 to 28, are used for finding the direction of helix. The arrows at the end bearings of the gears indicate the direction of the reaction against the thrust caused by the tooth pressure. The direction of the thrust depends on the direction of helix, the relative positions of driver and driven gear, and the direction of rotation. If the exact

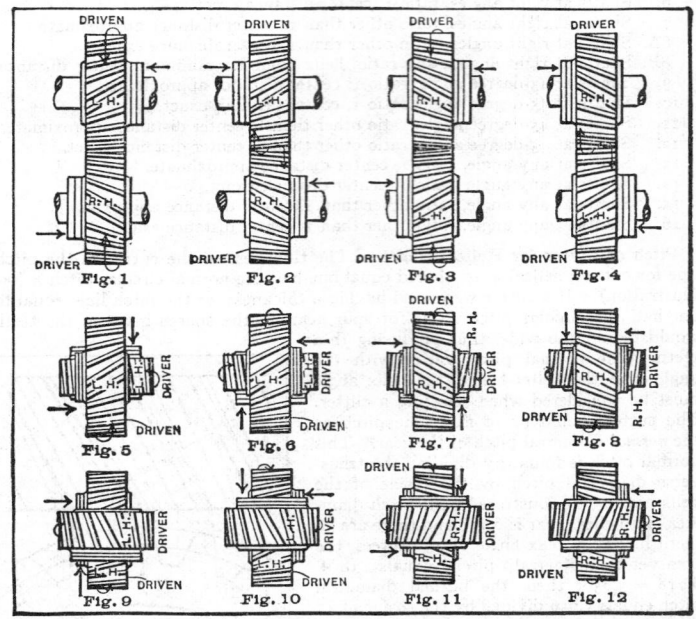

Figs. 1 to 12. Thrust Diagrams for Helical Gears — Direction of Thrust depends upon Direction of Rotation, Relative Position of Driver and Driven Gear, and Direction of Helix

condition with regard to thrust is not found in the diagrams, it may be obtained by changing any one of these three conditions; that is, in Fig. 1 the thrust may be changed to the opposite direction by interchanging driver or driven gear, by reversing the direction of rotation or by changing the direction of helix. Any one of these alterations will produce a thrust in the opposite direction.

The conditions of the design determine the nature of the center distance, whether it must be exact or approximate. The number of teeth in each gear is, of course, determined by the required speed ratios of the shafts. The angle of helix depends upon the conditions of the design and the relative position of the shafts. For parallel shafts the helix angle should not exceed 20 degrees in order to avoid excessive end thrust. In order to obtain smooth running gears, the helix angle should

be such that one end of the tooth remains in contact until the opposite end of the following tooth has found a bearing.

As far as the calculations are concerned, the formulas are the same for a 135-degree shaft angle as for an angle of 45 degrees. The following general rule relative to gears having a shaft angle of 45 degrees should be observed: When the helix angle of each gear is less than 45 degrees, then the helix angles are of the same hand and one helix angle is 45 degrees minus the other. When the helix angle of either gear is greater than 45 degrees, then the helix angles are of opposite hand, and the helix angle of one gear is 45 degrees plus the helix angle of the other.

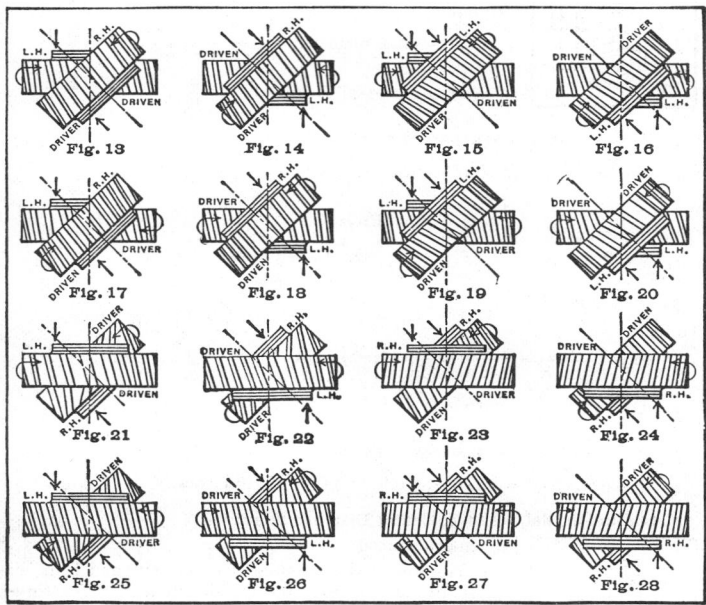

Figs. 13 to 28. Thrust Diagrams for Helical Gears — Direction of Thrust depends upon Direction of Rotation, Relative Position of Driver and Driven Gear, and Direction of Helix

The following rules should be observed for helical gears with shafts at any given angle. If each helix angle is less than the shaft angle, then the sum of the helix angles of the two gears will equal the angle between the shafts, and the helix is of the same hand in both gears; if the helix angle of one of the gears is greater than the shaft angle, then the difference between the helix angles of the two gears will be equal to the shaft angle, and the gears will be of opposite hand.

In some cases of helical gear design, particularly where trial-and-error methods of calculation must be used, it may be of considerable advantage to employ a graphical method for obtaining an approximate solution. (Such a method may be found in *Manual Of Gear Design*, Section Three, by Earle Buckingham.) The approximate solution thus obtained can then be modified to suit the problem at hand by making use of the standard helical gear formulas.

1. Shafts Parallel, Ratio 1 (or "1 to 1"), Center Distance Approximate. —

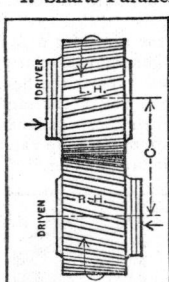

Given or assumed:

1. Hand of helix on driver or driven gear depending on rotation and direction in which thrust is to be received.
2. C_a = approximate center distance.
3. P_n = normal diametral pitch (pitch of cutter).
4. N = number of teeth.
5. α = angle of helix (usually less than 20 degrees for single gears to avoid excessive end thrust).

To find:

1. D = pitch diameter = $\dfrac{N}{P_n \cos \alpha}$

2. O = outside diameter = $D + \dfrac{2}{P_n}$

3. T = number of teeth marked on cutter = $\dfrac{N}{\cos^3 \alpha}$

4. L = lead of helix = $\pi D \cot \alpha$.

Example

Given or assumed:

1. See illustration.
2. C_a = 3 inches
3. P_n = 8.
4. N = 24.
5. α = 15 degrees.

To find:

1. $D = \dfrac{N}{P_n \cos \alpha} = \dfrac{24}{8 \times 0.9659} = 3.106$ inches.

2. $O = 3.106 + \dfrac{2}{8} = 3.356$ inches.

3. $T = \dfrac{N}{\cos^3 \alpha} = \dfrac{24}{0.9} = 26.6$, say 27 teeth.

4. $L = \pi D \cot 15° = 3.1416 \times 3.106 \times 3.732 = 36.416$ inches.

2. Shafts Parallel, Ratio 1, Center Distance Exact. —

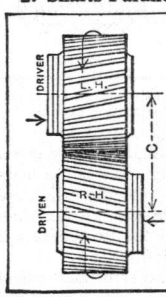

Given or assumed:

1. Position of gear having right- or left-hand helix, depending on rotation and direction in which thrust is to be received.
2. C = exact center distance = pitch diameter D.
3. P_n = normal diametral pitch (pitch of cutter).
4. N = number of teeth in each gear.

To find:

1. $\cos \alpha = \dfrac{N}{P_n D}$

2. O = outside diameter = $D + \dfrac{2}{P_n}$

3. T = number of teeth marked on formed milling cutter = $\dfrac{N}{\cos^3 \alpha}$

4. L = lead of helix = $\pi D \cot \alpha$.

α is usually less than 20 degrees to avoid excessive end thrust.

Example

Given or assumed:

1. See illustration.   2. $C = 3$ inches.
3. $P_n = 8$.   4. $N = 22$.

To find:

1. $\cos \alpha = \dfrac{N}{P_n D} = \dfrac{22}{8 \times 3} = 0.9166$, or $\alpha = 23° 34'$.

2. $O = D + \dfrac{2}{P_n} = 3 + \dfrac{2}{8} = 3\tfrac{1}{4}$ inches.

3. $T = \dfrac{N}{\cos^3 \alpha} = \dfrac{22}{(0.92)^3} = 28.2$, say 28 teeth.

4. $L = \pi D \cot \alpha = 3.1416 \times 3 \times 2.29 = 21.58$ inches.

3. Shafts Parallel, Ratio Other Than 1, Center Distance Approximate. —

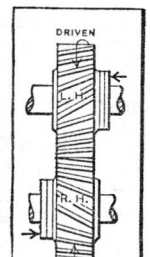

Given or assumed:

1. Position of gear having right- or left-hand helix, depending upon rotation and direction in which thrust is to be received.
2. C_a = approximate center distance.
3. P_n = normal diametral pitch.
4. N = number of teeth in large gear.
5. n = number of teeth in small gear.
6. α = angle of helix.

To find:

1. D = pitch diameter of large gear = $\dfrac{N}{P_n \cos \alpha}$

2. d = pitch diameter of small gear = $\dfrac{n}{P_n \cos \alpha}$

3. O = outside diameter of large gear = $D + \dfrac{2}{P_n}$

4. o = outside diameter of small gear = $d + \dfrac{2}{P_n}$

5. T = number of teeth marked on formed cutter (large gear) = $\dfrac{N}{\cos^3 \alpha}$

6. t = number of teeth marked on formed cutter (small gear) = $\dfrac{n}{\cos^3 \alpha}$

7. L = lead of helix on large gear = $\pi D \cot \alpha$.
8. l = lead of helix on small gear = $\pi d \cot \alpha$.
9. C = center distance (if not right vary α) = $\tfrac{1}{2} (D + d)$.

Example

Given or assumed:

1. See illustration.   2. $C_a = 17$ inches.   3. $P_n = 2$.   4. $N = 48$.
5. $n = 20$.   6. $\alpha = 20$ degrees.

To find:

1. $D = \dfrac{N}{P_n \cos \alpha} = \dfrac{48}{2 \times 0.9397} = 25.541$ inches.

2. $d = \dfrac{n}{P_n \cos \alpha} = \dfrac{20}{2 \times 0.9397} = 10.642$ inches.

3. $O = D + \dfrac{2}{P_n} = 25.541 + \dfrac{2}{2} = 26.541$ inches.

4. $o = d + \dfrac{2}{P_n} = 10.642 + \dfrac{2}{2} = 11.642$ inches.

5. $T = \dfrac{N}{\cos^3 \alpha} = \dfrac{48}{(0.9397)^3} = 57.8$, say 58 teeth.

6. $t = \dfrac{n}{\cos^3 \alpha} = \dfrac{20}{(0.9397)^3} = 24.1$, say 24 teeth.

7. $L = \pi D \cot \alpha = 3.1416 \times 25.541 \times 2.747 = 220.42$ inches.

8. $l = \pi d \cot \alpha = 3.1416 \times 10.642 \times 2.747 = 91.84$ inches.

9. $C = \tfrac{1}{2}(D + d) = \tfrac{1}{2}(25.541 + 10.642) = 18.091$ inches.

4. Shafts Parallel, Ratio Other Than 1, Center Distance Exact. —

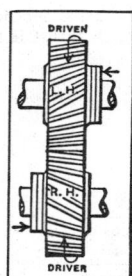

Given or assumed:

    1. Position of gear having right- or left-hand helix, depending upon rotation and direction in which thrust is to be received.

    2. C = exact center distance.

    3. P_n = normal diametral pitch (pitch of cutter).

    4. N = number of teeth in large gear.

    5. n = number of teeth in small gear.

To find:

    1. $\cos \alpha = \dfrac{N + n}{2 P_n C}$

    2. $\quad D$ = pitch diameter of large gear $= \dfrac{N}{P_n \cos \alpha}$

    3. $\quad d$ = pitch diameter of small gear $= \dfrac{n}{P_n \cos \alpha}$

4. O = outside diameter of large gear $= D + \dfrac{2}{P_n}$

5. o = outside diameter of small gear $= d + \dfrac{2}{P_n}$

6. T = number of teeth marked on formed milling cutter (large gear) $= \dfrac{N}{\cos^3 \alpha}$

7. t = number of teeth marked on formed milling cutter (small gear) $= \dfrac{n}{\cos^3 \alpha}$

8. L = lead of helix (large gear) $= \pi D \cot \alpha$

9. l = lead of helix (small gear) $= \pi d \cot \alpha$

Example

Given or assumed:

    1. See illustration.   2. $C = 18.75$ inches.   3. $P_n = 4$.   4. $N = 96$.   5. $n = 48$.

To find:

    1. $\cos \alpha = \dfrac{N + n}{2 P_n C} = \dfrac{96 + 48}{2 \times 4 \times 18.75} = 0.96$, or $\alpha = 16° \ 16'$.

    2. $\quad D = \dfrac{N}{P_n \cos \alpha} = \dfrac{96}{4 \times 0.96} = 25$ inches.

3. $\quad d = \dfrac{n}{P_n \cos \alpha} = \dfrac{48}{4 \times 0.96} = 12.5$ inches.

4. $\quad O = D + \dfrac{2}{P_n} = 25 + \dfrac{2}{4} = 25.5$ inches.

5. $\quad o = d + \dfrac{2}{P_n} = 12.5 + \dfrac{2}{4} = 13$ inches.

6. $\quad T = \dfrac{N}{\cos^3 \alpha} = \dfrac{96}{(0.96)^3} = 108$ teeth.

7. $\quad t = \dfrac{n}{\cos^3 \alpha} = \dfrac{48}{(0.96)^3} = 54$ teeth.

8. $\quad L = \pi D \cot \alpha = 3.1416 \times 25 \times 3.427 = 269.15$ inches.

9. $\quad l = \pi d \cot \alpha = 3.1416 \times 12.5 \times 3.427 = 134.57$ inches.

5. Shafts at Right Angles, Ratio 1, Center Distance Approximate. —

When the helix angles are 45 degrees, the gears are exactly alike; when other than 45 degrees, the sum of the helix angles must equal 90 degrees.

Given or assumed:

1. Position of gear having right- or left-hand helix, depending on the rotation and direction in which the thrust is to be received.
2. C_a = approximate center distance.
3. P_n = normal diametral pitch (pitch of cutter).
4. N = number of teeth.
5. α = angle of helix.

To find:

(a) When helix angles are 45 degrees.

1. D = pitch diameter = $\dfrac{N}{0.70711\, P_n}$ 2. O = outside diameter = $D + \dfrac{2}{P_n}$

3. T = number of teeth marked on formed milling cutter = $\dfrac{N}{0.353}$

4. L = lead of helix = πD.   5. C = center distance = D.

(b) When helix angles are other than 45 degrees.

1. D = pitch diameter = $\dfrac{N}{P_n \cos \alpha}$

2. T = number of teeth marked on formed milling cutter = $\dfrac{N}{\cos^3 \alpha}$

3. C = center distance = sum of pitch radii.

4. L = lead of helix = $\pi D \cot \alpha$.

Example

Given or assumed:

1. See illustration.   2. $C_a = 2.5$ inches.   3. $P_n = 10$.
          4. $N = 18$ teeth.   5. $\alpha = 45$ degrees.

To find:

1. $D = \dfrac{N}{0.70711\, P_n} = \dfrac{18}{0.70711 \times 10} = 2.546$ inches.

2. $O = D + \dfrac{2}{P_n} = 2.546 + \dfrac{2}{10} = 2.746$ inches.

3. $T = \dfrac{N}{\cos^3 \alpha} = \dfrac{18}{0.353} = 51$ teeth.

4. $L = \pi D \times 1 = 3.1416 \times 2.546 = 7.999$ inches.

6. Shafts at Right Angles, Ratio 1, Center Distance Exact. —

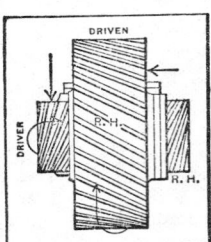

Gears have same direction of helix but probably different pitch diameters and helix angles; the sum of the latter must be 90 degrees. Given or assumed:

1. Position of gear having right- or left-hand helix depending on rotation and direction of thrust.
2. P_n = normal diametral pitch (pitch of cutter).
3. ϕ = approximate helix angle of one gear.
4. C = center distance.
5. N = number of teeth = nearest whole number to $CP_n \times \cos \phi$ for approximately 45°; and

$$CP_n \times \frac{\sin 2\phi}{\sqrt{1 + \sin 2\phi}} \text{ for any angle.}$$

To find:

1. α = helix angle of one gear

$$\sin 2\alpha = \frac{N^2}{2\,C^2 P_n^2} \pm \sqrt{\frac{N^2}{C^2 P_n^2} + \left(\frac{N^2}{2\,C^2 P_n^2}\right)^2}$$

2. β = helix angle of other gear = $90° - \alpha$.

3. D = pitch diameter of one gear = $\dfrac{N}{P_n \cos \alpha}$

4. d = pitch diameter of other gear = $\dfrac{N}{P_n \cos \beta}$

5. O = outside diameter of one gear = $D + \dfrac{2}{P_n}$

6. o = outside diameter of other gear = $d + \dfrac{2}{P_n}$

7. T = number of teeth marked on formed cutter for one gear = $\dfrac{N}{\cos^3 \alpha}$

8. t = number of teeth marked on formed cutter for other gear = $\dfrac{N}{\cos^3 \beta}$

9. L = lead of helix for one gear = $\pi D \cot \alpha$.
10. l = lead of helix for other gear = $\pi d \cot \beta$.

Example

Given or assumed:

1. See illustration.　　2. $P_n = 10$.　　3. $\phi = 45$ degrees.　　4. $C = 4$ inches.
5. $N = CP_n \cos \phi = 4 \times 10 \times 0.70711 = 28.28$, say 28 teeth.

To find:

1. $\sin 2\alpha = 0.98664$, or $\alpha = 40°\ 19'$.　　2. $\beta = 90° - \alpha = 49°\ 41'$.

3. $D = \dfrac{N}{P_n \cos \alpha} = \dfrac{28}{10 \times 0.76248} = 3.672$ inches.

4. $d = \dfrac{N}{P_n \cos \beta} = \dfrac{28}{10 \times 0\ 64701} = 4.328$ inches

5. $O = 3.672 + 0.2 = 3.872$ inches.

6. $o = 4.328 + 0.2 = 4.528$ inches.

7. $T = \dfrac{N}{\cos^3 \alpha} = \dfrac{28}{(0.762)^3} = 63.6$, say 64 teeth.

8. $t = \dfrac{N}{\cos^3 \beta} = \dfrac{28}{(0.647)^3} = 103.8$, say 104 teeth.

9. $L = \pi D \cot \alpha = 3.1416 \times 3.672 \times 1.1787 = 13.597$ inches.

10. $l = \pi d \cot \beta = 3.1416 \times 4.328 \times 0.84841 = 11.536$ inches.

7. Shafts at Right Angles, Ratio Other Than 1, Center Distance Approx. — Sum of helix angles of gear and pinion must equal 90 degrees.

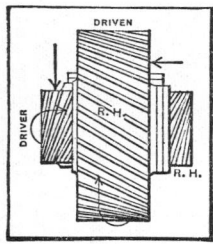

Given or assumed:

1. Position of gear having right- or left-hand helix, depending on rotation and direction in which thrust is to be received.
2. C_a = approximate center distance.
3. P_n = normal diametral pitch (pitch of cutter).
4. R = ratio of gear to pinion size.
5. n = number of teeth in pinion = $\dfrac{1.41\, C_a P_n}{R + 1}$ for 45 degrees; and $\dfrac{2\, C_a P_n \cos \alpha \cos \beta}{R \cos \beta + \cos \alpha}$ for any angle.
6. N = number of teeth in gear = nR.
7. α = angle of helix of gear.
8. β = angle of helix of pinion.

To find:

(a) When helix angles are 45 degrees.

1. D = pitch diameter of gear = $\dfrac{N}{0.70711\, P_n}$

2. d = pitch diameter of pinion = $\dfrac{n}{0.70711\, P_n}$

3. O = outside diameter of gear = $D + \dfrac{2}{P_n}$

4. o = outside diameter of pinion = $d + \dfrac{2}{P_n}$

5. T = number of formed cutter (gear) = $\dfrac{N}{0.353}$

6. t = number of formed cutter (pinion) = $\dfrac{n}{0.353}$

7. L = lead of helix of gear = πD.

8. l = lead of helix of pinion = πd.

9. C = center distance (exact) = $\dfrac{D + d}{2}$

(b) When helix angles are other than 45 degrees.

1. $D = \dfrac{N}{P_n \cos \alpha}$ 2. $d = \dfrac{n}{P_n \cos \beta}$ 3. $T = \dfrac{N}{\cos^3 \alpha}$

4. $t = \dfrac{n}{\cos^3 \beta}$ 5. $L = \pi D \cot \alpha$ 6. $l = \pi d \cot \beta$

Example

Given or assumed:

1. See illustration.   2. $C_a = 3.2$ inches.   3. $P_n = 10$.   4. $R = 1.5$.

5. $n = \dfrac{1.41\, C_a P_n}{R + 1} = \dfrac{1.41 \times 3.2 \times 10}{1.5 + 1} =$ say 18 teeth.

6. $N = nR = 18 \times 1.5 = 27$ teeth.

7. $\alpha = 45$ degrees.   8. $\beta = 45$ degrees.

To find:

1. $D = \dfrac{N}{0.70711\, P_n} = \dfrac{27}{0.70711 \times 10} = 3.818$ inches.

2. $d = \dfrac{n}{0.70711\, P_n} = \dfrac{18}{0.70711 \times 10} = 2.545$ inches.

3. $O = D + \dfrac{2}{P_n} = 3.818 + \dfrac{2}{10} = 4.018$ inches.

4. $o = d + \dfrac{2}{P_n} = 2.545 + \dfrac{2}{10} = 2.745$ inches.

5. $T = \dfrac{N}{0.353} = \dfrac{27}{0.353} = 76.5$, say 76 teeth.

6. $t = \dfrac{n}{0.353} = \dfrac{18}{0.353} = 51$ teeth.

7. $L = \pi D = 3.1416 \times 3.818 = 12$ inches.

8. $l = \pi d = 3.1416 \times 2.545 = 8$ inches.

9. $C = \dfrac{D + d}{2} = \dfrac{3.818 + 2.545}{2} = 3.182$ inches.

8A. Shafts at Right Angles, Ratios Other Than 1, Center Distance Exact.—

Gears have same direction of helix. The sum of the helix angles will equal 90 degrees.

Given or assumed:

1. Position of gear having right- or left-hand helix depending on rotation and direction in which thrust is to be received.
2. P_n = normal diametral pitch (pitch of cutter).
3. R = ratio of number of teeth in large gear to number of teeth in small gear.
4. α_a = approximate helix angle of large gear.
5. C = exact center distance.

To find:

1. n = number of teeth in small gear nearest

$$\frac{2\, C P_n \sin \alpha_a}{1 + R \tan \alpha_a}$$

2. N = number of teeth in large gear = Rn.

3. α = exact helix angle of large gear, found by trial from

$$R \sec \alpha + \operatorname{cosec} \alpha = \frac{2\, C P_n}{n}$$

4. β = exact helix angle of small gear = $90° - \alpha$.

5. D = pitch diameter of large gear = $N \div P_n \cos \alpha$.

6. d = pitch diameter of small gear = $n \div P_n \cos \beta$.

7. O = outside diameter of large gear = $D + (2 \div P_n)$.

8. o = outside diameter of small gear = $d + (2 \div P_n)$.

9. N' and n' = numbers of teeth marked on cutters for large and small gears (see page 1927).

10. L = lead of helix on large gear = $\pi D \cot \alpha$.

11. l = lead of helix on small gear = $\pi d \cot \beta$.

Example

Given or assumed:

1. See illustration. 2. $P_n = 8$. 3. $R = 3$. 4. $\alpha_a = 45$ degrees. 5. $C = 10$ in.

To find:

1. $n = \dfrac{2 \, C P_n \sin \alpha_a}{1 + R \tan \alpha_a} = \dfrac{2 \times 10 \times 8 \times 0.70711}{1 + 3} = 28.25$, say 28 teeth.

2. $N = Rn = 3 \times 28 = 84$ teeth.

3. $R \sec \alpha + \operatorname{cosec} \alpha = \dfrac{2 \, C P_n}{n} = \dfrac{2 \times 10 \times 8}{28} = 5.714$, or $\alpha = 46° 6'$.

4. $\beta = 90° - \alpha = 90° - 46° 6' = 43° 54'$.

5. $D = N \div P_n \cos \alpha = 84 \div (8 \times 0.6934) = 15.143$ inches.

6. $d = n \div P_n \cos \beta = 28 \div (8 \times 0.72055) = 4.857$ inches.

7. $O = D + \dfrac{2}{P_n} = 15.143 + 0.25 = 15.393$ inches.

8. $o = d + \dfrac{2}{P_n} = 4.857 + 0.25 = 5.107$ inches.

9. $N' = 275$; $n' = 94$ (see page 1927).

10. $L = \pi D \cot \alpha = 3.1416 \times 15.143 \times 0.96232 = 45.78$ inches.

11. $l = \pi d \cot \beta = 3.1416 \times 4.857 \times 1.0392 = 15.857$ inches.

8B. Shafts at Right Angles, Any Ratio, Helix Angle for Minimum Center Distance. — Diagram similar to 8A. Gears have same direction of helix. The sum of the helix angles will equal 90 degrees.

For any given ratio of gearing R there is a helix angle α for the larger gear and a helix angle $\beta = 90° - \alpha$ for the smaller gear that will make the center distance C a minimum. Helix angle α is found from the formula $\cot \alpha = \sqrt[3]{R}$. As an example, using the data found in Case 8A, helix angles α and β for minimum center distance would be: $\cot \alpha = \sqrt[3]{3} = 1.4422$; $\alpha = 34° 44'$ and $\beta = 90° - 34° 44' = 55° 16'$. Using these helix angles, $D = 12.777$; $d = 6.143$; and $C = 9.460$ from the formulas for D and d given under Case 8A.

9. Shafts at 45-Degree Angle, Ratio 1, Center Distance Approximate. —

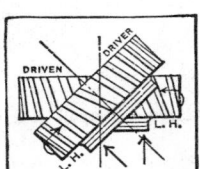

The sum of the helix angles of the two gears equals 45 degrees, and the gears are of the same hand, if each angle is less than 45 degrees. The difference between the helix angles equals 45 degrees, and the gears are of opposite hand, if either angle is greater than 45 degrees.

Given or assumed:

1. Hand of helix, depending on rotation and direction in which thrust is to be received.
2. C_a = approximate center distance.
3. P_n = normal diametral pitch (pitch of cutter).
4. α = angle of helix of driving gear.
5. β = angle of helix of driven gear.
6. N = number of teeth nearest $\dfrac{2\,C_a P_n \cos\alpha \cos\beta}{\cos\alpha + \cos\beta}$

To find:

(a) When helix angles are 22½ degrees.

1. D = pitch diameter = $\dfrac{N}{0.9239\,P_n}$ 2. O = outside diameter = $D + \dfrac{2}{P_n}$
3. T = number of teeth marked on formed cutter = $N \div 0.788$.
4. L = lead of helix = $7.584\,D$.
5. C = center distance = D.

(b) When helix angles are other than 22½ degrees.

1. D = pitch diameter of driver = $\dfrac{N}{P_n \cos\alpha}$

2. d = pitch diameter of driven gear = $\dfrac{N}{P_n \cos\beta}$

3. O = outside diameter of driver = $D + \dfrac{2}{P_n}$

4. o = outside diameter of driven gear = $d + \dfrac{2}{P_n}$

5. T = number of teeth marked on cutter for driver = $N \div \cos^3\alpha$.
6. t = number of teeth marked on cutter for driven gear = $N \div \cos^3\beta$.
7. L = lead of helix for driver = $\pi D \cot\alpha$.
8. l = lead of helix for driven gear = $\pi d \cot\beta$.
9. C = actual center distance = sum of pitch radii.

Example

Given or assumed:

1. See illustration.   2. C_a = 4 inches.   3. P_n = 10.
4 and 5. $\alpha = \beta = 22\tfrac{1}{2}$ deg.   6. N = 37.

To find:

1. $D = \dfrac{N}{0.9239\,P_n} = \dfrac{37}{0.9239 \times 10} = 4.005$ inches.

2. $O = D + \dfrac{2}{P_n} = 4.005 + \dfrac{2}{10} = 4.205$ inches.

3. $T = N \div 0.788 = 37 \div 0.788 = 47$ teeth.

4. $L = 7.584 D = 7.584 \times 4.005 = 30.374$ inches.

10. Shafts at 45-Degree Angle, Ratio 1, Center Distance Exact. —

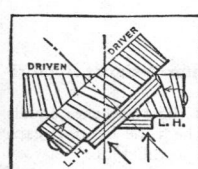

The sum of the helix angles of the two gears equals 45 degrees, and the gears are of the same hand, if each angle is less than 45 degrees. The difference between the helix angles equals 45 degrees, and the gears are of opposite hand, if either angle is greater than 45 degrees.

Given or assumed:

    1. Hand of helix, depending on rotation and direction. in which thrust is to be received.

    2. P_n = normal diametral pitch (pitch of cutter).

    3. C = center distance.

4. α_a = approximate helix angle of one gear.

5. β_a = approximate helix angle of the other gear.

6. N = number of teeth nearest $\dfrac{2\,CP_n \cos \alpha_a \cos \beta_a}{\cos \alpha_a + \cos \beta_a}$

To find:

1. α and β = exact helix angles found by trial from $\sec \alpha + \sec \beta = \dfrac{2\,CP_n}{N}$

2. D = pitch diameter of one gear = $\dfrac{N}{P_n \cos \alpha}$

3. d = pitch diameter of the other gear = $\dfrac{N}{P_n \cos \beta}$

4. O = outside diameter of one gear = $D + \dfrac{2}{P_n}$

5. o = outside diameter of other gear = $d + \dfrac{2}{P_n}$

6. T = number of teeth marked on formed cutter for one gear = $N \div \cos^3 \alpha$.

7. t = number of teeth marked on cutter for other gear = $N \div \cos^3 \beta$.

8. L = lead of helix for one gear = $\pi D \cot \alpha$.

9. l = lead of helix for other gear = $\pi d \cot \beta$.

Example

Given or assumed:

    1. See illustration.   2. $P_n = 8$.   3. $C = 10$ inches.   4. $\alpha_a = 15°$.   5. $\beta_a = 30°$.

6. $N = \dfrac{2\,CP_n \cos \alpha_a \cos \beta_a}{\cos \alpha_a + \cos \beta_a} = \dfrac{2 \times 10 \times 8 \times 0.96593 \times 0.86603}{0.96593 + 0.86603} = 73$ teeth.

To find:

1. α and β from $\sec \alpha + \sec \beta = \dfrac{2\,CP_n}{N} = \dfrac{2 \times 10 \times 8}{73} = 2.1918$; by trial α and β, respectively, $= 14° \, 44'$ and $30° \, 16'$.

2. $D = \dfrac{N}{P_n \cos \alpha} = \dfrac{73}{8 \times 0.96712} = 9.435$ inches.

3. $d = \dfrac{N}{P_n \cos \beta} = \dfrac{73}{8 \times 0.86369} = 10.565$ inches.

4. $O = D + \dfrac{2}{P_n} = 9.435 + \dfrac{2}{8} = 9.685$ inches.

5. $o = d + \dfrac{2}{P_n} = 10.565 + \dfrac{2}{8} = 10.815$ inches.

6. $T = N \div \cos^3 \alpha = 73 \div 0.904 = 81$ teeth.

7. $t = N \div \cos^3 \beta = 73 \div 0.645 = 113$ teeth.

8   $L = \pi D \cot \alpha = \pi \times 9.435 \times 3.803 = 112.72$ inches.

9. $l = \pi d \cot \beta = \pi \times 10.565 \times 1.714 = 56.889$ inches.

11. Shafts at 45-Degree Angle, Ratio Other Than 1, Center Distance Approximate. —

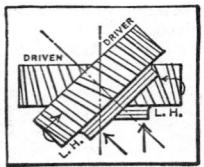

The sum of the helix angles of the two gears equals 45 degrees, and the gears are of the same hand, if each angle is less than 45 degrees. The difference between the helix angles equals 45 degrees, and the gears are of opposite hand, if either angle is greater than 45 degrees.

Given or assumed:

    1. Hand of helix, depending on rotation and direction in which thrust is to be received.

    2. C_a = center distance.

    3. P_n = normal diametral pitch (pitch of cutter).

4. R = ratio of gear to pinion size, $N \div n$.

5. α = angle of helix, gear.

6. β = angle of helix, pinion.

7. n = number of pinion teeth nearest $\dfrac{2\,C_a P_n \cos \alpha \cos \beta}{R \cos \beta + \cos \alpha}$

8. N = number of gear teeth = Rn.

To find:

   (*a*) When $\alpha = \beta = 22\frac{1}{2}$ degrees.

    1. D = pitch diameter of gear = $\dfrac{N}{0.9239\,P_n}$

    2. d = pitch diameter of pinion = $\dfrac{n}{0.9239\,P_n}$

    3. O = outside diameter of gear = $D + \dfrac{2}{P_n}$

    4. o = outside diameter of pinion = $d + \dfrac{2}{P_n}$

    5. T = number of teeth marked on formed cutter for gear = $N \div 0.788$.

    6. t = number of teeth marked on formed cutter for pinion = $n \div 0.788$.

    7. L = lead of helix on gear = $7.584\,D$.

    8. l = lead of helix on pinion = $7.584\,d$.

    9. C = actual center distance = $\dfrac{D + d}{2}$

   (*b*) When α and β are any angles.

    1. D = pitch diameter of gear = $\dfrac{N}{P_n \cos \alpha}$

    2. d = pitch diameter of pinion = $\dfrac{n}{P_n \cos \beta}$

    3. O = outside diameter of gear = $D + \dfrac{2}{P_n}$

4. o = outside diameter of pinion = $d + \dfrac{2}{P_n}$

5. T = number of teeth marked on formed cutter for gear = $N \div \cos^3 \alpha$.

6. t = number of teeth marked on formed cutter for pinion = $n \div \cos^3 \beta$.

7. L = lead of helix on gear = $\pi D \cot \alpha$.

8. l = lead of helix on pinion = $\pi d \cot \beta$.

9. C = actual center distance = $\dfrac{D + d}{2}$

Example

Given or assumed:

  1. See illustration.   2. $C = 12$ inches.   3. $P_n = 6$.   4. $R = 3$.

  5. $\alpha = 20$ deg.   6. $\beta = 25$ deg.

  7. $n = \dfrac{2\,C P_n \cos \alpha \cos \beta}{R \cos \beta + \cos \alpha} = \dfrac{2 \times 12 \times 6 \times 0.93969 \times 0.90631}{(3 \times 0.90631) + 0.93969}$ = 34 teeth, approx.

  8. $N = Rn = 3 \times 34 = 102$ teeth.

To find:

1. $D = \dfrac{N}{P_n \cos \alpha} = \dfrac{102}{6 \times 0.93969} = 18.091$ inches.

2. $d = \dfrac{n}{P_n \cos \beta} = \dfrac{34}{6 \times 0.90631} = 6.252$ inches.

3. $O = D + \dfrac{2}{P_n} = 18.091 + \dfrac{2}{6} = 18.424$ inches.

4. $o = d + \dfrac{2}{P_n} = 6.252 + \dfrac{2}{6} = 6.585$ inches.

5. $T = N \div \cos^3 \alpha = 102 \div 0.83 = 123$ teeth.

6. $t = n \div \cos^3 \beta = 34 \div 0.744 = 46$ teeth.

7. $L = \pi D \cot \alpha = \pi \times 18.091 \times 2.747 = 156.12$ inches.

8. $l = \pi d \cot \beta = \pi \times 6.252 \times 2.145 = 42.13$ inches.

9. $C = \dfrac{D + d}{2} = \dfrac{18.091 + 6.252}{2} = 12.1715$ inches.

12. Shafts at 45-Degree Angle, Ratio Other Than 1, Center Distance Exact.—

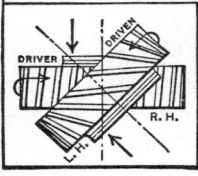

The sum of the helix angles of the two gears equals 45 degrees, and the gears are of the same hand, if each angle is less than 45 degrees. The difference between the helix angles equals 45 degrees, and the gears are of opposite hand, if either angle is greater than 45 degrees.

Given or assumed:

  1. Hand of helix, depending on rotation and direction in which thrust is to be received.

  2. P_n = normal diametral pitch (pitch of cutter).

3. R = ratio of large to small gear size = $N \div n$.

4. α_a = approximate helix angle of large gear.

5. β_a = approximate helix angle of small gear.

6. C = center distance.

7. n = number of teeth in small gear nearest $\dfrac{2\,CP_n \cos \alpha_a \cos \beta_a}{R \cos \beta_a + \cos \alpha_a}$

8. N = number of teeth, large gear = Rn.

To find:

1. α and β, exact helix angles, by trial from $R \sec \alpha + \sec \beta = \dfrac{2\,CP_n}{n}$

2. D = pitch diameter of large gear = $\dfrac{N}{P_n \cos \alpha}$

3. d = pitch diameter of small gear = $\dfrac{n}{P_n \cos \beta}$

4. O = outside diameter of large gear = $D + \dfrac{2}{P_n}$

5. o = outside diameter of small gear = $d + \dfrac{2}{P_n}$

6. T = number of teeth marked on formed cutter for large gear = $N \div \cos^3 \alpha$.

7. t = number of teeth marked on formed cutter for small gear = $n \div \cos^3 \beta$.

8. L = lead of helix for large gear = $\pi D \cot \alpha$.

9. l = lead of helix for small gear = $\pi d \cot \beta$.

Example

Given or assumed:

1. See illustration. 2. $P_n = 4$. 3. $R = 4$.

4. $\alpha_a = 50$ degrees. 5. $\beta_a = 5$ degrees. 6. $C = 30$ inches.

7. $n = \dfrac{2\,CP_n \cos \alpha_a \cos \beta_a}{R \cos \beta_a + \cos \alpha_a} = \dfrac{2 \times 30 \times 4 \times 0.643 \times 0.996}{(4 \times 0.996) + 0.643} = 33$ teeth.

8. $N = Rn = 4 \times 33 = 132$ teeth.

To find:

1. α and β from $R \sec \alpha + \sec \beta = \dfrac{2\,CP_n}{n} = \dfrac{2 \times 30 \times 4}{33} = 7.273$. By trial $\alpha = 50° 21'$, and $\beta = 5° 21'$.

2. $D = \dfrac{N}{P_n \cos \alpha} = \dfrac{132}{4 \times 0.63810} = 51.716$ inches.

3. $d = \dfrac{n}{P_n \cos \beta} = \dfrac{33}{4 \times 0.99564} = 8.286$ inches.

4. $O = D + \dfrac{2}{P_n} = 51.716 + \dfrac{2}{4} = 52.216$ inches.

5. $o = d + \dfrac{2}{P_n} = 8.286 + \dfrac{2}{4} = 8.786$ inches.

6. $T = N \div \cos^3 \alpha = 132 \div 0.26 = 508$ teeth.

7. $t = n \div \cos^3 \beta = 33 \div 0.987 = 33$ teeth.

8. $L = \pi D \cot \alpha = \pi \times 51.716 \times 0.82874 = 134.6$ inches.

9. $l = \pi d \cot \beta = \pi \times 8.286 \times 10.678 = 278$ inches.

13. Shafts at any Angle, Ratio 1, Center Distance Approximate. —

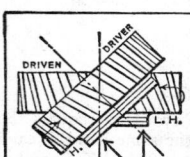

The sum of the helix angles of the two gears equals the shaft angle, and the gears are of the same hand, if each angle is less than the shaft angle. The difference between the helix angles equals the shaft angle, and the gears are of opposite hand, if either angle is greater than the shaft angle.

Given or assumed:

    1. Hand of helix, depending on rotation and direction in which thrust is to be received.

2. C_a = approximate center distance.
3. P_n = normal diametral pitch (pitch of cutter).
4. α = angle of helix of one gear.
5. β = angle of helix of other gear.

6. N = number of teeth nearest $\dfrac{2\,C_a P_n \cos \alpha \cos \beta}{\cos \alpha + \cos \beta}$

To find:

1. D = pitch diameter of one gear = $\dfrac{N}{P_n \cos \alpha}$

2. d = pitch diameter of other gear = $\dfrac{N}{P_n \cos \beta}$

3. O = outside diameter of one gear = $D + \dfrac{2}{P_n}$

4. o = outside diameter of other gear = $d + \dfrac{2}{P_n}$

5. T = number of teeth marked on formed cutter for one gear = $N \div \cos^3 \alpha$.
6. t = number of teeth marked on formed cutter for other gear = $N \div \cos^3 \beta$.
7. L = lead of helix for one gear = $\pi D \cot \alpha$.
8. l = lead of helix for other gear = $\pi d \cot \beta$.

9. C = actual center distance = $\dfrac{D + d}{2}$

Example

Given or assumed (angle of shafts, 30 degrees):

  1. See illustration.   2. C_a = 5 inches.   3. P_n = 10.
  4. α = 20 degrees.   5. β = 10 degrees.   6. N = 48.

To find:

1. $D = \dfrac{N}{P_n \cos \alpha} = \dfrac{48}{10 \times 0.9397} = 5.108$ inches.

2. $d = \dfrac{N}{P_n \cos \beta} = \dfrac{48}{10 \times 0.9848} = 4.874$ inches.

3. $O = D + \dfrac{2}{P_n} = 5.108 + \dfrac{2}{10} = 5.308$ inches.

4. $o = d + \dfrac{2}{P_n} = 4.874 + \dfrac{2}{10} = 5.074$ inches.

5. $T = N \div \cos^3 \alpha = 48 \div 0.83 = 58$ teeth.
6. $t = N \div \cos^3 \beta = 48 \div 0.96 = 50$ teeth.

7. $L = \pi D \cot \alpha = \pi \times 5.108 \times 2.747 = 44.08$ inches.
8. $l = \pi d \cot \beta = \pi \times 4.874 \times 5.671 = 86.84$ inches.
9. $C = \dfrac{D + d}{2} = \dfrac{5.108 + 4.874}{2} = 4.991$ inches.

14. Shafts at Any Angle, Ratio 1, Center Distance Exact. —

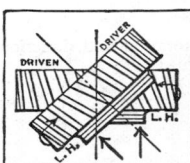

The sum of the helix angles of the two gears equals the shaft angle, and the gears are of the same hand, if each angle is less than the shaft angle. The difference between the helix angles equals the shaft angle, and the gears are of opposite hand, if either angle is greater than the shaft angle.

Given or assumed:

1. Hand of helix, depending on rotation and direction in which thrust is to be received.

2. C = center distance.
3. P_n = normal diametral pitch (pitch of cutter).
4. α_a = approximate helix angle of one gear.
5. β_a = approximate helix angle of other gear.
6. N = number of teeth nearest $\dfrac{2\,CP_n \cos \alpha_a \cos \beta_a}{\cos \alpha_a + \cos \beta_a}$

To find:

1. α and β = exact helix angles, found by trial from $\sec \alpha + \sec \beta = \dfrac{2\,CP_n}{N}$

2. D = pitch diameter of one gear = $\dfrac{N}{P_n \cos \alpha}$

3. d = pitch diameter of other gear = $\dfrac{N}{P_n \cos \beta}$

4. O = outside diameter of one gear = $D + \dfrac{2}{P_n}$

5. o = outside diameter of other gear = $d + \dfrac{2}{P_n}$

6. T = number of teeth marked on formed cutter for one gear = $N \div \cos^3 \alpha$.
7. t = number of teeth marked on formed cutter for other gear = $N \div \cos^3 \beta$.
8. L = lead of helix for one gear = $\pi D \cot \alpha$.
9. l = lead of helix for other gear = $\pi d \cot \beta$.

Example

Given or assumed (angle of shafts, 50 degrees):

1. See illustration. 2. $C = 10$ inches. 3. $P_n = 10$. 4. $\alpha_a = 20$ deg.
5. $\beta_a = 30$ deg.
6. $N = \dfrac{2\,CP_n \cos \alpha_a \cos \beta_a}{\cos \alpha_a + \cos \beta_a} = \dfrac{2 \times 10 \times 10 \times 0.93969 \times 0.86603}{0.93969 + 0.86603} = 90$ teeth.

To find:

1. α and β from $\sec \alpha + \sec \beta = \dfrac{2\,CP_n}{N} = \dfrac{2 \times 10 \times 10}{90} = 2.222$. By trial α and β, respectively, = $19° 20'$ and $30° 40'$.

2. $D = \dfrac{N}{P_n \cos \alpha} = \dfrac{90}{10 \times 0.94361} = 9.537$ inches.

3. $d = \dfrac{N}{P_n \cos \beta} = \dfrac{90}{10 \times 0.86015} = 10.463$ inches.

4. $O = D + \dfrac{2}{P_n} = 9.537 + \dfrac{2}{10} = 9.737$ inches.

5. $o = d + \dfrac{2}{P_n} = 10.463 + \dfrac{2}{10} = 10.663$ inches.

6. $T = N \div \cos^3 \alpha = 90 \div 0.84 = 107$ teeth.

7. $t = N \div \cos^3 \beta = 90 \div 0.64 = 141$ teeth.

8. $L = \pi D \cot \alpha = \pi \times 9.537 \times 2.85 = 85.39$ inches.

9. $l = \pi d \cot \beta = \pi \times 10.463 \times 1.686 = 55.42$ inches.

15. Shafts at Any Angle, Ratio Other Than 1, Center Distance Approx. —

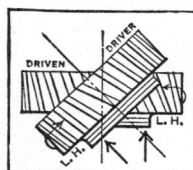

The sum of the helix angles of the two gears equals the shaft angle, and the gears are of the same hand, if each angle is less than the shaft angle. The difference between the helix angles equals the shaft angle, and the gears are of opposite hand, if either angle is greater than the shaft angle.

Given or assumed:

1. Hand of helix, depending on rotation and direction in which thrust is to be received.

2. C_a = center distance.

3. P_n = normal diametral pitch (pitch of cutter).

4. R = ratio of gear to pinion = $N \div n$.

5. α = angle of helix, gear.

6. β = angle of helix, pinion.

7. n = number of teeth in pinion nearest $\dfrac{2 C_a P_n \cos \alpha \cos \beta}{R \cos \beta + \cos \alpha}$ for any angle,

and $\dfrac{2 C_a P_n \cos \alpha}{R + 1}$ when both angles are equal.

8. N = number of teeth in gear = Rn.

To find:

1. D = pitch diameter of gear = $\dfrac{N}{P_n \cos \alpha}$

2. d = pitch diameter of pinion = $\dfrac{n}{P_n \cos \beta}$

3. O = outside diameter of gear = $D + \dfrac{2}{P_n}$

4. o = outside diameter of pinion = $d + \dfrac{2}{P_n}$

5. T = number of teeth marked on cutter for gear = $N \div \cos^3 \alpha$.

6. t = number of teeth marked on cutter for pinion = $n + \cos^3 \beta$.

7. L = lead of helix on gear = $\pi D \cot \alpha$.

8. l = lead of helix on pinion = $\pi d \cot \beta$.

9. C = actual center distance = $\dfrac{D + d}{2}$

Example

Given or assumed (angle of shafts, 60 degrees):

1. See illustration. 2. $C_a = 12$ inches. 3. $P_n = 8$.
4. $R = 4$. 5. $\alpha = 30$ degrees. 6. $\beta = 30$ degrees

7. $n = \dfrac{2 C_a P_n \cos \alpha}{R + 1} = \dfrac{2 \times 12 \times 8 \times 0.86603}{4 + 1} = 33$ teeth.

8. $N = 4 \times 33 = 132$ teeth.

To find:

1. $D = \dfrac{N}{P_n \cos \alpha} = \dfrac{132}{8 \times 0.86603} = 19.052$ inches.

2. $d = \dfrac{n}{P_n \cos \beta} = \dfrac{33}{8 \times 0.86603} = 4.763$ inches.

3. $O = D + \dfrac{2}{P_n} = 19.052 + \dfrac{2}{8} = 19.302$ inches.

4. $o = d + \dfrac{2}{P_n} = 4.763 + \dfrac{2}{8} = 5.013$ inches.

5. $T = N \div \cos^3 \alpha = 132 \div 0.65 = 203$ teeth.

6. $t = n \div \cos^3 \beta = 33 \div 0.65 = 51$ teeth.

7. $L = \pi D \cot \alpha = \pi \times 19.052 \times 1.732 = 103.66$ inches.

8. $l = \pi d \cot \beta = \pi \times 4.763 \times 1.732 = 25.92$ inches.

9. $C = \dfrac{D + d}{2} = \dfrac{19.052 + 4.763}{2} = 11.9075$ inches.

16. Shafts at Any Angle, Ratio Other Than 1, Center Distance Exact. —

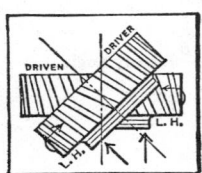

The sum of the helix angles of the two gears equals the shaft angle, and the gears are of the same hand, if each angle is less than the shaft angle. The difference between the helix angles equals the shaft angle, and the gears are of opposite hand, if either angle is greater than the shaft angle.

Given or assumed:

1. Hand of helix, depending on rotation and direction in which thrust is to be received.
2. C = center distance.
3. P_n = normal diametral pitch (pitch of cutter).
4. α_a = approximate helix angle of gear.
5. β_a = approximate helix angle of pinion.
6. R = ratio of gear to pinion size = $N \div n$.
7. n = number of pinion teeth nearest $\dfrac{2 C P_n \cos \alpha_a \cos \beta_a}{R \cos \beta_a + \cos \alpha_a}$
8. N = number of gear teeth = Rn.

To find:

1. α and β, exact helix angles, found by trial from $R \sec \alpha + \sec \beta = \dfrac{2 C P_n}{n}$

2. D = pitch diameter of gear = $\dfrac{N}{P_n \cos \alpha}$

3. d = pitch diameter of pinion = $\dfrac{n}{P_n \cos \beta}$

4. O = outside diameter of gear = $D + \dfrac{2}{P_n}$

5. o = outside diameter of pinion = $d + \dfrac{2}{P_n}$

6. N' = number of teeth marked on formed cutter for gear (see below).
7. n' = number of teeth marked on formed cutter for pinion (see below).
8. L = lead of helix on gear = $\pi D \cot \alpha$.
9. l = lead of helix on pinion = $\pi d \cot \beta$.

Selecting Cutter for Milling Helical Gears. — The proper milling cutter to use for *spur* gears depends upon the pitch of the teeth and also upon the number of teeth as explained on page 1794 but a cutter for milling helical gears is not selected with reference to the actual number of teeth in the gear, as in spur gearing, but rather with reference to a calculated number N' that takes into account the effect on the tooth profile of lead angle, normal diametral pitch, and cutter diameter.

In the helical gearing examples on pages 1910–1927 the number of teeth N' on which to base the selection of the cutter has been determined using the approximate formula $N' = N \div \cos^3 \alpha$ or $N' = N \sec^3 \alpha$, where N = the actual number of teeth in the helical gear and α = the helix angle. However, the use of this formula may,

Factors for Selecting Cutters for Milling Helical Gears

Helix Angle, α	K	K'	Helix Angle, α	K	K'	Helix Angle, α	K	K'	Helix Angle, α	K	K'
0	1.000	0	16	1.127	0.082	32	1.640	0.390	48	3.336	1.233
1	1.001	0	17	1.145	0.093	33	1.695	0.422	49	3.540	1.323
2	1.002	0.001	18	1.163	0.106	34	1.755	0.455	50	3.767	1.420
3	1.004	0.003	19	1.182	0.119	35	1.819	0.490	51	4.012	1.525
4	1.007	0.005	20	1.204	0.132	36	1.889	0.528	52	4.284	1.638
5	1.011	0.008	21	1.228	0.147	37	1.963	0.568	53	4.586	1.761
6	1.016	0.011	22	1.254	0.163	38	2.044	0.610	54	4.925	1.894
7	1.022	0.015	23	1.282	0.180	39	2.130	0.656	55	5.295	2.039
8	1.030	0.020	24	1.312	0.198	40	2.225	0.704	56	5.710	2.198
9	1.038	0.025	25	1.344	0.217	41	2.326	0.756	57	6.190	2.371
10	1.047	0.031	26	1.377	0.238	42	2.436	0.811	58	6.720	2.561
11	1.057	0.038	27	1.414	0.260	43	2.557	0.870	59	7.321	2.770
12	1.068	0.045	28	1.454	0.283	44	2.687	0.933	60	8.000	3.000
13	1.080	0.053	29	1.495	0.307	45	2.828	1	61	8.780	3.254
14	1.094	0.062	30	1.540	0.333	46	2.983	1.072	62	9.658	3.537
15	1.110	0.072	31	1.588	0.361	47	3.152	1.150	63	10.687	3.852

$K = 1 \div \cos^3 \alpha = \sec^3 \alpha$; $K' = \tan^2 \alpha$

Outside and Pitch Diameters of Standard Involute-form Milling Cutters*

Normal Diametral Pitch, P_n	Outside Diam., D_o	Pitch Diam., D_c	$Q = P_n D_c$	Normal Diametral Pitch, P_n	Outside Diam., D_o	Pitch Diam., D_c	$Q = P_n D_c$	Normal Diametral Pitch, P_n	Outside Diam., D_o	Pitch Diam., D_c	$Q = P_n D_c$
1	8.500	6.18	6.18	6	3.125	2.76	16.56	20	2.000	1.89	37.80
1¼	7.750	5.70	7.12	7	2.875	2.54	17.78	24	1.750	1.65	39.60
1½	7.000	5.46	8.19	8	2.875	2.61	20.88	28	1.750	1.67	46.76
1¾	6.500	5.04	8.82	9	2.750	2.50	22.50	32	1.750	1.68	53.76
2	5.750	4.60	9.20	10	2.375	2.14	21.40	36	1.750	1.69	60.84
2½	5.750	4.83	12.08	12	2.250	2.06	24.72	40	1.750	1.70	68.00
3	4.750	3.98	11.94	14	2.125	1.96	27.44	48	1.750	1.70	81.60
4	4.250	3.67	14.68	16	2.125	1.98	31.68	..	.....	....	.....
5	3.750	3.29	16.45	18	2.000	1.87	33.66	..	.....	....	.....

* The pitch diameters shown in the table are computed from the formula: $D_c = D_o - 2(1.157 \div P_n)$. This same formula may be used to compute the pitch diameter of a non-standard outside diameter cutter when the normal diametral pitch P_n and the outside diameter D_o are known.

where a combination of high helix angle and low tooth number is involved, result in the selection of a higher number of cutter than should actually be used for greatest accuracy. This condition is most likely to occur when the afore-mentioned formula is used to calculate N' for gears of high helix angle and low number of teeth.

To avoid the possibility of error in choice of cutter number, the following formula, which gives theoretically correct results for all combinations of helix angle and tooth numbers, is to be preferred:

$$N' = N \sec^3 \alpha + P_n D_c \tan^2 \alpha \qquad (1)$$

where: N' = number of teeth on which to base selection of cutter number from table on page 1794; N = actual number of teeth in helical gear; α = helix angle; P_n = normal diametral pitch of gear and cutter; and D_c = pitch diameter of cutter.

To simplify calculations, Formula (1) may be written as follows:

$$N' = NK + QK' \qquad (2)$$

In this formula, K, K' and Q are constants obtained from the tables on page 1927.

Example: Helix angle = 30 degrees; number of teeth in helical gear = 15; and normal diametral pitch = 20. From the tables on page 1927 K, K', and Q are, respectively, 1.540, 0.333, and 37.80.

$$N' = (15 \times 1.540) + (37.80 \times 0.333) = 23.10 + 12.60$$
$$= 35.70, \text{ say, } 36.$$

Hence, from page 1794 select a number 3 cutter. Had the approximate formula been used, then a number 5 cutter would have been selected on the basis of $N' = 23$.

Milling the Helical Teeth. — The teeth of a helical gear are proportioned from the normal pitch and not the circular pitch. The whole depth of the tooth can be found by dividing 2.157 by the normal diametral pitch of the gear, which corresponds to the pitch of the cutter. The thickness of the tooth at the pitch line equals 1.571 divided by the normal diametral pitch. After a tooth space has been milled, the cutter should be prevented from dragging through it when being returned for another cut. This can be done by lowering the blank slightly, or by stopping the machine and turning the cutter to such a position that the teeth will not touch the work. If the gear has teeth coarser than 10 or 12 diametral pitch, it is well to take a roughing and a finishing cut. When pressing a helical gear blank on the arbor, it should be remembered that it is more likely to slip when being milled than a spur gear, because the pressure of the cut, being at an angle, tends to rotate the blank on the arbor.

Angular Position of Table: When cutting a helical gear on a milling machine, the table is set to the helix angle of the gear. If the lead of the helical gear is known, but not the helix angle, the helix angle is determined by multiplying the pitch diameter of the gear by 3.1416 and dividing this product by the lead; the result is the tangent of the lead angle which may be obtained from trigonometric tables or a calculator.

American National Standard Fine-Pitch Teeth for Helical Gears. — This Standard, ANSI B6.7-1977, provides a 20-degree tooth form for both spur and helical gears of 20 diametral pitch and finer. Formulas for tooth parts are given on page 1775.

Enlargement of Helical Pinions of 20-Degree Normal Pressure Angle: To avoid undercutting of the teeth and to provide more favorable contact conditions near the base of the tooth, it is recommended that helical pinions with less than 24 teeth be enlarged in accordance with the following graph and formulas. As in the case of enlarged spur pinions, when an enlarged helical pinion is used it is necessary either to reduce the diameter of the mating gear or to increase the center distance. In the formulas that follow, ϕ_n = normal pressure angle; ϕ_t = transverse pressure angle;

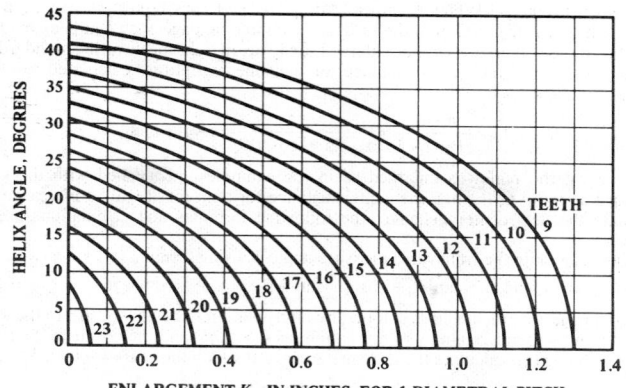

ENLARGEMENT K_h, IN INCHES, FOR 1 DIAMETRAL PITCH,
20° PRESSURE ANGLE PINIONS

ψ = helix angle of pinion; P_n = normal diametral pitch; P_t = transverse diametral pitch; d = pitch diameter of pinion; d_o = outside diameter of enlarged pinion; K_h = enlargement for full depth pinions of 1 normal diametral pitch; and n = number of teeth in pinion.

$$P_t = P_n \cos \psi \tag{1}$$

$$d = n \div P_t \tag{2}$$

$$\tan \phi_t = \tan \phi_n \div \cos \psi \tag{3}$$

$$K_h = 2.1 - \frac{n}{\cos \psi} (\sin \phi_t - \cos \phi_t \tan 5°) \sin \phi_t \tag{4}$$

$$d_o = d + \frac{2 + K_h}{P_n} \tag{5}$$

Formula (4) and the accompanying graph are based on the use of hobs having sharp corners at their top lands. Pinions cut by shaper cutters may not require as much modification as indicated by formula (4) or the graph. The number 2.1 appearing in (4) results from the use of a standard tooth thickness rack having an addendum of $1.05/P_n$ which will start contact at a roll angle 5 degrees above the base radius. The roll angle of 5 degrees is also reflected in formula (4).

To eliminate the need for making the calculations indicated in formulas (3) and (4), the accompanying graph may be used to obtain the value of K_h directly for full-depth pinions of 20-degree normal pressure angle.

Example: Find the outside diameter of a helical pinion having 12 teeth, 32 normal diametral pitch, 20-degree pressure angle, and 18-degree helix angle.

$$P_t = P_n \cos \psi = 32 \cos 18° = 32 \times 0.95106 = 30.4339$$

$$d = n \div P_t = 12 \div 30.4339 = 0.3943 \text{ inch}$$

$$K_h = 0.851 \text{ (from graph)}$$

$$d_o = 0.3943 + \frac{2 + 0.851}{32} = 0.4834$$

Center Distance at Which Modified Mating Helical Gears Will Mesh with no Backlash. — If the helical pinion in the previous example on page 1929 had been made to standard dimensions, that is, not enlarged, and was in tight mesh with a standard 24-tooth mating gear, the center distance for tight mesh could be calculated from the formula on page 1775:

$$C = \frac{n + N}{2P_n \cos \psi} = \frac{12 + 24}{2 \times 32 \times \cos 18°} = 0.5914 \text{ inch} \qquad (1)$$

However, if the pinion is enlarged as in the example and meshed with the same standard 24-tooth gear, then the center distance for tight mesh will be increased. To calculate the new center distance, the following formulas and calculations are required:

First, calculate the transverse pressure angle ϕ_t using formula (2):

$$\tan \phi_t = \tan \phi_n \div \cos \psi = \tan 20° \div \cos 18° = 0.38270 \qquad (2)$$

and from page 102, the angle ϕ_t is found by interpolation to be $20° \, 56' \, 30''$. In the same table, inv ϕ_t is found, again by interpolation, as 0.017196, and the cosine as 0.93394.

Next, using (3), calculate the pressure angle ϕ at which the gears are in tight mesh:

$$\text{inv } \phi = \text{inv } \phi_t + \frac{(t_{nP} + t_{nG}) - \pi}{n + N} \qquad (3)$$

In this formula, the value for t_{nP} for 1 diametral pitch is that found in Table 3c, page 1800, for a 12-tooth pinion, in the fourth column: 1.94703. The value of t_{nG} for 1 diametral pitch for a standard gear is always 1.5708.

$$\text{inv } \phi = 0.017196 + \frac{(1.94703 + 1.5708) - \pi}{12 + 24} = 0.027647.$$

From the table on page 106, 0.027647 is the involute of $24° \, 22' \, 7''$ and the cosine corresponding to this angle is 0.91091.

Finally, using formula (4), the center distance for tight mesh, C' is found:

$$C' = \frac{C \cos \phi_t}{\cos \phi} = \frac{0.5914 \times 0.93394}{0.91091} = 0.606 \text{ inch} \qquad (4)$$

Change-gears for Helical Gear Hobbing. — If a gear-hobbing machine is not equipped with a differential, there is a fixed relation between the index and feed gears and it is necessary to compensate for even slight errors in the index gear ratio, to avoid excessive lead errors. This may be done readily (as shown by the example to follow) by modifying the ratio of the feed gears slightly, thus offsetting the index gear error and making very accurate leads possible.

Machine Without Differential: The formulas which follow may be applied in computing the index gear ratio.

R = index-gear ratio
L = lead of gear, inches
F = feed per gear revolution, inch
K = machine constant
T = number of threads on hob

N = number of teeth on gear
P_n = normal diametral pitch
P_{nc} = normal circular pitch
A = helix angle, relative to axis
M = feed gear constant

$$R = \frac{L \div F}{(L \div F) \pm 1} \times \frac{KT}{N} = \frac{L}{L \pm F} \times \frac{KT}{N} = \frac{\text{Driving gear sizes}}{\text{Driven gear sizes}} \qquad (1)$$

Use minus (−) sign in formulas (1) and (2) when gear and hob are the same "hand" and plus (+) sign when they are of opposite hand; when *climb* hobbing is to be used, reverse this rule.

$$R = \frac{KT}{N \pm \dfrac{P_n \times \sin A \times F}{\pi}} = \frac{KT}{N \pm \dfrac{\sin A \times F}{P_{nc}}} \qquad (2)$$

$$\text{Ratio of feed gears} = \frac{F}{M}; \quad F = \frac{L(NR - KT)}{NR} \qquad (3)$$

$$L = \frac{FNR}{(NR - KT)} = \text{lead obtained with available index and feed gears} \qquad (4)$$

Note: If gear and hob are of opposite hand, then in Formulas (3) and (4) change $(NR - KT)$ to $(KT - NR)$. This change is also made if gear and hob are of same hand but *climb* hobbing is used.

Example: A right-hand helical gear with 48 teeth of 10 normal diametral pitch, has a lead of 44.0894 inches. The feed is to be 0.035 inch, with whatever slight adjustment may be necessary to compensate for the error in available index gears. $K = 30$ and $M = 0.075$. A single-thread right-hand hob is to be used.

$$R = \frac{44.0894}{44.0894 - 0.035} \times \frac{30 \times 1}{48} = 0.62549654;$$

Log $0.62549654 = \bar{1}.7962249$ (seven-place table).

Since the ratio is less than 1 in this case, eliminate the minus characteristic ($\bar{1}$) by subtracting log of desired ratio, from log of 1 which is 0.00000, thus obtaining log of reciprocal of ratio. Later, the fractions representing gear sizes will be inverted to obtain the actual ratio. If ratio is greater than 1, use log of desired ratio.

$$\begin{aligned}
0.000000 &= \text{Log of 1} \\
- \bar{1}.796225 &= \text{Log of ratio desired (rounded to six places)} \\
\hline
0.203775 &= \text{Log of reciprocal of ratio}
\end{aligned}$$

Note that $0.000000 - \bar{1}.796225 = 0.000000 - (-1 + .796225) = 0.000000 + 1 - .796225 = 0.203775.$

Select trial log from table of logarithms of gear ratios (see pages 1679–1702) that is equal approximately to one-half log of reciprocal. Then proceed as follows:

1. Subtract from log of reciprocal, trial log selected from table.
2. Compare difference between logs in step 1, with nearest log in table.
   Repeat step 1 until difference is restricted to at least the fifth place.

Assume that 0.101873 is selected as trial log.

$$\begin{aligned}
0.203775 &= \text{log of reciprocal} \\
- 0.101873 &= \text{trial log} \\
\hline
0.101902 &= \text{difference (not close to any table log)}
\end{aligned}$$

$$\begin{aligned}
0.203775 &= \text{log of reciprocal} \\
- 0.101990 &= \text{second trial log} \\
\hline
0.101785 &= \text{difference (0.101799 nearest log in table)}
\end{aligned}$$

Second trial log $0.101990 = \log \dfrac{43}{34}$ log nearest final difference or $0.101799 = \log \dfrac{67}{53}$. Inverting, obtainable index ratio $= \dfrac{34}{43} \times \dfrac{53}{67} = 0.62547726.$

(If desired index ratio is greater than 1 and log of index ratio is used instead of log of reciprocal of this ratio, the fractions obtained are *not* inverted. The procedure otherwise is the same as here outlined.)

Index ratio error = 0.62549654 − 0.62547726 = 0.00001928.

Now use formula (3) to find slight change required in rate of feed. This change compensates sufficiently for the error in available index gears.

Change in Feed Rate: Insert in formula (3) obtainable index ratio.

$$F = \frac{44.0894 \times (48 \times 0.62547726 - 30)}{48 \times 0.62547726} = 0.0336417$$

Modified feed gear ratio $= \dfrac{F}{M} = \dfrac{0.0336417}{0.075} = 0.448556$

Log 0.448556 = $\bar{1}$.651817; log of reciprocal = 0.348183.

To find close approximation to modified feed gear ratio, proceed as in finding suitable gears for index ratio, thus obtaining $\dfrac{106}{71} \times \dfrac{112}{75}$. Inverting, modified feed gear

ratio $= \dfrac{71}{106} \times \dfrac{75}{112} = 0.448534.$

Modified feed F = obtainable modified feed ratio × M = 0.448534 × 0.075 = 0.03364 inch. If the feed rate is not modified, even a small error in the index gear ratio may result in an excessive lead error.

Checking Accuracy of Lead: The modified feed and obtainable index ratio are inserted in formula (4). Desired lead = 44.0894 inches. Lead obtained = 44.087196 inches; hence the computed error = 44.0894 − 44.087196 = 0.002204 inch or about 0.00005 inch per inch of lead.

Machine with Differential: If a machine is equipped with a differential, the *lead gears* are computed in order to obtain the required helix angle and lead. The instructions of the hobbing machine manufacturer should be followed in computing the lead gears, because the ratio formula is affected by the location of the differential gears. If these gears are *ahead* of the index gears, the lead gear ratio is not affected by a change in the number of teeth to be cut (see formula 5); hence, the same lead gears are used when, for example, a gear and pinion are cut on the same machine. In the formulas which follow, the notation is the same as previously given, with these exceptions: R_d = lead gear ratio for machine with differential; P_a = axial or linear pitch of helical gear = distance from center of one tooth to center of next tooth measured parallel to gear axis = total lead L ÷ number of teeth N.

$$R_d = \frac{P_a \times T}{K} = \frac{L \times T}{N \times K} = \frac{\pi \times \operatorname{cosec} A \times T}{P_n \times K} = \frac{\text{Driven gear sizes}}{\text{Driving gear sizes}} \qquad (5)$$

The number of hob threads T is included in the formula because double-thread hobs are used sometimes, especially for roughing in order to reduce the hobbing time. Lead gears having a ratio sufficiently close to the required ratio, may be determined by using the table of gear ratio logarithms as previously described in connection with the non-differential type of machine. When using a machine equipped with a differential, the effect of a lead-gear ratio error upon the lead of the gear, is small in comparison with the effect of an index gear error when using a non-differential type of machine. The lead obtained with a given or obtainable lead gear ratio may be determined by the following formula: $L = (R_d N K) \div T$. In this formula, R_d represents the ratio obtained with available gears. If the given lead is 44.0894 inches, as in the preceding example, then the desired ratio as obtained with formula (5) would be 0.9185292 if $K = 1$. Assume that the lead gears selected by using logs of ratios have a ratio of 0.9184704; then this ratio error of 0.0000588 would result in a computed lead error of only 0.000065 inch per inch.

Formula (5), as mentioned, applies to machines having the differential located *ahead* of the index gears. If the differential is located after the index gears, it is necessary to change lead gears whenever the index gears are changed for hobbing a different number of teeth, as indicated by the following formula which gives the lead gear ratio. In this formula, D = pitch diameter.

$$R_d = \frac{L \times T}{K} = \frac{D \times \pi \times T}{K \times \tan A} = \frac{\text{Driven gear sizes}}{\text{Driving gear sizes}} \tag{6}$$

General Remarks on Helical Gear Hobbing. — In cutting teeth having large angles, it is desirable to have the direction of helix of the hob the same as the direction of helix of the gear, or in other words, the gear and the hob of the same "hand." Then the direction of the cut will come against the movement of the blank. At ordinary angles, however, one hob will cut both right- and left-hand gears. In setting up the hobbing machine for helical gears, care should be taken to see that the vertical feed does not trip until the machine has been stopped or the hob has fed down past the finished gear.

Herringbone Gears

Double helical or herringbone gears are commonly used in parallel-shaft transmissions, especially when a smooth, continuous action (due to the gradual overlapping engagement of the teeth) is essential, as in high-speed drives where the pitch-line velocity may range from about 1000 to 3000 feet per minute in commercial gearing and up to 12,000 feet per minute or higher in more specialized installations. These relatively high speeds are encountered in marine reduction gears, in certain speed-reducing and speed-increasing units, and in various other transmissions, particularly in connection with steam turbine and electric motor drives.

Causes of Herringbone Gear Failures. — Where failure occurs in a herringbone gear transmission, it is rarely due to tooth breakage but usually to excessive wear or sub-surface failures, such as pitting and spalling; hence, it is common practice to base the design of such gears upon durability, or upon tooth pressures which are within the allowable limits for wear. In this connection, it seems to have been well established by tests of both spur gears and herringbone gears, that there is a critical surface pressure value for teeth having given physical properties and coefficient of friction. According to these tests, pressures above the critical value result in rapid wear and a short gear life, whereas when pressures are below the critical, wear is negligible. The yield point or endurance limit of the material marks the critical loading point, and in practical designing a reasonable factor of safety would, of course, be employed.

General Classes of Helical Gear Problems. — There are two general classes of problems. In one case, the problem is to design gears capable of transmitting a given amount of power at a given speed, safely and without excessive wear; hence, in this case the required proportions must be determined. In the second case, the proportions and speed are known and the power-transmitting capacity is required. The first case is the more difficult and also the more common. In establishing the proportions of the gearing, there are numerous possible combinations of pinion diameter and face width which, theoretically at least, will meet the requirements. The speed of the driver and the ratio of the gearing ordinarily are known.

A.G.M.A. Horsepower Rating Formula. — Equation (1) which follows is the standard of the American Gear Manufacturers Association for determining the horsepower ratings of helical and herringbone gears. These ratings for wear or surface durability normally represent tooth loads that are well within the allowable limits for strength. In the equations which follow:

P_w = horsepower rating based upon wear or surface durability;

F_i = combined factor relating to face width and pinion location in gearing designed for either single, double, or triple speed reduction (see Table 1);

K_r = combined factor for hardness of pinion and gear, and ratio of gear to pinion size (K_r also takes into account tooth form) — see Table 2;

D_s = combined factor relating to pitch diameter and speed of pinion;

d = pitch diameter of pinion, inches; n = revolutions per minute of pinion; W = face width in inches; V = pitch line velocity in feet per minute; C_v = velocity factor. (In equation (3) for trial calculations, the constant 1.5 is an assumed mean value of the reciprocal of C_v.)

$$\text{Horsepower rating } P_w = F_i \times K_r \times D_s \qquad (1)$$

$$D_s = \frac{d^2 \times C_v \times n}{126,000} \quad \text{where} \quad C_v = \frac{78}{78 + \sqrt{V}} \qquad (2)$$

Load Capacity of Gearing is Based Upon Size of Pinion. — In designing herringbone gears it is the general practice except with low ratios, to base the load or power-transmitting capacity upon the pinion size which is assumed to be weaker than the gear and subject to greater wear. In preliminary calculations, one plan which has been applied quite extensively is to assume that the load-carrying capacity of the transmission is directly proportional to the product of the face width and the square of the pinion pitch diameter. The product of the face width and the square of the center distance has also been used as a power capacity factor. A third method which may be utilized in conjunction with the A.G.M.A. horsepower Equation (1) is to use the cube of the pinion diameter in connection with the preliminary calculations for establishing the proportions of gears having a given power capacity. Regardless of the method, the ratio of face width to pinion diameter should agree with established practice.

Ratio of Face Width to Pinion Diameter. — The face width is generally established with reference to the pinion diameter. The pinion width must be kept within certain limits to prevent excessive deflection between the supporting bearings. According to some authorities, the face width for ordinary applications should be limited to from 1½ to 2 times the pinion diameter. According to another source, 2 to 2½ times the pinion diameter represents approved practice. In some cases, the ratio of face width to pinion diameter may be as high as 3; hence, it is evident that quite a number of different combinations of pinion width and diameter might be employed in transmitting a given amount of power.

Table 1. F_i Factor for Given Face Width

Face Width (W) in Inches								
2	4	6	8	10	12	14	16	18
F_i factors for high-speed pinions of single, double and triple reduction units (See Note 1)								
1.25	2.45	3.55	4.50	5.35	6.05	6.70	7.20	7.65
F_i factors for low-speed pinions of double, and intermediate gears of triple reduction units (See Note 2)								
1.70	3.25	4.70	6.00	7.10	8.10	8.95	9.65	10.20
F_i factors for low-speed pinions of triple reduction units (See Note 3)								
1.85	3.50	4.95	6.30	7.50	8.50	9.45	10.20	10.80

For face widths greater than 18 inches:
- Note 1. $F_i = 0.425 \times W$
- Note 2. $F_i = 0.570 \times W$
- Note 3. $F_i = 0.600 \times W$

Table 2. K_r Factor for Given Ratio of Gear to Pinion — External Gears

Ratio of Gear to Pinion Size	Brinell Hardness of Gear (G) and Pinion (P) (Note: Both gear and pinion are cut after hardening)							
	(G)180 (P)210	(G)210 (P)245	(G)225 (P)265	(G)245 (P)285	(G)255 (P)300	(G)270 (P)315	(G)285 (P)335	(G)300 (P)350
	K_r Factor							
I	204	240	261	294	311	339	369	403
1.2	221	262	284	319	338	367	399	436
1.4	236	280	304	341	361	391	425	465
1.6	250	296	322	360	382	413	450	490
1.8	262	310	338	378	400	432	471	513
2	272	321	350	391	415	450	490	533
2.25	283	334	364	407	431	469	509	555
2.5	291	345	376	420	445	484	526	575
3	306	361	393	440	467	509	553	604
3.5	318	375	408	457	483	529	575	628
4	326	385	420	470	497	545	590	644
4.5	333	394	430	481	509	558	603	658
5	340	401	437	490	518	568	614	669
6	350	414	449	503	530	581	630	685
8	361	429	465	521	550	601	654	710
10	370	439	477	533	563	615	670	727
15	382	452	492	550	584	635	690	751
20	388	460	500	559	591	645	700	762

Formula for Preliminary Calculations Based upon Cube of Pinion Diameter. — The following equation will be found convenient to use in preliminary or trial calculations for determining the size of a pinion having a given power-transmitting capacity (constant 1.5 is assumed mean value of C_v reciprocal):

$$d^3 = \frac{1.5 \times P_w \times 126,000}{n \times K_r} \qquad (3)$$

The procedure in using this equation in conjunction with the A.G.M.A. Formula (1) for horsepower rating will be demonstrated by an example.

Example: Herringbone gears are to be designed for transmitting about 900 H.P. at 2400 R.P.M. Ratio of gear to pinion size is 4 to 1. Diametral pitch (to be selected) is pitch in transverse plane or plane of rotation. Determine the pinion diameter and face width. Assume for trial purposes that pinion hardness is 265 Brinell.

Find Factor K_r. — Table 2 shows that $K_r = 420$ for a pinion and gear hardness of 265 and 225 Brinell, respectively, and a ratio of gear to pinion size of 4. A trial pinion pitch diameter will now be determined.

$$d^3 = \frac{1.5 \times 900 \times 126,000}{2400 \times 420} = 169; \quad d = \sqrt[3]{169} = 5.53 \text{ inches}$$

Selecting Pitch Diameter Corresponding to Standard Diametral Pitch. — This approximate or trial pitch diameter of 5.53 is next changed to some near value corresponding to a standard diametral pitch and a number of teeth within the usual range for herringbone pinions (ordinarily from 14 to 34 teeth). Since diametral pitch is in the plane of rotation, pitch diameter = number of teeth ÷ diametral pitch; hence spur gear table on page 630 may be used in selecting a suitable diametral pitch and number of teeth. The following combinations have pitch diameters which are close to the trial value of 5.53: for 22 teeth of 4 D.P., pitch diameter = 5.5; for 28 teeth of 5 D.P., pitch diameter = 5.6; for 33 teeth of 6 D.P., pitch diameter = 5.5. Assume that we select for trial 28 teeth of 5 D.P. or a pitch diameter of 5.6 inches. (Note: Since pitch is in plane of rotation, hob or cutter must conform to desired helix angle.)

Find Trial Factor F_i from Table 1. — Assume by way of trial that F_i is equal to the pinion diameter, or 5.6 in this case. Table 1 shows that this value lies between F_i

Table 3. Additional K_r Factors — External Gears

Note: In the following formulas for K_r $C_r = \dfrac{\text{gear ratio}}{\text{gear ratio} + 1}$ where gear ratio $= \dfrac{\text{No. teeth in gear}}{\text{No. teeth in pinion}}$
Gear and Pinion Hardened after Cutting:—Case-hardened or through-hardened steel: Gear and pinion 575 Brinell, $K_r = 1530 \times 0.9 \times C_r$; gear and pinion 500 Brinell, $K_r = 1350 \times 0.9 \times C_r$; gear 350 Brinell, pinion 450 Brinell, $K_r = 1060 \times 0.9 \times C_r$.   Surface-hardened steel: gear and pinion 440 Brinell, $K_r = 0.890 \times 0.9 \times C_r$.
Gear Cut after Hardening, Pinion Hardened after Cutting: — Case-hardened or through-hardened steel: Gear 335 Brinell, pinion 380 Brinell, $K_r = 973 \times 0.95 \times C_r$; gear 315 Brinell, pinion 360 Brinell, $K_r = 870 \times 0.95 \times C_r$; gear 225 Brinell, pinion 450 Brinell, $K_r = 608 \times 0.95 \times C_r$.
Cast Iron: — Gear 200 Brinell, pinion 210 Brinell, $K_r = 344 \times C_r$.
Bronze: — Gear 40,000 pounds per square inch tensile strength, pinion 180 Brinell, $K_r = 274 \times C_r$.

factors for face widths of 10 and 12 inches. By interpolation, the face width W corresponding to a F_i value of 5.6 is found to be 10.7 inches.

Find Factor D_s. — This factor relating to pitch diameter and speed of pinion is found by Formula (2). First calculate V and C_v.

$$V = \frac{\pi \times d \times n}{12} = \frac{3.1416 \times 5.6 \times 2400}{12} = 3518 \text{ ft. per min.}$$

$$C_v = \frac{78}{78 + \sqrt{3518}} = 0.568; \quad D_s = \frac{5.6^2 \times 0.568 \times 2400}{126,000} = 0.34$$

Inserting these factors F_i, K_r, and D_s in Equation (1)

$$P_w = 5.6 \times 420 \times 0.34 = 800 \text{ horsepower}$$

Changing Trial Values to Obtain Given Power Capacity. — Since the horsepower capacity specified in this case is 900, or 100 more than shown by trial solution, we may either (1) increase face width and factor F_i; (2) increase pinion hardness and factor K_r; or (3) increase pinion diameter and factor D_s. Assume that face width is increased. Then,

$$F_i = 5.6 \times \frac{900}{800} = 6.3$$

If $F_i = 6.3$, then by interpolating in Table 1, we obtain a corresponding face width of 12.8. The ratio of this increased face width to pinion diameter $= 12.8 \div 5.6 = 2.28$, which is within the range of approved practice. When these revised figures are inserted in Equation (1), it will be seen that the capacity has been increased to 900 H.P. Thus,

$$P_w = 6.3 \times 420 \times 0.340 = 900 \text{ horsepower approximately.}$$

Gear Ratios. — A single gear train generally is used if the ratio of gear to pinion size is not over 10 or 12. Double or triple reductions would be used for higher ratios.

Helix Angles. — For herringbone gears, helix angles usually range from 20 to 45 degrees. Angles of 23 to 30 degrees have been used extensively. The higher angles are for precision gears and comparatively low tooth pressures.

Pitch of Pinion Teeth. — Comparatively fine pitches are used for herringbone gears in turbine-driven or other high-speed transmissions. Coarse pitches would not be satisfactory for such applications because of the reduction in contact in the axial plane. Where heavy shock loads are encountered, as in rolling mill drives, for example, large pitches are commonly employed. The number of teeth in a herring-

bone pinion of given size, does not affect materially the load-carrying capacity, provided the number is somewhere between 14 and 34. According to Farrel-Birmingham Co., the number of teeth should be related to the peripheral velocity as indicated by the following figures which are intended as a general guide. For velocities below 500 feet per minute, 14 to 25 teeth; for velocities between 500 and 1000 feet per minute, 17 to 27 teeth; for velocities higher than 1000 feet per minute, 19 to 33 teeth. Although the number of teeth selected for a given pitch diameter also fixes the pitch, any one of several combinations may meet the requirements.

Formula for Checking Diametral Pitch. — According to Buckingham, the pitch should not be finer than indicated by the following equation in which K varies according to pinion hardness as follows: $K = .036$ for 225 Brinell, .040 for 245 Brinell; .045 for 280 Brinell; .050 for 315 Brinell, and .054 for 350 Brinell.

Minimum tooth size (diametral pitch) = $(WdnK) \div$ horsepower capacity.

According to this formula, the diametral pitch of the herringbone gear referred to in the preceding example may be as fine as 8.

Replacement of Spur Gears by Helical Gears. — If spur gears are to be replaced either by single helical or herringbone gears without changing the center distance, the procedure is as follows:

Rule: Select a spur gear hob (or cutter) for generating slightly smaller teeth on the helical gearing. For example, if diametral pitch of spur gearing is 6, make normal diametral pitch of herringbone gearing 7. Then, $6 \div 7 =$ cosine of herringbone gear helix angle required to obtain 6 diametral pitch *in plane of rotation,* thus retaining spur gear center distance.

Note: If special hob is available having the same diametral pitch *in plane of rotation* as spur gearing to be replaced, merely cut helical or herringbone to whatever helix angle the hob (or other cutter) is intended for.

Planetary Gearing

Planetary or epicyclic gearing provides means of obtaining a compact design of transmission, with driving and driven shafts in line, and a large speed reduction when required. Typical arrangements of planetary gearing are shown by the following diagrams which are accompanied by speed ratio formulas. When planetary gears are arranged as shown by Figs. 5, 6, 9 and 12, the speed of the follower relative to the driver is increased, whereas Figs. 7, 8, 10 and 11 illustrate speed-reducing mechanisms.

Direction of Rotation. — In using the following formulas, if the final result is preceded by a minus sign (negative), this indicates that the driver and follower will rotate in opposite directions; otherwise, both will rotate in the same direction.

Compound Drive. — The formulas accompanying Figs. 19 to 22, inclusive, are for obtaining the speed ratios when there are *two* driving members rotating at different speeds. For example, in Fig. 19, the central shaft with its attached link is one driver. The internal gear z, instead of being fixed, is also rotated. In Fig. 22, if $z = 24$, $B = 60$ and $S = 3\frac{1}{2}$, with both drivers rotating in the same direction, then $F = 0$, thus indicating, in this case, the point where a larger value of S will reverse follower rotation.

Planetary Bevel Gears. — Two forms of planetary gears of the bevel type are shown in Figs. 23 and 24. The planet gear in Fig. 23 rotates about a fixed bevel gear at the center of which is the driven shaft. Figure 24 illustrates the Humpage reduction gear. This is sometimes referred to as cone-pulley back-gearing because of its use within the cone pulleys of certain types of machine tools.

Ratios of Planetary or Epicyclic Gearing

D = rotation of *driver* per revolution of follower or driven member.

F = rotation of *follower* or driven member per revolution of driver. (In Figs. 1 to 4, inclusive, F = rotation of planet type follower about its axis.)

A = size of driving gear (use either number of teeth or pitch diameter). Note: When follower derives its motion both from A and from a secondary driving member, A = size of *initial* driving gear, and formula gives speed relationship between A and follower.

B = size of *driven gear or follower* (use either pitch diameter or number of teeth).

C = size of *fixed gear* (use either pitch diameter or number of teeth).

x = size of *planet gear* as shown by diagram (use either pitch diameter or number of teeth).

y = size of *planet gear* as shown by diagram (use either pitch diameter or number of teeth).

z = size of secondary or *auxiliary driving gear*, when follower derives its motion from two driving members.

S = rotation of *secondary driver*, per revolution of *initial* driver. S is negative when secondary and initial drivers rotate in opposite directions. (Formulas in which S is used, give speed relationship between follower and the initial driver.)

Note: In all cases, if D is known, $F = 1 \div D$, or, if F is known, $D = 1 \div F$.

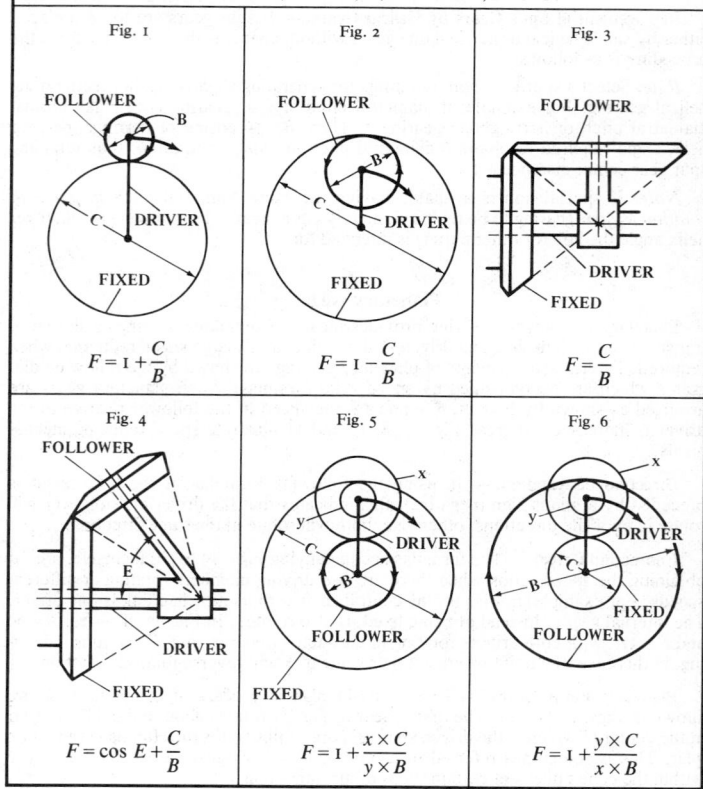

Fig. 1	Fig. 2	Fig. 3
$F = 1 + \dfrac{C}{B}$	$F = 1 - \dfrac{C}{B}$	$F = \dfrac{C}{B}$

Fig. 4	Fig. 5	Fig. 6
$F = \cos E + \dfrac{C}{B}$	$F = 1 + \dfrac{x \times C}{y \times B}$	$F = 1 + \dfrac{y \times C}{x \times B}$

Ratios of Planetary or Epicyclic Gearing

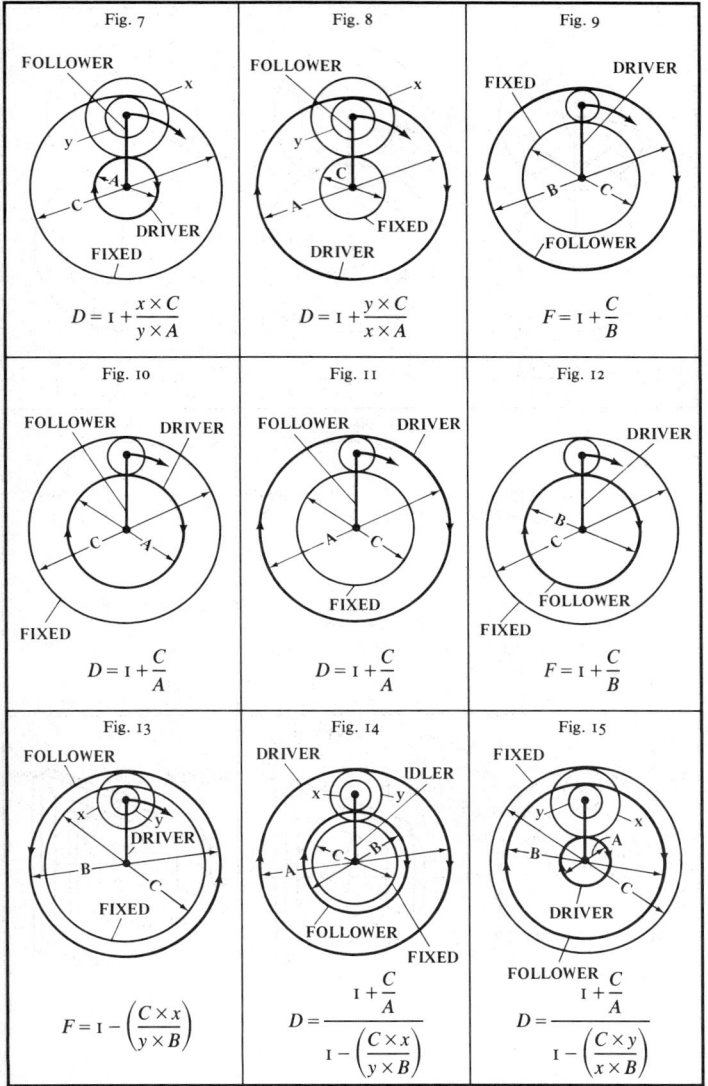

Fig. 7

$$D = 1 + \frac{x \times C}{y \times A}$$

Fig. 8

$$D = 1 + \frac{y \times C}{x \times A}$$

Fig. 9

$$F = 1 + \frac{C}{B}$$

Fig. 10

$$D = 1 + \frac{C}{A}$$

Fig. 11

$$D = 1 + \frac{C}{A}$$

Fig. 12

$$F = 1 + \frac{C}{B}$$

Fig. 13

$$F = 1 - \left(\frac{C \times x}{y \times B} \right)$$

Fig. 14

$$D = \frac{1 + \dfrac{C}{A}}{1 - \left(\dfrac{C \times x}{y \times B} \right)}$$

Fig. 15

$$D = \frac{1 + \dfrac{C}{A}}{1 - \left(\dfrac{C \times y}{x \times B} \right)}$$

Ratios of Planetary or Epicyclic Gearing

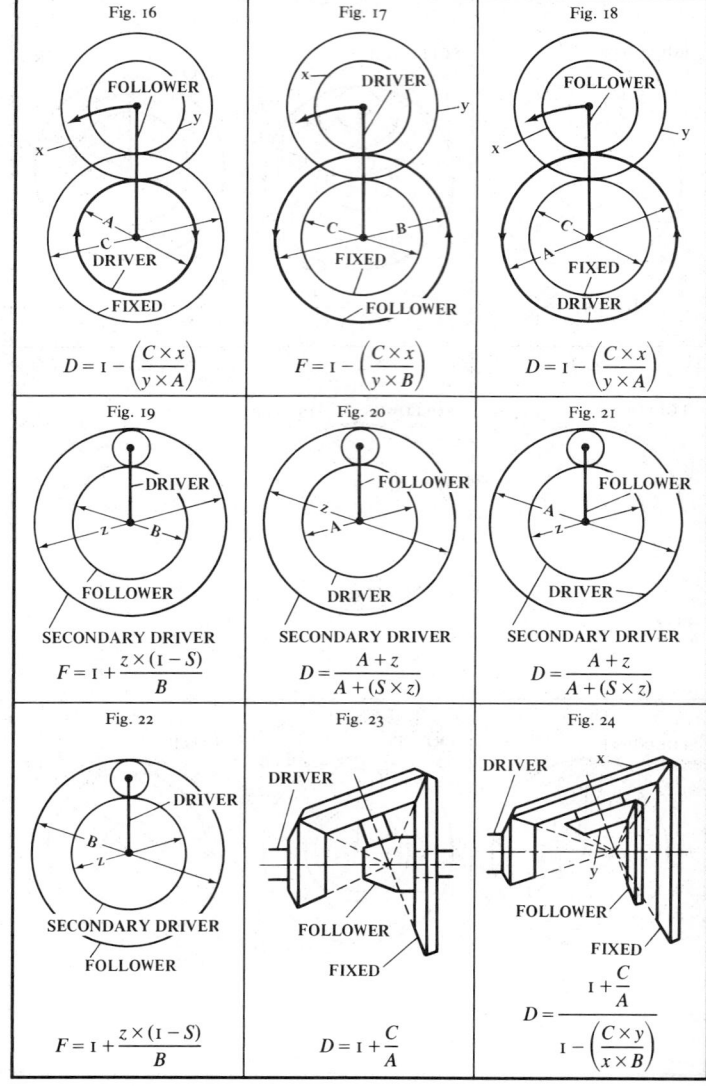

Fig. 16

$$D = 1 - \left(\frac{C \times x}{y \times A}\right)$$

Fig. 17

$$F = 1 - \left(\frac{C \times x}{y \times B}\right)$$

Fig. 18

$$D = 1 - \left(\frac{C \times x}{y \times A}\right)$$

Fig. 19

$$F = 1 + \frac{z \times (1 - S)}{B}$$

Fig. 20

$$D = \frac{A + z}{A + (S \times z)}$$

Fig. 21

$$D = \frac{A + z}{A + (S \times z)}$$

Fig. 22

$$F = 1 + \frac{z \times (1 - S)}{B}$$

Fig. 23

$$D = 1 + \frac{C}{A}$$

Fig. 24

$$D = \frac{1 + \dfrac{C}{A}}{1 - \left(\dfrac{C \times y}{x \times B}\right)}$$

Ratchet Gearing

Ratchet gearing may be used to transmit intermittent motion, or its only function may be to prevent the ratchet wheel from rotating backward. Ratchet gearing of this latter form is commonly used in connection with hoisting mechanisms of various kinds, to prevent the hoisting drum or shaft from rotating in a reverse direction under the action of the load.

Ratchet gearing in its simplest form consists of a toothed ratchet wheel *a* (see diagram *A*), and a pawl or detent *b*, and it may be used to transmit intermittent motion or to prevent relative motion between two parts except in one direction. The pawl *b* is pivoted to lever *c* which, when given an oscillating movement, imparts an intermittent rotary movement to ratchet wheel *a*. Diagram *B* illustrates another application of the ordinary ratchet and pawl mechanism. In this instance, the pawl is pivoted to a stationary member and its only function is to prevent the ratchet wheel from rotating backward. With the stationary design, illustrated at *C*, the pawl prevents the ratchet wheel from rotating in either direction, so long as it is in engagement with the wheel.

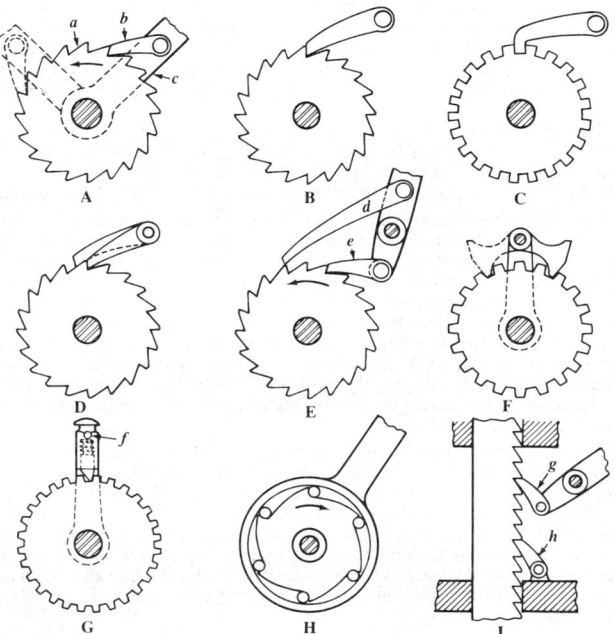

The principle of *multiple-pawl ratchet gearing* is illustrated at *D*, which shows the use of two pawls. One of these pawls is longer than the other, by an amount equal to one-half the pitch of the ratchet-wheel teeth, so that the practical effect is that of reducing the pitch one-half. By placing a number of driving pawls side by

side and proportioning their lengths according to the pitch of the teeth, a very fine feed can be obtained with a ratchet wheel of comparatively coarse pitch.

This method of obtaining a fine feed from relatively coarse-pitch ratchets may be preferable to the use of single ratchets of fine pitch which, although providing the feed required, may have considerably weaker teeth.

The type of ratchet gearing shown at E is sometimes employed to impart a rotary movement to the ratchet wheel for both the forward and backward motions of the lever to which the two pawls are attached.

A simple form of *reversing ratchet* is illustrated at F. The teeth of the wheel are so shaped that either side may be used for driving by simply changing the position of the double-ended pawl, as indicated by the full and dotted lines.

Another form of reversible ratchet gearing for shapers is illustrated at G. The pawl, in this case, instead of being a pivoted latch, is in the form of a plunger which is free to move in the direction of its axis, but is normally held into engagement with the ratchet wheel by a small spring. When the pawl is lifted and turned one-half revolution, the driving face then engages the opposite sides of the teeth and the ratchet wheel is given an intermittent rotary motion in the opposite direction.

The *frictional type* of ratchet gearing differs from the designs previously referred to, in that there is no positive engagement between the driving and driven members of the ratchet mechanism, the motion being transmitted by frictional resistance. One type of frictional ratchet gearing is illustrated at H. Rollers or balls are placed between the ratchet wheel and an outer ring which, when turned in one direction, causes the rollers or balls to wedge between the wheel and ring as they move up the inclined edges of the teeth.

Diagram I illustrates one method of utilizing ratchet gearing for moving the driven member in a straight line, as in the case of a lifting jack. The pawl g is pivoted to the operating lever of the jack and does the lifting, whereas the pawl h holds the load while the lifting pawl g is being returned preparatory to another lifting movement.

Shape of Ratchet Wheel Teeth. — When designing ratchet gearing, it is important to so shape the teeth that the pawl will remain in engagement when a load is applied. The faces of the teeth which engage the end of the pawl should be in such relation with the center of the pawl pivot that a line perpendicular to the face of the engaging tooth will pass somewhere between the center of the ratchet wheel and the center of the pivot about which the pawl swings. This is true if the pawl *pushes* the ratchet wheel, or if the ratchet wheel *pushes* the pawl. However, if the pawl *pulls* the ratchet wheel or if the ratchet wheel *pulls* the pawl, the perpendicular from the face of the ratchet teeth should fall outside the pawl pivot center. Ratchet teeth may be either cut by a milling cutter having the correct angle, or hobbed in a gear-hobbing machine by the use of a special hob.

Pitch of Ratchet Wheel Teeth. — The pitch of ratchet wheels used for holding suspended loads may be calculated by the following formula, in which P = circular pitch, in inches, measured at the outside circumference; M = turning moment acting upon the ratchet wheel shaft, in inch-pounds; L = length of tooth face, in inches (thickness of ratchet gear); S = safe stress (for steel, 2500 pounds per square inch when subjected to shock, and 4000 pounds per square inch when not subjected to shock); N = number of teeth in ratchet wheel; F = a factor the value of which is 50 for ratchet gears with 12 teeth or less, 35 for gears having from 12 to 20 teeth, and 20 for gears having over 20 teeth:

$$P = \sqrt{\frac{FM}{LSN}}$$

This formula has been used in the calculation of ratchet gears for crane design.

Gear Design Based upon Module System. — The *module* of a gear equals the pitch diameter divided by the number of teeth, whereas *diametral pitch* equals the number of teeth divided by the pitch diameter. The module system is in general use in countries which have adopted the metric system; hence the term module is usually understood to mean the pitch diameter *in millimeters* divided by the number of teeth. The module system may, however, also be based upon inch measurements and then it is known as English module to avoid confusion with the metric module. Module is an actual dimension, whereas diametral pitch is only a ratio. Thus, if the pitch diameter of a gear is 50 millimeters and the number of teeth 25, the module is 2 which means that there are 2 millimeters of pitch diameter for each tooth. The table "Tooth Dimensions Based Upon Module System" shows the relation between module, diametral pitch, and circular pitch.

German Standard Tooth Form for Spur and Bevel Gears (DIN — 867)

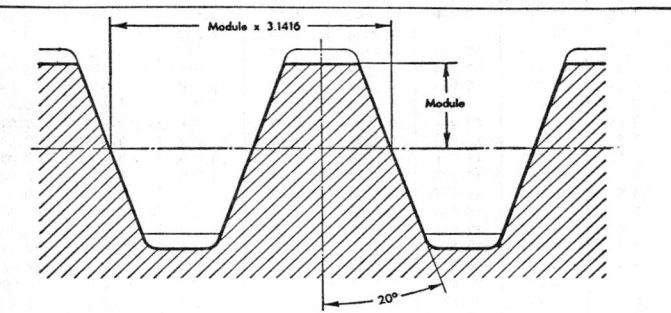

The flanks or sides are straight (involute system) and the pressure angle is 20 degrees. The shape of the root clearance space and the amount of clearance depend upon the method of cutting and special requirements. The amount of clearance may vary from 0.1 × module to 0.3 × module.

To Find	Module Known	Circular Pitch Known
Addendum	Equals module	0.3183 × Circular pitch
Dedendum	1.157 × module* 1.167 × module**	0.3683 × Circular pitch* 0.3714 × Circular pitch**
Working Depth	2 × module	0.6366 × Circular pitch
Total Depth	2.157 × module* 2.167 × module**	0.6866 × Circular pitch* 0.6898 × Circular pitch**
Tooth Thickness on Pitch Line	1.5708 × module	0.5 × Circular pitch

Formulas for dedendum and total depth, marked (*) are used when clearance equals 0.157 × module. Formulas marked (**) are used when clearance equals one-sixth module. It is the common practice among American cutter manufacturers to make the clearance of metric or module cutters equal to 0.157 × module.

Tooth Dimensions Based Upon Module System

Module, DIN Standard Series	Equivalent Diametral Pitch	Circular Pitch		Addendum, Millimeters	Dedendum, Millimeters*	Whole Depth,* Millimeters	Whole Depth,† Millimeters
		Millimeters	Inches				
0.3	84.667	0.943	0.0371	0.30	0.35	0.650	0.647
0.4	63.500	1.257	0.0495	0.40	0.467	0.867	0.863
0.5	50.800	1.571	0.0618	0.50	0.583	1.083	1.079
0.6	42.333	1.885	0.0742	0.60	0.700	1.300	1.294
0.7	36.286	2.199	0.0865	0.70	0.817	1.517	1.510
0.8	31.750	2.513	0.0989	0.80	0.933	1.733	1.726
0.9	28.222	2.827	0.1113	0.90	1.050	1.950	1.941
1	25.400	3.142	0.1237	1.00	1.167	2.167	2.157
1.25	20.320	3.927	0.1546	1.25	1.458	2.708	2.697
1.5	16.933	4.712	0.1855	1.50	1.750	3.250	3.236
1.75	14.514	5.498	0.2164	1.75	2.042	3.792	3.774
2	12.700	6.283	0.2474	2.00	2.333	4.333	4.314
2.25	11.289	7.069	0.2783	2.25	2.625	4.875	4.853
2.5	10.160	7.854	0.3092	2.50	2.917	5.417	5.392
2.75	9.236	8.639	0.3401	2.75	3.208	5.958	5.932
3	8.466	9.425	0.3711	3.00	3.500	6.500	6.471
3.25	7.815	10.210	0.4020	3.25	3.791	7.041	7.010
3.5	7.257	10.996	0.4329	3.50	4.083	7.583	7.550
3.75	6.773	11.781	0.4638	3.75	4.375	8.125	8.089
4	6.350	12.566	0.4947	4.00	4.666	8.666	8.628
4.5	5.644	14.137	0.5566	4.50	5.25	9.750	9.707
5	5.080	15.708	0.6184	5.00	5.833	10.833	10.785
5.5	4.618	17.279	0.6803	5.50	6.416	11.916	11.864
6	4.233	18.850	0.7421	6.00	7.000	13.000	12.942
6.5	3.908	20.420	0.8035	6.50	7.583	14.083	14.021
7	3.628	21.991	0.8658	7.	8.166	15.166	15.099
8	3.175	25.132	0.9895	8.	9.333	17.333	17.256
9	2.822	28.274	1.1132	9.	10.499	19.499	19.413
10	2.540	31.416	1.2368	10.	11.666	21.666	21.571
11	2.309	34.558	1.3606	11.	12.833	23.833	23.728
12	2.117	37.699	1.4843	12.	14.000	26.000	25.884
13	1.954	40.841	1.6079	13.	15.166	28.166	28.041
14	1.814	43.982	1.7317	14.	16.332	30.332	30.198
15	1.693	47.124	1.8541	15.	17.499	32.499	32.355
16	1.587	50.266	1.9790	16.	18.666	34.666	34.512
18	1.411	56.549	2.2263	18.	21.000	39.000	38.826
20	1.270	62.832	2.4737	20.	23.332	43.332	43.142
22	1.155	69.115	2.7210	22.	25.665	47.665	47.454
24	1.058	75.398	2.9685	24.	28.000	52.000	51.768
27	0.941	84.823	3.339	27.	31.498	58.498	58.239
30	0.847	94.248	3.711	30.	35.000	65.000	64.713
33	0.770	103.673	4.082	33.	38.498	71.498	71.181
36	0.706	113.097	4.453	36.	41.998	77.998	77.652
39	0.651	122.522	4.824	39.	45.497	84.497	84.123
42	0.605	131.947	5.195	42.	48.997	90.997	90.594
45	0.564	141.372	5.566	45.	52.497	97.497	97.065
50	0.508	157.080	6.184	50.	58.330	108.330	107.855
55	0.462	172.788	6.803	55.	64.163	119.163	118.635
60	0.423	188.496	7.421	60.	69.996	129.996	129.426
65	0.391	204.204	8.040	65.	75.829	140.829	140.205
70	0.363	219.911	8.658	70.	81.662	151.662	150.997
75	0.339	235.619	9.276	75.	87.495	162.495	161.775

* Dedendum and total depth when clearance = 0.1666 × module, or one-sixth module.
† Total depth equivalent to American standard full-depth teeth. (Clearance = 0.157 × module.)

Rules for Module System of Gearing

To Find	Rule
Metric Module	*Rule* 1: To find the metric module, divide the pitch diameter in millimeters by the number of teeth. *Example* 1: The pitch diameter of a gear is 200 millimeters and the number of teeth, 40; then $$\text{module} = \frac{200}{40} = 5$$ *Rule* 2: Multiply circular pitch in millimeters by 0.3183. *Example* 2: (Same as Example 1. Circular pitch of this gear equals 15.708 millimeters). $$\text{module} = 15.708 \times 0.3183 = 5$$ *Rule* 3: Divide outside diameter in millimeters by the number of teeth plus 2.
English Module	*Note:* The module system is usually applied when gear dimensions are expressed in millimeters, but module may also be based upon inch measurements. *Rule:* To find the English module, divide pitch diameter in inches by number of teeth. *Example:* A gear has 48 teeth and a pitch diameter of 12 inches. $$\text{module} = \frac{12}{48} = \frac{1}{4} \text{ module or 4 diametral pitch}$$
Metric Module Equivalent to Diametral Pitch	*Rule:* To find the metric module equivalent to a given diametral pitch, divide 25.4 by the diametral pitch. *Example:* Determine metric module equivalent to 10 diametral pitch. $$\text{Equivalent module} = \frac{25.4}{10} = 2.54$$ *Note:* The nearest standard module is 2.5.
Diametral Pitch Equivalent to Metric Module	*Rule:* To find the diametral pitch equivalent to a given module, divide 25.4 by the module. (25.4 = number of millimeters per inch.) *Example:* The module is 12; determine equivalent diametral pitch. $$\text{Equivalent diametral pitch} = \frac{25.4}{12} = 2.117$$ *Note:* A diametral pitch of 2 is the nearest *standard* equivalent.
Pitch Diameter	*Rule:* Multiply number of teeth by module. *Example:* The metric module is 8 and gear has 40 teeth; then $$D = 40 \times 8 = 320 \text{ millimeters} = 12.598 \text{ inches}$$
Outside Diameter	*Rule:* Add 2 to the number of teeth and multiply sum by the module. *Example:* A gear has 40 teeth and module is 6. Find outside or blank diameter. $$\text{Outside diameter} = (40 + 2) \times 6 = 252 \text{ millimeters}$$

For tooth dimensions, see table Tooth Dimensions Based Upon Module System; also formulas below German Standard Tooth Form.

Equivalent Diametral Pitches, Circular Pitches, and Metric Modules

Commonly Used Pitches and Modules in Bold Type

Diametral Pitch	Circular Pitch, Inches	Module Millimeters	Diametral Pitch	Circular Pitch, Inches	Module Millimeters	Diametral Pitch	Circular Pitch, Inches	Module Millimeters
½	6.2832	50.8000	2.2848	1⅜	11.1170	10.0531	5/16	2.5266
0.5080	6.1842	50	2.3091	1.3605	11	10.1600	0.3092	2½
0.5236	6	48.5104	2½	1.2566	10.1600	11	0.2856	2.3091
0.5644	5.5658	45	2.5133	1¼	10.1063	12	0.2618	2.1167
0.5712	5½	44.4679	2.5400	1.2368	10	12.5664	¼	2.0213
0.6283	5	40.4253	2¾	1.1424	9.2364	12.7000	0.2474	2
0.6350	4.9474	40	2.7925	1⅛	9.0957	13	0.2417	1.9538
0.6981	4½	36.3828	2.8222	1.1132	9	14	0.2244	1.8143
0.7257	4.3290	35	3	1.0472	8.4667	15	0.2094	1.6933
¾	4.1888	33.8667	3.1416	1	8.0851	16	0.1963	1.5875
0.7854	4	32.3403	3.1750	0.9895	8	16.7552	3/16	1.5160
0.8378	3¾	30.3190	3.3510	15/16	7.5797	16.9333	0.1855	1½
0.8467	3.7105	30	3½	0.8976	7.2571	17	0.1848	1.4941
0.8976	3½	28.2977	3.5904	⅞	7.0744	18	0.1745	1.4111
0.9666	3¼	26.2765	3.6286	0.8658	7	19	0.1653	1.3368
1	3.1416	25.4000	3.8666	13/16	6.5691	20	0.1571	1.2700
1.0160	3.0921	25	3.9078	0.8040	6½	22	0.1428	1.1545
1.0472	3	24.2552	4	0.7854	6.3500	24	0.1309	1.0583
1.1424	2¾	22.2339	4.1888	¾	6.0638	25	0.1257	1.0160
1¼	2.5133	20.3200	4.2333	0.7421	6	25.1328	⅛	1.0106
1.2566	2½	20.2127	4.5696	11/16	5.5585	25.4000	0.1237	1
1.2700	2.4737	20	4.6182	0.6803	5½	26	0.1208	0.9769
1.3963	2¼	18.1914	5	0.6283	5.0800	28	0.1122	0.9071
1.4111	2.2263	18	5.0265	⅝	5.0532	30	0.1047	0.8467
1½	2.0944	16.9333	5.0800	0.6184	5	32	0.0982	0.7937
1.5708	2	16.1701	5.5851	9/16	4.5478	34	0.0924	0.7470
1.5875	1.9790	16	5.6443	0.5566	4½	36	0.0873	0.7056
1.6755	1⅞	15.1595	6	0.5236	4.2333	38	0.0827	0.6684
1.6933	1.8553	15	6.2832	½	4.0425	40	0.0785	0.6350
1¾	1.7952	14.5143	6.3500	0.4947	4	42	0.0748	0.6048
1.7952	1¾	14.1489	7	0.4488	3.6286	44	0.0714	0.5773
1.8143	1.7316	14	7.1808	7/16	3.5372	46	0.0683	0.5522
1.9333	1⅝	13.1382	7.2571	0.4329	3½	48	0.0654	0.5292
1.9538	1.6079	13	8	0.3927	3.1750	50	0.0628	0.5080
2	1.5708	12.7000	8.3776	⅜	3.0319	50.2656	1/16	0.5053
2.0944	1½	12.1276	8.4667	0.3711	3	50.8000	0.0618	½
2.1167	1.4842	12	9	0.3491	2.8222	56	0.0561	0.4536
2¼	1.3963	11.2889	10	0.3142	2.5400	60	0.0524	0.4233

The module of a gear is the pitch diameter divided by the number of teeth. The module may be expressed in any units; but when no units are stated, it is understood to be in millimeters. The metric module, therefore, equals the pitch diameter in millimeters divided by the number of teeth. To find the metric module equivalent to a given diametral pitch, divide 25.4 by the diametral pitch. To find the diametral pitch equivalent to a given module, divide 25.4 by the module. (25.4 = number of millimeters per inch.)

British Standard for Spur and Helical Gears. — This revised standard No. 436-1940, amended in 1941, 1943, and 1956, applies to machine cut or ground spur gears and to single or double helical gears connecting parallel shafts. Internal as well as external gears are included. The pressure angle is 20 degrees and the working depth equals twice the module (whether English or metric should be stated). The tooth form represents a well-balanced compromise between strength, wear resistance, and quietness of operation. Gears are divided into five general classes.

Class A1, Precision Ground Gears (nominal proportions of basic rack tooth for this class are the same as shown by accompanying diagram for circular pitch of 1, except that fillet radius at the root is 0.0938 instead of 0.124 and the dedendum is 0.4583 instead of 0.3979).

Class A2, Precision Cut Gears for peripheral speeds above 2000 feet per minute.

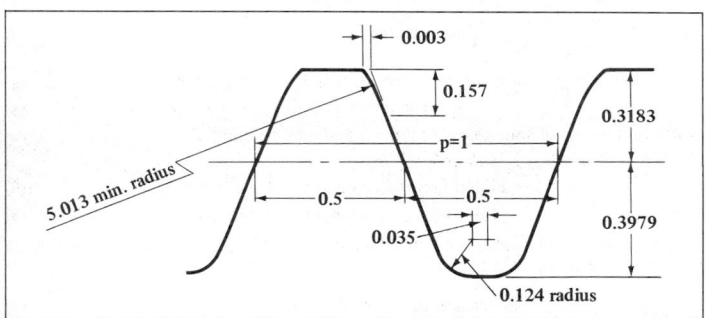

Class B, High-class Cut Gears for peripheral speeds between 750 and 3000 feet per minute (nominal proportions of basic rack for Classes A2 and B are shown by accompanying diagram).

Class C, Commercial Cut Gears for peripheral speeds below 1200 feet per minute

Class D, Large Internal Gears (basic rack for Classes C and D same as diagram except tip radius is 4.098, easing 0.006 and depth, 0.200.

The range of speeds specified for gears of Classes A2, B and C permits considerable overlap between the classes. The notation follows:

A = Gear addendum	p_n = Normal pitch
a = Pinion addendum	T = Number of gear teeth
k_p = Pinion correcton factor	t = Number of pinion teeth
k_w = Gear correction factor	Δ = Center distance extension factor
P_n = Normal diametral pitch	σ = Helix angle
p = Circular pitch	

Easing or Tip Relief. — The form is involute except for a slight easing at the point. The maximum amount of this easing or tip relief (e) is as follows:

For Classes A1, A2, and B (Precision Ground, Precision Cut and High-class Cut Gears)

$$e = 0.003p \text{ extending } 0.157p \text{ in depth}$$

For Classes C and D (Commercial Cut and Large Internal Gears)

$$e = 0.006p \text{ extending } 0.20p \text{ in depth}.$$

Helical Gear Teeth. — The shape and proportions of helical gear teeth in the normal section corresponds to the basic rack tooth forms, normal pitch being substituted for circular pitch. (Note that $p_n = p \cos \sigma$; also $P_n = \pi \div p_n$.)

Addendum — Gear and Pinion. — The *recommended* addendum values vary according to pitch and numbers of teeth in mating gears, in order to obtain full involute action, avoid undercutting in some cases, and obtain better zone and strength factors (factors used in calculating the horsepower rating).

$$\text{Pinion addendum } a = \frac{p_n}{\pi}(1 + k_p) = \frac{1 + k_p}{P_n} \tag{1}$$

$$\text{Gear addendum } A = \frac{p_n}{\pi}(1 + k_w) = \frac{1 + k_w}{P_n} \tag{2}$$

The correction factors k_p and k_w for pinion and gear, are determined as follows: *When $(t + T)$ sec³ σ is 60 or greater:*

$$\text{Pinion factor } k_p = 0.4\left(1 - \frac{t}{T}\right) \tag{3}$$

$$\text{or } 0.02(30 - t \sec^3 \sigma) \text{ whichever is greater} \tag{4}$$

$$\text{Gear factor } k_w = -k_p \tag{5}$$

Note: In the case of spur gears, $\sec^3 \sigma = 1$ and may be omitted. *When $(t + T)$ sec³ σ is less than 60:*

$$\text{Pinion factor } k_p = 0.02(30 - t \sec^3 \sigma) \tag{6}$$

$$\text{Gear factor } k_w = 0.02(30 - T \sec^3 \sigma) \tag{7}$$

The center distance also is extended an amount indicated by the following formula when $(t + T)$ sec³ σ is less than 60.

$$\text{Extension of center distance} = \frac{\Delta p_n}{\pi} = \frac{\Delta}{P_n} \tag{8}$$

Factors used in Center Distance Extension Formula No. 8

Sum of Correction Factors $k_p + k_w$	Factor Δ	Sum of Correction Factors $k_p + k_w$	Factor Δ	Sum of Correction Factors $k_p + k_w$	Factor Δ	Sum of Correction Factors $k_p + k_w$	Factor Δ
0.025	0.025	0.225	0.218	0.425	0.400	0.625	0.555
0.050	0.050	0.250	0.243	0.450	0.420	0.650	0.575
0.075	0.075	0.275	0.267	0.475	0.444	0.675	0.588
0.100	0.100	0.300	0.288	0.500	0.462	0.700	0.606
0.125	0.122	0.325	0.313	0.525	0.480	0.725	0.623
0.150	0.146	0.350	0.332	0.550	0.500	0.750	0.636
0.175	0.170	0.375	0.356	0.575	0.516	0.775	0.650
0.200	0.196	0.400	0.376	0.600	0.536	0.800	0.663

After finding the sum of k_p and k_w, the value of Δ is obtained either directly, or by interpolation, from the accompanying table (based upon chart in the British standard).

For internal gears (irrespective of numbers of teeth)

$$k_p = 0.4 \text{ and } k_w = -k_p \tag{9}$$

Outside Diameter of Pinion. — If number of pinion teeth is such that $t \sec^3 \sigma$ is less than 17, outside diameter is reduced but the pitch diameter and root diameter are not changed.

$$\text{Pinion diam. reduction} = \frac{p_n}{\pi} \times 0.04(17 - t \sec^3 \sigma) = \frac{0.04(17 - t \sec^3 \sigma)}{P_n} \tag{10}$$

Example 1. — Find the addendum values for a pair of spur gears. The pinion has 26 teeth, the gear 73 teeth, and the circular pitch is 0.5 inch. In this case, $t + T = 26 + 73 = 99$. Since this sum is larger than 60, pinion correction factor k_p is determined either by formula (3) or (4), whichever yields the greater value.

Applying formula (3), $k_p = 0.4\left(1 - \dfrac{26}{73}\right) = 0.258$.

Applying formula (4), $k_p = 0.02(30 - 26) = 0.08$.

Hence, pinion correction factor $k_p = 0.258$ and gear correction factor $k_w = -0.258$.

Applying formula (1), pinion addendum $a = \dfrac{0.5}{\pi} \times 1.258 = 0.200$ inch.

Applying formula (2), gear addendum $A = \dfrac{0.5}{\pi} \times 0.742 = 0.118$ inch.

Note: The regular unmodified addendum in this case would equal $0.3183 \times 0.5 = 0.159$ inch for pinion and gear.

Example 2. — Find the addendum values for a pair of helical gears. The pinion has 11 teeth, the gear 22 teeth. The normal diametral pitch is 4 and the helix angle is $22°30'$ ($\sec^3 22.5° = 1.268$).

First determine whether Formulas (3), (4) and (5) or Formulas (6) and (7) are to be used for finding the pinion and gear correction factors k_p and k_w.

$$(t + T)\sec^3 \sigma = (11 + 22) \times 1.268 = 41.8$$

Since 41.8 is less than 60, Formulas (6) and (7) should be used.

$$k_p = 0.02(30 - 11 \times 1.268) = 0.321$$
$$k_w = 0.02(30 - 22 \times 1.268) = 0.042$$

Next, determine the extension of the center distance using Formula (8).

$$k_p + k_w = 0.321 + 0.042 = 0.363$$

Factor Δ obtained from the accompanying table by interpolation is about 0.345; hence, using Formula (8)

Extension of center distance $= \dfrac{0.345}{4} = 0.086$ inch

Center distance $= \dfrac{t + T}{2P_n \cos \sigma} + 0.086 = \dfrac{11 + 22}{2 \times 4 \times 0.9239} + 0.086 = 4.551$ inches

Gear addendum $= \dfrac{1 + 0.042}{4} = 0.260$ inch

Finally, check to see if the outside diameter of the pinion should be reduced. In this example, $t \sec^3 \sigma = 11 \times 1.268 = 13.948$. Since this is less than 17, the pinion addendum is first obtained by Formula (1) and then it is reduced.

By Formula (1), $a = \dfrac{1 + 0.321}{4} = 0.330$ inch.

By Formula (10), diam. reduction $= \dfrac{0.04(17 - 13.948)}{4} = 0.030$ inch.

Actual pinion addendum $= 0.330 - \dfrac{0.030}{2} = 0.315$ inch.

British Standard Spur and Helical Gears. — Metric Modules (B.S. 436:Part 2: 1970, Amended 1977). — This British Standard is a metric-unit specification for external and internal spur and helical gears for use with parallel shafts. Preferred and second-choice modules are given, and the requirements for basic rack tooth profile, and accuracy are covered. Any of twelve different grades of accuracy may be applied to each gear element. Thus, gear requirements are met ranging from coarse commercial to high-speed and high-load precision applications. Tolerances on gear blanks are included in the specification. The Standard is a companion specification to B.S. 436: Part 1:1967, which covers spur and helical gears in the inch system.

Terminology, Definitions, and Notation. — For the purpose of this British Standard, the definitions and notation given in BS 2519, Part 1 (ISO/R 1122) and Part 2 (ISO 701) apply. This notation is as follows:

a = Center distance
d = Reference circle diameter
    d_1 Reference circle diameter, pinion
    d_2 Reference circle diameter, wheel
d_a = Tip diameter
    d_{a1} Tip diameter, pinion
    d_{a2} Tip diameter, wheel
y = Center distance modification coefficient
b = Face width
    b_1 Face width, pinion
    b_2 Face width, wheel
x = Addendum modification coefficient
    x_1 Addendum modification coefficient, pinion
    x_2 Addendum modification coefficient, wheel
l = Length of arc
m = module
m_n = Normal module
p_t = Transverse pitch
z = Number of teeth
    z_1 Number of teeth, pinion
    z_2 Number of teeth, wheel
β = Helix angle at reference cylinder
α = Pressure angle at reference cylinder
α_n = Normal pressure angle at reference cylinder
α_t = Transverse pressure angle at reference cylinder
α_{tw} = Transverse pressure angle, working

Basic Rack Tooth Profile. — The basic rack is generally in agreement with ISO 53, "Cylindrical gears for general and heavy engineering — Basic rack," and the modules are selected from ISO 54, "Cylindrical gears for general and heavy engineering — Modules and diametral pitches." The accuracy requirements are based on ISO 1328, "Parallel involute Gears — ISO system of accuracy." Grades 1 and 2 are intended for use as master gears, and in certain circumstances, Grades 3, 4, and 5 may also be used as master gears. In Fig. 1 is shown the profile of the basic rack for unit normal metric module, and this tooth profile has been adopted for the purposes of the Standard. The values shown are proportions of the module.

In practice, the basic rack tooth is usually modified, and the extent of modification shall be in accordance with the following: (*a*) The total depth may vary within the limits 2.25 to 2.40, which permits an increase in root clearance within the same limits to allow for the use of different manufacturing processes. (*b*) The root radius may vary within the limits 0.25 to 0.39. (*c*) Tip relief may be applied within the limits shown in Fig. 1.

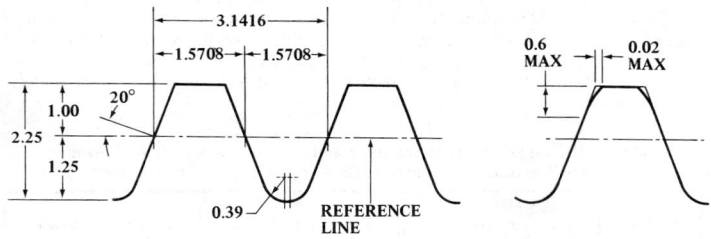

Fig. 1. (Left) British Standard Basic Rack Tooth Profile for Unit Normal Metric Module, and (Right) Limits of Tip Relief

Standard Normal Modules. — The modules, as shown in Table 1, are taken from ISO 54. Wherever possible, the preferred modules should be selected, rather than the second choice modules.

Accuracy Grades. — The accuracy grades, numbered 1 to 12, are based on the requirements of draft ISO 1328. The grade of a gear depends on the limits of tolerance for pitch, tooth profile, and tooth alignment, which are given in Table 2, and shown graphically for grades 3 to 12 in Charts 1, 3 and 4. When these three tolerances are used, the highest grade number selected for any of the three is the grade assigned to the particular gear. The tolerances on the other elements, which are also shown in the table, and in Charts 6, 7, and 8, may be subject to agreement between the purchaser and manufacturer. Generally, the elements of a pair of mating gears are of identical accuracy grades, but different grades may be applied to individual elements by agreement between user and manufacturer. It should be appreciated that the choice of particular element tolerances may depend on the procedure employed for accuracy testing, the choice of disposition of the fundamental tolerance, and assembly requirements, for example.

Tolerance on Pitch. — The limits of tolerance on transverse pitch, both adjacent and cumulative, can be obtained from Chart 1 for grades 3 to 12 when the length of arc l is known. This arc may be of any selected length in millimeters, less than $\pi d/2$.

Table 1. British Standard Spur and Helical Gears — Standard Normal Metric Modules (B.S. 436:Part 2:1970, Amended 1977)

Preferred Modules*	1	1.25	1.5	2	2.5	3	4	5	6
Second Choice Modules	1.125	1.375	1.75	2.25	2.75	3.5	4.5	5.5	7
Preferred Modules*	8	10	12	16	20	25	32	40	50
Second Choice Modules	9	11	14	18	22	28	36	45	. . .

The values are in millimeters.
* Wherever possible, the preferred modules should be applied rather than those of second choice.

Example: The pitch tolerance is required on an arc length of 40 mm for a gear of grade 6 accuracy. The graph given in Chart 1 is entered on the horizontal scale at the 40 mm length position, and the limit of tolerance is read off in relation to the curve for grade 6 accuracy. The figure obtained on the vertical scale is 22 micrometers, which is 0.022 mm.

Table 2. British Standard Metric Spur and Helical Gears — Basic Formulas for Limits of Tolerance on Elements* (B.S. 436: Part 2:1970, Amended 1977)

Gear Accuracy Grade	Limits of Tolerance on Pitch	Limits of Tolerance on Tooth Profile	Limits of Tolerance on Tooth Alignment
1	$0.25\sqrt{l} + 0.63$	$0.063\phi_f + 2$	$0.315\sqrt{b} + 1.6$
2	$0.4\ \sqrt{l} + 1$	$0.10\ \phi_f + 2.5$	$0.40\ \sqrt{b} + 2$
3	$0.63\sqrt{l} + 1.6$	$0.16\ \phi_f + 3.15$	$0.5\ \sqrt{b} + 2.5$
4	$1.0\ \sqrt{l} + 2.5$	$0.25\ \phi_f + 4.0$	$0.63\ \sqrt{b} + 3.15$
5	$1.6\ \sqrt{l} + 4.0$	$0.40\ \phi_f + 5.0$	$0.80\ \sqrt{b} + 4.0$
6	$2.5\ \sqrt{l} + 6.3$	$0.63\ \phi_f + 6.3$	$1.0\ \sqrt{b} + 5.0$
7	$3.55\sqrt{l} + 9.0$	$1.0\ \phi_f + 8.0$	$1.25\ \sqrt{b} + 6.3$
8	$5.0\ \sqrt{l} + 12.5$	$1.6\ \phi_f + 10.0$	$2.0\ \sqrt{b} + 10.0$
9	$7.1\ \sqrt{l} + 18.0$	$2.5\ \phi_f + 16.0$	$3.15\ \sqrt{b} + 16.0$
10	$10.0\ \sqrt{l} + 25.0$	$4.0\ \phi_f + 25.0$	$5.0\ \sqrt{b} + 25.0$
11	$14.0\ \sqrt{l} + 35.5$	$6.3\ \phi_f + 40.0$	$8.0\ \sqrt{b} + 40.0$
12	$20.0\ \sqrt{l} + 50.0$	$10.0\ \phi_f + 63.0$	$12.5\ \sqrt{b} + 63.0$

Gear Accuracy Grade	Limits of Tolerance on Radial Runout of Teeth	Limits of Tolerance on Tooth-to-Tooth Composite Error	Limits of Tolerance on Total Composite Error
1	$0.224\phi_p + 2.8$	$0.16\ \phi_p + 2.0$	$0.315\phi_p + 4.0$
2	$0.355\phi_p + 4.5$	$0.224\phi_p + 2.8$	$0.5\ \phi_p + 6.3$
3	$0.56\ \phi_p + 7.1$	$0.32\ \phi_p + 4.0$	$0.8\ \phi_p + 10$
4	$0.90\ \phi_p + 11.2$	$0.45\ \phi_p + 5.6$	$1.25\ \phi_p + 16.0$
5	$1.40\ \phi_p + 18.0$	$0.63\ \phi_p + 8.0$	$2.0\ \phi_p + 25.0$
6	$2.24\ \phi_p + 28.0$	$0.9\ \phi_p + 11.2$	$3.15\ \phi_p + 40.0$
7	$3.15\ \phi_p + 40.0$	$1.25\ \phi_p + 16.0$	$4.5\ \phi_p + 56.0$
8	$4.0\ \phi_p + 50.0$	$1.8\ \phi_p + 22.4$	$5.6\ \phi_p + 71.0$
9	$5.0\ \phi_p + 63.0$	$2.24\ \phi_p + 28.0$	$7.1\ \phi_p + 90.0$
10	$6.3\ \phi_p + 80.0$	$2.8\ \phi_p + 35.5$	$9.0\ \phi_p + 112.0$
11	$8.0\ \phi_p + 100.0$	$3.55\ \phi_p + 45.0$	$11.2\ \phi_p + 140.0$
12	$10.0\ \phi_p + 125.0$	$4.5\ \phi_p + 56.0$	$14.0\ \phi_p + 180.0$

The limits of tolerance are in micrometers.

* To simplify application, the limits of tolerance, and tolerance factors for Grades 3 to 12 are given graphically in Charts 1 through 8.

The values of symbols given in the above formulas are:

l = any selected length of arc in millimeters, less than $\pi d/2$.

$\phi_f = m_n + 0.1\ \sqrt{d}$, where m_n = normal module, and d = reference circle diameter in mm.

b = face width in mm, up to a maximum of 150 mm.

$\phi_p = m_n + 0.25\ \sqrt{d}$, where m_n = normal module, and d = reference circle diameter in mm.

This tolerance can also be calculated using the appropriate formula given in the pitch tolerance sub-table in Table 2. Thus, for gear of grade 6 accuracy, the formula is $2.5\ \sqrt{l} + 6.3$. Substituting 40 mm arc length, the calculation is $2.5\ \sqrt{40} + 6.3 = 2.5 \times 6.32 + 6.3 = 22.1$ micrometers, which rounded down is 0.022 mm.

Tolerance on Tooth Profile. — The limits of tolerance on tooth profile for Grades 3 to 12 may be obtained from Chart 3, when the tolerance factor ϕ_f has been arrived at for a particular application using Chart 2.

Example: The tolerance on tooth profile is required for a gear of 5 module, with a reference circle diameter of 100 mm. Entering Chart 2 on the vertical scale at 100 mm, the tolerance factor is read off in relation to the curve for 5 module, and the figure is 6, obtained on the horizontal scale. This factor is then applied in Chart 3. If the gear grade accuracy is 9, for example, the graph is entered on the horizontal ordinate at tolerance factor 6, and the profile tolerance is read off in relation to the curve for grade 9. The figure obtained on the vertical ordinate is approximately 31 micrometers, which is 0.031 mm.

Chart 1. British Standard Metric Spur and Helical Gears — Limits of Tolerance on Transverse Pitch (Adjacent and Cumulative)

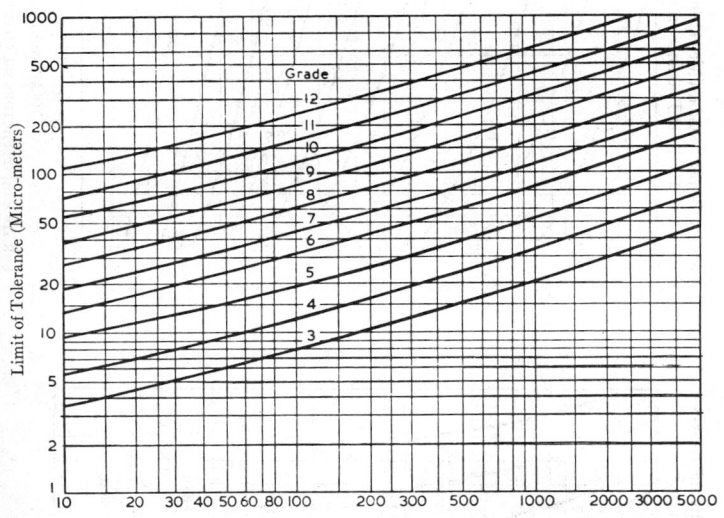

Selected Length of Arc l* (Millimeters)

* l = Any selected length of arc in millimeters, less than $\pi d/2$.

The tolerance factor is based on the formula $\phi_f = m_n + 0.1\sqrt{d}$, where m_n = normal module, and d = reference circle diameter in millimeters. When a particular tolerance factor is known, either by calculation or from Chart 2, the tooth profile tolerance may be calculated as an alternative to obtaining it from Chart 3, using the appropriate formula given in Table 2. Thus, for a gear of grade 9 accuracy, the formula is 2.5 ϕ_f + 16.0. If the tolerance factor is 6, as in the earlier example, then the calculation is 2.5 × 6 + 16 = 31 micrometers, or 0.031 mm.

The spread of the tolerance zone related to the design profile is shown in Fig. 2, left-hand view. Surface irregularities of geometrical form on any one flank of the actual profile, shall be within the tolerance zone contained by the parallel curves A and B. For most applications, positive departure from the design profile should not occur outside the central third of the working depth of the tooth flank as is shown diagrammatically in Fig. 2, right-hand view.

Chart 2. British Standard Metric Spur and Helical Gears — Tooth Profile Tolerance Factors ϕ_f

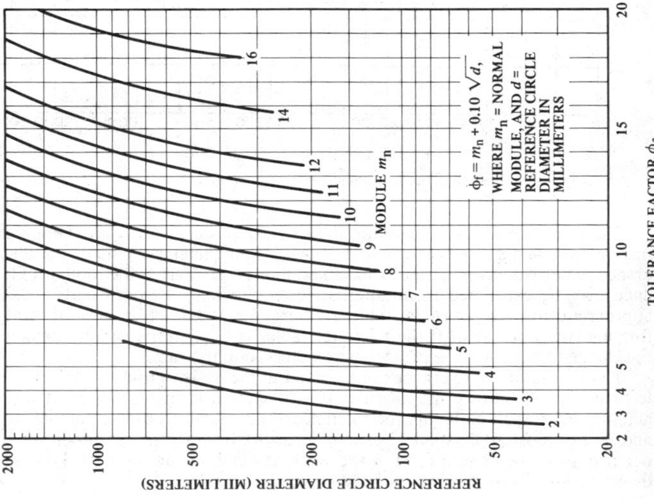
REFERENCE CIRCLE DIAMETER (MILLIMETERS)
2000
1000
500
200
100
50
20
TOLERANCE FACTOR ϕ_f
2
3
4
5
10
15
20
MODULE m_n
2
3
4
5
6
7
8
9
10
11
12
14
16
$\phi_f = m_n + 0.10\sqrt{d}$,
WHERE m_n = NORMAL MODULE, AND d = REFERENCE CIRCLE DIAMETER IN MILLIMETERS

Chart 3. British Standard Metric Spur and Helical Gears — Limits of Tolerance on Tooth Profile

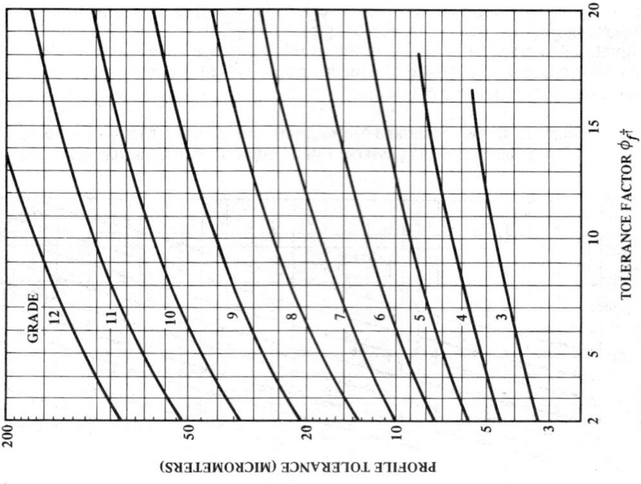
PROFILE TOLERANCE (MICROMETERS)
200
50
20
10
5
3
TOLERANCE FACTOR ϕ_f†
2
5
10
15
20
GRADE
12
11
10
9
8
7
6
5
4
3

† FOR TOLERANCE FACTORS SEE CHART 2.

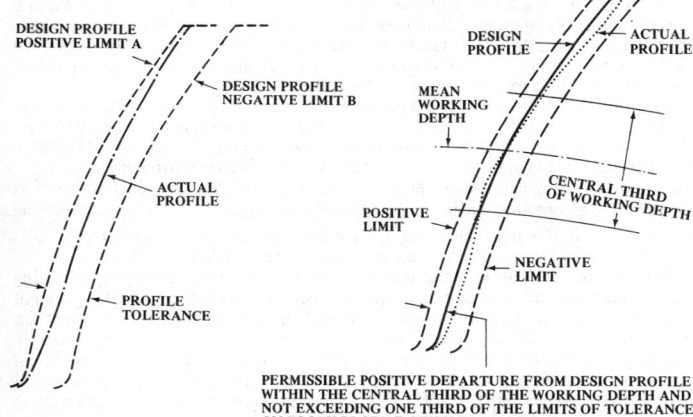

Fig. 2. British Standard Metric Spur and Helical Gears — (Left) Tolerance Zone of Tooth Profile Error, and (Right) Control of Positive Departures from Design Profile

Chart 4. British Standard Metric Spur and Helical Gears — Limits of Tolerance on Tooth Alignment

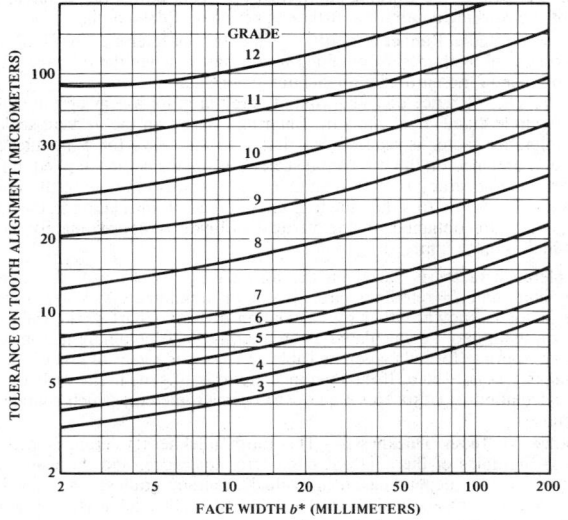

*b = FACE WIDTH UP TO A MAXIMUM OF 150 MILLIMETERS.

Tolerance on Tooth Alignment, and Accuracy of Meshing. — Tooth alignment limits of tolerance are related to a proportion or to the whole of the tooth face width b of a gear, and apply up to a maximum of $b = 150$ mm. Errors are measured as departures from nominal position in the transverse plane. The tolerance can be obtained from Chart 4 when the face width and accuracy grade are known, or may be calculated using the appropriate formula in Table 2.

Example: The tolerance on tooth alignment is required on a face width of 10 mm, for a gear of grade 7 accuracy. Chart 4 is entered on the horizontal scale at 10 mm, and the tolerance is read off in relation to the curve for grade 7 accuracy. The figure on the vertical scale is approximately 10 micrometers which is 0.010 mm.

If the tolerance is calculated, then the formula for a gear of grade 7 accuracy is $1.25\sqrt{b} + 6.3$ as obtained from the appropriate sub-table in Table 2. Substituting the face width, in this instance 10 mm, in the formula the calculation is $1.25\sqrt{10} + 6.3 = 1.25 \times 3.16 + 6.3 = 10.25$ micrometers, which rounded down is 0.010 mm.

Accuracy of meshing is determined by contact marking using the following method (excepting the case of master gears which are subject to special agreement between purchaser and manufacturer). The teeth of either the pinion or wheel are coated with a thin film of toolmaker's blue, and the pair accurately and suitably mounted to permit slow rotation. The mounting is arranged to provide just sufficient pressure between the gears to ensure contact of the tooth surfaces. The accuracy of meshing is considered satisfactory if the contact marking of the tooth flank area is not less than the following percentages:

Gear accuracy grades 3, 4, and 5: At least 40 per cent of the working depth for 50 per cent of the length, and at least 20 per cent of the working depth for a further 40 per cent of the length. *Gear accuracy grades 6, 7, and 8:* At least 40 per cent of the working depth for 35 per cent of the length, and at least 20 per cent of the working depth for a further 35 per cent of the length. *Gear accuracy grades 9, 10, 11, and 12:* At least 40 per cent of the working depth for 25 per cent of the length and at least 20 per cent of the working depth for a further 25 per cent of the length.

Tolerance on Radial Runout of Teeth. — The limits of tolerance for Grades 3 to 12 on radial runout of teeth may be obtained from Chart 6, when the tolerance factor ϕ_p has been obtained for a particular application from Chart 5.

Example: The tolerance on radial runout is required for a gear of 7 module, with a reference circle diameter of 200 mm. Entering Chart 5 on the vertical scale at 200 mm, the tolerance factor is read off in relation to the curve for 7 module, and the figure is 10.5, obtained on the horizontal scale. This factor is then applied in Chart 6. If the gear grade accuracy is 7, for example, the graph is entered on the horizontal scale at tolerance factor 10.5, and the tolerance is read off in relation to the curve for grade 7. The figure obtained on the vertical ordinate is approximately 73 micrometers, which is 0.073 mm.

The tolerance factor is based on the formula $\phi_p = m_n + 0.25\sqrt{d}$, where $m_n =$ normal module, and $d =$ reference circle diameter in millimeters. When a particular tolerance factor is known, either by calculation or from Chart 5, the radial runout tolerance on teeth may be calculated as an alternative to obtaining it from Chart 6, using the appropriate formula given in Table 2. Thus, for a gear of grade 7 accuracy, the formula is $3.15\,\phi_p + 40.0$. If the tolerance factor is 10.5, as in the earlier example, then the calculation is $3.15 \times 10.5 + 40.0 = 73.07$ micrometers, which rounded down is 0.073 mm.

Tolerance on Tooth Thickness. — The tooth thickness tolerance shall be determined as multiples of the adjacent pitch error tolerance, and the values are obtained from Chart 1, depending on individual accuracy grades. The magnitude of the values selected for application to the designed tooth thickness, will depend on functional considerations, with particular regard to the magnitude of the pitch

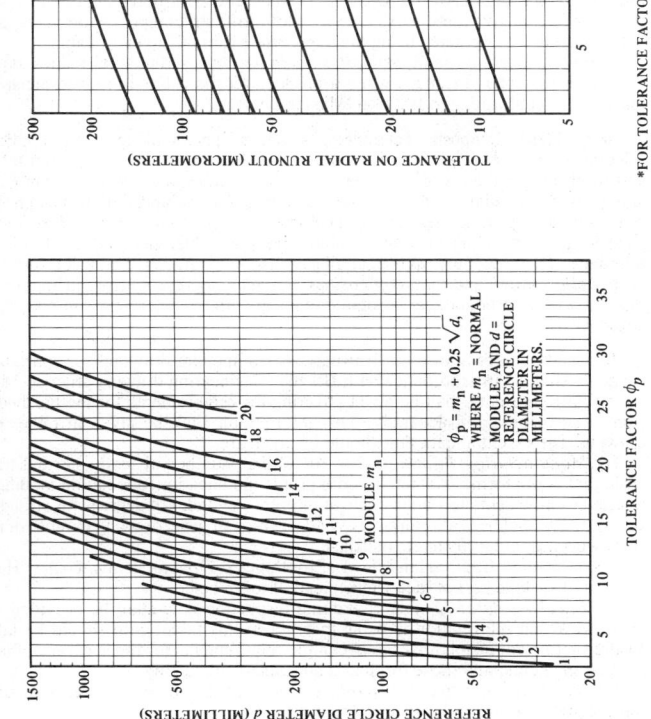

Chart 6. British Standard Metric Spur and Helical Gears — Limits of Tolerance on Radial Runout of Teeth

*FOR TOLERANCE FACTORS SEE CHART 5.

Chart 5. British Standard Metric Spur and Helical Gears — Tolerance Factors ϕ_p

$\phi_p = m_n + 0.25 \sqrt{d}$, WHERE m_n = NORMAL MODULE, AND d = REFERENCE CIRCLE DIAMETER IN MILLIMETERS.

error, tooth profile error, and radial runout error, since these features will have a direct effect on variations in tooth thickness around the gear.

Dual-Flank Composite Tolerance. — When dual-flank testing is applied, the *tooth-to-tooth composite tolerance* can be obtained from Chart 7. The tolerance factor ϕ_p is required when using the chart, and it is the same factor referred to earlier in connection with radial runout tolerance, and is obtained from Chart 5.

Example: The tooth-to-tooth composite tolerance is required for a gear of 5 module, with a reference circle diameter of 90 mm. Entering Chart 5 on the vertical scale at 90 mm, the tolerance factor is read off in relation to the curve for 5 module, and the figure is 7.5, obtained on the horizontal scale. This factor is then applied in Chart 7. If the gear grade accuracy is 8, for example, the graph is entered on the horizontal scale at tolerance factor 7.5, and the tooth-to-tooth composite tolerance is read off in relation to the curve for grade 8. The figure obtained on the vertical scale is approximately 36 micrometers, which is 0.036 mm.

If the tolerance is calculated as an alternative to obtaining it from Chart 7, then the appropriate formula in Table 2 is used.

When dual-flank testing is being applied to measure *total composite error*, the limits of tolerance may be obtained from Chart 8. Again, the tolerance factor ϕ_p is required, and is obtained from Chart 5. The procedure for obtaining the final figure is the same as that given in the example for tooth-to-tooth composite tolerance, and the figure may also be obtained using the appropriate formula given in Table 2.

For both *tooth-to-tooth*, and *total composite error*, the total values represent variations in center distances when a product gear is rotated in close mesh with a master gear conforming to BS 3696 'Master Gears.'

Single-Flank Composite Tolerances. — When single-flank testing is applied, the tolerance for *tooth-to-tooth composite error* is the sum of the single pitch tolerance and tooth profile tolerance, and the values for addition are obtained from Charts 1 and 3, or by calculation, as described earlier under the appropriate headings. This procedure applies to all gear grade accuracies. For *total composite errors*, the tolerance is the sum of the maximum cumulative pitch tolerance and the tooth profile tolerance. The procedure applies to all gear accuracy grades. The values derived for both *tooth-to-tooth* and *total composite tolerances*, represent errors in angular transmission when a product gear is rotated in mesh with a master gear conforming to BS 3696 'Master gears.'

Information to be Shown on Drawings. — Component drawings for spur and helical gears shall be in accordance with BS 308 'Engineering drawing practice,' and the appropriate gear data shall be given covering the requirements for the finished tooth form, dimensions, and accuracy. This information shall be given in tables on the drawing. The essential data that should be given are:

(1) *Manufacturing Information — All Gears:* number of teeth; normal module; basic rack tooth form; axial pitch; tooth profile modification; addendum modification; reference circle diameter; helix angle at reference cylinder (0 degrees for straight spur gears); tooth thickness at reference cylinder; grade of gear; drawing number of mating gear; working center distance; backlash.

Single Helical Gears: Hand; lead of tooth helix. *Double Helical Gears:* Hand (in relation to specific part of face width); lead of tooth helix.

(2) *Checking data:* A table containing inspection data shall be provided on the drawing. When information is being inserted in this table, care should be taken to ensure that no confusion arises between the requirements for the accuracy of individual gear elements, and those for dual- and single-flank testing.

(3) *Supplementary and Other Information* may be added to meet particular design, manufacturing, and inspection requirements.

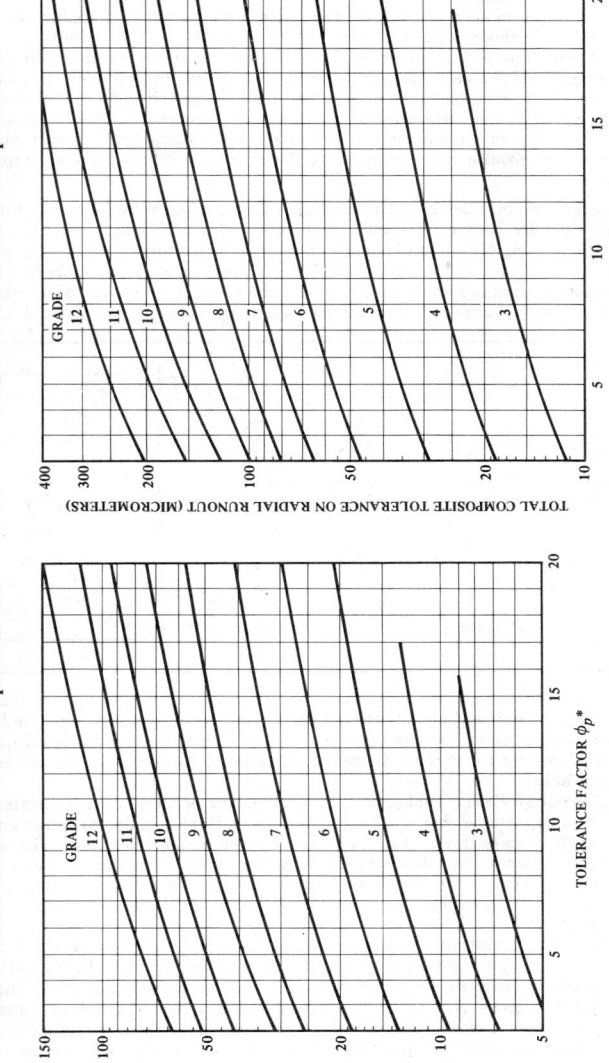

Chart 8. British Standard Metric Spur and Helical Gears — Limits of Tolerance on Total Composite Error

†FOR TOLERANCE FACTORS SEE CHART 5.

Chart 7. British Standard Metric Spur and Helical Gears — Limits of Tolerance on Tooth-to-Tooth Composite Error

*FOR TOLERANCE FACTORS SEE CHART 5.

Checking Gear Size by Measurement Over Wires or Pins

The wire or pin method of checking gear sizes is accurate, easily applied, and especially useful in shops with limited inspection equipment.   Two cylindrical wires or pins of predetermined diameter are placed in diametrically opposite tooth spaces (see diagram).   If the gear has an odd number of teeth, the wires are located as nearly opposite as possible, as shown by the diagram at the right.   The over-all measurement M is checked by using any sufficiently accurate method of measurement.   The value of measurement M when the pitch diameter is correct can be determined easily and quickly by means of the calculated values in the accompanying tables.

Measurements for Checking External Spur Gears when Wire Diameter Equals 1.728 Divided by Diametral Pitch.   Tables 1 and 2 give measurements M, in inches, for checking the pitch diameters of external spur gears of 1 diametral pitch.   For any other diametral pitch, divide the measurement given in the table by whatever diametral pitch is required.   The result shows what measurement M should be when the pitch diameter is correct *and there is no allowance for backlash*.   The

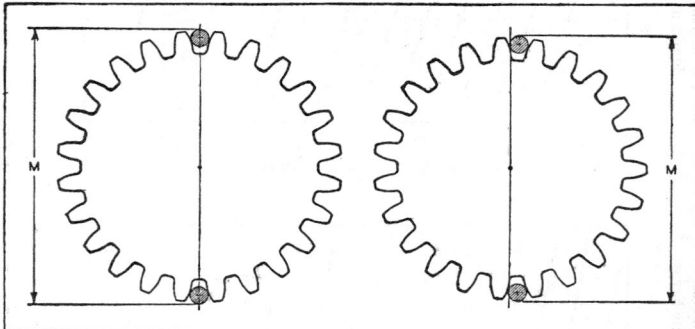

procedure for obtaining a given amount of backlash will be explained later.   Tables 1 to 4 inclusive are based upon wire sizes conforming to the Van Keuren standard.   For external spur gears the wire size equals 1.728 divided by the diametral pitch.   The wire diameters for various diametral pitches will be found in the left-hand section of Table 5.

Even Number of Teeth: Table 1 is for even numbers of teeth.   To illustrate the use of the table, assume that a spur gear has 32 teeth of 4 diametral pitch and a pressure angle of 20 degrees.   Table 1 shows that the measurement for 1 diametral pitch is 34.4130; hence, for 4 diametral pitch, the measurement equals 34.4130 ÷ 4 = 8.6032 inches.   This is the actual measurement over the wires when the pitch diameter is correct, provided there is no allowance for backlash.   The wire diameter in this case equals 1.728 ÷ 4 = .432 inch (Table 5).

Measurement for even numbers of teeth above 170 and not in the Table 1, may be determined as shown by the following example: Assume that number of teeth = 240 and pressure angle 14½ degrees; then, for 1 diametral pitch, figure at left of decimal point = given No. of teeth + 2 = 240 + 2 = 242.   Figure at right of decimal point lies between decimal values given in table for 200 teeth and 300 teeth and is obtained by interpolation.   Thus, 240 − 200 = 40 (change to .40); .5395 − .5321 = .0074 = difference between decimal values for 300 and 200 teeth;

Table 1. Checking External Spur Gear Sizes by Measurement Over Wires

EVEN NUMBERS OF TEETH

Dimensions in table are for 1 diametral pitch and Van Keuren standard wire sizes. For any other diametral pitch, divide dimension in table by given pitch.

$$\text{Wire or pin diameter} = \frac{1.728}{\text{Diametral Pitch}}$$

No. of Teeth	Pressure Angle				
	$14\frac{1}{2}°$	$17\frac{1}{2}°$	$20°$	$25°$	$30°$
6	8.2846	8.2927	8.3032	8.3340	8.3759
8	10.3160	10.3196	10.3271	10.3533	10.3919
10	12.3399	12.3396	12.3445	12.3667	12.4028
12	14.3590	14.3552	14.3578	14.3768	14.4108
14	16.3746	16.3677	16.3683	16.3846	16.4169
16	18.3877	18.3780	18.3768	18.3908	18.4217
18	20.3989	20.3866	20.3840	20.3959	20.4256
20	22.4087	22.3940	22.3900	22.4002	22.4288
22	24.4172	24.4004	24.3952	24.4038	24.4315
24	26.4247	26.4060	26.3997	26.4069	26.4339
26	28.4314	28.4110	28.4036	28.4096	28.4358
28	30.4374	30.4154	30.4071	30.4120	30.4376
30	32.4429	32.4193	32.4102	32.4141	32.4391
32	34.4478	34.4228	34.4130	34.4159	34.4405
34	36.4523	36.4260	36.4155	36.4176	36.4417
36	38.4565	38.4290	38.4178	38.4191	38.4428
38	40.4603	40.4317	40.4198	40.4205	40.4438
40	42.4638	42.4341	42.4217	42.4217	42.4447
42	44.4671	44.4364	44.4234	44.4228	44.4455
44	46.4701	46.4385	46.4250	46.4239	46.4463
46	48.4729	48.4404	48.4265	48.4248	48.4470
48	50.4756	50.4422	50.4279	50.4257	50.4476
50	52.4781	52.4439	52.4292	52.4265	52.4482
52	54.4804	54.4454	54.4304	54.4273	54.4487
54	56.4826	56.4469	56.4315	56.4280	56.4492
56	58.4847	58.4483	58.4325	58.4287	58.4497
58	60.4866	60.4496	60.4335	60.4293	60.4501
60	62.4884	62.4509	62.4344	62.4299	62.4506
62	64.4902	64.4520	64.4352	64.4304	64.4510
64	66.4918	66.4531	66.4361	66.4309	66.4513
66	68.4933	68.4542	68.4369	68.4314	68.4517
68	70.4948	70.4552	70.4376	70.4319	70.4520
70	72.4963	72.4561	72.4383	72.4323	72.4523
72	74.4977	74.4570	74.4390	74.4327	74.4526
74	76.4990	76.4578	76.4396	76.4331	76.4529
76	78.5002	78.4586	78.4402	78.4335	78.4532
78	80.5014	80.4594	80.4408	80.4339	80.4534
80	82.5026	82.4601	82.4413	82.4342	82.4536
82	84.5037	84.4608	84.4418	84.4345	84.4538
84	86.5047	86.4615	86.4423	86.4348	86.4540
86	88.5057	88.4621	88.4428	88.4351	88.4542
88	90.5067	90.4627	90.4433	90.4354	90.4544

Table 1.   Checking External Spur Gear Sizes by Measurement Over Wires

EVEN NUMBERS OF TEETH					
No. of Teeth	14½°	17½°	20°	25°	30°
90	92.5076	92.4633	92.4437	92.4357	92.4546
92	94.5085	94.4639	94.4441	94.4359	94.4548
94	96.5094	96.4644	96.4445	96.4362	96.4550
96	98.5102	98.4649	98.4449	98.4364	98.4552
98	100.5110	100.4655	100.4453	100.4367	100.4554
100	102.5118	102.4660	102.4456	102.4369	102.4555
102	104.5125	104.4665	104.4460	104.4370	104.4557
104	106.5132	106.4669	106.4463	106.4372	106.4558
106	108.5139	108.4673	108.4466	108.4374	108.4560
108	110.5146	110.4678	110.4469	110.4376	110.4561
110	112.5152	112.4682	112.4472	112.4378	112.4562
112	114.5159	114.4686	114.4475	114.4380	114.4563
114	116.5165	116.4690	116.4478	116.4382	116.4564
116	118.5171	118.4693	118.4481	118.4384	118.4565
118	120.5177	120.4697	120.4484	120.4385	120.4566
120	122.5182	122.4701	122.4486	122.4387	122.4567
122	124.5188	124.4704	124.4489	124.4388	124.4568
124	126.5193	126.4708	126.4491	126.4390	126.4569
126	128.5198	128.4711	128.4493	128.4391	128.4570
128	130.5203	130.4714	130.4496	130.4393	130.4571
130	132.5208	132.4717	132.4498	132.4394	132.4572
132	134.5213	134.4720	134.4500	134.4395	134.4573
134	136.5217	136.4723	136.4502	136.4397	136.4574
136	138.5221	138.4725	138.4504	138.4398	138.4575
138	140.5226	140.4728	140.4506	140.4399	140.4576
140	142.5230	142.4730	142.4508	142.4400	142.4577
142	144.5234	144.4733	144.4510	144.4401	144.4578
144	146.5238	146.4736	146.4512	146.4402	146.4578
146	148.5242	148.4738	148.4513	148.4403	148.4579
148	150.5246	150.4740	150.4515	150.4404	150.4580
150	152.5250	152.4742	152.4516	152.4405	152.4580
152	154.5254	154.4745	154.4518	154.4406	154.4581
154	156.5257	156.4747	156.4520	156.4407	156.4581
156	158.5261	158.4749	158.4521	158.4408	158.4582
158	160.5264	160.4751	160.4523	160.4409	160.4582
160	162.5267	162.4753	162.4524	162.4410	162.4583
162	164.5270	164.4755	164.4526	164.4411	164.4584
164	166.5273	166.4757	166.4527	166.4411	166.4584
166	168.5276	168.4759	168.4528	168.4412	168.4585
168	170.5279	170.4760	170.4529	170.4413	170.4585
170	172.5282	172.4761	172.4531	172.4414	172.4586
180	182.5297	182.4771	182.4537	182.4418	182.4589
190	192.5310	192.4780	192.4542	192.4421	192.4591
200	202.5321	202.4786	202.4548	202.4424	202.4593
300	302.5395	302.4831	302.4579	302.4443	302.4606
400	402.5434	402.4854	402.4596	402.4453	402.4613
500	502.5458	502.4868	502.4606	502.4458	502.4619

Table 2. Checking External Spur Gear Sizes by Measurement Over Wires

ODD NUMBERS OF TEETH

Dimensions in table are for 1 diametral pitch and Van Keuren standard wire sizes. For any other diametral pitch, divide dimension in table by given pitch.

$$\text{Wire or pin diameter} = \frac{1.728}{\text{Diametral Pitch}}$$

No. of Teeth	Pressure Angle				
	14½°	17½°	20°	25°	30°
7	9.1116	9.1172	9.1260	9.1536	9.1928
9	11.1829	11.1844	11.1905	11.2142	11.2509
11	13.2317	13.2296	13.2332	13.2536	13.2882
13	15.2677	15.2617	15.2639	15.2814	15.3142
15	17.2957	17.2873	17.2871	17.3021	17.3329
17	19.3182	19.3072	19.3053	19.3181	19.3482
19	21.3368	21.3233	21.3200	21.3310	21.3600
21	23.3524	23.3368	23.3321	23.3415	23.3696
23	25.3658	25.3481	25.3423	25.3502	25.3775
25	27.3774	27.3579	27.3511	27.3576	27.3842
27	29.3876	29.3664	29.3586	29.3640	29.3899
29	31.3966	31.3738	31.3652	31.3695	31.3948
31	33.4047	33.3804	33.3710	33.3743	33.3991
33	35.4119	35.3863	35.3761	35.3786	35.4029
35	37.4185	37.3916	37.3807	37.3824	37.4063
37	39.4245	39.3964	39.3849	39.3858	39.4094
39	41.4299	41.4007	41.3886	41.3889	41.4120
41	43.4348	43.4047	43.3920	43.3917	43.4145
43	45.4394	45.4083	45.3951	45.3942	45.4168
45	47.4437	47.4116	47.3980	47.3965	47.4188
47	49.4477	49.4147	49.4007	49.3986	49.4206
49	51.4514	51.4175	51.4031	51.4006	51.4223
51	53.4547	53.4202	53.4053	53.4024	53.4239
53	55.4579	55.4227	55.4074	55.4041	55.4254
55	57.4609	57.4249	57.4093	57.4056	57.4267
57	59.4637	59.4271	59.4111	59.4071	59.4280
59	61.4664	61.4291	61.4128	61.4084	61.4292
61	63.4689	63.4310	63.4144	63.4097	63.4303
63	65.4712	65.4328	65.4159	65.4109	65.4313
65	67.4734	67.4344	67.4173	67.4120	67.4323
67	69.4755	69.4360	69.4186	69.4130	69.4332
69	71.4775	71.4375	71.4198	71.4140	71.4341
71	73.4795	73.4389	73.4210	73.4150	73.4349
73	75.4813	75.4403	75.4221	75.4159	75.4357
75	77.4830	77.4416	77.4232	77.4167	77.4364
77	79.4847	79.4428	79.4242	79.4175	79.4371
79	81.4863	81.4440	81.4252	81.4183	81.4378
81	83.4877	83.4451	83.4262	83.4190	83.4384
83	85.4892	85.4462	85.4271	85.4196	85.4390
85	87.4906	87.4472	87.4279	87.4203	87.4395
87	89.4919	89.4481	89.4287	89.4209	89.4400
89	91.4932	91.4490	91.4295	91.4215	91.4405

Table 2. Checking External Spur Gear Sizes by Measurement Over Wires

ODD NUMBERS OF TEETH					
No. of Teeth	14½°	17½°	20°	25°	30°
91	93.4944	93.4499	93.4303	93.4221	93.4410
93	95.4956	95.4508	95.4310	95.4227	95.4415
95	97.4967	97.4516	97.4317	97.4232	97.4420
97	99.4978	99.4524	99.4323	99.4237	99.4424
99	101.4988	101.4532	101.4329	101.4242	101.4428
101	103.4998	103.4540	103.4335	103.4247	103.4432
103	105.5008	105.4546	105.4341	105.4252	105.4436
105	107.5017	107.4553	107.4346	107.4256	107.4440
107	109.5026	109.4559	109.4352	109.4260	109.4443
109	111.5035	111.4566	111.4357	111.4264	111.4447
111	113.5044	113.4572	113.4362	113.4268	113.4450
113	115.5052	115.4578	115.4367	115.4272	115.4453
115	117.5060	117.4584	117.4372	117.4275	117.4456
117	119.5068	119.4589	119.4376	119.4279	119.4459
119	121.5075	121.4594	121.4380	121.4282	121.4462
121	123.5082	123.4599	123.4384	123.4285	123.4465
123	125.5089	125.4604	125.4388	125.4288	125.4468
125	127.5096	127.4609	127.4392	127.4291	127.4471
127	129.5103	129.4614	129.4396	129.4294	129.4473
129	131.5109	131.4619	131.4400	131.4297	131.4476
131	133.5115	133.4623	133.4404	133.4300	133.4478
133	135.5121	135.4628	135.4408	135.4302	135.4480
135	137.5127	137.4632	137.4411	137.4305	137.4483
137	139.5133	139.4636	139.4414	139.4307	139.4485
139	141.5139	141.4640	141.4418	141.4310	141.4487
141	143.5144	143.4644	143.4421	143.4312	143.4489
143	145.5149	145.4648	145.4424	145.4315	145.4491
145	147.5154	147.4651	147.4427	147.4317	147.4493
147	149.5159	149.4655	149.4430	149.4319	149.4495
149	151.5164	151.4658	151.4433	151.4321	151.4497
151	153.5169	153.4661	153.4435	153.4323	153.4498
153	155.5174	155.4665	155.4438	155.4325	155.4500
155	157.5179	157.4668	157.4440	157.4327	157.4502
157	159.5183	159.4671	159.4443	159.4329	159.4504
159	161.5188	161.4674	161.4445	161.4331	161.4505
161	163.5192	163.4677	163.4448	163.4333	163.4507
163	165.5196	165.4680	165.4450	165.4335	165.4508
165	167.5200	167.4683	167.4453	167.4337	167.4510
167	169.5204	169.4686	169.4455	169.4338	169.4511
169	171.5208	171.4688	171.4457	171.4340	171.4513
171	173.5212	173.4691	173.4459	173.4342	173.4514
181	183.5230	183.4704	183.4469	183.4350	183.4520
191	193.5246	193.4715	193.4478	193.4357	193.4526
201	203.5260	203.4725	203.4487	203.4363	203.4532
301	303.5355	303.4790	303.4538	303.4402	303.4565
401	403.5404	403.4823	403.4565	403.4422	403.4582
501	503.5433	503.4843	503.4581	503.4434	503.4592

Table 3. Checking Internal Spur Gear Sizes by Measurement Between Wires

EVEN NUMBERS OF TEETH

Dimensions in table are for 1 diametral pitch and Van Keuren standard wire sizes. For any other diametral pitch, divide dimensions in table by given pitch.

$$\text{Wire or pin diameter} = \frac{1.44}{\text{Diametral Pitch}}$$

No. of Teeth	Pressure Angle				
	14½°	17½°	20°	25°	30°
10	8.8337	8.7383	8.6617	8.5209	8.3966
12	10.8394	10.7404	10.6623	10.5210	10.3973
14	12.8438	12.7419	12.6627	12.5210	12.3978
16	14.8474	14.7431	14.6630	14.5210	14.3982
18	16.8504	16.7441	16.6633	16.5210	16.3985
20	18.8529	18.7449	18.6635	18.5211	18.3987
22	20.8550	20.7456	20.6636	20.5211	20.3989
24	22.8569	22.7462	22.6638	22.5211	22.3991
26	24.8585	24.7467	24.6639	24.5211	24.3992
28	26.8599	26.7471	26.6640	26.5211	26.3993
30	28.8612	28.7475	28.6641	28.5211	28.3994
32	30.8623	30.7478	30.6642	30.5211	30.3995
34	32.8633	32.7481	32.6641	32.5211	32.3995
36	34.8642	34.7483	34.6643	34.5212	34.3996
38	36.8650	36.7486	36.6642	36.5212	36.3996
40	38.8658	38.7488	38.6644	38.5212	38.3997
42	40.8665	40.7490	40.6644	40.5212	40.3997
44	42.8672	42.7492	42.6645	42.5212	42.3998
46	44.8678	44.7493	44.6645	44.5212	44.3998
48	46.8683	46.7495	46.6646	46.5212	46.3999
50	48.8688	48.7496	48.6646	48.5212	48.3999
52	50.8692	50.7497	50.6646	50.5212	50.3999
54	52.8697	52.7499	52.6647	52.5212	52.4000
56	54.8701	54.7500	54.6647	54.5212	54.4000
58	56.8705	56.7501	56.6648	56.5212	56.4001
60	58.8709	58.7502	58.6648	58.5212	58.4001
62	60.8712	60.7503	60.6648	60.5212	60.4001
64	62.8715	62.7504	62.6648	62.5212	62.4001
66	64.8718	64.7505	64.6649	64.5212	64.4001
68	66.8721	66.7505	66.6649	66.5212	66.4001
70	68.8724	68.7506	68.6649	68.5212	68.4001
72	70.8727	70.7507	70.6649	70.5212	70.4002
74	72.8729	72.7507	72.6649	72.5212	72.4002
76	74.8731	74.7508	74.6649	74.5212	74.4002
78	76.8734	76.7509	76.6649	76.5212	76.4002
80	78.8736	78.7509	78.6649	78.5212	78.4002
82	80.8738	80.7510	80.6649	80.5212	80.4002
84	82.8740	82.7510	82.6649	82.5212	82.4002
86	84.8742	84.7511	84.6650	84.5212	84.4002
88	86.8743	86.7511	86.6650	86.5212	86.4003
90	88.8745	88.7512	88.6650	88.5212	88.4003

Table 3. Checking Internal Spur Gear Sizes by Measurement Between Wires

EVEN NUMBERS OF TEETH					
No. of Teeth	14½°	17½°	20°	25°	30°
92	90.8747	90.7512	90.6650	90.5212	90.4003
94	92.8749	92.7513	92.6650	92.5212	92.4003
96	94.8750	94.7513	94.6650	94.5212	94.4003
98	96.8752	96.7513	96.6650	96.5212	96.4003
100	98.8753	98.7514	98.6650	98.5212	98.4003
102	100.8754	100.7514	100.6650	100.5212	100.4003
104	102.8756	102.7514	102.6650	102.5212	102.4003
106	104.8757	104.7515	104.6650	104.5212	104.4003
108	106.8758	106.7515	106.6650	106.5212	106.4003
110	108.8759	108.7515	108.6651	108.5212	108.4004
112	110.8760	110.7516	110.6651	110.5212	110.4004
114	112.8761	112.7516	112.6651	112.5212	112.4004
116	114.8762	114.7516	114.6651	114.5212	114.4004
118	116.8763	116.7516	116.6651	116.5212	116.4004
120	118.8764	118.7517	118.6651	118.5212	118.4004
122	120.8765	120.7517	120.6651	120.5212	120.4004
124	122.8766	122.7517	122.6651	122.5212	122.4004
126	124.8767	124.7517	124.6651	124.5212	124.4004
128	126.8768	126.7518	126.6651	126.5212	126.4004
130	128.8769	128.7518	128.6652	128.5212	128.4004
132	130.8769	130.7518	130.6652	130.5212	130.4004
134	132.8770	132.7518	132.6652	132.5212	132.4004
136	134.8771	134.7519	134.6652	134.5212	134.4004
138	136.8772	136.7519	136.6652	136.5212	136.4004
140	138.8773	138.7519	138.6652	138.5212	138.4004
142	140.8773	140.7519	140.6652	140.5212	140.4004
144	142.8774	142.7519	142.6652	142.5212	142.4004
146	144.8774	144.7520	144.6652	144.5212	144.4004
148	146.8775	146.7520	146.6652	146.5212	146.4004
150	148.8775	148.7520	148.6652	148.5212	148.4005
152	150.8776	150.7520	150.6652	150.5212	150.4005
154	152.8776	152.7520	152.6652	152.5212	152.4005
156	154.8777	154.7520	154.6652	154.5212	154.4005
158	156.8778	156.7520	156.6652	156.5212	156.4005
160	158.8778	158.7520	158.6652	158.5212	158.4005
162	160.8779	160.7520	160.6652	160.5212	160.4005
164	162.8779	162.7521	162.6652	162.5212	162.4005
166	164.8780	164.7521	164.6652	164.5212	164.4005
168	166.8780	166.7521	166.6652	166.5212	166.4005
170	168.8781	168.7521	168.6652	168.5212	168.4005
180	178.8783	178.7522	178.6652	178.5212	178.4005
190	188.8785	188.7522	188.6652	188.5212	188.4005
200	198.8788	198.7523	198.6652	198.5212	198.4005
300	298.8795	298.7525	298.6654	298.5212	298.4005
400	398.8803	398.7527	398.6654	398.5212	398.4006
500	498.8810	498.7528	498.6654	498.5212	498.4006

Table 4. Checking Internal Spur Gear Sizes by Measurement Between Wires

	ODD NUMBERS OF TEETH				

Dimensions in table are for 1 diametral pitch and Van Keuren standard wire sizes. For any other diametral pitch, divide dimensions in table by given pitch.

$$\text{Wire or pin diameter} = \frac{1.44}{\text{Diametral Pitch}}$$

No. of Teeth	Pressure Angle				
	14½°	17½°	20°	25°	30°
7	5.6393	5.5537	5.4823	5.3462	5.2232
9	7.6894	7.5976	7.5230	7.3847	7.2618
11	9.7219	9.6256	9.5490	9.4094	9.2867
13	11.7449	11.6451	11.5669	11.4265	11.3040
15	13.7620	13.6594	13.5801	13.4391	13.3167
17	15.7752	15.6703	15.5902	15.4487	15.3265
19	17.7858	17.6790	17.5981	17.4563	17.3343
21	19.7945	19.6860	19.6045	19.4625	19.3405
23	21.8017	21.6918	21.6099	21.4676	21.3457
25	23.8078	23.6967	23.6143	23.4719	23.3501
27	25.8130	25.7009	25.6181	25.4755	25.3538
29	27.8176	27.7045	27.6214	27.4787	27.3571
31	29.8216	29.7076	29.6242	29.4814	29.3599
33	31.8251	31.7104	31.6267	31.4838	31.3623
35	33.8282	33.7128	33.6289	33.4860	33.3645
37	35.8311	35.7150	35.6310	35.4879	35.3665
39	37.8336	37.7169	37.6327	37.4896	37.3682
41	39.8359	39.7187	39.6343	39.4911	39.3698
43	41.8380	41.7203	41.6357	41.4925	41.3712
45	43.8399	43.7217	43.6371	43.4938	43.3725
47	45.8416	45.7231	45.6383	45.4950	45.3737
49	47.8432	47.7243	47.6394	47.4960	47.3748
51	49.8447	49.7254	49.6404	49.4970	49.3758
53	51.8461	51.7265	51.6414	51.4979	51.3768
55	53.8474	53.7274	53.6422	53.4988	53.3776
57	55.8486	55.7283	55.6431	55.4996	55.3784
59	57.8497	57.7292	57.6438	57.5003	57.3792
61	59.8508	59.7300	59.6445	59.5010	59.3799
63	61.8517	61.7307	61.6452	61.5016	61.3806
65	63.8526	63.7314	63.6458	63.5022	63.3812
67	65.8535	65.7320	65.6464	65.5028	65.3818
69	67.8543	67.7327	67.6469	67.5033	67.3823
71	69.8551	69.7332	69.6475	69.5038	69.3828
73	71.8558	71.7338	71.6480	71.5043	71.3833
75	73.8565	73.7343	73.6484	73.5048	73.3838
77	75.8572	75.7348	75.6489	75.5052	75.3842
79	77.8573	77.7352	77.6493	77.5056	77.3846
81	79.8584	79.7357	79.6497	79.5060	79.3850
83	81.8590	81.7361	81.6501	81.5064	81.3854
85	83.8595	83.7365	83.6505	83.5067	83.3858
87	85.8600	85.7369	85.6508	85.5071	85.3861
89	87.8605	87.7373	87.6511	87.5074	87.3864

Table 4. Checking Internal Spur Gear Sizes by Measurement Between Wires

	ODD NUMBERS OF TEETH				
No. of Teeth	14½°	17½°	20°	25°	30°
91	89.8610	89.7376	89.6514	89.5077	89.3867
93	91.8614	91.7379	91.6517	91.5080	91.3870
95	93.8619	93.7383	93.6520	93.5082	93.3873
97	95.8623	95.7386	95.6523	95.5085	95.3876
99	97.8627	97.7389	97.6526	97.5088	97.3879
101	99.8631	99.7391	99.6528	99.5090	99.3881
103	101.8635	101.7394	101.6531	101.5093	101.3883
105	103.8638	103.7397	103.6533	103.5095	103.3886
107	105.8642	105.7399	105.6535	105.5097	105.3888
109	107.8645	107.7402	107.6537	107.5099	107.3890
111	109.8648	109.7404	109.6539	109.5101	109.3893
113	111.8651	111.7406	111.6541	111.5103	111.3895
115	113.8654	113.7409	113.6543	113.5105	113.3897
117	115.8657	115.7411	115.6545	115.5107	115.3899
119	117.8660	117.7413	117.6547	117.5109	117.3900
121	119.8662	119.7415	119.6548	119.5110	119.3902
123	121.8663	121.7417	121.6550	121.5112	121.3904
125	123.8668	123.7418	123.6552	123.5114	123.3905
127	125.8670	125.7420	125.6554	125.5115	125.3907
129	127.8672	127.7422	127.6556	127.5117	127.3908
131	129.8675	129.7424	129.6557	129.5118	129.3910
133	131.8677	131.7425	131.6559	131.5120	131.3911
135	133.8679	133.7427	133.6560	133.5121	133.3913
137	135.8681	135.7428	135.6561	135.5123	135.3914
139	137.8683	137.7430	137.6563	137.5124	137.3916
141	139.8685	139.7431	139.6564	139.5125	139.3917
143	141.8687	141.7433	141.6565	141.5126	141.3918
145	143.8689	143.7434	143.6566	143.5127	143.3919
147	145.8691	145.7436	145.6568	145.5128	145.3920
149	147.8693	147.7437	147.6569	147.5130	147.3922
151	149.8694	149.7438	149.6570	149.5131	149.3923
153	151.8696	151.7439	151.6571	151.5132	151.3924
155	153.8698	153.7441	153.6572	153.5133	153.3925
157	155.8699	155.7442	155.6573	155.5134	155.3926
159	157.8701	157.7443	157.6574	157.5135	157.3927
161	159.8702	159.7444	159.6575	159.5136	159.3928
163	161.8704	161.7445	161.6576	161.5137	161.3929
165	163.8705	163.7446	163.6577	163.5138	163.3930
167	165.8707	165.7447	165.6578	165.5139	165.3931
169	167.8708	167.7448	167.6579	167.5139	167.3932
171	169.8710	169.7449	169.6580	169.5140	169.3933
181	179.8716	179.7453	179.6584	179.5144	179.3937
191	189.8721	189.7458	189.6588	189.5148	189.3940
201	199.8727	199.7461	199.6591	199.5151	199.3944
301	299.8759	299.7485	299.6612	299.5171	299.3965
401	399.8776	399.7496	399.6623	399.5182	399.3975
501	499.8786	499.7504	499.6629	499.5188	499.3981

hence, decimal required = = .5321 + (.40 × .0074) = .53506. Total dimension = 242.53506 divided by the diametral pitch required.

Odd Number of Teeth: Table 2 is for odd numbers of teeth. Measurement for odd numbers above 171 and not in Table 2, may be determined as shown by the following example: Assume that number of teeth = 335 and pressure angle 20 degrees; then, for 1 diametral pitch, figure at left of decimal point = given No. of teeth + 2 = 335 + 2 = 337. Figure at right of decimal point lies between decimal values given in table for 301 and 401 teeth. Thus, 335 − 301 = 34 (change to .34); .4565 − .4538 = .0027; hence, decimal required = .4538 + (.34 × .0027) = .4547. Total dimension = 337.4547.

Table 5. Van Keuren Wire Diameters for Gears

External Gears Wire Diam. = 1.728 ÷ D.P.				Internal Gears Wire Diam. = 1.44 ÷ D.P.			
D.P.	Diam.	D.P.	Diam.	D.P.	Diam.	D.P.	Diam.
2	.86400	16	.10800	2	.72000	16	.09000
2½	.69120	18	.09600	2½	.57600	18	.08000
3	.57600	20	.08640	3	.48000	20	.07200
4	.43200	22	.07855	4	.36000	22	.06545
5	.34560	24	.07200	5	.28800	24	.06000
6	.28800	28	.06171	6	.24000	28	.05143
7	.24686	32	.05400	7	.20571	32	.04500
8	.21600	36	.04800	8	.18000	36	.04000
9	.19200	40	.04320	9	.16000	40	.03600
10	.17280	48	.03600	10	.14400	48	.03000
11	.15709	64	.02700	11	.13091	64	.02250
12	.14400	72	.02400	12	.12000	72	.02000
14	.12343	80	.02160	14	.10286	80	.01800

Measurements for Checking Internal Gears when Wire Diameter Equals 1.44 Divided by Diametral Pitch. Tables 3 and 4 give measurements between wires for checking internal gears of 1 diametral pitch. For any other diametral pitch, divide the measurement given in the table by the diametral pitch required. These measurements are based upon the Van Keuren standard wire size, which, for internal spur gears, equals 1.44 divided by the diametral pitch (see Table 5).

Even Number of Teeth: For an even number of teeth above 170 and not in Table 3, proceed as shown by the following example: Assume that the number of teeth = 380 and pressure angle is 14½ degrees; then, for 1 diametral pitch, figure at left of decimal point = given number of teeth − 2 = 380 − 2 = 378. Figure at right of decimal point lies between decimal values given in table for 300 and 400 teeth and is obtained by interpolation. Thus, 380 − 300 = 80 (change to .80); .8803 − .8795 = .0008; hence, decimal required = .8795 + (.80 × .0008) = .88014. Total dimension = 378.88014.

Odd Number of Teeth: Table 4 is for internal gears having odd numbers of teeth. For tooth numbers above 171 and not in the table, proceed as shown by the following example: Assume that number of teeth = 337 and pressure angle is 14½ degrees; then, for 1 diametral pitch, figure at left of decimal point = given No. of teeth − 2 = 337 − 2 = 335. Figure at right of decimal point lies between decimal values given in table for 301 and 401 teeth and is obtained by interpolation. Thus, 337 − 301 = 36 (change to .36); .8776 − .8759 = .0017; hence, decimal required = .8759 + (.36 × .0017) = .8765. Total dimension = 335.8765.

Measurements for Checking External Spur Gears when Wire Diameter Equals 1.68 Divided by Diametral Pitch. — Tables 7 and 8 give measurements M, in inches, for checking the pitch diameters of external spur gears of 1 diametral pitch. For any other diametral pitch, divide the measurement given in the table by whatever diametral pitch is required. The result shows what measurement M should be when the pitch diameter is correct and there is no allowance for backlash. The procedure for checking for a given amount of backlash when the diameter of the measuring wires equals 1.68 divided by the diametral pitch is explained under a subsequent heading. Tables 7 and 8 are based upon wire sizes equal to 1.68 divided by the diametral pitch. The corresponding wire diameters for various diametral pitches are given in Table 9.

To find measurement M of an external spur gear using wire sizes equal to 1.68 inches divided by the diametral pitch, the same method is followed as in using Tables 7 and 8 as that outlined for Tables 1 and 2.

Allowance for Backlash. — Tables 1, 2, 7 and 8 give measurements over wires when the pitch diameters are correct and there is no allowance for backlash or play between meshing teeth. Backlash is obtained by cutting the teeth somewhat deeper than standard, thus reducing the thickness. Usually, the teeth of both mating gears are reduced in thickness an amount equal to one-half of the total backlash desired. However, if the pinion is small, it is common practice to reduce the gear teeth the full amount of backlash and the pinion is made to standard size. The changes in measurements M over wires, for obtaining backlash in external spur gears, are listed in Table 6.

Table 6. Backlash Allowances for External and Internal Spur Gears

External Gears: For each 0.001 inch reduction in pitch-line tooth thickness, *reduce* measurement over wires obtained from Tables 1, 2, 7, or 8 by the amount shown below.

Internal Gears: For each 0.001 inch reduction in pitch-line tooth thickness, *increase* measurement between wires obtained from Tables 3 or 4 by the amounts shown below.

Backlash on pitch line equals double tooth thickness reduction when teeth of *both* mating gears are reduced. If teeth of *one* gear only are reduced, backlash on pitch line equals amount of reduction.

Example: For a 30-tooth, 10-diametral pitch, 20-degree pressure angle, external gear the measurement over wires from Table 1 is 32.4102 ÷ 10. For a backlash of 0.002 this measurement must be reduced by 2 × 0.0024 to 3.2362 or (3.2410 − 0.0048).

No. of Teeth	14½°		17½°		20°		25°		30°	
	Ext.	Int.	Ext.	Int.	Ext.	Int.	Ext.	Int.	Ext.	Int.
5	.0019	.0024	.0018	.0024	.0017	.0023	.0015	.0021	.0013	.0019
10	.0024	.0029	.0022	.0027	.0020	.0026	.0017	.0022	.0015	.0018
20	.0028	.0032	.0025	.0029	.0023	.0027	.0019	.0022	.0016	.0018
30	.0030	.0034	.0026	.0030	.0024	.0027	.0020	.0022	.0016	.0018
40	.0031	.0035	.0027	.0030	.0025	.0027	.0020	.0022	.0017	.0018
50	.0032	.0036	.0028	.0031	.0025	.0027	.0020	.0022	.0017	.0018
100	.0035	.0037	.0030	.0031	.0026	.0027	.0021	.0022	.0017	.0017
200	.0036	.0038	.0031	.0031	.0027	.0027	.0021	.0022	.0017	.0017

Measurements for Checking Helical Gears using Wires or Balls. — Helical gears may be checked for size by using one wire, or ball; two wires, or balls; and three wires, depending on the case at hand. Three wires may be used for measurement of either even or odd tooth numbers provided that the face width and helix angle of the gear permit the arrangement of two wires in adjacent tooth spaces on one side of the gear and a third wire on the opposite side. The wires should be held between flat, parallel plates. The measurement between these plates, and perpendicular to the gear axis, will be the same for both even and odd numbers of teeth because the axial displacement of the wires with the odd numbers of teeth does not affect the perpendicular measurement between the plates. The calculation of measurements over three wires is the same as described for measurements over two wires for even numbers of teeth.

Table 7. Checking External Spur Gear Sizes by Measurement Over Wires

EVEN NUMBERS OF TEETH

Dimensions in table are for 1 diametral pitch and 1.68-inch series wire sizes (a Van Keuren standard). For any other diametral pitch, divide dimension in table by given pitch.

$$\text{Wire or pin diameter} = \frac{1.68}{\text{Diametral Pitch}}$$

No. of Teeth	Pressure Angle				
	14½°	17½°	20°	25°	30°
6	8.1298	8.1442	8.1600	8.2003	8.2504
8	10.1535	10.1647	10.1783	10.2155	10.2633
10	12.1712	12.1796	12.1914	12.2260	12.2722
12	14.1851	14.1910	14.2013	14.2338	14.2785
14	16.1964	16.2001	16.2091	16.2397	16.2833
16	18.2058	18.2076	18.2154	18.2445	18.2871
18	20.2137	20.2138	20.2205	20.2483	20.2902
20	22.2205	22.2190	22.2249	22.2515	22.2927
22	24.2265	24.2235	24.2286	24.2542	24.2949
24	26.2317	26.2275	26.2318	26.2566	26.2967
26	28.2363	28.2309	28.2346	28.2586	28.2982
28	30.2404	30.2339	30.2371	30.2603	30.2996
30	32.2441	32.2367	32.2392	32.2619	32.3008
32	34.2475	34.2391	34.2412	34.2632	34.3017
34	36.2505	36.2413	36.2430	36.2644	36.3026
36	38.2533	38.2433	38.2445	38.2655	38.3035
38	40.2558	40.2451	40.2460	40.2666	40.3044
40	42.2582	42.2468	42.2473	42.2675	42.3051
42	44.2604	44.2483	44.2485	44.2683	44.3057
44	46.2624	46.2497	46.2496	46.2690	46.3063
46	48.2642	48.2510	48.2506	48.2697	48.3068
48	50.2660	50.2522	50.2516	50.2704	50.3073
50	52.2676	52.2534	52.2525	52.2710	52.3078
52	54.2691	54.2545	54.2533	54.2716	54.3082
54	56.2705	56.2555	56.2541	56.2721	56.3086
56	58.2719	58.2564	58.2548	58.2726	58.3089
58	60.2731	60.2572	60.2555	60.2730	60.3093
60	62.2743	62.2580	62.2561	62.2735	62.3096
62	64.2755	64.2587	64.2567	64.2739	64.3099
64	66.2765	66.2594	66.2572	66.2742	66.3102
66	68.2775	68.2601	68.2577	68.2746	68.3104
68	70.2785	70.2608	70.2582	70.2749	70.3107
70	72.2794	72.2615	72.2587	72.2752	72.3109
72	74.2803	74.2620	74.2591	74.2755	74.3111
74	76.2811	76.2625	76.2596	76.2758	76.3113
76	78.2819	78.2631	78.2600	78.2761	78.3115
78	80.2827	80.2636	80.2604	80.2763	80.3117
80	82.2834	82.2641	82.2607	82.2766	82.3119
82	84.2841	84.2646	84.2611	84.2768	84.3121
84	86.2847	86.2650	86.2614	86.2771	86.3123
86	88.2854	88.2655	88.2617	88.2773	88.3124
88	90.2860	90.2659	90.2620	90.2775	90.3126

Table 7. Checking External Spur Gear Sizes by Measurement Over Wires

	EVEN NUMBERS OF TEETH				
No. of Teeth	14½°	17½°	20°	25°	30°
90	92.2866	92.2662	92.2624	92.2777	92.3127
92	94.2872	94.2666	94.2627	94.2779	94.3129
94	96.2877	96.2670	96.2630	96.2780	96.3130
96	98.2882	98.2673	98.2632	98.2782	98.3131
98	100.2887	100.2677	100.2635	100.2784	100.3132
100	102.2892	102.2680	102.2638	102.2785	102.3134
102	104.2897	104.2683	104.2640	104.2787	104.3135
104	106.2901	106.2685	106.2642	106.2788	106.3136
106	108.2905	108.2688	108.2644	108.2789	108.3137
108	110.2910	110.2691	110.2645	110.2791	110.3138
110	112.2914	112.2694	112.2647	112.2792	112.3139
112	114.2918	114.2696	114.2649	114.2793	114.3140
114	116.2921	116.2699	116.2651	116.2794	116.3141
116	118.2925	118.2701	118.2653	118.2795	118.3142
118	120.2929	120.2703	120.2655	120.2797	120.3142
120	122.2932	122.2706	122.2656	122.2798	122.3143
122	124.2936	124.2708	124.2658	124.2799	124.3144
124	126.2939	126.2710	126.2660	126.2800	126.3145
126	128.2941	128.2712	128.2661	128.2801	128.3146
128	130.2945	130.2714	130.2663	130.2802	130.3146
130	132.2948	132.2716	132.2664	132.2803	132.3147
132	134.2951	134.2718	134.2666	134.2804	134.3147
134	136.2954	136.2720	136.2667	136.2805	136.3148
136	138.2957	138.2722	138.2669	138.2806	138.3149
138	140.2960	140.2724	140.2670	140.2807	140.3149
140	142.2962	142.2725	142.2671	142.2808	142.3150
142	144.2965	144.2727	144.2672	144.2808	144.3151
144	146.2967	146.2729	146.2674	146.2809	146.3151
146	148.2970	148.2730	148.2675	148.2810	148.3152
148	150.2972	150.2732	150.2676	150.2811	150.3152
150	152.2974	152.2733	152.2677	152.2812	152.3153
152	154.2977	154.2735	154.2678	154.2812	154.3153
154	156.2979	156.2736	156.2679	156.2813	156.3154
156	158.2981	158.2737	158.2680	158.2813	158.3155
158	160.2983	160.2739	160.2681	160.2814	160.3155
160	162.2985	162.2740	162.2682	162.2815	162.3155
162	164.2987	164.2741	164.2683	164.2815	164.3156
164	166.2989	166.2742	166.2684	166.2816	166.3156
166	168.2990	168.2744	168.2685	168.2816	168.3157
168	170.2992	170.2745	170.2686	170.2817	170.3157
170	172.2994	172.2746	172.2687	172.2818	172.3158
180	182.3003	182.2752	182.2691	182.2820	182.3160
190	192.3011	192.2757	192.2694	192.2823	192.3161
200	202.3018	202.2761	202.2698	202.2825	202.3163
300	302.3063	302.2790	302.2719	302.2839	302.3173
400	402.3087	402.2804	402.2730	402.2845	402.3178
500	502.3101	502.2813	502.2736	502.2850	502.3181

Table 8. Checking External Spur Gear Sizes by Measurement Over Wires

ODD NUMBERS OF TEETH

Dimensions in table are for 1 diametral pitch and 1.68-inch series wire sizes (a VanKeuren standard). For any other diametral pitch, divide dimension in table by given pitch.

$$\text{Wire or pin diameter} = \frac{1.68}{\text{Diametral Pitch}}$$

No. of Teeth	Pressure Angle				
	14½°	17½°	20°	25°	30°
5	6.8485	6.8639	6.8800	6.9202	6.9691
7	8.9555	8.9679	8.9822	9.0199	9.0675
9	11.0189	11.0285	11.0410	11.0762	11.1224
11	13.0615	13.0686	13.0795	13.1126	13.1575
13	15.0925	15.0973	15.1068	15.1381	15.1819
15	17.1163	17.1190	17.1273	17.1570	17.1998
17	19.1351	19.1360	19.1432	19.1716	19.2136
19	21.1505	21.1498	21.1561	21.1832	21.2245
21	23.1634	23.1611	23.1665	23.1926	23.2334
23	25.1743	25.1707	25.1754	25.2005	25.2408
25	27.1836	27.1788	27.1828	27.2071	27.2469
27	29.1918	29.1859	29.1892	29.2128	29.2522
29	31.1990	31.1920	31.1948	31.2177	31.2568
31	33.2053	33.1974	33.1997	33.2220	33.2607
33	35.2110	35.2021	35.2041	35.2258	35.2642
35	37.2161	37.2065	37.2079	37.2292	37.2674
37	39.2208	39.2104	39.2115	39.2323	39.2702
39	41.2249	41.2138	41.2147	41.2349	41.2726
41	43.2287	43.2170	43.2174	43.2374	43.2749
43	45.2323	45.2199	45.2200	45.2396	45.2769
45	47.2355	47.2226	47.2224	47.2417	47.2788
47	49.2385	49.2251	49.2246	49.2435	49.2805
49	51.2413	51.2273	51.2266	51.2452	51.2820
51	53.2439	53.2294	53.2284	53.2468	53.2835
53	55.2463	55.2313	55.2302	55.2483	55.2848
55	57.2485	57.2331	57.2318	57.2497	57.2861
57	59.2506	59.2348	59.2333	59.2509	59.2872
59	61.2526	61.2363	61.2347	61.2521	61.2883
61	63.2545	63.2378	63.2360	63.2532	63.2893
63	65.2562	65.2392	65.2372	65.2543	65.2902
65	67.2579	67.2406	67.2383	67.2553	67.2911
67	69.2594	69.2419	69.2394	69.2562	69.2920
69	71.2609	71.2431	71.2405	71.2571	71.2928
71	73.2623	73.2442	73.2414	73.2579	73.2935
73	75.2636	75.2452	75.2423	75.2586	75.2942
75	77.2649	77.2462	77.2432	77.2594	77.2949
77	79.2661	79.2472	79.2440	79.2601	79.2955
79	81.2673	81.2481	81.2448	81.2607	81.2961
81	83.2684	83.2490	83.2456	83.2614	83.2967
83	85.2694	85.2498	85.2463	85.2620	85.2972
85	87.2704	87.2506	87.2470	87.2625	87.2977
87	89.2714	89.2514	89.2476	89.2631	89.2982
89	91.2723	91.2521	91.2482	91.2636	91.2987

Table 8. Checking External Spur Gear Sizes by Measurement Over Wires

	ODD NUMBERS OF TEETH				
No. of Teeth	14½°	17½°	20°	25°	30°
91	93.2732	93.2528	93.2489	93.2641	93.2991
93	95.2741	95.2534	95.2495	95.2646	95.2996
95	97.2749	97.2541	97.2500	97.2650	97.3000
97	99.2757	99.2547	99.2506	99.2655	99.3004
99	101.2764	101.2553	101.2511	101.2659	101.3008
101	103.2771	103.2558	103.2516	103.2663	103.3011
103	105.2778	105.2563	105.2520	105.2667	105.3015
105	107.2785	107.2568	107.2525	107.2671	107.3018
107	109.2791	109.2573	109.2529	109.2674	109.3021
109	111.2798	111.2578	111.2533	111.2678	111.3024
111	113.2804	113.2583	113.2537	113.2681	113.3027
113	115.2809	115.2588	115.2541	115.2684	115.3030
115	117.2815	117.2592	117.2544	117.2687	117.3033
117	119.2821	119.2596	119.2548	119.2690	119.3036
119	121.2826	121.2601	121.2552	121.2693	121.3038
121	123.2831	123.2605	123.2555	123.2696	123.3041
123	125.2836	125.2608	125.2558	125.2699	125.3043
125	127.2841	127.2612	127.2562	127.2702	127.3046
127	129.2846	129.2615	129.2565	129.2704	129.3048
129	131.2851	131.2619	131.2568	131.2707	131.3050
131	133.2855	133.2622	133.2571	133.2709	133.3053
133	135.2859	135.2626	135.2574	135.2712	135.3055
135	137.2863	137.2629	137.2577	137.2714	137.3057
137	139.2867	139.2632	139.2579	139.2716	139.3059
139	x41.2871	141.2635	141.2582	141.2718	141.3060
141	143.2875	143.2638	143.2584	143.2720	143.3062
143	145.2879	145.2641	145.2587	145.2722	145.3064
145	147.2883	147.2644	147.2589	147.2724	147.3066
147	149.2887	149.2647	149.2591	149.2726	149.3068
149	151.2890	151.2649	151.2594	151.2728	151.3069
151	153.2893	153.2652	153.2596	153.2730	153.3071
153	155.2897	155.2654	155.2598	155.2732	155.3073
155	157.2900	157.2657	157.2600	157.2733	157.3074
157	159.2903	159.2659	159.2602	159.2735	159.3076
159	161.2906	161.2661	161.2604	161.2736	161.3077
161	163.2909	163.2663	163.2606	163.2738	163.3078
163	165.2912	165.2665	165.2608	165.2740	165.3080
165	167.2915	167.2668	167.2610	167.2741	167.3081
167	169.2917	169.2670	169.2611	169.2743	169.3083
169	171.2920	171.2672	171.2613	171.2744	171.3084
171	173.2922	173.2674	173.2615	173.2746	173.3085
181	183.2936	183.2684	183.2623	183.2752	183.3091
191	193.2947	193.2692	193.2630	193.2758	193.3097
201	203.2957	203.2700	203.2636	203.2764	203.3101
301	303.3022	303.2749	303.2678	303.2798	303.3132
401	403.3056	403.2774	403.2699	403.2815	403.3147
501	503.3076	503.2789	503.2711	503.2825	503.3156

Table 9. Wire Diameters for Spur and Helical Gears Based upon 1.68 Constant*

Diametral or Normal Diametral Pitch	Wire Diameter	Diametral or Normal Diametral Pitch	Wire Diameter	Diametral or Normal Diametral Pitch	Wire Diameter	Diametral or Normal Diametral Pitch	Wire Diameter
2	.840	8	.210	18	.09333	40	.042
2½	.672	9	.18666	20	.084	48	.035
3	.560	10	.168	22	.07636	64	.02625
4	.420	11	.15273	24	.070	72	.02333
5	.336	12	.140	28	.060	80	.021
6	.280	14	.120	32	.0525	. . .	. . .
7	.240	16	.105	36	.04667	. . .	. . .

* Pin diameter = 1.68 ÷ diametral pitch for spur gears and 1.68 ÷ normal diametral pitch for helical gears.

Measurements over One Wire or One Ball for Even or Odd Numbers of Teeth: This measurement is calculated by the method for measurement over two wires for even numbers of teeth and the result divided by two to obtain the measurement from over the wire or ball to the center of the gear mounted on an arbor.

Measurement over Two Wires or Two Balls for Even Numbers of Teeth: The measurement over two wires (or two balls kept in the same plane by holding them against a surface parallel to the face of the gear) is calculated as follows: First, calculate the pitch diameter of the helical gear from the formula D = Number of teeth divided by the product of the normal diametral pitch and the cosine of the helix angle, $D = N \div (P_n \times \cos \psi)$. Next, calculate the number of teeth, N_e, there would be in a spur gear for it to have the same tooth curvature as the helical gear has in the normal plane: $N_e = N/\cos^3 \psi$. Next, refer to Table 7 for spur gears with even tooth numbers and find, by interpolation, the *decimal* value of the constant for this number of teeth under the given *normal* pressure angle. Finally, add 2 to this decimal value and divide the sum by the normal diametral pitch P_n. The result of this calculation, added to the pitch diameter D, is the measurement over two wires or balls.

Example: A helical gear has 32 teeth of 6 normal diametral pitch, 20 degree pressure angle, and 23 degree helix angle. Determine the measurement over two wires, M, without allowance for backlash.

$D = 32 \div 6 \times \cos 23° = 5.7939$; $N_e = 32 \div \cos^3 23° = 41.027$; and in Table 7, fourth column, the decimal part of the measurement for 40 teeth is .2473 and that for 42 teeth is .2485. The decimal part for 41.027 teeth is, by interpolation, $\dfrac{(41.027 - 40)}{(42 - 40)}$ $\times (.2485 - .2473) + .2473 = 0.2479$; $(0.2479 + 2) \div 6 = 0.3747$; and $M = 0.3747 + 5.7939 = 6.1686$.

This measurement over wires or balls is based upon the use of $1.68/P_n$ wires or balls. If measurements over $1.728/P_n$ diameter wires or balls are preferred, use Table 1 to find the decimal part described above instead of Table 7.

Measurement over Two Wires or Two Balls for Odd Numbers of Teeth: The procedure is similar to that for two wire or two ball measurement for even tooth numbers except that a correction is made in the final M value to account for the wires or balls not being diametrically opposite by one-half tooth interval. In addition, care must be taken to ensure that the balls or wires are kept in a plane of the gear's rotation as described previously.

Example: A helical gear has 13 teeth of 8 normal diametral pitch, 14½ degree pressure angle, and 45 degree helix angle. Determine measurement M without allowance for backlash based upon the use of $1.728/P_n$ balls or wires.

As before, $D = 13/8 \times \cos 45° = 2.2981$; $N_e = 13/\cos^3 45° = 36.770$; and in the second column of Table 1 the *decimal* part of the measurement for 36 teeth is .4565 and that for 38 teeth is .4603. The decimal part for 36.770 teeth is, by interpolation, $\frac{(36.770 - 36)}{(38 - 36)} \times (.4603 - .4565) + .4565 = 0.4580$; $(0.4580 + 2)/8 = 0.3073$; and $M = 0.3073 + 2.2981 = 2.6054$. This measurement is correct for three-wire measurements but, for two balls or wires held in the plane of rotation of the gear, M must be corrected as follows:

$$M \text{ corrected} = (M - \text{Ball Diam.}) \times \cos (90°/N) + \text{Ball Diam.}$$

$$= (2.6054 - 1.728/8) \times \cos (90°/13) + 1.728/8 = 2.5880$$

Checking Spur Gear Size by Chordal Measurement Over Two or More Teeth. — Another method of checking gear sizes, that is generally available, is illustrated by the diagram accompanying Table 10. A vernier caliper is used to measure the distance M over two or more teeth. The diagram illustrates the measurement over two teeth (or with one intervening tooth space), but three or more teeth might be included, depending upon the pitch. The jaws of the caliper are merely held in contact with the sides or profiles of the teeth and perpendicular to the axis of the gear. Measurement M for involute teeth of the correct size, is determined as follows:

Table for Determining the Chordal Dimension: Table 10 gives the chordal dimensions for one diametral pitch when measuring over the number of teeth indicated in Table 11. To obtain any chordal dimension, it is simply necessary to divide chord M in the table (opposite the given number of teeth) by the diametral pitch of the gear to be measured and then subtract from the quotient one-half the total backlash between the mating pair of gears. In cases where a small pinion is used with a large gear and all of the backlash is to be obtained by reducing the gear teeth, the total amount of backlash is subtracted from the chordal dimension of the gear and nothing from the chordal dimension of the pinion. The application of the tables will be illustrated by an example.

Example — Determine the chordal dimension for checking the size of a gear having 30 teeth of 5 diametral pitch and a pressure angle of 20 degrees. A total backlash of 0.008 inch is to be obtained by reducing equally the teeth of both mating gears.

Table 10 shows that the chordal distance for 30 teeth of one diametral pitch and a pressure angle of 20 degrees is 10.7526 inches; one-half of the backlash equals 0.004 inch; hence,

$$\text{Chordal dimension} = \frac{10.7526}{5} - 0.004 = 2.1465 \text{ inches}$$

Table 11 shows that this is the chordal dimension when the vernier caliper spans four teeth, this being the number of teeth to gage over whenever gears of 20-degree pressure angle have any number of teeth from 28 to 36, inclusive.

If it is considered necessary to leave enough stock on the gear teeth for a shaving or finishing cut, this allowance is simply added to the chordal dimension of the finished teeth to obtain the required measurement over the teeth for the roughing

Table 10. Chordal Measurements over Spur Gear Teeth of 1 Diametral Pitch

Find value of M under pressure angle and opposite number of teeth; divide M by diametral pitch of gear to be measured and then subtract one-half total backlash to obtain a measurement M equivalent to given pitch and backlash. The number of teeth to gage or measure over is shown by Table 11.

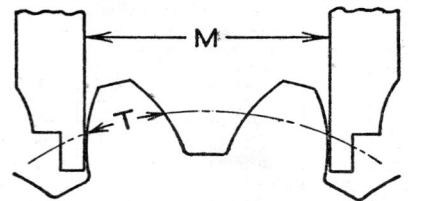

Number of Gear Teeth	M in Inches for 1 D.P.	Number of Gear Teeth	M in Inches for 1 D.P.	Number of Gear Teeth	M in Inches for 1 D.P.	Number of Gear Teeth	M in Inches for 1 D.P.
\multicolumn{8}{Pressure Angle, 14½ Degrees}							
12	4.6267	37	7.8024	62	14.0197	87	20.2370
13	4.6321	38	10.8493	63	17.0666	88	23.2838
14	4.6374	39	10.8547	64	17.0720	89	23.2892
15	4.6428	40	10.8601	65	17.0773	90	23.2946
16	4.6482	41	10.8654	66	17.0827	91	23.2999
17	4.6536	42	10.8708	67	17.0881	92	23.3053
18	4.6589	43	10.8762	68	17.0934	93	23.3107
19	7.7058	44	10.8815	69	17.0988	94	23.3160
20	7.7112	45	10.8869	70	17.1042	95	23.3214
21	7.7166	46	10.8923	71	17.1095	96	23.3268
22	7.7219	47	10.8976	72	17.1149	97	23.3322
23	7.7273	48	10.9030	73	17.1203	98	23.3375
24	7.7326	49	10.9084	74	17.1256	99	23.3429
25	7.7380	50	10.9137	75	17.1310	100	23.3483
26	7.7434	51	13.9606	76	20.1779	101	26.3952
27	7.7488	52	13.9660	77	20.1833	102	26.4005
28	7.7541	53	13.9714	78	20.1886	103	26.4059
29	7.7595	54	13.9767	79	20.1940	104	26.4113
30	7.7649	55	13.9821	80	20.1994	105	26.4166
31	7.7702	56	13.9875	81	20.2047	106	26.4220
32	7.7756	57	13.9929	82	20.2101	107	26.4274
33	7.7810	58	13.9982	83	20.2155	108	26.4327
34	7.7863	59	14.0036	84	20.2208	109	26.4381
35	7.7917	60	14.0090	85	20.2262	110	26.4435
36	7.7971	61	14.0143	86	20.2316	...	....
\multicolumn{8}{Pressure Angle, 20 Degrees}							
12	4.5963	30	10.7526	48	16.9090	66	23.0653
13	4.6103	31	10.7666	49	16.9230	67	23.0793
14	4.6243	32	10.7806	50	16.9370	68	23.0933
15	4.6383	33	10.7946	51	16.9510	69	23.1073
16	4.6523	34	10.8086	52	16.9650	70	23.1214
17	4.6663	35	10.8226	53	16.9790	71	23.1354
18	4.6803	36	10.8366	54	16.9930	72	23.1494
19	7.6464	37	13.8028	55	19.9591	73	26.1155
20	7.6604	38	13.8168	56	19.9731	74	26.1295
21	7.6744	39	13.8307	57	19.9872	75	26.1435
22	7.6884	40	13.8447	58	20.0012	76	26.1575
23	7.7024	41	13.8587	59	20.0152	77	26.1715
24	7.7165	42	13.8727	60	20.0292	78	26.1855
25	7.7305	43	13.8867	61	20.0432	79	26.1995
26	7.7445	44	13.9007	62	20.0572	80	26.2135
27	7.7585	45	13.9147	63	20.0712	81	26.2275
28	10.7246	46	16.8810	64	23.0373	...	....
29	10.7386	47	16.8950	65	23.0513	...	....

Table 11. Number of Teeth Included in Chordal Measurement

This table shows the number of teeth included between the jaws of the vernier caliper in measuring dimension M which is obtained as explained in connection with Table 10.

Tooth Range for 14½° Pressure Angle	Tooth Range for 20° Pressure Angle	Number of Teeth to Gage Over	Tooth Range for 14½° Pressure Angle	Tooth Range for 20° Pressure Angle	Number of Teeth to Gage Over
12 to 18	12 to 18	2	63 to 75	46 to 54	6
19 to 37	19 to 27	3	76 to 87	55 to 63	7
38 to 50	28 to 36	4	88 to 100	64 to 72	8
51 to 62	37 to 45	5	101 to 110	73 to 81	9

operation. It may be advisable to place this chordal dimension for roughing on the detail drawing.

Formula for Chordal Dimension M. — The required measurement M over spur gear teeth may be obtained by the following formula in which R = pitch radius of gear, A = pressure angle, T = tooth thickness along pitch circle, N = number of gear teeth, S = number of tooth *spaces* between caliper jaws, F = a factor depending upon the pressure angle = 0.01109 for 14½°; = 0.01973 for 17½°; = 0.0298 for 20°; = 0.04303 for 22½°; = 0.05995 for 25°. This factor F equals twice the involute function of the pressure angle.

$$M = R \times \cos A \times \left(\frac{T}{R} + \frac{6.2832 \times S}{N} + F \right)$$

Example — A spur gear has 30 teeth of 6 diametral pitch and a pressure angle of 14½ degrees. Determine measurement M over three teeth, there being two intervening tooth spaces.

The pitch radius = 2½ inches, the arc tooth thickness equivalent to 6 diametral pitch is 0.2618 inch (if no allowance is made for backlash) and factor F for 14½ degrees = 0.01109 inch.

$$M = 2.5 \times 0.96815 \times \left(\frac{0.2618}{2.5} + \frac{6.2832 \times 2}{30} + 0.01109 \right) = 1.2941 \text{ inches}$$

Checking Enlarged Pinions by Measuring Over Pins or Wires. — When the teeth of small spur gears or pinions would be undercut if generated by an unmodified straight-sided rack cutter or hob, it is common practice to make the outside diameter larger than standard. The amount of increase in outside diameter varies with the pressure angle and number of teeth, as shown by the table on page 1797. In all cases, the teeth are cut to standard depth on a generating type of machine such as a gear hobber or gear shaper; and since the number of teeth and pitch are not changed, the pitch diameter also remains unchanged. The tooth thickness on the pitch circle, however, is increased and wire sizes suitable for standard gears are not large enough to extend above the tops of these enlarged gears or pinions; hence the Van Keuren wire size recommended for these enlarged pinions equals 1.92 ÷ diametral pitch. Table 12 gives measurements over wires of this size, for checking full-depth involute gears of 1 diametral pitch. For any other pitch, merely divide the measurement given in the table by the diametral pitch. Table 12 applies to pinions which have been enlarged by the same amounts as given in the tables on pages 1797 and 1798. These enlarged pinions will mesh with standard gears; but if the standard center distance is to be maintained, reduce the gear diameter below the standard size as much as the pinion diameter is increased.

Table 12. Checking Enlarged Spur Pinions by Measurement Over Wires

Measurements over wires are given in table for 1 diametral pitch. For any other diametral pitch, divide measurement in table by given pitch. Wire size equals 1.92 ÷ diametral pitch.

Number of Teeth	Outside or Major Diameter (Note 1)	Circular Tooth Thickness (Note 2)	Measurement Over Wires	Number of Teeth	Outside or Major Diameter (Note 1)	Circular Tooth Thickness (Note 2)	Measurement Over Wires
14½-degree full-depth involute teeth:				20-degree full-depth involute teeth:			
10	13.3731	1.9259	13.6186	10	12.936	1.912	13.5039
11	14.3104	1.9097	14.4966	11	13.818	1.868	14.3299
12	15.2477	1.8935	15.6290	12	14.702	1.826	15.4086
13	16.1850	1.8773	16.5211	13	15.584	1.783	16.2473
14	17.1223	1.8611	17.6244	14	16.468	1.741	17.2933
15	18.0597	1.8449	18.5260	15	17.350	1.698	18.1383
16	18.9970	1.8286	19.6075	16	18.234	1.656	19.1596
17	19.9343	1.8124	20.5156	17	19.116	1.613	20.0080
18	20.8716	1.7962	21.5806				
19	21.8089	1.7800	22.4934				
20	22.7462	1.7638	23.5451				
21	23.6835	1.7476	24.4611				
22	24.6208	1.7314	25.5018				
23	25.5581	1.7151	26.4201				
24	26.4954	1.6989	27.4515				
25	27.4328	1.6827	28.3718				
26	28.3701	1.6665	29.3952				
27	29.3074	1.6503	30.3168				
28	30.2447	1.6341	31.3333				
29	31.1820	1.6179	32.2558				
30	32.1193	1.6017	33.2661				
31	33.0566	1.5854	34.1889				

Note 1: These enlargements, which are to improve the tooth form and avoid undercut, conform to those given in the tables on pages 1797 and 1798 where data will be found on the minimum number of teeth in the mating gear.

Note 2: The circular or arc thickness is at the standard pitch diameter. The corresponding chordal thickness may be found as follows: Multiply arc thickness by 90 and then divide product by 3.1416 × pitch radius; find sine of angle thus obtained and multiply it by pitch diameter.

General Formula for Checking External and Internal Spur Gears by Measurement Over Wires. — The following formulas may be used for pressure angles or wire sizes not covered by the tables. In these formulas, M = measurement *over* wires for external gears or measurement *between* wires for internal gears; D = pitch diameter; T = arc tooth thickness on pitch circle; W = wire diameter; N = number of gear teeth; A = pressure angle of gear; a = angle, the cosine of which is required in Formulas (2) and (3).

First determine the involute function of angle a (inv a); then the corresponding angle a is found by referring to the tables of involute functions beginning on page 82,

$$\text{inv } a = \text{inv } A \pm \frac{T}{D} \pm \frac{W}{D\cos A} \mp \frac{\pi}{N} \tag{1}$$

For even numbers of teeth, $M = \dfrac{D\cos A}{\cos a} \pm W$ (2)

For odd numbers of teeth, $M = \left(\dfrac{D\cos A}{\cos a}\right)\left(\cos\dfrac{90°}{N}\right) \pm W$ (3)

Note: In Formulas (1), (2), and (3), use the upper sign for *external* and the lower sign for *internal* gears wherever a ± or ∓ appears in the formulas.

GEAR SELECTION

Gear Selection. — To aid users and specifiers in the proper selection of gears for various applications, the American Gear Manufacturers Association has prepared a Gear Classification Manual, AGMA 390.03, which covers coarse- and fine-pitch spur, helical, herringbone, bevel and hypoid gears. The data in Tables 1 through 9 and information in the accompanying text have been extracted from this manual with the permission of the publisher, the American Gear Manufacturers Association, 1500 King St., Arlington, Va. 22314.

For convenience and simplification, gear selections are made according to assigned AGMA class numbers which combine a dimensional Quality Number, identifying tooth tolerances and a Material Treatment Number. Examples of the way these class numbers are used to denote or specify gears are given in Tables 1a and 1b.

Gear Tolerance Classifications. — Table 2 gives the American Gear Manufacturers Association tolerances for runout, pitch, and profile for coarse-pitch, spur, helical and herringbone gears. Table 3 gives lead tolerances; Table 4, total composite tolerances and Table 5, tooth-to-tooth composite tolerances, all for coarse-pitch, spur, helical and herringbone gears. Table 6 gives bevel gear tolerances. Table 7 gives the correlation of Quality Numbers with former fine-pitch classes.

Axial runout is the total variation in a direction parallel to the axis of rotation of a reference surface from a surface of revolution. Generally, the reference surface is a plane perpendicular to the axis of rotation.

Radial runout is the total variation in a direction perpendicular to the axis of rotation of a reference surface from a surface of revolution. Generally, the reference surface is the pitch surface of the gear. Radial runout includes eccentricity and out-of-roundness, and is approximately equal to twice the eccentricity.

Runout tolerance is the total allowable runout. In the case of gear teeth, it is measured by a specified probe such as a cylinder, ball, cone, rack or gear tooth. For spur, helical and herringbone gear teeth, the measurement is made perpendicular to the surface of revolution. On bevel gears, both axial and radial runout are included in the one measurement.

Tooth-to-tooth spacing tolerance is the allowable variation in spacing between corresponding sides of adjacent teeth. Normally, the measurements are made at or near the pitch circle.

Profile is that portion of the tooth flank between the specified form circle and the outside circle or start of tip chamfer. When the form circle is not specified, it should be that for a meshing rack.

Profile tolerance is the allowable deviation of the actual tooth form from the theoretical involute profile with the measurement made in the plane of rotation. Tip relief and any portion of the tooth surface below the active profile are not to be considered.

Lead is the axial advance of a helix for one complete turn, as in the threads of cylindrical worms and teeth of helical gears.

Lead tolerance as applied to spur, helical and herringbone gears and racks is the allowable deviation across the face width of a tooth surface measured normal to the surface established by the specified lead.

Composite action is the variation in center distance when a gear is rolled in tight mesh (double flank contact) with a specified gear.

Total composite tolerance is the allowance in center-distance variation in one complete revolution of the gear being inspected. This includes the effects of variations in active profile, lead, pitch, tooth thickness and runout.

Pitch variation (formerly pitch error) is the amount by which the circular arc between corresponding points on adjacent teeth differs from the value obtained by dividing the circumference of the corresponding circle by the number of teeth.

Pitch tolerance is the allowable amount of pitch variation. In general, measurements are made in a plane of rotation at or near the pitch surface. For bevel and hypoid gears it is customary to make measurements at or near the middle of the face width.

Spacing variation (also tooth-to-tooth spacing) is the difference in measured spacing between corresponding sides of adjacent teeth.

Spacing tolerance is the allowable amount of spacing variation. In general, measurements are made in a plane of rotation at or near the pitch surface. For bevel and hypoid gears, it is customary to make measurements at or near the middle of the face width.

Index variation (formerly accumulated pitch error) is the greatest displacement of any tooth profile from its theoretical position with respect to any other corresponding tooth profile measured at or near the pitch surface.

Index tolerance is the allowable amount of index variation. In general, measurements are made in a plane of rotation at or near the pitch surface. For bevel and hypoid gears, it is customary to make the measurements at or near the middle of the face width.

Tooth-to-tooth composite variation is the center-distance variation (on bevel and hypoid gears, movement of the pinion in a direction at right angles to the pinion axis) obtained as the gear is rotated from tooth to tooth (360 degrees divided by the number of teeth) in tight mesh with a specified mating gear for one complete revolution of the gear being checked. It includes the effects of variations in circular pitch, tooth thickness and profile on both tooth flanks.

Tooth-to-tooth composite tolerance is the allowable amount of tooth-to-tooth variation.

Total composite variation is the total variation in center distance (on bevel and hypoid gears, movement of the pinion in a direction at right angles to the pinion axis) obtained when a gear is rotated through one complete revolution in tight mesh with a specified mating gear. This includes the effect of all tooth and runout variations.

Total composite tolerance for bevel and hypoid gears is the allowable amount of total composite variation.

Changes in Gear Tolerances. — Any changes in gear element tolerances from those indicated in a specified AGMA Quality Number should be by mutual agreement between the manufacturer and the purchaser. It is recognized that there are precision gear applications that require closer tolerances than are indicated in Tables 2, 3, and 4. In such cases the designer or purchaser may wish to consult with the manufacturer to assure that manufacturing facilities and inspection equipment will be available to permit production and inspection to the closer tolerances specified.

In some applications requiring high accuracy gearing it is necessary to match pinion and gear tooth surfaces to obtain specified tooth contact. Matched sets can be provided but usually involve extra cost. In such a case, the manufacturer and purchaser should agree on details of the additional specifications covering how the matching is to be done and checked.

Unusual conditions may require that one or more of the individual element tolerances be of a lower or higher Quality Number than the other element tolerances. In such cases it is possible to modify the AGMA Class Number to include the Quality Number for each gear element tolerance. This should be done only when it results in a more satisfactory gear. Examples of such designations are shown in Table 1.

Certain control gearing applications require gears having a high degree of accuracy in the spacing of teeth. For such applications, a specification of the tooth spacing will be required in addition to the accuracy class specifications shown in Tables 2–4.

(*Continued on page* 1996)

Table 1a. Designation of Coarse-Pitch Gears and Fine-Pitch Bevel and Hypoid Gears and Racks by AGMA Class Numbers — Typical Examples

When All Gear Element Tolerances are the **Same** Quality Number		
Gear Selection Based On		
AGMA Quality Number*	AGMA Material Number† Material Treatment Hardness	AGMA Class Number
Runout 8 T-T Spacing 8 Profile 8 Lead 8	H-14 Alloy Steel Quench and Temper 285 to 325 Bhn	8-H-14
Runout 5 T-T Spacing 5 Profile 5	IN-3 Carbon Steel Induction Harden 52 Rc Min.	5-IN-3
Runout 8 T-T Spacing 8 Profile 8 Lead 8	NI-7 Nodular Iron Quench and Temper 270 Bhn Min.	8-NI-7
Runout 12 T-T Spacing 12 Profile 12 Lead 12	S-2 Stainless Steel No Treatment Hardness as Received	12-S-2
When All Gear Element Tolerances are **not the Same** Quality Number		
Runout 8 T-T Spacing 10 Profile 8 Lead 8	CH-15 Alloy Steel Carburize 60 Rc Min.	8-CH-15 (Except T-T Spacing Quality 10)
Runout 11 T-T Spacing 11 Profile 11 Lead 10	H-5 Carbon Steel Quench and Temper 285 to 325 Bhn	11-H-5 (Except Lead Quality 10)
Runout 8 T-T Spacing 9 Profile 9 Lead 8	F-12 Alloy Steel Flame Harden 48 Rc Min.	8-F-12 (Except T-T Spacing and Profile Quality 9)

* See Tables 2 and 3.   † See Table 8.

Table 1b. Designation of Fine-Pitch Spur, Helical and Herringbone Gears by AGMA Quality Numbers — Typical Example

Gear Selection Based On			
AGMA Quality Number*	AGMA Backlash** Designation	AGMA Material Number† Material Treatment Hardness	AGMA Class Number
10	C	A-4 Aluminum Bar 2024-T4 Heat Treated 120 Bhn (500 Kg)	10C-A-4

* See Table 4.   ** See page 1832.   † See Table 8.

Table 2. AGMA Coarse-Pitch Spur, Helical and Herringbone Gear Tolerances* — 1
(In ten-thousandths of an inch)

Runout Tolerance

AGMA Quality Number	Normal Diametral Pitch	Pitch Diameter, inches									
		¾	1½	3	6	12	25	50	100	200	400
3	½					788.2	938.6	1106.9	1305.5	1539.6	1815.7
	1				477.8	563.5	671.1	791.4	933.4	1100.8	1298.1
	2			289.7	341.6	402.9	479.8	565.9	667.4	787.1	928.2
	4			207.1	244.3	288.1	343.1	404.6	477.2	562.7	663.7
	8			148.1	174.7	206.0	245.3	289.3	341.2	402.4	474.5
4	½					563.0	670.4	790.7	932.5	1099.7	1297.0
	1				341.3	402.5	479.3	565.3	666.7	786.3	927.3
	2			206.9	244.0	287.8	342.7	404.2	476.7	562.2	663.0
	4			147.9	174.4	205.8	245.0	289.0	340.8	402.0	474.1
	8			105.8	124.8	147.1	175.2	206.6	243.7	287.4	338.9
5	½					402.1	478.9	564.8	666.1	785.5	926.4
	1				243.8	287.5	342.4	403.8	476.2	561.6	662.4
	2			147.8	174.3	205.6	244.8	288.7	340.5	401.6	473.6
	4		89.6	105.7	124.6	147.0	175.0	206.4	243.5	287.1	338.6
	8		64.1	75.6	89.1	105.1	125.1	147.6	174.1	205.3	242.1
6	½					287.2	342.1	403.4	475.8	561.1	661.7
	1				174.1	205.4	244.6	288.4	340.2	401.2	473.1
	2			105.6	124.5	146.8	174.9	206.2	243.2	286.8	338.3
	4		64.0	75.5	89.0	105.0	125.0	147.4	173.9	205.1	241.9
	8	38.8	45.8	54.0	63.6	75.1	89.4	105.4	124.3	146.6	172.9
	12	31.9	37.6	44.4	52.3	61.7	73.5	86.6	102.2	120.5	142.1
	20	24.9	29.4	34.6	40.8	48.2	57.4	67.7	79.8	94.1	111.0
7	½					205.2	244.3	288.1	339.8	400.8	472.7
	1				124.4	146.7	174.7	206.0	243.0	286.6	337.9
	2			75.4	88.9	104.9	124.9	147.3	173.7	204.9	241.6
	4		45.7	53.9	63.6	75.0	89.3	105.3	124.2	146.5	172.8
	8	27.7	32.7	38.5	45.5	53.6	63.9	75.3	88.8	104.7	123.5
	12	22.8	26.9	31.7	37.4	44.1	52.5	61.9	73.0	86.1	101.5
	20	17.8	21.0	24.7	29.2	34.4	41.0	48.3	57.0	67.2	79.3

Pitch Tolerance

AGMA Quality Number	Normal Diametral Pitch	Pitch Diameter, inches									
		¾	1½	3	6	12	25	50	100	200	400
6	½					38.4	43.7	49.4	55.9	63.2	71.4
	1				29.1	32.9	37.4	42.3	47.8	54.1	61.1
	2			22.0	24.9	28.1	32.0	36.2	41.0	46.3	52.3
	4		16.7	18.9	21.3	24.1	27.4	31.0	35.1	39.6	44.8
	8	12.6	14.3	16.1	18.2	20.6	23.5	26.6	30.0	33.9	38.4
	12	11.5	13.0	14.7	16.7	18.8	21.5	24.3	27.4	31.0	35.0
	20	10.3	11.6	13.1	14.9	16.8	19.1	21.6	24.5	27.6	31.3
7	½					27.0	30.8	34.8	39.3	44.5	50.3
	1				20.5	23.1	26.4	29.8	33.7	38.1	43.1
	2			15.5	17.5	19.8	22.6	25.5	28.8	32.6	36.9
	4		11.7	13.3	15.0	17.0	19.3	21.8	24.7	29.9	31.6
	8	8.9	10.1	11.4	12.9	14.5	16.5	18.7	21.1	23.9	27.0
	12	8.1	9.2	10.4	11.7	13.3	15.1	17.1	19.3	21.8	24.7
	20	7.2	8.2	9.3	10.5	11.8	13.5	15.2	17.2	19.5	22.0

* Extracted from AGMA Handbook for Unassembled Gears — Volume 1 — Gear Classifications, Materials, and Inspection (AGMA 390.03), with permission of the publisher, the American Gear Manufacturers Association, 1901 North Fort Myer Drive, Arlington, Va. 22209.

Table 2. AGMA Coarse-Pitch Spur, Helical and Herringbone Gear Tolerances — 2
(In ten-thousandths of an inch)

AGMA Quality Number	Normal Diametral Pitch	Pitch Diameter, inches									
		¾	1½	3	6	12	25	50	100	200	400
		Runout Tolerance									
8	½					146.5	174.5	205.8	242.7	286.3	337.6
	1				88.8	104.8	124.8	147.2	173.6	204.7	241.4
	2			53.9	63.5	74.9	89.2	105.2	124.1	146.3	172.6
	4		32.7	38.5	45.4	53.6	63.8	75.2	88.7	104.6	123.4
	8	19.8	23.3	27.5	32.5	38.3	45.6	53.8	63.4	74.8	88.2
	12	16.3	19.2	22.6	26.7	31.5	37.5	44.2	52.1	61.5	72.5
	20	12.7	15.0	17.7	20.8	24.6	29.3	34.5	40.7	48.0	56.6
9	½					104.7	124.7	147.0	173.4	204.5	241.2
	1				63.5	74.8	89.1	105.1	124.0	146.2	172.4
	2			38.5	45.4	53.5	63.7	75.2	88.6	104.5	123.3
	4		23.3	27.5	32.4	38.3	45.6	53.7	63.4	74.7	88.1
	8	14.1	16.7	19.7	23.2	27.4	32.6	38.4	45.3	53.4	63.0
	12	11.6	13.7	16.2	19.1	22.5	26.8	31.6	37.2	43.9	51.8
	20	9.1	10.7	12.6	14.9	17.6	20.9	24.7	29.1	34.3	40.4
10	½					74.8	89.0	105.0	123.8	146.1	172.3
	1				45.3	53.5	63.7	75.1	88.5	104.4	123.2
	2			27.5	32.4	38.2	45.5	53.7	63.3	74.7	88.1
	4		16.7	19.6	23.2	27.3	32.5	38.4	45.3	53.4	63.0
	8	10.1	11.9	14.0	16.6	19.5	23.3	27.4	32.4	38.2	45.0
	12	8.3	9.8	11.5	13.6	16.1	19.1	22.6	26.6	31.4	37.0
	20	6.5	7.6	9.0	10.6	12.5	14.9	17.6	20.8	24.5	28.9
11	½					53.4	63.6	75.0	88.5	104.3	123.0
	1				32.4	38.2	45.5	53.6	63.2	74.6	88.0
	2			19.6	23.1	27.3	32.5	38.3	45.2	53.3	62.9
	4		11.9	14.0	16.6	19.5	23.2	27.4	32.3	38.1	45.0
	8	7.2	8.5	10.0	11.8	14.0	16.6	19.6	23.1	27.3	32.2
	12	5.9	7.0	8.2	9.7	11.5	13.7	16.1	19.0	22.4	26.4
	20	4.6	5.5	6.4	7.6	9.0	10.7	12.6	14.8	17.5	20.6
12	½					38.1	45.4	53.6	63.2	74.5	87.9
	1				23.1	27.3	32.5	38.3	45.2	53.3	62.8
	2			14.0	16.5	19.5	23.2	27.4	32.3	38.1	44.9
	4		8.5	10.0	11.8	13.9	16.6	19.6	23.1	27.2	32.1
	8	5.2	6.1	7.2	8.5	10.0	11.9	14.0	16.5	19.5	23.0
	12	4.2	5.0	5.9	6.9	8.2	9.8	11.5	13.6	16.0	18.9
	20	3.3	3.9	4.6	5.4	6.4	7.6	9.0	10.6	12.5	14.7
13	½					27.2	32.4	38.3	45.1	53.2	62.8
	1				16.5	19.5	23.2	27.4	32.3	38.1	44.9
	2			10.0	11.8	13.9	16.6	19.6	23.1	27.2	32.1
	4		6.1	7.2	8.4	10.0	11.9	14.0	16.5	19.5	22.9
	8	3.7	4.3	5.1	6.0	7.1	8.5	10.0	11.8	13.9	16.4
	12	3.0	3.6	4.2	5.0	5.9	7.0	8.2	9.7	11.4	13.5
	20	2.4	2.8	3.3	3.9	4.6	5.4	6.4	7.6	8.9	10.5
14	½					19.5	23.2	27.3	32.2	38.0	44.8
	1				11.8	13.9	16.6	19.5	23.0	27.2	32.1
	2			7.2	8.4	9.9	11.8	14.0	16.5	19.4	22.9
	4		4.3	5.1	6.0	7.1	8.5	10.0	11.8	13.9	16.4
	8	2.6	3.1	3.7	4.3	5.1	6.1	7.1	8.4	9.9	11.7
	12	2.2	2.5	3.0	3.5	4.2	5.0	5.9	7.0	8.2	9.6
	20	1.7	2.0	2.3	2.8	3.3	3.9	4.6	5.4	6.4	7.5
15	½					13.9	16.6	19.5	23.0	27.2	32.0
	1				8.4	9.9	11.8	14.0	16.5	19.4	22.9
	2			5.1	6.0	7.1	8.5	10.0	11.8	13.9	16.4
	4		3.1	3.7	4.3	5.1	6.1	7.1	8.4	9.9	11.7
	8	1.9	2.2	2.6	3.1	3.6	4.3	5.1	6.0	7.1	8.4
	12	1.5	1.8	2.1	2.5	3.0	3.6	4.2	4.9	5.8	6.9
	20	1.2	1.4	1.7	2.0	2.3	2.8	3.3	3.9	4.6	5.4

Table 2. AGMA Coarse-Pitch Spur, Helical and Herringbone Gear Tolerances — 3
(In ten-thousandths of an inch)

AGMA Quality Number	Normal Diametral Pitch	Pitch Diameter, inches									
		¾	1½	3	6	12	25	50	100	200	400
		Pitch Tolerance									
8	½	....	....	....	....	19.0	21.7	24.5	27.7	31.3	35.4
	1	....	....	....	14.4	16.3	18.6	21.0	23.7	26.8	30.3
	2	....	....	10.9	12.3	14.0	15.9	18.0	20.3	23.0	26.0
	4	....	8.3	9.3	10.6	11.9	13.6	15.4	17.4	19.7	22.2
	8	6.3	7.1	8.0	9.0	10.2	11.7	13.2	14.9	16.8	19.0
	12	5.7	6.5	7.3	8.3	9.3	10.6	12.0	13.6	15.4	17.4
	20	5.1	5.8	6.5	7.4	8.3	9.5	10.7	12.1	13.7	15.5
9	½	....	....	....	....	13.4	15.3	17.3	19.5	22.1	24.9
	1	....	....	....	10.2	11.5	13.1	14.8	16.7	18.9	21.4
	2	....	....	7.7	8.7	9.8	11.2	12.7	14.3	16.2	18.3
	4	....	5.8	6.6	7.4	8.4	9.6	10.8	12.2	13.8	15.7
	8	4.4	5.0	5.6	6.4	7.2	8.2	9.3	10.5	11.9	13.4
	12	4.0	4.6	5.1	5.8	6.6	7.5	8.5	9.6	10.8	12.2
	20	3.6	4.1	4.6	5.2	5.9	6.7	7.6	8.5	9.7	10.9
10	½	....	....	....	....	9.4	10.8	12.2	13.7	15.5	17.6
	1	....	....	....	7.2	8.1	9.2	10.4	11.8	13.3	15.0
	2	....	....	5.4	6.1	6.9	7.9	8.9	10.1	11.4	12.9
	4	....	4.1	4.6	5.2	5.9	6.7	7.6	8.6	9.8	11.0
	8	3.1	3.5	4.0	4.5	5.1	5.8	6.5	7.4	8.3	9.4
	12	2.8	3.2	3.6	4.1	4.6	5.3	6.0	6.7	7.6	8.6
	20	2.5	2.9	3.2	3.7	4.1	4.7	5.3	6.0	6.8	7.7
11	½	....	....	....	....	6.6	7.6	8.6	9.7	10.9	12.4
	1	....	....	....	5.0	5.7	6.5	7.3	8.3	9.4	10.6
	2	....	....	3.8	4.3	4.9	5.6	6.3	7.1	8.0	9.1
	4	....	2.9	3.3	3.7	4.2	4.8	5.4	6.1	6.9	7.8
	8	2.2	2.5	2.8	3.2	3.6	4.1	4.6	5.2	5.9	6.6
	12	2.0	2.3	2.6	2.9	3.3	3.7	4.2	4.7	5.4	6.1
	20	1.8	2.0	2.3	2.6	2.9	3.3	3.7	4.2	4.8	5.4
12	½	....	....	....	....	4.7	5.3	6.0	6.8	7.7	8.7
	1	....	....	....	3.5	4.0	4.6	5.2	5.8	6.6	7.5
	2	....	....	2.7	3.0	3.4	3.9	4.4	5.0	5.6	6.4
	4	....	2.0	2.3	2.6	2.9	3.3	3.8	4.3	4.8	5.5
	8	1.5	1.7	2.0	2.2	2.5	2.9	3.2	3.7	4.1	4.7
	12	1.4	1.6	1.8	2.0	2.3	2.6	3.0	3.3	3.8	4.3
	20	1.3	1.4	1.6	1.8	2.0	2.3	2.6	3.0	3.4	3.8
13	½	....	....	....	....	3.3	3.8	4.2	4.8	5.4	6.1
	1	....	....	....	2.5	2.8	3.2	3.6	4.1	4.6	5.3
	2	....	....	1.9	2.1	2.4	2.8	3.1	3.5	4.0	4.5
	4	....	1.4	1.6	1.8	2.1	2.4	2.7	3.0	3.4	3.8
	8	1.1	1.3	1.4	1.6	1.8	2.0	2.3	2.6	2.9	3.3
	12	1.0	1.1	1.3	1.4	1.6	1.8	2.1	2.4	2.7	3.0
	20	0.9	1.0	1.1	1.3	1.4	1.6	1.9	2.1	2.4	2.7
14	½	....	....	....	....	2.3	2.6	3.0	3.4	3.8	4.3
	1	....	....	....	1.8	2.0	2.3	2.6	2.9	3.3	3.7
	2	....	....	1.3	1.5	1.7	1.9	2.2	2.5	2.8	3.2
	4	....	1.0	1.1	1.3	1.5	1.7	1.9	2.1	2.4	2.7
	8	0.8	0.9	1.0	1.1	1.2	1.4	1.6	1.8	2.1	2.3
	12	0.7	0.8	0.9	1.0	1.1	1.3	1.5	1.7	1.9	2.1
	20	0.6	0.7	0.8	0.9	1.0	1.2	1.3	1.5	1.7	1.9
15	½	....	....	....	....	1.6	1.9	2.1	2.4	2.7	3.0
	1	....	....	....	1.2	1.4	1.6	1.8	2.0	2.3	2.6
	2	....	....	0.9	1.1	1.2	1.4	1.5	1.7	2.0	2.2
	4	....	0.7	0.8	0.9	1.0	1.2	1.3	1.5	1.7	1.9
	8	0.5	0.6	0.7	0.8	0.9	1.0	1.1	1.3	1.4	1.6
	12	0.5	0.6	0.6	0.7	0.8	0.9	1.0	1.2	1.3	1.5
	20	0.4	0.5	0.6	0.6	0.7	0.8	0.9	1.0	1.2	1.3

Table 2. AGMA Coarse-Pitch Spur, Helical and Herringbone Gear Tolerances — 4
(In ten-thousandths of an inch)

AGMA Quality Number	Normal Diametral Pitch	Pitch Diameter, inches									
		3/4	1½	3	6	12	25	50	100	200	400
		Profile Tolerance									
8	½	....	....	....	....	42.6	47.7	53.1	59.1	65.7	73.1
	1	....	....	....	28.3	31.5	35.3	39.3	43.7	48.6	54.1
	2	....	....	18.8	21.0	23.3	26.1	29.0	32.3	36.0	40.0
	4	....	12.5	13.9	15.5	17.2	19.3	21.5	23.9	26.6	29.6
	8	8.3	9.3	10.3	11.5	12.8	14.3	15.9	17.7	19.7	21.9
	12	7.0	7.8	8.6	9.6	10.7	12.0	13.3	14.8	16.5	18.4
	20	5.6	6.2	6.9	7.7	8.6	9.6	10.7	11.9	13.2	14.7
9	½	....	....	....	....	30.4	34.1	37.9	42.2	46.9	52.2
	1	....	....	....	20.2	22.5	25.2	28.1	31.2	34.7	38.6
	2	....	....	13.5	15.0	16.7	18.6	20.7	23.1	25.7	28.6
	4	....	8.9	10.0	11.1	12.3	13.8	15.3	17.1	19.0	21.1
	8	5.9	6.6	7.4	8.2	9.1	10.2	11.4	12.6	14.1	15.6
	12	5.0	5.5	6.2	6.9	7.6	8.6	9.5	10.6	11.8	13.1
	20	4.0	4.4	4.9	5.5	6.1	6.8	7.6	8.5	9.4	10.5
10	½	....	....	....	....	21.7	24.3	27.1	30.1	33.5	37.3
	1	....	....	....	14.5	16.1	18.0	20.0	22.3	24.8	27.6
	2	....	....	9.6	10.7	11.9	13.3	14.8	16.5	18.3	20.4
	4	....	6.4	7.1	7.9	8.8	9.9	11.0	12.2	13.6	15.1
	8	4.2	4.7	5.3	5.9	6.5	7.3	8.1	9.0	10.0	11.2
	12	3.6	4.0	4.4	4.9	5.5	6.1	6.8	7.6	8.4	9.4
	20	2.9	3.2	3.5	3.9	4.4	4.9	5.4	6.1	6.7	7.5
11	½	....	....	....	....	15.5	17.4	19.3	21.5	24.0	26.7
	1	....	....	....	10.3	11.5	12.9	14.3	15.9	17.7	19.7
	2	....	....	6.9	7.6	8.5	9.5	10.6	11.8	13.1	14.6
	4	....	4.6	5.1	5.6	6.3	7.0	7.8	8.7	9.7	10.8
	8	3.0	3.4	3.8	4.2	4.6	5.2	5.8	6.4	7.2	8.0
	12	2.5	2.8	3.1	3.5	3.9	4.4	4.9	5.4	6.0	6.7
	20	2.0	2.3	2.5	2.8	3.1	3.5	3.9	4.3	4.8	5.4
12	½	....	....	....	....	11.1	12.4	13.8	15.4	17.1	19.0
	1	....	....	....	7.4	8.2	9.2	10.2	11.4	12.7	14.1
	2	....	....	4.9	5.5	6.1	6.8	7.6	8.4	9.4	10.4
	4	....	3.3	3.6	4.0	4.5	5.0	5.6	6.2	6.9	7.7
	8	2.2	2.4	2.7	3.0	3.3	3.7	4.1	4.6	5.1	5.7
	12	1.8	2.0	2.2	2.5	2.8	3.1	3.5	3.9	4.3	4.8
	20	1.5	1.6	1.8	2.0	2.2	2.5	2.8	3.1	3.4	3.8
13	½	....	....	....	....	7.9	8.9	9.9	11.0	12.2	13.6
	1	....	....	....	5.3	5.9	6.6	7.3	8.1	9.0	10.1
	2	....	....	3.5	3.9	4.3	4.9	5.4	6.0	6.7	7.4
	4	....	2.3	2.6	2.9	3.2	3.6	4.0	4.4	4.9	5.5
	8	1.5	1.7	1.9	2.1	2.4	2.7	3.0	3.3	3.7	4.1
	12	1.3	1.4	1.6	1.8	2.0	2.2	2.5	2.8	3.1	3.4
	20	1.0	1.2	1.3	1.4	1.6	1.8	2.0	2.2	2.5	2.7
14	½	....	....	....	....	5.7	6.3	7.1	7.8	8.7	9.7
	1	....	....	....	3.8	4.2	4.7	5.2	5.8	6.5	7.2
	2	....	....	2.5	2.8	3.1	3.5	3.9	4.3	4.8	5.3
	4	....	1.7	1.9	2.1	2.3	2.6	2.9	3.2	3.5	3.9
	8	1.1	1.2	1.4	1.5	1.7	1.9	2.1	2.3	2.6	2.9
	12	0.9	1.0	1.1	1.3	1.4	1.6	1.8	2.0	2.2	2.4
	20	0.7	0.8	0.9	1.0	1.1	1.3	1.4	1.6	1.8	2.0
15	½	....	....	....	....	4.0	4.5	5.0	5.6	6.2	6.9
	1	....	....	....	2.7	3.0	3.3	3.7	4.1	4.6	5.1
	2	....	....	1.8	2.0	2.2	2.5	2.8	3.1	3.4	3.8
	4	....	1.2	1.3	1.5	1.6	1.8	2.0	2.3	2.5	2.8
	8	0.8	0.9	1.0	1.1	1.2	1.4	1.5	1.7	1.9	2.1
	12	0.7	0.7	0.8	0.9	1.0	1.1	1.3	1.4	1.6	1.7
	20	0.5	0.6	0.7	0.7	0.8	0.9	1.0	1.1	1.3	1.4

Table 3. AGMA Coarse-Pitch Spur, Helical and Herringbone Gear Lead Tolerances*
(In ten-thousandths of an inch)

AGMA Quality Number	Face Width, inches				
	1 to 2	2	3	4	5
			Lead Tolerance		
8	5	8	11	13	16
9	4	7	9	11	13
10	3	5	7	9	10
11	3	4	6	7	8
12	2	3	5	6	7
13	2	3	4	4	5
14	1	2	3	4	4
15	1	2	2	3	3

* See footnote to table below.

Table 4. AGMA Coarse-Pitch Spur, Helical and Herringbone Gear Total Composite Tolerances* — I
(In ten-thousandths of an inch)

Pitch Diameter, inches — Total Composite Tolerance

AGMA Quality Number	Normal Diametral Pitch†	0.040	0.063	0.100	0.160	0.250	0.400	0.630	1.0	1.6	2.5	4.0	6.3	10	16	25	50	100	200
5	12.0							94.8	95.0	95.4	98.4	106.4	114.7	123.8	133.8	144.0	161.5	181.2	
	20.0						74.3	74.3	74.3	77.0	82.5	88.6	95.0	102.0	109.6	117.3	130.5		
6	0.5														434.3	447.5	480.3	577.7	694.8
	1.0												268.5	273.8	282.3	293.0	346.1	408.7	482.7
	2.0											176.3	179.4	184.4	196.9	216.8	251.9	292.6	340.0
	4.0									118.4	119.4	121.2	123.6	135.2	148.2	161.7	185.0	211.8	242.5
	8.0								82.6	83.1	83.9	87.9	95.2	103.3	112.2	121.3	137.1	154.8	
	12.0							67.5	67.7	68.1	70.3	76.0	81.9	88.4	95.6	102.9	115.4	129.4	
	20.0						53.0	53.0	53.0	55.0	58.9	63.3	67.9	72.8	78.3	83.8	93.2		
7	0.5														300.2	313.4	343.1	412.6	496.3
	1.0												186.2	191.5	201.6	209.3	247.2	291.9	344.8
	2.0											123.6	126.7	131.7	140.6	154.9	179.9	209.0	242.8
	4.0									83.4	84.5	86.3	88.3	96.6	105.8	115.5	132.2	151.3	173.2
	8.0								58.6	59.1	59.9	62.8	68.0	73.8	80.1	86.7	97.9	110.6	
	12.0							48.0	48.3	48.6	50.2	54.3	58.5	63.2	68.3	73.5	82.4	92.4	
	20.0						37.9	37.9	37.9	39.3	42.1	45.2	48.5	52.0	55.9	59.9	66.6		

* Extracted from AGMA Handbook for Unassembled Gears — Volume 1 — Gear Classifications, Materials, and Inspection (AGMA 390.03), with the permission of the publisher, the American Gear Manufacturers Association, 1901 North Fort Myer Drive, Arlington, Va. 22209.
† For diametral pitches above 20, see above mentioned Handbook.

Table 4. AGMA Coarse-Pitch Spur, Helical and Herringbone Gear Total Composite Tolerances* — 2
(In ten-thousandths of an inch)

Column group heading: **Pitch Diameter, inches** — values under **Total Composite Tolerance**.

AGMA Quality Number	Normal Diametral Pitch†	0.040	0.063	0.100	0.160	0.250	0.400	0.630	1.0	1.6	2.5	4.0	6.3	10	16	25	50	100	200
8	0.5														204.4	217.6	245.1	294.7	354.5
	1.0												127.4	132.7	141.2	149.5	176.6	208.5	246.3
	2.0											86.0	89.1	94.1	100.4	110.6	128.5	149.3	173.5
	4.0									58.4	59.5	61.3	63.0	69.0	75.6	82.5	94.4	108.1	123.7
	8.0								41.5	42.0	42.8	44.9	48.6	52.7	57.2	61.9	69.9	79.0	
	12.0							34.1	34.4	34.7	35.9	38.8	41.8	45.1	48.8	52.5	58.9	66.0	
	20.0						27.1	27.1	27.1	28.1	30.1	32.3	34.6	37.2	39.9	42.8	47.6		
9	0.5														136.0	149.1	175.0	210	253.2
	1.0												85.4	90.7	99.3	106.8	126.1	148.9	175.9
	2.0											59.1	62.2	67.2	71.7	79.0	91.8	106.7	123.9
	4.0									40.5	41.6	43.4	45.0	49.3	54.0	58.9	67.4	77.2	88.4
	8.0								29.2	29.8	30.6	32.0	34.7	37.6	40.9	44.2	49.9	56.4	
	12.0							24.2	24.4	24.8	25.6	27.7	27.7	32.2	34.8	37.5	42.0	47.1	
	20.0						19.3	19.3	19.2	20.1	21.5	23.1	24.7	26.5	28.5	30.5	34.0		
10	0.5														87.1	100.3	125.0	150.4	180.9
	1.0												55.5	60.7	69.3	76.3	90.1	106.4	125.6
	2.0											39.9	43.0	48.0	51.2	56.4	65.6	76.2	88.5
	4.0									27.8	28.9	30.7	32.2	35.2	38.6	42.1	48.2	55.1	63.1
	8.0								20.5	21.0	21.8	22.9	24.8	26.9	29.2	31.6	35.7	40.3	
	12.0							17.1	17.3	17.7	18.3	19.8	21.3	23.0	24.9	26.8	30.0	33.7	
	20.0						13.8	13.8	13.8	14.3	15.3	16.5	17.7	19.0	20.4	21.8	24.3		
11	0.5														52.2	65.3	89.3	107.4	129.2
	1.0												34.0	39.3	47.9	54.5	64.3	76.3	89.7
	2.0											26.2	29.3	34.3	36.6	40.3	46.8	54.4	63.2
	4.0									18.7	19.8	21.7	23.0	25.1	27.6	30.1	34.4	39.4	45.1
	8.0								14.3	14.8	15.6	17.7	17.7	19.2	20.9	22.6	25.5	28.8	
	12.0							12.0	12.3	12.6	13.1	14.1	15.2	16.4	17.8	19.1	21.5	24.1	
	20.0						9.9	9.9	9.9	10.2	11.0	11.8	12.7	13.5	14.6	15.6	17.3		

*† See footnotes on first page of table.

Table 4. AGMA Coarse-Pitch Spur, Helical and Herringbone Gear Total Composite Tolerances*† — 3
(In ten-thousandths of an inch)

AGMA Quality Number	Normal Diametral Pitch†	Pitch Diameter, inches (Total Composite Tolerance)																	
		0.040	0.063	0.100	0.160	0.250	0.400	0.630	1.0	1.6	2.5	4.0	6.3	10	16	25	50	100	200
12	0.5	…	…	…	…	…	…	…	…	…	…	…	…	…	27.2	40.4	63.8	76.7	92.3
	1.0	…	…	…	…	…	…	…	…	…	…	…	18.7	24.0	32.6	38.9	46.0	54.3	64.1
	2.0	…	…	…	…	…	…	…	…	…	…	16.4	19.5	24.5	26.1	28.8	33.5	38.9	45.2
	4.0	…	…	…	…	…	7.0	8.4	9.8	12.2	13.3	15.1	16.4	18.0	19.7	21.5	24.6	28.1	32.2
	8.0	…	…	…	…	…	…	7.0	8.6	10.3	11.1	11.7	12.6	13.7	14.9	16.1	18.2	20.6	…
	12.0	…	…	…	…	…	…	…	7.0	9.0	9.3	10.1	10.9	11.7	12.7	13.7	15.3	17.2	…
	20.0	…	…	…	…	…	…	…	…	7.3	7.8	8.4	9.0	9.7	10.4	11.1	13.4	…	…
13	0.5	…	…	…	…	…	…	…	…	…	…	…	…	…	9.4	22.6	45.6	54.8	65.9
	1.0	…	…	…	…	…	…	…	…	…	…	…	7.8	13.1	21.6	27.8	32.8	38.8	45.8
	2.0	…	…	…	…	…	…	…	…	…	…	9.4	12.5	17.5	18.7	20.6	23.9	27.8	32.3
	4.0	…	…	…	…	…	5.0	5.8	6.6	7.5	8.6	10.4	11.7	12.8	14.1	15.3	17.6	20.1	23.0
	8.0	…	…	…	…	…	…	5.0	6.1	7.1	8.0	8.3	9.0	9.8	10.6	11.5	13.0	14.7	…
	12.0	…	…	…	…	…	…	…	5.0	6.4	6.7	7.2	7.8	8.4	9.1	9.8	10.9	12.3	…
	20.0	…	…	…	…	…	…	…	…	5.2	5.6	6.0	6.4	6.9	7.4	8.0	8.8	…	…
14	0.5	…	…	…	…	…	…	…	…	…	…	…	…	…	…	9.9	32.5	39.1	47.1
	1.0	…	…	…	…	…	…	…	…	…	…	…	…	5.3	13.8	19.9	23.5	27.7	32.7
	2.0	…	…	…	…	…	…	…	…	…	…	4.4	7.5	12.5	13.3	14.7	17.1	19.8	23.0
	4.0	…	…	…	…	…	3.6	4.0	4.3	4.2	5.3	7.1	8.4	9.2	10.6	11.0	12.5	14.4	16.4
	8.0	…	…	…	…	…	…	3.6	4.2	4.9	5.7	6.0	6.5	7.0	8.2	8.2	9.3	10.5	…
	12.0	…	…	…	…	…	…	…	3.6	4.6	4.8	5.1	5.6	6.0	6.5	7.0	7.8	8.8	…
	20.0	…	…	…	…	…	…	…	…	3.7	4.0	4.3	4.6	4.9	5.3	5.7	6.3	…	…
15	20.0	…	…	…	…	…	2.6	2.6	2.6	2.7	2.9	3.1	3.3	3.5	3.8	4.1	4.5	…	…
16	20.0	…	…	…	…	…	1.8	1.8	1.8	1.9	2.0	2.2	2.3	2.5	2.7	2.9	3.2	…	…

*† See footnotes on first page of table.

Table 5. AGMA Coarse-Pitch Spur, Helical and Herringbone Gear Tooth-to-Tooth Composite Tolerances* — 1
(In ten-thousandths of an inch)

Pitch Diameter, inches — Tooth-to-Tooth Composite Tolerance

AGMA Quality Number	Normal Diametral Pitch	0.04	0.06	0.10	0.16	0.25	0.40	0.63	1.0	1.6	2.5	4.0	6.3	10	16	25	50	100	200
5	20.0						44.4	39.8	35.6	33.5	33.4	33.4	33.4	33.4	33.4	33.4	33.4		
6	0.5														76.9	69.1	57.8	57.9	57.9
	1.0											68.9	61.7	55.1	49.3	49.0	49.0	49.0
	2.0										55.1	49.4	44.2	40.7	41.5	41.5	41.5	41.5
	4.0								49.2	44.2	39.5	35.8	35.1	35.1	35.1	35.1	35.1	35.1
	8.0							39.5	35.3	31.7	29.7	29.7	29.7	29.7	29.7	29.7	29.7	
	12.0						36.3	32.5	29.1	27.2	27.0	27.0	27.0	27.0	27.0	27.0	27.0	
	20.0						31.7	28.4	25.5	24.0	23.9	23.9	23.9	23.9	23.9	23.9		
7	0.5														54.9	49.3	41.3	41.3	41.3
	1.0											49.2	44.1	39.4	35.2	35.0	35.0	35.0
	2.0										39.4	35.3	31.6	29.1	29.6	29.6	29.6	29.6
	4.0								35.2	31.6	28.2	25.6	25.1	25.1	25.1	25.1	25.1	25.1
	8.0							28.2	25.2	22.7	21.1	21.2	21.2	21.2	21.2	21.2	21.2	
	12.0						26.0	23.2	20.8	19.4	19.3	19.3	19.3	19.3	19.3	19.3	19.3	
	20.0						22.7	20.3	18.2	17.1	17.1	17.1	17.1	17.1	17.1	17.1		
8	0.5														39.2	35.2	29.5	29.5	29.5
	1.0											35.2	31.5	28.1	28.1	25.0	25.0	25.0
	2.0										28.1	25.2	22.6	20.8	21.2	21.2	21.2	21.2
	4.0								25.1	22.6	20.2	18.3	17.9	17.9	17.9	17.9	17.9	17.9
	8.0							20.2	18.0	16.2	15.1	15.2	15.2	15.2	15.2	15.2	15.2	
	12.0						18.5	16.6	14.8	13.9	13.8	13.8	13.8	13.8	13.8	13.8	13.8	
	20.0						16.2	14.5	13.0	12.2	12.2	12.2	12.2	12.2	12.2	12.2		

* Extracted from AGMA Handbook for Unassembled Gears — Volume 1 — Gear Classifications, Materials, and Inspection (AGMA 390.03), with the permission of the publisher, the American Gear Manufacturers Association, 1901 North Fort Myer Drive, Arlington, Va. 22209.

Table 5. AGMA Coarse-Pitch Spur, Helical and Herringbone Gear Tooth-to-Tooth Composite Tolerances* — 2
(In ten-thousandths of an inch)

AGMA Quality Number	Normal Diametral Pitch	Pitch Diameter, inches																	
		0.04	0.06	0.10	0.16	0.25	0.40	0.63	1.0	1.6	2.5	4.0	6.3	10	16	25	50	100	200
		Tooth-to-Tooth Composite Tolerance																	
9	0.5														28.0	25.2	21.1	21.1	21.1
	1.0												25.1	22.5	20.1	18.0	17.9	17.9	17.9
	2.0											20.1	18.0	16.1	14.8	15.1	15.1	15.1	15.1
	4.0									17.9	16.1	14.4	13.0	12.8	12.8	12.8	12.8	12.8	12.8
	8.0								14.4	12.9	11.6	10.8	10.8	10.8	10.8	10.8	10.8	10.8	
	12.0							13.2	11.9	10.6	9.9	9.8	9.8	9.8	9.8	9.8	9.8	9.8	
	20.0						11.6	10.4	9.3	8.7	8.7	8.7	8.7	8.7	8.7	8.7	8.7		
10	0.5														20.0	18.0	15.0	15.1	15.1
	1.0												17.9	16.1	14.3	12.8	12.8	12.8	12.8
	2.0											14.3	12.9	11.5	10.6	10.8	10.8	10.8	10.8
	4.0									12.8	11.5	10.3	9.3	9.1	9.1	9.1	9.1	9.1	9.1
	8.0								10.3	9.2	8.3	7.7	7.7	7.7	7.7	7.7	7.7	7.7	
	12.0							9.5	8.5	7.6	7.6	7.0	7.0	7.0	7.0	7.0	7.0	7.0	
	20.0						8.3	7.4	6.6	6.2	6.2	6.2	6.2	6.2	6.2	6.2	6.2		
11	0.5														14.3	12.8	10.7	10.8	10.8
	1.0												12.8	11.5	10.2	9.2	9.1	9.1	9.7
	2.0											10.2	9.2	8.2	7.6	7.7	7.7	7.7	7.7
	4.0									9.2	8.2	7.3	6.7	6.5	6.5	6.5	6.5	6.5	6.5
	8.0								7.3	6.6	5.9	5.5	5.5	5.5	5.5	5.5	5.5	5.5	
	12.0							6.8	6.0	5.4	5.1	5.0	5.0	5.0	5.0	5.0	5.0	5.0	
	20.0						5.9	5.3	4.7	4.5	4.4	4.4	4.4	4.4	4.4	4.4	4.4		
12	0.5														10.2	9.2	7.7	7.7	7.7
	1.0												9.2	8.2	7.3	6.5	6.5	6.5	6.5
	2.0											7.3	6.6	5.9	5.4	5.5	5.5	5.5	5.5
	4.0									6.5	5.9	5.1	4.8	4.7	4.7	4.7	4.7	4.7	4.7
	8.0								5.2	4.7	4.2	3.9	4.0	4.0	4.0	4.0	4.0	4.0	
	12.0							4.8	4.3	3.9	3.9	3.6	3.6	3.6	3.6	3.6	3.6	3.6	
	20.0						4.2	3.8	3.4	3.2	3.2	3.2	3.2	3.2	3.2	3.2	3.2		

* See footnote on first page of table.

Table 5. AGMA Coarse-Pitch Spur, Helical and Herringbone Gear Tooth-to-Tooth Composite Tolerances* — 3

(In ten-thousandths of an inch)

AGMA Quality Number	Normal Diametral Pitch	Pitch Diameter, inches																	
		0.04	0.06	0.10	0.16	0.25	0.40	0.63	1.0	1.6	2.5	4.0	6.3	10	16	25	50	100	200
		Tooth-to-Tooth Composite Tolerance																	
13	0.5														7.3	6.6	5.5	5.5	5.5
	1.0												6.5	5.9	5.2	4.7	4.6	4.6	4.6
	2.0											5.2	4.7	4.2	3.9	3.9	3.9	3.9	3.9
	4.0									4.7	4.2	3.8	3.4	3.3	3.3	3.3	3.3	3.3	3.3
	8.0								3.7	3.3	3.0	2.8	2.8	2.8	2.8	2.8	2.8	2.8	
	12.0							3.4	3.1	2.8	2.6	2.6	2.6	2.6	2.6	2.6	2.6	2.6	
	20.0						3.0	2.7	2.4	2.3	2.3	2.3	2.3	2.3	2.3	2.3	2.3		
14	0.5														5.2	4.7	3.9	3.9	3.9
	1.0												4.7	4.2	3.7	3.3	3.3	3.3	3.3
	2.0											3.7	3.3	3.0	2.8	2.8	2.8	2.8	2.8
	4.0									3.3	3.0	2.7	2.4	2.4	2.4	2.4	2.4	2.4	2.4
	8.0								2.7	2.4	2.1	2.0	2.0	2.0	2.0	2.0	2.0	2.0	
	12.0							2.5	2.2	1.8	1.8	1.8	1.8	1.8	1.8	1.8	1.8	1.8	
	20.0						2.1	1.9	1.7	1.6	1.6	1.6	1.6	1.6	1.6	1.6	1.6		
15	20.0						1.5	1.4	1.2	1.6	1.2	1.2	1.2	1.2	1.2	1.2	1.2		
16	20.0						1.1	1.0	0.9	0.8	0.8	0.8	0.8	0.8	0.8	0.8	0.8		

* See footnote on first page of table.

AGMA Bevel Gear Tolerances — 1
(In ten-thousandths of an inch)

AGMA Quality Number	Diametral Pitch	Pitch Diameter (Inches)								
		¾	1½	3	5	12	25	50	100	200
		Pitch Tolerance								
6	½	....	....	....	....	....	50	55	62	70
	1	....	....	....	....	31	33	38	42	50
	2	....	....	....	26	27	29	33	37	42
	4	....	....	22	22	24	26	28	32	37
	8	....	18	18	19	21	23	25	28	....
	16–19.99	16	16	16	17	18	19	21	23	....
7	½	....	....	....	....	....	37	40	45	50
	1	....	....	....	....	23	25	28	32	37
	2	....	....	....	19	20	22	24	27	31
	4	....	....	16	16	17	19	21	24	28
	8	....	13½	14	14½	15½	17	19	21	....
	16–19.99	11	11½	12	12½	13	14	15½	17½	....
8	½	....	....	....	....	....	26	28	31	....
	1	....	....	....	....	16	18	19	22	....
	2	....	....	....	14	14	15	17	19	....
	4	....	....	11	11	12	13	15	....	....
	8	....	9	10	10	11	12	....	....	....
	16–19.99	8	8	8	9	9	....	....	....	....
9	½	....	....	....	....	....	19	20	22	....
	1	....	....	....	....	11	12	14	16	....
	2	....	....	....	10	10	11	12	14	....
	4	....	....	8	8	9	9½	11	....	....
	8	....	7	7	7½	8	8½	....	....	....
	16–19.99	6	6	6	6½	7	....	....	....	....
10	1	....	....	....	....	8½	9	....	....	....
	2	....	....	....	7	7½	8	....	....	....
	4	....	....	6	6	6½	7	....	....	....
	8	....	5	5	5½	6	6½	....	....	....
	16–19.99	4½	4½	4½	4½	5	....	....	....	....
11	1	....	....	....	....	6	6	....	....	....
	2	....	....	....	5	5	6	....	....	....
	4	....	....	4	4	4½	5	....	....	....
	8	....	3½	3½	4	4	4½	....	....	....
	16–19.99	3	3	3	3½	3½	....	....	....	....
12	2	....	....	....	3½	4	4	....	....	....
	4	....	....	3	3	3½	3½	....	....	....
	8	....	2½	2½	3	3	3½	....	....	....
	16–19.99	2½	2½	2½	2½	2½	....	....	....	....
13	2	....	....	....	2½	3	3	....	....	....
	4	....	....	2	2	2½	2½	....	....	....
	8	....	2	2	2	2	2½	....	....	....
	16–19.99	2	2	2	2	2	....	....	....	....

Table 6. AGMA Bevel Gear Tolerances — 2
(In ten-thousandths of an inch)

AGMA Quality Number	Diametral Pitch	Pitch Diameter (Inches)								
		3/4	1½	3	6	12	25	50	100	200
		Runout Tolerance								
3	½	....	....	....	....	....	770	1010	1360	....
	1	....	....	....	....	540	710	930	1250	....
	2	....	....	....	382	498	660	860	1150	....
	4	....	....	280	355	460	608	800	....	....
4	½	....	....	....	....	....	540	700	940	....
	1	....	....	....	....	378	496	640	860	....
	2	....	....	....	272	348	452	590	790	....
	4	....	....	198	250	320	419	542	720	....
5	½	....	....	....	....	....	396	510	665	880
	1	....	....	....	....	270	350	450	582	775
	2	....	....	....	184	233	302	390	510	680
	4	....	....	130	160	203	262	340	440	590
	8	....	91	112	140	177	228	290	380	....
6	½	....	....	....	....	....	280	350	450	600
	1	....	....	....	....	188	235	295	378	508
	2	....	....	....	131	160	200	250	322	425
	4	....	....	92	110	135	170	210	270	360
	8	....	64	76	93	114	143	180	230	....
	16–19.99	46	55	66	80	98	122	152	193	....
7	½	....	....	....	....	....	209	260	335	445
	1	....	....	....	....	132	165	205	261	350
	2	....	....	....	84	103	130	163	210	280
	4	....	....	55	67	82	103	130	167	225
	8	....	37	44	54	66	82	103	132	....
	16–19.99	26	31	37	45	55	69	86	110	....
8	½	....	....	....	....	....	160	200	255	....
	1	....	....	....	....	95	115	140	180	....
	2	....	....	....	58	68	82	100	125	....
	4	....	....	36	41	47	56	67	....	....
	8	....	25	28	32	36	42	....	....	....
	16–19.99	19	21	26	30	....	....	....	....	....
	20–120	....	....	....	....	....	....	....	....	....
9	½	....	....	....	....	....	113	140	180	....
	1	....	....	....	....	68	81	100	125	....
	2	....	....	....	40	48	58	70	88	....
	4	....	....	26	29	34	40	48	....	....
	8	....	18	20	22	26	30	....	....	....
	16–19.99	14	15	16	18	21	....	....	....	....
	20–120	....	....	....	....	....	....	....	....	....
10	1	....	....	....	....	50	58	....	....	....
	2	....	....	....	30	34	40	....	....	....
	4	....	....	18	21	24	28	....	....	....
	8	....	13	14	16	18	21	....	....	....
	16–19.99	10	11	12	13	15	....	....	....	....
	20–120	....	....	....	....	....	....	....	....	....
11	1	....	....	....	....	34	41	....	....	....
	2	....	....	....	21	24	28	....	....	....
	4	....	....	13	15	17	20	....	....	....
	8	....	9	10	11	13	15	....	....	....
	16–19.99	7	8	8	9	11	....	....	....	....
	20–120	....	....	....	....	....	....	....	....	....
12	2	....	....	....	15	18	21	....	....	....
	4	....	....	9	11	12	14	....	....	....
	8	....	6½	7	8	9	11	....	....	....
	16–19.99	....	5½	6	7	8	....	....	....	....
	20–120	....	....	....	....	....	....	....	....	....
13	2	....	....	....	11	13	15	....	....	....
	4	....	....	7	7½	9	10½	....	....	....
	8	....	5	5½	6	7	8	....	....	....
	16–19.99	5	5	5	5	5½	....	....	....	....
	20–120	....	....	....	....	....	....	....	....	....

Table 6. AGMA Bevel Gear Tolerances — 3
(In ten-thousandths of an inch)

AGMA Quality Number	Diametral Pitch	Tooth to Tooth Composite Tolerance — Pitch Diameter (Inches)							Total Composite Tolerance — Pitch Diameter (Inches)						
		3/16	3/8	3/4	1½	3	6	12	3/16	3/8	3/4	1½	3	6	12
5	20–24	27	27	27	27	27	27	27	52	52	52	52	61	72	72
	24–32	27	27	27	27	27	..	..	52	52	52	52	61	..	..
	32–48	27	27	27	27	..	..	..	52	52	52	52	..	..	..
6	20–32	19	19	19	19	19	19	19	37	37	37	37	44	52	52
	32–48	19	19	19	19	19	..	..	37	37	37	37	44	..	..
	48–64	19	19	19	19	..	..	..	37	37	37	37	..	..	..
7	20–48	14	14	14	14	14	14	14	27	27	27	27	32	37	37
	48–64	14	14	14	14	14	..	..	27	27	27	27	32	..	..
	64–96	14	14	14	14	..	..	..	27	27	27	27	..	..	..
8	2	..	..	..	..	..	..	..	..	..	..	..	..	..	80
	4	..	..	..	..	..	..	..	..	..	..	..	46	52	58
	8	..	..	..	..	..	..	..	..	..	..	35	38	42	46
	16–19.99	..	..	..	..	..	..	..	..	..	..	27	30	34	37
	20–64	10	10	10	10	10	10	10	19	19	19	19	23	27	27
	64–96	10	10	10	10	10	..	..	19	19	19	19	23	..	..
	96–120	10	10	10	10	..	..	..	19	19	19	19	..	..	..
9	2	..	..	..	..	..	..	..	..	..	..	..	..	..	57
	4	..	..	..	..	..	..	..	..	..	..	..	33	37	42
	8	..	..	..	..	..	..	..	..	..	..	24	27	30	33
	16–19.99	..	..	..	..	..	..	..	..	..	..	19	22	24	27
	20–120	7	7	7	7	7	7	..	14	14	14	14	16	19	..
10	2	..	..	..	..	..	..	..	..	..	..	..	..	26	40
	4	..	..	..	..	..	..	..	..	..	..	..	23	26	29
	8	..	..	..	..	..	..	..	..	..	..	17	19	21	24
	16–19.99	..	..	..	..	..	..	..	..	..	..	14	15	17	19
	20–120	5	5	5	5	5	5	..	10	10	10	10	12	14	..
11	4	..	..	..	..	..	..	..	..	..	..	..	17	18	21
	8	..	..	..	..	..	..	..	..	..	..	13	14	15	17
	16–19.99	..	..	..	..	..	..	..	..	..	..	10	11	12	14
	20–120	4	4	4	4	4	4	..	7	7	7	7	9	10	..
12	4	..	..	..	..	..	..	..	..	..	..	..	12	13	15
	8	..	..	..	..	..	..	..	..	..	..	9	10	11	12
	16–19.99	..	..	..	..	..	..	..	..	..	..	7	8	9	10
	20–120	3	3	3	3	3	3	..	5	5	5	5	6	7	..
13	4	..	..	..	..	..	..	..	..	..	..	..	10	11	12
	8	..	..	..	..	..	..	..	..	..	..	8	9	10	11
	16–19.99	..	..	..	..	..	..	..	..	..	..	6	7	8	..
	20–120	2	2	2	2	2	2	..	4	4	4	4	4	5	..

Table 7. Correlation of Former Fine-Pitch Spur and Helical Gear Classes
(AGMA 236.04) With the Quality Numbering System of the
Gear Classification Manual (AGMA 390.03)

Previous AGMA Standard — AGMA 236.04	Correlation of Quality Numbers of AGMA 390.03
Commercial 1	5 or 6
Commercial 2	6 or 7
Commercial 3	8
Commercial 4	9
Precision 1	10 or 11
Precision 2	12
Precision 3	13 or 14

Gear Material and Treatment. — Table 8 gives AGMA designations for materials and heat treatments for coarse-pitch and fine-pitch gears. Because several of the listed material specifications could be selected for a given gearing application, there are other considerations that will help to determine a proper material selection.

Some of the fundamental factors to consider when making material and treatment selections are as follows:

a. The gear designer will have information as to the factors of safety, loading and operating conditions which will dictate the use of carbon steels or alloy steels and what hardness is required.

b. For replacement gearing, the life obtained from the previous gearing should be checked. If satisfactory, a similar material may be used for replacement. If longer life is required, the selection of a better material and treatment specification with higher hardness may provide the desired improvement.

c. Annealed carbon steels, bar stock, forgings, or castings are usually satisfactory for pinions and gears subjected to uniform or moderate shock loads when the size of the gearing is not an important factor.

d. Annealed carbon steel pinions with cast iron gears are sometimes used for the same reason mentioned in item *c.*

e. Alloy steel pinions are used for increased loads or where greater life is desired. They may be used with cast iron, or annealed (forged or cast) steel gears, usually where the ratio is about 6 to 1 or higher.

f. Alloy steel pinions and gears, heat treated, should be used with the higher hardness ranges when space limitation is a factor, *i.e.*, where smaller centers and face widths may be necessary.

g. For steel pinions and gears having ratios of 2 to 1 up to about 8 to 1, with both heat treated before machining and cutting, specify the minimum hardness on the pinion to be 40 Bhn higher than the minimum hardness on the gear.   A higher hardness on the pinion than this specification (in relation to the gear hardness) will provide increased wear resistance when the ratio is about 8 to 1 or higher.

h. When steel pinions and gears with ratios below 2 to 1 are both heat treated before cutting, they are generally made to the same hardness.

i. Steel pinions and gears hardened after cutting to 400 Bhn or higher are generally specified with the same hardness, unless extremely high hardness is desired for the pinion.

j. For the convenience of the heat treater, a range of 40 Bhn should be specified. For example, a pinion might be specified with a range from 265 to 305 Bhn and the mating gear 225 to 265 Bhn.

(*Continued on page* 1999)

Table 8. AGMA Materials and Treatments for Coarse-Pitch and Fine-Pitch Gears*

Designation Number	Material	Treatment	Hardness Range (See Notes 1 and 4)
UC-1	Carbon Steel	Anneal or as rolled	Equivalent to 179 Bhn for AGMA Durability Rating
UA-11	Alloy Steel	Anneal or as rolled	Equivalent to 179 Bhn for AGMA Durability Rating
HC-1	Carbon Steel	Normalize and Temper or Quench and Temper	212 to 248 Bhn
HC-2	Carbon Steel	Quench and Temper	223 to 262 Bhn
HC-3	Carbon Steel	Quench and Temper	248 to 285 Bhn
HC-4	Carbon Steel	Quench and Temper	262 to 302 Bhn
HC-5	Carbon Steel	Quench and Temper	285 to 321 Bhn
HC-6	Carbon Steel	Quench and Temper	302 to 351 Bhn
HA-11	Alloy Steel	Normalize and Temper or Quench and Temper	223 to 262 Bhn
HA-12	Alloy Steel	Quench and Temper	248 to 285 Bhn
HA-13	Alloy Steel	Quench and Temper	262 to 302 Bhn
HA-14	Alloy Steel	Quench and Temper	285 to 321 Bhn
HA-15	Alloy Steel	Quench and Temper	302 to 351 Bhn
HA-16	Alloy Steel	Quench and Temper	331 to 388 Bhn
HA-17	Alloy Steel	Quench and Temper	351 to 402 Bhn
HA-18	Alloy Steel	Quench and Temper	402 to 461 Bhn (42 to 49 Rc)
FC-1	Carbon Steel	Flame Harden	43 Rc Min.
FC-2	Carbon Steel	Flame Harden	48 Rc Min.
FC-3	Carbon Steel	Flame Harden	52 Rc Min.
FC-4	Carbon Steel	Flame Harden	55 Rc Min.
FA-11	Alloy Steel	Flame Harden	43 Rc Min.
FA-12	Alloy Steel	Flame Harden	48 Rc Min.
FA-13	Alloy Steel	Flame Harden	52 Rc Min.
FA-14	Alloy Steel	Flame Harden	55 Rc Min.
IC-1	Carbon Steel	Induction Harden	43 Rc Min.
IC-2	Carbon Steel	Induction Harden	48 Rc Min.
IC-3	Carbon Steel	Induction Harden	52 Rc Min.
IC-4	Carbon Steel	Induction Harden	55 Rc Min.
IA-11	Alloy Steel	Induction Harden	43 Rc Min.
IA-12	Alloy Steel	Induction Harden	48 Rc Min.
IA-13	Alloy Steel	Induction Harden	52 Rc Min.
IA-14	Alloy Steel	Induction Harden	55 Rc Min.
CC-1	Carbon Steel	Cyanide	55 Rc Min.
CA-11	Alloy Steel	Cyanide	55 Rc Min.
CN-1	Carbon Steel	Carbonitride	55 Rc Min.
CN-11	Alloy Steel	Carbonitride	55 Rc Min.
NA-11	Alloy Steel (4140-4340-4640)	Nitride	48 Rc Min.
NA-12	Alloy Steel	Nitride	64 Rc Min.
CH-1	Carbon Steel	Carburize	48 Rc Min.
CH-2	Carbon Steel	Carburize	50 Rc Min.
CH-3	Carbon Steel	Carburize	55 Rc Min.
CH-4	Carbon Steel	Carburize	58 Rc Min.
CH-5	Carbon Steel	Carburize	60 Rc Min.
CH-11	Alloy Steel	Carburize	48 Rc Min.
CH-12	Alloy Steel	Carburize	50 Rc Min.
CH-13	Alloy Steel	Carburize	55 Rc Min.
CH-14	Alloy Steel	Carburize	58 Rc Min.
CH-15	Alloy Steel	Carburize	60 Rc Min.
CI-20	Cast Iron	As Required	
CI-30	Cast Iron	As Required	174 Bhn Min.
CI-35	Cast Iron	As Required	183 Bhn Min.

* Extracted from AGMA Handbook for Unassembled Gears — Volume 1 — Gear Classifications, Materials, and Inspection (AMA 390.03), with permission of the publisher, the American Gear Manufacturers Association, 1500 King St., Arlington, Va. 22314.

Table 8. *(Continued).* **AGMA Materials and Treatments for Coarse-Pitch and Fine-Pitch Gears**

Desig-nation Number	Material	Treatment	Hardness Range (See Notes 1 and 4)
CI-40	Cast Iron	As Required	202 Bhn Min.
CI-50	Cast Iron	As Required	217 Bhn Min.
CI-60	Cast Iron	As Required	223 Bhn Min.
NI-1	Nodular Iron	Anneal	179 Bhn Min.
NI-2	Nodular Iron	Anneal or Normalize and Temper	212 Bhn Min.
NI-3	Nodular Iron	Anneal or Normalize and Temper	223 Bhn Min.
NI-4	Nodular Iron	Anneal or Normalize and Temper	248 Bhn Min.
NI-5	Nodular Iron	Quench and Temper	255 Bhn Min.
NI-6	Nodular Iron	Quench and Temper	262 Bhn Min.
NI-7	Nodular Iron	Quench and Temper	269 Bhn Min.
NI-8	Nodular Iron	Quench and Temper	277 Bhn Min.
NI-9	Nodular Iron	Quench and Temper	285 Bhn Min.
NI-10	Nodular Iron	Quench and Temper	302 Bhn Min.
NI-11	Nodular Iron	Quench and Temper	311 Bhn Min.
NI-12	Nodular Iron	Quench and Temper	331 Bhn Min.
NI-13	Nodular Iron	Quench and Temper	351 Bhn Min.
NI-14	Nodular Iron	Flame Harden or Induction Harden	48 Rc Min. (Note 3)
SN-1	Stainless Steel Non Magnetic (300 Series)	None Required	....
SM-2	Stainless Steel Magnetic (400 Series)	None Required	....
SM-3	Stainless Steel Magnetic (410, 416, 440)	Quench and Temper Induction Harden or Bright Harden	As Specified
SM-4	Stainless Steel Magnetic (440)	Harden and Temper Induction Harden or Bright Harden	As Specified
AB-1	Aluminum Bronze	As Cast	116 Bhn Min.
AB-2	Aluminum Bronze	As Cast	116 Bhn Min.
AB-2	Aluminum Bronze	Heat Treated	121 Bhn Min.
AB-3	Aluminum Bronze	As Cast	140 Bhn Min.
AB-3	Aluminum Bronze	Heat Treated	190 Bhn Min.
AB-4	Aluminum Bronze	As Cast	175 Bhn Min.
AB-4	Aluminum Bronze	Heat Treated	202 Bhn Min.
AB-5	Aluminum Bronze	Wrought, Heat Treated	180 Bhn Min.
AB-6	Aluminum Bronze	Wrought, Heat Treated	180 Bhn Min.
MB-1	Manganese Bronze	As Cast	85 Bhn (500 Kg)
MB-2	Manganese Bronze	As Cast	125 Bhn (500 Kg)
MB-3	Manganese Bronze	As Cast	200 Bhn (500 Kg)
MB-4	Manganese Bronze	As Cast	210 Bhn (500 Kg)
MB-5	Manganese Bronze	Wrought — Soft	150 Bhn (500 Kg)
MB-6	Manganese Bronze	Wrought — Half Hard	190 Bhn (500 Kg)
MB-7	Manganese Bronze	Wrought — Hard	210 Bhn (500 Kg)
BZ-1	Tin Bronze	As Cast	70 Bhn (500 Kg)
BZ-2	Tin Bronze	Chill Cast	80 Bhn (500 Kg)
BZ-3	Tin Bronze	As Cast	70 Bhn (500 Kg)

See notes at end of table.

Table 8. *(Concluded.)*   **AGMA Materials and Treatments for Coarse Pitch and Fine Pitch Gears**

Desig-nation Number	Material	Treatment	Hardness Range (See Notes 1 and 3)
BZ-4	Tin Bronze	Chill Cast	85 Bhn (500 Kg)
BZ-5	Tin Bronze	As Cast	
BZ-6	Tin Bronze	Chill Cast	95 Bhn (1500 Kg)
BZ-7	Tin Bronze	Centrifugal Cast	
BZ-8	Tin Bronze	Chill Cast	95 Bhn (1500 Kg)
AL-1	Aluminum 2017 T3 Sheet	Heat Treated	. . . .
AL-2	Aluminum 2017 T4 Sheet	Heat Treated	105 Bhn (500 Kg)
AL-3	Aluminum 2024 T3 Sheet	Heat Treated	120 Bhn (500 Kg)
AL-4	Aluminum 2024 T4 Sheet	Heat Treated	120 Bhn (500 Kg)
AL-5	Aluminum 6061 T6 Bar or Sheet	Heat Treated	95 Bhn (500 Kg)
BR-1	HH Brass	As Rolled	. . . .
NM	Nonmetallic	. . . .	(Note 2)

Note 1: All Brinell hardness numbers are those obtained with a 3000 kg load unless otherwise specified. For routine acceptance tests, the diameter of the impression should be read to 0.05 mm. The range of Brinell hardness should be at least 40 points, up to 285 Bhn and 50 points, over 302 Bhn.

Note 2: The use of and specifications for nonmetallic materials shall be established by agreement between gear manufacturer and buyer.

Note 3: Where a hardness of Rc Min. is shown, a designer may prefer to provide a range of hardness. In such cases, the range should provide at least 5 points Rc above the minimum indicated in the table, and should be by agreement between the gear manufacturer and the buyer. This includes carbon and alloy steels, for all surface-hardening methods, except nitriding.

For all surface-hardened treatments where Rockwell C measurements cannot be used, the hardnesses as specified are conversion readings in accordance with ASTM specification E140.

See also footnote on first page of table.

k. When core hardness is requested, consideration should be given to the section involved.

l. Where considerable impact loads exist, the use of alloys, plus a lowering of the hardness to a range of 50 to 56 Rockwell C, is recommended for carburized gears and pinions.

m. When accelerated wear is encountered in service, heat treated gearing to provide higher hardness will, in most cases, help to alleviate these conditions. The gear manufacturer should be consulted for appropriate recommendations.

AGMA Applications and Quality Numbers. — Table 9 includes a tabulation of many industrial and end use applications for spur, helical, herringbone, bevel and hypoid gearing, racks, and fine-pitch worms and worm gearing. A typical AGMA Quality Number range is shown for each of the many industries and applications. When selecting a Quality Number for an industry or an application which is not shown, use a similar industry or application as a guide.

The AGMA Quality Number shown opposite each item of equipment identifies the quality of gearing generally used. There may be certain designs or operating conditions that would justify specifying gears to a lower or higher Quality Number.

Table 9. AGMA Applications and Quality Numbers for Racks and Gears — 1

Gearing Application	Quality No.*	Gearing Application	Quality No.*
AEROSPACE		**CEMENT INDUSTRY**	
Actuators	7–11	(Continued)	
Control Gearing	7–11	Overhead Crane	5–6
Engine Accessories	10–13	Pug, Rod and Tube Mills	5–6
Engine Power	10–13	Pulverizer	5–6
Engine Starting	10–13	Raw and Finish Mill	5–6
Loading Hoist	7–11	Rotary Dryer	5–6
Propeller Feathering	10–13	Slurry Agitator	5–6
Small Engines	12–13	**CHEWING GUM INDUSTRY**	
AGRICULTURE		Chicle Grinder	6–8
Baler	3–7	Coater	6–8
Beet Harvester	5–7	Mixer-Kneader	6–8
Combine	5–7	Molder-Roller	6–8
Corn Picker	5–7	Wrapper	6–8
Cotton Picker	5–7	**CHOCOLATE INDUSTRY**	
Farm Elevator	3–7	Glazer, Finisher	6–8
Field Harvester	5–7	Mixer, Mill	6–8
Peanut Harvester	3–7	Molder	6–8
Potato Digger	5–7	Presser, Refiner	6–8
AIR COMPRESSOR	10–11	Tampering	6–8
AUTOMOTIVE INDUSTRY	10–11	Wrapper	6–8
BALING MACHINE	5–7	**CLAY WORKING MACHINERY**	5–7
BOTTLING INDUSTRY		**CONSTRUCTION EQUIPMENT**	
Capping	6–7	Backhoe	6–8
Filling	6–7	Cranes	
Labeling	6–7	Open Gearing	3–6
Washer, Sterilizer	6–7	Enclosed Gearing	6–8
BREWING INDUSTRY		Ditch Digger	3–8
Agitator	6–8	Transmission	6–8
Barrel Washer	6–8	Drag Line	5–8
Cookers	6–8	Dumpster	6–8
Filling Machine	6–8	Paver Loader	3
Mash Tubs	6–8	Transmission	8
Pasteurizer	6–8	Mixer	3–5
Racking Machine	6–8	Swing Gear	3–5
BRICK MAKING MACHINERY	5–7	Mixing Bucket	3
BRIDGE MACHINERY	5–7	Shaker	8
BRIQUETTE MACHINES	5–7	Shovels	
CEMENT INDUSTRY		Open Gearing	3–6
(Quarry Operation)		Enclosed Gearing	6–8
Conveyor	5–6	Stationary Mixer	
Crusher	5–6	Transmission	8
Diesel Electric		Drum Gears	3–5
Locomotive	8–9	Stone Crusher	
Electric Dragline		Transmission	8
(cast gear)	3	Conveyor	6
(cut gear)	6–8	Truck Mixer	
Electric Locomotive		Transfer Case	9
Electric Shovel		Drum Gears	3–5
(cast gear)	3	**COMMERCIAL METERS**	
(cut gear)	6–8	Gas	7–9
Elevator	5–6	Liquid, Water, Milk	7–9
Locomotive Crane		Parking	7–9
(cast gear)	3	**COMPUTING AND ACCOUNTING MACHINES**	
(cut gear)	5–6	Accounting-Billing	9–10
(Plant Operation)		Adding Machine-Calculator	7–9
Air Separator	5–6	Addressograph	7
Ball Mill	5–7	Bookkeeping	9–10
Compeb Mill	5–6	Cash Register	7
Conveyor	5–6	Comptometer	6–8
Cooler	5–6	Computing	10–11
Elevator	5–6	Data Processing	7–9
Feeder	5–6	Dictating Machine	9
Filter	5–6	Typewriter	8
Kiln	5–6		
Kiln Slurry Agitator	5–6		

* Quality Numbers are inclusive, from the lowest to highest numbers shown.

Table 9. AGMA Applications and Quality Numbers for Racks and Gears — 2

Gearing Application	Quality No.*	Gearing Application	Quality No.*
CRANES		FOUNDRY INDUSTRY	
Boom Hoist	5–6	Conveyor	5–6
Gantry	5–6	Elevator	5–6
Load Hoist	5–7	Ladle	5–6
Overhead	5–6	Molding Machine	5–6
Ship	5–7	Overhead Cranes	5–6
CRUSHERS		Sand Mixer	5–6
Ice, Feed	6–8	Sand Slinger	5–6
Portable and Stationary	6–8	Tumbling Mill	5–6
Rock, Ore, Coal	6–8	HOME APPLIANCES	
DAIRY INDUSTRY		Blender	6–8
Bottle Washer	6–7	Mixer	7–9
Homogenizer	7–9	Timer	8–10
Separator	7–9	Washing Machine	8–10
DISH WASHER		MACHINE TOOL INDUSTRY	
Commercial	5–7	Hand Motion (but not	
DISTILLERY INDUSTRY		Indexing and	
Agitator	5–7	Positioning)	6–9
Bottle Filler	5–7	Power Drives	
Conveyor, Elevator	6–7	0–800 FPM	6–8
Grain Pulverizer	6–8	800–2000 FPM	8–10
Mash Tub	5–7	2000–4000 FPM	10–12
Mixer	5–7	Over 4000 FPM	12 & Up
Yeast Tub	5–7	Indexing and Positioning	
ELECTRIC FURNACE		Approximate Positioning	6–10
Tilting Gears	5–7	Accurate Indexing and	
ELECTRONIC INSTRUMENT		Positioning	12 & Up
CONTROL AND		MARINE INDUSTRY	
GUIDANCE SYSTEMS		Anchor Hoist	6–8
Accelerometer	10–12	Cargo Hoist	7–8
Airborne Temperature		Conveyor	5–7
Recorder	12–14	Davit Gearing	5–7
Aircraft Instrument	12	Elevator	6–7
Altimeter-Stabilizer	9–11	Small Propulsion	10–12
Analog Computer	10–12	Steering Gear	8
Antenna Assembly	7–9	Winch	5–8
Antiaircraft Detector	12	METAL WORKING	
Automatic Pilot	9–11	Bending Roll	5–7
Digital Computer	10–12	Draw Bench	6–8
Gun Data Computer	12–14	Forge Press	5–7
Gyro Caging Mechanism	10–12	Punch Press	5–7
Gyroscope-Computer	12–14	Roll Lathe	5–7
Pressure Transducer	12–14	MINING AND PREPARATION	
Radar, Sonar, Tuner	10–12	Breaker	5–6
Recorder, Telemeter	10–12	Car Dump	5–6
Servo System Component	9–11	Concentrator	5–6
Sound Detector	9	Continuous Miner	6–7
Transmitter, Receiver	10–12	Conveyor	5–7
ENGINES		Cutting Machine	6–10
Combustion		Drag Line	
Engine Accessories	10–12	Open Gearing	3–6
Supercharger	10–12	Enclosed Gearing	6–8
Timing Gearings	10–12	Drills	5–6
Transmission	8–10	Drier	5–6
FARM EQUIPMENT		Electric Locomotive	6–8
Milking Machine	6–8	Elevator	5–6
Separator	8–10	Feeder	6–8
Sweeper	4–6	Flotation	5–6
FLOUR MILL INDUSTRY		Grizzly	5–7
Bleacher	7–8	Hoists, Skips	7–8
Grain Cleaner	7–8	Loader (Underground)	5–8
Grinder	7–8	Rock Drill	5–6
Hulling	7–8	Rotary Car Dump	6–8
Milling, Scouring	7–8	Screen (Rotary)	7–8
Polisher	7–8	Screen (Shaking)	7–8
Separator	7–8	Separator	5–6

* Quality Numbers are inclusive, from the lowest to highest numbers shown.

Table 9. AGMA Applications and Quality Numbers for Racks and Gears — 3

Gearing Application	Quality No.*	Gearing Application	Quality No.*
MINING AND PREPARATION		Mixer, Tuber	6–8
(Continued)		Refiner, Calender	5–7
Sedimentation	5–6	Rubber Mill, Scrap Cutter	5–7
Shaker	6–8	Tire Building	6–8
Shovel	3–8	Tire Chopper	5–7
Tipple Gearing	5–7	Washer, Banbury Mixer	5–7
Washer	6–8		
PAPER AND PULP		**SMALL POWER TOOLS**	
Bag Machines	6–8	Bench Grinder	6–8
Box Machines	6–8	Drills-Saws	7–9
Building Paper	6–8	Hair Clipper	7–9
Calendar	6–8	Hedge Clipper	7–9
Chipper	6–8	Sander, Polisher	8–10
Coating	6–8	Sprayer	6–8
Envelope Machines	6–8		
Food Container	6–8	**SPACE NAVIGATION**	
Glazing	6–8	Sextant and Star Tracker	14 & Up
Log Conveyor-Elevator	5–7		
Mixer, Agitator	6–8	**STEEL INDUSTRY**	
Paper Machine		Miscellaneous Drives	
Auxiliary	8–9	Bessemer Tilt-Car Dump	5–6
Main Drive	10–12	Coke Pusher, Distributor	5–6
Press, Couch, Drier Rolls	6–8	Conveyor, Door Lift	5–6
Slitting	10–12	Electric Furnace Tilt	5–6
Steam Drum	6–8	Hot Metal Car Tilt	5–6
Varnishing	6–8	Hot Metal Charger	5–6
Wall Paper Machines	6–8	Jib Hoist, Dolomite	
PAVING INDUSTRY		Machine	5–6
Aggregate Drier	5–7	Larry Car, Mud Gun	5–6
Aggregate Spreader	5–7	Mixing Bin, Mixer Tilt	5–6
Asphalt Mixer	5–7	Ore Crusher, Pig Machine	5–6
Asphalt Spreader	5–7	Pulverizer, Quench Car	5–6
Concrete Batch Mixer	5–7	Shaker, Sinter Conveyor	5–6
PHOTOGRAPHIC EQUIPMENT		Sinter Machine Skip Hoist	5–6
Aerial	10–12	Slag Crusher, Slag Shovel	5–6
Commercial	8–10		
PRINTING INDUSTRY		Primary and Secondary	
Press		Rolling Mill Drives	
Book	9–11	Blooming and Plate Mill	5–6
Flat	9–11	Heavy Duty Hot Mill	
Magazine	9–11	Drives	5–6
Newspaper	9–11	Slabbing and Strip Mill	5–6
Roll Reels	6–7		
Rotary	9–11	Hot Mill Drives	
PUMP INDUSTRY		Sendzimer-Stekel	7–8
Liquid	10–12	Tandem-Temper-Skin	6–7
Rotary	6–8		
Slush-Duplex-Triplex	6–8	Cold Mill Drives	
Vacuum	6–8	Bar, Merchant, Rail, Rod	5–6
QUARRY INDUSTRY		Structural, Tube	5–6
Conveyor-Elevator	6–7		
Crusher	5–7	Auxiliary and	
Rotary Screen	7–8	Miscellaneous Drives	
RADAR AND MISSILE		Annealing Furnace Car	5–6
Antenna Elevating	8–10	Bending Roll	5–6
Data Gear	10–12	Blooming Mill	
Launch Pad Azimuth	8	Manipulator	5–6
Ring Gear	9–12	Blooming Mill Rack and	
Rotating Drive	10–12	Pinion	5–6
RAILROADS		Blooming Mill Side	
Construction Hoist	5–7	Guard	5–6
Wrecking Crane	6–8	Car Haul	5–6
RUBBER AND PLASTICS		Coil Conveyor	5–6
Boot and Shoe Machines	6–8	Coil Dump	5–6
Drier, Press	6–8	Crop Conveyor	5–6
Extruder, Strainer	6–8	Edger Drives	5–6
		Electrolytic Line	6–7

* Quality Numbers are inclusive, from the lowest to highest numbers shown.

Table 9. AGMA Applications and Quality Numbers for Racks and Gears — 4

Gearing Application	Quality No.*	Gearing Application	Quality No.*
STEEL INDUSTRY (Continued)		STEEL INDUSTRY (Continued)	
Flange Machine Ingot		Turbine	9–10
Buggy	5–6	Overhead Cranes	
Leveler	6–7	Billet Charger, Cold Mill	5–6
Magazine Pusher	6–7	Bucket Handling	5–6
Mill Shear Drives	6–7	Car Repair Shop	5–6
Mill Table Drives		Cast House, Coil Storage	5–6
(under 800 ft/Min)	5–6	Charging Machine	5–6
Mill Table Drives		Cinder Yard, Hot Top	5–6
(over 800–1800 ft/Min)	6–7	Coal and Ore Bridges	5–6
Mill Table Drives		Electric Furnace Charger	5–6
(over 1800 ft/Min)	8	Hot Metal, Ladle	5–6
Nail and Spike Machine	5–6	Hot Mill, Ladle House	5–6
Piler	5–6	Jib Crane, Motor Room	5–6
Plate Mill Rack and		Mold Yard, Rod Mill	5–6
Pinion	5–6	Ore Unloader, Stripper	5–6
Plate Mill Side Guards	5–6	Overhead Hoist	5–6
Plate Turnover	5–6	Pickler Building	5–6
Preheat Furnace Pusher	5–6	Pig Machine, Sand House	5–6
Processor	6–7	Portable Hoist	5–6
Pusher Rack and Pinion	5–6	Scale Pit, Shipping	5–6
Rotary Furnace	5–6	Scrap Balers and Shears	5–6
Shear Depress Table	5–6	Scrap Preparation	5–6
Slab Squeezer	5–6	Service Shops	5–6
Slab Squeezer Rack and		Skull Cracker	5–6
Pinion	5–6	Slab Handling	5–6
Slitter, Side Trimmer	6–7		
Tension Reel	6–7	MISCELLANEOUS	
Tilt Table, Upcoiler	5–6	Clocks	6
Transfer Car	5–6	Counters	7–9
Wire Drawing Machine	6–7	Fishing Reel	6
Precision Gear Drives		Gauges	8–10
Diesel Electric Gearing	8–9	IBM Card Puncher, Sorter	8
Flying Shear	9–10	Metering Pumps	7–8
Shear Timing Gears	9–10	Motion-Picture Equipment	8
High Speed Reels	8–9	Popcorn Machine, Comm.	6–7
Locomotive Timing		Pumps	5–7
Gears	9–10	Sewing Machine	8
Pump Gears	8–9	Slicer	7–8
Tube Reduction Gearing	8–9	Vending Machines	6–7

Extracted from AGMA Handbook for Unassembled Gears — Volume 1 — Gear Classifications, Materials, and Inspection (AGMA 390.03), with permission of the publisher, the American Gear Manufacturers Association.
  * Quality Numbers are inclusive, from the lowest to highest number shown.

AGMA Measuring Methods and Practices. — Process control and inspection procedures for individual element checks of coarse- and fine-pitch, spur, helical, herringbone, bevel and hypoid gears are presented in AGMA Gear Handbook, Volume 1 (AGMA 390.03). It also covers composite inspection of coarse- and fine-pitch, spur, helical, herringbone, bevel and hypoid gears. In addition, backlash and tooth-contact pattern evaluation of bevel and hypoid gearing is discussed.

The quality level of a gear is determined during its manufacture by the specific sequence of steps followed and the degree of care employed at each step. This AGMA Handbook formalizes the long-established concept of process control as a quality determinant. Process control is a method by which gear accuracy is achieved and maintained through control of manufacturing equipment, methods and processes, without resort to inspection of individual elements of every gear produced.

This Gear Handbook is published by the American Gear Manufacturers Association, 1500 King St., Arlington, Va. 22314.

Gear Materials

Classification of Gear Steels. — Gear steels may be divided into two general classes — the plain carbon and the alloy steels. Alloy steels are used to some extent in the industrial field, but heat-treated plain carbon steels are far more common. The use of untreated alloy steels for gears is seldom, if ever, justified, and then, only when heat-treating facilities are lacking. The points to be considered in determining whether to use heat-treated plain carbon steels or heat-treated alloy steels are: Does the service condition or design require the superior characteristics of the alloy steels, or, if alloy steels are not required, will the advantages to be derived offset the additional cost? For most applications, plain carbon steels, heat-treated to obtain the best of their qualities for the service intended, are satisfactory and quite economical. The advantages obtained from using heat-treated alloy steels in place of heat-treated plain carbon steels are as follows:

1. Increased surface hardness and depth of hardness penetration for the same carbon content and quench.

2. Ability to obtain the same surface hardness with a less drastic quench and, in the case of some of the alloys, a lower quenching temperature, thus giving less distortion.

3. Increased toughness, as indicated by the higher values of yield point, elongation, and reduction of area.

4. Finer grain size, with the resulting higher impact toughness and increased wear resistance.

5. In the case of some of the alloys, better machining qualities or the possibility of machining at higher hardnesses.

Use of Casehardening Steels. — Each of the two general classes of gear steels may be further subdivided as follows: (1) Casehardening steels; (2) full-hardening steels; and (3) steels that are heat-treated and drawn to a hardness that will permit machining. The first two — casehardening and full-hardening steels — are interchangeable for some kinds of service, and the choice is often a matter of personal opinion. Casehardening steels with their extremely hard, fine-grained (when properly treated) case and comparatively soft and ductile core are generally used when resistance to wear is desired. Casehardening alloy steels have a fairly tough core, but not as tough as that of the full-hardening steels. In order to realize the greatest benefits from the core properties, casehardened steels should be double-quenched. This is particularly true of the alloy steels, because the benefits derived from their use seldom justify the additional expense, unless the core is refined and toughened by a second quench. The penalty that must be paid for the additional refinement is increased distortion, which may be excessive if the shape or design does not lend itself to the casehardening process.

Use of "Thru-Hardening" Steels. — Thru-hardening steels are used when great strength, high endurance limit, toughness, and resistance to shock are required. These qualities are governed by the kind of steel and treatment used. Fairly high surface hardnesses are obtainable in this group, though not so high as those of the casehardening steels. For that reason, the resistance to wear is not so great as might be obtained, but when wear resistance combined with great strength and toughness is required, this type of steel is superior to the others. Thru-hardening steels become distorted to some extent when hardened, the amount depending upon the steel and quenching medium used. For that reason, thru-hardening steels are not suitable for high-speed gearing where noise is a factor, or for gearing where

Steels for Industrial Gearing

CASE HARDENING STEELS

Material Specification	Hardness		Typical Heat Treatment, Characteristics and Uses
	Case R_c	Core Bhn	
AISI 1020 AISI 1116	55–60	160–230	Carburize, harden, temper at 350°F. For gears that must be wear resistant. Normalized material is easily machined. Core is ductile but has little strength.
AISI 4130 AISI 4140	50–55	270–370	Harden, temper at 900°F, Nitride. For parts requiring greater wear resistance than that of thru hardened steels but cannot tolerate the distortion of carburizing. Case is shallow, core is tough.
AISI 4615 AISI 4620 AISI 8615 AISI 8620	55–60 55–60	170–260 200–300	Carburize, harden, temper at 350°F. For gears requiring high fatigue resistance and strength. The 86xx series has better machinability. The 20 point steels are used for coarser teeth.
AISI 9310	58–63	250–350	Carburize, harden, temper at 300°F. Primarily for aerospace gears that are highly loaded and operate at high pitch line velocity and for other gears requiring high reliability under extreme operating conditions. This material is not used at high temperature.
Nitralloy N and Type 135 Mod. (15-N)	90–94	300–370	Harden, temper at 1200°F, Nitride. For gears requiring high strength and wear resistance that cannot tolerate the distortion of the carburizing process or that operate at high temperatures. Gear teeth are usually finished before nitriding. Care must be exercised in running nitrided gears together to avoid crazing of case.

THRU-HARDENING STEELS

Material Specification	Hardness		Typical Heat Treatment, Characteristics and Uses
AISI 1045 AISI 1140	24–40	....	Harden and temper to required hardness. Oil quench for lower hardness and water quench for higher hardness. For gears of medium and large size requiring moderate strength and wear resistance. Gears must have consistent, solid sections to withstand quenching.
AISI 4140 AISI 4340	24–40	....	Harden (oil quench), temper to required hardness. For gears requiring high strength and wear resistance, and high shock loading resistance. Use 41xx series for moderate sections and 43xx series for heavy sections. Gears must have consistent, solid sections to withstand quenching.

accuracy is of paramount importance, except, of course, in cases where grinding of the teeth is practicable. The medium and high-carbon percentages require an oil quench, but a water quench may be necessary for the lower carbon contents, in order to obtain the highest physical properties and hardness. The distortion, however, will be greater with the water quench.

Heat-Treatment that Permits Machining. — When the grinding of gear teeth is not practicable and a high degree of accuracy is required, hardened steels may be drawn or tempered to a hardness that will permit the cutting of the teeth. This treatment gives a highly refined structure, great toughness, and, in spite of the low hardness, excellent wearing qualities. The lower strength is somewhat compensated for by the elimination of the increment loads due to the impacts which are caused by inaccuracies. When steels that have a low degree of hardness penetration from surface to core are treated in this manner, the design cannot be based on the physical properties corresponding to the hardness at the surface. Since the physical properties are determined by the hardness, the drop in hardness from surface to core will give lower physical properties at the root of the tooth, where the stress is greatest. The quenching medium may be either oil, water, or brine, depending on the steel used and hardness penetration desired. The amount of distortion, of course, is immaterial, because the machining is done after heat-treating.

Making Pinion Harder than Gear to Equalize Wear. — Beneficial results from a wear standpoint are obtained by making the pinion harder than the gear. The pinion, having a lesser number of teeth than the gear, naturally does more work per tooth, and the differential in hardness between the pinion and the gear (the amount being dependent on the ratio) serves to equalize the rate of wear. The harder pinion teeth correct the errors in the gear teeth to some extent by the initial wear and then seem to burnish the teeth of the gear and increase its ability to withstand wear by the greater hardness due to the cold-working of the surface. In applications where the gear ratio is high and there are no severe shock loads, a casehardened pinion running with an oil-treated gear, treated to a Brinell hardness at which the teeth may be cut after treating, is an excellent combination. The pinion, being relatively small, is distorted but little, and distortion in the gear is circumvented by cutting the teeth after treatment.

Forged and Rolled Carbon Steels for Gears. — These compositions cover steel for gears in three groups, according to heat treatment, as follows: (a) case-hardened gears, (b) unhardened gears, not heat treated after machining, and (c) hardened and tempered gears.

Forged and rolled carbon gear steels are purchased on the basis of the requirements as to chemical composition specified in Table 1. Class N steel will normally be ordered in ten point carbon ranges within these limits. Requirements as to physical properties have been omitted, but when they are called for the requirements as to carbon shall be omitted. The steels may be made by either or both the open hearth and electric furnace processes.

Forged and Rolled Alloy Steels for Gears. — These compositions cover alloy steel for gears, in two classes according to heat treatment, as follows: (a) casehardened gears, and (b) hardened and tempered gears. Forged and rolled alloy gear steels are purchased on the basis of the requirements as to chemical composition specified in Table 2. Requirements as to physical properties have been omitted. The steel shall be made by either or both the open hearth and electric furnace process.

Steel Castings for Gears. — It is recommended that steel castings for cut gears be purchased on the basis of chemical analysis and that only two types of analysis be used, one for case-hardened gears and the other for both untreated gears and

those which are to be hardened and tempered. The steel is to be made by the open hearth, crucible or electric furnace processes. The converter process is not recognized. Sufficient risers shall be provided to secure soundness and freedom from undue segregation. Risers shall not be broken off the unannealed castings by force. Where risers are cut off with a torch the cut shall be at least one-half inch above the surface of the castings, and the remaining metal removed by chipping, grinding, or other non-injurious method.

Steel for use in gears shall conform to the requirements, as to chemical composition as indicated in Table 3. All steel castings for gears must be thoroughly normalized or annealed, using such temperature and time as will entirely eliminate the characteristic structure of unannealed castings.

Effect of Alloying Metals on Gear Steels. — The effect of the various alloying elements on steel will be summarized in order to assist engineers in deciding upon the particular kind of alloy steel to use for specific purposes. The characteristics, outlined apply only to heat-treated steels. When the effect of the addition of an alloying element is stated, it is understood that reference is made to alloy steels of a given carbon content, compared with a plain carbon steel of the same carbon content.

Nickel — The addition of nickel tends to increase the hardness and strength, with but little sacrifice of ductility. The hardness penetration is somewhat greater than that of plain carbon steels. Its use as an alloying element lowers the critical points and produces less distortion, due to the lower quenching temperature. The nickel steels of the case-hardening group carburize more slowly, but the grain growth is less.

Chromium — Chromium increases the hardness and strength over that obtained by the use of nickel, though the loss of ductility is greater. Chromium refines the grain and imparts a greater depth of hardness. Chromium steels have a high degree of wear resistance and are easily machined in spite of the fine grain.

Manganese — When present in sufficient amounts to warrant the use of the term alloy, the addition of manganese is very effective. It gives greater strength than nickel, and a higher degree of toughness than chromium. Owing to its susceptibility to cold-working, it is likely to flow under severe unit pressures. Up to the present time, it has never been used to any great extent for heat-treated gears, but is now receiving an increasing amount of attention.

Vanadium — Vanadium has a similar effect to that of manganese — increasing the hardness, strength, and toughness. The loss of ductility is somewhat more than that due to manganese, but the hardness penetration is greater than for any of the other alloying elements. Owing to the extremely fine-grained structure, the impact strength is high; but vanadium tends to make machining difficult.

Molybdenum — Molybdenum has the property of increasing the strength without affecting the ductility. For the same hardness, steels containing molybdenum are more ductile than any other alloy steels, and having nearly the same strength, are tougher; in spite of the increased toughness, the presence of molybdenum does not make machining more difficult. In fact, such steels can be machined at a higher hardness than any of the other alloy steels. The impact strength is nearly as great as that of the vanadium steels.

Chrome-Nickel Steels — The combination of the two alloying elements chromium and nickel adds the beneficial qualities of both. The high degree of ductility present in nickel steels is complemented by the high strength, finer grain size, deep hardening and wear-resistant properties imparted by the addition of chromium. The increased toughness makes these steels more difficult to machine than the plain carbon steels, and they are more difficult to heat-treat. The distortion increases with the amount of chromium and nickel.

Table 1. Compositions of Forged and Rolled Carbon Steels for Gears

Heat-treatment	Class	Carbon	Manganese	Phosphorus	Sulphur
Case-hardened........	C	0.15–0.25	0.40–0.70	0.045 max	0.055 max
Untreated.............	N	0.25–0.50	0.50–0.80	0.045 max	0.055 max
Hardened............. (or untreated)	H	0.40–0.50	0.40–0.70	0.045 max	0.055 max

Table 2. Compositions of Forged and Rolled Alloy Steels for Gears

Steel Specification	Chemical Composition*					
	C	Mn	Si	Ni	Cr	Mo
AISI 4130	.28–.30	.40– .60	.20–.35	....	.80–1.1	.15–.25
AISI 4140	.38–.43	.75–1.0	.20–.35	....	.80–1.1	.15–.25
AISI 4340	.38–.43	.60– .80	.20–.35	1.65–2.0	.70– .90	.20–.30
AISI 4615	.13–.18	.45– .65	.20–.35	1.65–2.0	....	.20–.30
AISI 4620	.17–.22	.46– .65	.20–.35	1.65–2.0	....	.20–.30
AISI 8615	.13–.18	.70– .90	.20–.35	.40– .70	.40– .60	.15–.25
AISI 8620	.18–.23	.70– .90	.20–.35	.40– .70	.40– .60	.15–.25
AISI 9310	.08–.13	.45– .65	.20–.35	3.0 –3.5	1.0 –1.4	.08–.15
Nitralloy Type N†	.20–.27	.40– .70	.20–.40	3.2 –3.8	1.0 –1.3	.20–.30
135 Mod.†	.38–.45	.40– .70	.20–.40	....	1.4 –1.8	.30–.45

* C = carbon; Mn = manganese; Si = silicon; Ni = nickel; Cr = chromium, and Mo = Molybdenum.
† Both Nitralloy alloys contain Aluminum .85–1.2%.

Table 3. Compositions of Cast Steels for Gears

Steel Specification	Chemical Composition*			
	C	Mn	Si	
SAE-0022	.12–.22	.50–.90	.60 Max.	May be carburized
SAE-0050	.40–.50	.50–.90	.80 Max.	Hardenable 210–250

* C = carbon; Mn = manganese; and Si = silicon.

Chrome-Vanadium Steels — Chrome-vanadium steels have practically the same tensile properties as the chrome-nickel steels, but the hardening power, impact strength, and wear resistance are increased by the finer grain size. They are difficult to machine and become distorted more easily than the other alloy steels.

Chrome-Molybdenum Steels — This group has the same qualities as the straight molybdenum steels, but the hardening depth and wear resistance are increased by the addition of chromium. This steel is very easily heat-treated and machined.

Nickel-Molybdenum Steels — Nickel-molybdenum steels have qualities similar to chrome-molybdenum steel. The toughness is said to be greater, but the steel is somewhat more difficult to machine.

Sintered Materials. — For high production of low and moderately loaded gears, significant production cost savings may be effected by the use of a sintered metal powder. With this material, the gear is formed in a die under high pressure and then sintered in a furnace. The primary cost saving comes from the great reduction in labor cost of machining the gear teeth and other gear blank surfaces. The volume of production must be high enough to amortize the cost of the die and the gear blank must be of such a configuration that it may be formed and readily ejected from the die.

Bronze and Brass Gear Castings. — These specifications cover non-ferrous metals for spur, bevel, and worm gears, bushings and flanges for composition gears. This material shall be purchased on the basis of chemical composition. The alloys may be made by any approved method.

Spur and Bevel Gears: For spur and bevel gears, hard cast bronze is recommended (A.S.T.M. B–10–18; S.A.E. No. 62; and the well-known 88–10–2 mixture) with the following limits as to composition: Copper, 86 to 89; tin, 9 to 11; zinc, 1 to 3; lead (max), 0.20; iron (max), 0.06 per cent. Good castings made from this bronze should have the following minimum physical characteristics: Ultimate strength, 30,000 pounds per square inch; yield point, 15,000 pounds per square inch; elongation in 2 inches, 14 per cent.

Worm Gears: For bronze worm gears, two alternative analyses of phosphor bronze are recommended, S.A.E. No. 65 and No. 63.

S.A.E. No. 65 (called phosphor gear bronze) has the following composition: Copper, 88 to 90; tin, 10 to 12; phosphorus, 0.1 to 0.3; lead, zinc and impurities (max), 0.5 per cent. Good castings made of this alloy should have the following minimum physical characteristics: Ultimate strength, 35,000 pounds per square inch; yield point, 20,000 pounds per square inch; elongation in 2 inches, 10 per cent.

The composition of S.A.E. No. 63 (called leaded gun metal) follows: Copper, 86 to 89; tin, 9 to 11; lead, 1 to 2.5; phosphorus (max), 0.25; zinc and impurities (max), 0.50 per cent.

Good castings made of this alloy should have the following minimum physical characteristics: Ultimate strength, 30,000 pounds per square inch; yield point, 12,000 pounds per square inch; elongation in 2 inches, 10 per cent.

These alloys, especially No. 65, are adapted to chilling for hardness and refinement of grain. No. 65 is to be preferred for use with worms of great hardness and fine accuracy. No. 63 is to be preferred for use with unhardened worms.

Gear Bushings: For bronze bushings for gears, S.A.E. No. 64 is recommended of the following analysis: Copper, 78.5 to 81.5; tin, 9 to 11; lead, 9 to 11; phosphorus, 0.05 to 0.25; zinc (max), 0.75; other impurities (max), 0.25 per cent. Good castings of this alloy should have the following minimum physical characteristics: Ultimate strength, 25,000 pounds per square inch; yield point, 12,000 pounds per square inch; elongation in 2 inches, 8 per cent.

Flanges for Composition Pinions: For brass flanges for composition pinions A.S.T.M. B–30–32T, and S.A.E. No. 40 are recommended. This is a good cast red brass of sufficient strength and hardness to take its share of load and wear when the design is such that the flanges mesh with the mating gear. The composition is as follows: Copper, 83 to 86; tin, 4.5 to 5.5; lead, 4.5 to 5.5; zinc, 4.5 to 5.5 iron (max) 0.35; antimony (max), 0.25 per cent; aluminum, none. Good castings made from this alloy should have the following minimum physical characteristics: Ultimate strength, 27,000 pounds per square inch; yield point, 12,000 pounds per square inch; elongation in 2 inches, 16 per cent.

Materials for Worm Gearing. — The Hamilton Gear & Machine Co. conducted an extensive series of tests on a variety of materials that might be used for worm gears, to ascertain which material is the most suitable. According to these tests chill-cast nickel-phosphor-bronze ranks first in resistance to wear and deformation. This bronze is composed of approximately 87.5 per cent copper, 11 per cent tin, 1.5 per cent nickel, with from 0.1 to 0.2 per cent phosphorus. The worms used in these tests were made from S.A.E.–2315, 3½ per cent nickel steel, casehardened, ground, and polished. The Shore scleroscope hardness of the worms was between 80 and 90. This nickel alloy steel was adopted after numerous tests of a variety of steels, because it provided the necessary strength, together with the degree of hardness required.

The material that showed up second best in these tests was a No. 65 S.A.E. bronze. Navy bronze (88-10-2) containing 2 per cent of zinc, with no phosphorus, and not chilled, performed satisfactorily at speeds of 600 revolutions per minute, but was not sufficiently strong at lower speeds. Red brass (85-5-5) proved slightly better at from 1500 to 1800 revolutions per minute, but would bend at lower speeds, before it would show actual wear.

Non-metallic Gearing. — Non-metallic or composition gearing is used primarily where quietness of operation at high speed is the first consideration. Non-metallic materials are also applied very generally to timing gears and numerous other classes of gearing. Rawhide was used originally for non-metallic gears, but other materials have been introduced which have important advantages. These later materials are sold by different firms under various trade names, such as Micarta, Textolite, Formica, Dilecto, Spauldite, Phenolite, Fibroc, Fabroil, Synthane, Celoron, etc. Most of these gear materials consist of layers of canvas which is impregnated with bakelite and forced together under hydraulic pressure, which, in conjunction with the application of heat, forms a dense rigid mass.

Although phenol resin gears in general are resilient, they are self-supporting and require no side plates or shrouds unless subjected to a heavy starting torque. The phenol resinoid element makes these gears proof against vermin and rodents.

The non-metallic gear materials referred to are generally assumed to have the power-transmitting capacity of cast iron. While the tensile strength may be considerably less than that of cast iron, the resiliency of these materials enables them to withstand impact and abrasion to a degree that might result in excessive wear of cast-iron teeth. Thus in many cases, composition gearing of impregnated canvas has proved to be more durable than cast iron and much more durable than rawhide.

Application of Non-metallic Gears. — The most effective field of use for these non-metallic materials is for high-speed duty. At low speeds, where the starting torque may be high, or where the load may fluctuate widely, or when high shock loads may be encountered, these non-metallic materials do not always prove satisfactory. In general, non-metallic materials should not be used for pitch-line velocities below 600 feet per minute.

Tooth Form: The best tooth form for non-metallic materials is the 20-degree stub-tooth system. When only a single pair of gears is involved and the center distance can be varied, the best results will be obtained by making the non-metallic driving pinion of all-addendum form, while the driven metal gear is made with standard tooth proportions. Such a drive will carry from 50 to 75 per cent greater loads than one of standard tooth proportions.

Material for Mating Gear: For durability under load, the use of hardened steel (over 400 Brinell) for the mating metal gear appears to give the best results. A good second choice for the material of the mating member is cast iron. The use of brass, bronze, or soft steel (under 400 Brinell) as a material for the mating member of phenolic laminated gears leads to excessive abrasive wear.

Power-transmitting Capacity of Non-metallic Gears. — The characteristics of gears made of phenolic laminated materials are so different from those of metal gears that they should be considered in a class by themselves. Because of the low modulus of elasticity, most of the effects of small errors in tooth form and spacing are absorbed at the tooth surfaces by the elastic deformation, and have but little effect on the strength of the gears.

If

S = safe working stress for a given velocity

S_s = allowable static stress

V = pitch-line velocity in feet per minute

then, according to the recommended practice of the American Gear Manufacturers' Association,

$$S = S_s \times \left(\frac{150}{200 + V} + 0.25 \right)$$

The value of S_s for phenolic laminated materials is given as 6000 pounds per square inch. The accompanying table gives the safe working stresses S for different pitch-line velocities. When the value of S is known, the horsepower capacity is determined by substituting the value of S for s_t in Equation 3 on page 1813 and solving for torque T_1. Then H.P. = $T_1 \times$ R.P.M. $\div$ 63,000.

Safe Working Stresses for Non-metallic Gears

Pitch-line Velocity Feet per Minute V	Safe Working Stresses	Pitch-line Velocity Feet per Minute V	Safe Working Stresses	Pitch-line Velocity Feet per Minute V	Safe Working Stresses
600	2625	1800	1950	4000	1714
700	2500	2000	1909	4500	1691
800	2400	2200	1875	5000	1673
900	2318	2400	1846	5500	1653
1000	2250	2600	1821	6000	1645
1200	2143	2800	1800	6500	1634
1400	2063	3000	1781	7000	1622
1600	2000	3500	1743	7500	1617

The tensile strength of the phenolic laminated materials used for gears, is slightly less than that of cast iron. These materials are far softer than any metal, and the modulus of elasticity is about one-thirtieth that of steel. In other words, if the tooth load on a steel gear which causes a deformation of 0.001 inch were applied to the tooth of a similar gear made of phenolic laminated material, the tooth of the non-metallic gear would be deformed about ½₂ inch. Under these conditions, several things will happen. With all gears, regardless of the theoretical duration of contact, one tooth only will carry the load until the load is sufficient to deform the tooth the amount of the error that may be present. On metal gears, when the tooth has been deformed the amount of the error, the stresses set up in the materials may approach or exceed the elastic limit of the material. Hence for standard tooth forms and those generated from standard basic racks, it is dangerous to calculate their strength as very much greater than that which can safely be carried on a single tooth. On gears made of phenolic laminated materials, on the other hand, the teeth will be deformed the amount of this normal error without setting up any appreciable stresses in the material, so that the load is actually supported by several teeth.

All materials have their own peculiar and distinct characteristics, so that under certain specific conditions, each material has a field of its own where it is superior to any other. Such fields may overlap to some extent, and only in such overlapping fields are different materials directly competitive. For example, steel is more or less ductile, has a high tensile strength, and a high modulus of elasticity. Cast iron, on the other hand, is not ductile, has a low tensile strength, but a high compressive strength, and a low modulus of elasticity. Hence when stiffness and high tensile strength are essential, steel is far superior to cast iron. On the other hand,

when these two characteristics are unimportant, but high compressive strength and a moderate amount of elasticity are essential, cast iron is superior to steel.

Preferred Pitch for Non-metallic Gears. — The pitch of the gear or pinion should bear a reasonable relation either to the horsepower or speed or to the applied torque, as shown by the accompanying table. The upper half of this table is based upon horsepower transmitted at a given pitch-line velocity. The lower half gives the torque in pounds-feet or the torque at a 1-foot radius. This torque T for any given horsepower and speed can be obtained from the following formula:

$$T = \frac{5252 \times \text{H.P.}}{\text{R.P.M.}}$$

Bore Sizes for Non-metallic Gears. — For plain phenolic laminated pinions, that is, pinions without metal end plates, a drive fit of 0.001 inch per inch of shaft diameter should be used. For shafts above 2.5 inches in diameter, the fit should be constant at 0.0025 to 0.003 inch. When metal reinforcing end plates are used, the drive fit should conform to the same standards as used for metal.

Preferred Pitches for Non-metallic Gears*

Diametral Pitch for Given Horsepower and Pitch Line Velocities			
Horsepower Transmitted	Pitch Line Velocity up to 1000 Feet per Minute	Pitch Line Velocity from 1000 to 2000 Feet per Minute	Pitch Line Velocity over 2000 Feet per Minute
¼–1	8–10	10–12	12–16
1–2	7–8	8–10	10–12
2–3	6–7	7–8	8–10
3–7½	5–6	6–7	7–8
7½–10	4–5	5–6	6–7
10–15	3–4	4–5	5–6
15–25	2½–3	3–4	4–5
25–60	2–2½	2½–3	3–4
60–100	1¾–2	2–2½	2½–3
100–150	1½–1¾	1¾–2	2–2½

Torque in Pounds-feet for Given Diametral Pitch					
Diametral Pitch	Torque in Pounds-feet		Diametral Pitch	Torque in Pounds-feet	
	Minimum	Maximum		Minimum	Maximum
16	1	2	4	50	100
12	2	4	3	100	200
10	4	8	2½	200	450
8	8	15	2	450	900
6	15	30	1½	900	1800
5	30	50	1	1800	3500

* These preferred pitches are applicable both to rawhide and the phenolic laminated types of materials.

The root diameter of a pinion of phenolic laminated type should be such that the minimum distance from the edge of the keyway to the root diameter will be at least equal to the depth of tooth.

Keyway Stresses for Non-metallic Gears. — The keyway stress should not exceed 3000 pounds per square inch on a plain phenolic laminated gear or pinion. The keyway stress is calculated by the formula:

$$S = \frac{33,000 \times \text{H.P.}}{V \times A}$$

in which

S = unit stress in pounds per square inch;

H.P. = horsepower transmitted;

V = peripheral speed of shaft in feet per minute; and

A = square inch area of keyway in pinion (length $\times$ height).

If the keyway stress formula is expressed in terms of shaft radius r and revolutions per minute, it will read:

$$S = \frac{63,000 \times \text{H.P.}}{\text{R.P.M.} \times r \times A}$$

When the design is such that the keyway stresses exceed 3000 pounds, metal reinforcing end plates may be used. Such end plates should not extend beyond the root diameter of the teeth. The distance from the outer edge of the retaining bolt to the root diameter of the teeth shall not be less than a full tooth depth. The use of drive keys should be avoided, but if required, metal end plates should be used on the pinion to take the wedging action of the key.

For phenolic laminated pinions, the face of the mating gear should be the same or slightly greater than the pinion face.

Invention of Gear Teeth. — The invention of gear teeth represents a gradual evolution from gearing of primitive form. The earliest evidence we have of an investigation of the problem of *uniform motion* from toothed gearing and the successful solution of that problem, dates from the time of Olaf Roemer, the celebrated Danish astronomer, who, in the year 1674, proposed the epicycloidal form to obtain uniform motion. Evidently Robert Willis, professor in the University of Cambridge, was the first to make a practical application of the epicycloidal curve so as to provide for an interchangeable series of gears. Willis gives credit to Camus for conceiving the idea of interchangeable gears, but claims for himself its first application. The involute tooth was suggested as a theory by early scientists and mathematicians, but it remained for Willis to present it in a practical form. Perhaps the earliest conception of the application of this form of teeth to gears was by Philippe de Lahire, a Frenchman, who considered it, in theory, equally suitable with the epicycloidal for tooth outlines. This was about 1695 and not long after Roemer had first demonstrated the epicycloidal form. The applicability of the involute had been further elucidated by Leonard Euler, a Swiss mathematician, born at Basel, 1707, who is credited by Willis with being the first to suggest it. Willis devised the Willis odontograph for laying out involute teeth.

A pressure angle of 14½ degrees was selected for three different reasons. First, because the sine of 14½ degrees is nearly ¼, making it convenient in calculation; second, because this angle coincided closely with the pressure angle resulting from the usual construction of epicycloidal gear teeth; third, because the angle of the straight-sided involute rack is the same as the 29-degree worm thread.

Calculating Replacement-Gear Dimensions from Simple Measurements.—The following tables provide formulas with which to calculate the dimensions needed to produce replacement spur, bevel, and helical gears when only the number of teeth, the outside diameter, and the tooth depth of the gear to be replaced are known.

For helical gears, exact helix angles can be obtained by the following procedure.

1. Using a common protractor, measure the approximate helix angle A at the approximate pitch line.

2. Place sample or its mating gear on the arbor of a gear hobbing machine.

3. Calculate the index and lead gears differentially for the angle obtained by the measurements, and set up the machine as though a gear is to be cut.

4. Attach a dial indicator on an adjustable arm to the vertical swivel head, with

SPUR GEARS					
Tooth Form and Pressure Angle	Diametral Pitch P	Pitch Diameter D	Circular Pitch P_c	Outside Diameter O	Addendum J
American Standard 14½- and 20-degree full depth	$\dfrac{N+2}{O}$	$\dfrac{N}{P}$	$\dfrac{3.1416}{P}$	$\dfrac{N+2}{P}$	$\dfrac{1}{P}$
American Standard 20-degree stub	$\dfrac{N+1.6}{O}$	$\dfrac{N}{P}$	$\dfrac{3.1416}{P}$	$\dfrac{N+1.6}{P}$	$\dfrac{0.8}{P}$
Fellows 20-degree stub	Note	$\dfrac{N}{P_N}$	$\dfrac{3.1416}{P_N}$	$\dfrac{N}{P_N}+\dfrac{2}{P_D}$	$\dfrac{1}{P_D}$
Tooth Form and Pressure Angle	Dedendum K	Whole Tooth Depth W	Clearance $K-J$	Tooth Thickness on Pitch Circle	
American Standard 14½- and 20-degree full depth	$\dfrac{1.157}{P}$	$\dfrac{2.157}{P}$	$\dfrac{0.157}{P}$	$\dfrac{1.5708}{P}$	
American Standard 20-degree stub	$\dfrac{1}{P}$	$\dfrac{1.8}{P}$	$\dfrac{0.2}{P}$	$\dfrac{1.5708}{P}$	
Fellows 20-degree stub	$\dfrac{1.25}{P_D}$	$\dfrac{2.25}{P_D}$	$\dfrac{0.25}{P_D}$	$\dfrac{1.5708}{P_N}$	

N = number of teeth.

In the Fellows stub-tooth system, P_N = diametral pitch in numerator of stub-tooth designation and is used to determine circular pitch and number of teeth, and P_D = diametral pitch in the diameter of stub-tooth designation and is used to determine tooth depth.

MILLED BEVEL GEARS — 90 DEGREE SHAFTS††				
Tooth Form and Pressure Angle	Tangent of Pitch-Cone Angle of Gear, tan A	Tangent of Pitch-Cone Angle of Pinion, tan a	Diametral Pitch of Both Gear and Pinion, P*	Outside Diameter of Gear, O, or Pinion, o
American Standard $14\frac{1}{2}$- and 20-degree full depth	$\dfrac{N_G}{N_P}$	$\dfrac{N_P}{N_G}$	$\dfrac{N_a + 2\cos A}{O}$ or $\dfrac{N_P + 2\cos a}{o}$	$\dfrac{N_a + 2\cos A}{P}$ or $\dfrac{N_P + 2\cos a}{P}$
American Standard 20-degree stub	$\dfrac{N_G}{N_P}$	$\dfrac{N_P}{N_G}$	$\dfrac{N_a + 1.6\cos A}{O}$ or $\dfrac{N_P + 1.6\cos a}{o}$	$\dfrac{N_a + 1.6\cos A}{P}$ or $\dfrac{N_P + 1.6\cos a}{P}$
Fellows 20-degree stub	$\dfrac{N_G}{N_P}$	$\dfrac{N_P}{N_G}$	. . .	$\dfrac{N_G}{P_N} + \dfrac{2\cos A}{P_D}$ or $\dfrac{N_P}{P_N} + \dfrac{2\cos a}{P_D}$

Tooth Form and Pressure Angle	Pitch-Cone Radius or Cone Distance, E*	Tangent of Addendum Angle*	Tangent of Dedendum Angle*	Cosine of Pitch-Cone Angle of Gear, cos A†
American Standard $14\frac{1}{2}$- and 20-degree full depth	$\dfrac{D}{2\sin A}$ or $\dfrac{d}{2\sin a}$	$\dfrac{2\sin A}{N_a}$ or $\dfrac{2\sin a}{N_P}$	$\dfrac{2.314\sin A}{N_G}$ or $\dfrac{2.314\sin a}{N_P}$	$\dfrac{(P \times O) - N_G}{2}$
American Standard 20-degree stub	$\dfrac{D}{2\sin A}$ or $\dfrac{d}{2\sin a}$	$\dfrac{1.6\sin A}{N_G}$ or $\dfrac{1.6\sin a}{N_P}$	$\dfrac{2\sin A}{N_G}$ or $\dfrac{2\sin a}{N_P}$	$\dfrac{(P \times O) - N_G}{1.6}$
Fellows 20-degree stub	$\dfrac{D}{2\sin A}$ or $\dfrac{d}{2\sin a}$	$\dfrac{2 P_N \sin A}{N_G \times P_D}$ or $\dfrac{2 P_N \sin a}{N_P \times P_D}$	$\dfrac{2.5 P_N \sin A}{N_G \times P_D}$ or $\dfrac{2.5 P_N \sin a}{N_P \times P_D}$	$\dfrac{P_D[(O \times P_N) - N_G]}{2 P_N}$

N_G = number of teeth in gear; N_P = number of teeth in pinion; O = outside diameter of gear; o = outside diameter of pinion; D = pitch diameter of gear = $N_G \div P$; d = pitch diameter of pinion = $N_P \div P$; P_c = circular pitch; J = addendum; K = dedendum; W = whole depth. See footnote in spur gear table for meaning of P_N and P_D. The tooth thickness on the pitch circle is found by means of the formulas in the last column under spur gears.

* These values are the same for both gear and pinion.

† The same formulas apply to the pinion, substituting N_P for N_G and o for O.

†† These formulas do not apply to Gleason System Gearing.

HELICAL GEARS

Tooth Form and Pressure Angle	Normal Diametral Pitch P_N	Diametral Pitch P	Outside Diameter of Blank O	Pitch Diameter D
American Standard 14½- and 20-degree full depth	$\dfrac{N+2\cos A}{O \times \cos A}$ or $\dfrac{P}{\cos A}$	$P_N \cos A$ or $\dfrac{N+2\cos A}{O}$	$\dfrac{N+2\cos A}{P_N \cos A}$ or $\dfrac{N+2\cos A}{P}$	$\dfrac{N}{P_N \cos A}$ or $\dfrac{N}{P}$
American Standard 20-degree stub	$\dfrac{N+1.6\cos A}{O \times \cos A}$ or $\dfrac{P}{\cos A}$	$P_n \cos A$ or $\dfrac{N+1.6\cos A}{O}$	$\dfrac{N+1.6\cos A}{P_N \cos A}$ or $\dfrac{N+1.6\cos A}{P}$	$\dfrac{N}{P_N \cos A}$ or $\dfrac{N}{P}$
Fellows 20-degree stub	$\cdots$	$\cdots$	$\dfrac{N}{(P_N)_N \cos A} + \dfrac{2}{(P_N)_D}$	$\dfrac{N}{(P_N)_N \cos A}$

Tooth Form and Pressure Angle	Cosine of Helix Angle A	Addendum	Dedendum	Whole Depth
American Standard 14½- and 20-degree full depth	$\dfrac{P}{P_N}$ or $\dfrac{N}{O \times P_N - 2}$	$\dfrac{1}{P_N}$ or $\dfrac{\cos A}{P}$	$\dfrac{1.157}{P_N}$ or $\dfrac{1.157\cos A}{P}$	$\dfrac{2.157}{P_N}$ or $\dfrac{2.157\cos A}{P}$
American Standard 20-degree stub	$\dfrac{P}{P_N}$ or $\dfrac{N}{O \times P_N - 1.6}$	$\dfrac{0.8}{P_N}$ or $\dfrac{0.8\cos A}{P}$	$\dfrac{1}{P_N}$ or $\dfrac{\cos A}{P}$	$\dfrac{1.8}{P_N}$ or $\dfrac{1.8\cos A}{P}$
Fellows 20-degree stub	$\dfrac{N}{(P_N)_N\left(O - \dfrac{2}{(P_N)_D}\right)}$	$\dfrac{1}{(P_N)_D}$	$\dfrac{1.25}{(P_N)_D}$	$\dfrac{2.25}{(P_N)_D}$

P_N = normal diametral pitch; = normal diametral pitch of cutter or hob used to cut teeth
P = diametral pitch;
O = outside diameter of blank;
D = pitch diameter;
A = helix angle;
N = number of teeth;

$(P_N)_N$ = normal diametral pitch in numerator of stub-tooth designation, which determines thickness of tooth and number of teeth;
$(P_N)_D$ = normal diametral pitch in denominator of stub-tooth designation, which determines the addendum, dedendum, and whole depth.

the indicator plunger in a plane perpendicular to the gear axis and in contact with the tooth face. Contact may be anywhere between the top and the root of the tooth.

5. With the power shut off, engage the starting lever and traverse the indicator plunger axially by means of the handwheel.

6. If angle A is correct, the indicator plunger will not move as it traverses the face width of the gear. If it does move from 0, note the amount. Divide the amount of movement by the width of the gear to obtain the tangent of the angle by which to correct angle A, plus or minus, depending on the direction of indicator movement.

SPLINES AND SERRATIONS

A splined shaft is one having a series of parallel keys formed integrally with the shaft and mating with corresponding grooves cut in a hub or fitting; this is in contrast to a shaft having a series of keys or feathers fitted into slots cut into the shaft. This latter construction weakens the shaft to a considerable degree because of the slots cut into it and, as a consequence, reduces its torque-transmitting capacity.

Splined shafts are most generally used in three types of applications: (1) for coupling shafts when relatively heavy torques are to be transmitted without slippage; (2) for transmitting power to slidably-mounted or permanently-fixed gears, pulleys, and other rotating members; and (3) for attaching parts that may require removal for indexing or change in angular position.

Splines having straight-sided teeth have been used in many applications (see SAE Parallel Side Splines for Soft Broached Holes in Fittings); however, the use of splines with teeth of involute profile has steadily increased since (1) involute spline couplings have greater torque-transmitting capacity than any other type; (2) they can be produced by the same techniques and equipment as is used to cut gears; and (3) they have a self-centering action under load even when there is backlash between mating members.

Involute Splines

American National Standard Involute Splines.* — These splines or multiple keys are similar in form to internal and external involute gears. The general practice is to form the external splines either by hobbing, rolling, or on a gear shaper, and internal splines either by broaching or on a gear shaper. The internal spline is held to basic dimensions and the external spline is varied to control the fit. Involute splines have maximum strength at the base, they can be accurately spaced and are self-centering, thus equalizing the bearing and stresses, and they can be measured and fitted accurately.

In American National Standard ANSI B92.1-1970 most of the features of the 1960 standard are retained; plus the addition of three tolerance classes, for a total of four. The term "involute serration," formerly applied to involute splines with 45-degree pressure angle, has been deleted and the standard now includes involute splines with 30-, 37.5- and 45-degree pressure angles. Tables for these splines have been rearranged accordingly. The term "serration" will no longer apply to splines covered by this Standard.

The Standard has only one fit class for all side fit splines; the former Class 2 fit. Class 1 fit has been deleted because of its infrequent use. The major diameter of the flat root side fit spline has been changed and a tolerance applied to include the range of the 1950 and the 1960 standards. The interchangeability limitations with splines made to previous standards are given later in the section entitled "Interchangeability."

There have been no tolerance nor fit changes to the major diameter fit section.

The Standard recognizes the fact that proper assembly between mating splines is dependent only on the spline being within effective specifications from the tip of the tooth to the form diameter. Therefore, on side fit splines, the internal spline major diameter now is shown as a maximum dimension and the external spline minor diameter is shown as a minimum dimension. The minimum internal major diameter and the maximum external minor diameter must clear the specified form diameter and thus do not need any additional control.

The spline specification tables now include a greater number of tolerance level selections. These tolerance classes were added for greater selection to suit end product needs. The selections differ only in the tolerance as applied to space width

* See American National Standard ANSI B92.2M-1980, Metric Module Involute Splines; also p. 2039.

and tooth thickness. The tolerance class which was used in ASA B5.15-1960 is the basis and is now designated as tolerance Class 5. The new tolerance classes are based on the following formulas:

$$\text{Tolerance Class } 4 = \text{Tolerance Class } 5 \times 0.71$$
$$\text{Tolerance Class } 6 = \text{Tolerance Class } 5 \times 1.40$$
$$\text{Tolerance Class } 7 = \text{Tolerance Class } 5 \times 2.00$$

All dimensions, listed in this standard, are for the finished part. Therefore, any compensation that must be made for operations which take place during processing, such as heat treatment, must be taken into account when selecting the tolerance level for manufacturing.

The standard has the same internal minimum effective space width and external maximum effective tooth thickness for all tolerance classes and has two types of fit. For tooth side fits, the minimum effective space width and the maximum effective tooth thickness are of equal value. This basic concept makes it possible to have interchangeable assembly between mating splines where made to this standard regardless of the tolerance class of the individual members. This allows a tolerance class "mix" of mating members which often is an advantage where one member is considerably less difficult to produce than its mate, and the "average" tolerance applied to the two units is such that it satisfies the design need. For instance, this can be the result of specification of Class 5 tolerance to one member and Class 7 to its mate, thus providing an assembly tolerance in the Class 6 range. The maximum effective tooth thickness is less than the minimum effective space width for major diameter fits to allow for eccentricity variations.

In the event the fit as provided in this standard does not satisfy a particular design need and a specific amount of effective clearance or press fit is desired, the change should be made only to the external spline by a reduction or an increase in effective tooth thickness and a like change in actual tooth thickness. The minimum effective space width, in this standard, is always basic. This basic minimum effective space width should always be retained when special designs are derived from the concept of this standard.

Terms Applied to Involute Splines. — The following definitions of involute spline terms, here listed in alphabetical order, are given in the American National Standard:

Active Spline Length (L_a) is the length of spline which contacts the mating spline. On sliding splines it exceeds the length of engagement.

Actual Space Width (s) is the circular width on the pitch circle of any single space considering an infinitely thin increment of axial spline length.

Actual Tooth Thickness (t) is the circular thickness on the pitch circle of any single tooth considering an infinitely thin increment of axial spline length.

Alignment Variation is the variation of the effective spline axis with respect to the reference axis (see Fig. 1).

Base Circle is the circle from which involute spline tooth profiles are constructed.

Base Diameter (D_b) is the diameter of the base circle.

Basic Space Width is the basic space width for 30-degree pressure angle splines; half the circular pitch. The basic space width for 37.5- and 45-degree pressure angle splines, however, is greater than half the circular pitch. The teeth are proportioned so that the external tooth, at its base, has about the same thickness as the internal tooth at the form diameter. This results in greater minor diameters than those of comparable involute splines of 30-degree pressure angle.

Circular Pitch (p) is the distance along the pitch circle between corresponding points of adjacent spline teeth.

Depth of Engagement is the radial distance from the minor circle of the internal spline to the major circle of the external spline, minus corner clearance and/or chamfer depth.

Diametral Pitch (P) is the number of spline teeth per inch of pitch diameter. The diametral pitch determines the circular pitch and the basic space width or tooth thickness. In conjunction with the number of teeth, it also determines the pitch diameter. (See also *Pitch*.)

Effective Clearance (c_v) is the effective space width of the internal spline minus the effective tooth thickness of the mating external spline.

Effective Space Width, (s_v) of an internal spline is equal to the circular tooth thickness on the pitch circle of an imaginary perfect external spline which would fit the internal spline without looseness or interference considering engagement of the entire axial length of the spline. The minimum effective space width of the internal spline is always basic, as shown in Table 3. Fit variations may be obtained by adjusting the tooth thickness of the external spline.

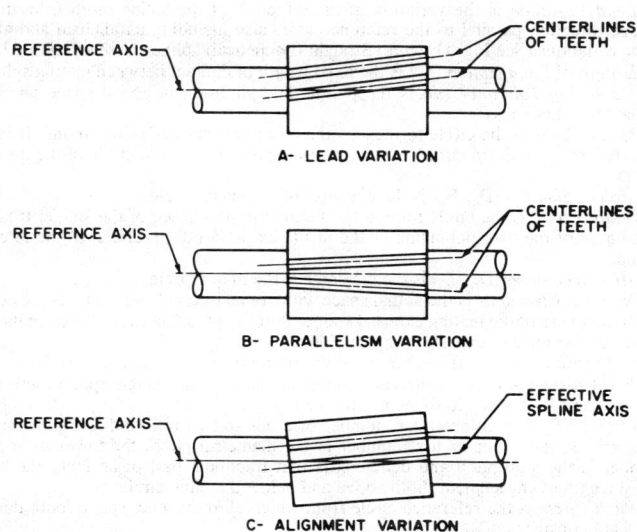

Fig. 1.   Three types of involute spline variations.

Effective Tooth Thickness (t_v) is the effective tooth thickness of an external spline, equal to the circular space width on the pitch circle of an imaginary perfect internal spline which would fit the external spline without looseness or interference, considering engagement of the entire axial length of the spline.

Effective Variation is the accumulated effect of the spline variations on the fit with the mating part.

External Spline is a spline formed on the outer surface of a cylinder.

Fillet is the concave portion of the tooth profile which joins the sides to the bottom of the space.

Fillet Root Splines are those in which a single fillet in the general form of an arc joins the sides of adjacent teeth.

Flat Root Splines are those in which fillets join the arcs of major or minor circles to the tooth sides.

Form Circle is the circle which defines the deepest points of involute form control of the tooth profile. This circle along with the tooth tip circle (or start of chamfer circle) determines the limits of tooth profile requiring control. It is located near the major circle on the internal spline and near the minor circle on the external spline.

Form Clearance (c_F) is the radial depth of involute profile beyond the depth of engagement with the mating part. It allows for looseness between mating splines and for eccentricities between the minor circle (internal), the major circle (external), and their respective pitch circles.

Form Diameter (D_{Fe}, D_{Fi}) the diameter of the form circle.

Internal Spline is a spline formed on the inner surface of a cylinder.

Involute Spline is one having teeth with involute profiles.

Lead Variation is the variation of the direction of the spline tooth from its intended direction parallel to the reference axis, also including parallelism and alignment variations (See Fig. 1). *Note:* Straight (nonhelical) splines have an infinite lead.

Length of Engagement (L_q) is the axial length of contact between mating splines.

Machining Tolerance (m) is the permissible variation in actual space width or actual tooth thickness.

Major Circle is the circle formed by the outermost surface of the spline. It is the outside circle (tooth tip circle) of the external spline or the root circle of the internal spline.

Major Diameter (D_o, D_{ri}) is the diameter of the major circle.

Minor Circle is the circle formed by the innermost surface of the spline. It is the root circle of the external spline or the inside circle (tooth tip circle) of the internal spline.

Minor Diameter (D_{re}, D_i) is the diameter of the minor circle.

Nominal Clearance is the actual space width of an internal spline minus the actual tooth thickness of the mating external spline. It does not define the fit between mating members, because of the effect of variations.

Out of Roundness is the variation of the spline from a true circular configuration.

Parallelism Variation is the variation of parallelism of a single spline tooth with respect to any other single spline tooth (See Fig. 1).

Pitch (P/P_s) is a combination number of a one-to-two ratio indicating the spline proportions; the upper or first number is the diametral pitch, the lower or second number is the stub pitch and denotes, as that fractional part of an inch, the basic radial length of engagement, both above and below the pitch circle.

Pitch Circle is the reference circle from which all transverse spline tooth dimensions are constructed.

Pitch Diameter (D) is the diameter of the pitch circle.

Pitch Point is the intersection of the spline tooth profile with the pitch circle.

Pressure Angle (ϕ) is the angle between a line tangent to an involute and a radial line through the point of tangency. Unless otherwise specified, it is the standard pressure angle.

Profile Variation is any variation from the specified tooth profile normal to the flank.

Spline is a machine element consisting of integral keys (spline teeth) or keyways (spaces) equally spaced around a circle or portion thereof.

Standard (Main) Pressure Angle (ϕ_D) is the pressure angle at the specified pitch diameter.

Stub Pitch (P_s) is a number used to denote the radial distance from the pitch circle to the major circle of the external spline and from the pitch circle to the minor circle

of the internal spline. The stub pitch for splines in this standard is twice the diametral pitch.

Total Index Variation is the greatest difference in any two teeth (adjacent or otherwise) between the actual and the perfect spacing of the tooth profiles.

Total Tolerance $(m + \lambda)$ is the machining tolerance plus the variation allowance.

Variation Allowance (λ) is the permissible effective variation.

Many of these terms and their associated symbols are illustrated in the diagrams in Tables 6 to 10, incl.

Tooth Proportions. — There are 17 pitches: 2.5/5, 3/6, 4/8, 5/10, 6/12, 8/16, 10/20, 12/24, 16/32, 20/40, 24/48, 32/64, 40/80, 48/96, 64/128, 80/160, and 128/256. The numerator in this fractional designation is known as the diametral pitch and controls the pitch diameter; the denominator, which is always double the numerator, is known as the stub pitch and controls the tooth depth. For convenience in calculation, only the numerator is used in the formulas given and is designated as P. Diametral pitch, as in gears, means the number of teeth per inch of pitch diameter.

Table 1 shows the symbols and Table 2 the formulas for basic tooth dimensions of involute spline teeth of various pitches. Basic dimensions are given in Table 3.

Table 1. American National Standard Involute Spline Symbols (ANSI B92.1-1970)

c_v	effective clearance	M_i	measurement between pins, internal spline
c_F	form clearance		
D	pitch diameter	N	number of teeth
D_b	base diameter	P	diametral pitch
D_{ci}	pin contact diameter, internal spline	P_s	stub pitch
		p	circular pitch
D_{ce}	pin contact diameter, external spline	r_f	fillet radius
		s	actual space width, circular
D_{Fe}	form diameter, external spline	s_v	effective space width, circular
D_{Fi}	form diameter, internal spline	s_c	allowable compressive stress, psi
D_i	minor diameter, internal spline	s_s	allowable shear stress, psi
D_o	major diameter, external spline	t	actual tooth thickness, circular
D_{re}	minor diameter, external spline (root)	t_v	effective tooth thickness, circular
		λ	variation allowance
D_{ri}	major diameter, internal spline (root)	ϵ	involute roll angle
		ϕ	pressure angle
d_e	diameter of measuring pin for external spline	ϕ_D	standard pressure angle
		ϕ_{ci}	pressure angle at pin contact diameter, internal spline
d_i	diameter of measuring pin for internal spline	ϕ_{ce}	pressure angle at pin contact diameter, external spline
K_e	change factor for external spline		
K_i	change factor for internal spline	ϕ_i	pressure angle at pin center, internal spline
L	spline length		
L_a	active spline length	ϕ_e	pressure angle at pin center, external spline
L_g	length of engagement		
m	machining tolerance	ϕ_F	pressure angle at form diameter
M_e	measurement over pins, external spline		

Tooth Numbers. — The American National Standard covers involute splines having tooth numbers ranging from 6 to 60 with a 30- or 37.5-degree pressure angle and from 6 to 100 with a 45-degree pressure angle. In selecting the number of teeth

Table 2. Formulas for Basic Dimensions (ANSI B92.1-1970)

Term	Symbol	30 deg φD			37.5 deg φD	45 deg φD
		Flat Root Side Fit	Flat Root Major Dia Fit	Fillet Root Side Fit	Fillet Root Side Fit	Fillet Root Side Fit
		2.5/5-32/64 Pitch	3/6-16/32 Pitch	2.5/5-48/96 Pitch	2.5/5-48/96 Pitch	10/20-128/256 Pitch
Stub Pitch	P_s	$2P$	$2P$	$2P$	$2P$	$2P$
Pitch Diameter	D	$\dfrac{N}{P}$	$\dfrac{N}{P}$	$\dfrac{N}{P}$	$\dfrac{N}{P}$	$\dfrac{N}{P}$
Base Diameter	D_b	$D\cos\phi D$	$D\cos\phi D$	$D\cos\phi D$	$D\cos\phi D$	$D\cos\phi D$
Circular Pitch	p	$\dfrac{\pi}{P}$	$\dfrac{\pi}{P}$	$\dfrac{\pi}{P}$	$\dfrac{\pi}{P}$	$\dfrac{\pi}{P}$
Minimum Effective Space Width	s_v	$\dfrac{\pi}{2P}$	$\dfrac{\pi}{2P}$	$\dfrac{\pi}{2P}$	$\dfrac{0.5\pi + 0.1}{P}$	$\dfrac{0.5\pi + 0.2}{P}$
Major Diameter, Internal	D_{ri}	$\dfrac{N+1.35}{P}$	$\dfrac{N+1}{P}$	$\dfrac{N+1.8}{P}$	$\dfrac{N+1.6}{P}$	$\dfrac{N+1.4}{P}$
Major Diameter, External	D_o	$\dfrac{N+1}{P}$	$\dfrac{N+1}{P}$	$\dfrac{N+1}{P}$	$\dfrac{N+1}{P}$	$\dfrac{N+1}{P}$
Minor Diameter, Internal	D_i	$\dfrac{N-1}{P}$	$\dfrac{N-1}{P}$	$\dfrac{N-1}{P}$	$\dfrac{N-0.8}{P}$	$\dfrac{N-0.6}{P}$
Minor Dia. Ext. — 2.5/5 thru 12/24 pitch	D_{re}	$\dfrac{N-1.35}{P}$	$\dfrac{N-1.35}{P}$	$\dfrac{N-1.8}{P}$	$\dfrac{N-1.3}{P}$	
Minor Dia. Ext. — 16/32 pitch and finer	D_{re}			$\dfrac{N-2}{P}$		
Minor Dia. Ext. — 10/20 pitch and finer	D_{re}					$\dfrac{N-1}{P}$
Form Diameter, Internal	D_{Fi}	$\dfrac{N+1}{P} + 2cF$	$\dfrac{N+0.8}{P} - 0.004 + 2cF$	$\dfrac{N+1}{P} + 2cF$	$\dfrac{N+1}{P} + 2cF$	$\dfrac{N+1}{P} + 2cF$
Form Diameter, External	D_{Fe}	$\dfrac{N-1}{P} - 2cF$	$\dfrac{N-1}{P} - 2cF$	$\dfrac{N-1}{P} - 2cF$	$\dfrac{N-0.8}{P} - 2cF$	$\dfrac{N-0.6}{P} - 2cF$
Form Clearance (Radial)	cF	$0.001D$, with max of 0.010, min of 0.002				

$\pi = 3.1415927$. NOTE: All spline specification table dimensions in the standard are derived from these basic formulas by application of tolerances.

Table 3. Basic Dimensions for Involute Splines (ANSI B92.1-1970)

Pitch, P/P_s	Circular Pitch, p	Min Effective Space Width (BASIC), S_v min			Pitch, P/P_s	Circular Pitch, p	Min Effective Space Width (BASIC), S_v min		
		30 deg φ	37.5 deg φ	45 deg φ			30 deg φ	37.5 deg φ	45 deg φ
2.5/5	1.2566	0.6283	0.6683	...	20/40	0.1571	0.0785	0.0835	0.0885
3/6	1.0472	0.5236	0.5569	...	24/48	0.1309	0.0654	0.0696	0.0738
4/8	0.7854	0.3927	0.4177	...	32/64	0.0982	0.0491	0.0522	0.0553
5/10	0.6283	0.3142	0.3342	...	40/80	0.0785	0.0393	0.0418	0.0443
6/12	0.5236	0.2618	0.2785	...	48/96	0.0654	0.0327	0.0348	0.0369
8/16	0.3927	0.1963	0.2088	...	64/128	0.0491	...	...	0.0277
10/20	0.3142	0.1571	0.1671	0.1771	80/160	0.0393	...	...	0.0221
12/24	0.2618	0.1309	0.1392	0.1476	128/256	0.0246	...	...	0.0138
16/32	0.1963	0.0982	0.1044	0.1107	...	...	...	...	...

for a given spline application, it is well to keep in mind that there are no advantages to be gained by using odd numbers of teeth and that the diameters of splines with odd tooth numbers, particularly internal splines, are troublesome to measure with pins since no two tooth spaces are diametrically opposite each other.

Types and Classes of Involute Spline Fits. — Two types of fits are covered by the American National Standard for involute splines, the side fit and the major diameter fit. Dimensional data for flat root side fit, flat root major diameter fit, and fillet root side fit splines are tabulated in this standard for 30-degree pressure angle splines; but for only the fillet root side fit for 37.5- and 45-degree pressure angle splines.

Side Fit: In the side fit, the mating members contact only on the sides of the teeth; major and minor diameters are clearance dimensions. The tooth sides act as drivers and centralize the mating splines.

Major Diameter Fit: Mating parts for this fit contact at the major diameter for centralizing. The sides of the teeth act as drivers. The minor diameters are clearance dimensions.

The major diameter fit provides a minimum effective clearance that will allow for contact and location at the major diameter with a minimum amount of location or centralizing effect by the sides of the teeth. The major diameter fit has only one space width and tooth thickness tolerance which is the same as side fit Class 5.

A fillet root may be specified for an external spline, even though it is otherwise designed to the flat root side fit or major diameter fit standard. An internal spline with a fillet root can be used only for the side fit.

Classes of Tolerances. — This standard includes four classes of tolerances on space width and tooth thickness. This has been done to provide a range of tolerances for selection to suit a design need. The classes are variations of the former single tolerance which is now Class 5 and are based on the formulas shown in the footnote of Table 4. All tolerance classes have the same minimum effective space width and maximum effective tooth thickness limits so that a mix of classes between mating parts is possible.

Table 4. Maximum Tolerances for Space Width and Tooth Thickness of Tolerance Class 5 Splines* (ANSI B92.1-1970)
(Values shown in ten thousandths (20 = 0.0020))

No. of Teeth	Pitch, P/P_s							
	2.5/5 and 3/6	4/8 and 5/10	6/12 and 8/16	10/20 and 12/24	16/32 and 20/40	24/48 thru 48/96	64/128 and 80/160	128/256
N	Machining Tolerance, *m*							
10	15.8	14.5	12.5	12.0	11.7	11.7	9.6	9.5
20	17.6	16.0	14.0	13.0	12.4	12.4	10.2	10.0
30	18.4	17.5	15.5	14.0	13.1	13.1	10.8	10.5
40	21.8	19.0	17.0	15.0	13.8	13.8	11.4	—
50	23.0	20.5	18.5	16.0	14.5	14.5	—	—
60	24.8	22.0	20.0	17.0	15.2	15.2	—	—
70	—	—	—	18.0	15.9	15.9	—	—
80	—	—	—	19.0	16.6	16.6	—	—
90	—	—	—	20.0	17.3	17.3	—	—
100	—	—	—	21.0	18.0	18.0	—	—
N	Variation Allowance, λ							
10	23.5	20.3	17.0	15.7	14.2	12.2	11.0	9.8
20	27.0	22.6	19.0	17.4	15.4	13.4	12.0	10.6
30	30.5	24.9	21.0	19.1	16.6	14.6	13.0	11.4
40	34.0	27.2	23.0	21.6	17.8	15.8	14.0	—
50	37.5	29.5	25.0	22.5	19.0	17.0	—	—
60	41.0	31.8	27.0	24.2	20.2	18.2	—	—
70	—	—	—	25.9	21.4	19.4	—	—
80	—	—	—	27.6	22.6	20.6	—	—
90	—	—	—	29.3	23.8	21.8	—	—
100	—	—	—	31.0	25.0	23.0	—	—
N	Total Index Variation							
10	20	17	15	15	14	12	11	10
20	24	20	18	17	15	13	12	11
30	28	22	20	19	16	15	14	13
40	32	25	22	20	18	16	15	—
50	36	27	25	22	19	17	—	—
60	40	30	27	24	20	18	—	—
70	—	—	—	26	21	20	—	—
80	—	—	—	28	22	21	—	—
90	—	—	—	29	24	23	—	—
100	—	—	—	31	25	24	—	—
N	Profile Variation							
All	+7 −10	+6 −8	+5 −7	+4 −6	+3 −5	+2 −4	+2 −4	+2 −4
	Lead Variation							
L_g, in.	0.3	0.5	1	2	3	4	5	6	7	8	9	10
Variation	2	3	4	5	6	7	8	9	10	11	12	13

* For other tolerance classes: Class 4 = 0.71 × Tabulated value
Class 5 = As tabulated in table
Class 6 = 1.40 × Tabulated value
Class 7 = 2.00 × Tabulated value

Fillets and Chamfers. — Spline teeth may have either a flat root or a rounded fillet root.

Flat Root Splines are suitable for most applications. The fillet which joins the sides to the bottom of the tooth space, if generated, has a varying radius of curvature. Specification of this fillet is usually not required. It is controlled by the form diameter which is the diameter at the deepest point of the desired true involute form (sometimes designated as TIF).

When flat root splines are used for heavily loaded couplings which are not suitable for fillet root spline application, it may be desirable to minimize the stress concentration in the flat root type by specifying an approximate radius for the fillet.

Since internal splines are stronger than external splines because of their broad bases and high pressure angles at the major diameter, broaches for flat root internal splines are normally made with the involute profile extending to the major diameter.

Fillet Root Splines are recommended for heavy loads because the larger fillets provided reduce the stress concentrations. The curvature along any generated fillet varies and cannot be specified by a radius of any given value.

External splines may be produced by generating with a pinion type shaper cutter or with a hob, or by cutting with no generating motion using a tool formed to the contour of a tooth space. External splines are also made of cold forming and in these cases are usually of the fillet root design. Internal splines are usually produced by broaching, by form cutting, or by generating with a shaper cutter. Even when full tip radius tools are used, each of these cutting methods produces a fillet contour with individual characteristics. Generated spline fillets are curves related to the prolate epicycloid for external splines and the prolate hypocycloid for internal splines. These fillets have a minimum radius of curvature at the point where the fillet is tangent to the external spline minor diameter circle or the internal spline major diameter circle and a rapidly increasing radius of curvature up to the point where the fillet comes tangent to the involute profile.

Chamfers and Corner Clearance: In major diameter fits, it is always necessary to provide corner clearance at the major diameter of the spline coupling. This is usually effected by providing a chamfer on the top corners of the external member. This method may not be possible or feasible because:

(a) If the external member is roll formed by plastic deformation, a chamfer cannot be provided by the process.

(b) A semitopping cutter may not be available.

(c) When cutting external splines with small numbers of teeth, a semitopping cutter may reduce the width of the top land to a prohibitive point.

In such cases the corner clearance can be provided on the internal spline as shown in Fig. 2.

When this option is used, the form diameter may fall in the protuberance area.

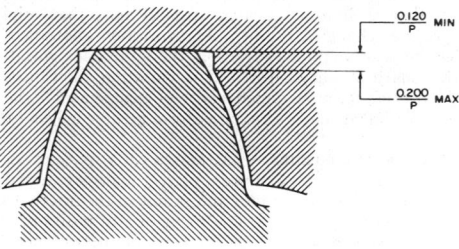

Fig. 2. Internal corner clearance

Spline Variations. — The maximum allowable variations for involute splines are listed in Table 4.

Profile Variation: The reference profile, from which variations occur, passes through the point which is used to determine the actual space width or tooth thickness. This is either the pitch point or the contact point of the standard measuring pins.

Profile variation is positive in the direction of the space and negative in the direction of the tooth. Profile variations which may occur at any point on the profile for establishing effective fits are shown in Table 4.

Lead Variations: The lead tolerance for the total spline length applies also to any portion thereof unless otherwise specified.

Out of Roundness: This condition may appear merely as a result of index and profile variations given in Table 4 and requires no further allowance. However, heat treatment and deflection of thin sections may cause out of roundness, which increases index and profile variations. Tolerances for such conditions depend on many variables and are therefore not tabulated. Additional tooth and/or space width tolerance must allow for such conditions.

Eccentricity: Eccentricity of major and minor diameters in relation to the effective diameter of side fit splines, should not cause contact beyond the form diameters of the mating splines, even under conditions of maximum effective clearance. This standard does not establish specific tolerances.

Eccentricity of major diameters in relation to the effective diameters of major diameter fit splines should be absorbed within the maximum material limits established by the tolerances on major diameter and effective space width or effective tooth thickness.

If the alignment of mating splines is affected by eccentricity of locating surfaces relative to each other and/or the splines, it may be necessary to decrease the effective and actual tooth thickness of the external splines in order to maintain the desired fit condition. This standard does not include allowances for eccentric location.

Effect of Spline Variations. — These can be classified as index variations, profile variations and lead variations.

Index Variations: These variations cause the clearance to vary from one set of mating tooth sides to another. Since the fit depends on the areas with minimum clearance, index variations reduce the effective clearance.

Profile Variations: Positive profile variations affect the fit by reducing effective clearance. Negative profile variations do not affect the fit but reduce the contact area.

Lead variations: These variations will cause clearance variations and therefore reduce the effective clearance.

Variation Allowance: The effect of individual spline variations on the fit (effective variation) is less than their total, because areas of more than minimum clearance can be altered without changing the fit. The variation allowance is 60 per cent of the sum of twice the positive profile variation, the total index variation and the lead variation for the length of engagement. The variation allowances in Table 4 are based on a lead variation for an assumed length of engagement equal to one-half the pitch diameter. Adjustment may be required for a greater length of engagement.

Effective and Actual Dimensions. — Although each space of an internal spline may have the same width as each tooth of a perfect mating external spline, the two may not fit because of variations of index and profile in the internal spline. In such a case, to allow the perfect external spline to fit in any position, all spaces of the internal spline must be widened by the amount of interference. The resulting

width of these tooth spaces is the *actual* space width of the internal spline. The *effective* space width is the tooth thickness of the perfect mating external spline. The same reasoning applied to an external spline which has variations of index and profile when mated with a perfect internal spline leads to the concept of effective tooth thickness which exceeds the actual tooth thickness by the effective variation.

The effective space width of the internal spline minus the effective tooth thickness of the external spline is the effective clearance. This defines the fit of the mating parts. (This is strictly true only if high points of mating parts come into contact.) Positive effective clearance represents looseness or backlash. Negative effective clearance represents tightness or interference.

Space Width and Tooth Thickness Limits. — The variation of actual space width and actual tooth thickness within the machining tolerance causes corresponding variations of effective dimensions, so that there are four limit dimensions for each component part.

These are shown diagrammatically in Table 5.

The minimum effective space width is always basic. The maximum effective tooth thickness is the same as the minimum effective space width except for the major diameter fit. The major diameter fit maximum effective tooth thickness is less than the minimum effective space width by an amount which allows for eccentricity between the effective spline and the major diameter. The permissible variation of the effective clearance is divided between the internal and external splines to arrive at the maximum effective space width and the minimum effective tooth thickness. Limits of the actual space width and actual tooth thickness are constructed from suitable variation allowances.

Use of Effective and Actual Dimensions. — Each of the four dimensions for space width and tooth thickness shown in Table 5, has a definite function.

Minimum Effective Space Width and Maximum Effective Tooth Thickness: These dimensions control the minimum effective clearance, and must always be specified.

Minimum Actual Space Width and Maximum Actual Tooth Thickness: These dimensions cannot be used for acceptance or rejection of parts. If the actual space width is less than the minimum without causing the effective space width to be undersized, or if the actual tooth thickness is more than the maximum without causing the effective tooth thickness to be oversized, the effective variation is less than anticipated; such parts are desirable and not defective. The specification of these actual dimensions as processing reference dimensions is optional. They are also used to analyze undersize effective space width or oversize effective tooth thickness conditions to determine whether or not these conditions are caused by excessive effective variation.

Maximum Actual Space Width and Minimum Actual Tooth Thickness: These dimensions control machining tolerance and also limit the effective variation. The spread between these dimensions, reduced by the effective variation of the internal and external spline, is the maximum effective clearance. Where the effective variation obtained in machining is appreciably less than the variation allowance, these dimensions must be adjusted in order to maintain the desired fit.

Maximum Effective Space Width and Minimum Effective Tooth Thickness: These dimensions define the maximum effective clearance but they do not limit the effective variation. They may be used, in addition to the maximum actual space width and minimum actual tooth thickness, to prevent the increase of maximum effective clearance due to reduction of effective variations. The notation "inspection optional" may be added where maximum effective clearance is an assembly requirement, but does not need absolute control. It will indicate, without necessarily

adding inspection time and equipment, that the actual space width of the internal spline must be held below the maximum, or the actual tooth thickness of the external spline above the minimum, if machining methods result in less than the allowable variations. Where effective variation needs no control or is controlled by laboratory inspection, these limits may be substituted for maximum actual space width and minimum actual tooth thickness.

Table 5.   Specification Guide for Space Width and Tooth Thickness
(ANSI B92.1-1970)

Dimension of Variations, Clearances, and Tolerances on Part				Dimensioning Method		
				Standard	Alternatives	
Dimension	Effective		Actual		A	B
Space Width of Internal Spline		c_v Max	Max	Required	Required	Ref.
	Max		Min	Ref.	Ref.	Ref.
				Ref.	Required	Required
(Basic)	$\frac{\pi}{2P}$ Min			Required	Required	Required
	Max					
Tooth Thickness of External Spline	Min		Max	Ref.	Required	Required
	c_v Min = 0			Ref.	Ref.	Ref.
			Min	Required	Required	Ref.

Note: The minimum effective clearance, c_v min, is greater than zero for major diameter fits. The maximum effective tooth thickness is less than the minimum effective space width for major diameter fits to allow for eccentricity variations.

Combinations of Involute Spline Types. — Flat root side fit internal splines may be used with fillet root external splines where the larger radius is desired on the external spline for control of stress concentrations. This combination of fits may also be permitted as a design option by specifying for the minimum root diameter of the external, the value of the minimum root diameter of the fillet root external spline and noting this as "optional root."

A design option may also be permitted to provide either flat root internal or fillet root internal by specifying for the maximum major diameter, the value of the maximum major diameter of the fillet root internal spline and noting this as "optional root."

Interchangeability. — Splines made to this standard may interchange with splines made to older standards. Exceptions are listed below.

External Splines: These will mate with older internal splines as follows:

Year	Major Dia. Fit	Flat Root Side Fit	Fillet Root Side Fit
1946	Yes	No (A)[c]	No (A)
1950[a]	Yes (B)	Yes (B)	Yes(C)
1950[b]	Yes (B)	No (A)	Yes (C)
1957 SAE	Yes	No (A)	Yes (C)
1960	Yes	No (A)	Yes (C)

[a]Full dedendum.
[b]Short dedendum.
[c]For exceptions A, B, C, see paragraph below.

Internal Splines: These will mate with older external splines as follows:

Year	Major Dia. Fit	Flat Root Side Fit	Fillet Root Side Fit
1946	No (D)[a]	No (E)	No (D)
1950	Yes (F)	Yes	Yes (C)
1957 SAE	Yes (G)	Yes	Yes
1960	Yes (G)	Yes	Yes

[a] For exceptions C, D, E, F, G, see paragraph below.

Exceptions:

A. The external major diameter, unless chamfered or reduced, may interfere with the internal form diameter on flat root side fit splines. Internal splines made to the 1957 and 1960 standards had the same dimensions as shown for the major diameter fit splines in this standard.

B. For 15 teeth or less, the minor diameter of the internal spline, unless chamfered, will interfere with the form diameter of the external spline.

C. For 9 teeth or less, the minor diameter of the internal spline, unless chamfered, will interfere with form diameter of the external spline.

D. The internal minor diameter, unless chamfered, will interfere with the external form diameter.

E. The internal minor diameter, unless chamfered, will interfere with the external form diameter.

F. For 10 teeth or less, the minimum chamfer on the major diameter of the external spline may not clear the internal form diameter.

G. Depending upon the pitch of the spline, the minimum chamfer on the major diameter may not clear the internal form diameter.

Drawing Data. — It is important that uniform specifications be used to show complete information on detail drawings of splines. Much misunderstanding will be avoided by following the suggested arrangement of dimensions and data as given in Table 6. The number of x's indicates the number of decimal places normally used. With this tabulated type of spline specifications, it is usually not necessary to show a graphic illustration of the spline teeth.

Spline Data and Reference Dimensions. — Spline data are used for engineering and manufacturing purposes. Pitch and pressure angle are not subject to individual inspection.

As used in this standard, *reference* is an added notation or modifier to a dimension, specification, or note when that dimension, specification, or note is:

1. Repeated for drawing clarification.
2. Needed to define a nonfeature datum or basis from which a form or feature is generated.
3. Needed to define a nonfeature dimension from which other specifications or dimensions are developed.
4. Needed to define a nonfeature dimension at which toleranced sizes of a feature are specified.
5. Needed to define a nonfeature dimension from which control tolerances or sizes are developed or added as useful information.

Any dimension, specification, or note that is noted "REF" should not be used as a criterion for part acceptance or rejection.

Table 6. Spline Terms, Symbols and Drawing Data, 30-Degree Pressure Angle, Flat Root Side Fit (ANSI B92.1-1970)

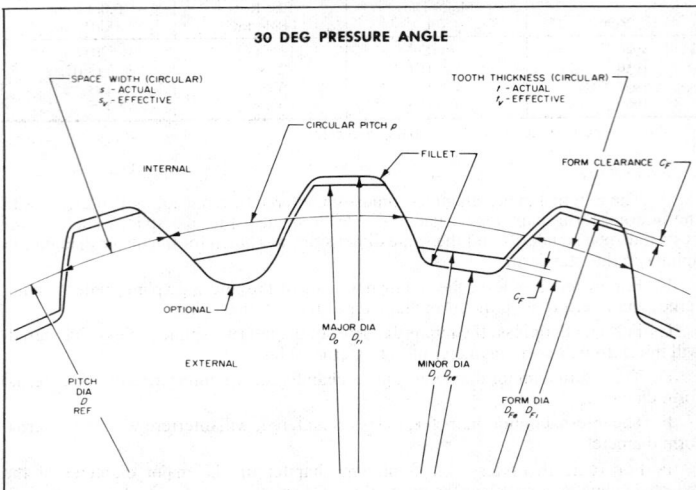

This fit is used in restricted areas (as with tubular parts with wall thickness too small to permit use of fillet roots, and to allow hobbing closer to shoulders, etc.) and for economy (when hobbing, shaping, etc. and shorter broaches for the internal member).

Press fits are not tabulated because their design depends on the degree of tightness desired and must allow for such factors as the shape of the blank, wall thickness, material, hardness, thermal expansion, etc. Close tolerances or selective size grouping may be required to limit fit variations.

DRAWING DATA

INTERNAL INVOLUTE SPLINE DATA		EXTERNAL INVOLUTE SPLINE DATA	
FLAT ROOT SIDE FIT		FLAT ROOT SIDE FIT	
NUMBER OF TEETH	xx	NUMBER OF TEETH	xx
PITCH	xx/xx	PITCH	xx/xx
PRESSURE ANGLE	30°	PRESSURE ANGLE	30°
BASE DIAMETER	x.xxxxxx REF	BASE DIAMETER	x.xxxxxx REF
PITCH DIAMETER	x.xxxxxx REF	PITCH DIAMETER	x.xxxxxx REF
MAJOR DIAMETER	x.xxx max	MAJOR DIAMETER	x.xxx/x.xxx
FORM DIAMETER	x.xxx	FORM DIAMETER	x.xxx
MINOR DIAMETER	x.xxx/x.xxx	MINOR DIAMETER	x.xxx min
CIRCULAR SPACE WIDTH		CIRCULAR TOOTH THICKNESS	
MAX ACTUAL	x.xxxx	MAX EFFECTIVE	x.xxxx
MIN EFFECTIVE	x.xxxx	MIN ACTUAL	x.xxxx
The following information may be added as required:		The following information may be added as required:	
MAX MEASUREMENT BETWEEN PINS	x.xxxx REF	MIN MEASUREMENT OVER PINS	x.xxxx REF
PIN DIAMETER	x.xxxx	PIN DIAMETER	x.xxxx

The above drawing data is required for the spline specifications.   The standard system is shown; for alternate systems, see Table 5.   Number of x's indicates number of decimal places normally used.

Estimating Key and Spline Sizes and Lengths. — Figure 1 may be used to estimate the size of American Standard involute splines required to transmit a given torque. It also may be used to find the outside diameter of shafts used with single keys. After the size of the shaft is found, the proportions of the key can be determined from Table 2 on page 2236.

Curve A is for flexible splines with teeth hardened to Rockwell C 55–65. For these splines, lengths are generally made equal to or somewhat greater than the pitch diameter for diameters below 1¼ inches; on larger diameters, the length is generally one-third to two-thirds the pitch diameter. Curve A also applies for a single key used as a fixed coupling, the length of the key being one to one and one-quarter times the shaft diameter. The stress in the shaft, neglecting stress concentration at the keyway, is about 7500 pounds per square inch. For the effect of keyways on shaft strength, see pages 312 and 313.

Curve B represents high-capacity single keys used as fixed couplings for stresses of 9500 pounds per square inch, neglecting stress concentration. Key-length is one to one and one-quarter times shaft diameter and both shaft and key are of moderately hard heat-treated steel. This type of connection is commonly used to key commercial flexible couplings to motor or generator shafts.

Curve C is for multiple-key fixed splines with lengths of three-quarters to one and one-quarter times pitch diameter and shaft hardness of 200–300 BHN.

Curve D is for high-capacity splines with lengths one-half to one times the pitch diameter. Hardnesses up to Rockwell C 58 are common and in aircraft applications the shaft is generally hollow to reduce weight.

Curve E represents a solid shaft with 65,000 pounds per square inch shear stress. For hollow shafts with inside diameter equal to three-quarters of the outside diameter the shear stress would be 95,000 pounds per square inch.

Length of Splines: Fixed splines with lengths of one-third the pitch diameter will have the same shear strength as the shaft, assuming uniform loading of the teeth; however, errors in spacing of teeth result in only half the teeth being fully loaded. Therefore, for balanced strength of teeth and shaft the length should be two-thirds the pitch diameter. If weight is not important, however, this may be increased to equal the pitch diameter. In the case of flexible splines, long lengths do not contribute to load carrying capacity when there is misalignment to be accommodated. Maximum effective length for flexible splines may be approximated from Fig. 2.

Formulas for Torque Capacity of Involute Splines. — The formulas for torque capacity of 30-degree involute splines given in the following paragraphs are derived largely from an article "When Splines Need Stress Control" by D. W. Dudley, *Product Engineering,* Dec. 23, 1957.

In the formulas that follow the symbols used are as defined on page 2021 with the following additions: D_h = inside diameter of hollow shaft, inches; K_a = application factor from Table 1; K_m = load distribution factor from Table 2; K_f = fatigue life factor from Table 3; K_w = wear life factor from Table 4; L_e = maximum effective length from Fig. 2, to be used in stress formulas even though the actual length may be greater; T = transmitted torque, pound-inches.

Definitions: A *fixed* spline is one which is either shrink fitted or loosely fitted but piloted with rings at each end to prevent rocking of the spline which results in small axial movements that cause wear. A *flexible* spline permits some rocking motion such as occurs when the shafts are not perfectly aligned. This flexing or rocking motion causes axial movement and consequently wear of the teeth. Straight-toothed flexible splines can accommodate only small angular misalignments (less than 1 deg.) before wear becomes a serious problem. For greater amounts of misalignment (up to about 5 deg.), crowned splines are preferable to reduce wear and end-loading of the teeth.

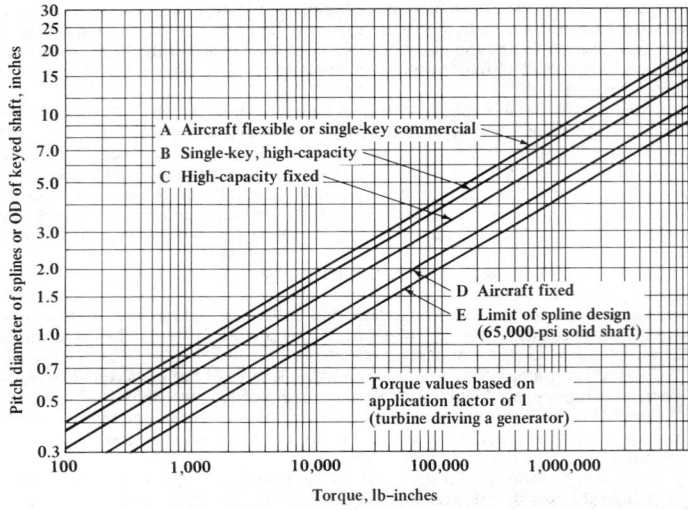

Fig. 1. Chart for Estimating Involute Spline Size Based on Diameter-Torque Relationships

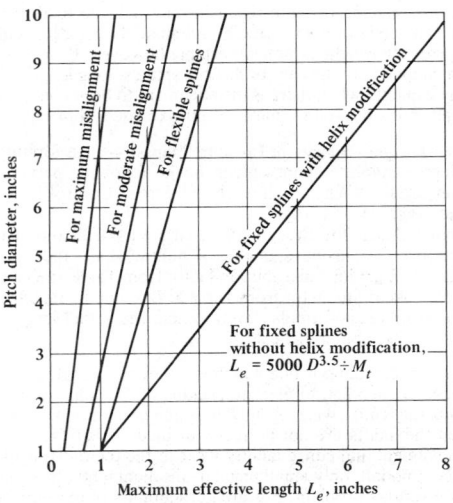

Fig. 2. Maximum Effective Length for Fixed and Flexible Splines

Table 1. Spline Application Factors, K_a

Power Source	Type of Load			
	Uniform (Generator-Fan)	Light Shock (Oscillating Pumps, etc.)	Intermittent Shock (Actuating Pumps, etc.)	Heavy Shock (Punches, Shears, etc.)
	Application Factor, K_a			
Uniform (Turbine, Motor)	1.0	1.2	1.5	1.8
Light Shock (Hydraulic Motor)	1.2	1.3	1.8	2.1
Medium Shock (Internal Combustion, Engine)	2.0	2.2	2.4	2.8

Table 2. Load Distribution Factors, K_m, for Misalignment of Flexible Splines*

Misalignment, inches per inch	Load Distribution Factor, K_m*			
	½-in. Face Width	1-in. Face Width	2-in. Face Width	4-in. Face Width
0.001	1	1	1	1½
0.002	1	1	1½	2
0.004	1	1½	2	2½
0.008	1½	2	2½	3

* For fixed splines, $K_m = 1$.

Table 3. Fatigue-Life Factors, K_f, for Splines

No. of Torque Cycles*	Fatigue-Life Factor, K_f	
	Unidirectional	Fully-reversed
1,000	1.8	1.8
10,000	1.0	1.0
100,000	0.5	0.4
1,000,000	0.4	0.3
10,000,000	0.3	0.2

* A torque cycle consists of one start and one stop, not the number of revolutions.

Shear Stress Under Roots of External Teeth: For a transmitted torque T, the torsional shear stress induced in the shaft under the root diameter of an external spline is:

$$S_s = \frac{16TK_a}{\pi D_{re}^3 K_f} \qquad \text{for a solid shaft} \qquad (1)$$

$$S_s = \frac{16TD_{re} K_a}{\pi (D_{re}^4 - D_h^4) K_f} \qquad \text{for a hollow shaft} \qquad (2)$$

The computed stress should not exceed the values in Table 5.

Table 4. Wear Life Factors, K_w, for Flexible Splines*

Number of Revolutions of Spline	Life Factor, K_w	Number of Revolutions of Spline	Life Factor, K_w
10,000	4.0	100,000,000	1.0
100,000	2.8	1,000,000,000	0.7
1,000,000	2.0	10,000,000,000	0.5
10,000,000	1.4	. . .	. . .

* Wear life factors, unlike fatigue life factors given in Table 3, are based on the total number of revolutions of the spline, since each revolution of a flexible spline results in a complete cycle of rocking motion which contributes to spline wear.

Table 5. Allowable Shear Stresses for Splines

Material	Hardness		Max. Allowable Shear Stress, psi
	Brinell	Rockwell C	
Steel	160–200	—	20,000
Steel	230–260	—	30,000
Steel	302–351	33–38	40,000
Surface-hardened Steel	—	48–53	40,000
Case-hardened Steel	—	58–63	50,000
Through-hardened Steel (Aircraft Quality)	—	42–46	45,000

Shear Stress at the Pitch Diameter of Teeth: The shear stress at the pitch line of the teeth for a transmitted torque T is:

$$S_s = \frac{4TK_aK_m}{DNL_etK_f} \qquad (3)$$

The factor of 4 in (3) assumes that only half the teeth will carry the load because of spacing errors. For poor manufacturing accuracies, change the factor to 6.

The computed stress should not exceed the values in Table 5.

Compressive Stresses on Sides of Spline Teeth: Allowable compressive stresses on splines are very much lower than for gear teeth since non-uniform load distribution and misalignment result in unequal load sharing and end loading of the teeth.

$$\text{For flexible splines, } S_c = \frac{2TK_mK_a}{DNL_ehK_w} \qquad (4)$$

$$\text{For fixed splines, } S_c = \frac{2TK_mK_a}{9DNL_ehK_f} \qquad (5)$$

In these formulas, h is the depth of engagement of the teeth, which for flat root splines is $0.9/P$ and for fillet root splines is $1/P$, approximately.

The stresses computed from Formulas (4) and (5) should not exceed the values in Table 6.

Table 6. Allowable Compressive Stresses for Splines

Material	Hardness		Max. Allowable Compressive Stress, psi	
	Brinell	Rockwell C	Straight	Crowned
Steel	160–200	—	1,500	6,000
Steel	230–260	—	2,000	8,000
Steel	302–351	33–38	3,000	12,000
Surface-hardened Steel	—	48–53	4,000	16,000
Case-hardened Steel	—	58–63	5,000	20,000

Bursting Stresses on Splines: Internal splines may burst due to three kinds of tensile stress: (1) tensile stress due to the radial component of the transmitted load; (2) centrifugal tensile stress; and (3) tensile stress due to the tangential force at the pitch line causing bending of the teeth.

$$\text{Radial load tensile stress, } S_1 = \frac{T \tan \phi}{\pi D t_w L} \tag{6}$$

where t_w = wall thickness of internal spline = outside diameter of spline sleeve minus spline major diameter, all divided by 2. L = full length of spline.

$$\text{Centrifugal tensile stress, } S_2 = \frac{1.656 \times (\text{rpm})^2 \, (D_{oi}^2 + 0.212 D_{ri}^2)}{1,000,000} \tag{7}$$

Where D_{oi} = outside diameter of spline sleeve.

$$\text{Beam loading tensile stress, } S_3 = \frac{4T}{D^2 L_e Y} \tag{8}$$

In this equation, Y is the Lewis form factor obtained from a tooth layout. For internal splines of 30-deg. pressure angle a value of $Y = 1.5$ is a satisfactory estimate. The factor 4 in (8) assumes that only half the teeth are carrying the load.

Table 7. Allowable Tensile Stresses for Splines

Material	Hardness		Max. Allowable Stress, psi
	Brinell	Rockwell C	
Steel	160–200	—	22,000
Steel	230–260	—	32,000
Steel	302–351	33–38	45,000
Surface-hardened Steel	—	48–53	45,000
Case-hardened Steel	—	58–63	55,000
Through-hardened Steel	—	42–46	50,000

The total tensile stress tending to burst the rim of the external member is:
$S_t = [K_a K_m (S_1 + S_3) + S_2]/K_f$; and should be less than those in Table 7.

Crowned Splines for Large Misalignments. — As mentioned on page 2031, crowned splines can accommodate misalignments of up to about 5 degrees. Crowned splines

have considerably less capacity than straight splines of the same size if both are operating with precise alignment. However, when large misalignments exist, the crowned spline has greater capacity.

American Standard tooth forms may be used for crowned external members so that they may be mated with straight internal members of Standard form.

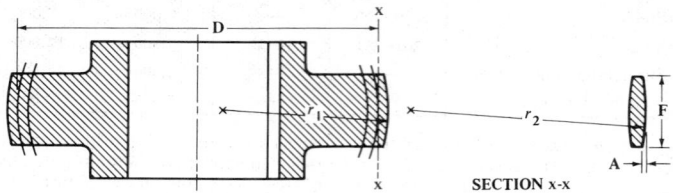

The accompanying diagram of a crowned spline shows the radius of the crown r_1; the radius of curvature of the crowned tooth, r_2; the pitch diameter of the spline, D; the face width, F; and the relief or crown height A at the ends of the teeth. The crown height A should always be made somewhat greater than one-half the face width multiplied by the tangent of the misalignment angle. For a crown height A, the approximate radius of curvature r_2 is $F^2 \div 8A$, and $r_1 = r_2 \tan \phi$, where ϕ is the pressure angle of the spline.

For a torque T, the compressive stress on the teeth is:

$$S_c = 2290 \sqrt{2T \div DNhr_2} \; ;$$

and should be less than the value in Table 6.

Fretting Damage to Splines and Other Machine Elements. — Fretting is wear that occurs when cyclic loading, such as vibration, causes two surfaces in intimate contact to undergo small oscillatory motions with respect to each other. During fretting, high points or asperities of the mating surfaces adhere to each other and small particles are pulled out, leaving minute, shallow pits and a powdery debris. In steel parts exposed to air, the metallic debris oxidizes rapidly and forms a red, rustlike powder or sludge; hence, the coined designation "fretting corrosion."

Fretting is mechanical in origin and has been observed in most materials, including those that do not oxidize, such as gold, platinum, and nonmetallics; hence, the corrosion accompanying fretting of steel parts is a secondary factor.

Fretting can occur in the operation of machinery subject to motion or vibration or both. It can destroy close fits; the debris may clog moving parts; and fatigue failure may be accelerated because stress levels to initiate fatigue in fretted parts are much lower than for undamaged material. Sites for fretting damage include interference fits; splined, bolted, keyed, pinned, and riveted joints; between wires in wire rope; flexible shafts and tubes; between leaves in leaf springs; friction clamps; small amplitude-of-oscillation bearings; and electrical contacts.

Vibration or cyclic loadings are the main causes of fretting. If these factors cannot be eliminated, greater clamping force may reduce movement but, if not effective, may actually worsen the damage. Lubrication may delay the onset of damage; hard plating or surface hardening methods may be effective, not by reducing fretting, but by increasing the fatigue strength of the material. Plating soft materials having inherent lubricity onto contacting surfaces is effective until the plating wears through.

Involute Spline Inspection Methods. — Spline gages are used for routine inspection of production parts.

Analytical inspection, which is the measurement of individual dimensions and variations, may be required:

A. To supplement inspection by gages, for example, where NOT GO composite gages are used in place of NOT GO sector gages and variations must be controlled.

B. To evaluate parts rejected by gages.

C. For prototype parts or short runs where spline gages are not used.

D. To supplement inspection by gages where each individual variation must be restrained from assuming too great a portion of the tolerance between the minimum material actual and the maximum material effective dimensions.

Inspection with Gages. — A variety of gages is used in the inspection of involute splines.

Types of Gages: A composite spline gage is a gage having a full complement of teeth. A sector spline gage is a gage having two diametrically opposite groups of teeth. A sector plug gage with only two teeth per sector is also known as a "paddle gage." A sector ring gage with only two teeth per sector is also known as a "snap ring gage." A progressive gage is a gage consisting of two or more adjacent sections with different inspection functions. Progressive GO gages are physical combinations of GO gage members which check consecutively first one feature or one group of features, then their relationship to other features. GO and NOT GO gages may also be combined physically to form a progressive gage.

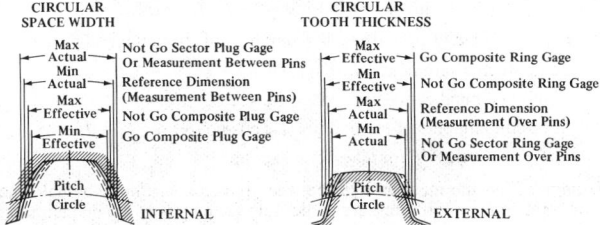

Fig. 3. Space width and tooth-thickness inspection.

GO and NOT GO Gages: GO gages are used to inspect maximum material conditions (maximum external, minimum internal dimensions). They may be used to inspect an individual dimension or the relationship between two or more functional dimensions. They control the minimum looseness or maximum interference.

NOT GO gages are used to inspect minimum material conditions (minimum external, maximum internal dimensions), thereby controlling the maximum looseness or minimum interference. Unless otherwise agreed upon, a product is acceptable only if the NOT GO gage does not enter or go on the part. A NOT GO gage can be used to inspect only one dimension. An attempt at simultaneous NOT GO inspection of more than one dimension could result in failure of such a gage to enter or go on (acceptance of part), even though all but one of the dimensions were outside product limits. In the event all dimensions are outside the limits, their relationship could be such as to allow acceptance.

Effective and Actual Dimensions: The effective space width and tooth thickness are inspected by means of an accurate mating member in the form of a composite spline gage.

The actual space width and tooth thickness are inspected with sector plug and ring gages, or by measurements with pins.

Measurements with Pins. — The actual space width of internal splines, and the actual tooth thickness of external splines, may be measured with pins. These measurements do not determine the fit between mating parts, but may be used as part of the analytic inspection of splines to evaluate the effective space width or effective tooth thickness by approximation.

Formulas for 2-Pin Measurement: For measurement *between* pins of internal splines using the symbols given on page 2021:

1. Find involute of pressure angle at pin center:

$$\text{inv } \phi_i = s/D + \text{inv } \phi_d - d_i/D_b$$

2. Find the value of ϕ_i, in degrees, in the table beginning on page 82. Find sec ϕ_i in the same table, using interpolation to obtain higher accuracy.

3. Compute measurement, M_i, between pins:

For even numbers of teeth: $M_i = D_b \text{ sec } \phi_i - d_i$

For odd numbers of teeth: $M_i = (D_b \cos 90°/N) \text{ sec } \phi_i - d_i$

where: $d_i = 1.7280/P$ for 30° and 37.5° standard pressure angle (ϕ_D) splines;

$d_i = 1.9200/P$ for 45° pressure angle splines.

For measurement *over* pins of external splines:

1a. Find involute of pressure angle at pin center:

$$\text{inv } \phi_e = t/D + \text{inv } \phi_D + d_e/D_b - \pi/N$$

2a. Find the value of ϕ_e and sec ϕ_e from the table beginning on page 82.

3a. Compute measurement, M_e, over pins:

For even numbers of teeth: $M_e = D_b \text{ sec } \phi_e + d_e$

For odd numbers of teeth: $M_e = (D_b \cos 90°/N) \text{ sec } \phi_e + d_e$

where $d_e = 1.9200/P$ for all external splines.

Example: Find the measurement between pins for *maximum* actual space width of an internal spline of 30° pressure angle, tolerance class 4, 3/6 diametral pitch, and 20 teeth.

The maximum actual space width to be substituted for s in Step 1 above is obtained as follows: In Table 5, page 2028, the maximum actual space width is the sum of the minimum effective space width (second column) and $\lambda + m$ (third column). The minimum effective space width s_v from Table 2, page 2022, is $\pi/2P = \pi/(2 \times 3)$. The values of λ and m from Table 4, page 2024, are, for a class 4 fit, 3/6 diametral pitch, 20-tooth spline: $\lambda = 0.0027 \times 0.71 = 0.00192$; and $m = 0.00176 \times 0.71 = 0.00125$, so that $s = 0.52360 + 0.00192 + 0.00125 = 0.52677$.

Other values required for Step 1 are:

$$D = N/P = 20/3 = 6.66666$$
$$\text{inv } \phi_D = \text{inv } 30° = 0.053751 \text{ from page } 112.$$
$$d_i = 1.7280/3 = 0.57600$$
$$D_b = D \cos \phi_D = 6.66666 \times 0.86603 = 5.77353$$

The computation is made as follows:

1. $\text{inv } \phi_i = 0.52677/6.66666 + 0.053751 - 0.57600/5.77353 = 0.03300$
2. By interpolation on page 107, $\phi_i = 25°46.18'$ and sec $\phi_i = 1.11044$
3. $M_i = 5.77353 \times 1.11044 - 0.57600 = 5.8352$ inches

American National Standard Metric Module Splines. — ANSI B92.2M-1980 is the American National Standards Institute version of the International Standards Organization involute spline standard. It is not a "soft metric" conversion of any previous, inch-based, standard,* and splines made to this hard metric version are not intended for use with components made to the B92.1 or other, previous standards. The ISO 4156 Standard from which this one is derived is the result of a cooperative effort between the ANSI B92 committee and other members of the ISO/TC 32 involute spline committee.

Many of the features of the previous standard, ANSI B92.1-1970, have been retained such as: 30-, 37.5-, and 45-degree pressure angles; flat root and fillet root side fits; the four tolerance classes 4, 5, 6, and 7; tables for a single class of fit; and the effective fit concept.

Among the major differences are: use of modules of from 0.25 through 10 mm in place of diametral pitch; dimensions in millimeters instead of inches; the "basic rack"; removal of the major diameter fit; and use of ISO symbols in place of those used previously. Also, provision is made for calculating three defined clearance fits.

The Standard recognizes that proper assembly between mating splines is dependent only on the spline being within effective specifications from the tip of the tooth to the form diameter. Therefore, the internal spline major diameter is shown as a maximum dimension and the external spline minor diameter is shown as a minimum dimension. The minimum internal major diameter and the maximum external minor diameter must clear the specified form diameter and thus require no additional control. All dimensions are for the finished part; any compensation that must be made for operations which take place during processing, such as heat treatment, must be considered when selecting the tolerance level for manufacturing.

The Standard provides the same internal minimum effective space width and external maximum effective tooth thickness for all tolerance classes. This basic concept makes possible interchangeable assembly between mating splines regardless of the tolerance class of the individual members, and permits a tolerance class "mix" of mating members. This is often an advantage when one member is considerably less difficult to produce than its mate, and the "average" tolerance applied to the two units is such that it satisfies the design need. For example, this can be the result of specifying Class 5 tolerance for one member and Class 7 for its mate, thus providing an assembly tolerance in the Class 6 range.

If a fit provided in this Standard does not satisfy a particular design need, and a specific clearance or press fit is desired, the change shall be made only to the external spline by a reduction of, or an increase in, the effective tooth thickness and a like change in the actual tooth thickness. The minimum effective space width is always basic and this basic width should always be retained when special designs are derived from the concept of this Standard.

Spline Terms and Definitions: The spline terms and definitions given for American National Standard ANSI B92.1-1970, described in the preceding section, may be used in regard to ANSI B92.2M-1980. The 1980 Standard utilizes ISO symbols in place of those used in the 1970 Standard; these differences are shown in Table 1.

Dimensions and Tolerances: Dimensions and tolerances of splines made to the 1980 Standard may be calculated using the formulas given in Table 2. These formulas are for metric module splines in the range of from 0.25 to 10 mm metric module of side-fit design and having pressure angles of 30-, 37.5-, and 45-degrees. The standard modules in the system are: 0.25; 0.5; 0.75; 1; 1.25; 1.5; 1.75; 2; 2.5; 3; 4; 5; 6; 8;

---

* A "soft" conversion is one in which dimensions in inches, when multiplied by 25.4 will, after being appropriately rounded off, provide equivalent dimensions in millimeters. In a "hard" system the tools of production, such as hobs, do not bear a usable relation to the tools in another system; i.e., a 10 diametral pitch hob calculates to be equal to a 2.54 module hob in the metric module system, a hob that does not exist in the metric standard.

and 10. The range of from 0.5 to 10 module applies to all splines except 45-degree fillet root splines; for these, the range of from 0.25 to 2.5 module applies.

Fit Classes: Four classes of side fit splines are provided: spline fit class H/h having a minimum effective clearance, $c_v = es = 0$; classes H/f, H/e, and H/d having tooth thickness modifications, *es*, of f, e, and d, respectively, to provide progressively greater effective clearance c_v. The tooth thickness modifications h, f, e, and d in Table 3 are fundamental deviations selected from ISO R286, "ISO System of Limits and Fits." They are applied to the external spline by shifting the tooth thickness total tolerance below the basic tooth thickness by the amount of the tooth thickness modification to provide a prescribed minimum effective clearance c_v.

Basic Rack Profiles: The basic rack profile for the standard pressure angle splines are shown in Figs. 1a, b, c, and d. The dimensions shown are for maximum material condition and for fit class H/h.

Table 1. Comparison of Symbols Used in ANSI B92.2M-1980 and Those in ANSI B92.1-1970

Symbol			Symbol		
B92.2M	B92.1	Meaning of Symbol	M92.2M	B92.1	Meaning of Symbol
c	...	theoretical clearance	E_{max}	s	max. actual circular space width
c_v	c_v	effective clearance	E_{min}	s	min. actual circular space width
c_F	c_F	form clearance			
D	D	pitch diameter	EV	s_v	effective circular space width
DB	D_b	base diameter			
d_{ce}	D_{ce}	pin contact diameter, external spline	S_{bsc}	t_v max	basic circular tooth thickness
d_{ci}	D_{ci}	internal spline			
DEE	D_o	major diam., ext. spline	S_{max}	t	max. actual circular tooth thick.
DEI	D_{ri}	major diam., int. spline			
DFE	D_{Fe}	form diam., ext. spline	S_{min}	t	min. actual circular tooth thick.
DFI	D_{Fi}	form diam., int. spline			
DIE	D_{re}	minor diam., ext. spline	SV	t_v	effective circular tooth thick.
DII	D_i	minor diam., int. spline			
DRE	d_e	pin diam., ext. spline	α	ϕ	pressure angle
DRI	d_i	pin diam., int. spline	α_D	ϕ_D	standard pressure angle
h_s	...	see Fig. 1a, b, c, and d	α_{ci}	ϕ_{ci}	press. angle at pin contact diameter, internal spline
λ	λ	effective variation			
inv α	...	involute $\alpha = \tan \alpha - \text{arc } \alpha$	α_{ce}	ϕ_{ce}	press. angle at pin contact diameter, external spline
KE	K_e	change factor, ext. spline			
KI	K_i	change factor, int. spline	α_i	ϕ_i	press. angle at pin center, internal spline
g	L	spline length			
g_w	L_a	active spline length	α_e	ϕ_e	press. angle at pin center, external spline
g_γ	L_g	length of engagement			
T	m	machining tolerance	α_{Fe}	ϕ_F	press. angle at form diameter, external spline
MRE	M_e	meas. over 2 pins, ext. spline			
			α_{Fi}	ϕ_F	press. angle at form diameter, internal spline
MRI	M_i	meas. bet. 2 pins, int. spline			
Z	N	number of teeth	es	...	ext. spline cir. tooth thick.
m	...	module			modification for required fit
...	P	diametral pitch			class = c_v min (Table 3)
...	P_s	stub pitch = $2P$	h, f, e,	...	tooth thick. size modifiers
P_b	...	base pitch	or d		(called fundamental deviation in ISO R286), Table 3
p	p	circular pitch			
π	π	3.141592654	H	...	space width size modifier
rfe	r_f	fillet rad., ext. spline			(called fundamental deviation in ISO R286), Table 3
rfi	r_f	fillet rad., int. spline			
E_{bsc}	s_v min	basic circular space width			

Table 2. Formulas for Dimensions and Tolerances for All Fit Classes — Metric Module Involute Splines (ANSI B92.2M-1980)

Term	Symbol	Formula			
		30-Degree Flat Root	30-Degree Fillet Root	37.5-Degree Fillet Root	45-Degree Fillet Root
Pitch Diameter	D	0.5 to 10 module	0.5 to 10 module	0.5 to 10 module	0.25 to 2.5 module
		mZ			
Base Diameter	DB	$mZ \cos \alpha_D$			
Circular Pitch	p	πm			
Base Pitch	p_b	$\pi m \cos \alpha_D$			
Tooth Thick Mod	es	According to selected fit class, H/h, H/f, H/e, or H/d (see Table 3)			
Min Maj Diam, Int	DEI min	$m(Z + 1.5)$	$m(Z + 1.8)$	$m(Z + 1.4)$	$m(Z + 1.2)$
Max Maj Diam, Int	DEI max	DEI min $+ (T + \lambda)/\tan \alpha_D$ (see Note 1)			
Form Diam, Int	DFI	$m(Z + 1) + 2c_F$	$m(Z + 1) + 2c_F$	$m(Z + 0.9) + 2c_F$	$m(Z + 0.8) + 2c_F$
Min Minor Diam, Int	DII min	$DFE + 2c_F$ (see Note 2)			
Max Minor Diam, Int	DII max	DII min $+ (0.2m^{0.667} - 0.01m^{-0.5})*$			
Cir Space Width, Basic	E_{bsc}	$0.5\pi m$			
Min Effective	EV min	$0.5\pi m$			
Max Actual	E max	EV min $+ (T + \lambda)$ for classes 4, 5, 6, and 7 (see Table 4 for $T + \lambda$)			
Min Actual	E min	EV min $+ \lambda$ (see text on page 2043 for λ)			
Max Effective	EV max	E max $- \lambda$ (see text on page 2043 for λ)			
Max Major Diam, Ext	DEE max	$m(Z + 1) - es/\tan \alpha_D$**	$m(Z + 1) - es/\tan \alpha_D$**	$m(Z + 0.9) - es/\tan \alpha_D$**	$m(Z + 0.8) - es/\tan \alpha_D$**
Min Major Diam, Ext	DEE min	DEE max $- (0.2m^{0.667} - 0.01m^{-0.5})*$			
Form Diam, External	DFE	$2 \times \sqrt{(0.5DB)^2 + \left[0.5D \sin \alpha_D - \dfrac{h_s + \left(\dfrac{0.5\ es}{\tan \alpha_D}\right)}{\sin \alpha_D}\right]^2}$			

See footnotes at end of table.

Table 2. *(Concluded)* **Formulas for Dimensions and Tolerances for All Fit Classes — Metric Module Involute Splines (ANSI B92.2M-1980)**

Term	Symbol	Formula			
		30-Degree Flat Root	30-Degree Fillet Root	37.5-Degree Fillet Root	45-Degree Fillet Root
Max Minor Diam, Ext	DIE max	0.5 to 10 module	0.5 to 10 module	0.5 to 10 module	0.25 to 2.5 module
		$m(Z - 1.5) - es/\tan \alpha_D$**	$m(Z - 1.8) - es/\tan \alpha_D$**	$m(Z - 1.4) - es/\tan \alpha_D$**	$m(Z - 1.2) - es/\tan \alpha_D$**
Min Minor Diam, Ext	DIE min	DIE max $- (T + \lambda)/\tan \alpha_D$ (see Note 1)			
Cir Tooth Thick, Basic	S_{bsc}	$0.5\pi m$			
Max Effective	SV max	$S_{bsc} - es$			
Min Actual	S min	SV max $- (T + \lambda)$ for classes 4, 5, 6, and 7 (see Table 4 for $T + \lambda$)			
Max Actual	S max	SV max $- \lambda$ (see text on page 2043 for λ)			
Min Effective	SV min	S min $+ \lambda$ (see text on page 2043 for λ)			
Total Tolerance on Circular Space Width or Tooth Thickness	$(T + \lambda)$	$(T + \lambda)$ from Table 4	See formulas in Table 4		
Machining Tolerance on Circular Space Width or Tooth Thickness	T	$T = (T + \lambda)$ from Table 4 from text on page 2043.			
Effective Variation Allowed on Circular Space Width or Tooth Thickness	λ	See text on page 2043.			
Form Clearance	c_F	0.1m			
Rack Dimension	h_s	0.6m (see Fig. 1a)	0.6m (see Fig. 1b)	0.55m (see Fig. 1c)	0.5m (see Fig. 1d)

Note 1: Use $(T + \lambda)$ for class 7 from Table 4.

Note 2: For all types of fit, always use the DFE value corresponding to the H/h fit.

* Values of $(0.2m^{0.667} - 0.01m^{-0.5})$ are as follows: for 10 module, 0.93; for 8 module, 0.80; for 6 module, 0.66; for 5 module, 0.58; for 4 module, 0.50; for 3 module, 0.41; for 2.5 module, 0.36; for 2 module, 0.31; for 1.75 module, 0.28; for 1.5 module, 0.25; for 1.25 module, 0.22; for 1 module, 0.19; for 0.75 module, 0.15; for 0.5 module, 0.11; and for 0.25 module, 0.06.

** See Table 6 for values of $es/\tan \alpha_D$.

**Table 3. Tooth Thickness Modification, *es*,
for Selected Spline Fit Classes**

Pitch Diameter in mm, D	External Splines*				Pitch Diameter in mm, D	External Splines*			
	Selected Fit Class					Selected Fit Class			
	d	e	f	h		d	e	f	h
	Tooth Thickness Modification (Reduction) Relative to Basic Tooth Thickness at Pitch Diameter, *es*, in mm					Tooth Thickness Modification (Reduction) Relative to Basic Tooth Thickness at Pitch Diameter, *es*, in mm			
≤ 3	0.020	0.014	0.006	0	> 120 to 180	0.145	0.085	0.043	0
> 3 to 6	0.030	0.020	0.010	0	> 180 to 250	0.170	0.100	0.050	0
> 6 to 10	**0.040**	0.025	0.013	0	> 250 to 315	0.190	0.110	0.056	0
> 10 to 18	0.050	0.032	0.016	0	> 315 to 400	0.210	**0.125**	**0.062**	0
> 18 to 30	0.065	0.040	0.020	0	> 400 to 500	0.230	0.135	**0.068**	0
> 30 to 50	0.080	0.050	0.025	0	> 500 to 630	0.260	0.145	**0.076**	0
> 50 to 80	0.100	0.060	0.030	0	> 630 to 800	0.290	0.160	**0.080**	0
> 80 to 120	0.120	**0.072**	0.036	0	> 800 to 1000	0.320	**0.170**	**0.086**	0

* Internal splines are fit class H and have space width modification from basic space width equal to zero; thus, an H/h fit class has effective clearance $c_v = 0$.

Note: The values listed in this table are taken from ISO R286 and have been computed on the basis of the geometrical mean of the size ranges shown. Values in **boldface** type do not comply with any documented rule for rounding but are those used by ISO R286; they are used in this table to comply with established international practice.

**Table 4. Space Width and Tooth Thickness Total Tolerance,
$(T + \lambda)$, in Millimeters**

Spline Tolerance Class	Formula for Total Tolerance, $(T + \lambda)$	Spline Tolerance Class	Formula for Total Tolerance, $(T + \lambda)$	In these formulas, i* and i** are tolerance units based upon pitch diameter and tooth thickness, respectively:
4	10i* + 40i**	6	25i* + 100i**	$i* = 0.001 (0.45 \sqrt[3]{D} + 0.001D)$ for D≤500 mm
5	16i* + 64i**	7	40i* + 160i**	$= 0.001 (0.004D + 2.1)$ for $D > 500$ mm
				$i** = 0.001 (0.45 \sqrt[3]{S_{bsc}} + 0.001S_{bsc})$

Spline Machining Tolerances and Variations. — The total tolerance $(T + \lambda)$, Table 4, is the sum of Effective Variation, λ, and a Machining Tolerance, T.

Effective Variation: The effective variation, λ, is the combined effect that total index variation, positive profile variation, and tooth alignment variation has on the effective fit of mating involute splines. The effect of these individual variations is less than the sum of these allowable variations because areas of more than minimum clearance can have profile, tooth alignment, or index variations without changing the fit. It is also unlikely that these variations would occur in their maximum amounts simultaneously on the same spline. For this reason, total index variation,

Table 5. Formulas for F_p, f_f, and F_β used to calculate λ

Spline Tolerance Class	Total Index Variation, in mm, F_p	Total Profile Variation, in mm, f_f	Total Lead Variation, in mm, F_β
4	$0.001(2.5\sqrt{mZ\pi/2} + 6.3)$	$0.001[1.6m(1 + 0.0125Z) + 10]$	$0.001(0.8\sqrt{g} + 4)$
5	$0.001(3.55\sqrt{mZ\pi/2} + 9)$	$0.001[2.5m(1 + 0.0125Z) + 16]$	$0.001(1.0\sqrt{g} + 5)$
6	$0.001(5\sqrt{mZ\pi/2} + 12.5)$	$0.001[4m(1 + 0.0125Z) + 25]$	$0.001(1.25\sqrt{g} + 6.3)$
7	$0.001(7.1\sqrt{mZ\pi/2} + 18)$	$0.001[6.3m(1 + 0.0125Z) + 40]$	$0.001(2\sqrt{g} + 10)$

g = length of spline in millimeters.

total profile variation, and tooth alignment variation are used to calculate the combined effect by the following formula:

$$\lambda = 0.6\sqrt{(F_p)^2 + (f_f)^2 + (F_\beta)^2} \text{ millimeters}$$

The above variation is based upon a length of engagement equal to one-half the pitch diameter of the spline; adjustment of λ may be required for a greater length of engagement. Formulas for values of F_p, f_f, and F_β used in the above formula are given in Table 5.

Table 6. Reduction, $es/\tan \alpha_D$, of External Spline Major and Minor Diameters Required for Selected Fit Classes*

Pitch Diameter D in mm	Standard Pressure Angle, in Degrees									All
	30	37.5	45	30	37.5	45	30	37.5	45	
	Classes of Fit									
	d			e			f			h
	$es/\tan \alpha_D$ in millimeters									
≤ 3	0.035	0.026	0.020	0.024	0.018	0.014	0.010	0.008	0.006	0
> 3 to 6	0.052	0.039	0.030	0.035	0.026	0.020	0.017	0.013	0.010	0
> 6 to 10	0.069	0.052	0.040	0.043	0.033	0.025	0.023	0.017	0.013	0
> 10 to 18	0.087	0.065	0.050	0.055	0.042	0.032	0.028	0.021	0.016	0
> 18 to 30	0.113	0.085	0.065	0.069	0.052	0.040	0.035	0.026	0.020	0
> 30 to 50	0.139	0.104	0.080	0.087	0.065	0.050	0.043	0.033	0.025	0
> 50 to 80	0.173	0.130	0.100	0.104	0.078	0.060	0.052	0.039	0.030	0
> 80 to 120	0.208	0.156	0.120	0.125	0.094	0.072	0.062	0.047	0.036	0
> 120 to 180	0.251	0.189	0.145	0.147	0.111	0.085	0.074	0.056	0.043	0
> 180 to 250	0.294	0.222	0.170	0.173	0.130	0.100	0.087	0.065	0.050	0
> 250 to 315	0.329	0.248	0.190	0.191	0.143	0.110	0.097	0.073	0.056	0
> 315 to 400	0.364	0.274	0.210	0.217	0.163	0.125	0.107	0.081	0.062	0
> 400 to 500	0.398	0.300	0.230	0.234	0.176	0.135	0.118	0.089	0.068	0
> 500 to 630	0.450	0.339	0.260	0.251	0.189	0.145	0.132	0.099	0.076	0
> 630 to 800	0.502	0.378	0.290	0.277	0.209	0.160	0.139	0.104	0.080	0
> 800 to 1000	0.554	0.417	0.320	0.294	0.222	0.170	0.149	0.112	0.086	0

* These values are used with the applicable formulas in Table 2.

Machining Tolerance: A value for machining tolerance may be obtained by subtracting the effective variation, λ, from the total tolerance $(T + \lambda)$. Design requirements or specific processes used in spline manufacture may require a different amount of machining tolerance in relation to the total tolerance.

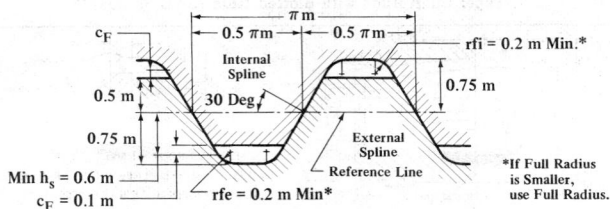

Fig. 1a. Profile of Basic Rack for 30° Flat Root Spline

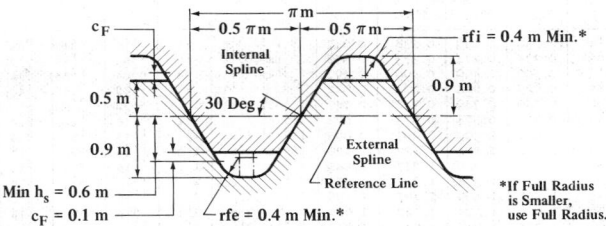

Fig. 1b. Profile of Basic Rack for 30° Fillet Root Spline

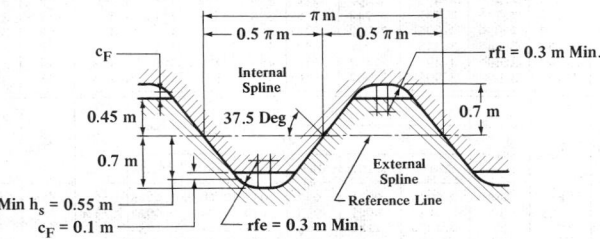

Fig. 1c. Profile of Basic Rack for 37.5° Fillet Root Spline

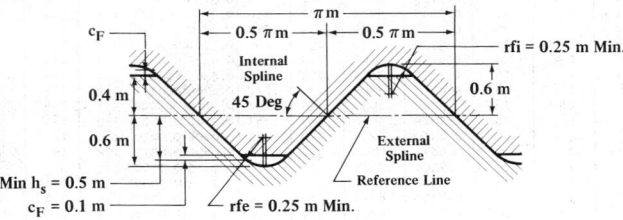

Fig. 1d. Profile of Basic Rack for 45° Fillet Root Spline

Taper Shaft Ends with Slotted Nuts (SAE Standard)

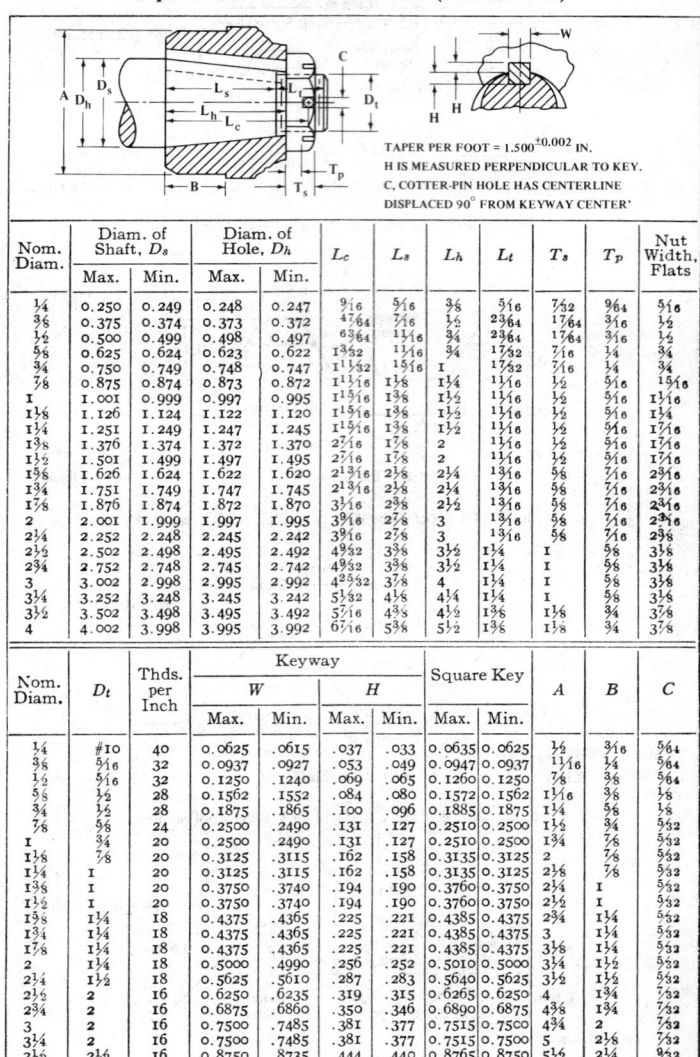

TAPER PER FOOT = 1.500$^{\pm0.002}$ IN.

H IS MEASURED PERPENDICULAR TO KEY.

C, COTTER-PIN HOLE HAS CENTERLINE DISPLACED 90° FROM KEYWAY CENTER

Nom. Diam.	Diam. of Shaft, Ds		Diam. of Hole, Dh		Lc	Ls	Lh	Lt	Ts	Tp	Nut Width, Flats
	Max.	Min.	Max.	Min.							
1/4	0.250	0.249	0.248	0.247	9/16	5/16	3/8	5/16	7/32	9/64	5/16
3/8	0.375	0.374	0.373	0.372	47/64	7/16	1/2	23/64	17/64	3/16	1/2
1/2	0.500	0.499	0.498	0.497	63/64	11/16	3/4	1 7/64	17/64	3/16	1/2
5/8	0.625	0.624	0.623	0.622	1 3/32	1 1/16	3/4	1 7/32	7/16	1/4	3/4
3/4	0.750	0.749	0.748	0.747	1 11/32	15/16	1	1 7/32	7/16	1/4	3/4
7/8	0.875	0.874	0.873	0.872	1 11/16	1 1/8	1 1/4	1 1/16	1/2	5/16	15/16
1	1.001	0.999	0.997	0.995	1 15/16	1 3/8	1 1/2	1 1/16	1/2	5/16	1 1/16
1 1/8	1.126	1.124	1.122	1.120	1 15/16	1 3/8	1 1/2	1 1/16	1/2	9/16	1 1/4
1 1/4	1.251	1.249	1.247	1.245	1 15/16	1 3/8	1 1/2	1 1/16	1/2	9/16	1 7/16
1 3/8	1.376	1.374	1.372	1.370	2 7/16	1 7/8	2	1 1/16	1/2	9/16	1 7/16
1 1/2	1.501	1.499	1.497	1.495	2 7/16	1 7/8	2	1 1/16	1/2	9/16	1 7/16
1 5/8	1.626	1.624	1.622	1.620	2 13/16	2 1/8	2 1/4	1 3/16	5/8	7/16	2 3/16
1 3/4	1.751	1.749	1.747	1.745	2 13/16	2 1/8	2 1/4	1 3/16	5/8	7/16	2 3/16
1 7/8	1.876	1.874	1.872	1.870	3 1/16	2 3/8	2 1/2	1 3/16	5/8	7/16	2 3/16
2	2.001	1.999	1.997	1.995	3 9/16	2 7/8	3	1 3/16	5/8	7/16	2 3/16
2 1/4	2.252	2.248	2.245	2.242	3 9/16	2 7/8	3	1 3/16	5/8	7/16	2 3/8
2 1/2	2.502	2.498	2.495	2.492	4 9/32	3 3/8	3 1/2	1 1/4	1	9/16	3 1/8
2 3/4	2.752	2.748	2.745	2.742	4 9/32	3 3/8	3 1/2	1 1/4	1	9/16	3 1/8
3	3.002	2.998	2.995	2.992	4 25/32	3 7/8	4	1 1/4	1	9/16	3 1/8
3 1/4	3.252	3.248	3.245	3.242	5 1/32	4 1/8	4 1/4	1 1/4	1	5/8	3 1/8
3 1/2	3.502	3.498	3.495	3.492	5 7/16	4 3/8	4 1/2	1 3/8	1 1/8	3/4	3 7/8
4	4.002	3.998	3.995	3.992	6 7/16	5 3/8	5 1/2	1 3/8	1 1/8	3/4	3 7/8

Nom. Diam.	Dt	Thds. per Inch	Keyway W		Keyway H		Square Key		A	B	C
			Max.	Min.	Max.	Min.	Max.	Min.			
1/4	#10	40	0.0625	.0615	.037	.033	0.0635	0.0625	1/2	3/16	5/64
3/8	5/16	32	0.0937	.0927	.053	.049	0.0947	0.0937	11/16	1/4	5/64
1/2	5/16	32	0.1250	.1240	.069	.065	0.1260	0.1250	7/8	3/8	5/64
5/8	1/2	28	0.1562	.1552	.084	.080	0.1572	0.1562	1 1/16	3/8	1/8
3/4	1/2	28	0.1875	.1865	.100	.096	0.1885	0.1875	1 1/4	5/8	1/8
7/8	5/8	24	0.2500	.2490	.131	.127	0.2510	0.2500	1 1/2	3/4	5/32
1	3/4	20	0.2500	.2490	.131	.127	0.2510	0.2500	1 3/4	7/8	5/32
1 1/8	7/8	20	0.3125	.3115	.162	.158	0.3135	0.3125	2	7/8	5/32
1 1/4	1	20	0.3125	.3115	.162	.158	0.3135	0.3125	2 1/2	7/8	5/32
1 3/8	1	20	0.3750	.3740	.194	.190	0.3760	0.3750	2 1/4	1	5/32
1 1/2	1	20	0.3750	.3740	.194	.190	0.3760	0.3750	2 1/2	1	5/32
1 5/8	1 1/4	18	0.4375	.4365	.225	.221	0.4385	0.4375	2 3/4	1 1/4	5/32
1 3/4	1 1/4	18	0.4375	.4365	.225	.221	0.4385	0.4375	3	1 1/4	5/32
1 7/8	1 1/4	18	0.4375	.4365	.225	.221	0.4385	0.4375	3 1/8	1 1/4	5/32
2	1 1/4	18	0.5000	.4990	.256	.252	0.5010	0.5000	3 1/4	1 1/2	5/32
2 1/4	1 1/2	18	0.5625	.5610	.287	.283	0.5640	0.5625	3 1/2	1 1/2	5/32
2 1/2	2	16	0.6250	.6235	.319	.315	0.6265	0.6250	4	1 3/4	7/32
2 3/4	2	16	0.6875	.6860	.350	.346	0.6890	0.6875	4 3/8	1 3/4	7/32
3	2	16	0.7500	.7485	.381	.377	0.7515	0.7500	4 3/4	2	7/32
3 1/4	2	16	0.7500	.7485	.381	.377	0.7515	0.7500	5	2 1/8	7/32
3 1/2	2 1/2	16	0.8750	.8735	.444	.440	0.8765	0.8750	5 1/2	2 1/4	9/32
4	2 1/2	16	1.0000	.9985	.506	.502	1.0015	1.0000	6 1/4	2 3/4	9/32

All dimensions in inches except where otherwise noted.

Polygon-Type Shaft Connections.—Involute-form and straight-sided splines are used for both fixed and sliding connections between machine members such as shafts and gears. Polygon-type connections, so called because they resemble regular polygons but with curved sides, may be used similarly. German DIN Standards 32711 and 32712 include data for three- and four-sided metric polygon connections. Data for 11 of the sizes shown in those Standards, but converted to inch dimensions by Stoffel Polygon Systems, are given in the accompanying table.

Dimensions of Three- and Four-Sided Polygon-type Shaft Connections

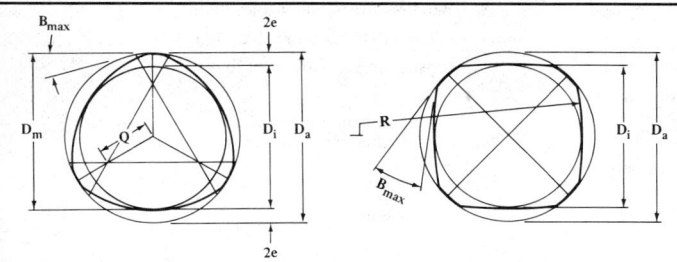

Three-Sided Designs				Four-Sided Designs			
Nominal Sizes			Design Data	Nominal Sizes			Design Data
D_A (in.)	D_I (in.)	e (in.)	Area (in.²)	Z_P (in.³)	D_A (in.)	D_I (in.)	e (in.)	Area (in.²)	Z_P (in.³)
0.530	0.470	0.015	0.194	0.020	0.500	0.415	0.075	0.155	0.014
0.665	0.585	0.020	0.302	0.039	0.625	0.525	0.075	0.250	0.028
0.800	0.700	0.025	0.434	0.067	0.750	0.625	0.125	0.350	0.048
0.930	0.820	0.027	0.594	0.108	0.875	0.725	0.150	0.470	0.075
1.080	0.920	0.040	0.765	0.153	1.000	0.850	0.150	0.650	0.12
1.205	1.045	0.040	0.977	0.224	1.125	0.950	0.200	0.810	0.17
1.330	1.170	0.040	1.208	0.314	1.250	1.040	0.200	0.980	0.22
1.485	1.265	0.055	1.450	0.397	1.375	1.135	0.225	1.17	0.29
1.610	1.390	0.055	1.732	0.527	1.500	1.260	0.225	1.43	0.39
1.870	1.630	0.060	2.378	0.850	1.750	1.480	0.250	1.94	0.64
2.140	1.860	0.070	3.090	1.260	2.000	1.700	0.250	2.60	0.92

Dimensions Q and R shown on the diagrams are approximate and used only for drafting purposes: $Q \approx 7.5e$; $R \approx D_I/2 + 16e$.

Dimension $D_M = D_I + 2e$. Pressure angle B_{max} is approximately $344e/D_M$ degrees for three sides, and $229e/D_M$ degrees for four sides.

Tolerances: ISO H7 tolerances apply to bore dimensions. For shafts, g6 tolerances apply for sliding fits; k7 tolerances for tight fits.

Choosing Between Three- and Four-Sided Designs: Three-sided designs are best for applications in which no relative movement between mating components is allowed while torque is transmitted. If a hub is to slide on a shaft while under torque, four-sided designs, which have larger pressure angles B_{max} than those of three-sided designs, are better suited to sliding even though the axial force needed to move the sliding member is approximately 50 per cent greater than for comparable involute spline connections.

Strength of Polygon Connections: In the formulas that follow,

$$H_w = \text{hub width, inches}$$
$$H_t = \text{hub wall thickness, inches}$$
$$M_b = \text{bending moment, lb-in.}$$
$$M_t = \text{torque, lb-in.}$$

Z = section modulus, bending, in.^{3}

    = $0.098 D_M{}^4/D_A$ for three sides

    = $0.15 D_I{}^3$ for four sides

Z_P = polar section modulus, torsion, in.^{3}

    = $0.196 D_M{}^4/D_A$ for three sides

    = $0.196 D_I{}^3$ for four sides

D_A and D_M. See table footnotes.

S_b = bending stress, allowable, lb/in.^{2}

S_s = shearing stress, allowable, lb/in.^{2}

S_t = tensile stress, allowable, lb/in.^{2}

For shafts,

$$M_t \text{ (maximum)} = S_s Z_P;$$
$$M_b \text{ (maximum)} = S_b Z$$

For bores,

$$H_t \text{ (minimum)} = K \sqrt{\frac{M_t}{S_t H_w}}$$

in which $K = 1.44$ for three sides except that if D_M is greater than 1.375 inches, then $K = 1.2$; $K = 0.7$ for four sides.

Failure may occur in the hub of a polygon connection if the hoop stresses in the hub exceed the allowable tensile stress for the material used. The radial force tending to expand the rim and cause tensile stresses is calculated from

$$\text{Radial Force, lb} = \frac{2 M_t}{D_I \, n \, \tan (B_{max} + 11.3)}$$

This radial force acting at n points may be used to calculate the tensile stress in the hub wall using formulas from strength of materials.

Manufacturing: Polygon shaft profiles may be produced using conventional machining processes such as hobbing, shaping, contour milling, copy turning, and numerically controlled milling and grinding. Bores are produced using broaches, spark erosion, gear shapers with generating cutters of appropriate form, and, in some instances, internal grinders of special design. Regardless of the production methods used, points on both of the mating profiles may be calculated from the following equations:

$$X = (D_I/2 + e) \cos \alpha - e \cos n\alpha \cos \alpha - ne \sin n\alpha \sin \alpha$$
$$Y = (D_I/2 + e) \sin \alpha - e \cos n\alpha \sin \alpha + ne \sin n\alpha \cos \alpha$$

In these equations, α is the angle of rotation of the workpiece from any selected reference position; n is the number of polygon sides, either 3 or 4; D_I is the diameter of the inscribed circle shown on the diagram in the table; and e is the dimension shown on the diagram in the table and which may be used as a setting on special polygon grinding machines. The value of e determines the shape of the profile. A value of 0, for example, results in a circular shaft having a diameter of D_I. The value of e in the table were selected arbitrarily to provide suitable proportions for the size shown.

CAMS AND CAM DESIGN

Classes of Cams. — Cams may, in general, be divided into two classes: uniform motion cams and accelerated motion cams. The uniform motion cam moves the follower at the same rate of speed from the beginning to the end of the stroke; but as the movement is started from zero to the full speed of the uniform motion and stops in the same abrupt way, there is a distinct shock at the beginning and end of the stroke, if the movement is at all rapid. In machinery working at a high rate of speed, therefore, it is important that cams are so constructed that sudden shocks are avoided when starting the motion or when reversing the direction of motion of the follower.

The uniformly accelerated motion cam is suitable for moderate speeds, but it has the disadvantage of sudden changes in acceleration at the beginning, middle and end of the stroke. A cycloidal motion curve cam produces no abrupt changes in acceleration and is often used in high-speed machinery because it results in low noise, vibration and wear. The cycloidal motion displacement curve is so called because it can be generated from a cycloid which is the locus of a point of a circle rolling on a straight line.*

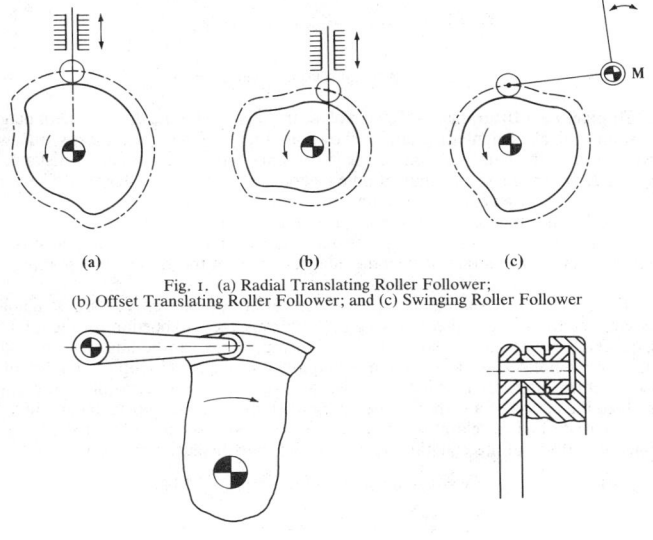

(a)   (b)   (c)

Fig. 1. (a) Radial Translating Roller Follower;
(b) Offset Translating Roller Follower; and (c) Swinging Roller Follower

(a)   (b)

Fig. 2. (a) Closed-Track Cam; (b) Closed-Track Cam With Two Rollers

Cam Follower Systems. — The three most used cam and follower systems are radial and offset translating roller follower, Figs. 1a and 1b; and the swinging roller follower, Fig. 1c. When the cam rotates, it imparts a translating motion to the roller followers in (a) and (b) and a swinging motion to the roller follower in (c). The motion

* Jensen, P. W., *Cam Design and Manufacture,* Industrial Press Inc.

of the follower is, of course, dependent on the shape of the cam; and the following section on displacement diagrams explains how a favorable motion is obtained so that the cam can rotate at high speed without shock.

The arrangements in Fig. 1 show open-track cams. In Fig. 2 the roller is forced to move in a closed track. Open-track cams build smaller than closed-track cams but, in general, springs are necessary to keep the roller in contact with the cam at all times. Closed-track cams do not require a spring and have the advantage of positive drive throughout the rise and return cycle. The positive drive is sometimes required as in the case where a broken spring would cause serious damage to a machine.

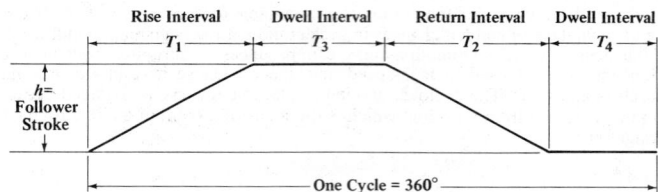

Fig. 3. A Simple Displacement Diagram

Displacement Diagrams. — Design of a cam begins with the displacement diagram. A simple displacement diagram is shown in Fig. 3. One cycle means one whole revolution of the cam; i.e., one cycle represents 360°. The horizontal distances T_1, T_2, T_3, T_4 are expressed in units of time (seconds); or radians or degrees. The vertical distance, h, represents the maximum "rise" or stroke of the follower.

The displacement diagram of Fig. 3 is not a very favorable one because the motion from rest (the horizontal lines) to constant velocity takes place instantaneously and this means that accelerations become infinitely large at these transition points.

Types of Cam Displacement Curves: A variety of cam curves are available for moving the follower. In the following sections only the rise portions of the total time-displacement diagram are studied. The return portions can be analyzed in a similar manner. Complex cams are frequently employed which may involve a number of rise-dwell-return intervals in which the rise and return aspects are quite different. To analyze the action of a cam it is necessary to study its time-displacement and associated velocity and acceleration curves. The latter are based on the first and second time-derivatives of the equation describing the time-displacement curve:

$$y = \text{displacement} = f(t); \quad \text{or} \quad y = f(\phi)$$

$$v = \frac{dy}{dt} = \text{velocity} = \omega \frac{dy}{d\phi}$$

$$a = \frac{d^2y}{dt^2} = \text{acceleration} = \omega^2 \frac{d^2y}{d\phi^2}$$

Meaning of Symbols and Equivalent Relations:

  y = displacement of follower, in.
  h = maximum displacement of follower, in.
  t = time for cam to rotate through angle ϕ, sec, = ϕ/ω, sec
  T = time for cam to rotate through angle β, sec, = β/ω, or $\beta/6N$, sec

ϕ = cam angle rotation for follower displacement y, degrees
β = cam angle rotation for total rise h, degrees
v = velocity of follower, in./sec
a = follower acceleration, in./sec^2
$t/T = \phi/\beta$
N = cam speed, rpm
ω = angular velocity of cam, degrees/sec = $\beta/T = \phi/t = d\phi/dt = 6N$
ω_R = angular velocity of cam, radians/sec = $\pi\omega/180$
W = effective weight, lbs
g = gravitational constant = 386 in./sec^2
$f(t)$ means a function of t
$f(\phi)$ means a function of ϕ
R_{min} = minimum radius to the cam pitch curve, in.
R_{max} = maximum radius to the cam pitch curve, in.
r_f = radius of cam follower roller, in.
ρ = radius of curvature of cam pitch curve (path of center of roller follower), in.
R_c = radius of curvature of actual cam surface, in., = $\rho - r_f$ for convex surface; = $\rho + r_f$ for concave surface.

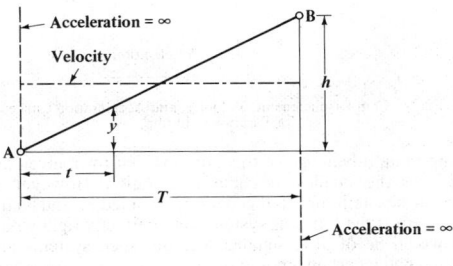

Fig. 4. Cam Displacement, Velocity, and Acceleration Curves for Constant Velocity Motion

Four displacement curves are of the greatest utility in cam design:
1. *Constant-Velocity Motion* (Fig. 4)

$$y = h\,\frac{t}{T}; \quad \text{or} \quad y = \frac{h\phi}{\beta} \right\} \tag{1a}$$

$$v = \frac{dy}{dt} = \frac{h}{T}; \quad \text{or} \quad v = \frac{h\omega}{\beta} \left.\right\} \; 0 < t < T \tag{1b}$$

$$a = \frac{d^2y}{dt^2} = 0^* \right\} \tag{1c}$$

* Except at $t = 0$ and $t = T$ where the acceleration is theoretically infinite.

This motion and its disadvantages were mentioned previously. While in the unaltered form shown it is rarely used except in very crude devices, nevertheless, the advantage of uniform velocity is an important one and by modifying the start and finish of the follower stroke this form of cam motion can be utilized. Such modification is explained in the section Displacement Diagram Synthesis.

2. *Parabolic Motion* (Fig. 5)

For $0 \leq t \leq T/2$ and $0 \leq \phi \leq \beta/2$

$$y = 2h(t/T)^2 = 2h(\phi/\beta)^2 \quad (2a)$$

$$v = 4ht/T^2 = 4h\omega\phi/\beta^2 \quad (2b)$$

$$a = 4h/T^2 = 4h(\omega/\beta)^2 \quad (2c)$$

For $T/2 \leq t \leq T$ and $\beta/2 \leq \phi \leq \beta$

$$y = h[1 - 2(1 - t/T)^2] = h[1 - 2(1 - \phi/\beta)^2] \quad (2d)$$

$$v = 4h/T(1 - t/T) = (4h\omega/\beta)(1 - \phi/\beta) \quad (2e)$$

$$a = -4h/T^2 = -4h(\omega/\beta)^2 \quad (2f)$$

Examination of the above formulas shows that the velocity is zero when $t = 0$ and $y = 0$; and when $t = T$ and $y = h$.

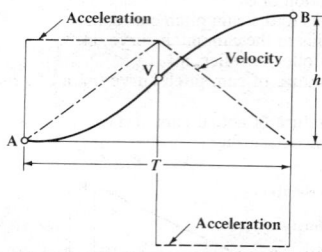

Fig. 5. Cam Displacement, Velocity, and Acceleration Curves for Parabolic Motion

The most important advantage of this curve is that for a given angle of rotation and rise it produces the smallest possible acceleration. However, because of the sudden changes in acceleration at the beginning, middle, and end of the stroke, shocks are produced. If the follower system were perfectly rigid with no backlash or flexibility, this would be of little significance. But such systems are mechanically impossible to build and a certain amount of impact is caused at each of these change-over points.

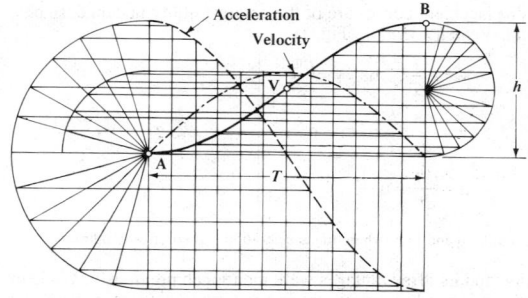

Fig. 6. Cam Displacement, Velocity, and Acceleration Curves for Simple Harmonic Motion

3. *Simple Harmonic Motion* (Fig. 6)

$$y = \frac{h}{2}\left[1 - \cos\left(\frac{180°t}{T}\right)\right]; \quad \text{or} \quad y = \frac{h}{2}\left[1 - \cos\left(\frac{180°\phi}{\beta}\right)\right] \tag{3a}$$

$$v = \frac{h}{2} \cdot \frac{\pi}{T} \sin\left(\frac{180°t}{T}\right); \quad \text{or} \quad v = \frac{h}{2} \cdot \frac{\pi\omega}{\beta} \sin\left(\frac{180°\phi}{\beta}\right) \quad \Bigg\} \quad 0 \leq t \leq T \tag{3b}$$

$$a = \frac{h}{2} \cdot \frac{\pi^2}{T^2} \cos\left(\frac{180°t}{T}\right); \quad \text{or} \quad a = \frac{h}{2} \cdot \left(\frac{\pi\omega}{\beta}\right)^2 \cos\left(\frac{180°\phi}{\beta}\right) \tag{3c}$$

Smoothness in velocity and acceleration during the stroke is the advantage inherent in this curve. However, the instantaneous changes in acceleration at the beginning and end of the stroke tend to cause vibration, noise, and wear. As can be seen from Fig. 6, the maximum acceleration values occur at the ends of the stroke. Thus, if inertia loads are to be overcome by the follower, the resulting forces cause stresses in the members. These forces are in many cases much larger than the externally applied loads.

4. *Cycloidal Motion* (Fig. 7)

$$y = h\left[\frac{t}{T} - \frac{1}{2\pi}\sin\left(\frac{360°t}{T}\right)\right]; \quad \text{or} \quad y = h\left[\frac{\phi}{\beta} - \frac{1}{2\pi}\sin\left(\frac{360°\phi}{\beta}\right)\right] \tag{4a}$$

$$v = \frac{h}{T}\left[1 - \cos\left(\frac{360°t}{T}\right)\right]; \quad \text{or} \quad v = \frac{h\omega}{\beta}\left[1 - \cos\left(\frac{360°\phi}{\beta}\right)\right] \quad \Bigg\} \quad 0 \leq t \leq T \tag{4b}$$

$$a = \frac{2\pi h}{T^2}\sin\left(\frac{360°t}{T}\right); \quad \text{or} \quad a = \frac{2\pi h\omega^2}{\beta^2}\sin\left(\frac{360°\phi}{\beta}\right) \tag{4c}$$

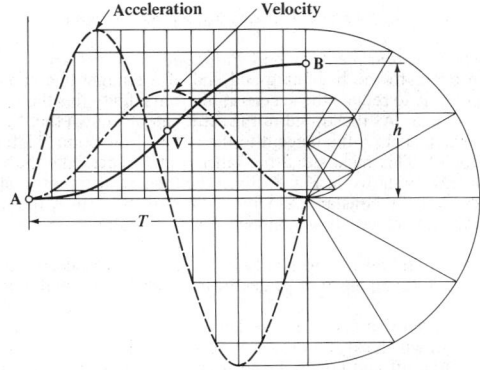

Fig. 7. Cam Displacement, Velocity, and Acceleration Curves for Cycloidal Motion

This time-displacement curve has excellent acceleration characteristics; there are no abrupt changes in its associated acceleration curve. The maximum value of the acceleration of the follower for a given rise and time is somewhat higher than that of the simple harmonic motion curve. In spite of this, the cycloidal curve is used often

as a basis for designing cams for high-speed machinery because it results in low levels of noise, vibration, and wear.

Displacement Diagram Synthesis. — The straight-line graph shown in Fig. 3 has the important advantage of uniform velocity. This is so desirable that many cams based on this graph are used. To avoid impact at the beginning and end of the stroke, a modification is introduced at these points. There are many different types of modifications possible, ranging from a simple circular arc to much more complicated curves. One of the better curves used for this purpose is the parabolic curve given by Eq. (2a). As seen from the derived time graphs, this curve causes the follower to begin a stroke with zero velocity but having a finite and constant acceleration. We must accept the necessity of acceleration, but effort should be made to hold it to a minimum.

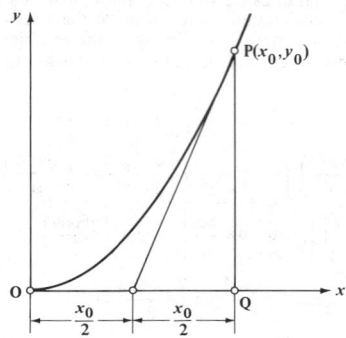

Fig. 8. The Tangent at P Bisects OQ, When Curve is a Parabola

Matching of Constant Velocity and Parabolic Motion Curves: By matching a parabolic cam curve to the beginning and end of a straight-line cam displacement diagram it is possible to reduce the acceleration from infinity to a finite constant value to avoid impact loads. As illustrated in Fig. 8, it can be shown that for any parabola the vertex of which is at O, the tangent to the curve at the point P intersects the line OQ at its midpoint. This means that the tangent at P represents the velocity of the follower at time X_0 as shown in Fig. 8. Since the tangent also represents the velocity of the follower over the constant velocity portion of the stroke, the transition from rest to the maximum velocity is accomplished with smoothness.

Example: A cam follower is to rise $\frac{1}{4}$ in. with constant acceleration; $1\frac{1}{4}$ in. with constant velocity, over an angle of 50 degrees; and then $\frac{1}{2}$ in. with constant deceleration.

In Fig. 9 the three rise distances are laid out, $y_1 = \frac{1}{4}$ in., $y_2 = 1\frac{1}{4}$ in., $y_3 = \frac{1}{2}$ in., and horizontals drawn. Next, an arbitrary horizontal distance ϕ_2 proportional to 50 degrees is measured off and points A and B are located. The line AB is extended to M_1 and M_2. By remembering that a tangent to a parabola, Fig. 8, will cut the abscissa axis at point $(X_0/2, 0)$ where X_0 is the abscissa of the point of tangency, the two values $\phi_1 = 20°$ and $\phi_3 = 40°$ will be found. Analytically,

$$\frac{M_1E}{\phi_2} = \frac{y_1}{y_2} \; ; \quad \frac{\frac{1}{2}\phi_1}{50°} = \frac{0.25}{1.25} \quad \therefore \phi_1 = 20°$$

$$\frac{FM_2}{\phi_2} = \frac{y_3}{y_2} \; ; \quad \frac{\frac{1}{2}\phi_3}{50°} = \frac{0.50}{1.25} \quad \therefore \; \phi_3 = 40°$$

In Fig. 9, the portions of the parabola have been drawn in; the details of this operation are as follows:

Assume that accuracy to the nearest thousandth of one inch is desired, and it is decided to plot values for every 5 degrees of cam rotation.

The formula for the acceleration portion of the parabolic curve is:

$$y = \frac{2h}{T^2} t^2 = 2h \left(\frac{\phi}{\beta} \right)^2 \tag{2a}$$

Two different parabolas are involved in this example; one for accelerating the follower during a cam rotation of 20 degrees, the other for decelerating it in 40 degrees, these two being tangent, to opposite ends of the same line AB.

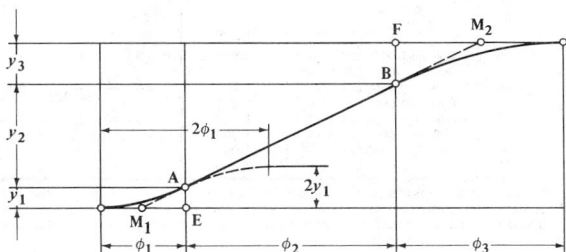

Fig. 9. Matching a Parabola at Each End of Straight Line Displacement Curve AB to Provide More Acceptable Acceleration and Deceleration

In Fig. 9 only the first half of a complete acceleration-deceleration parabolic curve is used to blend with the left end of the straight line AB. Therefore, in using the formula (2a) substitute $2y_1$ for h and $2\phi_1$ for β so that

$$y = \frac{2h\phi^2}{\beta^2} = \frac{(2)(2y_1)}{(2\phi_1)^2} \phi^2$$

For the right end of the straight line AB, the calculations are similar but, in using formula (2a), calculated y values are *subtracted* from the *total* rise of the cam ($y_1 + y_2 + y_3$) to obtain the follower displacement.

Table I shows the computations and resulting values for the cam displacement diagram described. The calculations are shown in detail so that if equations are programmed for a digital computer, the results can be verified easily. Obviously, the intermediate points are not needed to draw the straight line, but when the cam profile is later to be drawn or cut, these values will be needed since they are to be measured on radial lines.

The matching procedure when using cycloidal motion is exactly the same as for parabolic motion, because parabolic and cycloidal motion have the same maximum velocity for equal rise (or return) and lift angle (or return angle).

Cam Profile Determination. — In the cam constructions that follow an artificial device called an *inversion* is used. This represents a mental concept which is very helpful in performing the graphical work. The construction of a cam profile requires the drawing of many positions of the cam with the follower in each case in its related

**Table 1. Development of Modified Constant Velocity Cam
with Parabolic Matching**

Rise Angle	ϕ Degrees	Computation	Follower Displacement y	Explanation
$\phi_1 = 20°$	0	0	0	$\beta = 40°$, $h = 0.500$
	5	0.000625×5^2	0.016	$y = \dfrac{(2)\,(0.500)}{(40)^2}\phi^2$
	10	0.000625×10^2	0.063	
	15	0.000625×15^2	0.141	$= 0.000625\phi^2$
	20	0.000625×20^2	0.250	
$\phi_2 = 50°$	25		0.375	
	30		0.500	
	35		0.625	
	40		0.750	
	45		0.875	1.250 in. divided into
	50		1.000	10 uniform divisions
	55		1.125	
	60		1.250	
	65		1.375	
	70		1.500	
$\phi_3 = 40°$	75	$2.000 - (0.0003125 \times 35^2)$	1.617	$\beta = 80°$, $h = 1.000$
	80	$2.000 - (0.0003125 \times 30^2)$	1.719	$y = 2 - \dfrac{(2)\,(1.000)}{(80°)^2}(110° - \phi)^2$
	85	$2.000 - (0.0003125 \times 25^2)$	1.805	
	90	$2.000 - (0.0003125 \times 20^2)$	1.875	$= 2 - 0.0003125\,(110° - \phi)^2$
	95	$2.000 - (0.0003125 \times 15^2)$	1.930	(see * footnote)
	100	$2.000 - (0.0003125 \times 10^2)$	1.969	
	105	$2.000 - (0.0003125 \times \ 5^2)$	1.992	
	110	$2.000 - (0.0003125 \times \ 0^2)$	2.000	

* Since the deceleration portion of a parabolic cam is the same shape as the acceleration portion, but inverted, formula (2a) may be used to calculate the y values by substituting $2y_3$ for h and $2\phi_3$ for β and the result subtracted from the total cam rise ($y_1 + y_2 + y_3$) to obtain the follower displacement.

location. However, instead of revolving the cam, it is assumed that the follower rotates around the *fixed* cam. It requires the drawing of many follower positions, but since this is done more or less diagrammatically, it is relatively simple.

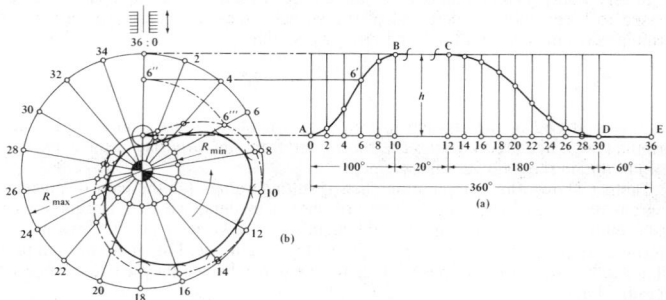

Fig. 10. (a) Time-Displacement Diagram for Cam to be Laid Out; (b) Construction of Contour of Cam With Radial Translating Roller Follower

As part of the inversion process, the direction of rotation is important. In order to preserve the correct sequence of events, the artificial rotation of the follower must be the reverse of the cam's prescribed rotation. Thus, in Fig. 10 the cam rotation is counterclockwise, whereas the artificial rotation of the follower is clockwise.

Radial Translating Roller Follower: The time-displacement diagram for a cam

with a radial translating roller follower is shown in Fig. 10(a). This diagram is read from left to right as follows: For 100 degrees of cam shaft rotation the follower rises *h* inches (*AB*), dwells in its upper position for 20 degrees (*BC*), returns over 180 degrees (*CD*), and finally dwells in its lowest position for 60 degrees (*DE*). Then the entire cycle is repeated.

Figure 10(b) shows the cam construction layout with the cam pitch curve as a dot and dash line. To locate a point on this curve, take a point on the displacement curve, as 6′ at the 60-degree position of the cam construction diagram, and project this horizontally to point 6″ on the 0-degree position of the cam construction diagram. Using the center of cam rotation, an arc is struck from point 6″ to intercept the 60-degree position radial line which gives point 6‴ on the cam pitch curve. It will be seen that the smaller circle in the cam construction layout has a radius R_{min} equal to the smallest distance from the center of cam rotation to the pitch curve and, similarly, the larger circle has a radius R_{max} equal to the largest distance to the pitch curve. Thus, the difference in radii of these two circles is equal to the maximum rise *h* of the follower.

The cam pitch curve is also the actual profile or working surface when a knife-edged follower is used. To get the profile or working surface for a cam with a roller follower, a series of arcs with centers on the pitch curve and radii equal to the radius of the roller are drawn and the inner envelope drawn tangent to these arcs is the cam working surface or profile shown as a solid line in Fig. 10(b).

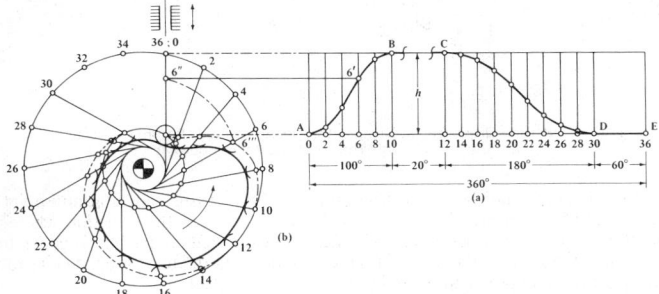

Fig. 11. (a) Time-Displacement Diagram for Cam to be Laid Out; (b) Construction of Contour of Cam With Offset Translating Roller Follower

Offset Translating Roller Follower: Given the time-displacement diagram Fig. 11(a) and an offset follower. The construction of the cam in this case is very similar to the foregoing case and is shown in Fig. 11(b). In this construction it will be noted that the angular position lines are not drawn radially from the cam shaft center but tangent to a circle having a radius equal to the amount of offset of the center line of the cam follower from the center of the cam shaft. For counterclockwise rotation of the cam, points 6′, 6″, and 6‴ are located in succession as indicated.

Swinging Roller Follower: Given the time-displacement diagram Fig. 12(a) and the length of the swinging follower arm L_f, it is required that the displacement of the follower center along the circular arc that it describes be equal to the corresponding displacements in the time-displacement diagram. If ϕ_o is known, the displacement *h* of Fig. 12(a) would be found from the formula $h = \pi \phi_o L_f / 180°$: otherwise the maximum rise *h* of the follower is stepped off on the arc drawn with *M* as a center and starting at a point on the R_{min} circle. Point *M* is the actual position of the pivot center of the swinging follower with respect to the cam shaft center. It is again required that

the rotation of the cam be counterclockwise and therefore M is considered to have been rotated clockwise around the cam shaft center, whereby the points 2, 4, 6, etc., are obtained as shown in Fig. 12(b). Around each of the pivot points, 2, 4, 6, etc., circular arcs whose radii equal L_f are drawn between the R_{min} and R_{max} circles giving the points 2′, 4′, 6′, etc. The R_{min} circle with center at the cam shaft center is drawn through the lowest position of the center of the roller follower and the R_{max} circle through the highest position as shown. The different points on the pitch curve are now located. Point 6‴, for instance, is found by stepping off the y_6 ordinate of the displacement diagram on arc 6′ starting at the R_{min} circle.

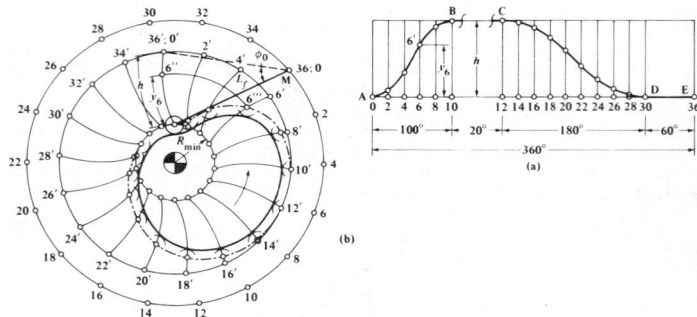

Fig. 12. (a) Time-Displacement Diagram for Cam to be Laid Out; (b) Construction of Contour of Cam With Swinging Roller Follower

Pressure Angle and Radius of Curvature. — The pressure angle at any point on the profile of a cam may be defined as the angle between the direction where the follower wants to go at that point and where the cam wants to push it. It is the angle between the tangent to the path of follower motion and the line perpendicular to the tangent of the cam profile at the point of cam-roller contact.

The size of the pressure angle is important because:

1. Increasing the pressure angle increases the side thrust and this increases the forces exerted on cam and follower.

2. Reducing the pressure angle increases the cam size and often this is not desirable because:

   a. The size of the cam determines, to a certain extent, the size of the machine.

   b. Larger cams require more precise cutting points in manufacturing and, therefore, an increase in cost.

   c. Larger cams have higher circumferential speed and small deviations from the theoretical path of the follower cause additional acceleration, the size of which increases with the square of the cam size.

   d. Larger cams mean more revolving weight and in high-speed machines this leads to increased vibrations in the machine.

   e. The inertia of a large cam may interfere with quick starting and stopping.

The maximum pressure angle α_m should, in general, be kept at or below 30 degrees for translating-type followers and at or below 45 degrees for swinging-type followers.

These values are on the conservative side and in many cases may be increased considerably, but beyond these limits trouble could develop and an analysis is necessary.

In the following, graphical methods are described by which a cam mechanism can be designed with translating or swinging roller followers having specified maximum pressure angles for rise and return. These methods are applicable to any kind of time-displacement diagram.

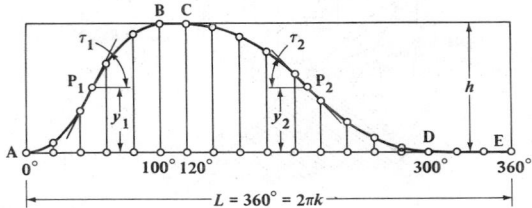

Fig. 13. Displacement Diagram

Determination of Cam Size for a Radial or an Offset Translating Follower. — Figure 13 shows a time-displacement diagram. The maximum displacement is preferably made to scale, but the length of the abscissa, L, can be chosen arbitrarily. The distance L from 0 to 360 degrees is measured and is set equal to $2\pi k$ from which

$$k = \frac{L}{2\pi}$$

k is calculated and laid out as length E to M in Fig. 14.

In Fig. 13 the two points P_1 and P_2 having the maximum angles of slope, τ_1 and τ_2, are located by inspection. In this example y_1 and y_2 are of equal length.

Angles τ_1 and τ_2 are laid out as shown in Fig. 14, and the points of intersection with a perpendicular to EM at M determine Q_1 and Q_2. The measured distances

$$MQ_1 = k \tan \tau_1 \quad \text{and} \quad MQ_2 = k \tan \tau_2$$

are laid out in Fig. 15, which is constructed as follows:

Draw a vertical line $R_u R_o$ of length h equal to the stroke of the roller follower, R_u being the lowest position and R_o the highest position of the center of the roller follower. From R_u lay out $R_u R_{y1} = y_1$ and $R_u R_{y2} = y_2$; these are equal lengths in this example. Next, if the rotation of the cam is counterclockwise, lay out $k \tan \tau_1$, to the left, $k \tan \tau_2$ to the right from points R_{y1} and R_{y2}, respectively, R_{y1} and R_{y2} being the same point in this case.

The specified maximum pressure angle α_1 is laid out at E_1 as shown, and a ray (line) $E_1 F_1$ is determined. Any point on this ray chosen as the cam shaft center will

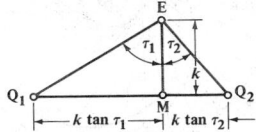

Fig. 14. Construction to Find $k \tan \tau_1$ and $k \tan \tau_2$

proportion the cam so that the pressure angle at a point on the cam profile corresponding to point P_1 of the displacement diagram will be exactly α_1.

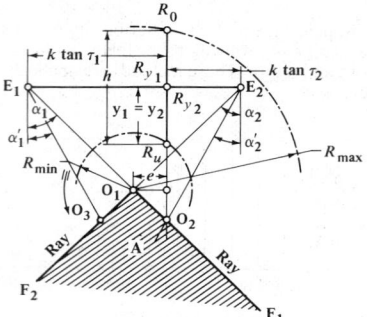

Fig. 15. Finding Proportions of Cam; Offset Translating Follower

The angle α_2 is laid out at E_2 as shown, and another ray E_2F_2 is determined.
Similarly, any point on this ray chosen as the cam shaft center will proportion the
cam so that the pressure angle at a point on the cam profile corresponding to point P_2
of the displacement diagram will be exactly α_2.

Any point chosen within the cross-hatched area A as the cam center will yield a
cam whose pressure angles at points corresponding to P_1 and P_2 will not exceed the
specified values α_1 and α_2 respectively. If O_1 is chosen as the cam shaft center, the
pressure angles on the cam profile corresponding to points P_1 and P_2 are exactly α_1
and α_2, respectively. Selection of point O_1 also yields the smallest possible cam for
the given requirements and requires an offset follower in which e is the offset distance.

If O_2 is chosen as the cam shaft center, a radial translating follower is obtained
(zero offset). In that case, the pressure angle α_1 for the rise is unchanged, whereas
the pressure angle for the return is changed from α_2 to α'_2. That is, the pressure angle
on the return stroke is reduced at the point P_2. If point O_3 had been selected, then α_2
would remain unchanged but α_1 would be decreased and the offset, e, increased.

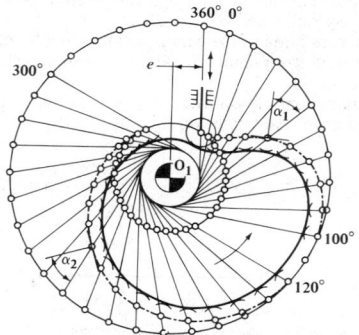

Fig. 16. Construction of Cam Contour; Offset Translating Follower

Figure 16 shows the shape of the cam when O_1 from Fig. 15 is chosen as the cam shaft center, and it is seen that the pressure angle at a point on the cam profile corresponding to point P_1 is α_1 and at a point corresponding to point P_2 is α_2.

In the foregoing, a cam mechanism has been so proportioned that the pressure angles α_1 and α_2 at points on the cam corresponding to points P_1 and P_2 were obtained. Even though P_1 and P_2 are the points of greatest slope on the displacement diagram, the pressure angles produced at some other points on the actual cam may be slightly greater.

However, if the pressure angles α_1 and α_2 are not to be exceeded at any point — i.e., they are to be maximum pressure angles — then P_1 and P_2 must be selected to be at the locations where these maximum pressure angles occur. If these locations are not known, then the graphical procedure described must be repeated, letting P_1 take various positions on the curve for rise (AB) and P_2 various positions on the return curve (CD) and then setting R_{min} equal to the largest of the values determined from the various positions.

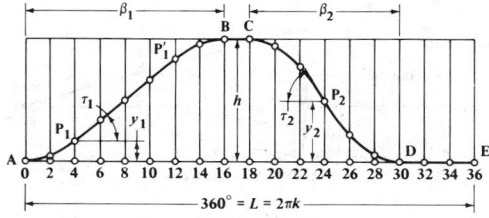

Fig. 17. Displacement Diagram

Determination of Cam Size for Swinging Roller Follower. — The proportioning of a cam with swinging roller follower having specific pressure angles at selected points follows the same procedure as that for a translating follower.

Example: Given the diagram for the roller displacement along its circular arc, Fig. 17 with $h = 1.95$ in., the periods of rise and fall, respectively, $\beta_1 = 160°$ and $\beta_2 = 120°$, the length of the swinging follower arm $L_f = 3.52$ in., rotation of the cam away from pivot point M, and pressure angles $\alpha_1 = \alpha_2 = 45°$ (corresponding to the points P_1 and P_2 in the displacement diagram). Find the cam proportions.

Solution: Distances $k \tan \tau_1$ and $k \tan \tau_2$ are determined as in the previous example, Fig. 14. In Fig. 18, R_{y1} is determined by making the distance $R_u R_{y1} = y_1$ along the arc $R_u R_o$ and R_{y2} by making $R_u R_{y2} = y_2$. The arc $R_u R_o = h$ and R_u indicates the

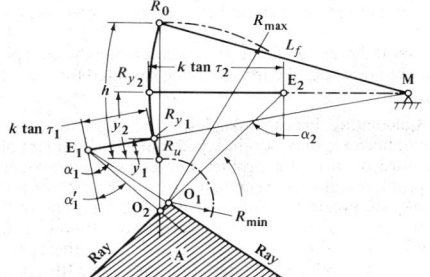

Fig. 18. Finding Proportions of Cam; Swinging Roller Follower
(CCW Rotation)

lowest position of the center of the swinging roller follower and R_o the highest position.

Because the cam (i.e., the surface of the cam as it passes under the follower roller) rotates away from pivot point M, $k \tan \tau_1$ is laid out away from M, that is, from R_{y1} to E_1 and $k \tan \tau_2$ is laid out toward M from R_{y2} to E_2. Angle α_1 at E_1 determines one ray and α_2 at E_2 another ray, which together subtend an area A having the property that if the cam shaft center is chosen inside this area, the pressure angles at the points of the cam corresponding to P_1 and P_2 in the displacement diagram will not exceed the given values α_1 and α_2, respectively. If the cam shaft center is chosen on the ray drawn from E_1 at an angle $\alpha_1 = 45°$, the pressure angle α_1 on the cam profile corresponding to point P_1 will be exactly 45°, and if chosen on the ray from E_2, the pressure angle α_2 corresponding to P_2 will be exactly 45°. If another point, O_2 for example, is chosen as the cam shaft center, the pressure angle corresponding to P_1 will be α'_1 and that corresponding to P_2 will be α_2.

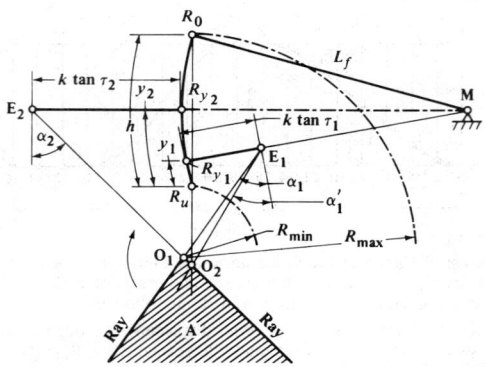

Fig. 19. Finding Proportions of Cam; Swinging Roller Follower (CW Rotation)

Figure 19 shows the construction for rotation toward pivot point M (clockwise rotation of the cam in this case). The layout of the cam curve is made in a manner similar to that shown previously in Fig. 12.

In this example, the cam mechanism was so proportioned that the pressure angles at certain points (corresponding to P_1 and P_2) do not exceed certain specified values (namely α_1 and α_2).

To make sure that the pressure angle at *no point* along the cam profile exceeds the specified value, the previous procedure should be repeated for a series of points along the profile.

Formulas for Calculating Pressure Angles. — The graphical methods described previously are useful because they permit layout and measurement of pressure angles and radii of curvature of *any* cam profile. For cams of complicated profiles, and especially if the profile cannot be represented by a simple formula, the graphical method may be the only practical solution. However, for some of the standard cam profiles utilizing *radial* translating roller followers, the following formulas may be used to determine key cam dimensions before laying out the cam. These formulas enable the designer to specify the maximum pressure angle (usually 30° or less) and, using the specified value, to calculate the minimum cam size that will satisfy the requirement.

The following symbols are in addition to those on pages 2050–2051.

α_{max} = specified maximum pressure angle, degrees;
$R_{\alpha\ max}$ = radius from cam center to point on pitch curve where α_{max} is located, inches;
ϕ_p = rise angle, in degrees, corresponding to α_{max} and $R_{\alpha\ max}$;
α = pressure angle at any selected point, degrees;
R_α = radius from cam center to pitch curve at α, inches;
ϕ = rise angle, in degrees, corresponding to α and R_α.

For Uniform Velocity Motion:

$$\alpha = \arctan\left[\frac{180°h}{\pi\beta R_\alpha}\right] \quad \text{at radius } R_\alpha \text{ to the pitch curve.} \tag{5a}$$

$$\alpha_{max} = \arctan\left[\frac{180°h}{\pi\beta R_{min}}\right] \quad \text{at radius } R_{min} \text{ of the pitch curve } (\phi = 0°). \tag{5b}$$

If α_{max} is specified, then the minimum radius to the lowest point on the pitch curve, R_{min}, is:

$$R_{min} = \frac{180°h}{\pi\beta \tan \alpha_{max}} \quad \text{which corresponds to } \phi = 0°. \tag{5c}$$

For Parabolic Motion:

$$\alpha = \arctan\left[\frac{720°h\phi}{\pi\beta^2 R_\alpha}\right] \quad \begin{array}{l}\text{at radius } R_\alpha \text{ to the pitch curve at angle } \phi, \text{ where}\\ 0 \le \phi \le \beta/2\end{array} \tag{6a}$$

and $\alpha = \arctan[720°h(1 - \phi/\beta)/(\pi\beta R_\alpha)]$ at radius R_α to the pitch curve at angle ϕ, where $\beta/2 \le \phi \le \beta$.

$$\alpha_{max} = \arctan\left[\frac{360°h}{\pi\beta R_\alpha}\right] \quad \text{which occurs at } \phi = \beta/2 \text{ and } R_\alpha = R_{min} + h/2. \tag{6b}$$

If α_{max} is specified, then the minimum radius to the lowest point of the pitch curve is:

$$R_{min} = \left[\frac{360°h}{\pi\beta \tan \alpha_{max}} - \frac{h}{2}\right] \quad \text{which corresponds to } \phi = 0°. \tag{6c}$$

For Simple Harmonic Motion:

$$\alpha = \arctan\left[\frac{90°h}{\beta R_\alpha} \sin\left(\frac{180°\phi}{\beta}\right)\right] \quad \begin{array}{l}\text{at radius } R_\alpha \text{ to the pitch curve at}\\ \text{angle } \phi.\end{array} \tag{7a}$$

$$\phi_p = \left(\frac{\beta}{180°}\right)\left[\operatorname{arccot}\left(\frac{\beta}{180°} \tan \alpha_{max}\right)\right] \quad \begin{array}{l}\phi_p = \text{value of } \phi \text{ where specified}\\ \text{pressure angle } \alpha_{max} \text{ occurs.}\end{array} \tag{7b}$$

$$R_{\alpha\ max} = \frac{h \left[\sin (180°\phi_p/\beta)\right]^2}{2 \cos (180°\phi_p/\beta)} \quad \text{at point where } \alpha = \alpha_{max} \text{ and } \phi = \phi_p. \tag{7c}$$

$$R_{min} = R_{\alpha\ max} - \frac{h}{2}\left[1 - \cos\left(\frac{180°\phi_p}{\beta}\right)\right] \tag{7d}$$

For Cycloidal Motion:

$$\alpha = \arctan\left[\frac{180°h}{\pi\beta R_\alpha}\left[1 - \cos\left(\frac{360°\phi}{\beta}\right)\right]\right] \quad \begin{array}{l}\text{at radius } R_\alpha \text{ to the pitch curve} \\ \text{at angle } \phi.\end{array} \quad (8a)$$

$$\phi_p = \frac{\beta}{180°}\left[\text{arccot}\left(\frac{\beta \tan \alpha_{\max}}{360°}\right)\right] \quad \begin{array}{l}\phi_p = \text{value of } \phi \text{ where specified} \\ \text{pressure angle } \alpha_{\max} \text{ occurs.}\end{array} \quad (8b)$$

$$R_{\alpha\max} = \frac{h}{2\pi}\frac{[1 - \cos(360°\phi_p/\beta)]^2}{\sin(360°\phi_p/\beta)} \quad \begin{array}{l}\text{at point where } \alpha = \alpha_{\max} \\ \text{and } \phi = \phi_p.\end{array} \quad (8c)$$

$$R_{\min} = R_{\alpha\max} - h\left[\frac{\phi_p}{\beta} - \frac{1}{2\pi}\sin\left(\frac{360°\phi_p}{\beta}\right)\right] \quad (8d)$$

Radius of Curvature. — The minimum radius of curvature of a cam should be kept as large as possible (1) to prevent undercutting of the convex portion of the cam and (2) to prevent too high surface stresses. Figures 20(a), (b) and (c) illustrate how undercutting occurs.

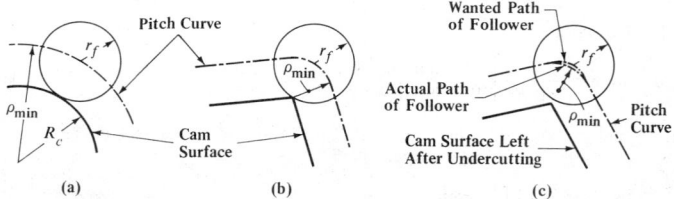

Fig. 20. (a) No Undercutting. (b) Sharp Corner on Cam. (c) Undercutting

In Fig. 20(a) the radius of curvature of the path of the follower is $\rho_{\min}$ and the cam will at that point have a radius of curvature $R_c = \rho_{\min} - r_f$.

In Fig. 20(b) $\rho_{\min} = r_f$ and $R_c = 0$. Therefore, the actual cam will have a sharp corner which in most cases will result in too high surface stresses.

In Fig. 20(c) is shown the case where $\rho_{\min} < r_f$. This case is not possible because undercutting will occur and the actual motion of the roller follower will deviate from the desired one as shown.

Undercutting cannot occur at the *concave* portion of the cam profile (working surface), but caution should be exerted in not making the radius of curvature equal to the radius of the roller follower. This condition would occur if there is a cusp on the displacement diagram which, of course, should always be avoided. To enable milling or grinding of *concave* portions of a cam profile, the radius of curvature of concave portions of the cam, $R_c = \rho_{\min} + r_f$, must be larger than the radius of the cutter to be used.

The radius of curvature is used in calculating surface stresses (see following section), and may be determined by measurement on the cam layout or, in the case of radial translating followers, may be calculated using the formulas that follow. Although these formulas are exact for radial followers, they may be used for offset and swinging followers to obtain an approximation.

Based upon polar coordinates, the radius of curvature is:

$$\rho = \frac{\left[r^2 + \left(\dfrac{dr}{d\phi} \right)^2 \right]^{3/2}}{r^2 + 2 \left(\dfrac{dr}{d\phi} \right)^2 - r \left(\dfrac{d^2r}{d\phi^2} \right)} \qquad (9)^*$$

In Eq. (9), $r = (R_{min} + y)$, where R_{min} is the smallest radius to the pitch curve (see Fig. 12) and y is the displacement of the follower from its lowest position given in terms of ϕ, the angle of cam rotation. The following formulas for r, $dr/d\phi$, and $d^2r/d\phi^2$ may be substituted into Eq. (9) to calculate the radius of curvature at any point of the cam pitch curve; however, to determine the possibility of undercutting of the convex portion of the cam, it is the minimum radius of curvature on the convex portion, ρ_{min}, that is needed. The minimum radius of curvature occurs, generally, at the point of maximum *negative* acceleration.

Parabolic motion:

$$r = R_{min} + h - 2h \left(1 - \frac{\phi}{\beta} \right)^2 \qquad (10a)\dagger$$

$$\frac{dr}{d\phi} = \frac{720^\circ h}{\pi\beta} \left(1 - \frac{\phi}{\beta} \right) \qquad \frac{\beta}{2} \leq \phi \leq \beta \qquad (10b)\dagger$$

$$\frac{d^2r}{d\phi^2} = \frac{-4(180^\circ)^2 h}{\pi^2\beta^2} \qquad (10c)\dagger$$

The minimum radius of curvature can occur at either $\phi = \beta/2$ or at $\phi = \beta$, depending on the magnitudes of h, R_{min}, and β. Therefore, to determine which is the case, make two calculations using formula (9), one for $\phi = \beta/2$, and the other for $\phi = \beta$.

Simple harmonic motion:

$$r = R_{min} + \frac{h}{2} \left[1 - \cos \left(\frac{180^\circ \phi}{\beta} \right) \right] \qquad (11a)$$

$$\frac{dr}{d\phi} = \frac{180^\circ h}{2\beta} \sin \left(\frac{180^\circ \phi}{\beta} \right) \qquad 0 \leq \phi \leq \beta \qquad (11b)$$

$$\frac{d^2r}{d\phi^2} = \frac{(180^\circ)^2 h}{2\beta^2} \cos \left(\frac{180^\circ \phi}{\beta} \right) \qquad (11c)$$

The minimum radius of curvature can occur at either $\phi = \beta/2$ or at $\phi = \beta$, depending on the magnitudes of h, R_{min}, and β. Therefore, to determine which is the case, make two calculations using formula (9), one for $\phi = \beta/2$, and the other for $\phi = \beta$.

---

* Positive values (+) indicate convex curve; negative values (−), concave.
† These equations are for the deceleration portion of the curve as explained in the footnote to Table 1.

Cycloidal motion:

$$r = R_{\min} + h \left[\frac{\phi}{\beta} - \frac{1}{2\pi} \sin\left(\frac{360°\phi}{\beta}\right) \right] \tag{12a}$$

$$\frac{dr}{d\phi} = \frac{180°h}{\pi\beta} \left[1 - \cos\left(\frac{360°\phi}{\beta}\right) \right] \qquad 0 \leqq \phi \leqq \beta \tag{12b}$$

$$\frac{d^2r}{d\phi^2} = \frac{2(180°)^2h}{\pi\beta^2} \sin\left(\frac{360°\phi}{\beta}\right) \tag{12c}$$

$$\rho_{\min} = \frac{[(R_{\min} + 0.91h)^2 + (180°h/\pi\beta)^2]^{3/2}}{(R_{\min} + 0.91h)^2 + 2(180°h/\pi\beta)^2 + (R_{\min} + 0.91h)\,[2(180°)^2h/\pi\beta^2]} \tag{12d}$$

($\rho_{\min}$ occurs near $\phi = 0.75\beta$.)

Example: Given $h = 1$ in., $R_{\min} = 2.9$ in., and $\beta = 60°$. Find $\rho_{\min}$ for parabolic motion, simple harmonic motion, and cycloidal motion.

Solution:

$\rho_{\min} = 2.02$ in. for parabolic motion, from Eq. (9).
$\rho_{\min} = 1.8$ in. for simple harmonic motion, from Eq. (9).
$\rho_{\min} = 1.6$ in. for cycloidal motion, from Eq. (12d).

The value of $\rho_{\min}$ on *any* cam may also be obtained by measurement on the layout of the cam using a compass.

Cam Forces, Contact Stresses, and Materials. — After a cam and follower configuration has been determined, the forces acting on the cam may be calculated or otherwise determined. Next, the stresses at the cam surface are calculated and suitable materials to withstand the stress are selected. If the calculated maximum stress is too great, it will be necessary to change the cam design. Such changes may include: (1) increasing the cam size to decrease pressure angle and increase the radius of curvature; (2) changing to an offset or swinging follower to reduce the pressure angle; (3) reducing the cam rotation speed to reduce inertia forces; (4) increasing the cam rise angle, β, during which the rise, h, occurs; (5) increasing the thickness of the cam, provided that deflections of the follower are small enough to maintain uniform loading across the width of the cam; (6) using a more suitable cam curve or modifying the cam curve at critical points.

Although parabolic motion seems to be the best with respect to minimizing the calculated maximum acceleration and, therefore, also the maximum acceleration forces, nevertheless, in the case of high speed cams, cycloidal motion yields the lower maximum acceleration forces. Thus, it can be shown that owing to the sudden change in acceleration (called *jerk* or *pulse*) in the case of parabolic motion, the actual forces acting on the cam are doubled and sometimes even tripled at high speed, whereas with cycloidal motion, owing to the gradually changing acceleration, the actual dynamic forces are only slightly higher than the theoretical. Therefore, the calculated force due to acceleration should be multiplied by at least a factor of 2 for parabolic and 1.05 for cycloidal motion to provide an allowance for the load-increasing effects of elasticity and backlash.

The main factors influencing cam forces are: (1) displacement and cam speed (forces due to acceleration); (2) dynamic forces due to backlash and flexibility; (3) linkage dimensions which affect weight and weight distribution; (4) pressure angle and friction forces; and (5) spring forces.

The main factors influencing stresses in cams are: (1) radius of curvature for cam and roller; and (2) materials.

Acceleration Forces: The formula for the force acting on a translating body given an acceleration a is:

$$R = \frac{Wa}{g} = \frac{Wa}{386} \qquad (13)$$

In this formula, $g = 386$ inches/second squared; $a =$ acceleration of W in inches/second squared; $R =$ resultant of all the external forces (except friction) acting on the weight W. For cam analysis purposes, W, in pounds, consists of the weight of the follower, a portion of the weight of the return spring ($\frac{1}{3}$), and the weight of the members of the external mechanism against which the follower pushes, for example, the weight of a piston:

$$W = W_f + \tfrac{1}{3} W_s + W_e \qquad (14)$$

where $W =$ equivalent single weight; $W_f =$ follower weight; $W_s =$ spring weight; and $W_e =$ external weight, all in pounds.

Spring Forces: The return spring, K_s, shown in Fig. 21a must be strong enough to hold the follower against the cam at all times. At high cam speeds the main force attempting to separate the follower from the cam surface is the acceleration force R at the point of maximum *negative* acceleration. Thus, at that point the spring must exert a force F_s,

$$F_s = R - W_f - F_e - F_f \qquad (15)$$

where $F_e =$ external force resisting motion of follower, and $F_f =$ friction force from follower guide bushings and other sources.

When the follower is at its lowest position (R_{min} in Fig. 21a), it is usual practice to have the spring provide some estimated preload to account for "set" that takes place in a spring after repeated use and to prevent roller sliding at the start of movement.

The required spring constant, K_s, in pounds per inch of spring deflection is:

$$K_s = \frac{F_s - \text{preload}}{y_a} \qquad (16)$$

where $y_a =$ rise of cam from R_{min} to height at which maximum negative acceleration takes place.

The force, F_y, that the spring exerts at any height y above R_{min} is:

$$F_y = yK_s + \text{preload} \qquad (17)$$

Pressure Angle and Friction Forces: As shown in Fig. 21b, the pressure angle of the cam causes a sideways component $F_n \sin \alpha$ which produces friction forces μF_1 and μF_2 in the guide bushing. If the follower rod is too flexible, bending of the follower will increase these friction forces. The effect of the friction forces and the pressure angle are taken into account in the formula,

$$F_n = \frac{P}{\cos \alpha - \dfrac{\mu \sin \alpha}{l_2}(2l_1 + l_2 - \mu d)} \qquad (18)$$

where $\mu =$ coefficient of friction in bushing; l_1, l_2, and d are as shown in Fig. 21; and $P =$ the sum of all the forces acting down against the upward motion of the follower (acceleration force + spring force + follower weight + external force)

$$P = \frac{W \times a}{386} + (yK_s + \text{preload}) + W_f + F_e \qquad (18a)$$

Cam Torque: The follower pressing against the cam causes resisting torques during the rise period and assisting torques during the return period. The maximum value of the resisting torque determines the cam drive requirements. Instantaneous torque values may be calculated from

$$T_o = \frac{30 v F_n \cos \alpha}{\pi N} = (R_{min} + y) F_n \sin \alpha \qquad (19)$$

in which T_o = instantaneous torque in pound-inches.

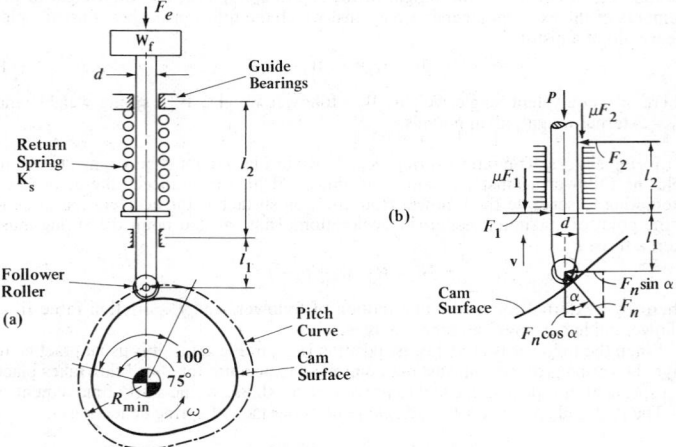

Fig. 21. (a) Radial Translating Follower and Cam System.
(b) Force Acting on a Translating Follower

Example of Force Analysis: A radial translating follower system is shown in Fig. 21a. The follower is moved with cycloidal motion over a distance of 1 in. and an angle of lift $\beta = 100°$. Cam speed $N = 900$ rpm. The weight of the follower mass, W_f, is 2 pounds. Both the spring weight W_s and the external weight W_e are negligible. The follower stem diameter is 0.75 in., $l_1 = 1.5$ in., $l_2 = 4$ in., coefficient of friction $\mu = 0.05$, external force $F_e = 10$ lbs, and the pressure angle is not to exceed 30°.

(a) What is the smallest radius R_{min} to the pitch curve?

From formula (8b) the rise angle ϕ_p to where the maximum pressure angle α_{max} exists is:

$$\phi_p = \frac{\beta}{180°} \left[\operatorname{arccot} \left(\frac{\beta \tan \alpha_{max}}{360°} \right) \right] = \frac{100°}{180°} \left[\operatorname{arccot} \left(\frac{100° \times \tan 30°}{360°} \right) \right]$$

$$= 44.94° = 45°$$

From formula (8c) the radius, $R_{\alpha\, max}$, at which the angle of rise is ϕ_p is:

$$R_{\alpha\, max} = \frac{h}{2\pi} \frac{[1 - \cos (360°\phi_p/\beta)]^2}{\sin (360°\phi_p/\beta)} = \frac{1}{2\pi} \frac{[1 - \cos [(360° \times 45°)/100°]]^2}{\sin [(360° \times 45°)/100°]} = 1.96 \text{ in.}$$

From formula (8d), R_{min} is given by

$$R_{min} = R_{\alpha\ max} - h \left[\frac{\phi_p}{\beta} - \frac{1}{2\pi} \sin \left(\frac{360° \ \phi_p}{\beta} \right) \right]$$

$$= 1.96 - 1 \times \left[\frac{45°}{100°} - \frac{1}{2\pi} \sin \left(\frac{360° \times 45°}{100°} \right) \right] = 1.560 \text{ in.}$$

The same results could have been obtained graphically. If this R_{min} is too small, i.e., if the cam bore and hub require a larger cam, then R_{min} can be increased, in which case the maximum pressure angle will be less than 30°.

(b) If the return spring K_s is specified to provide a preload of 36 lbs when the follower is at R_{min}, what is the spring constant required to hold the follower on the cam throughout the cycle?

The follower tends to leave the cam at the point of maximum *negative* acceleration. Figure 7 shows this to be at $\phi = \frac{3}{4}\beta = 75°$.

From formula (4c),

$$a = \frac{2\pi h \omega^2}{\beta^2} \sin \left(\frac{360° \phi}{\beta} \right) = \frac{2\pi \times 1 \times (6 \times 900)^2}{(100°)^2} \sin \left(\frac{360° \times 75°}{100°} \right) = -18,300 \text{ in./sec}^2$$

From formulas (13) and (14),

$$R = \frac{Wa}{386} = \frac{(W_f + \frac{1}{3}W_s + W_e)\ a}{386} = \frac{(2 + 0 + 0)\ (-18,300)}{386} = 95 \text{ lbs (upward)}$$

Using formula (15) to determine the spring force F_s to hold the follower on the cam,

$$F_s = R - W_f - F_e - F_f$$

as stated on page 2067, the value of R in the above formula should be multiplied by 1.05 for cycloidal motion to provide a factor of safety for dynamic pulses. Thus,

$$F_s = 1.05\ R - W_f - F_e - F_f = 1.05 \times 95 - 2 - 10 - 0 = 88 \text{ lbs (downward)}$$

The spring constant from formula (16) is:

$$K_s = \frac{F_s - \text{preload}}{y_a} = \frac{88 - 36}{y_a}$$

and, from formula (4a) y_a is:

$$y_a = h \left[\frac{\phi}{\beta} - \frac{1}{2\pi} \sin \left(\frac{360° \phi}{\beta} \right) \right] = 1 \times \left[\frac{75°}{100°} - \frac{1}{2\pi} \sin \left(\frac{360° \times 75°}{100°} \right) \right] = 0.909 \text{ in.}$$

so that $K_s = (88 - 36)/0.909 = 57$ lb/in.

(c) At the point where the pressure angle α_{max} is 30° ($\phi = 45°$) the rise of the follower is $1.96 - 1.56 = 0.40$ in. What is the normal force, F_n, on the cam? From formulas (18) and (18a)

$$F_n = \frac{Wa/386 + yK_s + \text{preload} + W_f + F_e}{\cos \alpha - \dfrac{\mu \sin \alpha}{l_2} (2l_1 + l_2 - \mu d)}$$

using $\phi = 45°$, $h = 1$ in., $\beta = 100°$, and $\omega = 6 \times 900$ in formula (4c) gives $a = 5660$ in./sec^2. So that, with $W = 2$ lbs, $y = 0.4$, $K_s = 57$, preload = 36 lbs, $W_f = 2$ lbs, $F_e = 10$ lbs, $\alpha = 30°$, $\mu = 0.05$, $l_1 = 1.5$, $l_2 = 4$, and $d = 0.75$,

$$F_n = \frac{(2 \times 5660)/386 + 0.4 \times 57 + 36 + 2}{\cos 30° - \dfrac{0.05 \times \sin 30°}{4}(2 \times 1.5 + 4 - 0.05 \times 0.75)} = 110 \text{ lbs}$$

Note: If the coefficient of friction had been assumed to be 0, then $F_n = 104$; on the other hand, if the follower is too flexible, so that sidewise bending occurs causing jamming in the bushing, the coefficient of friction may increase to, say, 0.5, in which case the calculated $F_n = 200$ lbs.

(d) Assuming that in the manufacture of this cam that an error or "bump" resulting from a chattermark or as a result of poor blending occurred, and that this "bump" rose to a height of 0.001 in. in a 1° rise of the cam in the vicinity of $\phi = 45°$. What effect would this bump have on the acceleration force R?

One formula that may be used to calculate the change in acceleration caused by such a cam error is:

$$\Delta a = \pm 2e \left(\frac{6N}{\Delta\phi}\right)^2 \qquad (20)$$

where Δa = change in acceleration,
$\qquad e$ = error in inches,
$\qquad \Delta\phi$ = width of error in degrees. The plus $(+)$ sign is used for a "bump" and the minus $(-)$ sign for a dent or hollow in the surface.

For $e = 0.001$, $\Delta\phi = 1°$, and $N = 900$ rpm,

$$\Delta a = +2 \times 0.001 \left(\frac{6 \times 900}{1°}\right)^2 = 58{,}320 \text{ in./sec}^2$$

which is 10 times the acceleration calculated for a perfect cam and would cause sufficient force F_n to damage the cam surface. On high speed cams, therefore, accuracy is of considerable importance.

(e) What is the cam torque at $\phi = 45°$?
From formula (19),

$$T_o = (R_{\min} + y)\, F_n \sin \alpha$$

$$= (1.56 + 0.4) \times 110 \times \sin 30° = 108 \text{ in.-lbs}$$

(f) What is the radius of curvature at $\phi = 45°$?
From formula (9),

$$\rho = \frac{\left[r^2 + \left(\dfrac{dr}{d\phi}\right)^2\right]^{3/2}}{r^2 + 2\left(\dfrac{dr}{d\phi}\right)^2 - r\left(\dfrac{d^2r}{d\phi^2}\right)}$$

$$r = R_{\min} + y = 1.56 + 0.4 = 1.96$$

From formula (12b),

$$\frac{dr}{d\phi} = \frac{180°h}{\pi\beta}\left[1 - \cos\left(\frac{360°\phi}{\beta}\right)\right] = \frac{180° \times 1}{\pi \times 100°}\left[1 - \cos\left(\frac{360° \times 45°}{100°}\right)\right]$$

$$= 1.12$$

From formula (12c),

$$\frac{d^2r}{d\phi^2} = \frac{2(180°)^2 h}{\pi\beta^2} \sin\left(\frac{360°\phi}{\beta}\right) = \frac{2 \times (180°)^2 \times 1}{\pi \times (100°)^2} \sin\left(\frac{360° \times 45°}{100°}\right)$$

$$= 0.64$$

$$\rho = \frac{[(1.96)^2 + (1.12)^2]^{3/2}}{(1.96)^2 + 2(1.12)^2 - 1.96 \times 0.64} = 2.26 \text{ in.}$$

Calculation of Contact Stresses. — When a roller follower is loaded against a cam, the compressive stress developed at the surface of contact may be calculated from

$$S_c = 2290 \sqrt{\frac{F_n}{b}\left(\frac{1}{r_f} \pm \frac{1}{R_c}\right)} \qquad (21)$$

for a steel roller against a steel cam. For a steel roller on a cast iron cam, use 1850 instead of 2290 in Eq. (21).

S_c = maximum calculated compressive stress, psi
F_n = normal load, lb
b = width of cam, in.
R_c = radius of curvature of cam surface, in.
r_f = radius of roller follower, in.

The plus sign in (21) is used in calculating the maximum compressive stress when the roller is in contact with the convex portion of the cam profile and the minus sign is used when the roller is in contact with the concave portion. When the roller is in contact with the straight (flat) portion of the cam profile, $R_c = \infty$ and $1/R_c = 0$. In practice, the greatest compressive stress is most apt to occur when the roller is in contact with that part of the cam profile which is convex and has the smallest radius of curvature.

Example: Given the previous cam example, the radius of the roller $r_f = 0.25$ in., the convex radius of the cam $R_c = (2.26-0.25)$ in., the width of contact $b = 0.3$ in., and the normal load $F_n = 110$ lbs. Find the maximum surface compressive stress. From (21),

$$S_c = 2290 \sqrt{\frac{110}{0.3}\left(\frac{1}{0.25} + \frac{1}{2.01}\right)} = 93,000 \text{ psi}$$

This calculated stress should be less than the allowable stress for the material selected from Table 2.

Cam Materials: In considering materials for cams it is difficult to select any single material as being the best for every application. Often the choice is based on custom or the machinability of the material rather than its strength. However, the failure of a cam or roller is commonly due to fatigue, so that an important factor to be considered is the limiting wear load, which depends on the surface endurance limits of the materials used and the relative hardnesses of the mating surfaces.

In Table 2 are given maximum permissible compressive stresses (surface endurance limits) for various cam materials when in contact with a roller of hardened steel. The stress values shown are based on 100,000,000 cycles or repetitions of stress for pure rolling. Where the repetitions of stress are considerably greater than 100,000,000, where there is appreciable misalignment, or where there is sliding, more conservative stress figures must be used.

Table 2. Cam Materials*

Cam Material for Use with Roller of Hardened Steel	Maximum Allowable Compressive Stress, psi
Gray-iron casting, ASTM A 48-48, Class 20, 160–190 Bhn, phosphate coated	58,000
Gray-iron casting, ASTM A 339-51T, Grade 20, 140–160 Bhn	51,000
Nodular-iron casting, ASTM A 339-51T, Grade 80-60-03, 207–241 Bhn	72,000
Gray-iron casting, ASTM A 48-48, Class 30, 200–220 Bhn	65,000
Gray-iron casting, ASTM A 48-48, Class 35, 225–255 Bhn	78,000
Gray-iron casting, ASTM A 48-48, Class 30, heat treated (Austempered), 255–300 Bhn	90,000
SAE 1020 steel, 130–150 Bhn	82,000
SAE 4150 steel, heat treated to 270–300 Bhn, phosphate coated	220,000
SAE 4150 steel, heat treated to 270–300 Bhn	188,000
SAE 1020 steel, carburized to 0.045 in. depth of case, 50–58 Rc	226,000
SAE 1340 steel, induction hardened to 45–55 Rc	198,000
SAE 4340 steel, induction hardened to 50–55 Rc	226,000

* Based on United Shoe Machinery Corp. data by Guy J. Talbourdet.

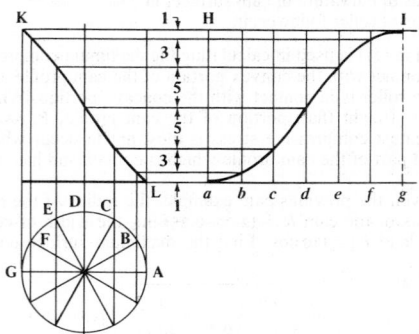

Fig. 22. Development of Cylindrical Cam

Layout of Cylinder Cams. — In Fig. 22 is shown the development of a uniformly accelerated motion cam curve laid out on the surface of a cylindrical cam. This development is necessary for finding the projection on the cylindrical surface, as shown at *KL*. To construct the developed curve, first divide the base circle of the cylinder into, say, twelve equal parts. Set off these parts along line *ag*. Only one-half of the layout has been shown, as the other half is constructed in the same manner, except that the curve is here falling instead of rising. Divide line *aH* into the same number of divisions as the half circle, the divisions being in the proportion 1 : 3 : 5 : 5 : 3 : 1. Draw horizontal lines from these division points and vertical lines from *a*, *b*, *c*, etc. The intersections between the two sets of lines are points on the developed cam curve. These points are transferred to the cylindrical surface at the left by projection in the usual manner.

AFBMA Standard Needle Roller Cam Followers*

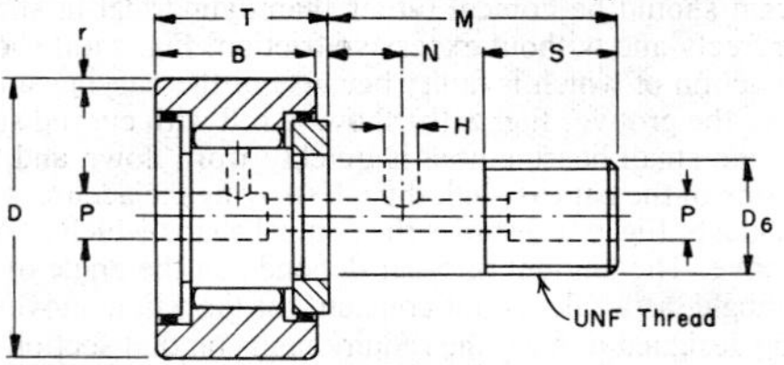

Code No.	Stud Diam. +0.001 −0.000 D_6	Roller O.D. +0.000 −0.001 D	Overall Width T	Roller Width +0.000 −0.005 B	Stud Length M	Thread Length S	Corner Radius r
001CTA	0.1900	0.5000	0.375	0.344	0.5000	0.2500	0.01
003CTA	0.1900	0.5000	0.406	0.375	0.6250	0.2500	0.01
005CTA	0.1900	0.5625	0.406	0.375	0.6250	0.2500	0.01
009CTA	0.2500	0.6250	0.438	0.406	0.6250	0.3125	0.02
013CTA	0.2500	0.6250	0.469	0.438	0.7500	0.3125	0.02
015CTA	0.2500	0.6875	0.469	0.438	0.7500	0.3125	0.02
021CTA	0.3750	0.7500	0.531	0.500	0.8750	0.3750	0.03
027CTA	0.3750	0.8750	0.531	0.500	0.8750	0.3750	0.03
031CTA	0.4375	1.0000	0.656	0.625	1.0000	0.5000	0.05
035CTA	0.4375	1.1250	0.656	0.625	1.0000	0.5000	0.05
045CTA	0.5000	1.2500	0.781	0.750	1.2500	0.6250	0.07
047CTA	0.5000	1.3750	0.781	0.750	1.2500	0.6250	0.07
051CTA	0.6250	1.5000	0.906	0.875	1.5000	0.7500	0.09
057CTA	0.6250	1.6250	0.906	0.875	1.5000	0.7500	0.09
065CTA	0.7500	1.7500	1.031	1.000	1.7500	0.8750	0.10
069CTA	0.7500	1.8750	1.031	1.000	1.7500	0.8750	0.10
075CTA	0.8750	2.0000	1.281	1.250	2.0000	1.0000	0.12
079CTA	0.8750	2.2500	1.281	1.250	2.0000	1.0000	0.12
085CTA	1.0000	2.5000	1.531	1.500	2.2500	1.1250	0.15
093CTA	1.0000	2.7500	1.531	1.500	2.2500	1.1250	0.15
101CTA	1.2500	3.0000	1.781	1.750	2.5000	1.2500	0.18
109CTA	1.2500	3.2500	1.781	1.750	2.5000	1.2500	0.18

Code No.	UNF Thread	N	H (Max)	Hole Diameter P	Bore Diameter for Stud	
					Max	Min
001CTA	10 x 32	.....	.....	0.1250†	0.1905	0.1900
003CTA	10 x 32	.....	.....	0.1250†	0.1905	0.1900
005CTA	10 x 32	.....	.....	0.1250†	0.1905	0.1900
009CTA	¼ x 28	.....	.....	0.1250†	0.2505	0.2500
013CTA	¼ x 28	.....	.....	0.1250†	0.2505	0.2500
015CTA	¼ x 28	.....	.....	0.1250†	0.2505	0.2500
021CTA	⅜ x 24	0.2500	0.0938	0.1875	0.3755	0.3750
027CTA	⅜ x 24	0.2500	0.0938	0.1875	0.3755	0.3750
031CTA	⁷⁄₁₆ x 20	0.2500	0.1250	0.1875	0.4380	0.4375
035CTA	⁷⁄₁₆ x 20	0.2500	0.1250	0.1875	0.4380	0.4375
045CTA	½ x 20	0.3125	0.1250	0.1875	0.5005	0.5000
047CTA	½ x 20	0.3125	0.1250	0.1875	0.5005	0.5000
051CTA	⅝ x 18	0.3750	0.1562	0.1875	0.6255	0.6250
057CTA	⅝ x 18	0.3750	0.1562	0.1875	0.6255	0.6250
065CTA	¾ x 16	0.4375	0.1562	0.1875	0.7505	0.7500
069CTA	¾ x 16	0.4375	0.1562	0.1875	0.7505	0.7500
075CTA	⅞ x 14	0.5000	0.1875	0.1875	0.8755	0.8750
079CTA	⅞ x 14	0.5000	0.1875	0.1875	0.8755	0.8750
085CTA	1 x 14	0.5625	0.1875	0.1875	1.0005	1.0000
093CTA	1 x 14	0.5625	0.1875	0.1875	1.0005	1.0000
101CTA	1¼ x 12	0.6250	0.2500	0.2500	1.2505	1.2500
109CTA	1¼ x 12	0.6250	0.2500	0.2500	1.2505	1.2500

* Anti-Friction Bearing Manufacturers Assn. All dimensions are given in inches.
† Grease fitting hole in head only.
   For larger sizes see AFBMA Standards, Section 2.

Shape of Rolls for Cylinder Cams. — The rolls for cylindrical cams working in a groove in the cam should be conical rather than cylindrical in shape, in order that they may rotate freely and without excessive friction. Fig. 23(a) shows a straight roll and groove, the action of which is faulty because of the varying surface speed at the top and bottom of the groove. Fig. 23(b) shows a roll with curved surface. For heavy work, however, the small bearing area is quickly worn down and the roll presses a groove into the side of the cam as well, thus destroying the accuracy of the movement and creating backlash. Fig. 23(c) shows the conical shape which permits a true rolling action in the groove. The amount of taper depends on the angle of spiral of the cam groove. As this angle, as a rule, is not constant for the whole movement, the roll and groove should be designed to meet the requirements on that section of the cam where the heaviest duty is performed. Frequently the cam groove is of a nearly even spiral angle for a considerable length. The method for determining the angle of the roll and groove to work correctly during the important part of the cycle is as follows:

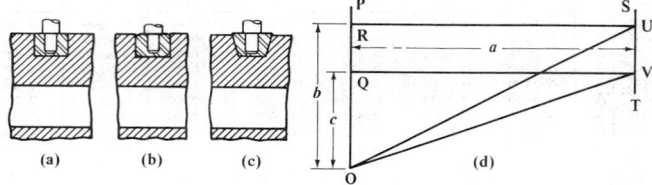

Fig. 23. Shape of Rolls for Cylinder Cams

In Fig. 23(d), b is the circumferential distance on the surface of the cam that includes the section of the groove for which correct rolling action is required. The throw of the cam for this circumferential movement is a. Line OU is the development of the movement of the cam roll during the given part of the cycle, and c is the movement corresponding to b, but on a circle the diameter of which is equal to that of the cam at the bottom of the groove. With the same throw a as before, the line OV will be the development of the cam at the bottom of the groove. OU then is the length of the helix traveled by the top of the roll, while OV is the travel at the bottom of the groove. If, then, the top width and bottom width of the groove be made proportional to OU and OV, the groove will be properly proportioned.

Cam Milling. — Plate cams having a constant rise, such as are used on automatic screw machines, can be cut in a universal milling machine, with the spiral head set at an angle α, as shown by the illustration. When the spiral head is set vertical, the "lead" of the cam (or its rise for one complete revolution) is the same as the lead for which the machine is geared; but when the spiral head and cutter are inclined, any lead or rise of the cam can be obtained, provided it is less than the lead for which the machine is geared, that is, less than the forward feed of the table for one turn of the spiral-head spindle. The cam lead, then, can be varied within certain limits by simply changing the inclination α of the spiral head and cutter. In the following formulas for determining this angle of inclination, for a given rise of cam and with the machine geared for a lead, L, selected from the tables beginning on page 1718, let

> $\alpha =$ angle to which index head and milling attachment are set from horizontal as shown in the accompanying diagram;
> $r =$ rise of cam in given part of circumference;
> $L =$ spiral lead for which milling machine is geared;
> $\phi =$ angle in which rise is required, expressed in degrees.

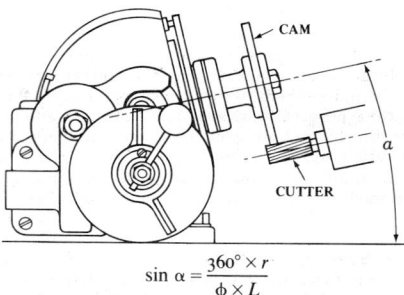

$$\sin \alpha = \frac{360° \times r}{\phi \times L}$$

For example, suppose a cam is to be milled having a rise of 0.125 inch in 300 degrees and that the machine is geared for the smallest possible lead, or 0.670 inch; then:

$$\sin \alpha = \frac{360° \times 0.125}{300° \times 0.670} = 0.2239$$

which is the sine of 12° 56′. Therefore, to secure a rise of 0.125 inch with the machine geared for 0.670 inch lead, the spiral head is elevated to an angle of 12° 56′ and the vertical milling attachment is also swiveled around to locate the cutter in line with the spiral-head spindle, so that the edge of the finished cam will be parallel to its axis of rotation. In the example given, the lead used was 0.670. A larger lead, say 0.930, could have been selected from the table on page 1470. In that case, α = 9° 17′.

When there are several lobes on a cam, having different leads, the machine can be geared for a lead somewhat in excess of the greatest lead on the cam, and then all the lobes can be milled without changing the spiral head gearing, by simply varying the angle of the spiral head and cutter to suit the different cam leads. Whenever possible, it is advisable to mill on the under side of the cam, as there is less interference from chips; moreover, it is easier to see any lines that may be laid out on the cam face. To set the cam for a new cut, it is first turned back by operating the handle of the table feed screw, after which the index crank is disengaged from the plate and turned the required amount.

Simple Method for Cutting Uniform Motion Cams. — Some cams are laid out with dividers, machined and filed to the line; but for a cam that must advance a certain number of thousandths per revolution of spindle this method is not accurate. Cams are easily and accurately cut in the following manner. Let it be required to make the heart cam shown in the illustration. The throw of this cam is 1.1 inch. Now, by setting the index on the milling machine to cut 200 teeth and also dividing 1.1 inch by 100, we find that we have 0.011 inch to recede from or advance towards the cam center for each cut across the cam. Placing the cam securely on an arbor, and the latter between the centers of the milling machine, and using a convex cutter set the proper distance from the center of the arbor, make the first cut across the cam. Then, by lowering the milling machine knee 0.011 inch and turning the index pin the proper number of holes on the index plate, take the next cut and so on.

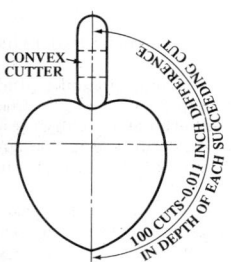

PLAIN BEARINGS

On the pages that follow are given data and procedures for designing full-film or hydrodynamically lubricated bearings of the journal and thrust types. However, before proceeding to these design methods, it may prove useful to review first those bearing aspects concerning the types of bearings available; lubricants and lubrication methods; hardness and surface finish; machining methods; seals; retainers; and typical length-to-diameter ratios for various applications.

The following paragraphs preceding the design sections provide guidance in these matters and suggest modifications in allowable loads when other than full-film operating conditions exist in a bearing.

Classes of Plain Bearings. — Bearings that provide sliding contact between mating surfaces fall into three general classes: *radial bearings* that support rotating shafts or journals; *thrust bearings* that support axial loads on rotating members; and *guide* or *slipper bearings* that guide moving parts in a straight line. Radial sliding bearings, more commonly called sleeve bearings, may be of several types, the most usual being the plain full journal bearing which has 360-degree contact with its mating journal, and the partial journal bearing, which has less than 180-degree contact. This latter type is used when the load direction is constant and has the advantages of simplicity, ease of lubrication, and reduced frictional loss.

The relative motions between the parts of plain bearings may take place: (1) As pure sliding without the benefit of a liquid or gaseous lubricating medium between the moving surfaces such as with the dry operation of nylon or Teflon; (2) With hydrodynamic lubrication in which a wedge or film build-up of lubricating medium is produced, with either whole or partial separation of the bearing surfaces; (3) With hydrostatic lubrication in which a lubricating medium is introduced under pressure between the mating surfaces causing a force opposite to the applied load and a lifting or separation of these surfaces; and (4) With a hybrid form or combination of hydrodynamic and hydrostatic lubrication.

Listed below are some of the advantages and disadvantages of sliding contact (plain) bearings as compared with rolling contact (anti-friction) bearings.

Advantages: 1. Require less space; 2. Are quieter in operation; 3. Are lower in cost, particularly in high volume production; 4. Have greater rigidity; and 5. Their life is generally not limited by fatigue.

Disadvantages: 1. Have higher frictional properties resulting in higher power consumption; 2. Are more susceptible to damage from foreign material in lubrication system; 3. Have more stringent lubrication requirements; and 4. Are more susceptible to damage from interrupted lubrication supply.

Types of Journal Bearings. — Many types of journal bearing configurations have been developed; some of these are shown in Fig. 1.

Circumferential-groove bearings, (a), have an oil groove extending circumferentially around the bearing. The oil is maintained under pressure in the groove. The groove divides the bearing into two shorter bearings which tend to run at a slightly greater eccentricity. However, the advantage in terms of stability is slight, and this design is most commonly used in reciprocating-load main and connecting-rod bearings because of the uniformity of oil distribution.

Short cylindrical bearings are a better solution than the circumferential-groove bearing for high-speed, low-load service. In many cases, the bearing can be shortened enough to increase the unit loading to a substantial value, causing the shaft to ride at a position of substantial eccentricity in the bearing. Experience has shown that instability rarely results when the shaft eccentricity is greater than 0.6. Very short bearings are not often used for this type of application, because they do not provide a high temporary

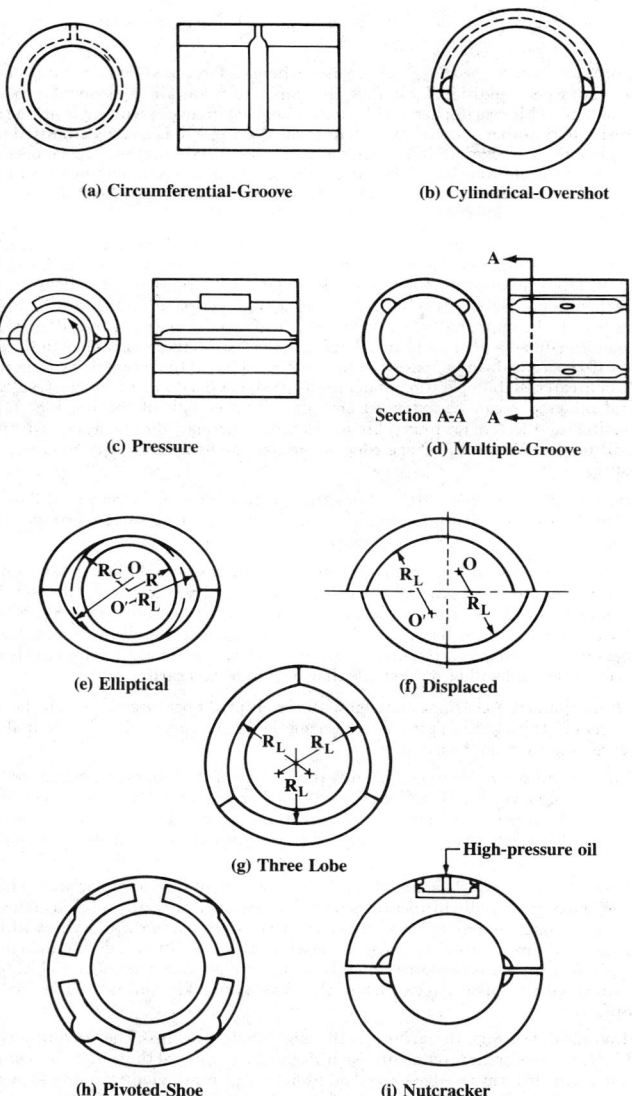

(a) Circumferential-Groove

(b) Cylindrical-Overshot

(c) Pressure

(d) Multiple-Groove

Section A-A

(e) Elliptical

(f) Displaced

(g) Three Lobe

High-pressure oil

(h) Pivoted-Shoe

(i) Nutcracker

Fig. 1. Typical shapes of several types of pressure-fed bearings.

rotating-load capacity in the event some unbalance should be created in the rotor during service.

Cylindrical-overshot bearings, (b), are used where surface speeds of 10,000 fpm or more exist, and where additional oil flow is desired to maintain a reasonable bearing temperature. This bearing has a wide circumferential groove extending from one axial oil groove to the other over the upper half of the bearing. Oil is usually admitted to the trailing-edge oil groove. An inlet orifice is used to control the oil flow. Cooler operation results from the elimination of shearing action over a large section of the upper half of the bearing and, to a great extent, from the additional flow of cool oil over the top half of the bearing.

Pressure bearings, (c), employ a groove over the top half of the bearing. The groove terminates at a sharp dam about 45 degrees beyond the vertical in the direction of shaft rotation. Oil is pumped into this groove by shear action from the rotation of the shaft and is then stopped by the dam. In high-speed operation, this situation creates a high oil pressure over the upper half of the bearing. The pressure created in the oil groove and surrounding upper half of the bearing increases the load on the lower half of the bearing. This self-generated load increases the shaft eccentricity. If the eccentricity is increased to 0.6 or greater, stable operation under high-speed, low-load conditions can result. The central oil groove can be extended around the lower half of the bearing, further increasing the effective loading. This design has one primary disadvantage: Dirt in the oil will tend to abrade the sharp edge of the dam and impair ability to create high pressures.

Multiple-groove bearings, (d), are sometimes used to provide increased oil flow. The interruptions in the oil film also appear to give this bearing some merit as a stable design.

Elliptical bearings, (e), are not truly elliptical, but are formed from two sections of a cylinder. This two-piece bearing has a large clearance in the direction of the split and a smaller clearance in the load direction at right angles to the split. At light loads, the shaft runs eccentric to both halves of the bearing, and hence, the elliptical bearing has a higher oil flow than the corresponding cylindrical bearing. Thus, the elliptical bearing will run cooler and will be more stable than a cylindrical bearing.

Elliptical-overshot bearings, (not shown), are elliptical bearings in which the upper half is relieved by a wide oil groove connecting the axial oil grooves. They are analogous to cylindrical-overshot bearings.

Displaced elliptical bearings, (f), shift the centers of the two bearing arcs in both a horizontal and a vertical direction. This design has greater stiffness than a cylindrical bearing, in both horizontal and vertical directions, with substantially higher oil flow. It has not been extensively used, but offers the prospect of high stability and cool operation.

Three-lobe bearings, (g), are made up in cross section of three circular arcs. They are most effective as anti-oilwhip bearings when the centers of curvature of each of the three lobes lie well outside the clearance circle, which the shaft center can describe within the bearing. Three axial oil-feed grooves are used. It is a more difficult design to manufacture, since it is almost necessary to make it in three parts instead of two. The bore is machined with shims between each of the three parts. The shims are removed after manufacture.

Pivoted-shoe bearings, (h), are one of the most stable bearings. The bearing surface is divided into three or more segments, each of which is pivoted at the center. In operation, each shoe tilts to form a wedge-shaped oil film, thus creating a force tending to push the shaft toward the center of the bearing. For single-direction rotation, the shoes are sometimes pivoted near one end and forced toward the shaft by springs.

Nutcracker bearings, (i), consist of two cylindrical half-bearings. The upper half-bearing is free to move in a vertical direction and is forced toward the shaft by a hydraulic cylinder. External oil pressure may be used to create load on the upper half of the bearing through the hydraulic cylinder. Or, the high-pressure oil may be obtained from the lower half of the bearing by tapping a hole into the high-pressure oil film, thus creating a self-loading bearing. Either type can increase bearing eccentricity to the point where stable operation can be achieved.

Hydrostatic Bearings. — Hydrostatic bearings are used when operating conditions require full film lubrication that cannot be developed hydrodynamically. The hydrostatically lubricated bearing, either thrust or radial, is supplied with lubricant under pressure from an external source. Some advantages of the hydrostatic bearing over bearings of other types are: low friction; high load capacity; high reliability; high stiffness; and long life.

Hydrostatic bearings are used successfully in many applications among which are machine tools, rolling mills, and other heavily loaded slow-moving machinery. However, specialized techniques, including a thorough understanding of hydraulic components external to the bearing package is required. The designer is cautioned against use of this type of bearing without a full knowledge of all aspects of the problem. Determination of the operating performance of hydrostatic bearings is a specialized area of the lubrication field and is described in specialized reference books.

Design. — The design of a sliding bearing is generally accomplished in one of two ways: (1) a bearing operating under similar conditions is used as a model or basis from which the new bearing is designed; (2) in the absence of any previous experience with similar bearings in similar environments, certain assumptions concerning operating conditions and requirements are made and a tentative design prepared based on general design parameters or rules of thumb. Detailed lubrication analysis is then performed to establish design and operating details and requirements.

Modes of Bearing Operation. — The load carrying ability of a sliding bearing depends upon the kind of fluid film which is formed between its moving surfaces. The formation of this film is dependent, in part, on the design of the bearing and, in part, on the speed of rotation. It results in three modes or regions of operation designated as *full film, mixed film* and *boundary* lubrication with effects on bearing friction as shown in Fig. 2.

In terms of physical bearing operation these three modes may be further described as follows:

1. Full film, or hydrodynamic lubrication produces a complete physical separation of the sliding surfaces. This results in low friction and long wear-free service life.

To promote full film lubrication in hydrodynamic operation the following parameters should be satisfied: 1. Lubricant selected has the correct viscosity for the proposed operation; 2. Proper lubricant flow rates are maintained; 3. Proper design methods and considerations have been utilized; and 4. Surface velocity in excess of 25 feet per minute is maintained.

When full film lubrication is achieved, a coefficient of friction between 0.001 and 0.005 can be expected.

2. Boundary lubrication takes place when the sliding surfaces are rubbing together with only an extremely thin film of lubricant present. This type of operation is acceptable only in applications with oscillating or slow rotary motion. In complete boundary lubrication the oscillatory or rotary motion is usually less than 10 feet per minute with resulting coefficients of friction of 0.08 to 0.14. These bearings are usually grease lubricated or periodically oil lubricated.

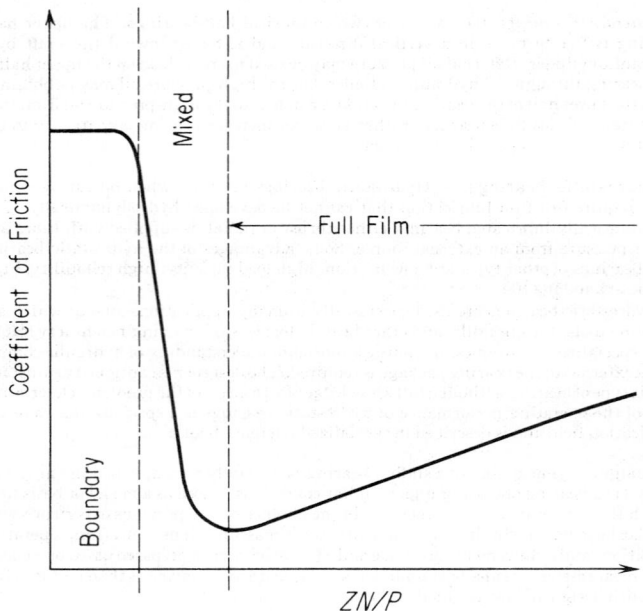

Fig. 2. Three modes of bearing operation.

3. Mixed-film lubrication is a mode of operation between the full film and boundary modes. With this mode there is a partial separation of the sliding surfaces by the lubricant film; however, as in boundary lubrication, limitations on surface speed and wear will result. With this type of lubrication a surface velocity in excess of 10 feet per minute is required with resulting coefficients of friction of 0.02 to 0.08.

A journal bearing in starting up and accelerating to its operating point passes through all three modes of operation. At rest, the journal and bearing are in contact and thus when starting, the operation is in the boundary lubrication region. As the shaft begins to rotate more rapidly and the hydrodynamic film starts to build up, bearing operation enters the region of mixed-film lubrication. When design speeds and loads are reached, the hydrodynamic action in a properly designed bearing will now promote full film lubrication.

Methods of Retaining Bearings. — A number of methods are available to insure that a bearing remain in place within a housing. Which method to use depends upon the particular application but requires first that the unit lends itself to convenient assembly and disassembly; additionally, the bearing wall should be of uniform thickness to avoid introduction of weak points in the construction which may lead to elastic or thermal distortion.

Press or Shrink Fit: One common and satisfactory technique for retaining the bearing is to press or shrink the bearing in the housing with an interference fit. This method permits the use of bearings having uniform wall thickness over the entire bearing length.

Standard bushings with finished inside and outside diameters are available in sizes up to approximately 5-inches inside diameter. Stock bushings are commonly provided 0.002- to 0.003-inch over nominal on outside diameter sizes of 3 inches or less. For diameters greater than 3 inches, actual outside diameters are 0.003 to 0.005 inch over nominal. Since these tolerances are built into standard bushings, the amount of press fit is controlled by the housing bore size.

As a result of a press or shrink fit, the bore of the bearing material "closes in" by some amount. In general, this diameter decrease is approximately 70 to 100 per cent of the amount of the interference fit. Any attempt to accurately predict the amount of close-in, in an effort to avoid final clearance machining, should be avoided.

Shrink fits may be accomplished by chilling the bearing in a mixture of dry-ice and alcohol, or in liquid air. These methods are easier than heating the housing and are preferred. Dry ice in alcohol has a temperature of −110 degrees F and liquid air boils at −310 degrees F.

When a bearing is pressed into the housing, the driving force should be uniformly applied to the end of the bearing to avoid upsetting or peening of the bearing. Of equal importance, the mating surfaces must be clean, smoothly finished, and free of machining imperfections.

Keying Methods: A variety of methods can be used to fix the position of the bearing with respect to its housing by "keying" the two together. Possible keying methods are

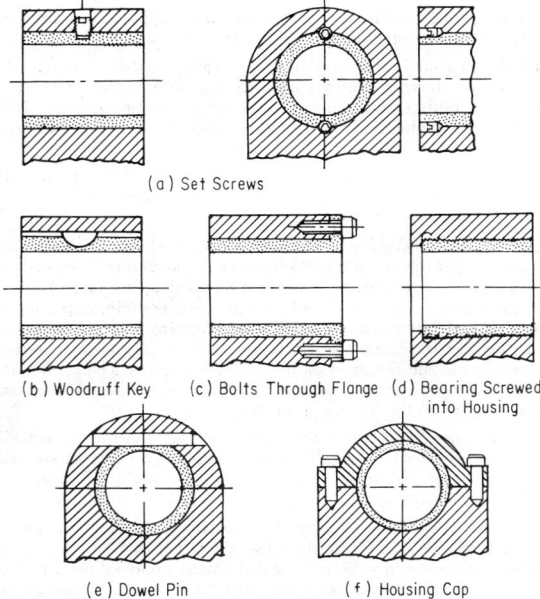

(a) Set Screws

(b) Woodruff Key   (c) Bolts Through Flange   (d) Bearing Screwed into Housing

(e) Dowel Pin   (f) Housing Cap

Fig. 3. Methods of bearing retention.

shown in Fig. 3 including: (a) Set screws; (b) Woodruff keys; (c) Bolted bearing flanges; (d) Threaded bearings; (e) Dowel pins; and (f) Housing caps.

Factors to be considered when selecting one of these methods are:

1. Maintaining uniform wall thickness of the bearing material, if possible, especially in the load-carrying region of the bearing.

2. Providing as much contact area as possible between bearing and housing. Mating surfaces should be clean, smooth, and free from imperfections to facilitate heat transfer.

3. Preventing any local deformation of the bearing that might result from the keying method. Machining after keying is recommended.

4. Considering the possibility of bearing distortion resulting from the effect of temperature changes on the particular keying method.

Methods of Sealing. — In applications where lubricants or process fluids are utilized in operation, provision must be made normally to prevent leakage to other areas. This is accomplished by the use of static and dynamic type sealing devices. In general three terms are used to describe the devices used for sealing:

Seal: A means of preventing migration of fluids, gases, or particles across a joint or opening in a container.

Packing: A dynamic seal, used where some form of relative motion occurs between rigid members of an assembly.

Gaskets: A static seal, used where there is no relative motion between joined parts.

Two major functions must be achieved by all sealing applications: prevent escape of fluid; and prevent migration of foreign matter from the outside.

The first determination in selecting the proper seal is whether the application is static or dynamic. To meet the requirements of a static application, there must be no relative motion betwen the joining parts or between the seal and the mating part. If there is any relative motion, the application must be considered dynamic, and the seal selected accordingly.

Dynamic sealing requires control of fluids leaking between parts with relative motion. Two primary methods are used to this end: positive contact or rubbing seals; and controlled clearance non-contact seals.

Positive Contact or Rubbing Seals: These are utilized where positive containment of liquids or gases is required, or where the seal area is continuously flooded. If properly selected and applied, contact seals can provide zero leakage for most fluids. However, because they are sensitive to temperature, pressure, and speed, improper application can result in early failure. These seals are applicable to rotating and reciprocating shafts. In many cases, the positive-contact seals are available as off-the-shelf items. In other instances, they are custom-designed to the special demands of a particular application. Custom design is offered by many seal manufacturers and, for extreme cases, probably offers the best solution to the sealing problem.

Controlled Clearance Non-Contact Seals: Representative of the controlled-clearance seals, which includes all seals in which there is no rubbing contact between the rotating and stationary members, are throttling bushings and labyrinths. Both of these types operate by fluid-throttling action in narrow annular or radial passages.

Clearance seals are frictionless and very insensitive to temperature and speed. They are chiefly effective as devices for limiting leakage rather than stopping it completely. Although they are employed as primary seals in many applications, the clearance seal also finds use as auxiliary protection in contact-seal applications. These seals are usually designed into the equipment by the designer himself, and they can take on many different forms.

Advantages of this seal are that friction is kept to an absolute minimum and that there is no wear or distortion during the life of the equipment. However, there are two significant disadvantages: The seal has limited use when leakage rates are critical; and it becomes quite costly as the configuration becomes elaborate.

Static Seals: Static seals such as gaskets, "O" rings, and molded packings cover very broad ranges of both design and materials. Some of the typical types are listed as follows: 1. Molded packings: A. Lip type, B. Squeeze-molded; 2. Simple compression packings; 3. Diaphragm seals; 4. Non-metallic gaskets; 5. "O" rings; and 6. Metallic gaskets and "O" rings.

Detailed design information for specific products should be obtained directly from manufacturers.

Hardness and Surface Finish. — Even in well-lubricated full-film sleeve bearings, momentary contact between journal and bearing may occur under such conditions as starting, stopping, or overloading. In mixed-film and boundary-film lubricated sleeve bearings, continuous metal-to-metal contact occurs. Hence, to allow for any necessary wearing-in, the journal is usually made harder than the bearing material. This allows the effects of scoring or wearing to take place on the bearing, which is more easily replaced, rather than on the more expensive shaft. As a general rule, recommended Brinnell hardness of the journal is at least 100 points harder than the bearing material.

The softer cast bronzes used for bearings are those with high lead content and very little tin. Such bronzes give adequate service in boundary and mixed-film applications where full advantage is taken of their excellent "bearing" characteristics.

High-tin, low-lead content cast bronzes are the harder bronzes and these have high ultimate load-carrying capacity: higher journal hardnesses are required with these bearing bronzes. Aluminum bronze, for example, requires a journal hardness in the range of 550 to 600 Bhn.

In general, harder bearing materials require better alignment and more reliable lubrication to minimize local heat generation if and when the journal touches the shaft. Also, abrasives which find their way into the bearing are a problem for the harder bearing and greater care should be taken to exclude them.

Surface Finish: Whether bearing operation is complete boundary, mixed film, or fluid film, surface finish of journal and bearing must receive careful attention. In applications where operation is hydrodynamic or full film, peak surface variations should be less than the expected minimum film thickness; otherwise, peaks on the journal surface will contact peaks on the bearing surface, with resulting high friction and temperature rise. Ranges of surface roughness obtained by various finishing methods are: boring, broaching, and reaming, 32 to 64 microinches, rms; grinding, 16 to 64 microinches, rms; and fine grinding, 4 to 16 microinches, rms.

In general, the better surface finishes are required for full-film bearings operating at high eccentricity ratios since full-film lubrication must be maintained with small clearances, and metal-to-metal contact must be avoided. Also, the harder the material the better the surface finish required. For boundary and mixed-film applications surface finish requirements may be somewhat relaxed since bearing wear-in will in time smooth the surfaces.

Figure 4 is a general guide to the ranges required for bearing and journal surface finishes. Selecting a particular surface finish in each range can be simplified by observing the general rule that smoother finishes are required for the harder materials, for high loads, and for high speeds.

Machining. — The methods most commonly used in finishing journal bearing bores are: boring, broaching, reaming, and burnishing.

Boring of journal bearings provides the best concentricity, alignment, and size control; this being the finishing method of choice when close tolerances and clear-

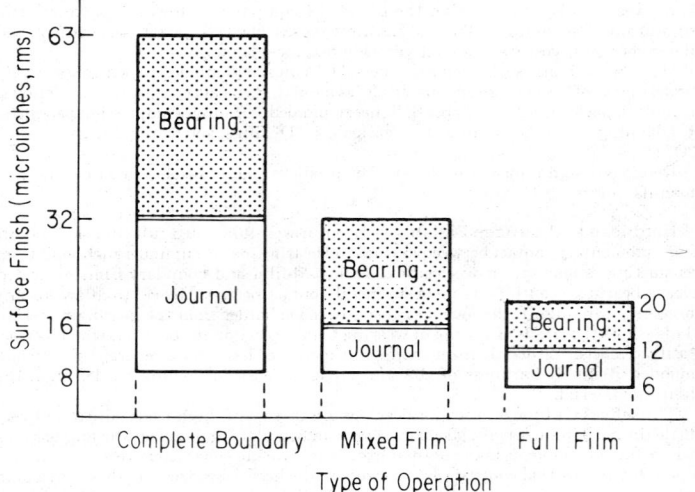

Fig. 4.   Recommended ranges of surface finish for the three types of
sleeve bearing operations.

ances are desirable. Broaching is a rapid finishing method providing good size and alignment control when adequate piloting is possible. Soft babbitt materials are particularly compatible with the broaching method. A third finishing method, reaming, facilitates good size and alignment control when piloting is utilized. Reaming can be accomplished both manually or by machine; the machine method being preferred. Burnishing is a fast sizing operation which gives good alignment control, but does not give as good size control as the cutting methods. It is not recommended for soft materials such as babbitt. Burnishing has an ironing effect which gives added seating of the bushing outside diameter in the housing bore; consequently, it is often used for this purpose, especially on a 1/32 inch wall bushing, even if a further sizing operation is to be used subsequently.

Methods of Lubrication. — There are numerous ways to supply lubricant to bearings. The more common of these are described as follows:

Pressure lubrication, in which an abundance of oil is fed to the bearing from a central groove, single or multiple holes, or axial grooves, is effective and efficient. The moving oil assists in flushing dirt from the bearing and helps keep the bearing cool. In fact it removes heat faster than other lubricating methods and, therefore, permits thinner oil films and unimpaired actual load capacities. The oil supply pressure needed for bushings carrying the basic load is directly proportional to the shaft speed, but for most installations 50 psi will be adequate.

Splash fed is a term applied to a variety of intermittently lubricated bushings. It includes everything from bearings spattered with oil from the action of other moving parts to bearings regularly dipped in oil. Like oil bath lubrication, splash feeding is practical when the housing can be made oiltight and when the moving parts do not churn the oil. The fluctuating nature of the load and the intermittent oil supply in

splash fed applications requires the designer to use experience and judgment when determining the probable load capacity of bearings lubricated in this way.

Oil bath lubrication, in which the bushing is submerged in oil, is the most reliable of all methods except pressure lubrication. It is practical if the housing can be made oiltight, and if the shaft speed is not so great as to cause excessive churning of the oil.

Oil ring lubrication, in which oil is supplied to the bearing by a ring in contact with the shaft will, within reasonable limits, bring enough oil to the bearing to maintain hydrodynamic lubrication. If the shaft speed is too low, little oil will follow the ring to the bearing; and, if the speed is too high, the ring speed will not keep pace with the shaft. Also, a ring revolving at high speed will lose oil by centrifugal force. For best results, the peripheral speed of the shaft should be between 200 and 2000 feet per minute. Safe load to achieve hydrodynamic lubrication should be one-half of that for pressure fed bearings.

Wick or waste pack lubrication delivers oil to a bushing by the capillary action of a wick or waste pack; the amount delivered being proportional to the size of the wick or pack. Unless the load is light, hydrodynamic lubrication is doubtful. The safe load, then, to

Z centistoke $= \left(.22t - \dfrac{180}{t}\right)$ where $t = $ secs Saybolt.

V $m^2/s = Z \times 10^{-6}$

Table 1. Oil Viscosity Unit Conversion

Convert from	Convert to					
	Poise (P)	Centipoise (Z)	Reyn (μ)	Stoke (S)	Centistoke (ν)	Secs Saybolt (t)
	Multiplying Factors					
Poise (P) $\dfrac{\text{dyne-sec}}{\text{cm}^2}$ or $\dfrac{\text{gram mass}}{\text{sec-cm}}$	1	100	1.45×10^{-5}	$\dfrac{1}{\rho}$	$\dfrac{100}{\rho}$	
Centipoise (Z) $\dfrac{\text{dyne-sec}}{100 \text{ cm}^2}$ or $\dfrac{\text{gram mass}}{100 \text{ sec-cm}}$	0.01	1	1.45×10^{-7}	$\dfrac{0.01}{\rho}$	$\dfrac{1}{\rho}$	
Reyn (μ) $\dfrac{\text{lb force-sec}}{\text{in.}^2}$	6.9×10^4	6.9×10^6	1	$\dfrac{6.9 \times 10^4}{\rho}$	$\dfrac{6.9 \times 10^6}{\rho}$	
Stoke (S) $\dfrac{\text{cm}^2}{\text{sec}}$	ρ	100ρ	$1.45 \times 10^{-5} \rho$	1	100	
Centistoke (ν) $\dfrac{\text{cm}^2}{100 \text{ sec}}$	0.01ρ	ρ	$1.45 \times 10^{-7} \rho$	0.01	1	$\left(.22t - \dfrac{180}{t}\right)$

ρ = Specific gravity of the oil.

To convert from a value in the "Convert from" column to a value in a "Convert to" column, multiply the "Convert from" column by the figure in the intersecting block, e.g. to change from Centipoise to Reyn, multiply Centipoise value by 1.45×10^{-7}.

achieve hydrodynamic lubrication, should be one-quarter of that of pressure fed
bearings.

Grease packed in a cavity surrounding the bushing is less adequate than an oil system,
but it has the advantage of being more or less permanent. Although hydrodynamic
lubrication is possible under certain very favorable circumstances, boundary lubrication is the usual state.

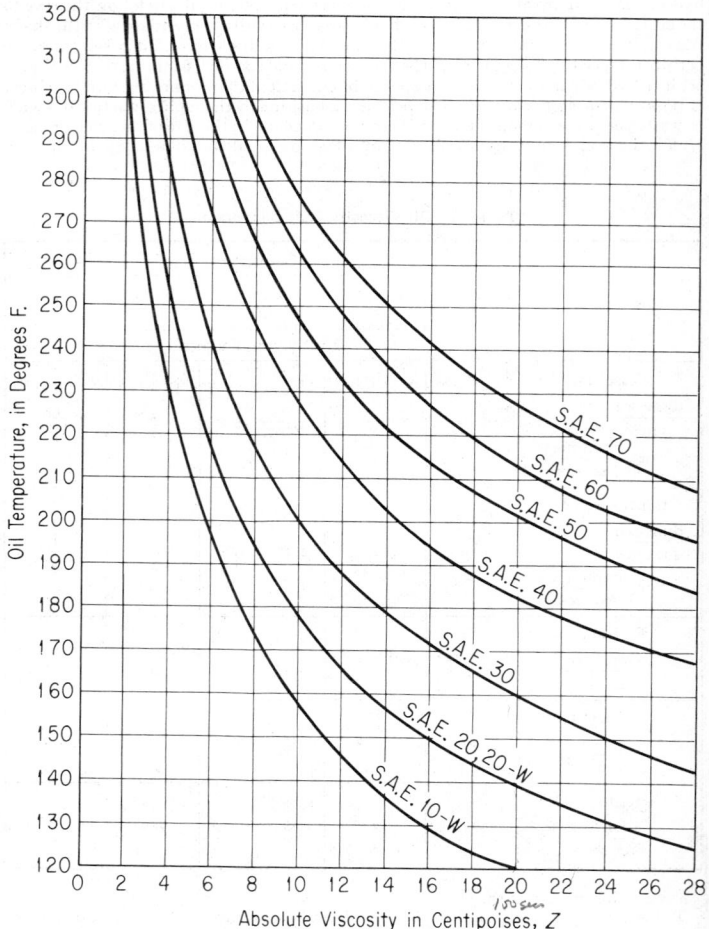

Fig. 5. Viscosity vs. temperature — SAE oils.

Lubricants. — The value of an oil as a lubricant depends mainly upon its film-forming capacity; that is, its capability of maintaining a film of oil between the bearing surfaces. The film-forming capacity depends to a large extent on the viscosity of the oil, but this should not be understood to mean that oil of the highest viscosity is in every case the most suitable lubricant. For practical reasons, an oil of the lowest viscosity which will retain an unbroken oil film between the bearing surfaces is the most suitable for purposes of lubrication. This is because a higher viscosity than that necessary to maintain the oil film results in a waste of power due to the expenditure of energy necessary to overcome the internal friction of the oil itself.

Figure 5 provides representative values of viscosity in centipoises for SAE mineral oils. Table 1 is provided as a means of converting viscosities of other units to centipoises.

Lubricant Selection. — In selecting lubricants for journal bearing operation several factors must be considered: 1. Type of operation (Full-, mixed-, or boundary film) anticipated; 2. Surface speed; and 3. Bearing loading.

Figure 6 combines these factors and facilitates general selection of the proper lubricant viscosity range.

As an example of using these curves, consider a lightly loaded bearing operating at 2000 rpm. At the bottom of the figure locate 2000 rpm and move vertically to intersect the light-load full-film lubrication curve which would indicate an SAE 5 oil.

As a general rule-of-thumb heavier oils are recommended for high loads and lighter oils for high speeds.

In addition, other than using conventional lubricating oils, journal bearings may be lubricated with either greases or solid lubricants. Some of the reasons for use of these lubricants are to:

1. Lengthen the period between relubrication;

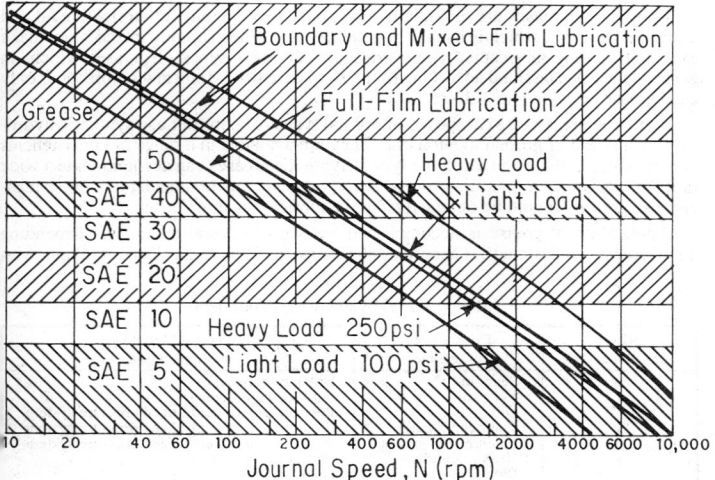

Fig. 6.   Lubricant selection guide.

Table 2. Commonly Used Greases and Solid Lubricants

Type	Operating Temperature, Degrees F	Load	Comments
Greases			
Calcium or lime soap	160	Moderate	. . .
Sodium soap	300	Wide	For wide speed range
Aluminum soap	180	Moderate	. . .
Lithium soap	300	Moderate	Good low temperature
Barium soap	350	Wide	. . .
Solid Lubricants			
Graphite	1000	Wide	. . .
Molybdenum disulfide	− 100 to 750	Wide	. . .

2. Avoid contaminating surrounding equipment or material with "leaking" lubricating oil;

3. Provide effective lubrication under extreme temperature ranges;

4. Provide effective lubrication in the presence of contaminating atmospheres; and

5. Prevent intimate metal-to-metal contact under conditions of high unit pressure which might destroy boundary lubricating films.

Greases: Where full-film lubrication is not possible or is impractical for slow-speed fairly high-load applications, greases are widely used as bearing lubricants. Although full-film lubrication with grease is possible, it is not normally considered since an elaborate pumping system is required to continuously supply a prescribed amount of grease to the bearing. Bearings supplied with grease are usually lubricated periodically. Grease lubrication, therefore, implies that the bearing will operate under conditions of complete boundary lubrication and should be designed accordingly.

Lubricating greases are essentially a combination of a mineral lubricating oil and a thickening agent, which is usually a metallic soap. When suitably mixed, they make excellent bearing lubricants. There are many different types of greases which, in general, may be classified according to the soap base used. Information on commonly used greases is shown in Table 2.

Synthetic greases are composed of normal types of soaps but use synthetic hydrocarbons instead of normal mineral oils. They are available in a range of consistencies in both water-soluble and insoluble types. Synthetic greases can accommodate a wide range of variation in operating temperature; however, recommendations on special-purpose greases should be obtained from the lubricant manufacturer.

Application of grease is accomplished by one of several techniques depending upon grease consistency. These classifications are shown in Table 3 along with

Table 3. NLGI* Consistency Numbers

NLGI Consistency No.	Consistency of Grease	Typical Method of Application
0	Semifluid	Brush or gun
1	Very soft	Pin-type cup or gun
2	Soft	Pressure gun or centralized pressure system
3	Light cup grease	Pressure gun or centralized pressure system
4	Medium cup grease	Pressure gun or centralized pressure system
5	Heavy cup grease	Pressure gun or hand
6	Block grease	Hand, cut to fit

* National Lubricating Grease Institute

typical methods of application. Grooves for grease are generally greater in width, up to 1.5 times, than for oil.

Coefficients of friction for grease-lubricated bearings range from 0.08 to 0.16, depending upon consistency of the grease, frequency of lubrication, and type of grease. An average value of 0.12 may be used for design purposes.

Solid Lubricants: The need for effective high-temperature lubricants led to the development of several solid lubricants. Essentially, solid lubricants may be described as low-shear-strength solid materials. Their function within a bronze bearing is to act as an intermediary material between sliding surfaces. Since these solids have very low shear strength, they shear more readily than the bearing material and thereby allow relative motion. So long as solid lubricant remains between the moving surfaces, effective lubrication is provided and friction and wear are reduced to acceptable levels.

Solid lubricants provide the most effective boundary films in terms of reduced friction, wear, and transfer of metal from one sliding component to the other. However, there is a significant deterioration in these desirable properties as the operating temperature of the boundary film approaches the melting point of the solid film. At this temperature the friction may increase by a factor of 5 to 10 and the rate of metal transfer may increase by as much as 1000. What occurs is that the molecules of the lubricant lose their orientation to the surface that exists when the lubricant is solid. As the temperature further increases, additional deterioration sets in with the friction increasing by some additional small amount but the transfer of metal accelerates by an additional factor of 20 or more. The final effect of too high temperature is the same as metal-to-metal contact without benefit of lubricant. These changes, which are due to the physical state of the lubricant, are reversed when cooling takes place.

The effects just described also partially explain why fatty acid lubricants are superior to paraffin base lubricants. The fatty acid lubricants react chemically with the metallic surfaces to form a metallic soap that has a higher melting point than the lubricant itself, the result being that the breakdown temperature of the film, now in the form of a metallic soap is raised so that it acts more like a solid film lubricant than a fluid film lubricant.

Journal or Sleeve Bearings

Although this type of bearing may take many shapes and forms, there are always three basic components: journal or shaft, bushing or bearing, and lubricant. Figure 7 shows these components with the nomenclature generally used to describe a journal bearing: W = applied load, N = revolution, e = eccentricity of journal center to bearing center, θ = attitude angle, which is the angle between the applied load and the point of minimum film thickness, d = diameter of the shaft, c_d = bearing clearance, $d + c_d$ = diameter of the bearing and h_o = minimum film thickness.

Grooving and Oil Feeding. — Grooving in a journal bearing has two purposes: (1) to establish and maintain an efficient film of lubricant between the bearing moving surfaces and (2) to provide adequate bearing cooling. The obvious and only practical location for introducing lubricant to the bearing is in a region of low pressure. A typical pressure profile of a bearing is shown by Fig. 8. The arrow W shows the applied load. Typical grooving configurations used for journal bearings are shown in Fig. 9.

Heat Radiating Capacity. — In a self-contained lubrication system for a journal bearing, the heat generated by bearing friction must be removed to prevent continued temperature rise to an unsatisfactory level. The heat-radiating capacity H_R of the bearing in foot-pounds per minute may be calculated from the formula $H_R = Ld\,Ct_R$ in which C is a constant determined by O. Lasche, and t_R is temperature rise in degrees Fahrenheit. Values for the product Ct_R may be found from

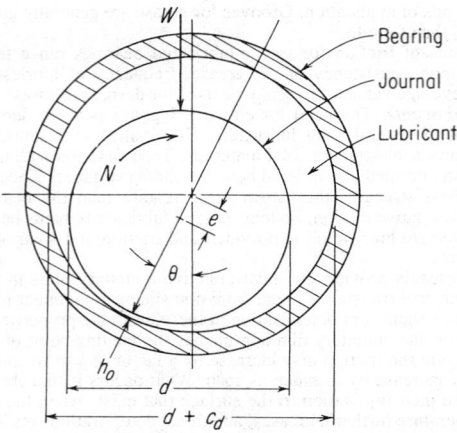

Fig. 7.   Basic components of a journal bearing.

the curves in Fig.10 for various values of bearing temperature rise t_R and for three operating conditions.   In this equation L = the total length of the bearing in inches and d = the bearing diameter in inches.

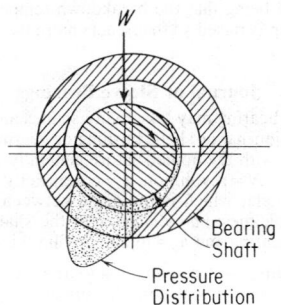

Fig. 8.   Typical pressure profile of journal bearing.

Journal Bearing Design Notation. — The symbols used in the following step-by-step procedure for lubrication analysis and design of a plain sleeve or journal bearing are listed below:

   c = specific heat of lubricant, Btu/lb/°F
   c_d = diametral clearance, inches
   C_n = bearing capacity number
   d = journal diameter, inches
   e = eccentricity, inches

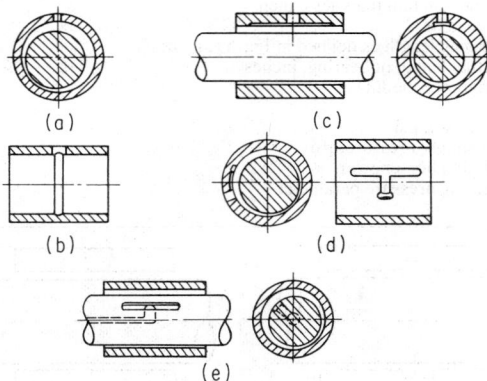

Fig. 9. Types of journal bearing oil grooving: (a) Single inlet hole. (b) Circular groove. (c) Straight axial groove. (d) Straight axial groove with feeder groove. (e) Straight axial groove in shaft.

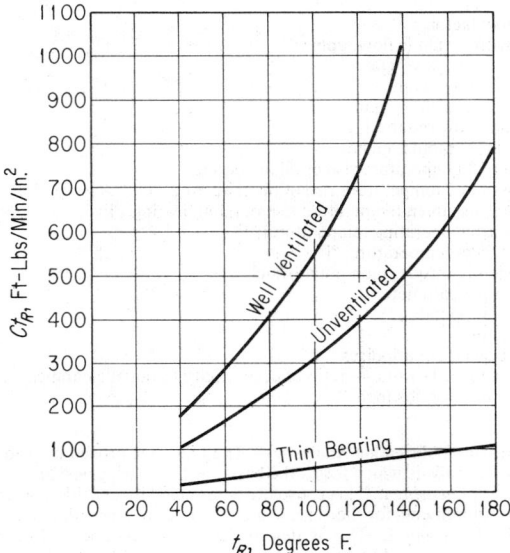

Fig. 10. Heat radiating capacity factor, Ct_R, vs. bearing temperature rise, t_R — journal bearings.

h_o = minimum film thickness, inch
K = constants
l = bearing length as defined in Fig. 11, inches
L = actual length of bearing, inches
m = clearance modulus
N = rpm
p_b = unit load, psi
p_s = oil supply pressure, psi
P_f = friction horsepower
P' = bearing pressure parameter

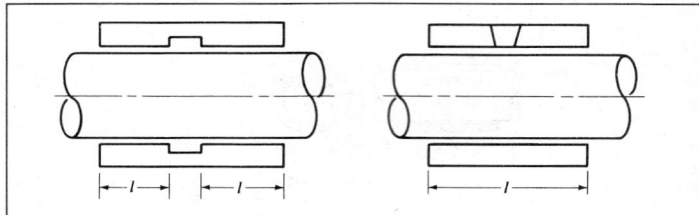

Fig. 11. Length, l, of bearing for circular groove type (left) and single inlet hole type (right).

q = flow factor
Q_1 = hydrodynamic flow, gpm
Q_2 = pressure flow, gpm
Q = total flow, gpm
Q_{new} = new total flow, gpm
Q_R = total flow required, gpm
r = journal radius, inches
Δt = actual temperature rise of oil in bearing, °F
Δt_a = assumed temperature rise of oil in bearing, °F
Δt_{new} = new assumed temperature rise of oil in bearing, °F
t_b = bearing operating temperature, °F
t_{in} = oil inlet temperature, °F
T_f = friction torque, inch-pounds/inch
T' = torque parameter
W = load, pounds
X = factor
Z = viscosity, centipoises
ϵ = eccentricity ratio — ratio of eccentricity to radial clearance
α = oil density, lbs/inch³

Journal Bearing Lubrication Analysis. — The following procedure leads to a complete lubrication analysis which forms the basis for the bearing design.

1. *Diameter of bearing d.* This is usually determined by considering strength and/or deflection requirements for the shaft using principles of strength of materials.

2. *Length of bearing L.* This is determined by an assumed l/d ratio in which l may or may not be equal to the overall length, L (See Step 6). Bearing pressure and the possibility of edge loading due to shaft deflection and misalignment are factors to be considered. In general, shaft misalignment resulting from location

tolerances and/or shaft deflections should be maintained below 0.0003 inch per inch of length.

3. *Bearing pressure p_b.* The unit load in pound per square inch is calculated from the formula:

$$p_b = \frac{W}{Kld}$$

where $K = 1$ for single oil hole
$K = 2$ for central groove
W = load, pounds
l = bearing length as defined in Fig. 11, inches
d = journal diameter, inches

Typical unit loads in service are shown in Table 4. These pressures can be used as a safe guide in selection. However, if space limitations impose a higher limit of loading, the complete lubrication analysis and evaluation of material properties will determine acceptability.

4. *Diametral clearance c_d.* This is selected on a trial basis from Fig. 12 which shows suggested diametral clearance ranges for various shaft sizes and for two speed ranges. These are *hot* or *operating* clearances so that thermal expansion of journal and bearing to these temperatures must be taken into consideration in establishing machining dimensions. The optimum operating clearance should be determined on the basis of a complete lubrication analysis (See paragraph following Step 23).

5. *Clearance modulus m.* This is calculated from the formula:

$$m = \frac{c_d}{d}$$

6. *Length to diameter ratio l/d.* This is usually between 1 and 2; however, with the modern trend toward higher speeds and more compact units, lower ratios down to 0.3 are used. In shorter bearings there is a consequent reduction in load carrying capacity due to excessive end or side leakage of lubricant. In longer bearings there may be a tendency towards edge loading. Length l for a single oil feed hole is taken as the total length of the bearing as shown in Fig. 11. For a central oil groove length, l is taken as one-half the total length.

Typical l/d ratio's use for various types of applications are given in Table 5.

Table 4. Allowable Sleeve Bearing Pressures for Various Classes of Bearings*

Types of Bearing or Kind of Service	Pressure, Lbs. per Sq. In.	Types of Bearing or Kind of Service	Pressure, Lbs. per Sq. In.
Electric Motor & Generator Bearings (General)	100–200	Diesel Engine Rod	1000–2000
Turbine & Reduction Gears	100–250	Wrist Pins	1800–2000
Heavy Line Shafting	100–150	Automotive, Main Bearings	500–700
Locomotive Axles	300–350	Rod Bearings	1500–2500
Light Line Shafting	15–35	Centrifugal Pumps	80–100
Diesel Engine, Main	800–1500	Aircraft Rod Bearings	700–3000

* These pressures in pounds per square inch of area equal to length times diameter are intended as a general guide only. The allowable unit pressure depends upon operating conditions, especially in regard to lubrication, design of bearings, workmanship, velocity, and nature of load.

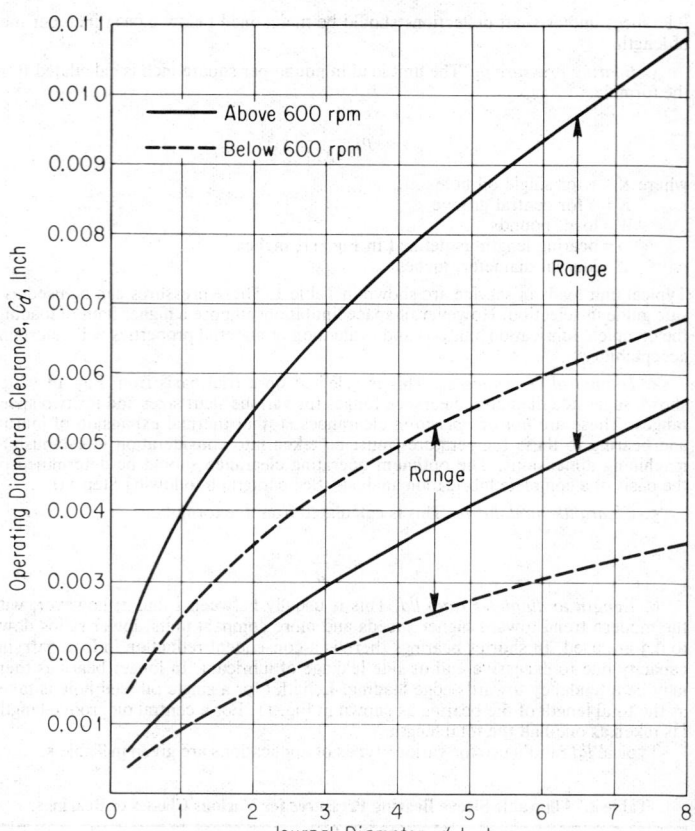

Fig. 12. Operating diametral clearance, c_d vs. journal diameter, d.

Table 5.   Representative l/d Ratios

Type of Service	l/d	Type of Service	l/d
Gasoline and diesel engine		Light shafting	2.5 to 3.5
Main bearings and crankpins	0.3 to 1.0	Heavy shafting	2.0 to 3.0
Generators and motors	1.2 to 2.5	Steam engine	
Turbogenerators	0.8 to 1.5	Main bearings	1.5 to 2.5
Machine tools	2.0 to 3.0	Crank and wrist pins	1.0 to 1.3

7. *Assumed operating temperature t_b.* A temperature rise of the lubricant as it passes through the bearing is assumed and the consequent operating temperature in degrees F. is calculated from the formula:

$$t_b = t_{in} + \Delta t_a$$

where t_{in} = inlet temperature of oil in °F

Δt_a = assumed temperature rise of oil in bearing in °F

An initial assumption of 20°F is usually made.

    8. *Viscosity of lubricant Z.* The viscosity in centipoises at the assumed bearing operating temperature is found from the curve in Fig. 5 which shows the viscosity of SAE grade oils versus temperature.

    9. *Bearing pressure parameter P′.* This value is required to find the eccentricity ratio and is calculated from the formula:

$$P' = \frac{6.9(1000m)^2 p_b}{ZN}$$

where N = rpm

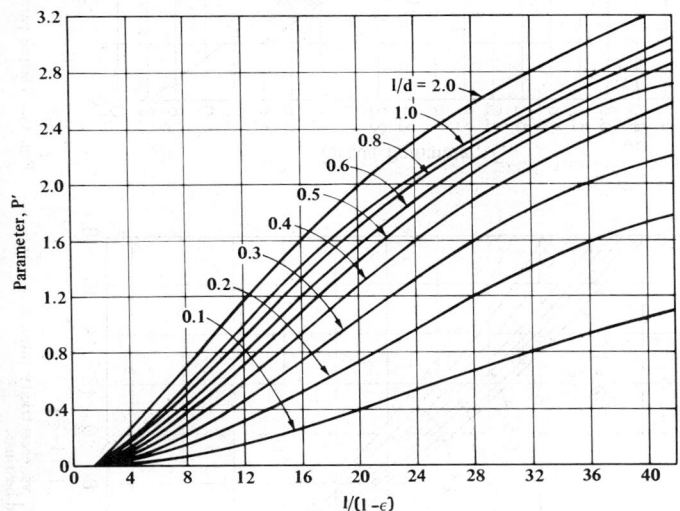

Fig. 13. Bearing parameter, P', vs. eccentricity ratio, $1/(1 - \epsilon)$ — journal bearings.

    10. *Eccentricity ratio ε.* Using P' and l/d, the value of $1/(1 - \epsilon)$ is determined from Fig. 13 and from this ε can be determined.

    11. *Torque parameter T′.* This value is obtained from Fig. 14 or Fig. 15 using $1/(1 - \epsilon)$ and l/d.

    12. *Friction torque T.* This value is calculated from the formula:

$$T = \frac{T' r^2 ZN}{6900(1000m)}$$

where r = journal radius, inches

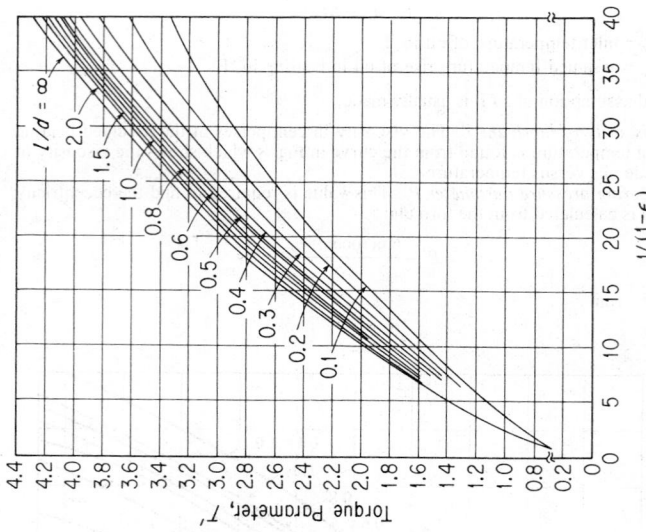

Fig. 15. Torque parameter, T', vs. eccentricity ratio, $1/(1-\epsilon)$ — journal bearings.

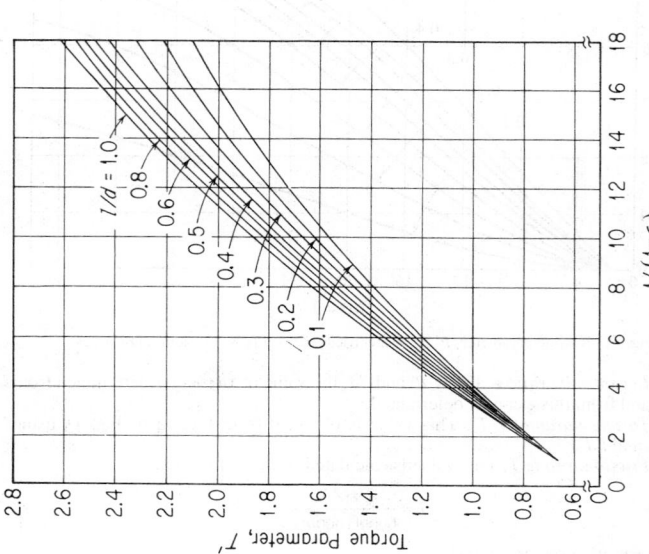

Fig. 14. Torque parameter, T', vs. eccentricity ratio, $1/(1-\epsilon)$ — journal bearings.

Table 6.   X Factor vs. Temperature of Mineral Oils

Temperature	X Factor
100	12.9
150	12.4
200	12.1
250	11.8
300	11.5

13. *Friction horsepower P_f.*   This is calculated from the formula:

$$P_f = \frac{KTNl}{63,000}$$

Where $K = 1$ for single oil hole, 2 for central groove

14. *Factor X.*   This factor is used in the calculation of the lubricant flow and can either be obtained from Table 6 or calculated from the formula:

$$X = 0.1837/\alpha c$$

where α = oil density in pounds per cubic inch
c = specific heat of lubricant in BTU/lb./°F

15. *Total flow of lubricant required Q_R.*   This is calculated from the formula:

$$Q_R = \frac{X(P_f)}{\Delta t_a}$$

16. *Bearing capacity number C_n.*   This is needed to obtain the flow factor and is calculated from the formula:

$$C_n = \left(\frac{l}{d}\right)^2 \Big/ 60P'$$

17. *Flow factor q.*   This is obtained from the curve in Fig. 16.

18. *Actual hydrodynamic flow of lubricant Q_1.*   This flow in gallons per minute is calculated from the formula:

$$Q_1 = \frac{Nlc_d qd}{294}$$

19. *Actual pressure flow of lubricant Q_2.*   This flow in gallons per minute is calculated from the formula:

$$Q_2 = \frac{Kp_s c_d^3 d(1 + 1.5\epsilon^2)}{Zl}$$

where $K = 1.64 \times 10^5$ for single oil hole
$K = 2.35 \times 10^5$ for central groove
p_s = oil supply pressure

20. *Actual total flow of lubricant Q.*   This is obtained by adding the hydrodynamic flow and the pressure flow.

$$Q = Q_1 + Q_2$$

21. *Actual bearing temperature rise Δt.*   This temperature rise in degrees F is obtained from the formula:

$$\Delta t = \frac{X(P_f)}{Q}$$

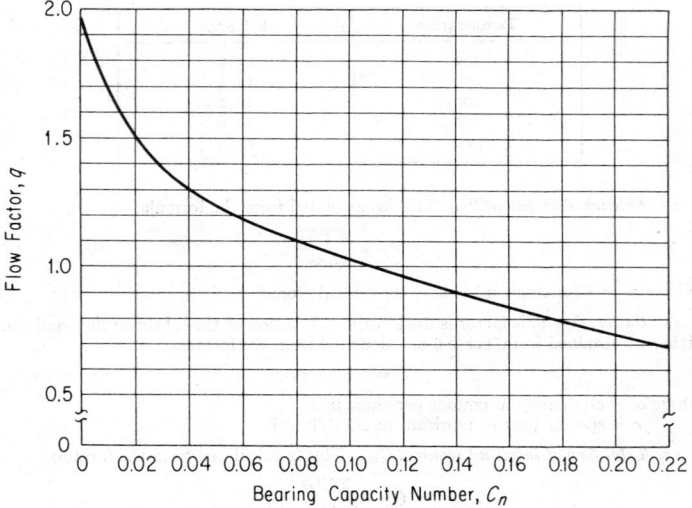

Fig. 16. Flow factor, q, vs. bearing capacity number, C_n — journal bearings.

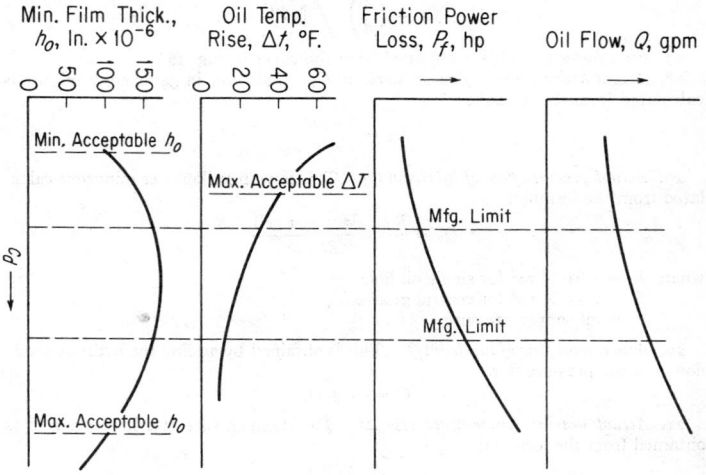

Fig. 17. Example of lubrication analysis curves for journal bearing.

22. *Comparison of actual and assumed temperature rises.* At this point if Δt_a and Δt differ by more than 5 degrees F, Steps 7 through 22 repeated using a Δt_{new} halfway between the former Δt_a and Δt.

23. *Minimum film thickness h_o.* When Step 22 has been satisfied, the minimum film thickness in inches is calculated from the formula: $h_o = \frac{1}{2} c_d (1 - \epsilon)$.

A new diametral clearance c_d is now assumed and Steps 5 through 23 are repeated. When this repetition has been done for a sufficient number of values for c_d, the full lubrication study is plotted as shown in Fig. 17. From this chart a working range of diametral clearance can be determined that optimizes film thickness, differential temperature, friction horsepower and oil flow.

Use of Lubrication Analysis. — Once the lubrication analysis has been completed and plotted as shown in Fig. 17, the following steps lead to the optimum bearing design, taking into consideration both basic operating requirements and requirements peculiar to the application.

1. Examine the curve (Fig. 17) for minimum film thickness and determine the acceptable range of diametral clearance, c_d, based on (a) a minimum of 200×10^{-6} inches for small bearings under 1 inch diameter; (b) a minimum of 500×10^{-6} inches for bearings from 1 to 4 inches diameter; (c) a minimum of 750×10^{-6} inches for larger bearings. More conservative designs would increase these requirements.

2. Determine the minimum acceptable c_d based on a maximum Δt of 40°F from the oil temperature rise curve (Fig. 17).

3. If there are no requirements for maintaining low friction horsepower and oil flow, the possible limits of diametral clearance are now defined.

4. The required manufacturing tolerances can now be placed within this band to optimize h_o as shown by Fig. 17.

5. If oil flow and power loss are a consideration, the manufacturing tolerances may then be shifted, within the range permitted by the requirements for h_o and Δt.

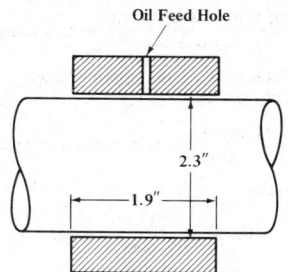

Oil Feed Hole

2.3"

1.9"

Fig. 18. Full journal bearing example design.

Example: A full journal bearing, Fig. 18, 2.3 inches in diameter and 1.9 inches long is to carry a load of 6000 pounds at 4800 rpm, using SAE 30 oil supplied at 200°F through a single oil hole at 30 psi. Determine the operating characteristics of this bearing as a function of diametral clearance.

1. *Diameter of bearing.* Given as 2.3 inches.
2. *Length of bearing.* Given as 1.9 inches.
3. *Bearing pressure.*

$$p_b = \frac{6000}{1 \times 1.9 \times 2.3} = 1372 \text{ lbs. per sq. in.}$$

4. *Diametral clearance.* Assume c_d is equal to 0.003 inch from Fig.12 for first calculation.

5. *Clearance modulus.*

$$m = \frac{0.003}{2.3} = 0.0013 \text{ inch}$$

6. *Length to diameter ratio.*

$$\frac{l}{d} = \frac{1.9}{2.3} = 0.83$$

7. *Assumed operating temperature.* If the temperature rise Δt_a is assumed to be 20°F,

$$t_b = 200 + 20 = 220°F$$

8. *Viscosity of lubricant.* From Fig. 5, $Z = 7.7$ centipoises

9. *Bearing pressure parameter.*

$$P' = \frac{6.9 \times 1.3^2 \times 1372}{7.7 \times 4800} = 0.43$$

10. *Eccentricity ratio.* From Fig.13, $\frac{1}{1-\epsilon} = 6.8$ and $\epsilon = 0.85$

11. *Torque parameter.* From Fig. 14, $T' = 1.46$

12. *Friction torque.*

$$T_f = \frac{1.46 \times 1.15^2 \times 7.7 \times 4800}{6900 \times 1.3} = 7.96 \text{ inch-pounds per inch}$$

13. *Friction horsepower.*

$$P_f = \frac{1 \times 7.96 \times 4800 \times 1.9}{63,000} = 1.15 \text{ horsepower}$$

14. *Factor X.* From Table 6, $X = 12$, approximately

15. *Total flow of lubricant required.*

$$Q_R = \frac{12 \times 1.15}{20} = 0.69 \text{ gallon per minute}$$

16. *Bearing capacity number*

$$C_n = \frac{0.83^2}{60 \times 0.43} = 0.027$$

17. *Flow factor.* From Fig. 16, $q = 1.43$

18. *Actual hydrodynamic flow of lubricant.*

$$Q_1 = \frac{4800 \times 1.9 \times 0.003 \times 1.43 \times 2.3}{294} = 0.306 \text{ gallon per minute}$$

19. *Actual pressure flow of lubricant.*

$$Q_2 = \frac{1.64 \times 10^5 \times 30 \times 0.003^3 \times 2.3 \times (1 + 1.5 \times 0.85^2)}{7.7 \times 1.9} = 0.044 \text{ gallon per minute}$$

20. *Actual total flow of lubricant.*

$$Q = 0.306 + 0.044 = 0.350 \text{ gallon per minute}$$

21. *Actual bearing temperature rise.*

$$\Delta t = \frac{12 \times 1.15}{0.350} = 39.4°F$$

22. *Comparison of actual and assumed temperature rises.* Since Δt_a and Δt differ by more than 5°F, a new Δt_a, midway between these two, of 30°F is assumed and Steps 7 through 22 are repeated.

7a. *Assumed operating temperature.*

$$t_b = 200 + 30 = 230°F$$

8a. *Viscosity of lubricant.* From Fig. 5, $Z = 6.8$ centipoises

9a. *Bearing pressure parameter.*

$$P' = \frac{6.9 \times 1.3^2 \times 1372}{6.8 \times 4800} = 0.49$$

10a. *Eccentricity ratio.* From Fig. 13, $\frac{1}{1-\epsilon} = 7.4$ and $\epsilon = 0.86$

11a. *Torque parameter.* From Fig. 14, $T' = 1.53$

12a. *Friction torque.*

$$T_f = \frac{1.53 \times 1.15^2 \times 6.8 \times 4800}{6900 \times 1.3} = 7.36 \text{ inch-pounds per inch}$$

13a. *Friction horsepower.*

$$P_f = \frac{1 \times 7.36 \times 4800 \times 1.9}{63,000} = 1.07 \text{ horsepower}$$

14a. *Factor X.* From Table 6, $X = 11.9$, approximately

15a. *Total flow of lubricant required.*

$$Q_R = \frac{11.9 \times 1.07}{30} = 0.42 \text{ gallon per minute}$$

16a. *Bearing capacity number.*

$$C_n = \frac{0.83^2}{60 \times 0.49} = 0.023$$

17a. *Flow factor.* From Fig. 16, $q = 1.48$

18a. *Actual hydrodynamic flow of lubricant.*

$$Q_1 = \frac{4800 \times 1.9 \times 0.003 \times 1.48 \times 2.3}{294} = 0.317 \text{ gallon per minute}$$

19a. *Pressure flow.*

$$Q_2 = \frac{1.64 \times 10^5 \times 30 \times 0.003^3 \times 2.3 \times (1 + 1.5 \times 0.86^2)}{6.8 \times 1.9} = 0.050 \text{ gallon per minute}$$

20a. *Actual flow of lubricant.*

$$Q_{new} = 0.317 + 0.050 = 0.367 \text{ gallon per minute}$$

21a. *Actual bearing temperature rise.*

$$\Delta t = \frac{11.9 \times 1.06}{0.367} = 34.4°F$$

22a. Comparison of actual and assumed temperature rises. Now Δt and Δt_a are within 5°F.

23. Minimum film thickness.

$$h_o = \frac{0.003}{2}(1 - 0.86) = 0.00021 \text{ inch.}$$

This analysis may now be repeated for other values of c_d determined from Fig. 12 and a complete lubrication analysis performed and plotted as shown in Fig. 17. An operating range for c_d can then be determined to optimize minimum clearance, friction horsepower loss, lubricant flow and temperature rise.

Thrust Bearings

Thrust bearings, as the name implies, are used to either absorb axial shaft loads or to position shafts. Brief descriptions of the normal designs for these bearings follow with approximate design methods for each. The generally accepted load ranges for these types of bearings are given in Table 7 and the schematic configurations are shown in Fig. 19.

The parallel or flat plate thrust bearing is probably the most frequently used type. It is the simplest and lowest in cost of those considered; however, it is also the least capable of absorbing load as can be seen from Table 7. It is most generally used as a positioning device where loads are either light or occasional.

The step bearing is, like the parallel plate, also a relatively simple design. This type of bearing will accept the normal range of thrust loads and lends itself to low-cost high-volume production. However, this bearing becomes sensitive to alignment as its size increases.

The tapered land thrust bearing, as shown in Table 7, is capable of high load capacity. Where the step bearing is generally used for small sizes, this type can be used in larger sizes. It is, however, more costly to manufacture and does require good alignment as size is increased.

The tilting pad or Kingsbury thrust bearing (as it is commonly referred to), is also capable of high thrust capacity. Because of its construction it is more costly, but it has the inherent advantage of being able to absorb significant amounts of misalignment.

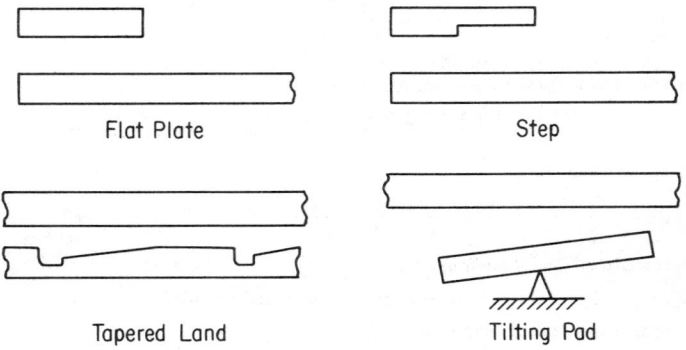

Flat Plate   Step

Tapered Land   Tilting Pad

Fig. 19. Types of thrust bearings.

Table 7. Thrust Bearing Loads*

Type	Normal Unit Loads, Lbs. per Sq. In.	Maximum Unit Loads, Lbs. per Sq. In.
Parallel Surface	<75	<150
Step	200	500
Tapered Land	200	500
Tilting Pad	200	500

Thrust Bearing Design Notation. — The symbols used in the design procedures which follow for flat plate, step, tapered land and tilting pad thrust bearings are listed below:

a = radial width of pad, inches
b = circumferential length of pad at pitch line, inches
b_2 = pad step length
B = circumference of pitch circle, inches
c = specific heat of oil, Btu/lb/°F
D = diameter, inches
e = depth of step, inch
f = coefficient of friction
g = depth of 45° chamfer, inches
h = film thickness, inch
i = number of pads
J = power loss coefficient
K = film thickness factor
K_g = fraction of circumference occupied by the pads; usually 0.8
l = length of chamfer, inches
M = horsepower per square inch
N = revolutions per minute
O = operating number
p = bearing unit load, psi
p_s = oil supply pressure, psi
P_f = friction horsepower
Q = total flow, gpm
Q_c = required flow per chamfer, gpm
$Q_c{}^0$ = uncorrected required flow per chamfer, gpm
Q_F = film flow, gpm
s = oil groove width
Δt = temperature rise, °F
U = velocity, feet per minute
V = effective width-to-length ratio for one pad
W = applied load, pounds
Y_G = oil flow factor
Y_L = leakage factor
Y_S = shape factor
Z = viscosity, centipoises
α = dimensionless film thickness factor
δ = taper
ξ = kinetic energy correction factor

Note: Subscript 1 denotes inside diameter and subscript 2 denotes outside diameter. Subscript i denotes inlet and subscript o denotes outlet.

* See footnote on page 2120.

Flat Plate Thrust Bearing Design. — The following steps define the performance of a flat plate thrust bearing, one section of which is shown in Fig. 20. Although each bearing section is wedge shaped, as shown at (2), for the purposes of design calculation it is considered to be a rectangle with a length b equal to the circumferential length along the pitch line of the actual section and a width a equal to the difference in the external and internal radii.

General Parameters: (a) From Table 7 the maximum unit load is between 75 and 100 pounds per square inch. (b) The outside diameter is usually between 1.5 and 2.5 times the inside diameter.

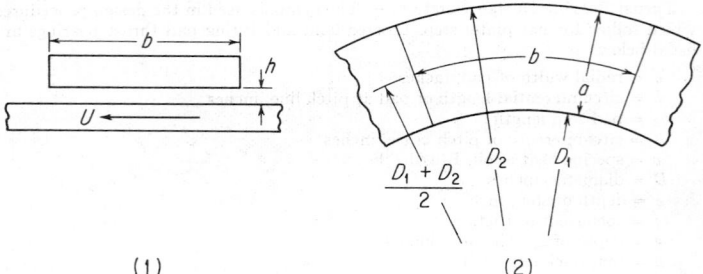

(1)   (2)

Fig. 20.   Basic elements of flat plate thrust bearing.

1. *Inside diameter, D_1.* This is determined by shaft size and clearance.
2. *Outside diameter, D_2.* This is calculated by the formula:

$$D_2 = \left(\frac{4W}{\pi K_g p} + D_1^2 \right)^{1/2}$$

where: W = applied load, pounds
K_g = fraction of circumference occupied by pads, usually 0.8
p = bearing unit load, psi

3. *Radial pad width, a.* This is equal to one-half the difference between the inside and outside diameters.

$$a = \frac{D_2 - D_1}{2}$$

4. *Pitch line circumference, B.* This is found from the pitch diameter.

$$B = (\pi)(D_2 - a)$$

5. *Number of pads, i.* Assume an oil groove width, s. If the length of pad is assumed to be optimum, i.e., equal to its width,

$$i_{app} = \frac{B}{a + s}$$

Take i as nearest even number.

6. *Length of pad, b.* If number of pads and oil groove width are known

$$b = \frac{B - (i \times s)}{i}$$

7. *Actual unit load, p.* This is calculated in pounds per square inch based on pad dimensions.

$$p = \frac{W}{iab}$$

8. *Pitch line velocity, U.* This is found in feet per minute.

$$U = \frac{BN}{12}$$

where N = rpm

9. *Friction power loss, P_f.* This is difficult to calculate for this type of bearing since there is no theoretical method of determining the operating film thickness. However, a good approximation can be made using Fig. 21. From this curve the value of M, horsepower loss per square inch of bearing surface, can be obtained. The total power loss P_f, is then calculated from

$$P_f = iabM$$

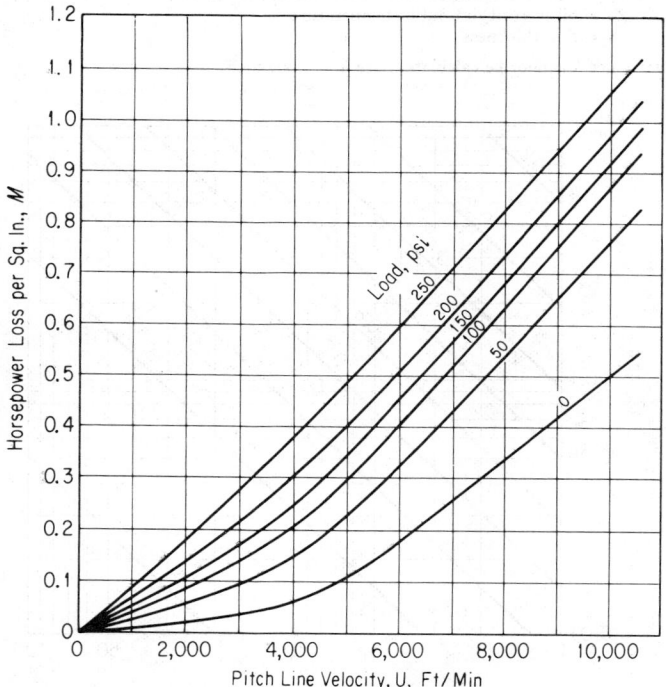

Fig. 21. Friction power loss, M, vs. peripheral speed, U — thrust bearings.*

* See footnote on page 2120.

10. *Oil flow required, Q.* This may be estimated in gallons per minute for a given temperature rise from:

$$Q = \frac{42.4 P_f}{c \Delta t}$$

where c = specific heat of oil in Btu/lb/°F
Δt = temperature rise of the oil in °F

Note: A Δt of 50°F is an acceptable maximum.

Since there is no theoretical method of predicting the minimum film thickness in this type of bearing, only an approximation, based on experience, of the film flow can be made. For this reason and based on practical experience, it is desirable to have a minimum of one-half of the desired oil flow pass through the chamfer.

11. *Film flow, Q_F.* This is calculated in gallons per minute from:

$$Q_F = \frac{(1.5)(10^6) i V h^3 p_s}{Z_2}$$

where V = effective width-to-length ratio for one pad, a/b
Z_2 = oil viscosity at outlet temperature
h = film thickness

Note: since h cannot be calculated, use h = 0.002 inch.

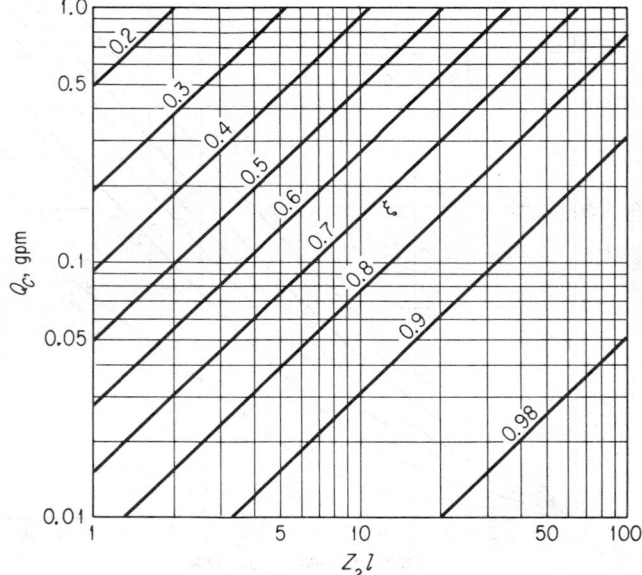

Fig. 22. Kinetic energy correction factor, ξ — thrust bearings.*

* See footnote on page 2120.

12. *Required flow per chamfer, Q_c.* This is readily found from the formula:

$$Q_c = \frac{Q}{i}$$

13. *Kinetic energy correction factor, ξ.* This is found by assuming a chamfer length l and entering Fig. 22 with a value $Z_2 l$ and Q_c.

14. *Uncorrected required flow per chamfer, $Q_c{}^0$.* This is found from the formula:

$$Q_c{}^0 = \frac{Q_c}{\xi}$$

15. *Depth of chamfer, g.* This is found from the formula:

$$g = \sqrt[4]{\frac{Q_c{}^0 l Z_2}{4.74 \times 10^4 p_s}}$$

Example: Design a flat plate thrust bearing to carry 900 pounds load at 4000 rpm using an SAE 10 oil with a specific heat of 3.5 Btu/lb/°F at 120°F and 30 psi inlet conditions. The shaft is 2¾ inches in diameter and the temperature rise is not to exceed 40°F. Figure 23 shows the final design of this bearing.

1. *Inside diameter.* Assumed to be 3 inches to clear shaft.

2. *Outside diameter.* Assuming a unit bearing load of 75 pounds per square inch from Table 7,

$$D_2 = \sqrt{\frac{4 \times 900}{\pi \times 0.8 \times 75} + 3^2} = 5.30 \text{ inches}$$

Use 5½ inches.

3. *Radial pad width.*

$$a = \frac{5.5 - 3}{2} = 1.25 \text{ inches}$$

4. *Pitch line circumference.*

$$B = \pi \times 4.25 = 13.3 \text{ inches}$$

5. *Number of pads.* Assume an oil groove width of 3/16 inch. If length of pad is assumed to be equal to width of pad, then

$$i_{app} = \frac{13.3}{1.25 + 0.1875} = 9 +$$

If the number of pads, i, is taken as 10, then

6. *Length of pad.* $b = \dfrac{13.3 - (10 \times 0.1875)}{10} = 1.14 \text{ inches}$

7. *Actual unit load.*

$$p = \frac{900}{10 \times 1.25 \times 1.14} = 63 \text{ psi}$$

8. *Pitch line velocity.*

$$U = \frac{13.3 \times 4000}{12} = 4,430 \text{ ft. per min.}$$

9. *Friction power loss.* From Fig. 21, $M = 0.19$

$$P_f = 10 \times 1.25 \times 1.14 \times 0.19 = 2.7 \text{ horsepower}$$

10. *Oil flow required.*

$$Q = \frac{42.4 \times 2.7}{3.5 \times 40} = 0.82 \text{ gallon per minute}$$

(Assuming a temperature rise of 40°F. — the maximum allowable according to the given condition — then the assumed operating temperature will be 120°F + 40°F = 160°F and the oil viscosity Z_2 is found from Fig. 5 to be 9.6 centipoises.)

11. *Film flow.*

$$Q_F = \frac{1.5 \times 10^6 \times 10 \times 1 \times (.002)^3 \times 30}{9.6} = 0.038 \text{ gpm}$$

Since this is a very small part of the required flow of 0.82 gpm, the bulk of the flow must be carried through the chamfers.

12. *Required flow per chamfer.* Assume that all of the oil flow is to be carried through the chamfers

$$Q_c = \frac{0.82}{10} = 0.082 \text{ gpm}$$

13. *Kinetic energy correction factor.* If l, the length of chamfer is made ⅛ inch, then $Z_2 l = 9.6 \times ⅛ = 1.2$. Entering Fig. 22 with this value and $Q_c = 0.082$,

$$\xi = 0.44$$

14. *Uncorrected required oil flow per chamfer.*

$$Q_c^0 = \frac{0.082}{0.44} = 0.186 \text{ gpm}$$

15. *Depth of chamfer.*

$$g = \sqrt[4]{\frac{0.186 \times 0.125 \times 9.6}{4.74 \times 10^4 \times 30}}$$

$$g = 0.02 \text{ inch}$$

A schematic drawing of this bearing is shown in Fig. 23.

Step Thrust Bearing Design. — The following steps define the performance of a step thrust bearing, one section of which is shown in Fig. 24. Although each bearing section is wedge shaped, as shown at (2), for the purposes of design calculation it is considered to be a rectangle with a length b equal to the circumferential length along the pitch line of the actual section and a width a equal to the difference in the external and internal radii.

General Parameters: For optimum proportions $a = b$, $b_2 = 1.2b_1$, and $e = 0.7h$.

1. *Internal diameter, D_1.* An internal diameter is assumed that is sufficient to clear the shaft.

2. *External diameter, D_2.* A unit bearing pressure is assumed from Table 7 and the external diameter is then found from the formula

$$D_2 = \sqrt{\frac{4W}{\pi K_g p} + D_1{}^2}$$

3. *Radial pad width, a.* This is equal to the difference between the external and internal radii.

$$a = \frac{D_2 - D_1}{2}$$

4. *Pitch line circumference, B.* This is found from the formula

$$B = \frac{\pi(D_1 + D_2)}{2}$$

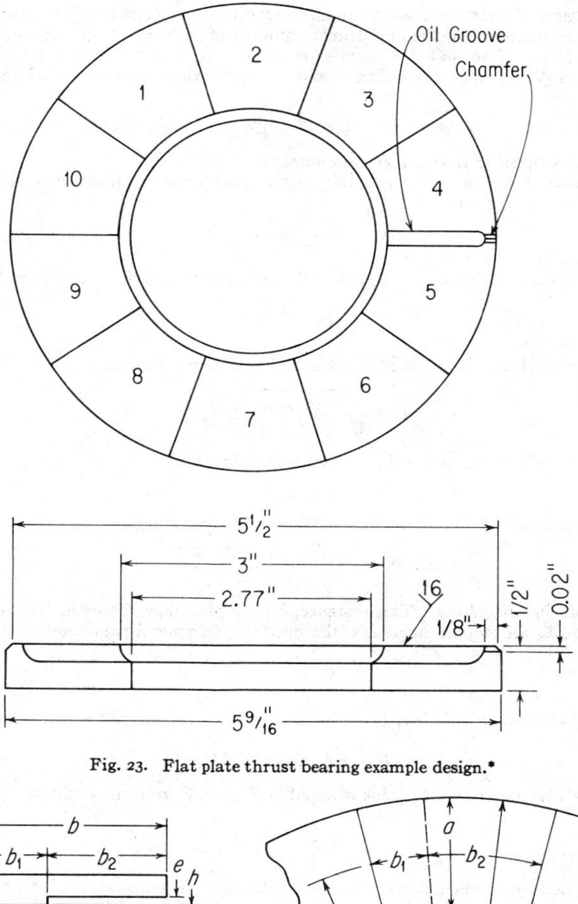

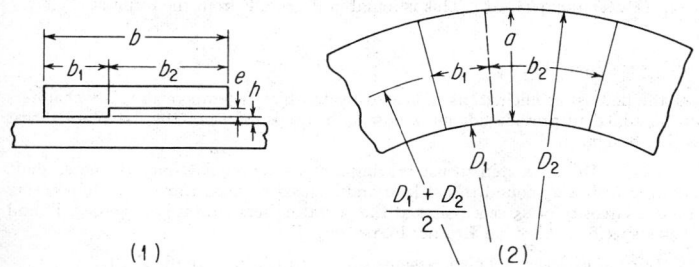

Fig. 23. Flat plate thrust bearing example design.*

Fig. 24. Basic elements of step thrust bearing.*

* See footnote on page 2120.

5. *Number of pads, i.* Assume an oil groove width, *s* (0.062 inch may be taken as a minimum), and find the approximate number of pads, assuming the pad length is equal to *a*. Note that if a chamfer is found necessary to increase the oil flow (see Step 13), the oil groove width should be greater than the chamfer width.

$$i_{app} = \frac{B}{a+s}$$

Then *i* is taken as the nearest even number.

6. *Length of pad, b.* This is readily determined since the number of pads and groove width are known.

$$b = \frac{B}{i} - s$$

7. *Pitch line velocity, U.* This is found in feet per minute from the formula

$$U = \frac{BN}{12}$$

8. *Film thickness, h.* This is found in inches from the formula

$$h = \sqrt{\frac{2.09 \times 10^{-9}ia^3UZ}{W}}$$

9. *Depth of step, e.* According to the general parameter

$$e = 0.7h$$

10. *Friction power loss, P_f.* This is found from the formula

$$P_f = \frac{7.35 \times 10^{-13}ia^2U^2Z}{h}$$

11. *Pad step length, b_2.* This distance, on the pitch line, from the leading edge of the pad to the step in inches is determined by the general parameters

$$b_2 = \frac{1.2b}{2.2}$$

12. *Hydrodynamic oil flow, Q.* This is found in gallons per minute from the formula

$$Q = 6.65 \times 10^{-4}iahU$$

13. *Temperature rise, Δt.* This is found in degrees F from the formula

$$\Delta t = \frac{42.4P_f}{cQ}$$

If the flow is insufficient, as indicated by too high a temperature rise, chamfers can be added to provide adequate flow as in Steps 12–15 of the flat plate thrust bearing design.

Example: Design a step thrust bearing for positioning a ⅞-inch diameter shaft operating with a 25-pound thrust load and a speed of 5,000 rpm. The lubricating oil has a viscosity of 25 centipoises at the operating temperature of 160 deg. F. and it has a specific heat of 3.4 Btu per lb. per deg. F.

1. *Internal diameter.* This is assumed to be 1 inch to clear the shaft.

2. *External diameter.* Since this is a positioning bearing with low total load, unit load will be negligible and the external diameter is not established by using

the formula given in Step 2 of the procedure but a convenient size is taken to give the desired overall bearing proportions.

$$D_2 = 3 \text{ inches}$$

3. *Radial pad width.*

$$a = \frac{3 - 1}{2} = 1 \text{ inch}$$

4. *Pitch line circumference.*

$$B = \frac{\pi(3 + 1)}{2} = 6.28 \text{ inches}$$

5. *Number of pads.* Assuming a minimum groove width of 0.062 inch

$$i_{app} = \frac{6.28}{1 + 0.062} = 5.9$$

Take $i = 6$

6. *Length of pad.*

$$b = \frac{6.28}{6} - 0.062 = 0.985$$

7. *Pitch line velocity.*

$$U = \frac{6.28 \times 5,000}{12} = 2,620 \text{ fpm}$$

8. *Film thickness.*

$$h = \sqrt{\frac{2.09 \times 10^{-9} \times 6 \times 1^3 \times 2,620 \times 25}{25}} = 0.0057 \text{ inch}$$

9. *Depth of step.*

$$e = 0.7 \times 0.0057 = 0.004 \text{ inch}$$

10. *Power loss.*

$$P_f = \frac{7.35 \times 10^{-13} \times 6 \times 1^2 \times 2,620^2 \times 25}{0.0057} = 0.133 \text{ hp}$$

11. *Pad step length.*

$$b_2 = \frac{1.2 \times 0.985}{2.2} = 0.537 \text{ inch}$$

12. *Total hydrodynamic oil flow.*

$$Q = 6.65 \times 10^{-4} \times 6 \times 1 \times 0.0057 \times 2,620 = 0.060 \text{ gpm}$$

13. *Temperature rise.*

$$\Delta t = \frac{42.4 \times 0.133}{3.4 \times 0.060} = 28°F.$$

Tapered Land Thrust Bearing Design. — The following steps define the performance of a tapered land thrust bearing, one section of which is shown in Fig. 25. Although each bearing section is wedge shaped, as shown at (2), for the purposes of design calculation it is considered to be a rectangle with a length b equal to the circumferential length along the pitch line of the actual section and a width a equal to the difference in the external and internal radii.

General Parameters: Usually the taper extends to only 80 per cent of the pad length with the remainder being flat, thus: $b_2 = 0.8b$ and $b_1 = 0.2b$.

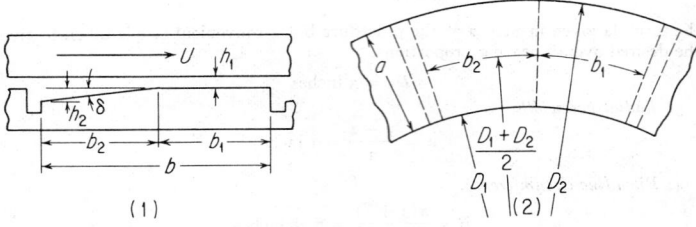

Fig. 25.   Basic elements of tapered-land thrust bearing.[*]

1. *Inside diameter, D_1.*   This is determined by shaft size and clearance.
2. *Outside diameter, D_2.*   This is calculated by the formula:

$$D_2 = \left(\frac{4W}{\pi K_g p_a} + D_1^2 \right)^{1/2}$$

where: K_g = 0.8 or 0.9 and W = applied load, pounds
       p_a = assumed unit load from Table 7, page 2103.

3. *Radial pad width, a.*   This is equal to one half the difference between the inside and outside diameters.

$$a = \frac{D_2 - D_1}{2}$$

4. *Pitch line circumference, B.*   This is found from the mean diameter:

$$B = \frac{\pi(D_1 + D_2)}{2}$$

5. *Number of pads, i.*   Assume an oil groove width, s, and find the approximate number of pads, assuming the pad length is equal to a.

$$i_{app} = \frac{B}{a + s}$$

Then i is taken as the nearest even number.
6. *Length of pad, b.*   This is readily determined since the number of pads and groove width are known.

$$b = \frac{B - is}{i}$$

7. *Taper values, δ_1 and δ_2.*   These can be taken from Table 8.
8. *Actual bearing unit load, p.*   This is calculated in pounds per square inch from the formula:

$$p = \frac{W}{iab}$$

9. *Pitch line velocity, U.*   This is found in feet per minute at the pitch circle from the formula:

$$U = \frac{BN}{12}$$

where N = rpm

* See footnote on page 2120.

Table 8. Taper Values for Tapered Land Thrust Bearing

Pad Dimensions, Inches	Taper, Inches	
$a \times b$	$\delta_1 = h_2 - h_1$ (at ID)	$\delta_2 = h_2 - h_1$ (at OD)
½ × ½	0.0025	0.0015
1 × 1	0.005	0.003
3 × 3	0.007	0.004
7 × 7	0.009	0.006

10. *Oil leakage factor, Y_L.* This is found either from Fig. 26 which shows curves for Y_L as functions of the pad width a and length of land b or from the formula:

$$Y_L = \frac{b}{1 + (\pi^2 b^2 / 12 a^2)}$$

11. *Film thickness factor, K.* This is calculated using the formula:

$$K = \frac{5.75 \times 10^6 p}{U Y_L Z}$$

12. *Minimum film thickness, h.* Using the value of K just determined and the selected taper values δ_1 and δ_2, h is found from Fig. 27. In general, h should be 0.001 inch for small bearings and 0.002 inch for larger and high-speed bearings.

13. *Friction power loss, P_f.* Using the film thickness h, the coefficient J can be obtained from Fig. 28. The friction loss in horsepower is then calculated from the formula:

$$P_f = 8.79 \times 10^{-13} i a b J U^2 Z$$

14. *Required oil flow, Q.* This may be estimated in gallons per minute for a given temperature rise Δt from the formula:

$$Q = \frac{42.4 \, P_f}{c \Delta t}$$

where c = specific heat of the oil in Btu/lb/°F

Note: A Δt of 50°F is an acceptable maximum.

15. *Shape factor, Y_S.* This is needed to compute the actual oil flow and is calculated from the formula:

$$Y_S = \frac{8ab}{D_2^2 - D_1^2}$$

16. *Oil flow factor, Y_G.* This is found from Fig. 29 using Y_S and D_1/D_2.

17. *Actual oil film flow, Q_F.* The amount of oil in gallons per minute that the bearing film will pass is calculated from the formula:

$$Q_F = \frac{8.9 \times 10^{-4} i \delta_2 D_2^3 N Y_G Y_S^2}{D_2 - D_1}$$

18. If the flow is insufficient, the tapers can be increased or chamfers calculated to provide adequate flow, as in Steps 12–15 of the flat plate thrust bearing design procedure.

Example: Design a tapered land thrust bearing for 70,000 pounds at 3600 rpm. The shaft diameter is 6.5 inches. The oil inlet temperature is 110°F @ 20 psi.

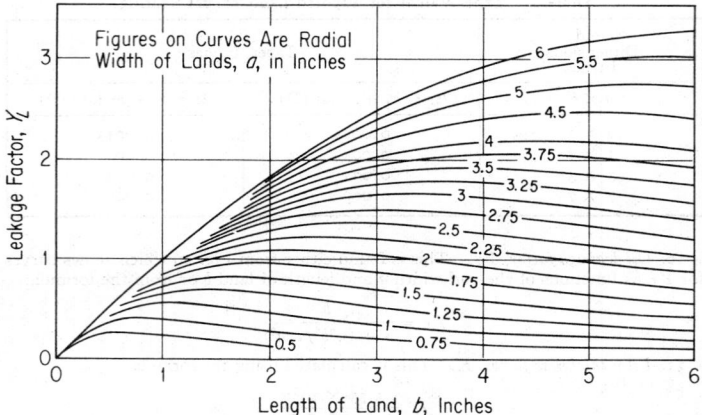

Fig. 26. Leakage factor, Y_L, vs. pad dimensions a and b — tapered-land thrust bearings.*

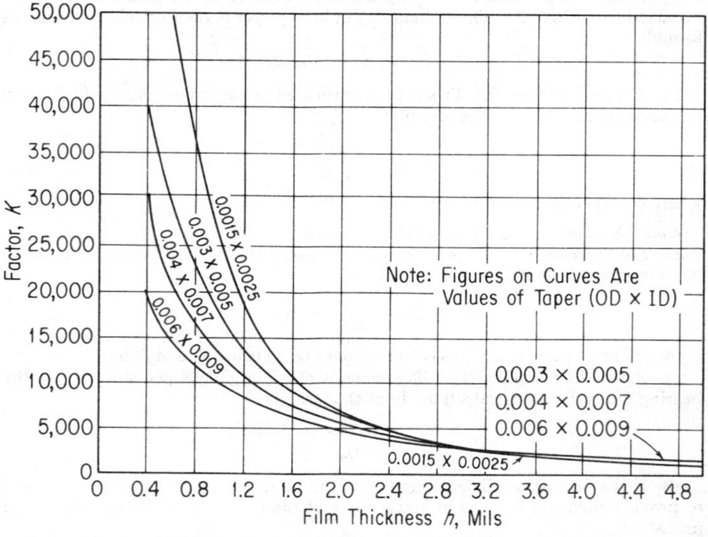

Fig. 27. Thickness, h, vs. factor K — tapered-land thrust bearings.*

* See footnote on page 2120.

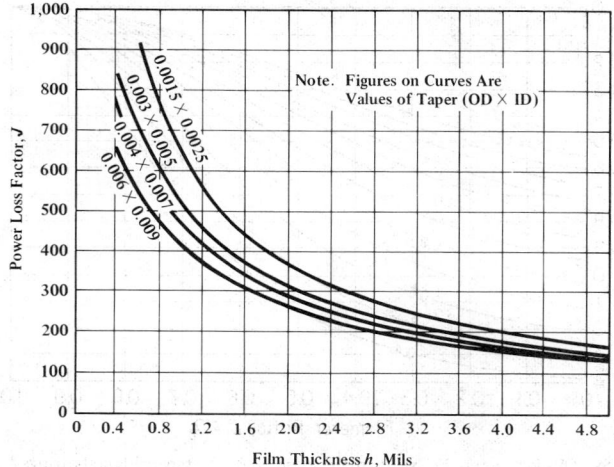

Fig. 28. Power-loss coefficient, *J*, vs. film thickness *h* — tapered-land thrust bearings.*

A maximum temperature rise of 50°F is acceptable and results in a viscosity of 18 centipoises. Use a value of $K_g = 0.9$ and a $c = 3.5$ Btu/lb/°F.

1. *Internal diameter.* Assume $D_1 = 7$ inches to clear shaft.

2. *External diameter.* Assume a unit bearing load p_a of 400 pounds per square inch from Table 7, then

$$D_2 = \sqrt{\frac{4 \times 70,000}{3.14 \times 0.9 \times 400} + 7^2} = 17.2 \text{ inches}$$

Round off to 17 inches.

3. *Radial pad width.*

$$a = \frac{17 - 7}{2} = 5 \text{ inches}$$

4. *Pitch line circumference.*

$$B = \frac{3.14(17 + 7)}{2} = 37.7 \text{ inches}$$

5. *Number of pads.* Assume groove width of 0.5 inch, then

$$i_{app} = \frac{37.7}{5 + 0.5} = 6.85$$

Take $i = 6$

6. *Length of pad.*

$$b = \frac{37.7 - 6 \times 0.5}{6} = 5.78 \text{ inches}$$

* See footnote on page 2120.

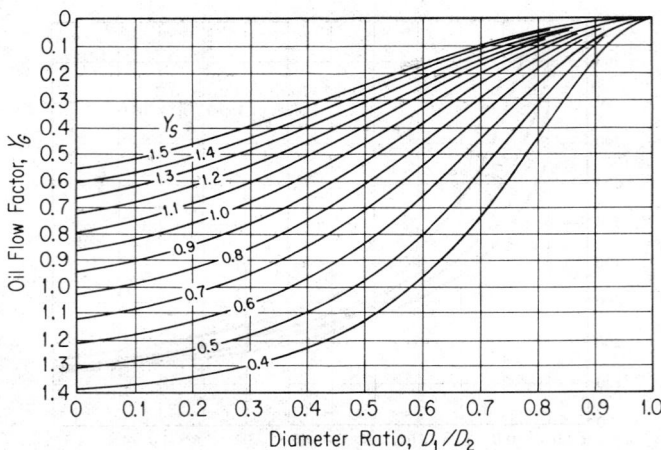

Fig. 29.   Oil-flow factor, Y_G, vs. diameter ratio D_1/D_2 — tapered-land bearings.*

7. *Taper values.*   Interpolate in Table 8 to obtain

$$\delta_1 = 0.008 \text{ inch and } \delta_2 = 0.005 \text{ inch}$$

8. *Actual bearing unit load.*

$$p = \frac{70,000}{6 \times 5 \times 5.78} = 404 \text{ psi}$$

9. *Pitch line velocity.*

$$U = \frac{37.7 \times 3600}{12} = 11,300 \text{ ft. per min.}$$

10. *Oil leakage factor.*

From Fig. 26, $Y_L = 2.75$

11. *Film thickness factor.*

$$K = \frac{5.75 \times 10^6 \times 404}{11,300 \times 2.75 \times 18} = 4150$$

12. *Minimum film thickness.*

From Fig. 27, $h = 2.2$ mils

13. *Friction power loss.*   From Fig. 28, $J = 260$ then

$$P_f = 8.79 \times 10^{-13} \times 6 \times 5 \times 5.78 \times 260 \times 11,300^2 \times 18 = 91 \text{ hp.}$$

14. *Required oil flow.*

$$Q = \frac{42.4 \times 91}{3.5 \times 50} = 22.0 \text{ gpm}$$

* See footnote on page 2120.

15. *Shape factor.*

$$Y_S = \frac{8 \times 5 \times 5.78}{17^2 - 7^2} = 0.684$$

16. *Oil flow factor.*

From Fig. 29, $Y_G = 0.087$

where $D_1/D_2 = 0.41$

17. *Actual oil film flow.*

$$Q_F = \frac{8.9 \times 10^{-4} \times 6 \times 0.005 \times 17^3 \times 3600 \times 0.087 \times (0.684)^2}{17 - 7} = 19.2 \text{ gpm}$$

Since film flow is less than required oil flow, either the tapers can be increased, Step 7, or chamfers calculated, Steps 12–15 of the flat plate thrust bearing design procedure.

Tilting Pad Thrust Bearing Design. — The following steps define the performance of a tilting pad thrust bearing, one section of which is shown in Fig. 30. Although each bearing section is wedge shaped, as shown at (2), for the purposes of design calculation it is considered to be a rectangle with a length b equal to the circumferential length along the pitch line of the actual section and a width a equal to the difference in the external and internal radii. The location of the pivot shown in Fig. 30 is optimum. If shaft rotation in both directions is required, however, the pivot must be at the midpoint which results in little or no detrimental effect on the performance.

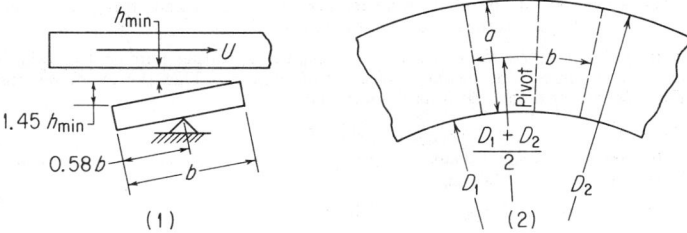

Fig. 30. Basic elements of tilting pad thrust bearing.*

1. *Inside diameter, D_1.* This is determined by shaft size and clearance.
2. *Outside diameter, D_2.* This is calculated from the formula:

$$D_2 = \left(\frac{4W}{\pi K_g p} + D_1^2 \right)^{\frac{1}{2}}$$

where W = applied load, pounds
K_g = 0.8
p = unit load from Table 7

3. *Radial pad width, a.* This is equal to one-half the difference between the inside and outside diameters.

$$a = \frac{D_2 - D_1}{2}$$

4. *Pitch line circumference, B.* This is found from the mean diameter:

$$B = \pi \left(\frac{D_1 + D_2}{2} \right)$$

* See footnote on page 2120.

5. *Number of pads, i.* The number of pads may be estimated from the formula:

$$i = \frac{BK_g}{a}$$

Select the nearest even number.

6. *Length of pad, b.* This can be found from the formula:

$$b \cong \frac{BK_g}{i}$$

7. *Pitch line velocity, U.* This is calculated in feet per minute from the formula:

$$U = \frac{BN}{12}$$

8. *Bearing unit load, p.* This is calculated from the formula:

$$p = \frac{W}{iab}$$

9. *Operating number, O.* This is calculated from the formula:

$$O = \frac{1.45 \times 10^{-7} Z_2 U}{5pb}$$

where Z_2 = viscosity of oil at outlet temperature (inlet temperature plus assumed temperature rise through the bearing).

10. *Minimum film thickness, h_{min}.* Using the operating number the value of α = dimensionless film thickness is found from Fig. 31. Then the actual minimum film thickness is calculated from the formula:

$$h_{min} = \alpha b$$

In general, this value should be 0.001 inch for small bearings and 0.002 inch for larger and high-speed bearings.

11. *Coefficient of friction, f.* This is found from Fig. 32.

12. *Friction power loss, P_f.* This horsepower loss now is calculated by the formula:

$$P_f = \frac{fWU}{33,000}$$

13. *Actual oil flow, Q.* This flow over the pad in gallons per minute is calculated from the formula:

$$Q = 0.0591 \alpha iab U$$

14. *Temperature rise, Δt.* This is found from the formula:

$$\Delta t = 0.0217 \frac{fp}{\alpha c}$$

where c = specific heat of oil in Btu/lb/°F

If the flow is insufficient, as indicated by too high a temperature rise, chamfers can be added to provide adequate flow as in Steps 12–15 of the flat plate thrust bearing design.

Example: Design a tilting pad thrust bearing for 70,000 pounds thrust at 3600 rpm. The shaft diameter is 6.5 inches and a maximum OD of 15 inches is available. The oil inlet temperature is 110°F and the supply pressure is 20 pounds per square

inch. A maximum temperature rise of 50°F is acceptable and results in a viscosity of 18 centipoises. Use a value of 3.5 Btu/lb/°F for c.

1. *Inside diameter.* Assume $D_1 = 7$ inches to clear shaft.

2. *Outside diameter.* Given maximum $D_2 = 15$ inches

3. *Radial pad width.*

$$a = \frac{15 - 7}{2} = 4 \text{ inches}$$

4. *Pitch line circumference.*

$$B = \pi \left(\frac{7 + 15}{2} \right) = 34.6 \text{ inches}$$

5. *Number of pads.*

$$i = \frac{34.6 \times 0.8}{4} = 6.9$$

Select 6 pads: $i = 6$

6. *Length of pad.*

$$b = \frac{34.6 \times 0.8}{6} = 4.61 \text{ inches}$$

Make $b = 4.75$ inches

7. *Pitch line velocity.*

$$U = \frac{34.6 \times 3600}{12} = 10,400 \text{ ft/min}$$

8. *Bearing unit load.*

$$p = \frac{70,000}{6 \times 4 \times 4.75} = 614 \text{ psi}$$

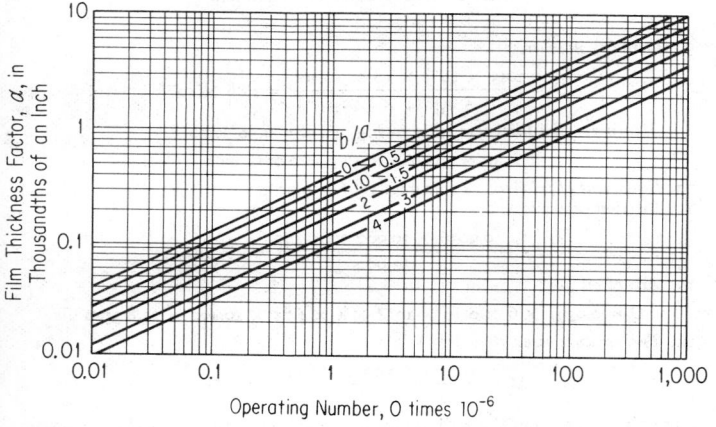

Fig. 31. Dimensionless minimum film thickness, α, vs. operating number, O — tilting pad thrust bearings.*

* See footnote on page 2120.

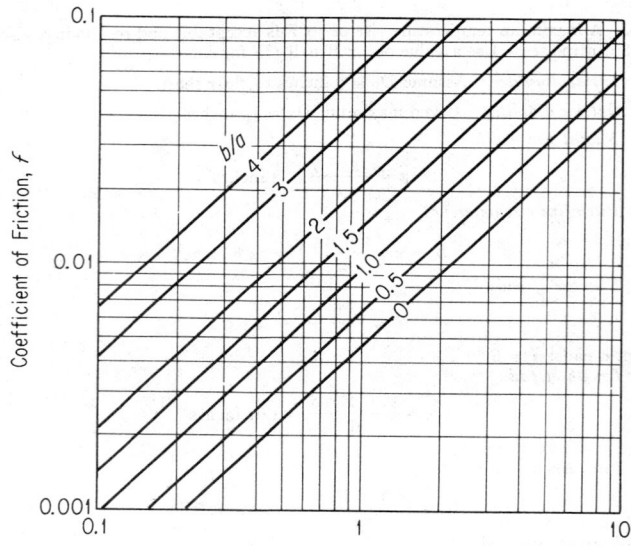

Fig. 32.   Coefficient of friction, f, vs. dimensionless film thickness α — tilting pad thrust bearings with optimum pivot location.*

9. *Operating number.*

$$O = \frac{1.45 \times 10^{-7} \times 18 \times 10,400}{5 \times 614 \times 4.75} = 1.86 \times 10^{-6}$$

10. *Minimum film thickness.*   From Fig. 31, $\alpha = 0.30 \times 10^{-3}$

$$h_{\min} = 0.00030 \times 4.75 = 0.0014 \text{ inch}$$

11. *Coefficient of friction.*   From Fig. 32, $f = 0.0036$
12. *Friction power loss.*

$$P_f = \frac{0.0036 \times 70,000 \times 10,400}{33,000} = 79.4 \text{ hp}$$

13. *Actual oil flow.*

$$Q = 0.0591 \times 6 \times 0.30 \times 10^{-3} \times 4 \times 4.75 \times 10,400 = 23.8 \text{ gpm}$$

14. *Temperature rise.*

$$\Delta t = \frac{0.0217 \times 0.0036 \times 614}{0.30 \times 10^{-3} \times 3.5} = 45.7 °F$$

Since this is less than the 50°F which is considered as the acceptable maximum, the design is satisfactory.

* Table 7 and Figs. 21-32 inclusive reproduced with permission from Wilcock and Booser, *Bearing Design and Application*, McGraw Hill Book Co., Copyright © 1957.

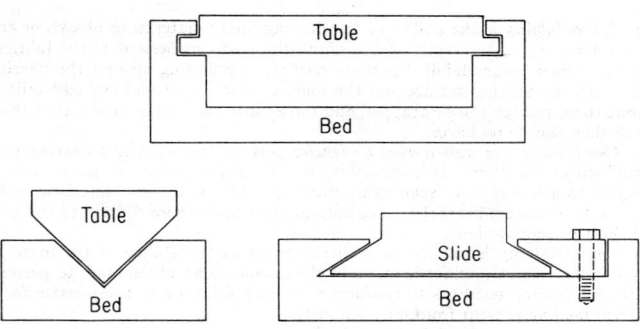

Fig. 33.   Types of guide bearings.

Guide Bearings

This type of bearing is generally used as a positioning device or as a guide to linear motion such as in machine tools.   Figure 33 shows several examples of this type of bearing.   It is normal for this type of bearing to operate in the boundary lubrication region with either dry, dry film (MoS_2 etc.), grease, oil or gaseous lubrication.   In order to improve performance, reduce wear, and increase stability, the use of hydrostatic lubrication is often resorted to.   This type of design provides oil or gas under pressure to pocket designs which provides a film and complete separation of the sliding surfaces.

Plain Bearing Materials

Materials used for sliding bearings cover a wide range of metals and non-metals. To make the optimum selection requires a complete analysis of the specific application.   The important general categories are: Babbitts, alkali hardened lead, cadmium alloys, copper lead, bronze aluminum, silver, sintered metals, plastic, wood, rubber and carbon graphite.

Properties of Bearing Materials. — For a material to be used as a plain bearing it must possess certain physical and chemical properties which permit it to operate properly.   If a material does not possess all of these characteristics to some degree it will not function long as a bearing.   It should be noted, however, that there are few, if any, materials which are outstanding in all of these characteristics.   Therefore, the selection of the optimum bearing material for a given application is at best a compromise to secure the most desirable combination of properties required for that particular application.   The seven properties generally acknowledged to be the most significant are: 1. Fatigue resistance; 2. Embeddability; 3. Compatibility; 4. Conformability; 5. Thermal conductivity; 6. Corrosion resistance; and 7. Load capacity.

These properties are described as follows:

1. *Fatigue resistance* is the ability of the bearing lining material to withstand repeated applications of stress and strain without cracking, flaking or otherwise being destroyed.

2. *Embeddability* is the ability of the bearing lining material to absorb or embed within itself any of the larger of the slight dirt particles present in the lubrication systems. Poor embeddability permits particles circulating around the bearing to score both the bearing surface and the journal or shaft. Good embeddability will permit these particles to be trapped and forced into the surface and out of the way where they can do no harm.

3. *Compatability or anti-scoring tendencies* permits the shaft and bearing to "get along" with each other. It is the ability to resist galling or seizing under conditions of metal-to-metal contact such as at start-up. This is the characteristic which is most truly a bearing property, since only at start up is there contact of the bearing and shaft in good designs.

4. *Conformability* is defined as malleability or as the ability of the material to creep or flow slightly under load, as in the initial stages of running, to permit the shaft and bearing contours to conform with each other or to compensate for non-uniform loading present from misalignment.

5. *High thermal conductivity* is required to absorb and carry away the heat generated in the bearing. This is most important, not in removing frictional heat generated in the oil film, but in preventing seizures due to local hot spots caused by local asperity break-throughs or foreign particles.

6. *Corrosion resistance* is required to resist attack by organic acids that are sometimes formed in oils at operating conditions.

7. *Load capacity or strength* is the ability of the material to withstand the hydrodynamic pressures exerted upon it during operation.

Babbitt or White Metal Alloys. — Many different bearing metal compositions are referred to as babbitt metals. The exact composition of the original babbitt metal is not known; however, the ingredients were probably tin, copper, and antimony in approximately the following percentages: 89.3, 3.6 and 7.1. Tin and lead-base babbitts are probably the best known of all bearing materials. With their excellent embeddability and compatibility characteristics under boundary lubrication, babbitt bearings are used in a wide range of applications including household appliances, automobile and diesel engines, railroad cars, electric motors, generators, steam and gas turbines, and industrial and marine gear units.

Both the Society of Automotive Engineers and American Society for Testing and Materials have classified white metal bearing alloys. Tables 9 and 10 give compositions and properties or characteristics for the two classifications.

In small bushings for fractional-horsepower motors and in automotive engine bearings the babbitt is generally used as a thin coating over a flat steel strip. After forming oil distribution grooves and drilling any required holes, the strip is cut to size, then rolled and shaped into the finished bearing. These are available for shaft diameters from 0.5 to 5 inches. Strip bearings are turned out by the millions yearly from automated factories and offer an excellent combination of low cost with good bearing properties.

For larger bearings in heavy-duty equipment, a thicker babbitt is cast on a rigid backing of steel or cast iron. Chemical and electrolytic cleaning of the bearing shell, thorough rinsing, tinning, and then centrifugal casting of the babbitt are desirable for sound bonding of the babbitt to the bearing shell. After machining, the babbitt layer is usually ⅛ to ¼ inch thick.

Compared to other bearing materials, babbitts generally have lower load-carrying capacity and fatigue strength, are a little higher in cost, and require a more complicated design. Also, their strength decreases rapidly with increasing temperature. These shortcomings can be avoided by using an intermediate layer of high-strength, fatigue resistant material which is placed between a steel backing and a thin babbitt

Table 9. Bearing and Bushing Alloys — Composition, Forms, Characteristics and Applications (SAE General Information)

SAE No. & Alloy Grouping		Nominal Composition, Per Cent	Form of Use (1), Characteristics (2), and Applications (3)
Sn-Base Alloys	11	Sn, 87.5; Sb, 6.75; Cu, 5.75.	(1) Cast on steel, bronze or brass backs, or directly in the bearing housing. (2) Soft, corrosion resistant with fair fatigue resistance. (3) Main and connecting-rod bearings; motor bushings. Operates with either hard or soft journal.
	12	Sn, 89; Sb, 7.5; Cu, 3.5.	
Pb-Base Alloys	13	Pb, 84; Sb, 10; Sn, 6.	(1)SAE 13 and 14 are cast on steel, bronze or brass, or in the bearing housing; SAE 15 is cast on steel; and SAE 16 is cast into and on a porous sintered matrix, usually copper-nickel bonded to steel. (2) Soft, moderately fatigue resistant, corrosion resistant. (3) Main and connecting-rod bearings. Operates with hard or soft journal with good finish.
	14	Pb, 75; Sb, 15; Sn, 10.	
	15	Pb, 83; Sb, 15; Sn, 1; As, 1.	
	16	Pb, 92; Sb, 3.5; Sn, 4.5.	
Pb-Sn Overlays	19	Pb, 90; Sn, 10.	(1) Electrodeposited as a thin layer on copper-lead or silver bearing faces. (2) Soft, corrosion resistant. Bearings so coated run satisfactorily against soft shafts throughout the life of the coating. (3) Heavy-duty, high-speed main and connecting-rod bearings.
	190	Pb, 93; Sn, 7.	
Cu-Pb Alloys	49	Cu, 76; Pb, 24.	(1) Cast or sintered on steel back with the exception of SAE 481 which is cast on steel back only. (2) Moderately hard. Somewhat subject to oil corrosion. Some oils minimize this; protection with overlay may be desirable. Fatigue resistance good to fairly good. Listed in order of decreasing hardness and fatigue resistance. (3) Main and connecting-rod bearings. The higher lead alloys can be used unplated against a soft shaft, although an overlay is helpful. The lower lead alloys may be used against a hard shaft, or with an overlay against a soft one.
	48	Cu, 70; Pb, 30.	
	480	Cu, 65; Pb, 35.	
	481	Cu, 60; Pb, 40.	
Cu-Pb-Sn Alloys	482	Cu, 67; Pb, 28; Sn, 5.	(1) Steel backed and lined with a structure combining sintered copper alloy matrix with corrosion-resistant lead alloy. (2) Moderately hard. Corrosion resistance improved over copper-leads of equal lead content without tin. Fatigue resistance fairly good. Listed in order of decreasing hardness and fatigue resistance. (3) Main and connecting-rod bearings. Generally used without overlay. SAE 484 and 485 may be used with hard or soft shaft, while a hardened or cast shaft is recommended for SAE 482.
	484	Cu, 55; Pb, 42; Sn, 3.	
	485	Cu, 46; Pb, 51; Sn, 3.	
Al-Base Alloys	770	Al, 91.75; Sn, 6.25; Cu, 1; Ni, 1.	(1) SAE 770 cast in permanent molds; work-hardened to improve physical properties. SAE 780 and 782 usually bonded to steel back but is procurable in strip form without steel backing. SAE 781 usually bonded to steel back but can be produced as castings or wrought strip without steel back. (2) Hard, extremely fatigue resistant, resistant to oil corrosion. (3) Main and connecting-rod bearings. Generally used with suitable overlay. SAE 781 and 782 also used for bushings and thrust bearings with or without overlay.
	780	Al, 91; Sn, 6; Si, 1.5; Cu, 1; Ni, 0.5.	
	781	Al, 95; Si, 4; Cd, 1.	
	782	Al, 95; Cu, 1; Ni, 1; Cd, 2.	
Other Cu-Base Alloys	795	Cu, 90; Zn, 9.5; Sn, 0.5.	(1) Wrought solid bronze. (2) Hard, strong, good fatigue resistance. (3) Intermediate-load oscillating motion such as tie-rods, brake shafts, and so forth.
	791	Cu, 88; Zn, 4; Sn, 4; Pb, 4.	(1) SAE 791, wrought solid bronze; SAE 793, cast on steel back; SAE 798, sintered on steel back. (2) General-purpose bearing material, good shock and load capacity. Resistant to high temperatures. Hard shaft desirable. Less score resistant than higher lead alloys. (3) Medium to high loads. Transmission bushings and thrust washers. SAE 791 also used for piston-pin and 793 and 798 for chassis bushings.
	793	Cu, 84; Pb, 8; Sn, 4; Zn, 4.	
	798	Cu, 84; Pb, 8; Sn, 4; Zn, 4.	

Table 9. (*Continued*) **Bearing and Bushing Alloys — Composition, Forms, Characteristics and Applications** (SAE General Information)

SAE No. & Alloy Grouping	Nominal Composition, Per Cent	Form of Use (1), Characteristics (2), and Applications (3)
Other Cu-Base Alloys	792 — Cu, 80; Sn, 10; Pb, 10.	(1) SAE 792, cast on steel back. SAE 797, sintered on steel back. (2) Has maximum shock and load carrying capacity of conventional cast bearing alloys; hard, both fatigue and corrosion resistant. Hard shaft desirable. (3) Heavy loads with oscillating or rotating motion. Used for piston pins, steering knuckles, differential axles, thrust washers, and wear plates.
	797 — Cu, 80; Sn, 10; Pb, 10.	
	794 — Cu, 73.5; Pb, 23; Sn, 3.5.	(1) SAE 794, cast on steel back; SAE 799, sintered on steel back. (2) Higher lead content gives improved surface action for higher speeds but results in somewhat less corrosion resistance. (3) Intermediate load application for both oscillating and rotating shafts, that is, rocker-arm bushings, transmissions, and farm implements.
	799 — Cu, 73.5; Pb, 23; Sn, 3.5.	

Table 10. **White Metal Bearing Alloys — Composition and Properties** (ASTM B23-49)

ASTM Alloy Grade Number	Nominal Composition, Per Cent				Compressive Yield Point,[b] psi		Ultimate Compressive Strength,[c] psi		Brinell Hardness, 10 mm ball-500 kg load-30 seconds		Melting Point, Deg. F	Proper Pouring Temperature, Deg. F
	Tin	Antimony	Lead	Copper	68° F	212° F	68° F	212° F	68° F	212° F		
1	91.0	4.5	...	4.5	4400	2650	12,850	6950	17.0	8.0	433	825
2	89.0	7.5	...	3.5	6100	3000	14,900	8700	24.5	12.0	466	795
3	83.33	8.33	...	8.33	6600	3150	17,600	9900	27.0	14.5	464	915
4	75.0	12.0	10.0	3.0	5550	2150	16,150	6900	24.5	12.0	363	710
5	65.0	15.0	18.0	2.0	5050	2150	15,050	6750	22.5	10.0	358	690
6	20.0	15.0	63.5	1.5	3800	2050	14,550	8050	21.0	10.5	358	655
7	10.0	15.0	75.0	...	3550	1600	15,650	6150	22.5	10.5	464	640
8	5.0	15.0	80.0	...	3400	1750	15,600	6150	20.0	9.5	459	645
10	2.0	15.0	83.0	...	3350	1850	15,450	5750	17.5	9.0	468	630
11	...	15.0	85.0	...	3050	1400	12,800	5100	15.0	7.0	471	630
12	...	10.0	90.0	...	2800	1250	12,900	5100	14.5	6.5	473	625
15[d]	1.0	15.0	82.5	0.5	...	...	...	...	21.0	13.0	479	662
16	10.0	12.5	77.0	0.5	...	...	...	...	27.5	13.6	471	620
19	5.0	9.0	86.0	...	...	...	15,600	6100	17.7	8.0	462	620

a The compression test specimens were cylinders 1.5 inches in length and 0.5 inch in diameter, machined from chill castings 2 inches in length and 0.75 inch in diameter. The Brinell tests were made on the bottom face of parallel machined specimens cast in a 2-inch diameter by 0.625-inch deep steel mold at room temperature.

b The values for yield point were taken from stress-strain curves at a deformation of 0.125 per cent reduction of gage length.

c The ultimate strength values were taken as the unit load necessary to produce a deformation of 25 per cent of the length of the specimen.

d Also nominal arsenic, 1 per cent.

surface layer. Such composite bearings frequently eliminate any need for using alternate materials with poorer bearing characteristics.

Tin babbitt is composed of 80 to 90 per cent tin to which is added about 3 to 8 per cent copper and 4 to 14 per cent antimony. An increase in copper or antimony

produces increased hardness and tensile strength and decreased ductility. However, if the percentage of these alloys is increased above those shown in Table 10, the resulting alloy will have decreased fatigue resistance. These alloys have very little tendency to cause wear to their journals because of their ability to embed dirt. They resist the corrosive effects of acids, are not prone to oil-film failure and are easily bonded and cast. Two drawbacks are encountered in their use. They have a low fatigue resistance and their hardness and strength drop appreciably at low temperatures.

Lead babbitt compositions generally range from 10 to 15 per cent antimony and up to 10 per cent tin in combination with the lead. These alloys, like tin-base babbitts, have little tendency to cause wear to their journals, embed dirt well, resist the corrosive effects of acids, are not prone to oil film failure and are easily bonded and cast. Their chief disadvantage when compared to the tin-base alloy is a rather lower strength and a susceptibility to corrosion.

Cadmium Base. — Cadmium alloy bearings have a greater resistance to fatigue than babbitt bearings; however, use of these materials is very limited due to their poor corrosion resistance. These alloys contain 1 to 15 per cent nickel, or 0.4 to 0.75 per cent copper, and 0.5 to 2.0 per cent silver. Their prime attribute is their high temperature capability. The load carrying capacity and relative basic bearing properties are shown in Tables 11 and 12.

Copper Lead. — Copper lead bearings are a binary mixture of copper and lead containing from 20 to 40 per cent lead. Since the lead is practically insoluble in copper, a cast microstructure consists of lead pockets in a copper matrix. A steel backing is commonly used with this material and high volume is achieved either by continuous casting or by powder metallurgy techniques. This material is very often used with an overplate such as lead tin and lead tin copper to increase basic bearing properties. Tables 11 and 12 provide comparisons of material properties.

The combination of good fatigue strength, high-load capacity, and high temperature performance has resulted in extensive use of this material for heavy-duty main and connecting-rod bearing as well as moderate load and speed applications in turbines and electric motors.

Leaded Bronze and Tin Bronze. — Leaded and tin bronzes contain up to 25 per cent lead or approximately 10 per cent tin respectively. Cast leaded bronze bearings offer good compatibility, excellent casting and easy machining character-

Table 11. Properties of Bearing Alloys

Material	Recommended Shaft Hardness, Brinell	Load Carrying Capacity, psi	Maximum Operating Temp., °F
Tin Base Babbitt	150 or less	800–1500	300
Lead Base Babbitt	150 or less	800–1200	300
Cadmium Base	200–250	1200–2000	500
Copper Lead	300	1500–2500	350
Tin Bronze	300–400	4000+	500+
Lead Bronze	300	3000–4500	450–500
Aluminum	300	4000+	225–300
Silver-Overplated	300	4000+	500
Tri Metal-Overplate	230 or less	2000–4000+	225–300

Table 12. Bearing Characteristics Ratings

Material	Compatibility	Conformability and Embeddability	Corrosion Resistance	Fatigue Strength
Tin Base Babbitt	1	1	1	5
Lead Base Babbitt	1	1	3	5
Cadmium Base	1	2	5	4
Copper Lead	2	2	5	3
Tin Bronze	3	5	2	1
Lead Bronze	3	4	4	2
Aluminum	5	3	1	2
Silver Overplated	2	3	1	1
Tri Metal-Overplated	1	2	2	3

Note: 1 is best; 5 is worst.

istics, low cost, good structural properties and high-load capacity, usefulness as a single material which requires neither a separate overlay nor a steel backing. Bronzes are available in standard bar stock, sand or permanent molds, investment, centrifugal or continuous casting. Leaded bronzes have better compatibility than tin bronzes because the spheroids of lead smear over the bearing surface under conditions of inadequate lubrication. These alloys are generally a first choice at intermediate loads and speeds. Tables 11 and 12 provide comparisons of basic bearing properties of these materials.

Aluminum. — Aluminum bearings are either cast solid aluminum, aluminum with a steel backing or aluminum with a suitable overlay. The aluminum is usually alloyed with small amounts of tin, silicon, cadmium, nickel, or copper, as shown in Table 9. An aluminum bearing alloy with 20 to 30 per cent tin alloy and up to 3 per cent copper has shown promise as a substitute for bronzes in some industrial applications.

These bearings are best suited for operation with hard journals. Owing to the high thermal expansion of the metal (resulting in diametral contraction when it is confined as a bearing in a rigid housing), large clearances are required, which tend to make the bearing noisy, especially on starting. Overlays of lead-tin, lead, or lead-tin-copper may be applied to aluminum bearings to facilitate their use with soft shafts.

Aluminum alloys are available with properties specifically designed for bearing applications, such as high load-carrying capacity, fatigue strength and thermal conductivity, in addition to excellent corrosion resistance and low cost.

Silver. — Silver bearings were developed for and have an excellent record in heavy duty applications such as aircraft master rod and diesel engine main bearings. Silver has a higher fatigue rating than any of the other bearing materials; in fact there have been cases where the steel backing used with this material has shown evidences of fatigue before the silver. The advent of overlays, or more commonly called overplates, made it possible for silver to be used as a bearing material. Silver by itself does not possess any of the desirable bearing qualities except high fatigue resistance and high thermal conductivity. The overlays such as lead, lead tin or lead indium improve the embeddability and antiscore properties of silver. The relative basic properties of this material, when used as an overplate, are shown in Tables 11 and 12.

Cast Iron. — Cast Iron is an inexpensive bearing material capable of operation at light loads and low speeds, i.e., to 130 feet per minute and 150 pounds per square inch. These bearings must be well lubricated and have a rather large clearance so as to avoid scoring from particles torn from the cast iron that ride between bearing and journal. A journal hardness of between 150 to 250 Brinell has been found to be best when using cast iron bearings.

Porous Metals. — Porous metal self-lubricating bearings are usually made by sintering metals such as plain or leaded bronze, iron, and stainless steel. The sintering produces a spongelike structure capable of absorbing fairly large quantities of oil, usually 10–35 per cent of the total volume. These bearings are used where lubrication supply is difficult, inadequate or infrequent. This type of bearing should be flooded from time to time to resaturate the material. Another use of these materials is to meter a small quantity of oil to the bearings such as in drop feed systems. The general design operating characteristics of this class of materials are shown in Table 13.

Table 13. Application Limits — Semi-Lubricated Sintered-Metal and Nonmetallic Bearings

Type of Bearing	Load Capacity Psi	Max. Temp. °F	Max. Surface Speed Ft/Min	PV Limit P = Psi Load V = Surface Ft/Min
Porous Metals	4000/8000	150	1500	50,000
Rubber	50	150	1000	15,000
Graphitic Materials	600	700	2500	15,000 Dry 150,000 Lubricated
Laminated Phenolics	6000	200	2500	15,000
Nylon	1000	200	1000	3,000
Wood (Maple & Lignum Vitae)	2000	150	2000	15,000
TFE	500	500	50	1,000
Reinforced TFE	2500	500	1000	10,000
TFE Fabric	60,000	500	150	25,000

The accompanying Table 14 gives the chemical compositions, permissible loads, interference fits, running clearances and size specifications of bronze base and iron base metal powder sintered bearings that are specified in the ASTM specifications for oil impregnated metal powder sintered bearings (B438-70 and B439-70).

Plastics (Phenolic, Nylon, TFE). — Plastics are finding increased usage as bearing materials because of their resistance to corrosion, quiet operation, ability to be molded into many configurations, and their excellent compatibility which minimizes or eliminates the need for lubrication. There are many plastics capable of operating as bearings, however, they are mostly phenolic, tetrafluoroethylene (TFE) or polyamide (nylon) resins. The general application limits for these materials are shown in Table 13.

Laminated Phenolics: These composite materials consist of cotton fabric, asbestos, or other fillers bonded with phenolic resin. They have excellent compatibility with various fluids as well as strength and shock resistance. However, precautions must be taken to maintain adequate bearing cooling since the thermal conductivity of these materials is low.

Nylon: This material has the widest use for small lightly loaded applications. It has low frictional properties and requires no lubrication.

Table 14. Copper- and Iron-Base Sintered Bearings (Oil Impregnated) —
(ASTM B438-70, B439-70, and Appendices)

CHEMICAL REQUIREMENTS

Alloying Elements[a]	Copper Base		Iron Base			
	Grade 1—Types 1 & 2	Grade 2—Type 1	Grade 1	Grade 2	Grade 3	Grade 4
			Composition, Per Cent			
Cu	87.5–90.5	82.6–88.5	...	...	7.0–11.0	18.0–22.0
Fe	1.0 max.	1.0 max.	96.25 min.	95.9 min.	Remainder[d]	Remainder[d]
Sn	9.5–10.5	9.5–10.5	...	...	...	...
Pb	...	2.0–4.0	...	...	...	...
Zn, max.	...	0.75	...	...	...	...
Ni, max.	...	0.35	...	...	...	...
Sb, max.	...	0.25	...	...	...	...
Si, max.	...	...	0.3	0.3	...	...
Al, max.	...	...	0.2	0.2	...	...
C, max.	1.75[c]	1.75[c]	...	...	...	...
Other, max.	0.5	0.5	3.0	3.0	3.0	3.0
Comb. C[b]	...	...	0.25 max.	0.25–0.60	...	...

PERMISSIBLE LOADS

Copper-Base Bearings			Iron-Base Bearings		
Shaft Velocity, fpm	Grade 1-Type 1	Grade 1-Type 2 Grade 2-Type 1	Shaft Velocity, fpm	Grade 1, Grade 2	Grade 3, Grade 4
	Permissible Load, psi			Permissible Load, psi	
Slow & intermittent	3200	4000	Slow & intermittent	3600	8000
25	2000	2000	25	1800	3000
50 to 100 incl.	550	500	50 to 100 incl.	450	700
Over 100-150 incl.	365	325	Over 100-150 incl.	300	400
Over 150-200 incl.	280	250	Over 150-200 incl.	225	300
Over 200	e	e	Over 200	e	e

PRESS FITS AND RUNNING CLEARANCES[f]

Outside Diameter of Bearing	Press Fit Min.	Max.	Shaft Size	Min. Recommended Clearance
Up to 0.760	.001	.003	Up to 0.760	.0005
0.761 to 1.510	.0015	.004	0.761 to 1.510	.001
1.511 to 2.510	.002	.005	1.511 to 2.510	.0015
2.511 to 3.010	.002	.006	Over 2.510	.002
Over 3.010	.002	.007	...	...

STANDARD COPPER-BASE SLEEVE BEARING SIZE SPECIFICATIONS

Inside Diam.[g] Fractional	Decimal	Outside Diam. Fractional	Decimal	Wall Thickness	Length	Outside Diam. Fractional	Decimal	Wall Thickness	Length
1/8	0.127	3/16	0.1905	1/32	0.250	1/4	0.253	1/16	0.250
5/32	0.158	1/4	0.253	3/64	0.312	...	...	...	...
3/16	0.1895	1/4	0.253	1/32	0.375	5/16	0.3155	1/16	0.375
1/4	0.252	3/8	0.378	1/16	0.500	7/16	0.4405	3/32	0.500
5/16	0.3145	7/16	0.4405	1/16	0.562	1/2	0.503	3/32	0.562
3/8	0.377	1/2	0.503	1/16	0.625	9/16	0.5655	3/32	0.625
1/2	0.502	5/8	0.628	1/16	0.750	3/4	0.753	1/8	0.750
5/8	0.627	3/4	0.753	1/16	0.750	7/8	0.879	1/8	0.937
3/4	0.752	7/8	0.879	1/16	1.125	1	1.004	1/8	1.125
1	1.003	1 1/4	1.254	1/8	1.500	1 3/8	1.379	3/16	1.500
1 1/4	1.2535	1 1/2	1.504	1/8	1.500	1 5/8	1.630	3/16	1.875
1 1/2	1.504	1 3/4	1.755	1/8	1.500	1 7/8	1.880	3/16	2.250
2	2.004	2 1/2	2.505	1/4	2.000	...	...	...	...
2 1/2	2.505	3	3.006	1/4	2.500	...	...	...	...
3	3.006	3 1/2	3.507	1/4	3.000	...	...	...	...

Note. Footnotes pertaining to the letter references can be found on the bottom of the continued table on the following page.

Table 14 *(Continued).* **Copper- and Iron-Base Sintered Bearings (Oil Impregnated) —** (ASTM B438-70, B439-70, and Appendices)

COMMERCIAL DIMENSIONAL TOLERANCES

Diameter Tolerance[h]			Length Tolerance[h]		
Inside Diameter or Outside Diameter	Total Tolerance for Inside or Outside Diameter		Length	Total Length Tolerance	
	Copper Base	Iron Base		Copper Base	Iron Base
Up to 0.760	−.001	−.001	Up to 1.495	±.010	±.010
0.761 to 1.510	−.001	−.0015	1.496 to 1.990	±.015	±.015
1.511 to 2.510	−.0015	−.002	1.991 to 2.990	±.015	±.020
2.511 to 3.010	−.002	−.003	2.991 to 4.985	±.020	±.030
3.011 to 4.010	−.003	−.004	...	...	...
4.011 to 5.010	−.004	−.005	...	...	...
5.011 to 6.010	−.005	−.006	...	...	...

Concentricity Tolerance[h,i]					
Outside Diameter	Copper Base		Iron Base		
	Wall Thickness, max.	Concentricity Tolerance	Wall Thickness, max.	Concentricity Tolerance	
Up to 1.010	Up to 0.255	.003	Up to 0.355	.003	
1.011 to 1.510	Up to 0.355	.003	Up to 0.355	.003	
1.511 to 2.010	Up to 0.505	.004	Up to 0.505	.004	
2.011 to 3.010	Up to 0.760	.005	Up to 1.010	.005	
3.011 to 4.010	Up to 1.010	.005	Up to 1.010	.005	
4.011 to 5.010	Up to 1.510	.006	Up to 1.510	.006	
5.011 to 6.010	Up to 2.010	.007	Up to 2.010	.007	

Flange and Thrust Bearings, Diameter and Thickness Tolerances[j]				
Diameter Range	Flange Diameter Tolerance		Flange Thickness Tolerance	
	Standard	Special	Standard	Special
0 to 1½	±.005	±.0025	±.005	±.0025
Over 1½ to 3	±.010	±.005	±.010	±.007
Over 3 to 6	±.025	±.010	±.015	±.010

	Parallelism on Faces, max.			
Diameter Range	Copper Base		Iron Base	
	Standard	Special	Standard	Special
0 to 1½	.003	.002	.005	.003
Over 1½ to 3	.004	.003	.007	.005
Over 3 to 6	.005	.004	.010	.007

All dimensions in inches except where otherwise noted.

[a] Abbreviations used for the alloying elements are as follows: Cu, copper; Fe, iron; Sn, tin; Pb, lead; Zn, zinc; Ni, nickel; Sb, antimony; Si, silicon; Al, aluminum; and C, carbon.

[b] Combined carbon (on basis of iron only) may be a metallographic estimate of the carbon in the iron.

[c] Commonly graphite. A minimum of 1.5 per cent of another type of solid lubricant may be substituted when authorized by the purchaser.

[d] Total of iron plus copper shall be 97 per cent, min.

[e] For shaft velocities over 200 fpm the permissible loads may be calculated as follows: $P = 50,000/V$; where P = safe load, psi of projected area, and V = shaft velocity, fpm.

[f] Only *minimum* recommended clearances are listed. It is assumed that ground steel shafting will be used and that all bearings will be oil impregnated.

[g] For some of the inside diameter sizes the standard provides for the selection of two possible sets of outside diameter, wall thickness and length dimensions as shown.

[h] Values given here are intended for copper-base bearings with a 4 to 1 maximum length to inside diameter ratio and a 24 to 1 maximum length to wall thickness ratio; also for iron-base bearings with a 3 to 1 maximum length to inside diameter ratio and a 20 to 1 maximum length to wall thickness ratio. Bearings having greater ratios than these are not covered here.

[i] Total indicator reading.

[j] Standard and special tolerances are specified for diameters, thickness and parallelism. Special tolerances should not be specified unless required since they require additional or secondary operations and, therefore are costlier. Thrust bearings (¼ inch thickness, max.) have a standard thickness tolerance of ±0.005 inch and a special thickness tolerance of ±0.0025 inch for all diameters. Thrust bearing outside diameter tolerances are the same as for flange bearings.

Teflon: This material, with its exceptional low coefficient of friction, self-lubricating characteristics, resistance to attack by almost any chemicals, and its wide temperature range, is one of the most interesting of the plastics for bearing use. High cost combined with low load capacity cause Teflon to be selected mostly in modified form, where other less expensive materials have proved inadequate for design requirements.

Bearings made of laminated phenolics, nylon, or Teflon are all unaffected by acids and alkalies except if highly concentrated and therefore can be used with lubricants containing dilute acids or alkalies. Water is used to lubricate most phenolic laminate bearings but oil, grease, and emulsions of grease and water are also used. Water and oil are used as lubricants for nylon and Teflon bearings. Almost all types of plastic bearings absorb water and oil to some extent. In some the dimensional change caused by the absorption may be as much as three per cent in one direction. This means that bearings have to be treated before use so that proper clearances will be kept. This may be done by boiling in water, for water lubricated bearings. Boiling in water makes bearings swell the maximum amount. Clearances for phenolic bearings are kept at about 0.001 inch per inch of diameter on treated bearings. Partially lubricated or dry nylon bearings are given a clearance of 0.004 to 0.006 inches for a one-inch diameter bearing.

Wood: Bearings made from such woods as lignum vitae, rock maple, or oak offer self-lubricating properties, low cost, and clean operation. However, they have frequently been displaced in recent years by various plastics, rubber and sintered-metal bearings. General applications are shown in Table 10.

Rubber: Rubber bearings give excellent performance on propeller shafts and rudders of ships, hydraulic turbines, pumps, sand and gravel washers, dredges and other industrial equipment that handle water or slurries. The resilience of rubber helps to isolate vibration and provide quiet operation, allows running with relatively large clearances and helps to compensate for misalignment. In these bearings a fluted rubber structure is supported by a metal shell. The flutes or scallops in the rubber form a series of grooves through which lubricant or, as generally used, water and foreign material such as sand may pass through the bearing.

Carbon-Graphite. — Bearings of molded and machined carbon-graphite are used where regular maintenance and lubrication cannot be given. They are dimensionally stable over a wide range of temperatures, may be lubricated if desired, and are not affected by chemicals. These bearings may be used up to temperatures of 700 to 750 degrees F. in air or 1200 degrees F. in a non-oxidizing atmosphere, and generally are operated at a maximum load of 20 pounds per square inch. In some instances a metal or metal alloy is added to the carbon-graphite composition to improve such properties as compressive strength and density. The temperature limitation depends upon the melting point of the metal or alloy and the maximum load is generally 350 pounds per square inch when used with no lubrication or 600 pounds per square inch when used with lubrication.

Normal running clearances for both types of carbon-graphite bearings used with steel shafts and operating at a temperature of less than 200 degrees F. are as follows: 0.001 inch for bearings of 0.187 to 0.500-inch inside diameter, 0.002 inch for bearings of 0.501 to 1.000-inch inside diameter, 0.003 inch for bearings of 1.001 to 1.250-inch inside diameter, 0.004 inch for bearings of 1.251 to 1.500-inch diameter, and 0.005 inch for bearings of 1.501 to 2.000-inch inside diameter. Speeds depend upon too many variables to list specifically so it can only be stated here that high loads require a low number of rpm and low loads permit a high number of rpm. Smooth journals are necessary in these bearings as rough ones tend to abrade the bearings quickly. Cast iron and hard chromium-plate steel shafts of 400 Brinell and over, and phosphor-bronze shafts over 135 Brinell are recommended.

Babbitting. — Babbitt metal is extensively used as a lining for bearings, not only for its anti-frictional qualities, but because it is much cheaper than a machined box. Prior to pouring the babbitt, the bearing should be heated to prevent the molten metal from becoming chilled and sluggish. Bronze shells which are to have babbitt linings should first be tinned by immersing in a pot of molten solder. Use solder of "half and half" composition, and zinc chloride as a flux. The shell should be babbitted immediately after tinning. Babbitt should not be used for tinning, because it has a much higher melting point, which makes it difficult to maintain a molten film on the surface to be tinned. Cast-iron shells are rarely, if ever, tinned.

If much work of this kind is being done, babbitting jigs should be made. These are simply fixtures which bear against or fit into any finished surface or hole with which the mandrel must be aligned, and which hold the latter in the correct position while the babbitt is being poured. Whenever practicable, the bearing should be placed in a vertical position while pouring. The ladle should preferably have a rounded spout rather than one which is sharp or broad. A broad, thin stream or one that is intermittent tends to produce porous areas or blow-holes. Putty is preferable for luting or sealing the ends of the bearings, as moisture in clay tends to produce sputtering.

Coatings for Babbitting Mandrels. — For babbitting solid bearings, the surface of the mandrel which comes into contact with the metal, when the latter is poured into the bearing, should be coated with some substance to facilitate the removal of the mandrel from the bearing. One method of coating the mandrel is to hold it near an oil flame so that the smoke will come into contact with it and cover the surface with carbon. Instead of smoking the mandrel, the surface is sometimes covered with a coat of thin white lead. Another method is to wrap a piece of paper about the mandrel. In babbitting two-part bearings a coating for the mandrel will not be necessary.

Temperature of Molten Babbitt. — Babbitt metal used for bearings is melted in iron pots or kettles, and the molten metal should be kept at a constant temperature of about 870 degrees F. A constant temperature is very important. The temperature should be increased slowly and the babbitt thoroughly stirred, especially when new babbitt is being melted, or when old babbitt which has been allowed to solidify in the pot is being re-melted. This is necessary in order to prevent certain of the constituent metals from rising to the top and becoming oxidized, as well as to prevent the heavier metals from sinking to the bottom of the pot, thus producing a non-uniform alloy. The babbitt may be melted in the pouring ladle over an open fire. The temperature is about right when a pine stick used for stirring chars but does not ignite. The mandrel temperature is right when water dries rapidly on its surface without spluttering.

Preheating Bearings and Mandrels. — All iron shells must be preheated to a temperature of from 200 to 300 degrees F. before pouring the babbitt. The higher temperature is preferable, as a rule, except where a lining is being poured in a very heavy shell, when it may be necessary to use the lower temperature to prevent the babbitt from cooling too slowly. The mandrels should be preheated to a temperature of from 200 to 300 degrees F. when pouring babbitt into the shells, but, for bronze shells, a somewhat lower temperature should be used. Oil-holes in the bearing shells are filled with asbestos or wood driven against the mandrel. Joints are made tight with clay.

Pouring Babbitt Metal. — To secure good results, the babbitt must be poured at the correct temperature. If the babbitt is poured at too high a temperature, extreme shrinkage will occur, resulting in porous areas in the lining and in broken anchors; furthermore, the babbitt will be oxidized, softened, and dirty, and its antifrictional qualities will be lowered. If the babbitt is poured at too low a temperature, a lining having a coarse granular formation is the result. If the shells and mandrels are too cold, blow-holes and similar defects will form, and the lining will shrink away from the shell in cooling. If the temperature of the shell or mandrel is too high, the babbitt will cool too slowly, and the heavier metals will have time to settle, producing a bearing which will be soft at one place and brittle at another.

BALL, ROLLER, AND NEEDLE BEARINGS
(ROLLING-CONTACT BEARINGS)

Rolling contact bearings substitute a rolling element, ball or roller, for a hydrodynamic or hydrostatic fluid film to carry an impressed load without wear and with reduced friction. Because of their greatly reduced starting friction, when compared to the conventional journal bearing, they have acquired the common designation of "anti-friction" bearings. Though normally made with hardened rolling elements and races, and usually utilizing a separator to space the rolling elements and reduce friction, many variations are in use throughout the mechanical and electrical industries. The most common anti-friction bearing application is that of the deep-groove ball bearing with ribbon-type separator and sealed-grease lubrication used to support a shaft with radial and thrust loads in rotating equipment. This shielded or sealed bearing has become a standard and commonplace item ordered from a supplier's catalogue in much the same manner as nuts and bolts. Because of the simple design approach and the elimination of a separate lubrication system or device, this bearing is found in as many installations as the wick-fed or impregnated porous plain bushing.

Currently, a number of manufacturers produce a complete range of ball and roller bearings in a fully interchangeable series with standard dimensions, tolerances and fits as specified in Anti-Friction Bearing Manufacturers Association (AFBMA) Standards. Except for deep-groove ball bearings, performance standards are not so well defined and sizing and selection must be done in close conformance with the specific manufacturer's catalogue requirements. In general, desired functional features should be carefully gone over with the vendor's representatives.

Rolling contact bearings are made to high standards of accuracy and with close metallurgical control. Balls and rollers are normally held to diametral tolerances of .0001 inch or less within one bearing and are often used as "gage" blocks in routine toolroom operations. This accuracy is essential to the performance and durability of rolling-contact bearings and also in limiting runout, providing proper radial and axial clearances, and insuring smoothness of operation.

Because of their low friction, both starting and running, rolling-contact bearings are utilized to reduce the complexity of many systems normally functioning with journal bearings. Aside from this advantage and that of precise radial and axial location of rotating elements, however, they also are desirable because of their reduced lubrication requirements and their ability to function during brief interruptions in normal lubrication.

In applying rolling-contact bearings it is well to appreciate that they have a life which is limited by the fatigue life of the material from which they are made and as modified by the lubricant used. In rolling-contact fatigue, precise relationships between life, load, and design characteristics are not predictable, but a statistical function described as the "probability of survival," is used to relate them according to equations recommended by the AFBMA. Deviations from these formulae result when certain extremes in applications such as speed, deflection, temperature, lubrication, and internal geometry must be dealt with.

Types of Anti-friction Bearings. — The general types are usually determined by the shape of the rolling element, but many variations have been developed which apply conventional elements in unique ways. Thus it is well to know that special bearings can be procured with races adapted to specific applications, though this is not practical for other than high volume configurations or where the requirements cannot be met in a more economical manner. "Special" races are appreciably more expensive. Quite often, in such situations, races are made to incorporate other functions of the mechanism, or are "submerged" in the surrounding structure, with the rolling elements supported by shaft or housing which has been hardened and finished in a suitable manner. Typical anti-friction bearing types are shown in Fig. 1.

Types of Ball Bearings. — Most types of ball bearings originate from three basic designs: the single-row radial, the single-row angular contact and the double-row angular contact.

BALL BEARINGS, SINGLE ROW, RADIAL CONTACT			
Symbol	Description	Symbol	Description
BC	Non-filling slot assembly 	BH	Non-separable counter-bore assembly 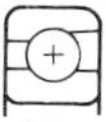
BL	Filling slot assembly 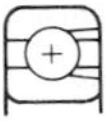	BM	Separable assembly 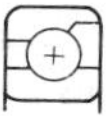
BALL BEARINGS, SINGLE ROW, ANGULAR CONTACT			
Symbol	Description	Symbol	Description
BN	Non-separable Nominal contact angle: from above 10° to and including 22° 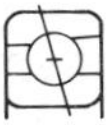	BAS	Separable inner ring Nominal contact angle: from above 22° to and including 32° 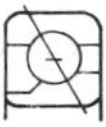
BNS	Separable outer ring Nominal contact angle: from above 10° to and including 22° 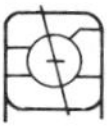	BT	Non-separable Nominal contact angle: from above 32° to and including 45° 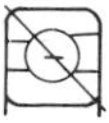
BNT	Separable inner ring Nominal contact angle: from above 10° to and including 22° 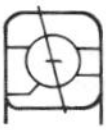	BY	Two-piece outer ring 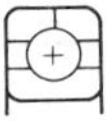
BA	Non-separable Nominal contact angle: from above 22° to and including 32° 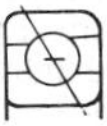	BZ	Two-piece inner ring 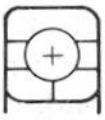
BALL BEARINGS, SINGLE ROW, RADIAL CONTACT, SPHERICAL OUTSIDE SURFACE			
Symbol	Description	Symbol	Description
BCA	Non-filling slot assembly 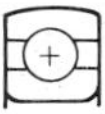	BLA	Filling slot assembly 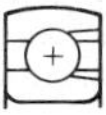

Fig. 1. Types of rolling element bearings and their symbols.

BALL BEARINGS, DOUBLE ROW, RADIAL CONTACT				
Symbol	Description		Symbol	Description
BF	Filling slot assembly 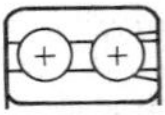		BHA	Non-separable two-piece outer ring 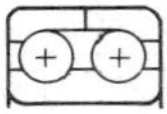
BK	Non-filling slot assembly 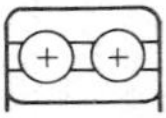			

BALL BEARINGS, DOUBLE ROW, ANGULAR CONTACT				
Symbol	Description		Symbol	Description
BD	Filling slot assembly Vertex of contact angles inside bearing 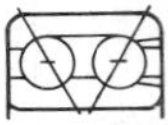		BG	Non-filling slot assembly Vertex of contact angles outside bearing 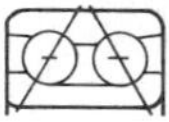
BE	Filling slot assembly Vertex of contact angles outside bearing 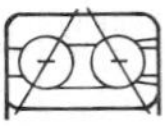		BAA	Non-separable Vertex of contact angles inside bearing Two-piece outer ring 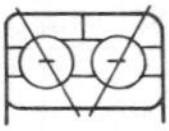
BJ	Non-filling slot assembly Vertex of contact angles inside bearing 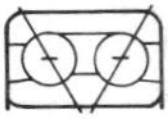		BVV	Separable Vertex of contact angles outside bearing Two-piece inner ring 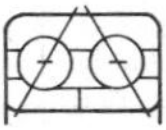

BALL BEARINGS, DOUBLE ROW, SELF-ALIGNING			
	Symbol	Description	
	BS	Raceway of outer ring spherical	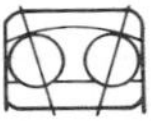

Fig. 1. (*Continued*). Types of rolling element bearings and their symbols

Single-row Radial, Non-filling Slot: This is probably the most widely used ball bearing and is employed in many modified forms. It is also known as the "Conrad" type or "Deep-groove" type. It is a symmetrical unit capable of taking combined radial and thrust loads in which the thrust component is relatively high, but is not intended for pure thrust loads, however. Because this type is not self-aligning, accurate alignment between shaft and housing bore is required.

Single-row Radial, Filling Slot: This type is designed primarily to carry radial loads. Bearings of this type are assembled with as many balls as can be introduced by eccentric displacement of the rings, as in the non-filling slot type, and then several more balls are inserted through the loading slot, aided by a slight spreading of the rings and heat expansion of the outer ring, if necessary. This type of bearing will take a certain degree of thrust when in combination with a radial load but is not recommended where thrust loads exceed 60 per cent of the radial load.

Single-row Angular-contact: This type is designed for combined radial and thrust loads where the thrust component may be large and axial deflection must be confined

CYLINDRICAL ROLLER BEARING, SINGLE ROW, NON-LOCATING TYPE			
Symbol	Description	Symbol	Description
RU	Inner ring without ribs Double-ribbed outer ring Inner ring separable 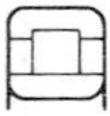	RNS	Double-ribbed inner ring Outer ring without ribs Outer ring separable Spherical outside surface 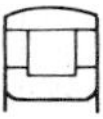
RUP	Inner ring without ribs Double-ribbed outer ring with one loose rib Both rings separable 	RAB	Inner ring without ribs Single-ribbed outer ring Both rings separable 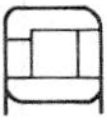
RUA	Inner ring without ribs Double-ribbed outer ring Inner ring separable Spherical outside surface 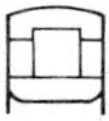	RM	Inner ring without ribs Rollers located by cage, end-rings or internal snap rings recesses in outer ring Inner ring separable 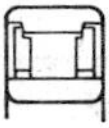
RN	Double-ribbed inner ring Outer ring without ribs Outer ring separable 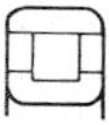	RNU	Inner ring without ribs Outer ring without ribs Both rings separable 

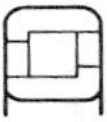

CYLINDRICAL ROLLER BEARINGS, SINGLE ROW, ONE-DIRECTION-LOCATING TYPE			
Symbol	Description	Symbol	Description
RR	Single-ribbed inner-ring Outer ring with two internal snap rings Inner ring separable 	RF	Double-ribbed inner ring Single-ribbed outer ring Outer ring separable 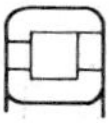
RJ	Single-ribbed inner ring Double-ribbed outer ring Inner ring separable 	RS	Single-ribbed inner ring Outer ring with one rib and one internal snap ring Inner ring separable 
RJP	Single-ribbed inner ring Double-ribbed outer ring with one loose rib Both rings separable 	RAA	Single-ribbed inner ring Single-ribbed outer ring

Fig. 1 (*Continued*). Types of rolling element bearings and their symbols.

within very close limits. A high shoulder on one side of the outer ring is provided to take the thrust, while the shoulder on the other side is only high enough to make the bearing non-separable. Except where used for a pure thrust load in one direction, this type is applied either in pairs (duplex) or one at each end of the shaft, opposed.

Double-row Bearings: These are, in effect, two single-row angular-contact bearings built as a unit with the internal fit between balls and raceway fixed at the time of

CYLINDRICAL ROLLER BEARINGS, SINGLE ROW, TWO-DIRECTION-LOCATING TYPE				
Symbol	Description	Symbol	Description	
RK	Double-ribbed inner ring Outer ring with two internal snap rings Non-separable	RY	Double-ribbed inner ring Outer ring with one rib and one internal snap ring Non-separable	
RC	Double-ribbed inner ring Double-ribbed outer ring Non-separable	RCS	Double-ribbed inner ring Double-ribbed outer ring Non-separable Spherical outside surface	
RG	Inner ring, with one rib and one snap ring Double-ribbed outer ring Non-separable			
RP	Double-ribbed inner ring Double-ribbed outer ring with one loose rib Outer ring separable	RT	Double-ribbed inner ring with one loose rib Double-ribbed outer ring Inner ring separable	

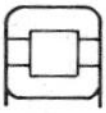

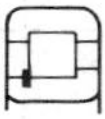

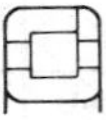

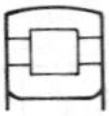

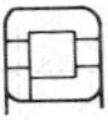

CYLINDRICAL ROLLER BEARINGS				
Double Row Non-Locating Type		Double Row Two-Direction-Locating Type		
Symbol	Description	Symbol	Description	
RA	Inner ring without ribs Three integral ribs on outer ring Inner ring separable	RB	Three integral ribs on inner ring Outer ring without ribs, with two internal snap rings Non-separable	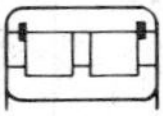
RD	Three integral ribs on inner ring Outer ring without ribs Outer ring separable	Multi-Row Non-Locating Type		
		Symbol	Description	
RE	Inner ring without ribs Outer rings without ribs, with two internal snap rings Inner ring separable	RV	Inner ring without ribs Double-ribbed outer ring (loose ribs) Both rings separable	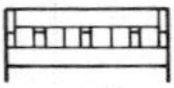

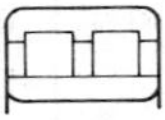

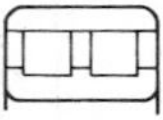

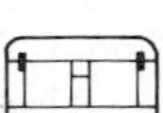

Fig. 1. (*Continued*). Types of rolling element bearings and their symbols.

bearing assembly. This fit is therefore not dependent upon mounting methods for internal rigidity. These bearings usually have a known amount of internal preload built in for maximum resistance to deflection under combined loads with thrust from either direction. Thus, with balls and races under compression before an external load is applied, due to this internal preload, the bearings are very effective for radial loads where bearing deflection must be minimized.

SELF-ALIGNING ROLLER BEARINGS, DOUBLE ROW			
Symbol	Description	Symbol	Description
SD	Three integral ribs on inner ring Raceway of outer ring spherical	SL	Raceway of outer ring spherical Rollers guided by the cage Two integral ribs on inner ring
SE	Raceway of outer ring spherical Rollers guided by separate center guide ring in outer ring	SELF-ALIGNING ROLLER BEARINGS, SINGLE ROW	
		Symbol	Description
SW	Raceway of inner ring spherical	SR	Inner ring with ribs Raceway of outer ring spherical Radial contact
		SA	Raceway of outer ring spherical Angular contact
SC	Raceway of outer ring spherical Rollers guided by separate axially floating guide ring on inner ring	SB	Raceway of inner ring spherical Angular contact
THRUST BALL BEARINGS		THRUST ROLLER BEARINGS	
Symbol	Description	Symbol	Description
TA	Single direction, grooved raceways, flat seats	TS	Self-aligning, single-direction, flat seats, asymmetrical barrel-shaped rollers
TAA	Single row, angular contact where a line through the ball contact points forms an angle from above 45° to and including 75° with a perpendicular to the bearing axis of rotation	TSA	Self-aligning, single-direction, flat seats, asymmetrical barrel-shaped rollers
		TP	Single-direction, flat seats, cylindrical rollers

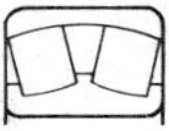

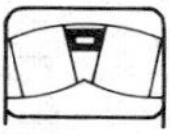

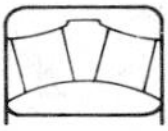

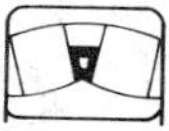

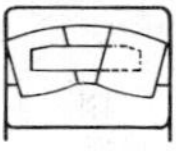

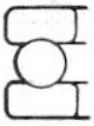

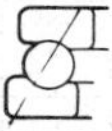

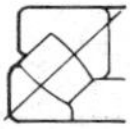

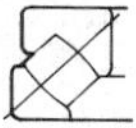

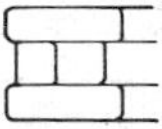

Fig. 1 (*Continued*). Types of rolling element bearings and their symbols.

Other Types: Modifications of these basic types provide arrangements for self-sealing, location by snap ring, shielding, etc., but the fundamentals of mounting are not changed. A special type is the *self-aligning* ball bearing which can be used to compensate for an appreciable degree of misalignment between shaft and housing due to shaft deflections, mounting inaccuracies, or other causes commonly encountered. With a single row of balls, alignment is provided by a spherical outer surface on the outer ring; with a double row of balls, alignment is provided by a spherical raceway on the outer ring. Bearings in the wide series have a considerable amount of thrust capacity.

Types of Roller Bearings. — Types of roller bearings are distinguished by the

TAPERED ROLLER BEARINGS	
Symbol	Description
TS	Single row 
TDI	Two row, double-cone single cups 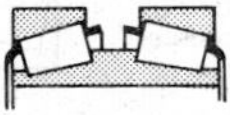
TDO	Two row, double-cup single-cone adjustable 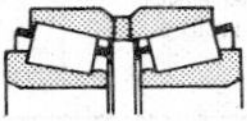
TNA	Two row, double-cup single cone nonadjustable 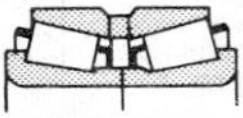
TQD	Four row, cup adjusted 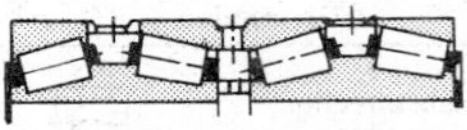
TQI	Four row, cup adjusted 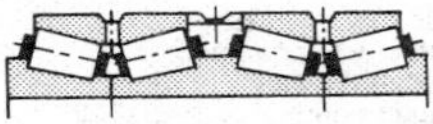
THRUST TAPERED ROLLER BEARINGS	
Symbol	Description
TT	Thrust bearings 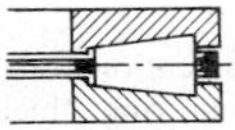

Fig. 1 (*Continued*). Types of rolling element bearings and their symbols.

design of rollers and raceways to handle axial, combined axial and thrust, or thrust loads.

Cylindrical Roller: These have solid or helically wound hollow cylindrical rollers. The free ring may have a restraining flange to provide some restraint to endwise movement in one direction or may be without a flange so that the bearing rings may be displaced axially with respect to each other. Either rolls or roller path on the races

NEEDLE ROLLER BEARINGS, DRAWN CUP			
Symbol*	Description	Symbol*	Description
NIB NB	Needle roller bearing, full complement, drawn cup, without inner ring	NIYM NYM	Needle roller bearing, full complement, rollers retained by lubricant, drawn cup, closed end, without inner ring
NIBM NBM	Needle roller bearing, full complement, drawn cup, closed end, without inner ring	NIH NH	Needle roller bearing, with cage, drawn cup, without inner ring
NIY NY	Needle roller bearing, full complement, rollers retained by lubricant, drawn cup, without inner ring	NIHM NHM	Neeedle roller bearing, with cage, drawn cup, closed end, without inner ring

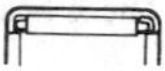

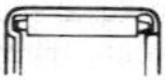

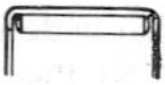

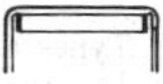

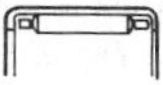

NEEDLE ROLLER BEARINGS	
Symbol*	Description
NIA NA	Needle roller bearing, with cage, machined ring, lubrication hole and groove in OD, without inner ring 

NEEDLE ROLLER AND CAGE ASSEMBLIES

Symbol*	Description
NIM NM	Needle roller and cage assembly

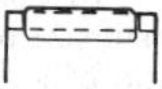

NEEDLE ROLLER BEARING INNER RINGS	
Symbol*	Description
NIR NR	Needle roller bearing inner ring, lubrication hole and groove in bore 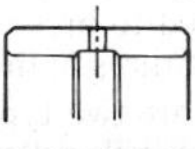

Machined Ring Needle Roller Bearings Type NIA may be used with inch dimensioned inner rings, Type NIR, and Type NA may be used with metric dimensioned inner rings, Type NR.

* Symbols with I, as NIB, are inch-dimensioned, and those without the I, as NB, are metric dimensioned.

Fig. 1 (Concluded). Types of rolling element bearings and their symbols.

may be slightly crowned to prevent edge loading under slight shaft misalignment. Low friction makes this type suitable for relatively high speeds.

Barrel Roller: These have rollers that are barrel-shaped and symmetrical. They are furnished in both single- and double-row mountings. As with cylindrical roller bearings, the single-row mounting type has a low thrust capacity, but angular mounting of rolls in the double-row type permits its use for combined axial and thrust loads.

Spherical Roller: These are usually furnished in a double-row, self-aligning mounting. Both rows of rollers have a common spherical outer raceway. The rollers are barrel-shaped with one end smaller than the other to provide a small thrust to keep the rollers in contact with the center guide flange. This type of roller bearing has a high radial and thrust load carrying capacity with the ability to maintain this capacity

under some degree of misalignment of shaft and bearing housing.

Tapered Roller: In this type, straight tapered rollers are held in accurate alignment by means of a guide flange on the inner ring. The basic characteristic of these bearings is that the apexes of the tapered working surfaces of both rollers and races, if extended, would coincide on the bearing axis. These bearings are separable. They have a high radial and thrust carrying capacity.

Types of Ball and Roller Thrust Bearings. — These are designed to take thrust loads alone or, in some cases, in combination with radial loads.

One-direction Ball Thrust: These consist of a shaft ring and a flat or spherical housing ring with a single row of balls between. They are capable of carrying pure thrust loads in one direction only. They cannot carry any radial load.

Two-direction Ball Thrust: These consist of a shaft ring with a ball groove in either side, two sets of balls, and two housing rings so arranged that thrust loads in either direction can be supported. No radial loads can be carried.

Spherical Roller Thrust: This type is similar in design to the radial spherical roller bearing except that it has a much larger contact angle. The rollers are barrel shaped with one end smaller than the other. This type bearing has a very high thrust load carrying capacity and can also carry radial loads.

Tapered Roller Thrust: In this type the rollers are tapered and several different arrangements of housing and shaft are used.

Roller Thrust: In this type the rollers are straight and several different arrangements of housing and shaft are used.

Types of Needle Bearings. — Needle bearings are characterized by their relatively small size rollers, usually not ranging above ¼ inch in diameter, and a relatively high ratio of length to diameter, usually ranging from about 3 to 1 and 10 to 1. Another feature that is characteristic of several types of needle bearings is the absence of a cage or separator for retaining the individual rollers. Needle bearings may be divided into three classes: loose-roller, outer race and retained roller, and non-separable units.

Loose-roller: This type of bearing has no integral races or retaining members, the needles being located directly between the shaft and the outer bearing bore. Usually both shaft and outer bore bearing surfaces are hardened and retaining members that have smooth unbroken surfaces are provided to prevent endwise movement. Compactness and high radial load capacity are features of this type.

Outer Race and Retained Roller: There are two types of outer race and retained roller bearings. In the *Drawn Shell* type, the needle rollers are enclosed by a hardened shell that acts as a retaining member and also as a hardened outer race. The needles roll directly on the shaft, the bearing surface of which should be hardened. The capacity for given roller length and shaft diameter is about two-thirds of the loose roller type. It is mounted in the housing with a press fit.

In the *Machined Race* type, the outer race consists of a heavy machined member. Various modifications of this type provide heavy ends or faces for end location of the needle rollers, or open end construction with end washers for roller retention, or a cage which maintains alignment of the rollers and is itself held in place by retaining rings. An auxiliary outer member with spherical seat which holds the outer race may be provided for self alignment. This type is applicable where split housings occur or where a press fit of the bearing into the housing is not possible.

Non-separable: This type consists of a non-separable unit of outer race, rollers and inner race. These bearings are used where high static or oscillating motion loads are expected as in certain aircraft components, and where both outer and inner races are necessary.

Special or Unconventional Types. — Rolling contact bearings have been developed for many highly specialized applications. They may be constructed of non-corrosive materials, non-magnetic materials, plastics, ceramics, and even wood. Though the materials are chosen to adapt more conventional configurations to difficult applications or environments, even greater ingenuity has been applied in utilizing rolling-contact for solving particular problems. Thus, linear or recirculating bearings are available to provide low friction, accurate location, and simplified lubrication features to such applications as machine ways, axial motion devices, jack-screws, steering linkages, collets, and chucks. This type of bearing utilizes the "full-complement" style of loading the rolling elements between "races" or ways without a cage and with each element advancing by the action of "races" in the loaded areas and by contact with the adjacent element in the unloaded areas. The "races" may not be cylindrical or bodies of revolution but plane surfaces, with suitable interruptions to free the rolling elements so that they can then follow a return trough or slot back to the entry-point at the start of the "race" contact area. Combinations of radial and thrust bearings are available for the user with special requirements.

Plastic Bearings. — A more recent development has been the use of Acetal Resin rollers and balls in applications where abrasive, corrosive and difficult-to-lubricate conditions exist. Though these bearings do not have the load carrying capacity nor the low friction factor of their hard steel counterparts, they do offer freedom from indentation, wear, and corrosion, while at the same time providing significant weight savings. Of additional value are: (1) their resistance to indentation from shock loads or oscillation and (2) their self-lubricating properties. Usually these bearings are not available in stock, but must be designed and produced in accordance with the data made available by the plastics processor.

Pillow Block and Flanged Housing Bearings. — Of great interest to the shop man and particularly adaptable to "line-shafting" applications are a series of ball and roller bearings supplied with their own housings, adapters, and seals. Often called pre-mounted bearings, they come with a wide variety of flange mountings permitting location on faces parallel to or perpendicular to the shaft axis.

Inner races can be mounted directly on ground shafts, or can be adapter-mounted to "drill-rod" or to commercial-shafting. For installations sensitive to unbalance and vibration, the use of accurately ground shaft seats is recommended.

Since most pillow block designs incorporate self-aligning bearing types, they do not require the precision mountings utilized with more normal bearing installations.

Conventional Bearing Materials. — Most rolling contact bearings are made with all load carrying members of full hard steel, either through- or case-hardened. For greater reliability this material is controlled and selected for cleanliness and alloying practices in conformity with rigid specifications in order to reduce the incidence of anomalies and inclusions which could limit the useful fatigue life. Magnaflux inspection is employed to ensure that elements are free from both material defects and cracks. Likewise, a light etch is employed between rough and finish grinding to ensure that burns due to heavy stock removal and associated decarburization will be eliminated from finished pieces.

Cage Materials. — Standard bearings are normally made with cages of free-machining brass or low carbon sulfurized steel. In high-speed applications or where lubrication may be intermittent or marginal, special materials may be employed. Iron-silicon-bronze, laminated phenolics, silver-plating, over-lays, solid-

film baked-on coatings, carbon-graphite inserts, and in extreme cases, sintered or even impregnated materials are used in separators.

Commercial bearings usually rely on stamped steel with or without a phosphate treatment; some economical varieties are found with snap-in plastic or metallic cages.

So long as lubrication is adequate and speeds are both reasonable and steady, the materials and design of the cage are of secondary importance when compared with those of the rolling elements and their contacts with the races. In spite of this tolerance however, a good portion of all rolling bearing failures encountered can be traced to cage failures resulting from inadequate lubrication. It can never be overemphasized that *no bearing can be designed to run continuously without lubrication!*

Standard Method of Bearing Designation. — The Anti-Friction Bearing Manufacturers Association has adopted a standard identification code that provides a specific designation for each different ball, roller, and needle bearing. Thus, for any given bearing, a uniform designation is provided for manufacturer and user alike, so that the confusion of different company designations can be avoided.

In this identification code there is a "basic number" for each bearing which consists of three elements: a one- to four-digit number which indicates the size of the bore in numbers of millimeters (metric series); a two- or three-letter symbol which indicates the type of bearing; and a two-digit number which indicates the dimension series to which the bearing belongs.

In addition to this "basic number" other numbers and letters are added to designate type of tolerance, cage, lubrication, fit up, ring modification, the addition of shields, seals, mounting accessories, etc. Thus, a complete designating symbol might be *50BC02JPXE0A10*, for example. The basic number is *50BC02* and the remainder is the supplementary number. For a radial bearing, this latter consists of up to four letters to indicate modification of design, one or two digits to indicate internal fit and tolerances, a letter to indicate lubricants and preservatives, and up to three digits to indicate special requirements.

For a thrust bearing the supplementary number would consist of two letters to indicate modifications of design, one digit to indicate tolerances, one letter to indicate lubricants and preservatives, and up to three digits to indicate special requirements.

For a needle bearing the supplementary number would consist of up to three letters indicating cage material or integral seal information or whether the outer ring has a crowned outside surface and one letter to indicate lubricants or preservatives.

Dimension Series: Annular ball, cylindrical roller, and self-aligning roller bearings are made in a series of different outside diameters for every given bore diameter and in a series of different widths for every given outside diameter. Thus, each of these bearings belongs to a dimension series which is designated by a two-digit number such as 01, 23, 93, etc. The first digit (8, 0, 1, 2, 3, 4, 5, 6 and 9) indicates the *width series* and the second digit (8, 9, 0 1, 2, 3, and 4) the *diameter series* to which the bearing belongs. In the case of ball and roller thrust bearings and needle roller bearings, similar types of identification codes are used.

Bearing Tolerances. — In order to provide standards of precision for proper application of ball or roller bearings in all types of equipment, four classes of tolerances have been established by the Anti-Friction Bearing Manufacturers Association for ball bearings, two for cylindrical roller bearings and one for spherical roller bearings. These tolerances are given in Tables 1, 2, 3 and 4. They are designated as ABEC-1, ABEC-5, ABEC-7 and ABEC-9 for ball bearings, the ABEC-9 being the most precise, RBEC-1 and RBEC-5 for roller bearings. In general, bearings to specifications closer than ABEC-1 or RBEC-1 are required because of the need for very precise fits on shaft or housing, to reduce eccentricity or runout of shaft or supported part, or to permit operation at very high speeds. All four classes

(Continued on page 2147)

Table 1. ABEC-1 and RBEC-1 Tolerance Limits for Metric Ball and Roller Bearings (ANSI/AFBMA Std 20-1977[1])

Basic Inner Ring Bore Diameter, d				Bore Tolerance Limits,[2] Inch				Radial Runout,[4] K_i Inch
mm		Inches		d_{mp}		d_s[3]		
Over	Incl.	Over	Incl.	Low	High	Low	High	
2.5	10	0.0984	0.3937	−.0003	+0	−.0004	+.0001	.0003
10	18	0.3937	0.7087	−.0003	+0	−.0004	+.0001	.0004
18	30	0.7087	1.1811	−.0004	+0	−.0005	+.0001	.0005
30	50	1.1811	1.9685	−.0005	+0	−.0006	+.0001	.0006
50	80	1.9685	3.1496	−.0006	+0	−.0008	+.0002	.0008
80	120	3.1496	4.7244	−.0008	+0	−.0010	+.0002	.0010
120	180	4.7244	7.0866	−.0010	+0	−.0012	+.0003	.0012

Basic Outer Ring Outside Diameter, D				Outside Diameter Tolerance Limits,[5] Inch						Radial Run-out,[4] K_e Inch
mm		Inches		D_{mp}		D_s[6]		D_s[7]		
Over	Incl.	Over	Incl.	High	Low	High	Low	High	Low	
6	18	0.2362	0.7087	+0	−.0003	+.0001	−.0004	+.0002	−.0005	.0006
18	30	0.7087	1.1811	+0	−.0004	+.0001	−.0005	+.0002	−.0006	.0006
30	50	1.1811	1.9685	+0	−.0005	+.0002	−.0007	+.0003	−.0008	.0008
50	80	1.9685	3.1496	+0	−.0005	+.0002	−.0007	+.0004	−.0009	.0010
80	120	3.1496	4.7244	+0	−.0006	+.0002	−.0008	+.0005	−.0011	.0014
120	150	4.7244	5.9055	+0	−.0008	+.0002	−.0010	+.0005	−.0013	.0016
150	180	5.9055	7.0866	+0	−.0010	+.0003	−.0013	+.0006	−.0016	.0018
180	250	7.0866	9.8425	+0	−.0012	+.0003	−.0015	+.0008	−.0020	.0020
250	315	9.8425	12.4015	+0	−.0014	+.0004	−.0018	+.0008	−.0022	.0024
315	400	12.4015	15.7480	+0	−.0016	+.0004	−.0020	+.0009	−.0025	.0028

Width Tolerances							
Normal Single Bearings				Modified Single Bearings[8]			
Basic Inner Ring Bore, d, mm		Width, B_s Tolerance Limits, Inch		Basic Inner Ring Bore, d, mm		Width, B_s Tolerance Limits, Inch	
Over	Incl.	High	Low	Over	Incl.	High	Low
2.5	50	+0	−.0050	2.5	50	+0	−.0100
50	80	+0	−.0060	50	80	+0	−.0150
80	120	+0	−.0080	80	120	+0	−.0150
120	180	+0	−.0100	120	180	+0	−.0200

For sizes beyond range of this table, see Standard.
[1] Does not cover tapered roller bearings. [2] The amounts by which the largest and smallest single diameter, d_s, of the bore and the mean, d_{mp}, of these two vary from the basic. Bore tolerance limits do not apply to tapered bore inner rings. [3] d_s applies only to: metric diameter series 0, up to and including $d = 40$ mm and in diameter series 2, up to and including 180 mm. For larger sizes in series 0 and 2 and all sizes of series 1, 8 and 9, d_s is not restricted. [4] Total indicator reading. [5] The amounts by which the largest and smallest outside diameter, D_s and the mean, D_{mp} of these two vary from the basic. [6] D_s applies only to open bearings of metric diameter series 0 up to and including $D = 80$ mm and diameter series 2, up to and including $D = 315$ mm. For larger sizes in series 0 and 2 and all sizes of series 1, 8 and 9, D_s is not restricted. [7] D_s applies only to bearings with shields or seals in metric diameter series 0 up to and including $D = 80$ mm and in diameter series 2 up to and including $D = 315$ mm. For larger sizes in series 0 and 2 and for all sizes of series 1, 8 and 9, D_s is not restricted. [8] This refers to a ball bearing on which one or both sides are so modified that two or more bearings can be mounted side by side as a unit.

Table 2. ABEC-5 and RBEC-5 Tolerance Limits for Metric Ball and Roller Bearings (ANSI/AFBMA Std 20-1977[1])

INNER RING

Inner Ring Bore Basic Diam., d — mm Over	Incl	Inches Over	Incl	All Series, d_{mp} High	Low	4, 3, 2, 1, 0 d_s High	Low	9, 8 d_s High	Low	Radial Runout K_i Max.	Ref. Side Runout with Bore[3] Max.	Raceway Runout with Ref. Side[3] Max.	Normal Single Bearings, B_s High	Low	Modified Single Bearings,[4] B_s High	Low	Width Variation Indiv. Ring Max.
0.6	10	0.0236	0.3937	+0	−.0002	+0	−.0002	+0	−.0002	.00015	.0003	.0003	+0	−.0016	+0	−.0100	.0002
10	18	0.3937	0.7087	+0	−.0002	+0	−.0002	+0	−.00025	.00015	.0003	.0003	+0	−.0032	+0	−.0100	.0002
18	30	0.7087	1.1811	+0	−.0002	+0	−.0002	+0	−.0003	.00015	.0003	.0003	+0	−.0050	+0	−.0100	.0002
30	50	1.1811	1.9685	+0	−.0002	+0	−.0002	+0	−.00035	.0002	.0003	.0003	+0	−.0050	+0	−.0100	.0002
50	80	1.9685	3.1496	+0	−.0002	+0	−.0003	+0	−.00045	.0002	.0003	.0003	+0	−.0060	+0	−.0100	.0002
80	120	3.1496	4.7244	+0	−.0003	+.00005	−.00035	+.00005	−.0005	.00025	.0003	.0003	+0	−.0080	+0	−.0150	.0003
120	180	4.7244	7.0866	+0	−.0004	+.0001	−.0005	+.0001	−.0006	.0003	.0004	.0004	+0	−.0100	+0	−.0150	.0003

OUTER RING

Basic Outside Diameter, D — mm Over	Incl	Inches Over	Incl	All Series, D_{mp} High	Low	Open Bearings 4, 3, 2, 1, 0 D_s High	Low	9, 8 D_s High	Low	Diam. Series[5] 4, 3, 2, 1, 0 D_s High	Low	9, 8 D_s High	Low	Radial Runout K_e Max.	O.D. Runout with Ref. Side[6] Max.	Raceway Runout with Ref. Side[3] Max.	Width Tol. Limits	Width Variation Indiv. Ring Max.
2.5	6	0.0984	0.2362	+0	−.0002	+0	−.0002	+0	−.0002	+.0001	−.0001	+.0001	−.0001	.0002	.0003	.0003	Same as those for Inner Ring of the same bearing	.0002
6	18	0.2362	0.7087	+0	−.0002	+0	−.0002	+.00005	−.00005	+.0001	−.0001	+.00015	−.00015	.0002	.0003	.0003		.0002
18	30	0.7087	1.1811	+0	−.0002	+0	−.0002	+.0001	−.0001	+.0001	−.0001	+.0002	−.0002	.0002	.0003	.0003		.0002
30	50	1.1811	1.9685	+0	−.0002	+0	−.0002	+.00015	−.00015	+.0002	−.0002	+.00035	−.00035	.0002	.0003	.0003		.0002
50	80	1.9685	3.1496	+0	−.0003	+0	−.0003	+.00015	−.00015	+.0003	−.0003	+.0004	−.0004	.0003	.0004	.0004		.0003
80	120	3.1496	4.7244	+0	−.0003	+.0005	−.00035	+.0002	−.0002	+.0003	−.0003	+.0005	−.0005	.0004	.0005	.0005		.0003
120	150	4.7244	5.9055	+0	−.0004	+.001	−.0005	+.00065	−.00065	+.0003	−.0003	+.0006	−.0006	.0004	.0005	.0005		.0003
150	180	5.9055	7.0866	+0	−.0005	+.0015	−.00065	+.0002	−.0002	+.0003	−.0003	+.0006	−.0006	.0005	.0005	.0006		.0003
180	250	7.0866	9.8425	+0	−.0005	+.0015	−.00065	+.00075	−.00075	+.0004	−.0004	+.0009	−.0009	.0005	.0006	.0006		.0004

All dimensions are in inches unless otherwise indicated. For sizes beyond range of this table see Standard.   [2] Bore tolerance limits do not apply to tapered bore innerrings.   [3] Does not apply to roller bearings.   [5] For bearings with shields or seals; D_s is not restricted for diameter series 9 and 8.   [6] Applies to bearings of width series 1 and narrower.   [1] Does not cover instrument bearings and tapered roller bearings.   [4] A bearing on which one or both sides are so modified that two or more bearings can be mounted side by side as a unit.

Table 3. ABEC-7 Tolerance Limits for Metric Radial and Angular Contact Ball Bearings (ANSI/AFBMA Std 20-1977[1])

INNER RING[2]

Inner Ring Bore Basic Diam.. d (mm)		Inches		Bore Tolerance Limits, Inch All Series, d_{mp}		Diam. Series 4, 3, 2, 1, 0 d_s		Diam. Series 9, 8 d_s		Radial Runout, K_i	Ref. Side Runout with Bore	Raceway Runout with Ref. Side[3]	Width Normal Single Bearings B_s		Modified Single Bearings[4] B_s		Width Variation Indiv. Ring
Over	Incl.	Over	Incl.	Low	High	Low	High	Low	High	Max.	Max.	Max.	High	Low	High	Low	Max.
0.6	10	0.0236	0.3937	− .00015	+0	− .00015	+0	− .00015	+0	.0001	.0001	.0001	+0	− .0016	+0	− .0100	.0001
10	18	0.3937	0.7087	− .00015	+0	− .00015	+0	− .0002	+.00005	.0001	.0001	.0001	+0	− .0032	+0	− .0100	.0001
18	30	0.7087	1.1811	− .00015	+0	− .00015	+0	− .00025	+.0001	.0001	.00015	.00015	+0	− .0050	+0	− .0100	.0001
30	50	1.1811	1.9685	− .0002	+0	− .0002	+0	− .0003	+.0001	.00015	.00015	.00015	+0	− .0050	+0	− .0100	.0001
50	80	1.9685	3.1496	− .0002	+0	− .0002	+0	− .0003	+.0001	.00015	.0002	.00015	+0	− .0060	+0	− .0100	.00015
80	120	3.1496	4.7244	− .00025	+0	− .0003	+.00005	− .00035	+.0001	.0002	.0002	.0002	+0	− .0080	+0	− .0150	.00015
120	180	4.7244	7.0866	− .0003	+0	− .00035	+.00005	− .00045	+.00015	.0003	.0003	.0003	+0	− .0100	+0	− .0150	.0002

OUTER RING[2]

Basic Outside Diameter, D (mm)		Inches		Outside Diameter Tolerance Limits, Inch All Series, D_{mp}		Open Bearings – Diam. Series 4, 3, 2, 1, 0 D_s		Diam. Series 9, 8 D_s		Diam. Series[5] 4, 3, 2, 1, 0 D_s		Radial Runout K_e	O. D. Runout with Ref. Side[6]	Raceway Runout with Ref. Side[3]	Width Tol. Limits	Width Variation Indiv. Ring
Over	Incl.	Over	Incl.	High	Low	High	Low	High	Low	High	Low	Max.	Max.	Max.		Max.
2.5	6	0.0984	0.2362	+0	− .0002	+0	− .0002	+0	− .0002	+.0001	− .0001	.00015	.00015	.0002	Same as those for Inner Ring of the same bearing	.0001
6	18	0.2362	0.7087	+0	− .0002	+0	− .0002	+.00005	− .00025	+.0001	− .0001	.00015	.00015	.0002		.0001
18	30	0.7087	1.1811	+0	− .0002	+0	− .0002	+.0001	− .0003	+.0001	− .0001	.00015	.00015	.0002		.0001
30	50	1.1811	1.9685	+0	− .0002	+0	− .0002	+.0001	− .0003	+.0002	− .0002	.0002	.00015	.0002		.0001
50	80	1.9685	3.1496	+0	− .0002	+0	− .0002	+.0001	− .0004	+.0002	− .0003	.0002	.00015	.0002		.0001
80	120	3.1496	4.7244	+0	− .0003	+.00005	− .00035	+.0001	− .0004	+.0003	− .0003	.0002	.0002	.0002		.0002
120	150	4.7244	5.9055	+0	− .0004	+.0001	− .0005	+.0002	− .0005	+.0003	− .0005	.0003	.0002	.0003		.0002
150	180	5.9055	7.0866	+0	− .0004	+.0001	− .0006	+.0002	− .0006	+.0003	− .0006	.0003	.0002	.0003		.0002
180	250	7.0866	9.8425	+0	− .0004	+.0001	− .0005	+.0002	− .0006	+.0003	− .0008	.0003	.0002	.0004		.0002

All dimensions are in inches unless otherwise indicated. For sizes beyond range of this table see Standard. tolerance limits are in inches. [1] Does not cover instrument bearings. [2] All tolerance limits are in inches. [3] Does not apply to self-aligning ball bearings. [4] A bearing on which one or both sides are so modified that two or more bearings can be mounted side by side as a unit. [5] For bearings with shields or seals; D_s is not restricted for dimension series 9 and 8. [6] Applies to bearings of width series 1, and narrower.

Table 4.   ABEC-9 Tolerance Limits for Metric Radial and Angular Contact Ball Bearings (ANSI/AFBMA Std 20-1977[1])

Inner Ring								
Basic Bore Diameter, d				Bore Tolerance Limits, Inch		Radial Run-out,[2] K_i Inch	Width Vari-ation, Inch	Ref. Side Runout with Bore, Inch	Race-way Runout with Side,[3] Inch
mm		Inches		d_s					
Over	Incl.	Over	Incl.	Low	High				
0.6	10	0.0236	0.3937	−.0001	+0	.00005	.00005	.00005	.00005
10	18	0.3937	0.7087	−.0001	+0	.00005	.00005	.00005	.00005
18	30	0.7087	1.1811	−.0001	+0	.0001	.00005	.00005	.0001
30	50	1.1811	1.9685	−.0001	+0	.0001	.00005	.00005	.0001
50	80	1.9685	3.1496	−.00015	+0	.0001	.00005	.00005	.0001
80	120	3.1496	4.7244	−.0002	+0	.0001	.0001	.0001	.0001
120	150	4.7244	5.9055	−.00025	+0	.0001	.0001	.0001	.0001
150	180	5.9055	7.0866	−.00025	+0	.0002	.00015	.00015	.0002

Outer Ring								
Basic Outside Diameter, D				Outside Diameter Tolerance Limits,[4] Inch		Radial Run-out,[5] K_e Inch	Width Vari-ations, Inch	Outside Cylin. Surface Runout with Side,[6] Inch	Race-way Runout with Side,[3] Inch
mm		Inches		D_s					
Over	Incl.	Over	Incl.	High	Low				
2.5	18	0.0984	0.7087	+0	−.0001	.00005	.00005	.00005	.00005
18	30	0.7087	1.1811	+0	−.00015	.0001	.00005	.00005	.0001
30	50	1.1811	1.9685	+0	−.00015	.0001	.00005	.00005	.0001
50	80	1.9685	3.1496	+0	−.00015	.00015	.00005	.00005	.00015
80	120	3.1496	4.7244	+0	−.0002	.0002	.0001	.0001	.0002
120	150	4.7244	5.9055	+0	−.0002	.0002	.0001	.0001	.0002
150	180	5.9055	7.0866	+0	−.00025	.0002	.0001	.0001	.0002
180	250	7.0866	9.8425	+0	−.0003	.00025	.00015	.00015	.00025
250	315	9.8425	12.4015	+0	−.0003	.00025	.00015	.00015	.00025

Width Tolerance Limits							
Normal Single Bearings				Modified Single Bearings[7]			
Basic Bore Diam., mm		Width, B_s Tolerance Limits, Inch		Basic Bore Diam., mm		Width, B_s Tolerance Limits, Inch	
Over	Incl.	High	Low	Over	Incl.	High	Low
0.6	10	+0	−.0010	0.6	80	+0	−.0100
10	18	+0	−.0032	80	150	+0	−.0150
18	50	+0	−.0050	150	250	+0	−.0200
50	80	+0	−.0060	....	....	....	....
80	120	+0	−.0080	....	....	....	....
120	150	+0	−.0100	....	....	....	....
150	180	+0	−.0120	....	....	....	....

For sizes beyond the range of this table see Standard.
[1] Does not cover instrument bearings.   [2] Difference between greatest and smallest radial distance between bore surface and middle of a raceway on outside of the ring.   [3] Does not apply to self-aligning ball bearings. "Side" is reference side.   [4] These tolerance limits apply before seals or shields are inserted.   [5] Difference between greatest and smallest radial distance between outside surface and middle of a raceway on inside of the ring.   [6] Applies to bearings of metric width series 1 or narrower.   [7] This refers to a ball bearing on which one or both sides are so modified that two or more bearings can be mounted side by side as a unit.

Table 5. AFBMA and American National Standard Tolerance Limits for Metric Single Direction Thrust Ball and Roller Bearings (ANSI/AFBMA Std 20-1977)

BASIC PLAN METRIC DIMENSIONED†									
Basic Bore Diam., d		Shaft Washer		Both Indiv. Washers	Bearing Height Tolerance Limits, H_m		Basic Outside Diameter, D		Housing Washer
		Bore Tolerance Limits, d_s		Race-way Runout with Seat Face					Outside Diam. Tolerance Limits, D_s
mm		Inch		Inch	Inch		mm		Inch
Over	Incl.	Low	High	Max.	High	Low	Over	Incl.	High	Low
0	18	−.0003	+.0002	.0004	0	−.0030	10	18	0	−.0004
18	30	−.0004	+.0003	.0004	0	−.0030	18	30	0	−.0005
30	50	−.0005	+.0004	.0004	0	−.0039	30	50	0	−.0006
50	80	−.0006	+.0005	.0004	0	−.0049	50	80	0	−.0007
80	120	−.0008	+.0006	.0006	0	−.0059	80	120	0	−.0009
120	180	−.0010	+.0007	.0006	0	−.0069	120	180	0	−.0010
180	250	−.0012	+.0009	.0008	0	−.0079	180	250	0	−.0012
250	315	−.0014	+.0010	.0010	0	−.0089	250	315	0	−.0014
315	400	−.0016	+.0011	.0012	0	−.0118	315	400	0	−.0016
400	500	−.0018	+.0013	.0012	....	....	400	500	0	−.0018

† For thrust bearings of types TS and TSA, tolerances in Table 1 apply.
For sizes beyond the range of this table see Standard.

include tolerances for bore, outside diameter, ring width, and radial runouts of inner and outer rings. ABEC-5, ABEC-7 and ABEC-9 provide added tolerances for parallelism of sides, side runout and groove parallelism with sides.

Thrust Bearings: Anti-Friction Bearing Manufacturers Association and American National Standard tolerance limits for metric single direction thrust ball and roller bearings are given in Table 5. Tolerance limits for single direction thrust ball bearings, inch dimensioned are given in Table 6, and for cylindrical thrust roller bearings, inch dimensioned in Table 7.

There is only one class of tolerance limits established for metric thrust bearings.

Radial Needle Roller Bearings: Tolerance limits for needle roller bearings, drawn cup, without inner ring, inch types NIB, NIBM, NIY, NIYM, NIH, and NIHM are given in Table 8 and for metric types NB, NBM, NY, NYM, NH and NHM are given in Table 9. Standard tolerance limits for needle roller bearings, with cage, machined ring, without inner ring, inch type NIA are given in Table 10 and for needle roller bearings inner rings, inch type NIR are given in Table 11.

Table 6. Tolerance Limits for Single Direction Ball Thrust Bearings—Inch Design (ANSI/AFBMA Std 21.2-1977)

Bore Diameter† d, Inches		Height Tolerance Limits, Inch		Outside Diameter D, Inches		Outside Diameter Tolerance, Inch	
Over	Incl.	High	Low	Over	Incl.	High	Low
0	1.8125	+.005	−.005	0	5.3125	+0	−.002
1.8125	12.0000	+.010	−.010	5.3125	17.3750	+0	−.003
12.0000	20.0000	+.015	−.015	17.3750	39.3701	+0	−.004

† Bore tolerance limits are: For bore diameters over 0 to 6.7500 inches, inclusive, +.005, −0; for bore diameters over 6.7500 to 20.0000 inches, inclusive, +.007, −0.

Table 7. Tolerance Limits for Cylindrical Roller Thrust Bearings — Inch Design (ANSI/AFBMA Std 21.2-1977)

Basic Bore Diam., d		Bore Tolerance Limits		Height Tolerance Limits		Basic Outside Diam., D		Outside Diam. Tolerance Limits	
Over	Incl.	d_{min}	d_{max}	High	Low	Over	Incl.	D_{max}	D_{min}

EXTRA LIGHT SERIES — TYPE TP

Over	Incl.	d_{min}	d_{max}	High	Low	Over	Incl.	D_{max}	D_{min}
0	0.9375	+.0040	+.0060	+.0050	−.0050	0	4.7188	+0	−.0030
0.9375	1.9375	+.0050	+.0070	+.0050	−.0050	4.7188	5.2188	+0	−.0030
1.9375	3.0000	+.0060	+.0080	+.0050	−.0050	....	....	....	....
3.0000	3.5000	+.0080	+.0100	+.0100	−.0100	....	....	....	....

Basic Bore Diameter, d		Bore Tolerance Limits		Basic Outside Diameter,[1] D		Outside Diam., D Tolerance Limits		Basic Outside Diameter,[2] D_1		Outside Diam., D_1 Tolerance Limits	
Over	Incl.	d_{max}	d_{min}	Over	Incl.	D_{max}	D_{min}	Over	Incl.	D_{max}	D_{min}

LIGHT SERIES — TYPES TP AND TR

Over	Incl.	d_{max}	d_{min}	Over	Incl.	D_{max}	D_{min}	Over	Incl.	D_{max}	D_{min}
0	1.1875	+0	−.0005	0	2.8750	+.0005	−0	0	3.0000	+.0007	−0
1.1875	1.3750	+0	−.0006	2.8750	3.3750	+.0007	−0	3.0000	3.3750	+.0009	−0
1.3750	1.5620	+0	−.0007	3.3750	3.7500	+.0009	−0	3.3750	3.6250	+.0011	−0
1.5620	1.7500	+0	−.0008	3.7500	4.1250	+.0011	−0	3.6250	3.8750	+.0013	−0
1.7500	1.9370	+0	−.0009	4.1250	4.7180	+.0013	−0	3.8750	4.5312	+.0015	−0
1.9370	2.1250	+0	−.0010	4.7180	5.0000	+.0015	−0	4.5312	5.0000	+.0017	−0
2.1250	2.5000	+0	−.0011	....	....	....	....	....	....	....	....
2.2500	3.0000	+0	−.0012	....	....	....	....	....	....	....	....
3.0000	3.5000	+0	−.0013	....	....	....	....	....	....	....	....

HEAVY SERIES — TYPES TP AND TR

Over	Incl.	d_{max}	d_{min}	Over	Incl.	D_{max}	D_{min}	Over	Incl.	D_{max}	D_{min}
2.0000	3.0000	+0	−.0010	5.0000	10.0000	+.0015	−0	5.0000	10.5000	+.0019	−0
3.0000	3.5000	+0	−.0012	10.0000	18.0000	+.0020	−0	10.5000	12.7500	+.0021	−0
3.5000	9.0000	+0	−.0015	18.0000	26.0000	+.0025	−0	12.7500	17.0000	+.0023	−0
9.0000	12.0000	+0	−.0018	26.0000	34.0000	+.0030	−0	17.0000	27.0000	+.0025	−0
12.0000	18.0000	+0	−.0020	34.0000	44.0000	+.0040	−0	27.0000	35.0000	+.0030	−0
18.0000	22.0000	+0	−.0025	....	....	....	....	....	....	....	....
22.0000	30.0000	+0	−.003	....	....	....	....	....	....	....	....

HEIGHT TOLERANCES — TYPES TP AND TR

Basic Bore Diameter, d		Height,[3] H Tolerance Limits		Height,[4] H_1 Tolerance Limits		Basic Bore Diameter, d		Height,[3] H Tolerance Limits		Height,[4] H_1 Tolerance Limits	
Over	Incl.	High	Low	High	Low	Over	Incl.	High	Low	High	Low
0	2.0000	+0	−.0060	+0	−.0080	6.0000	10.0000	+0	−.0150	+0	−.0200
2.0000	3.0000	+0	−.0080	+0	−.0100	10.0000	18.0000	+0	−.0200	+0	−.0250
3.0000	6.0000	+0	−.0100	+0	−.0150	18.0000	30.0000	+0	−.0250	+0	−.0300

All dimensions are in inches.
[1] D is outside diameter over bearing seat.
[2] D_1 is outside diameter over aligning washer.
[3] H is height excluding aligning washer.
[4] H_1 is height including aligning washer.

Table 8. AFBMA and American National Standard Tolerance Limits for Needle Roller Bearings, Drawn Cup, Without Inner Ring — Inch Types NIB, NIBM, NIY, NIYM, NIH, and NIHM
(ANSI/AFBMA Std 18.2-1982)

Ring Gage Bore Diameter*			Basic Bore Diameter under Needle Rollers, F_w		Allowable Deviation from F_w*		Allowable Deviation from Width, B	
Basic Outside Diameter, D Inch		Deviation from D Inch	Inch		Inch		Inch	
Over	Incl.		Over	Incl.	Low	High	High	Low
0.1875	0.9375	+ 0.0005						
0.9375	4.0000	− 0.0005	0.1875	0.6875	+ 0.0015	+ 0.0024	+ 0	− 0.0100
			0.6875	1.2500	+ 0.0005	1.0014	+ 0	− 0.0100
			1.2500	1.3750	+ 0.0005	+ 0.0015	+ 0	− 0.0100
			1.3750	1.6250	+ 0.0005	+ 0.0016	+ 0	− 0.0100
			1.6250	1.8750	+ 0.0005	+ 0.0017	+ 0	− 0.0100
			1.8750	2.0000	+ 0.0006	+ 0.0018	+ 0	− 0.0100
			2.0000	2.5000	+ 0.0006	+ 0.0020	+ 0	− 0.0100
			2.5000	3.5000	+ 0.0010	+ 0.0024	+ 0	− 0.0100

For fitting and mounting practice see Table 17.
* The bore diameter under needle rollers can be measured only when bearing is pressed into a ring gage which rounds and sizes the bearing.

Table 9. AFBMA and American National Standard Tolerance Limits for Needle Roller Bearings, Drawn Cup, Without Inner Ring — Metric Types NB, NBM, NY, NYM, NH, and NHM
(ANSI/AFBMA Std 18.1-1982)

Ring Gage Bore Diameter*			Basic Bore Diameter under Needle Rollers, F_w		Allowable Deviation from F_w*		Allowable Deviation from Width, B	
Basic Outside Diameter, D mm		Deviation from D Micrometers	mm		Micrometers		Micrometers	
Over	Incl.		Over	Incl.	Low	High	High	Low
6	10	− 16	3	6	+ 10	+ 28	+ 0	− 250
10	18	− 20	6	10	+ 13	+ 31	+ 0	− 250
18	30	− 24	10	18	+ 16	+ 34	+ 0	− 250
30	50	− 28	18	30	+ 20	+ 41	+ 0	− 250
50	80	− 33	30	50	+ 25	+ 50	+ 0	− 250
. . .	. . .	. . .	50	70	+ 30	+ 60	+ 0	− 250

For fitting and mounting practice, see Table 18.
* The bore diameter under needle rollers can be measured only when bearing is pressed into a ring gage which rounds and sizes the bearing.

Table 10. AFBMA and American National Standard Tolerance Limits for Needle Roller Bearings, With Cage, Machined Ring, Without Inner Ring — Inch Type NIA
(ANSI/AFBMA Std 18.2-1982)

Basic Outside Diameter, D		Allowable Deviation From D of Single Mean Diameter, D_{mp}		Basic Bore Diameter under Needle Rollers, F_w		Allowable Deviation from F_w		Allowable Deviation from Width, B	
Inch		Inch		Inch		Inch		Inch	
Over	Incl.	High	Low	Over	Incl.	Low	High	High	Low
0.7500	2.0000	+ 0	− 0.0005	0.3150	0.7087	+ 0.0008	+ 0.0017	+ 0	− 0.0050
2.0000	3.2500	+ 0	− 0.0006	0.7087	1.1811	+ 0.0009	+ 0.0018	+ 0	− 0.0050
3.2500	4.7500	+ 0	− 0.0008	1.1811	1.6535	+ 0.0010	+ 0.0019	+ 0	− 0.0050
4.7500	7.2500	+ 0	− 0.0010	1.6535	1.9685	+ 0.0010	+ 0.0020	+ 0	− 0.0050
				1.9685	2.7559	+ 0.0011	+ 0.0021	+ 0	− 0.0050
7.2500	10.2500	+ 0	− 0.0012	2.7559	3.1496	+ 0.0011	+ 0.0023	+ 0	− 0.0050
10.2500	11.1250	+ 0	− 0.0014	3.1496	4.0157	+ 0.0012	+ 0.0024	+ 0	− 0.0050
. . .	. . .	. . .	. . .	4.0157	4.7244	+ 0.0012	+ 0.0026	+ 0	− 0.0050
. . .	. . .	. . .	. . .	4.7244	6.2992	+ 0.0013	+ 0.0027	+ 0	− 0.0050
. . .	. . .	. . .	. . .	6.2992	7.0866	+ 0.0013	+ 0.0029	+ 0	− 0.0050
. . .	. . .	. . .	. . .	7.0866	7.8740	+ 0.0014	+ 0.0030	+ 0	− 0.0050
. . .	. . .	. . .	. . .	7.8740	9.2520	+ 0.0014	+ 0.0032	+ 0	− 0.0050

For fitting and mounting practice, see Table 19.

Table 11. AFBMA and American National Standard Tolerance Limits for Needle Roller Bearing Inner Rings — Inch Type NIR
(ANSI/AFBMA Std 18.2-1982)

Basic Outside Diameter, F		Allowable Deviation From F of Single Mean Diameter, F_{mp}		Basic Bore Diameter d		Allowable Deviation from d of Single Mean Diameter, d_{mp}		Allowable Deviation from Width, B	
Inch		Inch		Inch		Inch		Inch	
Over	Incl.	High	Low	Over	Incl.	High	Low	High	Low
0.3937	0.7087	− 0.0005	− 0.0009	0.3125	0.7500	+ 0	− 0.0004	+ 0.0100	+ 0.0050
0.7087	1.0236	− 0.0007	− 0.0012	0.7500	2.0000	+ 0	− 0.0005	+ 0.0100	+ 0.0050
1.0236	1.1811	− 0.0009	− 0.0014	2.0000	3.2500	+ 0	− 0.0006	+ 0.0100	+ 0.0050
1.1811	1.3780	− 0.0009	− 0.0015	3.2500	4.2500	+ 0	− 0.0008	+ 0.0100	+ 0.0050
1.3780	1.9685	− 0.0010	− 0.0016	4.2500	4.7500	+ 0	− 0.0008	+ 0.0150	+ 0.0100
1.9685	3.1496	− 0.0011	− 0.0018	4.7500	7.0000	+ 0	− 0.0010	+ 0.0150	+ 0.0100
3.1496	3.9370	− 0.0013	− 0.0022	7.0000	8.0000	+ 0	− 0.0012	+ 0.0150	+ 0.0100
3.9370	4.7244	− 0.0015	− 0.0024	. . .	. . .	. . .	. . .	. . .	. . .
4.7244	5.5118	− 0.0015	− 0.0025	. . .	. . .	. . .	. . .	. . .	. . .
5.5118	7.0866	− 0.0017	− 0.0027	. . .	. . .	. . .	. . .	. . .	. . .
7.0866	8.2677	− 0.0019	− 0.0031	. . .	. . .	. . .	. . .	. . .	. . .
8.2677	9.2520	− 0.0020	− 0.0032	. . .	. . .	. . .	. . .	. . .	. . .

For fitting and mounting practice, see Table 20.

Metric Radial Ball and Roller Bearing Shaft and Housing Fits. — To select the proper fits, it is necessary to consider the type and extent of the load, bearing type, and certain other design and performance requirements.

The required shaft and housing fits are indicated in Tables 12 and 15. The terms "Light," "Normal," and "Heavy" loads refer to radial loads that are generally within the following limits (*C* being the Basic Load Rating computed in accordance with AFBMA-ANSI Standards):

Radial Load	Ball Bearings	Roller Bearings
Light	up to 0.07*C*	up to 0.08*C*
Normal	from 0.07*C* to 0.15*C*	from 0.08*C* to 0.18*C*
Heavy	over 0.15*C*	over 0.18*C*

Shaft Fits: Table 12 indicates the initial approach to shaft fit selection. Note that for most normal applications where the shaft rotates and the radial load direction is constant, an interference fit should be used. Also, the heavier the load, the greater is the required interference. For stationary shaft conditions and constant radial load direction, the inner ring may be moderately loose on the shaft.

Note that for pure thrust (axial) loading, heavy interference fits are not necessary; only a moderately loose to tight shaft fit is needed.

The upper part of Table 13 shows how the shaft diameters for various ANSI shaft limit classifications deviate from the basic bore diameters.

Table 14 shows the actual diameter limits of bearing bores, and of shaft seats according to ANSI shaft limits g6, h6, h5, etc.

Housing Fits: Table 15 indicates the initial approach to housing fit selection. Note that the use of clearance or interference fits is mainly dependent upon which bearing ring rotates in relation to the radial load. For indeterminate or varying load directions, avoid clearance fits. Clearance fits are preferred in axially split housings to avoid distorting bearing outer rings. The extent of the radial load also influences the choice of fit.

The lower part of Table 13 shows how housing bores for various ANSI hole limit classifications deviate from the basic shaft outside diameters.

Table 16 shows the actual diameter limits of bearing outside diameters, and housing seats according to ANSI hole limits G7, H7, H6, etc.

Design and Installation Considerations. — Since interference fitting will reduce bearing radial internal clearance, it is recommended that prospective users consult bearing manufacturers to make certain that the required bearings are correctly specified to satisfy all mounting, environmental and other operating conditions and requirements. This is particularly necessary in those cases where heat sources in associated parts may further diminish bearing clearances in operation.

Standard values of radial internal clearances of radial bearings are listed in AFBMA-ANSI Standards.

Allowance for Axial Displacement. — Consideration should be given to axial displacement of bearing components due to thermal expansion or contraction of associated parts. Displacement may be accommodated either by the internal construction of the bearing or by allowing one of the bearing rings to be axially displaceable. For unusual applications consult bearing manufacturers.

Table 12. Selection of Shaft Tolerance Classifications for Metric Radial Ball and Roller Bearings of ABEC-1 and RBEC-1 Tolerance Classes

Operating Conditions		Ball Bearings		Cylindrical Roller Bearings		Spherical Roller Bearings		Tolerance Symbol[1]
		mm	Inch	mm	Inch	mm	Inch	
Inner ring stationary in relation to the direction of the load. — All loads — Inner ring has to be easily displaceable		All diameters	All diameters	All diameters	All diameters	All diameters	All diameters	g6
Inner ring does not have to be easily displaceable		All diameters	All diameters	All diameters	All diameters	All diameters	All diameters	h6
Direction of load indeterminate or the inner ring rotating in relation to the direction of the load. — Radial load: — LIGHT		≤18 >18	≤0.71 >0.71	≤40 (40)−140 (140)−320	≤1.57 (1.57)−5.52 (5.52)−12.6	≤40 (40)−100 (100)−200	≤1.57 (1.57)−3.94 (3.94)−7.88	h5 j6² k6² m6²
NORMAL		≤18 >18	≤0.71 >0.71	≤40 (40)−100 (100)−140 (140)−320	≤1.57 (1.57)−3.94 (3.94)−5.52 (5.52)−12.6	≤40 (40)−65 (65)−100 (100)−140 (140)−280 (286)−500 >500	≤1.57 (1.57)−2.56 (2.56)−3.94 (3.94)−5.52 (5.52)−11.10 (11.10)−19.7 >19.7	j5 k5 m5 m6 n6 p6 r7
HEAVY		(18)−100 >100	(0.71)−3.94 >3.94	≤40 (40)−65 (65)−140 (140)−320	≤1.57 (1.57)−2.56 (2.56)−5.52 (5.52)−12.6	≤40 (40)−65 (65)−100 (100)−140 (140)−200 >200	≤1.57 (1.57)−2.56 (2.56)−3.94 (3.94)−5.52 (5.52)−7.88 >7.88	k5 m5 m6² n6² p6² r6² r7²
Pure Thrust Load		All diams.	All diams.	All diams.	All diams.	All diams.	All diams.	j6

[1] For solid steel shafts. For hollow or nonferrous shafts, tighter fits may be needed. [2] When greater accuracy is required use j5, k5 and m5 instead of j6, k6 and m6, respectively.
Numerical values are given in Tables 13 and 14.

Table 13. AFBMA and American National Standard Shaft Diameter and Housing Bore Tolerance Limits* (ANSI B3.17-1973)

Allowable Deviations of Shaft Diameter from Basic Bore Diameter, Inch

Basic Bore Diameter, Inches (Over / Incl.)	mm (Over / Incl.)	g6	h5	h6	js	j6	k5	k6	m5	m6	n6	p6	r6	r7
.1181 / .2362	3 / 6	-.0002 / -.0005	0 / -.0002	0 / -.0003	+.0001 / -.0001	+.0003 / -.0001				+.0005 / +.0002				
.2362 / .3937	6 / 10	-.0002 / -.0006	0 / -.0003	0 / -.0004	+.0002 / -.0001	+.0003 / -.0001				+.0006 / +.0002				
.3937 / .7087	10 / 18	-.0003 / -.0007	0 / -.0003	0 / -.0004	+.0002 / -.0001	+.0003 / -.0002	+.0004 / +.0001	+.0005 / +.0001	+.0006 / +.0003	+.0007 / +.0003				
.7087 / 1.1811	18 / 30	-.0003 / -.0008	0 / -.0004	0 / -.0005	+.0002 / -.0002	+.0004 / -.0002	+.0005 / +.0001	+.0006 / +.0001	+.0007 / +.0003	+.0008 / +.0003	+.0011 / +.0006			
1.1811 / 1.9685	30 / 50	-.0004 / -.0010	0 / -.0004	0 / -.0006	+.0002 / -.0002	+.0004 / -.0002	+.0005 / +.0001	+.0007 / +.0001	+.0008 / +.0004	+.0010 / +.0004	+.0013 / +.0007	+.0016 / +.0010		
1.9685 / 3.1496	50 / 80	-.0004 / -.0011	0 / -.0005	0 / -.0007	+.0002 / -.0003	+.0004 / -.0003	+.0006 / +.0001	+.0008 / +.0001	+.0010 / +.0005	+.0012 / +.0005	+.0015 / +.0008	+.0021 / +.0014	+.0023 / +.0016	
3.1496 / 4.7244	80 / 120	-.0005 / -.0014	0 / -.0006	0 / -.0009	+.0002 / -.0004	+.0005 / -.0004	+.0007 / +.0001	+.0010 / +.0001	+.0011 / +.0005	+.0014 / +.0005	+.0019 / +.0010	+.0025 / +.0016	+.0029 / +.0020	

Allowable Deviations of Housing Bore from Basic Outside Diameter of Shaft, Inch

Basic Outside Diameter, Inches (Over / Incl.)	mm (Over / Incl.)	G7	H7	H6	J7	J6	K6	K7	M6	M7	N6	N7	P6	P7
.3937 / .7086	10 / 18	+.0003 / +.0011	0 / +.0007	0 / +.0004	-.0003 / +.0004	-.0002 / +.0002	-.0004 / +.0001	-.0005 / +.0002	-.0006 / -.0002	-.0007 / 0	-.0008 / -.0004	-.0009 / -.0002	-.0010 / -.0006	-.0011 / -.0004
.7086 / 1.1811	18 / 30	+.0003 / +.0011	0 / +.0008	0 / +.0005	-.0003 / +.0005	-.0002 / +.0003	-.0004 / +.0001	-.0006 / +.0002	-.0007 / -.0002	-.0008 / 0	-.0010 / -.0005	-.0011 / -.0003	-.0012 / -.0007	-.0013 / -.0005
1.1811 / 1.9685	30 / 50	+.0004 / +.0014	0 / +.0010	0 / +.0006	-.0004 / +.0006	-.0002 / +.0004	-.0005 / +.0001	-.0007 / +.0003	-.0008 / -.0002	-.0010 / 0	-.0011 / -.0005	-.0013 / -.0003	-.0014 / -.0008	-.0016 / -.0006
1.9685 / 3.1496	50 / 80	+.0004 / +.0016	0 / +.0012	0 / +.0007	-.0004 / +.0008	-.0002 / +.0005	-.0006 / +.0001	-.0008 / +.0004	-.0008 / -.0002	-.0012 / 0	-.0013 / -.0006	-.0015 / -.0003	-.0019 / -.0012	-.0021 / -.0009
3.1496 / 4.7244	80 / 120	+.0005 / +.0019	0 / +.0014	0 / +.0009	-.0005 / +.0009	-.0002 / +.0007	-.0007 / +.0002	-.0010 / +.0004	-.0010 / -.0003	-.0014 / 0	-.0016 / -.0007	-.0018 / -.0004	-.0022 / -.0013	-.0025 / -.0011
4.7244 / 7.0866	120 / 180	+.0006 / +.0022	0 / +.0016	0 / +.0010	-.0006 / +.0010	-.0003 / +.0007	-.0008 / +.0002	-.0011 / +.0005	-.0013 / -.0003	-.0016 / 0	-.0019 / -.0009	-.0022 / -.0006	-.0025 / -.0015	-.0028 / -.0012
7.0866 / 9.8425	180 / 250	+.0006 / +.0024	0 / +.0018	0 / +.0012	-.0007 / +.0011	-.0003 / +.0009	-.0010 / +.0002	-.0013 / +.0005	-.0015 / -.0003	-.0018 / 0	-.0022 / -.0010	-.0026 / -.0008	-.0028 / -.0016	-.0032 / -.0014

* Based on ANSI B4.1-1967, R1974 Preferred Limits and Fits for Cylindrical Parts. Symbols g6, h6, etc., are shaft and G7, H7, etc., hole limits designations. For larger diameters see AFBMA Standard 7.

Table 14. AFBMA and American National Standard Shaft Diameter Limits for Metric Radial Ball and Roller Bearings of ABEC-1 and RBEC-1 Tolerance Classes (ANSI B3.17-1973)

	Bearing Bore Diam. Inches		g6 Shaft Diam., Inch		h6 Shaft Diam., Inch		b5 Shaft Diam., Inch		j5 Shaft Diam., Inch		js Shaft Diam., Inch		j6 Shaft Diam., Inch	
mm.	Max.	Min	Max.	Min.	Max.	Min.	Max.	Min.	Max.	Min.	Max.	Min.	Max.	Min.
4	.1575	.1572	.1573	.1570	.1575	.1572	.1575	.1573	.1576	.1574	.1576	.1574		
5	.1969	.1966	.1967	.1964	.1969	.1966	.1969	.1967	.1970	.1968	.1970	.1968		
6	.2362	.2359	.2360	.2357	.2362	.2359	.2362	.2360	.2363	.2361	.2363	.2361		
7	.2756	.2753	.2754	.2750	.2756	.2752	.2756	.2753	.2758	.2755	.2758	.2755		
8	.3150	.3147	.3148	.3144	.3150	.3146	.3150	.3147	.3152	.3149	.3152	.3149		
9	.3543	.3540	.3541	.3537	.3543	.3539	.3543	.3540	.3545	.3542	.3545	.3542		
10	.3937	.3934	.3935	.3931	.3937	.3933	.3937	.3934	.3939	.3936	.3939	.3936		
12	.4724	.4721	.4721	.4717	.4724	.4720	.4724	.4721	.4726	.4723	.4726	.4723	.4727	.4723
15	.5906	.5903	.5903	.5899	.5906	.5902	.5906	.5903	.5908	.5905	.5908	.5905	.5909	.5905
17	.6693	.6690	.6690	.6686	.6693	.6689	.6693	.6690	.6695	.6692	.6695	.6692	.6696	.6692
20	.7874	.7870	.7871	.7866	.7874	.7869	.7874	.7870	.7876	.7872	.7876	.7872	.7877	.7872
25	.9843	.9839	.9840	.9835	.9843	.9838	.9843	.9839	.9845	.9841	.9845	.9841	.9846	.9841
30	1.1811	1.1807	1.1808	1.1803	1.1811	1.1806	1.1811	1.1807	1.1813	1.1809	1.1813	1.1809	1.1814	1.1809
35	1.3780	1.3775	1.3776	1.3770	1.3780	1.3774	1.3780	1.3776	1.3782	1.3778	1.3782	1.3778	1.3784	1.3778
40	1.5748	1.5743	1.5744	1.5738	1.5748	1.5742	1.5748	1.5744	1.5750	1.5746	1.5750	1.5746	1.5752	1.5746
45	1.7717	1.7712	1.7713	1.7707	1.7717	1.7711	1.7717	1.7713	1.7719	1.7715	1.7719	1.7715	1.7721	1.7715
50	1.9685	1.9680	1.9681	1.9675	1.9685	1.9679	1.9685	1.9681	1.9687	1.9683	1.9687	1.9683	1.9689	1.9683
55	2.1654	2.1648	2.1650	2.1643	2.1654	2.1647	2.1654	2.1649	2.1656	2.1651	2.1656	2.1651	2.1658	2.1651
60	2.3622	2.3616	2.3618	2.3611	2.3622	2.3615	2.3622	2.3617	2.3624	2.3619	2.3624	2.3619	2.3626	2.3619
65	2.5591	2.5585	2.5587	2.5580	2.5591	2.5584	2.5591	2.5586	2.5593	2.5588	2.5593	2.5588	2.5595	2.5588
70	2.7559	2.7553	2.7555	2.7548	2.7559	2.7552	2.7559	2.7554	2.7561	2.7556	2.7561	2.7556	2.7563	2.7556
75	2.9528	2.9522	2.9524	2.9517	2.9528	2.9521	2.9528	2.9523	2.9530	2.9525	2.9530	2.9525	2.9532	2.9525
80	3.1496	3.1490	3.1492	3.1485	3.1496	3.1489	3.1496	3.1491	3.1498	3.1493	3.1498	3.1493	3.1500	3.1493
85	3.3465	3.3457	3.3460	3.3451	3.3465	3.3456	3.3465	3.3459	3.3467	3.3461	3.3467	3.3461	3.3470	3.3461
90	3.5433	3.5425	3.5428	3.5419	3.5433	3.5424	3.5433	3.5427	3.5435	3.5429	3.5435	3.5429	3.5438	3.5429
95	3.7402	3.7394	3.7397	3.7388	3.7402	3.7393	3.7402	3.7396	3.7404	3.7398	3.7404	3.7398	3.7407	3.7398
100	3.9370	3.9362	3.9365	3.9356	3.9370	3.9361	3.9370	3.9364	3.9372	3.9366	3.9372	3.9366	3.9375	3.9366
105	4.1339	4.1331	4.1334	4.1325	4.1339	4.1330	4.1339	4.1333	4.1341	4.1335	4.1341	4.1335	4.1344	4.1335
110	4.3307	4.3299	4.3302	4.3293	4.3307	4.3298	4.3307	4.3301	4.3309	4.3303	4.3309	4.3303	4.3312	4.3303
115	4.5276	4.5268	4.5271	4.5262	4.5276	4.5267	4.5276	4.5270	4.5278	4.5272	4.5278	4.5272	4.5281	4.5272
120	4.7244	4.7236	4.7239	4.7230	4.7244	4.7235	4.7244	4.7238	4.7246	4.7240	4.7246	4.7240	4.7249	4.7240

For larger diameters see AFBMA Standard 7. See also Tables 12 and 13.

Table 14 (*Continued*).   AFBMA and American National Standard Shaft Diameter Limits for Metric Radial Ball and Roller Bearings of ABEC-1 and RBEC-1 Tolerance Classes (ANSI B3.17-1973)

Mm.	Bearing Bore Diam. Inches		k5 Shaft Diam., Inch		k6 Shaft Diam., Inch		m5 Shaft Diam., Inch		m6 Shaft Diam., Inch	
	Max.	Min.	Max.	Min.	Max.	Min.	Max.	Min.	Max.	Min.
12	.4724	.4721	.4728	.4725			.4730	.4727		
15	.5906	.5903	.5910	.5907			.5912	.5909		
17	.6693	.6690	.6697	.6694			.6699	.6696		
20	.7874	.7870	.7879	.7875	.7880	.7875	.7881	.7877		
25	.9843	.9839	.9848	.9844	.9849	.9844	.9850	.9846		
30	1.1811	1.1807	1.1816	1.1812	1.1817	1.1812	1.1818	1.1814		
35	1.3780	1.3775	1.3785	1.3781	1.3787	1.3781	1.3788	1.3784	1.3790	1.3784
40	1.5748	1.5743	1.5753	1.5749	1.5755	1.5749	1.5756	1.5752	1.5758	1.5752
45	1.7717	1.7712	1.7722	1.7718	1.7724	1.7718	1.7725	1.7721	1.7727	1.7721
50	1.9685	1.9680	1.9690	1.9686	1.9692	1.9686	1.9693	1.9689	1.9695	1.9689
55	2.1654	2.1648	2.1660	2.1655	2.1662	2.1655	2.1664	2.1659	2.1666	2.1659
60	2.3622	2.3616	2.3628	2.3623	2.3630	2.3623	2.3632	2.3627	2.3634	2.3627
65	2.5591	2.5585	2.5597	2.5592	2.5599	2.5592	2.5601	2.5596	2.5603	2.5596
70	2.7559	2.7553	2.7565	2.7560	2.7567	2.7560	2.7569	2.7564	2.7571	2.7564
75	2.9528	2.9522	2.9534	2.9529	2.9536	2.9529	2.9538	2.9533	2.9540	2.9533
80	3.1496	3.1490	3.1502	3.1497	3.1504	3.1497	3.1506	3.1501	3.1508	3.1501
85	3.3465	3.3457	3.3472	3.3466	3.3475	3.3466	3.3476	3.3470	3.3479	3.3470
90	3.5433	3.5425	3.5440	3.5434	3.5443	3.5434	3.5444	3.5438	3.5447	3.5438
95	3.7402	3.7394	3.7409	3.7403	3.7412	3.7403	3.7413	3.7407	3.7416	3.7407
100	3.9370	3.9362	3.9377	3.9371	3.9380	3.9371	3.9381	3.9375	3.9384	3.9375
105	4.1339	4.1331	4.1346	4.1340	4.1349	4.1340	4.1350	4.1344	4.1353	4.1344
110	4.3307	4.3299	4.3314	4.3308	4.3317	4.3308	4.3318	4.3312	4.3321	4.3312
115	4.5276	4.5268	4.5283	4.5277	4.5286	4.5277	4.5287	4.5281	4.5290	4.5281
120	4.7244	4.7236	4.7251	4.7245	4.7254	4.7245	4.7255	4.7249	4.7258	4.7249
125	4.9213	4.9203	4.9221	4.9214	4.9224	4.9214	4.9226	4.9219	4.9229	4.9219
130	5.1181	5.1171	5.1189	5.1182	5.1192	5.1182	5.1194	5.1187	5.1197	5.1187
140	5.5118	5.5108	5.5126	5.5119	5.5129	5.5119	5.5131	5.5124	5.5134	5.5124
150	5.9055	5.9045	5.9063	5.9056	5.9066	5.9056	5.9068	5.9061	5.9071	5.9061
160	6.2992	6.2982	6.3000	6.2993	6.3003	6.2993	6.3005	6.2998	6.3008	6.2998
170	6.6929	6.6919	6.6937	6.6930	6.6940	6.6930	6.6942	6.6935	6.6945	6.6935
180	7.0866	7.0856	7.0874	7.0867	7.0877	7.0867	7.0879	7.0872	7.0882	7.0872

For larger diameters see AFBMA Standard 7.   See also Tables 12 and 13.

Table 14 (*Concluded*). AFBMA and American National Standard Shaft Diameter Limits for Metric Radial Ball and Roller Bearings of ABEC-1 and RBEC-1 Tolerance Classes (ANSI B3.17-1973)

Mm.	Bearing Bore Diam. Inches		n6 Shaft Diam., Inch		p6 Shaft Diam., Inch		r6 Shaft Diam., Inch		r7 Shaft Diam., Inch	
	Max.	Min.	Max.	Min.	Max.	Min.	Max.	Min.	Max.	Min.
35	1.3780	1.3775	1.3793	1.3787						
40	1.5748	1.5743	1.5761	1.5755						
45	1.7717	1.7712	1.7730	1.7724						
50	1.9685	1.9680	1.9698	1.9692						
55	2.1654	2.1648	2.1669	2.1662	2.1675	2.1668				
60	2.3622	2.3616	2.3637	2.3630	2.3643	2.3636				
65	2.5591	2.5585	2.5606	2.5599	2.5612	2.5605				
70	2.7559	2.7553	2.7574	2.7567	2.7580	2.7573				
75	2.9528	2.9522	2.9543	2.9536	2.9549	2.9542				
80	3.1496	3.1490	3.1511	3.1504	3.1517	3.1510				
85	3.3465	3.3457	3.3484	3.3475	3.3490	3.3481	3.3494	3.3485		
90	3.5433	3.5425	3.5452	3.5443	3.5458	3.5449	3.5462	3.5453		
95	3.7402	3.7394	3.7421	3.7412	3.7427	3.7418	3.7431	3.7422		
100	3.9370	3.9362	3.9389	3.9380	3.9395	3.9386	3.9399	3.9390		
105	4.1339	4.1331	4.1358	4.1349	4.1364	4.1355	4.1368	4.1359		
110	4.3307	4.3299	4.3326	4.3317	4.3332	4.3323	4.3336	4.3327		
115	4.5276	4.5268	4.5295	4.5286	4.5301	4.5292	4.5305	4.5296		
120	4.7244	4.7236	4.7263	4.7254	4.7269	4.7260	4.7273	4.7264		
125	4.9213	4.9203	4.9235	4.9225	4.9241	4.9231	4.9248	4.9238		
130	5.1181	5.1171	5.1203	5.1193	5.1209	5.1199	5.1216	5.1206		
140	5.5118	5.5108	5.5140	5.5130	5.5146	5.5136	5.5153	5.5143		
150	5.9055	5.9045	5.9077	5.9067	5.9083	5.9073	5.9090	5.9080		
160	6.2992	6.2982	6.3014	6.3004	6.3020	6.3010	6.3027	6.3017		
170	6.6929	6.6919	6.6951	6.6941	6.6957	6.6947	6.6964	6.6954		
180	7.0866	7.0856	7.0888	7.0878	7.0894	7.0884	7.0901	7.0891		
190	7.4803	7.4791	7.4829	7.4817	7.4835	7.4823	7.4845	7.4833	7.4851	7.4833
200	7.8740	7.8728	7.8766	7.8754	7.8772	7.8760	7.8782	7.8770	7.8788	7.8770
220	8.6614	8.6602	8.6640	8.6628	8.6646	8.6634	8.6656	8.6644	8.6662	8.6644
240	9.4488	9.4476	9.4514	9.4502	9.4520	9.4508	9.4530	9.4518	9.4536	9.4518

For larger diameters see AFBMA Standard 7. See also Tables 12 and 13.

Table 15. Selection of Housing Tolerance Classifications for Metric Radial Ball and Roller Bearings of ABEC-1 and RBEC-1 Tolerance Classes

Rotational Conditions	Design and Operating Conditions			Tolerance Classification*
	Loading	Outer Ring Axial Displacement Limitations	Other Conditions	
Outer ring stationary in relation to load direction	Light, Normal and Heavy	Outer ring must be easily axially displaceable	Heat input through shaft	G7
			Housing split axially	H7
			Housing not split axially	H6†
Load direction is indeterminate	Shock with temporary complete unloading	Outer ring need not be axially displaceable	....	J6†
	Light and Normal			K6†
	Normal and Heavy		Split housing not recommended	M6†
	Heavy Shock			
Outer ring rotating in relation to load direction	Light			N6†
	Normal and Heavy			
	Heavy		Thin wall housing not split	P6†

* For cast iron or steel housings. For housings of nonferrous alloys tighter fits may be needed.

† Where wider tolerances are permissible, use tolerance classifications P7, N7, M7, K7, J7, and H7, in place of P6, N6, M6, K6, J6, and H6, respectively.

Numerical values are given in Tables 13 and 16.

Table 16. AFBMA and American National Standard Housing Bore Limits for Metric Radial Ball and Roller Bearings of ABEC-1 and RBEC-1 Tolerance Classes (ANSI B3.17-1973)

Mm.	Bearing Outside Diameter, Inches		G7 Housing Bore, Inch		H7 Housing Bore, Inch		H6 Housing Bore, Inch		J7 Housing Bore, Inch		J6 Housing Bore, Inch		K6 Housing Bore, Inch	
	Max.	Min.	Min.	Max.	Min.	Max.	Min.	Max.	Min.	Max.	Min.	Max.	Min.	Max.
16	.6299	.6296	.6302	.6309	.6299	.6306	.6299	.6303	.6296	.6303	.6297	.6301	.6295	.6299
19	.7480	.7476	.7483	.7491	.7480	.7488	.7480	.7485	.7477	.7485	.7478	.7483	.7475	.7480
21	.8268	.8264	.8271	.8279	.8268	.8276	.8268	.8273	.8265	.8273	.8266	.8271	.8263	.8268
22	.8661	.8657	.8664	.8672	.8661	.8669	.8661	.8666	.8658	.8666	.8659	.8664	.8656	.8661
24	.9449	.9445	.9452	.9460	.9449	.9457	.9449	.9454	.9446	.9454	.9447	.9452	.9444	.9449
26	1.0236	1.0232	1.0239	1.0247	1.0236	1.0244	1.0236	1.0241	1.0233	1.0241	1.0234	1.0239	1.0231	1.0236
28	1.1024	1.1020	1.1027	1.1035	1.1024	1.1032	1.1024	1.1029	1.1021	1.1029	1.1022	1.1027	1.1019	1.1024
30	1.1811	1.1807	1.1814	1.1822	1.1811	1.1819	1.1811	1.1816	1.1808	1.1816	1.1809	1.1814	1.1806	1.1811
32	1.2598	1.2593	1.2602	1.2612	1.2598	1.2608	1.2598	1.2604	1.2594	1.2604	1.2596	1.2602	1.2593	1.2599
35	1.3780	1.3775	1.3784	1.3794	1.3780	1.3790	1.3780	1.3786	1.3776	1.3786	1.3778	1.3784	1.3775	1.3781
37	1.4567	1.4562	1.4571	1.4581	1.4567	1.4577	1.4567	1.4573	1.4563	1.4573	1.4565	1.4571	1.4562	1.4568
40	1.5748	1.5743	1.5752	1.5762	1.5748	1.5758	1.5748	1.5754	1.5744	1.5754	1.5746	1.5752	1.5743	1.5749
42	1.6535	1.6530	1.6539	1.6549	1.6535	1.6545	1.6535	1.6541	1.6531	1.6541	1.6533	1.6539	1.6530	1.6536
47	1.8504	1.8499	1.8508	1.8518	1.8504	1.8514	1.8504	1.8510	1.8500	1.8510	1.8502	1.8508	1.8499	1.8505
52	2.0472	2.0467	2.0476	2.0488	2.0472	2.0484	2.0472	2.0479	2.0468	2.0480	2.0470	2.0477	2.0466	2.0473
55	2.1654	2.1649	2.1658	2.1670	2.1654	2.1666	2.1654	2.1661	2.1650	2.1662	2.1652	2.1659	2.1648	2.1655
58	2.2835	2.2830	2.2839	2.2851	2.2835	2.2847	2.2835	2.2842	2.2831	2.2843	2.2833	2.2840	2.2829	2.2836
62	2.4409	2.4404	2.4413	2.4425	2.4409	2.4421	2.4409	2.4416	2.4405	2.4417	2.4407	2.4414	2.4403	2.4410
65	2.5591	2.5586	2.5595	2.5607	2.5591	2.5603	2.5591	2.5598	2.5587	2.5599	2.5589	2.5596	2.5585	2.5592
68	2.6772	2.6767	2.6776	2.6788	2.6772	2.6784	2.6772	2.6779	2.6768	2.6780	2.6770	2.6777	2.6766	2.6773
72	2.8346	2.8341	2.8350	2.8362	2.8346	2.8358	2.8346	2.8353	2.8342	2.8354	2.8344	2.8351	2.8340	2.8347
75	2.9528	2.9523	2.9532	2.9544	2.9528	2.9540	2.9528	2.9535	2.9524	2.9536	2.9526	2.9533	2.9522	2.9529
78	3.0709	3.0704	3.0713	3.0725	3.0709	3.0721	3.0709	3.0716	3.0705	3.0717	3.0707	3.0714	3.0703	3.0710
80	3.1496	3.1491	3.1500	3.1512	3.1496	3.1508	3.1496	3.1503	3.1492	3.1504	3.1494	3.1501	3.1490	3.1497
85	3.3465	3.3459	3.3470	3.3484	3.3465	3.3479	3.3465	3.3474	3.3460	3.3474	3.3463	3.3472	3.3458	3.3467
90	3.5433	3.5427	3.5438	3.5452	3.5433	3.5447	3.5433	3.5442	3.5428	3.5442	3.5431	3.5440	3.5426	3.5435
95	3.7402	3.7396	3.7407	3.7421	3.7402	3.7416	3.7402	3.7411	3.7397	3.7411	3.7400	3.7409	3.7395	3.7404
100	3.9370	3.9364	3.9375	3.9389	3.9370	3.9384	3.9370	3.9379	3.9365	3.9379	3.9368	3.9377	3.9363	3.9372
105	4.1339	4.1333	4.1344	4.1358	4.1339	4.1353	4.1339	4.1348	4.1334	4.1348	4.1337	4.1346	4.1332	4.1341
110	4.3307	4.3301	4.3312	4.3326	4.3307	4.3321	4.3307	4.3316	4.3302	4.3316	4.3305	4.3314	4.3300	4.3309
115	4.5276	4.5270	4.5281	4.5295	4.5276	4.5290	4.5276	4.5285	4.5271	4.5285	4.5274	4.5283	4.5269	4.5278
120	4.7244	4.7238	4.7249	4.7263	4.7244	4.7258	4.7244	4.7253	4.7239	4.7253	4.7242	4.7251	4.7237	4.7246

For larger diameters see AFBMA Standard 7.   See also Tables 13 and 15.

Table 16 (Concluded). AFBMA and American National Standard Housing Bore Limits for Metric Radial Ball and Roller Bearings of ABEC-1 and RBEC-1 Tolerance Classes (ANSI B3.17-1973)

Bearing Outside Diameter			K7		M6		M7		N6		N7		P6		P7	
Mm	Inches		Housing Bore, Inch		Housing Bore, Inch		Housing Bore, Inch		Housing Bore, Inch		Housing Bore, Inch		Housing Bore, Inch		Housing Bore, Inch	
	Max.	Min.	Min.	Max.	Min.	Max.	Min.	Max.	Min.	Max.	Min.	Max.	Min.	Max.	Min.	Max.
16	.6299	.6295	.6294	.6301	.6293	.6297	.6292	.6299	.6291	.6295	.6290	.6297	.6289	.6293	.6288	.6295
19	.7480	.7476	.7474	.7482	.7473	.7478	.7472	.7480	.7470	.7475	.7469	.7477	.7468	.7473	.7467	.7475
21	.8268	.8264	.8262	.8270	.8261	.8266	.8260	.8268	.8258	.8263	.8257	.8265	.8256	.8261	.8255	.8263
22	.8661	.8657	.8655	.8663	.8654	.8659	.8653	.8661	.8651	.8656	.8650	.8658	.8649	.8654	.8648	.8656
24	.9449	.9445	.9443	.9451	.9442	.9447	.9441	.9449	.9439	.9444	.9438	.9446	.9437	.9442	.9436	.9444
26	1.0236	1.0232	1.0230	1.0238	1.0229	1.0234	1.0228	1.0236	1.0226	1.0231	1.0225	1.0233	1.0224	1.0229	1.0223	1.0231
28	1.1024	1.1020	1.1018	1.1026	1.1017	1.1022	1.1016	1.1024	1.1014	1.1019	1.1013	1.1021	1.1012	1.1017	1.1011	1.1019
30	1.1811	1.1807	1.1805	1.1813	1.1804	1.1809	1.1803	1.1811	1.1801	1.1806	1.1800	1.1808	1.1799	1.1804	1.1798	1.1806
32	1.2598	1.2593	1.2591	1.2601	1.2590	1.2596	1.2588	1.2598	1.2587	1.2593	1.2585	1.2595	1.2584	1.2590	1.2582	1.2592
35	1.3780	1.3775	1.3773	1.3783	1.3772	1.3778	1.3770	1.3780	1.3769	1.3775	1.3767	1.3777	1.3766	1.3772	1.3764	1.3774
37	1.4567	1.4562	1.4560	1.4570	1.4559	1.4565	1.4557	1.4567	1.4556	1.4562	1.4554	1.4564	1.4553	1.4559	1.4551	1.4561
40	1.5748	1.5743	1.5741	1.5751	1.5740	1.5746	1.5738	1.5748	1.5737	1.5743	1.5735	1.5745	1.5734	1.5740	1.5732	1.5742
42	1.6535	1.6530	1.6528	1.6538	1.6527	1.6533	1.6525	1.6535	1.6524	1.6530	1.6522	1.6532	1.6521	1.6527	1.6519	1.6529
47	1.8504	1.8499	1.8497	1.8507	1.8496	1.8502	1.8494	1.8504	1.8493	1.8499	1.8491	1.8501	1.8490	1.8496	1.8488	1.8498
52	2.0472	2.0467	2.0464	2.0476	2.0462	2.0469	2.0460	2.0472	2.0459	2.0466	2.0457	2.0469	2.0453	2.0460	2.0451	2.0463
55	2.1654	2.1649	2.1646	2.1658	2.1644	2.1651	2.1642	2.1654	2.1641	2.1648	2.1639	2.1651	2.1635	2.1642	2.1633	2.1645
58	2.2835	2.2830	2.2827	2.2839	2.2825	2.2832	2.2823	2.2835	2.2822	2.2829	2.2820	2.2832	2.2816	2.2823	2.2814	2.2826
62	2.4409	2.4404	2.4401	2.4413	2.4399	2.4406	2.4397	2.4409	2.4396	2.4403	2.4394	2.4406	2.4390	2.4397	2.4388	2.4400
65	2.5591	2.5586	2.5583	2.5595	2.5581	2.5588	2.5579	2.5591	2.5578	2.5585	2.5576	2.5588	2.5572	2.5579	2.5570	2.5582
68	2.6772	2.6767	2.6764	2.6776	2.6762	2.6769	2.6760	2.6772	2.6759	2.6766	2.6757	2.6769	2.6753	2.6760	2.6751	2.6763
72	2.8346	2.8341	2.8338	2.8350	2.8336	2.8343	2.8334	2.8346	2.8333	2.8340	2.8331	2.8343	2.8327	2.8334	2.8325	2.8337
75	2.9528	2.9523	2.9520	2.9532	2.9518	2.9525	2.9516	2.9528	2.9515	2.9522	2.9513	2.9525	2.9509	2.9516	2.9507	2.9519
78	3.0709	3.0704	3.0701	3.0713	3.0699	3.0706	3.0697	3.0709	3.0696	3.0703	3.0694	3.0706	3.0690	3.0697	3.0688	3.0700
80	3.1496	3.1491	3.1488	3.1500	3.1486	3.1493	3.1484	3.1496	3.1483	3.1490	3.1481	3.1493	3.1477	3.1484	3.1475	3.1487
85	3.3465	3.3459	3.3455	3.3469	3.3453	3.3462	3.3451	3.3465	3.3449	3.3458	3.3447	3.3461	3.3443	3.3452	3.3440	3.3454
90	3.5433	3.5427	3.5423	3.5437	3.5421	3.5430	3.5419	3.5433	3.5417	3.5426	3.5415	3.5429	3.5411	3.5420	3.5408	3.5422
95	3.7402	3.7396	3.7392	3.7406	3.7390	3.7399	3.7388	3.7402	3.7386	3.7395	3.7384	3.7398	3.7380	3.7389	3.7377	3.7391
100	3.9370	3.9364	3.9360	3.9374	3.9358	3.9367	3.9356	3.9370	3.9354	3.9363	3.9352	3.9366	3.9348	3.9357	3.9345	3.9359
105	4.1339	4.1333	4.1329	4.1343	4.1327	4.1336	4.1325	4.1339	4.1323	4.1332	4.1321	4.1335	4.1317	4.1326	4.1314	4.1328
110	4.3307	4.3301	4.3297	4.3311	4.3295	4.3304	4.3293	4.3307	4.3291	4.3300	4.3289	4.3303	4.3285	4.3294	4.3282	4.3296
115	4.5276	4.5270	4.5266	4.5280	4.5264	4.5273	4.5262	4.5276	4.5260	4.5269	4.5258	4.5272	4.5254	4.5263	4.5251	4.5265
120	4.7244	4.7238	4.7234	4.7248	4.7232	4.7241	4.7230	4.7244	4.7228	4.7237	4.7226	4.7240	4.7222	4.7231	4.7219	4.7233

For larger diameters see AFBMA Standard 7. See also Tables 13 and 15.

Needle Roller Bearing Fitting and Mounting Practice. — The tolerance limits required for shaft and housing seat diameters for needle roller bearings with inner and outer rings as well as limits for raceway diameters where inner or outer rings or both are omitted and rollers operate directly upon these surfaces are given in Tables 17 to 20, inclusive. Unusual design and operating conditions may require a departure from these practices. In such cases, bearing manufacturers should be consulted.

Needle Roller Bearings, Drawn Cup: These bearings without inner ring, Types NIB, NB, NIBM, NBM, NIY, NY, NIYM, NYM, NIH, NH, NIHM, NHM, and Inner Rings, Type NIR depend on the housings into which they are pressed for their size and shape. Therefore, the housings must not only have the proper bore dimensions but also must have sufficient strength. Tables 17 and 18, inclusive, show the bore tolerance limits for rigid housings such as those made from cast iron or steel of heavy radial section equal to or greater than the ring gage section given in AFBMA Standard 4 (ANSI B3.4-1984). The bearing manufacturers should be consulted for recommendations if the housings must be of lower strength materials such as aluminum or even of thin radial section. The shape of the housing bores should be such that when the mean bore diameter of a housing is measured in each of several radial planes, the maximum difference between these mean diameters should not exceed 0.0005 inch (0.013 mm) or one-half the housing bore tolerance limit, if smaller. Also, the radial deviation from circular form should not exceed 0.00025 inch (0.006 mm). The housing bore surface finish should not exceed 125 micro-inches (3.2 micrometers) arithmetical average.

Most needle roller bearings do not use inner rings, they operate directly on the surfaces of shafts. When shafts are used as inner raceways, they should be made of bearing quality steel hardened to Rockwell C 58 minimum. Tables 17 and 18 show the shaft raceway tolerance limits and Table 20 shows the shaft seat tolerance limits when inner rings are used. However, whether the shaft surfaces are used as inner raceways or as seats for inner rings, the mean outside diameter of the shaft surface in each of several radial planes should be determined. The difference between these mean diameters should not exceed 0.0003 inch (0.008 mm) or one-half the diameter tolerance limit, if smaller. The radial deviation from circular form should not exceed 0.0001 inch (0.0025 mm), for diameters up to and including 1 in. (25.4 mm). Above one inch the allowable deviation is 0.0001 times the shaft diameter. The surface finish should not exceed 16 micro-inches (0.4 micrometer) arithmetical average. The housing bore and shaft diameter tolerance limits depend upon whether the load rotates relative to the shaft or the housing.

Needle Roller Bearing With Cage, Machined Ring, Without Inner Ring: The following covers needle roller bearings Type NIA and inner rings Type NIR. The shape of the housing bores should be such that when the mean bore diameter of a housing is measured in each of several radial planes, the maximum difference between these mean diameters does not exceed 0.0005 inch (0.013 mm) or one-half the housing bore tolerance limit, if smaller. Also, the radial deviation from circular form should not exceed 0.00025 inch (0.006 mm). The housing bore surface finish should not exceed 125 micro-inches (3.2 micrometers) arithmetical average. Table 19 shows the housing bore tolerance limits.

When shafts are used as inner raceways their requirements are the same as those given above for Needle Roller Bearings, Drawn Cup. Table 19 shows the shaft raceway tolerance limits and Table 20 shows the shaft seat tolerance limits when inner rings are used.

Needle Roller and Cage Assemblies, Types NIM and NM. — For information concerning boundary dimensions, tolerance limits, and fitting and mounting practice, reference should be made to ANSI/AFBMA Std 18.1-1982 and ANSI/AFBMA Std 18.2-1982.

Table 17. AFBMA and American National Standard Tolerance Limits for Shaft Raceway and Housing Bore Diameters — Needle Roller Bearings, Drawn Cup, Without Inner Ring, Inch Types NIB, NIBM, NIY, NIYM, NIH, and NIHM (ANSI/AFBMA Std 18.2-1982)

Basic Bore Diameter under Needle Rollers, F_w		Shaft Raceway Diameter*		Basic Outside Diameter, D		Housing Bore Diameter*	
		Allowable Deviation from F_w				Allowable Deviation from D	
Inch		Inch		Inch		Inch	
Over	Incl.	High	Low	Over	Incl.	Low	High
OUTER RING STATIONARY RELATIVE TO LOAD							
0.1875	1.8750	+ 0	− 0.0005	0.3750	4.0000	− 0.0005	+ 0.0005
1.8750	3.5000	+ 0	− 0.0006	...	...	...	...
OUTER RING ROTATING RELATIVE TO LOAD							
0.1875	1.8750	− 0.0005	− 0.0010	0.3750	4.0000	− 0.0010	+ 0
1.8750	3.5000	− 0.0005	− 0.0011	...	...	...	...

For bearing tolerances, see Table 8.
* See text for additional requirements.

Table 18. AFBMA and American National Standard Tolerance Limits for Shaft Raceway and Housing Bore Diameters — Needle Roller Bearings, Drawn Cup, Without Inner Ring, Metric Types NB, NBM, NY, NYM, NH, and NHM (ANSI/AFBMA Std 18.1-1982)

Basic Bore Diameter under Needle Rollers, F_w				Shaft Raceway Diameter*		Basic Outside Diameter, D				Housing Bore Diameter*	
				Allowable Deviation from F_w						Allowable Deviation from D	
OUTER RING STATIONARY RELATIVE TO LOAD											
mm		Inch		ANSI h6, Inch		mm		Inch		ANSI N7, Inch	
Over	Incl.	Over	Incl.	High	Low	Over	Incl.	Over	Incl.	Low	High
3	6	0.1181	0.2362	+ 0	− 0.0003	6	10	0.2362	0.3937	− 0.0007	− 0.0002
6	10	0.2362	0.3937	+ 0	− 0.0004	10	18	0.3937	0.7087	− 0.0009	− 0.0002
10	18	0.3937	0.7087	+ 0	− 0.0004	18	30	0.7087	1.1811	− 0.0011	− 0.0003
18	30	0.7087	1.1811	+ 0	− 0.0005	30	50	1.1811	1.9685	− 0.0013	− 0.0003
30	50	1.1811	1.9685	+ 0	− 0.0006	50	80	1.9685	3.1496	− 0.0015	− 0.0004
50	80	1.9685	3.1496	+ 0	− 0.0007	...	...	...	...	...	...
OUTER RING ROTATING RELATIVE TO LOAD											
mm		Inch		ANSI f6, Inch		mm		Inch		ANSI R7, Inch	
Over	Incl.	Over	Incl.	High	Low	Over	Incl.	Over	Incl.	Low	High
3	6	0.1181	0.2362	− 0.0004	− 0.0007	6	10	0.2362	0.3937	− 0.0011	− 0.0005
6	10	0.2362	0.3937	− 0.0005	− 0.0009	10	18	0.3937	0.7087	− 0.0013	− 0.0006
10	18	0.3937	0.7087	− 0.0006	− 0.0011	18	30	0.7087	1.1811	− 0.0016	− 0.0008
18	30	0.7087	1.1811	− 0.0008	− 0.0013	30	50	1.1811	1.9685	− 0.0020	− 0.0010
30	50	1.1811	1.9685	− 0.0010	− 0.0016	50	65	1.9685	2.5591	− 0.0024	− 0.0012
50	80	1.9685	3.1496	− 0.0012	− 0.0019	65	80	2.5591	3.1496	− 0.0024	− 0.0013

For bearing tolerances, see Table 9.
* See text for additional requirements.

**Table 19. AFBMA and American National Standard Tolerance Limits for
Shaft Raceway and Housing Bore Diameters — Needle Roller Bearings,
With Cage, Machined Ring, Without Inner Ring, Inch Type NIA**
(ANSI/AFBMA Std 18.2-1982)

Basic Bore Diameter under Needle Rollers, F_w		Shaft Raceway Diameter* Allowable Deviation from F_w		Basic Outside Diameter, D		Housing Bore Diameter* Allowable Deviation from D	
OUTER RING STATIONARY RELATIVE TO LOAD							
Inch		ANSI h6, Inch		Inch		ANSI H7, Inch	
Over	Incl.	High	Low	Over	Incl.	Low	High
0.2362	0.3937	+ 0	− 0.0004	0.3937	0.7087	+ 0	+ 0.0007
0.3937	0.7087	+ 0	− 0.0004	0.7087	1.1811	+ 0	+ 0.0008
0.7087	1.1811	+ 0	− 0.0005	1.1811	1.9685	+ 0	+ 0.0010
1.1811	1.9685	+ 0	− 0.0006	1.9685	3.1496	+ 0	+ 0.0012
1.9685	3.1496	+ 0	− 0.0007	3.1496	4.7244	+ 0	+ 0.0014
3.1496	4.7244	+ 0	− 0.0009	4.7244	7.0866	+ 0	+ 0.0016
4.7244	7.0866	+ 0	− 0.0010	7.0866	9.8425	+ 0	+ 0.0018
7.0866	9.8425	+ 0	− 0.0011	9.8425	12.4016	+ 0	+ 0.0020
OUTER RING ROTATING RELATIVE TO LOAD							
Inch		ANSI f6, Inch		Inch		ANSI N7, Inch	
Over	Incl.	High	Low	Over	Incl.	Low	High
0.2362	0.3937	− 0.0005	− 0.0009	0.3937	0.7087	− 0.0009	− 0.0002
0.3937	0.7087	− 0.0006	− 0.0010	0.7087	1.1811	− 0.0011	− 0.0003
0.7087	1.1811	− 0.0008	− 0.0013	1.1811	1.9685	− 0.0013	− 0.0003
1.1811	1.9685	− 0.0010	− 0.0016	1.9685	3.1496	− 0.0015	− 0.0003
1.9685	3.1496	− 0.0012	− 0.0019	3.1496	4.7244	− 0.0019	− 0.0005
3.1496	4.7244	− 0.0014	− 0.0023	4.7244	7.0866	− 0.0022	− 0.0006
4.7244	7.0866	− 0.0016	− 0.0026	7.0866	9.8425	− 0.0024	− 0.0006
7.0866	9.8425	− 0.0020	− 0.0032	9.8425	11.2205	− 0.0026	− 0.0006

For bearing tolerances, see Table 10.
* See text for additional requirements.

**Table 20. AFBMA and American National Standard Tolerance Limits for
Shaft Diameters — Needle Roller Bearing Inner Rings, Inch Type NIR
(Used with Bearing Type NIA) (ANSI/AFBMA Std 18.2-1982)**

Basic Bore, d		Shaft Diameter*			
		Shaft Rotating Relative to Load, Outer Ring Stationary Relative to Load Allowable Deviation from d		Shaft Stationary Relative to Load, Outer Ring Rotating Relative to Load Allowable Deviation from d	
Inch		ANSI m5, Inch		ANSI g6, Inch	
Over	Incl.	High	Low	High	Low
0.2362	0.3937	+ 0.0005	+ 0.0002	− .0002	− 0.0006
0.3937	0.7087	+ 0.0006	+ 0.0003	− .0002	− 0.0007
0.7087	1.1811	+ 0.0007	+ 0.0003	− .0003	− 0.0008
1.1811	1.9685	+ 0.0008	+ 0.0004	− .0004	− 0.0010
1.9685	3.1496	+ 0.0009	+ 0.0004	− .0004	− 0.0011
3.1496	4.7244	+ 0.0011	+ 0.0005	− .0005	− 0.0013
4.7244	7.0866	+ 0.0013	+ 0.0006	− .0006	− 0.0015
7.0866	9.8425	+ 0.0015	+ 0.0007	− .0006	− 0.0017

For inner ring tolerance limits, see Table 11.
* See text for additional requirements.

Table 21. AFBMA Standard Lockwashers (Series W-00) for Ball Bearings and Cylindrical and Spherical Roller Bearings and (Series TW-100) for Tapered Roller Bearings. Inch Design.

Type W No.	Q	Type TW No.	Q	Tangs No.	Width T	Project V	Width S Min	Width S Max	X Min	X Max	X' Min	X' Max	Bore R Min	Bore R Max	Diameter E	Diameter Tol	Diam. Over Tangs, Max. B	Diam. Over Tangs, Max. B'
W-00	.032	TW-100	.032	9	.120	.031	.110	.120	.334	.359	.334	.359	.406	.421	.625	+.015	.875	.891
W-01	.032	TW-101	.032	9	.120	.031	.110	.120	.412	.437	.412	.437	.484	.499	.719	+.015	1.016	1.031
W-02	.032	TW-102	.048	11	.120	.031	.110	.120	.529	.554	.513	.538	.601	.616	.813	+.015	1.156	1.156
W-03	.032	TW-103	.048	11	.120	.031	.110	.120	.607	.632	.591	.616	.679	.694	.938	+.015	1.328	1.344
W-04	.032	TW-104	.048	11	.166	.031	.156	.176	.729	.754	.713	.738	.801	.816	1.125	+.015	1.531	1.563
W-05	.040	TW-105	.052	13	.166	.047	.156	.176	.909	.928	.897	.927	.989	1.009	1.281	+.015	1.719	1.703
W-06	.040	TW-106	.052	13	.166	.047	.156	.176	1.093	1.128	1.081	1.116	1.193	1.213	1.500	+.015	1.922	1.953
		TW-065	.052	15	.166	.047	.156	.176	...	...	1.221	1.256	1.333	1.353	1.813	+.015	2.234	2.234
W-07	.040	TW-107	.052	15	.166	.047	.156	.176	1.296	1.331	1.284	1.319	1.396	1.416	1.813	+.030	2.250	2.250
W-08	.048	TW-108	.062	15	.234	.062	.250	.290	1.475	1.510	1.461	1.496	1.583	1.603	2.000	+.030	2.469	2.484
W-09	.048	TW-109	.062	17	.234	.062	.250	.290	1.684	1.724	1.670	1.710	1.792	1.817	2.281	+.030	2.734	2.719
W-10	.048	TW-110	.062	17	.234	.062	.250	.290	1.884	1.924	1.870	1.910	1.992	2.017	2.438	+.030	2.922	2.922
W-11	.053	TW-111	.062	17	.234	.062	.250	.290	2.069	2.109	2.060	2.100	2.182	2.207	2.656	+.030	3.109	3.094
W-12	.053	TW-112	.072	19	.234	.062	.250	.290	2.267	2.307	2.248	2.288	2.400	2.425	2.844	+.030	3.344	3.328
W-13	.053	TW-113	.072	19	.234	.062	.250	.290	2.455	2.495	2.436	2.476	2.588	2.613	3.063	+.030	3.578	3.563
W-14	.053	TW-114	.072	19	.234	.094	.250	.290	2.658	2.698	2.639	2.679	2.791	2.816	3.313	+.030	3.828	3.813
W-15	.062	TW-115	.085	19	.328	.094	.313	.353	2.831	2.876	2.808	2.853	2.973	3.003	3.563	+.030	4.109	4.047
W-16	.062	TW-116	.085	19	.328	.094	.313	.353	3.035	3.080	3.012	3.057	3.177	3.207	3.844	+.030	4.375	4.391

All dimensions in inches. For dimensions in millimeters, multiply inch values by 25.4 and round result to two decimal places.

* *Tolerances:* On width T, − .010 inch for Types W-00 to W-03 and TW-100 to TW-103; − .020 inch for W-04 to W-07 and TW-104 to TW-107; − .030 inch for all others shown. On Projection V, + .031 inch for all sizes up through W-13 and TW-113; + .062 inch for all others shown.

Data for sizes larger than shown are given in ANSI/AFBMA Standard 8.2-1978.

Table 22. AFBMA Standard Locknuts (Series N-00) for Ball Bearings and Cylindrical and Spherical Roller Bearings and (Series TN-00) for Tapered Roller Bearings. Inch Design.

Runout and parallelism of faces measured on a tight fitting threaded arbor.

N-00 to N-06 = .002 Max.
N-07 to AN-15 = .004 Max.

TN-065 to TAN-15 = .002 Max.

Surface Finish Note

TN-065 to TN-11, 100μ in., max.
TN-12 to TAN-15, 120μ in., max.

N-00 through AN-15
TN-065 through TAN-15

BB & RB Nut No.	TRB Nut No.	Thds. per Inch	Thread Minor Diam.		Thread Pitch Diam.		Thd. Major Diam. d	Outside Diam. C	Face Diam. E		Slot Width G		Slot Height H	Thickness D	
			Min	Max	Min	Max	Min	Max	Min	Max	Min	Max	Max	Min	Max
N-00	—	32	0.3572	0.3606	0.3707	0.3733	0.391	0.755	0.605	0.625	.120	.130	.073	.209	.229
N-01	—	32	0.4352	0.4386	0.4487	0.4513	0.469	0.880	0.699	0.719	.120	.130	.073	.303	.323
N-02	—	32	0.5522	0.5556	0.5657	0.5687	0.586	1.005	0.793	0.813	.120	.130	.104	.303	.323
N-03	—	32	0.6302	0.6336	0.6437	0.6467	0.664	1.130	0.918	0.938	.120	.130	.104	.334	.354
N-04	—	32	0.7472	0.7506	0.7607	0.7641	0.781	1.386	1.105	1.125	.178	.198	.104	.365	.385
N-05	—	32	0.9352	0.9386	0.9487	0.9521	0.959	1.568	1.261	1.281	.178	.198	.104	.396	.416
N-06	—	18	1.1159	1.1189	1.1369	1.1409	1.173	1.755	1.480	1.500	.178	.198	.104	.396	.416
	TN-065	18	1.2524	1.2584	1.2764	1.2804	1.312	2.068	1.793	1.813	.178	.198	.104	.428	.448
N-07	TN-07	18	1.3159	1.3219	1.3399	1.3439	1.376	2.068	1.793	1.813	.178	.198	.104	.428	.448
N-08	TN-08	18	1.5029	1.5089	1.5269	1.5314	1.563	2.255	1.980	2.000	.240	.260	.104	.428	.448
N-09	TN-09	18	1.7069	1.7129	1.7309	1.7354	1.797	2.536	2.261	2.281	.240	.260	.104	.428	.448
N-10	TN-10	18	1.9069	1.9129	1.9309	1.9354	1.967	2.693	2.418	2.438	.240	.260	.104	.490	.510
N-11	TN-11	18	2.0999	2.1029	2.1209	2.1260	2.157	2.974	2.636	2.656	.240	.260	.135	.490	.510
N-12	TN-12	18	2.2999	2.3059	2.3239	2.3390	2.360	3.161	2.824	2.844	.240	.260	.135	.521	.541
N-13	TN-13	18	2.4879	2.4949	2.5119	2.5170	2.548	3.380	3.043	3.063	.240	.260	.135	.553	.573
N-14	TN-14	18	2.6909	2.6969	2.7149	2.7200	2.751	3.630	3.283	3.313	.240	.260	.135	.553	.573
AN-15	TAN-15	12	2.8428	2.8818	2.8789	2.8843	2.933	3.880	3.533	3.563	.360	.385	.135	.584	.604

All dimensions in inches. For dimensions in millimeters, multiply inch values, except thread diameters, by 25.4 and round result to two decimal places.

Threads are American National form, Class 3.

Typical steels for locknuts are: AISI, C1015, C1020, C1025, C1035, C1117, C1118, C1212, C1213, and C1215. Minimum hardness, tensile strength, yield strength and elongation are given in ANSI/AFBMA Std 8.2-1978 which also lists larger sizes of locknuts.

Table 23. AFBMA Standard for Shafts for Locknuts (Series N-00) for Ball Bearings and Cylindrical and Spherical Roller Bearings. Inch Design.

Locknut No.	Bearing Bore	V_2 Max.	No. per inch	Threads*			Length L Max.	Relief		Keyway		
				Major Diam. Max.	Pitch Diam. Max.	Minor Diam. Max.		Diam. A Max.	Width W Max.	Depth H Min.	Width S Min.	M Min.
N-00	0.3937	0.312	32	0.391	0.3707	0.3527	0.297	0.3421	0.078	0.062	0.125	0.094
N-01	0.4724	0.406	32	0.469	0.4487	0.4307	0.391	0.4201	0.078	0.062	0.125	0.094
N-02	0.5906	0.500	32	0.586	0.5657	0.5477	0.391	0.5371	0.078	0.078	0.125	0.094
N-03	0.6693	0.562	32	0.664	0.6437	0.6257	0.422	0.6151	0.078	0.078	0.125	0.094
N-04	0.7874	0.719	32	0.781	0.7607	0.7427	0.453	0.7321	0.078	0.078	0.188	0.094
N-05	0.9843	0.875	32	0.969	0.9487	0.9307	0.484	0.9201	0.078	0.094	0.188	0.125
N-06	1.1811	1.062	18	1.173	1.1369	1.1048	0.484	1.0942	0.109	0.094	0.188	0.125
N-07	1.3780	1.250	18	1.376	1.3399	1.3078	0.516	1.2972	0.109	0.094	0.188	0.125
N-08	1.5748	1.469	18	1.563	1.5269	1.4948	0.547	1.4842	0.109	0.094	0.312	0.125
N-09	1.7717	1.688	18	1.767	1.7399	1.6988	0.547	1.6882	0.141	0.094	0.312	0.156
N-10	1.9685	1.875	18	1.967	1.9309	1.8988	0.609	1.8882	0.141	0.094	0.312	0.156
N-11	2.1654	2.062	18	2.157	2.1209	2.0888	0.609	2.0782	0.141	0.125	0.312	0.156
N-12	2.3622	2.250	18	2.360	2.3239	2.2918	0.641	2.2812	0.141	0.125	0.312	0.156
N-13	2.5591	2.438	18	2.548	2.5119	2.4798	0.672	2.4692	0.141	0.125	0.312	0.156
N-14	2.7559	2.625	18	2.751	2.7149	2.6828	0.672	2.6722	0.141	0.125	0.312	0.250
AN-15	2.9528	2.781	12	2.933	2.8789	2.8308	0.703	2.8095	0.172	0.125	0.312	0.250
AN-16	3.1496	3.000	12	3.137	3.0829	3.0348	0.703	3.0135	0.172	0.125	0.375	0.250

All dimensions in inches. For dimensions in millimeters, multiply inch values, except thread diameters, by 25.4 and round result to two decimal places.
See footnote to Table 24 for material other than steel.
* Threads are American National form Class 3;
For sizes larger than shown, see ANSI/AFBMA Std 8.2-1978.

Table 24. AFBMA Standard for Shafts for Tapered Roller Bearing Locknuts. Inch Design.

Locknut Number	Bearing Bore	V_2 Max.	Threads*				Length		Relief		Keyway			
			No. per inch	Major Diam. Max.	Pitch Diam. Max.	Minor Diam. Max.	L_1 Max.	L_2 Max.	Diam. A Max.	Width W Max.	Depth H Min.	Width S Min.	M Min.	U Min.
N-00	0.3937	0.312	32	0.391	0.3707	0.3527	0.609	0.391	0.3421	0.078	0.094	0.125	0.094	0.469
N-01	0.4724	0.406	32	0.469	0.4487	0.4307	0.797	0.484	0.4201	0.078	0.094	0.125	0.094	0.562
N-02	0.5906	0.500	32	0.586	0.5657	0.5477	0.828	0.516	0.5371	0.078	0.094	0.125	0.094	0.594
N-03	0.6693	0.562	32	0.664	0.6437	0.6257	0.891	0.547	0.6151	0.078	0.094	0.125	0.094	0.625
N-04	0.7874	0.703	32	0.781	0.7607	0.7427	0.922	0.547	0.7331	0.078	0.094	0.188	0.094	0.625
N-05	0.9843	0.875	32	0.969	0.9487	0.9307	1.016	0.609	0.9201	0.078	0.125	0.188	0.125	0.719
N-06	1.1811	1.062	18	1.173	1.1369	1.1048	1.016	0.609	1.0942	0.109	0.125	0.188	0.125	0.719
TN-065	1.2598	1.188	18	1.312	1.2764	1.2443	1.078	0.641	1.2337	0.109	0.125	0.188	0.125	0.750
TN-07	1.3780	1.250	18	1.376	1.3399	1.3078	1.078	0.641	1.2972	0.109	0.125	0.188	0.125	0.750
TN-08	1.5748	1.438	18	1.563	1.5269	1.4948	1.078	0.641	1.4842	0.109	0.125	0.312	0.125	0.750
TN-09	1.7717	1.656	18	1.767	1.7309	1.6988	1.078	0.641	1.6882	0.141	0.125	0.312	0.156	0.781
TN-10	1.9685	1.859	18	1.967	1.9309	1.8988	1.203	0.703	1.8882	0.141	0.125	0.312	0.156	0.844
TN-11	2.1654	2.047	18	2.157	2.1209	2.0888	1.203	0.703	2.0782	0.141	0.125	0.312	0.156	0.844
TN-12	2.3622	2.250	18	2.360	2.3239	2.2918	1.297	0.766	2.2812	0.141	0.156	0.312	0.156	0.906
TN-13	2.5591	2.422	18	2.548	2.5119	2.4798	1.359	0.797	2.4692	0.141	0.156	0.312	0.156	0.938
TN-14	2.7559	2.625	18	2.751	2.7149	2.6828	1.359	0.797	2.6722	0.141	0.156	0.312	0.250	1.000
TAN-15	2.9528	2.781	12	2.933	2.8789	2.8308	1.422	0.828	2.8095	0.172	0.188	0.312	0.250	1.031
TAN-16	3.1496	3.000	12	3.137	3.0829	3.0348	1.422	0.828	3.0135	0.172	0.188	0.375	0.250	1.031

All dimensions in inches. For dimensions in millimeters, multiply inch values, except thread diameters, by 25.4 and round results to two decimal places.

* Threads are American National form Class 3.

These data apply to steel. When either the nut or the shaft is made of stainless steel, aluminum, or other material having a tendency to seize, it is recommended that the maximum thread diameter of the shaft, both major and pitch, be reduced by 20 per cent of the pitch diameter tolerance listed in the Standard.

For sizes larger than shown, see ANSI/AFBMA Std 8.2-1978.

Bearing Mounting Practice. — Because of their inherent design and material rigidity, rolling contact bearings must be mounted with careful control of their alignment and runout. Medium-speed or slower (400,000 DN values or less where D is the bearing bore in millimeters and N is the bearing speed in revolutions per minute), and medium to light load (C/P values of 7 or greater where C is the bearing specific dynamic capacity in pounds and P is the average bearing load in pounds) applications can endure misalignments equivalent to those acceptable for high-capacity, precision journal bearings utilizing hard bearing materials such as silver, copper-lead, or aluminum. In no case, however, should the maximum shaft deflection exceed .001 inch per inch for well-crowned roller bearings, and .003 inch per inch for deep-groove ball-bearings. Except for self-aligning ball-bearings and spherical or barrel roller bearings, all other types require shaft alignments with deflections no greater than .0002 inch per inch. With preloaded ball bearings, this same limit is recommended as a maximum. Close-clearance tapered bearings or thrust bearings of most types require the same shaft alignment also.

Of major importance for all bearings requiring good reliability, is the location of the races on the shaft and in the housing. Assembly methods must insure: (1) that the faces are square, before the cavity is closed; (2) that the cover face is square to the shoulder and pulled in evenly; and (3) that it will be located by a face parallel to it when finally seated against the housing. These requirements are shown in the accompanying figure. In applications not controlled by automatic tooling with closely controlled fixtures and bolt torquing mechanisms, races should be checked for squareness by sweeping with a dial indicator mounted as shown below. For

COMMERCIAL APPLICATION ALIGNMENT TOLERANCES

1. **Housing Face Runout** — Square to shaft center within .0004 inch/inch of radius full indicator reading.

2. **Outer Race Face Runout** — Square to shaft center within .0004 inch/inch of radius full indicator reading and complementary to the housing runout (not opposed).

3. **Inner Race Face Runout** — Square to shaft center within .0003 inch/inch of radius full indicator reading.

4. and 5. **Cover and Closure Mounting Face Parallelism** — Parallel within .001.

6. **Housing Mounting Face Parallelism** — Parallel within .001.

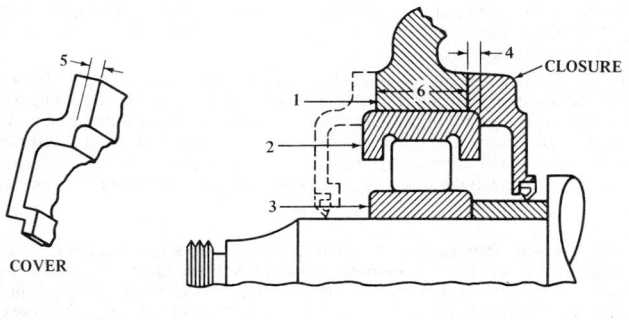

commercial applications with moderate life and reliability requirements, outer race runouts should be held to .0005 inch per inch of radius and inner race runout to .0004 inch per inch of radius. In preloaded and precision applications, these tolerances must be cut in half. In regard to the question of alignment, it must be recognized that rolling-contact bearings, being made of fully-hardened steel, do not wear in as may certain journal bearings when carefully applied and initially operated. Likewise, rolling contact bearings absorb relatively little deflection when loaded to C/P values of 6 or less. At such stress levels the rolling element-race deformation is generally not over .0002 inch. Consequently, proper mounting and control of shaft deflections are imperative for reliable bearing performance. Aside from inadequate lubrication, these factors are the most frequent causes of premature bearing failures.

Mountings for Precision and Quiet-running Applications. — In applications of rolling-element bearings where vibration or smoothness of operation is critical, special precautions must be taken to eliminate those conditions which can serve to initiate radial and axial motions. These exciting forces can result in shaft excursions which are in resonance with shaft or housing components over a range of frequencies from well below shaft speed to as much as 100 times above it. The more sensitive the configuration, the greater is the need for precision bearings and mountings to be used.

Precision bearings are normally made to much closer tolerances than standard and therefore benefit from better finishing techniques. Special inspection operations are required, however, to provide races and rolling elements with smoothness and runouts compatible with the needs of the application. Similarly, shafts and housings must be carefully controlled.

Among the important elements to be controlled are shaft, race, and housing roundness; squareness of faces, diameters, shoulders, and rolling paths. Though not readily appreciated, grinding chatter, lobular and compensating out-of-roundness, waviness, and flats of less than .0005 inch deviation from the average or mean diameter can cause significant roughness. To detect these and insure the selection of good pieces, three-point electronic indicator inspection must be made. For ultra-precise or quiet applications, pieces are often checked on a "Talyrond" or a similar continuous recording instrument capable of measuring to within a few millionths of an inch. Though this may seem extreme, it has been found that shaft deformities will be reflected through inner races shrunk onto them. Similarly, tight-fit outer races pick up significant deviations in housings. In many instrument and in missile guidance applications, such deviations and deformities may have to be limited to less than .00002 inch.

In most of these precision applications, bearings are used with rolling elements controlled to less than 5 millionths of an inch deviation from roundness and within the same range for diameter.

Special attention is required both in housing design and in assembly of the bearing to shaft and housing. Housing response to axial excursions forced by bearing wobble (which in itself is a result of out-of-square mounting) has been found to be a major source of small electric and other rotating equipment noise and howl. Stiffer, more massive housings and careful alignment of bearing races can make significant improvements in applications where noise or vibration has been found to be objectionable.

Squareness and Alignment. — In addition to the limits for roundness and wall variation of the races and their supports, squareness of end faces and shoulders must be closely controlled. Tolerances of .0001 inch full indicator reading per inch of diameter are normally required for end faces and shoulders, with appropriately

selected limits for fillet eccentricities. The latter must also fall within specified limits for radii tolerances to prevent interference and the resulting cocking of the race. Reference should be made to the bearing dimension tables which list corner radii for typical bearings. Shoulders must also be of a sufficient height to insure proper support for the races, since they are of hardened steel and are less capable of absorbing shock loads and abuse. The general subject of squareness and alignment is of primary importance to the life of rolling element bearings.

The following recommendations for shaft and housing design are given by the New Departure Division of General Motors Corporation:*

"As a rule, there is little trouble experienced with inaccuracies in shafts. Bearings seats and locating shoulders are turned and ground to size with the shaft held on centers and, with ordinary care, there is small chance for serious out-of-roundness or taper. Shaft shoulders should present sufficient surface in contact with the bearing face to assure positive and accurate location.

"Where an undercut must be made for wheel runout in grinding a bearing seat, care should be exercised that no sharp corners are left, for it is at such points that fatigue is most likely to result in shaft breakage. It is best to undercut as little as possible and to have the undercut end in a fillet instead of a sharp corner.

"Where clamping nuts are to be used, it is important to cut the threads as true and square as possible in order to insure even pressure at all points on the bearing inner ring faces when the nuts are set up tight. It is also important not to cut threads so far into the bearing seat as to leave part of the inner ring unsupported or carried on the threads. Excessive deflection is usually the result of improperly designed or undersized machine parts. With a weak shaft, it is possible to seriously affect bearing operation through misalignment due to shaft deflection. Where shafts are comparatively long, the diameter between bearings must be great enough to properly resist bending. In general, the use of more than two bearings on a single shaft should be avoided, owing to the difficulty of securing accurate alignment. With bearings mounted close to each other, this can result in extremely heavy bearing loads.

"Design is as important as careful machining in construction of accurate bearing housings. There should be plenty of metal in the wall sections and large, thin areas should be avoided as much as possible, since they are likely to permit deflection of the boring tool when the housing is being finish-machined.

"Wherever possible, it is best to design a housing so that the radial load placed on the bearing is transmitted as directly as possible to the wall or rib supporting the housing. Diaphragm walls connecting an offset housing to the main wall or side of a machine are apt to deflect unless made thick and well braced.

"When two bearings are to be mounted opposed, but in separate housings, the housings should be so reinforced with fins or webs as to prevent deflection due to the axial load under which the bearings are opposed.

"Where housings are deep and considerable overhang of the boring tool is required, there is a tendency to produce out-of-roundness and taper, unless the tool is very rigid and light finishing cuts are taken. In a too roughly bored housing there is a possibility for the ridges of metal to peen down under load, thus eventually resulting in too loose a fit for the bearing outer ring."

Soft Metal and Resilient Housings. — In applications relying on bearing housings made of soft materials (aluminum, magnesium, light sheet metal, etc.) or those which lose their fit because of differential thermal expansion, outer race mounting must be approached in a cautious manner. Of first importance is the determination of the

* New Departure Handbook, Vol. II — 1951.

possible consequences of race loosening and turning. In conjunction with this, the type of loading must be considered for it may serve to magnify the effect of race loosening. It must be remembered that generally, balancing processes do not insure zero unbalance at operating speeds, but rather an "acceptable" maximum. This force exerted by the rotating element on the outer race can initiate a precession which will aggravate the race loosening problem by causing further attrition through wear, pounding, and abrasion. Since this force is generally of an order greater than the friction forces in effect between the outer race, housing, and closures (retaining nuts also), no foolproof method can be recommended for securing outer races in housings which deform significantly under load or after appreciable service wear. Though many such "fixes" are offered, the only sure solution is to press the race into a housing of sufficient stiffness with the heaviest fit consistent with the installed and operating clearances. In many cases, inserts, or liners of cast iron or steel are provided to maintain the desired fit and increase useful life of both bearing and housing.

Quiet or Vibration-free Mountings. — In seeming contradiction is the approach to bearing mountings in which all shaft or rotating element excursions must be isolated from the frame, housing, or supporting structure. Here bearing outer races are often supported on elastomeric or metallic springs. Fundamentally, this is an isolation problem and must be approached with caution to insure solution of the primary bearing objective — location and restraint of the rotating body, as well as the reduction or elimination of the dynamic problem. Again, the danger of skidding rolling elements must be considered and reference to the resident engineers or sales engineers of the numerous bearing companies is recommended, as this problem generally develops requirements for special, or non-catalog-type bearings.

General Mounting Precautions. — Since the last operations involving the bearing application — mounting and closing — have such important effects on bearing performance, durability, and reliability, it must be cautioned that more bearings are abused or "killed" in this early stage of their life than wear out or "die" under conditions for which they were designed. Hammer and chisel "mechanics" invariably handle bearings as though no blow could be too hard, no dirt too abrasive, and no misalignment of any consequence. Proper tools, fixtures, and techniques are a must for rolling bearing application, and it is the responsibility of the design engineer to provide for this in his design, advisory notes, mounting instructions, and service manuals. Nicks, dents, scores, scratches, corrosion staining, and dirt must be avoided if reliability, long life, and smooth running are to be expected of rolling bearings. All manufacturers have pertinent service instructions available for the bearing user. These should be followed for best performance. In a later section, methods for inspecting bearings and descriptions of most common bearing deficiencies will be given.

Seating Fits for Bearings. — Anti-Friction Bearing Manufacturers Association (AFBMA) standard shaft and housing bearing seat tolerances are given in Tables 12 to 16, inclusive.

Clamping and Retaining Methods. — Various methods of clamping bearings to prevent axial movement on the shaft are employed, one of the most common being a nut screwed on the end of the shaft and held in place by a tongued lock washer (see Table 21). The shaft thread for the clamping nut (see Table 22) should be cut in accurate relation to bearing seats and shoulders if bearing stresses are to be avoided. The threads used are of American National Form, Class 3; special diameters and data for these are given in Tables 23 and 24. Where somewhat

closer than average accuracy is required, the washers and locknut faces may be obtained ground for closer alignment with the threads. For a high degree of accuracy the shaft threads are ground and a more precise clamping means is employed. Where a bearing inner ring is to be clamped, it is important to provide a sufficiently high shoulder on the shaft to locate the bearing positively and accurately. If the difference between bearing bore and maximum shaft diameter gives a low shoulder which would enter the corner of the radius of the bearing, a shoulder ring that extends above the shoulder and well into the shaft corner is employed. A shoulder ring with snap wire fitting into a groove in the shaft is sometimes used where no locating shaft shoulder is present. A snap ring fitting into a groove is frequently employed to prevent endwise movement of the bearing away from the locating shoulder where tight clamping is not required. Such a retaining ring should not be used where a slot in the shaft surface might lead to fatigue failure. Snap rings are also used to locate the outer bearing ring in the housing. Dimensions of snap rings used for this latter purpose are given in AFBMA and ANSI standards.

Bearing Closures. — Shields, seals, labyrinths, and slingers are employed to retain the lubricant in the bearing and to prevent the entry of dirt, moisture, or other harmful substances. The type selected for a given application depends upon the lubricant, shaft, speed, and the atmospheric conditions in which the unit is to operate. The shields or seals may be located in the bearing itself. Shields differ from seals in that they are attached to one bearing race but there is a definite clearance between the shield and the other, usually the inner, race. When a shielded bearing is placed in a housing in which the grease space has been filled, the bearing in running will tend to expel excess grease past the shields or to accept grease from the housing when the amount in the bearing itself is low.

Seals of leather, rubber, cork, felt, or plastic composition may be used. Since they must bear against the rotating member, excessive pressure should be avoided and some lubricant must be allowed to flow into the area of contact in order to prevent seizing and burning of the seal and scoring of the rotating member. Some seals are made up in the form of cartridges which can be pressed into the end of the bearing housing.

Leather seals may be used over a wide range of speeds. Although lubricant is best retained with a leather cupped inward toward the bearing, this arrangement is not suitable at high speeds due to danger of burning the leather. At high speeds where abrasive dust is present, the seal should be arranged with the leather cupped outward to lead some lubricant into the contact area. Only light pressure of leather against the shaft should be maintained.

Bearing Fits. — The slipping or creeping of a bearing ring on a rotating shaft or in a rotating housing occurs when the fit of the ring on the shaft or in the housing is loose. Such slipping or creeping action may cause rapid wear of both shaft and bearing ring when the surfaces are dry and highly loaded. To prevent this action the bearing is customarily mounted with the rotating ring a press fit and the stationary ring a push fit, the tightness or looseness depending upon the service intended. Thus, where shock or vibratory loads are to be encountered, fits should be made somewhat tighter than for ordinary service. The stationary ring, if correctly fitted, is allowed to creep very slowly so that prolonged stressing of one part of the raceway is avoided.

To facilitate the assembly of a bearing on a shaft it may become necessary to expand the inner ring by heating. This should be done in clean oil or in a temperature-controlled furnace at a temperature of between 200 and 250°F. The utmost care must be used to make sure that the temperature does not exceed 250°F. as overheating will tend to reduce the hardness of the rings. Prelubricated bearings should not be mounted by this method.

Friction Losses in Rolling Element Bearings. — The static and kinematic torques of rolling element bearings are generally small and in many applications are not significant. Bearing torque is a measure of the frictional resistance of the bearing to rotation and is the sum of three components: the torque due to the applied load; the torque due to viscous forces in lubricated rolling element bearings; and the torque due to roller end motions, for example, thrust loads against flanges. The friction or torque data may be used to calculate power absorption or heat generation within the bearing and can be utilized in efficiency or system-cooling studies.

Empirical equations have been developed for each of the torque components. These equations are influenced by such factors as bearing load, lubrication environment, and bearing design parameters. These design parameters include sliding friction from contact between the rolling elements and separator surfaces or between adjacent rolling elements; rolling friction from material deformations during the passage of the rolling elements over the race path; skidding or sliding of the Hertzian contact; and windage friction as a function of speed.

Starting or breakaway torques are also of interest in some situations. Breakaway torques tend to be between 1.5 and 1.8 times the running or kinetic torques.

When evaluating the torque requirements of a system under design, it should be noted that other components of the bearing package, such as seals and closures, can increase the overall system torque significantly. Seal torques have been shown to vary from a fraction of the bearing torque to several times that torque. In addition, the torque values given can vary significantly when load, speed of rotation, temperature, or lubrication are outside normal ranges.

For small instrument bearings friction torque has implications more critical than for larger types of bearings. These bearings have three operating friction torques to consider: starting torque, normal running torque, and peak running torque. These torque levels may vary between manufacturers and among lots from a given manufacturer.

Instrument bearings are even more critically dependent on design features — radial play, retainer type, and race conformity — than larger bearings. Typical starting torque values for small bearings are given in the accompanying table, extracted from the New Departure General Catalog.

Finally, if accurate control of friction torque is critical to a particular application, tests of the selected bearings should be conducted to evaluate performance.

Starting Torque — ABEC7

Bearing Bore (in.)	Max. Starting Torque (g cm)	Thrust Load (g)	Minimum Radial Play Range (inches)	
			High Carbon Chrome Steel and All Miniatures	Stainless Steel Except Miniatures
0.125	0.10	75	0.0003–0.0005	—
	0.14	75	0.0002–0.0004	0.0004–0.0006
	0.18	75	0.0001–0.0003	0.0003–0.0005
	0.22	75	0.0001–0.0003	0.0001–0.0003
0.1875–0.312	0.40	400	0.0005–0.0008	—
	0.45	400	0.0004–0.0006	0.0005–0.0008
	0.50	400	0.0003–0.0005	0.0003–0.0005
	0.63	400	0.0001–0.0003	0.0002–0.0004
0.375	0.50	400	0.0005–0.0008	0.0008–0.0011
	0.63	400	0.0004–0.0006	0.0005–0.0008
	0.75	400	0.0003–0.0005	0.0004–0.0006
	0.95	400	0.0002–0.0004	0.0003–0.0005

Selection of Ball and Roller Bearings. — As compared with sleeve bearings, ball and roller bearings offer the following advantages: (1) Starting friction is low; (2) Less axial space is required; (3) Relatively accurate shaft alignment can be maintained; (4) Both radial and axial loads can be carried by certain types; (5) Angle of load application is not restricted; (6) Replacement is relatively easy; (7) Comparatively heavy overloads can be carried momentarily; (8) Lubrication is simple, and (9) Design and application can be made with the assistance of bearing supplier engineers.

In selecting a ball or roller bearing for a specific application five choices must be made: (1) the bearing series, (2) the type of bearing, (3) the size of bearing, (4) the method of lubrication, and (5) the type of mounting. Naturally these considerations are modified or affected by the anticipated operating conditions, expected life, cost, and overhaul philosophy. It is well to review the possible history of the bearing and its function in the machine it will be applied to, thus: (1) Will it be expected to endure removal and reapplication? (2) Must it be free from maintenance attention during its useful life? (3) Can wear of the housing or shaft be tolerated during the overhaul period? (4) Must it be adjustable to take up wear, or to change shaft location? (5) How accurately can the load spectrum be estimated? and (6) Will it be relatively free from abuse in operation? Though many cautions could be pointed out, it should always be remembered that inadequate design approaches limit the utilization of rolling element bearings, reduce customer satisfaction, and reduce reliability. Time spent in this stage of design is the most rewarding effort of the bearing engineer, and here again he can depend on the bearing manufacturers' field organization for assistance.

Type: Where loads are low, ball bearings are usually less expensive than roller bearings in terms of unit-carrying capacity. Where loads are high, the reverse is usually true.

For a purely radial load, almost any type of radial bearing can be used, the actual choice being determined by other factors. To support a combination of thrust and radial loads, several types of bearings may be considered. If the thrust load component is large, it may be most economical to provide a separate thrust bearing. When a separate thrust bearing cannot be used due to high speed, lack of space, or other factors, the following types may be considered: angular contact ball bearing, deep groove ball bearing without filling slot, tapered roller bearing with steep contact angle, and self-aligning bearing of the wide type. If movement or deflection in an axial direction must be held to a minimum, then a separate thrust bearing or a preloaded bearing capable of taking considerable thrust load is required. To minimize deflection due to a moment in an axial plane, a rigid bearing such as a double row angular contact type with outwardly converging load lines is required. In such cases, the resulting stresses must be taken into consideration in determining the proper size of the bearing.

For shock loads or heavy loads of short duration, roller bearings are usually preferred.

Special bearing designs may be required where accelerations are usually high as in planetary or crank motions.

Where the problem of excessive shaft deflection or misalignment between shaft and housing is present, a self-aligning type of bearing may be a satisfactory solution.

It should be kept in mind that a great deal of difficulty can be avoided if standard types of bearings are used in preference to special designs, wherever possible.

Size: The size of bearing required for a given application is determined by the loads that are to be carried and, in some cases, by the amount of rigidity that is necessary to limit deflection to some specified amount.

The forces to which a bearing will be subjected can be calculated by the laws of engineering mechanics from the known loads, power, operating pressure, etc. Where

loads are irregular, varying, or of unknown magnitude, it may be difficult to determine the actual forces. In such cases, empirical determination of such forces, based on extensive experience in bearing design, may be needed to attack the problem successfully. Where such experience is lacking, the bearing manufacturer should be consulted or the services of a bearing expert obtained.

If a ball or roller bearing is to be subjected to a combination of radial and thrust loads, an *equivalent radial load* is computed in the case of radial or angular type bearings and an *equivalent thrust load* is computed in the case of thrust bearings.

Method of Lubrication. — If speeds are high, relubrication difficult, the shaft angle other than horizontal, the application environment incompatible with normal lubrication, leakage cannot be tolerated; if other elements of the mechanism establish the lubrication requirements, bearing selection must be made with these criteria as controlling influences. Modern bearing types cover a wide selection of lubrication means. Though the most popular type is the "cartridge" type of sealed grease ball bearing, many applications have requirements which dictate against them. Often, operating environments may subject bearings to temperatures too high for seals utilized in the more popular designs. If minute leakage or the accumulation of traces of dirt at seal lips cannot be tolerated by the application (as in baking industry machinery), then the selections of bearings must be made with other sealing and lubrication systems in mind.

High shaft speeds generally dictate bearing selection based on the need for cooling, the suppression of churning or aeration of conventional lubricants, and most important of all, the inherent speed limitations of certain bearing types. An example of the latter is the effect of cage design and of the roller-end thrust-flange contact on the lubrication requirements in commercial taper roller bearings, which limit the speed they can endure and the thrust load they can carry. Reference to the manufacturers' catalog and application-design manuals is recommended before making bearing selections.

Type of Mounting. — Many bearing installations are complicated because the best adapted type was not selected. Similarly, performance, reliability, and maintenance operations are restricted because the mounting was not thoroughly considered. There is no universally adaptable bearing for all needs. Careful reviews of the machine requirements should be made before designs are implemented. In many cases complicated machining, redundant shaft and housings, and use of an oversize bearing can be eliminated if the proper bearing in a well-thought-out mounting is chosen.

Advantage should be taken of the many race variations available in "standard" series of bearings. Puller grooves, tapered sleeves, flanged outer races, split races, fully demountable rolling-element and cage assemblies, flexible mountings, hydraulic removal features, relubrication holes and grooves, and many other innovations are available beyond the obvious advantages which are inherent in the basic bearing types.

Radial and Axial Clearance. — In designing the bearing mounting, a major consideration is to provide running clearances consistent with the requirements of the application. Race fits must be expected to absorb some of the original bearing clearance so that allowance should be made for approximately 80 per cent of the actual interference showing up in the diameter of the race. This will increase for heavy, stiff housings or for extra light series races shrunk onto solid shafts, while light metal housings (aluminum, magnesium, or sheet metal) and tubular shafts with wall sections less than the race wall thickness will cause a lesser change in the race diameter.

Where the application will impose heat losses through housing or shaft, or where a temperature differential may be expected, allowances must be made in the proper direction to insure proper operating clearance. Some compromises are required in applications where the indicated modification cannot be fully accommodated without endangering the bearing performance at lower speeds, during starting, or under lower temperature conditions than anticipated. Some leeway can be relied on with ball bearings since they can run with moderate preloads (.0005 inch, max.) without affecting bearing life or temperature rise. Roller bearings, however, have a lesser tolerance for preloading, and must be carefully controlled to avoid overheating and resulting self-destruction.

In all critical applications axial and radial clearances should be checked with feeler gages or dial indicators to insure mounted clearances within tolerances established by the design engineer. Since chips, scores, race misalignment, shaft or housing denting, housing distortion, end cover (closure) off-squareness, and mismatch of rotor and housing axial dimensions can rob the bearing of clearance, careful checks of running clearance is recommended.

For precision applications, taper-sleeve mountings, opposed ball or tapered-roller bearings with adjustable or shimmed closures are employed to provide careful control of radial and/or axial clearances. This practice requires skill and experience as well as the initial assistance of the bearing manufacturer's field engineer.

Tapered bore bearings are often used in applications such as these, again requiring careful and well worked-out assembly procedures. They can be assembled on either tapered shafts or on adapter sleeves. Advancement of the inner race over the tapered shaft can be done either by controlled heating (to expand the race as required) or by the use of a hydraulic jack. The adapter sleeve is supplied with a lock-nut which is used to advance the race on the tapered sleeve. With the heavier fits normally required to effect the clearance changes compatible with such mountings, hydraulic removal devices are normally recommended.

For the conventional application, with standard fits, clearances provided in the standard bearing are suitable for normal operation. To insure that the design conditions are "normal," a careful review of the application requirements, environments, operating speed range, anticipated abuses, and design parameters must be made.

General Bearing Handling Precautions. — To insure that rolling element bearings are capable of achieving their design life and that they perform without objectionable noise, temperature rise, or shaft excursions, the following precautions are recommended:

1. Use the best bearing available for the application, consistent with the value of the application. Remember, the cost of the best bearing is generally small compared to the replacement costs of the rotating components that can be destroyed if a bearing fails or malfunctions.

2. If questions arise in designing the bearing application, seek out the assistance of the bearing manufacturer's representative.

3. Handle bearings with care, keeping them in the sealed, original container until ready to use.

4. Follow the manufacturer's instructions in handling and assembling the bearings.

5. Work with clean tools, clean dry hands, and in clean surroundings.

6. Do not wash or wipe bearings prior to installation unless special instructions or requirements have been established to do so.

7. Place unwrapped bearings on clean paper and keep them similarly covered until applied, if they cannot be kept in the original container.

8. Don't use wooden mallets, brittle or chipped tools, or dirty fixtures and tools in mounting bearings.

9. Don't spin uncleaned bearings, nor spin *any* bearing with an air blast.

10. Use care not to scratch or nick bearings.

11. Don't strike or press on race flanges.

12. Use adapters for mounting which provide uniform steady pressure rather than hammering on a drift or sleeve.

13. Insure that races are started onto shafts and into housings evenly so as to prevent cocking.

14. Inspect shafts and housings before mounting bearing to insure that proper fits will be maintained.

15. When removing bearings, clean housings, covers, and shafts before exposing the bearings. All dirt can be considered an abrasive, dangerous to the reuse of any rolling bearing.

16. Treat used bearings, which may be reused, as new ones.

17. Protect dismantled bearings from dirt and moisture.

18. Use clean, lint-free rags if bearings are wiped.

19. Wrap bearings in clean, oil-proof paper when not in use.

20. Use clean filtered, water-free Stoddard's solvent or flushing oil to clean bearings.

21. In heating bearings for mounting onto shafts, follow manufacturer's instructions.

22. In assembling bearings onto shafts *never* strike the outer race, or press on it to force the inner race. Apply the pressure on the inner race only. In dismantling follow the same precautions.

23. Do not press, strike, or otherwise force the seal or shield on factory-sealed bearings.

Bearing Failures, Deficiencies, and Their Origins. — The general classifications of failures and deficiencies requiring bearing removal are:

(a) Overheating: 1. Inadequate or insufficient lubrication; 2. Excessive lubrication; 3. Grease liquefaction or aeration; 4. Oil foaming; 5. Abrasive or corrosive action due to contaminants in bearing; 6. Distortion of housing due to warping, or out-of-round; 7. Seal rubbing or failure; 8. Inadequate or blocked scavenge oil passages; 9. Inadequate bearing-clearance or bearing-preload; 10. Race turning; 11. Cage wear; and 12. Shaft expansion — loss of bearing or seal clearance.

(b) Vibration: 1. Dirt or chips in bearing; 2. Fatigued race or rolling elements; 3. Race turning; 4. Rotor unbalance; 5. Out-of-round shaft; 6. Race misalignment; 7. Housing resonance; 8. Cage wear; 9. Flats on races or rolling elements; 10. Excessive clearance; 11. Corrosion; 12. False-brinelling or indentation of races; 13. Electrical discharge (similar to corrosion effects); 14. Mixed rolling element diameters; and 15. Out-of-square rolling paths in races.

(c) Turning on shaft: 1. Growth of race due to overheating; 2. Fretting wear; 3. Improper initial fit; 4. Excessive shaft deflection; 5. Initially coarse shaft finish; and 6. Seal rub on inner race.

(d) Binding of the shaft: 1. Lubricant breakdown; 2. Contamination by abrasive or corrosive matter; 3. Housing distortion or out-of-round pinching bearing; 4. Uneven shimming of housing with loss of clearance; 5. Tight rubbing seals; 6. Preloaded bearings; 7. Cocked races; 8. Loss of clearance due to excessive tightening of adapter; 9. Thermal expansion of shaft or housing; and 10. Cage failure.

(e) Noisy bearing. 1. Lubrication breakdown, inadequate lubrication, stiff grease; 2. Contamination; 3. Pinched bearing; 4. Seal rubbing; 5. Loss of clearance and preloading; 6. Bearing slipping on shaft or in housing; 7. Flatted roller or ball; 8. Brinelling due to assembly abuse, handling, or shock loads; 9. Variation in size of rolling elements; 10. Out-of-round or lobular shaft; 11. Housing bore waviness; and 12. Chips or scores under bearing race seat.

(f) Displaced shaft: 1. Bearing wear; 2. Improper housing or closure assembly

3. Overheated and shifted bearing; 4. Inadequate shaft or housing shoulder; 5. Lubrication and cage failure permitting rolling elements to bunch; 6. Loosened retainer nut or adapter; 7. Excessive heat application in assembling inner race, causing growth and shifting on shaft; and 8. Housing pounding out.

(g) Lubricant leakage: 1. Overfilling of lubricant; 2. Grease churning due to use of too soft a consistency; 3. Grease deterioration due to excessive operating temperature; 4. Operating life longer than grease life (grease breakdown, aeration, and purging); 5. Seal wear; 6. Wrong shaft attitude (bearing seals designed for horizontal mounting only); 7. Seal failure; 8. Clogged breather; 9. Oil foaming due to churning or air flow through housing; 10. Gasket (O-ring) failure or misapplication; 11. Porous housing or closure; and 12. Lubricator set at wrong flow rate.

Load Ratings and Fatigue Life

Ball and Roller Bearing Life. — The performance of ball and roller bearings is a function of many variables. These include the bearing design, the characteristics of the material from which the bearings are made, the way in which they are manufactured, as well as many variables associated with their application. The only sure way to establish the satisfactory operation of a bearing selected for a specific application is by actual performance in the application. As this is often impractical, another basis is required to estimate the suitability of a particular bearing for a given application. Two factors are taken into consideration: the bearing fatigue life, and its ability to withstand static loading.

Life Criterion: Even if a ball or roller bearing is properly mounted, adequately lubricated, protected from foreign matter and not subjected to extreme operating conditions, it can ultimately fatigue. Under ideal conditions, the repeated stresses developed in the contact areas between the balls or rollers and the raceways eventually can result in the fatigue of the material which manifests itself with the spalling of the load-carrying surfaces. In most applications the fatigue life is the maximum useful life of a bearing.

Static Load Criterion: A static load is a load acting on a non-rotating bearing. Permanent deformations appear in balls or rollers and raceways under a static load of moderate magnitude and increase gradually with increasing load. The permissible static load is, therefore, dependent upon the permissible magnitude of permanent deformation. It has been found that for ball and roller bearings suitably manufactured from hardened alloy steel, deformations occurring under maximum contact stress of 4,000 megapascals (580,000 pounds per square inch) acting at the center of contact (in the case of roller bearings, of a uniformly loaded roller) do not greatly impair smoothness or friction. Depending on requirements for smoothness of operation, friction, or sound level, higher or lower static load limits may be tolerated.

Ball Bearing Types Covered. — AFBMA and American National Standard ANSI/AFBMA Std 9-1978 sets forth the method of determining ball bearing Rating Life and Static Load Rating and covers the following types:

1. *Radial, deep groove and angular contact ball bearings* whose inner ring raceways have a cross-sectional radius not larger than 52 percent of the ball diameter and whose outer ring raceways have a cross-sectional radius not larger than 53 percent of the ball diameter.

2. *Radial, self-aligning ball bearings* whose inner ring raceways have cross-sectional radii not larger than 53 percent of the ball diameter.

3. *Thrust ball bearings* whose washer raceways have cross-sectional radii not larger than 54 percent of the ball diameter.

4. *Double row, radial and angular contact ball bearings* and double direction thrust ball bearings are presumed to be symmetrical.

Limitations for Ball Bearings. — The following limitations apply:

1. *Truncated contact area.* This standard[1] may not be safely applied to ball bearings subjected to loading which causes the contact area of the ball with the raceway to be truncated by the raceway shoulder. This limitation depends strongly on details of bearing design which are not standardized.

2. *Material.* This standard applies only to ball bearings fabricated from hardened good quality steel.

3. *Types.* The f_c factors specified in the basic load rating formulas are valid only for those ball bearing types specified above.

4. *Lubrication.* The Rating Life calculated according to this standard is based on the assumption that the bearing is adequately lubricated. The determination of adequate lubrication depends upon the bearing application.

5. *Ring support and alignment.* The Rating Life calculated according to this standard assumes that the bearing inner and outer rings are rigidly supported and the inner and outer ring axes are properly aligned.

6. *Internal clearance.* The radial ball bearing Rating Life calculated according to this standard is based on the assumption that only a nominal interior clearance occurs in the mounted bearing at operating speed, load and temperature.

7. *High speed effects.* The Rating Life calculated according to this standard does not account for high speed effects such as ball centrifugal forces and gyroscopic moments. These effects tend to diminish fatigue life. Analytical evaluation of these effects frequently requires the use of high speed digital computation devices and hence is not covered in the standard.

8. *Groove radii.* If groove radii are smaller than those specified in the bearing types covered, the ability of a bearing to resist fatigue is not improved; however, it is diminished by the use of larger radii.

Ball Bearing Rating Life. — According to the Anti-Friction Bearing Manufacturers Association standards the Rating Life L_{10} of a group of apparently identical ball bearings is the life in millions of revolutions that 90 percent of the group will complete or exceed. For a single bearing, L_{10} also refers to the life associated with 90 percent reliability.

Radial and Angular Contact Ball Bearings: The magnitude of the Rating Life L_{10} in millions of revolutions, for a radial or angular contact ball bearing application is given by the formula:

$$L_{10} = \left(\frac{C}{P}\right)^3 \tag{1}$$

where C = basic load rating, newtons (pounds). See formulas (2) and (3).

P = equivalent radial load, newtons (pounds). See formula (4).

For radial and angular contact ball bearings with balls not larger than 25.4 mm (1 inch) in diameter, C is found by the formula:

$$C = f_c \, (i \cos \alpha)^{0.7} \, Z^{2/3} \, D^{1.8} \tag{2}$$

and with balls larger than 25.4 mm (1 inch) in diameter C is found by the formula:

$$C = 3.647 f_c \, (i \cos \alpha)^{0.7} \, Z^{2/3} \, D^{1.4} \quad \text{(metric)} \tag{3a}$$

$$C = f_c \, (i \cos \alpha)^{0.7} \, Z^{2/3} \, D^{1.4} \quad \text{(inch)} \tag{3b}$$

---

[1] All references to "standard" are to AFBMA and American National Standard "Load Ratings and Fatigue Life for Ball Bearings" (ANSI/AFBMA Std 9-1978).

where f_c = a factor which depends on the geometry of the bearing components, the accuracy to which the various bearing parts are made and the material. Values of f_c are given in Table 25.

i = number of rows of balls in the bearing

α = nominal contact angle, degrees

Z = number of balls per row in a radial or angular contact bearing

D = ball diameter, mm (inches)

The magnitude of the equivalent radial load, P, in newtons (pounds) for radial and angular contact ball bearings, under combined constant radial and constant thrust loads is given by the formula:

$$P = XF_r + YF_a \qquad (4)$$

where F_r = the applied radial load in newtons (pounds)

F_a = the applied axial load in newtons (pounds)

X = radial load factor as given in Table 26.

Y = axial load factor as given in Table 26.

Table 25. Values of f_c for Radial and Angular Contact Ball Bearings

$\dfrac{D \cos \alpha}{d_m}$	Single Row Radial Contact; Single and Double Row Angular Contact, Groove Type*		Double Row Radial Contact Groove Type		Self-Aligning	
	Values of f_c					
	Metric†	Inch‡	Metric†	Inch‡	Metric†	Inch‡
0.05	46.7	3550	44.2	3360	17.3	1310
0.06	49.1	3730	46.5	3530	18.6	1420
0.07	51.1	3880	48.4	3680	19.9	1510
0.08	52.8	4020	50.0	3810	21.1	1600
0.09	54.3	4130	51.4	3900	22.3	1690
0.10	55.5	4220	52.6	4000	23.4	1770
0.12	57.5	4370	54.5	4140	25.6	1940
0.14	58.8	4470	55.7	4230	27.7	2100
0.16	59.6	4530	56.5	4290	29.7	2260
0.18	59.9	4550	56.8	4310	31.7	2410
0.20	59.9	4550	56.8	4310	33.5	2550
0.22	59.6	4530	56.5	4290	35.2	2680
0.24	59.0	4480	55.9	4250	36.8	2790
0.26	58.2	4420	55.1	4190	38.2	2910
0.28	57.1	4340	54.1	4110	39.4	3000
0.30	56.0	4250	53.0	4030	40.3	3060
0.32	54.6	4160	51.8	3950	40.9	3110
0.34	53.2	4050	50.4	3840	41.2	3130
0.36	51.7	3930	48.9	3730	41.3	3140
0.38	50.0	3800	47.4	3610	41.0	3110
0.40	48.4	3670	45.8	3480	40.4	3070

* a. When calculating the basic load rating for a unit consisting of two similar, single row, radial contact ball bearings, in a duplex mounting, the pair is considered as one, double row, radial contact ball bearing.

b. When calculating the basic load rating for a unit consisting of two, similar, single row, angular contact ball bearings in a duplex mounting, "face-to-face" or "back-to-back," the pair is considered as one, double row, angular contact ball bearing.

c. When calculating the basic load rating for a unit consisting of two or more similar, single angular contact ball bearings mounted "in tandem," properly manufactured and mounted for equal load distribution, the rating of the combination is the number of bearings to the 0.7 power times the rating of a single row ball bearing. If the unit may be treated as a number of individually interchangeable single row bearings, this footnote c does not apply.

† Use to obtain C in newtons when D is given in mm.

‡ Use to obtain C in pounds when D is given in inches.

Table 26. Values of X and Y for Computing Equivalent Radial Load P of Radial and Angular Contact Ball Bearings

Contact Angle, α	Table Entering Factors*			Single Row Bearings† $\dfrac{F_a}{F_r} > e$			Double Row Bearings			
							$\dfrac{F_a}{F_r} \leq e$		$\dfrac{F_a}{F_r} > e$	
	F_a/C_o	F_a/iZD^2 Metric Units	Inch Units	e	X	Y	X	Y	X	Y
RADIAL CONTACT GROOVE BEARINGS										
0°	0.014	0.172	25	0.19		2.30				2.30
	0.028	0.345	50	0.22		1.99				1.99
	0.056	0.689	100	0.26		1.71				1.71
	0.084	1.03	150	0.28		1.56				1.55
	0.11	1.38	200	0.30	0.56	1.45	1	0	0.56	1.45
	0.17	2.07	300	0.34		1.31				1.31
	0.28	3.45	500	0.38		1.15				1.15
	0.42	5.17	750	0.42		1.04				1.04
	0.56	6.89	1000	0.44		1.00				1.00
ANGULAR CONTACT GROOVE BEARINGS (iF_a/C_o , F_a/ZD^2)										
5°	0.014	0.172	25	0.23	For this type use the X, Y, and e values applicable to single row radial contact bearings			2.78		3.74
	0.028	0.345	50	0.26				2.40		3.23
	0.056	0.689	100	0.30				2.07		2.78
	0.085	1.03	150	0.34				1.87		2.52
	0.11	1.38	200	0.36			1	1.75	0.78	2.36
	0.17	2.07	300	0.40				1.58		2.13
	0.28	3.45	500	0.45				1.39		1.87
	0.42	5.17	750	0.50				1.26		1.69
	0.56	6.89	1000	0.52				1.21		1.63
10°	0.014	0.172	25	0.29		1.88		2.18		3.06
	0.029	0.345	50	0.32		1.71		1.98		2.78
	0.057	0.689	100	0.36		1.52		1.76		2.47
	0.086	1.03	150	0.38		1.41		1.63		2.20
	0.11	1.38	200	0.40	0.46	1.34	1	1.55	0.75	2.18
	0.17	2.07	300	0.44		1.23		1.42		2.00
	0.29	3.45	500	0.49		1.10		1.27		1.79
	0.43	5.17	750	0.54		1.01		1.17		1.64
	0.57	6.89	1000	0.54		1.00		1.16		1.63
15°	0.015	0.172	25	0.38		1.47		1.65		2.39
	0.029	0.345	50	0.40		1.40		1.57		2.28
	0.058	0.689	100	0.43		1.30		1.46		2.11
	0.087	1.03	150	0.46		1.23		1.38		2.00
	0.12	1.38	200	0.47	0.44	1.19	1	1.34	0.72	1.93
	0.17	2.07	300	0.50		1.12		1.26		1.82
	0.29	3.45	500	0.55		1.02		1.14		1.66
	0.44	5.17	750	0.56		1.00		1.12		1.63
	0.58	6.89	1000	0.56		1.00		1.12		1.63
20°	...	...	...	0.57	0.43	1.00	1	1.09	0.70	1.63
25°	...	...	...	0.68	0.41	0.87	1	0.92	0.67	1.41
30°	...	...	...	0.80	0.39	0.76	1	0.78	0.63	1.24
35°	...	...	...	0.95	0.37	0.66	1	0.66	0.60	1.07
40°	...	...	...	1.14	0.35	0.57	1	0.55	0.57	0.98
Self-aligning Ball Bearings				$1.5 \tan\alpha$	0.40	$0.4 \cot\alpha$	1	$0.42 \cot\alpha$	0.65	$0.65 \cot\alpha$

* Symbol definitions are given on the following page. † For single row bearings when $F_a/F_r \leqq e$, use $X = 1$, $Y = 0$. Two similar, single row, angular contact ball bearings mounted face-to-face or back-to-back are considered as one double row, angular contact bearing.

Values of X, Y, and e for a load or contact angle other than shown are obtained by linear interpolation.

Values of X, Y, and e do not apply to filling slot bearings for applications in which ball-raceway contact areas project substantially into the filling slot under load.

Table 26 Symbol Definitions: F_a is the applied axial load in newtons (pounds); C_o is the static load rating in newtons (pounds) of the bearing under consideration and is found by formula (20); i is the number of rows of balls in the bearing; Z is the number of balls per row in a radial or angular contact bearing or the number of balls in a single row, single direction thrust bearing; D is the ball diameter in millimeters (inches); and F_r is the applied radial load in newtons (pounds).

Thrust Ball Bearings: The magnitude of the Rating Life L_{10} in millions of revolutions for a thrust ball bearing application is given by the formula:

$$L_{10} = \left(\frac{C_a}{P_a}\right)^3 \tag{5}$$

where C_a = the basic load rating, newtons (pounds). See formulas (6) to (10).
P_a = equivalent thrust load, newtons (pounds). See formula (11).

For single row, single and double direction, thrust ball bearing with balls not larger than 25.4 mm (1 inch) in diameter, C_a is found by the formulas:

for $\alpha = 90$ degrees, $\quad C_a = f_c Z^{2/3} D^{1.8}$ (6)

for $\alpha \neq 90$ degrees, $\quad C_a = f_c (\cos \alpha)^{0.7} Z^{2/3} D^{1.8} \tan \alpha$ (7)

and with balls larger than 25.4 mm (1 inch) in diameter, C_a is found by the formulas:

for $\alpha = 90$ degrees, $\quad C_a = 3.647 f_c Z^{2/3} D^{1.4}$ (metric) (8a)

$\qquad\qquad\qquad\qquad C_a = f_c Z^{2/3} D^{1.4}$ (inch) (8b)

for $\alpha \neq 90$ degrees, $\quad C_a = 3.647 f_c (\cos \alpha)^{0.7} Z^{2/3} D^{1.4} \tan \alpha$ (metric) (9a)

$\qquad\qquad\qquad\qquad C_a = f_c (\cos \alpha)^{0.7} Z^{2/3} D^{1.4} \tan \alpha$ (inch) (9b)

where f_c = a factor which depends on the geometry of the bearing components, the accuracy to which the various bearing parts are made, and the material. Values of f_c are given in Table 27.

Table 27. Values of f_c for Thrust Ball Bearings

$\dfrac{D}{d_m}$	$\alpha = 90°$		$D \cos \alpha / d_m$	$\alpha = 45°$		$\alpha = 60°$		$\alpha = 75°$	
	Metric†	Inch‡		Metric†	Inch‡	Metric†	Inch‡	Metric†	Inch‡
0.01	36.7	2790	0.01	42.1	3200	39.2	2970	37.3	2840
0.02	45.2	3430	0.02	51.7	3930	48.1	3650	45.9	3490
0.03	51.1	3880	0.03	58.2	4430	54.2	4120	51.7	3930
0.04	55.7	4230	0.04	63.3	4810	58.9	4470	56.1	4260
0.05	59.5	4520	0.05	67.3	5110	62.6	4760	59.7	4540
0.06	62.9	4780	0.06	70.7	5360	65.8	4990	62.7	4760
0.07	65.8	5000	0.07	73.5	5580	68.4	5190	65.2	4950
0.08	68.5	5210	0.08	75.9	5770	70.7	5360	67.3	5120
0.09	71.0	5390	0.09	78.0	5920	72.6	5510	69.2	5250
0.10	73.3	5570	0.10	79.7	6050	74.2	5630	70.7	5370
0.12	77.4	5880	0.12	82.3	6260	76.6	5830	. . .	. . .
0.14	81.1	6160	0.14	84.1	6390	78.3	5950	. . .	. . .
0.16	84.4	6410	0.16	85.1	6470	79.2	6020	. . .	. . .
0.18	87.4	6640	0.18	85.5	6500	79.6	6050	. . .	. . .
0.20	90.2	6854	0.20	85.4	6490	79.5	6040	. . .	. . .
0.22	92.8	7060	0.22	84.9	6450	. . .	. . .	. . .	. . .
0.24	95.3	7240	0.24	84.0	6380	. . .	. . .	. . .	. . .
0.26	97.6	7410	0.26	82.8	6290	. . .	. . .	. . .	. . .
0.28	99.8	7600	0.28	81.3	6180	. . .	. . .	. . .	. . .
0.30	101.9	7750	0.30	79.6	6040	. . .	. . .	. . .	. . .
0.32	103.9	7900		. . .	. . .	. . .	. . .	. . .	. . .
0.34	105.8	8050		. . .	. . .	. . .	. . .	. . .	. . .

† Use to obtain C_a in newtons when D is given in mm.
‡ Use to obtain C_a in pounds when D is given in inches.

Z = number of balls per row in a single row, single direction thrust ball bearing

D = ball diameter, mm (inches)

α = nominal contact angle, degrees

For thrust ball bearings with two or more rows of similar balls carrying loads in the same direction, the basic load rating, C_a, in newtons (pounds) is found by the formula:

$$C_a = (Z_1 + Z_2 + \ldots Z_n) \left[\left(\frac{Z_1}{C_{a1}} \right)^{10/3} + \left(\frac{Z_2}{C_{a2}} \right)^{10/3} + \ldots \left(\frac{Z_n}{C_{an}} \right)^{10/3} \right]^{-0.3} \quad (10)$$

where $Z_1, Z_2 \ldots Z_n$ = number of balls in respective rows of a single-direction multi-row thrust ball bearing.

$C_{a1}, C_{a2} \ldots C_{an}$ = basic load rating per row of a single-direction, multi-row thrust ball bearing, each calculated as a single-row bearing with $Z_1, Z_2 \ldots Z_n$ balls, respectively.

The magnitude of the equivalent thrust load, P_a, in newtons (pounds) for thrust ball bearings with $\alpha \neq 90$ degrees under combined constant thrust and constant radial loads is found by the formula:

$$P_a = XF_r + YF_a \quad (11)$$

where F_r = the applied radial load in newtons (pounds).

F_a = the applied axial load in newtons (pounds).

X = radial load factor as given in Table 28.

Y = axial load factor as given in Table 28.

Table 28. Values of X and Y for Computing Equivalent Thrust Load P_a for Thrust Ball Bearings

Contact Angle α	e	Single Direction Bearings		Double Direction Bearings			
		$\frac{F_a}{F_r} > e$		$\frac{F_a}{F_r} \leq e$		$\frac{F_a}{F_r} > e$	
		X	Y	X	Y	X	Y
45°	1.25	0.66	1	1.18	0.59	0.66	1
60°	2.17	0.92	1	1.90	0.54	0.92	1
75°	4.67	1.66	1	3.89	0.52	1.66	1

For $\alpha = 90°$, $F_r = 0$ and $Y = 1$.

Roller Bearing Types Covered. — This standard[1] applies to *cylindrical, tapered and self-aligning radial and thrust roller bearings* and to *needle roller bearings*. These bearings are presumed to be within the size ranges shown in the AFBMA dimensional standards, of good quality and produced in accordance with good manufacturing practice.

Roller bearings vary considerably in design and execution. Since small differences in relative shape of contacting surfaces may account for distinct differences in load carrying ability, this standard does not attempt to cover all design variations, rather it applies to basic roller bearing designs.

[1] All references to "standard" are to AFBMA and American National Standard "Load Ratings and Fatigue Life for Roller Bearings" (ANSI/AFBMA Std 11-1978).

Limitations for Roller Bearings. — The following limitations apply:

1. *Truncated contact area.* This standard may not be safely applied to roller bearings subjected to application conditions which cause the contact area of the roller with the raceway to be severely truncated by the edge of the raceway or roller.

2. *Stress concentrations.* A cylindrical, tapered or self-aligning roller bearing must be expected to have a basic load rating less than that obtained using a value of f_c taken from Table 29 or Table 31 if, under load, a stress concentration is present in some part of the roller-raceway contact. Such stress concentrations occur in the center of nominal point contacts, at the contact extremities for line contacts and at inadequately blended junctions of a rolling surface profile. Stress concentrations can also occur if the rollers are not accurately guided such as in bearings without cages and bearings not having rigid integral flanges. Values of f_c given in Tables 29 and 31 are based upon bearings manufactured to achieve optimized contact. For no bearing type or execution will the factor f_c be greater than that obtained in Tables 29 and 31.

3. *Material.* This standard applies only to roller bearings fabricated from hardened, good quality steel.

4. *Lubrication.* Rating Life calculated according to this standard is based on the assumption that the bearing is adequately lubricated. Determination of adequate lubrication depends upon the bearing application.

5. *Ring support and alignment.* Rating Life calculated according to this standard assumes that the bearing inner and outer rings are rigidly supported, and that the inner and outer ring axes are properly aligned.

6. *Internal clearance.* Radial roller bearing Rating Life calculated according to this standard is based on the assumption that only a nominal internal clearance occurs in the mounted bearing at operating speed, load, and temperature.

7. *High speed effects.* The Rating Life calculated according to this standard does not account for high speed effects such as roller centrifugal forces and gyroscopic moments: These effects tend to diminish fatigue life. Analytical evaluation of these effects frequently requires the use of high speed digital computation devices and hence, cannot be included.

Roller Bearing Rating Life. — The Rating Life L_{10} of a group of apparently identical roller bearings is the life in millions of revolutions that 90 percent of the group will complete or exceed. For a single bearing, L_{10} also refers to the life associated with 90 percent reliability.

Radial Roller Bearings: The magnitude of the Rating Life, L_{10}, in millions of revolutions, for a radial roller bearing application is given by the formula:

$$L_{10} = \left(\frac{C}{P}\right)^{10/3} \tag{12}$$

where C = the basic load rating in newtons (pounds). See formula (13).

P = equivalent radial load in newtons (pounds). See formula (14).

For radial roller bearings, C is found by the formula:

$$C = f_c \, (il_{eff} \cos \alpha)^{7/9} \, Z^{3/4} \, D^{29/27} \tag{13}$$

where f_c = a factor which depends on the geometry of the bearing components, the accuracy to which the various bearing parts are made and the material. Maximum values of f_c are given in Table 29.

i = number of rows of rollers in the bearing.

l_{eff} = effective length, mm (inches).

α = nominal contact angle, degrees.

Z = number of rollers per row in a radial roller bearing.

D = roller diameter, mm (inches) (mean diameter for a tapered roller, major diameter for a spherical roller).

Table 29. Values of f_c for Radial Roller Bearings

$\dfrac{D \cos \alpha}{d_m}$	f_c Metric†	f_c Inch‡	$\dfrac{D \cos \alpha}{d_m}$	f_c Metric†	f_c Inch‡	$\dfrac{D \cos \alpha}{d_m}$	f_c Metric†	f_c Inch‡
0.01	52.1	4680	0.18	88.8	7980	0.35	79.5	7140
0.02	60.8	5460	0.19	88.8	7980	0.36	78.6	7060
0.03	66.5	5970	0.20	88.7	7970	0.37	77.6	6970
0.04	70.7	6350	0.21	88.5	7950	0.38	76.7	6890
0.05	74.1	6660	0.22	88.2	7920	0.39	75.7	6800
0.06	76.9	6910	0.23	87.9	7890	0.40	74.6	6700
0.07	79.2	7120	0.24	87.5	7850	0.41	73.6	6610
0.08	81.2	7290	0.25	87.0	7810	0.42	72.5	6510
0.09	82.8	7440	0.26	86.4	7760	0.43	71.4	6420
0.10	84.2	7570	0.27	85.8	7710	0.44	70.3	6320
0.11	85.4	7670	0.28	85.2	7650	0.45	69.2	6220
0.12	86.4	7760	0.29	84.5	7590	0.46	68.1	6120
0.13	87.1	7830	0.30	83.8	7520	0.47	67.0	6010
0.14	87.7	7880	0.31	83.0	7450	0.48	65.8	5910
0.15	88.2	7920	0.32	82.2	7380	0.49	64.6	5810
0.16	88.5	7950	0.33	81.3	7300	0.50	63.5	5700
0.17	88.7	7970	0.34	80.4	7230	...	...	...

† Use to obtain C in newtons when l_{eff} and D are given in mm.
‡ Use to obtain C in pounds when l_{eff} and D are given in inches.

When rollers are longer than $2.5D$, a reduction in the f_c value must be anticipated. In this case, the bearing manufacturer may be expected to establish load ratings accordingly.

In applications where rollers operate directly on a shaft surface or a housing surface, such a surface must be equivalent in all respects to the raceway it replaces to achieve the basic load rating of the bearing.

When calculating the basic load rating for a unit consisting of two or more similar single-row bearings mounted "in tandem," properly manufactured and mounted for equal load distribution, the rating of the combination is the number of bearings to the 7/9 power times the rating of a single-row bearing. If, for some technical

Table 30. Values of X and Y for Computing Equivalent Radial Load P for Radial Roller Bearing

Bearing Type	$\dfrac{F_a}{F_r} \leq e^*$ X	$\dfrac{F_a}{F_r} \leq e^*$ Y	$\dfrac{F_a}{F_r} > e^*$ X	$\dfrac{F_a}{F_r} > e^*$ Y
Self-Aligning and Tapered Roller Bearings† $\alpha \neq 0°$	Single Row Bearings			
	1	0	0.4	0.4 cot α
	Double Row Bearings†			
	1	0.45 cot α	0.67	0.67 cot α

* $e = 1.5 \tan \alpha$.
† For $\alpha = 0°$, $F_a = 0$ and $X = 1$.

reason, the unit may be treated as a number of individually interchangeable single-row bearings, this consideration does not apply.

The magnitude of the equivalent radial load, P, in newtons (pounds), for radial roller bearings, under combined constant radial and constant thrust loads is given by the formula:

$$P = XF_r + YF_a \tag{14}$$

where F_r = the applied radial load in newtons (pounds).
F_a = the applied axial load in newtons (pounds).
X = radial load factor as given in Table 30.
Y = axial load factor as given in Table 30.

Roller bearings are generally designed to achieve optimized contact; however, they usually support loads other than the loading at which optimized contact is maintained. The 10/3 exponent in Rating Life formulas (12) and (15) was selected to yield satisfactory Rating Life estimates for a broad spectrum from light to heavy loading. When loading exceeds that which develops optimized contact, e.g., loading greater than $C/4$ to $C/2$ or $C_a/4$ to $C_a/2$, the user should consult the bearing manufacturer to establish the adequacy of the Rating Life formulas for the particular application.

Thrust Roller Bearings: The magnitude of the Rating Life, L_{10}, in millions of revolutions for a thrust roller bearing application is given by the formula:

$$L_{10} = \left(\frac{C_a}{P_a}\right)^{10/3} \tag{15}$$

where C_a = basic load rating, newtons (pounds). See formulas (16) to (18).
P_a = equivalent thrust load, newtons (pounds). See formula (19).

For single row, single and double direction, thrust roller bearings, the magnitude of the basic load rating, C_a, in newtons (pounds), is found by the formulas:

$$\text{for } \alpha = 90°, \ C_a = f_c l_{eff}^{7/9} \ Z^{3/4} \ D^{29/27} \tag{16}$$

$$\text{for } \alpha \neq 90°, \ C_a = f_c \ (l_{eff} \cos \alpha)^{7/9} \ Z^{3/4} \ D^{29/27} \tan \alpha \tag{17}$$

where f_c = a factor which depends on the geometry of the bearing components, the accuracy to which the various parts are made, and the material. Values of f_c are given in Table 31.
l_{eff} = effective length, mm (inches)
Z = number of rollers in a single row, single direction, thrust roller bearing
D = roller diameter, mm (inches) (mean diameter for a tapered roller, major diameter for a spherical roller)
α = nominal contact angle, degrees

For thrust roller bearings with two or more rows of rollers carrying loads in the same direction the magnitude of C_a is found by the formula:

$$C_a = (Z_1 l_{eff1} + Z_2 l_{eff2} \ldots Z_n l_{effn}) \left\{ \left[\frac{Z_1 l_{eff1}}{C_{a1}}\right]^{9/2} + \left[\frac{Z_2 l_{eff2}}{C_{a2}}\right]^{9/2} + \ldots \right.$$

$$\left. \left[\frac{Z_n l_{effn}}{C_{an}}\right]^{9/2} \right\}^{-2/9} \tag{18}$$

Where $Z_1, Z_2 \ldots Z_n$ = the number of rollers in respective rows of a single direction, multi-row bearing

Table 31. Values of f_c for Thrust Roller Bearings

$\dfrac{D \cos \alpha}{d_m}$	45° < α < 60°		60° < α < 75°		75° ≤ α < 90°		$\dfrac{D}{d_m}$	α = 90°	
	f_c							f_c	
	Metric†	Inch‡	Metric†	Inch‡	Metric†	Inch‡		Metric†	Inch‡
0.01	109.7	9840	107.1	9610	105.6	9470	0.01	105.4	9500
0.02	127.8	11460	124.7	11180	123.0	11030	0.02	122.9	11000
0.03	139.5	12510	136.2	12220	134.3	12050	0.03	134.5	12100
0.04	148.3	13300	144.7	12980	142.8	12810	0.04	143.4	12800
0.05	155.2	13920	151.5	13590	149.4	13400	0.05	150.7	13200
0.06	160.9	14430	157.0	14080	154.9	13890	0.06	156.9	14100
0.07	165.6	14850	161.6	14490	159.4	14300	0.07	162.4	14500
0.08	169.5	15200	165.5	14840	163.2	14640	0.08	167.2	15100
0.09	172.8	15500	168.7	15130	166.4	14930	0.09	171.7	15400
0.10	175.5	15740	171.4	15370	169.0	15160	0.10	175.7	15900
0.12	179.7	16120	175.4	15730	173.0	15520	0.12	183.0	16300
0.14	182.3	16350	177.9	15960	175.5	15740	0.14	189.4	17000
0.16	183.7	16480	179.3	16080	...	...	0.16	195.1	17500
0.18	184.1	16510	179.7	16120	...	...	0.18	200.3	18000
0.20	183.7	16480	179.3	16080	...	...	0.20	205.0	18500
0.22	182.6	16380	...	...	...	...	0.22	209.4	18800
0.24	180.9	16230	...	...	...	...	0.24	213.5	19100
0.26	178.7	16030	...	...	...	...	0.26	217.3	19600
0.28	...	...	...	...	...	...	0.28	220.9	19900
0.30	...	...	...	...	...	...	0.30	224.3	20100

† Use to obtain C_a in newtons when l_{eff} and D are given in mm.
‡ Use to obtain C_a in pounds when l_{eff} and D are given in inches.

$C_{a1}, C_{a2} \ldots C_{an}$ = the basic load rating per row of a single direction, multi-row, thrust roller bearing, each calculated as a single row bearing with $Z_1, Z_2 \ldots Z_n$ rollers respectively.

$l_{eff1}, l_{eff2} \ldots l_{effn}$ = effective length, mm (inches), or rollers in the respective rows.

In applications where rollers operate directly on a surface supplied by the user, such a surface must be equivalent in all respects to the washer raceway it replaces to achieve the basic load rating of the bearing.

In case the bearing is so designed that several rollers are located on a common axis, these rollers are considered as one roller of a length equal to the total effective length of contact of the several rollers. Rollers as defined above, or portions thereof which contact the same washer-raceway area, belong to one row.

When the ratio of the individual roller effective length to the pitch diameter (at which this roller operates) is too large, a reduction of the f_c value must be anticipated due to excessive slip in the roller-raceway contact.

When calculating the basic load rating for a unit consisting of two or more similar single row bearings mounted "in tandem," properly manufactured and mounted for equal load distribution, the rating of the combination is defined by formula (18). If, for some technical reason, the unit may be treated as a number of individually interchangeable single-row bearings, this consideration does not apply.

The magnitude of the equivalent thrust load, P_a, in pounds, for thrust roller bearings with α not equal to 90 degrees under combined constant thrust and constant radial loads is given by the formula:

$$P_a = XF_r + YF_a \qquad (19)$$

where F_r = applied radial load, newtons (pounds).
F_a = applied axial load, newtons (pounds).
X = radial load factor as given in Table 32.
Y = axial load factor as given in Table 32.

Table 32. Values of X and Y for Computing Equivalent Thrust Load P_a for Thrust Roller Bearings

| Bearing Type | Single Direction Bearings | | Double Direction Bearings | | | |
| | $\frac{F_a}{F_r} > e*$ | | $\frac{F_a}{F_r} \leqq e*$ | | $\frac{F_a}{F_r} > e*$ | |
	X	Y	X	Y	X	Y
Self-Aligning Tapered Thrust Roller Bearings† $\alpha \neq 0$	tan α	1	1.5 tan α	0.67	tan α	1

* $e = 1.5 \tan \alpha$.
† For $\alpha = 90°$, $F_r = 0$ and $Y = 1$.

Life Adjustment Factors. — In certain applications of ball or roller bearings it is desirable to specify life for a reliability other than 90 per cent. In other cases the bearings may be fabricated from special bearing steels such as vacuum-degassed and vacuum-melted steels, and improved processing techniques. Finally, application conditions may indicate other than normal lubrication, load distribution, or temperature. For such conditions a series of life adjustment factors may be applied to the fatigue life formula. This is fully explained in AFBMA and American National Standard "Load Ratings and Fatigue Life for Ball Bearings" (ANSI/AFBMA Std 9-1978) and AFBMA and American National Standard "Load Ratings and Fatigue Life for Roller Bearings" (ANSI/AFBMA Std 11-1978). In addition to consulting these standards it may be advantageous to also obtain information from the bearing manufacturer.

Life Adjustment Factor for Reliability. — For certain applications, it is desirable to specify life for a reliability greater than 90 per cent which is the basis of the Rating Life.

To determine the bearing life of ball or roller bearings for reliability greater than 90 per cent, the Rating Life must be adjusted by a factor a_1 such that $L_n = a_1 L_{10}$. For a reliability of 95 per cent, designated as L_5, the life adjustment factor a_1 is 0.62; for 96 per cent, L_4, a_1 is 0.53; for 97 per cent, L_3, a_1 is 0.44; for 98 per cent, L_2, a_1 is 0.33; and for 99 per cent, L_1, a_1 is 0.21.

Life Adjustment Factor for Material. — For certain types of ball or roller bearings which incorporate improved materials and processing, the Rating Life can be adjusted by a factor a_2 such that $L_{10}' = a_2 L_{10}$. Factor a_2 depends upon steel analysis, metallurgical processes, forming methods, heat treatment, and manufacturing methods in general. Ball and roller bearings fabricated from consumable vacuum remelted steels and certain other special analysis steels, have demonstrated extraordinarily long endurance. These steels are of exceptionally high quality, and bearings fabricated from these are usually considered special manufacture. Generally, a_2 values for such steels can be obtained from the bearing manufacturer. However, all of the specified limitations and qualifications for the application of the Rating Life formulas still apply.

Life Adjustment Factor for Application Condition. — Application conditions which affect ball or roller bearing life include: 1. lubrication; 2. load distribution (including effects of clearance, misalignment, housing and shaft stiffness, type of loading, and thermal gradients); and 3. temperature. Items 2 and 3 require special analytical and experimental techniques, therefore the user should consult the bearing manufacturer for evaluations and recommendations.

Operating conditions where the factor a_3 might be less than 1 include: (a) exceptionally low values of Nd_m (rpm times pitch diameter, in mm); e.g., $Nd_m < 10,000$; (b) lubricant viscosity at less than 70 SSU for ball bearings and 100 SSU for roller bearings at operating temperature; (c) excessively high operating temperatures. When a_3 is less than 1 it may not be assumed that the deficiency in lubrication can be overcome by using an improved steel. When this factor is applied, $L_{10}' = a_3 L_{10}$.

In most ball and roller bearing applications, lubrication is required to separate the rolling surfaces, i.e., rollers and raceways, to reduce the retainer-roller and retainer-land friction and sometimes to act as a coolant to remove heat generated by the bearing.

Factor Combinations. — A fatigue life formula embodying the foregoing life adjustment factors is $L_{10}' = a_1 a_2 a_3 L_{10}$. Indiscriminate application of the life adjustment factors in this formula may lead to serious overestimation of bearing endurance, since fatigue life is only one criterion for bearing selection. Care must be exercised to select bearings which are of sufficient size for the application.

Ball Bearing Static Load Rating. — For ball bearings suitably manufactured from hardened alloy steels, the static radial load rating is that uniformly distributed static radial bearing load which produces a maximum contact stress of 4,000 megapascals (580,000 pounds per square inch). In the case of a single row, angular contact ball bearing, the static radial load rating refers to the radial component of that load which causes a purely radial displacement of the bearing rings in relation to each other. The static axial load rating is that uniformly distributed static centric axial load which produces a maximum contact stress of 4,000 megapascals (580,000 pounds per square inch).

Radial and Angular Contact Groove Ball Bearings: The magnitude of the static load rating C_o in newtons (pounds) for radial ball bearings is found by the formula:

$$C_o = f_o \, i \, Z \, D^2 \cos \alpha \qquad (20)$$

where f_o = a factor for different kinds of ball bearings given in Table 33.
$\quad i$ = number of rows of balls in bearing.
$\quad Z$ = number of balls per row.
$\quad D$ = ball diameter, mm (inches).
$\quad \alpha$ = nominal contact angle, degrees.

This formula applies to bearings with a cross sectional raceway groove radius not larger than $0.52 \, D$ in radial and angular contact groove ball bearing inner rings and $0.53 \, D$ in radial and angular contact groove ball bearing outer rings and self-aligning ball bearing inner rings.

The load carrying ability of a ball bearing is not necessarily increased by the use of a smaller groove radius but is reduced by the use of a larger radius than those indicated above.

Radial or Angular Contact Ball Bearing Combinations: The basic static load rating for two similar single row radial or angular contact ball bearings mounted side by side on the same shaft such that they operate as a unit (duplex mounting) in "back-to-back" or "face-to-face" arrangement is two times the rating of one single row bearing.

Table 33. Values of f_o for Calculating Static Load Rating for Ball Bearings

$\dfrac{D \cos \alpha}{d_m}$	Radial and Angular Contact Groove Type		Radial Self-Aligning		Thrust	
	Metric†	Inch‡	Metric†	Inch‡	Metric†	Inch‡
0.00	12.7	1850	1.3	187	51.9	7730
0.01	13.0	1880	1.3	191	52.6	7620
0.02	13.2	1920	1.3	195	51.7	7500
0.03	13.5	1960	1.4	198	50.9	7380
0.04	13.7	1990	1.4	202	50.2	7280
0.05	14.0	2030	1.4	206	49.6	7190
0.06	14.3	2070	1.5	210	48.9	7090
0.07	14.5	2100	1.5	214	48.3	7000
0.08	14.7	2140	1.5	218	47.6	6900
0.09	14.5	2110	1.5	222	46.9	6800
0.10	14.3	2080	1.6	226	46.4	6730
0.11	14.1	2050	1.6	231	45.9	6660
0.12	13.9	2020	1.6	235	45.5	6590
0.13	13.6	1980	1.7	239	44.7	6480
0.14	13.4	1950	1.7	243	44.0	6380
0.15	13.2	1920	1.7	247	43.3	6280
0.16	13.0	1890	1.7	252	42.6	6180
0.17	12.7	1850	1.8	256	41.9	6070
0.18	12.5	1820	1.8	261	41.2	5970
0.19	12.3	1790	1.8	265	40.4	5860
0.20	12.1	1760	1.9	269	39.7	5760
0.21	11.9	1730	1.9	274	39.0	5650
0.22	11.6	1690	1.9	278	38.3	5550
0.23	11.4	1660	2.0	283	37.5	5440
0.24	11.2	1630	2.0	288	37.0	5360
0.25	11.0	1600	2.0	293	36.4	5280
0.26	10.8	1570	2.1	297	35.8	5190
0.27	10.6	1540	2.1	302	35.0	5080
0.28	10.4	1510	2.1	307	34.4	4980
0.29	10.3	1490	2.1	311	33.7	4890
0.30	10.1	1460	2.2	316	33.2	4810
0.31	9.9	1440	2.2	321	32.7	4740
0.32	9.7	1410	2.3	326	32.0	4640
0.33	9.5	1380	2.3	331	31.2	4530
0.34	9.3	1350	2.3	336	30.5	4420
0.35	9.1	1320	2.4	341	30.0	4350
0.36	8.9	1290	2.4	346	29.5	4270
0.37	8.7	1260	2.4	351	28.8	4170
0.38	8.5	1240	2.5	356	28.0	4060
0.39	8.3	1210	2.5	361	27.2	3950
0.40	8.1	1180	2.5	367	26.8	3880
0.41	8.0	1160	2.6	372	26.2	3800
0.42	7.8	1130	2.6	377	25.7	3720
0.43	7.6	1100	2.6	383	25.1	3640
0.44	7.4	1080	2.7	388	24.6	3560
0.45	7.2	1050	2.7	393	24.0	3480
0.46	7.1	1030	2.8	399	23.5	3400
0.47	6.9	1000	2.8	404	22.9	3320
0.48	6.7	977	2.8	410	22.4	3240
0.49	6.6	952	2.9	415	21.8	3160
0.50	6.4	927	2.9	421	21.2	3080

Note: Based on modulus of elasticity = 2.07×10^5 megapascals (30×10^6 pounds per square inch) and Poisson's ratio = 0.3.   † Use to obtain C_o or C_{oa} in newtons when D is given in mm.   ‡ Use to obtain C_o or C_{oa} in pounds when D is given in inches.

The basic static radial load rating for two or more single row radial or angular contact ball bearings mounted side by side on the same shaft such that they operate as a unit (duplex or stack mounting) in "tandem" arrangement, properly manufactured and mounted for equal load distribution, is the number of bearings times the rating of one single row bearing.

Thrust Ball Bearings: The magnitude of the static load rating C_{oa} for thrust ball bearings is found by the formula:

$$C_{oa} = f_o \, Z \, D^2 \sin \alpha \qquad (21)$$

where f_o = a factor given in Table 33.

Z = number of balls carrying the load in one direction.

D = ball diameter, mm (inches).

α = nominal contact angle, degrees.

This formula applies to thrust ball bearings with a cross sectional raceway radius not larger than $0.54\,D$. The load carrying ability of a bearing is not necessarily increased by use of a smaller radius, but is reduced by use of a larger radius.

Roller Bearing Static Load Rating. — For roller bearings suitably manufactured from hardened alloy steels, the static radial load rating is that uniformly distributed static radial bearing load which produces a maximum contact stress of 4,000 megapascals (580,000 pounds per square inch) acting at the center of contact of the most heavily loaded rolling element. The static axial load rating is that uniformly distributed static centric axial load which produces a maximum contact stress of 4,000 megapascals (580,000 pounds per square inch) acting at the center of contact of each rolling element.

Radial Roller Bearings: The magnitude of the static load rating C_o in newtons (pounds) for radial roller bearings is found by the formulas:

$$C_o = 44 \left(1 - \frac{D \cos \alpha}{d_m} \right) i \, Z \, l_{eff} \, D \cos \alpha \quad \text{(metric)} \qquad (22a)$$

$$C_o = 6430 \left(1 - \frac{D \cos \alpha}{d_m} \right) i \, Z \, l_{eff} \, D \cos \alpha \quad \text{(inch)} \qquad (22b)$$

where D = roller diameter, mm (inches); mean diameter for a tapered roller and major diameter for a spherical roller.

d_m = mean pitch diameter of the roller complement, mm (inches).

i = number of rows of rollers in bearing.

Z = number of rollers per row.

l_{eff} = effective length, mm (inches); overall roller length minus roller chamfers or minus grinding undercuts at the ring where contact is shortest.

α = nominal contact angle, degrees.

Radial Roller Bearing Combinations: The static load rating for two similar single row roller bearings mounted side by side on the same shaft such that they operate as a unit is two times the rating of one single row bearing.

The static radial load rating for two or more similar single row roller bearings mounted side by side on the same shaft such that they operate as a unit (duplex or stack mounting) in "tandem" arrangement, properly manufactured and mounted for equal load distribution, is the number of bearings times the rating of one single row bearing.

Thrust Roller Bearings: The magnitude of the static load rating C_{oa} in newtons (pounds) for thrust roller bearings is found by the formulas:

$$C_{oa} = 220 \left(1 - \frac{D \cos \alpha}{d_m} \right) Z \, l_{eff} \, D \sin \alpha \quad \text{(metric)} \qquad (23a)$$

$$C_{oa} = 32150 \left(1 - \frac{D \cos \alpha}{d_m} \right) Z \, l_{eff} \, D \sin \alpha \quad \text{(inch)} \qquad (23b)$$

where the symbol definitions are the same as for formulas (22a) and (22b).

Thrust Roller Bearing Combination: The static axial load rating for two or more similar single direction thrust roller bearings mounted side by side on the same shaft such that they operate as a unit (duplex or stack mounting) in "tandem" arrangement, properly manufactured and mounted for equal load distribution, is the number of bearings times the rating of one single direction bearing. The accuracy of this formula decreases in the case of single direction bearings when $F_r > 0.44 F_a \cot \alpha$ where F_r is the applied radial load in newtons (pounds) and F_a is the applied axial load in newtons (pounds).

Ball Bearing Static Equivalent Load. — For ball bearings the static equivalent radial load is that calculated static radial load which produces a maximum contact stress equal in magnitude to the maximum contact stress in the actual condition of loading. The static equivalent axial load is that calculated static centric axial load which produces a maximum contact stress equal in magnitude to the maximum contact stress in the actual condition of loading.

Radial and Angular Contact Ball Bearings: The magnitude of the static equivalent radial load P_o in newtons (pounds) for radial and angular contact ball bearings under combined thrust and radial loads is the greater of:

$$P_o = X_o F_r + Y_o F_a \qquad (24)$$

$$P_o = F_r \qquad (25)$$

where X_o = radial load factor given in Table 34.
Y_o = axial load factor given in Table 34.
F_r = applied radial load, newtons (pounds).
F_a = applied axial load, newtons (pounds).

**Table 34. Values of X_o and Y_o for Computing Static
Equivalent Radial Load P_o of Ball Bearings**

Contact Angle	Single Row Bearings[2]		Double Row Bearings	
	X_o	Y_o[3]	X_o	Y_o[3]
RADIAL CONTACT GROOVE BEARINGS[1,2]				
$\alpha = 0°$	0.6	0.5	0.6	0.5
ANGULAR CONTACT GROOVE BEARINGS				
$\alpha = 15°$	0.5	0.47	I	0.94
$\alpha = 20°$	0.5	0.42	I	0.84
$\alpha = 25°$	0.5	0.38	I	0.76
$\alpha = 30°$	0.5	0.33	I	0.66
$\alpha = 35°$	0.5	0.29	I	0.58
$\alpha = 40°$	0.5	0.26	I	0.52
SELF-ALIGNING BEARINGS				
. . .	0.5	0.22 $\cot \alpha$	I	0.44 $\cot \alpha$

[1] Permissible maximum value of F_a/C_o (where F_a is applied axial load and C_o is static radial load rating) depends on the bearing design (groove depth and internal clearance).
[2] P_o is always $\geq F_r$.
[3] Values of Y_o for intermediate contact angles are obtained by linear interpolation.

Thrust Ball Bearings: The magnitude of the static equivalent axial load P_{oa} in newtons (pounds) for thrust ball bearings with contact angle $\alpha \neq 90°$ under combined radial and thrust loads is found by the formula:

$$P_{oa} = F_a + 2.3\, F_r \tan \alpha \qquad (26)$$

where the symbol definitions are the same as for formulas (24) and (25). This formula is valid for all load directions in the case of double direction ball bearings. For single direction ball bearings, it is valid where $F_r/F_a \leq 0.44 \cot \alpha$ and gives a satisfactory but less conservative value of P_{oa} for F_r/F_a up to 0.67 cot α.

Thrust ball bearings with $\alpha = 90°$ can support axial loads only. The static equivalent load for this type of bearing is $P_{oa} = F_a$.

Roller Bearing Static Equivalent Load. — The static equivalent radial load for roller bearings is that calculated, static radial load which produces a maximum contact stress acting at the center of contact of a uniformly loaded rolling element equal in magnitude to the maximum contact stress in the actual condition of loading. The static equivalent axial load is that calculated, static centric axial load which produces a maximum contact stress acting at the center of contact of a uniformly loaded rolling element equal in magnitude to the maximum contact stress in the actual condition of loading.

Radial Roller Bearings: The magnitude of the static equivalent radial load P_o in newtons (pounds) for radial roller bearings under combined radial and thrust loads is the greater of:

$$P_o = X_o F_r + Y_o F_a \qquad (27)$$

$$P_o = F_r \qquad (28)$$

where X_o = radial factor given in Table 35.
$\quad Y_o$ = axial factor given in Table 35.
$\quad F_r$ = applied radial load, newtons (pounds).
$\quad F_a$ = applied axial load, newtons (pounds).

Table 35. Values of X_o and Y_o for Computing Static Equivalent Radial Load P_o for Self-Aligning and Tapered Roller Bearings

	Single Row[1]		Double Row	
Bearing Type	X_o	Y_o	X_o	Y_o
Self-Aligning and Tapered $\alpha \neq 0$	0.5	0.22 cot α	1	0.44 cot α

[1] P_o is always $\geq F_r$.

The static equivalent radial load for radial roller bearings with $\alpha = 0°$ and subjected to radial load only is $P_{or} = F_r$.

Note: The ability of radial roller bearings with $\alpha = 0°$ to support axial loads varies considerably with bearing design and execution. The bearing user should therefore consult the bearing manufacturer for recommendations regarding the evaluation of equivalent load in cases where bearings with $\alpha = 0°$ are subjected to axial load.

Radial Roller Bearing Combinations: When calculating the static equivalent radial load for two similar single row angular contact roller bearings mounted side by side on the same shaft such that they operate as a unit (duplex mounting) in ''back-to-back'' or ''face-to-face'' arrangement, use the X_o and Y_o values for a double row bearing and the F_r and F_a values for the total loads on the arrangement.

When calculating the static equivalent radial load for two or more similar single row angular contact roller bearings mounted side by side on the same shaft such that they operate as a unit (duplex or stack mounting) in "tandem" arrangement, use the X_o and Y_o values for a single row bearing and the F_r and F_a values for the total loads on the arrangement.

Thrust Roller Bearings: The magnitude of the static equivalent axial load P_{oa} in newtons (pounds) for thrust roller bearings with contact angle $\alpha \neq 90°$, under combined radial and thrust loads is found by the formula:

$$P_{oa} = F_a + 2.3 \, F_r \tan \alpha \qquad (29)$$

where F_a = applied axial load, newtons (pounds).
F_r = applied radial load, newtons (pounds).
α = nominal contact angle, degrees.

The accuracy of this formula decreases in the case of single direction thrust roller bearings when $F_r > 0.44 \, F_a \cot \alpha$.

Thrust Roller Bearing Combinations: When calculating the static equivalent axial load for two or more thrust roller bearings mounted side by side on the same shaft such that they operate as a unit (duplex or stack mounting) in "tandem" arrangement, use the F_r and F_a values for the total loads acting on the arrangement.

Standard Metal Balls. — American National Standard ANSI/AFBMA Std 10-1983 provides information for the user of metal balls permitting them to be described readily and accurately. It also covers certain measurable characteristics affecting ball quality.

On the following pages, tables taken from this Standard cover standard balls for bearings and other purposes by type of material, grade, and size range; preferred ball sizes; ball hardness corrections for curvature; various tolerances, marking increments, and maximum surface roughnesses by grades; total hardness ranges for various materials; and minimum case depths for carbon steel balls. The numbers of balls per pound and per kilogram for ferrous and non-ferrous metals are also shown.

Definitions and Symbols. — The following definitions and symbols apply to American National Standard metal balls.

Nominal Ball Diameter, D: The diameter value that is used for the general identification of a ball size, e.g., ¼ inch, 6 mm, etc.

Single Diameter of a Ball, D_s: The distance between two parallel planes tangent to the surface of a ball.

Mean Diameter of a Ball, D_m: The arithmetical mean of the largest and smallest single diameters of a ball.

Ball Diameter Variation, V_{Ds}: The difference between the largest and smallest single diameters of one ball.

Deviation from Spherical Form, W: The greatest radial distance in any radial plane between a sphere circumscribed around the ball surface and any point on the ball surface.

Lot: A definite quantity of balls manufactured under conditions which are presumed uniform and which is considered and identified as an entirety.

Lot Mean Diameter, D_{mL}: The arithmetical mean of the mean diameter of the largest ball and that of the smallest ball in the lot.

Lot Diameter Variation, V_{DL}: The difference between the mean diameter of the largest ball and that of the smallest ball in the lot.

Basic Diameter: The size ordered which is the basis to which the Basic Diameter Tolerance is applied.

Basic Diameter Tolerance: The maximum allowable deviation of any ball mean diameter from the Basic Diameter in any shipment to fill orders for that Basic Diameter.

Container Marking Increment: The Standard unit steps in millionths of an inch or in micrometers used to express the Specific Diameter.

Specific Diameter: The diameter marked on the unit container, expressed in the grade's Standard marking increment nearest to the mean diameter of the balls in that unit container.

Ball Gage Deviation, GDE: The difference between the lot mean diameter and the sum of the nominal mean diameter and the ball gage.

Surface Roughness: Surface roughness consists of all those irregularities which form surface relief and which are conventionally defined within the area where deviations of form and waviness are eliminated. (See Handbook Surface Texture Section.)

Ordering Specifications. — Unless otherwise agreed between producer and user, orders for metal balls should provide the following information: quantity, material, nominal ball diameter, grade, and ball gage. A *ball grade* embodies a specific combination of dimensional, form, and surface roughness tolerances. A *ball gage* is the prescribed small amount by which the lot mean diameter (arithmetic mean of the mean diameters of the largest and smallest balls in the lot) should differ from the nominal diameter, this amount being one of an established series of amounts as shown in the table below. The 0 ball gage is referred to as "OK".

Preferred Ball Gages for Grades 3 to 200

Grade	Ball Gages (in 0.0001-in. units)			Ball Gages (in 1-μm units)		
	Minus	OK	Plus	Minus	OK	Plus
3, 5	$-3 -2 -1$	0	$+1 +2 +3$	$-8 -7 -6 -5$ $-4 -3 -2 -1$	0	$+1 +2 +3 +4$ $+5 +6 +7 +8$
10, 16	$-4 -3 -2 -1$	0	$+1 +2 +3 +4$	$-10 -8 -6$ $-4 -2$	0	$+2 +4 +6 +8$ $+10$
24	$-5 -4 -3 -2 -1$	0	$+1 +2 +3 +4 +5$	$-12 -10 -8$ $-6 -4 -2$	0	$+2 +4 +6 +8$ $+10 +12$
48	$-6 -4 -2$	0	$+2 +4 +6$	$-16 -12 -8$ -4	0	$+4 +8 +12$ $+16$
100		0			0	
200		0			0	

Examples: A typical order, in inch units, might read as follows: 80,000 balls, chrome alloy steel, ¼-inch Nominal Diameter, Grade 16, and Ball Gage to be − 0.0002 inch.

A typical order, in metric units, might read as follows: 80,000 balls, chrome alloy steel, 6 mm Nominal Diameter, Grade 16, and Ball Gage to be − 4 μm.

Package Marking. — The ball manufacturer or supplier will identify packages containing each lot with information provided on the orders, as given above. In addition, the specific diameter of the contents shall be stated. Container marking increments are listed in Table 3.

Examples: Balls supplied to the order of the first of the previous examples would, if perfect size, be $D_{mL} = 0.249800$ inch. In grade 16 these balls would be acceptable with D_{mL} from 0.249760 to 0.249850 inch. If they actually measured 0.249823 (which would be rounded off to 0.249820), each package would be marked: 5,000 Balls, Chrome Alloy Steel, ¼″ Nominal Diameter, Grade 16, − 0.0002 inch Ball Gage, and − 0.000180 inch Specific Diameter.

Balls supplied to the order of the second of the two previous examples would, if perfect size, be $D_{mL} = 5.99600$ mm. In Grade 16 these balls would be acceptable with

Table 1. AFBMA Standard Balls — Typical Nominal Size Ranges by Material and Grade

Steel Balls†				Non-Ferrous Balls†			
		Size Range*				Size Range*	
Material	Grade	Inch	mm	Material	Grade	Inch	mm
Chrome Alloy	3	1/32–1	0.8–25	Aluminum	200	1/16–1	1.5–25
	5,10, 16,24	1/64–1 1/2	0.3–38	Aluminum Bronze	200	13/16–4	20–100
	48,100, 200, 500	1/32–2 7/8	0.8–75	Beryllium Copper	10	1/32–1/8	0.8–3
					16	1/32–3/16	0.8–5
	1000	3/8–4 1/2	10–115		24–100	1/32–1/4	0.8–6.5
AISI M-50	3	1/32–1/2	0.8–12	Brass	100,200, 500, 1000	1/16–3/4	1.5–19
	5,10,16	1/32–1 5/8	0.8–40				
	24,48						
Corrosion-Resisting Hardened	3,5,10,16	1/64–3/4	0.3–19	Bronze	200,500, 1000	1/16–3/4	1.5–19
	24	1/32–1	0.8–25				
	48	1/32–2	0.8–50	Monel Metal 400	100,200, 500	1/16–3/4	1.5–19
	100,200	1/32–4 1/2	0.8–115				
Corrosion-Resisting Unhardened	100,200, 500	1/16–3/4	1.5–19	K-Monel Metal 500	100	1/16–3/4	1.5–19
					200	1/16–1 11/16	1.5–45
Carbon Steel‡	100,200, 500, 1000	1/16–1 1/2	1.5–38	Tungsten Carbide	5	3/64–1/2	1.2–12
					10	3/64–3/4	1.2–19
					16	3/64–1	1.2–25
Silicon Molybdenum	200	1/4–1 1/8	6.5–28		24	3/64–1 1/4	1.2–32

* For tolerances see Table 4.
† For hardness ranges see Table 5.
‡ For minimum case depths see Table 6.

Table 2. Ball Hardness Corrections for Curvatures*

Hardness Reading, Rockwell C	Ball Diameters, Inch						
	1/4	5/16	3/8	1/2	5/8	3/4	1
	Correction — Rockwell C						
20	12.1	9.3	7.7	6.1	4.9	4.1	3.1
25	11.0	8.4	7.0	5.5	4.4	3.7	2.7
30	9.8	7.5	6.2	4.9	3.9	3.2	2.4
35	8.6	6.6	5.5	4.3	3.4	2.8	2.1
40	7.5	5.7	4.7	3.6	2.9	2.4	1.7
45	6.3	4.9	4.0	3.0	2.4	1.9	1.4
50	5.2	4.0	3.2	2.4	1.9	1.5	1.1
55	4.1	3.1	2.5	1.8	1.4	1.1	0.8
60	2.9	2.2	1.8	1.2	0.9	0.7	0.4
65	1.8	1.3	1.0	0.5	0.3	0.2	0.1

* Corrections to be added to Rockwell C readings obtained on spherical surfaces of chrome alloy steel, corrosion resisting hardened and unhardened steel, and carbon steel balls. For other ball sizes and hardness readings, interpolate between correction values shown.

a D_{mL} from 5.99500 to 5.99725 mm. If they actually measured 5.99627 mm (which would be rounded off to 5.99625 mm), each package would be marked: 5,000 Balls,

Table 2 (*Concluded*). **Ball Hardness Corrections for Curvatures***

Hardness Reading, Rockwell C	Ball Diameters, mm						
	6	8	10	12	15	20	25
	Correction — Rockwell C						
20	12.8	9.3	7.6	6.6	5.2	4.0	3.2
25	11.7	8.4	6.9	5.9	4.6	3.5	2.8
30	10.5	7.5	6.1	5.2	4.1	3.1	2.4
35	9.4	6.6	5.4	4.6	3.6	2.7	2.1
40	8.0	5.7	4.5	3.8	3.0	2.2	1.8
45	6.7	4.9	3.8	3.2	2.5	1.8	1.4
50	5.5	4.0	3.0	2.6	2.0	1.4	1.1
55	4.3	3.1	2.3	1.9	1.5	1.0	0.8
60	3.0	2.2	1.7	1.2	1.0	0.6	0.4
65	1.9	1.3	0.9	0.6	0.4	0.2	0.1

See footnote on preceding page.

Table 3. **AFBMA Standard Balls — Tolerances for Individual Balls and for Lots of Balls**

Grade	Allowable Ball Diameter Variation	Allowable Deviation from Spherical Form	Maximum Surface Roughness AA†	Allowable Lot Diameter Variation	Basic Diameter Tolerance (±)	Container Marking Increments
	For Individual Balls			For Lots of Balls		
	Millionths of an Inch					
3	3	3	0.5	5	30	10
5	5	5	0.8	10	50	10
10	10	10	1	20	100	10
16	16	16	1	32	100	10
24	24	24	2	48	100	10
48	48	48	3	96	200	50
100	100	100	5	200	500	*
200	200	200	8	400	1000	*
500	500	500	*	1000	2000	*
1000	1000	1000	*	2000	5000	*
	Micrometers					
3	0.08	0.08	0.012	0.13	0.75	0.25
5	0.13	0.13	0.02	0.25	1.25	0.25
10	0.25	0.25	0.025	0.5	0.35	0.25
16	0.4	0.4	0.025	0.8	2.5	0.25
24	0.6	0.6	0.05	1.2	2.5	0.25
48	1.2	1.2	0.08	2.4	5	1.25
100	2.5	2.5	0.125	5	12.5	*
200	5	5	0.2	10	25	*
500	13	13	*	25	50	*
1000	25	25	*	50	125	*

* Not applicable.
† AA — Arithmetical average.
Allowable ball gage (see text) deviation is for Grade 3: + 0.000030, − 0.000030 inch (+ 0.75, − 0.75 μm); for Grades 5, 10, and 16: + 0.000050, − 0.000040 inch (+ 1.25, − 1 μm); and for Grade 24: + 0.000100, − 0.000100 inch (+ 2.5, − 2.5 μm). Other grades not given.

Chrome Alloy Steel, 6 mm Nominal Diameter, Grade 16, − 4 μm Ball Gage, and − 3.75 μm Specific Diameter.

For complete details as to material requirements, quality specifications, quality assurance provisions, and methods of hardness testing, reference should be made to the Standard.

Table 4. AFBMA Standard Balls — Typical Hardness Ranges

Material	Common Standard	SAE Unified Number	Rockwell Value[1,2]
Steel —			
Alloy tool	AISI/SAE M50	K-88165	60–65 "C"[3,5]
Carbon[7]	AISI/SAE 1008	G-10080	60 Minimum "C"[2]
	AISI/SAE 1013	G-10130	60 Minimum "C"[2]
	AISI/SAE 1018	G-10180	60 Minimum "C"[2]
	AISI/SAE 1022	G-10220	60 Minimum "C"[2]
Chrome alloy	AISI/SAE E52100	G-52986	60–67 "C"[3,5]
Corrosion-resisting			
hardened	AISI/SAE 440C	S-44004	58–65 "C"[4,5]
	AISI/SAE 440B	S-44003	55–62 "C"[4,5]
	AISI/SAE 420	S-42000	52 Minimum "C"[4,5]
	AISI/SAE 410	S-41000	97 "B"; 41 "C"[4,5]
	AISI/SAE 329	S-32900	45 Minimum "C"[4,5]
Corrosion-resisting			
unhardened	AISI/SAE 302	S-30200	25–39 "C"[5,6]
	AISI/SAE 304	S-30400	25–39 "C"[5,6]
	AISI/SAE 305	S-30500	25–39 "C"[5,6]
	AISI/SAE 316	S-31600	25–39 "C"[5,6]
	AISI/SAE 430	S-43000	48–63 "A"[5]
Silicon molybdenum	AISI/SAE S2	T-41902	52–60 "C"[3]
Aluminum	AA-2017	A-92017	54–72 "B"
Aluminum bronze	CDA-624	C-62400	15–20 "C"
	CDA-630	C-63000	15–20 "C"
Beryllium copper	CDA-172	C-17200	38 Minimum "C"[5]
Brass	CDA-260	C-26000	75–87 "B"
Bronze	CDA-464	C-46400	75–98 "B"
Monel 400	AMS-4730	N-04400	85–95 "B"
Monel K-500	QQN-286	N-05500	27 Minimum "C"
Tungsten carbide	JIC Carbide Classification	. . .	84–91.5 "A"

[1] Rockwell Hardness Tests shall be conducted on parallel flats in accordance with ASTM Standard E-18 unless otherwise specified.

[2] Hardness readings taken on spherical surfaces are subject to the corrections shown in Table 3. Hardness readings for carbon steel balls smaller than 6 mm (¼ inch) shall be taken by the microhardness method (detailed in ANSI/AFBMA Std 10-1983) or as agreed between manufacturer and purchaser.

[3] Hardness of balls in any one lot shall be within 3 points on Rockwell C scale.

[4] Hardness of balls in any one lot shall be within 4 points on Rockwell C scale.

[5] When microhardness method (see ANSI/AFBMA Std 10-1983) is used, the Rockwell hardness values given are converted to DPH in accordance with ASTM Standard E 140-58, "Standard Hardness Conversion Tables for Metals."

[6] Annealed hardness of Rb 75–90 is available when specified.

[7] Choice of carbon steels shown to be at ball manufacturer's option.

Table 5. Preferred Ball Sizes

Nominal Ball Sizes Metric	Diameter mm	Diameter Inches	Nominal Ball Sizes Inch	Nominal Ball Sizes Metric	Diameter mm	Diameter Inches	Nominal Ball Sizes Inch
0.3	0.300 00	0.011 810			0.793 75	0.031 250	1/32
	0.396 88	0.015 625	1/64	0.8	0.800 00	0.031 496	
0.4	0.400 00	0.015 750		1	1.000 00	0.039 370	
0.5	0.500 00	0.019 680			1.190 63	0.046 875	3/64
	0.508 00	0.020 000	0.020	1.2	1.200 00	0.047 240	
0.6	0.600 00	0.023 620		1.5	1.500 00	0.059 060	
	0.635 00	0.025 000	0.025		1.587 50	0.062 500	1/16
0.7	0.700 00	0.027 560			1.984 38	0.078 125	5/64

Table 5 (Concluded). Preferred Ball Sizes

Nominal Ball Sizes Metric	Diameter mm	Diameter Inches	Nominal Ball Sizes Inch	Nominal Ball Sizes Metric	Diameter mm	Diameter Inches	Nominal Ball Sizes Inch
2	2.000 00	0.078 740			21.431 25	0.843 750	27/32
	2.381 25	0.093 750	3/32	21	21.000 00	0.826 770	
2.5	2.500 00	0.098 420		22	22.000 00	0.866 140	
	2.778 00	0.109 375	7/64		22.225 00	0.875 000	7/8
3	3.000 00	0.118 110			23.018 75	0.906 250	29/32
	3.175 00	0.125 000	1/8	23	23.000 00	0.905 510	
3.5	3.500 00	0.137 800			23.812 50	0.937 500	15/16
	3.571 87	0.140 625	9/64	24	24.000 00	0.944 880	
	3.968 75	0.156 250	5/32		24.606 25	0.968 750	31/32
4	4.000 00	0.157 480		25	25.000 00	0.984 250	
	4.365 63	0.171 875	11/64		25.400 00	1.000 000	1
4.5	4.500 00	0.177 160		26	26.000 00	1.023 620	
	4.762 50	0.187 500	3/16		26.987 50	1.062 500	1 1/16
5	5.000 00	0.196 850		28	28.000 00	1.102 360	
5.5	5.500 00	0.216 540			28.575 00	1.125 000	1 1/8
	5.556 25	0.218 750	7/32	30	30.000 00	1.181 100	
	5.953 12	0.234 375	15/64		30.162 50	1.187 500	1 3/16
6	6.000 00	0.236 220			31.750 00	1.250 000	1 1/4
	6.350 00	0.250 000	1/4	32	32.000 00	1.259 840	
6.5	6.500 00	0.255 900			33.337 50	1.312 500	1 5/16
	6.746 88	0.265 625	17/64	34	34.000 00	1.338 580	
7	7.000 00	0.275 590			34.925 00	1.375 000	1 3/8
	7.143 75	0.281 250	9/32	35	35.000 00	1.377 950	
7.5	7.500 00	0.295 280		36	36.000 00	1.417 320	
	7.540 63	0.296 875	19/64		36.512 50	1.437 500	1 7/16
	7.937 50	0.312 500	5/16	38	38.000 00	1.496 060	
8	8.000 00	0.314 960			38.100 00	1.500 000	1 1/2
8.5	8.500 00	0.334 640			39.687 50	1.562 500	1 9/16
	8.731 25	0.343 750	11/32	40	40.000 00	1.574 800	
9	9.000 00	0.354 330			41.275 00	1.625 000	1 5/8
	9.128 12	0.359 375	23/64		42.862 50	1.687 500	1 11/16
	9.525 00	0.375 000	3/8		44.450 00	1.750 000	1 3/4
	9.921 87	0.390 625	25/64	45	45.000 00	1.771 650	
10	10.000 00	0.393 700			46.037 50	1.812 500	1 13/16
	10.318 75	0.406 250	13/32		47.625 00	1.875 000	1 7/8
11	11.000 00	0.433 070			49.212 50	1.937 500	1 15/16
	11.112 50	0.437 500	7/16	50	50.000 00	1.968 500	
11.5	11.500 00	0.452 756			50.800 00	2.000 000	2
	11.509 38	0.453 125	29/64		53.975 00	2.125 000	2 1/8
	11.906 25	0.468 750	15/32	55	55.000 00	2.165 354	
	12.303 12	0.484 375	31/64		57.150 00	2.250 000	2 1/4
12	12.000 00	0.472 440		60	60.000 00	2.362 205	
	12.700 00	0.500 000	1/2		60.325 00	2.375 000	2 3/8
13	13.000 00	0.511 810			63.500 00	2.500 000	2 1/2
	13.493 75	0.531 250	17/32	65	65.000 00	2.559 055	
14	14.000 00	0.551 180			66.675 00	2.625 000	2 5/8
	14.287 50	0.562 500	9/16		69.850 00	2.750 000	2 3/4
15	15.000 00	0.590 550			73.025 00	2.875 000	2 7/8
	15.081 25	0.593 750	19/32		76.200 00	3.000 000	3
	15.875 00	0.625 000	5/8		79.375 00	3.125 000	3 1/8
16	16.000 00	0.629 920			82.550 00	3.250 000	3 1/4
	16.668 75	0.656 250	21/32		85.725 00	3.375 000	3 3/8
17	17.000 00	0.669 290			88.900 00	3.500 000	3 1/2
	17.462 50	0.687 500	11/16		92.075 00	3.625 000	3 5/8
18	18.000 00	0.708 660			95.250 00	3.750 000	3 3/4
	18.256 25	0.718 750	23/32		98.425 00	3.875 000	3 7/8
19	19.000 00	0.748 030			101.600 00	4.000 000	4
	19.050 00	0.750 000	3/4		104.775 00	4.125 000	4 1/8
	19.843 75	0.781 250	25/32		107.950 00	4.250 000	4 1/4
20	20.000 00	0.787 400			111.125 00	4.375 000	4 3/8
	20.637 50	0.812 500	13/16		114.300 00	4.500 000	4 1/2

Table 6. Number of Metal Balls per Pound

Nom. Diam.* Inches	Material Density, Pounds per Cubic Inch												
	.101	.274	.277	.279	.283	.284	.286	.288	.301	.304	.306	.319	.540
1/32	620 000	228 000	226 000	224 000	221 000	220 000	219 000	217 000	208 000	206 000	205 000	196 000	116 000
1/16	77 500	28 600	28 200	28 000	27 600	27 500	27 400	27 200	26 000	25 700	25 600	24 500	14 500
3/32	22 900	8 460	8 370	8 310	8 190	8 160	8 100	8 050	7 700	7 620	7 570	7 270	4 290
1/8	9 680	3 570	3 530	3 500	3 460	3 440	3 420	3 400	3 250	3 220	3 200	3 070	1 810
5/32	4 960	1 830	1 810	1 790	1 770	1 760	1 750	1 740	1 660	1 650	1 640	1 570	927
3/16	2 870	1 060	1 050	1 040	1 020	1 020	1 010	1 010	963	953	947	908	537
7/32	1 810	666	659	654	645	642	638	634	606	600	596	572	338
1/4	1 210	446	441	438	432	430	427	424	406	402	399	383	226
9/32	850	313	310	308	303	302	300	298	285	282	281	269	159
5/16	620	228	226	224	221	220	219	217	208	206	205	196	116
11/32	466	172.	170.	169.	166.	166.	164.	163.	156.	155.	154.	147.	87.1
3/8	359	132.	131.	130.	128.	128.	127.	126.	120.	119.	118.	114.	67.1
13/32	282	104.	103.	102.	101.	100.	99.6	98.9	94.6	93.7	93.1	89.3	52.8
7/16	226	83.2	82.3	81.7	80.6	80.3	79.7	79.2	75.8	75.0	74.5	71.5	42.2
15/32	184	67.7	66.9	66.5	65.5	65.3	64.8	64.4	61.6	61.0	60.6	58.1	34.3
1/2	151.	55.8	55.2	54.8	54.0	53.8	53.4	53.1	50.8	50.3	49.9	47.9	28.3
17/32	126.	46.5	46.0	45.7	45.0	44.9	44.5	44.2	42.3	41.9	41.6	39.9	23.6
9/16	106.	39.2	38.7	38.5	37.9	37.8	37.5	37.3	35.7	35.3	35.1	33.6	19.9
19/32	90.3	33.3	32.9	32.7	32.2	32.1	31.9	31.7	30.3	30.0	29.8	28.6	16.9
5/8	77.5	28.6	28.2	28.0	27.6	27.5	27.4	27.2	26.0	25.7	25.7	24.5	14.5
21/32	66.9	24.7	24.4	24.2	23.9	23.8	23.6	23.5	22.5	22.2	22.1	21.2	12.5
11/16	58.2	21.5	21.2	21.1	20.8	20.7	20.6	20.4	19.5	19.3	19.2	18.4	10.9
23/32	50.9	18.8	18.6	18.4	18.2	18.1	18.0	17.9	17.1	16.9	16.8	16.1	9.53
3/4	44.8	16.5	16.3	16.2	16.0	15.9	15.8	15.7	15.0	14.9	14.8	14.2	8.38
25/32	39.7	14.6	14.5	14.4	14.2	14.1	14.0	13.9	13.3	13.2	13.1	12.6	7.42
13/16	35.3	13.0	12.9	12.8	12.6	12.5	12.5	12.4	11.8	11.7	11.6	11.2	6.59
27/32	31.5	11.6	11.5	11.4	11.2	11.2	11.1	11.0	10.6	10.5	10.4	9.97	5.89
7/8	28.2	10.4	10.3	10.2	10.1	10.0	9.97	9.90	9.47	9.38	9.32	8.94	5.28
29/32	25.4	9.37	9.26	9.20	9.07	9.04	8.97	8.91	8.53	8.44	8.39	8.04	4.75
15/16	22.9	8.46	8.37	8.31	8.19	8.16	8.10	8.05	7.70	7.62	7.57	7.27	4.29
31/32	20.8	7.67	7.58	7.53	7.42	7.40	7.35	7.29	6.98	6.91	6.87	6.59	3.89
1	18.9	6.97	6.89	6.85	6.75	6.72	6.68	6.63	6.35	6.28	6.24	5.99	3.54

Ball material densities in pounds per cubic inch: aluminum .101; aluminum bronze .274; corrosion resisting hardened steel .277; AISI M-50 and silicon molybdenum steels .279; chrome alloy steel .283; carbon steel .284; AISI 302 corrosion resisting unhardened steel .286; AISI 316 corrosion resisting unhardened steel .288; beryllium copper .301; bronze .304; brass and K-Monel metal .306; Monel metal .319; and tungsten carbide .540. * For sizes above 1 in. diameter, use the following formula: No. balls per pound = 1.91 ÷ [(nom. diam., in.)^3 × (material density, lbs. per cubic in.)].

Table 7. Number of Metal Balls per Kilogram

Nom. Diam.,* mm	Material Density, Grams per Cubic Centimeter												
	14.947	8.830	8.470	8.415	8.332	7.972	7.916	7.861	7.833	7.723	7.667	7.584	2.796
0.3	4 730 000	8 010 000	8 350 000	8 410 000	8 490 000	8 870 000	8 940 000	9 000 000	9 030 000	9 160 000	9 230 000	9 330 000	25 300 000
0.4	2 000 000	3 380 000	3 520 000	3 550 000	3 580 000	3 740 000	3 770 000	3 800 000	3 810 000	3 860 000	3 890 000	3 930 000	10 670 000
0.5	1 020 000	1 730 000	1 800 000	1 820 000	1 830 000	1 920 000	1 930 000	1 940 000	1 950 000	1 980 000	1 990 000	2 010 000	5 470 000
0.7	373 000	631 000	657 000	662 000	668 000	698 000	703 000	708 000	711 000	721 000	726 000	734 000	1 990 000
0.8	250 000	422 000	440 000	443 000	448 000	468 000	471 000	475 000	476 000	483 000	487 000	492 000	1 330 000
1.0	128 000	216 000	225 000	227 000	229 000	240 000	241 000	243 000	244 000	247 000	249 000	252 000	683 000
1.2	73 900	125 000	130 000	131 000	133 000	139 000	140 000	141 000	141 000	143 000	144 000	146 000	395 000
1.5	37 900	64 100	66 800	67 200	67 900	71 000	71 500	72 000	72 200	73 300	73 800	74 600	202 000
2.0	16 000	28 200	28 200	28 400	28 700	29 900	30 200	30 400	30 500	30 900	31 100	31 500	85 400
2.5	8 180	13 800	14 400	14 500	14 700	15 300	15 400	15 500	15 600	15 800	15 900	16 100	43 700
3.0	4 730	8 010	8 350	8 410	8 490	8 870	8 940	9 000	9 030	9 160	9 230	9 330	25 300
3.5	2 980	5 040	5 260	5 290	5 350	5 590	5 630	5 670	5 690	5 770	5 810	5 870	15 900
4.0	2 000	3 386	3 520	3 550	3 580	3 740	3 770	3 800	3 810	3 866	3 890	3 930	10 700
4.5	1 400	2 370	2 470	2 490	2 520	2 630	2 650	2 670	2 680	2 710	2 730	2 760	7 500
5.0	1 020	1 730	1 800	1 820	1 830	1 920	1 930	1 940	1 950	1 980	1 990	2 010	5 470
5.5	768	1 300	1 360	1 360	1 380	1 440	1 450	1 460	1 470	1 490	1 500	1 510	4 110
6.0	592	1 000	1 040	1 050	1 060	1 110	1 120	1 120	1 130	1 140	1 150	1 170	3 160
6.5	465	788	821	826	835	872	878	885	888	901	907	917	2 490
7.0	373	631	657	662	668	698	703	708	711	721	726	734	1 990
7.5	303	513	534	538	543	568	572	576	578	586	590	597	1 620
8.0	250	422	440	443	448	468	471	475	476	483	487	492	1 330
8.5	208	352	367	370	373	390	393	396	397	403	406	410	1 110
9.0	175	297	309	311	314	329	331	333	334	339	342	345	937
10.0	128	216	225	227	229	240	241	243	244	247	249	252	683
11.0	96.0	163.0	169.0	171.0	172.0	180.0	181.0	183.0	183.0	186.0	187.0	189.0	513.0
11.5	84.0	142.0	148.0	149.0	151.0	158.0	159.0	160.0	160.0	163.0	164.0	166.0	449.0
12.0	73.9	125.0	130.0	131.0	133.0	139.0	140.0	141.0	141.0	143.0	144.0	146.0	395.0
13.0	58.2	98.5	103.0	103.0	104.0	109.0	110.0	111.0	111.0	113.0	113.0	115.0	311.0
14.0	46.6	78.8	82.2	82.7	83.5	87.3	87.9	88.5	88.9	90.1	90.8	91.8	249.0
15.0	37.9	64.1	66.8	67.2	67.9	71.0	71.5	72.0	72.2	73.3	73.8	74.6	202.0
16.0	31.2	52.8	55.1	55.4	56.0	58.5	58.9	59.3	59.5	60.4	60.8	61.5	167.0
17.0	26.0	44.0	45.9	46.2	46.7	48.8	49.1	49.5	49.6	50.3	50.7	51.3	139.0

Ball material densities in grams per cubic centimeter: aluminum, 2.796; aluminum bronze, 7.584; corrosion-resisting hardened steel, 7.677; AISI M-50 and silicon molybdenum steel, 7.723; chrome alloy steel, 7.833; carbon steel, 7.861; AISI 302 corrosion-resisting unhardened steel, 7.916; AISI 316 corrosion-resisting unhardened steel, 7.972; beryllium copper, 8.332; brass and K-Monel metal, 8.470; Monel metal, 8.830; tungsten carbide, 14.947. * For sizes above 17 mm diameter, use the following formula: No. balls per kilogram = $1{,}910{,}000 \div [(\text{nom. diam., mm})^3 \times (\text{material density, grams per cu. cm})]$.

FRICTION

Friction is the resistance to motion which takes place when one body is moved upon another, and is generally defined as "that force which acts between two bodies at their surface of contact, so as to resist their sliding on each other." The force of friction, F, bears — according to the conditions under which sliding occurs — a certain relation to the force between the two bodies; this force is called the normal force N. The relation between force of friction and normal force is given by the *coefficient of friction*, generally denoted by the Greek letter μ. Thus:

$$F = \mu \times N, \text{ and } \mu = \frac{F}{N}$$

Example: — A body weighing 28 pounds rests on a horizontal surface. The force required to keep it in motion along the surface is 7 pounds. Find the coefficient of friction.

$$\mu = \frac{F}{N} = \frac{7}{28} = 0.25$$

If a body is placed on an inclined plane, the friction between the body and the plane will prevent it from sliding down the inclined surface, provided the angle of the plane with the horizontal is not too great. There will be a certain angle, however, at which the body will just barely be able to remain stationary, the frictional resistance being very nearly overcome by the tendency of the body to slide down. This angle is termed the angle of repose, and the tangent of this angle equals the coefficient of friction. The angle of repose is frequently denoted by the Greek letter θ. Thus, $\mu =$ tan θ.

A greater force is required to start a body from a state of rest than to merely keep it in motion, because the *friction of rest* is greater than the *friction of motion*.

Laws of Friction. — The laws of friction for unlubricated or dry surfaces are summarized in the following statements.

1. For low pressures (normal force per unit area) the friction is directly proportional to the normal force between the two surfaces. As the pressure increases, the friction does not rise proportionally; but when the pressure becomes abnormally high, the friction increases at a rapid rate until seizing takes place.

2. The friction both in its total amount and its coefficient is independent of the areas in contact, so long as the normal force remains the same. This is true for moderate pressures only. For high pressures, this law is modified in the same way as in the first case.

3. At very low velocities the friction is independent of the velocity of rubbing. As the velocities increase, the friction decreases.

Lubricated Surfaces: For well lubricated surfaces, the laws of friction are considerably different from those governing dry or poorly lubricated surfaces.

1. The frictional resistance is almost independent of the pressure (normal force per unit area) if the surfaces are flooded with oil.

2. The friction varies directly as the speed, at low pressures; but for high pressures the friction is very great at low velocities, approaching a minimum at about two feet per second linear velocity, and afterwards increasing approximately as the square root of the speed.

3. For well lubricated surfaces the frictional resistance depends, to a very great extent, on the temperature, partly because of the change in the viscosity of the oil and partly because, for a journal bearing, the diameter of the bearing increases with the rise of temperature more rapidly than the diameter of the shaft, thus relieving the bearing of side pressure.

4. If the bearing surfaces are flooded with oil, the friction is almost independent of the nature of the material of the surfaces in contact. As the lubrication becomes less ample, the coefficient of friction becomes more dependent upon the material of the surfaces.

Influence of Friction on the Efficiency of Small Machine Elements. — Friction between machine parts lowers the efficiency of a machine. In the following are given average values of the efficiency, in per cent, of the most common machine elements when carefully made. Ordinary bearings, 95 to 98; roller bearings, 98; ball bearings, 99; spur gears with cut teeth, including bearings, 99; bevel gears with cut teeth, including bearings, 98; belting, from 96 to 98; high-class silent power transmission chain, 97 to 99; roller chains, 95 to 97.

Coefficients of Friction. — Tables 1 and 2 provide representative values of static friction for various combinations of materials with dry (clean, unlubricated) and lubricated surfaces. The static or breakaway friction shown in these tables will generally be higher than the subsequent or sliding friction. Typically, the steel-on-steel static coefficient of 0.8 unlubricated will drop to 0.4 when sliding has been initiated; with oil lubrication, the value will drop from 0.16 to 0.03.

Many factors affect friction, and even slight deviations from normal or test conditions can produce wide variations. Accordingly, when using friction coefficients in design calculations, due allowance or factors of safety should be considered and in critical cases, specific tests conducted to provide actual coefficients for material, geometry, and/or lubricant combinations.

Table 1.   Coefficients of Static Friction for Steel on Various Materials*

Material	Coefficient of Friction, μ	
	Clean	Lubricated
Steel..	0.8	0.16
Copper-lead alloy...........................	0.22	....
Phosphor bronze............................	0.35	....
Aluminum bronze...........................	0.45	....
Brass.......................................	0.35	0.19
Cast iron...................................	0.4	0.21
Bronze.....................................	....	0.16
Sintered bronze............................	....	0.13
Hard carbon...............................	0.14	0.11–0.14
Graphite...................................	0.1	0.1
Tungsten carbide...........................	0.4–0.6	0.1–0.2
Plexiglas...................................	0.4–0.5	0.4–0.5
Polystyrene.................................	0.3–0.35	0.3–0.35
Polythene..................................	0.2	0.2
Teflon.....................................	0.04	0.04

* With permission from *The Friction and Lubrication of Solids*, Vol. I, by Bowden and Tabor, Clarendon Press, Oxford, 1950.

Rolling Friction. — When a body rolls on a surface, the force resisting the motion is termed *rolling friction* or *rolling resistance.* Let W = total weight of rolling body or load on wheel, in pounds; r = radius of wheel, in inches; f = coefficient of rolling resistance, in inches. Then: Resistance to rolling, in pounds = $(W \times f) \div r$.

The coefficient of rolling resistance varies with the conditions. For wood on

Table 2. Coefficients of Static Friction for Various Material Combinations*

Material Combination	Coefficient of Friction, μ	
	Clean	Lubricated
Aluminum — aluminum	1.35	0.30
Cadmium — cadmium	0.5	0.05
Chromium — chromium	0.41	0.34
Copper — copper	1.0	0.08
Iron — iron	1.0	0.15–0.20
Magnesium — magnesium	0.6	0.08
Nickel — nickel	0.7	0.28
Platinum — platinum	1.2	0.25
Silver — silver	1.4	0.55
Zinc — zinc	0.6	0.04
Glass — glass	0.9–1.0	0.1–0.6
Glass — metal	0.5–0.7	0.2–0.3
Diamond — diamond	0.1	0.05–0.1
Diamond — metal	0.1–0.15	0.1
Sapphire — sapphire	0.2	0.2
Hard carbon on carbon	0.16	0.12–0.14
Graphite — graphite (in vacuum)	0.5–0.8	....
Graphite — graphite	0.1	0.1
Tungsten carbide — tungsten carbide	0.2–0.25	0.12
Plexiglas — plexiglas	0.8	0.8
Polystyrene — polystyrene	0.5	0.5
Teflon —Teflon	0.04	0.04
Nylon — nylon	0.15–0.25	....
Solids on rubber	1–4	....
Wood on wood (clean)	0.25–0.5	....
Wood on wood (wet)	0.2	....
Wood on metals (clean)	0.2–0.6	....
Wood on metals (wet)	0.2	....
Brick on wood	0.6	....
Leather on wood	0.3–0.4	....
Leather on metal (clean)	0.6	....
Leather on metal (wet)	0.4	....
Leather on metal (greasy)	0.2	....
Brake material on cast iron	0.4	....
Brake material on cast iron (wet)	0.2	....

* With permission from *The Friction and Lubrication of Solids*, Vol. I, by Bowden and Tabor, Clarendon Press, Oxford, 1950.

wood it may be assumed as 0.06 inch; for iron on iron, 0.02 inch; iron on granite, 0.085 inch; iron on asphalt, 0.15 inch; and iron on wood, 0.22 inch.

The coefficient of rolling resistance, f, is in inches and is not the same as the ordinary, or sliding coefficient of friction which is a dimensionless ratio between frictional resistance and normal load. As various investigators are not in close agreement on the true values for these coefficients, the foregoing values should only be used for approximate calculation of rolling resistance.

Lubricants. —The concept of supporting a load being moved on a friction-reducing film is called lubrication. The material of which the film is made up is defined as a lubricant. Lubricants may be in the form of fluids, such as conventional oils; semisolids, such as grease; and true solids, such as graphite and Teflon. Water, alcohol, and even air have been used as lubricants. Factors affecting the selection of the type of lubricant to be used are load, speed, temperature (both ambient

and generated) and environmental contamination such as in the food processing industry. Liquid lubricants usually are matched to the application by proper selection, of viscosity and load-carrying ability. (See "Lubricants and Lubrication," page 2205.)

Wear. — There is no apparent consistent relationship between friction and wear; friction may be high and wear low, or wear may be high and friction low. The principal types of wear are: adhesive, abrasive, and pitting.

Adhesive wear: As two surfaces slide over each other, wear occurs because of the shearing, deformation, and plucking away of material at points of adhesion; these points of adhesion occur at roughness peaks. The amount of wear is generally proportional to the load and to the distance over which the surfaces have slid, and inversely proportional to the hardness of the surface on which wear occurs.

Abrasive wear: This occurs when a hard rough surface slides against a softer one, ploughing a series of grooves and removing material; it also occurs when abrasive particles are introduced between sliding surfaces, or when a part is moved through an abrasive medium. Erosion abrasion takes place when abrasive particles impact on surfaces. These particles may be suspended in liquids, carried by air, or flow of their own weight such as sand particles down a chute.

Pitting: This is due to the surface fatigue failure of a material as a result of repeated surface or sub-surface stresses that exceed the endurance limit of the material. Pitting is a common mode of failure for gear teeth.

Pitting can be non-progressive or destructive. Non-progressive pitting occurs during the initial operation of a machine because of surface fatigue at roughness peaks. As the peaks wear, the surface becomes smoother and hence the load becomes more uniformly distributed. Pitting stops when the stress becomes less than the endurance limit of the material. Destructive pitting, however, is progressive and continues until the surface disintegrates.

Designing for Wear Resistance. — In adhesive wear situations the best results are usually obtained if parts have hard surfaces. Hard surfaces may be obtained by making parts from hard materials or by the application of appropriate surface treatments. Some of the hardest materials are diamond, brittle carbides, and martensitic steels. For less severe applications pearlitic and austentitic steels can be used.

Some common surface treatments of metal surfaces for the prevention of wear are: chromium and nickel plating, flame plating with tungsten carbide, phosphate treatment, carburizing, nitriding, cyaniding, carbonitriding, siliconizing, and chromizing. Aluminum and magnesium can be given hard anodic treatments. For moderate cost applications in which normal wear can be tolerated, low or medium carbon steels can be run against cast iron, brass, bronze, aluminum bronze, or nylon. Adequate lubricants should be used. In general it is best if unlike combinations of materials slide against each other. This is often accomplished by employing different surface treatments; for example, having a nitrided steel part run against one that is carburized.

Abrasive wear can be limited in several ways. Surfaces can be given a higher hardness than the abrasive particles. If a circulating lubricant is used, the abrasive particles can be removed by filtering. A combination of a hard and a soft surface can be used so that the abrasive particles can imbed themselves into the softer material. Typical softer materials used for this purpose are babbitt metals, copper-lead alloys, and aluminum alloys.

Pitting is controlled by keeping contact stresses within allowable limits. In the case of gears a hard pinion is often run with a softer gear. As the softer gear wears, there will be better conformity between gear teeth profiles, hence less localized high stresses.

Lubricants and Lubrication

A lubricant is used for one or more of the following purposes: 1. To reduce friction; 2. To prevent wear; 3. To prevent adhesion; 4. To aid in distributing the load; 5. To cool the moving elements; and 6. To prevent corrosion.

The range of materials used as lubricants has been greatly broadened over the years so that in addition to oils and greases many plastics and solids and even gases are now being applied in this role. The only limitations on many of these materials are their ability to replenish themselves, to dissipate frictional heat, their reaction to high environmental temperatures, and their stability in combined environments. Because of the wide selection of lubricating materials available, great care is advisable in choosing the material and the method of application. The following types of lubricants are available: 1. petroleum fluids; 2. synthetic fluids; 3. greases; 4. solid films; 5. working fluids; 6. gases; 7. plastics; 8. animal fat; 9. metallic and mineral films; and 10. vegetable oils.

Lubricating Oils. — The most versatile and best-known lubricant is mineral oil. When applied in well-designed applications which provide for the limitations of both mechanical and hydraulic elements, oil is recognized as the most reliable lubricant. Concurrently, it is offered in a wide selection of stocks, carefully developed to meet the requirements of the specific application.

Lubricating oils are seldom marketed without additives blended for a narrow range of applications. Since these "additive-packages" are developed for particular applications it is advisable to consult the sales-engineering representatives of a reputable petroleum company on the proper selection for the conditions under consideration. The following are the most common types of additives: 1. wear preventive; 2. oxidation inhibitor; 3. rust inhibitor; 4. detergent-dispersant; 5. viscosity index improver; 6. defoaming agent; and 7. pour point depressant.

A more recent development in the field of additives, is a series of organic compounds which leave no ash when heated to a temperature high enough to evaporate or burn off the base oil. Initially produced for internal combustion engine applications, they have found ready acceptance in those other applications where metallic or mineral trace elements would promote catalytic, corrosive, deposition, or degradation effects on mechanism materials.

Additives usually are not stable over the entire temperature and shear-rate ranges considered acceptable for the base stock oil application. Because of this, additive type oils must be carefully monitored to insure that they are not continued in service after their principal capabilities have been diminished or depleted. Of primary importance in this regard, is the action of the detergent-dispersant additives which function so well to reduce and control degradation products which would otherwise deposit on the operating parts and oil cavity walls. Since the materials cause the oil to carry a higher than normal amount of the breakdown products in a fine suspension, they may cause an accelerated deposition rate or foaming when they have been depleted or degenerated by thermal or contamination action. In this latter case, the ingestion of water by condensation or leaking can cause markedly harmful effects.

Viscosity index improvers serve to modify oils so that their change in viscosity is reduced over the operating temperature range. These materials may be used to improve both a heavy or a light oil; however, the original stock will tend to revert to its natural state when the additive has been depleted or degraded due to exposure to high temperatures or to the high shear rates normally encountered in the load carrying zones of bearings and gears. In heavy duty installations it is generally advisable to select a heavier or a more highly refined oil (and one that is generally more costly) rather than to rely on a less stable viscosity index improver product.

These oils are generally used in applications where the shear rate is well **below** 1,000,000 reciprocal seconds as determined in the following:

$$\text{Shear rate (sec.}^{-1}) = \frac{DN}{60t}$$

where D is the journal diameter in inches, N is the journal speed in RPM, and t is the film thickness in inches.

Types of Oils. — Aside from being aware of the many additives which can be obtained to satisfy particular application requirements and improve the performance of fluids, the designer must also be acquainted with the wide variety of oils, natural and synthetic, which are available. Each has its own special features which make it suitable for specific applications and which limit its utility in others. Though a complete description of each oil and its application feasibility cannot be given in the following paragraphs, reference to major petroleum and chemical company sales engineers will provide full descriptions and sound recommendations. In many applications, however, it must be accepted that the interrelation of many variables, including shear rate, load, and temperature variations, prohibit precise recommendations or predictions of fluid durability and performance. Thus, prototype and rig testing are often required to insure the final selection of the most satisfactory fluid.

The following table lists the major classifications, and properties of available commercial petroleum oils.

Properties of Commercial Petroleum Oils*

Group A			Group B		
Type	Viscosity, Centistokes		Type	Viscosity, Centistokes	Density, g/cc at 60°F.
	100°F.	210°F.	Density, g/cc at 60°F.		100°F.	210°F.
SAE 10W	41	6.0	0.870		22	3.9	0.880
SAE 20W	71	8.5	0.885	General Purpose	44	6.0	0.898
SAE 30	114	11.2	0.890		66	7.0	0.915
SAE 40	173	14.5	0.890		110	9.9	0.915
SAE 50	270	19.5	0.900		200	15.5	0.890

Group C			Group D		
SAE 75	47	7.0		Turbine Light	32	5.5	0.871
SAE 80	69	8.0	0.930, approx.	Medium	65	8.1	0.876
SAE 90	285	20.5		Heavy	99	10.7	0.885
SAE 140	725	34.0					
SAE 250	1,220	47.0					

Group E			Group F		
Aviation	5	1.5	0.858	Aviation	76	9.3	0.875
	10	2.5	0.864		268	20.0	0.891
					369	25.0	0.892

*Applications:
  Group A. Automotive. With increased additives, diesel and marine reciprocating engines.
  Group B. Gear trains and transmissions. With E. P. additives, hypoid gears.
  Group C. Machine tools and other industrial applications.
  Group D. Marine propulsion and stationary power turbines.
  Group E. Turbojet engines.
  Group F. Reciprocating engines.

Viscosity. — As noted above, fluids used as lubricants are generally categorized by their viscosity at 100 and 210 deg. F. Absolute viscosity is defined as a fluid's resistance to shear or motion — its internal friction in other words. This property is described in several ways, but basically it is the force required to move a plane surface of unit area with unit speed parallel to a second plane and at unit distance from it. In the metric system the unit of viscosity is called the "poise" and in the English system is called the "reyn." One reyn is equal to 68,950 poise. One poise is the viscosity of a fluid, such that one dyne force is required to move a surface of one square centimeter with a speed of one centimeter per second, the distance between surfaces being one centimeter. The range of kinematic viscosity for a series of typical fluids is shown in the table on page 559. Kinematic viscosity is related directly to the flow time of a fluid through the viscosimeter capillary. By multiplying the kinematic viscosity by the density of the fluid at the test temperature, one can determine the absolute viscosity. Since, in the metric system, the mass density is equal to the specific gravity, the conversion from kinematic to absolute viscosity is generally made in this system and then converted to English units where required. The densities of typical lubricating fluids with comparable viscosities at 100 deg. F. and 210 deg. F. are shown in this same table.

The following conversion table may be found helpful.

Multiply	By	To Get
Centipoises, Z, $\dfrac{\text{dyne-sec}}{100\ \text{cm}^2}$	1.45×10^{-7}	Reyns, μ, $\dfrac{\text{lbs. force-sec}}{\text{in.}^2}$
Centistokes, v, $\dfrac{\text{cm.}^2}{100\ \text{sec}}$	Density in g/cc	Centipoises, Z, $\dfrac{\text{dyne-sec}}{100\ \text{cm.}^2}$
Saybolt Universal Seconds, t_s	$.22t_s - \dfrac{180}{t_s}$	Centistokes, v, $\dfrac{\text{cm.}^2}{100\ \text{sec}}$

Finding Specific Gravity of Oils at Different Temperatures. — The standard practice in the oil industry is to obtain a measure of specific gravity at 60 deg. F. on an arbitrary scale, in degrees API, as specified by the American Petroleum Institute. As an example, API gravity, ρ_{API}, may be expressed as 27.5 degrees at 60 deg. F.

The relation between gravity in API degrees and specific gravity (grams of mass per cubic centimeter) at 60 deg. F., ρ_{60}, is:

$$\rho_{60} = \frac{141.5}{131.5 + \rho_{API}}$$

The specific gravity, ρ_T, at some other temperature, T, is found from the equation:

$$\rho_T = \rho_{60} - 0.00035(T - 60)$$

Normal values of specific gravity for sleeve bearing lubricants range from 0.75 to 0.95 at 60 deg. F. If the API rating is not known, an assumed value of 0.85 may be used.

Application of Lubricating Oils. — In the selection and application of lubricating oils careful attention must be given to the temperature in the critical operating area and its effect on oil properties. Analysis of each application should be made with detailed attention given to cooling, friction losses, shear rates, and contaminants.

Many oil selections are found to result in excessive operating temperatures because of a viscosity that is initially too high, which raises the friction losses. As a general rule, the lightest weight oil which can carry the maximum load should be

LUBRICANTS

used. Where it is felt that the load carrying capacity is borderline, lubricity improvers may be employed rather than an arbitrarily higher viscosity fluid. It is well to remember that in many mechanisms the thicker fluid may increase friction losses sufficiently to lower the operating viscosity into the range provided by an initially lighter fluid. In such situations also, improved cooling, such as may be accomplished by increasing the oil flow, can improve the fluid properties in the load zone.

Similar improvements can be accomplished in many gear trains and other mechanisms by reducing churning and aeration through improved scavenging, direction of oil jets, and elimination of obstacles to the flow of the fluid. Many devices, such as journal bearings, are extremely sensitive to the effects of cooling flow and can be improved by greater flow rates with a lighter fluid. In other cases it is well to remember that the load carrying capacity of a petroleum oil is affected by pressure, shear rate, and bearing surface finish as well as initial viscosity and therefore these must be considered in the selection of the fluid. Detailed explanation of these factors is not within the scope of this text; however the technical representatives of the petroleum companies can supply practical guides for most applications.

Other factors to consider in the selection of an oil include the following:

1. Compatibility with system materials
2. Water absorption properties
3. Break-in requirements
4. Detergent requirements
5. Corrosion protection
6. Low temperature properties
7. Foaming tendencies
8. Boundary lubrication properties
9. Oxidation resistance (high temperature properties)
10. Viscosity/temperature stability (Viscosity Temperature Index).

Generally, the factors listed above are those which are usually modified by additives as described earlier. Since additives are used in limited amounts in most petroleum products, blended oils are not as durable as the base stock and must therefore be used in carefully worked-out systems. Maintenance procedures must be established to monitor the oil so that it may be replaced when the effect of the additive is noted or expected to degrade. In large systems supervised by a lubricating engineer, sampling and associated laboratory analysis can be relied on, while in customer-maintained systems as in automobiles and reciprocating engines, the design engineer must specify a safe replacement period which takes into account any variation in type of service or utilization.

Some large systems, such as turbine-power units, have complete oil systems which are designed to filter, cool, monitor, meter, and replenish the oil automatically. In such facilities, much larger oil quantities are used and they are maintained by regularly assigned lubricating personnel. Here reliance is placed on conservatively chosen fluids with the expectation that they will endure many months or even years of service.

Centralized Lubrication Systems. — Various forms of centralized lubrication systems are used to simplify and render more efficient the task of lubricating machines. In general, a central reservoir provides the supply of oil, which is conveyed to each bearing either through individual lines of tubing or through a single line of tubing that has branches extending to each of the different bearings. Oil is pumped into the lines either manually by a single movement of a lever or handle, or automatically by mechanical drive from some revolving shaft or other

part of the machine. In either case, all bearings in the central system are lubricated simultaneously. Centralized force-feed lubrication is adaptable to various classes of machine tools such as lathes, planers, and milling machines and to many other types of machines. It permits the use of a lighter grade of oil, especially where complete coverage of the moving parts is assured.

Gravity Lubrication Systems. — Gravity systems of lubrication usually consist of a small number of distributing centers or manifolds from which oil is taken by piping as directly as possible to the various surfaces to be lubricated, each bearing point having its own independent pipe and set of connections. The aim of the gravity system, as of all lubrication systems, is to provide a reliable means of supplying the bearing surfaces with the proper amount of lubricating oil. The means employed to maintain this steady supply of oil include drip feeds, wick feeds, and the wiping type of oiler. Most manifolds are adapted to use either or both drip and wick feeds.

Drip-feed Lubricators: A drip feed consists of a simple cup or manifold mounted in a convenient position for filling and connected by a pipe or duct to each bearing to be oiled. The rate of feed in each pipe is regulated by a needle or conical valve. A loose-fitting cover is usually fitted to the manifold in order to prevent cinders or other foreign matter from becoming mixed with the oil. When a cylinder or other chamber operating under pressure is to be lubricated, the oil-cup takes the form of a lubricator having a tight-fitting screw cover and a valve in the oil line. To fill a lubricator of this kind, it is only necessary to close the valve and unscrew the cover.

Operation of Wick Feeds: For a wick feed, the siphoning effect of strands of worsted yarn is employed. The worsted wicks give a regular and reliable supply of oil and at the same time act as filters and strainers. A wick composed of the proper number of strands is fitted into each oil-tube. In order to insure using the proper sizes of wicks, a study should be made of the oil requirements of each installation, and the number of strands necessary to meet the demands of bearings at different rates of speed should be determined. When the necessary data have been obtained, a table should be prepared showing the size of wick or the number of strands to be used for each bearing of the machine.

Oil-conducting Capacity of Wicks: With the oil level maintained at a point ⅛ to ¾ inch below the top of an oil-tube, each strand of a clean worsted yarn will carry slightly more than one drop of oil a minute. A twenty-four-strand wick will feed approximately thirty drops a minute, which is ordinarily sufficient for operating a large bearing at high speed. The wicks should be removed from the oil-tubes when the machinery is idle. If left in place, they will continue to deliver oil to the bearings until the supply in the cup is exhausted, thus wasting a considerable quantity of oil, as well as flooding the bearing. When bearings require an extra supply of oil temporarily, it may be supplied by dipping the wicks or by pouring oil down the tubes from an oil-can or, in the case of drip feeds, by opening the needle valves. When equipment that has remained idle for some time is to be started up, the wicks should be dipped and the moving parts oiled by hand to insure an ample initial supply of oil. The oil should be kept at about the same level in the cup, as otherwise the rate of flow will be affected. Wicks should be lifted periodically to prevent dirt accumulations at the ends from obstructing the flow of oil.

How Lubricating Wicks are Made: Wicks for lubricating purposes are made by cutting worsted yarn into lengths about twice the height of the top of the oil-tube above the bottom of the oil-cup, plus 4 inches. Half the required number of strands are then assembled and doubled over a piece of soft copper wire, laid across the middle of the strands. The free ends are then caught together by a small piece of folded sheet lead, and the copper wire twisted together throughout its length.

The lead serves to hold the lower end of the wick in place, and the wire assists in forcing the other end of the wick several inches into the tube. When the wicks are removed, the free end of the copper wire may be hooked over the tube end to indicate which tube the wick belongs to. Dirt from the oil causes the wick to become gummy and to lose its filtering effect. Wicks that have thus become clogged with dirt should be cleaned or replaced by new ones. The cleaning is done by boiling the wicks in soda water and then rinsing them thoroughly to remove all traces of the soda. Oil-pipes are sometimes fitted with openings through which the flow of oil can be observed. In some installations, a short glass tube is substituted for such an opening.

Wiper-type Lubricating Systems: Wiper-type lubricators are used for out-of-the-way oscillating parts. A wiper consists of an oil-cup with a central blade or plate extending above the cup, and is attached to a moving part. A strip of fibrous material fed with oil from a source of supply is placed on a stationary part in such a position that the cup in its motion scrapes along the fibrous material and wipes off the oil, which then passes to the bearing surfaces.

Oil manifolds, cups, and pipes should be cleaned occasionally with steam conducted through a hose or with boiling soda water. When soda water is used, the pipes should be disconnected, so that no soda water can reach the bearings.

Oil Mist Systems. — A very effective system for both lubricating and cooling many elements which require a limited quantity of fluid is found in a device which generates a mist of oil, separates out the denser and larger (wet) oil particles, and then distributes the mist through a piping or conduit system. The mist is delivered into the bearing, gear, or lubricated element cavity through a condensing or spray nozzle, which also serves to meter the flow. In applications which do not encounter low temperatures or which permit the use of visual devices to monitor the accumulation of solid oil, oil mist devices offer advantages in providing cooling, clean lubricant, pressurized cavities which prevent entrance of contaminants, efficient application of limited lubricant quantities, and near-automatic performance. These devices are supplied with fluid reservoirs holding from a few ounces up to several gallons of oil and with accommodations for either accepting shop air or working from a self-contained compressor powered by electricity. With proper control of the fluid temperature, these units can atomize and dispense most motor and many gear oils.

Lubricating Greases. — In many applications, fluid lubricants cannot be used because of the difficulty of retention, relubrication, or the danger of churning. To satisfy these and other requirements such as simplification, greases are applied. These formulations are usually petroleum oils thickened by dispersions of soap, but may consist of synthetic oils with soap or inorganic thickeners, or oil with silaceous dispersions. In all cases, the thickener, which must be carefully prepared and mixed with the fluid, is used to immobilize the oil, serving as a storehouse from which the oil bleeds at a slow rate. Though the thickener very often has lubricating properties itself, the oil bleeding from the bulk of the grease is the determining lubricating function. Thus, it has been shown that when the oil has been depleted to the level of 50 per cent of the total weight of the grease, the lubricating ability of the material is no longer reliable. In some applications requiring an initially softer and wetter material, however, this level may be as high as 60 per cent.

Grease Consistency Classifications. — To classify greases as to mobility and oil content, they are divided into Grades by the NLGI (National Lubricating Grease Institute). These grades, ranging from 0, the softest, up through 6, the stiffest, are determined by testing in a penetrometer, with the depth of penetration of a specific

cone and weight being the controlling criterion. To insure proper averaging of specimen resistance to the cone, most specifications include a requirement that the specimen be worked in a sieve-like device before being packed into the penetrometer cup for the penetration test. Since many greases exhibit thixotropic properties (they soften with working, as they often do in an application with agitation of the bulk of the grease by the working elements or accelerations), this penetration of the worked specimen should be used as a guide to compare the material to the original manufactured condition of it and other greases, rather than to the exact condition in which it will be found in the application. Conversely, many greases are found to stiffen when exposed to high shear rates at moderate loads as in automatic grease dispensing equipment. The application of a grease, therefore must be determined by a carefully planned cut-and-try procedure. Most often this is done by the original equipment manufacturer with the aid of the petroleum company representatives, but in many cases it is advisable to include the bearing engineer as well. In this general area it is well to remember that shock loads, axial or thrust movement within or on the grease cavity can cause the grease to contact the moving parts and initiate shearing due to the shearing or working thus induced. To limit this action, grease-lubricated bearing assemblies often utilize dams or dividers to keep the bulk of the grease contained and unchanged by this working. Successful application of a grease depends however, on a relatively small amount of mobile lubricant (the oil bled out of the bulk) to replenish that small amount of lubricant in the element to be lubricated. If the space between the bulk of the mobile grease and the bearing is too large, then a critical delay period (which will be regulated by the grease bleed rate and the temperature at which it is held) will ensue before lubricant in the element can be resupplied. Since most lubricants undergo some attrition due to thermal degradation, evaporation, shearing, or decomposition in the bearing area to which applied, this delay can be fatal.

To prevent this from leading to failure, grease is normally applied so that the material in the cavity contacts the bearing in the lower quadrants, insuring that the excess orginally packed into it impinges on the material in the reservoir. With the proper selection of a grease which does not slump excessively, and a reservoir construction to prevent churning, the initial action of the bearing when started into operation will be to purge itself of excess grease, and to establish a flow path for bleed oil to enter the bearing. For this purpose, most greases selected will be of a grade 2 or 3 consistency, falling into the "channelling" variety or designation.

Types of Grease. — Greases are made with a variety of soaps and are chosen for many particular characteristics. Most popular today, however, are the lithium, or soda-soap grease and the modified-clay thickened materials. For high temperature applications (250 deg. F. and above) certain finely divided dyes and other synthetic thickeners are applied. For all-around use the lithium soap greases are best for moderate temperature applications (up to 225 deg. F.) while a number of soda-soap greases have been found to work well up to 285 deg. F. Since the major suppliers offer a number of different formulations for these temperature ranges it is recommended that the user contact the engineering representatives of a reputable petroleum company before choosing a grease. Greases also vary in volatility and viscosity according to the oil used. Since the former will affect the useful life of the bulk applied to the bearing and the latter will affect the load carrying capacity of the grease, they must both be considered in selecting a grease.

For application to certain gears and slow-speed journal bearings, a variety of greases are thickened with carbon, graphite, molybdenum disulfide, lead, or zinc oxide. Some of these materials are likewise used to inhibit fretting corrosion or wear in sliding or oscillating mechanisms and in screw or thread applications. One

material used as a "gear grease" is a residual asphaltic compound which is known as a "Crater Compound." Being extremely stiff and having an extreme temperature-viscosity relationship, its application must also be made with careful consideration of its limitations and only after careful evaluation in the actual application. Its oxidation resistance is limited and its low mobility in winter temperature ranges make it a material to be used with care. However, it is used extensively in the railroad industry and in other applications where containment and application of lubricants is difficult. In such conditions its ability to adhere to gear and chain contact surfaces far out-weighs its limitations and in some extremes it is "painted" onto the elements at regular intervals.

Temperature Effects on Grease Life. — Since most grease applications are made where long life is important and relubrication is not too practical, operating temperatures must be carefully considered and controlled. Being a hydro-carbon, and normally susceptible to oxidation, grease is subject to the general rule that: Above a critical threshold temperature, each 15- to 18-deg. F. rise in temperature reduces the oxidation life of the lubricant by half. For this reason, it is vital that all elements affecting the operating temperature of the application be considered, correlated, and controlled. With sealed-for-life bearings, in particular, grease life must be determined for representative bearings and limits must be established for all subsequent applications.

Most satisfactory control can be established by measuring bearing temperature rise during a controlled test, at a consistent measuring point or location. Once a base line and limiting range are determined, all deviating bearings should be dismantled, inspected, and reassembled with fresh lubricant for retest. In this manner mavericks or faulty assemblies will be ferreted out and the reliability of the application established. Generally, a well lubricated grease packed bearing will have a temperature rise above ambient, as measured at the outer race, of from 10 to 50 deg. F. In applications where heat is introduced into the bearing through the shaft or housing, a temperature rise must be added to that of the frame or shaft temperature.

In bearing applications care must be taken not to fill the cavity too full. The bearing should have a practical quantity of grease worked into it with the rolling elements thoroughly coated and the cage covered, but the housing (cap and cover) should be no more than 75 per cent filled; with softer greases, this should be no more than 50 per cent. Excessive packing is evidenced by overheating, churning, aerating, and eventual purging with final failure due to insufficient lubrication. In grease lubrication, *never* add a bit more for good luck — hold to the prescribed amount and determine this with care on a number of representative assemblies.

Relubricating with Grease. — In some applications, sealed-grease methods are not applicable and addition of grease at regular intervals is required. Where this is recommended by the manufacturer of the equipment, or where the method has been worked out as part of a development program, the procedure must be carefully followed. *First,* use the proper lubricant — the same as recommended by the manufacturer or as originally applied (grease performance can be drastically impaired if contaminated with another lubricant). *Second,* clean the lubrication fitting thoroughly with materials which will not affect the mechanism or penetrate into the grease cavity. *Third,* remove the cap (and if applicable, the drain or purge plug). *Fourth,* clean and inspect the drain or scavenge cavity. *Fifth,* weigh the grease gun or calibrate it to determine delivery rate. *Sixth,* apply the directed quantity or fill until grease is detected coming out the drain or purge hole. *Seventh,* operate the mechanism with the drain open so that excess grease is purged. *Last,* continue to operate the mechanism while determining the temperature rise and insure that it is

within limits. Where there is access to a laboratory, samples of the purged material may be analyzed to determine the deterioration of the lubricant and to search for foreign material which may be evidence of contamination or of bearing failure.

Normally, with modern types of grease and bearings, lubrication need only be considered at overhaul periods or over intervals of three to ten years.

Solid Film Lubricants. — Solids such as graphite, molybdenum disulfide, polytetrafluoroethylene, lead, babbit, silver, or metallic oxides are used to provide dry film lubrication in high-load, slow-speed or oscillating load conditions. Though most are employed in conjunction with fluid or grease lubricants, they are often applied as the primary or sole lubricant where their inherent limitations are acceptable. Of foremost importance is their inability to carry away heat. Second, they cannot replenish themselves, though they generally do lay down an oriented film on the contacting interface. Third, they are relatively immobile and must be bonded to the substrate by a carrier, by plating, fusing, or by chemical or thermal deposition.

Though these materials do not provide the low coefficient of friction associated with fluid lubrication, they do provide coefficients in the range of 0.4 down to 0.02, depending on the method of application and the material against which they rub. Polytetrafluoroethylene, in normal atmospheres and after establishing a film on both surfaces has been found to exhibit a coefficient of friction down to 0.02. However, this material is subject to cold flow and must be supported by a filler or on a matrix to continue its function. Since it can now be cemented in thin sheets and is often supplied with a fine glass fiber filler, it is practical in a number of installations where the speed and load do not combine to melt the bond or cause the material to sublime.

Bonded films of molybdenum disulfide, using various resins and ceramic combinations as binders, are deposited over phosphate treated steel, aluminum, or other metals with good success. Since its action produces a gradual wear of the lubricant, its life is limited by the thickness which can be applied (not over a thousandth or two in the conventional application). In most applications this is adequate if the material is used to promote break-in, prevent galling or pick-up, and to reduce fretting or abrasion in contacts otherwise impossible to separate.

In all applications of solid film lubricants, the performance of the film is limited by the care and preparation of the surface to which they are applied. If they can't adhere properly, they cannot perform, coming off in flakes and often jamming under flexible components. The best advice is to seek the assistance of the supplier's field engineer and set up a close control of the surface preparation and solid film application procedure. It should be noted that the functions of a good solid film lubricant cannot overcome the need for better surface finishing. Contacting surfaces should be smooth and flat to insure long life and minimum friction forces. Generally, surfaces should be finished to no more than 24 micro-inches AA with waviness no greater than 0.00002 inch.

Anti-friction Bearing Lubrication. — The limiting factors in bearing lubrication are the load and the linear velocity of the centers of the balls or rollers. Since these are difficult to evaluate, a speed factor which consists of the inner race bore diameter × RPM is used as a criterion. This factor will be referred to as S_i where the bore diameter is in inches and S_m where it is in millimeters.

In order to be suitable for use in anti-friction bearings, grease must have the following properties:

1. Freedom from chemically or mechanically active ingredients such as uncombined metals or oxides, and similar mineral or solid contaminants.

2. The slightest possible tendency of change in consistency, such as thickening, separation of oil, evaporation or hardening.

3. A melting point considerably higher than the operating temperatures.

The choice of lubricating oils is easier. They are more uniform in their characteristics and if resistant to oxidation, gumming and evaporation, can be selected primarily with regard to a suitable viscosity.

Grease Lubrication: Anti-friction bearings are normally grease lubricated, both because grease is much easier than oil to retain in the housing over a long period and because it acts to some extent as a seal against the entry of dirt and other contaminants into the bearings. For almost all applications, a No. 2 soda-base grease or a mixed-base grease with up to 5 per cent calcium soap to give a smoother consistency, blended with an oil of around 250 to 300 SSU (Saybolt Universal Seconds) at 100 degrees F. is suitable. In cases where speeds are high, say S_i is 5000 or over, a grease made with an oil of about 150 SSU at 100 degrees F. may be more suitable especially if temperatures are also high. In many cases where bearings are exposed to large quantities of water, it has been found that a standard soda-base ball-bearing grease, although classed as water soluble gives better results than water-insoluble types. Greases are available that will give satisfactory lubrication over a temperature range of −40 degrees to +250 degrees F.

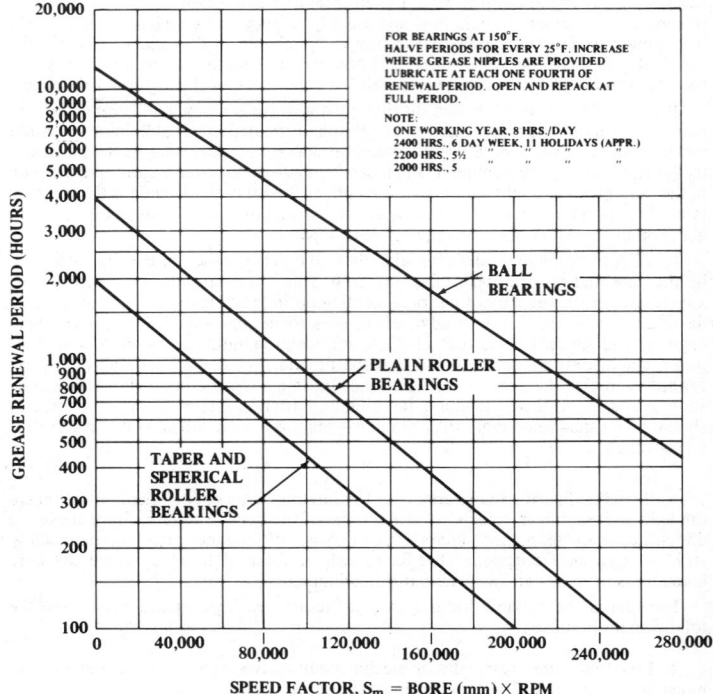

SPEED FACTOR, S_m = BORE (mm) × RPM

Conservative grease renewal periods will be found in the accompanying chart. Grease should not be allowed to remain in a bearing for longer than 48 months or if the service is very light and temperatures low, 60 months, irrespective of the number of hours' operation during that period as separation of the oil from the soap and oxidation continue whether the bearing is in operation or not.

Before renewing the grease in a hand-packed bearing, the bearing assembly should be removed and washed in clean kerosene, degreasing fluid or other solvent. As soon as the bearing is quite clean it should be washed at once in clean light mineral oil, preferably rust-inhibited. The bearing should *not be spun* before or while it is being oiled. Caustic solutions may be used if the old grease is hard and difficult to remove, but the best method is to soak the bearing for a few hours in light mineral oil, preferably warmed to about 130 degrees F., and then wash in cleaning fluid as described above. The use of chlorinated solvents is best avoided.

When replacing the grease, it should be forced with the fingers between the balls or rollers, dismantling the bearing, if convenient. The available space inside the bearing should be filled completely and the bearing then spun by hand. Any grease thrown out should be wiped off. The space on each side of the bearing in the housing should be not more than half-filled. Too much grease will result in considerable churning, high bearing temperatures and the possibility of early failure. Unlike any other kind of bearing, anti-friction bearings more often give trouble due to over-rather than to under-lubrication.

Oil Viscosities and Temperature Ranges for Ball Bearing Lubrication*

Maximum Temperature Range Degrees F.	Optimum Temperature Range, Degrees F.	Speed Factor, S_i†	
		Under 1000	Over 1000
		Viscosity	
− 40 to + 100	− 40 to − 10	80 to 90 SSU**	70 to 80 SSU**
− 10 to + 100	− 10 to + 30	100 to 115 SSU**	80 to 100 SSU**
+ 30 to + 150	+ 30 to + 150	SAE 20	SAE 10
+ 30 to + 200	+ 150 to + 200	SAE 40	SAE 30
+ 50 to + 300	+ 200 to + 300	SAE 70	SAE 60

* Not applicable to air-distributed oil mist lubrication.
** At 100 deg. F.
† Inner race bore diameter (inches) × RPM.

Grease is usually not very suitable for speed factors over 12,000 for S_i or 300,000 for S_m (although successful applications have been made up to an S_i of 50,000) or for temperatures much over 210 degrees F., 300 degrees F. being the extreme practical upper limit, even if synthetics are used. For temperatures above 210 degrees F., the grease renewal periods are very short.

Oil Lubrication: Oil lubrication is usually adopted when speeds and temperatures are high or when it is desired to adopt a central oil supply for the machine as a whole. Oil for anti-friction bearing lubrication should be well refined with high film strength and good resistance to oxidation and good corrosion protection. Anti-oxidation additives do no harm but are not really necessary at temperatures below about 200 degrees F. Anti-corrosion additives are always desirable. The accompanying table gives recommended viscosities of oil for ball bearing lubrication other than by an air-distributed oil mist. Within a given temperature and speed range, an oil towards the lighter end of the grade should be used, if convenient, as speeds increase. Roller bearings usually require an oil one grade heavier than do ball bearings for a given speed and temperature range. Cooled oil is sometimes circulated through an anti-friction bearing to carry off excess heat resulting from high speeds and heavy loads.

COUPLINGS AND CLUTCHES

Connecting Shafts: — For couplings to transmit up to about 150 horsepower, simple flange-type couplings of appropriate size, as shown in the table, are commonly used. The design shown is known as a safety flange coupling because the bolt heads and nuts are shrouded by the flange, but such couplings today are normally shielded by a sheet metal or other cover.

Safety Flange Couplings

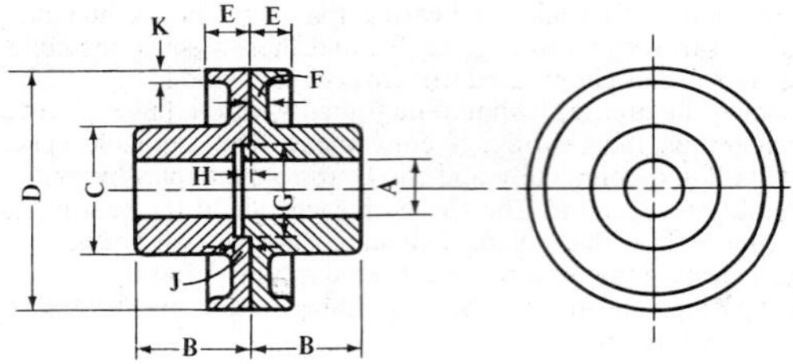

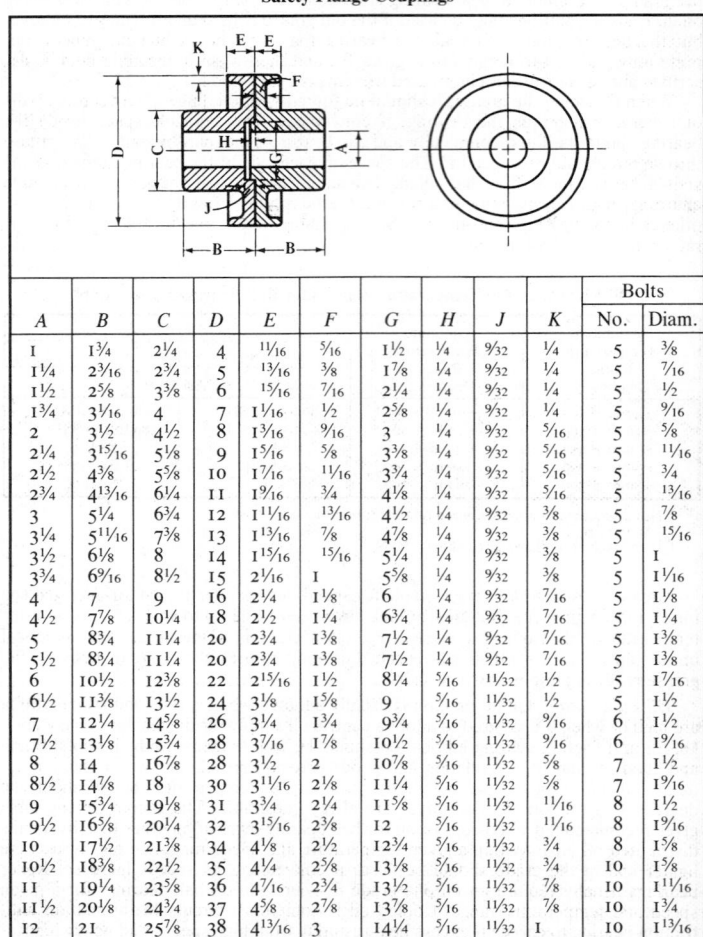

										Bolts	
A	*B*	*C*	*D*	*E*	*F*	*G*	*H*	*J*	*K*	No.	Diam.
1	1 3/4	2 1/4	4	11/16	5/16	1 1/2	1/4	9/32	1/4	5	3/8
1 1/4	2 3/16	2 3/4	5	13/16	3/8	1 7/8	1/4	9/32	1/4	5	7/16
1 1/2	2 5/8	3 3/8	6	15/16	7/16	2 1/4	1/4	9/32	1/4	5	1/2
1 3/4	3 1/16	4	7	1 1/16	1/2	2 5/8	1/4	9/32	1/4	5	9/16
2	3 1/2	4 1/2	8	1 3/16	9/16	3	1/4	9/32	5/16	5	5/8
2 1/4	3 15/16	5 1/8	9	1 5/16	5/8	3 3/8	1/4	9/32	5/16	5	11/16
2 1/2	4 3/8	5 5/8	10	1 7/16	11/16	3 3/4	1/4	9/32	5/16	5	3/4
2 3/4	4 13/16	6 1/4	11	1 9/16	3/4	4 1/8	1/4	9/32	5/16	5	13/16
3	5 1/4	6 3/4	12	1 11/16	13/16	4 1/2	1/4	9/32	3/8	5	7/8
3 1/4	5 11/16	7 3/8	13	1 13/16	7/8	4 7/8	1/4	9/32	3/8	5	15/16
3 1/2	6 1/8	8	14	1 15/16	15/16	5 1/4	1/4	9/32	3/8	5	1
3 3/4	6 9/16	8 1/2	15	2 1/16	1	5 5/8	1/4	9/32	3/8	5	1 1/16
4	7	9	16	2 1/4	1 1/8	6	1/4	9/32	7/16	5	1 1/8
4 1/2	7 7/8	10 1/4	18	2 1/2	1 1/4	6 3/4	1/4	9/32	7/16	5	1 1/4
5	8 3/4	11 1/4	20	2 3/4	1 3/8	7 1/2	1/4	9/32	7/16	5	1 3/8
5 1/2	8 3/4	11 1/4	20	2 3/4	1 3/8	7 1/2	1/4	9/32	7/16	5	1 3/8
6	10 1/2	12 3/8	22	2 15/16	1 1/2	8 1/4	5/16	11/32	1/2	5	1 7/16
6 1/2	11 3/8	13 1/2	24	3 1/8	1 5/8	9	5/16	11/32	1/2	5	1 1/2
7	12 1/4	14 5/8	26	3 1/4	1 3/4	9 3/4	5/16	11/32	9/16	6	1 1/2
7 1/2	13 1/8	15 3/4	28	3 7/16	1 7/8	10 1/2	5/16	11/32	9/16	6	1 9/16
8	14	16 7/8	28	3 1/2	2	10 7/8	5/16	11/32	5/8	7	1 1/2
8 1/2	14 7/8	18	30	3 11/16	2 1/8	11 1/4	5/16	11/32	5/8	7	1 9/16
9	15 3/4	19 1/8	31	3 3/4	2 1/4	11 5/8	5/16	11/32	11/16	8	1 1/2
9 1/2	16 5/8	20 1/4	32	3 15/16	2 3/8	12	5/16	11/32	11/16	8	1 9/16
10	17 1/2	21 3/8	34	4 1/8	2 1/2	12 3/4	5/16	11/32	3/4	8	1 5/8
10 1/2	18 3/8	22 1/2	35	4 1/4	2 5/8	13 1/8	5/16	11/32	3/4	10	1 5/8
11	19 1/4	23 5/8	36	4 7/16	2 3/4	13 1/2	5/16	11/32	7/8	10	1 11/16
11 1/2	20 1/8	24 3/4	37	4 5/8	2 7/8	13 7/8	5/16	11/32	7/8	10	1 3/4
12	21	25 7/8	38	4 13/16	3	14 1/4	5/16	11/32	1	10	1 13/16

For small sizes and low power applications, a setscrew may provide the connection between the hub and the shaft, but higher power usually requires a key and perhaps two setscrews, one of them above the key. A flat on the shaft and some means of locking the setscrew(s) in position are advisable. In the AGMA Class I and II fits the shaft tolerances are −0.0005 inch from ½ to 1½ inches diameter and −0.001 inch on larger diameters up to 7 inches.

Class I coupling bore tolerances are +0.001 inch up to 1½ inches diameter, then +0.0015 inch to 7 inches diameter. Class II coupling bore tolerances are +0.002 inch on sizes up to 3 inches diameter, +0.003 inch on sizes from 3¼ through 3¾ inches diameter, and +0.004 inch on larger diameters up to 7 inches.

Interference Fits. — Components of couplings transmitting over 150 horsepower often are made an interference fit on the shafts, which may reduce fretting corrosion. These couplings may or may not use keys, depending on the degree of interference. Keys may range in size from ⅛ inch wide by 1/16 inch high for ½-inch diameter shafts to 1¾ inches wide by ⅞ inch high for 7-inch diameter shafts. Couplings transmitting high torque or operating at high speeds or both may use two keys. Keys must be a good fit in their keyways to ensure good transmission of torque and prevent failure. AGMA standards provide recommendations for square parallel, rectangular section, and plain tapered keys, for shafts of 5/16 through 7 inches diameter, in three classes designated commercial, precision, and fitted. These standards also cover keyway offset, lead, parallelism, finish and radii, and face keys and splines. (See also ANSI and other Standards in Keys and Keyways section of this Handbook.)

Double-cone Clamping Couplings. — As shown in the table, double-cone clamping couplings are made in a range of sizes for shafts from 1 7/16 to 6 inches in diameter, and are easily assembled to shafts. These couplings provide an interference fit, but they usually cost more and have larger overall dimensions than regular flanged couplings.

Double-cone Clamping Couplings

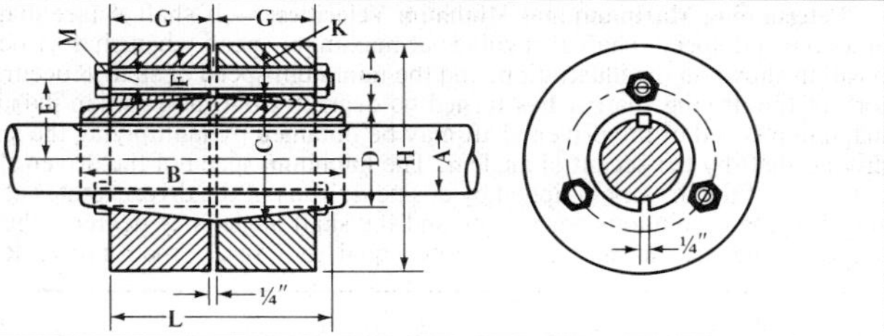

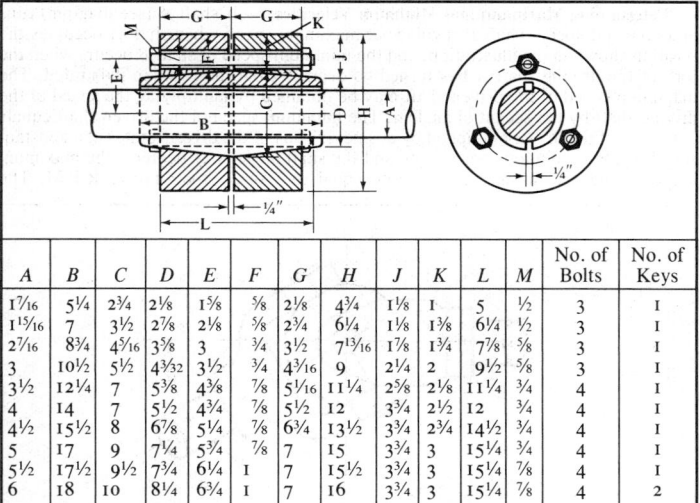

A	*B*	*C*	*D*	*E*	*F*	*G*	*H*	*J*	*K*	*L*	*M*	No. of Bolts	No. of Keys
1 7/16	5¼	2¾	2⅛	1⅝	⅝	2⅛	4¾	1⅛	1	5	½	3	1
1 15/16	7	3½	2⅞	2⅛	⅝	2¾	6¼	1⅛	1⅜	6¼	½	3	1
2 7/16	8¾	4 5/16	3⅝	3	¾	3½	7 13/16	1⅞	1¾	7⅞	⅝	3	1
3	10½	5½	4 3/32	3½	¾	4 3/16	9	2¼	2	9½	⅝	3	1
3½	12¼	7	5⅜	4⅜	⅞	5 1/16	11¼	2⅝	2⅛	11¼	¾	4	1
4	14	7	5½	4¾	⅞	5½	12	3¾	2½	12	¾	4	1
4½	15½	8	6⅞	5¼	⅞	6¾	13½	3¾	2¾	14½	¾	4	1
5	17	9	7¼	5¾	⅞	7	15	3¾	3	15¼	¾	4	1
5½	17½	9½	7¾	6¼	1	7	15½	3¾	3	15¼	⅞	4	1
6	18	10	8¼	6¾	1	7	16	3¾	3	15¼	⅞	4	2

Flexible Couplings. — Shafts that are out of alignment laterally or angularly can be connected by any of several designs of flexible couplings. Such couplings also permit some degree of axial movement in one or both shafts. Some couplings use disks or diaphragms to transmit the torque. Another simple form of flexible coupling consists of two flanges connected by links or endless belts made of leather or other strong, pliable material. Alternatively, the flanges may have projections that engage spacers of molded rubber or other flexible materials that accommodate uneven motion between the shafts. More highly developed flexible couplings use toothed flanges engaged by correspondingly toothed elements, permitting relative movement. These couplings require lubrication unless one or more of the elements is made of a self-lubricating material. Other couplings use diaphragms or bellows that can flex to accommodate relative movement between the shafts.

The Universal Joint. — This form of coupling, originally known as a Cardan or Hooke's coupling, is used for connecting two shafts the axes of which are not in line with each other, but which merely intersect at a point. There are many different designs of universal joints or couplings, which are based on the principle embodied in the original design. One well-known type is shown by the accompanying diagram.

As a rule, a universal joint does not work well if the angle α (see illustration) is more than 45 degrees, and the angle should preferably be limited to about 20 degrees or 25 degrees, excepting when the speed of rotation is slow and little power is transmitted.

Variation in Angular Velocity of Driven Shaft. — Owing to the angularity between two shafts connected by a universal joint, there is a variation in the angular velocity of one shaft during a single revolution, and because of this, the use of universal couplings is sometimes prohibited. Thus, the angular velocity of the driven shaft will not be the same at all points of the revolution as the angular velocity of the driving shaft. In other words, if the driving shaft moves with a uniform motion, then the driven shaft will have a variable motion and, therefore, the universal joint should not be used when absolute uniformity of motion is essential for the driven shaft.

Determining Maximum and Minimum Velocities. — If shaft A (see diagram) runs at a constant speed, shaft B revolves at maximum speed when shaft A occupies the position shown in the illustration, and the minimum speed of shaft B occurs when the fork of the driving shaft A has turned 90 degrees from the position illustrated. The maximum speed of the driven shaft may be obtained by multiplying the speed of the driving shaft by the secant of angle α. The minimum speed of the driven shaft equals the speed of the driver multiplied by cosine α. Thus, if the driver rotates at a constant speed of 100 revolutions per minute and the shaft angle is 25 degrees, the maximum speed of the driven shaft is at a rate equal to $1.1034 \times 100 = 110.34$ R.P.M. The

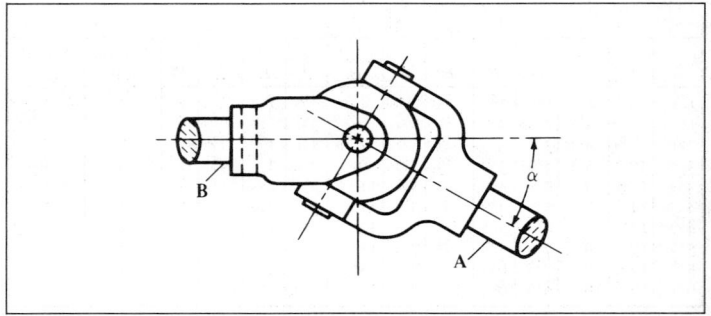

minimum speed rate equals $0.9063 \times 100 = 90.63$; hence, the extreme variation equals $110.34 - 90.63 = 19.71$ R.P.M.

Use of Intermediate Shaft between Two Universal Joints. — The lack of uniformity in the speed of the driven shaft resulting from the use of a universal coupling, as previously explained, is objectionable for some forms of mechanisms. This variation may be avoided if the two shafts are connected with an intermediate shaft and two universal joints, provided the latter are properly arranged or located. Two conditions are necessary to obtain a constant speed ratio between the driving and driven shafts. First, the shafts must make the same angle with the intermediate shaft; second, the universal joint forks (assuming that the fork design is employed) on the intermediate shaft must be placed relatively so that when the plane of the fork at the left end coincides with the center lines of the intermediate shaft and the shaft attached to the left-hand coupling, the plane of the right-hand fork must also coincide with the center lines of the intermediate shaft and the shaft attached to the right-hand coupling; therefore the driving and the driven shafts may be placed in a variety of positions. One of the most common arrangements is with the driving and driven shafts parallel. The forks on the intermediate shafts should then be placed in the same plane.

This intermediate connecting shaft is frequently made telescoping, and then the driving and driven shafts can be moved independently of each other within certain limits in longitudinal and lateral directions. The telescoping intermediate shaft consists of a rod which enters a sleeve and is provided with a suitable spline, to prevent rotation between the rod and sleeve and permit a sliding movement. This arrangement is applied to various machine tools.

Knuckle Joints. — Movement at the joint between two rods may be provided by knuckle joints, for which typical proportions are seen in the accompanying table.

Friction Clutches. — Clutches which transmit motion from the driving to the driven member by the friction between the engaging surfaces are built in many different designs, although practically all of them can be classified under four general types, namely, conical clutches; radially expanding clutches; contracting-band clutches; and friction disk clutches in single and multiple types. There are many modifications of these general classes, some of which combine the features of different types. The proportions of various sizes of cone clutches are given in the table "Cast-iron Friction Clutches." The multicone friction clutch is a further development of the cone clutch. Instead of having a single cone-shaped surface, there is a series of concentric conical rings which engage annular grooves formed by corresponding rings on the opposite clutch member. The internal-expanding type is provided with shoes which are forced outward against an enclosing drum by the action of levers connecting with a collar free to slide along the shaft. The engaging shoes are commonly lined with wood or other material to increase the coefficient of friction. Disk clutches are based on the principle of multiple-plane friction, and use alternating plates or disks so arranged that one set engages with an outside cylindrical case and the other set with the shaft. When these plates are pressed together by spring pressure, or by other means, motion is transmitted from the driving to the driven members connected to the clutch. Some disk clutches have a few rather heavy or thick plates and others a relatively large number of thinner plates. Clutches of the latter type are common in automobile transmissions. One set of disks may be of soft steel and the other set of phosphor-bronze, or some other combination may be employed. For instance, disks are sometimes provided with cork inserts, or one set or series of disks may be faced with a special friction material such as asbestos-wire fabric, as in "dry plate" clutches, the disks of which are not lubricated like the disks of a clutch having, for example, the steel and phosphor-bronze combination. It is common practice to hold the driving and driven members of friction clutches in engagement by means of spring pressure, although pneumatic or hydraulic pressure may be employed.

Proportions of Knuckle Joints

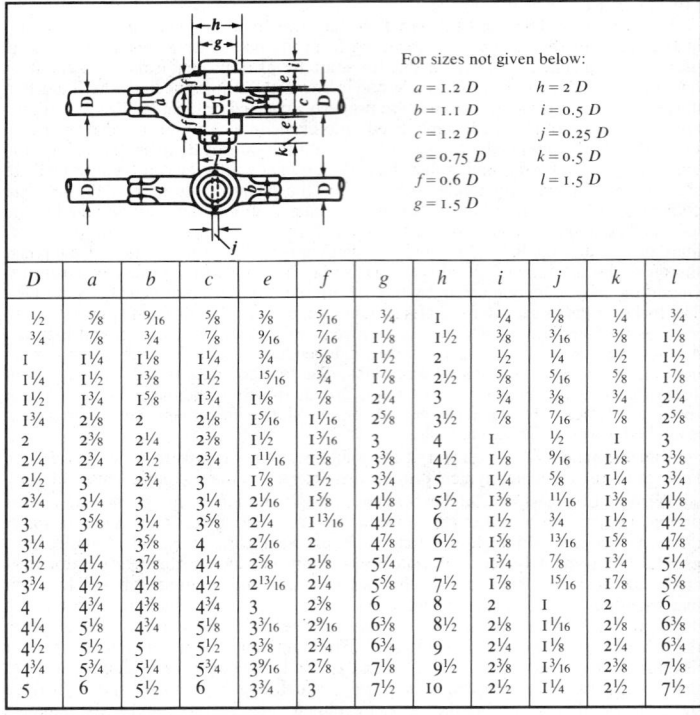

For sizes not given below:

$a = 1.2\,D$	$h = 2\,D$
$b = 1.1\,D$	$i = 0.5\,D$
$c = 1.2\,D$	$j = 0.25\,D$
$e = 0.75\,D$	$k = 0.5\,D$
$f = 0.6\,D$	$l = 1.5\,D$
$g = 1.5\,D$	

D	a	b	c	e	f	g	h	i	j	k	l
½	⅝	9/16	⅝	⅜	5/16	¾	1	¼	⅛	¼	¾
¾	⅞	¾	⅞	9/16	7/16	1⅛	1½	⅜	3/16	⅜	1⅛
1	1¼	1⅛	1¼	¾	⅝	1½	2	½	¼	½	1½
1¼	1½	1⅜	1½	15/16	¾	1⅞	2½	⅝	5/16	⅝	1⅞
1½	1¾	1⅝	1¾	1⅛	⅞	2¼	3	¾	⅜	¾	2¼
1¾	2⅛	2	2⅛	1 5/16	1 1/16	2⅝	3½	⅞	7/16	⅞	2⅝
2	2⅜	2¼	2⅜	1½	1 3/16	3	4	1	½	1	3
2¼	2¾	2½	2¾	1 11/16	1⅜	3⅜	4½	1⅛	9/16	1⅛	3⅜
2½	3	2¾	3	1⅞	1½	3¾	5	1¼	⅝	1¼	3¾
2¾	3¼	3	3¼	2 1/16	1⅝	4⅛	5½	1⅜	11/16	1⅜	4⅛
3	3⅝	3¼	3⅝	2¼	1 13/16	4½	6	1½	¾	1½	4½
3¼	4	3⅝	4	2 7/16	2	4⅞	6½	1⅝	13/16	1⅝	4⅞
3½	4¼	3⅞	4¼	2⅝	2⅛	5¼	7	1¾	⅞	1¾	5¼
3¾	4½	4⅛	4½	2 13/16	2¼	5⅝	7½	1⅞	15/16	1⅞	5⅝
4	4¾	4⅜	4¾	3	2⅜	6	8	2	1	2	6
4¼	5⅛	4¾	5⅛	3 3/16	2 9/16	6⅜	8½	2⅛	1 1/16	2⅛	6⅜
4½	5½	5	5½	3⅜	2¾	6¾	9	2¼	1⅛	2¼	6¾
4¾	5¾	5¼	5¾	3 9/16	2⅞	7⅛	9½	2⅜	1 3/16	2⅜	7⅛
5	6	5½	6	3¾	3	7½	10	2½	1¼	2½	7½

Power Transmitting Capacity of Friction Clutches. — When selecting a clutch for a given class of service, it is advisable to consider any overloads that may be encountered and base the power transmitting capacity of the clutch upon such overloads. When the load varies or is subject to frequent release or engagement, the clutch capacity should be greater than the actual amount of power transmitted. If the power is derived from a gas or gasoline engine, the horsepower rating of the clutch should be 75 or 100 per cent greater than that of the engine.

Power Transmitted by Disk Clutches. — The approximate amount of power that a disk clutch will transmit may be determined from the following formula, in which H = horsepower transmitted by the clutch: μ = coefficient of friction; r = mean radius of engaging surfaces; F = axial force in pounds (spring pressure) holding disks in contact; N = number of frictional surfaces; S = speed of shaft in revolutions per minute:

$$H = \frac{\mu r F N S}{63,000}$$

Cast-iron Friction Clutches

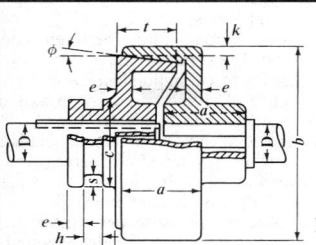

For sizes not given below:

$a = 2\,D$
$b = 4$ to $8\,D$
$c = 2\tfrac{1}{4}\,D$
$t = 1\tfrac{1}{2}\,D$
$e = \tfrac{3}{8}\,D$
$h = \tfrac{1}{2}\,D$
$s = \tfrac{5}{16}\,D$, nearly
$k = \tfrac{1}{4}\,D$

Note: — The angle ϕ of the cone may be from 4 to 10 degrees.

D	a	b	c	t	e	h	s	k
1	2	4– 8	2¼	1½	⅜	½	5⁄16	¼
1¼	2½	5–10	2⅞	1⅞	½	⅝	⅜	5⁄16
1½	3	6–12	3⅜	2¼	⅝	¾	½	⅜
1¾	3½	7–14	4	2⅝	⅝	⅞	⅝	7⁄16
2	4	8–16	4½	3	¾	1	⅝	½
2¼	4½	9–18	5	3⅜	⅞	1⅛	⅝	9⁄16
2½	5	10–20	5⅝	3¾	1	1¼	¾	⅝
2¾	5½	11–22	6¼	4⅛	1	1⅜	⅞	11⁄16
3	6	12–24	6¾	4½	1⅛	1½	⅞	¾
3¼	6½	13–26	7⅜	4⅞	1¼	1⅝	1	13⁄16
3½	7	14–28	7⅞	5¼	1⅜	1¾	1	⅞
3¾	7½	15–30	8½	5⅝	1⅜	1⅞	1¼	15⁄16
4	8	16–32	9	6	1½	2	1¼	1
4¼	8½	17–34	9½	6⅜	1⅝	2⅛	1⅜	1 1⁄16
4½	9	18–36	10¼	6¾	1¾	2¼	1⅜	1⅛
4¾	9½	19–38	10¾	7⅛	1¾	2⅜	1½	1 3⁄16
5	10	20–40	11¼	7½	1⅞	2½	1½	1¼
5¼	10½	21–42	11¾	7⅞	2	2⅝	1⅝	1 5⁄16
5½	11	22–44	12⅜	8¼	2	2¾	1¾	1⅜
5¾	11½	23–46	13	8⅝	2¼	2⅞	1¾	1 7⁄16
6	12	24–48	13½	9	2¼	3	1⅞	1½

Frictional Coefficients for Clutch Calculations. — While the frictional coefficients used by designers of clutches differ somewhat and depend upon variable factors, the following values may be used in clutch calculations: For greasy leather on cast iron about 0.20 or 0.25, leather on metal that is quite oily 0.15; metal and cork on oily metal 0.32; the same on dry metal 0.35; metal on dry metal 0.15; disk clutches having lubricated surfaces 0.10.

Formulas for Cone Clutches. — In cone clutch design, different formulas have been developed for determining the horsepower transmitted. These formulas, at first sight, do not seem to agree, there being a variation due to the fact that in some of the formulas the friction clutch surfaces are assumed to engage without slip, whereas, in others, some allowance is made for slip. The following formulas include both of these conditions:

H.P. = horsepower transmitted;
  N = revolutions per minute;
  r = mean radius of friction cone, in inches;
  r_1 = large radius of friction cone, in inches;
  r_2 = small radius of friction cone, in inches;
  R_1 = outside radius of leather band, in inches;
  R_2 = inside radius of leather band, in inches;
  V = velocity of a point at distance r from the center, in feet per minute;

F = tangential force acting at radius r, in pounds;
  P_n = total normal force between cone surfaces, in pounds;
  P_s = spring force, in pounds;
  α = angle of clutch surface with axis of shaft = 7 to 13 degrees;
  β = included angle of clutch leather, when developed, in degrees;
  f = coefficient of friction = 0.20 to 0.25 for greasy leather on iron;
  p = allowable pressure per square inch of leather band = 7 to 8 pounds;
  W = width of clutch leather, in inches.

$$R_1 = \frac{r_1}{\sin \alpha} \qquad R_2 = \frac{r_2}{\sin \alpha}$$

$$\beta = \sin \alpha \times 360 \qquad r = \frac{r_1 + r_2}{2}$$

$$V = \frac{2\pi r N}{12}$$

$$F = \frac{\text{H.P.} \times 33{,}000}{V}$$

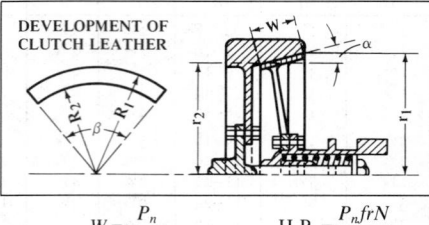

DEVELOPMENT OF CLUTCH LEATHER

$$W = \frac{P_n}{2\pi r p} \qquad \text{H.P.} = \frac{P_n f r N}{63{,}025}$$

For engagement with some slip:

$$P_n = \frac{P_s}{\sin \alpha} \qquad P_s = \frac{\text{H.P.} \times 63{,}025 \sin \alpha}{f r N}$$

For engagement without slip:

$$P_n = \frac{P_s}{\sin \alpha + f \cos \alpha} \qquad P_s = \frac{\text{H.P.} \times 63{,}025 \, (\sin \alpha + f \cos \alpha)}{f r N}$$

Angle of Cone. — If the angle of the conical surface of the cone type of clutch is too small, it may be difficult to release the clutch on account of the wedging effect, whereas, if the angle is too large, excessive spring force will be required to prevent slipping. The minimum angle for a leather-faced cone is about 8 or 9 degrees and the maximum angle about 13 degrees. An angle of 12½ degrees appears to be the most common and is generally considered good practice. These angles are given with relation to the clutch axis and are one-half the included angle.

Magnetic Clutches. — Many disk and other clutches are operated electromagnetically with the magnetic force used only to move the friction disk(s) and the clutch disk(s) into or out of engagement against spring or other pressure. On the other hand, in a magnetic particle clutch, transmission of power is accomplished by magnetizing a quantity of metal particles enclosed between the driving and the driven components, forming a bond between them. Such clutches can be controlled to provide either a rigid coupling or uniform slip, useful in wire drawing and manufacture of cables.

Another type of magnetic clutch uses eddy currents induced in the input member which interact with the field in the output rotor. Torque transmitted is proportional to the coil current, so precise control of torque is provided. A third type of magnetic clutch relies on the hysteresis loss between magnetic fields generated by a coil in an

input drum and a close-fitting cup on the output shaft, to transmit torque. Torque transmitted with this type of clutch also is proportional to coil current, so close control is possible.

Permanent-magnet types of clutches also are available, in which the engagement force is exerted by permanent magnets when the electrical supply to the disengagement coils is cut off. These types of clutches have capacities up to five times the torque-to-weight ratio of spring-operated clutches. In addition, if the controls are so arranged as to permit the coil polarity to be reversed instead of being cut off, the combined permanent magnet and electromagnetic forces can transmit even greater torque.

Centrifugal and Free-wheeling Clutches. — Centrifugal clutches have driving members that expand outward to engage a surrounding drum when speed is sufficient to generate centrifugal force. Free-wheeling clutches are made in many different designs and use balls, cams or sprags, ratchets, and fluids to transmit motion from one member to the other. These types of clutches are designed to transmit torque in only one direction and to take up the drive with various degrees of gradualness up to instantaneously.

Slipping Clutch/Couplings. — Where high shock loads are likely to be experienced, a slipping clutch or coupling or both should be used. The most common design uses a clutch plate that is clamped between the driving and driven plates by spring pressure that can be adjusted. When excessive load causes the driven member to slow, the clutch plate surfaces slip, allowing reduction of the torque transmitted. When the overload is removed, the drive is taken up automatically. Switches can be provided to cut off current supply to the driving motor when the driven shaft slows to a preset limit or to signal a warning or both. The slip or overload torque is calculated by taking 150 per cent of the normal running torque.

Wrapped-spring Clutches. — For certain applications, a simple steel spring sized so that its internal diameter is a snug fit on both driving and driven shafts will transmit adequate torque in one direction. The tightness of grip of the spring on the shafts increases as the torque transmitted increases. Disengagement can be effected by slight rotation of the spring, through a projecting tang, using electrical or mechanical means, to wind up the spring to a larger internal diameter, allowing one of the shafts to run free within the spring.

Normal running torque T_r in lb-ft = (required horsepower $\times$ 5250) $\div$ rpm. For heavy shock load applications, multiply by a 200 per cent or greater overload factor. (See Motors, factors governing selection.)

The clutch starting torque T_c, in lb-ft, required to accelerate a given inertia in a specific time is calculated from the formula:

$$T_c = \frac{WR^2 \times \Delta N}{308t}$$

where WR^2 = total inertia encountered by clutch in lb-ft^2 (W = weight and R = radius of gyration of rotating part); ΔN = final rpm $-$ initial rpm; 308 = constant (see Motors, factors governing selection); t = time to required speed in seconds.

Example: If the inertia is 80 lb-ft^2, and the speed of the driven shaft is to be increased from 0 to 1500 rpm in 3 seconds, find the clutch starting torque in lb-ft.

$$T_c = \frac{80 \times 1500}{308 \times 3} = 130 \text{ lb-ft}$$

The heat E, in BTU, generated in one engagement of a clutch can be calculated from the formula:

$$E = \frac{T_c \times WR^2 \times (N_1^2 - N_2^2)}{(T_c - T_1) \times 4.7 \times 10^6}$$

where: WR^2 = total inertia encountered by clutch in lb-ft.^{2}
N_1 = final rpm
N_2 = initial rpm
T_c = clutch torque in lb-ft.
T_1 = torque load in lb-ft.

Example: Calculate the heat generated for each engagement under the conditions cited for the first example.

$$E = \frac{130 \times 80 \times (1500)^2}{(130 - 10) \times 4.7 \times 10^6} = 41.5 \text{ BTU}$$

The preferred location for a clutch is on the high- rather than on the low-speed shaft because a smaller-capacity unit, of lower cost and with more rapid dissipation of heat, can be used. However, the heat generated may also be more because of the greater slippage at higher speeds, and the clutch may have a shorter life. For light-duty applications, such as to a machine tool, where cutting occurs after the spindle has reached operating speed, the calculated torque should be multiplied by a safety factor of 1.5 to arrive at the capacity of the clutch to be used. Heavy-duty applications such as frequent starting of a heavily loaded vibratory-finishing barrel require a safety factor of 3 or more.

Positive Clutches. — When the driving and driven members of a clutch are connected by the engagement of interlocking teeth or projecting lugs, the clutch is said to be "positive" to distinguish it from the type in which the power is transmitted by frictional contact. The positive clutch is employed when a sudden starting action is not objectionable and when the inertia of the driven parts is relatively small. The various forms of positive clutches differ merely in the angle or shape of the engaging surfaces. The least positive form is one having planes of engagement which incline backward, with respect to the direction of motion. The tendency of such a clutch is to disengage under load, in which case it must be held in position by axial pressure.

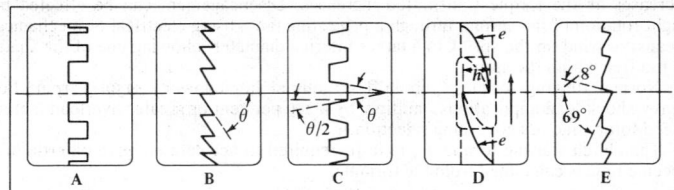

Fig. 1. Types of Clutch Teeth

This pressure may be regulated to perform normal duty, permitting the clutch to slip and disengage when over-loaded. Positive clutches, with the engaging planes parallel to the axis of rotation, are held together to obviate the tendency to jar out of engagement, but they provide no safety feature against over-load. So-called "under-cut" clutches engage more tightly the heavier the load, and are designed to be disengaged only when free from load. The teeth of positive clutches are made in a variety of forms, a few of the more common styles being shown in Fig. 1. Clutch A is a straight-toothed type, and B has angular or saw-shaped teeth. The driving member of the former can be rotated in either direction; the latter is adapted to the transmission of motion in one direction only, but is more readily engaged. The angle θ of the cutter for a saw-tooth clutch B is ordinarily 60 degrees. Clutch C is similar to A, except that the sides of the teeth are inclined to facilitate engagement and disengagement. Teeth of this shape are sometimes used when a clutch is required to run in either direction

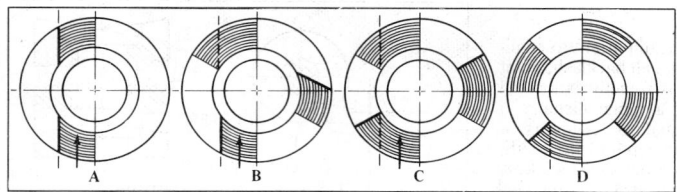

Fig. 2. Diagrammatic View Showing Method of Cutting Clutch Teeth

without backlash. Angle θ is varied to suit requirements and should not exceed 16 or 18 degrees. The straight-tooth clutch *A* is also modified to make the teeth engage more readily, by rounding the corners of the teeth at the top and bottom. Clutch *D* (commonly called a "spiral-jaw" clutch) differs from *B* in that the surfaces *e* are helicoidal. The driving member of this clutch can transmit motion in only one direction. Clutches

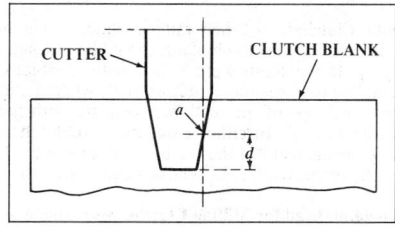

Fig. 3

of this type are known as right- and left-hand, the former driving when turning to the right, as indicated by the arrow in the illustration. Clutch *E* is the form used on the back-shaft of the Brown & Sharpe automatic screw machines. The faces of the teeth are radial and incline at an angle of 8 degrees with the axis, so that the clutch can readily be disengaged. This type of clutch is easily operated, with little jar or noise. The 2-inch diameter size has 10 teeth. Height of working face, ⅛ inch.

Cutting Clutch Teeth. — A common method of cutting a straight-tooth clutch is indicated by the diagrams *A*, *B* and *C*, Fig. 2, which show the first, second and third cuts required for forming the three teeth. The work is held in the chuck of a dividing-head, the latter being set at right angles to the table. A plain milling cutter may be used (unless the corners of the teeth are rounded), the side of the cutter being set to exactly coincide with the center-line. When the number of teeth in the clutch is odd, the cut can be taken clear across the blank as shown, thus finishing the sides of two teeth with one passage of the cutter. When the number of teeth is even, as at *D*, it is necessary to mill all the teeth on one side and then set the cutter for finishing the opposite side. Therefore, clutches of this type commonly have an odd number of teeth. The maximum width of the cutter depends upon the width of the space at the narrow ends of the teeth. If the cutter must be quite narrow in order to pass the narrow ends, some stock may be left in the tooth spaces, which must be removed by a separate cut. If the tooth is of the modified form shown at *C*, Fig. 1, the cutter should be set as indicated in Fig. 3; that is, so that a point *a* on the cutter at a radial distance *d* equal to one-half the depth of the clutch teeth lies in a radial plane. When

it is important to elimi-
nate all backlash, point
a is sometimes located
at a radial distance d
equal to six-tenths of
the depth of the tooth,
in order to leave clear-
ance spaces at the bot-
toms of the teeth; the
two clutch members
will then fit together
tightly. Clutches of this
type must be held in
mesh.

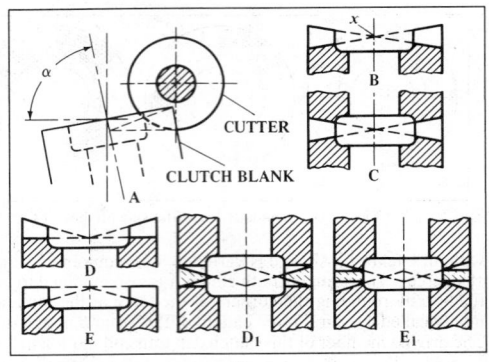

Fig. 4

Cutting Saw-tooth Clutches. — When milling clutches having angular teeth as shown at B, Fig. 1, the axis of the clutch blank should be inclined a certain angle α as shown at A in Fig. 4. If the teeth were milled with the blank vertical, the tops of the teeth would incline towards the center as at D, whereas, if the blank were set to such an angle that the tops of the teeth were square with the axis, the bottoms would incline upwards as at E. In either case, the two clutch members would not mesh completely; the engagement of the teeth cut as shown at D and E would be as indicated at D_1 and E_1 respectively. As will be seen, when the outer points of the

Angle of Dividing-head for Milling Clutches with Single-angle Cutter

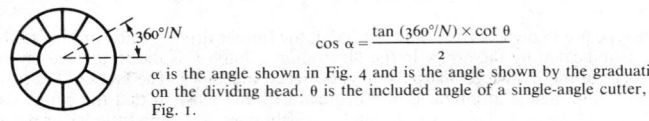

$$\cos \alpha = \frac{\tan (360°/N) \times \cot \theta}{2}$$

α is the angle shown in Fig. 4 and is the angle shown by the graduations on the dividing head. θ is the included angle of a single-angle cutter, see Fig. 1.

No. of Teeth, N	Angle of Single-angle Cutter, θ		
	60°	70°	80°
	Dividing Head Angle, α		
5	27° 19.2'	55° 56.3'	74° 15.4'
6	60	71 37.6	81 13
7	68 46.7	76 48.5	83 39.2
8	73 13.3	79 30.9	84 56.5
9	75 58.9	81 13	85 45.4
10	77 53.6	82 24.1	86 19.6
11	79 18.5	83 17	86 45.1
12	80 24.4	83 58.1	87 4.94
13	81 17.1	84 31.1	87 20.9
14	82 .536	84 58.3	87 34
15	82 36.9	85 21.2	87 45
16	83 7.95	85 40.6	87 54.4
17	83 34.7	85 57.4	88 2.56
18	83 58.1	86 12.1	88 9.67
19	84 18.8	86 25.1	88 15.9
20	84 37.1	86 36.6	88 21.5
21	84 53.5	86 46.9	88 26.5
22	85 8.26	86 56.2	88 31
23	85 21.6	87 4.63	88 35.1
24	85 33.8	87 12.3	88 38.8
25	85 45	87 19.3	88 42.2
26	85 55.2	87 25.7	88 45.3
27	86 4.61	87 31.7	88 48.2
28	86 13.3	87 37.2	88 50.8
29	86 21.4	87 42.3	88 53.3
30	86 28.9	87 47	88 55.6

teeth at D_1 are at the bottom of the grooves in the opposite member, the inner ends are not together, the contact area being represented by the dotted lines. At E_1 the inner ends of the teeth strike first and spaces are left between the teeth around the outside of the clutch. To overcome this objectionable feature, the clutch teeth should be cut as indicated at B, or so that the bottoms and tops of the teeth have the same inclination, converging at a central point x. The teeth of both members will then engage across the entire width as shown at C. The angle α required for cutting a clutch as at B can be determined by the following formula in which α equals the required angle, $N =$ number of teeth, $\theta =$ cutter angle, and $360°/N =$ angle between teeth:

$$\cos \alpha = \frac{\tan (360°/N) \times \cot \theta}{2}$$

The angles α for various numbers of teeth and for 60-, 70- or 80-degree single-angle cutters are given in the table on page 2226. The following table is for double-angle cutters used to cut V-shaped teeth.

Angle of Dividing-head for Milling V-shaped Teeth with Double-angle Cutter

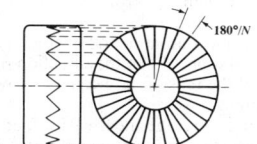

$$\cos \alpha = \frac{\tan (180°/N) \times \cot (\theta/2)}{2}$$

This is the angle (α, Fig. 4) shown by graduations on the dividing-head. θ is the included angle of a double-angle cutter, see Fig. 1.

No. of Teeth, N	Included Angle of Cutter, θ 60°	Included Angle of Cutter, θ 90°	No. of Teeth, N	Included Angle of Cutter, θ 60°	Included Angle of Cutter, θ 90°
	Dividing Head Angle, α			Dividing Head Angle, α	
10	73° 39.4'	80° 39'	31	84° 56.9'	87° 5.13'
11	75 16.1	81 33.5	32	85 6.42	87 10.6
12	76 34.9	82 18	33	85 15.4	87 15.8
13	77 40.5	82 55.3	34	85 23.8	87 20.7
14	78 36	83 26.8	35	85 31.8	87 25.2
15	79 23.6	83 54	36	85 39.3	87 29.6
16	80 4.83	84 17.5	37	85 46.4	87 33.7
17	80 41	84 38.2	38	85 53.1	87 37.5
18	81 13	84 56.5	39	85 59.5	87 41.2
19	81 41.5	85 12.8	40	86 5.51	87 44.7
20	82 6.97	85 27.5	41	86 11.3	87 48
21	82 30	85 40.7	42	86 16.7	87 51.2
22	82 50.8	85 52.6	43	86 22	87 54.2
23	83 9.82	86 3.56	44	86 26.9	87 57
24	83 27.2	86 13.5	45	86 31.7	87 59.8
25	83 43.1	86 22.7	46	86 36.2	88 2.4
26	83 57.8	86 31.2	47	86 40.6	88 4.91
27	84 11.4	86 39	48	86 44.8	88 7.32
28	84 24	86 46.2	49	86 48.8	88 9.63
29	84 35.7	86 53	50	86 52.6	88 11.8
30	84 46.7	86 59.3	51	86 56.3	88 14

The angles given in the table above are applicable to the milling of V-shaped grooves in brackets, etc., which must have toothed surfaces to prevent the two members from turning relative to each other, except when unclamped for angular adjustment.

FRICTION BRAKES

Formulas for Band Brakes. — In any band brake, such as shown in Fig. 1, in the tabulation of formulas, where the brake wheel rotates in a clockwise direction, the

tension in that part of the band marked x equals $P\dfrac{1}{e^{\mu\theta} - 1}$

The tension in that part marked y equals $P\dfrac{e^{\mu\theta}}{e^{\mu\theta} - 1}$

P = tangential force in pounds at rim of brake wheel;

e = base of natural logarithms = 2.71828;

μ = coefficient of friction between the brake band and the brake wheel;

θ = angle of contact of the brake band with the brake wheel expressed in

radians (one radian = $\dfrac{180 \text{ deg.}}{\pi \text{ radians}}$ = 57.296 $\dfrac{\text{deg.}}{\text{radian}}$).

For simplicity in the formulas presented, the tensions at x and y (Fig. 1) are denoted by T_1 and T_2 respectively, for clockwise rotation. When the direction of the rotation is reversed, the tension in x equals T_2, and the tension in y equals T_1, which is the reverse of the tension in the clockwise direction.

The value of the expression $e^{\mu\theta}$ in these formulas may be most easily found by using a hand-held calculator of the scientific type; that is, one capable of raising 2.71828 to the power $\mu\theta$. The following example outlines the steps in the calculations.

Table of Values of $e^{\mu\theta}$

Proportion of Contact to Whole Circumference, $\dfrac{\theta}{2\pi}$	Steel Band on Cast Iron, $\mu = 0.18$	Leather Belt on			
		Wood	Cast Iron		
		Slightly Greasy; $\mu = 0.47$	Very Greasy; $\mu = 0.12$	Slightly Greasy; $\mu = 0.28$	Damp; $\mu = 0.38$
0.1	1.12	1.34	1.08	1.19	1.27
0.2	1.25	1.81	1.16	1.42	1.61
0.3	1.40	2.43	1.25	1.69	2.05
0.4	1.57	3.26	1.35	2.02	2.60
0.425	1.62	3.51	1.38	2.11	2.76
0.45	1.66	3.78	1.40	2.21	2.93
0.475	1.71	4.07	1.43	2.31	3.11
0.5	1.76	4.38	1.46	2.41	3.30
0.525	1.81	4.71	1.49	2.52	3.50
0.55	1.86	5.07	1.51	2.63	3.72
0.6	1.97	5.88	1.57	2.81	4.19
0.7	2.21	7.90	1.66	3.43	5.32
0.8	2.47	10.60	1.83	4.09	6.75
0.9	2.77	14.30	1.97	4.87	8.57
1.0	3.10	19.20	2.12	5.81	10.90

Formulas for Simple and Differential Band Brakes

F = force in pounds at end of brake handle; P = tangential force in pounds at rim of brake wheel; e = base of natural logarithms = 2.71828; μ = coefficient of friction between the brake band and the brake wheel; θ = angle of contact of the brake band with the brake wheel, expressed in radians (one radian = 57.296 degrees).

$$T_1 = P \frac{1}{e^{\mu\theta} - 1} \qquad T_2 = P \frac{e^{\mu\theta}}{e^{\mu\theta} - 1}$$

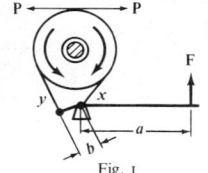 Fig. 1	Simple band brake. For clockwise rotation: $$F = \frac{bT_2}{a} = \frac{Pb}{a}\left(\frac{e^{\mu\theta}}{e^{\mu\theta} - 1}\right)$$ For counter clockwise rotation: $$F = \frac{bT_1}{a} = \frac{Pb}{a}\left(\frac{1}{e^{\mu\theta} - 1}\right)$$
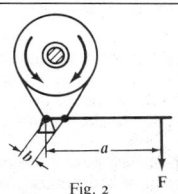 Fig. 2	Simple band brake. For clockwise rotation: $$F = \frac{bT_1}{a} = \frac{Pb}{a}\left(\frac{1}{e^{\mu\theta} - 1}\right)$$ For counter clockwise rotation: $$F = \frac{bT_2}{a} = \frac{Pb}{a}\left(\frac{e^{\mu\theta}}{e^{\mu\theta} - 1}\right)$$
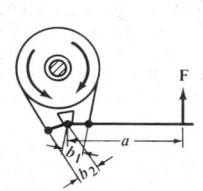 Fig. 3	Differential band brake. For clockwise rotation: $$F = \frac{b_2T_2 - b_1T_1}{a} = \frac{P}{a}\left(\frac{b_2e^{\mu\theta} - b_1}{e^{\mu\theta} - 1}\right)$$ For counter clockwise rotation: $$F = \frac{b_2T_1 - b_1T_2}{a} = \frac{P}{a}\left(\frac{b_2 - b_1e^{\mu\theta}}{e^{\mu\theta} - 1}\right)$$ In this case, if b_2 is equal to, or less than, $b_1e^{\mu\theta}$, the force F will be 0 or negative and the band brake works automatically.
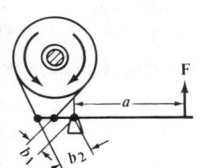 Fig. 4	Differential band brake. For clockwise rotation: $$F = \frac{b_2T_2 + b_1T_1}{a} = \frac{P}{a}\left(\frac{b_2e^{\mu\theta} + b_1}{e^{\mu\theta} - 1}\right)$$ For counter clockwise rotation: $$F = \frac{b_1T_2 + b_2T_1}{a} = \frac{P}{a}\left(\frac{b_1e^{\mu\theta} + b_2}{e^{\mu\theta} - 1}\right)$$ If $b_2 = b_1$, both of the above formulas reduce to $$F = \frac{Pb_1}{a}\left(\frac{e^{\mu\theta} + 1}{e^{\mu\theta} - 1}\right).$$ In this case, the same force F is required for rotation in either direction.

In a band brake of the type in Fig. 1, dimension $a = 24$ inches, and $b = 4$ inches; force $P = 100$ pounds; coefficient $\mu = 0.2$, and angle of contact = 240 degrees, or

$$\theta = \frac{240}{180} \times \pi = 4.18.$$

The rotation is clockwise. Find force F required.

$$F = \frac{Pb}{a}\left(\frac{e^{\mu\theta}}{e^{\mu\theta} - 1}\right)$$

$$= \frac{100 \times 4}{24}\left(\frac{2.71828^{0.2 \times 4.18}}{2.71828^{0.2 \times 4.18} - 1}\right)$$

$$= \frac{400}{24} \times \frac{2.71828^{0.836}}{2.71828^{0.836} - 1}$$

$$= 16.66 \times \frac{2.31}{2.31 - 1} = 29.4.$$

If a hand-held calculator is not used, determining the value of $e^{\mu\theta}$ is rather tedious, and the accompanying table will save calculations.

Coefficient of Friction in Brakes. — The coefficients of friction that may be assumed for friction brake calculations are as follows: Iron on iron, 0.25 to 0.3; leather on iron, 0.3; cork on iron, 0.35. Values somewhat lower than these should be assumed when the velocities exceed 400 feet per minute at the beginning of the braking operation.

For brakes where wooden brake blocks are used on iron drums, poplar has proved the best brake-block material. The best material for the brake drum is wrought iron. Poplar gives a high coefficient of friction, and is little affected by oil. The average coefficient of friction for poplar brake blocks and wrought-iron drums is 0.6; for poplar on cast iron, 0.35; for oak on wrought iron; 0.5; for oak on cast iron, 0.3; for beech on wrought iron, 0.5; for beech on cast iron, 0.3; for elm on wrought iron, 0.6; and for elm on cast iron, 0.35. The objection to elm is that the friction decreases rapidly if the friction surfaces are oily. The coefficient of friction for elm and wrought iron, if oily, is less than 0.4.

Calculating Horsepower from Dynamometer Tests. — When a dynamometer is arranged for obtaining the horsepower transmitted by a shaft, as indicated by the diagrammatic view in the illustration on page 2232, the horsepower may be obtained by the formula:

$$\text{H.P.} = \frac{2\pi LPN}{33,000}$$

in which H.P. = horsepower transmitted; N = number of revolutions per minute; L = distance (as shown in illustration) from center of pulley to point of action of weight P, in feet; P = weight hung on brake arm or read on scale.

By adopting a length of brake arm equal to 5 feet 3 inches, the formula may be reduced to the simple form:

$$\text{H.P.} = \frac{NP}{1000}$$

If a length of brake arm equal to 2 feet 7½ inches is adopted as a standard, the formula takes the form:

$$\text{H.P.} = \frac{NP}{2000}$$

Formulas for Block Brakes

F = force in pounds at end of brake handle;
P = tangential force in pounds at rim of brake wheel;
μ = coefficient of friction between the brake block and brake wheel.

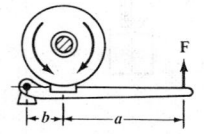

Fig. 1

Block brake.
For rotation in either direction:

$$F = P\,\frac{b}{a+b} \times \frac{1}{\mu} = \frac{Pb}{a+b}\left(\frac{1}{\mu}\right)$$

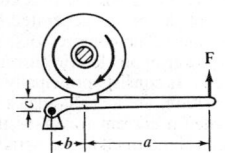

Fig. 2

Block brake.
For clockwise rotation:

$$F = \frac{\dfrac{Pb}{\mu} - Pc}{a+b} = \frac{Pb}{a+b}\left(\frac{1}{\mu} - \frac{c}{b}\right)$$

For counter clockwise rotation:

$$F = \frac{\dfrac{Pb}{\mu} + Pc}{a+b} = \frac{Pb}{a+b}\left(\frac{1}{\mu} + \frac{c}{b}\right)$$

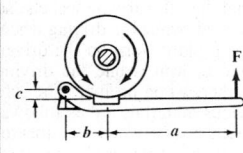

Fig. 3

Block brake.
For clockwise rotation:

$$F = \frac{\dfrac{Pb}{\mu} + Pc}{a+b} = \frac{Pb}{a+b}\left(\frac{1}{\mu} + \frac{c}{b}\right)$$

For counter clockwise rotation:

$$F = \frac{\dfrac{Pb}{\mu} - Pc}{a+b} = \frac{Pb}{a+b}\left(\frac{1}{\mu} - \frac{c}{b}\right)$$

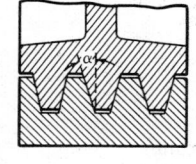

Fig. 4

The brake wheel and friction block of the block brake are often grooved as shown in Fig. 4. In this case, substitute for μ in the above equations the value $\dfrac{\mu}{\sin\alpha + \mu\cos\alpha}$ where α is one-half the angle included by the faces of the grooves.

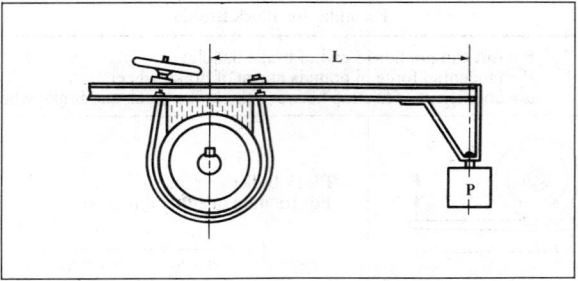

The *transmission* type of dynamometer measures the power by transmitting it through the mechanism of the dynamometer from the apparatus in which it is generated, or to the apparatus in which it is to be utilized. Dynamometers known as *indicators* operate by simultaneously measuring the pressure and volume of a confined fluid. This type may be used for the measurement of the power generated by steam or gas engines or absorbed by refrigerating machinery, air compressors, or pumps. An electrical dynamometer is for measuring the power of an electric current, based on the mutual action of currents flowing in two coils. It consists principally of one fixed and one movable coil, which, in the normal position, are at right angles to each other. Both coils are connected in series, and, when a current traverses the coils, the fields produced are at right angles; hence, the coils tend to take up a parallel position. The movable coil with an attached pointer will be deflected, the deflection measuring directly the electric current.

Friction Wheels for Power Transmission

When a rotating member is driven intermittently and the rate of driving does not need to be positive, friction wheels are frequently used, especially when the amount of power to be transmitted is comparatively small. The driven wheels in a pair of friction disks should always be made of a harder material than the driving wheels, so that if the driven wheel should be held stationary by the load, while the driving wheel revolves under its own pressure, a flat spot may not be rapidly worn on the driven wheel. The driven wheels, therefore, are usually made of iron, while the driving wheels are made of or covered with, rubber, paper, leather, wood or fiber. The safe working force per inch of face of contact for various materials are as follows: Straw fiber, 150; leather fiber, 240; tarred fiber, 240; leather, 150; wood, 100 to 150; paper, 150. Coefficients of friction for different combinations of materials are given in the following table. Smaller values should be used for exceptionally high speeds, or when the transmission must be started while under load.

Horsepower of Friction Wheels. — Let D = diameter of friction wheel in inches; N = Number of revolutions per minute; W = width of face in inches; f = coefficient of friction; P = force in pounds, per inch width of face. Then:

$$\text{H.P.} = \frac{3.1416 \times D \times N \times P \times W \times f}{33,000 \times 12}$$

Assume

$$\frac{3.1416 \times P \times f}{33,000 \times 12} = C;$$

then,

for $P = 100$ and $f = 0.20$, $\quad C = 0.00016$;
for $P = 150$ and $f = 0.20$, $\quad C = 0.00024$;
for $P = 200$ and $f = 0.20$, $\quad C = 0.00032$.

Working Values of Coefficient of Friction

Materials	Coefficient of Friction
Straw fiber and cast iron..............................	0.26
Straw fiber and aluminum..............................	0.27
Leather fiber and cast iron...........................	0.31
Leather fiber and aluminum...........................	0.30
Tarred fiber and cast iron............................	0.15
Paper and cast iron...................................	0.20
Tarred fiber and aluminum............................	0.18
Leather and cast iron.................................	0.14
Leather and aluminum.................................	0.22
Leather and typemetal................................	0.25
Wood and metal.......................................	0.25

The horsepower transmitted is then:

$$H.P. = D \times N \times W \times C.$$

Example: — Find the horsepower transmitted by a pair of friction wheels; the diameter of the driving wheel is 10 inches, and it revolves at 200 revolutions per minute. The width of the wheel is 2 inches. The force per inch width of face is 150 pounds, and the coefficient of friction 0.20.

$$H.P. = 10 \times 200 \times 2 \times 0.00024 = 0.96 \text{ horsepower.}$$

Horsepower Which May be Transmitted by Means of a Clean Paper Friction Wheel of One-inch Face when Run Under a Force of 150 Pounds
(Rockwood Mfg. Co.)

Diam. of Friction Wheel	Revolutions per Minute										
	25	50	75	100	150	200	300	400	600	800	1000
4	0.023	0.047	0.071	0.095	0.142	0.190	0.285	0.380	0.571	0.76	0.95
6	0.035	0.071	0.107	0.142	0.214	0.285	0.428	0.571	0.856	1.14	1.42
8	0.047	0.095	0.142	0.190	0.285	0.380	0.571	0.761	1.142	1.52	1.90
10	0.059	0.119	0.178	0.238	0.357	0.476	0.714	0.952	1.428	1.90	2.38
14	0.083	0.166	0.249	0.333	0.499	0.666	0.999	1.332	1.999	2.66	3.33
16	0.095	0.190	0.285	0.380	0.571	0.761	1.142	1.523	2.284	3.04	3.80
18	0.107	0.214	0.321	0.428	0.642	0.856	1.285	1.713	2.570	3.42	4.28
24	0.142	0.285	0.428	0.571	0.856	1.142	1.713	2.284	3.427	4.56	5.71
30	0.178	0.357	0.535	0.714	1.071	1.428	2.142	2.856	4.284	5.71	7.14
36	0.214	0.428	0.642	0.856	1.285	1.713	2.570	3.427	5.140	6.85	8.56
42	0.249	0.499	0.749	0.999	1.499	1.999	2.998	3.998	5.997	7.99	9.99
48	0.285	0.571	0.856	1.142	1.713	2.284	3.427	4.569	6.854	9.13	11.42
50	0.297	0.595	0.892	1.190	1.785	2.380	3.570	4.760	7.140	9.52	11.90

KEYS AND KEYSEATS

ANSI Standard Keys and Keyseats. — American National Standard, B17.1 Keys and Keyseats, based on current industry practice, was approved in 1967, and reaffirmed in 1973. This standard establishes a uniform relationship between shaft sizes and key sizes for parallel and taper keys as shown in Table 2. Other data in this standard are given in Tables 1 and 3–7. The sizes and tolerances shown are for single key applications only.

The following definitions are given in the standard:

Key: A demountable machinery part which, when assembled into keyseats, provides a positive means for transmitting torque between the shaft and hub.

Keyseat: An axially located rectangular groove in a shaft or hub.

This standard recognizes that there are two classes of stock for parallel keys used by industry. One is a close, plus toleranced key stock and the other is a broad, negative toleranced bar stock. Based on the use of two types of stock, two classes of fit are shown:

Class 1: A clearance of metal-to-metal side fit obtained by using bar stock keys and keyseat tolerances as given in Table 4. This is a relatively free fit and applies only to parallel keys.

Class 2: A side fit, with possible interference or clearance, obtained by using key stock and keyseat tolerances as given in Table 4. This is a relatively tight fit.

Class 3: This is an interference side fit and is not tabulated in Table 4 since the degree of interference has not been standardized. However, it is suggested that the top and bottom fit range given under Class 2 in Table 4, for parallel keys be used.

Key Size vs. Shaft Diameter: Shaft diameters are listed in Table 2 for identification of various key sizes and are not intended to establish shaft dimensions, tolerances or selections. For a stepped shaft, the size of a key is determined by the diameter of the shaft at the point of location of the key. Up through 6½-inch diameter shafts square keys are preferred; rectangular keys for larger shafts.

If special considerations dictate the use of a keyseat in the hub shallower than the preferred nominal depth shown, it is recommended that the tabulated preferred nominal standard keyseat be used in the shaft in all cases.

Keyseat Alignment Tolerances: A tolerance of 0.010 inch, max is provided for offset (due to parallel displacement of keyseat centerline from centerline of shaft or bore) of keyseats in shaft and bore. The following tolerances for maximum lead (due to angular displacement of keyseat centerline from centerline of shaft or bore and measured at right angles to the shaft or bore centerline) of keyseats in shaft and bore are specified: 0.002 inch for keyseat length up to and including 4 inches; 0.0005 inch per inch of length for keyseat lengths above 4 inches to and including 10 inches; and 0.005 inch for keyseat lengths above 10 inches. For the effect of keyways on shaft strength, see pages 312–313.

ANSI Standard Woodruff Keys and Keyseats. — American National Standard B17.2 was approved in 1967, and reaffirmed in 1978. Data from this standard are shown in Tables 9, 10, and 11.

The following definitions are given in this standard:

Woodruff Key: A demountable machinery part which, when assembled into keyseats, provides a positive means for transmitting torque between the shaft and hub.

Woodruff Key Number: An identification number by which the size of key may be readily determined.

Woodruff Keyseat — Shaft: The circular pocket in which the key is retained.

Woodruff Keyseat — Hub: An axially located rectangular groove in a hub. (This has been referred to as a keyway.)

Woodruff Keyseat Milling Cutter: An arbor type or shank type milling cutter normally used for milling Woodruff keyseats in shafts.

Table 1. ANSI Standard Plain and Gib Head Keys (ANSI B17.1-1967, R1973)

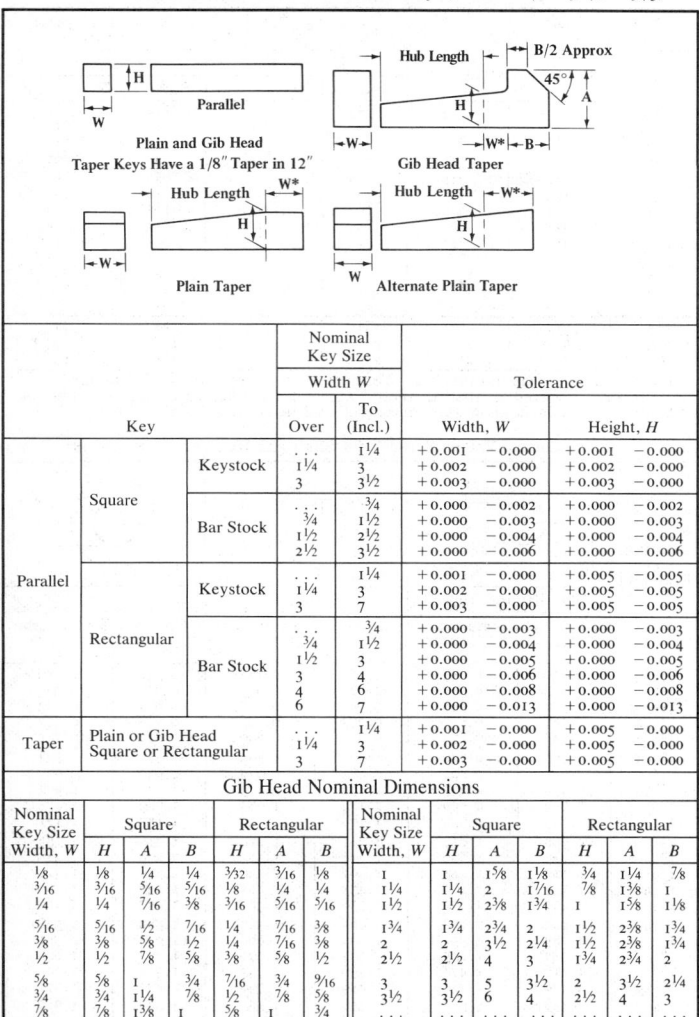

Key			Nominal Key Size Width W Over	To (Incl.)	Tolerance Width, W +	Width, W −	Height, H +	Height, H −
Parallel	Square	Keystock	. . .	1¼	+ 0.001	− 0.000	+ 0.001	− 0.000
			1¼	3	+ 0.002	− 0.000	+ 0.002	− 0.000
			3	3½	+ 0.003	− 0.000	+ 0.003	− 0.000
		Bar Stock	. . .	¾	+ 0.000	− 0.002	+ 0.000	− 0.002
			¾	1½	+ 0.000	− 0.003	+ 0.000	− 0.003
			1½	2½	+ 0.000	− 0.004	+ 0.000	− 0.004
			2½	3½	+ 0.000	− 0.006	+ 0.000	− 0.006
	Rectangular	Keystock	. . .	1¼	+ 0.001	− 0.000	+ 0.005	− 0.005
			1¼	3	+ 0.002	− 0.000	+ 0.005	− 0.005
			3	7	+ 0.003	− 0.000	+ 0.005	− 0.005
		Bar Stock	. . .	¾	+ 0.000	− 0.003	+ 0.000	− 0.003
			¾	1½	+ 0.000	− 0.004	+ 0.000	− 0.004
			1½	3	+ 0.000	− 0.005	+ 0.000	− 0.005
			3	4	+ 0.000	− 0.006	+ 0.000	− 0.006
			4	6	+ 0.000	− 0.008	+ 0.000	− 0.008
			6	7	+ 0.000	− 0.013	+ 0.000	− 0.013
Taper	Plain or Gib Head Square or Rectangular		. . .	1¼	+ 0.001	− 0.000	+ 0.005	− 0.000
			1¼	3	+ 0.002	− 0.000	+ 0.005	− 0.000
			3	7	+ 0.003	− 0.000	+ 0.005	− 0.000

Gib Head Nominal Dimensions

Nominal Key Size Width, W	Square H	A	B	Rectangular H	A	B	Nominal Key Size Width, W	Square H	A	B	Rectangular H	A	B
⅛	⅛	¼	¼	3/32	3/16	⅛	1	1	1⅝	1⅛	¾	1¼	⅞
3/16	3/16	5/16	5/16	⅛	¼	¼	1¼	1¼	2	1 7/16	⅞	1⅜	1
¼	¼	7/16	⅜	3/16	5/16	5/16	1½	1½	2⅜	1¾	1	1⅝	1⅛
5/16	5/16	½	7/16	¼	7/16	⅜	1¾	1¾	2¾	2	1½	2⅜	1¾
⅜	⅜	⅝	½	¼	7/16	⅜	2	2	3½	2¼	1½	2⅜	1¾
½	½	⅞	⅝	⅜	⅝	½	2½	2½	4	3	1¾	2¾	2
⅝	⅝	1	¾	7/16	¾	9/16	3	3	5	3½	2	3½	2¼
¾	¾	1¼	⅞	½	⅞	⅝	3½	3½	6	4	2½	4	3
⅞	⅞	1⅜	1	⅝	1	¾	. . .	. . .	. . .	. . .	. . .	. . .	. . .

All dimensions are given in inches.
* For locating position of dimension H. Tolerance does not apply.
For larger sizes the following relationships are suggested as guides for establishing A and B:
$A = 1.8H$ and $B = 1.2H$.

Table 2. Key Size Versus Shaft Diameter (ANSI B17.1-1967, R1973)

Nominal Shaft Diameter		Nominal Key Size			Nominal Keyseat Depth	
			Height, H		$H/2$	
Over	To (Incl.)	Width, W	Square	Rectangular	Square	Rectangular
5/16	7/16	3/32	3/32	....	3/64	....
7/16	9/16	1/8	1/8	3/32	1/16	3/64
9/16	7/8	3/16	3/16	1/8	3/32	1/16
7/8	1 1/4	1/4	1/4	3/16	1/8	3/32
1 1/4	1 3/8	5/16	5/16	1/4	5/32	1/8
1 3/8	1 3/4	3/8	3/8	1/4	3/16	1/8
1 3/4	2 1/4	1/2	1/2	3/8	1/4	3/16
2 1/4	2 3/4	5/8	5/8	7/16	5/16	7/32
2 3/4	3 1/4	3/4	3/4	1/2	3/8	1/4
3 1/4	3 3/4	7/8	7/8	5/8	7/16	5/16
3 3/4	4 1/2	1	1	3/4	1/2	3/8
4 1/2	5 1/2	1 1/4	1 1/4	7/8	5/8	7/16
5 1/2	6 1/2	1 1/2	1 1/2	1	3/4	1/2
6 1/2	7 1/2	1 3/4	1 3/4	1 1/2*	7/8	3/4
7 1/2	9	2	2	1 1/2	1	3/4
9	11	2 1/2	2 1/2	1 3/4	1 1/4	7/8

All dimensions are given in inches. For larger shaft sizes, see ANSI Standard.
Square keys preferred for shaft diameters above heavy line; rectangular keys, below.
* Some key standards show 1 1/4 inches; preferred height is 1 1/2 inches.

Table 3. Depth Control Values S and T for Shaft and Hub (ANSI B17.1-1967, R1973)

Nominal Shaft Diameter	Parallel and Taper		Parallel		Taper	
	Square	Rectangular	Square	Rectangular	Square	Rectangular
	S	S	T	T	T	T
1/2	0.430	0.445	0.560	0.544	0.535	0.519
9/16	0.493	0.509	0.623	0.607	0.598	0.582
5/8	0.517	0.548	0.709	0.678	0.684	0.653
11/16	0.581	0.612	0.773	0.742	0.748	0.717
3/4	0.644	0.676	0.837	0.806	0.812	0.781
13/16	0.708	0.739	0.900	0.869	0.875	0.844
7/8	0.771	0.802	0.964	0.932	0.939	0.907
15/16	0.796	0.827	1.051	1.019	1.026	0.994
1	0.859	0.890	1.114	1.083	1.089	1.058
1 1/16	0.923	0.954	1.178	1.146	1.153	1.121
1 1/8	0.986	1.017	1.241	1.210	1.216	1.185
1 3/16	1.049	1.080	1.304	1.273	1.279	1.248
1 1/4	1.112	1.144	1.367	1.336	1.342	1.311
1 5/16	1.137	1.169	1.455	1.424	1.430	1.399
1 3/8	1.201	1.232	1.518	1.487	1.493	1.462
1 7/16	1.225	1.288	1.605	1.543	1.580	1.518
1 1/2	1.289	1.351	1.669	1.606	1.644	1.581
1 9/16	1.352	1.415	1.732	1.670	1.707	1.645
1 5/8	1.416	1.478	1.796	1.733	1.771	1.708
1 11/16	1.479	1.541	1.859	1.796	1.834	1.771
1 3/4	1.542	1.605	1.922	1.860	1.897	1.835
1 13/16	1.527	1.590	2.032	1.970	2.007	1.945
1 7/8	1.591	1.654	2.096	2.034	2.071	2.009
1 15/16	1.655	1.717	2.160	2.097	2.135	2.072
2	1.718	1.781	2.223	2.161	2.198	2.136

All dimensions are given in inches. See Table 4 for tolerances.

Table 3. *(Concluded).* **Depth Control Values S and T for Shaft and Hub** (ANSI B17.1-1967, R1973)

Nominal Shaft Diameter	Parallel and Taper Square S	Parallel and Taper Rectangular S	Parallel Square T	Parallel Rectangular T	Taper Square T	Taper Rectangular T
2 1/16	1.782	1.844	2.287	2.224	2.262	2.199
2 1/8	1.845	1.908	2.350	2.288	2.325	2.263
2 3/16	1.909	1.971	2.414	2.351	2.389	2.326
2 1/4	1.972	2.034	2.477	2.414	2.452	2.389
2 5/16	1.957	2.051	2.587	2.493	2.562	2.468
2 3/8	2.021	2.114	2.651	2.557	2.626	2.532
2 7/16	2.084	2.178	2.714	2.621	2.689	2.596
2 1/2	2.148	2.242	2.778	2.684	2.753	2.659
2 9/16	2.211	2.305	2.841	2.748	2.816	2.723
2 5/8	2.275	2.369	2.905	2.811	2.880	2.786
2 11/16	2.338	2.432	2.968	2.874	2.943	2.849
2 3/4	2.402	2.495	3.032	2.938	3.007	2.913
2 13/16	2.387	2.512	3.142	3.017	3.117	2.992
2 7/8	2.450	2.575	3.205	3.080	3.180	3.055
2 15/16	2.514	2.639	3.269	3.144	3.244	3.119
3	2.577	2.702	3.332	3.207	3.307	3.182
3 1/16	2.641	2.766	3.396	3.271	3.371	3.246
3 1/8	2.704	2.829	3.459	3.334	3.434	3.309
3 3/16	2.768	2.893	3.523	3.398	3.498	3.373
3 1/4	2.831	2.956	3.586	3.461	3.561	3.436
3 5/16	2.816	2.941	3.696	3.571	3.671	3.546
3 3/8	2.880	3.005	3.760	3.635	3.735	3.610
3 7/16	2.943	3.068	3.823	3.698	3.798	3.673
3 1/2	3.007	3.132	3.887	3.762	3.862	3.737
3 9/16	3.070	3.195	3.950	3.825	3.925	3.800
3 5/8	3.134	3.259	4.014	3.889	3.989	3.864
3 11/16	3.197	3.322	4.077	3.952	4.052	3.927
3 3/4	3.261	3.386	4.141	4.016	4.116	3.991
3 13/16	3.246	3.371	4.251	4.126	4.226	4.101
3 7/8	3.309	3.434	4.314	4.189	4.289	4.164
3 15/16	3.373	3.498	4.378	4.253	4.353	4.228
4	3.436	3.561	4.441	4.316	4.416	4.291
4 3/16	3.627	3.752	4.632	4.507	4.607	4.482
4 1/4	3.690	3.815	4.695	4.570	4.670	4.545
4 3/8	3.817	3.942	4.822	4.697	4.797	4.672
4 7/16	3.880	4.005	4.885	4.760	4.860	4.735
4 1/2	3.944	4.069	4.949	4.824	4.924	4.799
4 3/4	4.041	4.229	5.296	5.109	5.271	5.084
4 7/8	4.169	4.356	5.424	5.236	5.399	5.211
4 15/16	4.232	4.422	5.487	5.300	5.462	5.275
5	4.296	4.483	5.551	5.363	5.526	5.338
5 3/16	4.486	4.674	5.741	5.554	5.716	5.529
5 1/4	4.550	4.737	5.805	5.617	5.780	5.592
5 7/16	4.740	4.927	5.995	5.807	5.970	5.782
5 1/2	4.803	4.991	6.058	5.871	6.033	5.846
5 3/4	4.900	5.150	6.405	6.155	6.380	6.130
5 15/16	5.091	5.341	6.596	6.346	6.571	6.321
6	5.155	5.405	6.660	6.410	6.635	6.385
6 1/4	5.409	5.659	6.914	6.664	6.889	6.639
6 1/2	5.662	5.912	7.167	6.917	7.142	6.892
6 3/4	5.760	*5.885	7.515	*7.390	7.490	*7.365
7	6.014	*6.139	7.769	*7.644	7.744	*7.619
7 1/4	6.268	*6.393	8.023	*7.898	7.998	*7.873
7 1/2	6.521	*6.646	8.276	*8.151	8.251	*8.126
7 3/4	6.619	6.869	8.624	8.374	8.599	8.349
8	6.873	7.123	8.878	8.628	8.853	8.603
9	7.887	8.137	9.892	9.642	9.867	9.617
10	8.591	8.966	11.096	10.721	11.071	10.696
11	9.606	9.981	12.111	11.736	12.086	11.711
12	10.309	10.809	13.314	12.814	13.289	12.789
13	11.325	11.825	14.330	13.830	14.305	13.805
14	12.028	12.528	15.533	15.033	15.508	15.008
15	13.043	13.543	16.548	16.048	16.523	16.023

All dimensions given in inches. See Table 4 for tolerances.
* 1 3/4 × 1 1/2 inch key.

Table 4. ANSI Standard Fits for Parallel and Taper Keys (ANSI B17.1-1967, R1973)

Type of Key	Key Width		Side Fit			Top and Bottom Fit			
			Width Tolerance			Depth Tolerance			
	Over	To (Incl.)	Key	Key-Seat	Fit Range*	Key	Shaft Key-Seat	Hub Key-Seat	Fit Range*
Class 1 Fit for Parallel Keys									
Square	...	½	+0.000 / −0.002	+0.002 / −0.000	0.004 CL / 0.000	+0.000 / −0.002	+0.000 / −0.015	+0.010 / −0.000	0.032 CL / 0.005 CL
	½	¾	+0.000 / −0.002	+0.003 / −0.000	0.005 CL / 0.000	+0.000 / −0.002	+0.000 / −0.015	+0.010 / −0.000	0.032 CL / 0.005 CL
	¾	1	+0.000 / −0.003	+0.003 / −0.000	0.006 CL / 0.000	+0.000 / −0.003	+0.000 / −0.015	+0.010 / −0.000	0.033 CL / 0.005 CL
	1	1½	+0.000 / −0.003	+0.004 / −0.000	0.007 CL / 0.000	+0.000 / −0.003	+0.000 / −0.015	+0.010 / −0.000	0.033 CL / 0.005 CL
	1½	2½	+0.000 / −0.004	+0.004 / −0.000	0.008 CL / 0.000	+0.000 / −0.004	+0.000 / −0.015	+0.010 / −0.000	0.034 CL / 0.005 CL
	2½	3½	+0.000 / −0.006	+0.004 / −0.000	0.010 CL / 0.000	+0.000 / −0.006	+0.000 / −0.015	+0.010 / −0.000	0.036 CL / 0.005 CL
Rectangular	...	½	+0.000 / −0.003	+0.002 / −0.000	0.005 CL / 0.000	+0.000 / −0.003	+0.000 / −0.015	+0.010 / −0.000	0.033 CL / 0.005 CL
	½	¾	+0.000 / −0.003	+0.003 / −0.000	0.006 CL / 0.000	+0.000 / −0.003	+0.000 / −0.015	+0.010 / −0.000	0.033 CL / 0.005 CL
	¾	1	+0.000 / −0.004	+0.003 / −0.000	0.007 CL / 0.000	+0.000 / −0.004	+0.000 / −0.015	+0.010 / −0.000	0.034 CL / 0.005 CL
	1	1½	+0.000 / −0.004	+0.004 / −0.000	0.008 CL / 0.000	+0.000 / −0.004	+0.000 / −0.015	+0.010 / −0.000	0.034 CL / 0.005 CL
	1½	3	+0.000 / −0.005	+0.004 / −0.000	0.009 CL / 0.000	+0.000 / −0.005	+0.000 / −0.015	+0.010 / −0.000	0.035 CL / 0.005 CL
	3	4	+0.000 / −0.006	+0.004 / −0.000	0.010 CL / 0.000	+0.000 / −0.006	+0.000 / −0.015	+0.010 / −0.000	0.036 CL / 0.005 CL
	4	6	+0.000 / −0.008	+0.004 / −0.000	0.012 CL / 0.000	+0.000 / −0.008	+0.000 / −0.015	+0.010 / −0.000	0.038 CL / 0.005 CL
	6	7	+0.000 / −0.013	+0.004 / −0.000	0.017 CL / 0.000	+0.000 / −0.013	+0.000 / −0.015	+0.010 / −0.000	0.043 CL / 0.005 CL
Class 2 Fit for Parallel and Taper Keys									
Parallel Square	...	1¼	+0.001 / −0.000	+0.002 / −0.000	0.002 CL / 0.001 INT	+0.001 / −0.000	+0.000 / −0.015	+0.010 / −0.000	0.030 CL / 0.004 CL
	1¼	3	+0.002 / −0.000	+0.002 / −0.000	0.002 CL / 0.002 INT	+0.002 / −0.000	+0.000 / −0.015	+0.010 / −0.000	0.030 CL / 0.003 CL
	3	3½	+0.003 / −0.000	+0.002 / −0.000	0.002 CL / 0.003 INT	+0.003 / −0.000	+0.000 / −0.015	+0.010 / −0.000	0.030 CL / 0.002 CL
Parallel Rectangular	...	1¼	+0.001 / −0.000	+0.002 / −0.000	0.002 CL / 0.001 INT	+0.005 / −0.005	+0.000 / −0.015	+0.010 / −0.000	0.035 CL / 0.000 CL
	1¼	3	+0.002 / −0.000	+0.002 / −0.000	0.002 CL / 0.002 INT	+0.005 / −0.005	+0.000 / −0.015	+0.010 / −0.000	0.035 CL / 0.000 CL
	3	7	+0.003 / −0.000	+0.002 / −0.000	0.002 CL / 0.003 INT	+0.005 / −0.005	+0.000 / −0.015	+0.010 / −0.000	0.035 CL / 0.000 CL
Taper	...	1¼	+0.001 / −0.000	+0.002 / −0.000	0.002 CL / 0.001 INT	+0.005 / −0.000	+0.000 / −0.015	+0.010 / −0.000	0.005 CL / 0.025 INT
	1¼	3	+0.002 / −0.000	+0.002 / −0.000	0.002 CL / 0.002 INT	+0.005 / −0.000	+0.000 / −0.015	+0.010 / −0.000	0.005 CL / 0.025 INT
	3	§	+0.003 / −0.000	+0.002 / −0.000	0.002 CL / 0.003 INT	+0.005 / −0.000	+0.000 / −0.015	+0.010 / −0.000	0.005 CL / 0.025 INT

All dimensions are given in inches.　See also text on page 2234.
* Limits of variation. CL = Clearance; INT = Interference.
§ To (Incl.) 3½-inch Square and 7-inch Rectangular key widths.

Chamfered Keys and Filleted Keyseats. — In general practice, chamfered keys and filleted keyseats are not used. However, it is recognized that fillets in keyseats decrease stress concentrations at corners. When used, fillet radii should be as large as possible without causing excessive bearing stresses due to reduced contact area between the key and its mating parts. Keys must be chamfered or rounded to clear fillet radii. Values in Table 5 assume general conditions and should be used only as a guide when critical stresses are encountered.

Table 5. Suggested Keyseat Fillet Radius and Key Chamfer (ANSI B17.1-1967, R1973)

Keyseat Depth, $H/2$		Fillet Radius	45 deg. Chamfer	Keyseat Depth, $H/2$		Fillet Radius	45 deg. Chamfer
Over	To (Incl.)			Over	To (Incl.)		
1/8	1/4	1/32	3/64	7/8	1 1/4	3/16	7/32
1/4	1/2	1/16	5/64	1 1/4	1 3/4	1/4	9/32
1/2	7/8	1/8	5/32	1 3/4	2 1/2	3/8	13/32

All dimensions are given in inches.

Table 6. ANSI Standard Keyseat Tolerances for Electric Motor and Generator Shaft Extensions (ANSI B17.1-1967, R1973)

Keyseat Width		Width Tolerance	Depth Tolerance
Over	To (Incl.)		
...	1/4	+0.001 −0.001	+0.000 −0.015
1/4	3/4	+0.000 −0.002	+0.000 −0.015
3/4	1 1/4	+0.000 −0.003	+0.000 −0.015

All dimensions are given in inches.

Table 7. Set Screws for Use Over Keys* (ANSI B17.1-1967, R1973)

Nom. Shaft Diam.		Nom. Key Width	Set Screw Diam.	Nom. Shaft Diam.		Nom. Key Width	Set Screw Diam.
Over	To (Incl.)			Over	To (Incl.)		
5/16	7/16	3/32	No. 10	2 1/4	2 3/4	5/8	1/2
7/16	9/16	1/8	No. 10	2 3/4	3 1/4	3/4	5/8
9/16	7/8	3/16	1/4	3 1/4	3 3/4	7/8	3/4
7/8	1 1/4	1/4	5/16	3 3/4	4 1/2	1	3/4
1 1/4	1 3/8	5/16	3/8	4 1/2	5 1/2	1 1/4	7/8
1 3/8	1 3/4	3/8	3/8	5 1/2	6 1/2	1 1/2	1
1 3/4	2 1/4	1/2	1/2	...	...	...	...

All dimensions are given in inches.
* These set screw diameter selections are offered as a guide but their use should be dependent upon design considerations.

Table 8. Finding Depth of Keyseat and Distance from Top of Key to Bottom of Shaft

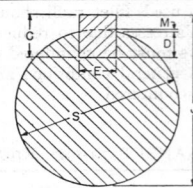

For milling keyseats, the total depth to feed cutter in from outside of shaft to bottom of keyseat is $M + D$, where D is depth of keyseat.

For checking an assembled key and shaft, caliper measurement J between top of key and bottom of shaft is used. $J = S - (M + D) + C$, where C is depth of key. For Woodruff keys, dimensions C and D can be found in Tables 9 to 11. Assuming shaft diameter S is nominal size, the tolerances on dimension J for Woodruff keys in keyslots are +0.000, −0.010 inch.

Diam. of Shaft S, Inches	Width of Keyseat, E — Dimension M, Inch														
	1/16	3/32	1/8	5/32	3/16	7/32	1/4	5/16	3/8	7/16	1/2	9/16	5/8	11/16	3/4
0.3125	.0032	...	...	...	...	...	...	...	...	...	...	...	...	...	...
0.3437	.0029	.0065	...	...	...	...	...	...	...	...	...	...	...	...	...
0.3750	.0026	.0060	.0107	...	...	...	...	...	...	...	...	...	...	...	...
0.4060	.0024	.0055	.0099	...	...	...	...	...	...	...	...	...	...	...	...
0.4375	.0022	.0051	.0091	...	...	...	...	...	...	...	...	...	...	...	...
0.4687	.0021	.0047	.0085	.0134	...	...	...	...	...	...	...	...	...	...	...
0.5000	.0020	.0044	.0079	.0125	...	...	...	...	...	...	...	...	...	...	...
0.5625	...	.0039	.0070	.0111	.0161	...	...	...	...	...	...	...	...	...	...
0.6250	...	.0035	.0063	.0099	.0144	.0198	...	...	...	...	...	...	...	...	...
0.6875	...	.0032	.0057	.0090	.0130	.0179	.0235	...	...	...	...	...	...	...	...
0.7500	...	.0029	.0052	.0082	.0119	.0163	.0214	.0341	...	...	...	...	...	...	...
0.8125	...	.0027	.0048	.0076	.0110	.0150	.0197	.0312	...	...	...	...	...	...	...
0.8750	...	.0025	.0045	.0070	.0102	.0139	.0182	.0288	...	...	...	...	...	...	...
0.9375	...	...	.0042	.0066	.0095	.0129	.0170	.0268	.0391	...	...	...	...	...	...
1.0000	...	...	.0039	.0061	.0089	.0121	.0159	.0250	.0365	...	...	...	...	...	...
1.0625	...	...	.0037	.0058	.0083	.0114	.0149	.0235	.0342	...	...	...	...	...	...
1.1250	...	...	.0035	.0055	.0079	.0107	.0141	.0221	.0322	.0443	...	...	...	...	...
1.1875	...	...	.0033	.0052	.0074	.0102	.0133	.0209	.0304	.0418	...	...	...	...	...
1.2500	...	...	.0031	.0049	.0071	.0097	.0126	.0198	.0288	.0395	...	...	...	...	...
1.3750	...	...	...	.0045	.0064	.0088	.0115	.0180	.0261	.0357	.0471	...	...	...	...
1.5000	...	...	...	.0041	.0059	.0080	.0105	.0165	.0238	.0326	.0429	...	...	...	...
1.6250	...	...	...	.0038	.0054	.0074	.0097	.0152	.0219	.0300	.0394	.0502	...	...	...
1.7500	...	...	...	...	.0050	.0069	.0090	.0141	.0203	.0278	.0365	.0464	...	...	...
1.8750	...	...	...	...	.0047	.0064	.0084	.0131	.0189	.0259	.0340	.0432	.0536	...	...
2.0000	...	...	...	...	.0044	.0060	.0078	.0123	.0177	.0242	.0318	.0404	.0501	...	...
2.1250	...	...	...	...	...	.0056	.0074	.0116	.0167	.0228	.0298	.0379	.0470	.0572	.0684
2.2500	...	...	...	...	...	...	.0070	.0109	.0157	.0215	.0281	.0357	.0443	.0538	.0643
2.3750	...	...	...	...	...	...	...	.0103	.0149	.0203	.0266	.0338	.0419	.0509	.0608
2.5000	...	...	...	...	...	...	...	...	.0141	.0193	.0253	.0321	.0397	.0482	.0576
2.6250	...	...	...	...	...	...	...	...	.0135	.0184	.0240	.0305	.0377	.0457	.0547
2.7500	...	...	...	...	...	...	...	...	...	.0175	.0229	.0291	.0360	.0437	.0521
2.8750	...	...	...	...	...	...	...	...	...	.0168	.0219	.0278	.0344	.0417	.0498
3.0000	...	...	...	...	...	...	...	...	...	...	.0210	.0266	.0329	.0399	.0476

Depths for Milling Keyseats. — The above table has been compiled to facilitate the accurate milling of keyseats. This table gives the distance M (see illustration accompanying table) between the top of the shaft and a line passing through the upper corners or edges of the keyseat. Dimension M is calculated by the formula: $M = \frac{1}{2}(S - \sqrt{S^2 - E^2})$ where S is diameter of shaft and E is width of keyseat. A simple approximate formula that gives M to within 0.001 inch is: $M = E^2 \div 4S$.

Table 9. ANSI Standard Woodruff Keys (ANSI B17.2-1967, R1978)

Key No.	Nominal Key Size $W \times B$	Actual Length F +0.000 −0.010	Height of Key				Distance Below Center E
			C		D		
			Max.	Min.	Max.	Min.	
202	1/16 × 1/4	0.248	0.109	0.104	0.109	0.104	1/64
202.5	1/16 × 5/16	0.311	0.140	0.135	0.140	0.135	1/64
302.5	3/32 × 5/16	0.311	0.140	0.135	0.140	0.135	1/64
203	1/16 × 3/8	0.374	0.172	0.167	0.172	0.167	1/64
303	3/32 × 3/8	0.374	0.172	0.167	0.172	0.167	1/64
403	1/8 × 3/8	0.374	0.172	0.167	0.172	0.167	1/64
204	1/16 × 1/2	0.491	0.203	0.198	0.194	0.188	3/64
304	3/32 × 1/2	0.491	0.203	0.198	0.194	0.188	3/64
404	1/8 × 1/2	0.491	0.203	0.198	0.194	0.188	3/64
305	3/32 × 5/8	0.612	0.250	0.245	0.240	0.234	1/16
405	1/8 × 5/8	0.612	0.250	0.245	0.240	0.234	1/16
505	5/32 × 5/8	0.612	0.250	0.245	0.240	0.234	1/16
605	3/16 × 5/8	0.612	0.250	0.245	0.240	0.234	1/16
406	1/8 × 3/4	0.740	0.313	0.308	0.303	0.297	1/16
506	5/32 × 3/4	0.740	0.313	0.308	0.303	0.297	1/16
606	3/16 × 3/4	0.740	0.313	0.308	0.303	0.297	1/16
806	1/4 × 3/4	0.740	0.313	0.308	0.303	0.297	1/16
507	5/32 × 7/8	0.866	0.375	0.370	0.365	0.359	1/16
607	3/16 × 7/8	0.866	0.375	0.370	0.365	0.359	1/16
707	7/32 × 7/8	0.866	0.375	0.370	0.365	0.359	1/16
807	1/4 × 7/8	0.866	0.375	0.370	0.365	0.359	1/16
608	3/16 × 1	0.992	0.438	0.433	0.428	0.422	1/16
708	7/32 × 1	0.992	0.438	0.433	0.428	0.422	1/16
808	1/4 × 1	0.992	0.438	0.433	0.428	0.422	1/16
1008	5/16 × 1	0.992	0.438	0.433	0.428	0.422	1/16
1208	3/8 × 1	0.992	0.438	0.433	0.428	0.422	1/16
609	3/16 × 1 1/8	1.114	0.484	0.479	0.475	0.469	5/64
709	7/32 × 1 1/8	1.114	0.484	0.479	0.475	0.469	5/64
809	1/4 × 1 1/8	1.114	0.484	0.479	0.475	0.469	5/64
1009	5/16 × 1 1/8	1.114	0.484	0.479	0.475	0.469	5/64
610	3/16 × 1 1/4	1.240	0.547	0.542	0.537	0.531	5/64
710	7/32 × 1 1/4	1.240	0.547	0.542	0.537	0.531	5/64
810	1/4 × 1 1/4	1.240	0.547	0.542	0.537	0.531	5/64
1010	5/16 × 1 1/4	1.240	0.547	0.542	0.537	0.531	5/64
1210	3/8 × 1 1/4	1.240	0.547	0.542	0.537	0.531	5/64
811	1/4 × 1 3/8	1.362	0.594	0.589	0.584	0.578	3/32
1011	5/16 × 1 3/8	1.362	0.594	0.589	0.584	0.578	3/32
1211	3/8 × 1 3/8	1.362	0.594	0.589	0.584	0.578	3/32
812	1/4 × 1 1/2	1.484	0.641	0.636	0.631	0.625	7/64
1012	5/16 × 1 1/2	1.484	0.641	0.636	0.631	0.625	7/64
1212	3/8 × 1 1/2	1.484	0.641	0.636	0.631	0.625	7/64

All dimensions are given in inches.

The key numbers indicate nominal key dimensions. The last two digits give the nominal diameter B in eighths of an inch and the digits preceding the last two give the nominal width W in thirty-seconds of an inch.

Table 10. ANSI Standard Woodruff Keys (ANSI B17.2-1967, R1978)

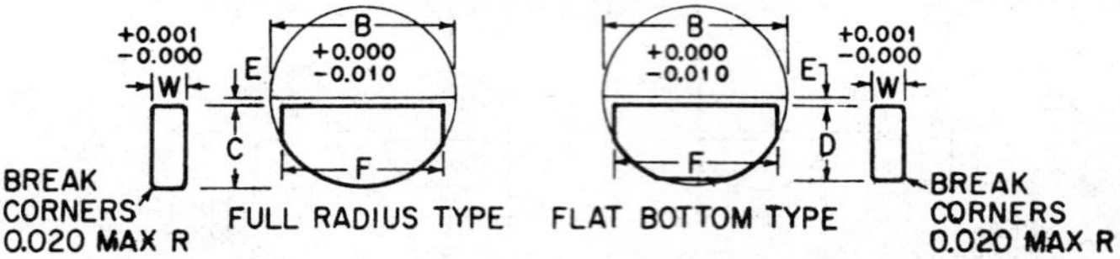

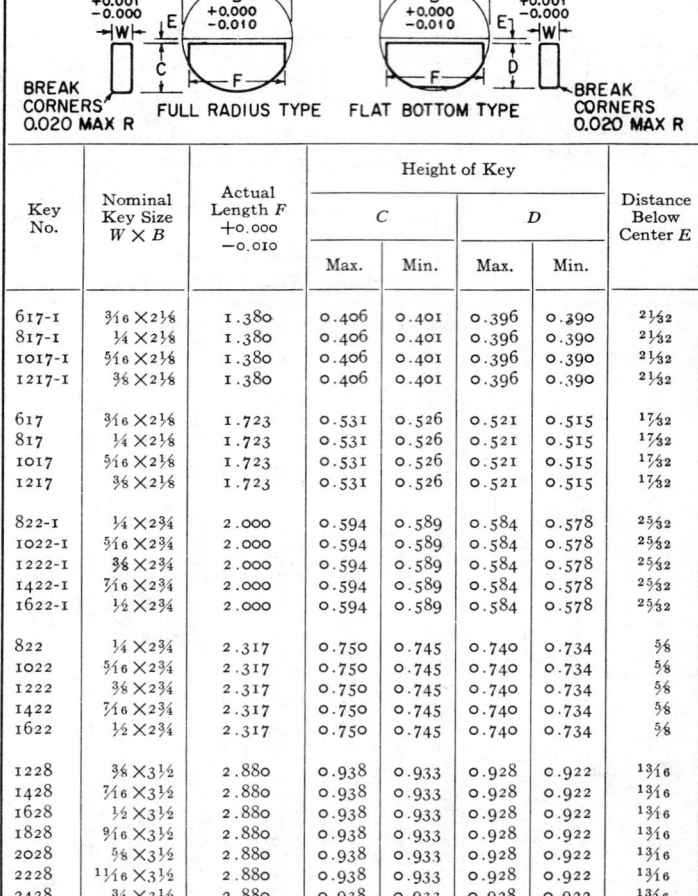

Key No.	Nominal Key Size $W \times B$	Actual Length F +0.000 −0.010	Height of Key				Distance Below Center E
			C		D		
			Max.	Min.	Max.	Min.	
617-1	3/16 × 2 1/8	1.380	0.406	0.401	0.396	0.390	21/32
817-1	1/4 × 2 1/8	1.380	0.406	0.401	0.396	0.390	21/32
1017-1	5/16 × 2 1/8	1.380	0.406	0.401	0.396	0.390	21/32
1217-1	3/8 × 2 1/8	1.380	0.406	0.401	0.396	0.390	21/32
617	3/16 × 2 1/8	1.723	0.531	0.526	0.521	0.515	17/32
817	1/4 × 2 1/8	1.723	0.531	0.526	0.521	0.515	17/32
1017	5/16 × 2 1/8	1.723	0.531	0.526	0.521	0.515	17/32
1217	3/8 × 2 1/8	1.723	0.531	0.526	0.521	0.515	17/32
822-1	1/4 × 2 3/4	2.000	0.594	0.589	0.584	0.578	25/32
1022-1	5/16 × 2 3/4	2.000	0.594	0.589	0.584	0.578	25/32
1222-1	3/8 × 2 3/4	2.000	0.594	0.589	0.584	0.578	25/32
1422-1	7/16 × 2 3/4	2.000	0.594	0.589	0.584	0.578	25/32
1622-1	1/2 × 2 3/4	2.000	0.594	0.589	0.584	0.578	25/32
822	1/4 × 2 3/4	2.317	0.750	0.745	0.740	0.734	5/8
1022	5/16 × 2 3/4	2.317	0.750	0.745	0.740	0.734	5/8
1222	3/8 × 2 3/4	2.317	0.750	0.745	0.740	0.734	5/8
1422	7/16 × 2 3/4	2.317	0.750	0.745	0.740	0.734	5/8
1622	1/2 × 2 3/4	2.317	0.750	0.745	0.740	0.734	5/8
1228	3/8 × 3 1/2	2.880	0.938	0.933	0.928	0.922	13/16
1428	7/16 × 3 1/2	2.880	0.938	0.933	0.928	0.922	13/16
1628	1/2 × 3 1/2	2.880	0.938	0.933	0.928	0.922	13/16
1828	9/16 × 3 1/2	2.880	0.938	0.933	0.928	0.922	13/16
2028	5/8 × 3 1/2	2.880	0.938	0.933	0.928	0.922	13/16
2228	11/16 × 3 1/2	2.880	0.938	0.933	0.928	0.922	13/16
2428	3/4 × 3 1/2	2.880	0.938	0.933	0.928	0.922	13/16

All dimensions are given in inches.

The key numbers indicate nominal key dimensions. The last two digits give the nominal diameter B in eighths of an inch and the digits preceding the last two give the nominal width W in thirty-seconds of an inch.

The key numbers with the -1 designation, while representing the nominal key size have a shorter length F and due to a greater distance below center E are less in height than the keys of the same number without the -1 designation.

Table 11. ANSI Keyseat Dimensions for Woodruff Keys (ANSI B17.2-1967, R1978)

KEYSEAT–SHAFT   KEY ABOVE SHAFT   KEYSEAT–HUB

| Key No. | Nominal Size Key | Keyseat — Shaft | | | | | Key Above Shaft | Keyseat — Hub | |
| | | Width A* | | Depth B | Diameter F | | Height C | Width D | Depth E |
		Min.	Max.	+0.005 / −0.000	Min.	Max.	+0.005 / −0.005	+0.002 / −0.000	+0.005 / −0.000
202	1/16 × 1/4	0.0615	0.0630	0.0728	0.250	0.268	0.0312	0.0635	0.0372
202.5	1/16 × 5/16	0.0615	0.0630	0.1038	0.312	0.330	0.0312	0.0635	0.0372
302.5	3/32 × 5/16	0.0928	0.0943	0.0882	0.312	0.330	0.0469	0.0948	0.0529
203	1/16 × 3/8	0.0615	0.0630	0.1358	0.375	0.393	0.0312	0.0635	0.0372
303	3/32 × 3/8	0.0928	0.0943	0.1202	0.375	0.393	0.0469	0.0948	0.0529
403	1/8 × 3/8	0.1240	0.1255	0.1045	0.375	0.393	0.0625	0.1260	0.0685
204	1/16 × 1/2	0.0615	0.0630	0.1668	0.500	0.518	0.0312	0.0635	0.0372
304	3/32 × 1/2	0.0928	0.0943	0.1511	0.500	0.518	0.0469	0.0948	0.0529
404	1/8 × 1/2	0.1240	0.1255	0.1355	0.500	0.518	0.0625	0.1260	0.0685
305	3/32 × 5/8	0.0928	0.0943	0.1981	0.625	0.643	0.0469	0.0948	0.0529
405	1/8 × 5/8	0.1240	0.1255	0.1825	0.625	0.643	0.0625	0.1260	0.0685
505	5/32 × 5/8	0.1553	0.1568	0.1669	0.625	0.643	0.0781	0.1573	0.0841
605	3/16 × 5/8	0.1863	0.1880	0.1513	0.625	0.643	0.0937	0.1885	0.0997
406	1/8 × 3/4	0.1240	0.1255	0.2455	0.750	0.768	0.0625	0.1260	0.0685
506	5/32 × 3/4	0.1553	0.1568	0.2299	0.750	0.768	0.0781	0.1573	0.0841
606	3/16 × 3/4	0.1863	0.1880	0.2143	0.750	0.768	0.0937	0.1885	0.0997
806	1/4 × 3/4	0.2487	0.2505	0.1830	0.750	0.768	0.1250	0.2510	0.1310
507	5/32 × 7/8	0.1553	0.1568	0.2919	0.875	0.895	0.0781	0.1573	0.0841
607	3/16 × 7/8	0.1863	0.1880	0.2763	0.875	0.895	0.0937	0.1885	0.0997
707	7/32 × 7/8	0.2175	0.2193	0.2607	0.875	0.895	0.1093	0.2198	0.1153
807	1/4 × 7/8	0.2487	0.2505	0.2450	0.875	0.895	0.1250	0.2510	0.1310
608	3/16 × 1	0.1863	0.1880	0.3393	1.000	1.020	0.0937	0.1885	0.0997
708	7/32 × 1	0.2175	0.2193	0.3237	1.000	1.020	0.1093	0.2198	0.1153
808	1/4 × 1	0.2487	0.2505	0.3080	1.000	1.020	0.1250	0.2510	0.1310
1008	5/16 × 1	0.3111	0.3130	0.2768	1.000	1.020	0.1562	0.3135	0.1622
1208	3/8 × 1	0.3735	0.3755	0.2455	1.000	1.020	0.1875	0.3760	0.1935
609	3/16 × 1 1/8	0.1863	0.1880	0.3853	1.125	1.145	0.0937	0.1885	0.0997
709	7/32 × 1 1/8	0.2175	0.2193	0.3697	1.125	1.145	0.1093	0.2198	0.1153
809	1/4 × 1 1/8	0.2487	0.2505	0.3540	1.125	1.145	0.1250	0.2510	0.1310
1009	5/16 × 1 1/8	0.3111	0.3130	0.3228	1.125	1.145	0.1562	0.3135	0.1622
610	3/16 × 1 1/4	0.1863	0.1880	0.4483	1.250	1.273	0.0937	0.1885	0.0997
710	7/32 × 1 1/4	0.2175	0.2193	0.4327	1.250	1.273	0.1093	0.2198	0.1153
810	1/4 × 1 1/4	0.2487	0.2505	0.4170	1.250	1.273	0.1250	0.2510	0.1310
1010	5/16 × 1 1/4	0.3111	0.3130	0.3858	1.250	1.273	0.1562	0.3135	0.1622
1210	3/8 × 1 1/4	0.3735	0.3755	0.3545	1.250	1.273	0.1875	0.3760	0.1935
811	1/4 × 1 3/8	0.2487	0.2505	0.4640	1.375	1.398	0.1250	0.2510	0.1310
1011	5/16 × 1 3/8	0.3111	0.3130	0.4328	1.375	1.398	0.1562	0.3135	0.1622
1211	3/8 × 1 3/8	0.3735	0.3755	0.4015	1.375	1.398	0.1875	0.3760	0.1935

All dimensions are given in inches.
* See footnote at end of table.

Table 11 (Concluded).   ANSI Standard Keyseat Dimensions for Woodruff Keys
(ANSI B17.2-1967, R1978)

Key No.	Nominal Size Key	Keyseat — Shaft					Key Above Shaft	Keyseat — Hub	
		Width A*		Depth B	Diameter F		Height C	Width D	Depth E
		Min.	Max.	+0.005 −0.000	Min.	Max.	+0.005 −0.005	+0.002 −0.000	+0.005 −0.000
812	¼ × 1½	0.2487	0.2505	0.5110	1.500	1.523	0.1250	0.2510	0.1310
1012	⁵⁄₁₆ × 1½	0.3111	0.3130	0.4798	1.500	1.523	0.1562	0.3135	0.1622
1212	⅜ × 1½	0.3735	0.3755	0.4485	1.500	1.523	0.1875	0.3760	0.1935
617-1	³⁄₁₆ × 2⅛	0.1863	0.1880	0.3073	2.125	2.160	0.0937	0.1885	0.0997
817-1	¼ × 2⅛	0.2487	0.2505	0.2760	2.125	2.160	0.1250	0.2510	0.1310
1017-1	⁵⁄₁₆ × 2⅛	0.3111	0.3130	0.2448	2.125	2.160	0.1562	0.3135	0.1622
1217-1	⅜ × 2⅛	0.3735	0.3755	0.2135	2.125	2.160	0.1875	0.3760	0.1935
617	³⁄₁₆ × 2⅛	0.1863	0.1880	0.4323	2.125	2.160	0.0937	0.1885	0.0997
817	¼ × 2⅛	0.2487	0.2505	0.4010	2.125	2.160	0.1250	0.2510	0.1310
1017	⁵⁄₁₆ × 2⅛	0.3111	0.3130	0.3698	2.125	2.160	0.1562	0.3135	0.1622
1217	⅜ × 2⅛	0.3735	0.3755	0.3385	2.125	2.160	0.1875	0.3760	0.1935
822-1	¼ × 2¾	0.2487	0.2505	0.4640	2.750	2.785	0.1250	0.2510	0.1310
1022-1	⁵⁄₁₆ × 2¾	0.3111	0.3130	0.4328	2.750	2.785	0.1562	0.3135	0.1622
1222-1	⅜ × 2¾	0.3735	0.3755	0.4015	2.750	2.785	0.1875	0.3760	0.1935
1422-1	⁷⁄₁₆ × 2¾	0.4360	0.4380	0.3703	2.750	2.785	0.2187	0.4385	0.2247
1622-1	½ × 2¾	0.4985	0.5005	0.3390	2.750	2.785	0.2500	0.5010	0.2560
822	¼ × 2¾	0.2487	0.2505	0.6200	2.750	2.785	0.1250	0.2510	0.1310
1022	⁵⁄₁₆ × 2¾	0.3111	0.3130	0.5888	2.750	2.785	0.1562	0.3135	0.1622
1222	⅜ × 2¾	0.3735	0.3755	0.5575	2.750	2.785	0.1875	0.3760	0.1935
1422	⁷⁄₁₆ × 2¾	0.4360	0.4380	0.5263	2.750	2.785	0.2187	0.4385	0.2247
1622	½ × 2¾	0.4985	0.5005	0.4950	2.750	2.785	0.2500	0.5010	0.2560
1228	⅜ × 3½	0.3735	0.3755	0.7455	3.500	3.535	0.1875	0.3760	0.1935
1428	⁷⁄₁₆ × 3½	0.4360	0.4380	0.7143	3.500	3.535	0.2187	0.4385	0.2247
1628	½ × 3½	0.4985	0.5005	0.6830	3.500	3.535	0.2500	0.5010	0.2560
1828	⁹⁄₁₆ × 3½	0.5610	0.5630	0.6518	3.500	3.535	0.2812	0.5635	0.2872
2028	⅝ × 3½	0.6235	0.6255	0.6205	3.500	3.535	0.3125	0.6260	0.3185
2228	¹¹⁄₁₆ × 3½	0.6860	0.6880	0.5893	3.500	3.535	0.3437	0.6885	0.3497
2428	¾ × 3½	0.7485	0.7505	0.5580	3.500	3.535	0.3750	0.7510	0.3810

All dimensions are given in inches.

* These Width A values were set with the maximum keyseat (shaft) width as that figure which will receive a key with the greatest amount of looseness consistent with assuring the key's sticking in the keyseat (shaft). Minimum keyseat width is that figure permitting the largest shaft distortion acceptable when assembling maximum key in minimum keyseat.

Dimensions A, B, C, D are taken at side intersection.

Cotters. — A cotter is a form of key that is used to connect rods, etc., that are subjected either to tension or compression or both, the cotter being subjected to shearing stresses at two transverse cross-sections. When taper cotters are used for drawing and holding parts together, if the cotter is held in place by the friction between the bearing surfaces, the taper should not be too great. Ordinarily a taper varying from ¼ to ½ inch per foot is used for plain cotters. When a set-screw or other device is used to prevent the cotter from backing out of its slot, the taper may vary from 1½ to 2 inches per foot.

British Standard Metric Keys and Keyways. — This British Standard (BS 4235:Part 1:1972) covers square and rectangular parallel keys and keyways, and square and rectangular taper keys and keyways. Plain and gib-head taper keys are specified. There are three classes of fit for the square and rectangular parallel keys and keyways, designated free, normal, and close. A *free fit* is applied when the hub of an assembly is required to slide over the key when in use; a *normal fit* is employed when the key is to be inserted in the keyway with the minimum amount of fitting, as may be required in mass production assembly work; and a *close fit* is applied when accurate fitting of the key is required under maximum material conditions, which may involve selection of components.

The Standard does not provide for misalignment or offset greater than can be accommodated within the dimensional tolerances. If an assembly is going to be heavily stressed, a check should be made to ensure that the cumulative effect of misalignment or offset, or both, does not prevent satisfactory bearing on the key. Radii and chamfers are not normally provided on keybar and keys as supplied, but they can be produced during manufacture by agreement between the user and supplier.

Unless otherwise specified, keys in compliance with this Standard are manufactured from steel complying with BS 970 having a tensile strength of not less than 550 MN/m^2 in the finished condition. BS 970, Part 1, lists the following steels and maximum section sizes, respectively, that meet this tensile strength requirement: 070M20, 25 × 14mm; 070M26; 36 × 20mm; 080M30, 90 × 45mm; and 080M40, 100 × 50mm.

At the time of publication of this Standard, the demand for metric keys was not sufficient to enable standard ranges of lengths to be established. The lengths given in the accompanying table are those shown as standard in ISO Recommendations R 773, "Rectangular or Square Parallel Keys and their Corresponding Keyways (Dimensions in millimeters)" and R 774, "Taper Keys and their Corresponding Keyways — with or without Gib Head (Dimensions in millimeters)."

Tables 1 through 4 on the following pages cover the dimensions and tolerances of square and rectangular keys and keyways, and square and rectangular taper keys and keyways.

British Standard Preferred Lengths of Metric Keys (BS 4235: Part 1: 1972)

Length	Type of key				Length	Type of key			
	Squ.	Rect.	Squ. taper	Rect. taper		Squ.	Rect.	Squ. taper	Rect. taper
6	x		x		63	x	x	x	x
8	x		x		70	x	x	x	x
10	x		x		80		x		x
12	x		x		90		x		x
14	x		x		100		x		x
16	x		x		110		x		x
18	x	x	x	x	125		x		x
20	x	x	x	x	140		x		x
22	x	x	x	x	160		x		x
25	x	x	x	x	180		x		x
28	x	x	x	x	200		x		x
32	x	x	x	x	220		x		x
36	x	x	x	x	250		x		x
40	x	x	x	x	280		x		x
45	x	x	x	x	320		x		x
50	x	x	x	x	360		x		x
56	x	x	x	x	400		x		x

Table 1. British Standard Metric Keyways for Square and Rectangular Parallel Keys (BS 4235: Part I: 1972)

Enlarged detail of key and keyways

Section X-X

Shaft		Key	Keyway										Depth				Radius r	
Nominal Diameter d		Size, b × h	Width b										Shaft t₁		Hub t₂			
				Free Fit		Normal Fit		Close Fit										
Over	Up to and Incl.		Nom.	Shaft (H9)	Hub (D10)	Shaft (N9)	Hub (Js9)*	Shaft and Hub (P9)					Nom.	Tol.	Nom.	Tol.	Max.	Min.
											Tolerances							
						Keyways for Square Parallel Keys												
6	8	2 × 2	2	+0.025 / 0	+0.060 / +0.020	−0.004 / −0.029	+0.012 / −0.012	−0.006 / −0.031					1.2	+0.1 / 0	1	+0.1 / 0	0.16	0.08
8	10	3 × 3	3	+0.025 / 0	+0.060 / +0.020	−0.004 / −0.029	+0.012 / −0.012	−0.006 / −0.031					1.8	+0.1 / 0	1.4	+0.1 / 0	0.16	0.08
10	12	4 × 4	4	+0.030 / 0	+0.078 / +0.030	0 / −0.030	+0.015 / −0.015	−0.012 / −0.042					2.5	+0.1 / 0	1.8	+0.1 / 0	0.16	0.08
12	17	5 × 5	5	+0.030 / 0	+0.078 / +0.030	0 / −0.030	+0.015 / −0.015	−0.012 / −0.042					3	+0.1 / 0	2.3	+0.1 / 0	0.25	0.16
17	22	6 × 6	6	+0.030 / 0	+0.078 / +0.030	0 / −0.030	+0.015 / −0.015	−0.012 / −0.042					3.5	+0.1 / 0	2.8	+0.1 / 0	0.25	0.16

All dimensions in millimeters.   * Tolerance limits Js9 are quoted from BS 4500, "ISO Limits and Fits," to three significant figures.

Table 1 (*Continued*). British Standard Metric Keyways for Square and Rectangular Parallel Keys (BS 4235: Part 1: 1972)

Keyways for Rectangular Parallel Keys — Tolerances

Shaft Nominal Diameter d Over	Up to and Incl.	Key Size, b × h	Width b Nom.	Free Fit Shaft (H9)	Free Fit Hub (D10)	Normal Fit Shaft (N9)	Normal Fit Hub (Js9)*	Close Fit Shaft and Hub (P9)	Depth Shaft t₁ Nom.	Shaft t₁ Tol.	Hub t₂ Nom.	Hub t₂ Tol.	Radius r Max.	Radius r Min.
22	30	8 × 7	8	+0.036 / 0	+0.098 / +0.040	0 / −0.036	+0.018 / −0.018	−0.015 / −0.051	4	+0.2 / 0	3.3	+0.2 / 0	0.25	0.16
30	38	10 × 8	10	+0.036 / 0	+0.098 / +0.040	0 / −0.036	+0.018 / −0.018	−0.015 / −0.051	5	+0.2 / 0	3.3	+0.2 / 0	0.40	0.25
38	44	12 × 8	12	+0.043 / 0	+0.120 / +0.050	0 / −0.043	+0.021 / −0.021	−0.018 / −0.061	5	+0.2 / 0	3.3	+0.2 / 0	0.40	0.25
44	50	14 × 9	14	+0.043 / 0	+0.120 / +0.050	0 / −0.043	+0.021 / −0.021	−0.018 / −0.061	5.5	+0.2 / 0	3.8	+0.2 / 0	0.40	0.25
50	58	16 × 10	16	+0.043 / 0	+0.120 / +0.050	0 / −0.043	+0.021 / −0.021	−0.018 / −0.061	6	+0.2 / 0	4.3	+0.2 / 0	0.40	0.25
58	65	18 × 11	18	+0.043 / 0	+0.120 / +0.050	0 / −0.043	+0.021 / −0.021	−0.018 / −0.061	7	+0.2 / 0	4.4	+0.2 / 0	0.40	0.25
65	75	20 × 12	20	+0.052 / 0	+0.149 / +0.065	0 / −0.052	+0.026 / −0.026	−0.022 / −0.074	7.5	+0.2 / 0	4.9	+0.2 / 0	0.60	0.40
75	85	22 × 14	22	+0.052 / 0	+0.149 / +0.065	0 / −0.052	+0.026 / −0.026	−0.022 / −0.074	9	+0.2 / 0	5.4	+0.2 / 0	0.60	0.40
85	95	25 × 14	25	+0.052 / 0	+0.149 / +0.065	0 / −0.052	+0.026 / −0.026	−0.022 / −0.074	9	+0.2 / 0	5.4	+0.2 / 0	0.60	0.40
95	110	28 × 16	28	+0.052 / 0	+0.149 / +0.065	0 / −0.052	+0.026 / −0.026	−0.022 / −0.074	10	+0.2 / 0	6.4	+0.2 / 0	0.60	0.40
110	130	32 × 18	32	+0.062 / 0	+0.180 / +0.080	0 / −0.062	+0.031 / −0.031	−0.026 / −0.088	11	+0.2 / 0	7.4	+0.2 / 0	0.60	0.40
130	150	36 × 20	36	+0.062 / 0	+0.180 / +0.080	0 / −0.062	+0.031 / −0.031	−0.026 / −0.088	12	+0.2 / 0	8.4	+0.2 / 0	1.00	0.70
150	170	40 × 22	40	+0.062 / 0	+0.180 / +0.080	0 / −0.062	+0.031 / −0.031	−0.026 / −0.088	13	+0.2 / 0	9.4	+0.2 / 0	1.00	0.70
170	200	45 × 25	45	+0.062 / 0	+0.180 / +0.080	0 / −0.062	+0.031 / −0.031	−0.026 / −0.088	15	+0.2 / 0	10.4	+0.2 / 0	1.00	0.70
200	230	50 × 28	50	+0.062 / 0	+0.180 / +0.080	0 / −0.062	+0.031 / −0.031	−0.026 / −0.088	17	+0.2 / 0	11.4	+0.2 / 0	1.00	0.70
230	260	56 × 32	56	+0.074 / 0	+0.220 / +0.100	0 / −0.074	+0.037 / −0.037	−0.032 / −0.106	20	+0.3 / 0	12.4	+0.3 / 0	1.60	1.20
260	290	63 × 32	63	+0.074 / 0	+0.220 / +0.100	0 / −0.074	+0.037 / −0.037	−0.032 / −0.106	20	+0.3 / 0	12.4	+0.3 / 0	1.60	1.20
290	330	70 × 36	70	+0.074 / 0	+0.220 / +0.100	0 / −0.074	+0.037 / −0.037	−0.032 / −0.106	22	+0.3 / 0	14.4	+0.3 / 0	1.60	1.20
330	380	80 × 40	80	+0.074 / 0	+0.220 / +0.100	0 / −0.074	+0.037 / −0.037	−0.032 / −0.106	25	+0.3 / 0	15.4	+0.3 / 0	2.50	2.00
380	440	90 × 45	90	+0.087 / 0	+0.260 / +0.120	0 / −0.087	+0.043 / −0.043	−0.037 / −0.124	28	+0.3 / 0	17.4	+0.3 / 0	2.50	2.00
440	500	100 × 50	100	+0.087 / 0	+0.260 / +0.120	0 / −0.087	+0.043 / −0.043	−0.037 / −0.124	31	+0.3 / 0	19.5	+0.3 / 0	2.50	2.00

All dimensions in millimeters.   * Tolerance limits Js9 are quoted from BS 4500, "ISO Limits and Fits," to three significant figures.

Table 2.　British Standard Metric Keyways for Square and Rectangular Taper Keys
(BS 4235: Part I: 1972)

Shaft		Key	Keyway							
Nominal Diameter d		Size, $b \times h$	Width b, Shaft and Hub		Depth				Corner Radius of Keyway	
					Shaft t_1		Hub t_2			
Over	Up to and Incl.		Nom.	Tol. (D10)	Nom.	Tol.	Nom.	Tol.	Max.	Min.
Keyways for Square Taper Keys										
6	8	2 × 2	2	+0.060 +0.020	1.2	+0.1 0	0.5	+0.1 0	0.16	0.08
8	10	3 × 3	3	+0.060 +0.020	1.8		0.9		0.16	0.08
10	12	4 × 4	4		2.5		1.2		0.16	0.08
12	17	5 × 5	5	+0.078 +0.030	3		1.7		0.25	0.16
17	22	6 × 6	6		3.5	+0.2 0	2.2	+0.2 0	0.25	0.16
Keyways for Rectangular Taper Keys										
22	30	8 × 7	8	+0.098 +0.040	4		2.4		0.25	0.16
30	38	10 × 8	10	+0.098 +0.040	5		2.4		0.40	0.25
38	44	12 × 8	12	+0.120 +0.050	5		2.4		0.40	0.25
44	50	14 × 9	14		5.5		2.9		0.40	0.25
50	58	16 × 10	16		6	+0.2 0	3.4	+0.2 0	0.40	0.25
58	65	18 × 11	18		7		3.4		0.40	0.25
65	75	20 × 12	20	+0.149 +0.065	7.5		3.9		0.60	0.40
75	85	22 × 14	22		9		4.4		0.60	0.40
85	95	25 × 14	25		9		4.4		0.60	0.40
95	110	28 × 16	28		10		5.4		0.60	0.40
110	130	32 × 18	32	+0.180 +0.080	11		6.4		0.60	0.40
130	150	36 × 20	36		12		7.1		1.00	0.70
150	170	40 × 22	40		13		8.1		1.00	0.70
170	200	45 × 25	45		15		9.1		1.00	0.70
200	230	50 × 28	50		17		10.1		1.00	0.70
230	260	56 × 32	56	+0.220 +0.120	20	+0.3 0	11.1	+0.3 0	1.60	1.20
260	290	63 × 32	63		20		11.1		1.60	1.20
290	330	70 × 36	70		22		13.1		1.60	1.20
330	380	80 × 40	80		25		14.1		2.50	2.00
380	440	90 × 45	90	+0.260 +0.120	28		16.1		2.50	2.00
440	500	100 × 50	100		31		18.1		2.50	2.00

Table 3. British Standard Metric Square and Rectangular Parallel Keys
(BS 4235: Part 1: 1972)

Width b Nom.	Width b Tol.*	Thickness h Nom.	Thickness h Tol.*	Chamfer s Min.	Chamfer s Max.	Length Range l From	Length Range l To
Square Parallel Keys
2	0 / − 0.025	2	0 / − 0.025	0.16	0.25	6	20
3		3		0.16	0.25	6	36
4	0 / − 0.030	4	0 / − 0.030	0.16	0.25	8	45
5		5		0.25	0.40	10	56
6		6		0.25	0.40	14	70
Rectangular Parallel Keys
8	0 / − 0.036	7		0.25	0.40	18	90
10		8		0.40	0.60	22	110
12		8	0 / − 0.090	0.40	0.60	28	140
14		9		0.40	0.60	36	160
16	0 / − 0.043	10		0.40	0.60	45	180
18		11		0.40	0.60	50	200
20		12		0.60	0.80	56	220
22	0 / − 0.052	14		0.60	0.80	63	250
25		14	0 / − 0.110	0.60	0.80	70	280
28		16		0.60	0.80	80	320
32		18		0.60	0.80	90	360
36	0 / − 0.062	20		1.00	1.20	100	400
40		22		1.00	1.20	...	...
45		25	0 / − 0.130	1.00	1.20	...	...
50		28		1.00	1.20	...	...
56		32		1.60	2.00	...	...
63	0 / − 0.074	32		1.60	2.00	...	...
70		36		1.60	2.00	...	...
80		40	0 / − 0.160	2.50	3.00	...	...
90	0 / − 0.087	45		2.50	3.00	...	...
100		50		2.50	3.00	...	...

* The tolerance on the width and thickness of square taper keys is h9, and on the width and thickness of rectangular keys, h9 and h11 respectively, in accordance with ISO metric limits and fits.

KEYS AND KEYWAYS

Table 4. British Standard Metric Square and Rectangular Taper Keys
(BS 4235: Part 1: 1972)

Width b		Thickness h		Chamfer s		Length Range l		Gib head h₁	Radius r
Nom.	Tol.*	Nom.	Tol.*	Min.	Max.	From	To	Nom.	Nom.
Square Taper Keys									
2	⎫ 0	2	⎫ 0	0.16	0.25	6	20	…	…
3	⎭ − 0.025	3	⎭ − 0.025	0.16	0.25	6	36	…	…
4	⎫	4	⎫	0.16	0.25	8	45	7	0.25
5	⎬ 0 − 0.030	5	⎬ 0 − 0.030	0.25	0.40	10	56	8	0.25
6	⎭	6	⎭	0.25	0.40	14	70	10	0.25
Rectangular Taper Keys									
8	⎫ 0	7	⎫	0.25	0.40	18	90	11	1.5
10	⎭ −0.036	8	⎬	0.40	0.60	22	110	12	1.5
12	⎫	8	⎬ 0 − 0.090	0.40	0.60	28	140	12	1.5
14	⎬ 0 − 0.043	9	⎬	0.40	0.60	36	160	14	1.5
16	⎭	10	⎭	0.40	0.60	45	180	16	3.2
18	⎭	11		0.40	0.60	50	200	18	3.2
20	⎫	12	⎫	0.60	0.80	56	220	20	3.2
22	⎬ 0 − 0.052	14	⎬ 0 − 0.110	0.60	0.80	63	250	22	3.2
25	⎬	14	⎬	0.60	0.80	70	280	22	3.2
28	⎭	16	⎭	0.60	0.80	80	320	25	3.2
32	⎭	18		0.60	0.80	90	360	28	6.4
36	⎫	20	⎫	1.00	1.20	100	400	32	6.4
40	⎬ 0 − 0.062	22	⎬ 0 − 0.130	1.00	1.20	…	…	36	6.4
45	⎬	25	⎬	1.00	1.20	…	…	40	6.4
50	⎭	28	⎭	1.00	1.20	…	…	45	6.4
56	⎫	32	⎫	1.60	2.00	…	…	50	9.5
63	⎬ 0 − 0.074	32	⎬	1.60	2.00	…	…	50	9.5
70	⎬	36	⎬	1.60	2.00	…	…	56	9.5
80	⎭	40	⎬ 0 − 0.160	2.50	3.00	…	…	63	9.5
90	⎫ 0 − 0.087	45	⎬	2.50	3.00	…	…	70	9.5
100	⎭	50	⎭	2.50	3.00	…	…	80	9.5

* The tolerance on the width and thickness of square taper keys is h9, and on the width and thickness of rectangular taper keys, h9 and h11 respectively, in accordance with ISO metric limits and fits. Does not apply to gib head dimensions.

British Standard Keys and Keyways. — Tables 1 through 6 (B.S. 46: Part 1: 1958) provide data for rectangular parallel keys and keyways, square parallel keys and keyways, plain and gib head rectangular taper keys and keyways, plain and gib head square taper keys and keyways, and Woodruff keys and keyways.

Parallel Keys: These keys are used for transmitting unidirectional torques in transmissions not subject to heavy starting loads and where periodic withdrawal or sliding of the hub member may be required. In many instances, particularly couplings, a gib-head cannot be accommodated, and there is insufficient room to drift out the key from behind. In these cases it is necessary to withdraw the component over the key and a parallel key is essential. Parallel square and rectangular keys are normally side fitting with top clearance and are usually retained in the shaft rather more securely than in the hub. The rectangular key is the general purpose key for shafts greater than 1 inch in diameter; the square key is intended for use with shafts up to and including 1-inch diameter or for shafts up to 6-inch diameter where it is desirable to have a greater key depth than is provided by rectangular keys. In cases of stepped shafts the larger diameters are usually required by considerations other than torque, e.g. resistance to bending. Where components such as fans, gears, impellers, etc., are attached to the larger shaft diameter, the use of a key smaller than standard for that diameter may be permissible. As this results in unequal disposition of the key in the shaft and its related hub the dimensions H and h must be recalculated to maintain the $T/2$ relationship.

Taper Keys: These keys are used for transmitting heavy unidirectional, reversing, or vibrating torques and in applications where periodic withdrawal of the key may be necessary. Taper keys are usually top fitting, but may be top and side fitting where required; in this case the keyway in the hub should have the same width value as the keyway in the shaft. Taper keys of rectangular section are used for general purposes and are of less depth than square keys; square sections are for use with shafts up to and including 1-inch diameter or for shafts up to 6-inch diameter where it is desirable to have greater key depth.

Woodruff Keys: These keys are used for light applications or the angular location of associated parts on tapered shaft ends. They are not recommended for other applications, but if so used, corner radii in the shaft and hub keyways are advisable to reduce stress concentration.

Dimensions and Tolerances for British Parallel and Taper Keys and Keyways. — Dimensions and tolerances for key and keyway widths given in Tables 1, 2, 3 and 4 are based on the width of key W and provide a fitting allowance. The fitting allowance is designed to permit an interference between the key and the shaft keyway and a slightly easier condition between the key and the hub keyway. In the case of shrink and heavy force fits it may be found necessary to depart from the width and depth tolerances specified. Any variation in the width of the keyway should be such that the greatest width is at the end from which the key enters and any variation in the depth of the keyway should be such that the greatest depth is at the end from which the key enters.

Keys and keybar normally are not chamfered or radiused as supplied, but this may be done at the time of fitting. Radii and chamfers are given in Tables 1, 2, 3 and 4. Corner radii are recommended for keyways to alleviate stress concentration.

Dimensions and Tolerances of British Woodruff Keys and Keyways. — Dimensions and tolerances are shown in Table 5. An optional alternative design of Woodruff key which differs from the normal form in its depth is given in the illustration accompanying the table. The method of designating British Woodruff Keys is the same as the American method explained in footnote on page 2241.

Table 1. British Standard Rectangular Parallel Keys, Keyways and Keybars (B.S. 46: Part 1: 1958)

All dimensions in inches

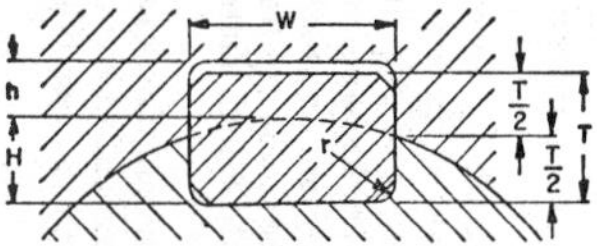

Diameter of Shaft		Size, W × T	Key				Keyway in Shaft				Keyway in Hub				Nominal Keyway Radius, r*	Keybar			
Over	Up to and Including		Width, W		Thickness, T		Width, Ws		Depth, H		Width, Wh		Depth, h			Width, W		Thickness, T	
			Max.	Min.	Max.	Min.	Min.	Max.	Min.	Max.	Min.	Max.	Min.	Max.		Max.	Min.	Max.	Min.
	1¼	5/16 × ¼	0.314	0.312	0.253	0.250	0.311	0.312	0.146	0.152	0.312	0.313	0.112	0.118	0.010	0.314	0.312	0.253	0.250
1¼	1½	3/8 × ¼	0.377	0.375	0.253	0.250	0.374	0.375	0.150	0.156	0.375	0.376	0.108	0.114	0.010	0.377	0.375	0.253	0.250
1½	1¾	7/16 × 5/16	0.440	0.438	0.315	0.312	0.437	0.438	0.186	0.192	0.438	0.439	0.135	0.141	0.020	0.440	0.438	0.315	0.312
1¾	2	½ × 5/16	0.502	0.500	0.315	0.312	0.499	0.500	0.190	0.196	0.500	0.501	0.131	0.137	0.020	0.502	0.500	0.315	0.312
2	2½	5/8 × 7/16	0.627	0.625	0.441	0.438	0.624	0.625	0.260	0.266	0.625	0.626	0.185	0.191	0.020	0.627	0.625	0.441	0.438
2½	3	¾ × ½	0.752	0.750	0.503	0.500	0.749	0.750	0.299	0.305	0.750	0.751	0.209	0.215	0.020	0.752	0.750	0.503	0.500
3	3½	⅞ × 5/8	0.877	0.875	0.629	0.625	0.874	0.875	0.370	0.376	0.875	0.876	0.264	0.270	0.062	0.877	0.875	0.629	0.625
3½	4	1 × ¾	1.003	1.000	0.754	0.750	0.999	1.000	0.441	0.447	1.000	1.001	0.318	0.324	0.062	1.003	1.000	0.754	0.750
4	5	1¼ × ⅞	1.253	1.250	0.879	0.875	1.248	1.250	0.518	0.524	1.250	1.252	0.366	0.372	0.062	1.253	1.250	0.879	0.875
5	6	1½ × 1	1.504	1.500	1.006	1.000	1.498	1.500	0.599	0.605	1.500	1.502	0.412	0.418	0.062	1.504	1.500	1.006	1.000
6	7	1¾ × 1¼	1.754	1.750	1.256	1.250	1.748	1.750	0.740	0.746	1.750	1.752	0.526	0.532	0.062	Bright keybar is not normally available in sections larger than the above.			
7	8	2 × 1⅜	2.005	2.000	1.381	1.375	1.998	2.000	0.818	0.824	2.000	2.002	0.573	0.579	0.125				
8	9	2¼ × 1½	2.255	2.250	1.506	1.500	2.248	2.250	0.897	0.905	2.250	2.252	0.619	0.627	0.125				
9	10	2½ × 1⅝	2.505	2.500	1.631	1.625	2.498	2.500	0.975	0.983	2.500	2.502	0.666	0.674	0.187				
10	11	2¾ × 1⅞	2.755	2.750	1.881	1.875	2.748	2.750	1.114	1.122	2.750	2.752	0.777	0.785	0.187				
11	12	3 × 2	3.006	3.000	2.008	2.000	2.998	3.000	1.195	1.203	3.000	3.002	0.823	0.831	0.187				
12	13	3¼ × 2⅛	3.256	3.250	2.133	2.125	3.248	3.250	1.273	1.281	3.250	3.252	0.870	0.878	0.187				
13	14	3½ × 2⅜	3.506	3.500	2.383	2.375	3.498	3.500	1.413	1.421	3.500	3.502	0.980	0.988	0.250				
14	15	3¾ × 2½	3.756	3.750	2.508	2.500	3.748	3.750	1.492	1.502	3.750	3.752	1.026	1.036	0.250				
15	16	4 × 2⅝	4.008	4.000	2.633	2.625	3.998	4.000	1.571	1.581	4.000	4.002	1.072	1.082	0.250				
16	17	4¼ × 2⅞	4.258	4.250	2.883	2.875	4.248	4.250	1.711	1.721	4.250	4.252	1.182	1.192	0.250				
17	18	4½ × 3	4.508	4.500	3.010	3.000	4.498	4.500	1.791	1.801	4.500	4.502	1.229	1.239	0.312				
18	19	4¾ × 3⅛	4.758	4.750	3.135	3.125	4.748	4.750	1.868	1.878	4.750	4.752	1.277	1.287	0.312				
19	20	5 × 3⅜	5.008	5.000	3.385	3.375	4.998	5.000	2.010	2.020	5.000	5.002	1.385	1.395	0.312				

* The key chamfer shall be the minimum to clear the keyway radius. Nominal values are given.

Table 2.   British Standard Square Parallel Keys, Keyways and Keybars (B.S. 46: Part 1: 1958)

All dimensions in inches

| Diameter of Shaft | | Key | | | Keyway in Shaft | | | | Keyway in Hub | | | | Nominal Keyway Radius, r | Bright Keybar | |
| Over | Up to and Including | Size, W × T | Width, W and Thickness, T | | Width, W_s | | Depth, H | | Width, W_h | | Depth, h | | | Width, W and Thickness, T | |
			Max.	Min.	Max.	Min.	Max.	Min.	Max.	Min.	Max.	Min.		Max.	Min.
1/4	1/2	1/8 × 1/8	0.127	0.125	0.125	0.124	0.078	0.072	0.126	0.125	0.066	0.060	0.010	0.127	0.125
1/2	3/4	3/16 × 3/16	0.190	0.188	0.188	0.187	0.113	0.107	0.189	0.188	0.094	0.088	0.010	0.190	0.188
3/4	1	1/4 × 1/4	0.252	0.250	0.250	0.249	0.148	0.142	0.251	0.250	0.121	0.115	0.010	0.252	0.250
1	1 1/4	5/16 × 5/16	0.314	0.312	0.312	0.311	0.183	0.177	0.313	0.312	0.148	0.142	0.010	0.314	0.312
1 1/4	1 1/2	3/8 × 3/8	0.377	0.375	0.375	0.374	0.219	0.213	0.376	0.375	0.175	0.169	0.010	0.377	0.375
1 1/2	1 3/4	7/16 × 7/16	0.440	0.438	0.438	0.437	0.254	0.248	0.439	0.438	0.203	0.197	0.020	0.440	0.438
1 3/4	2	1/2 × 1/2	0.502	0.500	0.500	0.499	0.289	0.283	0.501	0.500	0.230	0.224	0.020	0.502	0.500
2	2 1/2	5/8 × 5/8	0.627	0.625	0.625	0.624	0.360	0.354	0.626	0.625	0.284	0.278	0.020	0.627	0.625
2 1/2	3	3/4 × 3/4	0.752	0.750	0.750	0.749	0.430	0.424	0.751	0.750	0.339	0.333	0.020	0.752	0.750
3	3 1/2	7/8 × 7/8	0.877	0.875	0.875	0.874	0.501	0.495	0.876	0.875	0.393	0.387	0.062	0.877	0.875
3 1/2	4	1 × 1	1.003	1.000	1.000	0.999	0.572	0.566	1.001	1.000	0.448	0.442	0.062	1.003	1.000
4	5	1 1/4 × 1 1/4	1.253	1.250	1.250	1.248	0.713	0.707	1.252	1.250	0.557	0.551	0.062	1.253	1.250
5	6	1 1/2 × 1 1/2	1.504	1.500	1.500	1.498	0.854	0.848	1.502	1.500	0.667	0.661	0.062	1.504	1.500

• The key chamfer shall be the minimum to clear the keyway radius.   Nominal values are given.

Table 3. British Standard Rectangular Taper Keys and Keyways, Gib-head and Plain (B.S. 46: Part 1: 1958)

All dimensions in inches

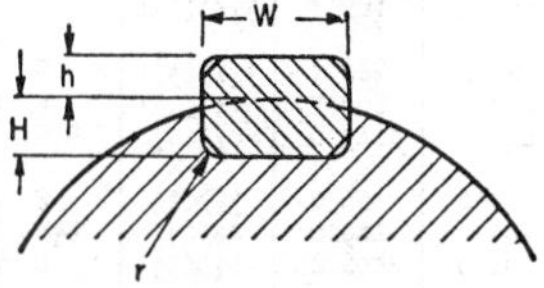
W
h
H
r

SECTION AT DEEP END OF KEYWAY IN HUB

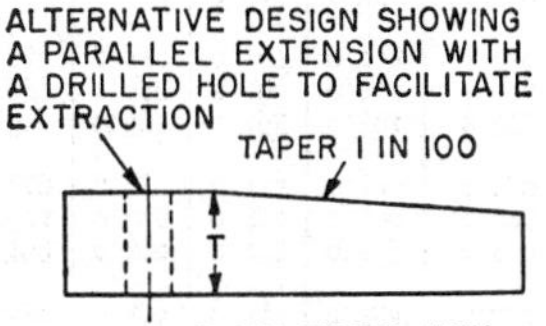
ALTERNATIVE DESIGN SHOWING A PARALLEL EXTENSION WITH A DRILLED HOLE TO FACILITATE EXTRACTION
TAPER 1 IN 100
T

PLAIN TAPER KEY

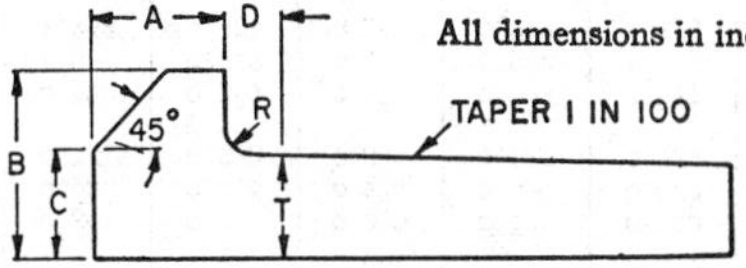
A
D
45°
R
B
C
T
TAPER 1 IN 100

GIB-HEAD KEY

Diameter of Shaft		Key					Keyway in Shaft and Hub								Nominal Keyway Radius, r*	Gib-head†				
Over	Up to and Including	Size, $W \times T$	Width, W		Thickness, T		Keyway in Shaft Width, W_s		Keyway in Hub Width, W_h		Depth in Shaft, H		Depth in Hub at Deep End of Keyway, h			A	B	C	D	Radius, R
			Max.	Min.	Max.	Min.	Min.	Max.	Min.	Max.	Min.	Max.	Min.	Max.						
1	1¼	5/16 × ¼	0.314	0.312	0.254	0.249	0.311	0.312	0.312	0.313	0.146	0.152	0.090	0.096	0.010	3/8	7/16	1/4	0.3	1/16
1¼	1½	3/8 × ¼	0.377	0.375	0.254	0.249	0.374	0.375	0.375	0.376	0.150	0.156	0.086	0.092	0.010	7/16	7/16	9/32	0.3	1/16
1½	1¾	7/16 × 5/16	0.440	0.438	0.316	0.311	0.437	0.438	0.438	0.439	0.186	0.192	0.112	0.118	0.020	1/2	9/16	5/16	0.4	1/16
1¾	2	½ × 5/16	0.502	0.500	0.316	0.311	0.499	0.500	0.500	0.501	0.190	0.196	0.108	0.114	0.020	9/16	5/8	3/8	0.4	1/16
2	2½	5/8 × 7/16	0.627	0.625	0.442	0.437	0.624	0.625	0.625	0.626	0.260	0.266	0.162	0.168	0.020	11/16	3/4	7/16	0.5	1/8
2½	3	¾ × ½	0.752	0.750	0.504	0.499	0.749	0.750	0.750	0.751	0.299	0.305	0.185	0.191	0.020	13/16	7/8	17/32	0.5	1/8
3	3½	7/8 × 5/8	0.877	0.875	0.630	0.624	0.874	0.875	0.875	0.876	0.370	0.376	0.239	0.245	0.062	15/16	1	21/32	0.6	1/8
3½	4	1 × ¾	1.003	1.000	0.755	0.749	0.999	1.000	1.000	1.001	0.441	0.447	0.293	0.299	0.062	1 1/16	1¼	23/32	0.6	1/8
4	5	1¼ × 7/8	1.253	1.250	0.880	0.874	1.248	1.250	1.250	1.252	0.518	0.524	0.340	0.346	0.062	1 5/16	1½	27/32	0.7	1/4
5	6	1½ × 1	1.504	1.500	1.007	0.999	1.498	1.500	1.500	1.502	0.599	0.605	0.384	0.390	0.062	1 9/16	1 5/8	1 1/32	0.7	1/4
6	7	1¾ × 1¼	1.754	1.750	1.257	1.249	1.748	1.750	1.750	1.752	0.740	0.746	0.493	0.499	0.125	1 13/16	2	1 7/32	0.8	1/4
7	8	2 × 1 3/8	2.005	2.000	1.382	1.374	1.998	2.000	2.000	2.002	0.818	0.824	0.539	0.545	0.125	2 1/16	2¼	1 13/32	0.8	1/4
8	9	2¼ × 1½	2.255	2.250	1.509	1.499	2.248	2.250	2.250	2.252	0.897	0.905	0.581	0.589	0.125	2 5/16	2½	1 9/16	0.9	3/8
9	10	2½ × 1 5/8	2.505	2.500	1.634	1.624	2.498	2.500	2.500	2.502	0.975	0.983	0.628	0.636	0.187	2 9/16	2¾	1 11/16	0.9	3/8
10	11	2¾ × 1 7/8	2.755	2.750	1.884	1.874	2.748	2.750	2.750	2.752	1.114	1.122	0.738	0.746	0.187	2 13/16	3	1 15/16	1.0	3/8
11	12	3 × 2	3.006	3.000	2.014	1.999	2.998	3.000	3.000	3.002	1.195	1.203	0.782	0.790	0.187	3 1/16	3¼	2 1/16	1.0	3/8

* The key chamfer shall be the minimum to clear the keyway radius. Nominal values shall be given.

† Dimensions A, B, C, D and R pertain to gib-head keys only.

Table 4. British Standard Square Taper Keys and Keyways, Gib-head or Plain (B.S. 46: Part 1: 1958)

All dimensions in inches

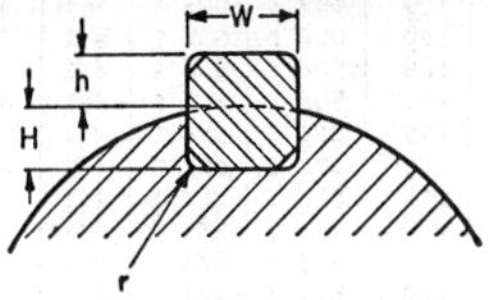

SECTION AT DEEP END OF KEYWAY IN HUB

ALTERNATIVE DESIGN SHOWING A PARALLEL EXTENSION WITH A DRILLED HOLE TO FACILITATE EXTRACTION.

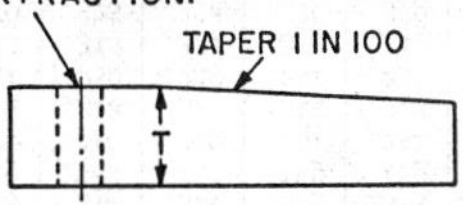

PLAIN TAPER KEY

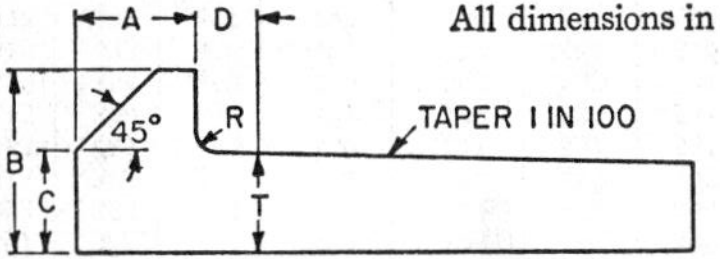

GIB-HEAD KEY

Diameter of Shaft		Key					Keyway in Shaft and Hub								Nominal Keyway Radius, r*	Gib-head†				
Over	Up to and Including	Size, W × T	Width, W		Thickness, T		Keyway in Shaft Width, W_s		Keyway in Hub Width, W_h		Depth in Shaft, H		Depth in Hub at Deep End of Keyway, h			A	B	C	D	Radius, R
			Max.	Min.	Max.	Min.	Min.	Max.	Min.	Max.	Min.	Max.	Min.	Max.						
1/4	1/2	1/8×1/8	0.127	0.125	0.129	0.124	0.124	0.125	0.125	0.126	0.072	0.078	0.039	0.045	0.010	3/16	1/4	5/32	0.1	1/32
1/2	3/4	3/16×3/16	0.190	0.188	0.192	0.187	0.187	0.188	0.188	0.189	0.107	0.113	0.067	0.073	0.010	1/4	3/8	7/32	0.2	1/32
3/4	1	1/4×1/4	0.252	0.250	0.254	0.249	0.249	0.250	0.250	0.251	0.142	0.148	0.094	0.100	0.010	5/16	7/16	9/32	0.2	1/16
1	1 1/4	5/16×5/16	0.314	0.312	0.316	0.311	0.311	0.312	0.312	0.313	0.177	0.183	0.121	0.127	0.010	3/8	9/16	11/32	0.3	1/16
1 1/4	1 1/2	3/8×3/8	0.377	0.375	0.379	0.374	0.374	0.375	0.375	0.376	0.213	0.219	0.148	0.154	0.010	7/16	5/8	13/32	0.3	1/16
1 1/2	1 3/4	7/16×7/16	0.440	0.438	0.442	0.437	0.437	0.438	0.438	0.439	0.248	0.254	0.175	0.181	0.020	1/2	3/4	15/32	0.4	1/16
1 3/4	2	1/2×1/2	0.502	0.500	0.504	0.499	0.499	0.500	0.500	0.501	0.283	0.289	0.202	0.208	0.020	9/16	7/8	17/32	0.4	1/16
2	2 1/2	5/8×5/8	0.627	0.625	0.630	0.624	0.624	0.625	0.625	0.626	0.354	0.360	0.256	0.262	0.020	11/16	1	21/32	0.5	1/8
2 1/2	3	3/4×3/4	0.752	0.750	0.755	0.749	0.749	0.750	0.750	0.751	0.424	0.430	0.310	0.316	0.020	13/16	1 1/4	25/32	0.5	1/8
3	3 1/2	7/8×7/8	0.877	0.875	0.880	0.874	0.874	0.875	0.875	0.876	0.495	0.501	0.364	0.370	0.062	15/16	1 3/8	29/32	0.6	1/8
3 1/2	4	1×1	1.003	1.000	1.007	0.999	0.999	1.000	1.000	1.001	0.566	0.572	0.418	0.424	0.062	1 1/16	1 5/8	1 1/32	0.6	1/8
4	5	1 1/4×1 1/4	1.253	1.250	1.257	1.249	1.248	1.250	1.250	1.252	0.707	0.713	0.526	0.532	0.062	1 5/16	2	1 9/32	0.7	1/4
5	6	1 1/2×1 1/2	1.504	1.500	1.509	1.499	1.498	1.500	1.500	1.502	0.848	0.854	0.635	0.641	0.062	1 9/16	2 1/2	1 17/32	0.7	1/4

* The key chamfer shall be the minimum to clear the keyway radius. Nominal values shall be given.
† Dimensions A, B, C, D and R pertain to gib-head keys only.

Table 5. British Standard Woodruff Keys and Keyways (B.S. 46: Part 1: 1958)

All dimensions are in inches

Key and Cutter No.	Nom. Width	Nom. Dia.	Dia. of Key, A Max.	A Min.	Depth of Key, B Max.	B Min.	Thickness of Key, C Max.	C Min.	Keyway Width in Shaft, D Min.	D Max.	Width of Keyway in Hub, E Min.	E Max.	Keyway Depth in Shaft, F Min.	F Max.	Keyway Depth in Hub at Center Line, G Min.	G Max.	Depth of Key (Optional Design), H Max.	H Min.	Dim. J Nom.
203	1/16	3/8	0.375	0.370	.171	.166	.063	.062	.061	.063	.063	.065	.135	.140	.042	.047	.162	.156	1/64
303	3/32	3/8	0.375	0.370	.171	.166	.095	.094	.093	.095	.095	.097	.119	.124	.057	.062	.162	.156	1/64
403	1/8	3/8	0.375	0.370	.171	.166	.126	.125	.124	.126	.126	.128	.104	.109	.073	.078	.162	.156	1/64
204	1/16	1/2	0.500	0.490	.203	.198	.063	.062	.061	.063	.063	.065	.167	.172	.042	.047	.194	.188	3/64
304	3/32	1/2	0.500	0.490	.203	.198	.095	.094	.093	.095	.095	.097	.151	.156	.057	.062	.194	.188	3/64
404	1/8	1/2	0.500	0.490	.203	.198	.126	.125	.124	.126	.126	.128	.136	.141	.073	.078	.194	.188	3/64
305	3/32	5/8	0.625	0.615	.250	.245	.095	.094	.093	.095	.095	.097	.199	.204	.057	.062	.240	.234	1/16
405	1/8	5/8	0.625	0.615	.250	.245	.126	.125	.124	.126	.126	.128	.182	.187	.073	.078	.240	.234	1/16
505	5/32	5/8	0.625	0.615	.250	.245	.157	.156	.155	.157	.157	.159	.167	.172	.089	.094	.240	.234	1/16
406	1/8	3/4	0.750	0.740	.313	.308	.126	.125	.124	.126	.126	.128	.246	.251	.073	.078	.303	.297	1/16
506	5/32	3/4	0.750	0.740	.313	.308	.157	.156	.155	.157	.157	.159	.230	.235	.089	.094	.303	.297	1/16
606	3/16	3/4	0.750	0.740	.313	.308	.189	.188	.187	.189	.189	.191	.214	.219	.104	.109	.303	.297	1/16
507	5/32	7/8	0.875	0.865	.375	.370	.157	.156	.155	.157	.157	.159	.292	.297	.089	.094	.365	.359	1/16
607	3/16	7/8	0.875	0.865	.375	.370	.189	.188	.187	.189	.189	.191	.276	.281	.104	.109	.365	.359	1/16
807	1/4	7/8	0.875	0.865	.375	.370	.251	.250	.249	.251	.251	.253	.245	.250	.136	.141	.365	.359	1/16
608	3/16	1	1.000	0.990	.438	.433	.189	.188	.187	.189	.189	.191	.339	.344	.104	.109	.428	.422	1/16
808	1/4	1	1.000	0.990	.438	.433	.251	.250	.249	.251	.251	.253	.308	.313	.136	.141	.428	.422	1/16
1008	5/16	1	1.000	0.990	.438	.433	.313	.312	.311	.313	.313	.315	.277	.282	.167	.172	.428	.422	1/16
609	3/16	1 1/8	1.125	1.115	.484	.479	.189	.188	.187	.189	.189	.191	.385	.390	.104	.109	.475	.469	5/64
809	1/4	1 1/8	1.125	1.115	.484	.479	.251	.250	.249	.251	.251	.253	.354	.359	.136	.141	.475	.469	5/64
1009	5/16	1 1/8	1.125	1.115	.484	.479	.313	.312	.311	.313	.313	.315	.323	.328	.167	.172	.475	.469	5/64
810	1/4	1 1/4	1.250	1.240	.547	.542	.251	.250	.249	.251	.251	.253	.417	.422	.136	.141	.537	.531	5/64
1010	5/16	1 1/4	1.250	1.240	.547	.542	.313	.312	.311	.313	.313	.315	.386	.391	.167	.172	.537	.531	5/64
1210	3/8	1 1/4	1.250	1.240	.547	.542	.376	.375	.374	.376	.376	.378	.354	.359	.198	.203	.537	.531	5/64
1011	5/16	1 3/8	1.375	1.365	.594	.589	.313	.312	.311	.313	.313	.315	.433	.438	.167	.172	.584	.578	3/32
1211	3/8	1 3/8	1.375	1.365	.594	.589	.376	.375	.374	.376	.376	.378	.402	.407	.198	.203	.584	.578	3/32
812	1/4	1 1/2	1.500	1.490	.641	.636	.251	.250	.249	.251	.251	.253	.511	.516	.136	.141	.631	.625	7/64
1012	5/16	1 1/2	1.500	1.490	.641	.636	.313	.312	.311	.313	.313	.315	.480	.485	.167	.172	.631	.625	7/64
1212	3/8	1 1/2	1.500	1.490	.641	.636	.376	.375	.374	.376	.376	.378	.448	.453	.198	.203	.631	.625	7/64

Table 6. British Preferred Lengths of Plain (Parallel or Taper) and Gib-head Keys, Rectangular and Square Section (B.S. 46: Part 1: 1958 Appendix)

All dimensions are in inches

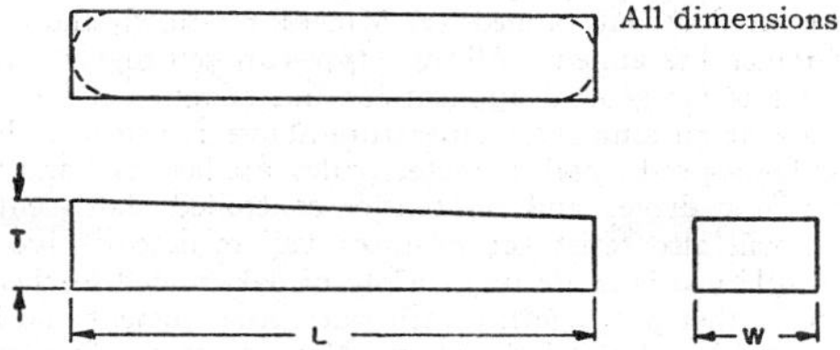

Plain Key Size W × T	Overall Length, L														
	¾	1	1¼	1½	1¾	2	2¼	2½	2¾	3	3½	4	4½	5	6
⅛ × ⅛	X	X													
3/16 × 3/16	X	X	X	X	X	X									
¼ × ¼	X	X	X	X	X	X	X	X	X	X	X				
5/16 × ¼	X	X	X	X	X	X	X	X	X	X	X				
5/16 × 5/16	X		X	X	X	X	X	X	X	X	X				
⅜ × ¼		X	X	X	X	X	X	X	X	X	X	X			
⅜ × ⅜		X	X	X	X	X	X	X	X	X	X	X	X		
7/16 × 5/16				X	X	X	X	X	X	X	X	X	X		
7/16 × 7/16					X	X	X	X	X	X	X	X	X		
½ × 5/16						X	X	X	X	X	X	X	X	X	
½ × ½						X	X	X	X	X	X	X	X	X	
⅝ × 7/16							X	X	X	X	X	X	X	X	
⅝ × ⅝								X	X	X	X	X	X	X	X
¾ × ½										X	X	X	X	X	X
¾ × ¾											X	X	X	X	X
⅞ × ⅝											X	X	X	X	X

All dimensions are in inches

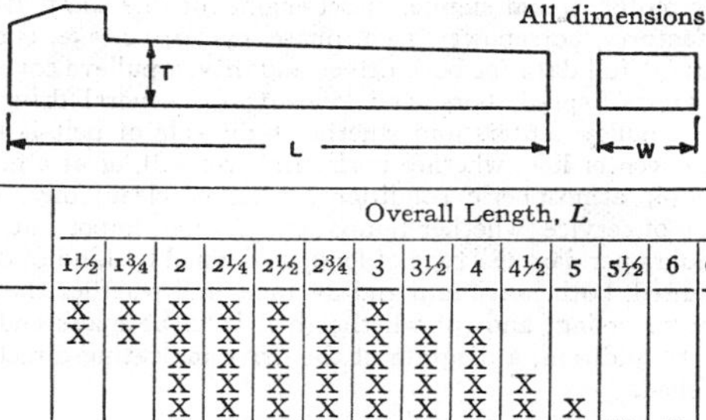

Gib-head Key Size, W × T	Overall Length, L																
	1½	1¾	2	2¼	2½	2¾	3	3½	4	4½	5	5½	6	6½	7	7½	8
3/16 × 3/16	X	X	X	X	X												
¼ × ¼	X	X	X	X	X												
5/16 × ¼			X	X	X	X	X	X	X								
5/16 × 5/16			X	X	X	X	X	X	X								
⅜ × ¼			X	X	X	X	X	X	X	X							
⅜ × ⅜			X	X	X	X	X	X	X	X	X						
7/16 × 5/16					X	X	X	X	X	X	X	X					
7/16 × 7/16						X	X	X	X	X	X	X	X				
½ × 5/16							X	X	X	X	X	X	X				
½ × ½							X	X	X	X	X	X	X	X			
⅝ × 7/16								X	X	X	X	X	X	X			
⅝ × ⅝								X	X	X	X	X	X	X	X		
¾ × ½									X	X	X	X	X	X	X		
¾ × ¾									X	X	X	X	X	X	X	X	X
⅞ × ⅝											X	X	X	X	X	X	X
⅞ × ⅞											X	X	X	X	X	X	X
1 × ¾												X	X	X	X	X	X
1 × 1												X		X	X		X

FLAT BELTS AND PULLEYS

Flat Leather Belting. — Three principal types of leather belting produced in the United States are: (1) Oak tanned; (2) Mineral retanned; and (3) Combination oak tanned and mineral retanned. All three types are put together with waterproof cement. The first is for general applications but should not be used where the drive is to operate at an ambient temperature above 120 degrees F. The second type is used for high-speed small-diameter pulley applications and is also suitable for high-speed motor drives and on tension-controlled short-center motor-base applications. It will also resist the mists or vapors of corrosive acids for long periods. The third type is made up of a ply of oak-tanned leather with a ply of mineral retanned leather put together with waterproof cement and is used on step cone pulleys and hard-pull shifting drives. None of these three types should be used, however, when a possibility exists of liquid acid coming in contact with the belt or when the ambient temperature is above 140 degrees F.

Specifying Flat Leather Belting. — When specifying a leather belt, the following information should be supplied: (1) Shortest steel tape length around pulleys. If measurement is stated as made with steel tape, manufacturer will make proper deductions for initial stretch. (2) The type, width, and thickness, (3) Whether the belt is to be supplied: (a) made endless at the factory; (b) with laps prepared for cementing on the job, or (c) with ends cut square for lacing.

Slide-rail mounted motors should be moved towards the driven pulley at least three-quarters of the total travel of the slide-rail base before taking measurements. It should be noted that the measurement of an old belt is not a dependable guide for the length of a new one since, being elastic, the new belt will let out somewhat when put under driving tension.

For important drives, where the belting manufacturer is called upon to recommend belt type, width, and thickness, the National Industrial Leather Association lists the following information to be given to the supplier: (1) source of power, whether electric motor, steam engine, diesel engine, or line shaft; if electric motor, then (2) manufacturer, horsepower, rpm, phase, cycles, voltage, current, and type of starting used; (3) full data for both driver and driven pulleys covering diameters, face widths, material, speed, bore, and keyway; (4) general drive data, such as distance between pulley centers and whether tight side of belt is top or bottom; (5) angle of drive center line, whether horizontal vertical, or at a given angle from the horizontal; (6) atmospheric conditions, whether clean, oily, wet, dusty, or normal; (7) type of service, whether temporary, normal, important, or continuous, together with hours per day; (8) type of load, as defined by kind of driven machine, horsepower required, both maximum and average, and whether the load is steady, jerky, shock, or reversing; and (9) whether belt is to be made endless on pulleys or at factory. In addition, a diagram of the drive indicating direction of rotation should be furnished.

Thicknesses of Flat Leather Belting. — The following thicknesses have been approved and adopted by the National Industrial Leather Association: Medium Single Ply, 11/64 inch, average; Heavy Single Ply, 13/64 inch, average; Light Double Ply, 18/64 inch, average; Medium Double Ply, 20/64 inch, average; Heavy Double Ply, 23/64 inch, average; Medium Triple Ply, 30/64 inch, average; and Heavy Triple Ply, 34/64 inch, average.

Measurement: All of the above thicknesses are average thicknesses and should be determined by measuring 20 coils and dividing the total by the number of coils measured. In rolls containing less than 20 coils, the average thickness should be determined by measuring all of the coils in the roll.

Table 1. Belt Capacity Factor, K — Flat Leather Belting

Officially adopted by the National Industrial Leather Association

Belt Speed, Feet per Minute	SINGLE PLY		DOUBLE PLY			TRIPLE PLY	
	*11/64″	*13/64″	*18/64″	*29/64″	*23/64″	*30/64″	*34/64″
	Medium	Heavy	Light	Medium	Heavy	Medium	Heavy
	Horsepower per Inch of Width, K — to be Corrected by Factors from Table 2						
600	1.1	1.2	1.5	1.8	2.2	2.5	2.8
800	1.4	1.7	2.0	2.4	2.9	3.3	3.6
1000	1.8	2.1	2.6	3.1	3.6	4.1	4.5
1200	2.1	2.5	3.1	3.7	4.3	4.9	5.4
1400	2.5	2.9	3.5	4.3	4.9	5.7	6.3
1600	2.8	3.3	4.0	4.9	5.6	6.5	7.1
1800	3.2	3.7	4.5	5.4	6.2	7.3	8.0
2000	3.5	4.1	4.9	6.0	6.9	8.1	8.9
2200	3.9	4.5	5.4	6.6	7.6	8.8	9.7
2400	4.2	4.9	5.9	7.1	8.2	9.5	10.5
2600	4.5	5.3	6.3	7.7	8.9	10.3	11.4
2800	4.9	5.6	6.8	8.2	9.5	11.0	12.1
3000	5.2	5.9	7.2	8.7	10.0	11.6	12.8
3200	5.4	6.3	7.6	9.2	10.6	12.3	13.5
3400	5.7	6.6	7.9	9.7	11.2	12.9	14.2
3600	5.9	6.9	8.3	10.1	11.7	13.4	14.8
3800	6.2	7.1	8.7	10.5	12.2	14.0	15.4
4000	6.4	7.4	9.0	10.9	12.6	14.5	16.0
4200	6.7	7.7	9.3	11.3	13.0	15.0	16.5
4400	6.9	7.9	9.6	11.7	13.4	15.4	16.9
4600	7.1	8.1	9.8	12.0	13.8	15.8	17.4
4800	7.2	8.3	10.1	12.3	14.1	16.2	17.8
5000	7.4	8.4	10.3	12.5	14.3	16.5	18.2
5200	7.5	8.6	10.5	12.8	14.6	16.8	18.5
5400	7.6	8.7	10.6	12.9	14.8	17.1	18.8
5600	7.7	8.8	10.8	13.1	15.0	17.3	19.0
5800	7.7	8.9	10.9	13.2	15.1	17.5	19.2
†6000	7.8	8.9	10.9	13.2	15.2	17.6	19.3
Belt Speed, fpm	Minimum Allowable Pulley Diameter, Inch, For Belt Thicknesses Listed Above						
Up to 2500	2½	3	4	5[1]	8[1]	16[2]	20[2]
2500 to 4000	3	3½	4½	6[1]	9[1]	18[2]	22[2]
4000 to 6000	3½	4	5	7[1]	10[1]	20[2]	24[2]

* The belt thicknesses are average thicknesses. See paragraph on Thicknesses of Flat Leather Belting.

† For belt speeds over 6000 feet per minute, consult a leather belting manufacturer.

[1] For belts 8 inches wide and over, add 2 inches to minimum pulley diameter shown.

[2] For belts 8 inches wide and over, add 4 inches to minimum pulley diameter shown.

Tolerances: Allowable tolerances for thicknesses of single and double ply belts are plus or minus 1/64 inch, based on the nominal thickness. At no point shall single-ply belting be more than 3/64 inch thicker or 2/64 inch thinner than the average thickness. For double-ply belting, the variation in thickness shall not be greater than 2/64 inch thicker or thinner than the average.

Triple-ply Belts: Most triple-ply belts are constructed for particular drive conditions. The manufacturer should be consulted for specific information concerning thickness and construction.

Pulleys. — On step-cone pulleys use a narrow, thick belt rather than a wide, thin belt of the same horsepower capacity. It is good practice to use pulleys with faces from ½ to 2 inches (depending upon their diameters) wider than the belt required for the drive.

Use with Tension-Controlling Motor Base. — When used with an electric motor drive, a flat leather belt will give best results if the motor is mounted on some kind of a tension controlling base. Three types are generally available. Two of these are pivoted; one uses the weight of the motor to maintain the proper belt tension; the other utilizes the reaction torque of the motor to accomplish this. The third type has a sliding action and controls the belt tension by means of springs. The effect of all three types is to cause the belt to maintain a uniform pull around and across the pulleys.

**Table 2.   Service Correction Factors *M*, *P* and *F*
Used in Determining Horsepower Rating**

Select the one appropriate factor from each of the three divisions in this table.

		M
Motor type and Starting Method	Squirrel cage, compensator starting	1.5
	Squirrel cage, line starting	2.0
	Slip ring and high starting torque	2.5
		P
Diameter of Small Pulley	4 inches and under	0.5
	4½ to 8 inches	0.6
	9 to 12 inches	0.7
	13 to 16 inches	0.8
	17 to 30 inches	0.9
	Over 30 inches	1.0
		F
Operating Conditions	Oily, wet, or dusty atmosphere	1.35
	Vertical drives	1.2
	Jerky loads	1.2
	Shock and reversing loads	1.4

Horsepower Ratings. — In Table 1 are given belt capacity factors for various types and thicknesses of flat leather belting and for various belt speeds in feet per minute. These factors are expressed in terms of horsepower per inch of width and are modified by service correction factors given in Table 2 as shown in the formula below. Where the pulley speed is known in terms of revolutions per minute, the corresponding belt speed in feet per minute can be found from Table 3.

The following formula is used to obtain the horsepower rating, *H*, of flat leather belt:

$$H = \frac{W \times K \times P}{M \times F} \tag{1}$$

Table 3. Conversion of Pulley Speeds in Revolutions per Minute into Feet per Minute

Pulley Diam. in Inches	Revolutions per Minute																				
	100	200	300	400	435	490	500	600	690	700	800	900	1000	1150	1200	1400	1500	1600	1750	1800	3600
	Velocity in Feet per Minute*																				
1	26	52	79	105	114	128	131	157	181	183	209	236	262	301	314	367	393	419	458	471	942
2	52	105	157	209	228	257	262	314	361	367	419	471	524	602	628	733	785	838	916	942	1885
3	79	157	236	314	342	385	393	471	542	550	628	707	785	903	942	1100	1178	1257	1374	1414	2827
4	105	209	314	419	456	513	524	628	722	733	838	942	1047	1204	1257	1466	1571	1676	1833	1885	3770
5	131	262	393	524	570	641	654	785	903	916	1047	1178	1309	1505	1571	1833	1964	2094	2291	2356	4712
6	157	314	471	628	683	770	785	942	1084	1100	1257	1414	1571	1806	1885	2199	2356	2513	2749	2827	5655
7	183	367	550	733	797	898	916	1100	1264	1283	1466	1649	1833	2107	2199	2566	2749	2932	3207	3299	…
8	209	419	628	838	911	1026	1047	1257	1445	1466	1676	1885	2094	2409	2513	2932	3142	3351	3665	3770	…
9	236	471	707	942	1025	1154	1178	1414	1625	1649	1885	2121	2356	2710	2827	3299	3534	3770	4123	4241	…
10	262	524	785	1048	1139	1283	1309	1571	1806	1833	2094	2356	2618	3011	3142	3665	3927	4189	4582	4712	…
20	524	1047	1571	2094	2278	2566	2618	3142	3612	3665	4189	4712	5236	…	…	…	…	…	…	…	…
30	785	1571	2356	3142	3416	3848	3927	4712	5418	5498	…	…	…	…	…	…	…	…	…	…	…
0.1	3	5	8	10	11	13	13	16	18	18	21	24	26	30	31	37	39	42	46	47	94
0.2	5	10	16	21	23	26	26	31	36	37	42	47	52	60	63	73	79	84	92	94	188
0.3	8	16	24	31	34	38	39	47	54	55	63	71	79	90	94	110	118	126	137	141	283
0.4	10	21	31	42	46	51	52	63	72	73	84	94	105	120	126	147	157	168	183	188	377
0.5	13	26	39	52	57	64	65	79	90	92	105	118	131	151	157	183	196	209	229	236	471
0.6	16	31	47	63	68	77	78	94	108	110	126	141	157	181	188	220	236	251	275	283	565
0.7	18	37	55	73	80	90	92	110	126	128	147	165	183	211	220	257	275	293	321	330	660
0.8	21	42	63	84	91	103	105	126	144	147	168	189	209	241	251	293	314	335	367	377	754
0.9	24	47	71	94	102	115	118	141	163	165	188	212	236	271	283	330	353	377	412	424	848

* Based on: Velocity in fpm = Diam. in inches × rpm × 0.2618. For velocities above 6,000 fpm, consult belt manufacturer.

Example: Find velocity in fpm of a 28.3-inch diameter pulley rotating at a speed of 600 rpm. *Solution:* Add the velocities equivalent to the 20-, 8-, and 0.3-inch diameters, thus: $V = 3142 + 1257 + 47 = 4446$ fpm.

where W = width of belt in inches.

     K = theoretical belt capacity factor taken from Table 1.

     P = correction factor for diameter of smaller pulley taken from Table 2.

     M = correction factor if type of motor and starting method is one of those given in Table 2. If not, use $M = 1$.

     F = special factor if operating condition is one of those shown in Table 2. If not, use $F = 1$.

If the belt is used on an electric motor drive and the horsepower rating of the motor is known at a given speed, then the required width of belt is found from the formula:

$$W = \frac{R \times M \times F}{K \times P} \tag{2}$$

where R = nameplate horsepower rating of the electric motor.

     M, F, K, and P are as in the previous formula.

Example: Find the proper width of flat leather belting for a drive employing a squirrel cage induction motor rated at 15 horsepower at 1750 rpm with across-the-line starting. Pulley diameters are: motor, 8 inches and driven, 16 inches.

    1. For 8-inch pulley running at 1750 rpm, Table 3 gives belt speed of 3665 fpm.

    2. The bottom section of Table 1 shows that for a belt speed of 2500 to 4000 fpm, and an 8-inch pulley, a Medium Double Ply belt is suitable.

    3. Table 1 gives the belt capacity factor for a Medium Double Ply belt running at 3665 fpm as 10.2 (interpolating between 10.1 for 3600 fpm and 10.5 for 3800 fpm).

    4. Table 2 gives the motor correction factor for squirrel cage motor, line starting as 2.0.

    5. The pulley correction factor for 8-inch pulley is given as 0.6 in Table 2.

Hence
$$W = \frac{R \times M}{K \times P} = \frac{15 \times 2.0}{10.2 \times 0.6} = 4.9 \text{ in.}$$

Use a 5-inch wide Medium Double Ply belt.

If the motor was being used on a vertical drive without a tension-controlled motor base, the belt width, as determined, should be multiplied by a factor of 1.2 taken from Table 2. The proper belt for use in this case would be 4.9×1.2 or a 6-inch Medium Double Ply belt.

Speed Limitation. — The use of flat leather belting for speeds in excess of 6000 feet per minute is not generally recommended. At these speeds the amount of horsepower which the belt can transmit decreases due to increased slippage because of the action of centrifugal force in holding the belt off the pulleys and thus reducing the areas of contact and the contact pressure.

Installation of Leather Belting. — An easy method of aligning shafting and pulleys is to first check shafts with a level and then place a taut string across the shafts and check each shaft against it with a large square. The alignment of pulleys can be checked by using a taut string along their edges. If the pulleys are the same width, the string should touch lightly at two opposite points on the rim of each pulley. If the pulleys are of different widths, the distance from the string to the pulley rim at opposite points on its circumference should be the same. If possible, each pulley should be given a half-turn and then rechecked. When pulleys are installed one above the other, a string and plumb-bob can be used to check alignment.

Belt Tension: For best results a leather belt should be run with the least tension needed to transmit the load without slipping. If the belt is too slack it will slip, causing its surface to glaze, then crack and peel. If it is too tight, it may put excessive load on the bearings.

Wherever possible, flat leather belts should be operated with the slack side on top. This arrangement will provide a greater arc of contact between belt and faces of the pulleys permitting lower tension. On short-center or vertical drives, a tension-controlled motor base should be used.

The maximum pulling power is obtained by running the grain or hair side of the belt next to the pulley faces.

Endless Belts: An endless belt should be forced over the pulleys with care to avoid putting a crook in it. Belts, particularly those six inches and wider, should either be made endless on the job as shown in the 19th edition of Machinery's Handbook or slipped on after temporarily shortening the center distance between pulleys, by moving the motor on slide rails, loosening hanger bolts, etc.

Running Direction: Care should be taken to have the outside feather edge of the lap face away from the direction in which the belt runs as shown in the accompanying

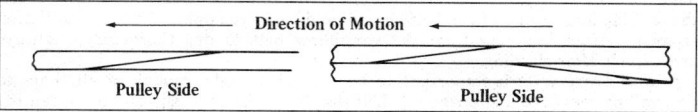

illustration. This mode tends to protect the outside points of the lap if they should strike guards, guides, or shifters. The lap is also protected from being opened by windage.

Belt Splicing: Wherever possible leather belts should be made endless on the job to avoid possibility of damage to the belt which frequently occurs when an endless belt, that is a tight fit, is forced over the pulleys.

The first step should be to shorten the pulley centers to the minimum. Then, a steel tape measurement is taken around the pulleys. The usual practice is to reduce this measured length by about $\frac{1}{8}$ inch per foot to provide adequate belt tension. To the resulting dimension, add the proper allowance as given in Table 4. The result will be the length of the belting to be cut from the roll.

Table 4. Lap and Break Lengths for Splicing Flat Leather Belting

Single Lap Belt			Double Lap Belt			
Belt Width	Lap Length	Allowance for Splice	Belt Width	Lap Length	Break Length	Allowance for Splice
Under 4	4	4	Under 8	4	8	12
4 to 10	5	5	8 to 12	4	10	14
Over 10	6	6	12 ro 18	4	16	20
. . .	. . .	. . .	18 and up	4	20	24

All dimensions in inches.

Maintenance of Leather Belting. — It is important to establish a system of inspection at regular intervals, at which times the following points should be checked.

1. *Dryness of Belt:* When pulley faces begin to polish, it is a sign that dressing is needed on the belt. Under normal conditions, dress belts every three to six months. Best results are usually obtained by using belt dressing sparingly, but frequently. Use a belt dressing approved by the manufacturer and designed to supply the necessary oils which were lost in use or during cleaning.

2. *Saturation with Oil:* Belts should be kept as clean as possible for best results. Oil or grease thrown from machine bearings will reduce belt life and pulling power. If the leak cannot be stopped at its source, the installation of deflectors or throwing discs will be helpful.

A small amount of oil on a belt can sometimes be removed by ordinary wiping. If this does not do the job, give it a thorough scrubbing with a solution of carbon tetrachloride and naphtha, using a stiff jute brush and working in the direction of lap joints so as not to lift them.

Another method is to remove the belt and soak it for five or six hours in a degreasing solution consisting of one part carbon tetrachloride to three parts naphtha. If carbon tetrachloride is not available, the belt can be soaked in any of the cleaning fluids used by dry cleaning establishments. Due to the fire hazard and toxic effect, the soaking and drying of the belt should be done in the open or where ventilation is good. After removing from the vat, allow belt to dry thoroughly. Always dress a belt after cleaning.

3. *Belt Tension:* It is important to keep the belt tight enough at all times to transmit power without slippage. A belt that is too slack will slip and burn, causing excessive wear. If the belt is too tight, it places undue strain on the bearings and the belt life will be shortened.

4. *Alignment of Shafting and Pulleys:* Belting cannot give good service if the pulleys or shafting are out of alignment. Indications of misalignment are: (a) belt running off the pulley at one side or (b) rubbing or climbing on flanged or step-cone pulleys. A simple test to determine whether the fault is the alignment or a crooked belt, is to turn the belt inside out or end for end. If it still runs to the same side of the pulley as before, the fault is in the alignment, not in the belt.

It is important to check drive alignment at least once a year. In multiple-story buildings, shifting of loads on floors above the shaft may cause it to be distorted or thrown out of alignment.

Some common faults in drive alignment are: (a) Shafts carrying driving and driven pulleys may not be parallel. (b) Shafting may be sprung out of line. (Hangers should always be located near the pulleys, the points of maximum load.) (c) Driving and driven pulleys may be offset. (d) Pulley may be eccentric with shafting.

5. *Laps and Plys:* When cementing laps, if first sizing coat does not dry with a shine, a second coat should be applied and allowed to dry. If sizing coat turns cloudy, apply a second coat and continue brushing until cloudiness disappears. If sizing coat does not dry in thirty minutes, wipe off and apply another sizing coat (this condition caused by an excessive amount of oil in the belt). If cemented laps show signs of opening, re-cement them immediately.

If belt guards, shifters, guides, or pulley flanges rub against edge of belt, laps and plies may open up. This condition should be corrected immediately. A good belt shifter has broad and well-rounded surfaces so as to spread thrust over a large, belt edge area.

Another cause of ply and lap separation is running too thick a belt on a small pulley.

Shortening of Leather Belts: When a belt becomes loose, it will slip excessively causing loss of power and undue wear on the belt. A loose belt should be shortened immediately.

Table 5.   Trouble-shooting Chart for Flat Leather Transmission Belts

Trouble	Cause	Remedy
1. Belt slips and squeals	(a) Belt too loose (b) Insufficient belt capacity (c) Pulley crown too high, causing increased wear of narrow center section of belt (d) Leather surface too dry and shiny	(a) Increase belt tension (b) Use thicker or wider belt (c) Decrease crown taper to ⅛ inch per foot (d) Apply suitable dressing
2. Excessive belt stretch	Belt capacity too low	Use thicker or wider belt
3. Belt runs crooked	(a) Belt stretched on one side by forcing over pulley (b) Belt ends not squared when joining (c) Belt unevenly stretched by running on misaligned pulleys (d) Loose belt unevenly stretched by running up on flanged or step-cone pulley	(a) (b) (c) (d) Repair damaged belt section or replace belt. Eliminate physical cause when installing
4. Belt runs off pulleys	(a) Misalignment of pulleys or shafting (if belt continues to run off same side when belt is turned end for end) (b) Crooked belt (c) Pulley crown too high	(a) Eliminate cause (b) Repair belt (c) Decrease crown taper to ⅛ inch per foot
5. Belt runs to one side of driven pulley	(a) Belt too slack (b) Load too great (c) Crooked belt (if it runs to opposite side when turned end for end) (d) Misalignment of pulleys or shafting	(a) Increase belt tension (b) Use thicker or wider belt (c) Repair belt (d) Eliminate cause
6. Belt whips and flaps	(a) Pulsating load or power source (b) Shaft, motor, or machine not rigidly supported (c) Lopsided pulley (d) Bent shaft (e) Too much or too little belt tension	(a) (b) (c) (d) (e) Eliminate cause where possible. Try change of speed or addition of flywheel to smooth out load
7. Belt weaves back and forth across pulley	(a) Wobbly pulley (b) High spot on pulley (c) Belt extremely crooked	(a) (b) Correct faulty condition (c) Repair or replace belt
8. Cracked outside ply	(a) Excessive belt tension (b) Pulley diameter too small	(a) Reduce tension (b) Provide proper pulley for belt thickness
9. Cracked inside ply	Burning caused by excessive slip	See Item 1.
10. Peeling grain	(a) Excessive slip (b) Improper belt dressing (c) Chemical fumes or oil	(a) See Item 1. (b) Clean belt with commercial solvent, scrape off any loose grain and use suitable dressing (c) Provide guards if possible, and use type of belt best suited for condition

Source: National Industrial Leather Association.

Angular Drive. — In laying out an angular drive — one in which the pulley shafts are not parallel — there is one fundamental rule to be followed: the belt must leave each pulley in the plane of the pulley toward which it is running. By "plane of the pulley" is meant that plane passing through the center of the face of the pulley and at right angles to its axis or shaft.

If the plane of one pulley of an open belt drive in which the shafts are parallel is rotated through an angle of 90 degrees and the pulleys are placed so that the above fundamental rule will hold, a "quarter-turn drive" will result.

Quarter-turn Drive. — To overcome the belt distortion encountered in a quarter-turn drive, and to distribute wear evenly on both of its sides, give one end of the belt a one-half turn (180 degrees) before making the splice. In a single-ply belt, made up in this way, the grain side at the joint is adjacent to the flesh side. This half-turn method of construction is particularly adaptable to double-ply belts, which have the grain exposed on both sides. The turning of the belt side to side also turns it edge to edge, with the result that not only is the wear distributed on the face of the belt, but the tension is also kept equalized on the edges of the belt.

When installing quarter-turn drives, it is particularly advantageous to make the belts endless with clamps and rods right on the job. Fewer shutdowns will result.

Rubber Belting. — Rubber belts are used in places exposed to the weather or the action of steam, as they do not absorb moisture or stretch as readily as leather belts, under like conditions. The quality of rubber belting depends on the mixture (containing more or less rubber) that forms the coating, the cotton duck that gives strength to the belt and the method of manufacture. The best grades of rubber belting contain nothing but new rubber; the cheapest grades are composed largely of reclaimed rubber. The weight of the cotton duck is an important consideration. High-grade belts contain what is known as a 32-ounce cotton duck, and the cheaper grades have either a 30-ounce or 28-ounce duck. If the proper weight of duck is used, a 3- or 4-ply rubber belt is equal in strength to a single leather belt; a 5- or 6-ply rubber belt is equal to a double leather belt, and a 7- or 8-ply rubber belt is equal to a triple leather belt.

Pulleys. — Flat belt pulleys are usually made of cast iron, fabricated steel, paper, fiber, or various kinds of wood. They may be solid or split and in either case the hub may be split for clamping to the shaft.

Pulley face widths are nominally the same as the widths of the belts they are to carry. Actually, however, the pulley face should be approximately one inch more than the belt width for belts under 12 inches wide, 2 inches more for belts from 12 to 24 inches wide, and 3 inches more for belts over 24 inches in width.

Belts may be made to center themselves on their pulleys by the use of crowned pulleys. The usual figure for the amount of crowning is ⅛ inch per foot of pulley width. Thus, the difference in maximum and minimum radii of a crowned 6-inch wide pulley would be 1/16 inch. Crowned pulleys have a rim section either with a convex curve or a flat V form. Flanges on the sides of flat belt pulleys are in general undesirable as the belt tends to crawl against them. Too much crown is undesirable because of the tendency to "break the belt's back." This is particularly true in the case of riding idlers close to driving pulleys where the curvature of the belt changes rapidly from one pulley to the other. In such cases the idler should under no circumstances be crowned and the adjacent pulley should have very little crown. Pulleys carrying shifting belts are not crowned.

Open belt drives connecting pulleys on short centers with one pulley considerably larger than the other may be unsatisfactory on account of the small angle of wrap on the smaller pulley. This angle may be increased by the use of idler pulleys on one or both sides of the belt.

Cast-Iron Pulleys. — Cast-iron pulleys formed of one solid casting may or may not have a split or divided hub. The solid-hub pulley is held to its shaft either by a key, a key and one or two set-screws, or by simply using one or more set-screws without a key as in the case of small pulleys, especially on low-grade machinery where there is little power to transmit. When the hub is split or divided, it is provided with clamping bolts, and when these are tightened, the split hub grips the shaft tightly. In addition to clamping bolts, a key or a key and set-screws may be used. Pulleys of this kind are known as the clamp-hub type.

Most pulleys have six arms. For diameters less than 15 or 20 inches, there may be four arms, and pulleys 5 feet or larger in diameter often have eight arms.

Split Cast-iron Pulleys. — The split pulley which is formed of two separate sections bolted together both at the hub and on opposite sides of the rim, can be placed between other pulleys on a shaft without removing either the pulleys or the shaft. These pulleys often have interchangeable hub bushings to fit shafts of different diameter. It is good practice to make pulleys having a face width of 10 inches or over, either of the clamp-hub or split form, because shrinkage strains are either greatly reduced or practically eliminated, and the hub of the pulley can be firmly clamped to a shaft even though the bore is not an accurate fit. If the face width of a cast-iron pulley is greater than from 20 to 24 inches, there should be two sets of arms to provide better support for the rim.

Wood Pulleys. — Wood pulleys are not only much lighter than cast-iron pulleys but they are superior as transmitters of power; in fact it is claimed that they will transmit from 35 to 50 per cent more power for the same belt tension. Wood pulleys should not be used where they are exposed to excessive moisture. Ordinarily the rims are built up of segments, and the arrangement of the arms varies on different sizes and makes. Some wood pulleys intended for unusually severe duty have a rim which is joined to an iron center or hub by a solid web of wood. Other pulleys of the iron-center type have cast-iron hubs and arms and a wood rim. Internal shrinkage strains are thus eliminated and the pulleys are adapted to unusually high speeds. Well-seasoned maple is adapted to wood pulleys. Wood bushings are often inserted in the hubs to permit using the pulleys on shafts of different size.

Steel Pulleys. — Pulleys formed of sheet steel combine lightness with strength and they are free from the initial stresses which are such an uncertain factor in many cast-iron pulleys. The weight is ordinarily from 45 to 55 per cent less than the weight of a cast-iron pulley of equal power-transmitting capacity, which lessens the weight on the lineshaft and reduces the frictional losses. A series of tests showed that the percentage of slip was from 2.35 to 2.70 per cent less for steel pulleys than for cast-iron pulleys. Steel pulleys are ordinarily of the split type.

Safe Speeds for Pulleys. — The maximum safe rim speeds for solid cast-iron pulleys is as a general rule about 5000 feet per minute. If the pulley is split or formed of separate sections which are bolted together at the rim, the maximum speed should be limited to about 55 or 60 per cent of the maximum speed for solid pulleys. While the safe speeds of built-up steel pulleys are subject to some variation on account of differences in design or construction, in general such pulleys may be run at about 6000 feet per minute. The safe speeds recommended for wood pulleys vary considerably according to the type; thus, the maximum speeds recommended may be 5000 feet per minute for some pulleys and 10,000 feet per minute for others of different construction. A pulley having a cast-iron hub and arms, with a wood rim, has been operated under test at a rim speed of five and one-half miles per minute. For additional information on speeds see pages 193 and 194 in the flywheel section.

Dimensions of Pulleys

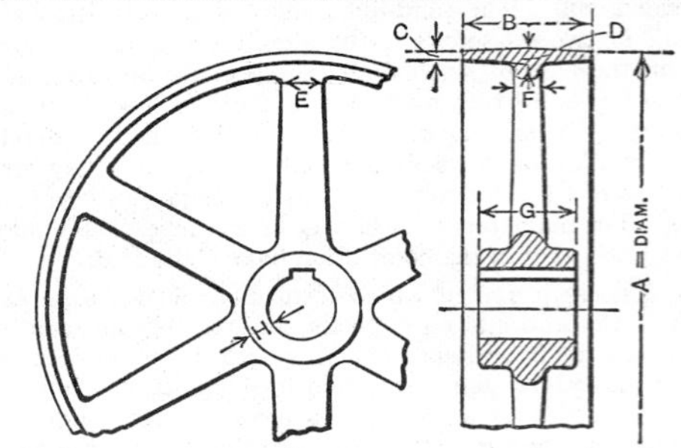

In all cases, the number of arms is 6. The arms increase in size towards the hub, the taper being ½ inch per foot. For safe speeds and descriptions of various pulley constructions, see pages 193 and 194.

Diam. *A*	Face *B*	*C*	*D*	*E*	*F*	*G*	*H*
6	4	1/8	3/16	3/4	7/16	3	3/8
6	6	1/8	3/16	3/4	7/16	3½	½
6	8	1/8	3/16	3/4	7/16	3½	½
6	12	1/8	3/16	3/4	7/16	4	½
8	4	1/8	3/16	13/16	7/16	3	3/8
8	6	1/8	3/16	13/16	7/16	3½	½
8	8	5/32	1/4	1 1/16	9/16	4½	½
8	12	5/32	1/4	1 1/16	9/16	5½	½
10	4	1/8	3/16	15/16	9/16	3	½
10	6	5/32	1/4	1 1/16	9/16	3½	½
10	8	5/32	1/4	1 1/16	9/16	4½	½
10	12	5/32	1/4	1 5/16	5/8	5½	5/8
12	4	5/32	1/4	1	7/16	3¼	½
12	6	5/32	1/4	1 3/4	½	4	½
12	8	5/32	1/4	1 3/4	½	5	5/8
12	12	3/16	5/16	1½	3/4	6½	5/8
14	4	5/32	1/4	1 1/8	½	3½	½
14	6	5/32	1/4	1 1/8	½	4½	5/8
14	8	3/16	5/16	1 5/16	9/16	5	5/8
14	12	3/16	5/16	1 11/16	13/16	6½	5/8
16	4	5/32	1/4	1 3/8	9/16	3½	½
16	8	3/16	5/16	1 7/16	5/8	5	5/8
16	12	7/32	11/32	1 7/16	5/8	6½	3/4
16	16	7/32	11/32	1 7/8	15/16	8¼	7/8
18	4	3/16	5/16	1 5/16	9/16	4	5/8
18	8	7/32	11/32	1½	11/16	5½	3/4
18	12	7/32	11/32	1½	11/16	7¼	7/8
18	20	1/4	3/8	2¼	1¼	9	7/8
20	4	3/16	5/16	1 3/8	5/8	4	5/8
20	8	3/16	5/16	1 3/8	5/8	5	3/4
20	12	7/32	11/32	1 5/8	3/4	7	3/4
20	20	9/32	7/16	2¼	1 1/8	10	1
22	4	3/16	5/16	1½	5/8	4	5/8
22	8	3/16	5/16	1½	5/8	5	3/4
22	12	7/32	11/32	1 3/4	13/16	6½	7/8
22	20	9/32	7/16	2½	1¼	11	1 1/8
24	4	7/32	11/32	1 9/16	11/16	4	5/8
24	8	7/32	11/32	1 9/16	11/16	5½	3/4

Rules for Calculating Diameters and Speeds of Pulleys

Speed of Driven Pulley Required. — Diameter and speed of driving pulley, and diameter of driven pulley are known. *Rule:* Multiply the diameter of the driving pulley by its speed in revolutions per minute, and divide the product by the diameter of the driven pulley.

Example: — If the diameter of the driving pulley is 15 inches and its speed, 180 revolutions per minute, and the diameter of the driven pulley, 9 inches, then the speed of the driven pulley $= \dfrac{15 \times 180}{9} = 300$ revolutions per minute.

Diameter of Driven Pulley Required. — Diameter and speed of driving pulley, and revolutions per minute of driven pulley are known. *Rule:* Multiply the diameter of the driving pulley by its speed in revolutions per minute, and divide the product by the required speed of the driven pulley.

Example: — If the diameter of the driving pulley is 24 inches and its speed, 100 revolutions per minute, and the driven pulley is to rotate 600 revolutions per minute, then the diameter of the driven pulley $= \dfrac{24 \times 100}{600} = 4$ inches.

Diameter of Driving Pulley Required. — Diameter and speed of driven pulley, and speed of driving pulley are known. *Rule:* Multiply the diameter of the driven pulley by its speed in revolutions per minute, and divide the product by the speed of the driving pulley.

Example: — If the diameter of the driven pulley is 36 inches and its required speed, 150 revolutions per minute, and the speed of the driving pulley is 600 revolutions per minute, then the diameter of the driving pulley $= \dfrac{36 \times 150}{600} = 9$ inches.

Speed of Driving Pulley Required. — Diameters of driving and driven pulleys, and speed of driven pulley are known. *Rule:* Multiply the diameter of the driven pulley by its speed, and divide the product by the diameter of the driving pulley.

Example: — If the diameter of driven pulley is 4 inches, its required speed, 800 revolutions per minute, and the diameter of the driver, 26 inches, then the required speed of the driver $= \dfrac{4 \times 800}{26} = 123$ revolutions per minute, approximately.

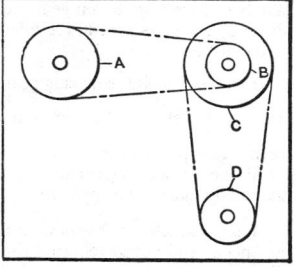

Speed of Driven Pulley in Compound Drive Required. — Diameters of pulleys *A*, *B*, *C* and *D* (see illustration), and speed of pulley *A* are known; find speed of pulley *D*. *Rule:* Divide product of diameters of driving pulleys by product of diameters of driven pulleys, and multiply quotient by speed of first driving pulley.

Example: — If the diameters of the driving pulleys *A* and *C* are 18 and 24 inches; the diameters of the driven pulleys *B* and *D*, 12 and 13 inches; and the speed of the driver *A*, 260 revolutions per minute; then the speed of the driven pulley $D = \dfrac{18 \times 24}{12 \times 13} \times 260 = 720$ revolutions per minute.

FLAT BELTS AND PULLEYS

Pulley Diameters in Compound Drive Required. — Speeds of driving and driven pulleys are known; find diameters of the four pulleys *A, B, C* and *D*. *Rule:* Place the speed of the driving pulley as the numerator of a fraction, and the speed of driven pulley as the denominator, and reduce this fraction to its lowest terms; then resolve both the numerator and denominator into two factors, and multiply each "pair" of factors (a pair being one factor in the numerator and one in the denominator) by a trial number which will give pulleys of suitable diameters.

Example: — If the speed of pulley *A* is 260 revolutions per minute, and the required speed of pulley *D* is 720 revolutions per minute, find the diameters of the four pulleys. The fraction $\dfrac{260}{720}$ reduced to its lowest terms is $\dfrac{13}{36}$, which represents the required speed ratio. Resolve $\dfrac{13}{36}$ into two factors; $\dfrac{13}{36} = \dfrac{1 \times 13}{2 \times 18}$. Multiply by trial numbers 12 and 1:

$$\frac{(1 \times 12) \times (13 \times 1)}{(2 \times 12) \times (18 \times 1)} = \frac{12 \times 13}{24 \times 18}$$

The values 12 and 13 in the numerator represent the diameters of the *driven* pulleys *B* and *D* and values 24 and 18 in the denominator, the diameters of the *driving* pulleys.

Lengths of Open and Crossed Belts. — In the following formulas for determining the lengths of belts on pulleys, *R* and *r* are radii of large and small pulleys, respectively; α = calculated belt angle used to calculate *L* and L_c; *C* = center distance; and *t* = belt thickness.

For open belts, $\sin \alpha = \dfrac{R - r}{C}$, and $L = \pi \,(R + r + t) + \dfrac{\pi\alpha}{90°}\,(R - r) + 2C \cos \alpha$

For crossed belts, $\sin \alpha = \dfrac{R + r + t}{C}$, and $L_c = \pi \,(R + r + t)\left(1 + \dfrac{\alpha}{90°}\right) + 2C \cos \alpha$

Rules for Calculating Speeds of Gearing

Speed of Driven Gear Required. — Number of teeth in driving gear, its speed, and number of teeth in driven gear are known. *Rule:* Multiply the number of teeth in the driving gear by its speed in revolutions per minute, and divide by the number of teeth in the driven gear.

Example: — If the driving gear has 20 teeth and rotates 80 revolutions per minute, and the driven gear has 40 teeth, then the speed of the driven gear $= \dfrac{20 \times 80}{40} = 40$ revolutions per minute.

If one or more intermediate gears are placed in a direct train between the driving and driven gears, the speed ratio will remain the same.

Pitch Diameter of Driven Gear Required. — The pitch diameter of the driving gear, its speed, and speed required for driven gear are known. *Rule:* Multiply the pitch diameter of the driving gear by its speed in revolutions per minute, and divide by the required speed of the driven gear.

Example: — If the pitch diameter of the driver is 8 inches, its speed, 75 revolutions per minute, and the speed required for the driven gear, 20 revolutions per minute, then the pitch diameter of the driven gear $= \dfrac{8 \times 75}{20} = 30$ inches.

V-BELTS AND SHEAVES

V-Belt Drives. — Belts of the V type, commonly manufactured of fabric, cord, or a combination of these, treated with natural or synthetic rubber compound and vulcanized together, provide a quiet, compact, and resilient form of power transmission. They are used extensively in single and multiple form for automotive, home and commercial equipment and in industrial drives for a wide range of horse-powers extending upwards from fractional values.

The tapered cross-sectional shape of a V-belt causes it to wedge firmly into the sheave groove during operation so that the driving action takes place through the sides of the belt rather than the bottom, which normally is not in contact with the sheave at all.

Light-duty or Fractional Horsepower V-belts. — These are typically used as single belts with fractional horsepower motors or engines and are intended for service that is commonly infrequent or intermittent rather than continuous. The Rubber Manufacturers Association with the cooperation of V-belt manufacturers has provided *Standards for Light Duty or Fractional Horsepower V-Belts* in which appear a description of the sizes and load capacities of light-duty V-belts together with sheave dimensions and horsepower ratings.

Belt Sizes: Cross-sectional dimensions as given in this standard are shown in Table 1. Standard lengths are given in Table 2. It should be noted that these are *outside* lengths and not pitch lengths as in the case of Multiple V-belts.

Size Designation: The size designation of the light-duty belt may show the cross section and the nominal outside length. For example, a 2L belt of 8-inch nominal length would be designated 2L080; a 3L belt of 52-inch nominal length would be designated 3L520; a 4L belt of 100-inch nominal length would be designated 4L1000. Other methods of designation may be used, however.

Sheave Dimensions: Dimensions of sheave grooves are given in Table 3. The minimum sheave outside diameters recommended for use with light-duty or fractional horsepower belts are: for 2L belt, 0.8 inch O.D.; 3L belt, 1.5 inch O.D.; 4L belt, 2.5 inch O.D.; and 5L belt, 3.5 inch O.D.

Speed Ratios: The sheave diameter used in calculating speed ratios and belt speeds is obtained by subtracting the value "2X" in Table 3 from the effective outside diameter of the sheave also given in this table.

Belt Length and Center Distance: The belt length required by a light-duty V-belt drive is computed from the effective *outside* diameters (Table 3) using the following formula:

$$L = 2C + 1.57(D + d) + \frac{(D - d)^2}{4C}$$

where: L = effective outside length of belt in inches
 C = distance between centers of sheaves in inches
 D = effective outside diameter of large sheave in inches
 = (for flat pulleys) the outside diameter of the pulley plus twice the nominal belt thickness
 d = effective outside diameter of small sheave in inches.

If this calculation results in a length which is not standard, the next longer standard length should be used and the necessary correction made in center distance.

If sheave effective diameters and belt length are known, the center distance between sheaves may be calculated as follows:

$$C = \frac{b + \sqrt{b^2 - 32(D - d)^2}}{16}, \text{ where: } b = 4L - 6.28(D + d)$$

Table 1.   Light-duty V-belt Cross-section Dimensions

Cross Section	2L*	3L	4L	5L
Nominal Top Width	¼	⅜	½	21/32
Nominal Thickness	5/32	7/32	5/16	⅜

* The 2L cross section is in limited usage and is not made by all manufacturers. All dimensions in inches.

Table 2.   Light-duty V-belt Standard Outside Lengths

Nom. Out-side Length, Inches	Standard Outside Lengths			Effect. Out-side Length Variations, Inches	Nom. Out-side Length, Inches	Standard Outside Lengths			Effect. Out-side Length Variations, Inches	Nom. Out-side Length, Inches	Standard Outside Lengths			Effect. Out-side Length Variations, Inches
	2L	3L	4L			3L	4L	5L			3L	4L	5L	
8	*	..	..	+1/8, −3/8	34	*	*	*	+1/4, −5/8	63	..	*	*	+5/16, −11/16
9	*	..	..	+1/8, −3/8	35	*	*	*	+1/4, −5/8	64	..	*	*	+5/16, −11/16
10	*	..	..	+1/8, −3/8	36	*	*	*	+1/4, −5/8	65	..	*	*	+5/16, −11/16
11	*	..	..	+1/8, −3/8	37	*	*	*	+1/4, −5/8	66	..	*	*	+5/16, −11/16
12	*	..	..	+1/8, −3/8	38	*	*	*	+1/4, −5/8	67	..	*	*	+5/16, −11/16
13	*	..	..	+1/8, −3/8	39	*	*	*	+1/4, −5/8	68	..	*	*	+5/16, −11/16
14	*	*	..	+1/8, −3/8	40	*	*	*	+1/4, −5/8	69	..	*	*	+5/16, −11/16
15	*	*	..	+1/8, −3/8	41	*	*	*	+1/4, −5/8	70	..	*	*	+5/16, −11/16
16	*	*	..	+1/8, −3/8	42	*	*	*	+1/4, −5/8	71	..	*	*	+5/16, −11/16
17	*	*	..	+1/8, −3/8	43	*	*	*	+1/4, −5/8	72	..	*	*	+5/16, −11/16
18	*	*	*	+1/8, −3/8	44	*	*	*	+1/4, −5/8	73	..	*	*	+5/16, −11/16
19	*	*	*	+1/8, −3/8	45	*	*	*	+1/4, −5/8	74	..	*	*	+5/16, −11/16
20	*	*	*	+1/8, −3/8	46	*	*	*	+1/4, −5/8	75	..	*	*	+5/16, −11/16
	3L	4L	5L		47	*	*	*	+1/4, −5/8	76	..	*	*	+5/16, −11/16
21	*	*	..	+1/4, −5/8	48	*	*	*	+1/4, −5/8	77	..	*	*	+5/16, −11/16
22	*	*	..	+1/4, −5/8	49	*	*	*	+1/4, −5/8	78	..	*	*	+5/16, −11/16
23	*	*	..	+1/4, −5/8	50	*	*	*	+1/4, −5/8	79	..	*	*	+5/16, −11/16
24	*	*	*	+1/4, −5/8	51	..	*	*	+1/4, −5/8	80	..	*	*	+5/8, −7/8
25	*	*	*	+1/4, −5/8	52	*	*	*	+1/4, −5/8	82	..	*	*	+5/8, −7/8
26	*	*	*	+1/4, −5/8	53	*	..	*	+1/4, −5/8	84	..	*	*	+5/8, −7/8
27	*	*	*	+1/4, −5/8	54	*	..	*	+1/4, −5/8	86	..	*	*	+5/8, −7/8
28	*	*	*	+1/4, −5/8	55	..	*	*	+1/4, −5/8	88	..	*	*	+5/8, −7/8
29	*	*	*	+1/4, −5/8	56	..	*	*	+1/4, −5/8	90	..	*	*	+5/8, −7/8
30	*	*	*	+1/4, −5/8	57	..	*	*	+1/4, −5/8	92	..	*	*	+5/8, −7/8
31	*	*	*	+1/4, −5/8	58	*	*	*	+1/4, −5/8	94	..	*	*	+5/8, −7/8
32	*	*	*	+1/4, −5/8	59	..	*	*	+1/4, −5/8	96	..	*	*	+5/8, −7/8
33	*	*	*	+1/4, −5/8	60	*	*	*	+1/4, −5/8	98	..	*	*	+5/8, −7/8
					61	..	*	*	+5/16, −11/16	100	..	*	*	+5/8, −7/8
					62	..	*	*	+5/16, −11/16					

If not dictated by other considerations, the recommended center distance is as follows: (a) if the speed ratio is less than 3, the recommended center distance is one-half of the sum of the two sheave diameters plus the diameter of the small sheave; (b) if the speed ratio is 3 or more, the recommended center distance is equal to the diameter of the large sheave.

The center distance obtained either from the formula mentioned above or by the rules given are for nominal length belts under operating tension. Provision must be made to move the sheave centers closer together so that any belt within the length tolerances given in Table 2 can be installed without being pried over the sheave. Also, the centers must be adjustable beyond the calculated distance to compensate for belt stretch and wear of belt and grooves. These required installation and take-up allowances in terms of subtractions from or additions to the nominal center distance are given in Table 4.

Table 3. Light-duty V-belt Sheave Dimensions

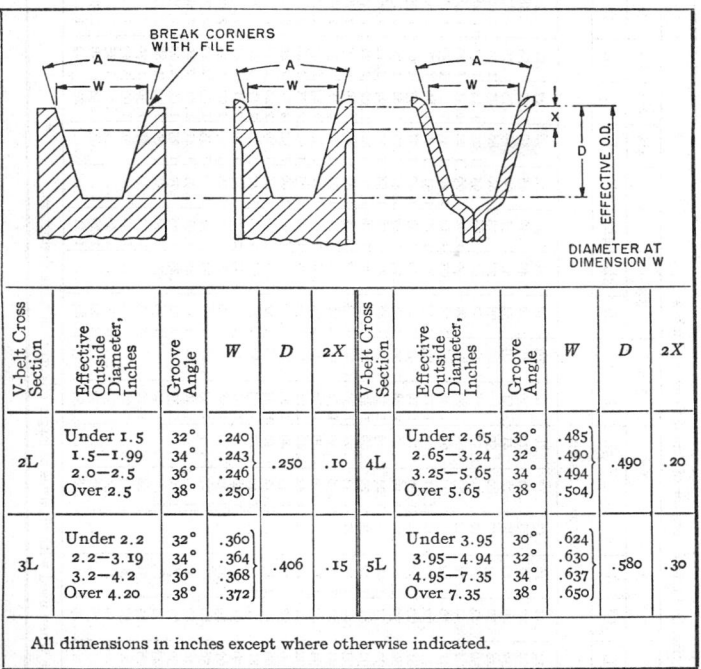

V-belt Cross Section	Effective Outside Diameter, Inches	Groove Angle	W	D	2X	V-belt Cross Section	Effective Outside Diameter, Inches	Groove Angle	W	D	2X
2L	Under 1.5 1.5—1.99 2.0—2.5 Over 2.5	32° 34° 36° 38°	.240 .243 .246 .250	.250	.10	4L	Under 2.65 2.65—3.24 3.25—5.65 Over 5.65	30° 32° 34° 38°	.485 .490 .494 .504	.490	.20
3L	Under 2.2 2.2—3.19 3.2—4.2 Over 4.20	32° 34° 36° 38°	.360 .364 .368 .372	.406	.15	5L	Under 3.95 3.95—4.94 4.95—7.35 Over 7.35	30° 32° 34° 38°	.624 .630 .637 .650	.580	.30

All dimensions in inches except where otherwise indicated.

Table 4. Minimum Center Distance Allowances for Installation and Take-up of Light-duty V-belts

Lengths	Minimum Allowance Below (−) and Above (+) Standard Center Distance, Inches			
	2L	3L	4L	5L
8 to 18	−⅜,+½	−⅝,+½	....	....
18 to 25	−⅜,+½	−⅝,+½	−¾,+½	....
25 to 38	....	−¾,+½	−¾,+½	−1,+½
38 to 61	....	−¾,+¾	−⅞,+¾	−1,+¾
61 to 80	....	....	−1,+1⅛	−1⅛,+1⅛
80 to 100, incl.	....	....	−1⅛,+1½	−1¼,+1½

Note: Minus values are for shortening center distance for installation. Plus values are for lengthening center distance to compensate for stretch and wear.

Table 5. Horsepower Ratings for Light-duty V-belts

Effective Outside Diameter of Small Sheave, Inches

Horsepower Ratings

Belt Speed, Ft. per Min.	5	4¾	4½	4¼	4	4	3¾	3¾	3½	3½	3¼	3¼	3	3	2¾	2¾	2½	2½	2¼	2¼	2	2	1¾	1½
	5L	5L	5L	5L	5L	4L*	5L	4L	5L	4L	5L	4L	4L	3L*	4L	3L	4L	3L	4L	3L	4L	3L	3L	3L
200	.30	.29	.27	.26	.24	.21	.21	.19	.19	.18	.16	.17	.16	.11	.14	.10	.13	.10	.09	.09	.07	.08	.07	.05
400	.58	.55	.52	.49	.45	.40	.40	.38	.35	.36	.29	.33	.31	.20	.27	.19	.23	.18	.17	.16	.12	.14	.12	.08
600	.83	.79	.75	.69	.64	.57	.57	.54	.49	.51	.41	.48	.43	.29	.38	.27	.32	.25	.24	.23	.15	.20	.16	.11
800	1.07	1.01	.95	.88	.80	.73	.72	.69	.62	.64	.50	.61	.54	.38	.48	.33	.41	.31	.30	.30	.17	.24	.19	.13
1000	1.28	1.21	1.14	1.05	.95	.88	.84	.83	.72	.78	.57	.74	.65	.43	.56	.40	.46	.37	.33	.33	.18	.28	.22	.14
1200	1.48	1.40	1.31	1.20	1.09	1.01	.96	.96	.82	.89	.63	.84	.74	.50	.64	.47	.51	.43	.36	.38	.17	.32	.24	.15
1400	1.67	1.57	1.46	1.34	1.20	1.14	1.05	1.08	.89	1.01	.67	.95	.82	.56	.71	.52	.56	.48	.40	.42	.16	.35	.27	.15
1600	1.84	1.72	1.60	1.46	1.31	1.26	1.13	1.19	.93	1.11	.69	1.04	.90	.62	.77	.58	.60	.52	.40	.46	.14	.38	.28	.15
1800	1.99	1.87	1.73	1.57	1.40	1.37	1.20	1.29	.97	1.19	.70	1.12	.96	.67	.81	.63	.62	.57	.39	.49	.11	.41	.29	.14
2000	2.13	1.99	1.84	1.67	1.47	1.47	1.25	1.38	1.00	1.28	.70	1.20	1.02	.72	.86	.67	.64	.60	.39	.50	.08	.43	.30	.13
2200	2.26	2.11	1.94	1.75	1.54	1.58	1.29	1.48	1.01	1.37	.69	1.28	1.08	.77	.90	.71	.66	.64	.37	.53	.04	.44	.30	.12
2400	2.37	2.21	2.02	1.81	1.58	1.66	1.32	1.55	1.01	1.43	.66	1.33	1.12	.81	.92	.74	.68	.66	.34	.55		.45	.29	.10
2600	2.47	2.29	2.09	1.87	1.62	1.75	1.33	1.63	1.00	1.50	.62	1.38	1.16	.84	.93	.77	.66	.69	.29	.57		.46	.28	.07
2800	2.56	2.36	2.15	1.91	1.63	1.81	1.33	1.68	.97	1.54	.56	1.41	1.18	.87	.94	.79	.65	.70	.25	.58		.46	.26	.04
3000	2.63	2.42	2.19	1.93	1.63	1.87	1.32	1.74	.93	1.58	.49	1.43	1.19	.89	.94	.81	.63	.72	.19	.59		.45	.23	.01
3200	2.68	2.46	2.21	1.94	1.62	1.92	1.31	1.78	.87	1.61	.40	1.43	1.20	.91	.92	.82	.60	.72	.11	.59		.44	.20	
3400	2.72	2.48	2.22	1.94	1.60	1.96	1.27	1.80	.79	1.63	.29	1.42	1.19	.92	.90	.83	.55	.72	.04	.59		.42	.16	
3600	2.73	2.48	2.22	1.92	1.54	1.98	1.22	1.82	.69	1.64	.16	1.40	1.16	.92	.86	.83	.50	.71		.57		.39	.12	
3800	2.72	2.46	2.20	1.89	1.47	2.00	1.15	1.83	.57	1.63		1.33	1.13	.92	.81	.82	.43	.69		.54		.36	.06	
4000	2.69	2.44	2.17	1.84	1.37	2.00	1.05	1.82	.43	1.61		1.27	1.09	.91	.76	.80	.35	.67		.51		.31		
4200	2.64	2.40	2.11	1.76	1.26	1.98	.93	1.79	.26	1.58		1.19	1.03	.89	.67	.77	.24	.64		.47		.26		
4400	2.57	2.35	2.03	1.67	1.12	1.95	.80	1.75	.08	1.53		1.09	.96	.85	.58	.74	.14	.60		.42		.21		
4600	2.48	2.27	1.93	1.55	.97	1.91	.64	1.70		1.46		.97	.87	.82	.48	.70	.01	.55		.37		.14		
4800	2.36	2.16	1.81	1.41	.78	1.85	.46	1.64		1.39		.83	.76	.77	.36	.64		.49		.30		.06		
5000	2.22	2.03	1.66	1.25	.58	1.78	.25	1.56		1.30		.68	.65	.71	.23	.58		.42		.22				
5200	2.06	1.88	1.49	1.06	.34	1.70	.02	1.45		1.19		.51	.51	.65	.08	.51		.34		.14				
5400	1.86	1.70	1.30	.85	.08	1.59		1.35		1.07		.33	.36	.58		.43		.26		.04				
5600	1.64	1.49	1.09	.61		1.46		1.20		.91		.14	.18	.49		.34		.16						
5800	1.39	1.25	.85	.34		1.29		1.03		.72				.39		.24		.05						
6000	1.11	.99	.54	.04		1.11		.84		.54				.29		.12								

* These horsepower ratings also hold for this belt size when used with sheaves of larger effective outside diameters.

Belt Horsepower Capacity: The horsepower ratings for Light-duty V-belts shown in Table 5 are the basic maximum ratings which take into consideration the degree and rate of belt flexing and the tensile pull, factors common to all V-belt drives. Where the arc of contact on the small sheave is less than 180 degrees as computed by the following formula:

$$\text{Arc of contact} = 180° - \frac{(D-d)60°}{C} \qquad (1)$$

in which D = effective outside diameter of large sheave (Table 3), d = effective outside diameter of smaller sheave (Table 3), and C = center distance of drive, all measured in inches; or where the drive is composed of a small sheave and a large diameter flat pulley (V to flat) in which D is the effective diameter of the flat pulley, the maximum ratings must be reduced by multiplying them by the correction factor for the arc of contact shown in Table 6.

Table 6.   Light-duty V-belt Arc of Contact Correction Factors

Arc of Contact, Deg.	Type of Drive		Arc of Contact	Type of Drive		Arc of Contact	Type of Drive	
	V to V	V to Flat*		V to V	V to Flat*		V to V	V to Flat*
	Correction Factor			Correction Factor			Correction Factor	
180	1.00	.75	145	.91	.83	110	.78	.78
175	.99	.76	140	.89	.84	105	.76	.76
170	.98	.77	135	.88	.85	100	.74	.74
165	.96	.78	130	.86	.86	95	.72	.72
160	.95	.80	125	.84	.84	90	.69	.69
155	.94	.81	120	.82	.82	...	...	...
150	.92	.82	115	.80	.80	...	...	...

* A V to Flat drive is one comprised of a small sheave and a large diameter flat pulley.

In finding the horsepower capacity required of the drive, it is good practice to start with the horsepower rating of the motor and multiply it by the proper service factor.

For *light service* such as small fans or blowers with light rotors, where the load comes on gradually as they come up to speed and where the final load is relatively steady, a service factor of 1.0 to 1.2 may be applied.

For *medium service* where the driven machine is started or stopped rather frequently or where the starting load is somewhat heavy, as for example, a fan or blower with a large relatively heavy rotor, a compressor with good flywheel effect to smooth out vibrations and with infrequent starting, or light-duty machines used in continuous production service, a service factor of 1.2 to 1.4 may be applied.

For *heavy service* where the machines are subjected to one or more of such severe factors as heavy starting loads, peak or shock loads, reciprocating loads, frequent starting and stopping, or industrial production service (included are refrigerator compressors, air compressors, reciprocating pumps, metalworking and woodworking machinery, stokers, drill presses or grinders, and other machinery subject to similar heavy-duty conditions), a service factor of 1.4 to 1.6 may be applied.

Grades of Multiple V-Belts. — Two grades of V-belts for multiple use are recognized in the *Engineering Standards for Multiple V-Belt Drives* sponsored by the Multiple V-Belt Drive and Mechanical Power Transmission Association and the Rubber Manufacturers Association.   The *Standard Quality* of V-belt is intended for the great majority of industrial drives with normal loads, speeds, center distances, sheave diameters and operating conditions.   Various types of V-belts which fall in the class of *Premium Quality* are designed to meet special drive conditions

such as repeated heavy shock loads, pulsating or vibrating loads, substandard sheave diameters, high speed operation, or extremes of temperature, humidity, etc.

Standard Multiple V-Belt and Sheave Dimensions. — Five sizes of V-belts are designated in the *Engineering Standards for Multiple V-Belt Drives*. Nominal width and thickness dimensions are as shown in Table 7. However, actual dimensions of V-belts of various manufacturers may vary somewhat from these nominal dimensions. Because of this fact, it is recommended that belts of different makes should never be mixed on the same drive. Standard V-belt *pitch* lengths and permissible pitch-length tolerances are given in Table 8. Groove dimensions and tolerances for multiple V-belt sheaves are given in Table 9.

Table 7. Standard Multiple V-belt Dimensions

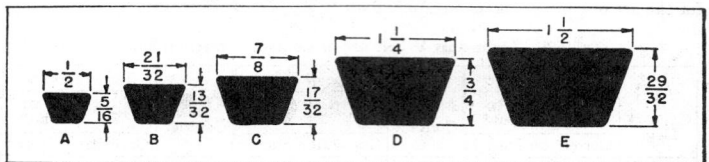

Measuring a Multiple V-Belt: The pitch length of a multiple V-belt is determined by placing the belt on a measuring fixture consisting of two equal diameter sheaves having standard dimensions and with a total tension of 50 pounds for an A V-belt, 65 pounds for a B V-belt, 165 pounds for a C V-belt, 300 pounds for a D V-belt, and 400 pounds for an E V-belt. One of the sheaves is fixed in position, while the other is movable along a graduated scale with the specified tension applied to it.

The sheaves should be rotated at least two revolutions to seat the belt properly in the sheave grooves and to equally divide the total tension between the two strands of the belt. The pitch length is the length obtained by adding the pitch circumference of one of the measuring sheaves to twice the measured center distance between them. Deviation of the measured pitch length from the standard pitch length shown in Table 8 should be within the tolerance limits also given in this table.

The grooves of the measuring sheaves should be machined and maintained to the following tolerances: pitch diameter, ± 0.002 inch; groove angle, ± 0 degrees, 20 minutes; and groove top width, ± 0.002 inch.

Belt Length and Center Distance. — The relation between center distance and belt *pitch* length is given by the following formula:

$$L = 2C + 1.57(D + d) + \frac{(D - d)^2}{4C} \qquad (2)$$

This formula can be rearranged to solve for center distance, as follows:

$$C = \frac{b + \sqrt{b^2 - 32(D - d)^2}}{16}, \qquad (3)$$

where: $b = 4L - 6.28(D + d)$
D = pitch diameter of large sheave, in inches
d = pitch diameter of small sheave, in inches
L = pitch length of belt in inches
C = center distance in inches.

Table 8. Standard Pitch Lengths for Multiple V-belts

Standard Length Designation*	Standard V-Belt Cross Sections					Permissible Deviations from Std. Pitch Length, Inches	Matching Limits for One Set†, Inch
	A	B	C	D	E		
	Standard Pitch Lengths, Inches						
26	27.3	.....	.....	.....	.....	+ .7,— .3	.10
31	32.3	.....	.....	.....	.....	+ .7,— .3	.10
33	34.3	.....	.....	.....	.....	+ .8,— .4	.10
35	36.3	36.8	.....	.....	.....	+ .8,— .4	.10
38	39.3	39.8	.....	.....	.....	+ .8,— .4	.10
42	43.3	43.8	.....	.....	.....	+ .8,— .4	.10
46	47.3	47.8	.....	.....	.....	+ .8,— .4	.10
48	49.3	49.8	.....	.....	.....	+ .9,— .5	.10
51	52.3	52.8	53.9	.....	.....	+ .9,— .5	.10
53	54.3	54.8	.....	.....	.....	+ .9,— .5	.10
55	56.3	56.8	.....	.....	.....	+ .9,— .5	.10
60	61.3	61.8	62.9	.....	.....	+ .9,— .5	.20
62	63.3	63.8	.....	.....	.....	+ .9,— .5	.20
64	65.3	65.8	.....	.....	.....	+ .9,— .5	.20
66	67.3	67.8	.....	.....	.....	+ .9,— .5	.20
68	69.3	69.8	70.9	.....	.....	+ .9,— .5	.20
71	72.3	72.8	.....	.....	.....	+ .9,— .5	.20
75	76.3	76.8	77.9	.....	.....	+ .9,— .5	.20
78	79.3	79.8	.....	.....	.....	+1.0,— .5	.30
80	81.3	.....	.....	.....	.....	+1.0,— .5	.30
81	.....	82.8	83.9	.....	.....	+1.0,— .5	.30
83	.....	84.8	.....	.....	.....	+1.0,— .5	.30
85	86.3	86.8	87.9	.....	.....	+1.0,— .5	.30
90	91.3	91.8	92.9	.....	.....	+1.0,— .5	.30
96	97.3	.....	98.9	.....	.....	+1.0,— .5	.30
97	.....	98.8	.....	.....	.....	+1.0,— .5	.30
105	106.3	106.8	107.9	.....	.....	+1.1,— .5	.40
112	113.3	113.8	114.9	.....	.....	+1.1,— .5	.40
120	121.3	121.8	122.9	123.3	.....	+1.2,— .5	.40
128	129.3	129.8	130.9	131.3	.....	+1.3,— .6	.40
136	.....	137.8	138.9	.....	.....	+1.3,— .6	.40
144	.....	145.8	146.9	147.3	.....	+1.4,— .6	.40
158	.....	159.8	160.9	161.3	.....	+1.5,— .6	.40
162	.....	.....	164.9	165.3	.....	+1.6,— .6	.40
173	.....	174.8	175.9	176.3	.....	+1.7,— .7	.50
180	.....	181.8	182.9	183.3	184.5	+1.7,— .7	.50
195	.....	196.8	197.9	198.3	199.5	+1.8,— .8	.50
210	.....	211.8	212.9	213.3	214.5	+2.0,— .8	.50
240	.....	240.3	240.9	240.8	241.0	+2.2,— .9	.50
270	.....	270.3	270.9	270.8	271.0	+2.4,—1.0	.50
300	.....	300.3	300.9	300.8	301.0	+2.5,—1.2	.60
330	.....	.....	330.9	330.8	331.0	+2.5,—1.2	.60
360	.....	.....	360.9	360.8	361.0	+2.5,—1.2	.60
390	.....	.....	390.9	390.8	391.0	+3.0,—1.5	.70
420	.....	.....	420.9	420.8	421.0	+3.5,—2.0	.70
480	.....	.....	.....	480.8	481.0	+4.0,—2.5	.70
540	.....	.....	.....	540.8	541.0	+4.5,—3.0	.70
600	.....	.....	.....	600.8	601.0	+5.0,—3.5	.70
660	.....	.....	.....	660.8	661.0	+6.0,—4.0	.70

* To specify belt size use the Standard Length Designation prefixed by the letter indicating cross section, for example: B90.

† Maximum allowable difference in actual pitch lengths of longest and shortest V-belts in a given set.

Table 9. Groove Dimensions and Tolerances for
Multiple V-belt Sheaves

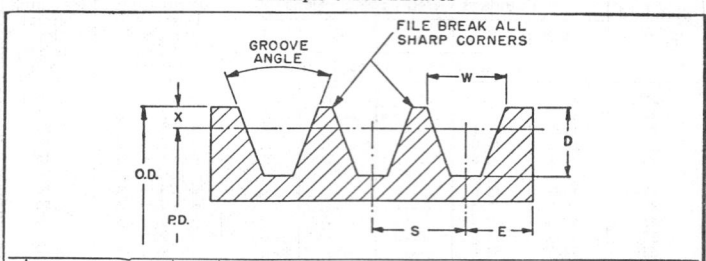

Belt	Pitch Diameter		Groove Angle	Standard Groove Dimensions					Deep Groove Dimensions				
	Minimum Recommended	Range		W	D	X	S¹	E	W	D	X	S¹	E
			±½°	(2)	±.031	....	±.031	(3)	(2)	±.031	....	±.031	(3)
A	3.0	2.6 to 5.4 Over 5.4	34° 38°	.494 .504	.490	.125	⅝	⅜	.589 .611	.645	.280	¾	⁷⁄₁₆
B	5.4	4.6 to 7.0 Over 7.0	34° 38°	.637 .650	.580	.175	¾	½	.747 .774	.760	.355	⅞	⁹⁄₁₆
C	9.0	7.0 to 7.99 8.0 to 12.0 Over 12.0	34° 36° 38°	.879 .887 .895	.780	.200	I	1¹⁄₁₆	1.066 1.085 1.105	1.085	.505	1¼	1³⁄₁₆
D	13.0	12.0 to 12.99 13.0 to 17.0 Over 17.0	34° 36° 38°	1.259 1.271 1.283	1.050	.300	1⁷⁄₁₆	⅞	1.513 1.541 1.569	1.465	.715	1¾	1¹⁄₁₆
E	21.0	18.0 to 24.0 Over 24.0	36° 38°	1.527 1.542	1.300	.400	1¾	1⅛	1.816 1.849	1.745	.845	2¹⁄₁₆	1⁵⁄₁₆

All dimensions in inches except groove angles are in degrees.
¹ Summation of the deviations from S for all grooves in any one sheave shall not exceed ±0.063 inch.
² Tolerances for W: for A and B belts are ±.005 inch; for C and D belts, ±.007 inch; and for E belt, ±.010 inch.
³ Tolerances for E: for A belts are, +.070, −.000 inch; for B and C belts, +.150, −.000 inch; and for D and E belts, +.250, −.000 inch.
Outside Diameter Tolerances: Under 12 inches, ±.020 inch; 12.0 up to 24.0 inches, ±.040 inch; 24 up to 58 inches, .060 inch; 58.0 up to 72.0 inches, ±.120 inch; and for 72 inches and above, ±.250 inch.
Outside Diameter Eccentricity: For 10.0-inch pitch diameter and under, .010 inch. Add .0005 inch for each additional inch of pitch diameter up to and including 60.0-inch pitch diameter. Add .001 inch for each additional inch of pitch diameter above 60 inches.
Side Wobble and Runout: .001 inch per inch of pitch diameter up to 20 inches. Add .0005 inch for each additional inch of pitch diameter up to and including 60.0 inches. Add .001 inch for each additional inch of pitch diameter above 60.0 inches.
For standard key and keyway dimensions, see pages 2235 and 2236.

Installation and Take-up Allowance. — After calculating a center distance from a standard pitch length, provision should be made for moving the centers together by an amount, as shown by the minus values in Table 10, to permit installing the belts over the sheaves without injury. Also shown in Table 10 is the minimum allowance above the standard center distance (plus values) for which the centers should be adjustable to take up any slack in the belts due to stretch and wear.

Table 10. Minimum Center Distance Allowances
for Installation and Take-up of Multiple V-belts

Range of Standard Lengths	Minimum Allowance Below (−) and Above (+) Standard Center Distance				
	A	B	C	D	E
26 to 38	−¾, +1	−1, +1	....	....	....
38 to 60	−¾, +1½	−1, +1½	−1½, +1½	....	....
60 to 90	−¾, +2	−1¼, +2	−1½, +2	....	....
90 to 120	−1, +2½	−1¼, +2½	−1½, +2½	....	....
120 to 158	−1, +3	−1¼, +3	−1½, +3	−2, +3	....
158 to 195	....	−1¼, +3½	−2, +3½	−2, +3½	−2½, +3½
195 to 240	....	−1½, +4	−2, +4	−2, +4	−2½, +4
240 to 270	....	....	−2, +4½	−2½, +4½	−2½, +4½
270 to 330	....	....	−2, +5	−2½, +5	−3, +5
330 to 420	....	....	−2, +6	−2½, +6	−3, +6
420 and over	....	....	....	−3,*	−3½,*

All dimensions in inches.
* For this belt size and lengths, 1.5 per cent of belt length above standard center distance for stretch and wear.

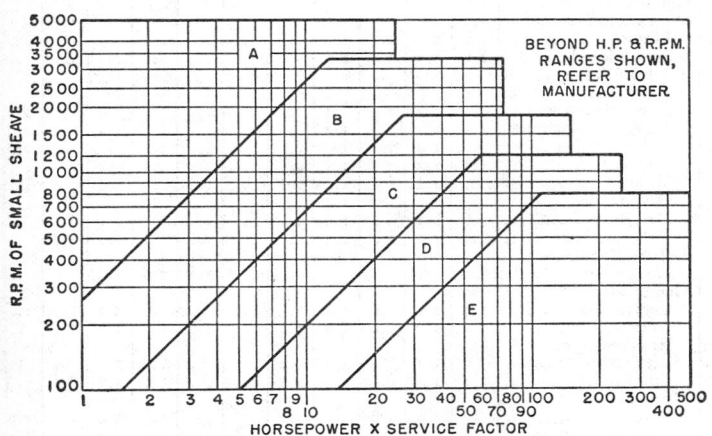

Chart for Selection of V-Belt for Given Drive

Selection of Multiple V-Belts. — The chart on page 2279 which appears in Engineering Standards for Multiple V-Belt Drives enables a V-belt of appropriate cross-section to be selected for a given drive if the revolutions per minute of the small sheave, the transmitted horsepower of the driving unit, and the service factor are known.   The selection procedure is as follows:

1. Multiply the horsepower to be transmitted by the drive by the proper service factor (Table 11) to obtain the "design horsepower."

2. Enter the chart at the rpm of the small sheave and proceed horizontally to a point in vertical line with the design horsepower.

3. If this point falls in the area marked A, then an A size belt is required or,

Table 11. Service Factors for Multiple V-belt Applications

Applications	Electric Motors										
	A.C.								D.C.		
	Squirrel Cage				Synchronous		Single Phase				
	Normal Torque Line Start	Normal Torque Compensator Start	High Torque	Wound Rotor (Slip Ring)	Normal Torque	High Torque	Repulsion and Split-Phase	Capacitor	Shunt Wound	Compound Wound	Line Shaft and Clutch Shifting
	Service Factors										
Agitators —											
Paddle-Propeller											
Liquid..............	1.0	1.0	1.2	...	...	...	...	...	...	...	...
Semi-Liquid.........	1.2	1.0	1.4	1.2	...	...	...	...	...	...	...
Brick and Clay											
Machinery											
Auger Machines......	...	1.2	1.4	1.4	...	...	...	...	1.4	...	2.0
De-Airing Machines.	...	1.2	1.4	1.4	...	...	...	...	1.4	...	2.0
Cutting Table.......	...	1.2	1.4	1.4	...	...	...	...	...	...	2.0
Pug Mill..............	1.5	1.3	1.8	1.5	...	...	...	...	...	...	...
Mixer.................	...	1.2	1.6	1.4	...	...	...	...	...	...	...
Granulator...........	...	1.2	1.4	1.4	...	...	...	...	...	...	...
Dry Press............	...	1.2	1.6	1.4	...	...	...	...	...	...	...
Rolls.................	...	1.2	1.4	1.4	...	...	...	...	...	...	...
Bakery Machinery											
Dough Mixer........	1.2	...	...	...	...	...	1.2	1.0	...	...	...
Compressors											
Centrifugal..........	1.2	1.2	...	1.4	1.4	...	...	...	1.2	...	...
Rotary...............	1.2	1.2	...	1.4	1.4	...	1.2	1.2	1.2	...	...
Reciprocating —											
3 or More Cyl......	1.2	1.2	...	1.4	1.4	...	...	...	1.2	...	...
1 or 2 Cyl..........	1.4	1.4	...	1.5	1.5	...	...	...	1.2	...	...
Conveyors											
Apron................	...	1.4	1.6	...	...	...	...	...	1.4	...	1.6
Belt(Ore, Coal, Sand)	...	1.2	1.4	...	...	...	...	...	1.2	...	1.4
Belt(Light Package).	...	1.0	1.1	...	...	...	...	...	1.0	...	1.2

Table 11. (*Continued*). Service Factors for Multiple V-Belt Applications

Applications	Electric Motors										
	A.C.								D.C.		
	Squirrel Cage				Synchronous		Single Phase				
	Normal Torque Line Start	Normal Torque Compensator Start	High Torque	Wound Rotor (Slip Ring)	Normal Torque	High Torque	Repulsion and Split-Phase	Capacitor	Shunt Wound	Compound Wound	Line Shaft and Clutch Shifting
	Service Factors										
Oven	...	1.0	1.1	...	...	...	...	...	1.0	...	1.2
Screw	...	1.6	1.8	...	...	...	...	...	1.6	...	1.8
Bucket	...	1.4	1.6	...	...	...	...	...	1.4	...	1.6
Pan	...	1.4	1.6	...	...	...	...	...	1.4	...	1.6
Flight	...	1.6	1.8	...	...	...	...	...	1.6	...	1.8
Elevator	...	1.4	1.6	...	...	...	...	...	1.4	...	1.6
Crushing Machinery											
Jaw Crushers	...	1.4	1.6	1.4	...	...	...	...	...	1.4	1.6
Gyratory Crushers	...	1.4	1.6	1.4	1.4	1.6	...	...	...	1.4	1.6
Cone Crushers	...	1.4	1.6	1.4	...	...	...	...	...	1.6	1.6
Crushing Rolls	...	1.4	1.6	1.4	...	...	...	...	...	1.4	1.6
Ball-Pebble and	...	1.4	1.6	1.4	1.4	1.6	...	...	...	1.4	1.6
Tube Mills	...	1.4	1.6	1.4	1.4	...	...	...	...	1.4	1.6
Fan and Blowers											
Centrifugal	1.2	1.2	...	1.4	...	...	...	...	1.2	...	...
Propeller	1.4	1.4	2.0	1.6	...	2.0	...	...	1.4	...	...
Induced Draft	1.2	1.2	...	1.4	...	...	...	...	1.4	...	...
Positive Blowers	1.6	1.6	...	2.0	2.0	2.0	...	...	...	...	...
Exhausters	1.2	1.2	...	1.4	...	...	...	...	1.4	...	1.5
Line Shafts	1.4	1.4	...	1.4	1.4	2.0	1.4	1.4	1.4	1.4	1.6
Machine Tools											
Grinders	1.2	...	...	1.4	...	...	1.2	1.0	1.2	1.2	...
Boring Mills	1.2	...	...	1.4	...	...	...	...	1.2	1.2	...
Lathes	1.0	...	...	1.2	...	...	1.0	1.0	1.0	1.0	...
Milling Machines	1.2	...	...	1.4	...	...	...	...	1.2	1.2	...
Screw Machines	1.0	...	...	1.0	...	...	1.0	1.0	1.0	1.0	...
Cam Cutters	1.0	...	...	1.0	...	...	...	...	1.0	1.0	...
Planers	1.2	...	...	1.4	...	...	1.2	1.0	1.2	1.2	...
Shapers	1.0	...	...	1.0	...	...	1.0	1.0	1.0	1.0	...
Drill Press	1.0	...	...	1.0	...	...	1.0	1.0	1.0	1.0	...
Drop Hammers	1.0	...	...	1.0	...	...	1.0	1.0	1.0	1.0	...
Shears	1.2	...	...	1.4	...	...	1.2	1.2	1.2	1.0	...
Mills											
Pebble	...	1.4	1.6	1.4	...	...	...	...	...	1.4	1.6
Rod	...	1.4	1.6	1.4	...	...	...	...	...	1.4	1.6
Ball	...	1.4	1.6	1.4	...	...	...	...	...	1.4	1.6
Roller Mills	...	1.4	1.6	1.4	...	...	...	...	...	1.4	1.6
Flaking Mills	...	1.6	1.6	1.4	...	...	...	...	...	1.4	1.6
Tumbling Barrels	...	1.6	1.6	1.4	...	...	...	...	...	1.4	1.6

Table 11 (*Continued*). Service Factors for Multiple V-Belt Applications

Applications	Squirrel Cage — Normal Torque Line Start	Squirrel Cage — Normal Torque Compensator Start	Squirrel Cage — High Torque	Wound Rotor (Slip Ring)	Synchronous — Normal Torque	Synchronous — High Torque	Single Phase — Repulsion and Split-Phase	Single Phase — Capacitor	D.C. — Shunt Wound	D.C. — Compound Wound	Line Shaft and Clutch Shifting
Paper Machinery											
Jordan Engines	1.5	1.3	1.8	1.5	1.6	1.8	...	...	1.5	1.5	...
Beaters	1.4	1.4	...	1.4	...	...	...	...	1.4	1.4	1.8
Calenders	1.2	1.2	...	1.2	...	...	...	...	1.2	1.2	...
Agitators	1.2	1.0	1.4	1.2	...	...	...	...	1.2	1.2	1.6
Dryers	1.2	1.2	...	1.2	...	...	...	...	1.2	1.2	...
Paper Machines	1.4	1.4	...	1.5	...	...	...	...	1.5	1.5	1.6
Pumps											
Centrifugal	1.2	1.2	1.4	1.4	...	...	1.2	1.2	...	...	...
Gear	1.2	1.2	1.4	1.4	...	...	1.2	1.2	...	...	...
Rotary	1.2	1.2	1.4	1.4	...	...	1.2	1.2	1.2	...	...
Reciprocating —											
3 or more Cyl	1.2	1.2	...	1.4	1.4	1.6	...	...	...	...	...
1 or 2 Cyl	1.4	1.4	...	1.6	1.6	1.8	...	...	...	...	...

similarly, a B, C, D, or E size belt. For example, for a small sheave rotating at 1750 rpm and a design horsepower of 5, a size A belt would be used.

If this point falls near the line of separation between two belt size areas, then both sizes may be considered as suitable for use. For example, a design horsepower of 40 to be transmitted at a small sheave speed of 800 rpm would call for a multiple drive of either C or D size V-belts.

Horsepower Rating for Multiple V-Belts. — The following formula and accompanying table of constants may be used to determine the general horsepower rating of a single V-belt.

$$\text{H.P.} = XS^{.91} - \frac{YS}{d_e} - ZS^3 \qquad (4)$$

where, X, Y, and Z are constants as given in the accompanying table; H.P. = the recommended horsepower which must be multiplied by the appropriate correction factors for length (see Table 12) and arc of contact (see Table 13).

S = belt speed in thousands of feet per minute. This is found by the formula:

$$S = \frac{3.14 \times \text{P.D. (inches)} \times \text{R.P.M.}}{12 \times 1000}$$

d_e = equivalent diameter of small sheave which is equal to pitch diameter (in inches) multiplied by small diameter factor (Table 14). This provides ratings that compensate for the flexing effect of the small and large sheaves of the drive. The maximum value of d_e to be used in the formula is: 5 for A belts; 7 for B belts; 12 for C belts; 17 for D belts; and 28 for E belts.

Factors X, Y, and Z for Use in Formula 4

Factor	Regular Quality Belts					Premium Quality Belts				
	Belt Cross Section									
	A	B	C	D	E	A	B	C	D	E
	Values of X, Y, and Z to be Used in H.P. Formula									
X	1.945	3.434	6.372	13.616	19.914	2.684	4.737	8.792	18.788	24.478
Y	3.801	9.830	26.948	93.899	177.74	5.326	13.962	38.819	137.70	263.04
Z	0.0136	0.0234	0.0416	0.0848	0.1222	0.0136	0.0234	0.0416	0.0848	0.1222

Example: Find the horsepower capacity of a standard quality A60 size V-belt for a drive in which the pitch diameter of the small sheave is 3 inches and that of the large, 9 inches.

The small sheave is to rotate at 1750 R.P.M.

1. Find center distance C (Formula 3, page 2276)

$$C = \frac{b + \sqrt{b^2 - 32(D - d)^2}}{16}$$

where $b = 4L - 6.28(D + d)$
$L = 61.3$ inches (Table 8)
$b = 4 \times 61.3 - 6.28(9 + 3) = 169.8$

$$C = \frac{169.8 + \sqrt{169.8^2 - 32(9 - 3)^2}}{16} = 21.0 \text{ inches}$$

2. Find arc of contact, A (Formula 5, page 2284)

$$A = 180° - \frac{(D - d)60°}{C} = 180° - \frac{(9 - 3)60°}{21.0} = 163°$$

3. Find correction factors
   Length correction factor = 0.98 (Table 12)
   Arc of contact correction factor = 0.96 (Table 13)
   Small diameter factor (Speed ratio = $9 \div 3 = 3$) = 1.14 (Table 14)
4. Compute belt speed in thousands of feet per minute:

$$S = \frac{3.14 \times \text{P.D.} \times \text{R.P.M.}}{12 \times 1000} = \frac{3.14 \times 3 \times 1750}{12 \times 1000} = 1.38$$

5. Compute equivalent diameter of small sheave

$$d_e = 3 \times 1.14 = 3.42 \text{ inches}$$

6. Compute belt H.P. using Formula 4:

$$\text{H.P.} = 1.945S^{.91} - \frac{3.801S}{d_e} - .0136S^3$$

$$= 1.945 \times 1.38^{.91} - \frac{3.801 \times 1.38}{3.42} - .0136 \times 1.38^3$$

$$= 1.945 \times 1.34 - 1.535 - .034 = 1.04$$

7. Apply length and arc of contact correction factors to get horsepower capacity:

$$1.04 \times 0.98 \times 0.96 = 0.98 \text{ H.P.}$$

8. Divide horsepower capacity into horsepower to be transmitted to obtain number of belts required for drive.

Table 12. Length Correction Factors

Standard Length Designation	Belt Cross Section			Standard Length Designation	Belt Cross Section				
	A	B	C		A	B	C	D	E
	Correction Factor				Correction Factor				
26	0.81	....	....	97	....	1.02	....	....	....
31	0.84	....	....	105	1.10	1.04	0.94	....	....
33	0.86	....	....	112	1.11	1.05	0.95	....	....
35	0.87	0.81	....	120	1.13	1.07	0.97	0.86	....
38	0.88	0.83	....	128	1.14	1.08	0.98	0.87	....
42	0.90	0.85	....	136	....	1.09	0.99	....	....
46	0.92	0.87	....	144	....	1.11	1.00	0.90	....
48	0.93	0.88	....	158	....	1.13	1.02	0.92	....
51	0.94	0.89	0.80	162	....	....	1.03	0.92	....
53	0.95	0.90	....	173	....	1.15	1.04	0.93	....
55	0.96	0.90	....	180	....	1.16	1.05	0.94	0.91
60	0.98	0.92	0.82	195	....	1.18	1.07	0.96	0.92
62	0.99	0.93	....	210	....	1.19	1.08	0.96	0.94
64	0.99	0.93	....	240	....	1.22	1.11	1.00	0.96
66	1.00	0.94	....	270	....	1.25	1.14	1.03	0.99
68	1.00	0.95	0.85	300	....	1.27	1.16	1.05	1.01
71	1.01	0.95	....	330	....	....	1.19	1.07	1.03
75	1.02	0.97	0.87	360	....	....	1.21	1.09	1.05
78	1.03	0.98	....	390	....	....	1.23	1.11	1.07
80	1.04	....	....	420	....	....	1.24	1.12	1.09
81	....	0.98	0.89	480	....	....	....	1.16	1.12
83	....	0.99	....	540	....	....	....	1.18	1.14
85	1.05	0.99	0.90	600	....	....	....	1.20	1.17
90	1.06	1.00	0.91	660	....	....	....	1.23	1.19
96	1.08	....	0.92	...	....	....	....	....	....

Arc of Contact. — The arc of contact made by the V-belt on the small sheave is of importance when computing the horsepower rating of a V-belt for a given drive. This may be found by the formula:

$$\text{Arc of Contact (degrees)} = 180° - \frac{(D - d)60°}{C} \qquad (5)$$

where D, d and C are as noted above. Correction factors, for various arcs of contact, used in finding horsepower capacities of mutiple V-belt drives (see example under Horsepower Rating for Multiple V-Belts) are given in Table 13.

Table 13. Arc of Contact Correction Factors

Arc of Contact on Small Sheave	Type of Drive		Arc of Contact on Small Sheave	Type of Drive	
	V to V	V to Flat*		V to V	V to Flat*
	Correction Factor			Correction Factor	
180°	1.00	.75	130°	.86	.86
170°	.98	.77	120°	.82	.82
160°	.95	.80	110°	.78	.78
150°	.92	.82	100°	.74	.74
140°	.89	.84	90°	.69	.69

* A V-Flat drive is one using a small sheave and a larger diameter flat pulley. .

Table 14. Small Diameter Factors

Speed Ratio Range	Small Diameter Factor	Speed Ratio Range	Small Diameter Factor	Speed Ratio Range	Small Diameter Factor
1.000–1.019	1.00	1.110–1.142	1.05	1.341–1.429	1.10
1.020–1.032	1.01	1.143–1.178	1.06	1.430–1.562	1.11
1.033–1.055	1.02	1.179–1.222	1.07	1.563–1.814	1.12
1.056–1.081	1.03	1.223–1.274	1.08	1.815–2.948	1.13
1.082–1.109	1.04	1.275–1.340	1.09	2.949 and over	1.14

Speed of Operation. — V-belts operate most efficiently at speeds of about 4500 feet per minute. For belt speeds of 5000 feet per minute and more the sheave should be both statically and dynamically balanced. Special design and materials may also be called for and the manufacturer should be consulted. Equivalent belt speed for given sheave pitch diameter and revolutions per minute can be found in Table 3 in Flat Leather Belt section.

Use of Idlers. — According to the B. F. Goodrich Company, the most successful drives are those where an idler is not necessary and where proper tension can be had by adjusting the position of either the driver unit or the driven unit. Where these units are not adjustable, a grooved idler can be used on the inside of the drive. Such an idler should have a diameter larger than the recommended minimum sheave diameter for the belt cross section. If the idler has the smallest diameter on the drive, the idler diameter should be used in determining the horsepower per belt. The drive should be designed taking into account the smallest arc of contact whether it be on the driver or on the driven sheaves.

Some fixed-center drives do not leave enough space for a grooved idler inside and for these there is no choice but to install a flat back bend idler. Such idlers are not recommended because they are inefficient and cause trouble. The belts have a tendency to turn over; the belt life is reduced from 20 to 50 per cent; and often a second (grooved) idler must be added to the drive ahead of the flat idler to make it workable. Where a flat back bend idler is used, the belt life will be improved if in the design computations an additional 0.2 service factor (see Table 11) is employed.

Quarter-turn Drives. — V-belt quarter-turn drives are used to transmit power from a horizontal shaft to a vertical shaft or vice versa. According to the Engineering Standards for Multiple V-Belt Drives, certain precautions must be taken in setting up this type of drive: (a) Direction of rotation must be such that the tight side of the drive is on the bottom; (b) The axis of the vertical shaft should lie in a plane perpendicular to the horizontal shaft, and intersecting it at the center of the

Table 15. "Y" Dimensions for Quarter-turn Drives

Center Distance	60	80	100	120	140	160	180	200	220	240
"Y" Dimension	2½	2¾	3	4	5¼	6½	7¾	9	10½	12

All dimensions in inches.

face of the sheave on the horizontal shaft; (c) the center of the face of the sheave of the vertical shaft should be below the axis of the horizontal shaft by an amount "Y" which depends on the center distance and is shown in Table 15.

Deep grooved sheaves (see Table 9) should always be used. The drive should have a minimum center distance of $5.5\,(D + (N - 1)S + w)$ where, D = pitch diameter of large sheave; N = number of belts; S = deep groove spacing (see Table 9) and w = nominal belt top width (see Table 7).

Open-end V-Belts. — V-belts in long lengths which can be cut to the desired length and used with metal V-belt fasteners or connectors are especially designed to permit firm anchorage of these fasteners. The most efficient speed of open-end belts is under 3000 feet per minute and the maximum safe speed is given by one manufacturer as 3500 feet per minute and as 4000 feet per minute by another. Open-end belts are not recommended for either V-flat or quarter-turn drives.

V-flat Drives. — When a combination of a V-grooved driver sheave and a flat driven pulley is used, it is called a V-flat drive. V-flat drives are frequently recommended as changeovers from existing flat belt drives where the larger flat pulley can be incorporated into the drive. Although V-flat drives are entirely practical as a means of power transmission, they have definite limitations since the V-belt cannot pull as effectively on a flat pulley as it can in a grooved sheave. V-belts manufacturers recommend that V-flat drives should be used only where the speed ratio is at least 3 to 1 or greater and where the center distance is approximately equal to or slightly less than the diameter of the flat pulley. If the center distance is increased beyond the recommended maximum, or if the speed ratio is reduced below 3 to 1, the efficiency of the V-flat drive drops off sharply. The pulley should be really flat and any crown face should be removed.

The effective pitch diameter of a flat pulley carrying a V-belt is equal to the outside diameter of the pulley plus an amount which depends upon the cross section of the V-belt used. These amounts for a given cross section may vary slightly from one manufacturer to another but the following values may be taken as more or less typical. For an A belt add 0.4 inch; for a B belt, 0.5 inch; for a C belt, 0.7 inch; for a D belt, 0.9 inch; and for an E belt, 1.0 inch.

The preferred minimum face widths of flat pulleys used in V-flat drives is as follows: for an A V-belt, 1¾ inch for one belt and ⅝ inch for each additional belt; for a B V-belt, 2¼ inches for one belt and ¾ inch for each additional belt; for a C V-belt, 2¾ inches for one belt and 1 inch for each additional belt; for a D V-belt, 3¾ inches for one belt and 1⁷⁄₁₆ inch for each additional belt; and for an E V-belt 4¾ inches for one belt and 1¾ inches for each additional belt.

V-belt Maintenance. — In obtaining the proper tension on V-belts, it is not necessary to pull them exceedingly taut. They should be tightened only enough to take out slack and undue sag. A good method for checking the proper tension of a V-belt drive is by "striking" the belt with the fist. Slack V-belts feel dead under this test, while properly adjusted V-belts vibrate and feel alive. Another simple test which can be made is to press down firmly on each individual belt in a multi-belt drive. When the top can be depressed so that it is in line with the bottom of other belts on the drive, the correct amount of tension has been applied.

Destructive Elements: Belts should be kept clean, free of oil, and protected from sunlight as much as possible. Mineral oil is especially destructive. To clean belts they should be wiped with a dry cloth. The safest way to remove dirt and grime is to wash with soap and water and rinse well. If by accident the belts become grease or oil spattered, remove with carbon tetrachloride. Belt dressing should never be used on a V-belt drive.

Proper Fit. — The V-belt should ride in the sheave groove so that the top surface is just above the highest point of the sheave. If the belt rides too high, it loses contact area. A low-riding belt may "bottom" in the sheave groove, reducing the wedging action on the sides and resulting in slipping and burring.

Double Angle V-Belts. — For drive applications where power must be transmitted to sheaves from both sides of the belt, a double angle V-belt is available which is hexagonal in cross section. Both sides provide natural V-belt wedging action.

SAE Standard V-Belts. — The data for V-belts and pulleys shown in Table 16 cover six sizes of V-belts and the corresponding pulleys for accommodating these. In this standard, as revised in 1954, no change has been made in the 0.380 and the 0.500 inch sizes. The ⅝ inch (28-degree groove) size has been dropped. For the other sizes (1¹⁄₁₆-, ¾-, ⅞-, and 1-inch), the new dimensions shown in the accompanying table are designed to permit the same actual grooves as in the previous standard.

The diagram and dimensions given permit checking of the pulley by means of measurement over balls or rods of specified diameter placed in the pulley groove as shown.

Standard belt lengths are in increments of one inch, without fractions, up to and including 60 inches. Above 60 inches, the increments are 2 inches without fractions. Standard belt length tolerances are based on the center distance and are as follows: For belt length of 40 inches or less, $+\frac{1}{8}$, $-\frac{5}{32}$ inch; belt lengths over 40 to 50 inches, inclusive, $+\frac{1}{8}$, $-\frac{3}{16}$ inch; for belt lengths over 50 to 60 inches, inclusive, $+\frac{5}{32}$, $-\frac{7}{32}$ inch; for belt lengths over 60 to 80 inches, inclusive, $+\frac{3}{16}$, $-\frac{9}{32}$ inch; for belt lengths over 80 to 100 inches, inclusive, $+\frac{7}{32}$, $-\frac{11}{32}$ inch.

Table 16. SAE V-belt and Pulley Dimensions

SAE Belt and Pulley Size	Nom. Belt Thickness, In.	Nom. Width of Groove, In., W	Pulley Nominal Diameters, In.	Groove Angle, Deg. ±30 Min., A	Ball or Rod Diameter, In., d	Ball Extension		Min. Depth of Groove, In., D
						K	2K	
0.380	⁵⁄₁₆	0.380	2.75 and up	36	0.3125	0.077	0.154	⁷⁄₁₆
0.500	¹³⁄₃₂	0.500	3 and up	36	0.4375	0.157	0.314	⁹⁄₁₆
1¹⁄₁₆	¹³⁄₃₂	0.597	3 to 4, incl Over 4 to 6, incl Over 6	34 36 38	0.500	0.129 0.140 0.151	0.258 0.280 0.302	⁹⁄₁₆
¾	⁷⁄₁₆	0.660	3 to 4, incl Over 4 to 6, incl Over 6	34 36 38	0.5625	0.164 0.176 0.187	0.328 0.352 0.374	⅝
⅞	½	0.785	3.5 to 4.5, incl Over 4.5 to 6, incl Over 6	34 36 38	0.6875	0.236 0.248 0.260	0.472 0.496 0.520	1¹⁄₁₆
1	⁹⁄₁₆	0.910	4 to 6, incl Over 6 to 8, incl Over 8	34 36 38	0.8125	0.308 0.321 0.333	0.616 0.642 0.666	1³⁄₁₆

TRANSMISSION CHAINS

In addition to the standard roller and inverted tooth types, a wide variety of drive chains of different construction is available. Such chains are manufactured to various degrees of precision ranging from unfinished castings or forgings to chains having certain machined parts. Practically all of these chains as well as standard roller chains can be equipped with attachments to fit them for conveyor use. A few such types are briefly described in the following paragraphs. Detailed information about them can be obtained from the manufacturers.

Detachable Chains: The links of this type of chain, which are identical, are easily detachable. Each has a hook-shaped end in which the bar of the adjacent link articulates. These chains are available in malleable iron or pressed steel. The chief advantage is the ease with which any link can be removed.

Cast Roller Chains: Cast roller chains are constructed, wholly or partly, of cast metal parts and are available in various styles. In general the rollers and side bars are accurately made castings without machine finish. The links are usually connected by means of forged pins secured by nuts or cotters. Such chains are used for slow speeds and moderate loads, or where the precision of standard roller chains is not required.

Pintle Chains: Unlike the roller chain, the pintle chain is composed of hollow-cored cylinders cast or forged integrally with two offset side bars and each link identical. The links are joined by pins inserted in holes in the ends of the side bars and through the cored holes in the adjacent links. Lugs prevent turning of the pins in the side bars and thus ensure articulation of the chain between the pin and the cored cylinder.

Standard Roller Transmission Chains

A roller chain is made up of two kinds of links: roller links and pin links alternately spaced throughout the length of the chain as shown in Table 1.

Roller chains are manufactured in several types, each designed for the particular service required. All roller chains are so constructed that the rollers are evenly spaced throughout the chain. The outstanding advantage of this type of chain is the ability of the rollers to rotate when contacting the teeth of the sprocket. Two arrangements of roller chains are in common use: the single-strand type and the multiple-strand type. In the latter type, two or more chains are joined side by side by means of common pins which maintain the alignment of the rollers in the different strands.

Standard roller chains are manufactured to the specifications in the American National Standard for precision power transmission roller chains, attachments, and sprockets (ANSI B29.1-1975) and, where indicated, the data in the subsequent tables have been taken from this standard. These roller chains and sprockets are commonly used for the transmission of power in industrial machinery, machine tools, motor trucks, motorcycles and tractors and similar applications.

Non-standard roller chains, developed individually by various manufacturers prior to the adoption of the ANSI standard, are similar in form and construction to standard roller chains but do not conform dimensionally to standard chains. Some sizes are still available from the originating manufacturers for replacement on existing equipment. They are not recommended for new installations, since their manufacture is being discontinued as rapidly as possible.

Standard double-pitch roller chains are like standard roller chains, except that their link plates have twice the pitch of the corresponding standard-pitch chain. Their design conforms to specifications in the ANSI Standard for double-pitch power transmission roller chains and sprockets (ANSI B29.3M-1985). They are especially useful for low speeds, moderate loads or long center distances.

Standard Roller Chain Nomenclature, Dimensions and Loads. — Standard nomenclature for roller chain parts are given in Table 1. Dimensions for Standard Series roller chain are given in Table 2.

Chain Pitch: Distance in inches between centers of adjacent joint members. Other dimensions are proportional to the pitch.

Tolerances for Chain Length: New chains, under standard measuring load, must not be under-length. Over-length tolerance is $0.001/(\text{pitch in inches})^2 + 0.015$ inch per foot. Length measurements are to be taken over a length of at least 12 inches.

Measuring Load: This is the load in pounds under which a chain should be measured for length. It is equal to one per cent of the ultimate tensile strength, with a minimum of 18 pounds and a maximum of 1000 pounds for both single and multiple-strand chain.

Minimum Ultimate Tensile Strength: For single-strand chain this is equal to or greater than $12,500 \ (\text{pitch in inches})^2$ pounds. The minimum tensile strength or breaking strength of a multiple-strand chain is equal to that of a single-strand chain multiplied by the number of strands.

Standard Roller Chain Numbers. — The right-hand figure in the chain number is zero for roller chains of the usual proportions, 1 for a light-weight chain and 5 for a rollerless bushing chain. The numbers to the left of the right-hand figure denote the number of ⅛ inches in the pitch. The letter H following the chain number denotes the heavy series; thus the number 80 H denotes a 1-inch pitch heavy chain. The hyphenated number 2 suffixed to the chain number denotes a double strand, 3 a triple strand, 4 a quadruple strand chain and so on.

Heavy Series: These chains, made in ¾-inch and larger pitches, have thicker link plates than those of the regular standard. Their value is only in the acceptance of higher loads at lower speeds.

Light-weight Machinery Chain: This chain is designated as No. 41. It is ½ inch pitch; ¼ inch wide; has 0.306-inch diameter rollers and a 0.141-inch pin diameter. The minimum ultimate tensile strength is 1500 pounds.

Multiple-strand Chain: This is essentially an assembly of two or more single-strand chains placed side by side with pins that extend through the entire width to maintain alignment of the different strands.

Types of Sprockets. — Four different designs or types of roller-chain sprockets are shown by the sectional views, Fig. 1. Type A is a plain plate; type B has a hub on one side only; type C, a hub on both sides; and type D, a detachable hub. Also used are shear pin and slip clutch sprockets designed to prevent damage to the drive or to other equipment caused by overloads or stalling.

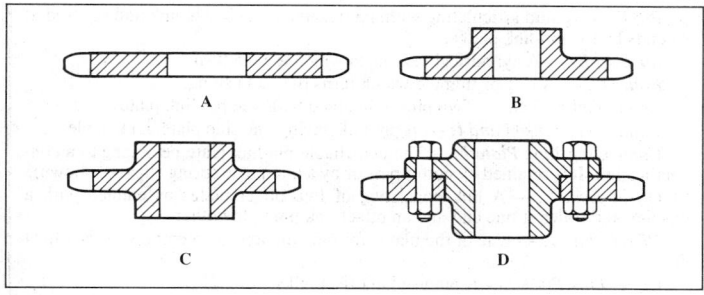

Fig. 1. Types of Sprockets

Table 1. ANSI Nomenclature for Roller Chain Parts (ANSI B29.1-1975)

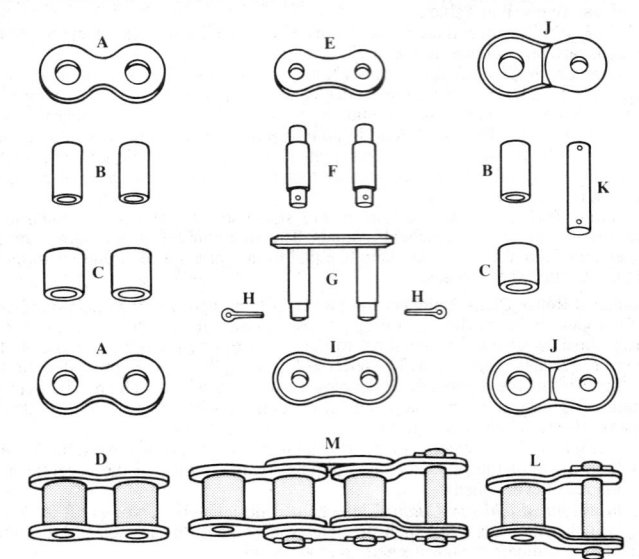

Roller Link D. — An inside link consisting of two inside plates, two bushings, and two rollers.

Pin Link G and E. — An outside link consisting of two pin-link plates assembled with two pins.

Inside Plate A. — One of the plates forming the tension members of a roller link.

Pin Link Plate E. — One of the plates forming the tension members of a pin link.

Pin F. — A stud articulating within a bushing of an inside link and secured at its ends by the pin-link plates.

Bushing B. — A cylindrical bearing in which the pin turns.

Roller C. — A ring or thimble which turns over a bushing.

Assembled Pins G. — Two pins assembled with one pin-link plate.

Connecting-Link G and I. — A pin link having one side plate detachable.

Connecting-Link Plate I. — The detachable pin-link plate belonging to a connecting link. It is retained by cotter pins or by a one-piece spring clip (not shown).

Offset Link L. — A link consisting of two offset plates assembled with a bushing and roller at one end and an offset link pin at the other.

Offset Plate J. — One of the plates forming the tension members of the offset link.

Offset Link Pin K. — A pin used in offset links.

Table 2. ANSI Roller Chain Dimensions (ANSI B29.1-1975)

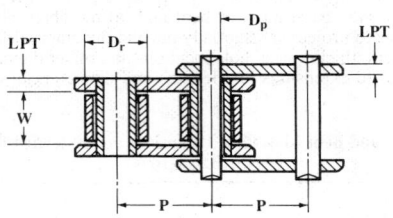

Roller Diameters D_r are approximately ⅝ P.
The *width W* is defined as the distance between the link plates. It is approximately ⅝ of the chain pitch.
Pin Diameters D_p are approximately 5/16 P or ½ of the roller diameter.
Thickness LPT of Inside and Outside Link Plates for the standard series is approximately ⅛ P.
Thickness of Link Plates for the heavy series of any pitch is approximately that of the next larger pitch Standard Series chain.
Maximum Height of Roller Link Plates = 0.95 Pitch.
Maximum Height of Pin Link Plates = 0.82 Pitch.
Maximum Pin Diameter = nominal pin diameter + 0.0005 inch.
Minimum Hole in Bushing = nominal pin diameter + 0.0015 inch.
Maximum Width of Roller Link = nominal width of chain + (2.12 × nominal link plate thickness.)
Minimum Distance between Pin Link Plates = maximum width of roller link + 0.005 inch.

		Standard Series					Heavy Series
Pitch P	Max. Roller Diameter D_r	Standard Chain No.	Width W	Pin Diameter D_p	Thickness of Link Plates LPT	Measuring Load,† Lb.	Thickness of Link Plates LPT
¼	*0.130	25	⅛	0.0905	0.030	18	...
⅜	*0.200	35	3/16	0.141	0.050	18	...
½	0.306	41	¼	0.141	0.050	18	...
½	0.312	40	5/16	0.156	0.060	31	...
⅝	0.400	50	⅜	0.200	0.080	49	...
¾	0.469	60	½	0.234	0.094	70	0.125
1	0.625	80	⅝	0.312	0.125	125	0.156
1¼	0.750	100	¾	0.375	0.156	195	0.187
1½	0.875	120	1	0.437	0.187	281	0.219
1¾	1	140	1	0.500	0.219	383	0.250
2	1.125	160	1¼	0.562	0.250	500	0.281
2¼	1.406	180	1 13/32	0.687	0.281	633	0.312
2½	1.562	200	1½	0.781	0.312	781	0.375
3	1.875	240	1⅞	0.937	0.375	1000	0.500

* Bushing diameter. This size chain has no rollers.   † For single-strand chain.

Attachments. — Modifications to standard chain components to adapt the chain for use in conveying, elevating, and timing operations are known as "attachments." The components commonly modified are: (1) the link plates, which are provided with extended lugs which may be straight or bent and (2) the chain pins, which are extended in length so as to project substantially beyond the outer surface of the pin link plates. Hole diameters, thicknesses, hole locations and offset dimensions for straight link and bent link plate extensions and lengths and diameters of extended pins are given in Table 3.

Table 3. Straight and Bent Link Plate Extensions and Extended Pin Dimensions
(ANSI B29.1-1975)

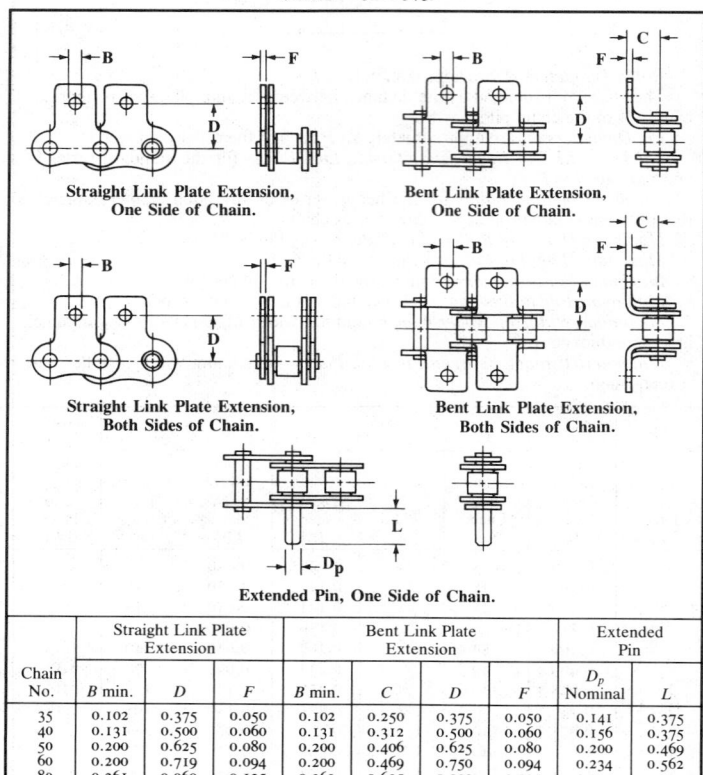

Straight Link Plate Extension,
One Side of Chain.

Bent Link Plate Extension,
One Side of Chain.

Straight Link Plate Extension,
Both Sides of Chain.

Bent Link Plate Extension,
Both Sides of Chain.

Extended Pin, One Side of Chain.

Chain No.	Straight Link Plate Extension			Bent Link Plate Extension				Extended Pin	
	B min.	D	F	B min.	C	D	F	D_p Nominal	L
35	0.102	0.375	0.050	0.102	0.250	0.375	0.050	0.141	0.375
40	0.131	0.500	0.060	0.131	0.312	0.500	0.060	0.156	0.375
50	0.200	0.625	0.080	0.200	0.406	0.625	0.080	0.200	0.469
60	0.200	0.719	0.094	0.200	0.469	0.750	0.094	0.234	0.562
80	0.261	0.969	0.125	0.261	0.625	1.000	0.125	0.312	0.750
100	0.323	1.250	0.156	0.323	0.781	1.250	0.156	0.375	0.938
120	0.386	1.438	0.188	0.386	0.906	1.500	0.188	0.437	1.125
140	0.448	1.750	0.219	0.448	1.125	1.750	0.219	0.500	1.312
160	0.516	2.000	0.250	0.516	1.250	2.000	0.250	0.562	1.500
200	0.641	2.500	0.312	0.641	1.688	2.500	0.312	0.781	1.875

All dimensions are in inches.

Sprocket Classes. — The American National Standard ANSI B29.1-1975 provides for two classes of sprockets designated as Commercial and Precision. The selection of either is a matter of drive application judgment. The usual moderate to slow speed commercial drive is adequately served by Commercial sprockets. Where extreme high speed in combination with high load is involved, or where the drive involves fixed centers, critical timing, or register problems, or close clearance with outside interference, then the use of Precision sprockets may be more appropriate.

As a general guide, drives requiring Type A or Type B lubrication (see page 2306) would be served by Commercial sprockets. Drives requiring Type C lubrication may require Precision sprockets; the manufacturer should be consulted.

Keys, Keyways, and Set Screws. — To secure sprockets to the shaft, both keys and set screws should be used. The key is used to prevent rotation of the sprocket on the shaft. Keys should be fitted carefully in the shaft and sprocket keyways to eliminate all backlash, especially on the fluctuating loads. A set screw should be located over a flat key to secure it against longitudinal displacement.

Where a set screw is to be used with a parallel key, the following sizes are recommended by the American Chain Association. For a sprocket bore and shaft diameter in the range of ½ through ⅞ inch, a ¼-inch set screw; for a range of 15/16 through 1¼ inches, a ⅜-inch set screw; for a range of 1 13/16 through 2¼ inches, a ½-inch set screw; for a range of 2 5/16 through 3¼ inches, a ⅝-inch set screw; for a range of 3⅜ through 4½ inches, a ¾-inch set screw; for a range of 4¾ through 5½ inches, a ⅞-inch set screw; for a range of 5¾ through 7⅜ inches, a 1-inch set screw; and for 7½ through 12½ inches, a 1¼-inch set screw.

Sprocket Diameters. — The various diameters of roller chain sprockets are shown in Figure 2. These are defined as follows.

Pitch Diameter: The pitch diameter is the diameter of the pitch circle that passes through the centers of the link pins as the chain is wrapped on the sprocket. Since

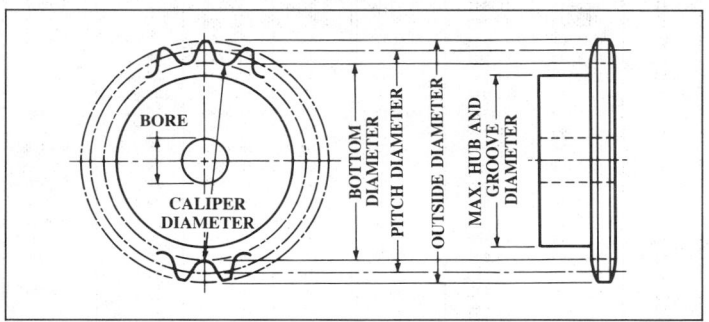

Fig. 2. Sprocket Diameters

the chain pitch is measured on a straight line between the centers of adjacent pins, the chain pitch lines form a series of chords of the sprocket pitch circle. Sprocket pitch diameters for one-inch pitch and for 9 to 108 teeth are given in Table 4. For lower (5 to 8) or higher (109 to 200) numbers of teeth use the following formula in which P = pitch, N = number of teeth: Pitch Diameter = $P ÷ \sin (180° ÷ N)$.

Bottom Diameter: The bottom diameter is the diameter of a circle tangent to the curve (called the seating curve) at the bottom of the tooth gap. It equals the pitch diameter minus the diameter of the roller.

Table 4. ANSI Roller Chain Sprocket Diameters (ANSI B29.1-1975)

These diameters and caliper factors apply only to chain of 1-inch pitch. For any other pitch, multiply the values given below by the pitch.

Caliper Diam. (even teeth) = Pitch Diameter − Roller Diam.
Caliper Diam. (odd teeth) = Caliper factor × Pitch − Roller Diam.
See Table 5 for tolerances on Caliper Diameters.

No. Teeth*	Pitch Diameter	Outside Diameter	Caliper Factor	No. Teeth*	Pitch Diameter	Outside Diameter	Caliper Factor
9	2.9238	3.348	2.8794	59	18.7892	19.363	18.7825
10	3.2361	3.678		60	19.1073	19.681	
11	3.5495	4.006	3.5133	61	19.4255	20.000	19.4190
12	3.8637	4.332		62	19.7437	20.318	
13	4.1786	4.657	4.1481	63	20.0618	20.637	20.0556
14	4.4940	4.981		64	20.3800	20.956	
15	4.8097	5.304	4.7834	65	20.6982	21.274	20.6921
16	5.1258	5.627		66	21.0164	21.593	
17	5.4422	5.949	5.4190	67	21.3346	21.911	21.3287
18	5.7588	6.271		68	21.6528	22.230	
19	6.0755	6.593	6.0548	69	21.9710	22.548	21.9653
20	6.3924	6.914		70	22.2892	22.867	
21	6.7095	7.235	6.6907	71	22.6074	23.185	22.6018
22	7.0267	7.555		72	22.9256	23.504	
23	7.3439	7.876	7.3268	73	23.2438	23.822	23.2384
24	7.6613	8.196		74	23.5620	24.141	
25	7.9787	8.516	7.9630	75	23.8802	24.459	23.8750
26	8.2962	8.836		76	24.1984	24.778	
27	8.6138	9.156	8.5992	77	24.5166	25.096	24.5116
28	8.9314	9.475		78	24.8349	25.415	
29	9.2491	9.795	9.2355	79	25.1531	25.733	25.1481
30	9.5668	10.114		80	25.4713	26.052	
31	9.8845	10.434	9.8718	81	25.7896	26.370	25.7847
32	10.2023	10.753		82	26.1078	26.689	
33	10.5201	11.073	10.5082	83	26.4260	27.007	26.4213
34	10.8379	11.392		84	26.7443	27.326	
35	11.1558	11.711	11.1446	85	27.0625	27.644	27.0579
36	11.4737	12.030		86	27.3807	27.962	
37	11.7916	12.349	11.7810	87	27.6990	28.281	27.6945
38	12.1095	12.668		88	28.0172	28.599	
39	12.4275	12.987	12.4174	89	28.3354	28.918	28.3310
40	12.7455	13.306		90	28.6537	29.236	
41	13.0635	13.625	13.0539	91	28.9719	29.555	28.9676
42	13.3815	13.944		92	29.2902	29.873	
43	13.6995	14.263	13.6904	93	29.6084	30.192	29.6042
44	14.0175	14.582		94	29.9267	30.510	
45	14.3355	14.901	14.3269	95	30.2449	30.828	30.2408
46	14.6536	15.219		96	30.5632	31.147	
47	14.9717	15.538	14.9634	97	30.8815	31.465	30.8774
48	15.2898	15.857		98	31.1997	31.784	
49	15.6079	16.176	15.5999	99	31.5180	32.102	31.5140
50	15.9260	16.495		100	31.8362	32.421	
51	16.2441	16.813	16.2364	101	32.1545	32.739	32.1506
52	16.5622	17.132		102	32.4727	33.057	
53	16.8803	17.451	16.8729	103	32.7910	33.376	32.7872
54	17.1984	17.769		104	33.1093	33.694	
55	17.5165	18.088	17.5094	105	33.4275	34.013	33.4238
56	17.8347	18.407		106	33.7458	34.331	
57	18.1528	18.725	18.1459	107	34.0641	34.649	34.0604
58	18.4710	19.044		108	34.3823	34.968	

* For 5-8 and 109-200 teeth see text, pages 2293, 2295.

Table 5. Minus Tolerances on the Caliper Diameters of Precision Sprockets*
(ANSI B29.1-1975)

Pitch	Number of Teeth				
	Up to 16	16–24	25–35	36–48	49–63
1/4	0.004	0.004	0.004	0.005	0.005
3/8	0.004	0.004	0.004	0.005	0.005
1/2	0.004	0.005	0.0055	0.006	0.0065
5/8	0.005	0.0055	0.006	0.007	0.008
3/4	0.005	0.006	0.007	0.008	0.009
1	0.006	0.007	0.008	0.009	0.010
1 1/4	0.007	0.008	0.009	0.010	0.012
1 1/2	0.007	0.009	0.0105	0.012	0.013
1 3/4	0.008	0.010	0.012	0.013	0.015
2	0.009	0.011	0.013	0.015	0.017
2 1/4	0.010	0.012	0.014	0.016	0.018
2 1/2	0.010	0.013	0.015	0.018	0.020
3	0.012	0.015	0.018	0.021	0.024

Pitch	Number of Teeth				
	64–80	81–99	100–120	121–143	144 up
1/4	0.005	0.005	0.006	0.006	0.006
3/8	0.006	0.006	0.006	0.007	0.007
1/2	0.007	0.0075	0.008	0.0085	0.009
5/8	0.009	0.009	0.009	0.010	0.011
3/4	0.010	0.010	0.011	0.012	0.013
1	0.011	0.012	0.013	0.014	0.015
1 1/4	0.013	0.014	0.016	0.017	0.018
1 1/2	0.015	0.016	0.018	0.019	0.021
1 3/4	0.017	0.019	0.020	0.022	0.024
2	0.019	0.021	0.023	0.025	0.027
2 1/4	0.021	0.023	0.025	0.028	0.030
2 1/2	0.023	0.025	0.028	0.030	0.033
3	0.027	0.030	0.033	0.036	0.039

* Minus tolerances for Commercial sprockets are twice those shown in this table.

Caliper Diameter: The caliper diameter is the same as the bottom diameter for a sprocket with an even number of teeth. For a sprocket with an odd number of teeth, it is defined as the distance from the bottom of one tooth gap to that of the nearest opposite tooth gap. The caliper diameter for an even tooth sprocket is equal to pitch diameter − roller diameter. The caliper diameter for an odd tooth sprocket is equal to caliper factor − roller diameter. Here, the caliper factor = $PD[\cos(90° ÷ N)]$, where PD = pitch diameter and N = number of teeth. Caliper factors for one-inch pitch and sprockets having 9 to 108 teeth are given in Table 4. For other tooth numbers use above formula. Caliper diameter tolerances are minus only and are equal to $0.002P \sqrt{N} + 0.006$ inch for the Commercial sprockets and $0.001P \sqrt{N} + 0.003$ inch for Precision sprockets. Tolerances are given in Table 5.

Outside Diameter: The outside diameter is the diameter over the tips of teeth. Sprocket outside diameters for one-inch pitch and 9 to 108 teeth are given in Table 4. For other tooth numbers the outside diameter may be determined by the following formula in which O = approximate outside diameter; P = pitch of chain; N = number of sprocket teeth: $O = P [0.6 + \cot(180° ÷ N)]$.

Table 6. American National Standard Roller Chain Sprocket Flange Thickness and Tooth Section Profile Dimensions (ANSI B29.1-1975)

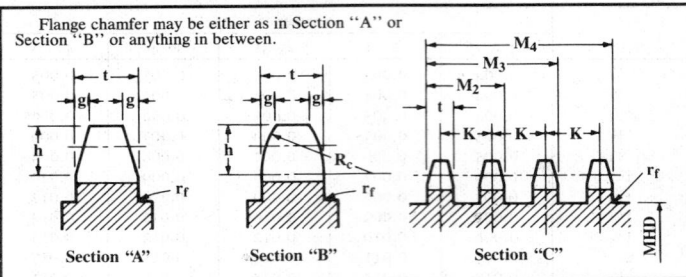

Flange chamfer may be either as in Section "A" or Section "B" or anything in between.

Section "A"   Section "B"   Section "C"

Sprocket Flange Thickness

Std. Chain No.	Width of Chain, W	Maximum Sprocket Flange Thickness, t			Minus Tolerance on t		Tolerance on M		Max. Variation of t on Each Flange	
		Single	Double & Triple	Quad. & Over	Commercial	Precision	Commercial Plus or Minus	Precision Minus Only	Commercial	Precision
25	0.125	0.110	0.106	0.096	0.021	0.007	0.007	0.007	0.021	0.004
35	0.188	0.169	0.163	0.150	0.027	0.008	0.008	0.008	0.027	0.004
41	0.250	0.226	...	...	0.032	0.009	...	...	0.032	0.004
40	0.312	0.284	0.275	0.256	0.035	0.009	0.009	0.009	0.035	0.004
50	0.375	0.343	0.332	0.310	0.036	0.010	0.010	0.010	0.036	0.005
60	0.500	0.459	0.444	0.418	0.036	0.011	0.011	0.011	0.036	0.006
80	0.625	0.575	0.556	0.526	0.040	0.012	0.012	0.012	0.040	0.006
100	0.750	0.692	0.669	0.633	0.046	0.014	0.014	0.014	0.046	0.007
120	1.000	0.924	0.894	0.848	0.057	0.016	0.016	0.016	0.057	0.008
140	1.000	0.924	0.894	0.848	0.057	0.016	0.016	0.016	0.057	0.008
160	1.250	1.156	1.119	1.063	0.062	0.018	0.018	0.018	0.062	0.009
180	1.406	1.302	1.259	1.198	0.068	0.020	0.020	0.020	0.068	0.010
200	1.500	1.389	1.344	1.278	0.072	0.021	0.021	0.021	0.072	0.010
240	1.875	1.738	1.682	1.602	0.087	0.025	0.025	0.025	0.087	0.012

Sprocket Tooth Section Profile Dimensions

Std. Chain No.	Chain Pitch P	Depth of Chamfer h	Width of Chamfer g	Minimum Radius R_c	Transverse Pitch K	
					Standard Series	Heavy Series
25	1/4	1/8	1/32	0.265	0.252	...
35	3/8	3/16	3/64	0.398	0.399	...
41	1/2	1/4	1/16	0.531	...	...
40	1/2	1/4	1/16	0.531	0.566	...
50	5/8	5/16	5/64	0.664	0.713	...
60	3/4	3/8	3/32	0.796	0.897	1.028
80	1	1/2	1/8	1.062	1.153	1.283
100	1 1/4	5/8	5/32	1.327	1.408	1.539
120	1 1/2	3/4	3/16	1.593	1.789	1.924
140	1 3/4	7/8	7/32	1.858	1.924	2.055
160	2	1	1/4	2.124	2.305	2.437
180	2 1/4	1 1/8	9/32	2.392	2.592	2.723
200	2 1/2	1 1/4	5/16	2.654	2.817	3.083
240	3	1 1/2	3/8	3.187	3.458	3.985

All dimensions are in inches. r_f max = 0.04 P for max. hub diameter.

Table 7. Typical Proportions of Single-Strand and Multiple-Strand Cast Roller Chain Sprockets

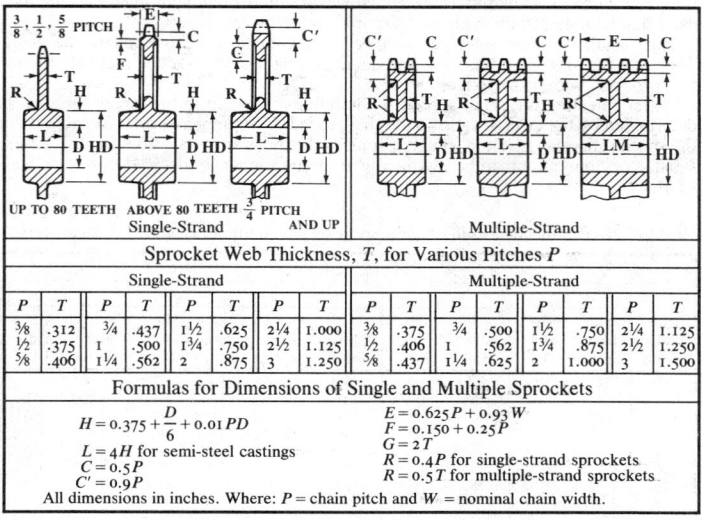

Sprocket Web Thickness, T, for Various Pitches P																	
Single-Strand								Multiple-Strand									
P	T	P	T	P	T	P	T	P	T	P	T	P	T	P	T
3/8	.312	3/4	.437	1½	.625	2¼	1.000	3/8	.375	3/4	.500	1½	.750	2¼	1.125
½	.375	1	.500	1¾	.750	2½	1.125	½	.406	1	.562	1¾	.875	2½	1.250
5/8	.406	1¼	.562	2	.875	3	1.250	5/8	.437	1¼	.625	2	1.000	3	1.500

Formulas for Dimensions of Single and Multiple Sprockets

$H = 0.375 + \dfrac{D}{6} + 0.01\,PD$

$L = 4H$ for semi-steel castings
$C = 0.5P$
$C' = 0.9P$

$E = 0.625P + 0.93W$
$F = 0.150 + 0.25P$
$G = 2T$
$R = 0.4P$ for single-strand sprockets
$R = 0.5T$ for multiple-strand sprockets

All dimensions in inches. Where: P = chain pitch and W = nominal chain width.

Table 8. Typical Proportions of Roller Chain Bar-steel Sprockets

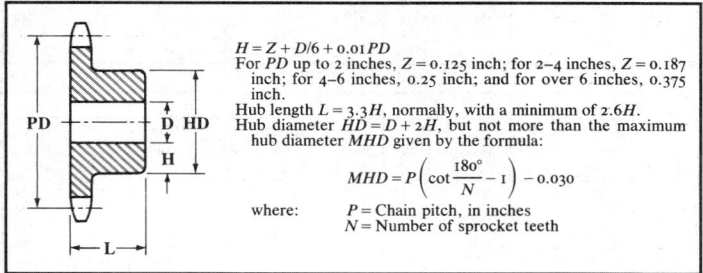

$H = Z + D/6 + 0.01\,PD$

For PD up to 2 inches, $Z = 0.125$ inch; for 2–4 inches, $Z = 0.187$ inch; for 4–6 inches, 0.25 inch; and for over 6 inches, 0.375 inch.

Hub length $L = 3.3H$, normally, with a minimum of $2.6H$.

Hub diameter $HD = D + 2H$, but not more than the maximum hub diameter MHD given by the formula:

$$MHD = P\left(\cot\frac{180°}{N} - 1\right) - 0.030$$

where: P = Chain pitch, in inches
N = Number of sprocket teeth

Proportions of Sprockets. — Typical proportions of single-strand and multiple-strand cast roller chain sprockets, as provided by the American Chain Association, are shown in Table 7. Typical proportions of roller chain bar-steel sprockets, also provided by this association, are shown in Table 8.

When sprocket wheels are designed with spokes, the usual assumptions made in order to determine suitable proportions are as follows: (1) That the maximum torque load acting on a sprocket is the chain tensile strength times the sprocket pitch radius; (2) That the torque load is equally divided between the arms by the rim; and (3) That each arm acts as a cantilever beam. The arms are generally elliptical in cross section, with the major axis twice the minor axis.

Selection of Chain and Sprockets. — The smallest applicable pitch of roller chain is desirable for quiet operation and high speed. The horsepower capacity varies with the chain pitch as shown in Table 15. However, short pitch with high working load can often be obtained by the use of multiple-strand chain.

The small sprocket selected must be large enough to accommodate the shaft. Table 9 gives maximum bore and hub diameters consistent with commercial practice for sprockets with up to 25 teeth.

After selecting the small sprocket, the number of teeth in the larger sprocket is determined by the desired ratio of the shaft speed. Overemphasis on the exactness in the speed ratio may result in a cumbersome and expensive installation. In most cases, satisfactory operation can be obtained with a minor change in speed of one or both shafts.

Table 9. Recommended Roller Chain Sprocket Maximum Bore and Hub Diameters*

	Roller Chain Pitch									
	3/8		1/2		5/8		3/4		1	
No. of Teeth	Max. Bore	Max. Hub Dia.	Max. Bore	Max. Hub Dia.	Max. Bore	Max. Hub Dia.	Max. Bore	Max. Hub Dia.	Max. Bore	Max. Hub Dia.
11	19/32	55/64	25/32	1 11/64	31/32	1 15/32	1 1/4	1 49/64	1 5/8	2 3/8
12	5/8	63/64	7/8	1 21/64	1 5/32	1 43/64	1 9/32	2 1/64	1 25/32	2 45/64
13	3/4	1 7/64	1	1 1/2	1 9/32	1 7/8	1 1/2	2 1/4	2	3 1/64
14	27/32	1 15/64	1 5/32	1 21/32	1 5/16	2 5/64	1 3/4	2 1/2	2 9/32	3 11/32
15	7/8	1 23/64	1 1/4	1 13/16	1 17/32	2 9/32	1 25/32	2 3/4	2 13/32	3 43/64
16	31/32	1 15/32	1 9/32	1 63/64	1 11/16	2 31/64	1 31/32	2 63/64	2 23/32	3 63/64
17	1 3/32	1 19/32	1 3/8	2 9/64	1 25/32	2 11/16	2 7/32	3 7/32	2 13/16	4 5/16
18	1 7/32	1 47/64	1 17/32	2 19/64	1 7/8	2 7/8	2 9/32	3 15/32	3 1/8	4 41/64
19	1 1/4	1 27/32	1 11/16	2 29/64	2 1/16	3 5/64	2 7/16	3 45/64	3 5/16	4 61/64
20	1 9/32	1 61/64	1 25/32	2 5/8	2 1/4	3 9/32	2 11/16	3 61/64	3 1/2	5 9/32
21	1 5/16	2 5/64	1 25/32	2 25/32	2 9/32	3 31/64	2 13/16	4 3/16	3 3/4	5 19/32
22	1 7/16	2 13/16	1 15/16	2 15/16	2 7/16	3 11/16	2 15/16	4 7/16	3 7/8	5 59/64
23	1 9/16	2 9/32	2 3/32	3 3/32	2 5/8	3 57/64	3 1/8	4 43/64	4 3/16	6 15/64
24	1 11/16	2 7/16	2 1/4	3 17/64	2 13/16	4 5/64	3 1/4	4 29/32	4 9/16	6 9/16
25	1 3/4	2 9/16	2 9/32	3 27/64	2 27/32	4 9/32	3 3/8	5 9/32	4 11/16	6 7/8

	Roller Chain Pitch									
	1 1/4		1 1/2		1 3/4		2		2 1/2	
No. of Teeth	Max. Bore	Max. Hub Dia.	Max. Bore	Max. Hub Dia.	Max. Bore	Max. Hub Dia.	Max. Bore	Max. Hub Dia.	Max. Bore	Max. Hub Dia.
11	1 31/32	2 31/32	2 5/16	3 37/64	2 13/16	4 11/64	3 9/32	4 25/32	3 15/16	5 63/64
12	2 9/32	3 3/8	2 3/4	4 1/16	3 1/4	4 3/4	3 5/8	5 27/64	4 23/32	6 51/64
13	2 17/32	3 25/32	3 1/16	4 35/64	3 9/16	5 5/16	4 1/16	6 5/64	5 3/32	7 39/64
14	2 11/16	4 3/64	3 5/16	5 1/32	3 7/8	5 7/8	4 11/16	6 23/32	5 23/32	8 27/64
15	3 3/32	4 19/32	3 3/4	5 33/64	4 7/16	6 29/64	4 7/8	7 3/8	6 1/4	9 7/32
16	3 9/32	5	4	6	4 11/16	7 1/64	5 1/2	8 1/64	7	10 1/32
17	3 21/32	5 13/32	4 15/32	6 31/64	5 1/16	7 37/64	5 11/16	8 21/32	7 7/16	10 27/32
18	3 25/32	5 51/64	4 21/32	6 31/32	5 5/8	8 9/64	6 1/4	9 5/16	8 1/8	11 41/64
19	4 3/16	6 13/64	4 15/16	7 29/64	5 13/16	8 45/64	6 7/8	9 61/64	9	12 7/16
20	4 19/32	6 9/64	5 7/16	7 15/16	6 1/4	9 17/64	7	10 19/32	9 3/4	13 1/4
21	4 11/16	7	5 11/16	8 27/64	6 13/16	9 53/64	7 3/4	11 15/64	10	14 3/64
22	4 7/8	7 13/32	5 7/8	8 57/64	7 1/4	10 25/64	8 3/8	11 7/8	10 7/8	14 27/32
23	5 5/16	7 13/16	6 3/8	9 3/8	7 7/16	10 15/16	9	12 33/64	11 5/8	15 21/32
24	5 11/16	8 13/64	6 13/16	9 55/64	8	11 1/2	9 5/8	13 5/32	13	16 29/64
25	5 23/32	8 39/64	7 1/4	10 11/32	8 9/16	12 1/16	10 1/4	13 51/64	13 1/2	17 1/4

* American Chain Association.
All dimensions in inches.
For standard key dimensions see pages 2234 to 2236.

Center Distance between Sprockets. — The center-to-center distance between sprockets, as a general rule, should not be less than 1½ times the diameter of the larger sprocket and not less than thirty times the pitch nor more than about 50 times the pitch, although much depends upon the speed and other conditions. A center distance equivalent to 80 pitches may be considered an approved maximum. Very long center distances result in catenary tension in the chain. If roller-chain drives are designed correctly, the center-to-center distance for some transmissions may be so short that the sprocket teeth nearly touch each other, assuming that the load is not too great and the number of teeth is not too small. To avoid interference of the sprocket teeth, the center distance must, of course, be somewhat greater than one-half the sum of the outside diameters of the sprockets. The chain should extend around at least 120 degrees of the pinion circumference, and this minimum amount of contact is obtained for all center distances provided the ratio is less than 3½ to 1. Other things being equal, a fairly long chain is recommended in preference to the shortest one allowed by the sprocket diameters, because the rate of chain elongation due to natural wear is inversely proportional to the length, and also because the greater elasticity of the longer strand tends to absorb irregularities of motion and to decrease the effect of shocks.

If possible, the center distance should be adjustable in order to take care of slack due to elongation from wear and this range of adjustment should be at least one and one-half pitches. A little slack is desirable as it allows the chain links to take the best position on the sprocket teeth and reduces the wear on the bearings. Too much sag or an excessive distance between the sprockets may cause the chain to whip up and down — a condition detrimental to smooth running and very destructive to the chain. The sprockets should run in a vertical plane, the sprocket axes being approximately horizontal, unless an idler is used on the slack side to keep the chain in position. The most satisfactory results are obtained when the slack side of the chain is on the bottom.

Center Distance for a Given Chain Length. — When the distance between the driving and driven sprockets can be varied to suit the length of the chain, this center distance for a tight chain may be determined by the following formula, in which $c =$ center-to-center distance in inches; $L =$ chain length in pitches; $P =$ pitch of chain; $N =$ number of teeth in large sprocket; $n =$ number of teeth in small sprocket.

$$c = \frac{P}{8}[2L - N - n + \sqrt{(2L - N - n)^2 - 0.810(N - n)^2}]$$

This formula is approximate, but the error is less than the variation in the length of the best chains. The length L in pitches should be an even number for a roller chain, so that the use of an offset connecting link will not be necessary. .

Idler Sprockets. — When sprockets have a fixed center distance or are non-adjustable, it may be advisable to use an idler sprocket for taking up the slack. The idler should preferably be placed against the slack side between the two strands of the chain. When a sprocket is applied to the tight side of the chain to reduce vibration, it should be on the lower side and so located that the chain will run in a straight line between the two main sprockets. A sprocket will wear excessively if the number of teeth is too small and the speed too high, because there is impact between the teeth and rollers even though the idler carries practically no load.

Length of Driving Chain. — The total length of a block chain should be given in multiples of the pitch, whereas for a roller chain, the length should be in multiples of twice the pitch, because the ends must be connected with an outside and inside link. The length of a chain can be calculated accurately enough for ordinary

practice by the use of the following formula, in which L = chain length in pitches; C = center distance in pitches; N = number of teeth in large sprocket; n = number of teeth in small sprocket:

$$L = 2C + \frac{N}{2} + \frac{n}{2} + \left(\frac{N-n}{2\pi}\right)^2 \times \frac{1}{C}$$

To the length obtained by this formula, add enough to make a whole number (and for a roller chain, an even number) of pitches. If a roller chain has an odd number of pitches, it will be necessary to use an offset connecting link.

Another formula for obtaining chain length in which D = distance between centers of shafts; R = pitch radius of large sprocket; r = pitch radius of small sprocket; N = number of teeth in large sprocket; n = number of teeth in small sprocket; P = pitch of chain and sprockets; and l = required chain length in inches, is:

$$l = \frac{180° + 2\alpha}{360°}NP + \frac{180° - 2\alpha}{360°}nP + 2D\cos\alpha; \quad \text{where } \sin\alpha = \frac{R-r}{D}$$

Cutting Standard Sprocket Tooth Form. — The proportions and seating curve data for the standard sprocket tooth form for roller chain are given in Table 10. Either formed or generating types of sprocket cutters may be employed.

Hobs: Only one hob will be required to cut any number of teeth for a given pitch and roller diameter. All hobs should be marked with pitch and roller diameter to be cut. Formulas and data for standard hob design are given in Table 11.

Space Cutters: Five cutters of this type will be required to cut from 7 teeth up for any given roller diameter. The ranges are, respectively, 7–8, 9–11, 12–17, 18–34, and 35 teeth and over. If less than 7 teeth is necessary, special cutters conforming to the required number of teeth should be used.

The regular cutters are based upon an intermediate number of teeth N_a equal to $2N_1N_2 \div (N_1 + N_2)$ in which N_1 = minimum number of teeth and N_2 = maximum number of teeth for which cutter is intended; but the topping curve radius F (see diagram in Table 12) is designed to produce adequate tooth height on a sprocket of N_2 teeth. The values of N_a for the several cutters are, respectively, 7.47, 9.9, 14.07, 23.54, and 56. Formulas and construction data for space cutter layout are given in Table 12 and recommended cutter sizes are given in Table 13.

Shaper Cutters: Only one will be required to cut any number of teeth for a given pitch and roller diameter. The manufacturer should be referred to for information concerning the cutter form design to be used.

Sprocket Manufacture. — Cast sprockets have cut teeth, and the rim, hub face, and bore are machined. The smaller sprockets are generally cut from steel bar stock, and are finished all over. Sprockets are often made from forgings or forged bars. The extent of finishing depends on the particular specifications that are applicable. Many sprockets are made by welding a steel hub to a steel plate. This process produces a one-piece sprocket of desired proportions and one that can be heat-treated.

Sprocket Materials. — For large sprockets, cast iron is commonly used, especially in drives with large speed ratios, since the teeth of the larger sprocket are subjected to fewer chain engagements in a given time. For severe service, cast steel or steel plate is preferred.

The smaller sprockets of a drive are usually made of steel. With this material the body of the sprocket can be heat-treated to produce toughness for shock resistance, and the tooth surfaces can be hardened to resist wear.

Stainless steel or bronze may be used for corrosion resistance and Formica, nylon or other suitable plastic materials for special applications.

Table 10. ANSI Sprocket Tooth Form for Roller Chain (ANSI B29.1-1975)

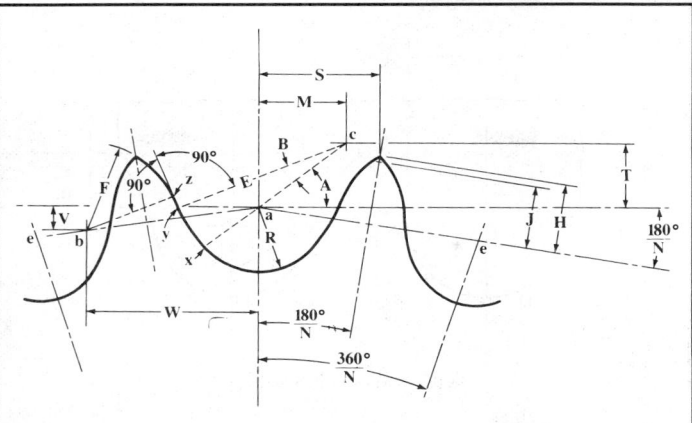

P = pitch (ae); N = number of teeth; D_r = nominal roller diameter

D_s = seating curve diameter = 1.005 D_r + 0.003 (in inches); R = ½ D_s

D_s has only plus tolerance

$A = 35° + (60° \div N)$; $B = 18° - (56° \div N)$; $ac = 0.8\ D_r$

$M = 0.8\ D_r \cos (35° + (60° \div N))$

$T = 0.8\ D_r \sin (35° + (60° \div N))$; $E = 1.3025\ D_r + 0.0015$ (in inches)

Chord $xy = (2.605\ D_r + 0.003) \sin (9° - (28° \div N))$ (in inches)

$yz = D_r\ [1.4 \sin (17° - (64° \div N)) - 0.8 \sin (18° - (56° \div N))]$

Length of a line between a and b = 1.4 D_r

$W = 1.4\ D_r \cos (180° \div N)$; $V = 1.4\ D_r \sin (180° \div N)$

$F = D_r\ [0.8 \cos (18° - (56° \div N)) + 1.4 \cos (17° - (64° \div N)) - 1.3025] - 0.0015$ in.

$H = \sqrt{F^2 - (1.4\ D_r - 0.5\ P)^2}$

$S = 0.5\ P \cos (180° \div N) + H \sin (180° \div N)$

Approximate O.D. of sprocket when J is 0.3 $P = P\ [0.6 + \cot (180° \div N)]$

O.D. of sprocket when tooth is pointed = $P \cot (180° \div N) + \cos (180°N)\ (D_s - D_r) + 2\ H$

Pressure angle for new chain = $xab = 35° - (120° \div N)$

Minimum pressure angle = $xab - B = 17° - (64° \div N)$; Average pressure angle = $26° - (92° \div N)$

				Seating Curve Data—Inches					
P	D_r	Min. R	Min. D_s	D_s Tol.*	P	D_r	Min. R	Min. D_s	D_s Tol.*
¼	0.130	0.0670	0.134	0.0055	1¼	0.750	0.3785	0.757	0.0070
⅜	0.200	0.1020	0.204	0.0055	1½	0.875	0.4410	0.882	0.0075
½	0.306	0.1585	0.317	0.0060	1¾	1	0.5040	1.008	0.0080
½	0.312	0.1585	0.317	0.0060	2	1.125	0.5670	1.134	0.0085
⅝	0.400	0.2025	0.405	0.0060	2¼	1.406	0.7080	1.416	0.0090
¾	0.469	0.2370	0.474	0.0065	2½	1.562	0.7870	1.573	0.0095
1	0.625	0.3155	0.631	0.0070	3	1.875	0.9435	1.887	0.0105

* Plus tolerance only.

Table 11. Standard Hob Design for Roller Chain Sprockets

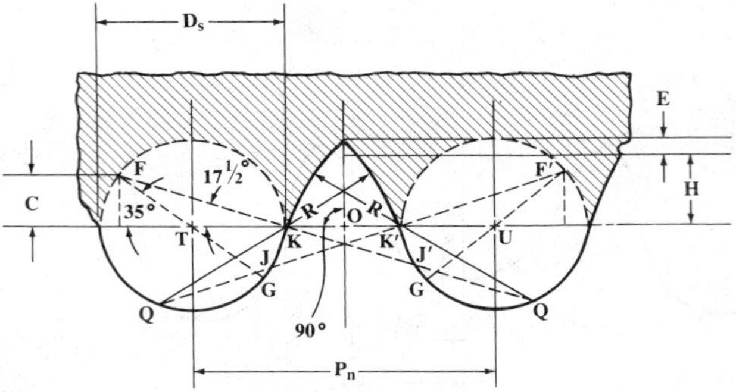

Section Normal to Hob Teeth

Hobs designed for a given roller diameter (D_r) and chain pitch (P) will cut any number of teeth.

P = Pitch of Chain; P_n = Normal Pitch of Hob = 1.011 P inches

D_s = Minimum Diameter of Seating Curve = $1.005\ D_r + 0.003$ inches

F = Radius Center for Arc GK; $TO = OU = P_n \div 2$

$H = 0.27\ P$; $E = 0.03\ P$ = Radius of Fillet Circle

Q is located on line passing through F and J. Point J is intersection of line XY with circle of diameter D_s. R is found by trial and the arc of this radius is tangent to arc KG at K and to fillet radius.

OD = Outside Diameter = 1.7 (Bore + D_r + 0.7 P) approx.

D_h = Pitch Diameter = $OD - D_s$; M = Helix Angle; $\sin M = P_n \div \pi D_h$

L = Lead = $P_n \div \cos M$; W = Width = Not less than 2 × Bore, or 6 D_r, or 3.2 P

Data for Laying Out Hob Outlines — Inches

P	P_n	H	E	O.D.	W	Bore	Keyway	No. Gashes
¼	0.2527	0.0675	0.0075	2⅝	2½	1.250	¼ × ⅛	13
⅜	0.379	0.101	0.012	3⅛	2½	1.250	¼ × ⅛	13
½	0.506	0.135	0.015	3⅜	2½	1.250	¼ × ⅛	12
⅝	0.632	0.170	0.018	3⅝	2½	1.250	¼ × ⅛	12
¾	0.759	0.202	0.023	3¾	2⅞	1.250	¼ × ⅛	11
1	1.011	0.270	0.030	4⅜	3¾	1.250	¼ × ⅛	11
1¼	1.264	0.337	0.038	4¾	4½	1.250	¼ × ⅛	10
1½	1.517	0.405	0.045	5⅜	5¼	1.250	¼ × ⅛	10
1¾	1.770	0.472	0.053	6⅜	6	1.500	⅜ × 3/16	9
2	2.022	0.540	0.060	6⅞	6¾	1.500	⅜ × 3/16	9
2¼	2.275	0.607	0.068	8	8½	1.750	⅜ × 3/16	8
2½	2.528	0.675	0.075	8⅝	9⅜	1.750	⅜ × 3/16	8
3	3.033	0.810	0.090	9¾	11¼	2.000	½ × 3/16	8

Table 12. Standard Space Cutters for Roller-Chain Sprockets

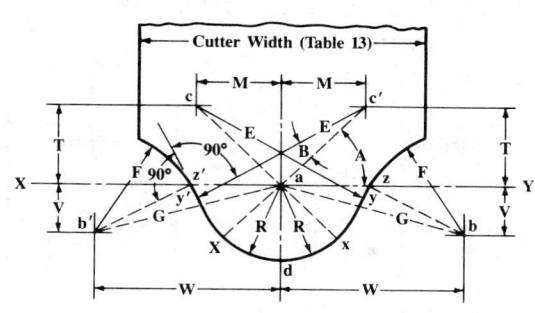

Angle Yab is equal to $180° \div N$ when the cutter is made for a specific number of teeth. For the design of cutters covering a range of teeth, angle Yab was determined by layout to assure chain roller clearance and to avoid pointed teeth on the larger sprockets of each range. It has values as given below for cutters covering the range of teeth shown. The following formulas are for cutters covering the standard ranges of teeth where N_a equals intermediate values given on page 2300.

$$W = 1.4 D_r \cos Yab; \qquad V = 1.4 D_r \sin Yab$$

$$yz = D_r \left[1.4 \sin \left(17° + \frac{116°}{N_a} - Yab \right) - 0.8 \sin \left(18° - \frac{56°}{N_a} \right) \right]$$

$$F = D_r \left[0.8 \cos \left(18° - \frac{56°}{N_a} \right) + 1.4 \cos \left(17° + \frac{116°}{N_a} - Yab \right) - 1.3025 \right] - 0.0015 \text{ in.}$$

For other points, use the value of N_a for N in the standard formulas in Table 10.

Data for Laying Out Space Cutter				
Range of Teeth	M	T	W	V
7–8	$0.5848\,D_r$	$0.5459\,D_r$	$1.2790\,D_r$	$0.5694\,D_r$
9–11	$0.6032\,D_r$	$0.5255\,D_r$	$1.3302\,D_r$	$0.4365\,D_r$
12–17	$0.6194\,D_r$	$0.5063\,D_r$	$1.3694\,D_r$	$0.2911\,D_r$
18–34	$0.6343\,D_r$	$0.4875\,D_r$	$1.3947\,D_r$	$0.1220\,D_r$
35 up	$0.6466\,D_r$	$0.4710\,D_r$	$1.4000\,D_r$	0
Range of Teeth	F	Chord xy	yz	Angle Yab
7–8	$0.8686\,D_r - 0.0015$	$0.2384\,D_r + 0.0003$	$0.0618\,D_r$	$24°$
9–11	$0.8554\,D_r - 0.0015$	$0.2800\,D_r + 0.0003$	$0.0853\,D_r$	$18°10'$
12–17	$0.8364\,D_r - 0.0015$	$0.3181\,D_r + 0.0004$	$0.1269\,D_r$	$12°$
18–34	$0.8073\,D_r - 0.0015$	$0.3540\,D_r + 0.0004$	$0.1922\,D_r$	$5°$
35 up	$0.7857\,D_r - 0.0015$	$0.3850\,D_r + 0.0004$	$0.2235\,D_r$	$0°$
E (same for all ranges) $= 1.3025\,D_r + 0.0015$; G (same for all ranges) $= 1.4\,D_r$				

Table 13. Recommended Space Cutter Sizes for Roller-Chain Sprockets

Pitch	Roller Diam.	Number of Teeth 6	7–8	9–11	12–17	18–34	35 up
		Cutter Diameter (Minimum)					
1/4	0.130	2 3/4	2 3/4	2 3/4	2 3/4	2 3/4	2 3/4
3/8	0.200	2 3/4	2 3/4	2 3/4	2 3/4	2 3/4	2 3/4
1/2	0.312	3	3	3 1/8	3 1/8	3 1/8	3 1/8
5/8	0.400	3 1/8	3 1/8	3 1/4	3 1/4	3 1/4	3 1/4
3/4	0.469	3 1/4	3 1/4	3 3/8	3 3/8	3 3/8	3 3/8
1	0.625	3 7/8	4	4 1/8	4 1/8	4 1/4	4 1/4
1 1/4	0.750	4 1/4	4 3/8	4 1/2	4 1/2	4 5/8	4 5/8
1 1/2	0.875	4 3/8	4 1/2	4 5/8	4 5/8	4 3/4	4 3/4
1 3/4	1.000	5	5 1/8	5 1/4	5 3/8	5 1/2	5 1/2
2	1.125	5 3/8	5 1/2	5 5/8	5 3/4	5 7/8	5 7/8
2 1/4	1.406	5 7/8	6	6 1/4	6 3/8	6 1/2	6 1/2
2 1/2	1.563	6 3/8	6 5/8	6 3/4	6 7/8	7	7 1/8
3	1.875	7 1/2	7 3/4	7 7/8	8	8	8 1/4
Pitch	Roller Diam.	Cutter Width (Minimum)					
1/4	0.130	5/16	5/16	5/16	5/16	9/32	9/32
3/8	0.200	15/32	15/32	15/32	7/16	7/16	13/32
1/2	0.312	3/4	3/4	3/4	3/4	23/32	11/16
5/8	0.400	3/4	3/4	3/4	3/4	23/32	11/16
3/4	0.469	29/32	29/32	29/32	7/8	27/32	13/16
1	0.625	1 1/2	1 1/2	1 15/32	1 15/32	1 13/32	1 11/32
1 1/4	0.750	1 13/16	1 13/16	1 25/32	1 3/4	1 11/16	1 5/8
1 1/2	0.875	1 13/16	1 13/16	1 25/32	1 3/4	1 11/16	1 5/8
1 3/4	1.000	2 3/32	2 3/32	2 1/16	2 1/32	1 31/32	1 7/8
2	1.125	2 13/32	2 13/32	2 3/8	2 5/16	2 1/4	2 5/32
2 1/4	1.406	2 11/16	2 11/16	2 21/32	2 19/32	2 15/32	2 13/32
2 1/2	1.563	3	3	2 15/16	2 29/32	2 3/4	2 11/16
3	1.875	3 19/32	3 19/32	3 17/32	3 15/32	3 11/32	3 7/32

Where the same roller diameter is commonly used with chains of two different pitches it is recommended that stock cutters be made wide enough to cut sprockets for both chains.

Marking of Cutters. — All cutters are to be marked, giving pitch, roller diameter and range of teeth to be cut.

Bores for Sprocket Cutters (recommended practice) are approximately as calculated from the formula:

$$\text{Bore} = 0.7\sqrt{(\text{Width of Cutter} + \text{Roller Diameter} + 0.7\ \text{Pitch})}$$

and are equal to 1 inch for 1/4- through 3/4-inch pitches; 1 1/4 inches for 1- through 1 1/2-inch pitches; 1 1/2 inches for 1 3/4- through 2 1/4-inch pitches; 1 3/4 inches for 2 1/2-inch pitch; and 2 inches for 3-inch pitch.

Minimum Outside Diameters of Space Cutters for 35 teeth and over (recommended practice) are approximately as calculated from the formula:

$$\text{Outside Diameter} = 1.2\ (\text{Bore} + \text{Roller Diameter} + 0.7\ \text{Pitch}) + 1\ \text{in.}$$

Roller Chain Drive Ratings. — In 1961, under auspices of The American Sprocket Chain Manufacturers Association (now called American Chain Association), a joint research program was begun to study pin-bushing interaction at high speeds and to gain further data on the phenomenon of chain joint galling. The objectives of this program were: (1) To determine how far roller chain could be operated safely into the galling area; (2) To investigate the factors influencing galling; and (3) To learn how to inhibit galling in future chain designs. This study resulted in 1966 in newly revised horsepower ratings of roller chains. Subsequent studies resulted in further revisions which appear in ANSI B29.1-1975. The ratings shown in Table 15 are below the galling range.

The horsepower ratings in Table 14 apply to lubricated, single-pitch, single-strand roller chains, both ANSI Standard and Heavy series. To obtain ratings of multiple-strand chains, a multiple-strand factor is applied. The ratings in Table 14 are based upon: (1) A service factor of 1; (2) A chain length of approximately 100 pitches; (3) Use of recommended lubrication methods; and (4) A drive arrangement where two aligned sprockets are mounted on parallel shafts in a horizontal plane. Under these conditions, approximately 15,000 hours of service life at full load operation may be expected.

Substantial increases in rated speed loads can be utilized, as when a service life of less than 15,000 hours is satisfactory, or when full load operation is encountered only during a portion of the required service life. Chain manufacturers should be consulted for assistance with any special application requirements.

The horsepower ratings shown in Table 15 relate to the speed of the smaller sprocket and drive selections are made on this basis, whether the drive is speed reducing or speed increasing. Drives with more than two sprockets, idlers, composite duty cycles, or other unusual conditions often require special consideration. Where quietness or extra smooth operation are of special importance, a small-pitch chain operating over large diameter sprockets will minimize noise and vibration.

When making drive selection, consideration is given to the loads imposed on the chain by the type of input power and the type of equipment to be driven. Service factors are used to compensate for these loads and the *required* horsepower rating of the chain is determined by the following formula:

$$\text{Required hp Table Rating} = \frac{\text{hp to be Transmitted} \times \text{Service Factor}}{\text{Multiple-Strand Factor}}$$

Service Factors: The service factors in Table 14 are for normal chain loading. For unusual or extremely severe operating conditions not shown in this table, it is desirable to use larger service factors.

Multiple-Strand factors: The horsepower ratings for multiple-strand chains equal single-strand ratings multiplied by these factors: for two strands, a factor of 1.7; for three strands, 2.5; and for four strands, 3.3.

Table 14. Roller Chain Drive Service Factors

Type of Driven Load	Type of Input Power		
	Internal Combustion Engine with Hydraulic Drive	Electric Motor or Turbine	Internal Combustion Engine with Mechanical Drive
Smooth	1.0	1.0	1.2
Moderate Shock	1.2	1.3	1.4
Heavy Shock	1.4	1.5	1.7

Lubrication. — It has been shown that a separating wedge of fluid lubricant is formed in operating chain joints much like that formed in journal bearings. Therefore, fluid lubricant must be applied to assure an oil supply to the joints and minimize metal-to-metal contact. Lubrication, if supplied in sufficient volume, also provides effective cooling and impact damping at higher speeds. For this reason, it is important that lubrication recommendations be followed. *The ratings in Table 15 apply only to drives lubricated in the manner specified in this table.*

Chain drives should be protected against dirt and moisture and the oil supply kept free of contamination. Periodic oil change is desirable. A good grade of non-detergent petroleum base oil is recommended. Heavy oils and greases are generally too stiff to enter and fill the chain joints. The following lubricant viscosities are recommended: For temperatures of 20° to 40° F, use SAE 20 lubricant; for 40° to 100°, use SAE 30; for 100° to 120°, use SAE 40; and for 120° to 140°, use SAE 50.

There are three basic types of lubrication for roller chain drives. The recommended type shown in Table 15 as Type A, Type B, or Type C is influenced by the chain speed and the amount of power transmitted. These are *minimum* lubrication requirements and the use of a better type (for example, Type C instead of Type B) is acceptable and may be beneficial. Chain life can vary appreciably depending upon the way the drive is lubricated. The better the chain lubrication, the longer the chain life. For this reason, it is important that the lubrication recommendations be followed when using the ratings given in Table 15. The types of lubrication are as follows:

Type A—Manual or Drip Lubrication: In manual lubrication, oil is applied copiously with a brush or spout can at least once every eight hours of operation. Volume and frequency should be sufficient to prevent overheating of the chain or discoloration of the chain joints. In drip lubrication, oil drops from a drip lubricator are directed between the link plate edges. The volume and frequency should be sufficient to prevent discoloration of the lubricant in the chain joints. Precaution must be taken against misdirection of the drops by windage.

Type B—Bath or Disc Lubrication: In bath lubrication, the lower strand of the chain runs through a sump of oil in the drive housing. The oil level should reach the pitch line of the chain at its lowest point while operating. In disc lubrication, the chain operates above the oil level. The disc picks up oil from the sump and deposits it onto the chain, usually by means of a trough. The diameter of the disc should be such as to produce rim speeds of between 600 and 8000 feet per minute.

Type C—Oil Stream Lubrication: The lubricant is usually supplied by a circulating pump capable of supplying each chain drive with a continuous stream of oil. The oil should be applied inside the chain loop evenly across the chain width, and directed at the slack strand.

The chain manufacturer should be consulted when it appears desirable to use a type of lubricant other than that recommended.

Installation and Alignment. — Sprockets should have the tooth form, thickness, profile, and diameters conforming to the ANSI B29.1 Standard. For maximum service life small sprockets operating at moderate to high speeds, or near the rated horsepower, should have hardened teeth. Normally, large sprockets should not exceed 120 teeth.

In general a center distance of 30 to 50 chain pitches is most desirable. The distance between sprocket centers should provide at least a 120 degree chain wrap on the smaller sprocket. Drives may be installed with either adjustable or fixed center distances. Adjustable centers simplify the control of chain slack. Sufficient housing clearance must always be provided for the chain slack to obtain full chain life.

Accurate alignment of shafts and sprocket tooth faces provides uniform distribution of the load across the entire chain width and contributes substantially to optimum drive life. Shafting, bearings, and foundations should be suitable to maintain the initial alignment. Periodic maintenance should include an inspection of alignment.

Table 15. Horsepower Ratings for Roller Chain — 1975

To properly use this table the following factors must be taken into consideration:

1. Service Factors
2. Multiple Strand Factors
3. Lubrication.

Service Factors: See Table 14.

Multiple Strand Factors: For two strands, the multiple strand factor is 1.7; for three strands, it is 2.5; and for four strands, it is 3.3.

Lubrication: Type A—Manual or Drip Lubrication
　　　　　　　 Type B—Bath or Disc Lubrication
　　　　　　　 Type C—Oil Stream Lubrication

Required type of lubrication is indicated at the bottom of each roller chain size section of the table.

For a description of each type of lubrication, see page 2306.

To find the required horsepower table rating, use the following formula:

$$\text{Required hp Table Rating} = \frac{\text{hp to be transmitted} \times \text{Service Factor}}{\text{Multiple Strand Factor}}$$

¼-inch Pitch Standard Single-Strand Roller Chain—No. 25

No. of Teeth Small Spkt.	Revolutions per Minute—Small Sprocket*												
	50	100	300	500	700	900	1200	1500	1800	2100	2500	3000	3500
	Horsepower Rating												
9	0.02	0.04	0.12	0.18	0.25	0.31	0.41	0.50	0.58	0.67	0.79	0.93	1.06
10	0.03	0.05	0.13	0.21	0.28	0.35	0.45	0.56	0.65	0.75	0.88	1.04	1.19
11	0.03	0.05	0.14	0.23	0.31	0.39	0.50	0.62	0.73	0.83	0.98	1.15	1.32
12	0.03	0.06	0.16	0.25	0.34	0.43	0.55	0.68	0.80	0.92	1.07	1.26	1.45
13	0.04	0.06	0.17	0.27	0.37	0.47	0.60	0.74	0.87	1.00	1.17	1.38	1.58
14	0.04	0.07	0.19	0.30	0.40	0.50	0.65	0.80	0.94	1.08	1.27	1.49	1.71
15	0.04	0.08	0.20	0.32	0.43	0.54	0.70	0.86	1.01	1.17	1.36	1.61	1.85
16	0.04	0.08	0.22	0.34	0.47	0.58	0.76	0.92	1.09	1.25	1.46	1.72	1.98
17	0.05	0.09	0.23	0.37	0.50	0.62	0.81	0.99	1.16	1.33	1.56	1.84	2.11
18	0.05	0.09	0.25	0.39	0.53	0.66	0.86	1.05	1.24	1.42	1.66	1.96	2.25
19	0.05	0.10	0.26	0.41	0.56	0.70	0.91	1.11	1.31	1.50	1.76	2.07	2.38
20	0.06	0.10	0.28	0.44	0.59	0.74	0.96	1.17	1.38	1.59	1.86	2.19	2.52
21	0.06	0.11	0.29	0.46	0.62	0.78	1.01	1.24	1.46	1.68	1.96	2.31	2.66
22	0.06	0.11	0.31	0.48	0.66	0.82	1.07	1.30	1.53	1.76	2.06	2.43	2.79
23	0.06	0.12	0.32	0.51	0.69	0.86	1.12	1.37	1.61	1.85	2.16	2.55	2.93
24	0.07	0.13	0.34	0.53	0.72	0.90	1.17	1.43	1.69	1.94	2.27	2.67	3.07
25	0.07	0.13	0.35	0.56	0.75	0.94	1.22	1.50	1.76	2.02	2.37	2.79	3.21
26	0.07	0.14	0.37	0.58	0.79	0.98	1.28	1.56	1.84	2.11	2.47	2.91	3.34
28	0.08	0.15	0.40	0.63	0.85	1.07	1.38	1.69	1.99	2.29	2.68	3.15	3.62
30	0.08	0.16	0.43	0.68	0.92	1.15	1.49	1.82	2.15	2.46	2.88	3.40	3.90
32	0.09	0.17	0.46	0.73	0.98	1.23	1.60	1.95	2.30	2.64	3.09	3.64	4.18
35	0.10	0.19	0.51	0.80	1.08	1.36	1.76	2.15	2.53	2.91	3.41	4.01	4.61
40	0.12	0.22	0.58	0.92	1.25	1.57	2.03	2.48	2.93	3.36	3.93	4.64	5.32
45	0.13	0.25	0.66	1.05	1.42	1.78	2.31	2.82	3.32	3.82	4.47	5.26	6.05
	Type A						Type B						

* For rpm above 3500, see ANSI B29.1-1975.

2308 TRANSMISSION ROLLER CHAIN

Table 15 *(Continued).* **Horsepower Ratings for Roller Chain—1975**

⅜-inch Pitch Standard Single-Strand Roller Chain—No. 35

No. of Teeth Small Spkt.	50	100	300	500	700	900	1200	1500	1800	2100	2500	3000	3500
	Revolutions per Minute—Small Sprocket*												
	Horsepower Rating												
9	0.08	0.15	0.39	0.62	0.84	1.06	1.37	1.68	1.98	2.27	2.65	2.17	1.73
10	0.09	0.16	0.44	0.70	0.95	1.19	1.54	1.88	2.21	2.54	2.97	2.55	2.02
11	0.10	0.18	0.49	0.77	1.05	1.31	1.70	2.08	2.45	2.82	3.30	2.94	2.33
12	0.11	0.20	0.54	0.85	1.15	1.44	1.87	2.29	2.70	3.10	3.62	3.35	2.66
13	0.12	0.22	0.59	0.93	1.26	1.57	2.04	2.49	2.94	3.38	3.95	3.77	3.00
14	0.13	0.24	0.63	1.01	1.36	1.71	2.21	2.70	3.18	3.66	4.28	4.22	3.35
15	0.14	0.25	0.68	1.08	1.47	1.84	2.38	2.91	3.43	3.94	4.61	4.68	3.71
16	0.15	0.27	0.73	1.16	1.57	1.97	2.55	3.12	3.68	4.22	4.94	5.15	4.09
17	0.16	0.29	0.78	1.24	1.68	2.10	2.73	3.33	3.93	4.51	5.28	5.64	4.48
18	0.17	0.31	0.83	1.32	1.78	2.24	2.90	3.54	4.18	4.80	5.61	6.15	4.88
19	0.18	0.33	0.88	1.40	1.89	2.37	3.07	3.76	4.43	5.09	5.95	6.67	5.29
20	0.19	0.35	0.93	1.48	2.00	2.51	3.25	3.97	4.68	5.38	6.29	7.20	5.72
21	0.20	0.37	0.98	1.56	2.11	2.64	3.42	4.19	4.93	5.67	6.63	7.75	6.15
22	0.21	0.38	1.03	1.64	2.22	2.78	3.60	4.40	5.19	5.96	6.97	8.21	6.59
23	0.22	0.40	1.08	1.72	2.33	2.92	3.78	4.62	5.44	6.25	7.31	8.62	7.05
24	0.23	0.42	1.14	1.80	2.44	3.05	3.96	4.84	5.70	6.55	7.66	9.02	7.51
25	0.24	0.44	1.19	1.88	2.55	3.19	4.13	5.05	5.95	6.84	8.00	9.43	7.99
26	0.25	0.46	1.24	1.96	2.66	3.33	4.31	5.27	6.21	7.14	8.35	9.84	8.47
28	0.27	0.50	1.34	2.12	2.88	3.61	4.67	5.71	6.73	7.73	9.05	10.7	9.47
30	0.29	0.54	1.45	2.29	3.10	3.89	5.03	6.15	7.25	8.33	9.74	11.5	10.5
32	0.31	0.58	1.55	2.45	3.32	4.17	5.40	6.77	7.77	8.93	10.4	12.3	11.6
35	0.34	0.64	1.71	2.70	3.66	4.59	5.95	7.27	8.56	9.84	11.5	13.6	13.2
40	0.39	0.73	1.97	3.12	4.23	5.30	6.87	8.40	9.89	11.4	13.3	15.7	16.2
45	0.45	0.83	2.24	3.55	4.80	6.02	7.80	9.53	11.2	12.9	15.1	17.8	19.3
	Type A		Type B								Type C		

½-inch Pitch Standard Single-Strand Roller Chain—No. 40

No. of Teeth Small Spkt.	50	100	200	300	400	500	700	900	1000	1200	1400	1600	1800
	Revolutions per Minute—Small Sprocket*												
	Horsepower Rating												
9	0.19	0.35	0.65	0.93	1.21	1.48	2.00	2.51	2.75	3.25	3.73	4.12	3.45
10	0.21	0.39	0.73	1.04	1.35	1.65	2.24	2.81	3.09	3.64	4.18	4.71	4.04
11	0.23	0.43	0.80	1.16	1.50	1.83	2.48	3.11	3.42	4.03	4.63	5.22	4.66
12	0.25	0.47	0.88	1.27	1.65	2.01	2.73	3.42	3.76	4.43	5.09	5.74	5.31
13	0.28	0.52	0.96	1.39	1.80	2.20	2.97	3.73	4.10	4.83	5.55	6.26	5.99
14	0.30	0.56	1.04	1.50	1.95	2.38	3.22	4.04	4.44	5.23	6.01	6.78	6.70
15	0.32	0.60	1.12	1.62	2.10	2.56	3.47	4.35	4.78	5.64	6.47	7.30	7.43
16	0.35	0.65	1.20	1.74	2.25	2.75	3.72	4.66	5.13	6.04	6.94	7.83	8.18
17	0.37	0.69	1.29	1.85	2.40	2.93	3.97	4.98	5.48	6.45	7.41	8.36	8.96
18	0.39	0.73	1.37	1.97	2.55	3.12	4.22	5.30	5.82	6.86	7.88	8.89	9.76
19	0.42	0.78	1.45	2.09	2.71	3.31	4.48	5.62	6.17	7.27	8.36	9.42	10.5
20	0.44	0.82	1.53	2.21	2.86	3.50	4.73	5.94	6.53	7.69	8.83	9.96	11.1
21	0.46	0.87	1.62	2.33	3.02	3.69	4.99	6.26	6.88	8.11	9.31	10.5	11.7
22	0.49	0.91	1.70	2.45	3.17	3.88	5.25	6.58	7.23	8.52	9.79	11.0	12.3
23	0.51	0.96	1.78	2.57	3.33	4.07	5.51	6.90	7.59	8.94	10.3	11.6	12.9
24	0.54	1.00	1.87	2.69	3.48	4.26	5.76	7.23	7.95	9.36	10.8	12.1	13.5
25	0.56	1.05	1.95	2.81	3.64	4.45	6.02	7.55	8.30	9.78	11.2	12.7	14.1
26	0.58	1.09	2.04	2.93	3.80	4.64	6.28	7.88	8.66	10.2	11.7	13.2	14.7
28	0.63	1.18	2.20	3.18	4.11	5.03	6.81	8.54	9.39	11.1	12.7	14.3	15.9
30	0.68	1.27	2.38	3.42	4.43	5.42	7.33	9.20	10.1	11.9	13.7	15.4	17.2
32	0.73	1.36	2.55	3.67	4.75	5.81	7.86	9.86	10.8	12.8	14.7	16.5	18.4
35	0.81	1.50	2.81	4.04	5.24	6.40	8.66	10.9	11.9	14.1	16.2	18.2	20.3
40	0.93	1.74	3.24	4.67	6.05	7.39	10.0	12.5	13.8	16.3	18.7	21.1	23.4
45	1.06	1.97	3.68	5.30	6.87	8.40	11.4	14.2	15.7	18.5	21.2	23.9	26.6
	Type A		Type B								Type C		

For use of table see page 2307.
*For lower or higher rpms see ANSI B29.1-1975.

Table 15 *(Continued).* **Horsepower Ratings for Roller Chain—1975**

½-inch Pitch Light Weight Machinery Roller Chain—No. 41

No. of Teeth Small Spkt.	Revolutions per Minute—Small Sprocket*												
	10	25	50	100	200	300	400	500	700	900	1000	1200	1400
	Horsepower Rating												
9	0.02	0.05	0.10	0.19	0.36	0.51	0.66	0.81	1.10	1.38	1.52	1.27	1.01
10	0.03	0.06	0.11	0.21	0.40	0.57	0.74	0.91	1.23	1.54	1.70	1.49	1.18
11	0.03	0.07	0.13	0.24	0.44	0.64	0.82	1.01	1.37	1.71	1.88	1.71	1.36
12	0.03	0.07	0.14	0.26	0.49	0.70	0.91	1.11	1.50	1.88	2.07	1.95	1.55
13	0.04	0.08	0.15	0.28	0.53	0.76	0.99	1.21	1.63	2.05	2.25	2.20	1.75
14	0.04	0.09	0.16	0.31	0.57	0.83	1.07	1.31	1.77	2.22	2.44	2.46	1.95
15	0.04	0.09	0.18	0.33	0.62	0.89	1.15	1.41	1.91	2.39	2.63	2.73	2.17
16	0.04	0.10	0.19	0.36	0.66	0.95	1.24	1.51	2.05	2.57	2.82	3.01	2.39
17	0.05	0.11	0.20	0.38	0.71	1.02	1.32	1.61	2.18	2.74	3.01	3.29	2.61
18	0.05	0.12	0.22	0.40	0.75	1.08	1.40	1.72	2.32	2.91	3.20	3.59	2.85
19	0.05	0.12	0.23	0.43	0.80	1.15	1.49	1.82	2.46	3.09	3.40	3.89	3.09
20	0.06	0.13	0.24	0.45	0.84	1.21	1.57	1.92	2.60	3.26	3.59	4.20	3.33
21	0.06	0.14	0.26	0.48	0.89	1.28	1.66	2.03	2.74	3.44	3.78	4.46	3.59
22	0.06	0.14	0.27	0.50	0.93	1.35	1.74	2.13	2.89	3.62	3.98	4.69	3.85
23	0.06	0.15	0.28	0.53	0.98	1.41	1.83	2.24	3.03	3.80	4.17	4.92	4.11
24	0.07	0.16	0.29	0.55	1.03	1.48	1.92	2.34	3.17	3.97	4.37	5.15	4.38
25	0.07	0.17	0.31	0.57	1.07	1.55	2.00	2.45	3.31	4.15	4.57	5.38	4.66
26	0.07	0.17	0.32	0.60	1.12	1.61	2.09	2.55	3.46	4.33	4.76	5.61	4.94
28	0.08	0.19	0.35	0.65	1.21	1.75	2.26	2.77	3.74	4.69	5.16	6.08	5.52
30	0.08	0.20	0.38	0.70	1.31	1.88	2.44	2.98	4.03	5.06	5.56	6.55	6.13
32	0.09	0.22	0.40	0.75	1.40	2.02	2.61	3.20	4.33	5.42	5.96	7.03	6.75
35	0.10	0.24	0.44	0.83	1.54	2.22	2.88	3.52	4.76	5.97	6.57	7.74	7.72
40	0.12	0.27	0.51	0.96	1.78	2.57	3.33	4.07	5.50	6.90	7.59	8.94	9.43
45	0.14	0.31	0.58	1.08	2.02	2.92	3.78	4.62	6.25	7.84	8.62	10.2	11.3
	Type A				Type B								T'pC

⅝-inch Pitch Standard Single-Strand Roller Chain—No. 50

No. of Teeth Small Spkt.	Revolutions per Minute—Small Sprocket*												
	25	50	100	200	300	400	500	700	900	1000	1200	1400	1600
	Horsepower Rating												
9	0.19	0.36	0.67	1.26	1.81	2.35	2.87	3.89	4.88	5.36	6.32	6.02	4.92
10	0.22	0.41	0.76	1.41	2.03	2.63	3.22	4.36	5.46	6.01	7.08	7.05	5.77
11	0.24	0.45	0.84	1.56	2.25	2.92	3.57	4.83	6.06	6.66	7.85	8.13	6.65
12	0.26	0.49	0.92	1.72	2.47	3.21	3.92	5.31	6.65	7.31	8.62	9.26	7.58
13	0.29	0.54	1.00	1.87	2.70	3.50	4.27	5.78	7.25	7.97	9.40	10.4	8.55
14	0.31	0.58	1.09	2.03	2.92	3.79	4.63	6.27	7.86	8.64	10.2	11.7	9.55
15	0.34	0.63	1.17	2.19	3.15	4.08	4.99	6.75	8.47	9.31	11.0	12.6	10.6
16	0.36	0.67	1.26	2.34	3.38	4.37	5.35	7.24	9.08	9.98	11.8	13.5	11.7
17	0.39	0.72	1.34	2.50	3.61	4.67	5.71	7.73	9.69	10.7	12.6	14.4	12.8
18	0.41	0.76	1.43	2.66	3.83	4.97	6.07	8.22	10.3	11.3	13.4	15.3	13.9
19	0.43	0.81	1.51	2.82	4.07	5.27	6.44	8.72	10.9	12.0	14.2	16.3	15.1
20	0.46	0.86	1.60	2.98	4.30	5.57	6.80	9.21	11.5	12.7	15.0	17.2	16.3
21	0.48	0.90	1.69	3.14	4.53	5.87	7.17	9.71	12.2	13.4	15.8	18.1	17.6
22	0.51	0.95	1.77	3.31	4.76	6.17	7.54	10.2	12.8	14.1	16.6	19.1	18.8
23	0.53	1.00	1.86	3.47	5.00	6.47	7.91	10.7	13.4	14.8	17.4	20.0	20.1
24	0.56	1.04	1.95	3.63	5.23	6.78	8.29	11.2	14.1	15.5	18.2	20.9	21.4
25	0.58	1.09	2.03	3.80	5.47	7.08	8.66	11.7	14.7	16.2	19.0	21.9	22.8
26	0.61	1.14	2.12	3.96	5.70	7.39	9.03	12.2	15.3	16.9	19.9	22.8	24.2
28	0.66	1.23	2.30	4.29	6.18	8.01	9.79	13.2	16.6	18.3	21.5	24.7	27.0
30	0.71	1.33	2.48	4.62	6.66	8.63	10.5	14.3	17.9	19.7	23.2	26.6	30.0
32	0.76	1.42	2.66	4.96	7.14	9.25	11.3	15.3	19.2	21.1	24.9	28.6	32.2
35	0.84	1.57	2.93	5.46	7.86	10.2	12.5	16.9	21.1	23.2	27.4	31.5	35.5
40	0.97	1.81	3.38	6.31	9.08	11.8	14.4	19.5	24.4	26.8	31.6	36.3	41.0
45	1.10	2.06	3.84	7.16	10.3	13.4	16.3	22.1	27.7	30.5	35.9	41.3	46.5
	Type A				Type B				Type C				

For use of table see page 2307.
*For lower or higher rpms see ANSI B29.1-1975.

Table 15 *(Continued)*. **Horsepower Ratings for Roller Chain—1975**

¾-inch Pitch Standard Single-Strand Roller Chain—No. 60

No. of Teeth Small Spkt.	Revolutions per Minute—Small Sprocket*												
	25	50	100	150	200	300	400	500	600	700	800	900	1000
	Horsepower Rating												
9	0.33	0.62	1.16	1.67	2.16	3.12	4.04	4.94	5.82	6.68	7.54	8.38	9.21
10	0.37	0.70	1.30	1.87	2.43	3.49	4.53	5.53	6.52	7.49	8.44	9.39	10.3
11	0.41	0.77	1.44	2.07	2.69	3.87	5.02	6.13	7.23	8.30	9.36	10.4	11.4
12	0.45	0.85	1.58	2.28	2.95	4.25	5.51	6.74	7.94	9.12	10.3	11.4	12.6
13	0.50	0.92	1.73	2.49	3.22	4.64	6.01	7.34	8.65	9.94	11.2	12.5	13.7
14	0.54	1.00	1.87	2.69	3.49	5.02	6.51	7.96	9.37	10.8	12.1	13.5	14.8
15	0.58	1.08	2.01	2.90	3.76	5.41	7.01	8.57	10.1	11.6	13.1	14.5	16.0
16	0.62	1.16	2.16	3.11	4.03	5.80	7.52	9.19	10.8	12.4	14.0	15.6	17.1
17	0.66	1.24	2.31	3.32	4.30	6.20	8.03	9.81	11.6	13.3	15.0	16.7	18.3
18	0.70	1.31	2.45	3.53	4.58	6.59	8.54	10.4	12.3	14.1	15.9	17.7	19.5
19	0.75	1.39	2.60	3.74	4.85	6.99	9.05	11.1	13.0	15.0	16.9	18.8	20.6
20	0.79	1.47	2.75	3.96	5.13	7.38	9.57	11.7	13.8	15.8	17.9	19.8	21.8
21	0.83	1.55	2.90	4.17	5.40	7.78	10.1	12.3	14.5	16.7	18.8	20.9	23.0
22	0.87	1.63	3.05	4.39	5.68	8.19	10.6	13.0	15.3	17.5	19.8	22.0	24.2
23	0.92	1.71	3.19	4.60	5.96	8.59	11.1	13.6	16.0	18.4	20.8	23.1	25.4
24	0.96	1.79	3.35	4.82	6.24	8.99	11.6	14.2	16.8	19.3	21.7	24.2	26.6
25	1.00	1.87	3.50	5.04	6.52	9.40	12.2	14.9	17.5	20.1	22.7	25.3	27.8
26	1.05	1.95	3.65	5.25	6.81	9.80	12.7	15.5	18.3	21.0	23.7	26.4	29.0
28	1.13	2.12	3.95	5.69	7.37	10.6	13.8	16.8	19.8	22.8	25.7	28.5	31.4
30	1.22	2.28	4.26	6.13	7.94	11.4	14.8	18.1	21.4	24.5	27.7	30.8	33.8
32	1.31	2.45	4.56	6.57	8.52	12.3	15.9	19.4	22.9	26.3	29.7	33.0	36.3
35	1.44	2.69	5.03	7.24	9.38	13.5	17.5	21.4	25.2	29.0	32.7	36.3	39.9
40	1.67	3.11	5.81	8.37	10.8	15.6	20.2	24.7	29.1	33.5	37.7	42.0	46.1
45	1.89	3.53	6.60	9.50	12.3	17.7	23.0	28.1	33.1	38.0	42.9	47.7	52.4
	Type A			Type B							Type C		

1-inch Pitch Standard Single-Strand Roller Chain—No. 80

No. of Teeth Small Spkt.	Revolutions per Minute—Small Sprocket*												
	25	50	100	150	200	300	400	500	600	700	800	900	1000
	Horsepower Rating												
9	0.78	1.45	2.71	3.90	5.05	7.28	9.43	11.5	13.6	15.6	17.6	17.0	14.5
10	0.87	1.63	3.03	4.37	5.66	8.16	10.6	12.9	15.2	17.5	19.7	19.9	17.0
11	0.97	1.80	3.36	4.84	6.28	9.04	11.7	14.3	16.9	19.4	21.9	23.0	19.6
12	1.06	1.98	3.69	5.32	6.89	9.93	12.9	15.7	18.5	21.3	24.0	26.2	22.3
13	1.16	2.16	4.03	5.80	7.52	10.8	14.0	17.1	20.2	23.2	26.2	29.1	25.2
14	1.25	2.34	4.36	6.29	8.14	11.7	15.2	18.6	21.9	25.1	28.4	31.5	28.2
15	1.35	2.52	4.70	6.77	8.77	12.6	16.4	20.0	23.6	27.1	30.6	34.0	31.2
16	1.45	2.70	5.04	7.26	9.41	13.5	17.6	21.5	25.3	29.0	32.8	36.4	34.4
17	1.55	2.88	5.38	7.75	10.0	14.5	18.7	22.9	27.0	31.0	35.0	38.9	37.7
18	1.64	3.07	5.72	8.25	10.7	15.4	19.9	24.4	28.7	33.0	37.2	41.4	41.1
19	1.74	3.25	6.07	8.74	11.3	16.3	21.1	25.8	30.4	35.0	39.4	43.8	44.5
20	1.84	3.44	6.41	9.24	12.0	17.2	22.3	27.3	32.2	37.0	41.7	46.3	48.1
21	1.94	3.62	6.76	9.74	12.6	18.2	23.5	28.8	33.9	39.0	43.9	48.9	51.7
22	2.04	3.81	7.11	10.2	13.3	19.1	24.8	30.3	35.7	41.0	46.2	51.4	55.5
23	2.14	4.00	7.46	10.7	13.9	20.1	26.0	31.8	37.4	43.0	48.5	53.9	59.3
24	2.24	4.19	7.81	11.3	14.6	21.0	27.2	33.2	39.2	45.0	50.8	56.4	62.0
25	2.34	4.37	8.16	11.8	15.2	21.9	28.4	34.7	40.9	47.0	53.0	59.0	64.8
26	2.45	4.56	8.52	12.3	15.9	22.9	29.7	36.2	42.7	49.1	55.3	61.5	67.6
28	2.65	4.94	9.23	13.3	17.2	24.8	32.1	39.3	46.3	53.2	59.9	66.7	73.3
30	2.85	5.33	9.94	14.3	18.5	26.7	34.6	42.3	49.9	57.3	64.6	71.8	78.9
32	3.06	5.71	10.7	15.3	19.9	28.6	37.1	45.4	53.5	61.4	69.2	77.0	84.6
35	3.37	6.29	11.7	16.9	21.9	31.6	40.9	50.0	58.9	67.6	76.3	84.8	93.3
40	3.89	7.27	13.6	19.5	25.3	36.4	47.2	57.7	68.0	78.1	88.1	99.0	108
45	4.42	8.25	15.4	22.2	28.7	41.4	53.6	65.6	77.2	88.7	100	111	122
	T'p A	Type B							Type C				

For use of table see page 2307.
*For lower or higher rpms see ANSI B29.1-1975.

Table 15 *(Concluded).* **Horsepower Ratings for Roller Chain—1975**

1¼-inch Pitch Standard Single-Strand Roller Chain—No. 100

No. of Teeth Small Spkt.	Revolutions per Minute—Small Sprocket*												
	10	25	50	100	150	200	300	400	500	600	700	800	900
	Horsepower Rating												
9	0.65	1.49	2.78	5.19	7.47	9.68	13.9	18.1	22.1	26.0	29.6	24.2	20.3
10	0.73	1.67	3.11	5.81	8.37	10.8	15.6	20.2	24.7	29.2	33.5	28.4	23.8
11	0.81	1.85	3.45	6.44	9.28	12.0	17.3	22.4	27.4	32.3	37.1	32.8	27.5
12	0.89	2.03	3.79	7.08	10.2	13.2	19.0	24.6	30.1	35.5	40.8	37.3	31.3
13	0.97	2.22	4.13	7.72	11.1	14.4	20.7	26.9	32.8	38.7	44.5	42.1	35.3
14	1.05	2.40	4.48	8.36	12.0	15.6	22.5	29.1	35.6	41.9	48.2	47.0	39.4
15	1.13	2.59	4.83	9.01	13.0	16.8	24.2	31.4	38.3	45.2	51.9	52.2	43.7
16	1.22	2.77	5.17	9.66	13.9	18.0	26.0	33.6	41.1	48.4	55.6	57.5	48.2
17	1.30	2.96	5.52	10.3	14.8	19.2	27.7	35.9	43.9	51.7	59.4	63.0	52.8
18	1.38	3.15	5.88	11.0	15.8	20.5	29.5	38.2	46.7	55.0	63.2	68.6	57.5
19	1.46	3.34	6.23	11.6	16.7	21.7	31.2	40.5	49.5	58.3	67.0	74.4	62.3
20	1.55	3.53	6.58	12.3	17.7	22.9	33.0	42.8	52.3	61.6	70.8	79.8	67.3
21	1.63	3.72	6.94	13.0	18.7	24.2	34.8	45.1	55.1	65.0	74.6	84.2	72.4
22	1.71	3.91	7.30	13.6	19.6	25.4	36.6	47.4	58.0	68.3	78.5	88.5	77.7
23	1.80	4.10	7.66	14.3	20.6	26.7	38.4	49.8	60.8	71.7	82.3	92.8	83.0
24	1.88	4.30	8.02	15.0	21.5	27.9	40.2	52.1	63.7	75.0	86.2	97.2	88.5
25	1.97	4.49	8.38	15.6	22.5	29.2	42.0	54.4	66.6	78.4	90.1	102	94.1
26	2.05	4.68	8.74	16.3	23.5	30.4	43.8	56.8	69.4	81.8	94.0	106	99.8
28	2.22	5.07	9.47	17.7	25.5	33.0	47.5	61.5	75.2	88.6	102	115	112
30	2.40	5.47	10.2	19.0	27.4	35.5	51.2	66.3	81.0	95.5	110	124	124
32	2.57	5.86	10.9	20.4	29.4	38.1	54.9	71.1	86.9	102	118	133	136
35	2.83	6.46	12.0	22.5	32.4	42.0	60.4	78.3	95.7	113	130	146	156
40	3.27	7.46	13.9	26.0	37.4	48.5	69.8	90.4	111	130	150	169	188
45	3.71	8.47	15.8	29.5	42.5	55.0	79.3	103	126	148	170	192	213
	Type A		Type B					Type C				

1½-inch Pitch Standard Single-Strand Roller Chain—No. 120

No. of Teeth Small Spkt.	Revolutions per Minute—Small Sprocket*												
	10	25	50	100	150	200	300	400	500	600	700	800	900
	Horsepower Rating												
9	1.10	2.52	4.69	8.76	12.6	16.3	23.5	30.5	37.3	43.2	34.3	28.1	23.5
10	1.24	2.82	5.26	9.81	14.1	18.3	26.4	34.2	41.8	49.2	40.1	32.9	27.5
11	1.37	3.12	5.83	10.9	15.7	20.3	29.2	37.9	46.3	54.6	46.3	37.9	31.8
12	1.50	3.43	6.40	11.9	17.2	22.3	32.1	41.6	50.9	59.9	52.8	43.2	36.2
13	1.64	3.74	6.98	13.0	18.8	24.3	35.0	45.4	55.5	65.3	59.5	48.7	40.8
14	1.78	4.05	7.56	14.1	20.3	26.3	37.9	49.1	60.1	70.8	66.5	54.4	45.6
15	1.91	4.37	8.15	15.2	21.9	28.4	40.9	53.0	64.7	76.3	73.8	60.4	50.6
16	2.05	4.68	8.74	16.3	23.5	30.4	43.8	56.8	69.4	81.8	81.3	66.5	55.7
17	2.19	5.00	9.33	17.4	25.1	32.5	46.8	60.6	74.1	87.3	89.0	72.8	61.0
18	2.33	5.32	9.92	18.5	26.7	34.6	49.8	64.5	78.8	92.9	97.0	79.4	66.5
19	2.47	5.64	10.5	19.6	28.3	36.6	52.8	68.4	83.6	98.5	105	86.1	72.1
20	2.61	5.96	11.1	20.7	29.9	38.7	55.8	72.2	88.3	104	114	92.9	77.9
21	2.75	6.28	11.7	21.9	31.5	40.8	58.8	76.2	93.1	110	122	100	83.8
22	2.90	6.60	12.3	23.0	33.1	42.9	61.8	80.1	97.9	115	131	107	89.9
23	3.04	6.93	12.9	24.1	34.8	45.0	64.9	84.0	103	121	139	115	96.1
24	3.18	7.25	13.5	25.3	36.4	47.1	67.9	88.0	108	127	146	122	102
25	3.32	7.58	14.1	26.4	38.0	49.3	71.0	91.9	112	132	152	130	109
26	3.47	7.91	14.8	27.5	39.7	51.4	74.0	95.9	117	138	159	138	115
28	3.76	8.57	16.0	29.8	43.0	55.7	80.2	104	127	150	172	154	129
30	4.05	9.23	17.2	32.1	46.3	60.0	86.4	112	137	161	185	171	143
32	4.34	9.90	18.5	34.5	49.6	64.3	92.6	120	147	173	199	188	158
35	4.78	10.9	20.3	38.0	54.7	70.9	102	132	162	190	219	215	180
40	5.52	12.6	23.5	43.9	63.2	81.8	118	153	187	220	253	...	...
45	6.27	14.3	26.7	49.8	71.7	92.9	134	173	212	250	287	...	...
	T'p A		Type B					Type C				

*For higher rpms and larger chain sizes see ANSI B29.1-1975.

Example of Roller Chain Drive Design Procedure — The selection of a roller chain and sprockets for a specific design requirement is best accomplished by a systematic step-by-step procedure such as is used in the following example.

Example: Select a roller chain drive to transmit 10 horsepower from a countershaft to the main shaft of a wire drawing machine. The countershaft is 1¹⁵⁄₁₆-inches diameter and operates at 1000 rpm. The main shaft is also 1¹⁵⁄₁₆-inches diameter and must operate between 378 and 382 rpm. Shaft centers, once established, are fixed and by initial calculations must be approximately 22½ inches. The load on the main shaft is uneven and presents "peaks," which place it in the heavy shock load category. The input power is supplied by an electric motor. The driving head is fully enclosed and all parts are lubricated from a central system.

Step 1. Service Factor: From Table 14 the service factor for heavy shock load and an electric motor drive is 1.5.

Step 2. Design Horsepower: The horsepower upon which the chain selection is based (design horsepower) is equal to the specified horsepower multiplied by the service factor, 10 × 1.5 = 15 hp.

Step 3. Chain Pitch and Small Sprocket Size for Single-Strand Drive: In Table 15 under 1000 rpm, a ⅝-inch pitch chain with a 24-tooth sprocket or a ¾-inch pitch chain with a 15-tooth sprocket are possible choices.

Step 4. Check of Chain Pitch and Sprocket Selection: From Table 9 it is seen that only the 24-tooth sprocket in Step 3 can be bored to fit the 1¹⁵⁄₁₆-inch diameter main shaft. In Table 15 a ⅝-pitch chain at a small sprocket speed of 1000 rpm is rated at 15.5 hp for a 24-tooth sprocket.

Step 5. Selection of Large Sprocket: Since the driver is to operate at 1000 rpm and the driven at a minimum of 378 rpm, the speed ratio 1000/378 = 2.646. Therefore the large sprocket should have 24 × 2.646 = 63.5 (use 63) teeth.

This combination of 24 and 63 teeth will produce a main drive shaft speed of 381 rpm which is within the limitation of 378 to 382 rpm established in the original specification.

Step 6. Computation of Chain Length: Since the 24- and 63-tooth sprockets are to be placed on 22½-inch centers, the chain length is determined from the formula:

$$L = 2C + \frac{N}{2} + \frac{n}{2} + \left(\frac{N-n}{2\pi}\right)^2 \times \frac{1}{C}$$

where L = chain length in pitches; C = shaft center distance in pitches; N = number of teeth in large sprocket; and n = number of teeth in small sprocket.

$$L = 2 \times 36 + \frac{63+24}{2} + \left(\frac{63-24}{6.28}\right)^2 \times \frac{1}{36} = 116.57 \text{ pitches}$$

Step 7. Correction of Center Distance: Since the chain is to couple at a whole number of pitches, 116 pitches will be used and the center distance recomputed based on this figure using the formula on page 2299 where c is the center distance in inches and P is the pitch.

$$c = \frac{P}{8}\left(2L - N - n + \sqrt{(2L-N-n)^2 - 0.810(N-n)^2}\right)$$

$$c = \frac{5}{64}\left(2 \times 116 - 63 - 24 + \sqrt{(2 \times 116 - 63 - 24)^2 - 0.810(63-24)^2}\right)$$

$$c = \frac{5}{64}(145 + 140.69) = 22.32 \text{ inches, say } 22\frac{3}{8} \text{ inches.}$$

Silent or Inverted Tooth Chain

Silent or inverted tooth chain consists of a series of toothed links alternately assembled either with pins or with a combination of joint components in such a way that the joints articulate between adjoining pitches. *Side Guide* chain has guide links which straddle the sprocket sides to control the chain laterally. *Center Guide* chain has guide links that run within a circumferential groove or grooves for lateral control.

Characteristics of Silent Chain Drives. — The silent or "inverted-tooth" driving chain has the following characteristics: The chain passes over the face of the wheel like a belt and the wheel teeth do not project through it; the chain engages the wheel by means of teeth extending across the full width of the under side, with the exception of those chains having a central guide link; the chain teeth and wheel teeth are of such a shape that as the chain pitch increases through wear at the joints, the chain shifts outward upon the teeth, thus engaging the wheel on a pitch circle of increasing diameter; the result of this action is that the pitch of the wheel teeth increases at the same rate as the chain pitch. The accompanying illustration shows an unworn chain to the left, and a worn chain to the right, which has moved outward as the result of

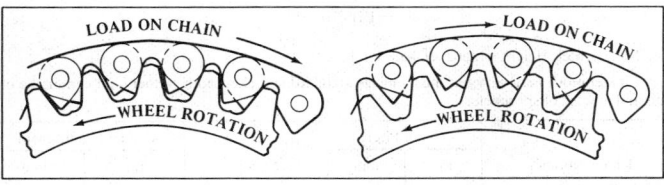

wear. Another distinguishing feature of the silent chain is that the power is transmitted by and to all the teeth in the arc of contact, irrespective of the increasing pitch due to elongation. The links have no sliding action either on or off the teeth, which results in a smooth and practically noiseless action, the chain being originally designed for the transmission of power at higher speeds than are suitable for roller chains. The efficiency of the silent chain itself may be as high as 99 per cent, and for the complete drive, from 96 to 97 per cent, under favorable conditions; from 94 to 96 per cent can be secured with well-designed drives under average conditions.

The life and upkeep of silent chains depend largely upon the design of the entire drive, including the provision for adjustment. If there is much slack, the whipping of the chain will greatly increase the wear, and means of adjustment may double the life of the chain. A slight amount of play is necessary for satisfactory operation. The minimum amount of sag should be about ⅛ inch. Although the silent chain shifts outward from the teeth and adjusts itself for an increase of pitch, it cannot take up the increased pitch in that portion of the chain between the wheels; therefore, the wheel must lag to the extent of the increased pitch in the straight portion of the chain.

Standard Silent Chain Designation. — The standard chain number or designation for ⅜-inch pitch or larger consists of: (1) a two letter symbol SC; (2) one or two numerical digits indicating the pitch in eighths of an inch; and (3) two or three numerical digits indicating the chain width in quarter-inches. Thus, SC302 designates a silent chain of ⅜-inch pitch and ½-inch width, while SC1012 designates a silent chain of 1¼-inch pitch and 3-inch width.

The standard chain number or designation for $\frac{3}{16}$-inch pitch consists of: (a) a two letter symbol SC; (b) a zero followed by a numerical digit indicating pitch in sixteenths of an inch; and (c) two numerical digits indicating the chain width in thirty-seconds of an inch. Thus, SC0309 designates a silent chain of $\frac{3}{16}$-inch pitch and $\frac{9}{32}$-inch width.

Silent Chain Links. — The joint components and link contours vary with each manufacturer's design. As shown in Table 1, minimum crotch height and pitch have been standardized for interchangeability. Chain link designations for $\frac{3}{8}$-inch and larger pitch are given in Table 1.

Table 1. American National Standard Silent Chain Links* (ANSI B29.2M-1982)

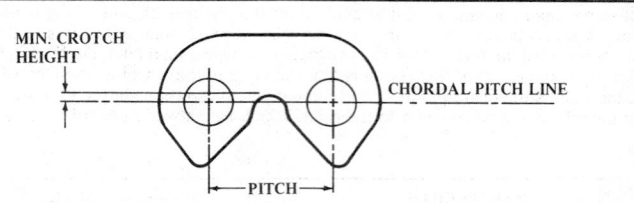

Min. Crotch Height = 0.062 × Chain Pitch.

Link contour may vary but must engage standard sprocket tooth so that joint centers lie on pitch diameter of sprockets.

Chain Number	Chain Pitch		Stamp	Crotch Height, Min.	
	in.	mm		in.	mm
SC3 (Width in ¼ in.)	0.375	9.52	SC3 or 3	0.0232	0.589
SC4 "	0.500	12.70	SC4 or 4	0.0310	0.787
SC5 "	0.625	15.88	SC5 or 5	0.0388	0.985
SC6 "	0.750	19.05	SC6 or 6	0.0465	1.181
SC8 "	1.000	25.40	SC8 or 8	0.0620	1.574
SC10 "	1.250	31.76	SC10 or 10	0.0775	1.968
SC12 "	1.500	38.10	SC12 or 12	0.0930	2.302
SC16 "	2.000	50.80	SC16 or 16	0.1240	3.149

* For $\frac{3}{8}$-inch and larger pitch chains.

Silent Chain Sprocket Diameters. — The important sprocket diameters are: (1) outside diameter; (2) pitch diameter; (3) maximum guide groove diameter; and (4) over-pin diameter. These are shown in the diagram in Table 2 and the symbols and formulas for each are also given in this table. Table 3A gives values of outside diameters for sprockets with rounded teeth and with square teeth, pitch diameters, and over-pin diameters for chains of 1-inch pitch and sprockets of various tooth numbers. Values for chains of other pitches ($\frac{3}{8}$ inch and larger) are found by multiplying the values shown by the pitch. Table 3B gives this information for $\frac{3}{16}$-in. pitch chains. Note that the over-pin diameter is measured over gage pins having a diameter $D_p = 0.625P$ in. for $\frac{3}{8}$-in. and larger pitch and $D_p = 0.667P$ in. for $\frac{3}{16}$-in. pitch chains. Over-pin diameter tolerances are given in Tables 4A and 4B.

Silent Chain Sprocket Profiles and Chain Widths. — Sprocket tooth face profiles for side guide chain, center guide chain and double guide chain are shown in Table 5

(Continued on page 2317)

Table 2. ANSI Silent Chain Sprocket Diameters (ANSI B29.2M-1982)

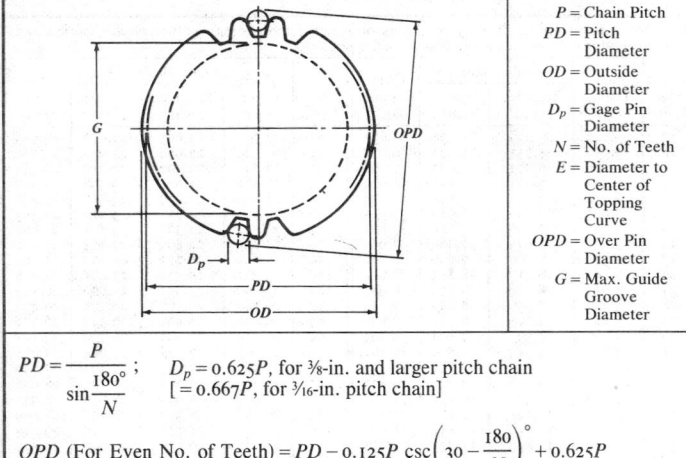

P = Chain Pitch
PD = Pitch Diameter
OD = Outside Diameter
D_p = Gage Pin Diameter
N = No. of Teeth
E = Diameter to Center of Topping Curve
OPD = Over Pin Diameter
G = Max. Guide Groove Diameter

$$PD = \frac{P}{\sin\frac{180°}{N}} \; ; \qquad D_p = 0.625P, \text{ for } \tfrac{3}{8}\text{-in. and larger pitch chain} \\ [= 0.667P, \text{ for } \tfrac{3}{16}\text{-in. pitch chain}]$$

$$OPD \text{ (For Even No. of Teeth)} = PD - 0.125P \csc\left(30 - \frac{180}{N}\right)° + 0.625P$$

$$\left[= PD - 0.160P \csc\left(35 - \frac{180}{N}\right)° + 0.667P\right]*$$

$$OPD \text{ (For Odd No. of Teeth)} = \cos\frac{90°}{N}\left[PD - 0.125P \csc\left(30 - \frac{180}{N}\right)°\right] + 0.625P$$

$$\left[= \cos\frac{90°}{N}\left[PD - 0.160P \csc\left(35 - \frac{180}{N}\right)°\right] + 0.667P\right]*$$

$$OD \text{ (For Rounded Teeth)} = P\left(\cot\frac{180°}{N} + 0.08\right)$$

$$\left[OD \text{ (For Nominal Rounded Teeth)} = P\left(\cot\frac{180°}{N} - 0.032\right)\right]*$$

$$OD \text{ (For Square Teeth)} = 2\sqrt{X^2 + L^2 - 2XL \cos\alpha}$$

$$\text{Where} \begin{cases} X = Y\cos\alpha - \sqrt{(0.15P)^2 - (Y\sin\alpha)^2} \\ Y = P(0.500 - 0.375\sec\alpha)\cot\alpha + 0.11P \\ L = Y + \frac{E}{2}\text{(See Table 8A for E)}; \; \alpha = \left(30 - \frac{360}{N}\right)° \end{cases}$$

$$G(\text{max.}) = P\left(\cot\frac{180°}{N} - 1.16\right) \qquad \left[= P\left(\cot\frac{180°}{N} - 1.20\right)\right]*$$

Tolerance = + 0, − 0.030 in. (0.76 mm) [= + 0, − 0.015 in. (0.38 mm)]*

* Applies to $\tfrac{3}{16}$-inch pitch chain. All other equations apply to $\tfrac{3}{8}$-inch and larger pitch chains.

Table 3A. American National Standard Silent Chain Sprocket Diameters
(ANSI B29.2M-1982)

These diameters apply only to chains of 1-inch pitch. For any other pitch (⅜ inch and larger) multiply the values given below by the pitch.

No. Teeth	Pitch Diameter	Outside Diameter		Over-Pin Dia.‡	No. Teeth	Pitch Diameter	Outside Diameter		Over-Pin Dia.‡
		Rounded Teeth*	Square Teeth†				Rounded Teeth*	Square Teeth†	
17	5.442	5.429	5.298	5.669	71	22.607	22.665	22.622	22.955
18	5.759	5.751	5.623	6.018	72	22.926	22.984	22.941	23.280
19	6.076	6.072	5.947	6.324	73	23.244	23.302	23.259	23.593
20	6.393	6.393	6.271	6.669	74	23.562	23.621	23.578	23.917
21	6.710	6.714	6.595	6.974	75	23.880	23.939	23.897	24.230
22	7.027	7.036	6.919	7.315	76	24.198	24.257	24.216	24.553
23	7.344	7.356	7.243	7.621	77	24.517	24.577	24.535	24.868
24	7.661	7.675	7.568	7.960	78	24.835	24.895	24.853	25.191
25	7.979	7.996	7.890	8.266	79	25.153	25.213	25.172	25.504
26	8.296	8.315	8.213	8.602	80	25.471	25.531	25.491	25.828
27	8.614	8.636	8.536	8.909	81	25.790	25.851	25.809	26.141
28	8.932	8.956	8.859	9.244	82	26.108	26.169	26.128	26.465
29	9.249	9.275	9.181	9.551	83	26.426	26.487	26.447	26.778
30	9.567	9.595	9.504	9.884	84	26.744	26.805	26.766	27.101
31	9.885	9.913	9.828	10.192	85	27.063	27.125	27.084	27.415
32	10.202	10.233	10.150	10.524	86	27.381	27.443	27.403	27.739
33	10.520	10.553	10.471	10.833	87	27.699	27.761	27.722	28.052
34	10.838	10.872	10.793	11.164	88	28.017	28.079	28.040	28.375
35	11.156	11.191	11.115	11.472	89	28.335	28.397	28.359	28.689
36	11.474	11.510	11.437	11.803	90	28.654	28.716	28.678	29.013
37	11.792	11.829	11.757	12.112	91	28.972	29.035	28.997	29.327
38	12.110	12.149	12.077	12.442	92	29.290	29.353	29.315	29.649
39	12.428	12.468	12.397	12.751	93	29.608	29.671	29.634	29.963
40	12.746	12.787	12.717	13.080	94	29.926	29.989	29.953	30.285
41	13.064	13.106	13.037	13.390	95	30.245	30.308	30.271	30.601
42	13.382	13.425	13.357	13.718	96	30.563	30.627	30.590	30.923
43	13.700	13.743	13.677	14.028	97	30.881	30.945	30.909	31.237
44	14.018	14.062	13.997	14.356	98	31.199	31.263	31.228	31.559
45	14.336	14.381	14.317	14.667	99	31.518	31.582	31.546	31.874
46	14.654	14.700	14.637	14.994	100	31.836	31.900	31.865	32.196
47	14.972	15.018	14.957	15.305	101	32.154	32.218	32.183	32.511
48	15.290	15.337	15.277	15.632	102	32.473	32.537	32.502	32.834
49	15.608	15.656	15.597	15.943	103	32.791	32.856	32.820	33.148
50	15.926	15.975	15.917	16.270	104	33.109	33.174	33.139	33.470
51	16.244	16.293	16.236	16.581	105	33.427	33.492	33.457	33.784
52	16.562	16.612	16.556	16.906	106	33.746	33.811	33.776	34.107
53	16.880	16.930	16.876	17.218	107	34.064	34.129	34.094	34.422
54	17.198	17.249	17.196	17.544	108	34.382	34.447	34.413	34.744
55	17.517	17.568	17.515	17.857	109	34.701	34.767	34.731	35.059
56	17.835	17.887	17.834	18.183	110	35.019	35.084	35.050	35.381
57	18.153	18.205	18.154	18.494	111	35.337	35.403	35.368	35.695
58	18.471	18.524	18.473	18.820	112	35.655	35.721	35.687	36.017
59	18.789	18.842	18.793	19.131	113	35.974	36.040	36.005	36.333
60	19.107	19.161	19.112	19.457	114	36.292	36.358	36.324	36.654
61	19.426	19.480	19.431	19.769	115	36.610	36.676	36.642	36.969
62	19.744	19.799	19.750	20.095	116	36.929	36.995	36.961	37.292
63	20.062	20.117	20.070	20.407	117	37.247	37.313	37.279	37.606
64	20.380	20.435	20.388	20.731	118	37.565	37.632	37.598	37.928
65	20.698	20.754	20.708	21.044	119	37.883	37.950	37.916	38.243
66	21.016	21.072	21.027	21.368	120	38.201	38.268	38.235	38.564
67	21.335	21.391	21.346	21.682	121	38.519	38.586	38.553	38.879
68	21.653	21.710	21.665	22.006	122	38.837	38.904	38.872	39.200
69	21.971	22.028	21.984	22.319	123	39.156	39.223	39.190	39.516
70	22.289	22.347	22.303	22.643	124	39.475	39.542	39.508	39.839

All dimensions in inches.

* Blank diameters are 0.020 inch larger and maximum guide groove diameters G are 1.240 inches smaller than these outside diameters.

† These diameters are maximum; tolerance is $+ 0$, $- 0.50 \times$ pitch, inches.

‡ For tolerances on over-pin diameters, see Table 4.

Tolerance for maximum eccentricity (total indicator reading) of pitch diameter with respect to bore is $0.001 \times PD$, but not less than 0.006 nor more than 0.032 inch.

Table 3A (*Concluded*). **ANSI Silent Chain Sprocket Diameters**
(ANSI B29.2M-1982)

These diameters apply only to chains of 1-inch pitch. For any other pitch (⅜ inch and larger) multiply the values given below by the pitch.

No. Teeth	Pitch Diameter	Outside Diameter		Over-Pin Dia.‡	No. Teeth	Pitch Diameter	Outside Diameter		Over-Pin Dia.‡
		Rounded Teeth*	Square Teeth†				Rounded Teeth*	Square Teeth†	
125	39.794	39.861	39.827	40.154	138	43.930	43.998	43.966	44.295
126	40.112	40.180	40.145	40.476	139	44.249	44.317	44.284	44.611
127	40.430	40.497	40.464	40.790	140	44.567	44.636	44.603	44.932
128	40.748	40.816	40.782	41.112	141	44.885	44.954	44.922	45.247
129	41.066	41.134	41.100	41.427	142	45.203	45.271	45.240	45.568
130	41.384	41.452	41.419	41.748	143	45.521	45.590	45.558	45.883
131	41.702	41.770	41.738	42.063	144	45.840	45.909	45.877	46.205
132	42.020	42.088	42.056	42.384	145	46.158	46.227	46.195	46.520
133	42.338	42.406	42.374	42.699	146	46.477	46.546	46.514	46.842
134	42.656	42.724	42.693	43.020	147	46.796	46.865	46.832	47.159
135	42.975	43.043	43.011	43.336	148	47.114	47.183	47.151	47.479
136	43.293	43.362	43.329	43.657	149	47.432	47.501	47.469	47.795
137	43.611	43.679	43.647	43.972	150	47.750	47.819	47.787	48.116

All dimensions in inches. See page 2316 for footnotes.

Table 3B. American National Standard Silent Chain Sprocket Diameters for ³⁄₁₆-in. Pitch Chain (ANSI B29.2M-1982)

No. Teeth	Pitch Diameter	Outside Diameter*·†	Over-Pin Diameter*·†	Max. Groove Diameter*	No. teeth	Pitch Diameter	Outside Diameter*·†	Over-Pin Diameter*·†	Max. Groove Diameter*
11	0.665	0.632	0.691	0.413	36	2.151	2.136	2.216	1.918
12	0.724	0.694	0.761	0.429	37	2.211	2.196	2.274	1.978
13	0.783	0.755	0.821	0.536	38	2.271	2.256	2.336	2.038
14	0.843	0.815	0.888	0.596	39	2.330	2.315	2.394	2.098
15	0.902	0.876	0.946	0.657	40	2.390	2.376	2.456	2.158
16	0.961	0.937	1.012	0.718	41	2.449	2.435	2.513	2.217
17	1.020	0.996	1.069	0.778	42	2.509	2.495	2.575	2.277
18	1.080	1.057	1.134	0.838	43	2.569	2.555	2.633	2.337
19	1.139	1.116	1.191	0.899	44	2.628	2.614	2.695	2.397
20	1.199	1.177	1.256	0.959	45	2.688	2.674	2.753	2.456
21	1.258	1.237	1.312	1.019	46	2.748	2.735	2.815	2.516
22	1.318	1.298	1.377	1.079	47	2.807	2.794	2.872	2.576
23	1.377	1.357	1.433	1.139	48	2.867	2.854	2.934	2.636
24	1.436	1.417	1.497	1.199	49	2.926	2.913	2.992	2.696
25	1.496	1.477	1.554	1.259	50	2.986	2.973	3.053	2.755
26	1.556	1.538	1.617	1.319	51	3.046	3.033	3.111	2.815
27	1.615	1.597	1.674	1.379	52	3.105	3.092	3.173	2.875
28	1.675	1.657	1.737	1.439	53	3.165	3.152	3.231	2.934
29	1.734	1.717	1.795	1.499	54	3.225	3.213	3.293	2.993
30	1.794	1.777	1.857	1.559	55	3.284	3.272	3.351	3.054
31	1.853	1.836	1.914	1.619	56	3.344	3.332	3.412	3.114
32	1.913	1.897	1.977	1.679	57	3.404	3.392	3.471	3.173
33	1.973	1.957	2.035	1.739	58	3.463	3.451	3.531	3.233
34	2.032	2.016	2.096	1.799	59	3.523	3.511	3.590	3.293
35	2.092	2.077	2.155	1.858	60	3.583	3.571	3.651	3.353

All dimensions in inches.
* Diameters given are maximum; all tolerances must be negative. (See Table 4B.)
† For rounded teeth.
‡ Gage pin diameter = 0.1250 in. Tolerance for guide groove diameter $G = +0, -0.015$ in. (-0.38 mm).
Tolerance for maximum eccentricity (total indicator reading) pitch diameter with respect to bore is 0.004 in. up to and including 4 in. diameter; and 0.008 in., over 4 in. diameter.

together with important dimensions for chains of various pitches and widths. Maximum over-all width M of the three types of chain are also given in this table for

(Continued on page 2319)

Table 3B (*Concluded*). **American National Standard Silent Chain Sprocket Diameters for 3⁄16-in. Pitch Chain** (ANSI B29.2M-1982)

No. Teeth	Pitch Diameter	Outside Diameter*,†	Over-Pin Diameter*,†	Max. Groove Diameter*	No. teeth	Pitch Diameter	Outside Diameter*,†	Over-Pin Diameter*,†	Max. Groove Diameter*
61	3.642	3.630	3.709	3.413	91	5.432	5.422	5.501	5.204
62	3.702	3.690	3.771	3.472	92	5.492	5.482	5.562	5.264
63	3.762	3.750	3.830	3.532	93	5.552	5.542	5.621	5.323
64	3.821	3.809	3.890	3.592	94	5.611	5.601	5.681	5.383
65	3.881	3.869	3.949	3.651	95	5.671	5.661	5.740	5.443
66	3.941	3.930	4.010	3.711	96	5.731	5.721	5.801	5.503
67	4.000	3.989	4.068	3.771	97	5.790	5.780	5.859	5.562
68	4.060	4.049	4.129	3.831	98	5.850	5.840	5.920	5.622
69	4.120	4.109	4.188	3.890	99	5.910	5.900	5.979	5.682
70	4.179	4.168	4.248	3.950	100	5.969	5.959	6.039	5.741
71	4.239	4.228	4.307	4.010	101	6.029	6.019	6.098	5.801
72	4.299	4.288	4.368	4.070	102	6.089	6.079	6.159	5.861
73	4.358	4.347	4.426	4.129	103	6.148	6.138	6.217	5.921
74	4.418	4.407	4.487	4.189	104	6.207	6.197	6.277	5.980
75	4.478	4.467	4.546	4.249	105	6.268	6.258	6.337	6.040
76	4.537	4.526	4.606	4.308	106	6.328	6.318	6.398	6.100
77	4.597	4.586	4.665	4.368	107	6.388	6.378	6.457	6.159
78	4.657	4.646	4.726	4.428	108	6.447	6.437	6.518	6.219
79	4.716	4.705	4.785	4.487	109	6.508	6.498	6.576	6.279
80	4.776	4.765	4.846	4.547	110	6.566	6.556	6.637	6.338
81	4.836	4.825	4.905	4.607	111	6.625	6.615	6.695	6.398
82	4.895	4.884	4.965	4.667	112	6.685	6.675	6.755	6.458
83	4.955	4.944	5.024	4.726	113	6.745	6.735	6.815	6.518
84	5.015	5.004	5.085	4.786	114	6.805	6.795	6.876	6.577
85	5.074	5.063	5.143	4.846	115	6.866	6.856	6.935	6.637
86	5.134	5.124	5.204	4.906	116	6.924	6.914	6.995	6.697
87	5.194	5.184	5.263	4.965	117	6.984	6.974	7.054	6.756
88	5.253	5.243	5.323	5.045	118	7.044	7.034	7.114	6.816
89	5.313	5.303	5.382	5.084	119	7.103	7.094	7.174	6.876
90	5.373	5.363	5.443	5.144	120	7.162	7.153	7.233	6.935

All dimensions in inches. See footnotes on page 2317.

Table 4A. Over-Pin Diameter Tolerances for American National Standard 3⁄8-in. Pitch and Larger Silent Chain Sprocket Measurement (ANSI B29.2M-1982)

Pitch	Number of Teeth									
	Up to 15	16–24	25–35	36–48	49–63	64–80	81–99	100–120	121–143	144 up
TOLERANCE,* INCHES										
0.375	0.005	0.005	0.005	0.006	0.006	0.007	0.007	0.007	0.008	0.008
0.500	0.005	0.006	0.006	0.007	0.007	0.008	0.008	0.009	0.009	0.010
0.625	0.006	0.006	0.007	0.008	0.009	0.010	0.010	0.010	0.011	0.012
0.750	0.006	0.007	0.008	0.009	0.010	0.011	0.011	0.012	0.013	0.014
1.000	0.007	0.008	0.009	0.010	0.011	0.012	0.013	0.014	0.015	0.016
1.250	0.008	0.009	0.010	0.011	0.013	0.014	0.015	0.017	0.018	0.019
1.500	0.008	0.010	0.011	0.013	0.014	0.016	0.017	0.019	0.020	0.022
2.000	0.010	0.012	0.014	0.016	0.018	0.020	0.022	0.024	0.026	0.028
TOLERANCE,* MILLIMETERS										
9.52	0.13	0.13	0.13	0.15	0.15	0.18	0.18	0.18	0.20	0.20
12.70	0.13	0.15	0.15	0.18	0.18	0.20	0.20	0.23	0.23	0.25
15.88	0.15	0.15	0.18	0.20	0.23	0.25	0.25	0.25	0.28	0.30
19.05	0.15	0.18	0.20	0.23	0.25	0.28	0.28	0.30	0.33	0.36
25.40	0.18	0.20	0.23	0.25	0.28	0.30	0.33	0.36	0.38	0.40
31.75	0.20	0.23	0.25	0.28	0.33	0.36	0.38	0.43	0.46	0.48
38.10	0.20	0.25	0.28	0.33	0.36	0.40	0.43	0.48	0.51	0.56
50.80	0.25	0.30	0.36	0.40	0.46	0.51	0.56	0.61	0.66	0.71

* All tolerances are *negative*. Tolerance = $(0.004 + 0.001P\sqrt{N})$, where P = chain pitch, N = number of teeth. See Table 3A for over-pin diameters.

**Table 4B. Over-Pin Diameter Tolerances for American National Standard
³⁄₁₆-in. Silent Chain Sprocket Measurement** (ANSI B29.2M-1982)

					Number of Teeth					
Pitch	Up to 15	16–24	25–35	36–48	49–63	64–80	81–99	100–120	121–143	144 up
					TOLERANCE,* INCHES					
0.1875	0.004	0.004	0.004	0.004	0.004	0.005	0.005	0.005	0.005	0.005
					TOLERANCE,* MILLIMETERS					
4.76	0.10	0.10	0.10	0.10	0.10	0.13	0.13	0.13	0.13	0.13

* All tolerances are negative.

**Table 5A. American National Standard ³⁄₈-in. Pitch and Larger
Silent Chain Widths and Sprocket Face Dimensions** (ANSI B29.2M-1982)

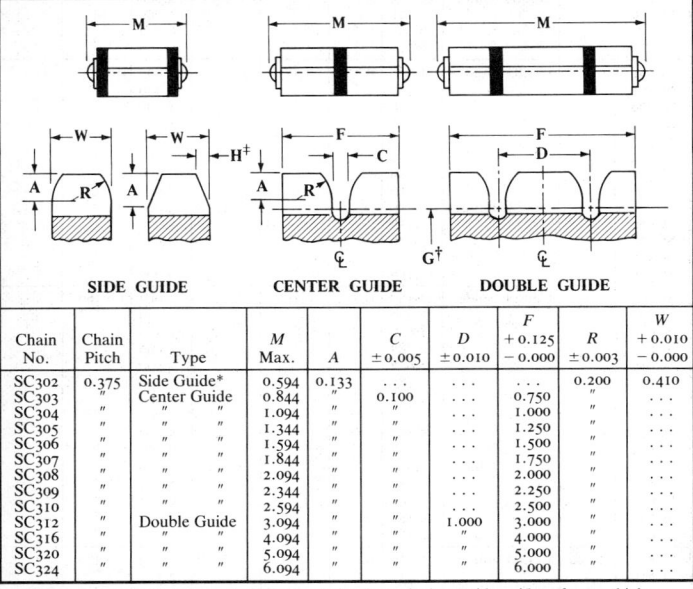

SIDE GUIDE   CENTER GUIDE   DOUBLE GUIDE

Chain No.	Chain Pitch	Type	M Max.	A	C ±0.005	D ±0.010	F +0.125 −0.000	R ±0.003	W +0.010 −0.000
SC302	0.375	Side Guide*	0.594	0.133	. . .	. . .	. . .	0.200	0.410
SC303	"	Center Guide	0.844	"	0.100	. . .	0.750	"	. . .
SC304	"	" "	1.094	"	"	. . .	1.000	"	. . .
SC305	"	" "	1.344	"	"	. . .	1.250	"	. . .
SC306	"	" "	1.594	"	"	. . .	1.500	"	. . .
SC307	"	" "	1.844	"	"	. . .	1.750	"	. . .
SC308	"	" "	2.094	"	"	. . .	2.000	"	. . .
SC309	"	" "	2.344	"	"	. . .	2.250	"	. . .
SC310	"	" "	2.594	"	"	. . .	2.500	"	. . .
SC312	"	Double Guide	3.094	"	"	1.000	3.000	"	. . .
SC316	"	" "	4.094	"	"	"	4.000	"	. . .
SC320	"	" "	5.094	"	"	"	5.000	"	. . .
SC324	"	" "	6.094	"	"	"	6.000	"	. . .

All dimensions in inches. * Side Guide chains have single outside guides of same thickness as toothed links. † Grooving tool may be either square or round end but groove must be full width down to diameter of G. For values of G (max.) see footnote to Table 3A. ‡ Values of H (± 0.003 in.) = 0.051 in. are given only for chain numbers SC302 and SC402. M = Max. overall width of chain. The maximum radius over a new chain engaged on a sprocket will not exceed the sprocket pitch radius plus 75 per cent of the chain pitch. To obtain the chain widths and sprocket face dimensions in millimeters, multiply each entry by 25.4.

various pitches and widths. It should be noted that the sprocket tooth width W for the side guide chain is given in Table 5 for one-half-inch wide chains of ³⁄₈-inch and ½-inch pitches. No values of W for other chain sizes are specified in American National Standard B29.2M-1982.

Table 5A (*Continued*). **American National Standard ⅜ in. Pitch and Larger Silent Chain Widths and Sprocket Face Dimensions** (ANSI B29.2M-1982)

Chain No.	Chain Pitch	Type	M^* Max.	A	C ±0.005	D ±0.010	F +0.125 −0.000	R ±0.003	W +0.010 −0.000
SC402	0.500	Side Guide*	0.750	0.133	. . .	. . .	. . .	0.200	0.410
SC403	"	Center Guide	0.875	"	0.100	. . .	0.750	"	. . .
SC404	"	" "	1.125	"	"	. . .	1.000	"	. . .
SC405	"	" "	1.375	"	"	. . .	1.250	"	. . .
SC406	"	" "	1.625	"	"	. . .	1.500	"	. . .
SC407	"	" "	1.875	"	"	. . .	1.750	"	. . .
SC408	"	" "	2.125	"	"	. . .	2.000	"	. . .
SC409	"	" "	2.375	"	"	. . .	2.250	"	. . .
SC410	"	" "	2.625	"	"	. . .	2.500	"	. . .
SC411	"	" "	2.875	"	"	. . .	2.750	"	. . .
SC412	"	" "	3.125	"	"	. . .	3.000	"	. . .
SC414	"	" "	3.625	"	"	. . .	3.500	"	. . .
SC416	"	Double Guide	4.125	"	"	1.000	4.000	"	. . .
SC420	"	" "	5.125	"	"	"	5.000	"	. . .
SC424	"	" "	6.125	"	"	"	6.000	"	. . .
SC432	"	" "	8.125	"	"	"	8.000	"	. . .
SC504	0.625	Center Guide	1.156	0.177	0.125	. . .	1.000	0.250	. . .
SC505	"	" "	1.406	"	"	. . .	1.250	"	. . .
SC506	"	" "	1.656	"	"	. . .	1.500	"	. . .
SC507	"	" "	1.906	"	"	. . .	1.750	"	. . .
SC508	"	" "	2.156	"	"	. . .	2.000	"	. . .
SC510	"	" "	2.656	"	"	. . .	2.500	"	. . .
SC512	"	" "	3.156	"	"	. . .	3.000	"	. . .
SC516	"	" "	4.156	"	"	. . .	4.000	"	. . .
SC520	"	Double Guide	5.156	"	"	2.000	5.000	"	. . .
SC524	"	" "	6.156	"	"	"	6.000	"	. . .
SC528	"	" "	7.156	"	"	"	7.000	"	. . .
SC532	"	" "	8.156	"	"	"	8.000	"	. . .
SC540	"	" "	10.156	"	"	"	10.000	"	. . .
SC604	0.750	Center Guide	1.187	0.274	0.180	. . .	1.000	0.360	. . .
SC605	"	" "	1.437	"	"	. . .	1.250	"	. . .
SC606	"	" "	1.687	"	"	. . .	1.500	"	. . .
SC608	"	" "	2.187	"	"	. . .	2.000	"	. . .
SC610	"	" "	2.687	"	"	. . .	2.500	"	. . .
SC612	"	" "	3.187	"	"	. . .	3.000	"	. . .
SC614	"	" "	3.687	"	"	. . .	3.500	"	. . .
SC616	"	" "	4.187	"	"	. . .	4.000	"	. . .
SC620	"	" "	5.187	"	"	. . .	5.000	"	. . .
SC624	"	" "	6.187	"	"	. . .	6.000	"	. . .
SC628	"	Double Guide	7.187	"	"	4.000	7.000	"	. . .
SC632	"	" "	8.187	"	"	"	8.000	"	. . .
SC636	"	" "	9.187	"	"	"	9.000	"	. . .
SC640	"	" "	10.187	"	"	"	10.000	"	. . .
SC648	"	" "	12.187	"	"	"	12.000	"	. . .
SC808	1.000	Center Guide	2.250	0.274	0.180	. . .	2.000	0.360	. . .
SC810	"	" "	2.750	"	"	. . .	2.500	"	. . .
SC812	"	" "	3.250	"	"	. . .	3.000	"	. . .
SC816	"	" "	4.250	"	"	. . .	4.000	"	. . .
SC820	"	" "	5.250	"	"	. . .	5.000	"	. . .
SC824	"	" "	6.250	"	"	. . .	6.000	"	. . .
SC828	"	Double Guide	7.250	"	"	4.000	7.000	"	. . .
SC832	"	" "	8.250	"	"	"	8.000	"	. . .
SC836	"	" "	9.250	"	"	"	9.000	"	. . .
SC840	"	" "	10.250	"	"	"	10.000	"	. . .
SC848	"	" "	12.250	"	"	"	12.000	"	. . .
SC856	"	" "	14.250	"	"	"	14.000	"	. . .
SC864	"	" "	16.250	"	"	"	16.000	"	. . .

* See footnotes on page 2321.

Table 5A (*Concluded*). **American National Standard ⅜ in. Pitch and Larger Silent Chain Widths and Sprocket Face Dimensions** (ANSI B29.2-M-1982)

Chain No.	Chain Pitch	Type	M* Max.	A	C ±0.005	D ±0.010	F +0.125 -0.000	R ±0.003	W +0.010 -0.000
SC1010	1.25	Center Guide	2.812	0.274	0.180	. . .	2.500	0.360	. . .
SC1012	"	" "	3.312	"	"	. . .	3.000	"	. . .
SC1016	"	" "	4.312	"	"	. . .	4.000	"	. . .
SC1020	"	" "	5.312	"	"	. . .	5.000	"	. . .
SC1024	"	" "	6.312	"	"	. . .	6.000	"	. . .
SC1028	"	" "	7.312	"	"	. . .	7.000	"	. . .
SC1032	"	Double Guide	8.312	"	"	4.000	8.000	"	. . .
SC1036	"	" "	9.312	"	"	"	9.000	"	. . .
SC1040	"	" "	10.312	"	"	"	10.000	"	. . .
SC1048	"	" "	12.312	"	"	"	12.000	"	. . .
SC1056	"	" "	14.312	"	"	"	14.000	"	. . .
SC1064	"	" "	16.312	"	"	"	16.000	"	. . .
SC1072	"	" "	18.312	"	"	"	18.000	"	. . .
SC1080	"	" "	20.312	"	"	"	20.000	"	. . .
SC1212	1.500	Center Guide	3.375	0.274	0.180	. . .	3.000	0.360	. . .
SC1216	"	" "	4.375	"	"	. . .	4.000	"	. . .
SC1220	"	" "	5.375	"	"	. . .	5.000	"	. . .
SC1224	"	" "	6.375	"	"	. . .	6.000	"	. . .
SC1228	"	" "	7.375	"	"	. . .	7.000	"	. . .
SC1232	"	Double Guide	8.375	"	"	4.000	8.000	"	. . .
SC1236	"	" "	9.375	"	"	"	9.000	"	. . .
SC1240	"	" "	10.375	"	"	"	10.000	"	. . .
SC1248	"	" "	12.375	"	"	"	12.000	"	. . .
SC1256	"	" "	14.375	"	"	"	14.000	"	. . .
SC1264	"	" "	16.375	"	"	"	16.000	"	. . .
SC1272	"	" "	18.375	"	"	"	18.000	"	. . .
SC1280	"	" "	20.375	"	"	"	20.000	"	. . .
SC1288	"	" "	22.375	"	"	"	22.000	"	. . .
SC1296	"	" "	24.375	"	"	"	24.000	"	. . .
SC1616	2.000	Center Guide	4.500	0.274	0.218	. . .	4.000	0.360	. . .
SC1620	"	" "	5.500	"	"	. . .	5.000	"	. . .
SC1624	"	" "	6.500	"	"	. . .	6.000	"	. . .
SC1628	"	" "	7.500	"	"	. . .	7.000	"	. . .
SC1632	"	Double Guide	8.500	"	"	4.000	8.000	"	. . .
SC1640	"	" "	10.500	"	"	"	10.000	"	. . .
SC1648	"	" "	12.500	"	"	"	12.000	"	. . .
SC1656	"	" "	14.500	"	"	"	14.000	"	. . .
SC1664	"	" "	16.500	"	"	"	16.000	"	. . .
SC1672	"	" "	18.500	"	"	"	18.000	"	. . .
SC1680	"	" "	20.500	"	"	"	20.000	"	. . .
SC1688	"	" "	22.500	"	"	"	22.000	"	. . .
SC1696	"	" "	24.500	"	"	"	24.000	"	. . .
SC16120	"	" "	30.500	"	"	"	30.000	"	. . .

All dimensions in inches. *M* = Max. overall width of chain. * Specify side guide or center guide type. † Grooving tool may be either square or round end but groove must be full width down to diameter *G*. For values of *G*, see Table 3B. ‡Values of *H* = 0.025 in. are given for chain numbers SC0305 through SC0315. To obtain chain width and sprocket face dimensions in millimeters, multiply each entry by 25.4.

Sprocket Hub Dimensions. — The important hub dimensions are the outside diameter, the bore, and the length. The maximum hub diameter is limited by the need to clear the chain guides and is of particular importance for sprockets with low numbers of teeth. The American National Standard for inverted tooth chains and sprocket teeth ANSI B29.2M-1982 provides the following formulas for calculating maximum hub diameters, *MHD*.

$$MHD \text{ (for hobbed teeth)} = P[\cot(180°/N) - 1.33]$$

$$MHD \text{ (for straddle cut teeth)} = P[\cot(180°/N) - 1.25]$$

Table 5B. American National Standard ³⁄₁₆ in. Pitch Silent Chain Widths and Sprocket Face Dimensions (B29.2M-1982)

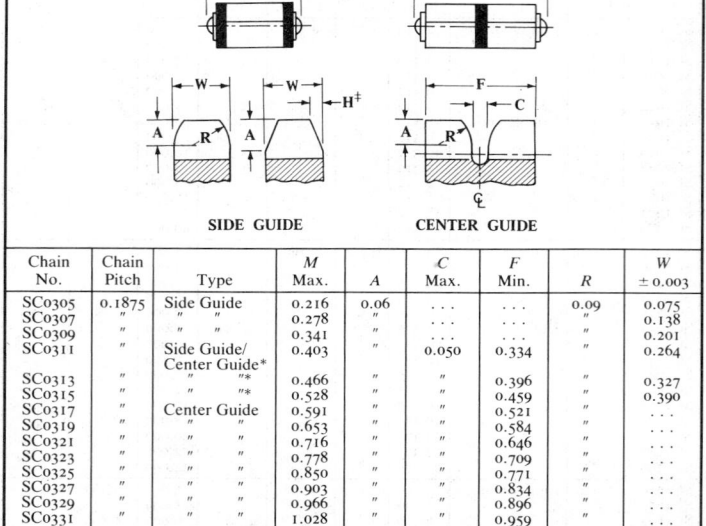

SIDE GUIDE   CENTER GUIDE

Chain No.	Chain Pitch	Type	M Max.	A	C Max.	F Min.	R	W ± 0.003
SC0305	0.1875	Side Guide	0.216	0.06	. . .	. . .	0.09	0.075
SC0307	"	" "	0.278	"	. . .	. . .	"	0.138
SC0309	"	" "	0.341	"	. . .	. . .	"	0.201
SC0311	"	Side Guide/ Center Guide*	0.403	"	0.050	0.334	"	0.264
SC0313	"	" " "*	0.466	"	"	0.396	"	0.327
SC0315	"	" " "*	0.528	"	"	0.459	"	0.390
SC0317	"	Center Guide	0.591	"	"	0.521	"	. . .
SC0319	"	" "	0.653	"	"	0.584	"	. . .
SC0321	"	" " "	0.716	"	"	0.646	"	. . .
SC0323	"	" " "	0.778	"	"	0.709	"	. . .
SC0325	"	" " "	0.850	"	"	0.771	"	. . .
SC0327	"	" " "	0.903	"	"	0.834	"	. . .
SC0329	"	" " "	0.966	"	"	0.896	"	. . .
SC0331	"	" " "	1.028	"	"	0.959	"	. . .

All dimensions in inches. M = Max. overall width of chain.
* Specify side guide or center guide type.
† Grooving tool may be either square or round end but groove must be full width down to diameter G. For values of G, see Table 3B.
‡Values of H = 0.025 in. are given for chain numbers SC0305 through SC0315.
To obtain chain width and sprocket face dimensions in millimeters, multiply each entry by 25.4.

Maximum hub diameters for sprockets with from 17 to 31 teeth are given in Table 6. Maximum hub diameters for other methods of cutting teeth may differ from these values. Recommended maximum bores are given in Table 7.

Sprocket Design and Tooth Form. — Except for tooth form, silent chain sprocket design parallels the general design practice of roller chain sprockets as covered in the previous section.

As shown in Tables 8A and 8B, sprockets for American National Standard silent chains have teeth with straight-line working faces. The tops of teeth for ⅜-in. and larger pitch chains may be rounded or square. Bottom clearance below the working face is not specified but must be sufficient to clear the chain teeth. The standard tooth form for ⅜-in. and larger pitch chains is designed to mesh with link plate contours having an included angle of 60 degrees as shown in the diagram of Table 8A. The standard tooth form for ³⁄₁₆-in. pitch chains has an included angle of 70 degrees as shown in Table 8B. It will be seen from these tables that the angle between the faces of a given tooth [$60° - 720°/N$ for ⅜-in. pitch and larger; $70° - 720°/N$ for ³⁄₁₆-in. pitch] becomes smaller as the number of teeth decreases. Therefore, for a ⅜-in. pitch or larger 12-tooth sprocket it will be zero. In other words the tooth faces will be parallel.

Table 6. American National Standard Minimum Hub Diameters for Silent Chain Sprockets (17 to 31 teeth)
(ANSI B29.2M-1982)

Values shown are for 1-inch pitch chain. For other pitches (3/8-inch and larger) multiply the values given by the pitch.

No. Teeth	Hob Cut	Straddle Cut	No. Teeth	Hob Cut	Straddle Cut	No. Teeth	Hob Cut	Straddle Cut
	Min. Hub Diam.			Min. Hub Diam.			Min. Hub Diam.	
17	4.019	4.099	22	5.626	5.706	27	7.226	7.306
18	4.341	4.421	23	5.946	6.026	28	7.546	7.626
19	4.662	4.742	24	6.265	6.345	29	7.865	7.945
20	4.983	5.063	25	6.586	6.666	30	8.185	8.265
21	5.304	5.384	26	6.905	6.985	31	8.503	8.583

All dimensions in inches.
Good practice indicates that teeth of sprockets up to and including 31 teeth should have a Rockwell hardness of C50 min.

For smaller tooth numbers the teeth would be undercut. For best results, 21 or more teeth are recommended; less than 17 should not be used.

Cutting Silent Chain Sprocket Teeth. — Sprocket teeth may be cut by either a straddle cutter or a hob. Essential dimensions for straddle cutters are given in Table 9 and for hobs in Tables 10 and 11. American National Standard silent chain hobs are stamped for identification as shown on page 2326.

Design of Silent Chain Drives. — The design of silent chain transmissions must be based not only upon the power to be transmitted and the ratio between driving and driven shafts, but also upon such factors as the speed of the faster running shaft, the available space, assuming that it affects the sprocket diameters, the character of the load and certain other factors. Determining the pitch of the chain and the number of teeth on the smallest sprocket are the important initial steps. Usually any one of several combinations of pitches and sprocket sizes may be employed for a given installation. In attempting to select the best combination, it is advisable to consult with the manufacturer of the chain to be used. Some of the more important fundamental points governing the design of silent chain transmissions will be summarized.

The design of a silent chain drive consists, primarily, of the selection of the chain size, sprockets, determination of chain length, center distance, lubrication method, and arrangement of casings.

Table 7. Recommended Maximum Sprocket Bores for Silent Chains*

Number of Teeth	Chain Pitch, Inches							
	3/8	1/2	5/8	3/4	1	1 1/4	1 1/2	2
	Max. Sprocket Bore, Inches							
17	1	1 3/8	1 3/4	2	2 3/4	3 3/8	4 1/8	5 1/2
19	1 1/4	1 5/8	2	2 3/8	3 1/4	4	4 7/8	6 3/4
21	1 3/8	1 7/8	2 1/4	2 3/4	3 3/4	4 1/2	5 1/2	7 3/4
23	1 5/8	2 1/8	2 5/8	3 1/4	4 3/8	5 1/2	6 1/2	9
25	1 3/4	2 3/8	3	3 5/8	4 3/4	6	7 1/4	10
27	2	2 5/8	3 3/8	3 7/8	5 3/8	6 3/4	8 1/8	11 1/4
29	2 1/8	2 13/16	3 5/8	4 3/8	5 3/4	7 3/8	9 1/8	12 1/2
31	2 5/16	3 1/16	3 7/8	4 5/8	6 3/8	8	9 7/8	13 1/2

* American Chain Association.

Table 8A. Tooth Form for ANSI ⅜-in. Pitch and Larger Silent Tooth Sprocket
(ANSI B29.2M-1982)

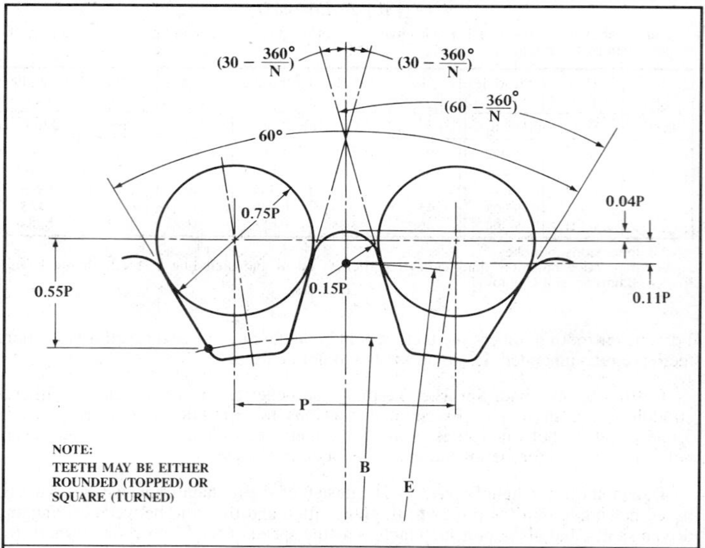

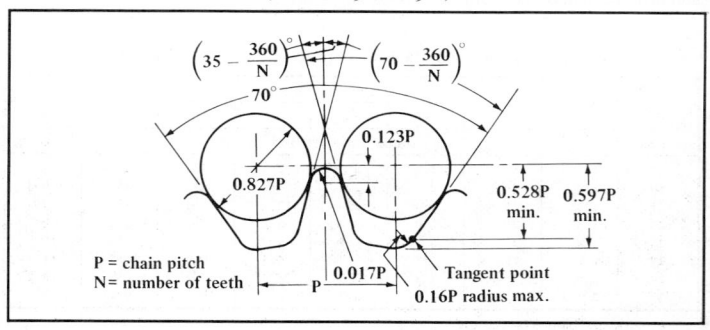

P = Chain Pitch
N = Number of Teeth
E = Diameter to Center of Topping Curve
B = Diameter to Base of Working Face

$$E = P\left(\cot\frac{180°}{N} - 0.22\right)$$

$$B = P\sqrt{1.515213 + \left(\cot\frac{180°}{N} - 1.1\right)^2}$$

Note: Shape of root line below working face may vary with type of cutter.

Table 8B. Tooth Form for ANSI 3⁄16-in. Pitch Silent Tooth Sprocket
(ANSI B29.2M-1982)

P = chain pitch
N = number of teeth

Table 9. Straddle Cutters for American National Standard ⅜-in. Pitch and Larger Silent Chain Sprocket Teeth

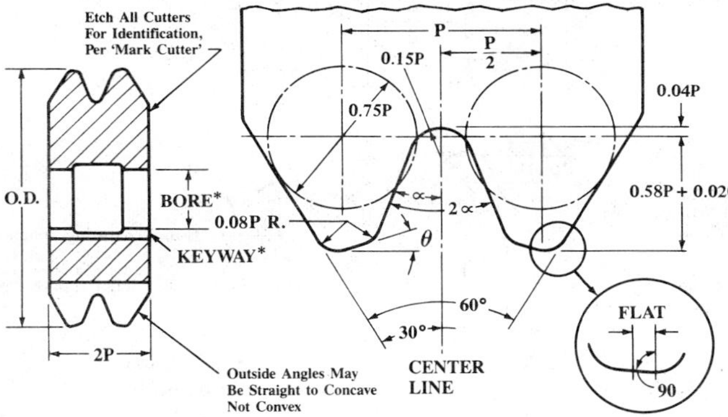

Etch All Cutters For Identification, Per 'Mark Cutter'

P
P/2
0.15P
0.75P
0.04P

O.D.
BORE*
0.08P R.
KEYWAY*

0.58P + 0.020

α
2 α
θ

60°
30°

FLAT
90

2P

Outside Angles May Be Straight to Concave Not Convex

CENTER LINE

Chain Pitch P	Mark Cutter†	Outside Diam.	0.75P	α	θ	Bore*
0.375	SC3-15 thru 35 SC3-36 up	3.625	0.2813	22°-30' 27°-30'	12° 5°	1.250
0.500	SC4-15 thru 35 SC4-36 up	3.875	0.3750	22°-30' 27°-30'	12° 5°	1.250
0.625	SC5-15 thru 35 SC5-36 up	4.250	0.4688	22°-30' 27°-30'	12° 5°	1.250
0.750	SC6-15 thru 35 SC6-36 up	4.625	0.5625	22°-30' 27°-30'	12° 5°	1.250
1.000	SC8-15 thru 35 SC8-36 up	5.250	0.7500	22°-30' 27°-30'	12° 5°	1.500
1.250	SC10-15 thru 35 SC10-36 up	5.750	0.9375	22°-30' 27°-30'	12° 5°	1.500
1.500	SC12-15 thru 35 SC12-36 up	6.250	1.1250	22°-30' 27°-30'	12° 5°	1.750
2.000	SC16-15 thru 35 SC16-36 up	6.500	1.5000	22°-30' 27°-30'	12° 5°	1.750

All dimensions in inches. To obtain values in millimeters, multiply inch values by 25.4.
* Suggested standard. Bores other than standard must be specified.
† Range of teeth is indicated in the cutter marking.
These data are given as supplementary information in ANSI B29.2M-1982 and are made available by the American Chain Association.

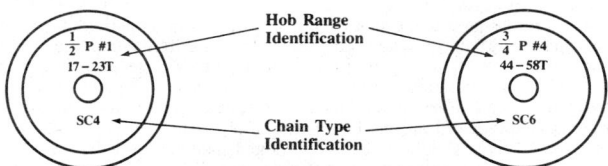

Hob Range Identification

½ P #1
17 – 23T
SC4

¾ P #4
44 – 58T
SC6

Chain Type Identification

Fig. 1. Identification of Inverted Tooth Chain Hobs

Table 10. Hobs for ANSI ⅜-in. Pitch and Larger Silent Chain Sprocket Teeth*

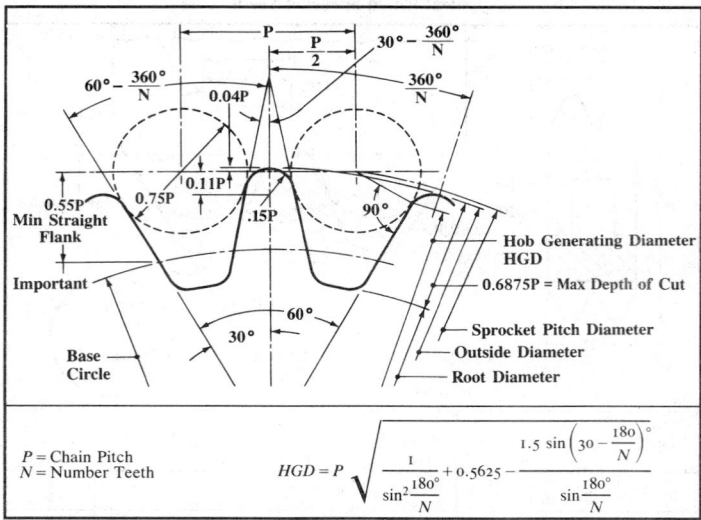

P = Chain Pitch
N = Number Teeth

$$HGD = P \sqrt{\dfrac{1}{\sin^2\dfrac{180°}{N}} + 0.5625 - \dfrac{1.5\sin\left(30 - \dfrac{180}{N}\right)°}{\sin\dfrac{180°}{N}}}$$

* Source: American Chain Association.

Pitch of Silent Chain. — The pitch is selected with reference to the speed of the faster running shaft which ordinarily is the driver and holds the smaller sprocket. The following pitches are recommended: for a faster running shaft of 2000 to 5000 rpm, ⅜-inch pitch; for 1500 to 2000 rpm, ½-inch pitch; for 1200 to 1500 rpm, ⅝-inch pitch, for 1000 to 1200 rpm, ¾-inch pitch; for 800 to 1000 rpm, 1-inch pitch; for 650 to 800 rpm, 1¼-inch pitch; for 500 to 600 rpm, 1½-inch pitch; for 300 to 500 rpm, 2-inch pitch; and for below 300 rpm, 2½-inch pitch. As the normal operating speeds increase, the allowable pitch decreases. Recommendations relating to the relationship between pitch and operating speed are intended for normal or average conditions. Speeds for a given pitch may be exceeded under favorable conditions and may have to be reduced when conditions are unfavorable. In general, smoother or quieter operation will result from using the smallest pitch suitable for a given speed and load. However, a larger pitch which might be applicable under the same conditions, will result in a narrower chain and a less expensive transmission. This relationship usually is true when there is a small speed reduction and comparatively long center distance. If there is a large speed reduction and short center distance, drives having the smaller pitches may be less expensive.

Maximum Ratios for Silent Chain Drives. — The maximum permissible ratios between driving and driven sprockets vary somewhat for different conditions and usually range from 6- or 7-to-1 up to 10-to-1. Some drives have even higher ratios, especially when the operating conditions are exceptionally favorable. When a large speed reduction is necessary, it is preferable as a general rule to use a double reduction or compound type of transmission instead of obtaining the entire reduction with two sprockets. Drives should be so proportioned that the angle between the two strands of a tight chain does not exceed 45 degrees. When the angle is larger, the chain does not have sufficient contact with the driving sprocket.

Table 11. Hobs for American National Standard ⅜-in Pitch and Larger Silent Chain Sprocket Teeth*

Chain Pitch	Hob Number	Basic Number of Teeth	Tooth Range of Hob	Generating Diam.	
				inches	millimeters
SC3 = 0.375 in. = 9.52 mm	1	20	17–23	2.311	58.70
	2	28	24–32	3.247	82.47
	3	38	33–43	4.428	112.47
	4	51	44–58	5.971	151.66
	5	69	59–79	8.114	206.10
	6	95	80–110	11.212	284.78
	7	130	111–150	15.385	390.78
SC4 = 0.500 in. = 12.70 mm	1	20	17–23	3.082	78.28
	2	28	24–32	4.329	109.96
	3	38	33–43	5.904	149.96
	4	51	44–58	7.962	202.23
	5	69	59–79	10.818	274.78
	6	95	80–110	14.950	379.73
	7	130	111–150	20.513	521.03
SC5 = 0.625 in. = 15.88 mm	1	20	17–23	3.852	97.84
	2	28	24–32	5.412	137.46
	3	38	33–43	7.381	187.48
	4	51	44–58	9.952	252.78
	5	69	59–79	13.522	343.46
	6	95	80–110	18.687	474.65
	7	130	111–150	25.641	651.28
SC6 = 0.750 in. = 19.05 mm	1	20	17–23	4.623	117.42
	2	28	24–32	6.494	164.95
	3	38	33–43	8.857	224.97
	4	51	44–58	11.943	303.35
	5	69	59–79	16.227	412.17
	6	95	80–110	22.424	569.57
	7	130	111–150	30.770	781.56
SC8 = 1.000 in. = 25.40 mm	1	20	17–23	6.163	156.54
	2	28	24–32	8.659	219.94
	3	38	33–43	11.809	299.95
	4	51	44–58	15.924	404.47
	5	69	59–79	21.636	549.55
	6	95	80–110	29.899	759.43
	7	130	111–150	41.026	1042.06
SC10 = 1.250 in. = 31.75 mm	1	20	17–23	7.704	195.68
	2	28	24–32	10.823	274.90
	3	38	33–43	14.761	374.93
	4	51	44–58	19.905	505.59
	5	69	59–79	27.045	686.94
	6	95	80–110	37.374	949.30
	7	130	111–150	51.283	1302.59
SC12 = 1.500 in. = 38.10 mm	1	20	17–23	9.245	234.82
	2	28	24–32	12.988	329.90
	3	38	33–43	17.713	449.91
	4	51	44–58	23.886	606.70
	5	69	59–79	32.454	824.33
	6	95	80–110	44.849	1139.16
	7	130	111–150	61.539	1563.09
SC16 = 2.000 in. = 50.80 mm	1	20	17–23	12.327	313.11
	2	28	24–32	17.317	439.85
	3	38	33–43	23.618	599.90
	4	51	44–58	31.848	808.94
	5	69	59–79	43.272	1099.11
	6	95	80–110	59.798	1518.87
	7	130	111–150	82.052	2084.12

* Source: American Chain Association.

Sprocket Size and Chain Speed: A driving sprocket with not less than 17 teeth is generally recommended. For the driven sprocket, one manufacturer recommends 127 teeth as a maximum limit and less than 100 as preferable. If practicable, the sprocket sizes should be small enough to limit the chain speed to from 1200 to 1400 feet per minute. If the chain speed exceeds these figures, this may indicate that the pitch is too large or that a smaller pitch, and, consequently, a reduction in sprocket diameters (and chain speed) will result in better operating conditions. Both sprockets should preferably have a "hunting tooth ratio" relative to the number of chain links for uniform wear. See "Hunting Tooth Ratios," page 1828.

If there is a small reduction in speed between the driving and driven shafts, both sprockets may be made as small as is consistent with satisfactory operation, either to obtain a compact drive or possibly to avoid excessive chain speed in cases where the rotative speed is high for a given horsepower. Under such conditions, one manufacturer recommends driving sprockets ranging from 17 to 30 teeth, and driven sprockets ranging from 19 to 33 teeth. If the number of revolutions per minute is low for a given horsepower and the center distance comparatively long, then the recommended range for driving sprockets is from 23 to 111 teeth, and driven sprockets from 27 to 129 teeth. The preferable range is from 17 to 75 teeth for the driving sprockets, and 19 to 102 teeth for the driven sprockets.

Center Distance for Silent Chain Drives. — If the ratio of the drive is small, it is possible to locate the sprockets so close that the teeth just clear; however, as a general rule, the minimum center-to-center distance should equal the sum of the diameters of both sprockets. According to the Whitney Chain & Mfg. Co., if the speed ratio is not over 2½-to-1, the center distance may be equal to one-half the sum of the sprocket diameters plus tooth clearance, providing this distance is not less than the minimum given in Table 12. If the speed ratio is greater than 2½-to-1, the center distance should not be less than the sum of the sprocket diameters.

Table 12. Minimum Center Distances for Various Pitches

Pitch, in.	3⁄8	1⁄2	5⁄8	3⁄4	1	1¼	1½
Min. Center Distance, in.	6	9	12	15	21	27	33

When the chain length in pitches is known, the equivalent center distance for a tight chain may be determined by the formula for roller chain found on page 2299.

In selecting chain length, factors determining length should be adjusted so that the use of offset links may be avoided wherever possible. Chain lengths of an uneven number of pitches are also to be avoided.

Silent Tooth Chain Horsepower Capacity. — The horsepower ratings given in Table 13 have been established on a life expectancy of approximately 15,000 hours under optimum drive conditions, i.e. for a uniform rate of work where there is relatively little shock or load variation throughout a single revolution of a driven sprocket. Using these horsepower ratings as a basis, engineering judgment should be exercised as to the severity of the operating conditions for the intended installation, taking into consideration the source of power, the nature of the load, and the resulting effects of inertia, strain, and shock. Thus, for other than optimum drive conditions, the specified horsepower must be multiplied by the applicable service factor to obtain a "design" horsepower value. This is the value used to enter Table 13 to obtain the required size of chain.

Service Factors: For a uniform type of load, a service factor of 1.0 for a 10-hour day and 1.3 for a 24-hour day are recommended. For a moderate shock load, service factors of 1.4 for a 10-hour day and 1.7 for a 24-hour day are recommended. For heavy shock loads, service factors of 1.7 for a 10-hour day and 2.0 for a 24-hour day are recommended. For extensive table of service factor applications, see supplementary information in ANSI B29.2M-1982.

Installation of Silent Chain Drives. — In installing chain transmissions of any kind, horizontal drives are those having driving and driven shafts in a horizontal plane. These are always preferable to vertical drives, which have a vertical center line intersecting the driving and driven shafts. If one sprocket must be higher than the other, avoid a vertical drive if possible by so locating the two sprockets that the common center line inclines from the vertical as far as is permitted by other conditions which might govern the installation. If practicable, an adjustment should be provided for the center distance between the driving and driven shafts.

Slack Side of Chain: As a general rule, the slack strand of a chain should be on the lower side of a horizontal drive. If the drive is not horizontal but angular or at some angle less than 90 degrees from the vertical, the slack should preferably be on that side which causes the strand to curve outward or away from the center line of the driving and driven shafts. Whenever the slack strand is on the upper side of either a horizontal or inclined drive, adjustment for the center distance is especially important to compensate for possible chain elongation.

Lubrication: The life of a silent chain subjected to conditions such as are common to automobile drives, depends largely upon the wear of the joints. On account of the high speed and whipping action, it is important to have the chains well oiled. When splash lubrication is employed, the supply pipe should be placed so that the oil will be directed against the inside of the chain. It is preferable that silent chains be operated in an oil-retaining casing with provisions for lubrication. Avoid using greases of any kind. The viscosity of the oil depends on temperature, as follows:

Ambient Temp., °F	Chain Pitch		Ambient Temp., °F	Chain Pitch	
	3/16 & 3/8 in.	1/2 in. & Larger		3/16 & 3/8 in.	1/2 in. & Larger
	Recommended Lubricant			Recommended Lubricant	
20–40	SAE 10	SAE 20	100–120	SAE 30	SAE 40
40–100	SAE 20	SAE 30	120–140	SAE 40	SAE 50

Double-Flexure Silent Chain. — In double-flexure chain, the teeth of the link plates project on both sides of the chain and the chain flexes in both directions. This chain is used where the drive arrangements require that sprockets contact both sides of the chain. Neither double-flexure chain nor sprockets are covered in American National Standard ANSI 29.2M-1982.

Table 13. Horsepower Ratings Per Inch of Chain Width for Silent Chain Drives — 1982

The following industrial standard horsepower ratings for silent chain drives have been supplied by the American Chain Association. These ratings are for American National Standard silent chain as covered by ANSI B29.2M-1982. These values may require modification by using the appropriate service factors (see page 2330). These factors, which apply to typical drives, are intended as a general guide only, and engineering judgment and experience may indicate different modifications to suit the nature of the load.

$$\text{Horsepower capacity of chain per inch of width} = \frac{\text{Rating in Table 13}}{\text{Service factor}}$$

$$\text{Chain width for given total hp capacity} = \frac{\text{hp} \times \text{Service factor}}{\text{Rating per inch, Table 13}}$$

Lubrication: The horsepower established from the sprocket and speed combinations of the drive under consideration will indicate a method of lubrication. This method or a better one must be used to obtain optimum chain life. The types of lubrication as indicated on the tables are: Type I, manual, brush, or oil cup; Type II, bath or disk; Type III, circulating pump.

Table 13 (Continued). Horsepower Ratings Per Inch at Chain Width for Silent Chain Drives — 1982

No. of Teeth Small Sprkt.	3/16-Inch Pitch Chain											
	Revolutions per Minute — Small Sprocket											
	500	600	700	800	900	1200	1800	2000	3500	5000	7000	9000
15	0.28	0.33	0.38	0.43	0.47	0.60	0.80	0.90	1.33	1.66	1.94	1.96
17	0.33	0.39	0.44	0.50	0.55	0.70	0.96	1.05	1.60	2.00	2.40	2.52
19	0.37	0.43	0.50	0.55	0.61	0.80	1.10	1.20	1.80	2.30	2.76	2.92
21	0.41	0.48	0.55	0.62	0.68	0.87	1.22	1.33	2.03	2.58	3.12	3.35
23	0.45	0.53	0.60	0.68	0.75	0.96	1.35	1.47	2.25	2.88	3.50	3.78
25	0.49	0.58	0.66	0.74	0.82	1.05	1.47	1.60	2.45	3.13	3.80	4.10
27	0.53	0.62	0.71	0.80	0.88	1.15	1.58	1.72	2.63	3.35	4.06	4.37
29	0.57	0.67	0.76	0.86	0.95	1.21	1.70	1.85	2.83	3.61	4.40	4.72
31	0.60	0.72	0.81	0.91	1.01	1.30	1.81	1.97	3.02	3.84	4.66	5.00
33	0.64	0.75	0.86	0.97	1.07	1.37	1.90	2.08	3.17	4.02	4.85	...
35	0.68	0.80	0.92	1.03	1.14	1.45	2.03	2.21	3.41	4.27	5.16	...
37	0.71	0.84	0.96	1.08	1.19	1.52	2.11	2.30	3.48	4.39	5.24	...
40	0.77	0.91	1.04	1.16	1.29	1.64	2.28	2.50	3.77	4.76	...	...
45	0.86	1.02	1.15	1.30	1.43	1.83	2.53	2.75	4.15	5.21	...	...
50	0.95	1.12	1.27	1.37	1.58	2.00	2.78	3.02	4.52	5.65	...	...
	TYPE I						TYPE II			TYPE III		

No. of Teeth Small Sprkt.	3/8-Inch Pitch Chain												
	Revolutions per Minute — Small Sprocket												
	100	500	1000	1200	1500	1800	2000	2500	3000	3500	4000	5000	6000
*17	0.46	2.1	4.6	4.9	5.3	6.5	6.9	7.9	8.5	8.8	8.8	...	...
*19	0.53	2.5	4.8	5.4	6.5	7.4	7.9	9.1	9.9	10	11	9.8	...
21	0.58	2.8	5.1	6.0	7.3	8.3	10	11	11	12	12	12	10
23	0.63	3.0	5.6	6.6	8.0	9.3	10	12	13	14	14	14	12
25	0.69	3.3	6.1	7.3	8.8	10	11	13	14	15	15	15	14
27	0.74	3.5	6.8	7.9	9.5	11	12	14	15	16	18	18	16
29	0.80	3.8	7.3	8.5	10	12	13	15	16	18	19	19	18
31	0.85	4.1	7.8	9.1	11	13	14	16	18	19	20	20	19
33	0.90	4.4	8.3	9.8	12	14	15	18	19	21	21	21	20
35	0.96	4.6	8.8	10	13	15	16	19	20	23	23	23	21
37	1.0	4.9	9.1	11	14	15	16	20	21	24	24	24	...
40	1.1	5.3	10	12	15	16	18	21	24	25	26	26	...
45	1.3	6.0	11	13	16	19	20	24	26	28	29	...	...
50	1.4	6.6	13	15	18	20	23	26	29	30	...	...	...
	TYPE I			TYPE II				TYPE III					

No. of Teeth Small Sprkt.	1/2-Inch Pitch Chain										
	Revolutions per Minute — Small Sprocket										
	100	500	700	1000	1200	1800	2000	2500	3000	3500	4000
*17	0.83	3.8	5.0	6.3	7.5	10	11	11	11	11	...
*19	0.93	3.8	5.0	7.5	8.8	11	13	14	14	14	...
21	1.0	5.0	6.3	8.8	10	14	14	15	16	16	...
23	1.1	5.0	7.5	10	11	15	16	18	19	19	18
25	1.2	5.0	7.5	10	13	16	18	20	21	21	20
27	1.3	6.3	8.8	11	13	18	19	21	24	24	23
29	1.4	6.3	8.8	13	14	19	21	24	25	25	25
31	1.5	7.5	10	13	15	21	23	25	28	28	28
33	1.6	7.5	10	14	16	23	24	28	29	30	29
35	1.8	7.5	11	15	18	24	25	29	31	31	30
37	1.9	8.8	11	16	19	25	26	30	33	33	...
40	2.0	8.8	13	18	20	28	29	33	35	35	...
45	2.5	10	14	19	23	30	30	36	39	...	...
50	2.5	10	15	21	25	34	36	40	...	...	...
	TYPE I			TYPE II				TYPE III			

* For best results, smaller sprocket should have at least 21 teeth.

Table 13 (*Continued*). Horsepower Ratings per Inch of Chain Width for Silent Chain Drives — 1982

No. of Teeth Small Sprkt.	⅝-Inch Pitch Chain Revolutions per Minute — Small Sprocket									
	100	500	700	1000	1200	1800	2000	2500	3000	3500
*17	1.3	6.3	7.5	10	11	14	15	14	...	...
*19	1.4	6.3	8.8	13	14	16	18	18	...	...
21	1.6	7.5	10	13	15	19	20	20	20	...
23	1.8	7.5	11	15	16	21	23	24	23	...
25	1.9	8.8	11	16	19	24	25	26	26	24
27	2.0	10	13	18	20	26	28	29	29	26
29	2.1	10	14	19	21	28	30	31	31	29
31	2.4	11	15	20	23	30	31	34	34	31
33	2.5	11	16	21	25	33	34	36	36	34
35	2.6	13	16	23	26	34	36	39	39	35
37	2.8	13	18	24	28	36	39	43	41	...
40	3.0	14	19	26	30	39	41	44	...	...
45	3.4	16	21	29	34	44	46	...	...	...
50	3.8	18	24	33	38	48	50	...	...	...
	TYPE I			TYPE II			TYPE III			

No. of Teeth Small Sprkt.	¾-Inch Pitch Chain Revolutions per Minute — Small Sprocket								
	100	500	700	1000	1200	1500	1800	2000	2500
*17	1.9	8.1	11	14	15	16	18	18	...
*19	2.0	9.3	13	15	18	20	21	21	...
21	2.3	10	14	18	20	23	24	25	24
23	2.5	11	15	20	23	25	28	28	28
25	2.8	13	16	21	25	29	31	31	30
27	2.9	14	18	24	28	31	34	35	35
29	3.1	15	20	26	30	34	36	38	38
31	3.4	15	21	28	31	36	40	41	41
33	3.6	16	23	30	34	39	43	44	44
35	3.8	18	24	31	36	41	45	46	46
37	4.0	19	25	34	39	44	48	49	49
40	4.4	20	28	36	41	48	51	53	53
45	4.9	23	30	40	46	53	56	58	...
50	5.4	25	34	45	51	58	61	...	...
	TYPE I			TYPE II			TYPE III		

No. of Teeth Small Sprkt.	1-Inch Pitch Chain Revolutions per Minute — Small Sprocket										
	100	200	300	400	500	700	1000	1200	1500	1800	2000
*17	3.8	6.3	8.8	11	14	18	21	23	...	...	...
*19	3.8	7.5	10	13	15	20	25	26	28	...	...
21	3.8	7.5	11	15	18	23	29	31	33	33	...
23	3.8	8.8	13	16	19	25	31	35	38	38	...
25	5.0	8.8	14	18	21	28	35	39	41	41	41
27	5.0	10	15	19	24	30	39	43	46	46	45
29	5.0	11	16	20	25	33	41	46	50	51	50
31	6.3	11	16	23	28	35	45	50	54	55	54
33	6.3	13	18	24	29	38	49	54	59	59	58
35	6.3	13	19	25	30	40	51	56	61	63	61
37	6.8	14	20	26	33	43	54	60	65	66	...
40	7.5	15	23	29	35	45	59	65	70	...	...
45	8.8	16	25	31	39	51	65	71	76	...	...
50	10	19	28	35	43	56	71	78	...	...	...
	TYPE I			TYPE II			TYPE III			

* For best results, smaller sprocket should have at least 21 teeth.

Table 13 (*Concluded*). **Horsepower Ratings Per Inch of Chain Width for Silent Chain Drives — 1982**

No. of Teeth Small Sprkt.	1¼-Inch Pitch Chain Revolutions per Minute — Small Sprocket										
	100	200	300	400	500	600	700	800	1000	1200	1500
*19	5.6	10	15	20	24	26	29	31	34	35	...
21	6.3	11	18	23	26	30	33	36	40	41	...
23	6.9	13	19	24	29	34	36	40	45	46	46
25	7.5	14	20	26	31	36	40	44	50	53	53
27	8.0	15	23	29	35	40	44	49	54	58	58
29	8.6	16	24	31	38	43	48	53	59	63	64
31	9.3	18	26	34	40	46	51	56	64	68	69
33	9.9	19	28	35	43	49	55	60	69	73	74
35	11	20	29	38	45	53	59	64	73	78	78
37	11	21	30	40	48	55	63	68	76	81	...
40	12	24	34	44	53	60	68	74	83	88	...
45	13	26	38	49	59	68	75	81	91	...	...
50	15	29	43	54	65	74	83	90	100	...	...
	TYPE I			TYPE II				TYPE III			

No. of Teeth Small Sprkt.	1½-Inch Pitch Chain Revolutions per Minute — Small Sprocket										
	100	200	300	400	500	600	700	800	900	1000	1200
*19	8.0	15	21	28	31	35	39	40	41	43	...
21	8.8	16	24	30	36	40	44	46	49	49	...
23	10	19	26	34	40	45	49	53	55	56	55
25	10	20	29	38	44	50	55	59	61	65	64
27	11	23	31	40	48	54	60	64	68	70	70
29	13	24	34	44	51	58	65	70	74	75	76
31	14	25	36	46	55	64	70	75	79	81	83
33	14	28	39	50	59	68	75	80	85	88	89
35	15	29	41	53	63	71	79	85	90	93	94
37	16	30	44	59	66	76	84	90	96	99	...
40	18	33	48	66	73	83	90	98	105	...	...
45	19	38	54	68	81	93	101	108	113	...	...
50	21	41	59	75	89	101	111	118	...	...	...
	TYPE I		TYPE II				TYPE III				

No. of Teeth Small Sprkt.	2-Inch Pitch Chain Revolutions per Minute — Small Sprocket								
	100	200	300	400	500	600	700	800	900
*19	14	26	36	44	50	54	56	...	...
21	16	29	40	50	53	63	65	...	...
23	17	33	45	55	64	70	74	75	...
25	18	35	49	61	70	78	83	85	85
27	20	38	54	66	78	85	91	94	94
29	21	41	58	73	84	93	99	103	103
31	23	44	63	78	90	100	106	110	110
33	25	46	66	83	96	106	114	118	118
35	26	50	71	88	103	114	121	125	125
37	28	53	75	93	110	124	128	131	...
40	30	58	81	101	118	129	138	141	...
45	34	64	90	113	131	144	151	...	...
50	38	71	100	125	144	156	...	...	...
	TYPE I		TYPE II				TYPE III		

* For best results, smaller sprocket should have at least 21 teeth.

Sprocket Wheels for Ordinary Link Chain

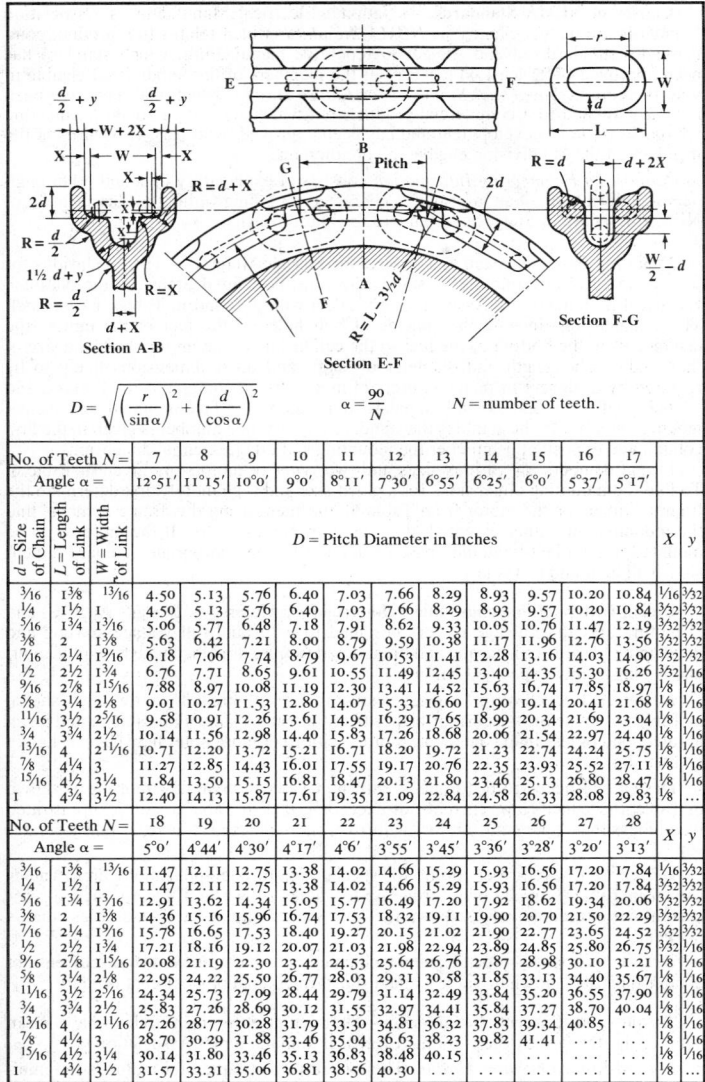

$$D = \sqrt{\left(\frac{r}{\sin \alpha}\right)^2 + \left(\frac{d}{\cos \alpha}\right)^2} \qquad \alpha = \frac{90}{N} \qquad N = \text{number of teeth.}$$

No. of Teeth N =	7	8	9	10	11	12	13	14	15	16	17		
Angle α =	12°51'	11°15'	10°0'	9°0'	8°11'	7°30'	6°55'	6°25'	6°0'	5°37'	5°17'		
d = Size of Chain / L = Length of Link / W = Width of Link			D = Pitch Diameter in Inches										X	y
3/16, 1 3/8, 13/16	4.50	5.13	5.76	6.40	7.03	7.66	8.29	8.93	9.57	10.20	10.84	1/16	3/32
1/4, 1 1/2, 1	4.50	5.13	5.76	6.40	7.03	7.66	8.29	8.93	9.57	10.20	10.84	3/32	3/32
5/16, 1 3/4, 1 3/16	5.06	5.77	6.48	7.18	7.91	8.62	9.33	10.05	10.76	11.47	12.19	3/32	3/32
3/8, 2, 1 3/8	5.63	6.42	7.21	8.00	8.79	9.59	10.38	11.17	11.96	12.76	13.56	3/32	3/32
7/16, 2 1/4, 1 9/16	6.18	7.06	7.74	8.79	9.67	10.53	11.41	12.28	13.16	14.03	14.90	3/32	3/32
1/2, 2 1/2, 1 3/4	6.76	7.71	8.65	9.61	10.55	11.49	12.45	13.40	14.35	15.30	16.26	3/32	1/16
9/16, 2 7/8, 1 15/16	7.88	8.97	10.08	11.19	12.30	13.41	14.52	15.63	16.74	17.85	18.97	1/8	1/16
5/8, 3 1/4, 2 1/8	9.01	10.27	11.53	12.80	14.07	15.33	16.60	17.90	19.14	20.41	21.68	1/8	1/16
11/16, 3 1/2, 2 5/16	9.58	10.91	12.26	13.61	14.95	16.29	17.65	18.99	20.34	21.69	23.04	1/8	1/16
3/4, 3 3/4, 2 1/2	10.14	11.56	12.98	14.40	15.83	17.26	18.68	20.06	21.54	22.97	24.40	1/8	1/16
13/16, 4, 2 11/16	10.71	12.20	13.72	15.21	16.71	18.20	19.72	21.23	22.74	24.24	25.75	1/8	1/16
7/8, 4 1/4, 3	11.27	12.85	14.43	16.01	17.55	19.17	20.76	22.35	23.93	25.52	27.11	1/8	1/16
15/16, 4 1/2, 3 1/4	11.84	13.50	15.15	16.81	18.47	20.13	21.80	23.46	25.13	26.80	28.47	1/8	1/16
1, 4 3/4, 3 1/2	12.40	14.13	15.87	17.61	19.35	21.09	22.84	24.58	26.33	28.08	29.83	1/8	...

No. of Teeth N =	18	19	20	21	22	23	24	25	26	27	28		
Angle α =	5°0'	4°44'	4°30'	4°17'	4°6'	3°55'	3°45'	3°36'	3°28'	3°20'	3°13'	X	y
3/16, 1 3/8, 13/16	11.47	12.11	12.75	13.38	14.02	14.66	15.29	15.93	16.56	17.20	17.84	1/16	3/32
1/4, 1 1/2, 1	11.47	12.11	12.75	13.38	14.02	14.66	15.29	15.93	16.56	17.20	17.84	3/32	3/32
5/16, 1 3/4, 1 3/16	12.91	13.62	14.34	15.05	15.77	16.49	17.20	17.92	18.62	19.34	20.06	3/32	3/32
3/8, 2, 1 3/8	14.36	15.16	15.96	16.74	17.53	18.32	19.11	19.90	20.70	21.50	22.29	3/32	3/32
7/16, 2 1/4, 1 9/16	15.78	16.65	17.53	18.40	19.27	20.15	21.02	21.90	22.77	23.65	24.52	3/32	3/32
1/2, 2 1/2, 1 3/4	17.21	18.16	19.12	20.07	21.03	21.98	22.94	23.89	24.85	25.80	26.75	3/32	1/16
9/16, 2 7/8, 1 15/16	20.08	21.19	22.30	23.42	24.53	25.64	26.76	27.87	28.98	30.10	31.21	1/8	1/16
5/8, 3 1/4, 2 1/8	22.95	24.22	25.50	26.77	28.03	29.31	30.58	31.85	33.13	34.40	35.67	1/8	1/16
11/16, 3 1/2, 2 5/16	24.34	25.73	27.09	28.44	29.79	31.14	32.49	33.84	35.20	36.55	37.90	1/8	1/16
3/4, 3 3/4, 2 1/2	25.83	27.26	28.69	30.12	31.55	32.97	34.41	35.84	37.27	38.70	40.04	1/8	1/16
13/16, 4, 2 11/16	27.26	28.77	30.28	31.79	33.30	34.81	36.32	37.83	39.34	40.85	...	1/8	1/16
7/8, 4 1/4, 3	28.70	30.29	31.88	33.46	35.04	36.63	38.23	39.82	41.41	...	...	1/8	1/16
15/16, 4 1/2, 3 1/4	30.14	31.80	33.46	35.13	36.83	38.48	40.15	...	...	...	...	1/8	1/16
1, 4 3/4, 3 1/2	31.57	33.31	35.06	36.81	38.56	40.30	...	...	...	...	...	1/8	...

STANDARDS FOR ELECTRIC MOTORS

Classes of NEMA Standards. — National Electrical Manufacturers Association Standards are of two classes: 1. *NEMA Standard*, which relates to a product commercially standardized and subject to repetitive manufacture, which standard has been approved by at least 90 per cent of the members of the Subdivision eligible to vote thereon; 2. *Suggested Standard for Future Design*, which may not have been regularly applied to a commercial product, but which suggests a sound engineering approach to future development and has been approved by at least two-thirds of the members of the Subdivision eligible to vote thereon.

Authorized Engineering Information consists of explanatory data and other engineering information of an informative character not falling within the classification of NEMA Standard or Suggested Standard for Future Design.

Mounting Dimensions and Frame Sizes for Electric Motors. — The dimensions for foot-mounted electric motors as standardized in the United States by the National Electrical Manufacturers Association (NEMA) will be found in Tables 1 to 3, incl. These dimensions include the spacing of bolt holes in the feet of the motor, the distance from the bottom of the feet to the center-line of the motor shaft, the size of the conduit, the length and diameter of shaft, and other dimensions likely to be required by designers or manufacturers of motor-driven equipment. In Tables 4 and 5, incl., will be found NEMA standard dimensions for face-mounted and flange-mounted motors. In these tables the standard motor frame number is given in the first column and opposite this number the mounting and other essential dimensions.

Frame numbers for various sizes and two types of motors are given in Table 6. To find the mounting dimensions for a given size and type motor, first determine the frame number for the motor from Table 6, and then, using this frame number, find the mounting and other essential dimensions in Tables 1 to 5. If the motor is to be mounted upon a belt-tightening base or upon rails, the appropriate standard dimensions will be found in Table 3.

Design Letters of Polyphase Integral-horsepower Motors. — Designs A, B, C, and D motors are squirrel-cage motors designed to withstand full voltage starting and developing locked-rotor torque and breakdown torque, drawing locked-rotor current, and having a slip as specified below:

Design A: Locked-rotor torque as shown in Table 9, breakdown torque as shown in Table 10, locked-rotor current higher than the values shown in Table 8, and a slip at rated load of less than 5 per cent. Motors with 10 or more poles may have a slightly greater slip.

Design B: Locked-rotor torque as shown in Table 9, breakdown torque as shown in Table 10, locked-rotor current not exceeding that in Table 8, and a slip at rated load of less than 5 per cent. Motors with 10 or more poles may have a slightly greater slip.

Design C: Locked-rotor torque for special high-torque applications up to values shown in Table 9, breakdown torque up to values shown in Table 10, locked-rotor current not exceeding values shown in Table 8 and a slip at rated load of less than 5 per cent.

Design D: Locked-rotor torque as indicated in Table 9, locked-rotor current not greater than that shown in Table 8 and a slip at rated load of 5 per cent or more.

Torque and Current Definitions. — The definitions which follow have been adopted as standard by the National Electrical Manufacturers Association.

Locked-Rotor or Static Torque: The locked-rotor torque of a motor is the mini-

mum torque which it will develop at rest for all angular positions of the rotor, with rated voltage applied at rated frequency.

Breakdown Torque: The breakdown torque of a motor is the maximum torque which the motor will develop, with rated voltage applied at rated frequency, without an abrupt drop in speed.

Full-Load Torque: The full-load torque of a motor is the torque necessary to produce its rated horsepower at full load speed. In pounds at 1-foot radius, it is equal to the horsepower times 5252 divided by the full-load speed.

Pull-Out Torque: The pull-out torque of a synchronous motor is the maximum sustained torque which the motor will develop at synchronous speed with rated voltage applied at rated frequency and with normal excitation.

Pull-In Torque: The pull-in torque of a synchronous motor is the maximum constant torque under which the motor will pull its connected inertia load into synchronism at rated voltage and frequency, when its field excitation is applied.

Pull-Up Torque: The pull-up torque of an alternating current motor is the minimum torque developed by the motor during the period of acceleration from rest to the speed at which breakdown torque occurs. For motors which do not have a definite breakdown torque, the pull-up torque is the minimum torque developed up to rated speed.

Locked Rotor Current: The locked rotor current of a motor is the steady-state current taken from the line with the rotor locked and with rated voltage (and rated frequency in the case of alternating-current motors) applied to the motor.

Standard Direction of Motor Rotation. — The standard direction of rotation for all non-reversing direct-current motors, all alternating-current single-phase motors, all synchronous motors, and all universal motors, is *counterclockwise* when facing that end of the motor opposite the drive.

This rule does not apply to two- and three-phase induction motors, as in most applications the phase sequence of the power lines is rarely known.

Motor Types According to Variability of Speed. — Five types of motors classified according to variability of speed are:

Constant-speed Motors: In this type of motor the normal operating speed is constant or practically constant; for example, a synchronous motor, an induction motor with small slip, or a direct-current shunt-wound motor.

Varying-speed Motor: In this type of motor, the speed varies with the load, ordinarily decreasing when the load increases; such as a series-wound or repulsion motor.

Adjustable-speed Motor: In this type of motor, the speed can be varied gradually over a considerable range, but when once adjusted remains practically unaffected by the load; such as a direct-current shunt-wound motor with field resistance control designed for a considerable range of speed adjustment.

The base speed of an adjustable-speed motor is the lowest rated speed obtained at rated load and rated voltage at the temperature rise specified in the rating.

Adjustable Varying-speed Motor: This type of motor is one in which the speed can be adjusted gradually, but when once adjusted for a given load will vary in considerable degree with the change in load; such as a direct-current compound-wound motor adjusted by field control or a wound-rotor induction motor with rheostatic speed control.

Multispeed Motor: This type of motor is one which can be operated at any one of two or more definite speeds, each being practically independent of the load; such as a direct-current motor with two armature windings or an induction motor with windings capable of various pole groupings. In the case of multispeed permanent-split capacitor and shaded pole motors, the speeds are dependent upon the load.

Table I. NEMA Standard Dimensions for Alternating-current Foot-mounted Motors with Single Straight-shaft Extension

Frame No.	A Max	B Max	D*	E†	2F†	BA	H†	U	N-W	V Min	Keyseat ES Min	S
42	...	...	2.62	1.75	1.69	2.06	0.28	0.3750	1.12	...	...	flat
48	...	...	3.00	2.12	2.75	2.50	0.34	0.5000	1.50	...	...	flat
48H	...	...	3.00	2.12	4.75	2.50	0.34	0.5000	1.50	...	...	flat
56	...	...	3.50	2.44	3.00	2.75	0.34	0.6250	1.88	...	1.41	0.188
56H	...	...	3.50	2.44	5.00	2.75	0.34	0.6250	1.88	...	1.41	0.188
143T	7.0	6.0	3.50	2.75	4.00	2.25	0.34	0.8750	2.25	2.00	1.41	0.188
145T	7.0	6.0	3.50	2.75	5.00	2.25	0.34	0.8750	2.25	2.00	1.41	0.188
182T	9.0	6.5	4.50	3.75	4.50	2.75	0.41	1.1250	2.75	2.50	1.78	0.250
184T	9.0	7.5	4.50	3.75	5.50	2.75	0.41	1.1250	2.75	2.50	1.78	0.250
213T	10.5	7.5	5.25	4.25	5.50	3.50	0.41	1.3750	3.38	3.12	2.41	0.312
215T	10.5	9.0	5.25	4.25	7.00	3.50	0.41	1.3750	3.38	3.12	2.41	0.312
254T	12.5	10.8	6.25	5.00	8.25	4.25	0.53	1.625	4.00	3.75	2.91	0.375
256T	12.5	12.5	6.25	5.00	10.00	4.25	0.53	1.625	4.00	3.75	2.91	0.375
284T	14.0	12.5	7.00	5.50	9.50	4.75	0.53	1.875	4.62	4.38	3.28	0.500
284TS	14.0	12.5	7.00	5.50	9.50	4.75	0.53	1.625	3.25	3.00	1.91	0.375
286T	14.0	14.0	7.00	5.50	11.00	4.75	0.53	1.875	4.62	4.38	3.28	0.500
286TS	14.0	14.0	7.00	5.50	11.00	4.75	0.53	1.625	3.25	3.00	1.91	0.375
324T	16.0	14.0	8.00	6.25	10.50	5.25	0.66	2.125	5.25	5.00	3.91	0.500
324TS	16.0	14.0	8.00	6.25	10.50	5.25	0.66	1.875	3.75	3.50	2.03	0.500
326T	16.0	15.5	8.00	6.25	12.00	5.25	0.66	2.125	5.25	5.00	3.91	0.500
326TS	16.0	15.5	8.00	6.25	12.00	5.25	0.66	1.875	3.75	3.50	2.03	0.500
364T	18.0	15.2	9.00	7.00	11.25	5.88	0.66	2.375	5.88	5.62	4.28	0.625
364TS	18.0	15.2	9.00	7.00	11.25	5.88	0.66	1.875	3.75	3.50	2.03	0.500
365T	18.0	16.2	9.00	7.00	12.25	5.88	0.66	2.375	5.88	5.62	4.28	0.625
365TS	18.0	16.2	9.00	7.00	12.25	5.88	0.66	1.875	3.75	3.50	2.03	0.500
404T	20.0	16.2	10.00	8.00	12.25	6.62	0.81	2.875	7.25	7.00	5.65	0.750
404TS	20.0	16.2	10.00	8.00	12.25	6.62	0.81	2.125	4.25	4.00	2.78	0.500
405T	20.0	17.8	10.00	8.00	13.75	6.62	0.81	2.875	7.25	7.00	5.65	0.750
405TS	20.0	17.8	10.00	8.00	13.75	6.62	0.81	2.125	4.25	4.00	2.78	0.500
444T	22.0	18.5	11.00	9.00	14.50	7.50	0.81	3.375	8.50	8.25	6.91	0.875
444TS	22.0	18.5	11.00	9.00	14.50	7.50	0.81	2.375	4.75	4.50	3.03	0.625
445T	22.0	20.5	11.00	9.00	16.50	7.50	0.81	3.375	8.50	8.25	6.91	0.875
445TS	22.0	20.5	11.00	9.00	16.50	7.50	0.81	2.375	4.75	4.50	3.03	0.625

All dimensions are in inches. See Fig. 1 for diagram showing letter symbols.

* Dimension D will never be greater than the above values for rigid-base motors. However, it may be less, so that shims are usually required for coupled or geared motors. When the exact dimension is required, shims up to 0.03 inch may be necessary on frame sizes whose D dimension is 8.00 inches or less; on larger frames, shims up to 0.06 inch may be necessary. No tolerances have been established for the D dimension of resilient mounted motors.

† Frame Nos. 42, 48, 48H, 56 and 56H have a tolerance for the 2F dimension of ± 0.03 inch and for the H dimension (width of slot) + 0.02, − 0 inch. For frame Nos. 143T to 445T, inclusive, the tolerance for the 2E and 2F dimensions is ± 0.03 inch and for the H dimension (diameter of hole) is + 0.05, − 0 inch.

The minimum size of the threaded or clearance hole, AA, for external conduit entrance (expressed in conduit size) to the terminal housing for frame Nos. 143T through 184T, ¾ inch; for frame Nos. 213T and 215T, 1 inch; for frame Nos. 254T and 256T, 1¼ inches; for frame Nos. 284T through 286TS, 1½ inches; for frame Nos. 324T through 326TS, 2 inches; and for frame Nos. 364 through 445TS, 3 inches.

For larger frame sizes see NEMA Standards.

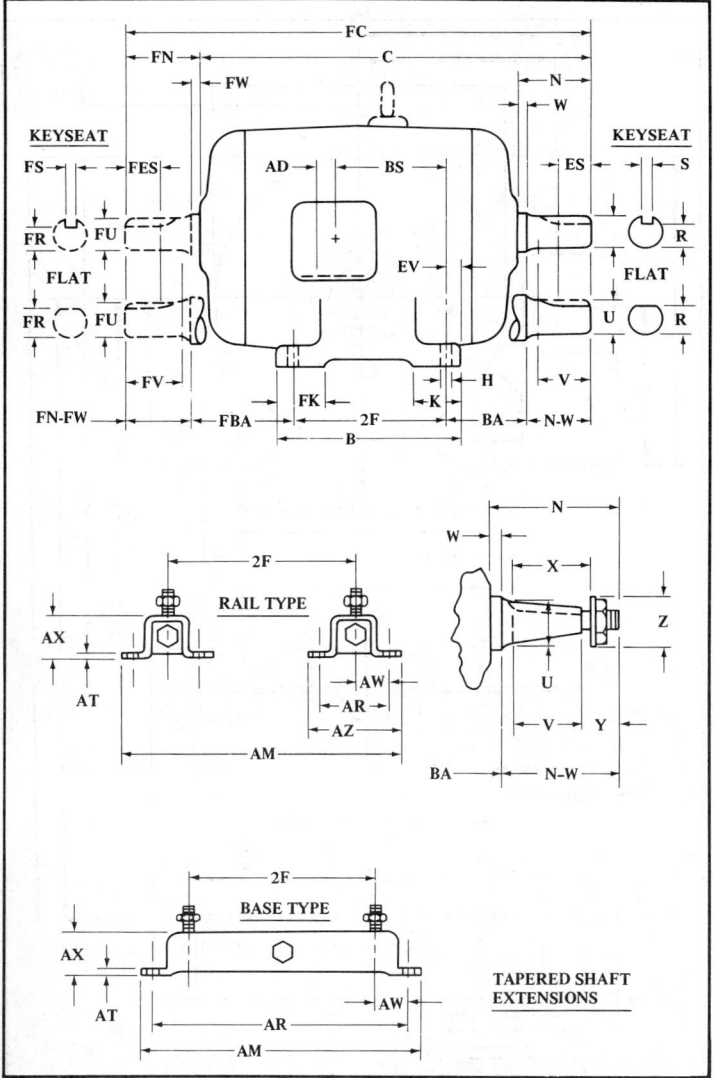

Fig. 1a.   NEMA Standard dimensional designations for alternating-current and direct-current foot-mounted motors—side view.

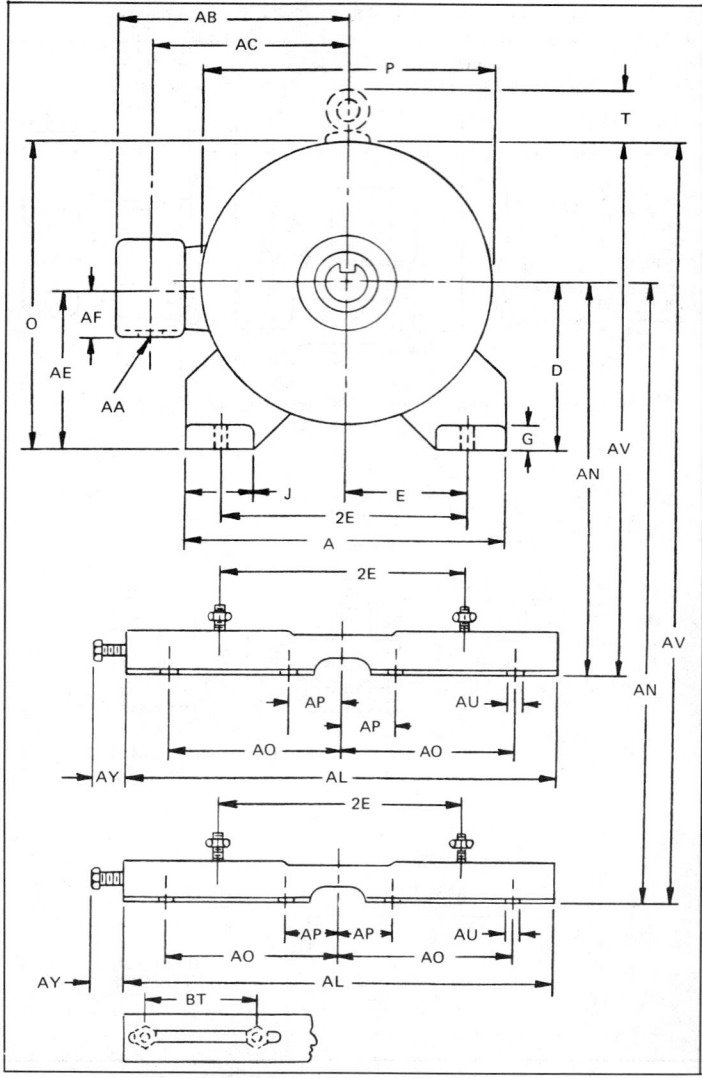

Fig. 1b. NEMA Standard dimensional designations for alternating-current and direct-current foot-mounted motors—drive end view.

Table 2. NEMA Standard Shaft Extension and Keyseat Dimensions for Alternating-current, Foot-mounted Motors with Single Tapered or Double Straight/Tapered Shaft Extension

Frame No.	Drive End—Tapered Shaft Extension†										
	BA	U	N–W	V	X	Y	Z max	Shaft Threads	Keyseat		Key Length‡
									Width	Depth	
143TR and 145TR	2.25	0.8750	2.62	1.75	1.88	0.75	1.38	⅝–18	0.188	0.094	1.50
182TR and 184TR	2.75	1.1250	3.38	2.25	2.38	0.88	1.50	¾–16	0.250	0.125	2.00
213TR and 215TR	3.50	1.3750	4.12	2.62	2.75	1.25	2.00	1 –14	0.312	0.156	2.38
254TR and 256TR	4.25	1.625	4.50	2.88	3.00	1.25	2.00	1 –14	0.375	0.188	2.62
284TR and 286TR	4.75	1.875	4.75	3.12	3.25	1.25	2.38	1¼–12	0.500	0.250	2.88
324TR and 326TR	5.25	2.125	5.25	3.62	3.62	1.38	2.75	1½–8	0.500	0.250	3.25
364TR and 365TR	5.88	2.375	5.75	3.75	3.88	1.50	3.25	1¾–8	0.625	0.312	3.50
404TR and 405TR	6.62	2.875	6.62	4.38	4.50	1.75	3.62	2 –8	0.750	0.375	4.12
444TR and 445TR	7.50	3.375	7.50	5.00	5.12	2.00	4.12	2¼–8	0.875	0.438	4.75

Opposite Drive End—Straight Shaft Extension**

Frame No. Series	FU	FN–FW	FV Min	Keyseat*		
				R	ES Min	S
140	0.6250	1.62	1.38	0.517	0.91	0.188
180	0.8750	2.25	2.00	0.771	1.41	0.188
210	1.1250	2.75	2.50	0.986	1.78	0.250
250	1.3750	3.38	3.12	1.201	2.41	0.312
280	1.625	4.00	3.75	1.416	2.91	0.375
280 Short Shaft	1.625	3.25	3.00	1.416	1.91	0.375
320	1.875	4.62	4.38	1.591	3.28	0.500
320 Short Shaft	1.875	3.75	3.50	1.591	2.03	0.500
360	1.875	4.62	4.38	1.591	3.28	0.500
360 Short Shaft	1.875	3.75	3.50	1.591	2.03	0.500
400	2.125	5.25	5.00	1.845	3.91	0.500
400 Short Shaft	2.125	4.25	4.00	1.845	2.78	0.500
440	2.375	5.88	5.62	2.021	4.28	0.625
440 Short Shaft	2.375	4.75	4.50	2.021	3.03	0.625

Opposite Drive End—Tapered Shaft Extension†**

Frame No. Series	FU	FN–FW	FV	FX	FY	FZ Max	Shaft Threads	Keyseat		Key Length‡
								Width	Depth	
140	0.6250	2.00	1.38	1.50	0.50	1.12	⅜–24	0.188	0.094	1.12
180	0.8750	2.62	1.75	1.88	0.75	1.38	⅝–18	0.188	0.094	1.50
210	1.1250	3.38	2.25	2.38	0.88	1.50	¾–16	0.250	0.125	2.00
250	1.3750	4.12	2.62	2.75	1.25	2.00	1 –14	0.312	0.156	2.38
280	1.6250	4.50	2.88	3.00	1.25	2.00	1 –14	0.375	0.188	2.62
320	1.8750	4.75	3.12	3.25	1.25	2.38	1¼–12	0.500	0.250	2.88
360	1.8750	4.75	3.12	3.25	1.25	2.38	1¼–12	0.500	0.250	2.88
400	2.1250	5.25	3.50	3.62	1.38	2.75	1½–8	0.500	0.250	3.25
440	2.3750	5.75	3.75	3.88	1.50	3.25	1¾–8	0.625	0.312	3.50

All dimensions are in inches.
See Fig. 1 for diagram showing letter dimensions.
* Motors furnished with keyseats cut in the shaft extension for pulley, coupling, pinion, etc., are usually furnished with a key.
** For drive applications other than direct connected, the motor manufacturer should be consulted.
† The threaded end of the tapered shaft is furnished with a nut and suitable locking device. The taper of the shaft is at the rate of 1.25 inches in diameter per foot of length.
‡ The tolerance on the length of key is ±0.03 inch.

Table 3. NEMA Standard Dimensions for Foot-mounted Industrial Direct-current Motors

Frame No.	A Max	B Max	D*	E†	2F†	BA	H† Hole
182AT	9.00	6.50	4.50	3.75	4.50	2.75	0.41
183AT	9.00	7.00	4.50	3.75	5.00	2.75	0.41
184AT	9.00	7.50	4.50	3.75	5.50	2.75	0.41
185AT	9.00	8.25	4.50	3.75	6.25	2.75	0.41
186AT	9.00	9.00	4.50	3.75	7.00	2.75	0.41
187AT	9.00	10.00	4.50	3.75	8.00	2.75	0.41
188AT	9.00	11.00	4.50	3.75	9.00	2.75	0.41
189AT	9.00	12.00	4.50	3.75	10.00	2.75	0.41
1810AT	9.00	13.00	4.50	3.75	11.00	2.75	0.41
213AT	10.50	7.50	5.25	4.25	5.50	3.50	0.41
214AT	10.50	8.25	5.25	4.25	6.25	3.50	0.41
215AT	10.50	9.00	5.25	4.25	7.00	3.50	0.41
216AT	10.50	10.00	5.25	4.25	8.00	3.50	0.41
217AT	10.50	11.00	5.25	4.25	9.00	3.50	0.41
218AT	10.50	12.00	5.25	4.25	10.00	3.50	0.41
219AT	10.50	13.00	5.25	4.25	11.00	3.50	0.41
2110AT	10.50	14.50	5.25	4.25	12.50	3.50	0.41
253AT	12.50	9.50	6.25	5.00	7.00	4.25	0.53
254AT	12.50	10.75	6.25	5.00	8.25	4.25	0.53
255AT	12.50	11.50	6.25	5.00	9.00	4.25	0.53
256AT	12.50	12.50	6.25	5.00	10.00	4.25	0.53
257AT	12.50	13.50	6.25	5.00	11.00	4.25	0.53
258AT	12.50	15.00	6.25	5.00	12.50	4.25	0.53
259AT	12.50	16.50	6.25	5.00	14.00	4.25	0.53
283AT	14.00	11.00	7.00	5.50	8.00	4.75	0.53
284AT	14.00	12.50	7.00	5.50	9.50	4.75	0.53
285AT	14.00	13.00	7.00	5.50	10.00	4.75	0.53
286AT	14.00	14.00	7.00	5.50	11.00	4.75	0.53
287AT	14.00	15.50	7.00	5.50	12.50	4.75	0.53
288AT	14.00	17.00	7.00	5.50	14.00	4.75	0.53
289AT	14.00	19.00	7.00	5.50	16.00	4.75	0.53
323AT	16.00	12.50	8.00	6.25	9.00	5.25	0.66
324AT	16.00	14.00	8.00	6.25	10.50	5.25	0.66
325AT	16.00	14.50	8.00	6.25	11.00	5.25	0.66
326AT	16.00	15.50	8.00	6.25	12.00	5.25	0.66
327AT	16.00	17.50	8.00	6.25	14.00	5.25	0.66
328AT	16.00	19.50	8.00	6.25	16.00	5.25	0.66
329AT	16.00	21.50	8.00	6.25	18.00	5.25	0.66
363AT	18.00	14.00	9.00	7.00	10.00	5.88	0.81
364AT	18.00	15.25	9.00	7.00	11.25	5.88	0.81
365AT	18.00	16.25	9.00	7.00	12.25	5.88	0.81
366AT	18.00	18.00	9.00	7.00	14.00	5.88	0.81
367AT	18.00	20.00	9.00	7.00	16.00	5.88	0.81
368AT	18.00	22.00	9.00	7.00	18.00	5.88	0.81
369AT	18.00	24.00	9.00	7.00	20.00	5.88	0.81
403AT	20.00	15.00	10.00	8.00	11.00	6.62	0.94
404AT	20.00	16.25	10.00	8.00	12.25	6.62	0.94
405AT	20.00	17.75	10.00	8.00	13.75	6.62	0.94
406AT	20.00	20.00	10.00	8.00	16.00	6.62	0.94
407AT	20.00	22.00	10.00	8.00	18.00	6.62	0.94
408AT	20.00	24.00	10.00	8.00	20.00	6.62	0.94
409AT	20.00	26.00	10.00	8.00	22.00	6.62	0.94
443AT	22.00	16.50	11.00	9.00	12.50	7.50	1.06
444AT	22.00	18.50	11.00	9.00	15.00	7.50	1.06
445AT	22.00	20.50	11.00	9.00	16.50	7.50	1.06
446AT	22.00	22.00	11.00	9.00	18.00	7.50	1.06

All dimensions are in inches. See Fig. 1 for diagram showing letter symbols.

* Dimension D will never be greater than the values shown in this table, but it may be less so that shims are usually required for coupled or geared motors. When the exact dimension is required, shims up to 0.03 inch may be necessary on frame sizes whose dimension D is 8 inches or less; on larger frame sizes, shims up to 0.06 inch may be necessary.

† The tolerance for the $2E$ and $2F$ dimensions is ±0.03 inch and for the H dimensions, +0.05, −0 inch.

For larger frame sizes see NEMA Standards.

Table 3. (*Continued*). **NEMA Standard Dimensions for Foot-mounted Industrial Direct-current Motors**

Frame No.	AL	AM	AO	AR	AU	AX	AY Max Bases	BT
182AT	12.75	9.50	4.50	4.25	0.50	1.50	0.50	3.00
183AT	12.75	10.00	4.50	4.50	0.50	1.50	0.50	3.00
184AT	12.75	10.50	4.50	4.75	0.50	1.50	0.50	3.00
185AT	12.75	11.25	4.50	5.12	0.50	1.50	0.50	3.00
186AT	12.75	12.00	4.50	5.50	0.50	1.50	0.50	3.00
187AT	12.75	13.00	4.50	6.00	0.50	1.50	0.50	3.00
188AT	12.75	14.00	4.50	6.50	0.50	1.50	0.50	3.00
189AT	12.75	15.00	4.50	7.00	0.50	1.50	0.50	3.00
1810AT	12.75	16.00	4.50	7.50	0.50	1.50	0.50	3.00
213AT	15.00	11.00	5.25	4.75	0.50	1.75	0.50	3.50
214AT	15.00	11.75	5.25	5.12	0.50	1.75	0.50	3.50
215AT	15.00	12.50	5.25	5.50	0.50	1.75	0.50	3.50
216AT	15.00	13.50	5.25	6.00	0.50	1.75	0.50	3.50
217AT	15.00	14.50	5.25	6.50	0.50	1.75	0.50	3.50
218AT	15.00	15.50	5.25	7.00	0.50	1.75	0.50	3.50
219AT	15.00	16.50	5.25	7.50	0.50	1.75	0.50	3.50
2110AT	15.00	18.00	5.25	8.25	0.50	1.75	0.50	3.50
253AT	17.75	13.88	6.25	6.00	0.62	2.00	0.62	4.00
254AT	17.75	15.12	6.25	6.62	0.62	2.00	0.62	4.00
255AT	17.75	15.88	6.25	7.00	0.62	2.00	0.62	4.00
256AT	17.75	16.88	6.25	7.50	0.62	2.00	0.62	4.00
257AT	17.75	17.88	6.25	8.00	0.62	2.00	0.62	4.00
258AT	17.75	19.38	6.25	8.75	0.62	2.00	0.62	4.00
259AT	17.75	20.88	6.25	9.50	0.62	2.00	0.62	4.00
283AT	19.75	15.38	7.00	6.75	0.62	2.00	0.62	4.00
284AT	19.75	16.88	7.00	7.50	0.62	2.00	0.62	4.00
285AT	19.75	17.38	7.00	7.75	0.62	2.00	0.62	4.00
286AT	19.75	18.38	7.00	8.25	0.62	2.00	0.62	4.00
287AT	19.75	19.88	7.00	9.00	0.62	2.00	0.62	4.00
288AT	19.75	21.38	7.00	9.75	0.62	2.00	0.62	4.00
289AT	19.75	23.38	7.00	10.75	0.62	2.00	0.62	4.00
323AT	22.75	17.75	8.00	7.75	0.75	2.50	0.75	5.25
324AT	22.75	19.25	8.00	8.50	0.75	2.50	0.75	5.25
325AT	22.75	19.75	8.00	8.75	0.75	2.50	0.75	5.25
326AT	22.75	20.75	8.00	9.25	0.75	2.50	0.75	5.25
327AT	22.75	22.75	8.00	10.25	0.75	2.50	0.75	5.25
328AT	22.75	24.75	8.00	11.25	0.75	2.50	0.75	5.25
329AT	22.75	26.75	8.00	12.25	0.75	2.50	0.75	5.25
363AT	25.50	19.25	9.00	8.25	0.88	2.50	0.75	6.00
364AT	25.50	20.50	9.00	9.12	0.88	2.50	0.75	6.00
365AT	25.50	21.50	9.00	9.62	0.88	2.50	0.75	6.00
366AT	25.50	23.25	9.00	10.50	0.88	2.50	0.75	6.00
367AT	25.50	25.25	9.00	11.50	0.88	2.50	0.75	6.00
368AT	25.50	27.25	9.00	12.50	0.88	2.50	0.75	6.00
369AT	25.50	29.25	9.00	13.50	0.88	2.50	0.75	6.00
403AT	28.75	21.12	10.00	9.25	1.00	3.00	0.88	7.00
404AT	28.75	22.38	10.00	9.88	1.00	3.00	0.88	7.00
405AT	28.75	23.88	10.00	10.62	1.00	3.00	0.88	7.00
406AT	28.75	26.12	10.00	11.75	1.00	3.00	0.88	7.00
407AT	28.75	28.12	10.00	12.75	1.00	3.00	0.88	7.00
408AT	28.75	30.12	10.00	13.75	1.00	3.00	0.88	7.00
409AT	28.75	32.12	10.00	14.75	1.00	3.00	0.88	7.00
443AT	31.25	22.62	11.00	10.00	1.12	3.00	0.88	7.50
444AT	31.25	24.62	11.00	11.00	1.12	3.00	0.88	7.50
445AT	31.25	26.62	11.00	12.00	1.12	3.00	0.88	7.50
446AT	31.25	28.12	11.00	12.75	1.12	3.00	0.88	7.50

All dimensions are in inches.
See Fig. 1 for diagram showing letter symbols.
For larger frame sizes, see NEMA Standards.

Table 3. *(Concluded).* **NEMA Standard Dimensions for Foot-mounted Industrial Direct-current Motors**

Frame No.‡	Drive End—For Belt Drive					End Opposite Drive—Straight				
	U	N–W	V Min	Keyseat		FU	FN–FW	FV Min	Keyseat	
				ES Min	S				FES Min	FS
182AT-1810AT	1.1250	2.25	2.00	1.41	0.250	0.8750	1.75	1.50	0.91	0.188
213AT-2110AT	1.3750	2.75	2.50	1.78	0.312	1.1250	2.25	2.00	1.41	0.250
253AT-259AT	1.625	3.25	3.00	2.28	0.375	1.3750	2.75	2.50	1.78	0.312
283AT-289AT	1.875	3.75	3.50	2.53	0.500	1.625	3.25	3.00	2.28	0.375
323AT-329AT	2.125	4.25	4.00	3.03	0.500	1.875	3.75	3.50	2.53	0.500
363AT-369AT	2.375	4.75	4.50	3.53	0.625	2.125	4.25	4.00	3.03	0.500
403AT-409AT	2.625	5.25	5.00	4.03	0.625	2.375	4.75	4.50	3.53	0.625
443AT-449AT	2.875	5.75	5.50	4.53	0.750	2.625	5.25	5.00	4.03	0.625

All dimensions are in inches. See Fig. 1 for diagram showing letter symbols.

Table 4. **NEMA Standard Dimensions for Face-mounted and Flange-mounted Industrial Direct-current Motors**

Type C Face Mounted											
Frame No.	AJ	AK	BA	BB Min	BC	BD Max	BF Tap Size*	U	AH	Keyseat	
										ES Min	S
182ATC-1810ATC	7.250	8.500	2.75	0.25	0.12	9.00	½-13	1.1250	2.12	1.41	0.250
213ATC-2110ATC	7.250	8.500	3.50	0.25	0.25	9.00	½-13	1.3750	2.50	1.78	0.312
253ATC-259ATC	7.250	8.500	4.25	0.25	0.25	10.00	½-13	1.625	3.00	2.28	0.375
283ATC-289ATC	9.000	10.500	4.75	0.25	0.25	11.25	½-13	1.875	3.50	2.53	0.500
323ATC-329ATC	11.000	12.500	5.25	0.25	0.25	14.00	⅝-11	2.125	4.00	3.03	0.500

Type D Flange Mounted										
Frame No.	AJ	AK	BB Max	BC	BD Max	BE Nom	BF Size†	U	AH	Keyseat
										S
182ATD-1810ATD	10.00	9.000	0.25	o	11.00	0.50	0.53	1.1250	2.25	0.250
213ATD-2110ATD	12.50	11.000	0.25	o	14.00	0.75	0.81	1.3750	2.75	0.312
253ATD-259ATD	16.00	14.000	0.25	o	18.00	0.75	0.81	1.625	3.25	0.375
283ATD-289ATD	16.00	14.000	0.25	o	18.00	0.75	0.81	1.875	3.75	0.500
323ATD-329ATD	16.00	14.000	0.25	o	18.00	0.75	0.81	2.125	4.25	0.500
363ATD-369ATD	20.00	18.000	0.25	o	22.00	1.00	0.81	2.375	4.75	0.625
403ATD-409ATD	22.00	18.000	0.25	o	24.00	1.00	0.81	2.625	5.25	0.625
443ATD-449ATD	22.00	18.000	0.25	o	24.00	1.00	0.81	2.875	5.75	0.750

All dimensions are in inches.
See Figs. 2 and 3 for diagrams showing letter symbols.
* The number of holes is 4. The bolt penetration allowance is 0.75 inch for frame Nos. 182ATC through 289ATC and 0.94 inch for frame Nos. 323ATC through 329ATC.
† The number of holes is 4 for frame Nos. 182ATD through 329ATD and 8 for frame Nos. 363ATD through 449ATD. Recommended bolt lengths are 1.25 inches for frame Nos. 182ATD through 1810ATD; 2.00 inches for frame Nos. 213ATD through 329ATD; and 2.25 inches for frame Nos. 363ATD through 449ATD. For larger frame sizes, see NEMA Standards.

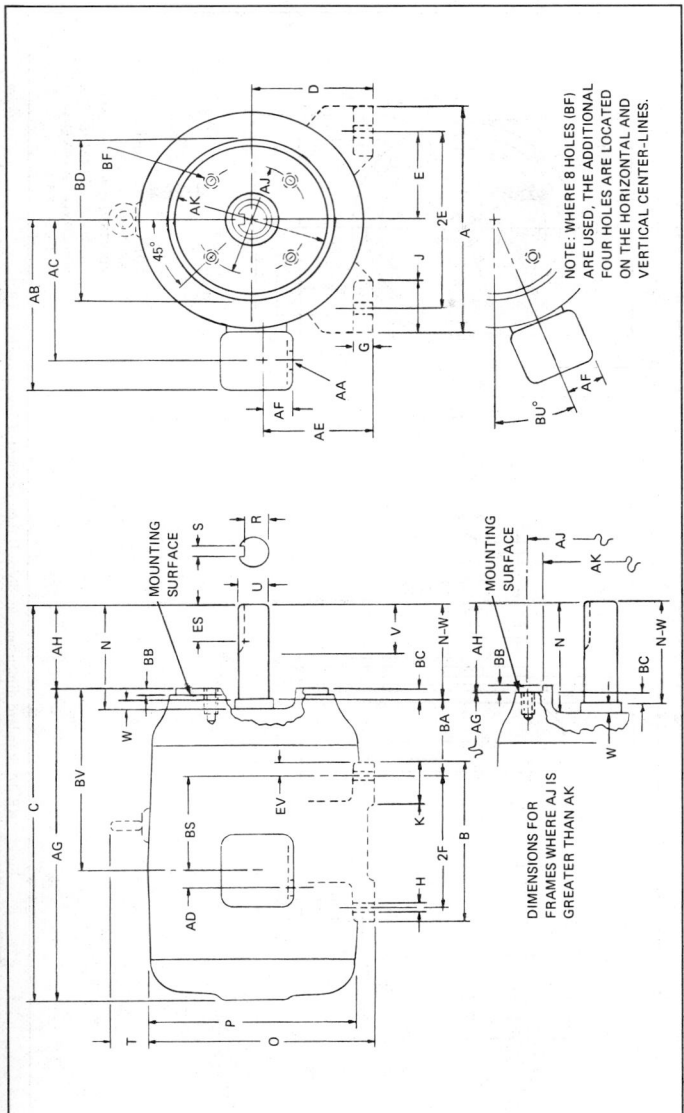

NOTE: WHERE 8 HOLES (8F) ARE USED, THE ADDITIONAL FOUR HOLES ARE LOCATED ON THE HORIZONTAL AND VERTICAL CENTER-LINES.

Fig. 2 NEMA Standard dimensions for Type C face-mounted foot or footless motors.

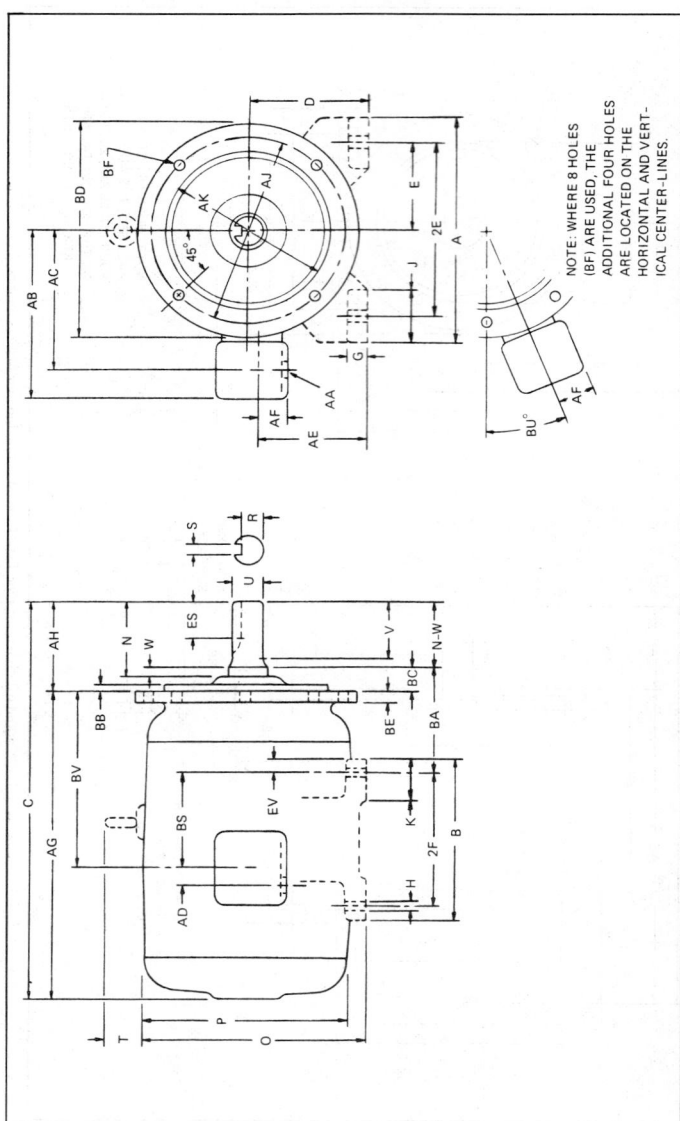

NOTE: WHERE 8 HOLES (BF) ARE USED, THE ADDITIONAL FOUR HOLES ARE LOCATED ON THE HORIZONTAL AND VERTICAL CENTER-LINES.

Fig. 3   NEMA Standard dimensions for Type D flange-mounted foot or footless motors.

Table 5. NEMA Standard Dimensions for Face-mounted and Flange-mounted Foot and Footless Alternating-current Motors

Type C Face Mounted

Frame No.	AJ**	AK	BA	BB Min	BC	BD Max	BF Tap Size§	U	AH	ES Min	S
42C	3.750	3.000	2.062	0.16†	−0.19	5.00‡	¼−20	0.3750	1.312*	...	flat
48C	3.750	3.000	2.50	0.16†	−0.19	5.625‡	¼−20	0.500	1.69*	...	flat
56C	5.875	4.500	2.75	0.16†	−0.19	6.50‡	⅜−16	0.6250	2.06*	1.41	0.188
143TC and 145TC	5.875	4.500	2.75	0.16†	+0.12	6.50‡	⅜−16	0.8750	2.12	1.41	0.188
182TC and 184TC	7.250	8.500	3.50	0.25	+0.12	9.00	½−13	1.1250	2.62	1.78	0.250
182TCH and 184TCH	5.875	4.500	3.50	0.16†	+0.12	6.50‡	⅜−16	1.1250	2.62	1.78	0.250
213TC and 215TC	7.250	8.500	4.25	0.25	+0.25	9.00	½−13	1.3750	3.12	2.41	0.312
254TC and 256TC	7.250	8.500	4.75	0.25	+0.25	10.00	½−13	1.625	3.75	2.91	0.375
284TC and 286TC	9.000	10.500	4.75	0.25	+0.25	11.25	½−13	1.875	4.38	3.28	0.500
284TSC and 286TSC	9.000	10.500	4.75	0.25	+0.25	11.25	½−13	1.625	3.00	1.91	0.375
324TC and 326TC	11.000	12.500	5.25	0.25	+0.25	14.00	⅝−11	2.125	5.00	3.91	0.500
324TSC and 326TSC	11.000	12.500	5.25	0.25	+0.25	14.00	⅝−11	1.875	3.50	2.03	0.500
364TC and 365TC	11.000	12.500	5.88	0.25	+0.25	14.00	⅝−11	2.375	5.62	4.28	0.625
364TSC and 365TSC	11.000	12.500	5.88	0.25	+0.25	14.00	⅝−11	1.875	3.50	2.03	0.500
404TC and 405TC	11.000	12.500	6.62	0.25	+0.25	15.50	⅝−11	2.875	7.00	5.65	0.750
404TSC and 405TSC	11.000	12.500	6.62	0.25	+0.25	15.50	⅝−11	2.125	4.00	2.78	0.500
444TC and 445TC	14.000	16.000	7.50	0.25	+0.25	18.00	⅝−11	3.375	8.25	6.91	0.875
444TSC and 445TSC	14.000	16.000	7.50	0.25	+0.25	18.00	⅝−11	2.375	4.50	3.03	0.625

Type D Flange Mounted

Frame No.	AJ	AK	BA	BB**	BC	BD Max	BE Nom	BF Size§	U	AH	ES Min	S
143TD and 145TD	10.00	9.000	2.75	0.25	0.00	11.00	0.50	0.53	0.8750	2.25	1.41	0.188
182TD and 184TD	10.00	9.000	3.50	0.25	0.00	11.00	0.50	0.53	1.1250	2.75	1.78	0.250
213TD and 215TD	10.00	9.000	4.25	0.25	0.00	11.00	0.50	0.53	1.3750	3.38	2.41	0.312
254TD and 256TD	12.50	11.000	4.75	0.25	0.00	14.00	0.75	0.81	1.625	4.00	2.91	0.375
284TD and 286TD	12.50	11.000	4.75	0.25	0.00	14.00	0.75	0.81	1.875	4.62	3.28	0.500
284TSD and 286TSD	12.50	11.000	4.75	0.25	0.00	14.00	0.75	0.81	1.625	3.25	1.91	0.375
324TD and 326TD	16.00	14.000	5.25	0.25	0.00	18.00	0.75	0.81	2.125	5.25	3.91	0.500
324TSD and 326TSD	16.00	14.000	5.25	0.25	0.00	18.00	0.75	0.81	1.875	3.75	2.03	0.500
364TD and 365TD	16.00	14.000	5.88	0.25	0.00	18.00	0.75	0.81	2.375	5.88	4.28	0.625
364TSD and 365TSD	16.00	14.000	5.88	0.25	0.00	18.00	0.75	0.81	1.875	3.75	2.03	0.500
404TD and 405TD	20.00	18.000	6.62	0.25	0.00	22.00	1.00	0.81	2.875	7.25	5.65	0.750
404TSD and 405TSD	20.00	18.000	6.62	0.25	0.00	22.00	1.00	0.81	2.125	4.25	2.78	0.500
444TD and 445TD	20.00	18.000	7.50	0.25	0.00	22.00	1.00	0.81	3.375	8.50	6.91	0.875
444TSD and 445TSD	20.00	18.000	7.50	0.25	0.00	22.00	1.00	0.81	2.375	4.75	3.03	0.625

All dimensions are in inches,
See Figs. 2 and 3 for diagrams showing letter dimensions.
*If the shaft extension length of the motor is not suitable for the application, it is recommended that deviations in length be in increments of 0.25 inch.
**For frames 182TC and 184TC, and 213TC through 500TC, the centerline of the bolt holes shall be within 0.025 inch of true location. True location is defined as angular and diametral location with reference to the centerline of the AK dimension.
†These BB dimensions are maximum dimensions.
‡These BD dimensions are nominal dimensions.
**The tolerance is +0.00, −0.06 inch.
§For Type C face-mounted motors, the number of holes is 4 for frame Nos. 42C through 326TSC and 8 for frame Nos. 364TC through 445TSC. The bolt penetration allowance is 0.56 inch for frame Nos. 143TC, 145TC, 182TCH, and 184TCH; 0.75 inch for frame Nos. 182TC, 184TC, and 213TC through 286TSC; and 0.94 inch for frame Nos. 324TC through 445TSC. For Type D flange-mounted motors, the number of holes is 4 for frame Nos. 143TD through 365TSD and 8 for frame Nos. 404TD through 445TSD. The recommended bolt length is 1.25 inches for frame Nos. 143TD through 215TD; 2.00 inches for frame Nos. 254TD through 365TSD; and 2.25 inches for frame Nos. 404TD through 445TSD.

Table 6.   NEMA Standard Frame Numbers for Polyphase, Squirrel-cage, Designs A and B, Horizontal and Vertical Motors, 60 Hertz, Class B Insulation System—575 Volts and Less

Hp.	Totally Enclosed Fan-cooled Type*				Open Type**			
	Speed, rpm				Speed, rpm			
	3600	1800	1200	900	3600	1800	1200	900
½	...	...	...	143T	...	...	...	143T
¾	...	...	143T	145T	...	...	143T	145T
1	...	143T	145T	182T	...	143T	145T	182T
1½	143T	145T	182T	184T	143T	145T	182T	184T
2	145T	145T	184T	213T	145T	145T	184T	213T
3	182T	182T	213T	215T	145T	182T	213T	215T
5	184T	184T	215T	254T	182T	184T	215T	254T
7½	213T	213T	254T	256T	184T	213T	254T	256T
10	215T	215T	256T	284T	213T	215T	256T	284T
15	254T	254T	284T	286T	215T	254T	284T	286T
20	256T	256T	286T	324T	254T	256T	286T	324T
25	284TS	284T	324T	326T	256T	284T	324T	326T
30	286TS	286T	326T	364T	284TS	286T	326T	364T
40	324TS	324T	364T	365T	286TS	324T	364T	365T
50	326TS	326T	365T	404T	324TS	326T	365T	404T
60	364TS	364TS†	404T	405T	326TS	364TS†	404T	405T
75	365TS	365TS†	405T	444T	364TS	365TS†	405T	444T
100	405TS	405TS†	444T	445T	365TS	404TS†	444T	445T
125	444TS	444TS†	445T	...	404TS	405TS†	445T	...
150	445TS	445TS†	...	...	405TS	444TS†	...	...
200	...	...	...	...	444TS†	...	...	...

The voltage rating of 115 volts applies only to motors rated 15 hp and smaller.
*1.00 Service Factor.   **1.15 Service Factor.
†When motors are to be used with V-belt or chain drives, the correct frame size is the frame size shown but with the suffix letter S omitted.

Table 7.   NEMA Standard Synchronous Speed Ratings of Polyphase Integral-horsepower Induction Motors

Hp.	60 hertz							50 hertz			
	Synchronous Speed, rpm										
	3600	1800	1200	900	720	600	514	3000	1500	1000	750
½	...	...	...	900	720	600	514	...	...	...	750
¾	...	...	1200	900	720	600	514	...	...	1000	750
1	...	1800	1200	900	720	600	514	...	1500	1000	750
1½	3600*	1800	1200	900	720	600	514	3000*	1500	1000	750
2	3600*	1800	1200	900	720	600	514	3000*	1500	1000	750
3	3600*	1800	1200	900	720	600	514	3000*	1500	1000	750
5	3600*	1800	1200	900	720	600	514	3000*	1500	1000	750
7½	3600*	1800	1200	900	720	600	514	3000*	1500	1000	750
10	3600*	1800	1200	900	720	600	514	3000*	1500	1000	750
15	3600*	1800	1200	900	720	600	514	3000*	1500	1000	750
20	3600*	1800	1200	900	720	600	514	3000*	1500	1000	750
25	3600*	1800	1200	900	720	600	514	3000*	1500	1000	750
30	3600*	1800	1200	900	720	600	514	3000*	1500	1000	750
40	3600*	1800	1200	900	720	600	514	3000*	1500	1000	750
50	3600*	1800	1200	900	720	600	514	3000*	1500	1000	750
60	3600*	1800	1200	900	720	600	514	3000*	1500	1000	750
75	3600*	1800	1200	900	720	600	514	3000*	1500	1000	750
100	3600*	1800	1200	900	720	600	514	3000*	1500	1000	750
125	3600*	1800	1200	900	720	600	514	3000*	1500	1000	750
150	3600*	1800	1200	900	720	600	...	3000*	1500	1000	750
200	3600*	1800	1200	900	720	...	...	3000*	1500	1000	750

*Applies to squirrel-cage motors only.
For motors above 200 horsepower, see NEMA Standard MG1-10.32.

Table 8. NEMA Standard Locked-rotor Current of 3-phase 60-hertz Integral-horsepower Squirrel-cage Induction Motors Rated at 230 Volts*

Horse-power	Locked-rotor Current, Amps.	Design Letters	Horse-power	Locked-rotor Current, Amps.	Design Letters	Horse-power	Locked-rotor Current, Amps.	Design Letters
½	20	B, D	7½	127	B, C, D	50	725	B, C, D
¾	25	B, D	10	162	B, C, D	60	870	B, C, D
1	30	B, D	15	232	B, C, D	75	1085	B, C, D
1½	40	B, D	20	290	B, C, D	100	1450	B, C, D
2	50	B, D	25	365	B, C, D	125	1815	B, C, D
3	64	B, C, D	30	435	B, C, D	150	2170	B, C, D
5	92	B, C, D	40	580	B, C, D	200	2900	B, C

Note: The locked-rotor current of a motor is the steady-state current taken from the line with the rotor locked and with rated voltage and frequency applied to the motor.

*For motors designed for voltages other than 230 volts, the locked-rotor current is inversely proportional to the voltages. For motors larger than 200 hp, see NEMA Standard MG 1-12.34.

Table 9. NEMA Standard Locked-rotor Torque of Single-speed Polyphase 60- and 50-hertz Squirrel-cage Integral-horsepower Motors with Continuous Ratings

Hp	Designs A and B							Design C		
60 hertz	3600	1800	1200	900	720	600	514	1800	1200	900
50 hertz	3000	1500	1000	750	...	...	...	1500	1000	750
½	...	...	...	140	140	115	110	...	...	...
¾	...	...	175	135	135	115	110	...	...	...
1	...	275	170	135	135	115	110	...	...	...
1½	175	250	165	130	130	115	110	...	...	...
2	170	235	160	130	125	115	110	...	...	...
3	160	215	155	130	125	115	110	...	250	225
5	150	185	150	130	125	115	110	250	250	225
7½	140	175	150	125	120	115	110	250	225	200
10	135	165	150	125	120	115	110	250	225	200
15	130	160	140	125	120	115	110	225	200	200
20	130	150	135	125	120	115	110			
25	130	150	135	125	120	115	110			
30	130	150	135	125	120	115	110	200 for all sizes above 15 hp.		
40	125	140	135	125	120	115	110			
50	120	140	135	125	120	115	110			
60	120	140	135	125	120	115	110			
75	105	140	135	125	120	115	110			
100	105	125	125	125	120	115	110			
125	100	110	125	120	115	115	110	For Design D motors, see footnote.		
150	100	110	120	120	115	115	...			
200	100	100	120	120	115	...	...			

Note: The locked-rotor torque of a motor is the minimum torque which it will develop at rest for all angular positions of the rotor, with rated voltage applied at rated frequency.

*These values represent the upper limit of application for these motors.

The locked-rotor torque of Design D, 60- and 50-hertz 4-, 6-, and 8-pole single-speed, polyphase squirrel-cage motors rated 150 hp and smaller, with rated voltage and frequency applied is 275 per cent of full-load torque, which represents the upper limit of application for these motors.

For motors larger than 200 hp, see NEMA Standard MG 1-12.37.

Table 10. NEMA Standard Breakdown Torque of Single-speed Polyphase Squirrel-cage, Integral-horsepower Motors with Continuous Ratings

Horse-power	Synchronous Speed, rpm						
	60 hertz 3600	1800	1200	900	720	600	514
	50 hertz 3000	1500	1000	750	...	...	...
	Per Cent of Full Load Torque						
	Designs A and B*						
½	...	...	...	225	200	200	200
¾	...	...	275	220	200	200	200
1	...	300	265	215	200	200	200
1½	250	280	250	210	200	200	200
2	240	270	240	210	200	200	200
3	230	250	230	205	200	200	200
5	215	225	215	205	200	200	200
7½	200	215	205	200	200	200	200
10-125, incl.	200	200	200	200	200	200	200
150	200	200	200	200	200	200	...
200	200	200	200	200	200	...	...
	Design C						
3	...	...	225	200	...	...	...
5	...	200	200	200	...	...	...
7½-200, incl.	...	190	190	190	...	...	...

*Design A values are in excess of those shown.
These values represent the upper limit of the range of application for these motors. For above 200 hp, see NEMA Standard MG1-12.38.

Table 11. NEMA Standard Breakdown Torque of Polyphase Wound-rotor Motors with Continuous Ratings — 60- and 50-hertz

Horse-power	Speed, rpm			Horse-power	Speed, rpm		
	1800	1200	900		1800	1200	900
	Per cent of Full-load Torque				Per cent of Full-load Torque		
1	...	...	250	7½	275	250	225
1½	...	...	250	10	275	250	225
2	275	275	250	15	250	225	225
3	275	275	250	20–200, incl.	225	225	225
5	275	275	250	...	...	...	...

These values represent the upper limit of the range of application for these motors.

Pull-up Torque. — NEMA Standard pull-up torques for single-speed, polyphase, squirrel-cage integral-horsepower motors, Designs A and B, with continuous ratings and with rated voltage and frequency applied are as follows: When the locked-rotor torque given in Table 9 is 110 per cent or less, the pull-up torque is 90 per cent of the locked-rotor torque; when the locked-rotor torque is greater than 110 per cent but less than 145 per cent, the pull-up torque is 100 per cent of full-load torque; and when the locked-rotor torque is 145 per cent or more, the pull-up torque is 70 per cent of the locked-rotor torque. For Design C motors, with rated voltage and frequency applied, the pull-up torque is not less than 70 per cent of the locked-rotor torque as given in Table 9.

Types and Characteristics
of Electric Motors

Types of Direct-Current Motors. — Direct-current motors may be grouped into three general classes: series-wound, shunt-wound and compound-wound. In the *series-wound motor* the field windings, which are fixed in the stator frame, and the armature windings, which are placed around the rotor, are connected in series so that all current passing through the armature also passes through the field. In the *shunt-wound motor*, both armature and field are connected across the main power supply so that the armature and field currents are separate. In the *compound-wound motor*, both series and shunt field windings are provided and these may be connected so that the currents in both are flowing in the same direction, called *cumulative compounding*, or so that the currents in each are flowing in opposite directions, called *differential compounding*.

Characteristics of Series-wound Direct-Current Motors. — In the series-wound motor, any increase in load results in more current passing through the armature and the field windings. As the field is strengthened by this increased current, the motor speed decreases. Conversely, as the load is decreased the field is weakened and the speed increases and at very light loads may become excessive. For this reason, series-wound direct-current motors are usually directly connected or geared to the load to prevent "runaway." (A series-wound motor designated as series-shunt wound, is sometimes provided with a light shunt field winding to prevent dangerously high speeds at light loads.) The increase in armature current with increasing load produces increased torque, so that the series-wound motor is particularly suited to heavy starting duty and where severe overloads may be expected. Its speed may be adjusted by means of a variable resistance placed in series with the motor, but due to variation with load, the speed cannot be held at any constant value. This variation of speed with load becomes greater as the speed is reduced. Series-wound motors are used where the load is practically constant and can easily be controlled by hand. They are usually limited to traction and lifting service.

Shunt-wound Direct-Current Motors. — In the shunt-wound motor, the strength of the field is not affected appreciably by change in the load, so that a fairly constant speed (about 10 to 12 per cent drop from no load to full load speed) is obtainable. This type of motor may be used for the operation of machines requiring an approximately constant speed and imposing low starting torque and light overload on the motor.

The shunt-wound motor becomes an adjustable-speed motor by means of field control or by armature control. If a variable resistance is placed in the field circuit, the amount of current in the field windings and hence the speed of rotation can be controlled. As the speed increases, the torque decreases proportionately, resulting in nearly constant horsepower. A speed range of 6 to 1 is possible using field control, but 4 to 1 is more common. Speed regulation is somewhat greater than in the constant-speed shunt-wound motors, ranging from about 15 to 22 per cent. If a variable resistance is placed in the armature circuit, the voltage applied to the armature can be reduced and hence the speed of rotation can be reduced over a range of about 2 to 1. With armature control, speed regulation becomes poorer as speed is decreased, and is about 100 per cent for a 2 to 1 speed range. Since the current through the field remains unchanged, the torque remains constant.

Machine Tool Applications: The adjustable-speed shunt-wound motors are useful on larger machines of the boring mill, lathe, and planer type and are particularly adapted to spindle drives because constant horsepower characteristics permit heavy

cuts at low speed and light or finishing cuts at high speed. They have long been used for planer drives because they can provide an adjustable low speed for the cutting stroke and a high speed for the return stroke. Their application has been limited, however, to plants in which direct-current power is available.

Adjustable-voltage Shunt-wound Motor Drive. — More extensive use of the shunt-wound motor has been made possible by a combination drive that includes a means of converting alternating current to direct current. This conversion may be effected by a self-contained unit consisting of a separately excited direct-current generator driven by a constant speed alternating-current motor connected to the regular alternating-current line, or by an electronic rectifier with suitable controls connected to the regular alternating-current supply lines. The latter has the advantage of causing no vibration when mounted directly on the machine tool, an important factor in certain types of grinders.

In this type of adjustable-speed, shunt-wound motor drive, speed control is effected by varying the voltage applied to the armature while supplying constant voltage to the field. In addition to providing for the adjustment of the voltage supplied by the conversion unit to the armature of the shunt-wound motor, the amount of current passing through the motor field may also be controlled. In fact, a single control may be provided to vary the motor speed from minimum to base speed (speed of the motor at full load with rated voltage on armature and field) by varying the voltage applied to the armature and from base speed to maximum speed by varying the current flowing through the field. When so controlled, the motor operates at constant torque up to base speed and at constant horsepower above base speed.

Speed Range: Speed ranges of at least 20 to 1 below base speed and 4 or 5 to 1 above base speed (a total range of 100 to 1, or more) are obtainable as compared with about 2 to 1 below normal speed and 3 or 4 to 1 above normal speed for the conventional type of control. Speed regulation may be as great as 25 per cent at high speeds. Special electronic controls, when used with this type shunt motor drive, make possible maintenance of motor speeds with as little variation as ½ to 1 per cent of full load speed from full load to no load over a line voltage variation of ± 10 per cent and over any normal variation in motor temperature and ambient temperature.

Applications: These direct-current, adjustable-voltage drives, as they are sometimes called, have been applied successfully to such machine tools as planers, milling machines, boring mills and lathes, as well as to other industrial machines where wide, stepless speed control, uniform speed under all operating conditions, constant torque acceleration and adaptability to automatic operation are required.

Compound-wound Motors. — In the compound-wound motor, the speed variation due to load changes is much less than in the series-wound motor, but greater than in the shunt-wound motor (ranging up to 25 per cent from full load to no load). It has a greater starting torque than the shunt-wound motor, is able to withstand heavier overloads, but has a narrower adjustable speed range. Standard motors of this type have a cumulative-compound winding, the differential-compound winding being limited to special applications. They are used where the starting load is very heavy or where the load changes suddenly and violently as with reciprocating pumps, printing presses and punch presses.

Types of Polyphase Alternating-Current Motors. — The most widely used polyphase motors are of the induction type. The *"squirrel cage" induction motor* consists of a wound stator which is connected to an external source of alternating-current power and a laminated steel core rotor with a number of heavy aluminum or copper conductors set into the core around its periphery and parallel to its axis. These conductors are connected together at each end of the rotor by a heavy ring,

which provides closed paths for the currents induced in the rotor to circulate. This forms, in effect, a "squirrel-cage" from which the motor takes its name.

Wound-rotor type of *Induction motor:* This type has in addition to a squirrel cage, a series of coils set into the rotor which are connected through slip-rings to external variable resistors. By varying the resistance of the wound-rotor circuits, the amount of current flowing in these circuits and hence the speed of the motor can be controlled. Since the rotor of an induction motor is not connected to the power supply, the motor is said to operate by transfer action and is analogous to a transformer with a short-circuited secondary that is free to rotate. Induction motors are built with a wide range of speed and torque characteristics which are discussed under "Operating Characteristics of Polyphase Induction Motors."

Synchronous Motor: The other type of polyphase alternating-current motor used industrially is the *synchronous motor.* In contrast to the induction motor, the rotor of the synchronous motor is connected to a direct-current supply which provides a field that rotates in step with the alternating-current field in the stator. After having been brought up to synchronous speed, which is governed by the frequency of the power supply and the number of poles in the rotor, the synchronous motor operates at this constant speed throughout its entire load range.

Operating Characteristics of Squirrel-cage Induction Motors. — In general, squirrel-cage induction motors are simple in design and construction and offer rugged service. They are essentially constant-speed motors, their speed changing very little with load and not being subject to adjustment. They are used for a wide range of industrial applications calling for integral horsepower ratings. According to the NEMA (National Electrical Manufacturers Association) Standards, there are four classes of squirrel-cage induction motors designated respectively as *A, B, C,* and *D.*

Design A motors are not commonly used since Design *B* has similar characteristics with the advantage of lower starting current.

Design B motors may be designated as a general purpose type suitable for the majority of polyphase alternating-current applications such as blowers, compressors, drill presses, grinders, hammer mills, lathes, planers, polishers, saws, screw machines, shakers, stokers, etc. The starting torque at 1800 R.P.M. is 250 to 275 per cent of full load torque for 3 H.P. and below; for 5 H.P. to 75 H.P. ratings the starting torque ranges from 185 to 150 per cent of full load torque. They have low starting current requirements, usually no more than 5 to 6 times full load current and can be started at full voltage. Their slip (difference between synchronous speed and actual speed at rated load) is relatively low.

Design C motors have high starting torque (up to 250 per cent of full load torque) but low starting current. They can be started at full voltage. Slip at rated load is relatively low. They are used for compressors requiring a loaded start, heavy conveyors, reciprocating pumps and other applications requiring high starting torque.

Design D motors have high slip at rated load, that is the motor speed drops off appreciably as the load increases, permitting use of the stored energy of a flywheel. They provide heavy starting torque, up to 275 per cent of full load torque, are quiet in operation and have relatively low starting current. Applications are for impact, shock and other high peak loads or flywheel drives such as trains, elevators, hoists, punch and drawing presses, shears, etc.

Design F motors are no longer standard. They had low starting torque, about 125 per cent of full-load torque, and low starting current. They were used to drive machines which required infrequent starting at no load or at very light load.

Multiple-Speed Induction Motors. — This type has a number of windings in the stator so arranged and connected that the number of effective poles and hence the speed can be changed. These motors are for the same types of starting conditions as the conventional squirrel-cage induction motors and are available in designs that provide constant horsepower at all rated speeds and in designs that provide constant torque at all rated speeds. Typical speed combinations obtainable in these motors are 600, 900, 1200 and 1800 R.P.M.; 450, 600, 900 and 1200 R.P.M.; and 600, 720, 900 and 1200 R.P.M. Where a gradual change in speed is called for, a wound rotor may be provided in addition to the multiple stator windings.

Wound-Rotor Induction Motors. — These motors are designed for applications where extremely low starting current with high starting torque are called for, such as in blowers, conveyors, compressors, fans and pumps. They may be employed for adjustable-varying speed service where the speed range does not extend below 50 per cent of synchronous speed, as for steel plate-forming rolls, printing presses, cranes, blowers, stokers, lathes and milling machines of certain types. The speed regulation of a wound rotor induction motor ranges from 5 to 10 per cent at maximum speed and from 18 to 30 per cent at low speed. They are also employed for reversing service as in cranes, gates, hoists and elevators.

High-Frequency Induction Motors. — This type is used in conjunction with frequency changers when very high speeds are desired, as on grinders, drills, routers, portable tools or woodworking machinery. These motors have an advantage over the series-wound or universal type of high speed motor in that they operate at a relatively constant speed over the entire load range. A motor-generator set, a two-unit frequency converter or a single unit inductor frequency converter may be used to supply three-phase power at the frequency required. The single unit frequency converter may be obtained for delivering any one of a number of frequencies ranging from 360 to 2160 cycles and it is self-driven and self-excited from the general polyphase power supply.

Synchronous Motors. — These are widely used in electric timing devices; to drive machines that must operate in synchronism; and also to operate compressors, rolling mills, crushers which are started without load, paper mill screens, shredders, vacuum pumps and motor-generator sets. Synchronous motors have an inherently high power factor and are often employed to make corrections for the low power factor of other types of motors on the same system.

Types of Single-Phase Alternating-Current Motors. — Most of the single-phase alternating-current motors are basically induction motors distinguished by different arrangements for starting. (A single-phase induction motor with only a squirrel-cage rotor has no starting torque.) In the *capacitor-start* single-phase motor, an auxiliary winding in the stator is connected in series with a capacitor and a centrifugal switch. During the starting and accelerating period the motor operates as a two-phase induction motor. At about two-thirds full-load speed, the auxiliary circuit is disconnected by the switch and the motor then runs as a single-phase induction motor. In the *capacitor-start, capacitor-run* motor, the auxiliary circuit is arranged to provide high effective capacity for high starting torque and to remain connected to the line but with reduced capacity during the running period. In the *single-value capacitor* or *capacitor split-phase* motor, a relatively small continuously-rated capacitor is permanently connected in one of the two stator windings and the motor both starts and runs like a two-phase motor.

In the *repulsion-start* single-phase motor, a drum-wound rotor circuit is connected to a commutator with a pair of short-circuited brushes set so that the magnetic axis of the rotor winding is inclined to the magnetic axis of the stator winding.

The current flowing in this rotor circuit reacts with the field to produce starting and accelerating torques. At about two-thirds full load speed the brushes are lifted, the commutator is short circuited and the motor runs as a single-phase squirrel-cage motor. The *repulsion* motor employs a repulsion winding on the rotor for both starting and running. The *repulsion-induction* motor has an outer winding on the rotor acting as a repulsion winding and an inner squirrel-cage winding. As the motor comes up to speed, the induced rotor current partially shifts from the repulsion winding to the squirrel-cage winding and the motor runs partly as an induction motor.

In the *split-phase* motor, an auxiliary winding in the stator is used for starting with either a resistance connected in series with the auxiliary winding (*resistance-start*) or a reactor in series with the main winding (*reactor-start*).

The *series-wound* single-phase motor has a rotor winding in series with the stator winding as in the series-wound direct-current motor. Since this motor may also be operated on direct current, it is called a *universal* motor.

Characteristics of Single-Phase Alternating-Current Motors. — Single-phase motors are used in sizes up to about $7\frac{1}{2}$ horsepower for heavy starting duty chiefly in home and commercial appliances for which polyphase power is not available. The *capacitor-start* motor is available in normal starting torque designs for such applications as centrifugal pumps, fans, and blowers and in high-starting torque designs for reciprocating compressors, pumps, loaded conveyors, or belts. The *capacitor-start, capacitor-run* motor is exceptionally quiet in operation when loaded to at least 50 per cent of capacity. It is available in low-torque designs for fans and centrifugal pumps and in high-torque designs for applications similar to those of the capacitor-start motor.

The *capacitor split-phase* motor requires the least maintenance of all single-phase motors, but has very low starting torque. Its high maximum torque makes it potentially useful in floor sanders or in grinders where momentary overloads due to excessive cutting pressure are experienced. It is also used for slow-speed direct connected fans.

The *repulsion-start, induction-run* motor has higher starting torque than the capacitor motors, although for the same current, the capacitor motors have equivalent pull-up and maximum torque. Electrical and mechanical noise and the extra maintenance sometimes required are disadvantages. These motors are used for compressors, conveyors and stokers starting under full load. The *repulsion-induction* motor has relatively high starting torque and low starting current. It also has a smooth speed-torque curve with no break and a greater ability to withstand long accelerating periods than capacitor type motors. It is particularly suitable for severe starting and accelerating duty and for high inertia loads such as laundry extractors. Brush noise is, however, continuous.

The *repulsion* motor has no limiting synchronous speed and the speed changes with the load. At certain loads, slight changes in load cause wide changes in speed. A brush shifting arrangement may be provided to adjust the speed which may have a range of 4 to 1 if full rated constant torque is applied but a decreasing range as the torque falls below this value. This type of motor may be reversed by shifting the brushes beyond the neutral point. These motors are suitable for machines requiring constant-torque and adjustable speed.

The *split-phase* and *universal* motors are limited to about $\frac{1}{3}$ H.P. ratings and are used chiefly for small appliance and office machine applications.

Motors with Built-in Speed Reducers. — Electric motors having built-in speed-changing units are compact and the design of these motorized speed reducers tends to improve the appearance of the machines which they drive. There are

several types of these speed reducers; they may be classified according to whether they are equipped with worm gearing, a regular gear train with parallel shafts, or planetary gearing.

The claims made for the worm gearing type of reduction unit are that the drive is quiet in operation and well adapted for use where the slow-speed shaft must be at right angles to the motor shaft and where a high speed ratio is essential.

For very low speeds, the double reduction worm gearing units are suitable. In these units two sets of worm gearing form the gear train, and both the slow-speed shaft and the armature shaft are parallel. The intermediate wormgear shaft can be built to extend from the housing, if required, so as to make two countershaft speeds available on the same unit.

In the parallel-shaft type of speed reducer, the slow-speed shaft is parallel with the armature shaft. The slow-speed shaft is rotated by a pinion on the armature shaft, this pinion meshing with a larger gear on the slow-speed shaft.

Geared motors having built-in speed-changing units are available with constant-mesh change gears for varying the speed ratio.

Planetary gearing permits a large speed reduction with few parts; hence, it is well adapted for geared-head motor units where economy and compactness are essential. The slow-speed shaft is in line with the armature shaft.

Factors Governing Motor Selection

Speed, Horsepower, Torque and Inertia Requirements. — Where more than one speed or a range of speeds are called for, one of the following types of motors may be selected, depending upon other requirements: For direct-current, the standard shunt-wound motor with field control has a 2 to 1 range in some designs; the adjustable speed motor may have a range of from 3 to 1 up to 6 to 1; the shunt motor with adjustable voltage supply has a range up to 20 to 1 or more below base speed and 4 or 5 to 1 above base speed, making a total range of up to 100 to 1 or more. For polyphase alternating current, multi-speed squirrel-cage induction motors have 2, 3 or 4 fixed speeds; the wound-rotor motor has a 2 to 1 range. The two-speed wound-rotor motor has a 4 to 1 range. The brush-shifting shunt motor has a 4 to 1 range. The brush-shifting series motor has a 3 to 1 range; and the squirrel-cage motor with a variable-frequency supply has a very wide range. For single-phase alternating current, the brush-shifting repulsion motor has a 2½ to 1 range; the capacitor motor with tapped winding has a 2 to 1 range and the multi-speed capacitor motor has 2 or 3 fixed speeds. Speed regulation (variation in speed from no load to full load) is greatest with motors having series field windings and entirely absent with synchronous motors.

Horsepower: Where the load to be carried by the motor is not constant but follows a definite cycle, a horsepower-time curve enables the peak horsepower to be determined as well as the root-mean-square-average horsepower, which indicates the proper motor rating from a heating standpoint. Where the load is maintained at a constant value for a period of from 15 minutes to 2 hours depending on the size, the horsepower rating required will usually not be less than this constant value. When selecting the size of an induction motor, it should be kept in mind that this type of motor operates at maximum efficiency when it is loaded to full capacity. Where operation is to be at several speeds, the horsepower requirement for each speed should be considered.

Torque: Starting torque requirements may vary from 10 per cent of full load to 250 per cent of full load torque depending upon the type of machine being driven. Starting torque may vary for a given machine because of frequency of start, temperature, type and amount of lubricant, etc. and such variables should be taken into account. The motor torque supplied to the machine must be well above that

required by the driven machine at all points up to full speed. The greater the excess torque, the more rapid the acceleration. The approximate time required for acceleration from rest to full speed is given by the formula:

$$\text{Time} = \frac{N \times WR^2}{T_a \times 308} \text{ seconds}$$

where N = Full load speed in R.P.M.

T_a = Torque = average foot-pounds available for acceleration.

WR^2 = Inertia of rotating part in pounds feet squared (W = weight and R = radius of gyration of rotating part).

308 = Combined constant converting minutes into seconds, weight into mass and radius into circumference.

If the time required for acceleration is greater than 20 seconds, special motors or starters may be required to avoid overheating.

The running torque T_r is found by the formula:

$$T_r = \frac{5250 \times \text{H.P.}}{N} \text{ foot pounds}$$

where H.P. = Horsepower being supplied to the driven machine

N = Running speed in R.P.M.

5250 = Combined constant converting horsepower to foot-pounds per minute and work per revolution into torque.

The peak horsepower determines the maximum torque required by the driven machine and the motor must have a maximum running torque in excess of this value.

Inertia: The inertia or flywheel effect of the rotating parts of a driven machine will, if large, appreciably affect the accelerating time and hence, the amount of heating in the motor. If synchronous motors are used, the inertia (WR^2) of both the motor rotor and the rotating parts of the machine must be known since the pull-in torque (torque required to bring the driven machine up to synchronous speed) varies approximately as the square root of the total inertia of motor and load.

Space Limitations in Motor Selection. — If the motor is to become an integral part of the machine which it drives and space is at a premium, a partial motor may be called for. A complete motor is one made up of a stator, a rotor, a shaft, and two end shields with bearings. A *partial motor* is without one or more of these elements. One common type is furnished without drive-end end shield and bearing and is directly connected to the end or side of the machine which it drives, such as the headstock of a lathe. A so-called *shaftless type of motor* is supplied without shaft, end shields or bearings and is intended for built-in application in such units as multiple drilling machines, precision grinders, deep well pumps, compressors and hoists where the rotor is actually made a part of the driven machine. Where a partial motor is used, however, proper ventilation, mounting, alignment and bearings must be arranged for by the designer of the machine to which it is applied.

Sometimes it is possible to use a motor having a smaller frame size and wound with Class *B* insulation, permitting it to be subjected to a higher temperature rise than the larger-frame Class *A* insulated motor having the same horsepower rating.

Temperatures. — The applicability of a given motor is limited not only by its load starting and carrying ability, but also by the temperature which it reaches under load. Motors are given temperature ratings which are based upon the type

of insulation (Class A or Class B are the most common) used in their construction and their type of frame (open, semi-enclosed or enclosed).

Insulating Materials: Class A materials are: (1) Cotton, silk, paper and similar organic materials when either impregnated or immersed in a liquid dielectric; (2) molded and laminated materials with cellulose filler, phenolic resins and other resins of similar properties; (3) films and sheets of cellulose acetate and other cellulose derivatives of similar properties; (4) varnishes (enamel) as applied to conductors.

Class B insulating materials are: Materials or combinations of materials such as mica, glass fiber, asbestos, etc. with suitable bonding substances. Other materials shown capable of operation at Class B temperatures may be included.

Ambient Temperature and Allowable Temperature Rise: Normal ambient temperature is taken to be 40° C. (104° F.). For open general-purpose motors with Class A insulation, the normal temperature rise on which the performance guarantees are based is 40° C. (72° F.).

Motors with Class A insulation having protected, semi-protected, drip-proof, splash-proof, or drip-proof protected enclosures have a 50° C. (90° F.) rise rating.

Motors with Class A insulation and having totally enclosed, totally enclosed fan-cooled, explosion-proof, water-proof, dust-tight, submersible or dust-explosion-proof enclosures have a 55° C. (99° F.) rise rating.

Motors with Class B insulation are permissible for total temperatures up to 110 degrees C. (230° F.) for open motors and 115° C. (239° F.) for enclosed motors.

Motors Exposed to Injurious Conditions. — Where motors are to be used in locations imposing unusual operating conditions, the manufacturer should be consulted, especially where any of the following conditions apply: (1) exposed to chemical fumes; (2) operated in damp places; (3) operated at speeds in excess of specified overspeed; (4) exposed to combustible or explosive dust; (5) exposed to gritty or conducting dust; (6) exposed to lint; (7) exposed to steam; (8) operated in poorly ventilated rooms; (9) operated in pits, or where entirely enclosed in boxes; (10) exposed to inflammable or explosive gases; (11) exposed to temperatures below 10° C. (50° F.); (12) exposed to oil vapor; (13) exposed to salt air; (14) exposed to abnormal shock or vibration from external sources; (15) where the departure from rated voltage is excessive; (16) where the alternating-current supply voltage is unbalanced.

Improved insulating materials and processes and greater mechanical protection against falling materials and liquids make it possible to use general-purpose motors in many locations where special purpose motors were previously considered necessary. *Splash-proof motors* having well protected ventilated openings and specially treated windings are used where they are to be subjected to falling and splashing water or are to be washed down as with a hose. Where climatic conditions are not severe, this type of motor is also successfully used in unprotected out-of-door installations.

If the surrounding atmosphere carries abnormal quantities of metallic, abrasive or non-explosive dust or acid or alkali fumes, a *totally enclosed fan-cooled motor* may be called for. In this type, the motor proper is completely enclosed but air is blown through an outer shell which completely or partially surrounds the inner case. Where the dust in the atmosphere is of a kind which tends to pack or solidify and close the air passages of open splash-proof or totally enclosed fan-cooled motors, *totally enclosed (non-ventilated) motors* are used. This type, which is limited to low horsepower ratings, is also used for outdoor service in mild or severe climates.

In addition to these special-purpose motors there are two types *of explosion-proof motors* designed for hazardous locations. One type is for operation in hazardous dust locations (Class II, Group *G* of the National Electrical Code) while the other is for atmospheres containing explosive vapors and fumes classified as Class I, Group *D* (gasoline, naptha, alcohols, acetone, lacquer-solvent vapors, natural gas).

Table 1. Characteristics and Applications of D.C. Motors, 1–300 H.P.

Type	Starting Duty	Maximum Momentary Running Torque	Speed Regulation	Speed Control†	Applications
Shunt-wound, constant-speed	Medium starting torque. Varies with voltage supplied to armature, and is limited by starting resistor to 125 to 200 per cent full-load torque	125 to 200 per cent. Limited by commutation	8 to 12 per cent	Basic speed to 200 per cent basic speed by field control	Drives where starting requirements are not severe. Use constant-speed or adjustable-speed, depending on speed required. Centrifugal pumps, fans, blowers, conveyors, elevators, wood- and metal-working machines
Shunt-wound, adjustable-speed			10 to 20 per cent, increases with weak fields	Basic speed to 60 per cent basic speed (lower for some ratings) by field control	
Shunt-wound, adjustable voltage control			Up to 25 per cent. Less than 5 per cent obtainable with special rotating regulator	Basic speed to 2 per cent basic speed and basic speed to 200 per cent basic speed	Drives where wide, stepless speed control, uniform speed, constant-torque acceleration and adaptability to automatic operation are required. Planers, milling machines, boring machines, lathes, etc.
Compound-wound, constant-speed	Heavy starting torque. Limited by starting resistor to 130 to 260 per cent of full-load torque	130 to 260 per cent. Limited by commutation	Standard compounding 25 per cent. Depends on amount of series winding	Basic speed to 125 per cent basic speed by field control	Drives requiring high starting torque and fairly constant speed. Pulsating loads. Shears, bending rolls, pumps, conveyors, crushers, etc.
Series-wound, varying-speed	Very heavy starting torque. Limited to 300 per cent to 350 per cent full-load torque	300 to 350 per cent. Limited by commutation	Very high. Infinite no-load speed	From zero to maximum speed, depending on control and load	Drives where very high starting torque is required and speed can be regulated. Cranes, hoists, gates, bridges, car dumpers, etc.

† Minimum speed below basic speed by armature control limited by heating.

Table 2. Characteristics and Applications of Polyphase A.C. Motors

Polyphase Type	Ratings Hp	Speed Regulation	Speed Control	Starting Torque	Breakdown Torque	Applications
General-purpose squirrel cage, normal stg current, normal stg torque, Design B	0.5 to 200 hp	Less than 5%	None, except multi-speed types, designed for 2 to 4 fixed speeds	100 to 250% of full-load	200 to 300% of full-load	Constant-speed service where starting torque is not excessive. Fans, blowers, rotary compressors, centrifugal pumps, wood-working machines, machine tools, line shafts
Full-voltage starting, high stg torque, normal stg current, squirrel-cage, Design C	3 to 150 hp	Less than 5%	None, except multi-speed types, designed for 2 to 4 fixed speeds	200 to 250% of full-load	190 to 225% of full-load	Constant-speed service where fairly high starting torque is required at infrequent intervals with starting current of about 300% full load. Reciprocating pumps and compressors, conveyors, crushers, pulverizers, agitators, etc.
Full-voltage starting, high stg torque, high-slip squirrel-cage, Design D	0.5 to 150 hp	Drops about 7 to 12% from no load to full load	None, except multi-speed types, designed for 2 to 4 fixed speeds	275% of full load, depending upon speed and rotor resistance	275%. This motor will usually not stall until loaded to its maximum torque, which occurs at standstill	Constant-speed service and high starting torque if starting not too frequent, and for taking high-peak loads with or without flywheels. Punch presses, die stamping, shears, bulldozers, bailers, hoists, cranes, elevators, etc.
Wound-rotor, external-resistance starting	0.5 to several thousand	With rotor rings short-circuited drops about 3% normal for large sizes, to 5% for small sizes	Speed can be reduced to 50% of normal by rotor resistance. Speed varies inversely as the load	Up to 300% depending upon external resistance and how distributed	200% when rotor slip rings are short circuited	Where high-starting torque with low-starting current or where limited speed control is required. Fans, centrifugal and plunger pumps, compressors, conveyors, hoists, cranes, gate hoists, etc.
Synchronous	25 to several thousand	Constant	None, except special motors designed for 2 fixed speeds	40% for slow speed to 160% for medium speed 80% p-f designs. Special high torque designs	Pull-out torque of unity-p-f motors 170%; 80%—p-f motors 225%. Special designs up to 300%	For constant-speed service, direct connection to slow-speed machines and where power-factor correction is required.

General Electric Co.

ELECTRIC MOTOR MAINTENANCE

Electric Motor Inspection Schedule. — Frequency and thoroughness of inspection depend upon such factors as; (1) importance of the motor in the production scheme; (2) percentage of days the motor operates; (3) nature of service; (4) winding conditions. The following schedules, recommended by the General Electric Company, and covering both A.C. and D.C. motors are based on average conditions in so far as duty and dirt are concerned.

Weekly Inspection. — (1) *Surroundings.* Check to see if the windings are exposed to any dripping water, acid or alcoholic fumes; also, check for any unusual amount of dust, chips or lint on or about the motor. See if any boards, covers, canvas, etc., have been left about that might interfere with the motor ventilation or jam moving parts.

(2) *Lubrication of sleeve bearing motors.* In sleeve-bearing motors check oil level, if a gage is used, and fill to the specified line. If the journal diameter is less than 2 inches, the motor should be stopped before checking the oil level. For special lubricating systems, such as wool-packed, forced lubrication, flood and disk lubrication, follow instruction book. Oil should be added to bearing housing only when motor is at rest. A check should be made to see if oil is creeping along the shaft toward windings where it may harm the insulation.

(3) *Mechanical condition.* — Note any unusual noise which may be caused by metal to metal contact or any odor as from scorching insulation varnish.

(4) *Ball or roller bearings.* Feel ball- or roller-bearing housings for evidence of vibration, and listen for any unusual noise. Inspect for creepage of grease on inside of motor.

(5) *Commutators and brushes.* Check brushes and commutator for sparking. If the motor is on cyclic duty it should be observed through several cycles. Note color and surface condition of the commutator. A stable copper oxide-carbon film (as distinguished from a pure copper surface) on the commutator is an essential requirement for good commutation. Such a film may vary in color all the way from copper to straw, chocolate to black. It should be clean and smooth and have a high polish. All brushes should be checked for wear and pigtail connections for looseness. The commutator surface may be cleaned by using a piece of dry canvas or other hard, nonlinting material which is wound around and securely fastened to a wooden stick, and held against the rotating commutator.

(6) *Rotors and armatures.* The air gap on sleeve bearing motors should be checked, especially if they have been recently overhauled. After installing new bearings, make sure that the average reading is within 10 per cent, provided reading should be less than 0.020 inch. Check air passages through punchings and make sure they are free of foreign matter.

(7) *Windings.* If necessary clean windings by suction or mild blowing. After making sure that the motor is dead, wipe off windings with dry cloth, note evidence of moisture and see if any water has accumulated in the bottom of frame. Check also and see if any oil or grease has worked its way up to the rotor or armature windings. Clean with carbon tetrachloride in a well-ventilated room.

(8) *General.* This is a good time to check the belt, gears, flexible couplings, chain and sprockets for excessive wear or improper location. The motor starting should be checked to make sure that it comes up to proper speed each time power is applied.

Monthly or Bi-Monthly Inspection. — (1) *Windings.* Check shunt, series and commutating field windings for tightness. Try to move field spools on the poles, as drying out may have caused some play. If this condition exists, a service shop should be consulted. The motor cable connections should be checked for tightness.

(2) *Brushes.* Check brushes in holders for fit and free play. Also check the brush-spring pressure. Tighten brush studs in holders to take up slack from drying out of washers, making sure that studs are not displaced, particularly on D.C. motors. Replace brushes that are worn down almost to the brush rivet, examine brush faces for chipped toes or heels, and for heat cracks. Damaged brushes should be replaced immediately.

(3) *Commutators.* Examine commutator surface for high bars and high mica, or evidence of scratches or roughness. See that the risers are clean and have not been damaged in any way.

(4) *Ball or roller bearings.* On hard-driven, 24-hour service ball- or roller-bearing motors, purge out old grease through drain hole and apply new grease. Check to make sure grease or oil is not leaking out of the bearing housing. If any leakage is present, correct the condition before continuing to operate.

(5) *Sleeve bearings.* Check sleeve bearings for wear, including end-play bearing surfaces. Clean out oil wells if there is evidence of dirt or sludge. Flush with lighter oil before refilling.

(6) *Enclosed gears.* For motors with enclosed gears, open drain plug and check oil flow for presence of metal scale, sand or water. If condition of oil is bad, drain, flush and refill as directed. Rock rotor to see if slack or backlash is increasing.

(7) *Loads.* Check loads for changed conditions, bad adjustment, poor handling or control.

(8) *Couplings and other drive details.* Note if belt-tightening adjustment is all used up. Shorten belt if this condition exists. See if belt runs steadily and close to inside (motor edge) of pulley. Chain should be checked for evidence of wear and stretch. Clean inside of chain housing. Check chain-lubricating system. Note incline of slanting base to make sure it does not cause oil rings to rub on housing.

Annual or Bi-Annual Inspection. — (1) *Windings.* Check insulation resistance by using either a megohmmeter or a volt meter having a resistance of about 100 ohms per volt. Check insulation surfaces for dry cracks and other evidence of need for coatings of insulating material. Clean surfaces and ventilating passages thoroughly if inspection shows accumulation of dust. Check for mold or water standing in frame to determine if windings need to be dried out, varnished and baked.

(2) *Air gap and bearings.* Check air gap to make sure that average reading is within 10 per cent, provided reading should be less than 0.020 inch. All bearings, ball, roller and sleeve should be thoroughly checked and defective ones replaced. Waste-packed and wick-oiled bearings should have waste or wicks renewed, if they have become glazed or filled with metal or dirt, making sure that new waste bears well against shaft.

(3) *Rotors (squirrel-cage).* Check squirrel-cage rotors for broken or loose bars and evidence of local heating. If fan blades are not cast in place, check for loose blades. Look for marks on rotor surface indicating foreign matter in air gap or a worn bearing.

(4) *Rotors (wound).* Clean wound rotors thoroughly around collector rings, washers and connections. Tighten connections if necessary. If rings are rough, spotted or eccentric, refer to service shop for refinishing. See that all top sticks or wedges are tight. If any are loose, refer to service shop.

(5) *Armatures.* Clean all armature air passages thoroughly if any are obstructed. Look for oil or grease creeping along shaft, checking back to bearing. Check commutator for surface condition, high bars, high mica or eccentricity. If necessary, turn down the commutator to secure a smooth fresh surface.

(6) *Loads.* Read load on motor with instruments at no load, full load or through an entire cycle, as a check on the mechanical condition of the driven machine.

PRESSURES AND FLOW OF WATER

Water Pressures. — Water is composed of two gases, hydrogen and oxygen, in the ratio of two volumes of the former to one of the latter. In the English System of measure, water boils under atmospheric pressure at 212 degrees F. and freezes at 32 degrees F. Its greatest density is 62.425 pounds per cubic foot, at 39.1 degrees F. In metric SI measure, water boils under atmospheric pressure at 100°C (Celsius) and freezes at 0°C. Its density is equal to one kilogram per liter, where one liter is one cubic decimeter. Also in metric SI, pressure is given in pascals (Pa) or the equivalent Newton per square meter. See page 2411 for additional information on metric SI.

For higher temperatures, the pressure slightly decreases in the proportion indicated by the table "Weight of Water per Cubic Foot at Different Temperatures." The pressure per square inch is equal in all directions, downwards, upwards and sideways. Water can be compressed only in a very slight degree, the compressibility being so slight that even at the depth of a mile, a cubic foot of water weighs only about one-half pound more than at the surface.

Pressure in Pounds per Square Inch for Different Heads of Water

Head, Feet	0	1	2	3	4	5	6	7	8	9
0	....	0.43	0.87	1.30	1.73	2.16	2.60	3.03	3.46	3.90
10	4.33	4.76	5.20	5.63	6.06	6.49	6.93	7.36	7.79	8.23
20	8.66	9.09	9.53	9.96	10.39	10.82	11.26	11.69	12.12	12.56
30	12.99	13.42	13.86	14.29	14.72	15.15	15.59	16.02	16.45	16.89
40	17.32	17.75	18.19	18.62	19.05	19.48	19.92	20.35	20.78	21.22
50	21.65	22.08	22.52	22.95	23.38	23.81	24.25	24.68	25.11	25.55
60	25.98	26.41	26.85	27.28	27.71	28.14	28.58	29.01	29.44	29.88
70	30.31	30.74	31.18	31.61	32.04	32.47	32.91	33.34	33.77	34.21
80	34.64	35.07	35.51	35.94	36.37	36.80	37.24	37.67	38.10	38.54
90	38.97	39.40	39.84	40.27	40.70	41.13	41.57	42.00	42.43	42.87

Heads of Water in Feet Corresponding to Certain Pressures in Pounds per Square Inch

Pressure, Lbs.	0	1	2	3	4	5	6	7	8	9
0	....	2.3	4.6	6.9	9.2	11.5	13.9	16.2	18.5	20.8
10	23.1	25.4	27.7	30.0	32.3	34.6	36.9	39.3	41.6	43.9
20	46.2	48.5	50.8	53.1	55.4	57.7	60.0	62.4	64.7	67.0
30	69.3	71.6	73.9	76.2	78.5	80.8	83.1	85.4	87.8	90.1
40	92.4	94.7	97.0	99.3	101.6	103.9	106.2	108.5	110.8	113.2
50	115.5	117.8	120.1	122.4	124.7	127.0	129.3	131.6	133.9	136.3
60	138.6	140.9	143.2	145.5	147.8	150.1	152.4	154.7	157.0	159.3
70	161.7	164.0	166.3	168.6	170.9	173.2	175.5	177.8	180.1	182.4
80	184.8	187.1	189.4	191.7	194.0	196.3	198.6	200.9	203.2	205.5
90	207.9	210.2	212.5	214.8	217.1	219.4	221.7	224.0	226.3	228.6

Comparison of Different Methods of Measuring Pressures

Ounces and Pounds per Square Inch, and Inches of Water and Mercury

Ounces per Square Inch	Pounds per Square Inch	Inches of Water	Inches of Mercury	Ounces per Square Inch	Pounds per Square Inch	Inches of Water	Inches of Mercury
0.25	0.016	0.433	0.0319	8	0.500	13.856	1.020
0.50	0.031	0.866	0.0638	9	0.562	15.588	1.148
1	0.062	1.732	0.1275	10	0.625	17.320	1.275
2	0.125	3.464	0.2551	11	0.687	19.052	1.403
3	0.187	5.196	0.3826	12	0.750	20.784	1.531
4	0.250	6.928	0.5102	13	0.812	22.516	1.658
5	0.312	8.660	0.6377	14	0.875	24.248	1.786
6	0.375	10.392	0.7653	15	0.937	25.980	1.913
7	0.437	12.124	0.8928	16	1.000	27.712	2.041

Pounds per Square Inch, Inches and Feet of Water and Inches of Mercury

Pounds per Square Inch	Inches of Water	Feet of Water	Inches of Mercury	Pounds per Square Inch	Inches of Water	Feet of Water	Inches of Mercury
1	27.71	2.31	2.041	14	387.97	32.33	28.57
2	55.42	4.62	4.081	14.7	407.37	33.95	30.00
3	83.14	6.93	6.122	15	415.68	34.64	30.61
4	110.85	9.24	8.163	16	443.40	36.95	32.65
5	138.56	11.55	10.20	17	471.11	39.26	34.69
6	166.27	13.86	12.24	18	498.82	41.57	36.73
7	193.99	16.17	14.28	19	526.53	43.88	38.77
8	221.70	18.47	16.33	20	554.25	46.19	40.81
9	249.41	20.78	18.37	21	581.96	48.50	42.85
10	277.12	23.09	20.41	22	609.67	50.81	44.89
11	304.84	25.40	22.45	23	637.38	53.12	46.94
12	332.55	27.71	24.49	24	665.10	55.42	48.98
13	360.26	30.02	26.53	25	692.81	57.73	51.02

Volume of Water at Different Temperatures

Degrees Fahr.	Volume	Degrees Fahr.	Volume	Degrees Fahr.	Volume	Degrees Fahr.	Volume
39.1	1.00000	86	1.00425	131	1.01423	176	1.02872
50	1.00025	95	1.00586	140	1.01678	185	1.03213
59	1.00083	104	1.00767	149	1.01951	194	1.03570
68	1.00171	113	1.00967	158	1.02241	203	1.03943
77	1.00286	122	1.01186	167	1.02548	212	1.04332

Flow of Water in Pipes. — The quantity of water that will be discharged through a pipe depends primarily on the head and also upon the diameter of the pipe, the character of the interior surface, and the number and shape of the bends. The head may be either the actual distance between the levels of the surface of water in a reservoir and the point of discharge, or it may be caused by mechanically applied pressure, as by pumping, in which case the head is calculated as the vertical distance corresponding to the pressure. One pound per square inch is equal to 2.309 feet head, or 1 foot head is equal to a pressure of 0.433 pound per square inch.

Weight of Water per Cubic Foot at Different Temperatures

Temperature, Degrees F.	Weight per Cubic Foot, Pounds	Temperature, Degrees F.	Weight per Cubic Foot, Pounds	Temperature, Degrees F.	Weight per Cubic Foot, Pounds	Temperature, Degrees F.	Weight per Cubic Foot, Pounds	Temperature, Degrees F.	Weight per Cubic Foot, Pounds	Temperature, Degrees F.	Weight per Cubic Foot, Pounds
32	62.42	130	61.56	220	59.63	320	56.66	420	52.6	520	47.6
40	62.42	140	61.37	230	59.37	330	56.30	430	52.2	530	47.0
50	62.41	150	61.18	240	59.11	340	55.94	440	51.7	540	46.3
60	62.37	160	60.98	250	58.83	350	55.57	450	51.2	550	45.6
70	62.31	170	60.77	260	58.55	360	55.18	460	50.7	560	44.9
80	62.23	180	60.55	270	58.26	370	54.78	470	50.2	570	44.1
90	62.13	190	60.32	280	57.96	380	54.36	480	49.7	580	43.3
100	62.02	200	60.12	290	57.65	390	53.94	490	49.2	590	42.6
110	61.89	210	59.88	300	57.33	400	53.50	500	48.7	600	41.8
120	61.74	212	59.83	310	57.00	410	53.00	510	48.1	...	...

Table of Horsepower Due to Certain Head of Water

The table gives the horsepower of 1 cubic foot of water per minute, and is based on an efficiency of 85 per cent.

Heads in Feet	Horsepower	Heads in Feet	Horsepower	Heads in Feet	Horsepower	Heads in Feet	Horsepower	Heads in Feet	Horsepower
1	0.0016	170	0.274	340	0.547	520	0.837	1250	2.012
10	0.0161	180	0.290	350	0.563	540	0.869	1300	2.093
20	0.0322	190	0.306	360	0.580	560	0.901	1350	2.173
30	0.0483	200	0.322	370	0.596	580	0.934	1400	2.254
40	0.0644	210	0.338	380	0.612	600	0.966	1450	2.334
50	0.0805	220	0.354	390	0.628	650	1.046	1500	2.415
60	0.0966	230	0.370	400	0.644	700	1.127	1550	2.495
70	0.1127	240	0.386	410	0.660	750	1.207	1600	2.576
80	0.1288	250	0.402	420	0.676	800	1.288	1650	2.656
90	0.1449	260	0.418	430	0.692	850	1.368	1700	2.737
100	0.1610	270	0.435	440	0.708	900	1.449	1750	2.818
110	0.1771	280	0.451	450	0.724	950	1.529	1800	2.898
120	0.1932	290	0.467	460	0.740	1000	1.610	1850	2.978
130	0.2093	300	0.483	470	0.757	1050	1.690	1900	3.059
140	0.2254	310	0.499	480	0.773	1100	1.771	1950	3.139
150	0.2415	320	0.515	490	0.789	1150	1.851	2000	3.220
160	0.2576	330	0.531	500	0.805	1200	1.932	2100	3.381

All formulas for finding the amount of water that will flow through a pipe in a given time are approximate. The formula below will give results within 5 or 10 per cent of actual results, if applied to pipe lines carefully laid and in a fair condition.

$$V = C \sqrt{\frac{hD}{L + 54\,D}}$$

in which V = approximate mean velocity in feet per second;
C = coefficient from the accompanying table;
D = diameter of pipe in feet;
h = total head in feet;
L = total length of pipe line in feet.

Values of Coefficient C

Diam. of Pipe		C	Diam. of Pipe		C	Diam. of Pipe		C
Feet	Inches		Feet	Inches		Feet	Inches	
0.1	1.2	23	0.8	9.6	46	3.5	42	64
0.2	2.4	30	0.9	10.8	47	4.0	48	66
0.3	3.6	34	1.0	12.0	48	5.0	60	68
0.4	4.8	37	1.5	18.0	53	6.0	72	70
0.5	6.0	39	2.0	24.0	57	7.0	84	72
0.6	7.2	42	2.5	30.0	60	8.0	96	74
0.7	8.4	44	3.0	36.0	62	10.0	120	77

Example. — A pipe line, 1 mile long, 12 inches in diameter, discharges water under a head of 100 feet. Find the velocity and quantity of discharge.

From the table, the coefficient C is found to be 48 for a pipe 1 foot in diameter, hence:

$$V = 48 \sqrt{\frac{100 \times 1}{5280 + 54 \times 1}} = 6.57 \text{ feet per second.}$$

To find the discharge in cubic feet per second, multiply the velocity found by the area of cross-section of the pipe in square feet:

$$6.57 \times 0.7854 = 5.16 \text{ cubic feet per second.}$$

The loss of head due to a bend in the pipe is most frequently given in the equivalent length of straight pipe, which would cause the same loss in head as the bend. Experiments show that a right-angle bend should have a radius of about three times the diameter of the pipe. Assuming this curvature, then, if D is the diameter of the pipe in inches and L is the length of straight pipe in feet, which causes the same loss of head as the bend in the pipe, the following formula gives the equivalent length of straight pipe that should be added to compensate for a right-angle bend:

$$L = 4\,D \div 3.$$

Thus the loss of head due to a right-angle bend in a six-inch pipe would be equal to that in 8 feet of straight pipe. Experiments undertaken to determine the losses due to valves in pipe lines indicate that a fully open gate valve in a pipe causes a loss of head corresponding to that in a length of pipe equal to six diameters.

Flow of Water Through Nozzles in Cubic Feet per Second

Head in Feet, at Nozzle	Pressure, Pounds per Square Inch	Theoretical Velocity, Feet per Second	Diameter of Nozzle, Inches							
			1	1½	2	2½	3	3½	4	4½
5	2.17	17.93	0.10	0.22	0.39	0.61	0.88	1.20	1.56	2.04
10	4.33	25.36	0.14	0.31	0.55	0.86	1.24	1.69	2.21	2.87
20	8.66	35.86	0.19	0.44	0.78	1.22	1.76	2.39	3.13	4.07
30	12.99	43.92	0.24	0.54	0.96	1.50	2.16	2.93	3.83	4.98
40	17.32	50.72	0.28	0.62	1.10	1.73	2.49	3.39	4.43	5.75
50	21.65	56.71	0.31	0.70	1.24	1.93	2.78	3.79	4.95	6.43
60	25.99	62.12	0.34	0.76	1.35	2.12	3.05	4.15	5.42	7.04
70	30.32	67.10	0.37	0.82	1.46	2.29	3.29	4.48	5.86	7.61
80	34.65	71.73	0.39	0.88	1.56	2.44	3.52	4.79	6.26	8.13
90	38.98	76.08	0.42	0.94	1.66	2.59	3.73	5.08	6.64	8.63
100	43.31	80.20	0.44	0.99	1.75	2.73	3.94	5.38	7.00	9.09
120	51.97	87.88	0.49	1.08	1.87	3.00	4.31	5.87	7.67	9.96
140	60.63	94.89	0.52	1.17	2.07	3.23	4.66	6.35	8.28	10.76
160	69.29	101.45	0.56	1.25	2.21	3.46	4.98	6.78	8.86	11.50
180	77.96	107.59	0.59	1.32	2.34	3.67	5.28	7.19	9.39	12.20
200	86.62	113.41	0.62	1.39	2.47	3.87	5.57	7.57	9.90	12.86
250	108.50	126.80	0.70	1.56	2.76	4.32	6.22	8.47	11.07	14.38
300	130.20	138.91	0.76	1.71	3.03	4.74	6.82	9.27	12.13	15.75
350	151.90	150.04	0.82	1.84	3.27	5.12	7.37	10.02	13.10	17.01
400	173.60	160.40	0.88	1.97	3.50	5.47	7.87	10.71	14.00	18.19
450	195.30	170.12	0.93	2.09	3.71	5.80	8.35	11.36	14.85	19.39
500	216.00	179.33	0.99	2.21	3.91	6.11	8.80	11.98	15.65	20.34

Head in Feet, at Nozzle	Pressure, Pounds per Square Inch	Theoretical Velocity, Feet per Second	Diameter of Nozzle, Inches							
			5	6	7	8	9	10	11	12
5	2.17	17.93	2.44	3.52	4.81	6.3	7.9	9.8	12.8	14.1
10	4.33	25.36	3.46	4.98	6.78	8.8	11.2	13.8	16.7	19.9
20	8.66	35.86	4.88	7.04	9.58	12.5	15.8	19.6	23.7	28.2
30	12.99	43.92	5.99	8.62	11.74	15.3	19.4	23.9	29.0	34.5
40	17.32	50.72	6.92	9.96	13.56	17.7	22.4	27.7	33.5	39.8
50	21.65	56.71	7.73	11.13	15.16	19.8	25.0	30.9	37.4	44.5
60	25.99	62.12	8.44	12.19	16.60	21.7	27.4	33.9	41.0	48.8
70	30.32	67.10	9.15	13.17	17.93	23.4	29.6	36.6	44.3	52.7
80	34.65	71.73	9.78	14.08	19.17	25.0	31.7	39.1	47.3	56.4
90	38.98	76.08	10.38	14.93	20.35	26.6	33.6	41.5	50.2	59.7
100	43.31	80.20	10.94	15.74	21.44	28.0	35.4	43.7	52.9	63.0
120	51.97	87.88	11.99	17.25	23.49	30.7	38.8	47.9	58.0	69.0
140	60.63	94.89	12.94	18.63	25.36	33.1	41.9	51.7	62.6	74.5
160	69.29	101.45	13.84	19.91	27.12	35.4	44.8	55.3	67.0	79.7
180	77.96	107.59	14.67	21.12	28.76	37.6	47.5	58.7	71.0	84.5
200	86.62	113.41	15.47	22.26	30.31	39.6	50.1	61.8	74.8	89.1
250	108.50	126.80	17.29	24.86	33.89	44.3	56.0	69.2	83.7	99.6
300	130.20	138.91	18.90	27.27	37.13	48.5	61.4	75.8	91.7	109.1
350	151.90	150.04	20.46	29.45	40.10	52.4	66.3	81.8	99.0	117.8
400	173.60	160.40	21.88	31.49	42.87	56.0	70.9	87.5	105.9	126.0
450	195.30	170.12	23.20	33.39	45.26	59.4	75.2	92.8	112.2	133.6
500	216.00	179.33	24.46	35.20	47.93	62.6	79.2	97.8	118.4	140.8

Theoretical Velocity of Water Due to Head in Feet

Head in Feet	Theoretical Velocity, Feet per Second	Theoretical Velocity, Feet per Minute	Head in Feet	Theoretical Velocity, Feet per Second	Theoretical Velocity, Feet per Minute	Head in Feet	Theoretical Velocity, Feet per Second	Theoretical Velocity, Feet per Minute
1	8.02	481	48	55.60	3336	95	78.22	4693
2	11.34	682	49	56.17	3370	96	78.63	4718
3	13.90	834	50	56.74	3405	97	79.04	4742
4	16.05	963	51	57.31	3438	98	79.44	4767
5	17.94	1077	52	57.87	3472	99	79.85	4791
6	19.66	1179	53	58.42	3505	100	80.25	4815
7	21.23	1274	54	58.97	3538	105	82.23	4934
8	22.70	1362	55	59.51	3571	110	84.17	5050
9	24.07	1445	56	60.05	3603	115	86.06	5163
10	25.38	1523	57	60.59	3635	120	87.91	5274
11	26.61	1597	58	61.12	3667	125	89.72	5383
12	27.80	1668	59	61.64	3698	130	91.50	5490
13	28.93	1736	60	62.16	3730	135	93.24	5594
14	30.03	1802	61	62.68	3761	140	94.95	5697
15	31.08	1865	62	63.19	3791	145	96.63	5798
16	32.10	1926	63	63.70	3822	150	98.28	5897
17	33.09	1985	64	64.20	3852	155	99.91	5994
18	34.05	2043	65	64.70	3882	160	101.50	6090
19	34.98	2099	66	65.19	3912	165	103.08	6185
20	35.89	2153	67	65.69	3941	170	104.63	6278
21	36.77	2206	68	66.17	3970	175	106.16	6370
22	37.64	2258	69	66.66	4000	180	107.66	6460
23	38.49	2309	70	67.14	4028	185	109.15	6549
24	39.31	2359	71	67.62	4057	190	110.61	6637
25	40.12	2407	72	68.09	4086	195	112.06	6724
26	40.92	2455	73	68.56	4114	200	113.49	6809
27	41.70	2502	74	69.03	4142	205	114.90	6894
28	42.46	2548	75	69.50	4170	210	116.29	6978
29	43.21	2593	76	69.96	4198	215	117.66	7060
30	43.95	2637	77	70.42	4225	220	119.03	7142
31	44.68	2681	78	70.87	4252	225	120.38	7222
32	45.40	2724	79	71.33	4280	230	121.70	7302
33	46.10	2766	80	71.78	4307	235	123.02	7381
34	46.79	2783	81	72.22	4333	240	124.32	7459
35	47.48	2848	82	72.67	4360	245	125.60	7537
36	48.15	2889	83	73.11	4387	250	126.88	7613
37	48.81	2929	84	73.55	4413	255	128.15	7649
38	49.47	2968	85	73.99	4439	260	129.39	7764
39	50.12	3007	86	74.42	4465	270	131.86	7912
40	50.75	3045	87	74.85	4491	280	134.28	8057
41	51.38	3083	88	75.28	4517	290	136.66	8200
42	52.01	3120	89	75.71	4542	300	138.99	8340
43	52.62	3157	90	76.13	4568	310	141.29	8478
44	53.23	3194	91	76.55	4593	320	143.55	8613
45	53.83	3230	92	76.97	4618	330	145.78	8761
46	54.43	3266	93	77.39	4643	340	147.97	8878
47	55.02	3301	94	77.80	4668	350	150.13	9008

PROPERTIES, COMPRESSION AND FLOW OF AIR

Properties of Air. — Air is a mechanical mixture composed of 78 per cent, by volume, of nitrogen, 21 per cent of oxygen and 1 per cent of argon. The density of dry air at 32 degrees F and atmospheric pressure (29.92 inches of mercury or 14.70 pounds per square inch) is 0.08073 pound per cubic foot. The density of air at any other temperature or pressure is

$$\rho = \frac{1.325 \times B}{T}$$

in which ρ = density in pounds per cubic foot; B = height of barometric pressure in inches of mercury; T = absolute temperature Rankine. (When using pounds as a unit, here and elsewhere, care must be exercised to differentiate between pounds mass and pounds force. See pages 131–134.)

Volume and Weight of Air at Different Temperatures, at Atmospheric Pressure

Temperature, Degrees Fahr.	Volume of 1 Lb. of Air in Cubic Feet	Density, Pounds per Cubic Foot	Temperature, Degrees Fahr.	Volume of 1 Lb. of Air in Cubic Feet	Density, Pounds per Cubic Foot	Temperature, Degrees Fahr.	Volume of 1 Lb. of Air in Cubic Feet	Density, Pounds per Cubic Foot
0	11.57	0.0864	172	15.92	0.0628	800	31.75	0.0315
12	11.88	0.0842	182	16.18	0.0618	900	34.25	0.0292
22	12.14	0.0824	192	16.42	0.0609	1000	37.31	0.0268
32	12.39	0.0807	202	16.67	0.0600	1100	39.37	0.0254
42	12.64	0.0791	212	16.92	0.0591	1200	41.84	0.0239
52	12.89	0.0776	230	17.39	0.0575	1300	44.44	0.0225
62	13.14	0.0761	250	17.89	0.0559	1400	46.95	0.0213
72	13.39	0.0747	275	18.52	0.0540	1500	49.51	0.0202
82	13.64	0.0733	300	19.16	0.0522	1600	52.08	0.0192
92	13.89	0.0720	325	19.76	0.0506	1700	54.64	0.0183
102	14.14	0.0707	350	20.41	0.0490	1800	57.14	0.0175
112	14.41	0.0694	375	20.96	0.0477	2000	62.11	0.0161
122	14.66	0.0682	400	21.69	0.0461	2200	67.11	0.0149
132	14.90	0.0671	450	22.94	0.0436	2400	72.46	0.0138
142	15.17	0.0659	500	24.21	0.0413	2600	76.92	0.0130
152	15.41	0.0649	600	26.60	0.0376	2800	82.64	0.0121
162	15.67	0.0638	700	29.59	0.0338	3000	87.72	0.0114

The absolute zero from which all temperatures must be counted when dealing with the weight and volume of gases is assumed to be -459.7 degrees F. Hence, to obtain the absolute temperature T used in the formula above, add to the temperature observed on a regular Fahrenheit thermometer the value 459.7.

In obtaining the value of B, 1 inch of mercury at 32 degrees F may be taken as equal to a pressure of 0.491 pound per square inch.

Example: — What would be the weight of a cubic foot of air at atmospheric pressure (29.92 inches of mercury) at 100 degrees F? [The weight, W, is given by $W = \rho V$.]

$$W = \rho V = \frac{1.325 \times 29.92}{100 + 459.7} \times 1 = 0.0708 \text{ pound.}$$

Density of Air at Different Pressures and Temperatures

Temp. of Air, Degrees Fahr.	Gage Pressure, Pounds														
	0	5	10	20	30	40	50	60	80	100	120	150	200	250	300
	Density in Pounds per Cubic Foot														
-20	0.0900	0.1205	0.1515	0.2125	0.274	0.336	0.397	0.458	0.580	0.702	0.825	1.010	1.318	1.625	1.930
-10	0.0882	0.1184	0.1485	0.2090	0.268	0.328	0.388	0.448	0.567	0.687	0.807	0.989	1.288	1.588	1.890
0	0.0864	0.1160	0.1455	0.2040	0.263	0.321	0.380	0.438	0.555	0.672	0.790	0.968	1.260	1.553	1.850
10	0.0846	0.1136	0.1425	0.1995	0.257	0.314	0.372	0.429	0.543	0.658	0.774	0.947	1.233	1.520	1.810
20	0.0828	0.1112	0.1395	0.1955	0.252	0.307	0.364	0.420	0.533	0.645	0.757	0.927	1.208	1.489	1.770
30	0.0811	0.1088	0.1366	0.1916	0.246	0.301	0.357	0.412	0.522	0.632	0.742	0.908	1.184	1.460	1.735
40	0.0795	0.1067	0.1338	0.1876	0.241	0.295	0.350	0.404	0.511	0.619	0.727	0.890	1.161	1.431	1.701
50	0.0780	0.1045	0.1310	0.1839	0.237	0.290	0.343	0.396	0.501	0.607	0.713	0.873	1.139	1.403	1.668
60	0.0764	0.1025	0.1283	0.1803	0.232	0.284	0.336	0.388	0.493	0.596	0.700	0.856	1.116	1.376	1.636
80	0.0736	0.0988	0.1239	0.1738	0.224	0.274	0.324	0.374	0.473	0.572	0.673	0.824	1.074	1.325	1.573
100	0.0710	0.0954	0.1197	0.1676	0.215	0.264	0.312	0.360	0.455	0.551	0.648	0.794	1.035	1.276	1.517
120	0.0686	0.0921	0.1155	0.1618	0.208	0.255	0.302	0.348	0.440	0.533	0.626	0.767	1.001	1.234	1.465
140	0.0663	0.0889	0.1115	0.1565	0.201	0.246	0.291	0.336	0.426	0.516	0.606	0.742	0.968	1.194	1.416
150	0.0652	0.0874	0.1096	0.1541	0.198	0.242	0.286	0.331	0.419	0.508	0.596	0.730	0.953	1.175	1.392
175	0.0626	0.0840	0.1054	0.1482	0.191	0.233	0.275	0.318	0.403	0.488	0.573	0.701	0.914	1.128	1.337
200	0.0603	0.0809	0.1014	0.1427	0.184	0.225	0.265	0.305	0.388	0.470	0.552	0.674	0.879	1.084	1.287
225	0.0581	0.0779	0.0976	0.1373	0.177	0.216	0.255	0.295	0.374	0.452	0.531	0.649	0.846	1.043	1.240
250	0.0560	0.0751	0.0941	0.1323	0.170	0.208	0.247	0.284	0.360	0.436	0.513	0.627	0.817	1.007	1.197
275	0.0541	0.0726	0.0910	0.1278	0.164	0.201	0.238	0.274	0.348	0.421	0.494	0.605	0.789	0.972	1.155
300	0.0523	0.0707	0.0881	0.1237	0.159	0.194	0.230	0.265	0.336	0.407	0.478	0.585	0.762	0.940	1.118
350	0.0491	0.0658	0.0825	0.1160	0.149	0.183	0.216	0.249	0.316	0.382	0.449	0.549	0.715	0.883	1.048
400	0.0463	0.0621	0.0779	0.1090	0.140	0.172	0.203	0.235	0.297	0.360	0.423	0.517	0.674	0.831	0.987
450	0.0437	0.0586	0.0735	0.1033	0.133	0.163	0.192	0.222	0.281	0.340	0.399	0.488	0.637	0.786	0.934
500	0.0414	0.0555	0.0696	0.0978	0.126	0.154	0.182	0.210	0.266	0.322	0.379	0.463	0.604	0.746	0.885
550	0.0394	0.0528	0.0661	0.0930	0.120	0.146	0.173	0.200	0.253	0.306	0.359	0.440	0.573	0.709	0.841
600	0.0376	0.0504	0.0631	0.0885	0.114	0.139	0.165	0.190	0.241	0.292	0.343	0.419	0.547	0.675	0.801

Relation between Pressure, Temperature and Volume of Air. — This relationship is expressed by the formula:

$$\frac{P \times V}{T} = 53.3,$$

in which P = absolute pressure in pounds per square foot; V = volume in cubic feet of one pound of air at the given pressure and temperature; T = absolute temperature in degrees R.

Example. — What is the volume of one pound of air at a pressure of 24.7 pounds per square inch and at a temperature of 210 degrees F?

$$\frac{24.7 \times 144 \times V}{210 + 459.7} = 53.3, \text{ or } V = \frac{53.3 \times 669.7}{24.7 \times 144} = 10.04 \text{ cubic feet.}$$

Relation Between Barometric Pressure, and Pressures in Pounds per Square Inch and Square Foot

Barometer, Inches	Pressure in Pounds per Square Inch	Pressure in Pounds per Square Foot	Barometer, Inches	Pressure in Pounds per Square Inch	Pressure in Pounds per Square Foot	Barometer, Inches	Pressure in Pounds per Square Inch	Pressure in Pounds per Square Foot
28.00	13.75	1980	29.25	14.36	2068	30.50	14.98	2156
28.25	13.87	1997	29.50	14.48	2086	30.75	15.10	2174
28.50	13.99	2015	29.75	14.61	2103	31.00	15.22	2192
28.75	14.12	2033	30.00	14.73	2121	31.25	15.34	2210
29.00	14.24	2050	30.25	14.85	2139	.....	.....	......

Expansion and Compression of Air. — The formula for the relationship of pressure, temperature and volume of air just given indicates that when the pressure remains constant the volume is directly proportional to the absolute temperature. If the temperature remains constant, the volume is inversely proportional to the absolute pressure. Theoretically, air (as well as other gases) can be expanded or compressed according to different laws. *Adiabatic* expansion or compression takes place when the air is expanded or compressed without transmission of heat to or from it; as for example, if the air could be expanded or compressed in a cylinder of an absolutely non-conducting material. Let:

P_1 = initial absolute pressure in pounds per square foot;
V_1 = initial volume in cubic feet;
T_1 = initial absolute temperature in degrees R;
P_2 = absolute pressure in pounds per square foot, after compression;
V_2 = volume in cubic feet, after compression;
T_2 = absolute temperature in degrees R, after compression.

Then:

$$\frac{V_2}{V_1} = \left(\frac{P_1}{P_2}\right)^{0.71} \qquad \frac{P_2}{P_1} = \left(\frac{V_1}{V_2}\right)^{1.41} \qquad \frac{T_2}{T_1} = \left(\frac{V_1}{V_2}\right)^{0.41}$$

$$\frac{V_2}{V_1} = \left(\frac{T_1}{T_2}\right)^{2.46} \qquad \frac{P_2}{P_1} = \left(\frac{T_2}{T_1}\right)^{3.46} \qquad \frac{T_2}{T_1} = \left(\frac{P_2}{P_1}\right)^{0.29}$$

These formulas are also applicable if all pressures are in pounds per square inch; if all volumes are in cubic inches; or if any other consistent set of units is used for pressure or volume.

Isothermal expansion or compression takes place when the gas is expanded or compressed with an addition or transmission of sufficient heat to maintain a constant temperature. Let:

P_1 = initial absolute pressure in pounds per square foot;

V_1 = initial volume in cubic feet;

P_2 = absolute pressure in pounds per square foot, after compression;

V_2 = volume in cubic feet, after compression;

R = 53.3

T = temperature in degrees Rankine maintained during isothermal expansion or contraction.

Then:

$$P_1 \times V_1 = P_2 \times V_2 = RT.$$

Example. — A volume of 165 cubic feet of air, at a pressure of 15 pounds per square inch, is compressed adiabatically to a pressure of 80 pounds per square inch. What will be the volume at this pressure?

$$V_2 = V_1 \left(\frac{P_1}{P_2}\right)^{0.71} = 165 \left(\frac{15}{80}\right)^{0.71} = 50 \text{ cubic feet, approx.}$$

Example. — The same volume of air is compressed isothermally from 15 to 80 pounds per square inch. What will be the volume after compression?

$$V_2 = \frac{P_1 \times V_1}{P_2} = \frac{15 \times 165}{80} = 31 \text{ cubic feet.}$$

Foot-pounds of Work Required in Compression of Air
Initial Pressure = 1 atmosphere = 14.7 pounds per square inch

Gage Pressure in Pounds per Square Inch	Isothermal Compression	Adiabatic Compression	Actual Compression	Gage Pressure in Pounds per Square Inch	Isothermal Compression	Adiabatic Compression	Actual Compression
	Foot-pounds Required per Cubic Foot of Air at Initial Pressure				Foot-pounds Required per Cubic Foot of Air at Initial Pressure		
5	619.6	649.5	637.5	55	3393.7	4188.9	3870.8
10	1098.2	1192.0	1154.6	60	3440.4	4422.8	4029.8
15	1488.3	1661.2	1592.0	65	3577.6	4645.4	4218.2
20	1817.7	2074.0	1971.4	70	3706.3	4859.6	4398.1
25	2102.6	2451.6	2312.0	75	3828.0	5063.9	4569.5
30	2353.6	2794.0	2617.8	80	3942.9	5259.7	4732.9
35	2578.0	3111.0	2897.8	85	4051.5	5450.0	4890.1
40	2780.8	3405.5	3155.6	90	4155.7	5633.1	5042.1
45	2966.0	3681.7	3395.4	95	4254.3	5819.3	5187.3
50	3136.2	3942.3	3619.8	100	4348.1	5981.2	5327.9

Work Required in Compression of Air. — The total work required for compression and expulsion of air, adiabatically compressed, is:

$$\text{Total work in foot-pounds} = 3.46 \, P_1 V_1 \left[\left(\frac{P_2}{P_1}\right)^{0.29} - 1 \right]$$

in which P_1 = initial absolute pressure in pounds per square foot;
P_2 = absolute pressure in pounds per square foot, after compression;
V_1 = initial volume in cubic feet.

The total work required for isothermal compression is:

$$\text{Total work in foot-pounds} = P_1 V_1 \log_e \frac{V_1}{V_2}$$

in which P_1, P_2 and V_1 denote the same quantities as in the previous equation, and V_2 = volume of air in cubic feet, after compression.

The work required to compress air isothermally, that is, when the heat of compression is removed as rapidly as produced, is considerably less than the work required for compressing air adiabatically, or when all the heat is retained. In practice, neither of these two theoretical extremes is obtainable, but the power required for air compression is about the median between the powers that would be required for each. The accompanying table gives the average number of foot-pounds of work required to compress air.

Horsepower Required to Compress Air. — In the accompanying tables is given the horsepower required for compressing one cubic foot of free air per minute (isothermally and adiabatically) from atmospheric pressure (14.7 pounds per square inch) to various gage pressures, for one-, two- and three-stage compression. The formula for calculating the horsepower required to compress, adiabatically, a given volume of free air to a given pressure is:

$$\text{H. P.} = \frac{144\,NPVn}{33000\,(n-1)} \left[\left(\frac{P_2}{P}\right)^{\frac{n-1}{Nn}} - 1 \right]$$

in which N = number of stages in which compression is accomplished;
P = atmospheric pressure in pounds per square inch;
P_2 = absolute terminal pressure in pounds per square inch;
V = volume of air, in cubic feet, compressed per minute, at atmospheric pressure;
n = exponent of the compression curve = 1.41 for adiabatic compression.

For different methods of compression and for one cubic foot of air per minute, this formula may be simplified as follows:

For one-stage compression: H. P. = $0.015\,P(R^{0.29} - 1)$
For two-stage compression: H. P. = $0.030\,P(R^{0.145} - 1)$
For three-stage compression: H. P. = $0.045\,P(R^{0.0975} - 1)$
For four-stage compression: H. P. = $0.060\,P(R^{0.0725} - 1)$

In these latter formulas $R = \dfrac{P_2}{P}$ = number of atmospheres to be compressed.

The formula for calculating the horsepower required to compress isothermally a given volume of free air to a given pressure is:

$$\text{H. P.} = \frac{144\,PV}{33000} \left(\log_e \frac{P_2}{P} \right)$$

Natural logarithms are obtained by multiplying common logarithms by 2.30259 or by using a handheld calculator.

Horsepower Required to Compress Air

Horsepower Required for Compressing One Cubic Foot of Free Air per Minute (Isothermally and Adiabatically) from Atmospheric Pressure (14.7 pounds per square inch) to Various Gage Pressures. — Single-stage Compression

(Initial Temperature of Air, 60° F. — Jacket-cooling not considered)

Gage Pressure, Pounds	Absolute Pressure, Pounds	Number of Atmospheres	Isothermal Compression		Adiabatic Compression			
			Mean Effective Pressure *	Horsepower	Mean Effective Pressure,* Theoretical	Mean Eff. Pressure plus 15 per cent Friction	Horsepower, Theoretical	Horsepower plus 15 per cent Friction
5	19.7	1.34	4.13	0.018	4.46	5.12	0.019	0.022
10	24.7	1.68	7.57	0.033	8.21	9.44	0.036	0.041
15	29.7	2.02	11.02	0.048	11.46	13.17	0.050	0.057
20	34.7	2.36	12.62	0.055	14.30	16.44	0.062	0.071
25	39.7	2.70	14.68	0.064	16.94	19.47	0.074	0.085
30	44.7	3.04	16.30	0.071	19.32	22.21	0.084	0.096
35	49.7	3.38	17.90	0.078	21.50	24.72	0.094	0.108
40	54.7	3.72	19.28	0.084	23.53	27.05	0.103	0.118
45	59.7	4.06	20.65	0.090	25.40	29.21	0.111	0.127
50	64.7	4.40	21.80	0.095	27.23	31.31	0.119	0.136
55	69.7	4.74	22.95	0.100	28.90	33.23	0.126	0.145
60	74.7	5.08	23.90	0.104	30.53	35.10	0.133	0.153
65	79.7	5.42	24.80	0.108	32.10	36.91	0.140	0.161
70	84.7	5.76	25.70	0.112	33.57	38.59	0.146	0.168
75	89.7	6.10	26.62	0.116	35.00	40.25	0.153	0.175
80	94.7	6.44	27.52	0.120	36.36	41.80	0.159	0.182
85	99.7	6.78	28.21	0.123	37.63	43.27	0.164	0.189
90	104.7	7.12	28.93	0.126	38.89	44.71	0.169	0.195
95	109.7	7.46	29.60	0.129	40.11	46.12	0.175	0.201
100	114.7	7.80	30.30	0.132	41.28	47.46	0.180	0.207
110	124.7	8.48	31.42	0.137	43.56	50.09	0.190	0.218
120	134.7	9.16	32.60	0.142	45.69	52.53	0.199	0.229
130	144.7	9.84	33.75	0.147	47.72	54.87	0.208	0.239
140	154.7	10.52	34.67	0.151	49.64	57.08	0.216	0.249
150	164.7	11.20	35.59	0.155	51.47	59.18	0.224	0.258
160	174.7	11.88	36.30	0.158	53.70	61.80	0.234	0.269
170	184.7	12.56	37.20	0.162	55.60	64.00	0.242	0.278
180	194.7	13.24	38.10	0.166	57.20	65.80	0.249	0.286
190	204.7	13.92	38.80	0.169	58.80	67.70	0.256	0.294
200	214.7	14.60	39.50	0.172	60.40	69.50	0.263	0.303

* Mean Effective Pressure (MEP) is defined as that single pressure rise, above atmospheric, which would require the same horsepower as the actual varying pressures during compression.

Horsepower Required to Compress Air

Horsepower Required for Compressing One Cubic Foot of Free Air per Minute (Isothermally and Adiabatically) from Atmospheric Pressure (14.7 pounds per square inch) to Various Gage Pressures. — Two-stage Compression

(Initial Temperature of Air, 60° F. — Jacket-cooling not considered)

Gage Pressure, Pounds	Absolute Pressure, Pounds	Number of Atmospheres	Correct Ratio of Cylinder Volumes	Intercooler Gage Pressure	Isothermal Compression Mean Effective Pressure *	Isothermal Compression Horsepower	Adiabatic Compression Mean Eff. Pressure, Theoretical *	Adiabatic Compression Mean Eff. Pressure plus 15 per cent Friction	Adiabatic Compression Horsepower, Theoretical	Adiabatic Compression H.P. plus 15 per cent Friction	Percentage of Saving over One-stage Compression
50	64.7	4.40	2.10	16.2	21.80	0.095	24.30	27.90	0.106	0.123	10 9
60	74.7	5.08	2.25	18.4	23.90	0.104	27.20	31.30	0.118	0.136	11.3
70	84.7	5.76	2.40	20.6	25.70	0.112	29.31	33.71	0.128	0.147	12.3
80	94.7	6.44	2.54	22.7	27.52	0.120	31.44	36.15	0.137	0.158	13.8
90	104.7	7.12	2.67	24.5	28.93	0.126	33.37	38.36	0.145	0.167	14.2
100	114.7	7.80	2.79	26.3	30.30	0.132	35.20	40.48	0.153	0.176	15.0
110	124.7	8.48	2.91	28.1	31.42	0.137	36.82	42.34	0.161	0.185	15.2
120	134.7	9.16	3.03	29.8	32.60	0.142	38.44	44.20	0.168	0.193	15.6
130	144.7	9.84	3.14	31.5	33.75	0.147	39.86	45.83	0.174	0.200	16.3
140	154.7	10.52	3.24	32.9	34.67	0.151	41.28	47.47	0.180	0.207	16.7
150	164.7	11.20	3.35	34.5	35.59	0.155	42.60	48.99	0.186	0.214	16.9
160	174.7	11.88	3.45	36.1	36.30	0.158	43.82	50.39	0.191	0.219	18.4
170	184.7	12.56	3.54	37.3	37.20	0.162	44.93	51.66	0.196	0.225	19.0
180	194.7	13.24	3.64	38.8	38.10	0.166	46.05	52.95	0.201	0.231	19.3
190	204.7	13.92	3.73	40.1	38.80	0.169	47.16	54.22	0.206	0.236	19.5
200	214.7	14.60	3.82	41.4	39.50	0.172	48.18	55.39	0.210	0.241	20.1
210	224.7	15.28	3.91	42.8	40.10	0.174	49.35	56.70	0.216	0.247	.....
220	234.7	15.96	3.99	44.0	40.70	0.177	50.30	57.70	0.220	0.252	.....
230	244.7	16.64	4.08	45.3	41.30	0.180	51.30	59.10	0.224	0.257	.....
240	254.7	17.32	4.17	46.6	41.90	0.183	52.25	60.10	0.228	0.262	.....
250	264.7	18.00	4.24	47.6	42.70	0.186	52.84	60.76	0.230	0.264	.....
260	274.7	18.68	4.32	48.8	43.00	0.188	53.85	62.05	0.235	0.270	.....
270	284.7	19.36	4.40	50.0	43.50	0.190	54.60	62.90	0.238	0.274	.....
280	294.7	20.04	4.48	51.1	44.00	0.192	55.50	63.85	0.242	0.278	.....
290	304.7	20.72	4.55	52.2	44.50	0.194	56.20	64.75	0.246	0.282	.....
300	314.7	21.40	4.63	53.4	45.80	0.197	56.70	65.20	0.247	0.283	.....
350	364.7	24.80	4.98	58.5	47.30	0.206	60.15	69.16	0.262	0.301	.....
400	414.7	28.20	5.31	63.3	49.20	0.214	63.19	72.65	0.276	0.317	.....
450	464.7	31.60	5.61	67.8	51.20	0.223	65.93	75.81	0.287	0.329	.....
500	514.7	35.01	5.91	72.1	52.70	0.229	68.46	78.72	0.298	0.342	.....

* See footnote on page 2372.

Horsepower Required to Compress Air

Horsepower Required for Compressing One Cubic Foot of Free Air per Minute (Isothermally and Adiabatically) from Atmospheric Pressure (14.7 pounds per square inch) to Various Gage Pressures. — Three-stage Compression

(Initial Temperature of Air, 60° F. — Jacket-cooling not considered)

Gage Pressure, Pounds	Absolute Pressure, Pounds	Number of Atmospheres	Correct Ratio of Cylinder Volumes	Intercooler Gage Pressure, First and Second Stages	Isothermal Compression		Adiabatic Compression				Percentage of Saving over Two-stage Compression
					Mean Effective Pressure *	Horsepower	Mean Eff. Pressure,* Theoretical	Mean Eff. Pressure plus 15 per cent Friction	Horsepower, Theoretical	H.P. plus 15 per cent Friction	
100	114.7	7.8	1.98	14.4– 42.9	30.30	0.132	33.30	38.30	0.145	0.167	5.23
150	164.7	11.2	2.24	18.2– 59.0	35.59	0.155	40.30	46.50	0.175	0.202	5.92
200	214.7	14.6	2.44	21.2– 73.0	39.50	0.172	45.20	52.00	0.196	0.226	6.67
250	264.7	18.0	2.62	23.8– 86.1	42.70	0.186	49.20	56.60	0.214	0.246	6.96
300	314.7	21.4	2.78	26.1– 98.7	45.30	0.197	52.70	60.70	0.229	0.264	7.28
350	364.7	24.8	2.92	28.2–110.5	47.30	0.206	55.45	63.80	0.242	0.277	7.64
400	414.7	28.2	3.04	30.0–121.0	49.20	0.214	58.25	66.90	0.253	0.292	8.33
450	464.7	31.6	3.16	31.8–132.3	51.20	0.223	60.40	69.40	0.263	0.302	8.36
500	514.7	35.0	3.27	33.4–142.4	52.70	0.229	62.30	71.70	0.273	0.314	8.38
550	564.7	38.4	3.38	35.0–153.1	53.75	0.234	65.00	74.75	0.283	0.326	8.80
600	614.7	41.8	3.47	36.3–162.3	54.85	0.239	66.85	76.90	0.291	0.334	8.86
650	664.7	45.2	3.56	37.6–171.5	56.00	0.244	67.90	78.15	0.296	0.340	9.02
700	714.7	48.6	3.65	38.9–180.8	57.15	0.249	69.40	79.85	0.303	0.348	9.18
750	764.7	52.0	3.73	40.1–189.8	58.10	0.253	70.75	81.40	0.309	0.355	....
800	814.7	55.4	3.82	41.4–199.5	59.00	0.257	72.45	83.25	0.315	0.362	....
850	864.7	58.8	3.89	42.5–207.8	60.20	0.262	73.75	84.90	0.321	0.369	....
900	914.7	62.2	3.95	43.4–214.6	60.80	0.265	74.80	86.00	0.326	0.375	....
950	964.7	65.6	4.03	44.6–224.5	61.72	0.269	76.10	87.50	0.331	0.381	....
1000	1014.7	69.0	4.11	45.7–233.3	62.40	0.272	77.20	88.80	0.336	0.383	....
1050	1064.7	72.4	4.15	46.3–238.3	63.10	0.275	78.10	90.10	0.340	0.391	....
1100	1114.7	75.8	4.23	47.5–248.3	63.80	0.278	79.10	91.10	0.344	0.396	....
1150	1164.7	79.2	4.30	48.5–256.8	64.40	0.281	80.15	92.20	0.349	0.401	....
1200	1214.7	82.6	4.33	49.0–261.3	65.00	0.283	81.00	93.15	0.353	0.405	....
1250	1264.7	86.0	4.42	50.3–272.3	65.60	0.286	82.00	94.30	0.357	0.411	....
1300	1314.7	89.4	4.48	51.3–280.8	66.30	0.289	82.90	95.30	0.362	0.416	....
1350	1364.7	92.8	4.53	52.0–287.3	66.70	0.291	84.00	96.60	0.366	0.421	....
1400	1414.7	96.2	4.58	52.6–293.5	67.00	0.292	84.60	97.30	0.368	0.423	....
1450	1464.7	99.6	4.64	53.5–301.5	67.70	0.295	85.30	98.20	0.371	0.426	....
1500	1514.7	103.0	4.69	54.3–309.3	68.30	0.298	85.80	98.80	0.374	0.430	....
1550	1564.7	106.4	4.74	55.0–317.3	68.80	0.300	86.80	99.85	0.378	0.434	....
1600	1614.7	109.8	4.79	55.8–323.3	69.10	0.302	87.60	100.80	0.382	0.438	....

* See footnote on page 2372.

Flow of Air in Pipes. — The following formulas are used:

$$v = \sqrt{\frac{25{,}000\, dp}{L}} \qquad p = \frac{Lv^2}{25{,}000\, d}$$

in which v = velocity of air in feet per second;

p = loss of pressure due to flow through the pipes in ounces per square inch;

d = inside diameter of pipe in inches;

L = length of pipe in feet.

The quantity of air discharged in cubic feet per second is the product of the velocity as obtained from the formula above and the area of the pipe in square feet. The horsepower required to drive air through a pipe equals the volume of air in cubic feet per second multiplied by the pressure in pounds per square foot, and this product divided by 550.

Volume of Air Transmitted, in Cubic Feet per Minute, Through Pipes

Velocity of Air in Feet per Second	Actual Inside Diameter of Pipe, Inches									
	1	2	3	4	6	8	10	12	16	24
1	0.33	1.31	2.95	5.2	11.8	20.9	32.7	47.1	83.8	188
2	0.65	2.62	5.89	10.5	23.6	41.9	65.4	94.2	167.5	377
3	0.98	3.93	8.84	15.7	35.3	62.8	98.2	141.4	251.3	565
4	1.31	5.24	11.78	20.9	47.1	83.8	131.0	188.0	335.0	754
5	1.64	6.55	14.7	26.2	59.0	104.0	163.0	235.0	419.0	942
6	1.96	7.85	17.7	31.4	70.7	125.0	196.0	283.0	502.0	1131
7	2.29	9.16	20.6	36.6	82.4	146.0	229.0	330.0	586.0	1319
8	2.62	10.50	23.5	41.9	94.0	167.0	262.0	377.0	670.0	1508
9	2.95	11.78	26.5	47.0	106.0	188.0	294.0	424.0	754.0	1696
10	3.27	13.1	29.4	52.0	118.0	209.0	327.0	471.0	838.0	1885
12	3.93	15.7	35.3	63.0	141.0	251.0	393.0	565.0	1005.0	2262
15	4.91	19.6	44.2	78.0	177.0	314.0	491.0	707.0	1256.0	2827
18	5.89	23.5	53.0	94.0	212.0	377.0	589.0	848.0	1508.0	3393
20	6.55	26.2	59.0	105.0	235.0	419.0	654.0	942.0	1675.0	3770
24	7.86	31.4	71.0	125.0	283.0	502.0	785.0	1131.0	2010.0	4524
25	8.18	32.7	73.0	131.0	294.0	523.0	818.0	1178.0	2094.0	4712
28	9.16	36.6	82.0	146.0	330.0	586.0	916.0	1319.0	2346.0	5278
30	9.80	39.3	88.0	157.0	353.0	628.0	982.0	1414.0	2513.0	5655

Flow of Compressed Air in Pipes. — When there is a comparatively small difference of pressure at the two ends of the pipe, the volume of flow in cubic feet per minute is found by the formula:

$$V = 58 \sqrt{\frac{p d^5}{WL}}$$

in which V = volume of air in cubic feet per minute;

p = difference in pressure at the two ends of the pipe in pounds per square inch;

d = inside diameter of pipe in inches;

W = weight in pounds of one cubic foot of entering air;

L = length of pipe in feet.

Velocity of Escaping Compressed Air. — If air, or gas, flows from one chamber to another, as from a chamber or tank through an orifice or nozzle into the open air, large changes in velocity may take place owing to the difference in pressures. Since the change takes place almost instantly little heat can escape from the fluid and the flow may be assumed to be adiabatic.

For a large container with a small orifice or hole from which the air escapes, the velocity of escape (theoretical) may be calculated from the formula:

$$v_2 = \sqrt{2g \cdot \frac{k}{k-1} \cdot 53.3(459.7 + F)\left[1 - \left(\frac{p_2}{p_1}\right)^{\frac{k-1}{k}}\right]}$$

In this formula, v_2 = velocity of escaping air in feet per second; g = acceleration due to gravity, 32.16 feet per second squared; $k = 1.41$ for adiabatic expansion or compression of air; F = temperature, degrees F; p_2 = atmospheric pressure = 14.7 pounds per square inch; and p_1 = pressure of air in container, pounds per square inch. It should be noted, in applying the above formula, that when the ratio p_2/p_1 equals 0.53, approximately, under normal temperature conditions at sea level, the escape velocity v_2 will be equal to the velocity of sound. Increasing the pressure p_1 will not increase the velocity of escaping air beyond this limiting velocity unless a special converging diverging nozzle design is used rather than an orifice.

The accompanying table provides velocity of escaping air for various values of p_1. These were calculated from the formula above simplified by substituting the appropriate constants:

$$v_2 = 108.58\sqrt{(459.7 + F)\left[1 - \left(\frac{14.7}{p_1}\right)^{0.29}\right]}$$

Velocity of Escaping Air at 70-Degrees F

Pressure Above Atmospheric Pressure			Theoretical Velocity, Feet per Second	Pressure Above Atmospheric Pressure			Theoretical Velocity, Feet per Second
In Atmospheres	In Inches Mercury	In Lbs. per Sq. In.		In Atmospheres	In Inches Mercury	In Lbs. per Sq. In.	
0.010	0.30	0.147	134	0.408	12.24	6.00	769
0.068	2.04	1.00	344	0.500	15.00	7.35	833
0.100	3.00	1.47	413	0.544	16.33	8.00	861
0.136	4.08	2.00	477	0.612	18.37	9.00	900
0.204	6.12	3.00	573	0.680	20.41	10.0	935
0.272	8.16	4.00	650	0.816	24.49	12.0	997
0.340	10.20	5.00	714	0.884	26.53	13.0	1025

The theoretical velocities in the table above must be reduced by multiplying by a "coefficient of discharge," which varies with the orifice and the pressure. The following coefficients are used for orifices in thin plate and short tubes.

Type of Orifice	Pressures in Atmospheres Above Atmospheric Pressure			
	0.01	0.1	0.5	1
Orifice in thin plate	0.65	0.64	0.57	0.54
Orifice in short tube	0.83	0.82	0.71	0.67

PIPE AND PIPE FITTINGS

Wrought Steel Pipe. — ANSI B36.10-1979 covers dimensions of welded and seamless wrought steel pipe, for high or low temperatures or pressures.

The word *pipe* as distinguished from *tube* is used to apply to tubular products of dimensions commonly used for pipelines and piping systems. Pipe dimensions of sizes 12 inches and smaller have outside diameters numerically larger than the corresponding nominal sizes whereas outside diameters of tubes are identical to nominal sizes.

Size: The size of all pipe is identified by the nominal pipe size. The manufacture of pipe in the nominal sizes of ⅛ inch to 12 inches, inclusive, is based on a standardized outside diameter (OD). This OD was originally selected so that pipe with a standard OD and having a wall thickness which was typical of the period would have an inside diameter (ID) approximately equal to the nominal size. Although there is now no such relation between the existing standard thicknesses, ODs and nominal sizes, these nominal sizes and standard ODs continue in use as "standard."

The manufacture of pipe in nominal sizes of 14-inch OD and larger proceeds on the basis of an OD corresponding to the nominal size.

Weight: The nominal weights of steel pipe are calculated values and are tabulated in Table 1. They are based on the following formula:

$$W_{pe} = 10.68(D - t)t$$

where W_{pe} = nominal plain end weight to the nearest 0.01 lb/ft.
D = outside diameter to the nearest 0.001 in.
t = specified wall thickness rounded to the nearest 0.001 in.

Wall thickness: The nominal wall thicknesses are given in Table 1 which also indicates the wall thicknesses in API Standards 5L and 5LX. Thicknesses listed in API Standard 5LS are not indicated but may be found in that Standard or in ANSI B36.10-1979.

The wall thickness designations "Standard," "Extra-Strong," and "Double Extra-Strong" have been commercially used designations for many years. The Schedule Numbers were subsequently added as a convenient designation for use in ordering pipe. "Standard" and Schedule 40 are identical for nominal pipe sizes up to 10 inches, inclusive. All larger sizes of "Standard" have ⅜-inch wall thickness. "Extra-Strong" and Schedule 80 are identical for nominal pipe sizes up to 8 inch, inclusive. All larger sizes of "Extra-Strong" have ½-inch wall thickness.

Wall Thickness Selection: When the selection of wall thickness depends primarily on capacity to resist internal pressure under given conditions, the designer shall compute the exact value of wall thickness suitable for conditions for which the pipe is required as prescribed in the "ASME Boiler and Pressure Vessel Code," "ANSI B31 Code for Pressure Piping," or other similar codes, whichever governs the construction. A thickness can then be selected from Table 1 to suit the value computed to fulfill the conditions for which the pipe is desired.

Metric Weights and Mass: Standard SI metric dimensions in millimeters for outside diameters and wall thicknesses may be found by multiplying the inch dimensions by 25.4. Outside diameters converted from those shown in Table 1 should be rounded to the nearest 0.1 mm and wall thicknesses to the nearest 0.01 mm.

The following formula may be used to calculate the SI metric plain end mass in kg/m using the converted metric diameters and thicknesses:

$$W_{pe} = 0.02466(D - t)t$$

where W_{pe} = nominal plain end mass rounded to the nearest 0.01 kg/m.
D = outside diameter to the nearest 0.1 mm for sizes shown in Table 1.
t = specified wall thickness rounded to the nearest 0.01 mm.

Table I. American National Standard Weights and Dimensions of Welded and Seamless Wrought Steel Pipe
(ANSI B36.10-1979)

Nom. Size and (O.D.), in.	Wall Thick., in.	Plain End Wgt., lb/ft	Sch. No.	Other*
1/8 (0.405)	0.068	0.24	40	5L STD
	0.095	0.31	80	5L XS
1/4 (0.540)	0.088	0.42	40	5L STD
	0.119	0.54	80	5L XS
3/8 (0.675)	0.091	0.57	40	5L STD
	0.126	0.74	80	5L XS
1/2 (0.840)	0.109	0.85	40	5L STD
	0.147	1.09	80	5L XS
	0.188	1.31	160	...
	0.294	1.71	...	5L XXS
3/4 (1.050)	0.113	1.13	40	5L STD
	0.154	1.47	80	5L XS
	0.219	1.94	160	...
	0.308	2.44	...	5L XXS
1 (1.315)	0.133	1.68	40	5L STD
	0.179	2.17	80	5L XS
	0.250	2.84	160	...
	0.358	3.66	...	5L XXS
1¼ (1.660)	0.140	2.27	40	5L STD
	0.191	3.00	80	5L XS
	0.250	3.76	160	...
	0.382	5.21	...	5L XXS
1½ (1.900)	0.145	2.72	40	5L STD
	0.200	3.63	80	5L XS
	0.281	4.86	160	...
	0.400	6.41	...	5L XXS
2 (2.375)	0.083	2.03	...	5L, 5LX ...
	0.109	2.64	...	5L, 5LX ...
	0.125	3.00	...	5L, 5LX ...
	0.141	3.36	...	5L, 5LX ...
	0.154	3.65	40	5L, 5LX STD
	0.172	4.05	...	5L, 5LX ...
	0.188	4.39	...	5L, 5LX ...
	0.218	5.02	80	5L, 5LX XS
	0.250	5.67	...	5L, 5LX ...
	0.281	6.28	...	5L, 5LX ...
	0.344	7.46	160	...
	0.436	9.03	...	5L, 5LX XXS
2½ (2.875)	0.083	2.47	...	5L, 5LX ...
	0.109	3.22	...	5L, 5LX ...
	0.125	3.67	...	5L, 5LX ...
	0.141	4.12	...	5L, 5LX ...
	0.156	4.53	...	5L, 5LX ...
	0.172	4.97	...	5L, 5LX ...
	0.188	5.40	...	5L, 5LX ...
	0.203	5.79	40	5L, 5LX STD
	0.216	6.13	...	5L, 5LX ...
	0.250	7.01	...	5L, 5LX ...
	0.276	7.66	80	5L, 5LX XS
	0.375	10.01	160	...
	0.552	13.69	...	5L, 5LX XXS

Nom. Size and (O.D.), in.	Wall Thick., in.	Plain End Wgt., lb/ft	Sch. No.	Other*
3 (3.500)	0.083	3.03	...	5L, 5LX ...
	0.109	3.95	...	5L, 5LX ...
	0.125	4.51	...	5L, 5LX ...
	0.141	5.06	...	5L, 5LX ...
	0.156	5.57	...	5L, 5LX ...
	0.172	6.11	...	5L, 5LX ...
	0.188	6.65	...	5L, 5LX ...
	0.216	7.58	40	5L, 5LX STD
	0.250	8.68	...	5L, 5LX ...
	0.281	9.66	...	5L, 5LX ...
	0.300	10.25	80	5L, 5LX XS
	0.438	14.32	160	...
	0.600	18.58	...	5L, 5LX XXS
3½ (4.000)	0.083	3.47	...	5L, 5LX
	0.109	4.53	...	5L, 5LX
	0.125	5.17	...	5L, 5LX
	0.141	5.81	...	5L, 5LX
	0.156	6.40	...	5L, 5LX
	0.172	7.03	...	5L, 5LX
	0.188	7.65	...	5L, 5LX
	0.226	9.11	40	5L, 5LX STD
	0.250	10.01	...	5L, 5LX
	0.281	11.16	...	5L, 5LX
	0.318	12.50	80	5L, 5LX XS
4 (4.500)	0.083	3.92	...	5L, 5LX
	0.109	5.11	...	5L
	0.125	5.84	...	5L, 5LX
	0.141	6.56	...	5L, 5LX
	0.156	7.24	...	5L, 5LX
	0.172	7.95	...	5L, 5LX
	0.188	8.66	...	5L, 5LX
	0.203	9.32	...	5L, 5LX
	0.219	10.01	...	5L, 5LX
	0.237	10.79	40	5L, 5LX STD
	0.250	11.35	...	5L, 5LX
	0.281	12.66	...	5L, 5LX
	0.312	13.96	...	5L, 5LX
	0.337	14.98	80	5L, 5LX XS
	0.438	19.00	120	5L, 5LX ...
	0.531	22.51	160	5L, 5LX ...
	0.674	27.54	...	5L, 5LX XXS
5 (5.563)	0.083	4.86	...	5L ...
	0.125	7.26	...	5L
	0.156	9.01	...	5L
	0.188	10.79	...	5L
	0.219	12.50	...	5L
	0.258	14.62	40	5L STD
	0.281	15.85	...	5L
	0.312	17.50	...	5L
	0.344	19.17	...	5L
	0.375	20.78	80	5L XS
	0.500	27.04	120	5L ...
	0.625	32.96	160	5L ...
	0.750	38.55	...	5L XXS

* Wall thicknesses listed in American Petroleum Institute (API) Standards 5L and 5LX are indicated but wall thicknesses listed in API Standard 5LS are not indicated. For these see ANSI B36.10-1979 or API 5LS Standard. Commercial designations are: STD = Standard; XS = Extra Strong; and XXS = Double Extra Strong.

Table 1 (*Concluded*). **American National Standard Weights and Dimensions of Welded and Seamless Wrought Steel Pipe (ANSI B36.10-1979)**

Nom. Size and (O.D.), in.†	Wall Thick., in.	Plain End Wgt., lb/ft	Sch. No.	Other*
6 (6.625)	0.083	5.80	...	5L, 5LX ...
	0.109	7.59	...	5L, 5LX ...
	0.125	8.68	...	5L, 5LX ...
	0.141	9.76	...	5L, 5LX ...
	0.156	10.78	...	5L, 5LX ...
	0.172	11.85	...	5L, 5LX ...
	0.188	12.92	...	5L, 5LX ...
	0.203	13.92	...	5L, 5LX ...
	0.219	14.98	...	5L, 5LX ...
	0.250	17.02	...	5L, 5LX ...
	0.280	18.97	40	5L, 5LX STD
	0.312	21.04	...	5L, 5LX ...
	0.344	23.08	...	5L, 5LX ...
	0.375	25.03	...	5L, 5LX ...
	0.432	28.57	80	5L, 5LX XS
	0.500	32.71	...	5L, 5LX ...
	0.562	36.39	120	5L, 5LX ...
	0.625	40.05	...	5L, 5LX ...
	0.719	45.35	160	5L, 5LX ...
	0.864	53.16	...	5L XXS
8 (8.625)	0.125	11.35	...	5L, 5LX ...
	0.156	14.11	...	5L, 5LX ...
	0.188	16.94	...	5L, 5LX ...
	0.203	18.26	...	5LX ...
	0.219	19.66	...	5L, 5LX ...
	0.250	22.36	20	5L, 5LX ...
	0.277	24.70	30	5L, 5LX ...
	0.312	27.70	...	5L, 5LX ...
	0.322	28.55	40	5L, 5LX STD
	0.344	30.42	...	5L, 5LX ...
	0.375	33.04	...	5L, 5LX ...
	0.406	35.64	60	...
	0.438	38.30	...	5L, 5LX ...
	0.500	43.39	80	5L, 5LX XS
	0.562	48.40	...	5L, 5LX ...
	0.594	50.95	100	...
	0.625	53.40	...	5L, 5LX ...
	0.719	60.71	120	5L, 5LX ...
	0.812	67.76	140	...
	0.875	72.42	...	5L XXS
	0.906	74.69	160	...
10 (10.750)	0.156	17.65	...	5L, 5LX ...
	0.188	21.21	...	5L, 5LX ...
	0.203	22.87	...	5LX ...
	0.219	24.63	...	5L, 5LX ...
	0.250	28.04	20	5L, 5LX ...
	0.279	31.20	...	5L, 5LX ...
	0.307	34.24	30	5L, 5LX ...
	0.344	38.23	...	5L, 5LX ...
	0.365	40.48	40	5L, 5LX STD
	0.438	48.24	...	5L, 5LX ...
	0.500	54.74	60	5L, 5LX XS
	0.562	61.15	...	5L, 5LX ...
	0.594	64.43	80	...
	0.625	67.58	...	5L, 5LX ...
10 (10.750)	0.719	77.03	100	5L, 5LX ...
	0.812	86.18	...	5L ...
	0.844	89.29	120	...
	1.000	104.13	140	... XXS
	1.125	115.64	160	...
12 (12.750)	0.172	23.11	...	5L, 5LX ...
	0.188	25.22	...	5L, 5LX ...
	0.203	27.20	...	5LX ...
	0.219	29.31	...	5L, 5LX ...
	0.250	33.38	20	5L, 5LX ...
	0.281	37.42	...	5L, 5LX ...
	0.312	41.45	...	5L, 5LX ...
	0.330	43.77	30	5L, 5LX ...
	0.344	45.58	...	5L, 5LX ...
	0.375	49.56	...	5L, 5LX STD
	0.406	53.52	40	5LX ...
	0.438	57.59	...	5L, 5LX ...
	0.500	65.42	...	5L, 5LX XS
	0.562	73.15	60	5L, 5LX ...
	0.625	80.93	...	5L, 5LX ...
	0.688	88.63	80	5L, 5LX ...
	0.750	96.12	...	5L, 5LX ...
	0.812	103.53	...	5L, 5LX ...
	0.844	107.32	100	...
	0.875	110.97	...	5L, 5LX ...
	1.000	125.49	120	... XXS
	1.125	139.67	140	...
	1.312	160.27	160	...
14 (14.000)	0.188	27.73	...	5L, 5LX ...
	0.203	29.91	...	5L ...
	0.210	30.93	...	5LX ...
	0.219	32.23	...	5LX ...
	0.250	36.71	10	5L, 5LX ...
	0.281	41.17	...	5L, 5LX ...
	0.312	45.61	20	5L, 5LX ...
	0.344	50.17	...	5L, 5LX ...
	0.375	54.57	30	5L, 5LX STD
	0.406	58.94	...	5LX ...
	0.438	63.44	40	5L, 5LX ...
	0.469	67.78	...	5LX ...
	0.500	72.09	...	5L, 5LX XS
	0.562	80.66	...	5L, 5LX ...
	0.594	85.05	60	...
	0.625	89.28	...	5L, 5LX ...
	0.688	97.81	...	5L, 5LX ...
	0.750	106.13	80	5L, 5LX ...
	0.812	114.37	...	5L, 5LX ...
	0.875	122.65	...	5L, 5LX ...
	0.938	130.85	100	5L, 5LX ...
	1.094	150.79	120	...
	1.250	170.21	140	...
	1.406	189.11	160	...
	2.000	256.32	...	...
	2.125	269.50	...	...
	2.200	277.25	...	...
	2.500	307.05	...	...

* Wall thicknesses listed in American Petroleum Institute (API) Standards 5L and 5LX are indicated but wall thicknesses listed in API Standard 5LS are not indicated. For these see ANSI B36.10-1979 or API 5LS Standard. Commercial Designations are: STD = Standard; XS = Extra Strong; and XXS = Double Extra Strong.

† For sizes larger than 14 inches see ANSI B36.10-1979 Standard.

Table 2. Properties of American National Standard Schedule 40 Welded and Seamless Wrought Steel Pipe

Diameter, Inches			Wall Thickness, Inches	Cross-Sectional Area of Metal	Weight per Foot, Pounds		Capacity per Foot of Length		Length of Pipe in Feet to Contain		Properties of Sections		
Nominal	Actual Inside	Actual Outside			Of Pipe	Of Water in Pipe	In Cubic Inches	In Gallons	One Cubic Foot	One Gallon	Moment of Inertia	Radius of Gyration	Section Modulus
1/8	0.269	0.405	0.068	0.072	0.24	0.025	0.682	0.003	2532.	338.7	0.00106	0.122	0.00525
1/4	0.364	0.540	0.088	0.125	0.42	0.045	1.249	0.005	1384.	185.0	0.00331	0.163	0.01227
3/8	0.493	0.675	0.091	0.167	0.57	0.083	2.291	0.010	754.4	100.8	0.00729	0.209	0.02160
1/2	0.622	0.840	0.109	0.250	0.85	0.132	3.646	0.016	473.9	63.35	0.01709	0.261	0.04070
3/4	0.824	1.050	0.113	0.333	1.13	0.231	6.399	0.028	270.0	36.10	0.03704	0.334	0.07055
1	1.049	1.315	0.133	0.494	1.68	0.374	10.37	0.045	166.6	22.27	0.08734	0.421	0.1328
1 1/4	1.380	1.660	0.140	0.669	2.27	0.648	17.95	0.078	96.28	12.87	0.1947	0.539	0.2346
1 1/2	1.610	1.900	0.145	0.799	2.72	0.882	24.43	0.106	70.73	9.456	0.3099	0.623	0.3262
2	2.067	2.375	0.154	1.075	3.65	1.454	40.27	0.174	42.91	5.737	0.6658	0.787	0.5607
2 1/2	2.469	2.875	0.203	1.704	5.79	2.072	57.45	0.249	30.08	4.021	1.530	0.947	1.064
3	3.068	3.500	0.216	2.228	7.58	3.202	88.71	0.384	19.48	2.604	3.017	1.163	1.724
3 1/2	3.548	4.000	0.226	2.680	9.11	4.283	118.6	0.514	14.56	1.947	4.788	1.337	2.394
4	4.026	4.500	0.237	3.174	10.79	5.515	152.8	0.661	11.31	1.512	7.233	1.510	3.215
5	5.047	5.563	0.258	4.300	14.62	8.666	240.1	1.04	7.198	0.9622	15.16	1.878	5.451
6	6.065	6.625	0.280	5.581	18.97	12.52	346.7	1.50	4.984	0.6663	28.14	2.245	8.496
8	7.981	8.625	0.322	8.399	28.55	21.67	600.3	2.60	2.878	0.3848	72.49	2.938	16.81
10	10.020	10.750	0.365	11.91	40.48	34.16	946.3	4.10	1.826	0.2441	160.7	3.674	29.91
12	11.938	12.750	0.406	15.74	53.52	48.49	1343.	5.81	1.286	0.1720	300.2	4.364	47.09
16	15.000	16.000	0.500	24.35	82.77	76.55	2121.	9.18	0.8149	0.1089	732.0	5.484	91.50
18	16.876	18.000	0.562	30.79	104.7	96.90	2684.	11.62	0.6438	0.0861	1172.	6.168	130.2
20	18.812	20.000	0.594	36.21	123.1	120.4	3335.	14.44	0.5181	0.0693	1706.	6.864	170.6
24	22.624	24.000	0.688	50.39	171.3	174.1	4824.	20.88	0.3582	0.0479	3426.	8.246	285.5
32	30.624	32.000	0.688	67.68	230.1	319.1	8839.	38.26	0.1955	0.0261	8299.	11.07	518.7

Note: Torsional section modulus equals twice section modulus.

Table 3. Properties of American National Standard Schedule 80 Welded and Seamless Wrought Steel Pipe

Diameter, Inches			Wall Thickness, Inches	Cross-Sectional Area of Metal	Weight per Foot, Pounds		Capacity per Foot of Length		Length of Pipe in Feet to Contain		Properties of Sections		
Nominal	Inside Actual	Outside Actual			Of Pipe	Of Water in Pipe	In Cubic Inches	In Gallons	One Cubic Foot	One Gallon	Moment of Inertia	Radius of Gyration	Section Modulus
⅛	0.215	0.405	0.095	0.093	0.315	0.016	0.436	0.0019	3966.	530.2	0.00122	0.115	0.00600
¼	0.302	0.540	0.119	0.157	0.537	0.031	0.860	0.0037	2010.	268.7	0.00377	0.155	0.01395
⅜	0.423	0.675	0.126	0.217	0.739	0.061	1.686	0.0073	1025.	137.0	0.00862	0.199	0.02554
½	0.546	0.840	0.147	0.320	1.088	0.101	2.810	0.0122	615.0	82.22	0.02008	0.250	0.04780
¾	0.742	1.050	0.154	0.433	1.474	0.187	5.189	0.0225	333.0	44.52	0.04479	0.321	0.08531
1	0.957	1.315	0.179	0.639	2.172	0.312	8.632	0.0374	200.2	26.76	0.1056	0.407	0.1606
1¼	1.278	1.660	0.191	0.881	2.997	0.556	15.39	0.0667	112.3	15.01	0.2418	0.524	0.2913
1½	1.500	1.900	0.200	1.068	3.631	0.766	21.21	0.0918	81.49	10.89	0.3912	0.605	0.4118
2	1.939	2.375	0.218	1.477	5.022	1.279	35.43	0.1534	48.77	6.519	0.8680	0.766	0.7309
2½	2.323	2.875	0.276	2.254	7.661	1.836	50.86	0.2202	33.98	4.542	1.924	0.924	1.339
3	2.900	3.500	0.300	3.016	10.25	2.861	79.26	0.3431	21.80	2.914	3.895	1.136	2.225
3½	3.364	4.000	0.318	3.678	12.50	3.850	106.7	0.4617	16.20	2.166	6.280	1.307	3.140
4	3.826	4.500	0.337	4.407	14.98	4.980	138.0	0.5972	12.53	1.674	9.611	1.477	4.272
5	4.813	5.563	0.375	6.112	20.78	7.882	218.3	0.9451	7.915	1.058	20.67	1.839	7.432
6	5.761	6.625	0.432	8.405	28.57	11.29	312.8	1.354	5.524	0.738	40.49	2.195	12.22
8	7.625	8.625	0.500	12.76	43.39	19.78	548.0	2.372	3.153	0.422	105.7	2.878	24.52
10	9.562	10.750	0.594	18.95	64.42	31.11	861.7	3.730	2.005	0.268	245.2	3.597	45.62
12	11.374	12.750	0.688	26.07	88.63	44.02	1219.	5.278	1.417	0.189	475.7	4.271	74.62
14	12.500	14.000	0.750	31.22	106.1	53.16	1473.	6.375	1.173	0.157	687.4	4.692	98.19
16	14.312	16.000	0.844	40.19	136.6	69.69	1931.	8.357	0.895	0.120	1158.	5.366	144.7
18	16.124	18.000	0.938	50.28	170.9	88.46	2450.	10.61	0.705	0.094	1835.	6.041	203.9
20	17.938	20.000	1.031	61.44	208.9	109.5	3033.	13.13	0.570	0.076	2772.	6.716	277.2
22	19.750	22.000	1.125	73.78	250.8	132.7	3676.	15.91	0.470	0.063	4031.	7.391	366.4

Note: Torsional section modulus equals twice section modulus.

Volume of Flow at 1 Foot Per-Minute Velocity in Pipe and Tube*

Nominal Diam, Inches	Schedule 40 Pipe			Schedule 80 Pipe			Type K Copper Tube			Type L Copper Tube		
	Cu. Ft. per Minute	Gallons per Minute	Pounds 60 F Water per Min.	Cu. Ft. per Minute	Gallons per Minute	Pounds 60 F Water per Min.	Cu. Ft. per Minute	Gallons per Minute	Pounds 60 F Water per Min.	Cu. Ft. per Minute	Gallons per Minute	Pounds 60 F Water per Min.
1/8	0.0004	0.003	0.025	0.0003	0.002	0.016	0.0002	0.0014	0.012	0.0002	0.002	0.014
1/4	0.0007	0.005	0.044	0.0005	0.004	0.031	0.0005	0.0039	0.033	0.0005	0.004	0.034
3/8	0.0013	0.010	0.081	0.0010	0.007	0.061	0.0009	0.0066	0.055	0.0010	0.008	0.063
1/2	0.0021	0.016	0.132	0.0016	0.012	0.102	0.0015	0.0113	0.094	0.0016	0.012	0.101
3/4	0.0037	0.028	0.232	0.0030	0.025	0.213	0.0030	0.0267	0.189	0.0034	0.025	0.210
1	0.0062	0.046	0.387	0.0050	0.037	0.312	0.0054	0.0404	0.338	0.0057	0.043	0.358
1 1/4	0.0104	0.078	0.649	0.0088	0.067	0.555	0.0085	0.0632	0.53	0.0087	0.065	0.545
1 1/2	0.0141	0.106	0.882	0.0123	0.092	0.765	0.0196	0.1465	1.22	0.0124	0.093	0.770
2	0.0233	0.174	1.454	0.0206	0.154	1.280	0.0209	0.1565	1.31	0.0215	0.161	1.34
2 1/2	0.0332	0.248	2.073	0.0294	0.220	1.830	0.0323	0.2418	2.02	0.0331	0.248	2.07
3	0.0514	0.383	3.201	0.0460	0.344	2.870	0.0461	0.3446	2.88	0.0473	0.354	2.96
3 1/2	0.0682	0.513	4.287	0.0617	0.458	3.720	0.0625	0.4675	3.91	0.0640	0.479	4.00
4	0.0884	0.660	5.516	0.0800	0.597	4.970	0.0811	0.6068	5.07	0.0841	0.622	5.20
5	0.1390	1.040	8.674	0.1260	0.947	7.940	0.1259	0.9415	7.87	0.1296	0.969	8.10
6	0.2010	1.500	12.52	0.1820	1.355	11.300	0.1797	1.3440	11.2	0.1882	1.393	11.6
8	0.3480	2.600	21.68	0.3180	2.380	19.800	0.3135	2.3446	19.6	0.3253	2.434	20.3
10	0.5476	4.10	34.18	0.5560	4.165	31.130	0.4867	3.4405	30.4	0.5050	3.777	21.6
12	0.7773	5.81	48.52	0.7060	5.280	44.040	0.6978	5.2194	43.6	0.7291	5.454	45.6
14	0.9396	7.03	58.65	0.8520	6.380	53.180	—	—	—	—	—	—
16	1.227	9.18	76.60	1.1170	8.360	69.730	—	—	—	—	—	—
18	1.553	11.62	96.95	1.4180	10.610	88.500	—	—	—	—	—	—
20	1.931	14.44	120.5	1.7550	13.130	109.510	—	—	—	—	—	—

* To obtain volume of flow at any other velocity, multiply values in table by velocity in feet per minute.

Seamless Drawn Brass and Copper Pipe

Made to correspond with iron pipe and to fit iron pipe fittings (American Tube Works).

Diameter			Approximate Weight per Foot, Pounds		Diameter			Approximate Weight per Foot, Pounds	
Iron Pipe Size	Approx. Outside Diam.	Exact Outside Diam.	Brass	Copper	Iron Pipe Size	Approx. Outside Diam.	Exact Outside Diam.	Brass	Copper
1/8	3/8	0.405	0.25	0.26	2½	2⅞	2.875	5.75	6.05
¼	9/16	0.540	0.43	0.45	3	3½	3.500	8.30	8.74
3/8	11/16	0.675	0.62	0.65	3½	4	4.000	10.90	11.47
½	13/16	0.840	0.90	0.95	4	4½	4.500	12.70	13.37
¾	1 1/16	1.050	1.25	1.32	4½	5	5.000	13.90	14.63
1	1 5/16	1.315	1.70	1.79	5	5 9/16	5.563	15.75	16.58
1¼	1 5/8	1.660	2.50	2.63	6	6 5/8	6.625	18.31	19.27
1½	1 7/8	1.900	3.00	3.16	7	7 5/8	7.625	23.73	24.98
2	2 3/8	2.375	4.00	4.21	.....	.....	.....	.....	....

Threading Pipe. — Clean, smooth pipe threads are essential to a good joint and depend largely upon the rake or lip angle and lead of the chasers, and the clearance, chip space and number of chasers in the die-head. The lip angle should vary from 15 to 25 degrees, depending upon the style and condition of the chasers and chaser holders. The chip space in front of the chasers should be large enough to allow room for accumulation of chips and at the same time provide means of

Length of Thread on Pipe Required to Make a Tight Joint
(Crane Co.)

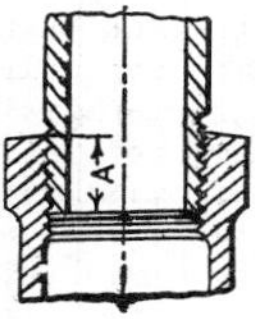

	Size of Pipe, Inches	Dimension A, Inches	Size of Pipe, Inches	Dimension A, Inches	Size of Pipe, Inches	Dimension A, Inches
	1/8	¼	1½	1 1/16	5	1¼
	¼	3/8	2	¾	6	1 5/16
	3/8	3/8	2½	15/16	7	1 3/8
	½	½	3	1	8	1 7/16
	¾	9/16	3½	1 1/16	9	1½
	1	11/16	4	1 1/8	10	1 5/8
	1¼	11/16	4½	1 3/16	12	1 3/4

Dimensions do not allow for variation in tapping or threading.

lubricating the chasers. This is an important point, as insufficient chip space will cause the chips to clog and tear the threads. The lead of the chaser is the angle which is machined or ground on the leading or front side, to enable the die to start readily on the pipe, and also to distribute the work of cutting over a number of threads. To secure a good thread, the lead should cover the first three threads. As the heaviest cutting is done by this beveled part, it should have a slightly greater clearance angle than the rest of the threads on the chaser. When re-grinding chasers which have become dull on the lead, care should be taken to give each chaser the same length of lead, as otherwise the work will be unevenly distributed.

(Continued on page 2388)

Sizes and Weights in Pounds per Foot of Seamless Brass Tubes*

Outside Diam. of Tube, Inches	Thickness — Stub's or Birmingham Gage											
	3	4	5	6	7	8	9	10	11	12	13	14
	Decimal Equivalent of Gage Number, Inch											
	0.259	0.238	0.220	0.203	0.180	0.165	0.148	0.134	0.120	0.109	0.095	0.083
1/8	...	...	...	...	...	...	...	...	...	...	...	...
3/16	...	...	...	...	...	...	...	...	...	...	...	...
1/4	...	...	...	...	...	...	...	...	0.18	0.177	0.170	0.160
5/16	...	...	...	...	...	...	...	...	0.27	0.256	0.238	0.220
3/8	...	...	...	...	...	0.40	0.39	0.37	0.35	0.335	0.307	0.280
7/16	...	...	...	...	...	0.52	0.49	0.47	0.44	0.413	0.376	0.340
1/2	...	...	...	0.70	0.66	0.64	0.60	0.57	0.53	0.492	0.444	0.400
9/16	...	...	...	0.84	0.79	0.76	0.71	0.66	0.61	0.571	0.513	0.460
5/8	1.09	1.06	1.03	0.99	0.92	0.88	0.81	0.76	0.70	0.649	0.581	0.520
11/16	1.28	1.23	1.19	1.13	1.05	0.99	0.92	0.86	0.79	0.728	0.650	0.580
3/4	1.47	1.41	1.35	1.28	1.18	1.11	1.03	0.95	0.87	0.807	0.718	0.640
13/16	1.65	1.58	1.50	1.43	1.31	1.23	1.13	1.05	0.96	0.885	0.787	0.700
7/8	1.84	1.75	1.66	1.57	1.44	1.35	1.24	1.15	1.04	0.964	0.855	0.759
15/16	2.03	1.92	1.82	1.72	1.57	1.47	1.35	1.24	1.13	1.042	0.924	0.819
1	2.22	2.09	1.98	1.87	1.70	1.59	1.45	1.34	1.22	1.12	0.99	0.88
1 1/8	2.60	2.44	2.30	2.16	1.96	1.83	1.67	1.53	1.39	1.28	1.13	1.00
1 1/4	2.97	2.78	2.61	2.45	2.22	2.07	1.88	1.73	1.56	1.44	1.27	1.12
1 3/8	3.35	3.12	2.93	2.75	2.48	2.30	2.10	1.92	1.74	1.59	1.40	1.24
1 1/2	3.72	3.47	3.25	3.04	2.74	2.54	2.31	2.11	1.91	1.75	1.54	1.36
1 5/8	4.09	3.81	3.57	3.33	3.00	2.78	2.52	2.31	2.08	1.91	1.68	1.48
1 3/4	4.47	4.15	3.88	3.62	3.26	3.02	2.74	2.50	2.26	2.06	1.82	1.60
1 7/8	4.84	4.50	4.20	3.92	3.52	3.26	2.95	2.69	2.43	2.22	1.95	1.72
2	5.21	4.84	4.52	4.21	3.78	3.50	3.16	2.89	2.60	2.38	2.09	1.84
2 1/8	5.59	5.18	4.84	4.50	4.04	3.73	3.38	3.08	2.78	2.54	2.23	1.96
2 1/4	5.96	5.53	5.15	4.80	4.30	3.97	3.59	3.27	2.95	2.69	2.36	2.08
2 3/8	6.34	5.87	5.47	5.09	4.56	4.21	3.80	3.47	3.12	2.85	2.50	2.20
2 1/2	6.71	6.21	5.79	5.38	4.82	4.45	4.02	3.66	3.30	3.01	2.64	2.32
2 5/8	7.08	6.56	6.11	5.67	5.08	4.69	4.23	3.85	3.47	3.17	2.77	2.44
2 3/4	7.46	6.90	6.42	5.97	5.34	4.92	4.44	4.05	3.64	3.32	2.91	2.56
2 7/8	7.83	7.24	6.74	6.26	5.60	5.16	4.66	4.24	3.81	3.48	3.05	2.68
3	8.20	7.59	7.06	6.55	5.86	5.40	4.87	4.43	3.99	3.64	3.19	2.79
3 1/8	8.58	7.93	7.38	6.85	6.12	5.64	5.08	4.63	4.16	3.79	3.32	2.91
3 1/4	8.95	8.27	7.69	7.14	6.38	5.88	5.30	4.82	4.33	3.95	3.46	3.03
3 3/8	9.33	8.62	8.01	7.43	6.64	6.11	5.51	5.01	4.51	4.11	3.60	3.15
3 1/2	9.70	8.96	8.33	7.72	6.90	6.35	5.72	5.21	4.68	4.27	3.73	3.27
3 5/8	10.07	9.30	8.65	8.02	7.16	6.59	5.94	5.40	4.85	4.42	3.87	3.39
3 3/4	10.45	9.65	8.96	8.31	7.42	6.83	6.15	5.59	5.03	4.58	4.01	3.51
3 7/8	10.82	9.99	9.28	8.60	7.68	7.07	6.37	5.79	5.20	4.74	4.15	3.63

To determine weight per foot of a tube of a given *inside diameter*, add to weights in above list the weights in pounds per foot given below under corresponding gage numbers.

Gage No.	3	4	5	6	7	8	9	10	11	12	13	14
Weight Added	1.549	1.308	1.117	0.951	0.748	0.628	0.506	0.414	0.332	0.274	0.208	0.159

* Bridgeport Brass Co.

Sizes and Weights in Pounds per Foot of Seamless Brass Tubes

Outside Diam. of Tube, Inches	Thickness — Stub's or Birmingham Gage										
	15	16	17	18	19	20	21	22	23	24	25
	Decimal Equivalent of Gage Number, Inch										
	0.072	0.065	0.058	0.049	0.042	0.035	0.032	0.028	0.025	0.022	0.020
⅛	....	0.045	0.045	0.043	0.040	0.036	0.034	0.031	0.029	0.026	0.024
³⁄₁₆	0.096	0.092	0.087	0.078	0.070	0.062	0.057	0.051	0.047	0.042	0.039
¼	0.148	0.139	0.129	0.114	0.101	0.087	0.080	0.072	0.065	0.058	0.053
⁵⁄₁₆	0.200	0.186	0.170	0.149	0.131	0.112	0.104	0.092	0.083	0.074	0.067
⅜	0.252	0.233	0.212	0.184	0.161	0.137	0.127	0.112	0.101	0.090	0.082
⁷⁄₁₆	0.304	0.279	0.254	0.220	0.192	0.163	0.150	0.132	0.119	0.106	0.096
½	0.356	0.326	0.296	0.255	0.222	0.188	0.173	0.152	0.137	0.121	0.111
⁹⁄₁₆	0.408	0.373	0.338	0.290	0.252	0.213	0.196	0.173	0.155	0.137	0.125
⅝	0.460	0.420	0.380	0.326	0.283	0.238	0.219	0.193	0.173	0.153	0.140
¹¹⁄₁₆	0.511	0.467	0.421	0.361	0.313	0.264	0.242	0.213	0.191	0.169	0.154
¾	0.563	0.514	0.463	0.396	0.343	0.289	0.265	0.233	0.209	0.185	0.169
¹³⁄₁₆	0.615	0.561	0.505	0.432	0.373	0.314	0.288	0.253	0.227	0.201	0.183
⅞	0.667	0.608	0.547	0.467	0.404	0.339	0.311	0.274	0.245	0.217	0.197
¹⁵⁄₁₆	0.719	0.655	0.589	0.502	0.434	0.365	0.334	0.294	0.263	0.232	0.211
1	0.77	0.70	0.63	0.54	0.46	0.389	0.358	0.314	0.281	0.248	0.226
1⅛	0.87	0.79	0.71	0.61	0.52	0.439	0.404	0.354	0.317	0.280	0.255
1¼	0.98	0.89	0.80	0.68	0.59	0.490	0.450	0.395	0.354	0.312	0.284
1⅜	1.08	0.98	0.88	0.75	0.65	0.540	0.496	0.435	0.390	0.343	0.313
1½	1.19	1.08	0.96	0.82	0.71	0.591	0.542	0.476	0.426	0.375	0.342
1⅝	1.29	1.17	1.05	0.89	0.77	0.641	0.588	0.516	0.462	0.407	0.371
1¾	1.39	1.26	1.13	0.96	0.83	0.692	0.635	0.556	0.498	0.439	0.399
1⅞	1.50	1.36	1.22	1.03	0.89	0.742	0.681	0.597	0.534	0.470	0.428
2	1.60	1.45	1.30	1.10	0.95	0.793	0.727	0.637	0.570	0.502	0.457
2⅛	1.71	1.55	1.38	1.17	1.01	0.843	0.773	0.678	0.606	0.534	0.486
2¼	1.81	1.64	1.47	1.24	1.07	0.894	0.819	0.718	0.642	0.566	0.515
2⅜	1.91	1.73	1.55	1.32	1.13	0.944	0.866	0.758	0.678	0.597	0.544
2½	2.02	1.83	1.63	1.39	1.19	0.995	0.912	0.799	0.714	0.629	0.573
2⅝	2.12	1.92	1.72	1.46	1.25	1.045	0.958	0.839	0.750	0.661	.....
2¾	2.23	2.01	1.80	1.53	1.31	1.096	1.004	0.880	0.786	0.693	.....
2⅞	2.33	2.11	1.89	1.60	1.37	1.146	1.050	0.920	0.822	0.724	.....
3	2.43	2.20	1.97	1.67	1.43	1.197	1.096	0.960	0.859	0.756	.....
3⅛	2.54	2.30	2.05	1.74	1.49	1.247	1.143	1.001	0.895	0.788	.....
3¼	2.64	2.39	2.14	1.81	1.55	1.298	1.189	1.041	0.931	0.820	.....
3⅜	2.74	2.48	2.22	1.88	1.62	1.348	1.235	1.082	0 967	0.851	.....
3½	2.85	2.58	2.30	1.95	1.68	1.399	1.281	1.122	1.003	0.883	.....
3⅝	2.95	2.67	2.39	2.02	1.74	1.449	1.327	1.162	1.039	0.915	.....
3¾	3.06	2.76	2.47	2.09	1.80	1.50	1.373	1.203	1.075	0.946	.....
3⅞	3.16	2.86	2.56	2.16	1.86	1.55	1.42	1.243	1.111	0.978	.....

To determine weight per foot of a tube of a given *inside diameter*, add to weights in above list the weights in pounds per foot given below under corresponding gage numbers.

Gage No.	15	16	17	18	19	20	21	22	23	24	25
Weight Added	0.120	0.097	0.078	0.055	0.041	0.028	0.024	0.018	0.014	0.011	0.009

Sizes and Weights in Pounds per Foot of Seamless Brass Tubes

Outside Diam. of Tube, Inches	Thickness — Stub's or Birmingham Gage									
	3	4	5	6	7	8	9	10	11	12
	Decimal Equivalent of Gage Number, Inch									
	0.259	0.238	0.220	0.203	0.180	0.165	0.148	0.134	0.120	0.109
4	11.19	10.33	9.60	8.90	7.94	7.31	6.58	5.98	5.37	4.89
4⅛	11.57	10.68	9.91	9.19	8.20	7.54	6.79	6.17	5.55	5.05
4¼	11.94	11.02	10.23	9.48	8.46	7.78	7.01	6.37	5.72	5.21
4⅜	12.32	11.36	10.55	9.77	8.72	8.02	7.22	6.56	5.89	5.37
4½	12.69	11.71	10.87	10.07	8.98	8.26	7.43	6.75	6.06	5.52
4⅝	13.06	12.05	11.18	10.36	9.24	8.50	7.65	6.94	6.24	5.68
4¾	13.44	12.39	11.50	10.65	9.50	8.73	7.86	7.14	6.41	5.84
4⅞	13.81	12.74	11.82	10.95	9.76	8.97	8.07	7.33	6.58	6.00
5	14.18	13.08	12.14	11.24	10.02	9.21	8.29	7.53	6.76	6.15
5⅛	14.56	13.42	12.45	11.53	10.28	9.45	8.50	7.72	6.93	6.31
5¼	14.93	13.77	12.77	11.82	10.53	9.69	8.71	7.91	7.10	6.47
5⅜	15.31	14.11	13.09	12.12	10.79	9.92	8.93	8.11	7.28	6.62
5½	15.68	14.45	13.41	12.41	11.05	10.16	9.14	8.30	7.45	6.78
5⅝	16.05	14.80	13.72	12.70	11.31	10.40	9.35	8.49	7.62	6.94
5¾	16.43	15.14	14.04	13.00	11.57	10.64	9.57	8.69	7.80	7.10
5⅞	16.80	15.48	14.36	13.29	11.83	10.88	9.78	8.88	7.97	7.25
6	17.17	15.83	14.67	13.58	12.09	11.12	9.99	9.07	8.14	7.41
6⅛	17.55	16.17	14.99	13.87	12.35	11.35	10.21	9.27	8.32	7.57
6¼	17.92	16.51	15.31	14.17	12.61	11.59	10.42	9.46	8.49	7.72
6⅜	18.30	16.86	15.63	14.46	12.87	11.83	10.64	9.65	8.66	7.88
6½	18.67	17.20	15.94	14.75	13.13	12.07	10.85	9.85	8.84	8.04
6⅝	19.04	17.54	16.26	15.05	13.39	12.31	11.06	10.04	9.01	8.20
6¾	19.42	17.89	16.58	15.34	13.65	12.54	11.28	10.23	9.18	8.35
6⅞	19.79	18.23	16.90	15.63	13.91	12.78	11.49	10.43	9.35	8.51
7	20.16	18.57	17.21	15.92	14.17	13.02	11.70	10.62	9.53	8.67
7⅛	20.54	18.92	17.53	16.22	14.43	13.26	11.92	10.81	9.70	8.83
7¼	20.91	19.26	17.85	16.51	14.69	13.50	12.13	11.01	9.87	8.98
7⅜	21.29	19.60	18.17	16.80	14.95	13.73	12.34	11.20	10.05	9.14
7½	21.66	19.95	18.48	17.10	15.21	13.97	12.56	11.39	10.22	9.30
7⅝	22.03	20.29	18.80	17.39	15.47	14.21	12.77	11.59	10.39	9.45
7¾	22.41	20.64	19.12	17.68	15.73	14.45	12.98	11.78	10.57	9.61
7⅞	22.78	20.98	19.44	17.98	15.99	14.69	13.20	11.97	10.74	9.77
8	23.15	21.32	19.75	18.27	16.25	14.93	13.41	12.17	10.91	9.93

To determine weight per foot of a tube of a given *inside diameter*, add to weights in above list the weights in pounds per foot given below under corresponding gage numbers.

Gage No.	3	4	5	6	7	8	9	10	11	12
Weight Added	1.549	1.308	1.117	0.951	0.748	0.628	0.506	0.414	0.332	0.274

Sizes and Weights in Pounds per Foot of Seamless Brass Tubes

Outside Diam. of Tube, Inches	Thickness — Stub's or Birmingham Gage											
	13	14	15	16	17	18	19	20	21	22	23	24
	Decimal Equivalent of Gage Number, Inch											
	0.095	0.083	0.072	0.065	0.058	0.049	0.042	0.035	0.032	0.028	0.025	0.022
4	4.28	3.75	3.26	2.95	2.64	2.23	1.92	1.601	1.466	1.284	1.147	1.010
4⅛	4.42	3.87	3.37	3.05	2.72	2.30	1.98	1.651	1.512	1.324	1.183	....
4¼	4.56	3.99	3.47	3.14	2.81	2.38	2.04	1.702	1.558	1.364	1.219	....
4⅜	4.69	4.11	3.58	3.23	2.89	2.45	2.10	1.752	1.604	1.405	1.255	....
4½	4.83	4.23	3.68	3.33	2.97	2.52	2.16	1.803	1.650	1.445	1.291	....
4⅝	4.97	4.35	3.78	3.42	3.06	2.59	2.22	1.853	1.697	1.486	...	
4¾	5.11	4.47	3.89	3.52	3.14	2.66	2.28	1.904	1.743	1.526	...	
4⅞	5.24	4.59	3.99	3.61	3.22	2.73	2.34	1.954	1.789	1.566	...	
5	5.38	4.71	4.09	3.70	3.31	2.80	2.40	2.005	1.835	1.607	...	
5⅛	5.52	4.83	4.20	3.79	3.39	2.87	2.46	2.055	1.881	....		
5¼	5.65	4.95	4.30	3.89	3.48	2.94	2.52	2.106	1.928	....	...	...
5⅜	5.79	5.07	4.41	3.98	3.56	3.01	2.58	2.156	1.974	....	...	...
5½	5.93	5.19	4.51	4.08	3.64	3.08	2.65	2.207	2.02	....	...	...
5⅝	6.07	5.31	4.61	4.17	3.73	3.15	2.71	2.257	....			
5¾	6.20	5.43	4.72	4.26	3.81	3.22	2.77	2.308	....			
5⅞	6.34	5.55	4.82	4.36	3.89	3.29	2.83	2.358		....		
6	6.48	5.67	4.93	4.45	3.98	3.37	2.89	2.409		....		
6⅛	6.61	5.79	5.03	4.54	4.06	3.44	....	....				
6¼	6.75	5.91	5.13	4.64	4.15	3.51	....	....				
6⅜	6.89	6.03	5.24	4.73	4.23	3.58	....	....				
6½	7.03	6.15	5.34	4.83	4.31	3.65	....					
6⅝	7.16	6.27	5.45	4.92	4.40	3.72	....					
6¾	7.30	6.39	5.55	5.01	4.48	3.79	....					
6⅞	7.44	6.51	5.65	5.11	4.56	3.86	....					
7	7.57	6.63	5.76	5.20	4.65	3.93	....					
7⅛	7.71	6.75	5.86	5.29	....							
7¼	7.85	6.87	5.96	5.39	....							
7⅜	7.99	6.99	6.07	5.48	....							
7½	8.12	7.11	6.17	5.58	....							
7⅝	8.26	7.23	6.28	5.67	....							
7¾	8.40	7.35	6.38	5.76	....	....	....	....				
7⅞	8.53	7.47	6.48	5.86	....	....	....	....				
8	8.67	7.58	6.59	5.95	....	....	....	....				

To determine weight per foot of a tube of a given *inside diameter*, add to weights in above list the weights in pounds per foot given below under corresponding gage numbers.

Gage No.	13	14	15	16	17	18	19	20	21	22	23	24
Weight Added	0.208	0.159	0.120	0.097	0.078	0.055	0.041	0.028	0.024	0.018	0.014	0.011

The number of chasers with which a die should be equipped depends upon the size of the die. The number recommended for different sizes is as follows:

Size of Die	Number of Chasers	Size of Die	Number of Chasers
Up to 1¼ inch	4	10 to 12 inches	12
1¼ to 4 inches	6	12 to 14 inches	14
4 to 7 inches	8	14 to 18 inches	16
7 to 10 inches	10	18 to 20 inches	18

Pipe threading dies should be lubricated with a good quality of lard oil or crude cotton-seed oil, the lubricant being used in liberal quantities.

Pipe and Tube Bending. — In bending a pipe or tube, the outer part of the bend is stretched and the inner section compressed, and as the result of opposite and unequal stresses, the pipe or tube tends to flatten or collapse. To prevent such distortion, the common practice is to support the wall of the pipe or tube in some manner during the bending operation. This support may be in the form of a filling material, or, when a bending machine or fixture is used, an internal mandrel or ball-shaped member may support the inner wall when required. If a filling material is used, it is melted and poured into the pipe or tube. One filler material (a commercial alloy known as "Bendalloy") has a melting point of only 160 degrees F. and is composed of bismuth, lead, tin, and cadmium. With this material, tubes having very thin walls have been bent to small radii. The metal filler conforms to the inside of the tube so closely that the tube can be bent just as though it were a solid rod. The filler is removed readily by melting. This method has been applied to the bending of copper, brass, duralumin, plain steel, and stainless steel tubes with uniform success. Tubes plated with chromium or nickel can be bent without danger of the plate flaking off.

Other filling materials such as resin, tar, lead, and dry sand have also been used.

Pipes are often bent to avoid the use of fittings, thus eliminating joints, providing a smooth unobstructed passage for fluids, and resulting in certain other advantages.

Minimum Radius: The safe minimum radius for a given diameter, material, and method of bending depends upon the thickness of the pipe wall, it being possible, for example, to bend extra heavy pipe to a smaller radius than pipe of standard weight. As a general rule, wrought iron or steel pipe of standard weight may readily be bent to a radius equal to five or six times the nominal pipe diameter. The minimum radius for standard weight pipe should, as a rule, be three and one-half to four times the diameter. It will be understood, however, that the minimum radius may vary considerably, depending upon the method of bending. Extra heavy pipe may be bent to radii varying from two and one-half times the diameter for smaller sizes to three and one-half to four times the diameter for larger sizes.

Rules for Finding Lengths of Bends: In determining the required length of a pipe or tube before bending, the lengths of the straight sections are, of course, added to the lengths required for the curved sections in order to make the proper allowance for bends. The following rules are for finding the lengths of the curved sections.

Rule for 90-Degree Bend: To find the length of a 90-degree or right-angle bend, multiply the radius of the bend by 1.57 (the radius is measured to the center of the pipe or to a point midway between the inner and outer walls).

Rule for 180-Degree Bend: To find the length of a 180-degree or U bend, multiply the radius of the bend by 3.14.

General Rule: A general rule for finding the lengths of sections having degrees of curvature other than 90 and 180 is as follows: Multiply the radius of the bend by the included angle, and then multiply the product by the constant 0.01745. The result is the length of the curved section.

Plastic Pipe. — Shortly after World War II, plastic pipe became an acceptable substitute, under certain service conditions, for other piping materials. Now, however, plastic pipe is specified on the basis of its own special capabilities and limitations. The largest volume of application has been for water piping systems.

Besides being light in weight, plastic pipe performs well in resisting deterioration from corrosive or caustic fluids. Even if the fluid borne is harmless, the chemical resistance of plastic pipe offers protection against a harmful exterior environment, such as when buried in a corrosive soil.

Generally, plastic pipe is limited by its temperature and pressure capacities. The higher the operating pressure of the pipe system, the less will be its temperature capability. The reverse is true, also. Since it is formed from organic resins, plastic pipe will burn. For various piping compositions, ignition temperatures vary from 700° to 800°F (370° to 430°C).

The following are accepted methods for joining plastic pipe:

Solvent Welding is usually accomplished by brushing a solvent cement on the end of the length of pipe and into the socket end of a fitting or the flange of the next pipe section. A chemical weld then joins and seals the pipe after connection.

Threading is a procedure not recommended for thin-walled plastic pipe or for specific grades of plastic. During connection of thicker-walled pipe, strap wrenches are used to avoid damaging and weakening the plastic.

Heat Fusion involves the use of heated air and plastic filler rods to weld plastic pipe assemblies. A properly welded joint can have a tensile strength equal to 90 percent that of the pipe material.

Elastomeric Sealing is used with bell-end piping. It is a recommended procedure for large diameter piping and for underground installations. The joints are set quickly and have good pressure capabilities.

Table 1. Dimensions and Weights of Thermoplastic Pipe*

Nominal Pipe Size		Outside Diameter		Schedule 40				Schedule 80			
				Nom. Wall Thickness		Nominal Weight*		Nom. Wall Thickness		Nominal Weight*	
in.	cm	in.	cm	in.	cm	lb/100'	kg/m	in.	cm	lb/100'	kg/m
1/8	0.3	0.405	1.03	0.072	0.18	3.27	0.05	0.101	0.256	4.18	0.06
1/4	0.6	0.540	1.37	0.093	0.24	5.66	0.08	0.126	0.320	7.10	0.11
3/8	1.0	0.675	1.71	0.096	0.24	7.57	0.11	0.134	0.340	9.87	0.15
1/2	1.3	0.840	2.13	0.116	0.295	11.4	0.17	0.156	0.396	14.5	0.22
3/4	2.0	1.050	2.67	0.120	0.305	15.2	0.23	0.163	0.414	19.7	0.29
1	2.5	1.315	3.34	0.141	0.358	22.5	0.33	0.190	0.483	29.1	0.43
1 1/4	3.2	1.660	4.22	0.148	0.376	30.5	0.45	0.202	0.513	40.1	0.60
1 1/2	3.8	1.900	4.83	0.154	0.391	36.6	0.54	0.212	0.538	48.7	0.72
2	5.1	2.375	6.03	0.163	0.414	49.1	0.73	0.231	0.587	67.4	1.00
2 1/2	6.4	2.875	7.30	0.215	0.546	77.9	1.16	0.293	0.744	103	1.5
3	7.6	3.500	8.89	0.229	0.582	102	1.5	0.318	0.808	138	2.1
3 1/2	8.9	4.000	10.16	0.240	0.610	123	1.8	0.337	0.856	168	2.5
4	10.2	4.500	11.43	0.251	0.638	145	2.2	0.357	0.907	201	3.0
5	12.7	5.563	14.13	0.273	0.693	197	2.9	0.398	1.011	280	4.2
6	15.2	6.625	16.83	0.297	0.754	256	3.8	0.458	1.163	385	5.7
8	20.3	8.625	21.91	0.341	0.866	385	5.7	0.530	1.346	584	8.7
10	25.4	10.75	27.31	0.387	0.983	546	8.1	0.629	1.598	867	12.9
12	30.5	12.75	32.39	0.430	1.09	722	10.7	0.728	1.849	1192	17.7

* The nominal weights of plastic pipe given in this table are based on an empirically chosen material density of 1.00 g/cm³. The nominal unit weight for a specific plastic pipe formulation can be obtained by multiplying the weight values from the table by the density in g/cm³ or by the specific gravity of the particular plastic composition.

The following are ranges of density factors for various plastic pipe materials: PE, 0.93 to 0.96; PVC, 1.35 to 1.40; CPVC, 1.55; ABS, 1.04 to 1.08; SR, 1.05; PB, 0.91 to 0.92; and PP, 0.91. For meanings of abbreviations see Table 2.

Information supplied by the Plastics Pipe Institute.

Insert Fitting is particularly useful for PE and PB pipe. For joining pipe sections, insert fittings are pushed into the pipe and secured by stainless steel clamps.

Transition Fitting involves specially designed connectors to join plastic pipe with other materials, such as cast iron, steel, copper, clay, and concrete.

Plastic pipe can be specified by means of Schedules 40, 80, and 120, which conform dimensionally to metal pipe, or through a Standard Dimension Ratio (SDR). The SDR is a rounded value obtained by dividing the average outside diameter of the pipe by the wall thickness. Within an individual SDR series of pipe, pressure ratings are uniform, regardless of pipe diameter.

Table 1 provides the weights and dimensions for Schedule 40 and 80 thermoplastic pipe, while Table 3 gives ranges of water pressure ratings and pipe support centers for a variety of pipe grades under three generic descriptions: PE (polyethylene); PVC (polyvinyl chloride); and ABS (acrylonitrile-butadiene styrene).

For more detailed information concerning the properties of a particular plastic pipe formulation, consult the pipe manufacturer or The Plastics Pipe Institute, 250 Park Ave., New York, N. Y. 10017.

Table 2.   General Properties and Uses of Plastic Pipe*

Plastic Pipe Material	Properties	Common Uses	Operating Temperature†		Joining Methods
			With Pressure	Without Pressure	
ABS (Acrylonitrile-butadiene styrene)	Rigid; excellent impact strength at low temperatures; maintains rigidity at higher temperatures.	Water, Drain, Waste, Vent, Sewage.	100°F (38°C)	180°F (82°C)	Solvent cement, Threading, Transition fitting.
PE (Polyethylene)	Flexible; excellent impact strength; good performance at low temperatures.	Water, Gas, Chemical, Irrigation.	100°F (38°C)	180°F (82°C)	Heat fusion, Insert and Transition fitting.
PVC (Polyvinyl chloride)	Rigid; fire self-extinguishing; high impact and tensile strength.	Water, Gas, Sewage, Industrial process, Irrigation.	100°F (38°C)	180°F (82°C)	Solvent cement, Elastomeric seal, Mechanical coupling, Transition fitting.
CPVC (Chlorinated polyvinyl chloride)	Rigid; fire self-extinguishing; high impact and tensile strength.	Hot and cold water, Chemical.	180°F (82°C) at 100 psig (690kPa) for SDR-11		Solvent cement, Threading, Mechanical coupling, Transition fitting.
PB (Polybutylene)	Flexible; good performance at elevated temperatures.	Water, Gas, Irrigation.	180°F (82°C)	200°F (93°C)	Insert fitting, Heat fusion, Transition fitting.
PP (Polypropylene)	Rigid; very light; high chemical resistance, particularly to sulfur-bearing compounds.	Chemical waste and processing.	100°F (38°C)	180°F (82°C)	Mechanical coupling, Heat fusion, Threading.
SR (Styrene rubber plastic)	Rigid; moderate chemical resistance; fair impact strength.	Drainage, Septic fields.	150°F (66°C)	.....	Solvent cement, Transition fitting, Elastomeric seal.

* From information supplied by the Plastics Pipe Institute.
† The operating temperatures shown are general guide points. For specific operating temperature and pressure data for various grades of the types of plastic pipe given, please consult the pipe manufacturer or the Plastics Pipe Institute.

Table 3. Thermoplastic Pipe Water Pressure Ratings at 73°F (23°C) and Horizontal Support-Center Distances — 1†

Nominal Pipe Size		Schedule 40											
		PE				PVC				ABS			
		Pressure* Rating		Support§ Center Dist.		Pressure* Rating		Support§ Center Dist.		Pressure* Rating		Support§ Center Dist.	
in.	cm	psi	kPa	ft	m	psi	kPa	ft	m	psi	kPa	ft	m
⅛	0.3	....	....	...	...	400 to 810	2760 to 5580	...	...	....	....	...	...
¼	0.6	....	....	1.0 to 1.5	0.3 to 0.5	390 to 780	2690 to 5380	...	...	....	....	...	...
⅜	1.0	....	....	1.0 to 1.5	0.3 to 0.5	310 to 620	2140 to 4270	...	...	....	....	...	...
½	1.3	119 to 188	820 to 1300	1.2 to 2.0	0.4 to 0.6	300 to 600	2070 to 4140	3.5	1.1	298 to 476	2050 to 3280	4.0	1.2
¾	2.0	96 to 152	660 to 1050	1.2 to 2.0	0.4 to 0.6	240 to 480	1650 to 3310	3.5	1.1	241 to 385	1660 to 2650	4.0	1.2
1	2.5	90 to 142	620 to 980	1.2 to 2.0	0.4 to 0.6	220 to 450	1520 to 3100	4.0	1.2	225 to 360	1550 to 2480	4.5	1.4
1¼	3.2	74 to 116	510 to 800	1.5 to 2.2	0.5 to 0.7	180 to 370	1240 to 2550	4.5	1.4	184 to 294	1270 to 2030	4.5	1.4
1½	3.8	66 to 104	460 to 717	1.5 to 2.2	0.5 to 0.7	170 to 330	1170 to 2280	4.5	1.4	165 to 264	1140 to 1820	5.0	1.5
2	5.1	55 to 87	380 to 600	1.7 to 2.7	0.5 to 0.8	140 to 280	970 to 1930	4.5	1.4	139 to 222	960 to 1530	5.0	1.5
2½	6.4	61 to 96	420 to 660	1.7 to 2.7	0.5 to 0.8	150 to 300	1030 to 2070	...	...	152 to 243	1050 to 1680	...	...
3	7.6	53 to 83	370 to 570	2.0 to 3.0	0.6 to 0.9	130 to 260	900 to 1790	5.5	1.7	132 to 211	910 to 1450	6	1.8
3½	8.9	50 to 75	340 to 520			120 to 240	830 to 1650	...	...	....	....	...	...
4	10.2	55 to 70	380 to 480	...	...	110 to 220	760 to 1520	6.2	1.9	111 to 177	765 to 1220	6.2	1.9
5	12.7	50 to 61	340 to 420	...	...	100 to 190	690 to 1310	...	...	....	....	...	...
6	15.2	55	380	...	...	90 to 180	620 to 1240	6.7	2.0	88 to 141	610 to 972	6.7	2.0
8	20.3	50	340	...	...	80 to 160	550 to 1100	7.5	2.3	....	....	...	...
10	25.4	....	....	...	...	70 to 140	480 to 970	7.7	2.3	....	....	...	...
12	30.5	....	....	...	...	70 to 130	480 to 900	8.0	2.4	....	....	...	...

† From information provided by the Plastics Pipe Institute.
* The pressure ratings given apply only to unthreaded pipe. Threading is not recommended for: (1) all PE; (2) Schedule 40 ABS; and (3) Schedule 40 PVC, 6 in. or less in nominal pipe diameter. § Support-center values for PE are based on water being carried at 70°F (21°C). From 70° to 100°F (21° to 38°C) closer support spacing is required. Above 100°F (38°C) support should be continuous. All support centers given for PE pipe apply only to pipe grades having water pressure ratings of 80 psi (550 kPa) or more.

Table 3. Thermoplastic Pipe Water Pressure Ratings at 73°F (23°C) and Horizontal Support-Center Distances — 2†

Nominal Pipe Size		Schedule 80											
		PE				PVC				ABS			
		Pressure* Rating		Support§ Center Dist.		Pressure* Rating		Support§ Center Dist.		Pressure* Rating		Support§ Center Dist.	
in.	cm	psi	kPa	ft	m	psi	kPa	ft	m	psi	kPa	ft	m
1/8	0.3	....	....	...	...	610 to 1230	4200 to 8480	...	...	....	....	...	...
1/4	0.6	....	....	1.0 to 1.5	0.3 to 0.5	570 to 1130	3900 to 7790	...	...	....	....	...	...
3/8	1.0	....	....	1.0 to 1.5	0.3 to 0.5	460 to 920	3200 to 6300	...	...	....	....	...	...
1/2	1.3	170 to 267	1170 to 1840	1.2 to 2.0	0.4 to 0.6	420 to 850	2900 to 5900	3.5	1.1	424 to 678	2920 to 4670	5.0	1.5
3/4	2.0	137 to 217	945 to 1500	1.2 to 2.0	0.4 to 0.6	340 to 690	2300 to 4800	4.0	1.2	344 to 550	2370 to 3790	5.0	1.5
1	2.5	126 to 199	869 to 1370	1.2 to 2.0	0.4 to 0.6	320 to 630	2200 to 4300	4.5	1.4	315 to 504	2170 to 3470	5.5	1.7
1¼	3.2	104 to 164	717 to 1130	1.5 to 2.2	0.5 to 0.7	260 to 520	1800 to 3600	5.0	1.5	260 to 416	1790 to 2870	5.5	1.7
1½	3.8	94 to 148	650 to 1020	1.5 to 2.2	0.5 to 0.7	240 to 470	1700 to 3200	5.0	1.5	235 to 376	1620 to 2590	6.0	1.8
2	5.1	81 to 127	560 to 876	1.7 to 2.7	0.5 to 0.8	200 to 400	1400 to 2800	5.0	1.5	202 to 323	1390 to 2230	6.0	1.8
2½	6.4	85 to 134	590 to 924	1.7 to 2.7	0.5 to 0.8	210 to 420	1500 to 2900	...	...	212 to 340	1460 to 2340	...	...
3	7.6	75 to 118	520 to 814	2.0 to 3.0	0.6 to 0.9	190 to 370	1300 to 2600	6.0	1.8	187 to 297	1290 to 2050	7	2.1
3½	8.9	69 to 109	480 to 752	...	...	170 to 350	1200 to 2400	...	...	....	....	...	...
4	10.2	65 to 102	450 to 703	2.2 to 3.5	0.7 to 1.1	160 to 320	1100 to 2200	7.5	2.3	162 to 259	1120 to 1790	7.5	2.3
5	12.7	58 to 91	400 to 630	...	...	140 to 290	970 to 2000	...	...	....	....	...	...
6	15.2	56 to 88	390 to 610	2.7 to 4.2	0.8 to 1.3	140 to 280	970 to 1900	8.5	2.6	139 to 222	958 to 1530	8.5	2.6
8	20.3	....	....	...	...	120 to 250	830 to 1720	9.0	2.7	....	....	...	...
10	25.4	....	....	...	...	120 to 230	830 to 1600	9.5	2.9	....	....	...	...
12	30.5	....	....	...	...	110 to 230	760 to 1600	10.0	3.0	....	....	...	...

† From information provided by the Plastics Pipe Institute.
* The pressure ratings given apply only to unthreaded pipe. Threading is not recommended for all PE. However, if threaded pipe is Schedule 80 PVC, one-half the pressure ratings shown in the table can be used. § Support-center values for PE are based on water being carried at 70°F (21°C). From 70° to 100°F (21° to 38°C) closer support spacing is required. Above 100°F (38°C) support should be continuous. All support centers given for PE pipe apply only to pipe grades having water pressure ratings of 80 psi (550 kPa) or more.

Definitions of Pipe Fittings

The following definitions for various pipe fittings are given by the National Tube Co.:

Armstrong Joint. — A two-bolt, flanged or lugged connection for high pressures. The ends of the pipes are peculiarly formed to properly hold a gutta-percha ring. It was originally made for cast-iron pipe. The two-bolt feature has much to commend it. There are various substitutes for this joint, many of which employ rubber in place of gutta-percha; others use more bolts in order to reduce the cost.

Bell and Spigot Joint. — (1) The usual term for the joint in cast-iron pipe. Each piece is made with an enlarged diameter or bell at one end into which the plain or spigot end of another piece is inserted when laying. The joint is then made tight by cement, oakum, lead, rubber or other suitable substance, which is driven in or calked into the bell and around the spigot. When a similar joint is made in wrought pipe by means of a cast bell (or hub), it is at times called hub and spigot joint (poor usage). Matheson joint is the name applied to a similar joint in wrought pipe which has the bell formed from the pipe. (2) Applied to fittings or valves, means that one end of the run is a "bell," and the other end is a "spigot," similar to those used on regular cast-iron pipe.

Bonnet. — (1) A cover used to guide and enclose the tail end of a valve spindle. (2) A cap over the end of a pipe (poor usage).

Branch. — The outlet or inlet of a fitting not in line with the run, but which may make any angle.

Branch Ell. — (1) Used to designate an elbow having a back outlet in line with one of the outlets of the "run." It is also called a heel outlet elbow. (2) Incorrectly used to designate side outlet or back outlet elbow.

Branch Pipe. — A very general term used to signify a pipe either cast or wrought, that is equipped with one or more branches. Such pipes are used so frequently that they have acquired common names such as tees, crosses, side or back outlet elbows, manifolds, double-branch elbows, etc. The term branch pipe is generally restricted to such as do not conform to usual dimensions.

Branch Tee (Header). — A tee having many side branches. (See Manifold.)

Bull Head Tee. — A tee the branch of which is larger than the run.

Bushing. — A pipe fitting for the purpose of connecting a pipe with a fitting of larger size, being a hollow plug with internal and external threads to suit the different diameters.

Card Weight Pipe. — A term used to designate standard cr full weight pipe, which is the Briggs' standard thickness of pipe.

Close Nipple. — One the length of which is about twice the length of a standard pipe thread and is without any shoulder.

Coupling. — A threaded sleeve used to connect two pipes. Commercial couplings are threaded inside to suit the exterior thread of the pipe. The term coupling is occasionally used to mean any jointing device and may be applied to either straight or reducing sizes.

Cross. — A pipe fitting with four branches arranged in pairs, each pair on one axis, and the axes at right angles. When the outlets are otherwise arranged the fittings are branch pipes or specials.

Cross-over. — A small fitting with a double offset, or shaped like the letter *U* with the ends turned out. It is only made in small sizes and used to pass the flow of one pipe past another when the pipes are in the same plane.

Cross-over Tee. — A fitting made along lines similar to the cross-over, but having at one end two openings in a tee-head the plane of which is at right angles to the plane of the cross-over bend.

Cross Valve. — (1) A valve fitted on a transverse pipe so as to open communi-

cation at will between two parallel lines of piping. Much used in connection with oil and water pumping arrangements, especially on ship board. (2) Usually considered as an angle valve with a back outlet in the same plane as the other two openings.

Crotch. — A fitting that has the general shape of the letter *Y*. Caution should be exercised not to confuse the crotch and wye.

Double-branch Elbow. — A fitting that, in a manner, looks like a tee, or as if two elbows had been shaved and then placed together, forming a shape something like the letter *Y* or a crotch.

Double Sweep Tee. — A tee made with easy curves between body and branch, *i.e.*, the center of the curve between run and branch lies outside the body.

Drop Elbow. — A small sized ell that is frequently used where gas is put into a building. These fittings have wings cast on each side. The wings have small countersunk holes so that they may be fastened by wood screws to a ceiling or wall or framing timbers.

Drop Tee. — One having the same peculiar wings as the drop elbow.

Dry Joint. — One made without gasket or packing or smear of any kind, as a ground joint.

Elbow (*Ell*). — A fitting that makes an angle between adjacent pipes. The angle is always 90 degrees, unless another angle is stated. (See Branch, Service, and Union Ell.)

Extra Heavy. — When applied to pipe, means pipe thicker than standard pipe; when applied to valves and fittings, indicates goods suitable for a working pressure of 250 pounds per square inch.

Header. — A large pipe into which one set of boilers is connected by suitable nozzles or tees, or similar large pipes from which a number of smaller ones lead to consuming points. Headers are often used for other purposes — for heaters or in refrigeration work. Headers are essentially branch pipes with many outlets, which are usually parallel. Largely used for tubes of water-tube boilers.

Hydrostatic Joint. — Used in large water mains, in which sheet lead is forced tightly into the bell of a pipe by means of the hydrostatic pressure of a liquid.

Kewanee Union. — A patented pipe union having one pipe end of brass and the other of malleable iron, with a ring or nut of malleable iron, in which the arrangement and finish of the several parts is such as to provide a non-corrosive ball-and-socket joint at the junction of the pipe ends, and a non-corrosive connection between the ring and brass pipe end.

Lead Joint. — (1) Generally used to signify the connection between pipes which is made by pouring molten lead into the annular space between a bell and spigot, and then making the lead tight by calking. (2) Rarely used to mean the joint made by pressing the lead between adjacent pieces, as when a lead gasket is used between flanges.

Lead Wool. — A material used in place of molten lead for making pipe joints. It is lead fiber, about as coarse as fine excelsior, and when made in a strand, it can be calked into the joints, making them very solid.

Line Pipe. — Special brand of pipe that employs recessed and taper thread couplings, and usually greater length of thread than Briggs' standard. The pipe is also subjected to higher test.

Lip Union. — (1) A special form of union characterized by the lip that prevents the gasket from being squeezed into the pipe so as to obstruct the flow. (2) A ring union, unless flange is specified.

Manifold. — (1) A fitting with numerous branches used to convey fluids between a large pipe and several smaller pipes. (See Branch Tee.) (2) A header for a coil.

Matheson Joint. — A wrought pipe joint made by enlarging one end of the pipe to form a suitable lead recess, similar to the bell end of a cast-iron pipe, and which receives the male or spigot end of the next length. Practically the same style of a joint as used for cast-iron pipe.

Medium Pressure. — When applied to valves and fittings, means suitable for a working pressure of from 125 to 175 pounds per square inch.

Needle Valve. — A valve provided with a long tapering point in place of the ordinary valve disk. The tapering point permits fine graduation of the opening. At times called a needle point valve.

Nipple. — (1) A tubular pipe fitting usually threaded on both ends and under 12 inches in length. Pipe over 12 inches long is regarded as cut pipe. (See Close, Short, Shoulder and Space Nipple.)

Reducer. — (1) A fitting having a larger size at one end than at the other. Some have tried to establish the term "increaser" — thinking of direction of flow — but this has been due to a misunderstanding of the trade custom of always giving the largest size of run of a fitting first; hence, all fittings having more than one size are reducers. They are always threaded inside, unless specified flanged or for some special joint. (2) Threaded type, made with abrupt reduction. (3) Flanged pattern with taper body. (4) Flanged eccentric pattern with taper body, but flanges at 90 degrees to one side of body. (5) Misapplied at times, to a reducing coupling.

Run. — (1) A length of pipe that is made of more than one piece of pipe. (2) The portion of any fitting having its ends "in line" or nearly so, in contradistinction to the branch or side opening, as of a tee. The two main openings of an ell also indicate its run, and when there is a third opening on an ell, the fitting is a "side outlet" or "back outlet" elbow, except that when all three openings are in one plane and the back outlet is in line with one of the run openings, the fitting is a "heel outlet elbow" or a "single sweep tee" or sometimes a "branch tee."

Rust Joint. — Employed to secure rigid connection. The joint is made by packing an intervening space tightly with a stiff paste which oxidizes the iron, the whole rusting together and hardening into a solid mass. It generally cannot be separated except by destroying some of the pieces. One recipe is 80 pounds cast-iron borings or filings, 1 pound sal-ammoniac, 2 pounds flowers of sulphur, mixed to a paste with water.

Service Ell. — An elbow having an outside thread on one end. Also known as street ell.

Service Pipe. — A pipe connecting mains with a dwelling.

Service Tee. — A tee having inside thread on one end and on branch, but outside thread on other end of run. Also known as street tee.

Short Nipple. — One whose length is a little greater than that of two threaded lengths or somewhat longer than a close nipple. It always has some unthreaded portion between the two threads.

Shoulder Nipple. — A nipple of any length, which has a portion of pipe between two pipe threads. As generally used, however, it is a nipple about halfway between the length of a close nipple and a short nipple.

Space Nipple. — A nipple with a portion of pipe or shoulder between the two threads. It may be of any length long enough to allow a shoulder.

Standard Pressure. — A term applied to valves and fittings suitable for a working steam pressure of 125 pounds per square inch.

Tee. — A fitting, either cast or wrought, that has one side outlet at right angles to the run. A single outlet branch pipe. (See Branch, Bull Head, Cross-over, Double Sweep, Drop, Service and Union Tee.)

Union. — (1) The usual trade term for a device used to connect pipes. It

commonly consists of three pieces which are, first, the thread end fitted with exterior and interior threads; second, the bottom end fitted with interior threads and a small exterior shoulder; and third, the ring which has an inside flange at one end while the other end has an inside thread like that on the exterior of the thread end. A gasket is placed between the thread and bottom ends, which are drawn together by the ring. Unions are very extensively used, because they permit of connections with little disturbance of the pipe positions.

Union Ell. — An ell with a male or female union at one end.

Union Joint. — A pipe coupling, usually threaded, which permits disconnection without disturbing other sections.

Union Tee. — A tee with male or female union at connection on one end of run.

Wiped Joint. — A lead joint in which the molten solder is poured upon the desired place, after scraping and fitting the parts together, and the joint is wiped up by hand with a moleskin or cloth pad while the metal is in a plastic condition.

Wye (Y). — A fitting either cast or wrought that has one side outlet at any angle other than 90 degrees. The angle is usually 45 degrees, unless another angle is specified. The fitting is usually indicated by the letter Y.

Adhesives and Sealants

Adhesives Bonding. — By strict definition, an adhesive is any substance that fastens or bonds materials to be joined (adherends) by means of surface attachment. However, besides bonding a joint, an adhesive may serve as a seal against attack by or passage of foreign materials. When an adhesive performs both bonding and sealing functions, it is usually called an adhesive sealant.

Where the design of an assembly permits, bonding with adhesives can replace bolting, welding, and riveting. When considering other fastening methods for thin cross-sections, the joint loads might be of such an unacceptable concentration that adhesives bonding may provide the only viable alternative. Too, properly designed adhesive joints can minimize or eliminate irregularities and breaks in the contour of an assembly. Adhesives can also serve as dielectric insulation. An adhesive with dielectric properties can act as a barrier against galvanic corrosion when two dissimilar metals such as aluminum and magnesium are joined together. Conversely, adhesive products are available which also conduct electricity.

An adhesive can be classified as structural or non-structural. Agreement is not universal on the exact separation between both classifications. But, in a general way, an adhesive can be considered structural when it is capable of supporting heavy loads; non-structural when it cannot. Most adhesives are found in liquid, paste, or granular form, though film and fabric-backed tape varieties are available. Adhesive formulations are applied by brush, roller, trowel, or spatula. If application surfaces are particularly large or if high rates of production are required, power-fed flow guns, brushes, or sprays can be used.

The hot-melt adhesives are relatively new to the assembly field. In general, they permit fastening speeds that are much greater than water- or solvent-based adhesives. Supplied in solid form, the hot-melts liquefy when heated. After application, they cool quickly, solidifying and forming the adhesive bond. They have been used successfully for a wide variety of adherends, and can greatly reduce the need for clamping and lengths of time for curing storage.

If an adhesive bonding agent is to give the best results, time restrictions recommended by the manufacturer, such as shelf life and working life must be observed. The shelf life is considered as the period of time an adhesive can be stored after its manufacture. Working or "pot" life is the span of time between the mixing or making ready of an adhesive, on the job, and when it is no longer usable.

The actual performance of an adhesive-bonded joint depends on a wide range of

factors, many of them quite complex. They include: the size and nature of the applied loads; environmental conditions such as moisture or contact with other fluids or vapors; the nature of prior surface treatment of adherends; temperatures, pressures and curing times in the bonding process.

A great number of adhesives, under various brand names, may be available for a particular bonding task. However, there can be substantial differences in the cost of purchase and difficulties in application. Therefore, it is always best to check with manufacturers' information before making a proper choice. Also, testing under conditions approximating those required of the assembly in service will help assure that joints meet expected performance.

Though not meant to be all-inclusive, the information which follows correlates classes of adherends and some successful adhesive compositions from the many that can be readily purchased.

Bonding Metal: Epoxy resin adhesives perform well in bonding metallic adherends. One type of epoxy formulation is a two-part adhesive which can be applied at room temperature. It takes, however, seven days at room temperature for full curing, achieving shear strengths as high as 2500 psi (17.2 MPa). Curing times for this adhesive can be greatly accelerated by elevating the bonding temperature. For example, curing takes only one hour at 160°F (71°C).

A structural adhesive-filler is available for metals which is composed of aluminum powder and epoxy resin. It is made ready by adding a catalyst to the base components, and can be used to repair structural defects. At a temperature of 140°F (60°C) it cures in approximately one hour. Depending on service temperatures and design of the joint, this adhesive-filler is capable of withstanding flexural stresses above 10,000 psi (69 MPa), tension above 5,000 psi (34 MPa), and compression over 30,000 psi (207 MPa).

Many non-structural adhesives for metal-to-metal bonding are also suitable for fastening combinations of types of materials. Polysulfide, neoprene, or rubber-based adhesives are used to bond metal foils. Ethylene cellulose cements, available in a selection of colors, are used to plug machined recesses in metal surfaces, such as with screw insets. They harden within 24 hours. Other, stronger adhesive fillers are available for the non-structural patching of defects in metallic parts. One variety, used for iron and steel castings, is a cement that combines powdered iron with water-activated binding agents. The consistency of the prepared mix is such that it can be applied with a trowel and sets within 24 hours at room temperature. The filler comes in types that can be applied to both dry and wet castings, and is able to resist the quick changes of temperature during quenching operations.

Polyester cement can replace lead and other fillers for dents and openings in sheet metal. One type, used successfully on truck and auto bodies, is a two-part cement consisting of a paste resin that can be combined with a paste or powder extender. It is brushed or trowelled on, and is ready for finishing operations in one hour.

Adhesives can be used for both structural and non-structural applications which combine metals with non-metals. Structural polyester-based adhesives can bond reinforced plastic laminates to metal surfaces. One type has produced joints, between glass reinforced epoxy and stainless steel, that have tensile strengths of over 3000 psi (21 MPa). Elevated temperature service is not recommended for this adhesive. However, it is easily brushed on and bonds under slight pressure at room temperature, requiring several days for curing. The curing process accelerates when heat is added in a controlled environment, but there results a moderate reduction in tensile strength.

Low-density epoxy adhesives are successful in structurally adhering light plastics, such as polyurethane foam, to various metals. Applied by brush or spatula, the bonds cure within 24 hours at room temperatures.

Metals can be bonded structurally to wood with a liquid adhesive made up of neoprene and synthetic resin. For the best surface coverage, the adhesive should be applied in a minimum of two coats. The joints formed are capable of reaching shear stresses of 125 psi, and can gain an additional 25 percent in shear strength with the passage of time. This adhesive also serves as a strong, general purpose bonding agent for other adherend combinations, including fabrics and ceramics.

For bonding strengths in shear over 500 psi (3.4 MPa) and at service temperatures slightly above 160°F (71°C), one- and two-part powder and jelly forms of metal-to-wood types are available.

Besides epoxy formulations, there are general purpose rubber, cellulose, and vinyl adhesives suitable for the non-structural bonding of metals to other adherends, which include glass and leather. These adhesives, however, are not limited only to applications in which one of the adherends is metal. The vinyl and cellulose types have similar bonding properties. But while the vinyls are less flammable, they are weaker in resistance to moisture than the comparable cellulosics. Rubber-based adhesives, in turn, have good resistance to moisture and lubricating oil. They can form non-structural bonds between metal and rubber.

One manufacturer has produced an acrylic-based adhesive that is highly suitable for rapidly bonding metal with other adherends at room temperature. For some applications it can be used as a structural adhesive, in the absence of moisture and high temperature. It cures within 24 hours and can be purchased in small bottles with dispenser tips.

A two-part epoxy adhesive is commercially available for non-structural bonding of joints or for patchwork in which one of the adherends is metal. Supplied in small tubes, it performs well even when temperatures vary between −50° to 200°F (−46° to 93°C). However, it is not recommended for use on assemblies that may experience heavy vibrations.

Bonding Plastic: Depending on the type of resin compound used in its manufacture, a plastic material can be classified as one of two types: a thermoplastic or a thermoset.

Thermoplastic materials have the capability of being repeatedly softened by heat and hardened by cooling. Common thermoplastics are nylon, polyethylene, acetal, polycarbonate, polyvinyl chloride, cellulose nitrate and cellulose acetate. Also, solvents can easily dissolve a number of thermoplastic materials. Because of these physical and chemical characteristics of thermoplastics, heat or solvent welding may in many instances offer a better bonding alternative than adhesives.

Thermoplastics commonly require temperatures between 200° and 400°F (93° and 204°C) for successful heat welding. However, if the maximum temperature limit for a particular thermoplastic formulation is exceeded, the plastic material will experience permanent damage. Heat can be applied directly to thermoplastic adherends, as in hot-air welding. More sophisticated joining techniques employ processes in which the heat generated for fusing thermoplastics is activated by electrical, sonic, or frictional means.

In the solvent welding of thermoplastics, solvent is applied to the adherend surfaces with the bond forming as the solvent dries. Some common solvents for thermoplastics are: a solution of phenol and formic acid for nylon; methylene chloride for polycarbonate; and methyl alcohol for the cellulosics.

Many adhesive bonding agents for thermoplastics are "dope" cements. Dope or solvent cements combine solvent with a base material that is the same thermoplastic as the adherend. One type is used successfully on polyvinyl chloride water (PVC) pipe. This liquid adhesive, with a polyvinyl chloride base, is applied in at least two coats. The pipe joint, however, must be closed in less than a minute after the adhesive is applied. Resulting joint bonds can resist hydrostatic pressures over 400 psi (28 MPa), for limited periods, and also have good resistance to impact.

Previously mentioned general purpose adhesives, such as the cellulosics, vinyls, rubber cements, and epoxies are also used successfully on thermoplastics.

Thermoset plastics lack the fusibility and solubility of the thermoplastics and are usually joined by adhesive bonding. The phenolics, epoxies, and alkyds are common thermoset plastics. Epoxy-based adhesives can join most thermoset materials, as can neoprene, nitrile rubber, and polyester-based cements. Again, these adhesives are of a general purpose nature, and can bond both thermoplastics and thermosets to other materials which include ceramics, fabric, wood, and metal.

Bonding Rubber: Adhesives are available commercially which can bond natural, butyl, nitrile, neoprene, and silicone rubbers. Natural and synthetic rubber cements will provide flexible joints; some types resist lubricating and other oils. Certain general purpose adhesives, such as the acrylics or epoxies, can bond rubber to almost anything else, though joints will be rigid. Depending on the choice of adhesive as well as adherend types, the bonds can carry loadings that vary from weak non-structural to mild structural in description. One type of natural rubber with a benzene-naphtha solvent can resist shear stresses to 12.5 psi (83 kPa).

Bonding Wood: Animal glues, available in liquid and powder form, are familiar types of wood-to-wood adhesives, commonly used in building laminated assemblies. Both forms, however, require heavy bonding pressures for joints capable of resisting substantial loadings. Also, animal glues are very sensitive to variations in temperature and moisture.

Casein types of adhesive offer moderate resistance to moisture and high temperature, but also require heavy bonding pressures, as much as 200 psi, for strong joints. Urea resin adhesives also offer moderate weather resistance, but are good for bonding wood to laminated plastics as well as to other wooden adherends. For outdoor service, under severe weather conditions, phenol-resorcinol adhesives are recommended.

Vinyl-acetate emulsions are excellent for bonding wood to other materials that have especially non-porous surfaces, such as metal and certain plastic laminates. These adhesives, too, tend to be sensitive to temperature and moisture, but are recommended for wooden patternmaking.

Rubber, acrylic, and epoxy general-purpose adhesives also perform well with wood and other adherend combinations. Specific rubber-based formulations resist attack by oil.

Fabric and Paper Bonding: The general purpose adhesives, which include the rubber cements and epoxies previously mentioned, are capable of bonding fabrics together and fabrics with other adherend materials. A butadiene-acrylonitrile adhesive, suitable also for fastening metals, glass, plastic, and rubber, forms joints in fabric that are highly resistant to oil and which maintain bonding strength at temperatures up to 160°F (70°C). This adhesive, however, requires a long curing period, the first few hours of which are at an elevated temperature.

Commonly, when coated fabric materials must be joined, the base material forming the suitable adhesive is of the same type as that protecting the fabric. For example, a polyvinyl chloride-based adhesive is acceptable for vinyl-coated fabrics; and neoprene-based cements for neoprene-coated materials.

Rubber cements, gum mucilages, wheat pastes, and wood rosin adhesive can join paper as well as fabric assemblies. Solvent-based rosins can be used on glass and wood also. Rosin adhesives can also be treated as hot-melt adhesives for rapid curing. Generally, the rosins are water resistant, but usually weak against attack by organic solvents.

Sealants. — Normally, the primary role of a sealant composition is the prevention of leakage or access by dust, fluid, and other materials in assembly structures. Nevertheless, many products are currently being manufactured that are capable of performing additional functions. For example, though a sealant is normally not an adhesive, there exists a family of adhesive sealants which in varying degrees can bond structural joints as well. Besides resisting chemical attack, some sealant surface coatings can protect against physical wear. Sealants can also dampen noise and vibration, or restrict the flow of heat or electricity. Many sealant products are available in decorative tints that can help improve the appearance of an assembly.

Most sealants tend to be limited by the operating temperatures and pressures under which they are capable of sustained performance. Also, before a suitable choice of sealant formulation is made, other properties have to be examined; these include: strength of the sealant; its degree of rigidity; ease of repair; curing characteristics; and even shelf and working life.

Dozens of manufacturers supply hundreds of sealant compounds, a number of which may fill the requirements for a particular application. The following information, however, lists common uses for sealants, along with types of compositions that have been employed successfully within each category.

Gasket Materials: Silicone rubber gasket compositions are supplied in tubes in a semiliquid form ready for manual application. They can also be obtained in larger containers for power-fed applications. Suppliers offer a silicone rubber-based composition that can replace preformed paper, cork, and rubber gaskets for many manufacturing operations. This composition has performed successfully in sealing water pumps, engine filter housings, and oil pans. It can also seal gear housings and other joints that require a flexible gasket material that besides resisting shock can sustain large temperature changes. Silicone rubber compositions can withstand temperatures that vary from $-100°F$ to $450°F$ ($-73°C$ to $232°C$).

Gasket tapes, ropes, and strips can also be readily purchased to fit many assembly applications. One type of sealant tape combines a pressure-sensitive adhesive with a strip of silicone-rubber sponge. This tape has good cushioning properties for vibration damping and can stick to metal, plastic, ceramic, and glass combinations.

TFE-based gasketing strips are also available. This non-stick gasketing material can perform at pressures up to 200 psi (1.4 MPa) and temperatures to $250°F$ ($120°C$). Because of the TFE base, the strip does not adhere to or gum joint surfaces.

Sealing Pipe Joints: Phenolic-based sealants can seal threaded joints on high-pressure steam lines. One type, that is available in liquid or paste form, resists pressure up to 1200 psi (8.3 MPa) and temperatures to $950°F$ ($510°C$). This compound is brushed on and the joint closed and tightened to a torque of 135 in.-lb. (15.3 N · m). The connection is then subjected to a 24-hour cure with superheated steam.

The joining and sealing of plastic pipe is covered under the previous adhesives bonding section.

Sulfur-based compounds, though lacking the durability of caulking lead, can be used on bell and spigot sewer pipe. Available in a formulation that can resist temperatures up to $200°F$ ($93°C$), one sulfur-based sealant is applied as a hot-melt and allowed to flow into the bell and spigot connection. It quickly solidifies at room temperature, and can develop a joint tensile strength over 300 psi (2.1 MPa).

There are asphalt, coal-tar and plastic-based compositions that can be used on both cast-iron and ceramic bell and spigot pipe. Portland cement mortars also seal ceramic piping.

THERMAL ENERGY OR HEAT

Thermometer Scales. — There are two thermometer scales in general use: the Fahrenheit (F), which is used in the United States and in other countries still using the English system of units, and the Celsius (C) or Centigrade used throughout the rest of the world.

In the Fahrenheit thermometer, the freezing point of water is marked at 32 degrees on the scale and the boiling point, at atmospheric pressure, at 212 degrees. The distance between these two points is divided into 180 degrees. On the Celsius scale, the freezing point of water is at 0 degrees and the boiling point at 100 degrees. The following formulas may be used for converting temperatures given on any one of the scales to the other scale:

$$\text{Degrees Fahrenheit} = \frac{9 \times \text{degrees C}}{5} + 32$$

$$\text{Degrees Celsius} = \frac{5 \times (\text{degrees F} - 32)}{9}$$

Tables appear on the pages which follow that can be used to convert degrees Celsius into degrees Fahrenheit or vice versa. In the event that the conversions are not covered in the tables use those applicable portions of the formulas given above for converting.

Absolute Temperature and Absolute Zero. — A point has been determined on the thermometer scale, by theoretical considerations, which is called the absolute zero and beyond which a further decrease in temperature is inconceivable. This point is located at − 273.2 degrees Celsius or − 459.7 degrees F. A temperature reckoned from this point, instead of from the zero on the ordinary thermometers, is called absolute temperature. Absolute temperature in degrees C is known as "degrees Kelvin" or the "Kelvin scale" (K) and absolute temperature in degrees F is known as "degrees Rankine" or the "Rankine scale" (R).

$$\text{Degrees Kelvin} = \text{degrees C} + 273.2$$

$$\text{Degrees Rankine} = \text{degrees F} + 459.7$$

Measures of the Quantity of Thermal Energy. — The unit of quantity of thermal energy used in the United States is the British thermal unit, which is the quantity of heat or thermal energy required to raise the temperature of one pound of pure water one degree F. (American National Standard abbreviation, Btu; conventional British symbol, B.Th.U.) The French thermal unit or *kilogram calorie*, is the quantity of heat or thermal energy required to raise the temperature of one kilogram of pure water one degree C. One kilogram calorie = 3.968 British thermal units = 1000 gram calories. The number of foot-pounds of mechanical energy equivalent to one British thermal unit is called the *mechanical equivalent of heat*, and equals 778 foot-pounds.

In the modern metric or SI system of units, the unit for thermal energy is the *joule* (J); a commonly used multiple being the kilojoule (kJ) or 1000 joules. See page 2411 for an explanation of the SI System. One kilojoule = 0.9478 Btu. Also in the SI System, the *watt* (W), equal to joule per second (J/s), is used for power, where one watt = 3.412 Btu per hour.

Fahrenheit — Celsius (Centigrade) Conversion. — A simple way to convert a Fahrenheit temperature reading into a Celsius temperature reading or vice versa

is to enter the accompanying table in the center or boldface column of figures. These figures refer to the temperature in either Fahrenheit or Celsius degrees. If it is desired to convert from Fahrenheit to Celsius degrees, consider the center column as a table of Fahrenheit temperatures and read the corresponding Celsius temperature in the column at the left. If it is desired to convert from Celsius to Fahrenheit degrees, consider the center column as a table of Celsius values, and read the corresponding Fahrenheit temperature on the right.

Interpolation Factors

deg C		deg F	deg C		deg F
0.56	1	1.8	3.33	6	10.8
1.11	2	3.6	3.89	7	12.6
1.67	3	5.4	4.44	8	14.4
2.22	4	7.2	5.00	9	16.2
2.78	5	9.0	5.56	10	18.0

Interpolation factors are given for use with that portion of the table in which the center column advances in increments of 10. To illustrate, suppose it is desired to find the Fahrenheit equivalent of 314 degrees C. The equivalent of 310 degrees C, found in the body of the main table, is seen to be 590.0 degrees F. The Fahrenheit equivalent of a 4-degree C difference is seen to be 7.2, as read in the table of interpolating factors. The answer is the sum or 597.2 degrees F.

Fahrenheit — Celsius (Centigrade) Conversion Table

deg C		deg F.	deg C		deg F	deg C		deg F	degC		deg.F
−273	−459.4	...	−101	−150	−238	− 8.3	17	62.6	9.4	49	120.2
−268	−450	...	− 96	−140	−220	− 7.8	18	64.4	10.0	50	122.0
−262	−440	...	− 90	−130	−202	− 7.2	19	66.2	10.6	51	123.8
−257	−430	...	− 84	−120	−184	− 6.7	20	68.0	11.1	52	125.6
−251	−420	...	− 79	−110	−166	− 6.1	21	69.8	11.7	53	127.4
−246	−410	...	− 73	−100	−148	− 5.6	22	71.6	12.2	54	129.2
−240	−400	...	− 68	− 90	−130	− 5.0	23	73.4	12.8	55	131.0
−234	−390	...	− 62	− 80	−112	− 4.4	24	75.2	13.3	56	132.8
−229	−380	...	− 57	− 70	− 94	− 3.9	25	77.0	13.9	57	134.6
−223	−370	...	− 51	− 60	− 76	− 3.3	26	78.8	14.4	58	136.4
−218	−360	...	−46	−50	−58	− 2.8	27	80.6	15.0	59	138.2
−212	−350	...	−40	−40	−40	− 2.2	28	82.4	15.6	60	140.0
−207	−340	...	−34	−30	−22	− 1.7	29	84.2	16.1	61	141.8
−201	−330	...	−29	−20	− 4	− 1.1	30	86.0	16.7	62	143.6
−196	−320	...	−23	−10	14	− 0.6	31	87.8	17.2	63	145.4
−190	−310	...	−17.8	0	32—	0—	32	89.6	17.8	64	147.2
−184	−300	...	−17.2	1	33.8	0.6	33	91.4	18.3	65	149.0
−179	−290	...	−16.7	2	35.6	1.1	34	93.2	18.9	66	150.8
−173	−280	...	−16.1	3	37.4	1.7	35	95.0	19.4	67	152.6
−169	−273	−459.4	−15.6	4	39.2	2.2	36	96.8	20.0	68	154.4
−168	−270	−454	−15.0	5	41.0	2.7	37	98.6	20.6	69	156.2
−162	−260	−436	−14.4	6	42.8	3.3	38	100.4	21.1	70	158.0
−157	−250	−418	−13.9	7	44.6	3.9	39	102.2	21.7	71	159.8
−151	−240	−400	−13.3	8	46.4	4.4	40	104.0	22.2	72	161.6
−146	−230	−382	−12.8	9	48.2	5.0	41	105.8	22.8	73	163.4
−140	−220	−364	−12.2	10	50.0	5.6	42	107.6	23.3	74	165.2
−134	−210	−346	−11.7	11	51.8	6.1	43	109.4	23.9	75	167.0
−129	−200	−328	−11.1	12	53.6	6.7	44	111.2	24.4	76	168.8
−123	−190	−310	−10.6	13	55.4	7.2	45	113.0	25.0	77	170.6
−118	−180	−292	−10.0	14	57.2	7.8	46	114.8	25.6	78	172.4
−112	−170	−274	− 9.4	15	59.0	8.3	47	116.6	26.1	79	174.2
−107	−160	−256	− 8.9	16	60.8	8.9	48	118.4	26.7	80	176.0

Fahrenheit — Celsius (Centigrade) Conversion Table (Continued)

deg C		deg F	deg C		deg F	deg C		deg F	deg C		deg F
27.2	81	177.8	58.3	137	278.6	89.4	193	379.4	304.4	580	1076
27.8	82	179.6	58.9	138	280.4	90.0	194	381.2	310.0	590	1094
28.3	83	181.4	59.4	139	282.2	90.6	195	383.0	315.6	600	1112
28.9	84	183.2	60.0	140	284.0	91.1	196	384.8	321.1	610	1130
29.4	85	185.0	60.6	141	285.8	91.7	197	386.6	326.7	620	1148
30.0	86	186.8	61.1	142	287.6	92.2	198	388.4	332.2	630	1166
30.6	87	188.6	61.7	143	289.4	92.8	199	390.2	337.8	640	1184
31.1	88	190.4	62.2	144	291.2	93.3	200	392.0	343.3	650	1202
31.7	89	192.2	62.8	145	293.0	93.9	201	393.8	348.9	660	1220
32.2	90	194.0	63.3	146	294.8	94.4	202	395.6	354.4	670	1238
32.8	91	195.8	63.9	147	296.6	95.0	203	397.4	360.0	680	1256
33.3	92	197.6	64.4	148	298.4	95.6	204	399.2	365.6	690	1274
33.9	93	199.4	65.0	149	300.2	96.1	205	401.0	371.1	700	1292
34.4	94	201.2	65.6	150	302.0	96.7	206	402.8	376.7	710	1310
35.0	95	203.0	66.1	151	303.8	97.2	207	404.6	382.2	720	1328
35.6	96	204.8	66.7	152	305.6	97.8	208	406.4	387.8	730	1346
36.1	97	206.6	67.2	153	307.4	98.3	209	408.2	393.3	740	1364
36.7	98	208.4	67.8	154	309.2	98.9	210	410.0	398.9	750	1382
37.2	99	210.2	68.3	155	311.0	99.4	211	411.8	404.4	760	1400
37.8	100	212.0	68.9	156	312.8	100.0	212	413.6	410.0	770	1418
38.3	101	213.8	69.4	157	314.6	104.4	220	428.0	415.6	780	1436
38.9	102	215.6	70.0	158	316.4	110.0	230	446.0	421.1	790	1454
39.4	103	217.4	70.6	159	318.2	115.6	240	464.0	426.7	800	1472
40.0	104	219.2	71.1	160	320.0	121.1	250	482.0	432.2	810	1490
40.6	105	221.0	71.7	161	321.8	126.7	260	500.0	437.8	820	1508
41.1	106	222.8	72.2	162	323.6	132.2	270	518.0	443.3	830	1526
41.7	107	224.6	72.8	163	325.4	137.8	280	536.0	448.9	840	1544
42.2	108	226.4	73.3	164	327.2	143.3	290	554.0	454.4	850	1562
42.8	109	228.2	73.9	165	329.0	148.9	300	572.0	460.0	860	1580
43.3	110	230.0	74.4	166	330.8	154.4	310	590.0	465.6	870	1598
43.9	111	231.8	75.0	167	332.6	160.0	320	608.0	471.1	880	1616
44.4	112	233.6	75.6	168	334.4	165.6	330	626.0	476.7	890	1634
45.0	113	235.4	76.1	169	336.2	171.1	340	644.0	482.2	900	1652
45.6	114	237.2	76.7	170	338.0	176.7	350	662.0	487.8	910	1670
46.1	115	239.0	77.2	171	339.8	182.2	360	680.0	493.3	920	1688
46.7	116	240.8	77.8	172	341.6	187.8	370	698.0	498.9	930	1706
47.2	117	242.6	78.3	173	343.4	193.3	380	716.0	504.4	940	1724
47.8	118	244.4	78.9	174	345.2	198.9	390	734.0	510.0	950	1742
48.3	119	246.2	79.4	175	347.0	204.4	400	752.0	515.6	960	1760
48.9	120	248.0	80.0	176	348.8	210	410	770.0	521.1	970	1778
49.4	121	249.8	80.6	177	350.6	215.6	420	788	526.7	980	1796
50.0	122	251.6	81.1	178	352.4	221.1	430	806	532.2	990	1814
50.6	123	253.4	81.7	179	354.2	226.7	440	824	537.8	1000	1832
51.1	124	255.2	82.2	180	356.0	232.2	450	842	565.6	1050	1922
51.7	125	257.0	82.8	181	357.8	237.8	460	860	593.3	1100	2012
52.2	126	258.8	83.3	182	359.6	243.3	470	878	621.1	1150	2102
52.8	127	260.6	83.9	183	361.4	248.9	480	896	648.9	1200	2192
53.3	128	262.4	84.4	184	363.2	254.4	490	914	676.7	1250	2282
53.9	129	264.2	85.0	185	365.0	260.0	500	932	704.4	1300	2372
54.4	130	266.0	85.6	186	366.8	265.6	510	950	732.2	1350	2462
55.0	131	267.8	86.1	187	368.6	271.1	520	968	760.0	1400	2552
55.6	132	269.6	86.7	188	370.4	276.7	530	986	787.8	1450	2642
56.1	133	271.4	87.2	189	372.2	282.2	540	1004	815.6	1500	2732
56.7	134	273.2	87.8	190	374.0	287.8	550	1022	1093.9	2000	3632
57.2	135	275.0	88.3	191	375.8	293.3	560	1040	1648.9	3000	5432
57.8	136	276.8	88.9	192	377.6	298.9	570	1058	2760.0	5000	9032

Above 1000 in the center column, the table increases in increments of 50. To convert 1462 degrees F to Celsius, for instance, add to the Celsius equivalent of 1400 degrees F ten times the interpolation factor for 6 and the interpolation factor for 2 or 760.0 + 33.3 + 1.11, which equals 794.4.

Coefficients of Heat Transmission

Heat transmitted, in British thermal units, per second, through metal 1 inch thick, per square inch of surface, for a temperature difference of 1° F.

Metal	Btu per Second	Metal	Btu per Second	Metal	Btu per Second
Aluminum	0.00203	German silver	0.00050	Steel, soft	0.00062
Antimony	0.00022	Iron	0.00089	Silver	0.00610
Brass, yellow	0.00142	Lead	0.00045	Tin	0.00084
Brass, red	0.00157	Mercury	0.00011	Zinc	0.00170
Copper	0.00404	Steel, hard	0.00034		

Coefficients of Heat Radiation

Heat radiated, in British thermal units, per square foot of surface per hour, for a temperature difference of 1° F.

Surface	Btu per Hour	Surface	Btu per Hour
Cast-iron, new	0.6480	Sawdust	0.7215
Cast-iron, rusted	0.6868	Sand, fine	0.7400
Copper, polished	0.0327	Silver, polished	0.0266
Glass	0.5948	Tin, polished	0.0439
Iron, ordinary	0.5662	Tinned iron, polished	0.0858
Iron, sheet-, polished	0.0920	Water	1.0853
Oil	1.4800		

Freezing Mixtures

Mixture	Temperature Change, Degrees F.	
	From	To
Common salt (NaCl), 1 part; snow, 3 parts	32	±0
Common salt (NaCl), 1 part; snow, 1 part	32	−0.4
Calcium chloride (CaCl₂), 3 parts; snow, 2 parts	32	−27
Calcium chloride (CaCl₂), 2 parts; snow, 1 part	32	−44
Sal ammoniac (NH₄Cl), 5 parts; saltpeter (KNO₃), 5 parts; water, 16 parts	50	+10
Sal ammoniac (NH₄Cl), 1 part; saltpeter (KNO₃), 1 part; water, 1 part	46	−11
Ammonium nitrate (NH₄NO₃), 1 part; water, 1 part	50	+ 3
Potassium hydrate (KOH), 4 parts; snow, 3 parts	32	−35

Ignition Temperatures. — The following temperatures are required to ignite the different substances specified: Phosphorus, transparent, 120 degrees F.; bisulphide of carbon, 300 degrees F.; gun cotton, 430 degrees F.; nitro-glycerine, 490 degrees F.; phosphorus, amorphous, 500 degrees F.; rifle powder, 550 degrees F.; charcoal, 660 degrees F.; dry pine wood, 800 degrees F.; dry oak wood, 900 degrees F.

Latent Heat. — When a body changes from the solid to the liquid state or from the liquid to the gaseous state, a certain amount of heat is used to accomplish this change. This heat does not raise the temperature of the body and is called latent heat. When the body changes again from the gaseous to the liquid, or from the liquid to the solid state, this quantity of heat is given out by it. The *latent heat of fusion* is the heat supplied to a solid body at the melting point; this heat is absorbed by the body although its temperature remains nearly stationary during the whole operation of melting. The *latent heat of evaporation* is the heat that must be supplied to a liquid at the boiling point to transform the liquid into a vapor. The latent heat is generally given in British thermal units per pound. When it is said that the latent heat of evaporation of water is 966.6, this means that it takes 966.6 heat units to evaporate one pound of water after it has been raised to the boiling point, 212 degrees F.

Latent Heat of Fusion

Substance	Btu per Pound	Substance	Btu per Pound	Substance	Btu per Pound
Bismuth.........	22.75	Paraffine.........	63.27	Sulphur.........	16.86
Beeswax.........	76.14	Phosphorus......	9.06	Tin.............	25.65
Cast iron, gray...	41.40	Lead.............	10.00	Zinc............	50.63
Cast iron, white..	59.40	Silver	37.92	Ice.............	144.00

Latent Heat of Evaporation

Liquid	Btu per Pound	Liquid	Btu per Pound	Liquid	Btu per Pound
Alcohol, ethyl....	371.0	Bisulphide of		Sulphur dioxide	164.0
Alcohol, methyl..	481.0	carbon.........	160.0	Turpentine.....	133.0
Ammonia........	529.0	Ether...........	162.8	Water..........	966.6

Boiling Points of Various Substances at Atmospheric Pressure

Substance	Boiling Point, Degrees F.	Substance	Boiling Point, Degrees F.	Substance	Boiling Point, Degrees F.
Aniline..........	363	Chloroform......	140	Saturated brine	226
Alcohol.........	173	Ether...........	100	Sulphur.........	833
Ammonia........	−28	Linseed oil......	597	Sulphuric acid...	590
Benzine.........	176	Mercury.........	676	Water, pure....	212
Bromine.........	145	Napthaline......	428	Water, sea......	213.2
Carbon bisul-		Nitric acid......	248	Wood alcohol...	150
phide..........	118	Oil of turpentine..	315	....	......

Specific Heat. — The specific heat of a substance is the ratio of the heat required to raise the temperature of a certain weight of the given substance one degree F. to that required to raise the temperature of the same weight of water one degree. As the specific heat is not constant at all temperatures, it is generally assumed that it is determined by raising the temperature from 62 to 63 degrees F. For most substances, however, it is practically constant for temperatures up to 212 degrees F.

Average Specific Heats (Btu/lb-F) of Various Substances

Substance	Specific Heat	Substance	Specific Heat
Alcohol (absolute)............	0.700	Kerosene.....................	0.500
Alcohol (density o.8).........	0.622	Lead........................	0.031
Aluminum....................	0.214	Limestone...................	0.217
Antimony....................	0.051	Magnesia....................	0.222
Benzine......................	0.450	Marble......................	0.210
Brass........................	0.094	Masonry, brick..............	0.200
Brickwork...................	0.200	Mercury.....................	0.033
Cadmium....................	0.057	Naphtha....................	0.310
Charcoal....................	0.200	Nickel......................	0.109
Chalk.......................	0.215	Oil, machine................	0.400
Coal........................	0.240	Oil, olive...................	0.350
Coke........................	0.203	Phosphorus.................	0.189
Copper, 32° to 212° F.......	0.094	Platinum....................	0.032
Copper, 32° to 572° F........	0.101	Quartz.....................	0.188
Corundum...................	0.198	Sand.......................	0.195
Ether.......................	0.503	Silica......................	0.191
Fusel oil....................	0.564	Silver......................	0.056
Glass.......................	0.194	Soda.......................	0.231
Gold........................	0.031	Steel, mild.................	0.116
Graphite....................	0.201	Steel, high carbon..........	0.117
Ice.........................	0.504	Stone (generally)............	0.200
Iron, cast..................	0.130	Sulphur....................	0.178
Iron, wrought, 32° to 212° F..	0.110	Sulphuric acid..............	0.330
32° to 392° F..............	0.115	Tin........................	0.056
32° to 572° F..............	0.122	Turpentine.................	0.472
32° to 662° F..............	0.126	Water......................	1.000
Iron, at high temperatures:		Wood, fir..................	0.650
1382° to 1832° F...........	0.213	Wood, oak.................	0.570
1750° to 1840° F...........	0.218	Wood, pine................	0.467
1920° to 2190° F...........	0.199	Zinc.......................	0.095

Specific Heat of Gases (Btu/lb-F)

Gas	Constant Pressure	Constant Volume	Gas	Constant Pressure	Constant Volume
Acetic acid.........	0.412	....	Chloroform.........	0.157	....
Air.................	0.238	0.168	Hydrogen..........	3.409	2.412
Alcohol............	0.453	0.399	Nitrogen...........	0.244	0.173
Ammonia..........	0.508	0.399	Oxygen............	0.217	0.155
Carbonic acid......	0.217	0.171	Ethylene...........	0.404	0.332
Carbonic oxide.....	0.245	0.176	Steam..............	0.480	0.346
Chlorine...........	0.121	....	....	....	....

Heat Loss from Uncovered Steam Pipes. — The loss of heat from a bare steam or hot water pipe varies with the difference between the temperature inside the pipe and that of the surrounding air. The loss is 2.15 Btu per hour, per square foot of pipe surface, per degree F. of temperature difference when the latter is 100 degrees; for a difference of 200 degrees, the loss is 2.66 Btu; for 300 degrees, 3.26 Btu; for 400 degrees, 4.03 Btu; for 500 degrees, 5.18 Btu. Thus, if the pipe area is 1.18 square feet per foot of length, and the temperature difference 300 degrees F., the loss per hour per foot of length = 1.18 × 300 × 3.26 = 1154 Btu.

Values of Thermal Conductivity (k) and of Conductance (C) of Common Building and Insulating Materials

Type of Material	Thickness, in.	k or C*	Type of Material	Thickness, in.	k or C*	Max. Temp., °F	Density, Lb per cu. ft.	k*
BUILDING			BUILDING (Continued)					
Batt:	...	...	Siding:					
Mineral Fiber	2–2¾	0.14	Metal‡	Avg.	1.61	...	...	...
Mineral Fiber	3–3½	0.09	Wood, Med. Density	7/16	1.49	...	...	...
Mineral Fiber	3½–6½	0.05	Stone:					
Mineral Fiber	6–7	0.04	Lime or Sand	1	12.50	...	...	...
Mineral Fiber	8½	0.03	Wall Tile:					
Block:	...	...	Hollow Clay, 1-Cell	4	0.9	...	...	...
Cinder	4	0.90	Hollow Clay, 2-Cell	8	0.54	...	...	...
Cinder	8	0.58	Hollow Clay, 3-Cell	12	0.40	...	...	...
Cinder	12	0.53	Hollow Gypsum	Avg.	0.7	...	...	...
Block:	...	...	INSULATING					
Concrete	4	1.40	Blanket, Mineral Fiber:	...	...	...	...	...
Concrete	8	0.90	Felt	...	...	400	3 to 8	0.26
Concrete	12	0.78	Rock or Slag	...	...	1200	6 to 12	0.26†
Board:	...	...	Glass	...	...	350	0.65	0.33
Asbestos Cement	¼	16.5	Textile	...	...	350	0.65	0.31
Plaster	½	2.22	Blanket, Hairfelt	...	...	180	10	0.29
Plywood	¾	1.07	Board, Block and Pipe	...	...	...	...	...
Brick:	...	...	Insulation:	...	...	...	...	...
Common	1	5.0	Amosite	...	...	1500	15 to 18	0.32†
Face	1	9.0	Asbestos Paper	...	...	700	30	0.40†
Concrete (poured)	1	12.0	Glass or Slag (for Pipe)	...	...	350	3 to 4	0.23
Floor:	...	...	Glass or Slag (for Pipe)	...	...	1000	10 to 15	0.33†
Wood Subfloor	¾	1.06	Glass, Cellular	...	...	800	9	0.40
Hardwood Finish	¾	1.47	Magnesia (85%)	...	...	600	11 to 12	0.35†
Tile	Avg.	20.0	Mineral Fiber	...	...	100	15	0.29
Glass:	...	...	Polystyrene, Beaded	...	...	170	1	0.28
Architectural	...	10.00	Polystyrene, Rigid	...	...	170	1.8	0.25
Mortar:	...	...	Rubber, Rigid Foam	...	...	150	4.5	0.22
Cement	1	5.0	Wood Felt	...	...	180	20	0.31
Plaster:	...	...	Loose Fill:	...	...	...	...	...
Sand	⅜	13.30	Cellulose	...	...	...	2.5 to 3	0.27
Sand and Gypsum	½	11.10	Mineral Fiber	...	...	...	2 to 5	0.28
Stucco	1	5.0	Perlite	...	...	...	5 to 8	0.37
Roofing:	...	...	Silica Aerogel	...	...	...	7.6	0.17
Asphalt Roll	Avg.	6.50	Vermiculite	...	...	...	7 to 8.2	0.47
Shingle, asb. cem.	Avg.	4.76	Mineral Fiber Cement	...	...	...	...	...
Shingle, asphalt	Avg.	2.27	Clay Binder	...	...	1800	24 to 30	0.49†
Shingle, wood	Avg.	1.06	Hydraulic Binder	...	...	1200	30 to 40	0.75†

* Units are in Btu/hr-ft²-°F. Where thickness is given as 1 inch, the value given is thermal conductivity (k); for other thicknesses the value given is thermal conductance (C). All values are for a test mean temperature of 75°F, except those designated with (†), which are for 100°F. † See * footnote. ‡ Over hollowback sheathing. *Source:* American Society of Heating, Refrigerating and Air-Conditioning Engineers, Inc.: HANDBOOK OF FUNDAMENTALS.

Linear Expansion of Various Substances between 32 and 212 Deg. Fahr.

(For linear expansion of metals see "Specific Gravity and Properties of Metals")
Expansion of volume = 3 × linear expansion.

Substance	Linear Expansion for 1 Deg. Fahr.	Substance	Linear Expansion for 1 Deg. Fahr.
Brick..................	0.0000030	Masonry, brick from......	0.0000026
Cement, Portland	0.0000060	to	0.0000050
Concrete	0.0000080	Plaster..................	0.0000092
Ebonite................	0.0000428	Porcelain................	0.0000020
Glass, thermometer	0.0000050	Quartz, from	0.0000043
Glass, hard	0.0000040	to	0.0000079
Granite.................	0.0000044	Slate	0.0000058
Marble, from............	0.0000031	Sandstone...............	0.0000065
to	0.0000079	Wood (pine)	0.0000028

WEIGHTS AND MEASURES

Measures of Length

1 mile = 1760 yards = 5280 feet.
1 yard = 3 feet = 36 inches.   1 foot = 12 inches.
1 mil = 0.001 inch.   1 fathom = 2 yards = 6 feet.
1 rod = 5.5 yards = 16.5 feet.   1 hand = 4 inches.   1 span = 9 inches.
1 micro-inch = one millionth inch or 0.000001 inch.   (1 micrometer or micron = one
     millionth meter = 0.00003937 inch.)

Surveyor's Measure

1 mile = 8 furlongs = 80 chains.
1 furlong = 10 chains = 220 yards.
1 chain = 4 rods = 22 yards = 66 feet = 100 links.
1 link = 7.92 inches.

Nautical Measure

1 league = 3 nautical miles.
1 nautical mile = 6076.11549 feet = 1.1508 statute miles. (The *knot,* which is a nautical
     unit of speed, is equivalent to a speed of 1 nautical mile per hour.)
One degree at the equator = 60 nautical miles = 69.047 statute miles. 360 degrees =
     21,600 nautical miles = 24,856.8 statute miles = circumference at equator.

Square Measure

1 square mile = 640 acres = 6400 square chains.
1 acre = 10 square chains = 4840 square yards = 43,560 square feet.
1 square chain = 16 square rods = 484 square yards = 4356 square feet.
1 square rod = 30.25 square yards = 272.25 square feet = 625 square links.
1 square yard = 9 square feet.
1 square foot = 144 square inches.
    An acre is equal to a square, the side of which is 208.7 feet.

Measure used for Diameters and Areas of Electric Wires

1 circular inch = area of circle 1 inch in diameter = 0.7854 square inch.
1 circular inch = 1,000,000 circular mils.
1 square inch = 1.2732 circular inch = 1,273,239 circular mils.
    A circular mil is the area of a circle 0.001 inch in diameter.

Cubic Measure

1 cubic yard = 27 cubic feet.
1 cubic foot = 1728 cubic inches.
    The following measures are also used for wood and masonry:
1 cord of wood = 4 × 4 × 8 feet = 128 cubic feet.
1 perch of masonry = 16½ × 1½ × 1 foot = 24¾ cubic feet.

Shipping Measure

    For measuring entire internal capacity of a vessel:
1 register ton = 100 cubic feet.
    For measurement of cargo:
    Approximately 40 cubic feet of merchandise is considered a shipping ton, unless
that bulk would weigh more than 2000 pounds, in which case the freight charge may
be based upon weight.
40 cubic feet = 32.143 U.S. bushels = 31.16 Imperial bushels.

Dry Measure

1 bushel (U.S. or Winchester struck bushel) = 1.2445 cubic feet = 2150.42 cubic inches.
1 bushel = 4 pecks = 32 quarts = 64 pints.
1 peck = 8 quarts = 16 pints.
1 quart = 2 pints.
1 heaped bushel = 1¼ struck bushel.
1 cubic foot = 0.8036 struck bushel.
1 British Imperial bushel = 8 Imperial gallons = 1.2837 cubic feet = 2218.19 cubic inches.

Liquid Measure

1 U.S. gallon = 0.1337 cubic foot = 231 cubic inches = 4 quarts = 8 pints.
1 quart = 2 pints = 8 gills.
1 pint = 4 gills.
1 British Imperial gallon = 1.2009 U.S. gallon = 277.42 cubic inches.
1 cubic foot = 7.48 U.S. gallons.

Old Liquid Measure

1 tun = 2 pipes = 3 puncheons.
1 pipe or butt = 2 hogsheads = 4 barrels = 126 gallons.
1 puncheon = 2 tierces = 84 gallons.
1 hogshead = 2 barrels = 63 gallons.
1 tierce = 42 gallons.
1 barrel = 31½ gallons.

Apothecaries' Fluid Measure

1 U.S. fluid ounce = 8 drachms = 1.805 cubic inch = $\frac{1}{128}$ U.S. gallon.
1 fluid drachm = 60 minims.
1 British fluid ounce = 1.732 cubic inch.

Measures of Weight

Avoirdupois or Commercial Weight

1 gross or long ton = 2240 pounds.
1 net or short ton = 2000 pounds.
1 pound = 16 ounces = 7000 grains.
1 ounce = 16 drachms = 437.5 grains.
   The following measures for weight are now seldom used in the United States:
1 hundred-weight = 4 quarters = 112 pounds (1 gross or long ton = 20 hundred-weights); 1 quarter = 28 pounds; 1 stone = 14 pounds; 1 quintal = 100 pounds.

Troy Weight, used for Weighing Gold and Silver

1 pound = 12 ounces = 5760 grains.
1 ounce = 20 pennyweights = 480 grains.
1 pennyweight = 24 grains.
1 carat (used in weighing diamonds) = 3.086 grains.
1 grain Troy = 1 grain avoirdupois = 1 grain apothecaries' weight.

Apothecaries' Weight

1 pound = 12 ounces = 5760 grains.
1 ounce = 8 drachms = 480 grains.
1 drachm = 3 scruples = 60 grains.
1 scruple = 20 grains.

Measures of Pressure

1 pound per square inch = 144 pounds per square foot = 0.068 atmosphere = 2.042 inches of mercury at 62 degrees F. = 27.7 inches of water at 62 degrees F. = 2.31 feet of water at 62 degrees F.

1 atmosphere = 30 inches of mercury at 62 degrees F. = 14.7 pounds per square inch = 2116.3 pounds per square foot = 33.95 feet of water at 62 degrees F.

1 foot of water at 62 degrees F. = 62.355 pounds per square foot = 0.433 pound per square inch.

1 inch of mercury at 62 degrees F. = 1.132 foot of water = 13.58 inches of water = 0.491 pound per square inch.

Miscellaneous

1 great gross = 12 gross = 144 dozen.
1 gross = 12 dozen = 144 units.
1 dozen = 12 units.
1 score = 20 units.

1 quire = 24 sheets.
1 ream = 20 quires = 480 sheets.
1 ream printing paper = 500 sheets.

Decimal Equivalents of Fractions of an Inch

1/64	0.015 625	11/32	0.343 75	43/64	0.671 875
1/32	0.031 25	23/64	0.359 375	11/16	0.687 5
3/64	0.046 875	3/8	0.375	45/64	0.703 125
1/16	0.062 5	25/64	0.390 625	23/32	0.718 75
5/64	0.078 125	13/32	0.406 25	47/64	0.734 375
3/32	0.093 75	27/64	0.421 875	3/4	0.750
7/64	0.109 375	7/16	0.437 5	49/64	0.765 625
1/8	0.125	29/64	0.453 125	25/32	0.781 25
9/64	0.140 625	15/32	0.468 75	51/64	0.796 875
5/32	0.156 25	31/64	0.484 375	13/16	0.812 5
11/64	0.171 875	1/2	0.500	53/64	0.828 125
3/16	0.187 5	33/64	0.515 625	27/32	0.843 75
13/64	0.203 125	17/32	0.531 25	55/64	0.859 375
7/32	0.218 75	35/64	0.546 875	7/8	0.875
15/64	0.234 375	9/16	0.562 5	57/64	0.890 625
1/4	0.250	37/64	0.578 125	29/32	0.906 25
17/64	0.265 625	19/32	0.593 75	59/64	0.921 875
9/32	0.281 25	39/64	0.609 375	15/16	0.937 5
19/64	0.296 875	5/8	0.625	61/64	0.953 125
5/16	0.312 5	41/64	0.640 625	31/32	0.968 75
21/64	0.328 125	21/32	0.656 25	63/64	0.984 375

METRIC SYSTEMS OF MEASUREMENT

A metric system of measurement was first established in France in the years following the French Revolution, and various systems of metric units have been developed since that time. All metric unit systems are based, at least in part, on the International Metric Standards which are the meter and kilogram, or decimal multiples or sub-multiples of these standards.

In 1795, a metric system called the centimeter-gram-second (cgs) system was proposed, and was adopted in France in 1799. In 1873, the British Association for the Advancement of Science recommended the use of the cgs system, and since then it has been widely used in all branches of science throughout the world. From the base units in the cgs system are derived:

    Unit of velocity = 1 centimeter per second.
    Acceleration due to gravity (at Paris) = 981 centimeters per sec. per sec.
    Unit of force = 1 dyne = ⅟₉₈₁ gram.
    Unit of work = 1 erg = 1 dyne-centimeter.
    Unit of power = 1 watt = 10,000,000 ergs per second.

Another metric system called the MKS (meter-kilogram-second) system of units was proposed by Professor G. Giorgi in 1902. In 1935, the International Electrotechnical Commission (IEC) accepted his recommendation that this system of units of mechanics should be linked with the electro-magnetic units by the adoption of a fourth base unit. In 1950, the IEC adopted the ampere, the unit of electric current, as the fourth unit, and the MKSA system thus came into being.

A gravitational system of metric units, known as the technical system, is based on the meter, the kilogram as a force, and the second. It has been widely used in engineering. Because the standard of force is defined as the weight of the mass of the standard kilogram, the fundamental unit of force varies due to the difference in gravitational pull at different locations around the earth. By international agreement, a standard value for acceleration due to gravity was chosen (9.81 meters per second squared) which for all practical measurements is approximately the same as the local value at the point of measurement.

The International System of Units (SI). — The Conference Generale des Poids et Mesures (CGPM) which is the body responsible for all international matters concerning the metric system, adopted in 1954, a rationalized and coherent system of units, based on the four MKSA units (see above), and including the *kelvin* as the unit of temperature and the *candela* as the unit of luminous intensity. In 1960, the CGPM formerly named this system the Systeme International d'Unites, for which the abbreviation is SI in all languages. In 1971, the 14th CGPM adopted a seventh base unit, the *mole* which is the unit of quantity ("amount of substance")

In the period since the first metric system was established in France towards the end of the 18th century, most of the countries of the world have adopted a metric system. At the present time most of the industrially advanced metric-using countries are changing from their traditional metric system to SI. Those countries which are currently changing or considering change from the English system of measurement to metric, have the advantage that they can convert directly to the modernized system. The United Kingdom, which can be said to have led the now worldwide move to change from the English system, went straight to SI.

The use of SI units instead of the traditional metric units has little effect on everyday life or trade. The units of linear measurement, mass, volume, and time remain the same, viz. meter, kilogram, liter, and second.

The SI, like the traditional metric system, is based on decimal arithmetic. For each physical quantity, units of different sizes are formed by multiplying or dividing a single base value by powers of 10. Thus, changes can be made very simply by adding zeros or shifting decimal points. For example, the meter is the basic unit of length; the kilometer is a multiple (1000 meters); and the millimeter is a sub-multiple (one-thousandth of a meter).

In the older metric systems, the simplicity of a series of units linked by powers of ten is an advantage for plain quantities such as length, but this simplicity is lost as soon as more complex units are encountered. For example, in different branches of science and engineering, energy may appear as the erg, the calorie, the kilogram meter, the liter atmosphere or the horsepower hour. In contrast, the SI provides only one basic unit for each physical quantity, and universality is thus achieved.

As mentioned above, there are seven base-units, which are for the basic quantities of length, mass, time, electric current, thermodynamic temperature, amount of substance and luminous intensity, expressed as the meter (m), the kilogram (kg), the second (s), the ampere (A), the kelvin (K), the mole (mol) and the candela (cd). The units are defined in the accompanying table.

The SI is a coherent system. A system is said to be coherent if the product or quotient of any two unit quantities in the system is the unit of the resultant quantity. For example, in a coherent system in which the foot is the unit of length, the square foot is the unit of area, whereas the acre is not.

Other physical quantities are derived from the base units. For example, the unit of velocity is the meter per second (m/s), which is a combination of the base units of length and time. The unit of acceleration is the meter per second squared (m/s^2). By applying Newton's second law of motion — force is proportional to mass multiplied by acceleration — the unit of force is obtained which is the kilogram meter per second squared (kgm/s^2). This unit is known as the newton or N. Work, or force times distance is the kilogram meter squared per second squared (kgm^2/s^2), which is the joule, (1 joule = 1 newton-meter), and energy is also expressed in these terms. The abbreviation for joule is J. Power or work per unit time is the kilogram meter squared per second cubed (kgm^2/s^3), which is the watt (1 watt = 1 joule per second = 1 newton-meter per second.) The abbreviation for watt is W. The term horsepower is not used in the SI and is replaced by the watt which together with multiples and sub-multiples — kilowatt and milliwatt, for example — is the same unit as that used in electrical work.

The use of the newton as the unit of force is of particular interest to engineers. In practical work using the English or traditional metric systems of measurements, it is a common practice to apply weight units as force units. Thus, the unit of force in those systems is that force which when applied to unit mass produces an acceleration g, rather than unit acceleration. The value of gravitational acceleration g varies around the earth, and thus the weight of a given mass also varies. In an effort to account for this minor error, the kilogram-force and pound-force were introduced, which are defined as the forces due to "standard gravity" acting on bodies of one kilogram or one pound mass respectively. The standard gravitational acceleration is taken as 9.80665 meters per second squared or 32.174 feet per second squared. The newton is defined as "that force which when applied to a body having a mass of one kilogram, gives it an acceleration of one meter per second squared." It is independent of g. As a result, the factor g disappears from a wide range of formulas in dynamics. However, in some formulas in statics, where the weight of a body is important rather than its mass, g does appear where it was formerly absent (the weight of a mass of W kilograms is equal to a force of Wg newtons, where g = approximately 9.81 meter per second squared). Details concerning the use of SI units in mechanics calculations are given on page 132, and throughout the Mechanics section in this Handbook. The use of SI units in strength of materials

calculations is covered in the section on that subject.

Decimal multiples and sub-multiples of the SI units are formed by means of the prefixes given in the following table, which represent the numerical factors shown.

Factors and Prefixes for Forming Decimal Multiples and Sub-multiples of the SI Units

Factor by which the unit is multiplied	Prefix	Symbol	Factor by which the unit is multiplied	Prefix	Symbol
10^{12}	tera	T	10^{-2}	centi	c
10^{9}	giga	G	10^{-3}	milli	m
10^{6}	mega	M	10^{-6}	micro	μ
10^{3}	kilo	k	10^{-9}	nano	n
10^{2}	hecto	h	10^{-12}	pico	p
10	deka	da	10^{-15}	femto	f
10^{-1}	deci	d	10^{-18}	atto	a

Standard of Length. — In 1866 the United States, by act of Congress, passed a law making legal the meter, the only measure of length that has been legalized by the United States Government. The United States yard is defined by the relation: 1 yard = $3600/3937$ meter. The legal equivalent of the meter for commercial purposes was fixed as 39.37 inches, by law, in July, 1866, and experience having shown that this value was exact within the error of observation, the United States Office of Standard Weights and Measures was, in 1893, authorized to derive the yard from the meter by the use of this relation. The United States prototype meters Nos. 27 and 21 were received from the International Bureau of Weights and Measures in 1889. Meter No. 27, sealed in its metal case, is preserved in a fireproof vault at the Bureau of Standards.

Comparisons made prior to 1893 indicated that the relation of the yard to the meter, fixed by the Act of 1866, was by chance the exact relation between the international meter and the British imperial yard, within the error of observation. A subsequent comparison made between the standards just mentioned indicates that the legal relation adopted by Congress is in error 0.0001 inch; but, in view of the fact that certain comparisons made by the English Standards Office between the imperial yard and its authentic copies show variations as great if not greater than this, it cannot be said with certainty that there is a difference between the imperial yard of Great Britain and the United States yard derived from the meter. The bronze yard No. 11, which was an exact copy of the British imperial yard both in form and material, had shown changes when compared with the imperial yard in 1876 and 1888, which could not reasonably be said to be entirely due to changes in Bronze No. 11. On the other hand, the new meters represented the most advanced ideas of standards, and it therefore seemed that greater stability as well as higher accuracy would be secured by accepting the international meter as a fundamental standard of length.

For more information on SI practice, the reader is referred to the following publications:

Metric Practice Guide, published by the American Society for Testing and Materials, 1916 Race St., Philadelphia, PA 19103.

ISO International Standard 1000. This publication covers the rules for use of SI units, their multiples and submultiples. It can be obtained from the American National Standards Institute, 1430 Broadway, New York, NY 10018.

The International System of Units, Special Publication 330 of the National Bureau of Standards — available from the Superintendent of Documents, U.S. Government Printing Office, Washington, DC 20402.

Traditional Metric Measures

Measures of Length

10 millimeters (mm)	= 1 centimeter (cm).
10 centimeters	= 1 decimeter (dm).
10 decimeters	= 1 meter (m).
1000 meters	= 1 kilometer (km.)

Square Measure

100 square millimeters (mm^2)	= 1 square centimeter (cm^2).
100 square centimeters	= 1 square decimeter (dm^2).
100 square decimeters	= 1 square meter (m^2).

Surveyor's Square Measure

100 square meters (m^2)	= 1 are (a).
100 ares	= 1 hectare (ha).
100 hectares	= 1 square kilometer (km^2).

Cubic Measure

1000 cubic millimeters (mm^3)	= 1 cubic centimeter (cm^3).
1000 cubic centimeters	= 1 cubic decimeter (dm^3).
1000 cubic decimeters	= 1 cubic meter (m^3).

Dry and Liquid Measure

10 milliliters (ml)	= 1 centiliter (cl).
10 centiliters	= 1 deciliter (dl).
10 deciliters	= 1 liter (l).
100 liters	= 1 hectoliter (hl).

1 liter = 1 cubic decimeter = the volume of 1 kilogram of pure water at a temperature of 39.2 degrees F.

Measures of Weight

10 milligrams (mg)	= 1 centigram (cg).
10 centigrams	= 1 decigram (dg).
10 decigrams	= 1 gram (g).
10 grams	= 1 dekagram (dag).
10 dekagrams	= 1 hectogram (hg).
10 hectograms	= 1 kilogram (kg).
1000 kilograms	= 1 (metric) ton (t).

Greek Letters

The Greek letters are frequently used in mathematical expressions and formulas. The Greek alphabet is given below.

A	α	Alpha	H	η	Eta	N	ν	Nu	T	τ	Tau
B	β	Beta	Θ	ϑ θ	Theta	Ξ	ξ	Xi	Υ	υ	Upsilon
Γ	γ	Gamma	I	ι	Iota	O	o	Omicron	Φ	ϕ	Phi
Δ	δ	Delta	K	κ	Kappa	Π	π	Pi	X	χ	Chi
E	ϵ	Epsilon	Λ	λ	Lambda	P	ρ	Rho	Ψ	ψ	Psi
Z	ζ	Zeta	M	μ	Mu	Σ	σ ς	Sigma	Ω	ω	Omega

Table 1.   International System (SI) Units

PHYSICAL QUANTITY	NAME OF UNIT	UNIT SYMBOL	DEFINITION
Basic SI Units			
Length	metre	m	Distance traveled by light in vacuo during 1/299,792,458 of a second.
Mass	kilogram	kg	Mass of the international prototype which is in the custody of the Bureau International des Poids et Mesures (BIPM) at Sèvres, near Paris.
Time	second	s	The duration of 9,192,631,770 periods of the radiation corresponding to the transition between the two hyperfine levels of the ground state of the cesium-133 atom.
Electric Current	ampere	A	The constant current which, if maintained in two parallel rectilinear conductors of infinite length, of negligible circular cross section, and placed at a distance of one metre apart in a vacuum, would produce between these conductors a force equal to 2×10^{-7} N/m length.
Thermodynamic Temperature	degree kelvin	K	The fraction 1/273.16 of the thermodynamic temperature of the triple point of water.
Amount of Substance	mole	mol	The amount of substance of a system which contains as many elementary entities as there are atoms in 0.012 kilogram of carbon 12.
Luminous Intensity	candela	cd	The luminous intensity, in the perpendicular direction, of a surface of 1/600,000 square metre of a black body at the temperature of freezing platinum under a pressure of 101,325 newtons per square metre.
SI Units Having Special Names			
Force	newton	$N = kg \cdot m/s^2$	That force which, when applied to a body having a mass of one kilogramme, gives it an acceleration of one metre per second squared.
Work, Energy, Quantity of Heat	joule	$J = N \cdot m$	The work done when the point of application of a force of one newton is displaced through a distance of one metre in the direction of the force.
Power	watt	$W = J/s$	One joule per second.
Electric Charge	coulomb	$C = A \cdot s$	The quantity of electricity transported in one second by a current of one ampere.
Electric Potential	volt	$V = W/A$	The difference of potential between two points of a conducting wire carrying a constant current of one ampere, when the power dissipated between these points is equal to one watt.
Electric Capacitance	farad	$F = C/V$	The capacitance of a capacitor between the plates of which there appears a difference of potential of one volt when it is charged by a quantity of electricity equal to one coulomb.

Table 1 (*Continued*). **International System (SI) Units**

PHYSICAL QUANTITY	NAME OF UNIT	UNIT SYMBOL	DEFINITION
SI Units Having Special Names			
Electric Resistance	ohm	$\Omega = V/A$	The resistance between two points of a conductor when a constant difference of potential of one volt, applied between these two points, produces in this conductor a current of one ampere, this conductor not being the source of any electromotive force.
Magnetic Flux	weber	$Wb = V\ s$	The flux which, linking a circuit of one turn produces in it an electromotive force of one volt as it is reduced to zero at a uniform rate in one second.
Inductance	henry	$H = V\ s/A$	The inductance of a closed circuit in which an electromotive force of one volt is produced when the electric current in the circuit varies uniformly at the rate of one ampere per second.
Luminous Flux	lumen	$lm = cd\ sr$	The flux emitted within a unit solid angle of one steradian by a point source having a uniform intensity of one candela.
Illumination	lux	$lx = lm/m^2$	An illumination of one lumen per square metre.

Table 2. **International System (SI) Units with Complex Names**

PHYSICAL QUANTITY	SI UNIT	UNIT SYMBOL
SI Units Having Complex Names		
Area	square metre	m^2
Volume	cubic metre	m^3
Frequency	hertz*	Hz
Density (Mass Density)	kilogram per cubic metre	kg/m^3
Velocity	metre per second	m/s
Angular Velocity	radian per second	rad/s
Acceleration	metre per second squared	m/s^2
Angular Acceleration	radian per second squared	rad/s^2
Pressure	pascal‡	Pa
Surface Tension	newton per metre	N/m
Dynamic Viscosity	newton second per metre squared	$N\ s/m^2$
Kinematic Viscosity ⎰ Diffusion Coefficient ⎱	metre squared per second	m^2/s
Thermal Conductivity	watt per metre degree Kelvin	$W/(m\ °K)$
Electric Field Strength	volt per metre	V/m
Magnetic Flux Density	tesla†	T
Magnetic Field Strength	ampere per metre	A/m
Luminance	candela per square metre	cd/m^2

* Hz = cycle/second.
† T = weber/metre².
‡ Pa = newton/metre².

Metric Conversion Factors

(Symbols of SI units, multiples and submultiples are
given in parentheses in the right-hand column)

Multiply	By	To Obtain
LENGTH		
centimetre	0.03280840	foot
centimetre	0.3937008	inch
fathom	1.8288*	metre (m)
foot	0.3048*	metre (m)
foot	30.48*	centimetre (cm)
foot	304.8*	millimetre (mm)
inch	0.0254*	metre (m)
inch	2.54*	centimetre (cm)
inch	25.4*	millimetre (mm)
kilometre	0.6213712	mile [U. S. statute]
metre	39.37008	inch
metre	0.5468066	fathom
metre	3.280840	foot
metre	0.1988388	rod
metre	1.093613	yard
metre	0.0006213712	mile [U. S. statute]
microinch	0.0254*	micrometre [micron] (μm)
micrometre [micron]	39.37008	microinch
mile [U. S. statute]	1609.344*	metre (m)
mile [U. S. statute]	1.609344*	kilometre (km)
millimetre	0.003280840	foot
millimetre	0.03937008	inch
rod	5.0292*	metre (m)
yard	0.9144*	metre (m)
AREA		
acre	4046.856	metre^2 (m^2)
acre	0.4046856	hectare
centimetre^2	0.1550003	inch^2
centimetre^2	0.001076391	foot^2
foot^2	0.09290304*	metre^2 (m^2)
foot^2	929.0304*	centimetre^2 (cm^2)
foot^2	92,903.04*	millimetre^2 (mm^2)
hectare	2.471054	acre
inch^2	645.16*	millimetre^2 (mm^2)
inch^2	6.4516*	centimetre^2 (cm^2)
inch^2	0.00064516*	metre^2 (m^2)
metre^2	1550.003	inch^2
metre^2	10.763910	foot^2
metre^2	1.195990	yard^2
metre^2	0.0002471054	acre
millimetre^2	0.00001076391	foot^2
millimetre^2	0.001550003	inch^2
yard^2	0.8361274	metre^2 (m^2)

* Where an asterisk is shown, the figure is exact.

Metric Conversion Factors (*Continued*)

Multiply	By	To Obtain
VOLUME (including CAPACITY)		
centimetre³	0.06102376	inch³
foot³	0.02831685	metre³ (m³)
foot³	28.31685	litre
gallon [U. K. liquid]	0.004546092	metre³ (m³)
gallon [U. K. liquid]	4.546092	litre
gallon [U. S. liquid]	0.003785412	metre³ (m³)
gallon [U. S. liquid]	3.785412	litre
inch³	16,387.06	millimetre³ (mm³)
inch³	16.38706	centimetre³ (cm³)
inch³	0.00001638706	metre³ (m³)
litre	0.001*	metre³ (m³)
litre	0.2199692	gallon [U. K. liquid]
litre	0.2641720	gallon [U. S. liquid]
litre	0.03531466	foot³
metre³	219.9692	gallon [U. K. liquid]
metre³	264.1720	gallon [U. S. liquid]
metre³	35.31466	foot³
metre³	1.307951	yard³
metre³	1000.*	litre
metre³	61,023.76	inch³
millimetre³	0.00006102376	inch³
yard³	0.7645549	metre³ (m³)
VELOCITY, ACCELERATION, and FLOW		
centimetre/second	1.968504	foot/minute
centimetre/second	0.03280840	foot/second
centimetre/minute	0.3937008	inch/minute
foot/hour	0.00008466667	metre/second (m/s)
foot/hour	0.00508*	metre/minute
foot/hour	0.3048*	metre/hour
foot/minute	0.508*	centimetre/second
foot/minute	18.288*	metre/hour
foot/minute	0.3048*	metre/minute
foot/minute	0.00508*	metre/second (m/s)
foot/second	30.48*	centimetre/second
foot/second	18.288*	metre/minute
foot/second	0.3048*	metre/second (m/s)
foot/second²	0.3048*	metre/second² (m/s²)
foot³/minute	28.31685	litre/minute
foot³/minute	0.0004719474	metre³/second (m³/s)
gallon [U. S. liquid]/min.	0.003785412	metre³/minute
gallon [U. S. liquid]/min.	0.00006309020	metre³/second (m³/s)
gallon [U. S. liquid]/min.	0.06309020	litre/second
gallon [U. S. liquid]/min.	3.785412	litre/minute
gallon [U. K. liquid]/min.	0.004546092	metre³/minute
gallon [U. K. liquid]/min.	0.00007576820	metre³/second (m³/s)
inch/minute	25.4*	millimetre/minute
inch/minute	2.54*	centimetre/minute
inch/minute	0.0254*	metre/minute
inch/second²	0.0254*	metre/second² (m/s²)

* Where an asterisk is shown, the figure is exact.

Metric Conversion Factors (*Continued*)

Multiply	By	To Obtain
VELOCITY, ACCELERATION, and FLOW (*Continued*)		
kilometre/hour	0.6213712	mile/hour [U. S. statute]
litre/minute	0.03531466	foot^3/minute
litre/minute	0.2641720	gallon [U. S. liquid]/minute
litre/second	15.85032	gallon [U. S. liquid]/minute
mile/hour	1.609344*	kilometre/hour
millimetre/minute	0.03937008	inch/minute
metre/second	11,811.02	foot/hour
metre/second	196.8504	foot/minute
metre/second	3.280840	foot/second
metre/second^2	3.280840	foot/second^2
metre/second^2	39.37008	inch/second^2
metre/minute	3.280840	foot/minute
metre/minute	0.05468067	foot/second
metre/minute	39.37008	inch/minute
metre/hour	3.280840	foot/hour
metre/hour	0.05468067	foot/minute
metre^3/second	2118.880	foot^3/minute
metre^3/second	13,198.15	gallon [U. K. liquid]/minute
metre^3/second	15,850.32	gallon [U. S. liquid]/minute
metre^3/minute	219.9692	gallon [U. K. liquid]/minute
metre^3/minute	264.1720	gallon [U. S. liquid]/minute
MASS and DENSITY		
grain [1/7000 lb avoirdupois]	0.06479891	gram (g)
gram	15.43236	grain
gram	0.001*	kilogram (kg)
gram	0.03527397	ounce [avoirdupois]
gram	0.03215074	ounce [troy]
gram/centimetre^3	0.03612730	pound/inch^3
hundredweight [long]	50.80235	kilogram (kg)
hundredweight [short]	45.35924	kilogram (kg)
kilogram	1000.*	gram (g)
kilogram	35.27397	ounce [avoirdupois]
kilogram	32.15074	ounce [troy]
kilogram	2.204622	pound [avoirdupois]
kilogram	0.06852178	slug
kilogram	0.0009842064	ton [long]
kilogram	0.001102311	ton [short]
kilogram	0.001*	ton [metric]
kilogram	0.001*	tonne
kilogram	0.01968413	hundredweight [long]
kilogram	0.02204622	hundredweight [short]
kilogram/metre^3	0.06242797	pound/foot^3
kilogram/metre^3	0.01002242	pound/gallon [U. K. liquid]
kilogram/metre^3	0.008345406	pound/gallon [U. S. liquid]
ounce [avoirdupois]	28.34952	gram (g)
ounce [avoirdupois]	0.02834952	kilogram (kg)

* Where an asterisk is shown, the figure is exact.

Metric Conversion Factors (*Continued*)

Multiply	By	To Obtain
MASS and DENSITY (*Continued*)		
ounce [troy]	31.10348	gram (g)
ounce [troy]	0.03110348	kilogram (kg)
pound [avoirdupois]	0.4535924	kilogram (kg)
pound/foot^3	16.01846	kilogram/metre^3 (kg/m^3)
pound/inch^3	27.67990	gram/centimetre^3 (g/cm^3)
pound/gal [U. S. liquid]	119.8264	kilogram/metre^3 (kg/m^3)
pound/gal [U. K. liquid]	99.77633	kilogram/metre^3 (kg/m^3)
slug	14.59390	kilogram (kg)
ton [long 2240 lb]	1016.047	kilogram (kg)
ton [short 2000 lb]	907.1847	kilogram (kg)
ton [metric]	1000.*	kilogram (kg)
tonne	1000.*	kilogram (kg)
FORCE and FORCE/LENGTH		
dyne	0.00001*	newton (N)
kilogram-force	9.806650*	newton (N)
kilopond	9.806650*	newton (N)
newton	0.1019716	kilogram-force
newton	0.1019716	kilopond
newton	0.2248089	pound-force
newton	100,000.*	dyne
newton	7.23301	poundal
newton	3.596942	ounce-force
newton/metre	0.005710148	pound/inch
newton/metre	0.06852178	pound/foot
ounce-force	0.2780139	newton (N)
pound-force	4.448222	newton (N)
poundal	0.1382550	newton (N)
pound/inch	175.1268	newton/metre (N/m)
pound/foot	14.59390	newton/metre (N/m)
BENDING MOMENT or TORQUE		
dyne-centimetre	0.0000001*	newton-metre (N · m)
kilogram-metre	9.806650*	newton-metre (N · m)
ounce-inch	7.061552	newton-millimetre
ounce-inch	0.007061552	newton-metre (N · m)
newton-metre	0.7375621	pound-foot
newton-metre	10,000,000.*	dyne-centimetre
newton-metre	0.1019716	kilogram-metre
newton-metre	141.6119	ounce-inch
newton-millimetre	0.1416119	ounce-inch
pound-foot	1.355818	newton-metre (N · m)

* Where an asterisk is shown, the figure is exact.

Metric Conversion Factors (*Continued*)

Multiply	By	To Obtain
MOMENT OF INERTIA and SECTION MODULUS		
moment of inertia [kg · m²]	23.73036	pound-foot²
moment of inertia [kg · m²]	3417.171	pound-inch²
moment of inertia [lb · ft²]	0.04214011	kilogram-metre² (kg · m²)
moment of inertia [lb · inch²]	0.0002926397	kilogram-metre² (kg · m²)
moment of section [foot⁴]	0.008630975	metre⁴ (m⁴)
moment of section [inch⁴]	41.62314	centimetre⁴
moment of section [metre⁴]	115.8618	foot⁴
moment of section [centimetre⁴]	0.02402510	inch⁴
section modulus [foot³]	0.02831685	metre³ (m³)
section modulus [inch³]	0.00001638706	metre³ (m³)
section modulus [metre³]	35.31466	foot³
section modulus [metre³]	61,023.76	inch³
MOMENTUM		
kilogram-metre/second	7.233011	pound-foot/second
kilogram-metre/second	86.79614	pound-inch/second
pound-foot/second	0.1382550	kilogram-metre/second (kg · m/s)
pound-inch/second	0.01152125	kilogram-metre/second (kg · m/s)
PRESSURE and STRESS		
atmosphere [14.6959 lb/inch²]	101,325.	pascal (Pa)
bar	100,000.*	pascal (Pa)
bar	14.50377	pound/inch²
bar	100,000.*	newton/metre² (N/m²)
hectobar	0.6474898	ton [long]/inch²
kilogram/centimetre²	14.22334	pound/inch²
kilogram/metre²	9.806650*	newton/metre² (N/m²)
kilogram/metre²	9.806650*	pascal (Pa)
kilogram/metre²	0.2048161	pound/foot²
kilonewton/metre²	0.1450377	pound/inch²
newton/centimetre²	1.450377	pound/inch²
newton/metre²	0.00001*	bar
newton/metre²	1.0*	pascal (Pa)
newton/metre²	0.0001450377	pound/inch²
newton/metre²	0.1019716	kilogram/metre²
newton/millimetre²	145.0377	pound/inch²
pascal	0.00000986923	atmosphere
pascal	0.00001*	bar
pascal	0.1019716	kilogram/metre²
pascal	1.0*	newton/metre² (N/m²)
pascal	0.02088543	pound/foot²
pascal	0.0001450377	pound/inch²

* Where an asterisk is shown, the figure is exact.

Metric Conversion Factors (*Continued*)

Multiply	By	To Obtain
PRESSURE and STRESS (*Continued*)		
pound/foot²	4.882429	kilogram/metre²
pound/foot²	47.88026	pascal (Pa)
pound/inch²	0.06894757	bar
pound/inch²	0.07030697	kilogram/centimetre²
pound/inch²	0.6894757	newton/centimetre²
pound/inch²	6.894757	kilonewton/metre²
pound/inch²	6894.757	newton/metre² (N/m²)
pound/inch²	0.006894757	newton/millimetre² (N/mm²)
pound/inch²	6894.757	pascal (Pa)
ton [long]/inch²	1.544426	hectobar
ENERGY and WORK		
Btu [International Table]	1055.056	joule (J)
Btu [mean]	1055.87	joule (J)
calorie [mean]	4.19002	joule (J)
foot-pound	1.355818	joule (J)
foot-poundal	0.04214011	joule (J)
joule	0.0009478170	Btu [International Table]
joule	0.0009470863	Btu [mean]
joule	0.2386623	calorie [mean]
joule	0.7375621	foot-pound
joule	23.73036	foot-poundal
joule	0.9998180	joule [International U. S.]
joule	0.9999830	joule [U. S. legal, 1948]
joule [International U. S.]	1.000182	joule (J)
joule [U. S. legal, 1948]	1.000017	joule (J)
joule	.0002777778	watt-hour
watt-hour	3600.*	joule (J)
POWER		
Btu [International Table]/hour	0.2930711	watt (W)
foot-pound/hour	0.0003766161	watt (W)
foot-pound/minute	0.02259697	watt (W)
horsepower [550 ft-lb/s]	0.7456999	kilowatt (kW)
horsepower [550 ft-lb/s]	745.6999	watt (W)
horsepower [electric]	746.*	watt (W)
horsepower [metric]	735.499	watt (W)
horsepower [U. K.]	745.70	watt (W)
kilowatt	1.341022	horsepower [550 ft-lb/s]
watt	2655.224	foot-pound/hour
watt	44.25372	foot-pound/minute
watt	0.001341022	horsepower [550 ft-lb/s]
watt	0.001340483	horsepower [electric]
watt	0.001359621	horsepower [metric]
watt	0.001341022	horsepower [U. K.]
watt	3.412141	Btu [International Table]/hour

* Where an asterisk is shown, the figure is exact.

Metric Conversion Factors (*Concluded*)

Multiply	By	To Obtain
VISCOSITY		
centipoise	0.001*	pascal-second (Pa · s)
centistoke	0.000001*	metre^2/second (m^2/s)
metre^2/second	1,000,000.*	centistoke
metre^2/second	10,000.*	stoke
pascal-second	1000.*	centipoise
pascal-second	10.*	poise
poise	0.1*	pascal-second (Pa · s)
stoke	0.0001*	metre^2/second (m^2/s)

To Convert From	To	Use Formula
TEMPERATURE		
temperature Celsius, t_C	temperature Kelvin, t_K	$t_K = t_C + 273.15$
temperature Fahrenheit, t_F	temperature Kelvin, t_K	$t_K = (t_F + 459.67)/1.8$
temperature Celsius, t_C	temperature Fahrenheit, t_F	$t_F = 1.8\, t_C + 32$
temperature Fahrenheit, t_F	temperature Celsius, t_C	$t_C = (t_F - 32)/1.8$
temperature Kelvin, t_K	temperature Celsius, t_C	$t_C = t_K - 273.15$
temperature Kelvin, t_K	temperature Fahrenheit, t_F	$t_F = 1.8\, t_K - 459.67$
temperature Kelvin, t_K	temperature Rankine, t_R	$t_R = 9/5\, t_K$
temperature Rankine, t_R	temperature Kelvin, t_K	$t_K = 5/9\, t_R$

* Where an asterisk is shown, the figure is exact.

Miscellaneous Conversion Factors
(English Units)

Multiply	By	To Obtain
atmospheres	29.92	inches of mercury (32 deg. F.)
atmospheres	14.70	pounds/inch^2
British thermal units/hour	12.96	foot-pounds/minute
circular mils	0.7854	square mils
feet of water (60 deg. F.)	0.8843	inches of mercury (60 deg. F.)
feet of water (60 deg. F.)	0.4331	pounds/inch^2
feet/minute	0.01136	miles/hour
foot-pounds/second	0.07716	British thermal units/minute
gallons (U.S.) of water (60 deg. F.)	8.337	pounds of water (60 deg. F.)
gallons (U.S.)/second	8.021	feet^3/minute
inches of mercury (32 deg. F.)	0.03342	atmospheres
inches of mercury (60 deg. F.)	1.131	feet of water (60 deg. F.)
inches of mercury (60 deg. F.)	0.4898	pounds/inch^2
inches of water (60 deg. F.)	0.03609	pounds/inch^2
knots (International)	1.151	miles (statute)/hour
miles/hour	88	feet/minute
miles (statute)/hour	0.8690	knots (International)
ounces (avoirdupois)	0.9115	ounces (troy)
ounces (troy)	1.097	ounces (avoirdupois)
ounces (troy)	0.06857	pounds (avoirdupois)
pounds (avoirdupois)	14.58	ounces (troy)
pounds of water (60 deg. F.)	0.01603	feet^3
pounds of water (60 deg. F.)	0.1199	gallons (U.S.)
pounds/inch^2	0.06805	atmospheres
pounds/inch^2	2.309	feet of water (60 deg. F.)
pounds/inch^2	2.042	inches of mercury (60 deg. F.)
pounds/inch^2	27.71	inches of water (60 deg. F.)
square mils	1.273	circular mils

Metric and English Equivalents

Linear Measure

I kilometer = 0.6214 mile.

I meter = $\begin{cases} 39.37 \text{ inches.} \\ 3.2808 \text{ feet.} \\ 1.0936 \text{ yards.} \end{cases}$

I centimeter = 0.3937 inch.
I millimeter = 0.03937 inch.

I mile = 1.609 kilometers.
I yard = 0.9144 meter.
I foot = 0.3048 meter.
I foot = 304.8 millimeters.
I inch = 2.54 centimeters.
I inch = 25.4 millimeters.

Square Measure

I square kilometer = 0.3861 square mile = 247.1 acres.
I hectare = 2.471 acres = 107,639 square feet.
I are = 0.0247 acre = 1076.4 square feet.
I square meter = 10.764 square feet = 1.196 square yards.
I square centimeter = 0.155 square inch.
I square millimeter = 0.00155 square inch.

I square mile = 2.5899 square kilometers.
I acre = 0.4047 hectare = 40.47 ares.
I square yard = 0.836 square meter.
I square foot = 0.0929 square meter = 929 square centimeters.
I square inch = 6.452 square centimeters = 645.2 square millimeters.

Cubic Measure

I cubic meter = 35.315 cubic feet = 1.308 cubic yards.
I cubic meter = 264.2 U.S. gallons.
I cubic centimeter = 0.061 cubic inch.
I liter (cubic decimeter) = 0.0353 cubic foot = 61.023 cubic inches.
I liter = 0.2642 U.S. gallon = 1.0567 U.S. quarts.

I cubic yard = 0.7646 cubic meter.
I cubic foot = 0.02832 cubic meter = 28.317 liters.
I cubic inch = 16.38706 cubic centimeters.
I U.S. gallon = 3.785 liters.
I U.S. quart = 0.946 liter.

Weight

I metric ton = 0.9842 ton (of 2240 pounds) = 2204.6 pounds.
I kilogram = 2.2046 pounds = 35.274 ounces avoirdupois.
I gram = 0.03215 ounce troy = 0.03527 ounce avoirdupois.
I gram = 15.432 grains.

I ton (of 2240 pounds) = 1.016 metric ton = 1016 kilograms.
I pound = 0.4536 kilogram = 453.6 grams.
I ounce avoirdupois = 28.35 grams.
I ounce troy = 31.103 grams.
I grain = 0.0648 gram.

I kilogram per square millimeter = 1422.32 pounds per square inch.
I kilogram per square centimeter = 14.223 pounds per square inch.
I kilogram-meter = 7.233 foot-pounds.
I pound per square inch = 0.0703 kilogram per square centimeter.
I calorie (kilogram calorie) = 3.968 Btu (British thermal unit).

Use of Conversion Tables. — On this and following pages tables are given which permit conversion from English to metric units and vice versa over a wide range of values. Where the desired value cannot be obtained directly from these tables, a simple addition of two or more values taken directly from the table will suffice as shown in the following examples:

Example 1: Find the millimeter equivalent of 0.4476 inch.

$$
\begin{aligned}
.4 \quad &\text{in.} = 10.16000 \text{ mm} \\
.04 \quad &\text{in.} = 1.01600 \text{ mm} \\
.007 \quad &\text{in.} = .17780 \text{ mm} \\
\underline{.0006} \quad &\text{in.} = \underline{.01524 \text{ mm}} \\
.4476 \quad &\text{in.} = 11.36904 \text{ mm}
\end{aligned}
$$

Example 2: Find the inch equivalent of 84.9 mm.

$$
\begin{aligned}
80. \quad &\text{mm} = 3.14961 \text{ in.} \\
4. \quad &\text{mm} = 0.15748 \text{ in.} \\
\underline{0.9} \quad &\text{mm} = \underline{0.03543 \text{ in.}} \\
84.9 \quad &\text{mm} = 3.34252 \text{ in.}
\end{aligned}
$$

Inch—Millimeter and Inch—Centimeter Conversion Table*
(Based on 1 inch = 25.4 millimeters, exactly)

INCHES TO MILLIMETERS											
in.	mm	in.	mm	in.	mm	in.	mm	in.	mm	in.	mm
10	254.00000	1	25.40000	.1	2.54000	.01	.25400	.001	.02540	.0001	.00254
20	508.00000	2	50.80000	.2	5.08000	.02	.50800	.002	.05080	.0002	.00508
30	762.00000	3	76.20000	.3	7.62000	.03	.76200	.003	.07620	.0003	.00762
40	1,016.00000	4	101.60000	.4	10.16000	.04	1.01600	.004	.10160	.0004	.01016
50	1,270.00000	5	127.00000	.5	12.70000	.05	1.27000	.005	.12700	.0005	.01270
60	1,524.00000	6	152.40000	.6	15.24000	.06	1.52400	.006	.15240	.0006	.01524
70	1,778.00000	7	177.80000	.7	17.78000	.07	1.77800	.007	.17780	.0007	.01778
80	2,032.00000	8	203.20000	.8	20.32000	.08	2.03200	.008	.20320	.0008	.02032
90	2,286.00000	9	228.60000	.9	22.86000	.09	2.28600	.009	.22860	.0009	.02286
100	2,540.00000	10	254.00000	1.0	25.40000	.10	2.54000	.010	.25400	.0010	.02540

MILLIMETERS TO INCHES											
mm	in.	mm	in.	mm.	in.	mm	in.	mm	in.	mm	in.
100	3.93701	10	.39370	1	.03937	.1	.00394	.01	.00039	.001	.00004
200	7.87402	20	.78740	2	.07874	.2	.00787	.02	.00079	.002	.00008
300	11.81102	30	1.18110	3	.11811	.3	.01181	.03	.00118	.003	.00012
400	15.74803	40	1.57480	4	.15748	.4	.01575	.04	.00157	.004	.00016
500	19.68504	50	1.96850	5	.19685	.5	.01969	.05	.00197	.005	.00020
600	23.62205	60	2.36220	6	.23622	.6	.02362	.06	.00236	.006	.00024
700	27.55906	70	2.75591	7	.27559	.7	.02756	.07	.00276	.007	.00028
800	31.49606	80	3.14961	8	.31496	.8	.03150	.08	.00315	.008	.00031
900	35.43307	90	3.54331	9	.35433	.9	.03543	.09	.00354	.009	.00035
1,000	39.37008	100	3.93701	10	.39370	1.0	.03937	.10	.00394	.010	.00039

* For inches to centimeters, shift decimal point in mm column one place to left and read centimeters, thus:

$$40 \text{ in.} = 1016 \text{ mm} = 101.6 \text{ cm}$$

For centimeters to inches, shift decimal point of centimeter value one place to right and enter mm column, thus:

$$70 \text{ cm} = 700 \text{ mm} = 27.55906 \text{ inches}$$

Decimals of an Inch to Millimeters
(Based on 1 inch = 25.4 millimeters, exactly)

Inches	0.000	0.001	0.002	0.003	0.004	0.005	0.006	0.007	0.008	0.009
	Millimeters									
0.000	...	0.0254	0.0508	0.0762	0.1016	0.1270	0.1524	0.1778	0.2032	0.2286
0.010	0.2540	0.2794	0.3048	0.3302	0.3556	0.3810	0.4064	0.4318	0.4572	0.4826
0.020	0.5080	0.5334	0.5588	0.5842	0.6096	0.6350	0.6604	0.6858	0.7112	0.7366
0.030	0.7620	0.7874	0.8128	0.8382	0.8636	0.8890	0.9144	0.9398	0.9652	0.9906
0.040	1.0160	1.0414	1.0668	1.0922	1.1176	1.1430	1.1684	1.1938	1.2192	1.2446
0.050	1.2700	1.2954	1.3208	1.3462	1.3716	1.3970	1.4224	1.4478	1.4732	1.4986
0.060	1.5240	1.5494	1.5748	1.6002	1.6256	1.6510	1.6764	1.7018	1.7272	1.7526
0.070	1.7780	1.8034	1.8288	1.8542	1.8796	1.9050	1.9304	1.9558	1.9812	2.0066
0.080	2.0320	2.0574	2.0828	2.1082	2.1336	2.1590	2.1844	2.2098	2.2352	2.2606
0.090	2.2860	2.3114	2.3368	2.3622	2.3876	2.4130	2.4384	2.4638	2.4892	2.5146
0.100	2.5400	2.5654	2.5908	2.6162	2.6416	2.6670	2.6924	2.7178	2.7432	2.7686
0.110	2.7940	2.8194	2.8448	2.8702	2.8956	2.9210	2.9464	2.9718	2.9972	3.0226
0.120	3.0480	3.0734	3.0988	3.1242	3.1496	3.1750	3.2004	3.2258	3.2512	3.2766
0.130	3.3020	3.3274	3.3528	3.3782	3.4036	3.4290	3.4544	3.4798	3.5052	3.5306
0.140	3.5560	3.5814	3.6068	3.6322	3.6576	3.6830	3.7084	3.7338	3.7592	3.7846
0.150	3.8100	3.8354	3.8608	3.8862	3.9116	3.9370	3.9624	3.9878	4.0132	4.0386
0.160	4.0640	4.0894	4.1148	4.1402	4.1656	4.1910	4.2164	4.2418	4.2672	4.2926
0.170	4.3180	4.3434	4.3688	4.3942	4.4196	4.4450	4.4704	4.4958	4.5212	4.5466
0.180	4.5720	4.5974	4.6228	4.6482	4.6736	4.6990	4.7244	4.7498	4.7752	4.8006
0.190	4.8260	4.8514	4.8768	4.9022	4.9276	4.9530	4.9784	5.0038	5.0292	5.0546
0.200	5.0800	5.1054	5.1308	5.1562	5.1816	5.2070	5.2324	5.2578	5.2832	5.3086
0.210	5.3340	5.3594	5.3848	5.4102	5.4356	5.4610	5.4864	5.5118	5.5372	5.5626
0.220	5.5880	5.6134	5.6388	5.6642	5.6896	5.7150	5.7404	5.7658	5.7912	5.8166
0.230	5.8420	5.8674	5.8928	5.9182	5.9436	5.9690	5.9944	6.0198	6.0452	6.0706
0.240	6.0960	6.1214	6.1468	6.1722	6.1976	6.2230	6.2484	6.2738	6.2992	6.3246
0.250	6.3500	6.3754	6.4008	6.4262	6.4516	6.4770	6.5024	6.5278	6.5532	6.5786
0.260	6.6040	6.6294	6.6548	6.6802	6.7056	6.7310	6.7564	6.7818	6.8072	6.8326
0.270	6.8580	6.8834	6.9088	6.9342	6.9596	6.9850	7.0104	7.0358	7.0612	7.0866
0.280	7.1120	7.1374	7.1628	7.1882	7.2136	7.2390	7.2644	7.2898	7.3152	7.3406
0.290	7.3660	7.3914	7.4168	7.4422	7.4676	7.4930	7.5184	7.5438	7.5692	7.5946
0.300	7.6200	7.6454	7.6708	7.6962	7.7216	7.7470	7.7724	7.7978	7.8232	7.8486
0.310	7.8740	7.8994	7.9248	7.9502	7.9756	8.0010	8.0264	8.0518	8.0772	8.1026
0.320	8.1280	8.1534	8.1788	8.2042	8.2296	8.2550	8.2804	8.3058	8.3312	8.3566
0.330	8.3820	8.4074	8.4328	8.4582	8.4836	8.5090	8.5344	8.5598	8.5852	8.6106
0.340	8.6360	8.6614	8.6868	8.7122	8.7376	8.7630	8.7884	8.8138	8.8392	8.8646
0.350	8.8900	8.9154	8.9408	8.9662	8.9916	9.0170	9.0424	9.0678	9.0932	9.1186
0.360	9.1440	9.1694	9.1948	9.2202	9.2456	9.2710	9.2964	9.3218	9.3472	9.3726
0.370	9.3980	9.4234	9.4488	9.4742	9.4996	9.5250	9.5504	9.5758	9.6012	9.6266
0.380	9.6520	9.6774	9.7028	9.7282	9.7536	9.7790	9.8044	9.8298	9.8552	9.8806
0.390	9.9060	9.9314	9.9568	9.9822	10.0076	10.0330	10.0584	10.0838	10.1092	10.1346
0.400	10.1600	10.1854	10.2108	10.2362	10.2616	10.2870	10.3124	10.3378	10.3632	10.3886
0.410	10.4140	10.4394	10.4648	10.4902	10.5156	10.5410	10.5664	10.5918	10.6172	10.6426
0.420	10.6680	10.6934	10.7188	10.7442	10.7696	10.7950	10.8204	10.8458	10.8712	10.8966
0.430	10.9220	10.9474	10.9728	10.9982	11.0236	11.0490	11.0744	11.0998	11.1252	11.1506
0.440	11.1760	11.2014	11.2268	11.2522	11.2776	11.3030	11.3284	11.3538	11.3792	11.4046
0.450	11.4300	11.4554	11.4808	11.5062	11.5316	11.5570	11.5824	11.6078	11.6332	11.6586
0.460	11.6840	11.7094	11.7348	11.7602	11.7856	11.8110	11.8364	11.8618	11.8872	11.9126
0.470	11.9380	11.9634	11.9888	12.0142	12.0396	12.0650	12.0904	12.1158	12.1412	12.1666
0.480	12.1920	12.2174	12.2428	12.2682	12.2936	12.3190	12.3444	12.3698	12.3952	12.4206
0.490	12.4460	12.4714	12.4968	12.5222	12.5476	12.5730	12.5984	12.6238	12.6492	12.6746

Use previous table to obtain whole inch equivalents to add to decimal equivalents above. All values given in this table are exact; figures to the right of the last place figures are all zeros.

Decimals of an Inch to Millimeters
(Based on 1 inch = 25.4 millimeters, exactly)

Inches	0.000	0.001	0.002	0.003	0.004	0.005	0.006	0.007	0.008	0.009
	Millimeters									
0.500	12.7000	12.7254	12.7508	12.7762	12.8016	12.8270	12.8524	12.8778	12.9032	12.9286
0.510	12.9540	12.9794	13.0048	13.0302	13.0556	13.0810	13.1064	13.1318	13.1572	13.1826
0.520	13.2080	13.2334	13.2588	13.2842	13.3096	13.3350	13.3604	13.3858	13.4112	13.4366
0.530	13.4620	13.4874	13.5128	13.5382	13.5636	13.5890	13.6144	13.6398	13.6652	13.6906
0.540	13.7160	13.7414	13.7668	13.7922	13.8176	13.8430	13.8684	13.8938	13.9192	13.9446
0.550	13.9700	13.9954	14.0208	14.0462	14.0716	14.0970	14.1224	14.1478	14.1732	14.1986
0.560	14.2240	14.2494	14.2748	14.3002	14.3256	14.3510	14.3764	14.4018	14.4272	14.4526
0.570	14.4780	14.5034	14.5288	14.5542	14.5796	14.6050	14.6304	14.6558	14.6812	14.7066
0.580	14.7320	14.7574	14.7828	14.8082	14.8336	14.8590	14.8844	14.9098	14.9352	14.9606
0.590	14.9860	15.0114	15.0368	15.0622	15.0876	15.1130	15.1384	15.1638	15.1892	15.2146
0.600	15.2400	15.2654	15.2908	15.3162	15.3416	15.3670	15.3924	15.4178	15.4432	15.4686
0.610	15.4940	15.5194	15.5448	15.5702	15.5956	15.6210	15.6464	15.6718	15.6972	15.7226
0.620	15.7480	15.7734	15.7988	15.8242	15.8496	15.8750	15.9004	15.9258	15.9512	15.9766
0.630	16.0020	16.0274	16.0528	16.0782	16.1036	16.1290	16.1544	16.1798	16.2052	16.2306
0.640	16.2560	16.2814	16.3068	16.3322	16.3576	16.3830	16.4084	16.4338	16.4592	16.4846
0.650	16.5100	16.5354	16.5608	16.5862	16.6116	16.6370	16.6624	16.6878	16.7132	16.7386
0.660	16.7640	16.7894	16.8148	16.8402	16.8656	16.8910	16.9164	16.9418	16.9672	16.9926
0.670	17.0180	17.0434	17.0688	17.0942	17.1196	17.1450	17.1704	17.1958	17.2212	17.2466
0.680	17.2720	17.2974	17.3228	17.3482	17.3736	17.3990	17.4244	17.4498	17.4752	17.5006
0.690	17.5260	17.5514	17.5768	17.6022	17.6276	17.6530	17.6784	17.7038	17.7292	17.7546
0.700	17.7800	17.8054	17.8308	17.8562	17.8816	17.9070	17.9324	17.9578	17.9832	18.0086
0.710	18.0340	18.0594	18.0848	18.1102	18.1356	18.1610	18.1864	18.2118	18.2372	18.2626
0.720	18.2880	18.3134	18.3388	18.3642	18.3896	18.4150	18.4404	18.4658	18.4912	18.5166
0.730	18.5420	18.5674	18.5928	18.6182	18.6436	18.6690	18.6944	18.7198	18.7452	18.7706
0.740	18.7960	18.8214	18.8468	18.8722	18.8976	18.9230	18.9484	18.9738	18.9992	19.0246
0.750	19.0500	19.0754	19.1008	19.1262	19.1516	19.1770	19.2024	19.2278	19.2532	19.2786
0.760	19.3040	19.3294	19.3548	19.3802	19.4056	19.4310	19.4564	19.4818	19.5072	19.5326
0.770	19.5580	19.5834	19.6088	19.6342	19.6596	19.6850	19.7104	19.7358	19.7612	19.7866
0.780	19.8120	19.8374	19.8628	19.8882	19.9136	19.9390	19.9644	19.9898	20.0152	20.0406
0.790	20.0660	20.0914	20.1168	20.1422	20.1676	20.1930	20.2184	20.2438	20.2692	20.2946
0.800	20.3200	20.3454	20.3708	20.3962	20.4216	20.4470	20.4724	20.4978	20.5232	20.5486
0.810	20.5740	20.5994	20.6248	20.6502	20.6756	20.7010	20.7264	20.7518	20.7772	20.8026
0.820	20.8280	20.8534	20.8788	20.9042	20.9296	20.9550	20.9804	21.0058	21.0312	21.0566
0.830	21.0820	21.1074	21.1328	21.1582	21.1836	21.2090	21.2344	21.2598	21.2852	21.3106
0.840	21.3360	21.3614	21.3868	21.4122	21.4376	21.4630	21.4884	21.5138	21.5392	21.5646
0.850	21.5900	21.6154	21.6408	21.6662	21.6916	21.7170	21.7424	21.7678	21.7932	21.8186
0.860	21.8440	21.8694	21.8948	21.9202	21.9456	21.9710	21.9964	22.0218	22.0472	22.0726
0.870	22.0980	22.1234	22.1488	22.1742	22.1996	22.2250	22.2504	22.2758	22.3012	22.3266
0.880	22.3520	22.3774	22.4028	22.4282	22.4536	22.4790	22.5044	22.5298	22.5552	22.5806
0.890	22.6060	22.6314	22.6568	22.6822	22.7076	22.7330	22.7584	22.7838	22.8092	22.8346
0.900	22.8600	22.8854	22.9108	22.9362	22.9616	22.9870	23.0124	23.0378	23.0632	23.0886
0.910	23.1140	23.1394	23.1648	23.1902	23.2156	23.2410	23.2664	23.2918	23.3172	23.3426
0.920	23.3680	23.3934	23.4188	23.4442	23.4696	23.4950	23.5204	23.5458	23.5712	23.5966
0.930	23.6220	23.6474	23.6728	23.6982	23.7236	23.7490	23.7744	23.7998	23.8252	23.8506
0.940	23.8760	23.9014	23.9268	23.9522	23.9776	24.0030	24.0284	24.0538	24.0792	24.1046
0.950	24.1300	24.1554	24.1808	24.2062	24.2316	24.2570	24.2824	24.3078	24.3332	24.3586
0.960	24.3840	24.4094	24.4348	24.4602	24.4856	24.5110	24.5364	24.5618	24.5872	24.6126
0.970	24.6380	24.6634	24.6888	24.7142	24.7396	24.7650	24.7904	24.8158	24.8412	24.8666
0.980	24.8920	24.9174	24.9428	24.9682	24.9936	25.0190	25.0444	25.0698	25.0952	25.1206
0.990	25.1460	25.1714	25.1968	25.2222	25.2476	25.2730	25.2984	25.3238	25.3492	25.3746
1.000	25.4000	...	...	...	...	...	...	...	...	...

Use previous table to obtain whole inch equivalents to add to decimal equivalents above. All values given in this table are exact; figures to the right of the last place figures are all zeros.

Millimeters to Inches
(Based on 1 inch = 25.4 millimeters, exactly)

Milli-meters	0	1	2	3	4	5	6	7	8	9
					Inches					
0	...	0.03937	0.07874	0.11811	0.15748	0.19685	0.23622	0.27559	0.31496	0.35433
10	0.39370	0.43307	0.47244	0.51181	0.55118	0.59055	0.62992	0.66929	0.70866	0.74803
20	0.78740	0.82677	0.86614	0.90551	0.94488	0.98425	1.02362	1.06299	1.10236	1.14173
30	1.18110	1.22047	1.25984	1.29921	1.33858	1.37795	1.41732	1.45669	1.49606	1.53543
40	1.57480	1.61417	1.65354	1.69291	1.73228	1.77165	1.81102	1.85039	1.88976	1.92913
50	1.96850	2.00787	2.04724	2.08661	2.12598	2.16535	2.20472	2.24409	2.28346	2.32283
60	2.36220	2.40157	2.44094	2.48031	2.51969	2.55906	2.59843	2.63780	2.67717	2.71654
70	2.75591	2.79528	2.83465	2.87402	2.91339	2.95276	2.99213	3.03150	3.07087	3.11024
80	3.14961	3.18898	3.22835	3.26772	3.30709	3.34646	3.38583	3.42520	3.46457	3.50394
90	3.54331	3.58268	3.62205	3.66142	3.70079	3.74016	3.77953	3.81890	3.85827	3.89764
100	3.93701	3.97638	4.01575	4.05512	4.09449	4.13386	4.17323	4.21260	4.25197	4.29134
110	4.33071	4.37008	4.40945	4.44882	4.48819	4.52756	4.56693	4.60630	4.64567	4.68504
120	4.72441	4.76378	4.80315	4.84252	4.88189	4.92126	4.96063	5.00000	5.03937	5.07874
130	5.11811	5.15748	5.19685	5.23622	5.27559	5.31496	5.35433	5.39370	5.43307	5.47244
140	5.51181	5.55118	5.59055	5.62992	5.66929	5.70866	5.74803	5.78740	5.82677	5.86614
150	5.90551	5.94488	5.98425	6.02362	6.06299	6.10236	6.14173	6.18110	6.22047	6.25984
160	6.29921	6.33858	6.37795	6.41732	6.45669	6.49606	6.53543	6.57480	6.61417	6.65354
170	6.69291	6.73228	6.77165	6.81102	6.85039	6.88976	6.92913	6.96850	7.00787	7.04724
180	7.08661	7.12598	7.16535	7.20472	7.24409	7.28346	7.32283	7.36220	7.40157	7.44094
190	7.48031	7.51969	7.55906	7.59843	7.63780	7.67717	7.71654	7.75591	7.79528	7.83465
200	7.87402	7.91339	7.95276	7.99213	8.03150	8.07087	8.11024	8.14961	8.18898	8.22835
210	8.26772	8.30709	8.34646	8.38583	8.42520	8.46457	8.50394	8.54331	8.58268	8.62205
220	8.66142	8.70079	8.74016	8.77953	8.81890	8.85827	8.89764	8.93701	8.97638	9.01575
230	9.05512	9.09449	9.13386	9.17323	9.21260	9.25197	9.29134	9.33071	9.37008	9.40945
240	9.44882	9.48819	9.52756	9.56693	9.60630	9.64567	9.68504	9.72441	9.76378	9.80315
250	9.84252	9.88189	9.92126	9.96063	10.0000	10.0394	10.0787	10.1181	10.1575	10.1969
260	10.2362	10.2756	10.3150	10.3543	10.3937	10.4331	10.4724	10.5118	10.5512	10.5906
270	10.6299	10.6693	10.7087	10.7480	10.7874	10.8268	10.8661	10.9055	10.9449	10.9843
280	11.0236	11.0630	11.1024	11.1417	11.1811	11.2205	11.2598	11.2992	11.3386	11.3780
290	11.4173	11.4567	11.4961	11.5354	11.5748	11.6142	11.6535	11.6929	11.7323	11.7717
300	11.8110	11.8504	11.8898	11.9291	11.9685	12.0079	12.0472	12.0866	12.1260	12.1654
310	12.2047	12.2441	12.2835	12.3228	12.3622	12.4016	12.4409	12.4803	12.5197	12.5591
320	12.5984	12.6378	12.6772	12.7165	12.7559	12.7953	12.8346	12.8740	12.9134	12.9528
330	12.9921	13.0315	13.0709	13.1102	13.1496	13.1890	13.2283	13.2677	13.3071	13.3465
340	13.3858	13.4252	13.4646	13.5039	13.5433	13.5827	13.6220	13.6614	13.7008	13.7402
350	13.7795	13.8189	13.8583	13.8976	13.9370	13.9764	14.0157	14.0551	14.0945	14.1339
360	14.1732	14.2126	14.2520	14.2913	14.3307	14.3701	14.4094	14.4488	14.4882	14.5276
370	14.5669	14.6063	14.6457	14.6850	14.7244	14.7638	14.8031	14.8425	14.8819	14.9213
380	14.9606	15.0000	15.0394	15.0787	15.1181	15.1575	15.1969	15.2362	15.2756	15.3150
390	15.3543	15.3937	15.4331	15.4724	15.5118	15.5512	15.5906	15.6299	15.6693	15.7087
400	15.7480	15.7874	15.8268	15.8661	15.9055	15.9449	15.9843	16.0236	16.0630	16.1024
410	16.1417	16.1811	16.2205	16.2598	16.2992	16.3386	16.3780	16.4173	16.4567	16.4961
420	16.5354	16.5748	16.6142	16.6535	16.6929	16.7323	16.7717	16.8110	16.8504	16.8898
430	16.9291	16.9685	17.0079	17.0472	17.0866	17.1260	17.1654	17.2047	17.2441	17.2835
440	17.3228	17.3622	17.4016	17.4409	17.4803	17.5197	17.5591	17.5984	17.6378	17.6772
450	17.7165	17.7559	17.7953	17.8346	17.8740	17.9134	17.9528	17.9921	18.0315	18.0709
460	18.1102	18.1496	18.1890	18.2283	18.2677	18.3071	18.3465	18.3858	18.4252	18.4646
470	18.5039	18.5433	18.5827	18.6220	18.6614	18.7008	18.7402	18.7795	18.8189	18.8583
480	18.8976	18.9370	18.9764	19.0157	19.0551	19.0945	19.1339	19.1732	19.2126	19.2520
490	19.2913	19.3307	19.3701	19.4094	19.4488	19.4882	19.5276	19.5669	19.6063	19.6457

Millimeters to Inches
(Based on 1 inch = 25.4 millimeters, exactly)

Milli-meters	0	1	2	3	4	5	6	7	8	9
					Inches					
500	19.6850	19.7244	19.7638	19.8031	19.8425	19.8819	19.9213	19.9606	20.0000	20.0394
510	20.0787	20.1181	20.1575	20.1969	20.2362	20.2756	20.3150	20.3543	20.3937	20.4331
520	20.4724	20.5118	20.5512	20.5906	20.6299	20.6693	20.7087	20.7480	20.7874	20.8268
530	20.8661	20.9055	20.9449	20.9843	21.0236	21.0630	21.1024	21.1417	21.1811	21.2205
540	21.2598	21.2992	21.3386	21.3780	21.4173	21.4567	21.4961	21.5354	21.5748	21.6142
550	21.6535	21.6929	21.7323	21.7717	21.8110	21.8504	21.8898	21.9291	21.9685	22.0079
560	22.0472	22.0866	22.1260	22.1654	22.2047	22.2441	22.2835	22.3228	22.3622	22.4016
570	22.4409	22.4803	22.5197	22.5591	22.5984	22.6378	22.6772	22.7165	22.7559	22.7953
580	22.8346	22.8740	22.9134	22.9528	22.9921	23.0315	23.0709	23.1102	23.1496	23.1890
590	23.2283	23.2677	23.3071	23.3465	23.3858	23.4252	23.4646	23.5039	23.5433	23.5827
600	23.6220	23.6614	23.7008	23.7402	23.7795	23.8189	23.8583	23.8976	23.9370	23.9764
610	24.0157	24.0551	24.0945	24.1339	24.1732	24.2126	24.2520	24.2913	24.3307	24.3701
620	24.4094	24.4488	24.4882	24.5276	24.5669	24.6063	24.6457	24.6850	24.7244	24.7638
630	24.8031	24.8425	24.8819	24.9213	24.9606	25.0000	25.0394	25.0787	25.1181	25.1575
640	25.1969	25.2362	25.2756	25.3150	25.3543	25.3937	25.4331	25.4724	25.5118	25.5512
650	25.5906	25.6299	25.6693	25.7087	25.7480	25.7874	25.8268	25.8661	25.9055	25.9449
660	25.9843	26.0236	26.0630	26.1024	26.1417	26.1811	26.2205	26.2598	26.2992	26.3386
670	26.3780	26.4173	26.4567	26.4961	26.5354	26.5748	26.6142	26.6535	26.6929	26.7323
680	26.7717	26.8110	26.8504	26.8898	26.9291	26.9685	27.0079	27.0472	27.0866	27.1260
690	27.1654	27.2047	27.2441	27.2835	27.3228	27.3622	27.4016	27.4409	27.4803	27.5197
700	27.5591	27.5984	27.6378	27.6772	27.7165	27.7559	27.7953	27.8346	27.8740	27.9134
710	27.9528	27.9921	28.0315	28.0709	28.1102	28.1496	28.1890	28.2283	28.2677	28.3071
720	28.3465	28.3858	28.4252	28.4646	28.5039	28.5433	28.5827	28.6220	28.6614	28.7008
730	28.7402	28.7795	28.8189	28.8583	28.8976	28.9370	28.9764	29.0157	29.0551	29.0945
740	29.1339	29.1732	29.2126	29.2520	29.2913	29.3307	29.3701	29.4094	29.4488	29.4882
750	29.5276	29.5669	29.6063	29.6457	29.6850	29.7244	29.7638	29.8031	29.8425	29.8819
760	29.9213	29.9606	30.0000	30.0394	30.0787	30.1181	30.1575	30.1969	30.2362	30.2756
770	30.3150	30.3543	30.3937	30.4331	30.4724	30.5118	30.5512	30.5906	30.6299	30.6693
780	30.7087	30.7480	30.7874	30.8268	30.8661	30.9055	30.9449	30.9843	31.0236	31.0630
790	31.1024	31.1417	31.1811	31.2205	31.2598	31.2992	31.3386	31.3780	31.4173	31.4567
800	31.4961	31.5354	31.5748	31.6142	31.6535	31.6929	31.7323	31.7717	31.8110	31.8504
810	31.8898	31.9291	31.9685	32.0079	32.0472	32.0866	32.1260	32.1654	32.2047	32.2441
820	32.2835	32.3228	32.3622	32.4016	32.4409	32.4803	32.5197	32.5591	32.5984	32.6378
830	32.6772	32.7165	32.7559	32.7953	32.8346	32.8740	32.9134	32.9528	32.9921	33.0315
840	33.0709	33.1102	33.1496	33.1890	33.2283	33.2677	33.3071	33.3465	33.3858	33.4252
850	33.4646	33.5039	33.5433	33.5827	33.6220	33.6614	33.7008	33.7402	33.7795	33.8189
860	33.8583	33.8976	33.9370	33.9764	34.0157	34.0551	34.0945	34.1339	34.1732	34.2126
870	34.2520	34.2913	34.3307	34.3701	34.4094	34.4488	34.4882	34.5276	34.5669	34.6063
880	34.6457	34.6850	34.7244	34.7638	34.8031	34.8425	34.8819	34.9213	34.9606	35.0000
890	35.0394	35.0787	35.1181	35.1575	35.1969	35.2362	35.2756	35.3150	35.3543	35.3937
900	35.4331	35.4724	35.5118	35.5512	35.5906	35.6299	35.6693	35.7087	35.7480	35.7874
910	35.8268	35.8661	35.9055	35.9449	35.9843	36.0236	36.0630	36.1024	36.1417	36.1811
920	36.2205	36.2598	36.2992	36.3386	36.3780	36.4173	36.4567	36.4961	36.5354	36.5748
930	36.6142	36.6535	36.6929	36.7323	36.7717	36.8110	36.8504	36.8898	36.9291	36.9685
940	37.0079	37.0472	37.0866	37.1260	37.1654	37.2047	37.2441	37.2835	37.3228	37.3622
950	37.4016	37.4409	37.4803	37.5197	37.5591	37.5984	37.6378	37.6772	37.7165	37.7559
960	37.7953	37.8346	37.8740	37.9134	37.9528	37.9921	38.0315	38.0709	38.1102	38.1496
970	38.1890	38.2283	38.2677	38.3071	38.3465	38.3858	38.4252	38.4646	38.5039	38.5433
980	38.5827	38.6220	38.6614	38.7008	38.7402	38.7795	38.8189	38.8583	38.8976	38.9370
990	38.9764	39.0157	39.0551	39.0945	39.1339	39.1732	39.2126	39.2520	39.2913	39.3307
1000	39.3701	...	...	...	...	...	...	...	...	...

Fractional Inch—Millimeter and Foot—Millimeter Conversion Tables

(Based on 1 inch = 25.4 millimeters, exactly)

FRACTIONAL INCH TO MILLIMETERS

in.	mm	in.	mm	in.	mm	in.	mm
1/64	0.397	17/64	6.747	33/64	13.097	49/64	19.447
1/32	0.794	9/32	7.144	17/32	13.494	25/32	19.844
3/64	1.191	19/64	7.541	35/64	13.891	51/64	20.241
1/16	1.588	5/16	7.938	9/16	14.288	13/16	20.638
5/64	1.984	21/64	8.334	37/64	14.684	53/64	21.034
3/32	2.381	11/32	8.731	19/32	15.081	27/32	21.431
7/64	2.778	23/64	9.128	39/64	15.478	55/64	21.828
1/8	3.175	3/8	9.525	5/8	15.875	7/8	22.225
9/64	3.572	25/64	9.922	41/64	16.272	57/64	22.622
5/32	3.969	13/32	10.319	21/32	16.669	29/32	23.019
11/64	4.366	27/64	10.716	43/64	17.066	59/64	23.416
3/16	4.762	7/16	11.112	11/16	17.462	15/16	23.812
13/64	5.159	29/64	11.509	45/64	17.859	61/64	24.209
7/32	5.556	15/32	11.906	23/32	18.256	31/32	24.606
15/64	5.953	31/64	12.303	47/64	18.653	63/64	25.003
1/4	6.350	1/2	12.700	3/4	19.050	1	25.400

INCHES TO MILLIMETERS

in.	mm	in.	mm	in.	mm	in.	mm	in.	mm	in.	mm
1	25.4	3	76.2	5	127.0	7	177.8	9	228.6	11	279.4
2	50.8	4	101.6	6	152.4	8	203.2	10	254.0	12	304.8

FEET TO MILLIMETERS

ft	mm	ft	mm	ft	mm	ft	mm	ft	mm
100	30,480	10	3,048	1	304.8	0.1	30.48	0.01	3.048
200	60,960	20	6,096	2	609.6	0.2	60.96	0.02	6.096
300	91,440	30	9,144	3	914.4	0.3	91.44	0.03	9.144
400	121,920	40	12,192	4	1,219.2	0.4	121.92	0.04	12.192
500	152,400	50	15,240	5	1,524.0	0.5	152.40	0.05	15.240
600	182,880	60	18,288	6	1,828.8	0.6	182.88	0.06	18.288
700	213,360	70	21,336	7	2,133.6	0.7	213.36	0.07	21.336
800	243,840	80	24,384	8	2,438.4	0.8	243.84	0.08	24.384
900	274,320	90	27,432	9	2,743.2	0.9	274.32	0.09	27.432
1,000	304,800	100	30,480	10	3,048.0	1.0	304.80	0.10	30.480

Example 1: Find millimeter equivalent of 293 feet, 5 47/64 inches.

$$
\begin{aligned}
200 \text{ ft} &= 60{,}960. \text{ mm} \\
90 \text{ ft} &= 27{,}432. \text{ mm} \\
3 \text{ ft} &= 914.4 \text{ mm} \\
5 \text{ in.} &= 127.0 \text{ mm} \\
47/64 \text{ in.} &= 18.653 \text{ mm} \\
\hline
293 \text{ ft, } 5\,47/64 \text{ in.} &= 89{,}452.053 \text{ mm}
\end{aligned}
$$

Example 2: Find millimeter equivalent of 71.86 feet.

$$
\begin{aligned}
70. \text{ ft} &= 21{,}336. \text{ mm} \\
1. \text{ ft} &= 304.8 \text{ mm} \\
.80 \text{ ft} &= 243.84 \text{ mm} \\
.06 \text{ ft} &= 18.288 \text{ mm} \\
\hline
71.86 \text{ ft} &= 21{,}902.928 \text{ mm}
\end{aligned}
$$

Microinches to Micrometers (microns)
(Based on 1 microinch = 0.0254 micrometers, exactly)

Micro-inches	0	1	2	3	4	5	6	7	8	9
					Micrometers (microns)					
0	.....	0.025	0.051	0.076	0.102	0.127	0.152	0.178	0.203	0.229
10	0.254	0.279	0.305	0.330	0.356	0.381	0.406	0.432	0.457	0.483
20	0.508	0.533	0.559	0.584	0.610	0.635	0.660	0.686	0.711	0.737
30	0.762	0.787	0.813	0.838	0.864	0.889	0.914	0.940	0.965	0.991
40	1.016	1.041	1.067	1.092	1.118	1.143	1.168	1.194	1.219	1.245
50	1.270	1.295	1.321	1.346	1.372	1.397	1.422	1.448	1.473	1.499
60	1.524	1.549	1.575	1.600	1.626	1.651	1.676	1.702	1.727	1.753
70	1.778	1.803	1.829	1.854	1.880	1.905	1.930	1.956	1.981	2.007
80	2.032	2.057	2.083	2.108	2.134	2.159	2.184	2.210	2.235	2.261
90	2.286	2.311	2.337	2.362	2.388	2.413	2.438	2.464	2.489	2.515
100	2.540	2.565	2.591	2.616	2.642	2.667	2.692	2.718	2.743	2.769
110	2.794	2.819	2.845	2.870	2.896	2.921	2.946	2.972	2.997	3.023
120	3.048	3.073	3.099	3.124	3.150	3.175	3.200	3.226	3.251	3.277
130	3.302	3.327	3.353	3.378	3.404	3.429	3.454	3.480	3.505	3.531
140	3.556	3.581	3.607	3.632	3.658	3.683	3.708	3.734	3.759	3.785
150	3.810	3.835	3.861	3.886	3.912	3.937	3.962	3.988	4.013	4.039
160	4.064	4.089	4.115	4.140	4.166	4.191	4.216	4.242	4.267	4.293
170	4.318	4.343	4.369	4.394	4.420	4.445	4.470	4.496	4.521	4.547
180	4.572	4.597	4.623	4.648	4.674	4.699	4.724	4.750	4.775	4.801
190	4.826	4.851	4.877	4.902	4.928	4.953	4.978	5.004	5.029	5.055
200	5.080	5.105	5.131	5.156	5.182	5.207	5.232	5.258	5.283	5.309
210	5.334	5.359	5.385	5.410	5.436	5.461	5.486	5.512	5.537	5.563
220	5.588	5.613	5.639	5.664	5.690	5.715	5.740	5.766	5.791	5.817
230	5.842	5.867	5.893	5.918	5.944	5.969	5.994	6.020	6.045	6.071
240	6.096	6.121	6.147	6.172	6.198	6.223	6.248	6.274	6.299	6.325
250	6.350	6.375	6.401	6.426	6.452	6.477	6.502	6.528	6.553	6.579
260	6.604	6.629	6.655	6.680	6.706	6.731	6.756	6.782	6.807	6.833
270	6.858	6.883	6.909	6.934	6.960	6.985	7.010	7.036	7.061	7.087
280	7.112	7.137	7.163	7.188	7.214	7.239	7.264	7.290	7.315	7.341
290	7.366	7.391	7.417	7.442	7.468	7.493	7.518	7.544	7.569	7.595

The following short table permits conversion of microinches to micrometers for ranges higher than in the main table given above. Appropriate quantities chosen from both tables are simply added to obtain the higher converted value:

μin.	μm	μin.	μm	μin.	μm	μin.	μm	μin.	μm
300	7.620	900	22.860	1500	38.100	2100	53.340	2700	68.580
600	15.240	1200	30.480	1800	45.720	2400	60.960	3000	76.200

Example: Convert 1375 μin. to μm:
From above table: 1200 μin. = 30.480 μm
From main table: 175 μin. = 4.445 μm
1375 μin. = 34.925 μm

Micrometers (microns) to Microinches — 1
(Based on 1 microinch = 0.0254 micrometers, exactly)

Micro-meters (microns)	0	0.01	0.02	0.03	0.04	0.05	0.06	0.07	0.08	0.09
					Microinches					
0	....	0.4	0.8	1.2	1.6	2.0	2.4	2.8	3.1	3.5
0.10	3.9	4.3	4.7	5.1	5.5	5.9	6.3	6.7	7.1	7.5
0.20	7.9	8.3	8.7	9.1	9.4	9.8	10.2	10.6	11.0	11.4
0.30	11.8	12.2	12.6	13.0	13.4	13.8	14.2	14.6	15.0	15.4
0.40	15.7	16.1	16.5	16.9	17.3	17.7	18.1	18.5	18.9	19.3

Micrometers (microns) to Microinches — 2

Micrometers (microns)	0	0.01	0.02	0.03	0.04	0.05	0.06	0.07	0.08	0.09
	Microinches									
0.50	19.7	20.1	20.5	20.9	21.3	21.7	22.0	22.4	22.8	23.2
0.60	23.6	24.0	24.4	24.8	25.2	25.6	26.0	26.4	26.8	27.2
0.70	27.6	28.0	28.3	28.7	29.1	29.5	29.9	30.3	30.7	31.1
0.80	31.5	31.9	32.3	32.7	33.1	33.5	33.9	34.3	34.6	35.0
0.90	35.4	35.8	36.2	36.6	37.0	37.4	37.8	38.2	38.6	39.0
1.00	39.4	39.8	40.2	40.6	40.9	41.3	41.7	42.1	42.5	42.9
1.10	43.3	43.7	44.1	44.5	44.9	45.3	45.7	46.1	46.5	46.9
1.20	47.2	47.6	48.0	48.4	48.8	49.2	49.6	50.0	50.4	50.8
1.30	51.2	51.6	52.0	52.4	52.8	53.1	53.5	53.9	54.3	54.7
1.40	55.1	55.5	55.9	56.3	56.7	57.1	57.5	57.9	58.3	58.7
1.50	59.1	59.4	59.8	60.2	60.6	61.0	61.4	61.8	62.2	62.6
1.60	63.0	63.4	63.8	64.2	64.6	65.0	65.4	65.7	66.1	66.5
1.70	66.9	67.3	67.7	68.1	68.5	68.9	69.3	69.7	70.1	70.5
1.80	70.9	71.3	71.7	72.0	72.4	72.8	73.2	73.6	74.0	74.4
1.90	74.8	75.2	75.6	76.0	76.4	76.8	77.2	77.6	78.0	78.3
2.00	78.7	79.1	79.5	79.9	80.3	80.7	81.1	81.5	81.9	82.3
2.10	82.7	83.1	83.5	83.9	84.3	84.6	85.0	85.4	85.8	86.2
2.20	86.6	87.0	87.4	87.8	88.2	88.6	89.0	89.4	89.8	90.2
2.30	90.6	90.9	91.3	91.7	92.1	92.5	92.9	93.3	93.7	94.1
2.40	94.5	94.9	95.3	95.7	96.1	96.5	96.9	97.2	97.6	98.0
2.50	98.4	98.8	99.2	99.6	100.0	100.4	100.8	101.2	101.6	102.0
2.60	102.4	102.8	103.1	103.5	103.9	104.3	104.7	105.1	105.5	105.9
2.70	106.3	106.7	107.1	107.5	107.9	108.3	108.7	109.1	109.4	109.8
2.80	110.2	110.6	111.0	111.4	111.8	112.2	112.6	113.0	113.4	113.8
2.90	114.2	114.6	115.0	115.4	115.7	116.1	116.5	116.9	117.3	117.7
3.00	118.1	118.5	118.9	119.3	119.7	120.1	120.5	120.9	121.3	121.7
3.10	122.0	122.4	122.8	123.2	123.6	124.0	124.4	124.8	125.2	125.6
3.20	126.0	126.4	126.8	127.2	127.6	128.0	128.3	128.7	129.1	129.5
3.30	129.9	130.3	130.7	131.1	131.5	131.9	132.3	132.7	133.1	133.5
3.40	133.9	134.3	134.6	135.0	135.4	135.8	136.2	136.6	137.0	137.4
3.50	137.8	138.2	138.6	139.0	139.4	139.8	140.2	140.6	140.9	141.3
3.60	141.7	142.1	142.5	142.9	143.3	143.7	144.1	144.5	144.9	145.3
3.70	145.7	146.1	146.5	146.9	147.2	147.6	148.0	148.4	148.8	149.2
3.80	149.6	150.0	150.4	150.8	151.2	151.6	152.0	152.4	152.8	153.1
3.90	153.5	153.9	154.3	154.7	155.1	155.5	155.9	156.3	156.7	157.1
4.00	157.5	157.9	158.3	158.7	159.1	159.4	159.8	160.2	160.6	161.0
4.10	161.4	161.8	162.2	162.6	163.0	163.4	163.8	164.2	164.6	165.0
4.20	165.4	165.7	166.1	166.5	166.9	167.3	167.7	168.1	168.5	168.9
4.30	169.3	169.7	170.1	170.5	170.9	171.3	171.7	172.0	172.4	172.8
4.40	173.2	173.6	174.0	174.4	174.8	175.2	175.6	176.0	176.4	176.8
4.50	177.2	177.6	178.0	178.3	178.7	179.1	179.5	179.9	180.3	180.7
4.60	181.1	181.5	181.9	182.3	182.7	183.1	183.5	183.9	184.3	184.6
4.70	185.0	185.4	185.8	186.2	186.6	187.0	187.4	187.8	188.2	188.6
4.80	189.0	189.4	189.8	190.2	190.6	190.9	191.3	191.7	192.1	192.5
4.90	192.9	193.3	193.7	194.1	194.5	194.9	195.3	195.7	196.1	196.5
5.00	196.9	197.2	197.6	198.0	198.4	198.8	199.2	199.6	200.0	200.4

The table given below can be used with the preceding main table to obtain higher converted values, simply by adding appropriate quantities chosen from each table:

μm	μin.	μm	μin.	μm	μin.	μm	μin.	μm	μin.
10	393.7	20	787.4	30	1,181.1	40	1,574.8	50	1,968.5
15	590.6	25	984.3	35	1,378.0	45	1,771.7	55	2,165.4

Example: Convert 23.55 μm to μin.:
From above table: 20.00 μm = 787.4 μin.
From main table: 3.55 μm = 139.8 μin.
            23.55 μm = 927.2 μin.

Foot — Meter and Mile — Kilometer Conversion Tables
(Based on 1 foot = 0.3048 meter, exactly)

Feet to Meters
(1 ft = 0.3048 m, exactly)

feet	meters	feet	meters	feet	meters	feet	meters	feet	meters
100	30.480	10	3.048	1	0.305	0.1	0.030	0.01	0.003
200	60.960	20	6.096	2	0.610	0.2	0.061	0.02	0.006
300	91.440	30	9.144	3	0.914	0.3	0.091	0.03	0.009
400	121.920	40	12.192	4	1.219	0.4	0.122	0.04	0.012
500	152.400	50	15.240	5	1.524	0.5	0.152	0.05	0.015
600	182.880	60	18.288	6	1.829	0.6	0.183	0.06	0.018
700	213.360	70	21.336	7	2.134	0.7	0.213	0.07	0.021
800	243.840	80	24.384	8	2.438	0.8	0.244	0.08	0.024
900	274.320	90	27.432	9	2.743	0.9	0.274	0.09	0.027
1,000	304.800	100	30.480	10	3.048	1.0	0.305	0.10	0.030

Meters to Feet
(1 m = 3.280840 ft)

meters	feet	meters	feet	meters	feet	meters	feet	meters	feet
100	328.084	10	32.808	1	3.281	0.1	0.328	0.01	0.033
200	656.168	20	65.617	2	6.562	0.2	0.656	0.02	0.066
300	984.252	30	98.425	3	9.843	0.3	0.984	0.03	0.098
400	1,312.336	40	131.234	4	13.123	0.4	1.312	0.04	0.131
500	1,640.420	50	164.042	5	16.404	0.5	1.640	0.05	0.164
600	1,968.504	60	196.850	6	19.685	0.6	1.969	0.06	0.197
700	2,296.588	70	229.659	7	22.966	0.7	2.297	0.07	0.230
800	2,624.672	80	262.467	8	26.247	0.8	2.625	0.08	0.262
900	2,952.756	90	295.276	9	29.528	0.9	2.953	0.09	0.295
1,000	3,280.840	100	328.084	10	32.808	1.0	3.281	0.10	0.328

Miles to Kilometers
(1 mile = 1.609344 km, exactly)

miles	km	miles	km	miles	km	miles	km	miles	km
1,000	1,609.34	100	160.93	10	16.09	1	1.61	0.1	0.16
2,000	3,218.69	200	321.87	20	32.19	2	3.22	0.2	0.32
3,000	4,828.03	300	482.80	30	48.28	3	4.83	0.3	0.48
4,000	6,437.38	400	643.74	40	64.37	4	6.44	0.4	0.64
5,000	8,046.72	500	804.67	50	80.47	5	8.05	0.5	0.80
6,000	9,656.06	600	965.61	60	96.56	6	9.66	0.6	0.97
7,000	11,265.41	700	1,126.54	70	112.65	7	11.27	0.7	1.13
8,000	12,874.75	800	1,287.48	80	128.75	8	12.87	0.8	1.29
9,000	14,484.10	900	1,448.41	90	144.84	9	14.48	0.9	1.45
10,000	16,093.44	1,000	1,609.34	100	160.93	10	16.09	1.0	1.61

Kilometers to Miles
(1 km = 0.6213712 mile)

km	miles	km	miles	km	miles	km	miles	km	miles
1,000	621.37	100	62.14	10	6.21	1	0.62	0.1	0.06
2,000	1,242.74	200	124.27	20	12.43	2	1.24	0.2	0.12
3,000	1,864.11	300	186.41	30	18.64	3	1.86	0.3	0.19
4,000	2,485.48	400	248.55	40	24.85	4	2.49	0.4	0.25
5,000	3,106.86	500	310.69	50	31.07	5	3.11	0.5	0.31
6,000	3,728.23	600	372.82	60	37.28	6	3.73	0.6	0.37
7,000	4,349.60	700	434.96	70	43.50	7	4.35	0.7	0.43
8,000	4,970.97	800	497.10	80	49.71	8	4.97	0.8	0.50
9,000	5,592.34	900	559.23	90	55.92	9	5.59	0.9	0.56
10,000	6,213.71	1,000	621.37	100	62.14	10	6.21	1.0	0.62

Square Inch — Square Centimeter and Square Foot — Square Meter Conversion Tables
(Based on 1 inch = 2.54 centimeters, exactly)

Square Inches to Square Centimeters (1 in.² = 6.4516 cm², exactly)									
in.²	cm²	in.²	cm²	in.²	cm²	in.²	cm²	in.²	cm²
100	645.16	10	64.52	1	6.45	0.1	0.65	0.01	0.06
200	1,290.32	20	129.03	2	12.90	0.2	1.29	0.02	0.13
300	1,935.48	30	193.55	3	19.35	0.3	1.94	0.03	0.19
400	2,580.64	40	258.06	4	25.81	0.4	2.58	0.04	0.26
500	3,225.80	50	322.58	5	32.26	0.5	3.23	0.05	0.32
600	3,870.96	60	387.10	6	38.71	0.6	3.87	0.06	0.39
700	4,516.12	70	451.61	7	45.16	0.7	4.52	0.07	0.45
800	5,161.28	80	516.13	8	51.61	0.8	5.16	0.08	0.52
900	5,806.44	90	580.64	9	58.06	0.9	5.81	0.09	0.58
1,000	6,451.60	100	645.16	10	64.52	1.0	6.45	0.10	0.65

Square Centimeters to Square Inches (1 cm² = 0.1550003 in.²)									
cm²	in.²	cm²	in.²	cm²	in.²	cm²	in.²	cm²	in.²
100	15.500	10	1.550	1	0.155	0.1	0.016	0.01	0.002
200	31.000	20	3.100	2	0.310	0.2	0.031	0.02	0.003
300	46.500	30	4.650	3	0.465	0.3	0.047	0.03	0.005
400	62.000	40	6.200	4	0.620	0.4	0.062	0.04	0.006
500	77.500	50	7.750	5	0.775	0.5	0.078	0.05	0.008
600	93.000	60	9.300	6	0.930	0.6	0.093	0.06	0.009
700	108.500	70	10.850	7	1.085	0.7	0.109	0.07	0.011
800	124.000	80	12.400	8	1.240	0.8	0.124	0.08	0.012
900	139.500	90	13.950	9	1.395	0.9	0.140	0.09	0.014
1,000	155.000	100	15.500	10	1.550	1.0	0.155	0.10	0.016

Square Feet to Square Meters (1 ft² = 0.09290304 m², exactly)									
ft²	m²	ft²	m²	ft²	m²	ft²	m²	ft²	m²
1,000	92.903	100	9.290	10	0.929	1	0.093	0.1	0.009
2,000	185.806	200	18.581	20	1.858	2	0.186	0.2	0.019
3,000	278.709	300	27.871	30	2.787	3	0.279	0.3	0.028
4,000	371.612	400	37.161	40	3.716	4	0.372	0.4	0.037
5,000	464.515	500	46.452	50	4.645	5	0.465	0.5	0.046
6,000	557.418	600	55.742	60	5.574	6	0.557	0.6	0.056
7,000	650.321	700	65.032	70	6.503	7	0.650	0.7	0.065
8,000	743.224	800	74.322	80	7.432	8	0.743	0.8	0.074
9,000	836.127	900	83.613	90	8.361	9	0.836	0.9	0.084
10,000	929.030	1,000	92.903	100	9.290	10	0.929	1.0	0.093

Square Meters to Square Feet (1 m² = 10.76391 ft²)									
m²	ft²	m²	ft²	m²	ft²	m²	ft²	m²	ft²
100	1,076.39	10	107.64	1	10.76	0.1	1.08	0.01	0.11
200	2,152.78	20	215.28	2	21.53	0.2	2.15	0.02	0.22
300	3,229.17	30	322.92	3	32.29	0.3	3.23	0.03	0.32
400	4,305.56	40	430.56	4	43.06	0.4	4.31	0.04	0.43
500	5,381.96	50	538.20	5	53.82	0.5	5.38	0.05	0.54
600	6,458.35	60	645.83	6	64.58	0.6	6.46	0.06	0.65
700	7,534.74	70	753.47	7	75.35	0.7	7.53	0.07	0.75
800	8,611.13	80	861.11	8	86.11	0.8	8.61	0.08	0.86
900	9,687.52	90	968.75	9	96.88	0.9	9.69	0.09	0.97
1,000	10,763.91	100	1,076.39	10	107.64	1.0	10.76	0.10	1.08

Acre — Hectare and U.K. Gallon — Liter Conversion Tables

Acres to Hectares
(1 acre = 0.4046856 hectare)

acres	0	10	20	30	40	50	60	70	80	90
					hectares					
0	...	4.047	8.094	12.141	16.187	20.234	24.281	28.328	32.375	36.422
100	40.469	44.515	48.562	52.609	56.656	60.703	64.750	68.797	72.843	76.890
200	80.937	84.984	89.031	93.078	97.125	101.171	105.218	109.265	113.312	117.359
300	121.406	125.453	129.499	133.546	137.593	141.640	145.687	149.734	153.781	157.827
400	161.874	165.921	169.968	174.015	178.062	182.109	186.155	190.202	194.249	198.296
500	202.343	206.390	210.437	214.483	218.530	222.577	226.624	230.671	234.718	238.765
600	242.811	246.858	250.905	254.952	258.999	263.046	267.092	271.139	275.186	279.233
700	283.280	287.327	291.374	295.420	299.467	303.514	307.561	311.608	315.655	319.702
800	323.748	327.795	331.842	335.889	339.936	343.983	348.030	352.076	356.123	360.170
900	364.217	368.264	372.311	376.358	380.404	384.451	388.498	392.545	396.592	400.639
1000	404.686	...	...	...	...	...	...	...	...	...

Hectares to Acres
(1 hectare = 2.471054 acres)

hectares	0	10	20	30	40	50	60	70	80	90
					acres					
0	...	24.71	49.42	74.13	98.84	123.55	148.26	172.97	197.68	222.39
100	247.11	271.82	296.53	321.24	345.95	370.66	395.37	420.08	444.79	469.50
200	494.21	518.92	543.63	568.34	593.05	617.76	642.47	667.18	691.90	716.61
300	741.32	766.03	790.74	815.45	840.16	864.87	889.58	914.29	939.00	963.71
400	988.42	1013.13	1037.84	1062.55	1087.26	1111.97	1136.68	1161.40	1186.11	1210.82
500	1235.53	1260.24	1284.95	1309.66	1334.37	1359.08	1383.79	1408.50	1433.21	1457.92
600	1482.63	1507.34	1532.05	1556.76	1581.47	1606.19	1630.90	1655.61	1680.32	1705.03
700	1729.74	1754.45	1779.16	1803.87	1828.58	1853.29	1878.00	1902.71	1927.42	1952.13
800	1976.84	2001.55	2026.26	2050.97	2075.69	2100.40	2125.11	2149.82	2174.53	2199.24
900	2223.95	2248.66	2273.37	2298.08	2322.79	2347.50	2372.21	2396.92	2421.63	2446.34
1000	2471.05	...	...	...	...	...	...	...	...	...

U.K. Gallons to Liters
(1 U.K. gallon = 4.546092 liters)

Imp. gals	0	1	2	3	4	5	6	7	8	9
					liters					
0	...	4.546	9.092	13.638	18.184	22.730	27.277	31.823	36.369	40.915
10	45.461	50.007	54.553	59.099	63.645	68.191	72.737	77.284	81.830	86.376
20	90.922	95.468	100.014	104.560	109.106	113.652	118.198	122.744	127.291	131.837
30	136.383	140.929	145.475	150.021	154.567	159.113	163.659	168.205	172.751	177.298
40	181.844	186.390	190.936	195.482	200.028	204.574	209.120	213.666	218.212	222.759
50	227.305	231.851	236.397	240.943	245.489	250.035	254.581	259.127	263.673	268.219
60	272.766	277.312	281.858	286.404	290.950	295.496	300.042	304.588	309.134	313.680
70	318.226	322.773	327.319	331.865	336.411	340.957	345.503	350.049	354.595	359.141
80	363.687	368.233	372.780	377.326	381.872	386.418	390.964	395.510	400.056	404.602
90	409.148	413.694	418.240	422.787	427.333	431.879	436.425	440.971	445.517	450.063
100	454.609	459.155	463.701	468.247	472.794	477.340	481.886	486.432	490.978	495.524

Liters to U.K. Gallons
(1 liter = 0.2199692 U.K. gallons)

liters	0	1	2	3	4	5	6	7	8	9
					Imperial gallons					
0	...	0.220	0.440	0.660	0.880	1.100	1.320	1.540	1.760	1.980
10	2.200	2.420	2.640	2.860	3.080	3.300	3.520	3.739	3.959	4.179
20	4.399	4.619	4.839	5.059	5.279	5.499	5.719	5.939	6.159	6.379
30	6.599	6.819	7.039	7.259	7.479	7.699	7.919	8.139	8.359	8.579
40	8.799	9.019	9.239	9.459	9.679	9.899	10.119	10.339	10.559	10.778
50	10.998	11.218	11.438	11.658	11.878	12.098	12.318	12.538	12.758	12.978
60	13.198	13.418	13.638	13.858	14.078	14.298	14.518	14.738	14.958	15.178
70	15.398	15.618	15.838	16.058	16.278	16.498	16.718	16.938	17.158	17.378
80	17.598	17.818	18.037	18.257	18.477	18.697	18.917	19.137	19.357	19.577
90	19.797	20.017	20.237	20.457	20.677	20.897	21.117	21.337	21.557	21.777
100	21.997	22.217	22.437	22.657	22.877	23.097	23.317	23.537	23.757	23.977

Cubic Inch — Cubic Centimeter and Cubic Foot — Cubic Meter Conversion Tables
(Based on 1 inch = 2.54 centimeters, exactly)

Cubic Inches to Cubic Centimeters
(1 in.^{3} = 16.38706 cm^3)

in.^{3}	cm^3	in.^{3}	cm^3	in.^{3}	cm^3	in.^{3}	cm^3	in.^{3}	cm^3
100	1,638.71	10	163.87	1	16.39	0.1	1.64	0.01	0.16
200	3,277.41	20	327.74	2	32.77	0.2	3.28	0.02	0.33
300	4,916.12	30	491.61	3	49.16	0.3	4.92	0.03	0.49
400	6,554.82	40	655.48	4	65.55	0.4	6.55	0.04	0.66
500	8,193.53	50	819.35	5	81.94	0.5	8.19	0.05	0.82
600	9,832.24	60	983.22	6	98.32	0.6	9.83	0.06	0.98
700	11,470.94	70	1,147.09	7	114.71	0.7	11.47	0.07	1.15
800	13,109.65	80	1,310.96	8	131.10	0.8	13.11	0.08	1.31
900	14,748.35	90	1,474.84	9	147.48	0.9	14.75	0.09	1.47
1,000	16,387.06	100	1,638.71	10	163.87	1.0	16.39	0.10	1.64

Cubic Centimeters to Cubic Inches
(1 cm^3 = 0.06102376 in.^{3})

cm^3	in.^{3}	cm^3	in.^{3}	cm^3	in.^{3}	cm^3	in.^{3}	cm^3	in.^{3}
1,000	61.024	100	6.102	10	0.610	1	0.061	0.1	0.006
2,000	122.048	200	12.205	20	1.220	2	0.122	0.2	0.012
3,000	183.071	300	18.307	30	1.831	3	0.183	0.3	0.018
4,000	244.095	400	24.410	40	2.441	4	0.244	0.4	0.024
5,000	305.119	500	30.512	50	3.051	5	0.305	0.5	0.031
6,000	366.143	600	36.614	60	3.661	6	0.366	0.6	0.037
7,000	427.166	700	42.717	70	4.272	7	0.427	0.7	0.043
8,000	488.190	800	48.819	80	4.882	8	0.488	0.8	0.049
9,000	549.214	900	54.921	90	5.492	9	0.549	0.9	0.055
10,000	610.238	1,000	61.024	100	6.102	10	0.610	1.0	0.061

Cubic Feet to Cubic Meters
(1 ft^3 = 0.02831685 m^3)

ft^3	m^3	ft^3	m^3	ft^3	m^3	ft^3	m^3	ft^3	m^3
1,000	28.317	100	2.832	10	0.283	1	0.028	0.1	0.003
2,000	56.634	200	5.663	20	0.566	2	0.057	0.2	0.006
3,000	84.951	300	8.495	30	0.850	3	0.085	0.3	0.008
4,000	113.267	400	11.327	40	1.133	4	0.113	0.4	0.011
5,000	141.584	500	14.158	50	1.416	5	0.142	0.5	0.014
6,000	169.901	600	16.990	60	1.699	6	0.170	0.6	0.017
7,000	198.218	700	19.822	70	1.982	7	0.198	0.7	0.020
8,000	226.535	800	22.653	80	2.265	8	0.227	0.8	0.023
9,000	254.852	900	25.485	90	2.549	9	0.255	0.9	0.025
10,000	283.168	1,000	28.317	100	2.832	10	0.283	1.0	0.028

Cubic Meters to Cubic Feet
(1 m^3 = 35.31466 ft^3)

m^3	ft^3	m^3	ft^3	m^3	ft^3	m^3	ft^3	m^3	ft^3
100	3,531.47	10	353.15	1	35.31	0.1	3.53	0.01	0.35
200	7,062.93	20	706.29	2	70.63	0.2	7.06	0.02	0.71
300	10,594.40	30	1,059.44	3	105.94	0.3	10.59	0.03	1.06
400	14,125.86	40	1,412.59	4	141.26	0.4	14.13	0.04	1.41
500	17,657.33	50	1,765.73	5	176.57	0.5	17.66	0.05	1.77
600	21,188.80	60	2,118.88	6	211.89	0.6	21.19	0.06	2.12
700	24,720.26	70	2,472.03	7	247.20	0.7	24.72	0.07	2.47
800	28,251.73	80	2,825.17	8	282.52	0.8	28.25	0.08	2.83
900	31,783.19	90	3,178.32	9	317.83	0.9	31.78	0.09	3.18
1,000	35,314.66	100	3,531.47	10	353.15	1.0	35.31	0.10	3.53

Cubic Foot — Liter and Gallon — Liter Conversion Tables
(Based on 1 liter = 1000 cubic centimeters)

Cubic Feet to Liters
($1 \text{ ft}^3 = 28.31685$ liters)

ft³	liters	ft³	liters	ft³	liters	ft³	liters	ft³	liters
100	2,831.68	10	283.17	1	28.32	0.1	2.83	0.01	0.28
200	5,663.37	20	566.34	2	56.63	0.2	5.66	0.02	0.57
300	8,495.06	30	849.51	3	84.95	0.3	8.50	0.03	0.85
400	11,326.74	40	1,132.67	4	113.27	0.4	11.33	0.04	1.13
500	14,158.42	50	1,415.84	5	141.58	0.5	14.16	0.05	1.42
600	16,990.11	60	1,699.01	6	169.90	0.6	16.99	0.06	1.70
700	19,821.80	70	1,982.18	7	198.22	0.7	19.82	0.07	1.98
800	22,653.48	80	2,263.35	8	226.53	0.8	22.65	0.08	2.27
900	25,485.16	90	2,548.52	9	254.85	0.9	25.49	0.09	2.55
1,000	28,316.85	100	2,831.68	10	283.17	1.0	28.32	0.10	2.83

Liters to Cubic Feet
($1 \text{ liter} = 0.03531466 \text{ ft}^3$)

liters	ft³	liters	ft³	liters	ft³	liters	ft³	liters	ft³
1,000	35.315	100	3.531	10	0.353	1	0.035	0.1	0.004
2,000	70.629	200	7.063	20	0.706	2	0.071	0.2	0.007
3,000	105.944	300	10.594	30	1.059	3	0.106	0.3	0.011
4,000	141.259	400	14.126	40	1.413	4	0.141	0.4	0.014
5,000	176.573	500	17.657	50	1.766	5	0.177	0.5	0.018
6,000	211.888	600	21.189	60	2.119	6	0.212	0.6	0.021
7,000	247.203	700	24.720	70	2.472	7	0.247	0.7	0.025
8,000	282.517	800	28.252	80	2.825	8	0.283	0.8	0.028
9,000	317.832	900	31.783	90	3.178	9	0.318	0.9	0.032
10,000	353.147	1,000	35.315	100	3.531	10	0.353	1.0	0.035

U.S. Gallons to Liters
($1 \text{ U.S. gallon} = 3.785412$ liters)

gals	liters	gals	liters	gals	liters	gals	liters	gals	liters
1,000	3,785.41	100	378.54	10	37.85	1	3.79	0.1	0.38
2,000	7,570.82	200	757.08	20	75.71	2	7.57	0.2	0.76
3,000	11,356.24	300	1,135.62	30	113.56	3	11.36	0.3	1.14
4,000	15,141.65	400	1,514.16	40	151.42	4	15.14	0.4	1.51
5,000	18,927.06	500	1,892.71	50	189.27	5	18.93	0.5	1.89
6,000	22,712.47	600	2,271.25	60	227.12	6	22.71	0.6	2.27
7,000	26,497.88	700	2,649.79	70	264.98	7	26.50	0.7	2.65
8,000	30,283.30	800	3,028.33	80	302.83	8	30.28	0.8	3.03
9,000	34,068.71	900	3,406.87	90	340.69	9	34.07	0.9	3.41
10,000	37,854.12	1,000	3,785.41	100	378.54	10	37.85	1.0	3.79

Liters to U.S. Gallons
($1 \text{ liter} = 0.2641720 \text{ U.S. gallon}$)

liters	gals	liters	gals	liters	gals	liters	gals	liters	gals
1,000	264.17	100	26.42	10	2.64	1	0.26	0.1	0.03
2,000	528.34	200	52.83	20	5.28	2	0.53	0.2	0.05
3,000	792.52	300	79.25	30	7.93	3	0.79	0.3	0.08
4,000	1,056.69	400	105.67	40	10.57	4	1.06	0.4	0.11
5,000	1,320.86	500	132.09	50	13.21	5	1.32	0.5	0.13
6,000	1,585.03	600	158.50	60	15.85	6	1.59	0.6	0.16
7,000	1,849.20	700	184.92	70	18.49	7	1.85	0.7	0.18
8,000	2,113.38	800	211.34	80	21.13	8	2.11	0.8	0.21
9,000	2,377.55	900	237.75	90	23.78	9	2.38	0.9	0.24
10,000	2,641.72	1,000	264.17	100	26.42	10	2.64	1.0	0.26

Pound — Kilogram and Ounce — Gram
Conversion Tables

Pounds to Kilograms
(1 pound = 0.4535924 kilogram)

lb	kg	lb	kg	lb	kg	lb	kg	lb	kg
1,000	453.59	100	45.36	10	4.54	1	0.45	0.1	0.05
2,000	907.18	200	90.72	20	9.07	2	0.91	0.2	0.09
3,000	1,360.78	300	136.08	30	13.61	3	1.36	0.3	0.14
4,000	1,814.37	400	181.44	40	18.14	4	1.81	0.4	0.18
5,000	2,267.96	500	226.80	50	22.68	5	2.27	0.5	0.23
6,000	2,721.55	600	272.16	60	27.22	6	2.72	0.6	0.27
7,000	3,175.15	700	317.51	70	31.75	7	3.18	0.7	0.32
8,000	3,628.74	800	362.87	80	36.29	8	3.63	0.8	0.36
9,000	4,082.33	900	408.23	90	40.82	9	4.08	0.9	0.41
10,000	4,535.92	1,000	453.59	100	45.36	10	4.54	1.0	0.45

Kilograms to Pounds
(1 kilogram = 2.204622 pounds)

kg	lb	kg	lb	kg	lb	kg	lb	kg	lb
1,000	2,204.62	100	220.46	10	22.05	1	2.20	0.1	0.22
2,000	4,409.24	200	440.92	20	44.09	2	4.41	0.2	0.44
3,000	6,613.87	300	661.39	30	66.14	3	6.61	0.3	0.66
4,000	8,818.49	400	881.85	40	88.18	4	8.82	0.4	0.88
5,000	11,023.11	500	1,102.31	50	110.23	5	11.02	0.5	1.10
6,000	13,227.73	600	1,322.77	60	132.28	6	13.23	0.6	1.32
7,000	15,432.35	700	1,543.24	70	154.32	7	15.43	0.7	1.54
8,000	17,636.98	800	1,763.70	80	176.37	8	17.64	0.8	1.76
9,000	19,841.60	900	1,984.16	90	198.42	9	19.84	0.9	1.98
10,000	22,046.22	1,000	2,204.62	100	220.46	10	22.05	1.0	2.20

Ounces to Grams
(1 ounce = 28.34952 grams)

oz	g	oz	g	oz	g	oz	g	oz	g
10	283.50	1	28.35	0.1	2.83	0.01	0.28	0.001	0.03
20	566.99	2	56.70	0.2	5.67	0.02	0.57	0.002	0.06
30	850.49	3	85.05	0.3	8.50	0.03	0.85	0.003	0.09
40	1,133.98	4	113.40	0.4	11.34	0.04	1.13	0.004	0.11
50	1,417.48	5	141.75	0.5	14.17	0.05	1.42	0.005	0.14
60	1,700.97	6	170.10	0.6	17.01	0.06	1.70	0.006	0.17
70	1,984.47	7	198.45	0.7	19.84	0.07	1.98	0.007	0.20
80	2,267.96	8	226.80	0.8	22.68	0.08	2.27	0.008	0.23
90	2,551.46	9	255.15	0.9	25.51	0.09	2.55	0.009	0.26
100	2,834.95	10	283.50	1.0	28.35	0.10	2.83	0.010	0.28

Grams to Ounces
(1 gram = 0.03527397 ounce)

g	oz	g	oz	g	oz	g	oz	g	oz
100	3.527	10	0.353	1	0.035	0.1	0.004	0.01	0.000
200	7.055	20	0.705	2	0.071	0.2	0.007	0.02	0.001
300	10.582	30	1.058	3	0.106	0.3	0.011	0.03	0.001
400	14.110	40	1.411	4	0.141	0.4	0.014	0.04	0.001
500	17.637	50	1.764	5	0.176	0.5	0.018	0.05	0.002
600	21.164	60	2.116	6	0.212	0.6	0.021	0.06	0.002
700	24.692	70	2.469	7	0.247	0.7	0.025	0.07	0.002
800	28.219	80	2.822	8	0.282	0.8	0.028	0.08	0.003
900	31.747	90	3.175	9	0.317	0.9	0.032	0.09	0.003
1,000	35.274	100	3.527	10	0.353	1.0	0.035	0.10	0.004

**Pounds per Square Inch —'Kilograms per Square Centimeter and
Pounds per Square Foot — Kilograms per Square Meter Conversion Tables**

Pounds per Square Inch to Kilograms per Square Centimeter
(1 lb/in.^{2} = 0.07030697 kg/cm^2)

lb/in.^{2}	kg/cm^2	lb/in.^{2}	kg/cm^2	lb/in.^{2}	kg/cm^2	lb/in.^{2}	kg/cm^2	lb/in.^{2}	kg/cm^2
1,000	70.307	100	7.031	10	0.703	1	0.070	0.1	0.007
2,000	140.614	200	14.061	20	1.406	2	0.141	0.2	0.014
3,000	210.921	300	21.092	30	2.109	3	0.211	0.3	0.021
4,000	281.228	400	28.123	40	2.812	4	0.281	0.4	0.028
5,000	351.535	500	35.153	50	3.515	5	0.352	0.5	0.035
6,000	421.842	600	42.184	60	4.218	6	0.422	0.6	0.042
7,000	492.149	700	49.215	70	4.921	7	0.492	0.7	0.049
8,000	562.456	800	56.246	80	5.625	8	0.562	0.8	0.056
9,000	632.763	900	63.276	90	6.328	9	0.633	0.9	0.063
10,000	703.070	1,000	70.307	100	7.031	10	0.703	1.0	0.070

Kilograms per Square Centimeter to Pounds per Square Inch
(1 kg/cm^2 = 14.22334 lb/in.^{2})

kg/cm^2	lb/in.^{2}	kg/cm^2	lb/in.^{2}	kg/cm^2	lb/in.^{2}	kg/cm^2	lb/in.^{2}	kg/cm^2	lb/in.^{2}
100	1,422.33	10	142.23	1	14.22	0.1	1.42	0.01	0.14
200	2,844.67	20	284.47	2	28.45	0.2	2.84	0.02	0.28
300	4,267.00	30	426.70	3	42.67	0.3	4.27	0.03	0.43
400	5,689.34	40	568.93	4	56.89	0.4	5.69	0.04	0.57
500	7,111.67	50	711.17	5	71.12	0.5	7.11	0.05	0.71
600	8,534.00	60	853.40	6	85.34	0.6	8.53	0.06	0.85
700	9,956.34	70	995.63	7	99.56	0.7	9.96	0.07	1.00
800	11,378.67	80	1,137.87	8	113.79	0.8	11.38	0.08	1.14
900	12,801.01	90	1,280.10	9	128.01	0.9	12.80	0.09	1.28
1,000	14,223.34	100	1,422.33	10	142.23	1.0	14.22	0.10	1.42

Pounds per Square Foot to Kilograms per Square Meter
(1 lb/ft^2 = 4.882429 kg/m^2)

lb/ft^2	kg/m^2	lb/ft^2	kg/m^2	lb/ft^2	kg/m^2	lb/ft^2	kg/m^2	lb/ft^2	kg/m^2
1,000	4,882.43	100	488.24	10	48.82	1	4.88	0.1	0.49
2,000	9,764.86	200	976.49	20	97.65	2	9.76	0.2	0.98
3,000	14,647.29	300	1,464.73	30	146.47	3	14.65	0.3	1.46
4,000	19,529.72	400	1,952.97	40	195.30	4	19.53	0.4	1.95
5,000	24,412.14	500	2,441.21	50	244.12	5	24.41	0.5	2.44
6,000	29,294.57	600	2,929.46	60	292.95	6	29.29	0.6	2.93
7,000	34,177.00	700	3,417.70	70	341.77	7	34.18	0.7	3.42
8,000	39,059.43	800	3,905.94	80	390.59	8	39.06	0.8	3.91
9,000	43,941.86	900	4,394.19	90	439.42	9	43.94	0.9	4.39
10,000	48,824.28	1,000	4,882.43	100	488.24	10	48.82	1.0	4.88

Kilograms per Square Meter to Pounds per Square Foot
(1 kg/m^2 = 0.2048161 lb/ft^2)

kg/m^2	lb/ft^2	kg/m^2	lb/ft^2	kg/m^2	lb/ft^2	kg/m^2	lb/ft^2	kg/m^2	lb/ft^2
1,000	204.82	100	20.48	10	2.05	1	0.20	0.1	0.02
2,000	409.63	200	40.96	20	4.10	2	0.41	0.2	0.04
3,000	614.45	300	61.44	30	6.14	3	0.61	0.3	0.06
4,000	819.26	400	81.93	40	8.19	4	0.82	0.4	0.08
5,000	1,024.08	500	102.41	50	10.24	5	1.02	0.5	0.10
6,000	1,228.90	600	122.89	60	12.29	6	1.23	0.6	0.12
7,000	1,433.71	700	143.37	70	14.34	7	1.43	0.7	0.14
8,000	1,638.53	800	163.85	80	16.39	8	1.64	0.8	0.16
9,000	1,843.34	900	184.33	90	18.43	9	1.84	0.9	0.18
10,000	2,048.16	1,000	204.82	100	20.48	10	2.05	1.0	0.20

Pounds per Square Inch — Kilopascals Conversion Table*

| | | | | | Pounds per Square Inch to Kilopascals | | | | | |
| | | | | | (1 lb/in² = 6.894757 kPa) | | | | | |
lb/in.²	0	1	2	3	4	5	6	7	8	9
						kilopascals				
0	...	6.895	13.790	20.684	27.579	34.474	41.369	48.263	55.158	62.053
10	68.948	75.842	82.737	89.632	96.527	103.421	110.316	117.211	124.106	131.000
20	137.895	144.790	151.685	158.579	165.474	172.369	179.264	186.158	193.053	199.948
30	206.843	213.737	220.632	227.527	234.422	241.316	248.211	255.106	262.001	268.896
40	275.790	282.685	289.580	296.475	303.369	310.264	317.159	324.054	330.948	337.843
50	344.738	351.633	358.527	365.422	372.317	379.212	386.106	393.001	399.896	406.791
60	413.685	420.580	427.475	434.370	441.264	448.159	455.054	461.949	468.843	475.738
70	482.633	489.528	496.423	503.317	510.212	517.107	524.002	530.896	537.791	544.686
80	551.581	558.475	565.370	572.265	579.160	586.054	592.949	599.844	606.739	613.633
90	620.528	627.423	634.318	641.212	648.107	655.002	661.897	668.791	675.686	682.581
100	689.476	696.370	703.265	710.160	717.055	723.949	730.844	737.739	744.634	751.529

| | | | | | Kilopascals to Pounds Per Square Inch | | | | | |
| | | | | | (1 kPa = 0.1450377 lb/in²) | | | | | |
kPa	0	1	2	3	4	5	6	7	8	9
						lb/in.²				
0	...	0.145	0.290	0.435	0.580	0.725	0.870	1.015	1.160	1.305
10	1.450	1.595	1.740	1.885	2.031	2.176	2.321	2.466	2.611	2.756
20	2.901	3.046	3.191	3.336	3.481	3.626	3.771	3.916	4.061	4.206
30	4.351	4.496	4.641	4.786	4.931	5.076	5.221	5.366	5.511	5.656
40	5.802	5.947	6.092	6.237	6.382	6.527	6.672	6.817	6.962	7.107
50	7.252	7.397	7.542	7.687	7.832	7.977	8.122	8.267	8.412	8.557
60	8.702	8.847	8.992	9.137	9.282	9.427	9.572	9.718	9.863	10.008
70	10.153	10.298	10.443	10.588	10.733	10.878	11.023	11.168	11.313	11.458
80	11.603	11.748	11.893	12.038	12.183	12.328	12.473	12.618	12.763	12.908
90	13.053	13.198	13.343	13.489	13.634	13.779	13.924	14.069	14.214	14.359
100	14.504	14.649	14.794	14.939	15.084	15.229	15.374	15.519	15.664	15.809

*Note: 1 kilopascal = 1 kilonewton/meter².

Power and Heat Equivalents

1 horsepower-hour = 0.746 kilowatt-hour = 1,980,000 foot-pounds = 2545 Btu (British thermal units) = 2.64 pounds of water evaporated at 212°F = 17 pounds of water raised from 62° to 212°F.

1 kilowatt-hour = 1000 watt-hours = 1.34 horsepower-hour = 2,655,200 foot-pounds = 3,600,000 joules = 3415 Btu = 3.54 pounds of water evaporated at 212°F = 22.8 pounds of water raised from 62° to 212°F.

1 horsepower = 746 watts = 0.746 kilowatt = 33,000 foot-pounds per minute = 550 foot-pounds per second = 2545 Btu per hour = 42.4 Btu per minute = 0.71 Btu per second = 2.64 lbs. of water evaporated per hour at 212°F.

1 kilowatt = 1000 watts = 1.34 horsepower = 2,655,200 foot-pounds per hour = 44,200 foot-pounds per minute = 737 foot-pounds per second = 3415 Btu per hour = 57 Btu per minute = 0.95 Btu per second = 3.54 pounds of water evaporated per hour at 212°F.

1 watt = 1 joule per second = 0.00134 horsepower = 0.001 kilowatt = 3.42 Btu per hour = 44.22 foot-pounds per minute = 0.74 foot-pounds per second = 0.0035 pound of water evaporated per hour at 212°F.

1 Btu (British thermal unit) = 1052 watt-seconds = 778 foot-pounds = 0.252 kilogram-calorie = 0.000292 kilowatt-hour = 0.000393, horsepower-hour = 0.00104 pound of water evaporated at 212°F.

1 foot-pound = 1.36 joules = 0.000000377 kilowatt-hour = 0.00129 Btu = 0.0000005 horsepower-hour.

1 joule = 1 watt-second = 0.000000278 kilowatt-hour = 0.00095 Btu = 0.74 foot-pound.

Pounds per Cubic Inch — Grams per Cubic Centimeter and Pounds per Cubic Foot — Kilograms per Cubic Meter Conversion Tables

Pounds per Cubic Inch to Grams per Cubic Centimeter
($1 \text{ lb/in.}^3 = 27.67990 \text{ g/cm}^3$)

lb/in.^{3}	g/cm^3	lb/in.^{3}	g/cm^3	lb/in.^{3}	g/cm^3	lb/in.^{3}	g/cm^3	lb/in.^{3}	g/cm^3
100	2,767.99	10	276.80	1	27.68	0.1	2.77	0.01	0.28
200	5,535.98	20	553.60	2	55.36	0.2	5.54	0.02	0.55
300	8,303.97	30	830.40	3	83.04	0.3	8.30	0.03	0.83
400	11,071.96	40	1,107.20	4	110.72	0.4	11.07	0.04	1.11
500	13,839.95	50	1,384.00	5	138.40	0.5	13.84	0.05	1.38
600	16,607.94	60	1,660.79	6	166.08	0.6	16.61	0.06	1.66
700	19,375.93	70	1,937.59	7	193.76	0.7	19.38	0.07	1.94
800	22,143.92	80	2,214.39	8	221.44	0.8	22.14	0.08	2.21
900	24,911.91	90	2,491.19	9	249.12	0.9	24.91	0.09	2.49
1,000	27,679.90	100	2,767.99	10	276.80	1.0	27.68	0.10	2.77

Grams per Cubic Centimeter to Pounds per Cubic Inch
($1 \text{ g/cm}^3 = 0.03612730 \text{ lb/in.}^3$)

g/cm^3	lb/in.^{3}	g/cm^3	lb/in.^{3}	g/cm^3	lb/in.^{3}	g/cm^3	lb/in.^{3}	g/cm^3	lb/in.^{3}
1,000	36.127	100	3.613	10	0.361	1	0.036	0.1	0.004
2,000	72.255	200	7.225	20	0.723	2	0.072	0.2	0.007
3,000	108.382	300	10.838	30	1.084	3	0.108	0.3	0.011
4,000	144.509	400	14.451	40	1.445	4	0.145	0.4	0.014
5,000	180.636	500	18.064	50	1.806	5	0.181	0.5	0.018
6,000	216.764	600	21.676	60	2.168	6	0.217	0.6	0.022
7,000	252.891	700	25.289	70	2.529	7	0.253	0.7	0.025
8,000	289.018	800	28.902	80	2.890	8	0.289	0.8	0.029
9,000	325.146	900	32.515	90	3.251	9	0.325	0.9	0.033
10,000	361.273	1,000	36.127	100	3.613	10	0.361	1.0	0.036

Pounds per Cubic Foot to Kilograms per Cubic Meter
($1 \text{ lb/ft}^3 = 16.01846 \text{ kg/m}^3$)

lb/ft^3	kg/m^3	lb/ft^3	kg/m^3	lb/ft^3	kg/m^3	lb/ft^3	kg/m^3	lb/ft^3	kg/m^3
100	1,601.85	10	160.18	1	16.02	0.1	1.60	0.01	0.16
200	3,203.69	20	320.37	2	32.04	0.2	3.20	0.02	0.32
300	4,805.54	30	480.55	3	48.06	0.3	4.81	0.03	0.48
400	6,407.38	40	640.74	4	64.07	0.4	6.41	0.04	0.64
500	8,009.23	50	800.92	5	80.09	0.5	8.01	0.05	0.80
600	9,611.08	60	961.11	6	96.11	0.6	9.61	0.06	0.96
700	11,212.92	70	1,121.29	7	112.13	0.7	11.21	0.07	1.12
800	12,814.77	80	1,281.48	8	128.15	0.8	12.81	0.08	1.28
900	14,416.61	90	1,441.66	9	144.17	0.9	14.42	0.09	1.44
1,000	16,018.46	100	1,601.85	10	160.18	1.0	16.02	0.10	1.60

Kilograms per Cubic Meter to Pounds per Cubic Foot
($1 \text{ kg/m}^3 = 0.06242797 \text{ lb/ft}^3$)

kg/m^3	lb/ft^3	kg/m^3	lb/ft^3	kg/m^3	lb/ft^3	kg/m^3	lb/ft^3	kg/m^3	lb/ft^3
1,000	62.428	100	6.243	10	0.624	1	0.062	0.1	0.006
2,000	124.856	200	12.486	20	1.249	2	0.125	0.2	0.012
3,000	187.284	300	18.728	30	1.873	3	0.187	0.3	0.019
4,000	249.712	400	24.971	40	2.497	4	0.250	0.4	0.025
5,000	312.140	500	31.214	50	3.121	5	0.312	0.5	0.031
6,000	374.568	600	37.457	60	3.746	6	0.375	0.6	0.037
7,000	436.996	700	43.700	70	4.370	7	0.437	0.7	0.044
8,000	499.424	800	49.942	80	4.994	8	0.499	0.8	0.050
9,000	561.852	900	56.185	90	5.619	9	0.562	0.9	0.056
10,000	624.280	1,000	62.428	100	6.243	10	0.624	1.0	0.062

British Thermal Unit — Foot-Pound and Horsepower — Kilowatt Conversion Tables

British Thermal Units to Foot-Pounds
(1 Btu = 778.26 ft.-lb.)*

Btu	Ft.-lb.	Btu	Ft.-lb.	Btu	Ft.-lb.	Btu	Ft.-lb.	Btu	Ft.-lb.
100	77,826	10	7,783	1	778	0.1	78	0.01	8
200	155,652	20	15,565	2	1,557	0.2	156	0.02	16
300	233,478	30	23,348	3	2,335	0.3	233	0.03	23
400	311,304	40	31,130	4	3,113	0.4	311	0.04	31
500	389,130	50	38,913	5	3,891	0.5	389	0.05	39
600	466,956	60	46,696	6	4,670	0.6	467	0.06	47
700	544,782	70	54,478	7	5,448	0.7	545	0.07	54
800	622,608	80	62,261	8	6,226	0.8	623	0.08	62
900	700,434	90	70,043	9	7,004	0.9	700	0.09	70
1,000	778,260	100	77,826	10	7,783	1.0	778	0.10	78

Foot-Pounds to British Thermal Units
(1 ft.-lb. = 0.00128492 Btu)*

Ft.-lb.	Btu	Ft.-lb.	Btu	Ft.-lb.	Btu	Ft.-lb.	Btu	Ft.-lb.	Btu
10,000	12.849	1,000	1.285	100	0.128	10	0.013	1	0.001
20,000	25.698	2,000	2.570	200	0.257	20	0.026	2	0.003
30,000	38.548	3,000	3.855	300	0.385	30	0.039	3	0.004
40,000	51.397	4,000	5.140	400	0.514	40	0.051	4	0.005
50,000	64.246	5,000	6.425	500	0.642	50	0.064	5	0.006
60,000	77.095	6,000	7.710	600	0.771	60	0.077	6	0.008
70,000	89.944	7,000	8.994	700	0.899	70	0.090	7	0.009
80,000	102.794	8,000	10.279	800	1.028	80	0.103	8	0.010
90,000	115.643	9,000	11.564	900	1.156	90	0.116	9	0.012
100,000	128.492	10,000	12.849	1,000	1.285	100	0.128	10	0.013

Horsepower to Kilowatts
(1 hp = 0.7456999 kW)†

hp	kW	hp	kW	hp	kW	hp	kW	hp	kW
1,000	745.7	100	74.6	10	7.5	1	0.7	0.1	0.07
2,000	1,491.4	200	149.1	20	14.9	2	1.5	0.2	0.15
3,000	2,237.1	300	223.7	30	22.4	3	2.2	0.3	0.22
4,000	2,982.8	400	298.3	40	29.8	4	3.0	0.4	0.30
5,000	3,728.5	500	372.8	50	37.3	5	3.7	0.5	0.37
6,000	4,474.2	600	447.4	60	44.7	6	4.5	0.6	0.45
7,000	5,219.9	700	522.0	70	52.2	7	5.2	0.7	0.52
8,000	5,965.6	800	596.6	80	59.7	8	6.0	0.8	0.60
9,000	6,711.3	900	671.1	90	67.1	9	6.7	0.9	0.67
10,000	7,457.0	1,000	745.7	100	74.6	10	7.5	1.0	0.75

Kilowatts to Horsepower
(1 kW = 1.341022 hp)†

kW	hp	kW	hp	kW	hp	kW	hp	kW	hp
1,000	1,341.0	100	134.1	10	13.4	1	1.3	0.1	0.13
2,000	2,682.0	200	268.2	20	26.8	2	2.7	0.2	0.27
3,000	4,023.1	300	402.3	30	40.2	3	4.0	0.3	0.40
4,000	5,364.1	400	536.4	40	53.6	4	5.4	0.4	0.54
5,000	6,705.1	500	670.5	50	67.1	5	6.7	0.5	0.67
6,000	8,046.1	600	804.6	60	80.5	6	8.0	0.6	0.80
7,000	9,387.2	700	938.7	70	93.9	7	9.4	0.7	0.94
8,000	10,728.2	800	1,072.8	80	107.3	8	10.7	0.8	1.07
9,000	12,069.2	900	1,206.9	90	120.7	9	12.1	0.9	1.21
10,000	13,410.2	1,000	1,341.0	100	134.1	10	13.4	1.0	1.34

*Conversion factor defined by International Steam Table Conference, 1929.
†Based on 1 horsepower = 550 foot-pounds per second.

British Thermal Unit — Kilojoule and Foot-Pound — Joule Conversion Tables

British Thermal Units to Kilojoules
(1 Btu = 1055.056 joules.)

Btu	0	100	200	300	400	500	600	700	800	900
					kilojoules					
0	...	105.51	211.01	316.52	422.02	527.53	633.03	738.54	844.04	949.55
1000	1055.06	1160.56	1266.07	1371.57	1477.08	1582.58	1688.09	1793.60	1899.10	2004.61
2000	2110.11	2215.62	2321.12	2426.63	2532.13	2637.64	2743.15	2848.65	2954.16	3059.66
3000	3165.17	3270.67	3376.18	3481.68	3587.19	3692.70	3798.20	3903.71	4009.21	4114.72
4000	4220.22	4325.73	4431.24	4536.74	4642.25	4747.75	4853.26	4958.76	5064.27	5169.77
5000	5275.28	5380.79	5486.29	5591.80	5697.30	5802.81	5908.31	6013.82	6119.32	6224.83
6000	6330.34	6435.84	6541.35	6646.85	6752.36	6857.86	6963.37	7068.88	7174.38	7279.89
7000	7385.39	7490.90	7596.40	7701.91	7807.41	7912.92	8018.43	8123.93	8229.44	8334.94
8000	8440.45	8545.95	8651.46	8756.96	8862.47	8967.98	9073.48	9178.99	9284.49	9390.00
9000	9495.50	9601.01	9706.52	9812.02	9917.53	10023.0	10128.5	10234.0	10339.5	10445.1
10000	10550.6	...	...	...	...	...	...	...	...	...

Kilojoules to British Thermal Units
1 joule = 0.0009478170 Btu.)

kJ	0	100	200	300	400	500	600	700	800	900
					British Thermal Units					
0	...	94.78	189.56	284.35	379.13	473.91	568.69	663.47	758.25	853.04
1000	947.82	1042.60	1137.38	1232.16	1326.94	1421.73	1516.51	1611.29	1706.07	1800.85
2000	1895.63	1990.42	2085.20	2179.98	2274.76	2369.54	2464.32	2559.11	2653.89	2748.67
3000	2843.45	2938.23	3033.01	3127.80	3222.58	3317.36	3412.14	3506.92	3601.70	3696.49
4000	3791.27	3886.05	3980.83	4075.61	4170.39	4265.18	4359.96	4454.74	4549.52	4644.30
5000	4739.08	4833.87	4928.65	5023.43	5118.21	5212.99	5307.78	5402.56	5497.34	5592.12
6000	5686.90	5781.68	5876.47	5971.25	6066.03	6160.81	6255.59	6350.37	6445.16	6539.94
7000	6634.72	6729.50	6824.28	6919.06	7013.85	7108.63	7203.41	7298.19	7392.97	7487.75
8000	7582.54	7677.32	7772.10	7866.88	7961.66	8056.44	8151.23	8246.01	8340.79	8435.57
9000	8530.35	8625.13	8719.92	8814.70	8909.48	9004.26	9099.04	9193.82	9288.61	9383.39
10000	9478.17	...	...	...	...	...	...	...	...	...

Foot-Pounds to Joules
(1 foot-pound = 1.355818 joules)

ft-lb	0	1	2	3	4	5	6	7	8	9
					joules					
0	...	1.356	2.712	4.067	5.423	6.779	8.135	9.491	10.847	12.202
10	13.558	14.914	16.270	17.626	18.981	20.337	21.693	23.049	24.405	25.761
20	27.116	28.472	29.828	31.184	32.540	33.895	35.251	36.607	37.963	39.319
30	40.675	42.030	43.386	44.742	46.098	47.454	48.809	50.165	51.521	52.877
40	54.233	55.589	56.944	58.300	59.656	61.012	62.368	63.723	65.079	66.435
50	67.791	69.147	70.503	71.858	73.214	74.570	75.926	77.282	78.637	79.993
60	81.349	82.705	84.061	85.417	86.772	88.128	89.484	90.840	92.196	93.551
70	94.907	96.263	97.619	98.975	100.331	101.686	103.042	104.398	105.754	107.110
80	108.465	109.821	111.177	112.533	113.889	115.245	116.600	117.956	119.312	120.668
90	122.024	123.379	124.735	126.091	127.447	128.803	130.159	131.514	132.870	134.226
100	135.582	136.938	138.293	139.649	141.005	142.361	143.717	145.073	146.428	147.784

Joules to Foot-Pounds
(1 joule = 0.7375621 foot-pound)

J	0	1	2	3	4	5	6	7	8	9
					foot-pounds					
0	...	0.7376	1.4751	2.2127	2.9502	3.6878	4.4254	5.1629	5.9005	6.6381
10	7.3756	8.1132	8.8507	9.5883	10.3259	11.0634	11.8010	12.5761	13.2761	14.0137
20	14.7512	15.4888	16.2264	16.9639	17.7015	18.4391	19.1766	19.9142	20.6517	21.3893
30	22.1269	22.8644	23.6020	24.3395	25.0771	25.8147	26.5522	27.2898	28.0274	28.7649
40	29.5025	30.2400	30.9776	31.7152	32.4527	33.1903	33.9279	34.6654	35.4030	36.1405
50	36.8781	37.6157	38.3532	39.0908	39.8284	40.5659	41.3035	42.0410	42.7786	43.5162
60	44.2537	44.9913	45.7289	46.4664	47.2040	47.9415	48.6791	49.4167	50.1542	50.8918
70	51.6293	52.3669	53.1045	53.8420	54.5796	55.3172	56.0547	56.7923	57.5298	58.2674
80	59.0050	59.7425	60.4801	61.2177	61.9552	62.6928	63.4303	64.1679	64.9055	65.6430
90	66.3806	67.1182	67.8557	68.5933	69.3308	70.0684	70.8060	71.5435	72.2811	73.0186
100	73.7562	74.4938	75.2313	75.9689	76.7065	77.4440	78.1816	78.9191	79.6567	80.3943

Pounds-force — Newtons and Pound-Inches to Newton-Meters Conversion Tables

Pounds-Force to Newtons
(1 pound-force = 4.448222 newtons.)

lbf	0	1	2	3	4	5	6	7	8	9
	newtons									
0	...	4.448	8.896	13.345	17.793	22.241	26.689	31.138	35.586	40.034
10	44.482	48.930	53.379	57.827	62.275	66.723	71.172	75.620	80.068	84.516
20	88.964	93.413	97.861	102.309	106.757	111.206	115.654	120.102	124.550	128.998
30	133.447	137.895	142.343	146.791	151.240	155.688	160.136	164.584	169.032	173.481
40	177.929	182.377	186.825	191.274	195.722	200.170	204.618	209.066	213.515	217.963
50	222.411	226.859	231.308	235.756	240.204	244.652	249.100	253.549	257.997	262.445
60	266.893	271.342	275.790	280.238	284.686	289.134	293.583	298.031	302.479	306.927
70	311.376	315.824	320.272	324.720	329.168	333.617	338.065	342.513	346.961	351.410
80	355.858	360.306	364.754	369.202	373.651	378.099	382.547	386.995	391.444	395.892
90	400.340	404.788	409.236	413.685	418.133	422.581	427.029	431.478	435.926	440.374
100	444.822	449.270	453.719	458.167	462.615	467.063	471.512	475.960	480.408	484.856

Newtons to Pounds-Force
(1 newton = 0.2248089 pound-force.)

N	0	1	2	3	4	5	6	7	8	9
	pounds-force									
0	...	0.22481	0.44962	0.67443	0.89924	1.12404	1.34885	1.57366	1.79847	2.02328
10	2.24809	2.47290	2.69771	2.92252	3.14732	3.37213	3.59694	3.82175	4.04656	4.27137
20	4.49618	4.72099	4.94580	5.17060	5.39541	5.62022	5.84503	6.06984	6.29465	6.51946
30	6.74427	6.96908	7.19388	7.41869	7.64350	7.86831	8.09312	8.31793	8.54274	8.76755
40	8.99236	9.21716	9.44197	9.66678	9.89159	10.1164	10.3412	10.5660	10.7908	11.0156
50	11.2404	11.4653	11.6901	11.9149	12.1397	12.3645	12.5893	12.8141	13.0389	13.2637
60	13.4885	13.7133	13.9382	14.1630	14.3878	14.6126	14.8374	15.0622	15.2870	15.5118
70	15.7366	15.9614	16.1862	16.4110	16.6359	16.8607	17.0855	17.3103	17.5351	17.7599
80	17.9847	18.2095	18.4343	18.6591	18.8839	19.1088	19.3336	19.5584	19.7832	20.0080
90	20.2328	20.4576	20.6824	20.9072	21.1320	21.3568	21.5817	21.8065	22.0313	22.2561
100	22.4809	22.7057	22.9305	23.1553	23.3801	23.6049	23.8297	24.0546	24.2794	24.5042

Pound-Inches to Newton-Meters
(1 pound-force-inch = 0.1129848 newton-meter)

lbf-in.	N·m	lbf-in.	N·m	lbf-in.	N·m	lbf-in.	N·m	lbf-in.	N·m
100	11.298	10	1.130	1	0.113	0.1	0.011	0.01	0.001
200	22.597	20	2.260	2	0.226	0.2	0.023	0.02	0.002
300	33.895	30	3.390	3	0.339	0.3	0.034	0.03	0.003
400	45.194	40	4.519	4	0.452	0.4	0.045	0.04	0.005
500	56.492	50	5.649	5	0.565	0.5	0.056	0.05	0.006
600	67.791	60	6.779	6	0.678	0.6	0.068	0.06	0.007
700	79.089	70	7.909	7	0.791	0.7	0.079	0.07	0.008
800	90.388	80	9.039	8	0.904	0.8	0.090	0.08	0.009
900	101.686	90	10.169	9	1.017	0.9	0.102	0.09	0.010
1000	112.985	100	11.298	10	1.130	1.0	0.113	0.10	0.011

Newton-Meters to Pound-Inches
(1 newton meter = 8.850748 pound-force-inches)

N·m	lbf-in.	N·m	lbf-in.	N·m	lbf-in.	N·m	lbf-in.	N·m	lbf-in.
100	885.07	10	88.51	1	8.85	0.1	0.89	0.01	0.09
200	1770.15	20	177.01	2	17.70	0.2	1.77	0.02	0.18
300	2655.22	30	265.52	3	26.55	0.3	2.66	0.03	0.27
400	3540.30	40	354.03	4	35.40	0.4	3.54	0.04	0.35
500	4425.37	50	442.54	5	44.25	0.5	4.43	0.05	0.44
600	5310.45	60	531.04	6	53.10	0.6	5.31	0.06	0.53
700	6195.52	40	619.55	7	61.96	0.7	6.20	0.07	0.62
800	7080.60	80	708.06	8	70.81	0.8	7.08	0.08	0.71
900	7965.67	90	796.57	9	79.66	0.9	7.97	0.09	0.80
1000	8850.75	100	885.07	10	88.51	1.0	8.85	0.10	0.89

American National Standard Abbreviations for Scientific and Engineering Terms (ANSI Y1.1-1972, R 1984) — 1

Only the most commonly used terms have been included. These forms are recommended for those whose familiarity with the terms used makes possible a maximum of abbreviations. For others, less contracted combinations made up from this list may be used. For example, the list gives the abbreviation of the term "feet per second" as "fps." To some, however, ft per sec will be more easily understood.

Absolute	abs
Alternating current	ac
Ampere	amp
Ampere-hour	amp hr
Angstrom unit	A
Antilogarithm	antilog
Arithmetical average	aa
Atmosphere	atm
Atomic weight	at wt
Avoirdupois	avdp
Barometer	baro
Board feet (feet board measure)	fbm
Boiler pressure	bopress
Boiling point	bp
Brinell hardness number	Bhn
British thermal unit	Btu or B
Bushel	bu
Calorie	cal
Candle	cd
Center to center	c to c
Centimeter	cm
Centimeter-gram-second (system)	cgs
Chemical	chem
Chemically pure	cp
Circular	circ
Circular mil	cmil
Coefficient	coef
Cologarithm	colog
Concentrate	conc
Conductivity	cndct
Constant	const
Cord	cd
Cosecant	csc
Cosine	cos
Cost, insurance, and freight	cif
Cotangent	ctn
Counter electromotive force	cemf
Cubic	cu
Cubic centimeter	cm^3 or cc
Cubic foot	ft^3 or cu ft
Cubic feet per second	ft^3/s or cfs
Cubic inch	in^3 or cu in
Cubic meter	m^3 or cu m
Cubic millimeter	mm^3 or cu mm
Cubic yard	yd^3 or cu yd
Current density	cd
Cylinder	cyl
Decibel	dB
Degree	deg or °
Degree Centigrade	°C
Degree Fahrenheit	°F
Degree Kelvin	K
Diameter	dia
Direct current	dc
Dozen	doz
Dram	dr
Efficiency	eff
Electric	elec
Electromotive force	emf
Elevation	el
Engine	eng
Engineer	engr
Engineering	engrg
Equation	eq
External	ext
Fluid	fl
Foot	ft
Foot-candle	fc
Foot-Lambert	fL or fl
Foot per minute	fpm
Foot per second	fps
Foot-pound	ft lb
Foot-pound-second (system)	fps
Free on board	fob
Freezing point	fp
Frequency	freq
Fusion point	fnpt
Gallon	gal
Gallon per minute	gpm
Gallon per second	gps
Grain	gr
Gram	g
Greatest common divisor	gcd
High pressure	hp
Horsepower	hp
Horsepower-hour	hp hr
Hour	h or hr
Hyperbolic cosine	cosh
Hyperbolic sine	sinh
Hyperbolic tangent	tanh
Inch	in
Inch per second	in/s or ips
Inch-pound	in lb

American National Standard Abbreviations for Scientific and Engineering Terms (ANSI Y1.1-1972, R 1984) — 2

Indicated horsepower-hour	ihph
Intermediate pressure	ip
Internal	intl
Kilovolt-ampere/hour	KVA-h or kVah
Kilowatt-hour meter	kwhm
Latitude	lat
Least common multiple	lcm
Liquid	liq
Logarithm (common)	log
Logarithm (natural)	ln
Low pressure	lp
Lumen per watt	lm/W or lpw
Magnetomotive force	mmf
Mathematics (ical)	math
Maximum	max
Mean effective pressure	mep
Melting point	mp
Meter	m
Meter-kilogram-second	mks
Microfarad	μF
Mile	mi
Mile per hour	mi/h or mph
Milliampere	mA
Minimum	min
Molecular weight	mol wt
Molecule	mo
National Electrical Code	NEC
Ounce	oz
Ounce-inch	oz in
Pennyweight	dwt
Pint	pt
Potential	pot
Potential difference	pd
Pound	lb
pound-force foot	$lb_f \cdot ft$ or lb ft
pound-force inch	$lb_f \cdot in$ or lb in

pound-force per square foot	lb_f/ft^2 or psf
pound-force per square inch	lb_f/in^2 or psi
pound per horsepower	lb/hp or php
Power factor	pf
Quart	qt
Reactive volt-ampere meter	rva
Revolution per minute	r/min or rpm
Revolution per second	r/s or rps
Root mean square	rms
Round	rnd
Secant	sec
Second	s or sec
Sine	sin
Specific gravity	sp gr
Specific heat	sp ht
Square	sq
Square centimeter	cm^2 or sq cm
Square foot	ft^2 or sq ft
Square inch	in^2 or sq in
Square kilometer	km^2 or sq km
Square root of mean square	rms
Standard	std
Tangent	tan
Temperature	temp
Tensile strength	ts
Versed sine	vers
Volt	V
Watt	W
Watthour	Wh
Week	wk
Weight	wt
Yard	yd

Alternative abbreviations conforming to the practice of the International Electrotechnical Commission.

Ampere	A	Kilowatthour	kWh	Millivolt	mV
Ampere-hour	Ah	Megawatt	MW	Ohm	Ω
Coulomb	C	Megohm	MΩ	Volt	V
Farad	F	Microampere	μA	Volt-ampere	VA
Henry	H	Microfarad	μF	Volt-coulomb	VC
Joule	J	Microwatt	μW	Watt	W
Kilovolt	kV	Milliampere	mA	Watthour	Wh
Kilovolt-ampere	kVA	Millifarad	mF		
Kilowatt	kW	Millihenry	mH		

Abbreviations should be used sparingly and only where their meaning will be clear. If there is any doubt, then spell out the term or unit of measurement.

The following points are good practice when preparing engineering documentation. Terms denoting units of measurement should be abbreviated in text only when preceded by the amounts indicated in numerals: "several inches," "one inch," "12 in." A sentence should not begin with a numeral followed by an abbreviation. The use of conventional signs for abbreviations in text should be avoided: use "lb," not "#" or "in," not ".

Symbols for the chemical elements are listed in the table on page 394.

INDEX